UNIT CONVERSION FACTORS

Length
1 m = 100 cm = 1000 mm = 10^6 μm = 10^9 nm
1 km = 1000 m = 0.6214 mi
1 m = 3.281 ft = 39.37 in.
1 cm = 0.3937 in.
1 in. = 2.540 cm
1 ft = 30.48 cm
1 yd = 91.44 cm
1 mi = 5280 ft = 1.609 km
1 Å = 10^{-10} m = 10^{-8} cm = 10^{-1} nm
1 nautical mile = 6080 ft
1 light-year = 9.461×10^{15} m

Area
1 cm^2 = 0.155 $in.^2$
1 m^2 = 10^4 cm^2 = 10.76 ft^2
1 $in.^2$ = 6.452 cm^2
1 ft^2 = 144 $in.^2$ = 0.0929 m^2

Volume
1 liter = 1000 cm^3 = 10^{-3} m^3 = 0.03531 ft^3 = 61.02 $in.^3$
1 ft^3 = 0.02832 m^3 = 28.32 liters = 7.477 gallons
1 gallon = 3.788 liters

Time
1 min = 60 s
1 h = 3600 s
1 d = 86,400 s
1 y = 365.24 d = 3.156×10^7 s

Angle
1 rad = 57.30° = 180°/π
1° = 0.01745 rad = π/180 rad
1 revolution = 360° = 2π rad
1 rev/min (rpm) = 0.1047 rad/s

Speed
1 m/s = 3.281 ft/s
1 ft/s = 0.3048 m/s
1 mi/min = 60 mi/h = 88 ft/s
1 km/h = 0.2778 m/s = 0.6214 mi/h
1 mi/h = 1.466 ft/s = 0.4470 m/s = 1.609 km/h
1 furlong/fortnight = 1.662×10^{-4} m/s

Acceleration
1 m/s^2 = 100 cm/s^2 = 3.281 ft/s^2
1 cm/s^2 = 0.01 m/s^2 = 0.03281 ft/s^2
1 ft/s^2 = 0.3048 m/s^2 = 30.48 cm/s^2
1 mi/h·s = 1.467 ft/s^2

Mass
1 kg = 10^3 g = 0.0685 slug
1 g = 6.85×10^{-5} slug
1 slug = 14.59 kg
1 u = 1.661×10^{-27} kg
1 kg has a weight of 2.205 lb when g = 9.80 m/s^2

Force
1 N = 10^5 dyn = 0.2248 lb
1 lb = 4.448 N = 4.448×10^5 dyn

Pressure
1 Pa = 1 N/m^2 = 1.450×10^{-4} $lb/in.^2$ = 0.0209 lb/ft^2
1 bar = 10^5 Pa
1 $lb/in.^2$ = 6895 Pa
1 lb/ft^2 = 47.88 Pa
1 atm = 1.013×10^5 Pa = 1.013 bar
    = 14.7 $lb/in.^2$ = 2117 lb/ft^2
1 mm Hg = 1 torr = 133.3 Pa

Energy
1 J = 10^7 ergs = 0.239 cal
1 cal = 4.186 J (based on 15° calorie)
1 ft·lb = 1.356 J
1 Btu = 1055 J = 252 cal = 778 ft·lb
1 eV = 1.602×10^{-19} J
1 kWh = 3.600×10^6 J

Mass-Energy Equivalence
1 kg $\leftrightarrow$ 8.988×10^{16} J
1 u $\leftrightarrow$ 931.5 MeV
1 eV $\leftrightarrow$ 1.074×10^{-9} u

Power
1 W = 1 J/s
1 hp = 746 W = 550 ft·lb/s
1 Btu/h = 0.293 W

PhET SIMULATIONS

Available in the Pearson eText and in the Study Area of MasteringPhysics

Extended Edition includes Chapters 1–44. Standard Edition includes Chapters 1–37.
Three-volume edition: Volume 1 includes Chapters 1–20, Volume 2 includes Chapters 21–37,
and Volume 3 includes Chapters 37–44.

*Indicates an associated tutorial is available in the MasteringPhysics Item Library.

APPLICATIONS

SEARS AND ZEMANSKY'S

UNIVERSITY PHYSICS

WITH MODERN PHYSICS

14TH EDITION

HUGH D. YOUNG

ROGER A. FREEDMAN
University of California, Santa Barbara

CONTRIBUTING AUTHOR
A. LEWIS FORD
Texas A&M University

PEARSON

Editor in Chief, Physical Sciences: Jeanne Zalesky
Executive Editor: Nancy Whilton
Project Manager: Beth Collins
Program Manager: Katie Conley
Executive Development Editor: Karen Karlin
Assistant Editor: Sarah Kaubisch
Rights and Permissions Project Manager: William Opaluch
Rights and Permissions Management: Lumina Datamatics
Development Manager: Cathy Murphy
Program and Project Management Team Lead: Kristen Flathman
Production Management: Cindy Johnson Publishing Services

Copyeditor: Carol Reitz
Compositor: Cenveo Publisher Services
Design Manager: Mark Ong
Interior and Cover Designer: Cadence Design Studio
Illustrators: Rolin Graphics, Inc.
Photo Permissions Management: Maya Melenchuk and Eric Schrader
Photo Researcher: Eric Schrader
Senior Manufacturing Buyer: Maura Zaldivar-Garcia
Marketing Manager: Will Moore

Cover Photo Credit: Knut Bry
About the Cover Image: www.leonardobridgeproject.org

The Leonardo Bridge Project is a project to build functional interpretations of Leonardo da Vinci's Golden Horn Bridge design, conceived and built first in Norway by artist Vebjørn Sand as a global public art project, linking people and cultures in communities in every continent.

Library of Congress Cataloging-in-Publication Data
Young, Hugh D.
    Sears and Zemansky's university physics / Hugh D. Young, Roger A. Freedman, University of California, Santa Barbara ; contributing author, A. Lewis Ford, Texas A&M University. -- 14th edition.
      pages cm
    Includes index.
    ISBN 978-0-321-97361-0 (student ed. : alk. paper) -- ISBN 978-0-13-397798-1 (balc : alk. paper)
1. Physics--Textbooks. 2. Physics--Study and teaching (Higher) I. Freedman, Roger A. II. Ford, A. Lewis (Albert Lewis) III. Sears, Francis Weston, 1898-1975. University physics. IV. Title. V. Title: University physics.
    QC21.3.Y68 2016
    530--dc23

2014040348

7 18

PEARSON
www.pearsonhighered.com

ISBN 10: 0-321-97361-5; ISBN 13: 978-0-321-97361-0 (Student edition)
ISBN 10: 0-13-397798-6; ISBN 13: 978-0-13-397798-1 (BALC)

BRIEF CONTENTS

THE BENCHMARK FOR CLARITY AND RIGOR

Since its first edition, *University Physics* has been renowned for its emphasis on fundamental principles and how to apply them. This text is known for its clear and thorough narrative and for its uniquely broad, deep, and thoughtful set of worked examples—key tools for developing both conceptual understanding and problem-solving skills.

The **Fourteenth Edition** improves the defining features of the text while adding new features influenced by physics education research. A focus on visual learning, new problem types, and pedagogy informed by MasteringPhysics metadata headline the improvements designed to create the best learning resource for today's physics students.

A FOCUS ON PROBLEM SOLVING

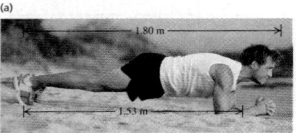

EXAMPLE 11.2   LOCATING YOUR CENTER OF GRAVITY WHILE YOU WORK OUT

The *plank* (**Fig. 11.8a**) is a great way to strengthen abdominal, back, and shoulder muscles. You can also use this exercise position to locate your center of gravity. Holding plank position with a scale under his toes and another under his forearms, one athlete measured that 66.0% of his weight was supported by his forearms and 34.0% by his toes. (That is, the total normal forces on his forearms and toes were 0.660w and 0.340w, respectively, where w is the athlete's weight.) He is 1.80 m tall, and in plank position

11.8 An athlete in plank position.

(a)

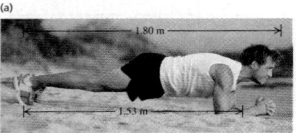

(b)

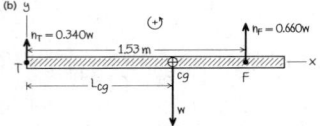

the distance from his toes to the middle of his forearms is 1.53 m. How far from his toes is his center of gravity?

SOLUTION

IDENTIFY and SET UP: We can use the two conditions for equilibrium, Eqs. (11.6), for an athlete at rest. So both the net force and net torque on the athlete are zero. Figure 11.8b shows a free-body diagram, including x- and y-axes and our convention that counterclockwise torques are positive. The weight w acts at the center of gravity, which is between the two supports (as it must be; see Section 11.2). Our target variable is the distance L_{cg}, the lever arm of the weight with respect to the toes T, so it is wise to take torques with respect to T. The torque due to the weight is negative (it tends to cause a clockwise rotation around T), and the upward normal force at the forearms F is po... a counterclockwise rotation around T).

EXECUTE: The first condition for equilibrium... $\Sigma F_x = 0$ because there are no x-comp... because $0.340w + 0.660w + (-w) = 0$... equation and solve for L_{cg}:

$$\Sigma \tau_R = 0.340w(0) - wL_{cg} + 0.660...$$
$$L_{cg} = 1.01 \text{ m}$$

EVALUATE: The center of gravity is slight... navel (as it is for most people) and closer t... his toes, which is why his forearms suppo... You can check our result by writing the tor... forearms F. You'll find that his center of gra... forearms, or $(1.53 \text{ m}) - (0.52 \text{ m}) = 1.01$ ...

◀ A research-based **PROBLEM-SOLVING APPROACH—IDENTIFY, SET UP, EXECUTE, EVALUATE**—is used in every Example and throughout the Student's and Instructor's Solutions Manuals and the Study Guide. This consistent approach teaches students to tackle problems thoughtfully rather than cutting straight to the math.

PROBLEM-SOLVING STRATEGY 3.1   PROJECTILE MOTION

NOTE: The strategies we used in Sections 2.4 and 2.5 for straight-line, constant-acceleration problems are also useful here.

IDENTIFY *the relevant concepts:* The key concept is that throughout projectile motion, the acceleration is downward and has a constant magnitude g. Projectile-motion equations don't apply to *throwing* a ball, because during the throw the ball is acted on by both the thrower's hand and gravity. These equations apply only *after* the ball leaves the thrower's hand.

SET UP *the problem* using the following steps:
1. Define your coordinate system and make a sketch showing your axes. It's almost always best to make the x-axis horizontal and the y-axis vertical, and to choose the origin to be where the body first becomes a projectile (for example, where a ball leaves the thrower's hand). Then the components of acceleration are $a_x = 0$ and $a_y = -g$, as in Eq. (3.13); the initial position is $x_0 = y_0 = 0$; and you can use Eqs. (3.19) through (3.22). (If you choose a different origin or axes, you'll have to modify these equations.)
2. List the unknown and known quantities, and decide which unknowns are your target variables. For example, you might be given the initial velocity (either the components or the magnitude and direction) and asked to find the coordinates and velocity components at some later time. Make sure that

you have as many equations as there are target variables to be found. In addition to Eqs. (3.19) through (3.22), Eqs. (3.23) through (3.26) may be useful.
3. State the problem in words and then translate those words into symbols. For example, *when* does the particle arrive at a certain point? (That is, at what value of t?) *Where* is the particle when its velocity has a certain value? (That is, what are the values of x and y when v_x or v_y has the specified value?) Since $v_y = 0$ at the highest point in a trajectory, the question "When does the projectile reach its highest point?" translates into "What is the value of t when $v_y = 0$?" Similarly, "When does the projectile return to its initial elevation?" translates into "What is the value of t when $y = y_0$?"

EXECUTE *the solution:* Find the target variables using the equations you chose. Resist the temptation to break the trajectory into segments and analyze each segment separately! You don't have to start all over when the projectile reaches its highest point! It's almost always easier to use the same axes and time scale throughout the problem. If you need numerical values, use $g = 9.80 \text{ m/s}^2$. Remember that g is positive!

EVALUATE *your answer:* Do your results make sense? Do the numerical values seem reasonable?

PROBLEM-SOLVING STRATEGIES ▶ coach students in how to approach specific types of problems.

BRIDGING PROBLEM   HOW LONG TO DRAIN?

A large cylindrical tank with diameter D is open to the air at the top. The tank contains water to a height H. A small circular hole with diameter d, where $d \ll D$, is then opened at the bottom of the tank (**Fig. 12.32**). Ignore any effects of viscosity. (a) Find y, the height of water in the tank a time t after the hole is opened, as a function of t. (b) How long does it take to drain the tank completely? (c) If you double height H, by what factor does the time to drain the tank increase?

SOLUTION GUIDE

IDENTIFY and SET UP
1. Draw a sketch of the situation that shows all of the relevant dimensions.
2. List the unknown quantities, and decide which of these are the target variables.
3. At what speed does water flow out of the bottom of the tank? How is this related to the volume flow rate of water out of the tank? How is the volume flow rate related to the rate of change of y?

EXECUTE
4. Use your results from step 3 to write an equation for dy/dt.
5. Your result from step 4 is a relatively simple differential equation. With your knowledge of calculus, you can integrate it to find y as a function of t. (*Hint:* Once you've done the integration, you'll still have to do a little algebra.)

12.32 A water tank that is open at the top and has a hole at the bottom.

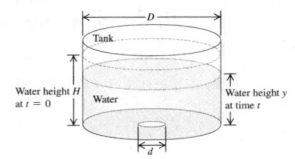

Tank

Water height H at t = 0

Water

Water height y at time t

6. Use your result from step 5 to find the time when the tank is empty. How does your result depend on the initial height H?

EVALUATE
7. Check whether your answers are reasonable. A good check is to draw a graph of y versus t. According to your graph, what is the algebraic sign of dy/dt at different times? Does this make sense?

◀ **BRIDGING PROBLEMS**, which help students move from single-concept worked examples to multi-concept problems at the end of the chapter, have been revised, based on reviewer feedback, ensuring that they are effective and at the appropriate difficulty level.

PEDAGOGY INFORMED BY DATA AND RESEARCH

DATA *SPEAKS*

Gravitation

When students were given a problem about superposition of gravitational forces, more than 60% gave an incorrect response. Common errors:

- Assuming that equal-mass objects *A* and *B* must exert equally strong gravitational attraction on an object *C* (which is not true when *A* and *B* are different distances from *C*).

- Neglecting to account for the vector nature of force. (To add two forces that point in different directions, you can't just add the force magnitudes.)

◀ **DATA SPEAKS SIDEBARS**, based on MasteringPhysics metadata, alert students to the statistically most common mistakes made in solving problems on a given topic.

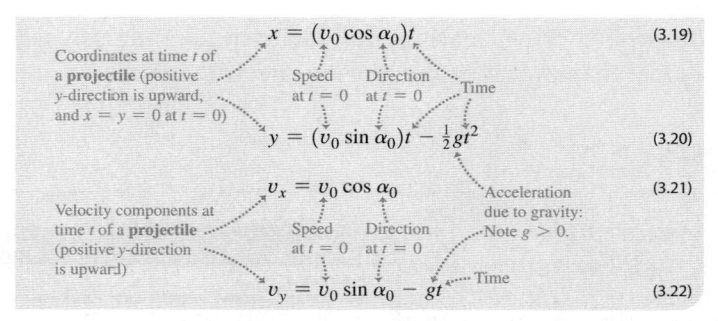

Coordinates at time *t* of a **projectile** (positive *y*-direction is upward, and $x = y = 0$ at $t = 0$)

$$x = (v_0 \cos \alpha_0)t \quad (3.19)$$

Speed at $t = 0$ Direction at $t = 0$ Time

$$y = (v_0 \sin \alpha_0)t - \tfrac{1}{2}gt^2 \quad (3.20)$$

Velocity components at time *t* of a **projectile** (positive *y*-direction is upward)

$$v_x = v_0 \cos \alpha_0 \quad (3.21)$$

Speed at $t = 0$ Direction at $t = 0$ Acceleration due to gravity: Note $g > 0$.

$$v_y = v_0 \sin \alpha_0 - gt \quad (3.22)$$

Time

▲ All **KEY EQUATIONS ARE NOW ANNOTATED** to help students make a connection between a conceptual and a mathematical understanding of physics.

PASSAGE PROBLEMS

BIO **PRESERVING CELLS AT COLD TEMPERATURES.** In cryopreservation, biological materials are cooled to a very low temperature to slow down chemical reactions that might damage the cells or tissues. It is important to prevent the materials from forming ice crystals during freezing. One method for preventing ice formation is to place the material in a protective solution called a *cryoprotectant*. Stated values of the thermal properties of one cryoprotectant are listed here:

Melting point	−20°C
Latent heat of fusion	2.80×10^5 J/kg
Specific heat (liquid)	4.5×10^3 J/kg · K
Specific	
Therma	
Therma	

DATA PROBLEMS appear in each chapter. These data-based reasoning problems, many of which are context rich, require students to use experimental evidence, presented in a tabular or graphical format, to formulate conclusions. ▼

17.117 Careful measurements show that the specific heat of the solid phase depends on temperature (**Fig. P17.117**). How will the actual time needed for this cryoprotectant to come to equilibrium with the cold plate compare with the time predicted by using the values in the table? Assume that all values other than the specific heat (solid) are correct. The actual time (a) will be shorter; (b) will be longer; (c) will be the same; (d) depends on the density of the cryoprotectant.

Figure **P17.117**

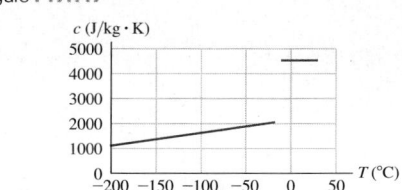

▲ Each chapter includes three to five **PASSAGE PROBLEMS**, which follow the format used in the MCATs. These problems require students to investigate multiple aspects of a real-life physical situation, typically biological in nature, as described in a reading passage.

9.89 •• **DATA** You are rebuilding a 1965 Chevrolet. To decide whether to replace the flywheel with a newer, lighter-weight one, you want to determine the moment of inertia of the original, 35.6-cm-diameter flywheel. It is not a uniform disk, so you can't use $I = \tfrac{1}{2}MR^2$ to calculate the moment of inertia. You remove the flywheel from the car and use low-friction bearings to mount it on a horizontal, stationary rod that passes through the center of the flywheel, which can then rotate freely (about 2 m above the ground). After gluing one end of a long piece of flexible fishing line to the rim of the flywheel, you wrap the line a number of turns around the rim and suspend a 5.60-kg metal block from the free end of the line. When you release the block from rest, it descends as the flywheel rotates. With high-speed photography you measure the distance *d* the block has moved downward as a function of the time since it was released. The equation for the graph shown in **Fig. P9.89** that gives a good fit to the data points is $d = (165 \text{ cm/s}^2)t^2$. (a) Based on the graph, does the block fall with constant acceleration? Explain. (b) Use the graph to calculate the speed of the block when it has descended 1.50 m. (c) Apply conservation of mechanical energy to the system of flywheel and block to calculate the moment of inertia of the flywheel. (d) You are relieved that the fishing line doesn't break. Apply Newton's second law to the block to find the tension in the line as the block descended.

PERSONALIZE LEARNING WITH MASTERINGPHYSICS

MasteringPhysics® from Pearson is the leading online homework, tutorial, and assessment system, designed to improve results by engaging students before, during, and after class with powerful content. Instructors can now ensure that students arrive ready to learn by assigning educationally effective content before class, and encourage critical thinking and retention with in-class resources such as Learning Catalytics. Students can further master concepts after class through traditional and adaptive homework assignments that provide hints and answer-specific feedback. The Mastering gradebook records scores for all automatically graded assignments in one place, while diagnostic tools give instructors access to rich data to assess student understanding and misconceptions.

Mastering brings learning full circle by continuously adapting to each student and making learning more personal than ever—before, during, and after class.

BEFORE CLASS

INTERACTIVE PRE-LECTURE VIDEOS address the rapidly growing movement toward pre-lecture teaching and flipped classrooms. These videos provide a conceptual introduction to key topics. Embedded assessment helps students to prepare before lecture and instructors to identify student misconceptions.

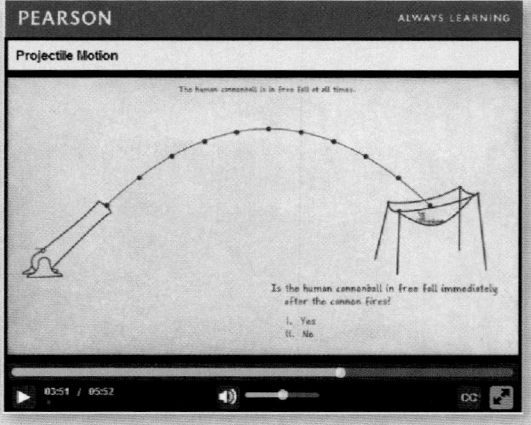

PRE-LECTURE CONCEPT QUESTIONS check familiarity with key concepts, prompting students to do their assigned reading prior to coming to class. These quizzes keep students on track, keep them more engaged in lecture, and help you spot the concepts with which they have the most difficulty. Open-ended essay questions help students identify what they find most difficult about a concept, better informing you and assisting with "just-in-time" teaching.

DURING CLASS

LEARNING CATALYTICS™ is a "bring your own device" student engagement, assessment, and classroom intelligence system. With Learning Catalytics you can:

- Assess students in real time, using open-ended tasks to probe student understanding.
- Understand immediately where students are and adjust your lecture accordingly.
- Improve your students' critical-thinking skills.
- Access rich analytics to understand student performance.
- Add your own questions to make Learning Catalytics fit your course exactly.
- Manage student interactions with intelligent grouping and timing.

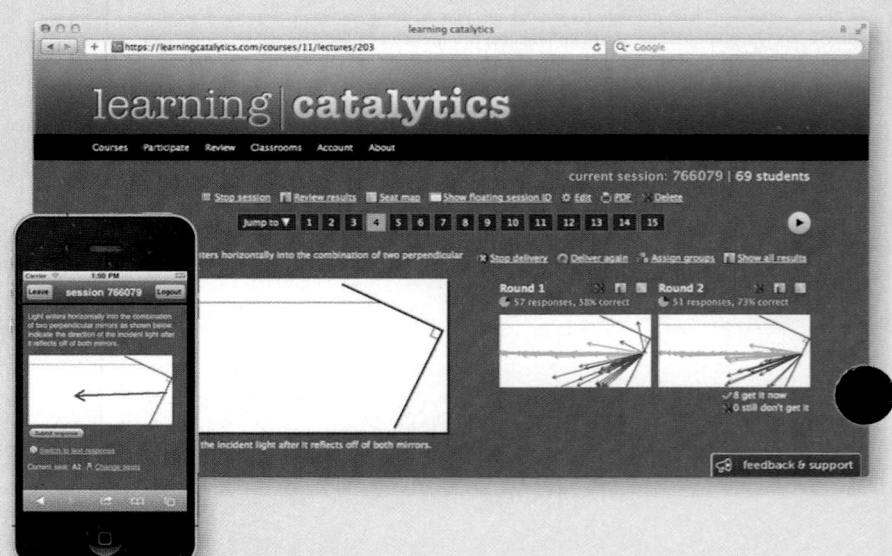

AFTER CLASS

TUTORIALS featuring specific wrong-answer feedback, hints, and a wide variety of educationally effective content guide your students through the toughest topics in physics. The hallmark Hints and Feedback offer instruction similar to what students would experience in an office hour, allowing them to learn from their mistakes without being given the answer.

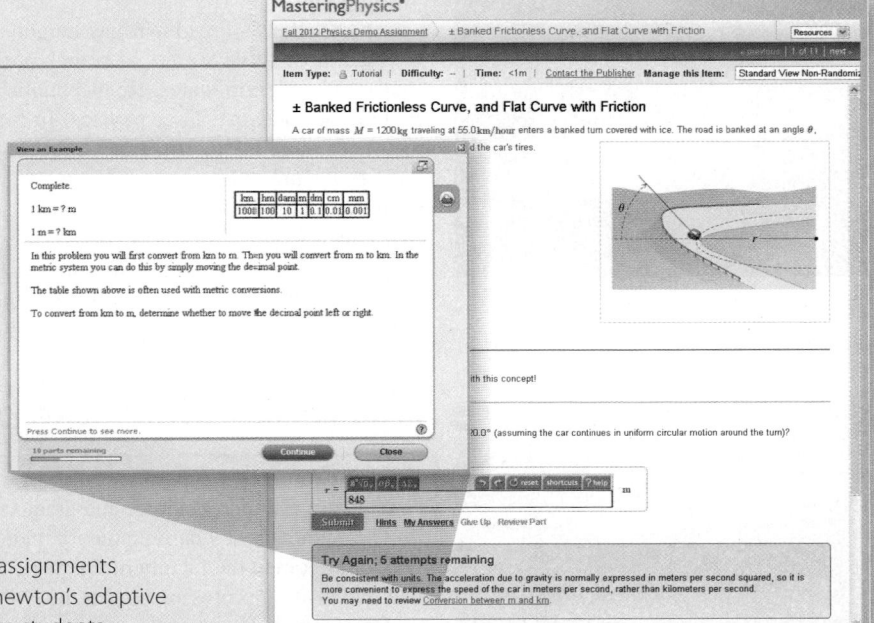

ADAPTIVE FOLLOW-UPS are personalized assignments that pair Mastering's powerful content with Knewton's adaptive learning engine to provide personalized help to students. These assignments address common student misconceptions and topics students struggled with on assigned homework, including core prerequisite topics. ▼

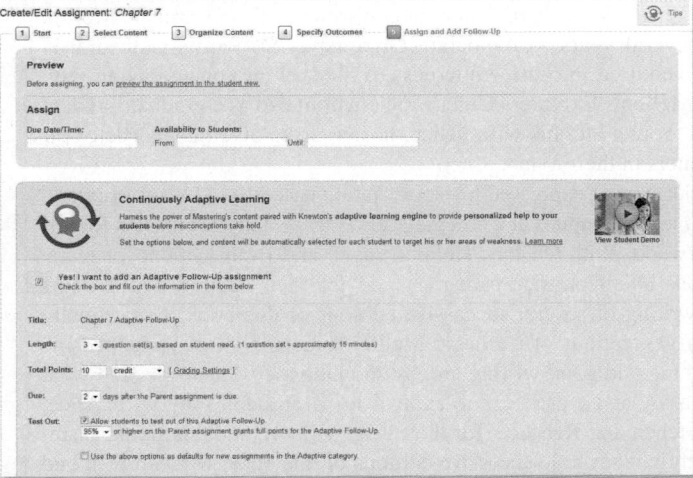

VIDEO TUTOR DEMONSTRATIONS, available in the Study Area and in the Item Library and accessible by QR code in the textbook, feature "pause-and-predict" demonstrations of key physics concepts as assessment to engage students actively in understanding key concepts. New VTDs build on the existing collection, adding new topics for a more robust set of demonstrations. ▼

◄ **VIDEO TUTOR SOLUTIONS** are tied to each worked example and Bridging Problem in the textbook and can be accessed through MasteringPhysics or from QR codes in the textbook. They walk students through the problem-solving process, providing a virtual teaching assistant on a round-the-clock basis.

ABOUT THE AUTHORS

Roger A. Freedman is a Lecturer in Physics at the University of California, Santa Barbara. He was an undergraduate at the University of California campuses in San Diego and Los Angeles and did his doctoral research in nuclear theory at Stanford University under the direction of Professor J. Dirk Walecka. Dr. Freedman came to UCSB in 1981 after three years of teaching and doing research at the University of Washington.

At UCSB, Dr. Freedman has taught in both the Department of Physics and the College of Creative Studies, a branch of the university intended for highly gifted and motivated undergraduates. He has published research in nuclear physics, elementary particle physics, and laser physics. In recent years, he has worked to make physics lectures a more interactive experience through the use of classroom response systems and pre-lecture videos.

In the 1970s Dr. Freedman worked as a comic book letterer and helped organize the San Diego Comic-Con (now the world's largest popular culture convention) during its first few years. Today, when not in the classroom or slaving over a computer, Dr. Freedman can be found either flying (he holds a commercial pilot's license) or with his wife, Caroline, cheering on the rowers of UCSB Men's and Women's Crew.

IN MEMORIAM: HUGH YOUNG (1930–2013)

Hugh D. Young was Emeritus Professor of Physics at Carnegie Mellon University. He earned both his undergraduate and graduate degrees from that university. He earned his Ph.D. in fundamental particle theory under the direction of the late Richard Cutkosky. Dr. Young joined the faculty of Carnegie Mellon in 1956 and retired in 2004. He also had two visiting professorships at the University of California, Berkeley.

Dr. Young's career was centered entirely on undergraduate education. He wrote several undergraduate-level textbooks, and in 1973 he became a coauthor with Francis Sears and Mark Zemansky for their well-known introductory textbooks. In addition to his role on Sears and Zemansky's *University Physics,* he was the author of Sears and Zemansky's *College Physics.*

Dr. Young earned a bachelor's degree in organ performance from Carnegie Mellon in 1972 and spent several years as Associate Organist at St. Paul's Cathedral in Pittsburgh. He often ventured into the wilderness to hike, climb, or go caving with students in Carnegie Mellon's Explorers Club, which he founded as a graduate student and later advised. Dr. Young and his wife, Alice, hosted up to 50 students each year for Thanksgiving dinners in their home.

Always gracious, Dr. Young expressed his appreciation earnestly: "I want to extend my heartfelt thanks to my colleagues at Carnegie Mellon, especially Professors Robert Kraemer, Bruce Sherwood, Ruth Chabay, Helmut Vogel, and Brian Quinn, for many stimulating discussions about physics pedagogy and for their support and encouragement during the writing of several successive editions of this book. I am equally indebted to the many generations of Carnegie Mellon students who have helped me learn what good teaching and good writing are, by showing me what works and what doesn't. It is always a joy and a privilege to express my gratitude to my wife, Alice, and our children, Gretchen and Rebecca, for their love, support, and emotional sustenance during the writing of several successive editions of this book. May all men and women be blessed with love such as theirs." We at Pearson appreciated his professionalism, good nature, and collaboration. He will be missed.

A. Lewis Ford is Professor of Physics at Texas A&M University. He received a B.A. from Rice University in 1968 and a Ph.D. in chemical physics from the University of Texas at Austin in 1972. After a one-year postdoc at Harvard University, he joined the Texas A&M physics faculty in 1973 and has been there ever since. Professor Ford has specialized in theoretical atomic physics—in particular, atomic collisions. At Texas A&M he has taught a variety of undergraduate and graduate courses, but primarily introductory physics.

HOW TO SUCCEED IN PHYSICS BY REALLY TRYING

Mark Hollabaugh, *Normandale Community College, Emeritus*

Physics encompasses the large and the small, the old and the new. From the atom to galaxies, from electrical circuitry to aerodynamics, physics is very much a part of the world around us. You probably are taking this introductory course in calculus-based physics because it is required for subsequent courses that you plan to take in preparation for a career in science or engineering. Your professor wants you to learn physics and to enjoy the experience. He or she is very interested in helping you learn this fascinating subject. That is part of the reason your professor chose this textbook for your course. That is also the reason Drs. Young and Freedman asked me to write this introductory section. We want you to succeed!

The purpose of this section of *University Physics* is to give you some ideas that will assist your learning. Specific suggestions on how to use the textbook will follow a brief discussion of general study habits and strategies.

PREPARATION FOR THIS COURSE

If you had high school physics, you will probably learn concepts faster than those who have not because you will be familiar with the language of physics. If English is a second language for you, keep a glossary of new terms that you encounter and make sure you understand how they are used in physics. Likewise, if you are further along in your mathematics courses, you will pick up the mathematical aspects of physics faster. Even if your mathematics is adequate, you may find a book such as Arnold D. Pickar's *Preparing for General Physics: Math Skill Drills and Other Useful Help (Calculus Version)* to be useful. Your professor may assign sections of this math review to assist your learning.

LEARNING TO LEARN

Each of us has a different learning style and a preferred means of learning. Understanding your own learning style will help you to focus on aspects of physics that may give you difficulty and to use those components of your course that will help you overcome the difficulty. Obviously you will want to spend more time on those aspects that give you the most trouble. If you learn by hearing, lectures will be very important. If you learn by explaining, then working with other students will be useful to you. If solving problems is difficult for you, spend more time learning how to solve problems. Also, it is important to understand and develop good study habits. Perhaps the most important thing you can do for yourself is set aside adequate, regularly scheduled study time in a distraction-free environment.

Answer the following questions for yourself:
- Am I able to use fundamental mathematical concepts from algebra, geometry, and trigonometry? (If not, plan a program of review with help from your professor.)
- In similar courses, what activity has given me the most trouble? (Spend more time on this.) What has been the easiest for me? (Do this first; it will build your confidence.)
- Do I understand the material better if I read the book before or after the lecture? (You may learn best by skimming the material, going to lecture, and then undertaking an in-depth reading.)

- Do I spend adequate time studying physics? (A rule of thumb for a class like this is to devote, on average, 2.5 hours out of class for each hour in class. For a course that meets 5 hours each week, that means you should spend about 10 to 15 hours per week studying physics.)
- Do I study physics every day? (Spread that 10 to 15 hours out over an entire week!) At what time of the day am I at my best for studying physics? (Pick a specific time of the day and stick to it.)
- Do I work in a quiet place where I can maintain my focus? (Distractions will break your routine and cause you to miss important points.)

WORKING WITH OTHERS

Scientists or engineers seldom work in isolation from one another but rather work cooperatively. You will learn more physics and have more fun doing it if you work with other students. Some professors may formalize the use of cooperative learning or facilitate the formation of study groups. You may wish to form your own informal study group with members of your class. Use e-mail to keep in touch with one another. Your study group is an excellent resource when you review for exams.

LECTURES AND TAKING NOTES

An important component of any college course is the lecture. In physics this is especially important, because your professor will frequently do demonstrations of physical principles, run computer simulations, or show video clips. All of these are learning activities that will help you understand the basic principles of physics. Don't miss lectures. If for some reason you do, ask a friend or member of your study group to provide you with notes and let you know what happened.

Take your class notes in outline form, and fill in the details later. It can be very difficult to take word-for-word notes, so just write down key ideas. Your professor may use a diagram from the textbook. Leave a space in your notes and add the diagram later. After class, edit your notes, filling in any gaps or omissions and noting things that you need to study further. Make references to the textbook by page, equation number, or section number.

Ask questions in class, or see your professor during office hours. Remember that the only "dumb" question is the one that is not asked. Your college may have teaching assistants or peer tutors who are available to help you with any difficulties.

EXAMINATIONS

Taking an examination is stressful. But if you feel adequately prepared and are well rested, your stress will be lessened. Preparing for an exam is a continuous process; it begins the moment the previous exam is over. You should immediately go over the exam to understand any mistakes you made. If you worked a problem and made substantial errors, try this: Take a piece of paper and divide it down the middle with a line from top to bottom. In one column, write the proper solution to the problem. In the other column, write what you did and why, if you know, and why your solution was incorrect. If you are uncertain why you made your mistake or how to avoid making it again, talk with your professor. Physics constantly builds on fundamental ideas, and it is important to correct any misunderstandings immediately. *Warning:* Although cramming at the last minute may get you through the present exam, you will not adequately retain the concepts for use on the next exam.

TO THE INSTRUCTOR

PREFACE

This book is the product of six and a half decades of leadership and innovation in physics education. When the first edition of *University Physics* by Francis W. Sears and Mark W. Zemansky was published in 1949, it was revolutionary among calculus-based physics textbooks in its emphasis on the fundamental principles of physics and how to apply them. The success of *University Physics* with generations of several million students and educators around the world is a testament to the merits of this approach and to the many innovations it has introduced subsequently.

In preparing this new Fourteenth Edition, we have further augmented and developed *University Physics* to assimilate the best ideas from education research with enhanced problem-solving instruction, pioneering visual and conceptual pedagogy, all-new categories of end-of-chapter problems, and the most pedagogically proven and widely used online homework and tutorial system in the world.

NEW TO THIS EDITION

- **All key equations now include annotations** that describe the equation and explain the meanings of the symbols in the equation. These annotations help promote in-depth processing of information and greater recall.
- **DATA SPEAKS sidebars** in each chapter, based on data captured from thousands of students, alert students to the statistically most common mistakes students make when working problems on related topics in MasteringPhysics.
- **Updated modern physics content** includes sections on quantum measurement (Chapter 40) and quantum entanglement (Chapter 41), as well as recent data on the Higgs boson and cosmic background radiation (Chapter 44).
- **Additional bioscience applications** appear throughout the text, mostly in the form of marginal photos with explanatory captions, to help students see how physics is connected to many breakthroughs and discoveries in the biosciences.
- The **text has been streamlined** with tighter and more focused language.
- **Based on data from MasteringPhysics, changes to the end-of-chapter content** include the following:
  - **25%–30% of problems are new or revised.**
  - Most chapters include **six to ten biosciences-related problems.**
  - The number of **context-rich problems** is increased to facilitate the greater learning gains that they can offer.
  - **Three new DATA problems** appear in each chapter. These typically context-rich, data-based reasoning problems require students to use experimental evidence, presented in a tabular or graphical format, to formulate conclusions.
  - Each chapter now includes **three to five new Passage Problems,** which follow the format that is used in the MCATs. These problems require students to investigate multiple aspects of a real-life physical situation, typically biological in nature, that is described in a reading passage.
- **Looking back at ...** essential past concepts are listed at the beginning of each chapter, so that students know what they need to have mastered before digging into the current chapter.

Standard, Extended,
and Three-Volume Editions

With MasteringPhysics:
- **Standard Edition:** Chapters 1–37
  (ISBN 978-0-13-409650-6)
- **Extended Edition:** Chapters 1–44
  (ISBN 978-0-321-98258-2)

Without MasteringPhysics:
- **Standard Edition:** Chapters 1–37
  (ISBN 978-0-13-396929-0)
- **Extended Edition:** Chapters 1–44
  (ISBN 978-0-321-97361-0)
- **Volume 1:** Chapters 1–20
  (ISBN 978-0-13-397804-9)
- **Volume 2:** Chapters 21–37
  (ISBN 978-0-13-397800-1)
- **Volume 3:** Chapters 37–44
  (ISBN 978-0-13-397802-5)

KEY FEATURES OF *UNIVERSITY PHYSICS*

DEMO

- More than 620 **QR codes** throughout the book allow students to use a mobile phone to watch an interactive video of a physics instructor giving a relevant physics demonstration (Video Tutor Demonstration) or showing a narrated and animated worked Example (Video Tutor Solution).

  All of these videos also play directly through links within the Pearson eText as well as the Study Area within MasteringPhysics.

- End-of-chapter **Bridging Problems,** many revised, provide a transition between the single-concept Examples and the more challenging end-of-chapter problems. Each Bridging Problem poses a difficult, multiconcept problem that typically incorporates physics from earlier chapters. A skeleton **Solution Guide,** consisting of questions and hints, helps train students to approach and solve challenging problems with confidence.

- Deep and extensive **problem sets** cover a wide range of difficulty (with blue dots to indicate relative difficulty level) and exercise both physical understanding and problem-solving expertise. Many problems are based on complex real-life situations.

- This textbook offers more **Examples** and **Conceptual Examples** than most other leading calculus-based textbooks, allowing students to explore problem-solving challenges that are not addressed in other textbooks.

- A research-based **problem-solving approach (Identify, Set Up, Execute, Evaluate)** is used in every Example as well as in the Problem-Solving Strategies, in the Bridging Problems, and throughout the Instructor's Solutions Manual and the Study Guide. This consistent approach teaches students to tackle problems thoughtfully rather than cutting straight to the math.

- **Problem-Solving Strategies** coach students in how to approach specific types of problems.

- The **figures** use a simplified graphical style to focus on the physics of a situation, and they incorporate more **explanatory annotations** than in the previous edition. Both techniques have been demonstrated to have a strong positive effect on learning.

- Many figures that illustrate Example solutions take the form of black-and-white **pencil sketches,** which directly represent what a student should draw in solving such problems themselves.

- The popular **Caution paragraphs** focus on typical misconceptions and student problem areas.

- End-of-section **Test Your Understanding** questions let students check their grasp of the material and use a multiple-choice or ranking-task format to probe for common misconceptions.

- **Visual Summaries** at the end of each chapter present the key ideas in words, equations, and thumbnail pictures, helping students review more effectively.

- **Approximately 70 PhET simulations** are linked to the Pearson eText and provided in the Study Area of the MasteringPhysics website (with icons in the printed book). These powerful simulations allow students to interact productively with the physics concepts they are learning. PhET clicker questions are also included on the Instructor's Resource DVD.

INSTRUCTOR'S SUPPLEMENTS

Note: *For convenience, all of the following instructor's supplements (except for the Instructor's Resource DVD) can be downloaded from the Instructor Resources Area accessed via MasteringPhysics (www.masteringphysics.com).*

The **Instructor's Solutions Manual,** prepared by A. Lewis Ford (Texas A&M University) and Wayne Anderson, contains complete and detailed solutions to all end-of-chapter problems. All solutions follow consistently the same Identify/Set Up/Execute/Evaluate problem-solving framework used in the textbook. Download only from the MasteringPhysics Instructor Area or from the Instructor Resource Center (www.pearsonhighered.com/irc).

The cross-platform **Instructor's Resource DVD** (978-0-13-393364-7) provides a comprehensive library of approximately 350 applets from ActivPhysics OnLine as well as all art and photos from the textbook in JPEG and PowerPoint formats. In addition, all of the key equations, problem-solving strategies, tables, and chapter summaries are provided in JPEGs and editable Word format, and all of the new Data Speaks boxes are offered in JPEGs. In-class weekly multiple-choice questions for use with various Classroom Response Systems (CRS) are also provided, based on the Test Your Understanding questions and chapter-opening questions in the text. Written by Roger Freedman, many new CRS questions that increase in difficulty level have been added. Lecture outlines and PhET clicker questions, both in PowerPoint format, are also included along with about 70 PhET simulations and the Video Tutor Demonstrations (interactive video demonstrations) that are linked to QR codes throughout the textbook.

MasteringPhysics® (www.masteringphysics.com) from Pearson is the leading online teaching and learning system designed to improve results by engaging students before, during, and after class with powerful content. Ensure that students arrive ready to learn by assigning educationally effective content before class, and encourage critical thinking and retention with in-class resources such as Learning Catalytics. Students can further master concepts after class through traditional homework assignments that provide hints and answer-specific feedback. The Mastering gradebook records scores for all automatically graded assignments, while diagnostic tools give instructors access to rich data to assess student understanding and misconceptions.

Mastering brings learning full circle by continuously adapting to each student and making learning more personal than ever—before, during, and after class.
- **NEW!** The **Mastering Instructor Resources Area** contains all of the contents of the Instructor's Resource DVD—lecture outlines; Classroom Response System questions; images, tables, key equations, problem-solving strategies, Data Speaks boxes, and chapter summaries from the textbook; access to the Instructor's Solutions Manual, Test Bank, ActivPhysics Online—and much more.
- **NEW! Pre-lecture Videos** are assignable interactive videos that introduce students to key topics before they come to class. Each one includes assessment that feeds to the gradebook and alerts the instructor to potential trouble spots for students.
- **Pre-lecture Concept Questions** check students' familiarity with key concepts, prompting students to do their assigned reading before they come to class. These quizzes keep students on track, keep them more engaged in lecture, and help you spot the concepts that students find the most difficult.
- **NEW! Learning Catalytics** is a "bring your own device" student engagement, assessment, and classroom intelligence system that allows you to assess students in real time, understand immediately where they are and adjust your lecture accordingly, improve their critical-thinking skills, access rich analytics to understand student performance, add your own questions to fit your course exactly, and manage student interactions with intelligent grouping and timing. Learning Catalytics can be used both during and after class.
- **NEW! Adaptive Follow-Ups** allow Mastering to adapt continuously to each student, making learning more personal than ever. These assignments pair Mastering's powerful content with Knewton's adaptive learning engine to provide personalized help to students before misconceptions take hold. They are based on each student's performance on homework assignments and on all work in the course to date, including core prerequisite topics.

- **Video Tutor Demonstrations,** linked to QR codes in the textbook, feature "Pause and predict" videos of key physics concepts that ask students to submit a prediction before they see the outcome. These interactive videos are available in the Study Area of Mastering and in the Pearson eText.
- **Video Tutor Solutions** are linked to QR codes in the textbook. In these videos, which are available in the Study Area of Mastering and in the Pearson eText, an instructor explains and solves each worked example and Bridging Problem.
- **NEW!** An **Alternative Problem Set** in the Item Library of Mastering includes hundreds of new end-of-chapter questions and problems to offer instructors a wealth of options.
- **NEW! Physics/Biology Tutorials for MasteringPhysics** are assignable, multipart tutorials that emphasize biological processes and structures but also teach the physics principles that underlie them. They contain assessment questions that are based on the core competencies outlined in the 2015 MCAT.
- **PhET Simulations** (from the PhET project at the University of Colorado) are interactive, research-based simulations of physical phenomena. These tutorials, correlated to specific topics in the textbook, are available in the Pearson eText and in the Study Area within www.masteringphysics.com.
- **ActivPhysics OnLine**™ (which is accessed through the Study Area and Instructor Resources within www.masteringphysics.com) provides a comprehensive library of approximately 350 tried and tested ActivPhysics applets updated for web delivery.
- Mastering's **powerful gradebook** records all scores for automatically graded assignments. Struggling students and challenging assignments are highlighted in red, giving you an at-a-glance view of potential hurdles in the course. With a single click, charts summarize the most difficult problems, identify vulnerable students, and show the grade distribution, allowing for just-in-time teaching to address student misconceptions.
- **Learning Management System (LMS) Integration** gives seamless access to modified Mastering. Having all of your course materials and communications in one place makes life less complicated for you and your students. We've made it easier to link from within your LMS to modified Mastering and provide solutions, regardless of your LMS platform. With seamless, single sign-on your students will gain access to the personalized learning resources that make studying more efficient and more effective. You can access modified Mastering assignments, rosters, and resources and synchronize grades from modified Mastering with LMS.
- The **Test Bank** contains more than 2000 high-quality problems, with a range of multiple-choice, true/false, short-answer, and regular homework-type questions. Test files are provided both in TestGen (an easy-to-use, fully networkable program for creating and editing quizzes and exams) and in Word format. Download only from the MasteringPhysics Instructor Resources Area or from the Instructor Resources Center (www.pearsonhighered.com/irc).

MasteringPhysics enables instructors to:
- Quickly build homework assignments that combine regular end-of-chapter problems and tutoring (through additional multistep tutorial problems that offer wrong-answer feedback and simpler problems upon request).
- Expand homework to include the widest range of automatically graded activities available—from numerical problems with randomized values, through algebraic answers, to free-hand drawing.
- Choose from a wide range of nationally pre-tested problems that provide accurate estimates of time to complete and difficulty.
- After an assignment is completed, quickly identify not only the problems that were the trickiest for students but also the individual problem types with which students had trouble.

- Compare class results against the system's worldwide average for each problem assigned, to identify issues to be addressed with just-in-time teaching.
- Check the work of an individual student in detail, including the time spent on each problem, what wrong answers were submitted at each step, how much help was asked for, and how many practice problems were worked.

STUDENT'S SUPPLEMENTS

The **Student's Study Guide** by Laird Kramer reinforces the textbook's emphasis on problem-solving strategies and student misconceptions. The *Study Guide for Volume 1* (978-0-13-398361-6) covers Chapters 1–20, and the *Study Guide for Volumes 2 and 3* (978-0-13-398360-9) covers Chapters 21–44.

The **Student's Solutions Manual** by A. Lewis Ford (Texas A&M University) and Wayne Anderson contains detailed, step-by-step solutions to more than half of the odd-numbered end-of-chapter problems from the textbook. All solutions follow consistently the same Identify/Set Up/Execute/Evaluate problem-solving framework used in the textbook. The *Student's Solutions Manual for Volume 1* (978-0-13-398171-1) covers Chapters 1–20, and the *Student's Solutions Manual for Volumes 2 and 3* (978-0-13-396928-3) covers Chapters 21–44.

MasteringPhysics® (www.masteringphysics.com) is a homework, tutorial, and assessment system based on years of research into how students work physics problems and precisely where they need help. Studies show that students who use MasteringPhysics compared to handwritten homework significantly increase their scores. MasteringPhysics achieves this improvement by providing students with instantaneous feedback specific to their wrong answers, simpler sub-problems upon request when they get stuck, and partial credit for their method(s). This individualized, 24/7 Socratic tutoring is recommended by nine out of ten students to their peers as the most effective and time-efficient way to study.

Pearson eText is available through MasteringPhysics either automatically, when MasteringPhysics is packaged with new books, or as a purchased upgrade online. Allowing students access to the text wherever they have access to the Internet, Pearson eText comprises the full text, including figures that can be enlarged for better viewing. With eText, students are also able to pop up definitions and terms to help with vocabulary and the reading of the material. Students can also take notes in eText by using the annotation feature at the top of each page.

Pearson Tutor Services (www.pearsontutorservices.com). Each student's subscription to MasteringPhysics also contains complimentary access to Pearson Tutor Services, powered by Smarthinking, Inc. By logging in with their MasteringPhysics ID and password, students are connected to highly qualified e-instructors who provide additional interactive online tutoring on the major concepts of physics. Some restrictions apply; the offer is subject to change.

TIPERs (Tasks Inspired by Physics Education Research) are workbooks that give students the practice they need to develop reasoning about physics and that promote a conceptual understanding of problem solving:
- **NEW! TIPERs: Sensemaking Tasks for Introductory Physics** (978-0-13-285458-0) by Curtis Hieggelke, Stephen Kanim, David Maloney, and Thomas O'Kuma
- **Newtonian Tasks Inspired by Physics Education Research: nTIPERs** (978-0-321-75375-5) by Curtis Hieggelke, David Maloney, and Stephen Kanim
- **E&M TIPERs: Electricity & Magnetism Tasks** (978-0-13-185499-4) by Curtis Hieggelke, David Maloney, Thomas O'Kuma, and Stephen Kanim

Tutorials in Introductory Physics (978-0-13-097069-5) by Lillian C. McDermott and Peter S. Schaffer presents a series of physics tutorials designed by a leading physics education research group. Emphasizing the development of concepts and scientific reasoning skills, the tutorials focus on the specific conceptual and reasoning difficulties that students tend to encounter.

ACKNOWLEDGMENTS

I would like to thank the hundreds of reviewers and colleagues who have offered valuable comments and suggestions over the life of this textbook. The continuing success of *University Physics* is due in large measure to their contributions.

Miah Adel (U. of Arkansas at Pine Bluff), Edward Adelson (Ohio State U.), Julie Alexander (Camosun C.), Ralph Alexander (U. of Missouri at Rolla), J. G. Anderson, R. S. Anderson, Wayne Anderson (Sacramento City C.), Sanjeev Arora (Fort Valley State U.), Alex Azima (Lansing Comm. C.), Dilip Balamore (Nassau Comm. C.), Harold Bale (U. of North Dakota), Arun Bansil (Northeastern U.), John Barach (Vanderbilt U.), J. D. Barnett, H. H. Barschall, Albert Bartlett (U. of Colorado), Marshall Bartlett (Hollins U.), Paul Baum (CUNY, Queens C.), Frederick Becchetti (U. of Michigan), B. Bederson, David Bennum (U. of Nevada, Reno), Lev I. Berger (San Diego State U.), Angela Biselli (Fairfield U.), Robert Boeke (William Rainey Harper C.), Bram Boroson (Clayton State U.), S. Borowitz, A. C. Braden, James Brooks (Boston U.), Nicholas E. Brown (California Polytechnic State U., San Luis Obispo), Tony Buffa (California Polytechnic State U., San Luis Obispo), Shane Burns (Colorado C.), A. Capecelatro, Michael Cardamone (Pennsylvania State U.), Duane Carmony (Purdue U.), Troy Carter (UCLA), P. Catranides, John Cerne (SUNY at Buffalo), Shinil Cho (La Roche C.), Tim Chupp (U. of Michigan), Roger Clapp (U. of South Florida), William M. Cloud (Eastern Illinois U.), Leonard Cohen (Drexel U.), W. R. Coker (U. of Texas, Austin), Malcolm D. Cole (U. of Missouri at Rolla), H. Conrad, David Cook (Lawrence U.), Gayl Cook (U. of Colorado), Hans Courant (U. of Minnesota), Carl Covatto (Arizona State U.), Bruce A. Craver (U. of Dayton), Larry Curtis (U. of Toledo), Jai Dahiya (Southeast Missouri State U.), Dedra Demaree (Georgetown U.), Steve Detweiler (U. of Florida), George Dixon (Oklahoma State U.), Steve Drasco (Grinnell C.), Donald S. Duncan, Boyd Edwards (West Virginia U.), Robert Eisenstein (Carnegie Mellon U.), Amy Emerson Missourn (Virginia Institute of Technology), Olena Erhardt (Richland C.), William Faissler (Northeastern U.), Gregory Falabella (Wagner C.), William Fasnacht (U.S. Naval Academy), Paul Feldker (St. Louis Comm. C.), Carlos Figueroa (Cabrillo C.), L. H. Fisher, Neil Fletcher (Florida State U.), Allen Flora (Hood C.), Robert Folk, Peter Fong (Emory U.), A. Lewis Ford (Texas A&M U.), D. Frantszog, James R. Gaines (Ohio State U.), Solomon Gartenhaus (Purdue U.), Ron Gautreau (New Jersey Institute of Technology), J. David Gavenda (U. of Texas, Austin), Dennis Gay (U. of North Florida), Elizabeth George (Wittenberg U.), James Gerhart (U. of Washington), N. S. Gingrich, J. L. Glathart, S. Goodwin, Rich Gottfried (Frederick Comm. C.), Walter S. Gray (U. of Michigan), Paul Gresser (U. of Maryland), Benjamin Grinstein (UC, San Diego), Howard Grotch (Pennsylvania State U.), John Gruber (San Jose State U.), Graham D. Gutsche (U.S. Naval Academy), Michael J. Harrison (Michigan State U.), Harold Hart (Western Illinois U.), Howard Hayden (U. of Connecticut), Carl Helrich (Goshen C.), Andrew Hirsch (Purdue U.), Linda Hirst (UC, Merced), Laurent Hodges (Iowa State U.), C. D. Hodgman, Elizabeth Holden (U. of Wisconsin, Platteville), Michael Hones (Villanova U.), Keith Honey (West Virginia Institute of Technology), Gregory Hood (Tidewater Comm. C.), John Hubisz (North Carolina State U.), Eric Hudson (Pennsylvania State U.), M. Iona, Bob Jacobsen (UC, Berkeley), John Jaszczak (Michigan Technical U.), Alvin Jenkins (North Carolina State U.), Charles Johnson (South Georgia State C.), Robert P. Johnson (UC, Santa Cruz), Lorella Jones (U. of Illinois), Manoj Kaplinghat (UC, Irvine), John Karchek (GMI Engineering & Management Institute), Thomas Keil (Worcester Polytechnic Institute), Robert Kraemer (Carnegie Mellon U.), Jean P. Krisch (U. of Michigan), Robert A. Kromhout, Andrew Kunz (Marquette U.), Charles Lane (Berry C.), Stewart Langton (U. of Victoria), Thomas N. Lawrence (Texas State U.), Robert J. Lee, Alfred Leitner (Rensselaer Polytechnic U.), Frederic Liebrand (Walla Walla U.), Gerald P. Lietz (DePaul U.), Gordon Lind (Utah State U.), S. Livingston (U. of Wisconsin, Milwaukee), Jorge Lopez (U. of Texas, El Paso), Elihu Lubkin (U. of Wisconsin, Milwaukee), Robert Luke (Boise State U.), David Lynch (Iowa State U.), Michael Lysak (San Bernardino Valley C.), Jeffrey Mallow (Loyola U.), Robert Mania (Kentucky State U.), Robert Marchina (U. of Memphis), David Markowitz (U. of Connecticut), Philip Matheson (Utah Valley U.), R. J. Maurer, Oren Maxwell (Florida International U.), Joseph L. McCauley (U. of Houston), T. K. McCubbin, Jr. (Pennsylvania State U.), Charles McFarland (U. of Missouri at Rolla), James Mcguire (Tulane U.), Lawrence McIntyre (U. of Arizona), Fredric Messing (Carnegie Mellon U.), Thomas Meyer (Texas A&M U.), Andre Mirabelli (St. Peter's C., New Jersey), Herbert Muether

(SUNY, Stony Brook), Jack Munsee (California State U., Long Beach), Lorenzo Narducci (Drexel U.), Van E. Neie (Purdue U.), Forrest Newman (Sacramento City C.), David A. Nordling (U.S. Naval Academy), Benedict Oh (Pennsylvania State U.), L. O. Olsen, Michael Ottinger (Missouri Western State U.), Russell Palma (Minnesota State U., Mankato), Jim Pannell (DeVry Institute of Technology), Neeti Parashar (Purdue U., Calumet), W. F. Parks (U. of Missouri), Robert Paulson (California State U., Chico), Jerry Peacher (U. of Missouri at Rolla), Arnold Perlmutter (U. of Miami), Lennart Peterson (U. of Florida), R. J. Peterson (U. of Colorado, Boulder), R. Pinkston, Ronald Poling (U. of Minnesota), Yuri Popov (U. of Michigan), J. G. Potter, C. W. Price (Millersville U.), Francis Prosser (U. of Kansas), Shelden H. Radin, Roberto Ramos (Drexel U.), Michael Rapport (Anne Arundel Comm. C.), R. Resnick, James A. Richards, Jr., John S. Risley (North Carolina State U.), Francesc Roig (UC, Santa Barbara), T. L. Rokoske, Richard Roth (Eastern Michigan U.), Carl Rotter (U. of West Virginia), S. Clark Rowland (Andrews U.), Rajarshi Roy (Georgia Institute of Technology), Russell A. Roy (Santa Fe Comm. C.), Desi Saludes (Hillsborough Comm. C.), Thomas Sandin (North Carolina A&T State U.), Dhiraj Sardar (U. of Texas, San Antonio), Tumer Sayman (Eastern Michigan U.), Bruce Schumm (UC, Santa Cruz), Melvin Schwartz (St. John's U.), F. A. Scott, L. W. Seagondollar, Paul Shand (U. of Northern Iowa), Stan Shepherd (Pennsylvania State U.), Douglas Sherman (San Jose State U.), Bruce Sherwood (Carnegie Mellon U.), Hugh Siefkin (Greenville C.), Christopher Sirola (U. of Southern Mississippi), Tomasz Skwarnicki (Syracuse U.), C. P. Slichter, Jason Slinker (U. of Texas, Dallas), Charles W. Smith (U. of Maine, Orono), Malcolm Smith (U. of Lowell), Ross Spencer (Brigham Young U.), Julien Sprott (U. of Wisconsin), Victor Stanionis (Iona C.), James Stith (American Institute of Physics), Chuck Stone (North Carolina A&T State U.), Edward Strother (Florida Institute of Technology), Conley Stutz (Bradley U.), Albert Stwertka (U.S. Merchant Marine Academy), Kenneth Szpara-DeNisco (Harrisburg Area Comm. C.), Devki Talwar (Indiana U. of Pennsylvania), Fiorella Terenzi (Florida International U.), Martin Tiersten (CUNY, City C.), David Toot (Alfred U.), Greg Trayling (Rochester Institute of Technology), Somdev Tyagi (Drexel U.), Matthew Vannette (Saginaw Valley State U.), Eswara Venugopal (U. of Detroit, Mercy), F. Verbrugge, Helmut Vogel (Carnegie Mellon U.), Aaron Warren (Purdue U., North Central), Robert Webb (Texas A&M U.), Thomas Weber (Iowa State U.), M. Russell Wehr (Pennsylvania State U.), Robert Weidman (Michigan Technical U.), Dan Whalen (UC, San Diego), Lester V. Whitney, Thomas Wiggins (Pennsylvania State U.), Robyn Wilde (Oregon Institute of Technology), David Willey (U. of Pittsburgh, Johnstown), George Williams (U. of Utah), John Williams (Auburn U.), Stanley Williams (Iowa State U.), Jack Willis, Suzanne Willis (Northern Illinois U.), Robert Wilson (San Bernardino Valley C.), L. Wolfenstein, James Wood (Palm Beach Junior C.), Lowell Wood (U. of Houston), R. E. Worley, D. H. Ziebell (Manatee Comm. C.), George O. Zimmerman (Boston U.)

In addition, I would like to thank my past and present colleagues at UCSB, including Rob Geller, Carl Gwinn, Al Nash, Elisabeth Nicol, and Francesc Roig, for their wholehearted support and for many helpful discussions. I owe a special debt of gratitude to my early teachers Willa Ramsay, Peter Zimmerman, William Little, Alan Schwettman, and Dirk Walecka for showing me what clear and engaging physics teaching is all about, and to Stuart Johnson for inviting me to become a coauthor of *University Physics* beginning with the Ninth Edition. Special acknowledgments go out to Lewis Ford for creating a wealth of new problems for this edition, including the new category of DATA problems; to Wayne Anderson, who carefully reviewed all of the problems and solved them, along with Forrest Newman and Michael Ottinger; and to Elizabeth George, who provided most of the new category of Passage Problems. A particular tip of the hat goes to Tom Sandin for his numerous contributions to the end-of-chapter problems, including carefully checking all problems and writing new ones. Hats off as well and a tremendous reception to Linda Hirst for contributing a number of ideas that became new Application features in this edition. I want to express special thanks to the editorial staff at Pearson: to Nancy Whilton for her editorial vision; to Karen Karlin for her keen eye and careful development of this edition; to Charles Hibbard for his careful reading of the page proofs; and to Beth Collins, Katie Conley, Sarah Kaubisch, Eric Schrader, and Cindy Johnson for keeping the editorial and production pipelines flowing. Most of all, I want to express my gratitude and love to my wife, Caroline, to whom I dedicate my contribution to this book. Hey, Caroline, the new edition's done at last—let's go flying!

PLEASE TELL ME WHAT YOU THINK!

I welcome communications from students and professors, especially concerning errors or deficiencies that you find in this edition. The late Hugh Young and I have devoted a lot of time and effort to writing the best book we know how to write, and I hope it will help as you teach and learn physics. In turn, you can help me by letting me know what still needs to be improved! Please feel free to contact me either electronically or by ordinary mail. Your comments will be greatly appreciated.

August 2014

Roger A. Freedman
Department of Physics
University of California, Santa Barbara
Santa Barbara, CA 93106-9530
airboy@physics.ucsb.edu
http://www.physics.ucsb.edu/~airboy/
Twitter: @RogerFreedman

DETAILED CONTENTS

MECHANICS

THERMODYNAMICS

ELECTROMAGNETISM

T-bacteriophage viruses

100 nm = 0.1 μm

Viral
DNA

? Tornadoes are spawned by severe thunderstorms, so being able to predict the path of thunderstorms is essential. If a thunderstorm is moving at 15 km/h in a direction 37° north of east, how far north does the thunderstorm move in 2.0 h? (i) 30 km; (ii) 24 km; (iii) 18 km; (iv) 12 km; (v) 9 km.

1 UNITS, PHYSICAL QUANTITIES, AND VECTORS

P hysics is one of the most fundamental of the sciences. Scientists of all disciplines use the ideas of physics, including chemists who study the structure of molecules, paleontologists who try to reconstruct how dinosaurs walked, and climatologists who study how human activities affect the atmosphere and oceans. Physics is also the foundation of all engineering and technology. No engineer could design a flat-screen TV, a prosthetic leg, or even a better mousetrap without first understanding the basic laws of physics.

The study of physics is also an adventure. You will find it challenging, sometimes frustrating, occasionally painful, and often richly rewarding. If you've ever wondered why the sky is blue, how radio waves can travel through empty space, or how a satellite stays in orbit, you can find the answers by using fundamental physics. You will come to see physics as a towering achievement of the human intellect in its quest to understand our world and ourselves.

In this opening chapter, we'll go over some important preliminaries that we'll need throughout our study. We'll discuss the nature of physical theory and the use of idealized models to represent physical systems. We'll introduce the systems of units used to describe physical quantities and discuss ways to describe the accuracy of a number. We'll look at examples of problems for which we can't (or don't want to) find a precise answer, but for which rough estimates can be useful and interesting. Finally, we'll study several aspects of vectors and vector algebra. We'll need vectors throughout our study of physics to help us describe and analyze physical quantities, such as velocity and force, that have direction as well as magnitude.

1.1 THE NATURE OF PHYSICS

Physics is an *experimental* science. Physicists observe the phenomena of nature and try to find patterns that relate these phenomena. These patterns are called physical theories or, when they are very well established and widely used, physical laws or principles.

CAUTION   **The meaning of "theory"**   A theory is *not* just a random thought or an unproven concept. Rather, a theory is an explanation of natural phenomena based on observation and accepted fundamental principles. An example is the well-established theory of biological evolution, which is the result of extensive research and observation by generations of biologists.

1.1 Two research laboratories.

(a) According to legend, Galileo investigated falling objects by dropping them from the Leaning Tower of Pisa, Italy, ...

... and he studied pendulum motion by observing the swinging chandelier in the adjacent cathedral.

(b) The Planck spacecraft is designed to study the faint electromagnetic radiation left over from the Big Bang 13.8 billion years ago.

These technicians are reflected in the spacecraft's light-gathering mirror during pre-launch testing.

To develop a physical theory, a physicist has to learn to ask appropriate questions, design experiments to try to answer the questions, and draw appropriate conclusions from the results. **Figure 1.1** shows two important facilities used for physics experiments.

Legend has it that Galileo Galilei (1564–1642) dropped light and heavy objects from the top of the Leaning Tower of Pisa (Fig. 1.1a) to find out whether their rates of fall were different. From examining the results of his experiments (which were actually much more sophisticated than in the legend), he made the inductive leap to the principle, or theory, that the acceleration of a falling object is independent of its weight.

The development of physical theories such as Galileo's often takes an indirect path, with blind alleys, wrong guesses, and the discarding of unsuccessful theories in favor of more promising ones. Physics is not simply a collection of facts and principles; it is also the *process* by which we arrive at general principles that describe how the physical universe behaves.

No theory is ever regarded as the final or ultimate truth. The possibility always exists that new observations will require that a theory be revised or discarded. It is in the nature of physical theory that we can disprove a theory by finding behavior that is inconsistent with it, but we can never prove that a theory is always correct.

Getting back to Galileo, suppose we drop a feather and a cannonball. They certainly do *not* fall at the same rate. This does not mean that Galileo was wrong; it means that his theory was incomplete. If we drop the feather and the cannonball *in a vacuum* to eliminate the effects of the air, then they do fall at the same rate. Galileo's theory has a **range of validity:** It applies only to objects for which the force exerted by the air (due to air resistance and buoyancy) is much less than the weight. Objects like feathers or parachutes are clearly outside this range.

1.2 SOLVING PHYSICS PROBLEMS

At some point in their studies, almost all physics students find themselves thinking, "I understand the concepts, but I just can't solve the problems." But in physics, truly understanding a concept *means* being able to apply it to a variety of problems. Learning how to solve problems is absolutely essential; you don't *know* physics unless you can *do* physics.

How do you learn to solve physics problems? In every chapter of this book you will find *Problem-Solving Strategies* that offer techniques for setting up and solving problems efficiently and accurately. Following each *Problem-Solving Strategy* are one or more worked *Examples* that show these techniques in action. (The *Problem-Solving Strategies* will also steer you away from some *incorrect* techniques that you may be tempted to use.) You'll also find additional examples that aren't associated with a particular *Problem-Solving Strategy*. In addition, at the end of each chapter you'll find a *Bridging Problem* that uses more than one of the key ideas from the chapter. Study these strategies and problems carefully, and work through each example for yourself on a piece of paper.

Different techniques are useful for solving different kinds of physics problems, which is why this book offers dozens of *Problem-Solving Strategies.* No matter what kind of problem you're dealing with, however, there are certain key steps that you'll always follow. (These same steps are equally useful for problems in math, engineering, chemistry, and many other fields.) In this book we've organized these steps into four stages of solving a problem.

All of the *Problem-Solving Strategies* and *Examples* in this book will follow these four steps. (In some cases we will combine the first two or three steps.) We encourage you to follow these same steps when you solve problems yourself. You may find it useful to remember the acronym *I SEE*—short for *Identify, Set up, Execute,* and *Evaluate.*

PROBLEM-SOLVING STRATEGY 1.1 SOLVING PHYSICS PROBLEMS

IDENTIFY *the relevant concepts:* Use the physical conditions stated in the problem to help you decide which physics concepts are relevant. Identify the **target variables** of the problem—that is, the quantities whose values you're trying to find, such as the speed at which a projectile hits the ground, the intensity of a sound made by a siren, or the size of an image made by a lens. Identify the known quantities, as stated or implied in the problem. This step is essential whether the problem asks for an algebraic expression or a numerical answer.

SET UP *the problem:* Given the concepts you have identified, the known quantities, and the target variables, choose the equations that you'll use to solve the problem and decide how you'll use them. Make sure that the variables you have identified correlate exactly with those in the equations. If appropriate, draw a sketch of the situation described in the problem. (Graph paper, ruler, protractor, and compass will help you make clear, useful sketches.)

As best you can, estimate what your results will be and, as appropriate, predict what the physical behavior of a system will be. The worked examples in this book include tips on how to make these kinds of estimates and predictions. If this seems challenging, don't worry—you'll get better with practice!

EXECUTE *the solution:* This is where you "do the math." Study the worked examples to see what's involved in this step.

EVALUATE *your answer:* Compare your answer with your estimates, and reconsider things if there's a discrepancy. If your answer includes an algebraic expression, assure yourself that it correctly represents what would happen if the variables in it had very large or very small values. For future reference, make note of any answer that represents a quantity of particular significance. Ask yourself how you might answer a more general or more difficult version of the problem you have just solved.

Idealized Models

In everyday conversation we use the word "model" to mean either a small-scale replica, such as a model railroad, or a person who displays articles of clothing (or the absence thereof). In physics a **model** is a simplified version of a physical system that would be too complicated to analyze in full detail.

For example, suppose we want to analyze the motion of a thrown baseball (**Fig. 1.2a**). How complicated is this problem? The ball is not a perfect sphere (it has raised seams), and it spins as it moves through the air. Air resistance and wind influence its motion, the ball's weight varies a little as its altitude changes, and so on. If we try to include all these things, the analysis gets hopelessly complicated. Instead, we invent a simplified version of the problem. We ignore the size and shape of the ball by representing it as a point object, or **particle.** We ignore air resistance by making the ball move in a vacuum, and we make the weight constant. Now we have a problem that is simple enough to deal with (Fig. 1.2b). We will analyze this model in detail in Chapter 3.

We have to overlook quite a few minor effects to make an idealized model, but we must be careful not to neglect too much. If we ignore the effects of gravity completely, then our model predicts that when we throw the ball up, it will go in a straight line and disappear into space. A useful model simplifies a problem enough to make it manageable, yet keeps its essential features.

1.2 To simplify the analysis of (a) a baseball in flight, we use (b) an idealized model.

(a) A real baseball in flight

Baseball spins and has a complex shape.

Air resistance and wind exert forces on the ball.

Direction of motion

Gravitational force on ball depends on altitude.

(b) An idealized model of the baseball

Treat the baseball as a point object (particle).

No air resistance.

Direction of motion

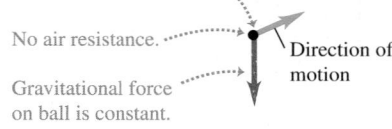

Gravitational force on ball is constant.

The validity of the predictions we make using a model is limited by the validity of the model. For example, Galileo's prediction about falling objects (see Section 1.1) corresponds to an idealized model that does not include the effects of air resistance. This model works fairly well for a dropped cannonball, but not so well for a feather.

Idealized models play a crucial role throughout this book. Watch for them in discussions of physical theories and their applications to specific problems.

1.3 STANDARDS AND UNITS

As we learned in Section 1.1, physics is an experimental science. Experiments require measurements, and we generally use numbers to describe the results of measurements. Any number that is used to describe a physical phenomenon quantitatively is called a **physical quantity.** For example, two physical quantities that describe you are your weight and your height. Some physical quantities are so fundamental that we can define them only by describing how to measure them. Such a definition is called an **operational definition.** Two examples are measuring a distance by using a ruler and measuring a time interval by using a stopwatch. In other cases we define a physical quantity by describing how to calculate it from other quantities that we *can* measure. Thus we might define the average speed of a moving object as the distance traveled (measured with a ruler) divided by the time of travel (measured with a stopwatch).

When we measure a quantity, we always compare it with some reference standard. When we say that a Ferrari 458 Italia is 4.53 meters long, we mean that it is 4.53 times as long as a meter stick, which we define to be 1 meter long. Such a standard defines a **unit** of the quantity. The meter is a unit of distance, and the second is a unit of time. When we use a number to describe a physical quantity, we must always specify the unit that we are using; to describe a distance as simply "4.53" wouldn't mean anything.

To make accurate, reliable measurements, we need units of measurement that do not change and that can be duplicated by observers in various locations. The system of units used by scientists and engineers around the world is commonly called "the metric system," but since 1960 it has been known officially as the **International System,** or **SI** (the abbreviation for its French name, *Système International*). Appendix A gives a list of all SI units as well as definitions of the most fundamental units.

1.3 The measurements used to determine **(a)** the duration of a second and **(b)** the length of a meter. These measurements are useful for setting standards because they give the same results no matter where they are made.

(a) Measuring the second

Microwave radiation with a frequency of exactly 9,192,631,770 cycles per second ...

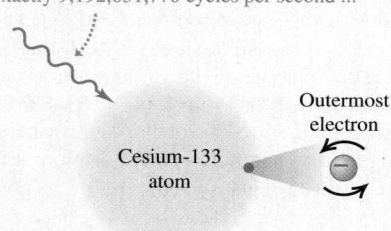

... causes the outermost electron of a cesium-133 atom to reverse its spin direction.

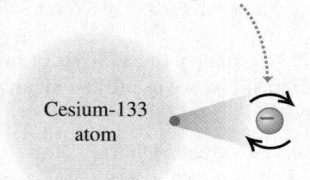

An atomic clock uses this phenomenon to tune microwaves to this exact frequency. It then counts 1 second for each 9,192,631,770 cycles.

(b) Measuring the meter

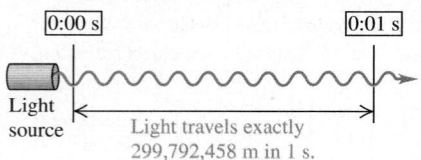

Time

From 1889 until 1967, the unit of time was defined as a certain fraction of the mean solar day, the average time between successive arrivals of the sun at its highest point in the sky. The present standard, adopted in 1967, is much more precise. It is based on an atomic clock, which uses the energy difference between the two lowest energy states of the cesium atom (^{133}Cs). When bombarded by microwaves of precisely the proper frequency, cesium atoms undergo a transition from one of these states to the other. One **second** (abbreviated s) is defined as the time required for 9,192,631,770 cycles of this microwave radiation (**Fig. 1.3a**).

Length

In 1960 an atomic standard for the meter was also established, using the wavelength of the orange-red light emitted by excited atoms of krypton (^{86}Kr). From this length standard, the speed of light in vacuum was measured to be 299,792,458 m/s. In November 1983, the length standard was changed again so that the speed of light in vacuum was *defined* to be precisely 299,792,458 m/s.

Hence the new definition of the **meter** (abbreviated m) is the distance that light travels in vacuum in 1/299,792,458 second (Fig. 1.3b). This modern definition provides a much more precise standard of length than the one based on a wavelength of light.

Mass

The standard of mass, the **kilogram** (abbreviated kg), is defined to be the mass of a particular cylinder of platinum–iridium alloy kept at the International Bureau of Weights and Measures at Sèvres, near Paris (**Fig. 1.4**). An atomic standard of mass would be more fundamental, but at present we cannot measure masses on an atomic scale with as much accuracy as on a macroscopic scale. The *gram* (which is not a fundamental unit) is 0.001 kilogram.

Other *derived units* can be formed from the fundamental units. For example, the units of speed are meters per second, or m/s; these are the units of length (m) divided by the units of time (s).

Unit Prefixes

Once we have defined the fundamental units, it is easy to introduce larger and smaller units for the same physical quantities. In the metric system these other units are related to the fundamental units (or, in the case of mass, to the gram) by multiples of 10 or $\frac{1}{10}$ Thus one kilometer (1 km) is 1000 meters, and one centimeter (1 cm) is $\frac{1}{100}$ meter. We usually express multiples of 10 or $\frac{1}{10}$ in exponential notation: $1000 = 10^3$, $\frac{1}{1000} = 10^{-3}$, and so on. With this notation, $1\text{ km} = 10^3\text{ m}$ and $1\text{ cm} = 10^{-2}\text{ m}$.

The names of the additional units are derived by adding a **prefix** to the name of the fundamental unit. For example, the prefix "kilo-," abbreviated k, always means a unit larger by a factor of 1000; thus

$$1\text{ kilometer} = 1\text{ km} = 10^3\text{ meters} = 10^3\text{ m}$$
$$1\text{ kilogram} = 1\text{ kg} = 10^3\text{ grams} = 10^3\text{ g}$$
$$1\text{ kilowatt} = 1\text{ kW} = 10^3\text{ watts} = 10^3\text{ W}$$

A table in Appendix A lists the standard SI units, with their meanings and abbreviations.

Table 1.1 gives some examples of the use of multiples of 10 and their prefixes with the units of length, mass, and time. **Figure 1.5** (next page) shows how these prefixes are used to describe both large and small distances.

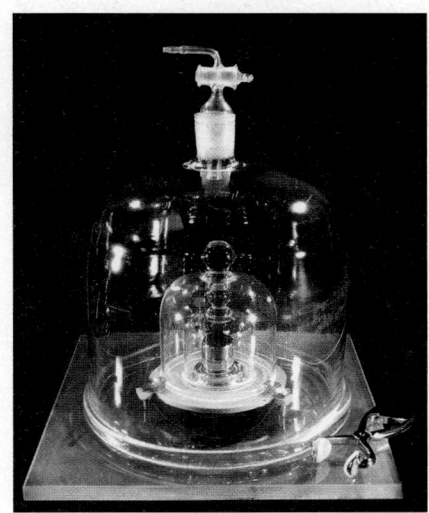

1.4 The international standard kilogram is the metal object carefully enclosed within these nested glass containers.

TABLE 1.1 Some Units of Length, Mass, and Time		
Length	**Mass**	**Time**
1 nanometer $= 1$ nm $= 10^{-9}$ m (a few times the size of the largest atom)	1 microgram $= 1\ \mu$g $= 10^{-6}$ g $= 10^{-9}$ kg (mass of a very small dust particle)	1 nanosecond $= 1$ ns $= 10^{-9}$ s (time for light to travel 0.3 m)
1 micrometer $= 1\ \mu$m $= 10^{-6}$ m (size of some bacteria and other cells)	1 milligram $= 1$ mg $= 10^{-3}$ g $= 10^{-6}$ kg (mass of a grain of salt)	1 microsecond $= 1\ \mu$s $= 10^{-6}$ s (time for space station to move 8 mm)
1 millimeter $= 1$ mm $= 10^{-3}$ m (diameter of the point of a ballpoint pen)	1 gram $= 1$ g $= 10^{-3}$ kg (mass of a paper clip)	1 millisecond $= 1$ ms $= 10^{-3}$ s (time for a car moving at freeway speed to travel 3 cm)
1 centimeter $= 1$ cm $= 10^{-2}$ m (diameter of your little finger)		
1 kilometer $= 1$ km $= 10^3$ m (distance in a 10-minute walk)		

1.5 Some typical lengths in the universe.

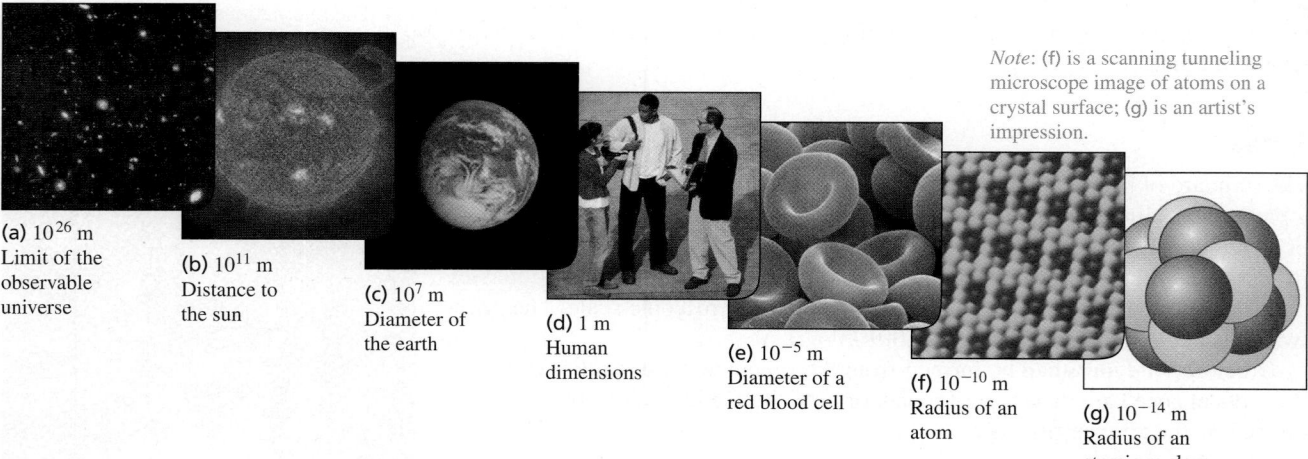

Note: (f) is a scanning tunneling microscope image of atoms on a crystal surface; (g) is an artist's impression.

(a) 10^{26} m
Limit of the observable universe

(b) 10^{11} m
Distance to the sun

(c) 10^7 m
Diameter of the earth

(d) 1 m
Human dimensions

(e) 10^{-5} m
Diameter of a red blood cell

(f) 10^{-10} m
Radius of an atom

(g) 10^{-14} m
Radius of an atomic nucleus

1.6 Many everyday items make use of both SI and British units. An example is this speedometer from a U.S.-built automobile, which shows the speed in both kilometers per hour (inner scale) and miles per hour (outer scale).

The British System

Finally, we mention the British system of units. These units are used in only the United States and a few other countries, and in most of these they are being replaced by SI units. British units are now officially defined in terms of SI units, as follows:

Length: 1 inch = 2.54 cm (exactly)

Force: 1 pound = 4.448221615260 newtons (exactly)

The newton, abbreviated N, is the SI unit of force. The British unit of time is the second, defined the same way as in SI. In physics, British units are used in mechanics and thermodynamics only; there is no British system of electrical units.

In this book we use SI units for all examples and problems, but we occasionally give approximate equivalents in British units. As you do problems using SI units, you may also wish to convert to the approximate British equivalents if they are more familiar to you (**Fig. 1.6**). But you should try to *think* in SI units as much as you can.

1.4 USING AND CONVERTING UNITS

We use equations to express relationships among physical quantities, represented by algebraic symbols. Each algebraic symbol always denotes both a number and a unit. For example, d might represent a distance of 10 m, t a time of 5 s, and v a speed of 2 m/s.

An equation must always be **dimensionally consistent.** You can't add apples and automobiles; two terms may be added or equated only if they have the same units. For example, if a body moving with constant speed v travels a distance d in a time t, these quantities are related by the equation

$$d = vt$$

If d is measured in meters, then the product vt must also be expressed in meters. Using the above numbers as an example, we may write

$$10 \text{ m} = \left(2\,\frac{\text{m}}{\cancel{\text{s}}}\right)(5\,\cancel{\text{s}})$$

Because the unit s in the denominator of m/s cancels, the product has units of meters, as it must. In calculations, units are treated just like algebraic symbols with respect to multiplication and division.

CAUTION **Always use units in calculations** Make it a habit to *always* write numbers with the correct units and carry the units through the calculation as in the example above. This provides a very useful check. If at some stage in a calculation you find that an equation or an expression has inconsistent units, you know you have made an error. In this book we will *always* carry units through all calculations, and we strongly urge you to follow this practice when you solve problems. ∎

PROBLEM-SOLVING STRATEGY 1.2) SOLVING PHYSICS PROBLEMS

IDENTIFY *the relevant concepts:* In most cases, it's best to use the fundamental SI units (lengths in meters, masses in kilograms, and times in seconds) in every problem. If you need the answer to be in a different set of units (such as kilometers, grams, or hours), wait until the end of the problem to make the conversion.

SET UP *the problem* and **EXECUTE** *the solution:* Units are multiplied and divided just like ordinary algebraic symbols. This gives us an easy way to convert a quantity from one set of units to another: Express the same physical quantity in two different units and form an equality.

For example, when we say that 1 min = 60 s, we don't mean that the number 1 is equal to the number 60; rather, we mean that 1 min represents the same physical time interval as 60 s. For this reason, the ratio (1 min)/(60 s) equals 1, as does its reciprocal, (60 s)/(1 min). We may multiply a quantity by either of these

factors (which we call *unit multipliers*) without changing that quantity's physical meaning. For example, to find the number of seconds in 3 min, we write

$$3 \text{ min} = (3 \text{ min})\left(\frac{60 \text{ s}}{1 \text{ min}}\right) = 180 \text{ s}$$

EVALUATE *your answer:* If you do your unit conversions correctly, unwanted units will cancel, as in the example above. If, instead, you had multiplied 3 min by (1 min)/(60 s), your result would have been the nonsensical $\frac{1}{20}$ min²/s. To be sure you convert units properly, include the units at *all* stages of the calculation.

Finally, check whether your answer is reasonable. For example, the result 3 min = 180 s is reasonable because the second is a smaller unit than the minute, so there are more seconds than minutes in the same time interval.

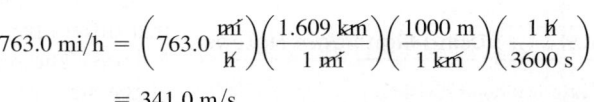

EXAMPLE 1.1   CONVERTING SPEED UNITS

The world land speed record of 763.0 mi/h was set on October 15, 1997, by Andy Green in the jet-engine car *Thrust SSC*. Express this speed in meters per second.

$$763.0 \text{ mi/h} = \left(763.0 \, \frac{\text{mi}}{\text{h}}\right)\left(\frac{1.609 \text{ km}}{1 \text{ mi}}\right)\left(\frac{1000 \text{ m}}{1 \text{ km}}\right)\left(\frac{1 \text{ h}}{3600 \text{ s}}\right)$$

$$= 341.0 \text{ m/s}$$

SOLUTION

IDENTIFY, SET UP, and EXECUTE: We need to convert the units of a speed from mi/h to m/s. We must therefore find unit multipliers that relate (i) miles to meters and (ii) hours to seconds. In Appendix E we find the equalities 1 mi = 1.609 km, 1 km = 1000 m, and 1 h = 3600 s. We set up the conversion as follows, which ensures that all the desired cancellations by division take place:

EVALUATE: This example shows a useful rule of thumb: A speed expressed in m/s is a bit less than half the value expressed in mi/h, and a bit less than one-third the value expressed in km/h. For example, a normal freeway speed is about 30 m/s = 67 mi/h = 108 km/h, and a typical walking speed is about 1.4 m/s = 3.1 mi/h = 5.0 km/h.

EXAMPLE 1.2   CONVERTING VOLUME UNITS

One of the world's largest cut diamonds is the First Star of Africa (mounted in the British Royal Sceptre and kept in the Tower of London). Its volume is 1.84 cubic inches. What is its volume in cubic centimeters? In cubic meters?

SOLUTION

IDENTIFY, SET UP, and EXECUTE: Here we are to convert the units of a volume from cubic inches (in.³) to both cubic centimeters (cm³) and cubic meters (m³). Appendix E gives us the equality 1 in. = 2.540 cm, from which we obtain 1 in.³ = (2.54 cm)³. We then have

$$1.84 \text{ in.}^3 = (1.84 \text{ in.}^3)\left(\frac{2.54 \text{ cm}}{1 \text{ in.}}\right)^3$$

$$= (1.84)(2.54)^3 \, \frac{\text{in.}^3 \, \text{cm}^3}{\text{in.}^3} = 30.2 \text{ cm}^3$$

Appendix E also gives us 1 m = 100 cm, so

$$30.2 \text{ cm}^3 = (30.2 \text{ cm}^3)\left(\frac{1 \text{ m}}{100 \text{ cm}}\right)^3$$

$$= (30.2)\left(\frac{1}{100}\right)^3 \frac{\text{cm}^3 \, \text{m}^3}{\text{cm}^3} = 30.2 \times 10^{-6} \text{ m}^3$$

$$= 3.02 \times 10^{-5} \text{ m}^3$$

EVALUATE: Following the pattern of these conversions, can you show that 1 in.³ ≈ 16 cm³ and that 1 m³ ≈ 60,000 in.³?

1.7 This spectacular mishap was the result of a very small percent error—traveling a few meters too far at the end of a journey of hundreds of thousands of meters.

TABLE 1.2 Using Significant Figures

Multiplication or division:
Result can have no more significant figures than the factor with the fewest significant figures:

$$\frac{0.745 \times 2.2}{3.885} = 0.42$$

$$1.32578 \times 10^7 \times 4.11 \times 10^{-3} = 5.45 \times 10^4$$

Addition or subtraction:
Number of significant figures is determined by the term with the largest uncertainty (i.e., fewest digits to the right of the decimal point):

$$27.153 + 138.2 - 11.74 = 153.6$$

1.8 Determining the value of π from the circumference and diameter of a circle.

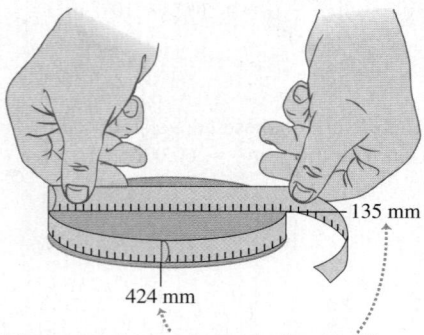

135 mm

424 mm

The measured values have only three significant figures, so their calculated ratio (π) also has only three significant figures.

1.5 UNCERTAINTY AND SIGNIFICANT FIGURES

Measurements always have uncertainties. If you measure the thickness of the cover of a hardbound version of this book using an ordinary ruler, your measurement is reliable to only the nearest millimeter, and your result will be 3 mm. It would be *wrong* to state this result as 3.00 mm; given the limitations of the measuring device, you can't tell whether the actual thickness is 3.00 mm, 2.85 mm, or 3.11 mm. But if you use a micrometer caliper, a device that measures distances reliably to the nearest 0.01 mm, the result will be 2.91 mm. The distinction between the measurements with a ruler and with a caliper is in their **uncertainty;** the measurement with a caliper has a smaller uncertainty. The uncertainty is also called the **error** because it indicates the maximum difference there is likely to be between the measured value and the true value. The uncertainty or error of a measured value depends on the measurement technique used.

We often indicate the **accuracy** of a measured value—that is, how close it is likely to be to the true value—by writing the number, the symbol $\pm$, and a second number indicating the uncertainty of the measurement. If the diameter of a steel rod is given as 56.47 $\pm$ 0.02 mm, this means that the true value is likely to be within the range from 56.45 mm to 56.49 mm. In a commonly used shorthand notation, the number 1.6454(21) means 1.6454 $\pm$ 0.0021. The numbers in parentheses show the uncertainty in the final digits of the main number.

We can also express accuracy in terms of the maximum likely **fractional error** or **percent error** (also called *fractional uncertainty* and *percent uncertainty*). A resistor labeled "47 ohms $\pm$ 10%" probably has a true resistance that differs from 47 ohms by no more than 10% of 47 ohms—that is, by about 5 ohms. The resistance is probably between 42 and 52 ohms. For the diameter of the steel rod given above, the fractional error is $(0.02 \text{ mm})/(56.47 \text{ mm})$, or about 0.0004; the percent error is $(0.0004)(100\%)$, or about 0.04%. Even small percent errors can be very significant (**Fig. 1.7**).

In many cases the uncertainty of a number is not stated explicitly. Instead, the uncertainty is indicated by the number of meaningful digits, or **significant figures,** in the measured value. We gave the thickness of the cover of the book as 2.91 mm, which has three significant figures. By this we mean that the first two digits are known to be correct, while the third digit is uncertain. The last digit is in the hundredths place, so the uncertainty is about 0.01 mm. Two values with the *same* number of significant figures may have *different* uncertainties; a distance given as 137 km also has three significant figures, but the uncertainty is about 1 km. A distance given as 0.25 km has two significant figures (the zero to the left of the decimal point doesn't count); if given as 0.250 km, it has three significant figures.

When you use numbers that have uncertainties to compute other numbers, the computed numbers are also uncertain. When numbers are multiplied or divided, the result can have no more significant figures than the factor with the fewest significant figures has. For example, $3.1416 \times 2.34 \times 0.58 = 4.3$. When we add and subtract numbers, it's the location of the decimal point that matters, not the number of significant figures. For example, $123.62 + 8.9 = 132.5$. Although 123.62 has an uncertainty of about 0.01, 8.9 has an uncertainty of about 0.1. So their sum has an uncertainty of about 0.1 and should be written as 132.5, not 132.52. **Table 1.2** summarizes these rules for significant figures.

To apply these ideas, suppose you want to verify the value of π, the ratio of the circumference of a circle to its diameter. The true value of this ratio to ten digits is 3.141592654. To test this, you draw a large circle and measure its circumference and diameter to the nearest millimeter, obtaining the values 424 mm and 135 mm (**Fig. 1.8**). You punch these into your calculator and obtain the quotient $(424 \text{ mm})/(135 \text{ mm}) = 3.140740741$. This may seem to disagree with the true value of π, but keep in mind that each of your measurements has three significant figures, so your measured value of π can have only three significant figures. It should be stated simply as 3.14. Within the limit of three significant figures, your value does agree with the true value.

In the examples and problems in this book we usually give numerical values with three significant figures, so your answers should usually have no more than three significant figures. (Many numbers in the real world have even less accuracy. An automobile speedometer, for example, usually gives only two significant figures.) Even if you do the arithmetic with a calculator that displays ten digits, a ten-digit answer would misrepresent the accuracy of the results. Always round your final answer to keep only the correct number of significant figures or, in doubtful cases, one more at most. In Example 1.1 it would have been wrong to state the answer as 341.01861 m/s. Note that when you reduce such an answer to the appropriate number of significant figures, you must *round,* not *truncate.* Your calculator will tell you that the ratio of 525 m to 311 m is 1.688102894; to three significant figures, this is 1.69, not 1.68.

When we work with very large or very small numbers, we can show significant figures much more easily by using **scientific notation,** sometimes called **powers-of-10 notation.** The distance from the earth to the moon is about 384,000,000 m, but writing the number in this form doesn't indicate the number of significant figures. Instead, we move the decimal point eight places to the left (corresponding to dividing by 10^8) and multiply by 10^8; that is,

$$384{,}000{,}000 \text{ m} = 3.84 \times 10^8 \text{ m}$$

In this form, it is clear that we have three significant figures. The number 4.00×10^{-7} also has three significant figures, even though two of them are zeros. Note that in scientific notation the usual practice is to express the quantity as a number between 1 and 10 multiplied by the appropriate power of 10.

When an integer or a fraction occurs in an algebraic equation, we treat that number as having no uncertainty at all. For example, in the equation $v_x^2 = v_{0x}^2 + 2a_x(x - x_0)$, which is Eq. (2.13) in Chapter 2, the coefficient 2 is *exactly* 2. We can consider this coefficient as having an infinite number of significant figures (2.000000 . . .). The same is true of the exponent 2 in v_x^2 and v_{0x}^2.

Finally, let's note that **precision** is not the same as *accuracy.* A cheap digital watch that gives the time as 10:35:17 A.M. is very *precise* (the time is given to the second), but if the watch runs several minutes slow, then this value isn't very *accurate.* On the other hand, a grandfather clock might be very accurate (that is, display the correct time), but if the clock has no second hand, it isn't very precise. A high-quality measurement is both precise *and* accurate.

EXAMPLE 1.3 **SIGNIFICANT FIGURES IN MULTIPLICATION**

The rest energy E of an object with rest mass m is given by Albert Einstein's famous equation $E = mc^2$, where c is the speed of light in vacuum. Find E for an electron for which (to three significant figures) $m = 9.11 \times 10^{-31}$ kg. The SI unit for E is the joule (J); $1 \text{ J} = 1 \text{ kg} \cdot \text{m}^2/\text{s}^2$.

SOLUTION

IDENTIFY and SET UP: Our target variable is the energy E. We are given the value of the mass m; from Section 1.3 (or Appendix F) the speed of light is $c = 2.99792458 \times 10^8$ m/s.

EXECUTE: Substituting the values of m and c into Einstein's equation, we find

$$E = (9.11 \times 10^{-31} \text{ kg})(2.99792458 \times 10^8 \text{ m/s})^2$$
$$= (9.11)(2.99792458)^2(10^{-31})(10^8)^2 \text{ kg} \cdot \text{m}^2/\text{s}^2$$
$$= (81.87659678)(10^{[-31+(2\times8)]}) \text{ kg} \cdot \text{m}^2/\text{s}^2$$
$$= 8.187659678 \times 10^{-14} \text{ kg} \cdot \text{m}^2/\text{s}^2$$

Since the value of m was given to only three significant figures, we must round this to

$$E = 8.19 \times 10^{-14} \text{ kg} \cdot \text{m}^2/\text{s}^2 = 8.19 \times 10^{-14} \text{ J}$$

EVALUATE: While the rest energy contained in an electron may seem ridiculously small, on the atomic scale it is tremendous. Compare our answer to 10^{-19} J, the energy gained or lost by a single atom during a typical chemical reaction. The rest energy of an electron is about 1,000,000 times larger! (We'll discuss the significance of rest energy in Chapter 37.)

PhET: Estimation

1.6 ESTIMATES AND ORDERS OF MAGNITUDE

We have stressed the importance of knowing the accuracy of numbers that represent physical quantities. But even a very crude estimate of a quantity often gives us useful information. Sometimes we know how to calculate a certain quantity, but we have to guess at the data we need for the calculation. Or the calculation might be too complicated to carry out exactly, so we make rough approximations. In either case our result is also a guess, but such a guess can be useful even if it is uncertain by a factor of two, ten, or more. Such calculations are called **order-of-magnitude estimates.** The great Italian-American nuclear physicist Enrico Fermi (1901–1954) called them "back-of-the-envelope calculations."

Exercises 1.17 through 1.23 at the end of this chapter are of the estimating, or order-of-magnitude, variety. Most require guesswork for the needed input data. Don't try to look up a lot of data; make the best guesses you can. Even when they are off by a factor of ten, the results can be useful and interesting.

EXAMPLE 1.4   AN ORDER-OF-MAGNITUDE ESTIMATE

You are writing an adventure novel in which the hero escapes with a billion dollars' worth of gold in his suitcase. Could anyone carry that much gold? Would it fit in a suitcase?

SOLUTION

IDENTIFY, SET UP, and EXECUTE: Gold sells for about $1400 an ounce, or about $100 for $\frac{1}{14}$ ounce. (The price per ounce has varied between $200 and $1900 over the past twenty years or so.) An ounce is about 30 grams, so $100 worth of gold has a mass of about $\frac{1}{14}$ of 30 grams, or roughly 2 grams. A billion (10^9) dollars' worth of gold has a mass 10^7 times greater, about 2×10^7 (20 million) grams or 2×10^4 (20,000) kilograms. A thousand kilograms has a weight in British units of about a ton, so the suitcase weighs roughly 20 tons! No human could lift it.

Roughly what is the *volume* of this gold? The density of water is 10^3 kg/m^3; if gold, which is much denser than water, has a density 10 times greater, then 10^4 kg of gold fit into a volume of 1 m^3. So 10^9 dollars' worth of gold has a volume of 2 m^3, many times the volume of a suitcase.

EVALUATE: Clearly your novel needs rewriting. Try the calculation again with a suitcase full of five-carat (1-gram) diamonds, each worth $500,000. Would this work?

Application Scalar Temperature, Vector Wind The comfort level on a wintry day depends on the temperature, a scalar quantity that can be positive or negative (say, +5°C or −20°C) but has no direction. It also depends on the wind velocity, a vector quantity with both magnitude and direction (for example, 15 km/h from the west).

1.7 VECTORS AND VECTOR ADDITION

Some physical quantities, such as time, temperature, mass, and density, can be described completely by a single number with a unit. But many other important quantities in physics have a *direction* associated with them and cannot be described by a single number. A simple example is the motion of an airplane: We must say not only how fast the plane is moving but also in what direction. The speed of the airplane combined with its direction of motion constitute a quantity called *velocity*. Another example is *force,* which in physics means a push or pull exerted on a body. Giving a complete description of a force means describing both how hard the force pushes or pulls on the body and the direction of the push or pull.

When a physical quantity is described by a single number, we call it a **scalar quantity.** In contrast, a **vector quantity** has both a **magnitude** (the "how much" or "how big" part) and a direction in space. Calculations that combine scalar quantities use the operations of ordinary arithmetic. For example, 6 kg + 3 kg = 9 kg, or 4 × 2 s = 8 s. However, combining vectors requires a different set of operations.

To understand more about vectors and how they combine, we start with the simplest vector quantity, **displacement.** Displacement is a change in the position of an object. Displacement is a vector quantity because we must state not only how far the object moves but also in what direction. Walking 3 km north from your front door doesn't get you to the same place as walking 3 km southeast; these two displacements have the same magnitude but different directions.

We usually represent a vector quantity such as displacement by a single letter, such as $\vec{A}$ in **Fig. 1.9a.** In this book we always print vector symbols in **boldface italic type with an arrow above them.** We do this to remind you that vector quantities have different properties from scalar quantities; the arrow is a reminder that vectors have direction. When you handwrite a symbol for a vector, *always* write it with an arrow on top. If you don't distinguish between scalar and vector quantities in your notation, you probably won't make the distinction in your thinking either, and confusion will result.

We always *draw* a vector as a line with an arrowhead at its tip. The length of the line shows the vector's magnitude, and the direction of the arrowhead shows the vector's direction. Displacement is always a straight-line segment directed from the starting point to the ending point, even though the object's actual path may be curved (Fig. 1.9b). Note that displacement is not related directly to the total *distance* traveled. If the object were to continue past P_2 and then return to P_1, the displacement for the entire trip would be *zero* (Fig. 1.9c).

If two vectors have the same direction, they are **parallel.** If they have the same magnitude *and* the same direction, they are *equal,* no matter where they are located in space. The vector $\vec{A}\,'$ from point P_3 to point P_4 in **Fig. 1.10** has the same length and direction as the vector $\vec{A}$ from P_1 to P_2. These two displacements are equal, even though they start at different points. We write this as $\vec{A}\,' = \vec{A}$ in Fig. 1.10; the boldface equals sign emphasizes that equality of two vector quantities is not the same relationship as equality of two scalar quantities. Two vector quantities are equal only when they have the same magnitude *and* the same direction.

Vector $\vec{B}$ in Fig. 1.10, however, is not equal to $\vec{A}$ because its direction is *opposite* that of $\vec{A}$. We define the **negative of a vector** as a vector having the same magnitude as the original vector but the *opposite* direction. The negative of vector quantity $\vec{A}$ is denoted as $-\vec{A}$, and we use a boldface minus sign to emphasize the vector nature of the quantities. If $\vec{A}$ is 87 m south, then $-\vec{A}$ is 87 m north. Thus we can write the relationship between $\vec{A}$ and $\vec{B}$ in Fig. 1.10 as $\vec{A} = -\vec{B}$ or $\vec{B} = -\vec{A}$. When two vectors $\vec{A}$ and $\vec{B}$ have opposite directions, whether their magnitudes are the same or not, we say that they are **antiparallel.**

We usually represent the *magnitude* of a vector quantity by the same letter used for the vector, but in *lightface italic type* with *no* arrow on top. For example, if displacement vector $\vec{A}$ is 87 m south, then $A = 87$ m. An alternative notation is the vector symbol with vertical bars on both sides:

$$(\text{Magnitude of } \vec{A}) = A = |\vec{A}| \qquad (1.1)$$

The magnitude of a vector quantity is a scalar quantity (a number) and is *always positive.* Note that a vector can never be equal to a scalar because they are different kinds of quantities. The expression "$\vec{A} = 6$ m" is just as wrong as "2 oranges = 3 apples"!

When we draw diagrams with vectors, it's best to use a scale similar to those used for maps. For example, a displacement of 5 km might be represented in a diagram by a vector 1 cm long, and a displacement of 10 km by a vector 2 cm long.

1.9 Displacement as a vector quantity.

(a) We represent a displacement by an arrow that points in the direction of displacement.

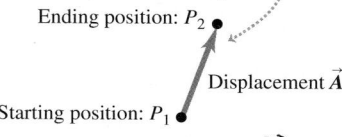

Ending position: P_2

Displacement $\vec{A}$

Starting position: P_1

Handwritten notation: $\vec{A}$

(b) A displacement is always a straight arrow directed from the starting position to the ending position. It does not depend on the path taken, even if the path is curved.

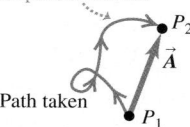

P_2

$\vec{A}$

Path taken

P_1

(c) Total displacement for a round trip is 0, regardless of the path taken or distance traveled.

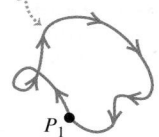

P_1

1.10 The meaning of vectors that have the same magnitude and the same or opposite direction.

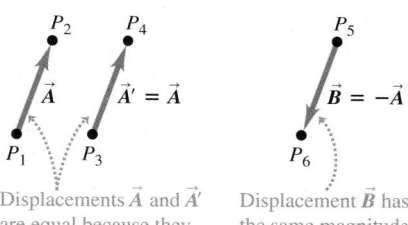

P_2 P_4 P_5

$\vec{A}$ $\vec{A}\,' = \vec{A}$ $\vec{B} = -\vec{A}$

P_1 P_3 P_6

Displacements $\vec{A}$ and $\vec{A}\,'$ are equal because they have the same length and direction.

Displacement $\vec{B}$ has the same magnitude as $\vec{A}$ but opposite direction; $\vec{B}$ is the negative of $\vec{A}$.

1.11 Three ways to add two vectors.

(a) We can add two vectors by placing them head to tail.

The vector sum $\vec{C}$ extends from the tail of vector $\vec{A}$ ...

... to the head of vector $\vec{B}$.

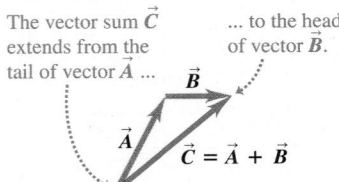

$\vec{C} = \vec{A} + \vec{B}$

(b) Adding them in reverse order gives the same result: $\vec{A} + \vec{B} = \vec{B} + \vec{A}$. The order doesn't matter in vector addition.

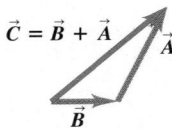

$\vec{C} = \vec{B} + \vec{A}$

(c) We can also add two vectors by placing them tail to tail and constructing a parallelogram.

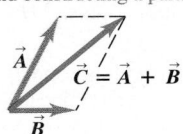

$\vec{C} = \vec{A} + \vec{B}$

1.12 Adding vectors that are (a) parallel and (b) antiparallel.

(a) Only when vectors $\vec{A}$ and $\vec{B}$ are parallel does the magnitude of their vector sum $\vec{C}$ equal the sum of their magnitudes: $C = A + B$.

$\vec{C} = \vec{A} + \vec{B}$

(b) When $\vec{A}$ and $\vec{B}$ are antiparallel, the magnitude of their vector sum $\vec{C}$ equals the *difference* of their magnitudes: $C = |A - B|$.

$\vec{C} = \vec{A} + \vec{B}$

Vector Addition and Subtraction

Suppose a particle undergoes a displacement $\vec{A}$ followed by a second displacement $\vec{B}$. The final result is the same as if the particle had started at the same initial point and undergone a single displacement $\vec{C}$ (**Fig. 1.11a**). We call displacement $\vec{C}$ the **vector sum,** or **resultant,** of displacements $\vec{A}$ and $\vec{B}$. We express this relationship symbolically as

$$\vec{C} = \vec{A} + \vec{B} \tag{1.2}$$

The boldface plus sign emphasizes that adding two vector quantities requires a geometrical process and is not the same operation as adding two scalar quantities such as $2 + 3 = 5$. In vector addition we usually place the *tail* of the *second* vector at the *head,* or tip, of the *first* vector (Fig. 1.11a).

If we make the displacements $\vec{A}$ and $\vec{B}$ in reverse order, with $\vec{B}$ first and $\vec{A}$ second, the result is the same (Fig. 1.11b). Thus

$$\vec{C} = \vec{B} + \vec{A} \quad \text{and} \quad \vec{A} + \vec{B} = \vec{B} + \vec{A} \tag{1.3}$$

This shows that the order of terms in a vector sum doesn't matter. In other words, vector addition obeys the *commutative* law.

Figure 1.11c shows another way to represent the vector sum: If we draw vectors $\vec{A}$ and $\vec{B}$ with their tails at the same point, vector $\vec{C}$ is the diagonal of a parallelogram constructed with $\vec{A}$ and $\vec{B}$ as two adjacent sides.

> CAUTION **Magnitudes in vector addition** It's a common error to conclude that if $\vec{C} = \vec{A} + \vec{B}$, then magnitude C equals magnitude A plus magnitude B. In general, this conclusion is *wrong;* for the vectors shown in Fig. 1.11, $C < A + B$. The magnitude of $\vec{A} + \vec{B}$ depends on the magnitudes of $\vec{A}$ and $\vec{B}$ *and* on the angle between $\vec{A}$ and $\vec{B}$. Only in the special case in which $\vec{A}$ and $\vec{B}$ are *parallel* is the magnitude of $\vec{C} = \vec{A} + \vec{B}$ equal to the sum of the magnitudes of $\vec{A}$ and $\vec{B}$ (**Fig. 1.12a**). When the vectors are *antiparallel* (Fig. 1.12b), the magnitude of $\vec{C}$ equals the *difference* of the magnitudes of $\vec{A}$ and $\vec{B}$. Be careful to distinguish between scalar and vector quantities, and you'll avoid making errors about the magnitude of a vector sum. ▌

Figure 1.13a shows *three* vectors $\vec{A}$, $\vec{B}$, and $\vec{C}$. To find the vector sum of all three, in Fig. 1.13b we first add $\vec{A}$ and $\vec{B}$ to give a vector sum $\vec{D}$; we then add vectors $\vec{C}$ and $\vec{D}$ by the same process to obtain the vector sum $\vec{R}$:

$$\vec{R} = (\vec{A} + \vec{B}) + \vec{C} = \vec{D} + \vec{C}$$

Alternatively, we can first add $\vec{B}$ and $\vec{C}$ to obtain vector $\vec{E}$ (Fig. 1.13c), and then add $\vec{A}$ and $\vec{E}$ to obtain $\vec{R}$:

$$\vec{R} = \vec{A} + (\vec{B} + \vec{C}) = \vec{A} + \vec{E}$$

We don't even need to draw vectors $\vec{D}$ and $\vec{E}$; all we need to do is draw $\vec{A}$, $\vec{B}$, and $\vec{C}$ in succession, with the tail of each at the head of the one preceding it. The sum vector $\vec{R}$ extends from the tail of the first vector to the head of the last vector

1.13 Several constructions for finding the vector sum $\vec{A} + \vec{B} + \vec{C}$.

(a) To find the sum of these three vectors ...

(b) ... add $\vec{A}$ and $\vec{B}$ to get $\vec{D}$ and then add $\vec{C}$ to $\vec{D}$ to get the final sum (resultant) $\vec{R}$ ...

(c) ... or add $\vec{B}$ and $\vec{C}$ to get $\vec{E}$ and then add $\vec{A}$ to $\vec{E}$ to get $\vec{R}$ ...

(d) ... or add $\vec{A}$, $\vec{B}$, and $\vec{C}$ to get $\vec{R}$ directly ...

(e) ... or add $\vec{A}$, $\vec{B}$, and $\vec{C}$ in any other order and still get $\vec{R}$.

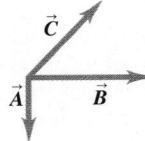

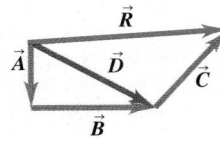

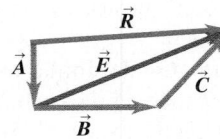

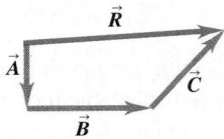

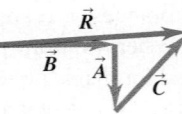

1.14 To construct the vector difference $\vec{A} - \vec{B}$, you can either place the tail of $-\vec{B}$ at the head of $\vec{A}$ or place the two vectors $\vec{A}$ and $\vec{B}$ head to head.

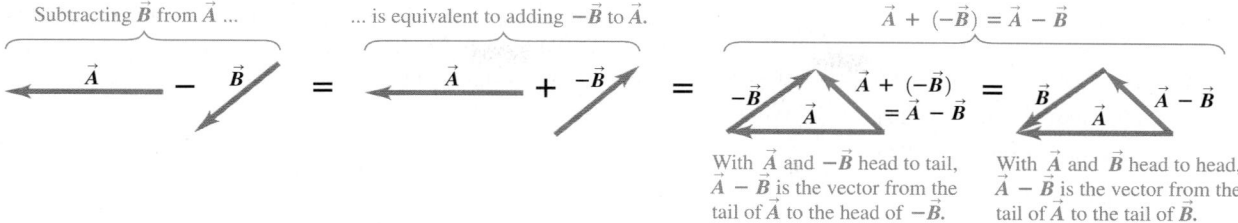

(Fig. 1.13d). The order makes no difference; Fig. 1.13e shows a different order, and you should try others. Vector addition obeys the *associative* law.

We can *subtract* vectors as well as add them. To see how, recall that vector $-\vec{A}$ has the same magnitude as $\vec{A}$ but the opposite direction. We define the difference $\vec{A} - \vec{B}$ of two vectors $\vec{A}$ and $\vec{B}$ to be the vector sum of $\vec{A}$ and $-\vec{B}$:

$$\vec{A} - \vec{B} = \vec{A} + (-\vec{B}) \tag{1.4}$$

Figure 1.14 shows an example of vector subtraction.

A vector quantity such as a displacement can be multiplied by a scalar quantity (an ordinary number). The displacement $2\vec{A}$ is a displacement (vector quantity) in the same direction as vector $\vec{A}$ but twice as long; this is the same as adding $\vec{A}$ to itself (**Fig. 1.15a**). In general, when we multiply a vector $\vec{A}$ by a scalar c, the result $c\vec{A}$ has magnitude $|c|A$ (the absolute value of c multiplied by the magnitude of vector $\vec{A}$). If c is positive, $c\vec{A}$ is in the same direction as $\vec{A}$; if c is negative, $c\vec{A}$ is in the direction opposite to $\vec{A}$. Thus $3\vec{A}$ is parallel to $\vec{A}$, while $-3\vec{A}$ is antiparallel to $\vec{A}$ (Fig. 1.15b).

A scalar used to multiply a vector can also be a physical quantity. For example, you may be familiar with the relationship $\vec{F} = m\vec{a}$; the net force $\vec{F}$ (a vector quantity) that acts on a body is equal to the product of the body's mass m (a scalar quantity) and its acceleration $\vec{a}$ (a vector quantity). The direction of $\vec{F}$ is the same as that of $\vec{a}$ because m is positive, and the magnitude of $\vec{F}$ is equal to the mass m multiplied by the magnitude of $\vec{a}$. The unit of force is the unit of mass multiplied by the unit of acceleration.

PhET: Vector Addition

1.15 Multiplying a vector by a scalar.

(a) Multiplying a vector by a positive scalar changes the magnitude (length) of the vector but not its direction.

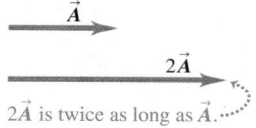

$2\vec{A}$ is twice as long as $\vec{A}$.

(b) Multiplying a vector by a negative scalar changes its magnitude and reverses its direction.

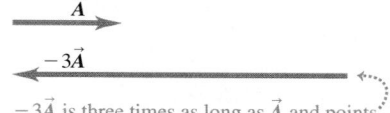

$-3\vec{A}$ is three times as long as $\vec{A}$ and points in the opposite direction.

EXAMPLE 1.5 **ADDING TWO VECTORS AT RIGHT ANGLES**

A cross-country skier skis 1.00 km north and then 2.00 km east on a horizontal snowfield. How far and in what direction is she from the starting point?

SOLUTION

IDENTIFY and SET UP: The problem involves combining two displacements at right angles to each other. This vector addition amounts to solving a right triangle, so we can use the Pythagorean theorem and simple trigonometry. The target variables are the skier's straight-line distance and direction from her starting point. **Figure 1.16** is a scale diagram of the two displacements and the resultant net displacement. We denote the direction from the starting point by the angle ϕ (the Greek letter phi). The displacement appears to be a bit more than 2 km. Measuring the angle with a protractor indicates that ϕ is about 63°.

1.16 The vector diagram, drawn to scale, for a ski trip.

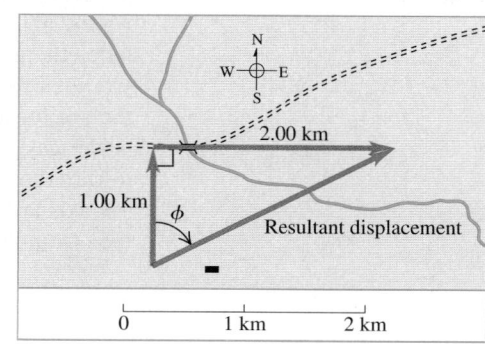

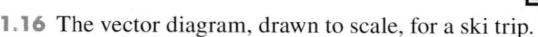

Continued

EXECUTE: The distance from the starting point to the ending point is equal to the length of the hypotenuse:

$$\sqrt{(1.00 \text{ km})^2 + (2.00 \text{ km})^2} = 2.24 \text{ km}$$

A little trigonometry (from Appendix B) allows us to find angle ϕ:

$$\tan \phi = \frac{\text{Opposite side}}{\text{Adjacent side}} = \frac{2.00 \text{ km}}{1.00 \text{ km}} = 2.00$$

$$\phi = \arctan 2.00 = 63.4°$$

We can describe the direction as 63.4° east of north or 90° − 63.4° = 26.6° north of east.

EVALUATE: Our answers (2.24 km and $\phi = 63.4°$) are close to our predictions. In Section 1.8 we'll learn how to easily add two vectors *not* at right angles to each other.

DATA *SPEAKS*

Vector Addition and Subtraction

When students were given a problem about adding or subtracting two vectors, more than 28% gave an incorrect answer. Common errors:

- When adding vectors, drawing vectors $\vec{A}$, $\vec{B}$, and $\vec{A} + \vec{B}$ incorrectly. The head-to-tail arrangement shown in Figs. 1.11a and 1.11b is easiest.
- When subtracting vectors, drawing vectors $\vec{A}$, $\vec{B}$, and $\vec{A} - \vec{B}$ incorrectly. Remember that subtracting $\vec{B}$ from $\vec{A}$ is the same as adding $-\vec{B}$ to $\vec{A}$ (Fig. 1.14).

TEST YOUR UNDERSTANDING OF SECTION 1.7 Two displacement vectors, $\vec{S}$ and $\vec{T}$, have magnitudes $S = 3$ m and $T = 4$ m. Which of the following could be the magnitude of the difference vector $\vec{S} - \vec{T}$? (There may be more than one correct answer.) (i) 9 m; (ii) 7 m; (iii) 5 m; (iv) 1 m; (v) 0 m; (vi) −1 m. ▌

1.8 COMPONENTS OF VECTORS

In Section 1.7 we added vectors by using a scale diagram and properties of right triangles. Making measurements of a diagram offers only very limited accuracy, and calculations with right triangles work only when the two vectors are perpendicular. So we need a simple but general method for adding vectors. This is called the method of *components*.

To define what we mean by the components of a vector $\vec{A}$, we begin with a rectangular (Cartesian) coordinate system of axes (**Fig. 1.17**). If we think of $\vec{A}$ as a displacement vector, we can regard $\vec{A}$ as the sum of a displacement parallel to the *x*-axis and a displacement parallel to the *y*-axis. We use the numbers A_x and A_y to tell us how much displacement there is parallel to the *x*-axis and how much there is parallel to the *y*-axis, respectively. For example, if the +*x*-axis points east and the +*y*-axis points north, $\vec{A}$ in Figure 1.17 could be the sum of a 2.00-m displacement to the east and a 1.00-m displacement to the north. Then $A_x = +2.00$ m and $A_y = +1.00$ m. We can use the same idea for any vectors, not just displacement vectors. The two numbers A_x and A_y are called the **components** of $\vec{A}$.

1.17 Representing a vector $\vec{A}$ in terms of its components A_x and A_y.

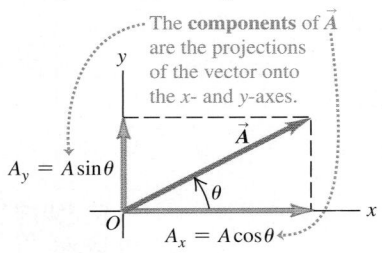

The **components** of $\vec{A}$ are the projections of the vector onto the *x*- and *y*-axes.

$A_y = A \sin \theta$

$A_x = A \cos \theta$

In this case, both A_x and A_y are positive.

CAUTION **Components are not vectors** The components A_x and A_y of a vector $\vec{A}$ are numbers; they are *not* vectors themselves. This is why we print the symbols for components in lightface italic type with *no* arrow on top instead of in boldface italic with an arrow, which is reserved for vectors. ▌

We can calculate the components of vector $\vec{A}$ if we know its magnitude A and its direction. We'll describe the direction of a vector by its angle relative to some reference direction. In Fig. 1.17 this reference direction is the positive *x*-axis, and the angle between vector $\vec{A}$ and the positive *x*-axis is θ (the Greek letter theta). Imagine that vector $\vec{A}$ originally lies along the +*x*-axis and that you then rotate it to its true direction, as indicated by the arrow in Fig. 1.17 on the arc for angle θ. If this rotation is from the +*x*-axis toward the +*y*-axis, as is the case in Fig. 1.17, then θ is *positive;* if the rotation is from the +*x*-axis toward the −*y*-axis, then θ is *negative*. Thus the +*y*-axis is at an angle of 90°, the −*x*-axis at 180°, and the −*y*-axis at 270° (or −90°). If θ is measured in this way, then from the definition of the trigonometric functions,

$$\frac{A_x}{A} = \cos \theta \qquad \text{and} \qquad \frac{A_y}{A} = \sin \theta$$

$$A_x = A \cos \theta \qquad \text{and} \qquad A_y = A \sin \theta \qquad (1.5)$$

(θ measured from the +*x*-axis, rotating toward the +*y*-axis)

In Fig. 1.17 A_x and A_y are positive. This is consistent with Eqs. (1.5); θ is in the first quadrant (between 0° and 90°), and both the cosine and the sine of an angle in this quadrant are positive. But in **Fig. 1.18a** the component B_x is negative and the component B_y is positive. (If the $+x$-axis points east and the $+y$-axis points north, $\vec{B}$ could represent a displacement of 2.00 m west and 1.00 m north. Since west is in the $-x$-direction and north is in the $+y$-direction, $B_x = -2.00$ m is negative and $B_y = +1.00$ m is positive.) Again, this is consistent with Eqs. (1.5); now θ is in the second quadrant, so $\cos\theta$ is negative and $\sin\theta$ is positive. In Fig. 1.18b both C_x and C_y are negative (both $\cos\theta$ and $\sin\theta$ are negative in the third quadrant).

1.18 The components of a vector may be positive or negative numbers.

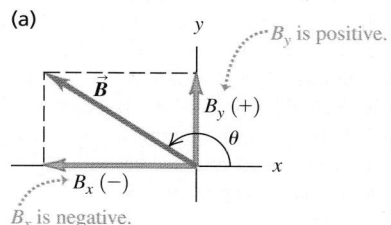

CAUTION **Relating a vector's magnitude and direction to its components** Equations (1.5) are correct *only* when the angle θ is measured from the positive x-axis. If the angle of the vector is given from a different reference direction or you use a different rotation direction, the relationships are different! Example 1.6 illustrates this point. ∎

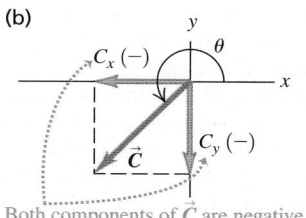

Both components of $\vec{C}$ are negative.

EXAMPLE 1.6 FINDING COMPONENTS

(a) What are the x- and y-components of vector $\vec{D}$ in **Fig. 1.19a**? The magnitude of the vector is $D = 3.00$ m, and angle $\alpha = 45°$. (b) What are the x- and y-components of vector $\vec{E}$ in Fig. 1.19b? The magnitude of the vector is $E = 4.50$ m, and angle $\beta = 37.0°$.

SOLUTION

IDENTIFY and SET UP: We can use Eqs. (1.5) to find the components of these vectors, but we must be careful: Neither angle α nor β in Fig. 1.19 is measured from the $+x$-axis toward the $+y$-axis. We estimate from the figure that the lengths of both

1.19 Calculating the x- and y-components of vectors.

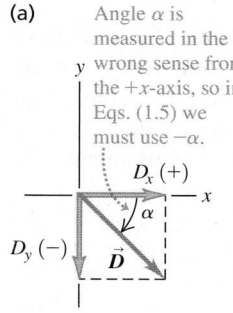
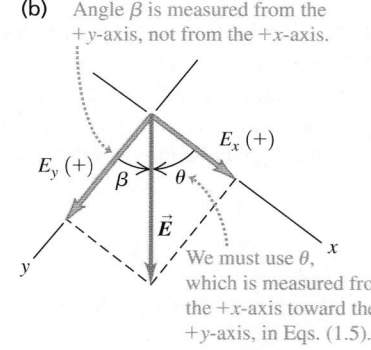

components in part (a) are roughly 2 m, and that those in part (b) are 3 m and 4 m. The figure indicates the signs of the components.

EXECUTE: (a) The angle α (the Greek letter alpha) between the positive x-axis and $\vec{D}$ is measured toward the *negative* y-axis. The angle we must use in Eqs. (1.5) is $\theta = -\alpha = -45°$. We then find

$$D_x = D \cos\theta = (3.00\text{ m})(\cos(-45°)) = +2.1\text{ m}$$
$$D_y = D \sin\theta = (3.00\text{ m})(\sin(-45°)) = -2.1\text{ m}$$

Had we carelessly substituted $+45°$ for θ in Eqs. (1.5), our result for D_y would have had the wrong sign.

(b) The x- and y-axes in Fig. 1.19b are at right angles, so it doesn't matter that they aren't horizontal and vertical, respectively. But we can't use the angle β (the Greek letter beta) in Eqs. (1.5), because β is measured from the $+y$-axis. Instead, we must use the angle $\theta = 90.0° - \beta = 90.0° - 37.0° = 53.0°$. Then we find

$$E_x = E \cos 53.0° = (4.50\text{ m})(\cos 53.0°) = +2.71\text{ m}$$
$$E_y = E \sin 53.0° = (4.50\text{ m})(\sin 53.0°) = +3.59\text{ m}$$

EVALUATE: Our answers to both parts are close to our predictions. But why do the answers in part (a) correctly have only two significant figures?

Using Components to Do Vector Calculations

Using components makes it relatively easy to do various calculations involving vectors. Let's look at three important examples: finding a vector's magnitude and direction, multiplying a vector by a scalar, and calculating the vector sum of two or more vectors.

CAUTION Finding the direction of a vector from its components There's one complication in using Eqs. (1.7) to find θ: Any two angles that differ by 180° have the same tangent. Suppose $A_x = 2$ m and $A_y = -2$ m as in **Fig. 1.20**; then $\tan\theta = -1$. But both 135° and 315° (or $-45°$) have tangents of -1. To decide which is correct, we have to look at the individual components. Because A_x is positive and A_y is negative, the angle must be in the fourth quadrant; thus $\theta = 315°$ (or $-45°$) is the correct value. Most pocket calculators give $\arctan(-1) = -45°$. In this case that is correct; but if instead we have $A_x = -2$ m and $A_y = 2$ m, then the correct angle is 135°. Similarly, when both A_x and A_y are negative, the tangent is positive, but the angle is in the third quadrant. *Always* draw a sketch like Fig. 1.20 to determine which of the two possibilities is correct. ▮

1.20 Drawing a sketch of a vector reveals the signs of its x- and y-components.

Suppose that $\tan\theta = \dfrac{A_y}{A_x} = -1$. What is θ?

Two angles have tangents of -1: 135° and 315°. The diagram shows that θ must be 315°.

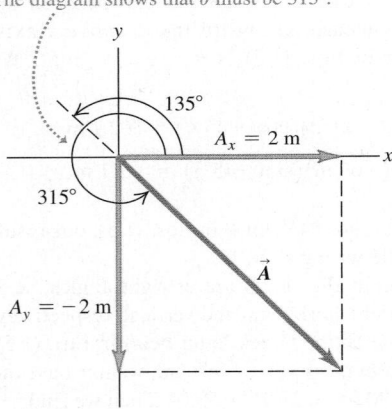

1.21 Finding the vector sum (resultant) of $\vec{A}$ and $\vec{B}$ using components.

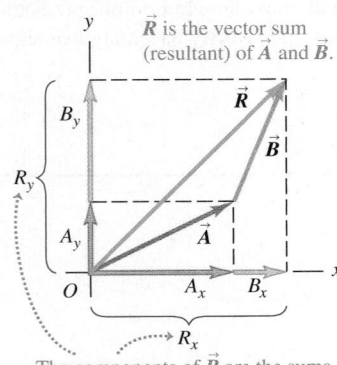

$\vec{R}$ is the vector sum (resultant) of $\vec{A}$ and $\vec{B}$.

The components of $\vec{R}$ are the sums of the components of $\vec{A}$ and $\vec{B}$:

$$R_y = A_y + B_y \qquad R_x = A_x + B_x$$

1. **Finding a vector's magnitude and direction from its components**. We can describe a vector completely by giving either its magnitude and direction or its x- and y-components. Equations (1.5) show how to find the components if we know the magnitude and direction. We can also reverse the process: We can find the magnitude and direction if we know the components. By applying the Pythagorean theorem to Fig. 1.17, we find that the magnitude of vector $\vec{A}$ is

$$A = \sqrt{A_x^2 + A_y^2} \tag{1.6}$$

(We always take the positive root.) Equation (1.6) is valid for any choice of x-axis and y-axis, as long as they are mutually perpendicular. The expression for the vector direction comes from the definition of the tangent of an angle. If θ is measured from the positive x-axis, and a positive angle is measured toward the positive y-axis (as in Fig. 1.17), then

$$\tan\theta = \frac{A_y}{A_x} \quad \text{and} \quad \theta = \arctan\frac{A_y}{A_x} \tag{1.7}$$

We will always use the notation arctan for the inverse tangent function (see Example 1.5 in Section 1.7). The notation $\tan^{-1}$ is also commonly used, and your calculator may have an INV or 2ND button to be used with the TAN button.

2. **Multiplying a vector by a scalar.** If we multiply a vector $\vec{A}$ by a scalar c, each component of the product $\vec{D} = c\vec{A}$ is the product of c and the corresponding component of $\vec{A}$:

$$D_x = cA_x, \qquad D_y = cA_y \qquad \text{(components of } \vec{D} = c\vec{A}) \tag{1.8}$$

For example, Eqs. (1.8) say that each component of the vector $2\vec{A}$ is twice as great as the corresponding component of $\vec{A}$, so $2\vec{A}$ is in the same direction as $\vec{A}$ but has twice the magnitude. Each component of the vector $-3\vec{A}$ is three times as great as the corresponding component of $\vec{A}$ but has the opposite sign, so $-3\vec{A}$ is in the opposite direction from $\vec{A}$ and has three times the magnitude. Hence Eqs. (1.8) are consistent with our discussion in Section 1.7 of multiplying a vector by a scalar (see Fig. 1.15).

3. **Using components to calculate the vector sum (resultant) of two or more vectors.** Figure **1.21** shows two vectors $\vec{A}$ and $\vec{B}$ and their vector sum $\vec{R}$, along with the x- and y-components of all three vectors. The x-component R_x of the vector sum is simply the sum $(A_x + B_x)$ of the x-components of the vectors being added. The same is true for the y-components. In symbols,

Each **component of $\vec{R} = \vec{A} + \vec{B}$** ...

$$R_x = A_x + B_x, \qquad R_y = A_y + B_y \tag{1.9}$$

... is the sum of the corresponding components of $\vec{A}$ and $\vec{B}$.

Figure 1.21 shows this result for the case in which the components A_x, A_y, B_x, and B_y are all positive. Draw additional diagrams to verify for yourself that Eqs. (1.9) are valid for *any* signs of the components of $\vec{A}$ and $\vec{B}$.

If we know the components of any two vectors $\vec{A}$ and $\vec{B}$, perhaps by using Eqs. (1.5), we can compute the components of the vector sum $\vec{R}$. Then if we need the magnitude and direction of $\vec{R}$, we can obtain them from Eqs. (1.6) and (1.7) with the A's replaced by R's.

We can use the same procedure to find the sum of any number of vectors. If $\vec{R}$ is the vector sum of $\vec{A}, \vec{B}, \vec{C}, \vec{D}, \vec{E}, \ldots$, the components of $\vec{R}$ are

$$R_x = A_x + B_x + C_x + D_x + E_x + \cdots$$
$$R_y = A_y + B_y + C_y + D_y + E_y + \cdots \tag{1.10}$$

We have talked about vectors that lie in the xy-plane only, but the component method works just as well for vectors having any direction in space. We can introduce a z-axis perpendicular to the xy-plane; then in general a vector $\vec{A}$ has components A_x, A_y, and A_z in the three coordinate directions. Its magnitude A is

$$A = \sqrt{A_x^2 + A_y^2 + A_z^2} \tag{1.11}$$

Again, we always take the positive root (**Fig. 1.22**). Also, Eqs. (1.10) for the vector sum $\vec{R}$ have a third component:

$$R_z = A_z + B_z + C_z + D_z + E_z + \cdots$$

We've focused on adding *displacement* vectors, but the method is applicable to all vector quantities. When we study the concept of force in Chapter 4, we'll find that forces are vectors that obey the same rules of vector addition.

1.22 A vector in three dimensions.

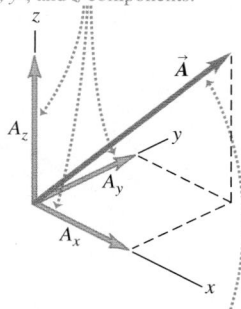

In three dimensions, a vector has x-, y-, and z-components.

The magnitude of vector $\vec{A}$ is $A = \sqrt{A_x^2 + A_y^2 + A_z^2}$.

PROBLEM-SOLVING STRATEGY 1.3 VECTOR ADDITION

IDENTIFY *the relevant concepts:* Decide what the target variable is. It may be the magnitude of the vector sum, the direction, or both.

SET UP *the problem:* Sketch the vectors being added, along with suitable coordinate axes. Place the tail of the first vector at the origin of the coordinates, place the tail of the second vector at the head of the first vector, and so on. Draw the vector sum $\vec{R}$ from the tail of the first vector (at the origin) to the head of the last vector. Use your sketch to estimate the magnitude and direction of $\vec{R}$. Select the mathematical tools you'll use for the full calculation: Eqs. (1.5) to obtain the components of the vectors given, if necessary, Eqs. (1.10) to obtain the components of the vector sum, Eq. (1.11) to obtain its magnitude, and Eqs. (1.7) to obtain its direction.

EXECUTE *the solution* as follows:
1. Find the x- and y-components of each individual vector and record your results in a table, as in Example 1.7. If a vector is described by a magnitude A and an angle θ, measured from

the $+x$-axis toward the $+y$-axis, then its components are given by Eqs. 1.5:

$$A_x = A \cos \theta \qquad A_y = A \sin \theta$$

If the angles of the vectors are given in some other way, perhaps using a different reference direction, convert them to angles measured from the $+x$-axis as in Example 1.6.
2. Add the individual x-components algebraically (including signs) to find R_x, the x-component of the vector sum. Do the same for the y-components to find R_y. See Example 1.7.
3. Calculate the magnitude R and direction θ of the vector sum by using Eqs. (1.6) and (1.7):

$$R = \sqrt{R_x^2 + R_y^2} \qquad \theta = \arctan \frac{R_y}{R_x}$$

EVALUATE *your answer:* Confirm that your results for the magnitude and direction of the vector sum agree with the estimates you made from your sketch. The value of θ that you find with a calculator may be off by 180°; your drawing will indicate the correct value.

EXAMPLE 1.7 USING COMPONENTS TO ADD VECTORS

Three players on a reality TV show are brought to the center of a large, flat field. Each is given a meter stick, a compass, a calculator, a shovel, and (in a different order for each contestant) the following three displacements:

$\vec{A}$: 72.4 m, 32.0° east of north
$\vec{B}$: 57.3 m, 36.0° south of west
$\vec{C}$: 17.8 m due south

The three displacements lead to the point in the field where the keys to a new Porsche are buried. Two players start measuring immediately, but the winner first *calculates* where to go. What does she calculate?

SOLUTION

IDENTIFY and SET UP: The goal is to find the sum (resultant) of the three displacements, so this is a problem in vector addition. See **Figure 1.23.** We have chosen the $+x$-axis as east and the $+y$-axis as north. We estimate from the diagram that the vector sum $\vec{R}$ is about 10 m, 40° west of north (so θ is about 90° plus 40°, or about 130°).

Continued

1.23 Three successive displacements $\vec{A}$, $\vec{B}$, and $\vec{C}$ and the resultant (vector sum) displacement $\vec{R} = \vec{A} + \vec{B} + \vec{C}$.

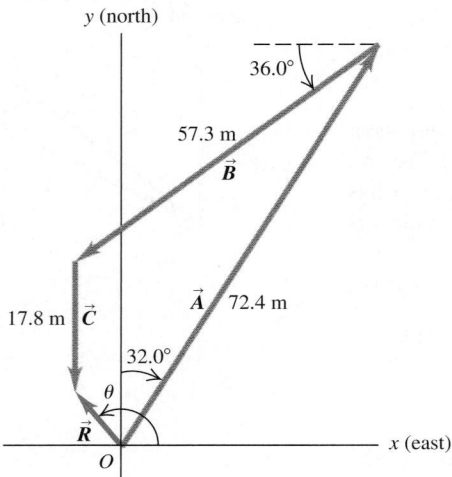

We've kept an extra significant figure in the components; we'll round to the correct number of significant figures at the end of our calculation. The table below shows the components of all the displacements, the addition of the components, and the other calculations from Eqs. (1.6) and (1.7).

Distance	Angle	x-component	y-component
A = 72.4 m	58.0°	38.37 m	61.40 m
B = 57.3 m	216.0°	−46.36 m	−33.68 m
C = 17.8 m	270.0°	0.00 m	−17.80 m
		$R_x = -7.99$ m	$R_y = 9.92$ m

$$R = \sqrt{(-7.99 \text{ m})^2 + (9.92 \text{ m})^2} = 12.7 \text{ m}$$

$$\theta = \arctan\frac{9.92 \text{ m}}{-7.99 \text{ m}} = -51°$$

Comparing to angle θ in Fig. 1.23 shows that the calculated angle is clearly off by 180°. The correct value is $\theta = 180° + (-51°) = 129°$, or 39° west of north.

EVALUATE: Our calculated answers for R and θ agree with our estimates. Notice how drawing the diagram in Fig. 1.23 made it easy to avoid a 180° error in the direction of the vector sum.

EXECUTE: The angles of the vectors, measured from the +x-axis toward the +y-axis, are $(90.0° - 32.0°) = 58.0°$, $(180.0° + 36.0°) = 216.0°$, and 270.0°, respectively. We may now use Eqs. (1.5) to find the components of $\vec{A}$:

$$A_x = A \cos\theta_A = (72.4 \text{ m})(\cos 58.0°) = 38.37 \text{ m}$$
$$A_y = A \sin\theta_A = (72.4 \text{ m})(\sin 58.0°) = 61.40 \text{ m}$$

TEST YOUR UNDERSTANDING OF SECTION 1.8 Two vectors $\vec{A}$ and $\vec{B}$ lie in the xy-plane. (a) Can $\vec{A}$ have the same magnitude as $\vec{B}$ but different components? (b) Can $\vec{A}$ have the same components as $\vec{B}$ but a different magnitude? ∎

1.9 UNIT VECTORS

A **unit vector** is a vector that has a magnitude of 1, with no units. Its only purpose is to *point*—that is, to describe a direction in space. Unit vectors provide a convenient notation for many expressions involving components of vectors. We will always include a caret, or "hat" (^), in the symbol for a unit vector to distinguish it from ordinary vectors whose magnitude may or may not be equal to 1.

In an xy-coordinate system we can define a unit vector $\hat{i}$ that points in the direction of the positive x-axis and a unit vector $\hat{j}$ that points in the direction of the positive y-axis (**Fig. 1.24a**). Then we can write a vector $\vec{A}$ in terms of its components as

$$\vec{A} = A_x\hat{i} + A_y\hat{j} \qquad (1.12)$$

Equation (1.12) is a vector equation; each term, such as $A_x\hat{i}$, is a vector quantity (Fig. 1.24b).

Using unit vectors, we can express the vector sum $\vec{R}$ of two vectors $\vec{A}$ and $\vec{B}$ as follows:

$$\vec{A} = A_x\hat{i} + A_y\hat{j}$$
$$\vec{B} = B_x\hat{i} + B_y\hat{j}$$
$$\vec{R} = \vec{A} + \vec{B} \qquad (1.13)$$
$$= (A_x\hat{i} + A_y\hat{j}) + (B_x\hat{i} + B_y\hat{j})$$
$$= (A_x + B_x)\hat{i} + (A_y + B_y)\hat{j}$$
$$= R_x\hat{i} + R_y\hat{j}$$

1.24 (a) The unit vectors $\hat{i}$ and $\hat{j}$. (b) Expressing a vector $\vec{A}$ in terms of its components.

(a)

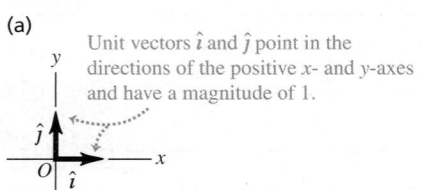

Unit vectors $\hat{i}$ and $\hat{j}$ point in the directions of the positive x- and y-axes and have a magnitude of 1.

(b)

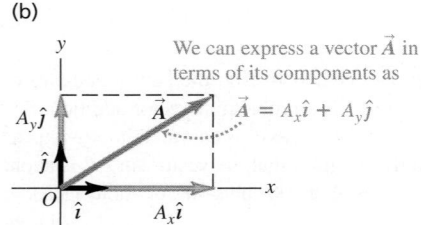

We can express a vector $\vec{A}$ in terms of its components as $\vec{A} = A_x\hat{i} + A_y\hat{j}$

Equation (1.13) restates the content of Eqs. (1.9) in the form of a single vector equation rather than two component equations.

If not all of the vectors lie in the *xy*-plane, then we need a third component. We introduce a third unit vector $\hat{k}$ that points in the direction of the positive *z*-axis (**Fig. 1.25**). Then Eqs. (1.12) and (1.13) become

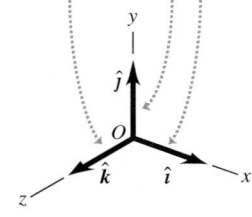

1.25 The unit vectors $\hat{\imath}$, $\hat{\jmath}$, and $\hat{k}$.

Unit vectors $\hat{\imath}$, $\hat{\jmath}$, and $\hat{k}$ point in the directions of the positive *x*-, *y*-, and *z*-axes and have a magnitude of 1.

Any vector can be expressed in terms of its *x*-, *y*-, and *z*-components ...

$$\vec{A} = A_x\hat{\imath} + A_y\hat{\jmath} + A_z\hat{k}$$
$$\vec{B} = B_x\hat{\imath} + B_y\hat{\jmath} + B_z\hat{k}$$

(1.14)

... and unit vectors $\hat{\imath}$, $\hat{\jmath}$, and $\hat{k}$.

$$\vec{R} = (A_x + B_x)\hat{\imath} + (A_y + B_y)\hat{\jmath} + (A_z + B_z)\hat{k}$$
$$= R_x\hat{\imath} + R_y\hat{\jmath} + R_z\hat{k}$$

(1.15)

EXAMPLE 1.8   USING UNIT VECTORS

Given the two displacements

$$\vec{D} = (6.00\,\hat{\imath} + 3.00\,\hat{\jmath} - 1.00\hat{k})\,\text{m} \quad \text{and}$$
$$\vec{E} = (4.00\,\hat{\imath} - 5.00\,\hat{\jmath} + 8.00\hat{k})\,\text{m}$$

find the magnitude of the displacement $2\vec{D} - \vec{E}$.

SOLUTION

IDENTIFY and SET UP: We are to multiply vector $\vec{D}$ by 2 (a scalar) and subtract vector $\vec{E}$ from the result, so as to obtain the vector $\vec{F} = 2\vec{D} - \vec{E}$. Equation (1.8) says that to multiply $\vec{D}$ by 2, we multiply each of its components by 2. We can use Eq. (1.15) to do the subtraction; recall from Section 1.7 that subtracting a vector is the same as adding the negative of that vector.

EXECUTE: We have

$$\vec{F} = 2(6.00\hat{\imath} + 3.00\,\hat{\jmath} - 1.00\hat{k})\,\text{m} - (4.00\hat{\imath} - 5.00\,\hat{\jmath} + 8.00\hat{k})\,\text{m}$$
$$= [(12.00 - 4.00)\hat{\imath} + (6.00 + 5.00)\,\hat{\jmath} + (-2.00 - 8.00)\hat{k}]\,\text{m}$$
$$= (8.00\hat{\imath} + 11.00\,\hat{\jmath} - 10.00\hat{k})\,\text{m}$$

From Eq. (1.11) the magnitude of $\vec{F}$ is

$$F = \sqrt{F_x^2 + F_y^2 + F_z^2}$$
$$= \sqrt{(8.00\,\text{m})^2 + (11.00\,\text{m})^2 + (-10.00\,\text{m})^2}$$
$$= 16.9\,\text{m}$$

EVALUATE: Our answer is of the same order of magnitude as the larger components that appear in the sum. We wouldn't expect our answer to be much larger than this, but it could be much smaller.

TEST YOUR UNDERSTANDING OF SECTION 1.9   Arrange the following vectors in order of their magnitude, with the vector of largest magnitude first. (i) $\vec{A} = (3\hat{\imath} + 5\hat{\jmath} - 2\hat{k})$ m; (ii) $\vec{B} = (-3\hat{\imath} + 5\hat{\jmath} - 2\hat{k})$ m; (iii) $\vec{C} = (3\hat{\imath} - 5\hat{\jmath} - 2\hat{k})$ m; (iv) $\vec{D} = (3\hat{\imath} + 5\hat{\jmath} + 2\hat{k})$ m. ∎

1.10 PRODUCTS OF VECTORS

We saw how vector addition develops naturally from the problem of combining displacements. It will prove useful for calculations with many other vector quantities. We can also express many physical relationships by using *products* of vectors. Vectors are not ordinary numbers, so we can't directly apply ordinary multiplication to vectors. We'll define two different kinds of products of vectors. The first, called the *scalar product,* yields a result that is a scalar quantity. The second, the *vector product,* yields another vector.

Scalar Product

We denote the **scalar product** of two vectors $\vec{A}$ and $\vec{B}$ by $\vec{A} \cdot \vec{B}$. Because of this notation, the scalar product is also called the **dot product.** Although $\vec{A}$ and $\vec{B}$ are vectors, the quantity $\vec{A} \cdot \vec{B}$ is a scalar.

1.26 Calculating the scalar product of two vectors, $\vec{A} \cdot \vec{B} = AB \cos \phi$.

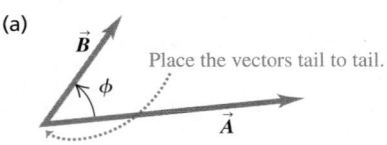

(a)

Place the vectors tail to tail.

(b) $\vec{A} \cdot \vec{B}$ equals $A(B \cos \phi)$.

(Magnitude of $\vec{A}$) $\times$ $\begin{pmatrix} \text{Component of } \vec{B} \\ \text{in direction of } \vec{A} \end{pmatrix}$

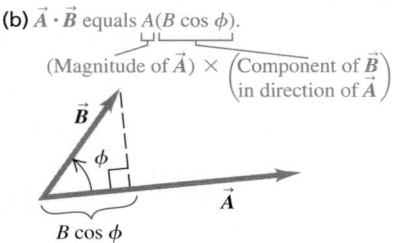

$B \cos \phi$

(c) $\vec{A} \cdot \vec{B}$ also equals $B(A \cos \phi)$.

(Magnitude of $\vec{B}$) $\times$ $\begin{pmatrix} \text{Component of } \vec{A} \\ \text{in direction of } \vec{B} \end{pmatrix}$

$A \cos \phi$

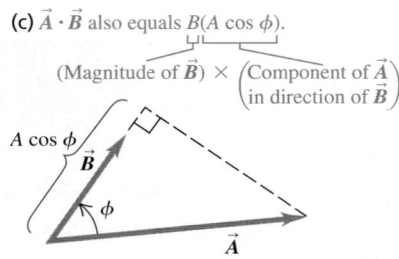

1.27 The scalar product $\vec{A} \cdot \vec{B} = AB \cos \phi$ can be positive, negative, or zero, depending on the angle between $\vec{A}$ and $\vec{B}$.

(a)

If ϕ is between 0° and 90°, $\vec{A} \cdot \vec{B}$ is positive ...

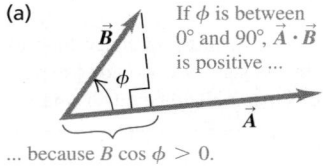

... because $B \cos \phi > 0$.

(b)

If ϕ is between 90° and 180°, $\vec{A} \cdot \vec{B}$ is negative ...

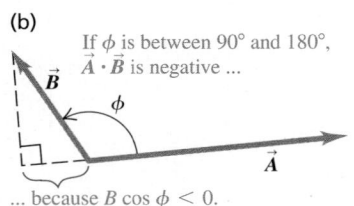

... because $B \cos \phi < 0$.

(c)

If $\phi = 90°$, $\vec{A} \cdot \vec{B} = 0$ because $\vec{B}$ has zero component in the direction of $\vec{A}$.

$\phi = 90°$

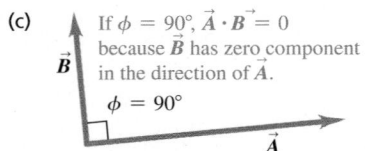

To define the scalar product $\vec{A} \cdot \vec{B}$ we draw the two vectors $\vec{A}$ and $\vec{B}$ with their tails at the same point (**Fig. 1.26a**). The angle ϕ (the Greek letter phi) between their directions ranges from 0° to 180°. Figure 1.26b shows the projection of vector $\vec{B}$ onto the direction of $\vec{A}$; this projection is the component of $\vec{B}$ in the direction of $\vec{A}$ and is equal to $B \cos \phi$. (We can take components along *any* direction that's convenient, not just the *x*- and *y*-axes.) We define $\vec{A} \cdot \vec{B}$ to be the magnitude of $\vec{A}$ multiplied by the component of $\vec{B}$ in the direction of $\vec{A}$, or

Scalar (dot) product of vectors $\vec{A}$ and $\vec{B}$ .... Magnitudes of $\vec{A}$ and $\vec{B}$

$$\vec{A} \cdot \vec{B} = AB \cos \phi = |\vec{A}||\vec{B}| \cos \phi \qquad (1.16)$$

Angle between $\vec{A}$ and $\vec{B}$ when placed tail to tail

Alternatively, we can define $\vec{A} \cdot \vec{B}$ to be the magnitude of $\vec{B}$ multiplied by the component of $\vec{A}$ in the direction of $\vec{B}$, as in Fig. 1.26c. Hence $\vec{A} \cdot \vec{B} = B(A \cos \phi) = AB \cos \phi$, which is the same as Eq. (1.16).

The scalar product is a scalar quantity, not a vector, and it may be positive, negative, or zero. When ϕ is between 0° and 90°, $\cos \phi > 0$ and the scalar product is positive (**Fig. 1.27a**). When ϕ is between 90° and 180° so $\cos \phi < 0$, the component of $\vec{B}$ in the direction of $\vec{A}$ is negative, and $\vec{A} \cdot \vec{B}$ is negative (Fig. 1.27b). Finally, when $\phi = 90°$, $\vec{A} \cdot \vec{B} = 0$ (Fig. 1.27c). *The scalar product of two perpendicular vectors is always zero.*

For any two vectors $\vec{A}$ and $\vec{B}$, $AB \cos \phi = BA \cos \phi$. This means that $\vec{A} \cdot \vec{B} = \vec{B} \cdot \vec{A}$. The scalar product obeys the commutative law of multiplication; the order of the two vectors does not matter.

We'll use the scalar product in Chapter 6 to describe work done by a force. In later chapters we'll use the scalar product for a variety of purposes, from calculating electric potential to determining the effects that varying magnetic fields have on electric circuits.

Using Components to Calculate the Scalar Product

We can calculate the scalar product $\vec{A} \cdot \vec{B}$ directly if we know the *x*-, *y*-, and *z*-components of $\vec{A}$ and $\vec{B}$. To see how this is done, let's first work out the scalar products of the unit vectors $\hat{\imath}$, $\hat{\jmath}$, and $\hat{k}$. All unit vectors have magnitude 1 and are perpendicular to each other. Using Eq. (1.16), we find

$$\hat{\imath} \cdot \hat{\imath} = \hat{\jmath} \cdot \hat{\jmath} = \hat{k} \cdot \hat{k} = (1)(1) \cos 0° = 1$$
$$\hat{\imath} \cdot \hat{\jmath} = \hat{\imath} \cdot \hat{k} = \hat{\jmath} \cdot \hat{k} = (1)(1) \cos 90° = 0 \qquad (1.17)$$

Now we express $\vec{A}$ and $\vec{B}$ in terms of their components, expand the product, and use these products of unit vectors:

$$\vec{A} \cdot \vec{B} = (A_x \hat{\imath} + A_y \hat{\jmath} + A_z \hat{k}) \cdot (B_x \hat{\imath} + B_y \hat{\jmath} + B_z \hat{k})$$

$$= A_x \hat{\imath} \cdot B_x \hat{\imath} + A_x \hat{\imath} \cdot B_y \hat{\jmath} + A_x \hat{\imath} \cdot B_z \hat{k}$$

$$+ A_y \hat{\jmath} \cdot B_x \hat{\imath} + A_y \hat{\jmath} \cdot B_y \hat{\jmath} + A_y \hat{\jmath} \cdot B_z \hat{k}$$

$$+ A_z \hat{k} \cdot B_x \hat{\imath} + A_z \hat{k} \cdot B_y \hat{\jmath} + A_z \hat{k} \cdot B_z \hat{k} \qquad (1.18)$$

$$= A_x B_x \hat{\imath} \cdot \hat{\imath} + A_x B_y \hat{\imath} \cdot \hat{\jmath} + A_x B_z \hat{\imath} \cdot \hat{k}$$

$$+ A_y B_x \hat{\jmath} \cdot \hat{\imath} + A_y B_y \hat{\jmath} \cdot \hat{\jmath} + A_y B_z \hat{\jmath} \cdot \hat{k}$$

$$+ A_z B_x \hat{k} \cdot \hat{\imath} + A_z B_y \hat{k} \cdot \hat{\jmath} + A_z B_z \hat{k} \cdot \hat{k}$$

From Eqs. (1.17) you can see that six of these nine terms are zero. The three that survive give

> **Scalar (dot) product**
> of vectors $\vec{A}$ and $\vec{B}$ .... Components of $\vec{A}$
>
> $$\vec{A} \cdot \vec{B} = A_x B_x + A_y B_y + A_z B_z \qquad (1.19)$$
>
> Components of $\vec{B}$

Thus *the scalar product of two vectors is the sum of the products of their respective components.*

The scalar product gives a straightforward way to find the angle ϕ between any two vectors $\vec{A}$ and $\vec{B}$ whose components are known. In this case we can use Eq. (1.19) to find the scalar product of $\vec{A}$ and $\vec{B}$. Example 1.10 shows how to do this.

EXAMPLE 1.9   CALCULATING A SCALAR PRODUCT

Find the scalar product $\vec{A} \cdot \vec{B}$ of the two vectors in **Fig. 1.28.** The magnitudes of the vectors are $A = 4.00$ and $B = 5.00$.

SOLUTION

IDENTIFY and SET UP: We can calculate the scalar product in two ways: using the magnitudes of the vectors and the angle between them (Eq. 1.16), and using the components of the vectors (Eq. 1.19). We'll do it both ways, and the results will check each other.

1.28 Two vectors $\vec{A}$ and $\vec{B}$ in two dimensions.

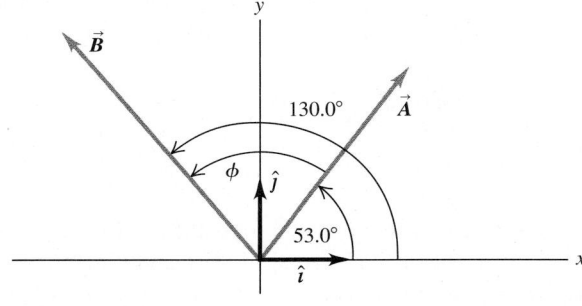

EXECUTE: The angle between the two vectors $\vec{A}$ and $\vec{B}$ is $\phi = 130.0° - 53.0° = 77.0°$, so Eq. (1.16) gives us

$$\vec{A} \cdot \vec{B} = AB \cos \phi = (4.00)(5.00) \cos 77.0° = 4.50$$

To use Eq. (1.19), we must first find the components of the vectors. The angles of $\vec{A}$ and $\vec{B}$ are given with respect to the $+x$-axis and are measured in the sense from the $+x$-axis to the $+y$-axis, so we can use Eqs. (1.5):

$$A_x = (4.00) \cos 53.0° = 2.407$$
$$A_y = (4.00) \sin 53.0° = 3.195$$
$$B_x = (5.00) \cos 130.0° = -3.214$$
$$B_y = (5.00) \sin 130.0° = 3.830$$

As in Example 1.7, we keep an extra significant figure in the components and round at the end. Equation (1.19) now gives us

$$\vec{A} \cdot \vec{B} = A_x B_x + A_y B_y + A_z B_z$$
$$= (2.407)(-3.214) + (3.195)(3.830) + (0)(0) = 4.50$$

EVALUATE: Both methods give the same result, as they should.

EXAMPLE 1.10   FINDING AN ANGLE WITH THE SCALAR PRODUCT

Find the angle between the vectors

$$\vec{A} = 2.00\hat{i} + 3.00\hat{j} + 1.00\hat{k}$$

and

$$\vec{B} = -4.00\hat{i} + 2.00\hat{j} - 1.00\hat{k}$$

SOLUTION

IDENTIFY and SET UP: We're given the x-, y-, and z-components of two vectors. Our target variable is the angle ϕ between them (**Fig. 1.29**). To find this, we'll solve Eq. (1.16), $\vec{A} \cdot \vec{B} = AB \cos \phi$, for ϕ in terms of the scalar product $\vec{A} \cdot \vec{B}$ and the magnitudes A and B. We can use Eq. (1.19) to evaluate the scalar product, $\vec{A} \cdot \vec{B} = A_x B_x + A_y B_y + A_z B_z$, and we can use Eq. (1.6) to find A and B.

1.29 Two vectors in three dimensions.

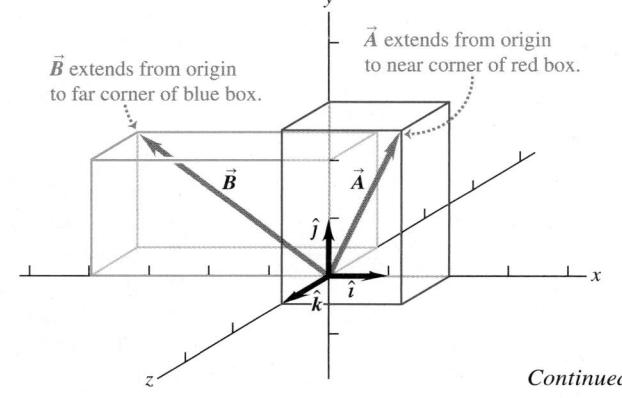

EXECUTE: We solve Eq. (1.16) for $\cos\phi$ and use Eq. (1.19) to write $\vec{A}\cdot\vec{B}$:

$$\cos\phi = \frac{\vec{A}\cdot\vec{B}}{AB} = \frac{A_xB_x + A_yB_y + A_zB_z}{AB}$$

We can use this formula to find the angle between *any* two vectors $\vec{A}$ and $\vec{B}$. Here we have $A_x = 2.00$, $A_y = 3.00$, and $A_z = 1.00$, and $B_x = -4.00$, $B_y = 2.00$, and $B_z = -1.00$. Thus

$$\vec{A}\cdot\vec{B} = A_xB_x + A_yB_y + A_zB_z$$

$$= (2.00)(-4.00) + (3.00)(2.00) + (1.00)(-1.00)$$

$$= -3.00$$

$$A = \sqrt{A_x^2 + A_y^2 + A_z^2} = \sqrt{(2.00)^2 + (3.00)^2 + (1.00)^2}$$
$$= \sqrt{14.00}$$

$$B = \sqrt{B_x^2 + B_y^2 + B_z^2} = \sqrt{(-4.00)^2 + (2.00)^2 + (-1.00)^2}$$
$$= \sqrt{21.00}$$

$$\cos\phi = \frac{A_xB_x + A_yB_y + A_zB_z}{AB} = \frac{-3.00}{\sqrt{14.00}\sqrt{21.00}} = -0.175$$

$$\phi = 100°$$

EVALUATE: As a check on this result, note that the scalar product $\vec{A}\cdot\vec{B}$ is negative. This means that ϕ is between 90° and 180° (see Fig. 1.27), which agrees with our answer.

1.30 The vector product of (a) $\vec{A}\times\vec{B}$ and (b) $\vec{B}\times\vec{A}$.

(a) Using the right-hand rule to find the direction of $\vec{A}\times\vec{B}$

① Place $\vec{A}$ and $\vec{B}$ tail to tail.

② Point fingers of right hand along $\vec{A}$, with palm facing $\vec{B}$.

③ Curl fingers toward $\vec{B}$.

④ Thumb points in direction of $\vec{A}\times\vec{B}$.

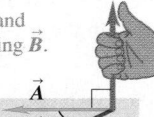

(b) Using the right-hand rule to find the direction of $\vec{B}\times\vec{A} = -\vec{A}\times\vec{B}$ (vector product is anticommutative)

① Place $\vec{B}$ and $\vec{A}$ tail to tail.

② Point fingers of right hand along $\vec{B}$, with palm facing $\vec{A}$.

③ Curl fingers toward $\vec{A}$.

④ Thumb points in direction of $\vec{B}\times\vec{A}$.

⑤ $\vec{B}\times\vec{A}$ has same magnitude as $\vec{A}\times\vec{B}$ but points in opposite direction.

Vector Product

We denote the **vector product** of two vectors $\vec{A}$ and $\vec{B}$, also called the **cross product,** by $\vec{A}\times\vec{B}$. As the name suggests, the vector product is itself a vector. We'll use this product in Chapter 10 to describe torque and angular momentum; in Chapters 27 and 28 we'll use it to describe magnetic fields and forces.

To define the vector product $\vec{A}\times\vec{B}$, we again draw the two vectors $\vec{A}$ and $\vec{B}$ with their tails at the same point (**Fig. 1.30a**). The two vectors then lie in a plane. We define the vector product to be a vector quantity with a direction perpendicular to this plane (that is, perpendicular to both $\vec{A}$ and $\vec{B}$) and a magnitude equal to $AB\sin\phi$. That is, if $\vec{C} = \vec{A}\times\vec{B}$, then

Magnitude of **vector (cross) product** of vectors $\vec{B}$ and $\vec{A}$

$$C = AB\sin\phi \qquad (1.20)$$

Magnitudes of $\vec{A}$ and $\vec{B}$ ⟋ Angle between $\vec{A}$ and $\vec{B}$ when placed tail to tail

We measure the angle ϕ from $\vec{A}$ toward $\vec{B}$ and take it to be the smaller of the two possible angles, so ϕ ranges from 0° to 180°. Then $\sin\phi \geq 0$ and C in Eq. (1.20) is never negative, as must be the case for a vector magnitude. Note that when $\vec{A}$ and $\vec{B}$ are parallel or antiparallel, $\phi = 0°$ or 180° and $C = 0$. That is, *the vector product of two parallel or antiparallel vectors is always zero.* In particular, *the vector product of any vector with itself is zero.*

CAUTION **Vector product vs. scalar product** Do not confuse the expression $AB\sin\phi$ for the magnitude of the vector product $\vec{A}\times\vec{B}$ with the similar expression $AB\cos\phi$ for the scalar product $\vec{A}\cdot\vec{B}$. To see the difference between these two expressions, imagine that we vary the angle between $\vec{A}$ and $\vec{B}$ while keeping their magnitudes constant. When $\vec{A}$ and $\vec{B}$ are parallel, the magnitude of the vector product will be zero and the scalar product will be maximum. When $\vec{A}$ and $\vec{B}$ are perpendicular, the magnitude of the vector product will be maximum and the scalar product will be zero. ▌

There are always *two* directions perpendicular to a given plane, one on each side of the plane. We choose which of these is the direction of $\vec{A}\times\vec{B}$ as follows. Imagine rotating vector $\vec{A}$ about the perpendicular line until $\vec{A}$ is aligned with $\vec{B}$, choosing the smaller of the two possible angles between $\vec{A}$ and $\vec{B}$. Curl the fingers of your right hand around the perpendicular line so that your fingertips point in the direction of rotation; your thumb will then point in the direction of $\vec{A}\times\vec{B}$. Figure 1.30a shows this **right-hand rule** and describes a second way to think about this rule.

Similarly, we determine the direction of $\vec{B} \times \vec{A}$ by rotating $\vec{B}$ into $\vec{A}$ as in Fig. 1.30b. The result is a vector that is *opposite* to the vector $\vec{A} \times \vec{B}$. The vector product is *not* commutative but instead is *anticommutative:* For any two vectors $\vec{A}$ and $\vec{B}$,

$$\vec{A} \times \vec{B} = -\vec{B} \times \vec{A} \qquad (1.21)$$

Just as we did for the scalar product, we can give a geometrical interpretation of the magnitude of the vector product. In **Fig. 1.31a,** $B \sin\phi$ is the component of vector $\vec{B}$ that is *perpendicular* to the direction of vector $\vec{A}$. From Eq. (1.20) the magnitude of $\vec{A} \times \vec{B}$ equals the magnitude of $\vec{A}$ multiplied by the component of $\vec{B}$ that is perpendicular to $\vec{A}$. Figure 1.31b shows that the magnitude of $\vec{A} \times \vec{B}$ also equals the magnitude of $\vec{B}$ multiplied by the component of $\vec{A}$ that is perpendicular to $\vec{B}$. Note that Fig. 1.31 shows the case in which ϕ is between $0°$ and $90°$; draw a similar diagram for ϕ between $90°$ and $180°$ to show that the same geometrical interpretation of the magnitude of $\vec{A} \times \vec{B}$ applies.

Using Components to Calculate the Vector Product

If we know the components of $\vec{A}$ and $\vec{B}$, we can calculate the components of the vector product by using a procedure similar to that for the scalar product. First we work out the multiplication table for unit vectors $\hat{\imath}$, $\hat{\jmath}$, and $\hat{k}$, all three of which are perpendicular to each other (**Fig. 1.32a**). The vector product of any vector with itself is zero, so

$$\hat{\imath} \times \hat{\imath} = \hat{\jmath} \times \hat{\jmath} = \hat{k} \times \hat{k} = \mathbf{0}$$

The boldface zero is a reminder that each product is a zero *vector*—that is, one with all components equal to zero and an undefined direction. Using Eqs. (1.20) and (1.21) and the right-hand rule, we find

$$\hat{\imath} \times \hat{\jmath} = -\hat{\jmath} \times \hat{\imath} = \hat{k}$$

$$\hat{\jmath} \times \hat{k} = -\hat{k} \times \hat{\jmath} = \hat{\imath} \qquad (1.22)$$

$$\hat{k} \times \hat{\imath} = -\hat{\imath} \times \hat{k} = \hat{\jmath}$$

You can verify these equations by referring to Fig. 1.32a.

Next we express $\vec{A}$ and $\vec{B}$ in terms of their components and the corresponding unit vectors, and we expand the expression for the vector product:

$$\begin{aligned}\vec{A} \times \vec{B} = {}& (A_x\hat{\imath} + A_y\hat{\jmath} + A_z\hat{k}) \times (B_x\hat{\imath} + B_y\hat{\jmath} + B_z\hat{k}) \\ = {}& A_x\hat{\imath} \times B_x\hat{\imath} + A_x\hat{\imath} \times B_y\hat{\jmath} + A_x\hat{\imath} \times B_z\hat{k} \\ & + A_y\hat{\jmath} \times B_x\hat{\imath} + A_y\hat{\jmath} \times B_y\hat{\jmath} + A_y\hat{\jmath} \times B_z\hat{k} \\ & + A_z\hat{k} \times B_x\hat{\imath} + A_z\hat{k} \times B_y\hat{\jmath} + A_z\hat{k} \times B_z\hat{k}\end{aligned} \qquad (1.23)$$

We can also rewrite the individual terms in Eq. (1.23) as $A_x\hat{\imath} \times B_y\hat{\jmath} = (A_xB_y)\hat{\imath} \times \hat{\jmath}$, and so on. Evaluating these by using the multiplication table for the unit vectors in Eqs. (1.22) and then grouping the terms, we get

$$\vec{A} \times \vec{B} = (A_yB_z - A_zB_y)\hat{\imath} + (A_zB_x - A_xB_z)\hat{\jmath} + (A_xB_y - A_yB_x)\hat{k} \qquad (1.24)$$

If you compare Eq. (1.24) with Eq. (1.14), you'll see that the components of $\vec{C} = \vec{A} \times \vec{B}$ are

> ····Components of vector (cross) product $\vec{A} \times \vec{B}$····
>
> $$C_x = A_yB_z - A_zB_y \qquad C_y = A_zB_x - A_xB_z \qquad C_z = A_xB_y - A_yB_x \qquad (1.25)$$
>
> $A_x, A_y, A_z = $ components of $\vec{A}$ $B_x, B_y, B_z = $ components of $\vec{B}$

1.31 Calculating the magnitude $AB \sin\phi$ of the vector product of two vectors, $\vec{A} \times \vec{B}$.

(a)

(Magnitude of $\vec{A} \times \vec{B}$) equals $A(B \sin\phi)$.

(Magnitude of $\vec{A}$) $\times$ (Component of $\vec{B}$ perpendicular to $\vec{A}$)

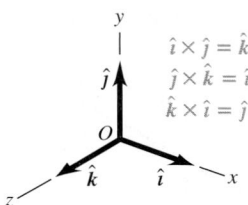

(b)

(Magnitude of $\vec{A} \times \vec{B}$) also equals $B(A \sin\phi)$.

(Magnitude of $\vec{B}$) $\times$ (Component of $\vec{A}$ perpendicular to $\vec{B}$)

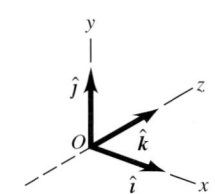

1.32 (a) We will always use a right-handed coordinate system, like this one. (b) We will never use a left-handed coordinate system (in which $\hat{\imath} \times \hat{\jmath} = -\hat{k}$, and so on).

(a) A right-handed coordinate system

$\hat{\imath} \times \hat{\jmath} = \hat{k}$
$\hat{\jmath} \times \hat{k} = \hat{\imath}$
$\hat{k} \times \hat{\imath} = \hat{\jmath}$

(b) A left-handed coordinate system; we will not use these.

With the axis system of Fig. 1.32a, if we reverse the direction of the z-axis, we get the system shown in Fig. 1.32b. Then, as you may verify, the definition of the vector product gives $\hat{\imath} \times \hat{\jmath} = -\hat{k}$ instead of $\hat{\imath} \times \hat{\jmath} = \hat{k}$. In fact, all vector products of unit vectors $\hat{\imath}$, $\hat{\jmath}$, and $\hat{k}$ would have signs opposite to those in Eqs. (1.22). So there are two kinds of coordinate systems, which differ in the signs of the vector products of unit vectors. An axis system in which $\hat{\imath} \times \hat{\jmath} = \hat{k}$, as in Fig. 1.32a, is called a **right-handed system.** The usual practice is to use *only* right-handed systems, and we'll follow that practice throughout this book.

EXAMPLE 1.11 CALCULATING A VECTOR PRODUCT

Vector $\vec{A}$ has magnitude 6 units and is in the direction of the $+x$-axis. Vector $\vec{B}$ has magnitude 4 units and lies in the xy-plane, making an angle of 30° with the $+x$-axis (**Fig. 1.33**). Find the vector product $\vec{C} = \vec{A} \times \vec{B}$.

SOLUTION

IDENTIFY and SET UP: We'll find the vector product in two ways, which will provide a check of our calculations. First we'll use Eq. (1.20) and the right-hand rule; then we'll use Eqs. (1.25) to find the vector product by using components.

1.33 Vectors $\vec{A}$ and $\vec{B}$ and their vector product $\vec{C} = \vec{A} \times \vec{B}$. Vector $\vec{B}$ lies in the xy-plane.

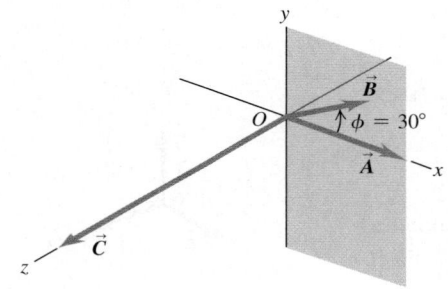

EXECUTE: From Eq. (1.20) the magnitude of the vector product is

$$AB \sin \phi = (6)(4)(\sin 30°) = 12$$

By the right-hand rule, the direction of $\vec{A} \times \vec{B}$ is along the $+z$-axis (the direction of the unit vector $\hat{k}$), so $\vec{C} = \vec{A} \times \vec{B} = 12\hat{k}$.

To use Eqs. (1.25), we first determine the components of $\vec{A}$ and $\vec{B}$. Note that $\vec{A}$ points along the x-axis, so its only nonzero component is A_x. For $\vec{B}$, Fig. 1.33 shows that $\phi = 30°$ is measured from the $+x$-axis toward the $+y$-axis, so we can use Eqs. (1.5):

$$A_x = 6 \qquad A_y = 0 \qquad A_z = 0$$
$$B_x = 4 \cos 30° = 2\sqrt{3} \qquad B_y = 4 \sin 30° = 2 \qquad B_z = 0$$

Then Eqs. (1.25) yield

$$C_x = (0)(0) - (0)(2) = 0$$
$$C_y = (0)(2\sqrt{3}) - (6)(0) = 0$$
$$C_z = (6)(2) - (0)(2\sqrt{3}) = 12$$

Thus again we have $\vec{C} = 12\hat{k}$.

EVALUATE: Both methods give the same result. Depending on the situation, one or the other of the two approaches may be the more convenient one to use.

TEST YOUR UNDERSTANDING OF SECTION 1.10 Vector $\vec{A}$ has magnitude 2 and vector $\vec{B}$ has magnitude 3. The angle ϕ between $\vec{A}$ and $\vec{B}$ is (i) 0°, (ii) 90°, or (iii) 180°. For each of the following situations, state what the value of ϕ must be. (In each situation there may be more than one correct answer.) (a) $\vec{A} \cdot \vec{B} = 0$; (b) $\vec{A} \times \vec{B} = 0$; (c) $\vec{A} \cdot \vec{B} = 6$; (d) $\vec{A} \cdot \vec{B} = -6$; (e) (magnitude of $\vec{A} \times \vec{B}$) = 6. ∎

Physical quantities and units: Three fundamental physical quantities are mass, length, and time. The corresponding fundamental SI units are the kilogram, the meter, and the second. Derived units for other physical quantities are products or quotients of the basic units. Equations must be dimensionally consistent; two terms can be added only when they have the same units. (See Examples 1.1 and 1.2.)

Significant figures: The accuracy of a measurement can be indicated by the number of significant figures or by a stated uncertainty. The significant figures in the result of a calculation are determined by the rules summarized in Table 1.2. When only crude estimates are available for input data, we can often make useful order-of-magnitude estimates. (See Examples 1.3 and 1.4.)

Significant figures in magenta

$$\pi = \frac{C}{2r} = \frac{0.424 \text{ m}}{2(0.06750 \text{ m})} = 3.14$$

$$123.62 + 8.9 = 132.5$$

Scalars, vectors, and vector addition: Scalar quantities are numbers and combine according to the usual rules of arithmetic. Vector quantities have direction as well as magnitude and combine according to the rules of vector addition. The negative of a vector has the same magnitude but points in the opposite direction. (See Example 1.5.)

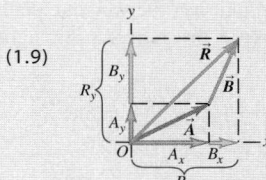

Vector components and vector addition: Vectors can be added by using components of vectors. The x-component of $\vec{R} = \vec{A} + \vec{B}$ is the sum of the x-components of $\vec{A}$ and $\vec{B}$, and likewise for the y- and z-components. (See Examples 1.6 and 1.7.)

$$R_x = A_x + B_x$$
$$R_y = A_y + B_y \qquad (1.9)$$
$$R_z = A_z + B_z$$

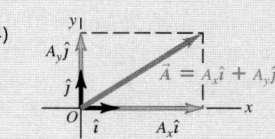

Unit vectors: Unit vectors describe directions in space. A unit vector has a magnitude of 1, with no units. The unit vectors $\hat{\imath}$, $\hat{\jmath}$, and $\hat{k}$, aligned with the x-, y-, and z-axes of a rectangular coordinate system, are especially useful. (See Example 1.8.)

$$\vec{A} = A_x\hat{\imath} + A_y\hat{\jmath} + A_z\hat{k} \qquad (1.14)$$

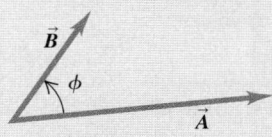

Scalar product: The scalar product $C = \vec{A} \cdot \vec{B}$ of two vectors $\vec{A}$ and $\vec{B}$ is a scalar quantity. It can be expressed in terms of the magnitudes of $\vec{A}$ and $\vec{B}$ and the angle ϕ between the two vectors, or in terms of the components of $\vec{A}$ and $\vec{B}$. The scalar product is commutative; $\vec{A} \cdot \vec{B} = \vec{B} \cdot \vec{A}$. The scalar product of two perpendicular vectors is zero. (See Examples 1.9 and 1.10.)

$$\vec{A} \cdot \vec{B} = AB\cos\phi = |\vec{A}||\vec{B}|\cos\phi \qquad (1.16)$$
$$\vec{A} \cdot \vec{B} = A_xB_x + A_yB_y + A_zB_z \qquad (1.19)$$

Scalar product $\vec{A} \cdot \vec{B} = AB\cos\phi$

Vector product: The vector product $\vec{C} = \vec{A} \times \vec{B}$ of two vectors $\vec{A}$ and $\vec{B}$ is a third vector $\vec{C}$. The magnitude of $\vec{A} \times \vec{B}$ depends on the magnitudes of $\vec{A}$ and $\vec{B}$ and the angle ϕ between the two vectors. The direction of $\vec{A} \times \vec{B}$ is perpendicular to the plane of the two vectors being multiplied, as given by the right-hand rule. The components of $\vec{C} = \vec{A} \times \vec{B}$ can be expressed in terms of the components of $\vec{A}$ and $\vec{B}$. The vector product is not commutative; $\vec{A} \times \vec{B} = -\vec{B} \times \vec{A}$. The vector product of two parallel or antiparallel vectors is zero. (See Example 1.11.)

$$C = AB \sin\phi \qquad (1.20)$$

$$C_x = A_y B_z - A_z B_y$$
$$C_y = A_z B_x - A_x B_z \qquad (1.25)$$
$$C_z = A_x B_y - A_y B_x$$

$\vec{A} \times \vec{B}$ is perpendicular to the plane of $\vec{A}$ and $\vec{B}$.

(Magnitude of $\vec{A} \times \vec{B}$) $= AB \sin\phi$

BRIDGING PROBLEM VECTORS ON THE ROOF

An air-conditioning unit is fastened to a roof that slopes at an angle of 35° above the horizontal (**Fig. 1.34**). Its weight is a force $\vec{F}$ on the air conditioner that is directed vertically downward. In order that the unit not crush the roof tiles, the component of the unit's weight perpendicular to the roof cannot exceed 425 N. (One newton, or 1 N, is the SI unit of force. It is equal to 0.2248 lb.) (a) What is the maximum allowed weight of the unit? (b) If the fasteners fail, the unit slides 1.50 m along the roof before it comes to a halt against a ledge. How much work does the weight force do on the unit during its slide if the unit has the weight calculated in part (a)? The work done by a force $\vec{F}$ on an object that undergoes a displacement $\vec{s}$ is $W = \vec{F} \cdot \vec{s}$.

1.34 An air-conditioning unit on a slanted roof.

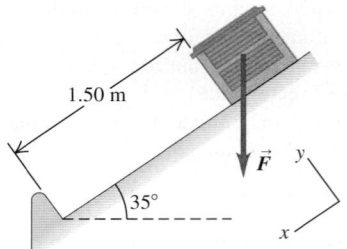

SOLUTION GUIDE

IDENTIFY and SET UP
1. This problem involves vectors and components. What are the known quantities? Which aspect(s) of the weight vector (magnitude, direction, and/or particular components) represent the target variable for part (a)? Which aspect(s) must you know to solve part (b)?
2. Make a sketch based on Fig. 1.34. Draw the x- and y-axes, choosing the positive direction for each. Your axes don't have to be horizontal and vertical, but they do have to be mutually perpendicular. Figure 1.34 shows a convenient choice of axes: The x-axis is parallel to the slope of the roof.
3. Choose the equations you'll use to determine the target variables.

EXECUTE
4. Use the relationship between the magnitude and direction of a vector and its components to solve for the target variable in part (a). Be careful: Is 35° the correct angle to use in the equation? (*Hint:* Check your sketch.)
5. Make sure your answer has the correct number of significant figures.
6. Use the definition of the scalar product to solve for the target variable in part (b). Again, use the correct number of significant figures.

EVALUATE
7. Did your answer to part (a) include a vector component whose absolute value is greater than the magnitude of the vector? Is that possible?
8. There are two ways to find the scalar product of two vectors, one of which you used to solve part (b). Check your answer by repeating the calculation, using the other way. Do you get the same answer?

Problems

•, ••, •••: Difficulty levels. CP: Cumulative problems incorporating material from earlier chapters. CALC: Problems requiring calculus.
DATA: Problems involving real data, scientific evidence, experimental design, and/or statistical reasoning. BIO: Biosciences problems.

DISCUSSION QUESTIONS

Q1.1 How many correct experiments do we need to disprove a theory? How many do we need to prove a theory? Explain.

Q1.2 Suppose you are asked to compute the tangent of 5.00 meters. Is this possible? Why or why not?

Q1.3 What is your height in centimeters? What is your weight in newtons?

Q1.4 The U.S. National Institute of Standards and Technology (NIST) maintains several accurate copies of the international standard kilogram. Even after careful cleaning, these national standard kilograms are gaining mass at an average rate of about 1 μg/y (y = year) when compared every 10 years or so to the standard international kilogram. Does this apparent increase have any importance? Explain.

Q1.5 What physical phenomena (other than a pendulum or cesium clock) could you use to define a time standard?

Q1.6 Describe how you could measure the thickness of a sheet of paper with an ordinary ruler.

Q1.7 The quantity $\pi = 3.14159\ldots$ is a number with no dimensions, since it is a ratio of two lengths. Describe two or three other geometrical or physical quantities that are dimensionless.

Q1.8 What are the units of volume? Suppose another student tells you that a cylinder of radius r and height h has volume given by $\pi r^3 h$. Explain why this cannot be right.

Q1.9 Three archers each fire four arrows at a target. Joe's four arrows hit at points 10 cm above, 10 cm below, 10 cm to the left, and 10 cm to the right of the center of the target. All four of Moe's arrows hit within 1 cm of a point 20 cm from the center, and Flo's four arrows hit within 1 cm of the center. The contest judge says that one of the archers is precise but not accurate, another archer is accurate but not precise, and the third archer is both accurate and precise. Which description applies to which archer? Explain.

Q1.10 Is the vector $(\hat{\imath} + \hat{\jmath} + \hat{k})$ a unit vector? Is the vector $(3.0\hat{\imath} - 2.0\hat{\jmath})$ a unit vector? Justify your answers.

Q1.11 A circular racetrack has a radius of 500 m. What is the displacement of a bicyclist when she travels around the track from the north side to the south side? When she makes one complete circle around the track? Explain.

Q1.12 Can you find two vectors with different lengths that have a vector sum of zero? What length restrictions are required for three vectors to have a vector sum of zero? Explain.

Q1.13 The "direction of time" is said to proceed from past to future. Does this mean that time is a vector quantity? Explain.

Q1.14 Air traffic controllers give instructions called "vectors" to tell airline pilots in which direction they are to fly. If these are the only instructions given, is the name "vector" used correctly? Why or why not?

Q1.15 Can you find a vector quantity that has a magnitude of zero but components that are not zero? Explain. Can the magnitude of a vector be less than the magnitude of any of its components? Explain.

Q1.16 (a) Does it make sense to say that a vector is *negative*? Why? (b) Does it make sense to say that one vector is the negative of another? Why? Does your answer here contradict what you said in part (a)?

Q1.17 If $\vec{C} = \vec{A} + \vec{B}$, what must be true about the directions and magnitudes of $\vec{A}$ and $\vec{B}$ if $C = A + B$? What must be true about the directions and magnitudes of $\vec{A}$ and $\vec{B}$ if $C = 0$?

Q1.18 If $\vec{A}$ and $\vec{B}$ are nonzero vectors, is it possible for *both* $\vec{A} \cdot \vec{B}$ and $\vec{A} \times \vec{B}$ to be zero? Explain.

Q1.19 What does $\vec{A} \cdot \vec{A}$, the scalar product of a vector with itself, give? What about $\vec{A} \times \vec{A}$, the vector product of a vector with itself?

Q1.20 Let $\vec{A}$ represent any nonzero vector. Why is $\vec{A}/A$ a unit vector, and what is its direction? If θ is the angle that $\vec{A}$ makes with the $+x$-axis, explain why $(\vec{A}/A) \cdot \hat{\imath}$ is called the *direction cosine* for that axis.

Q1.21 Figure 1.7 shows the result of an unacceptable error in the stopping position of a train. If a train travels 890 km from Berlin to Paris and then overshoots the end of the track by 10.0 m, what is the percent error in the total distance covered? Is it correct to write the total distance covered by the train as 890,010 m? Explain.

Q1.22 Which of the following are legitimate mathematical operations: (a) $\vec{A} \cdot (\vec{B} - \vec{C})$; (b) $(\vec{A} - \vec{B}) \times \vec{C}$; (c) $\vec{A} \cdot (\vec{B} \times \vec{C})$; (d) $\vec{A} \times (\vec{B} \times \vec{C})$; (e) $\vec{A} \times (\vec{B} \cdot \vec{C})$? In each case, give the reason for your answer.

Q1.23 Consider the vector products $\vec{A} \times (\vec{B} \times \vec{C})$ and $(\vec{A} \times \vec{B}) \times \vec{C}$. Give an example that illustrates the general rule that these two vector products do not have the same magnitude or direction. Can you choose vectors $\vec{A}$, $\vec{B}$, and $\vec{C}$ such that these two vector products *are* equal? If so, give an example.

Q1.24 Show that, no matter what $\vec{A}$ and $\vec{B}$ are, $\vec{A} \cdot (\vec{A} \times \vec{B}) = 0$. (*Hint:* Do not look for an elaborate mathematical proof. Consider the definition of the direction of the cross product.)

Q1.25 (a) If $\vec{A} \cdot \vec{B} = 0$, does it necessarily follow that $A = 0$ or $B = 0$? Explain. (b) If $\vec{A} \times \vec{B} = \mathbf{0}$, does it necessarily follow that $A = 0$ or $B = 0$? Explain.

Q1.26 If $\vec{A} = \mathbf{0}$ for a vector in the xy-plane, does it follow that $A_x = -A_y$? What *can* you say about A_x and A_y?

EXERCISES

Section 1.3 Standards and Units
Section 1.4 Using and Converting Units

1.1 • Starting with the definition 1 in. = 2.54 cm, find the number of (a) kilometers in 1.00 mile and (b) feet in 1.00 km.

1.2 •• According to the label on a bottle of salad dressing, the volume of the contents is 0.473 liter (L). Using only the conversions 1 L = 1000 cm³ and 1 in. = 2.54 cm, express this volume in cubic inches.

1.3 •• How many nanoseconds does it take light to travel 1.00 ft in vacuum? (This result is a useful quantity to remember.)

1.4 •• The density of gold is 19.3 g/cm³. What is this value in kilograms per cubic meter?

1.5 • The most powerful engine available for the classic 1963 Chevrolet Corvette Sting Ray developed 360 horsepower and had a displacement of 327 cubic inches. Express this displacement in liters (L) by using only the conversions 1 L = 1000 cm³ and 1 in. = 2.54 cm.

1.6 •• A square field measuring 100.0 m by 100.0 m has an area of 1.00 hectare. An acre has an area of 43,600 ft². If a lot has an area of 12.0 acres, what is its area in hectares?

1.7 • How many years older will you be 1.00 gigasecond from now? (Assume a 365-day year.)

1.8 • While driving in an exotic foreign land, you see a speed limit sign that reads 180,000 furlongs per fortnight. How many miles per hour is this? (One furlong is $\frac{1}{8}$ mile, and a fortnight is 14 days. A furlong originally referred to the length of a plowed furrow.)

1.9 • A certain fuel-efficient hybrid car gets gasoline mileage of 55.0 mpg (miles per gallon). (a) If you are driving this car in Europe and want to compare its mileage with that of other European cars, express this mileage in km/L (L = liter). Use the conversion factors in Appendix E. (b) If this car's gas tank holds 45 L, how many tanks of gas will you use to drive 1500 km?

1.10 • The following conversions occur frequently in physics and are very useful. (a) Use 1 mi = 5280 ft and 1 h = 3600 s to convert 60 mph to units of ft/s. (b) The acceleration of a freely falling object is 32 ft/s². Use 1 ft = 30.48 cm to express this acceleration in units of m/s². (c) The density of water is 1.0 g/cm³. Convert this density to units of kg/m³.

1.11 •• **Neptunium.** In the fall of 2002, scientists at Los Alamos National Laboratory determined that the critical mass of neptunium-237 is about 60 kg. The critical mass of a fissionable material is the minimum amount that must be brought together to start a nuclear chain reaction. Neptunium-237 has a density of 19.5 g/cm³. What would be the radius of a sphere of this material that has a critical mass?

1.12 • BIO (a) The recommended daily allowance (RDA) of the trace metal magnesium is 410 mg/day for males. Express this quantity in μg/day. (b) For adults, the RDA of the amino acid lysine is 12 mg per kg of body weight. How many grams per day should a 75-kg adult receive? (c) A typical multivitamin tablet can contain 2.0 mg of vitamin B_2 (riboflavin), and the RDA is 0.0030 g/day. How many such tablets should a person take each day to get the proper amount of this vitamin, if he gets none from other sources? (d) The RDA for the trace element selenium is 0.000070 g/day. Express this dose in mg/day.

1.13 •• BIO **Bacteria.** Bacteria vary in size, but a diameter of 2.0 μm is not unusual. What are the volume (in cubic centimeters) and surface area (in square millimeters) of a spherical bacterium of that size? (Consult Appendix B for relevant formulas.)

Section 1.5 Uncertainty and Significant Figures

1.14 • With a wooden ruler, you measure the length of a rectangular piece of sheet metal to be 12 mm. With micrometer calipers, you measure the width of the rectangle to be 5.98 mm. Use the correct number of significant figures: What is (a) the area of the rectangle; (b) the ratio of the rectangle's width to its length; (c) the perimeter of the rectangle; (d) the difference between length and the width; and (e) the ratio of the length to the width?

1.15 •• A useful and easy-to-remember approximate value for the number of seconds in a year is $\pi \times 10^7$. Determine the percent error in this approximate value. (There are 365.24 days in one year.)

1.16 • Express each approximation of π to six significant figures: (a) 22/7 and (b) 355/113. (c) Are these approximations accurate to that precision?

Section 1.6 Estimates and Orders of Magnitude

1.17 •• BIO A rather ordinary middle-aged man is in the hospital for a routine checkup. The nurse writes "200" on the patient's medical chart but forgets to include the units. Which of these quantities could the 200 plausibly represent? The patient's (a) mass in kilograms; (b) height in meters; (c) height in centimeters; (d) height in millimeters; (e) age in months.

1.18 • How many gallons of gasoline are used in the United States in one day? Assume that there are two cars for every three people, that each car is driven an average of 10,000 miles per year, and that the average car gets 20 miles per gallon.

1.19 • BIO How many times does a typical person blink her eyes in a lifetime?

1.20 • BIO Four astronauts are in a spherical space station. (a) If, as is typical, each of them breathes about 500 cm³ of air with each breath, approximately what volume of air (in cubic meters) do these astronauts breathe in a year? (b) What would the diameter (in meters) of the space station have to be to contain all this air?

1.21 • In Wagner's opera *Das Rheingold*, the goddess Freia is ransomed for a pile of gold just tall enough and wide enough to hide her from sight. Estimate the monetary value of this pile. The density of gold is 19.3 g/cm³, and take its value to be about $10 per gram.

1.22 • BIO How many times does a human heart beat during a person's lifetime? How many gallons of blood does it pump? (Estimate that the heart pumps 50 cm³ of blood with each beat.)

1.23 • You are using water to dilute small amounts of chemicals in the laboratory, drop by drop. How many drops of water are in a 1.0-L bottle? (*Hint:* Start by estimating the diameter of a drop of water.)

Section 1.7 Vectors and Vector Addition

1.24 •• For the vectors $\vec{A}$ and $\vec{B}$ in **Fig. E1.24**, use a scale drawing to find the magnitude and direction of (a) the vector sum $\vec{A} + \vec{B}$ and (b) the vector difference $\vec{A} - \vec{B}$. Use your answers to find the magnitude and direction of (c) $-\vec{A} - \vec{B}$ and (d) $\vec{B} - \vec{A}$. (See also Exercise 1.31 for a different approach.)

Figure **E1.24**

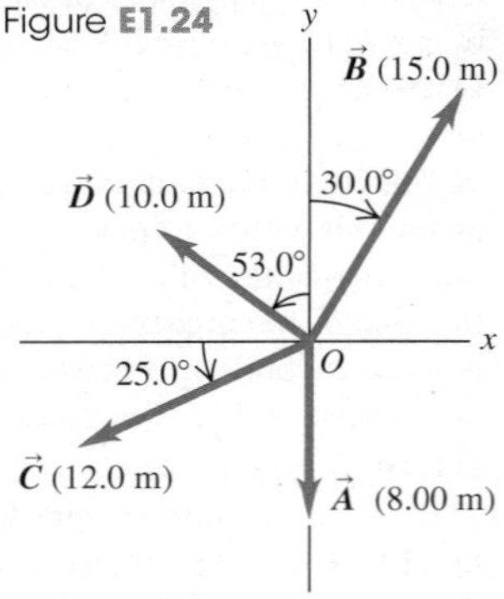

1.25 •• A postal employee drives a delivery truck along the route shown in **Fig. E1.25**. Determine the magnitude and direction of the resultant displacement by drawing a scale diagram. (See also Exercise 1.32 for a different approach.)

Figure **E1.25**

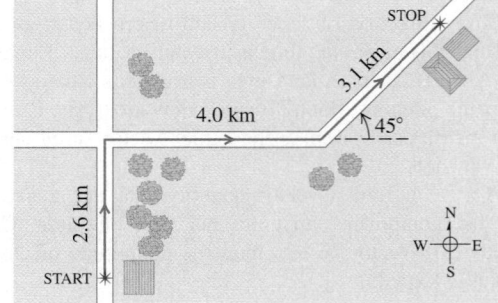

1.26 •• A spelunker is surveying a cave. She follows a passage 180 m straight west, then 210 m in a direction 45° east of south, and then 280 m at 30° east of north. After a fourth displacement, she finds herself back where she started. Use a scale drawing to determine the magnitude and direction of the fourth displacement. (See also Problem 1.61 for a different approach.)

Section 1.8 Components of Vectors

1.27 • Compute the x- and y-components of the vectors $\vec{A}, \vec{B}, \vec{C}$, and $\vec{D}$ in Fig. E1.24.

1.28 •• Let θ be the angle that the vector $\vec{A}$ makes with the $+x$-axis, measured counterclockwise from that axis. Find angle θ for a vector that has these components: (a) $A_x = 2.00$ m, $A_y = -1.00$ m; (b) $A_x = 2.00$ m, $A_y = 1.00$ m; (c) $A_x = -2.00$ m, $A_y = 1.00$ m; (d) $A_x = -2.00$ m, $A_y = -1.00$ m.

1.29 • Vector $\vec{A}$ has y-component $A_y = +9.60$ m. $\vec{A}$ makes an angle of $32.0°$ counterclockwise from the $+y$-axis. (a) What is the x-component of $\vec{A}$? (b) What is the magnitude of $\vec{A}$?

1.30 • Vector $\vec{A}$ is in the direction $34.0°$ clockwise from the $-y$-axis. The x-component of $\vec{A}$ is $A_x = -16.0$ m. (a) What is the y-component of $\vec{A}$? (b) What is the magnitude of $\vec{A}$?

1.31 • For the vectors $\vec{A}$ and $\vec{B}$ in Fig. E1.24, use the method of components to find the magnitude and direction of (a) the vector sum $\vec{A} + \vec{B}$; (b) the vector sum $\vec{B} + \vec{A}$; (c) the vector difference $\vec{A} - \vec{B}$; (d) the vector difference $\vec{B} - \vec{A}$.

1.32 •• A postal employee drives a delivery truck over the route shown in Fig. E1.25. Use the method of components to determine the magnitude and direction of her resultant displacement. In a vector-addition diagram (roughly to scale), show that the resultant displacement found from your diagram is in qualitative agreement with the result you obtained by using the method of components.

1.33 •• A disoriented physics professor drives 3.25 km north, then 2.20 km west, and then 1.50 km south. Find the magnitude and direction of the resultant displacement, using the method of components. In a vector-addition diagram (roughly to scale), show that the resultant displacement found from your diagram is in qualitative agreement with the result you obtained by using the method of components.

1.34 • Find the magnitude and direction of the vector represented by the following pairs of components: (a) $A_x = -8.60$ cm, $A_y = 5.20$ cm; (b) $A_x = -9.70$ m, $A_y = -2.45$ m; (c) $A_x = 7.75$ km, $A_y = -2.70$ km.

1.35 •• Vector $\vec{A}$ is 2.80 cm long and is $60.0°$ above the x-axis in the first quadrant. Vector $\vec{B}$ is 1.90 cm long and is $60.0°$ below the x-axis in the fourth quadrant (**Fig. E1.35**). Use components to find the magnitude and direction of (a) $\vec{A} + \vec{B}$; (b) $\vec{A} - \vec{B}$; (c) $\vec{B} - \vec{A}$. In each case, sketch the vector addition or subtraction and show that your numerical answers are in qualitative agreement with your sketch.

Figure **E1.35**

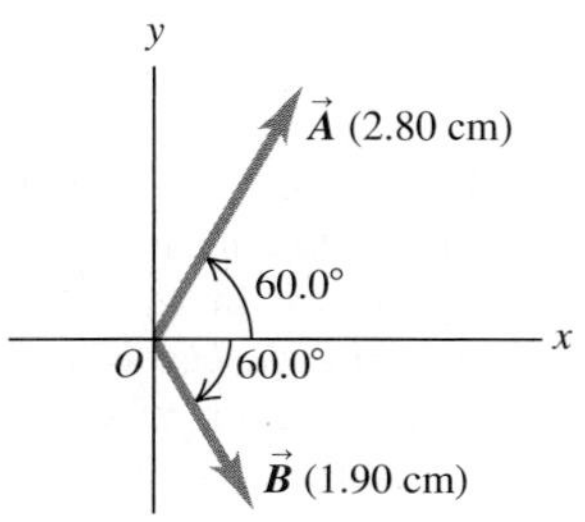

Section 1.9 Unit Vectors

1.36 • In each case, find the x- and y-components of vector $\vec{A}$: (a) $\vec{A} = 5.0\hat{i} - 6.3\hat{j}$; (b) $\vec{A} = 11.2\hat{j} - 9.91\hat{i}$; (c) $\vec{A} = -15.0\hat{i} + 22.4\hat{j}$; (d) $\vec{A} = 5.0\vec{B}$, where $\vec{B} = 4\hat{i} - 6\hat{j}$.

1.37 •• Write each vector in Fig. E1.24 in terms of the unit vectors $\hat{i}$ and $\hat{j}$.

1.38 •• Given two vectors $\vec{A} = 4.00\hat{i} + 7.00\hat{j}$ and $\vec{B} = 5.00\hat{i} - 2.00\hat{j}$, (a) find the magnitude of each vector; (b) use unit vectors

to write an expression for the vector difference $\vec{A} - \vec{B}$; and (c) find the magnitude and direction of the vector difference $\vec{A} - \vec{B}$. (d) In a vector diagram show $\vec{A}, \vec{B}$, and $\vec{A} - \vec{B}$, and show that your diagram agrees qualitatively with your answer to part (c).

1.39 •• (a) Write each vector in Fig. E1.39 in terms of the unit vectors $\hat{i}$ and $\hat{j}$. (b) Use unit vectors to express vector $\vec{C}$, where $\vec{C} = 3.00\vec{A} - 4.00\vec{B}$. (c) Find the magnitude and direction of $\vec{C}$.

Figure **E1.39**

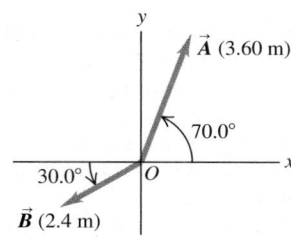

1.40 • You are given two vectors $\vec{A} = -3.00\hat{i} + 6.00\hat{j}$ and $\vec{B} = 7.00\hat{i} + 2.00\hat{j}$. Let counterclockwise angles be positive. (a) What angle does $\vec{A}$ make with the $+x$-axis? (b) What angle does $\vec{B}$ make with the $+x$-axis? (c) Vector $\vec{C}$ is the sum of $\vec{A}$ and $\vec{B}$, so $\vec{C} = \vec{A} + \vec{B}$. What angle does $\vec{C}$ make with the $+x$-axis?

1.41 • Given two vectors $\vec{A} = -2.00\hat{i} + 3.00\hat{j} + 4.00\hat{k}$ and $\vec{B} = 3.00\hat{i} + 1.00\hat{j} - 3.00\hat{k}$, (a) find the magnitude of each vector; (b) use unit vectors to write an expression for the vector difference $\vec{A} - \vec{B}$; and (c) find the magnitude of the vector difference $\vec{A} - \vec{B}$. Is this the same as the magnitude of $\vec{B} - \vec{A}$? Explain.

Section 1.10 Products of Vectors

1.42 •• (a) Find the scalar product of the vectors $\vec{A}$ and $\vec{B}$ given in Exercise 1.38. (b) Find the angle between these two vectors.

1.43 • For the vectors $\vec{A}, \vec{B}$, and $\vec{C}$ in Fig. E1.24, find the scalar products (a) $\vec{A} \cdot \vec{B}$; (b) $\vec{B} \cdot \vec{C}$; (c) $\vec{A} \cdot \vec{C}$.

1.44 •• Find the vector product $\vec{A} \times \vec{B}$ (expressed in unit vectors) of the two vectors given in Exercise 1.38. What is the magnitude of the vector product?

1.45 •• Find the angle between each of these pairs of vectors:

(a) $\vec{A} = -2.00\hat{i} + 6.00\hat{j}$ and $\vec{B} = 2.00\hat{i} - 3.00\hat{j}$

(b) $\vec{A} = 3.00\hat{i} + 5.00\hat{j}$ and $\vec{B} = 10.00\hat{i} + 6.00\hat{j}$

(c) $\vec{A} = -4.00\hat{i} + 2.00\hat{j}$ and $\vec{B} = 7.00\hat{i} + 14.00\hat{j}$

1.46 • For the two vectors in Fig. E1.35, find the magnitude and direction of (a) the vector product $\vec{A} \times \vec{B}$; (b) the vector product $\vec{B} \times \vec{A}$.

1.47 • For the two vectors $\vec{A}$ and $\vec{D}$ in Fig. E1.24, find the magnitude and direction of (a) the vector product $\vec{A} \times \vec{D}$; (b) the vector product $\vec{D} \times \vec{A}$.

1.48 • For the two vectors $\vec{A}$ and $\vec{B}$ in Fig. E1.39, find (a) the scalar product $\vec{A} \cdot \vec{B}$; (b) the magnitude and direction of the vector product $\vec{A} \times \vec{B}$.

PROBLEMS

1.49 •• **White Dwarfs and Neutron Stars.** Recall that density is mass divided by volume, and consult Appendix B as needed. (a) Calculate the average density of the earth in g/cm^3, assuming our planet is a perfect sphere. (b) In about 5 billion years, at the end of its lifetime, our sun will end up as a white dwarf that has about the same mass as it does now but is reduced to about 15,000 km in diameter. What will be its density at that stage? (c) A neutron star is the remnant of certain supernovae (explosions of giant stars). Typically, neutron stars are about 20 km in diameter and have about the same mass as our sun. What is a typical neutron star density in g/cm^3?

1.50 • An acre has a length of one furlong ($\frac{1}{8}$ mi) and a width one-tenth of its length. (a) How many acres are in a square mile? (b) How many square feet are in an acre? See Appendix E. (c) An acre-foot is the volume of water that would cover 1 acre of flat land to a depth of 1 foot. How many gallons are in 1 acre-foot?

1.51 •• **An Earthlike Planet.** In January 2006 astronomers reported the discovery of a planet, comparable in size to the earth, orbiting another star and having a mass about 5.5 times the earth's mass. It is believed to consist of a mixture of rock and ice, similar to Neptune. If this planet has the same density as Neptune (1.76 g/cm^3), what is its radius expressed (a) in kilometers and (b) as a multiple of earth's radius? Consult Appendix F for astronomical data.

1.52 •• **The Hydrogen Maser.** A maser is a laser-type device that produces electromagnetic waves with frequencies in the microwave and radio-wave bands of the electromagnetic spectrum. You can use the radio waves generated by a hydrogen maser as a standard of frequency. The frequency of these waves is 1,420,405,751.786 hertz. (A hertz is another name for one cycle per second.) A clock controlled by a hydrogen maser is off by only 1 s in 100,000 years. For the following questions, use only three significant figures. (The large number of significant figures given for the frequency simply illustrates the remarkable accuracy to which it has been measured.) (a) What is the time for one cycle of the radio wave? (b) How many cycles occur in 1 h? (c) How many cycles would have occurred during the age of the earth, which is estimated to be 4.6×10^9 years? (d) By how many seconds would a hydrogen maser clock be off after a time interval equal to the age of the earth?

1.53 • **BIO** **Breathing Oxygen.** The density of air under standard laboratory conditions is 1.29 kg/m^3, and about 20% of that air consists of oxygen. Typically, people breathe about $\frac{1}{2}$ L of air per breath. (a) How many grams of oxygen does a person breathe in a day? (b) If this air is stored uncompressed in a cubical tank, how long is each side of the tank?

1.54 ••• A rectangular piece of aluminum is 7.60 ± 0.01 cm long and 1.90 ± 0.01 cm wide. (a) Find the area of the rectangle and the uncertainty in the area. (b) Verify that the fractional uncertainty in the area is equal to the sum of the fractional uncertainties in the length and in the width. (This is a general result.)

1.55 ••• As you eat your way through a bag of chocolate chip cookies, you observe that each cookie is a circular disk with a diameter of 8.50 ± 0.02 cm and a thickness of 0.050 ± 0.005 cm. (a) Find the average volume of a cookie and the uncertainty in the volume. (b) Find the ratio of the diameter to the thickness and the uncertainty in this ratio.

1.56 • **BIO** Biological tissues are typically made up of 98% water. Given that the density of water is 1.0×10^3 kg/m^3, estimate the mass of (a) the heart of an adult human; (b) a cell with a diameter of 0.5 μm; (c) a honeybee.

1.57 • **BIO** Estimate the number of atoms in your body. (*Hint:* Based on what you know about biology and chemistry, what are the most common types of atom in your body? What is the mass of each type of atom? Appendix D gives the atomic masses of different elements, measured in atomic mass units; you can find the value of an atomic mass unit, or 1 u, in Appendix E.)

1.58 •• Two ropes in a vertical plane exert equal-magnitude forces on a hanging weight but pull with an angle of 72.0° between them. What pull does each rope exert if their resultant pull is 372 N directly upward?

1.59 ••• Two workers pull horizontally on a heavy box, but one pulls twice as hard as the other. The larger pull is directed at 21.0° west of north, and the resultant of these two pulls is 460.0 N directly northward. Use vector components to find the magnitude of each of these pulls and the direction of the smaller pull.

1.60 •• Three horizontal ropes pull on a large stone stuck in the ground, producing the vector forces $\vec{A}$, $\vec{B}$, and $\vec{C}$ shown in **Fig. P1.60.** Find the magnitude and direction of a fourth force on the stone that will make the vector sum of the four forces zero.

Figure **P1.60**

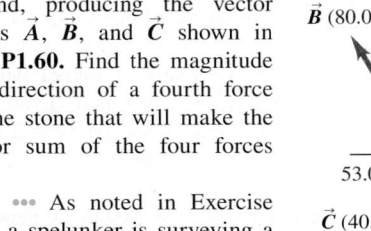

1.61 ••• As noted in Exercise 1.26, a spelunker is surveying a cave. She follows a passage 180 m straight west, then 210 m in a direction 45° east of south, and then 280 m at 30° east of north. After a fourth displacement, she finds herself back where she started. Use the method of components to determine the magnitude and direction of the fourth displacement. Draw the vector-addition diagram and show that it is in qualitative agreement with your numerical solution.

1.62 ••• **Emergency Landing.** A plane leaves the airport in Galisteo and flies 170 km at 68.0° east of north; then it changes direction to fly 230 km at 36.0° south of east, after which it makes an immediate emergency landing in a pasture. When the airport sends out a rescue crew, in which direction and how far should this crew fly to go directly to this plane?

1.63 ••• **BIO** **Dislocated Shoulder.** A patient with a dislocated shoulder is put into a traction apparatus as shown in **Fig. P1.63.** The pulls $\vec{A}$ and $\vec{B}$ have equal magnitudes and must combine to produce an outward traction force of 12.8 N on the patient's arm. How large should these pulls be?

Figure **P1.63**

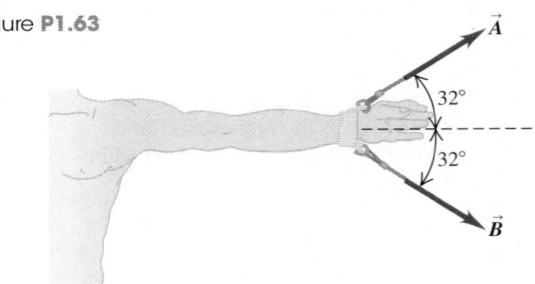

1.64 •• A sailor in a small sailboat encounters shifting winds. She sails 2.00 km east, next 3.50 km southeast, and then an additional distance in an unknown direction. Her final position is 5.80 km directly east of the starting point (**Fig. P1.64**). Find the magnitude and direction of the third leg of the journey. Draw the vector-addition diagram and show that it is in qualitative agreement with your numerical solution.

Figure **P1.64**

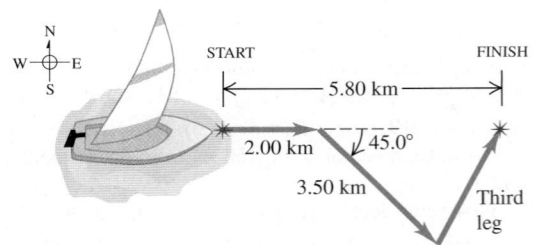

1.65 ·· You leave the airport in College Station and fly 23.0 km in a direction 34.0° south of east. You then fly 46.0 km due north. How far and in what direction must you then fly to reach a private landing strip that is 32.0 km due west of the College Station airport?

1.66 ··· On a training flight, a student pilot flies from Lincoln, Nebraska, to Clarinda, Iowa, next to St. Joseph, Missouri, and then to Manhattan, Kansas (**Fig. P1.66**). The directions are shown relative to north: 0° is north, 90° is east, 180° is south, and 270° is west. Use the method of components to find (a) the distance she has to fly from Manhattan to get back to Lincoln, and (b) the direction (relative to north) she must fly to get there. Illustrate your solutions with a vector diagram.

Figure **P1.66**

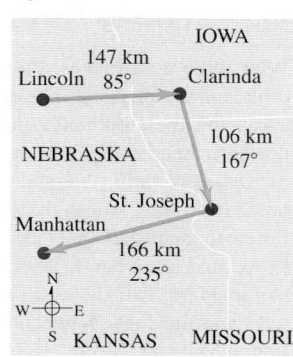

1.67 ·· As a test of orienteering skills, your physics class holds a contest in a large, open field. Each contestant is told to travel 20.8 m due north from the starting point, then 38.0 m due east, and finally 18.0 m in the direction 33.0° west of south. After the specified displacements, a contestant will find a silver dollar hidden under a rock. The winner is the person who takes the shortest time to reach the location of the silver dollar. Remembering what you learned in class, you run on a straight line from the starting point to the hidden coin. How far and in what direction do you run?

1.68 ··· **Getting Back.** An explorer in Antarctica leaves his shelter during a whiteout. He takes 40 steps northeast, next 80 steps at 60° north of west, and then 50 steps due south. Assume all of his steps are equal in length. (a) Sketch, roughly to scale, the three vectors and their resultant. (b) Save the explorer from becoming hopelessly lost by giving him the displacement, calculated by using the method of components, that will return him to his shelter.

1.69 ·· You are lost at night in a large, open field. Your GPS tells you that you are 122.0 m from your truck, in a direction 58.0° east of south. You walk 72.0 m due west along a ditch. How much farther, and in what direction, must you walk to reach your truck?

1.70 ··· A ship leaves the island of Guam and sails 285 km at 62.0° north of west. In which direction must it now head and how far must it sail so that its resultant displacement will be 115 km directly east of Guam?

1.71 ·· BIO **Bones and Muscles.** A physical therapy patient has a forearm that weighs 20.5 N and lifts a 112.0-N weight. These two forces are directed vertically downward. The only other significant forces on this forearm come from the biceps muscle (which acts perpendicular to the forearm) and the force at the elbow. If the biceps produces a pull of 232 N when the forearm is raised 43.0° above the horizontal, find the magnitude and direction of the force that the elbow exerts on the forearm. (The sum of the elbow force and the biceps force must balance the weight of the arm and the weight it is carrying, so their vector sum must be 132.5 N, upward.)

1.72 ··· You decide to go to your favorite neighborhood restaurant. You leave your apartment, take the elevator 10 flights down (each flight is 3.0 m), and then walk 15 m south to the apartment exit. You then proceed 0.200 km east, turn north, and walk 0.100 km to the entrance of the restaurant. (a) Determine the displacement from your apartment to the restaurant. Use unit vector notation for your answer, clearly indicating your choice of coordinates. (b) How far did you travel along the path you took from your apartment to the restaurant, and what is the magnitude of the displacement you calculated in part (a)?

1.73 ·· While following a treasure map, you start at an old oak tree. You first walk 825 m directly south, then turn and walk 1.25 km at 30.0° west of north, and finally walk 1.00 km at 32.0° north of east, where you find the treasure: a biography of Isaac Newton! (a) To return to the old oak tree, in what direction should you head and how far will you walk? Use components to solve this problem. (b) To see whether your calculation in part (a) is reasonable, compare it with a graphical solution drawn roughly to scale.

1.74 ·· A fence post is 52.0 m from where you are standing, in a direction 37.0° north of east. A second fence post is due south from you. How far are you from the second post if the distance between the two posts is 68.0 m?

1.75 ·· A dog in an open field runs 12.0 m east and then 28.0 m in a direction 50.0° west of north. In what direction and how far must the dog then run to end up 10.0 m south of her original starting point?

1.76 ··· Ricardo and Jane are standing under a tree in the middle of a pasture. An argument ensues, and they walk away in different directions. Ricardo walks 26.0 m in a direction 60.0° west of north. Jane walks 16.0 m in a direction 30.0° south of west. They then stop and turn to face each other. (a) What is the distance between them? (b) In what direction should Ricardo walk to go directly toward Jane?

1.77 ··· You are camping with Joe and Karl. Since all three of you like your privacy, you don't pitch your tents close together. Joe's tent is 21.0 m from yours, in the direction 23.0° south of east. Karl's tent is 32.0 m from yours, in the direction 37.0° north of east. What is the distance between Karl's tent and Joe's tent?

1.78 ·· **Bond Angle in Methane.** In the methane molecule, CH_4, each hydrogen atom is at a corner of a regular tetrahedron with the carbon atom at the center. In coordinates for which one of the C—H bonds is in the direction of $\hat{\imath} + \hat{\jmath} + \hat{k}$, an adjacent C—H bond is in the $\hat{\imath} - \hat{\jmath} - \hat{k}$ direction. Calculate the angle between these two bonds.

1.79 ·· Vectors $\vec{A}$ and $\vec{B}$ have scalar product −6.00, and their vector product has magnitude +9.00. What is the angle between these two vectors?

1.80 ·· A cube is placed so that one corner is at the origin and three edges are along the x-, y-, and z-axes of a coordinate system (**Fig. P1.80**). Use vectors to compute (a) the angle between the edge along the z-axis (line ab) and the diagonal from the origin to the opposite corner (line ad), and (b) the angle between line ac (the diagonal of a face) and line ad.

Figure **P1.80**

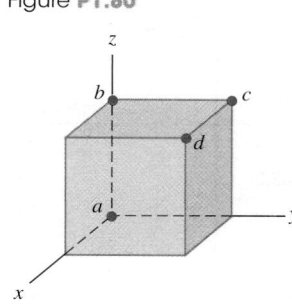

1.81 ·· Vector $\vec{A}$ has magnitude 12.0 m, and vector $\vec{B}$ has magnitude 16.0 m. The scalar product $\vec{A} \cdot \vec{B}$ is 112.0 m². What is the magnitude of the vector product between these two vectors?

1.82 ··· Obtain a *unit vector* perpendicular to the two vectors given in Exercise 1.41.

1.83 ·· The scalar product of vectors $\vec{A}$ and $\vec{B}$ is +48.0 m². Vector $\vec{A}$ has magnitude 9.00 m and direction 28.0° west of south. If vector $\vec{B}$ has direction 39.0° south of east, what is the magnitude of $\vec{B}$?

1.84 •• Two vectors $\vec{A}$ and $\vec{B}$ have magnitudes $A = 3.00$ and $B = 3.00$. Their vector product is $\vec{A} \times \vec{B} = -5.00\hat{k} + 2.00\hat{i}$. What is the angle between $\vec{A}$ and $\vec{B}$?

1.85 •• You are given vectors $\vec{A} = 5.0\hat{i} - 6.5\hat{j}$ and $\vec{B} = 3.5\hat{i} - 7.0\hat{j}$. A third vector, $\vec{C}$, lies in the xy-plane. Vector $\vec{C}$ is perpendicular to vector $\vec{A}$, and the scalar product of $\vec{C}$ with $\vec{B}$ is 15.0. From this information, find the components of vector $\vec{C}$.

1.86 •• Later in our study of physics we will encounter quantities represented by $(\vec{A} \times \vec{B}) \cdot \vec{C}$. (a) Prove that for any three vectors $\vec{A}$, $\vec{B}$, and $\vec{C}$, $\vec{A} \cdot (\vec{B} \times \vec{C}) = (\vec{A} \times \vec{B}) \cdot \vec{C}$. (b) Calculate $(\vec{A} \times \vec{B}) \cdot \vec{C}$ for vector $\vec{A}$ with magnitude $A = 5.00$ and angle $\theta_A = 26.0°$ (measured from the $+x$-axis toward the $+y$-axis), vector $\vec{B}$ with $B = 4.00$ and $\theta_B = 63.0°$, and vector $\vec{C}$ with magnitude 6.00 and in the $+z$-direction. Vectors $\vec{A}$ and $\vec{B}$ are in the xy-plane.

1.87 ••• DATA You are a team leader at a pharmaceutical company. Several technicians are preparing samples, and you want to compare the densities of the samples (density = mass/volume) by using the mass and volume values they have reported. Unfortunately, you did not specify what units to use. The technicians used a variety of units in reporting their values, as shown in the following table.

Sample ID	Mass	Volume
A	8.00 g	1.67×10^{-6} m³
B	6.00 µg	9.38×10^{6} µm³
C	8.00 mg	2.50×10^{-3} cm³
D	9.00×10^{-4} kg	2.81×10^{3} mm³
E	9.00×10^{4} ng	1.41×10^{-2} mm³
F	6.00×10^{-2} mg	1.25×10^{8} µm³

List the sample IDs in order of increasing density of the sample.

1.88 ••• DATA You are a mechanical engineer working for a manufacturing company. Two forces, $\vec{F_1}$ and $\vec{F_2}$, act on a component part of a piece of equipment. Your boss asked you to find the magnitude of the larger of these two forces. You can vary the angle between $\vec{F_1}$ and $\vec{F_2}$ from 0° to 90° while the magnitude of each force stays constant. And, you can measure the magnitude of the resultant force they produce (their vector sum), but you cannot directly measure the magnitude of each separate force. You measure the magnitude of the resultant force for four angles θ between the directions of the two forces as follows:

θ	Resultant force (N)
0.0°	8.00
45.0°	7.43
60.0°	7.00
90.0°	5.83

(a) What is the magnitude of the larger of the two forces? (b) When the equipment is used on the production line, the angle between the two forces is 30.0°. What is the magnitude of the resultant force in this case?

1.89 ••• DATA **Navigating in the Solar System.** The *Mars Polar Lander* spacecraft was launched on January 3, 1999. On December 3, 1999, the day *Mars Polar Lander* impacted the Martian surface at high velocity and probably disintegrated, the positions of the earth and Mars were given by these coordinates:

	x	y	z
Earth	0.3182 AU	0.9329 AU	0.0000 AU
Mars	1.3087 AU	−0.4423 AU	−0.0414 AU

With these coordinates, the sun is at the origin and the earth's orbit is in the xy-plane. The earth passes through the $+x$-axis once a year on the autumnal equinox, the first day of autumn in the northern hemisphere (on or about September 22). One AU, or *astronomical unit,* is equal to 1.496×10^{8} km, the average distance from the earth to the sun. (a) Draw the positions of the sun, the earth, and Mars on December 3, 1999. (b) Find these distances in AU on December 3, 1999: from (i) the sun to the earth; (ii) the sun to Mars; (iii) the earth to Mars. (c) As seen from the earth, what was the angle between the direction to the sun and the direction to Mars on December 3, 1999? (d) Explain whether Mars was visible from your current location at midnight on December 3, 1999. (When it is midnight, the sun is on the opposite side of the earth from you.)

CHALLENGE PROBLEMS

1.90 ••• **Completed Pass.** The football team at Enormous State University (ESU) uses vector displacements to record its plays, with the origin taken to be the position of the ball before the play starts. In a certain pass play, the receiver starts at $+1.0\hat{i} - 5.0\hat{j}$, where the units are yards, $\hat{i}$ is to the right, and $\hat{j}$ is downfield. Subsequent displacements of the receiver are $+9.0\hat{i}$ (he is in motion before the snap), $+11.0\hat{j}$ (breaks downfield), $-6.0\hat{i} + 4.0\hat{j}$ (zigs), and $+12.0\hat{i} + 18.0\hat{j}$ (zags). Meanwhile, the quarterback has dropped straight back to a position $-7.0\hat{j}$. How far and in which direction must the quarterback throw the ball? (Like the coach, you will be well advised to diagram the situation before solving this numerically.)

1.91 ••• **Navigating in the Big Dipper.** All of the stars of the Big Dipper (part of the constellation Ursa Major) may appear to be the same distance from the earth, but in fact they are very far from each other. **Figure P1.91** shows the distances from the earth to each of these stars. The distances are given in light-years (ly), the distance that light travels in one year. One light-year equals 9.461×10^{15} m. (a) Alkaid and Merak are 25.6° apart in the earth's sky. In a diagram, show the relative positions of Alkaid, Merak, and our sun. Find the distance in light-years from Alkaid to Merak. (b) To an inhabitant of a planet orbiting Merak, how many degrees apart in the sky would Alkaid and our sun be?

Figure **P1.91**

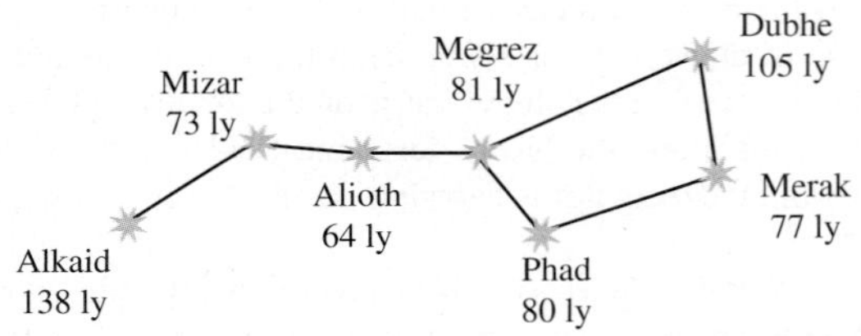

BIO **CALCULATING LUNG VOLUME IN HUMANS.** In humans, oxygen and carbon dioxide are exchanged in the blood within many small sacs called alveoli in the lungs. Alveoli provide a large surface area for gas exchange. Recent careful measurements show that the total number of alveoli in a typical pair of lungs is about 480×10^6 and that the average volume of a single alveolus is $4.2 \times 10^6 \ \mu m^3$. (The volume of a sphere is $V = \frac{4}{3}\pi r^3$, and the area of a sphere is $A = 4\pi r^2$.)

1.92 What is total volume of the gas-exchanging region of the lungs? (a) 2000 μm^3; (b) 2 m^3; (c) 2.0 L; (d) 120 L.

1.93 If we assume that alveoli are spherical, what is the diameter of a typical alveolus? (a) 0.20 mm; (b) 2 mm; (c) 20 mm; (d) 200 mm.

1.94 Individuals vary considerably in total lung volume. **Figure P1.94** shows the results of measuring the total lung volume and average alveolar volume of six individuals. From these data, what can you infer about the relationship among alveolar size, total lung volume, and number of alveoli per individual? As the total volume of the lungs increases, (a) the number and volume of individual alveoli increase; (b) the number of alveoli increases and the volume of individual alveoli decreases; (c) the volume of the individual alveolus remains constant and the number of alveoli increases; (d) both the number of alveoli and the volume of individual alveoli remain constant.

Figure **P1.94**

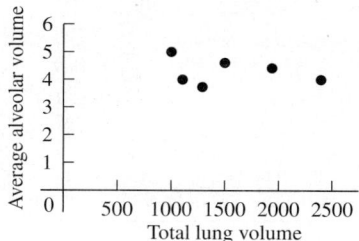

Answers

Chapter Opening Question **?**

(iii) Take the $+x$-axis to point east and the $+y$-axis to point north. Then we need to find the y-component of the velocity vector, which has magnitude $v = 15$ km/h and is at an angle $\theta = 37°$ measured from the $+x$-axis toward the $+y$-axis. From Eqs. (1.5) we have $v_y = v \sin\theta = (15 \text{ km/h}) \sin 37° = 9.0$ km/h. So the thunderstorm moves 9.0 km north in 1 h and 18 km north in 2 h.

Test Your Understanding Questions

1.5 (ii) Density $= (1.80 \text{ kg})/(6.0 \times 10^{-4} \text{ m}^3) = 3.0 \times 10^3 \text{ kg/m}^3$. When we multiply or divide, the number with the fewest significant figures controls the number of significant figures in the result.

1.6 The answer depends on how many students are enrolled at your campus.

1.7 (ii), (iii), and (iv) Vector $-\vec{T}$ has the same magnitude as vector $\vec{T}$, so $\vec{S} - \vec{T} = \vec{S} + (-\vec{T})$ is the *sum* of one vector of magnitude 3 m and one of magnitude 4 m. This sum has magnitude 7 m if $\vec{S}$ and $-\vec{T}$ are parallel and magnitude 1 m if $\vec{S}$ and $-\vec{T}$ are antiparallel. The magnitude of $\vec{S} - \vec{T}$ is 5 m if $\vec{S}$ and $-\vec{T}$ are perpendicular, when vectors $\vec{S}$, $\vec{T}$, and $\vec{S} - \vec{T}$ form a 3–4–5 right triangle. Answer (i) is impossible because the magnitude of the sum of two vectors cannot be greater than the sum of the magnitudes; answer (v) is impossible because the sum of two vectors can be zero only if the two vectors are antiparallel and have the same magnitude; and answer (vi) is impossible because the magnitude of a vector cannot be negative.

1.8 (a) yes, (b) no Vectors $\vec{A}$ and $\vec{B}$ can have the same magnitude but different components if they point in different directions. If they have the same components, however, they are the same vector $(\vec{A} = \vec{B})$ and so must have the same magnitude.

1.9 All have the same magnitude. Vectors $\vec{A}$, $\vec{B}$, $\vec{C}$, and $\vec{D}$ point in different directions but have the same magnitude:

$$A = B = C = D = \sqrt{(\pm 3 \text{ m})^2 + (\pm 5 \text{ m})^2 + (\pm 2 \text{ m})^2}$$
$$= \sqrt{9 \text{ m}^2 + 25 \text{ m}^2 + 4 \text{ m}^2} = \sqrt{38 \text{ m}^2} = 6.2 \text{ m}$$

1.10 (a) (ii) $\phi = 90°$, **(b) (i)** $\phi = 0°$ or **(iii)** $\phi = 180°$, **(c) (i)** $\phi = 0°$, **(d) (iii)** $\phi = 180°$, **(e) (ii)** $\phi = 90°$ (a) The scalar product is zero only if $\vec{A}$ and $\vec{B}$ are perpendicular. (b) The vector product is zero only if $\vec{A}$ and $\vec{B}$ are parallel or antiparallel. (c) The scalar product is equal to the product of the magnitudes $(\vec{A} \cdot \vec{B} = AB)$ only if $\vec{A}$ and $\vec{B}$ are parallel. (d) The scalar product is equal to the negative of the product of the magnitudes $(\vec{A} \cdot \vec{B} = -AB)$ only if $\vec{A}$ and $\vec{B}$ are antiparallel. (e) The magnitude of the vector product is equal to the product of the magnitudes $[$(magnitude of $\vec{A} \times \vec{B}) = AB]$ only if $\vec{A}$ and $\vec{B}$ are perpendicular.

Bridging Problem

(a) $5.2 \times 10^2 \text{ N}$
(b) $4.5 \times 10^2 \text{ N} \cdot \text{m}$

2 MOTION ALONG A STRAIGHT LINE

LEARNING GOALS

Looking forward at ...

2.1 How the ideas of displacement and average velocity help us describe straight-line motion.

2.2 The meaning of instantaneous velocity; the difference between velocity and speed.

2.3 How to use average acceleration and instantaneous acceleration to describe changes in velocity.

2.4 How to use equations and graphs to solve problems that involve straight-line motion with constant acceleration.

2.5 How to solve problems in which an object is falling freely under the influence of gravity alone.

2.6 How to analyze straight-line motion when the acceleration is not constant.

Looking back at ...

1.7 The displacement vector.

1.8 Components of a vector.

What distance must an airliner travel down a runway before it reaches takeoff speed? When you throw a baseball straight up in the air, how high does it go? When a glass slips from your hand, how much time do you have to catch it before it hits the floor? These are the kinds of questions you will learn to answer in this chapter. We begin our study of physics with *mechanics,* the study of the relationships among force, matter, and motion. In this chapter and the next we will study *kinematics,* the part of mechanics that enables us to describe motion. Later we will study *dynamics,* which helps us understand why objects move in different ways.

In this chapter we'll concentrate on the simplest kind of motion: a body moving along a straight line. To describe this motion, we introduce the physical quantities *velocity* and *acceleration.* In physics these quantities have definitions that are more precise and slightly different from the ones used in everyday language. Both velocity and acceleration are *vectors:* As you learned in Chapter 1, this means that they have both magnitude and direction. Our concern in this chapter is with motion along a straight line only, so we won't need the full mathematics of vectors just yet. But using vectors will be essential in Chapter 3 when we consider motion in two or three dimensions.

We'll develop simple equations to describe straight-line motion in the important special case when acceleration is constant. An example is the motion of a freely falling body. We'll also consider situations in which acceleration varies during the motion; in this case, it's necessary to use integration to describe the motion. (If you haven't studied integration yet, Section 2.6 is optional.)

2.1 DISPLACEMENT, TIME, AND AVERAGE VELOCITY

Suppose a drag racer drives her dragster along a straight track (**Fig. 2.1**). To study the dragster's motion, we need a coordinate system. We choose the *x*-axis to lie along the dragster's straight-line path, with the origin *O* at the starting line.

2.1 Positions of a dragster at two times during its run.

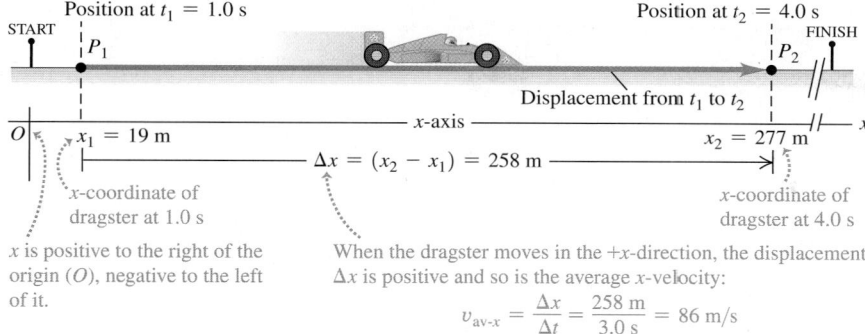

We also choose a point on the dragster, such as its front end, and represent the entire dragster by that point. Hence we treat the dragster as a **particle.**

A useful way to describe the motion of this particle is in terms of the change in its coordinate x over a time interval. Suppose that 1.0 s after the start the front of the dragster is at point P_1, 19 m from the origin, and 4.0 s after the start it is at point P_2, 277 m from the origin. The *displacement* of the particle is a vector that points from P_1 to P_2 (see Section 1.7). Figure 2.1 shows that this vector points along the x-axis. The x-component (see Section 1.8) of the displacement is the change in the value of x, $(277 \text{ m} - 19 \text{ m}) = 258 \text{ m}$, that took place during the time interval of $(4.0 \text{ s} - 1.0 \text{ s}) = 3.0 \text{ s}$. We define the dragster's **average velocity** during this time interval as a *vector* whose x-component is the change in x divided by the time interval: $(258 \text{ m})/(3.0 \text{ s}) = 86 \text{ m/s}$.

In general, the average velocity depends on the particular time interval chosen. For a 3.0-s time interval *before* the start of the race, the dragster is at rest at the starting line and has zero displacement, so its average velocity for this time interval is zero.

Let's generalize the concept of average velocity. At time t_1 the dragster is at point P_1, with coordinate x_1, and at time t_2 it is at point P_2, with coordinate x_2. The displacement of the dragster during the time interval from t_1 to t_2 is the vector from P_1 to P_2. The x-component of the displacement, denoted Δx, is the change in the coordinate x:

$$\Delta x = x_2 - x_1 \tag{2.1}$$

The dragster moves along the x-axis only, so the y- and z-components of the displacement are equal to zero.

CAUTION **The meaning of Δx** Note that Δx is *not* the product of Δ and x; it is a single symbol that means "the change in quantity x." We use the Greek capital letter Δ (delta) to represent a *change* in a quantity, equal to the *final* value of the quantity minus the *initial* value—never the reverse. Likewise, the time interval from t_1 to t_2 is Δt, the change in t: $\Delta t = t_2 - t_1$ (final time minus initial time). ▮

The x-component of average velocity, or the **average x-velocity,** is the x-component of displacement, Δx, divided by the time interval Δt during which the displacement occurs. We use the symbol $v_{\text{av-}x}$ for average x-velocity (the subscript "av" signifies average value, and the subscript x indicates that this is the x-component):

Average x-velocity of a particle in **straight-line motion** during time interval from t_1 to t_2 — x-component of the particle's displacement

$$v_{\text{av-}x} = \frac{\Delta x}{\Delta t} = \frac{x_2 - x_1}{t_2 - t_1} \tag{2.2}$$

Time interval — Final time minus initial time

Final x-coordinate minus initial x-coordinate

2.2 Positions of an official's truck at two times during its motion. The points P_1 and P_2 now indicate the positions of the truck, not the dragster, and so are the reverse of Fig. 2.1.

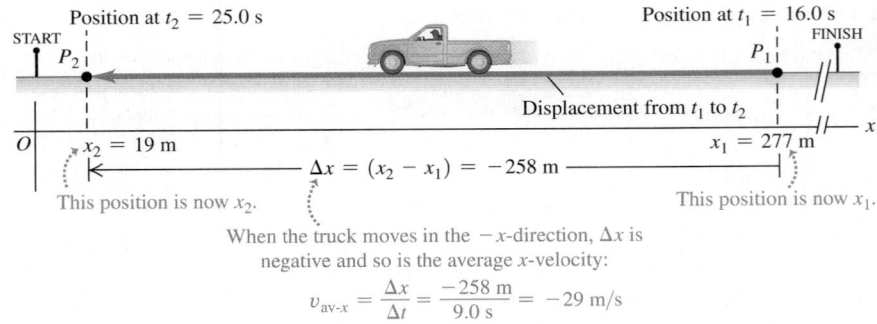

Position at $t_2 = 25.0$ s

Position at $t_1 = 16.0$ s

Displacement from t_1 to t_2

$x_2 = 19$ m

$x_1 = 277$ m

This position is now x_2.

$\Delta x = (x_2 - x_1) = -258$ m

This position is now x_1.

When the truck moves in the $-x$-direction, Δx is negative and so is the average x-velocity:

$$v_{\text{av-}x} = \frac{\Delta x}{\Delta t} = \frac{-258 \text{ m}}{9.0 \text{ s}} = -29 \text{ m/s}$$

As an example, for the dragster in Fig. 2.1, $x_1 = 19$ m, $x_2 = 277$ m, $t_1 = 1.0$ s, and $t_2 = 4.0$ s. So Eq. (2.2) gives

$$v_{\text{av-}x} = \frac{277 \text{ m} - 19 \text{ m}}{4.0 \text{ s} - 1.0 \text{ s}} = \frac{258 \text{ m}}{3.0 \text{ s}} = 86 \text{ m/s}$$

The average x-velocity of the dragster is positive. This means that during the time interval, the coordinate x increased and the dragster moved in the positive x-direction (to the right in Fig. 2.1).

If a particle moves in the *negative* x-direction during a time interval, its average velocity for that time interval is negative. For example, suppose an official's truck moves to the left along the track (**Fig. 2.2**). The truck is at $x_1 = 277$ m at $t_1 = 16.0$ s and is at $x_2 = 19$ m at $t_2 = 25.0$ s. Then $\Delta x = (19 \text{ m} - 277 \text{ m}) = -258$ m and $\Delta t = (25.0 \text{ s} - 16.0 \text{ s}) = 9.0$ s. The x-component of average velocity is $v_{\text{av-}x} = \Delta x/\Delta t = (-258 \text{ m})/(9.0 \text{ s}) = -29$ m/s. **Table 2.1** lists some simple rules for deciding whether the x-velocity is positive or negative.

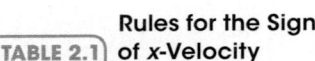

TABLE 2.1	Rules for the Sign of *x*-Velocity
If *x*-coordinate is:	**... *x*-velocity is:**
Positive & increasing (getting more positive)	Positive: Particle is moving in $+x$-direction
Positive & decreasing (getting less positive)	Negative: Particle is moving in $-x$-direction
Negative & increasing (getting less negative)	Positive: Particle is moving in $+x$-direction
Negative & decreasing (getting more negative)	Negative: Particle is moving in $-x$-direction

Note: These rules apply to both the average x-velocity $v_{\text{av-}x}$ and the instantaneous x-velocity v_x (to be discussed in Section 2.2).

CAUTION **The sign of average *x*-velocity** In our example positive $v_{\text{av-}x}$ means motion to the right, as in Fig. 2.1, and negative $v_{\text{av-}x}$ means motion to the left, as in Fig. 2.2. But that's *only* because we chose the $+x$-direction to be to the right. Had we chosen the $+x$-direction to be to the left, the average x-velocity $v_{\text{av-}x}$ would have been negative for the dragster moving to the right and positive for the truck moving to the left. In most problems the direction of the coordinate axis is yours to choose. Once you've made your choice, you *must* take it into account when interpreting the signs of $v_{\text{av-}x}$ and other quantities that describe motion! ▮

With straight-line motion we sometimes call Δx simply the displacement and $v_{\text{av-}x}$ simply the average velocity. But remember that these are the x-components of vector quantities that, in this special case, have *only* x-components. In Chapter 3, displacement, velocity, and acceleration vectors will have two or three nonzero components.

Figure 2.3 is a graph of the dragster's position as a function of time—that is, an **x-t graph.** The curve in the figure *does not* represent the dragster's path;

2.3 A graph of the position of a dragster as a function of time.

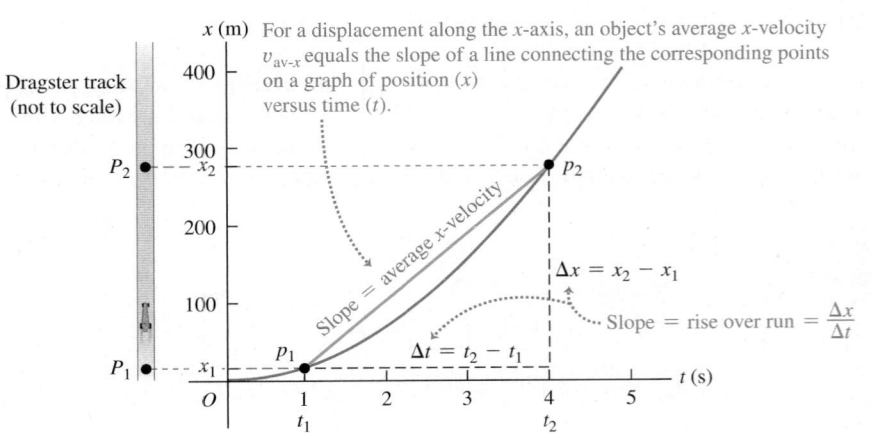

For a displacement along the x-axis, an object's average x-velocity $v_{\text{av-}x}$ equals the slope of a line connecting the corresponding points on a graph of position (x) versus time (t).

Dragster track (not to scale)

Slope = average x-velocity

$\Delta x = x_2 - x_1$

Slope = rise over run = $\dfrac{\Delta x}{\Delta t}$

$\Delta t = t_2 - t_1$

as Fig. 2.1 shows, the path is a straight line. Rather, the graph represents how the dragster's position changes with time. The points p_1 and p_2 on the graph correspond to the points P_1 and P_2 along the dragster's path. Line p_1p_2 is the hypotenuse of a right triangle with vertical side $\Delta x = x_2 - x_1$ and horizontal side $\Delta t = t_2 - t_1$. The average x-velocity $v_{\text{av-}x} = \Delta x/\Delta t$ of the dragster equals the *slope* of the line p_1p_2—that is, the ratio of the triangle's vertical side Δx to its horizontal side Δt. (The slope has units of meters divided by seconds, or m/s, the correct units for average x-velocity.)

The average x-velocity depends on only the total displacement $\Delta x = x_2 - x_1$ that occurs during the time interval $\Delta t = t_2 - t_1$, not on what happens during the time interval. At time t_1 a motorcycle might have raced past the dragster at point P_1 in Fig. 2.1, then slowed down to pass through point P_2 at the same time t_2 as the dragster. Both vehicles have the same displacement during the same time interval and so have the same average x-velocity.

If distance is given in meters and time in seconds, average velocity is measured in meters per second, or m/s (**Table 2.2**). Other common units of velocity are kilometers per hour (km/h), feet per second (ft/s), miles per hour (mi/h), and knots (1 knot = 1 nautical mile/h = 6080 ft/h).

TABLE 2.2	Typical Velocity Magnitudes
A snail's pace	10^{-3} m/s
A brisk walk	2 m/s
Fastest human	11 m/s
Freeway speeds	30 m/s
Fastest car	341 m/s
Random motion of air molecules	500 m/s
Fastest airplane	1000 m/s
Orbiting communications satellite	3000 m/s
Electron orbiting in a hydrogen atom	2×10^6 m/s
Light traveling in vacuum	3×10^8 m/s

TEST YOUR UNDERSTANDING OF SECTION 2.1 Each of the following five trips takes one hour. The positive x-direction is to the east. (i) Automobile A travels 50 km due east. (ii) Automobile B travels 50 km due west. (iii) Automobile C travels 60 km due east, then turns around and travels 10 km due west. (iv) Automobile D travels 70 km due east. (v) Automobile E travels 20 km due west, then turns around and travels 20 km due east. (a) Rank the five trips in order of average x-velocity from most positive to most negative. (b) Which trips, if any, have the same average x-velocity? (c) For which trip, if any, is the average x-velocity equal to zero? ▮

2.2 INSTANTANEOUS VELOCITY

Sometimes average velocity is all you need to know about a particle's motion. For example, a race along a straight line is really a competition to see whose average velocity, $v_{\text{av-}x}$, has the greatest magnitude. The prize goes to the competitor who can travel the displacement Δx from the start to the finish line in the shortest time interval, Δt (**Fig. 2.4**).

But the average velocity of a particle during a time interval can't tell us how fast, or in what direction, the particle was moving at any given time during the interval. For that we need to know the **instantaneous velocity,** or the velocity at a specific instant of time or specific point along the path.

CAUTION **How long is an instant?** You might use the phrase "It lasted just an instant" to refer to something that spanned a very short time interval. But in physics an instant has no duration at all; it refers to a single value of time. ▮

To find the instantaneous velocity of the dragster in Fig. 2.1 at point P_1, we move point P_2 closer and closer to point P_1 and compute the average velocity $v_{\text{av-}x} = \Delta x/\Delta t$ over the ever-shorter displacement and time interval. Both Δx and Δt become very small, but their ratio does not necessarily become small. In the language of calculus, the limit of $\Delta x/\Delta t$ as Δt approaches zero is called the **derivative** of x with respect to t and is written dx/dt. We use the symbol v_x, with no "av" subscript, for the instantaneous velocity along the x-axis, or the **instantaneous x-velocity:**

2.4 The winner of a 50-m swimming race is the swimmer whose average velocity has the greatest magnitude—that is, the swimmer who traverses a displacement Δx of 50 m in the shortest elapsed time Δt.

The **instantaneous** x-velocity of a particle in straight-line motion ...

$$v_x = \lim_{\Delta t \to 0} \frac{\Delta x}{\Delta t} = \frac{dx}{dt} \tag{2.3}$$

... equals the limit of the particle's average x-velocity as the time interval approaches zero ...   ... and equals the instantaneous rate of change of the particle's x-coordinate.

2.5 In any problem involving straight-line motion, the choice of which direction is positive and which is negative is entirely up to you.

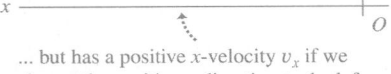

······ A bicyclist moving to the left ...

... has a negative x-velocity v_x if we choose the positive x-direction to the right ...

... but has a positive x-velocity v_x if we choose the positive x-direction to the left.

The time interval Δt is always positive, so v_x has the same algebraic sign as Δx. A positive value of v_x means that x is increasing and the motion is in the positive x-direction; a negative value of v_x means that x is decreasing and the motion is in the negative x-direction. A body can have positive x and negative v_x, or the reverse; x tells us where the body is, while v_x tells us how it's moving (**Fig. 2.5**). The rules that we presented in Table 2.1 (Section 2.1) for the sign of average x-velocity $v_{av\text{-}x}$ also apply to the sign of instantaneous x-velocity v_x.

Instantaneous velocity, like average velocity, is a vector; Eq. (2.3) defines its x-component. In straight-line motion, all other components of instantaneous velocity are zero. In this case we often call v_x simply the instantaneous velocity. (In Chapter 3 we'll deal with the general case in which the instantaneous velocity can have nonzero x-, y-, and z-components.) When we use the term "velocity," we will always mean instantaneous rather than average velocity.

"Velocity" and "speed" are used interchangeably in everyday language, but they have distinct definitions in physics. We use the term **speed** to denote distance traveled divided by time, on either an average or an instantaneous basis. Instantaneous *speed,* for which we use the symbol v with *no* subscripts, measures how fast a particle is moving; instantaneous *velocity* measures how fast *and* in what direction it's moving. Instantaneous speed is the magnitude of instantaneous velocity and so can never be negative. For example, a particle with instantaneous velocity $v_x = 25$ m/s and a second particle with $v_x = -25$ m/s are moving in opposite directions at the same instantaneous speed 25 m/s.

CAUTION **Average speed and average velocity** Average speed is *not* the magnitude of average velocity. When César Cielo set a world record in 2009 by swimming 100.0 m in 46.91 s, his average speed was $(100.0 \text{ m})/(46.91 \text{ s}) = 2.132$ m/s. But because he swam two lengths in a 50-m pool, he started and ended at the same point and so had zero total displacement and zero average *velocity!* Both average speed and instantaneous speed are scalars, not vectors, because these quantities contain no information about direction. ∎

EXAMPLE 2.1 AVERAGE AND INSTANTANEOUS VELOCITIES

A cheetah is crouched 20 m to the east of a vehicle (**Fig. 2.6a**). At time $t = 0$ the cheetah begins to run due east toward an antelope that is 50 m to the east of the vehicle. During the first 2.0 s of the chase, the cheetah's x-coordinate varies with time according to the equation $x = 20 \text{ m} + (5.0 \text{ m/s}^2)t^2$. (a) Find the cheetah's displacement between $t_1 = 1.0$ s and $t_2 = 2.0$ s. (b) Find its average velocity during that interval. (c) Find its instantaneous velocity at $t_1 = 1.0$ s by taking $\Delta t = 0.1$ s, then 0.01 s, then 0.001 s. (d) Derive an expression for the cheetah's instantaneous velocity as a function of time, and use it to find v_x at $t = 1.0$ s and $t = 2.0$ s.

2.6 A cheetah attacking an antelope from ambush. The animals are not drawn to the same scale as the axis.

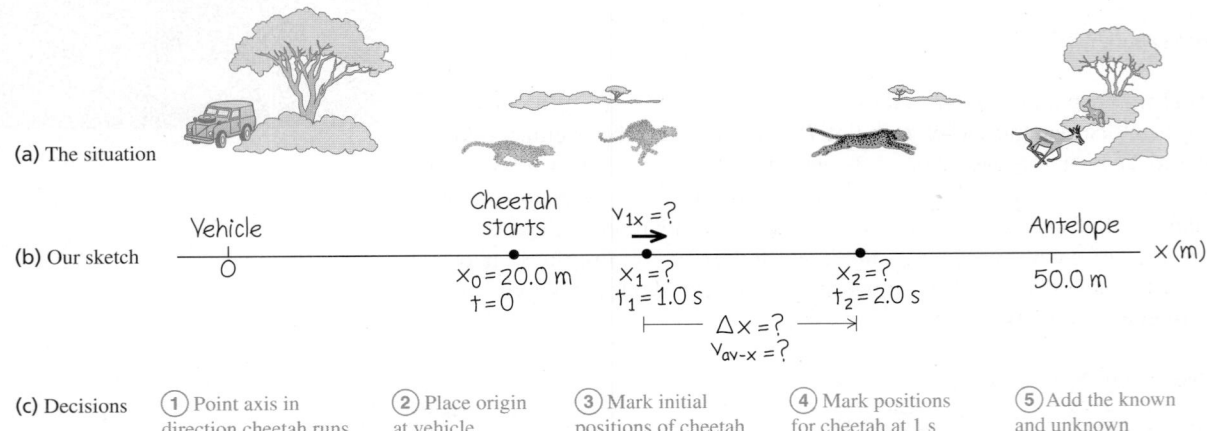

(a) The situation

(b) Our sketch

(c) Decisions
① Point axis in direction cheetah runs, so that all values will be positive.
② Place origin at vehicle.
③ Mark initial positions of cheetah and antelope.
④ Mark positions for cheetah at 1 s and 2 s.
⑤ Add the known and unknown quantities.

SOLUTION

IDENTIFY and SET UP: Figure 2.6b shows our sketch of the cheetah's motion. We use Eq. (2.1) for displacement, Eq. (2.2) for average velocity, and Eq. (2.3) for instantaneous velocity.

EXECUTE: (a) At $t_1 = 1.0$ s and $t_2 = 2.0$ s the cheetah's positions x_1 and x_2 are

$$x_1 = 20 \text{ m} + (5.0 \text{ m/s}^2)(1.0 \text{ s})^2 = 25 \text{ m}$$

$$x_2 = 20 \text{ m} + (5.0 \text{ m/s}^2)(2.0 \text{ s})^2 = 40 \text{ m}$$

The displacement during this 1.0-s interval is

$$\Delta x = x_2 - x_1 = 40 \text{ m} - 25 \text{ m} = 15 \text{ m}$$

(b) The average x-velocity during this interval is

$$v_{\text{av-}x} = \frac{x_2 - x_1}{t_2 - t_1} = \frac{40 \text{ m} - 25 \text{ m}}{2.0 \text{ s} - 1.0 \text{ s}} = \frac{15 \text{ m}}{1.0 \text{ s}}$$

$$= 15 \text{ m/s}$$

(c) With $\Delta t = 0.1$ s the time interval is from $t_1 = 1.0$ s to a new $t_2 = 1.1$ s. At t_2 the position is

$$x_2 = 20 \text{ m} + (5.0 \text{ m/s}^2)(1.1 \text{ s})^2 = 26.05 \text{ m}$$

The average x-velocity during this 0.1-s interval is

$$v_{\text{av-}x} = \frac{26.05 \text{ m} - 25 \text{ m}}{1.1 \text{ s} - 1.0 \text{ s}} = 10.5 \text{ m/s}$$

Following this pattern, you can calculate the average x-velocities for 0.01-s and 0.001-s intervals: The results are 10.05 m/s and 10.005 m/s. As Δt gets smaller, the average x-velocity gets closer to 10.0 m/s, so we conclude that the instantaneous x-velocity at $t = 1.0$ s is 10.0 m/s. (We suspended the rules for significant-figure counting in these calculations.)

(d) From Eq. (2.3) the instantaneous x-velocity is $v_x = dx/dt$. The derivative of a constant is zero and the derivative of t^2 is $2t$, so

$$v_x = \frac{dx}{dt} = \frac{d}{dt}[20 \text{ m} + (5.0 \text{ m/s}^2)t^2]$$

$$= 0 + (5.0 \text{ m/s}^2)(2t) = (10 \text{ m/s}^2)t$$

At $t = 1.0$ s, this yields $v_x = 10$ m/s, as we found in part (c); at $t = 2.0$ s, $v_x = 20$ m/s.

EVALUATE: Our results show that the cheetah picked up speed from $t = 0$ (when it was at rest) to $t = 1.0$ s ($v_x = 10$ m/s) to $t = 2.0$ s ($v_x = 20$ m/s). This makes sense; the cheetah covered only 5 m during the interval $t = 0$ to $t = 1.0$ s, but it covered 15 m during the interval $t = 1.0$ s to $t = 2.0$ s.

Finding Velocity on an x-t Graph

We can also find the x-velocity of a particle from the graph of its position as a function of time. Suppose we want to find the x-velocity of the dragster in Fig. 2.1 at point P_1. As point P_2 in Fig. 2.1 approaches point P_1, point p_2 in the x-t graphs of **Figs. 2.7a** and 2.7b approaches point p_1 and the average x-velocity is calculated over shorter time intervals Δt. In the limit that $\Delta t \rightarrow 0$, shown in Fig. 2.7c, the slope of the line p_1p_2 equals the slope of the line tangent to the curve at point p_1. Thus, *on a graph of position as a function of time for straight-line motion, the instantaneous x-velocity at any point is equal to the slope of the tangent to the curve at that point.*

If the tangent to the x-t curve slopes upward to the right, as in Fig. 2.7c, then its slope is positive, the x-velocity is positive, and the motion is in the positive x-direction. If the tangent slopes downward to the right, the slope of the x-t graph and the x-velocity are negative, and the motion is in the negative x-direction. When the tangent is horizontal, the slope and the x-velocity are zero. **Figure 2.8** (next page) illustrates these three possibilities.

2.7 Using an x-t graph to go from (a), (b) average x-velocity to (c) instantaneous x-velocity v_x. In (c) we find the slope of the tangent to the x-t curve by dividing any vertical interval (with distance units) along the tangent by the corresponding horizontal interval (with time units).

(a)

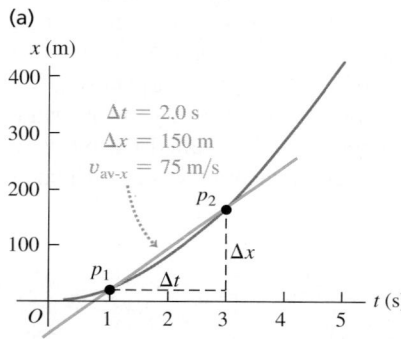

As the average x-velocity $v_{\text{av-}x}$ is calculated over shorter and shorter time intervals ...

(b)

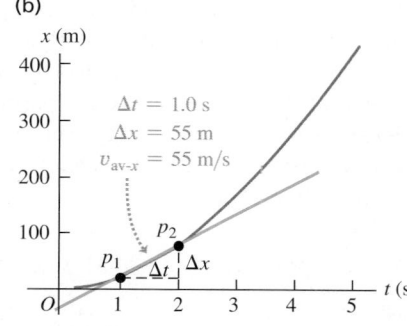

... its value $v_{\text{av-}x} = \Delta x/\Delta t$ approaches the instantaneous x-velocity.

(c)

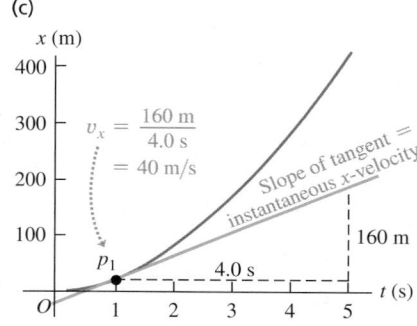

The instantaneous x-velocity v_x at any given point equals the slope of the tangent to the x-t curve at that point.

2.8 (a) The *x-t* graph of the motion of a particular particle. (b) A motion diagram showing the position and velocity of the particle at each of the times labeled on the *x-t* graph.

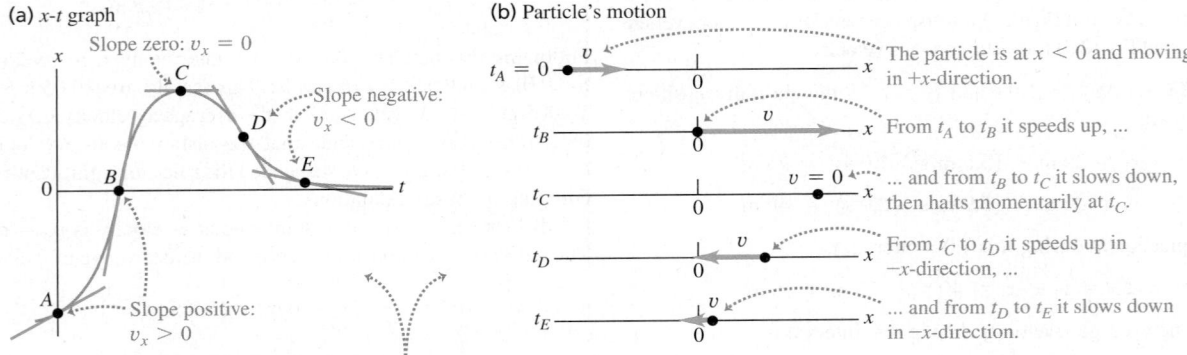

(a) *x-t* graph

Slope zero: $v_x = 0$

Slope negative: $v_x < 0$

Slope positive: $v_x > 0$

(b) Particle's motion

$t_A = 0$ — The particle is at $x < 0$ and moving in $+x$-direction.

t_B — From t_A to t_B it speeds up, ...

t_C — $v = 0$ ... and from t_B to t_C it slows down, then halts momentarily at t_C.

t_D — From t_C to t_D it speeds up in $-x$-direction, ...

t_E — ... and from t_D to t_E it slows down in $-x$-direction.

• On an *x-t* graph, the slope of the tangent at any point equals the particle's velocity at that point.
• The steeper the slope (positive or negative), the greater the particle's speed in the positive or negative *x*-direction.

Figure 2.8 depicts the motion of a particle in two ways: as (a) an *x-t* graph and (b) a **motion diagram** that shows the particle's position at various instants (like frames from a video of the particle's motion) as well as arrows to represent the particle's velocity at each instant. We will use both *x-t* graphs and motion diagrams in this chapter to represent motion. You will find it helpful to draw *both* an *x-t* graph and a motion diagram when you solve any problem involving motion.

2.9 An *x-t* graph for a particle.

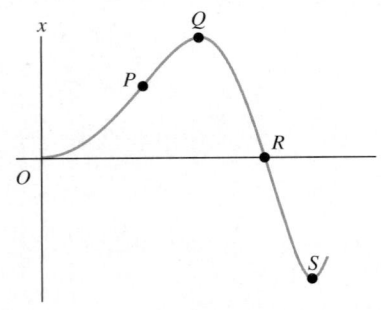

TEST YOUR UNDERSTANDING OF SECTION 2.2 **Figure 2.9** is an *x-t* graph of the motion of a particle. (a) Rank the values of the particle's *x*-velocity v_x at points P, Q, R, and S from most positive to most negative. (b) At which points is v_x positive? (c) At which points is v_x negative? (d) At which points is v_x zero? (e) Rank the values of the particle's *speed* at points P, Q, R, and S from fastest to slowest. ∎

2.3 AVERAGE AND INSTANTANEOUS ACCELERATION

Just as velocity describes the rate of change of position with time, *acceleration* describes the rate of change of velocity with time. Like velocity, acceleration is a vector quantity. When the motion is along a straight line, its only nonzero component is along that line. As we'll see, acceleration in straight-line motion can refer to either speeding up or slowing down.

Average Acceleration

Let's consider again a particle moving along the *x*-axis. Suppose that at time t_1 the particle is at point P_1 and has *x*-component of (instantaneous) velocity v_{1x}, and at a later time t_2 it is at point P_2 and has *x*-component of velocity v_{2x}. So the *x*-component of velocity changes by an amount $\Delta v_x = v_{2x} - v_{1x}$ during the time interval $\Delta t = t_2 - t_1$. As the particle moves from P_1 to P_2, its **average acceleration** is a vector quantity whose *x*-component $a_{\text{av-}x}$ (called the **average x-acceleration**) equals Δv_x, the change in the *x*-component of velocity, divided by the time interval Δt:

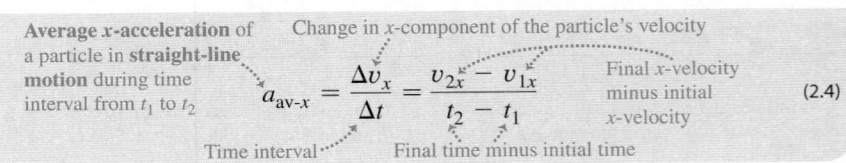

Average *x*-acceleration of a particle in **straight-line motion** during time interval from t_1 to t_2

Change in *x*-component of the particle's velocity

$$a_{\text{av-}x} = \frac{\Delta v_x}{\Delta t} = \frac{v_{2x} - v_{1x}}{t_2 - t_1}$$

Final *x*-velocity minus initial *x*-velocity

Time interval — Final time minus initial time

(2.4)

For straight-line motion along the *x*-axis we will often call $a_{\text{av-}x}$ simply the average acceleration. (We'll encounter the other components of the average acceleration vector in Chapter 3.)

If we express velocity in meters per second and time in seconds, then average acceleration is in meters per second per second. This is usually written as m/s^2 and is read "meters per second squared."

CAUTION **Don't confuse velocity and acceleration** Velocity describes how a body's position changes with time; it tells us how fast and in what direction the body moves. Acceleration describes how the velocity changes with time; it tells us how the speed and direction of motion change. To see the difference, imagine you are riding along with the moving body. If the body accelerates forward and gains speed, you feel pushed backward in your seat; if it accelerates backward and loses speed, you feel pushed forward. If the velocity is constant and there's no acceleration, you feel neither sensation. (We'll explain these sensations in Chapter 4.) ▮

EXAMPLE 2.2 AVERAGE ACCELERATION

An astronaut has left an orbiting spacecraft to test a new personal maneuvering unit. As she moves along a straight line, her partner on the spacecraft measures her velocity every 2.0 s, starting at time $t = 1.0$ s:

t	v_x	t	v_x
1.0 s	0.8 m/s	9.0 s	−0.4 m/s
3.0 s	1.2 m/s	11.0 s	−1.0 m/s
5.0 s	1.6 m/s	13.0 s	−1.6 m/s
7.0 s	1.2 m/s	15.0 s	−0.8 m/s

Find the average *x*-acceleration, and state whether the speed of the astronaut increases or decreases over each of these 2.0-s time intervals: (a) $t_1 = 1.0$ s to $t_2 = 3.0$ s; (b) $t_1 = 5.0$ s to $t_2 = 7.0$ s; (c) $t_1 = 9.0$ s to $t_2 = 11.0$ s; (d) $t_1 = 13.0$ s to $t_2 = 15.0$ s.

SOLUTION

IDENTIFY and SET UP: We'll use Eq. (2.4) to determine the average acceleration $a_{\text{av-}x}$ from the change in velocity over each time interval. To find the changes in speed, we'll use the idea that speed v is the magnitude of the instantaneous velocity v_x.

The upper part of **Fig. 2.10** is our graph of the *x*-velocity as a function of time. On this v_x-t graph, the slope of the line connecting the endpoints of each interval is the average *x*-acceleration $a_{\text{av-}x} = \Delta v_x / \Delta t$ for that interval. The four slopes (and thus the *signs* of the average accelerations) are, from left to right, positive, negative, negative, and positive. The third and fourth slopes (and thus the average accelerations themselves) have greater magnitude than the first and second.

EXECUTE: Using Eq. (2.4), we find:

(a) $a_{\text{av-}x} = (1.2 \text{ m/s} - 0.8 \text{ m/s})/(3.0 \text{ s} - 1.0 \text{ s}) = 0.2 \text{ m/s}^2$. The speed (magnitude of instantaneous *x*-velocity) increases from 0.8 m/s to 1.2 m/s.

(b) $a_{\text{av-}x} = (1.2 \text{ m/s} - 1.6 \text{ m/s})/(7.0 \text{ s} - 5.0 \text{ s}) = -0.2 \text{ m/s}^2$. The speed decreases from 1.6 m/s to 1.2 m/s.

(c) $a_{\text{av-}x} = [-1.0 \text{ m/s} - (-0.4 \text{ m/s})]/(11.0 \text{ s} - 9.0 \text{ s}) = -0.3 \text{ m/s}^2$. The speed increases from 0.4 m/s to 1.0 m/s.

2.10 Our graphs of *x*-velocity versus time (top) and average *x*-acceleration versus time (bottom) for the astronaut.

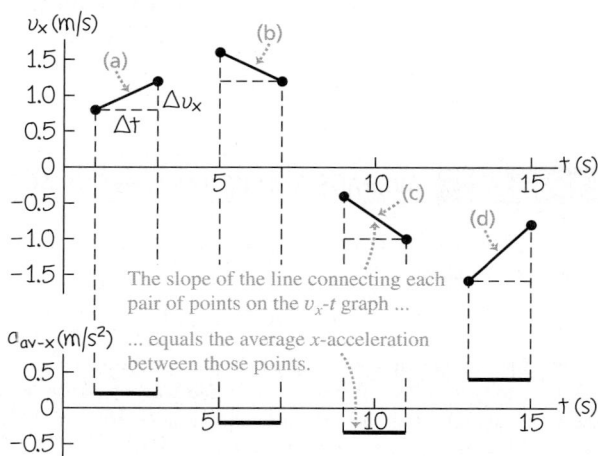

(d) $a_{\text{av-}x} = [-0.8 \text{ m/s} - (-1.6 \text{ m/s})]/(15.0 \text{ s} - 13.0 \text{ s}) = 0.4 \text{ m/s}^2$. The speed decreases from 1.6 m/s to 0.8 m/s.

In the lower part of Fig. 2.10, we graph the values of $a_{\text{av-}x}$.

EVALUATE: The signs and relative magnitudes of the average accelerations agree with our qualitative predictions. Notice that when the average *x*-acceleration has the *same* algebraic sign as the initial velocity, as in intervals (a) and (c), the astronaut goes faster. When $a_{\text{av-}x}$ has the *opposite* algebraic sign from the initial velocity, as in intervals (b) and (d), she slows down. Thus positive *x*-acceleration means speeding up if the *x*-velocity is positive [interval (a)] but slowing down if the *x*-velocity is negative [interval (d)]. Similarly, negative *x*-acceleration means speeding up if the *x*-velocity is negative [interval (c)] but slowing down if the *x*-velocity is positive [interval (b)].

2.11 A Grand Prix car at two points on the straightaway.

Instantaneous Acceleration

We can now define **instantaneous acceleration** by following the same procedure that we used to define instantaneous velocity. Suppose a race car driver is driving along a straightaway as shown in **Fig. 2.11**. To define the instantaneous acceleration at point P_1, we take point P_2 in Fig. 2.11 to be closer and closer to P_1 so that the average acceleration is computed over shorter and shorter time intervals. Thus

The **instantaneous** x-acceleration of a particle in **straight-line motion** ...

$$a_x = \lim_{\Delta t \to 0} \frac{\Delta v_x}{\Delta t} = \frac{dv_x}{dt} \qquad (2.5)$$

... equals the limit of the particle's average x-acceleration as the time interval approaches zero ...

... and equals the instantaneous rate of change of the particle's x-velocity.

In Eq. (2.5) a_x is the x-component of the acceleration vector, which we call the **instantaneous x-acceleration;** in straight-line motion, all other components of this vector are zero. From now on, when we use the term "acceleration," we will always mean instantaneous acceleration, not average acceleration.

EXAMPLE 2.3 **AVERAGE AND INSTANTANEOUS ACCELERATIONS**

Suppose the x-velocity v_x of the car in Fig. 2.11 at any time t is given by the equation

$$v_x = 60 \text{ m/s} + (0.50 \text{ m/s}^3)t^2$$

(a) Find the change in x-velocity of the car in the time interval $t_1 = 1.0$ s to $t_2 = 3.0$ s. (b) Find the average x-acceleration in this time interval. (c) Find the instantaneous x-acceleration at time $t_1 = 1.0$ s by taking Δt to be first 0.1 s, then 0.01 s, then 0.001 s. (d) Derive an expression for the instantaneous x-acceleration as a function of time, and use it to find a_x at $t = 1.0$ s and $t = 3.0$ s.

SOLUTION

IDENTIFY and SET UP: This example is analogous to Example 2.1 in Section 2.2. In that example we found the average x-velocity from the change in position over shorter and shorter time intervals, and we obtained an expression for the instantaneous x-velocity by differentiating the position as a function of time. In this example we have an exact parallel. Using Eq. (2.4), we'll find the average x-acceleration from the change in x-velocity over a time interval. Likewise, using Eq. (2.5), we'll obtain an expression for the instantaneous x-acceleration by differentiating the x-velocity as a function of time.

EXECUTE: (a) Before we can apply Eq. (2.4), we must find the x-velocity at each time from the given equation. At $t_1 = 1.0$ s and $t_2 = 3.0$ s, the velocities are

$$v_{1x} = 60 \text{ m/s} + (0.50 \text{ m/s}^3)(1.0 \text{ s})^2 = 60.5 \text{ m/s}$$

$$v_{2x} = 60 \text{ m/s} + (0.50 \text{ m/s}^3)(3.0 \text{ s})^2 = 64.5 \text{ m/s}$$

The change in x-velocity Δv_x between $t_1 = 1.0$ s and $t_2 = 3.0$ s is

$$\Delta v_x = v_{2x} - v_{1x} = 64.5 \text{ m/s} - 60.5 \text{ m/s} = 4.0 \text{ m/s}$$

(b) The average x-acceleration during this time interval of duration $t_2 - t_1 = 2.0$ s is

$$a_{\text{av-}x} = \frac{v_{2x} - v_{1x}}{t_2 - t_1} = \frac{4.0 \text{ m/s}}{2.0 \text{ s}} = 2.0 \text{ m/s}^2$$

During this time interval the x-velocity and average x-acceleration have the same algebraic sign (in this case, positive), and the car speeds up.

(c) When $\Delta t = 0.1$ s, we have $t_2 = 1.1$ s. Proceeding as before, we find

$$v_{2x} = 60 \text{ m/s} + (0.50 \text{ m/s}^3)(1.1 \text{ s})^2 = 60.605 \text{ m/s}$$

$$\Delta v_x = 0.105 \text{ m/s}$$

$$a_{\text{av-}x} = \frac{\Delta v_x}{\Delta t} = \frac{0.105 \text{ m/s}}{0.1 \text{ s}} = 1.05 \text{ m/s}^2$$

You should follow this pattern to calculate $a_{\text{av-}x}$ for $\Delta t = 0.01$ s and $\Delta t = 0.001$ s; the results are $a_{\text{av-}x} = 1.005 \text{ m/s}^2$ and $a_{\text{av-}x} = 1.0005 \text{ m/s}^2$, respectively. As Δt gets smaller, the average x-acceleration gets closer to 1.0 m/s^2, so the instantaneous x-acceleration at $t = 1.0$ s is 1.0 m/s^2.

(d) By Eq. (2.5) the instantaneous x-acceleration is $a_x = dv_x/dt$. The derivative of a constant is zero and the derivative of t^2 is $2t$, so

$$a_x = \frac{dv_x}{dt} = \frac{d}{dt}[60 \text{ m/s} + (0.50 \text{ m/s}^3)t^2]$$

$$= (0.50 \text{ m/s}^3)(2t) = (1.0 \text{ m/s}^3)t$$

When $t = 1.0$ s,

$$a_x = (1.0 \text{ m/s}^3)(1.0 \text{ s}) = 1.0 \text{ m/s}^2$$

When $t = 3.0$ s,

$$a_x = (1.0 \text{ m/s}^3)(3.0 \text{ s}) = 3.0 \text{ m/s}^2$$

EVALUATE: Neither of the values we found in part (d) is equal to the average x-acceleration found in part (b). That's because the car's instantaneous x-acceleration varies with time. The rate of change of acceleration with time is sometimes called the "jerk."

Finding Acceleration on a $v_x\text{-}t$ Graph or an $x\text{-}t$ Graph

In Section 2.2 we interpreted average and instantaneous x-velocity in terms of the slope of a graph of position versus time. In the same way, we can interpret average and instantaneous x-acceleration by using a graph of instantaneous velocity v_x versus time t—that is, a $v_x\text{-}t$ **graph** (**Fig. 2.12**). Points p_1 and p_2 on the graph correspond to points P_1 and P_2 in Fig. 2.11. The average x-acceleration $a_{\text{av-}x} = \Delta v_x / \Delta t$ during this interval is the slope of the line p_1p_2.

As point P_2 in Fig. 2.11 approaches point P_1, point p_2 in the $v_x\text{-}t$ graph of Fig. 2.12 approaches point p_1, and the slope of the line p_1p_2 approaches the slope of the line tangent to the curve at point p_1. Thus, *on a graph of x-velocity as a function of time, the instantaneous x-acceleration at any point is equal to the slope of the tangent to the curve at that point.* Tangents drawn at different points along the curve in Fig. 2.12 have different slopes, so the instantaneous x-acceleration varies with time.

CAUTION **Signs of x-acceleration and x-velocity** The algebraic sign of the x-acceleration does *not* tell you whether a body is speeding up or slowing down. **?** You must compare the signs of the x-velocity and the x-acceleration. When v_x and a_x have the *same* sign, the body is speeding up. If both are positive, the body is moving in the positive direction with increasing speed. If both are negative, the body is moving in the negative direction with an x-velocity that is becoming more negative, and again the speed is increasing. When v_x and a_x have *opposite* signs, the body is slowing down. If v_x is positive and a_x is negative, the body is moving in the positive direction with decreasing speed; if v_x is negative and a_x is positive, the body is moving in the negative direction with an x-velocity that is becoming less negative, and again the body is slowing down. **Table 2.3** summarizes these rules, and **Fig. 2.13** (next page) illustrates some of them. ❚

The term "deceleration" is sometimes used for a decrease in speed. Because it may mean positive or negative a_x, depending on the sign of v_x, we avoid this term.

We can also learn about the acceleration of a body from a graph of its *position* versus time. Because $a_x = dv_x/dt$ and $v_x = dx/dt$, we can write

$$a_x = \frac{dv_x}{dt} = \frac{d}{dt}\left(\frac{dx}{dt}\right) = \frac{d^2x}{dt^2} \qquad (2.6)$$

Rules for the Sign of x-Acceleration

TABLE 2.3	
If x-velocity is:	**. . . x-acceleration is:**
Positive & increasing (getting more positive)	Positive: Particle is moving in $+x$-direction & speeding up
Positive & decreasing (getting less positive)	Negative: Particle is moving in $+x$-direction & slowing down
Negative & increasing (getting less negative)	Positive: Particle is moving in $-x$-direction & slowing down
Negative & decreasing (getting more negative)	Negative: Particle is moving in $-x$-direction & speeding up

Note: These rules apply to both the average x-acceleration $a_{\text{av-}x}$ and the instantaneous x-acceleration a_x.

2.12 A $v_x\text{-}t$ graph of the motion in Fig. 2.11.

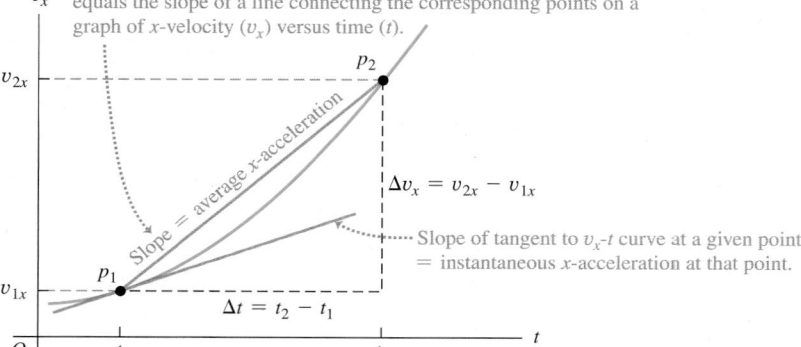

For a displacement along the x-axis, an object's average x-acceleration equals the slope of a line connecting the corresponding points on a graph of x-velocity (v_x) versus time (t).

Slope = average x-acceleration

$\Delta v_x = v_{2x} - v_{1x}$

Slope of tangent to $v_x\text{-}t$ curve at a given point = instantaneous x-acceleration at that point.

$\Delta t = t_2 - t_1$

2.13 (a) The v_x-t graph of the motion of a different particle from that shown in Fig. 2.8. (b) A motion diagram showing the position, velocity, and acceleration of the particle at each of the times labeled on the v_x-t graph.

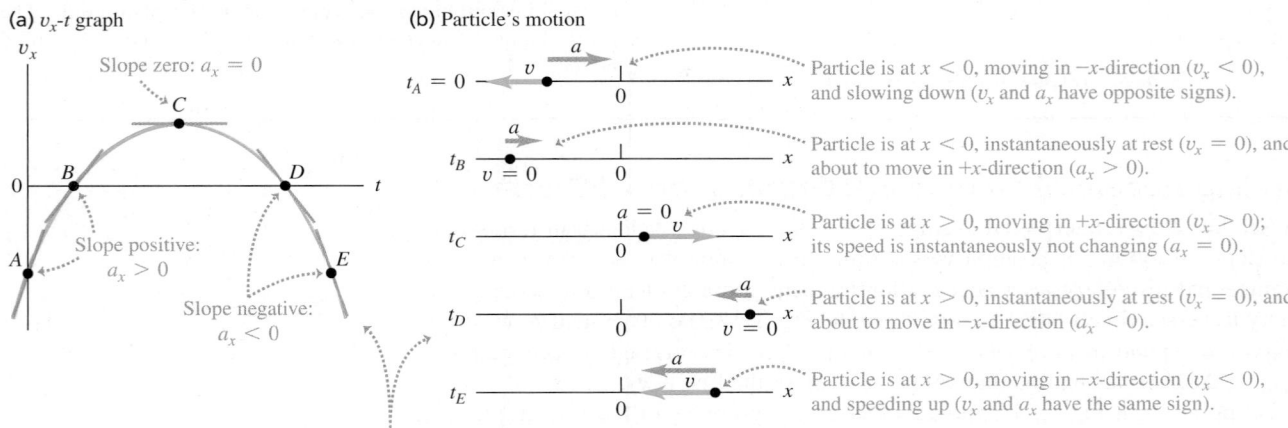

(a) v_x-t graph

(b) Particle's motion

Slope zero: $a_x = 0$

Slope positive: $a_x > 0$

Slope negative: $a_x < 0$

$t_A = 0$ — Particle is at $x < 0$, moving in $-x$-direction ($v_x < 0$), and slowing down (v_x and a_x have opposite signs).

t_B — Particle is at $x < 0$, instantaneously at rest ($v_x = 0$), and about to move in $+x$-direction ($a_x > 0$).

t_C — Particle is at $x > 0$, moving in $+x$-direction ($v_x > 0$); its speed is instantaneously not changing ($a_x = 0$).

t_D — Particle is at $x > 0$, instantaneously at rest ($v_x = 0$), and about to move in $-x$-direction ($a_x < 0$).

t_E — Particle is at $x > 0$, moving in $-x$-direction ($v_x < 0$), and speeding up (v_x and a_x have the same sign).

• On a v_x-t graph, the slope of the tangent at any point equals the particle's acceleration at that point.
• The steeper the slope (positive or negative), the greater the particle's acceleration in the positive or negative x-direction.

That is, a_x is the second derivative of x with respect to t. The second derivative of any function is directly related to the *concavity* or *curvature* of the graph of that function (**Fig. 2.14**). At a point where the x-t graph is concave up (curved upward), such as point A or E in Fig. 2.14a, the x-acceleration is positive and v_x is increasing. At a point where the x-t graph is concave down (curved downward), such as point C in Fig. 2.14a, the x-acceleration is negative and v_x is decreasing. At a point where the x-t graph has no curvature, such as the inflection points B and D in Fig. 2.14a, the x-acceleration is zero and the velocity is not changing.

Examining the curvature of an x-t graph is an easy way to identify the *sign* of acceleration. This technique is less helpful for determining numerical values of acceleration because the curvature of a graph is hard to measure accurately.

TEST YOUR UNDERSTANDING OF SECTION 2.3 Look again at the x-t graph in Fig. 2.9 at the end of Section 2.2. (a) At which of the points P, Q, R, and S is the x-acceleration a_x positive? (b) At which points is the x-acceleration negative? (c) At which points does the x-acceleration appear to be zero? (d) At each point state whether the velocity is increasing, decreasing, or not changing. ∎

2.14 (a) The same x-t graph as shown in Fig. 2.8a. (b) A motion diagram showing the position, velocity, and acceleration of the particle at each of the times labeled on the x-t graph.

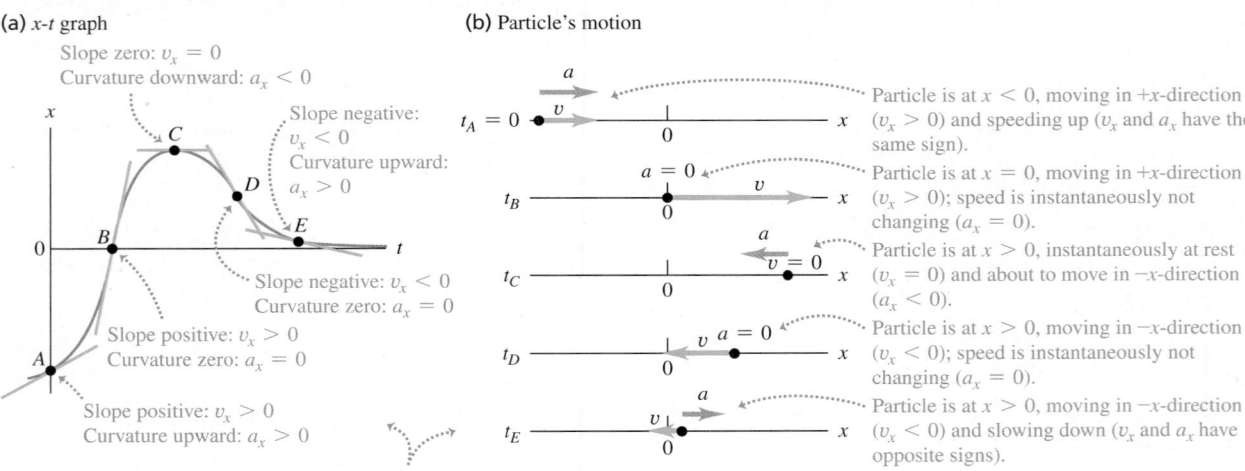

(a) x-t graph

(b) Particle's motion

Slope zero: $v_x = 0$
Curvature downward: $a_x < 0$

Slope negative:
$v_x < 0$
Curvature upward:
$a_x > 0$

Slope negative: $v_x < 0$
Curvature zero: $a_x = 0$

Slope positive: $v_x > 0$
Curvature zero: $a_x = 0$

Slope positive: $v_x > 0$
Curvature upward: $a_x > 0$

$t_A = 0$ — Particle is at $x < 0$, moving in $+x$-direction ($v_x > 0$) and speeding up (v_x and a_x have the same sign).

t_B — Particle is at $x = 0$, moving in $+x$-direction ($v_x > 0$); speed is instantaneously not changing ($a_x = 0$).

t_C — Particle is at $x > 0$, instantaneously at rest ($v_x = 0$) and about to move in $-x$-direction ($a_x < 0$).

t_D — Particle is at $x > 0$, moving in $-x$-direction ($v_x < 0$); speed is instantaneously not changing ($a_x = 0$).

t_E — Particle is at $x > 0$, moving in $-x$-direction ($v_x < 0$) and slowing down (v_x and a_x have opposite signs).

• On an x-t graph, the curvature at any point tells you the particle's acceleration at that point.
• The greater the curvature (positive or negative), the greater the particle's acceleration in the positive or negative x-direction.

2.4 MOTION WITH CONSTANT ACCELERATION

The simplest kind of accelerated motion is straight-line motion with *constant* acceleration. In this case the velocity changes at the same rate throughout the motion. As an example, a falling body has a constant acceleration if the effects of the air are not important. The same is true for a body sliding on an incline or along a rough horizontal surface, or for an airplane being catapulted from the deck of an aircraft carrier.

Figure 2.15 is a motion diagram showing the position, velocity, and acceleration of a particle moving with constant acceleration. **Figures 2.16** and **2.17** depict this same motion in the form of graphs. Since the x-acceleration is constant, the a_x-t **graph** (graph of x-acceleration versus time) in Figure 2.16 is a horizontal line. The graph of x-velocity versus time, or v_x-t graph, has a constant *slope* because the acceleration is constant, so this graph is a straight line (Fig. 2.17).

When the x-acceleration a_x is constant, the average x-acceleration $a_{av\text{-}x}$ for any time interval is the same as a_x. This makes it easy to derive equations for the position x and the x-velocity v_x as functions of time. To find an expression for v_x, we first replace $a_{av\text{-}x}$ in Eq. (2.4) by a_x:

$$a_x = \frac{v_{2x} - v_{1x}}{t_2 - t_1} \tag{2.7}$$

Now we let $t_1 = 0$ and let t_2 be any later time t. We use the symbol v_{0x} for the initial x-velocity at time $t = 0$; the x-velocity at the later time t is v_x. Then Eq. (2.7) becomes

$$a_x = \frac{v_x - v_{0x}}{t - 0} \qquad \text{or}$$

> **x-velocity** at time t of a particle with **constant x-acceleration** ⟶ Constant x-acceleration of the particle x-velocity of the particle at time 0   Time
>
> $$v_x = v_{0x} + a_x t \tag{2.8}$$

In Eq. (2.8) the term $a_x t$ is the product of the constant rate of change of x-velocity, a_x, and the time interval t. Therefore it equals the *total* change in x-velocity from $t = 0$ to time t. The x-velocity v_x at any time t then equals the initial x-velocity v_{0x} (at $t = 0$) plus the change in x-velocity $a_x t$ (Fig. 2.17).

Equation (2.8) also says that the change in x-velocity $v_x - v_{0x}$ of the particle between $t = 0$ and any later time t equals the *area* under the a_x-t graph between those two times. You can verify this from Fig. 2.16: Under this graph is a rectangle of vertical side a_x, horizontal side t, and area $a_x t$. From Eq. (2.8) the area $a_x t$ is indeed equal to the change in velocity $v_x - v_{0x}$. In Section 2.6 we'll show that even if the x-acceleration is not constant, the change in x-velocity during a time interval is still equal to the area under the a_x-t curve, although then Eq. (2.8) does not apply.

Next we'll derive an equation for the position x as a function of time when the x-acceleration is constant. To do this, we use two different expressions for the average x-velocity $v_{av\text{-}x}$ during the interval from $t = 0$ to any later time t. The first expression comes from the definition of $v_{av\text{-}x}$, Eq. (2.2), which is true whether or not the acceleration is constant. We call the position at time $t = 0$ the *initial position*, denoted by x_0. The position at time t is simply x. Thus for the time interval $\Delta t = t - 0$ the displacement is $\Delta x = x - x_0$, and Eq. (2.2) gives

$$v_{av\text{-}x} = \frac{x - x_0}{t} \tag{2.9}$$

To find a second expression for $v_{av\text{-}x}$, note that the x-velocity changes at a constant rate if the x-acceleration is constant. In this case the average x-velocity

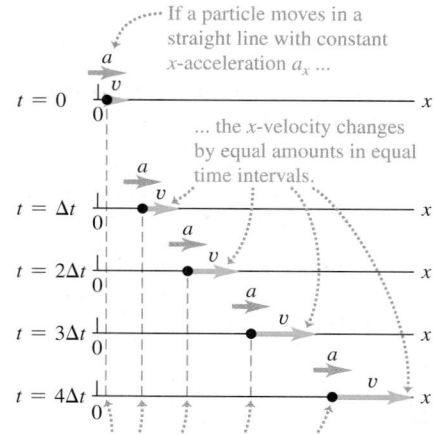

2.15 A motion diagram for a particle moving in a straight line in the positive x-direction with constant positive x-acceleration a_x.

If a particle moves in a straight line with constant x-acceleration a_x ...

... the x-velocity changes by equal amounts in equal time intervals.

However, the position changes by *different* amounts in equal time intervals because the velocity is changing.

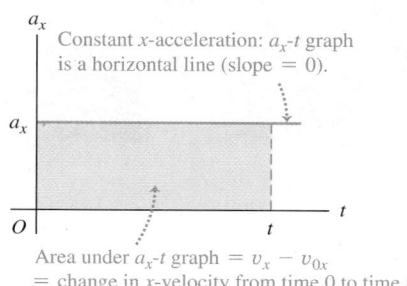

2.16 An acceleration-time (a_x-t) graph of straight-line motion with constant positive x-acceleration a_x.

Constant x-acceleration: a_x-t graph is a horizontal line (slope = 0).

Area under a_x-t graph $= v_x - v_{0x}$ = change in x-velocity from time 0 to time t.

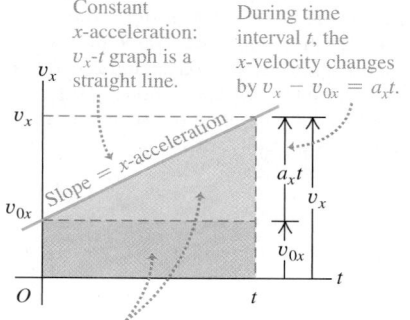

2.17 A velocity-time (v_x-t) graph of straight-line motion with constant positive x-acceleration a_x. The initial x-velocity v_{0x} is also positive in this case.

Constant x-acceleration: v_x-t graph is a straight line.

During time interval t, the x-velocity changes by $v_x - v_{0x} = a_x t$.

Slope = x-acceleration

Total area under v_x-t graph $= x - x_0$ = change in x-coordinate from time 0 to time t.

BIO **Application Testing Humans at High Accelerations** In experiments carried out by the U.S. Air Force in the 1940s and 1950s, humans riding a rocket sled could withstand accelerations as great as 440 m/s². The first three photos in this sequence show Air Force physician John Stapp speeding up from rest to 188 m/s (678 km/h = 421 mi/h) in just 5 s. Photos 4–6 show the even greater magnitude of acceleration as the rocket sled braked to a halt.

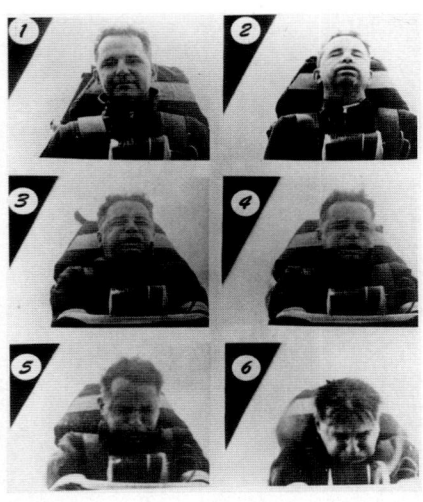

for the time interval from 0 to t is simply the average of the x-velocities at the beginning and end of the interval:

$$v_{\text{av-}x} = \tfrac{1}{2}(v_{0x} + v_x) \quad \text{(constant } x\text{-acceleration only)} \tag{2.10}$$

[Equation (2.10) is *not* true if the x-acceleration varies during the time interval.] We also know that with constant x-acceleration, the x-velocity v_x at any time t is given by Eq. (2.8). Substituting that expression for v_x into Eq. (2.10), we find

$$v_{\text{av-}x} = \tfrac{1}{2}(v_{0x} + v_{0x} + a_xt) \qquad \text{(constant}$$
$$= v_{0x} + \tfrac{1}{2}a_xt \qquad x\text{-acceleration only)} \tag{2.11}$$

Finally, we set Eqs. (2.9) and (2.11) equal to each other and simplify:

$$v_{0x} + \tfrac{1}{2}a_xt = \frac{x - x_0}{t} \qquad \text{or}$$

Position of the particle at time 0

Position at time t of a particle with **constant** x-acceleration

Time

$$x = x_0 + v_{0x}t + \tfrac{1}{2}a_xt^2 \tag{2.12}$$

x-velocity of the particle at time 0   Constant x-acceleration of the particle

Equation (2.12) tells us that the particle's position at time t is the sum of three terms: its initial position at $t = 0$, x_0, plus the displacement $v_{0x}t$ it would have if its x-velocity remained equal to its initial value, plus an additional displacement $\tfrac{1}{2}a_xt^2$ caused by the change in x-velocity.

A graph of Eq. (2.12)—that is, an x-t graph for motion with constant x-acceleration (**Fig. 2.18a**)—is always a *parabola*. Figure 2.18b shows such a graph. The curve intercepts the vertical axis (x-axis) at x_0, the position at $t = 0$. The slope of the tangent at $t = 0$ equals v_{0x}, the initial x-velocity, and the slope of the tangent at any time t equals the x-velocity v_x at that time. The slope and x-velocity are continuously increasing, so the x-acceleration a_x is positive and the graph in Fig. 2.18b is concave up (it curves upward). If a_x is negative, the x-t graph is a parabola that is concave down (has a downward curvature).

If there is zero x-acceleration, the x-t graph is a straight line; if there is a constant x-acceleration, the additional $\tfrac{1}{2}a_xt^2$ term in Eq. (2.12) for x as a function of t curves the graph into a parabola (**Fig. 2.19a**). Similarly, if there is zero x-acceleration, the v_x-t graph is a horizontal line (the x-velocity is constant). Adding a constant x-acceleration in Eq. (2.8) gives a slope to the graph (Fig. 2.19b).

Here's another way to derive Eq. (2.12). Just as the change in x-velocity of the particle equals the area under the a_x-t graph, the displacement (change in position) equals the area under the v_x-t graph. So the displacement $x - x_0$ of the particle between $t = 0$ and any later time t equals the area under the v_x-t graph between

2.18 (a) Straight-line motion with constant acceleration. (b) A position-time (x-t) graph for this motion (the same motion as is shown in Figs. 2.15, 2.16, and 2.17). For this motion the initial position x_0, the initial velocity v_{0x}, and the acceleration a_x are all positive.

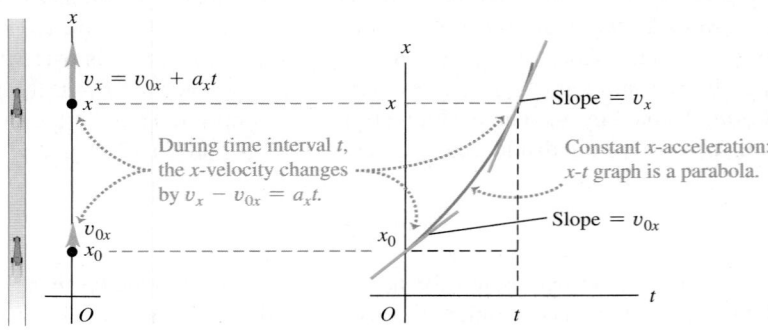

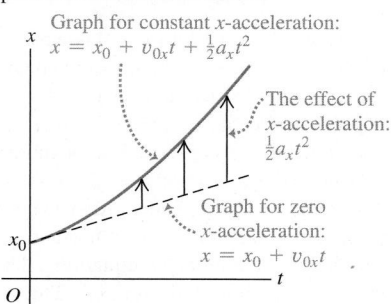

(a) An x-t graph for a particle moving with positive constant x-acceleration

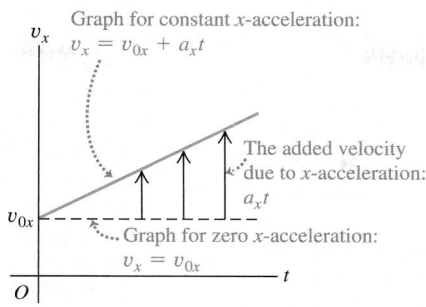

(b) The v_x-t graph for the same particle

2.19 **(a)** How a constant x-acceleration affects a particle's **(a)** x-t graph and **(b)** v_x-t graph.

those times. In Fig. 2.17 we divide the area under the graph into a dark-colored rectangle (vertical side v_{0x}, horizontal side t, and area $v_{0x}t$) and a light-colored right triangle (vertical side a_xt, horizontal side t, and area $\frac{1}{2}(a_xt)(t) = \frac{1}{2}a_xt^2$). The total area under the v_x-t graph is $x - x_0 = v_{0x}t + \frac{1}{2}a_xt^2$, in accord with Eq. (2.12).

The displacement during a time interval is always equal to the area under the v_x-t curve. This is true even if the acceleration is *not* constant, although in that case Eq. (2.12) does not apply. (We'll show this in Section 2.6.)

It's often useful to have a relationship for position, x-velocity, and (constant) x-acceleration that does not involve time. To obtain this, we first solve Eq. (2.8) for t and then substitute the resulting expression into Eq. (2.12):

$$t = \frac{v_x - v_{0x}}{a_x}$$

$$x = x_0 + v_{0x}\left(\frac{v_x - v_{0x}}{a_x}\right) + \frac{1}{2}a_x\left(\frac{v_x - v_{0x}}{a_x}\right)^2$$

We transfer the term x_0 to the left side, multiply through by $2a_x$, and simplify:

$$2a_x(x - x_0) = 2v_{0x}v_x - 2v_{0x}{}^2 + v_x{}^2 - 2v_{0x}v_x + v_{0x}{}^2$$

Finally,

x-velocity at time t of a particle with **constant x-acceleration** ..... x-velocity of the particle at time 0

$$v_x{}^2 = v_{0x}{}^2 + 2a_x(x - x_0) \tag{2.13}$$

Constant x-acceleration of the particle ..... Position of the particle at time t ..... Position of the particle at time 0

We can get one more useful relationship by equating the two expressions for $v_{\text{av-}x}$, Eqs. (2.9) and (2.10), and multiplying through by t:

Position at time t of a particle with **constant x-acceleration** ..... Position of the particle at time 0 ..... Time

$$x - x_0 = \frac{1}{2}(v_{0x} + v_x)t \tag{2.14}$$

x-velocity of the particle at time 0 ..... x-velocity of the particle at time t

Note that Eq. (2.14) does not contain the x-acceleration a_x. This equation can be handy when a_x is constant but its value is unknown.

Equations (2.8), (2.12), (2.13), and (2.14) are the *equations of motion with constant acceleration* (**Table 2.4**). By using these equations, we can solve *any* problem involving straight-line motion of a particle with constant acceleration.

For the particular case of motion with constant x-acceleration depicted in Fig. 2.15 and graphed in Figs. 2.16, 2.17, and 2.18, the values of x_0, v_{0x}, and a_x are all positive. We recommend that you redraw these figures for cases in which one, two, or all three of these quantities are negative.

Equations of Motion with
TABLE 2.4 Constant Acceleration

Equation		Includes Quantities			
$v_x = v_{0x} + a_xt$	(2.8)	t		v_x	a_x
$x = x_0 + v_{0x}t + \frac{1}{2}a_xt^2$	(2.12)	t	x		a_x
$v_x{}^2 = v_{0x}{}^2 + 2a_x(x - x_0)$	(2.13)		x	v_x	a_x
$x - x_0 = \frac{1}{2}(v_{0x} + v_x)t$	(2.14)	t	x	v_x	

PROBLEM-SOLVING STRATEGY 2.1 MOTION WITH CONSTANT ACCELERATION

IDENTIFY *the relevant concepts:* In most straight-line motion problems, you can use the constant-acceleration equations (2.8), (2.12), (2.13), and (2.14). If you encounter a situation in which the acceleration *isn't* constant, you'll need a different approach (see Section 2.6).

SET UP *the problem* using the following steps:

1. Read the problem carefully. Make a motion diagram showing the location of the particle at the times of interest. Decide where to place the origin of coordinates and which axis direction is positive. It's often helpful to place the particle at the origin at time $t = 0$; then $x_0 = 0$. Your choice of the positive axis direction automatically determines the positive directions for *x*-velocity and *x*-acceleration. If *x* is positive to the right of the origin, then v_x and a_x are also positive toward the right.

2. Identify the physical quantities (times, positions, velocities, and accelerations) that appear in Eqs. (2.8), (2.12), (2.13), and (2.14) and assign them appropriate symbols: x, x_0, v_x, v_{0x}, and a_x, or symbols related to those. Translate the prose into physics: "*When* does the particle arrive at its highest point" means "What is the value of *t* when *x* has its maximum value?" In Example 2.4, "Where is he when his speed is 25 m/s?" means "What is the value of *x* when $v_x = 25$ m/s?" Be alert for implicit information. For example, "A car sits at a stop light" usually means $v_{0x} = 0$.

3. List the quantities such as x, x_0, v_x, v_{0x}, a_x, and t. Some of them will be known and some will be unknown. Write down the values of the known quantities, and decide which of the unknowns are the target variables. Make note of the *absence* of any of the quantities that appear in the four constant-acceleration equations.

4. Use Table 2.4 to identify the applicable equations. (These are often the equations that don't include any of the absent quantities that you identified in step 3.) Usually you'll find a single equation that contains only one of the target variables. Sometimes you must find two equations, each containing the same two unknowns.

5. Sketch graphs corresponding to the applicable equations. The v_x-*t* graph of Eq. (2.8) is a straight line with slope a_x. The *x*-*t* graph of Eq. (2.12) is a parabola that's concave up if a_x is positive and concave down if a_x is negative.

6. On the basis of your experience with such problems, and taking account of what your sketched graphs tell you, make any qualitative and quantitative predictions you can about the solution.

EXECUTE *the solution:* If a single equation applies, solve it for the target variable, *using symbols only*; then substitute the known values and calculate the value of the target variable. If you have two equations in two unknowns, solve them simultaneously for the target variables.

EVALUATE *your answer:* Take a hard look at your results to see whether they make sense. Are they within the general range of values that you expected?

EXAMPLE 2.4 CONSTANT-ACCELERATION CALCULATIONS

SOLUTION

A motorcyclist heading east through a small town accelerates at a constant 4.0 m/s^2 after he leaves the city limits (**Fig. 2.20**). At time $t = 0$ he is 5.0 m east of the city-limits signpost while he moves east at 15 m/s. (a) Find his position and velocity at $t = 2.0$ s. (b) Where is he when his speed is 25 m/s?

SOLUTION

IDENTIFY and SET UP: The *x*-acceleration is constant, so we can use the constant-acceleration equations. We take the signpost as the origin of coordinates ($x = 0$) and choose the positive *x*-axis to point east (see Fig. 2.20, which is also a motion diagram). The known variables are the initial position and velocity, $x_0 = 5.0$ m and $v_{0x} = 15$ m/s, and the acceleration, $a_x = 4.0$ m/s^2. The unknown target variables in part (a) are the values of the position x and the *x*-velocity v_x at $t = 2.0$ s; the target variable in part (b) is the value of x when $v_x = 25$ m/s.

EXECUTE: (a) Since we know the values of x_0, v_{0x}, and a_x, Table 2.4 tells us that we can find the position x at $t = 2.0$ s by using

Eq. (2.12) and the *x*-velocity v_x at this time by using Eq. (2.8):

$$x = x_0 + v_{0x}t + \tfrac{1}{2}a_x t^2$$
$$= 5.0 \text{ m} + (15 \text{ m/s})(2.0 \text{ s}) + \tfrac{1}{2}(4.0 \text{ m/s}^2)(2.0 \text{ s})^2$$
$$= 43 \text{ m}$$
$$v_x = v_{0x} + a_x t$$
$$= 15 \text{ m/s} + (4.0 \text{ m/s}^2)(2.0 \text{ s}) = 23 \text{ m/s}$$

(b) We want to find the value of x when $v_x = 25$ m/s, but we don't know the time when the motorcycle has this velocity. Table 2.4 tells us that we should use Eq. (2.13), which involves x, v_x, and a_x but does *not* involve t:

$$v_x^2 = v_{0x}^2 + 2a_x(x - x_0)$$

Solving for x and substituting the known values, we find

$$x = x_0 + \frac{v_x^2 - v_{0x}^2}{2a_x}$$

$$= 5.0 \text{ m} + \frac{(25 \text{ m/s})^2 - (15 \text{ m/s})^2}{2(4.0 \text{ m/s}^2)} = 55 \text{ m}$$

EVALUATE: You can check the result in part (b) by first using Eq. (2.8), $v_x = v_{0x} + a_x t$, to find the time at which $v_x = 25$ m/s, which turns out to be $t = 2.5$ s. You can then use Eq. (2.12), $x = x_0 + v_{0x}t + \tfrac{1}{2}a_x t^2$, to solve for x. You should find $x = 55$ m, the same answer as above. That's the long way to solve the problem, though. The method we used in part (b) is much more efficient.

2.20 A motorcyclist traveling with constant acceleration.

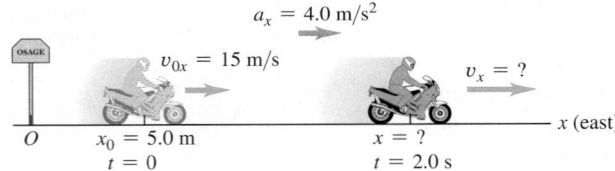

EXAMPLE 2.5   TWO BODIES WITH DIFFERENT ACCELERATIONS

A motorist traveling at a constant 15 m/s (about 34 mi/h) passes a school crossing where the speed limit is 10 m/s (about 22 mi/h). Just as the motorist passes the school-crossing sign, a police officer on a motorcycle stopped there starts in pursuit with constant acceleration 3.0 m/s² (**Fig. 2.21a**). (a) How much time elapses before the officer passes the motorist? At that time, (b) what is the officer's speed and (c) how far has each vehicle traveled?

SOLUTION

IDENTIFY and SET UP: Both the officer and the motorist move with constant acceleration (equal to zero for the motorist), so we can use the constant-acceleration formulas. We take the origin at the sign, so $x_0 = 0$ for both, and we take the positive direction to the right. Let x_P and x_M represent the positions of the police officer and the motorist at any time. Their initial velocities are $v_{P0x} = 0$ and $v_{M0x} = 15$ m/s, and their accelerations are $a_{Px} = 3.0$ m/s² and $a_{Mx} = 0$. Our target variable in part (a) is the time when the officer and motorist are at the same position x; Table 2.4 tells us that Eq. (2.12) is useful for this part. In part (b) we'll use Eq. (2.8) to find the officer's speed v (the magnitude of her velocity) at the time found in part (a). In part (c) we'll use Eq. (2.12) again to find the position of either vehicle at this same time.

Figure 2.21b shows an x-t graph for both vehicles. The straight line represents the motorist's motion, $x_M = x_{M0} + v_{M0x}t = v_{M0x}t$. The graph for the officer's motion is the right half of a parabola with upward curvature:

$$x_P = x_{P0} + v_{P0x}t + \tfrac{1}{2}a_{Px}t^2 = \tfrac{1}{2}a_{Px}t^2$$

A good sketch shows that the officer and motorist are at the same position $(x_P = x_M)$ at about $t = 10$ s, at which time both have traveled about 150 m from the sign.

EXECUTE: (a) To find the value of the time t at which the motorist and police officer are at the same position, we set $x_P = x_M$ by equating the expressions above and solving that equation for t:

$$v_{M0x}t = \tfrac{1}{2}a_{Px}t^2$$

$$t = 0 \quad \text{or} \quad t = \frac{2v_{M0x}}{a_{Px}} = \frac{2(15 \text{ m/s})}{3.0 \text{ m/s}^2} = 10 \text{ s}$$

Both vehicles have the same x-coordinate at *two* times, as Fig. 2.21b indicates. At $t = 0$ the motorist passes the officer; at $t = 10$ s the officer passes the motorist.

(b) We want the magnitude of the officer's x-velocity v_{Px} at the time t found in part (a). Substituting the values of v_{P0x} and a_{Px} into Eq. (2.8) along with $t = 10$ s from part (a), we find

$$v_{Px} = v_{P0x} + a_{Px}t = 0 + (3.0 \text{ m/s}^2)(10 \text{ s}) = 30 \text{ m/s}$$

The officer's speed is the absolute value of this, which is also 30 m/s.

(c) In 10 s the motorist travels a distance

$$x_M = v_{M0x}t = (15 \text{ m/s})(10 \text{ s}) = 150 \text{ m}$$

and the officer travels

$$x_P = \tfrac{1}{2}a_{Px}t^2 = \tfrac{1}{2}(3.0 \text{ m/s}^2)(10 \text{ s})^2 = 150 \text{ m}$$

This verifies that they have gone equal distances after 10 s.

EVALUATE: Our results in parts (a) and (c) agree with our estimates from our sketch. Note that when the officer passes the motorist, they do *not* have the same velocity: The motorist is moving at 15 m/s and the officer is moving at 30 m/s. You can also see this from Fig. 2.21b. Where the two x-t curves cross, their slopes (equal to the values of v_x for the two vehicles) are different.

Is it just coincidence that when the two vehicles are at the same position, the officer is going twice the speed of the motorist? Equation (2.14), $x - x_0 = \tfrac{1}{2}(v_{0x} + v_x)t$, gives the answer. The motorist has constant velocity, so $v_{M0x} = v_{Mx}$, and the motorist's displacement $x - x_0$ in time t is $v_{M0x}t$. Because $v_{P0x} = 0$, in the same time t the officer's displacement is $\tfrac{1}{2}v_{Px}t$. The two vehicles have the same displacement in the same amount of time, so $v_{M0x}t = \tfrac{1}{2}v_{Px}t$ and $v_{Px} = 2v_{M0x}$—that is, the officer has exactly twice the motorist's velocity. This is true no matter what the value of the officer's acceleration.

2.21 (a) Motion with constant acceleration overtaking motion with constant velocity. (b) A graph of x versus t for each vehicle.

(a)

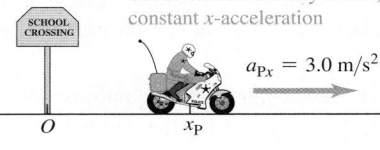

(b)

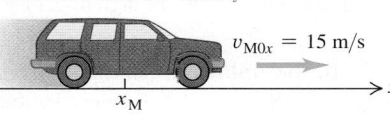

2.22 Multiflash photo of a freely falling ball.

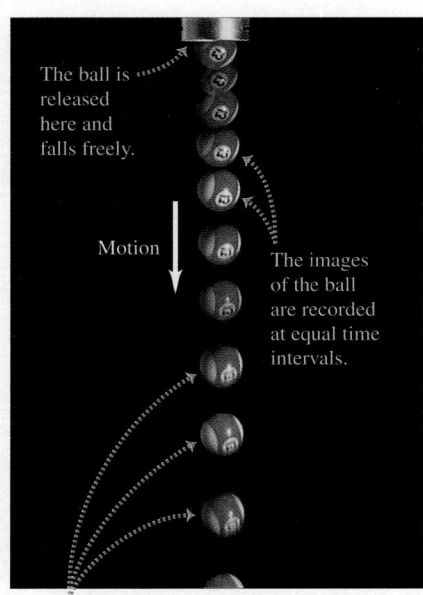

- The average velocity in each time interval is proportional to the distance between images.
- This distance continuously increases, so the ball's velocity is continuously changing; the ball is accelerating downward.

PhET: Lunar Lander

DATA *SPEAKS*

Free Fall

When students were given a problem about free fall, more than 20% gave an incorrect answer. Common errors:

- Confusing speed, velocity, and acceleration. Speed can never be negative; velocity can be positive or negative, depending on the direction of motion. In free fall, speed and velocity change continuously but acceleration (rate of change of velocity) is constant and downward.

- Not realizing that a freely falling body that moves upward at a certain speed past a point will pass that same point at the same speed as it moves downward (see Example 2.7).

TEST YOUR UNDERSTANDING OF SECTION 2.4 Four possible v_x-t graphs are shown for the two vehicles in Example 2.5. Which graph is correct?

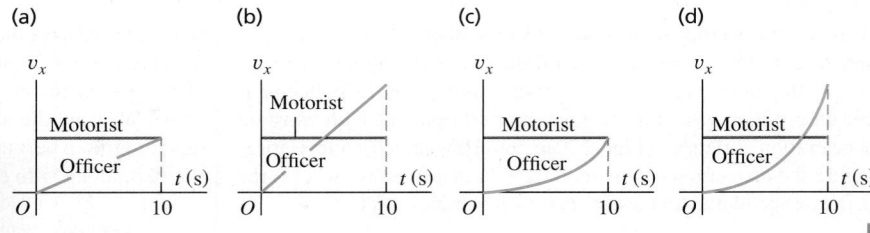

2.5 FREELY FALLING BODIES

The most familiar example of motion with (nearly) constant acceleration is a body falling under the influence of the earth's gravitational attraction. Such motion has held the attention of philosophers and scientists since ancient times. In the fourth century B.C., Aristotle thought (erroneously) that heavy bodies fall faster than light bodies, in proportion to their weight. Nineteen centuries later, Galileo (see Section 1.1) argued that a body should fall with a downward acceleration that is constant and independent of its weight.

Experiment shows that if the effects of the air can be ignored, Galileo is right; all bodies at a particular location fall with the same downward acceleration, regardless of their size or weight. If in addition the distance of the fall is small compared with the radius of the earth, and if we ignore small effects due to the earth's rotation, the acceleration is constant. The idealized motion that results under all of these assumptions is called **free fall,** although it includes rising as well as falling motion. (In Chapter 3 we will extend the discussion of free fall to include the motion of projectiles, which move both vertically and horizontally.)

Figure 2.22 is a photograph of a falling ball made with a stroboscopic light source that produces a series of short, intense flashes at equal time intervals. As each flash occurs, an image of the ball at that instant is recorded on the photograph. The increasing spacing between successive images in Fig. 2.22 indicates that the ball is accelerating downward. Careful measurement shows that the velocity change is the same in each time interval, so the acceleration of the freely falling ball is constant.

The constant acceleration of a freely falling body is called the **acceleration due to gravity,** and we denote its magnitude with the letter g. We will frequently use the approximate value of g at or near the earth's surface:

$$g = 9.80 \text{ m/s}^2 = 980 \text{ cm/s}^2 = 32.2 \text{ ft/s}^2 \quad \begin{array}{l}\text{(approximate value near the}\\\text{earth's surface)}\end{array}$$

The exact value varies with location, so we will often give the value of g at the earth's surface to only two significant figures as 9.8 m/s². On the moon's surface, the acceleration due to gravity is caused by the attractive force of the moon rather than the earth, and $g = 1.6$ m/s². Near the surface of the sun, $g = 270$ m/s².

CAUTION **g is always a positive number** Because g is the *magnitude* of a vector quantity, it is always a *positive* number. If you take the positive direction to be upward, as we do in most situations involving free fall, the acceleration is negative (downward) and equal to $-g$. Be careful with the sign of g, or you'll have trouble with free-fall problems. ▌

In the following examples we use the constant-acceleration equations developed in Section 2.4. Review Problem-Solving Strategy 2.1 in that section before you study the next examples.

EXAMPLE 2.6   A FREELY FALLING COIN

A one-euro coin is dropped from the Leaning Tower of Pisa and falls freely from rest. What are its position and velocity after 1.0 s, 2.0 s, and 3.0 s?

SOLUTION

IDENTIFY and SET UP: "Falls freely" means "falls with constant acceleration due to gravity," so we can use the constant-acceleration equations. The right side of **Fig. 2.23** shows our motion diagram

2.23 A coin freely falling from rest.

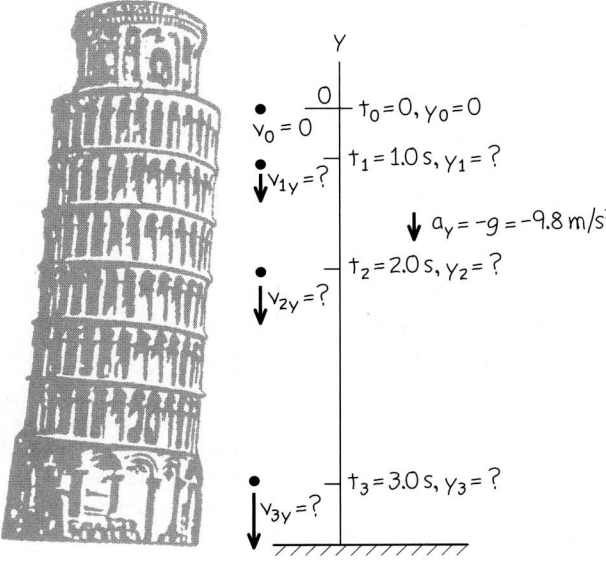

for the coin. The motion is vertical, so we use a vertical coordinate axis and call the coordinate y instead of x. We take the origin O at the starting point and the *upward* direction as positive. Both the initial coordinate y_0 and initial y-velocity v_{0y} are zero. The y-acceleration is downward (in the negative y-direction), so $a_y = -g = -9.8 \text{ m/s}^2$. (Remember that, by definition, g is a positive quantity.) Our target variables are the values of y and v_y at the three given times. To find these, we use Eqs. (2.12) and (2.8) with x replaced by y. Our choice of the upward direction as positive means that all positions and velocities we calculate will be negative.

EXECUTE: At a time t after the coin is dropped, its position and y-velocity are

$$y = y_0 + v_{0y}t + \tfrac{1}{2}a_y t^2 = 0 + 0 + \tfrac{1}{2}(-g)t^2 = (-4.9 \text{ m/s}^2)t^2$$

$$v_y = v_{0y} + a_y t = 0 + (-g)t = (-9.8 \text{ m/s}^2)t$$

When $t = 1.0$ s, $y = (-4.9 \text{ m/s}^2)(1.0 \text{ s})^2 = -4.9$ m and $v_y = (-9.8 \text{ m/s}^2)(1.0 \text{ s}) = -9.8$ m/s; after 1.0 s, the coin is 4.9 m below the origin (y is negative) and has a downward velocity (v_y is negative) with magnitude 9.8 m/s.

We can find the positions and y-velocities at 2.0 s and 3.0 s in the same way. The results are $y = -20$ m and $v_y = -20$ m/s at $t = 2.0$ s, and $y = -44$ m and $v_y = -29$ m/s at $t = 3.0$ s.

EVALUATE: All our answers are negative, as we expected. If we had chosen the positive y-axis to point downward, the acceleration would have been $a_y = +g$ and all our answers would have been positive.

EXAMPLE 2.7   UP-AND-DOWN MOTION IN FREE FALL

You throw a ball vertically upward from the roof of a tall building. The ball leaves your hand at a point even with the roof railing with an upward speed of 15.0 m/s; the ball is then in free fall. On its way back down, it just misses the railing. Find (a) the ball's position and velocity 1.00 s and 4.00 s after leaving your hand; (b) the ball's velocity when it is 5.00 m above the railing; (c) the maximum height reached; (d) the ball's acceleration when it is at its maximum height.

SOLUTION

IDENTIFY and SET UP: The words "in free fall" mean that the acceleration is due to gravity, which is constant. Our target variables are position [in parts (a) and (c)], velocity [in parts (a) and (b)], and acceleration [in part (d)]. We take the origin at the point where the ball leaves your hand, and take the positive direction to be upward (**Fig. 2.24**). The initial position y_0 is zero, the initial y-velocity v_{0y} is $+15.0$ m/s, and the y-acceleration is $a_y = -g = -9.80 \text{ m/s}^2$. In part (a), as in Example 2.6, we'll use Eqs. (2.12) and (2.8) to find the position and velocity as functions of time. In part (b) we must find the velocity at a given *position* (no time is given), so we'll use Eq. (2.13).

2.24 Position and velocity of a ball thrown vertically upward.

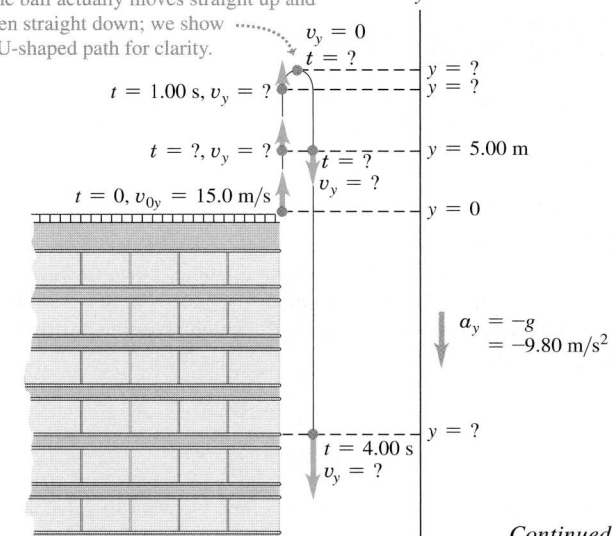

Continued

2.25 (a) Position and (b) velocity as functions of time for a ball thrown upward with an initial speed of 15.0 m/s.

(a) *y-t* graph (curvature is downward because $a_y = -g$ is negative)

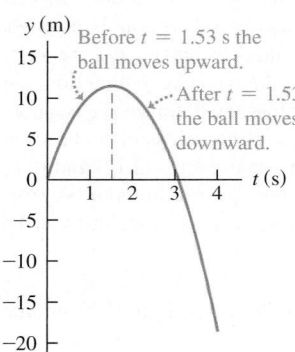

(b) v_y-*t* graph (straight line with negative slope because $a_y = -g$ is constant and negative)

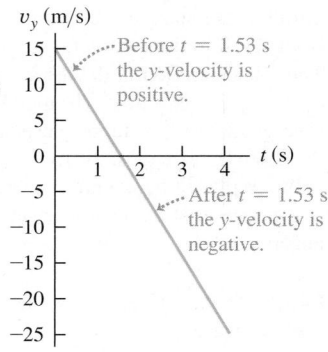

Figure 2.25 shows the *y-t* and v_y-*t* graphs for the ball. The *y-t* graph is a concave-down parabola that rises and then falls, and the v_y-*t* graph is a downward-sloping straight line. Note that the ball's velocity is zero when it is at its highest point.

EXECUTE: (a) The position and *y*-velocity at time *t* are given by Eqs. (2.12) and (2.8) with *x*'s replaced by *y*'s:

$$y = y_0 + v_{0y}t + \tfrac{1}{2}a_y t^2 = y_0 + v_{0y}t + \tfrac{1}{2}(-g)t^2$$
$$= (0) + (15.0 \text{ m/s})t + \tfrac{1}{2}(-9.80 \text{ m/s}^2)t^2$$
$$v_y = v_{0y} + a_y t = v_{0y} + (-g)t$$
$$= 15.0 \text{ m/s} + (-9.80 \text{ m/s}^2)t$$

When *t* = 1.00 s, these equations give *y* = +10.1 m and v_y = +5.2 m/s. That is, the ball is 10.1 m above the origin (*y* is positive) and moving upward (v_y is positive) with a speed of 5.2 m/s. This is less than the initial speed because the ball slows as it ascends. When *t* = 4.00 s, those equations give *y* = −18.4 m and v_y = −24.2 m/s. The ball has passed its highest point and is 18.4 m *below* the origin (*y* is negative). It is moving *downward*

(v_y is negative) with a speed of 24.2 m/s. Equation (2.13) tells us that the ball is moving at the initial 15.0-m/s speed as it moves downward past the launching point and continues to gain speed as it descends further.

(b) The *y*-velocity at any position *y* is given by Eq. (2.13) with *x*'s replaced by *y*'s:

$$v_y^2 = v_{0y}^2 + 2a_y(y - y_0) = v_{0y}^2 + 2(-g)(y - 0)$$
$$= (15.0 \text{ m/s})^2 + 2(-9.80 \text{ m/s}^2)y$$

When the ball is 5.00 m above the origin we have *y* = +5.00 m, so

$$v_y^2 = (15.0 \text{ m/s})^2 + 2(-9.80 \text{ m/s}^2)(5.00 \text{ m}) = 127 \text{ m}^2/\text{s}^2$$
$$v_y = \pm 11.3 \text{ m/s}$$

We get *two* values of v_y because the ball passes through the point *y* = +5.00 m twice, once on the way up (so v_y is positive) and once on the way down (so v_y is negative) (see Figs. 2.24 and 2.25a).

(c) At the instant at which the ball reaches its maximum height y_1, its *y*-velocity is momentarily zero: $v_y = 0$. We use Eq. (2.13) to find y_1. With $v_y = 0$, $y_0 = 0$, and $a_y = -g$, we get

$$0 = v_{0y}^2 + 2(-g)(y_1 - 0)$$
$$y_1 = \frac{v_{0y}^2}{2g} = \frac{(15.0 \text{ m/s})^2}{2(9.80 \text{ m/s}^2)} = +11.5 \text{ m}$$

(d) ▱CAUTION▱ **A free-fall misconception** It's a common misconception that at the highest point of free-fall motion, where the velocity is zero, the acceleration is also zero. If this were so, once the ball reached the highest point it would hang there suspended in midair! Remember that acceleration is the rate of change of velocity, and the ball's velocity is continuously changing. At every point, including the highest point, and at any velocity, including zero, the acceleration in free fall is always $a_y = -g = -9.80 \text{ m/s}^2$. ▮

EVALUATE: A useful way to check any free-fall problem is to draw the *y-t* and v_y-*t* graphs, as we did in Fig. 2.25. Note that these are graphs of Eqs. (2.12) and (2.8), respectively. Given the initial position, initial velocity, and acceleration, you can easily create these graphs by using a graphing calculator or an online mathematics program. ▮

EXAMPLE 2.8 **TWO SOLUTIONS OR ONE?**

At what time after being released has the ball in Example 2.7 fallen 5.00 m below the roof railing?

SOLUTION

IDENTIFY and SET UP: We treat this as in Example 2.7, so y_0, v_{0y}, and $a_y = -g$ have the same values as there. Now, however, the target variable is the time at which the ball is at *y* = −5.00 m. The best equation to use is Eq. (2.12), which gives the position *y* as a function of time *t*:

$$y = y_0 + v_{0y}t + \tfrac{1}{2}a_y t^2$$
$$= y_0 + v_{0y}t + \tfrac{1}{2}(-g)t^2$$

This is a *quadratic* equation for *t*, which we want to solve for the value of *t* when *y* = −5.00 m.

EXECUTE: We rearrange the equation so that it has the standard form of a quadratic equation for an unknown *x*, $Ax^2 + Bx + C = 0$:

$$\left(\tfrac{1}{2}g\right)t^2 + (-v_{0y})t + (y - y_0) = At^2 + Bt + C = 0$$

By comparison, we identify $A = \tfrac{1}{2}g$, $B = -v_{0y}$, and $C = y - y_0$. The quadratic formula (see Appendix B) tells us that this equation has *two* solutions:

$$t = \frac{-B \pm \sqrt{B^2 - 4AC}}{2A}$$
$$= \frac{-(-v_{0y}) \pm \sqrt{(-v_{0y})^2 - 4\left(\tfrac{1}{2}g\right)(y - y_0)}}{2\left(\tfrac{1}{2}g\right)}$$
$$= \frac{v_{0y} \pm \sqrt{v_{0y}^2 - 2g(y - y_0)}}{g}$$

Substituting the values $y_0 = 0$, $v_{0y} = +15.0$ m/s, $g = 9.80$ m/s^2, and $y = -5.00$ m, we find

$$t = \frac{(15.0 \text{ m/s}) \pm \sqrt{(15.0 \text{ m/s})^2 - 2(9.80 \text{ m/s}^2)(-5.00 \text{ m} - 0)}}{9.80 \text{ m/s}^2}$$

You can confirm that the numerical answers are $t = +3.36$ s and $t = -0.30$ s. The answer $t = -0.30$ s doesn't make physical sense, since it refers to a time *before* the ball left your hand at $t = 0$. So the correct answer is $t = +3.36$ s.

EVALUATE: Why did we get a second, fictitious solution? The explanation is that constant-acceleration equations like Eq. (2.12) are based on the assumption that the acceleration is constant for *all* values of time, whether positive, negative, or zero. Hence the solution $t = -0.30$ s refers to an imaginary moment when a freely falling ball was 5.00 m below the roof railing and rising to meet your hand. Since the ball didn't leave your hand and go into free fall until $t = 0$, this result is pure fiction.

Repeat these calculations to find the times when the ball is 5.00 m *above* the origin ($y = +5.00$ m). The two answers are $t = +0.38$ s and $t = +2.68$ s. Both are positive values of t, and both refer to the real motion of the ball after leaving your hand. At the earlier time the ball passes through $y = +5.00$ m moving upward; at the later time it passes through this point moving downward. [Compare this with part (b) of Example 2.7, and again refer to Fig. 2.25a.]

You should also solve for the times when $y = +15.0$ m. In this case, both solutions involve the square root of a negative number, so there are *no* real solutions. Again Fig. 2.25a shows why; we found in part (c) of Example 2.7 that the ball's maximum height is $y = +11.5$ m, so it *never* reaches $y = +15.0$ m. While a quadratic equation such as Eq. (2.12) always has two solutions, in some situations one or both of the solutions aren't physically reasonable.

TEST YOUR UNDERSTANDING OF SECTION 2.5   If you toss a ball upward with a certain initial speed, it falls freely and reaches a maximum height h a time t after it leaves your hand. (a) If you throw the ball upward with double the initial speed, what new maximum height does the ball reach? (i) $h\sqrt{2}$; (ii) $2h$; (iii) $4h$; (iv) $8h$; (v) $16h$. (b) If you throw the ball upward with double the initial speed, how long does it take to reach its new maximum height? (i) $t/2$; (ii) $t/\sqrt{2}$; (iii) t; (iv) $t\sqrt{2}$; (v) $2t$. ❚

2.6 VELOCITY AND POSITION BY INTEGRATION

This section is intended for students who have already learned a little integral calculus. In Section 2.4 we analyzed the special case of straight-line motion with constant acceleration. When a_x is not constant, as is frequently the case, the equations that we derived in that section are no longer valid (**Fig. 2.26**). But even when a_x varies with time, we can still use the relationship $v_x = dx/dt$ to find the x-velocity v_x as a function of time if the position x is a known function of time. And we can still use $a_x = dv_x/dt$ to find the x-acceleration a_x as a function of time if the x-velocity v_x is a known function of time.

In many situations, however, position and velocity are not known functions of time, while acceleration is (**Fig. 2.27**). How can we find the position and velocity in straight-line motion from the acceleration function $a_x(t)$?

2.26 When you push a car's accelerator pedal to the floorboard, the resulting acceleration is *not* constant: The greater the car's speed, the more slowly it gains additional speed. A typical car takes twice as long to accelerate from 50 km/h to 100 km/h as it does to accelerate from 0 to 50 km/h.

2.27 The inertial navigation system (INS) on board a long-range airliner keeps track of the airliner's acceleration. Given the airliner's initial position and velocity before takeoff, the INS uses the acceleration data to calculate the airliner's position and velocity throughout the flight.

2.28 An a_x-t graph for a body whose x-acceleration is not constant.

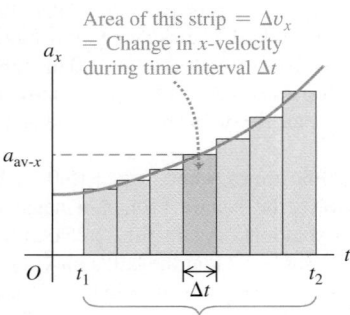

Area of this strip $= \Delta v_x$
$=$ Change in x-velocity during time interval Δt

Total area under the x-t graph from t_1 to t_2
$=$ Net change in x-velocity from t_1 to t_2

Figure 2.28 is a graph of x-acceleration versus time for a body whose acceleration is not constant. We can divide the time interval between times t_1 and t_2 into many smaller subintervals, calling a typical one Δt. Let the average x-acceleration during Δt be $a_{\text{av-}x}$. From Eq. (2.4) the change in x-velocity Δv_x during Δt is

$$\Delta v_x = a_{\text{av-}x}\, \Delta t$$

Graphically, Δv_x equals the area of the shaded strip with height $a_{\text{av-}x}$ and width Δt—that is, the area under the curve between the left and right sides of Δt. The total change in x-velocity from t_1 to t_2 is the sum of the x-velocity changes Δv_x in the small subintervals. So the total x-velocity change is represented graphically by the *total* area under the a_x-t curve between the vertical lines t_1 and t_2. (In Section 2.4 we showed this for the special case in which a_x is constant.)

In the limit that all the Δt's become very small and they become very large in number, the value of $a_{\text{av-}x}$ for the interval from any time t to $t + \Delta t$ approaches the instantaneous x-acceleration a_x at time t. In this limit, the area under the a_x-t curve is the *integral* of a_x (which is in general a function of t) from t_1 to t_2. If v_{1x} is the x-velocity of the body at time t_1 and v_{2x} is the velocity at time t_2, then

$$v_{2x} - v_{1x} = \int_{v_{1x}}^{v_{2x}} dv_x = \int_{t_1}^{t_2} a_x\, dt \qquad (2.15)$$

The change in the x-velocity v_x is the time integral of the x-acceleration a_x.

We can carry out exactly the same procedure with the curve of x-velocity versus time. If x_1 is a body's position at time t_1 and x_2 is its position at time t_2, from Eq. (2.2) the displacement Δx during a small time interval Δt is equal to $v_{\text{av-}x}\, \Delta t$, where $v_{\text{av-}x}$ is the average x-velocity during Δt. The total displacement $x_2 - x_1$ during the interval $t_2 - t_1$ is given by

$$x_2 - x_1 = \int_{x_1}^{x_2} dx = \int_{t_1}^{t_2} v_x\, dt \qquad (2.16)$$

The change in position x—that is, the displacement—is the time integral of x-velocity v_x. Graphically, the displacement between times t_1 and t_2 is the area under the v_x-t curve between those two times. [This is the same result that we obtained in Section 2.4 for the special case in which v_x is given by Eq. (2.8).]

If $t_1 = 0$ and t_2 is any later time t, and if x_0 and v_{0x} are the position and velocity, respectively, at time $t = 0$, then we can rewrite Eqs. (2.15) and (2.16) as

x-velocity of a particle at time t ⋯⋯⋯⋯⋯ x-velocity of the particle at time 0

$$v_x = v_{0x} + \int_0^t a_x\, dt \qquad (2.17)$$

Integral of the x-acceleration of the particle from time 0 to time t

Position of a particle at time t ⋯⋯⋯⋯ Position of the particle at time 0

$$x = x_0 + \int_0^t v_x\, dt \qquad (2.18)$$

Integral of the x-velocity of the particle from time 0 to time t

If we know the x-acceleration a_x as a function of time and we know the initial velocity v_{0x}, we can use Eq. (2.17) to find the x-velocity v_x at any time; in other words, we can find v_x as a function of time. Once we know this function, and given the initial position x_0, we can use Eq. (2.18) to find the position x at any time.

EXAMPLE 2.9 MOTION WITH CHANGING ACCELERATION

Sally is driving along a straight highway in her 1965 Mustang. At $t = 0$, when she is moving at 10 m/s in the positive x-direction, she passes a signpost at $x = 50$ m. Her x-acceleration as a function of time is

$$a_x = 2.0 \text{ m/s}^2 - (0.10 \text{ m/s}^3)t$$

(a) Find her x-velocity v_x and position x as functions of time. (b) When is her x-velocity greatest? (c) What is that maximum x-velocity? (d) Where is the car when it reaches that maximum x-velocity?

SOLUTION

IDENTIFY and SET UP: The x-acceleration is a function of time, so we *cannot* use the constant-acceleration formulas of Section 2.4. Instead, we use Eq. (2.17) to obtain an expression for v_x as a function of time, and then use that result in Eq. (2.18) to find an expression for x as a function of t. We'll then be able to answer a variety of questions about the motion.

EXECUTE: (a) At $t = 0$, Sally's position is $x_0 = 50$ m and her x-velocity is $v_{0x} = 10$ m/s. To use Eq. (2.17), we note that the integral of t^n (except for $n = -1$) is $\int t^n \, dt = \frac{1}{n+1} t^{n+1}$. Hence

$$v_x = 10 \text{ m/s} + \int_0^t [2.0 \text{ m/s}^2 - (0.10 \text{ m/s}^3)t] \, dt$$

$$= 10 \text{ m/s} + (2.0 \text{ m/s}^2)t - \tfrac{1}{2}(0.10 \text{ m/s}^3)t^2$$

Now we use Eq. (2.18) to find x as a function of t:

$$x = 50 \text{ m} + \int_0^t [10 \text{ m/s} + (2.0 \text{ m/s}^2)t - \tfrac{1}{2}(0.10 \text{ m/s}^3)t^2] \, dt$$

$$= 50 \text{ m} + (10 \text{ m/s})t + \tfrac{1}{2}(2.0 \text{ m/s}^2)t^2 - \tfrac{1}{6}(0.10 \text{ m/s}^3)t^3$$

Figure 2.29 shows graphs of a_x, v_x, and x as functions of time as given by the previous equations. Note that for any time t, the slope of the v_x-t graph equals the value of a_x and the slope of the x-t graph equals the value of v_x.

(b) The maximum value of v_x occurs when the x-velocity stops increasing and begins to decrease. At that instant, $dv_x/dt = a_x = 0$. So we set the expression for a_x equal to zero and solve for t:

$$0 = 2.0 \text{ m/s}^2 - (0.10 \text{ m/s}^3)t$$

$$t = \frac{2.0 \text{ m/s}^2}{0.10 \text{ m/s}^3} = 20 \text{ s}$$

(c) We find the maximum x-velocity by substituting $t = 20$ s, the time from part (b) when velocity is maximum, into the equation for v_x from part (a):

$$v_{\text{max-}x} = 10 \text{ m/s} + (2.0 \text{ m/s}^2)(20 \text{ s}) - \tfrac{1}{2}(0.10 \text{ m/s}^3)(20 \text{ s})^2$$

$$= 30 \text{ m/s}$$

(d) To find the car's position at the time that we found in part (b), we substitute $t = 20$ s into the expression for x from part (a):

$$x = 50 \text{ m} + (10 \text{ m/s})(20 \text{ s}) + \tfrac{1}{2}(2.0 \text{ m/s}^2)(20 \text{ s})^2$$

$$- \tfrac{1}{6}(0.10 \text{ m/s}^3)(20 \text{ s})^3$$

$$= 517 \text{ m}$$

EVALUATE: Figure 2.29 helps us interpret our results. The left-hand graph shows that a_x is positive between $t = 0$ and $t = 20$ s and negative after that. It is zero at $t = 20$ s, the time at which v_x is maximum (the high point in the middle graph). The car speeds up until $t = 20$ s (because v_x and a_x have the same sign) and slows down after $t = 20$ s (because v_x and a_x have opposite signs).

Since v_x is maximum at $t = 20$ s, the x-t graph (the right-hand graph in Fig. 2.29) has its maximum positive slope at this time. Note that the x-t graph is concave up (curved upward) from $t = 0$ to $t = 20$ s, when a_x is positive. The graph is concave down (curved downward) after $t = 20$ s, when a_x is negative.

2.29 The position, velocity, and acceleration of the car in Example 2.9 as functions of time. Can you show that if this motion continues, the car will stop at $t = 44.5$ s?

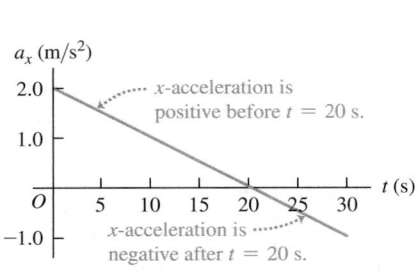

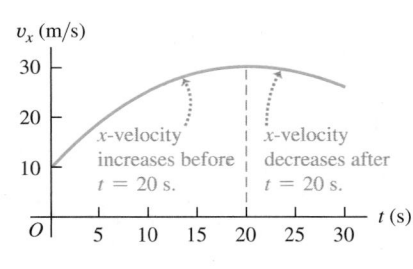

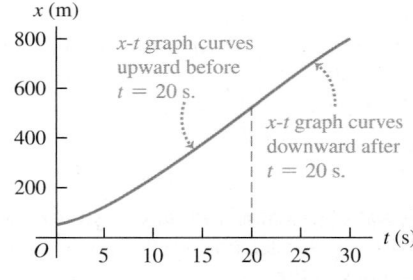

TEST YOUR UNDERSTANDING OF SECTION 2.6 If the x-acceleration a_x of an object moving in straight-line motion is increasing with time, will the v_x-t graph be (i) a straight line, (ii) concave up (i.e., with an upward curvature), or (iii) concave down (i.e., with a downward curvature)? ∎

CHAPTER 2 SUMMARY

SOLUTIONS TO ALL EXAMPLES

Straight-line motion, average and instantaneous x-velocity: When a particle moves along a straight line, we describe its position with respect to an origin O by means of a coordinate such as x. The particle's average x-velocity $v_{\text{av-}x}$ during a time interval $\Delta t = t_2 - t_1$ is equal to its displacement $\Delta x = x_2 - x_1$ divided by Δt. The instantaneous x-velocity v_x at any time t is equal to the average x-velocity over the time interval from t to $t + \Delta t$ in the limit that Δt goes to zero. Equivalently, v_x is the derivative of the position function with respect to time. (See Example 2.1.)

$$v_{\text{av-}x} = \frac{\Delta x}{\Delta t} = \frac{x_2 - x_1}{t_2 - t_1} \tag{2.2}$$

$$v_x = \lim_{\Delta t \to 0} \frac{\Delta x}{\Delta t} = \frac{dx}{dt} \tag{2.3}$$

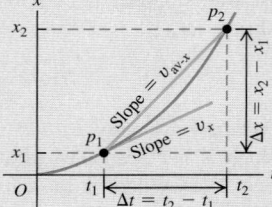

Average and instantaneous x-acceleration: The average x-acceleration $a_{\text{av-}x}$ during a time interval Δt is equal to the change in velocity $\Delta v_x = v_{2x} - v_{1x}$ during that time interval divided by Δt. The instantaneous x-acceleration a_x is the limit of $a_{\text{av-}x}$ as Δt goes to zero, or the derivative of v_x with respect to t. (See Examples 2.2 and 2.3.)

$$a_{\text{av-}x} = \frac{\Delta v_x}{\Delta t} = \frac{v_{2x} - v_{1x}}{t_2 - t_1} \tag{2.4}$$

$$a_x = \lim_{\Delta t \to 0} \frac{\Delta v_x}{\Delta t} = \frac{dv_x}{dt} \tag{2.5}$$

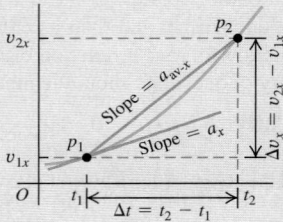

Straight-line motion with constant acceleration: When the x-acceleration is constant, four equations relate the position x and the x-velocity v_x at any time t to the initial position x_0, the initial x-velocity v_{0x} (both measured at time $t = 0$), and the x-acceleration a_x. (See Examples 2.4 and 2.5.)

Constant x-acceleration only:

$$v_x = v_{0x} + a_x t \tag{2.8}$$

$$x = x_0 + v_{0x}t + \tfrac{1}{2}a_x t^2 \tag{2.12}$$

$$v_x^2 = v_{0x}^2 + 2a_x(x - x_0) \tag{2.13}$$

$$x - x_0 = \tfrac{1}{2}(v_{0x} + v_x)t \tag{2.14}$$

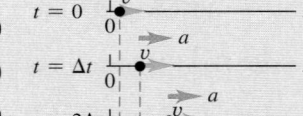

Freely falling bodies: Free fall is a case of motion with constant acceleration. The magnitude of the acceleration due to gravity is a positive quantity, g. The acceleration of a body in free fall is always downward. (See Examples 2.6–2.8.)

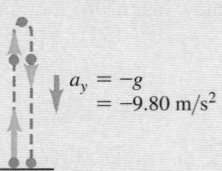

$a_y = -g$
$= -9.80 \text{ m/s}^2$

Straight-line motion with varying acceleration: When the acceleration is not constant but is a known function of time, we can find the velocity and position as functions of time by integrating the acceleration function. (See Example 2.9.)

$$v_x = v_{0x} + \int_0^t a_x \, dt \tag{2.17}$$

$$x = x_0 + \int_0^t v_x \, dt \tag{2.18}$$

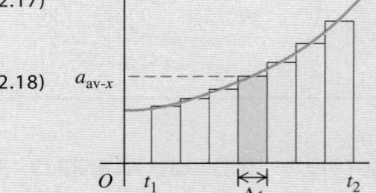

SOLUTION

BRIDGING PROBLEM   THE FALL OF A SUPERHERO

The superhero Green Lantern steps from the top of a tall building. He falls freely from rest to the ground, falling half the total distance to the ground during the last 1.00 s of his fall (**Fig. 2.30**). What is the height *h* of the building?

SOLUTION GUIDE

IDENTIFY and SET UP

1. You're told that Green Lantern falls freely from rest. What does this imply about his acceleration? About his initial velocity?
2. Choose the direction of the positive *y*-axis. It's easiest to make the same choice we used for freely falling objects in Section 2.5.
3. You can divide Green Lantern's fall into two parts: from the top of the building to the halfway point and from the halfway point to the ground. You know that the second part of the fall lasts 1.00 s. Decide what you would need to know about Green Lantern's motion at the halfway point in order to solve for the target variable *h*. Then choose two equations, one for the first part of the fall and one for the second part, that you'll use together to find an expression for *h*. (There are several pairs of equations that you could choose.)

EXECUTE

4. Use your two equations to solve for the height *h*. Heights are always positive numbers, so your answer should be positive.

2.30 Our sketch for this problem.

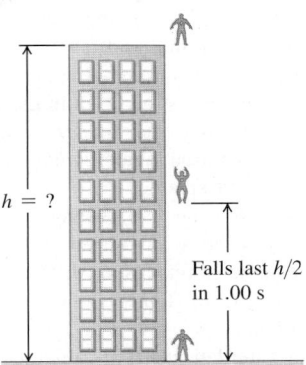

$h = ?$

Falls last $h/2$ in 1.00 s

EVALUATE

5. To check your answer for *h*, use one of the free-fall equations to find how long it takes Green Lantern to fall (i) from the top of the building to half the height and (ii) from the top of the building to the ground. If your answer for *h* is correct, time (ii) should be 1.00 s greater than time (i). If it isn't, go back and look for errors in how you found *h*.

Problems

For assigned homework and other learning materials, go to MasteringPhysics®.   **MP**

°, °°, °°°: Difficulty levels. CP: Cumulative problems incorporating material from earlier chapters. CALC: Problems requiring calculus. DATA: Problems involving real data, scientific evidence, experimental design, and/or statistical reasoning. BIO: Biosciences problems.

DISCUSSION QUESTIONS

Q2.1 Does the speedometer of a car measure speed or velocity? Explain.

Q2.2 The black dots at the top of **Fig. Q2.2** represent a series of high-speed photographs of an insect flying in a straight line from left to right (in the positive *x*-direction). Which of the graphs in Fig. Q2.2 most plausibly depicts this insect's motion?

Figure **Q2.2**

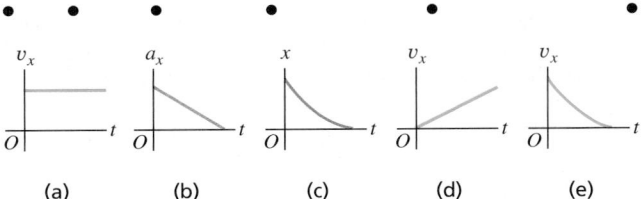

(a)   (b)   (c)   (d)   (e)

Q2.3 Can an object with constant acceleration reverse its direction of travel? Can it reverse its direction *twice*? In both cases, explain your reasoning.

Q2.4 Under what conditions is average velocity equal to instantaneous velocity?

Q2.5 Is it possible for an object to be (a) slowing down while its acceleration is increasing in magnitude; (b) speeding up while its acceleration is decreasing? In both cases, explain your reasoning.

Q2.6 Under what conditions does the magnitude of the average velocity equal the average speed?

Q2.7 When a Dodge Viper is at Elwood's Car Wash, a BMW Z3 is at Elm and Main. Later, when the Dodge reaches Elm and Main, the BMW reaches Elwood's Car Wash. How are the cars' average velocities between these two times related?

Q2.8 A driver in Massachusetts was sent to traffic court for speeding. The evidence against the driver was that a policewoman observed the driver's car alongside a second car at a certain moment, and the policewoman had already clocked the second car going faster than the speed limit. The driver argued, "The second car was passing me. I was not speeding." The judge ruled against the driver because, in the judge's words, "If two cars were side by side, both of you were speeding." If you were a lawyer representing the accused driver, how would you argue this case?

Q2.9 Can you have zero displacement and nonzero average velocity? Zero displacement and nonzero velocity? Illustrate your answers on an *x-t* graph.

Q2.10 Can you have zero acceleration and nonzero velocity? Use a v_x-*t* graph to explain.

Q2.11 Can you have zero velocity and nonzero average acceleration? Zero velocity and nonzero acceleration? Use a v_x-*t* graph to explain, and give an example of such motion.

Q2.12 An automobile is traveling west. Can it have a velocity toward the west and at the same time have an acceleration toward the east? Under what circumstances?

Q2.13 The official's truck in Fig. 2.2 is at $x_1 = 277$ m at $t_1 = 16.0$ s and is at $x_2 = 19$ m at $t_2 = 25.0$ s. (a) Sketch *two* different possible x-t graphs for the motion of the truck. (b) Does the average velocity $v_{\text{av-}x}$ during the time interval from t_1 to t_2 have the same value for both of your graphs? Why or why not?

Q2.14 Under constant acceleration the average velocity of a particle is half the sum of its initial and final velocities. Is this still true if the acceleration is *not* constant? Explain.

Q2.15 You throw a baseball straight up in the air so that it rises to a maximum height much greater than your height. Is the magnitude of the ball's acceleration greater while it is being thrown or after it leaves your hand? Explain.

Q2.16 Prove these statements: (a) As long as you can ignore the effects of the air, if you throw anything vertically upward, it will have the same speed when it returns to the release point as when it was released. (b) The time of flight will be twice the time it takes to get to its highest point.

Q2.17 A dripping water faucet steadily releases drops 1.0 s apart. As these drops fall, does the distance between them increase, decrease, or remain the same? Prove your answer.

Q2.18 If you know the initial position and initial velocity of a vehicle and have a record of the acceleration at each instant, can you compute the vehicle's position after a certain time? If so, explain how this might be done.

Q2.19 From the top of a tall building, you throw one ball straight up with speed v_0 and one ball straight down with speed v_0. (a) Which ball has the greater speed when it reaches the ground? (b) Which ball gets to the ground first? (c) Which ball has a greater displacement when it reaches the ground? (d) Which ball has traveled the greater distance when it hits the ground?

Q2.20 You run due east at a constant speed of 3.00 m/s for a distance of 120.0 m and then continue running east at a constant speed of 5.00 m/s for another 120.0 m. For the total 240.0-m run, is your average velocity 4.00 m/s, greater than 4.00 m/s, or less than 4.00 m/s? Explain.

Q2.21 An object is thrown straight up into the air and feels no air resistance. How can the object have an acceleration when it has stopped moving at its highest point?

Q2.22 When you drop an object from a certain height, it takes time T to reach the ground with no air resistance. If you dropped it from three times that height, how long (in terms of T) would it take to reach the ground?

EXERCISES

Section 2.1 Displacement, Time, and Average Velocity

2.1 • A car travels in the $+x$-direction on a straight and level road. For the first 4.00 s of its motion, the average velocity of the car is $v_{\text{av-}x} = 6.25$ m/s. How far does the car travel in 4.00 s?

2.2 •• In an experiment, a shearwater (a seabird) was taken from its nest, flown 5150 km away, and released. The bird found its way back to its nest 13.5 days after release. If we place the origin at the nest and extend the $+x$-axis to the release point, what was the bird's average velocity in m/s (a) for the return flight and (b) for the whole episode, from leaving the nest to returning?

2.3 •• **Trip Home.** You normally drive on the freeway between San Diego and Los Angeles at an average speed of 105 km/h (65 mi/h), and the trip takes 1 h and 50 min. On a Friday afternoon, however, heavy traffic slows you down and you drive the same distance at an average speed of only 70 km/h (43 mi/h). How much longer does the trip take?

2.4 •• **From Pillar to Post.** Starting from a pillar, you run 200 m east (the $+x$-direction) at an average speed of 5.0 m/s and then run 280 m west at an average speed of 4.0 m/s to a post. Calculate (a) your average speed from pillar to post and (b) your average velocity from pillar to post.

2.5 • Starting from the front door of a ranch house, you walk 60.0 m due east to a windmill, turn around, and then slowly walk 40.0 m west to a bench, where you sit and watch the sunrise. It takes you 28.0 s to walk from the house to the windmill and then 36.0 s to walk from the windmill to the bench. For the entire trip from the front door to the bench, what are your (a) average velocity and (b) average speed?

2.6 •• A Honda Civic travels in a straight line along a road. The car's distance x from a stop sign is given as a function of time t by the equation $x(t) = \alpha t^2 - \beta t^3$, where $\alpha = 1.50$ m/s^2 and $\beta = 0.0500$ m/s^3. Calculate the average velocity of the car for each time interval: (a) $t = 0$ to $t = 2.00$ s; (b) $t = 0$ to $t = 4.00$ s; (c) $t = 2.00$ s to $t = 4.00$ s.

Section 2.2 Instantaneous Velocity

2.7 • CALC A car is stopped at a traffic light. It then travels along a straight road such that its distance from the light is given by $x(t) = bt^2 - ct^3$, where $b = 2.40$ m/s^2 and $c = 0.120$ m/s^3. (a) Calculate the average velocity of the car for the time interval $t = 0$ to $t = 10.0$ s. (b) Calculate the instantaneous velocity of the car at $t = 0$, $t = 5.0$ s, and $t = 10.0$ s. (c) How long after starting from rest is the car again at rest?

2.8 • CALC A bird is flying due east. Its distance from a tall building is given by $x(t) = 28.0$ m $+ (12.4$ m/s$)t - (0.0450$ m/s$^3)t^3$. What is the instantaneous velocity of the bird when $t = 8.00$ s?

2.9 •• A ball moves in a straight line (the x-axis). The graph in **Fig. E2.9** shows this ball's velocity as a function of time. (a) What are the ball's average speed and average velocity during the first 3.0 s? (b) Suppose that the ball moved in such a way that the graph segment after 2.0 s was -3.0 m/s instead of $+3.0$ m/s. Find the ball's average speed and average velocity in this case.

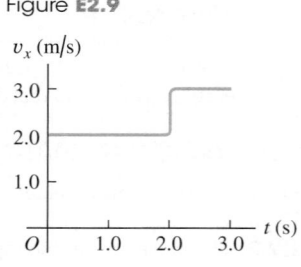

Figure **E2.9**

2.10 •• A physics professor leaves her house and walks along the sidewalk toward campus. After 5 min it starts to rain, and she returns home. Her distance from her house as a function of time is shown in **Fig. E2.10.** At which of the labeled points is her velocity (a) zero? (b) constant and positive? (c) constant and negative? (d) increasing in magnitude? (e) decreasing in magnitude?

Figure **E2.10**

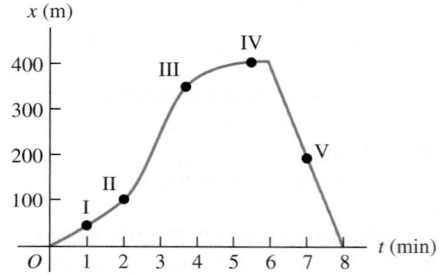

2.11 •• A test car travels in a straight line along the x-axis. The graph in **Fig. E2.11** shows the car's position x as a function of time. Find its instantaneous velocity at points A through G.

Figure **E2.11**

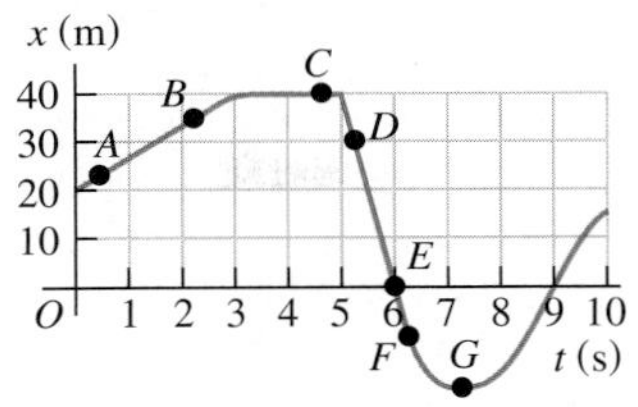

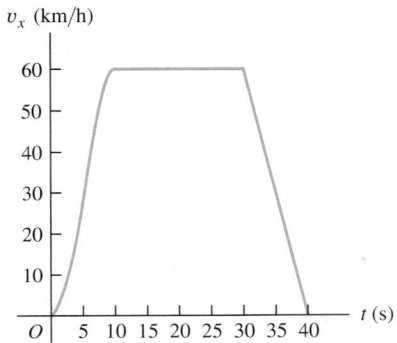

Section 2.3 Average and Instantaneous Acceleration

2.12 • **Figure E2.12** shows the velocity of a solar-powered car as a function of time. The driver accelerates from a stop sign, cruises for 20 s at a constant speed of 60 km/h, and then brakes to come to a stop 40 s after leaving the stop sign. (a) Compute the average acceleration during these time intervals: (i) $t = 0$ to $t = 10$ s; (ii) $t = 30$ s to $t = 40$ s; (iii) $t = 10$ s to $t = 30$ s; (iv) $t = 0$ to $t = 40$ s. (b) What is the instantaneous acceleration at $t = 20$ s and at $t = 35$ s?

Figure **E2.12**

v_x (km/h)

[graph with values 10, 20, 30, 40, 50, 60 on vertical axis and 5, 10, 15, 20, 25, 30, 35, 40 on horizontal axis labeled t (s)]

2.13 • **The Fastest (and Most Expensive) Car!** The table shows test data for the Bugatti Veyron Super Sport, the fastest street car made. The car is moving in a straight line (the x-axis).

Time (s)	0	2.1	20.0	53
Speed (mi/h)	0	60	200	253

(a) Sketch a v_x-t graph of this car's velocity (in mi/h) as a function of time. Is its acceleration constant? (b) Calculate the car's average acceleration (in m/s²) between (i) 0 and 2.1 s; (ii) 2.1 s and 20.0 s; (iii) 20.0 s and 53 s. Are these results consistent with your graph in part (a)? (Before you decide to buy this car, it might be helpful to know that only 300 will be built, it runs out of gas in 12 minutes at top speed, and it costs more than $1.5 million!)

2.14 •• CALC A race car starts from rest and travels east along a straight and level track. For the first 5.0 s of the car's motion, the eastward component of the car's velocity is given by $v_x(t) = (0.860 \text{ m/s}^3)t^2$. What is the acceleration of the car when $v_x = 12.0$ m/s?

2.15 • CALC A turtle crawls along a straight line, which we will call the x-axis with the positive direction to the right. The equation for the turtle's position as a function of time is $x(t) = 50.0 \text{ cm} + (2.00 \text{ cm/s})t - (0.0625 \text{ cm/s}^2)t^2$. (a) Find the turtle's initial velocity, initial position, and initial acceleration. (b) At what time t is the velocity of the turtle zero? (c) How long after

starting does it take the turtle to return to its starting point? (d) At what times t is the turtle a distance of 10.0 cm from its starting point? What is the velocity (magnitude and direction) of the turtle at each of those times? (e) Sketch graphs of x versus t, v_x versus t, and a_x versus t, for the time interval $t = 0$ to $t = 40$ s.

2.16 • An astronaut has left the International Space Station to test a new space scooter. Her partner measures the following velocity changes, each taking place in a 10-s interval. What are the magnitude, the algebraic sign, and the direction of the average acceleration in each interval? Assume that the positive direction is to the right. (a) At the beginning of the interval, the astronaut is moving toward the right along the x-axis at 15.0 m/s, and at the end of the interval she is moving toward the right at 5.0 m/s. (b) At the beginning she is moving toward the left at 5.0 m/s, and at the end she is moving toward the left at 15.0 m/s. (c) At the beginning she is moving toward the right at 15.0 m/s, and at the end she is moving toward the left at 15.0 m/s.

2.17 • CALC A car's velocity as a function of time is given by $v_x(t) = \alpha + \beta t^2$, where $\alpha = 3.00$ m/s and $\beta = 0.100$ m/s³. (a) Calculate the average acceleration for the time interval $t = 0$ to $t = 5.00$ s. (b) Calculate the instantaneous acceleration for $t = 0$ and $t = 5.00$ s. (c) Draw v_x-t and a_x-t graphs for the car's motion between $t = 0$ and $t = 5.00$ s.

2.18 •• CALC The position of the front bumper of a test car under microprocessor control is given by $x(t) = 2.17 \text{ m} + (4.80 \text{ m/s}^2)t^2 - (0.100 \text{ m/s}^6)t^6$. (a) Find its position and acceleration at the instants when the car has zero velocity. (b) Draw x-t, v_x-t, and a_x-t graphs for the motion of the bumper between $t = 0$ and $t = 2.00$ s.

Section 2.4 Motion with Constant Acceleration

2.19 •• An antelope moving with constant acceleration covers the distance between two points 70.0 m apart in 6.00 s. Its speed as it passes the second point is 15.0 m/s. What are (a) its speed at the first point and (b) its acceleration?

2.20 •• BIO **Blackout?** A jet fighter pilot wishes to accelerate from rest at a constant acceleration of 5g to reach Mach 3 (three times the speed of sound) as quickly as possible. Experimental tests reveal that he will black out if this acceleration lasts for more than 5.0 s. Use 331 m/s for the speed of sound. (a) Will the period of acceleration last long enough to cause him to black out? (b) What is the greatest speed he can reach with an acceleration of 5g before he blacks out?

2.21 • **A Fast Pitch.** The fastest measured pitched baseball left the pitcher's hand at a speed of 45.0 m/s. If the pitcher was in contact with the ball over a distance of 1.50 m and produced constant acceleration, (a) what acceleration did he give the ball, and (b) how much time did it take him to pitch it?

2.22 •• **A Tennis Serve.** In the fastest measured tennis serve, the ball left the racquet at 73.14 m/s. A served tennis ball is typically in contact with the racquet for 30.0 ms and starts from rest. Assume constant acceleration. (a) What was the ball's acceleration during this serve? (b) How far did the ball travel during the serve?

2.23 •• BIO **Automobile Air Bags.** The human body can survive an acceleration trauma incident (sudden stop) if the magnitude of the acceleration is less than 250 m/s². If you are in an automobile accident with an initial speed of 105 km/h (65 mi/h) and are stopped by an airbag that inflates from the dashboard, over what distance must the airbag stop you for you to survive the crash?

2.24 • BIO A pilot who accelerates at more than $4g$ begins to "gray out" but doesn't completely lose consciousness. (a) Assuming constant acceleration, what is the shortest time that a jet pilot starting from rest can take to reach Mach 4 (four times the speed of sound) without graying out? (b) How far would the plane travel during this period of acceleration? (Use 331 m/s for the speed of sound in cold air.)

2.25 • BIO **Air-Bag Injuries.** During an auto accident, the vehicle's air bags deploy and slow down the passengers more gently than if they had hit the windshield or steering wheel. According to safety standards, air bags produce a maximum acceleration of $60g$ that lasts for only 36 ms (or less). How far (in meters) does a person travel in coming to a complete stop in 36 ms at a constant acceleration of $60g$?

2.26 • BIO **Prevention of Hip Fractures.** Falls resulting in hip fractures are a major cause of injury and even death to the elderly. Typically, the hip's speed at impact is about 2.0 m/s. If this can be reduced to 1.3 m/s or less, the hip will usually not fracture. One way to do this is by wearing elastic hip pads. (a) If a typical pad is 5.0 cm thick and compresses by 2.0 cm during the impact of a fall, what constant acceleration (in m/s² and in g's) does the hip undergo to reduce its speed from 2.0 m/s to 1.3 m/s? (b) The acceleration you found in part (a) may seem rather large, but to assess its effects on the hip, calculate how long it lasts.

2.27 • BIO **Are We Martians?** It has been suggested, and not facetiously, that life might have originated on Mars and been carried to the earth when a meteor hit Mars and blasted pieces of rock (perhaps containing primitive life) free of the Martian surface. Astronomers know that many Martian rocks have come to the earth this way. (For instance, search the Internet for "ALH 84001.") One objection to this idea is that microbes would have had to undergo an enormous lethal acceleration during the impact. Let us investigate how large such an acceleration might be. To escape Mars, rock fragments would have to reach its escape velocity of 5.0 km/s, and that would most likely happen over a distance of about 4.0 m during the meteor impact. (a) What would be the acceleration (in m/s² and g's) of such a rock fragment, if the acceleration is constant? (b) How long would this acceleration last? (c) In tests, scientists have found that over 40% of *Bacillus subtilis* bacteria survived after an acceleration of 450,000g. In light of your answer to part (a), can we rule out the hypothesis that life might have been blasted from Mars to the earth?

2.28 • **Entering the Freeway.** A car sits on an entrance ramp to a freeway, waiting for a break in the traffic. Then the driver accelerates with constant acceleration along the ramp and onto the freeway. The car starts from rest, moves in a straight line, and has a speed of 20 m/s (45 mi/h) when it reaches the end of the 120-m-long ramp. (a) What is the acceleration of the car? (b) How much time does it take the car to travel the length of the ramp? (c) The traffic on the freeway is moving at a constant speed of 20 m/s. What distance does the traffic travel while the car is moving the length of the ramp?

2.29 •• At launch a rocket ship weighs 4.5 million pounds. When it is launched from rest, it takes 8.00 s to reach 161 km/h; at the end of the first 1.00 min, its speed is 1610 km/h. (a) What is the average acceleration (in m/s²) of the rocket (i) during the first 8.00 s and (ii) between 8.00 s and the end of the first 1.00 min? (b) Assuming the acceleration is constant during each time interval (but not necessarily the same in both intervals), what distance does the rocket travel (i) during the first 8.00 s and (ii) during the interval from 8.00 s to 1.00 min?

2.30 •• A cat walks in a straight line, which we shall call the x-axis, with the positive direction to the right. As an observant physicist, you make measurements of this cat's motion and construct a graph of the feline's velocity as a function of time (**Fig. E2.30**). (a) Find the cat's velocity at $t = 4.0$ s and at $t = 7.0$ s. (b) What is the cat's acceleration at $t = 3.0$ s? At $t = 6.0$ s? At $t = 7.0$ s? (c) What distance does the cat move during the first 4.5 s? From $t = 0$ to $t = 7.5$ s? (d) Assuming that the cat started at the origin, sketch clear graphs of the cat's acceleration and position as functions of time.

Figure **E2.30**

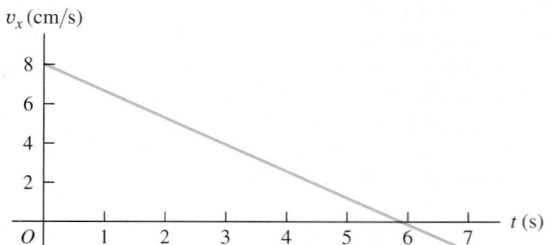

2.31 •• The graph in **Fig. E2.31** shows the velocity of a motorcycle police officer plotted as a function of time. (a) Find the instantaneous acceleration at $t = 3$ s, $t = 7$ s, and $t = 11$ s. (b) How far does the officer go in the first 5 s? The first 9 s? The first 13 s?

Figure **E2.31**

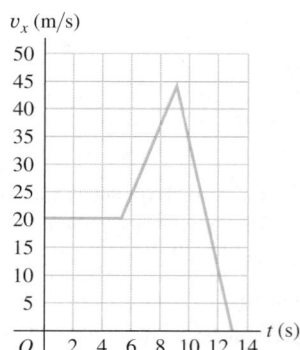

2.32 • Two cars, A and B, move along the x-axis. **Figure E2.32** is a graph of the positions of A and B versus time. (a) In motion diagrams (like Figs. 2.13b and 2.14b), show the position, velocity, and acceleration of each of the two cars at $t = 0$, $t = 1$ s, and $t = 3$ s. (b) At what time(s), if any, do A and B have the same position? (c) Graph velocity versus time for both A and B. (d) At what time(s), if any, do A and B have the same velocity? (e) At what time(s), if any, does car A pass car B? (f) At what time(s), if any, does car B pass car A?

Figure **E2.32**

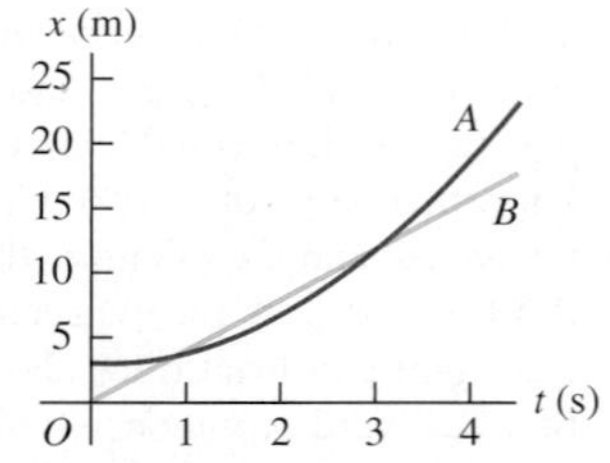

2.33 •• A small block has constant acceleration as it slides down a frictionless incline. The block is released from rest at the top of the incline, and its speed after it has traveled 6.80 m to the bottom of the incline is 3.80 m/s. What is the speed of the block when it is 3.40 m from the top of the incline?

2.34 • At the instant the traffic light turns green, a car that has been waiting at an intersection starts ahead with a constant acceleration of 2.80 m/s². At the same instant a truck, traveling with a constant speed of 20.0 m/s, overtakes and passes the car. (a) How far beyond its starting point does the car overtake the truck? (b) How fast is the car traveling when it overtakes the truck? (c) Sketch an *x-t* graph of the motion of both vehicles. Take *x* = 0 at the intersection. (d) Sketch a v_x-*t* graph of the motion of both vehicles.

Section 2.5 Freely Falling Bodies

2.35 •• (a) If a flea can jump straight up to a height of 0.440 m, what is its initial speed as it leaves the ground? (b) How long is it in the air?

2.36 •• A small rock is thrown vertically upward with a speed of 22.0 m/s from the edge of the roof of a 30.0-m-tall building. The rock doesn't hit the building on its way back down and lands on the street below. Ignore air resistance. (a) What is the speed of the rock just before it hits the street? (b) How much time elapses from when the rock is thrown until it hits the street?

2.37 • A juggler throws a bowling pin straight up with an initial speed of 8.20 m/s. How much time elapses until the bowling pin returns to the juggler's hand?

2.38 •• You throw a glob of putty straight up toward the ceiling, which is 3.60 m above the point where the putty leaves your hand. The initial speed of the putty as it leaves your hand is 9.50 m/s. (a) What is the speed of the putty just before it strikes the ceiling? (b) How much time from when it leaves your hand does it take the putty to reach the ceiling?

2.39 •• A tennis ball on Mars, where the acceleration due to gravity is 0.379*g* and air resistance is negligible, is hit directly upward and returns to the same level 8.5 s later. (a) How high above its original point did the ball go? (b) How fast was it moving just after it was hit? (c) Sketch graphs of the ball's vertical position, vertical velocity, and vertical acceleration as functions of time while it's in the Martian air.

2.40 •• **Touchdown on the Moon.** A lunar lander is making its descent to Moon Base I (**Fig. E2.40**). The lander descends slowly under the retro-thrust of its descent engine. The engine is cut off when the lander is 5.0 m above the surface and has a downward speed of 0.8 m/s. With the engine off, the lander is in free fall. What is the speed of the lander just before it touches the surface? The acceleration due to gravity on the moon is 1.6 m/s².

Figure **E2.40**

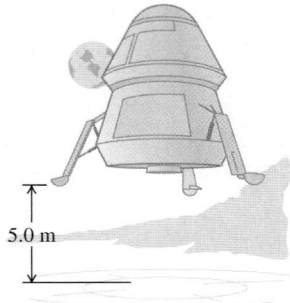

5.0 m

2.41 •• **A Simple Reaction-Time Test.** A meter stick is held vertically above your hand, with the lower end between your thumb and first finger. When you see the meter stick released, you grab it with those two fingers. You can calculate your reaction time from the distance the meter stick falls, read directly from the point where your fingers grabbed it. (a) Derive a relationship for your reaction time in terms of this measured distance, *d*. (b) If the measured distance is 17.6 cm, what is your reaction time?

2.42 •• A brick is dropped (zero initial speed) from the roof of a building. The brick strikes the ground in 1.90 s. You may ignore air resistance, so the brick is in free fall. (a) How tall, in meters, is the building? (b) What is the magnitude of the brick's velocity just before it reaches the ground? (c) Sketch a_y-*t*, v_y-*t*, and *y-t* graphs for the motion of the brick.

2.43 •• **Launch Failure.** A 7500-kg rocket blasts off vertically from the launch pad with a constant upward acceleration of 2.25 m/s² and feels no appreciable air resistance. When it has reached a height of 525 m, its engines suddenly fail; the only force acting on it is now gravity. (a) What is the maximum height this rocket will reach above the launch pad? (b) How much time will elapse after engine failure before the rocket comes crashing down to the launch pad, and how fast will it be moving just before it crashes? (c) Sketch a_y-*t*, v_y-*t*, and *y-t* graphs of the rocket's motion from the instant of blast-off to the instant just before it strikes the launch pad.

2.44 •• A hot-air balloonist, rising vertically with a constant velocity of magnitude 5.00 m/s, releases a sandbag at an instant when the balloon is 40.0 m above the ground (**Fig. E2.44**). After the sandbag is released, it is in free fall. (a) Compute the position and velocity of the sandbag at 0.250 s and 1.00 s after its release. (b) How many seconds after its release does the bag strike the ground? (c) With what magnitude of velocity does it strike the ground? (d) What is the greatest height above the ground that the sandbag reaches? (e) Sketch a_y-*t*, v_y-*t*, and *y-t* graphs for the motion.

Figure **E2.44**

v = 5.00 m/s

40.0 m to ground

2.45 • **BIO** The rocket-driven sled *Sonic Wind No. 2*, used for investigating the physiological effects of large accelerations, runs on a straight, level track 1070 m (3500 ft) long. Starting from rest, it can reach a speed of 224 m/s (500 mi/h) in 0.900 s. (a) Compute the acceleration in m/s², assuming that it is constant. (b) What is the ratio of this acceleration to that of a freely falling body (*g*)? (c) What distance is covered in 0.900 s? (d) A magazine article states that at the end of a certain run, the speed of the sled decreased from 283 m/s (632 mi/h) to zero in 1.40 s and that during this time the magnitude of the acceleration was greater than 40*g*. Are these figures consistent?

2.46 • An egg is thrown nearly vertically upward from a point near the cornice of a tall building. The egg just misses the cornice on the way down and passes a point 30.0 m below its starting point 5.00 s after it leaves the thrower's hand. Ignore air resistance. (a) What is the initial speed of the egg? (b) How high does it rise above its starting point? (c) What is the magnitude of its velocity at the highest point? (d) What are the magnitude and direction of its acceleration at the highest point? (e) Sketch a_y-*t*, v_y-*t*, and *y-t* graphs for the motion of the egg.

2.47 •• A 15-kg rock is dropped from rest on the earth and reaches the ground in 1.75 s. When it is dropped from the same height on Saturn's satellite Enceladus, the rock reaches the ground in 18.6 s. What is the acceleration due to gravity on Enceladus?

2.48 • A large boulder is ejected vertically upward from a volcano with an initial speed of 40.0 m/s. Ignore air resistance. (a) At what time after being ejected is the boulder moving at 20.0 m/s upward? (b) At what time is it moving at 20.0 m/s downward? (c) When is the displacement of the boulder from its initial position zero? (d) When is the velocity of the boulder zero? (e) What are the magnitude and direction of the acceleration while the boulder is (i) moving upward? (ii) Moving downward? (iii) At the highest point? (f) Sketch a_y-t, v_y-t, and y-t graphs for the motion.

2.49 •• You throw a small rock straight up from the edge of a highway bridge that crosses a river. The rock passes you on its way down, 6.00 s after it was thrown. What is the speed of the rock just before it reaches the water 28.0 m below the point where the rock left your hand? Ignore air resistance.

2.50 •• CALC A small object moves along the x-axis with acceleration $a_x(t) = -(0.0320 \text{ m/s}^3)(15.0 \text{ s} - t)$. At $t = 0$ the object is at $x = -14.0$ m and has velocity $v_{0x} = 8.00$ m/s. What is the x-coordinate of the object when $t = 10.0$ s?

Section 2.6 Velocity and Position by Integration

2.51 • CALC A rocket starts from rest and moves upward from the surface of the earth. For the first 10.0 s of its motion, the vertical acceleration of the rocket is given by $a_y = (2.80 \text{ m/s}^3)t$, where the $+y$-direction is upward. (a) What is the height of the rocket above the surface of the earth at $t = 10.0$ s? (b) What is the speed of the rocket when it is 325 m above the surface of the earth?

2.52 •• CALC The acceleration of a bus is given by $a_x(t) = \alpha t$, where $\alpha = 1.2 \text{ m/s}^3$. (a) If the bus's velocity at time $t = 1.0$ s is 5.0 m/s, what is its velocity at time $t = 2.0$ s? (b) If the bus's position at time $t = 1.0$ s is 6.0 m, what is its position at time $t = 2.0$ s? (c) Sketch a_y-t, v_y-t, and x-t graphs for the motion.

2.53 •• CALC The acceleration of a motorcycle is given by $a_x(t) = At - Bt^2$, where $A = 1.50 \text{ m/s}^3$ and $B = 0.120 \text{ m/s}^4$. The motorcycle is at rest at the origin at time $t = 0$. (a) Find its position and velocity as functions of time. (b) Calculate the maximum velocity it attains.

2.54 •• BIO **Flying Leap of the Flea.** High-speed motion pictures (3500 frames/second) of a jumping, 210-μg flea yielded the data used to plot the graph in **Fig. E2.54**. (See "The Flying Leap of the Flea" by M. Rothschild, Y. Schlein, K. Parker, C. Neville, and S. Sternberg in the November 1973 *Scientific American*.) This flea was about 2 mm long and jumped at a nearly vertical takeoff angle. Use the graph to answer these questions: (a) Is the acceleration of the flea ever zero? If so, when? Justify your answer. (b) Find the maximum height the flea reached in the first 2.5 ms. (c) Find the flea's acceleration at 0.5 ms, 1.0 ms, and 1.5 ms. (d) Find the flea's height at 0.5 ms, 1.0 ms, and 1.5 ms.

Figure **E2.54**

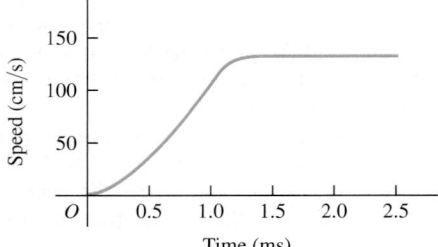

PROBLEMS

2.55 • BIO A typical male sprinter can maintain his maximum acceleration for 2.0 s, and his maximum speed is 10 m/s. After he reaches this maximum speed, his acceleration becomes zero, and then he runs at constant speed. Assume that his acceleration is constant during the first 2.0 s of the race, that he starts from rest, and that he runs in a straight line. (a) How far has the sprinter run when he reaches his maximum speed? (b) What is the magnitude of his average velocity for a race of these lengths: (i) 50.0 m; (ii) 100.0 m; (iii) 200.0 m?

2.56 • CALC A lunar lander is descending toward the moon's surface. Until the lander reaches the surface, its height above the surface of the moon is given by $y(t) = b - ct + dt^2$, where $b = 800$ m is the initial height of the lander above the surface, $c = 60.0$ m/s, and $d = 1.05 \text{ m/s}^2$. (a) What is the initial velocity of the lander, at $t = 0$? (b) What is the velocity of the lander just before it reaches the lunar surface?

2.57 ••• **Earthquake Analysis.** Earthquakes produce several types of shock waves. The most well known are the P-waves (P for *primary* or *pressure*) and the S-waves (S for *secondary* or *shear*). In the earth's crust, P-waves travel at about 6.5 km/s and S-waves move at about 3.5 km/s. The time delay between the arrival of these two waves at a seismic recording station tells geologists how far away an earthquake occurred. If the time delay is 33 s, how far from the seismic station did the earthquake occur?

2.58 •• A brick is dropped from the roof of a tall building. After it has been falling for a few seconds, it falls 40.0 m in a 1.00-s time interval. What distance will it fall during the next 1.00 s? Ignore air resistance.

2.59 ••• A rocket carrying a satellite is accelerating straight up from the earth's surface. At 1.15 s after liftoff, the rocket clears the top of its launch platform, 63 m above the ground. After an additional 4.75 s, it is 1.00 km above the ground. Calculate the magnitude of the average velocity of the rocket for (a) the 4.75-s part of its flight and (b) the first 5.90 s of its flight.

2.60 ••• A subway train starts from rest at a station and accelerates at a rate of 1.60 m/s^2 for 14.0 s. It runs at constant speed for 70.0 s and slows down at a rate of 3.50 m/s^2 until it stops at the next station. Find the *total* distance covered.

2.61 • A gazelle is running in a straight line (the x-axis). The graph in **Fig. P2.61** shows this animal's velocity as a function of time. During the first 12.0 s, find (a) the total distance moved and (b) the displacement of the gazelle. (c) Sketch an a_x-t graph showing this gazelle's acceleration as a function of time for the first 12.0 s.

Figure **P2.61**

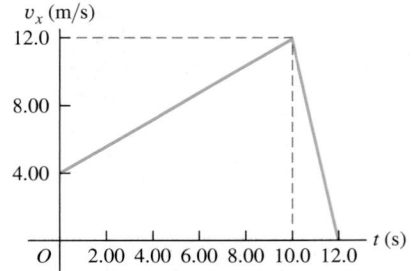

2.62 •• **Collision.** The engineer of a passenger train traveling at 25.0 m/s sights a freight train whose caboose is 200 m ahead on the same track (**Fig. P2.62**). The freight train is traveling at 15.0 m/s in the same direction as the passenger train. The engineer of the passenger train immediately applies the brakes,

Figure **P2.62**

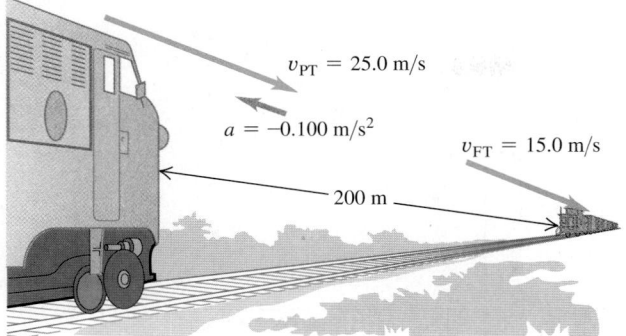

$v_{PT} = 25.0$ m/s

$a = -0.100$ m/s^2

$v_{FT} = 15.0$ m/s

200 m

causing a constant acceleration of 0.100 m/s^2 in a direction opposite to the train's velocity, while the freight train continues with constant speed. Take $x = 0$ at the location of the front of the passenger train when the engineer applies the brakes. (a) Will the cows nearby witness a collision? (b) If so, where will it take place? (c) On a single graph, sketch the positions of the front of the passenger train and the back of the freight train.

2.63 ••• A ball starts from rest and rolls down a hill with uniform acceleration, traveling 200 m during the second 5.0 s of its motion. How far did it roll during the first 5.0 s of motion?

2.64 •• Two cars start 200 m apart and drive toward each other at a steady 10 m/s. On the front of one of them, an energetic grasshopper jumps back and forth between the cars (he has strong legs!) with a constant horizontal velocity of 15 m/s relative to the ground. The insect jumps the instant he lands, so he spends no time resting on either car. What total distance does the grasshopper travel before the cars hit?

2.65 • A car and a truck start from rest at the same instant, with the car initially at some distance behind the truck. The truck has a constant acceleration of 2.10 m/s^2, and the car has an acceleration of 3.40 m/s^2. The car overtakes the truck after the truck has moved 60.0 m. (a) How much time does it take the car to overtake the truck? (b) How far was the car behind the truck initially? (c) What is the speed of each when they are abreast? (d) On a single graph, sketch the position of each vehicle as a function of time. Take $x = 0$ at the initial location of the truck.

2.66 •• You are standing at rest at a bus stop. A bus moving at a constant speed of 5.00 m/s passes you. When the rear of the bus is 12.0 m past you, you realize that it is your bus, so you start to run toward it with a constant acceleration of 0.960 m/s^2. How far would you have to run before you catch up with the rear of the bus, and how fast must you be running then? Would an average college student be physically able to accomplish this?

2.67 •• **Passing.** The driver of a car wishes to pass a truck that is traveling at a constant speed of 20.0 m/s (about 45 mi/h). Initially, the car is also traveling at 20.0 m/s, and its front bumper is 24.0 m behind the truck's rear bumper. The car accelerates at a constant 0.600 m/s^2, then pulls back into the truck's lane when the rear of the car is 26.0 m ahead of the front of the truck. The car is 4.5 m long, and the truck is 21.0 m long. (a) How much time is required for the car to pass the truck? (b) What distance does the car travel during this time? (c) What is the final speed of the car?

2.68 •• CALC An object's velocity is measured to be $v_x(t) = \alpha - \beta t^2$, where $\alpha = 4.00$ m/s and $\beta = 2.00$ m/s^3. At $t = 0$ the object is at $x = 0$. (a) Calculate the object's position and acceleration as functions of time. (b) What is the object's maximum *positive* displacement from the origin?

2.69 ••• CALC The acceleration of a particle is given by $a_x(t) = -2.00$ m/s$^2 + (3.00$ m/s$^3)t$. (a) Find the initial velocity v_{0x} such that the particle will have the same x-coordinate at $t = 4.00$ s as it had at $t = 0$. (b) What will be the velocity at $t = 4.00$ s?

2.70 • **Egg Drop.** You are on the roof of the physics building, 46.0 m above the ground (**Fig. P2.70**). Your physics professor, who is 1.80 m tall, is walking alongside the building at a constant speed of 1.20 m/s. If you wish to drop an egg on your professor's head, where should the professor be when you release the egg? Assume that the egg is in free fall.

Figure **P2.70**

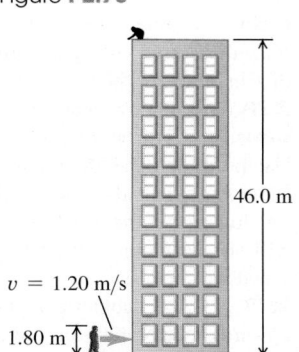

46.0 m

$v = 1.20$ m/s

1.80 m

2.71 • A certain volcano on earth can eject rocks vertically to a maximum height H. (a) How high (in terms of H) would these rocks go if a volcano on Mars ejected them with the same initial velocity? The acceleration due to gravity on Mars is 3.71 m/s^2; ignore air resistance on both planets. (b) If the rocks are in the air for a time T on earth, for how long (in terms of T) would they be in the air on Mars?

2.72 •• An entertainer juggles balls while doing other activities. In one act, she throws a ball vertically upward, and while it is in the air, she runs to and from a table 5.50 m away at an average speed of 3.00 m/s, returning just in time to catch the falling ball. (a) With what minimum initial speed must she throw the ball upward to accomplish this feat? (b) How high above its initial position is the ball just as she reaches the table?

2.73 ••• **Look Out Below.** Sam heaves a 16-lb shot straight up, giving it a constant upward acceleration from rest of 35.0 m/s^2 for 64.0 cm. He releases it 2.20 m above the ground. Ignore air resistance. (a) What is the speed of the shot when Sam releases it? (b) How high above the ground does it go? (c) How much time does he have to get out of its way before it returns to the height of the top of his head, 1.83 m above the ground?

2.74 ••• A flowerpot falls off a windowsill and passes the window of the story below. Ignore air resistance. It takes the pot 0.380 s to pass from the top to the bottom of this window, which is 1.90 m high. How far is the top of the window below the windowsill from which the flowerpot fell?

2.75 •• Two stones are thrown vertically upward from the ground, one with three times the initial speed of the other. (a) If the faster stone takes 10 s to return to the ground, how long will it take the slower stone to return? (b) If the slower stone reaches a maximum height of H, how high (in terms of H) will the faster stone go? Assume free fall.

2.76 ••• **A Multistage Rocket.** In the first stage of a two-stage rocket, the rocket is fired from the launch pad starting from rest but with a constant acceleration of 3.50 m/s^2 upward. At 25.0 s after launch, the second stage fires for 10.0 s, which boosts the rocket's velocity to 132.5 m/s upward at 35.0 s after launch. This firing uses up all of the fuel, however, so after the second stage has finished firing, the only force acting on the rocket is gravity. Ignore air resistance. (a) Find the maximum height that the stage-two rocket reaches above the launch pad. (b) How much time after the end of the stage-two firing will it take for the rocket to fall back to the launch pad? (c) How fast will the stage-two rocket be moving just as it reaches the launch pad?

2.77 ••• During your summer internship for an aerospace company, you are asked to design a small research rocket. The rocket is to be launched from rest from the earth's surface and is to reach a maximum height of 960 m above the earth's surface. The rocket's engines give the rocket an upward acceleration of 16.0 m/s² during the time T that they fire. After the engines shut off, the rocket is in free fall. Ignore air resistance. What must be the value of T in order for the rocket to reach the required altitude?

2.78 •• A physics teacher performing an outdoor demonstration suddenly falls from rest off a high cliff and simultaneously shouts "Help." When she has fallen for 3.0 s, she hears the echo of her shout from the valley floor below. The speed of sound is 340 m/s. (a) How tall is the cliff? (b) If we ignore air resistance, how fast will she be moving just before she hits the ground? (Her actual speed will be less than this, due to air resistance.)

2.79 ••• A helicopter carrying Dr. Evil takes off with a constant upward acceleration of 5.0 m/s². Secret agent Austin Powers jumps on just as the helicopter lifts off the ground. After the two men struggle for 10.0 s, Powers shuts off the engine and steps out of the helicopter. Assume that the helicopter is in free fall after its engine is shut off, and ignore the effects of air resistance. (a) What is the maximum height above ground reached by the helicopter? (b) Powers deploys a jet pack strapped on his back 7.0 s after leaving the helicopter, and then he has a constant downward acceleration with magnitude 2.0 m/s². How far is Powers above the ground when the helicopter crashes into the ground?

2.80 •• **Cliff Height.** You are climbing in the High Sierra when you suddenly find yourself at the edge of a fog-shrouded cliff. To find the height of this cliff, you drop a rock from the top; 8.00 s later you hear the sound of the rock hitting the ground at the foot of the cliff. (a) If you ignore air resistance, how high is the cliff if the speed of sound is 330 m/s? (b) Suppose you had ignored the time it takes the sound to reach you. In that case, would you have overestimated or underestimated the height of the cliff? Explain.

2.81 •• CALC An object is moving along the x-axis. At $t = 0$ it has velocity $v_{0x} = 20.0$ m/s. Starting at time $t = 0$ it has acceleration $a_x = -Ct$, where C has units of m/s³. (a) What is the value of C if the object stops in 8.00 s after $t = 0$? (b) For the value of C calculated in part (a), how far does the object travel during the 8.00 s?

2.82 •• A ball is thrown straight up from the ground with speed v_0. At the same instant, a second ball is dropped from rest from a height H, directly above the point where the first ball was thrown upward. There is no air resistance. (a) Find the time at which the two balls collide. (b) Find the value of H in terms of v_0 and g such that at the instant when the balls collide, the first ball is at the highest point of its motion.

2.83 • CALC Cars A and B travel in a straight line. The distance of A from the starting point is given as a function of time by $x_A(t) = \alpha t + \beta t^2$, with $\alpha = 2.60$ m/s and $\beta = 1.20$ m/s². The distance of B from the starting point is $x_B(t) = \gamma t^2 - \delta t^3$, with $\gamma = 2.80$ m/s² and $\delta = 0.20$ m/s³. (a) Which car is ahead just after the two cars leave the starting point? (b) At what time(s) are the cars at the same point? (c) At what time(s) is the distance from A to B neither increasing nor decreasing? (d) At what time(s) do A and B have the same acceleration?

2.84 •• DATA In your physics lab you release a small glider from rest at various points on a long, frictionless air track that is inclined at an angle θ above the horizontal. With an electronic photocell, you measure the time t it takes the glider to slide a distance x from the release point to the bottom of the track. Your measurements are given in **Fig. P2.84,** which shows a

Figure **P2.84**

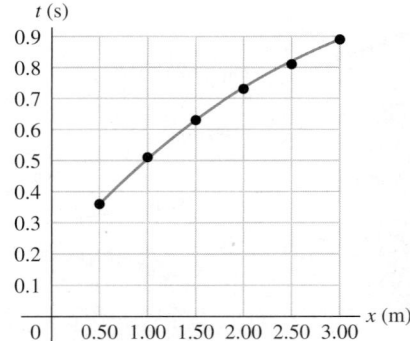

second-order polynomial (quadratic) fit to the plotted data. You are asked to find the glider's acceleration, which is assumed to be constant. There is some error in each measurement, so instead of using a single set of x and t values, you can be more accurate if you use graphical methods and obtain your measured value of the acceleration from the graph. (a) How can you re-graph the data so that the data points fall close to a straight line? (*Hint:* You might want to plot x or t, or both, raised to some power.) (b) Construct the graph you described in part (a) and find the equation for the straight line that is the best fit to the data points. (c) Use the straight-line fit from part (b) to calculate the acceleration of the glider. (d) The glider is released at a distance $x = 1.35$ m from the bottom of the track. Use the acceleration value you obtained in part (c) to calculate the speed of the glider when it reaches the bottom of the track.

2.85 •• DATA In a physics lab experiment, you release a small steel ball at various heights above the ground and measure the ball's speed just before it strikes the ground. You plot your data on a graph that has the release height (in meters) on the vertical axis and the square of the final speed (in m²/s²) on the horizontal axis. In this graph your data points lie close to a straight line. (a) Using $g = 9.80$ m/s² and ignoring the effect of air resistance, what is the numerical value of the slope of this straight line? (Include the correct units.) The presence of air resistance reduces the magnitude of the downward acceleration, and the effect of air resistance increases as the speed of the object increases. You repeat the experiment, but this time with a tennis ball as the object being dropped. Air resistance now has a noticeable effect on the data. (b) Is the final speed for a given release height higher than, lower than, or the same as when you ignored air resistance? (c) Is the graph of the release height versus the square of the final speed still a straight line? Sketch the qualitative shape of the graph when air resistance is present.

2.86 ••• DATA A model car starts from rest and travels in a straight line. A smartphone mounted on the car has an app that transmits the magnitude of the car's acceleration (measured by an accelerometer) every second. The results are given in the table:

Time (s)	Acceleration (m/s²)
0	5.95
1.00	5.52
2.00	5.08
3.00	4.55
4.00	3.96
5.00	3.40

Each measured value has some experimental error. (a) Plot acceleration versus time and find the equation for the straight line that gives the best fit to the data. (b) Use the equation for $a(t)$ that you found in part (a) to calculate $v(t)$, the speed of the car as a function of time. Sketch the graph of v versus t. Is this graph a straight line? (c) Use your result from part (b) to calculate the speed of the car at $t = 5.00$ s. (d) Calculate the distance the car travels between $t = 0$ and $t = 5.00$ s.

CHALLENGE PROBLEMS

2.87 ••• In the vertical jump, an athlete starts from a crouch and jumps upward as high as possible. Even the best athletes spend little more than 1.00 s in the air (their "hang time"). Treat the athlete as a particle and let y_{max} be his maximum height above the floor. To explain why he seems to hang in the air, calculate the ratio of the time he is above $y_{max}/2$ to the time it takes him to go from the floor to that height. Ignore air resistance.

2.88 ••• **Catching the Bus.** A student is running at her top speed of 5.0 m/s to catch a bus, which is stopped at the bus stop. When the student is still 40.0 m from the bus, it starts to pull away, moving with a constant acceleration of 0.170 m/s^2. (a) For how much time and what distance does the student have to run at 5.0 m/s before she overtakes the bus? (b) When she reaches the bus, how fast is the bus traveling? (c) Sketch an x-t graph for both the student and the bus. Take $x = 0$ at the initial position of the student. (d) The equations you used in part (a) to find the time have a second solution, corresponding to a later time for which the student and bus are again at the same place if they continue their specified motions. Explain the significance of this second solution. How fast is the bus traveling at this point? (e) If the student's top speed is 3.5 m/s, will she catch the bus? (f) What is the *minimum* speed the student must have to just catch up with the bus? For what time and what distance does she have to run in that case?

2.89 ••• A ball is thrown straight up from the edge of the roof of a building. A second ball is dropped from the roof 1.00 s later. Ignore air resistance. (a) If the height of the building is 20.0 m, what must the initial speed of the first ball be if both are to hit the ground at the same time? On the same graph, sketch the positions of both balls as a function of time, measured from when the first ball is thrown. Consider the same situation, but now let the initial speed v_0 of the first ball be given and treat the height h of the building as an unknown. (b) What must the height of the building be for both balls to reach the ground at the same time if (i) v_0 is 6.0 m/s and (ii) v_0 is 9.5 m/s? (c) If v_0 is greater than some value v_{max}, no value of h exists that allows both balls to hit the ground at the same time. Solve for v_{max}. The value v_{max} has a simple physical interpretation. What is it? (d) If v_0 is less than some value v_{min}, no value of h exists that allows both balls to hit the ground at the same time. Solve for v_{min}. The value v_{min} also has a simple physical interpretation. What is it?

BIO **BLOOD FLOW IN THE HEART.** The human circulatory system is closed—that is, the blood pumped out of the left ventricle of the heart into the arteries is constrained to a series of continuous, branching vessels as it passes through the capillaries and then into the veins as it returns to the heart. The blood in each of the heart's four chambers comes briefly to rest before it is ejected by contraction of the heart muscle.

2.90 If the contraction of the left ventricle lasts 250 ms and the speed of blood flow in the aorta (the large artery leaving the heart) is 0.80 m/s at the end of the contraction, what is the average acceleration of a red blood cell as it leaves the heart? (a) 310 m/s^2; (b) 31 m/s^2; (c) 3.2 m/s^2; (d) 0.32 m/s^2.

2.91 If the aorta (diameter d_a) branches into two equal-sized arteries with a combined area equal to that of the aorta, what is the diameter of one of the branches? (a) $\sqrt{d_a}$; (b) $d_a/\sqrt{2}$; (c) $2d_a$; (d) $d_a/2$.

2.92 The velocity of blood in the aorta can be measured directly with ultrasound techniques. A typical graph of blood velocity versus time during a single heartbeat is shown in **Fig. P2.92**. Which statement is the best interpretation of this graph? (a) The blood flow changes direction at about 0.25 s; (b) the speed of the blood flow begins to decrease at about 0.10 s; (c) the acceleration of the blood is greatest in magnitude at about 0.25 s; (d) the acceleration of the blood is greatest in magnitude at about 0.10 s.

Figure **P2.92**

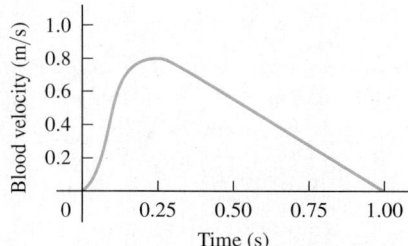

Answers

Chapter Opening Question ?

(iii) Acceleration refers to *any* change in velocity, including both speeding up and slowing down.

Test Your Understanding Questions

2.1 (a): (iv), (i) and (iii) (tie), (v), (ii); (b): (i) and (iii); (c): (v) In (a) the average x-velocity is $v_{\text{av-}x} = \Delta x/\Delta t$. For all five trips, $\Delta t = 1$ h. For the individual trips, (i) $\Delta x = +50$ km, $v_{\text{av-}x} = +50$ km/h; (ii) $\Delta x = -50$ km, $v_{\text{av-}x} = -50$ km/h; (iii) $\Delta x = 60$ km $- 10$ km $= +50$ km, $v_{\text{av-}x} = +50$ km/h; (iv) $\Delta x = +70$ km, $v_{\text{av-}x} = +70$ km/h; (v) $\Delta x = -20$ km $+ 20$ km $= 0$, $v_{\text{av-}x} = 0$. In (b) both have $v_{\text{av-}x} = +50$ km/h.

2.2 (a) P, Q and S (tie), R The x-velocity is **(b)** positive when the slope of the x-t graph is positive (**P**), **(c)** negative when the slope is negative (**R**), and **(d)** zero when the slope is zero (**Q and S**). **(e) R, P, Q and S (tie)** The speed is greatest when the slope of the x-t graph is steepest (either positive or negative) and zero when the slope is zero.

2.3 (a) S, where the x-t graph is curved upward (concave up). **(b) Q,** where the x-t graph is curved downward (concave down). **(c) P and R,** where the x-t graph is not curved either up or down. **(d)** At P, $a_x = 0$ (velocity is **not changing**); at Q, $a_x < 0$ (velocity is **decreasing,** i.e., changing from positive to zero to negative); at R, $a_x = 0$ (velocity is **not changing**); and at S, $a_x > 0$ (velocity is **increasing,** i.e., changing from negative to zero to positive).

2.4 (b) The officer's x-acceleration is constant, so her v_x-t graph is a straight line. The motorcycle is moving faster than the car when the two vehicles meet at $t = 10$ s.

2.5 (a) (iii) Use Eq. (2.13) with x replaced by y and $a_y = -g$; $v_y^2 = v_{0y}^2 - 2g(y - y_0)$. The starting height is $y_0 = 0$ and the y-velocity at the maximum height $y = h$ is $v_y = 0$, so $0 = v_{0y}^2 - 2gh$ and $h = v_{0y}^2/2g$. If the initial y-velocity is increased by a factor of 2, the maximum height increases by a factor of $2^2 = 4$ and the ball goes to height $4h$. **(b) (v)** Use Eq. (2.8) with x replaced by y and $a_y = -g$; $v_y = v_{0y} - gt$. The y-velocity at the maximum height is $v_y = 0$, so $0 = v_{0y} - gt$ and $t = v_{0y}/g$. If the initial y-velocity is increased by a factor of 2, the time to reach the maximum height increases by a factor of 2 and becomes $2t$.

2.6 (ii) The acceleration a_x is equal to the slope of the v_x-t graph. If a_x is increasing, the slope of the v_x-t graph is also increasing and the graph is concave up.

Bridging Problem

$h = 57.1$ m

? If a cyclist is going around a curve at constant speed, is he accelerating? If so, what is the direction of his acceleration? (i) No; (ii) yes, in the direction of his motion; (iii) yes, toward the inside of the curve; (iv) yes, toward the outside of the curve; (v) yes, but in some other direction.

3 MOTION IN TWO OR THREE DIMENSIONS

LEARNING GOALS

Looking forward at ...

3.1 How to use vectors to represent the position and velocity of a particle in two or three dimensions.

3.2 How to find the vector acceleration of a particle, why a particle can have an acceleration even if its speed is constant, and how to interpret the components of acceleration parallel and perpendicular to a particle's path.

3.3 How to solve problems that involve the curved path followed by a projectile.

3.4 How to analyze motion in a circular path, with either constant speed or varying speed.

3.5 How to relate the velocities of a moving body as seen from two different frames of reference.

Looking back at ...

2.1 Average x-velocity.

2.2 Instantaneous x-velocity.

2.3 Average and instantaneous x-acceleration.

2.4 Straight-line motion with constant acceleration.

2.5 The motion of freely falling bodies.

What determines where a batted baseball lands? How do you describe the motion of a roller coaster car along a curved track or the flight of a circling hawk? Which hits the ground first: a baseball that you simply drop or one that you throw horizontally?

We can't answer these kinds of questions by using the techniques of Chapter 2, in which particles moved only along a straight line. Instead, we need to extend our descriptions of motion to two- and three-dimensional situations. We'll still use the vector quantities displacement, velocity, and acceleration, but now these quantities will no longer lie along a single line. We'll find that several important kinds of motion take place in two dimensions only—that is, in a *plane*.

We also need to consider how the motion of a particle is described by different observers who are moving relative to each other. The concept of *relative velocity* will play an important role later in the book when we explore electromagnetic phenomena and when we introduce Einstein's special theory of relativity.

This chapter merges the vector mathematics of Chapter 1 with the kinematic language of Chapter 2. As before, we're concerned with describing motion, not with analyzing its causes. But the language you learn here will be an essential tool in later chapters when we study the relationship between force and motion.

3.1 POSITION AND VELOCITY VECTORS

Let's see how to describe a particle's motion in space. If the particle is at a point P at a certain instant, the **position vector** $\vec{r}$ of the particle at this instant is a vector that goes from the origin of the coordinate system to point P (**Fig. 3.1** on next page). The Cartesian coordinates x, y, and z of point P are the x-, y-, and z-components of vector $\vec{r}$. Using the unit vectors we introduced in Section 1.9, we can write

Position vector of a particle at a given instant ⋯⋯➤ $\vec{r} = x\hat{\imath} + y\hat{\jmath} + z\hat{k}$ Coordinates of particle's position

Unit vectors in x-, y-, and z-directions ⋯⋯ (3.1)

3.1 The position vector $\vec{r}$ from origin O to point P has components x, y, and z.

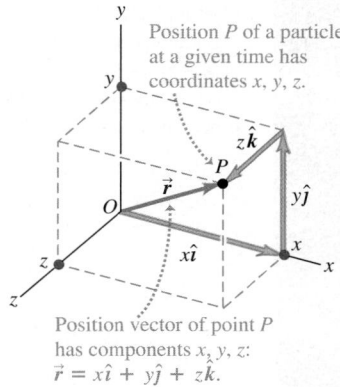

Position P of a particle at a given time has coordinates x, y, z.

Position vector of point P has components x, y, z:
$\vec{r} = x\hat{\imath} + y\hat{\jmath} + z\hat{k}$.

3.2 The average velocity $\vec{v}_{av}$ between points P_1 and P_2 has the same direction as the displacement $\Delta\vec{r}$.

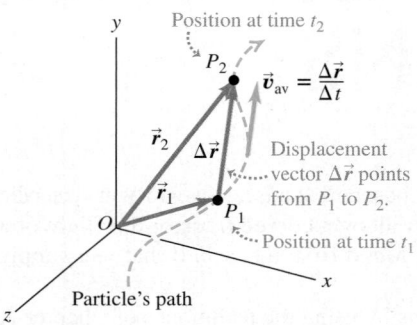

Position at time t_2

$\vec{v}_{av} = \dfrac{\Delta\vec{r}}{\Delta t}$

Displacement vector $\Delta\vec{r}$ points from P_1 to P_2.

Position at time t_1

Particle's path

3.3 The vectors $\vec{v}_1$ and $\vec{v}_2$ are the instantaneous velocities at the points P_1 and P_2 shown in Fig. 3.2.

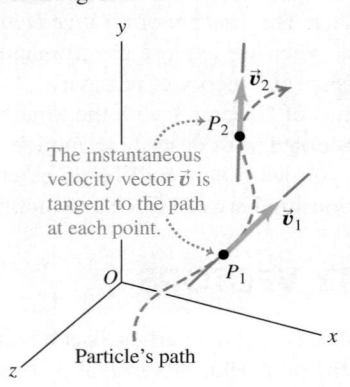

The instantaneous velocity vector $\vec{v}$ is tangent to the path at each point.

Particle's path

During a time interval Δt the particle moves from P_1, where its position vector is $\vec{r}_1$, to P_2, where its position vector is $\vec{r}_2$. The change in position (the displacement) during this interval is $\Delta\vec{r} = \vec{r}_2 - \vec{r}_1 = (x_2 - x_1)\hat{\imath} + (y_2 - y_1)\hat{\jmath} + (z_2 - z_1)\hat{k}$. We define the **average velocity** $\vec{v}_{av}$ during this interval in the same way we did in Chapter 2 for straight-line motion, as the displacement divided by the time interval (**Fig. 3.2**):

Change in the particle's position vector

Average velocity vector of a particle during time interval from t_1 to t_2

$$\vec{v}_{av} = \frac{\Delta\vec{r}}{\Delta t} = \frac{\vec{r}_2 - \vec{r}_1}{t_2 - t_1} \tag{3.2}$$

Final position minus initial position

Time interval   Final time minus initial time

Dividing a vector by a scalar is a special case of *multiplying* a vector by a scalar, described in Section 1.7; the average velocity $\vec{v}_{av}$ is equal to the displacement vector $\Delta\vec{r}$ multiplied by $1/\Delta t$. Note that the x-component of Eq. (3.2) is $v_{av-x} = (x_2 - x_1)/(t_2 - t_1) = \Delta x/\Delta t$. This is just Eq. (2.2), the expression for average x-velocity that we found in Section 2.1 for one-dimensional motion.

We now define **instantaneous velocity** just as we did in Chapter 2: It equals the instantaneous rate of change of position with time. The key difference is that both position $\vec{r}$ and instantaneous velocity $\vec{v}$ are now vectors:

The **instantaneous velocity** vector of a particle ...

$$\vec{v} = \lim_{\Delta t \to 0} \frac{\Delta\vec{r}}{\Delta t} = \frac{d\vec{r}}{dt} \tag{3.3}$$

... equals the limit of its average velocity vector as the time interval approaches zero ...   ... and equals the instantaneous rate of change of its position vector.

At any instant, the *magnitude* of $\vec{v}$ is the *speed* v of the particle at that instant, and the *direction* of $\vec{v}$ is the direction in which the particle is moving at that instant.

As $\Delta t \to 0$, points P_1 and P_2 in Fig. 3.2 move closer and closer together. In this limit, the vector $\Delta\vec{r}$ becomes tangent to the path. The direction of $\Delta\vec{r}$ in this limit is also the direction of $\vec{v}$. So *at every point along the path, the instantaneous velocity vector is tangent to the path at that point* (**Fig. 3.3**).

It's often easiest to calculate the instantaneous velocity vector by using components. During any displacement $\Delta\vec{r}$, the changes Δx, Δy, and Δz in the three coordinates of the particle are the *components* of $\Delta\vec{r}$. It follows that the components v_x, v_y, and v_z of the instantaneous velocity $\vec{v} = v_x\hat{\imath} + v_y\hat{\jmath} + v_z\hat{k}$ are simply the time derivatives of the coordinates x, y, and z:

Each **component of** a particle's **instantaneous velocity vector** ...

$$v_x = \frac{dx}{dt} \qquad v_y = \frac{dy}{dt} \qquad v_z = \frac{dz}{dt} \tag{3.4}$$

... equals the instantaneous rate of change of its corresponding coordinate.

The x-component of $\vec{v}$ is $v_x = dx/dt$, which is the same as Eq. (2.3) for straight-line motion (see Section 2.2). Hence Eq. (3.4) is a direct extension of instantaneous velocity to motion in three dimensions.

We can also get Eq. (3.4) by taking the derivative of Eq. (3.1). The unit vectors $\hat{\imath}$, $\hat{\jmath}$, and $\hat{k}$ don't depend on time, so their derivatives are zero and we find

$$\vec{v} = \frac{d\vec{r}}{dt} = \frac{dx}{dt}\hat{\imath} + \frac{dy}{dt}\hat{\jmath} + \frac{dz}{dt}\hat{k} \tag{3.5}$$

This shows again that the components of $\vec{v}$ are dx/dt, dy/dt, and dz/dt.

The magnitude of the instantaneous velocity vector $\vec{v}$—that is, the speed—is given in terms of the components v_x, v_y, and v_z by the Pythagorean relation:

$$|\vec{v}| = v = \sqrt{v_x{}^2 + v_y{}^2 + v_z{}^2} \tag{3.6}$$

Figure 3.4 shows the situation when the particle moves in the *xy*-plane. In this case, z and v_z are zero. Then the speed (the magnitude of $\vec{v}$) is

$$v = \sqrt{v_x^2 + v_y^2}$$

and the direction of the instantaneous velocity $\vec{v}$ is given by angle α (the Greek letter alpha) in the figure. We see that

$$\tan \alpha = \frac{v_y}{v_x} \qquad (3.7)$$

(We use α for the direction of the instantaneous velocity vector to avoid confusion with the direction θ of the *position* vector of the particle.)

From now on, when we use the word "velocity," we will always mean the *instantaneous* velocity vector $\vec{v}$ (rather than the average velocity vector). Usually, we won't even bother to call $\vec{v}$ a vector; it's up to you to remember that velocity is a vector quantity with both magnitude and direction.

3.4 The two velocity components for motion in the *xy*-plane.

The instantaneous velocity vector $\vec{v}$ is always tangent to the path.

Particle's path in the *xy*-plane

v_x and v_y are the *x*- and *y*- components of $\vec{v}$.

EXAMPLE 3.1 CALCULATING AVERAGE AND INSTANTANEOUS VELOCITY

A robotic vehicle, or rover, is exploring the surface of Mars. The stationary Mars lander is the origin of coordinates, and the surrounding Martian surface lies in the *xy*-plane. The rover, which we represent as a point, has *x*- and *y*-coordinates that vary with time:

$$x = 2.0\ \text{m} - (0.25\ \text{m/s}^2)t^2$$
$$y = (1.0\ \text{m/s})t + (0.025\ \text{m/s}^3)t^3$$

(a) Find the rover's coordinates and distance from the lander at $t = 2.0$ s. (b) Find the rover's displacement and average velocity vectors for the interval $t = 0.0$ s to $t = 2.0$ s. (c) Find a general expression for the rover's instantaneous velocity vector $\vec{v}$. Express $\vec{v}$ at $t = 2.0$ s in component form and in terms of magnitude and direction.

SOLUTION

IDENTIFY and SET UP: This problem involves motion in two dimensions, so we must use the vector equations obtained in this section. **Figure 3.5** shows the rover's path (dashed line). We'll use Eq. (3.1) for position $\vec{r}$, the expression $\Delta\vec{r} = \vec{r}_2 - \vec{r}_1$ for displacement, Eq. (3.2) for average velocity, and Eqs. (3.5), (3.6), and (3.7) for instantaneous velocity and its magnitude and direction.

3.5 At $t = 0.0$ s the rover has position vector $\vec{r}_0$ and instantaneous velocity vector $\vec{v}_0$. Likewise, $\vec{r}_1$ and $\vec{v}_1$ are the vectors at $t = 1.0$ s; $\vec{r}_2$ and $\vec{v}_2$ are the vectors at $t = 2.0$ s.

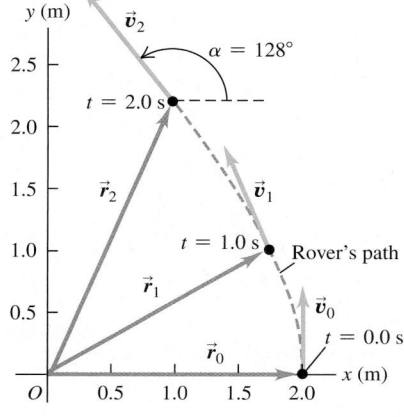

EXECUTE: (a) At $t = 2.0$ s the rover's coordinates are

$$x = 2.0\ \text{m} - (0.25\ \text{m/s}^2)(2.0\ \text{s})^2 = 1.0\ \text{m}$$
$$y = (1.0\ \text{m/s})(2.0\ \text{s}) + (0.025\ \text{m/s}^3)(2.0\ \text{s})^3 = 2.2\ \text{m}$$

The rover's distance from the origin at this time is

$$r = \sqrt{x^2 + y^2} = \sqrt{(1.0\ \text{m})^2 + (2.2\ \text{m})^2} = 2.4\ \text{m}$$

(b) To find the displacement and average velocity over the given time interval, we first express the position vector $\vec{r}$ as a function of time t. From Eq. (3.1) this is

$$\vec{r} = x\hat{\imath} + y\hat{\jmath}$$
$$= [2.0\ \text{m} - (0.25\ \text{m/s}^2)t^2]\hat{\imath}$$
$$\quad + [(1.0\ \text{m/s})t + (0.025\ \text{m/s}^3)t^3]\hat{\jmath}$$

At $t = 0.0$ s the position vector $\vec{r}_0$ is

$$\vec{r}_0 = (2.0\ \text{m})\hat{\imath} + (0.0\ \text{m})\hat{\jmath}$$

From part (a), the position vector $\vec{r}_2$ at $t = 2.0$ s is

$$\vec{r}_2 = (1.0\ \text{m})\hat{\imath} + (2.2\ \text{m})\hat{\jmath}$$

The displacement from $t = 0.0$ s to $t = 2.0$ s is therefore

$$\Delta\vec{r} = \vec{r}_2 - \vec{r}_0 = (1.0\ \text{m})\hat{\imath} + (2.2\ \text{m})\hat{\jmath} - (2.0\ \text{m})\hat{\imath}$$
$$= (-1.0\ \text{m})\hat{\imath} + (2.2\ \text{m})\hat{\jmath}$$

During this interval the rover moves 1.0 m in the negative *x*-direction and 2.2 m in the positive *y*-direction. From Eq. (3.2), the average velocity over this interval is the displacement divided by the elapsed time:

$$\vec{v}_{\text{av}} = \frac{\Delta\vec{r}}{\Delta t} = \frac{(-1.0\ \text{m})\hat{\imath} + (2.2\ \text{m})\hat{\jmath}}{2.0\ \text{s} - 0.0\ \text{s}}$$
$$= (-0.50\ \text{m/s})\hat{\imath} + (1.1\ \text{m/s})\hat{\jmath}$$

The components of this average velocity are $v_{\text{av-}x} = -0.50$ m/s and $v_{\text{av-}y} = 1.1$ m/s.

Continued

(c) From Eq. (3.4) the components of *instantaneous* velocity are the time derivatives of the coordinates:

$$v_x = \frac{dx}{dt} = (-0.25 \text{ m/s}^2)(2t)$$

$$v_y = \frac{dy}{dt} = 1.0 \text{ m/s} + (0.025 \text{ m/s}^3)(3t^2)$$

Hence the instantaneous velocity vector is

$$\vec{v} = v_x \hat{\imath} + v_y \hat{\jmath}$$
$$= (-0.50 \text{ m/s}^2)t\hat{\imath} + [1.0 \text{ m/s} + (0.075 \text{ m/s}^3)t^2]\hat{\jmath}$$

At $t = 2.0$ s the velocity vector $\vec{v}_2$ has components

$$v_{2x} = (-0.50 \text{ m/s}^2)(2.0 \text{ s}) = -1.0 \text{ m/s}$$

$$v_{2y} = 1.0 \text{ m/s} + (0.075 \text{ m/s}^3)(2.0 \text{ s})^2 = 1.3 \text{ m/s}$$

The magnitude of the instantaneous velocity (that is, the speed) at $t = 2.0$ s is

$$v_2 = \sqrt{v_{2x}{}^2 + v_{2y}{}^2} = \sqrt{(-1.0 \text{ m/s})^2 + (1.3 \text{ m/s})^2}$$
$$= 1.6 \text{ m/s}$$

Figure 3.5 shows the direction of velocity vector $\vec{v}_2$, which is at an angle α between 90° and 180° with respect to the positive *x*-axis. From Eq. (3.7) we have

$$\arctan\frac{v_y}{v_x} = \arctan\frac{1.3 \text{ m/s}}{-1.0 \text{ m/s}} = -52°$$

This is off by 180°; the correct value is $\alpha = 180° - 52° = 128°$, or 38° west of north.

EVALUATE: Compare the components of *average* velocity from part (b) for the interval from $t = 0.0$ s to $t = 2.0$ s ($v_{\text{av-}x} = -0.50$ m/s, $v_{\text{av-}y} = 1.1$ m/s) with the components of *instantaneous* velocity at $t = 2.0$ s from part (c) ($v_{2x} = -1.0$ m/s, $v_{2y} = 1.3$ m/s). Just as in one dimension, the average velocity vector $\vec{v}_{\text{av}}$ over an interval is in general *not* equal to the instantaneous velocity $\vec{v}$ at the end of the interval (see Example 2.1).

Figure 3.5 shows the position vectors $\vec{r}$ and instantaneous velocity vectors $\vec{v}$ at $t = 0.0$ s, 1.0 s, and 2.0 s. (Calculate these quantities for $t = 0.0$ s and $t = 1.0$ s.) Notice that $\vec{v}$ is tangent to the path at every point. The magnitude of $\vec{v}$ increases as the rover moves, which means that its speed is increasing.

TEST YOUR UNDERSTANDING OF SECTION 3.1 In which of these situations *would* the average velocity vector $\vec{v}_{\text{av}}$ over an interval be equal to the instantaneous velocity $\vec{v}$ at the end of the interval? (i) A body moving along a curved path at constant speed; (ii) a body moving along a curved path and speeding up; (iii) a body moving along a straight line at constant speed; (iv) a body moving along a straight line and speeding up. ∎

3.2 THE ACCELERATION VECTOR

Now let's consider the *acceleration* of a particle moving in space. Just as for motion in a straight line, acceleration describes how the velocity of the particle changes. But since we now treat velocity as a vector, acceleration will describe changes in the velocity magnitude (that is, the speed) *and* changes in the direction of velocity (that is, the direction in which the particle is moving).

In **Fig. 3.6a**, a car (treated as a particle) is moving along a curved road. Vectors $\vec{v}_1$ and $\vec{v}_2$ represent the car's instantaneous velocities at time t_1, when the car is

3.6 (a) A car moving along a curved road from P_1 to P_2. (b) How to obtain the change in velocity $\Delta\vec{v} = \vec{v}_2 - \vec{v}_1$ by vector subtraction. (c) The vector $\vec{a}_{\text{av}} = \Delta\vec{v}/\Delta t$ represents the average acceleration between P_1 and P_2.

(a)

This car accelerates by slowing while rounding a curve. (Its instantaneous velocity changes in both magnitude and direction.)

(b)

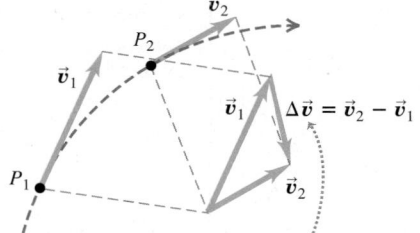

To find the car's average acceleration between P_1 and P_2, we first find the change in velocity $\Delta\vec{v}$ by subtracting $\vec{v}_1$ from $\vec{v}_2$. (Notice that $\vec{v}_1 + \Delta\vec{v} = \vec{v}_2$.)

(c)

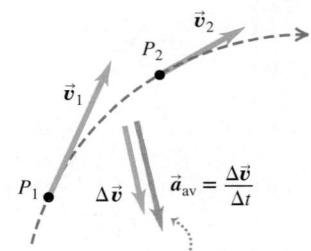

The average acceleration has the same direction as the change in velocity, $\Delta\vec{v}$.

at point P_1, and at time t_2, when the car is at point P_2. During the time interval from t_1 to t_2, the *vector change in velocity* is $\vec{v}_2 - \vec{v}_1 = \Delta\vec{v}$, so $\vec{v}_2 = \vec{v}_1 + \Delta\vec{v}$ (Fig. 3.6b). The **average acceleration** $\vec{a}_{av}$ of the car during this time interval is the velocity change divided by the time interval $t_2 - t_1 = \Delta t$:

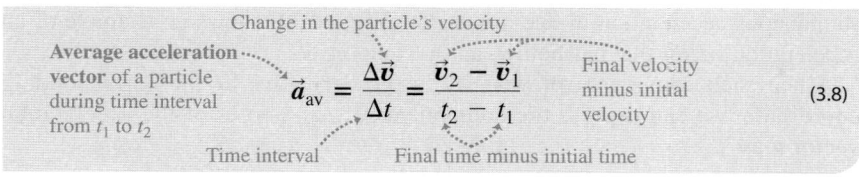

Average acceleration
vector of a particle
during time interval
from t_1 to t_2 $\qquad \vec{a}_{av} = \dfrac{\Delta\vec{v}}{\Delta t} = \dfrac{\vec{v}_2 - \vec{v}_1}{t_2 - t_1}$ $\qquad$ Final velocity
minus initial
velocity $\qquad$ (3.8)

Change in the particle's velocity

Time interval $\qquad$ Final time minus initial time

Average acceleration is a *vector* quantity in the same direction as $\Delta\vec{v}$ (Fig. 3.6c). The x-component of Eq. (3.8) is $a_{av-x} = (v_{2x} - v_{1x})/(t_2 - t_1) = \Delta v_x/\Delta t$, which is just Eq. (2.4) for average acceleration in straight-line motion.

As in Chapter 2, we define the **instantaneous acceleration** $\vec{a}$ (a *vector* quantity) at point P_1 as the limit of the average acceleration vector when point P_2 approaches point P_1, so both $\Delta\vec{v}$ and Δt approach zero (**Fig. 3.7**):

The **instantaneous**
acceleration vector
of a particle ... $\qquad \vec{a} = \lim\limits_{\Delta t \to 0} \dfrac{\Delta\vec{v}}{\Delta t} = \dfrac{d\vec{v}}{dt}$ $\qquad$ (3.9)

... equals the limit of its average acceleration
vector as the time interval approaches zero ... $\qquad$ ... and equals the instantaneous rate
of change of its velocity vector.

The velocity vector $\vec{v}$ is always tangent to the particle's path, but the instantaneous acceleration vector $\vec{a}$ does *not* have to be tangent to the path. If the path is curved, $\vec{a}$ points toward the concave side of the path—that is, toward the inside of any turn that the particle is making (Fig. 3.7a). The acceleration is tangent to the path only if the particle moves in a straight line (Fig. 3.7b).

CAUTION **Any particle following a curved path is accelerating** When a particle is moving in a curved path, it always has nonzero acceleration, even when it moves with constant speed. This conclusion is contrary to the everyday use of the word "acceleration" to mean that speed is increasing. The more precise definition given in Eq. (3.9) shows that there is a nonzero acceleration whenever the velocity vector changes in *any* way, whether there is a change of speed, direction, or both. ▮

To convince yourself that a particle is accelerating as it moves on a curved path with constant speed, think of your sensations when you ride in a car. When the car accelerates, you tend to move inside the car in a direction *opposite* to the car's acceleration. (In Chapter 4 we'll learn why this is so.) Thus you tend to slide toward the back of the car when it accelerates forward (speeds up) and toward the front of the car when it accelerates backward (slows down). If the car makes a turn on a level road, you tend to slide toward the outside of the turn; hence the car is accelerating toward the inside of the turn.

We'll usually be interested in instantaneous acceleration, not average acceleration. From now on, we'll use the term "acceleration" to mean the instantaneous acceleration vector $\vec{a}$.

Each component of the acceleration vector $\vec{a} = a_x\hat{\imath} + a_y\hat{\jmath} + a_z\hat{k}$ is the derivative of the corresponding component of velocity:

Each **component** of a particle's **instantaneous acceleration vector** ...

$$a_x = \frac{dv_x}{dt} \qquad a_y = \frac{dv_y}{dt} \qquad a_z = \frac{dv_z}{dt} \qquad (3.10)$$

... equals the instantaneous rate of change of its corresponding velocity component.

DEMO

3.7 (a) Instantaneous acceleration $\vec{a}$ at point P_1 in Fig. 3.6. (b) Instantaneous acceleration for motion along a straight line.

(a) Acceleration: curved trajectory

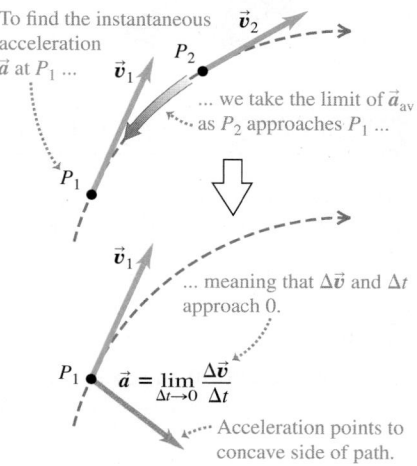

To find the instantaneous acceleration $\vec{a}$ at P_1 ... ... we take the limit of $\vec{a}_{av}$ as P_2 approaches P_1 ...

... meaning that $\Delta\vec{v}$ and Δt approach 0.

$\vec{a} = \lim\limits_{\Delta t \to 0} \dfrac{\Delta\vec{v}}{\Delta t}$

Acceleration points to concave side of path.

(b) Acceleration: straight-line trajectory

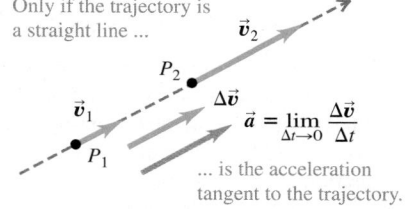

Only if the trajectory is a straight line ...

$\vec{a} = \lim\limits_{\Delta t \to 0} \dfrac{\Delta\vec{v}}{\Delta t}$

... is the acceleration tangent to the trajectory.

BIO **Application Horses on a Curved Path** By leaning to the side and hitting the ground with their hooves at an angle, these horses give themselves the sideways acceleration necessary to make a sharp change in direction.

3.8 When the fingers release the arrow, its acceleration vector has a horizontal component (a_x) and a vertical component (a_y).

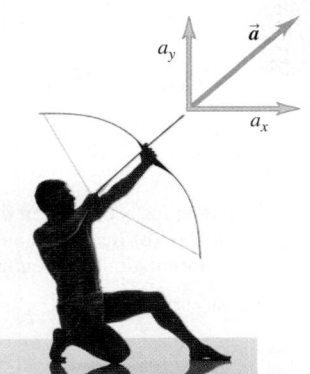

In terms of unit vectors,

$$\vec{a} = \frac{d\vec{v}}{dt} = \frac{dv_x}{dt}\hat{\imath} + \frac{dv_y}{dt}\hat{\jmath} + \frac{dv_z}{dt}\hat{k} \qquad (3.11)$$

The x-component of Eqs. (3.10) and (3.11), $a_x = dv_x/dt$, is just Eq. (2.5) for instantaneous acceleration in one dimension. **Figure 3.8** shows an example of an acceleration vector that has both x- and y-components.

Since each component of velocity is the derivative of the corresponding coordinate, we can express the components a_x, a_y, and a_z of the acceleration vector $\vec{a}$ as

$$a_x = \frac{d^2x}{dt^2} \qquad a_y = \frac{d^2y}{dt^2} \qquad a_z = \frac{d^2z}{dt^2} \qquad (3.12)$$

EXAMPLE 3.2 CALCULATING AVERAGE AND INSTANTANEOUS ACCELERATION

Let's return to the motions of the Mars rover in Example 3.1. (a) Find the components of the average acceleration for the interval $t = 0.0$ s to $t = 2.0$ s. (b) Find the instantaneous acceleration at $t = 2.0$ s.

SOLUTION

IDENTIFY and SET UP: In Example 3.1 we found the components of the rover's instantaneous velocity at any time t:

$$v_x = \frac{dx}{dt} = (-0.25 \text{ m/s}^2)(2t) = (-0.50 \text{ m/s}^2)t$$

$$v_y = \frac{dy}{dt} = 1.0 \text{ m/s} + (0.025 \text{ m/s}^3)(3t^2)$$

$$= 1.0 \text{ m/s} + (0.075 \text{ m/s}^3)t^2$$

We'll use the vector relationships among velocity, average acceleration, and instantaneous acceleration. In part (a) we determine the values of v_x and v_y at the beginning and end of the interval and then use Eq. (3.8) to calculate the components of the average acceleration. In part (b) we obtain expressions for the instantaneous acceleration components at any time t by taking the time derivatives of the velocity components as in Eqs. (3.10).

EXECUTE: (a) In Example 3.1 we found that at $t = 0.0$ s the velocity components are

$$v_x = 0.0 \text{ m/s} \qquad v_y = 1.0 \text{ m/s}$$

and that at $t = 2.0$ s the components are

$$v_x = -1.0 \text{ m/s} \qquad v_y = 1.3 \text{ m/s}$$

Thus the components of average acceleration in the interval $t = 0.0$ s to $t = 2.0$ s are

$$a_{\text{av-}x} = \frac{\Delta v_x}{\Delta t} = \frac{-1.0 \text{ m/s} - 0.0 \text{ m/s}}{2.0 \text{ s} - 0.0 \text{ s}} = -0.50 \text{ m/s}^2$$

$$a_{\text{av-}y} = \frac{\Delta v_y}{\Delta t} = \frac{1.3 \text{ m/s} - 1.0 \text{ m/s}}{2.0 \text{ s} - 0.0 \text{ s}} = 0.15 \text{ m/s}^2$$

(b) Using Eqs. (3.10), we find

$$a_x = \frac{dv_x}{dt} = -0.50 \text{ m/s}^2 \qquad a_y = \frac{dv_y}{dt} = (0.075 \text{ m/s}^3)(2t)$$

Hence the instantaneous acceleration vector $\vec{a}$ at time t is

$$\vec{a} = a_x\hat{\imath} + a_y\hat{\jmath} = (-0.50 \text{ m/s}^2)\hat{\imath} + (0.15 \text{ m/s}^3)t\hat{\jmath}$$

At $t = 2.0$ s the components of acceleration and the acceleration vector are

$$a_x = -0.50 \text{ m/s}^2 \qquad a_y = (0.15 \text{ m/s}^3)(2.0 \text{ s}) = 0.30 \text{ m/s}^2$$

$$\vec{a} = (-0.50 \text{ m/s}^2)\hat{\imath} + (0.30 \text{ m/s}^2)\hat{\jmath}$$

The magnitude of acceleration at this time is

$$a = \sqrt{a_x^2 + a_y^2}$$

$$= \sqrt{(-0.50 \text{ m/s}^2)^2 + (0.30 \text{ m/s}^2)^2} = 0.58 \text{ m/s}^2$$

A sketch of this vector (**Fig. 3.9**) shows that the direction angle β of $\vec{a}$ with respect to the positive x-axis is between $90°$ and $180°$. From Eq. (3.7) we have

$$\arctan\frac{a_y}{a_x} = \arctan\frac{0.30 \text{ m/s}^2}{-0.50 \text{ m/s}^2} = -31°$$

Hence $\beta = 180° + (-31°) = 149°$.

3.9 The path of the robotic rover, showing the velocity and acceleration at $t = 0.0$ s ($\vec{v}_0$ and $\vec{a}_0$), $t = 1.0$ s ($\vec{v}_1$ and $\vec{a}_1$), and $t = 2.0$ s ($\vec{v}_2$ and $\vec{a}_2$).

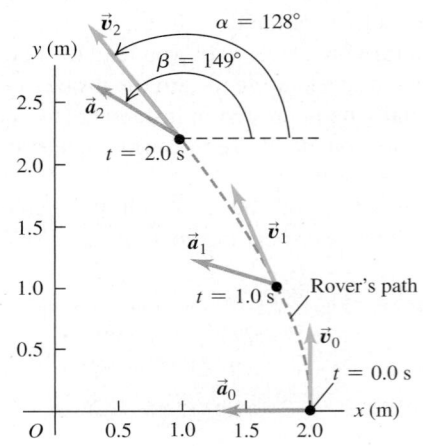

EVALUATE: Figure 3.9 shows the rover's path and the velocity and acceleration vectors at $t = 0.0$ s, 1.0 s, and 2.0 s. (Use the results of part (b) to calculate the instantaneous acceleration at $t = 0.0$ s and $t = 1.0$ s for yourself.) Note that $\vec{v}$ and $\vec{a}$ are *not* in the same direction at any of these times. The velocity vector $\vec{v}$ is tangent to the path at each point (as is always the case), and the acceleration vector $\vec{a}$ points toward the concave side of the path.

Parallel and Perpendicular Components of Acceleration

Equations (3.10) tell us about the components of a particle's instantaneous acceleration vector $\vec{a}$ along the x-, y-, and z-axes. Another useful way to think about $\vec{a}$ is in terms of one component *parallel* to the particle's path and to its velocity $\vec{v}$, and one component *perpendicular* to the path and to $\vec{v}$ (**Fig. 3.10**). That's because the parallel component $a_\parallel$ tells us about changes in the particle's *speed*, while the perpendicular component $a_\perp$ tells us about changes in the particle's *direction of motion*. To see why the parallel and perpendicular components of $\vec{a}$ have these properties, let's consider two special cases.

In **Fig. 3.11a** the acceleration vector is in the same direction as the velocity $\vec{v}_1$, so $\vec{a}$ has only a parallel component $a_\parallel$ (that is, $a_\perp = 0$). The velocity change $\Delta\vec{v}$ during a small time interval Δt is in the same direction as $\vec{a}$ and hence in the same direction as $\vec{v}_1$. The velocity $\vec{v}_2$ at the end of Δt is in the same direction as $\vec{v}_1$ but has greater magnitude. Hence during the time interval Δt the particle in Fig. 3.11a moved in a straight line with increasing speed (compare Fig. 3.7b).

In Fig. 3.11b the acceleration is *perpendicular* to the velocity, so $\vec{a}$ has only a perpendicular component $a_\perp$ (that is, $a_\parallel = 0$). In a small time interval Δt, the velocity change $\Delta\vec{v}$ is very nearly perpendicular to $\vec{v}_1$, and so $\vec{v}_1$ and $\vec{v}_2$ have different directions. As the time interval Δt approaches zero, the angle ϕ in the figure also approaches zero, $\Delta\vec{v}$ becomes perpendicular to *both* $\vec{v}_1$ and $\vec{v}_2$, and $\vec{v}_1$ and $\vec{v}_2$ have the same magnitude. In other words, the speed of the particle stays the same, but the direction of motion changes and the path of the particle curves.

In the most general case, the acceleration $\vec{a}$ has *both* components parallel and perpendicular to the velocity $\vec{v}$, as in Fig. 3.10. Then the particle's speed will change (described by the parallel component $a_\parallel$) *and* its direction of motion will change (described by the perpendicular component $a_\perp$).

Figure 3.12 shows a particle moving along a curved path for three situations: constant speed, increasing speed, and decreasing speed. If the speed is constant, $\vec{a}$ is perpendicular, or *normal,* to the path and to $\vec{v}$ and points toward the concave side of the path (Fig. 3.12a). If the speed is increasing, there is still a perpendicular component of $\vec{a}$, but there is also a parallel component with the same direction as $\vec{v}$ (Fig. 3.12b). Then $\vec{a}$ points ahead of the normal to the path. (This was the case in Example 3.2.) If the speed is decreasing, the parallel component has the direction opposite to $\vec{v}$, and $\vec{a}$ points behind the normal to the path (Fig. 3.12c; compare Fig. 3.7a). We will use these ideas again in Section 3.4 when we study the special case of motion in a circle.

3.10 The acceleration can be resolved into a component $a_\parallel$ parallel to the path (that is, along the tangent to the path) and a component $a_\perp$ perpendicular to the path (that is, along the normal to the path).

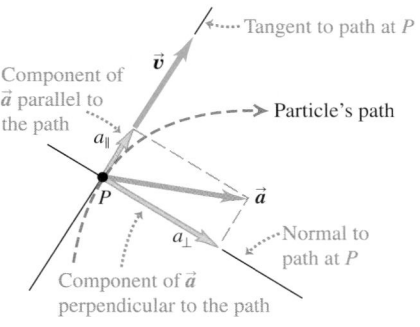

3.11 The effect of acceleration directed (a) parallel to and (b) perpendicular to a particle's velocity.

(a) Acceleration parallel to velocity

Changes only *magnitude* of velocity: speed changes; direction doesn't.

(b) Acceleration perpendicular to velocity

Changes only *direction* of velocity: particle follows curved path at constant speed.

(**MP**)

PhET: Maze Game

3.12 Velocity and acceleration vectors for a particle moving through a point P on a curved path with (a) constant speed, (b) increasing speed, and (c) decreasing speed.

(a) When speed is constant along a curved path ...

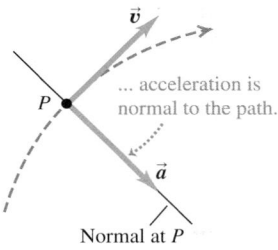

(b) When speed is increasing along a curved path ...

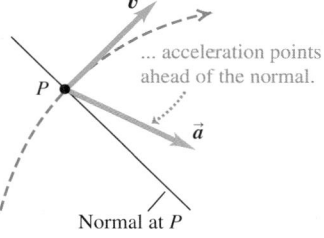

(c) When speed is decreasing along a curved path ...

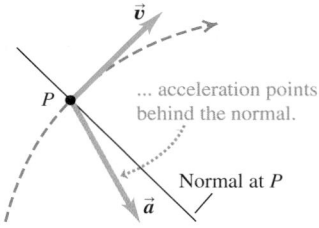

EXAMPLE 3.3   **CALCULATING PARALLEL AND PERPENDICULAR COMPONENTS OF ACCELERATION**

For the rover of Examples 3.1 and 3.2, find the parallel and perpendicular components of the acceleration at $t = 2.0$ s.

SOLUTION

IDENTIFY and SET UP: We want to find the components of the acceleration vector $\vec{a}$ that are parallel and perpendicular to velocity vector $\vec{v}$. We found the directions of $\vec{v}$ and $\vec{a}$ in Examples 3.1 and 3.2, respectively; Fig. 3.9 shows the results. From these directions we can find the angle between the two vectors and the components of $\vec{a}$ with respect to the direction of $\vec{v}$.

EXECUTE: From Example 3.2, at $t = 2.0$ s the particle has an acceleration of magnitude 0.58 m/s² at an angle of 149° with respect to the positive x-axis. In Example 3.1 we found that at this time the velocity vector is at an angle of 128° with respect to the positive x-axis. The angle between $\vec{a}$ and $\vec{v}$ is therefore 149° − 128° = 21° (**Fig. 3.13**). Hence the components of acceleration parallel and perpendicular to $\vec{v}$ are

$$a_\parallel = a \cos 21° = (0.58 \text{ m/s}^2)\cos 21° = 0.54 \text{ m/s}^2$$

$$a_\perp = a \sin 21° = (0.58 \text{ m/s}^2)\sin 21° = 0.21 \text{ m/s}^2$$

3.13 The parallel and perpendicular components of the acceleration of the rover at $t = 2.0$ s.

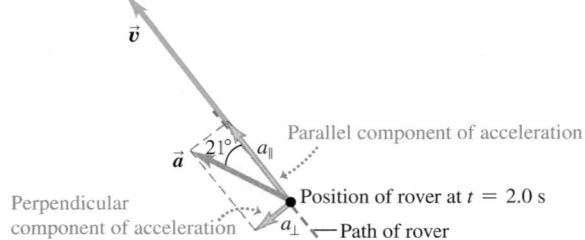

EVALUATE: The parallel component $a_\parallel$ is positive (in the same direction as $\vec{v}$), which means that the speed is increasing at this instant. The value $a_\parallel = +0.54$ m/s² tells us that the speed is increasing at this instant at a rate of 0.54 m/s per second. The perpendicular component $a_\perp$ is not zero, which means that at this instant the rover is turning—that is, it is changing direction and following a curved path.

CONCEPTUAL EXAMPLE 3.4   **ACCELERATION OF A SKIER**

A skier moves along a ski-jump ramp (**Fig. 3.14a**). The ramp is straight from point A to point C and curved from point C onward. The skier speeds up as she moves downhill from point A to point E, where her speed is maximum. She slows down after passing point E. Draw the direction of the acceleration vector at each of the points B, D, E, and F.

SOLUTION

Figure 3.14b shows our solution. At point B the skier is moving in a straight line with increasing speed, so her acceleration points downhill, in the same direction as her velocity. At points D, E, and F the skier is moving along a curved path, so her acceleration has a component perpendicular to the path (toward the concave side of the path) at each of these points. At point D there is also an acceleration component in the direction of her motion because she is speeding up. So the acceleration vector points *ahead* of the normal to her path at point D. At point E, the skier's speed is instantaneously not changing; her speed is maximum at this point, so its derivative is zero. There is therefore no parallel component of $\vec{a}$, and the acceleration is perpendicular to her motion. At point F there is an acceleration component *opposite to* the direction of her motion because she's slowing down. The acceleration vector therefore points *behind* the normal to her path.

In the next section we'll consider the skier's acceleration after she flies off the ramp.

3.14 (a) The skier's path. (b) Our solution.

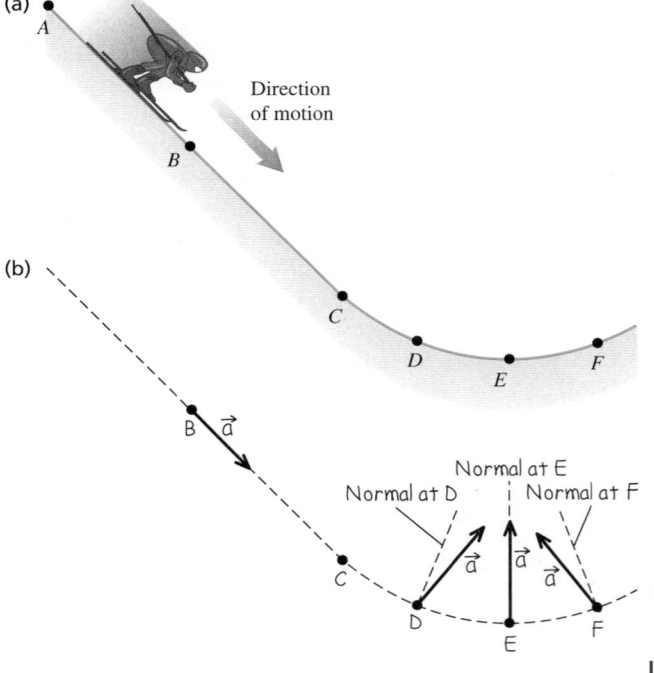

TEST YOUR UNDERSTANDING OF SECTION 3.2   A sled travels over the crest of a snow-covered hill. The sled slows down as it climbs up one side of the hill and gains speed as it descends on the other side. Which of the vectors (1 through 9) in the figure correctly shows the direction of the sled's acceleration at the crest? (Choice 9 is that the acceleration is zero.) ❙

or 9: acceleration = 0

3.3 PROJECTILE MOTION

A **projectile** is any body that is given an initial velocity and then follows a path determined entirely by the effects of gravitational acceleration and air resistance. A batted baseball, a thrown football, and a bullet shot from a rifle are all projectiles. The path followed by a projectile is called its **trajectory.**

To analyze the motion of a projectile, we'll use an idealized model. We'll represent the projectile as a particle with an acceleration (due to gravity) that is constant in both magnitude and direction. We'll ignore the effects of air resistance and the curvature and rotation of the earth. This model has limitations, however: We have to consider the earth's curvature when we study the flight of long-range missiles, and air resistance is of crucial importance to a sky diver. Nevertheless, we can learn a lot from analysis of this simple model. For the remainder of this chapter the phrase "projectile motion" will imply that we're ignoring air resistance. In Chapter 5 we'll see what happens when air resistance cannot be ignored.

Projectile motion is always confined to a vertical plane determined by the direction of the initial velocity (**Fig. 3.15**). This is because the acceleration due to gravity is purely vertical; gravity can't accelerate the projectile sideways. Thus projectile motion is *two-dimensional.* We will call the plane of motion the *xy*-coordinate plane, with the *x*-axis horizontal and the *y*-axis vertically upward.

The key to analyzing projectile motion is that we can treat the *x*- and *y*-coordinates separately. **Figure 3.16** illustrates this for two projectiles: a red ball dropped from rest and a yellow ball projected horizontally from the same height. The figure shows that the horizontal motion of the yellow projectile has *no* effect on its vertical motion. For both projectiles, the *x*-component of acceleration is zero and the *y*-component is constant and equal to $-g$. (By definition, g is always positive; with our choice of coordinate directions, a_y is negative.) So *we can analyze projectile motion as a combination of horizontal motion with constant velocity and vertical motion with constant acceleration.*

We can then express all the vector relationships for the projectile's position, velocity, and acceleration by separate equations for the horizontal and vertical components. The components of $\vec{a}$ are

$$a_x = 0 \qquad a_y = -g \qquad \text{(projectile motion, no air resistance)} \qquad (3.13)$$

Since both the *x*-acceleration and *y*-acceleration are constant, we can use Eqs. (2.8), (2.12), (2.13), and (2.14) directly. Suppose that at time $t = 0$ our particle is at the point (x_0, y_0) and its initial velocity at this time has components v_{0x} and v_{0y}. The components of acceleration are $a_x = 0$, $a_y = -g$. Considering the *x*-motion first, we substitute 0 for a_x in Eqs. (2.8) and (2.12). We find

$$v_x = v_{0x} \qquad (3.14)$$

$$x = x_0 + v_{0x}t \qquad (3.15)$$

For the *y*-motion we substitute y for x, v_y for v_x, v_{0y} for v_{0x}, and $a_y = -g$ for a_x:

$$v_y = v_{0y} - gt \qquad (3.16)$$

$$y = y_0 + v_{0y}t - \tfrac{1}{2}gt^2 \qquad (3.17)$$

3.15 The trajectory of an idealized projectile.

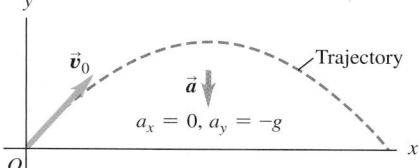

- A projectile moves in a vertical plane that contains the initial velocity vector $\vec{v}_0$.
- Its trajectory depends only on $\vec{v}_0$ and on the downward acceleration due to gravity.

3.16 The red ball is dropped from rest, and the yellow ball is simultaneously projected horizontally.

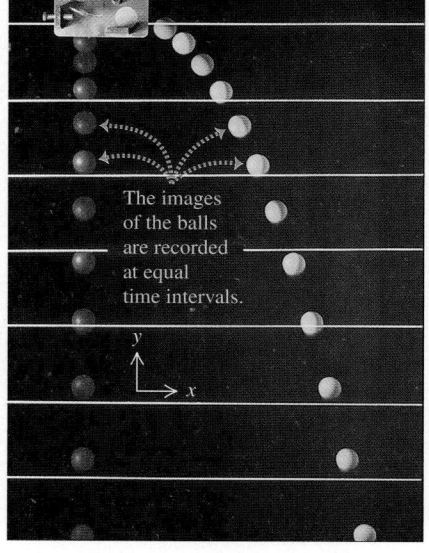

The images of the balls are recorded at equal time intervals.

- At any time the two balls have different *x*-coordinates and *x*-velocities but the same *y*-coordinate, *y*-velocity, and *y*-acceleration.
- The horizontal motion of the yellow ball has no effect on its vertical motion.

3.17 If air resistance is negligible, the trajectory of a projectile is a combination of horizontal motion with constant velocity and vertical motion with constant acceleration.

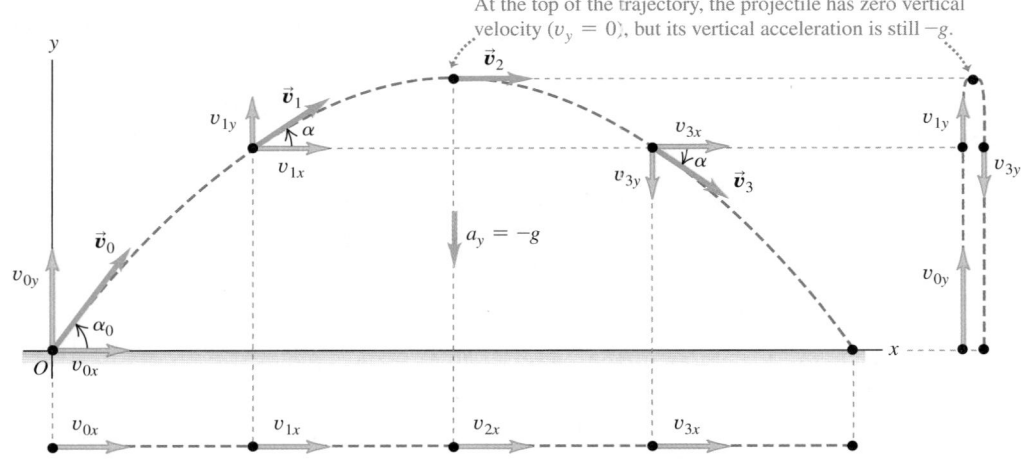

At the top of the trajectory, the projectile has zero vertical velocity ($v_y = 0$), but its vertical acceleration is still $-g$.

Vertically, the projectile is in constant-acceleration motion in response to the earth's gravitational pull. Thus its vertical velocity *changes* by equal amounts during equal time intervals.

Horizontally, the projectile is in constant-velocity motion: Its horizontal acceleration is zero, so it moves equal x-distances in equal time intervals.

3.18 The initial velocity components v_{0x} and v_{0y} of a projectile (such as a kicked soccer ball) are related to the initial speed v_0 and initial angle α_0.

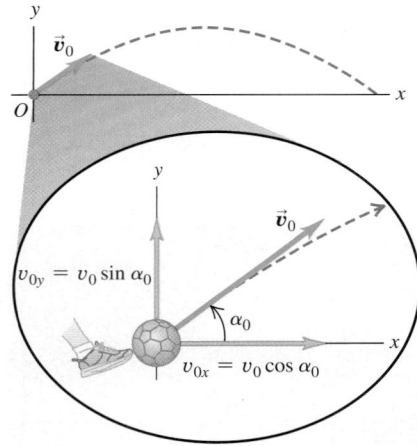

It's usually simplest to take the initial position (at $t = 0$) as the origin; then $x_0 = y_0 = 0$. This might be the position of a ball at the instant it leaves the hand of the person who throws it or the position of a bullet at the instant it leaves the gun barrel.

Figure 3.17 shows the trajectory of a projectile that starts at (or passes through) the origin at time $t = 0$, along with its position, velocity, and velocity components at equal time intervals. The x-velocity v_x is constant; the y-velocity v_y changes by equal amounts in equal times, just as if the projectile were launched vertically with the same initial y-velocity.

We can also represent the initial velocity $\vec{v}_0$ by its magnitude v_0 (the initial speed) and its angle α_0 with the positive x-axis (**Fig. 3.18**). In terms of these quantities, the components v_{0x} and v_{0y} of the initial velocity are

$$v_{0x} = v_0 \cos\alpha_0 \qquad v_{0y} = v_0 \sin\alpha_0 \qquad (3.18)$$

If we substitute Eqs. (3.18) into Eqs. (3.14) through (3.17) and set $x_0 = y_0 = 0$, we get the following equations. They describe the position and velocity of the projectile in Fig. 3.17 at any time t:

DEMO

DEMO

DEMO

PhET: Projectile Motion

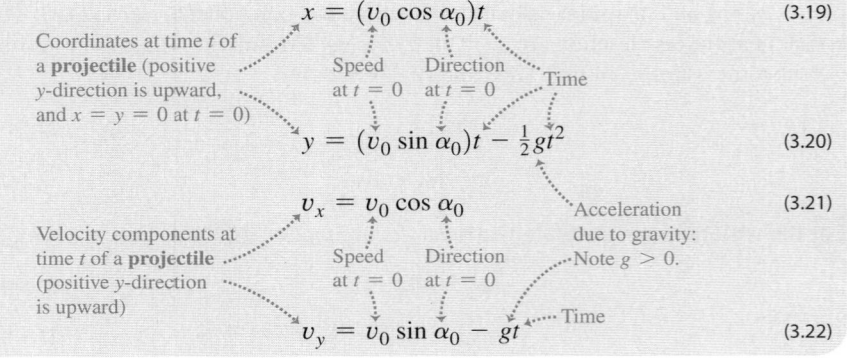

$$x = (v_0 \cos\alpha_0)t \qquad (3.19)$$

Coordinates at time t of a **projectile** (positive y-direction is upward, and $x = y = 0$ at $t = 0$)

Speed at $t = 0$ ⋯ Direction at $t = 0$ ⋯ Time

$$y = (v_0 \sin\alpha_0)t - \tfrac{1}{2}gt^2 \qquad (3.20)$$

$$v_x = v_0 \cos\alpha_0 \qquad (3.21)$$

Velocity components at time t of a **projectile** (positive y-direction is upward)

Speed at $t = 0$ ⋯ Direction at $t = 0$ ⋯ Acceleration due to gravity: Note $g > 0$. ⋯ Time

$$v_y = v_0 \sin\alpha_0 - gt \qquad (3.22)$$

We can get a lot of information from Eqs. (3.19) through (3.22). For example, the distance r from the origin to the projectile at any time t is

$$r = \sqrt{x^2 + y^2} \qquad (3.23)$$

The projectile's speed (the magnitude of its velocity) at any time is

$$v = \sqrt{v_x^2 + v_y^2} \qquad (3.24)$$

The *direction* of the velocity, in terms of the angle α it makes with the positive x-direction (see Fig. 3.17), is

$$\tan \alpha = \frac{v_y}{v_x} \qquad (3.25)$$

The velocity vector $\vec{v}$ is tangent to the trajectory at each point.

We can derive an equation for the trajectory's shape in terms of x and y by eliminating t. From Eqs. (3.19) and (3.20), we find $t = x/(v_0 \cos \alpha_0)$ and

$$y = (\tan \alpha_0)x - \frac{g}{2v_0^2 \cos^2 \alpha_0}x^2 \qquad (3.26)$$

Don't worry about the details of this equation; the important point is its general form. Since v_0, $\tan \alpha_0$, $\cos \alpha_0$, and g are constants, Eq. (3.26) has the form

$$y = bx - cx^2$$

where b and c are constants. This is the equation of a *parabola*. In our simple model of projectile motion, the trajectory is always a parabola (**Fig. 3.19**).

When air resistance *isn't* negligible and has to be included, calculating the trajectory becomes a lot more complicated; the effects of air resistance depend on velocity, so the acceleration is no longer constant. **Figure 3.20** shows a computer simulation of the trajectory of a baseball both without air resistance and with air resistance proportional to the square of the baseball's speed. We see that air resistance has a very large effect; the projectile does not travel as far or as high, and the trajectory is no longer a parabola.

DEMO

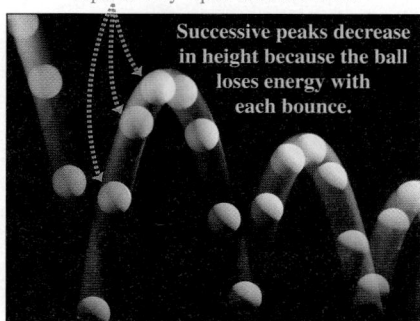
3.19 The nearly parabolic trajectories of a bouncing ball.

Successive images of the ball are separated by equal time intervals.

Successive peaks decrease in height because the ball loses energy with each bounce.

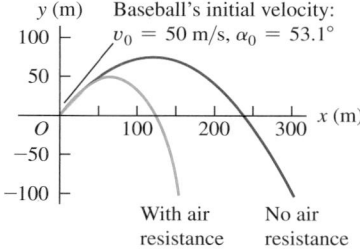
3.20 Air resistance has a large cumulative effect on the motion of a baseball. In this simulation we allow the baseball to fall below the height from which it was thrown (for example, the baseball could have been thrown from a cliff).

CONCEPTUAL EXAMPLE 3.5   ACCELERATION OF A SKIER, CONTINUED

SOLUTION

Let's consider again the skier in Conceptual Example 3.4. What is her acceleration at each of the points G, H, and I in **Fig. 3.21a** *after* she flies off the ramp? Neglect air resistance.

SOLUTION

Figure 3.21b shows our answer. The skier's acceleration changed from point to point while she was on the ramp. But as soon as she leaves the ramp, she becomes a projectile. So at points G, H, and I, and indeed at *all* points after she leaves the ramp, the skier's acceleration points vertically downward and has magnitude g. No matter how complicated the acceleration of a particle before it becomes a projectile, its acceleration as a projectile is given by $a_x = 0$, $a_y = -g$.

3.21 (a) The skier's path during the jump. (b) Our solution.

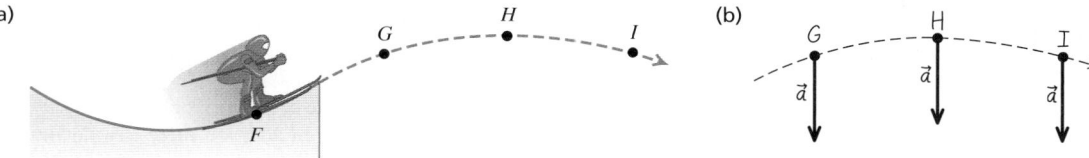

PROBLEM-SOLVING STRATEGY 3.1) PROJECTILE MOTION

NOTE: The strategies we used in Sections 2.4 and 2.5 for straight-line, constant-acceleration problems are also useful here.

IDENTIFY *the relevant concepts:* The key concept is that throughout projectile motion, the acceleration is downward and has a constant magnitude *g*. Projectile-motion equations don't apply to *throwing* a ball, because during the throw the ball is acted on by both the thrower's hand and gravity. These equations apply only *after* the ball leaves the thrower's hand.

SET UP *the problem* using the following steps:
1. Define your coordinate system and make a sketch showing your axes. It's almost always best to make the *x*-axis horizontal and the *y*-axis vertical, and to choose the origin to be where the body first becomes a projectile (for example, where a ball leaves the thrower's hand). Then the components of acceleration are $a_x = 0$ and $a_y = -g$, as in Eq. (3.13); the initial position is $x_0 = y_0 = 0$; and you can use Eqs. (3.19) through (3.22). (If you choose a different origin or axes, you'll have to modify these equations.)
2. List the unknown and known quantities, and decide which unknowns are your target variables. For example, you might be given the initial velocity (either the components or the magnitude and direction) and asked to find the coordinates and velocity components at some later time. Make sure that

you have as many equations as there are target variables to be found. In addition to Eqs. (3.19) through (3.22), Eqs. (3.23) through (3.26) may be useful.
3. State the problem in words and then translate those words into symbols. For example, *when* does the particle arrive at a certain point? (That is, at what value of *t*?) *Where* is the particle when its velocity has a certain value? (That is, what are the values of *x* and *y* when v_x or v_y has the specified value?) Since $v_y = 0$ at the highest point in a trajectory, the question "When does the projectile reach its highest point?" translates into "What is the value of *t* when $v_y = 0$?" Similarly, "When does the projectile return to its initial elevation?" translates into "What is the value of *t* when $y = y_0$?"

EXECUTE *the solution:* Find the target variables using the equations you chose. Resist the temptation to break the trajectory into segments and analyze each segment separately. You don't have to start all over when the projectile reaches its highest point! It's almost always easier to use the same axes and time scale throughout the problem. If you need numerical values, use $g = 9.80 \text{ m/s}^2$. Remember that *g* is positive!

EVALUATE *your answer:* Do your results make sense? Do the numerical values seem reasonable?

EXAMPLE 3.6   A BODY PROJECTED HORIZONTALLY

A motorcycle stunt rider rides off the edge of a cliff. Just at the edge his velocity is horizontal, with magnitude 9.0 m/s. Find the motorcycle's position, distance from the edge of the cliff, and velocity 0.50 s after it leaves the edge of the cliff.

SOLUTION

IDENTIFY and SET UP: Figure 3.22 shows our sketch of the trajectory of motorcycle and rider. He is in projectile motion as soon as he leaves the edge of the cliff, which we take to be the origin (so $x_0 = y_0 = 0$). His initial velocity $\vec{v}_0$ at the edge of the cliff is horizontal (that is, $\alpha_0 = 0$), so its components are $v_{0x} = v_0 \cos \alpha_0 = 9.0 \text{ m/s}$ and $v_{0y} = v_0 \sin \alpha_0 = 0$. To find the motorcycle's position

3.22 Our sketch for this problem.

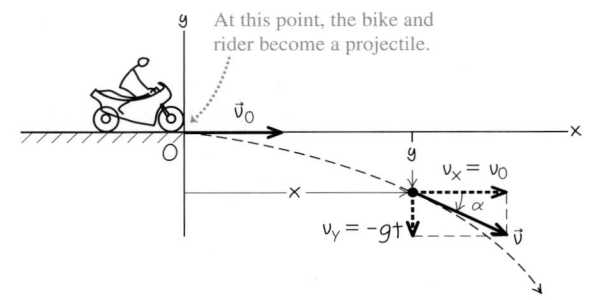

At this point, the bike and rider become a projectile.

at $t = 0.50$ s, we use Eqs. (3.19) and (3.20); we then find the distance from the origin using Eq. (3.23). Finally, we use Eqs. (3.21) and (3.22) to find the velocity components at $t = 0.50$ s.

EXECUTE: From Eqs. (3.19) and (3.20), the motorcycle's *x*- and *y*-coordinates at $t = 0.50$ s are

$$x = v_{0x}t = (9.0 \text{ m/s})(0.50 \text{ s}) = 4.5 \text{ m}$$
$$y = -\tfrac{1}{2}gt^2 = -\tfrac{1}{2}(9.80 \text{ m/s}^2)(0.50 \text{ s})^2 = -1.2 \text{ m}$$

The negative value of *y* shows that the motorcycle is below its starting point.

From Eq. (3.23), the motorcycle's distance from the origin at $t = 0.50$ s is

$$r = \sqrt{x^2 + y^2} = \sqrt{(4.5 \text{ m})^2 + (-1.2 \text{ m})^2} = 4.7 \text{ m}$$

From Eqs. (3.21) and (3.22), the velocity components at $t = 0.50$ s are

$$v_x = v_{0x} = 9.0 \text{ m/s}$$
$$v_y = -gt = (-9.80 \text{ m/s}^2)(0.50 \text{ s}) = -4.9 \text{ m/s}$$

The motorcycle has the same horizontal velocity v_x as when it left the cliff at $t = 0$, but in addition there is a downward (negative) vertical velocity v_y. The velocity vector at $t = 0.50$ s is

$$\vec{v} = v_x\hat{\imath} + v_y\hat{\jmath} = (9.0 \text{ m/s})\hat{\imath} + (-4.9 \text{ m/s})\hat{\jmath}$$

From Eqs. (3.24) and (3.25), at $t = 0.50$ s the velocity has magnitude v and angle α given by

$$v = \sqrt{v_x^2 + v_y^2} = \sqrt{(9.0 \text{ m/s})^2 + (-4.9 \text{ m/s})^2} = 10.2 \text{ m/s}$$

$$\alpha = \arctan \frac{v_y}{v_x} = \arctan\left(\frac{-4.9 \text{ m/s}}{9.0 \text{ m/s}}\right) = -29°$$

The motorcycle is moving at 10.2 m/s in a direction 29° below the horizontal.

EVALUATE: Just as in Fig. 3.17, the motorcycle's horizontal motion is unchanged by gravity; the motorcycle continues to move horizontally at 9.0 m/s, covering 4.5 m in 0.50 s. The motorcycle initially has zero vertical velocity, so it falls vertically just like a body released from rest and descends a distance $\frac{1}{2}gt^2 = 1.2$ m in 0.50 s.

EXAMPLE 3.7 HEIGHT AND RANGE OF A PROJECTILE I: A BATTED BASEBALL

A batter hits a baseball so that it leaves the bat at speed $v_0 = 37.0$ m/s at an angle $\alpha_0 = 53.1°$. (a) Find the position of the ball and its velocity (magnitude and direction) at $t = 2.00$ s. (b) Find the time when the ball reaches the highest point of its flight, and its height h at this time. (c) Find the *horizontal range* R—that is, the horizontal distance from the starting point to where the ball hits the ground—and the ball's velocity just before it hits.

SOLUTION

IDENTIFY and SET UP: As Fig. 3.20 shows, air resistance strongly affects the motion of a baseball. For simplicity, however, we'll ignore air resistance here and use the projectile-motion equations to describe the motion. The ball leaves the bat at $t = 0$ a meter or so above ground level, but we'll ignore this distance and assume that it starts at ground level ($y_0 = 0$). **Figure 3.23** shows our sketch of the ball's trajectory. We'll use the same coordinate system as in Figs. 3.17 and 3.18, so we can use Eqs. (3.19) through (3.22). Our target variables are (a) the position and velocity of the ball 2.00 s after it leaves the bat, (b) the time t when the ball is at its maximum height (that is, when $v_y = 0$) and the y-coordinate at this time, and (c) the x-coordinate when the ball returns to ground level ($y = 0$) and the ball's vertical component of velocity then.

EXECUTE: (a) We want to find x, y, v_x, and v_y at $t = 2.00$ s. The initial velocity of the ball has components

$$v_{0x} = v_0 \cos \alpha_0 = (37.0 \text{ m/s})\cos 53.1° = 22.2 \text{ m/s}$$

$$v_{0y} = v_0 \sin \alpha_0 = (37.0 \text{ m/s})\sin 53.1° = 29.6 \text{ m/s}$$

From Eqs. (3.19) through (3.22),

$$x = v_{0x}t = (22.2 \text{ m/s})(2.00 \text{ s}) = 44.4 \text{ m}$$

$$y = v_{0y}t - \tfrac{1}{2}gt^2$$
$$= (29.6 \text{ m/s})(2.00 \text{ s}) - \tfrac{1}{2}(9.80 \text{ m/s}^2)(2.00 \text{ s})^2 = 39.6 \text{ m}$$

3.23 Our sketch for this problem.

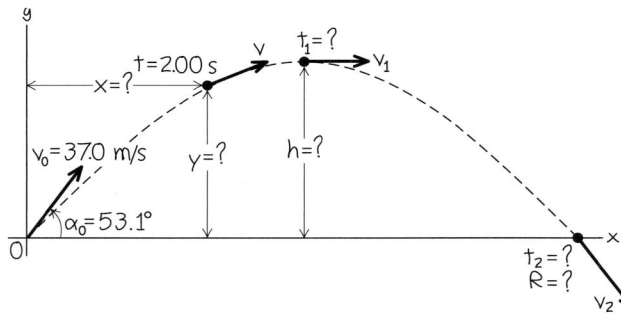

$$v_x = v_{0x} = 22.2 \text{ m/s}$$

$$v_y = v_{0y} - gt = 29.6 \text{ m/s} - (9.80 \text{ m/s}^2)(2.00 \text{ s}) = 10.0 \text{ m/s}$$

The y-component of velocity is positive at $t = 2.00$ s, so the ball is still moving upward (Fig. 3.23). From Eqs. (3.24) and (3.25), the magnitude and direction of the velocity are

$$v = \sqrt{v_x^2 + v_y^2} = \sqrt{(22.2 \text{ m/s})^2 + (10.0 \text{ m/s})^2} = 24.4 \text{ m/s}$$

$$\alpha = \arctan\left(\frac{10.0 \text{ m/s}}{22.2 \text{ m/s}}\right) = \arctan 0.450 = 24.2°$$

The ball is moving at 24.4 m/s in a direction 24.2° above the horizontal.

(b) At the highest point, the vertical velocity v_y is zero. Call the time when this happens t_1; then

$$v_y = v_{0y} - gt_1 = 0$$

$$t_1 = \frac{v_{0y}}{g} = \frac{29.6 \text{ m/s}}{9.80 \text{ m/s}^2} = 3.02 \text{ s}$$

The height h at the highest point is the value of y at time t_1:

$$h = v_{0y}t_1 - \tfrac{1}{2}gt_1^2$$
$$= (29.6 \text{ m/s})(3.02 \text{ s}) - \tfrac{1}{2}(9.80 \text{ m/s}^2)(3.02 \text{ s})^2 = 44.7 \text{ m}$$

(c) We'll find the horizontal range in two steps. First, we find the time t_2 when $y = 0$ (the ball is at ground level):

$$y = 0 = v_{0y}t_2 - \tfrac{1}{2}gt_2^2 = t_2\left(v_{0y} - \tfrac{1}{2}gt_2\right)$$

This is a quadratic equation for t_2. It has two roots:

$$t_2 = 0 \quad \text{and} \quad t_2 = \frac{2v_{0y}}{g} = \frac{2(29.6 \text{ m/s})}{9.80 \text{ m/s}^2} = 6.04 \text{ s}$$

The ball is at $y = 0$ at both times. The ball *leaves* the ground at $t_2 = 0$, and it hits the ground at $t_2 = 2v_{0y}/g = 6.04$ s.

The horizontal range R is the value of x when the ball returns to the ground at $t_2 = 6.04$ s:

$$R = v_{0x}t_2 = (22.2 \text{ m/s})(6.04 \text{ s}) = 134 \text{ m}$$

The vertical component of velocity when the ball hits the ground is

$$v_y = v_{0y} - gt_2 = 29.6 \text{ m/s} - (9.80 \text{ m/s}^2)(6.04 \text{ s})$$
$$= -29.6 \text{ m/s}$$

Continued

That is, v_y has the same magnitude as the initial vertical velocity v_{0y} but the opposite direction (down). Since v_x is constant, the angle $\alpha = -53.1°$ (below the horizontal) at this point is the negative of the initial angle $\alpha_0 = 53.1°$.

EVALUATE: It's often useful to check results by getting them in a different way. For example, we can also find the maximum height in part (b) by applying the constant-acceleration formula Eq. (2.13) to the y-motion:

$$v_y^2 = v_{0y}^2 + 2a_y(y - y_0) = v_{0y}^2 - 2g(y - y_0)$$

At the highest point, $v_y = 0$ and $y = h$. Solve this equation for h; you should get the answer that we obtained in part (b). (Do you?)

Note that the time to hit the ground, $t_2 = 6.04$ s, is exactly twice the time to reach the highest point, $t_1 = 3.02$ s. Hence the time of descent equals the time of ascent. This is *always* true if the starting point and endpoint are at the same elevation and if air resistance can be ignored.

Note also that $h = 44.7$ m in part (b) is comparable to the 61.0-m height above second base of the roof at Marlins Park in Miami, and the horizontal range $R = 134$ m in part (c) is greater than the 99.7-m distance from home plate to the right-field fence at Safeco Field in Seattle. In reality, due to air resistance (which we have ignored) a batted ball with the initial speed and angle we've used here won't go as high or as far as we've calculated (see Fig. 3.20).

EXAMPLE 3.8   HEIGHT AND RANGE OF A PROJECTILE II: MAXIMUM HEIGHT, MAXIMUM RANGE

Find the maximum height h and horizontal range R (see Fig. 3.23) of a projectile launched with speed v_0 at an initial angle α_0 between 0 and 90°. For a given v_0, what value of α_0 gives maximum height? What value gives maximum horizontal range?

SOLUTION

IDENTIFY and SET UP: This is almost the same as parts (b) and (c) of Example 3.7, except that now we want general expressions for h and R. We also want the values of α_0 that give the maximum values of h and R. In part (b) of Example 3.7 we found that the projectile reaches the high point of its trajectory (so that $v_y = 0$) at time $t_1 = v_{0y}/g$, and in part (c) we found that the projectile returns to its starting height (so that $y = y_0$) at time $t_2 = 2v_{0y}/g = 2t_1$. We'll use Eq. (3.20) to find the y-coordinate h at t_1 and Eq. (3.19) to find the x-coordinate R at time t_2. We'll express our answers in terms of the launch speed v_0 and launch angle α_0 by using Eqs. (3.18).

EXECUTE: From Eqs. (3.18), $v_{0x} = v_0 \cos\alpha_0$ and $v_{0y} = v_0 \sin\alpha_0$. Hence we can write the time t_1 when $v_y = 0$ as

$$t_1 = \frac{v_{0y}}{g} = \frac{v_0 \sin\alpha_0}{g}$$

Equation (3.20) gives the height $y = h$ at this time:

$$h = (v_0 \sin\alpha_0)\left(\frac{v_0 \sin\alpha_0}{g}\right) - \frac{1}{2}g\left(\frac{v_0 \sin\alpha_0}{g}\right)^2 = \frac{v_0^2 \sin^2\alpha_0}{2g}$$

For a given launch speed v_0, the maximum value of h occurs for $\sin\alpha_0 = 1$ and $\alpha_0 = 90°$—that is, when the projectile is launched straight up. (If it is launched horizontally, as in Example 3.6, $\alpha_0 = 0$ and the maximum height is zero!)

The time t_2 when the projectile hits the ground is

$$t_2 = \frac{2v_{0y}}{g} = \frac{2v_0 \sin\alpha_0}{g}$$

The horizontal range R is the value of x at this time. From Eq. (3.19), this is

$$R = (v_0 \cos\alpha_0)t_2 = (v_0 \cos\alpha_0)\frac{2v_0 \sin\alpha_0}{g} = \frac{v_0^2 \sin 2\alpha_0}{g}$$

(We used the trigonometric identity $2\sin\alpha_0\cos\alpha_0 = \sin 2\alpha_0$, found in Appendix B.) The maximum value of $\sin 2\alpha_0$ is 1; this occurs when $2\alpha_0 = 90°$, or $\alpha_0 = 45°$. This angle gives the maximum range for a given initial speed if air resistance can be ignored.

EVALUATE: Figure 3.24 is based on a composite photograph of three trajectories of a ball projected from a small spring gun at angles of 30°, 45°, and 60°. The initial speed v_0 is approximately the same in all three cases. The horizontal range is greatest for the 45° angle. The ranges are nearly the same for the 30° and 60° angles: Can you prove that for a given value of v_0 the range is the same for both an initial angle α_0 and an initial angle $90° - \alpha_0$? (This is not the case in Fig. 3.24 due to air resistance.)

CAUTION   **Height and range of a projectile** We don't recommend memorizing the above expressions for h and R. They are applicable only in the special circumstances we've described. In particular, you can use the expression for the range R *only* when launch and landing heights are equal. There are many end-of-chapter problems to which these equations do *not* apply. ▮

3.24 A launch angle of 45° gives the maximum horizontal range. The range is shorter with launch angles of 30° and 60°.

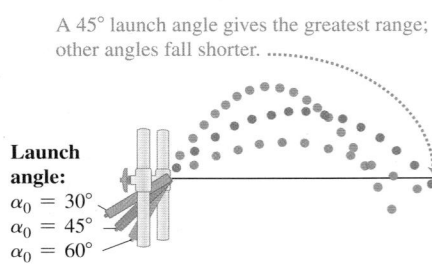

A 45° launch angle gives the greatest range; other angles fall shorter.

Launch angle:
$\alpha_0 = 30°$
$\alpha_0 = 45°$
$\alpha_0 = 60°$

SOLUTION

EXAMPLE 3.9   DIFFERENT INITIAL AND FINAL HEIGHTS

You throw a ball from your window 8.0 m above the ground. When the ball leaves your hand, it is moving at 10.0 m/s at an angle of 20° below the horizontal. How far horizontally from your window will the ball hit the ground? Ignore air resistance.

SOLUTION

IDENTIFY and SET UP: As in Examples 3.7 and 3.8, we want to find the horizontal coordinate of a projectile when it is at a given y-value. The difference here is that this value of y is *not* the same as the initial value. We again choose the x-axis to be horizontal and the y-axis to be upward, and place the origin of coordinates at the point where the ball leaves your hand (**Fig. 3.25**). We have $v_0 = 10.0$ m/s and $\alpha_0 = -20°$ (the angle is negative because the initial velocity is below the horizontal). Our target variable is the value of x when the ball reaches the ground at $y = -8.0$ m. We'll use Eq. (3.20) to find the time t when this happens and then use Eq. (3.19) to find the value of x at this time.

3.25 Our sketch for this problem.

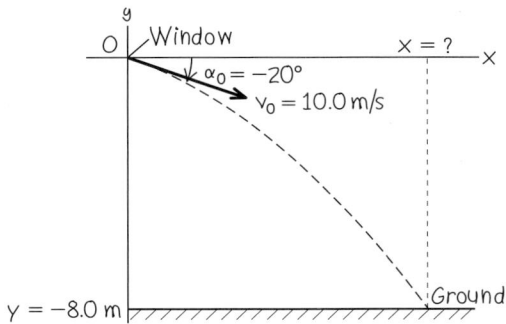

EXECUTE: To determine t, we rewrite Eq. (3.20) in the standard form for a quadratic equation for t:

$$\tfrac{1}{2}gt^2 - (v_0\sin\alpha_0)t + y = 0$$

The roots of this equation are

$$t = \frac{v_0\sin\alpha_0 \pm \sqrt{(-v_0\sin\alpha_0)^2 - 4\left(\tfrac{1}{2}g\right)y}}{2\left(\tfrac{1}{2}g\right)}$$

$$= \frac{v_0\sin\alpha_0 \pm \sqrt{v_0^2\sin^2\alpha_0 - 2gy}}{g}$$

$$= \frac{\left[\begin{array}{c}(10.0\text{ m/s})\sin(-20°) \\ \pm \sqrt{(10.0\text{ m/s})^2\sin^2(-20°) - 2(9.80\text{ m/s}^2)(-8.0\text{ m})}\end{array}\right]}{9.80\text{ m/s}^2}$$

$$= -1.7\text{ s}\quad\text{or}\quad 0.98\text{ s}$$

We discard the negative root, since it refers to a time before the ball left your hand. The positive root tells us that the ball reaches the ground at $t = 0.98$ s. From Eq. (3.19), the ball's x-coordinate at that time is

$$x = (v_0\cos\alpha_0)t = (10.0\text{ m/s})[\cos(-20°)](0.98\text{ s}) = 9.2\text{ m}$$

The ball hits the ground a horizontal distance of 9.2 m from your window.

EVALUATE: The root $t = -1.7$ s is an example of a "fictional" solution to a quadratic equation. We discussed these in Example 2.8 in Section 2.5; review that discussion.

SOLUTION

EXAMPLE 3.10   THE ZOOKEEPER AND THE MONKEY

A monkey escapes from the zoo and climbs a tree. After failing to entice the monkey down, the zookeeper fires a tranquilizer dart directly at the monkey (**Fig. 3.26**). The monkey lets go at the instant the dart leaves the gun. Show that the dart will *always* hit the monkey, provided that the dart reaches the monkey before he hits the ground and runs away.

SOLUTION

IDENTIFY and SET UP: We have *two* bodies in projectile motion: the dart and the monkey. They have different initial positions and initial velocities, but they go into projectile motion at the same time $t = 0$. We'll first use Eq. (3.19) to find an expression for the time t when the x-coordinates x_{monkey} and x_{dart} are equal. Then we'll use that expression in Eq. (3.20) to see whether y_{monkey} and y_{dart} are also equal at this time; if they are, the dart hits the monkey. We make the usual choice for the x- and y-directions, and place the origin of coordinates at the muzzle of the tranquilizer gun (Fig. 3.26).

EXECUTE: The monkey drops straight down, so $x_{\text{monkey}} = d$ at all times. From Eq. (3.19), $x_{\text{dart}} = (v_0\cos\alpha_0)t$. We solve for the time t when these x-coordinates are equal:

$$d = (v_0\cos\alpha_0)t\quad\text{so}\quad t = \frac{d}{v_0\cos\alpha_0}$$

We must now show that $y_{\text{monkey}} = y_{\text{dart}}$ at this time. The monkey is in one-dimensional free fall; its position at any time is given by Eq. (2.12), with appropriate symbol changes. Figure 3.26 shows that the monkey's initial height above the dart-gun's muzzle is $y_{\text{monkey}-0} = d\tan\alpha_0$, so

$$y_{\text{monkey}} = d\tan\alpha_0 - \tfrac{1}{2}gt^2$$

From Eq. (3.20),

$$y_{\text{dart}} = (v_0\sin\alpha_0)t - \tfrac{1}{2}gt^2$$

Continued

3.26 The tranquilizer dart hits the falling monkey.

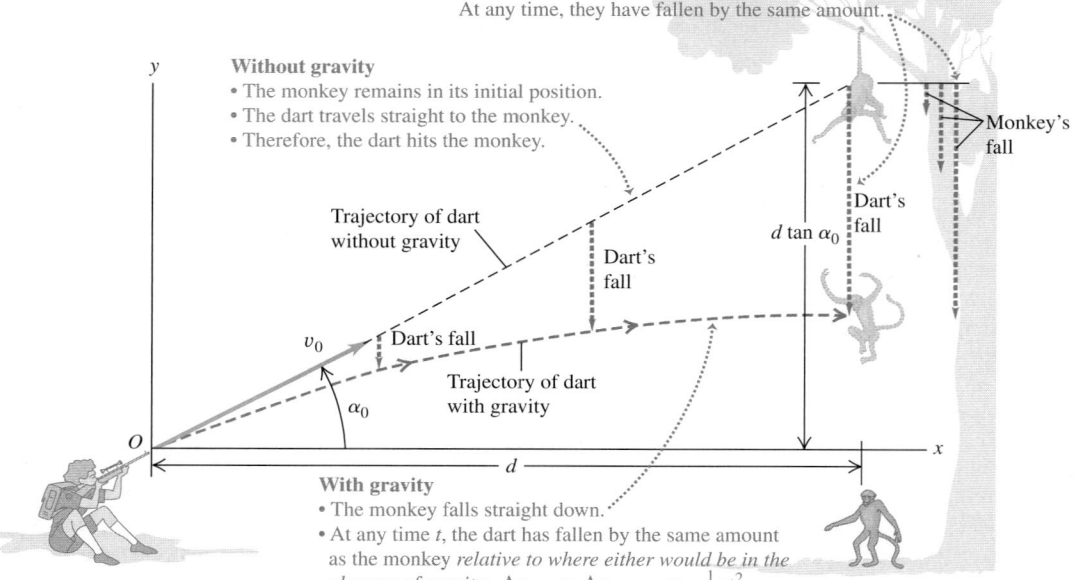

Dashed arrows show how far the dart and monkey have fallen at specific times relative to where they would be without gravity. At any time, they have fallen by the same amount.

Without gravity
• The monkey remains in its initial position.
• The dart travels straight to the monkey.
• Therefore, the dart hits the monkey.

Monkey's fall

Trajectory of dart without gravity

$d \tan \alpha_0$

Dart's fall

Dart's fall

v_0 Dart's fall

Trajectory of dart with gravity

α_0

O

d

x

With gravity
• The monkey falls straight down.
• At any time t, the dart has fallen by the same amount as the monkey *relative to where either would be in the absence of gravity:* $\Delta y_{\text{dart}} = \Delta y_{\text{monkey}} = -\frac{1}{2}gt^2$.
• Therefore, the dart always hits the monkey.

Comparing these two equations, we see that we'll have $y_{\text{monkey}} = y_{\text{dart}}$ (and a hit) if $d \tan \alpha_0 = (v_0 \sin \alpha_0)t$ when the two x-coordinates are equal. To show that this happens, we replace t with $d/(v_0 \cos \alpha_0)$, the time when $x_{\text{monkey}} = x_{\text{dart}}$. Sure enough,

$$(v_0 \sin \alpha_0)t = (v_0 \sin \alpha_0)\frac{d}{v_0 \cos \alpha_0} = d \tan \alpha_0$$

EVALUATE: We've proved that the y-coordinates of the dart and the monkey are equal at the same time that their x-coordinates are equal; a dart aimed at the monkey *always* hits it, no matter what v_0 is (provided the monkey doesn't hit the ground first). This result is independent of the value of g, the acceleration due to gravity. With no gravity ($g = 0$), the monkey would remain motionless, and the dart would travel in a straight line to hit him. With gravity, both fall the same distance $gt^2/2$ below their $t = 0$ positions, and the dart still hits the monkey (Fig. 3.26).

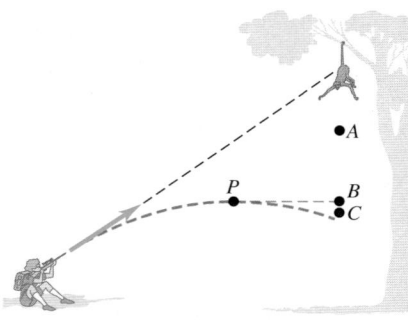

TEST YOUR UNDERSTANDING OF SECTION 3.3 In Example 3.10, suppose the tranquilizer dart has a relatively low muzzle velocity so that the dart reaches a maximum height at a point P before striking the monkey, as shown in the figure. When the dart is at point P, will the monkey be (i) at point A (higher than P), (ii) at point B (at the same height as P), or (iii) at point C (lower than P)? Ignore air resistance. ❚

3.4 MOTION IN A CIRCLE

When a particle moves along a curved path, the direction of its velocity changes. As we saw in Section 3.2, this means that the particle *must* have a component of acceleration perpendicular to the path, even if its speed is constant (see Fig. 3.11b). In this section we'll calculate the acceleration for the important special case of motion in a circle.

Uniform Circular Motion

When a particle moves in a circle with *constant speed,* the motion is called **uniform circular motion.** A car rounding a curve with constant radius at constant speed, a satellite moving in a circular orbit, and an ice skater skating in a circle

3.27 A car moving along a circular path. If the car is in uniform circular motion as in (a), the speed is constant and the acceleration is directed toward the center of the circular path (compare Fig. 3.12).

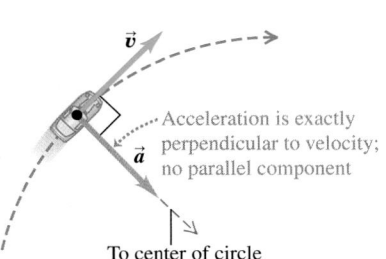

(a) Uniform circular motion: Constant speed along a circular path

Acceleration is exactly perpendicular to velocity; no parallel component

To center of circle

(b) Car speeding up along a circular path

Component of acceleration parallel to velocity: Changes car's speed

Component of acceleration perpendicular to velocity: Changes car's direction

(c) Car slowing down along a circular path

Component of acceleration perpendicular to velocity: Changes car's direction

Component of acceleration parallel to velocity: Changes car's speed

with constant speed are all examples of uniform circular motion (**Fig. 3.27a**; compare Fig. 3.12a). There is no component of acceleration parallel (tangent) to the path; otherwise, the speed would change. The acceleration vector is perpendicular (normal) to the path and hence directed inward (never outward!) toward the center of the circular path. This causes the direction of the velocity to change without changing the speed.

We can find a simple expression for the magnitude of the acceleration in uniform circular motion. We begin with **Fig. 3.28a**, which shows a particle moving with constant speed in a circular path of radius R with center at O. The particle moves a distance Δs from P_1 to P_2 in a time interval Δt. Figure 3.28b shows the vector change in velocity $\Delta \vec{v}$ during this interval.

The angles labeled $\Delta \phi$ in Figs. 3.28a and 3.28b are the same because $\vec{v}_1$ is perpendicular to the line OP_1 and $\vec{v}_2$ is perpendicular to the line OP_2. Hence the triangles in Figs. 3.28a and 3.28b are *similar*. The ratios of corresponding sides of similar triangles are equal, so

$$\frac{|\Delta \vec{v}|}{v_1} = \frac{\Delta s}{R} \quad \text{or} \quad |\Delta \vec{v}| = \frac{v_1}{R} \Delta s$$

The magnitude a_{av} of the average acceleration during Δt is therefore

$$a_{av} = \frac{|\Delta \vec{v}|}{\Delta t} = \frac{v_1}{R} \frac{\Delta s}{\Delta t}$$

The magnitude a of the *instantaneous* acceleration $\vec{a}$ at point P_1 is the limit of this expression as we take point P_2 closer and closer to point P_1:

$$a = \lim_{\Delta t \to 0} \frac{v_1}{R} \frac{\Delta s}{\Delta t} = \frac{v_1}{R} \lim_{\Delta t \to 0} \frac{\Delta s}{\Delta t}$$

If the time interval Δt is short, Δs is the distance the particle moves along its curved path. So the limit of $\Delta s / \Delta t$ is the speed v_1 at point P_1. Also, P_1 can be any point on the path, so we can drop the subscript and let v represent the speed at any point. Then

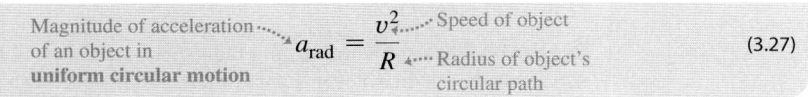

Magnitude of acceleration of an object in **uniform circular motion** $a_{rad} = \dfrac{v^2}{R}$ Speed of object · Radius of object's circular path   (3.27)

The subscript "rad" is a reminder that the direction of the instantaneous acceleration at each point is always along a radius of the circle (toward the center of the circle; see Figs. 3.27a and 3.28c). So *in uniform circular motion, the magnitude*

3.28 Finding the velocity change $\Delta \vec{v}$, average acceleration $\vec{a}_{av}$, and instantaneous acceleration $\vec{a}_{rad}$ for a particle moving in a circle with constant speed.

(a) A particle moves a distance Δs at constant speed along a circular path.

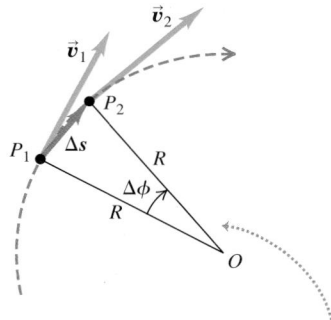

(b) The corresponding change in velocity and average acceleration

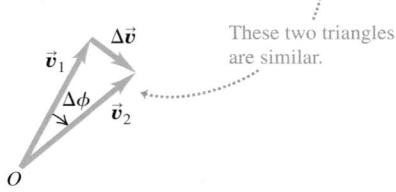

These two triangles are similar.

(c) The instantaneous acceleration

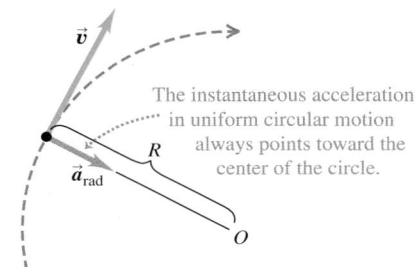

The instantaneous acceleration in uniform circular motion always points toward the center of the circle.

3.29 Acceleration and velocity **(a)** for a particle in uniform circular motion and **(b)** for a projectile with no air resistance.

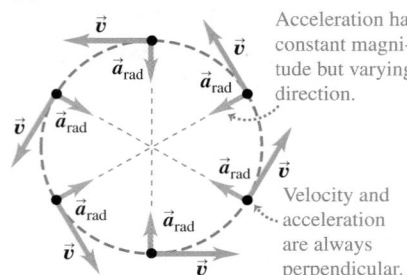

(a) Uniform circular motion

Acceleration has constant magnitude but varying direction.

Velocity and acceleration are always perpendicular.

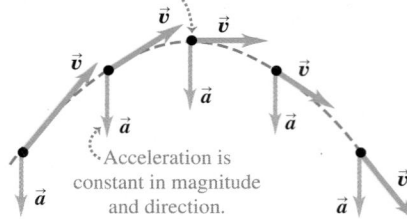

(b) Projectile motion

Velocity and acceleration are perpendicular only at the peak of the trajectory.

Acceleration is constant in magnitude and direction.

DATA *SPEAKS*

Acceleration on a Curved Path

When students were given a problem about an object following a curved path (not necessarily the parabolic path of a projectile), more than 46% gave an incorrect answer. Common errors:

- Confusion between the acceleration vector $\vec{a}$ and the velocity vector $\vec{v}$. Remember that $\vec{a}$ is the rate of change of $\vec{v}$, and on a curved path $\vec{a}$ and $\vec{v}$ cannot be in the same direction (see Fig. 3.12).

- Confusion about the direction of $\vec{a}$. If the path is curved, $\vec{a}$ always has a component toward the inside of the curve (see Fig. 3.12).

PhET: Ladybug Revolution
PhET: Motion in 2D

a_{rad} *of the instantaneous acceleration is equal to the square of the speed v divided by the radius R of the circle. Its direction is perpendicular to $\vec{v}$ and inward along the radius* (**Fig. 3.29a**).

Because the acceleration in uniform circular motion is always directed toward the center of the circle, it is sometimes called **centripetal acceleration.** The word "centripetal" is derived from two Greek words meaning "seeking the center."

CAUTION **Uniform circular motion vs. projectile motion** Notice the differences between acceleration in uniform circular motion (Fig. 3.29a) and acceleration in projectile motion (Fig. 3.29b). It's true that in both kinds of motion the *magnitude* of acceleration is the same at all times. However, in uniform circular motion the *direction* of $\vec{a}$ changes continuously—it always points toward the center of the circle. In projectile motion, the direction of $\vec{a}$ remains the same at all times. ▌

We can also express the magnitude of the acceleration in uniform circular motion in terms of the **period** T of the motion, the time for one revolution (one complete trip around the circle). In a time T the particle travels a distance equal to the circumference $2\pi R$ of the circle, so its speed is

$$v = \frac{2\pi R}{T} \tag{3.28}$$

When we substitute this into Eq. (3.27), we obtain the alternative expression

Magnitude of acceleration of an object in **uniform circular motion** $\cdots a_{\text{rad}} = \dfrac{4\pi^2 R}{T^2}$ $\cdots$ Radius of object's circular path $\cdots$ Period of motion $\tag{3.29}$

EXAMPLE 3.11 **CENTRIPETAL ACCELERATION ON A CURVED ROAD**

An Aston Martin V8 Vantage sports car has a "lateral acceleration" of $0.96g = (0.96)(9.8 \text{ m/s}^2) = 9.4 \text{ m/s}^2$. This is the maximum centripetal acceleration the car can sustain without skidding out of a curved path. If the car is traveling at a constant 40 m/s (about 89 mi/h, or 144 km/h) on level ground, what is the radius R of the tightest unbanked curve it can negotiate?

SOLUTION

IDENTIFY, SET UP, and EXECUTE: The car is in uniform circular motion because it's moving at a constant speed along a curve that is a segment of a circle. Hence we can use Eq. (3.27) to solve for the target variable R in terms of the given centripetal accelera-

tion a_{rad} and speed v:

$$R = \frac{v^2}{a_{\text{rad}}} = \frac{(40 \text{ m/s})^2}{9.4 \text{ m/s}^2} = 170 \text{ m (about 560 ft)}$$

This is the *minimum* turning radius because a_{rad} is the *maximum* centripetal acceleration.

EVALUATE: The minimum turning radius R is proportional to the *square* of the speed, so even a small reduction in speed can make R substantially smaller. For example, reducing v by 20% (from 40 m/s to 32 m/s) would decrease R by 36% (from 170 m to 109 m).

Another way to make the minimum turning radius smaller is to *bank* the curve. We'll investigate this option in Chapter 5.

EXAMPLE 3.12   **CENTRIPETAL ACCELERATION ON A CARNIVAL RIDE**

Passengers on a carnival ride move at constant speed in a horizontal circle of radius 5.0 m, making a complete circle in 4.0 s. What is their acceleration?

SOLUTION

IDENTIFY and SET UP: The speed is constant, so this is uniform circular motion. We are given the radius $R = 5.0$ m and the period $T = 4.0$ s, so we can use Eq. (3.29) to calculate the acceleration directly, or we can calculate the speed v by using Eq. (3.28) and then find the acceleration by using Eq. (3.27).

EXECUTE: From Eq. (3.29),

$$a_{rad} = \frac{4\pi^2(5.0 \text{ m})}{(4.0 \text{ s})^2} = 12 \text{ m/s}^2 = 1.3g$$

EVALUATE: We can check this answer by using the second, round-about approach. From Eq. (3.28), the speed is

$$v = \frac{2\pi R}{T} = \frac{2\pi(5.0 \text{ m})}{4.0 \text{ s}} = 7.9 \text{ m/s}$$

The centripetal acceleration is then

$$a_{rad} = \frac{v^2}{R} = \frac{(7.9 \text{ m/s})^2}{5.0 \text{ m}} = 12 \text{ m/s}^2$$

As in Fig. 3.29a, the direction of $\vec{a}$ is always toward the center of the circle. The magnitude of $\vec{a}$ is relatively mild as carnival rides go; some roller coasters subject their passengers to accelerations as great as $4g$.

Nonuniform Circular Motion

We have assumed throughout this section that the particle's speed is constant as it goes around the circle. If the speed varies, we call the motion **nonuniform circular motion.** In nonuniform circular motion, Eq. (3.27) still gives the *radial* component of acceleration $a_{rad} = v^2/R$, which is always *perpendicular* to the instantaneous velocity and directed toward the center of the circle. But since the speed v has different values at different points in the motion, the value of a_{rad} is not constant. The radial (centripetal) acceleration is greatest at the point in the circle where the speed is greatest.

In nonuniform circular motion there is also a component of acceleration that is *parallel* to the instantaneous velocity (see Figs. 3.27b and 3.27c). This is the component $a_{\parallel}$ that we discussed in Section 3.2; here we call this component a_{tan} to emphasize that it is *tangent* to the circle. The tangential component of acceleration a_{tan} is equal to the rate of change of *speed*. Thus

$$a_{rad} = \frac{v^2}{R} \quad \text{and} \quad a_{tan} = \frac{d|\vec{v}|}{dt} \quad \text{(nonuniform circular motion)} \quad (3.30)$$

The tangential component is in the same direction as the velocity if the particle is speeding up, and in the opposite direction if the particle is slowing down (**Fig. 3.30**). If the particle's speed is constant, $a_{tan} = 0$.

CAUTION   **Uniform vs. nonuniform circular motion**   The two quantities

$$\frac{d|\vec{v}|}{dt} \quad \text{and} \quad \left|\frac{d\vec{v}}{dt}\right|$$

are *not* the same. The first, equal to the tangential acceleration, is the rate of change of speed; it is zero whenever a particle moves with constant speed, even when its direction of motion changes (such as in *uniform* circular motion). The second is the magnitude of the vector acceleration; it is zero only when the particle's acceleration *vector* is zero—that is, when the particle moves in a straight line with constant speed. In *uniform* circular motion $|d\vec{v}/dt| = a_{rad} = v^2/r$; in *nonuniform* circular motion there is also a tangential component of acceleration, so $|d\vec{v}/dt| = \sqrt{a_{rad}^2 + a_{tan}^2}$.

TEST YOUR UNDERSTANDING OF SECTION 3.4   Suppose that the particle in Fig. 3.30 experiences four times the acceleration at the bottom of the loop as it does at the top of the loop. Compared to its speed at the top of the loop, is its speed at the bottom of the loop (i) $\sqrt{2}$ times as great; (ii) 2 times as great; (iii) $2\sqrt{2}$ times as great; (iv) 4 times as great; or (v) 16 times as great? ∎

Application Watch Out: Tight Curves Ahead!   These roller coaster cars are in nonuniform circular motion: They slow down and speed up as they move around a vertical loop. The large accelerations involved in traveling at high speed around a tight loop mean extra stress on the passengers' circulatory systems, which is why people with cardiac conditions are cautioned against going on such rides.

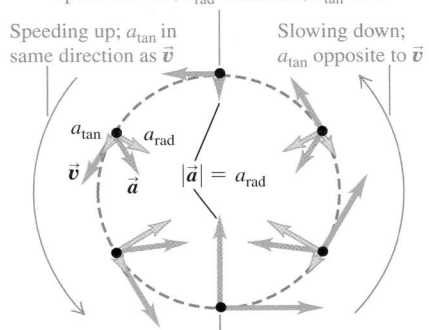

3.30 A particle moving in a vertical loop with a varying speed, like a roller coaster car.

Speed slowest, a_{rad} minimum, a_{tan} zero

Speeding up; a_{tan} in same direction as $\vec{v}$

Slowing down; a_{tan} opposite to $\vec{v}$

a_{tan} a_{rad}

$\vec{v}$ $\vec{a}$ $|\vec{a}| = a_{rad}$

Speed fastest, a_{rad} maximum, a_{tan} zero

3.31 Airshow pilots face a complicated problem involving relative velocities. They must keep track of their motion relative to the air (to maintain enough airflow over the wings to sustain lift), relative to each other (to keep a tight formation without colliding), and relative to their audience (to remain in sight of the spectators).

3.32 (a) A passenger walking in a train. (b) The position of the passenger relative to the cyclist's frame of reference and the train's frame of reference.

(a)

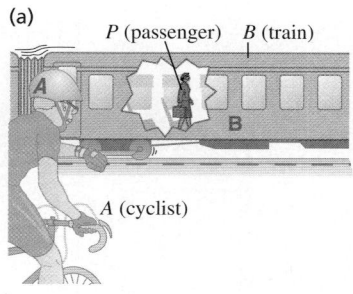

(b)

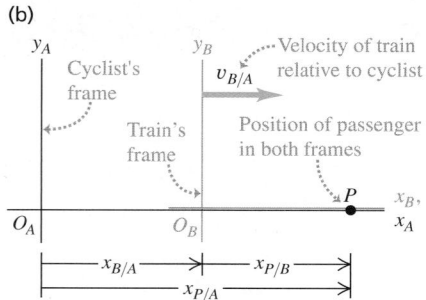

3.5 RELATIVE VELOCITY

If you stand next to a one-way highway, all the cars appear to be moving forward. But if you're driving in the fast lane on that highway, slower cars appear to be moving backward. In general, when two observers measure the velocity of the same body, they get different results if one observer is moving relative to the other. The velocity seen by a particular observer is called the velocity *relative* to that observer, or simply **relative velocity.** In many situations relative velocity is extremely important (**Fig. 3.31**).

We'll first consider relative velocity along a straight line and then generalize to relative velocity in a plane.

Relative Velocity in One Dimension

A passenger walks with a velocity of 1.0 m/s along the aisle of a train that is moving with a velocity of 3.0 m/s (**Fig. 3.32a**). What is the passenger's velocity? It's a simple enough question, but it has no single answer. As seen by a second passenger sitting in the train, she is moving at 1.0 m/s. A person on a bicycle standing beside the train sees the walking passenger moving at 1.0 m/s + 3.0 m/s = 4.0 m/s. An observer in another train going in the opposite direction would give still another answer. We have to specify which observer we mean, and we speak of the velocity *relative* to a particular observer. The walking passenger's velocity relative to the train is 1.0 m/s, her velocity relative to the cyclist is 4.0 m/s, and so on. Each observer, equipped in principle with a meter stick and a stopwatch, forms what we call a **frame of reference.** Thus a frame of reference is a coordinate system plus a time scale.

Let's use the symbol A for the cyclist's frame of reference (at rest with respect to the ground) and the symbol B for the frame of reference of the moving train. In straight-line motion the position of a point P relative to frame A is given by $x_{P/A}$ (the position of P with respect to A), and the position of P relative to frame B is given by $x_{P/B}$ (Fig. 3.32b). The position of the origin of B with respect to the origin of A is $x_{B/A}$. Figure 3.32b shows that

$$x_{P/A} = x_{P/B} + x_{B/A} \qquad (3.31)$$

In words, the coordinate of P relative to A equals the coordinate of P relative to B plus the coordinate of B relative to A.

The x-velocity of P relative to frame A, denoted by $v_{P/A\text{-}x}$, is the derivative of $x_{P/A}$ with respect to time. We can find the other velocities in the same way. So the time derivative of Eq. (3.31) gives us a relationship among the various velocities:

$$\frac{dx_{P/A}}{dt} = \frac{dx_{P/B}}{dt} + \frac{dx_{B/A}}{dt} \qquad \text{or}$$

Relative velocity along a line:
$$v_{P/A\text{-}x} = v_{P/B\text{-}x} + v_{B/A\text{-}x} \qquad (3.32)$$

x-velocity of P relative to A x-velocity of P relative to B x-velocity of B relative to A

Getting back to the passenger on the train in Fig. 3.32a, we see that A is the cyclist's frame of reference, B is the frame of reference of the train, and point P represents the passenger. Using the above notation, we have

$$v_{P/B\text{-}x} = +1.0 \text{ m/s} \qquad v_{B/A\text{-}x} = +3.0 \text{ m/s}$$

From Eq. (3.32) the passenger's velocity $v_{P/A\text{-}x}$ relative to the cyclist is

$$v_{P/A\text{-}x} = +1.0 \text{ m/s} + 3.0 \text{ m/s} = +4.0 \text{ m/s}$$

as we already knew.

In this example, both velocities are toward the right, and we have taken this as the positive x-direction. If the passenger walks toward the *left* relative to the train, then $v_{P/B\text{-}x} = -1.0$ m/s, and her x-velocity relative to the cyclist is $v_{P/A\text{-}x} = -1.0$ m/s $+ 3.0$ m/s $= +2.0$ m/s. The sum in Eq. (3.32) is always an algebraic sum, and any or all of the x-velocities may be negative.

When the passenger looks out the window, the stationary cyclist on the ground appears to her to be moving backward; we call the cyclist's velocity relative to her $v_{A/P\text{-}x}$. This is just the negative of the *passenger's* velocity relative to the cyclist, $v_{P/A\text{-}x}$. In general, if A and B are any two points or frames of reference,

$$v_{A/B\text{-}x} = -v_{B/A\text{-}x} \qquad (3.33)$$

PROBLEM-SOLVING STRATEGY 3.2 RELATIVE VELOCITY

IDENTIFY *the relevant concepts:* Whenever you see the phrase "velocity relative to" or "velocity with respect to," it's likely that the concepts of relative velocity will be helpful.

SET UP *the problem:* Sketch and label each frame of reference in the problem. Each moving body has its own frame of reference; in addition, you'll almost always have to include the frame of reference of the earth's surface. (Statements such as "The car is traveling north at 90 km/h" implicitly refer to the car's velocity relative to the surface of the earth.) Use the labels to help identify the target variable. For example, if you want to find the x-velocity of a car (C) with respect to a bus (B), your target variable is $v_{C/B\text{-}x}$.

EXECUTE *the solution:* Solve for the target variable using Eq. (3.32). (If the velocities aren't along the same direction, you'll need to use

the vector form of this equation, derived later in this section.) It's important to note the order of the double subscripts in Eq. (3.32): $v_{B/A\text{-}x}$ means "x-velocity of B relative to A." These subscripts obey a kind of algebra. If we regard each one as a fraction, then the fraction on the left side is the *product* of the fractions on the right side: $P/A = (P/B)(B/A)$. You can apply this rule to any number of frames of reference. For example, if there are three frames of reference A, B, and C, Eq. (3.32) becomes

$$v_{P/A\text{-}x} = v_{P/C\text{-}x} + v_{C/B\text{-}x} + v_{B/A\text{-}x}$$

EVALUATE *your answer:* Be on the lookout for stray minus signs in your answer. If the target variable is the x-velocity of a car relative to a bus ($v_{C/B\text{-}x}$), make sure that you haven't accidentally calculated the x-velocity of the *bus* relative to the *car* ($v_{B/C\text{-}x}$). If you've made this mistake, you can recover by using Eq. (3.33).

EXAMPLE 3.13 RELATIVE VELOCITY ON A STRAIGHT ROAD

You drive north on a straight two-lane road at a constant 88 km/h. A truck in the other lane approaches you at a constant 104 km/h (**Fig. 3.33**). Find (a) the truck's velocity relative to you and (b) your velocity relative to the truck. (c) How do the relative velocities change after you and the truck pass each other? Treat this as a one-dimensional problem.

3.33 Reference frames for you and the truck.

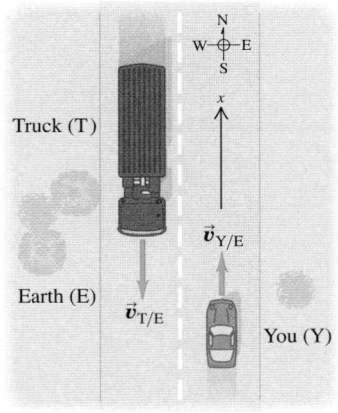

SOLUTION

IDENTIFY and SET UP: In this problem about relative velocities along a line, there are three reference frames: you (Y), the truck (T), and the earth's surface (E). Let the positive x-direction be north (Fig. 3.33). Then your x-velocity relative to the earth is $v_{Y/E\text{-}x} = +88$ km/h. The truck is initially approaching you, so it is moving south and its x-velocity with respect to the earth is $v_{T/E\text{-}x} = -104$ km/h. The target variables in parts (a) and (b) are $v_{T/Y\text{-}x}$ and $v_{Y/T\text{-}x}$, respectively. We'll use Eq. (3.32) to find the first target variable and Eq. (3.33) to find the second.

EXECUTE: (a) To find $v_{T/Y\text{-}x}$, we write Eq. (3.32) for the known $v_{T/E\text{-}x}$ and rearrange:

$$v_{T/E\text{-}x} = v_{T/Y\text{-}x} + v_{Y/E\text{-}x}$$
$$v_{T/Y\text{-}x} = v_{T/E\text{-}x} - v_{Y/E\text{-}x}$$
$$= -104 \text{ km/h} - 88 \text{ km/h} = -192 \text{ km/h}$$

The truck is moving at 192 km/h in the negative x-direction (south) relative to you.

Continued

(b) From Eq. (3.33),

$$v_{Y/T\text{-}x} = -v_{T/Y\text{-}x} = -(-192 \text{ km/h}) = +192 \text{ km/h}$$

You are moving at 192 km/h in the positive x-direction (north) relative to the truck.

(c) The relative velocities do *not* change after you and the truck pass each other. The relative *positions* of the bodies don't matter. After it passes you the truck is still moving at 192 km/h toward the south relative to you, even though it is now moving away from you instead of toward you.

EVALUATE: To check your answer in part (b), use Eq. (3.32) directly in the form $v_{Y/T\text{-}x} = v_{Y/E\text{-}x} + v_{E/T\text{-}x}$. (The x-velocity of the earth with respect to the truck is the opposite of the x-velocity of the truck with respect to the earth: $v_{E/T\text{-}x} = -v_{T/E\text{-}x}$.) Do you get the same result?

Relative Velocity in Two or Three Dimensions

Let's extend the concept of relative velocity to include motion in a plane or in space. Suppose that the passenger in Fig. 3.32a is walking not down the aisle of the railroad car but from one side of the car to the other, with a speed of 1.0 m/s (**Fig. 3.34a**). We can again describe the passenger's position P in two frames of reference: A for the stationary ground observer and B for the moving train. But instead of coordinates x, we use position vectors $\vec{r}$ because the problem is now two-dimensional. Then, as Fig. 3.34b shows,

$$\vec{r}_{P/A} = \vec{r}_{P/B} + \vec{r}_{B/A} \tag{3.34}$$

Just as we did before, we take the time derivative of this equation to get a relationship among the various velocities; the velocity of P relative to A is $\vec{v}_{P/A} = d\vec{r}_{P/A}/dt$ and so on for the other velocities. We get

Relative velocity in space:	$\vec{v}_{P/A} = \vec{v}_{P/B} + \vec{v}_{B/A}$	(3.35)
	Velocity of P relative to A Velocity of P relative to B Velocity of B relative to A	

Equation (3.35) is known as the *Galilean velocity transformation*. It relates the velocity of a body P with respect to frame A and its velocity with respect to frame B ($\vec{v}_{P/A}$ and $\vec{v}_{P/B}$, respectively) to the velocity of frame B with respect to frame A ($\vec{v}_{B/A}$). If all three of these velocities lie along the same line, then Eq. (3.35) reduces to Eq. (3.32) for the components of the velocities along that line.

If the train is moving at $v_{B/A} = 3.0$ m/s relative to the ground and the passenger is moving at $v_{P/B} = 1.0$ m/s relative to the train, then the passenger's velocity

3.34 (a) A passenger walking across a railroad car. (b) Position of the passenger relative to the cyclist's frame and the train's frame. (c) Vector diagram for the velocity of the passenger relative to the ground (the cyclist's frame), $\vec{v}_{P/A}$.

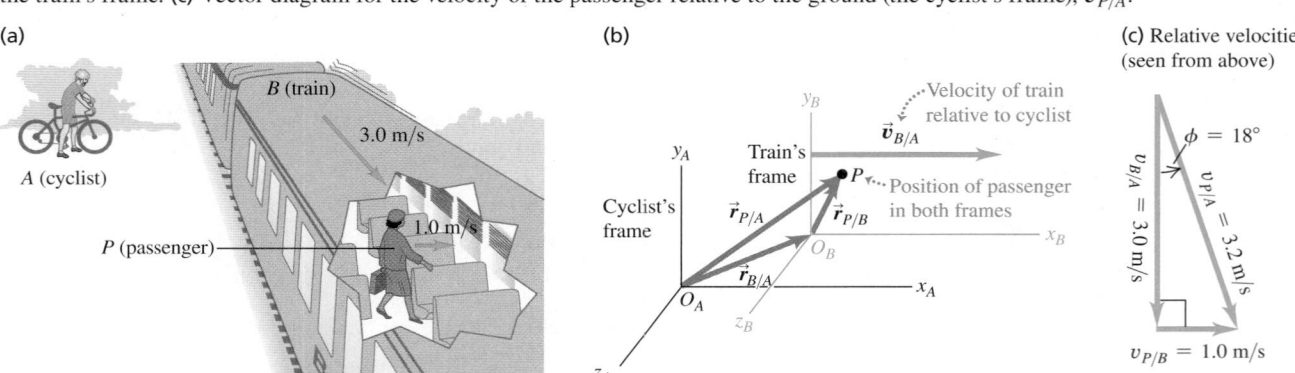

vector $\vec{v}_{P/A}$ relative to the ground is as shown in Fig. 3.34c. The Pythagorean theorem then gives us

$$v_{P/A} = \sqrt{(3.0 \text{ m/s})^2 + (1.0 \text{ m/s})^2} = \sqrt{10 \text{ m}^2/\text{s}^2} = 3.2 \text{ m/s}$$

Figure 3.34c also shows that the *direction* of the passenger's velocity vector relative to the ground makes an angle ϕ with the train's velocity vector $\vec{v}_{B/A}$, where

$$\tan \phi = \frac{v_{P/B}}{v_{B/A}} = \frac{1.0 \text{ m/s}}{3.0 \text{ m/s}} \quad \text{and} \quad \phi = 18°$$

As in the case of motion along a straight line, we have the general rule that if A and B are *any* two points or frames of reference,

$$\vec{v}_{A/B} = -\vec{v}_{B/A} \qquad (3.36)$$

The velocity of the passenger relative to the train is the negative of the velocity of the train relative to the passenger, and so on.

In the early 20th century Albert Einstein showed that Eq. (3.35) has to be modified when speeds approach the speed of light, denoted by c. It turns out that if the passenger in Fig. 3.32a could walk down the aisle at $0.30c$ and the train could move at $0.90c$, then her speed relative to the ground would be not $1.20c$ but $0.94c$; nothing can travel faster than light! We'll return to Einstein and his *special theory of relativity* in Chapter 37.

EXAMPLE 3.14   FLYING IN A CROSSWIND

An airplane's compass indicates that it is headed due north, and its airspeed indicator shows that it is moving through the air at 240 km/h. If there is a 100-km/h wind from west to east, what is the velocity of the airplane relative to the earth?

SOLUTION

IDENTIFY and SET UP: This problem involves velocities in two dimensions (northward and eastward), so it is a relative velocity problem using vectors. We are given the magnitude and direction of the velocity of the plane (P) relative to the air (A). We are also given the magnitude and direction of the wind velocity, which is the velocity of the air A with respect to the earth (E):

$$\vec{v}_{P/A} = 240 \text{ km/h} \quad \text{due north}$$
$$\vec{v}_{A/E} = 100 \text{ km/h} \quad \text{due east}$$

We'll use Eq. (3.35) to find our target variables: the magnitude and direction of velocity $\vec{v}_{P/E}$ of the plane relative to the earth.

EXECUTE: From Eq. (3.35) we have

$$\vec{v}_{P/E} = \vec{v}_{P/A} + \vec{v}_{A/E}$$

Figure 3.35 shows that the three relative velocities constitute a right-triangle vector addition; the unknowns are the speed $v_{P/E}$ and the angle α. We find

$$v_{P/E} = \sqrt{(240 \text{ km/h})^2 + (100 \text{ km/h})^2} = 260 \text{ km/h}$$

$$\alpha = \arctan\left(\frac{100 \text{ km/h}}{240 \text{ km/h}}\right) = 23° \text{ E of N}$$

3.35 The plane is pointed north, but the wind blows east, giving the resultant velocity $\vec{v}_{P/E}$ relative to the earth.

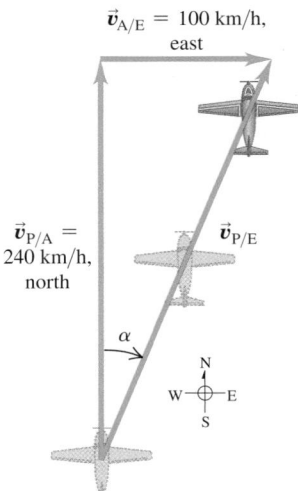

EVALUATE: You can check the results by taking measurements on the scale drawing in Fig. 3.35. The crosswind increases the speed of the airplane relative to the earth, but pushes the airplane off course.

EXAMPLE 3.15 CORRECTING FOR A CROSSWIND

With wind and airspeed as in Example 3.14, in what direction should the pilot head to travel due north? What will be her velocity relative to the earth?

SOLUTION

IDENTIFY and SET UP: Like Example 3.14, this is a relative velocity problem with vectors. **Figure 3.36** is a scale drawing of the situation. Again the vectors add in accordance with Eq. (3.35) and form a right triangle:

$$\vec{v}_{P/E} = \vec{v}_{P/A} + \vec{v}_{A/E}$$

As Fig. 3.36 shows, the pilot points the nose of the airplane at an angle β into the wind to compensate for the crosswind. This angle, which tells us the direction of the vector $\vec{v}_{P/A}$ (the velocity of the airplane relative to the air), is one of our target variables. The other target variable is the speed of the airplane over the ground, which is the magnitude of the vector $\vec{v}_{P/E}$ (the velocity of the airplane relative to the earth). The known and unknown quantities are

$\vec{v}_{P/E}$ = magnitude unknown   due north
$\vec{v}_{P/A}$ = 240 km/h   direction unknown
$\vec{v}_{A/E}$ = 100 km/h   due east

We'll solve for the target variables by using Fig. 3.36 and trigonometry.

EXECUTE: From Fig. 3.36 the speed $v_{P/E}$ and the angle β are

$$v_{P/E} = \sqrt{(240 \text{ km/h})^2 - (100 \text{ km/h})^2} = 218 \text{ km/h}$$

$$\beta = \arcsin\left(\frac{100 \text{ km/h}}{240 \text{ km/h}}\right) = 25°$$

3.36 The pilot must point the plane in the direction of the vector $\vec{v}_{P/A}$ to travel due north relative to the earth.

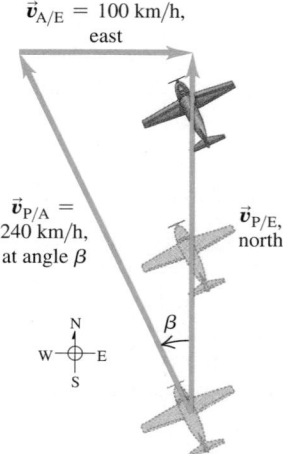

The pilot should point the airplane 25° west of north, and her ground speed is then 218 km/h.

EVALUATE: There were two target variables—the magnitude of a vector and the direction of a vector—in both this example and Example 3.14. In Example 3.14 the magnitude and direction referred to the *same* vector ($\vec{v}_{P/E}$); here they refer to *different* vectors ($\vec{v}_{P/E}$ and $\vec{v}_{P/A}$).

While we expect a *headwind* to reduce an airplane's speed relative to the ground, this example shows that a *crosswind* does, too. That's an unfortunate fact of aeronautical life.

TEST YOUR UNDERSTANDING OF SECTION 3.5   Suppose the nose of an airplane is pointed due east and the airplane has an airspeed of 150 km/h. Due to the wind, the airplane is moving due *north* relative to the ground and its speed relative to the ground is 150 km/h. What is the velocity of the air relative to the earth? (i) 150 km/h from east to west; (ii) 150 km/h from south to north; (iii) 150 km/h from southeast to northwest; (iv) 212 km/h from east to west; (v) 212 km/h from south to north; (vi) 212 km/h from southeast to northwest; (vii) there is no possible wind velocity that could cause this. ▮

Position, velocity, and acceleration vectors: The position vector $\vec{r}$ of a point P in space is the vector from the origin to P. Its components are the coordinates x, y, and z.

The average velocity vector $\vec{v}_{av}$ during the time interval Δt is the displacement $\Delta \vec{r}$ (the change in position vector $\vec{r}$) divided by Δt. The instantaneous velocity vector $\vec{v}$ is the time derivative of $\vec{r}$, and its components are the time derivatives of x, y, and z. The instantaneous speed is the magnitude of $\vec{v}$. The velocity $\vec{v}$ of a particle is always tangent to the particle's path. (See Example 3.1.)

The average acceleration vector $\vec{a}_{av}$ during the time interval Δt equals $\Delta \vec{v}$ (the change in velocity vector $\vec{v}$) divided by Δt. The instantaneous acceleration vector $\vec{a}$ is the time derivative of $\vec{v}$, and its components are the time derivatives of v_x, v_y, and v_z. (See Example 3.2.)

The component of acceleration parallel to the direction of the instantaneous velocity affects the speed, while the component of $\vec{a}$ perpendicular to $\vec{v}$ affects the direction of motion. (See Examples 3.3 and 3.4.)

$$\vec{r} = x\hat{\imath} + y\hat{\jmath} + z\hat{k} \qquad (3.1)$$

$$\vec{v}_{av} = \frac{\vec{r}_2 - \vec{r}_1}{t_2 - t_1} = \frac{\Delta \vec{r}}{\Delta t} \qquad (3.2)$$

$$\vec{v} = \lim_{\Delta t \to 0} \frac{\Delta \vec{r}}{\Delta t} = \frac{d\vec{r}}{dt} \qquad (3.3)$$

$$v_x = \frac{dx}{dt} \quad v_y = \frac{dy}{dt} \quad v_z = \frac{dz}{dt} \qquad (3.4)$$

$$\vec{a}_{av} = \frac{\vec{v}_2 - \vec{v}_1}{t_2 - t_1} = \frac{\Delta \vec{v}}{\Delta t} \qquad (3.8)$$

$$\vec{a} = \lim_{\Delta t \to 0} \frac{\Delta \vec{v}}{\Delta t} = \frac{d\vec{v}}{dt} \qquad (3.9)$$

$$a_x = \frac{dv_x}{dt}$$

$$a_y = \frac{dv_y}{dt} \qquad (3.10)$$

$$a_z = \frac{dv_z}{dt}$$

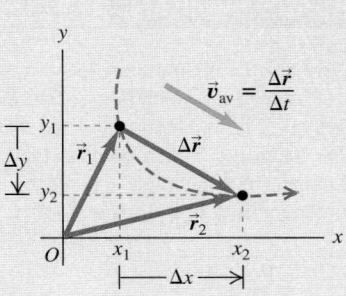

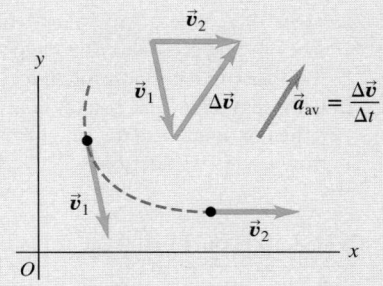

Projectile motion: In projectile motion with no air resistance, $a_x = 0$ and $a_y = -g$. The coordinates and velocity components are simple functions of time, and the shape of the path is always a parabola. We usually choose the origin to be at the initial position of the projectile. (See Examples 3.5–3.10.)

$$x = (v_0 \cos \alpha_0)t \qquad (3.19)$$

$$y = (v_0 \sin \alpha_0)t - \tfrac{1}{2}gt^2 \qquad (3.20)$$

$$v_x = v_0 \cos \alpha_0 \qquad (3.21)$$

$$v_y = v_0 \sin \alpha_0 - gt \qquad (3.22)$$

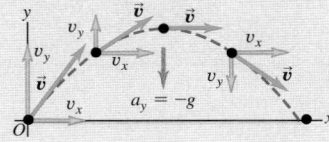

Uniform and nonuniform circular motion: When a particle moves in a circular path of radius R with constant speed v (uniform circular motion), its acceleration $\vec{a}$ is directed toward the center of the circle and perpendicular to $\vec{v}$. The magnitude a_{rad} of the acceleration can be expressed in terms of v and R or in terms of R and the period T (the time for one revolution), where $v = 2\pi R/T$. (See Examples 3.11 and 3.12.)

If the speed is not constant in circular motion (nonuniform circular motion), there is still a radial component of $\vec{a}$ given by Eq. (3.27) or (3.29), but there is also a component of $\vec{a}$ parallel (tangential) to the path. This tangential component is equal to the rate of change of speed, dv/dt.

$$a_{rad} = \frac{v^2}{R} \qquad (3.27)$$

$$a_{rad} = \frac{4\pi^2 R}{T^2} \qquad (3.29)$$

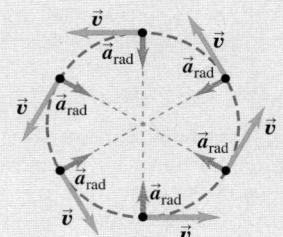

Relative velocity: When a body P moves relative to a body (or reference frame) B, and B moves relative to a body (or reference frame) A, we denote the velocity of P relative to B by $\vec{v}_{P/B}$, the velocity of P relative to A by $\vec{v}_{P/A}$, and the velocity of B relative to A by $\vec{v}_{B/A}$. If these velocities are all along the same line, their components along that line are related by Eq. (3.32). More generally, these velocities are related by Eq. (3.35). (See Examples 3.13–3.15.)

$$v_{P/A\text{-}x} = v_{P/B\text{-}x} + v_{B/A\text{-}x}$$
(relative velocity along a line) $\qquad (3.32)$

$$\vec{v}_{P/A} = \vec{v}_{P/B} + \vec{v}_{B/A}$$
(relative velocity in space) $\qquad (3.35)$

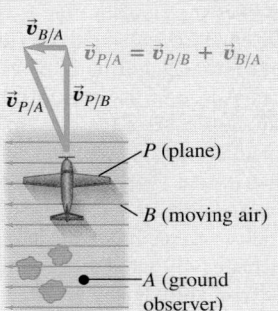

91

BRIDGING PROBLEM LAUNCHING UP AN INCLINE

You fire a ball with an initial speed v_0 at an angle ϕ above the surface of an incline, which is itself inclined at an angle θ above the horizontal (**Fig. 3.37**). (a) Find the distance, measured along the incline, from the launch point to the point when the ball strikes the incline. (b) What angle ϕ gives the maximum range, measured along the incline? Ignore air resistance.

SOLUTION GUIDE

IDENTIFY and SET UP

1. Since there's no air resistance, this is a problem in projectile motion. The goal is to find the point where the ball's parabolic trajectory intersects the incline.
2. Choose the x- and y-axes and the position of the origin. When in doubt, use the suggestions given in Problem-Solving Strategy 3.1 in Section 3.3.

3.37 Launching a ball from an inclined ramp.

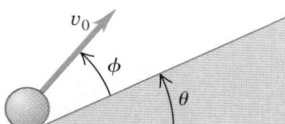

3. In the projectile equations in Section 3.3, the launch angle α_0 is measured from the horizontal. What is this angle in terms of θ and ϕ? What are the initial x- and y-components of the ball's initial velocity?
4. You'll need to write an equation that relates x and y for points along the incline. What is this equation? (This takes just geometry and trigonometry, not physics.)

EXECUTE

5. Write the equations for the x-coordinate and y-coordinate of the ball as functions of time t.
6. When the ball hits the incline, x and y are related by the equation that you found in step 4. Based on this, at what time t does the ball hit the incline?
7. Based on your answer from step 6, at what coordinates x and y does the ball land on the incline? How far is this point from the launch point?
8. What value of ϕ gives the *maximum* distance from the launch point to the landing point? (Use your knowledge of calculus.)

EVALUATE

9. Check your answers for the case $\theta = 0$, which corresponds to the incline being horizontal rather than tilted. (You already know the answers for this case. Do you know why?)

Problems

For assigned homework and other learning materials, go to MasteringPhysics®.

•, ••, •••: Difficulty levels. **CP**: Cumulative problems incorporating material from earlier chapters. **CALC**: Problems requiring calculus.
DATA: Problems involving real data, scientific evidence, experimental design, and/or statistical reasoning. **BIO**: Biosciences problems.

DISCUSSION QUESTIONS

Q3.1 A simple pendulum (a mass swinging at the end of a string) swings back and forth in a circular arc. What is the direction of the acceleration of the mass when it is at the ends of the swing? At the midpoint? In each case, explain how you obtained your answer.

Q3.2 Redraw Fig. 3.11a if $\vec{a}$ is antiparallel to $\vec{v}_1$. Does the particle move in a straight line? What happens to its speed?

Q3.3 A projectile moves in a parabolic path without air resistance. Is there any point at which $\vec{a}$ is parallel to $\vec{v}$? Perpendicular to $\vec{v}$? Explain.

Q3.4 A book slides off a horizontal tabletop. As it leaves the table's edge, the book has a horizontal velocity of magnitude v_0. The book strikes the floor in time t. If the initial velocity of the book is doubled to $2v_0$, what happens to (a) the time the book is in the air, (b) the horizontal distance the book travels while it is in the air, and (c) the speed of the book just before it reaches the floor? In particular, does each of these quantities stay the same, double, or change in another way? Explain.

Q3.5 At the instant that you fire a bullet horizontally from a rifle, you drop a bullet from the height of the gun barrel. If there is no air resistance, which bullet hits the level ground first? Explain.

Q3.6 A package falls out of an airplane that is flying in a straight line at a constant altitude and speed. If you ignore air resistance, what would be the path of the package as observed by the pilot? As observed by a person on the ground?

Q3.7 Sketch the six graphs of the x- and y-components of position, velocity, and acceleration versus time for projectile motion with $x_0 = y_0 = 0$ and $0 < \alpha_0 < 90°$.

Q3.8 If a jumping frog can give itself the same initial speed regardless of the direction in which it jumps (forward or straight up), how is the maximum vertical height to which it can jump related to its maximum horizontal range $R_{max} = v_0^2/g$?

Q3.9 A projectile is fired upward at an angle θ above the horizontal with an initial speed v_0. At its maximum height, what are its velocity vector, its speed, and its acceleration vector?

Q3.10 In uniform circular motion, what are the *average* velocity and *average* acceleration for one revolution? Explain.

Q3.11 In uniform circular motion, how does the acceleration change when the speed is increased by a factor of 3? When the radius is decreased by a factor of 2?

Q3.12 In uniform circular motion, the acceleration is perpendicular to the velocity at every instant. Is this true when the motion is not uniform—that is, when the speed is not constant?

Q3.13 Raindrops hitting the side windows of a car in motion often leave diagonal streaks even if there is no wind. Why? Is the explanation the same or different for diagonal streaks on the windshield?

Q3.14 In a rainstorm with a strong wind, what determines the best position in which to hold an umbrella?

Q3.15 You are on the west bank of a river that is flowing north with a speed of 1.2 m/s. Your swimming speed relative to the water is 1.5 m/s, and the river is 60 m wide. What is your path relative to the earth that allows you to cross the river in the shortest time? Explain your reasoning.

Q3.16 A stone is thrown into the air at an angle above the horizontal and feels negligible air resistance. Which graph in **Fig. Q3.16** best depicts the stone's *speed* v as a function of time t while it is in the air?

Figure **Q3.16**

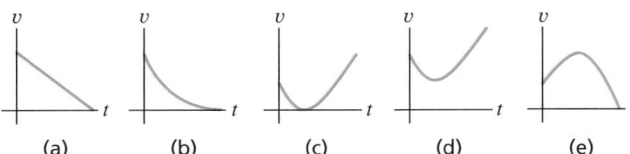

(a) (b) (c) (d) (e)

EXERCISES

Section 3.1 Position and Velocity Vectors

3.1 • A squirrel has x- and y-coordinates (1.1 m, 3.4 m) at time $t_1 = 0$ and coordinates (5.3 m, −0.5 m) at time $t_2 = 3.0$ s. For this time interval, find (a) the components of the average velocity, and (b) the magnitude and direction of the average velocity.

3.2 • A rhinoceros is at the origin of coordinates at time $t_1 = 0$. For the time interval from $t_1 = 0$ to $t_2 = 12.0$ s, the rhino's average velocity has x-component −3.8 m/s and y-component 4.9 m/s. At time $t_2 = 12.0$ s, (a) what are the x- and y-coordinates of the rhino? (b) How far is the rhino from the origin?

3.3 •• CALC A web page designer creates an animation in which a dot on a computer screen has position

$$\vec{r} = [4.0 \text{ cm} + (2.5 \text{ cm/s}^2)t^2]\hat{i} + (5.0 \text{ cm/s})t\hat{j}.$$

(a) Find the magnitude and direction of the dot's average velocity between $t = 0$ and $t = 2.0$ s.(b) Find the magnitude and direction of the instantaneous velocity at $t = 0$, $t = 1.0$ s, and $t = 2.0$ s. (c) Sketch the dot's trajectory from $t = 0$ to $t = 2.0$ s, and show the velocities calculated in part (b).

3.4 • CALC The position of a squirrel running in a park is given by $\vec{r} = [(0.280 \text{ m/s})t + (0.0360 \text{ m/s}^2)t^2]\hat{i} + (0.0190 \text{ m/s}^3)t^3\hat{j}$. (a) What are $v_x(t)$ and $v_y(t)$, the x- and y-components of the velocity of the squirrel, as functions of time? (b) At $t = 5.00$ s, how far is the squirrel from its initial position? (c) At $t = 5.00$ s, what are the magnitude and direction of the squirrel's velocity?

Section 3.2 The Acceleration Vector

3.5 • A jet plane is flying at a constant altitude. At time $t_1 = 0$, it has components of velocity $v_x = 90$ m/s, $v_y = 110$ m/s. At time $t_2 = 30.0$ s, the components are $v_x = -170$ m/s, $v_y = 40$ m/s. (a) Sketch the velocity vectors at t_1 and t_2. How do these two vectors differ? For this time interval calculate (b) the components of the average acceleration, and (c) the magnitude and direction of the average acceleration.

3.6 •• A dog running in an open field has components of velocity $v_x = 2.6$ m/s and $v_y = -1.8$ m/s at $t_1 = 10.0$ s. For the time interval from $t_1 = 10.0$ s to $t_2 = 20.0$ s, the average acceleration of the dog has magnitude 0.45 m/s^2 and direction 31.0° measured from the +x-axis toward the +y-axis. At $t_2 = 20.0$ s, (a) what are

the x- and y-components of the dog's velocity? (b) What are the magnitude and direction of the dog's velocity? (c) Sketch the velocity vectors at t_1 and t_2. How do these two vectors differ?

3.7 •• CALC The coordinates of a bird flying in the xy-plane are given by $x(t) = \alpha t$ and $y(t) = 3.0 \text{ m} - \beta t^2$, where $\alpha = 2.4$ m/s and $\beta = 1.2$ m/s^2. (a) Sketch the path of the bird between $t = 0$ and $t = 2.0$ s. (b) Calculate the velocity and acceleration vectors of the bird as functions of time. (c) Calculate the magnitude and direction of the bird's velocity and acceleration at $t = 2.0$ s. (d) Sketch the velocity and acceleration vectors at $t = 2.0$ s. At this instant, is the bird's speed increasing, decreasing, or not changing? Is the bird turning? If so, in what direction?

3.8 • CALC A remote-controlled car is moving in a vacant parking lot. The velocity of the car as a function of time is given by $\vec{v} = [5.00 \text{ m/s} - (0.0180 \text{ m/s}^3)t^2]\hat{i} + [2.00 \text{ m/s} + (0.550 \text{ m/s}^2)t]\hat{j}$. (a) What are $a_x(t)$ and $a_y(t)$, the x- and y-components of the car's velocity as functions of time? (b) What are the magnitude and direction of the car's velocity at $t = 8.00$ s? (b) What are the magnitude and direction of the car's acceleration at $t = 8.00$ s?

Section 3.3 Projectile Motion

3.9 • A physics book slides off a horizontal tabletop with a speed of 1.10 m/s. It strikes the floor in 0.480 s. Ignore air resistance. Find (a) the height of the tabletop above the floor; (b) the horizontal distance from the edge of the table to the point where the book strikes the floor; (c) the horizontal and vertical components of the book's velocity, and the magnitude and direction of its velocity, just before the book reaches the floor. (d) Draw x-t, y-t, v_x-t, and v_y-t graphs for the motion.

3.10 •• A daring 510-N swimmer dives off a cliff with a running horizontal leap, as shown in **Fig. E3.10.** What must her minimum speed be just as she leaves the top of the cliff so that she will miss the ledge at the bottom, which is 1.75 m wide and 9.00 m below the top of the cliff?

Figure **E3.10**

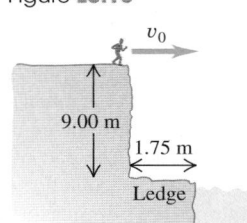

3.11 • Crickets Chirpy and Milada jump from the top of a vertical cliff. Chirpy drops downward and reaches the ground in 2.70 s, while Milada jumps horizontally with an initial speed of 95.0 cm/s. How far from the base of the cliff will Milada hit the ground? Ignore air resistance.

3.12 • A rookie quarterback throws a football with an initial upward velocity component of 12.0 m/s and a horizontal velocity component of 20.0 m/s. Ignore air resistance. (a) How much time is required for the football to reach the highest point of the trajectory? (b) How high is this point? (c) How much time (after it is thrown) is required for the football to return to its original level? How does this compare with the time calculated in part (a)? (d) How far has the football traveled horizontally during this time? (e) Draw x-t, y-t, v_x-t, and v_y-t graphs for the motion.

3.13 •• **Leaping the River I.** During a storm, a car traveling on a level horizontal road comes upon a bridge that has washed out. The driver must get to the other side, so he decides to try leaping the river with his car. The side of the road the car is on is 21.3 m above the river, while the opposite side is only 1.8 m above the river. The river itself is a raging torrent 48.0 m wide. (a) How fast should the car be traveling at the time it leaves the road in order just to clear the river and land safely on the opposite side? (b) What is the speed of the car just before it lands on the other side?

3.14 • BIO **The Champion Jumper of the Insect World.** The froghopper, *Philaenus spumarius,* holds the world record for insect jumps. When leaping at an angle of 58.0° above the horizontal, some of the tiny critters have reached a maximum height of 58.7 cm above the level ground. (See *Nature,* Vol. 424, July 31, 2003, p. 509.) (a) What was the takeoff speed for such a leap? (b) What horizontal distance did the froghopper cover for this world-record leap?

3.15 •• Inside a starship at rest on the earth, a ball rolls off the top of a horizontal table and lands a distance *D* from the foot of the table. This starship now lands on the unexplored Planet X. The commander, Captain Curious, rolls the same ball off the same table with the same initial speed as on earth and finds that it lands a distance 2.76*D* from the foot of the table. What is the acceleration due to gravity on Planet X?

3.16 • On level ground a shell is fired with an initial velocity of 40.0 m/s at 60.0° above the horizontal and feels no appreciable air resistance. (a) Find the horizontal and vertical components of the shell's initial velocity. (b) How long does it take the shell to reach its highest point? (c) Find its maximum height above the ground. (d) How far from its firing point does the shell land? (e) At its highest point, find the horizontal and vertical components of its acceleration and velocity.

3.17 • A major leaguer hits a baseball so that it leaves the bat at a speed of 30.0 m/s and at an angle of 36.9° above the horizontal. Ignore air resistance. (a) At what *two* times is the baseball at a height of 10.0 m above the point at which it left the bat? (b) Calculate the horizontal and vertical components of the baseball's velocity at each of the two times calculated in part (a). (c) What are the magnitude and direction of the baseball's velocity when it returns to the level at which it left the bat?

3.18 • A shot putter releases the shot some distance above the level ground with a velocity of 12.0 m/s, 51.0° above the horizontal. The shot hits the ground 2.08 s later. Ignore air resistance. (a) What are the components of the shot's acceleration while in flight? (b) What are the components of the shot's velocity at the beginning and at the end of its trajectory? (c) How far did she throw the shot horizontally? (d) Why does the expression for *R* in Example 3.8 *not* give the correct answer for part (c)? (e) How high was the shot above the ground when she released it? (f) Draw *x-t*, *y-t*, v_x-*t*, and v_y-*t* graphs for the motion.

3.19 •• **Win the Prize.** In a carnival booth, you can win a stuffed giraffe if you toss a quarter into a small dish. The dish is on a shelf above the point where the quarter leaves your hand and is a horizontal distance of 2.1 m from this point (**Fig. E3.19**). If you toss the coin with a velocity of 6.4 m/s at an angle of 60° above the horizontal, the coin will land in the dish. Ignore air resistance.

Figure **E3.19**

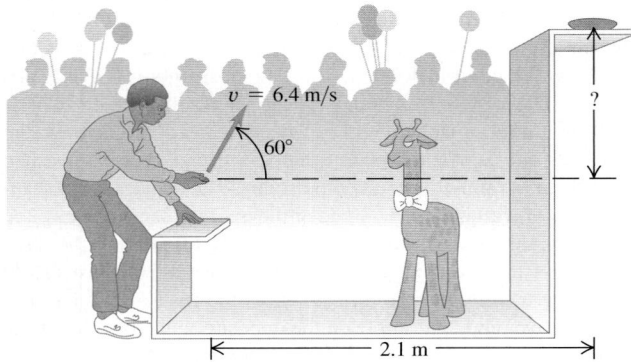

(a) What is the height of the shelf above the point where the quarter leaves your hand? (b) What is the vertical component of the velocity of the quarter just before it lands in the dish?

3.20 • Firemen use a high-pressure hose to shoot a stream of water at a burning building. The water has a speed of 25.0 m/s as it leaves the end of the hose and then exhibits projectile motion. The firemen adjust the angle of elevation α of the hose until the water takes 3.00 s to reach a building 45.0 m away. Ignore air resistance; assume that the end of the hose is at ground level. (a) Find α. (b) Find the speed and acceleration of the water at the highest point in its trajectory. (c) How high above the ground does the water strike the building, and how fast is it moving just before it hits the building?

3.21 •• A man stands on the roof of a 15.0-m-tall building and throws a rock with a speed of 30.0 m/s at an angle of 33.0° above the horizontal. Ignore air resistance. Calculate (a) the maximum height above the roof that the rock reaches; (b) the speed of the rock just before it strikes the ground; and (c) the horizontal range from the base of the building to the point where the rock strikes the ground. (d) Draw *x-t*, *y-t*, v_x-*t*, and v_y-*t* graphs for the motion.

3.22 •• A 124-kg balloon carrying a 22-kg basket is descending with a constant downward velocity of 20.0 m/s. A 1.0-kg stone is thrown from the basket with an initial velocity of 15.0 m/s perpendicular to the path of the descending balloon, as measured relative to a person at rest in the basket. That person sees the stone hit the ground 5.00 s after it was thrown. Assume that the balloon continues its downward descent with the same constant speed of 20.0 m/s. (a) How high is the balloon when the rock is thrown? (b) How high is the balloon when the rock hits the ground? (c) At the instant the rock hits the ground, how far is it from the basket? (d) Just before the rock hits the ground, find its horizontal and vertical velocity components as measured by an observer (i) at rest in the basket and (ii) at rest on the ground.

Section 3.4 Motion in a Circle

3.23 •• The earth has a radius of 6380 km and turns around once on its axis in 24 h. (a) What is the radial acceleration of an object at the earth's equator? Give your answer in m/s^2 and as a fraction of *g*. (b) If a_{rad} at the equator is greater than *g*, objects will fly off the earth's surface and into space. (We will see the reason for this in Chapter 5.) What would the period of the earth's rotation have to be for this to occur?

3.24 •• BIO **Dizziness.** Our balance is maintained, at least in part, by the endolymph fluid in the inner ear. Spinning displaces this fluid, causing dizziness. Suppose that a skater is spinning very fast at 3.0 revolutions per second about a vertical axis through the center of his head. Take the inner ear to be approximately 7.0 cm from the axis of spin. (The distance varies from person to person.) What is the radial acceleration (in m/s^2 and in *g*'s) of the endolymph fluid?

3.25 • BIO **Pilot Blackout in a Power Dive.** A jet plane comes in for a downward dive as shown in **Fig. E3.25**. The bottom part of the path is a quarter circle with a radius of curvature of 280 m. According to medical tests, pilots will lose consciousness when they pull out of a dive at an upward acceleration greater than 5.5*g*. At what speed (in m/s and in mph) will the pilot black out during this dive?

Figure **E3.25**

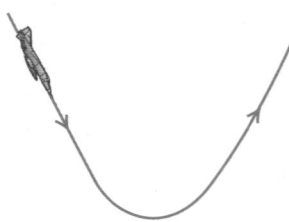

3.26 •• A model of a helicopter rotor has four blades, each 3.40 m long from the central shaft to the blade tip. The model is rotated in a wind tunnel at 550 rev/min. (a) What is the linear speed of the blade tip, in m/s? (b) What is the radial acceleration of the blade tip expressed as a multiple of g?

3.27 • A Ferris wheel with radius 14.0 m is turning about a horizontal axis through its center (**Fig. E3.27**). The linear speed of a passenger on the rim is constant and equal to 6.00 m/s. What are the magnitude and direction of the passenger's acceleration as she passes through (a) the lowest point in her circular motion and (b) the highest point in her circular motion? (c) How much time does it take the Ferris wheel to make one revolution?

Figure **E3.27**

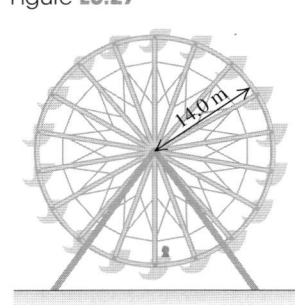

3.28 • The radius of the earth's orbit around the sun (assumed to be circular) is 1.50×10^8 km, and the earth travels around this orbit in 365 days. (a) What is the magnitude of the orbital velocity of the earth, in m/s? (b) What is the radial acceleration of the earth toward the sun, in m/s^2? (c) Repeat parts (a) and (b) for the motion of the planet Mercury (orbit radius $= 5.79 \times 10^7$ km, orbital period $= 88.0$ days).

3.29 •• BIO **Hypergravity.** At its Ames Research Center, NASA uses its large "20-G" centrifuge to test the effects of very large accelerations ("hypergravity") on test pilots and astronauts. In this device, an arm 8.84 m long rotates about one end in a horizontal plane, and an astronaut is strapped in at the other end. Suppose that he is aligned along the centrifuge's arm with his head at the outermost end. The maximum sustained acceleration to which humans are subjected in this device is typically 12.5g. (a) How fast must the astronaut's head be moving to experience this maximum acceleration? (b) What is the *difference* between the acceleration of his head and feet if the astronaut is 2.00 m tall? (c) How fast in rpm (rev/min) is the arm turning to produce the maximum sustained acceleration?

Section 3.5 Relative Velocity

3.30 • A railroad flatcar is traveling to the right at a speed of 13.0 m/s relative to an observer standing on the ground. Someone is riding a motor scooter on the flatcar (**Fig. E3.30**). What is the velocity (magnitude and direction) of the scooter relative to the flatcar if the scooter's velocity relative to the observer on the ground is (a) 18.0 m/s to the right? (b) 3.0 m/s to the left? (c) zero?

Figure **E3.30**

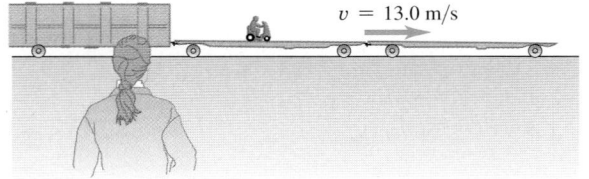

$v = 13.0$ m/s

3.31 • A "moving sidewalk" in an airport terminal moves at 1.0 m/s and is 35.0 m long. If a woman steps on at one end and walks at 1.5 m/s relative to the moving sidewalk, how much time does it take her to reach the opposite end if she walks (a) in the same direction the sidewalk is moving? (b) In the opposite direction?

3.32 • Two piers, A and B, are located on a river; B is 1500 m downstream from A (**Fig. E3.32**). Two friends must make round trips from pier A to pier B and return. One rows a boat at a constant speed of 4.00 km/h relative to the water; the other walks on the shore at a constant speed of 4.00 km/h. The velocity of the river is 2.80 km/h in the direction from A to B. How much time does it take each person to make the round trip?

Figure **E3.32**

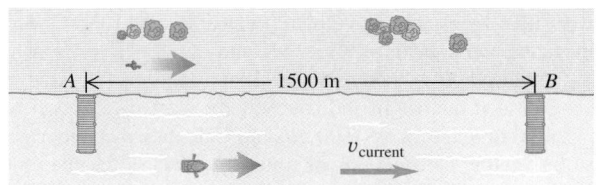

3.33 •• A canoe has a velocity of 0.40 m/s southeast relative to the earth. The canoe is on a river that is flowing 0.50 m/s east relative to the earth. Find the velocity (magnitude and direction) of the canoe relative to the river.

3.34 •• The nose of an ultralight plane is pointed due south, and its airspeed indicator shows 35 m/s. The plane is in a 10-m/s wind blowing toward the southwest relative to the earth. (a) In a vector-addition diagram, show the relationship of $\vec{v}_{P/E}$ (the velocity of the plane relative to the earth) to the two given vectors. (b) Let x be east and y be north, and find the components of $\vec{v}_{P/E}$. (c) Find the magnitude and direction of $\vec{v}_{P/E}$.

3.35 • **Crossing the River I.** A river flows due south with a speed of 2.0 m/s. You steer a motorboat across the river; your velocity relative to the water is 4.2 m/s due east. The river is 500 m wide. (a) What is your velocity (magnitude and direction) relative to the earth? (b) How much time is required to cross the river? (c) How far south of your starting point will you reach the opposite bank?

3.36 • **Crossing the River II.** (a) In which direction should the motorboat in Exercise 3.35 head to reach a point on the opposite bank directly east from your starting point? (The boat's speed relative to the water remains 4.2 m/s.) (b) What is the velocity of the boat relative to the earth? (c) How much time is required to cross the river?

3.37 •• BIO **Bird Migration.** Canada geese migrate essentially along a north–south direction for well over a thousand kilometers in some cases, traveling at speeds up to about 100 km/h. If one goose is flying at 100 km/h relative to the air but a 40-km/h wind is blowing from west to east, (a) at what angle relative to the north–south direction should this bird head to travel directly southward relative to the ground? (b) How long will it take the goose to cover a ground distance of 500 km from north to south? (*Note:* Even on cloudy nights, many birds can navigate by using the earth's magnetic field to fix the north–south direction.)

3.38 •• An airplane pilot wishes to fly due west. A wind of 80.0 km/h (about 50 mi/h) is blowing toward the south. (a) If the airspeed of the plane (its speed in still air) is 320.0 km/h (about 200 mi/h), in which direction should the pilot head? (b) What is the speed of the plane over the ground? Draw a vector diagram.

PROBLEMS

3.39 • **CALC** A rocket is fired at an angle from the top of a tower of height $h_0 = 50.0$ m. Because of the design of the engines, its position coordinates are of the form $x(t) = A + Bt^2$ and $y(t) = C + Dt^3$, where A, B, C, and D are constants. The acceleration of the rocket 1.00 s after firing is $\vec{a} = (4.00\hat{\imath} + 3.00\hat{\jmath})$ m/s². Take the origin of coordinates to be at the base of the tower. (a) Find the constants A, B, C, and D, including their SI units. (b) At the instant after the rocket is fired, what are its acceleration vector and its velocity? (c) What are the x- and y-components of the rocket's velocity 10.0 s after it is fired, and how fast is it moving? (d) What is the position vector of the rocket 10.0 s after it is fired?

3.40 ••• **CALC** A faulty model rocket moves in the xy-plane (the positive y-direction is vertically upward). The rocket's acceleration has components $a_x(t) = \alpha t^2$ and $a_y(t) = \beta - \gamma t$, where $\alpha = 2.50$ m/s⁴, $\beta = 9.00$ m/s², and $\gamma = 1.40$ m/s³. At $t = 0$ the rocket is at the origin and has velocity $\vec{v}_0 = v_{0x}\hat{\imath} + v_{0y}\hat{\jmath}$ with $v_{0x} = 1.00$ m/s and $v_{0y} = 7.00$ m/s. (a) Calculate the velocity and position vectors as functions of time. (b) What is the maximum height reached by the rocket? (c) What is the horizontal displacement of the rocket when it returns to $y = 0$?

3.41 •• **CALC** If $\vec{r} = bt^2\hat{\imath} + ct^3\hat{\jmath}$, where b and c are positive constants, when does the velocity vector make an angle of 45.0° with the x- and y-axes?

3.42 •• **CALC** The position of a dragonfly that is flying parallel to the ground is given as a function of time by $\vec{r} = [2.90 \text{ m} + (0.0900 \text{ m/s}^2)t^2]\hat{\imath} - (0.0150 \text{ m/s}^3)t^3\hat{\jmath}$. (a) At what value of t does the velocity vector of the dragonfly make an angle of 30.0° clockwise from the +x-axis? (b) At the time calculated in part (a), what are the magnitude and direction of the dragonfly's acceleration vector?

3.43 ••• **CP** A test rocket starting from rest at point A is launched by accelerating it along a 200.0-m incline at 1.90 m/s² (**Fig. P3.43**). The incline rises at 35.0° above the horizontal, and at the instant the rocket leaves it, the engines turn off and the rocket is subject to gravity only (ignore air resistance). Find (a) the maximum height above the ground that the rocket reaches, and (b) the rocket's greatest horizontal range beyond point A.

Figure **P3.43**

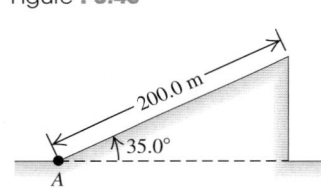

3.44 •• **CALC** A bird flies in the xy-plane with a velocity vector given by $\vec{v} = (\alpha - \beta t^2)\hat{\imath} + \gamma t\hat{\jmath}$, with $\alpha = 2.4$ m/s, $\beta = 1.6$ m/s³, and $\gamma = 4.0$ m/s². The positive y-direction is vertically upward. At $t = 0$ the bird is at the origin. (a) Calculate the position and acceleration vectors of the bird as functions of time. (b) What is the bird's altitude (y-coordinate) as it flies over $x = 0$ for the first time after $t = 0$?

3.45 •• A sly 1.5-kg monkey and a jungle veterinarian with a blow-gun loaded with a tranquilizer dart are 25 m above the ground in trees 70 m apart. Just as the veterinarian shoots horizontally at the monkey, the monkey drops from the tree in a vain attempt to escape being hit. What must the minimum muzzle velocity of the dart be for the dart to hit the monkey before the monkey reaches the ground?

3.46 ••• **BIO Spiraling Up.** Birds of prey typically rise upward on thermals. The paths these birds take may be spiral-like. You can model the spiral motion as uniform circular motion combined with a constant upward velocity. Assume that a bird completes a circle of radius 6.00 m every 5.00 s and rises vertically at

a constant rate of 3.00 m/s. Determine (a) the bird's speed relative to the ground; (b) the bird's acceleration (magnitude and direction); and (c) the angle between the bird's velocity vector and the horizontal.

3.47 •• In fighting forest fires, airplanes work in support of ground crews by dropping water on the fires. For practice, a pilot drops a canister of red dye, hoping to hit a target on the ground below. If the plane is flying in a horizontal path 90.0 m above the ground and has a speed of 64.0 m/s (143 mi/h), at what horizontal distance from the target should the pilot release the canister? Ignore air resistance.

3.48 ••• A movie stuntwoman drops from a helicopter that is 30.0 m above the ground and moving with a constant velocity whose components are 10.0 m/s upward and 15.0 m/s horizontal and toward the south. Ignore air resistance. (a) Where on the ground (relative to the position of the helicopter when she drops) should the stuntwoman have placed foam mats to break her fall? (b) Draw x-t, y-t, v_x-t, and v_y-t graphs of her motion.

3.49 •• An airplane is flying with a velocity of 90.0 m/s at an angle of 23.0° above the horizontal. When the plane is 114 m directly above a dog that is standing on level ground, a suitcase drops out of the luggage compartment. How far from the dog will the suitcase land? Ignore air resistance.

3.50 •• A cannon, located 60.0 m from the base of a vertical 25.0-m-tall cliff, shoots a 15-kg shell at 43.0° above the horizontal toward the cliff. (a) What must the minimum muzzle velocity be for the shell to clear the top of the cliff? (b) The ground at the top of the cliff is level, with a constant elevation of 25.0 m above the cannon. Under the conditions of part (a), how far does the shell land past the edge of the cliff?

3.51 • **CP CALC** A toy rocket is launched with an initial velocity of 12.0 m/s in the horizontal direction from the roof of a 30.0-m-tall building. The rocket's engine produces a horizontal acceleration of $(1.60 \text{ m/s}^3)t$, in the same direction as the initial velocity, but in the vertical direction the acceleration is g, downward. Ignore air resistance. What horizontal distance does the rocket travel before reaching the ground?

3.52 ••• An important piece of landing equipment must be thrown to a ship, which is moving at 45.0 cm/s, before the ship can dock. This equipment is thrown at 15.0 m/s at 60.0° above the horizontal from the top of a tower at the edge of the water, 8.75 m above the ship's deck (**Fig. P3.52**). For this equipment to land at the front of the ship, at what distance D from the dock should the ship be when the equipment is thrown? Ignore air resistance.

Figure **P3.52**

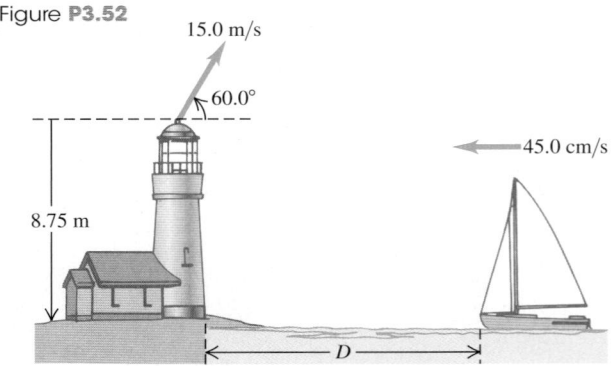

3.53 ••• **The Longest Home Run.** According to *Guinness World Records,* the longest home run ever measured was hit by Roy "Dizzy" Carlyle in a minor league game. The ball traveled 188 m (618 ft) before landing on the ground outside the ballpark. (a) If the ball's initial velocity was in a direction 45° above the horizontal, what did the initial speed of the ball need to be to produce such a home run if the ball was hit at a point 0.9 m (3.0 ft) above ground level? Ignore air resistance, and assume that the ground was perfectly flat. (b) How far would the ball be above a fence 3.0 m (10 ft) high if the fence was 116 m (380 ft) from home plate?

3.54 •• **An Errand of Mercy.** An airplane is dropping bales of hay to cattle stranded in a blizzard on the Great Plains. The pilot releases the bales at 150 m above the level ground when the plane is flying at 75 m/s in a direction 55° above the horizontal. How far in front of the cattle should the pilot release the hay so that the bales land at the point where the cattle are stranded?

3.55 •• A baseball thrown at an angle of 60.0° above the horizontal strikes a building 18.0 m away at a point 8.00 m above the point from which it is thrown. Ignore air resistance. (a) Find the magnitude of the ball's initial velocity (the velocity with which the ball is thrown). (b) Find the magnitude and direction of the velocity of the ball just before it strikes the building.

3.56 ••• A water hose is used to fill a large cylindrical storage tank of diameter D and height $2D$. The hose shoots the water at 45° above the horizontal from the same level as the base of the tank and is a distance $6D$ away (**Fig. P3.56**). For what *range* of launch speeds (v_0) will the water enter the tank? Ignore air resistance, and express your answer in terms of D and g.

Figure **P3.56**

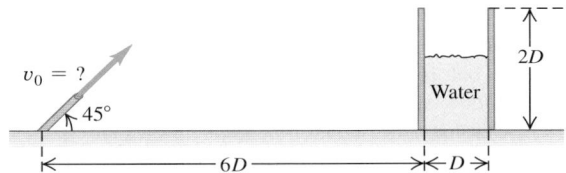

3.57 •• A grasshopper leaps into the air from the edge of a vertical cliff, as shown in **Fig. P3.57**. Find (a) the initial speed of the grasshopper and (b) the height of the cliff.

Figure **P3.57**

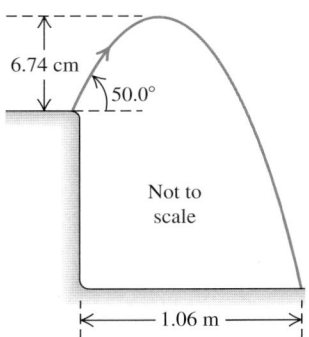

Not to scale

3.58 •• **Kicking an Extra Point.** In Canadian football, after a touchdown the team has the opportunity to earn one more point by kicking the ball over the bar between the goal posts. The bar is 10.0 ft above the ground, and the ball is kicked from ground level, 36.0 ft horizontally from the bar (**Fig. P3.58**). Football regulations are stated in English units, but convert them to SI units for this problem. (a) There is a minimum angle above the ground such that if the ball is launched below this angle, it can never clear the bar, no matter how fast it is kicked. What is this angle? (b) If the ball is kicked at 45.0° above the horizontal, what must its initial speed be if it is just to clear the bar? Express your answer in m/s and in km/h.

Figure **P3.58**

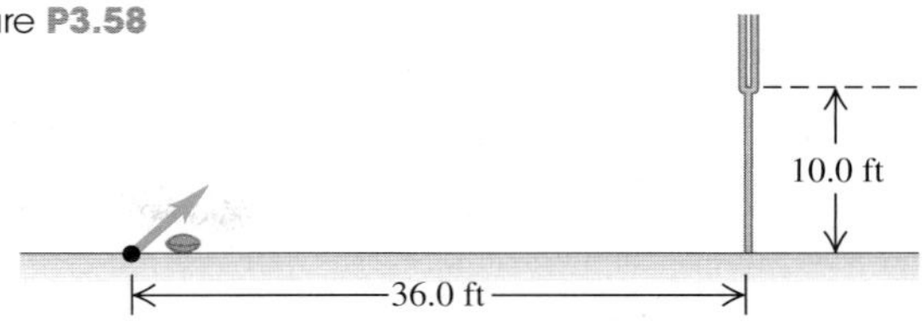

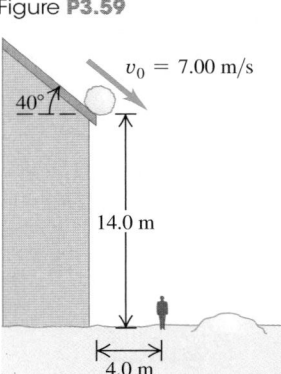

3.59 ••• **Look Out!** A snowball rolls off a barn roof that slopes downward at an angle of 40° (**Fig. P3.59**). The edge of the roof is 14.0 m above the ground, and the snowball has a speed of 7.00 m/s as it rolls off the roof. Ignore air resistance. (a) How far from the edge of the barn does the snowball strike the ground if it doesn't strike anything else while falling? (b) Draw x-t, y-t, v_x-t, and v_y-t graphs for the motion in part (a). (c) A man 1.9 m tall is standing 4.0 m from the edge of the barn. Will the snowball hit him?

Figure **P3.59**

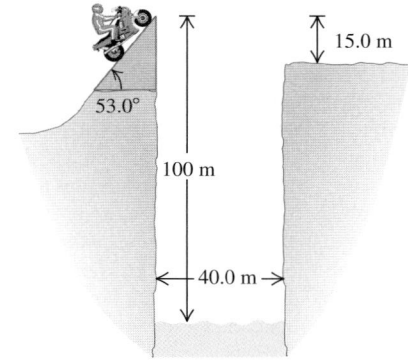

3.60 •• A boy 12.0 m above the ground in a tree throws a ball for his dog, who is standing right below the tree and starts running the instant the ball is thrown. If the boy throws the ball horizontally at 8.50 m/s, (a) how fast must the dog run to catch the ball just as it reaches the ground, and (b) how far from the tree will the dog catch the ball?

3.61 •• Suppose that the boy in Problem 3.60 throws the ball upward at 60.0° above the horizontal, but all else is the same. Repeat parts (a) and (b) of that problem.

3.62 •• A rock is thrown with a velocity v_0, at an angle of α_0 from the horizontal, from the roof of a building of height h. Ignore air resistance. Calculate the speed of the rock just before it strikes the ground, and show that this speed is independent of α_0.

3.63 •• **Leaping the River II.** A physics professor did daredevil stunts in his spare time. His last stunt was an attempt to jump across a river on a motorcycle (**Fig. P3.63**). The takeoff ramp was inclined at 53.0°, the river was 40.0 m wide, and the far bank was 15.0 m lower than the top of the ramp. The river itself was 100 m below the ramp. Ignore air resistance. (a) What should his speed have been at the top of the ramp to have just made it to the edge of the far bank? (b) If his speed was only half the value found in part (a), where did he land?

Figure **P3.63**

3.64 • A 2.7-kg ball is thrown upward with an initial speed of 20.0 m/s from the edge of a 45.0-m-high cliff. At the instant the ball is thrown, a woman starts running away from the base of the cliff with a constant speed of 6.00 m/s. The woman runs in a straight line on level ground. Ignore air resistance on the ball. (a) At what angle above the horizontal should the ball be thrown so that the runner will catch it just before it hits the ground, and how far does she run before she catches the ball? (b) Carefully sketch the ball's trajectory as viewed by (i) a person at rest on the ground and (ii) the runner.

3.65 • A 76.0-kg rock is rolling horizontally at the top of a vertical cliff that is 20 m above the surface of a lake (**Fig. P3.65**). The top of the vertical face of a dam is located 100 m from the foot of the cliff, with the top of the dam level with the surface of the water in the lake. A level plain is 25 m below the top of the dam. (a) What must be the minimum speed of the rock just as it leaves the cliff so that it will reach the plain without striking the dam? (b) How far from the foot of the dam does the rock hit the plain?

Figure **P3.65**

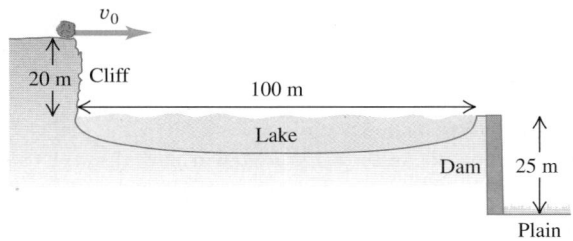

3.66 •• **Tossing Your Lunch.** Henrietta is jogging on the sidewalk at 3.05 m/s on the way to her physics class. Bruce realizes that she forgot her bag of bagels, so he runs to the window, which is 38.0 m above the street level and directly above the sidewalk, to throw the bag to her. He throws it horizontally 9.00 s after she has passed below the window, and she catches it on the run. Ignore air resistance. (a) With what initial speed must Bruce throw the bagels so that Henrietta can catch the bag just before it hits the ground? (b) Where is Henrietta when she catches the bagels?

3.67 •• A cart carrying a vertical missile launcher moves horizontally at a constant velocity of 30.0 m/s to the right. It launches a rocket vertically upward. The missile has an initial vertical velocity of 40.0 m/s relative to the cart. (a) How high does the rocket go? (b) How far does the cart travel while the rocket is in the air? (c) Where does the rocket land relative to the cart?

3.68 •• A firefighting crew uses a water cannon that shoots water at 25.0 m/s at a fixed angle of 53.0° above the horizontal. The firefighters want to direct the water at a blaze that is 10.0 m above ground level. How far from the building should they position their cannon? There are *two* possibilities; can you get them both? (*Hint:* Start with a sketch showing the trajectory of the water.)

3.69 ••• In the middle of the night you are standing a horizontal distance of 14.0 m from the high fence that surrounds the estate of your rich uncle. The top of the fence is 5.00 m above the ground. You have taped an important message to a rock that you want to throw over the fence. The ground is level, and the width of the fence is small enough to be ignored. You throw the rock from a height of 1.60 m above the ground and at an angle of 56.0° above the horizontal. (a) What minimum initial speed must the rock have as it leaves your hand to clear the top of the fence? (b) For the initial velocity calculated in part (a), what horizontal distance beyond the fence will the rock land on the ground?

3.70 ••• **CP Bang!** A student sits atop a platform a distance h above the ground. He throws a large firecracker horizontally with a speed v. However, a wind blowing parallel to the ground gives the firecracker a constant horizontal acceleration with magnitude a. As a result, the firecracker reaches the ground directly below the student. Determine the height h in terms of v, a, and g. Ignore the effect of air resistance on the vertical motion.

3.71 •• An airplane pilot sets a compass course due west and maintains an airspeed of 220 km/h. After flying for 0.500 h, she finds herself over a town 120 km west and 20 km south of her starting point. (a) Find the wind velocity (magnitude and direction). (b) If the wind velocity is 40 km/h due south, in what direction should the pilot set her course to travel due west? Use the same airspeed of 220 km/h.

3.72 •• **Raindrops.** When a train's velocity is 12.0 m/s eastward, raindrops that are falling vertically with respect to the earth make traces that are inclined 30.0° to the vertical on the windows of the train. (a) What is the horizontal component of a drop's velocity with respect to the earth? With respect to the train? (b) What is the magnitude of the velocity of the raindrop with respect to the earth? With respect to the train?

3.73 ••• In a World Cup soccer match, Juan is running due north toward the goal with a speed of 8.00 m/s relative to the ground. A teammate passes the ball to him. The ball has a speed of 12.0 m/s and is moving in a direction 37.0° east of north, relative to the ground. What are the magnitude and direction of the ball's velocity relative to Juan?

3.74 •• An elevator is moving upward at a constant speed of 2.50 m/s. A bolt in the elevator ceiling 3.00 m above the elevator floor works loose and falls. (a) How long does it take for the bolt to fall to the elevator floor? What is the speed of the bolt just as it hits the elevator floor (b) according to an observer in the elevator? (c) According to an observer standing on one of the floor landings of the building? (d) According to the observer in part (c), what distance did the bolt travel between the ceiling and the floor of the elevator?

3.75 •• Two soccer players, Mia and Alice, are running as Alice passes the ball to Mia. Mia is running due north with a speed of 6.00 m/s. The velocity of the ball relative to Mia is 5.00 m/s in a direction 30.0° east of south. What are the magnitude and direction of the velocity of the ball relative to the ground?

3.76 •• **DATA** A spring-gun projects a small rock from the ground with speed v_0 at an angle θ_0 above the ground. You have been asked to determine v_0. From the way the spring-gun is constructed, you know that to a good approximation v_0 is independent of the launch angle. You go to a level, open field, select a launch angle, and measure the horizontal distance the rock travels. You use $g = 9.80$ m/s^2 and ignore the small height of the end of the spring-gun's barrel above the ground. Since your measurement includes some uncertainty in values measured for the launch angle and for the horizontal range, you repeat the measurement for several launch angles and obtain the results given in **Fig. 3.76**. You

Figure **P3.76**

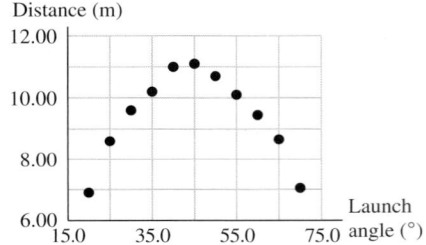

ignore air resistance because there is no wind and the rock is small and heavy. (a) Select a way to represent the data well as a straight line. (b) Use the slope of the best straight-line fit to your data from part (a) to calculate v_0. (c) When the launch angle is $36.9°$, what maximum height above the ground does the rock reach?

3.77 •• DATA You have constructed a hair-spray-powered potato gun and want to find the muzzle speed v_0 of the potatoes, the speed they have as they leave the end of the gun barrel. You use the same amount of hair spray each time you fire the gun, and you have confirmed by repeated firings at the same height that the muzzle speed is approximately the same for each firing. You climb on a microwave relay tower (with permission, of course) to launch the potatoes horizontally from different heights above the ground. Your friend measures the height of the gun barrel above the ground and the range R of each potato. You obtain the following data:

Launch height h	Horizontal range R
2.00 m	10.4 m
6.00 m	17.1 m
9.00 m	21.3 m
12.00 m	25.8 m

Each of the values of h and R has some measurement error: The muzzle speed is not precisely the same each time, and the barrel isn't precisely horizontal. So you use all of the measurements to get the best estimate of v_0. No wind is blowing, so you decide to ignore air resistance. You use $g = 9.80 \text{ m/s}^2$ in your analysis. (a) Select a way to represent the data well as a straight line. (b) Use the slope of the best-fit line from part (a) to calculate the average value of v_0. (c) What would be the horizontal range of a potato that is fired from ground level at an angle of $30.0°$ above the horizontal? Use the value of v_0 that you calculated in part (b).

3.78 ••• DATA You are a member of a geological team in Central Africa. Your team comes upon a wide river that is flowing east. You must determine the width of the river and the current speed (the speed of the water relative to the earth). You have a small boat with an outboard motor. By measuring the time it takes to cross a pond where the water isn't flowing, you have calibrated the throttle settings to the speed of the boat in still water. You set the throttle so that the speed of the boat relative to the river is a constant 6.00 m/s. Traveling due north across the river, you reach the opposite bank in 20.1 s. For the return trip, you change the throttle setting so that the speed of the boat relative to the water is 9.00 m/s. You travel due south from one bank to the other and cross the river in 11.2 s. (a) How wide is the river, and what is the current speed? (b) With the throttle set so that the speed of the boat relative to the water is 6.00 m/s, what is the shortest time in which you could cross the river, and where on the far bank would you land?

CHALLENGE PROBLEMS

3.79 ••• CALC A projectile thrown from a point P moves in such a way that its distance from P is always increasing. Find the maximum angle above the horizontal with which the projectile could have been thrown. Ignore air resistance.

3.80 ••• Two students are canoeing on a river. While heading upstream, they accidentally drop an empty bottle overboard. They then continue paddling for 60 minutes, reaching a point 2.0 km farther upstream. At this point they realize that the bottle is missing and, driven by ecological awareness, they turn around and head downstream. They catch up with and retrieve the bottle (which has been moving along with the current) 5.0 km downstream from the turnaround point. (a) Assuming a constant paddling effort throughout, how fast is the river flowing? (b) What would the canoe speed in a still lake be for the same paddling effort?

3.81 ••• CP A rocket designed to place small payloads into orbit is carried to an altitude of 12.0 km above sea level by a converted airliner. When the airliner is flying in a straight line at a constant speed of 850 km/h, the rocket is dropped. After the drop, the airliner maintains the same altitude and speed and continues to fly in a straight line. The rocket falls for a brief time, after which its rocket motor turns on. Once that motor is on, the combined effects of thrust and gravity give the rocket a constant acceleration of magnitude $3.00g$ directed at an angle of $30.0°$ above the horizontal. For safety, the rocket should be at least 1.00 km in front of the airliner when it climbs through the airliner's altitude. Your job is to determine the minimum time that the rocket must fall before its engine starts. Ignore air resistance. Your answer should include (i) a diagram showing the flight paths of both the rocket and the airliner, labeled at several points with vectors for their velocities and accelerations; (ii) an x-t graph showing the motions of both the rocket and the airliner; and (iii) a y-t graph showing the motions of both the rocket and the airliner. In the diagram and the graphs, indicate when the rocket is dropped, when the rocket motor turns on, and when the rocket climbs through the altitude of the airliner.

PASSAGE PROBLEMS

BIO **BALLISTIC SEED DISPERSAL.** Some plants disperse their seeds when the fruit splits and contracts, propelling the seeds through the air. The trajectory of these seeds can be determined with a high-speed camera. In an experiment on one type of plant, seeds are projected at 20 cm above ground level with initial speeds between 2.3 m/s and 4.6 m/s. The launch angle is measured from the horizontal, with $+90°$ corresponding to an initial velocity straight up and $-90°$ straight down.

3.82 The experiment is designed so that the seeds move no more than 0.20 mm between photographic frames. What minimum frame rate for the high-speed camera is needed to achieve this? (a) 250 frames/s; (b) 2500 frames/s; (c) 25,000 frames/s; (d) 250,000 frames/s.

3.83 About how long does it take a seed launched at $90°$ at the highest possible initial speed to reach its maximum height? Ignore air resistance. (a) 0.23 s; (b) 0.47 s; (c) 1.0 s; (d) 2.3 s.

3.84 If a seed is launched at an angle of $0°$ with the maximum initial speed, how far from the plant will it land? Ignore air resistance, and assume that the ground is flat. (a) 20 cm; (b) 93 cm; (c) 2.2 m; (d) 4.6 m.

3.85 A large number of seeds are observed, and their initial launch angles are recorded. The range of projection angles is found to be $-51°$ to $75°$, with a mean of $31°$. Approximately 65% of the seeds are launched between $6°$ and $56°$. (See W. J. Garrison et al., "Ballistic seed projection in two herbaceous species," *Amer. J. Bot.,* Sept. 2000, 87:9, 1257–64.) Which of these hypotheses is best supported by the data? Seeds are preferentially launched (a) at angles that maximize the height they travel above the plant; (b) at angles below the horizontal in order to drive the seeds into the ground with more force; (c) at angles that maximize the horizontal distance the seeds travel from the plant; (d) at angles that minimize the time the seeds spend exposed to the air.

Answers

Chapter Opening Question **?**

(iii) A cyclist going around a curve at constant speed has an acceleration directed toward the inside of the curve (see Section 3.2, especially Fig. 3.12a).

Test Your Understanding Questions

3.1 (iii)   If the instantaneous velocity $\vec{v}$ is constant over an interval, its value at any point (including the end of the interval) is the same as the average velocity $\vec{v}_{av}$ over the interval. In (i) and (ii) the direction of $\vec{v}$ at the end of the interval is tangent to the path at that point, while the direction of $\vec{v}_{av}$ points from the beginning of the path to its end (in the direction of the net displacement). In (iv) both $\vec{v}$ and $\vec{v}_{av}$ are directed along the straight line, but $\vec{v}$ has a greater magnitude because the speed has been increasing.

3.2 Vector 7   At the high point of the sled's path, the speed is minimum. At that point the speed is neither increasing nor decreasing, and the parallel component of the acceleration (that is, the horizontal component) is zero. The acceleration has only a perpendicular component toward the inside of the sled's curved path. In other words, the acceleration is downward.

3.3 (i)   If there were no gravity $(g = 0)$, the monkey would not fall and the dart would follow a straight-line path (shown as a dashed line). The effect of gravity is to make both the monkey and the dart fall the same distance $\frac{1}{2}gt^2$ below their $g = 0$ positions. Point A is the same distance below the monkey's initial position as point P is below the dashed straight line, so point A is where we would find the monkey at the time in question.

3.4 (ii)   At both the top and bottom of the loop, the acceleration is purely radial and is given by Eq. (3.27). Radius R is the same at both points, so the difference in acceleration is due purely to differences in speed. Since a_{rad} is proportional to the square of v, the speed must be twice as great at the bottom of the loop as at the top.

3.5 (vi)   The effect of the wind is to cancel the airplane's eastward motion and give it a northward motion. So the velocity of the air relative to the ground (the wind velocity) must have one 150-km/h component to the west and one 150-km/h component to the north. The combination of these is a vector of magnitude $\sqrt{(150 \text{ km/h})^2 + (150 \text{ km/h})^2} = 212 \text{ km/h}$ that points to the northwest.

Bridging Problem

(a) $R = \dfrac{2v_0^2}{g} \dfrac{\cos(\theta + \phi)\sin\phi}{\cos^2\theta}$ **(b)** $\phi = 45° - \dfrac{\theta}{2}$

? Under what circumstances does the barbell push on the weightlifter just as hard as he pushes on the barbell? (i) When he holds the barbell stationary; (ii) when he raises the barbell; (iii) when he lowers the barbell; (iv) two of (i), (ii), and (iii); (v) all of (i), (ii), and (iii); (vi) none of these.

4 NEWTON'S LAWS OF MOTION

LEARNING GOALS

Looking forward at ...

4.1 What the concept of force means in physics, why forces are vectors, and the significance of the net force on an object.

4.2 What happens when the net force on an object is zero, and the significance of inertial frames of reference.

4.3 How the acceleration of an object is determined by the net force on the object and the object's mass.

4.4 The difference between the mass of an object and its weight.

4.5 How the forces that two objects exert on each other are related.

4.6 How to use a free-body diagram to help analyze the forces on an object.

Looking back at ...

2.4 Straight-line motion with constant acceleration.

2.5 The motion of freely falling bodies.

3.2 Acceleration as a vector.

3.4 Uniform circular motion.

3.5 Relative velocity.

We've seen in the last two chapters how to use *kinematics* to describe motion in one, two, or three dimensions. But what *causes* bodies to move the way that they do? For example, why does a dropped feather fall more slowly than a dropped baseball? Why do you feel pushed backward in a car that accelerates forward? The answers to such questions take us into the subject of **dynamics,** the relationship of motion to the forces that cause it.

The principles of dynamics were clearly stated for the first time by Sir Isaac Newton (1642–1727); today we call them **Newton's laws of motion.** The first law states that when the net force on a body is zero, its motion doesn't change. The second law tells us that a body accelerates when the net force is *not* zero. The third law relates the forces that two interacting bodies exert on each other.

Newton did not *derive* the three laws of motion, but rather *deduced* them from a multitude of experiments performed by other scientists, especially Galileo Galilei (who died the year Newton was born). Newton's laws are the foundation of **classical mechanics** (also called **Newtonian mechanics**); using them, we can understand most familiar kinds of motion. Newton's laws need modification only for situations involving extremely high speeds (near the speed of light) or very small sizes (such as within the atom).

Newton's laws are very simple to state, yet many students find these laws difficult to grasp and to work with. The reason is that before studying physics, you've spent years walking, throwing balls, pushing boxes, and doing dozens of things that involve motion. Along the way, you've developed a set of "common sense" ideas about motion and its causes. But many of these "common sense" ideas don't stand up to logical analysis. A big part of the job of this chapter—and of the rest of our study of physics—is helping you recognize how "common sense" ideas can sometimes lead you astray, and how to adjust your understanding of the physical world to make it consistent with what experiments tell us.

4.1 Some properties of forces.

- A force is a push or a pull.
- A force is an interaction between two objects or between an object and its environment.
- A force is a vector quantity, with magnitude and direction.

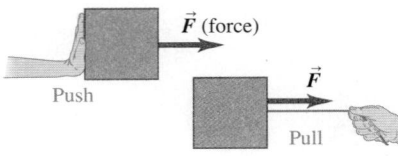

4.2 Four common types of forces.

(a) **Normal force** $\vec{n}$: When an object rests or pushes on a surface, the surface exerts a push on it that is directed perpendicular to the surface.

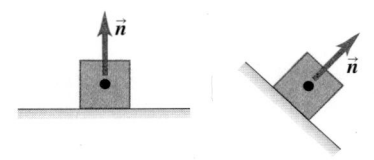

(b) **Friction force** $\vec{f}$: In addition to the normal force, a surface may exert a friction force on an object, directed parallel to the surface.

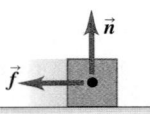

(c) **Tension force** $\vec{T}$: A pulling force exerted on an object by a rope, cord, etc.

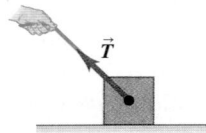

(d) **Weight** $\vec{w}$: The pull of gravity on an object is a long-range force (a force that acts over a distance).

4.1 FORCE AND INTERACTIONS

In everyday language, a **force** is a push or a pull. A better definition is that a force is an *interaction* between two bodies or between a body and its environment (**Fig. 4.1**). That's why we always refer to the force that one body *exerts* on a second body. When you push on a car that is stuck in the snow, you exert a force on the car; a steel cable exerts a force on the beam it is hoisting at a construction site; and so on. As Fig. 4.1 shows, force is a *vector* quantity; you can push or pull a body in different directions.

When a force involves direct contact between two bodies, such as a push or pull that you exert on an object with your hand, we call it a **contact force.** **Figures 4.2a**, 4.2b, and 4.2c show three common types of contact forces. The **normal force** (Fig. 4.2a) is exerted on an object by any surface with which it is in contact. The adjective *normal* means that the force always acts perpendicular to the surface of contact, no matter what the angle of that surface. By contrast, the **friction force** (Fig. 4.2b) exerted on an object by a surface acts *parallel* to the surface, in the direction that opposes sliding. The pulling force exerted by a stretched rope or cord on an object to which it's attached is called a **tension force** (Fig. 4.2c). When you tug on your dog's leash, the force that pulls on her collar is a tension force.

In addition to contact forces, there are **long-range forces** that act even when the bodies are separated by empty space. The force between two magnets is an example of a long-range force, as is the force of gravity (Fig. 4.2d); the earth pulls a dropped object toward it even though there is no direct contact between the object and the earth. The gravitational force that the earth exerts on your body is called your **weight.**

To describe a force vector $\vec{F}$, we need to describe the *direction* in which it acts as well as its *magnitude,* the quantity that describes "how much" or "how hard" the force pushes or pulls. The SI unit of the magnitude of force is the *newton,* abbreviated N. (We'll give a precise definition of the newton in Section 4.3.) **Table 4.1** lists some typical force magnitudes.

A common instrument for measuring force magnitudes is the *spring balance.* It consists of a coil spring enclosed in a case with a pointer attached to one end. When forces are applied to the ends of the spring, it stretches by an amount that depends on the force. We can make a scale for the pointer by using a number of identical bodies with weights of exactly 1 N each. When one, two, or more of these are suspended simultaneously from the balance, the total force stretching the spring is 1 N, 2 N, and so on, and we can label the corresponding positions of the pointer 1 N, 2 N, and so on. Then we can use this instrument to measure the magnitude of an unknown force. We can also make a similar instrument that measures pushes instead of pulls.

TABLE 4.1 Typical Force Magnitudes	
Sun's gravitational force on the earth	3.5×10^{22} N
Weight of a large blue whale	1.9×10^{6} N
Maximum pulling force of a locomotive	8.9×10^{5} N
Weight of a 250-lb linebacker	1.1×10^{3} N
Weight of a medium apple	1 N
Weight of the smallest insect eggs	2×10^{-6} N
Electric attraction between the proton and the electron in a hydrogen atom	8.2×10^{-8} N
Weight of a very small bacterium	1×10^{-18} N
Weight of a hydrogen atom	1.6×10^{-26} N
Weight of an electron	8.9×10^{-30} N
Gravitational attraction between the proton and the electron in a hydrogen atom	3.6×10^{-47} N

4.3 Using a vector arrow to denote the force that we exert when **(a)** pulling a block with a string or **(b)** pushing a block with a stick.

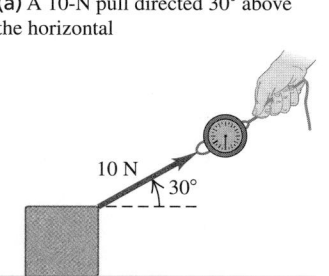
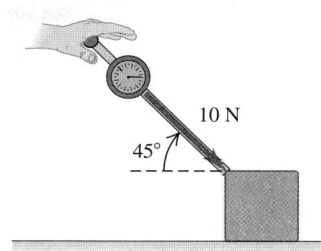

(a) A 10-N pull directed 30° above the horizontal

(b) A 10-N push directed 45° below the horizontal

Figure 4.3 shows a spring balance being used to measure a pull or push that we apply to a box. In each case we draw a vector to represent the applied force. The length of the vector shows the magnitude; the longer the vector, the greater the force magnitude.

Superposition of Forces

When you throw a ball, at least two forces act on it: the push of your hand and the downward pull of gravity. Experiment shows that when two forces $\vec{F}_1$ and $\vec{F}_2$ act at the same time at the same point on a body (**Fig. 4.4**), the effect on the body's motion is the same as if a single force $\vec{R}$ were acting equal to the vector sum, or resultant, of the original forces: $\vec{R} = \vec{F}_1 + \vec{F}_2$. More generally, *any number of forces applied at a point on a body have the same effect as a single force equal to the vector sum of the forces.* This important principle is called **superposition of forces.**

Since forces are vector quantities and add like vectors, we can use all of the rules of vector mathematics that we learned in Chapter 1 to solve problems that involve vectors. This would be a good time to review the rules for vector addition presented in Sections 1.7 and 1.8.

We learned in Section 1.8 that it's easiest to add vectors by using components. That's why we often describe a force $\vec{F}$ in terms of its x- and y-components F_x and F_y. Note that the x- and y-coordinate axes do *not* have to be horizontal and vertical, respectively. As an example, **Fig. 4.5** shows a crate being pulled up a ramp by a force $\vec{F}$. In this situation it's most convenient to choose one axis to be parallel to the ramp and the other to be perpendicular to the ramp. For the case shown in Fig. 4.5, both F_x and F_y are positive; in other situations, depending on your choice of axes and the orientation of the force $\vec{F}$, either F_x or F_y may be negative or zero.

CAUTION **Using a wiggly line in force diagrams** In Fig. 4.5 we draw a wiggly line through the force vector $\vec{F}$ to show that we have replaced it by its x- and y-components. Otherwise, the diagram would include the same force twice. We will draw such a wiggly line in any force diagram where a force is replaced by its components. Look for this wiggly line in other figures in this and subsequent chapters. ▌

We will often need to find the vector sum (resultant) of *all* forces acting on a body. We call this the **net force** acting on the body. We will use the Greek letter Σ (capital sigma, equivalent to the Roman S) as a shorthand notation for a sum. If the forces are labeled $\vec{F}_1$, $\vec{F}_2$, $\vec{F}_3$, and so on, we can write

The net force acting on a body ... $\vec{R} = \Sigma\vec{F} = \vec{F}_1 + \vec{F}_2 + \vec{F}_3 + \cdots$ (4.1)

... is the vector sum, or resultant, of all individual forces acting on that body.

4.4 Superposition of forces.

Two forces $\vec{F}_1$ and $\vec{F}_2$ acting on a body at point O have the same effect as a single force $\vec{R}$ equal to their vector sum.

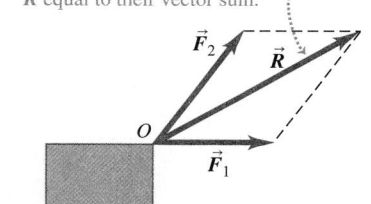

4.5 F_x and F_y are the components of $\vec{F}$ parallel and perpendicular to the sloping surface of the inclined plane.

We cross out a vector when we replace it by its components.

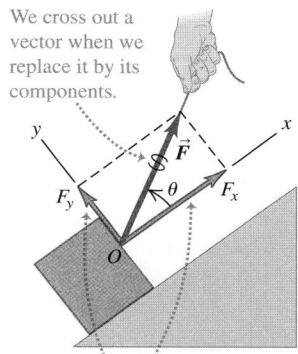

The x- and y-axes can have any orientation, just so they're mutually perpendicular.

4.6 Finding the components of the vector sum (resultant) $\vec{R}$ of two forces $\vec{F_1}$ and $\vec{F_2}$.

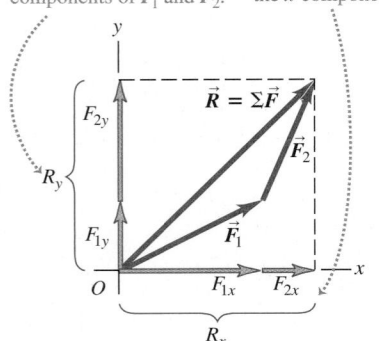

The y-component of $\vec{R}$ equals the sum of the y-components of $\vec{F_1}$ and $\vec{F_2}$.

The same goes for the x-components.

We read $\Sigma\vec{F}$ as "the vector sum of the forces" or "the net force." The x-component of the net force is the sum of the x-components of the individual forces, and likewise for the y-component (**Fig. 4.6**):

$$R_x = \Sigma F_x \qquad R_y = \Sigma F_y \qquad (4.2)$$

Each component may be positive or negative, so be careful with signs when you evaluate these sums.

Once we have R_x and R_y we can find the magnitude and direction of the net force $\vec{R} = \Sigma\vec{F}$ acting on the body. The magnitude is

$$R = \sqrt{R_x^2 + R_y^2}$$

and the angle θ between $\vec{R}$ and the +x-axis can be found from the relationship $\tan\theta = R_y/R_x$. The components R_x and R_y may be positive, negative, or zero, and the angle θ may be in any of the four quadrants.

In three-dimensional problems, forces may also have z-components; then we add the equation $R_z = \Sigma F_z$ to Eq. (4.2). The magnitude of the net force is then

$$R = \sqrt{R_x^2 + R_y^2 + R_z^2}$$

EXAMPLE 4.1 SUPERPOSITION OF FORCES

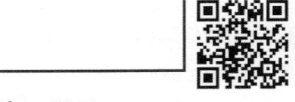

Three professional wrestlers are fighting over a champion's belt. **Figure 4.7a** shows the horizontal force each wrestler applies to the belt, as viewed from above. The forces have magnitudes $F_1 = 250$ N, $F_2 = 50$ N, and $F_3 = 120$ N. Find the x- and y-components of the net force on the belt, and find its magnitude and direction.

SOLUTION

IDENTIFY and SET UP: This is a problem in vector addition in which the vectors happen to represent forces. We want to find the x- and y-components of the net force $\vec{R}$, so we'll use the component method of vector addition expressed by Eqs. (4.2). Once we know the components of $\vec{R}$, we can find its magnitude and direction.

EXECUTE: From Fig. 4.7a the angles between the three forces $\vec{F_1}$, $\vec{F_2}$, and $\vec{F_3}$ and the +x-axis are $\theta_1 = 180° - 53° = 127°$, $\theta_2 = 0°$, and $\theta_3 = 270°$. The x- and y-components of the three forces are

$$F_{1x} = (250 \text{ N}) \cos 127° = -150 \text{ N}$$

$$F_{1y} = (250 \text{ N}) \sin 127° = 200 \text{ N}$$

$$F_{2x} = (50 \text{ N}) \cos 0° = 50 \text{ N}$$

$$F_{2y} = (50 \text{ N}) \sin 0° = 0 \text{ N}$$

$$F_{3x} = (120 \text{ N}) \cos 270° = 0 \text{ N}$$

$$F_{3y} = (120 \text{ N}) \sin 270° = -120 \text{ N}$$

From Eqs. (4.2) the net force $\vec{R} = \Sigma\vec{F}$ has components

$$R_x = F_{1x} + F_{2x} + F_{3x} = (-150 \text{ N}) + 50 \text{ N} + 0 \text{ N} = -100 \text{ N}$$

$$R_y = F_{1y} + F_{2y} + F_{3y} = 200 \text{ N} + 0 \text{ N} + (-120 \text{ N}) = 80 \text{ N}$$

The net force has a negative x-component and a positive y-component, as Fig. 4.7b shows.

The magnitude of $\vec{R}$ is

$$R = \sqrt{R_x^2 + R_y^2} = \sqrt{(-100 \text{ N})^2 + (80 \text{ N})^2} = 128 \text{ N}$$

To find the angle between the net force and the +x-axis, we use Eq. (1.7):

$$\theta = \arctan\frac{R_y}{R_x} = \arctan\left(\frac{80 \text{ N}}{-100 \text{ N}}\right) = \arctan(-0.80)$$

The arctangent of -0.80 is $-39°$, but Fig. 4.7b shows that the net force lies in the second quadrant. Hence the correct solution is $\theta = -39° + 180° = 141°$.

EVALUATE: The net force is *not* zero. Your intuition should suggest that wrestler 1 (who exerts the greatest force on the belt, $F_1 = 250$ N) will walk away with it when the struggle ends.

You should check the direction of $\vec{R}$ by adding the vectors $\vec{F_1}$, $\vec{F_2}$, and $\vec{F_3}$ graphically. Does your drawing show that $\vec{R} = \vec{F_1} + \vec{F_2} + \vec{F_3}$ points in the second quadrant as we found?

4.7 (a) Three forces acting on a belt. (b) The net force $\vec{R} = \Sigma\vec{F}$ and its components.

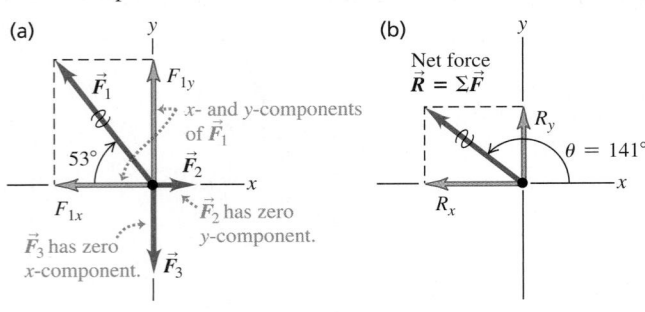

TEST YOUR UNDERSTANDING OF SECTION 4.1   Figure 4.5 shows a force $\vec{F}$ acting on a crate. With the x- and y-axes shown in the figure, which statement about the components of the *gravitational* force that the earth exerts on the crate (the crate's weight) is *correct*? (i) Both the x- and y-components are positive. (ii) The x-component is zero and the y-component is positive. (iii) The x-component is negative and the y-component is positive. (iv) Both the x- and y-components are negative. (v) The x-component is zero and the y-component is negative. (vi) The x-component is positive and the y-component is negative. ▍

4.2 NEWTON'S FIRST LAW

How do the forces that act on a body affect its motion? To begin to answer this question, let's first consider what happens when the net force on a body is *zero*. You would almost certainly agree that if a body is at rest, and if no net force acts on it (that is, no net push or pull), that body will remain at rest. But what if there is zero net force acting on a body in *motion*?

To see what happens in this case, suppose you slide a hockey puck along a horizontal tabletop, applying a horizontal force to it with your hand (**Fig. 4.8a**). After you stop pushing, the puck *does not* continue to move indefinitely; it slows down and stops. To keep it moving, you have to keep pushing (that is, applying a force). You might come to the "common sense" conclusion that bodies in motion naturally come to rest and that a force is required to sustain motion.

But now imagine pushing the puck across a smooth surface of ice (Fig. 4.8b). After you quit pushing, the puck will slide a lot farther before it stops. Put it on an air-hockey table, where it floats on a thin cushion of air, and it moves still farther (Fig. 4.8c). In each case, what slows the puck down is *friction,* an interaction between the lower surface of the puck and the surface on which it slides. Each surface exerts a friction force on the puck that resists the puck's motion; the difference in the three cases is the magnitude of the friction force. The ice exerts less friction than the tabletop, so the puck travels farther. The gas molecules of the air-hockey table exert the least friction of all. If we could eliminate friction completely, the puck would never slow down, and we would need no force at all to keep the puck moving once it had been started. Thus the "common sense" idea that a force is required to sustain motion is *incorrect.*

Experiments like the ones we've just described show that when *no* net force acts on a body, the body either remains at rest *or* moves with constant velocity in a straight line. Once a body has been set in motion, no net force is needed to keep it moving. We call this observation *Newton's first law of motion:*

> **NEWTON'S FIRST LAW OF MOTION:** A body acted on by no net force has a constant velocity (which may be zero) and zero acceleration.

The tendency of a body to keep moving once it is set in motion is called **inertia.** You use inertia when you try to get ketchup out of a bottle by shaking it. First you start the bottle (and the ketchup inside) moving forward; when you jerk the bottle back, the ketchup tends to keep moving forward and, you hope, ends up on your burger. Inertia is also the tendency of a body at rest to remain at rest. You may have seen a tablecloth yanked out from under the china without breaking anything. The force on the china isn't great enough to make it move appreciably during the short time it takes to pull the tablecloth away.

It's important to note that the *net* force is what matters in Newton's first law. For example, a physics book at rest on a horizontal tabletop has two forces acting on it: an upward supporting force, or normal force, exerted by the tabletop (see Fig. 4.2a) and the downward force of the earth's gravity (which acts even if the tabletop is elevated above the ground; see Fig. 4.2d). The upward push of the surface is just as great as the downward pull of gravity, so the *net* force acting

4.8 The slicker the surface, the farther a puck slides after being given an initial velocity. On an air-hockey table (c) the friction force is practically zero, so the puck continues with almost constant velocity.

(a) Table: puck stops short.

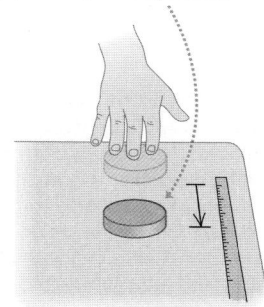

(b) Ice: puck slides farther.

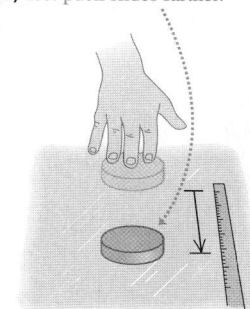

(c) Air-hockey table: puck slides even farther.

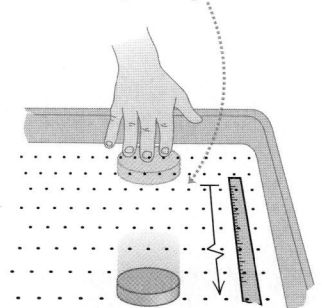

4.9 (a) A hockey puck accelerates in the direction of a net applied force $\vec{F}_1$. (b) When the net force is zero, the acceleration is zero, and the puck is in equilibrium.

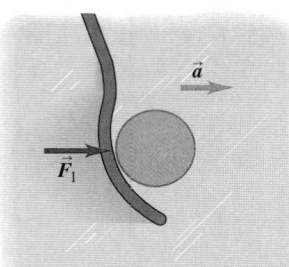

(a) A puck on a frictionless surface accelerates when acted on by a single horizontal force.

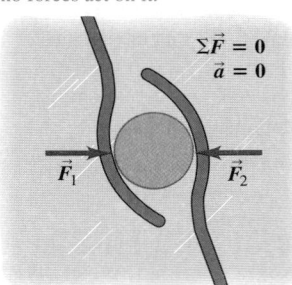

(b) This puck is acted on by two horizontal forces whose vector sum is zero. The puck behaves as though no forces act on it.

$$\Sigma \vec{F} = 0$$
$$\vec{a} = 0$$

on the book (that is, the vector sum of the two forces) is zero. In agreement with Newton's first law, if the book is at rest on the tabletop, it remains at rest. The same principle applies to a hockey puck sliding on a horizontal, frictionless surface: The vector sum of the upward push of the surface and the downward pull of gravity is zero. Once the puck is in motion, it continues to move with constant velocity because the *net* force acting on it is zero.

Here's another example. Suppose a hockey puck rests on a horizontal surface with negligible friction, such as an air-hockey table or a slab of wet ice. If the puck is initially at rest and a single horizontal force $\vec{F}_1$ acts on it (**Fig. 4.9a**), the puck starts to move. If the puck is in motion to begin with, the force changes its speed, its direction, or both, depending on the direction of the force. In this case the net force is equal to $\vec{F}_1$, which is *not* zero. (There are also two vertical forces: the earth's gravitational attraction and the upward normal force exerted by the surface. But as we mentioned earlier, these two forces cancel.)

Now suppose we apply a second force, $\vec{F}_2$ (Fig. 4.9b), equal in magnitude to $\vec{F}_1$ but opposite in direction. The two forces are negatives of each other, $\vec{F}_2 = -\vec{F}_1$, and their vector sum is zero:

$$\Sigma \vec{F} = \vec{F}_1 + \vec{F}_2 = \vec{F}_1 + (-\vec{F}_1) = 0$$

Again, we find that if the body is at rest at the start, it remains at rest; if it is initially moving, it continues to move in the same direction with constant speed. These results show that in Newton's first law, *zero net force is equivalent to no force at all.* This is just the principle of superposition of forces that we saw in Section 4.1.

When a body is either at rest or moving with constant velocity (in a straight line with constant speed), we say that the body is in **equilibrium.** For a body to be in equilibrium, it must be acted on by no forces, or by several forces such that their vector sum—that is, the net force—is zero:

Newton's first law: $\cdots\!\!\rightarrow \Sigma \vec{F} = 0 \leftarrow\!\!\cdots$ ... must be zero if body
Net force on a body ... $\qquad\qquad$ is in **equilibrium.** $\qquad$ (4.3)

We're assuming that the body can be represented adequately as a point particle. When the body has finite size, we also have to consider *where* on the body the forces are applied. We'll return to this point in Chapter 11.

Application Sledding with Newton's First Law The downward force of gravity acting on the child and sled is balanced by an upward normal force exerted by the ground. The adult's foot exerts a forward force that balances the backward force of friction on the sled. Hence there is no net force on the child and sled, and they slide with a constant velocity.

CONCEPTUAL EXAMPLE 4.2 **ZERO NET FORCE MEANS CONSTANT VELOCITY**

In the classic 1950 science-fiction film *Rocketship X-M*, a space-ship is moving in the vacuum of outer space, far from any star or planet, when its engine dies. As a result, the spaceship slows down and stops. What does Newton's first law say about this scene?

SOLUTION

No forces act on the spaceship after the engine dies, so according to Newton's first law it will *not* stop but will continue to move in a straight line with constant speed. Some science-fiction movies are based on accurate science; this is not one of them.

CONCEPTUAL EXAMPLE 4.3 **CONSTANT VELOCITY MEANS ZERO NET FORCE**

You are driving a Maserati GranTurismo S on a straight testing track at a constant speed of 250 km/h. You pass a 1971 Volkswagen Beetle doing a constant 75 km/h. On which car is the net force greater?

SOLUTION

The key word in this question is "net." Both cars are in equilibrium because their velocities are constant; Newton's first law therefore says that the *net* force on each car is *zero*.

This seems to contradict the "common sense" idea that the faster car must have a greater force pushing it. Thanks to your

Maserati's high-power engine, it's true that the track exerts a greater forward force on your Maserati than it does on the Volkswagen. But a *backward* force also acts on each car due to road friction and air resistance. When the car is traveling with constant velocity, the vector sum of the forward and backward forces is zero. There is more air resistance on the fast-moving Maserati than on the slow-moving Volkswagen, which is why the Maserati's engine must be more powerful than that of the Volkswagen.

Inertial Frames of Reference

In discussing relative velocity in Section 3.5, we introduced the concept of *frame of reference*. This concept is central to Newton's laws of motion. Suppose you are in a bus that is traveling on a straight road and speeding up. If you could stand in the aisle on roller skates, you would start moving *backward* relative to the bus as the bus gains speed. If instead the bus was slowing to a stop, you would start moving forward down the aisle. In either case, it looks as though Newton's first law is not obeyed; there is no net force acting on you, yet your velocity changes. What's wrong?

The point is that the bus is accelerating with respect to the earth and is *not* a suitable frame of reference for Newton's first law. This law is valid in some frames of reference and not valid in others. A frame of reference in which Newton's first law *is* valid is called an **inertial frame of reference.** The earth is at least approximately an inertial frame of reference, but the bus is not. (The earth is not a completely inertial frame, owing to the acceleration associated with its rotation and its motion around the sun. These effects are quite small, however; see Exercises 3.23 and 3.28.) Because Newton's first law is used to define what we mean by an inertial frame of reference, it is sometimes called the *law of inertia*.

Figure 4.10 helps us understand what you experience when riding in a vehicle that's accelerating. In Fig. 4.10a, a vehicle is initially at rest and then begins to

4.10 Riding in an accelerating vehicle.

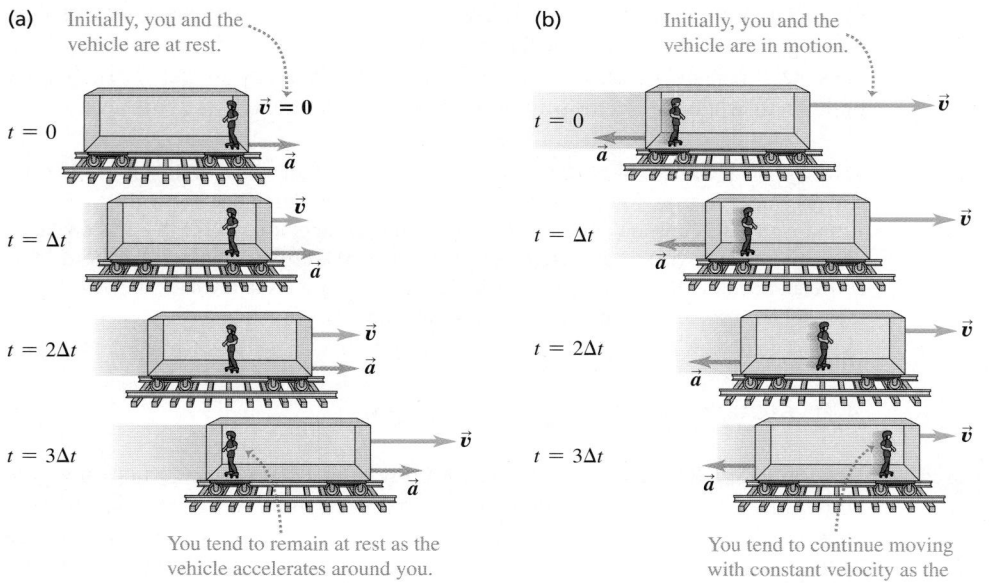

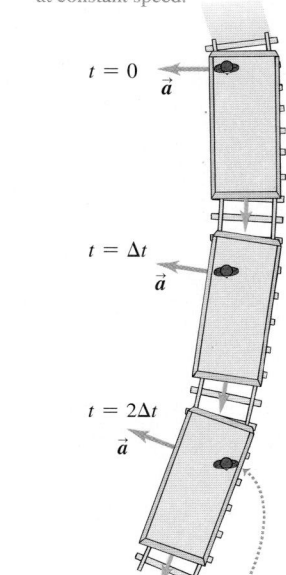

accelerate to the right. A passenger standing on roller skates (which nearly eliminate the effects of friction) has virtually no net force acting on her, so she tends to remain at rest relative to the inertial frame of the earth. As the vehicle accelerates around her, she moves backward relative to the vehicle. In the same way, a passenger in a vehicle that is slowing down tends to continue moving with constant velocity relative to the earth, and so moves forward relative to the vehicle (Fig. 4.10b). A vehicle is also accelerating if it moves at a constant speed but is turning (Fig. 4.10c). In this case a passenger tends to continue moving relative to the earth at constant speed in a straight line; relative to the vehicle, the passenger moves to the side of the vehicle on the outside of the turn.

In each case shown in Fig. 4.10, an observer in the vehicle's frame of reference might be tempted to conclude that there *is* a net force acting on the passenger, since the passenger's velocity *relative to the vehicle* changes in each case. This conclusion is simply wrong; the net force on the passenger is indeed zero. The vehicle observer's mistake is in trying to apply Newton's first law in the vehicle's frame of reference, which is *not* an inertial frame and in which Newton's first law isn't valid (**Fig. 4.11**). In this book we will use *only* inertial frames of reference.

We've mentioned only one (approximately) inertial frame of reference: the earth's surface. But there are many inertial frames. If we have an inertial frame of reference A, in which Newton's first law is obeyed, then *any* second frame of reference B will also be inertial if it moves relative to A with constant velocity $\vec{v}_{B/A}$. We can prove this by using the relative-velocity relationship Eq. (3.35) from Section 3.5:

$$\vec{v}_{P/A} = \vec{v}_{P/B} + \vec{v}_{B/A}$$

Suppose that P is a body that moves with constant velocity $\vec{v}_{P/A}$ with respect to an inertial frame A. By Newton's first law the net force on this body is zero. The velocity of P relative to another frame B has a different value, $\vec{v}_{P/B} = \vec{v}_{P/A} - \vec{v}_{B/A}$. But if the relative velocity $\vec{v}_{B/A}$ of the two frames is constant, then $\vec{v}_{P/B}$ is constant as well. Thus B is also an inertial frame; the velocity of P in this frame is constant, and the net force on P is zero, so Newton's first law is obeyed in B. Observers in frames A and B will disagree about the velocity of P, but they will agree that P has a constant velocity (zero acceleration) and has zero net force acting on it.

There is no single inertial frame of reference that is preferred over all others for formulating Newton's laws. If one frame is inertial, then every other frame moving relative to it with constant velocity is also inertial. Viewed in this light, the state of rest and the state of motion with constant velocity are not very different; both occur when the vector sum of forces acting on the body is zero.

4.11 From the frame of reference of the car, it seems as though a force is pushing the crash test dummies forward as the car comes to a sudden stop. But there is really no such force: As the car stops, the dummies keep moving forward as a consequence of Newton's first law.

TEST YOUR UNDERSTANDING OF SECTION 4.2   In which of the following situations is there zero net force on the body? (i) An airplane flying due north at a steady 120 m/s and at a constant altitude; (ii) a car driving straight up a hill with a 3° slope at a constant 90 km/h; (iii) a hawk circling at a constant 20 km/h at a constant height of 15 m above an open field; (iv) a box with slick, frictionless surfaces in the back of a truck as the truck accelerates forward on a level road at 5 m/s². ∎

4.3 NEWTON'S SECOND LAW

Newton's first law tells us that when a body is acted on by zero net force, the body moves with constant velocity and zero acceleration. In **Fig. 4.12a**, a hockey puck is sliding to the right on wet ice. There is negligible friction, so there are no horizontal forces acting on the puck; the downward force of gravity and the upward normal force exerted by the ice surface sum to zero. So the net force $\sum \vec{F}$ acting on the puck is zero, the puck has zero acceleration, and its velocity is constant.

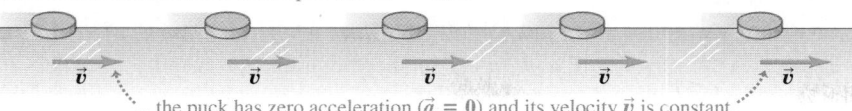

(a) If there is zero net force on the puck, so $\sum \vec{F} = 0$, ...

... the puck has zero acceleration ($\vec{a} = 0$) and its velocity $\vec{v}$ is constant.

(b) If a constant net force $\sum \vec{F}$ acts on the puck in the direction of its motion ...

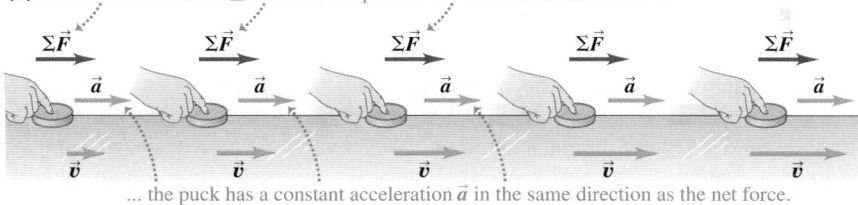

... the puck has a constant acceleration $\vec{a}$ in the same direction as the net force.

(c) If a constant net force $\sum \vec{F}$ acts on the puck opposite to the direction of its motion ...

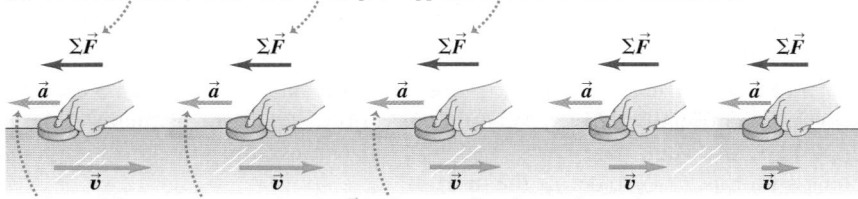

... the puck has a constant acceleration $\vec{a}$ in the same direction as the net force.

4.12 Using a hockey puck on a friction-less surface to explore the relationship between the net force $\sum \vec{F}$ on a body and the resulting acceleration $\vec{a}$ of the body.

4.13 A top view of a hockey puck in uniform circular motion on a frictionless horizontal surface.

Puck moves at constant speed around circle.

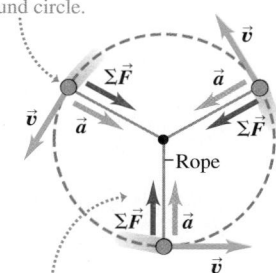

At all points, the acceleration $\vec{a}$ and the net force $\sum \vec{F}$ point in the same direction—always toward the center of the circle.

But what happens when the net force is *not* zero? In Fig. 4.12b we apply a constant horizontal force to a sliding puck in the same direction that the puck is moving. Then $\sum \vec{F}$ is constant and in the same horizontal direction as $\vec{v}$. We find that during the time the force is acting, the velocity of the puck changes at a constant rate; that is, the puck moves with constant acceleration. The speed of the puck increases, so the acceleration $\vec{a}$ is in the same direction as $\vec{v}$ and $\sum \vec{F}$.

In Fig. 4.12c we reverse the direction of the force on the puck so that $\sum \vec{F}$ acts opposite to $\vec{v}$. In this case as well, the puck has an acceleration; the puck moves more and more slowly to the right. The acceleration $\vec{a}$ in this case is to the left, in the same direction as $\sum \vec{F}$. As in the previous case, experiment shows that the acceleration is constant if $\sum \vec{F}$ is constant.

We conclude that *a net force acting on a body causes the body to accelerate in the same direction as the net force.* If the magnitude of the net force is constant, as in Figs. 4.12b and 4.12c, then so is the magnitude of the acceleration.

These conclusions about net force and acceleration also apply to a body moving along a curved path. For example, **Fig. 4.13** shows a hockey puck moving in a horizontal circle on an ice surface of negligible friction. A rope is attached to the puck and to a stick in the ice, and this rope exerts an inward tension force of constant magnitude on the puck. The net force and acceleration are both constant in magnitude and directed toward the center of the circle. The speed of the puck is constant, so this is uniform circular motion, as discussed in Section 3.4.

Figure 4.14a shows another experiment to explore the relationship between acceleration and net force. We apply a constant horizontal force to a puck on a frictionless horizontal surface, using the spring balance described in Section 4.1 with the spring stretched a constant amount. As in Figs. 4.12b and 4.12c, this horizontal force equals the net force on the puck. If we change the magnitude of the net force, the acceleration changes in the same proportion. Doubling the net force doubles the acceleration (Fig. 4.14b), halving the net force halves the acceleration (Fig. 4.14c), and so on. Many such experiments show that *for any given body, the magnitude of the acceleration is directly proportional to the magnitude of the net force acting on the body.*

4.14 The magnitude of a body's acceleration $\vec{a}$ is directly proportional to the magnitude of the net force $\sum \vec{F}$ acting on the body of mass m.

(a) A constant net force $\sum \vec{F}$ causes a constant acceleration $\vec{a}$.

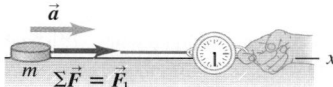

(b) Doubling the net force doubles the acceleration.

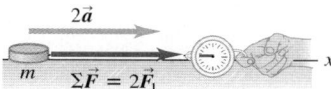

(c) Halving the force halves the acceleration.

Mass and Force

Our results mean that for a given body, the *ratio* of the magnitude $|\Sigma \vec{F}|$ of the net force to the magnitude $a = |\vec{a}|$ of the acceleration is constant, regardless of the magnitude of the net force. We call this ratio the *inertial mass,* or simply the **mass,** of the body and denote it by m. That is,

$$m = \frac{|\Sigma \vec{F}|}{a} \qquad \text{or} \qquad |\Sigma \vec{F}| = ma \qquad \text{or} \qquad a = \frac{|\Sigma \vec{F}|}{m} \qquad (4.4)$$

Mass is a quantitative measure of inertia, which we discussed in Section 4.2. The last of the equations in Eqs. (4.4) says that the greater a body's mass, the more the body "resists" being accelerated. When you hold a piece of fruit in your hand at the supermarket and move it slightly up and down to estimate its heft, you're applying a force and seeing how much the fruit accelerates up and down in response. If a force causes a large acceleration, the fruit has a small mass; if the same force causes only a small acceleration, the fruit has a large mass. In the same way, if you hit a table-tennis ball and then a basketball with the same force, the basketball has much smaller acceleration because it has much greater mass.

The SI unit of mass is the **kilogram.** We mentioned in Section 1.3 that the kilogram is officially defined to be the mass of a cylinder of platinum–iridium alloy kept in a vault near Paris (Fig. 1.4). We can use this standard kilogram, along with Eqs. (4.4), to define the **newton:**

> **One newton is the amount of net force that gives an acceleration of 1 meter per second squared to a body with a mass of 1 kilogram.**

This definition allows us to calibrate the spring balances and other instruments used to measure forces. Because of the way we have defined the newton, it is related to the units of mass, length, and time. For Eqs. (4.4) to be dimensionally consistent, it must be true that

$$1 \text{ newton} = (1 \text{ kilogram})(1 \text{ meter per second squared})$$

or

$$1 \text{ N} = 1 \text{ kg} \cdot \text{m/s}^2$$

We will use this relationship many times in the next few chapters, so keep it in mind.

We can also use Eqs. (4.4) to compare a mass with the standard mass and thus to *measure* masses. Suppose we apply a constant net force $\Sigma \vec{F}$ to a body having a known mass m_1 and we find an acceleration of magnitude a_1 (**Fig. 4.15a**). We then apply the same force to another body having an unknown mass m_2, and we find an acceleration of magnitude a_2 (Fig. 4.15b). Then, according to Eqs. (4.4),

$$m_1 a_1 = m_2 a_2$$

$$\frac{m_2}{m_1} = \frac{a_1}{a_2} \qquad \text{(same net force)} \qquad (4.5)$$

For the same net force, the ratio of the masses of two bodies is the inverse of the ratio of their accelerations. In principle we could use Eq. (4.5) to measure an unknown mass m_2, but it is usually easier to determine mass indirectly by measuring the body's *weight.* We'll return to this point in Section 4.4.

When two bodies with masses m_1 and m_2 are fastened together, we find that the mass of the composite body is always $m_1 + m_2$ (Fig. 4.15c). This additive property of mass may seem obvious, but it has to be verified experimentally. Ultimately, the mass of a body is related to the number of protons, electrons, and neutrons it contains. This wouldn't be a good way to *define* mass because there is no practical way to count these particles. But the concept of mass is the most fundamental way to characterize the quantity of matter in a body.

4.15 For a given net force $\Sigma \vec{F}$ acting on a body, the acceleration is inversely proportional to the mass of the body. Masses add like ordinary scalars.

(a) A known force $\Sigma \vec{F}$ causes an object with mass m_1 to have an acceleration $\vec{a}_1$.

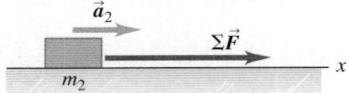

(b) Applying the same force $\Sigma \vec{F}$ to a second object and noting the acceleration allow us to measure the mass.

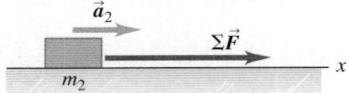

(c) When the two objects are fastened together, the same method shows that their composite mass is the sum of their individual masses.

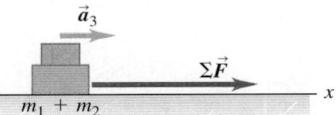

Stating Newton's Second Law

Experiment shows that the *net* force on a body is what causes that body to accelerate. If a combination of forces $\vec{F}_1$, $\vec{F}_2$, $\vec{F}_3$, and so on is applied to a body, the body will have the same acceleration vector $\vec{a}$ as when only a single force is applied, if that single force is equal to the vector sum $\vec{F}_1 + \vec{F}_2 + \vec{F}_3 + \cdots$. In other words, the principle of superposition of forces (see Fig. 4.4) also holds true when the net force is not zero and the body is accelerating.

Equations (4.4) relate the magnitude of the net force on a body to the magnitude of the acceleration that it produces. We have also seen that the direction of the net force is the same as the direction of the acceleration, whether the body's path is straight or curved. What's more, the forces that affect a body's motion are *external* forces, those exerted on the body by other bodies in its environment. Newton wrapped up all these results into a single concise statement that we now call *Newton's second law of motion:*

> **NEWTON'S SECOND LAW OF MOTION:** **If a net external force acts on a body, the body accelerates. The direction of acceleration is the same as the direction of the net force. The mass of the body times the acceleration vector of the body equals the net force vector.**

In symbols,

Newton's second law:
If there is a net force on a body ...
$$\sum \vec{F} = m\vec{a}$$
... the body accelerates in same direction as the net force.
Mass of body
(4.6)

An alternative statement is that the acceleration of a body is equal to the net force acting on the body divided by the body's mass:

$$\vec{a} = \frac{\sum \vec{F}}{m}$$

Newton's second law is a fundamental law of nature, the basic relationship between force and motion. Most of the remainder of this chapter and all of the next are devoted to learning how to apply this principle in various situations.

Equation (4.6) has many practical applications (**Fig. 4.16**). You've actually been using it all your life to measure your body's acceleration. In your inner ear, microscopic hair cells sense the magnitude and direction of the force that they must exert to cause small membranes to accelerate along with the rest of your body. By Newton's second law, the acceleration of the membranes—and hence that of your body as a whole—is proportional to this force and has the same direction. In this way, you can sense the magnitude and direction of your acceleration even with your eyes closed!

Using Newton's Second Law

There are at least four aspects of Newton's second law that deserve special attention. First, Eq. (4.6) is a *vector* equation. Usually we will use it in component form, with a separate equation for each component of force and the corresponding component of acceleration:

Newton's second law: Each component of net force on a body ...
$$\sum F_x = ma_x \qquad \sum F_y = ma_y \qquad \sum F_z = ma_z$$
... equals body's mass times the corresponding acceleration component.
(4.7)

This set of component equations is equivalent to the single vector Eq. (4.6).

Second, the statement of Newton's second law refers to *external* forces. It's impossible for a body to affect its own motion by exerting a force on itself; if it were possible, you could lift yourself to the ceiling by pulling up on your belt! That's why only external forces are included in the sum $\sum \vec{F}$ in Eqs. (4.6) and (4.7).

4.16 The design of high-performance motorcycles depends fundamentally on Newton's second law. To maximize the forward acceleration, the designer makes the motorcycle as light as possible (that is, minimizes the mass) and uses the most powerful engine possible (thus maximizing the forward force).

Lightweight body (small m)

Powerful engine (large F)

Application Blame Newton's Second Law This car stopped because of Newton's second law: The tree exerted an external force on the car, giving the car an acceleration that changed its velocity to zero.

Third, Eqs. (4.6) and (4.7) are valid only when the mass m is *constant*. It's easy to think of systems whose masses change, such as a leaking tank truck, a rocket ship, or a moving railroad car being loaded with coal. Such systems are better handled by using the concept of momentum; we'll get to that in Chapter 8.

DEMO

Finally, Newton's second law is valid in inertial frames of reference only, just like the first law. Thus it is not valid in the reference frame of any of the accelerating vehicles in Fig. 4.10; relative to any of these frames, the passenger accelerates even though the net force on the passenger is zero. We will usually assume that the earth is an adequate approximation to an inertial frame, although because of its rotation and orbital motion it is not precisely inertial.

CAUTION $m\vec{a}$ **is not a force** Even though the vector $m\vec{a}$ is equal to the vector sum $\Sigma\vec{F}$ of all the forces acting on the body, the vector $m\vec{a}$ is *not* a force. Acceleration is a *result* of a nonzero net force; it is not a force itself. It's "common sense" to think that there is a "force of acceleration" that pushes you back into your seat when your car accelerates forward from rest. But *there is no such force;* instead, your inertia causes you to tend to stay at rest relative to the earth, and the car accelerates around you (see Fig. 4.10a). The "common sense" confusion arises from trying to apply Newton's second law where it isn't valid—in the noninertial reference frame of an accelerating car. We will always examine motion relative to *inertial* frames of reference only. ▌

In learning how to use Newton's second law, we will begin in this chapter with examples of straight-line motion. Then in Chapter 5 we will consider more general cases and develop more detailed problem-solving strategies.

SOLUTION

EXAMPLE 4.4 DETERMINING ACCELERATION FROM FORCE

A worker applies a constant horizontal force with magnitude 20 N to a box with mass 40 kg resting on a level floor with negligible friction. What is the acceleration of the box?

SOLUTION

IDENTIFY and SET UP: This problem involves force and acceleration, so we'll use Newton's second law. In *any* problem involving forces, the first steps are to choose a coordinate system and to identify all of the forces acting on the body in question. It's usually convenient to take one axis either along or opposite the direction of the body's acceleration, which in this case is horizontal. Hence we take the $+x$-axis to be in the direction of the applied horizontal force (which is the direction in which the box accelerates) and the $+y$-axis to be upward (**Fig. 4.17**). In most force problems that you'll encounter (including this one), the force vectors all lie in a plane, so the z-axis isn't used.

4.17 Our sketch for this problem. The tiles under the box are freshly waxed, so we assume that friction is negligible.

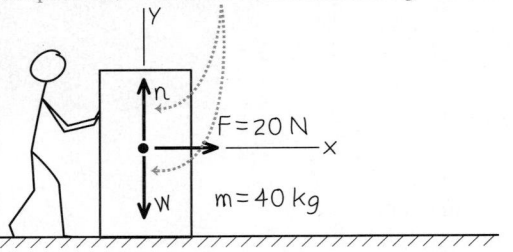

The box has no vertical acceleration, so the vertical components of the net force sum to zero. Nevertheless, for completeness, we show the vertical forces acting on the box.

The forces acting on the box are (i) the horizontal force $\vec{F}$ exerted by the worker, of magnitude 20 N; (ii) the weight $\vec{w}$ of the box—that is, the downward gravitational force exerted by the earth; and (iii) the upward supporting force $\vec{n}$ exerted by the floor. As in Section 4.2, we call $\vec{n}$ a *normal* force because it is normal (perpendicular) to the surface of contact. (We use an italic letter n to avoid confusion with the abbreviation N for newton.) Friction is negligible, so no friction force is present.

The box doesn't move vertically, so the y-acceleration is zero: $a_y = 0$. Our target variable is the x-acceleration, a_x. We'll find it by using Newton's second law in component form, Eqs. (4.7).

EXECUTE: From Fig. 4.17 only the 20-N force exerted by the worker has a nonzero x-component. Hence the first of Eqs. (4.7) tells us that

$$\Sigma F_x = F = 20\,\text{N} = ma_x$$

The x-component of acceleration is therefore

$$a_x = \frac{\Sigma F_x}{m} = \frac{20\,\text{N}}{40\,\text{kg}} = \frac{20\,\text{kg}\cdot\text{m/s}^2}{40\,\text{kg}} = 0.50\,\text{m/s}^2$$

EVALUATE: The acceleration is in the $+x$-direction, the same direction as the net force. The net force is constant, so the acceleration is also constant. If we know the initial position and velocity of the box, we can find its position and velocity at any later time from the constant-acceleration equations of Chapter 2.

To determine a_x, we didn't need the y-component of Newton's second law from Eqs. (4.7), $\Sigma F_y = ma_y$. Can you use this equation to show that the magnitude n of the normal force in this situation is equal to the weight of the box?

SOLUTION

EXAMPLE 4.5 DETERMINING FORCE FROM ACCELERATION

A waitress shoves a ketchup bottle with mass 0.45 kg to her right along a smooth, level lunch counter. The bottle leaves her hand moving at 2.8 m/s, then slows down as it slides because of a constant horizontal friction force exerted on it by the countertop. It slides for 1.0 m before coming to rest. What are the magnitude and direction of the friction force acting on the bottle?

SOLUTION

IDENTIFY and SET UP: This problem involves forces and acceleration (the slowing of the ketchup bottle), so we'll use Newton's second law to solve it. As in Example 4.4, we choose a coordinate system and identify the forces acting on the bottle (**Fig. 4.18**). We choose the $+x$-axis to be in the direction that the bottle slides, and take the origin to be where the bottle leaves the waitress's hand. The friction force $\vec{f}$ slows the bottle down, so its direction must be opposite the direction of the bottle's velocity (see Fig. 4.12c).

Our target variable is the magnitude f of the friction force. We'll find it by using the x-component of Newton's second law from Eqs. (4.7). We aren't told the x-component of the bottle's acceleration, a_x, but we know that it's constant because the friction force that causes the acceleration is constant. Hence we can use a constant-acceleration formula from Section 2.4 to calculate a_x. We know the bottle's initial and final x-coordinates ($x_0 = 0$ and $x = 1.0$ m) and its initial and final x-velocity ($v_{0x} = 2.8$ m/s and $v_x = 0$), so the easiest equation to use is Eq. (2.13), $v_x^2 = v_{0x}^2 + 2a_x(x - x_0)$.

EXECUTE: We solve Eq. (2.13) for a_x:

$$a_x = \frac{v_x^2 - v_{0x}^2}{2(x - x_0)} = \frac{(0 \text{ m/s})^2 - (2.8 \text{ m/s})^2}{2(1.0 \text{ m} - 0 \text{ m})} = -3.9 \text{ m/s}^2$$

The negative sign means that the bottle's acceleration is toward the *left* in Fig. 4.18, opposite to its velocity; this is as it must be, because the bottle is slowing down. The net force in the x-direction is the x-component $-f$ of the friction force, so

$$\Sigma F_x = -f = ma_x = (0.45 \text{ kg})(-3.9 \text{ m/s}^2)$$
$$= -1.8 \text{ kg} \cdot \text{m/s}^2 = -1.8 \text{ N}$$

The negative sign shows that the net force on the bottle is toward the left. The *magnitude* of the friction force is $f = 1.8$ N.

EVALUATE: As a check on the result, try repeating the calculation with the $+x$-axis to the *left* in Fig. 4.18. You'll find that ΣF_x is equal to $+f = +1.8$ N (because the friction force is now in the $+x$-direction), and again you'll find $f = 1.8$ N. The answers for the *magnitudes* of forces don't depend on the choice of coordinate axes!

4.18 Our sketch for this problem.

We draw one diagram for the bottle's motion and one showing the forces on the bottle.

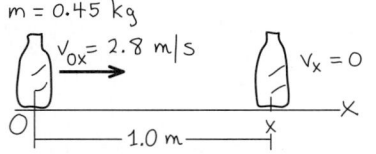
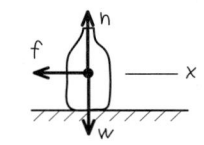

Some Notes on Units

A few words about units are in order. In the cgs metric system (not used in this book), the unit of mass is the gram, equal to 10^{-3} kg, and the unit of distance is the centimeter, equal to 10^{-2} m. The cgs unit of force is called the *dyne:*

$$1 \text{ dyne} = 1 \text{ g} \cdot \text{cm/s}^2 = 10^{-5} \text{ N}$$

In the British system, the unit of force is the *pound* (or pound-force) and the unit of mass is the *slug* (**Fig. 4.19**). The unit of acceleration is 1 foot per second squared, so

$$1 \text{ pound} = 1 \text{ slug} \cdot \text{ft/s}^2$$

The official definition of the pound is

$$1 \text{ pound} = 4.448221615260 \text{ newtons}$$

It is handy to remember that a pound is about 4.4 N and a newton is about 0.22 pound. Another useful fact: A body with a mass of 1 kg has a weight of about 2.2 lb at the earth's surface.

Table 4.2 lists the units of force, mass, and acceleration in the three systems.

4.19 Despite its name, the English unit of mass has nothing to do with the type of slug shown here. A common garden slug has a mass of about 15 grams, or about 10^{-3} slug.

Units of Force, Mass, and Acceleration

TABLE 4.2			
System of Units	Force	Mass	Acceleration
SI	newton (N)	kilogram (kg)	m/s^2
cgs	dyne (dyn)	gram (g)	cm/s^2
British	pound (lb)	slug	ft/s^2

TEST YOUR UNDERSTANDING OF SECTION 4.3   Rank the following situations in order of the magnitude of the object's acceleration, from lowest to highest. Are there any cases that have the same magnitude of acceleration? (i) A 2.0-kg object acted on by a 2.0-N net force; (ii) a 2.0-kg object acted on by an 8.0-N net force; (iii) an 8.0-kg object acted on by a 2.0-N net force; (iv) an 8.0-kg object acted on by an 8.0-N net force. ▮

4.4 MASS AND WEIGHT

One of the most familiar forces is the *weight* of a body, which is the gravitational force that the earth exerts on the body. (If you are on another planet, your weight is the gravitational force that planet exerts on you.) Unfortunately, the terms *mass* and *weight* are often misused and interchanged in everyday conversation. It is absolutely essential for you to understand clearly the distinctions between these two physical quantities.

Mass characterizes the *inertial* properties of a body. Mass is what keeps the china on the table when you yank the tablecloth out from under it. The greater the mass, the greater the force needed to cause a given acceleration; this is reflected in Newton's second law, $\sum \vec{F} = m\vec{a}$.

Weight, on the other hand, is a *force* exerted on a body by the pull of the earth. Mass and weight are related: Bodies that have large mass also have large weight. A large stone is hard to throw because of its large *mass,* and hard to lift off the ground because of its large *weight.*

To understand the relationship between mass and weight, note that a freely falling body has an acceleration of magnitude g (see Section 2.5). Newton's second law tells us that a force must act to produce this acceleration. If a 1-kg body falls with an acceleration of 9.8 m/s², the required force has magnitude

$$F = ma = (1 \text{ kg})(9.8 \text{ m/s}^2) = 9.8 \text{ kg} \cdot \text{m/s}^2 = 9.8 \text{ N}$$

The force that makes the body accelerate downward is its weight. Any body near the surface of the earth that has a mass of 1 kg *must* have a weight of 9.8 N to give it the acceleration we observe when it is in free fall. More generally,

$$\underset{\substack{\text{Magnitude of} \\ \text{weight of a body}}}{} \quad w = mg \quad \underset{\substack{\text{Mass of body} \\ \text{Magnitude of acceleration} \\ \text{due to gravity}}}{} \tag{4.8}$$

Hence the magnitude w of a body's weight is directly proportional to its mass m. The weight of a body is a force, a vector quantity, and we can write Eq. (4.8) as a vector equation (**Fig. 4.20**):

$$\vec{w} = m\vec{g} \tag{4.9}$$

Remember that g is the *magnitude* of $\vec{g}$, the acceleration due to gravity, so g is always a positive number, by definition. Thus w, given by Eq. (4.8), is the *magnitude* of the weight and is also always positive.

4.20 Relating the mass and weight of a body.

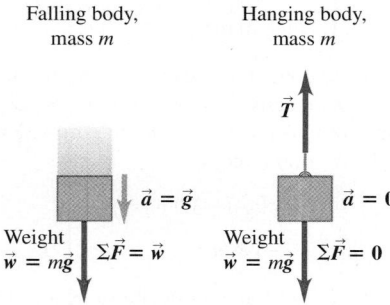

Falling body, mass m   Hanging body, mass m

$\vec{a} = \vec{g}$ $\vec{a} = 0$

Weight $\vec{w} = m\vec{g}$ $\sum\vec{F} = \vec{w}$ Weight $\vec{w} = m\vec{g}$ $\sum\vec{F} = 0$

- The relationship of mass to weight: $\vec{w} = m\vec{g}$.
- This relationship is the same whether a body is falling or stationary.

CAUTION **A body's weight acts at all times** The weight of a body acts on the body *all the time,* whether it is in free fall or not. If we suspend an object from a rope, it is in equilibrium and its acceleration is zero. But its weight, given by Eq. (4.9), is still pulling down on it (Fig. 4.20). In this case the rope pulls up on the object, applying an upward force. The *vector sum* of the forces is zero, but the weight still acts. ∎

CONCEPTUAL EXAMPLE 4.6   **NET FORCE AND ACCELERATION IN FREE FALL**

In Example 2.6 of Section 2.5, a one-euro coin was dropped from rest from the Leaning Tower of Pisa. If the coin falls freely, so that the effects of the air are negligible, how does the net force on the coin vary as it falls?

SOLUTION

In free fall, the acceleration $\vec{a}$ of the coin is constant and equal to $\vec{g}$. Hence by Newton's second law the net force $\sum\vec{F} = m\vec{a}$ is also constant and equal to $m\vec{g}$, which is the coin's weight $\vec{w}$ (**Fig. 4.21**). The coin's velocity changes as it falls, but the net force acting on it is constant. (If this surprises you, reread Conceptual Example 4.3.)

The net force on a freely falling coin is constant even if you initially toss it upward. The force that your hand exerts on the coin to toss it is a contact force, and it disappears the instant the coin

leaves your hand. From then on, the only force acting on the coin is its weight $\vec{w}$.

4.21 The acceleration of a freely falling object is constant, and so is the net force acting on the object.

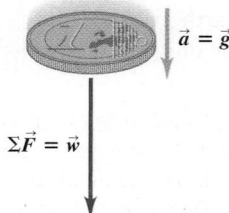

$\vec{a} = \vec{g}$

$\sum\vec{F} = \vec{w}$

Variation of *g* with Location

We will use $g = 9.80 \text{ m/s}^2$ for problems set on the earth (or, if the other data in the problem are given to only two significant figures, $g = 9.8 \text{ m/s}^2$). In fact, the value of *g* varies somewhat from point to point on the earth's surface—from about 9.78 to 9.82 m/s²—because the earth is not perfectly spherical and because of effects due to its rotation and orbital motion. At a point where $g = 9.80 \text{ m/s}^2$, the weight of a standard kilogram is $w = 9.80 \text{ N}$. At a different point, where $g = 9.78 \text{ m/s}^2$, the weight is $w = 9.78 \text{ N}$ but the mass is still 1 kg. The weight of a body varies from one location to another; the mass does not.

If we take a standard kilogram to the surface of the moon, where the acceleration of free fall (equal to the value of *g* at the moon's surface) is 1.62 m/s², its weight is 1.62 N but its mass is still 1 kg (**Fig. 4.22**). An 80.0-kg astronaut has a weight on earth of $(80.0 \text{ kg})(9.80 \text{ m/s}^2) = 784 \text{ N}$, but on the moon the astronaut's weight would be only $(80.0 \text{ kg})(1.62 \text{ m/s}^2) = 130 \text{ N}$. In Chapter 13 we'll see how to calculate the value of *g* at the surface of the moon or on other worlds.

Measuring Mass and Weight

In Section 4.3 we described a way to compare masses by comparing their accelerations when they are subjected to the same net force. Usually, however, the easiest way to measure the mass of a body is to measure its weight, often by comparing with a standard. Equation (4.8) says that two bodies that have the same weight at a particular location also have the same mass. We can compare weights very precisely; the familiar equal-arm balance (**Fig. 4.23**) can determine with great precision (up to 1 part in 10^6) when the weights of two bodies are equal and hence when their masses are equal.

The concept of mass plays two rather different roles in mechanics. The weight of a body (the gravitational force acting on it) is proportional to its mass; we call the property related to gravitational interactions *gravitational mass*. On the other hand, we call the inertial property that appears in Newton's second law the *inertial mass*. If these two quantities were different, the acceleration due to gravity might well be different for different bodies. However, extraordinarily precise experiments have established that in fact the two *are* the same to a precision of better than one part in 10^{12}.

CAUTION **Don't confuse mass and weight** The SI units for mass and weight are often misused in everyday life. For example, it's incorrect to say "This box weighs 6 kg"; what this really means is that the *mass* of the box, probably determined indirectly by *weighing*, is 6 kg. Avoid this sloppy usage in your own work! In SI units, weight (a force) is measured in newtons, while mass is measured in kilograms. ∎

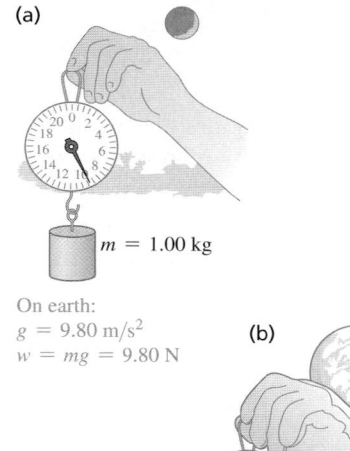

4.22 The weight of a 1-kilogram mass (a) on earth and (b) on the moon.

(a)

$m = 1.00 \text{ kg}$

On earth:
$g = 9.80 \text{ m/s}^2$
$w = mg = 9.80 \text{ N}$

(b)

On the moon:
$g = 1.62 \text{ m/s}^2$
$w = mg = 1.62 \text{ N}$
$m = 1.00 \text{ kg}$

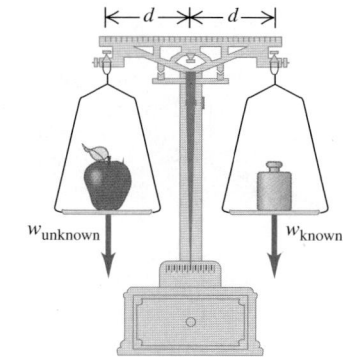

4.23 An equal-arm balance determines the mass of a body (such as an apple) by comparing its weight to a known weight.

w_{unknown} w_{known}

EXAMPLE 4.7   MASS AND WEIGHT

SOLUTION

A 2.49×10^4 N Rolls-Royce Phantom traveling in the $+x$-direction makes an emergency stop; the x-component of the net force acting on it is -1.83×10^4 N. What is its acceleration?

SOLUTION

IDENTIFY and SET UP: Our target variable is the x-component of the car's acceleration, a_x. We use the x-component portion of Newton's second law, Eqs. (4.7), to relate force and acceleration. To do this, we need to know the car's mass. The newton is a unit for force, however, so 2.49×10^4 N is the car's *weight*, not its

mass. Hence we'll first use Eq. (4.8) to determine the car's mass from its weight. The car has a positive x-velocity and is slowing down, so its x-acceleration will be negative.

EXECUTE: The mass of the car is

$$m = \frac{w}{g} = \frac{2.49 \times 10^4 \text{ N}}{9.80 \text{ m/s}^2} = \frac{2.49 \times 10^4 \text{ kg} \cdot \text{m/s}^2}{9.80 \text{ m/s}^2}$$

$$= 2540 \text{ kg}$$

Continued

Then $\sum F_x = ma_x$ gives

$$a_x = \frac{\sum F_x}{m} = \frac{-1.83 \times 10^4 \text{ N}}{2540 \text{ kg}} = \frac{-1.83 \times 10^4 \text{ kg} \cdot \text{m/s}^2}{2540 \text{ kg}}$$

$$= -7.20 \text{ m/s}^2$$

EVALUATE: The negative sign means that the acceleration vector points in the negative x-direction, as we expected. The magnitude of this acceleration is pretty high; passengers in this car will experience a lot of rearward force from their shoulder belts.

This acceleration equals $-0.735g$. The number -0.735 is also the ratio of -1.83×10^4 N (the x-component of the net force) to 2.49×10^4 N (the weight). In fact, the acceleration of a body, expressed as a multiple of g, is *always* equal to the ratio of the net force on the body to its weight. Can you see why?

TEST YOUR UNDERSTANDING OF SECTION 4.4   Suppose an astronaut landed on a planet where $g = 19.6 \text{ m/s}^2$. Compared to earth, would it be easier, harder, or just as easy for her to walk around? Would it be easier, harder, or just as easy for her to catch a ball that is moving horizontally at 12 m/s? (Assume that the astronaut's spacesuit is a lightweight model that doesn't impede her movements in any way.) ▌

4.5 NEWTON'S THIRD LAW

A force acting on a body is always the result of its interaction with another body, so forces always come in pairs. You can't pull on a doorknob without the doorknob pulling back on you. When you kick a football, the forward force that your foot exerts on the ball launches it into its trajectory, but you also feel the force the ball exerts back on your foot.

In each of these cases, the force that you exert on the other body is in the opposite direction to the force that body exerts on you. Experiments show that whenever two bodies interact, the two forces that they exert on each other are always *equal in magnitude* and *opposite in direction*. This fact is called *Newton's third law of motion:*

> **NEWTON'S THIRD LAW OF MOTION:** If body A exerts a force on body B (an "action"), then body B exerts a force on body A (a "reaction"). These two forces have the same magnitude but are opposite in direction. These two forces act on *different* bodies.

For example, in **Fig. 4.24** $\vec{F}_{A \text{ on } B}$ is the force applied *by* body A (first subscript) *on* body B (second subscript), and $\vec{F}_{B \text{ on } A}$ is the force applied *by* body B (first subscript) *on* body A (second subscript). In equation form,

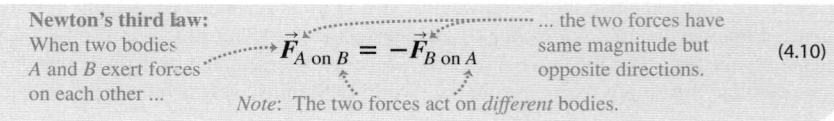

Newton's third law:
When two bodies
A and B exert forces
on each other ...
$$\vec{F}_{A \text{ on } B} = -\vec{F}_{B \text{ on } A}$$
... the two forces have
same magnitude but
opposite directions.   (4.10)
Note: The two forces act on *different* bodies.

It doesn't matter whether one body is inanimate (like the soccer ball in Fig. 4.24) and the other is not (like the kicker's foot): They necessarily exert forces on each other that obey Eq. (4.10).

In the statement of Newton's third law, "action" and "reaction" are the two opposite forces (in Fig. 4.24, $\vec{F}_{A \text{ on } B}$ and $\vec{F}_{B \text{ on } A}$); we sometimes refer to them as an **action–reaction pair.** This is *not* meant to imply any cause-and-effect relationship; we can consider either force as the "action" and the other as the "reaction." We often say simply that the forces are "equal and opposite," meaning that they have equal magnitudes and opposite directions.

CAUTION   **The two forces in an action–reaction pair act on different bodies**   We stress that the two forces described in Newton's third law act on *different* bodies. This is important in problems involving Newton's first or second law, which involve the forces that act on a single body. For instance, the net force on the soccer ball in Fig. 4.24 is the vector sum of the weight of the ball and the force $\vec{F}_{A \text{ on } B}$ exerted by the kicker. You wouldn't include the force $\vec{F}_{B \text{ on } A}$ because this force acts on the kicker, not on the ball. ▌

4.24 Newton's third law of motion.

If body A exerts force $\vec{F}_{A \text{ on } B}$ on body B
(for example, a foot kicks a ball) ...

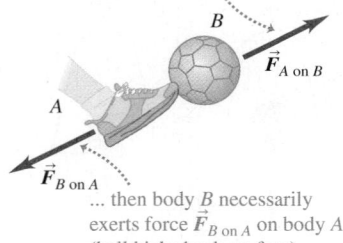

... then body B necessarily exerts force $\vec{F}_{B \text{ on } A}$ on body A (ball kicks back on foot).

The two forces have same magnitude but opposite directions: $\vec{F}_{A \text{ on } B} = -\vec{F}_{B \text{ on } A}$.

DEMO

In Fig. 4.24 the action and reaction forces are *contact* forces that are present only when the two bodies are touching. But Newton's third law also applies to *long-range* forces that do not require physical contact, such as the force of gravitational attraction. A table-tennis ball exerts an upward gravitational force on the earth that's equal in magnitude to the downward gravitational force the earth exerts on the ball. When you drop the ball, both the ball and the earth accelerate toward each other. The net force on each body has the same magnitude, but the earth's acceleration is microscopically small because its mass is so great. Nevertheless, it does move!

CONCEPTUAL EXAMPLE 4.8   WHICH FORCE IS GREATER?

After your sports car breaks down, you start to push it to the nearest repair shop. While the car is starting to move, how does the force you exert on the car compare to the force the car exerts on you? How do these forces compare when you are pushing the car along at a constant speed?

SOLUTION

Newton's third law says that in *both* cases, the force you exert on the car is equal in magnitude and opposite in direction to the force the car exerts on you. It's true that you have to push harder to get the car going than to keep it going. But no matter how hard you push on the car, the car pushes just as hard back on you. Newton's third law gives the same result whether the two bodies are at rest, moving with constant velocity, or accelerating.

You may wonder how the car "knows" to push back on you with the same magnitude of force that you exert on it. It may help to visualize the forces you and the car exert on each other as interactions between the atoms at the surface of your hand and the atoms at the surface of the car. These interactions are analogous to miniature springs between adjacent atoms, and a compressed spring exerts equally strong forces on both of its ends.

Fundamentally, though, the reason we know that objects of different masses exert equally strong forces on each other is that experiment tells us so. Physics isn't merely a collection of rules and equations; rather, it's a systematic description of the natural world based on experiment and observation.

CONCEPTUAL EXAMPLE 4.9   APPLYING NEWTON'S THIRD LAW: OBJECTS AT REST

An apple sits at rest on a table, in equilibrium. What forces act on the apple? What is the reaction force to each of the forces acting on the apple? What are the action–reaction pairs?

SOLUTION

Figure 4.25a shows the forces acting on the apple. $\vec{F}_{\text{earth on apple}}$ is the weight of the apple—that is, the downward gravitational force exerted *by* the earth *on* the apple. Similarly, $\vec{F}_{\text{table on apple}}$ is the upward normal force exerted *by* the table *on* the apple.

Figure 4.25b shows one of the action–reaction pairs involving the apple. As the earth pulls down on the apple, with force $\vec{F}_{\text{earth on apple}}$, the apple exerts an equally strong upward pull on the earth $\vec{F}_{\text{apple on earth}}$. By Newton's third law (Eq. 4.10) we have

$$\vec{F}_{\text{apple on earth}} = -\vec{F}_{\text{earth on apple}}$$

Also, as the table pushes up on the apple with force $\vec{F}_{\text{table on apple}}$, the corresponding reaction is the downward force $\vec{F}_{\text{apple on table}}$

4.25 The two forces in an action–reaction pair always act on different bodies.

(a) The forces acting on the apple

(b) The action–reaction pair for the interaction between the apple and the earth

(c) The action–reaction pair for the interaction between the apple and the table

(d) We eliminate one of the forces acting on the apple.

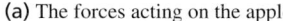
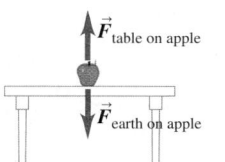
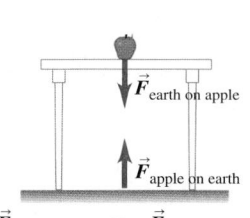
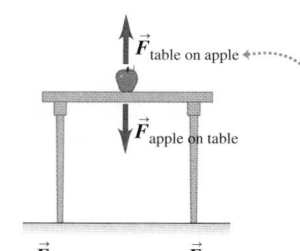
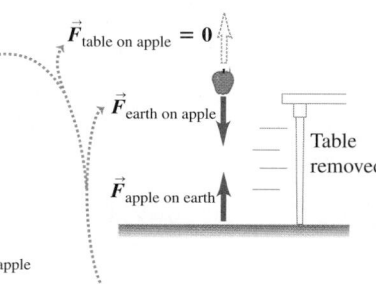

Continued

exerted by the apple on the table (Fig. 4.25c). For this action–reaction pair we have

$$\vec{F}_{\text{apple on table}} = -\vec{F}_{\text{table on apple}}$$

The two forces acting on the apple, $\vec{F}_{\text{table on apple}}$ and $\vec{F}_{\text{earth on apple}}$, are *not* an action–reaction pair, despite being equal in magnitude and opposite in direction. They do not represent the mutual interaction of two bodies; they are two different forces

acting on the *same* body. Figure 4.25d shows another way to see this. If we suddenly yank the table out from under the apple, the forces $\vec{F}_{\text{apple on table}}$ and $\vec{F}_{\text{table on apple}}$ suddenly become zero, but $\vec{F}_{\text{apple on earth}}$ and $\vec{F}_{\text{earth on apple}}$ are unchanged (the gravitational interaction is still present). Because $\vec{F}_{\text{table on apple}}$ is now zero, it can't be the negative of the nonzero $\vec{F}_{\text{earth on apple}}$, and these two forces can't be an action–reaction pair. *The two forces in an action–reaction pair **never** act on the same body.*

CONCEPTUAL EXAMPLE 4.10 APPLYING NEWTON'S THIRD LAW: OBJECTS IN MOTION

A stonemason drags a marble block across a floor by pulling on a rope attached to the block (**Fig. 4.26a**). The block is not necessarily in equilibrium. How are the various forces related? What are the action–reaction pairs?

SOLUTION

We'll use the subscripts B for the block, R for the rope, and M for the mason. In Fig. 4.26b the vector $\vec{F}_{\text{M on R}}$ represents the force exerted by the *mason* on the *rope*. The corresponding reaction is the force $\vec{F}_{\text{R on M}}$ exerted by the *rope* on the *mason*. Similarly, $\vec{F}_{\text{R on B}}$ represents the force exerted by the *rope* on the *block,* and the corresponding reaction is the force $\vec{F}_{\text{B on R}}$ exerted by the *block* on the *rope.* The forces in each action–reaction pair are equal and opposite:

$$\vec{F}_{\text{R on M}} = -\vec{F}_{\text{M on R}} \quad \text{and} \quad \vec{F}_{\text{B on R}} = -\vec{F}_{\text{R on B}}$$

Forces $\vec{F}_{\text{M on R}}$ and $\vec{F}_{\text{B on R}}$ (Fig. 4.26c) are *not* an action–reaction pair, because both of these forces act on the *same* body (the rope); an action and its reaction *must* always act on *different* bodies. Furthermore, the forces $\vec{F}_{\text{M on R}}$ and $\vec{F}_{\text{B on R}}$ are not necessarily equal in magnitude. Applying Newton's second law to the rope, we get

$$\Sigma\vec{F} = \vec{F}_{\text{M on R}} + \vec{F}_{\text{B on R}} = m_{\text{rope}}\vec{a}_{\text{rope}}$$

If the block and rope are accelerating (speeding up or slowing down), the rope is not in equilibrium, and $\vec{F}_{\text{M on R}}$ must have a

different magnitude than $\vec{F}_{\text{B on R}}$. By contrast, the action–reaction forces $\vec{F}_{\text{M on R}}$ and $\vec{F}_{\text{R on M}}$ are always equal in magnitude, as are $\vec{F}_{\text{R on B}}$ and $\vec{F}_{\text{B on R}}$. Newton's third law holds whether or not the bodies are accelerating.

In the special case in which the rope is in equilibrium, the forces $\vec{F}_{\text{M on R}}$ and $\vec{F}_{\text{B on R}}$ are equal in magnitude, and they are opposite in direction. But this is an example of Newton's *first* law, not his third; these are two forces on the same body, not forces of two bodies on each other. Another way to look at this is that in equilibrium, $\vec{a}_{\text{rope}} = 0$ in the previous equation. Then $\vec{F}_{\text{B on R}} = -\vec{F}_{\text{M on R}}$ because of Newton's first or second law.

Another special case is if the rope is accelerating but has negligibly small mass compared to that of the block or the mason. In this case, $m_{\text{rope}} = 0$ in the previous equation, so again $\vec{F}_{\text{B on R}} = -\vec{F}_{\text{M on R}}$. Since Newton's third law says that $\vec{F}_{\text{B on R}}$ *always* equals $-\vec{F}_{\text{R on B}}$ (they are an action–reaction pair), in this "massless-rope" case $\vec{F}_{\text{R on B}}$ also equals $\vec{F}_{\text{M on R}}$.

For both the "massless-rope" case and the case of the rope in equilibrium, the force of the rope on the block is equal in magnitude and direction to the force of the mason on the rope (Fig. 4.26d). Hence we can think of the rope as "transmitting" to the block the force the mason exerts on the rope. This is a useful point of view, but remember that it is valid *only* when the rope has negligibly small mass or is in equilibrium.

4.26 Identifying the forces that act when a mason pulls on a rope attached to a block.

(a) The block, the rope, and the mason

(b) The action–reaction pairs

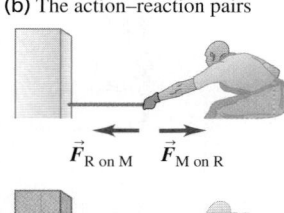

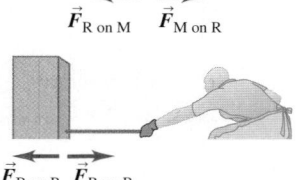

(c) *Not* an action–reaction pair

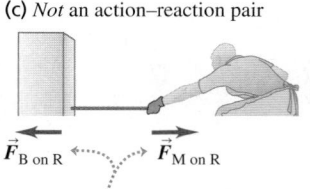

These forces cannot be an action–reaction pair because they act on the same object (the rope).

(d) Not necessarily equal

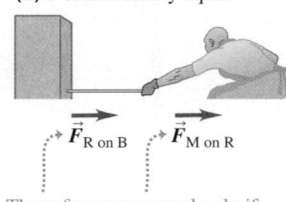

These forces are equal only if the rope is in equilibrium (or can be treated as massless).

SOLUTION

CONCEPTUAL EXAMPLE 4.11   A NEWTON'S THIRD LAW PARADOX?

We saw in Conceptual Example 4.10 that the stonemason pulls as hard on the rope–block combination as that combination pulls back on him. Why, then, does the block move while the stonemason remains stationary?

SOLUTION

To resolve this seeming paradox, keep in mind the difference between Newton's *second* and *third* laws. The only forces involved in Newton's second law are those that act *on* a given body. The vector sum of these forces determines the body's acceleration, if any. By contrast, Newton's third law relates the forces that two *different* bodies exert on *each other*. The third law alone tells you nothing about the motion of either body.

If the rope–block combination is initially at rest, it begins to slide if the stonemason exerts a force $\vec{F}_{\text{M on R}}$ that is *greater* in magnitude than the friction force that the floor exerts on the block (**Fig. 4.27**). (The block has a smooth underside, which minimizes friction.) Then there is a net force to the right on the rope–block combination, and it accelerates to the right. By contrast, the stonemason *doesn't* move because the net force acting on him is *zero*. His shoes have nonskid soles that don't slip on the floor, so the friction force that the floor exerts on him is strong enough to balance the pull of the rope on him, $\vec{F}_{\text{R on M}}$. (Both the block and the stonemason also experience a downward force of gravity and an upward normal force exerted by the floor. These forces balance each other, so we haven't included them in Fig. 4.27.)

Once the block is moving, the stonemason doesn't need to pull as hard; he must exert only enough force to balance the friction force on the block. Then the net force on the moving block is zero, and by Newton's first law the block continues to move toward the mason at a constant velocity.

4.27 The horizontal forces acting on the block–rope combination (left) and the mason (right). (The vertical forces are not shown.)

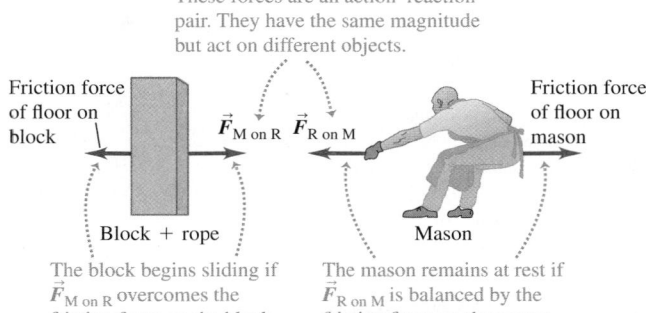

These forces are an action–reaction pair. They have the same magnitude but act on different objects.

Friction force of floor on block

$\vec{F}_{\text{M on R}}$ $\vec{F}_{\text{R on M}}$

Friction force of floor on mason

Block + rope

The block begins sliding if $\vec{F}_{\text{M on R}}$ overcomes the friction force on the block.

Mason

The mason remains at rest if $\vec{F}_{\text{R on M}}$ is balanced by the friction force on the mason.

So the block accelerates but the stonemason doesn't because different amounts of friction act on them. If the floor were freshly waxed, so that there was little friction between the floor and the stonemason's shoes, pulling on the rope might start the block sliding to the right *and* start him sliding to the left.

The moral of this example is that when analyzing the motion of a body, you must remember that only the forces acting *on* a body determine its motion. From this perspective, Newton's third law is merely a tool that can help you determine what those forces are.

A body that has pulling forces applied at its ends, such as the rope in Fig. 4.26, is said to be in *tension*. The **tension** at any point along the rope is the magnitude of the force acting at that point (see Fig. 4.2c). In Fig. 4.26b the tension at the right end of the rope is the magnitude of $\vec{F}_{\text{M on R}}$ (or of $\vec{F}_{\text{R on M}}$), and the tension at the left end equals the magnitude of $\vec{F}_{\text{B on R}}$ (or of $\vec{F}_{\text{R on B}}$). If the rope is in equilibrium and if no forces act except at its ends, the tension is the *same* at both ends and throughout the rope. Thus, if the magnitudes of $\vec{F}_{\text{B on R}}$ and $\vec{F}_{\text{M on R}}$ are 50 N each, the tension in the rope is 50 N (*not* 100 N). The *total* force vector $\vec{F}_{\text{B on R}} + \vec{F}_{\text{M on R}}$ acting on the rope in this case is zero!

We emphasize once again that the two forces in an action–reaction pair *never* act on the same body. Remembering this fact can help you avoid confusion about action–reaction pairs and Newton's third law.

TEST YOUR UNDERSTANDING OF SECTION 4.5   You are driving a car on a country road when a mosquito splatters on the windshield. Which has the greater magnitude: the force that the car exerted on the mosquito or the force that the mosquito exerted on the car? Or are the magnitudes the same? If they are different, how can you reconcile this fact with Newton's third law? If they are equal, why is the mosquito splattered while the car is undamaged? ∎

DATA *SPEAKS*

Force and Motion

When students were given a problem about forces acting on an object and how they affect the object's motion, more than 20% gave an incorrect answer. Common errors:

- Confusion about contact forces. If your fingers push on an object, the force you exert acts only when your fingers and the object are in contact. Once contact is broken, the force is no longer present even if the object is still moving.

- Confusion about Newton's third law. The third law relates the forces that two objects exert on each other. By itself, this law can't tell you anything about two forces that act on the same object.

4.6 FREE-BODY DIAGRAMS

DEMO   DEMO

4.28 The simple act of walking depends crucially on Newton's third law. To start moving forward, you push backward on the ground with your foot. As a reaction, the ground pushes forward on your foot (and hence on your body as a whole) with a force of the same magnitude. This *external* force provided by the ground is what accelerates your body forward.

——————
CAUTION **Forces in free-body diagrams** For a free-body diagram to be complete, you *must* be able to answer this question for each force: What other body is applying this force? If you can't answer that question, you may be dealing with a nonexistent force. Avoid nonexistent forces such as "the force of acceleration" or "the $m\vec{a}$ force," discussed in Section 4.3. ❚

Newton's three laws of motion contain all the basic principles we need to solve a wide variety of problems in mechanics. These laws are very simple in form, but the process of applying them to specific situations can pose real challenges. In this brief section we'll point out three key ideas and techniques to use in any problems involving Newton's laws. You'll learn others in Chapter 5, which also extends the use of Newton's laws to cover more complex situations.

1. *Newton's first and second laws apply to a specific body.* Whenever you use Newton's first law, $\sum \vec{F} = \mathbf{0}$, for an equilibrium situation or Newton's second law, $\sum \vec{F} = m\vec{a}$, for a nonequilibrium situation, you must decide at the beginning to which body you are referring. This decision may sound trivial, but it isn't.

2. *Only forces acting on the body matter.* The sum $\sum \vec{F}$ includes all the forces that act *on* the body in question. Hence, once you've chosen the body to analyze, you have to identify all the forces acting on it. Don't confuse the forces acting on a body with the forces exerted by that body on some other body. For example, to analyze a person walking, you would include in $\sum \vec{F}$ the force that the ground exerts on the person as he walks, but *not* the force that the person exerts on the ground (**Fig. 4.28**). These forces form an action–reaction pair and are related by Newton's third law, but only the member of the pair that acts on the body you're working with goes into $\sum \vec{F}$.

3. *Free-body diagrams are essential to help identify the relevant forces.* A **free-body diagram** shows the chosen body by itself, "free" of its surroundings, with vectors drawn to show the magnitudes and directions of all the forces that act on the body. We've already shown free-body diagrams in Figs. 4.17, 4.18, 4.20, and 4.25a. Be careful to include all the forces acting *on* the body, but be equally careful *not* to include any forces that the body exerts on any other body. In particular, the two forces in an action–reaction pair must *never* appear in the same free-body diagram because they never act on the same body. Furthermore, never include forces that a body exerts on itself, since these can't affect the body's motion.

When a problem involves more than one body, you have to take the problem apart and draw a separate free-body diagram for each body. For example, Fig. 4.26c shows a separate free-body diagram for the rope in the case in which the rope is considered massless (so that no gravitational force acts on it). Figure 4.27 also shows diagrams for the block and the mason, but these are *not* complete free-body diagrams because they don't show all the forces acting on each body. (We left out the vertical forces—the weight force exerted by the earth and the upward normal force exerted by the floor.)

In **Fig. 4.29** we present three real-life situations and the corresponding complete free-body diagrams. Note that in each situation a person exerts a force on something in his or her surroundings, but the force that shows up in the person's free-body diagram is the surroundings pushing back *on* the person.

TEST YOUR UNDERSTANDING OF SECTION 4.6 The buoyancy force shown in Fig. 4.29c is one half of an action–reaction pair. What force is the other half of this pair? (i) The weight of the swimmer; (ii) the forward thrust force; (iii) the backward drag force; (iv) the downward force that the swimmer exerts on the water; (v) the backward force that the swimmer exerts on the water by kicking. ❚

4.29 Examples of free-body diagrams. Each free-body diagram shows all of the external forces that act on the object in question.

(a)

$\vec{F}_y$ $\vec{F}_{\text{block on runner}}$

$\vec{F}_x$

$\vec{w}$

The force of the starting block on the runner has a vertical component that counteracts her weight and a large horizontal component that accelerates her.

(b)

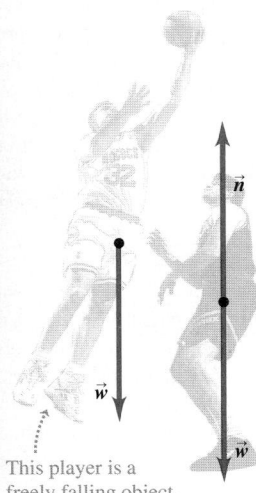

$\vec{n}$

To jump up, this player will push down against the floor, increasing the upward reaction force $\vec{n}$ of the floor on him.

$\vec{w}$

$\vec{w}$

This player is a freely falling object.

(c)

The water exerts a buoyancy force that counters the swimmer's weight.

$\vec{F}_{\text{buoyancy}}$

$\vec{F}_{\text{thrust}}$ $\vec{F}_{\text{drag}}$

Kicking causes the water to exert a forward reaction force, or thrust, on the swimmer.

$\vec{w}$

Thrust is countered by drag forces exerted by the water on the moving swimmer.

CHAPTER 4 **SUMMARY**

SOLUTIONS TO ALL EXAMPLES

Force as a vector: Force is a quantitative measure of the interaction between two bodies. It is a vector quantity. When several forces act on a body, the effect on its motion is the same as when a single force, equal to the vector sum (resultant) of the forces, acts on the body. (See Example 4.1.)

$$\vec{R} = \Sigma\vec{F} = \vec{F}_1 + \vec{F}_2 + \vec{F}_3 + \cdots \quad (4.1)$$

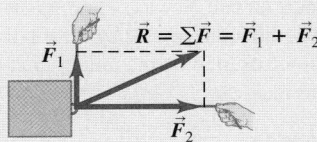

$\vec{F}_1$ $\vec{R} = \Sigma\vec{F} = \vec{F}_1 + \vec{F}_2$

$\vec{F}_2$

The net force on a body and Newton's first law: Newton's first law states that when the vector sum of all forces acting on a body (the *net force*) is zero, the body is in equilibrium and has zero acceleration. If the body is initially at rest, it remains at rest; if it is initially in motion, it continues to move with constant velocity. This law is valid in inertial frames of reference only. (See Examples 4.2 and 4.3.)

$$\Sigma\vec{F} = 0 \quad (4.3)$$

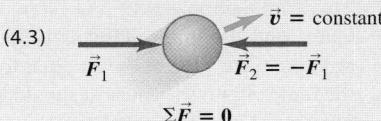

$\vec{v}$ = constant

$\vec{F}_1$ $\vec{F}_2 = -\vec{F}_1$

$\Sigma\vec{F} = 0$

Mass, acceleration, and Newton's second law: The inertial properties of a body are characterized by its *mass*. The acceleration of a body under the action of a given set of forces is directly proportional to the vector sum of the forces (the *net force*) and inversely proportional to the mass of the body. This relationship is Newton's second law. Like Newton's first law, this law is valid in inertial frames of reference only. The unit of force is defined in terms of the units of mass and acceleration. In SI units, the unit of force is the newton (N), equal to $1 \text{ kg} \cdot \text{m/s}^2$. (See Examples 4.4 and 4.5.)

$$\sum \vec{F} = m\vec{a} \qquad (4.6)$$

$$\sum F_x = ma_x$$
$$\sum F_y = ma_y \qquad (4.7)$$
$$\sum F_z = ma_z$$

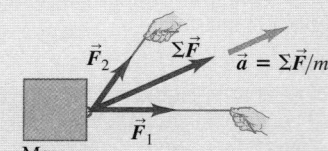

Weight: The weight $\vec{w}$ of a body is the gravitational force exerted on it by the earth. Weight is a vector quantity. The magnitude of the weight of a body at any specific location is equal to the product of its mass m and the magnitude of the acceleration due to gravity g at that location. While the weight of a body depends on its location, the mass is independent of location. (See Examples 4.6 and 4.7.)

$$w = mg \qquad (4.8)$$

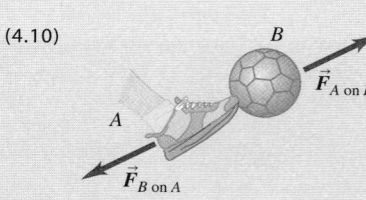

Newton's third law and action–reaction pairs: Newton's third law states that when two bodies interact, they exert forces on each other that are equal in magnitude and opposite in direction. These forces are called action and reaction forces. Each of these two forces acts on only one of the two bodies; they never act on the same body. (See Examples 4.8–4.11.)

$$\vec{F}_{A \text{ on } B} = -\vec{F}_{B \text{ on } A} \qquad (4.10)$$

BRIDGING PROBLEM LINKS IN A CHAIN

A student suspends a chain consisting of three links, each of mass $m = 0.250 \text{ kg}$, from a light rope. The rope is attached to the top link of the chain, which does not swing. She pulls upward on the rope, so that the rope applies an upward force of 9.00 N to the chain. (a) Draw a free-body diagram for the entire chain, considered as a body, and one for each of the three links. (b) Use the diagrams of part (a) and Newton's laws to find (i) the acceleration of the chain, (ii) the force exerted by the top link on the middle link, and (iii) the force exerted by the middle link on the bottom link. Treat the rope as massless.

SOLUTION GUIDE

IDENTIFY and SET UP

1. There are four objects of interest in this problem: the chain as a whole and the three individual links. For each of these four objects, make a list of the external forces that act on it. Besides the force of gravity, your list should include only forces exerted by other objects that *touch* the object in question.
2. Some of the forces in your lists form action–reaction pairs (one pair is the force of the top link on the middle link and the force of the middle link on the top link). Identify all such pairs.
3. Use your lists to help you draw a free-body diagram for each of the four objects. Choose the coordinate axes.

4. Use your lists to decide how many unknowns there are in this problem. Which of these are target variables?

EXECUTE

5. Write a Newton's second law equation for each of the four objects, and write a Newton's third law equation for each action–reaction pair. You should have at least as many equations as there are unknowns (see step 4). Do you?
6. Solve the equations for the target variables.

EVALUATE

7. You can check your results by substituting them back into the equations from step 6. This is especially important to do if you ended up with more equations in step 5 than you used in step 6.
8. Rank the force of the rope on the chain, the force of the top link on the middle link, and the force of the middle link on the bottom link in order from smallest to largest magnitude. Does this ranking make sense? Explain.
9. Repeat the problem for the case in which the upward force that the rope exerts on the chain is only 7.35 N. Is the ranking in step 8 the same? Does this make sense?

Problems

•, ••, •••: Difficulty levels. **CP**: Cumulative problems incorporating material from earlier chapters. **CALC**: Problems requiring calculus. **DATA**: Problems involving real data, scientific evidence, experimental design, and/or statistical reasoning. **BIO**: Biosciences problems.

DISCUSSION QUESTIONS

Q4.1 Can a body be in equilibrium when only one force acts on it? Explain.

Q4.2 A ball thrown straight up has zero velocity at its highest point. Is the ball in equilibrium at this point? Why or why not?

Q4.3 A helium balloon hovers in midair, neither ascending nor descending. Is it in equilibrium? What forces act on it?

Q4.4 When you fly in an airplane at night in smooth air, you have no sensation of motion, even though the plane may be moving at 800 km/h (500 mi/h). Why?

Q4.5 If the two ends of a rope in equilibrium are pulled with forces of equal magnitude and opposite directions, why isn't the total tension in the rope zero?

Q4.6 You tie a brick to the end of a rope and whirl the brick around you in a horizontal circle. Describe the path of the brick after you suddenly let go of the rope.

Q4.7 When a car stops suddenly, the passengers tend to move forward relative to their seats. Why? When a car makes a sharp turn, the passengers tend to slide to one side of the car. Why?

Q4.8 Some people say that the "force of inertia" (or "force of momentum") throws the passengers forward when a car brakes sharply. What is wrong with this explanation?

Q4.9 A passenger in a moving bus with no windows notices that a ball that has been at rest in the aisle suddenly starts to move toward the rear of the bus. Think of two possible explanations, and devise a way to decide which is correct.

Q4.10 Suppose you chose the fundamental physical quantities to be force, length, and time instead of mass, length, and time. What would be the units of mass in terms of those fundamental quantities?

Q4.11 Why is the earth only approximately an inertial reference frame?

Q4.12 Does Newton's second law hold true for an observer in a van as it speeds up, slows down, or rounds a corner? Explain.

Q4.13 Some students refer to the quantity $m\vec{a}$ as "the force of acceleration." Is it correct to refer to this quantity as a force? If so, what exerts this force? If not, what is a better description of this quantity?

Q4.14 The acceleration of a falling body is measured in an elevator that is traveling upward at a constant speed of 9.8 m/s. What value is obtained?

Q4.15 You can play catch with a softball in a bus moving with constant speed on a straight road, just as though the bus were at rest. Is this still possible when the bus is making a turn at constant speed on a level road? Why or why not?

Q4.16 Students sometimes say that the force of gravity on an object is 9.8 m/s². What is wrong with this view?

Q4.17 Why can it hurt your foot more to kick a big rock than a small pebble? *Must* the big rock hurt more? Explain.

Q4.18 "It's not the fall that hurts you; it's the sudden stop at the bottom." Translate this saying into the language of Newton's laws of motion.

Q4.19 A person can dive into water from a height of 10 m without injury, but a person who jumps off the roof of a 10-m-tall building and lands on a concrete street is likely to be seriously injured. Why is there a difference?

Q4.20 Why are cars designed to crumple in front and back for safety? Why not for side collisions and rollovers?

Q4.21 When a string barely strong enough lifts a heavy weight, it can lift the weight by a steady pull; but if you jerk the string, it will break. Explain in terms of Newton's laws of motion.

Q4.22 A large crate is suspended from the end of a vertical rope. Is the tension in the rope greater when the crate is at rest or when it is moving upward at constant speed? If the crate is traveling upward, is the tension in the rope greater when the crate is speeding up or when it is slowing down? In each case, explain in terms of Newton's laws of motion.

Q4.23 Which feels a greater pull due to the earth's gravity: a 10-kg stone or a 20-kg stone? If you drop the two stones, why doesn't the 20-kg stone fall with twice the acceleration of the 10-kg stone? Explain.

Q4.24 Why is it incorrect to say that 1.0 kg *equals* 2.2 lb?

Q4.25 A horse is hitched to a wagon. Since the wagon pulls back on the horse just as hard as the horse pulls on the wagon, why doesn't the wagon remain in equilibrium, no matter how hard the horse pulls?

Q4.26 True or false? You exert a push *P* on an object and it pushes back on you with a force *F*. If the object is moving at constant velocity, then *F* is equal to *P*, but if the object is being accelerated, then *P* must be greater than *F*.

Q4.27 A large truck and a small compact car have a head-on collision. During the collision, the truck exerts a force $\vec{F}_{\text{T on C}}$ on the car, and the car exerts a force $\vec{F}_{\text{C on T}}$ on the truck. Which force has the larger magnitude, or are they the same? Does your answer depend on how fast each vehicle was moving before the collision? Why or why not?

Q4.28 When a car comes to a stop on a level highway, what force causes it to slow down? When the car increases its speed on the same highway, what force causes it to speed up? Explain.

Q4.29 A small compact car is pushing a large van that has broken down, and they travel along the road with equal velocities and accelerations. While the car is speeding up, is the force it exerts on the van larger than, smaller than, or the same magnitude as the force the van exerts on it? Which vehicle has the larger net force on it, or are the net forces the same? Explain.

Q4.30 Consider a tug-of-war between two people who pull in opposite directions on the ends of a rope. By Newton's third law, the force that *A* exerts on *B* is just as great as the force that *B* exerts on *A*. So what determines who wins? (*Hint:* Draw a free-body diagram showing all the forces that act on each person.)

Q4.31 Boxes *A* and *B* are in contact on a horizontal, frictionless surface. You push on box *A* with a horizontal 100-N force (**Fig. Q4.31**). Box *A* weighs 150 N, and box *B* weighs 50 N. Is the force that box *A* exerts on box *B* equal to 100 N, greater than 100 N, or less than 100 N? Explain.

Figure **Q4.31**

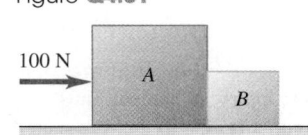

Q4.32 A manual for student pilots contains this passage: "When an airplane flies at a steady altitude, neither climbing nor descending, the upward lift force from the wings equals the plane's weight. When the plane is climbing at a steady rate, the upward lift is greater than the weight; when the plane is descending at a steady rate, the upward lift is less than the weight." Are these statements correct? Explain.

Q4.33 If your hands are wet and no towel is handy, you can remove some of the excess water by shaking them. Why does this work?

Q4.34 If you squat down (such as when you examine the books on a bottom shelf) and then suddenly get up, you may temporarily feel light-headed. What do Newton's laws of motion have to say about why this happens?

Q4.35 When a car is hit from behind, the occupants may experience whiplash. Use Newton's laws of motion to explain what causes this result.

Q4.36 In a head-on auto collision, passengers who are not wearing seat belts may be thrown through the windshield. Use Newton's laws of motion to explain why this happens.

Q4.37 In a head-on collision between a compact 1000-kg car and a large 2500-kg car, which one experiences the greater force? Explain. Which one experiences the greater acceleration? Explain why. Why are passengers in the small car more likely to be injured than those in the large car, even when the two car bodies are equally strong?

Q4.38 Suppose you are in a rocket with no windows, traveling in deep space far from other objects. Without looking outside the rocket or making any contact with the outside world, explain how you could determine whether the rocket is (a) moving forward at a constant 80% of the speed of light and (b) accelerating in the forward direction.

EXERCISES

Section 4.1 Force and Interactions

4.1 •• Two dogs pull horizontally on ropes attached to a post; the angle between the ropes is 60.0°. If Rover exerts a force of 270 N and Fido exerts a force of 300 N, find the magnitude of the resultant force and the angle it makes with Rover's rope.

4.2 • To extricate an SUV stuck in the mud, workmen use three horizontal ropes, producing the force vectors shown in **Fig. E4.2**. (a) Find the x- and y-components of each of the three pulls. (b) Use the components to find the magnitude and direction of the resultant of the three pulls.

Figure **E4.2**

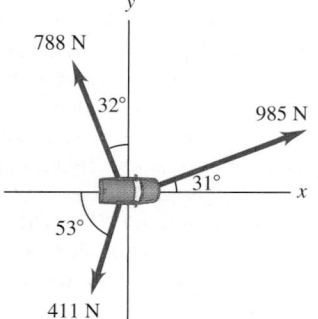

788 N
32°
985 N
31°
53°
411 N

4.3 • **BIO Jaw Injury.** Due to a jaw injury, a patient must wear a strap (**Fig. E4.3**) that produces a net upward force of 5.00 N on his chin. The tension is the same throughout the strap. To what tension must the strap be adjusted to provide the necessary upward force?

Figure **E4.3**

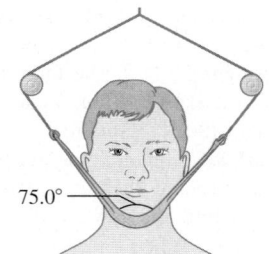

75.0°

4.4 • A man is dragging a trunk up the loading ramp of a mover's truck. The ramp has a slope angle of 20.0°, and the man pulls upward with a force $\vec{F}$ whose direction makes an angle of 30.0° with the ramp (**Fig. E4.4**). (a) How large a force $\vec{F}$ is necessary for the component F_x parallel to the ramp to be 90.0 N? (b) How large will the component F_y perpendicular to the ramp be then?

Figure **E4.4**

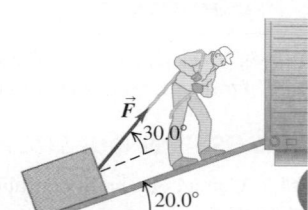

$\vec{F}$
30.0°
20.0°

4.5 • Forces $\vec{F}_1$ and $\vec{F}_2$ act at a point. The magnitude of $\vec{F}_1$ is 9.00 N, and its direction is 60.0° above the x-axis in the second quadrant. The magnitude of $\vec{F}_2$ is 6.00 N, and its direction is 53.1° below the x-axis in the third quadrant. (a) What are the x- and y-components of the resultant force? (b) What is the magnitude of the resultant force?

Section 4.3 Newton's Second Law

4.6 • An electron (mass = 9.11×10^{-31} kg) leaves one end of a TV picture tube with zero initial speed and travels in a straight line to the accelerating grid, which is 1.80 cm away. It reaches the grid with a speed of 3.00×10^6 m/s. If the accelerating force is constant, compute (a) the acceleration; (b) the time to reach the grid; and (c) the net force, in newtons. Ignore the gravitational force on the electron.

4.7 •• A 68.5-kg skater moving initially at 2.40 m/s on rough horizontal ice comes to rest uniformly in 3.52 s due to friction from the ice. What force does friction exert on the skater?

4.8 •• You walk into an elevator, step onto a scale, and push the "up" button. You recall that your normal weight is 625 N. Draw a free-body diagram. (a) When the elevator has an upward acceleration of magnitude 2.50 m/s², what does the scale read? (b) If you hold a 3.85-kg package by a light vertical string, what will be the tension in this string when the elevator accelerates as in part (a)?

4.9 • A box rests on a frozen pond, which serves as a frictionless horizontal surface. If a fisherman applies a horizontal force with magnitude 48.0 N to the box and produces an acceleration of magnitude 2.20 m/s², what is the mass of the box?

4.10 •• A dockworker applies a constant horizontal force of 80.0 N to a block of ice on a smooth horizontal floor. The frictional force is negligible. The block starts from rest and moves 11.0 m in 5.00 s. (a) What is the mass of the block of ice? (b) If the worker stops pushing at the end of 5.00 s, how far does the block move in the next 5.00 s?

4.11 • A hockey puck with mass 0.160 kg is at rest at the origin ($x = 0$) on the horizontal, frictionless surface of the rink. At time $t = 0$ a player applies a force of 0.250 N to the puck, parallel to the x-axis; she continues to apply this force until $t = 2.00$ s. (a) What are the position and speed of the puck at $t = 2.00$ s? (b) If the same force is again applied at $t = 5.00$ s, what are the position and speed of the puck at $t = 7.00$ s?

4.12 • A crate with mass 32.5 kg initially at rest on a warehouse floor is acted on by a net horizontal force of 14.0 N. (a) What acceleration is produced? (b) How far does the crate travel in 10.0 s? (c) What is its speed at the end of 10.0 s?

4.13 • A 4.50-kg experimental cart undergoes an acceleration in a straight line (the x-axis). The graph in **Fig. E4.13** shows this acceleration as a function of time. (a) Find the maximum net force on this cart. When does this maximum force occur? (b) During what times is the net force on the cart a constant? (c) When is the net force equal to zero?

Figure **E4.13**

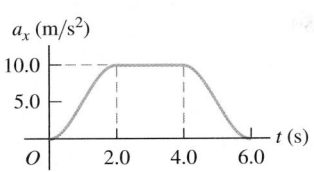

4.14 · A 2.75-kg cat moves in a straight line (the *x*-axis). **Figure E4.14** shows a graph of the *x*-component of this cat's velocity as a function of time. (a) Find the maximum net force on this cat. When does this force occur? (b) When is the net force on the cat equal to zero? (c) What is the net force at time 8.5 s?

Figure **E4.14**

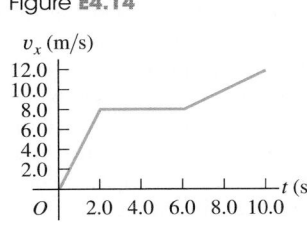

4.15 · A small 8.00-kg rocket burns fuel that exerts a time-varying upward force on the rocket (assume constant mass) as the rocket moves upward from the launch pad. This force obeys the equation $F = A + Bt^2$. Measurements show that at $t = 0$, the force is 100.0 N, and at the end of the first 2.00 s, it is 150.0 N. (a) Find the constants A and B, including their SI units. (b) Find the *net* force on this rocket and its acceleration (i) the instant after the fuel ignites and (ii) 3.00 s after the fuel ignites. (c) Suppose that you were using this rocket in outer space, far from all gravity. What would its acceleration be 3.00 s after fuel ignition?

Section 4.4 Mass and Weight

4.16 · An astronaut's pack weighs 17.5 N when she is on the earth but only 3.24 N when she is at the surface of a moon. (a) What is the acceleration due to gravity on this moon? (b) What is the mass of the pack on this moon?

4.17 · Superman throws a 2400-N boulder at an adversary. What horizontal force must Superman apply to the boulder to give it a horizontal acceleration of 12.0 m/s²?

4.18 · BIO (a) An ordinary flea has a mass of 210 μg. How many newtons does it weigh? (b) The mass of a typical froghopper is 12.3 mg. How many newtons does it weigh? (c) A house cat typically weighs 45 N. How many pounds does it weigh, and what is its mass in kilograms?

4.19 · At the surface of Jupiter's moon Io, the acceleration due to gravity is $g = 1.81$ m/s². A watermelon weighs 44.0 N at the surface of the earth. (a) What is the watermelon's mass on the earth's surface? (b) What would be its mass and weight on the surface of Io?

Section 4.5 Newton's Third Law

4.20 · A small car of mass 380 kg is pushing a large truck of mass 900 kg due east on a level road. The car exerts a horizontal force of 1600 N on the truck. What is the magnitude of the force that the truck exerts on the car?

4.21 · BIO World-class sprinters can accelerate out of the starting blocks with an acceleration that is nearly horizontal and has magnitude 15 m/s². How much horizontal force must a 55-kg sprinter exert on the starting blocks to produce this acceleration? Which body exerts the force that propels the sprinter: the blocks or the sprinter herself?

4.22 ·· The upward normal force exerted by the floor is 620 N on an elevator passenger who weighs 650 N. What are the reaction forces to these two forces? Is the passenger accelerating? If so, what are the magnitude and direction of the acceleration?

4.23 ·· Boxes *A* and *B* are in contact on a horizontal, frictionless surface (**Fig. E4.23**). Box *A* has mass 20.0 kg and box *B* has mass 5.0 kg. A horizontal force of 250 N is exerted on box *A*. What is the magnitude of the force that box *A* exerts on box *B*?

Figure **E4.23**

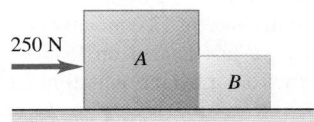

4.24 ·· A student of mass 45 kg jumps off a high diving board. What is the acceleration of the earth toward her as she accelerates toward the earth with an acceleration of 9.8 m/s²? Use 6.0×10^{24} kg for the mass of the earth, and assume that the net force on the earth is the force of gravity she exerts on it.

Section 4.6 Free-Body Diagrams

4.25 ·· Crates *A* and *B* sit at rest side by side on a frictionless horizontal surface. They have masses m_A and m_B, respectively. When a horizontal force $\vec{F}$ is applied to crate *A*, the two crates move off to the right. (a) Draw clearly labeled free-body diagrams for crate *A* and for crate *B*. Indicate which pairs of forces, if any, are third-law action–reaction pairs. (b) If the magnitude of $\vec{F}$ is less than the total weight of the two crates, will it cause the crates to move? Explain.

4.26 ·· You pull horizontally on block *B* in **Fig. E4.26**, causing both blocks to move together as a unit. For this moving system, make a carefully labeled free-body diagram of block *A* if (a) the table is frictionless and (b) there is friction between block *B* and the table and the pull is equal in magnitude to the friction force on block *B* due to the table.

Figure **E4.26**

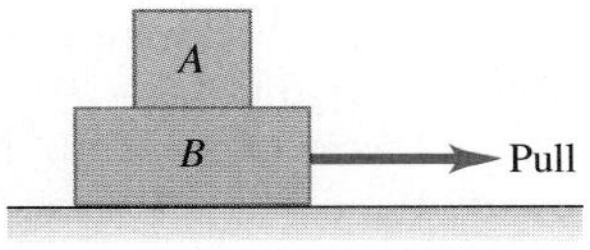

Horizontal table

4.27 · A ball is hanging from a long string that is tied to the ceiling of a train car traveling eastward on horizontal tracks. An observer inside the train car sees the ball hang motionless. Draw a clearly labeled free-body diagram for the ball if (a) the train has a uniform velocity and (b) the train is speeding up uniformly. Is the net force on the ball zero in either case? Explain.

4.28 ·· CP A .22-caliber rifle bullet traveling at 350 m/s strikes a large tree and penetrates it to a depth of 0.130 m. The mass of the bullet is 1.80 g. Assume a constant retarding force. (a) How much time is required for the bullet to stop? (b) What force, in newtons, does the tree exert on the bullet?

4.29 ·· A chair of mass 12.0 kg is sitting on the horizontal floor; the floor is not frictionless. You push on the chair with a force $F = 40.0$ N that is directed at an angle of 37.0° below the horizontal, and the chair slides along the floor. (a) Draw a clearly labeled free-body diagram for the chair. (b) Use your diagram and Newton's laws to calculate the normal force that the floor exerts on the chair.

PROBLEMS

4.30 ••• A large box containing your new computer sits on the bed of your pickup truck. You are stopped at a red light. When the light turns green, you stomp on the gas and the truck accelerates. To your horror, the box starts to slide toward the back of the truck. Draw clearly labeled free-body diagrams for the truck and for the box. Indicate pairs of forces, if any, that are third-law action–reaction pairs. (The horizontal truck bed is *not* frictionless.)

4.31 •• CP A 5.60-kg bucket of water is accelerated upward by a cord of negligible mass whose breaking strength is 75.0 N. If the bucket starts from rest, what is the minimum time required to raise the bucket a vertical distance of 12.0 m without breaking the cord?

4.32 •• CP You have just landed on Planet X. You release a 100-g ball from rest from a height of 10.0 m and measure that it takes 3.40 s to reach the ground. Ignore any force on the ball from the atmosphere of the planet. How much does the 100-g ball weigh on the surface of Planet X?

4.33 •• Two adults and a child want to push a wheeled cart in the direction marked *x* in **Fig. P4.33**. The two adults push with horizontal forces $\vec{F}_1$ and $\vec{F}_2$ as shown. (a) Find the magnitude and direction of the *smallest* force that the child should exert. Ignore the effects of friction. (b) If the child exerts the minimum force found in part (a), the cart accelerates at 2.0 m/s^2 in the +*x*-direction. What is the weight of the cart?

Figure **P4.33**

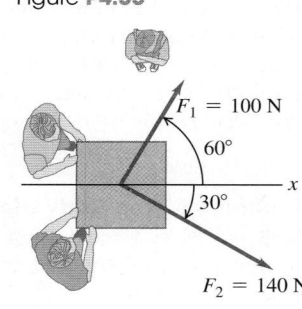

$F_1 = 100$ N

60°

30°

x

$F_2 = 140$ N

4.34 • CP An oil tanker's engines have broken down, and the wind is blowing the tanker straight toward a reef at a constant speed of 1.5 m/s (**Fig. P4.34**). When the tanker is 500 m from the reef, the wind dies down just as the engineer gets the engines going again. The rudder is stuck, so the only choice is to try to accelerate straight backward away from the reef. The mass of the tanker and cargo is 3.6 × 10^7 kg, and the engines produce a net horizontal force of 8.0 × 10^4 N on the tanker. Will the ship hit the reef? If it does, will the oil be safe? The hull can withstand an impact at a speed of 0.2 m/s or less. Ignore the retarding force of the water on the tanker's hull.

Figure **P4.34**

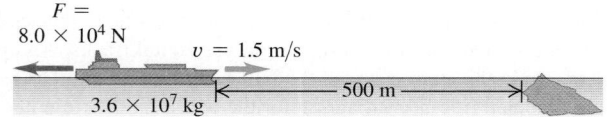

$F =$ 8.0 × 10^4 N

$v = 1.5$ m/s

500 m

3.6 × 10^7 kg

4.35 •• CP BIO **A Standing Vertical Jump.** Basketball player Darrell Griffith is on record as attaining a standing vertical jump of 1.2 m (4 ft). (This means that he moved upward by 1.2 m after his feet left the floor.) Griffith weighed 890 N (200 lb). (a) What was his speed as he left the floor? (b) If the time of the part of the jump before his feet left the floor was 0.300 s, what was his average acceleration (magnitude and direction) while he pushed against the floor? (c) Draw his free-body diagram. In terms of the forces on the diagram, what was the net force on him? Use Newton's laws and the results of part (b) to calculate the average force he applied to the ground.

4.36 ••• CP An advertisement claims that a particular automobile can "stop on a dime." What net force would be necessary to

stop a 850-kg automobile traveling initially at 45.0 km/h in a distance equal to the diameter of a dime, 1.8 cm?

4.37 •• BIO **Human Biomechanics.** The fastest pitched baseball was measured at 46 m/s. A typical baseball has a mass of 145 g. If the pitcher exerted his force (assumed to be horizontal and constant) over a distance of 1.0 m, (a) what force did he produce on the ball during this record-setting pitch? (b) Draw free-body diagrams of the ball during the pitch and just *after* it left the pitcher's hand.

4.38 •• BIO **Human Biomechanics.** The fastest served tennis ball, served by "Big Bill" Tilden in 1931, was measured at 73.14 m/s. The mass of a tennis ball is 57 g, and the ball, which starts from rest, is typically in contact with the tennis racquet for 30.0 ms. Assuming constant acceleration, (a) what force did Big Bill's tennis racquet exert on the ball if he hit it essentially horizontally? (b) Draw free-body diagrams of the ball during the serve and just after it moved free of the racquet.

4.39 • Two crates, one with mass 4.00 kg and the other with mass 6.00 kg, sit on the frictionless surface of a frozen pond, connected by a light rope (**Fig. P4.39**). A woman wearing golf shoes (for traction) pulls horizontally on the 6.00-kg crate with a force *F* that gives the crate an acceleration of 2.50 m/s^2. (a) What is the acceleration of the 4.00-kg crate? (b) Draw a free-body diagram for the 4.00-kg crate. Use that diagram and Newton's second law to find the tension *T* in the rope that connects the two crates. (c) Draw a free-body diagram for the 6.00-kg crate. What is the direction of the net force on the 6.00-kg crate? Which is larger in magnitude, *T* or *F*? (d) Use part (c) and Newton's second law to calculate the magnitude of *F*.

Figure **P4.39**

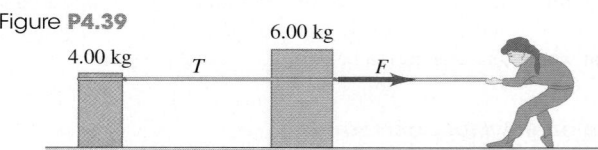

4.00 kg

T

6.00 kg

F

4.40 •• CP Two blocks connected by a light horizontal rope sit at rest on a horizontal, frictionless surface. Block *A* has mass 15.0 kg, and block *B* has mass *m*. A constant horizontal force $F = 60.0$ N is applied to block *A* (**Fig. P4.40**). In the first 5.00 s after the force is applied, block *A* moves 18.0 m to the right. (a) While the blocks are moving, what is the tension *T* in the rope that connects the two blocks? (b) What is the mass of block *B*?

Figure **P4.40**

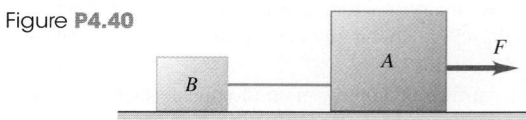

B

A

F

4.41 • CALC To study damage to aircraft that collide with large birds, you design a test gun that will accelerate chicken-sized objects so that their displacement along the gun barrel is given by $x = (9.0 \times 10^3 \text{ m/s}^2)t^2 - (8.0 \times 10^4 \text{ m/s}^3)t^3$. The object leaves the end of the barrel at $t = 0.025$ s. (a) How long must the gun barrel be? (b) What will be the speed of the objects as they leave the end of the barrel? (c) What net force must be exerted on a 1.50-kg object at (i) $t = 0$ and (ii) $t = 0.025$ s?

4.42 •• CP A 6.50-kg instrument is hanging by a vertical wire inside a spaceship that is blasting off from rest at the earth's surface. This spaceship reaches an altitude of 276 m in 15.0 s with constant acceleration. (a) Draw a free-body diagram for the instrument during this time. Indicate which force is greater. (b) Find the force that the wire exerts on the instrument.

4.43 •• BIO **Insect Dynamics.** The froghopper (*Philaenus spumarius*), the champion leaper of the insect world, has a mass of 12.3 mg and leaves the ground (in the most energetic jumps) at 4.0 m/s from a vertical start. The jump itself lasts a mere 1.0 ms before the insect is clear of the ground. Assuming constant acceleration, (a) draw a free-body diagram of this mighty leaper during the jump; (b) find the force that the ground exerts on the froghopper during the jump; and (c) express the force in part (b) in terms of the froghopper's weight.

4.44 • A loaded elevator with very worn cables has a total mass of 2200 kg, and the cables can withstand a maximum tension of 28,000 N. (a) Draw the free-body force diagram for the elevator. In terms of the forces on your diagram, what is the net force on the elevator? Apply Newton's second law to the elevator and find the maximum upward acceleration for the elevator if the cables are not to break. (b) What would be the answer to part (a) if the elevator were on the moon, where $g = 1.62$ m/s^2?

4.45 •• CP After an annual checkup, you leave your physician's office, where you weighed 683 N. You then get into an elevator that, conveniently, has a scale. Find the magnitude and direction of the elevator's acceleration if the scale reads (a) 725 N and (b) 595 N.

4.46 ••• CP A nail in a pine board stops a 4.9-N hammer head from an initial downward velocity of 3.2 m/s in a distance of 0.45 cm. In addition, the person using the hammer exerts a 15-N downward force on it. Assume that the acceleration of the hammer head is constant while it is in contact with the nail and moving downward. (a) Draw a free-body diagram for the hammer head. Identify the reaction force for each action force in the diagram. (b) Calculate the downward force $\vec{F}$ exerted by the hammer head on the nail while the hammer head is in contact with the nail and moving downward. (c) Suppose that the nail is in hardwood and the distance the hammer head travels in coming to rest is only 0.12 cm. The downward forces on the hammer head are the same as in part (b). What then is the force $\vec{F}$ exerted by the hammer head on the nail while the hammer head is in contact with the nail and moving downward?

4.47 •• CP **Jumping to the Ground.** A 75.0-kg man steps off a platform 3.10 m above the ground. He keeps his legs straight as he falls, but his knees begin to bend at the moment his feet touch the ground; treated as a particle, he moves an additional 0.60 m before coming to rest. (a) What is his speed at the instant his feet touch the ground? (b) If we treat the man as a particle, what is his acceleration (magnitude and direction) as he slows down, if the acceleration is assumed to be constant? (c) Draw his free-body diagram. In terms of the forces on the diagram, what is the net force on him? Use Newton's laws and the results of part (b) to calculate the average force his feet exert on the ground while he slows down. Express this force both in newtons and as a multiple of his weight.

4.48 •• The two blocks in **Fig. P4.48** are connected by a heavy uniform rope with a mass of 4.00 kg. An upward force of 200 N is applied as shown. (a) Draw three free-body diagrams: one for the 6.00-kg block, one for the 4.00-kg rope, and another one for the 5.00-kg block. For each force, indicate what body exerts that force. (b) What is the acceleration of the system? (c) What is the tension at the top of the heavy rope? (d) What is the tension at the midpoint of the rope?

Figure **P4.48**

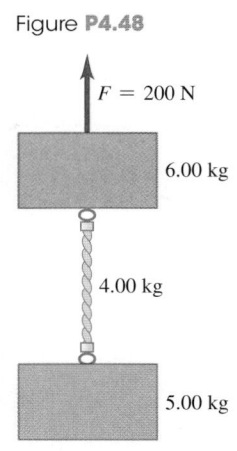

$F = 200$ N

6.00 kg

4.00 kg

5.00 kg

4.49 •• CP Boxes A and B are connected to each end of a light vertical rope (**Fig. P4.49**). A constant upward force $F = 80.0$ N is applied to box A. Starting from rest, box B descends 12.0 m in 4.00 s. The tension in the rope connecting the two boxes is 36.0 N. What are the masses of (a) box B, (b) box A?

Figure **P4.49**

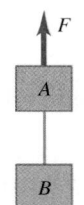

F

A

B

4.50 •• CP **Extraterrestrial Physics.** You have landed on an unknown planet, Newtonia, and want to know what objects weigh there. When you push a certain tool, starting from rest, on a frictionless horizontal surface with a 12.0-N force, the tool moves 16.0 m in the first 2.00 s. You next observe that if you release this tool from rest at 10.0 m above the ground, it takes 2.58 s to reach the ground. What does the tool weigh on Newtonia, and what does it weigh on Earth?

4.51 •• CP CALC A mysterious rocket-propelled object of mass 45.0 kg is initially at rest in the middle of the horizontal, frictionless surface of an ice-covered lake. Then a force directed east and with magnitude $F(t) = (16.8$ N/s$)t$ is applied. How far does the object travel in the first 5.00 s after the force is applied?

4.52 ••• CALC The position of a training helicopter (weight 2.75×10^5 N) in a test is given by $\hat{r} = (0.020$ m/s$^3)t^3\hat{i} + (2.2$ m/s$)t\hat{j} - (0.060$ m/s$^2)t^2\hat{k}$. Find the net force on the helicopter at $t = 5.0$ s.

4.53 •• DATA The table[*] gives automobile performance data for a few types of cars:

Make and Model (Year)	Mass (kg)	Time (s) to go from 0 to 60 mph
Alpha Romeo 4C (2013)	895	4.4
Honda Civic 2.0i (2011)	1320	6.4
Ferrari F430 (2004)	1435	3.9
Ford Focus RS500 (2010)	1468	5.4
Volvo S60 (2013)	1650	7.2

[*]Source: www.autosnout.com

(a) During an acceleration of 0 to 60 mph, which car has the largest average net force acting on it? The smallest? (b) During this acceleration, for which car would the average net force on a 72.0-kg passenger be the largest? The smallest? (c) When the Ferrari F430 accelerates from 0 to 100 mph in 8.6 s, what is the average net force acting on it? How does this net force compare with the average net force during the acceleration from 0 to 60 mph? Explain why these average net forces might differ. (d) Discuss why a car has a top speed. What is the net force on the Ferrari F430 when it is traveling at its top speed, 196 mph?

4.54 •• DATA An 8.00-kg box sits on a level floor. You give the box a sharp push and find that it travels 8.22 m in 2.8 s before coming to rest again. (a) You measure that with a different push the box traveled 4.20 m in 2.0 s. Do you think the box has a constant acceleration as it slows down? Explain your reasoning. (b) You add books to the box to increase its mass. Repeating the experiment, you give the box a push and measure how long it takes the box to come to rest and how far the box travels. The results, including the initial experiment with no added mass, are given in the table:

Added Mass (kg)	Distance (m)	Time (s)
0	8.22	2.8
3.00	10.75	3.2
7.00	9.45	3.0
12.0	7.10	2.6

In each case, did your push give the box the same initial speed? What is the ratio between the greatest initial speed and the smallest initial speed for these four cases? (c) Is the average horizontal force f exerted on the box by the floor the same in each case? Graph the magnitude of force f versus the total mass m of the box plus its contents, and use your graph to determine an equation for f as a function of m.

4.55 •• DATA You are a Starfleet captain going boldly where no man has gone before. You land on a distant planet and visit an engineering testing lab. In one experiment a short, light rope is attached to the top of a block and a constant upward force F is applied to the free end of the rope. The block has mass m and is initially at rest. As F is varied, the time for the block to move upward 8.00 m is measured. The values that you collected are given in the table:

F (N)	Time (s)
250	3.3
300	2.2
350	1.7
400	1.5
450	1.3
500	1.2

(a) Plot F versus the acceleration a of the block. (b) Use your graph to determine the mass m of the block and the acceleration of gravity g at the surface of the planet. Note that even on that planet, measured values contain some experimental error.

CHALLENGE PROBLEM

4.56 ••• CALC An object of mass m is at rest in equilibrium at the origin. At $t = 0$ a new force $\vec{F}(t)$ is applied that has components

$$F_x(t) = k_1 + k_2 y \qquad F_y(t) = k_3 t$$

where k_1, k_2, and k_3 are constants. Calculate the position $\vec{r}(t)$ and velocity $\vec{v}(t)$ vectors as functions of time.

BIO **FORCES ON A DANCER'S BODY.** Dancers experience large forces associated with the jumps they make. For example, when a dancer lands after a vertical jump, the force exerted on the head by the neck must exceed the head's weight by enough to cause the head to slow down and come to rest. The head is about 9.4% of a typical person's mass. Video analysis of a 65-kg dancer landing after a vertical jump shows that her head decelerates from 4.0 m/s to rest in a time of 0.20 s.

4.57 What is the magnitude of the average force that her neck exerts on her head during the landing? (a) 0 N; (b) 60 N; (c) 120 N; (d) 180 N.

4.58 Compared with the force her neck exerts on her head during the landing, the force her head exerts on her neck is (a) the same; (b) greater; (c) smaller; (d) greater during the first half of the landing and smaller during the second half of the landing.

4.59 While the dancer is in the air and holding a fixed pose, what is the magnitude of the force her neck exerts on her head? (a) 0 N; (b) 60 N; (c) 120 N; (d) 180 N.

4.60 The forces on a dancer can be measured directly when a dancer performs a jump on a force plate that measures the force between her feet and the ground. A graph of force versus time throughout a vertical jump performed on a force plate is shown in **Fig. P4.60**. What is happening at 0.4 s? The dancer is (a) bending her legs so that her body is accelerating downward; (b) pushing her body up with her legs and is almost ready to leave the ground; (c) in the air and at the top of her jump; (d) landing and her feet have just touched the ground.

Figure **P4.60**

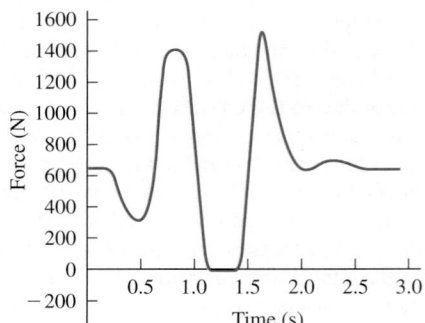

Answers

Chapter Opening Question ?

(v) Newton's third law tells us that the barbell pushes on the weightlifter just as hard as the weightlifter pushes on the barbell in *all* circumstances, no matter how the barbell is moving. However, the magnitude of the force that the weightlifter exerts is different in different circumstances. This force magnitude is equal to the weight of the barbell when the barbell is stationary, moving upward at a constant speed, or moving downward at a constant speed; it is greater than the weight of the barbell when the barbell accelerates upward; and it is less than the weight of the barbell when the barbell accelerates downward. But in each case the push of the barbell on the weightlifter has exactly the same magnitude as the push of the weightlifter on the barbell.

Test Your Understanding Questions

4.1 (iv) The gravitational force on the crate points straight downward. In Fig. 4.5 the *x*-axis points up and to the right, and the *y*-axis points up and to the left. Hence the gravitational force has both an *x*-component and a *y*-component, and both are negative.

4.2 (i), (ii), and (iv) In (i), (ii), and (iv) the body is not accelerating, so the net force on the body is zero. [In (iv), the box remains stationary as seen in the inertial reference frame of the ground as the truck accelerates forward, like the person on skates in Fig. 4.10a.] In (iii), the hawk is moving in a circle; hence it is accelerating and is *not* in equilibrium.

4.3 (iii), (i) and (iv) (tie), (ii) The acceleration is equal to the net force divided by the mass. Hence the magnitude of the acceleration in each situation is

  (i) $a = (2.0\ \text{N})/(2.0\ \text{kg}) = 1.0\ \text{m/s}^2$;

  (ii) $a = (8.0\ \text{N})/(2.0\ \text{N}) = 4.0\ \text{m/s}^2$;

  (iii) $a = (2.0\ \text{N})/(8.0\ \text{kg}) = 0.25\ \text{m/s}^2$;

  (iv) $a = (8.0\ \text{N})/(8.0\ \text{kg}) = 1.0\ \text{m/s}^2$.

4.4 It would take twice the effort for the astronaut to walk around because her weight on the planet would be twice as much as on the earth. But it would be just as easy to catch a ball moving horizontally. The ball's *mass* is the same as on earth, so the horizontal force the astronaut would have to exert to bring it to a stop (i.e., to give it the same acceleration) would also be the same as on earth.

4.5 By Newton's third law, the two forces have equal magnitude. Because the car has much greater mass than the mosquito, it undergoes only a tiny, imperceptible acceleration in response to the force of the impact. By contrast, the mosquito, with its minuscule mass, undergoes a catastrophically large acceleration.

4.6 (iv) The buoyancy force is an *upward* force that the *water* exerts on the *swimmer*. By Newton's third law, the other half of the action–reaction pair is a *downward* force that the *swimmer* exerts on the *water* and has the same magnitude as the buoyancy force. It's true that the weight of the swimmer is also downward and has the same magnitude as the buoyancy force; however, the weight acts on the same body (the swimmer) as the buoyancy force, and so these forces aren't an action–reaction pair.

Bridging Problem

(a) *See the Video Tutor Solution on MasteringPhysics*®

(b) (i) 2.20 m/s²; (ii) 6.00 N; (iii) 3.00 N

? Each of the seeds being blown off the head of a dandelion (genus *Taraxacum*) has a feathery structure called a pappus. The pappus acts like a parachute and enables the seed to be borne by the wind and drift gently to the ground. If a seed with its pappus descends straight down at a steady speed, which force acting on the seed has a greater magnitude? (i) The force of gravity; (ii) the upward force exerted by the air; (iii) both forces have the same magnitude; (iv) it depends on the speed at which the seed descends.

5 APPLYING NEWTON'S LAWS

W e saw in Chapter 4 that Newton's three laws of motion, the foundation of classical mechanics, can be stated very simply. But *applying* these laws to situations such as an iceboat skating across a frozen lake, a toboggan sliding down a hill, or an airplane making a steep turn requires analytical skills and problem-solving technique. In this chapter we'll help you extend the problem-solving skills you began to develop in Chapter 4.

We'll begin with equilibrium problems, in which we analyze the forces that act on a body that is at rest or moving with constant velocity. We'll then consider bodies that are not in equilibrium, for which we'll have to deal with the relationship between forces and motion. We'll learn how to describe and analyze the contact force that acts on a body when it rests on or slides over a surface. We'll also analyze the forces that act on a body that moves in a circle with constant speed. We close the chapter with a brief look at the fundamental nature of force and the classes of forces found in our physical universe.

5.1 USING NEWTON'S FIRST LAW: PARTICLES IN EQUILIBRIUM

We learned in Chapter 4 that a body is in *equilibrium* when it is at rest or moving with constant velocity in an inertial frame of reference. A hanging lamp, a kitchen table, an airplane flying straight and level at a constant speed—all are examples of equilibrium situations. In this section we consider only the equilibrium of a body that can be modeled as a particle. (In Chapter 11 we'll see how to analyze a body in equilibrium that can't be represented adequately as a particle, such as a bridge that's supported at various points along its span.) The essential physical principle is Newton's first law:

Newton's first law: $\cdots\cdots\rightarrow$ ... must be *zero* for a
Net force on a body ... $\quad\sum \vec{F} = 0 \leftarrow\cdots$ body in equilibrium.

| Sum of *x*-components of force on body must be zero. | Sum of *y*-components of force on body must be zero. | (5.1) |

$$\sum F_x = 0 \qquad \sum F_y = 0$$

This section is about using Newton's first law to solve problems dealing with bodies in equilibrium. Some of these problems may seem complicated, but remember that *all* problems involving particles in equilibrium are done in the same way. Problem-Solving Strategy 5.1 details the steps you need to follow for any and all such problems. Study this strategy carefully, look at how it's applied in the worked-out examples, and try to apply it when you solve assigned problems.

PROBLEM-SOLVING STRATEGY 5.1 | NEWTON'S FIRST LAW: EQUILIBRIUM OF A PARTICLE

IDENTIFY *the relevant concepts:* You must use Newton's *first* law, Eqs. (5.1), for any problem that involves forces acting on a body in equilibrium—that is, either at rest or moving with constant velocity. A car is in equilibrium when it's parked, but also when it's traveling down a straight road at a steady speed.

If the problem involves more than one body and the bodies interact with each other, you'll also need to use Newton's *third* law. This law allows you to relate the force that one body exerts on a second body to the force that the second body exerts on the first one.

Identify the target variable(s). Common target variables in equilibrium problems include the magnitude and direction (angle) of one of the forces, or the components of a force.

SET UP *the problem* by using the following steps:
1. Draw a very simple sketch of the physical situation, showing dimensions and angles. You don't have to be an artist!
2. Draw a free-body diagram for each body that is in equilibrium. For now, we consider the body as a particle, so you can represent it as a large dot. In your free-body diagram, *do not* include the other bodies that interact with it, such as a surface it may be resting on or a rope pulling on it.
3. Ask yourself what is interacting with the body by contact or in any other way. On your free-body diagram, draw a force vector for each interaction. Label each force with a symbol for the *magnitude* of the force. If you know the angle at which a force is directed, draw the angle accurately and label it. Include the body's weight, unless the body has negligible mass. If the mass is given, use $w = mg$ to find the weight. A surface in contact with the body exerts a normal force perpendicular to the surface and possibly a friction force parallel to the surface. A rope or chain exerts a pull (never a push) in a direction along its length.

4. *Do not* show in the free-body diagram any forces exerted *by* the body on any other body. The sums in Eqs. (5.1) include only forces that act *on* the body. For each force on the body, ask yourself "What other body causes that force?" If you can't answer that question, you may be imagining a force that isn't there.
5. Choose a set of coordinate axes and include them in your free-body diagram. (If there is more than one body in the problem, choose axes for each body separately.) Label the positive direction for each axis. If a body rests or slides on a plane surface, for simplicity choose axes that are parallel and perpendicular to this surface, even when the plane is tilted.

EXECUTE *the solution* as follows:
1. Find the components of each force along each of the body's coordinate axes. Draw a wiggly line through each force vector that has been replaced by its components, so you don't count it twice. The *magnitude* of a force is always positive, but its *components* may be positive or negative.
2. Set the sum of all *x*-components of force equal to zero. In a separate equation, set the sum of all *y*-components equal to zero. (*Never* add *x*- and *y*-components in a single equation.)
3. If there are two or more bodies, repeat all of the above steps for each body. If the bodies interact with each other, use Newton's third law to relate the forces they exert on each other.
4. Make sure that you have as many independent equations as the number of unknown quantities. Then solve these equations to obtain the target variables.

EVALUATE *your answer:* Look at your results and ask whether they make sense. When the result is a symbolic expression or formula, check to see that your formula works for any special cases (particular values or extreme cases for the various quantities) for which you can guess what the results ought to be.

EXAMPLE 5.1 ONE-DIMENSIONAL EQUILIBRIUM: TENSION IN A MASSLESS ROPE

A gymnast with mass $m_G = 50.0$ kg suspends herself from the lower end of a hanging rope of negligible mass. The upper end of the rope is attached to the gymnasium ceiling. (a) What is the gymnast's weight? (b) What force (magnitude and direction) does the rope exert on her? (c) What is the tension at the top of the rope?

SOLUTION

IDENTIFY and SET UP: The gymnast and the rope are in equilibrium, so we can apply Newton's first law to both bodies. We'll use Newton's third law to relate the forces that they exert on each

Continued

other. The target variables are the gymnast's weight, w_G; the force that the bottom of the rope exerts on the gymnast (call it $T_{R \text{ on } G}$); and the force that the ceiling exerts on the top of the rope (call it $T_{C \text{ on } R}$). **Figure 5.1** shows our sketch of the situation and free-body diagrams for the gymnast and for the rope. We take the positive y-axis to be upward in each diagram. Each force acts in the vertical direction and so has only a y-component.

The forces $T_{R \text{ on } G}$ (the upward force of the rope on the gymnast, Fig. 5.1b) and $T_{G \text{ on } R}$ (the downward force of the gymnast on the rope, Fig. 5.1c) form an action–reaction pair. By Newton's third law, they must have the same magnitude.

Note that Fig. 5.1c includes only the forces that act *on* the rope. In particular, it doesn't include the force that the *rope* exerts on the *ceiling* (compare the discussion of the apple in Conceptual Example 4.9 in Section 4.5).

5.1 Our sketches for this problem.

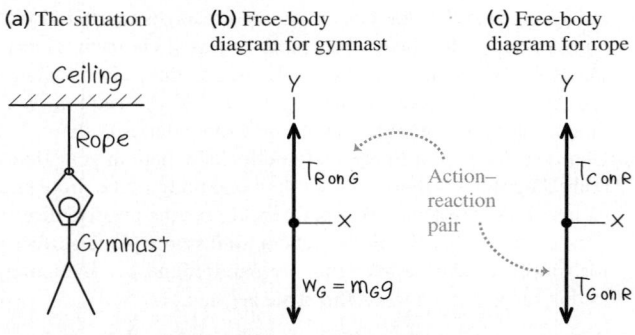

(a) The situation

(b) Free-body diagram for gymnast

(c) Free-body diagram for rope

EXECUTE: (a) The magnitude of her mass and the acceleration due to gravity, g:

$$w_G = m_G g = (50.0 \text{ kg})(9.80 \text{ m/s}^2) = 490 \text{ N}$$

(b) The gravitational force on the gymnast (her weight) points in the negative y-direction, so its y-component is $-w_G$. The upward force of the rope on the gymnast has unknown magnitude $T_{R \text{ on } G}$ and positive y-component $+T_{R \text{ on } G}$. We find this by using Newton's first law from Eqs. (5.1):

$$\text{Gymnast:} \quad \Sigma F_y = T_{R \text{ on } G} + (-w_G) = 0 \quad \text{so}$$
$$T_{R \text{ on } G} = w_G = 490 \text{ N}$$

The rope pulls *up* on the gymnast with a force $T_{R \text{ on } G}$ of magnitude 490 N. (By Newton's third law, the gymnast pulls *down* on the rope with a force of the same magnitude, $T_{G \text{ on } R} = 490 \text{ N}$.)

(c) We have assumed that the rope is weightless, so the only forces on it are those exerted by the ceiling (upward force of unknown magnitude $T_{C \text{ on } R}$) and by the gymnast (downward force of magnitude $T_{G \text{ on } R} = 490 \text{ N}$). From Newton's first law, the *net* vertical force on the rope in equilibrium must be zero:

$$\text{Rope:} \quad \Sigma F_y = T_{C \text{ on } R} + (-T_{G \text{ on } R}) = 0 \quad \text{so}$$
$$T_{C \text{ on } R} = T_{G \text{ on } R} = 490 \text{ N}$$

EVALUATE: The *tension* at any point in the rope is the magnitude of the force that acts at that point. For this weightless rope, the tension $T_{G \text{ on } R}$ at the lower end has the same value as the tension $T_{C \text{ on } R}$ at the upper end. For such an ideal weightless rope, the tension has the same value at any point along the rope's length. (See the discussion in Conceptual Example 4.10 in Section 4.5.)

EXAMPLE 5.2 **ONE-DIMENSIONAL EQUILIBRIUM: TENSION IN A ROPE WITH MASS**

Find the tension at each end of the rope in Example 5.1 if the weight of the rope is 120 N.

SOLUTION

IDENTIFY and SET UP: As in Example 5.1, the target variables are the magnitudes $T_{G \text{ on } R}$ and $T_{C \text{ on } R}$ of the forces that act at the bottom and top of the rope, respectively. Once again, we'll apply Newton's first law to the gymnast and to the rope, and use Newton's third law to relate the forces that the gymnast and rope exert on each other. Again we draw separate free-body diagrams for the gymnast (**Fig. 5.2a**) and the rope (Fig. 5.2b). There is now a *third* force acting on the rope, however: the weight of the rope, of magnitude $w_R = 120 \text{ N}$.

EXECUTE: The gymnast's free-body diagram is the same as in Example 5.1, so her equilibrium condition is also the same. From Newton's third law, $T_{R \text{ on } G} = T_{G \text{ on } R}$, and we again have

$$\text{Gymnast:} \quad \Sigma F_y = T_{R \text{ on } G} + (-w_G) = 0 \quad \text{so}$$
$$T_{R \text{ on } G} = T_{G \text{ on } R} = w_G = 490 \text{ N}$$

The equilibrium condition $\Sigma F_y = 0$ for the rope is now

$$\text{Rope:} \quad \Sigma F_y = T_{C \text{ on } R} + (-T_{G \text{ on } R}) + (-w_R) = 0$$

Note that the y-component of $T_{C \text{ on } R}$ is positive because it points in the $+y$-direction, but the y-components of both $T_{G \text{ on } R}$ and w_R are negative. We solve for $T_{C \text{ on } R}$ and substitute the values $T_{G \text{ on } R} = T_{R \text{ on } G} = 490 \text{ N}$ and $w_R = 120 \text{ N}$:

$$T_{C \text{ on } R} = T_{G \text{ on } R} + w_R = 490 \text{ N} + 120 \text{ N} = 610 \text{ N}$$

5.2 Our sketches for this problem, including the weight of the rope.

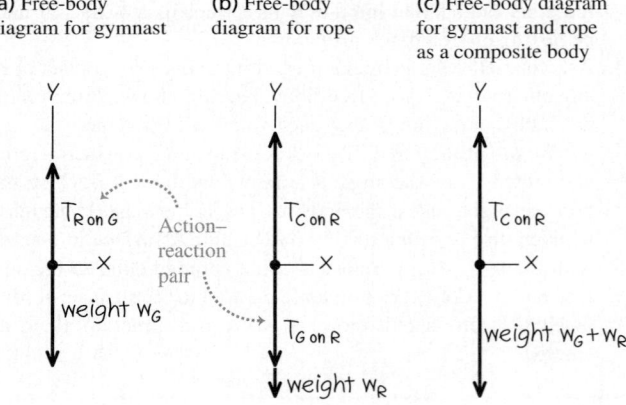

(a) Free-body diagram for gymnast

(b) Free-body diagram for rope

(c) Free-body diagram for gymnast and rope as a composite body

EVALUATE: When we include the weight of the rope, the tension is *different* at the rope's two ends: 610 N at the top and 490 N at the bottom. The force $T_{C \text{ on } R} = 610 \text{ N}$ exerted by the ceiling has to hold up both the 490-N weight of the gymnast and the 120-N weight of the rope.

To see this more clearly, we draw a free-body diagram for a composite body consisting of the gymnast and rope together (Fig. 5.2c). Only two external forces act on this composite body: the force $T_{C \text{ on } R}$ exerted by the ceiling and the total weight

$w_G + w_R = 490\ \text{N} + 120\ \text{N} = 610\ \text{N}$. (The forces $T_{\text{G on R}}$ and $T_{\text{R on G}}$ are *internal* to the composite body. Newton's first law applies only to *external* forces, so these internal forces play no role.) Hence Newton's first law applied to this composite body is

Composite body: $\sum F_y = T_{\text{C on R}} + [-(w_G + w_R)] = 0$

and so $T_{\text{C on R}} = w_G + w_R = 610\ \text{N}$.

Treating the gymnast and rope as a composite body is simpler, but we can't find the tension $T_{\text{G on R}}$ at the bottom of the rope by this method. *Moral: Whenever you have more than one body in a problem involving Newton's laws, the safest approach is to treat each body separately.*

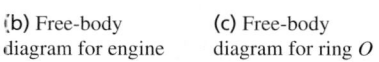

EXAMPLE 5.3   TWO-DIMENSIONAL EQUILIBRIUM

In **Fig. 5.3a**, a car engine with weight w hangs from a chain that is linked at ring O to two other chains, one fastened to the ceiling and the other to the wall. Find expressions for the tension in each of the three chains in terms of w. The weights of the ring and chains are negligible compared with the weight of the engine.

SOLUTION

IDENTIFY and SET UP: The target variables are the tension magnitudes T_1, T_2, and T_3 in the three chains (Fig. 5.3a). All the bodies are in equilibrium, so we'll use Newton's first law. We need three independent equations, one for each target variable. However, applying Newton's first law in component form to just one body gives only *two* equations [the x- and y-equations in Eqs. (5.1)]. So we'll have to consider more than one body in equilibrium. We'll look at the engine (which is acted on by T_1) and the ring (which is attached to all three chains and so is acted on by all three tensions).

Figures 5.3b and 5.3c show our free-body diagrams and choice of coordinate axes. Two forces act on the engine: its weight w and the upward force T_1 exerted by the vertical chain. Three forces act on the ring: the tensions from the vertical chain (T_1), the horizontal chain (T_2), and the slanted chain (T_3). Because the vertical chain has negligible weight, it exerts forces of the same magnitude T_1 at both of its ends (see Example 5.1). (If the weight of this chain were not negligible, these two forces would have different magnitudes; see Example 5.2.) The weight of the ring is also negligible, so it isn't included in Fig. 5.3c.

EXECUTE: The forces acting on the engine are along the y-axis only, so Newton's first law [Eqs. (5.1)] says

Engine: $\sum F_y = T_1 + (-w) = 0$ and $T_1 = w$

The horizontal and slanted chains don't exert forces on the engine itself because they are not attached to it. These forces do appear when we apply Newton's first law to the ring, however. In the free-body diagram for the ring (Fig. 5.3c), remember that T_1, T_2, and T_3 are the *magnitudes* of the forces. We resolve the force with magnitude T_3 into its x- and y-components. Applying Newton's first law in component form to the ring gives us the two equations

Ring: $\sum F_x = T_3 \cos 60° + (-T_2) = 0$

Ring: $\sum F_y = T_3 \sin 60° + (-T_1) = 0$

Because $T_1 = w$ (from the engine equation), we can rewrite the second ring equation as

$$T_3 = \frac{T_1}{\sin 60°} = \frac{w}{\sin 60°} = 1.2w$$

We can now use this result in the first ring equation:

$$T_2 = T_3 \cos 60° = w\frac{\cos 60°}{\sin 60°} = 0.58w$$

EVALUATE: The chain attached to the ceiling exerts a force on the ring with a *vertical* component equal to T_1, which in turn is equal to w. But this force also has a horizontal component, so its magnitude T_3 is somewhat greater than w. This chain is under the greatest tension and is the one most susceptible to breaking.

To get enough equations to solve this problem, we had to consider not only the forces on the engine but also the forces acting on a second body (the ring connecting the chains). Situations like this are fairly common in equilibrium problems, so keep this technique in mind.

5.3 Our sketches for this problem.

(a) Engine, chains, and ring

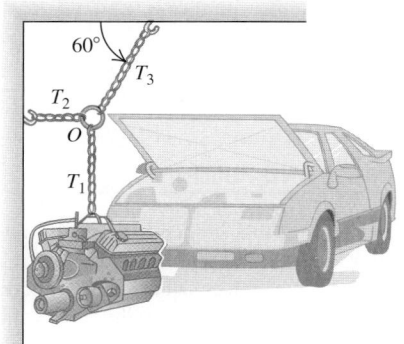

(b) Free-body diagram for engine

(c) Free-body diagram for ring O

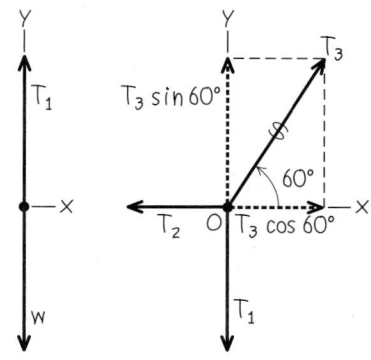

EXAMPLE 5.4 AN INCLINED PLANE

A car of weight w rests on a slanted ramp attached to a trailer (**Fig. 5.4a**). Only a cable running from the trailer to the car prevents the car from rolling off the ramp. (The car's brakes are off and its transmission is in neutral.) Find the tension in the cable and the force that the ramp exerts on the car's tires.

SOLUTION

IDENTIFY: The car is in equilibrium, so we use Newton's first law. The ramp exerts a separate force on each of the car's tires, but for simplicity we lump these forces into a single force. For a further simplification, we'll neglect any friction force the ramp exerts on the tires (see Fig. 4.2b). Hence the ramp exerts only a force on the car that is *perpendicular* to the ramp. As in Section 4.1, we call this force the *normal* force (see Fig. 4.2a). The two target variables are the magnitude T of the tension in the cable and the magnitude n of the normal force.

SET UP: Figure 5.4 shows the situation and a free-body diagram for the car. The three forces acting on the car are its weight (magnitude w), the tension in the cable (magnitude T), and the normal force (magnitude n). Note that the angle α between the ramp and the horizontal is equal to the angle α between the weight vector $\vec{w}$ and the downward normal to the plane of the ramp. Note also that we choose the x- and y-axes to be parallel and perpendicular to the ramp so that we need to resolve only one force (the weight) into x- and y-components. If we had chosen axes that were horizontal and vertical, we'd have to resolve both the normal force and the tension into components.

EXECUTE: To write down the x- and y-components of Newton's first law, we must first find the components of the weight. One complication is that the angle α in Fig. 5.4b is *not* measured from the $+x$-axis toward the $+y$-axis. Hence we *cannot* use Eqs. (1.5) directly to find the components. (You may want to review Section 1.8 to make sure that you understand this important point.)

5.4 A cable holds a car at rest on a ramp.

(a) Car on ramp

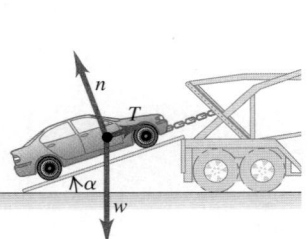

(b) Free-body diagram for car

We replace the weight by its components.

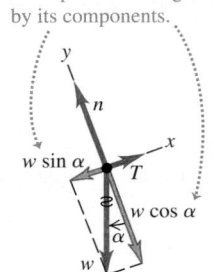

One way to find the components of $\vec{w}$ is to consider the right triangles in Fig. 5.4b. The sine of α is the magnitude of the x-component of $\vec{w}$ (that is, the side of the triangle opposite α) divided by the magnitude w (the hypotenuse of the triangle). Similarly, the cosine of α is the magnitude of the y-component (the side of the triangle adjacent to α) divided by w. Both components are negative, so $w_x = -w\sin\alpha$ and $w_y = -w\cos\alpha$.

Another approach is to recognize that one component of $\vec{w}$ must involve $\sin\alpha$ while the other component involves $\cos\alpha$. To decide which is which, draw the free-body diagram so that the angle α is noticeably smaller or larger than 45°. (You'll have to fight the natural tendency to draw such angles as being close to 45°.) We've drawn Fig. 5.4b so that α is smaller than 45°, so $\sin\alpha$ is less than $\cos\alpha$. The figure shows that the x-component of $\vec{w}$ is smaller than the y-component, so the x-component must involve $\sin\alpha$ and the y-component must involve $\cos\alpha$. We again find $w_x = -w\sin\alpha$ and $w_y = -w\cos\alpha$.

In Fig. 5.4b we draw a wiggly line through the original vector representing the weight to remind us not to count it twice. Newton's first law gives us

$$\sum F_x = T + (-w\sin\alpha) = 0$$
$$\sum F_y = n + (-w\cos\alpha) = 0$$

(Remember that T, w, and n are all *magnitudes* of vectors and are therefore all positive.) Solving these equations for T and n, we find

$$T = w\sin\alpha$$
$$n = w\cos\alpha$$

EVALUATE: Our answers for T and n depend on the value of α. To check this dependence, let's look at some special cases. If the ramp is horizontal ($\alpha = 0$), we get $T = 0$ and $n = w$: No cable tension T is needed to hold the car, and the normal force n is equal in magnitude to the weight. If the ramp is vertical ($\alpha = 90°$), we get $T = w$ and $n = 0$: The cable tension T supports all of the car's weight, and there's nothing pushing the car against the ramp.

CAUTION **Normal force and weight may not be equal** It's a common error to assume that the normal-force magnitude n equals the weight w. Our result shows that this is *not* always the case. Always treat n as a variable and solve for its value, as we've done here. ▌

How would the answers for T and n be affected if the car were being pulled up the ramp at a constant speed? This, too, is an equilibrium situation, since the car's velocity is constant. So the calculation is the same, and T and n have the same values as when the car is at rest. (It's true that T must be greater than $w\sin\alpha$ to *start* the car moving up the ramp, but that's not what we asked.)

EXAMPLE 5.5 EQUILIBRIUM OF BODIES CONNECTED BY CABLE AND PULLEY

Your firm needs to haul granite blocks up a 15° slope out of a quarry and to lower dirt into the quarry to fill the holes. You design a system in which a granite block on a cart with steel wheels (weight w_1, including both block and cart) is pulled uphill on steel rails by a dirt-filled bucket (weight w_2, including both dirt

and bucket) that descends vertically into the quarry (**Fig. 5.5a**). How must the weights w_1 and w_2 be related in order for the system to move with constant speed? Ignore friction in the pulley and wheels, and ignore the weight of the cable.

5.5 Our sketches for this problem.

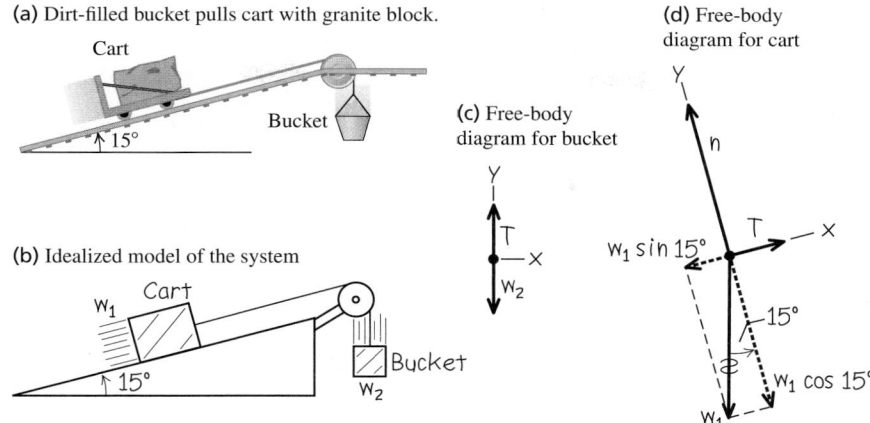

(a) Dirt-filled bucket pulls cart with granite block.

Cart

Bucket

15°

(b) Idealized model of the system

w_1 Cart

15°   Bucket w_2

(c) Free-body
diagram for bucket

Y

T

X

w_2

(d) Free-body
diagram for cart

Y

n

$w_1 \sin 15°$ T   X

15°

$w_1 \cos 15°$

w_1

SOLUTION

IDENTIFY and SET UP: The cart and bucket each move with a constant velocity (in a straight line at constant speed). Hence each body is in equilibrium, and we can apply Newton's first law to each. Our target is an expression relating the weights w_1 and w_2.

Figure 5.5b shows our idealized model for the system, and Figs. 5.5c and 5.5d show our free-body diagrams. The two forces on the bucket are its weight w_2 and an upward tension exerted by the cable. As for the car on the ramp in Example 5.4, three forces act on the cart: its weight w_1, a normal force of magnitude n exerted by the rails, and a tension force from the cable. Since we're assuming that the cable has negligible weight, the tension forces that the cable exerts on the cart and on the bucket have the same magnitude T. (We're ignoring friction, so we assume that the rails exert no force on the cart parallel to the incline.) Note that we orient the axes differently for each body; the choices shown are the most convenient ones. We find the components of the weight force in the same way that we did in Example 5.4. (Compare Fig. 5.5d with Fig. 5.4b.)

EXECUTE: Applying $\Sigma F_y = 0$ to the bucket in Fig. 5.5c, we find

$$\Sigma F_y = T + (-w_2) = 0 \qquad \text{so} \qquad T = w_2$$

Applying $\Sigma F_x = 0$ to the cart (and block) in Fig. 5.5d, we get

$$\Sigma F_x = T + (-w_1 \sin 15°) = 0 \qquad \text{so} \qquad T = w_1 \sin 15°$$

Equating the two expressions for T, we find

$$w_2 = w_1 \sin 15° = 0.26w_1$$

EVALUATE: Our analysis doesn't depend at all on the direction in which the cart and bucket move. Hence the system can move with constant speed in *either* direction if the weight of the dirt and bucket is 26% of the weight of the granite block and cart. What would happen if w_2 were greater than $0.26w_1$? If it were less than $0.26w_1$?

Notice that we didn't need the equation $\Sigma F_y = 0$ for the cart and block. Can you use this to show that $n = w_1 \cos 15°$?

TEST YOUR UNDERSTANDING OF SECTION 5.1   A traffic light of weight w hangs from two lightweight cables, one on each side of the light. Each cable hangs at a 45° angle from the horizontal. What is the tension in each cable? (i) $w/2$; (ii) $w/\sqrt{2}$; (iii) w; (iv) $w\sqrt{2}$; (v) $2w$. ∎

5.2 USING NEWTON'S SECOND LAW: DYNAMICS OF PARTICLES

We are now ready to discuss *dynamics* problems. In these problems, we apply Newton's second law to bodies on which the net force is *not* zero. These bodies are *not* in equilibrium and hence are accelerating:

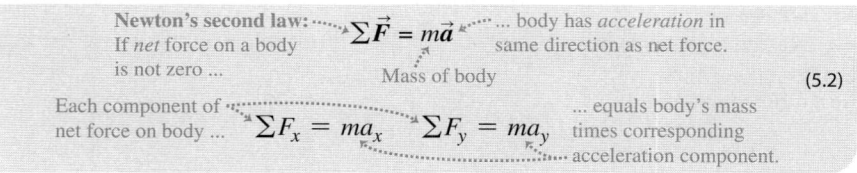

Newton's second law: ⋯⋯ $\Sigma \vec{F} = m\vec{a}$ ⋯⋯ body has *acceleration* in
If *net* force on a body   same direction as net force.
is not zero ...   Mass of body   (5.2)

Each component of ⋯⋯ $\Sigma F_x = ma_x$ $\Sigma F_y = ma_y$ ... equals body's mass
net force on body ...   times corresponding
                                                                   ⋯⋯ acceleration component.

5.6 Correct and incorrect free-body diagrams for a falling body.

(a)

Only the force of gravity acts on this falling fruit.

(b) Correct free-body diagram

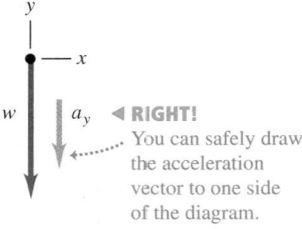

◄ **RIGHT!**
You can safely draw the acceleration vector to one side of the diagram.

(c) Incorrect free-body diagram

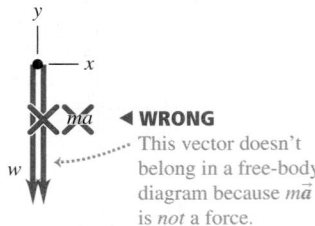

◄ **WRONG**
This vector doesn't belong in a free-body diagram because $m\vec{a}$ is *not* a force.

The following problem-solving strategy is very similar to Problem-Solving Strategy 5.1 for equilibrium problems in Section 5.1. Study it carefully, watch how we apply it in our examples, and use it when you tackle the end-of-chapter problems. You can use this strategy to solve *any* dynamics problem.

> CAUTION $m\vec{a}$ **doesn't belong in free-body diagrams** Remember that the quantity $m\vec{a}$ is the *result* of forces acting on a body, *not* a force itself. When you draw the free-body diagram for an accelerating body (like the fruit in **Fig. 5.6a**), *never* include the "$m\vec{a}$ force" because *there is no such force* (Fig. 5.6c). Review Section 4.3 if you're not clear on this point. Sometimes we draw the acceleration vector $\vec{a}$ *alongside* a free-body diagram, as in Fig. 5.6b. But we *never* draw the acceleration vector with its tail touching the body (a position reserved exclusively for forces that act on the body). ▌

PROBLEM-SOLVING STRATEGY 5.2) NEWTON'S SECOND LAW: DYNAMICS OF PARTICLES

IDENTIFY *the relevant concepts:* You have to use Newton's second law, Eqs. (5.2), for *any* problem that involves forces acting on an accelerating body.

Identify the target variable—usually an acceleration or a force. If the target variable is something else, you'll need to select another concept to use. For example, suppose the target variable is how fast a sled is moving when it reaches the bottom of a hill. Newton's second law will let you find the sled's acceleration; you'll then use the constant-acceleration relationships from Section 2.4 to find velocity from acceleration.

SET UP *the problem* by using the following steps:
1. Draw a simple sketch of the situation that shows each moving body. For each body, draw a free-body diagram that shows all the forces acting *on* the body. [The sums in Eqs. (5.2) include the forces that act on the body, *not* the forces that it exerts on anything else.] Make sure you can answer the question "What other body is applying this force?" for each force in your diagram. Never include the quantity $m\vec{a}$ in your free-body diagram; it's not a force!
2. Label each force with an algebraic symbol for the force's *magnitude*. Usually, one of the forces will be the body's weight; it's usually best to label this as $w = mg$.
3. Choose your *x*- and *y*-coordinate axes for each body, and show them in its free-body diagram. Indicate the positive direction for each axis. If you know the direction of the acceleration, it usually simplifies things to take one positive axis along that direction. If your problem involves two or more bodies that accelerate in different directions, you can use a different set of axes for each body.

4. In addition to Newton's second law, $\sum\vec{F} = m\vec{a}$, identify any other equations you might need. For example, you might need one or more of the equations for motion with constant acceleration. If more than one body is involved, there may be relationships among their motions; for example, they may be connected by a rope. Express any such relationships as equations relating the accelerations of the various bodies.

EXECUTE *the solution* as follows:
1. For each body, determine the components of the forces along each of the body's coordinate axes. When you represent a force in terms of its components, draw a wiggly line through the original force vector to remind you not to include it twice.
2. List all of the known and unknown quantities. In your list, identify the target variable or variables.
3. For each body, write a separate equation for each component of Newton's second law, as in Eqs. (5.2). Write any additional equations that you identified in step 4 of "Set Up." (You need as many equations as there are target variables.)
4. Do the easy part—the math! Solve the equations to find the target variable(s).

EVALUATE *your answer:* Does your answer have the correct units? (When appropriate, use the conversion $1\text{ N} = 1\text{ kg}\cdot\text{m/s}^2$.) Does it have the correct algebraic sign? When possible, consider particular values or extreme cases of quantities and compare the results with your intuitive expectations. Ask, "Does this result make sense?"

EXAMPLE 5.6 STRAIGHT-LINE MOTION WITH A CONSTANT FORCE

An iceboat is at rest on a frictionless horizontal surface (**Fig. 5.7a**). Due to the blowing wind, 4.0 s after the iceboat is released, it is moving to the right at 6.0 m/s (about 22 km/h, or 13 mi/h). What constant horizontal force F_W does the wind exert on the iceboat? The combined mass of iceboat and rider is 200 kg.

___
SOLUTION

IDENTIFY and SET UP: Our target variable is one of the forces (F_W) acting on the accelerating iceboat, so we need to use Newton's second law. The forces acting on the iceboat and rider (considered as a unit) are the weight w, the normal force n exerted by the surface, and the horizontal force F_W. Figure 5.7b shows the free-body diagram. The net force and hence the acceleration are to the right, so we chose the positive x-axis in this direction. The acceleration isn't given; we'll need to find it. Since the wind is assumed to exert a constant force, the resulting acceleration is constant and we can use one of the constant-acceleration formulas from Section 2.4.

5.7 Our sketches for this problem.

(a) Iceboat and rider on frictionless ice

(b) Free-body diagram for iceboat and rider

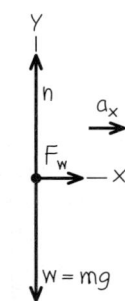

The iceboat starts at rest (its initial x-velocity is $v_{0x} = 0$) and it attains an x-velocity $v_x = 6.0$ m/s after an elapsed time $t = 4.0$ s. To relate the x-acceleration a_x to these quantities we use Eq. (2.8), $v_x = v_{0x} + a_x t$. There is no vertical acceleration, so we expect that the normal force on the iceboat is equal in magnitude to the iceboat's weight.

EXECUTE: The *known* quantities are the mass $m = 200$ kg, the initial and final x-velocities $v_{0x} = 0$ and $v_x = 6.0$ m/s, and the elapsed time $t = 4.0$ s. There are three *unknown* quantities: the acceleration a_x, the normal force n, and the horizontal force F_W. Hence we need three equations.

The first two equations are the x- and y-equations for Newton's second law, Eqs. (5.2). The force F_W is in the positive x-direction, while the forces n and $w = mg$ are in the positive and negative y-directions, respectively. Hence we have

$$\sum F_x = F_W = ma_x$$
$$\sum F_y = n + (-mg) = 0 \quad \text{so} \quad n = mg$$

The third equation is Eq. (2.8) for constant acceleration:

$$v_x = v_{0x} + a_x t$$

To find F_W, we first solve this third equation for a_x and then substitute the result into the $\sum F_x$ equation:

$$a_x = \frac{v_x - v_{0x}}{t} = \frac{6.0 \text{ m/s} - 0 \text{ m/s}}{4.0 \text{ s}} = 1.5 \text{ m/s}^2$$

$$F_W = ma_x = (200 \text{ kg})(1.5 \text{ m/s}^2) = 300 \text{ kg} \cdot \text{m/s}^2$$

Since 1 kg $\cdot$ m/s^2 = 1 N, the final answer is

$$F_W = 300 \text{ N (about 67 lb)}$$

EVALUATE: Our answers for F_W and n have the correct units for a force, and (as expected) the magnitude n of the normal force is equal to mg. Does it seem reasonable that the force F_W is substantially *less* than mg?

EXAMPLE 5.7 STRAIGHT-LINE MOTION WITH FRICTION

Suppose a constant horizontal friction force with magnitude 100 N opposes the motion of the iceboat in Example 5.6. In this case, what constant force F_W must the wind exert on the iceboat to cause the same constant x-acceleration $a_x = 1.5$ m/s^2?

___
SOLUTION

IDENTIFY and SET UP: Again the target variable is F_W. We are given the x-acceleration, so to find F_W all we need is Newton's second law. **Figure 5.8** shows our new free-body diagram. The only difference from Fig. 5.7b is the addition of the friction force $\vec{f}$, which points in the negative x-direction (opposite the motion). Because the wind must now overcome the friction force to yield the same acceleration as in Example 5.6, we expect our answer for F_W to be greater than the 300 N we found there.

5.8 Our free-body diagram for the iceboat and rider with friction force $\vec{f}$ opposing the motion.

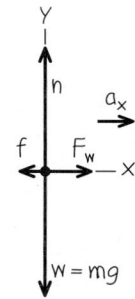

Continued

EXECUTE: Two forces now have x-components: the force of the wind (x-component $+F_W$) and the friction force (x-component $-f$). The x-component of Newton's second law gives

$$\Sigma F_x = F_W + (-f) = ma_x$$

$$F_W = ma_x + f = (200\text{ kg})(1.5\text{ m/s}^2) + (100\text{ N}) = 400\text{ N}$$

EVALUATE: The required value of F_W is 100 N greater than in Example 5.6 because the wind must now push against an additional 100-N friction force.

EXAMPLE 5.8 TENSION IN AN ELEVATOR CABLE

An elevator and its load have a combined mass of 800 kg (**Fig. 5.9a**). The elevator is initially moving downward at 10.0 m/s; it slows to a stop with constant acceleration in a distance of 25.0 m. What is the tension T in the supporting cable while the elevator is being brought to rest?

SOLUTION

IDENTIFY and SET UP: The target variable is the tension T, which we'll find by using Newton's second law. As in Example 5.6, we'll use a constant-acceleration formula to determine the acceleration. Our free-body diagram (Fig. 5.9b) shows two forces acting on the elevator: its weight w and the tension force T of the cable. The elevator is moving downward with decreasing speed, so its acceleration is upward; we chose the positive y-axis to be upward.

5.9 Our sketches for this problem.

(a) Descending elevator

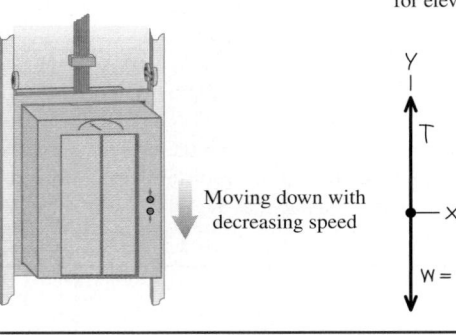

(b) Free-body diagram for elevator

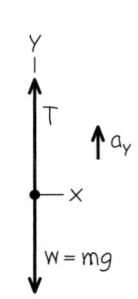

Moving down with decreasing speed

The elevator is moving in the negative y-direction, so both its initial y-velocity v_{0y} and its y-displacement $y - y_0$ are negative: $v_{0y} = -10.0\text{ m/s}$ and $y - y_0 = -25.0\text{ m}$. The final y-velocity is $v_y = 0$. To find the y-acceleration a_y from this information, we'll use Eq. (2.13) in the form $v_y^2 = v_{0y}^2 + 2a_y(y - y_0)$. Once we have a_y, we'll substitute it into the y-component of Newton's second law from Eqs. (5.2) and solve for T. The net force must be upward to give an upward acceleration, so we expect T to be greater than the weight $w = mg = (800\text{ kg})(9.80\text{ m/s}^2) = 7840\text{ N}$.

EXECUTE: First let's write out Newton's second law. The tension force acts upward and the weight acts downward, so

$$\Sigma F_y = T + (-w) = ma_y$$

We solve for the target variable T:

$$T = w + ma_y = mg + ma_y = m(g + a_y)$$

To determine a_y, we rewrite the constant-acceleration equation $v_y^2 = v_{0y}^2 + 2a_y(y - y_0)$:

$$a_y = \frac{v_y^2 - v_{0y}^2}{2(y - y_0)} = \frac{(0)^2 - (-10.0\text{ m/s})^2}{2(-25.0\text{ m})} = +2.00\text{ m/s}^2$$

The acceleration is upward (positive), just as it should be.

Now we can substitute the acceleration into the equation for the tension:

$$T = m(g + a_y) = (800\text{ kg})(9.80\text{ m/s}^2 + 2.00\text{ m/s}^2) = 9440\text{ N}$$

EVALUATE: The tension is greater than the weight, as expected. Can you see that we would get the same answers for a_y and T if the elevator were moving *upward* and *gaining* speed at a rate of 2.00 m/s²?

EXAMPLE 5.9 APPARENT WEIGHT IN AN ACCELERATING ELEVATOR

A 50.0-kg woman stands on a bathroom scale while riding in the elevator in Example 5.8. What is the reading on the scale?

SOLUTION

IDENTIFY and SET UP: The scale (**Fig. 5.10a**) reads the magnitude of the downward force exerted *by* the woman *on* the scale. By Newton's third law, this equals the magnitude of the upward normal force exerted *by* the scale *on* the woman. Hence our target variable is the magnitude n of the normal force. We'll find n by applying Newton's second law to the woman. We already know her acceleration; it's the same as the acceleration of the elevator, which we calculated in Example 5.8.

Figure 5.10b shows our free-body diagram for the woman. The forces acting on her are the normal force n exerted by the

5.10 Our sketches for this problem.

(a) Woman in a descending elevator

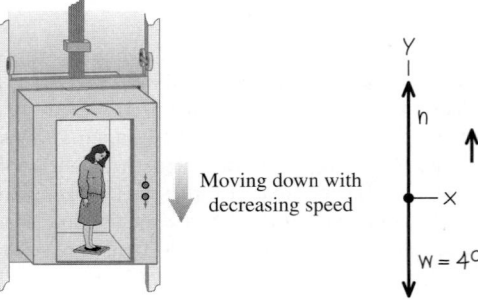

(b) Free-body diagram for woman

Moving down with decreasing speed

scale and her weight $w = mg = (50.0 \text{ kg})(9.80 \text{ m/s}^2) = 490$ N. (The tension force, which played a major role in Example 5.8, doesn't appear here because it doesn't act on the woman.) From Example 5.8, the y-acceleration of the elevator and of the woman is $a_y = +2.00 \text{ m/s}^2$. As in Example 5.8, the upward force on the body accelerating upward (in this case, the normal force on the woman) will have to be greater than the body's weight to produce the upward acceleration.

EXECUTE: Newton's second law gives

$$\Sigma F_y = n + (-mg) = ma_y$$

$$n = mg + ma_y = m(g + a_y)$$

$$= (50.0 \text{ kg})(9.80 \text{ m/s}^2 + 2.00 \text{ m/s}^2) = 590 \text{ N}$$

EVALUATE: Our answer for n means that while the elevator is stopping, the scale pushes up on the woman with a force of 590 N. By Newton's third law, she pushes down on the scale with the same force. So the scale reads 590 N, which is 100 N more than her actual weight. The scale reading is called the passenger's **apparent weight.** The woman *feels* the floor pushing up harder on her feet than when the elevator is stationary or moving with constant velocity.

What would the woman feel if the elevator were accelerating *downward*, so that $a_y = -2.00 \text{ m/s}^2$? This would be the case if the elevator were moving upward with decreasing speed or moving downward with increasing speed. To find the answer for this situation, we just insert the new value of a_y in our equation for n:

$$n = m(g + a_y) = (50.0 \text{ kg})[9.80 \text{ m/s}^2 + (-2.00 \text{ m/s}^2)]$$

$$= 390 \text{ N}$$

Now the woman would feel as though she weighs only 390 N, or 100 N *less* than her actual weight w.

You can feel these effects yourself; try taking a few steps in an elevator that is coming to a stop after descending (when your apparent weight is greater than w) or coming to a stop after ascending (when your apparent weight is less than w).

Apparent Weight and Apparent Weightlessness

Let's generalize the result of Example 5.9. When a passenger with mass m rides in an elevator with y-acceleration a_y, a scale shows the passenger's apparent weight to be

$$n = m(g + a_y)$$

When the elevator is accelerating upward, a_y is positive and n is greater than the passenger's weight $w = mg$. When the elevator is accelerating downward, a_y is negative and n is less than the weight. If the passenger doesn't know the elevator is accelerating, she may feel as though her weight is changing; indeed, this is just what the scale shows.

The extreme case occurs when the elevator has a downward acceleration $a_y = -g$—that is, when it is in free fall. In that case $n = 0$ and the passenger *seems* to be weightless. Similarly, an astronaut orbiting the earth with a spacecraft experiences *apparent weightlessness* (**Fig. 5.11**). In each case, the person is not truly weightless because a gravitational force still acts. But the person's *sensations* in this free-fall condition are exactly the same as though the person were in outer space with no gravitational force at all. In both cases the person and the vehicle (elevator or spacecraft) fall together with the same acceleration g, so nothing pushes the person against the floor or walls of the vehicle.

5.11 Astronauts in orbit feel "weightless" because they have the same acceleration as their spacecraft. They are *not* outside the pull of the earth's gravity. (We'll discuss the motions of orbiting bodies in detail in Chapter 12.)

EXAMPLE 5.10 ACCELERATION DOWN A HILL

A toboggan loaded with students (total weight w) slides down a snow-covered hill that slopes at a constant angle α. The toboggan is well waxed, so there is virtually no friction. What is its acceleration?

SOLUTION

IDENTIFY and SET UP: Our target variable is the acceleration, which we'll find by using Newton's second law. There is no friction, so only two forces act on the toboggan: its weight w and the normal force n exerted by the hill.

Figure 5.12 shows our sketch and free-body diagram. We take axes parallel and perpendicular to the surface of the hill, so that the acceleration (which is parallel to the hill) is along the positive x-direction.

5.12 Our sketches for this problem.

(a) The situation

(b) Free-body diagram for toboggan

Continued

EXECUTE: The normal force has only a y-component, but the weight has both x- and y-components: $w_x = w\sin\alpha$ and $w_y = -w\cos\alpha$. (In Example 5.4 we had $w_x = -w\sin\alpha$. The difference is that the positive x-axis was uphill in Example 5.4 but is downhill in Fig. 5.12b.) The wiggly line in Fig. 5.12b reminds us that we have resolved the weight into its components. The acceleration is purely in the $+x$-direction, so $a_y = 0$. Newton's second law in component form from Eqs. (5.2) then tells us that

$$\sum F_x = w\sin\alpha = ma_x$$
$$\sum F_y = n - w\cos\alpha = ma_y = 0$$

Since $w = mg$, the x-component equation gives $mg\sin\alpha = ma_x$, or

$$a_x = g\sin\alpha$$

Note that we didn't need the y-component equation to find the acceleration. That's part of the beauty of choosing the x-axis to lie along the acceleration direction! The y-equation tells us the magnitude of the normal force exerted by the hill on the toboggan:

$$n = w\cos\alpha = mg\cos\alpha$$

EVALUATE: Notice that the normal force n is not equal to the toboggan's weight (compare Example 5.4). Notice also that the mass m does not appear in our result for the acceleration. That's because the downhill force on the toboggan (a component of the weight) is proportional to m, so the mass cancels out when we use $\sum F_x = ma_x$ to calculate a_x. Hence *any* toboggan, regardless of its mass, slides down a frictionless hill with acceleration $g\sin\alpha$.

If the plane is horizontal, $\alpha = 0$ and $a_x = 0$ (the toboggan does not accelerate); if the plane is vertical, $\alpha = 90°$ and $a_x = g$ (the toboggan is in free fall).

CAUTION **Common free-body diagram errors Figure 5.13** shows both the correct way (Fig. 5.13a) and a common *incorrect* way (Fig. 5.13b) to draw the free-body diagram for the toboggan. The diagram in Fig. 5.13b is wrong for two reasons: The normal force must be drawn perpendicular to the surface (remember, "normal" means perpendicular), and there's no such thing as the "$m\vec{a}$ force." ▮

5.13 Correct and incorrect free-body diagrams for a toboggan on a frictionless hill.

(a) Correct free-body diagram for the sled

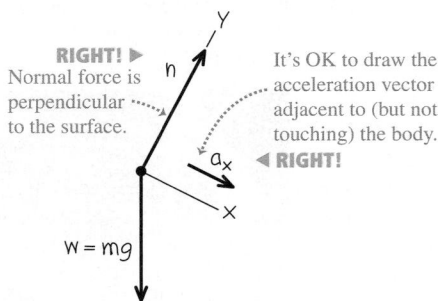

RIGHT! ▶
Normal force is perpendicular to the surface.

It's OK to draw the acceleration vector adjacent to (but not touching) the body. ◀ RIGHT!

(b) Incorrect free-body diagram for the sled

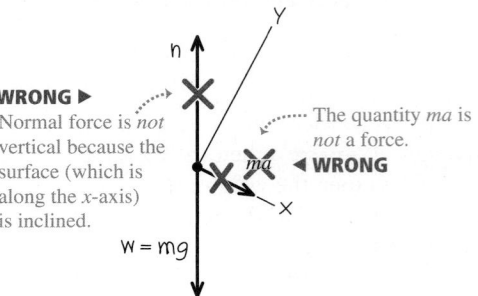

WRONG ▶
Normal force is *not* vertical because the surface (which is along the x-axis) is inclined.

The quantity ma is *not* a force. ◀ WRONG

EXAMPLE 5.11 **TWO BODIES WITH THE SAME ACCELERATION**

You push a 1.00-kg food tray through the cafeteria line with a constant 9.0-N force. The tray pushes a 0.50-kg milk carton (**Fig. 5.14a**). The tray and carton slide on a horizontal surface so greasy that friction can be ignored. Find the acceleration of the tray and carton and the horizontal force that the tray exerts on the carton.

SOLUTION

IDENTIFY and SET UP: Our *two* target variables are the acceleration of the tray–carton system and the force of the tray on the carton. We'll use Newton's second law to get two equations, one for each target variable. We set up and solve the problem in two ways.

Method 1: We treat the carton (mass m_C) and tray (mass m_T) as separate bodies, each with its own free-body diagram (Figs. 5.14b and 5.14c). The force F that you exert on the tray doesn't appear in the free-body diagram for the carton, which is accelerated by the force (of magnitude $F_{T\,on\,C}$) exerted on it by the tray. By Newton's third law, the carton exerts a force of equal magnitude on the tray: $F_{C\,on\,T} = F_{T\,on\,C}$. We take the acceleration to be in the positive x-direction; both the tray and milk carton move with the same x-acceleration a_x.

Method 2: We treat the tray and milk carton as a composite body of mass $m = m_T + m_C = 1.50\,kg$ (Fig. 5.14d). The only horizontal force acting on this body is the force F that you exert. The forces $F_{T\,on\,C}$ and $F_{C\,on\,T}$ don't come into play because they're *internal* to this composite body, and Newton's second law tells us that only *external* forces affect a body's acceleration (see Section 4.3). To find the magnitude $F_{T\,on\,C}$ we'll again apply Newton's second law to the carton, as in Method 1.

EXECUTE: *Method 1:* The x-component equations of Newton's second law are

Tray: $\quad \sum F_x = F - F_{C\,on\,T} = F - F_{T\,on\,C} = m_T a_x$

Carton: $\quad \sum F_x = F_{T\,on\,C} = m_C a_x$

These are two simultaneous equations for the two target variables a_x and $F_{T\,on\,C}$. (Two equations are all we need, which means that the y-components don't play a role in this example.) An easy way to solve the two equations for a_x is to add them; this eliminates $F_{T\,on\,C}$, giving

$$F = m_T a_x + m_C a_x = (m_T + m_C)a_x$$

5.14 Pushing a food tray and milk carton in the cafeteria line.

(a) A milk carton and a food tray

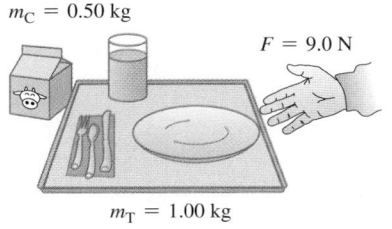

$m_C = 0.50$ kg

$F = 9.0$ N

$m_T = 1.00$ kg

(b) Free-body diagram for milk carton

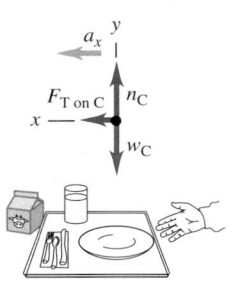

(c) Free-body diagram for food tray

(d) Free-body diagram for carton and tray as a composite body

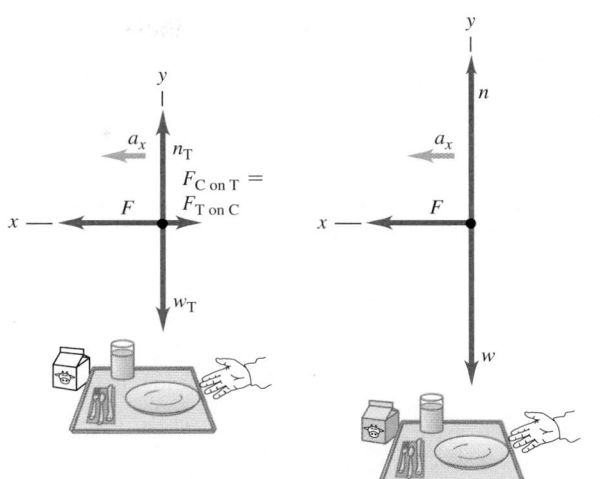

We solve this equation for a_x:

$$a_x = \frac{F}{m_T + m_C} = \frac{9.0 \text{ N}}{1.00 \text{ kg} + 0.50 \text{ kg}} = 6.0 \text{ m/s}^2 = 0.61g$$

Substituting this value into the carton equation gives

$$F_{T \text{ on } C} = m_C a_x = (0.50 \text{ kg})(6.0 \text{ m/s}^2) = 3.0 \text{ N}$$

Method 2: The x-component of Newton's second law for the composite body of mass m is

$$\sum F_x = F = ma_x$$

The acceleration of this composite body is

$$a_x = \frac{F}{m} = \frac{9.0 \text{ N}}{1.50 \text{ kg}} = 6.0 \text{ m/s}^2$$

Then, looking at the milk carton by itself, we see that to give it an acceleration of 6.0 m/s² requires that the tray exert a force

$$F_{T \text{ on } C} = m_C a_x = (0.50 \text{ kg})(6.0 \text{ m/s}^2) = 3.0 \text{ N}$$

EVALUATE: The answers are the same with both methods. To check the answers, note that there are different forces on the two sides of the tray: $F = 9.0$ N on the right and $F_{C \text{ on } T} = 3.0$ N on the left. The net horizontal force on the tray is $F - F_{C \text{ on } T} = 6.0$ N, exactly enough to accelerate a 1.00-kg tray at 6.0 m/s².

Treating two bodies as a single, composite body works *only* if the two bodies have the same magnitude *and* direction of acceleration. If the accelerations are different we must treat the two bodies separately, as in the next example.

EXAMPLE 5.12 **TWO BODIES WITH THE SAME MAGNITUDE OF ACCELERATION**

Figure 5.15a shows an air-track glider with mass m_1 moving on a level, frictionless air track in the physics lab. The glider is connected to a lab weight with mass m_2 by a light, flexible, nonstretching string that passes over a stationary, frictionless pulley. Find the acceleration of each body and the tension in the string.

SOLUTION

IDENTIFY and SET UP: The glider and weight are accelerating, so again we must use Newton's second law. Our three target variables are the tension T in the string and the accelerations of the two bodies.

The two bodies move in different directions—one horizontal, one vertical—so we can't consider them to be a single unit as we did the bodies in Example 5.11. Figures 5.15b and 5.15c show our free-body diagrams and coordinate systems. It's convenient to have both bodies accelerate in the positive axis directions, so we chose the positive y-direction for the lab weight to be downward.

We consider the string to be massless and to slide over the pulley without friction, so the tension T in the string is the same throughout and it applies a force of the same magnitude T to each

5.15 Our sketches for this problem.

(a) Apparatus

(b) Free-body diagram for glider

(c) Free-body diagram for weight

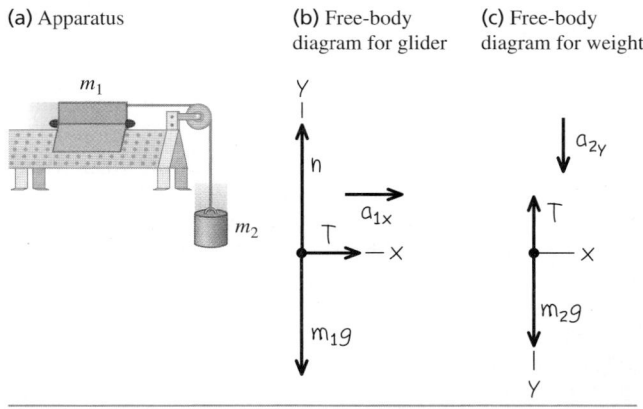

body. (You may want to review Conceptual Example 4.10, in which we discussed the tension force exerted by a massless rope.) The weights are m_1g and m_2g.

Continued

While the *directions* of the two accelerations are different, their *magnitudes* are the same. (That's because the string doesn't stretch, so the two bodies must move equal distances in equal times and their speeds at any instant must be equal. When the speeds change, they change at the same rate, so the accelerations of the two bodies must have the same magnitude a.) We can express this relationship as $a_{1x} = a_{2y} = a$, which means that we have only *two* target variables: a and the tension T.

What results do we expect? If $m_1 = 0$ (or, approximately, for m_1 much less than m_2) the lab weight will fall freely with acceleration g, and the tension in the string will be zero. For $m_2 = 0$ (or, approximately, for m_2 much less than m_1) we expect zero acceleration and zero tension.

EXECUTE: Newton's second law gives

$$\text{Glider:} \quad \sum F_x = T = m_1 a_{1x} = m_1 a$$
$$\text{Glider:} \quad \sum F_y = n + (-m_1 g) = m_1 a_{1y} = 0$$
$$\text{Lab weight:} \quad \sum F_y = m_2 g + (-T) = m_2 a_{2y} = m_2 a$$

(There are no forces on the lab weight in the x-direction.) In these equations we've used $a_{1y} = 0$ (the glider doesn't accelerate vertically) and $a_{1x} = a_{2y} = a$.

The x-equation for the glider and the equation for the lab weight give us two simultaneous equations for T and a:

$$\text{Glider:} \quad T = m_1 a$$
$$\text{Lab weight:} \quad m_2 g - T = m_2 a$$

We add the two equations to eliminate T, giving

$$m_2 g = m_1 a + m_2 a = (m_1 + m_2)a$$

and so the magnitude of each body's acceleration is

$$a = \frac{m_2}{m_1 + m_2} g$$

Substituting this back into the glider equation $T = m_1 a$, we get

$$T = \frac{m_1 m_2}{m_1 + m_2} g$$

EVALUATE: The acceleration is in general less than g, as you might expect; the string tension keeps the lab weight from falling freely. The tension T is *not* equal to the weight $m_2 g$ of the lab weight, but is *less* by a factor of $m_1/(m_1 + m_2)$. If T *were* equal to $m_2 g$, then the lab weight would be in equilibrium, and it isn't.

As predicted, the acceleration is equal to g for $m_1 = 0$ and equal to zero for $m_2 = 0$, and $T = 0$ for either $m_1 = 0$ or $m_2 = 0$.

CAUTION **Tension and weight may not be equal** It's a common mistake to assume that if an object is attached to a vertical string, the string tension must be equal to the object's weight. That was the case in Example 5.5, where the acceleration was zero, but it's not the case in this example! The only safe approach is *always* to treat the tension as a variable, as we did here. ▮

PhET: Lunar Lander

TEST YOUR UNDERSTANDING OF SECTION 5.2 Suppose you hold the glider in Example 5.12 so that it and the weight are initially at rest. You give the glider a push to the left in Fig. 5.15a and then release it. The string remains taut as the glider moves to the left, comes instantaneously to rest, then moves to the right. At the instant the glider has zero velocity, what is the tension in the string? (i) Greater than in Example 5.12; (ii) the same as in Example 5.12; (iii) less than in Example 5.12 but greater than zero; (iv) zero. ▮

5.3 FRICTION FORCES

We've seen several problems in which a body rests or slides on a surface that exerts forces on the body. Whenever two bodies interact by direct contact (touching) of their surfaces, we describe the interaction in terms of *contact forces*. The normal force is one example of a contact force; in this section we'll look in detail at another contact force, the force of friction.

Friction is important in many aspects of everyday life. The oil in a car engine minimizes friction between moving parts, but without friction between the tires and the road we couldn't drive or turn the car. Air drag—the friction force exerted by the air on a body moving through it—decreases automotive fuel economy but makes parachutes work. Without friction, nails would pull out and most forms of animal locomotion would be impossible (**Fig. 5.16**).

5.16 There is friction between the feet of this caterpillar (the larval stage of a butterfly of the family Papilionidae) and the surfaces over which it walks. Without friction, the caterpillar could not move forward or climb over obstacles.

Kinetic and Static Friction

When you try to slide a heavy box of books across the floor, the box doesn't move at all unless you push with a certain minimum force. Once the box starts moving, you can usually keep it moving with less force than you needed to get it started. If you take some of the books out, you need less force to get it started or keep it moving. What can we say in general about this behavior?

First, when a body rests or slides on a surface, we can think of the surface as exerting a single contact force on the body, with force components perpendicular and parallel to the surface (**Fig. 5.17**). The perpendicular component vector is the normal force, denoted by $\vec{n}$. The component vector parallel to the surface (and perpendicular to $\vec{n}$) is the **friction force**, denoted by $\vec{f}$. If the surface is frictionless, then $\vec{f}$ is zero but there is still a normal force. (Frictionless surfaces are an unattainable idealization, like a massless rope. But we can approximate a surface as frictionless if the effects of friction are negligibly small.) The direction of the friction force is always such as to oppose relative motion of the two surfaces.

The kind of friction that acts when a body slides over a surface is called a **kinetic friction force** $\vec{f}_k$. The adjective "kinetic" and the subscript "k" remind us that the two surfaces are moving relative to each other. The *magnitude* of the kinetic friction force usually increases when the normal force increases. This is why it takes more force to slide a full box of books across the floor than an empty one. Automotive brakes use the same principle: The harder the brake pads are squeezed against the rotating brake discs, the greater the braking effect. In many cases the magnitude of the kinetic friction force f_k is found experimentally to be approximately *proportional* to the magnitude n of the normal force:

$$\underset{\substack{\text{Magnitude of kinetic} \\ \text{friction force}}}{} f_k = \mu_k n \underset{\substack{\text{Coefficient of kinetic friction} \\ \text{Magnitude of normal force}}}{} \qquad (5.3)$$

Here μ_k (pronounced "mu-sub-k") is a constant called the **coefficient of kinetic friction.** The more slippery the surface, the smaller this coefficient. Because it is a quotient of two force magnitudes, μ_k is a pure number without units.

CAUTION **Friction and normal forces are always perpendicular** Remember that Eq. (5.3) is *not* a vector equation because $\vec{f}_k$ and $\vec{n}$ are always perpendicular. Rather, it is a scalar relationship between the magnitudes of the two forces. ∎

Equation (5.3) is only an approximate representation of a complex phenomenon. On a microscopic level, friction and normal forces result from the intermolecular forces (electrical in nature) between two rough surfaces at points where they come into contact (**Fig. 5.18**). As a box slides over the floor, bonds between the two surfaces form and break, and the total number of such bonds varies. Hence the kinetic friction force is not perfectly constant. Smoothing the surfaces can actually increase friction, since more molecules can interact and bond; bringing two smooth surfaces of the same metal together can cause a "cold weld." Lubricating oils work because an oil film between two surfaces (such as the pistons and cylinder walls in a car engine) prevents them from coming into actual contact.

Table 5.1 lists some representative values of μ_k. Although these values are given with two significant figures, they are only approximate, since friction forces can also depend on the speed of the body relative to the surface. For now we'll ignore this effect and assume that μ_k and f_k are independent of speed, in order to concentrate on the simplest cases. Table 5.1 also lists coefficients of static friction; we'll define these shortly.

Friction forces may also act when there is *no* relative motion. If you try to slide a box across the floor, the box may not move at all because the floor exerts an equal and opposite friction force on the box. This is called a **static friction force** $\vec{f}_s$.

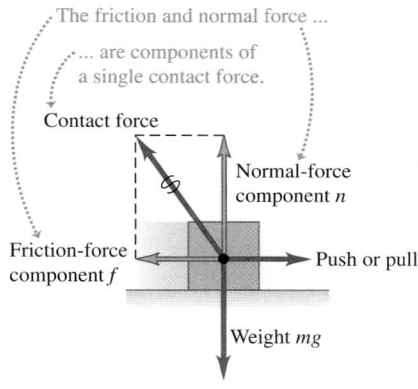

5.17 When a block is pushed or pulled over a surface, the surface exerts a contact force on it.

The friction and normal force ...

... are components of a single contact force.

Contact force

Normal-force component n

Friction-force component f

Push or pull

Weight mg

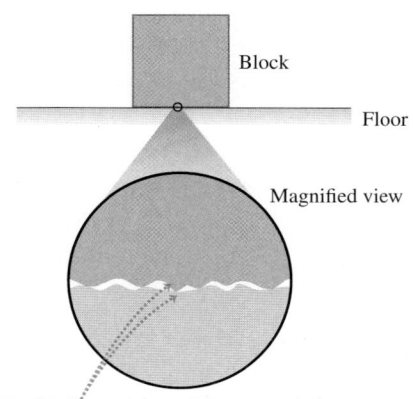

5.18 A microscopic view of the friction and normal forces.

Block

Floor

Magnified view

The friction and normal forces result from interactions between molecules in the block and in the floor where the two rough surfaces touch.

Approximate
TABLE 5.1 **Coefficients of Friction**

Materials	Coefficient of Static Friction, μ_s	Coefficient of Kinetic Friction, μ_k
Steel on steel	0.74	0.57
Aluminum on steel	0.61	0.47
Copper on steel	0.53	0.36
Brass on steel	0.51	0.44
Zinc on cast iron	0.85	0.21
Copper on cast iron	1.05	0.29
Glass on glass	0.94	0.40
Copper on glass	0.68	0.53
Teflon on Teflon	0.04	0.04
Teflon on steel	0.04	0.04
Rubber on concrete (dry)	1.0	0.8
Rubber on concrete (wet)	0.30	0.25

5.19 When there is no relative motion, the magnitude of the static friction force f_s is less than or equal to $\mu_s n$. When there is relative motion, the magnitude of the kinetic friction force f_k equals $\mu_k n$.

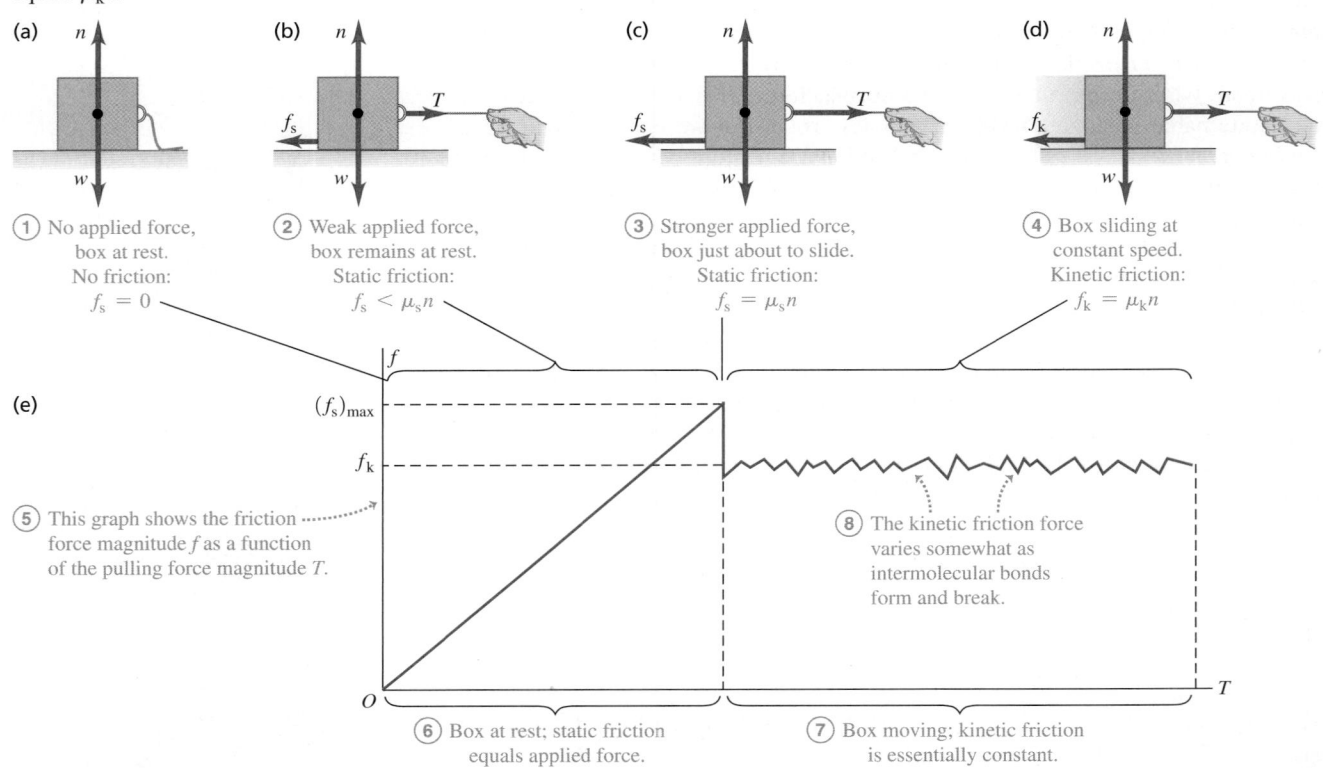

(a) n

(1) No applied force, box at rest. No friction:
$$f_s = 0$$

(b) n, T, f_s, w

(2) Weak applied force, box remains at rest. Static friction:
$$f_s < \mu_s n$$

(c) n, T, f_s, w

(3) Stronger applied force, box just about to slide. Static friction:
$$f_s = \mu_s n$$

(d) n, T, f_k, w

(4) Box sliding at constant speed. Kinetic friction:
$$f_k = \mu_k n$$

(e)

$(f_s)_{max}$

f_k

(5) This graph shows the friction force magnitude f as a function of the pulling force magnitude T.

(8) The kinetic friction force varies somewhat as intermolecular bonds form and break.

(6) Box at rest; static friction equals applied force.

(7) Box moving; kinetic friction is essentially constant.

In **Fig. 5.19a**, the box is at rest, in equilibrium, under the action of its weight $\vec{w}$ and the upward normal force $\vec{n}$. The normal force is equal in magnitude to the weight ($n = w$) and is exerted on the box by the floor. Now we tie a rope to the box (Fig. 5.19b) and gradually increase the tension T in the rope. At first the box remains at rest because the force of static friction f_s also increases and stays equal in magnitude to T.

At some point T becomes greater than the maximum static friction force f_s the surface can exert. Then the box "breaks loose" and starts to slide. Figure 5.19c shows the forces when T is at this critical value. For a given pair of surfaces the maximum value of f_s depends on the normal force. Experiment shows that in many cases this maximum value, called $(f_s)_{max}$, is approximately *proportional* to n; we call the proportionality factor μ_s the **coefficient of static friction.** Table 5.1 lists some representative values of μ_s. In a particular situation, the actual force of static friction can have any magnitude between zero (when there is no other force parallel to the surface) and a maximum value given by $\mu_s n$:

Magnitude of static friction force ....
Coefficient of static friction ....
$$f_s \leq (f_s)_{max} = \mu_s n \tag{5.4}$$
Maximum static friction force ....
.... Magnitude of normal force

Like Eq. (5.3), this is a relationship between magnitudes, *not* a vector relationship. The equality sign holds only when the applied force T has reached the critical value at which motion is about to start (Fig. 5.19c). When T is less than this value (Fig. 5.19b), the inequality sign holds. In that case we have to use the equilibrium conditions ($\Sigma \vec{F} = 0$) to find f_s. If there is no applied force ($T = 0$) as in Fig. 5.19a, then there is no static friction force either ($f_s = 0$).

As soon as the box starts to slide (Fig. 5.19d), the friction force usually *decreases* (Fig. 5.19e); it's easier to keep the box moving than to start it moving.

Hence the coefficient of kinetic friction is usually *less* than the coefficient of static friction for any given pair of surfaces, as Table 5.1 shows.

In some situations the surfaces will alternately stick (static friction) and slip (kinetic friction). This is what causes the horrible sound made by chalk held at the wrong angle on a blackboard and the shriek of tires sliding on as-phalt pavement. A more positive example is the motion of a violin bow against the string.

In the linear air tracks used in physics laboratories, gliders move with very little friction because they are supported on a layer of air. The friction force is velocity dependent, but at typical speeds the effective coefficient of friction is of the order of 0.001.

PhET: Forces in 1 Dimension
PhET: Friction
PhET: The Ramp

EXAMPLE 5.13 FRICTION IN HORIZONTAL MOTION

You want to move a 500-N crate across a level floor. To start the crate moving, you have to pull with a 230-N horizontal force. Once the crate starts to move, you can keep it moving at constant velocity with only 200 N. What are the coefficients of static and kinetic friction?

SOLUTION

IDENTIFY and SET UP: The crate is in equilibrium both when it is at rest and when it is moving at constant velocity, so we use Newton's first law, as expressed by Eqs. (5.1). We use Eqs. (5.3) and (5.4) to find the target variables μ_s and μ_k.

Figures 5.20a and 5.20b show our sketch and free-body diagram for the instant just before the crate starts to move, when the static friction force has its maximum possible value $(f_s)_{max} = \mu_s n$.

5.20 Our sketches for this problem.

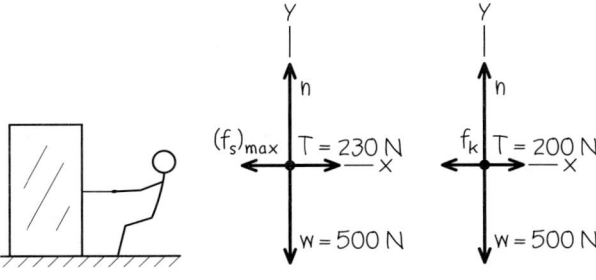

(a) Pulling a crate

(b) Free-body diagram for crate just before it starts to move

(c) Free-body diagram for crate moving at constant speed

Once the crate is moving, the friction force changes to its kinetic form (Fig. 5.20c). In both situations, four forces act on the crate: the downward weight (magnitude $w = 500$ N), the upward normal force (magnitude n) exerted by the floor, a tension force (magni-tude T) to the right exerted by the rope, and a friction force to the left exerted by the floor. Because the rope in Fig. 5.20a is in equi-librium, the tension is the same at both ends. Hence the tension force that the rope exerts on the crate has the same magnitude as the force you exert on the rope. Since it's easier to keep the crate moving than to start it moving, we expect that $\mu_k < \mu_s$.

EXECUTE: Just before the crate starts to move (Fig. 5.20b), we have from Eqs. (5.1)

$$\sum F_x = T + (-(f_s)_{max}) = 0 \quad so \quad (f_s)_{max} = T = 230 \text{ N}$$
$$\sum F_y = n + (-w) = 0 \quad\quad so \quad n = w = 500 \text{ N}$$

Now we solve Eq. (5.4), $(f_s)_{max} = \mu_s n$, for the value of μ_s:

$$\mu_s = \frac{(f_s)_{max}}{n} = \frac{230 \text{ N}}{500 \text{ N}} = 0.46$$

After the crate starts to move (Fig. 5.20c) we have

$$\sum F_x = T + (-f_k) = 0 \quad so \quad f_k = T = 200 \text{ N}$$
$$\sum F_y = n + (-w) = 0 \quad so \quad n = w = 500 \text{ N}$$

Using $f_k = \mu_k n$ from Eq. (5.3), we find

$$\mu_k = \frac{f_k}{n} = \frac{200 \text{ N}}{500 \text{ N}} = 0.40$$

EVALUATE: As expected, the coefficient of kinetic friction is less than the coefficient of static friction.

EXAMPLE 5.14 STATIC FRICTION CAN BE LESS THAN THE MAXIMUM

In Example 5.13, what is the friction force if the crate is at rest on the surface and a horizontal force of 50 N is applied to it?

SOLUTION

IDENTIFY and SET UP: The applied force is less than the maxi-mum force of static friction, $(f_s)_{max} = 230$ N. Hence the crate remains at rest and the net force acting on it is zero. The target variable is the magnitude f_s of the friction force. The free-body

diagram is the same as in Fig. 5.20b, but with $(f_s)_{max}$ replaced by f_s and $T = 230$ N replaced by $T = 50$ N.

EXECUTE: From the equilibrium conditions, Eqs. (5.1), we have

$$\sum F_x = T + (-f_s) = 0 \quad so \quad f_s = T = 50 \text{ N}$$

EVALUATE: The friction force can prevent motion for any horizon-tal applied force up to $(f_s)_{max} = \mu_s n = 230$ N. Below that value, f_s has the same magnitude as the applied force.

EXAMPLE 5.15 | MINIMIZING KINETIC FRICTION

In Example 5.13, suppose you move the crate by pulling upward on the rope at an angle of 30° above the horizontal. How hard must you pull to keep it moving with constant velocity? Assume that $\mu_k = 0.40$.

SOLUTION

IDENTIFY and SET UP: The crate is in equilibrium because its velocity is constant, so we again apply Newton's first law. Since the crate is in motion, the floor exerts a *kinetic* friction force. The target variable is the magnitude T of the tension force.

Figure 5.21 shows our sketch and free-body diagram. The kinetic friction force f_k is still equal to $\mu_k n$, but now the normal

5.21 Our sketches for this problem.

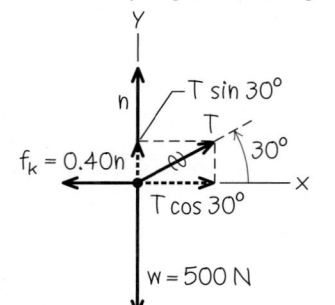

(a) Pulling a crate at an angle

(b) Free-body diagram for moving crate

force n is *not* equal in magnitude to the crate's weight. The force exerted by the rope has a vertical component that tends to lift the crate off the floor; this *reduces* n and so reduces f_k.

EXECUTE: From the equilibrium conditions and Eq. (5.3), $f_k = \mu_k n$, we have

$$\Sigma F_x = T\cos 30° + (-f_k) = 0 \quad \text{so} \quad T\cos 30° = \mu_k n$$

$$\Sigma F_y = T\sin 30° + n + (-w) = 0 \quad \text{so} \quad n = w - T\sin 30°$$

These are two equations for the two unknown quantities T and n. One way to find T is to substitute the expression for n in the second equation into the first equation and then solve the resulting equation for T:

$$T\cos 30° = \mu_k(w - T\sin 30°)$$

$$T = \frac{\mu_k w}{\cos 30° + \mu_k \sin 30°} = 188 \text{ N}$$

We can substitute this result into either of the original equations to obtain n. If we use the second equation, we get

$$n = w - T\sin 30° = (500 \text{ N}) - (188 \text{ N})\sin 30° = 406 \text{ N}$$

EVALUATE: As expected, the normal force is less than the 500-N weight of the box. It turns out that the tension required to keep the crate moving at constant speed is a little less than the 200-N force needed when you pulled horizontally in Example 5.13. Can you find an angle where the required pull is *minimum*?

EXAMPLE 5.16 | TOBOGGAN RIDE WITH FRICTION I

Let's go back to the toboggan we studied in Example 5.10. The wax has worn off, so there is now a nonzero coefficient of kinetic friction μ_k. The slope has just the right angle to make the toboggan slide with constant velocity. Find this angle in terms of w and μ_k.

SOLUTION

IDENTIFY and SET UP: Our target variable is the slope angle α. The toboggan is in equilibrium because its velocity is constant, so we use Newton's first law in the form of Eqs. (5.1).

Three forces act on the toboggan: its weight, the normal force, and the kinetic friction force. The motion is downhill, so the friction force (which opposes the motion) is directed uphill. **Figure 5.22** shows our sketch and free-body diagram (compare Fig. 5.12b in Example 5.10). From Eq. (5.3), the magnitude of the kinetic friction force is $f_k = \mu_k n$. We expect that the greater the value of μ_k, the steeper will be the required slope.

EXECUTE: The equilibrium conditions are

$$\Sigma F_x = w\sin\alpha + (-f_k) = w\sin\alpha - \mu_k n = 0$$

$$\Sigma F_y = n + (-w\cos\alpha) = 0$$

Rearranging these two equations, we get

$$\mu_k n = w\sin\alpha \quad \text{and} \quad n = w\cos\alpha$$

As in Example 5.10, the normal force is *not* equal to the weight. We eliminate n by dividing the first of these equations by the

5.22 Our sketches for this problem.

(a) The situation

(b) Free-body diagram for toboggan

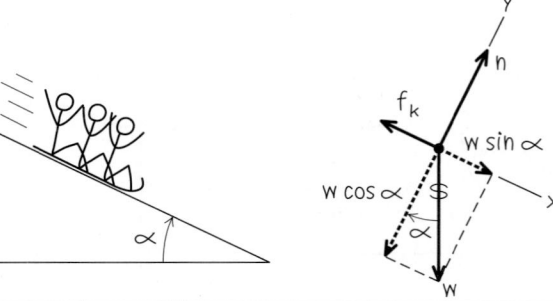

second, with the result

$$\mu_k = \frac{\sin\alpha}{\cos\alpha} = \tan\alpha \quad \text{so} \quad \alpha = \arctan\mu_k$$

EVALUATE: The weight w doesn't appear in this expression. *Any* toboggan, regardless of its weight, slides down an incline with constant speed if the coefficient of kinetic friction equals the tangent of the slope angle of the incline. The arctangent function increases as its argument increases, so it's indeed true that the slope angle α increases as μ_k increases.

EXAMPLE 5.17   TOBOGGAN RIDE WITH FRICTION II

The same toboggan with the same coefficient of friction as in Example 5.16 *accelerates* down a steeper hill. Derive an expression for the acceleration in terms of g, α, μ_k, and w.

SOLUTION

IDENTIFY and SET UP: The toboggan is accelerating, so we must use Newton's second law as given in Eqs. (5.2). Our target variable is the downhill acceleration.

Our sketch and free-body diagram (**Fig. 5.23**) are almost the same as for Example 5.16. The toboggan's y-component of acceleration a_y is still zero but the x-component a_x is not, so we've drawn $w \sin \alpha$, the downhill component of weight, as a longer vector than the (uphill) friction force.

EXECUTE: It's convenient to express the weight as $w = mg$. Then Newton's second law in component form says

$$\sum F_x = mg \sin \alpha + (-f_k) = ma_x$$
$$\sum F_y = n + (-mg \cos \alpha) = 0$$

5.23 Our sketches for this problem.

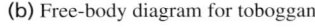

(a) The situation

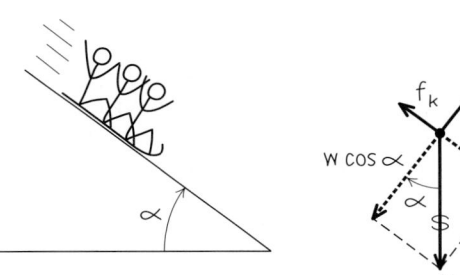

(b) Free-body diagram for toboggan

From the second equation and Eq. (5.3) we get an expression for f_k:

$$n = mg \cos \alpha$$
$$f_k = \mu_k n = \mu_k mg \cos \alpha$$

We substitute this into the x-component equation and solve for a_x:

$$mg \sin \alpha + (-\mu_k mg \cos \alpha) = ma_x$$
$$a_x = g(\sin \alpha - \mu_k \cos \alpha)$$

EVALUATE: As for the frictionless toboggan in Example 5.10, the acceleration doesn't depend on the mass m of the toboggan. That's because all of the forces that act on the toboggan (weight, normal force, and kinetic friction force) are proportional to m.

Let's check some special cases. If the hill is vertical ($\alpha = 90°$) so that $\sin \alpha = 1$ and $\cos \alpha = 0$, we have $a_x = g$ (the toboggan falls freely). For a certain value of α the acceleration is zero; this happens if

$$\sin \alpha = \mu_k \cos \alpha \qquad \text{and} \qquad \mu_k = \tan \alpha$$

This agrees with our result for the constant-velocity toboggan in Example 5.16. If the angle is even smaller, $\mu_k \cos \alpha$ is greater than $\sin \alpha$ and a_x is *negative;* if we give the toboggan an initial downhill push to start it moving, it will slow down and stop. Finally, if the hill is frictionless so that $\mu_k = 0$, we retrieve the result of Example 5.10: $a_x = g \sin \alpha$.

Notice that we started with a simple problem (Example 5.10) and extended it to more and more general situations. The general result we found in this example includes *all* the previous ones as special cases. Don't memorize this result, but do make sure you understand how we obtained it and what it means.

Suppose instead we give the toboggan an initial push *up* the hill. The direction of the kinetic friction force is now reversed, so the acceleration is different from the downhill value. It turns out that the expression for a_x is the same as for downhill motion except that the minus sign becomes plus. Can you show this?

Rolling Friction

It's a lot easier to move a loaded filing cabinet across a horizontal floor by using a cart with wheels than by sliding it. How much easier? We can define a **coefficient of rolling friction** μ_r, which is the horizontal force needed for constant speed on a flat surface divided by the upward normal force exerted by the surface. Transportation engineers call μ_r the *tractive resistance*. Typical values of μ_r are 0.002 to 0.003 for steel wheels on steel rails and 0.01 to 0.02 for rubber tires on concrete. These values show one reason railroad trains are generally much more fuel efficient than highway trucks.

Fluid Resistance and Terminal Speed

Sticking your hand out the window of a fast-moving car will convince you of the existence of **fluid resistance,** the force that a fluid (a gas or liquid) exerts on a body moving through it. The moving body exerts a force on the fluid to push it out of the way. By Newton's third law, the fluid pushes back on the body with an equal and opposite force.

The *direction* of the fluid resistance force acting on a body is always opposite the direction of the body's velocity relative to the fluid. The *magnitude* of the fluid resistance force usually increases with the speed of the body through the fluid.

5.24 Motion with fluid resistance.

(a) Metal ball falling through oil

(b) Free-body diagram for ball in oil

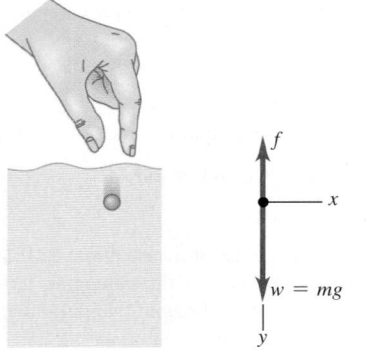

BIO Application Pollen and Fluid Resistance These spiky spheres are pollen grains from the ragweed flower (*Ambrosia artemisiifolia*) and a common cause of hay fever. Because of their small radius (about 10 μm = 0.01 mm), when they are released into the air the fluid resistance force on them is proportional to their speed. The terminal speed given by Eq. (5.8) is only about 1 cm/s. Hence even a moderate wind can keep pollen grains aloft and carry them substantial distances from their source.

This is very different from the kinetic friction force between two surfaces in contact, which we can usually regard as independent of speed. For small objects moving at very low speeds, the magnitude f of the fluid resistance force is approximately proportional to the body's speed v:

$$f = kv \qquad \text{(fluid resistance at low speed)} \qquad (5.5)$$

where k is a proportionality constant that depends on the shape and size of the body and the properties of the fluid. Equation (5.5) is appropriate for dust particles falling in air or a ball bearing falling in oil. For larger objects moving through air at the speed of a tossed tennis ball or faster, the resisting force is approximately proportional to v^2 rather than to v. It is then called **air drag** or simply *drag*. Airplanes, falling raindrops, and bicyclists all experience air drag. In this case we replace Eq. (5.5) by

$$f = Dv^2 \qquad \text{(fluid resistance at high speed)} \qquad (5.6)$$

Because of the v^2 dependence, air drag increases rapidly with increasing speed. The air drag on a typical car is negligible at low speeds but comparable to or greater than rolling resistance at highway speeds. The value of D depends on the shape and size of the body and on the density of the air. You should verify that the units of the constant k in Eq. (5.5) are N $\cdot$ s/m or kg/s, and that the units of the constant D in Eq. (5.6) are N $\cdot$ s^2/m^2 or kg/m.

Because of the effects of fluid resistance, an object falling in a fluid does *not* have a constant acceleration. To describe its motion, we can't use the constant-acceleration relationships from Chapter 2; instead, we have to start over with Newton's second law. As an example, suppose you drop a metal ball at the surface of a bucket of oil and let it fall to the bottom (**Fig. 5.24a**). The fluid resistance force in this situation is given by Eq. (5.5). What are the acceleration, velocity, and position of the metal ball as functions of time?

Figure 5.24b shows the free-body diagram. We take the positive y-direction to be downward and neglect any force associated with buoyancy in the oil. Since the ball is moving downward, its speed v is equal to its y-velocity v_y and the fluid resistance force is in the $-y$-direction. There are no x-components, so Newton's second law gives

$$\sum F_y = mg + (-kv_y) = ma_y \qquad (5.7)$$

When the ball first starts to move, $v_y = 0$, the resisting force is zero and the initial acceleration is $a_y = g$. As the speed increases, the resisting force also increases, until finally it is equal in magnitude to the weight. At this time $mg - kv_y = 0$, the acceleration is zero, and there is no further increase in speed. The final speed v_t, called the **terminal speed,** is given by $mg - kv_t = 0$, or

$$v_t = \frac{mg}{k} \qquad \text{(terminal speed, fluid resistance } f = kv) \qquad (5.8)$$

Figure 5.25 shows how the acceleration, velocity, and position vary with time. As time goes by, the acceleration approaches zero and the velocity approaches v_t

5.25 Graphs of the motion of a body falling without fluid resistance and with fluid resistance proportional to the speed.

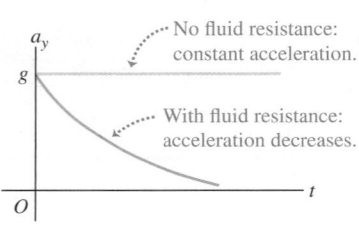

Acceleration versus time

No fluid resistance: constant acceleration.

With fluid resistance: acceleration decreases.

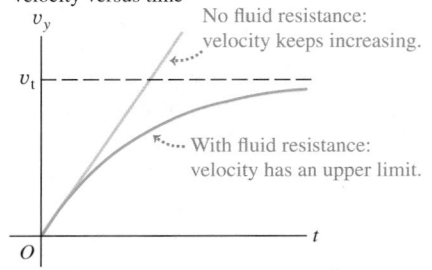

Velocity versus time

No fluid resistance: velocity keeps increasing.

With fluid resistance: velocity has an upper limit.

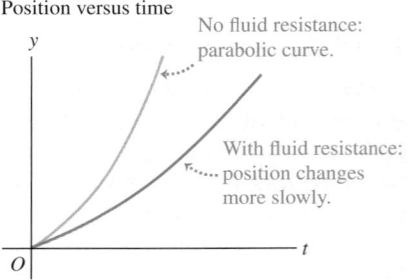

Position versus time

No fluid resistance: parabolic curve.

With fluid resistance: position changes more slowly.

(remember that we chose the positive y-direction to be down). The slope of the graph of y versus t becomes constant as the velocity becomes constant.

To see how the graphs in Fig. 5.25 are derived, we must find the relationship between velocity and time during the interval before the terminal speed is reached. We go back to Newton's second law for the falling ball, Eq. (5.7), which we rewrite with $a_y = dv_y/dt$:

$$m\frac{dv_y}{dt} = mg - kv_y$$

After rearranging terms and replacing mg/k by v_t, we integrate both sides, noting that $v_y = 0$ when $t = 0$:

$$\int_0^v \frac{dv_y}{v_y - v_t} = -\frac{k}{m}\int_0^t dt$$

which integrates to

$$\ln\frac{v_t - v_y}{v_t} = -\frac{k}{m}t \quad \text{or} \quad 1 - \frac{v_y}{v_t} = e^{-(k/m)t}$$

and finally

$$v_y = v_t\big[1 - e^{-(k/m)t}\big] \tag{5.9}$$

Note that v_y becomes equal to the terminal speed v_t only in the limit that $t \rightarrow \infty$; the ball cannot attain terminal speed in any finite length of time.

The derivative of v_y in Eq. (5.9) gives a_y as a function of time, and the integral of v_y gives y as a function of time. We leave the derivations for you to complete; the results are

$$a_y = ge^{-(k/m)t} \tag{5.10}$$

$$y = v_t\bigg[t - \frac{m}{k}(1 - e^{-(k/m)t})\bigg] \tag{5.11}$$

Now look again at Fig. 5.25, which shows graphs of these three relationships.

In deriving the terminal speed in Eq. (5.8), we assumed that the fluid resistance force is proportional to the speed. For an object falling through the air at high speeds, so that the fluid resistance is equal to Dv^2 as in Eq. (5.6), the terminal speed is reached when Dv^2 equals the weight mg (**Fig. 5.26a**). You can show that the terminal speed v_t is given by

$$v_t = \sqrt{\frac{mg}{D}} \quad \text{(terminal speed, fluid resistance } f = Dv^2\text{)} \tag{5.12}$$

This expression for terminal speed explains why heavy objects in air tend to fall faster than light objects. Two objects that have the same physical size but different mass (say, a table-tennis ball and a lead ball with the same radius) have the same value of D but different values of m. The more massive object has a higher terminal speed and falls faster. The same idea explains why a sheet of paper falls faster if you first crumple it into a ball; the mass m is the same, but the smaller size makes D smaller (less air drag for a given speed) and v_t larger. Skydivers use the same principle to control their descent (Fig. 5.26b).

Figure 5.27 shows the trajectories of a baseball with and without air drag, assuming a coefficient $D = 1.3 \times 10^{-3}$ kg/m (appropriate for a batted ball at sea level). Both the range of the baseball and the maximum height reached are substantially smaller than the zero-drag calculation would lead you to believe. Hence the baseball trajectory we calculated in Example 3.7 (Section 3.3) by ignoring air drag is unrealistic. Air drag is an important part of the game of baseball!

5.26 (a) Air drag and terminal speed. (b) By changing the positions of their arms and legs while falling, skydivers can change the value of the constant D in Eq. (5.6) and hence adjust the terminal speed of their fall [Eq. (5.12)].

(a) Free-body diagrams for falling with air drag

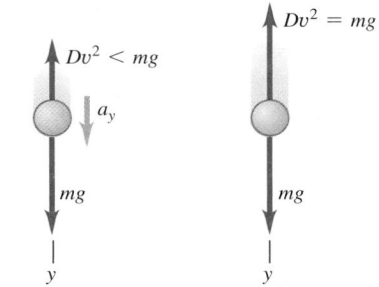

Before terminal speed: Object accelerating, drag force less than weight.

At terminal speed v_t: Object in equilibrium, drag force equals weight.

(b) A skydiver falling at terminal speed

5.27 Computer-generated trajectories of a baseball launched at 50 m/s at 35° above the horizontal. Note that the scales are different on the horizontal and vertical axes.

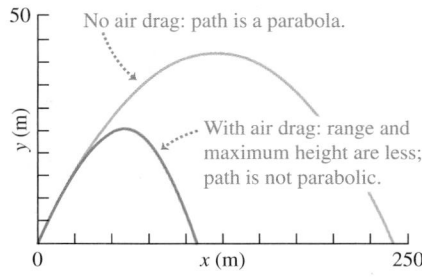

No air drag: path is a parabola.

With air drag: range and maximum height are less; path is not parabolic.

EXAMPLE 5.18   **TERMINAL SPEED OF A SKYDIVER**

For a human body falling through air in a spread-eagle position (Fig. 5.26b), the numerical value of the constant D in Eq. (5.6) is about 0.25 kg/m. Find the terminal speed for a 50-kg skydiver.

SOLUTION

IDENTIFY and SET UP: This example uses the relationship among terminal speed, mass, and drag coefficient. We use Eq. (5.12) to find the target variable v_t.

EXECUTE: We find for $m = 50$ kg:

$$v_t = \sqrt{\frac{mg}{D}} = \sqrt{\frac{(50 \text{ kg})(9.8 \text{ m/s}^2)}{0.25 \text{ kg/m}}}$$

$$= 44 \text{ m/s (about 160 km/h, or 99 mi/h)}$$

EVALUATE: The terminal speed is proportional to the square root of the skydiver's mass. A skydiver with the same drag coefficient D but twice the mass would have a terminal speed $\sqrt{2} = 1.41$ times greater, or 63 m/s. (A more massive skydiver would also have more frontal area and hence a larger drag coefficient, so his terminal speed would be a bit less than 63 m/s.) Even the 50-kg skydiver's terminal speed is quite high, so skydives don't last very long. A drop from 2800 m (9200 ft) to the surface at the terminal speed takes only $(2800 \text{ m})/(44 \text{ m/s}) = 64$ s.

When the skydiver deploys the parachute, the value of D increases greatly. Hence the terminal speed of the skydiver and parachute decreases dramatically to a much lower value.

5.28 Net force, acceleration, and velocity in uniform circular motion.

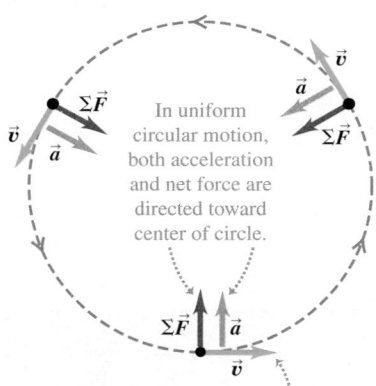

In uniform circular motion, both acceleration and net force are directed toward center of circle.

Velocity is tangent to circle.

5.29 What happens if the inward radial force suddenly ceases to act on a body in circular motion?

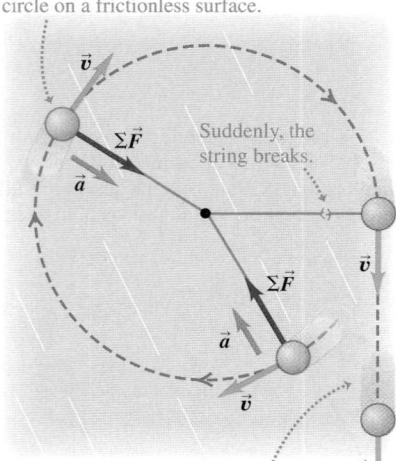

A ball attached to a string whirls in a circle on a frictionless surface.

Suddenly, the string breaks.

No net force now acts on the ball, so it obeys Newton's first law—it moves in a straight line at constant velocity.

TEST YOUR UNDERSTANDING OF SECTION 5.3   Consider a box that is placed on different surfaces. (a) In which situation(s) is *no* friction force acting on the box? (b) In which situation(s) is a *static* friction force acting on the box? (c) In which situation(s) is a *kinetic* friction force acting on the box? (i) The box is at rest on a rough horizontal surface. (ii) The box is at rest on a rough tilted surface. (iii) The box is on the rough-surfaced flat bed of a truck that is moving at a constant velocity on a straight, level road, and the box remains in place in the middle of the truck bed. (iv) The box is on the rough-surfaced flat bed of a truck that is speeding up on a straight, level road, and the box remains in place in the middle of the truck bed. (v) The box is on the rough-surfaced flat bed of a truck that is climbing a hill, and the box is sliding toward the back of the truck. ❙

5.4 DYNAMICS OF CIRCULAR MOTION

We talked about uniform circular motion in Section 3.4. We showed that when a particle moves in a circular path with constant speed, the particle's acceleration has a constant magnitude a_{rad} given by

Magnitude of acceleration of an object in **uniform circular motion** $a_{rad} = \dfrac{v^2}{R}$ Speed of object / Radius of object's circular path   (5.13)

The subscript "rad" is a reminder that at each point the acceleration points radially inward toward the center of the circle, perpendicular to the instantaneous velocity. We explained in Section 3.4 why this acceleration is often called *centripetal acceleration* or *radial acceleration*.

We can also express the centripetal acceleration a_{rad} in terms of the *period T*, the time for one revolution:

$$T = \frac{2\pi R}{v}$$ (5.14)

In terms of the period, a_{rad} is

Magnitude of acceleration of an object in **uniform circular motion** $a_{rad} = \dfrac{4\pi^2 R}{T^2}$ Radius of object's circular path / Period of motion   (5.15)

Uniform circular motion, like all other motion of a particle, is governed by Newton's second law. To make the particle accelerate toward the center of the circle, the net force $\sum \vec{F}$ on the particle must always be directed toward the center (**Fig. 5.28**). The magnitude of the acceleration is constant, so the magnitude F_{net} of the net force must also be constant. If the inward net force stops acting, the particle flies off in a straight line tangent to the circle (**Fig. 5.29**).

The magnitude of the radial acceleration is given by $a_{rad} = v^2/R$, so the magnitude F_{net} of the net force on a particle with mass m in uniform circular motion must be

$$F_{net} = ma_{rad} = m\frac{v^2}{R} \quad \text{(uniform circular motion)} \quad (5.16)$$

Uniform circular motion can result from *any* combination of forces, just so the net force $\sum \vec{F}$ is always directed toward the center of the circle and has a constant magnitude. Note that the body need not move around a complete circle: Equation (5.16) is valid for *any* path that can be regarded as part of a circular arc.

CAUTION **Avoid using "centrifugal force"** **Figure 5.30** shows a correct free-body diagram for uniform circular motion (Fig. 5.30a) and an *incorrect* diagram (Fig. 5.30b). Figure 5.30b is incorrect because it includes an extra outward force of magnitude $m(v^2/R)$ to "keep the body out there" or to "keep it in equilibrium." There are three reasons not to include such an outward force, called *centrifugal force* ("centrifugal" means "fleeing from the center"). First, the body does *not* "stay out there": It is in constant motion around its circular path. Because its velocity is constantly changing in direction, the body accelerates and is *not* in equilibrium. Second, if there *were* an outward force that balanced the inward force, the net force would be zero and the body would move in a straight line, not a circle (Fig. 5.29). Third, the quantity $m(v^2/R)$ is *not* a force; it corresponds to the $m\vec{a}$ side of $\sum \vec{F} = m\vec{a}$ and does not appear in $\sum \vec{F}$ (Fig. 5.30a). It's true that when you ride in a car that goes around a circular path, you tend to slide to the outside of the turn as though there was a "centrifugal force." But we saw in Section 4.2 that what happens is that you tend to keep moving in a straight line, and the outer side of the car "runs into" you as the car turns (Fig. 4.10c). *In an inertial frame of reference there is no such thing as "centrifugal force."* We won't mention this term again, and we strongly advise you to avoid it.

5.30 Right and wrong ways to depict uniform circular motion.

(a) Correct free-body diagram

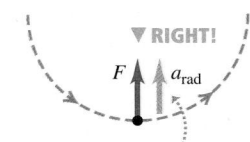

▼ RIGHT!

$F \uparrow \quad a_{rad}$

If you include the acceleration, draw it to one side of the body to show that it's not a force.

(b) Incorrect free-body diagram

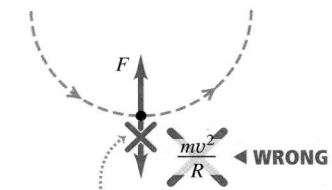

F

$\frac{mv^2}{R}$ ◄ WRONG

The quantity mv^2/R is *not* a force—it doesn't belong in a free-body diagram.

DEMO

EXAMPLE 5.19 FORCE IN UNIFORM CIRCULAR MOTION

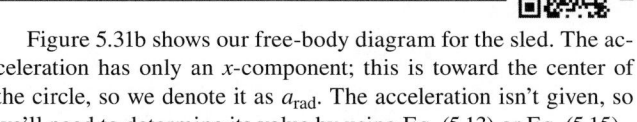

A sled with a mass of 25.0 kg rests on a horizontal sheet of essentially frictionless ice. It is attached by a 5.00-m rope to a post set in the ice. Once given a push, the sled revolves uniformly in a circle around the post (**Fig. 5.31a**). If the sled makes five complete revolutions every minute, find the force F exerted on it by the rope.

SOLUTION

IDENTIFY and SET UP: The sled is in uniform circular motion, so it has a constant radial acceleration. We'll apply Newton's second law to the sled to find the magnitude F of the force exerted by the rope (our target variable).

5.31 (a) The situation. (b) Our free-body diagram.

(a) A sled in uniform circular motion

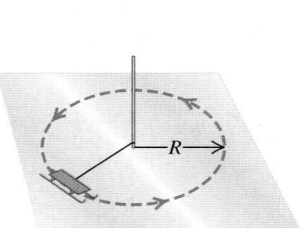

(b) Free-body diagram for sled

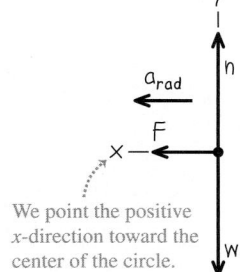

We point the positive x-direction toward the center of the circle.

Figure 5.31b shows our free-body diagram for the sled. The acceleration has only an x-component; this is toward the center of the circle, so we denote it as a_{rad}. The acceleration isn't given, so we'll need to determine its value by using Eq. (5.13) or Eq. (5.15).

EXECUTE: The force F appears in Newton's second law for the x-direction:

$$\sum F_x = F = ma_{rad}$$

We can find the centripetal acceleration a_{rad} by using Eq. (5.15). The sled moves in a circle of radius $R = 5.00$ m with a period $T = (60.0 \text{ s})/(5 \text{ rev}) = 12.0$ s, so

$$a_{rad} = \frac{4\pi^2 R}{T^2} = \frac{4\pi^2(5.00 \text{ m})}{(12.0 \text{ s})^2} = 1.37 \text{ m/s}^2$$

The magnitude F of the force exerted by the rope is then

$$F = ma_{rad} = (25.0 \text{ kg})(1.37 \text{ m/s}^2)$$
$$= 34.3 \text{ kg} \cdot \text{m/s}^2 = 34.3 \text{ N}$$

EVALUATE: You can check our value for a_{rad} by first using Eq. (5.14), $v = 2\pi R/T$, to find the speed and then using $a_{rad} = v^2/R$ from Eq. (5.13). Do you get the same result?

A greater force would be needed if the sled moved around the circle at a higher speed v. In fact, if v were doubled while R remained the same, F would be four times greater. Can you show this? How would F change if v remained the same but the radius R were doubled?

EXAMPLE 5.20 A CONICAL PENDULUM

An inventor designs a pendulum clock using a bob with mass m at the end of a thin wire of length L. Instead of swinging back and forth, the bob is to move in a horizontal circle at constant speed v, with the wire making a fixed angle β with the vertical direction (**Fig. 5.32a**). This is called a *conical pendulum* because the suspending wire traces out a cone. Find the tension F in the wire and the period T (the time for one revolution of the bob).

SOLUTION

IDENTIFY and SET UP: To find our target variables, the tension F and period T, we need two equations. These will be the horizontal and vertical components of Newton's second law applied to the bob. We'll find the radial acceleration of the bob from one of the circular motion equations.

Figure 5.32b shows our free-body diagram and coordinate system for the bob at a particular instant. There are just two forces on the bob: the weight mg and the tension F in the wire. Note that the

center of the circular path is in the same horizontal plane as the bob, *not* at the top end of the wire. The horizontal component of tension is the force that produces the radial acceleration a_{rad}.

EXECUTE: The bob has zero vertical acceleration; the horizontal acceleration is toward the center of the circle, which is why we use the symbol a_{rad}. Newton's second law, Eqs. (5.2), says

$$\sum F_x = F \sin\beta = ma_{rad}$$
$$\sum F_y = F \cos\beta + (-mg) = 0$$

These are two equations for the two unknowns F and β. The equation for $\sum F_y$ gives $F = mg/\cos\beta$; that's our target expression for F in terms of β. Substituting this result into the equation for $\sum F_x$ and using $\sin\beta/\cos\beta = \tan\beta$, we find

$$a_{rad} = g \tan\beta$$

To relate β to the period T, we use Eq. (5.15) for a_{rad}, solve for T, and insert $a_{rad} = g \tan\beta$:

$$a_{rad} = \frac{4\pi^2 R}{T^2} \quad \text{so} \quad T^2 = \frac{4\pi^2 R}{a_{rad}}$$

$$T = 2\pi\sqrt{\frac{R}{g\tan\beta}}$$

Figure 5.32a shows that $R = L\sin\beta$. We substitute this and use $\sin\beta/\tan\beta = \cos\beta$:

$$T = 2\pi\sqrt{\frac{L\cos\beta}{g}}$$

EVALUATE: For a given length L, as the angle β increases, $\cos\beta$ decreases, the period T becomes smaller, and the tension $F = mg/\cos\beta$ increases. The angle can never be 90°, however; this would require that $T = 0$, $F = \infty$, and $v = \infty$. A conical pendulum would not make a very good clock because the period depends on the angle β in such a direct way.

5.32 (a) The situation. (b) Our free-body diagram.

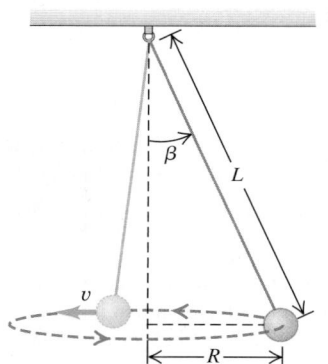
(a) The situation

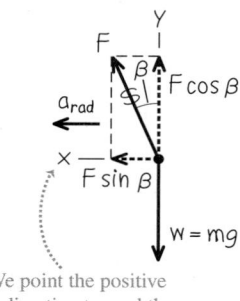
(b) Free-body diagram for pendulum bob

We point the positive *x*-direction toward the center of the circle.

EXAMPLE 5.21 ROUNDING A FLAT CURVE

The sports car in Example 3.11 (Section 3.4) is rounding a flat, unbanked curve with radius R (**Fig. 5.33a**). If the coefficient of static friction between tires and road is μ_s, what is the maximum speed v_{max} at which the driver can take the curve without sliding?

SOLUTION

IDENTIFY and SET UP: The car's acceleration as it rounds the curve has magnitude $a_{rad} = v^2/R$. Hence the maximum speed v_{max} (our target variable) corresponds to the maximum acceleration a_{rad} and to the maximum horizontal force on the car toward the center of its circular path. The only horizontal force acting on the car is the friction force exerted by the road. So to solve this problem we'll need Newton's second law, the equations of uniform circular motion, and our knowledge of the friction force from Section 5.3.

The free-body diagram in Fig. 5.33b includes the car's weight $w = mg$ and the two forces exerted by the road: the normal force n and the horizontal friction force f. The friction force must point toward the center of the circular path in order to cause the radial

5.33 (a) The situation. (b) Our free-body diagram.

(a) Car rounding flat curve

(b) Free-body diagram for car

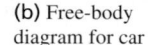

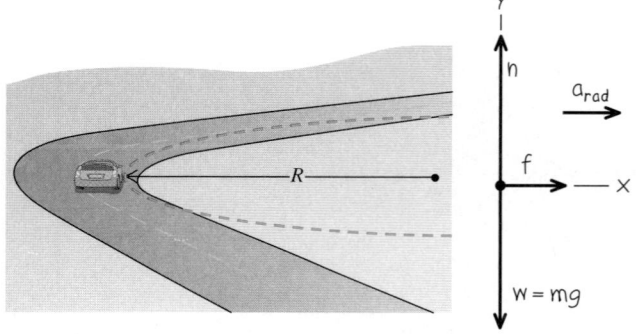

acceleration. The car doesn't slide toward or away from the center of the circle, so the friction force is *static* friction, with a maximum magnitude $f_{max} = \mu_s n$ [see Eq. (5.4)].

EXECUTE: The acceleration toward the center of the circular path is $a_{rad} = v^2/R$. There is no vertical acceleration. Thus

$$\Sigma F_x = f = ma_{rad} = m\frac{v^2}{R}$$

$$\Sigma F_y = n + (-mg) = 0$$

The second equation shows that $n = mg$. The first equation shows that the friction force *needed* to keep the car moving in its circular path increases with the car's speed. But the maximum friction force *available* is $f_{max} = \mu_s n = \mu_s mg$, and this determines the car's maximum speed. Substituting $\mu_s mg$ for f and v_{max} for v in the first equation, we find

$$\mu_s mg = m\frac{v_{max}^2}{R} \quad \text{so} \quad v_{max} = \sqrt{\mu_s gR}$$

As an example, if $\mu_s = 0.96$ and $R = 230$ m, we have

$$v_{max} = \sqrt{(0.96)(9.8 \text{ m/s}^2)(230 \text{ m})} = 47 \text{ m/s}$$

or about 170 km/h (100 mi/h). This is the maximum speed for this radius.

EVALUATE: If the car's speed is slower than $v_{max} = \sqrt{\mu_s gR}$, the required friction force is less than the maximum value $f_{max} = \mu_s mg$, and the car can easily make the curve. If we try to take the curve going *faster* than v_{max}, we will skid. We could still go in a circle without skidding at this higher speed, but the radius would have to be larger.

The maximum centripetal acceleration (called the "lateral acceleration" in Example 3.11) is equal to $\mu_s g$. That's why it's best to take curves at less than the posted speed limit if the road is wet or icy, either of which can reduce the value of μ_s and hence $\mu_s g$.

SOLUTION

EXAMPLE 5.22 ROUNDING A BANKED CURVE

For a car traveling at a certain speed, it is possible to bank a curve at just the right angle so that no friction is needed to maintain the car's turning radius. Then a car can safely round the curve even on wet ice. (Bobsled racing depends on this idea.) Your engineering firm plans to rebuild the curve in Example 5.21 so that a car moving at a chosen speed v can safely make the turn even with no friction (**Fig. 5.34a**). At what angle β should the curve be banked?

SOLUTION

IDENTIFY and SET UP: With no friction, the only forces acting on the car are its weight and the normal force. Because the road is banked, the normal force (which acts perpendicular to the road surface) has a horizontal component. This component causes the car's horizontal acceleration toward the center of the car's circular path. We'll use Newton's second law to find the target variable β.

Our free-body diagram (Fig. 5.34b) is very similar to the diagram for the conical pendulum in Example 5.20 (Fig. 5.32b). The normal force acting on the car plays the role of the tension force exerted by the wire on the pendulum bob.

EXECUTE: The normal force $\vec{n}$ is perpendicular to the roadway and is at an angle β with the vertical (Fig. 5.34b). Thus it has a

vertical component $n\cos\beta$ and a horizontal component $n\sin\beta$. The acceleration in the x-direction is the centripetal acceleration $a_{rad} = v^2/R$; there is no acceleration in the y-direction. Thus the equations of Newton's second law are

$$\Sigma F_x = n\sin\beta = ma_{rad}$$
$$\Sigma F_y = n\cos\beta + (-mg) = 0$$

From the ΣF_y equation, $n = mg/\cos\beta$. Substituting this into the ΣF_x equation and using $a_{rad} = v^2/R$, we get an expression for the bank angle:

$$\tan\beta = \frac{a_{rad}}{g} = \frac{v^2}{gR} \quad \text{so} \quad \beta = \arctan\frac{v^2}{gR}$$

EVALUATE: The bank angle depends on both the speed and the radius. For a given radius, no one angle is correct for all speeds. In the design of highways and railroads, curves are often banked for the average speed of the traffic over them. If $R = 230$ m and $v = 25$ m/s (equal to a highway speed of 88 km/h, or 55 mi/h), then

$$\beta = \arctan\frac{(25 \text{ m/s})^2}{(9.8 \text{ m/s}^2)(230 \text{ m})} = 15°$$

This is within the range of bank angles actually used in highways.

5.34 (a) The situation. (b) Our free-body diagram.

(a) Car rounding banked curve

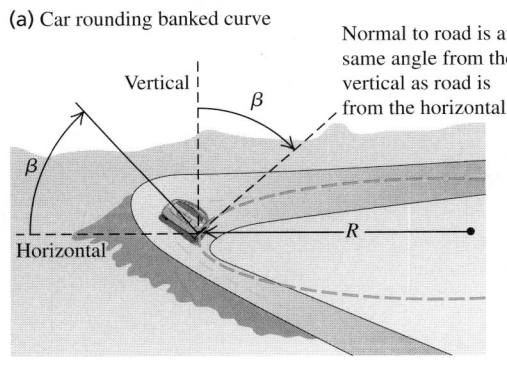

Normal to road is at same angle from the vertical as road is from the horizontal.

(b) Free-body diagram for car

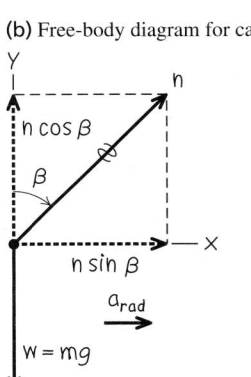

5.35 An airplane banks to one side in order to turn in that direction. The vertical component of the lift force $\vec{L}$ balances the force of gravity; the horizontal component of $\vec{L}$ causes the acceleration v^2/R.

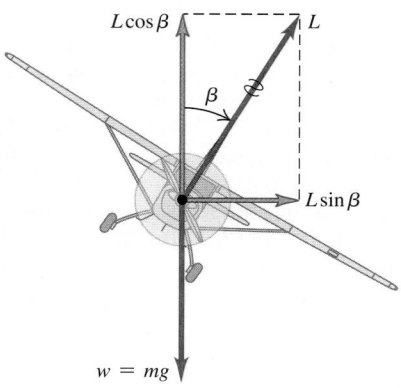

Banked Curves and the Flight of Airplanes

The results of Example 5.22 also apply to an airplane when it makes a turn in level flight (**Fig. 5.35**). When an airplane is flying in a straight line at a constant speed and at a steady altitude, the airplane's weight is exactly balanced by the lift force $\vec{L}$ exerted by the air. (The upward lift force that the air exerts on the wings is a reaction to the downward push the wings exert on the air as they move through it.) To make the airplane turn, the pilot banks the airplane to one side so that the lift force has a horizontal component, as Fig. 5.35 shows. (The pilot also changes the angle at which the wings "bite" into the air so that the vertical component of lift continues to balance the weight.) The bank angle is related to the airplane's speed v and the radius R of the turn by the same expression as in Example 5.22: $\tan\beta = v^2/gR$. For an airplane to make a tight turn (small R) at high speed (large v), $\tan\beta$ must be large and the required bank angle β must approach 90°.

We can also apply the results of Example 5.22 to the *pilot* of an airplane. The free-body diagram for the pilot of the airplane is exactly as shown in Fig. 5.34b; the normal force $n = mg/\cos\beta$ is exerted on the pilot by the seat. As in Example 5.9, n is equal to the apparent weight of the pilot, which is greater than the pilot's true weight mg. In a tight turn with a large bank angle β, the pilot's apparent weight can be tremendous: $n = 5.8mg$ at $\beta = 80°$ and $n = 9.6mg$ at $\beta = 84°$. Pilots black out in such tight turns because the apparent weight of their blood increases by the same factor, and the human heart isn't strong enough to pump such apparently "heavy" blood to the brain.

Motion in a Vertical Circle

In Examples 5.19, 5.20, 5.21, and 5.22 the body moved in a horizontal circle. Motion in a *vertical* circle is no different in principle, but the weight of the body has to be treated carefully. The following example shows what we mean.

SOLUTION

EXAMPLE 5.23 **UNIFORM CIRCULAR MOTION IN A VERTICAL CIRCLE**

A passenger on a carnival Ferris wheel moves in a vertical circle of radius R with constant speed v. The seat remains upright during the motion. Find expressions for the force the seat exerts on the passenger when at the top of the circle and when at the bottom.

SOLUTION

IDENTIFY and SET UP: The target variables are n_T, the upward normal force the seat applies to the passenger at the top of the circle, and n_B, the normal force at the bottom. We'll find these by using Newton's second law and the uniform circular motion equations.

Figure 5.36a shows the passenger's velocity and acceleration at the two positions. The acceleration always points toward the center of the circle—downward at the top of the circle and upward at the bottom of the circle. At each position the only forces acting are vertical: the upward normal force and the downward force of gravity. Hence we need only the vertical component of Newton's second law. Figures 5.36b and 5.36c show free-body diagrams for the two positions. We take the positive y-direction as upward in both cases (that is, *opposite* the direction of the acceleration at the top of the circle).

EXECUTE: At the top the acceleration has magnitude v^2/R, but its vertical component is negative because its direction is downward.

5.36 Our sketches for this problem.

(a) Sketch of two positions

(b) Free-body diagram for passenger at top

(c) Free-body diagram for passenger at bottom

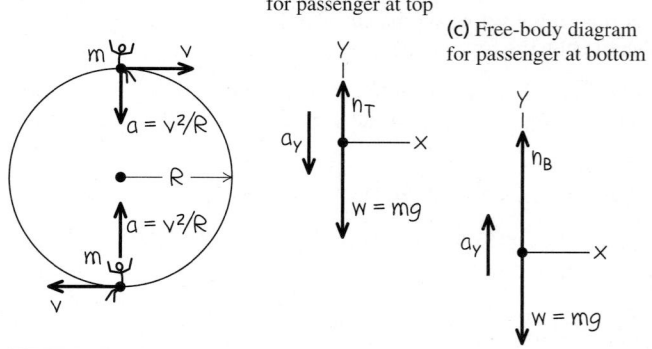

Hence $a_y = -v^2/R$ and Newton's second law tells us that

$$\text{Top:} \quad \sum F_y = n_T + (-mg) = -m\frac{v^2}{R} \quad \text{or}$$

$$n_T = mg\left(1 - \frac{v^2}{gR}\right)$$

At the bottom the acceleration is upward, so $a_y = +v^2/R$ and Newton's second law says

Bottom: $\quad \Sigma F_y = n_B + (-mg) = +m\dfrac{v^2}{R} \quad$ or

$$n_B = mg\left(1 + \frac{v^2}{gR}\right)$$

EVALUATE: Our result for n_T tells us that at the top of the Ferris wheel, the upward force the seat applies to the passenger is *smaller*

in magnitude than the passenger's weight $w = mg$. If the ride goes fast enough that $g - v^2/R$ becomes zero, the seat applies *no* force, and the passenger is about to become airborne. If v becomes still larger, n_T becomes negative; this means that a *downward* force (such as from a seat belt) is needed to keep the passenger in the seat. By contrast, the normal force n_B at the bottom is always *greater* than the passenger's weight. You feel the seat pushing up on you more firmly than when you are at rest. You can see that n_T and n_B are the values of the passenger's *apparent weight* at the top and bottom of the circle (see Section 5.2).

When we tie a string to an object and whirl it in a vertical circle, the analysis in Example 5.23 isn't directly applicable. The reason is that v is *not* constant in this case; except at the top and bottom of the circle, the net force (and hence the acceleration) does *not* point toward the center of the circle (**Fig. 5.37**). So both $\Sigma \vec{F}$ and $\vec{a}$ have a component tangent to the circle, which means that the speed changes. Hence this is a case of *nonuniform* circular motion (see Section 3.4). Even worse, we can't use the constant-acceleration formulas to relate the speeds at various points because *neither* the magnitude nor the direction of the acceleration is constant. The speed relationships we need are best obtained by using the concept of energy. We'll consider such problems in Chapter 7.

5.37 A ball moving in a vertical circle.

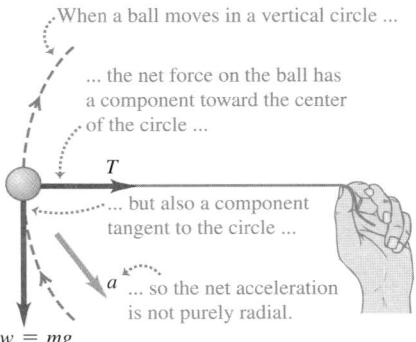

When a ball moves in a vertical circle ...

... the net force on the ball has a component toward the center of the circle ...

T

... but also a component tangent to the circle ...

a ... so the net acceleration is not purely radial.

$w = mg$

TEST YOUR UNDERSTANDING OF SECTION 5.4   Satellites are held in orbit by the force of our planet's gravitational attraction. A satellite in a small-radius orbit moves at a higher speed than a satellite in an orbit of large radius. Based on this information, what can you conclude about the earth's gravitational attraction for the satellite? (i) It increases with increasing distance from the earth. (ii) It is the same at all distances from the earth. (iii) It decreases with increasing distance from the earth. (iv) This information by itself isn't enough to answer the question. ❙

5.5 THE FUNDAMENTAL FORCES OF NATURE

We have discussed several kinds of forces—including weight, tension, friction, fluid resistance, and the normal force—and we will encounter others as we continue our study of physics. How many kinds of forces are there? Our best understanding is that all forces are expressions of just four distinct classes of *fundamental* forces, or interactions between particles (**Fig. 5.38**, next page). Two are familiar in everyday experience. The other two involve interactions between subatomic particles that we cannot observe with the unaided senses.

   Gravitational interactions include the familiar force of your *weight,* which results from the earth's gravitational attraction acting on you. The mutual gravitational attraction of various parts of the earth for each other holds our planet together, and likewise for the other planets (Fig. 5.38a). Newton recognized that the sun's gravitational attraction for the earth keeps our planet in its nearly circular orbit around the sun. In Chapter 13 we'll study gravitational interactions in more detail, including their vital role in the motions of planets and satellites.

   The second familiar class of forces, **electromagnetic interactions,** includes electric and magnetic forces. If you run a comb through your hair, the comb ends up with an electric charge; you can use the electric force exerted by this charge to pick up bits of paper. All atoms contain positive and negative electric charge, so atoms and molecules can exert electric forces on one another. Contact forces, including the normal force, friction, and fluid resistance, are the result of electrical interactions between atoms on the surface of an object and atoms in its surroundings (Fig. 5.38b). *Magnetic* forces, such as those between magnets or between a magnet and a piece of iron, are actually the result of electric charges in motion. For example, an electromagnet causes magnetic interactions because

BIO Application Circular Motion in a Centrifuge   An important tool in medicine and biological research is the ultracentrifuge, a device that makes use of the dynamics of circular motion. A tube is filled with a solvent that contains various small particles (for example, blood containing platelets and white and red blood cells). The tube is inserted into the centrifuge, which then spins at thousands of revolutions per minute. The solvent provides the inward force that keeps the particles in circular motion. The particles slowly drift away from the rotation axis within the solvent. Because the drift rate depends on the particle size and density, particles of different types become separated in the tube, making analysis much easier.

5.38 Examples of the fundamental interactions in nature.

(a) The gravitational interaction

Saturn is held together by the mutual gravitional attraction of all of its parts.

The particles that make up the rings are held in orbit by Saturn's gravitational force.

(b) The electromagnetic interaction

The contact forces between the microphone and the singer's hand are electrical in nature.

This microphone uses electric and magnetic effects to convert sound into an electrical signal that can be amplified and recorded.

(c) The strong interaction

The nucleus of a gold atom has 79 protons and 118 neutrons.

The strong interaction holds the protons and neutrons together and overcomes the electric repulsion of the protons.

(d) The weak interaction

Scientists find the age of this ancient skull by measuring its carbon-14—a form of carbon that is radioactive thanks to the weak interaction.

electric charges move through its wires. We will study electromagnetic interactions in detail in the second half of this book.

On the atomic or molecular scale, gravitational forces play no role because electric forces are enormously stronger: The electrical repulsion between two protons is stronger than their gravitational attraction by a factor of about 10^{35}. But in bodies of astronomical size, positive and negative charges are usually present in nearly equal amounts, and the resulting electrical interactions nearly cancel out. Gravitational interactions are thus the dominant influence in the motion of planets and in the internal structure of stars.

The other two classes of interactions are less familiar. One, the **strong interaction,** is responsible for holding the nucleus of an atom together (Fig. 5.38c). Nuclei contain electrically neutral neutrons and positively charged protons. The electric force between charged protons tries to push them apart; the strong attractive force between nuclear particles counteracts this repulsion and makes the nucleus stable. In this context the strong interaction is also called the *strong nuclear force*. It has much shorter range than electrical interactions, but within its range it is much stronger. Without the strong interaction, the nuclei of atoms essential to life, such as carbon (six protons, six neutrons) and oxygen (eight protons, eight neutrons), would not exist and you would not be reading these words!

Finally, there is the **weak interaction.** Its range is so short that it plays a role only on the scale of the nucleus or smaller. The weak interaction is responsible for a common form of radioactivity called beta decay, in which a neutron in a radioactive nucleus is transformed into a proton while ejecting an electron and a nearly massless particle called an antineutrino. The weak interaction between the antineutrino and ordinary matter is so feeble that an antineutrino could easily penetrate a wall of lead a million kilometers thick!

An important application of the weak interaction is *radiocarbon dating,* a technique that enables scientists to determine the ages of many biological specimens (Fig. 5.38d). Naturally occurring carbon includes atoms of both carbon-12 (with six protons and six neutrons in the nucleus) and carbon-14 (with two additional neutrons). Living organisms take in carbon atoms of both kinds from their environment but stop doing so when they die. The weak interaction makes carbon-14 nuclei unstable—one of the neutrons changes to a proton, an electron, and an antineutrino—and these nuclei decay at a known rate. By measuring the fraction of carbon-14 that is left in an organism's remains, scientists can determine how long ago the organism died.

In the 1960s physicists developed a theory that described the electromagnetic and weak interactions as aspects of a single *electroweak* interaction. This theory has passed every experimental test to which it has been put. Encouraged by this success, physicists have made similar attempts to describe the strong, electromagnetic, and weak interactions in terms of a single *grand unified theory* (GUT) and have taken steps toward a possible unification of all interactions into a *theory of everything* (TOE). Such theories are still speculative, and there are many unanswered questions in this very active field of current research.

CHAPTER 5 SUMMARY

SOLUTIONS TO ALL EXAMPLES

Using Newton's first law: When a body is in equilibrium in an inertial frame of reference—that is, either at rest or moving with constant velocity—the vector sum of forces acting on it must be zero (Newton's first law). Free-body diagrams are essential in identifying the forces that act on the body being considered.

Newton's third law (action and reaction) is also frequently needed in equilibrium problems. The two forces in an action–reaction pair *never* act on the same body. (See Examples 5.1–5.5.)

The normal force exerted on a body by a surface is *not* always equal to the body's weight. (See Example 5.4.)

Vector form:

$$\sum \vec{F} = 0$$

Component form:

$$\sum F_x = 0 \qquad \sum F_y = 0$$

(5.1)

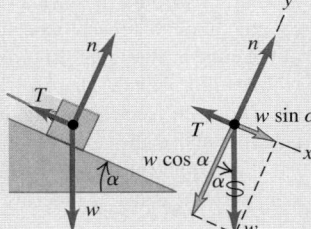

Using Newton's second law: If the vector sum of forces on a body is *not* zero, the body accelerates. The acceleration is related to the net force by Newton's second law.

Just as for equilibrium problems, free-body diagrams are essential for solving problems involving Newton's second law, and the normal force exerted on a body is not always equal to its weight. (See Examples 5.6–5.12.)

Vector form:

$$\sum \vec{F} = m\vec{a}$$

Component form:

$$\sum F_x = ma_x \qquad \sum F_y = ma_y$$

(5.2)

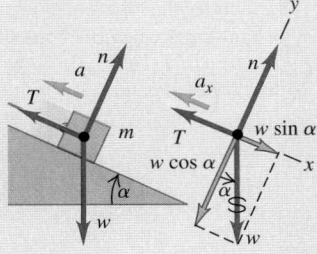

Friction and fluid resistance: The contact force between two bodies can always be represented in terms of a normal force $\vec{n}$ perpendicular to the surface of contact and a friction force $\vec{f}$ parallel to the surface.

Magnitude of kinetic friction force:

$$f_k = \mu_k n \qquad (5.3)$$

Magnitude of static friction force:

$$f_s \leq (f_s)_{max} = \mu_s n \qquad (5.4)$$

When a body is sliding over the surface, the friction force is called *kinetic* friction. Its magnitude f_k is approximately equal to the normal force magnitude n multiplied by the coefficient of kinetic friction μ_k.

When a body is *not* moving relative to a surface, the friction force is called *static* friction. The *maximum* possible static friction force is approximately equal to the magnitude n of the normal force multiplied by the coefficient of static friction μ_s. The *actual* static friction force may be anything from zero to this maximum value, depending on the situation. Usually μ_s is greater than μ_k for a given pair of surfaces in contact. (See Examples 5.13–5.17.)

Rolling friction is similar to kinetic friction, but the force of fluid resistance depends on the speed of an object through a fluid. (See Example 5.18.)

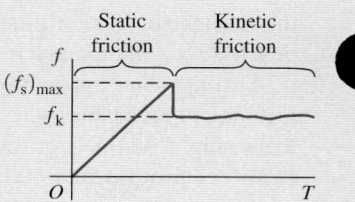

Forces in circular motion: In uniform circular motion, the acceleration vector is directed toward the center of the circle. The motion is governed by Newton's second law, $\Sigma\vec{F} = m\vec{a}$. (See Examples 5.19–5.23.)

Acceleration in uniform circular motion:

$$a_{rad} = \frac{v^2}{R} = \frac{4\pi^2 R}{T^2} \qquad (5.13), (5.15)$$

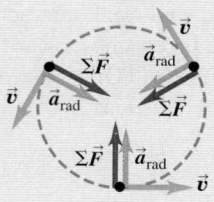

BRIDGING PROBLEM | IN A ROTATING CONE

A small block with mass m is placed inside an inverted cone that is rotating about a vertical axis such that the time for one revolution of the cone is T (**Fig. 5.39**). The walls of the cone make an angle β with the horizontal. The coefficient of static friction between the block and the cone is μ_s. If the block is to remain at a constant height h above the apex of the cone, what are (a) the maximum value of T and (b) the minimum value of T? (That is, find expressions for T_{max} and T_{min} in terms of β and h.)

SOLUTION GUIDE

IDENTIFY and SET UP

1. Although we want the block not to slide up or down on the inside of the cone, this is *not* an equilibrium problem. The block rotates with the cone and is in uniform circular motion, so it has an acceleration directed toward the center of its circular path.
2. Identify the forces on the block. What is the direction of the friction force when the cone is rotating as slowly as possible, so T

5.39 A block inside a spinning cone.

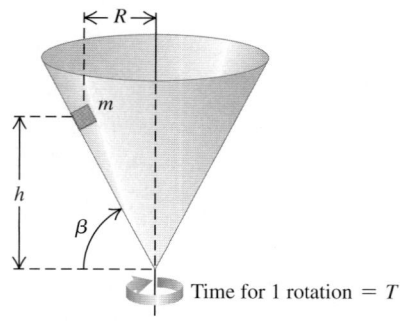

Time for 1 rotation $= T$

has its maximum value T_{max}? What is the direction of the friction force when the cone is rotating as rapidly as possible, so T has its minimum value T_{min}? In these situations does the static friction force have its *maximum* magnitude? Why or why not?

3. Draw a free-body diagram for the block when the cone is rotating with $T = T_{max}$ and a free-body diagram when the cone is rotating with $T = T_{min}$. Choose coordinate axes, and remember that it's usually easiest to choose one of the axes to be in the direction of the acceleration.
4. What is the radius of the circular path that the block follows? Express this in terms of β and h.
5. List the unknown quantities, and decide which of these are the target variables.

EXECUTE

6. Write Newton's second law in component form for the case in which the cone is rotating with $T = T_{max}$. Write the acceleration in terms of T_{max}, β, and h, and write the static friction force in terms of the normal force n.
7. Solve these equations for the target variable T_{max}.
8. Repeat steps 6 and 7 for the case in which the cone is rotating with $T = T_{min}$, and solve for the target variable T_{min}.

EVALUATE

9. You'll end up with some fairly complicated expressions for T_{max} and T_{min}, so check them over carefully. Do they have the correct units? Is the minimum time T_{min} less than the maximum time T_{max}, as it must be?
10. What do your expressions for T_{max} and T_{min} become if $\mu_s = 0$? Check your results by comparing them with Example 5.22 in Section 5.4.

Problems

•, ••, •••: Difficulty levels. CP: Cumulative problems incorporating material from earlier chapters. CALC: Problems requiring calculus. DATA: Problems involving real data, scientific evidence, experimental design, and/or statistical reasoning. BIO: Biosciences problems.

[Always assume that pulleys are frictionless and massless and that strings and cords are massless, unless otherwise noted.]

DISCUSSION QUESTIONS

Q5.1 A man sits in a seat that is hanging from a rope. The rope passes over a pulley suspended from the ceiling, and the man holds the other end of the rope in his hands. What is the tension in the rope, and what force does the seat exert on him? Draw a free-body force diagram for the man.

Q5.2 "In general, the normal force is not equal to the weight." Give an example in which these two forces are equal in magnitude, and at least two examples in which they are not.

Q5.3 A clothesline hangs between two poles. No matter how tightly the line is stretched, it sags a little at the center. Explain why.

Q5.4 You drive a car up a steep hill at constant speed. Discuss all of the forces that act on the car. What pushes it up the hill?

Q5.5 For medical reasons, astronauts in outer space must determine their body mass at regular intervals. Devise a scheme for measuring body mass in an apparently weightless environment.

Q5.6 To push a box up a ramp, which requires less force: pushing horizontally or pushing parallel to the ramp? Why?

Q5.7 A woman in an elevator lets go of her briefcase, but it does not fall to the floor. How is the elevator moving?

Q5.8 A block rests on an inclined plane with enough friction to prevent it from sliding down. To start the block moving, is it easier to push it up the plane or down the plane? Why?

Q5.9 A crate slides up an inclined ramp and then slides down the ramp after momentarily stopping near the top. There is kinetic friction between the surface of the ramp and the crate. Which is greater? (i) The crate's acceleration going up the ramp; (ii) the crate's acceleration going down the ramp; (iii) both are the same. Explain.

Q5.10 A crate of books rests on a level floor. To move it along the floor at a constant velocity, why do you exert less force if you pull it at an angle θ above the horizontal than if you push it at the same angle below the horizontal?

Q5.11 In a world without friction, which of the following activities could you do (or not do)? Explain your reasoning. (a) Drive around an unbanked highway curve; (b) jump into the air; (c) start walking on a horizontal sidewalk; (d) climb a vertical ladder; (e) change lanes while you drive.

Q5.12 When you stand with bare feet in a wet bathtub, the grip feels fairly secure, and yet a catastrophic slip is quite possible. Explain this in terms of the two coefficients of friction.

Q5.13 You are pushing a large crate from the back of a freight elevator to the front as the elevator is moving to the next floor. In which situation is the force you must apply to move the crate the least, and in which is it the greatest: when the elevator is accelerating upward, when it is accelerating downward, or when it is traveling at constant speed? Explain.

Q5.14 It is often said that "friction always opposes motion." Give at least one example in which (a) static friction *causes* motion, and (b) kinetic friction *causes* motion.

Q5.15 If there is a net force on a particle in uniform circular motion, why doesn't the particle's speed change?

Q5.16 A curve in a road has a bank angle calculated and posted for 80 km/h. However, the road is covered with ice, so you cautiously plan to drive slower than this limit. What might happen to your car? Why?

Q5.17 You swing a ball on the end of a lightweight string in a horizontal circle at constant speed. Can the string ever be truly horizontal? If not, would it slope above the horizontal or below the horizontal? Why?

Q5.18 The centrifugal force is not included in the free-body diagrams of Figs. 5.34b and 5.35. Explain why not.

Q5.19 A professor swings a rubber stopper in a horizontal circle on the end of a string in front of his class. He tells Caroline, in the front row, that he is going to let the string go when the stopper is directly in front of her face. Should Caroline worry?

Q5.20 To keep the forces on the riders within allowable limits, many loop-the-loop roller coaster rides are designed so that the loop is not a perfect circle but instead has a larger radius of curvature at the bottom than at the top. Explain.

Q5.21 A tennis ball drops from rest at the top of a tall glass cylinder—first with the air pumped out of the cylinder so that there is no air resistance, and again after the air has been readmitted to the cylinder. You examine multiflash photographs of the two drops. Can you tell which photo belongs to which drop? If so, how?

Q5.22 You throw a baseball straight upward with speed v_0. When the ball returns to the point from where you threw it, how does its speed compare to v_0 (a) in the absence of air resistance and (b) in the presence of air resistance? Explain.

Q5.23 You throw a baseball straight upward. If you do *not* ignore air resistance, how does the time required for the ball to reach its maximum height compare to the time required for it to fall from its maximum height back down to the height from which you threw it? Explain.

Q5.24 You have two identical tennis balls and fill one with water. You release both balls simultaneously from the top of a tall building. If air resistance is negligible, which ball will strike the ground first? Explain. What if air resistance is *not* negligible?

Q5.25 A ball is dropped from rest and feels air resistance as it falls. Which of the graphs in **Fig. Q5.25** best represents its acceleration as a function of time?

Figure **Q5.25**

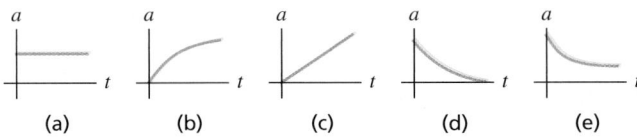

(a)   (b)   (c)   (d)   (e)

Q5.26 A ball is dropped from rest and feels air resistance as it falls. Which of the graphs in **Fig. Q5.26** best represents its vertical velocity component as a function of time?

Figure **Q5.26**

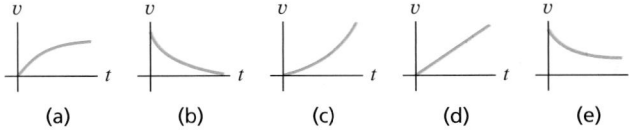

(a)   (b)   (c)   (d)   (e)

Q5.27 When a batted baseball moves with air drag, when does the ball travel a greater horizontal distance? (i) While climbing to its maximum height; (ii) while descending from its maximum height back to the ground; (iii) the same for both? Explain in terms of the forces acting on the ball.

Q5.28 "A ball is thrown from the edge of a high cliff. Regardless of the angle at which it is thrown, due to air resistance, the ball will eventually end up moving vertically downward." Justify this statement.

EXERCISES

Section 5.1 Using Newton's First Law: Particles in Equilibrium

5.1 • Two 25.0-N weights are suspended at opposite ends of a rope that passes over a light, frictionless pulley. The pulley is attached to a chain from the ceiling. (a) What is the tension in the rope? (b) What is the tension in the chain?

5.2 • In **Fig. E5.2** each of the suspended blocks has weight w. The pulleys are frictionless, and the ropes have negligible weight. In each case, draw a free-body diagram and calculate the tension T in the rope in terms of w.

Figure **E5.2**

(a)   (b)   (c)

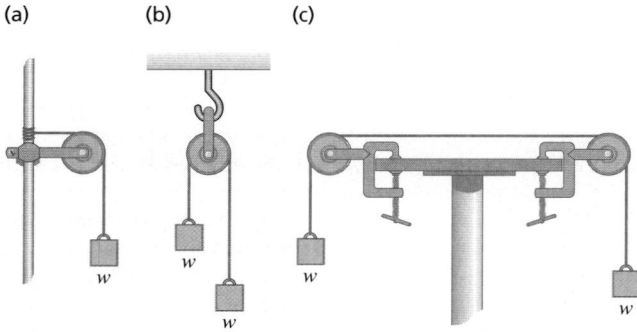

5.3 • A 75.0-kg wrecking ball hangs from a uniform, heavy-duty chain of mass 26.0 kg. (a) Find the maximum and minimum tensions in the chain. (b) What is the tension at a point three-fourths of the way up from the bottom of the chain?

5.4 •• BIO **Injuries to the Spinal Column.** In the treatment of spine injuries, it is often necessary to provide tension along the spinal column to stretch the backbone. One device for doing this is the Stryker frame (**Fig. E5.4a**). A weight W is attached to the patient (sometimes around a neck collar, Fig. E5.4b), and friction between the person's body and the bed prevents sliding. (a) If the coefficient of static friction between a 78.5-kg patient's body and the bed is 0.75, what is the maximum traction force along the spinal column that W can provide without causing the patient to slide? (b) Under the conditions of maximum traction, what is the tension in each cable attached to the neck collar?

Figure **E5.4**

(a)   (b)

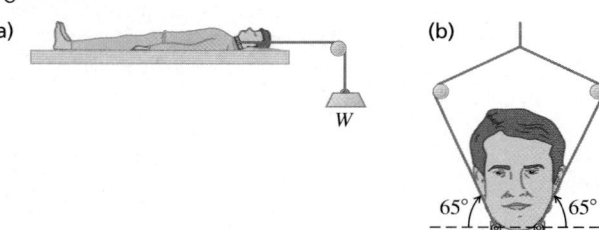

5.5 •• A picture frame hung against a wall is suspended by two wires attached to its upper corners. If the two wires make the same angle with the vertical, what must this angle be if the tension in each wire is equal to 0.75 of the weight of the frame? (Ignore any friction between the wall and the picture frame.)

5.6 •• A large wrecking ball is held in place by two light steel cables (**Fig. E5.6**). If the mass m of the wrecking ball is 3620 kg, what are (a) the tension T_B in the cable that makes an angle of 40° with the vertical and (b) the tension T_A in the horizontal cable?

Figure **E5.6**

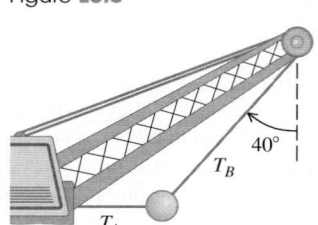

5.7 •• Find the tension in each cord in **Fig. E5.7** if the weight of the suspended object is w.

Figure **E5.7**

(a)   (b)

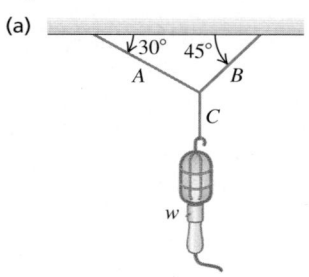
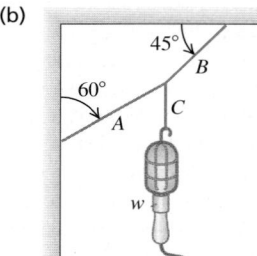

5.8 •• A 1130-kg car is held in place by a light cable on a very smooth (frictionless) ramp (**Fig. E5.8**). The cable makes an angle of 31.0° above the surface of the ramp, and the ramp itself rises at 25.0° above the horizontal. (a) Draw a free-body diagram for the car. (b) Find the tension in the cable. (c) How hard does the surface of the ramp push on the car?

Figure **E5.8**

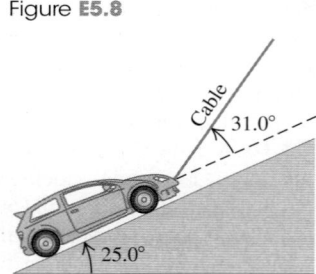

5.9 •• A man pushes on a piano with mass 180 kg; it slides at constant velocity down a ramp that is inclined at 19.0° above the horizontal floor. Neglect any friction acting on the piano. Calculate the magnitude of the force applied by the man if he pushes (a) parallel to the incline and (b) parallel to the floor.

5.10 •• In **Fig. E5.10** the weight w is 60.0 N. (a) What is the tension in the diagonal string? (b) Find the magnitudes of the horizontal forces $\vec{F}_1$ and $\vec{F}_2$ that must be applied to hold the system in the position shown.

Figure **E5.10**

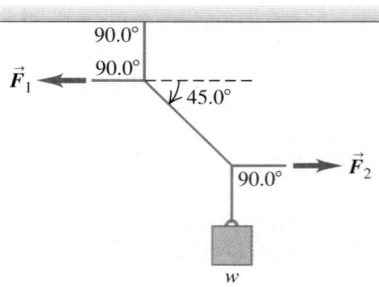

Section 5.2 Using Newton's Second Law: Dynamics of Particles

5.11 •• BIO **Stay Awake!** An astronaut is inside a 2.25×10^6 kg rocket that is blasting off vertically from the launch pad. You want this rocket to reach the speed of sound (331 m/s) as quickly as possible, but astronauts are in danger of blacking out at an acceleration greater than 4g. (a) What is the maximum initial thrust this rocket's engines can have but just barely avoid blackout? Start with a free-body diagram of the rocket. (b) What force, in terms of the astronaut's weight w, does the rocket exert on her? Start with a free-body diagram of the astronaut. (c) What is the shortest time it can take the rocket to reach the speed of sound?

5.12 •• A rocket of initial mass 125 kg (including all the contents) has an engine that produces a constant vertical force (the *thrust*) of 1720 N. Inside this rocket, a 15.5-N electrical power supply rests on the floor. (a) Find the initial acceleration of the rocket. (b) When the rocket initially accelerates, how hard does the floor push on the power supply? (*Hint:* Start with a free-body diagram for the power supply.)

5.13 •• CP **Genesis Crash.** On September 8, 2004, the *Genesis* spacecraft crashed in the Utah desert because its parachute did not open. The 210-kg capsule hit the ground at 311 km/h and penetrated the soil to a depth of 81.0 cm. (a) What was its acceleration (in m/s^2 and in g's), assumed to be constant, during the crash? (b) What force did the ground exert on the capsule during the crash? Express the force in newtons and as a multiple of the capsule's weight. (c) How long did this force last?

5.14 • Three sleds are being pulled horizontally on frictionless horizontal ice using horizontal ropes (**Fig. E5.14**). The pull is of magnitude 190 N. Find (a) the acceleration of the system and (b) the tension in ropes A and B.

Figure **E5.14**

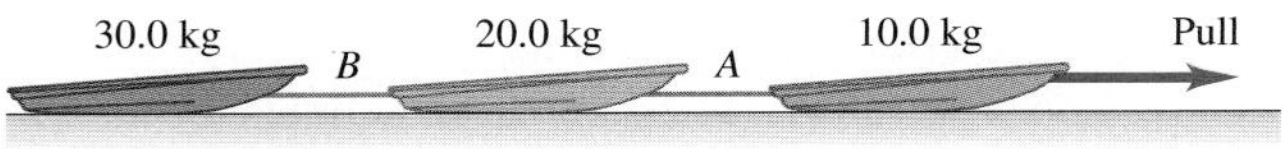

30.0 kg		20.0 kg		10.0 kg	Pull
	B		A		

5.15 •• **Atwood's Machine.** A 15.0-kg load of bricks hangs from one end of a rope that passes over a small, frictionless pulley. A 28.0-kg counterweight is suspended from the other end of the rope (**Fig. E5.15**). The system is released from rest. (a) Draw two free-body diagrams, one for the load of bricks and one for the counterweight. (b) What is the magnitude of the upward acceleration of the load of bricks? (c) What is the tension in the rope while the load is moving? How does the tension compare to the weight of the load of bricks? To the weight of the counterweight?

Figure **E5.15**

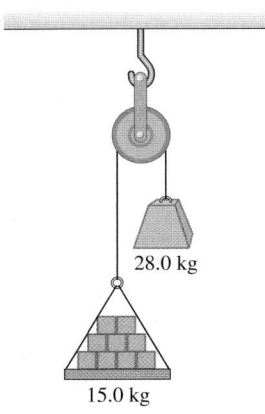

28.0 kg

15.0 kg

5.16 •• CP An 8.00-kg block of ice, released from rest at the top of a 1.50-m-long frictionless ramp, slides downhill, reaching a speed of 2.50 m/s at the bottom. (a) What is the angle between the ramp and the horizontal? (b) What would be the speed of the ice at the bottom if the motion were opposed by a constant friction force of 10.0 N parallel to the surface of the ramp?

5.17 •• A light rope is attached to a block with mass 4.00 kg that rests on a frictionless, horizontal surface. The horizontal rope passes over a frictionless, massless pulley, and a block with mass m is suspended from the other end. When the blocks are released, the tension in the rope is 15.0 N. (a) Draw two free-body diagrams: one for each block. (b) What is the acceleration of either block? (c) Find m. (d) How does the tension compare to the weight of the hanging block?

5.18 •• CP **Runway Design.** A transport plane takes off from a level landing field with two gliders in tow, one behind the other. The mass of each glider is 700 kg, and the total resistance (air drag plus friction with the runway) on each may be assumed constant and equal to 2500 N. The tension in the towrope between the transport plane and the first glider is not to exceed 12,000 N. (a) If a speed of 40 m/s is required for takeoff, what minimum length of runway is needed? (b) What is the tension in the towrope between the two gliders while they are accelerating for the takeoff?

5.19 •• CP A 750.0-kg boulder is raised from a quarry 125 m deep by a long uniform chain having a mass of 575 kg. This chain is of uniform strength, but at any point it can support a maximum tension no greater than 2.50 times its weight without breaking. (a) What is the maximum acceleration the boulder can have and still get out of the quarry, and (b) how long does it take to be lifted out at maximum acceleration if it started from rest?

5.20 •• **Apparent Weight.** A 550-N physics student stands on a bathroom scale in an elevator that is supported by a cable. The combined mass of student plus elevator is 850 kg. As the elevator starts moving, the scale reads 450 N. (a) Find the acceleration of the elevator (magnitude and direction). (b) What is the acceleration if the scale reads 670 N? (c) If the scale reads zero, should the student worry? Explain. (d) What is the tension in the cable in parts (a) and (c)?

5.21 •• CP BIO **Force During a Jump.** When jumping straight up from a crouched position, an average person can reach a maximum height of about 60 cm. During the jump, the person's body from the knees up typically rises a distance of around 50 cm. To keep the calculations simple and yet get a reasonable result, assume that the *entire body* rises this much during the jump. (a) With what initial speed does the person leave the ground to reach a height of 60 cm? (b) Draw a free-body diagram of the person during the jump. (c) In terms of this jumper's weight w, what force does the ground exert on him or her during the jump?

5.22 CP CALC A 2540-kg test rocket is launched vertically from the launch pad. Its fuel (of negligible mass) provides a thrust force such that its vertical velocity as a function of time is given by $v(t) = At + Bt^2$, where A and B are constants and time is measured from the instant the fuel is ignited. The rocket has an upward acceleration of 1.50 m/s^2 at the instant of ignition and, 1.00 s later, an upward velocity of 2.00 m/s. (a) Determine A and B, including their SI units. (b) At 4.00 s after fuel ignition, what is the acceleration of the rocket, and (c) what thrust force does the burning fuel exert on it, assuming no air resistance? Express the thrust in newtons and as a multiple of the rocket's weight. (d) What was the initial thrust due to the fuel?

5.23 •• CP CALC A 2.00-kg box is moving to the right with speed 9.00 m/s on a horizontal, frictionless surface. At $t = 0$ a horizontal force is applied to the box. The force is directed to the left and has magnitude $F(t) = (6.00 \text{ N/s}^2)t^2$. (a) What distance does the box move from its position at $t = 0$ before its speed is reduced to zero? (b) If the force continues to be applied, what is the speed of the box at $t = 3.00$ s?

5.24 •• CP CALC A 5.00-kg crate is suspended from the end of a short vertical rope of negligible mass. An upward force $F(t)$ is applied to the end of the rope, and the height of the crate above its initial position is given by $y(t) = (2.80 \text{ m/s})t + (0.610 \text{ m/s}^3)t^3$. What is the magnitude of F when $t = 4.00$ s?

Section 5.3 Friction Forces

5.25 • BIO **The Trendelenburg Position.** After emergencies with major blood loss, a patient is placed in the Trendelenburg position, in which the foot of the bed is raised to get maximum blood flow to the brain. If the coefficient of static friction between a typical patient and the bedsheets is 1.20, what is the maximum angle at which the bed can be tilted with respect to the floor before the patient begins to slide?

5.26 • In a laboratory experiment on friction, a 135-N block resting on a rough horizontal table is pulled by a horizontal wire. The pull gradually increases until the block begins to move and continues to increase thereafter. **Figure E5.26** shows a graph of the friction force on this block as a function of the pull. (a) Identify the regions of the graph where static friction and kinetic friction occur. (b) Find the coefficients of static friction and kinetic friction between the block and the table. (c) Why does the graph slant upward at first but then level out? (d) What would the graph look like if a 135-N brick were placed on the block, and what would the coefficients of friction be?

Figure **E5.26**

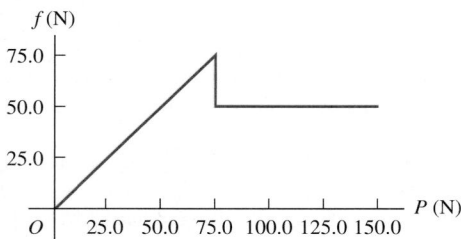

5.27 •• CP A stockroom worker pushes a box with mass 16.8 kg on a horizontal surface with a constant speed of 3.50 m/s. The coefficient of kinetic friction between the box and the surface is 0.20. (a) What horizontal force must the worker apply to maintain the motion? (b) If the force calculated in part (a) is removed, how far does the box slide before coming to rest?

5.28 •• A box of bananas weighing 40.0 N rests on a horizontal surface. The coefficient of static friction between the box and the surface is 0.40, and the coefficient of kinetic friction is 0.20. (a) If no horizontal force is applied to the box and the box is at rest, how large is the friction force exerted on it? (b) What is the magnitude of the friction force if a monkey applies a horizontal force of 6.0 N to the box and the box is initially at rest? (c) What minimum horizontal force must the monkey apply to start the box in motion? (d) What minimum horizontal force must the monkey apply to keep the box moving at constant velocity once it has been started? (e) If the monkey applies a horizontal force of 18.0 N, what is the magnitude of the friction force and what is the box's acceleration?

5.29 •• A 45.0-kg crate of tools rests on a horizontal floor. You exert a gradually increasing horizontal push on it, and the crate just begins to move when your force exceeds 313 N. Then you must reduce your push to 208 N to keep it moving at a steady 25.0 cm/s. (a) What are the coefficients of static and kinetic friction between the crate and the floor? (b) What push must you exert

to give it an acceleration of 1.10 m/s²? (c) Suppose you were performing the same experiment on the moon, where the acceleration due to gravity is 1.62 m/s². (i) What magnitude push would cause it to move? (ii) What would its acceleration be if you maintained the push in part (b)?

5.30 •• Some sliding rocks approach the base of a hill with a speed of 12 m/s. The hill rises at 36° above the horizontal and has coefficients of kinetic friction and static friction of 0.45 and 0.65, respectively, with these rocks. (a) Find the acceleration of the rocks as they slide up the hill. (b) Once a rock reaches its highest point, will it stay there or slide down the hill? If it stays, show why. If it slides, find its acceleration on the way down.

5.31 •• A box with mass 10.0 kg moves on a ramp that is inclined at an angle of 55.0° above the horizontal. The coefficient of kinetic friction between the box and the ramp surface is $\mu_k = 0.300$. Calculate the magnitude of the acceleration of the box if you push on the box with a constant force $F = 120.0$ N that is parallel to the ramp surface and (a) directed down the ramp, moving the box down the ramp; (b) directed up the ramp, moving the box up the ramp.

5.32 •• A pickup truck is carrying a toolbox, but the rear gate of the truck is missing. The toolbox will slide out if it is set moving. The coefficients of kinetic friction and static friction between the box and the level bed of the truck are 0.355 and 0.650, respectively. Starting from rest, what is the shortest time this truck could accelerate uniformly to 30.0 m/s without causing the box to slide? Draw a free-body diagram of the toolbox.

5.33 •• You are lowering two boxes, one on top of the other, down a ramp by pulling on a rope parallel to the surface of the ramp (**Fig. E5.33**). Both boxes move together at a constant speed of 15.0 cm/s. The coefficient of kinetic friction between the ramp and the lower box is 0.444, and the coefficient of static friction between the two boxes is 0.800. (a) What force do you need to exert to accomplish this? (b) What are the magnitude and direction of the friction force on the upper box?

Figure **E5.33**

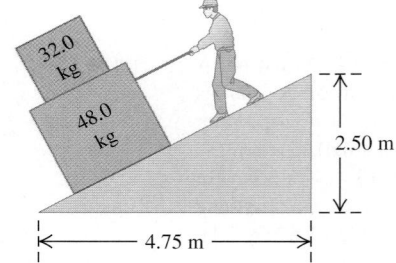

5.34 •• Consider the system shown in **Fig. E5.34**. Block A weighs 45.0 N, and block B weighs 25.0 N. Once block B is set into downward motion, it descends at a constant speed. (a) Calculate the coefficient of kinetic friction between block A and the tabletop. (b) A cat, also of weight 45.0 N, falls asleep on top of block A. If block B is now set into downward motion, what is its acceleration (magnitude and direction)?

Figure **E5.34**

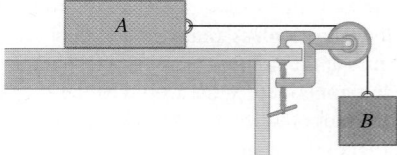

5.35 •• **CP Stopping Distance.** (a) If the coefficient of kinetic friction between tires and dry pavement is 0.80, what is the shortest distance in which you can stop a car by locking the brakes when the car is traveling at 28.7 m/s (about 65 mi/h)? (b) On wet pavement the coefficient of kinetic friction may be only 0.25. How fast should you drive on wet pavement to be able to stop in the same distance as in part (a)? (*Note:* Locking the brakes is *not* the safest way to stop.)

5.36 •• **CP** A 25.0-kg box of textbooks rests on a loading ramp that makes an angle α with the horizontal. The coefficient of kinetic friction is 0.25, and the coefficient of static friction is 0.35. (a) As α is increased, find the minimum angle at which the box starts to slip. (b) At this angle, find the acceleration once the box has begun to move. (c) At this angle, how fast will the box be moving after it has slid 5.0 m along the loading ramp?

5.37 • Two crates connected by a rope lie on a horizontal surface (Fig. E5.37). Crate A has mass m_A, and crate B has mass m_B. The coefficient of kinetic friction between each crate and the surface is μ_k. The crates are pulled to the right at constant velocity by a horizontal force $\vec{F}$. Draw one or more free-body diagrams to calculate the following in terms of m_A, m_B, and μ_k: (a) the magnitude of $\vec{F}$ and (b) the tension in the rope connecting the blocks.

Figure **E5.37**

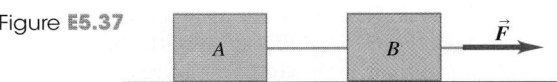

5.38 •• A box with mass m is dragged across a level floor with coefficient of kinetic friction μ_k by a rope that is pulled upward at an angle θ above the horizontal with a force of magnitude F. (a) In terms of m, μ_k, θ, and g, obtain an expression for the magnitude of the force required to move the box with constant speed. (b) Knowing that you are studying physics, a CPR instructor asks you how much force it would take to slide a 90-kg patient across a floor at constant speed by pulling on him at an angle of 25° above the horizontal. By dragging weights wrapped in an old pair of pants down the hall with a spring balance, you find that $\mu_k = 0.35$. Use the result of part (a) to answer the instructor's question.

5.39 •• **CP** As shown in Fig. E5.34, block A (mass 2.25 kg) rests on a tabletop. It is connected by a horizontal cord passing over a light, frictionless pulley to a hanging block B (mass 1.30 kg). The coefficient of kinetic friction between block A and the tabletop is 0.450. The blocks are released then from rest. Draw one or more free-body diagrams to find (a) the speed of each block after they move 3.00 cm and (b) the tension in the cord.

5.40 •• You throw a baseball straight upward. The drag force is proportional to v^2. In terms of g, what is the y-component of the ball's acceleration when the ball's speed is half its terminal speed and (a) it is moving up? (b) It is moving back down?

5.41 •• A large crate with mass m rests on a horizontal floor. The coefficients of friction between the crate and the floor are μ_s and μ_k. A woman pushes downward with a force $\vec{F}$ on the crate at an angle θ below the horizontal. (a) What magnitude of force $\vec{F}$ is required to keep the crate moving at constant velocity? (b) If μ_s is greater than some critical value, the woman cannot start the crate moving no matter how hard she pushes. Calculate this critical value of μ_s.

5.42 • (a) In Example 5.18 (Section 5.3), what value of D is required to make $v_t = 42$ m/s for the skydiver? (b) If the skydiver's daughter, whose mass is 45 kg, is falling through the air and has the same D (0.25 kg/m) as her father, what is the daughter's terminal speed?

Section 5.4 Dynamics of Circular Motion

5.43 • A stone with mass 0.80 kg is attached to one end of a string 0.90 m long. The string will break if its tension exceeds 60.0 N. The stone is whirled in a horizontal circle on a frictionless tabletop; the other end of the string remains fixed. (a) Draw a free-body diagram of the stone. (b) Find the maximum speed the stone can attain without the string breaking.

5.44 • **BIO Force on a Skater's Wrist.** A 52-kg ice skater spins about a vertical axis through her body with her arms horizontally outstretched; she makes 2.0 turns each second. The distance from one hand to the other is 1.50 m. Biometric measurements indicate that each hand typically makes up about 1.25% of body weight. (a) Draw a free-body diagram of one of the skater's hands. (b) What horizontal force must her wrist exert on her hand? (c) Express the force in part (b) as a multiple of the weight of her hand.

5.45 •• A small remote-controlled car with mass 1.60 kg moves at a constant speed of $v = 12.0$ m/s in a track formed by a vertical circle inside a hollow metal cylinder that has a radius of 5.00 m (Fig. E5.45). What is the magnitude of the normal force exerted on the car by the walls of the cylinder at (a) point A (bottom of the track) and (b) point B (top of the track)?

Figure **E5.45**

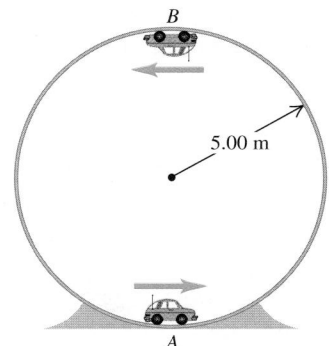

5.00 m

5.46 •• A small car with mass 0.800 kg travels at constant speed on the inside of a track that is a vertical circle with radius 5.00 m (Fig. E5.45). If the normal force exerted by the track on the car when it is at the top of the track (point B) is 6.00 N, what is the normal force on the car when it is at the bottom of the track (point A)?

5.47 • A small model car with mass m travels at constant speed on the inside of a track that is a vertical circle with radius 5.00 m (Fig. E5.45). If the normal force exerted by the track on the car when it is at the bottom of the track (point A) is equal to 2.50mg, how much time does it take the car to complete one revolution around the track?

5.48 • A flat (unbanked) curve on a highway has a radius of 170.0 m. A car rounds the curve at a speed of 25.0 m/s. (a) What is the minimum coefficient of static friction that will prevent sliding? (b) Suppose that the highway is icy and the coefficient of static friction between the tires and pavement is only one-third of what you found in part (a). What should be the maximum speed of the car so that it can round the curve safely?

5.49 •• A 1125-kg car and a 2250-kg pickup truck approach a curve on a highway that has a radius of 225 m. (a) At what angle should the highway engineer bank this curve so that vehicles traveling at 65.0 mi/h can safely round it regardless of the condition of their tires? Should the heavy truck go slower than the lighter car? (b) As the car and truck round the curve at 65.0 mi/h, find the normal force on each one due to the highway surface.

5.50 •• The "Giant Swing" at a county fair consists of a vertical central shaft with a number of horizontal arms attached at its upper end. Each arm supports a seat suspended from a cable 5.00 m long, and the upper end of the cable is fastened to the arm at a point 3.00 m from the central shaft (**Fig. E5.50**). (a) Find the time of one revolution of the swing if the cable supporting a seat makes an angle of 30.0° with the vertical. (b) Does the angle depend on the weight of the passenger for a given rate of revolution?

Figure **E5.50**

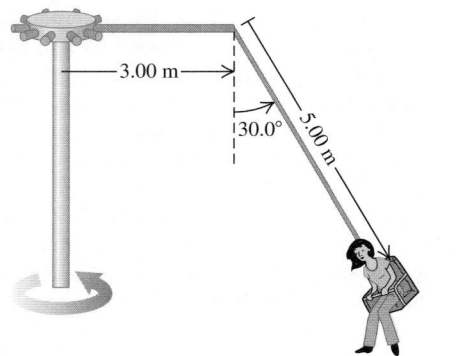

5.51 •• In another version of the "Giant Swing" (see Exercise 5.50), the seat is connected to two cables, one of which is horizontal (**Fig. E5.51**). The seat swings in a horizontal circle at a rate of 28.0 rpm (rev/min). If the seat weighs 255 N and an 825-N person is sitting in it, find the tension in each cable.

Figure **E5.51**

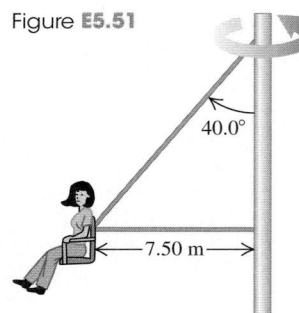

5.52 •• A small button placed on a horizontal rotating platform with diameter 0.520 m will revolve with the platform when it is brought up to a speed of 40.0 rev/min, provided the button is no more than 0.220 m from the axis. (a) What is the coefficient of static friction between the button and the platform? (b) How far from the axis can the button be placed, without slipping, if the platform rotates at 60.0 rev/min?

5.53 •• **Rotating Space Stations.** One problem for humans living in outer space is that they are apparently weightless. One way around this problem is to design a space station that spins about its center at a constant rate. This creates "artificial gravity" at the outside rim of the station. (a) If the diameter of the space station is 800 m, how many revolutions per minute are needed for the "artificial gravity" acceleration to be 9.80 m/s^2? (b) If the space station is a waiting area for travelers going to Mars, it might be desirable to simulate the acceleration due to gravity on the Martian surface (3.70 m/s^2). How many revolutions per minute are needed in this case?

5.54 • The Cosmo Clock 21 Ferris wheel in Yokohama, Japan, has a diameter of 100 m. Its name comes from its 60 arms, each of which can function as a second hand (so that it makes one revolution every 60.0 s). (a) Find the speed of the passengers when the Ferris wheel is rotating at this rate. (b) A passenger weighs 882 N at the weight-guessing booth on the ground. What is his apparent weight at the highest and at the lowest point on the Ferris wheel?

(c) What would be the time for one revolution if the passenger's apparent weight at the highest point were zero? (d) What then would be the passenger's apparent weight at the lowest point?

5.55 •• An airplane flies in a loop (a circular path in a vertical plane) of radius 150 m. The pilot's head always points toward the center of the loop. The speed of the airplane is not constant; the airplane goes slowest at the top of the loop and fastest at the bottom. (a) What is the speed of the airplane at the top of the loop, where the pilot feels weightless? (b) What is the apparent weight of the pilot at the bottom of the loop, where the speed of the airplane is 280 km/h? His true weight is 700 N.

5.56 •• A 50.0-kg stunt pilot who has been diving her airplane vertically pulls out of the dive by changing her course to a circle in a vertical plane. (a) If the plane's speed at the lowest point of the circle is 95.0 m/s, what is the minimum radius of the circle so that the acceleration at this point will not exceed 4.00g? (b) What is the apparent weight of the pilot at the lowest point of the pullout?

5.57 • **Stay Dry!** You tie a cord to a pail of water and swing the pail in a vertical circle of radius 0.600 m. What minimum speed must you give the pail at the highest point of the circle to avoid spilling water?

5.58 •• A bowling ball weighing 71.2 N (16.0 lb) is attached to the ceiling by a 3.80-m rope. The ball is pulled to one side and released; it then swings back and forth as a pendulum. As the rope swings through the vertical, the speed of the bowling ball is 4.20 m/s. At this instant, what are (a) the acceleration of the bowling ball, in magnitude and direction, and (b) the tension in the rope?

5.59 •• **BIO Effect on Blood of Walking.** While a person is walking, his arms swing through approximately a 45° angle in $\frac{1}{2}$ s. As a reasonable approximation, assume that the arm moves with constant speed during each swing. A typical arm is 70.0 cm long, measured from the shoulder joint. (a) What is the acceleration of a 1.0-g drop of blood in the fingertips at the bottom of the swing? (b) Draw a free-body diagram of the drop of blood in part (a). (c) Find the force that the blood vessel must exert on the drop of blood in part (a). Which way does this force point? (d) What force would the blood vessel exert if the arm were not swinging?

PROBLEMS

5.60 •• An adventurous archaeologist crosses between two rock cliffs by slowly going hand over hand along a rope stretched between the cliffs. He stops to rest at the middle of the rope (**Fig. P5.60**). The rope will break if the tension in it exceeds 2.50×10^4 N, and our hero's mass is 90.0 kg. (a) If the angle θ is 10.0°, what is the tension in the rope? (b) What is the smallest value θ can have if the rope is not to break?

Figure **P5.60**

5.61 ••• Two ropes are connected to a steel cable that supports a hanging weight (**Fig. P5.61**). (a) Draw a free-body diagram showing all of the forces acting at the knot that connects the two ropes to the steel cable. Based on your diagram, which of the two ropes will have the greater tension? (b) If the maximum tension either rope can sustain without breaking is 5000 N, determine the maximum value of the hanging weight that these ropes can safely support. Ignore the weight of the ropes and of the steel cable.

Figure **P5.61**

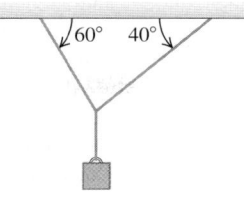

5.62 •• In **Fig. P5.62** a worker lifts a weight w by pulling down on a rope with a force $\vec{F}$. The upper pulley is attached to the ceiling by a chain, and the lower pulley is attached to the weight by another chain. Draw one or more free-body diagrams to find the tension in each chain and the magnitude of $\vec{F}$, in terms of w, if the weight is lifted at constant speed. Assume that the rope, pulleys, and chains have negligible weights.

Figure **P5.62**

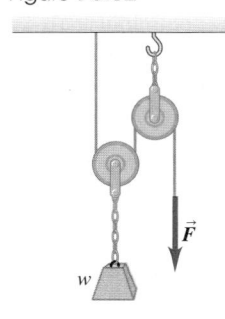

5.63 •• In a repair shop a truck engine that has mass 409 kg is held in place by four light cables (**Fig. P5.63**). Cable A is horizontal, cables B and D are vertical, and cable C makes an angle of $37.1°$ with a vertical wall. If the tension in cable A is 722 N, what are the tensions in cables B and C?

Figure **P5.63**

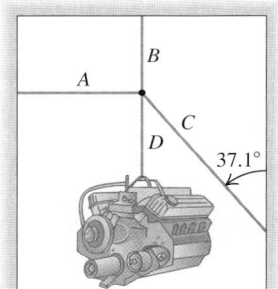

5.64 ••• A horizontal wire holds a solid uniform ball of mass m in place on a tilted ramp that rises $35.0°$ above the horizontal. The surface of this ramp is perfectly smooth, and the wire is directed away from the center of the ball (**Fig. P5.64**). (a) Draw a free-body diagram of the ball. (b) How hard does the surface of the ramp push on the ball? (c) What is the tension in the wire?

Figure **P5.64**

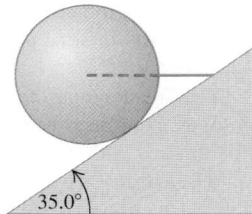

5.65 ••• A solid uniform 45.0-kg ball of diameter 32.0 cm is supported against a vertical, frictionless wall by a thin 30.0-cm wire of negligible mass (**Fig. P5.65**). (a) Draw a free-body diagram for the ball, and use the diagram to find the tension in the wire. (b) How hard does the ball push against the wall?

Figure **P5.65**

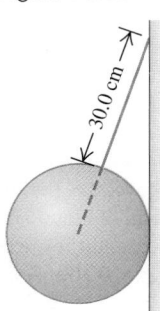

5.66 •• CP A box is sliding with a constant speed of 4.00 m/s in the $+x$-direction on a horizontal, frictionless surface. At $x = 0$ the box encounters a rough patch of the surface, and then the surface becomes even rougher. Between $x = 0$ and $x = 2.00$ m, the coefficient of kinetic friction between the box and the surface is 0.200; between $x = 2.00$ m and $x = 4.00$ m, it is 0.400. (a) What is the x-coordinate of the point where the box comes to rest? (b) How much time does it take the box to come to rest after it first encounters the rough patch at $x = 0$?

5.67 •• CP BIO **Forces During Chin-ups.** When you do a chin-up, you raise your chin just over a bar (the chinning bar), supporting yourself with only your arms. Typically, the body below the arms is raised by about 30 cm in a time of 1.0 s, starting from rest. Assume that the entire body of a 680-N person doing chin-ups is raised by 30 cm, and that half the 1.0 s is spent accelerating upward and the other half accelerating downward, uniformly in both cases. Draw a free-body diagram of the person's body, and use it to find the force his arms must exert on him during the accelerating part of the chin-up.

5.68 •• CP CALC A 2.00-kg box is suspended from the end of a light vertical rope. A time-dependent force is applied to the upper end of the rope, and the box moves upward with a velocity magnitude that varies in time according to $v(t) = (2.00 \text{ m/s}^2)t + (0.600 \text{ m/s}^3)t^2$. What is the tension in the rope when the velocity of the box is 9.00 m/s?

5.69 ••• CALC A 3.00-kg box that is several hundred meters above the earth's surface is suspended from the end of a short vertical rope of negligible mass. A time-dependent upward force is applied to the upper end of the rope and results in a tension in the rope of $T(t) = (36.0 \text{ N/s})t$. The box is at rest at $t = 0$. The only forces on the box are the tension in the rope and gravity. (a) What is the velocity of the box at (i) $t = 1.00$ s and (ii) $t = 3.00$ s? (b) What is the maximum distance that the box descends below its initial position? (c) At what value of t does the box return to its initial position?

5.70 •• CP A 5.00-kg box sits at rest at the bottom of a ramp that is 8.00 m long and is inclined at $30.0°$ above the horizontal. The coefficient of kinetic friction is $\mu_k = 0.40$, and the coefficient of static friction is $\mu_s = 0.43$. What constant force F, applied parallel to the surface of the ramp, is required to push the box to the top of the ramp in a time of 6.00 s?

5.71 •• Two boxes connected by a light horizontal rope are on a horizontal surface (Fig. E5.37). The coefficient of kinetic friction between each box and the surface is $\mu_k = 0.30$. Box B has mass 5.00 kg, and box A has mass m. A force F with magnitude 40.0 N and direction $53.1°$ above the horizontal is applied to the 5.00-kg box, and both boxes move to the right with $a = 1.50 \text{ m/s}^2$. (a) What is the tension T in the rope that connects the boxes? (b) What is m?

5.72 ••• A 6.00-kg box sits on a ramp that is inclined at 37.0° above the horizontal. The coefficient of kinetic friction between the box and the ramp is $\mu_k = 0.30$. What *horizontal* force is required to move the box up the incline with a constant acceleration of 3.60 m/s^2?

5.73 •• CP An 8.00-kg box sits on a ramp that is inclined at 33.0° above the horizontal. The coefficient of kinetic friction between the box and the surface of the ramp is $\mu_k = 0.300$. A constant *horizontal* force $F = 26.0 \text{ N}$ is applied to the box (**Fig. P5.73**), and the box moves down the ramp. If the box is initially at rest, what is its speed 2.00 s after the force is applied?

Figure **P5.73**

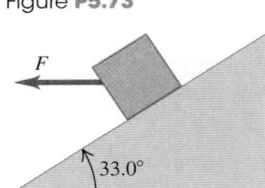

5.74 •• CP In **Fig. P5.74**, $m_1 = 20.0 \text{ kg}$ and $\alpha = 53.1°$. The coefficient of kinetic friction between the block of mass m_1 and the incline is $\mu_k = 0.40$. What must be the mass m_2 of the hanging block if it is to descend 12.0 m in the first 3.00 s after the system is released from rest?

Figure **P5.74**

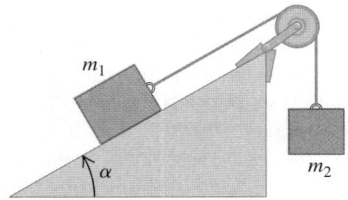

5.75 •• CP You place a book of mass 5.00 kg against a vertical wall. You apply a constant force $\vec{F}$ to the book, where $F = 96.0 \text{ N}$ and the force is at an angle of 60.0° above the horizontal (**Fig. P5.75**). The coefficient of kinetic friction between the book and the wall is 0.300. If the book is initially at rest, what is its speed after it has traveled 0.400 m up the wall?

Figure **P5.75**

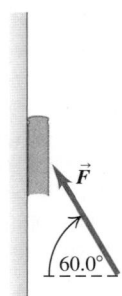

5.76 •• Block A in **Fig. P5.76** weighs 60.0 N. The coefficient of static friction between the block and the surface on which it rests is 0.25. The weight w is 12.0 N, and the system is in equilibrium. (a) Find the friction force exerted on block A. (b) Find the maximum weight w for which the system will remain in equilibrium.

Figure **P5.76**

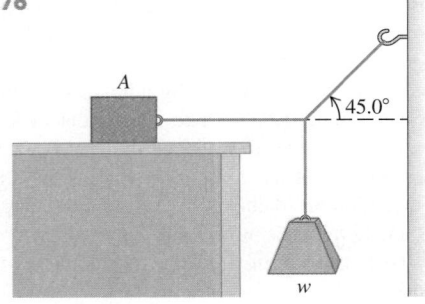

5.77 •• A block with mass m_1 is placed on an inclined plane with slope angle α and is connected to a hanging block with mass m_2 by a cord passing over a small, frictionless pulley (Fig. P5.74). The coefficient of static friction is μ_s, and the coefficient of kinetic friction is μ_k. (a) Find the value of m_2 for which the block of mass m_1 moves up the plane at constant speed once it is set in motion. (b) Find the value of m_2 for which the block of mass m_1 moves down the plane at constant speed once it is set in motion. (c) For what range of values of m_2 will the blocks remain at rest if they are released from rest?

5.78 •• BIO **The Flying Leap of a Flea.** High-speed motion pictures (3500 frames/second) of a jumping 210-μg flea yielded the data to plot the flea's acceleration as a function of time, as shown in **Fig. P5.78**. (See "The Flying Leap of the Flea," by M. Rothschild et al., *Scientific American*, November 1973.) This flea was about 2 mm long and jumped at a nearly vertical takeoff angle. Using the graph, (a) find the *initial* net external force on the flea. How does it compare to the flea's weight? (b) Find the *maximum* net external force on this jumping flea. When does this maximum force occur? (c) Use the graph to find the flea's maximum speed.

Figure **P5.78**

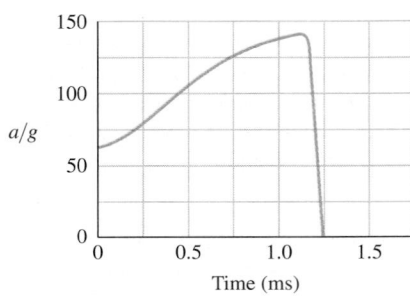

5.79 •• Block A in **Fig. P5.79** weighs 1.20 N, and block B weighs 3.60 N. The coefficient of kinetic friction between all surfaces is 0.300. Find the magnitude of the horizontal force $\vec{F}$ necessary to drag block B to the left at constant speed (a) if A rests on B and moves with it (Fig. P5.79a), (b) if A is held at rest (Fig. P5.79b).

Figure **P5.79**

(a)

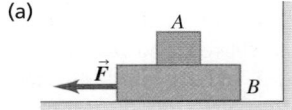

(b)

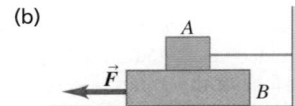

5.80 ••• CP **Elevator Design.** You are designing an elevator for a hospital. The force exerted on a passenger by the floor of the elevator is not to exceed 1.60 times the passenger's weight. The elevator accelerates upward with constant acceleration for a distance of 3.0 m and then starts to slow down. What is the maximum speed of the elevator?

5.81 ••• CP CALC You are standing on a bathroom scale in an elevator in a tall building. Your mass is 64 kg. The elevator starts from rest and travels upward with a speed that varies with time according to $v(t) = (3.0 \text{ m/s}^2)t + (0.20 \text{ m/s}^3)t^2$. When $t = 4.0 \text{ s}$, what is the reading on the bathroom scale?

5.82 ·· A hammer is hanging by a light rope from the ceiling of a bus. The ceiling is parallel to the roadway. The bus is traveling in a straight line on a horizontal street. You observe that the hammer hangs at rest with respect to the bus when the angle between the rope and the ceiling of the bus is 56.0°. What is the acceleration of the bus?

5.83 ·· A 40.0-kg packing case is initially at rest on the floor of a 1500-kg pickup truck. The coefficient of static friction between the case and the truck floor is 0.30, and the coefficient of kinetic friction is 0.20. Before each acceleration given below, the truck is traveling due north at constant speed. Find the magnitude and direction of the friction force acting on the case (a) when the truck accelerates at 2.20 m/s² northward and (b) when it accelerates at 3.40 m/s² southward.

5.84 ··· If the coefficient of static friction between a table and a uniform, massive rope is μ_s, what fraction of the rope can hang over the edge of the table without the rope sliding?

5.85 ··· Two identical 15.0-kg balls, each 25.0 cm in diameter, are suspended by two 35.0-cm wires (**Fig. P5.85**). The entire apparatus is supported by a single 18.0-cm wire, and the surfaces of the balls are perfectly smooth. (a) Find the tension in each of the three wires. (b) How hard does each ball push on the other one?

Figure **P5.85**

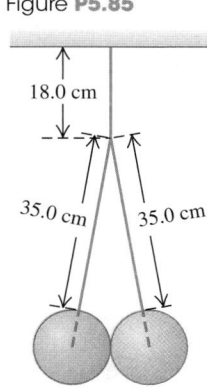

18.0 cm

35.0 cm   35.0 cm

5.86 · CP **Traffic Court.** You are called as an expert witness in a trial for a traffic violation. The facts are these: A driver slammed on his brakes and came to a stop with constant acceleration. Measurements of his tires and the skid marks on the pavement indicate that he locked his car's wheels, the car traveled 192 ft before stopping, and the coefficient of kinetic friction between the road and his tires was 0.750. He was charged with speeding in a 45-mi/h zone but pleads innocent. What is your conclusion: guilty or innocent? How fast was he going when he hit his brakes?

5.87 ··· Block A in **Fig. P5.87** weighs 1.90 N, and block B weighs 4.20 N. The coefficient of kinetic friction between all surfaces is 0.30. Find the magnitude of the horizontal force $\vec{F}$ necessary to drag block B to the left at constant speed if A and B are connected by a light, flexible cord passing around a fixed, frictionless pulley.

Figure **P5.87**

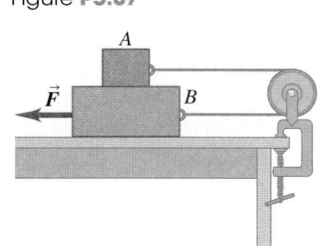

5.88 ·· CP **Losing Cargo.** A 12.0-kg box rests on the level bed of a truck. The coefficients of friction between the box and bed are $\mu_s = 0.19$ and $\mu_k = 0.15$. The truck stops at a stop sign and then starts to move with an acceleration of 2.20 m/s². If the box is 1.80 m from the rear of the truck when the truck starts, how much time elapses before the box falls off the truck? How far does the truck travel in this time?

5.89 ·· Block A in **Fig. P5.89** has mass 4.00 kg, and block B has mass 12.0 kg. The coefficient of kinetic friction between block B and the horizontal surface is 0.25. (a) What is the mass of block C

if block B is moving to the right and speeding up with an acceleration of 2.00 m/s²? (b) What is the tension in each cord when block B has this acceleration?

Figure **P5.89**

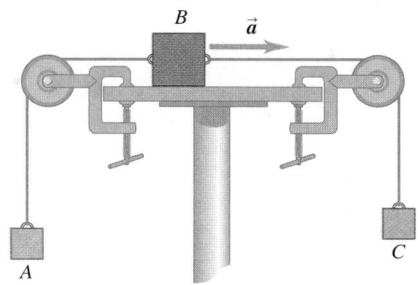

5.90 ·· Two blocks connected by a cord passing over a small, frictionless pulley rest on frictionless planes (**Fig. P5.90**). (a) Which way will the system move when the blocks are released from rest? (b) What is the acceleration of the blocks? (c) What is the tension in the cord?

Figure **P5.90**

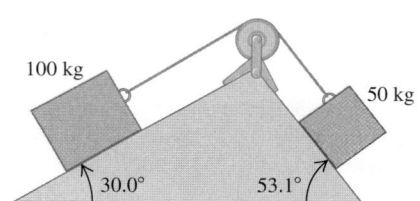

100 kg   50 kg

30.0°   53.1°

5.91 ·· In terms of m_1, m_2, and g, find the acceleration of each block in **Fig. P5.91**. There is no friction anywhere in the system.

Figure **P5.91**

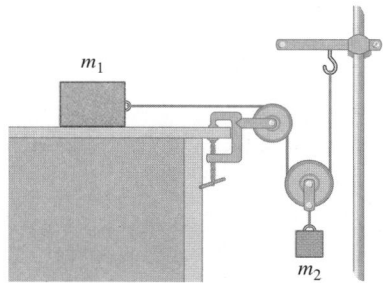

m_1

m_2

5.92 ··· Block B, with mass 5.00 kg, rests on block A, with mass 8.00 kg, which in turn is on a horizontal tabletop (**Fig. P5.92**). There is no friction between block A and the tabletop, but the coefficient of static friction between blocks A and B is 0.750. A light string attached to block A passes over a frictionless, massless pulley, and block C is suspended from the other end of the string. What is the largest mass that block C can have so that blocks A and B still slide together when the system is released from rest?

Figure **P5.92**

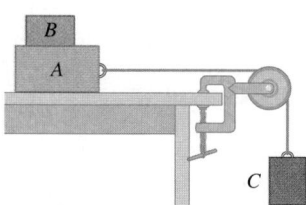

B

A

C

5.93 ••• Two objects, with masses 5.00 kg and 2.00 kg, hang 0.600 m above the floor from the ends of a cord that is 6.00 m long and passes over a frictionless pulley. Both objects start from rest. Find the maximum height reached by the 2.00-kg object.

5.94 •• **Friction in an Elevator.** You are riding in an elevator on the way to the 18th floor of your dormitory. The elevator is accelerating upward with $a = 1.90 \text{ m/s}^2$. Beside you is the box containing your new computer; the box and its contents have a total mass of 36.0 kg. While the elevator is accelerating upward, you push horizontally on the box to slide it at constant speed toward the elevator door. If the coefficient of kinetic friction between the box and the elevator floor is $\mu_k = 0.32$, what magnitude of force must you apply?

5.95 • A block is placed against the vertical front of a cart (**Fig. P5.95**). What acceleration must the cart have so that block A does not fall? The coefficient of static friction between the block and the cart is μ_s. How would an observer on the cart describe the behavior of the block?

Figure **P5.95**

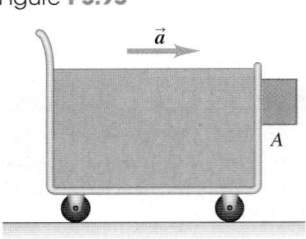

5.96 ••• Two blocks, with masses 4.00 kg and 8.00 kg, are connected by a string and slide down a 30.0° inclined plane (**Fig. P5.96**). The coefficient of kinetic friction between the 4.00-kg block and the plane is 0.25; that between the 8.00-kg block and the plane is 0.35. Calculate (a) the acceleration of each block and (b) the tension in the string. (c) What happens if the positions of the blocks are reversed, so that the 4.00-kg block is uphill from the 8.00-kg block?

Figure **P5.96**

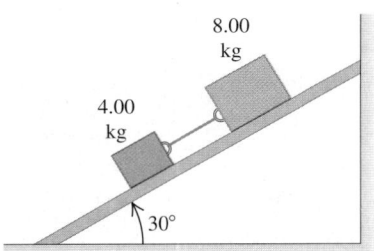

5.97 ••• Block A, with weight $3w$, slides down an inclined plane S of slope angle 36.9° at a constant speed while plank B, with weight w, rests on top of A. The plank is attached by a cord to the wall (**Fig. P5.97**). (a) Draw a diagram of all the forces acting on block A. (b) If the coefficient of kinetic friction is the same between A and B and between S and A, determine its value.

Figure **P5.97**

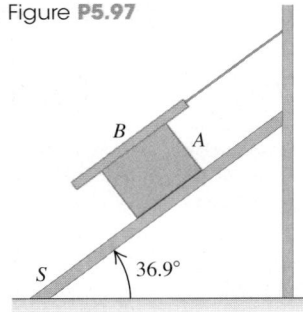

5.98 •• Jack sits in the chair of a Ferris wheel that is rotating at a constant 0.100 rev/s. As Jack passes through the highest point of his circular path, the upward force that the chair exerts on him is equal to one-fourth of his weight. What is the radius of the circle in which Jack travels? Treat him as a point mass.

5.99 ••• **Banked Curve I.** A curve with a 120-m radius on a level road is banked at the correct angle for a speed of 20 m/s. If an automobile rounds this curve at 30 m/s, what is the minimum coefficient of static friction needed between tires and road to prevent skidding?

5.100 •• **Banked Curve II.** Consider a wet roadway banked as in Example 5.22 (Section 5.4), where there is a coefficient of static friction of 0.30 and a coefficient of kinetic friction of 0.25 between the tires and the roadway. The radius of the curve is $R = 50$ m. (a) If the bank angle is $\beta = 25°$, what is the *maximum* speed the automobile can have before sliding *up* the banking? (b) What is the *minimum* speed the automobile can have before sliding *down* the banking?

5.101 ••• Blocks A, B, and C are placed as in **Fig. P5.101** and connected by ropes of negligible mass. Both A and B weigh 25.0 N each, and the coefficient of kinetic friction between each block and the surface is 0.35. Block C descends with constant velocity. (a) Draw separate free-body diagrams showing the forces acting on A and on B. (b) Find the tension in the rope connecting blocks A and B. (c) What is the weight of block C? (d) If the rope connecting A and B were cut, what would be the acceleration of C?

Figure **P5.101**

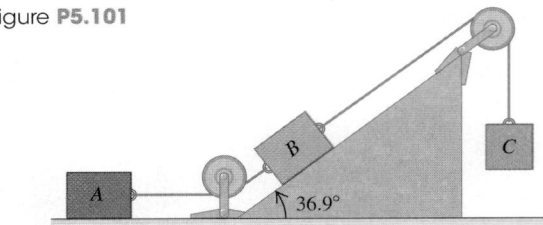

5.102 •• You are riding in a school bus. As the bus rounds a flat curve at constant speed, a lunch box with mass 0.500 kg, suspended from the ceiling of the bus by a string 1.80 m long, is found to hang at rest relative to the bus when the string makes an angle of 30.0° with the vertical. In this position the lunch box is 50.0 m from the curve's center of curvature. What is the speed v of the bus?

5.103 •• **CALC** You throw a rock downward into water with a speed of $3mg/k$, where k is the coefficient in Eq. (5.5). Assume that the relationship between fluid resistance and speed is as given in Eq. (5.5), and calculate the speed of the rock as a function of time.

5.104 ••• A 4.00-kg block is attached to a vertical rod by means of two strings. When the system rotates about the axis of the rod, the strings are extended as shown in **Fig. P5.104** and the tension in the upper string is 80.0 N. (a) What is the tension in the lower cord? (b) How many revolutions per minute does the system make? (c) Find the number of revolutions per minute at which the lower cord just goes slack. (d) Explain what happens if the number of revolutions per minute is less than that in part (c).

Figure **P5.104**

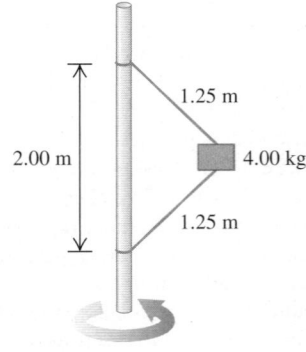

5.105 •• On the ride "Spindletop" at the amusement park Six Flags Over Texas, people stood against the inner wall of a hollow vertical cylinder with radius 2.5 m. The cylinder started to rotate, and when it reached a constant rotation rate of 0.60 rev/s, the floor dropped about 0.5 m. The people remained pinned against the wall without touching the floor. (a) Draw a force diagram for a person on this ride after the floor has dropped. (b) What minimum coefficient of static friction was required for the person not to slide downward to the new position of the floor? (c) Does your answer in part (b) depend on the person's mass? (*Note:* When such a ride is over, the cylinder is slowly brought to rest. As it slows down, people slide down the walls to the floor.)

5.106 •• A 70-kg person rides in a 30-kg cart moving at 12 m/s at the top of a hill that is in the shape of an arc of a circle with a radius of 40 m. (a) What is the apparent weight of the person as the cart passes over the top of the hill? (b) Determine the maximum speed that the cart can travel at the top of the hill without losing contact with the surface. Does your answer depend on the mass of the cart or the mass of the person? Explain.

5.107 •• A small bead can slide without friction on a circular hoop that is in a vertical plane and has a radius of 0.100 m. The hoop rotates at a constant rate of 4.00 rev/s about a vertical diameter (**Fig. P5.107**). (a) Find the angle β at which the bead is in vertical equilibrium. (It has a radial acceleration toward the axis.) (b) Is it possible for the bead to "ride" at the same elevation as the center of the hoop? (c) What will happen if the hoop rotates at 1.00 rev/s?

Figure **P5.107**

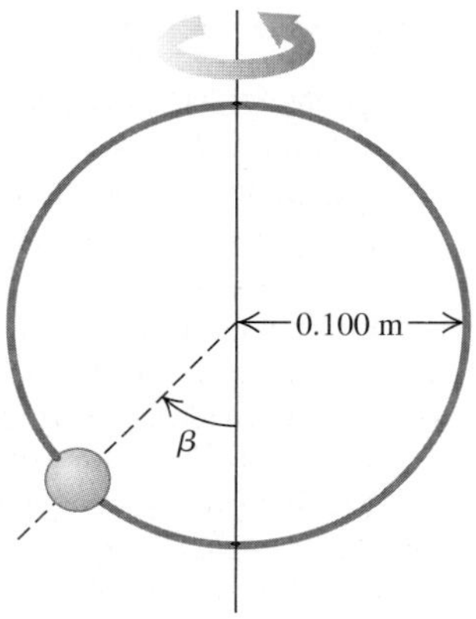

5.108 •• A physics major is working to pay her college tuition by performing in a traveling carnival. She rides a motorcycle inside a hollow, transparent plastic sphere. After gaining sufficient speed, she travels in a vertical circle with radius 13.0 m. She has mass 70.0 kg, and her motorcycle has mass 40.0 kg. (a) What minimum speed must she have at the top of the circle for the motorcycle tires to remain in contact with the sphere? (b) At the bottom of the circle, her speed is twice the value calculated in part (a). What is the magnitude of the normal force exerted on the motorcycle by the sphere at this point?

5.109 •• DATA In your physics lab, a block of mass m is at rest on a horizontal surface. You attach a light cord to the block and apply a horizontal force to the free end of the cord. You find that the block remains at rest until the tension T in the cord exceeds 20.0 N. For $T > 20.0$ N, you measure the acceleration of the block when T is maintained at a constant value, and you plot the results (**Fig. P5.109**). The equation for the straight line that best fits your data is $a = [0.182\ \text{m}/(\text{N} \cdot \text{s}^2)]T - 2.842\ \text{m/s}^2$. For this block and surface, what are (a) the coefficient of static friction and (b) the coefficient of kinetic friction? (c) If the experiment were done on the earth's moon, where g is much smaller than on the earth, would the graph of a versus T still be fit well by a straight line? If so, how would the slope and intercept of the line differ from the values in Fig. P5.109? Or, would each of them be the same?

Figure **P5.109**

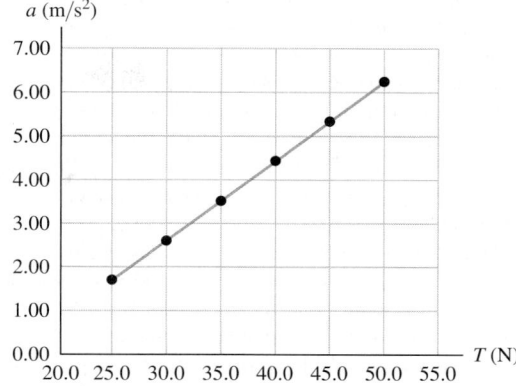

5.110 •• DATA A road heading due east passes over a small hill. You drive a car of mass m at constant speed v over the top of the hill, where the shape of the roadway is well approximated as an arc of a circle with radius R. Sensors have been placed on the road surface there to measure the downward force that cars exert on the surface at various speeds. The table gives values of this force versus speed for your car:

Speed (m/s)	6.00	8.00	10.0	12.0	14.0	16.0
Force (N)	8100	7690	7050	6100	5200	4200

Treat the car as a particle. (a) Plot the values in such a way that they are well fitted by a straight line. You might need to raise the speed, the force, or both to some power. (b) Use your graph from part (a) to calculate m and R. (c) What maximum speed can the car have at the top of the hill and still not lose contact with the road?

5.111 •• DATA You are an engineer working for a manufacturing company. You are designing a mechanism that uses a cable to drag heavy metal blocks a distance of 8.00 m along a ramp that is sloped at 40.0° above the horizontal. The coefficient of kinetic friction between these blocks and the incline is $\mu_k = 0.350$. Each block has a mass of 2170 kg. The block will be placed on the bottom of the ramp, the cable will be attached, and the block will then be given just enough of a momentary push to overcome static friction. The block is then to accelerate at a constant rate to move the 8.00 m in 4.20 s. The cable is made of wire rope and is parallel to the ramp surface. The table gives the breaking strength of the cable as a function of its diameter; the safe load tension, which is 20% of the breaking strength; and the mass per meter of the cable:

Cable Diameter (in.)	Breaking Strength (kN)	Safe Load (kN)	Mass per Meter (kg/m)
$\frac{1}{4}$	24.4	4.89	0.16
$\frac{3}{8}$	54.3	10.9	0.36
$\frac{1}{2}$	95.2	19.0	0.63
$\frac{5}{8}$	149	29.7	0.98
$\frac{3}{4}$	212	42.3	1.41
$\frac{7}{8}$	286	57.4	1.92
1	372	74.3	2.50

Source: www.engineeringtoolbox.com

(a) What is the minimum diameter of the cable that can be used to pull a block up the ramp without exceeding the safe load value of the tension in the cable? Ignore the mass of the cable, and select the diameter from those listed in the table. (b) You need to know safe load values for diameters that aren't in the table, so you hypothesize that the breaking strength and safe load limit are proportional to the cross-sectional area of the cable. Draw a graph that tests this hypothesis, and discuss its accuracy. What is your estimate of the safe load value for a cable with diameter $\frac{9}{16}$ in.? (c) The coefficient of static friction between the crate and the ramp is $\mu_s = 0.620$, which is nearly twice the value of the coefficient of kinetic friction. If the machinery jams and the block stops in the middle of the ramp, what is the tension in the cable? Is it larger or smaller than the value when the block is moving? (d) Is the actual tension in the cable, at its upper end, larger or smaller than the value calculated when you ignore the mass of the cable? If the cable is 9.00 m long, how accurate is it to ignore the cable's mass?

CHALLENGE PROBLEMS

5.112 ••• **Moving Wedge.** A wedge with mass M rests on a frictionless, horizontal tabletop. A block with mass m is placed on the wedge (**Fig. P5.112a**). There is no friction between the block and the wedge. The system is released from rest. (a) Calculate the acceleration of the wedge and the horizontal and vertical components of the acceleration of the block. (b) Do your answers to part (a) reduce to the correct results when M is very large? (c) As seen by a stationary observer, what is the shape of the trajectory of the block?

Figure **P5.112**

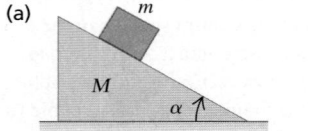
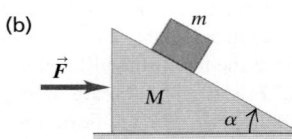

5.113 ••• A wedge with mass M rests on a frictionless, horizontal tabletop. A block with mass m is placed on the wedge, and a horizontal force $\vec{F}$ is applied to the wedge (Fig. P5.112b). What must the magnitude of $\vec{F}$ be if the block is to remain at a constant height above the tabletop?

5.114 ••• **Double Atwood's Machine.** In **Fig. P5.114** masses m_1 and m_2 are connected by a light string A over a light, frictionless pulley B. The axle of pulley B is connected by a light string C over a light, frictionless pulley D to a mass m_3. Pulley D is suspended from the ceiling by an attachment to its axle. The system is released from rest. In terms of m_1, m_2, m_3, and g, what are (a) the acceleration of block m_3; (b) the acceleration of pulley B; (c) the acceleration of block m_1; (d) the acceleration of block m_2; (e) the tension in string A; (f) the tension in string C? (g) What do your expressions give for the special case of $m_1 = m_2$ and $m_3 = m_1 + m_2$? Is this reasonable?

Figure **P5.114**

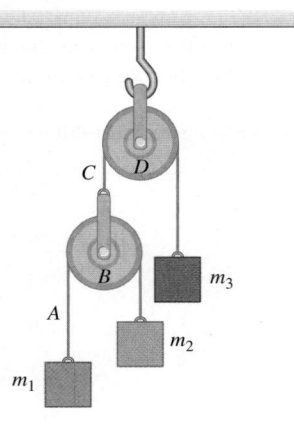

5.115 ••• A ball is held at rest at position A in **Fig. P5.115** by two light strings. The horizontal string is cut, and the ball starts swinging as a pendulum. Position B is the farthest to the right that the ball can go as it swings back and forth. What is the ratio of the tension in the supporting string at B to its value at A before the string was cut?

Figure **P5.115**

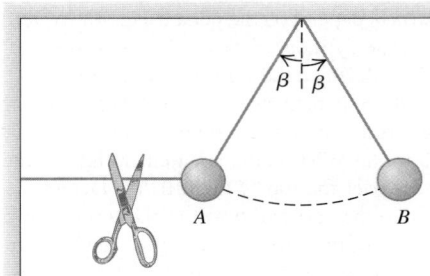

FRICTION AND CLIMBING SHOES. Shoes made for the sports of bouldering and rock climbing are designed to provide a great deal of friction between the foot and the surface of the ground. Such shoes on smooth rock might have a coefficient of static friction of 1.2 and a coefficient of kinetic friction of 0.90.

5.116 For a person wearing these shoes, what's the maximum angle (with respect to the horizontal) of a smooth rock that can be walked on without slipping? (a) 42°; (b) 50°; (c) 64°; (d) larger than 90°.

5.117 If the person steps onto a smooth rock surface that's inclined at an angle large enough that these shoes begin to slip, what will happen? (a) She will slide a short distance and stop; (b) she will accelerate down the surface; (c) she will slide down the surface at constant speed; (d) we can't tell what will happen without knowing her mass.

5.118 A person wearing these shoes stands on a smooth, horizontal rock. She pushes against the ground to begin running. What is the maximum horizontal acceleration she can have without slipping? (a) 0.20g; (b) 0.75g; (c) 0.90g; (d) 1.2g.

Answers

Chapter Opening Question **?**

(iii) The upward force exerted by the air has the same magnitude as the force of gravity. Although the seed and pappus are descending, their vertical velocity is constant, so their vertical acceleration is zero. According to Newton's first law, the net vertical force on the seed and pappus must also be zero. The individual vertical forces must balance.

Test Your Understanding Questions

5.1 (ii) The two cables are arranged symmetrically, so the tension in either cable has the same magnitude T. The vertical component of the tension from each cable is $T\sin 45°$ (or, equivalently, $T\cos 45°$), so Newton's first law applied to the vertical forces tells us that $2T\sin 45° - w = 0$. Hence $T = w/(2\sin 45°) = w/\sqrt{2} = 0.71w$. Each cable supports half of the weight of the traffic light, but the tension is greater than $w/2$ because only the vertical component of the tension counteracts the weight.

5.2 (ii) No matter what the instantaneous velocity of the glider, its acceleration is constant and has the value found in Example 5.12. In the same way, the acceleration of a body in free fall is the same whether it is ascending, descending, or at the high point of its motion (see Section 2.5).

5.3 (a): (i), (iii); (b): (ii), (iv); (c): (v) In situations (i) and (iii) the box is not accelerating (so the net force on it must be zero) and no other force is acting parallel to the horizontal surface; hence no friction force is needed to prevent sliding. In situations (ii) and (iv) the box would start to slide over the surface if no friction were present, so a static friction force must act to prevent this. In situation (v) the box is sliding over a rough surface, so a kinetic friction force acts on it.

5.4 (iii) A satellite of mass m orbiting the earth at speed v in an orbit of radius r has an acceleration of magnitude v^2/r, so the net force acting on it from the earth's gravity has magnitude $F = mv^2/r$. The farther the satellite is from the earth, the greater the value of r, the smaller the value of v, and hence the smaller the values of v^2/r and of F. In other words, the earth's gravitational force decreases with increasing distance.

Bridging Problem

(a) $T_{\max} = 2\pi\sqrt{\dfrac{h(\cos\beta + \mu_s\sin\beta)}{g\tan\beta(\sin\beta - \mu_s\cos\beta)}}$

(b) $T_{\min} = 2\pi\sqrt{\dfrac{h(\cos\beta - \mu_s\sin\beta)}{g\tan\beta(\sin\beta + \mu_s\cos\beta)}}$

? A baseball pitcher does work with his throwing arm to give the ball a property called kinetic energy, which depends on the ball's mass and speed. Which has the greatest kinetic energy? (i) A ball of mass 0.145 kg moving at 20.0 m/s; (ii) a smaller ball of mass 0.0145 kg moving at 200 m/s; (iii) a larger ball of mass 1.45 kg moving at 2.00 m/s; (iv) all three balls have the same kinetic energy; (v) it depends on the directions in which the balls move.

6 WORK AND KINETIC ENERGY

S uppose you try to find the speed of an arrow that has been shot from a bow. You apply Newton's laws and all the problem-solving techniques that we've learned, but you run across a major stumbling block: After the archer releases the arrow, the bow string exerts a *varying* force that depends on the arrow's position. As a result, the simple methods that we've learned aren't enough to calculate the speed. Never fear; we aren't by any means finished with mechanics, and there are other methods for dealing with such problems.

The new method that we're about to introduce uses the ideas of *work* and *energy*. The importance of the energy idea stems from the *principle of conservation of energy*: Energy is a quantity that can be converted from one form to another but cannot be created or destroyed. In an automobile engine, chemical energy stored in the fuel is converted partially to the energy of the automobile's motion and partially to thermal energy. In a microwave oven, electromagnetic energy obtained from your power company is converted to thermal energy of the food being cooked. In these and all other processes, the *total* energy—the sum of all energy present in all different forms—remains the same. No exception has ever been found.

We'll use the energy idea throughout the rest of this book to study a tremendous range of physical phenomena. This idea will help you understand how automotive engines work, how a camera's flash unit can produce a short burst of light, and the meaning of Einstein's famous equation $E = mc^2$.

In this chapter, though, our concentration will be on mechanics. We'll learn about one important form of energy called *kinetic energy,* or energy of motion, and how it relates to the concept of *work*. We'll also consider *power*, which is the time rate of doing work. In Chapter 7 we'll expand these ideas into a deeper understanding of the concepts of energy and the conservation of energy.

6.1 WORK

You'd probably agree that it's hard work to pull a heavy sofa across the room, to lift a stack of encyclopedias from the floor to a high shelf, or to push a stalled car off the road. Indeed, all of these examples agree with the everyday meaning of *work*—any activity that requires muscular or mental effort.

In physics, work has a much more precise definition. By making use of this definition we'll find that in any motion, no matter how complicated, the total work done on a particle by all forces that act on it equals the change in its *kinetic energy*—a quantity that's related to the particle's mass and speed. This relationship holds even when the forces acting on the particle aren't constant, a situation that can be difficult or impossible to handle with the techniques you learned in Chapters 4 and 5. The ideas of work and kinetic energy enable us to solve problems in mechanics that we could not have attempted before.

In this section we'll see how work is defined and how to calculate work in a variety of situations involving *constant* forces. Later in this chapter we'll relate work and kinetic energy, and then apply these ideas to problems in which the forces are *not* constant.

The three examples of work described above—pulling a sofa, lifting encyclopedias, and pushing a car—have something in common. In each case you do work by exerting a *force* on a body while that body *moves* from one place to another—that is, undergoes a *displacement* (**Fig. 6.1**). You do more work if the force is greater (you push harder on the car) or if the displacement is greater (you push the car farther down the road).

The physicist's definition of work is based on these observations. Consider a body that undergoes a displacement of magnitude s along a straight line. (For now, we'll assume that any body we discuss can be treated as a particle so that we can ignore any rotation or changes in shape of the body.) While the body moves, a constant force $\vec{F}$ acts on it in the same direction as the displacement $\vec{s}$ (**Fig. 6.2**). We define the **work** W done by this constant force under these circumstances as the product of the force magnitude F and the displacement magnitude s:

$$W = Fs \quad \text{(constant force in direction of straight-line displacement)} \quad (6.1)$$

The work done on the body is greater if either the force F or the displacement s is greater, in agreement with our observations above.

CAUTION **Work = W, weight = w** Don't confuse uppercase W (work) with lowercase w (weight). Though the symbols are similar, work and weight are different quantities. ▮

The SI unit of work is the **joule** (abbreviated J, pronounced "jool," and named in honor of the 19th-century English physicist James Prescott Joule). From Eq. (6.1) we see that in any system of units, the unit of work is the unit of force multiplied by the unit of distance. In SI units the unit of force is the newton and the unit of distance is the meter, so 1 joule is equivalent to 1 *newton-meter* (N·m):

$$1 \text{ joule} = (1 \text{ newton})(1 \text{ meter}) \quad \text{or} \quad 1 \text{ J} = 1 \text{ N·m}$$

If you lift an object with a weight of 1 N (about the weight of a medium-sized apple) a distance of 1 m at a constant speed, you exert a 1-N force on the object in the same direction as its 1-m displacement and so do 1 J of work on it.

As an illustration of Eq. (6.1), think of a person pushing a stalled car. If he pushes the car through a displacement $\vec{s}$ with a constant force $\vec{F}$ in the direction of motion, the amount of work he does on the car is given by Eq. (6.1): W = Fs.

6.1 These people are doing work as they push on the car because they exert a force on the car as it moves.

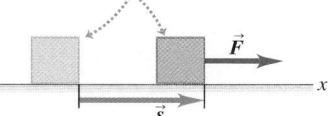
6.2 The work done by a constant force acting in the same direction as the displacement.

If a body moves through a displacement $\vec{s}$ while a constant force $\vec{F}$ acts on it in the same direction ...

... the work done by the force on the body is W = Fs.

BIO **Application Work and Muscle Fibers** Our ability to do work with our bodies comes from our skeletal muscles. The fiberlike cells of skeletal muscle, shown in this micrograph, can shorten, causing the muscle as a whole to contract and to exert force on the tendons to which it attaches. Muscle can exert a force of about 0.3 N per square millimeter of cross-sectional area: The greater the cross-sectional area, the more fibers the muscle has and the more force it can exert when it contracts.

6.3 The work done by a constant force acting at an angle to the displacement.

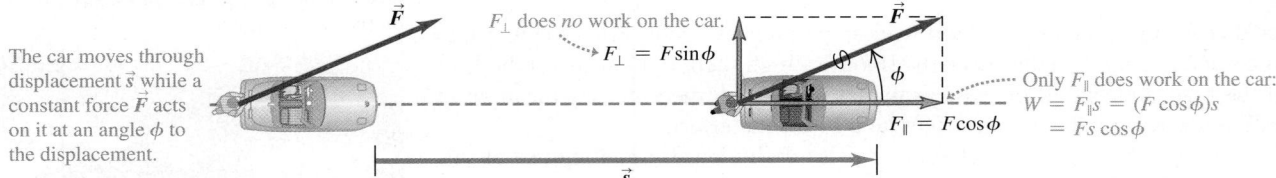

The car moves through displacement $\vec{s}$ while a constant force $\vec{F}$ acts on it at an angle ϕ to the displacement.

$F_\perp$ does *no* work on the car.
$F_\perp = F\sin\phi$

Only $F_\parallel$ does work on the car:
$W = F_\parallel s = (F\cos\phi)s$
$= Fs\cos\phi$

$F_\parallel = F\cos\phi$

But what if the person pushes at an angle ϕ to the car's displacement (**Fig. 6.3**)? Then $\vec{F}$ has a component $F_\parallel = F\cos\phi$ in the direction of the displacement $\vec{s}$ and a component $F_\perp = F\sin\phi$ that acts perpendicular to $\vec{s}$. (Other forces must act on the car so that it moves along $\vec{s}$, not in the direction of $\vec{F}$. We're interested in only the work that the person does, however, so we'll consider only the force he exerts.) Only the parallel component $F_\parallel$ is effective in moving the car, so we define the work as the product of this force component and the magnitude of the displacement. Hence $W = F_\parallel s = (F\cos\phi)s$, or

Work done on a particle by **constant force** $\vec{F}$ during **straight-line displacement** $\vec{s}$ ····· Magnitude of $\vec{F}$

$$W = Fs\cos\phi \longleftarrow \text{Angle between } \vec{F} \text{ and } \vec{s}$$ (6.2)

····· Magnitude of $\vec{s}$

If $\phi = 0$, so that $\vec{F}$ and $\vec{s}$ are in the same direction, then $\cos\phi = 1$ and we are back to Eq. (6.1).

Equation (6.2) has the form of the *scalar product* of two vectors, which we introduced in Section 1.10: $\vec{A} \cdot \vec{B} = AB\cos\phi$. You may want to review that definition. Hence we can write Eq. (6.2) more compactly as

Work done on a particle by **constant force** $\vec{F}$ during **straight-line displacement** $\vec{s}$

$$W = \vec{F} \cdot \vec{s}$$ (6.3)

Scalar product (dot product) of vectors $\vec{F}$ and $\vec{s}$

CAUTION **Work is a scalar** An essential point: Work is a *scalar* quantity, even though it's calculated from two vector quantities (force and displacement). A 5-N force toward the east acting on a body that moves 6 m to the east does the same amount of work as a 5-N force toward the north acting on a body that moves 6 m to the north. ▮

EXAMPLE 6.1 **WORK DONE BY A CONSTANT FORCE**

(a) Steve exerts a steady force of magnitude 210 N (about 47 lb) on the stalled car in Fig. 6.3 as he pushes it a distance of 18 m. The car also has a flat tire, so to make the car track straight Steve must push at an angle of 30° to the direction of motion. How much work does Steve do? (b) In a helpful mood, Steve pushes a second stalled car with a steady force $\vec{F} = (160\text{ N})\hat{\imath} - (40\text{ N})\hat{\jmath}$. The displacement of the car is $\vec{s} = (14\text{ m})\hat{\imath} + (11\text{ m})\hat{\jmath}$. How much work does Steve do in this case?

SOLUTION

IDENTIFY and SET UP: In both parts (a) and (b), the target variable is the work W done by Steve. In each case the force is constant and the displacement is along a straight line, so we can use Eq. (6.2) or (6.3). The angle between $\vec{F}$ and $\vec{s}$ is given in part (a), so we can apply Eq. (6.2) directly. In part (b) both $\vec{F}$ and $\vec{s}$ are given in terms

of components, so it's best to calculate the scalar product by using Eq. (1.19): $\vec{A} \cdot \vec{B} = A_xB_x + A_yB_y + A_zB_z$.

EXECUTE: (a) From Eq. (6.2),

$$W = Fs\cos\phi = (210\text{ N})(18\text{ m})\cos 30° = 3.3 \times 10^3\text{ J}$$

(b) The components of $\vec{F}$ are $F_x = 160\text{ N}$ and $F_y = -40\text{ N}$, and the components of $\vec{s}$ are $x = 14\text{ m}$ and $y = 11\text{ m}$. (There are no z-components for either vector.) Hence, using Eqs. (1.19) and (6.3), we have

$$W = \vec{F} \cdot \vec{s} = F_xx + F_yy$$
$$= (160\text{ N})(14\text{ m}) + (-40\text{ N})(11\text{ m})$$
$$= 1.8 \times 10^3\text{ J}$$

EVALUATE: In each case the work that Steve does is more than 1000 J. This shows that 1 joule is a rather small amount of work.

6.4 A constant force $\vec{F}$ can do positive, negative, or zero work depending on the angle between $\vec{F}$ and the displacement $\vec{s}$.

Direction of Force (or Force Component)	Situation	Force Diagram
(a) **Force $\vec{F}$ has a component in direction of displacement:** $W = F_{\parallel}s = (F\cos\phi)s$ Work is *positive*.		$F_{\parallel} = F\cos\phi$
(b) **Force $\vec{F}$ has a component opposite to direction of displacement:** $W = F_{\parallel}s = (F\cos\phi)s$ Work is *negative* (because $F\cos\phi$ is negative for $90° < \phi < 180°$).		$F_{\parallel} = F\cos\phi$
(c) **Force $\vec{F}$ (or force component $F_{\perp}$) is perpendicular to direction of displacement:** The force (or force component) does *no* work on the object.		$\phi = 90°$

Work: Positive, Negative, or Zero

In Example 6.1 the work done in pushing the cars was positive. But it's important to understand that work can also be negative or zero. This is the essential way in which work as defined in physics differs from the "everyday" definition of work. When the force has a component in the *same direction* as the displacement (ϕ between 0° and 90°), $\cos\phi$ in Eq. (6.2) is positive and the work W is *positive* (**Fig. 6.4a**). When the force has a component *opposite* to the displacement (ϕ between 90° and 180°), $\cos\phi$ is negative and the work is *negative* (Fig. 6.4b). When the force is *perpendicular* to the displacement, $\phi = 90°$ and the work done by the force is *zero* (Fig. 6.4c). The cases of zero work and negative work bear closer examination, so let's look at some examples.

There are many situations in which forces act but do zero work. You might think it's "hard work" to hold a barbell motionless in the air for 5 minutes (**Fig. 6.5**). But in fact, you aren't doing any work on the barbell because there is no displacement. (Holding the barbell requires you to keep the muscles of your arms contracted, and this consumes energy stored in carbohydrates and fat within your body. As these energy stores are used up, your muscles feel fatigued even though you do no work on the barbell.) Even when you carry a book while you walk with constant velocity on a level floor, you do no work on the book. It has a displacement, but the (vertical) supporting force that you exert on the book has no component in the direction of the (horizontal) motion. Then $\phi = 90°$ in Eq. (6.2), and $\cos\phi = 0$. When a body slides along a surface, the work done on the body by the normal force is zero; and when a ball on a string moves in uniform circular motion, the work done on the ball by the tension in the string is also zero. In both cases the work is zero because the force has no component in the direction of motion.

What does it mean to do *negative* work? The answer comes from Newton's third law of motion. When a weightlifter lowers a barbell as in **Fig. 6.6a** (next page), his hands and the barbell move together with the same displacement $\vec{s}$. The barbell exerts a force $\vec{F}_{\text{barbell on hands}}$ on his hands in the same direction as the hands' displacement, so the work done by the *barbell* on his *hands* is positive (Fig. 6.6b). But by Newton's third law the weightlifter's hands exert an equal and opposite force $\vec{F}_{\text{hands on barbell}} = -\vec{F}_{\text{barbell on hands}}$ on the barbell (Fig. 6.6c). This force, which keeps the barbell from crashing to the floor, acts opposite to the barbell's displacement. Thus the work done by his *hands* on the *barbell* is negative. Because the weightlifter's hands and the barbell have the same displacement, the

6.5 A weightlifter does no work on a barbell as long as he holds it stationary.

···The weightlifter exerts an upward force on the barbell ...

... but because the barbell is stationary (its displacement is zero), he does no work on it.

6.6 This weightlifter's hands do negative work on a barbell as the barbell does positive work on his hands.

(a) A weightlifter lowers a barbell to the floor.

(b) The barbell does *positive* work on the weightlifter's hands.

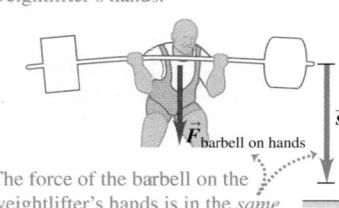

The force of the barbell on the weightlifter's hands is in the *same* direction as the hands' displacement.

(c) The weightlifter's hands do *negative* work on the barbell.

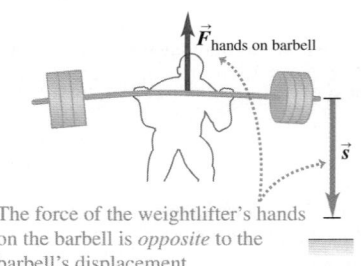

$\vec{F}_{\text{hands on barbell}}$

The force of the weightlifter's hands on the barbell is *opposite* to the barbell's displacement.

work that his hands do on the barbell is just the negative of the work that the barbell does on his hands. In general, when one body does negative work on a second body, the second body does an equal amount of *positive* work on the first body.

CAUTION **Keep track of who's doing the work** We always speak of work done *on* a particular body *by* a specific force. Always specify exactly what force is doing the work. When you lift a book, you exert an upward force on it and the book's displacement is upward, so the work done by the lifting force on the book is positive. But the work done by the *gravitational* force (weight) on a book being lifted is *negative* because the downward gravitational force is opposite to the upward displacement. ▮

Total Work

How do we calculate work when *several* forces act on a body? One way is to use Eq. (6.2) or (6.3) to compute the work done by each separate force. Then, because work is a scalar quantity, the *total* work W_{tot} done on the body by all the forces is the algebraic sum of the quantities of work done by the individual forces. An alternative way to find the total work W_{tot} is to compute the vector sum of the forces (that is, the net force) and then use this vector sum as $\vec{F}$ in Eq. (6.2) or (6.3). The following example illustrates both of these techniques.

DATA *SPEAKS*

Positive, Negative, and Zero Work

When students were given a problem that required them to find the work done by a constant force during a straight-line displacement, more than 59% gave an incorrect answer. Common errors:

- Forgetting that a force does negative work if it acts opposite to the direction of the object's displacement.
- Forgetting that, even if a force is present, it does zero work if it acts perpendicular to the direction of the displacement.

EXAMPLE 6.2 **WORK DONE BY SEVERAL FORCES**

A farmer hitches her tractor to a sled loaded with firewood and pulls it a distance of 20 m along level ground (**Fig. 6.7a**). The total weight of sled and load is 14,700 N. The tractor exerts a constant 5000-N force at an angle of 36.9° above the horizontal. A 3500-N friction force opposes the sled's motion. Find the work done by each force acting on the sled and the total work done by all the forces.

6.7 Calculating the work done on a sled of firewood being pulled by a tractor.

(a)

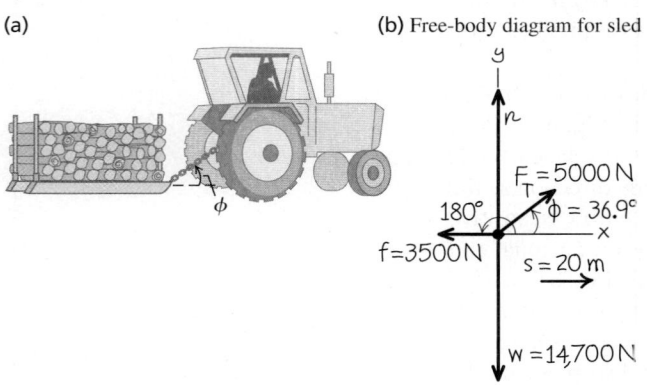

(b) Free-body diagram for sled

$F_T = 5000\,\text{N}$
$\phi = 36.9°$
$180°$
$f = 3500\,\text{N}$
$s = 20\,\text{m}$
$w = 14{,}700\,\text{N}$

SOLUTION

IDENTIFY and SET UP: Each force is constant and the sled's displacement is along a straight line, so we can use the ideas of this section to calculate the work. We'll find the total work in two ways: (1) by adding the work done on the sled by each force and (2) by finding the work done by the net force on the sled. We first draw a free-body diagram showing all of the forces acting on the sled, and we choose a coordinate system (Fig. 6.7b). For each force—weight, normal force, force of the tractor, and friction force—we know the angle between the displacement (in the positive *x*-direction) and the force. Hence we can use Eq. (6.2) to calculate the work each force does.

As in Chapter 5, we'll find the net force by adding the components of the four forces. Newton's second law tells us that because the sled's motion is purely horizontal, the net force can have only a horizontal component.

EXECUTE: (1) The work W_w done by the weight is zero because its direction is perpendicular to the displacement (compare Fig. 6.4c). For the same reason, the work W_n done by the normal force is also zero. (Note that we don't need to calculate the magnitude *n* to conclude this.) So $W_w = W_n = 0$.

That leaves the work W_T done by the force F_T exerted by the tractor and the work W_f done by the friction force f. From Eq. (6.2),

$$W_T = F_T s \cos 36.9° = (5000 \text{ N})(20 \text{ m})(0.800) = 80,000 \text{ N} \cdot \text{m}$$
$$= 80 \text{ kJ}$$

The friction force $\vec{f}$ is opposite to the displacement, so for this force $\phi = 180°$ and $\cos \phi = -1$. Again from Eq. (6.2),

$$W_f = fs \cos 180° = (3500 \text{ N})(20 \text{ m})(-1) = -70,000 \text{ N} \cdot \text{m}$$
$$= -70 \text{ kJ}$$

The total work W_{tot} done on the sled by all forces is the *algebraic* sum of the work done by the individual forces:

$$W_{tot} = W_w + W_n + W_T + W_f = 0 + 0 + 80 \text{ kJ} + (-70 \text{ kJ})$$
$$= 10 \text{ kJ}$$

(2) In the second approach, we first find the *vector* sum of all the forces (the net force) and then use it to compute the total work. It's easiest to find the net force by using components. From Fig. 6.7b,

$$\Sigma F_x = F_T \cos \phi + (-f) = (5000 \text{ N}) \cos 36.9° - 3500 \text{ N}$$
$$= 500 \text{ N}$$
$$\Sigma F_y = F_T \sin \phi + n + (-w)$$
$$= (5000 \text{ N}) \sin 36.9° + n - 14,700 \text{ N}$$

We don't need the second equation; we know that the y-component of force is perpendicular to the displacement, so it does no work. Besides, there is no y-component of acceleration, so ΣF_y must be zero anyway. The total work is therefore the work done by the total x-component:

$$W_{tot} = (\Sigma \vec{F}) \cdot \vec{s} = (\Sigma F_x)s = (500 \text{ N})(20 \text{ m}) = 10,000 \text{ J}$$
$$= 10 \text{ kJ}$$

EVALUATE: We get the same result for W_{tot} with either method, as we should. Note that the net force in the x-direction is *not* zero, and so the sled must accelerate as it moves. In Section 6.2 we'll return to this example and see how to use the concept of work to explore the sled's changes of speed.

TEST YOUR UNDERSTANDING OF SECTION 6.1 An electron moves in a straight line toward the east with a constant speed of 8×10^7 m/s. It has electric, magnetic, and gravitational forces acting on it. During a 1-m displacement, the total work done on the electron is (i) positive; (ii) negative; (iii) zero; (iv) not enough information is given. ❙

6.2 KINETIC ENERGY AND THE WORK-ENERGY THEOREM

The total work done on a body by external forces is related to the body's displacement—that is, to changes in its position. But the total work is also related to changes in the *speed* of the body. To see this, consider **Fig. 6.8**, which shows a block sliding on a frictionless table. The forces acting on the block are its weight $\vec{w}$, the normal force $\vec{n}$, and the force $\vec{F}$ exerted on it by the hand.

In Fig. 6.8a the net force on the block is in the direction of its motion. From Newton's second law, this means that the block speeds up; from Eq. (6.1), this also means that the total work W_{tot} done on the block is positive. The total work

PhET: The Ramp

6.8 The relationship between the total work done on a body and how the body's speed changes.

(a)

A block slides to the right on a frictionless surface.

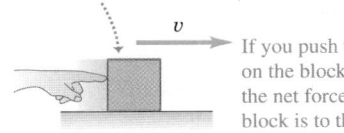

If you push to the right on the block as it moves, the net force on the block is to the right.

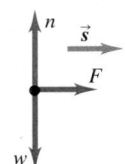

- The total work done on the block during a displacement $\vec{s}$ is positive: $W_{tot} > 0$.
- The block speeds up.

(b)

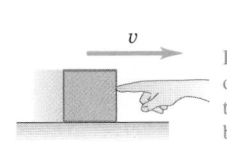

If you push to the left on the block as it moves, the net force on the block is to the left.

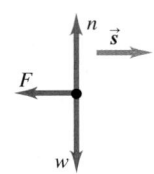

- The total work done on the block during a displacement $\vec{s}$ is negative: $W_{tot} < 0$.
- The block slows down.

(c)

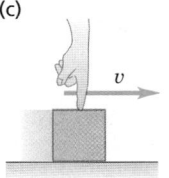

If you push straight down on the block as it moves, the net force on the block is zero.

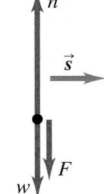

- The total work done on the block during a displacement $\vec{s}$ is zero: $W_{tot} = 0$.
- The block's speed stays the same.

6.9 A constant net force $\vec{F}$ does work on a moving body.

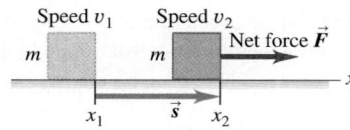

is *negative* in Fig. 6.8b because the net force opposes the displacement; in this case the block slows down. The net force is zero in Fig. 6.8c, so the speed of the block stays the same and the total work done on the block is zero. We can conclude that *when a particle undergoes a displacement, it speeds up if* $W_{tot} > 0$, *slows down if* $W_{tot} < 0$, *and maintains the same speed if* $W_{tot} = 0$.

Let's make this more quantitative. In **Fig. 6.9** a particle with mass m moves along the x-axis under the action of a constant net force with magnitude F that points in the positive x-direction. The particle's acceleration is constant and given by Newton's second law (Section 4.3): $F = ma_x$. As the particle moves from point x_1 to x_2, it undergoes a displacement $s = x_2 - x_1$ and its speed changes from v_1 to v_2. Using a constant-acceleration equation from Section 2.4, Eq. (2.13), and replacing v_{0x} by v_1, v_x by v_2, and $(x - x_0)$ by s, we have

$$v_2^2 = v_1^2 + 2a_x s$$

$$a_x = \frac{v_2^2 - v_1^2}{2s}$$

When we multiply this equation by m and equate ma_x to the net force F, we find

$$F = ma_x = m\frac{v_2^2 - v_1^2}{2s} \quad \text{and}$$

$$Fs = \tfrac{1}{2}mv_2^2 - \tfrac{1}{2}mv_1^2 \tag{6.4}$$

In Eq. (6.4) the product Fs is the work done by the net force F and thus is equal to the total work W_{tot} done by all the forces acting on the particle. The quantity $\frac{1}{2}mv^2$ is called the **kinetic energy** K of the particle:

Kinetic energy ····→ $K = \tfrac{1}{2}mv^2$ ←···· Mass of particle
of a particle ←···· Speed of particle $\tag{6.5}$

6.10 Comparing the kinetic energy $K = \frac{1}{2}mv^2$ of different bodies.

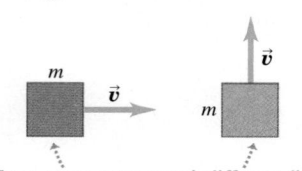

Same mass, same speed, different directions of motion: *same* kinetic energy

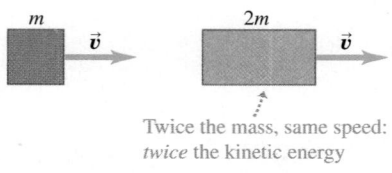

Twice the mass, same speed: *twice* the kinetic energy

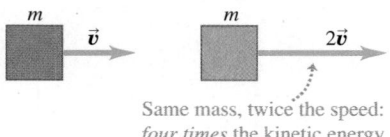

Same mass, twice the speed: *four times* the kinetic energy

Like work, the kinetic energy of a particle is a scalar quantity; it depends on only the particle's mass and speed, not its direction of motion (**Fig. 6.10**). Kinetic energy can never be negative, and it is zero only when the particle is at rest.

We can now interpret Eq. (6.4) in terms of work and kinetic energy. The first term on the right side of Eq. (6.4) is $K_2 = \frac{1}{2}mv_2^2$, the final kinetic energy of the particle (that is, after the displacement). The second term is the initial kinetic energy, $K_1 = \frac{1}{2}mv_1^2$, and the difference between these terms is the *change* in kinetic energy. So Eq. (6.4) says:

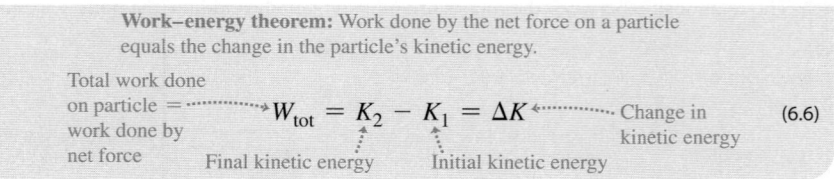

Work–energy theorem: Work done by the net force on a particle equals the change in the particle's kinetic energy.

Total work done
on particle = ····→ $W_{tot} = K_2 - K_1 = \Delta K$ ←···· Change in
work done by   kinetic energy $\tag{6.6}$
net force   Final kinetic energy   Initial kinetic energy

This **work–energy theorem** agrees with our observations about the block in Fig. 6.8. When W_{tot} is *positive*, the kinetic energy *increases* (the final kinetic energy K_2 is greater than the initial kinetic energy K_1) and the particle is going faster at the end of the displacement than at the beginning. When W_{tot} is *negative*, the kinetic energy *decreases* (K_2 is less than K_1) and the speed is less after the displacement. When $W_{tot} = 0$, the kinetic energy stays the same ($K_1 = K_2$) and the speed is unchanged. Note that the work–energy theorem by itself tells us only about changes in *speed*, not velocity, since the kinetic energy doesn't depend on the direction of motion.

From Eq. (6.4) or Eq. (6.6), kinetic energy and work must have the same units. Hence the joule is the SI unit of both work and kinetic energy (and, as we will see later, of all kinds of energy). To verify this, note that in SI the quantity $K = \frac{1}{2}mv^2$ has units $\text{kg} \cdot (\text{m/s})^2$ or $\text{kg} \cdot \text{m}^2/\text{s}^2$; we recall that $1 \text{ N} = 1 \text{ kg} \cdot \text{m/s}^2$, so

$$1 \text{ J} = 1 \text{ N} \cdot \text{m} = 1 \ (\text{kg} \cdot \text{m/s}^2) \cdot \text{m} = 1 \text{ kg} \cdot \text{m}^2/\text{s}^2$$

Because we used Newton's laws in deriving the work–energy theorem, we can use this theorem only in an inertial frame of reference. Note that the work–energy theorem is valid in *any* inertial frame, but the values of W_{tot} and $K_2 - K_1$ may differ from one inertial frame to another (because the displacement and speed of a body may be different in different frames).

We've derived the work–energy theorem for the special case of straight-line motion with constant forces, and in the following examples we'll apply it to this special case only. We'll find in the next section that the theorem is valid even when the forces are not constant and the particle's trajectory is curved.

PROBLEM-SOLVING STRATEGY 6.1 | WORK AND KINETIC ENERGY

IDENTIFY *the relevant concepts:* The work–energy theorem, $W_{\text{tot}} = K_2 - K_1$, is extremely useful when you want to relate a body's speed v_1 at one point in its motion to its speed v_2 at a different point. (It's less useful for problems that involve the *time* it takes a body to go from point 1 to point 2 because the work–energy theorem doesn't involve time at all. For such problems it's usually best to use the relationships among time, position, velocity, and acceleration described in Chapters 2 and 3.)

SET UP *the problem* using the following steps:
1. Identify the initial and final positions of the body, and draw a free-body diagram showing all the forces that act on the body.
2. Choose a coordinate system. (If the motion is along a straight line, it's usually easiest to have both the initial and final positions lie along one of the axes.)
3. List the unknown and known quantities, and decide which unknowns are your target variables. The target variable may be the body's initial or final speed, the magnitude of one of the forces acting on the body, or the body's displacement.

EXECUTE *the solution:* Calculate the work W done by each force. If the force is constant and the displacement is a straight line, you can use Eq. (6.2) or Eq. (6.3). (Later in this chapter we'll see how to handle varying forces and curved trajectories.) Be sure to check signs; W must be positive if the force has a component in the direction of the displacement, negative if the force has a component opposite to the displacement, and zero if the force and displacement are perpendicular.

Add the amounts of work done by each force to find the total work W_{tot}. Sometimes it's easier to calculate the vector sum of the forces (the net force) and then find the work done by the net force; this value is also equal to W_{tot}.

Write expressions for the initial and final kinetic energies, K_1 and K_2. Note that kinetic energy involves *mass,* not *weight;* if you are given the body's weight, use $w = mg$ to find the mass.

Finally, use Eq. (6.6), $W_{\text{tot}} = K_2 - K_1$, and Eq. (6.5), $K = \frac{1}{2}mv^2$, to solve for the target variable. Remember that the right-hand side of Eq. (6.6) represents the change of the body's kinetic energy between points 1 and 2; that is, it is the *final* kinetic energy minus the *initial* kinetic energy, never the other way around. (If you can predict the sign of W_{tot}, you can predict whether the body speeds up or slows down.)

EVALUATE *your answer:* Check whether your answer makes sense. Remember that kinetic energy $K = \frac{1}{2}mv^2$ can never be negative. If you come up with a negative value of K, perhaps you interchanged the initial and final kinetic energies in $W_{\text{tot}} = K_2 - K_1$ or made a sign error in one of the work calculations.

EXAMPLE 6.3   USING WORK AND ENERGY TO CALCULATE SPEED

Let's look again at the sled in Fig. 6.7 and our results from Example 6.2. Suppose the sled's initial speed v_1 is 2.0 m/s. What is the speed of the sled after it moves 20 m?

6.11 Our sketch for this problem.

SOLUTION

IDENTIFY and SET UP: We'll use the work–energy theorem, Eq. (6.6), $W_{\text{tot}} = K_2 - K_1$, since we are given the initial speed $v_1 = 2.0$ m/s and want to find the final speed v_2. **Figure 6.11** shows our sketch of the situation. The motion is in the positive x-direction. In Example 6.2 we calculated the total work done by all the forces: $W_{\text{tot}} = 10$ kJ. Hence the kinetic energy of the sled and its load must increase by 10 kJ, and the speed of the sled must also increase.

EXECUTE: To write expressions for the initial and final kinetic energies, we need the mass of the sled and load. The combined *weight* is 14,700 N, so the mass is

$$m = \frac{w}{g} = \frac{14{,}700 \text{ N}}{9.8 \text{ m/s}^2} = 1500 \text{ kg}$$

Continued

Then the initial kinetic energy K_1 is

$$K_1 = \tfrac{1}{2}mv_1^2 = \tfrac{1}{2}(1500 \text{ kg})(2.0 \text{ m/s})^2 = 3000 \text{ kg} \cdot \text{m}^2/\text{s}^2$$
$$= 3000 \text{ J}$$

The final kinetic energy K_2 is

$$K_2 = \tfrac{1}{2}mv_2^2 = \tfrac{1}{2}(1500 \text{ kg})v_2^2$$

The work–energy theorem, Eq. (6.6), gives

$$K_2 = K_1 + W_{\text{tot}} = 3000 \text{ J} + 10{,}000 \text{ J} = 13{,}000 \text{ J}$$

Setting these two expressions for K_2 equal, substituting $1 \text{ J} = 1 \text{ kg} \cdot \text{m}^2/\text{s}^2$, and solving for the final speed v_2, we find

$$v_2 = 4.2 \text{ m/s}$$

EVALUATE: The total work is positive, so the kinetic energy increases ($K_2 > K_1$) and the speed increases ($v_2 > v_1$).

This problem can also be solved without the work–energy theorem. We can find the acceleration from $\sum \vec{F} = m\vec{a}$ and then use the equations of motion for constant acceleration to find v_2. Since the acceleration is along the x-axis,

$$a = a_x = \frac{\sum F_x}{m} = \frac{500 \text{ N}}{1500 \text{ kg}} = 0.333 \text{ m/s}^2$$

Then, using Eq. (2.13),

$$v_2^2 = v_1^2 + 2as = (2.0 \text{ m/s})^2 + 2(0.333 \text{ m/s}^2)(20 \text{ m})$$
$$= 17.3 \text{ m}^2/\text{s}^2$$
$$v_2 = 4.2 \text{ m/s}$$

This is the same result we obtained with the work–energy approach, but there we avoided the intermediate step of finding the acceleration. You will find several other examples in this chapter and the next that *can* be done without using energy considerations but that are easier when energy methods are used. When a problem can be done by two methods, doing it by both methods (as we did here) is a good way to check your work.

EXAMPLE 6.4 FORCES ON A HAMMERHEAD

The 200-kg steel hammerhead of a pile driver is lifted 3.00 m above the top of a vertical I-beam being driven into the ground (**Fig. 6.12a**). The hammerhead is then dropped, driving the I-beam 7.4 cm deeper into the ground. The vertical guide rails exert a constant 60-N friction force on the hammerhead. Use the work–energy theorem to find (a) the speed of the hammerhead just as it hits the I-beam and (b) the average force the hammerhead exerts on the I-beam. Ignore the effects of the air.

SOLUTION

IDENTIFY: We'll use the work–energy theorem to relate the hammerhead's speed at different locations and the forces acting on it.

There are *three* locations of interest: point 1, where the hammerhead starts from rest; point 2, where it first contacts the I-beam; and point 3, where the hammerhead and I-beam come to a halt (Fig. 6.12a). The two target variables are the hammerhead's speed at point 2 and the average force the hammerhead exerts between points 2 and 3. Hence we'll apply the work–energy theorem twice: once for the motion from 1 to 2, and once for the motion from 2 to 3.

SET UP: Figure 6.12b shows the vertical forces on the hammerhead as it falls from point 1 to point 2. (We can ignore any horizontal forces that may be present because they do no work as the hammerhead moves vertically.) For this part of the motion, our target variable is the hammerhead's final speed v_2.

6.12 (a) A pile driver pounds an I-beam into the ground. (b), (c) Free-body diagrams. Vector lengths are not to scale.

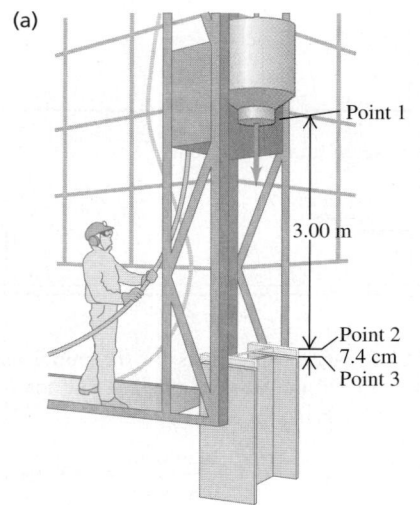

(a)

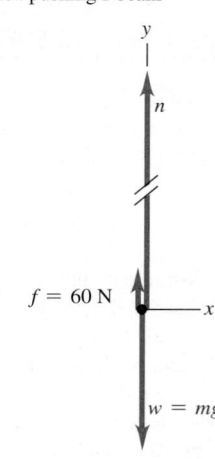

(b) Free-body diagram for falling hammerhead

(c) Free-body diagram for hammerhead when pushing I-beam

Figure 6.12c shows the vertical forces on the hammerhead during the motion from point 2 to point 3. In addition to the forces shown in Fig. 6.12b, the I-beam exerts an upward normal force of magnitude n on the hammerhead. This force actually varies as the hammerhead comes to a halt, but for simplicity we'll treat n as a constant. Hence n represents the *average* value of this upward force during the motion. Our target variable for this part of the motion is the force that the *hammerhead* exerts on the I-beam; it is the reaction force to the normal force exerted by the I-beam, so by Newton's third law its magnitude is also n.

EXECUTE: (a) From point 1 to point 2, the vertical forces are the downward weight $w = mg = (200 \text{ kg})(9.8 \text{ m/s}^2) = 1960 \text{ N}$ and the upward friction force $f = 60 \text{ N}$. Thus the net downward force is $w - f = 1900 \text{ N}$. The displacement of the hammerhead from point 1 to point 2 is downward and equal to $s_{12} = 3.00 \text{ m}$. The total work done on the hammerhead between point 1 and point 2 is then

$$W_{\text{tot}} = (w - f)s_{12} = (1900 \text{ N})(3.00 \text{ m}) = 5700 \text{ J}$$

At point 1 the hammerhead is at rest, so its initial kinetic energy K_1 is zero. Hence the kinetic energy K_2 at point 2 equals the total work done on the hammerhead between points 1 and 2:

$$W_{\text{tot}} = K_2 - K_1 = K_2 - 0 = \tfrac{1}{2}mv_2^2 - 0$$

$$v_2 = \sqrt{\frac{2W_{\text{tot}}}{m}} = \sqrt{\frac{2(5700 \text{ J})}{200 \text{ kg}}} = 7.55 \text{ m/s}$$

This is the hammerhead's speed at point 2, just as it hits the I-beam.

(b) As the hammerhead moves downward from point 2 to point 3, its displacement is $s_{23} = 7.4 \text{ cm} = 0.074 \text{ m}$ and the net downward force acting on it is $w - f - n$ (Fig. 6.12c). The total work done on the hammerhead during this displacement is

$$W_{\text{tot}} = (w - f - n)s_{23}$$

The initial kinetic energy for this part of the motion is K_2, which from part (a) equals 5700 J. The final kinetic energy is $K_3 = 0$ (the hammerhead ends at rest). From the work–energy theorem,

$$W_{\text{tot}} = (w - f - n)s_{23} = K_3 - K_2$$

$$n = w - f - \frac{K_3 - K_2}{s_{23}}$$

$$= 1960 \text{ N} - 60 \text{ N} - \frac{0 \text{ J} - 5700 \text{ J}}{0.074 \text{ m}} = 79{,}000 \text{ N}$$

The downward force that the hammerhead exerts on the I-beam has this same magnitude, 79,000 N (about 9 tons)—more than 40 times the weight of the hammerhead.

EVALUATE: The net change in the hammerhead's kinetic energy from point 1 to point 3 is zero; a relatively small net force does positive work over a large distance, and then a much larger net force does negative work over a much smaller distance. The same thing happens if you speed up your car gradually and then drive it into a brick wall. The very large force needed to reduce the kinetic energy to zero over a short distance is what does the damage to your car—and possibly to you.

The Meaning of Kinetic Energy

Example 6.4 gives insight into the physical meaning of kinetic energy. The hammerhead is dropped from rest, and its kinetic energy when it hits the I-beam equals the total work done on it up to that point by the net force. This result is true in general: To accelerate a particle of mass m from rest (zero kinetic energy) up to a speed v, the total work done on it must equal the change in kinetic energy from zero to $K = \tfrac{1}{2}mv^2$:

$$W_{\text{tot}} = K - 0 = K$$

So *the kinetic energy of a particle is equal to the total work that was done to accelerate it from rest to its present speed* (**Fig. 6.13**). The definition $K = \tfrac{1}{2}mv^2$, Eq. (6.5), wasn't chosen at random; it's the *only* definition that agrees with this interpretation of kinetic energy.

In the second part of Example 6.4 the kinetic energy of the hammerhead did work on the I-beam and drove it into the ground. This gives us another interpretation of kinetic energy: *The kinetic energy of a particle is equal to the total work that particle can do in the process of being brought to rest.* This is why you pull your hand and arm backward when you catch a ball. As the ball comes to rest, it does an amount of work (force times distance) on your hand equal to the ball's initial kinetic energy. By pulling your hand back, you maximize the distance over which the force acts and so minimize the force on your hand.

6.13 Imparting kinetic energy to a cue ball.

When a billiards player hits a cue ball at rest, the ball's kinetic energy after being hit is equal to the work that was done on it by the cue.

The greater the force exerted by the cue and the greater the distance the ball moves while in contact with it, the greater the ball's kinetic energy.

CONCEPTUAL EXAMPLE 6.5 **COMPARING KINETIC ENERGIES**

Two iceboats like the one in Example 5.6 (Section 5.2) hold a race on a frictionless horizontal lake (**Fig. 6.14**). The two iceboats have masses m and $2m$. The iceboats have identical sails, so the wind exerts the same constant force $\vec{F}$ on each iceboat. They start from rest and cross the finish line a distance s away. Which iceboat crosses the finish line with greater kinetic energy?

SOLUTION

If you use the definition of kinetic energy, $K = \frac{1}{2}mv^2$, Eq. (6.5), the answer to this problem isn't obvious. The iceboat of mass $2m$

6.14 A race between iceboats.

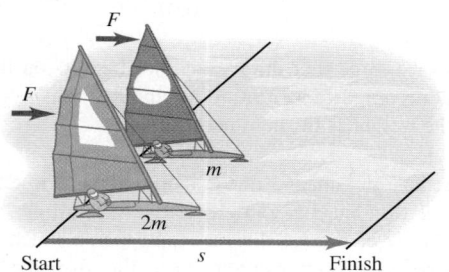

has greater mass, so you might guess that it has greater kinetic energy at the finish line. But the lighter iceboat, of mass m, has greater acceleration and crosses the finish line with a greater speed, so you might guess that *this* iceboat has the greater kinetic energy. How can we decide?

The key is to remember that *the kinetic energy of a particle is equal to the total work done to accelerate it from rest.* Both iceboats travel the same distance s from rest, and only the horizontal force F in the direction of motion does work on either iceboat. Hence the total work done between the starting line and the finish line is the *same* for each iceboat, $W_{tot} = Fs$. At the finish line, each iceboat has a kinetic energy equal to the work W_{tot} done on it, because each iceboat started from rest. So both iceboats have the *same* kinetic energy at the finish line!

You might think this is a "trick" question, but it isn't. If you really understand the meanings of quantities such as kinetic energy, you can solve problems more easily and with better insight.

Notice that we didn't need to know anything about how much time each iceboat took to reach the finish line. This is because the work–energy theorem makes no direct reference to time, only to displacement. In fact the iceboat of mass m has greater acceleration and so takes less time to reach the finish line than does the iceboat of mass $2m$.

Work and Kinetic Energy in Composite Systems

In this section we've been careful to apply the work–energy theorem only to bodies that we can represent as *particles*—that is, as moving point masses. New subtleties appear for more complex systems that have to be represented as many particles with different motions. We can't go into these subtleties in detail in this chapter, but here's an example.

Suppose a boy stands on frictionless roller skates on a level surface, facing a rigid wall (**Fig. 6.15**). He pushes against the wall, which makes him move to the right. The forces acting on him are his weight $\vec{w}$, the upward normal forces $\vec{n}_1$ and $\vec{n}_2$ exerted by the ground on his skates, and the horizontal force $\vec{F}$ exerted on him by the wall. There is no vertical displacement, so $\vec{w}$, $\vec{n}_1$, and $\vec{n}_2$ do no work. Force $\vec{F}$ accelerates him to the right, but the parts of his body where that force is applied (the boy's hands) do not move while the force acts. Thus the force $\vec{F}$ also does no work. Where, then, does the boy's kinetic energy come from?

The explanation is that it's not adequate to represent the boy as a single point mass. Different parts of the boy's body have different motions; his hands remain stationary against the wall while his torso is moving away from the wall. The various parts of his body interact with each other, and one part can exert forces and do work on another part. Therefore the *total* kinetic energy of this *composite* system of body parts can change, even though no work is done by forces applied by bodies (such as the wall) that are outside the system. In Chapter 8 we'll consider further the motion of a collection of particles that interact with each other. We'll discover that just as for the boy in this example, the total kinetic energy of such a system can change even when no work is done on any part of the system by anything outside it.

6.15 The external forces acting on a skater pushing off a wall. The work done by these forces is zero, but the skater's kinetic energy changes nonetheless.

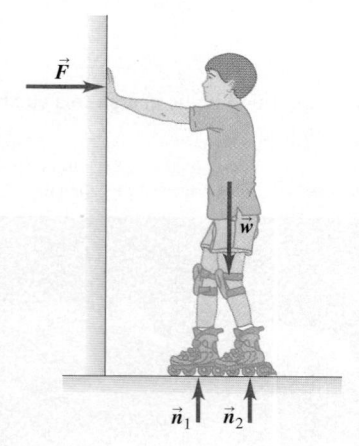

TEST YOUR UNDERSTANDING OF SECTION 6.2 Rank the following bodies in order of their kinetic energy, from least to greatest. (i) A 2.0-kg body moving at 5.0 m/s; (ii) a 1.0-kg body that initially was at rest and then had 30 J of work done on it; (iii) a 1.0-kg body that initially was moving at 4.0 m/s and then had 20 J of work done on it; (iv) a 2.0-kg body that initially was moving at 10 m/s and then did 80 J of work on another body. ∎

6.3 WORK AND ENERGY WITH VARYING FORCES

So far we've considered work done by *constant forces* only. But what happens when you stretch a spring? The more you stretch it, the harder you have to pull, so the force you exert is *not* constant as the spring is stretched. We've also restricted our discussion to *straight-line* motion. There are many situations in which a body moves along a curved path and is acted on by a force that varies in magnitude, direction, or both. We need to be able to compute the work done by the force in these more general cases. Fortunately, the work–energy theorem holds true even when forces are varying and when the body's path is not straight.

Work Done by a Varying Force, Straight-Line Motion

To add only one complication at a time, let's consider straight-line motion along the x-axis with a force whose x-component F_x may change as the body moves. (A real-life example is driving a car along a straight road with stop signs, so the driver has to alternately step on the gas and apply the brakes.) Suppose a particle moves along the x-axis from point x_1 to x_2 (**Fig. 6.16a**). Figure 6.16b is a graph of the x-component of force as a function of the particle's coordinate x. To find the work done by this force, we divide the total displacement into narrow segments Δx_a, Δx_b, and so on (Fig. 6.16c). We approximate the work done by the force during segment Δx_a as the average x-component of force F_{ax} in that segment multiplied by the x-displacement Δx_a. We do this for each segment and then add the results for all the segments. The work done by the force in the total displacement from x_1 to x_2 is approximately

$$W = F_{ax}\Delta x_a + F_{bx}\Delta x_b + \cdots$$

In the limit that the number of segments becomes very large and the width of each becomes very small, this sum becomes the *integral* of F_x from x_1 to x_2:

Work done on a particle by a varying x-component of force F_x during straight-line displacement along x-axis

$$W = \int_{x_1}^{x_2} F_x\,dx \qquad (6.7)$$

Upper limit = final position
Integral of x-component of force
Lower limit = initial position

Note that $F_{ax}\Delta x_a$ represents the *area* of the first vertical strip in Fig. 6.16c and that the integral in Eq. (6.7) represents the area under the curve of Fig. 6.16b between x_1 and x_2. *On such a graph of force as a function of position, the total work done by the force is represented by the area under the curve between the initial and final positions.* Alternatively, the work W equals the average force that acts over the entire displacement, multiplied by the displacement.

In the special case that F_x, the x-component of the force, is constant, we can take it outside the integral in Eq. (6.7):

$$W = \int_{x_1}^{x_2} F_x\,dx = F_x\int_{x_1}^{x_2} dx = F_x(x_2 - x_1) \qquad \text{(constant force)}$$

But $x_2 - x_1 = s$, the total displacement of the particle. So in the case of a constant force F, Eq. (6.7) says that $W = Fs$, in agreement with Eq. (6.1). The interpretation of work as the area under the curve of F_x as a function of x also holds for a constant force: $W = Fs$ is the area of a rectangle of height F and width s (**Fig. 6.17**).

Now let's apply these ideas to the stretched spring. To keep a spring stretched beyond its unstretched length by an amount x, we have to apply a force of equal

6.16 Calculating the work done by a varying force F_x in the x-direction as a particle moves from x_1 to x_2.

(a) A particle moves from x_1 to x_2 in response to a changing force in the x-direction.

(b) The force F_x varies with position x ...

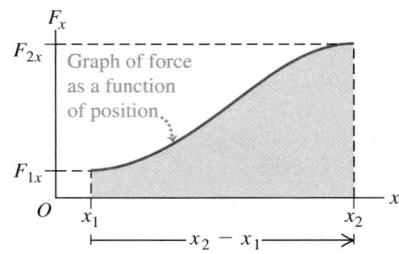

(c) ... but over a short displacement Δx, the force is essentially constant.

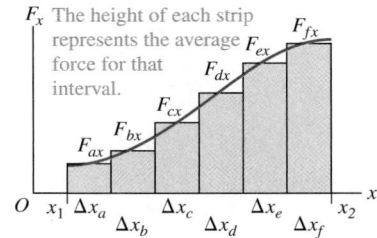

(MP)

PhET: Molecular Motors

PhET: Stretching DNA

6.17 The work done by a constant force F in the x-direction as a particle moves from x_1 to x_2.

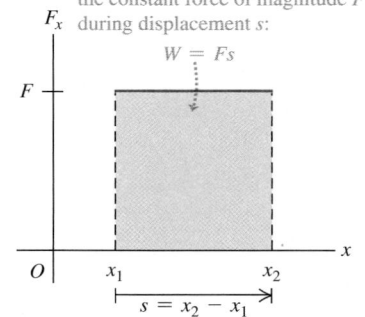

6.18 The force needed to stretch an ideal spring is proportional to the spring's elongation: $F_x = kx$.

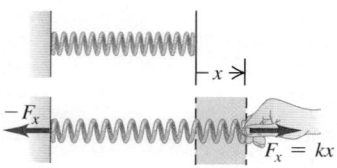

6.19 Calculating the work done to stretch a spring by a length X.

The area under the graph represents the work done on the spring as the spring is stretched from $x = 0$ to a maximum value X:

$$W = \tfrac{1}{2}kX^2$$

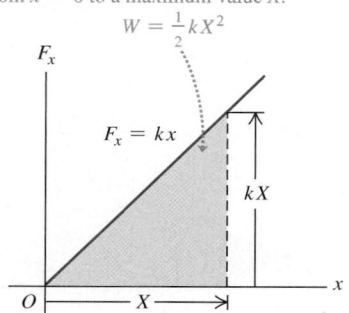

$F_x = kx$

magnitude at each end (**Fig. 6.18**). If the elongation x is not too great, the force we apply to the right-hand end has an x-component directly proportional to x:

$$F_x = kx \qquad \text{(force required to stretch a spring)} \qquad (6.8)$$

where k is a constant called the **force constant** (or spring constant) of the spring. The units of k are force divided by distance: N/m in SI units. A floppy toy spring such as a Slinky™ has a force constant of about 1 N/m; for the much stiffer springs in an automobile's suspension, k is about 10^5 N/m. The observation that force is directly proportional to elongation for elongations that are not too great was made by Robert Hooke in 1678 and is known as **Hooke's law.** It really shouldn't be called a "law," since it's a statement about a specific device and not a fundamental law of nature. Real springs don't always obey Eq. (6.8) precisely, but it's still a useful idealized model. We'll discuss Hooke's law more fully in Chapter 11.

To stretch a spring, we must do work. We apply equal and opposite forces to the ends of the spring and gradually increase the forces. We hold the left end stationary, so the force we apply at this end does no work. The force at the moving end *does* do work. **Figure 6.19** is a graph of F_x as a function of x, the elongation of the spring. The work done by this force when the elongation goes from zero to a maximum value X is

$$W = \int_0^X F_x \, dx = \int_0^X kx \, dx = \tfrac{1}{2}kX^2 \qquad (6.9)$$

We can also obtain this result graphically. The area of the shaded triangle in Fig. 6.19, representing the total work done by the force, is equal to half the product of the base and altitude, or

$$W = \tfrac{1}{2}(X)(kX) = \tfrac{1}{2}kX^2$$

This equation also says that the work is the *average* force $kX/2$ multiplied by the total displacement X. We see that the total work is proportional to the *square* of the final elongation X. To stretch an ideal spring by 2 cm, you must do four times as much work as is needed to stretch it by 1 cm.

Equation (6.9) assumes that the spring was originally unstretched. If initially the spring is already stretched a distance x_1, the work we must do to stretch it to a greater elongation x_2 (**Fig. 6.20a**) is

$$W = \int_{x_1}^{x_2} F_x \, dx = \int_{x_1}^{x_2} kx \, dx = \tfrac{1}{2}kx_2^2 - \tfrac{1}{2}kx_1^2 \qquad (6.10)$$

Use your knowledge of geometry to convince yourself that the trapezoidal area under the graph in Fig. 6.20b is given by the expression in Eq. (6.10).

6.20 Calculating the work done to stretch a spring from one elongation to a greater one.

(a) Stretching a spring from elongation x_1 to elongation x_2

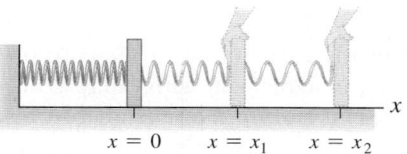

$x = 0 \qquad x = x_1 \qquad x = x_2$

(b) Force-versus-distance graph

The trapezoidal area under the graph represents the work done on the spring to stretch it from $x = x_1$ to $x = x_2$: $W = \tfrac{1}{2}kx_2^2 - \tfrac{1}{2}kx_1^2$.

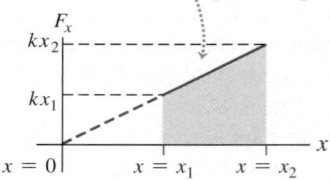

If the spring has spaces between the coils when it is unstretched, then it can also be compressed, and Hooke's law holds for compression as well as stretching. In this case the force and displacement are in the opposite directions from those shown in Fig. 6.18, so both F_x and x in Eq. (6.8) are negative. Since both F_x and x are reversed, the force again is in the same direction as the displacement, and the work done by F_x is again positive. So the total work is still given by Eq. (6.9) or (6.10), even when X is negative or either or both of x_1 and x_2 are negative.

CAUTION Work done *on* a spring vs. work done *by* a spring Equation (6.10) gives the work that *you* must do *on* a spring to change its length. If you stretch a spring that's originally relaxed, then $x_1 = 0, x_2 > 0$, and $W > 0$: The force you apply to one end of the spring is in the same direction as the displacement, and the work you do is positive. By contrast, the work that the *spring* does on whatever it's attached to is given by the *negative* of Eq. (6.10). Thus, as you pull on the spring, the spring does negative work on you. ▌

EXAMPLE 6.6 WORK DONE ON A SPRING SCALE

SOLUTION

A woman weighing 600 N steps on a bathroom scale that contains a stiff spring (**Fig. 6.21**). In equilibrium, the spring is compressed 1.0 cm under her weight. Find the force constant of the spring and the total work done on it during the compression.

SOLUTION

IDENTIFY and SET UP: In equilibrium the upward force exerted by the spring balances the downward force of the woman's weight. We'll use this principle and Eq. (6.8) to determine the force

6.21 Compressing a spring in a bathroom scale.

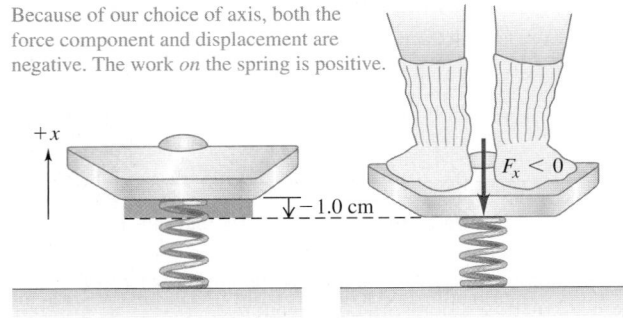

Because of our choice of axis, both the force component and displacement are negative. The work *on* the spring is positive.

$+x$

$F_x < 0$

-1.0 cm

constant k, and we'll use Eq. (6.10) to calculate the work W that the woman does on the spring to compress it. We take positive values of x to correspond to elongation (upward in Fig. 6.21), so that both the displacement of the end of the spring (x) and the x-component of the force that the woman exerts on it (F_x) are negative. The applied force and the displacement are in the same direction, so the work done on the spring will be positive.

EXECUTE: The top of the spring is displaced by $x = -1.0$ cm $= -0.010$ m, and the woman exerts a force $F_x = -600$ N on the spring. From Eq. (6.8) the force constant is then

$$k = \frac{F_x}{x} = \frac{-600\ \text{N}}{-0.010\ \text{m}} = 6.0 \times 10^4\ \text{N/m}$$

Then, using $x_1 = 0$ and $x_2 = -0.010$ m in Eq. (6.10), we have

$$W = \tfrac{1}{2}kx_2^2 - \tfrac{1}{2}kx_1^2$$
$$= \tfrac{1}{2}(6.0 \times 10^4\ \text{N/m})(-0.010\ \text{m})^2 - 0 = 3.0\ \text{J}$$

EVALUATE: The work done is positive, as expected. Our arbitrary choice of the positive direction has no effect on the answer for W. You can test this by taking the positive x-direction to be downward, corresponding to compression. Do you get the same values for k and W as we found here?

Work–Energy Theorem for Straight-Line Motion, Varying Forces

In Section 6.2 we derived the work–energy theorem, $W_{\text{tot}} = K_2 - K_1$, for the special case of straight-line motion with a constant net force. We can now prove that this theorem is true even when the force varies with position. As in Section 6.2, let's consider a particle that undergoes a displacement x while being acted on by a net force with x-component F_x, which we now allow to vary. Just as in Fig. 6.16, we divide the total displacement x into a large number of small segments Δx. We can apply the work–energy theorem, Eq. (6.6), to each segment because the value

BIO Application Tendons Are Nonideal Springs Muscles exert forces via the tendons that attach them to bones. A tendon consists of long, stiff, elastic collagen fibers. The graph shows how the tendon from the hind leg of a wallaby (a small kangaroo-like marsupial) stretches in response to an applied force. The tendon does not exhibit the simple, straight-line behavior of an ideal spring, so the work it does has to be found by integration [Eq. (6.7)]. The tendon exerts less force while relaxing than while stretching. As a result, the relaxing tendon does only about 93% of the work that was done to stretch it.

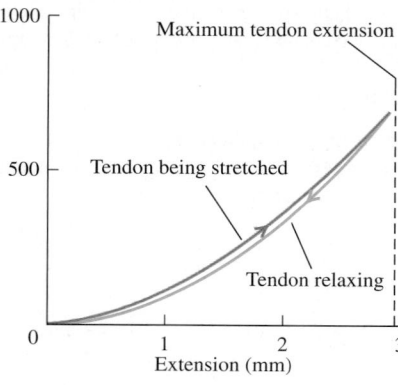

Force exerted by tendon (N)

of F_x in each small segment is approximately constant. The change in kinetic energy in segment Δx_a is equal to the work $F_{ax}\Delta x_a$, and so on. The total change of kinetic energy is the sum of the changes in the individual segments, and thus is equal to the total work done on the particle during the entire displacement. So $W_{tot} = \Delta K$ holds for varying forces as well as for constant ones.

Here's an alternative derivation of the work–energy theorem for a force that may vary with position. It involves making a change of variable from x to v_x in the work integral. Note first that the acceleration a of the particle can be expressed in various ways, using $a_x = dv_x/dt$, $v_x = dx/dt$, and the chain rule for derivatives:

$$a_x = \frac{dv_x}{dt} = \frac{dv_x}{dx}\frac{dx}{dt} = v_x\frac{dv_x}{dx} \tag{6.11}$$

From this result, Eq. (6.7) tells us that the total work done by the *net* force F_x is

$$W_{tot} = \int_{x_1}^{x_2} F_x\, dx = \int_{x_1}^{x_2} ma_x\, dx = \int_{x_1}^{x_2} mv_x\frac{dv_x}{dx}\, dx \tag{6.12}$$

Now $(dv_x/dx)dx$ is the change in velocity dv_x during the displacement dx, so we can make that substitution in Eq. (6.12). This changes the integration variable from x to v_x, so we change the limits from x_1 and x_2 to the corresponding x-velocities v_1 and v_2:

$$W_{tot} = \int_{v_1}^{v_2} mv_x\, dv_x$$

The integral of $v_x\, dv_x$ is just $v_x^2/2$. Substituting the upper and lower limits, we finally find

$$W_{tot} = \tfrac{1}{2}mv_2^2 - \tfrac{1}{2}mv_1^2 \tag{6.13}$$

This is the same as Eq. (6.6), so the work–energy theorem is valid even without the assumption that the net force is constant.

EXAMPLE 6.7 **MOTION WITH A VARYING FORCE**

An air-track glider of mass 0.100 kg is attached to the end of a horizontal air track by a spring with force constant 20.0 N/m (**Fig. 6.22a**). Initially the spring is unstretched and the glider is moving at 1.50 m/s to the right. Find the maximum distance d that the glider moves to the right (a) if the air track is turned on, so that there is no friction, and (b) if the air is turned off, so that there is kinetic friction with coefficient $\mu_k = 0.47$.

SOLUTION

IDENTIFY and SET UP: The force exerted by the spring is not constant, so we *cannot* use the constant-acceleration formulas of Chapter 2 to solve this problem. Instead, we'll use the work–energy theorem, since the total work done involves the distance moved (our target variable). In Figs. 6.22b and 6.22c we choose the positive x-direction to be to the right (in the direction of the glider's motion). We take $x = 0$ at the glider's initial position (where the spring is unstretched) and $x = d$ (the target variable) at the position where the glider stops. The motion is purely horizontal, so only the horizontal forces do work. Note that Eq. (6.10) gives the work done by the *glider* on the *spring* as it stretches; to use the work–energy theorem we need the work done by the

6.22 (a) A glider attached to an air track by a spring. (b), (c) Our free-body diagrams.

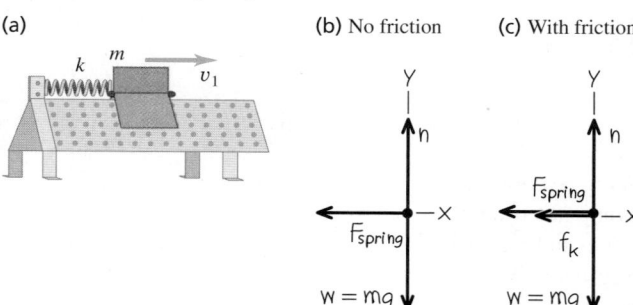

spring on the *glider*, which is the negative of Eq. (6.10). We expect the glider to move farther without friction than with friction.

EXECUTE: (a) Equation (6.10) says that as the glider moves from $x_1 = 0$ to $x_2 = d$, it does an amount of work $W = \frac{1}{2}kd^2 - \frac{1}{2}k(0)^2 = \frac{1}{2}kd^2$ on the spring. The amount of work that the *spring* does on the *glider* is the negative of this, $-\frac{1}{2}kd^2$. The spring stretches until the glider comes instantaneously to rest, so the final kinetic energy K_2 is zero. The initial kinetic energy is $\frac{1}{2}mv_1^2$, where $v_1 = 1.50$ m/s is the glider's initial speed. From the work–energy theorem,

$$-\tfrac{1}{2}kd^2 = 0 - \tfrac{1}{2}mv_1^2$$

We solve for the distance d the glider moves:

$$d = v_1\sqrt{\frac{m}{k}} = (1.50 \text{ m/s})\sqrt{\frac{0.100 \text{ kg}}{20.0 \text{ N/m}}}$$

$$= 0.106 \text{ m} = 10.6 \text{ cm}$$

The stretched spring subsequently pulls the glider back to the left, so the glider is at rest only instantaneously.

(b) If the air is turned off, we must include the work done by the kinetic friction force. The normal force n is equal in magnitude to the weight of the glider, since the track is horizontal and there are no other vertical forces. Hence the kinetic friction force has constant magnitude $f_k = \mu_k n = \mu_k mg$. The friction force is directed opposite to the displacement, so the work done by friction is

$$W_{\text{fric}} = f_k d \cos 180° = -f_k d = -\mu_k mgd$$

The total work is the sum of W_{fric} and the work done by the spring, $-\frac{1}{2}kd^2$. The work–energy theorem then says that

$$-\mu_k mgd - \tfrac{1}{2}kd^2 = 0 - \tfrac{1}{2}mv_1^2 \qquad \text{or}$$

$$\tfrac{1}{2}kd^2 + \mu_k mgd - \tfrac{1}{2}mv_1^2 = 0$$

This is a quadratic equation for d. The solutions are

$$d = -\frac{\mu_k mg}{k} \pm \sqrt{\left(\frac{\mu_k mg}{k}\right)^2 + \frac{mv_1^2}{k}}$$

We have

$$\frac{\mu_k mg}{k} = \frac{(0.47)(0.100 \text{ kg})(9.80 \text{ m/s}^2)}{20.0 \text{ N/m}} = 0.02303 \text{ m}$$

$$\frac{mv_1^2}{k} = \frac{(0.100 \text{ kg})(1.50 \text{ m/s})^2}{20.0 \text{ N/m}} = 0.01125 \text{ m}^2$$

so

$$d = -(0.02303 \text{ m}) \pm \sqrt{(0.02303 \text{ m})^2 + 0.01125 \text{ m}^2}$$

$$= 0.086 \text{ m} \quad \text{or} \quad -0.132 \text{ m}$$

The quantity d is a positive displacement, so only the positive value of d makes sense. Thus with friction the glider moves a distance $d = 0.086$ m = 8.6 cm.

EVALUATE: If we set $\mu_k = 0$, our algebraic solution for d in part (b) reduces to $d = v_1\sqrt{m/k}$, the zero-friction result from part (a). With friction, the glider goes a shorter distance. Again the glider stops instantaneously, and again the spring force pulls it toward the left; whether it moves or not depends on how great the *static* friction force is. How large would the coefficient of static friction μ_s have to be to keep the glider from springing back to the left?

Work–Energy Theorem for Motion Along a Curve

We can generalize our definition of work further to include a force that varies in direction as well as magnitude, and a displacement that lies along a curved path. **Figure 6.23a** shows a particle moving from P_1 to P_2 along a curve. We divide the curve between these points into many infinitesimal vector displacements, and we call a typical one of these $d\vec{l}$. Each $d\vec{l}$ is tangent to the path at its position. Let $\vec{F}$ be the force at a typical point along the path, and let ϕ be the angle between $\vec{F}$ and $d\vec{l}$ at this point. Then the small element of work dW done on the particle during the displacement $d\vec{l}$ may be written as

$$dW = \vec{F} \cdot d\vec{l} = F\cos\phi\, dl = F_\parallel\, dl$$

where $F_\parallel = F\cos\phi$ is the component of $\vec{F}$ in the direction parallel to $d\vec{l}$ (Fig. 6.23b). The work done by $\vec{F}$ on the particle as it moves from P_1 to P_2 is

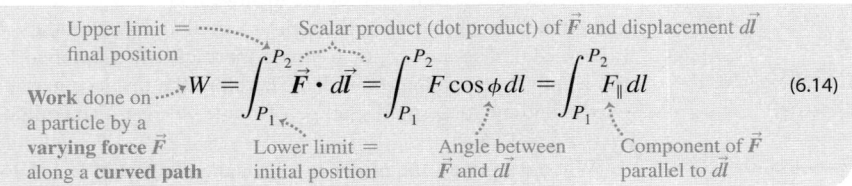

Upper limit = final position		Scalar product (dot product) of $\vec{F}$ and displacement $d\vec{l}$		

Work done on a particle by a **varying force** $\vec{F}$ along a **curved path**

$$W = \int_{P_1}^{P_2} \vec{F} \cdot d\vec{l} = \int_{P_1}^{P_2} F\cos\phi\, dl = \int_{P_1}^{P_2} F_\parallel\, dl \qquad (6.14)$$

Lower limit = initial position   Angle between $\vec{F}$ and $d\vec{l}$ Component of $\vec{F}$ parallel to $d\vec{l}$

The integral in Eq. (6.14) (shown in three versions) is called a *line integral*. We'll see shortly how to evaluate an integral of this kind.

6.23 A particle moves along a curved path from point P_1 to P_2, acted on by a force $\vec{F}$ that varies in magnitude and direction.

(a)

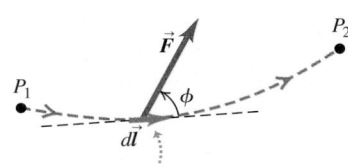

During an infinitesimal displacement $d\vec{l}$, the force $\vec{F}$ does work dW on the particle:
$$dW = \vec{F} \cdot d\vec{l} = F\cos\phi\, dl$$

(b)

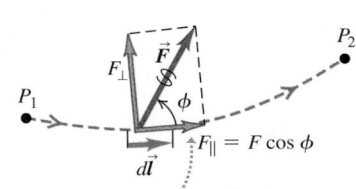

Only the component of $\vec{F}$ parallel to the displacement, $F_\parallel = F\cos\phi$, contributes to the work done by $\vec{F}$.

We can now show that the work–energy theorem, Eq. (6.6), holds true even with varying forces and a displacement along a curved path. The force $\vec{F}$ is essentially constant over any given infinitesimal segment $d\vec{l}$ of the path, so we can apply the work–energy theorem for straight-line motion to that segment. Thus the change in the particle's kinetic energy K over that segment equals the work $dW = F_\parallel dl = \vec{F} \cdot d\vec{l}$ done on the particle. Adding up these infinitesimal quantities of work from all the segments along the whole path gives the total work done, Eq. (6.14), which equals the total change in kinetic energy over the whole path. So $W_{\text{tot}} = \Delta K = K_2 - K_1$ is true *in general,* no matter what the path and no matter what the character of the forces. This can be proved more rigorously by using steps like those in Eqs. (6.11) through (6.13).

Note that only the component of the net force parallel to the path, $F_\parallel$, does work on the particle, so only this component can change the speed and kinetic energy of the particle. The component perpendicular to the path, $F_\perp = F \sin\phi$, has no effect on the particle's speed; it acts only to change the particle's direction.

To evaluate the line integral in Eq. (6.14) in a specific problem, we need some sort of detailed description of the path and of the way in which $\vec{F}$ varies along the path. We usually express the line integral in terms of some scalar variable, as in the following example.

EXAMPLE 6.8   MOTION ON A CURVED PATH

At a family picnic you are appointed to push your obnoxious cousin Throckmorton in a swing (**Fig. 6.24a**). His weight is w, the length of the chains is R, and you push Throcky until the chains make an angle θ_0 with the vertical. To do this, you exert a varying horizontal force $\vec{F}$ that starts at zero and gradually increases just enough that Throcky and the swing move very slowly and remain very nearly in equilibrium throughout the process. (a) What is the total work done on Throcky by all forces? (b) What is the work done by the tension T in the chains? (c) What is the work you do by exerting force $\vec{F}$? (Ignore the weight of the chains and seat.)

SOLUTION

IDENTIFY and SET UP: The motion is along a curve, so we'll use Eq. (6.14) to calculate the work done by the net force, by the tension force, and by the force $\vec{F}$. Figure 6.24b shows our free-body diagram and coordinate system for some arbitrary point in Throcky's motion. We have replaced the sum of the tensions in the two chains with a single tension T.

6.24 (a) Pushing cousin Throckmorton in a swing. (b) Our free-body diagram.

(a)

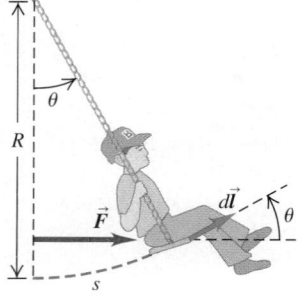

(b) Free-body diagram for Throckmorton (neglecting the weight of the chains and seat)

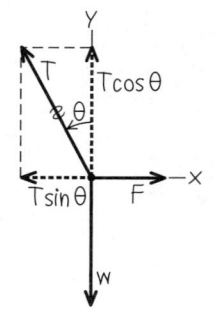

EXECUTE: (a) There are two ways to find the total work done during the motion: (1) by calculating the work done by each force and then adding those quantities, and (2) by calculating the work done by the net force. The second approach is far easier here because Throcky is nearly in equilibrium at every point. Hence the net force on him is zero, the integral of the net force in Eq. (6.14) is zero, and the total work done on him is zero.

(b) It's also easy to find the work done by the chain tension T because this force is perpendicular to the direction of motion at all points along the path. Hence at all points the angle between the chain tension and the displacement vector $d\vec{l}$ is 90° and the scalar product in Eq. (6.14) is zero. Thus the chain tension does zero work.

(c) To compute the work done by $\vec{F}$, we need to calculate the line integral in Eq. (6.14). Inside the integral is the quantity $F\cos\phi\,dl$; let's see how to express each term in this quantity.

Figure 6.24a shows that the angle between $\vec{F}$ and $d\vec{l}$ is θ, so we replace ϕ in Eq. (6.14) with θ. The value of θ changes as Throcky moves.

To find the magnitude F of force $\vec{F}$, note that the net force on Throcky is zero (he is nearly in equilibrium at all points), so $\Sigma F_x = 0$ and $\Sigma F_y = 0$. From Fig. 6.24b,

$$\Sigma F_x = F + (-T\sin\theta) = 0 \qquad \Sigma F_y = T\cos\theta + (-w) = 0$$

If you eliminate T from these two equations, you can show that $F = w\tan\theta$. As the angle θ increases, the tangent increases and F increases (you have to push harder).

To find the magnitude dl of the infinitesimal displacement $d\vec{l}$, note that Throcky moves through a circular arc of radius R (Fig. 6.24a). The arc length s equals the radius R multiplied by the length θ (in radians): $s = R\theta$. Therefore the displacement $d\vec{l}$ corresponding to a small change of angle $d\theta$ has a magnitude $dl = ds = R\,d\theta$.

When we put all the pieces together, the integral in Eq. (6.14) becomes

$$W = \int_{P_1}^{P_2} F\cos\phi\,dl = \int_0^{\theta_0} (w\tan\theta)\cos\theta(R\,d\theta) = \int_0^{\theta_0} wR\sin\theta\,d\theta$$

(Recall that $\tan\theta = \sin\theta/\cos\theta$, so $\tan\theta\cos\theta = \sin\theta$.) We've converted the *line* integral into an *ordinary* integral in terms of the angle θ. The limits of integration are from the starting position at $\theta = 0$ to the final position at $\theta = \theta_0$. The final result is

$$W = wR\int_0^{\theta_0}\sin\theta\,d\theta = -wR\cos\theta\big|_0^{\theta_0} = -wR(\cos\theta_0 - 1)$$

$$= wR(1 - \cos\theta_0)$$

EVALUATE: If $\theta_0 = 0$, there is no displacement; then $\cos\theta_0 = 1$ and $W = 0$, as we should expect. As θ_0 increases, $\cos\theta_0$ decreases and $W = wR(1 - \cos\theta_0)$ increases. So the farther along the arc you push Throcky, the more work you do. You can confirm that the quantity $R(1 - \cos\theta_0)$ is equal to h, the increase in Throcky's height during the displacement. So the work that you do to raise Throcky is just equal to his weight multiplied by the height that you raise him.

We can check our results by calculating the work done by the force of gravity $\vec{w}$. From part (a) the total work done on Throcky is zero, and from part (b) the work done by tension is zero. So gravity must do a negative amount of work that just balances the positive work done by the force $\vec{F}$ that we calculated in part (c).

For variety, let's calculate the work done by gravity by using the form of Eq. (6.14) that involves the quantity $\vec{F} \cdot d\vec{l}$, and express the force $\vec{w}$ and displacement $d\vec{l}$ in terms of their x- and y-components. The force of gravity has zero x-component and a y-component of $-w$. Figure 6.24a shows that $d\vec{l}$ has a magnitude of ds, an x-component of $ds\cos\theta$, and a y-component of $ds\sin\theta$.

So

$$\vec{w} = \hat{\jmath}(-w)$$
$$d\vec{l} = \hat{\imath}(ds\cos\theta) + \hat{\jmath}(ds\sin\theta)$$

Use Eq. (1.19) to calculate the scalar product $\vec{w} \cdot d\vec{l}$:

$$\vec{w} \cdot d\vec{l} = (-w)(ds\sin\theta) = -w\sin\theta\,ds$$

Using $ds = R\,d\theta$, we find the work done by the force of gravity:

$$\int_{P_1}^{P_2}\vec{w} \cdot d\vec{l} = \int_0^{\theta_0}(-w\sin\theta)R\,d\theta = -wR\int_0^{\theta_0}\sin\theta\,d\theta$$

$$= -wR(1 - \cos\theta_0)$$

The work done by gravity is indeed the negative of the work done by force $\vec{F}$ that we calculated in part (c). Gravity does negative work because the force pulls downward while Throcky moves upward.

As we saw earlier, $R(1 - \cos\theta_0)$ is equal to h, the increase in Throcky's height during the displacement. So the work done by gravity along the curved path is $-mgh$, the *same* work that gravity would have done if Throcky had moved *straight upward* a distance h. This is an example of a more general result that we'll prove in Section 7.1.

TEST YOUR UNDERSTANDING OF SECTION 6.3 In Example 5.20 (Section 5.4) we examined a conical pendulum. The speed of the pendulum bob remains constant as it travels around the circle shown in Fig. 5.32a. (a) Over one complete circle, how much work does the tension force F do on the bob? (i) A positive amount; (ii) a negative amount; (iii) zero. (b) Over one complete circle, how much work does the weight do on the bob? (i) A positive amount; (ii) a negative amount; (iii) zero. ▌

6.4 POWER

The definition of work makes no reference to the passage of time. If you lift a barbell weighing 100 N through a vertical distance of 1.0 m at constant velocity, you do $(100\,\text{N})(1.0\,\text{m}) = 100$ J of work whether it takes you 1 second, 1 hour, or 1 year to do it. But often we need to know how quickly work is done. We describe this in terms of *power*. In ordinary conversation the word "power" is often synonymous with "energy" or "force." In physics we use a much more precise definition: **Power** is the time *rate* at which work is done. Like work and energy, power is a scalar quantity.

The average work done per unit time, or **average power** P_{av}, is defined to be

Average power during time interval Δt ⸱⸱⸱⸱⸱▸$P_{av} = \dfrac{\Delta W}{\Delta t}$ ◂⸱⸱⸱⸱Work done during time interval ⸱⸱⸱⸱⸱Duration of time interval (6.15)

The rate at which work is done might not be constant. We define **instantaneous power** P as the quotient in Eq. (6.15) as Δt approaches zero:

Instantaneous ⸱⸱⸱⸱▸power $P = \lim\limits_{\Delta t\to 0}\dfrac{\Delta W}{\Delta t} = \dfrac{dW}{dt}$ ⸱⸱⸱Time rate of doing work (6.16)

Average power over infinitesimally short time interval

The SI unit of power is the **watt** (W), named for the English inventor James Watt. One watt equals 1 joule per second: $1\,\text{W} = 1\,\text{J/s}$ (**Fig. 6.25**). The kilowatt $(1\,\text{kW} = 10^3\,\text{W})$ and the megawatt $(1\,\text{MW} = 10^6\,\text{W})$ are also commonly used.

6.25 The same amount of work is done in both of these situations, but the power (the rate at which work is done) is different.

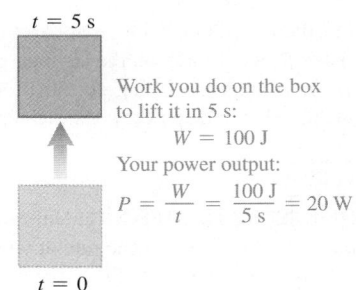

$t = 5$ s

Work you do on the box to lift it in 5 s:
$W = 100$ J
Your power output:
$P = \dfrac{W}{t} = \dfrac{100\ \text{J}}{5\ \text{s}} = 20$ W

$t = 0$

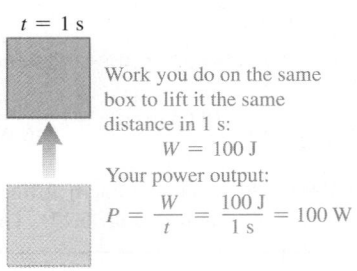

$t = 1$ s

Work you do on the same box to lift it the same distance in 1 s:
$W = 100$ J
Your power output:
$P = \dfrac{W}{t} = \dfrac{100\ \text{J}}{1\ \text{s}} = 100$ W

$t = 0$

6.26 A one-horsepower (746-W) propulsion system.

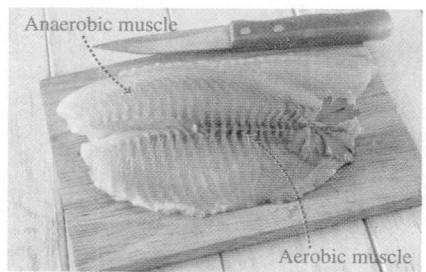

Anaerobic muscle

Aerobic muscle

Another common unit of power is the *horsepower* (hp) (**Fig. 6.26**). The value of this unit derives from experiments by James Watt, who measured that in one minute a horse could do an amount of work equivalent to lifting 33,000 pounds (lb) a distance of 1 foot (ft), or 33,000 ft · lb. Thus 1 hp = 33,000 ft · lb/min. Using 1 ft = 0.3048 m, 1 lb = 4.448 N, and 1 min = 60 s, we can show that

$$1 \text{ hp} = 746 \text{ W} = 0.746 \text{ kW}$$

The watt is a familiar unit of *electrical* power; a 100-W light bulb converts 100 J of electrical energy into light and heat each second. But there's nothing inherently electrical about a watt. A light bulb could be rated in horsepower, and an engine can be rated in kilowatts.

The *kilowatt-hour* (kW · h) is the usual commercial unit of electrical energy. One kilowatt-hour is the total work done in 1 hour (3600 s) when the power is 1 kilowatt (10^3 J/s), so

$$1 \text{ kW} \cdot \text{h} = (10^3 \text{ J/s})(3600 \text{ s}) = 3.6 \times 10^6 \text{ J} = 3.6 \text{ MJ}$$

The kilowatt-hour is a unit of *work* or *energy,* not power.

In mechanics we can also express power in terms of force and velocity. Suppose that a force $\vec{F}$ acts on a body while it undergoes a vector displacement $\Delta \vec{s}$. If $F_\parallel$ is the component of $\vec{F}$ tangent to the path (parallel to $\Delta \vec{s}$), then the work done by the force is $\Delta W = F_\parallel \Delta s$. The average power is

$$P_{av} = \frac{F_\parallel \Delta s}{\Delta t} = F_\parallel \frac{\Delta s}{\Delta t} = F_\parallel v_{av} \tag{6.17}$$

Instantaneous power P is the limit of this expression as $\Delta t \rightarrow 0$:

$$P = F_\parallel v \tag{6.18}$$

where v is the magnitude of the instantaneous velocity. We can also express Eq. (6.18) in terms of the scalar product:

Instantaneous power for a force doing work on a particle ⋯⋯→ $P = \vec{F} \cdot \vec{v}$ ←⋯ Force that acts on particle ⋯⋯ Velocity of particle (6.19)

EXAMPLE 6.9 FORCE AND POWER

Each of the four jet engines on an Airbus A380 airliner develops a thrust (a forward force on the airliner) of 322,000 N (72,000 lb). When the airplane is flying at 250 m/s (900 km/h, or roughly 560 mi/h), what horsepower does each engine develop?

SOLUTION

IDENTIFY, SET UP, and EXECUTE: Our target variable is the instantaneous power P, which is the rate at which the thrust does work. We use Eq. (6.18). The thrust is in the direction of motion, so $F_\parallel$ is just equal to the thrust. At $v = 250$ m/s, the power developed by each engine is

$$P = F_\parallel v = (3.22 \times 10^5 \text{ N})(250 \text{ m/s}) = 8.05 \times 10^7 \text{ W}$$

$$= (8.05 \times 10^7 \text{ W})\frac{1 \text{ hp}}{746 \text{ W}} = 108{,}000 \text{ hp}$$

EVALUATE: The speed of modern airliners is directly related to the power of their engines (**Fig. 6.27**). The largest propeller-driven

6.27 (a) Propeller-driven and (b) jet airliners.

(a)  (b)

airliners of the 1950s had engines that each developed about 3400 hp (2.5×10^6 W), giving them maximum speeds of about 600 km/h (370 mi/h). Each engine on an Airbus A380 develops more than 30 times more power, enabling it to fly at about 900 km/h (560 mi/h) and to carry a much heavier load.

If the engines are at maximum thrust while the airliner is at rest on the ground so that $v = 0$, the engines develop *zero* power. Force and power are not the same thing!

EXAMPLE 6.10 A "POWER CLIMB"

A 50.0-kg marathon runner runs up the stairs to the top of Chicago's 443-m-tall Willis Tower, the second tallest building in the United States (**Fig. 6.28**). To lift herself to the top in 15.0 minutes, what must be her average power output? Express your answer in watts, in kilowatts, and in horsepower.

SOLUTION

IDENTIFY and SET UP: We'll treat the runner as a particle of mass m. Her average power output P_{av} must be enough to lift her at constant speed against gravity.

We can find P_{av} in two ways: (1) by determining how much work she must do and dividing that quantity by the elapsed time, as in Eq. (6.15), or (2) by calculating the average upward force she must exert (in the direction of the climb) and multiplying that quantity by her upward velocity, as in Eq. (6.17).

EXECUTE: (1) As in Example 6.8, lifting a mass m against gravity requires an amount of work equal to the weight mg multiplied by the height h it is lifted. Hence the work the runner must do is

$$W = mgh = (50.0 \text{ kg})(9.80 \text{ m/s}^2)(443 \text{ m})$$
$$= 2.17 \times 10^5 \text{ J}$$

She does this work in a time 15.0 min = 900 s, so from Eq. (6.15) the average power is

$$P_{av} = \frac{2.17 \times 10^5 \text{ J}}{900 \text{ s}} = 241 \text{ W} = 0.241 \text{ kW} = 0.323 \text{ hp}$$

(2) The force exerted is vertical and the average vertical component of velocity is $(443 \text{ m})/(900 \text{ s}) = 0.492$ m/s, so from Eq. (6.17) the average power is

$$P_{av} = F_{\parallel}v_{av} = (mg)v_{av}$$
$$= (50.0 \text{ kg})(9.80 \text{ m/s}^2)(0.492 \text{ m/s}) = 241 \text{ W}$$

which is the same result as before.

EVALUATE: The runner's *total* power output will be several times greater than 241 W. The reason is that the runner isn't really a particle but a collection of parts that exert forces on each other and do work, such as the work done to inhale and exhale and to make her arms and legs swing. What we've calculated is only the part of her power output that lifts her to the top of the building.

6.28 How much power is required to run up the stairs of Chicago's Willis Tower in 15 minutes?

TEST YOUR UNDERSTANDING OF SECTION 6.4 The air surrounding an airplane in flight exerts a drag force that acts opposite to the airplane's motion. When the Airbus A380 in Example 6.9 is flying in a straight line at a constant altitude at a constant 250 m/s, what is the rate at which the drag force does work on it? (i) 432,000 hp; (ii) 108,000 hp; (iii) 0; (iv) −108,000 hp; (v) −432,000 hp. ∎

Work done by a force: When a constant force $\vec{F}$ acts on a particle that undergoes a straight-line displacement $\vec{s}$, the work done by the force on the particle is defined to be the scalar product of $\vec{F}$ and $\vec{s}$. The unit of work in SI units is 1 joule = 1 newton-meter (1 J = 1 N · m). Work is a scalar quantity; it can be positive or negative, but it has no direction in space. (See Examples 6.1 and 6.2.)

$$W = \vec{F} \cdot \vec{s} = Fs\cos\phi$$

$$\phi = \text{angle between } \vec{F} \text{ and } \vec{s}$$

(6.2), (6.3)

Kinetic energy: The kinetic energy K of a particle equals the amount of work required to accelerate the particle from rest to speed v. It is also equal to the amount of work the particle can do in the process of being brought to rest. Kinetic energy is a scalar that has no direction in space; it is always positive or zero. Its units are the same as the units of work: $1\text{ J} = 1\text{ N} \cdot \text{m} = 1\text{ kg} \cdot \text{m}^2/\text{s}^2$.

$$K = \tfrac{1}{2}mv^2$$

(6.5)

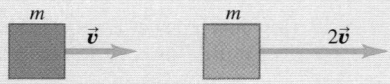

Doubling m doubles K.

Doubling v quadruples K.

The work–energy theorem: When forces act on a particle while it undergoes a displacement, the particle's kinetic energy changes by an amount equal to the total work done on the particle by all the forces. This relationship, called the work–energy theorem, is valid whether the forces are constant or varying and whether the particle moves along a straight or curved path. It is applicable only to bodies that can be treated as particles. (See Examples 6.3–6.5.)

$$W_{\text{tot}} = K_2 - K_1 = \Delta K$$

(6.6)

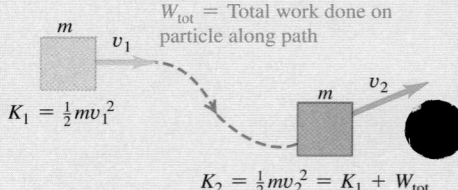

$W_{\text{tot}} = $ Total work done on particle along path

$K_1 = \tfrac{1}{2}mv_1^2$

$K_2 = \tfrac{1}{2}mv_2^2 = K_1 + W_{\text{tot}}$

Work done by a varying force or on a curved path: When a force varies during a straight-line displacement, the work done by the force is given by an integral, Eq. (6.7). (See Examples 6.6 and 6.7.) When a particle follows a curved path, the work done on it by a force $\vec{F}$ is given by an integral that involves the angle ϕ between the force and the displacement. This expression is valid even if the force magnitude and the angle ϕ vary during the displacement. (See Example 6.8.)

$$W = \int_{x_1}^{x_2} F_x \, dx$$

(6.7)

$$W = \int_{P_1}^{P_2} \vec{F} \cdot d\vec{l}$$

$$= \int_{P_1}^{P_2} F\cos\phi \, dl = \int_{P_1}^{P_2} F_\parallel \, dl$$

(6.14)

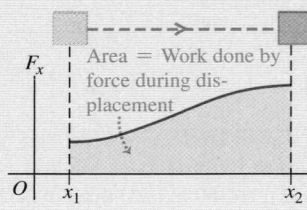

Area = Work done by force during displacement

Power: Power is the time rate of doing work. The average power P_{av} is the amount of work ΔW done in time Δt divided by that time. The instantaneous power is the limit of the average power as Δt goes to zero. When a force $\vec{F}$ acts on a particle moving with velocity $\vec{v}$, the instantaneous power (the rate at which the force does work) is the scalar product of $\vec{F}$ and $\vec{v}$. Like work and kinetic energy, power is a scalar quantity. The SI unit of power is 1 watt = 1 joule/second (1 W = 1 J/s). (See Examples 6.9 and 6.10.)

$$P_{\text{av}} = \frac{\Delta W}{\Delta t}$$

(6.15)

$$P = \lim_{\Delta t \to 0} \frac{\Delta W}{\Delta t} = \frac{dW}{dt}$$

(6.16)

$$P = \vec{F} \cdot \vec{v}$$

(6.19)

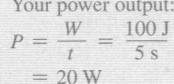

$t = 5$ s Work you do on the box to lift it in 5 s: $W = 100$ J

Your power output:

$P = \dfrac{W}{t} = \dfrac{100\text{ J}}{5\text{ s}}$

$t = 0$ $= 20$ W

BRIDGING PROBLEM | A SPRING THAT DISOBEYS HOOKE'S LAW

Consider a hanging spring of negligible mass that does *not* obey Hooke's law. When the spring is pulled downward by a distance x, the spring exerts an upward force of magnitude αx^2, where α is a positive constant. Initially the hanging spring is relaxed (not extended). We then attach a block of mass m to the spring and release the block. The block stretches the spring as it falls (**Fig. 6.29**). (a) How fast is the block moving when it has fallen a distance x_1? (b) At what rate does the spring do work on the block at this point? (c) Find the maximum distance x_2 that the spring stretches. (d) Will the block *remain* at the point found in part (c)?

SOLUTION GUIDE

IDENTIFY and SET UP

1. The spring force in this problem isn't constant, so you have to use the work–energy theorem. You'll also need Eq. (6.7) to find the work done by the spring over a given displacement.

6.29 The block is attached to a spring that does not obey Hooke's law.

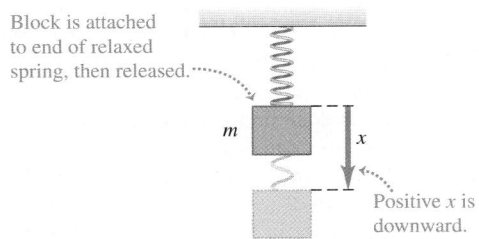

Block is attached to end of relaxed spring, then released.

m

x

Positive x is downward.

2. Draw a free-body diagram for the block, including your choice of coordinate axes. Note that x represents how far the spring is *stretched,* so choose the positive x-direction to be downward, as in Fig. 6.29. On your coordinate axis, label the points $x = x_1$ and $x = x_2$.

3. Make a list of the unknown quantities, and decide which of these are the target variables.

EXECUTE

4. Calculate the work done on the block by the spring as the block falls an arbitrary distance x. (The integral isn't a difficult one. Use Appendix B if you need a reminder.) Is the work done by the spring positive, negative, or zero?

5. Calculate the work done on the block by any other forces as the block falls an arbitrary distance x. Is this work positive, negative, or zero?

6. Use the work–energy theorem to find the target variables. (You'll also need an equation for power.) *Hint:* When the spring is at its maximum stretch, what is the speed of the block?

7. To answer part (d), consider the *net* force that acts on the block when it is at the point found in part (c).

EVALUATE

8. We learned in Section 2.5 that after an object dropped from rest has fallen freely a distance x_1, its speed is $\sqrt{2gx_1}$. Use this to decide whether your answer in part (a) makes sense. In addition, ask yourself whether the algebraic sign of your answer in part (b) makes sense.

9. Find the value of x where the net force on the block would be zero. How does this compare to your result for x_2? Is this consistent with your answer in part (d)?

Problems

For assigned homework and other learning materials, go to MasteringPhysics®. **MP**

•, ••, •••: Difficulty levels. CP: Cumulative problems incorporating material from earlier chapters. CALC: Problems requiring calculus. DATA: Problems involving real data, scientific evidence, experimental design, and/or statistical reasoning. BIO: Biosciences problems.

DISCUSSION QUESTIONS

Q6.1 The sign of many physical quantities depends on the choice of coordinates. For example, a_y for free-fall motion can be negative or positive, depending on whether we choose upward or downward as positive. Is the same true of work? In other words, can we make positive work negative by a different choice of coordinates? Explain.

Q6.2 An elevator is hoisted by its cables at constant speed. Is the total work done on the elevator positive, negative, or zero? Explain.

Q6.3 A rope tied to a body is pulled, causing the body to accelerate. But according to Newton's third law, the body pulls back on the rope with a force of equal magnitude and opposite direction. Is the total work done then zero? If so, how can the body's kinetic energy change? Explain.

Q6.4 If it takes total work W to give an object a speed v and kinetic energy K, starting from rest, what will be the object's speed (in terms of v) and kinetic energy (in terms of K) if we do twice as much work on it, again starting from rest?

Q6.5 If there is a net nonzero force on a moving object, can the total work done on the object be zero? Explain, using an example.

Q6.6 In Example 5.5 (Section 5.1), how does the work done on the bucket by the tension in the cable compare with the work done on the cart by the tension in the cable?

Q6.7 In the conical pendulum of Example 5.20 (Section 5.4), which of the forces do work on the bob while it is swinging?

Q6.8 For the cases shown in **Fig. Q6.8**, the object is released from rest at the top and feels no friction or air resistance. In which (if any) cases will the mass have (i) the greatest speed at the bottom and (ii) the most work done on it by the time it reaches the bottom?

Figure **Q6.8**

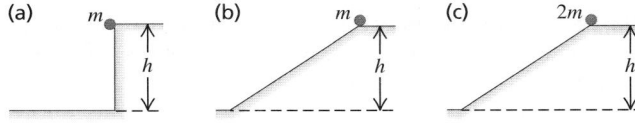

Q6.9 A force $\vec{F}$ is in the *x*-direction and has a magnitude that depends on *x*. Sketch a possible graph of *F* versus *x* such that the force does zero work on an object that moves from x_1 to x_2, even though the force magnitude is not zero at all *x* in this range.

Q6.10 Does a car's kinetic energy change more when the car speeds up from 10 to 15 m/s or from 15 to 20 m/s? Explain.

Q6.11 A falling brick has a mass of 1.5 kg and is moving straight downward with a speed of 5.0 m/s. A 1.5-kg physics book is sliding across the floor with a speed of 5.0 m/s. A 1.5-kg melon is traveling with a horizontal velocity component 3.0 m/s to the right and a vertical component 4.0 m/s upward. Do all of these objects have the same velocity? Do all of them have the same kinetic energy? For both questions, give your reasoning.

Q6.12 Can the *total* work done on an object during a displacement be negative? Explain. If the total work is negative, can its magnitude be larger than the initial kinetic energy of the object? Explain.

Q6.13 A net force acts on an object and accelerates it from rest to a speed v_1. In doing so, the force does an amount of work W_1. By what factor must the work done on the object be increased to produce three times the final speed, with the object again starting from rest?

Q6.14 A truck speeding down the highway has a lot of kinetic energy relative to a stopped state trooper but no kinetic energy relative to the truck driver. In these two frames of reference, is the same amount of work required to stop the truck? Explain.

Q6.15 You are holding a briefcase by the handle, with your arm straight down by your side. Does the force your hand exerts do work on the briefcase when (a) you walk at a constant speed down a horizontal hallway and (b) you ride an escalator from the first to second floor of a building? In both cases justify your answer.

Q6.16 When a book slides along a tabletop, the force of friction does negative work on it. Can friction ever do *positive* work? Explain. (*Hint:* Think of a box in the back of an accelerating truck.)

Q6.17 Time yourself while running up a flight of steps, and compute the average rate at which you do work against the force of gravity. Express your answer in watts and in horsepower.

Q6.18 Fractured Physics. Many terms from physics are badly misused in everyday language. In both cases, explain the errors involved. (a) A *strong* person is called *powerful*. What is wrong with this use of *power*? (b) When a worker carries a bag of concrete along a level construction site, people say he did a lot of *work*. Did he?

Q6.19 An advertisement for a portable electrical generating unit claims that the unit's diesel engine produces 28,000 hp to drive an electrical generator that produces 30 MW of electrical power. Is this possible? Explain.

Q6.20 A car speeds up while the engine delivers constant power. Is the acceleration greater at the beginning of this process or at the end? Explain.

Q6.21 Consider a graph of instantaneous power versus time, with the vertical *P*-axis starting at $P = 0$. What is the physical significance of the area under the *P*-versus-*t* curve between vertical lines at t_1 and t_2? How could you find the average power from the graph? Draw a *P*-versus-*t* curve that consists of two straight-line sections and for which the peak power is equal to twice the average power.

Q6.22 A nonzero net force acts on an object. Is it possible for any of the following quantities to be constant: the object's (a) speed; (b) velocity; (c) kinetic energy?

Q6.23 When a certain force is applied to an ideal spring, the spring stretches a distance *x* from its unstretched length and does work *W*. If instead twice the force is applied, what distance (in terms of *x*) does the spring stretch from its unstretched length, and how much work (in terms of *W*) is required to stretch it this distance?

Q6.24 If work *W* is required to stretch a spring a distance *x* from its unstretched length, what work (in terms of *W*) is required to stretch the spring an *additional* distance *x*?

EXERCISES

Section 6.1 Work

6.1 • You push your physics book 1.50 m along a horizontal table-top with a horizontal push of 2.40 N while the opposing force of friction is 0.600 N. How much work does each of the following forces do on the book: (a) your 2.40-N push, (b) the friction force, (c) the normal force from the tabletop, and (d) gravity? (e) What is the net work done on the book?

6.2 • Using a cable with a tension of 1350 N, a tow truck pulls a car 5.00 km along a horizontal roadway. (a) How much work does the cable do on the car if it pulls horizontally? If it pulls at 35.0° above the horizontal? (b) How much work does the cable do on the tow truck in both cases of part (a)? (c) How much work does gravity do on the car in part (a)?

6.3 • A factory worker pushes a 30.0-kg crate a distance of 4.5 m along a level floor at constant velocity by pushing horizontally on it. The coefficient of kinetic friction between the crate and the floor is 0.25. (a) What magnitude of force must the worker apply? (b) How much work is done on the crate by this force? (c) How much work is done on the crate by friction? (d) How much work is done on the crate by the normal force? By gravity? (e) What is the total work done on the crate?

6.4 •• Suppose the worker in Exercise 6.3 pushes downward at an angle of 30° below the horizontal. (a) What magnitude of force must the worker apply to move the crate at constant velocity? (b) How much work is done on the crate by this force when the crate is pushed a distance of 4.5 m? (c) How much work is done on the crate by friction during this displacement? (d) How much work is done on the crate by the normal force? By gravity? (e) What is the total work done on the crate?

6.5 •• A 75.0-kg painter climbs a ladder that is 2.75 m long and leans against a vertical wall. The ladder makes a 30.0° angle with the wall. (a) How much work does gravity do on the painter? (b) Does the answer to part (a) depend on whether the painter climbs at constant speed or accelerates up the ladder?

6.6 •• Two tugboats pull a disabled supertanker. Each tug exerts a constant force of 1.80×10^6 N, one 14° west of north and the other 14° east of north, as they pull the tanker 0.75 km toward the north. What is the total work they do on the supertanker?

6.7 • Two blocks are connected by a very light string passing over a massless and frictionless pulley (**Fig. E6.7**). Traveling at constant speed, the 20.0-N block moves 75.0 cm to the right and the 12.0-N block moves 75.0 cm downward. How much work is done

Figure **E6.7**

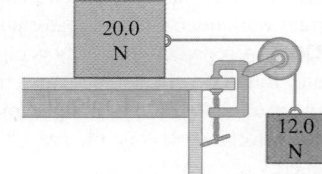

(a) on the 12.0-N block by (i) gravity and (ii) the tension in the string? (b) How much work is done on the 20.0-N block by

(i) gravity, (ii) the tension in the string, (iii) friction, and (iv) the normal force? (c) Find the total work done on each block.

6.8 •• A loaded grocery cart is rolling across a parking lot in a strong wind. You apply a constant force $\vec{F} = (30 \text{ N})\hat{\imath} - (40 \text{ N})\hat{\jmath}$ to the cart as it undergoes a displacement $\vec{s} = (-9.0 \text{ m})\hat{\imath} - (3.0 \text{ m})\hat{\jmath}$. How much work does the force you apply do on the grocery cart?

6.9 • A 0.800-kg ball is tied to the end of a string 1.60 m long and swung in a vertical circle. (a) During one complete circle, starting anywhere, calculate the total work done on the ball by (i) the tension in the string and (ii) gravity. (b) Repeat part (a) for motion along the semicircle from the lowest to the highest point on the path.

6.10 •• A 12.0-kg package in a mail-sorting room slides 2.00 m down a chute that is inclined at 53.0° below the horizontal. The coefficient of kinetic friction between the package and the chute's surface is 0.40. Calculate the work done on the package by (a) friction, (b) gravity, and (c) the normal force. (d) What is the net work done on the package?

6.11 • A 128.0-N carton is pulled up a frictionless baggage ramp inclined at 30.0° above the horizontal by a rope exerting a 72.0-N pull parallel to the ramp's surface. If the carton travels 5.20 m along the surface of the ramp, calculate the work done on it by (a) the rope, (b) gravity, and (c) the normal force of the ramp. (d) What is the net work done on the carton? (e) Suppose that the rope is angled at 50.0° above the horizontal, instead of being parallel to the ramp's surface. How much work does the rope do on the carton in this case?

6.12 •• A boxed 10.0-kg computer monitor is dragged by friction 5.50 m upward along a conveyor belt inclined at an angle of 36.9° above the horizontal. If the monitor's speed is a constant 2.10 cm/s, how much work is done on the monitor by (a) friction, (b) gravity, and (c) the normal force of the conveyor belt?

6.13 •• A large crate sits on the floor of a warehouse. Paul and Bob apply constant horizontal forces to the crate. The force applied by Paul has magnitude 48.0 N and direction 61.0° south of west. How much work does Paul's force do during a displacement of the crate that is 12.0 m in the direction 22.0° east of north?

6.14 •• You apply a constant force $\vec{F} = (-68.0 \text{ N})\hat{\imath} + (36.0 \text{ N})\hat{\jmath}$ to a 380-kg car as the car travels 48.0 m in a direction that is 240.0° counterclockwise from the $+x$-axis. How much work does the force you apply do on the car?

6.15 •• On a farm, you are pushing on a stubborn pig with a constant horizontal force with magnitude 30.0 N and direction 37.0° counterclockwise from the $+x$-axis. How much work does this force do during a displacement of the pig that is (a) $\vec{s} = (5.00 \text{ m})\hat{\imath}$; (b) $\vec{s} = -(6.00 \text{ m})\hat{\jmath}$; (c) $\vec{s} = -(2.00 \text{ m})\hat{\imath} + (4.00 \text{ m})\hat{\jmath}$?

Section 6.2 Kinetic Energy and the Work–Energy Theorem

6.16 •• A 1.50-kg book is sliding along a rough horizontal surface. At point A it is moving at 3.21 m/s, and at point B it has slowed to 1.25 m/s. (a) How much work was done on the book between A and B? (b) If -0.750 J of work is done on the book from B to C, how fast is it moving at point C? (c) How fast would it be moving at C if $+0.750$ J of work was done on it from B to C?

6.17 •• BIO **Animal Energy.** Adult cheetahs, the fastest of the great cats, have a mass of about 70 kg and have been clocked to run at up to 72 mi/h (32 m/s). (a) How many joules of kinetic energy does such a swift cheetah have? (b) By what factor would its kinetic energy change if its speed were doubled?

6.18 • **Some Typical Kinetic Energies.** (a) In the Bohr model of the atom, the ground-state electron in hydrogen has an orbital speed of 2190 km/s. What is its kinetic energy? (Consult Appendix F.) (b) If you drop a 1.0-kg weight (about 2 lb) from a height of 1.0 m, how many joules of kinetic energy will it have when it reaches the ground? (c) Is it reasonable that a 30-kg child could run fast enough to have 100 J of kinetic energy?

6.19 • **Meteor Crater.** About 50,000 years ago, a meteor crashed into the earth near present-day Flagstaff, Arizona. Measurements from 2005 estimate that this meteor had a mass of about 1.4×10^8 kg (around 150,000 tons) and hit the ground at a speed of 12 km/s. (a) How much kinetic energy did this meteor deliver to the ground? (b) How does this energy compare to the energy released by a 1.0-megaton nuclear bomb? (A megaton bomb releases the same amount of energy as a million tons of TNT, and 1.0 ton of TNT releases 4.184×10^9 J of energy.)

6.20 • A 4.80-kg watermelon is dropped from rest from the roof of an 18.0-m-tall building and feels no appreciable air resistance. (a) Calculate the work done by gravity on the watermelon during its displacement from the roof to the ground. (b) Just before it strikes the ground, what is the watermelon's (i) kinetic energy and (ii) speed? (c) Which of the answers in parts (a) and (b) would be *different* if there were appreciable air resistance?

6.21 •• Use the work–energy theorem to solve each of these problems. You can use Newton's laws to check your answers. Neglect air resistance in all cases. (a) A branch falls from the top of a 95.0-m-tall redwood tree, starting from rest. How fast is it moving when it reaches the ground? (b) A volcano ejects a boulder directly upward 525 m into the air. How fast was the boulder moving just as it left the volcano?

6.22 •• Use the work–energy theorem to solve each of these problems. You can use Newton's laws to check your answers. (a) A skier moving at 5.00 m/s encounters a long, rough horizontal patch of snow having a coefficient of kinetic friction of 0.220 with her skis. How far does she travel on this patch before stopping? (b) Suppose the rough patch in part (a) was only 2.90 m long. How fast would the skier be moving when she reached the end of the patch? (c) At the base of a frictionless icy hill that rises at 25.0° above the horizontal, a toboggan has a speed of 12.0 m/s toward the hill. How high vertically above the base will it go before stopping?

6.23 •• You are a member of an Alpine Rescue Team. You must project a box of supplies up an incline of constant slope angle α so that it reaches a stranded skier who is a vertical distance h above the bottom of the incline. The incline is slippery, but there is some friction present, with kinetic friction coefficient μ_k. Use the work–energy theorem to calculate the minimum speed you must give the box at the bottom of the incline so that it will reach the skier. Express your answer in terms of g, h, μ_k, and α.

6.24 •• You throw a 3.00-N rock vertically into the air from ground level. You observe that when it is 15.0 m above the ground, it is traveling at 25.0 m/s upward. Use the work–energy theorem to find (a) the rock's speed just as it left the ground and (b) its maximum height.

6.25 • A sled with mass 12.00 kg moves in a straight line on a frictionless, horizontal surface. At one point in its path, its speed is 4.00 m/s; after it has traveled 2.50 m beyond this point, its speed is 6.00 m/s. Use the work–energy theorem to find the force acting on the sled, assuming that this force is constant and that it acts in the direction of the sled's motion.

6.26 ·· A mass m slides down a smooth inclined plane from an initial vertical height h, making an angle α with the horizontal. (a) The work done by a force is the sum of the work done by the components of the force. Consider the components of gravity parallel and perpendicular to the surface of the plane. Calculate the work done on the mass by each of the components, and use these results to show that the work done by gravity is exactly the same as if the mass had fallen straight down through the air from a height h. (b) Use the work–energy theorem to prove that the speed of the mass at the bottom of the incline is the same as if the mass had been dropped from height h, independent of the angle α of the incline. Explain how this speed can be independent of the slope angle. (c) Use the results of part (b) to find the speed of a rock that slides down an icy frictionless hill, starting from rest 15.0 m above the bottom.

6.27 · A 12-pack of Omni-Cola (mass 4.30 kg) is initially at rest on a horizontal floor. It is then pushed in a straight line for 1.20 m by a trained dog that exerts a horizontal force with magnitude 36.0 N. Use the work–energy theorem to find the final speed of the 12-pack if (a) there is no friction between the 12-pack and the floor, and (b) the coefficient of kinetic friction between the 12-pack and the floor is 0.30.

6.28 ·· A soccer ball with mass 0.420 kg is initially moving with speed 2.00 m/s. A soccer player kicks the ball, exerting a constant force of magnitude 40.0 N in the same direction as the ball's motion. Over what distance must the player's foot be in contact with the ball to increase the ball's speed to 6.00 m/s?

6.29 · A little red wagon with mass 7.00 kg moves in a straight line on a frictionless horizontal surface. It has an initial speed of 4.00 m/s and then is pushed 3.0 m in the direction of the initial velocity by a force with a magnitude of 10.0 N. (a) Use the work–energy theorem to calculate the wagon's final speed. (b) Calculate the acceleration produced by the force. Use this acceleration in the kinematic relationships of Chapter 2 to calculate the wagon's final speed. Compare this result to that calculated in part (a).

6.30 ·· A block of ice with mass 2.00 kg slides 1.35 m down an inclined plane that slopes downward at an angle of 36.9° below the horizontal. If the block of ice starts from rest, what is its final speed? Ignore friction.

6.31 · **Stopping Distance.** A car is traveling on a level road with speed v_0 at the instant when the brakes lock, so that the tires slide rather than roll. (a) Use the work–energy theorem to calculate the minimum stopping distance of the car in terms of v_0, g, and the coefficient of kinetic friction μ_k between the tires and the road. (b) By what factor would the minimum stopping distance change if (i) the coefficient of kinetic friction were doubled, or (ii) the initial speed were doubled, or (iii) both the coefficient of kinetic friction and the initial speed were doubled?

6.32 ·· A 30.0-kg crate is initially moving with a velocity that has magnitude 3.90 m/s in a direction 37.0° west of north. How much work must be done on the crate to change its velocity to 5.62 m/s in a direction 63.0° south of east?

Section 6.3 Work and Energy with Varying Forces

6.33 · BIO **Heart Repair.** A surgeon is using material from a donated heart to repair a patient's damaged aorta and needs to know the elastic characteristics of this aortal material. Tests performed on a 16.0-cm strip of the donated aorta reveal that it stretches 3.75 cm when a 1.50-N pull is exerted on it. (a) What is the force constant of this strip of aortal material? (b) If the maximum distance it will be able to stretch when it replaces the aorta in the damaged heart is 1.14 cm, what is the greatest force it will be able to exert there?

6.34 ·· To stretch a spring 3.00 cm from its unstretched length, 12.0 J of work must be done. (a) What is the force constant of this spring? (b) What magnitude force is needed to stretch the spring 3.00 cm from its unstretched length? (c) How much work must be done to compress this spring 4.00 cm from its unstretched length, and what force is needed to compress it this distance?

6.35 · Three identical 8.50-kg masses are hung by three identical springs (**Fig. E6.35**). Each spring has a force constant of 7.80 kN/m and was 12.0 cm long before any masses were attached to it. (a) Draw a free-body diagram of each mass. (b) How long is each spring when hanging as shown? (*Hint:* First isolate only the bottom mass. Then treat the bottom two masses as a system. Finally, treat all three masses as a system.)

Figure **E6.35**

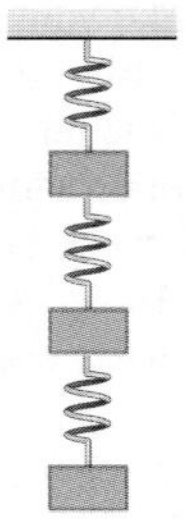

6.36 · A child applies a force $\vec{F}$ parallel to the x-axis to a 10.0-kg sled moving on the frozen surface of a small pond. As the child controls the speed of the sled, the x-component of the force she applies varies with the x-coordinate of the sled as shown in **Fig. E6.36**. Calculate the work done by $\vec{F}$ when the sled moves (a) from $x = 0$ to $x = 8.0$ m; (b) from $x = 8.0$ m to $x = 12.0$ m; (c) from $x = 0$ to 12.0 m.

Figure **E6.36**

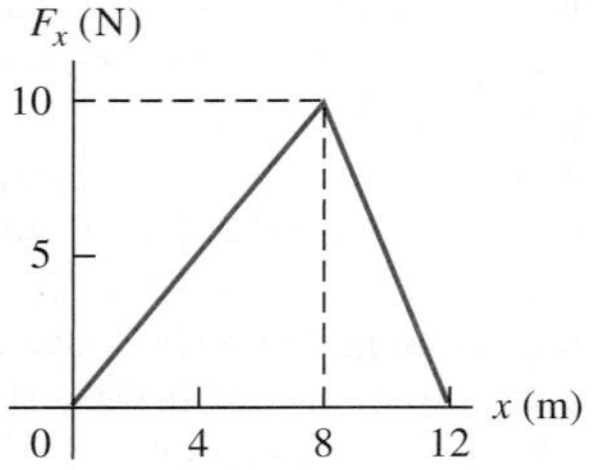

6.37 ·· Suppose the sled in Exercise 6.36 is initially at rest at $x = 0$. Use the work–energy theorem to find the speed of the sled at (a) $x = 8.0$ m and (b) $x = 12.0$ m. Ignore friction between the sled and the surface of the pond.

6.38 ·· A spring of force constant 300.0 N/m and unstretched length 0.240 m is stretched by two forces, pulling in opposite directions at opposite ends of the spring, that increase to 15.0 N. How long will the spring now be, and how much work was required to stretch it that distance?

6.39 ·· A 6.0-kg box moving at 3.0 m/s on a horizontal, frictionless surface runs into a light spring of force constant 75 N/cm. Use the work–energy theorem to find the maximum compression of the spring.

6.40 ·· **Leg Presses.** As part of your daily workout, you lie on your back and push with your feet against a platform attached to two stiff springs arranged side by side so that they are parallel to each other. When you push the platform, you compress the springs. You do 80.0 J of work when you compress the springs 0.200 m from their uncompressed length. (a) What magnitude of force must you apply to hold the platform in this position? (b) How much *additional* work must you do to move the platform 0.200 m *farther*, and what maximum force must you apply?

6.41 ·· (a) In Example 6.7 (Section 6.3) it was calculated that with the air track turned off, the glider travels 8.6 cm before it stops instantaneously. How large would the coefficient of static friction μ_s have to be to keep the glider from springing back to the left? (b) If the coefficient of static friction between the glider and the track is $\mu_s = 0.60$, what is the maximum initial speed v_1 that the glider can be given and still remain at rest after it stops instantaneously? With the air track turned off, the coefficient of kinetic friction is $\mu_k = 0.47$.

6.42 • A 4.00-kg block of ice is placed against a horizontal spring that has force constant $k = 200$ N/m and is compressed 0.025 m. The spring is released and accelerates the block along a horizontal surface. Ignore friction and the mass of the spring. (a) Calculate the work done on the block by the spring during the motion of the block from its initial position to where the spring has returned to its uncompressed length. (b) What is the speed of the block after it leaves the spring?

6.43 • A force $\vec{F}$ is applied to a 2.0-kg, radio-controlled model car parallel to the x-axis as it moves along a straight track. The x-component of the force varies with the x-coordinate of the car (**Fig. E6.43**). Calculate the work done by the force $\vec{F}$ when the car moves from (a) $x = 0$ to $x = 3.0$ m; (b) $x = 3.0$ m to $x = 4.0$ m; (c) $x = 4.0$ m to $x = 7.0$ m; (d) $x = 0$ to $x = 7.0$ m; (e) $x = 7.0$ m to $x = 2.0$ m.

Figure **E6.43**

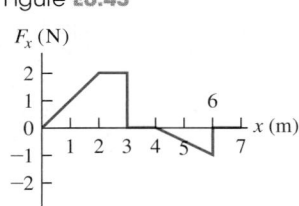

6.44 • Suppose the 2.0-kg model car in Exercise 6.43 is initially at rest at $x = 0$ and $\vec{F}$ is the net force acting on it. Use the work–energy theorem to find the speed of the car at (a) $x = 3.0$ m; (b) $x = 4.0$ m; (c) $x = 7.0$ m.

6.45 •• At a waterpark, sleds with riders are sent along a slippery, horizontal surface by the release of a large compressed spring. The spring, with force constant $k = 40.0$ N/cm and negligible mass, rests on the frictionless horizontal surface. One end is in contact with a stationary wall. A sled and rider with total mass 70.0 kg are pushed against the other end, compressing the spring 0.375 m. The sled is then released with zero initial velocity. What is the sled's speed when the spring (a) returns to its uncompressed length and (b) is still compressed 0.200 m?

6.46 • **Half of a Spring.** (a) Suppose you cut a massless ideal spring in half. If the full spring had a force constant k, what is the force constant of each half, in terms of k? (*Hint:* Think of the original spring as two equal halves, each producing the same force as the entire spring. Do you see why the forces must be equal?) (b) If you cut the spring into three equal segments instead, what is the force constant of each one, in terms of k?

6.47 •• A small glider is placed against a compressed spring at the bottom of an air track that slopes upward at an angle of 40.0° above the horizontal. The glider has mass 0.0900 kg. The spring has $k = 640$ N/m and negligible mass. When the spring is released, the glider travels a maximum distance of 1.80 m along the air track before sliding back down. Before reaching this maximum distance, the glider loses contact with the spring. (a) What distance was the spring originally compressed? (b) When the glider has traveled along the air track 0.80 m from its initial position against the compressed spring, is it still in contact with the spring? What is the kinetic energy of the glider at this point?

6.48 •• An ingenious bricklayer builds a device for shooting bricks up to the top of the wall where he is working. He places a brick on a vertical compressed spring with force constant $k = 450$ N/m and negligible mass. When the spring is released, the brick is propelled upward. If the brick has mass 1.80 kg and is to reach a maximum height of 3.6 m above its initial position on the compressed spring, what distance must the bricklayer compress the spring initially? (The brick loses contact with the spring when the spring returns to its uncompressed length. Why?)

6.49 •• CALC A force in the $+x$-direction with magnitude $F(x) = 18.0$ N $- (0.530$ N/m$)x$ is applied to a 6.00-kg box that is sitting on the horizontal, frictionless surface of a frozen lake. $F(x)$ is the only horizontal force on the box. If the box is initially at rest at $x = 0$, what is its speed after it has traveled 14.0 m?

Section 6.4 Power

6.50 •• A crate on a motorized cart starts from rest and moves with a constant eastward acceleration of $a = 2.80$ m/s². A worker assists the cart by pushing on the crate with a force that is eastward and has magnitude that depends on time according to $F(t) = (5.40$ N/s$)t$. What is the instantaneous power supplied by this force at $t = 5.00$ s?

6.51 • How many joules of energy does a 100-watt light bulb use per hour? How fast would a 70-kg person have to run to have that amount of kinetic energy?

6.52 •• BIO **Should You Walk or Run?** It is 5.0 km from your home to the physics lab. As part of your physical fitness program, you could run that distance at 10 km/h (which uses up energy at the rate of 700 W), or you could walk it leisurely at 3.0 km/h (which uses energy at 290 W). Which choice would burn up more energy, and how much energy (in joules) would it burn? Why does the more intense exercise burn up less energy than the less intense exercise?

6.53 •• **Magnetar.** On December 27, 2004, astronomers observed the greatest flash of light ever recorded from outside the solar system. It came from the highly magnetic neutron star SGR 1806-20 (a *magnetar*). During 0.20 s, this star released as much energy as our sun does in 250,000 years. If P is the average power output of our sun, what was the average power output (in terms of P) of this magnetar?

6.54 •• A 20.0-kg rock is sliding on a rough, horizontal surface at 8.00 m/s and eventually stops due to friction. The coefficient of kinetic friction between the rock and the surface is 0.200. What average power is produced by friction as the rock stops?

6.55 • A tandem (two-person) bicycle team must overcome a force of 165 N to maintain a speed of 9.00 m/s. Find the power required per rider, assuming that each contributes equally. Express your answer in watts and in horsepower.

6.56 •• When its 75-kW (100-hp) engine is generating full power, a small single-engine airplane with mass 700 kg gains altitude at a rate of 2.5 m/s (150 m/min, or 500 ft/min). What fraction of the engine power is being used to make the airplane climb? (The remainder is used to overcome the effects of air resistance and of inefficiencies in the propeller and engine.)

6.57 •• **Working Like a Horse.** Your job is to lift 30-kg crates a vertical distance of 0.90 m from the ground onto the bed of a truck. How many crates would you have to load onto the truck in 1 minute (a) for the average power output you use to lift the crates to equal 0.50 hp; (b) for an average power output of 100 W?

6.58 •• An elevator has mass 600 kg, not including passengers. The elevator is designed to ascend, at constant speed, a vertical distance of 20.0 m (five floors) in 16.0 s, and it is driven by a motor that can provide up to 40 hp to the elevator. What is the maximum number of passengers that can ride in the elevator? Assume that an average passenger has mass 65.0 kg.

6.59 •• A ski tow operates on a 15.0° slope of length 300 m. The rope moves at 12.0 km/h and provides power for 50 riders at one time, with an average mass per rider of 70.0 kg. Estimate the power required to operate the tow.

6.60 • You are applying a constant horizontal force $\vec{F} = (-8.00 \text{ N})\hat{\imath} + (3.00 \text{ N})\hat{\jmath}$ to a crate that is sliding on a factory floor. At the instant that the velocity of the crate is $\vec{v} = (3.20 \text{ m/s})\hat{\imath} + (2.20 \text{ m/s})\hat{\jmath}$, what is the instantaneous power supplied by this force?

6.61 • BIO While hovering, a typical flying insect applies an average force equal to twice its weight during each downward stroke. Take the mass of the insect to be 10 g, and assume the wings move an average downward distance of 1.0 cm during each stroke. Assuming 100 downward strokes per second, estimate the average power output of the insect.

PROBLEMS

6.62 ••• CALC A balky cow is leaving the barn as you try harder and harder to push her back in. In coordinates with the origin at the barn door, the cow walks from $x = 0$ to $x = 6.9$ m as you apply a force with x-component $F_x = -[20.0 \text{ N} + (3.0 \text{ N/m})x]$. How much work does the force you apply do on the cow during this displacement?

6.63 • A luggage handler pulls a 20.0-kg suitcase up a ramp inclined at 32.0° above the horizontal by a force $\vec{F}$ of magnitude 160 N that acts parallel to the ramp. The coefficient of kinetic friction between the ramp and the incline is $\mu_k = 0.300$. If the suitcase travels 3.80 m along the ramp, calculate (a) the work done on the suitcase by $\vec{F}$; (b) the work done on the suitcase by the gravitational force; (c) the work done on the suitcase by the normal force; (d) the work done on the suitcase by the friction force; (e) the total work done on the suitcase. (f) If the speed of the suitcase is zero at the bottom of the ramp, what is its speed after it has traveled 3.80 m along the ramp?

6.64 • BIO **Chin-ups.** While doing a chin-up, a man lifts his body 0.40 m. (a) How much work must the man do per kilogram of body mass? (b) The muscles involved in doing a chin-up can generate about 70 J of work per kilogram of muscle mass. If the man can just barely do a 0.40-m chin-up, what percentage of his body's mass do these muscles constitute? (For comparison, the *total* percentage of muscle in a typical 70-kg man with 14% body fat is about 43%.) (c) Repeat part (b) for the man's young son, who has arms half as long as his father's but whose muscles can also generate 70 J of work per kilogram of muscle mass. (d) Adults and children have about the same percentage of muscle in their bodies. Explain why children can commonly do chin-ups more easily than their fathers.

6.65 ••• Consider the blocks in Exercise 6.7 as they move 75.0 cm. Find the total work done on each one (a) if there is no friction between the table and the 20.0-N block, and (b) if $\mu_s = 0.500$ and $\mu_k = 0.325$ between the table and the 20.0-N block.

6.66 •• A 5.00-kg package slides 2.80 m down a long ramp that is inclined at 24.0° below the horizontal. The coefficient of kinetic friction between the package and the ramp is $\mu_k = 0.310$. Calculate (a) the work done on the package by friction; (b) the work done on the package by gravity; (c) the work done on the package by the normal force; (d) the total work done on the package. (e) If the package has a speed of 2.20 m/s at the top of the ramp, what is its speed after it has slid 2.80 m down the ramp?

6.67 •• CP BIO **Whiplash Injuries.** When a car is hit from behind, its passengers undergo sudden forward acceleration, which can cause a severe neck injury known as *whiplash*. During normal acceleration, the neck muscles play a large role in accelerating the head so that the bones are not injured. But during a very sudden acceleration, the muscles do not react immediately because they are flexible; most of the accelerating force is provided by the neck bones. Experiments have shown that these bones will fracture if they absorb more than 8.0 J of energy. (a) If a car waiting at a stoplight is rear-ended in a collision that lasts for 10.0 ms, what is the greatest speed this car and its driver can reach without breaking neck bones if the driver's head has a mass of 5.0 kg (which is about right for a 70-kg person)? Express your answer in m/s and in mi/h. (b) What is the acceleration of the passengers during the collision in part (a), and how large a force is acting to accelerate their heads? Express the acceleration in m/s² and in g's.

6.68 •• CALC A net force along the x-axis that has x-component $F_x = -12.0 \text{ N} + (0.300 \text{ N/m}^2)x^2$ is applied to a 5.00-kg object that is initially at the origin and moving in the $-x$-direction with a speed of 6.00 m/s. What is the speed of the object when it reaches the point $x = 5.00$ m?

6.69 • CALC **Varying Coefficient of Friction.** A box is sliding with a speed of 4.50 m/s on a horizontal surface when, at point P, it encounters a rough section. The coefficient of friction there is not constant; it starts at 0.100 at P and increases linearly with distance past P, reaching a value of 0.600 at 12.5 m past point P. (a) Use the work–energy theorem to find how far this box slides before stopping. (b) What is the coefficient of friction at the stopping point? (c) How far would the box have slid if the friction coefficient didn't increase but instead had the constant value of 0.100?

6.70 •• CALC Consider a spring that does not obey Hooke's law very faithfully. One end of the spring is fixed. To keep the spring stretched or compressed an amount x, a force along the x-axis with x-component $F_x = kx - bx^2 + cx^3$ must be applied to the free end. Here $k = 100 \text{ N/m}$, $b = 700 \text{ N/m}^2$, and $c = 12,000 \text{ N/m}^3$. Note that $x > 0$ when the spring is stretched and $x < 0$ when it is compressed. How much work must be done (a) to stretch this spring by 0.050 m from its unstretched length? (b) To *compress* this spring by 0.050 m from its unstretched length? (c) Is it easier to stretch or compress this spring? Explain why in terms of the dependence of F_x on x. (Many real springs behave qualitatively in the same way.)

6.71 •• CP A small block with a mass of 0.0600 kg is attached to a cord passing through a hole in a frictionless, horizontal surface (**Fig. P6.71**). The block is originally revolving at a distance of 0.40 m from the hole with a speed of 0.70 m/s. The cord is then pulled from below, shortening the radius of the circle in which the block revolves to 0.10 m. At this new distance, the speed of the block is 2.80 m/s. (a) What is the tension in the cord in the original situation, when the block has speed $v = 0.70$ m/s? (b) What is the tension in the cord in the final situation, when the block has speed $v = 2.80$ m/s? (c) How much work was done by the person who pulled on the cord?

Figure **P6.71**

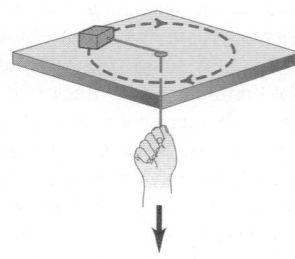

6.72 •• CALC **Proton Bombardment.** A proton with mass 1.67×10^{-27} kg is propelled at an initial speed of 3.00×10^5 m/s directly toward a uranium nucleus 5.00 m away. The proton is repelled by the uranium nucleus with a force of magnitude $F = \alpha/x^2$, where x is the separation between the two objects and $\alpha = 2.12 \times 10^{-26} \text{ N} \cdot \text{m}^2$. Assume that the uranium nucleus remains at rest. (a) What is the speed of the proton when it is 8.00×10^{-10} m from the uranium nucleus? (b) As the proton approaches the uranium nucleus, the repulsive force slows down the proton until it comes momentarily to rest, after which the proton moves away from the uranium nucleus. How close to the uranium nucleus does the proton get? (c) What is the speed of the proton when it is again 5.00 m away from the uranium nucleus?

6.73 •• You are asked to design spring bumpers for the walls of a parking garage. A freely rolling 1200-kg car moving at 0.65 m/s is to compress the spring no more than 0.090 m before stopping. What should be the force constant of the spring? Assume that the spring has negligible mass.

6.74 •• You and your bicycle have combined mass 80.0 kg. When you reach the base of a bridge, you are traveling along the road at 5.00 m/s (**Fig. P6.74**). At the top of the bridge, you have climbed a vertical distance of 5.20 m and slowed to 1.50 m/s. Ignore work done by friction and any inefficiency in the bike or your legs. (a) What is the total work done on you and your bicycle when you go from the base to the top of the bridge? (b) How much work have you done with the force you apply to the pedals?

Figure **P6.74**

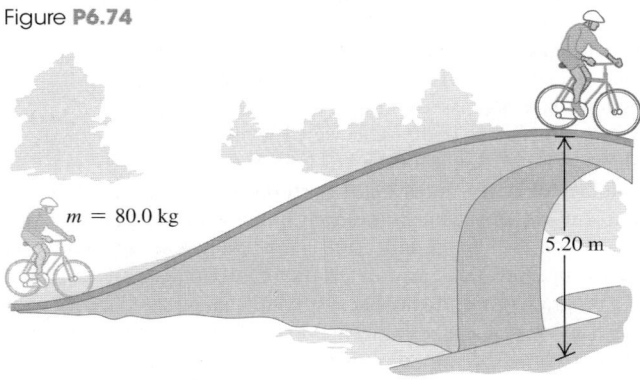

$m = 80.0$ kg

5.20 m

6.75 ••• A 2.50-kg textbook is forced against a horizontal spring of negligible mass and force constant 250 N/m, compressing the spring a distance of 0.250 m. When released, the textbook slides on a horizontal tabletop with coefficient of kinetic friction $\mu_k = 0.30$. Use the work–energy theorem to find how far the textbook moves from its initial position before it comes to rest.

6.76 •• The spring of a spring gun has force constant $k = 400$ N/m and negligible mass. The spring is compressed 6.00 cm, and a ball with mass 0.0300 kg is placed in the horizontal barrel against the compressed spring. The spring is then released, and the ball is propelled out the barrel of the gun. The barrel is 6.00 cm long, so the ball leaves the barrel at the same point that it loses contact with the spring. The gun is held so that the barrel is horizontal. (a) Calculate the speed with which the ball leaves the barrel if you can ignore friction. (b) Calculate the speed of the ball as it leaves the barrel if a constant resisting force of 6.00 N acts on the ball as it moves along the barrel. (c) For the situation in part (b), at what position along the barrel does the ball have the greatest speed, and what is that speed? (In this case, the maximum speed does not occur at the end of the barrel.)

6.77 •• One end of a horizontal spring with force constant 130.0 N/m is attached to a vertical wall. A 4.00-kg block sitting on the floor is placed against the spring. The coefficient of kinetic friction between the block and the floor is $\mu_k = 0.400$. You apply a constant force $\vec{F}$ to the block. $\vec{F}$ has magnitude $F = 82.0$ N and is directed toward the wall. At the instant that the spring is compressed 80.0 cm, what are (a) the speed of the block, and (b) the magnitude and direction of the block's acceleration?

6.78 •• One end of a horizontal spring with force constant 76.0 N/m is attached to a vertical post. A 2.00-kg block of frictionless ice is attached to the other end and rests on the floor. The spring is initially neither stretched nor compressed. A constant horizontal force of 54.0 N is then applied to the block, in the direction away from the post. (a) What is the speed of the block when the spring is stretched 0.400 m? (b) At that instant, what are the magnitude and direction of the acceleration of the block?

6.79 • A 5.00-kg block is moving at $v_0 = 6.00$ m/s along a frictionless, horizontal surface toward a spring with force constant $k = 500$ N/m that is attached to a wall (**Fig. P6.79**). The spring has negligible mass. (a) Find the maximum distance the spring will be compressed. (b) If the spring is to compress by no more than 0.150 m, what should be the maximum value of v_0?

Figure **P6.79**

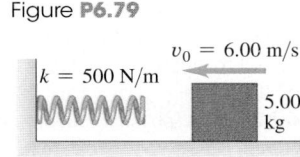

$v_0 = 6.00$ m/s

$k = 500$ N/m

5.00 kg

6.80 ••• A physics professor is pushed up a ramp inclined upward at 30.0° above the horizontal as she sits in her desk chair, which slides on frictionless rollers. The combined mass of the professor and chair is 85.0 kg. She is pushed 2.50 m along the incline by a group of students who together exert a constant horizontal force of 600 N. The professor's speed at the bottom of the ramp is 2.00 m/s. Use the work–energy theorem to find her speed at the top of the ramp.

6.81 •• Consider the system shown in **Fig. P6.81**. The rope and pulley have negligible mass, and the pulley is frictionless. Initially the 6.00-kg block is moving downward and the 8.00-kg block is moving to the right, both with a speed of 0.900 m/s. The blocks come to rest after moving 2.00 m. Use the work–energy theorem to calculate the coefficient of kinetic friction between the 8.00-kg block and the tabletop.

Figure **P6.81**

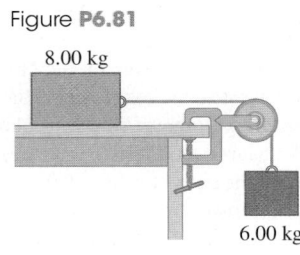

8.00 kg

6.00 kg

6.82 •• Consider the system shown in Fig. P6.81. The rope and pulley have negligible mass, and the pulley is frictionless. The coefficient of kinetic friction between the 8.00-kg block and the tabletop is $\mu_k = 0.250$. The blocks are released from rest. Use energy methods to calculate the speed of the 6.00-kg block after it has descended 1.50 m.

6.83 •• On an essentially frictionless, horizontal ice rink, a skater moving at 3.0 m/s encounters a rough patch that reduces her speed to 1.65 m/s due to a friction force that is 25% of her weight. Use the work–energy theorem to find the length of this rough patch.

6.84 •• BIO All birds, independent of their size, must maintain a power output of 10–25 watts per kilogram of body mass in order to fly by flapping their wings. (a) The Andean giant hummingbird (*Patagona gigas*) has mass 70 g and flaps its wings 10 times per second while hovering. Estimate the amount of work done by such a hummingbird in each wingbeat. (b) A 70-kg athlete can maintain a power output of 1.4 kW for no more than a few seconds; the *steady* power output of a typical athlete is only 500 W or so. Is it possible for a human-powered aircraft to fly for extended periods by flapping its wings? Explain.

6.85 •• A pump is required to lift 800 kg of water (about 210 gallons) per minute from a well 14.0 m deep and eject it with a speed of 18.0 m/s. (a) How much work is done per minute in lifting the water? (b) How much work is done in giving the water the kinetic energy it has when ejected? (c) What must be the power output of the pump?

6.86 ••• The Grand Coulee Dam is 1270 m long and 170 m high. The electrical power output from generators at its base is approximately 2000 MW. How many cubic meters of water must flow from the top of the dam per second to produce this amount of power if 92% of the work done on the water by gravity is converted to electrical energy? (Each cubic meter of water has a mass of 1000 kg.)

6.87 ••• A physics student spends part of her day walking between classes or for recreation, during which time she expends energy at an average rate of 280 W. The remainder of the day she is sitting in class, studying, or resting; during these activities, she expends energy at an average rate of 100 W. If she expends a total of 1.1×10^7 J of energy in a 24-hour day, how much of the day did she spend walking?

6.88 • CALC An object has several forces acting on it. One of these forces is $\vec{F} = \alpha xy\hat{\imath}$, a force in the x-direction whose magnitude depends on the position of the object, with $\alpha = 2.50$ N/m². Calculate the work done on the object by this force for the following displacements of the object: (a) The object starts at the point ($x = 0$, $y = 3.00$ m) and moves parallel to the x-axis to the point ($x = 2.00$ m, $y = 3.00$ m). (b) The object starts at the point ($x = 2.00$ m, $y = 0$) and moves in the y-direction to the point ($x = 2.00$ m, $y = 3.00$ m). (c) The object starts at the origin and moves on the line $y = 1.5x$ to the point ($x = 2.00$ m, $y = 3.00$ m).

6.89 • BIO **Power of the Human Heart.** The human heart is a powerful and extremely reliable pump. Each day it takes in and discharges about 7500 L of blood. Assume that the work done by the heart is equal to the work required to lift this amount of blood a height equal to that of the average American woman (1.63 m). The density (mass per unit volume) of blood is 1.05×10^3 kg/m³. (a) How much work does the heart do in a day? (b) What is the heart's power output in watts?

6.90 •• DATA **Figure P6.90** shows the results of measuring the force F exerted on both ends of a rubber band to stretch it a distance x from its unstretched position. (Source: www.sciencebuddies.org) The data points are well fit by the equation $F = 33.55x^{0.4871}$, where F is in newtons and x is in meters. (a) Does this rubber band obey Hooke's law over the range of x shown in the graph? Explain. (b) The stiffness of a spring that obeys Hooke's law is measured by the value of its force constant k, where $k = F/x$. This can be

written as $k = dF/dx$ to emphasize the quantities that are changing. Define $k_{\text{eff}} = dF/dx$ and calculate k_{eff} as a function of x for this rubber band. For a spring that obeys Hooke's law, k_{eff} is constant, independent of x. Does the stiffness of this band, as measured by k_{eff}, increase or decrease as x is increased, within the range of the data? (c) How much work must be done to stretch the rubber band from $x = 0$ to $x = 0.0400$ m? From $x = 0.0400$ m to $x = 0.0800$ m? (d) One end of the rubber band is attached to a stationary vertical rod, and the band is stretched horizontally 0.0800 m from its unstretched length. A 0.300-kg object on a horizontal, frictionless surface is attached to the free end of the rubber band and released from rest. What is the speed of the object after it has traveled 0.0400 m?

6.91 ••• DATA In a physics lab experiment, one end of a horizontal spring that obeys Hooke's law is attached to a wall. The spring is compressed 0.400 m, and a block with mass 0.300 kg is attached to it. The spring is then released, and the block moves along a horizontal surface. Electronic sensors measure the speed v of the block after it has traveled a distance d from its initial position against the compressed spring. The measured values are listed in the table. (a) The data show that the speed v of the block increases

d (m)	v (m/s)
0	0
0.05	0.85
0.10	1.11
0.15	1.24
0.25	1.26
0.30	1.14
0.35	0.90
0.40	0.36

and then decreases as the spring returns to its unstretched length. Explain why this happens, in terms of the work done on the block by the forces that act on it. (b) Use the work–energy theorem to derive an expression for v^2 in terms of d. (c) Use a computer graphing program (for example, Excel or Matlab) to graph the data as v^2 (vertical axis) versus d (horizontal axis). The equation that you derived in part (b) should show that v^2 is a quadratic function of d, so, in your graph, fit the data by a second-order polynomial (quadratic) and have the graphing program display the equation for this trendline. Use that equation to find the block's maximum speed v and the value of d at which this speed occurs. (d) By comparing the equation from the graphing program to the formula you derived in part (b), calculate the force constant k for the spring and the coefficient of kinetic friction for the friction force that the surface exerts on the block.

6.92 •• DATA For a physics lab experiment, four classmates run up the stairs from the basement to the top floor of their physics building—a vertical distance of 16.0 m. The classmates and their masses are: Tatiana, 50.2 kg; Bill, 68.2 kg; Ricardo, 81.8 kg; and Melanie, 59.1 kg. The time it takes each of them is shown in **Fig. P6.92**. (a) Considering only the work done against gravity, which person had the largest average power output? The smallest? (b) Chang is very fit and has mass 62.3 kg. If his average power output is 1.00 hp, how many seconds does it take him to run up the stairs?

Figure **P6.90**

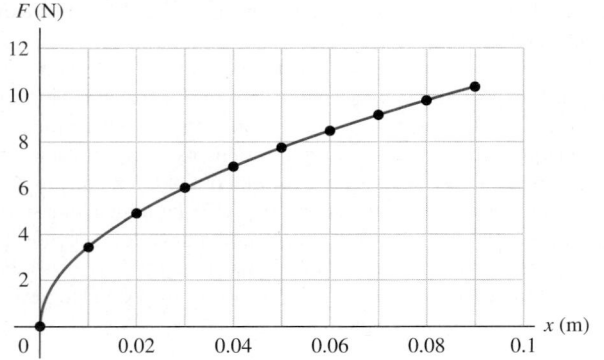

Figure **P6.92**

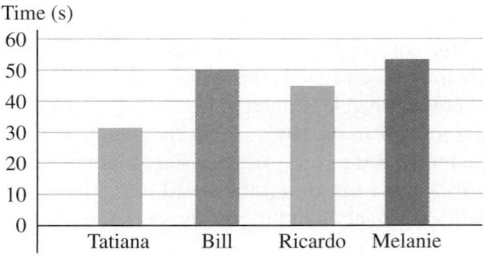

CHALLENGE PROBLEMS

6.93 ••• CALC **A Spring with Mass.** We usually ignore the kinetic energy of the moving coils of a spring, but let's try to get a reasonable approximation to this. Consider a spring of mass M, equilibrium length L_0, and force constant k. The work done to stretch or compress the spring by a distance L is $\frac{1}{2}kX^2$, where $X = L - L_0$. Consider a spring, as described above, that has one end fixed and the other end moving with speed v. Assume that the speed of points along the length of the spring varies linearly with distance l from the fixed end. Assume also that the mass M of the spring is distributed uniformly along the length of the spring. (a) Calculate the kinetic energy of the spring in terms of M and v. (*Hint:* Divide the spring into pieces of length dl; find the speed of each piece in terms of l, v, and L; find the mass of each piece in terms of dl, M, and L; and integrate from 0 to L. The result is *not* $\frac{1}{2}Mv^2$, since not all of the spring moves with the same speed.) In a spring gun, a spring of mass 0.243 kg and force constant 3200 N/m is compressed 2.50 cm from its unstretched length. When the trigger is pulled, the spring pushes horizontally on a 0.053-kg ball. The work done by friction is negligible. Calculate the ball's speed when the spring reaches its uncompressed length (b) ignoring the mass of the spring and (c) including, using the results of part (a), the mass of the spring. (d) In part (c), what is the final kinetic energy of the ball and of the spring?

6.94 ••• CALC An airplane in flight is subject to an air resistance force proportional to the square of its speed v. But there is an additional resistive force because the airplane has wings. Air flowing over the wings is pushed down and slightly forward, so from Newton's third law the air exerts a force on the wings and airplane that is up and slightly backward (**Fig. P6.94**). The upward force is the lift force that keeps the airplane aloft, and the backward force is called *induced drag*. At flying speeds, induced drag is inversely proportional to v^2, so the total air resistance force can be expressed by $F_{\text{air}} = \alpha v^2 + \beta/v^2$, where α and β are positive constants that depend on the shape and size of the airplane and the density of the air. For a Cessna 150, a small single-engine airplane, $\alpha = 0.30\ \text{N} \cdot \text{s}^2/\text{m}^2$ and $\beta = 3.5 \times 10^5\ \text{N} \cdot \text{m}^2/\text{s}^2$. In steady flight, the engine must provide a forward force that exactly balances the air resistance force. (a) Calculate the speed (in km/h) at which this airplane will have the maximum *range* (that is, travel the greatest distance) for a given quantity of fuel. (b) Calculate the speed (in km/h) for which the airplane will have the maximum *endurance* (that is, remain in the air the longest time).

Figure **P6.94**

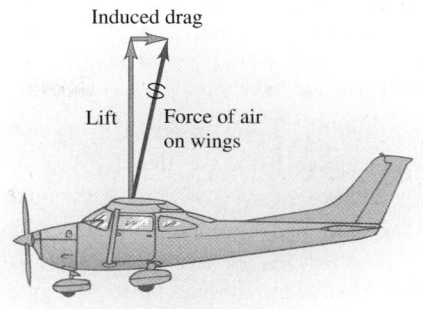

PASSAGE PROBLEMS

BIO **ENERGY OF LOCOMOTION.** On flat ground, a 70-kg person requires about 300 W of metabolic power to walk at a steady pace of 5.0 km/h(1.4 m/s). Using the same metabolic power output, that person can bicycle over the same ground at 15 km/h.

6.95 Based on the given data, how does the energy used in biking 1 km compare with that used in walking 1 km? Biking takes (a) $\frac{1}{3}$ of the energy of walking the same distance; (b) the same energy as walking the same distance; (c) 3 times the energy of walking the same distance; (d) 9 times the energy of walking the same distance.

6.96 A 70-kg person walks at a steady pace of 5.0 km/h on a treadmill at a 5.0% grade. (That is, the vertical distance covered is 5.0% of the horizontal distance covered.) If we assume the metabolic power required is equal to that required for walking on a flat surface plus the rate of doing work for the vertical climb, how much power is required? (a) 300 W; (b) 315 W; (c) 350 W; (d) 370 W.

6.97 How many times greater is the kinetic energy of the person when biking than when walking? Ignore the mass of the bike. (a) 1.7; (b) 3; (c) 6; (d) 9.

Answers

Chapter Opening Question ?

(ii) The expression for kinetic energy is $K = \frac{1}{2}mv^2$. If we calculate K for the three balls, we find (i) $K = \frac{1}{2}(0.145 \text{ kg}) \times (20.0 \text{ m/s})^2 = 29.0 \text{ kg} \cdot \text{m}^2/\text{s}^2 = 29.0 \text{ J}$, (ii) $K = \frac{1}{2}(0.0145 \text{ kg}) \times (200 \text{ m/s})^2 = 290 \text{ J}$, and (iii) $K = \frac{1}{2}(1.45 \text{ kg})(2.00 \text{ m/s})^2 = 2.90 \text{ J}$. The smaller ball has the least mass of all three, but it also has the greatest speed and so the most kinetic energy. Since kinetic energy is a scalar, it does not depend on the direction of motion.

Test Your Understanding Questions

6.1 (iii) The electron has constant velocity, so its acceleration is zero and (by Newton's second law) the net force on the electron is also zero. Therefore the total work done by all the forces (equal to the work done by the net force) must be zero as well. The individual forces may do nonzero work, but that's not what the question asks.

6.2 (iv), (i), (iii), (ii) Body (i) has kinetic energy $K = \frac{1}{2}mv^2 = \frac{1}{2}(2.0 \text{ kg})(5.0 \text{ m/s})^2 = 25 \text{ J}$. Body (ii) had zero kinetic energy initially and then had 30 J of work done on it, so its final kinetic energy is $K_2 = K_1 + W = 0 + 30 \text{ J} = 30 \text{ J}$. Body (iii) had initial kinetic energy $K_1 = \frac{1}{2}mv_1^2 = \frac{1}{2}(1.0 \text{ kg})(4.0 \text{ m/s})^2 = 8.0 \text{ J}$ and then had 20 J of work done on it, so its final kinetic energy is $K_2 = K_1 + W = 8.0 \text{ J} + 20 \text{ J} = 28 \text{ J}$. Body (iv) had initial kinetic energy $K_1 = \frac{1}{2}mv_1^2 = \frac{1}{2}(2.0 \text{ kg})(10 \text{ m/s})^2 = 100 \text{ J}$; when it did 80 J of work on another body, the other body did -80 J of work on body (iv), so the final kinetic energy of body (iv) is $K_2 = K_1 + W = 100 \text{ J} + (-80 \text{ J}) = 20 \text{ J}$.

6.3 (a) (iii), (b) (iii) At any point during the pendulum bob's motion, both the tension force and the weight act perpendicular to the motion—that is, perpendicular to an infinitesimal displacement $d\vec{l}$ of the bob. (In Fig. 5.32b, the displacement $d\vec{l}$ would be directed outward from the plane of the free-body diagram.) Hence for either force the scalar product inside the integral in Eq. (6.14) is $\vec{F} \cdot d\vec{l} = 0$, and the work done along any part of the circular path (including a complete circle) is $W = \int \vec{F} \cdot d\vec{l} = 0$.

6.4 (v) The airliner has a constant horizontal velocity, so the net horizontal force on it must be zero. Hence the backward drag force must have the same magnitude as the forward force due to the combined thrust of the four engines. This means that the drag force must do *negative* work on the airplane at the same rate that the combined thrust force does *positive* work. The combined thrust does work at a rate of $4(108,000 \text{ hp}) = 432,000 \text{ hp}$, so the drag force must do work at a rate of $-432,000$ hp.

Bridging Problem

(a) $v_1 = \sqrt{\dfrac{2}{m}\left(mgx_1 - \frac{1}{3}\alpha x_1^3\right)} = \sqrt{2gx_1 - \dfrac{2\alpha x_1^3}{3m}}$

(b) $P = -F_{\text{spring}-1}v_1 = -\alpha x_1^2 \sqrt{2gx_1 - \dfrac{2\alpha x_1^3}{3m}}$

(c) $x_2 = \sqrt{\dfrac{3mg}{\alpha}}$ **(d)** No

? As this sandhill crane (*Grus canadensis*) glides in to a landing, it descends along a straight-line path at a constant speed. During the glide, what happens to the mechanical energy (the sum of kinetic energy and gravitational potential energy)? (i) It stays the same; (ii) it increases due to the effect of gravity; (iii) it increases due to the effect of the air; (iv) it decreases due to the effect of gravity; (v) it decreases due to the effect of the air.

7 POTENTIAL ENERGY AND ENERGY CONSERVATION

W hen a diver jumps off a high board into a swimming pool, she hits the water moving pretty fast, with a lot of kinetic energy—energy of *motion*. Where does that energy come from? The answer we learned in Chapter 6 was that the gravitational force does work on the diver as she falls, and her kinetic energy increases by an amount equal to the work done.

However, there's a useful alternative way to think about work and kinetic energy. This new approach uses the idea of *potential energy*, which is associated with the *position* of a system rather than with its motion. In this approach, there is *gravitational potential energy* even when the diver is at rest on the high board. As she falls, this potential energy is *transformed* into her kinetic energy.

If the diver bounces on the end of the board before she jumps, the bent board stores a second kind of potential energy called *elastic potential energy*. We'll discuss elastic potential energy of simple systems such as a stretched or compressed spring. (An important third kind of potential energy is associated with the forces between electrically charged objects. We'll return to this in Chapter 23.)

We will prove that in some cases the sum of a system's kinetic and potential energies, called the *total mechanical energy* of the system, is constant during the motion of the system. This will lead us to the general statement of the *law of conservation of energy,* one of the most fundamental principles in all of science.

7.1 GRAVITATIONAL POTENTIAL ENERGY

In many situations it seems as though energy has been stored in a system, to be recovered later. For example, you must do work to lift a heavy stone over your head. It seems reasonable that in hoisting the stone into the air you are storing energy in the system, energy that is later converted into kinetic energy when you let the stone fall.

7.1 The greater the height of a basketball, the greater the associated gravitational potential energy. As the basketball descends, gravitational potential energy is converted to kinetic energy and the basketball's speed increases.

This example points to the idea of an energy associated with the *position* of bodies in a system. This kind of energy is a measure of the *potential* or *possibility* for work to be done; if you raise a stone into the air, there is a potential for the gravitational force to do work on it, but only if you allow the stone to fall to the ground. For this reason, energy associated with position is called **potential energy.** The potential energy associated with a body's weight and its height above the ground is called *gravitational potential energy* (**Fig. 7.1**).

We now have *two* ways to describe what happens when a body falls without air resistance. One way, which we learned in Chapter 6, is to say that a falling body's kinetic energy increases because the force of the earth's gravity does work on the body. The other way is to say that the kinetic energy increases as the gravitational potential energy decreases. Later in this section we'll use the work–energy theorem to show that these two descriptions are equivalent.

Let's derive the expression for gravitational potential energy. Suppose a body with mass m moves along the (vertical) y-axis, as in **Fig. 7.2**. The forces acting on it are its weight, with magnitude $w = mg$, and possibly some other forces; we call the vector sum (resultant) of all the other forces $\vec{F}_{\text{other}}$. We'll assume that the body stays close enough to the earth's surface that the weight is constant. (We'll find in Chapter 13 that weight decreases with altitude.) We want to find the work done by the weight when the body moves downward from a height y_1 above the origin to a lower height y_2 (Fig. 7.2a). The weight and displacement are in the same direction, so the work W_{grav} done on the body by its weight is positive:

$$W_{\text{grav}} = Fs = w(y_1 - y_2) = mgy_1 - mgy_2 \tag{7.1}$$

This expression also gives the correct work when the body moves *upward* and y_2 is greater than y_1 (Fig. 7.2b). In that case the quantity $(y_1 - y_2)$ is negative, and W_{grav} is negative because the weight and displacement are opposite in direction.

Equation (7.1) shows that we can express W_{grav} in terms of the values of the quantity mgy at the beginning and end of the displacement. This quantity is called the **gravitational potential energy,** U_{grav}:

7.2 When a body moves vertically from an initial height y_1 to a final height y_2, the gravitational force $\vec{w}$ does work and the gravitational potential energy changes.

(a) A body moves downward

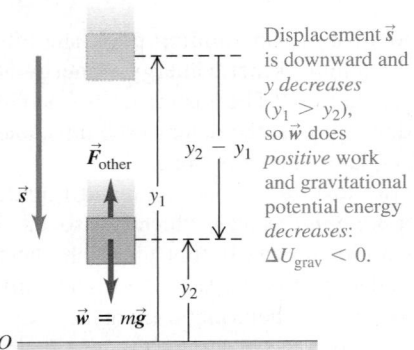

Displacement $\vec{s}$ is downward and y decreases ($y_1 > y_2$), so $\vec{w}$ does *positive* work and gravitational potential energy *decreases*: $\Delta U_{\text{grav}} < 0$.

Gravitational potential energy ⋯ associated with a particle

Vertical coordinate of particle (y increases if particle moves upward)

$$U_{\text{grav}} = mgy \tag{7.2}$$

Mass of particle ⋯ ⋯ Acceleration due to gravity

Its initial value is $U_{\text{grav, 1}} = mgy_1$ and its final value is $U_{\text{grav, 2}} = mgy_2$. The change in U_{grav} is the final value minus the initial value, or $\Delta U_{\text{grav}} = U_{\text{grav, 2}} - U_{\text{grav, 1}}$. Using Eq. (7.2), we can rewrite Eq. (7.1) for the work done by the gravitational force during the displacement from y_1 to y_2:

$$W_{\text{grav}} = U_{\text{grav, 1}} - U_{\text{grav, 2}} = -(U_{\text{grav, 2}} - U_{\text{grav, 1}}) = -\Delta U_{\text{grav}}$$

or

(b) A body moves upward

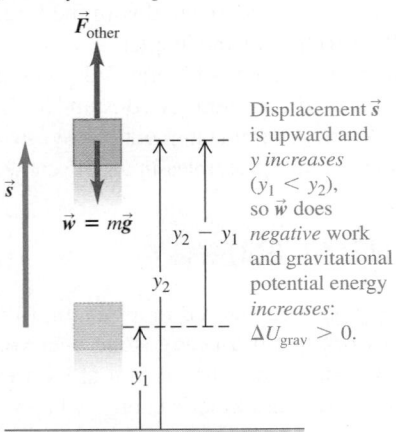

Displacement $\vec{s}$ is upward and y increases ($y_1 < y_2$), so $\vec{w}$ does *negative* work and gravitational potential energy *increases*: $\Delta U_{\text{grav}} > 0$.

Work done by the gravitational force on a particle ...

... equals the **negative of the change in** the gravitational potential energy.

$$W_{\text{grav}} = mgy_1 - mgy_2 = U_{\text{grav, 1}} - U_{\text{grav, 2}} = -\Delta U_{\text{grav}} \tag{7.3}$$

Mass of particle ⋯ Acceleration due to gravity

Initial and final vertical coordinates of particle

The negative sign in front of ΔU_{grav} is *essential*. When the body moves up, y increases, the work done by the gravitational force is negative, and the gravitational potential energy increases ($\Delta U_{\text{grav}} > 0$). When the body moves down, y decreases, the gravitational force does positive work, and the gravitational potential energy decreases ($\Delta U_{\text{grav}} < 0$). It's like drawing money out of the bank (decreasing U_{grav}) and spending it (doing positive work). The unit of potential energy is the joule (J), the same unit as is used for work.

CAUTION **To what body does gravitational potential energy "belong"?** It is *not* correct to call $U_{grav} = mgy$ the "gravitational potential energy of the body." The reason is that U_{grav} is a *shared* property of the body and the earth. The value of U_{grav} increases if the earth stays fixed and the body moves upward, away from the earth; it also increases if the body stays fixed and the earth is moved away from it. Notice that the formula $U_{grav} = mgy$ involves characteristics of both the body (its mass m) and the earth (the value of g). ▮

Conservation of Mechanical Energy (Gravitational Forces Only)

To see what gravitational potential energy is good for, suppose a body's weight is the *only* force acting on it, so $\vec{F}_{other} = 0$. The body is then falling freely with no air resistance and can be moving either up or down. Let its speed at point y_1 be v_1 and let its speed at y_2 be v_2. The work–energy theorem, Eq. (6.6), says that the total work done on the body equals the change in the body's kinetic energy: $W_{tot} = \Delta K = K_2 - K_1$. If gravity is the only force that acts, then from Eq. (7.3), $W_{tot} = W_{grav} = -\Delta U_{grav} = U_{grav,1} - U_{grav,2}$. Putting these together, we get

$$\Delta K = -\Delta U_{grav} \quad \text{or} \quad K_2 - K_1 = U_{grav,1} - U_{grav,2}$$

which we can rewrite as

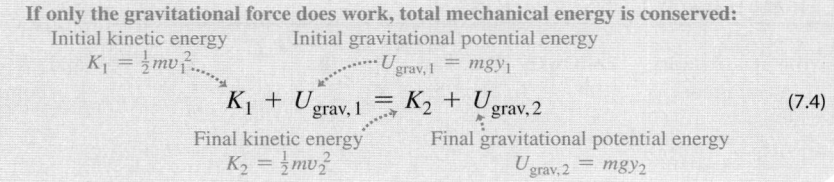

If only the gravitational force does work, total mechanical energy is conserved:

Initial kinetic energy
$K_1 = \frac{1}{2}mv_1^2$

Initial gravitational potential energy
$U_{grav,1} = mgy_1$

$$K_1 + U_{grav,1} = K_2 + U_{grav,2} \tag{7.4}$$

Final kinetic energy
$K_2 = \frac{1}{2}mv_2^2$

Final gravitational potential energy
$U_{grav,2} = mgy_2$

The sum $K + U_{grav}$ of kinetic and potential energies is called E, the **total mechanical energy of the system.** By "system" we mean the body of mass m and the earth considered together, because gravitational potential energy U is a shared property of both bodies. Then $E_1 = K_1 + U_{grav,1}$ is the total mechanical energy at y_1 and $E_2 = K_2 + U_{grav,2}$ is the total mechanical energy at y_2. Equation (7.4) says that when the body's weight is the only force doing work on it, $E_1 = E_2$. That is, E is constant; it has the same value at y_1 and y_2. But since positions y_1 and y_2 are arbitrary points in the motion of the body, the total mechanical energy E has the same value at *all* points during the motion:

$$E = K + U_{grav} = \text{constant} \quad \text{(if only gravity does work)}$$

A quantity that always has the same value is called a *conserved* quantity. *When only the force of gravity does work, the total mechanical energy is constant— that is, it is conserved* (**Fig. 7.3**). This is our first example of the **conservation of mechanical energy.**

DEMO

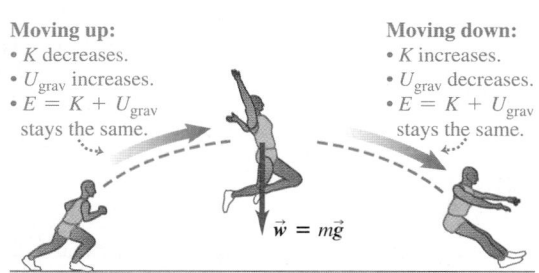

Moving up:
• K decreases.
• U_{grav} increases.
• $E = K + U_{grav}$ stays the same.

Moving down:
• K increases.
• U_{grav} decreases.
• $E = K + U_{grav}$ stays the same.

$\vec{w} = m\vec{g}$

7.3 While this athlete is in midair, only gravity does work on him (if we neglect the minor effects of air resistance). Mechanical energy E—the sum of kinetic and gravitational potential energy—is conserved.

When we throw a ball into the air, its speed decreases on the way up as kinetic energy is converted to potential energy: $\Delta K < 0$ and $\Delta U_{grav} > 0$. On the way back down, potential energy is converted back to kinetic energy and the ball's speed increases: $\Delta K > 0$ and $\Delta U_{grav} < 0$. But the *total* mechanical energy (kinetic plus potential) is the same at every point in the motion, provided that no force other than gravity does work on the ball (that is, air resistance must be negligible). It's still true that the gravitational force does work on the body as it moves up or down, but we no longer have to calculate work directly; keeping track of changes in the value of U_{grav} takes care of this completely.

Equation (7.4) is also valid if forces other than gravity are present but do *not* do work. We'll see a situation of this kind later, in Example 7.4.

CAUTION **Choose "zero height" to be wherever you like** When working with gravitational potential energy, we may choose any height to be $y = 0$. If we shift the origin for y, the values of y_1 and y_2 change, as do the values of $U_{grav,1}$ and $U_{grav,2}$. But this shift has no effect on the *difference* in height $y_2 - y_1$ or on the *difference* in gravitational potential energy $U_{grav,2} - U_{grav,1} = mg(y_2 - y_1)$. As Example 7.1 shows, the physically significant quantity is not the value of U_{grav} at a particular point but the *difference* in U_{grav} between two points. We can define U_{grav} to be zero at whatever point we choose. ∎

EXAMPLE 7.1 HEIGHT OF A BASEBALL FROM ENERGY CONSERVATION

You throw a 0.145-kg baseball straight up, giving it an initial velocity of magnitude 20.0 m/s. Find how high it goes, ignoring air resistance.

SOLUTION

IDENTIFY and SET UP: After the ball leaves your hand, only gravity does work on it. Hence mechanical energy is conserved, and we can use Eq. (7.4). We take point 1 to be where the ball leaves your hand and point 2 to be where it reaches its maximum height. As in Fig. 7.2, we take the positive y-direction to be upward. The ball's speed at point 1 is $v_1 = 20.0$ m/s; at its maximum height it is instantaneously at rest, so $v_2 = 0$. We take the origin at point 1, so $y_1 = 0$ (**Fig. 7.4**). Our target variable, the distance the ball moves vertically between the two points, is the displacement $y_2 - y_1 = y_2 - 0 = y_2$.

7.4 After a baseball leaves your hand, mechanical energy $E = K + U$ is conserved.

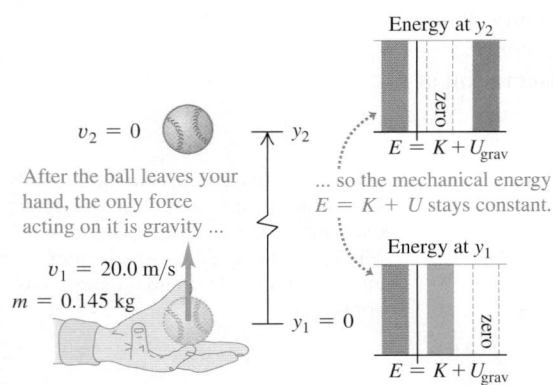

EXECUTE: We have $y_1 = 0$, $U_{grav,1} = mgy_1 = 0$, and $K_2 = \frac{1}{2}mv_2^2 = 0$. Then Eq. (7.4), $K_1 + U_{grav,1} = K_2 + U_{grav,2}$, becomes

$$K_1 = U_{grav,2}$$

As the energy bar graphs in Fig. 7.4 show, this equation says that the kinetic energy of the ball at point 1 is completely converted to gravitational potential energy at point 2. We substitute $K_1 = \frac{1}{2}mv_1^2$ and $U_{grav,2} = mgy_2$ and solve for y_2:

$$\frac{1}{2}mv_1^2 = mgy_2$$

$$y_2 = \frac{v_1^2}{2g} = \frac{(20.0 \text{ m/s})^2}{2(9.80 \text{ m/s}^2)} = 20.4 \text{ m}$$

EVALUATE: As a check, use the given value of v_1 and our result for y_2 to calculate the kinetic energy at point 1 and the gravitational potential energy at point 2. You should find that these are equal: $K_1 = \frac{1}{2}mv_1^2 = 29.0$ J and $U_{grav,2} = mgy_2 = 29.0$ J. Note that we could have found the result $y_2 = v_1^2/2g$ by using Eq. (2.13) in the form $v_{2y}^2 = v_{1y}^2 - 2g(y_2 - y_1)$.

What if we put the origin somewhere else—for example, 5.0 m below point 1, so that $y_1 = 5.0$ m? Then the total mechanical energy at point 1 is part kinetic and part potential; at point 2 it's still purely potential because $v_2 = 0$. You'll find that this choice of origin yields $y_2 = 25.4$ m, but again $y_2 - y_1 = 20.4$ m. In problems like this, you are free to choose the height at which $U_{grav} = 0$. The physics doesn't depend on your choice.

When Forces Other Than Gravity Do Work

If other forces act on the body in addition to its weight, then $\vec{F}_{\text{other}}$ in Fig. 7.2 is *not* zero. For the pile driver described in Example 6.4 (Section 6.2), the force applied by the hoisting cable and the friction with the vertical guide rails are examples of forces that might be included in $\vec{F}_{\text{other}}$. The gravitational work W_{grav} is still given by Eq. (7.3), but the total work W_{tot} is then the sum of W_{grav} and the work done by $\vec{F}_{\text{other}}$. We will call this additional work W_{other}, so the total work done by all forces is $W_{\text{tot}} = W_{\text{grav}} + W_{\text{other}}$. Equating this to the change in kinetic energy, we have

$$W_{\text{other}} + W_{\text{grav}} = K_2 - K_1 \qquad (7.5)$$

Also, from Eq. (7.3), $W_{\text{grav}} = U_{\text{grav},1} - U_{\text{grav},2}$, so Eq. (7.5) becomes

$$W_{\text{other}} + U_{\text{grav},1} - U_{\text{grav},2} = K_2 - K_1$$

which we can rearrange in the form

$$K_1 + U_{\text{grav},1} + W_{\text{other}} = K_2 + U_{\text{grav},2} \qquad \text{(if forces other than gravity do work)} \qquad (7.6)$$

We can use the expressions for the various energy terms to rewrite Eq. (7.6):

$$\tfrac{1}{2}mv_1^2 + mgy_1 + W_{\text{other}} = \tfrac{1}{2}mv_2^2 + mgy_2 \qquad \text{(if forces other than gravity do work)} \qquad (7.7)$$

The meaning of Eqs. (7.6) and (7.7) is this: *The work done by all forces other than the gravitational force equals the change in the total mechanical energy $E = K + U_{\text{grav}}$ of the system, where U_{grav} is the gravitational potential energy.* When W_{other} is positive, E increases and $K_2 + U_{\text{grav},2}$ is greater than $K_1 + U_{\text{grav},1}$. When W_{other} is negative, E decreases (**Fig. 7.5**). In the special case in which no forces other than the body's weight do work, $W_{\text{other}} = 0$. The total mechanical energy is then constant, and we are back to Eq. (7.4).

7.5 As this parachutist moves downward, the upward force of air resistance does negative work W_{other} on him. Hence the total mechanical energy $E = K + U$ decreases.

$\vec{F}_{\text{other}}$ (air resistance)

$\vec{s}$ (displacement)

- $\vec{F}_{\text{other}}$ and $\vec{s}$ are opposite, so $W_{\text{other}} < 0$.
- Hence $E = K + U_{\text{grav}}$ must decrease.
- The parachutist's speed remains constant, so K is constant.
- The parachutist descends, so U_{grav} decreases.

| PROBLEM-SOLVING STRATEGY 7.1 | **PROBLEMS USING MECHANICAL ENERGY I** |

IDENTIFY *the relevant concepts:* Decide whether the problem should be solved by energy methods, by using $\sum \vec{F} = m\vec{a}$ directly, or by a combination of these. The energy approach is best when the problem involves varying forces or motion along a curved path (discussed later in this section). If the problem involves elapsed time, the energy approach is usually *not* the best choice because it doesn't involve time directly.

SET UP *the problem* using the following steps:
1. When using the energy approach, first identify the initial and final states (the positions and velocities) of the bodies in question. Use the subscript 1 for the initial state and the subscript 2 for the final state. Draw sketches showing these states.
2. Define a coordinate system, and choose the level at which $y = 0$. Choose the positive y-direction to be upward. (The equations in this section require this.)
3. Identify any forces that do work on each body and that *cannot* be described in terms of potential energy. (So far, this means

any forces other than gravity. In Section 7.2 we'll see that the work done by an ideal spring can also be expressed as a change in potential energy.) Sketch a free-body diagram for each body.
4. List the unknown and known quantities, including the coordinates and velocities at each point. Identify the target variables.

EXECUTE *the solution:* Write expressions for the initial and final kinetic and potential energies K_1, K_2, $U_{\text{grav},1}$, and $U_{\text{grav},2}$. If no other forces do work, use Eq. (7.4). If there are other forces that do work, use Eq. (7.6). Draw bar graphs showing the initial and final values of K, $U_{\text{grav},1}$, and $E = K + U_{\text{grav}}$. Then solve to find your target variables.

EVALUATE *your answer:* Check whether your answer makes physical sense. Remember that the gravitational work is included in ΔU_{grav}, so do not include it in W_{other}.

EXAMPLE 7.2 | **WORK AND ENERGY IN THROWING A BASEBALL**

In Example 7.1 suppose your hand moves upward by 0.50 m while you are throwing the ball. The ball leaves your hand with an upward velocity of 20.0 m/s. (a) Find the magnitude of the force (assumed constant) that your hand exerts on the ball. (b) Find the speed of the ball at a point 15.0 m above the point where it leaves your hand. Ignore air resistance.

SOLUTION

IDENTIFY and SET UP: In Example 7.1 only gravity did work. Here we must include the nongravitational, "other" work done by your hand. **Figure 7.6** shows a diagram of the situation, including a free-body diagram for the ball while it is being thrown. We let point 1 be where your hand begins to move, point 2 be where the ball leaves your hand, and point 3 be where the ball is 15.0 m above point 2. The nongravitational force $\vec{F}$ of your hand acts only between points 1 and 2. Using the same coordinate system as in Example 7.1, we have $y_1 = -0.50$ m, $y_2 = 0$, and $y_3 = 15.0$ m. The ball starts at rest at point 1, so $v_1 = 0$, and the ball's speed as it leaves your hand is $v_2 = 20.0$ m/s. Our target variables are (a) the magnitude F of the force of your hand and (b) the magnitude of the ball's velocity v_{3y} at point 3.

EXECUTE: (a) To determine F, we'll first use Eq. (7.6) to calculate the work W_{other} done by this force. We have

$$K_1 = 0$$

$$U_{\text{grav},1} = mgy_1 = (0.145 \text{ kg})(9.80 \text{ m/s}^2)(-0.50 \text{ m}) = -0.71 \text{ J}$$

$$K_2 = \tfrac{1}{2}mv_2^2 = \tfrac{1}{2}(0.145 \text{ kg})(20.0 \text{ m/s})^2 = 29.0 \text{ J}$$

$$U_{\text{grav},2} = mgy_2 = (0.145 \text{ kg})(9.80 \text{ m/s}^2)(0) = 0$$

(Don't worry that $U_{\text{grav},1}$ is less than zero; all that matters is the *difference* in potential energy from one point to another.) From Eq. (7.6),

$$K_1 + U_{\text{grav},1} + W_{\text{other}} = K_2 + U_{\text{grav},2}$$

$$W_{\text{other}} = (K_2 - K_1) + (U_{\text{grav},2} - U_{\text{grav},1})$$

$$= (29.0 \text{ J} - 0) + [0 - (-0.71 \text{ J})] = 29.7 \text{ J}$$

But since $\vec{F}$ is constant and upward, the work done by $\vec{F}$ equals the force magnitude times the displacement: $W_{\text{other}} = F(y_2 - y_1)$. So

$$F = \frac{W_{\text{other}}}{y_2 - y_1} = \frac{29.7 \text{ J}}{0.50 \text{ m}} = 59 \text{ N}$$

This is more than 40 times the weight of the ball (1.42 N).

(b) To find v_{3y}, note that between points 2 and 3 only gravity acts on the ball. So between these points mechanical energy is

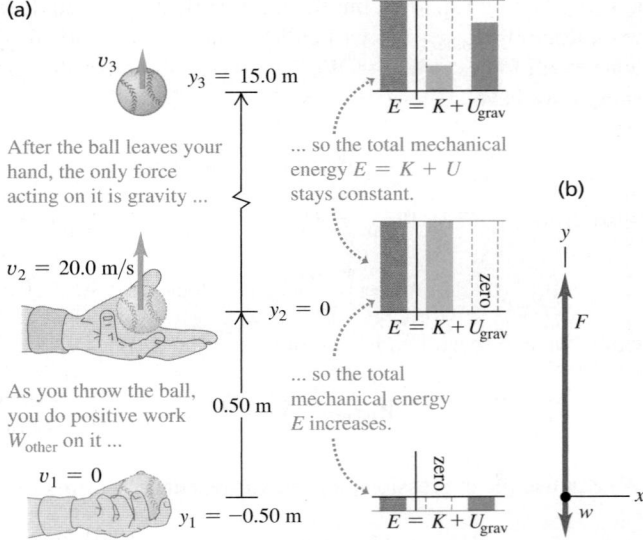

7.6 (a) Applying energy ideas to a ball thrown vertically upward. (b) Free-body diagram for the ball as you throw it.

(a)

v_3 $y_3 = 15.0$ m

After the ball leaves your hand, the only force acting on it is gravity ...

... so the total mechanical energy $E = K + U$ stays constant.

$E = K + U_{\text{grav}}$

(b)

$v_2 = 20.0$ m/s $y_2 = 0$

$E = K + U_{\text{grav}}$

As you throw the ball, you do positive work W_{other} on it ...

0.50 m

... so the total mechanical energy E increases.

$v_1 = 0$ $y_1 = -0.50$ m

$E = K + U_{\text{grav}}$

y

F

w

x

conserved and $W_{\text{other}} = 0$. From Eq. (7.4), we can solve for K_3 and from that solve for v_{3y}:

$$K_2 + U_{\text{grav},2} = K_3 + U_{\text{grav},3}$$

$$U_{\text{grav},3} = mgy_3 = (0.145 \text{ kg})(9.80 \text{ m/s}^2)(15.0 \text{ m}) = 21.3 \text{ J}$$

$$K_3 = (K_2 + U_{\text{grav},2}) - U_{\text{grav},3}$$

$$= (29.0 \text{ J} + 0 \text{ J}) - 21.3 \text{ J} = 7.7 \text{ J}$$

Since $K_3 = \tfrac{1}{2}mv_{3y}^2$, we find

$$v_{3y} = \pm\sqrt{\frac{2K_3}{m}} = \pm\sqrt{\frac{2(7.7 \text{ J})}{0.145 \text{ kg}}} = \pm 10 \text{ m/s}$$

The plus-or-minus sign reminds us that the ball passes point 3 on the way up and again on the way down. The ball's kinetic energy $K_3 = 7.7$ J at point 3, and hence its speed at that point, doesn't depend on the direction the ball is moving. The velocity v_{3y} is positive ($+10$ m/s) when the ball is moving up and negative (-10 m/s) when it is moving down; the speed v_3 is 10 m/s in either case.

EVALUATE: In Example 7.1 we found that the ball reaches a maximum height $y = 20.4$ m. At that point all of the kinetic energy it had when it left your hand at $y = 0$ has been converted to gravitational potential energy. At $y = 15.0$ m, the ball is about three-fourths of the way to its maximum height, so about three-fourths of its mechanical energy should be in the form of potential energy. Can you verify this from our results for K_3 and $U_{\text{grav},3}$?

Gravitational Potential Energy for Motion Along a Curved Path

In our first two examples the body moved along a straight vertical line. What happens when the path is slanted or curved (**Fig. 7.7a**)? The body is acted on by the gravitational force $\vec{w} = m\vec{g}$ and possibly by other forces whose resultant we call $\vec{F}_{\text{other}}$. To find the work W_{grav} done by the gravitational force during this displacement, we divide the path into small segments $\Delta\vec{s}$; Fig. 7.7b shows a typical segment. The work done by the gravitational force over this segment is the scalar product of the force and the displacement. In terms of unit vectors, the force is $\vec{w} = m\vec{g} = -mg\hat{j}$ and the displacement is $\Delta\vec{s} = \Delta x\hat{i} + \Delta y\hat{j}$, so

$$W_{\text{grav}} = \vec{w} \cdot \Delta\vec{s} = -mg\hat{j} \cdot (\Delta x\hat{i} + \Delta y\hat{j}) = -mg\Delta y$$

The work done by gravity is the same as though the body had been displaced vertically a distance Δy, with no horizontal displacement. This is true for every segment, so the *total* work done by the gravitational force is $-mg$ multiplied by the *total* vertical displacement $(y_2 - y_1)$:

$$W_{\text{grav}} = -mg(y_2 - y_1) = mgy_1 - mgy_2 = U_{\text{grav},1} - U_{\text{grav},2}$$

This is the same as Eq. (7.1) or (7.3), in which we assumed a purely vertical path. So even if the path a body follows between two points is curved, the total work done by the gravitational force depends on only the difference in height between the two points of the path. This work is unaffected by any horizontal motion that may occur. So *we can use the same expression for gravitational potential energy whether the body's path is curved or straight.*

7.7 Calculating the change in gravitational potential energy for a displacement along a curved path.

(a)

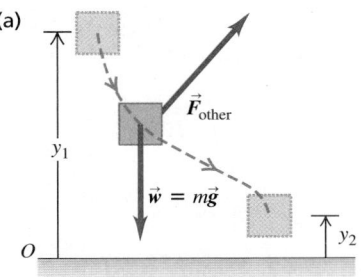

(b)

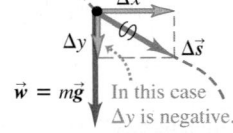

The work done by the gravitational force depends only on the vertical component of displacement Δy.

In this case Δy is negative.

CONCEPTUAL EXAMPLE 7.3   **ENERGY IN PROJECTILE MOTION**

A batter hits two identical baseballs with the same initial speed and from the same initial height but at different initial angles. Prove that both balls have the same speed at any height h if air resistance can be ignored.

SOLUTION

The only force acting on each ball after it is hit is its weight. Hence the total mechanical energy for each ball is constant. **Figure 7.8** shows the trajectories of two balls batted at the same height with the same initial speed, and thus the same total mechanical energy, but with different initial angles. At all points at the same height the potential energy is the same. Thus the kinetic energy at this height must be the same for both balls, and the speeds are the same.

7.8 For the same initial speed and initial height, the speed of a projectile at a given elevation h is always the same, if we ignore air resistance.

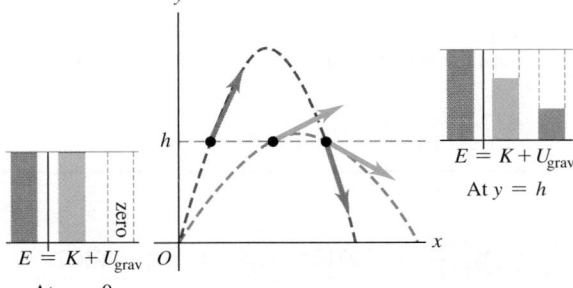

$E = K + U_{\text{grav}}$
At $y = h$

$E = K + U_{\text{grav}}$
At $y = 0$

EXAMPLE 7.4   **SPEED AT THE BOTTOM OF A VERTICAL CIRCLE**

Your cousin Throckmorton skateboards from rest down a curved, frictionless ramp. If we treat Throcky and his skateboard as a particle, he moves through a quarter-circle with radius $R = 3.00$ m (**Fig. 7.9**, next page). Throcky and his skateboard have a total mass of 25.0 kg. (a) Find his speed at the bottom of the ramp. (b) Find the normal force that acts on him at the bottom of the curve.

SOLUTION

IDENTIFY: We can't use the constant-acceleration equations of Chapter 2 because Throcky's acceleration isn't constant; the slope decreases as he descends. Instead, we'll use the energy approach. Throcky moves along a circular arc, so we'll also use what we learned about circular motion in Section 5.4.

SET UP: The only forces on Throcky are his weight and the normal force $\vec{n}$ exerted by the ramp (Fig. 7.9b). Although $\vec{n}$ acts all along the path, it does zero work because $\vec{n}$ is perpendicular to Throcky's displacement at every point. Hence $W_{\text{other}} = 0$ and mechanical energy is conserved. We treat Throcky as a particle located at the center of his body, take point 1 at the particle's starting point, and take point 2 (which we let be $y = 0$) at the particle's low point. We take the positive y-direction upward; then $y_1 = R$ and $y_2 = 0$. Throcky starts at rest at the top, so $v_1 = 0$. In part (a) our target variable is his speed v_2 at the bottom; in part (b) the target variable is the magnitude n of the normal force at point 2. To find n, we'll use Newton's second law and the relation $a = v^2/R$.

Continued

7.9 (a) Throcky skateboarding down a frictionless circular ramp. The total mechanical energy is constant.
(b) Free-body diagrams for Throcky and his skateboard at various points on the ramp.

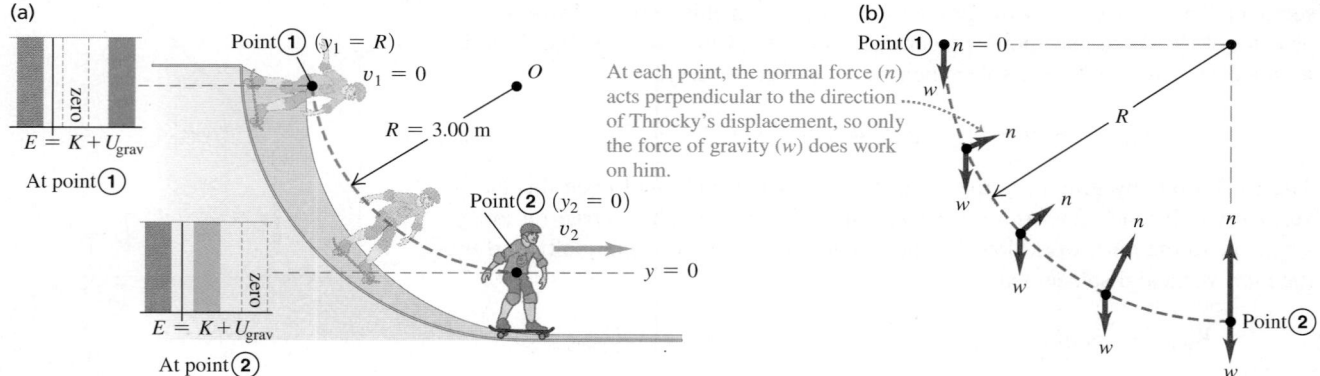

EXECUTE: (a) The various energy quantities are

$$K_1 = 0 \qquad U_{\text{grav},1} = mgR$$
$$K_2 = \tfrac{1}{2}mv_2^2 \qquad U_{\text{grav},2} = 0$$

From conservation of mechanical energy, Eq. (7.4),

$$K_1 + U_{\text{grav},1} = K_2 + U_{\text{grav},2}$$
$$0 + mgR = \tfrac{1}{2}mv_2^2 + 0$$
$$v_2 = \sqrt{2gR} = \sqrt{2(9.80 \text{ m/s}^2)(3.00 \text{ m})} = 7.67 \text{ m/s}$$

This answer doesn't depend on the ramp being circular; Throcky would have the same speed $v_2 = \sqrt{2gR}$ at the bottom of any ramp of height R, no matter what its shape.

(b) To use Newton's second law to find n at point 2, we need the free-body diagram at that point (Fig. 7.9b). At point 2, Throcky is moving at speed $v_2 = \sqrt{2gR}$ in a circle of radius R; his acceleration is toward the center of the circle and has magnitude

$$a_{\text{rad}} = \frac{v_2^2}{R} = \frac{2gR}{R} = 2g$$

The y-component of Newton's second law is

$$\Sigma F_y = n + (-w) = ma_{\text{rad}} = 2mg$$
$$n = w + 2mg = 3mg$$
$$= 3(25.0 \text{ kg})(9.80 \text{ m/s}^2) = 735 \text{ N}$$

At point 2 the normal force is three times Throcky's weight. This result doesn't depend on the radius R of the ramp. We saw in Examples 5.9 and 5.23 that the magnitude of n is the *apparent weight,* so at the bottom of the *curved part* of the ramp Throcky feels as though he weighs three times his true weight mg. But when he reaches the *horizontal* part of the ramp, immediately to the right of point 2, the normal force decreases to $w = mg$ and thereafter Throcky feels his true weight again. Can you see why?

EVALUATE: This example shows a general rule about the role of forces in problems in which we use energy techniques: What matters is not simply whether a force *acts,* but whether that force *does work.* If the force does no work, like the normal force $\vec{n}$ here, then it does not appear in Eqs. (7.4) and (7.6).

EXAMPLE 7.5 A VERTICAL CIRCLE WITH FRICTION

Suppose that the ramp of Example 7.4 is not frictionless and that Throcky's speed at the bottom is only 6.00 m/s, not the 7.67 m/s we found there. What work was done on him by the friction force?

SOLUTION

IDENTIFY and SET UP: The setup is the same as in Example 7.4. **Figure 7.10** shows that again the normal force does no work, but now there is a friction force $\vec{f}$ that *does* do work W_f. Hence the nongravitational work W_{other} done on Throcky between points 1 and 2 is equal to W_f and is not zero. Our target variable is $W_f = W_{\text{other}}$, which we'll find by using Eq. (7.6). Since $\vec{f}$ points opposite to Throcky's motion, W_f is negative.

EXECUTE: The energy quantities are

$$K_1 = 0$$
$$U_{\text{grav},1} = mgR = (25.0 \text{ kg})(9.80 \text{ m/s}^2)(3.00 \text{ m}) = 735 \text{ J}$$
$$K_2 = \tfrac{1}{2}mv_2^2 = \tfrac{1}{2}(25.0 \text{ kg})(6.00 \text{ m/s})^2 = 450 \text{ J}$$
$$U_{\text{grav},2} = 0$$

7.10 Energy bar graphs and free-body diagrams for Throcky skateboarding down a ramp with friction.

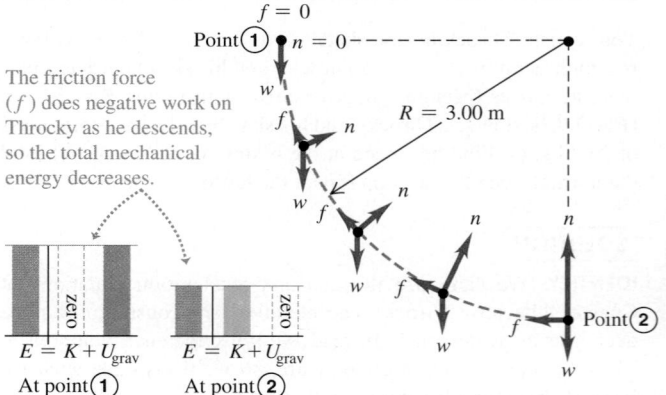

From Eq. (7.6),

$$W_f = W_{other}$$
$$= K_2 + U_{grav,2} - K_1 - U_{grav,1}$$
$$= 450\,J + 0 - 0 - 735\,J$$
$$= -285\,J$$

The work done by the friction force is $-285\,J$, and the total mechanical energy *decreases* by 285 J.

EVALUATE: Our result for W_f is negative. Can you see from the free-body diagrams in Fig. 7.10 why this must be so?

It would be very difficult to apply Newton's second law, $\sum \vec{F} = m\vec{a}$, directly to this problem because the normal and friction forces and the acceleration are continuously changing in both magnitude and direction as Throcky descends. The energy approach, by contrast, relates the motions at the top and bottom of the ramp without involving the details of the motion in between.

EXAMPLE 7.6 AN INCLINED PLANE WITH FRICTION

We want to slide a 12-kg crate up a 2.5-m-long ramp inclined at 30°. A worker, ignoring friction, calculates that he can do this by giving it an initial speed of 5.0 m/s at the bottom and letting it go. But friction is *not* negligible; the crate slides only 1.6 m up the ramp, stops, and slides back down (**Fig. 7.11a**). (a) Find the magnitude of the friction force acting on the crate, assuming that it is constant. (b) How fast is the crate moving when it reaches the bottom of the ramp?

SOLUTION

IDENTIFY and SET UP: The friction force does work on the crate as it slides from point 1, at the bottom of the ramp, to point 2, where the crate stops instantaneously ($v_2 = 0$). Friction also does work as the crate returns to the bottom of the ramp, which we'll call point 3 (Fig. 7.11a). We take the positive y-direction upward. We take $y = 0$ (and hence $U_{grav} = 0$) to be at ground level (point 1), so $y_1 = 0$, $y_2 = (1.6\,m)\sin 30° = 0.80\,m$, and $y_3 = 0$. We are given $v_1 = 5.0$ m/s. In part (a) our target variable is f, the magnitude of the friction force as the crate slides up; we'll find this by using the energy approach. In part (b) our target variable is v_3, the crate's speed at the bottom of the ramp. We'll calculate the work done by friction as the crate slides back down, then use the energy approach to find v_3.

EXECUTE: (a) The energy quantities are

$$K_1 = \tfrac{1}{2}(12\,kg)(5.0\,m/s)^2 = 150\,J$$
$$U_{grav,1} = 0$$
$$K_2 = 0$$
$$U_{grav,2} = (12\,kg)(9.8\,m/s^2)(0.80\,m) = 94\,J$$
$$W_{other} = -fs$$

Here $s = 1.6$ m. Using Eq. (7.6), we find

$$K_1 + U_{grav,1} + W_{other} = K_2 + U_{grav,2}$$
$$W_{other} = -fs = (K_2 + U_{grav,2}) - (K_1 + U_{grav,1})$$
$$= (0 + 94\,J) - (150\,J + 0) = -56\,J = -fs$$
$$f = \frac{W_{other}}{s} = \frac{56\,J}{1.6\,m} = 35\,N$$

The friction force of 35 N, acting over 1.6 m, causes the mechanical energy of the crate to decrease from 150 J to 94 J (Fig. 7.11b).

(b) As the crate moves from point 2 to point 3, the work done by friction has the same negative value as from point 1 to point 2. (Both the friction force and the displacement reverse direction, but their magnitudes don't change.) The total work done by friction between points 1 and 3 is therefore

$$W_{other} = W_{fric} = -2fs = -2(56\,J) = -112\,J$$

From part (a), $K_1 = 150\,J$ and $U_{grav,1} = 0$; in addition, $U_{grav,3} = 0$ since $y_3 = 0$. Equation (7.6) then gives

$$K_1 + U_{grav,1} + W_{other} = K_3 + U_{grav,3}$$
$$K_3 = K_1 + U_{grav,1} - U_{grav,3} + W_{other}$$
$$= 150\,J + 0 - 0 + (-112\,J) = 38\,J$$

The crate returns to the bottom of the ramp with only 38 J of the original 150 J of mechanical energy (Fig. 7.11b). Since $K_3 = \tfrac{1}{2}mv_3^2$,

$$v_3 = \sqrt{\frac{2K_3}{m}} = \sqrt{\frac{2(38\,J)}{12\,kg}} = 2.5\,m/s$$

EVALUATE: Energy is lost due to friction, so the crate's speed $v_3 = 2.5$ m/s when it returns to the bottom of the ramp is less than the speed $v_1 = 5.0$ m/s at which it left that point. In part (b) we applied Eq. (7.6) to points 1 and 3, considering the round trip as a whole. Alternatively, we could have considered the second part of the motion by itself and applied Eq. (7.6) to points 2 and 3. Try it; do you get the same result for v_3?

7.11 (a) A crate slides partway up the ramp, stops, and slides back down. (b) Energy bar graphs for points 1, 2, and 3.

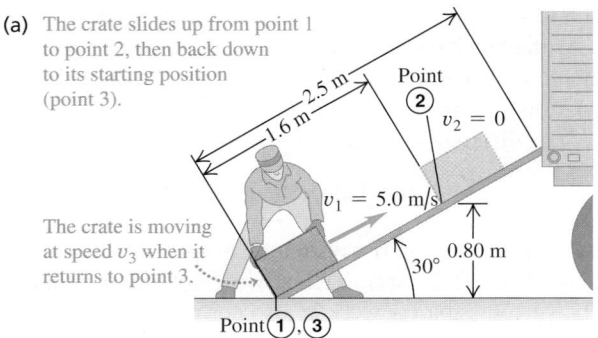

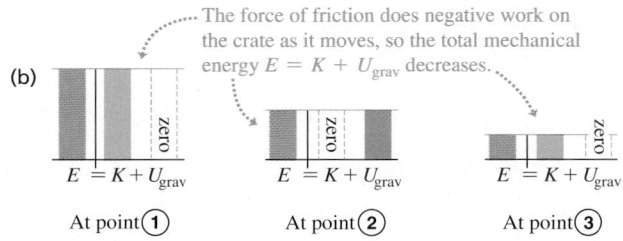

TEST YOUR UNDERSTANDING OF SECTION 7.1   The figure shows two friction-less ramps. The heights y_1 and y_2 are the same for both ramps. If a block of mass m is released from rest at the left-hand end of each ramp, which block arrives at the right-hand end with the greater speed? (i) Block I; (ii) block II; (iii) the speed is the same for both blocks.

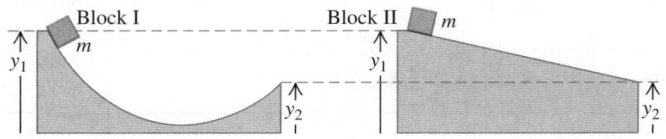

7.2 ELASTIC POTENTIAL ENERGY

In many situations we encounter potential energy that is not gravitational in nature. One example is a rubber-band slingshot. Work is done on the rubber band by the force that stretches it, and that work is stored in the rubber band until you let it go. Then the rubber band gives kinetic energy to the projectile.

This is the same pattern we saw with the baseball in Example 7.2: Do work on the system to store energy, which can later be converted to kinetic energy. We'll describe the process of storing energy in a deformable body such as a spring or rubber band in terms of *elastic potential energy* (**Fig. 7.12**). A body is called *elastic* if it returns to its original shape and size after being deformed.

To be specific, we'll consider storing energy in an ideal spring, like the ones we discussed in Section 6.3. To keep such an ideal spring stretched by a distance x, we must exert a force $F = kx$, where k is the force constant of the spring. Many elastic bodies show this same direct proportionality between force $\vec{F}$ and displacement x, provided that x is sufficiently small.

Let's proceed just as we did for gravitational potential energy. We begin with the work done by the elastic (spring) force and then combine this with the work–energy theorem. The difference is that gravitational potential energy is a shared property of a body and the earth, but elastic potential energy is stored in just the spring (or other deformable body).

Figure 7.13 shows the ideal spring from Fig. 6.18 but with its left end held stationary and its right end attached to a block with mass m that can move along the x-axis. In Fig. 7.13a the block is at $x = 0$ when the spring is neither stretched nor compressed. We move the block to one side, thereby stretching or compressing the spring, then let it go. As the block moves from a different position x_1 to a different position x_2, how much work does the elastic (spring) force do on the block?

We found in Section 6.3 that the work we must do *on* the spring to move one end from an elongation x_1 to a different elongation x_2 is

$$W = \tfrac{1}{2}kx_2^2 - \tfrac{1}{2}kx_1^2 \qquad \text{(work done } on \text{ a spring)} \qquad (7.8)$$

where k is the force constant of the spring. If we stretch the spring farther, we do positive work on the spring; if we let the spring relax while holding one end, we do negative work on it. This expression for work is also correct when the spring is compressed such that x_1, x_2, or both are negative. Now, from Newton's third law the work done *by* the spring is just the negative of the work done *on* the spring. So by changing the signs in Eq. (7.8), we find that in a displacement from x_1 to x_2 the spring does an amount of work W_{el} given by

$$W_{el} = \tfrac{1}{2}kx_1^2 - \tfrac{1}{2}kx_2^2 \qquad \text{(work done } by \text{ a spring)} \qquad (7.9)$$

The subscript "el" stands for *elastic*. When both x_1 and x_2 are positive and $x_2 > x_1$ (Fig. 7.13b), the spring does negative work on the block, which moves in the $+x$-direction while the spring pulls on it in the $-x$-direction. The spring stretches farther, and the block slows down. When both x_1 and x_2 are positive

7.12 The Achilles tendon, which runs along the back of the ankle to the heel bone, acts like a natural spring. When it stretches and then relaxes, this tendon stores and then releases elastic potential energy. This spring action reduces the amount of work your leg muscles must do as you run.

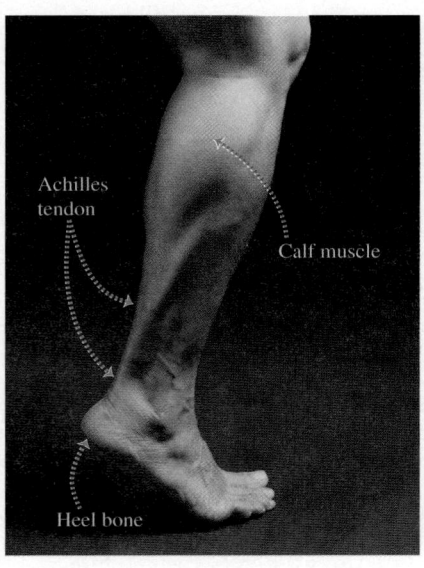

and $x_2 < x_1$ (Fig. 7.13c), the spring does positive work as it relaxes and the block speeds up. If the spring can be compressed as well as stretched, x_1, x_2, or both may be negative, but the expression for W_{el} is still valid. In Fig. 7.13d, both x_1 and x_2 are negative, but x_2 is less negative than x_1; the compressed spring does positive work as it relaxes, speeding the block up.

Just as for gravitational work, we can express Eq. (7.9) for the work done by the spring in terms of a quantity at the beginning and end of the displacement. This quantity is $\frac{1}{2}kx^2$, and we define it to be the **elastic potential energy:**

Elastic potential energy stored in a spring ⟶ $U_{el} = \frac{1}{2}kx^2$ Force constant of spring / Elongation of spring ($x > 0$ if stretched, $x < 0$ if compressed)   (7.10)

Figure 7.14 is a graph of Eq. (7.10). As for all other energy and work quantities, the unit of U_{el} is the joule (J); to see this from Eq. (7.10), recall that the units of k are N/m and that $1 \text{ N} \cdot \text{m} = 1 \text{ J}$. We can now use Eq. (7.10) to rewrite Eq. (7.9) for the work W_{el} done by the spring:

Work done by the elastic force ... ... equals the **negative of the change in elastic potential energy.**

$$W_{el} = \frac{1}{2}kx_1^2 - \frac{1}{2}kx_2^2 = U_{el,1} - U_{el,2} = -\Delta U_{el} \quad (7.11)$$

Force constant of spring   Initial and final elongations of spring

When a stretched spring is stretched farther, as in Fig. 7.13b, W_{el} is negative and U_{el} increases; more elastic potential energy is stored in the spring. When a stretched spring relaxes, as in Fig. 7.13c, x decreases, W_{el} is positive, and U_{el} decreases; the spring loses elastic potential energy. Figure 7.14 shows that U_{el} is positive for both positive and negative x values; Eqs. (7.10) and (7.11) are valid for both cases. The more a spring is compressed or stretched, the greater its elastic potential energy.

CAUTION **Gravitational potential energy vs. elastic potential energy** An important difference between gravitational potential energy $U_{grav} = mgy$ and elastic potential energy $U_{el} = \frac{1}{2}kx^2$ is that we *cannot* choose $x = 0$ to be wherever we wish. In Eq. (7.10), $x = 0$ *must* be the position at which the spring is neither stretched nor compressed. At that position, both its elastic potential energy and the force that it exerts are zero. |

The work–energy theorem says that $W_{tot} = K_2 - K_1$, no matter what kind of forces are acting on a body. If the elastic force is the *only* force that does work on the body, then

$$W_{tot} = W_{el} = U_{el,1} - U_{el,2}$$

and so

If only the elastic force does work, total mechanical energy is conserved:

Initial kinetic energy   Initial elastic potential energy
$K_1 = \frac{1}{2}mv_1^2$ $U_{el,1} = \frac{1}{2}kx_1^2$

$$K_1 + U_{el,1} = K_2 + U_{el,2} \quad (7.12)$$

Final kinetic energy   Final elastic potential energy
$K_2 = \frac{1}{2}mv_2^2$ $U_{el,2} = \frac{1}{2}kx_2^2$

In this case the total mechanical energy $E = K + U_{el}$—the sum of kinetic and *elastic* potential energies—is *conserved*. An example of this is the motion of the block in Fig. 7.13, provided the horizontal surface is frictionless so no force does work other than that exerted by the spring.

For Eq. (7.12) to be strictly correct, the ideal spring that we've been discussing must also be *massless*. If the spring has mass, it also has kinetic energy as the

7.13 Calculating the work done by a spring attached to a block on a horizontal surface. The quantity x is the extension or compression of the spring.

(a)

Here the spring is neither stretched nor compressed.
$x = 0$

(b)

As the spring stretches, it does negative work on the block.
$\vec{s}$
x_2
x_1
$\vec{F}_{spring}$

(c)

As the spring relaxes, it does positive work on the block. $\vec{s}$
x_1
x_2
$\vec{F}_{spring}$

(d)

$\vec{s}$
x_1
x_2
A compressed spring also does positive work on the block as it relaxes.
$\vec{F}_{spring}$

7.14 The graph of elastic potential energy for an ideal spring is a parabola: $U_{el} = \frac{1}{2}kx^2$, where x is the extension or compression of the spring. Elastic potential energy U_{el} is never negative.

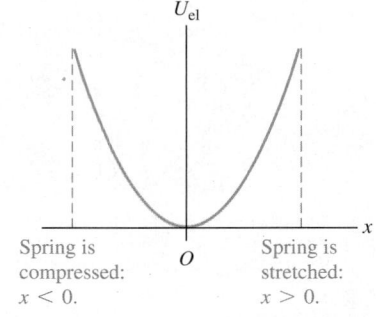

U_{el}

Spring is compressed: $x < 0$. O Spring is stretched: $x > 0$.

BIO Application Elastic Potential Energy of a Cheetah When a cheetah (*Acinonyx jubatus*) gallops, its back flexes and extends dramatically. Flexion of the back stretches tendons and muscles along the top of the spine and also compresses the spine, storing elastic potential energy. When the cheetah launches into its next bound, this energy is released, enabling the cheetah to run more efficiently.

Difference in nose-to-tail length

7.15 Trampoline jumping involves an interplay among kinetic energy, gravitational potential energy, and elastic potential energy. Due to air resistance and friction forces within the trampoline, mechanical energy is not conserved. That's why the bouncing eventually stops unless the jumper does work with his or her legs to compensate for the lost energy.

Gravitational potential energy increases as jumper ascends.

Kinetic energy increases as jumper moves faster.

Elastic potential energy increases when trampoline is stretched.

coils of the spring move back and forth. We can ignore the kinetic energy of the spring if its mass is much less than the mass m of the body attached to the spring. For instance, a typical automobile has a mass of 1200 kg or more. The springs in its suspension have masses of only a few kilograms, so their mass can be ignored if we want to study how a car bounces on its suspension.

Situations with Both Gravitational and Elastic Potential Energy

Equation (7.12) is valid when the only potential energy in the system is elastic potential energy. What happens when we have *both* gravitational and elastic forces, such as a block attached to the lower end of a vertically hanging spring? And what if work is also done by other forces that *cannot* be described in terms of potential energy, such as the force of air resistance on a moving block? Then the total work is the sum of the work done by the gravitational force (W_{grav}), the work done by the elastic force (W_{el}), and the work done by other forces (W_{other}): $W_{tot} = W_{grav} + W_{el} + W_{other}$. The work–energy theorem then gives

$$W_{grav} + W_{el} + W_{other} = K_2 - K_1$$

The work done by the gravitational force is $W_{grav} = U_{grav,1} - U_{grav,2}$ and the work done by the spring is $W_{el} = U_{el,1} - U_{el,2}$. Hence we can rewrite the work–energy theorem for this most general case as

$$K_1 + U_{grav,1} + U_{el,1} + W_{other} = K_2 + U_{grav,2} + U_{el,2} \quad \text{(valid in general)} \quad (7.13)$$

or, equivalently,

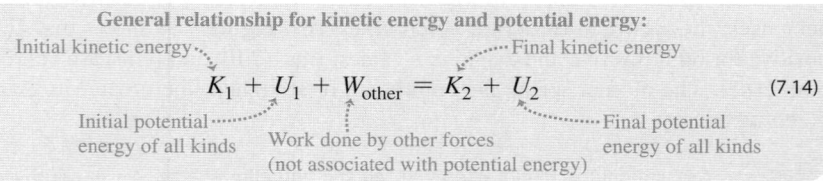

General relationship for kinetic energy and potential energy:
Initial kinetic energy ⋯ Final kinetic energy
$$K_1 + U_1 + W_{other} = K_2 + U_2 \quad (7.14)$$
Initial potential energy of all kinds ⋯ Work done by other forces (not associated with potential energy) ⋯ Final potential energy of all kinds

where $U = U_{grav} + U_{el} = mgy + \frac{1}{2}kx^2$ is the *sum* of gravitational potential energy and elastic potential energy. We call U simply "the potential energy."

Equation (7.14) is *the most general statement* of the relationship among kinetic energy, potential energy, and work done by other forces. It says:

> The work done by all forces other than the gravitational force or elastic force equals the change in the total mechanical energy $E = K + U$ of the system.

The "system" is made up of the body of mass m, the earth with which it interacts through the gravitational force, and the spring of force constant k.

If W_{other} is positive, $E = K + U$ increases; if W_{other} is negative, E decreases. If the gravitational and elastic forces are the *only* forces that do work on the body, then $W_{other} = 0$ and the total mechanical energy $E = K + U$ is conserved. [Compare Eq. (7.14) to Eqs. (7.6) and (7.7), which include gravitational potential energy but not elastic potential energy.]

Trampoline jumping (**Fig. 7.15**) involves transformations among kinetic energy, elastic potential energy, and gravitational potential energy. As the jumper descends through the air from the high point of the bounce, gravitational potential energy U_{grav} decreases and kinetic energy K increases. Once the jumper touches the trampoline, some of the mechanical energy goes into elastic potential energy U_{el} stored in the trampoline's springs. At the lowest point of the trajectory (U_{grav} is minimum), the jumper comes to a momentary halt ($K = 0$) and the springs are maximally stretched (U_{el} is maximum). The springs then convert their energy back into K and U_{grav}, propelling the jumper upward.

PROBLEM-SOLVING STRATEGY 7.2 PROBLEMS USING MECHANICAL ENERGY II

Problem-Solving Strategy 7.1 (Section 7.1) is useful in solving problems that involve elastic forces as well as gravitational forces. The only new wrinkle is that the potential energy U now includes the elastic potential energy $U_{el} = \frac{1}{2}kx^2$, where x is the displacement of the spring *from its unstretched length*. The work done by the gravitational and elastic forces is accounted for by their potential energies; the work done by other forces, W_{other}, must still be included separately.

EXAMPLE 7.7 MOTION WITH ELASTIC POTENTIAL ENERGY

A glider with mass $m = 0.200$ kg sits on a frictionless, horizontal air track, connected to a spring with force constant $k = 5.00$ N/m. You pull on the glider, stretching the spring 0.100 m, and release it from rest. The glider moves back toward its equilibrium position $(x = 0)$. What is its x-velocity when $x = 0.080$ m?

SOLUTION

IDENTIFY and SET UP: As the glider starts to move, elastic potential energy is converted to kinetic energy. The glider remains at the same height throughout the motion, so gravitational potential energy is not a factor and $U = U_{el} = \frac{1}{2}kx^2$. **Figure 7.16** shows our sketches. Only the spring force does work on the glider, so

7.16 Our sketches and energy bar graphs for this problem.

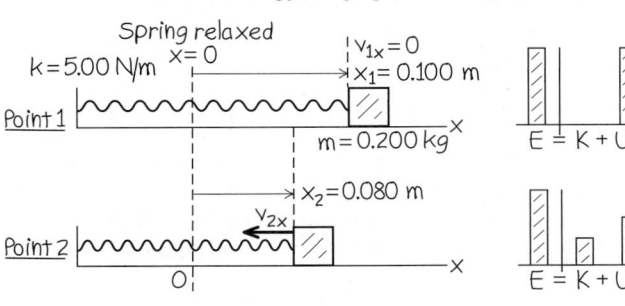

$W_{other} = 0$ in Eq. (7.14). We designate the point where the glider is released as point 1 (that is, $x_1 = 0.100$ m) and $x_2 = 0.080$ m as point 2. We are given $v_{1x} = 0$; our target variable is v_{2x}.

EXECUTE: The energy quantities are

$$K_1 = \tfrac{1}{2}mv_{1x}^2 = \tfrac{1}{2}(0.200 \text{ kg})(0)^2 = 0$$

$$U_1 = \tfrac{1}{2}kx_1^2 = \tfrac{1}{2}(5.00 \text{ N/m})(0.100 \text{ m})^2 = 0.0250 \text{ J}$$

$$K_2 = \tfrac{1}{2}mv_{2x}^2$$

$$U_2 = \tfrac{1}{2}kx_2^2 = \tfrac{1}{2}(5.00 \text{ N/m})(0.080 \text{ m})^2 = 0.0160 \text{ J}$$

We use Eq. (7.14) with $W_{other} = 0$ to solve for K_2 and then find v_{2x}:

$$K_2 = K_1 + U_1 - U_2 = 0 + 0.0250 \text{ J} - 0.0160 \text{ J} = 0.0090 \text{ J}$$

$$v_{2x} = \pm\sqrt{\frac{2K_2}{m}} = \pm\sqrt{\frac{2(0.0090 \text{ J})}{0.200 \text{ kg}}} = \pm 0.30 \text{ m/s}$$

We choose the negative root because the glider is moving in the $-x$-direction. Our answer is $v_{2x} = -0.30$ m/s.

EVALUATE: Eventually the spring will reverse the glider's motion, pushing it back in the $+x$-direction (see Fig. 7.13d). The solution $v_{2x} = +0.30$ m/s tells us that when the glider passes through $x = 0.080$ m on this return trip, its speed will be 0.30 m/s, just as when it passed through this point while moving to the left.

EXAMPLE 7.8 MOTION WITH ELASTIC POTENTIAL ENERGY AND WORK DONE BY OTHER FORCES

Suppose the glider in Example 7.7 is initially at rest at $x = 0$, with the spring unstretched. You then push on the glider with a constant force $\vec{F}$ (magnitude 0.610 N) in the $+x$-direction. What is the glider's velocity when it has moved to $x = 0.100$ m?

SOLUTION

IDENTIFY and SET UP: Although the force $\vec{F}$ you apply is constant, the spring force isn't, so the acceleration of the glider won't be constant. Total mechanical energy is not conserved because of the work done by force $\vec{F}$, so W_{other} in Eq. (7.14) is not zero. As in Example 7.7, we ignore gravitational potential energy because the glider's height doesn't change. Hence we again have $U = U_{el} = \frac{1}{2}kx^2$. This time, we let point 1 be at $x_1 = 0$, where the velocity is $v_{1x} = 0$, and let point 2 be at $x = 0.100$ m. The glider's

displacement is then $\Delta x = x_2 - x_1 = 0.100$ m. Our target variable is v_{2x}, the velocity at point 2.

EXECUTE: Force $\vec{F}$ is constant and in the same direction as the displacement, so the work done by this force is $F\Delta x$. Then the energy quantities are

$$K_1 = 0$$

$$U_1 = \tfrac{1}{2}kx_1^2 = 0$$

$$K_2 = \tfrac{1}{2}mv_{2x}^2$$

$$U_2 = \tfrac{1}{2}kx_2^2 = \tfrac{1}{2}(5.00 \text{ N/m})(0.100 \text{ m})^2 = 0.0250 \text{ J}$$

$$W_{other} = F\Delta x = (0.610 \text{ N})(0.100 \text{ m}) = 0.0610 \text{ J}$$

Continued

The initial total mechanical energy is zero; the work done by $\vec{F}$ increases the total mechanical energy to 0.0610 J, of which $U_2 = 0.0250$ J is elastic potential energy. The remainder is kinetic energy. From Eq. (7.14),

$$K_1 + U_1 + W_{other} = K_2 + U_2$$
$$K_2 = K_1 + U_1 + W_{other} - U_2$$
$$= 0 + 0 + 0.0610 \text{ J} - 0.0250 \text{ J} = 0.0360 \text{ J}$$
$$v_{2x} = \sqrt{\frac{2K_2}{m}} = \sqrt{\frac{2(0.0360 \text{ J})}{0.200 \text{ kg}}} = 0.60 \text{ m/s}$$

We choose the positive square root because the glider is moving in the $+x$-direction.

EVALUATE: What would be different if we disconnected the glider from the spring? Then only $\vec{F}$ would do work, there would be zero elastic potential energy at all times, and Eq. (7.14) would give us

$$K_2 = K_1 + W_{other} = 0 + 0.0610 \text{ J}$$
$$v_{2x} = \sqrt{\frac{2K_2}{m}} = \sqrt{\frac{2(0.0610 \text{ J})}{0.200 \text{ kg}}} = 0.78 \text{ m/s}$$

Our answer $v_{2x} = 0.60$ m/s is less than 0.78 m/s because the spring does negative work on the glider as it stretches (see Fig. 7.13b).

If you stop pushing on the glider when it reaches $x = 0.100$ m, only the spring force does work on it thereafter. Hence for $x > 0.100$ m, the total mechanical energy $E = K + U = 0.0610$ J is constant. As the spring continues to stretch, the glider slows down and the kinetic energy K decreases as the potential energy increases. The glider comes to rest at some point $x = x_3$, at which the kinetic energy is zero and the potential energy $U = U_{el} = \frac{1}{2}kx_3^2$ equals the total mechanical energy 0.0610 J. Can you show that $x_3 = 0.156$ m? (It moves an additional 0.056 m after you stop pushing.) If there is no friction, will the glider remain at rest?

EXAMPLE 7.9 MOTION WITH GRAVITATIONAL, ELASTIC, AND FRICTION FORCES

A 2000-kg (19,600-N) elevator with broken cables in a test rig is falling at 4.00 m/s when it contacts a cushioning spring at the bottom of the shaft. The spring is intended to stop the elevator, compressing 2.00 m as it does so (**Fig. 7.17**). During the motion a safety clamp applies a constant 17,000-N friction force to the elevator. What is the necessary force constant k for the spring?

SOLUTION

IDENTIFY and SET UP: We'll use the energy approach and Eq. (7.14) to determine k, which appears in the expression for elastic potential energy. This problem involves *both* gravitational and elastic potential energies. Total mechanical energy is not conserved because the friction force does negative work W_{other} on the elevator. We take point 1 as the position of the bottom of the elevator when it contacts the spring, and point 2 as its position when it stops. We choose the origin to be at point 1, so $y_1 = 0$

7.17 The fall of an elevator is stopped by a spring and by a constant friction force.

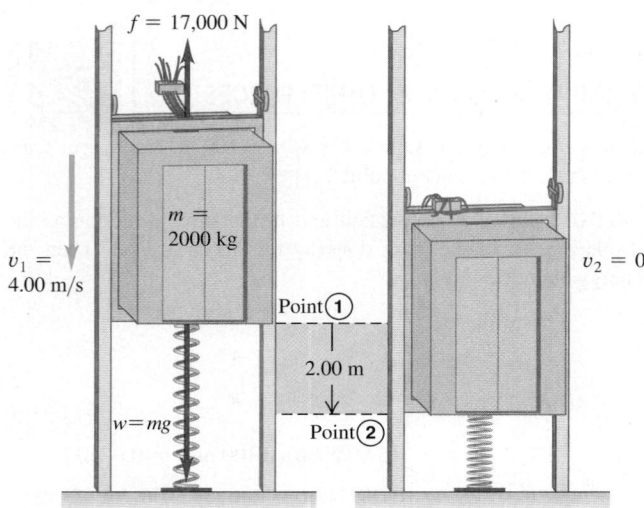

and $y_2 = -2.00$ m. With this choice the coordinate of the upper end of the spring after contact is the same as the coordinate of the elevator, so the elastic potential energy at any point between points 1 and 2 is $U_{el} = \frac{1}{2}ky^2$. The gravitational potential energy is $U_{grav} = mgy$ as usual. We know the initial and final speeds of the elevator and the magnitude of the friction force, so the only unknown is the force constant k (our target variable).

EXECUTE: The elevator's initial speed is $v_1 = 4.00$ m/s, so its initial kinetic energy is

$$K_1 = \tfrac{1}{2}mv_1^2 = \tfrac{1}{2}(2000 \text{ kg})(4.00 \text{ m/s})^2 = 16,000 \text{ J}$$

The elevator stops at point 2, so $K_2 = 0$. At point 1 the potential energy $U_1 = U_{grav} + U_{el}$ is zero; U_{grav} is zero because $y_1 = 0$, and $U_{el} = 0$ because the spring is uncompressed. At point 2 there are both gravitational and elastic potential energies, so

$$U_2 = mgy_2 + \tfrac{1}{2}ky_2^2$$

The gravitational potential energy at point 2 is

$$mgy_2 = (2000 \text{ kg})(9.80 \text{ m/s}^2)(-2.00 \text{ m}) = -39,200 \text{ J}$$

The "other" force is the constant 17,000-N friction force. It acts opposite to the 2.00-m displacement, so

$$W_{other} = -(17,000 \text{ N})(2.00 \text{ m}) = -34,000 \text{ J}$$

We put these terms into Eq. (7.14), $K_1 + U_1 + W_{other} = K_2 + U_2$:

$$K_1 + 0 + W_{other} = 0 + (mgy_2 + \tfrac{1}{2}ky_2^2)$$
$$k = \frac{2(K_1 + W_{other} - mgy_2)}{y_2^2}$$
$$= \frac{2[16,000 \text{ J} + (-34,000 \text{ J}) - (-39,200 \text{ J})]}{(-2.00 \text{ m})^2}$$
$$= 1.06 \times 10^4 \text{ N/m}$$

This is about one-tenth the force constant of a spring in an automobile suspension.

EVALUATE: There might seem to be a paradox here. The elastic potential energy at point 2 is

$$\tfrac{1}{2}ky_2^2 = \tfrac{1}{2}(1.06 \times 10^4 \text{ N/m})(-2.00 \text{ m})^2 = 21{,}200 \text{ J}$$

This is *more* than the total mechanical energy at point 1:

$$E_1 = K_1 + U_1 = 16{,}000 \text{ J} + 0 = 16{,}000 \text{ J}$$

But the friction force *decreased* the mechanical energy of the system by 34,000 J between points 1 and 2. Did energy appear from nowhere? No. At point 2, which is below the origin, there is also *negative* gravitational potential energy $mgy_2 = -39{,}200$ J. The total mechanical energy at point 2 is therefore not 21,200 J but

$$E_2 = K_2 + U_2 = 0 + \tfrac{1}{2}ky_2^2 + mgy_2$$
$$= 0 + 21{,}200 \text{ J} + (-39{,}200 \text{ J}) = -18{,}000 \text{ J}$$

This is just the initial mechanical energy of 16,000 J minus 34,000 J lost to friction.

Will the elevator stay at the bottom of the shaft? At point 2 the compressed spring exerts an upward force of magnitude $F_{\text{spring}} = (1.06 \times 10^4 \text{ N/m})(2.00 \text{ m}) = 21{,}200$ N, while the downward force of gravity is only $w = mg = (2000 \text{ kg})(9.80 \text{ m/s}^2) = 19{,}600$ N. If there were no friction, there would be a net upward force of 21,200 N − 19,600 N = 1600 N, and the elevator would rebound. But the safety clamp can exert a kinetic friction force of 17,000 N, and it can presumably exert a maximum static friction force greater than that. Hence the clamp will keep the elevator from rebounding.

TEST YOUR UNDERSTANDING OF SECTION 7.2 Consider the situation in Example 7.9 at the instant when the elevator is still moving downward and the spring is compressed by 1.00 m. Which of the energy bar graphs in the figure most accurately shows the kinetic energy K, gravitational potential energy U_{grav}, and elastic potential energy U_{el} at this instant?

(i) 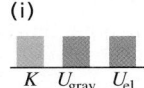 (ii) 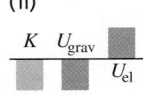 (iii) 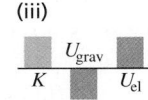 (iv)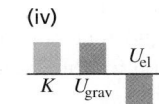

7.3 CONSERVATIVE AND NONCONSERVATIVE FORCES

In our discussions of potential energy we have talked about "storing" kinetic energy by converting it to potential energy, with the idea that we can retrieve it again as kinetic energy. For example, when you throw a ball up in the air, it slows down as kinetic energy is converted to gravitational potential energy. But on the way down the ball speeds up as potential energy is converted back to kinetic energy. If there is no air resistance, the ball is moving just as fast when you catch it as when you threw it.

Another example is a glider moving on a frictionless horizontal air track that runs into a spring bumper. The glider compresses the spring and then bounces back. If there is no friction, the glider ends up with the same speed and kinetic energy it had before the collision. Again, there is a two-way conversion from kinetic to potential energy and back. In both cases the total mechanical energy, kinetic plus potential, is constant or *conserved* during the motion.

Conservative Forces

A force that offers this opportunity of two-way conversion between kinetic and potential energies is called a **conservative force.** We have seen two examples of conservative forces: the gravitational force and the spring force. (Later in this book we'll study another conservative force, the electric force between charged objects.) An essential feature of conservative forces is that their work is always *reversible*. Anything that we deposit in the energy "bank" can later be withdrawn without loss. Another important aspect of conservative forces is that if a body follows different paths from point 1 to point 2, the work done by a conservative force is the same for all of these paths (**Fig. 7.18**). For example, if a body stays close to the surface of the earth, the gravitational force $m\vec{g}$ is independent of

7.18 The work done by a conservative force such as gravity depends on only the endpoints of a path, not the specific path taken between those points.

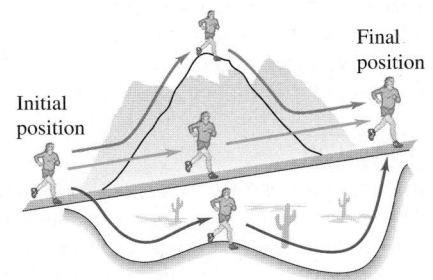

Because the gravitational force is conservative, the work it does is the same for all three paths.

Initial position

Final position

PhET: The Ramp

height, and the work done by this force depends on only the change in height. If the body moves around a closed path, ending at the same height where it started, the *total* work done by the gravitational force is always zero.

In summary, the work done by a conservative force has four properties:

1. It can be expressed as the difference between the initial and final values of a *potential-energy* function.

2. It is reversible.

3. It is independent of the path of the body and depends on only the starting and ending points.

4. When the starting and ending points are the same, the total work is zero.

When the *only* forces that do work are conservative forces, the total mechanical energy $E = K + U$ is constant.

Nonconservative Forces

Not all forces are conservative. Consider the friction force acting on the crate sliding on a ramp in Example 7.6 (Section 7.1). When the body slides up and then back down to the starting point, the total work done on it by the friction force is *not* zero. When the direction of motion reverses, so does the friction force, and friction does *negative* work in *both* directions. Friction also acts when a car with its brakes locked skids with decreasing speed (and decreasing kinetic energy). The lost kinetic energy can't be recovered by reversing the motion or in any other way, and mechanical energy is *not* conserved. So there is *no* potential-energy function for the friction force.

In the same way, the force of fluid resistance (see Section 5.3) is not conservative. If you throw a ball up in the air, air resistance does negative work on the ball while it's rising *and* while it's descending. The ball returns to your hand with less speed and less kinetic energy than when it left, and there is no way to get back the lost mechanical energy.

A force that is not conservative is called a **nonconservative force.** The work done by a nonconservative force *cannot* be represented by a potential-energy function. Some nonconservative forces, like kinetic friction or fluid resistance, cause mechanical energy to be lost or dissipated; a force of this kind is called a **dissipative force.** There are also nonconservative forces that *increase* mechanical energy. The fragments of an exploding firecracker fly off with very large kinetic energy, thanks to a chemical reaction of gunpowder with oxygen. The forces unleashed by this reaction are nonconservative because the process is not reversible. (The fragments never spontaneously reassemble themselves into a complete firecracker!)

EXAMPLE 7.10 **FRICTIONAL WORK DEPENDS ON THE PATH**

You are rearranging your furniture and wish to move a 40.0-kg futon 2.50 m across the room. A heavy coffee table, which you don't want to move, blocks this straight-line path. Instead, you slide the futon along a dogleg path; the doglegs are 2.00 m and 1.50 m long. How much more work must you do to push the futon along the dogleg path than along the straight-line path? The coefficient of kinetic friction is $\mu_k = 0.200$.

SOLUTION

IDENTIFY and SET UP: Here both you and friction do work on the futon, so we must use the energy relationship that includes "other" forces. We'll use this relationship to find a connection between the work that *you* do and the work that *friction* does. **Figure 7.19**

shows our sketch. The futon is at rest at both point 1 and point 2, so $K_1 = K_2 = 0$. There is no elastic potential energy (there are no springs), and the gravitational potential energy does not

7.19 Our sketch for this problem.

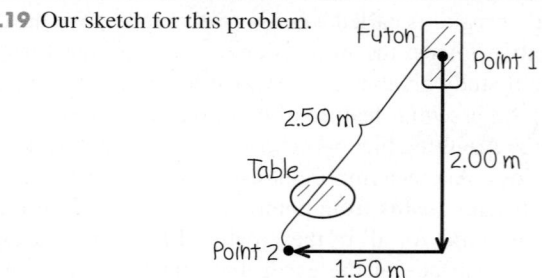

change because the futon moves only horizontally, so $U_1 = U_2$. From Eq. (7.14) it follows that $W_{\text{other}} = 0$. That "other" work done on the futon is the sum of the positive work you do, W_{you}, and the negative work done by friction, W_{fric}. Since the sum of these is zero, we have

$$W_{\text{you}} = -W_{\text{fric}}$$

So we can calculate the work done by friction to determine W_{you}.

EXECUTE: The floor is horizontal, so the normal force on the futon equals its weight mg and the magnitude of the friction force is $f_k = \mu_k n = \mu_k mg$. The work you do over each path is then

$$W_{\text{you}} = -W_{\text{fric}} = -(-f_k s) = +\mu_k mgs$$
$$= (0.200)(40.0 \text{ kg})(9.80 \text{ m/s}^2)(2.50 \text{ m})$$
$$= 196 \text{ J} \quad \text{(straight-line path)}$$

$$W_{\text{you}} = -W_{\text{fric}} = +\mu_k mgs$$
$$= (0.200)(40.0 \text{ kg})(9.80 \text{ m/s}^2)(2.00 \text{ m} + 1.50 \text{ m})$$
$$= 274 \text{ J} \quad \text{(dogleg path)}$$

The extra work you must do is $274 \text{ J} - 196 \text{ J} = 78 \text{ J}$.

EVALUATE: Friction does different amounts of work on the futon, -196 J and -274 J, on these different paths between points 1 and 2. Hence friction is a *nonconservative* force.

EXAMPLE 7.11 CONSERVATIVE OR NONCONSERVATIVE?

In a region of space the force on an electron is $\vec{F} = Cx\hat{j}$, where C is a positive constant. The electron moves around a square loop in the xy-plane (**Fig. 7.20**). Calculate the work done on the electron by force $\vec{F}$ during a counterclockwise trip around the square. Is this force conservative or nonconservative?

SOLUTION

IDENTIFY and SET UP: Force $\vec{F}$ is not constant and in general is not in the same direction as the displacement. To calculate the work done by $\vec{F}$, we'll use the general expression Eq. (6.14):

$$W = \int_{P_1}^{P_2} \vec{F} \cdot d\vec{l}$$

where $d\vec{l}$ is an infinitesimal displacement. We'll calculate the work done on each leg of the square separately, and add the results to find the work done on the round trip. If this round-trip work is zero, force $\vec{F}$ is conservative and can be represented by a potential-energy function.

EXECUTE: On the first leg, from $(0, 0)$ to $(L, 0)$, the force is everywhere perpendicular to the displacement. So $\vec{F} \cdot d\vec{l} = 0$,

and the work done on the first leg is $W_1 = 0$. The force has the same value $\vec{F} = CL\hat{j}$ everywhere on the second leg, from $(L, 0)$ to (L, L). The displacement on this leg is in the $+y$-direction, so $d\vec{l} = dy\hat{j}$ and

$$\vec{F} \cdot d\vec{l} = CL\hat{j} \cdot dy\hat{j} = CL \, dy$$

The work done on the second leg is then

$$W_2 = \int_{(L, 0)}^{(L, L)} \vec{F} \cdot d\vec{l} = \int_{y=0}^{y=L} CL \, dy = CL \int_0^L dy = CL^2$$

On the third leg, from (L, L) to $(0, L)$, $\vec{F}$ is again perpendicular to the displacement and so $W_3 = 0$. The force is zero on the final leg, from $(0, L)$ to $(0, 0)$, so $W_4 = 0$. The work done by $\vec{F}$ on the round trip is therefore

$$W = W_1 + W_2 + W_3 + W_4 = 0 + CL^2 + 0 + 0 = CL^2$$

The starting and ending points are the same, but the total work done by $\vec{F}$ is not zero. This is a *nonconservative* force; it *cannot* be represented by a potential-energy function.

EVALUATE: Because $W > 0$, the mechanical energy *increases* as the electron goes around the loop. This is actually what happens in an electric generating plant: A loop of wire is moved through a magnetic field, which gives rise to a nonconservative force similar to the one here. Electrons in the wire gain energy as they move around the loop, and this energy is carried via transmission lines to the consumer. (We'll discuss this in Chapter 29.)

If the electron went *clockwise* around the loop, $\vec{F}$ would be unaffected but the direction of each infinitesimal displacement $d\vec{l}$ would be reversed. Thus the sign of work would also reverse, and the work for a clockwise round trip would be $W = -CL^2$. This is a different behavior than the nonconservative friction force. The work done by friction on a body that slides in any direction over a stationary surface is always negative (see Example 7.6 in Section 7.1).

7.20 An electron moving around a square loop while being acted on by the force $\vec{F} = Cx\hat{j}$.

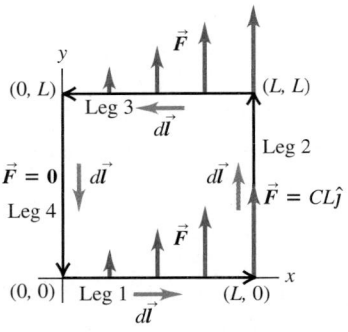

Application Nonconservative Forces and Internal Energy in a Tire An automobile tire deforms and flexes like a spring as it rolls, but it is not an ideal spring: Nonconservative internal friction forces act within the rubber. As a result, mechanical energy is lost and converted to internal energy of the tire. Thus the temperature of a tire increases as it rolls, which causes the pressure of the air inside the tire to increase as well. That's why tire pressures are best checked before the car is driven, when the tire is cold.

7.21 The battery pack in this radio-controlled helicopter contains 2.4×10^4 J of electric energy. When this energy is used up, the internal energy of the battery pack decreases by this amount, so $\Delta U_{int} = -2.4 \times 10^4$ J. This energy can be converted to kinetic energy to make the rotor blades and helicopter go faster, or to gravitational potential energy to make the helicopter climb.

The Law of Conservation of Energy

Nonconservative forces cannot be represented in terms of potential energy. But we can describe the effects of these forces in terms of kinds of energy other than kinetic or potential energy. When a car with locked brakes skids to a stop, both the tires and the road surface become hotter. The energy associated with this change in the state of the materials is called **internal energy.** Raising the temperature of a body increases its internal energy; lowering the body's temperature decreases its internal energy.

To see the significance of internal energy, let's consider a block sliding on a rough surface. Friction does *negative* work on the block as it slides, and the change in internal energy of the block and surface (both of which get hotter) is *positive*. Careful experiments show that the increase in the internal energy is *exactly* equal to the absolute value of the work done by friction. In other words,

$$\Delta U_{int} = -W_{other}$$

where ΔU_{int} is the change in internal energy. We substitute this into Eq. (7.14):

$$K_1 + U_1 - \Delta U_{int} = K_2 + U_2$$

Writing $\Delta K = K_2 - K_1$ and $\Delta U = U_2 - U_1$, we can finally express this as

Law of conservation of energy:

$$\Delta K + \Delta U + \Delta U_{int} = 0 \qquad (7.15)$$

Change in kinetic energy   Change in potential energy   Change in internal energy

This remarkable statement is the general form of the **law of conservation of energy.** In a given process, the kinetic energy, potential energy, and internal energy of a system may all change. But the *sum* of those changes is always zero. If there is a decrease in one form of energy, it is made up for by an increase in the other forms (**Fig. 7.21**). When we expand our definition of energy to include internal energy, Eq. (7.15) says: *Energy is never created or destroyed; it only changes form.* No exception to this rule has ever been found.

The concept of work has been banished from Eq. (7.15); instead, it suggests that we think purely in terms of the conversion of energy from one form to another. For example, when you throw a baseball straight up, you convert a portion of the internal energy of your molecules to kinetic energy of the baseball. This is converted to gravitational potential energy as the ball climbs and back to kinetic energy as the ball falls. If there is air resistance, part of the energy is used to heat up the air and the ball and increase their internal energy. Energy is converted back to the kinetic form as the ball falls. If you catch the ball in your hand, whatever energy was not lost to the air once again becomes internal energy; the ball and your hand are now warmer than they were at the beginning.

In Chapters 19 and 20, we will study the relationship of internal energy to temperature changes, heat, and work. This is the heart of the area of physics called *thermodynamics.*

CONCEPTUAL EXAMPLE 7.12 **WORK DONE BY FRICTION**

Let's return to Example 7.5 (Section 7.1), in which Throcky skateboards down a curved ramp. He starts with zero kinetic energy and 735 J of potential energy, and at the bottom he has 450 J of kinetic energy and zero potential energy; hence $\Delta K = +450$ J and $\Delta U = -735$ J. The work $W_{other} = W_{fric}$ done by the friction forces is -285 J, so the change in internal energy is $\Delta U_{int} = -W_{other} = +285$ J. The skateboard wheels and bearings

and the ramp all get a little warmer. In accordance with Eq. (7.15), the sum of the energy changes equals zero:

$$\Delta K + \Delta U + \Delta U_{int} = +450 \text{ J} + (-735 \text{ J}) + 285 \text{ J} = 0$$

The total energy of the system (including internal, nonmechanical forms of energy) is conserved.

TEST YOUR UNDERSTANDING OF SECTION 7.3   In a hydroelectric generating station, falling water is used to drive turbines ("water wheels"), which in turn run electric generators. Compared to the amount of gravitational potential energy released by the falling water, how much electrical energy is produced? (i) The same; (ii) more; (iii) less. ▌

7.4 FORCE AND POTENTIAL ENERGY

For the two kinds of conservative forces (gravitational and elastic) we have studied, we started with a description of the behavior of the *force* and derived from that an expression for the *potential energy*. For example, for a body with mass m in a uniform gravitational field, the gravitational force is $F_y = -mg$. We found that the corresponding potential energy is $U(y) = mgy$. The force that an ideal spring exerts on a body is $F_x = -kx$, and the corresponding potential-energy function is $U(x) = \frac{1}{2}kx^2$.

In studying physics, however, you'll encounter situations in which you are given an expression for the *potential energy* as a function of position and have to find the corresponding *force*. We'll see several examples of this kind when we study electric forces later in this book: It's often far easier to calculate the electric potential energy first and then determine the corresponding electric force afterward.

Here's how we find the force that corresponds to a given potential-energy expression. First let's consider motion along a straight line, with coordinate x. We denote the x-component of force, a function of x, by $F_x(x)$ and the potential energy as $U(x)$. This notation reminds us that both F_x and U are *functions* of x. Now we recall that in any displacement, the work W done by a conservative force equals the negative of the change ΔU in potential energy:

$$W = -\Delta U$$

Let's apply this to a small displacement Δx. The work done by the force $F_x(x)$ during this displacement is approximately equal to $F_x(x)\,\Delta x$. We have to say "approximately" because $F_x(x)$ may vary a little over the interval Δx. So

$$F_x(x)\,\Delta x = -\Delta U \qquad \text{and} \qquad F_x(x) = -\frac{\Delta U}{\Delta x}$$

You can probably see what's coming. We take the limit as $\Delta x \to 0$; in this limit, the variation of F_x becomes negligible, and we have the exact relationship

Force from potential energy:	... is the negative
In **one-dimensional motion,** ...	of the derivative at x
the value of a conservative $\qquad F_x(x) = -\dfrac{dU(x)}{dx}$	of the associated $\qquad$ (7.16)
force at point x ...	potential-energy function.

This result makes sense; in regions where $U(x)$ changes most rapidly with x (that is, where $dU(x)/dx$ is large), the greatest amount of work is done during a given displacement, and this corresponds to a large force magnitude. Also, when $F_x(x)$ is in the positive x-direction, $U(x)$ *decreases* with increasing x. So $F_x(x)$ and $dU(x)/dx$ should indeed have opposite signs. The physical meaning of Eq. (7.16) is that *a conservative force always acts to push the system toward lower potential energy.*

As a check, let's consider the function for elastic potential energy, $U(x) = \frac{1}{2}kx^2$. Substituting this into Eq. (7.16) yields

$$F_x(x) = -\frac{d}{dx}\left(\tfrac{1}{2}kx^2\right) = -kx$$

which is the correct expression for the force exerted by an ideal spring (**Fig. 7.22a**, next page). Similarly, for gravitational potential energy we have $U(y) = mgy$; taking care to change x to y for the choice of axis, we get $F_y = -dU/dy = -d(mgy)/dy = -mg$, which is the correct expression for gravitational force (Fig. 7.22b).

7.22 A conservative force is the negative derivative of the corresponding potential energy.

(a) Elastic potential energy and force as functions of x

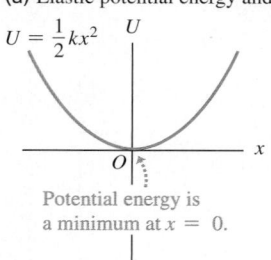

$U = \frac{1}{2}kx^2$

Potential energy is a minimum at $x = 0$.

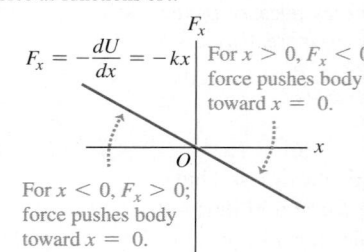

$F_x = -\dfrac{dU}{dx} = -kx$

For $x > 0$, $F_x < 0$; force pushes body toward $x = 0$.

For $x < 0$, $F_x > 0$; force pushes body toward $x = 0$.

(b) Gravitational potential energy and force as functions of y

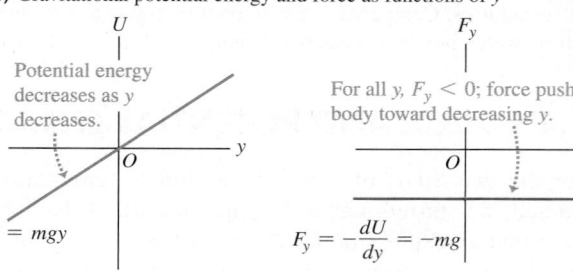

Potential energy decreases as y decreases.

$U = mgy$

For all y, $F_y < 0$; force pushes body toward decreasing y.

$F_y = -\dfrac{dU}{dy} = -mg$

EXAMPLE 7.13 AN ELECTRIC FORCE AND ITS POTENTIAL ENERGY

An electrically charged particle is held at rest at the point $x = 0$; a second particle with equal charge is free to move along the positive x-axis. The potential energy of the system is $U(x) = C/x$, where C is a positive constant that depends on the magnitude of the charges. Derive an expression for the x-component of force acting on the movable particle as a function of its position.

SOLUTION

IDENTIFY and SET UP: We are given the potential-energy function $U(x)$. We'll find the corresponding force function by using Eq. (7.16), $F_x(x) = -dU(x)/dx$.

EXECUTE: The derivative of $1/x$ with respect to x is $-1/x^2$. So for $x > 0$ the force on the movable charged particle is

$$F_x(x) = -\frac{dU(x)}{dx} = -C\left(-\frac{1}{x^2}\right) = \frac{C}{x^2}$$

EVALUATE: The x-component of force is positive, corresponding to a repulsion between like electric charges. Both the potential energy and the force are very large when the particles are close together (small x), and both get smaller as the particles move farther apart (large x). The force pushes the movable particle toward large positive values of x, where the potential energy is lower. (We'll study electric forces in detail in Chapter 21.)

Force and Potential Energy in Three Dimensions

We can extend this analysis to three dimensions, for a particle that may move in the x-, y-, or z-direction, or all at once, under the action of a conservative force that has components F_x, F_y, and F_z. Each component of force may be a function of the coordinates x, y, and z. The potential-energy function U is also a function of all three space coordinates. The potential-energy change ΔU when the particle moves a small distance Δx in the x-direction is again given by $-F_x \Delta x$; it doesn't depend on F_y and F_z, which represent force components that are perpendicular to the displacement and do no work. So we again have the approximate relationship

$$F_x = -\frac{\Delta U}{\Delta x}$$

We determine the y- and z-components in exactly the same way:

$$F_y = -\frac{\Delta U}{\Delta y} \qquad F_z = -\frac{\Delta U}{\Delta z}$$

To make these relationships exact, we take the limits $\Delta x \rightarrow 0$, $\Delta y \rightarrow 0$, and $\Delta z \rightarrow 0$ so that these ratios become derivatives. Because U may be a function of all three coordinates, we need to remember that when we calculate each of these derivatives, only one coordinate changes at a time. We compute the derivative of U with respect to x by assuming that y and z are constant and only x varies, and so on. Such a derivative is called a *partial derivative*. The usual notation for a partial derivative is $\partial U/\partial x$ and so on; the symbol ∂ is a modified d. So we write

> **Force from potential energy:** In **three-dimensional motion,** the value at a given point of each component of a conservative force ...
>
> $$F_x = -\frac{\partial U}{\partial x} \qquad F_y = -\frac{\partial U}{\partial y} \qquad F_z = -\frac{\partial U}{\partial z} \qquad (7.17)$$
>
> ... is the negative of the partial derivative at that point of the associated potential-energy function.

We can use unit vectors to write a single compact vector expression for the force $\vec{F}$:

> **Force from potential energy:** The vector value of a conservative force at a given point ...
>
> $$\vec{F} = -\left(\frac{\partial U}{\partial x}\hat{\imath} + \frac{\partial U}{\partial y}\hat{\jmath} + \frac{\partial U}{\partial z}\hat{k}\right) = -\vec{\nabla}U \qquad (7.18)$$
>
> ... is the negative of the gradient at that point of the associated potential-energy function.

In Eq. (7.18) we take the partial derivative of U with respect to each coordinate, multiply by the corresponding unit vector, and then take the vector sum. This operation is called the **gradient** of U and is often abbreviated as $\vec{\nabla}U$.

As a check, let's substitute into Eq. (7.18) the function $U = mgy$ for gravitational potential energy:

$$\vec{F} = -\vec{\nabla}(mgy) = -\left(\frac{\partial(mgy)}{\partial x}\hat{\imath} + \frac{\partial(mgy)}{\partial y}\hat{\jmath} + \frac{\partial(mgy)}{\partial z}\hat{k}\right) = (-mg)\hat{\jmath}$$

This is just the familiar expression for the gravitational force.

Application Topography and Potential Energy Gradient The greater the elevation of a hiker in Canada's Banff National Park, the greater the gravitational potential energy U_{grav}. Think of an x-axis that runs horizontally from west to east and a y-axis that runs horizontally from south to north. Then the function $U_{\text{grav}}(x, y)$ tells us the elevation as a function of position in the park. Where the mountains have steep slopes, $\vec{F} = -\vec{\nabla}U_{\text{grav}}$ has a large magnitude and there's a strong force pushing you along the mountain's surface toward a region of lower elevation (and hence lower U_{grav}). There's zero force along the surface of the lake, which is all at the same elevation. Hence U_{grav} is constant at all points on the lake surface, and $\vec{F} = -\vec{\nabla}U_{\text{grav}} = \mathbf{0}.$

EXAMPLE 7.14 FORCE AND POTENTIAL ENERGY IN TWO DIMENSIONS

A puck with coordinates x and y slides on a level, frictionless air-hockey table. It is acted on by a conservative force described by the potential-energy function

$$U(x, y) = \tfrac{1}{2}k(x^2 + y^2)$$

Note that $r = \sqrt{x^2 + y^2}$ is the distance on the table surface from the puck to the origin. Find a vector expression for the force acting on the puck, and find an expression for the magnitude of the force.

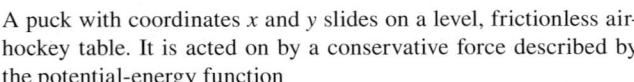

SOLUTION

IDENTIFY and SET UP: Starting with the function $U(x, y)$, we need to find the vector components and magnitude of the corresponding force $\vec{F}$. We'll use Eq. (7.18) to find the components. The function U doesn't depend on z, so the partial derivative of U with respect to z is $\partial U/\partial z = 0$ and the force has no z-component. We'll determine the magnitude F of the force by using $F = \sqrt{F_x^2 + F_y^2}$.

EXECUTE: The x- and y-components of $\vec{F}$ are

$$F_x = -\frac{\partial U}{\partial x} = -kx \qquad F_y = -\frac{\partial U}{\partial y} = -ky$$

From Eq. (7.18), the vector expression for the force is

$$\vec{F} = (-kx)\hat{\imath} + (-ky)\hat{\jmath} = -k(x\hat{\imath} + y\hat{\jmath})$$

The magnitude of the force is

$$F = \sqrt{(-kx)^2 + (-ky)^2} = k\sqrt{x^2 + y^2} = kr$$

EVALUATE: Because $x\hat{\imath} + y\hat{\jmath}$ is just the position vector $\vec{r}$ of the particle, we can rewrite our result as $\vec{F} = -k\vec{r}$. This represents a force that is opposite in direction to the particle's position vector— that is, a force directed toward the origin, $r = 0$. This is the force that would be exerted on the puck if it were attached to one end of a spring that obeys Hooke's law and has a negligibly small unstretched length compared to the other distances in the problem. (The other end is attached to the air-hockey table at $r = 0$.)

To check our result, note that $U = \tfrac{1}{2}kr^2$. We can find the force from this expression using Eq. (7.16) with x replaced by r:

$$F_r = -\frac{dU}{dr} = -\frac{d}{dr}\left(\tfrac{1}{2}kr^2\right) = -kr$$

As we found above, the force has magnitude kr; the minus sign indicates that the force is toward the origin (at $r = 0$).

7.23 (a) A glider on an air track. The spring exerts a force $F_x = -kx$. (b) The potential-energy function.

(a)

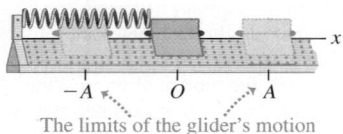

The limits of the glider's motion are at $x = A$ and $x = -A$.

(b)

On the graph, the limits of motion are the points where the U curve intersects the horizontal line representing total mechanical energy E.

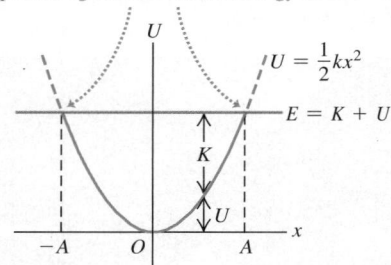

Application Acrobats in Equilibrium
Each of these acrobats is in *unstable* equilibrium. The gravitational potential energy is lower no matter which way an acrobat tips, so if she begins to fall she will keep on falling. Staying balanced requires the acrobats' constant attention.

TEST YOUR UNDERSTANDING OF SECTION 7.4 A particle moving along the x-axis is acted on by a conservative force F_x. At a certain point, the force is zero. (a) Which of the following statements about the value of the potential-energy function $U(x)$ at that point is correct? (i) $U(x) = 0$; (ii) $U(x) > 0$; (iii) $U(x) < 0$; (iv) not enough information is given to decide. (b) Which of the following statements about the value of the derivative of $U(x)$ at that point is correct? (i) $dU(x)/dx = 0$; (ii) $dU(x)/dx > 0$; (iii) $dU(x)/dx < 0$; (iv) not enough information is given to decide. ∎

7.5 ENERGY DIAGRAMS

When a particle moves along a straight line under the action of a conservative force, we can get a lot of insight into its possible motions by looking at the graph of the potential-energy function $U(x)$. **Figure 7.23a** shows a glider with mass m that moves along the x-axis on an air track. The spring exerts on the glider a force with x-component $F_x = -kx$. Figure 7.23b is a graph of the corresponding potential-energy function $U(x) = \frac{1}{2}kx^2$. If the elastic force of the spring is the *only* horizontal force acting on the glider, the total mechanical energy $E = K + U$ is constant, independent of x. A graph of E as a function of x is thus a straight horizontal line. We use the term **energy diagram** for a graph like this, which shows both the potential-energy function $U(x)$ and the energy of the particle subjected to the force that corresponds to $U(x)$.

The vertical distance between the U and E graphs at each point represents the difference $E - U$, equal to the kinetic energy K at that point. We see that K is greatest at $x = 0$. It is zero at the values of x where the two graphs cross, labeled A and $-A$ in Fig. 7.23b. Thus the speed v is greatest at $x = 0$, and it is zero at $x = \pm A$, the points of *maximum* possible displacement from $x = 0$ for a given value of the total energy E. The potential energy U can never be greater than the total energy E; if it were, K would be negative, and that's impossible. The motion is a back-and-forth oscillation between the points $x = A$ and $x = -A$.

From Eq. (7.16), at each point the force F_x on the glider is equal to the negative of the slope of the $U(x)$ curve: $F_x = -dU/dx$ (see Fig. 7.22a). When the particle is at $x = 0$, the slope and the force are zero, so this is an *equilibrium* position. When x is positive, the slope of the $U(x)$ curve is positive and the force F_x is negative, directed toward the origin. When x is negative, the slope is negative and F_x is positive, again directed toward the origin. Such a force is called a *restoring force;* when the glider is displaced to either side of $x = 0$, the force tends to "restore" it back to $x = 0$. An analogous situation is a marble rolling around in a round-bottomed bowl. We say that $x = 0$ is a point of **stable equilibrium.** More generally, *any minimum in a potential-energy curve is a stable equilibrium position.*

Figure 7.24a shows a hypothetical but more general potential-energy function $U(x)$. Figure 7.24b shows the corresponding force $F_x = -dU/dx$. Points x_1 and x_3 are stable equilibrium points. At both points, F_x is zero because the slope of the $U(x)$ curve is zero. When the particle is displaced to either side, the force pushes back toward the equilibrium point. The slope of the $U(x)$ curve is also zero at points x_2 and x_4, and these are also equilibrium points. But when the particle is displaced a little to the right of either point, the slope of the $U(x)$ curve becomes negative, corresponding to a positive F_x that tends to push the particle still farther from the point. When the particle is displaced a little to the left, F_x is negative, again pushing away from equilibrium. This is analogous to a marble rolling on the top of a bowling ball. Points x_2 and x_4 are called **unstable equilibrium** points; *any maximum in a potential-energy curve is an unstable equilibrium position.*

7.24 The maxima and minima of a potential-energy function $U(x)$ correspond to points where $F_x = 0$.

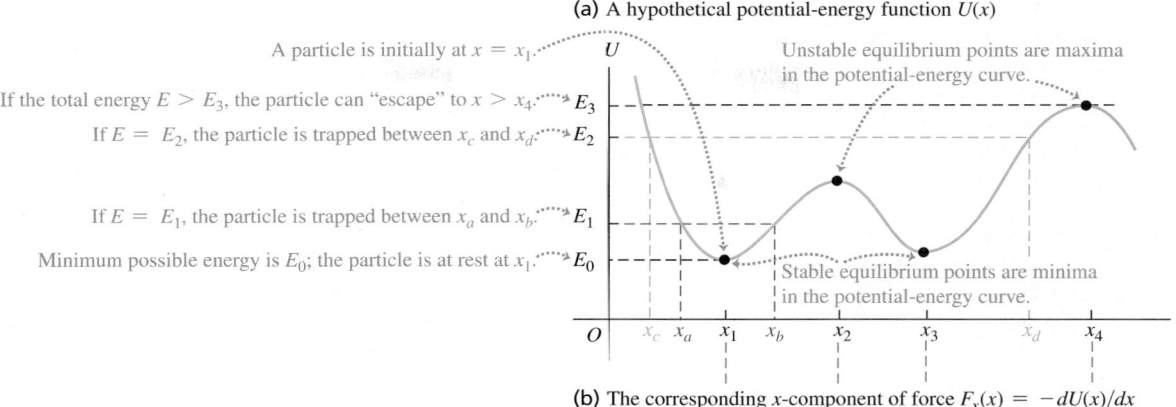

(a) A hypothetical potential-energy function $U(x)$

A particle is initially at $x = x_1$.

Unstable equilibrium points are maxima in the potential-energy curve.

If the total energy $E > E_3$, the particle can "escape" to $x > x_4$. → E_3

If $E = E_2$, the particle is trapped between x_c and x_d. → E_2

If $E = E_1$, the particle is trapped between x_a and x_b. → E_1

Minimum possible energy is E_0; the particle is at rest at x_1. → E_0

Stable equilibrium points are minima in the potential-energy curve.

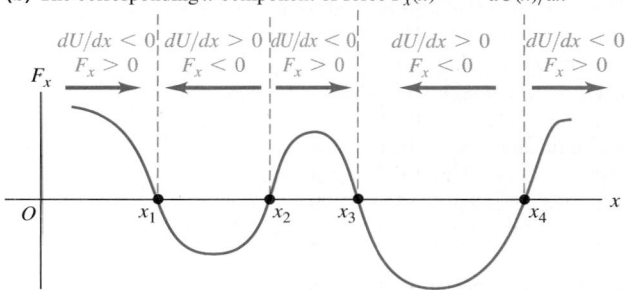

(b) The corresponding x-component of force $F_x(x) = -dU(x)/dx$

$dU/dx < 0$ $dU/dx > 0$ $dU/dx < 0$ $dU/dx > 0$ $dU/dx < 0$
$F_x > 0$ $F_x < 0$ $F_x > 0$ $F_x < 0$ $F_x > 0$

CAUTION **Potential energy and the direction of a conservative force** The direction of the force on a body is *not* determined by the sign of the potential energy U. Rather, it's the sign of $F_x = -dU/dx$ that matters. The physically significant quantity is the *difference* in the values of U between two points (Section 7.1), which is what the derivative $F_x = -dU/dx$ measures. You can always add a constant to the potential-energy function without changing the physics. ▌

MP

PhET: Energy Skate Park

If the total energy is E_1 and the particle is initially near x_1, it can move only in the region between x_a and x_b determined by the intersection of the E_1 and U graphs (Fig. 7.24a). Again, U cannot be greater than E_1 because K can't be negative. We speak of the particle as moving in a *potential well*, and x_a and x_b are the *turning points* of the particle's motion (since at these points, the particle stops and reverses direction). If we increase the total energy to the level E_2, the particle can move over a wider range, from x_c to x_d. If the total energy is greater than E_3, the particle can "escape" and move to indefinitely large values of x. At the other extreme, E_0 represents the minimum total energy the system can have.

TEST YOUR UNDERSTANDING OF SECTION 7.5 The curve in Fig. 7.24b has a maximum at a point between x_2 and x_3. Which statement correctly describes what happens to the particle when it is at this point? (i) The particle's acceleration is zero. (ii) The particle accelerates in the positive x-direction; the magnitude of the acceleration is less than at any other point between x_2 and x_3. (iii) The particle accelerates in the positive x-direction; the magnitude of the acceleration is greater than at any other point between x_2 and x_3. (iv) The particle accelerates in the negative x-direction; the magnitude of the acceleration is less than at any other point between x_2 and x_3. (v) The particle accelerates in the negative x-direction; the magnitude of the acceleration is greater than at any other point between x_2 and x_3. ▌

Gravitational potential energy and elastic potential energy: The work done on a particle by a constant gravitational force can be represented as a change in the gravitational potential energy, $U_{grav} = mgy$. This energy is a shared property of the particle and the earth. A potential energy is also associated with the elastic force $F_x = -kx$ exerted by an ideal spring, where x is the amount of stretch or compression. The work done by this force can be represented as a change in the elastic potential energy of the spring, $U_{el} = \frac{1}{2}kx^2$.

$$W_{grav} = mgy_1 - mgy_2$$
$$= U_{grav,1} - U_{grav,2}$$
$$= -\Delta U_{grav} \qquad (7.2), (7.3)$$

$$W_{el} = \frac{1}{2}kx_1^2 - \frac{1}{2}kx_2^2$$
$$= U_{el,1} - U_{el,2} = -\Delta U_{el} \qquad (7.10), (7.11)$$

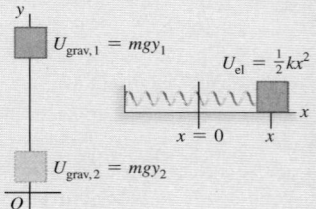

When total mechanical energy is conserved: The total potential energy U is the sum of the gravitational and elastic potential energies: $U = U_{grav} + U_{el}$. If no forces other than the gravitational and elastic forces do work on a particle, the sum of kinetic and potential energies is conserved. This sum $E = K + U$ is called the total mechanical energy. (See Examples 7.1, 7.3, 7.4, and 7.7.)

$$K_1 + U_1 = K_2 + U_2 \qquad (7.4), (7.12)$$

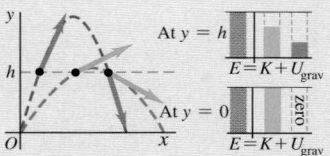

When total mechanical energy is not conserved: When forces other than the gravitational and elastic forces do work on a particle, the work W_{other} done by these other forces equals the change in total mechanical energy (kinetic energy plus total potential energy). (See Examples 7.2, 7.5, 7.6, 7.8, and 7.9.)

$$K_1 + U_1 + W_{other} = K_2 + U_2 \qquad (7.14)$$

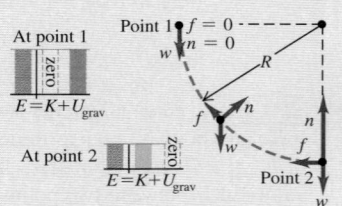

Conservative forces, nonconservative forces, and the law of conservation of energy: All forces are either conservative or nonconservative. A conservative force is one for which the work–kinetic energy relationship is completely reversible. The work of a conservative force can always be represented by a potential-energy function, but the work of a nonconservative force cannot. The work done by nonconservative forces manifests itself as changes in the internal energy of bodies. The sum of kinetic, potential, and internal energies is always conserved. (See Examples 7.10–7.12.)

$$\Delta K + \Delta U + \Delta U_{int} = 0 \qquad (7.15)$$

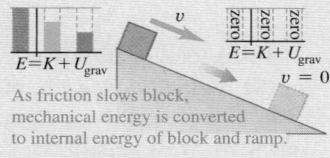

As friction slows block, mechanical energy is converted to internal energy of block and ramp.

Determining force from potential energy: For motion along a straight line, a conservative force $F_x(x)$ is the negative derivative of its associated potential-energy function U. In three dimensions, the components of a conservative force are negative partial derivatives of U. (See Examples 7.13 and 7.14.)

$$F_x(x) = -\frac{dU(x)}{dx} \qquad (7.16)$$

$$F_x = -\frac{\partial U}{\partial x} \qquad F_y = -\frac{\partial U}{\partial y}$$
$$F_z = -\frac{\partial U}{\partial z} \qquad (7.17)$$

$$\vec{F} = -\left(\frac{\partial U}{\partial x}\hat{i} + \frac{\partial U}{\partial y}\hat{j} + \frac{\partial U}{\partial z}\hat{k}\right) \qquad (7.18)$$
$$= -\vec{\nabla}U$$

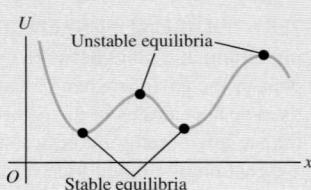

BRIDGING PROBLEM   A SPRING AND FRICTION ON AN INCLINE

A 2.00-kg package is released on a 53.1° incline, 4.00 m from a long spring with force constant 1.20×10^2 N/m that is attached at the bottom of the incline (**Fig. 7.25**). The coefficients of friction between the package and incline are $\mu_s = 0.400$ and $\mu_k = 0.200$. The mass of the spring is negligible. (a) What is the maximum compression of the spring? (b) The package rebounds up the incline. How close does it get to its original position? (c) What is the change in the internal energy of the package and incline from the point at which the package is released until it rebounds to its maximum height?

7.25 The initial situation.

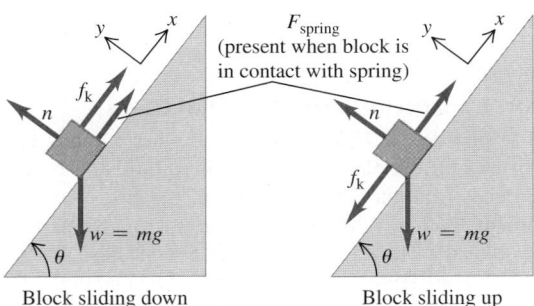

$m = 2.00$ kg

$D = 4.00$ m

$\theta = 53.1°$

SOLUTION GUIDE

IDENTIFY and SET UP

1. This problem involves the gravitational force, a spring force, and the friction force, as well as the normal force that acts on the package. Since the spring force isn't constant, you'll have to use energy methods. Is mechanical energy conserved during any part of the motion? Why or why not?
2. Draw free-body diagrams for the package as it is sliding down the incline and sliding back up the incline. Include your choice of coordinate axes (see below). (*Hint:* If you choose $x = 0$ to be at the end of the uncompressed spring, you'll be able to use $U_{el} = \frac{1}{2}kx^2$ for the elastic potential energy of the spring.)

y x F_{spring} (present when block is in contact with spring) y x

f_k n n f_k

$w = mg$ θ $w = mg$ θ

Block sliding down   Block sliding up

3. Label the three critical points in the package's motion: its starting position, its position when it comes to rest with the spring maximally compressed, and its position when it has rebounded as far as possible up the incline. (*Hint:* You can assume that the package is no longer in contact with the spring at the last of these positions. If this turns out to be incorrect, you'll calculate a value of x that tells you the spring is still partially compressed at this point.)
4. List the unknown quantities and decide which of these are the target variables.

EXECUTE

5. Find the magnitude of the friction force that acts on the package. Does the magnitude of this force depend on whether the package is moving up or down the incline, or on whether the package is in contact with the spring? Does the *direction* of the friction force depend on any of these?
6. Write the general energy equation for the motion of the package between the first two points you labeled in step 3. Use this equation to solve for the distance that the spring is compressed when the package is at its lowest point. (*Hint:* You'll have to solve a quadratic equation. To decide which of the two solutions of this equation is the correct one, remember that the distance the spring is compressed is positive.)
7. Write the general energy equation for the motion of the package between the second and third points you labeled in step 3. Use this equation to solve for how far the package rebounds.
8. Calculate the change in internal energy for the package's trip down and back up the incline. Remember that the amount the internal energy *increases* is equal to the amount the total mechanical energy *decreases*.

EVALUATE

9. Was it correct to assume in part (b) that the package is no longer in contact with the spring when it reaches its maximum rebound height?
10. Check your result for part (c) by finding the total work done by the force of friction over the entire trip. Is this in accordance with your result from step 8?

Problems

For assigned homework and other learning materials, go to MasteringPhysics®.

•, ••, •••: Difficulty levels. **CP**: Cumulative problems incorporating material from earlier chapters. **CALC**: Problems requiring calculus.
DATA: Problems involving real data, scientific evidence, experimental design, and/or statistical reasoning. **BIO**: Biosciences problems.

DISCUSSION QUESTIONS

Q7.1 A baseball is thrown straight up with initial speed v_0. If air resistance cannot be ignored, when the ball returns to its initial height its speed is less than v_0. Explain why, using energy concepts.

Q7.2 A projectile has the same initial kinetic energy no matter what the angle of projection. Why doesn't it rise to the same maximum height in each case?

Q7.3 An object is released from rest at the top of a ramp. If the ramp is frictionless, does the object's speed at the bottom of the ramp depend on the shape of the ramp or just on its height? Explain. What if the ramp is *not* frictionless?

Q7.4 An egg is released from rest from the roof of a building and falls to the ground. Its fall is observed by a student on the roof of the building, who uses coordinates with origin at the roof, and by a student on the ground, who uses coordinates with origin at the ground. Do the values the two students assign to the following quantities match each other: initial gravitational potential energy, final gravitational potential energy, change in gravitational potential energy, and kinetic energy of the egg just before it strikes the ground? Explain.

Q7.5 A physics teacher had a bowling ball suspended from a very long rope attached to the high ceiling of a large lecture hall.

To illustrate his faith in conservation of energy, he would back up to one side of the stage, pull the ball far to one side until the taut rope brought it just to the end of his nose, and then release it. The massive ball would swing in a mighty arc across the stage and then return to stop momentarily just in front of the nose of the stationary, unflinching teacher. However, one day after the demonstration he looked up just in time to see a student at the other side of the stage *push* the ball away from his nose as he tried to duplicate the demonstration. Tell the rest of the story, and explain the reason for the potentially tragic outcome.

Q7.6 Is it possible for a friction force to *increase* the mechanical energy of a system? If so, give examples.

Q7.7 A woman bounces on a trampoline, going a little higher with each bounce. Explain how she increases the total mechanical energy.

Q7.8 Fractured Physics. People often call their electric bill a *power* bill, yet the quantity on which the bill is based is expressed in *kilowatt-hours*. What are people really being billed for?

Q7.9 (a) A book is lifted upward a vertical distance of 0.800 m. During this displacement, does the gravitational force acting on the book do positive work or negative work? Does the gravitational potential energy of the book increase or decrease? (b) A can of beans is released from rest and falls downward a vertical distance of 2.00 m. During this displacement, does the gravitational force acting on the can do positive work or negative work? Does the gravitational potential energy of the can increase or decrease?

Q7.10 (a) A block of wood is pushed against a spring, which is compressed 0.080 m. Does the force on the block exerted by the spring do positive or negative work? Does the potential energy stored in the spring increase or decrease? (b) A block of wood is placed against a vertical spring that is compressed 6.00 cm. The spring is released and pushes the block upward. From the point where the spring is compressed 6.00 cm to where it is compressed 2.00 cm from its equilibrium length and the block has moved 4.00 cm upward, does the spring force do positive or negative work on the block? During this motion, does the potential energy stored in the spring increase or decrease?

Q7.11 A 1.0-kg stone and a 10.0-kg stone are released from rest at the same height above the ground. Ignore air resistance. Which of these statements about the stones are true? Justify each answer. (a) Both have the same initial gravitational potential energy. (b) Both will have the same acceleration as they fall. (c) Both will have the same speed when they reach the ground. (d) Both will have the same kinetic energy when they reach the ground.

Q7.12 Two objects with different masses are launched vertically into the air by placing them on identical compressed springs and then releasing the springs. The two springs are compressed by the same amount before launching. Ignore air resistance and the masses of the springs. Which of these statements about the masses are true? Justify each answer. (a) Both reach the same maximum height. (b) At their maximum height, both have the same gravitational potential energy, if the initial gravitational potential of each mass is taken to be zero.

Q7.13 When people are cold, they often rub their hands together to warm up. How does doing this produce heat? Where does the heat come from?

Q7.14 A box slides down a ramp and work is done on the box by the forces of gravity and friction. Can the work of each of these forces be expressed in terms of the change in a potential-energy function? For each force explain why or why not.

Q7.15 In physical terms, explain why friction is a nonconservative force. Does it store energy for future use?

Q7.16 Since only changes in potential energy are important in any problem, a student decides to let the elastic potential energy of a spring be zero when the spring is stretched a distance x_1. The student decides, therefore, to let $U = \frac{1}{2}k(x - x_1)^2$. Is this correct? Explain.

Q7.17 Figure 7.22a shows the potential-energy function for the force $F_x = -kx$. Sketch the potential-energy function for the force $F_x = +kx$. For this force, is $x = 0$ a point of equilibrium? Is this equilibrium stable or unstable? Explain.

Q7.18 Figure 7.22b shows the potential-energy function associated with the gravitational force between an object and the earth. Use this graph to explain why objects always fall toward the earth when they are released.

Q7.19 For a system of two particles we often let the potential energy for the force between the particles approach zero as the separation of the particles approaches infinity. If this choice is made, explain why the potential energy at noninfinite separation is positive if the particles repel one another and negative if they attract.

Q7.20 Explain why the points $x = A$ and $x = -A$ in Fig. 7.23b are called *turning points*. How are the values of E and U related at a turning point?

Q7.21 A particle is in *neutral equilibrium* if the net force on it is zero and remains zero if the particle is displaced slightly in any direction. Sketch the potential-energy function near a point of neutral equilibrium for the case of one-dimensional motion. Give an example of an object in neutral equilibrium.

Q7.22 The net force on a particle of mass m has the potential-energy function graphed in Fig. 7.24a. If the total energy is E_1, graph the speed v of the particle versus its position x. At what value of x is the speed greatest? Sketch v versus x if the total energy is E_2.

Q7.23 The potential-energy function for a force $\vec{F}$ is $U = \alpha x^3$, where α is a positive constant. What is the direction of $\vec{F}$?

EXERCISES

Section 7.1 Gravitational Potential Energy

7.1 • In one day, a 75-kg mountain climber ascends from the 1500-m level on a vertical cliff to the top at 2400 m. The next day, she descends from the top to the base of the cliff, which is at an elevation of 1350 m. What is her change in gravitational potential energy (a) on the first day and (b) on the second day?

7.2 • **BIO How High Can We Jump?** The maximum height a typical human can jump from a crouched start is about 60 cm. By how much does the gravitational potential energy increase for a 72-kg person in such a jump? Where does this energy come from?

7.3 •• **CP** A 90.0-kg mail bag hangs by a vertical rope 3.5 m long. A postal worker then displaces the bag to a position 2.0 m sideways from its original position, always keeping the rope taut. (a) What horizontal force is necessary to hold the bag in the new position? (b) As the bag is moved to this position, how much work is done (i) by the rope and (ii) by the worker?

7.4 •• **BIO Food Calories.** The *food calorie,* equal to 4186 J, is a measure of how much energy is released when the body metabolizes food. A certain fruit-and-cereal bar contains 140 food calories. (a) If a 65-kg hiker eats one bar, how high a mountain must he climb to "work off" the calories, assuming that all the food energy goes into increasing gravitational potential energy? (b) If, as is typical, only 20% of the food calories go into mechanical energy, what would be the answer to part (a)? (*Note:* In this and all other problems, we are assuming that 100% of the food calories that are eaten are absorbed and used by the body. This is not true. A person's "metabolic efficiency" is the percentage of calories eaten that are actually used; the body eliminates the rest. Metabolic efficiency varies considerably from person to person.)

7.5 • A baseball is thrown from the roof of a 22.0-m-tall building with an initial velocity of magnitude 12.0 m/s and directed at an angle of 53.1° above the horizontal. (a) What is the speed of the ball just before it strikes the ground? Use energy methods and ignore air resistance. (b) What is the answer for part (a) if the initial velocity is at an angle of 53.1° *below* the horizontal? (c) If the effects of air resistance are included, will part (a) or (b) give the higher speed?

7.6 •• A crate of mass M starts from rest at the top of a frictionless ramp inclined at an angle α above the horizontal. Find its speed at the bottom of the ramp, a distance d from where it started. Do this in two ways: Take the level at which the potential energy is zero to be (a) at the bottom of the ramp with y positive upward, and (b) at the top of the ramp with y positive upward. (c) Why didn't the normal force enter into your solution?

7.7 •• **BIO Human Energy vs. Insect Energy.** For its size, the common flea is one of the most accomplished jumpers in the animal world. A 2.0-mm-long, 0.50-mg flea can reach a height of 20 cm in a single leap. (a) Ignoring air drag, what is the takeoff speed of such a flea? (b) Calculate the kinetic energy of this flea at takeoff and its kinetic energy per kilogram of mass. (c) If a 65-kg, 2.0-m-tall human could jump to the same height compared with his length as the flea jumps compared with its length, how high could the human jump, and what takeoff speed would the man need? (d) Most humans can jump no more than 60 cm from a crouched start. What is the kinetic energy per kilogram of mass at takeoff for such a 65-kg person? (e) Where does the flea store the energy that allows it to make sudden leaps?

7.8 •• **BIO Bone Fractures.** The maximum energy that a bone can absorb without breaking depends on characteristics such as its cross-sectional area and elasticity. For healthy human leg bones of approximately 6.0 cm² cross-sectional area, this energy has been experimentally measured to be about 200 J. (a) From approximately what maximum height could a 60-kg person jump and land rigidly upright on both feet without breaking his legs? (b) You are probably surprised at how small the answer to part (a) is. People obviously jump from much greater heights without breaking their legs. How can that be? What else absorbs the energy when they jump from greater heights? (*Hint:* How did the person in part (a) land? How do people normally land when they jump from greater heights?) (c) Why might older people be much more prone than younger ones to bone fractures from simple falls (such as a fall in the shower)?

7.9 •• **CP** A small rock with mass 0.20 kg is released from rest at point A, which is at the top edge of a large, hemispherical bowl with radius $R = 0.50$ m (**Fig. E7.9**). Assume that the size of the rock is small compared to R, so that the rock can be treated as a particle, and assume that the rock slides rather than rolls. The work done by friction on the rock when it moves from point A to point B at the bottom of the bowl has magnitude 0.22 J. (a) Between points A and B, how much work is done on the rock by (i) the normal force and (ii) gravity? (b) What is the speed of the rock as it reaches point B? (c) Of the three forces acting on the rock as it slides down the bowl, which (if any) are constant and which are not? Explain. (d) Just as the rock reaches point B, what is the normal force on it due to the bottom of the bowl?

Figure **E7.9**

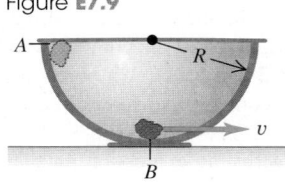

7.10 •• A 25.0-kg child plays on a swing having support ropes that are 2.20 m long. Her brother pulls her back until the ropes are 42.0° from the vertical and releases her from rest. (a) What is her potential energy just as she is released, compared with the potential energy at the bottom of the swing's motion? (b) How fast will she be moving at the bottom? (c) How much work does the tension in the ropes do as she swings from the initial position to the bottom of the motion?

7.11 •• You are testing a new amusement park roller coaster with an empty car of mass 120 kg. One part of the track is a vertical loop with radius 12.0 m. At the bottom of the loop (point A) the car has speed 25.0 m/s, and at the top of the loop (point B) it has speed 8.0 m/s. As the car rolls from point A to point B, how much work is done by friction?

7.12 • **Tarzan and Jane.** Tarzan, in one tree, sights Jane in another tree. He grabs the end of a vine with length 20 m that makes an angle of 45° with the vertical, steps off his tree limb, and swings down and then up to Jane's open arms. When he arrives, his vine makes an angle of 30° with the vertical. Determine whether he gives her a tender embrace or knocks her off her limb by calculating Tarzan's speed just before he reaches Jane. Ignore air resistance and the mass of the vine.

7.13 •• **CP** A 10.0-kg microwave oven is pushed 6.00 m up the sloping surface of a loading ramp inclined at an angle of 36.9° above the horizontal, by a constant force $\vec{F}$ with a magnitude 110 N and acting parallel to the ramp. The coefficient of kinetic friction between the oven and the ramp is 0.250. (a) What is the work done on the oven by the force $\vec{F}$? (b) What is the work done on the oven by the friction force? (c) Compute the increase in potential energy for the oven. (d) Use your answers to parts (a), (b), and (c) to calculate the increase in the oven's kinetic energy. (e) Use $\sum \vec{F} = m\vec{a}$ to calculate the oven's acceleration. Assuming that the oven is initially at rest, use the acceleration to calculate the oven's speed after the oven has traveled 6.00 m. From this, compute the increase in the oven's kinetic energy, and compare it to your answer for part (d).

Section 7.2 Elastic Potential Energy

7.14 •• An ideal spring of negligible mass is 12.00 cm long when nothing is attached to it. When you hang a 3.15-kg weight from it, you measure its length to be 13.40 cm. If you wanted to store 10.0 J of potential energy in this spring, what would be its *total* length? Assume that it continues to obey Hooke's law.

7.15 •• A force of 520 N keeps a certain spring stretched a distance of 0.200 m. (a) What is the potential energy of the spring when it is stretched 0.200 m? (b) What is its potential energy when it is compressed 5.00 cm?

7.16 • **BIO Tendons.** Tendons are strong elastic fibers that attach muscles to bones. To a reasonable approximation, they obey Hooke's law. In laboratory tests on a particular tendon, it was found that, when a 250-g object was hung from it, the tendon stretched 1.23 cm. (a) Find the force constant of this tendon in N/m. (b) Because of its thickness, the maximum tension this tendon can support without rupturing is 138 N. By how much can the tendon stretch without rupturing, and how much energy is stored in it at that point?

7.17 • A spring stores potential energy U_0 when it is compressed a distance x_0 from its uncompressed length. (a) In terms of U_0, how much energy does the spring store when it is compressed (i) twice as much and (ii) half as much? (b) In terms of x_0, how much must the spring be compressed from its uncompressed length to store (i) twice as much energy and (ii) half as much energy?

7.18 • A slingshot will shoot a 10-g pebble 22.0 m straight up. (a) How much potential energy is stored in the slingshot's rubber band? (b) With the same potential energy stored in the rubber band, how high can the slingshot shoot a 25-g pebble? (c) What physical effects did you ignore in solving this problem?

7.19 •• A spring of negligible mass has force constant $k = 800 \text{ N/m}$. (a) How far must the spring be compressed for 1.20 J of potential energy to be stored in it? (b) You place the spring vertically with one end on the floor. You then lay a 1.60-kg book on top of the spring and release the book from rest. Find the maximum distance the spring will be compressed.

7.20 • A 1.20-kg piece of cheese is placed on a vertical spring of negligible mass and force constant $k = 1800 \text{ N/m}$ that is compressed 15.0 cm. When the spring is released, how high does the cheese rise from this initial position? (The cheese and the spring are *not* attached.)

7.21 •• A spring of negligible mass has force constant $k = 1600 \text{ N/m}$. (a) How far must the spring be compressed for 3.20 J of potential energy to be stored in it? (b) You place the spring vertically with one end on the floor. You then drop a 1.20-kg book onto it from a height of 0.800 m above the top of the spring. Find the maximum distance the spring will be compressed.

7.22 •• (a) For the elevator of Example 7.9 (Section 7.2), what is the speed of the elevator after it has moved downward 1.00 m from point 1 in Fig. 7.17? (b) When the elevator is 1.00 m below point 1 in Fig. 7.17, what is its acceleration?

7.23 •• A 2.50-kg mass is pushed against a horizontal spring of force constant 25.0 N/cm on a frictionless air table. The spring is attached to the tabletop, and the mass is not attached to the spring in any way. When the spring has been compressed enough to store 11.5 J of potential energy in it, the mass is suddenly released from rest. (a) Find the greatest speed the mass reaches. When does this occur? (b) What is the greatest acceleration of the mass, and when does it occur?

7.24 •• A 2.50-kg block on a horizontal floor is attached to a horizontal spring that is initially compressed 0.0300 m. The spring has force constant 840 N/m. The coefficient of kinetic friction between the floor and the block is $\mu_k = 0.40$. The block and spring are released from rest, and the block slides along the floor. What is the speed of the block when it has moved a distance of 0.0200 m from its initial position? (At this point the spring is compressed 0.0100 m.)

7.25 •• You are asked to design a spring that will give a 1160-kg satellite a speed of 2.50 m/s relative to an orbiting space shuttle. Your spring is to give the satellite a maximum acceleration of $5.00g$. The spring's mass, the recoil kinetic energy of the shuttle, and changes in gravitational potential energy will all be negligible. (a) What must the force constant of the spring be? (b) What distance must the spring be compressed?

Section 7.3 Conservative and Nonconservative Forces

7.26 • A 75-kg roofer climbs a vertical 7.0-m ladder to the flat roof of a house. He then walks 12 m on the roof, climbs down another vertical 7.0-m ladder, and finally walks on the ground back to his starting point. How much work is done on him by gravity (a) as he climbs up; (b) as he climbs down; (c) as he walks on the roof and on the ground? (d) What is the total work done on him by gravity during this round trip? (e) On the basis of your answer to part (d), would you say that gravity is a conservative or nonconservative force? Explain.

7.27 • A 0.60-kg book slides on a horizontal table. The kinetic friction force on the book has magnitude 1.8 N. (a) How much work is done on the book by friction during a displacement of 3.0 m to the left? (b) The book now slides 3.0 m to the right, returning to its starting point. During this second 3.0-m displacement, how much work is done on the book by friction? (c) What is the total work done on the book by friction during the complete round trip? (d) On the basis of your answer to part (c), would you say that the friction force is conservative or nonconservative? Explain.

7.28 •• CALC In an experiment, one of the forces exerted on a proton is $\vec{F} = -\alpha x^2 \hat{\imath}$, where $\alpha = 12 \text{ N/m}^2$. (a) How much work does $\vec{F}$ do when the proton moves along the straight-line path from the point (0.10 m, 0) to the point (0.10 m, 0.40 m)? (b) Along the straight-line path from the point (0.10 m, 0) to the point (0.30 m, 0)? (c) Along the straight-line path from the point (0.30 m, 0) to the point (0.10 m, 0)? (d) Is the force $\vec{F}$ conservative? Explain. If $\vec{F}$ is conservative, what is the potential-energy function for it? Let $U = 0$ when $x = 0$.

7.29 •• A 62.0-kg skier is moving at 6.50 m/s on a frictionless, horizontal, snow-covered plateau when she encounters a rough patch 4.20 m long. The coefficient of kinetic friction between this patch and her skis is 0.300. After crossing the rough patch and returning to friction-free snow, she skis down an icy, frictionless hill 2.50 m high. (a) How fast is the skier moving when she gets to the bottom of the hill? (b) How much internal energy was generated in crossing the rough patch?

7.30 • While a roofer is working on a roof that slants at 36° above the horizontal, he accidentally nudges his 85.0-N toolbox, causing it to start sliding downward from rest. If it starts 4.25 m from the lower edge of the roof, how fast will the toolbox be moving just as it reaches the edge of the roof if the kinetic friction force on it is 22.0 N?

Section 7.4 Force and Potential Energy

7.31 •• CALC A force parallel to the x-axis acts on a particle moving along the x-axis. This force produces potential energy $U(x)$ given by $U(x) = \alpha x^4$, where $\alpha = 0.630 \text{ J/m}^4$. What is the force (magnitude and direction) when the particle is at $x = -0.800$ m?

7.32 •• CALC The potential energy of a pair of hydrogen atoms separated by a large distance x is given by $U(x) = -C_6/x^6$, where C_6 is a positive constant. What is the force that one atom exerts on the other? Is this force attractive or repulsive?

7.33 •• CALC A small block with mass 0.0400 kg is moving in the xy-plane. The net force on the block is described by the potential-energy function $U(x, y) = (5.80 \text{ J/m}^2)x^2 - (3.60 \text{ J/m}^3)y^3$. What are the magnitude and direction of the acceleration of the block when it is at the point $(x = 0.300 \text{ m}, y = 0.600 \text{ m})$?

7.34 •• CALC An object moving in the xy-plane is acted on by a conservative force described by the potential-energy function $U(x, y) = \alpha[(1/x^2) + (1/y^2)]$, where α is a positive constant. Derive an expression for the force expressed in terms of the unit vectors $\hat{\imath}$ and $\hat{\jmath}$.

Section 7.5 Energy Diagrams

7.35 • CALC The potential energy of two atoms in a diatomic molecule is approximated by $U(r) = (a/r^{12}) - (b/r^6)$, where r is the spacing between atoms and a and b are positive constants. (a) Find the force $F(r)$ on one atom as a function of r. Draw two graphs: one of $U(r)$ versus r and one of $F(r)$ versus r. (b) Find the equilibrium distance between the two atoms. Is this equilibrium stable? (c) Suppose the distance between the two atoms is equal to the equilibrium distance found in part (b). What minimum energy must be added to the molecule to *dissociate* it—that is, to separate the two atoms to an infinite distance apart? This is called the *dissociation energy* of the molecule. (d) For the molecule CO, the equilibrium distance between the carbon and oxygen atoms is 1.13×10^{-10} m and the dissociation energy is 1.54×10^{-18} J per molecule. Find the values of the constants a and b.

7.36 • A marble moves along the x-axis. The potential-energy function is shown in **Fig. E7.36.** (a) At which of the labeled x-coordinates is the force on the marble zero? (b) Which of the labeled x-coordinates is a position of stable equilibrium? (c) Which of the labeled x-coordinates is a position of unstable equilibrium?

Figure **E7.36**

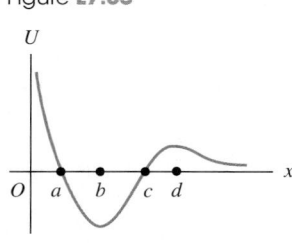

PROBLEMS

7.37 ••• At a construction site, a 65.0-kg bucket of concrete hangs from a light (but strong) cable that passes over a light, friction-free pulley and is connected to an 80.0-kg box on a horizontal roof (**Fig. P7.37**). The cable pulls horizontally on the box, and a 50.0-kg bag of gravel rests on top of the box. The coefficients of friction between the box and roof are shown. (a) Find the friction force on the bag of gravel and on the box. (b) Suddenly a worker picks up the bag of gravel. Use energy conservation to find the speed of the bucket after it has descended 2.00 m from rest. (Use Newton's laws to check your answer.)

Figure **P7.37**

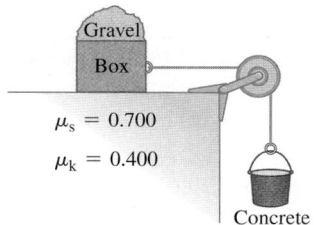

$\mu_s = 0.700$
$\mu_k = 0.400$

7.38 •• Two blocks with different masses are attached to either end of a light rope that passes over a light, frictionless pulley suspended from the ceiling. The masses are released from rest, and the more massive one starts to descend. After this block has descended 1.20 m, its speed is 3.00 m/s. If the total mass of the two blocks is 22.0 kg, what is the mass of each block?

7.39 • A block with mass 0.50 kg is forced against a horizontal spring of negligible mass, compressing the spring a distance of 0.20 m (**Fig. P7.39**). When released, the block moves on a horizontal tabletop for 1.00 m before coming to rest. The force constant k is 100 N/m. What is the coefficient of kinetic friction μ_k between the block and the tabletop?

Figure **P7.39**

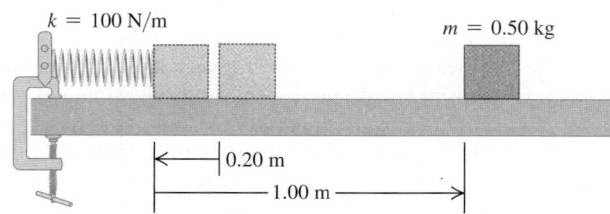

7.40 • A 2.00-kg block is pushed against a spring with negligible mass and force constant $k = 400$ N/m, compressing it 0.220 m. When the block is released, it moves along a frictionless, horizontal surface and then up a frictionless incline with slope 37.0° (**Fig. P7.40**). (a) What is the speed of the block as it slides along the horizontal surface after having left the spring? (b) How far does the block travel up the incline before starting to slide back down?

Figure **P7.40**

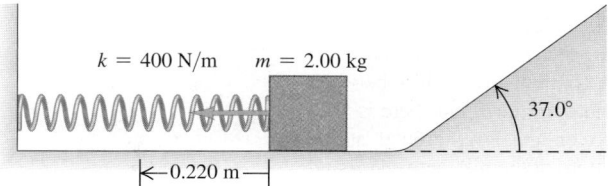

7.41 •• A 350-kg roller coaster car starts from rest at point A and slides down a frictionless loop-the-loop (**Fig. P7.41**). (a) How fast is this roller coaster car moving at point B? (b) How hard does it press against the track at point B?

Figure **P7.41**

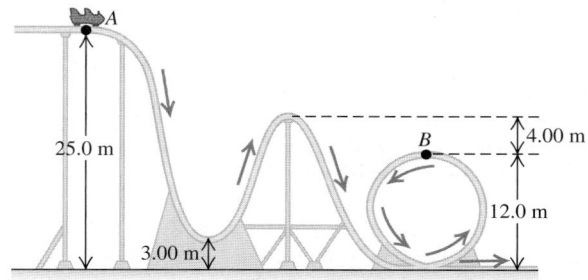

7.42 •• CP **Riding a Loop-the-Loop.** A car in an amusement park ride rolls without friction around a track (**Fig. P7.42**). The car starts from rest at point A at a height h above the bottom of the loop. Treat the car as a particle. (a) What is the minimum value of h (in terms of R) such that the car moves around the loop without falling off at the top (point B)? (b) If $h = 3.50R$ and $R = 14.0$ m, compute the speed, radial acceleration, and tangential acceleration of the passengers when the car is at point C, which is at the end of a horizontal diameter. Show these acceleration components in a diagram, approximately to scale.

Figure **P7.42**

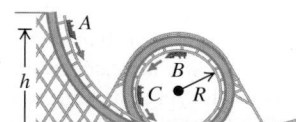

7.43 •• A 2.0-kg piece of wood slides on a curved surface (**Fig. P7.43**). The sides of the surface are perfectly smooth, but the rough horizontal bottom is 30 m long and has a kinetic friction coefficient of 0.20 with the wood. The piece of wood starts from rest 4.0 m above the rough bottom. (a) Where will this wood eventually come to rest? (b) For the motion from the initial release until the piece of wood comes to rest, what is the total amount of work done by friction?

Figure **P7.43**

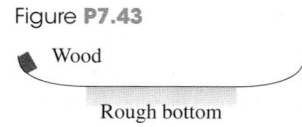

7.44 •• **Up and Down the Hill.** A 28-kg rock approaches the foot of a hill with a speed of 15 m/s. This hill slopes upward at a constant angle of 40.0° above the horizontal. The coefficients of static and kinetic friction between the hill and the rock are 0.75 and 0.20, respectively. (a) Use energy conservation to find the maximum height above the foot of the hill reached by the rock. (b) Will the rock remain at rest at its highest point, or will it slide back down the hill? (c) If the rock does slide back down, find its speed when it returns to the bottom of the hill.

7.45 •• A 15.0-kg stone slides down a snow-covered hill (**Fig. P7.45**), leaving point A at a speed of 10.0 m/s. There is no friction on the hill between points A and B, but there is friction on the level ground at the bottom of the hill, between B and the wall. After entering the rough horizontal region, the stone travels 100 m and then runs into a very long, light spring with force constant 2.00 N/m. The coefficients of kinetic and static friction between the stone and the horizontal ground are 0.20 and 0.80, respectively. (a) What is the speed of the stone when it reaches point B? (b) How far will the stone compress the spring? (c) Will the stone move again after it has been stopped by the spring?

Figure **P7.45**

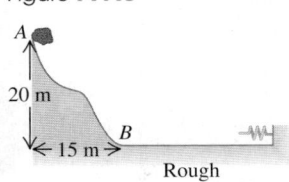

7.46 •• CP A 2.8-kg block slides over the smooth, icy hill shown in **Fig. P7.46**. The top of the hill is horizontal and 70 m higher than its base. What minimum speed must the block have at the base of the 70-m hill to pass over the pit at the far (right-hand) side of that hill?

Figure **P7.46**

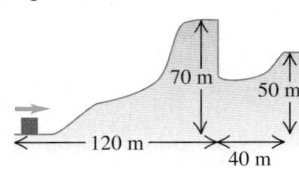

7.47 ••• **Bungee Jump.** A bungee cord is 30.0 m long and, when stretched a distance x, it exerts a restoring force of magnitude kx. Your father-in-law (mass 95.0 kg) stands on a platform 45.0 m above the ground, and one end of the cord is tied securely to his ankle and the other end to the platform. You have promised him that when he steps off the platform he will fall a maximum distance of only 41.0 m before the cord stops him. You had several bungee cords to select from, and you tested them by stretching them out, tying one end to a tree, and pulling on the other end with a force of 380.0 N. When you do this, what distance will the bungee cord that you should select have stretched?

7.48 ••• You are designing a delivery ramp for crates containing exercise equipment. The 1470-N crates will move at 1.8 m/s at the top of a ramp that slopes downward at 22.0°. The ramp exerts a 515-N kinetic friction force on each crate, and the maximum static friction force also has this value. Each crate will compress a spring at the bottom of the ramp and will come to rest after traveling a total distance of 5.0 m along the ramp. Once stopped, a crate must not rebound back up the ramp. Calculate the largest force constant of the spring that will be needed to meet the design criteria.

7.49 ••• The Great Sandini is a 60-kg circus performer who is shot from a cannon (actually a spring gun). You don't find many men of his caliber, so you help him design a new gun. This new gun has a very large spring with a very small mass and a force constant of 1100 N/m that he will compress with a force of 4400 N. The inside of the gun barrel is coated with Teflon, so the average friction force will be only 40 N during the 4.0 m he moves in the barrel. At what speed will he emerge from the end of the barrel, 2.5 m above his initial rest position?

7.50 •• A 1500-kg rocket is to be launched with an initial upward speed of 50.0 m/s. In order to assist its engines, the engineers will start it from rest on a ramp that rises 53° above the horizontal (**Fig. P7.50**). At the bottom, the ramp turns upward and launches the rocket vertically. The engines provide a constant forward thrust of 2000 N, and friction with the ramp surface is a constant 500 N. How far from the base of the ramp should the rocket start, as measured along the surface of the ramp?

Figure **P7.50**

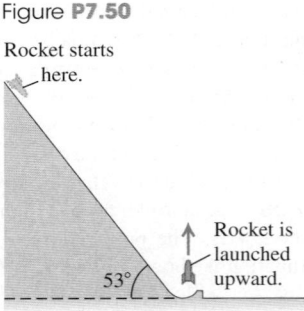

Rocket starts here.

Rocket is launched upward.

53°

7.51 •• A system of two paint buckets connected by a light-weight rope is released from rest with the 12.0-kg bucket 2.00 m above the floor (**Fig. P7.51**). Use the principle of conservation of energy to find the speed with which this bucket strikes the floor. Ignore friction and the mass of the pulley.

Figure **P7.51**

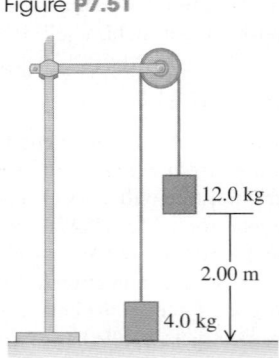

12.0 kg

2.00 m

4.0 kg

7.52 •• These results are from a computer simulation for a batted baseball with mass 0.145 kg, including air resistance:

t	x	y	v_x	v_y
0	0	0	30.0 m/s	40.0 m/s
3.05 s	70.2 m	53.6 m	18.6 m/s	0
6.59 s	124.4 m	0	11.9 m/s	−28.7 m/s

How much work did the air do on the baseball (a) as the ball moved from its initial position to its maximum height, and (b) as the ball moved from its maximum height back to the starting elevation? (c) Explain why the magnitude of the answer in part (b) is smaller than the magnitude of the answer in part (a).

7.53 •• CP A 0.300-kg potato is tied to a string with length 2.50 m, and the other end of the string is tied to a rigid support. The potato is held straight out horizontally from the point of support, with the string pulled taut, and is then released. (a) What is the speed of the potato at the lowest point of its motion? (b) What is the tension in the string at this point?

7.54 •• A 60.0-kg skier starts from rest at the top of a ski slope 65.0 m high. (a) If friction forces do −10.5 kJ of work on her as she descends, how fast is she going at the bottom of the slope? (b) Now moving horizontally, the skier crosses a patch of soft snow where $\mu_k = 0.20$. If the patch is 82.0 m wide and the average force of air resistance on the skier is 160 N, how fast is she going after crossing the patch? (c) The skier hits a snowdrift and penetrates 2.5 m into it before coming to a stop. What is the average force exerted on her by the snowdrift as it stops her?

7.55 • CP A skier starts at the top of a very large, frictionless snowball, with a very small initial speed, and skis straight down the side (**Fig. P7.55**). At what point does she lose contact with the snowball and fly off at a tangent? That is, at the instant she loses contact with the snowball, what angle α does a radial line from the center of the snowball to the skier make with the vertical?

Figure **P7.55**

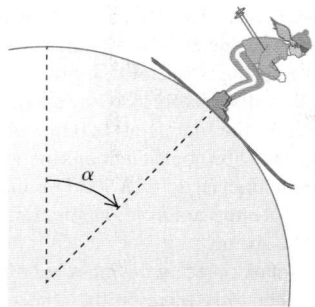

7.56 •• A ball is thrown upward with an initial velocity of 15 m/s at an angle of 60.0° above the horizontal. Use energy conservation to find the ball's greatest height above the ground.

7.57 •• In a truck-loading station at a post office, a small 0.200-kg package is released from rest at point A on a track that is one-quarter of a circle with radius 1.60 m (**Fig. P7.57**). The size of the package is much less than 1.60 m, so the package can be treated as a particle. It slides down the track and reaches point B with a speed of 4.80 m/s. From point B, it slides on a level surface a distance of 3.00 m to point C, where it comes to rest. (a) What is the coefficient of kinetic friction on the horizontal surface? (b) How much work is done on the package by friction as it slides down the circular arc from A to B?

Figure **P7.57**

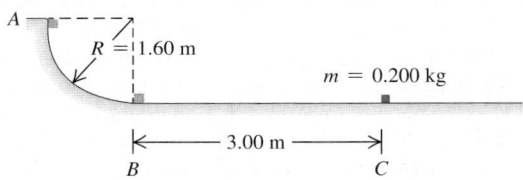

7.58 ••• A truck with mass m has a brake failure while going down an icy mountain road of constant downward slope angle α (**Fig. P7.58**). Initially the truck is moving downhill at speed v_0. After careening downhill a distance L with negligible friction, the truck driver steers the runaway vehicle onto a runaway truck ramp of constant upward slope angle β. The truck ramp has a soft sand surface for which the coefficient of rolling friction is μ_r. What is the distance that the truck moves up the ramp before coming to a halt? Solve by energy methods.

Figure **P7.58**

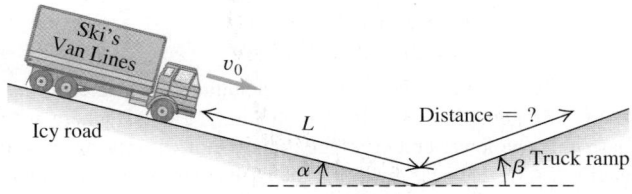

7.59 •• CALC A certain spring found *not* to obey Hooke's law exerts a restoring force $F_x(x) = -\alpha x - \beta x^2$ if it is stretched or compressed, where $\alpha = 60.0$ N/m and $\beta = 18.0$ N/m^2. The mass of the spring is negligible. (a) Calculate the potential-energy function $U(x)$ for this spring. Let $U = 0$ when $x = 0$. (b) An object with mass 0.900 kg on a frictionless, horizontal surface is attached to this spring, pulled a distance 1.00 m to the right (the $+x$-direction) to stretch the spring, and released. What is the speed of the object when it is 0.50 m to the right of the $x = 0$ equilibrium position?

7.60 •• CP A sled with rider having a combined mass of 125 kg travels over a perfectly smooth icy hill (**Fig. P7.60**). How far does the sled land from the foot of the cliff?

Figure **P7.60**

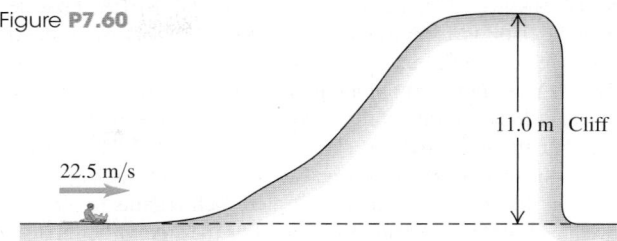

7.61 •• CALC A conservative force $\vec{F}$ is in the $+x$-direction and has magnitude $F(x) = \alpha/(x + x_0)^2$, where $\alpha = 0.800$ N·m^2 and $x_0 = 0.200$ m. (a) What is the potential-energy function $U(x)$ for this force? Let $U(x) \rightarrow 0$ as $x \rightarrow \infty$. (b) An object with mass $m = 0.500$ kg is released from rest at $x = 0$ and moves in the $+x$-direction. If $\vec{F}$ is the only force acting on the object, what is the object's speed when it reaches $x = 0.400$ m?

7.62 •• A 3.00-kg block is connected to two ideal horizontal springs having force constants $k_1 = 25.0$ N/cm and $k_2 = 20.0$ N/cm (**Fig. P7.62**).

Figure **P7.62**

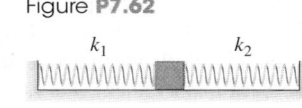

The system is initially in equilibrium on a horizontal, frictionless surface. The block is now pushed 15.0 cm to the right and released from rest. (a) What is the maximum speed of the block? Where in the motion does the maximum speed occur? (b) What is the maximum compression of spring 1?

7.63 •• A 0.150-kg block of ice is placed against a horizontal, compressed spring mounted on a horizontal tabletop that is 1.20 m above the floor. The spring has force constant 1900 N/m and is initially compressed 0.045 m. The mass of the spring is negligible. The spring is released, and the block slides along the table, goes off the edge, and travels to the floor. If there is negligible friction between the block of ice and the tabletop, what is the speed of the block of ice when it reaches the floor?

7.64 •• If a fish is attached to a vertical spring and slowly lowered to its equilibrium position, it is found to stretch the spring by an amount d. If the same fish is attached to the end of the unstretched spring and then allowed to fall from rest, through what maximum distance does it stretch the spring? (*Hint:* Calculate the force constant of the spring in terms of the distance d and the mass m of the fish.)

7.65 ••• CALC You are an industrial engineer with a shipping company. As part of the package-handling system, a small box with mass 1.60 kg is placed against a light spring that is compressed 0.280 m. The spring has force constant $k = 45.0$ N/m. The spring and box are released from rest, and the box travels along a horizontal surface for which the coefficient of kinetic friction with the box is $\mu_k = 0.300$. When the box has traveled 0.280 m and the spring has reached its equilibrium length, the box loses contact with the spring. (a) What is the speed of the box at the instant when it leaves the spring? (b) What is the maximum speed of the box during its motion?

7.66 •• A basket of negligible weight hangs from a vertical spring scale of force constant 1500 N/m. (a) If you suddenly put a 3.0-kg adobe brick in the basket, find the maximum distance that the spring will stretch. (b) If, instead, you release the brick from 1.0 m above the basket, by how much will the spring stretch at its maximum elongation?

7.67 ••• CALC A 3.00-kg fish is attached to the lower end of a vertical spring that has negligible mass and force constant 900 N/m. The spring initially is neither stretched nor compressed. The fish is released from rest. (a) What is its speed after it has descended 0.0500 m from its initial position? (b) What is the maximum speed of the fish as it descends?

7.68 •• You are designing an amusement park ride. A cart with two riders moves horizontally with speed $v = 6.00$ m/s. You assume that the total mass of cart plus riders is 300 kg. The cart hits a light spring that is attached to a wall, momentarily comes to rest as the spring is compressed, and then regains speed as it moves back in the opposite direction. For the ride to be thrilling but safe, the maximum acceleration of the cart during this motion should be 3.00g. Ignore friction. What is (a) the required force constant of the spring, (b) the maximum distance the spring will be compressed?

7.69 • A 0.500-kg block, attached to a spring with length 0.60 m and force constant 40.0 N/m, is at rest with the back of the block at point A on a frictionless, horizontal air table (**Fig. P7.69**). The mass of the spring is negligible. You move the block to the right along the surface by pulling with a constant 20.0-N horizontal force. (a) What is the block's speed when the back of the block reaches point B, which is 0.25 m to the right of point A? (b) When the back of the block reaches point B, you let go of the block. In the subsequent motion, how close does the block get to the wall where the left end of the spring is attached?

Figure **P7.69**

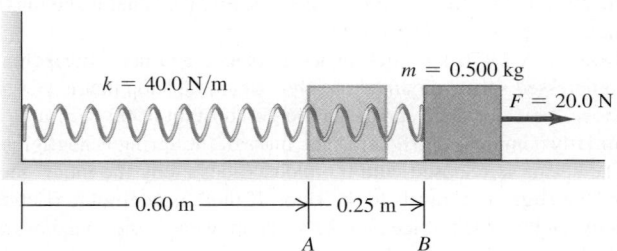

7.70 ••• CP A small block with mass 0.0400 kg slides in a vertical circle of radius $R = 0.500$ m on the inside of a circular track. During one of the revolutions of the block, when the block is at the bottom of its path, point A, the normal force exerted on the block by the track has magnitude 3.95 N. In this same revolution, when the block reaches the top of its path, point B, the normal force exerted on the block has magnitude 0.680 N. How much work is done on the block by friction during the motion of the block from point A to point B?

7.71 ••• CP A small block with mass 0.0500 kg slides in a vertical circle of radius $R = 0.800$ m on the inside of a circular track. There is no friction between the track and the block. At the bottom of the block's path, the normal force the track exerts on the block has magnitude 3.40 N. What is the magnitude of the normal force that the track exerts on the block when it is at the top of its path?

7.72 •• CP **Pendulum.** A small rock with mass 0.12 kg is fastened to a massless string with length 0.80 m to form a pendulum. The pendulum is swinging so as to make a maximum angle of 45° with the vertical. Air resistance is negligible. (a) What is the speed of the rock when the string passes through the vertical position? What is the tension in the string (b) when it makes an angle of 45° with the vertical, (c) as it passes through the vertical?

7.73 ••• A wooden block with mass 1.50 kg is placed against a compressed spring at the bottom of an incline of slope 30.0° (point A). When the spring is released, it projects the block up the incline. At point B, a distance of 6.00 m up the incline from A, the block is moving up the incline at 7.00 m/s and is no longer in contact with the spring. The coefficient of kinetic friction between the block and the incline is $\mu_k = 0.50$. The mass of the spring is negligible. Calculate the amount of potential energy that was initially stored in the spring.

7.74 •• CALC A small object with mass $m = 0.0900$ kg moves along the +x-axis. The only force on the object is a conservative force that has the potential-energy function $U(x) = -\alpha x^2 + \beta x^3$, where $\alpha = 2.00$ J/m² and $\beta = 0.300$ J/m³. The object is released from rest at small x. When the object is at $x = 4.00$ m, what are its (a) speed and (b) acceleration (magnitude and direction)? (c) What is the maximum value of x reached by the object during its motion?

7.75 ••• CALC A cutting tool under microprocessor control has several forces acting on it. One force is $\vec{F} = -\alpha xy^2\hat{j}$, a force in the negative y-direction whose magnitude depends on the position of the tool. For $\alpha = 2.50$ N/m³, consider the displacement of the tool from the origin to the point $(x = 3.00$ m, $y = 3.00$ m). (a) Calculate the work done on the tool by $\vec{F}$ if this displacement is along the straight line $y = x$ that connects these two points. (b) Calculate the work done on the tool by $\vec{F}$ if the tool is first moved out along the x-axis to the point $(x = 3.00$ m, $y = 0)$ and then moved parallel to the y-axis to the point $(x = 3.00$ m, $y = 3.00$ m). (c) Compare the work done by $\vec{F}$ along these two paths. Is $\vec{F}$ conservative or nonconservative? Explain.

7.76 • A particle moves along the x-axis while acted on by a single conservative force parallel to the x-axis. The force corresponds to the potential-energy function graphed in **Fig. P7.76**. The particle is released from rest at point A. (a) What is the direction of the force on the particle when it is at point A? (b) At point B? (c) At what value of x is the kinetic energy of the particle a maximum? (d) What is the force on the particle when it is at point C? (e) What is the largest value of x reached by the particle during its motion? (f) What value or values of x correspond to points of stable equilibrium? (g) Of unstable equilibrium?

Figure **P7.76**

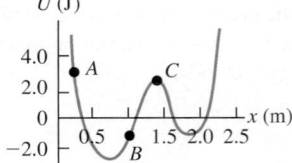

7.77 •• DATA You are designing a pendulum for a science museum. The pendulum is made by attaching a brass sphere with mass m to the lower end of a long, light metal wire of (unknown) length L. A device near the top of the wire measures the tension in the wire and transmits that information to your laptop computer. When the wire is vertical and the sphere is at rest, the sphere's center is 0.800 m above the floor and the tension in the wire is 265 N. Keeping the wire taut, you then pull the sphere to one side (using a ladder if necessary) and gently release it. You record the height h of the center of the sphere above the floor at the point where the sphere is released and the tension T in the wire as the sphere swings through its lowest point. You collect your results:

h (m)	0.800	2.00	4.00	6.00	8.00	10.0	12.0
T (N)	265	274	298	313	330	348	371

Assume that the sphere can be treated as a point mass, ignore the mass of the wire, and assume that mechanical energy is conserved through each measurement. (a) Plot T versus h, and use this graph to calculate L. (b) If the breaking strength of the wire is 822 N, from what maximum height h can the sphere be released if the tension in the wire is not to exceed half the breaking strength? (c) The pendulum is swinging when you leave at the end of the day. You lock the museum doors, and no one enters the building until you return the next morning. You find that the sphere is hanging at rest. Using energy considerations, how can you explain this behavior?

7.78 •• **DATA** A long ramp made of cast iron is sloped at a constant angle $\theta = 52.0°$ above the horizontal. Small blocks, each with mass 0.42 kg but made of different materials, are released from rest at a vertical height h above the bottom of the ramp. In each case the coefficient of static friction is small enough that the blocks start to slide down the ramp as soon as they are released. You are asked to find h so that each block will have a speed of 4.00 m/s when it reaches the bottom of the ramp. You are given these coefficients of sliding (kinetic) friction for different pairs of materials:

Material 1	Material 2	Coefficient of Sliding Friction
Cast iron	Cast iron	0.15
Cast iron	Copper	0.29
Cast iron	Lead	0.43
Cast iron	Zinc	0.85

Source: www.engineershandbook.com

(a) Use work and energy considerations to find the required value of h if the block is made from (i) cast iron; (ii) copper; (iii) zinc. (b) What is the required value of h for the copper block if its mass is doubled to 0.84 kg? (c) For a given block, if θ is increased while h is kept the same, does the speed v of the block at the bottom of the ramp increase, decrease, or stay the same?

7.79 •• **DATA** A single conservative force $F(x)$ acts on a small sphere of mass m while the sphere moves along the x-axis. You release the sphere from rest at $x = -1.50$ m. As the sphere moves, you measure its velocity as a function of position. You use the velocity data to calculate the kinetic energy K; **Fig. P7.79** shows your data. (a) Let $U(x)$ be the potential-energy function for $F(x)$. Is $U(x)$ symmetric about $x = 0$? [If so, then $U(x) = U(-x)$.] (b) If you set $U = 0$ at $x = 0$, what is the value of U at $x = -1.50$ m? (c) Sketch $U(x)$. (d) At what values of x (if any) is $F = 0$? (e) For

Figure **P7.79**

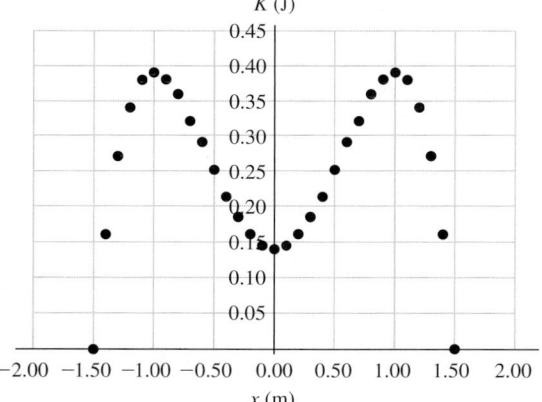

K (J)

what range of values of x between $x = -1.50$ m and $x = +1.50$ m is F positive? Negative? (f) If you release the sphere from rest at $x = -1.30$ m, what is the largest value of x that it reaches during its motion? The largest value of kinetic energy that it has during its motion?

CHALLENGE PROBLEM

7.80 ••• **CALC** A proton with mass m moves in one dimension. The potential-energy function is $U(x) = (\alpha/x^2) - (\beta/x)$, where α and β are positive constants. The proton is released from rest at $x_0 = \alpha/\beta$. (a) Show that $U(x)$ can be written as

$$U(x) = \frac{\alpha}{x_0^2}\left[\left(\frac{x_0}{x}\right)^2 - \frac{x_0}{x}\right]$$

Graph $U(x)$. Calculate $U(x_0)$ and thereby locate the point x_0 on the graph. (b) Calculate $v(x)$, the speed of the proton as a function of position. Graph $v(x)$ and give a qualitative description of the motion. (c) For what value of x is the speed of the proton a maximum? What is the value of that maximum speed? (d) What is the force on the proton at the point in part (c)? (e) Let the proton be released instead at $x_1 = 3\alpha/\beta$. Locate the point x_1 on the graph of $U(x)$. Calculate $v(x)$ and give a qualitative description of the motion. (f) For each release point $(x = x_0$ and $x = x_1)$, what are the maximum and minimum values of x reached during the motion?

PASSAGE PROBLEMS

BIO THE DNA SPRING. A DNA molecule, with its double-helix structure, can in some situations behave like a spring. Measuring the force required to stretch single DNA molecules under various conditions can provide information about the biophysical properties of DNA. A technique for measuring the stretching force makes use of a very small cantilever, which consists of a beam that is supported at one end and is free to move at the other end, like a tiny diving board. The cantilever is constructed so that it obeys Hooke's law—that is, the displacement of its free end is proportional to the force applied to it. Because different cantilevers have different force constants, the cantilever's response must first be calibrated by applying a known force and determining the resulting deflection of the cantilever. Then one end of a DNA molecule is attached to the free end of the cantilever, and the other end of the DNA molecule is attached to a small stage that can be moved away from the cantilever, stretching the DNA. The stretched DNA pulls on the cantilever, deflecting the end of the cantilever very slightly. The measured deflection is then used to determine the force on the DNA molecule.

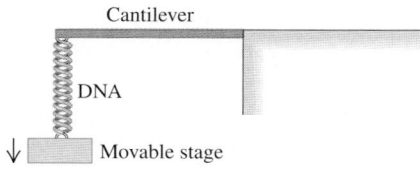

7.81 During the calibration process, the cantilever is observed to deflect by 0.10 nm when a force of 3.0 pN is applied to it. What deflection of the cantilever would correspond to a force of 6.0 pN? (a) 0.07 nm; (b) 0.14 nm; (c) 0.20 nm; (d) 0.40 nm.

7.82 A segment of DNA is put in place and stretched. **Figure P7.82** shows a graph of the force exerted on the DNA as a function of the displacement of the stage. Based on this graph, which statement is the best interpretation of the DNA's behavior over this range of displacements? The DNA (a) does not follow Hooke's law, because its force constant increases as the force on it increases; (b) follows Hooke's law and has a force constant of about 0.1 pN/nm; (c) follows Hooke's law and has a force constant of about 10 pN/nm; (d) does not follow Hooke's law, because its force constant decreases as the force on it increases.

Figure **P7.82**

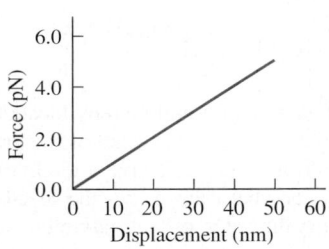

7.83 Based on Fig. P7.82, how much elastic potential energy is stored in the DNA when it is stretched 50 nm? (a) 2.5×10^{-19} J; (b) 1.2×10^{-19} J; (c) 5.0×10^{-12} J; (d) 2.5×10^{-12} J.

7.84 The stage moves at a constant speed while stretching the DNA. Which of the graphs in **Fig. P7.84** best represents the power supplied to the stage versus time?

Figure **P7.84**

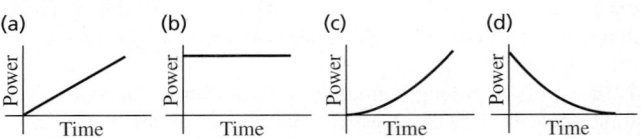

Answers

Chapter Opening Question ?

(v) As the crane descends, air resistance directed opposite to the bird's motion prevents its speed from increasing. Because the crane's speed stays the same, its kinetic energy K remains constant, but the gravitational potential energy U_{grav} decreases as the crane descends. Hence the total mechanical energy $E = K + U_{\text{grav}}$ decreases. The lost mechanical energy goes into warming the crane's skin (that is, an increase in the crane's internal energy) and stirring up the air through which the crane passes (an increase in the internal energy of the air). See Section 7.3.

Test Your Understanding Questions

7.1 (iii) The initial kinetic energy $K_1 = 0$, the initial potential energy $U_1 = mgy_1$, and the final potential energy $U_2 = mgy_2$ are the same for both blocks. Mechanical energy is conserved in both cases, so the final kinetic energy $K_2 = \frac{1}{2}mv_2^2$ is also the same for both blocks. Hence the speed at the right-hand end is the *same* in both cases!

7.2 (iii) The elevator is still moving downward, so the kinetic energy K is positive (remember that K can never be negative); the elevator is below point 1, so $y < 0$ and $U_{\text{grav}} < 0$; and the spring is compressed, so $U_{\text{el}} > 0$.

7.3 (iii) Because of friction in the turbines and between the water and turbines, some of the potential energy goes into raising the temperatures of the water and the mechanism.

7.4 (a) (iv), (b) (i) If $F_x = 0$ at a point, then the derivative of $U(x)$ must be zero at that point because $F_x = -dU(x)/dx$. However, this tells us absolutely nothing about the *value* of $U(x)$ at that point.

7.5 (iii) Figure 7.24b shows the x-component of force, F_x. Where this is maximum (most positive), the x-component of force and the x-acceleration have more positive values than at adjacent values of x.

Bridging Problem

(a) 1.06 m **(b)** 1.32 m **(c)** 20.7 J

? Which of the following three bullets (all of the same length and diameter) can do greater damage to this carrot? (i) A .22-caliber bullet moving at 220 m/s, as shown here; (ii) a bullet with half the mass moving at twice the speed; (iii) a bullet with double the mass moving at half the speed; (iv) all do the same amount of damage.

8 MOMENTUM, IMPULSE, AND COLLISIONS

Many questions involving forces can't be answered by directly applying Newton's second law, $\sum \vec{F} = m\vec{a}$. For example, when a truck collides head-on with a compact car, what determines which way the wreckage moves after the collision? In playing pool, how do you decide how to aim the cue ball in order to knock the eight ball into the pocket? And when a meteorite collides with the earth, how much of the meteorite's kinetic energy is released in the impact?

All of these questions involve forces about which we know very little: the forces between the car and the truck, between the two pool balls, or between the meteorite and the earth. Remarkably, we will find in this chapter that we don't have to know *anything* about these forces to answer questions of this kind!

Our approach uses two new concepts, *momentum* and *impulse*, and a new conservation law, *conservation of momentum*. This conservation law is every bit as important as the law of conservation of energy. The law of conservation of momentum is valid even in situations in which Newton's laws are inadequate, such as bodies moving at very high speeds (near the speed of light) or objects on a very small scale (such as the constituents of atoms). Within the domain of Newtonian mechanics, conservation of momentum enables us to analyze many situations that would be very difficult if we tried to use Newton's laws directly. Among these are *collision* problems, in which two bodies collide and can exert very large forces on each other for a short time. We'll also use momentum ideas to solve problems in which an object's mass changes as it moves, including the important special case of a rocket (which loses mass as it expends fuel).

8.1 MOMENTUM AND IMPULSE

In Section 6.2 we re-expressed Newton's second law for a particle, $\sum \vec{F} = m\vec{a}$, in terms of the work–energy theorem. This theorem helped us tackle a great number of problems and led us to the law of conservation of energy. Let's return to $\sum \vec{F} = m\vec{a}$ and see yet another useful way to restate this fundamental law.

Newton's Second Law in Terms of Momentum

Consider a particle of constant mass m. Because $\vec{a} = d\vec{v}/dt$, we can write Newton's second law for this particle as

$$\sum \vec{F} = m\frac{d\vec{v}}{dt} = \frac{d}{dt}(m\vec{v}) \tag{8.1}$$

We can move the mass m inside the derivative because it is constant. Thus Newton's second law says that the net force $\sum \vec{F}$ acting on a particle equals the time rate of change of the product of the particle's mass and velocity. We'll call this product the **momentum,** or **linear momentum,** of the particle:

Momentum of a particle ·····→ **Particle mass**
(a vector quantity) $\vec{p} = m\vec{v}$ ·····Particle velocity $\tag{8.2}$

The greater the mass m and speed v of a particle, the greater is its magnitude of momentum mv. Keep in mind that momentum is a *vector* quantity with the same direction as the particle's velocity (**Fig. 8.1**). A car driving north at 20 m/s and an identical car driving east at 20 m/s have the same *magnitude* of momentum (mv) but different momentum *vectors* ($m\vec{v}$) because their directions are different.

We often express the momentum of a particle in terms of its components. If the particle has velocity components v_x, v_y, and v_z, then its momentum components p_x, p_y, and p_z (which we also call the *x-momentum, y-momentum,* and *z-momentum*) are

$$p_x = mv_x \qquad p_y = mv_y \qquad p_z = mv_z \tag{8.3}$$

These three component equations are equivalent to Eq. (8.2).

The units of the magnitude of momentum are units of mass times speed; the SI units of momentum are kg · m/s. The plural of momentum is "momenta."

Let's now substitute the definition of momentum, Eq. (8.2), into Eq. (8.1):

Newton's second law
in terms of momentum: ·······→ $\sum \vec{F} = \dfrac{d\vec{p}}{dt}$ ←······· ... equals the rate of change
The net force acting on of the particle's momentum. $\tag{8.4}$
a particle ...

The net force (vector sum of all forces) acting on a particle equals the time rate of change of momentum of the particle. This, not $\sum \vec{F} = m\vec{a}$, is the form in which Newton originally stated his second law (although he called momentum the "quantity of motion"). This law is valid only in inertial frames of reference (see Section 4.2). As Eq. 8.4 shows, a rapid change in momentum requires a large net force, while a gradual change in momentum requires a smaller net force (**Fig. 8.2**).

The Impulse–Momentum Theorem

Both a particle's momentum $\vec{p} = m\vec{v}$ and its kinetic energy $K = \frac{1}{2}mv^2$ depend on the mass and velocity of the particle. What is the fundamental difference between these two quantities? A purely mathematical answer is that momentum

---

8.1 The velocity and momentum vectors of a particle.

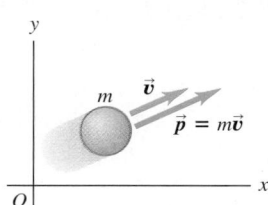

Momentum $\vec{p}$ is a vector quantity; a particle's momentum has the same direction as its velocity $\vec{v}$.

8.2 When you land after jumping upward, your momentum changes from a downward value to zero. It's best to land with your knees bent so that your legs can flex: You then take a relatively long time to stop, and the force that the ground exerts on your legs is small. If you land with your legs extended, you stop in a short time, the force on your legs is larger, and the possibility of injury is greater.

is a vector whose magnitude is proportional to speed, while kinetic energy is a scalar proportional to the speed squared. But to see the *physical* difference between momentum and kinetic energy, we must first define a quantity closely related to momentum called *impulse*.

Let's first consider a particle acted on by a *constant* net force $\Sigma\vec{F}$ during a time interval Δt from t_1 to t_2. The **impulse** of the net force, denoted by $\vec{J}$, is defined to be the product of the net force and the time interval:

$$\underset{\substack{\text{Impulse of a} \\ \text{constant net force}}}{} \vec{J} = \underset{\text{Constant net force}}{\Sigma\vec{F}}(t_2 - t_1) = \Sigma\vec{F}\underset{\substack{\text{Time interval over} \\ \text{which net force acts}}}{\Delta t} \qquad (8.5)$$

Impulse is a vector quantity; its direction is the same as the net force $\Sigma\vec{F}$. The SI unit of impulse is the newton-second (N·s). Because $1\text{ N} = 1\text{ kg}\cdot\text{m/s}^2$, an alternative set of units for impulse is $\text{kg}\cdot\text{m/s}$, the same as for momentum.

To see what impulse is good for, let's go back to Newton's second law as restated in terms of momentum, Eq. (8.4). If the net force $\Sigma\vec{F}$ is constant, then $d\vec{p}/dt$ is also constant. In that case, $d\vec{p}/dt$ is equal to the *total* change in momentum $\vec{p}_2 - \vec{p}_1$ during the time interval $t_2 - t_1$, divided by the interval:

$$\Sigma\vec{F} = \frac{\vec{p}_2 - \vec{p}_1}{t_2 - t_1}$$

Multiplying this equation by $(t_2 - t_1)$, we have

$$\Sigma\vec{F}(t_2 - t_1) = \vec{p}_2 - \vec{p}_1$$

Comparing with Eq. (8.5), we end up with

Impulse–momentum theorem: The impulse of the net force on a particle during a time interval equals the change in momentum of that particle during that interval:

$$\underset{\substack{\text{Impulse of net force over} \\ \text{a time interval}}}{} \vec{J} = \underset{\text{Final momentum}}{\vec{p}_2} - \underset{\text{Initial momentum}}{\vec{p}_1} = \underset{\substack{\text{Change in} \\ \text{momentum}}}{\Delta\vec{p}} \qquad (8.6)$$

The impulse–momentum theorem also holds when forces are not constant. To see this, we integrate both sides of Newton's second law $\Sigma\vec{F} = d\vec{p}/dt$ over time between the limits t_1 and t_2:

$$\int_{t_1}^{t_2} \Sigma\vec{F}\, dt = \int_{t_1}^{t_2} \frac{d\vec{p}}{dt}\, dt = \int_{\vec{p}_1}^{\vec{p}_2} d\vec{p} = \vec{p}_2 - \vec{p}_1$$

We see from Eq. (8.6) that the integral on the left is the impulse of the net force:

$$\underset{\substack{\text{Impulse of a} \\ \text{general net force} \\ \text{(either constant} \\ \text{or varying)}}}{} \vec{J} = \int_{t_1}^{t_2} \Sigma\vec{F}\, dt \overset{\substack{\text{Upper limit = final time}}}{\underset{\substack{\text{Time integral of} \\ \text{net force} \\ \text{Lower limit = initial time}}}{}} \qquad (8.7)$$

If the net force $\Sigma\vec{F}$ is constant, the integral in Eq. (8.7) reduces to Eq. (8.5). We can define an *average* net force $\vec{F}_{av}$ such that even when $\Sigma\vec{F}$ is not constant, the impulse $\vec{J}$ is given by

$$\vec{J} = \vec{F}_{av}(t_2 - t_1) \qquad (8.8)$$

When $\Sigma\vec{F}$ is constant, $\Sigma\vec{F} = \vec{F}_{av}$ and Eq. (8.8) reduces to Eq. (8.5).

Figure 8.3a shows the x-component of net force ΣF_x as a function of time during a collision. This might represent the force on a soccer ball that is in contact with a player's foot from time t_1 to t_2. The x-component of impulse during this interval is represented by the red area under the curve between t_1 and t_2. This area is equal to the green rectangular area bounded by t_1, t_2, and $(F_{av})_x$,

BIO Application Woodpecker Impulse The pileated woodpecker (*Dryocopus pileatus*) has been known to strike its beak against a tree up to 20 times a second and up to 12,000 times a day. The impact force can be as much as 1200 times the weight of the bird's head. Because the impact lasts such a short time, the impulse—the product of the net force during the impact multiplied by the duration of the impact—is relatively small. (The woodpecker has a thick skull of spongy bone as well as shock-absorbing cartilage at the base of the lower jaw, and so avoids injury.)

8.3 The meaning of the area under a graph of ΣF_x versus t.

(a)

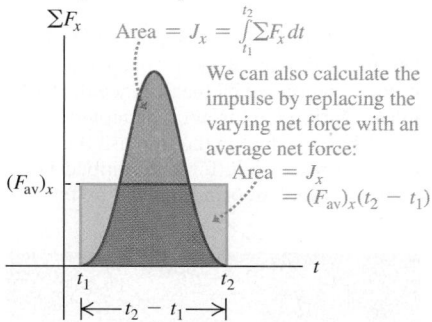

The area under the curve of net force versus time equals the impulse of the net force:

$$\text{Area} = J_x = \int_{t_1}^{t_2} \Sigma F_x\, dt$$

We can also calculate the impulse by replacing the varying net force with an average net force:

$$\text{Area} = J_x = (F_{av})_x(t_2 - t_1)$$

(b)

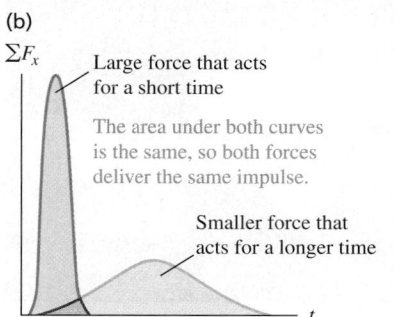

Large force that acts for a short time

The area under both curves is the same, so both forces deliver the same impulse.

Smaller force that acts for a longer time

8.4 The impulse–momentum theorem explains how air bags reduce the chance of injury by minimizing the force on an occupant of an automobile.

- Impulse–momentum theorem:
$$\vec{J} = \vec{p}_2 - \vec{p}_1 = \vec{F}_{av}\Delta t$$
- Impulse is the same no matter how the driver is brought to rest (so $\vec{p}_2 = \mathbf{0}$).
- Compared to striking the steering wheel, striking the air bag brings the driver to rest over a longer time interval Δt.
- Hence with an air bag, average force $\vec{F}_{av}$ on the driver is less.

so $(F_{av})_x(t_2 - t_1)$ is equal to the impulse of the actual time-varying force during the same interval. Note that a large force acting for a short time can have the same impulse as a smaller force acting for a longer time if the areas under the force–time curves are the same (Fig. 8.3b). We used this idea in Fig. 8.2: A small force acting for a relatively long time (as when you land with your legs bent) has the same effect as a larger force acting for a short time (as when you land stiff-legged). Automotive air bags use the same principle (**Fig. 8.4**).

Both impulse and momentum are vector quantities, and Eqs. (8.5)–(8.8) are vector equations. It's often easiest to use them in component form:

$$J_x = \int_{t_1}^{t_2} \Sigma F_x \, dt = (F_{av})_x(t_2 - t_1) = p_{2x} - p_{1x} = mv_{2x} - mv_{1x}$$

$$(8.9)$$

$$J_y = \int_{t_1}^{t_2} \Sigma F_y \, dt = (F_{av})_y(t_2 - t_1) = p_{2y} - p_{1y} = mv_{2y} - mv_{1y}$$

and similarly for the z-component.

Momentum and Kinetic Energy Compared

We can now see the fundamental difference between momentum and kinetic energy. The impulse–momentum theorem, $\vec{J} = \vec{p}_2 - \vec{p}_1$, says that changes in a particle's momentum are due to impulse, which depends on the *time* over which the net force acts. By contrast, the work–energy theorem, $W_{tot} = K_2 - K_1$, tells us that kinetic energy changes when work is done on a particle; the total work depends on the *distance* over which the net force acts.

Let's consider a particle that starts from rest at t_1 so that $\vec{v}_1 = \mathbf{0}$. Its initial momentum is $\vec{p}_1 = m\vec{v}_1 = \mathbf{0}$, and its initial kinetic energy is $K_1 = \frac{1}{2}mv_1^2 = 0$. Now let a constant net force equal to $\vec{F}$ act on that particle from time t_1 until time t_2. During this interval, the particle moves a distance s in the direction of the force. From Eq. (8.6), the particle's momentum at time t_2 is

$$\vec{p}_2 = \vec{p}_1 + \vec{J} = \vec{J}$$

8.5 The *kinetic energy* of a pitched base-ball is equal to the work the pitcher does on it (force multiplied by the distance the ball moves during the throw). The *momentum* of the ball is equal to the impulse the pitcher imparts to it (force multiplied by the time it took to bring the ball up to speed).

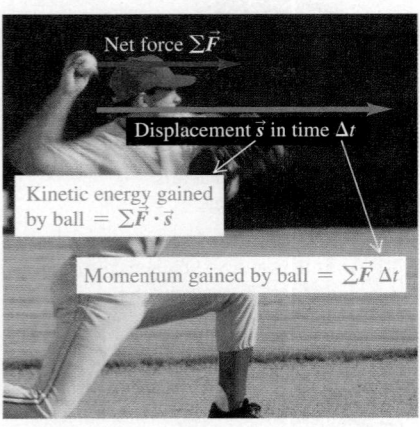

where $\vec{J} = \vec{F}(t_2 - t_1)$ is the impulse that acts on the particle. So *the momentum of a particle equals the impulse that accelerated it from rest to its present speed;* impulse is the product of the net force that accelerated the particle and the *time* required for the acceleration. By comparison, the kinetic energy of the particle at t_2 is $K_2 = W_{tot} = Fs$, the total *work* done on the particle to accelerate it from rest. The total work is the product of the net force and the *distance* required to accelerate the particle (**Fig. 8.5**).

Here's an application of the distinction between momentum and kinetic energy. Which is easier to catch: a 0.50-kg ball moving at 4.0 m/s or a 0.10-kg ball moving at 20 m/s? Both balls have the same magnitude of momentum, $p = mv = (0.50 \text{ kg})(4.0 \text{ m/s}) = (0.10 \text{ kg})(20 \text{ m/s}) = 2.0 \text{ kg} \cdot \text{m/s}$. However, the two balls have different values of kinetic energy $K = \frac{1}{2}mv^2$: The large, slow-moving ball has $K = 4.0 \text{ J}$, while the small, fast-moving ball has $K = 20 \text{ J}$. Since the momentum is the same for both balls, both require the same *impulse* to be brought to rest. But stopping the 0.10-kg ball with your hand requires five times more *work* than stopping the 0.50-kg ball because the smaller ball has five times more kinetic energy. For a given force that you exert with your hand, it takes the same amount of time (the duration of the catch) to stop either ball, but your hand and arm will be pushed back five times farther if you choose to catch the small, fast-moving ball. To minimize arm strain, you should choose to catch the 0.50-kg ball with its lower kinetic energy.

Both the impulse–momentum and work–energy theorems rest on the foundation of Newton's laws. They are *integral* principles, relating the motion at two different times separated by a finite interval. By contrast, Newton's second law itself (in either of the forms $\sum \vec{F} = m\vec{a}$ or $\sum \vec{F} = d\vec{p}/dt$) is a *differential* principle that concerns the rate of change of velocity or momentum at each instant.

CONCEPTUAL EXAMPLE 8.1 MOMENTUM VERSUS KINETIC ENERGY

Consider again the race described in Conceptual Example 6.5 (Section 6.2) between two iceboats on a frictionless frozen lake. The boats have masses m and $2m$, and the wind exerts the same constant horizontal force $\vec{F}$ on each boat (see Fig. 6.14). The boats start from rest and cross the finish line a distance s away. Which boat crosses the finish line with greater momentum?

SOLUTION

In Conceptual Example 6.5 we asked how the *kinetic energies* of the boats compare when they cross the finish line. We answered this by remembering that *a body's kinetic energy equals the total work done to accelerate it from rest.* Both boats started from rest, and the total work done was the same for both boats (because the net force and the displacement were the same for both). Hence both boats had the same kinetic energy at the finish line.

Similarly, to compare the *momenta* of the boats we use the idea that *the momentum of each boat equals the impulse that*

accelerated it from rest. As in Conceptual Example 6.5, the net force on each boat equals the constant horizontal wind force $\vec{F}$. Let Δt be the time a boat takes to reach the finish line, so that the impulse on the boat during that time is $\vec{J} = \vec{F} \Delta t$. Since the boat starts from rest, this equals the boat's momentum $\vec{p}$ at the finish line:

$$\vec{p} = \vec{F} \Delta t$$

Both boats are subjected to the same force $\vec{F}$, but they take different times Δt to reach the finish line. The boat of mass $2m$ accelerates more slowly and takes a longer time to travel the distance s; thus there is a greater impulse on this boat between the starting and finish lines. So the boat of mass $2m$ crosses the finish line with a greater magnitude of momentum than the boat of mass m (but with the same kinetic energy). Can you show that the boat of mass $2m$ has $\sqrt{2}$ times as much momentum at the finish line as the boat of mass m?

EXAMPLE 8.2 A BALL HITS A WALL

You throw a ball with a mass of 0.40 kg against a brick wall. It is moving horizontally to the left at 30 m/s when it hits the wall; it rebounds horizontally to the right at 20 m/s. (a) Find the impulse of the net force on the ball during its collision with the wall. (b) If the ball is in contact with the wall for 0.010 s, find the average horizontal force that the wall exerts on the ball during the impact.

SOLUTION

IDENTIFY and SET UP: We're given enough information to determine the initial and final values of the ball's momentum, so we can use the impulse–momentum theorem to find the impulse. We'll then use the definition of impulse to determine the average force. **Figure 8.6** shows our sketch. We need only a single axis because the motion is purely horizontal. We'll take the positive x-direction to be to the right. In part (a) our target variable is the x-component of impulse, J_x, which we'll find by using Eqs. (8.9). In part (b), our target variable is the average x-component of force $(F_{av})_x$; once we know J_x, we can also find this force by using Eqs. (8.9).

8.6 Our sketch for this problem.

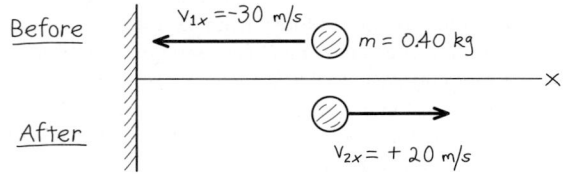

EXECUTE: (a) With our choice of x-axis, the initial and final x-components of momentum of the ball are

$$p_{1x} = mv_{1x} = (0.40 \text{ kg})(-30 \text{ m/s}) = -12 \text{ kg} \cdot \text{m/s}$$
$$p_{2x} = mv_{2x} = (0.40 \text{ kg})(+20 \text{ m/s}) = +8.0 \text{ kg} \cdot \text{m/s}$$

From the x-equation in Eqs. (8.9), the x-component of impulse equals the *change* in the x-momentum:

$$J_x = p_{2x} - p_{1x}$$
$$= 8.0 \text{ kg} \cdot \text{m/s} - (-12 \text{ kg} \cdot \text{m/s}) = 20 \text{ kg} \cdot \text{m/s} = 20 \text{ N} \cdot \text{s}$$

(b) The collision time is $t_2 - t_1 = \Delta t = 0.010$ s. From the x-equation in Eqs. (8.9), $J_x = (F_{av})_x(t_2 - t_1) = (F_{av})_x \Delta t$, so

$$(F_{av})_x = \frac{J_x}{\Delta t} = \frac{20 \text{ N} \cdot \text{s}}{0.010 \text{ s}} = 2000 \text{ N}$$

EVALUATE: The x-component of impulse J_x is positive—that is, to the right in Fig. 8.6. The impulse represents the "kick" that the wall imparts to the ball, and this "kick" is certainly to the right.

CAUTION **Momentum is a vector** Because momentum is a vector, we had to include the negative sign in writing $p_{1x} = -12 \text{ kg} \cdot \text{m/s}$. Had we omitted it, we would have calculated the impulse to be $8.0 \text{ kg} \cdot \text{m/s} - (12 \text{ kg} \cdot \text{m/s}) = -4 \text{ kg} \cdot \text{m/s}$. This would say that the wall had somehow given the ball a kick to the *left*! Remember the *direction* of momentum in your calculations. ▮

Continued

The force that the wall exerts on the ball must have such a large magnitude (2000 N, equal to the weight of a 200-kg object) to change the ball's momentum in such a short time. Other forces that act on the ball during the collision are comparatively weak; for instance, the gravitational force is only 3.9 N. Thus, during the short time that the collision lasts, we can ignore all other forces on the ball. **Figure 8.7** shows the impact of a tennis ball and racket.

Note that the 2000-N value we calculated is the *average* horizontal force that the wall exerts on the ball during the impact. It corresponds to the horizontal line $(F_{av})_x$ in Fig. 8.3a. The horizontal force is zero before impact, rises to a maximum, and then decreases to zero when the ball loses contact with the wall. If the ball is relatively rigid, like a baseball or golf ball, the collision lasts a short time and the maximum force is large, as in the blue curve in Fig. 8.3b. If the ball is softer, like a tennis ball, the collision time is longer and the maximum force is less, as in the orange curve in Fig. 8.3b.

8.7 Typically, a tennis ball is in contact with the racket for approximately 0.01 s. The ball flattens noticeably due to the tremendous force exerted by the racket.

 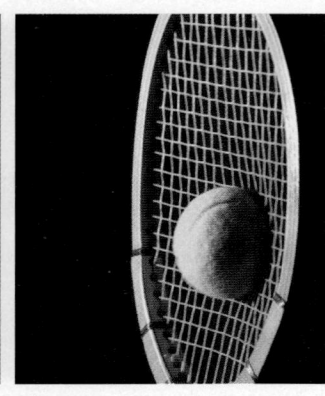

EXAMPLE 8.3   KICKING A SOCCER BALL

A soccer ball has a mass of 0.40 kg. Initially it is moving to the left at 20 m/s, but then it is kicked. After the kick it is moving at 45° upward and to the right with speed 30 m/s (**Fig. 8.8a**). Find the impulse of the net force and the average net force, assuming a collision time $\Delta t = 0.010$ s.

8.8 (a) Kicking a soccer ball. (b) Finding the average force on the ball from its components.

(a) Before-and-after diagram

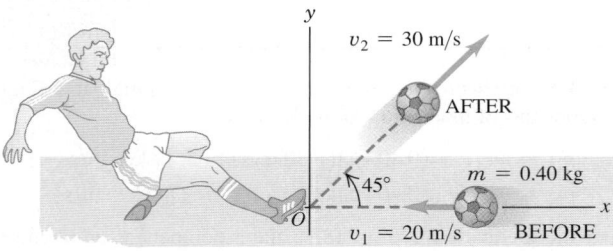

(b) Average force on the ball

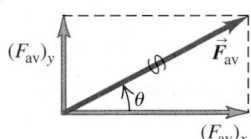

SOLUTION

IDENTIFY and SET UP: The ball moves in two dimensions, so we must treat momentum and impulse as vector quantities. We take the x-axis to be horizontally to the right and the y-axis to be vertically upward. Our target variables are the components of the net impulse on the ball, J_x and J_y, and the components of the average net force on the ball, $(F_{av})_x$ and $(F_{av})_y$. We'll find them by using the impulse–momentum theorem in its component form, Eqs. (8.9).

EXECUTE: Using $\cos 45° = \sin 45° = 0.707$, we find the ball's velocity components before and after the kick:

$$v_{1x} = -20 \text{ m/s} \qquad v_{1y} = 0$$
$$v_{2x} = v_{2y} = (30 \text{ m/s})(0.707) = 21.2 \text{ m/s}$$

From Eqs. (8.9), the impulse components are

$$J_x = p_{2x} - p_{1x} = m(v_{2x} - v_{1x})$$
$$= (0.40 \text{ kg})[21.2 \text{ m/s} - (-20 \text{ m/s})] = 16.5 \text{ kg} \cdot \text{m/s}$$

$$J_y = p_{2y} - p_{1y} = m(v_{2y} - v_{1y})$$
$$= (0.40 \text{ kg})(21.2 \text{ m/s} - 0) = 8.5 \text{ kg} \cdot \text{m/s}$$

From Eq. (8.8), the average net force components are

$$(F_{av})_x = \frac{J_x}{\Delta t} = 1650 \text{ N} \qquad (F_{av})_y = \frac{J_y}{\Delta t} = 850 \text{ N}$$

The magnitude and direction of the $\vec{F}_{av}$ vector (Fig. 8.8b) are

$$F_{av} = \sqrt{(1650 \text{ N})^2 + (850 \text{ N})^2} = 1.9 \times 10^3 \text{ N}$$
$$\theta = \arctan \frac{850 \text{ N}}{1650 \text{ N}} = 27°$$

The ball was not initially at rest, so its final velocity does *not* have the same direction as the average force that acted on it.

EVALUATE: $\vec{F}_{av}$ includes the force of gravity, which is very small; the weight of the ball is only 3.9 N. As in Example 8.2, the average force acting during the collision is exerted almost entirely by the object that the ball hit (in this case, the soccer player's foot).

TEST YOUR UNDERSTANDING OF SECTION 8.1 Rank the following situations according to the magnitude of the impulse of the net force, from largest value to smallest value. In each situation a 1000-kg automobile is moving along a straight east–west road. The automobile is initially (i) moving east at 25 m/s and comes to a stop in 10 s; (ii) moving east at 25 m/s and comes to a stop in 5 s; (iii) at rest, and a 2000-N net force toward the east is applied to it for 10 s; (iv) moving east at 25 m/s, and a 2000-N net force toward the west is applied to it for 10 s; (v) moving east at 25 m/s; over a 30-s period, the automobile reverses direction and ends up moving west at 25 m/s. ❚

8.2 CONSERVATION OF MOMENTUM

The concept of momentum is particularly important in situations in which we have two or more bodies that *interact*. To see why, let's consider first an idealized system of two bodies that interact with each other but not with anything else—for example, two astronauts who touch each other as they float freely in the zero-gravity environment of outer space (**Fig. 8.9**). Think of the astronauts as particles. Each particle exerts a force on the other; according to Newton's third law, the two forces are always equal in magnitude and opposite in direction. Hence, the *impulses* that act on the two particles are equal in magnitude and opposite in direction, as are the changes in momentum of the two particles.

Let's go over that again with some new terminology. For any system, the forces that the particles of the system exert on each other are called **internal forces.** Forces exerted on any part of the system by some object outside it are called **external forces.** For the system shown in Fig. 8.9, the internal forces are $\vec{F}_{B \text{ on } A}$, exerted by particle B on particle A, and $\vec{F}_{A \text{ on } B}$, exerted by particle A on particle B. There are *no* external forces; when this is the case, we have an **isolated system.**

The net force on particle A is $\vec{F}_{B \text{ on } A}$ and the net force on particle B is $\vec{F}_{A \text{ on } B}$, so from Eq. (8.4) the rates of change of the momenta of the two particles are

$$\vec{F}_{B \text{ on } A} = \frac{d\vec{p}_A}{dt} \qquad \vec{F}_{A \text{ on } B} = \frac{d\vec{p}_B}{dt} \tag{8.10}$$

The momentum of each particle changes, but these changes are related to each other by Newton's third law: Forces $\vec{F}_{B \text{ on } A}$ and $\vec{F}_{A \text{ on } B}$ are always equal in magnitude and opposite in direction. That is, $\vec{F}_{B \text{ on } A} = -\vec{F}_{A \text{ on } B}$, so $\vec{F}_{B \text{ on } A} + \vec{F}_{A \text{ on } B} = \mathbf{0}$. Adding together the two equations in Eq. (8.10), we have

$$\vec{F}_{B \text{ on } A} + \vec{F}_{A \text{ on } B} = \frac{d\vec{p}_A}{dt} + \frac{d\vec{p}_B}{dt} = \frac{d(\vec{p}_A + \vec{p}_B)}{dt} = \mathbf{0} \tag{8.11}$$

The rates of change of the two momenta are equal and opposite, so the rate of change of the vector sum $\vec{p}_A + \vec{p}_B$ is zero. We define the **total momentum** $\vec{P}$ of the system of two particles as the vector sum of the momenta of the individual particles; that is,

$$\vec{P} = \vec{p}_A + \vec{p}_B \tag{8.12}$$

Then Eq. (8.11) becomes

$$\vec{F}_{B \text{ on } A} + \vec{F}_{A \text{ on } B} = \frac{d\vec{P}}{dt} = \mathbf{0} \tag{8.13}$$

The time rate of change of the *total* momentum $\vec{P}$ is zero. Hence the total momentum of the system is constant, even though the individual momenta of the particles that make up the system can change.

If external forces are also present, they must be included on the left side of Eq. (8.13) along with the internal forces. Then the total momentum is, in general, not constant. But if the vector sum of the external forces is zero, as in **Fig. 8.10**, these forces have no effect on the left side of Eq. (8.13), and $d\vec{P}/dt$ is again zero. Thus we have the following general result:

> **If the vector sum of the external forces on a system is zero, the total momentum of the system is constant.**

This is the simplest form of the **principle of conservation of momentum.** This principle is a direct consequence of Newton's third law. What makes this principle useful is that it doesn't depend on the detailed nature of the internal forces

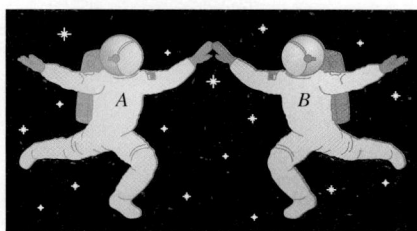

8.9 Two astronauts push each other as they float freely in the zero-gravity environment of space.

No external forces act on the two-astronaut system, so its total momentum is conserved.

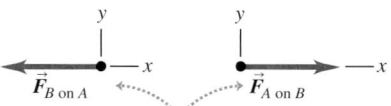

The forces the astronauts exert on each other form an action–reaction pair.

8.10 Two ice skaters push each other as they skate on a frictionless, horizontal surface. (Compare to Fig. 8.9.)

The forces the skaters exert on each other form an action–reaction pair.

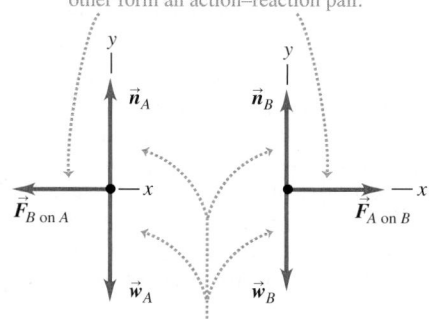

Although the normal and gravitational forces are external, their vector sum is zero, so the total momentum is conserved.

that act between members of the system. This means that we can apply conservation of momentum even if (as is often the case) we know very little about the internal forces. We have used Newton's second law to derive this principle, so we have to be careful to use it only in inertial frames of reference.

We can generalize this principle for a system that contains any number of particles A, B, C, . . . interacting only with one another, with total momentum

> Total momentum of a system of particles A, B, C, ...
> $$\vec{P} = \vec{p}_A + \vec{p}_B + \cdots = m_A\vec{v}_A + m_B\vec{v}_B + \cdots \tag{8.14}$$
> ... equals vector sum of momenta of all particles in the system.

We make the same argument as before: The total rate of change of momentum of the system due to each action–reaction pair of internal forces is zero. Thus the total rate of change of momentum of the entire system is zero whenever the vector sum of the external forces acting on it is zero. The internal forces can change the momenta of individual particles but not the *total* momentum of the system.

CAUTION **Conservation of momentum means conservation of its components** When you apply the conservation of momentum to a system, remember that momentum is a *vector* quantity. Hence you must use vector addition to compute the total momentum of a system (**Fig. 8.11**). Using components is usually the simplest method. If p_{Ax}, p_{Ay}, and p_{Az} are the components of momentum of particle A, and similarly for the other particles, then Eq. (8.14) is equivalent to the component equations

$$P_x = p_{Ax} + p_{Bx} + \cdots, \quad P_y = p_{Ay} + p_{By} + \cdots, \quad P_z = p_{Az} + p_{Bz} + \cdots \tag{8.15}$$

If the vector sum of the external forces on the system is zero, then P_x, P_y, and P_z are all constant.

In some ways the principle of conservation of momentum is more general than the principle of conservation of mechanical energy. For example, mechanical energy is conserved only when the internal forces are *conservative*—that is, when the forces allow two-way conversion between kinetic and potential energies. But conservation of momentum is valid even when the internal forces are *not* conservative. In this chapter we will analyze situations in which both momentum and mechanical energy are conserved, and others in which only momentum is conserved. These two principles play a fundamental role in all areas of physics, and we will encounter them throughout our study of physics.

8.11 When applying conservation of momentum, remember that momentum is a vector quantity!

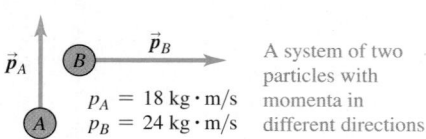

A system of two particles with momenta in different directions

$p_A = 18 \text{ kg} \cdot \text{m/s}$
$p_B = 24 \text{ kg} \cdot \text{m/s}$

You CANNOT find the magnitude of the total momentum by adding the magnitudes of the individual momenta!

$P = p_A + p_B = 42 \text{ kg} \cdot \text{m/s}$ ◄ **WRONG**

Instead, use vector addition:

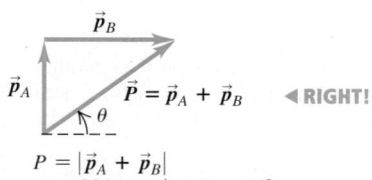

$\vec{P} = \vec{p}_A + \vec{p}_B$ ◄ **RIGHT!**

$P = |\vec{p}_A + \vec{p}_B|$
$= 30 \text{ kg} \cdot \text{m/s at } \theta = 37°$

PROBLEM-SOLVING STRATEGY 8.1 | **CONSERVATION OF MOMENTUM**

IDENTIFY *the relevant concepts:* Confirm that the vector sum of the external forces acting on the system of particles is zero. If it isn't zero, you can't use conservation of momentum.

SET UP *the problem* using the following steps:
1. Treat each body as a particle. Draw "before" and "after" sketches, including velocity vectors. Assign algebraic symbols to each magnitude, angle, and component. Use letters to label each particle and subscripts 1 and 2 for "before" and "after" quantities. Include any given values.
2. Define a coordinate system and show it in your sketches; define the positive direction for each axis.
3. Identify the target variables.

EXECUTE *the solution:*
4. Write an equation in symbols equating the total initial and final x-components of momentum, using $p_x = mv_x$ for each particle. Write a corresponding equation for the y-components. Components can be positive or negative, so be careful with signs!
5. In some problems, energy considerations (discussed in Section 8.4) give additional equations relating the velocities.
6. Solve your equations to find the target variables.

EVALUATE *your answer:* Does your answer make physical sense? If your target variable is a certain body's momentum, check that the direction of the momentum is reasonable.

EXAMPLE 8.4   RECOIL OF A RIFLE

A marksman holds a rifle of mass $m_R = 3.00$ kg loosely, so it can recoil freely. He fires a bullet of mass $m_B = 5.00$ g horizontally with a velocity relative to the ground of $v_{Bx} = 300$ m/s. What is the recoil velocity v_{Rx} of the rifle? What are the final momentum and kinetic energy of the bullet and rifle?

SOLUTION

IDENTIFY and SET UP: If the marksman exerts negligible horizontal forces on the rifle, then there is no net horizontal force on the system (the bullet and rifle) during the firing, and the total horizontal momentum of the system is conserved. **Figure 8.12** shows our sketch. We take the positive x-axis in the direction of aim. The rifle and the bullet are initially at rest, so the initial x-component of total momentum is zero. After the shot is fired, the bullet's x-momentum is $p_{Bx} = m_B v_{Bx}$ and the rifle's x-momentum is $p_{Rx} = m_R v_{Rx}$. Our target variables are v_{Rx}, p_{Bx}, p_{Rx}, and the final kinetic energies $K_B = \frac{1}{2} m_B v_{Bx}^2$ and $K_R = \frac{1}{2} m_R v_{Rx}^2$.

8.12 Our sketch for this problem.

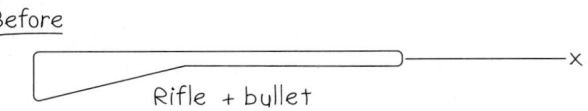

Before

Rifle + bullet

After
$v_{Rx} = ?$

$v_{Bx} = 300$ m/s

$m_R = 3.00$ kg

$m_B = 5.00$ g

EXECUTE: Conservation of the x-component of total momentum gives

$$P_x = 0 = m_B v_{Bx} + m_R v_{Rx}$$

$$v_{Rx} = -\frac{m_B}{m_R} v_{Bx} = -\left(\frac{0.00500 \text{ kg}}{3.00 \text{ kg}}\right)(300 \text{ m/s}) = -0.500 \text{ m/s}$$

The negative sign means that the recoil is in the direction opposite to that of the bullet.

The final momenta and kinetic energies are

$$p_{Bx} = m_B v_{Bx} = (0.00500 \text{ kg})(300 \text{ m/s}) = 1.50 \text{ kg} \cdot \text{m/s}$$

$$K_B = \tfrac{1}{2} m_B v_{Bx}^2 = \tfrac{1}{2}(0.00500 \text{ kg})(300 \text{ m/s})^2 = 225 \text{ J}$$

$$p_{Rx} = m_R v_{Rx} = (3.00 \text{ kg})(-0.500 \text{ m/s}) = -1.50 \text{ kg} \cdot \text{m/s}$$

$$K_R = \tfrac{1}{2} m_R v_{Rx}^2 = \tfrac{1}{2}(3.00 \text{ kg})(-0.500 \text{ m/s})^2 = 0.375 \text{ J}$$

EVALUATE: The bullet and rifle have equal and opposite final *momenta* thanks to Newton's third law: They experience equal and opposite interaction forces that act for the same *time,* so the impulses are equal and opposite. But the bullet travels a much greater *distance* than the rifle during the interaction. Hence the force on the bullet does more work than the force on the rifle, giving the bullet much greater *kinetic energy* than the rifle. The 600:1 ratio of the two kinetic energies is the inverse of the ratio of the masses; in fact, you can show that this always happens in recoil situations (see Exercise 8.26).

EXAMPLE 8.5   COLLISION ALONG A STRAIGHT LINE

Two gliders with different masses move toward each other on a frictionless air track (**Fig. 8.13a**). After they collide (Fig. 8.13b), glider B has a final velocity of $+2.0$ m/s (Fig. 8.13c). What is the final velocity of glider A? How do the changes in momentum and in velocity compare?

SOLUTION

IDENTIFY and SET UP: As for the skaters in Fig. 8.10, the total vertical force on each glider is zero, and the net force on each individual glider is the horizontal force exerted on it by the other glider. The net external force on the *system* of two gliders is zero, so their total momentum is conserved. We take the positive x-axis to be to the right. We are given the masses and initial velocities of both gliders and the final velocity of glider B. Our target variables are v_{A2x} (the final x-component of velocity of glider A), and the changes in momentum and in velocity of the two gliders (the value *after* the collision minus the value *before* the collision).

8.13 Two gliders colliding on an air track.

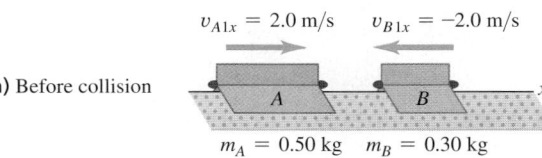

(a) Before collision

$v_{A1x} = 2.0$ m/s $v_{B1x} = -2.0$ m/s

$m_A = 0.50$ kg $m_B = 0.30$ kg

(b) Collision

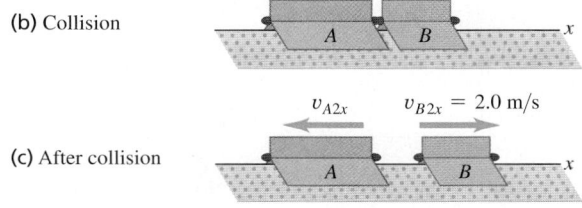

(c) After collision

v_{A2x} $v_{B2x} = 2.0$ m/s

Continued

EXECUTE: The x-component of total momentum before the collision is

$$P_x = m_A v_{A1x} + m_B v_{B1x}$$
$$= (0.50 \text{ kg})(2.0 \text{ m/s}) + (0.30 \text{ kg})(-2.0 \text{ m/s})$$
$$= 0.40 \text{ kg} \cdot \text{m/s}$$

This is positive (to the right in Fig. 8.13) because A has a greater magnitude of momentum than B. The x-component of total momentum has the same value after the collision, so

$$P_x = m_A v_{A2x} + m_B v_{B2x}$$

We solve for v_{A2x}:

$$v_{A2x} = \frac{P_x - m_B v_{B2x}}{m_A} = \frac{0.40 \text{ kg} \cdot \text{m/s} - (0.30 \text{ kg})(2.0 \text{ m/s})}{0.50 \text{ kg}}$$
$$= -0.40 \text{ m/s}$$

The changes in the x-momenta are

$$m_A v_{A2x} - m_A v_{A1x} = (0.50 \text{ kg})(-0.40 \text{ m/s})$$
$$- (0.50 \text{ kg})(2.0 \text{ m/s})$$
$$= -1.2 \text{ kg} \cdot \text{m/s}$$

$$m_B v_{B2x} - m_B v_{B1x} = (0.30 \text{ kg})(2.0 \text{ m/s})$$
$$- (0.30 \text{ kg})(-2.0 \text{ m/s})$$
$$= +1.2 \text{ kg} \cdot \text{m/s}$$

The changes in x-velocities are

$$v_{A2x} - v_{A1x} = (-0.40 \text{ m/s}) - 2.0 \text{ m/s} = -2.4 \text{ m/s}$$
$$v_{B2x} - v_{B1x} = 2.0 \text{ m/s} - (-2.0 \text{ m/s}) = +4.0 \text{ m/s}$$

EVALUATE: The gliders were subjected to equal and opposite interaction forces for the same time during their collision. By the impulse–momentum theorem, they experienced equal and opposite impulses and therefore equal and opposite changes in momentum. But by Newton's second law, the less massive glider (B) had a greater magnitude of acceleration and hence a greater velocity change.

EXAMPLE 8.6   COLLISION IN A HORIZONTAL PLANE

Figure 8.14a shows two battling robots on a frictionless surface. Robot A, with mass 20 kg, initially moves at 2.0 m/s parallel to the x-axis. It collides with robot B, which has mass 12 kg and is initially at rest. After the collision, robot A moves at 1.0 m/s in a direction that makes an angle $\alpha = 30°$ with its initial direction (Fig. 8.14b). What is the final velocity of robot B?

SOLUTION

IDENTIFY and SET UP: There are no horizontal external forces, so the x- and y-components of the total momentum of the system are conserved. Hence the sum of the x-components of momentum before the collision (subscript 1) must equal the sum after the collision (subscript 2), and similarly for the sums of the y-components. Our target variable is $\vec{v}_{B2}$, the final velocity of robot B.

EXECUTE: The momentum-conservation equations and their solutions for v_{B2x} and v_{B2y} are

$$m_A v_{A1x} + m_B v_{B1x} = m_A v_{A2x} + m_B v_{B2x}$$

$$v_{B2x} = \frac{m_A v_{A1x} + m_B v_{B1x} - m_A v_{A2x}}{m_B}$$

$$= \frac{\left[\begin{array}{l}(20 \text{ kg})(2.0 \text{ m/s}) + (12 \text{ kg})(0) \\ - (20 \text{ kg})(1.0 \text{ m/s})(\cos 30°)\end{array}\right]}{12 \text{ kg}}$$

$$= 1.89 \text{ m/s}$$

$$m_A v_{A1y} + m_B v_{B1y} = m_A v_{A2y} + m_B v_{B2y}$$

$$v_{B2y} = \frac{m_A v_{A1y} + m_B v_{B1y} - m_A v_{A2y}}{m_B}$$

$$= \frac{\left[\begin{array}{l}(20 \text{ kg})(0) + (12 \text{ kg})(0) \\ - (20 \text{ kg})(1.0 \text{ m/s})(\sin 30°)\end{array}\right]}{12 \text{ kg}}$$

$$= -0.83 \text{ m/s}$$

Figure 8.14b shows the motion of robot B after the collision. The magnitude of $\vec{v}_{B2}$ is

$$v_{B2} = \sqrt{(1.89 \text{ m/s})^2 + (-0.83 \text{ m/s})^2} = 2.1 \text{ m/s}$$

and the angle of its direction from the positive x-axis is

$$\beta = \arctan \frac{-0.83 \text{ m/s}}{1.89 \text{ m/s}} = -24°$$

8.14 Views from above of the robot velocities.

(a) Before collision

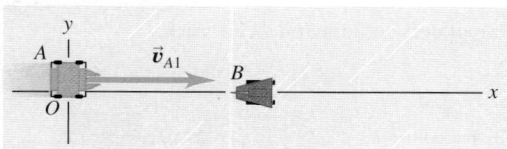

(b) After collision

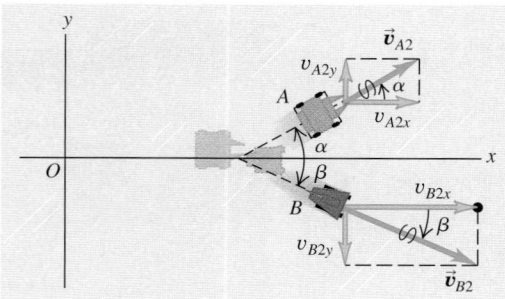

EVALUATE: Let's confirm that the components of total momentum before and after the collision are equal. Initially robot A has x-momentum $m_A v_{A1x} = (20\ \text{kg})(2.0\ \text{m/s}) = 40\ \text{kg}\cdot\text{m/s}$ and zero y-momentum; robot B has zero momentum. Afterward, the momentum components are $m_A v_{A2x} = (20\ \text{kg})(1.0\ \text{m/s})(\cos 30°) =$ 17 kg·m/s and $m_B v_{B2x} = (12\ \text{kg})(1.89\ \text{m/s}) = 23\ \text{kg}\cdot\text{m/s}$; the total x-momentum is 40 kg·m/s, the same as before the collision. The final y-components are $m_A v_{A2y} = (20\ \text{kg})(1.0\ \text{m/s})(\sin 30°) =$ 10 kg·m/s and $m_B v_{B2y} = (12\ \text{kg})(-0.83\ \text{m/s}) = -10\ \text{kg}\cdot\text{m/s}$; the total y-component of momentum is zero, as before the collision.

TEST YOUR UNDERSTANDING OF SECTION 8.2 A spring-loaded toy sits at rest on a horizontal, frictionless surface. When the spring releases, the toy breaks into equal-mass pieces A, B, and C, which slide along the surface. Piece A moves off in the negative x-direction, while piece B moves off in the negative y-direction. (a) What are the signs of the velocity components of piece C? (b) Which of the three pieces is moving the fastest? ❙

8.3 MOMENTUM CONSERVATION AND COLLISIONS

To most people the term *collision* is likely to mean some sort of automotive disaster. We'll broaden the meaning to include any strong interaction between bodies that lasts a relatively short time. So we include not only car accidents but also balls colliding on a billiard table, neutrons hitting atomic nuclei in a nuclear reactor, and a close encounter of a spacecraft with the planet Saturn.

If the forces between the colliding bodies are much larger than any external forces, as is the case in most collisions, we can ignore the external forces and treat the bodies as an *isolated* system. Then momentum is conserved and the total momentum of the system has the same value before and after the collision. Two cars colliding at an icy intersection provide a good example. Even two cars colliding on dry pavement can be treated as an isolated system during the collision if the forces between the cars are much larger than the friction forces of pavement against tires.

Elastic and Inelastic Collisions

If the forces between the bodies are also *conservative,* so no mechanical energy is lost or gained in the collision, the total *kinetic* energy of the system is the same after the collision as before. Such a collision is called an **elastic collision.** A collision between two marbles or two billiard balls is almost completely elastic. **Figure 8.15** shows a model for an elastic collision. When the gliders collide, their springs are momentarily compressed and some of the original kinetic energy is momentarily converted to elastic potential energy. Then the gliders bounce apart, the springs expand, and this potential energy is converted back to kinetic energy.

A collision in which the total kinetic energy after the collision is *less* than before the collision is called an **inelastic collision.** A meatball landing on a plate of spaghetti and a bullet embedding itself in a block of wood are examples of inelastic collisions. An inelastic collision in which the colliding bodies stick together and move as one body after the collision is called a **completely inelastic collision.** **Figure 8.16** shows an example; we have replaced the spring bumpers in Fig. 8.15 with Velcro®, which sticks the two bodies together.

8.15 Two gliders undergoing an elastic collision on a frictionless surface. Each glider has a steel spring bumper that exerts a conservative force on the other glider.

(a) Before collision

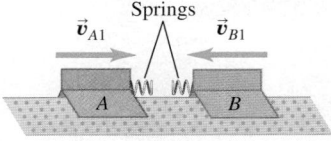

Springs

$\vec{v}_{A1}$ $\vec{v}_{B1}$

(b) Elastic collision

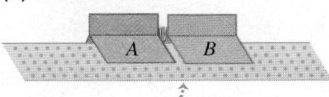

Kinetic energy is stored as potential energy in compressed springs.

(c) After collision

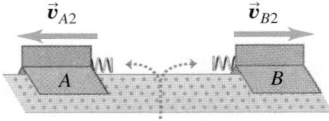

$\vec{v}_{A2}$ $\vec{v}_{B2}$

The system of the two gliders has the same kinetic energy after the collision as before it.

DEMO

8.16 Two gliders undergoing a completely inelastic collision. The spring bumpers on the gliders are replaced by Velcro®, so the gliders stick together after collision.

(a) Before collision

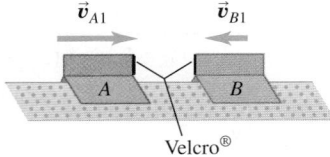

$\vec{v}_{A1}$ $\vec{v}_{B1}$

Velcro®

(b) Completely inelastic collision

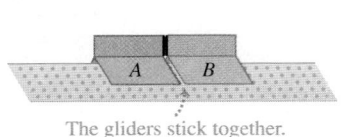

The gliders stick together.

(c) After collision

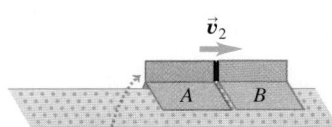

$\vec{v}_2$

The system of the two gliders has less kinetic energy after the collision than before it.

CAUTION An inelastic collision doesn't have to be *completely* inelastic Inelastic collisions include many situations in which the bodies do *not* stick. If two cars bounce off each other in a "fender bender," the work done to deform the fenders cannot be recovered as kinetic energy of the cars, so the collision is inelastic (**Fig. 8.17**). ▮

8.17 Cars are designed so that collisions are inelastic—the structure of the car absorbs as much of the energy of the collision as possible. This absorbed energy cannot be recovered, since it goes into a permanent deformation of the car.

Remember this rule: **In any collision in which external forces can be ignored, momentum is conserved and the total momentum before equals the total momentum after; in elastic collisions *only*, the total kinetic energy before equals the total kinetic energy after.**

Completely Inelastic Collisions

Let's look at what happens to momentum and kinetic energy in a *completely* inelastic collision of two bodies (A and B), as in Fig. 8.16. Because the two bodies stick together after the collision, they have the same final velocity $\vec{v}_2$:

$$\vec{v}_{A2} = \vec{v}_{B2} = \vec{v}_2$$

Conservation of momentum gives the relationship

$$m_A\vec{v}_{A1} + m_B\vec{v}_{B1} = (m_A + m_B)\vec{v}_2 \quad \text{(completely inelastic collision)} \quad (8.16)$$

If we know the masses and initial velocities, we can compute the common final velocity $\vec{v}_2$.

Suppose, for example, that a body with mass m_A and initial x-component of velocity v_{A1x} collides inelastically with a body with mass m_B that is initially at rest ($v_{B1x} = 0$). From Eq. (8.16) the common x-component of velocity v_{2x} of both bodies after the collision is

$$v_{2x} = \frac{m_A}{m_A + m_B}v_{A1x} \quad \begin{array}{l}\text{(completely inelastic collision,} \\ \text{B initially at rest)}\end{array} \quad (8.17)$$

Let's verify that the total kinetic energy after this completely inelastic collision is less than before the collision. The motion is purely along the x-axis, so the kinetic energies K_1 and K_2 before and after the collision, respectively, are

$$K_1 = \tfrac{1}{2}m_A v_{A1x}{}^2$$

$$K_2 = \tfrac{1}{2}(m_A + m_B)v_{2x}{}^2 = \tfrac{1}{2}(m_A + m_B)\left(\frac{m_A}{m_A + m_B}\right)^2 v_{A1x}{}^2$$

The ratio of final to initial kinetic energy is

$$\frac{K_2}{K_1} = \frac{m_A}{m_A + m_B} \quad \begin{array}{l}\text{(completely inelastic collision,} \\ \text{B initially at rest)}\end{array} \quad (8.18)$$

The right side is always less than unity because the denominator is always greater than the numerator. Even when the initial velocity of m_B is not zero, the kinetic energy after a completely inelastic collision is always less than before.

Please note: Don't memorize Eq. (8.17) or (8.18)! We derived them only to prove that kinetic energy is always lost in a completely inelastic collision.

EXAMPLE 8.7 **A COMPLETELY INELASTIC COLLISION**

We repeat the collision described in Example 8.5 (Section 8.2), but this time equip the gliders so that they stick together when they collide. Find the common final x-velocity, and compare the initial and final kinetic energies of the system.

8.18 Our sketch for this problem.

Before
$v_{A1x} = 2.0$ m/s $v_{B1x} = -2.0$ m/s
$A \longrightarrow$ $\longleftarrow B$
$m_A = 0.50$ kg $m_B = 0.30$ kg

After
$A\,B \longrightarrow$ $v_{2x} = ?$

SOLUTION

IDENTIFY and SET UP: There are no external forces in the x-direction, so the x-component of momentum is conserved. **Figure 8.18** shows our sketch. Our target variables are the final x-velocity, v_{2x}, and the initial and final kinetic energies, K_1 and K_2.

SOLUTION

EXECUTE: From conservation of momentum,

$$m_A v_{A1x} + m_B v_{B1x} = (m_A + m_B)v_{2x}$$

$$v_{2x} = \frac{m_A v_{A1x} + m_B v_{B1x}}{m_A + m_B}$$

$$= \frac{(0.50 \text{ kg})(2.0 \text{ m/s}) + (0.30 \text{ kg})(-2.0 \text{ m/s})}{0.50 \text{ kg} + 0.30 \text{ kg}}$$

$$= 0.50 \text{ m/s}$$

Because v_{2x} is positive, the gliders move together to the right after the collision. Before the collision, the kinetic energies are

$$K_A = \tfrac{1}{2}m_A v_{A1x}^2 = \tfrac{1}{2}(0.50 \text{ kg})(2.0 \text{ m/s})^2 = 1.0 \text{ J}$$

$$K_B = \tfrac{1}{2}m_B v_{B1x}^2 = \tfrac{1}{2}(0.30 \text{ kg})(-2.0 \text{ m/s})^2 = 0.60 \text{ J}$$

The total kinetic energy before the collision is $K_1 = K_A + K_B = 1.6 \text{ J}$. The kinetic energy after the collision is

$$K_2 = \tfrac{1}{2}(m_A + m_B)v_{2x}^2 = \tfrac{1}{2}(0.50 \text{ kg} + 0.30 \text{ kg})(0.50 \text{ m/s})^2$$

$$= 0.10 \text{ J}$$

EVALUATE: The final kinetic energy is only $\frac{1}{16}$ of the original; $\frac{15}{16}$ is converted from mechanical energy to other forms. If there is a wad of chewing gum between the gliders, it squashes and becomes warmer. If there is a spring between the gliders that is compressed as they lock together, the energy is stored as potential energy of the spring. In both cases the *total* energy of the system is conserved, although *kinetic* energy is not. In an isolated system, however, momentum is *always* conserved whether the collision is elastic or not.

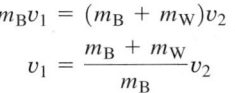

EXAMPLE 8.8 THE BALLISTIC PENDULUM

Figure 8.19 shows a ballistic pendulum, a simple system for measuring the speed of a bullet. A bullet of mass m_B makes a completely inelastic collision with a block of wood of mass m_W, which is suspended like a pendulum. After the impact, the block swings up to a maximum height h. In terms of h, m_B, and m_W, what is the initial speed v_1 of the bullet?

SOLUTION

IDENTIFY: We'll analyze this event in two stages: (1) the bullet embeds itself in the block, and (2) the block swings upward. The first stage happens so quickly that the block does not move appreciably. The supporting strings remain nearly vertical, so negligible external horizontal force acts on the bullet–block system, and the horizontal component of momentum is conserved. Mechanical energy is *not* conserved during this stage, however, because a nonconservative force does work (the force of friction between bullet and block).

In the second stage, the block and bullet move together. The only forces acting on this system are gravity (a conservative force) and the string tensions (which do no work). Thus, as the block swings,

mechanical energy is conserved. Momentum is *not* conserved during this stage, however, because there is a net external force (the forces of gravity and string tension don't cancel when the strings are inclined).

SET UP: We take the positive x-axis to the right and the positive y-axis upward. Our target variable is v_1. Another unknown quantity is the speed v_2 of the system just after the collision. We'll use momentum conservation in the first stage to relate v_1 to v_2, and we'll use energy conservation in the second stage to relate v_2 to h.

EXECUTE: In the first stage, all velocities are in the +x-direction. Momentum conservation gives

$$m_B v_1 = (m_B + m_W)v_2$$

$$v_1 = \frac{m_B + m_W}{m_B}v_2$$

At the beginning of the second stage, the system has kinetic energy $K = \tfrac{1}{2}(m_B + m_W)v_2^2$. The system swings up and comes to rest for an instant at a height h, where its kinetic energy is zero and the potential energy is $(m_B + m_W)gh$; it then swings back down. Energy conservation gives

$$\tfrac{1}{2}(m_B + m_W)v_2^2 = (m_B + m_W)gh$$

$$v_2 = \sqrt{2gh}$$

We substitute this expression for v_2 into the momentum equation:

$$v_1 = \frac{m_B + m_W}{m_B}\sqrt{2gh}$$

EVALUATE: Let's plug in the realistic numbers $m_B = 5.00 \text{ g} = 0.00500 \text{ kg}$, $m_W = 2.00 \text{ kg}$, and $h = 3.00 \text{ cm} = 0.0300 \text{ m}$:

$$v_1 = \frac{0.00500 \text{ kg} + 2.00 \text{ kg}}{0.00500 \text{ kg}}\sqrt{2(9.80 \text{ m/s}^2)(0.0300 \text{ m})}$$

$$= 307 \text{ m/s}$$

$$v_2 = \sqrt{2gh} = \sqrt{2(9.80 \text{ m/s}^2)(0.0300 \text{ m})} = 0.767 \text{ m/s}$$

The speed v_2 of the block after impact is *much* lower than the initial speed v_1 of the bullet. The kinetic energy of the bullet before impact is $\tfrac{1}{2}(0.00500 \text{ kg})(307 \text{ m/s})^2 = 236 \text{ J}$. Just after impact the kinetic energy of the system is $\tfrac{1}{2}(2.005 \text{ kg})(0.767 \text{ m/s})^2 = 0.590 \text{ J}$. Nearly all the kinetic energy disappears as the wood splinters and the bullet and block become warmer.

8.19 A ballistic pendulum.

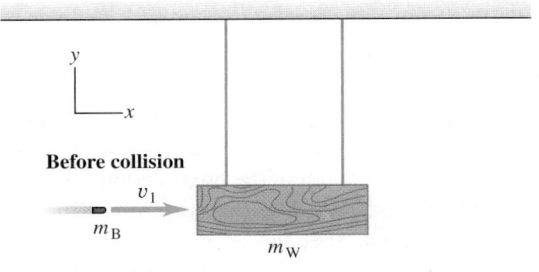

Before collision

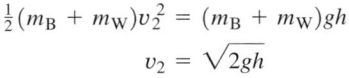

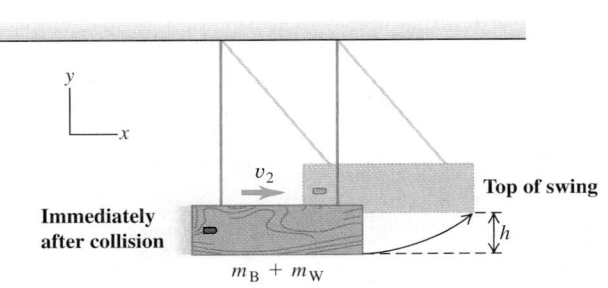

Immediately after collision

Top of swing

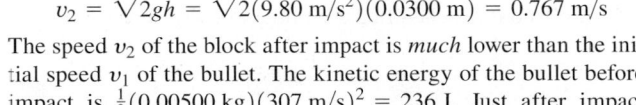

EXAMPLE 8.9 AN AUTOMOBILE COLLISION

A 1000-kg car traveling north at 15 m/s collides with a 2000-kg truck traveling east at 10 m/s. The occupants, wearing seat belts, are uninjured, but the two vehicles move away from the impact point as one. The insurance adjustor asks you to find the velocity of the wreckage just after impact. What is your answer?

SOLUTION

IDENTIFY and SET UP: Any horizontal external forces (such as friction) on the vehicles during the collision are very small compared with the forces that the colliding vehicles exert on each other. (We'll verify this below.) So we can treat the cars as an isolated system, and the momentum of the system is conserved. **Figure 8.20** shows our sketch and the x- and y-axes. We can use Eqs. (8.15) to find the total momentum $\vec{P}$ before the collision. The momentum has the same value just after the collision; hence we can find the velocity $\vec{V}$ just after the collision (our target variable) by using $\vec{P} = M\vec{V}$, where $M = m_C + m_T = 3000$ kg is the mass of the wreckage.

EXECUTE: From Eqs. (8.15), the components of $\vec{P}$ are

$$P_x = p_{Cx} + p_{Tx} = m_C v_{Cx} + m_T v_{Tx}$$
$$= (1000 \text{ kg})(0) + (2000 \text{ kg})(10 \text{ m/s})$$
$$= 2.0 \times 10^4 \text{ kg} \cdot \text{m/s}$$
$$P_y = p_{Cy} + p_{Ty} = m_C v_{Cy} + m_T v_{Ty}$$
$$= (1000 \text{ kg})(15 \text{ m/s}) + (2000 \text{ kg})(0)$$
$$= 1.5 \times 10^4 \text{ kg} \cdot \text{m/s}$$

The magnitude of $\vec{P}$ is

$$P = \sqrt{(2.0 \times 10^4 \text{ kg} \cdot \text{m/s})^2 + (1.5 \times 10^4 \text{ kg} \cdot \text{m/s})^2}$$
$$= 2.5 \times 10^4 \text{ kg} \cdot \text{m/s}$$

and its direction is given by the angle θ shown in Fig. 8.20:

$$\tan\theta = \frac{P_y}{P_x} = \frac{1.5 \times 10^4 \text{ kg} \cdot \text{m/s}}{2.0 \times 10^4 \text{ kg} \cdot \text{m/s}} = 0.75 \quad \theta = 37°$$

8.20 Our sketch for this problem.

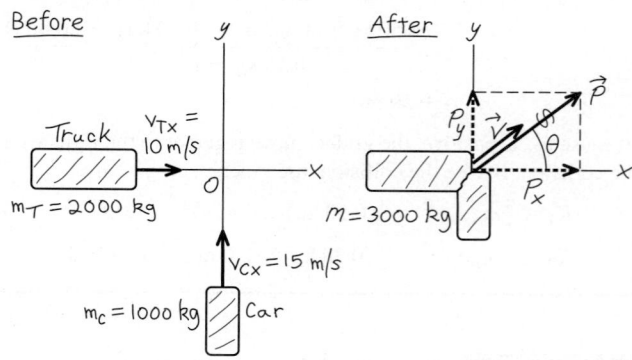

From $\vec{P} = M\vec{V}$, the direction of the velocity $\vec{V}$ just after the collision is also $\theta = 37°$. The velocity magnitude is

$$V = \frac{P}{M} = \frac{2.5 \times 10^4 \text{ kg} \cdot \text{m/s}}{3000 \text{ kg}} = 8.3 \text{ m/s}$$

EVALUATE: As you can show, the initial kinetic energy is 2.1×10^5 J and the final value is 1.0×10^5 J. In this inelastic collision, the total kinetic energy is less after the collision than before.

We can now justify our neglect of the external forces on the vehicles during the collision. The car's weight is about 10,000 N; if the coefficient of kinetic friction is 0.5, the friction force on the car during the impact is about 5000 N. The car's initial kinetic energy is $\frac{1}{2}(1000 \text{ kg})(15 \text{ m/s})^2 = 1.1 \times 10^5$ J, so -1.1×10^5 J of work must be done to stop it. If the car crumples by 0.20 m in stopping, a force of magnitude $(1.1 \times 10^5 \text{ J})/(0.20 \text{ m}) = 5.5 \times 10^5$ N would be needed; that's 110 times the friction force. So it's reasonable to treat the external force of friction as negligible compared with the internal forces the vehicles exert on each other.

Classifying Collisions

It's important to remember that we can classify collisions according to energy considerations (**Fig. 8.21**). A collision in which kinetic energy is conserved is called *elastic*. (We'll explore this type in more depth in the next section.) A collision in which the total kinetic energy decreases is called *inelastic*. When the two bodies have a common final velocity, we say that the collision is *completely inelastic*. There are also cases in which the final kinetic energy is *greater* than the initial value. Rifle recoil, discussed in Example 8.4 (Section 8.2), is an example.

8.21 Collisions are classified according to energy considerations.

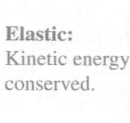

Elastic:
Kinetic energy conserved.

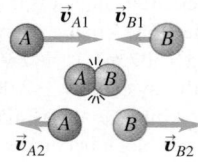

Inelastic:
Some kinetic energy lost.

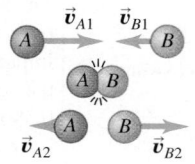

Completely inelastic:
Bodies have same final velocity.

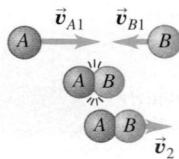

Finally, we emphasize again that we can sometimes use momentum conservation even when external forces are acting on the system, if the net external force acting on the colliding bodies is small in comparison with the internal forces during the collision (as in Example 8.9).

8.4 ELASTIC COLLISIONS

We saw in Section 8.3 that an *elastic collision* in an isolated system is one in which kinetic energy (as well as momentum) is conserved. Elastic collisions occur when the forces between the colliding bodies are *conservative*. When two billiard balls collide, they squash a little near the surface of contact, but then they spring back. Some of the kinetic energy is stored temporarily as elastic potential energy, but at the end it is reconverted to kinetic energy (**Fig. 8.22**).

Let's look at a *one-dimensional* elastic collision between two bodies A and B, in which all the velocities lie along the same line. We call this line the x-axis, so each momentum and velocity has only an x-component. We call the x-velocities before the collision v_{A1x} and v_{B1x}, and those after the collision v_{A2x} and v_{B2x}. From conservation of kinetic energy we have

$$\tfrac{1}{2}m_A v_{A1x}^2 + \tfrac{1}{2}m_B v_{B1x}^2 = \tfrac{1}{2}m_A v_{A2x}^2 + \tfrac{1}{2}m_B v_{B2x}^2$$

and conservation of momentum gives

$$m_A v_{A1x} + m_B v_{B1x} = m_A v_{A2x} + m_B v_{B2x}$$

If the masses m_A and m_B and the initial velocities v_{A1x} and v_{B1x} are known, we can solve these two equations to find the two final velocities v_{A2x} and v_{B2x}.

Elastic Collisions, One Body Initially at Rest

The general solution to the above equations is a little complicated, so we will concentrate on the particular case in which body B is at rest before the collision (so $v_{B1x} = 0$). Think of body B as a target for body A to hit. Then the kinetic energy and momentum conservation equations are, respectively,

$$\tfrac{1}{2}m_A v_{A1x}^2 = \tfrac{1}{2}m_A v_{A2x}^2 + \tfrac{1}{2}m_B v_{B2x}^2 \qquad (8.19)$$

$$m_A v_{A1x} = m_A v_{A2x} + m_B v_{B2x} \qquad (8.20)$$

We can solve for v_{A2x} and v_{B2x} in terms of the masses and the initial velocity v_{A1x}. This involves some fairly strenuous algebra, but it's worth it. No pain, no gain! The simplest approach is somewhat indirect, but along the way it uncovers an additional interesting feature of elastic collisions.

First we rearrange Eqs. (8.19) and (8.20) as follows:

$$m_B v_{B2x}^2 = m_A(v_{A1x}^2 - v_{A2x}^2) = m_A(v_{A1x} - v_{A2x})(v_{A1x} + v_{A2x}) \qquad (8.21)$$

$$m_B v_{B2x} = m_A(v_{A1x} - v_{A2x}) \qquad (8.22)$$

Now we divide Eq. (8.21) by Eq. (8.22) to obtain

$$v_{B2x} = v_{A1x} + v_{A2x} \qquad (8.23)$$

8.22 Billiard balls deform very little when they collide, and they quickly spring back from any deformation they do undergo. Hence the force of interaction between the balls is almost perfectly conservative, and the collision is almost perfectly elastic.

DATA *SPEAKS*

Conservation of Momentum

When students were given a problem about conservation of momentum, more than 29% gave an incorrect response. Common errors:

- Forgetting that momentum $\vec{p}$ is a vector. Its components can be positive or negative depending on the direction of $\vec{p}$.

- Adding momenta incorrectly. If two momentum vectors point in different directions, you cannot find the total momentum by simply adding the magnitudes of the two momenta.

8.23 One-dimensional elastic collisions between bodies with different masses.

(a) Moving Ping-Pong ball strikes initially stationary bowling ball.

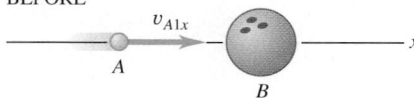

BEFORE

AFTER

(b) Moving bowling ball strikes initially stationary Ping-Pong ball.

BEFORE

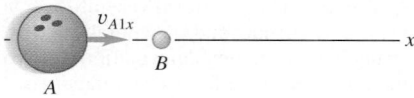

AFTER

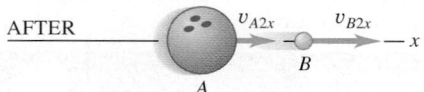

8.24 A one-dimensional elastic collision between bodies of equal mass.

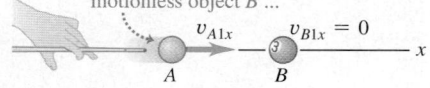

When a moving object A has a 1-D elastic collision with an equal-mass, motionless object B ...

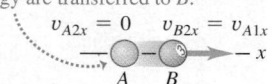

... all of A's momentum and kinetic energy are transferred to B.

We substitute this expression back into Eq. (8.22) to eliminate v_{B2x} and then solve for v_{A2x}:

$$m_B(v_{A1x} + v_{A2x}) = m_A(v_{A1x} - v_{A2x})$$

$$v_{A2x} = \frac{m_A - m_B}{m_A + m_B} v_{A1x} \tag{8.24}$$

Finally, we substitute this result back into Eq. (8.23) to obtain

$$v_{B2x} = \frac{2m_A}{m_A + m_B} v_{A1x} \tag{8.25}$$

Now we can interpret the results. Suppose A is a Ping-Pong ball and B is a bowling ball. Then we expect A to bounce off after the collision with a velocity nearly equal to its original value but in the opposite direction (**Fig. 8.23a**), and we expect B's velocity to be much less. That's just what the equations predict. When m_A is much smaller than m_B, the fraction in Eq. (8.24) is approximately equal to (-1), so v_{A2x} is approximately equal to $-v_{A1x}$. The fraction in Eq. (8.25) is much smaller than unity, so v_{B2x} is much less than v_{A1x}. Figure 8.23b shows the opposite case, in which A is the bowling ball and B the Ping-Pong ball and m_A is much larger than m_B. What do you expect to happen then? Check your predictions against Eqs. (8.24) and (8.25).

Another interesting case occurs when the masses are equal (**Fig. 8.24**). If $m_A = m_B$, then Eqs. (8.24) and (8.25) give $v_{A2x} = 0$ and $v_{B2x} = v_{A1x}$. That is, the body that was moving stops dead; it gives all its momentum and kinetic energy to the body that was at rest. This behavior is familiar to all pool players.

Elastic Collisions and Relative Velocity

Let's return to the more general case in which A and B have different masses. Equation (8.23) can be rewritten as

$$v_{A1x} = v_{B2x} - v_{A2x} \tag{8.26}$$

Here $v_{B2x} - v_{A2x}$ is the velocity of B relative to A *after* the collision; from Eq. (8.26), this equals v_{A1x}, which is the *negative* of the velocity of B relative to A *before* the collision. (We discussed relative velocity in Section 3.5.) The relative velocity has the same magnitude, but opposite sign, before and after the collision. The sign changes because A and B are approaching each other before the collision but moving apart after the collision. If we view this collision from a second coordinate system moving with constant velocity relative to the first, the velocities of the bodies are different but the *relative* velocities are the same. Hence our statement about relative velocities holds for *any* straight-line elastic collision, even when neither body is at rest initially. *In a straight-line elastic collision of two bodies, the relative velocities before and after the collision have the same magnitude but opposite sign.* This means that if B is moving before the collision, Eq. (8.26) becomes

$$v_{B2x} - v_{A2x} = -(v_{B1x} - v_{A1x}) \tag{8.27}$$

It turns out that a *vector* relationship similar to Eq. (8.27) is a general property of *all* elastic collisions, even when both bodies are moving initially and the velocities do not all lie along the same line. This result provides an alternative and equivalent definition of an elastic collision: *In an elastic collision, the relative velocity of the two bodies has the same magnitude before and after the collision.* Whenever this condition is satisfied, the total kinetic energy is also conserved.

When an elastic two-body collision isn't head-on, the velocities don't all lie along a single line. If they all lie in a plane, then each final velocity has two unknown components, and there are four unknowns in all. Conservation of energy and conservation of the x- and y-components of momentum give only three equations. To determine the final velocities uniquely, we need additional information, such as the direction or magnitude of one of the final velocities.

EXAMPLE 8.10 AN ELASTIC STRAIGHT-LINE COLLISION

We repeat the air-track collision of Example 8.5 (Section 8.2), but now we add ideal spring bumpers to the gliders so that the collision is elastic. What are the final velocities of the gliders?

SOLUTION

IDENTIFY and SET UP: The net external force on the system is zero, so the momentum of the system is conserved. **Figure 8.25** shows our sketch. We'll find our target variables, v_{A2x} and v_{B2x}, by using Eq. (8.27), the relative-velocity relationship for an elastic collision, and the momentum-conservation equation.

EXECUTE: From Eq. (8.27),

$$v_{B2x} - v_{A2x} = -(v_{B1x} - v_{A1x})$$
$$= -(-2.0 \text{ m/s} - 2.0 \text{ m/s}) = 4.0 \text{ m/s}$$

From conservation of momentum,

$$m_A v_{A1x} + m_B v_{B1x} = m_A v_{A2x} + m_B v_{B2x}$$
$$(0.50 \text{ kg})(2.0 \text{ m/s}) + (0.30 \text{ kg})(-2.0 \text{ m/s})$$
$$= (0.50 \text{ kg})v_{A2x} + (0.30 \text{ kg})v_{B2x}$$
$$0.50 v_{A2x} + 0.30 v_{B2x} = 0.40 \text{ m/s}$$

(To get the last equation we divided both sides of the equation just above it by 1 kg. This makes the units the same as in the first equation.) Solving these equations simultaneously, we find

$$v_{A2x} = -1.0 \text{ m/s} \qquad v_{B2x} = 3.0 \text{ m/s}$$

8.25 Our sketch for this problem.

Before: $v_{A1x} = 2.0$ m/s, A, $m_A = 0.50$ kg; $v_{B1x} = -2.0$ m/s, B, $m_B = 0.30$ kg

After: $v_{A2x} = ?$, A; $v_{B2x} = ?$, B

EVALUATE: Both bodies reverse their direction of motion; A moves to the left at 1.0 m/s and B moves to the right at 3.0 m/s. This is unlike the result of Example 8.5 because that collision was *not* elastic. The more massive glider A slows down in the collision and so loses kinetic energy. The less massive glider B speeds up and gains kinetic energy. The total kinetic energy before the collision (which we calculated in Example 8.7) is 1.6 J. The total kinetic energy after the collision is

$$\tfrac{1}{2}(0.50 \text{ kg})(-1.0 \text{ m/s})^2 + \tfrac{1}{2}(0.30 \text{ kg})(3.0 \text{ m/s})^2 = 1.6 \text{ J}$$

The kinetic energies before and after this elastic collision are equal. Kinetic energy is transferred from A to B, but none of it is lost.

CAUTION **Be careful with the elastic collision equations** You could *not* have solved this problem by using Eqs. (8.24) and (8.25), which apply only if body B is initially *at rest*. Always be sure that you use equations that are applicable!

EXAMPLE 8.11 MODERATING FISSION NEUTRONS IN A NUCLEAR REACTOR

The fission of uranium nuclei in a nuclear reactor produces high-speed neutrons. Before such neutrons can efficiently cause additional fissions, they must be slowed down by collisions with nuclei in the *moderator* of the reactor. The first nuclear reactor (built in 1942 at the University of Chicago) used carbon (graphite) as the moderator. Suppose a neutron (mass 1.0 u) traveling at 2.6×10^7 m/s undergoes a head-on elastic collision with a carbon nucleus (mass 12 u) initially at rest. Neglecting external forces during the collision, find the velocities after the collision. (1 u is the *atomic mass unit*, equal to 1.66×10^{-27} kg.)

SOLUTION

IDENTIFY and SET UP: We ignore external forces, so momentum is conserved in the collision. The collision is elastic, so kinetic energy is also conserved. **Figure 8.26** shows our sketch. We take the x-axis to be in the direction in which the neutron is moving initially. The collision is head-on, so both particles move along this same axis after the collision. The carbon nucleus is initially at rest, so we can use Eqs. (8.24) and (8.25); we replace A by n (for the neutron) and B by C (for the carbon nucleus). We have $m_n = 1.0$ u, $m_C = 12$ u, and $v_{n1x} = 2.6 \times 10^7$ m/s. The target variables are the final velocities v_{n2x} and v_{C2x}.

8.26 Our sketch for this problem.

Before: $v_{n1x} = 2.6 \times 10^7$ m/s, n, $m_n = 1.0$ u; C, $m_c = 12$ u

After: $v_{n2x} = ?$, n; $v_{C2x} = ?$, C

EXECUTE: You can do the arithmetic. (*Hint:* There's no reason to convert atomic mass units to kilograms.) The results are

$$v_{n2x} = -2.2 \times 10^7 \text{ m/s} \qquad v_{C2x} = 0.4 \times 10^7 \text{ m/s}$$

EVALUATE: The neutron ends up with $|(m_n - m_C)/(m_n + m_C)| = \tfrac{11}{13}$ of its initial speed, and the speed of the recoiling carbon nucleus is $|2m_n/(m_n + m_C)| = \tfrac{2}{13}$ of the neutron's initial speed. Kinetic energy is proportional to speed squared, so the neutron's final kinetic energy is $\left(\tfrac{11}{13}\right)^2 \approx 0.72$ of its original value. After a second head-on collision, its kinetic energy is $(0.72)^2$, or about half its original value, and so on. After a few dozen collisions (few of which are head-on), the neutron speed will be low enough that it can efficiently cause a fission reaction in a uranium nucleus.

EXAMPLE 8.12 A TWO-DIMENSIONAL ELASTIC COLLISION

Figure 8.27 shows an elastic collision of two pucks (masses $m_A = 0.500$ kg and $m_B = 0.300$ kg) on a frictionless air-hockey table. Puck A has an initial velocity of 4.00 m/s in the positive x-direction and a final velocity of 2.00 m/s in an unknown direction α. Puck B is initially at rest. Find the final speed v_{B2} of puck B and the angles α and β.

8.27 An elastic collision that isn't head-on.

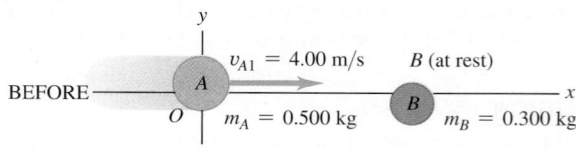

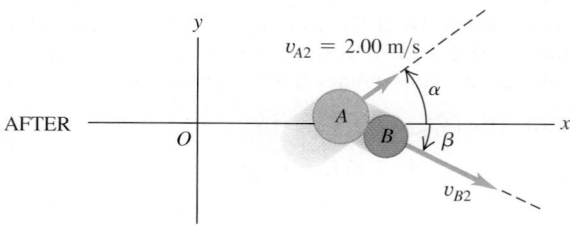

SOLUTION

IDENTIFY and SET UP: We'll use the equations for conservation of energy and conservation of x- and y-momentum. These three equations should be enough to solve for the three target variables.

EXECUTE: The collision is elastic, so the initial and final kinetic energies of the system are equal:

$$\tfrac{1}{2}m_A v_{A1}^2 = \tfrac{1}{2}m_A v_{A2}^2 + \tfrac{1}{2}m_B v_{B2}^2$$

$$v_{B2}^2 = \frac{m_A v_{A1}^2 - m_A v_{A2}^2}{m_B}$$

$$= \frac{(0.500 \text{ kg})(4.00 \text{ m/s})^2 - (0.500 \text{ kg})(2.00 \text{ m/s})^2}{0.300 \text{ kg}}$$

$$v_{B2} = 4.47 \text{ m/s}$$

Conservation of the x- and y-components of total momentum gives

$$m_A v_{A1x} = m_A v_{A2x} + m_B v_{B2x}$$

$$(0.500 \text{ kg})(4.00 \text{ m/s}) = (0.500 \text{ kg})(2.00 \text{ m/s})(\cos \alpha)$$
$$+ (0.300 \text{ kg})(4.47 \text{ m/s})(\cos \beta)$$

$$0 = m_A v_{A2y} + m_B v_{B2y}$$

$$0 = (0.500 \text{ kg})(2.00 \text{ m/s})(\sin \alpha)$$
$$- (0.300 \text{ kg})(4.47 \text{ m/s})(\sin \beta)$$

These are two simultaneous equations for α and β. You can supply the details of the solution. (*Hint:* Solve the first equation for $\cos \beta$ and the second for $\sin \beta$; square each equation and add. Since $\sin^2 \beta + \cos^2 \beta = 1$, this eliminates β and leaves an equation that you can solve for $\cos \alpha$ and hence for α. Substitute this value into either of the two equations and solve for β.) The results are

$$\alpha = 36.9° \qquad \beta = 26.6°$$

EVALUATE: To check the answers we confirm that the y-momentum, which was zero before the collision, is in fact zero after the collision. The y-momenta are

$$p_{A2y} = (0.500 \text{ kg})(2.00 \text{ m/s})(\sin 36.9°) = +0.600 \text{ kg} \cdot \text{m/s}$$
$$p_{B2y} = -(0.300 \text{ kg})(4.47 \text{ m/s})(\sin 26.6°) = -0.600 \text{ kg} \cdot \text{m/s}$$

and their sum is indeed zero.

TEST YOUR UNDERSTANDING OF SECTION 8.4 Most present-day nuclear reactors use water as a moderator (see Example 8.11). Are water molecules (mass $m_w = 18.0$ u) a better or worse moderator than carbon atoms? (One advantage of water is that it also acts as a coolant for the reactor's radioactive core.) ∎

8.5 CENTER OF MASS

We can restate the principle of conservation of momentum in a useful way by using the concept of **center of mass**. Suppose we have several particles with masses m_1, m_2, and so on. Let the coordinates of m_1 be (x_1, y_1), those of m_2 be (x_2, y_2), and so on. We define the center of mass of the system as the point that has coordinates (x_{cm}, y_{cm}) given by

$$x_{cm} = \frac{m_1 x_1 + m_2 x_2 + m_3 x_3 + \cdots}{m_1 + m_2 + m_3 + \cdots} = \frac{\sum_i m_i x_i}{\sum_i m_i}$$

$$y_{cm} = \frac{m_1 y_1 + m_2 y_2 + m_3 y_3 + \cdots}{m_1 + m_2 + m_3 + \cdots} = \frac{\sum_i m_i y_i}{\sum_i m_i}$$

(center of mass) (8.28)

We can express the position of the center of mass as a vector $\vec{r}_{cm}$:

$$\underset{\substack{\text{Position vector of} \\ \textbf{center of mass of} \\ \text{a system of particles}}}{\vec{r}_{cm}} = \frac{\overset{\text{Position vectors of individual particles}}{m_1\vec{r}_1 + m_2\vec{r}_2 + m_3\vec{r}_3 + \cdots}}{\underset{\text{Masses of individual particles}}{m_1 + m_2 + m_3 + \cdots}} = \frac{\sum_i m_i\vec{r}_i}{\sum_i m_i} \qquad (8.29)$$

We say that the center of mass is a *mass-weighted average* position of the particles.

EXAMPLE 8.13   CENTER OF MASS OF A WATER MOLECULE

Figure 8.28 shows a simple model of a water molecule. The oxygen–hydrogen separation is $d = 9.57 \times 10^{-11}$ m. Each hydrogen atom has mass 1.0 u, and the oxygen atom has mass 16.0 u. Find the position of the center of mass.

SOLUTION

IDENTIFY and SET UP: Nearly all the mass of each atom is concentrated in its nucleus, whose radius is only about 10^{-5} times the overall radius of the atom. Hence we can safely represent each atom as a point particle. Figure 8.28 shows our coordinate system, with the x-axis chosen to lie along the molecule's symmetry axis. We'll use Eqs. (8.28) to find x_{cm} and y_{cm}.

8.28 Where is the center of mass of a water molecule?

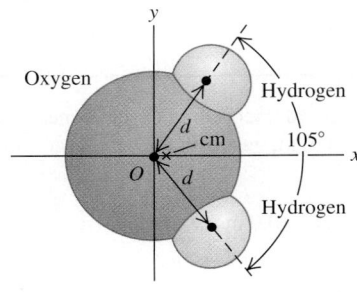

EXECUTE: The oxygen atom is at $x = 0$, $y = 0$. The x-coordinate of each hydrogen atom is $d\cos(105°/2)$; the y-coordinates are $\pm d\sin(105°/2)$. From Eqs. (8.28),

$$x_{cm} = \frac{\begin{bmatrix}(1.0\text{ u})(d\cos 52.5°) + (1.0\text{ u})(d\cos 52.5°) \\ + (16.0\text{ u})(0)\end{bmatrix}}{1.0\text{ u} + 1.0\text{ u} + 16.0\text{ u}} = 0.068d$$

$$y_{cm} = \frac{\begin{bmatrix}(1.0\text{ u})(d\sin 52.5°) + (1.0\text{ u})(-d\sin 52.5°) \\ + (16.0\text{ u})(0)\end{bmatrix}}{1.0\text{ u} + 1.0\text{ u} + 16.0\text{ u}} = 0$$

Substituting $d = 9.57 \times 10^{-11}$ m, we find

$$x_{cm} = (0.068)(9.57 \times 10^{-11}\text{ m}) = 6.5 \times 10^{-12}\text{ m}$$

EVALUATE: The center of mass is much closer to the oxygen atom (located at the origin) than to either hydrogen atom because the oxygen atom is much more massive. The center of mass lies along the molecule's *axis of symmetry*. If the molecule is rotated 180° around this axis, it looks exactly the same as before. The position of the center of mass can't be affected by this rotation, so it *must* lie on the axis of symmetry.

For solid bodies, in which we have (at least on a macroscopic level) a continuous distribution of matter, the sums in Eqs. (8.28) have to be replaced by integrals. The calculations can get quite involved, but we can say three general things about such problems (**Fig. 8.29**). First, whenever a homogeneous body has a geometric center, such as a billiard ball, a sugar cube, or a can of frozen orange juice, the center of mass is at the geometric center. Second, whenever a body has an axis of symmetry, such as a wheel or a pulley, the center of mass always lies on that axis. Third, there is no law that says the center of mass has to be within the body. For example, the center of mass of a donut is in the middle of the hole.

8.29 Locating the center of mass of a symmetric object.

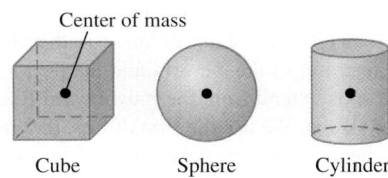

If a homogeneous object has a geometric center, that is where the center of mass is located.

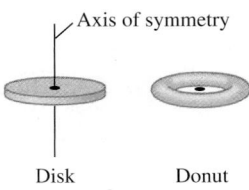

If an object has an axis of symmetry, the center of mass lies along it. As in the case of the donut, the center of mass may not be within the object.

Motion of the Center of Mass

To see the significance of the center of mass of a collection of particles, we must ask what happens to the center of mass when the particles move. The x- and y-components of velocity of the center of mass, $v_{cm\text{-}x}$ and $v_{cm\text{-}y}$, are the time derivatives of x_{cm} and y_{cm}. Also, dx_1/dt is the x-component of velocity of particle 1, so $dx_1/dt = v_{1x}$, and so on. Taking time derivatives of Eqs. (8.28), we get

$$v_{cm\text{-}x} = \frac{m_1v_{1x} + m_2v_{2x} + m_3v_{3x} + \cdots}{m_1 + m_2 + m_3 + \cdots}$$

$$v_{cm\text{-}y} = \frac{m_1v_{1y} + m_2v_{2y} + m_3v_{3y} + \cdots}{m_1 + m_2 + m_3 + \cdots} \qquad (8.30)$$

8.30 The net external force on this wrench is almost zero as it spins on a smooth, horizontal surface (seen from above). Hence the center of mass moves in a straight line with nearly constant velocity.

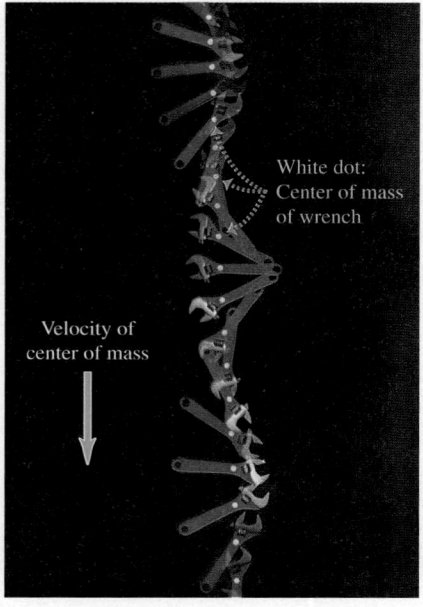

White dot: Center of mass of wrench

Velocity of center of mass

These equations are equivalent to the single vector equation obtained by taking the time derivative of Eq. (8.29):

$$\vec{v}_{cm} = \frac{m_1\vec{v}_1 + m_2\vec{v}_2 + m_3\vec{v}_3 + \cdots}{m_1 + m_2 + m_3 + \cdots} \qquad (8.31)$$

We denote the *total* mass $m_1 + m_2 + \cdots$ by M. We can then rewrite Eq. (8.31) as

Total mass of a system of particles
Momenta of individual particles
Velocity of center of mass

$$M\vec{v}_{cm} = m_1\vec{v}_1 + m_2\vec{v}_2 + m_3\vec{v}_3 + \cdots = \vec{P} \qquad (8.32)$$

Total momentum of system

So *the total momentum* $\vec{P}$ *of a system equals the total mass times the velocity of the center of mass.* When you catch a baseball, you are really catching a collection of a very large number of molecules of masses $m_1, m_2, m_3, \ldots$. The impulse you feel is due to the total momentum of this entire collection. But this impulse is the same as if you were catching a single particle of mass $M = m_1 + m_2 + m_3 + \cdots$ moving with $\vec{v}_{cm}$, the velocity of the collection's center of mass. So Eq. (8.32) helps us justify representing an extended body as a particle.

For a system of particles on which the net external force is zero, so that the total momentum $\vec{P}$ is constant, the velocity of the center of mass $\vec{v}_{cm} = \vec{P}/M$ is also constant. **Figure 8.30** shows an example. The overall motion of the wrench appears complicated, but the center of mass follows a straight line, as though all the mass were concentrated at that point.

EXAMPLE 8.14   A TUG-OF-WAR ON THE ICE

James (mass 90.0 kg) and Ramon (mass 60.0 kg) are 20.0 m apart on a frozen pond. Midway between them is a mug of their favorite beverage. They pull on the ends of a light rope stretched between them. When James has moved 6.0 m toward the mug, how far and in what direction has Ramon moved?

SOLUTION

IDENTIFY and SET UP: The surface is horizontal and (we assume) frictionless, so the net external force on the system of James, Ramon, and the rope is zero; their total momentum is conserved. Initially there is no motion, so the total momentum is zero. The velocity of the center of mass is therefore zero, and it remains at rest. Let's take the origin at the position of the mug and let the $+x$-axis extend from the mug toward Ramon. **Figure 8.31** shows our sketch. We use the first of Eqs. (8.28) to calculate the position of the center of mass; we ignore the mass of the light rope.

8.31 Our sketch for this problem.

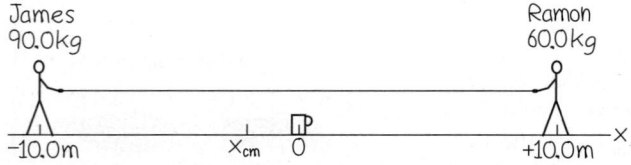

James
90.0kg

Ramon
60.0kg

-10.0m x_{cm} 0 $+10.0$m x

EXECUTE: The initial x-coordinates of James and Ramon are -10.0 m and $+10.0$ m, respectively, so the x-coordinate of the center of mass is

$$x_{cm} = \frac{(90.0\text{ kg})(-10.0\text{ m}) + (60.0\text{ kg})(10.0\text{ m})}{90.0\text{ kg} + 60.0\text{ kg}} = -2.0\text{ m}$$

When James moves 6.0 m toward the mug, his new x-coordinate is -4.0 m; we'll call Ramon's new x-coordinate x_2. The center of mass doesn't move, so

$$x_{cm} = \frac{(90.0\text{ kg})(-4.0\text{ m}) + (60.0\text{ kg})x_2}{90.0\text{ kg} + 60.0\text{ kg}} = -2.0\text{ m}$$

$$x_2 = 1.0\text{ m}$$

James has moved 6.0 m and is still 4.0 m from the mug, but Ramon has moved 9.0 m and is only 1.0 m from it.

EVALUATE: The ratio of the distances moved, $(6.0\text{ m})/(9.0\text{ m}) = \frac{2}{3}$, is the *inverse* ratio of the masses. Can you see why? Because the surface is frictionless, the two men will keep moving and collide at the center of mass; Ramon will reach the mug first. This is independent of how hard either person pulls; pulling harder just makes them move faster.

External Forces and Center-of-Mass Motion

If the net external force on a system of particles is not zero, then total momentum is not conserved and the velocity of the center of mass changes. Let's look at this situation in more detail.

Equations (8.31) and (8.32) give the *velocity* of the center of mass in terms of the velocities of the individual particles. We take the time derivatives of these equations to show that the *accelerations* are related in the same way. Let $\vec{a}_{cm} = d\vec{v}_{cm}/dt$ be the acceleration of the center of mass; then

$$M\vec{a}_{cm} = m_1\vec{a}_1 + m_2\vec{a}_2 + m_3\vec{a}_3 + \cdots \tag{8.33}$$

Now $m_1\vec{a}_1$ is equal to the vector sum of forces on the first particle, and so on, so the right side of Eq. (8.33) is equal to the vector sum $\Sigma\vec{F}$ of *all* the forces on *all* the particles. Just as we did in Section 8.2, we can classify each force as *external* or *internal*. The sum of all forces on all the particles is then

$$\Sigma\vec{F} = \Sigma\vec{F}_{ext} + \Sigma\vec{F}_{int} = M\vec{a}_{cm}$$

Because of Newton's third law, all of the internal forces cancel in pairs, and $\Sigma\vec{F}_{int} = \mathbf{0}$. What survives on the left side is the sum of only the *external* forces:

Net external force ⋯⋯ on a body or a collection of particles
$$\Sigma\vec{F}_{ext} = M\vec{a}_{cm} \tag{8.34}$$
⋯⋯ Total mass of body or collection of particles
⋯⋯ Acceleration of center of mass

When a body or a collection of particles is acted on by external forces, the center of mass moves as though all the mass were concentrated at that point and it were acted on by a net force equal to the sum of the external forces on the system.

This result is central to the whole subject of mechanics. In fact, we've been using this result all along; without it, we would not be able to represent an extended body as a point particle when we apply Newton's laws. It explains why only *external* forces can affect the motion of an extended body. If you pull upward on your belt, your belt exerts an equal downward force on your hands; these are *internal* forces that cancel and have no effect on the overall motion of your body.

As an example, suppose that a cannon shell traveling in a parabolic trajectory (ignoring air resistance) explodes in flight, splitting into two fragments with equal mass (**Fig. 8.32**). The fragments follow new parabolic paths, but the center of mass continues on the original parabolic trajectory, as though all the mass were still concentrated at that point.

This property of the center of mass is important when we analyze the motion of rigid bodies. In Chapter 10 we'll describe the motion of an extended body as a combination of translational motion of the center of mass and rotational motion about an axis through the center of mass. This property also plays an important role in the motion of astronomical objects. It's not correct to say that the moon orbits the earth; rather, both the earth and the moon move in orbits around their common center of mass.

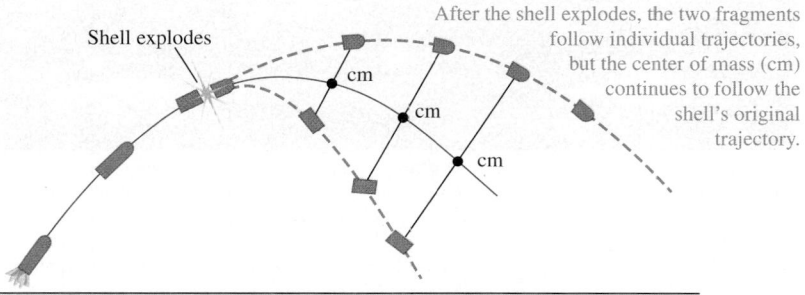

Shell explodes

After the shell explodes, the two fragments follow individual trajectories, but the center of mass (cm) continues to follow the shell's original trajectory.

8.32 A shell explodes into two fragments in flight. If air resistance is ignored, the center of mass continues on the same trajectory as the shell's path before the explosion.

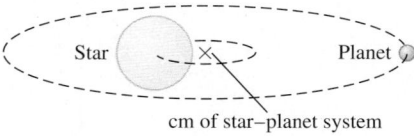

Star   × - - - - - -   Planet

cm of star–planet system

There's one more useful way to describe the motion of a system of particles. Using $\vec{a}_{cm} = d\vec{v}_{cm}/dt$, we can rewrite Eq. (8.33) as

$$M\vec{a}_{cm} = M\frac{d\vec{v}_{cm}}{dt} = \frac{d(M\vec{v}_{cm})}{dt} = \frac{d\vec{P}}{dt} \tag{8.35}$$

The total system mass M is constant, so we're allowed to move it inside the derivative. Substituting Eq. (8.35) into Eq. (8.34), we find

$$\sum \vec{F}_{ext} = \frac{d\vec{P}}{dt} \qquad \text{(extended body or system of particles)} \tag{8.36}$$

This equation looks like Eq. (8.4). The difference is that Eq. (8.36) describes a *system* of particles, such as an extended body, while Eq. (8.4) describes a single particle. The interactions between the particles that make up the system can change the individual momenta of the particles, but the *total* momentum $\vec{P}$ of the system can be changed only by external forces acting from outside the system.

If the net external force is zero, Eqs. (8.34) and (8.36) show that the center-of-mass acceleration $\vec{a}_{cm}$ is zero (so the center-of-mass velocity $\vec{v}_{cm}$ is constant) and the total momentum $\vec{P}$ is constant. This is just our statement from Section 8.3: If the net external force on a system is zero, momentum is conserved.

TEST YOUR UNDERSTANDING OF SECTION 8.5   Will the center of mass in Fig. 8.32 continue on the same parabolic trajectory even after one of the fragments hits the ground? Why or why not? ▮

8.6 ROCKET PROPULSION

Momentum considerations are particularly useful for analyzing a system in which the masses of parts of the system change with time. In such cases we can't use Newton's second law $\sum \vec{F} = m\vec{a}$ directly because m changes. Rocket propulsion is an important example of this situation. A rocket is propelled forward by rearward ejection of burned fuel that initially was in the rocket (which is why rocket fuel is also called *propellant*). The forward force on the rocket is the reaction to the backward force on the ejected material. The total mass of the system is constant, but the mass of the rocket itself decreases as material is ejected.

For simplicity, let's consider a rocket in outer space, where there is no gravitational force and no air resistance. Let m denote the mass of the rocket, which will change as it expends fuel. We choose our x-axis to be along the rocket's direction of motion. **Figure 8.33a** shows the rocket at a time t, when its mass is m and its x-velocity relative to our coordinate system is v. (To simplify, we will drop the subscript x in this discussion.) The x-component of total momentum at this instant is $P_1 = mv$. In a short time interval dt, the mass of the rocket changes by an amount dm. This is an inherently negative quantity because the rocket's mass m *decreases* with time. During dt, a *positive* mass $-dm$ of burned fuel is ejected from the rocket. Let v_{ex} be the exhaust *speed* of this material *relative*

(a)   (b)

8.33 A rocket moving in gravity-free outer space at **(a)** time t and **(b)** time $t + dt$.

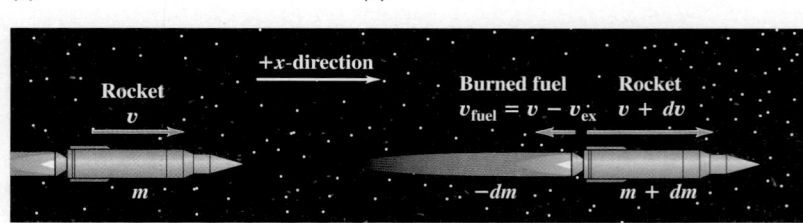

At **time** t, the rocket has mass m and x-component of velocity v.

At **time** $t + dt$, the rocket has mass $m + dm$ (where dm is inherently *negative*) and x-component of velocity $v + dv$. The burned fuel has x-component of velocity $v_{fuel} = v - v_{ex}$ and mass $-dm$. (The minus sign is needed to make $-dm$ *positive* because dm is negative.)

DEMO

to the rocket; the burned fuel is ejected opposite the direction of motion, so its x-component of *velocity* relative to the rocket is $-v_{ex}$. The x-velocity v_{fuel} of the burned fuel relative to our coordinate system is then

$$v_{fuel} = v + (-v_{ex}) = v - v_{ex}$$

and the x-component of momentum of the ejected mass $(-dm)$ is

$$(-dm)v_{fuel} = (-dm)(v - v_{ex})$$

Figure 8.33b shows that at the end of the time interval dt, the x-velocity of the rocket and unburned fuel has increased to $v + dv$, and its mass has decreased to $m + dm$ (remember that dm is negative). The rocket's momentum at this time is

$$(m + dm)(v + dv)$$

Thus the *total* x-component of momentum P_2 of the rocket plus ejected fuel at time $t + dt$ is

$$P_2 = (m + dm)(v + dv) + (-dm)(v - v_{ex})$$

According to our initial assumption, the rocket and fuel are an isolated system. Thus momentum is conserved, and the total x-component of momentum of the system must be the same at time t and at time $t + dt$: $P_1 = P_2$. Hence

$$mv = (m + dm)(v + dv) + (-dm)(v - v_{ex})$$

This can be simplified to

$$m\, dv = -dm\, v_{ex} - dm\, dv$$

We can ignore the term $(-dm\, dv)$ because it is a product of two small quantities and thus is much smaller than the other terms. Dropping this term, dividing by dt, and rearranging, we find

$$m\frac{dv}{dt} = -v_{ex}\frac{dm}{dt} \qquad (8.37)$$

Now dv/dt is the acceleration of the rocket, so the left side of Eq. (8.37) (mass times acceleration) equals the net force F, or *thrust,* on the rocket:

$$F = -v_{ex}\frac{dm}{dt} \qquad (8.38)$$

The thrust is proportional both to the relative speed v_{ex} of the ejected fuel and to the mass of fuel ejected per unit time, $-dm/dt$. (Remember that dm/dt is negative because it is the rate of change of the rocket's mass, so F is positive.)

The x-component of acceleration of the rocket is

$$a = \frac{dv}{dt} = -\frac{v_{ex}}{m}\frac{dm}{dt} \qquad (8.39)$$

This is positive because v_{ex} is positive (remember, it's the exhaust *speed*) and dm/dt is negative. The rocket's mass m decreases continuously while the fuel is being consumed. If v_{ex} and dm/dt are constant, the acceleration increases until all the fuel is gone.

Equation (8.38) tells us that an effective rocket burns fuel at a rapid rate (large $-dm/dt$) and ejects the burned fuel at a high relative speed (large v_{ex}), as in **Fig. 8.34**. In the early days of rocket propulsion, people who didn't understand conservation of momentum thought that a rocket couldn't function in outer space because "it doesn't have anything to push against." In fact, rockets work *best* in outer space, where there is no air resistance! The launch vehicle in Fig. 8.34 is *not* "pushing against the ground" to ascend.

If the exhaust speed v_{ex} is constant, we can integrate Eq. (8.39) to relate the velocity v at any time to the remaining mass m. At time $t = 0$, let the mass be m_0 and the velocity be v_0. Then we rewrite Eq. (8.39) as

$$dv = -v_{ex}\frac{dm}{m}$$

8.34 To provide enough thrust to lift its payload into space, this *Atlas V* launch vehicle ejects more than 1000 kg of burned fuel per second at speeds of nearly 4000 m/s.

We change the integration variables to v' and m', so we can use v and m as the upper limits (the final speed and mass). Then we integrate both sides, using limits v_0 to v and m_0 to m, and take the constant v_{ex} outside the integral:

$$\int_{v_0}^{v} dv' = -\int_{m_0}^{m} v_{ex} \frac{dm'}{m'} = -v_{ex} \int_{m_0}^{m} \frac{dm'}{m'}$$

$$v - v_0 = -v_{ex} \ln \frac{m}{m_0} = v_{ex} \ln \frac{m_0}{m} \qquad (8.40)$$

The ratio m_0/m is the original mass divided by the mass after the fuel has been exhausted. In practical spacecraft this ratio is made as large as possible to maximize the speed gain, which means that the initial mass of the rocket is almost all fuel. The final velocity of the rocket will be greater in magnitude (and is often *much* greater) than the relative speed v_{ex} if $\ln(m_0/m) > 1$—that is, if $m_0/m > e = 2.71828\ldots$.

We've assumed throughout this analysis that the rocket is in gravity-free outer space. However, gravity must be taken into account when a rocket is launched from the surface of a planet, as in Fig. 8.34.

EXAMPLE 8.15   ACCELERATION OF A ROCKET

The engine of a rocket in outer space, far from any planet, is turned on. The rocket ejects burned fuel at a constant rate; in the first second of firing, it ejects $\frac{1}{120}$ of its initial mass m_0 at a relative speed of 2400 m/s. What is the rocket's initial acceleration?

SOLUTION

IDENTIFY and SET UP: We are given the rocket's exhaust speed v_{ex} and the fraction of the initial mass lost during the first second of firing, from which we can find dm/dt. We'll use Eq. (8.39) to find the acceleration of the rocket.

EXECUTE: The initial rate of change of mass is

$$\frac{dm}{dt} = -\frac{m_0/120}{1 \text{ s}} = -\frac{m_0}{120 \text{ s}}$$

From Eq. (8.39),

$$a = -\frac{v_{ex}}{m_0} \frac{dm}{dt} = -\frac{2400 \text{ m/s}}{m_0} \left(-\frac{m_0}{120 \text{ s}} \right) = 20 \text{ m/s}^2$$

EVALUATE: The answer doesn't depend on m_0. If v_{ex} is the same, the initial acceleration is the same for a 120,000-kg spacecraft that ejects 1000 kg/s as for a 60-kg astronaut equipped with a small rocket that ejects 0.5 kg/s.

EXAMPLE 8.16   SPEED OF A ROCKET

Suppose that $\frac{3}{4}$ of the initial mass of the rocket in Example 8.15 is fuel, so the fuel is completely consumed at a constant rate in 90 s. The final mass of the rocket is $m = m_0/4$. If the rocket starts from rest in our coordinate system, find its speed at the end of this time.

SOLUTION

IDENTIFY, SET UP, and EXECUTE: We are given the initial velocity $v_0 = 0$, the exhaust speed $v_{ex} = 2400$ m/s, and the final mass m as a fraction of the initial mass m_0. We'll use Eq. (8.40) to find the final speed v:

$$v = v_0 + v_{ex} \ln \frac{m_0}{m} = 0 + (2400 \text{ m/s})(\ln 4) = 3327 \text{ m/s}$$

EVALUATE: Let's examine what happens as the rocket gains speed. (To illustrate our point, we use more figures than are significant.) At the start of the flight, when the velocity of the rocket is zero, the ejected fuel is moving backward at 2400 m/s relative to our frame of reference. As the rocket moves forward and speeds up, the fuel's speed relative to our system decreases; when the rocket speed reaches 2400 m/s, this relative speed is *zero*. [Knowing the rate of fuel consumption, you can solve Eq. (8.40) to show that this occurs at about $t = 75.6$ s.] After this time the ejected burned fuel moves *forward*, not backward, in our system. Relative to our frame of reference, the last bit of ejected fuel has a forward velocity of 3327 m/s − 2400 m/s = 927 m/s.

TEST YOUR UNDERSTANDING OF SECTION 8.6 (a) If a rocket in gravity-free outer space has the same thrust at all times, is its acceleration constant, increasing, or decreasing? (b) If the rocket has the same acceleration at all times, is the thrust constant, increasing, or decreasing? ∎

Momentum of a particle: The momentum $\vec{p}$ of a particle is a vector quantity equal to the product of the particle's mass m and velocity $\vec{v}$. Newton's second law says that the net force on a particle is equal to the rate of change of the particle's momentum.

$$\vec{p} = m\vec{v} \tag{8.2}$$

$$\sum \vec{F} = \frac{d\vec{p}}{dt} \tag{8.4}$$

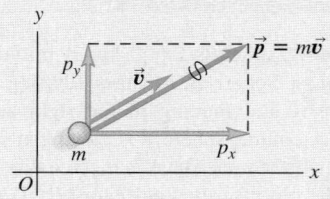

Impulse and momentum: If a constant net force $\sum \vec{F}$ acts on a particle for a time interval Δt from t_1 to t_2, the impulse $\vec{J}$ of the net force is the product of the net force and the time interval. If $\sum \vec{F}$ varies with time, $\vec{J}$ is the integral of the net force over the time interval. In any case, the change in a particle's momentum during a time interval equals the impulse of the net force that acted on the particle during that interval. The momentum of a particle equals the impulse that accelerated it from rest to its present speed. (See Examples 8.1–8.3.)

$$\vec{J} = \sum \vec{F}(t_2 - t_1) = \sum \vec{F} \, \Delta t \tag{8.5}$$

$$\vec{J} = \int_{t_1}^{t_2} \sum \vec{F} \, dt \tag{8.7}$$

$$\vec{J} = \vec{p}_2 - \vec{p}_1 \tag{8.6}$$

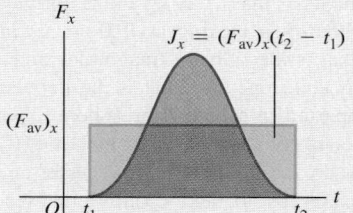

Conservation of momentum: An internal force is a force exerted by one part of a system on another. An external force is a force exerted on any part of a system by something outside the system. If the net external force on a system is zero, the total momentum of the system $\vec{P}$ (the vector sum of the momenta of the individual particles that make up the system) is constant, or conserved. Each component of total momentum is separately conserved. (See Examples 8.4–8.6.)

$$\vec{P} = \vec{p}_A + \vec{p}_B + \cdots$$
$$= m_A\vec{v}_A + m_B\vec{v}_B + \cdots \tag{8.14}$$

If $\sum \vec{F} = 0$, then $\vec{P}$ = constant.

Collisions: In collisions of all kinds, the initial and final total momenta are equal. In an elastic collision between two bodies, the initial and final total kinetic energies are also equal, and the initial and final relative velocities have the same magnitude. In an inelastic two-body collision, the total kinetic energy is less after the collision than before. If the two bodies have the same final velocity, the collision is completely inelastic. (See Examples 8.7–8.12.)

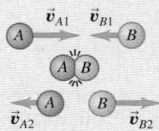

Center of mass: The position vector of the center of mass of a system of particles, $\vec{r}_{cm}$, is a weighted average of the positions $\vec{r}_1, \vec{r}_2, \ldots$ of the individual particles. The total momentum $\vec{P}$ of a system equals the system's total mass M multiplied by the velocity of its center of mass, $\vec{v}_{cm}$. The center of mass moves as though all the mass M were concentrated at that point. If the net external force on the system is zero, the center-of-mass velocity $\vec{v}_{cm}$ is constant. If the net external force is not zero, the center of mass accelerates as though it were a particle of mass M being acted on by the same net external force. (See Examples 8.13 and 8.14.)

$$\vec{r}_{cm} = \frac{m_1\vec{r}_1 + m_2\vec{r}_2 + m_3\vec{r}_3 + \cdots}{m_1 + m_2 + m_3 + \cdots}$$

$$= \frac{\sum_i m_i\vec{r}_i}{\sum_i m_i} \tag{8.29}$$

$$\vec{P} = m_1\vec{v}_1 + m_2\vec{v}_2 + m_3\vec{v}_3 + \cdots$$
$$= M\vec{v}_{cm} \tag{8.32}$$

$$\sum \vec{F}_{ext} = M\vec{a}_{cm} \tag{8.34}$$

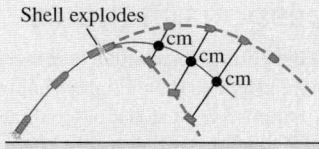

Rocket propulsion: In rocket propulsion, the mass of a rocket changes as the fuel is used up and ejected from the rocket. Analysis of the motion of the rocket must include the momentum carried away by the spent fuel as well as the momentum of the rocket itself. (See Examples 8.15 and 8.16.)

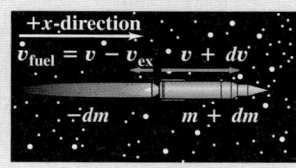

SOLUTION

BRIDGING PROBLEM ONE COLLISION AFTER ANOTHER

Sphere A, of mass 0.600 kg, is initially moving to the right at 4.00 m/s. Sphere B, of mass 1.80 kg, is initially to the right of sphere A and moving to the right at 2.00 m/s. After the two spheres collide, sphere B is moving at 3.00 m/s in the same direction as before. (a) What is the velocity (magnitude and direction) of sphere A after this collision? (b) Is this collision elastic or inelastic? (c) Sphere B then has an off-center collision with sphere C, which has mass 1.20 kg and is initially at rest. After this collision, sphere B is moving at 19.0° to its initial direction at 2.00 m/s. What is the velocity (magnitude and direction) of sphere C after this collision? (d) What is the impulse (magnitude and direction) imparted to sphere B by sphere C when they collide? (e) Is this second collision elastic or inelastic? (f) What is the velocity (magnitude and direction) of the center of mass of the system of three spheres (A, B, and C) after the second collision? No external forces act on any of the spheres in this problem.

SOLUTION GUIDE

IDENTIFY and SET UP

1. Momentum is conserved in these collisions. Can you explain why?
2. Choose the x- and y-axes, and use your choice of axes to draw three figures that show the spheres (i) before the first collision, (ii) after the first collision but before the second collision, and (iii) after the second collision. Assign subscripts to values in each of situations (i), (ii), and (iii).
3. Make a list of the target variables, and choose the equations that you'll use to solve for these.

EXECUTE

4. Solve for the velocity of sphere A after the first collision. Does A slow down or speed up in the collision? Does this make sense?
5. Now that you know the velocities of both A and B after the first collision, decide whether or not this collision is elastic. (How will you do this?)
6. The second collision is two-dimensional, so you'll have to demand that *both* components of momentum are conserved. Use this to find the speed and direction of sphere C after the second collision. (*Hint:* After the first collision, sphere B maintains the same velocity until it hits sphere C.)
7. Use the definition of impulse to find the impulse imparted to sphere B by sphere C. Remember that impulse is a vector.
8. Use the same technique that you employed in step 5 to decide whether the second collision is elastic.
9. Find the velocity of the center of mass after the second collision.

EVALUATE

10. Compare the directions of the vectors you found in steps 6 and 7. Is this a coincidence? Why or why not?
11. Find the velocity of the center of mass before and after the first collision. Compare to your result from step 9. Again, is this a coincidence? Why or why not?

Problems

For assigned homework and other learning materials, go to MasteringPhysics®. **MP**

•, ••, •••: Difficulty levels. **CP**: Cumulative problems incorporating material from earlier chapters. **CALC**: Problems requiring calculus.
DATA: Problems involving real data, scientific evidence, experimental design, and/or statistical reasoning. **BIO**: Biosciences problems.

DISCUSSION QUESTIONS

Q8.1 In splitting logs with a hammer and wedge, is a heavy hammer more effective than a lighter hammer? Why?

Q8.2 Suppose you catch a baseball and then someone invites you to catch a bowling ball with either the same momentum or the same kinetic energy as the baseball. Which would you choose? Explain.

Q8.3 When rain falls from the sky, what happens to its momentum as it hits the ground? Is your answer also valid for Newton's famous apple?

Q8.4 A car has the same kinetic energy when it is traveling south at 30 m/s as when it is traveling northwest at 30 m/s. Is the momentum of the car the same in both cases? Explain.

Q8.5 A truck is accelerating as it speeds down the highway. One inertial frame of reference is attached to the ground with its origin at a fence post. A second frame of reference is attached to a police car that is traveling down the highway at constant velocity. Is the momentum of the truck the same in these two reference frames? Explain. Is the rate of change of the truck's momentum the same in these two frames? Explain.

Q8.6 (a) If the momentum of a *single* point object is equal to zero, must the object's kinetic energy also be zero? (b) If the momentum of a *pair* of point objects is equal to zero, must the kinetic energy of those objects also be zero? (c) If the kinetic energy of a pair of point objects is equal to zero, must the momentum of those objects also be zero? Explain your reasoning in each case.

Q8.7 A woman holding a large rock stands on a frictionless, horizontal sheet of ice. She throws the rock with speed v_0 at an angle α above the horizontal. Consider the system consisting of the woman plus the rock. Is the momentum of the system conserved? Why or why not? Is any component of the momentum of the system conserved? Again, why or why not?

Q8.8 In Example 8.7 (Section 8.3), where the two gliders of Fig. 8.18 stick together after the collision, the collision is inelastic because $K_2 < K_1$. In Example 8.5 (Section 8.2), is the collision inelastic? Explain.

Q8.9 In a completely inelastic collision between two objects, where the objects stick together after the collision, is it possible for the final kinetic energy of the system to be zero? If so, give an example in which this would occur. If the final kinetic energy is zero, what must the initial momentum of the system be? Is the initial kinetic energy of the system zero? Explain.

Q8.10 Since for a particle the kinetic energy is given by $K = \frac{1}{2}mv^2$ and the momentum by $\vec{P} = m\vec{v}$, it is easy to show that $K = p^2/2m$. How, then, is it possible to have an event during which the total momentum of the system is constant but the total kinetic energy changes?

Q8.11 In each of Examples 8.10, 8.11, and 8.12 (Section 8.4), verify that the relative velocity vector of the two bodies has the same magnitude before and after the collision. In each case, what happens to the *direction* of the relative velocity vector?

Q8.12 A glass dropped on the floor is more likely to break if the floor is concrete than if it is wood. Why? (Refer to Fig. 8.3b.)

Q8.13 In Fig. 8.23b, the kinetic energy of the Ping-Pong ball is larger after its interaction with the bowling ball than before. From where does the extra energy come? Describe the event in terms of conservation of energy.

Q8.14 A machine gun is fired at a steel plate. Is the average force on the plate from the bullet impact greater if the bullets bounce off or if they are squashed and stick to the plate? Explain.

Q8.15 A net force of 4 N acts on an object initially at rest for 0.25 s and gives it a final speed of 5 m/s. How could a net force of 2 N produce the same final speed?

Q8.16 A net force with x-component ΣF_x acts on an object from time t_1 to time t_2. The x-component of the momentum of the object is the same at t_1 as it is at t_2, but ΣF_x is not zero at all times between t_1 and t_2. What can you say about the graph of ΣF_x versus t?

Q8.17 A tennis player hits a tennis ball with a racket. Consider the system made up of the ball and the racket. Is the total momentum of the system the same just before and just after the hit? Is the total momentum just after the hit the same as 2 s later, when the ball is in midair at the high point of its trajectory? Explain any differences between the two cases.

Q8.18 In Example 8.4 (Section 8.2), consider the system consisting of the rifle plus the bullet. What is the speed of the system's center of mass after the rifle is fired? Explain.

Q8.19 An egg is released from rest from the roof of a building and falls to the ground. As the egg falls, what happens to the momentum of the system of the egg plus the earth?

Q8.20 A woman stands in the middle of a perfectly smooth, frictionless, frozen lake. She can set herself in motion by throwing things, but suppose she has nothing to throw. Can she propel herself to shore *without* throwing anything?

Q8.21 At the highest point in its parabolic trajectory, a shell explodes into two fragments. Is it possible for *both* fragments to fall straight down after the explosion? Why or why not?

Q8.22 When an object breaks into two pieces (explosion, radioactive decay, recoil, etc.), the lighter fragment gets more kinetic energy than the heavier one. This is a consequence of momentum conservation, but can you also explain it by using Newton's laws of motion?

Q8.23 An apple falls from a tree and feels no air resistance. As it is falling, which of these statements about it are true? (a) Only its momentum is conserved; (b) only its mechanical energy is conserved; (c) both its momentum and its mechanical energy are conserved; (d) its kinetic energy is conserved.

Q8.24 Two pieces of clay collide and stick together. During the collision, which of these statements are true? (a) Only the momentum of the clay is conserved; (b) only the mechanical energy of the clay is conserved; (c) both the momentum and the

mechanical energy of the clay are conserved; (d) the kinetic energy of the clay is conserved.

Q8.25 Two objects of mass M and $5M$ are at rest on a horizontal, frictionless table with a compressed spring of negligible mass between them. When the spring is released, which of the following statements are true? (a) The two objects receive equal magnitudes of momentum. (b) The two objects receive equal amounts of kinetic energy from the spring. (c) The heavier object gains more kinetic energy than the lighter object. (d) The lighter object gains more kinetic energy than the heavier object. Explain your reasoning in each case.

Q8.26 A very heavy SUV collides head-on with a very light compact car. Which of these statements about the collision are correct? (a) The amount of kinetic energy lost by the SUV is equal to the amount of kinetic energy gained by the compact; (b) the amount of momentum lost by the SUV is equal to the amount of momentum gained by the compact; (c) the compact feels a considerably greater force during the collision than the SUV does; (d) both cars lose the same amount of kinetic energy.

EXERCISES

Section 8.1 Momentum and Impulse

8.1 • (a) What is the magnitude of the momentum of a 10,000-kg truck whose speed is 12.0 m/s? (b) What speed would a 2000-kg SUV have to attain in order to have (i) the same momentum? (ii) the same kinetic energy?

8.2 • In a certain track and field event, the shotput has a mass of 7.30 kg and is released with a speed of 15.0 m/s at 40.0° above the horizontal over a competitor's straight left leg. What are the initial horizontal and vertical components of the momentum of this shotput?

8.3 • Objects A, B, and C are moving as shown in **Fig. E8.3**. Find the x- and y-components of the net momentum of the particles if we define the system to consist of (a) A and C, (b) B and C, (c) all three objects.

Figure **E8.3**

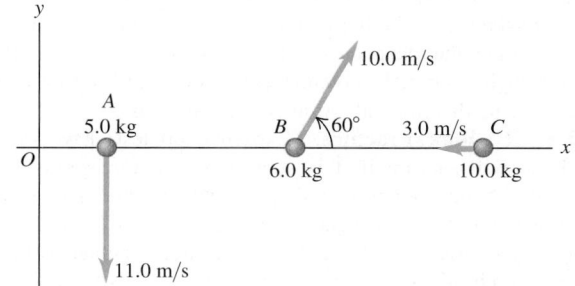

8.4 • Two vehicles are approaching an intersection. One is a 2500-kg pickup traveling at 14.0 m/s from east to west (the $-x$-direction), and the other is a 1500-kg sedan going from south to north (the $+y$-direction) at 23.0 m/s. (a) Find the x- and y-components of the net momentum of this system. (b) What are the magnitude and direction of the net momentum?

8.5 • One 110-kg football lineman is running to the right at 2.75 m/s while another 125-kg lineman is running directly toward him at 2.60 m/s. What are (a) the magnitude and direction of the net momentum of these two athletes, and (b) their total kinetic energy?

8.6 •• BIO **Biomechanics.** The mass of a regulation tennis ball is 57 g (although it can vary slightly), and tests have shown that the ball is in contact with the tennis racket for 30 ms. (This number can also vary, depending on the racket and swing.) We shall assume a 30.0-ms contact time. The fastest-known served tennis ball was served by "Big Bill" Tilden in 1931, and its speed was measured to be 73 m/s. (a) What impulse and what force did Big Bill exert on the tennis ball in his record serve? (b) If Big Bill's opponent returned his serve with a speed of 55 m/s, what force and what impulse did he exert on the ball, assuming only horizontal motion?

8.7 • **Force of a Golf Swing.** A 0.0450-kg golf ball initially at rest is given a speed of 25.0 m/s when a club strikes it. If the club and ball are in contact for 2.00 ms, what average force acts on the ball? Is the effect of the ball's weight during the time of contact significant? Why or why not?

8.8 • **Force of a Baseball Swing.** A baseball has mass 0.145 kg. (a) If the velocity of a pitched ball has a magnitude of 45.0 m/s and the batted ball's velocity is 55.0 m/s in the opposite direction, find the magnitude of the change in momentum of the ball and of the impulse applied to it by the bat. (b) If the ball remains in contact with the bat for 2.00 ms, find the magnitude of the average force applied by the bat.

8.9 • A 0.160-kg hockey puck is moving on an icy, frictionless, horizontal surface. At $t = 0$, the puck is moving to the right at 3.00 m/s. (a) Calculate the velocity of the puck (magnitude and direction) after a force of 25.0 N directed to the right has been applied for 0.050 s. (b) If, instead, a force of 12.0 N directed to the left is applied from $t = 0$ to $t = 0.050$ s, what is the final velocity of the puck?

8.10 •• A bat strikes a 0.145-kg baseball. Just before impact, the ball is traveling horizontally to the right at 40.0 m/s; when it leaves the bat, the ball is traveling to the left at an angle of 30° above horizontal with a speed of 52.0 m/s. If the ball and bat are in contact for 1.75 ms, find the horizontal and vertical components of the average force on the ball.

8.11 • CALC At time $t = 0$ a 2150-kg rocket in outer space fires an engine that exerts an increasing force on it in the $+x$-direction. This force obeys the equation $F_x = At^2$, where t is time, and has a magnitude of 781.25 N when $t = 1.25$ s. (a) Find the SI value of the constant A, including its units. (b) What impulse does the engine exert on the rocket during the 1.50-s interval starting 2.00 s after the engine is fired? (c) By how much does the rocket's velocity change during this interval? Assume constant mass.

8.12 •• BIO **Bone Fracture.** Experimental tests have shown that bone will rupture if it is subjected to a force density of 1.03×10^8 N/m². Suppose a 70.0-kg person carelessly roller-skates into an overhead metal beam that hits his forehead and completely stops his forward motion. If the area of contact with the person's forehead is 1.5 cm², what is the greatest speed with which he can hit the wall without breaking any bone if his head is in contact with the beam for 10.0 ms?

8.13 • A 2.00-kg stone is sliding to the right on a frictionless, horizontal surface at 5.00 m/s when it is suddenly struck by an object that exerts a large horizontal force on it for a short period of time. The graph in **Fig. E8.13** shows the magnitude of this force as a function of time. (a) What impulse does this

Figure **E8.13**

F (kN)

2.50 ┄┄┄

O   15.0   16.0 t (ms)

force exert on the stone? (b) Just after the force stops acting, find the magnitude and direction of the stone's velocity if the force acts (i) to the right or (ii) to the left.

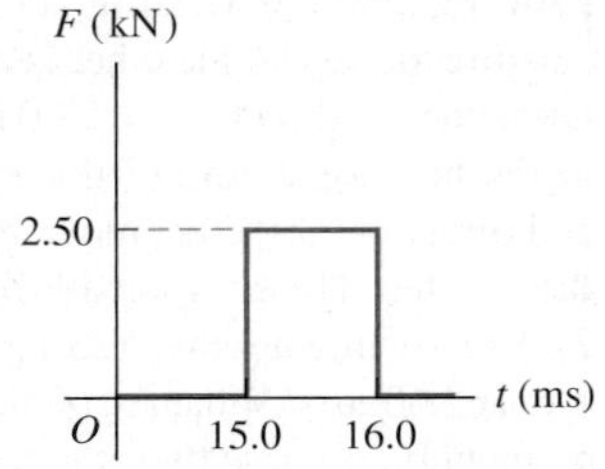

8.14 •• CALC Starting at $t = 0$, a horizontal net force $\vec{F} = (0.280 \text{ N/s})t\hat{i} + (-0.450 \text{ N/s}^2)t^2\hat{j}$ is applied to a box that has an initial momentum $\vec{p} = (-3.00 \text{ kg} \cdot \text{m/s})\hat{i} + (4.00 \text{ kg} \cdot \text{m/s})\hat{j}$. What is the momentum of the box at $t = 2.00$ s?

8.15 •• To warm up for a match, a tennis player hits the 57.0-g ball vertically with her racket. If the ball is stationary just before it is hit and goes 5.50 m high, what impulse did she impart to it?

Section 8.2 Conservation of Momentum

8.16 • A 68.5-kg astronaut is doing a repair in space on the orbiting space station. She throws a 2.25-kg tool away from her at 3.20 m/s relative to the space station. With what speed and in what direction will she begin to move?

8.17 •• The expanding gases that leave the muzzle of a rifle also contribute to the recoil. A .30-caliber bullet has mass 0.00720 kg and a speed of 601 m/s relative to the muzzle when fired from a rifle that has mass 2.80 kg. The loosely held rifle recoils at a speed of 1.85 m/s relative to the earth. Find the momentum of the propellant gases in a coordinate system attached to the earth as they leave the muzzle of the rifle.

8.18 • Two figure skaters, one weighing 625 N and the other 725 N, push off against each other on frictionless ice. (a) If the heavier skater travels at 1.50 m/s, how fast will the lighter one travel? (b) How much kinetic energy is "created" during the skaters' maneuver, and where does this energy come from?

8.19 • BIO **Animal Propulsion.** Squids and octopuses propel themselves by expelling water. They do this by keeping water in a cavity and then suddenly contracting the cavity to force out the water through an opening. A 6.50-kg squid (including the water in the cavity) at rest suddenly sees a dangerous predator. (a) If the squid has 1.75 kg of water in its cavity, at what speed must it expel this water suddenly to achieve a speed of 2.50 m/s to escape the predator? Ignore any drag effects of the surrounding water. (b) How much kinetic energy does the squid create by this maneuver?

8.20 •• You are standing on a sheet of ice that covers the football stadium parking lot in Buffalo; there is negligible friction between your feet and the ice. A friend throws you a 0.600-kg ball that is traveling horizontally at 10.0 m/s. Your mass is 70.0 kg. (a) If you catch the ball, with what speed do you and the ball move afterward? (b) If the ball hits you and bounces off your chest, so afterward it is moving horizontally at 8.0 m/s in the opposite direction, what is your speed after the collision?

8.21 •• On a frictionless, horizontal air table, puck A (with mass 0.250 kg) is moving toward puck B (with mass 0.350 kg), which is initially at rest. After the collision, puck A has a velocity of 0.120 m/s to the left, and puck B has a velocity of 0.650 m/s to the right. (a) What was the speed of puck A before the collision? (b) Calculate the change in the total kinetic energy of the system that occurs during the collision.

8.22 •• When cars are equipped with flexible bumpers, they will bounce off each other during low-speed collisions, thus causing less damage. In one such accident, a 1750-kg car traveling to the right at 1.50 m/s collides with a 1450-kg car going to the left at 1.10 m/s. Measurements show that the heavier car's speed just after the collision was 0.250 m/s in its original direction. Ignore any road friction during the collision. (a) What was the speed of the lighter car just after the collision? (b) Calculate the change in the combined kinetic energy of the two-car system during this collision.

8.23 •• Two identical 0.900-kg masses are pressed against opposite ends of a light spring of force constant 1.75 N/cm, compressing the spring by 20.0 cm from its normal length. Find the speed of each mass when it has moved free of the spring on a frictionless, horizontal table.

8.24 • Block A in **Fig. E8.24** has mass 1.00 kg, and block B has mass 3.00 kg. The blocks are forced together, compressing a spring S between them; then the system is released from rest on a level, frictionless surface. The spring, which has negligible mass, is not fastened to either block and drops to the surface after it has expanded. Block B acquires a speed of 1.20 m/s. (a) What is the final speed of block A? (b) How much potential energy was stored in the compressed spring?

Figure **E8.24** $m_A = 1.00$ kg $m_B = 3.00$ kg

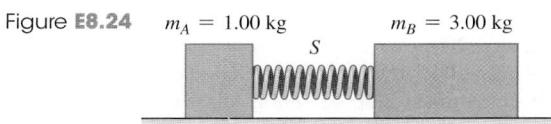

8.25 •• A hunter on a frozen, essentially frictionless pond uses a rifle that shoots 4.20-g bullets at 965 m/s. The mass of the hunter (including his gun) is 72.5 kg, and the hunter holds tight to the gun after firing it. Find the recoil velocity of the hunter if he fires the rifle (a) horizontally and (b) at 56.0° above the horizontal.

8.26 • An atomic nucleus suddenly bursts apart (fissions) into two pieces. Piece A, of mass m_A, travels off to the left with speed v_A. Piece B, of mass m_B, travels off to the right with speed v_B. (a) Use conservation of momentum to solve for v_B in terms of m_A, m_B, and v_A. (b) Use the results of part (a) to show that $K_A/K_B = m_B/m_A$, where K_A and K_B are the kinetic energies of the two pieces.

8.27 •• Two ice skaters, Daniel (mass 65.0 kg) and Rebecca (mass 45.0 kg), are practicing. Daniel stops to tie his shoelace and, while at rest, is struck by Rebecca, who is moving at 13.0 m/s before she collides with him. After the collision, Rebecca has a velocity of magnitude 8.00 m/s at an angle of 53.1° from her initial direction. Both skaters move on the frictionless, horizontal surface of the rink. (a) What are the magnitude and direction of Daniel's velocity after the collision? (b) What is the change in total kinetic energy of the two skaters as a result of the collision?

8.28 •• You are standing on a large sheet of frictionless ice and holding a large rock. In order to get off the ice, you throw the rock so it has velocity 12.0 m/s relative to the earth at an angle of 35.0° above the horizontal. If your mass is 70.0 kg and the rock's mass is 3.00 kg, what is your speed after you throw the rock? (See Discussion Question Q8.7.)

8.29 •• You (mass 55 kg) are riding a frictionless skateboard (mass 5.0 kg) in a straight line at a speed of 4.5 m/s. A friend standing on a balcony above you drops a 2.5-kg sack of flour straight down into your arms. (a) What is your new speed while you hold the sack? (b) Since the sack was dropped vertically, how can it affect your *horizontal* motion? Explain. (c) Now you try to rid yourself of the extra weight by throwing the sack straight up. What will be your speed while the sack is in the air? Explain.

8.30 • An astronaut in space cannot use a conventional means, such as a scale or balance, to determine the mass of an object. But she does have devices to measure distance and time accurately. She knows her own mass is 78.4 kg, but she is unsure of the mass of a large gas canister in the airless rocket. When this canister is approaching her at 3.50 m/s, she pushes against it, which slows it down to 1.20 m/s (but does not reverse it) and gives her a speed of 2.40 m/s. What is the mass of this canister?

8.31 •• **Asteroid Collision.** Two asteroids of equal mass in the asteroid belt between Mars and Jupiter collide with a glancing blow. Asteroid A, which was initially traveling at 40.0 m/s, is deflected 30.0° from its original direction, while asteroid B, which was initially at rest, travels at 45.0° to the original direction of A (**Fig. E8.31**). (a) Find the speed of each asteroid after the collision. (b) What fraction of the original kinetic energy of asteroid A dissipates during this collision?

Figure **E8.31**

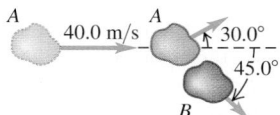

Section 8.3 Momentum Conservation and Collisions

8.32 • Two skaters collide and grab on to each other on frictionless ice. One of them, of mass 70.0 kg, is moving to the right at 4.00 m/s, while the other, of mass 65.0 kg, is moving to the left at 2.50 m/s. What are the magnitude and direction of the velocity of these skaters just after they collide?

8.33 •• A 15.0-kg fish swimming at 1.10 m/s suddenly gobbles up a 4.50-kg fish that is initially stationary. Ignore any drag effects of the water. (a) Find the speed of the large fish just after it eats the small one. (b) How much mechanical energy was dissipated during this meal?

8.34 • Two fun-loving otters are sliding toward each other on a muddy (and hence frictionless) horizontal surface. One of them, of mass 7.50 kg, is sliding to the left at 5.00 m/s, while the other, of mass 5.75 kg, is slipping to the right at 6.00 m/s. They hold fast to each other after they collide. (a) Find the magnitude and direction of the velocity of these free-spirited otters right after they collide. (b) How much mechanical energy dissipates during this play?

8.35 • **Deep Impact Mission.** In July 2005, NASA's "Deep Impact" mission crashed a 372-kg probe directly onto the surface of the comet Tempel 1, hitting the surface at 37,000 km/h. The original speed of the comet at that time was about 40,000 km/h, and its mass was estimated to be in the range $(0.10 - 2.5) \times 10^{14}$ kg. Use the smallest value of the estimated mass. (a) What change in the comet's velocity did this collision produce? Would this change be noticeable? (b) Suppose this comet were to hit the earth and fuse with it. By how much would it change our planet's velocity? Would this change be noticeable? (The mass of the earth is 5.97×10^{24} kg.)

8.36 • A 1050-kg sports car is moving westbound at 15.0 m/s on a level road when it collides with a 6320-kg truck driving east on the same road at 10.0 m/s. The two vehicles remain locked together after the collision. (a) What is the velocity (magnitude and direction) of the two vehicles just after the collision? (b) At what speed should the truck have been moving so that both it and the car are stopped in the collision? (c) Find the change in kinetic energy of the system of two vehicles for the situations of part (a) and part (b). For which situation is the change in kinetic energy greater in magnitude?

8.37 •• On a very muddy football field, a 110-kg linebacker tackles an 85-kg halfback. Immediately before the collision, the linebacker is slipping with a velocity of 8.8 m/s north and the halfback is sliding with a velocity of 7.2 m/s east. What is the velocity (magnitude and direction) at which the two players move together immediately after the collision?

8.38 •• **Accident Analysis.** Two cars collide at an intersection. Car A, with a mass of 2000 kg, is going from west to east, while car B, of mass 1500 kg, is going from north to south at 15 m/s. As a result, the two cars become enmeshed and move as one. As an expert witness, you inspect the scene and determine that, after the collision, the enmeshed cars moved at an angle of 65° south of east from the point of impact. (a) How fast were the enmeshed cars moving just after the collision? (b) How fast was car A going just before the collision?

8.39 •• Jack (mass 55.0 kg) is sliding due east with speed 8.00 m/s on the surface of a frozen pond. He collides with Jill (mass 48.0 kg), who is initially at rest. After the collision, Jack is traveling at 5.00 m/s in a direction 34.0° north of east. What is Jill's velocity (magnitude and direction) after the collision? Ignore friction.

8.40 •• BIO **Bird Defense.** To protect their young in the nest, peregrine falcons will fly into birds of prey (such as ravens) at high speed. In one such episode, a 600-g falcon flying at 20.0 m/s hit a 1.50-kg raven flying at 9.0 m/s. The falcon hit the raven at right angles to its original path and bounced back at 5.0 m/s. (These figures were estimated by the author as he watched this attack occur in northern New Mexico.) (a) By what angle did the falcon change the raven's direction of motion? (b) What was the raven's speed right after the collision?

8.41 • At the intersection of Texas Avenue and University Drive, a yellow subcompact car with mass 950 kg traveling east on University collides with a red pickup truck with mass 1900 kg that is traveling north on Texas and has run a red light (**Fig. E8.41**). The two vehicles stick together as a result of the collision, and the wreckage slides at 16.0 m/s in the direction 24.0° east of north.

Figure **E8.41**

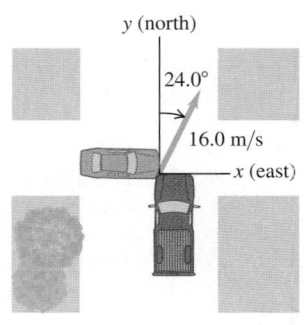

Calculate the speed of each vehicle before the collision. The collision occurs during a heavy rainstorm; ignore friction forces between the vehicles and the wet road.

8.42 •• A 5.00-g bullet is fired horizontally into a 1.20-kg wooden block resting on a horizontal surface. The coefficient of kinetic friction between block and surface is 0.20. The bullet remains embedded in the block, which is observed to slide 0.310 m along the surface before stopping. What was the initial speed of the bullet?

8.43 •• **A Ballistic Pendulum.** A 12.0-g rifle bullet is fired with a speed of 380 m/s into a ballistic pendulum with mass 6.00 kg, suspended from a cord 70.0 cm long (see Example 8.8 in Section 8.3). Compute (a) the vertical height through which the pendulum rises, (b) the initial kinetic energy of the bullet, and (c) the kinetic energy of the bullet and pendulum immediately after the bullet becomes embedded in the wood.

8.44 •• **Combining Conservation Laws.** A 15.0-kg block is attached to a very light horizontal spring of force constant 500.0 N/m and is resting on a frictionless horizontal table (**Fig. E8.44**).

Figure **E8.44**

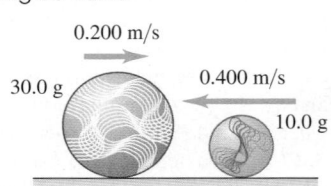

Suddenly it is struck by a 3.00-kg stone traveling horizontally at 8.00 m/s to the right, whereupon the stone rebounds at 2.00 m/s horizontally to the left. Find the maximum distance that the block will compress the spring after the collision.

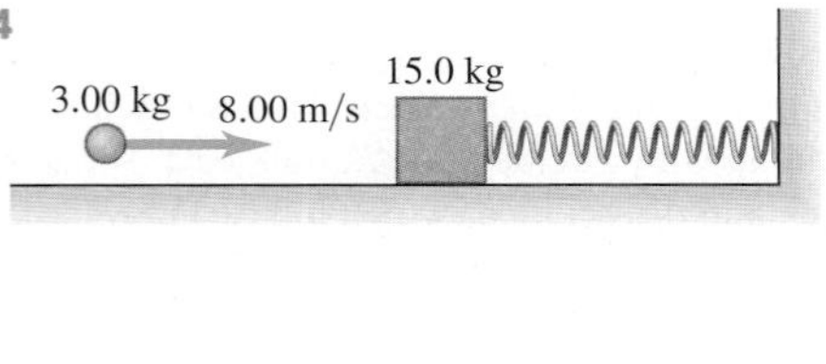

8.45 •• CP A 0.800-kg ornament is hanging by a 1.50-m wire when the ornament is suddenly hit by a 0.200-kg missile traveling horizontally at 12.0 m/s. The missile embeds itself in the ornament during the collision. What is the tension in the wire immediately after the collision?

Section 8.4 Elastic Collisions

8.46 •• A 0.150-kg glider is moving to the right with a speed of 0.80 m/s on a frictionless, horizontal air track. The glider has a head-on collision with a 0.300-kg glider that is moving to the left with a speed of 2.20 m/s. Find the final velocity (magnitude and direction) of each glider if the collision is elastic.

8.47 •• Blocks A (mass 2.00 kg) and B (mass 6.00 kg) move on a frictionless, horizontal surface. Initially, block B is at rest and block A is moving toward it at 2.00 m/s. The blocks are equipped with ideal spring bumpers, as in Example 8.10 (Section 8.4). The collision is head-on, so all motion before and after the collision is along a straight line. (a) Find the maximum energy stored in the spring bumpers and the velocity of each block at that time. (b) Find the velocity of each block after they have moved apart.

8.48 • A 10.0-g marble slides to the left at a speed of 0.400 m/s on the frictionless, horizontal surface of an icy New York sidewalk and has a head-on, elastic collision with a larger 30.0-g marble sliding to the right at a speed of 0.200 m/s (**Fig. E8.48**).

Figure **E8.48**

(a) Find the velocity of each marble (magnitude and direction) after the collision. (Since the collision is head-on, all motion is along a line.) (b) Calculate the *change in momentum* (the momentum after the collision minus the momentum before the collision) for each marble. Compare your values for each marble. (c) Calculate the *change in kinetic energy* (the kinetic energy after the collision minus the kinetic energy before the collision) for each marble. Compare your values for each marble.

8.49 •• **Moderators.** Canadian nuclear reactors use *heavy water* moderators in which elastic collisions occur between the neutrons and deuterons of mass 2.0 u (see Example 8.11 in Section 8.4). (a) What is the speed of a neutron, expressed as a fraction of its original speed, after a head-on, elastic collision with a deuteron that is initially at rest? (b) What is its kinetic energy, expressed as a fraction of its original kinetic energy? (c) How many such successive collisions will reduce the speed of a neutron to 1/59,000 of its original value?

8.50 •• You are at the controls of a particle accelerator, sending a beam of 1.50×10^7 m/s protons (mass m) at a gas target of an unknown element. Your detector tells you that some protons bounce straight back after a collision with one of the nuclei of the unknown element. All such protons rebound with a speed of 1.20×10^7 m/s. Assume that the initial speed of the target nucleus is negligible and the collision is elastic. (a) Find the mass of one nucleus of the unknown element. Express your answer in terms of the proton mass m. (b) What is the speed of the unknown nucleus immediately after such a collision?

Section 8.5 Center of Mass

8.51 • Three odd-shaped blocks of chocolate have the following masses and center-of-mass coordinates: (1) 0.300 kg, (0.200 m, 0.300 m); (2) 0.400 kg, (0.100 m, −0.400 m); (3) 0.200 kg, (−0.300 m, 0.600 m). Find the coordinates of the center of mass of the system of three chocolate blocks.

8.52 • Find the position of the center of mass of the system of the sun and Jupiter. (Since Jupiter is more massive than the rest of the solar planets combined, this is essentially the position of the center of mass of the solar system.) Does the center of mass lie inside or outside the sun? Use the data in Appendix F.

8.53 •• **Pluto and Charon.** Pluto's diameter is approximately 2370 km, and the diameter of its satellite Charon is 1250 km. Although the distance varies, they are often about 19,700 km apart, center to center. Assuming that both Pluto and Charon have the same composition and hence the same average density, find the location of the center of mass of this system relative to the center of Pluto.

8.54 • A 1200-kg SUV is moving along a straight highway at 12.0 m/s. Another car, with mass 1800 kg and speed 20.0 m/s, has its center of mass 40.0 m ahead of the center of mass of the SUV (**Fig. E8.54**). Find (a) the position of the center of mass of the system consisting of the two cars; (b) the magnitude of the system's total momentum, by using the given data; (c) the speed of the system's center of mass; (d) the system's total momentum, by using the speed of the center of mass. Compare your result with that of part (b).

Figure **E8.54**

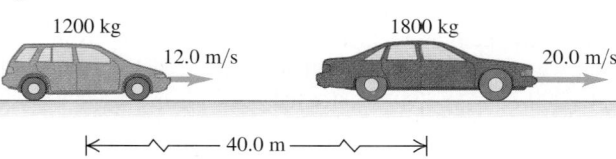

8.55 • A machine part consists of a thin, uniform 4.00-kg bar that is 1.50 m long, hinged perpendicular to a similar vertical bar of mass 3.00 kg and length 1.80 m. The longer bar has a small but dense 2.00-kg ball at one end (**Fig. E8.55**). By what distance will the center of mass of this part move horizontally and vertically if the vertical bar is pivoted counterclockwise through 90° to make the entire part horizontal?

Figure **E8.55**

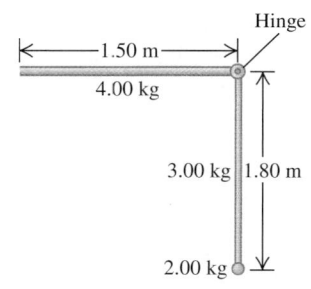

8.56 • At one instant, the center of mass of a system of two particles is located on the x-axis at $x = 2.0$ m and has a velocity of $(5.0 \text{ m/s})\hat{\imath}$. One of the particles is at the origin. The other particle has a mass of 0.10 kg and is at rest on the x-axis at $x = 8.0$ m. (a) What is the mass of the particle at the origin? (b) Calculate the total momentum of this system. (c) What is the velocity of the particle at the origin?

8.57 •• In Example 8.14 (Section 8.5), Ramon pulls on the rope to give himself a speed of 1.10 m/s. What is James's speed?

8.58 • CALC A system consists of two particles. At $t = 0$ one particle is at the origin; the other, which has a mass of 0.50 kg,

is on the y-axis at $y = 6.0$ m. At $t = 0$ the center of mass of the system is on the y-axis at $y = 2.4$ m. The velocity of the center of mass is given by $(0.75 \text{ m/s}^3)t^2\hat{\imath}$. (a) Find the total mass of the system. (b) Find the acceleration of the center of mass at any time t. (c) Find the net external force acting on the system at $t = 3.0$ s.

8.59 • CALC A radio-controlled model airplane has a momentum given by $[(-0.75 \text{ kg·m/s}^3)t^2 + (3.0 \text{ kg·m/s})]\hat{\imath} + (0.25 \text{ kg·m/s}^2)t\hat{\jmath}$. What are the x-, y-, and z-components of the net force on the airplane?

8.60 •• BIO **Changing Your Center of Mass.** To keep the calculations fairly simple but still reasonable, we model a human leg that is 92.0 cm long (measured from the hip joint) by assuming that the upper leg and the lower leg (which includes the foot) have equal lengths and are uniform. For a 70.0-kg person, the mass of the upper leg is 8.60 kg, while that of the lower leg (including the foot) is 5.25 kg. Find the location of the center of mass of this leg, relative to the hip joint, if it is (a) stretched out horizontally and (b) bent at the knee to form a right angle with the upper leg remaining horizontal.

Section 8.6 Rocket Propulsion

8.61 •• A 70-kg astronaut floating in space in a 110-kg MMU (manned maneuvering unit) experiences an acceleration of 0.029 m/s² when he fires one of the MMU's thrusters. (a) If the speed of the escaping N_2 gas relative to the astronaut is 490 m/s, how much gas is used by the thruster in 5.0 s? (b) What is the thrust of the thruster?

8.62 • A small rocket burns 0.0500 kg of fuel per second, ejecting it as a gas with a velocity relative to the rocket of magnitude 1600 m/s. (a) What is the thrust of the rocket? (b) Would the rocket operate in outer space where there is no atmosphere? If so, how would you steer it? Could you brake it?

8.63 •• Obviously, we can make rockets to go very fast, but what is a reasonable top speed? Assume that a rocket is fired from rest at a space station in deep space, where gravity is negligible. (a) If the rocket ejects gas at a relative speed of 2000 m/s and you want the rocket's speed eventually to be $1.00 \times 10^{-3}c$, where c is the speed of light in vacuum, what fraction of the initial mass of the rocket and fuel is *not* fuel? (b) What is this fraction if the final speed is to be 3000 m/s?

PROBLEMS

8.64 •• A steel ball with mass 40.0 g is dropped from a height of 2.00 m onto a horizontal steel slab. The ball rebounds to a height of 1.60 m. (a) Calculate the impulse delivered to the ball during impact. (b) If the ball is in contact with the slab for 2.00 ms, find the average force on the ball during impact.

8.65 •• Just before it is struck by a racket, a tennis ball weighing 0.560 N has a velocity of $(20.0 \text{ m/s})\hat{\imath} - (4.0 \text{ m/s})\hat{\jmath}$. During the 3.00 ms that the racket and ball are in contact, the net force on the ball is constant and equal to $-(380 \text{ N})\hat{\imath} + (110 \text{ N})\hat{\jmath}$. What are the x- and y-components (a) of the impulse of the net force applied to the ball; (b) of the final velocity of the ball?

8.66 • Three identical pucks on a horizontal air table have repelling magnets. They are held together and then released simultaneously. Each has the same speed at any instant. One puck moves due west. What is the direction of the velocity of each of the other two pucks?

8.67 •• Blocks A (mass 2.00 kg) and B (mass 10.00 kg, to the right of A) move on a frictionless, horizontal surface. Initially, block B is moving to the left at 0.500 m/s and block A is moving to the right at 2.00 m/s. The blocks are equipped with ideal spring bumpers, as in Example 8.10 (Section 8.4). The collision is head-on, so all motion before and after it is along a straight line. Find (a) the maximum energy stored in the spring bumpers and the velocity of each block at that time; (b) the velocity of each block after they have moved apart.

8.68 •• A railroad handcar is moving along straight, frictionless tracks with negligible air resistance. In the following cases, the car initially has a total mass (car and contents) of 200 kg and is traveling east with a velocity of magnitude 5.00 m/s. Find the *final velocity* of the car in each case, assuming that the handcar does not leave the tracks. (a) A 25.0-kg mass is thrown sideways out of the car with a velocity of magnitude 2.00 m/s relative to the car's initial velocity. (b) A 25.0-kg mass is thrown backward out of the car with a velocity of 5.00 m/s relative to the initial motion of the car. (c) A 25.0-kg mass is thrown into the car with a velocity of 6.00 m/s relative to the ground and opposite in direction to the initial velocity of the car.

8.69 • Spheres A (mass 0.020 kg), B (mass 0.030 kg), and C (mass 0.050 kg) are approaching the origin as they slide on a frictionless air table. The initial velocities of A and B are given in **Fig. P8.69**. All three spheres arrive at the origin at the same time and stick together. (a) What must the x- and y-components of the initial velocity of C be if all three objects are to end up moving at 0.50 m/s in the $+x$-direction after the collision? (b) If C has the velocity found in part (a), what is the change in the kinetic energy of the system of three spheres as a result of the collision?

Figure **P8.69**

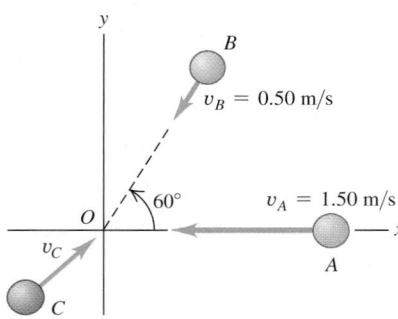

8.70 ••• You and your friends are doing physics experiments on a frozen pond that serves as a frictionless, horizontal surface. Sam, with mass 80.0 kg, is given a push and slides eastward. Abigail, with mass 50.0 kg, is sent sliding northward. They collide, and after the collision Sam is moving at 37.0° north of east with a speed of 6.00 m/s and Abigail is moving at 23.0° south of east with a speed of 9.00 m/s. (a) What was the speed of each person before the collision? (b) By how much did the total kinetic energy of the two people decrease during the collision?

8.71 •• CP An 8.00-kg block of wood sits at the edge of a frictionless table, 2.20 m above the floor. A 0.500-kg blob of clay slides along the length of the table with a speed of 24.0 m/s, strikes the block of wood, and sticks to it. The combined object leaves the edge of the table and travels to the floor. What horizontal distance has the combined object traveled when it reaches the floor?

8.72 ••• CP A small wooden block with mass 0.800 kg is suspended from the lower end of a light cord that is 1.60 m long. The

block is initially at rest. A bullet with mass 12.0 g is fired at the block with a horizontal velocity v_0. The bullet strikes the block and becomes embedded in it. After the collision the combined object swings on the end of the cord. When the block has risen a vertical height of 0.800 m, the tension in the cord is 4.80 N. What was the initial speed v_0 of the bullet?

8.73 •• **Combining Conservation Laws.** A 5.00-kg chunk of ice is sliding at 12.0 m/s on the floor of an ice-covered valley when it collides with and sticks to another 5.00-kg chunk of ice that is initially at rest (**Fig. P8.73**). Since the valley is icy, there is no friction. After the collision, how high above the valley floor will the combined chunks go?

Figure **P8.73**

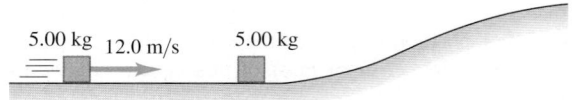

8.74 •• CP Block B (mass 4.00 kg) is at rest at the edge of a smooth platform, 2.60 m above the floor. Block A (mass 2.00 kg) is sliding with a speed of 8.00 m/s along the platform toward block B. A strikes B and rebounds with a speed of 2.00 m/s. The collision projects B horizontally off the platform. What is the speed of B just before it strikes the floor?

8.75 •• Two blocks have a spring compressed between them, as in Exercise 8.24. The spring has force constant 720 N/m and is initially compressed 0.225 m from its original length. For each block, what is (a) the acceleration just after the blocks are released; (b) the final speed after the blocks leave the spring?

8.76 •• **Automobile Accident Analysis.** You are called as an expert witness to analyze the following auto accident: Car B, of mass 1900 kg, was stopped at a red light when it was hit from behind by car A, of mass 1500 kg. The cars locked bumpers during the collision and slid to a stop with brakes locked on all wheels. Measurements of the skid marks left by the tires showed them to be 7.15 m long. The coefficient of kinetic friction between the tires and the road was 0.65. (a) What was the speed of car A just before the collision? (b) If the speed limit was 35 mph, was car A speeding, and if so, by how many miles per hour was it *exceeding* the speed limit?

8.77 •• **Accident Analysis.** A 1500-kg sedan goes through a wide intersection traveling from north to south when it is hit by a 2200-kg SUV traveling from east to west. The two cars become enmeshed due to the impact and slide as one thereafter. On-the-scene measurements show that the coefficient of kinetic friction between the tires of these cars and the pavement is 0.75, and the cars slide to a halt at a point 5.39 m west and 6.43 m south of the impact point. How fast was each car traveling just before the collision?

8.78 ••• CP A 0.150-kg frame, when suspended from a coil spring, stretches the spring 0.0400 m. A 0.200-kg lump of putty is dropped from rest onto the frame from a height of 30.0 cm (**Fig. P8.78**). Find the maximum distance the frame moves downward from its initial equilibrium position.

Figure **P8.78**

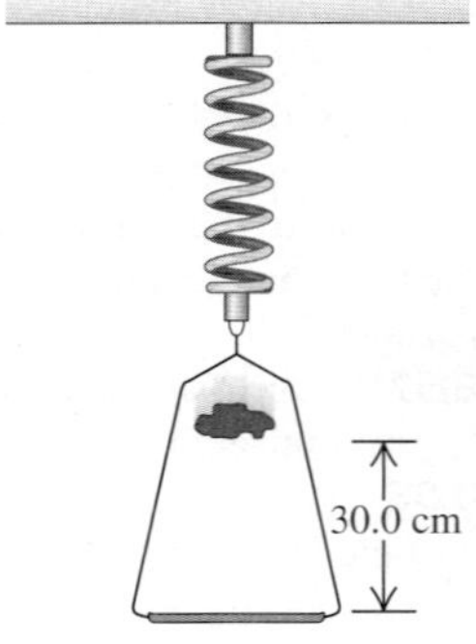

8.79 • A rifle bullet with mass 8.00 g strikes and embeds itself in a block with mass 0.992 kg that rests on a frictionless, horizontal surface and is attached to a coil spring (**Fig. P8.79**). The impact compresses the spring 15.0 cm. Calibration of the spring shows that a force of 0.750 N is required to compress the spring 0.250 cm. (a) Find the magnitude of the block's velocity just after impact. (b) What was the initial speed of the bullet?

Figure **P8.79**

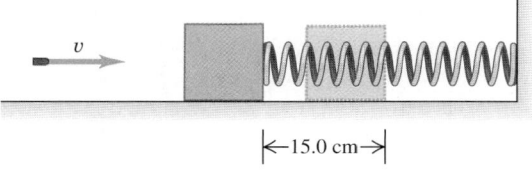

$\leftarrow$15.0 cm$\rightarrow$

8.80 •• **A Ricocheting Bullet.** A 0.100-kg stone rests on a frictionless, horizontal surface. A bullet of mass 6.00 g, traveling horizontally at 350 m/s, strikes the stone and rebounds horizontally at right angles to its original direction with a speed of 250 m/s. (a) Compute the magnitude and direction of the velocity of the stone after it is struck. (b) Is the collision perfectly elastic?

8.81 •• A movie stuntman (mass 80.0 kg) stands on a window ledge 5.0 m above the floor (**Fig. P8.81**). Grabbing a rope attached to a chandelier, he swings down to grapple with the movie's villain (mass 70.0 kg), who is standing directly under the chandelier. (Assume that the stuntman's center of mass moves downward 5.0 m. He releases the rope just as he reaches the villain.) (a) With what speed do the entwined foes start to slide across the floor? (b) If the coefficient of kinetic friction of their bodies with the floor is $\mu_k = 0.250$, how far do they slide?

Figure **P8.81**

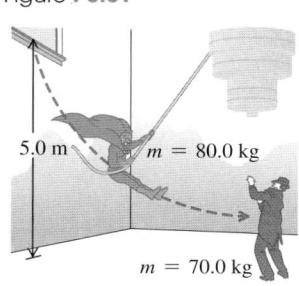

5.0 m $m = 80.0$ kg

$m = 70.0$ kg

8.82 •• CP Two identical masses are released from rest in a smooth hemispherical bowl of radius R from the positions shown in **Fig. P8.82**. Ignore friction between the masses and the surface of the bowl. If the masses stick together when they collide, how high above the bottom of the bowl will they go after colliding?

Figure **P8.82**

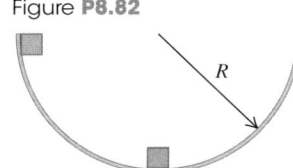

R

8.83 •• A ball with mass M, moving horizontally at 4.00 m/s, collides elastically with a block with mass $3M$ that is initially hanging at rest from the ceiling on the end of a 50.0-cm wire. Find the maximum angle through which the block swings after it is hit.

8.84 ••• CP A 20.00-kg lead sphere is hanging from a hook by a thin wire 2.80 m long and is free to swing in a complete circle. Suddenly it is struck horizontally by a 5.00-kg steel dart that embeds itself in the lead sphere. What must be the minimum initial speed of the dart so that the combination makes a complete circular loop after the collision?

8.85 •• A 4.00-g bullet, traveling horizontally with a velocity of magnitude 400 m/s, is fired into a wooden block with mass 0.800 kg, initially at rest on a level surface. The bullet passes through the block and emerges with its speed reduced to 190 m/s. The block slides a distance of 72.0 cm along the surface from its initial position. (a) What is the coefficient of kinetic friction between block and surface? (b) What is the decrease in kinetic energy of the bullet? (c) What is the kinetic energy of the block at the instant after the bullet passes through it?

8.86 •• A 5.00-g bullet is shot *through* a 1.00-kg wood block suspended on a string 2.00 m long. The center of mass of the block rises a distance of 0.38 cm. Find the speed of the bullet as it emerges from the block if its initial speed is 450 m/s.

8.87 •• CP In a shipping company distribution center, an open cart of mass 50.0 kg is rolling to the left at a speed of 5.00 m/s (**Fig. P8.87**). Ignore friction between the cart and the floor. A 15.0-kg package slides down a chute that is inclined at 37° from the horizontal and leaves the end of the chute with a speed of 3.00 m/s. The package lands in the cart and they roll together. If the lower end of the chute is a vertical distance of 4.00 m above the bottom of the cart, what are (a) the speed of the package just before it lands in the cart and (b) the final speed of the cart?

Figure **P8.87**

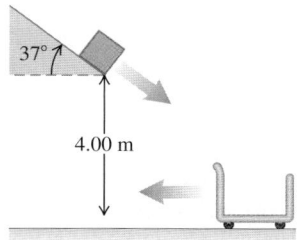

37°

4.00 m

8.88 ••• **Neutron Decay.** A neutron at rest decays (breaks up) to a proton and an electron. Energy is released in the decay and appears as kinetic energy of the proton and electron. The mass of a proton is 1836 times the mass of an electron. What fraction of the total energy released goes into the kinetic energy of the proton?

8.89 • **Antineutrino.** In beta decay, a nucleus emits an electron. A ^{210}Bi (bismuth) nucleus at rest undergoes beta decay to ^{210}Po (polonium). Suppose the emitted electron moves to the right with a momentum of 5.60×10^{-22} kg · m/s. The ^{210}Po nucleus, with mass 3.50×10^{-25} kg, recoils to the left at a speed of 1.14×10^3 m/s. Momentum conservation requires that a second particle, called an antineutrino, must also be emitted. Calculate the magnitude and direction of the momentum of the antineutrino that is emitted in this decay.

8.90 •• Jonathan and Jane are sitting in a sleigh that is at rest on frictionless ice. Jonathan's weight is 800 N, Jane's weight is 600 N, and that of the sleigh is 1000 N. They see a poisonous spider on the floor of the sleigh and immediately jump off. Jonathan jumps to the left with a velocity of 5.00 m/s at 30.0° above the horizontal (relative to the ice), and Jane jumps to the right at 7.00 m/s at 36.9° above the horizontal (relative to the ice). Calculate the sleigh's horizontal velocity (magnitude and direction) after they jump out.

8.91 •• Friends Burt and Ernie stand at opposite ends of a uniform log that is floating in a lake. The log is 3.0 m long and has mass 20.0 kg. Burt has mass 30.0 kg; Ernie has mass 40.0 kg. Initially, the log and the two friends are at rest relative to the shore. Burt then offers Ernie a cookie, and Ernie walks to Burt's end of the log to get it. Relative to the shore, what distance has the log moved by the time Ernie reaches Burt? Ignore any horizontal force that the water exerts on the log, and assume that neither friend falls off the log.

8.92 •• A 45.0-kg woman stands up in a 60.0-kg canoe 5.00 m long. She walks from a point 1.00 m from one end to a point 1.00 m from the other end (**Fig. P8.92**). If you ignore resistance to motion of the canoe in the water, how far does the canoe move during this process?

Figure **P8.92**

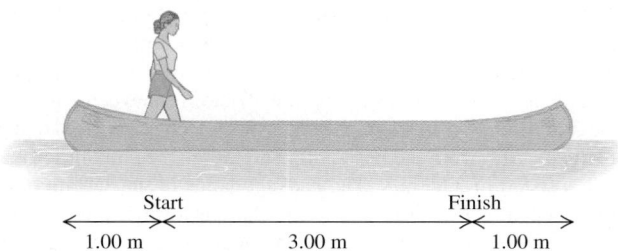

8.93 •• You are standing on a concrete slab that in turn is resting on a frozen lake. Assume there is no friction between the slab and the ice. The slab has a weight five times your weight. If you begin walking forward at 2.00 m/s relative to the ice, with what speed, relative to the ice, does the slab move?

8.94 •• CP In a fireworks display, a rocket is launched from the ground with a speed of 18.0 m/s and a direction of 51.0° above the horizontal. During the flight, the rocket explodes into two pieces of equal mass (see Fig. 8.32). (a) What horizontal distance from the launch point will the center of mass of the two pieces be after both have landed on the ground? (b) If one piece lands a horizontal distance of 26.0 m from the launch point, where does the other piece land?

8.95 •• A 7.0-kg shell at rest explodes into two fragments, one with a mass of 2.0 kg and the other with a mass of 5.0 kg. If the heavier fragment gains 100 J of kinetic energy from the explosion, how much kinetic energy does the lighter one gain?

8.96 •• CP A 20.0-kg projectile is fired at an angle of 60.0° above the horizontal with a speed of 80.0 m/s. At the highest point of its trajectory, the projectile explodes into two fragments with equal mass, one of which falls vertically with zero initial speed. Ignore air resistance. (a) How far from the point of firing does the other fragment strike if the terrain is level? (b) How much energy is released during the explosion?

8.97 ••• CP A fireworks rocket is fired vertically upward. At its maximum height of 80.0 m, it explodes and breaks into two pieces: one with mass 1.40 kg and the other with mass 0.28 kg. In the explosion, 860 J of chemical energy is converted to kinetic energy of the two fragments. (a) What is the speed of each fragment just after the explosion? (b) It is observed that the two fragments hit the ground at the same time. What is the distance between the points on the ground where they land? Assume that the ground is level and air resistance can be ignored.

8.98 ••• A 12.0-kg shell is launched at an angle of 55.0° above the horizontal with an initial speed of 150 m/s. At its highest point, the shell explodes into two fragments, one three times heavier than the other. The two fragments reach the ground at the same time. Ignore air resistance. If the heavier fragment lands back at the point from which the shell was launched, where will the lighter fragment land, and how much energy was released in the explosion?

8.99 • CP An outlaw cuts loose a wagon with two boxes of gold, of total mass 300 kg, when the wagon is at rest 50 m up a 6.0° slope. The outlaw plans to have the wagon roll down the slope and across the level ground, and then fall into a canyon where his accomplices wait. But in a tree 40 m from the canyon's cliff wait the

Lone Ranger (mass 75.0 kg) and Tonto (mass 60.0 kg). They drop vertically into the wagon as it passes beneath them (**Fig. P8.99**). (a) If they require 5.0 s to grab the gold and jump out, will they make it before the wagon goes over the cliff? The wagon rolls with negligible friction. (b) When the two heroes drop into the wagon, is the kinetic energy of the system of heroes plus wagon conserved? If not, does it increase or decrease, and by how much?

Figure **P8.99**

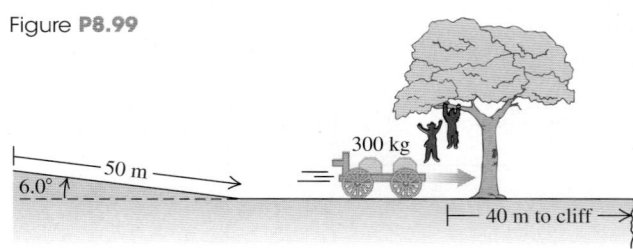

8.100 •• DATA A 2004 Prius with a 150-lb driver and no passengers weighs 3071 lb. The car is initially at rest. Starting at $t = 0$, a net horizontal force $F_x(t)$ in the $+x$-direction is applied to the car. The force as a function of time is given in **Fig. P8.100**. (a) For the time interval $t = 0$ to $t = 4.50$ s, what is the impulse applied to the car? (b) What is the speed of the car at $t = 4.50$ s? (c) At $t = 4.50$ s, the 3500-N net force is replaced by a constant net braking force $B_x = -5200$ N. Once the braking force is first applied, how long does it take the car to stop? (d) How much work must be done on the car by the braking force to stop the car? (e) What distance does the car travel from the time the braking force is first applied until the car stops?

Figure **P8.100**

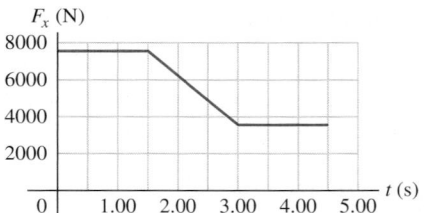

8.101 •• DATA In your job in a police lab, you must design an apparatus to measure the muzzle velocities of bullets fired from handguns. Your solution is to attach a 2.00-kg wood block that rests on a horizontal surface to a light horizontal spring. The other end of the spring is attached to a wall. Initially the spring is at its equilibrium length. A bullet is fired horizontally into the block and remains embedded in it. After the bullet strikes the block, the block compresses the spring a maximum distance d. You have measured that the coefficient of kinetic friction between the block and the horizontal surface is 0.38. The table lists some firearms that you will test:

Bullet ID	Type	Bullet Mass (grains)	Muzzle Velocity (ft/s)
A	.38Spec Glaser Blue	80	1667
B	.38Spec Federal	125	945
C	.44Spec Remington	240	851
D	.44Spec Winchester	200	819
E	0.45ACP Glaser Blue	140	1355

Source: www.chuckhawks.com

A grain is a unit of mass equal to 64.80 mg. (a) Of bullets A through E, which will produce the maximum compression of the spring? The minimum? (b) You want the maximum compression of the spring to be 0.25 m. What must be the force constant of the spring? (c) For the bullet that produces the minimum spring compression, what is the compression d if the spring has the force constant calculated in part (b)?

8.102 •• DATA For the Texas Department of Public Safety, you are investigating an accident that occurred early on a foggy morning in a remote section of the Texas Panhandle. A 2012 Prius traveling due north collided in a highway intersection with a 2013 Dodge Durango that was traveling due east. After the collision, the wreckage of the two vehicles was locked together and skidded across the level ground until it struck a tree. You measure that the tree is 35 ft from the point of impact. The line from the point of impact to the tree is in a direction 39° north of east. From experience, you estimate that the coefficient of kinetic friction between the ground and the wreckage is 0.45. Shortly before the collision, a highway patrolman with a radar gun measured the speed of the Prius to be 50 mph and, according to a witness, the Prius driver made no attempt to slow down. Four people with a total weight of 460 lb were in the Durango. The only person in the Prius was the 150-lb driver. The Durango with its passengers had a weight of 6500 lb, and the Prius with its driver had a weight of 3042 lb. (a) What was the Durango's speed just before the collision? (b) How fast was the wreckage traveling just before it struck the tree?

CHALLENGE PROBLEMS

8.103 • CALC **A Variable-Mass Raindrop.** In a rocket-propulsion problem the mass is variable. Another such problem is a raindrop falling through a cloud of small water droplets. Some of these small droplets adhere to the raindrop, thereby *increasing* its mass as it falls. The force on the raindrop is

$$F_{ext} = \frac{dp}{dt} = m\frac{dv}{dt} + v\frac{dm}{dt}$$

Suppose the mass of the raindrop depends on the distance x that it has fallen. Then $m = kx$, where k is a constant, and $dm/dt = kv$. This gives, since $F_{ext} = mg$,

$$mg = m\frac{dv}{dt} + v(kv)$$

Or, dividing by k,

$$xg = x\frac{dv}{dt} + v^2$$

This is a differential equation that has a solution of the form $v = at$, where a is the acceleration and is constant. Take the initial velocity of the raindrop to be zero. (a) Using the proposed solution for v, find the acceleration a. (b) Find the distance the raindrop has fallen in $t = 3.00$ s. (c) Given that $k = 2.00$ g/m, find the mass of the raindrop at $t = 3.00$ s. (For many more intriguing aspects of this problem, see K. S. Krane, *American Journal of Physics*, Vol. 49 (1981), pp. 113–117.)

8.104 •• CALC In Section 8.5 we calculated the center of mass by considering objects composed of a *finite* number of point masses or objects that, by symmetry, could be represented by a finite number of point masses. For a solid object whose mass distribution does not allow for a simple determination of the center of mass by symmetry, the sums of Eqs. (8.28) must be generalized to integrals

$$x_{cm} = \frac{1}{M}\int x\,dm \qquad y_{cm} = \frac{1}{M}\int y\,dm$$

where x and y are the coordinates of the small piece of the object that has mass dm. The integration is over the whole of the object. Consider a thin rod of length L, mass M, and cross-sectional area A. Let the origin of the coordinates be at the left end of the rod and the positive x-axis lie along the rod. (a) If the density $\rho = M/V$ of the object is uniform, perform the integration described above to show that the x-coordinate of the center of mass of the rod is at its geometrical center. (b) If the density of the object varies linearly with x—that is, $\rho = \alpha x$, where α is a positive constant—calculate the x-coordinate of the rod's center of mass.

8.105 •• CALC Use the methods of Challenge Problem 8.104 to calculate the x- and y-coordinates of the center of mass of a semicircular metal plate with uniform density ρ and thickness t. Let the radius of the plate be a. The mass of the plate is thus $M = \frac{1}{2}\rho\pi a^2 t$. Use the coordinate system indicated in **Fig. P8.105**.

Figure **P8.105**

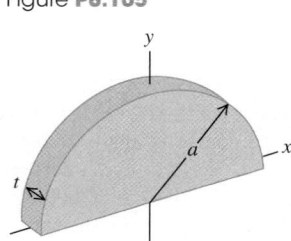

PASSAGE PROBLEMS

BIO **MOMENTUM AND THE ARCHERFISH.** Archerfish are tropical fish that hunt by shooting drops of water from their mouths at insects above the water's surface to knock them into the water, where the fish can eat them. A 65-g fish at rest just at the surface of the water can expel a 0.30-g drop of water in a short burst of 5.0 ms. High-speed measurements show that the water has a speed of 2.5 m/s just after the archerfish expels it.

8.106 What is the momentum of one drop of water immediately after it leaves the fish's mouth? (a) 7.5×10^{-4} kg·m/s; (b) 1.5×10^{-4} kg·m/s; (c) 7.5×10^{-3} kg·m/s; (d) 1.5×10^{-3} kg·m/s.

8.107 What is the speed of the archerfish immediately after it expels the drop of water? (a) 0.0025 m/s; (b) 0.012 m/s; (c) 0.75 m/s; (d) 2.5 m/s.

8.108 What is the average force the fish exerts on the drop of water? (a) 0.00015 N; (b) 0.00075 N; (c) 0.075 N; (d) 0.15 N.

8.109 The fish shoots the drop of water at an insect that hovers on the water's surface, so just before colliding with the insect, the drop is still moving at the speed it had when it left the fish's mouth. In the collision, the drop sticks to the insect, and the speed of the insect and water just after the collision is measured to be 2.0 m/s. What is the insect's mass? (a) 0.038 g; (b) 0.075 g; (c) 0.24 g; (d) 0.38 g.

Answers

Chapter Opening Question ?

(ii) All three bullets have the same magnitude of momentum $p = mv$ (the product of mass and speed), but the fast, lightweight bullet has twice the kinetic energy $K = \frac{1}{2}mv^2$ of the .22-caliber bullet and four times the kinetic energy of the heavyweight bullet. Hence, the lightweight bullet can do the most work on the carrot (and do the most damage) in the process of coming to a halt (see Section 8.1).

Test Your Understanding Questions

8.1 (v), (i) and (ii) (tied for second place), (iii) and (iv) (tied for third place) We use two interpretations of the impulse of the net force: (1) the net force multiplied by the time that the net force acts, and (2) the change in momentum of the particle on which the net force acts. Which interpretation we use depends on what information we are given. We take the positive x-direction to be to the east. (i) The force is not given, so we use interpretation 2: $J_x = mv_{2x} - mv_{1x} = (1000\ \text{kg})(0) - (1000\ \text{kg})(25\ \text{m/s}) = -25{,}000\ \text{kg} \cdot \text{m/s}$, so the magnitude of the impulse is $25{,}000\ \text{kg} \cdot \text{m/s} = 25{,}000\ \text{N} \cdot \text{s}$. (ii) For the same reason and values as in (i), we use interpretation 2, and the magnitude of the impulse is again $25{,}000\ \text{N} \cdot \text{s}$. (iii) The final velocity is not given, so we use interpretation 1: $J_x = (\Sigma F_x)_{\text{av}}(t_2 - t_1) = (2000\ \text{N})(10\ \text{s}) = 20{,}000\ \text{N} \cdot \text{s}$, so the magnitude of the impulse is $20{,}000\ \text{N} \cdot \text{s}$. (iv) For the same reason as in (iii), we use interpretation 1: $J_x = (\Sigma F_x)_{\text{av}}(t_2 - t_1) = (-2000\ \text{N})(10\ \text{s}) = -20{,}000\ \text{N} \cdot \text{s}$, so the magnitude of the impulse is $20{,}000\ \text{N} \cdot \text{s}$. (v) The force is not given, so we use interpretation 2: $J_x = mv_{2x} - mv_{1x} = (1000\ \text{kg})(-25\ \text{m/s}) - (1000\ \text{kg})(25\ \text{m/s}) = -50{,}000\ \text{kg} \cdot \text{m/s}$, so the magnitude of the impulse is $50{,}000\ \text{kg} \cdot \text{m/s} = 50{,}000\ \text{N} \cdot \text{s}$.

8.2 (a) $v_{C2x} > 0, v_{C2y} > 0$, **(b) piece C** There are no external horizontal forces, so the x- and y-components of the total momentum of the system are conserved. Both components of the total momentum are zero before the spring releases, so they must be zero after the spring releases. Hence,

$$P_x = 0 = m_A v_{A2x} + m_B v_{B2x} + m_C v_{C2x}$$
$$P_y = 0 = m_A v_{A2y} + m_B v_{B2y} + m_C v_{C2y}$$

We are given that $m_A = m_B = m_C$, $v_{A2x} < 0$, $v_{A2y} = 0$, $v_{B2x} = 0$, and $v_{B2y} < 0$. You can solve the above equations to show that $v_{C2x} = -v_{A2x} > 0$ and $v_{C2y} = -v_{B2y} > 0$, so both velocity components of piece C are positive. Piece C has speed $\sqrt{v_{C2x}^2 + v_{C2y}^2} = \sqrt{v_{A2x}^2 + v_{B2y}^2}$, which is greater than the speed of either piece A or piece B.

8.3 (a) elastic, (b) inelastic, (c) completely inelastic In each case gravitational potential energy is converted to kinetic energy as the ball falls, and the collision is between the ball and the ground. In (a) all of the initial energy is converted back to gravitational potential energy, so no kinetic energy is lost in the bounce and the collision is elastic. In (b) there is less gravitational potential energy at the end than at the beginning, so some kinetic energy is lost in the bounce. Hence the collision is inelastic. In (c) the ball loses all of its kinetic energy, the ball and the ground stick together, and the collision is completely inelastic.

8.4 worse After colliding with a water molecule initially at rest, the neutron has speed $\left|(m_{\text{n}} - m_{\text{w}})/(m_{\text{n}} + m_{\text{w}})\right| = \left|(1.0\ \text{u} - 18\ \text{u})/(1.0\ \text{u} + 18\ \text{u})\right| = \frac{17}{19}$ of its initial speed, and its kinetic energy is $\left(\frac{17}{19}\right)^2 = 0.80$ of the initial value. Hence a water molecule is a worse moderator than a carbon atom, for which the corresponding numbers are $\frac{11}{13}$ and $\left(\frac{11}{13}\right)^2 = 0.72$.

8.5 no If gravity is the only force acting on the system of two fragments, the center of mass will follow the parabolic trajectory of a freely falling object. Once a fragment lands, however, the ground exerts a normal force on that fragment. Hence the net force on the system has changed, and the trajectory of the center of mass changes in response.

8.6 (a) increasing, (b) decreasing From Eqs. (8.37) and (8.38), the thrust F is equal to $m(dv/dt)$, where m is the rocket's mass and dv/dt is its acceleration. Because m decreases with time, if the thrust F is constant, then the acceleration must increase with time (the same force acts on a smaller mass); if the acceleration dv/dt is constant, then the thrust must decrease with time (a smaller force is all that's needed to accelerate a smaller mass).

Bridging Problem

(a) $1.00\ \text{m/s}$ to the right **(b)** Elastic
(c) $1.93\ \text{m/s}$ at $-30.4°$
(d) $2.31\ \text{kg} \cdot \text{m/s}$ at $149.6°$ **(e)** Inelastic
(f) $1.67\ \text{m/s}$ in the positive x-direction

? Each blade on a rotating airplane propeller is like a long, thin rod. If each blade were stretched to double its length (while the mass of each blade and the propeller's angular speed stay the same), by what factor would the kinetic energy of the rotating blades increase? (i) 2; (ii) 4; (iii) 8; (iv) the kinetic energy would not change; (v) the kinetic energy would decrease, not increase.

9 ROTATION OF RIGID BODIES

LEARNING GOALS

Looking forward at …

9.1 How to describe the rotation of a rigid body in terms of angular coordinate, angular velocity, and angular acceleration.

9.2 How to analyze rigid-body rotation when the angular acceleration is constant.

9.3 How to relate the rotation of a rigid body to the linear velocity and linear acceleration of a point on the body.

9.4 The meaning of a body's moment of inertia about a rotation axis, and how it relates to rotational kinetic energy.

9.5 How to relate the values of a body's moment of inertia for two different but parallel rotation axes.

9.6 How to calculate the moment of inertia of bodies with various shapes.

Looking back at …

1.10 Vector product of two vectors.

2.2–2.4 Linear velocity, linear acceleration, and motion with constant acceleration.

3.4 Motion in a circle.

7.1 Using mechanical energy to solve problems.

What do the motions of an airplane propeller, a Blu-ray disc, a Ferris wheel, and a circular saw blade have in common? None of these can be represented adequately as a moving *point;* each involves a body that *rotates* about an axis that is stationary in some inertial frame of reference.

Rotation occurs at all scales, from the motions of electrons in atoms to the motions of entire galaxies. We need to develop some general methods for analyzing the motion of a rotating body. In this chapter and the next we consider bodies that have definite size and definite shape, and that in general can have rotational as well as translational motion.

Real-world bodies can be very complicated; the forces that act on them can deform them—stretching, twisting, and squeezing them. We'll ignore these deformations for now and assume that the body has a perfectly definite and unchanging shape and size. We call this idealized model a **rigid body.** This chapter and the next are mostly about rotational motion of a rigid body.

We begin with kinematic language for *describing* rotational motion. Next we look at the kinetic energy of rotation, the key to using energy methods for rotational motion. Then in Chapter 10 we'll develop dynamic principles that relate the forces on a body to its rotational motion.

9.1 ANGULAR VELOCITY AND ACCELERATION

In analyzing rotational motion, let's think first about a rigid body that rotates about a *fixed axis*—an axis that is at rest in some inertial frame of reference and does not change direction relative to that frame. The rotating rigid body might be a motor shaft, a chunk of beef on a barbecue skewer, or a merry-go-round.

Figure 9.1 (next page) shows a rigid body rotating about a fixed axis. The axis passes through point O and is perpendicular to the plane of the diagram, which we'll call the xy-plane. One way to describe the rotation of this body would be to choose a particular point P on the body and to keep track of the x- and

9.1 A speedometer needle (an example of a rigid body) rotating counterclockwise about a fixed axis.

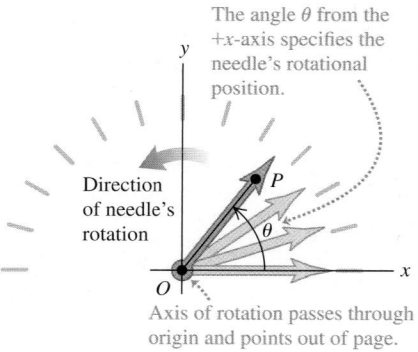

The angle θ from the +x-axis specifies the needle's rotational position.

Direction of needle's rotation

Axis of rotation passes through origin and points out of page.

9.2 Measuring angles in radians.

(a)

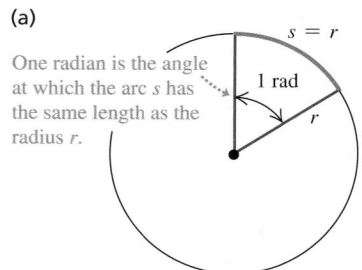

One radian is the angle at which the arc s has the same length as the radius r.

$s = r$
1 rad

(b)

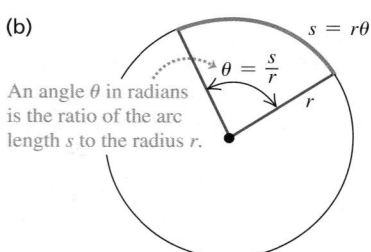

An angle θ in radians is the ratio of the arc length s to the radius r.

$s = r\theta$
$\theta = \dfrac{s}{r}$

y-coordinates of P. This isn't very convenient, since it takes two numbers (the two coordinates x and y) to specify the rotational position of the body. Instead, we notice that the line OP is fixed in the body and rotates with it. The angle θ that OP makes with the +x-axis is a single **angular cooordinate** that completely describes the body's rotational position.

The angular coordinate θ of a rigid body rotating around a fixed axis can be positive or negative. If we choose positive angles to be measured counterclockwise from the positive x-axis, then the angle θ in Fig. 9.1 is positive. If we instead choose the positive rotation direction to be clockwise, then θ in Fig. 9.1 is negative. When we considered the motion of a particle along a straight line, it was essential to specify the direction of positive displacement along that line; when we discuss rotation around a fixed axis, it's just as essential to specify the direction of positive rotation.

The most natural way to measure the angle θ is not in degrees but in **radians.** As **Fig. 9.2a** shows, one radian (1 rad) is the angle subtended at the center of a circle by an arc with a length equal to the radius of the circle. In Fig. 9.2b an angle θ is subtended by an arc of length s on a circle of radius r. The value of θ (in radians) is equal to s divided by r:

$$\theta = \frac{s}{r} \quad \text{or} \quad s = r\theta \quad (\theta \text{ in radians}) \tag{9.1}$$

An angle in radians is the ratio of two lengths, so it is a pure number, without dimensions. If $s = 3.0$ m and $r = 2.0$ m, then $\theta = 1.5$, but we will often write this as 1.5 rad to distinguish it from an angle measured in degrees or revolutions.

The circumference of a circle (that is, the arc length all the way around the circle) is 2π times the radius, so there are 2π (about 6.283) radians in one complete revolution (360°). Therefore

$$1 \text{ rad} = \frac{360°}{2\pi} = 57.3°$$

Similarly, $180° = \pi$ rad, $90° = \pi/2$ rad, and so on. If we had measured angle θ in degrees, we would have needed an extra factor of $(2\pi/360)$ on the right-hand side of $s = r\theta$ in Eq. (9.1). By measuring angles in radians, we keep the relationship between angle and distance along an arc as simple as possible.

Angular Velocity

The coordinate θ shown in Fig. 9.1 specifies the rotational position of a rigid body at a given instant. We can describe the rotational *motion* of such a rigid body in terms of the rate of change of θ. We'll do this in an analogous way to our description of straight-line motion in Chapter 2. In **Fig. 9.3a**, a reference line OP in a rotating body makes an angle θ_1 with the +x-axis at time t_1. At a later time t_2 the angle has changed to θ_2. We define the **average angular velocity** $\omega_{\text{av-}z}$ (the

9.3 (a) Angular displacement $\Delta\theta$ of a rotating body. (b) Every part of a rotating rigid body has the same average angular velocity $\Delta\theta/\Delta t$.

(a)

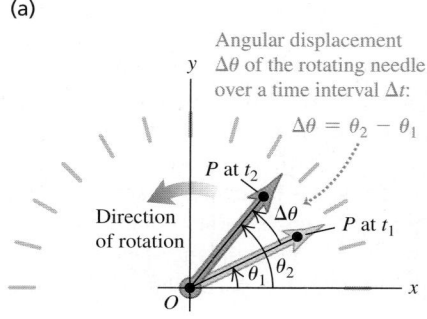

Angular displacement $\Delta\theta$ of the rotating needle over a time interval Δt:
$\Delta\theta = \theta_2 - \theta_1$

P at t_2

Direction of rotation

P at t_1

(b)

Greek letter omega) of the body in the time interval $\Delta t = t_2 - t_1$ as the ratio of the **angular displacement** $\Delta\theta = \theta_2 - \theta_1$ to Δt:

$$\omega_{\text{av-}z} = \frac{\theta_2 - \theta_1}{t_2 - t_1} = \frac{\Delta\theta}{\Delta t} \tag{9.2}$$

The subscript z indicates that the body in Fig. 9.3a is rotating about the z-axis, which is perpendicular to the plane of the diagram. The **instantaneous angular velocity** ω_z is the limit of $\omega_{\text{av-}z}$ as Δt approaches zero:

The **instantaneous angular velocity** of a rigid body rotating around the z-axis ...

$$\omega_z = \lim_{\Delta t \to 0} \frac{\Delta\theta}{\Delta t} = \frac{d\theta}{dt} \tag{9.3}$$

... equals the limit of the body's average angular velocity as the time interval approaches zero ...

... and equals the instantaneous rate of change of the body's angular coordinate.

When we refer simply to "angular velocity," we mean the instantaneous angular velocity, not the average angular velocity.

The angular velocity ω_z can be positive or negative, depending on the direction in which the rigid body is rotating (**Fig. 9.4**). The angular *speed* ω, which we'll use in Sections 9.3 and 9.4, is the magnitude of angular velocity. Like linear speed v, the angular speed is never negative.

CAUTION Angular velocity vs. linear velocity Keep in mind the distinction between angular velocity ω_z and *linear velocity* v_x (see Section 2.2). If an object has a linear velocity v_x, the object as a whole is *moving* along the x-axis. By contrast, if an object has an angular velocity ω_z, then it is *rotating* around the z-axis. We do *not* mean that the object is moving along the z-axis. ▌

Different points on a rotating rigid body move different distances in a given time interval, depending on how far each point lies from the rotation axis. But because the body is rigid, *all* points rotate through the same angle in the same time (Fig. 9.3b). Hence *at any instant, every part of a rotating rigid body has the same angular velocity.*

If angle θ is in radians, the unit of angular velocity is the radian per second (rad/s). Other units, such as the revolution per minute (rev/min or rpm), are often used. Since 1 rev = 2π rad, two useful conversions are

$$1 \text{ rev/s} = 2\pi \text{ rad/s} \quad \text{and} \quad 1 \text{ rev/min} = 1 \text{ rpm} = \frac{2\pi}{60} \text{rad/s}$$

That is, 1 rad/s is about 10 rpm.

9.4 A rigid body's average angular velocity (shown here) and instantaneous angular velocity can be positive or negative.

We choose the angle θ to increase in the counterclockwise rotation.

Counterclockwise rotation:	Clockwise rotation:
θ increases, so angular velocity is positive. $\Delta\theta > 0$, so $\omega_{\text{av-}z} = \Delta\theta/\Delta t > 0$	θ decreases, so angular velocity is negative. $\Delta\theta < 0$, so $\omega_{\text{av-}z} = \Delta\theta/\Delta t < 0$

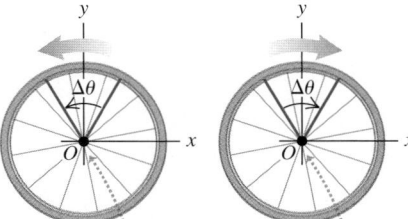

Axis of rotation (z-axis) passes through origin and points out of page.

EXAMPLE 9.1 CALCULATING ANGULAR VELOCITY

The angular position θ of a 0.36-m-diameter flywheel is given by

$$\theta = (2.0 \text{ rad/s}^3)t^3$$

(a) Find θ, in radians and in degrees, at $t_1 = 2.0$ s and $t_2 = 5.0$ s. (b) Find the distance that a particle on the flywheel rim moves from $t_1 = 2.0$ s to $t_2 = 5.0$ s. (c) Find the average angular velocity, in rad/s and in rev/min, over that interval. (d) Find the instantaneous angular velocities at $t_1 = 2.0$ s and $t_2 = 5.0$ s.

SOLUTION

IDENTIFY and SET UP: Our target variables are θ_1 and θ_2 (the angular positions at times t_1 and t_2) and the angular displacement $\Delta\theta = \theta_2 - \theta_1$. We'll find these from the given expression for θ as a function of time. Knowing $\Delta\theta$, we'll find the distance traveled

and the average angular velocity between t_1 and t_2 by using Eqs. (9.1) and (9.2), respectively. To find the instantaneous angular velocities ω_{1z} (at time t_1) and ω_{2z} (at time t_2), we'll take the derivative of the given equation for θ with respect to time, as in Eq. (9.3).

EXECUTE: (a) We substitute the values of t into the equation for θ:

$$\theta_1 = (2.0 \text{ rad/s}^3)(2.0 \text{ s})^3 = 16 \text{ rad}$$

$$= (16 \text{ rad})\frac{360°}{2\pi \text{ rad}} = 920°$$

$$\theta_2 = (2.0 \text{ rad/s}^3)(5.0 \text{ s})^3 = 250 \text{ rad}$$

$$= (250 \text{ rad})\frac{360°}{2\pi \text{ rad}} = 14{,}000°$$

Continued

(b) During the interval from t_1 to t_2 the flywheel's angular displacement is $\Delta\theta = \theta_2 - \theta_1 = 250 \text{ rad} - 16 \text{ rad} = 234 \text{ rad}$. The radius r is half the diameter, or 0.18 m. To use Eq. (9.1), the angles *must* be expressed in radians:

$$s = r\theta_2 - r\theta_1 = r\Delta\theta = (0.18 \text{ m})(234 \text{ rad}) = 42 \text{ m}$$

We can drop "radians" from the unit for s because θ is a dimensionless number; like r, s is measured in meters.

(c) From Eq. (9.2),

$$\omega_{\text{av-}z} = \frac{\theta_2 - \theta_1}{t_2 - t_1} = \frac{250 \text{ rad} - 16 \text{ rad}}{5.0 \text{ s} - 2.0 \text{ s}} = 78 \text{ rad/s}$$

$$= \left(78\frac{\text{rad}}{\text{s}}\right)\left(\frac{1 \text{ rev}}{2\pi \text{ rad}}\right)\left(\frac{60 \text{ s}}{1 \text{ min}}\right) = 740 \text{ rev/min}$$

(d) From Eq. (9.3),

$$\omega_z = \frac{d\theta}{dt} = \frac{d}{dt}[(2.0 \text{ rad/s}^3)t^3] = (2.0 \text{ rad/s}^3)(3t^2)$$

$$= (6.0 \text{ rad/s}^3)t^2$$

At times $t_1 = 2.0$ s and $t_2 = 5.0$ s we have

$$\omega_{1z} = (6.0 \text{ rad/s}^3)(2.0 \text{ s})^2 = 24 \text{ rad/s}$$

$$\omega_{2z} = (6.0 \text{ rad/s}^3)(5.0 \text{ s})^2 = 150 \text{ rad/s}$$

EVALUATE: The angular velocity $\omega_z = (6.0 \text{ rad/s}^3)t^2$ increases with time. Our results are consistent with this; the instantaneous angular velocity at the end of the interval ($\omega_{2z} = 150 \text{ rad/s}$) is greater than at the beginning ($\omega_{1z} = 24 \text{ rad/s}$), and the average angular velocity $\omega_{\text{av-}z} = 78 \text{ rad/s}$ over the interval is intermediate between these two values.

Angular Velocity As a Vector

As we have seen, our notation for the angular velocity ω_z about the z-axis is reminiscent of the notation v_x for the ordinary velocity along the x-axis (see Section 2.2). Just as v_x is the x-component of the velocity vector $\vec{v}$, ω_z is the z-component of an angular velocity *vector* $\vec{\omega}$ directed along the axis of rotation. As **Fig. 9.5a** shows, the direction of $\vec{\omega}$ is given by the right-hand rule that we used to define the vector product in Section 1.10. If the rotation is about the z-axis, then $\vec{\omega}$ has only a z-component. This component is positive if $\vec{\omega}$ is along the positive z-axis and negative if $\vec{\omega}$ is along the negative z-axis (Fig. 9.5b).

The vector formulation is especially useful when the direction of the rotation axis *changes*. We'll examine such situations briefly at the end of Chapter 10. In this chapter, however, we'll consider only situations in which the rotation axis is fixed. Hence throughout this chapter we'll use "angular velocity" to refer to ω_z, the component of $\vec{\omega}$ along the axis.

CAUTION **The angular velocity vector is perpendicular to the plane of rotation, not in it** It's a common error to think that an object's angular velocity vector $\vec{\omega}$ points in the direction in which some particular part of the object is moving. Another error is to think that $\vec{\omega}$ is a "curved vector" that points around the rotation axis in the direction of rotation (like the curved arrows in Figs. 9.1, 9.3, and 9.4). Neither of these is true! Angular velocity is an attribute of the *entire* rotating rigid body, not any one part, and there's no such thing as a curved vector. We choose the direction of $\vec{\omega}$ to be along the rotation axis—*perpendicular* to the plane of rotation—because that axis is common to every part of a rotating rigid body. ▌

9.5 (a) The right-hand rule for the direction of the angular velocity vector $\vec{\omega}$. Reversing the direction of rotation reverses the direction of $\vec{\omega}$. (b) The sign of ω_z for rotation along the z-axis.

(a)

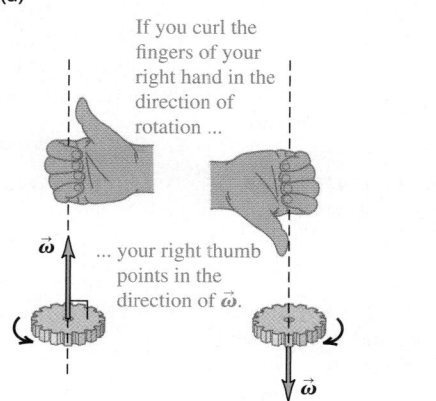

If you curl the fingers of your right hand in the direction of rotation ...

$\vec{\omega}$

... your right thumb points in the direction of $\vec{\omega}$.

$\vec{\omega}$

(b)

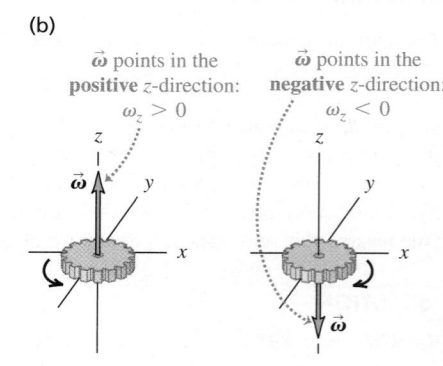

$\vec{\omega}$ points in the **positive** z-direction: $\omega_z > 0$

$\vec{\omega}$ points in the **negative** z-direction: $\omega_z < 0$

$\vec{\omega}$

$\vec{\omega}$

Angular Acceleration

A rigid body whose angular velocity changes has an *angular acceleration*. When you pedal your bicycle harder to make the wheels turn faster or apply the brakes to bring the wheels to a stop, you're giving the wheels an angular acceleration.

If ω_{1z} and ω_{2z} are the instantaneous angular velocities at times t_1 and t_2, we define the **average angular acceleration** $\alpha_{\text{av-}z}$ over the interval $\Delta t = t_2 - t_1$ as the change in angular velocity divided by Δt (**Fig. 9.6**):

$$\alpha_{\text{av-}z} = \frac{\omega_{2z} - \omega_{1z}}{t_2 - t_1} = \frac{\Delta\omega_z}{\Delta t} \tag{9.4}$$

The **instantaneous angular acceleration** α_z is the limit of $\alpha_{\text{av-}z}$ as $\Delta t \to 0$:

The **instantaneous angular acceleration** of a rigid body rotating around the z-axis ...

$$\alpha_z = \lim_{\Delta t \to 0} \frac{\Delta\omega_z}{\Delta t} = \frac{d\omega_z}{dt} \tag{9.5}$$

... equals the limit of the body's average angular acceleration as the time interval approaches zero ...

... and equals the instantaneous rate of change of the body's angular velocity.

The usual unit of angular acceleration is the radian per second per second, or rad/s². From now on we will use the term "angular acceleration" to mean the instantaneous angular acceleration rather than the average angular acceleration.

Because $\omega_z = d\theta/dt$, we can also express angular acceleration as the second derivative of the angular coordinate:

$$\alpha_z = \frac{d}{dt}\frac{d\theta}{dt} = \frac{d^2\theta}{dt^2} \tag{9.6}$$

You've probably noticed that we use Greek letters for angular kinematic quantities: θ for angular position, ω_z for angular velocity, and α_z for angular acceleration. These are analogous to x for position, v_x for velocity, and a_x for acceleration in straight-line motion. In each case, velocity is the rate of change of position with respect to time and acceleration is the rate of change of velocity with respect to time. We sometimes use the terms "*linear* velocity" for v_x and "*linear* acceleration" for a_x to distinguish clearly between these and the *angular* quantities introduced in this chapter.

If the angular acceleration α_z is positive, then the angular velocity ω_z is increasing; if α_z is negative, then ω_z is decreasing. The rotation is speeding up if α_z and ω_z have the same sign and slowing down if α_z and ω_z have opposite signs. (These are exactly the same relationships as those between *linear* acceleration a_x and *linear* velocity v_x for straight-line motion; see Section 2.3.)

9.6 Calculating the average angular acceleration of a rotating rigid body.

The average angular acceleration is the change in angular velocity divided by the time interval:

$$\alpha_{\text{av-}z} = \frac{\omega_{2z} - \omega_{1z}}{t_2 - t_1} = \frac{\Delta\omega_z}{\Delta t}$$

ω_{1z} ω_{2z}

At t_1 At t_2

EXAMPLE 9.2   CALCULATING ANGULAR ACCELERATION

For the flywheel of Example 9.1, (a) find the average angular acceleration between $t_1 = 2.0$ s and $t_2 = 5.0$ s. (b) Find the instantaneous angular accelerations at $t_1 = 2.0$ s and $t_2 = 5.0$ s.

SOLUTION

IDENTIFY and SET UP: We use Eqs. (9.4) and (9.5) for the average and instantaneous angular accelerations.

EXECUTE: (a) From Example 9.1, the values of ω_z at the two times are

$$\omega_{1z} = 24 \text{ rad/s} \qquad \omega_{2z} = 150 \text{ rad/s}$$

From Eq. (9.4), the average angular acceleration is

$$\alpha_{\text{av-}z} = \frac{150 \text{ rad/s} - 24 \text{ rad/s}}{5.0 \text{ s} - 2.0 \text{ s}} = 42 \text{ rad/s}^2$$

(b) We found in Example 9.1 that $\omega_z = (6.0 \text{ rad/s}^3)t^2$ for the flywheel. From Eq. (9.5), the value of α_z at any time t is

$$\alpha_z = \frac{d\omega_z}{dt} = \frac{d}{dt}[(6.0 \text{ rad/s}^3)(t^2)] = (6.0 \text{ rad/s}^3)(2t)$$

$$= (12 \text{ rad/s}^3)t$$

Hence

$$\alpha_{1z} = (12 \text{ rad/s}^3)(2.0 \text{ s}) = 24 \text{ rad/s}^2$$
$$\alpha_{2z} = (12 \text{ rad/s}^3)(5.0 \text{ s}) = 60 \text{ rad/s}^2$$

EVALUATE: The angular acceleration is *not* constant in this situation. The angular velocity ω_z is always increasing because α_z is always positive. Furthermore, the rate at which angular velocity increases is itself increasing, since α_z increases with time.

9.7 When the rotation axis is fixed, both the angular acceleration and angular velocity vectors lie along that axis.

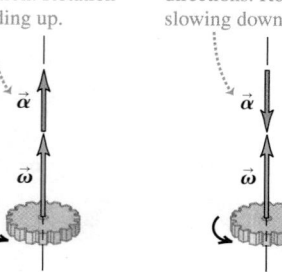

$\vec{\alpha}$ and $\vec{\omega}$ in the **same** direction: Rotation speeding up.

$\vec{\alpha}$ and $\vec{\omega}$ in the **opposite** directions: Rotation slowing down.

$\vec{\alpha}$

$\vec{\omega}$

$\vec{\alpha}$

$\vec{\omega}$

Angular Acceleration As a Vector

Just as we did for angular velocity, it's useful to define an angular acceleration *vector* $\vec{\alpha}$. Mathematically, $\vec{\alpha}$ is the time derivative of the angular velocity vector $\vec{\omega}$. If the object rotates around the fixed z-axis, then $\vec{\alpha}$ has only a z-component α_z. In this case, $\vec{\alpha}$ is in the same direction as $\vec{\omega}$ if the rotation is speeding up and opposite to $\vec{\omega}$ if the rotation is slowing down (**Fig. 9.7**).

The vector $\vec{\alpha}$ will be particularly useful in Chapter 10 when we discuss what happens when the rotation axis changes direction. In this chapter, however, the rotation axis will always be fixed and we need only the z-component α_z.

TEST YOUR UNDERSTANDING OF SECTION 9.1 The figure shows a graph of ω_z and α_z versus time for a particular rotating body. (a) During which time intervals is the rotation speeding up? (i) $0 < t < 2$ s; (ii) 2 s $< t < 4$ s; (iii) 4 s $< t < 6$ s. (b) During which time intervals is the rotation slowing down? (i) $0 < t < 2$ s; (ii) 2 s $< t < 4$ s; (iii) 4 s $< t < 6$ s.

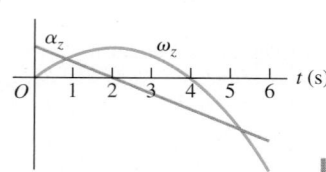

9.2 ROTATION WITH CONSTANT ANGULAR ACCELERATION

In Chapter 2 we found that straight-line motion is particularly simple when the acceleration is constant. This is also true of rotational motion about a fixed axis. When the angular acceleration is constant, we can derive equations for angular velocity and angular position by using the same procedure that we used for straight-line motion in Section 2.4. In fact, the equations we are about to derive are identical to Eqs. (2.8), (2.12), (2.13), and (2.14) if we replace x with θ, v_x with ω_z, and a_x with α_z. We suggest that you review Section 2.4 before continuing.

Let ω_{0z} be the angular velocity of a rigid body at time $t = 0$ and ω_z be its angular velocity at a later time t. The angular acceleration α_z is constant and equal to the average value for any interval. From Eq. (9.4) with the interval from 0 to t,

$$\alpha_z = \frac{\omega_z - \omega_{0z}}{t - 0} \quad \text{or}$$

Angular velocity at time t of a rigid body with **constant angular acceleration**

Angular velocity of body at time 0

$$\omega_z = \omega_{0z} + \alpha_z t \quad\quad \text{(9.7)}$$

Time

Constant angular acceleration of body

The product $\alpha_z t$ is the total change in ω_z between $t = 0$ and the later time t; angular velocity ω_z at time t is the sum of the initial value ω_{0z} and this total change.

With constant angular acceleration, the angular velocity changes at a uniform rate, so its average value between 0 and t is the average of the initial and final values:

$$\omega_{\text{av-}z} = \frac{\omega_{0z} + \omega_z}{2} \quad\quad \text{(9.8)}$$

We also know that $\omega_{\text{av-}z}$ is the total angular displacement $(\theta - \theta_0)$ divided by the time interval $(t - 0)$:

$$\omega_{\text{av-}z} = \frac{\theta - \theta_0}{t - 0} \quad\quad \text{(9.9)}$$

When we equate Eqs. (9.8) and (9.9) and multiply the result by t, we get

Angular position at time t of a rigid body with **constant angular acceleration**

Angular position of body at time 0

$$\theta - \theta_0 = \tfrac{1}{2}(\omega_{0z} + \omega_z)t \quad\quad \text{(9.10)}$$

Angular velocity of body at time 0

Time

Angular velocity of body at time t

BIO Application Rotational Motion in Bacteria *Escherichia coli* bacteria (about 2 μm by 0.5 μm) are found in the lower intestines of humans and other warm-blooded animals. The bacteria swim by rotating their long, corkscrew-shaped flagella, which act like the blades of a propeller. Each flagellum is powered by a remarkable motor (made of protein) located at the base of the bacterial cell. The motor can rotate the flagellum at angular speeds from 200 to 1000 rev/min (about 20 to 100 rad/s) and can vary its speed to give the flagellum an angular acceleration.

Flagella

To obtain a relationship between θ and t that doesn't contain ω_z, we substitute Eq. (9.7) into Eq. (9.10):

$$\theta - \theta_0 = \tfrac{1}{2}\left[\omega_{0z} + (\omega_{0z} + \alpha_z t)\right]t \qquad \text{or}$$

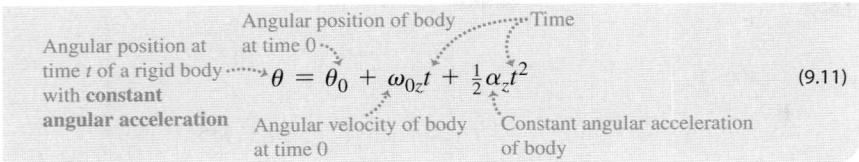

Angular position at time t of a rigid body with **constant angular acceleration** ·········→ Angular position of body at time 0 ····→ $\theta = \theta_0 + \omega_{0z}t + \tfrac{1}{2}\alpha_z t^2$ ···· Time

Angular velocity of body at time 0 · Constant angular acceleration of body

(9.11)

That is, if at the initial time $t = 0$ the body is at angular position θ_0 and has angular velocity ω_{0z}, then its angular position θ at any later time t is θ_0, plus the rotation $\omega_{0z}t$ it would have if the angular velocity were constant, plus an additional rotation $\tfrac{1}{2}\alpha_z t^2$ caused by the changing angular velocity.

Following the same procedure as for straight-line motion in Section 2.4, we can combine Eqs. (9.7) and (9.11) to obtain a relationship between θ and ω_z that does not contain t. We invite you to work out the details, following the same procedure we used to get Eq. (2.13). (See Exercise 9.12.) We get

Angular velocity at time t of a rigid body with **constant angular acceleration** ·······→ Angular velocity of body at time 0 ···→ $\omega_z^2 = \omega_{0z}^2 + 2\alpha_z(\theta - \theta_0)$

Constant angular ···· acceleration of body ···· Angular position of body at time t ···· Angular position of body at time 0

(9.12)

CAUTION Constant angular acceleration Keep in mind that all of these results are valid *only* when the angular acceleration α_z is *constant;* do not try to apply them to problems in which α_z is *not* constant. **Table 9.1** shows the analogy between Eqs. (9.7), (9.10), (9.11), and (9.12) for fixed-axis rotation with constant angular acceleration and the corresponding equations for straight-line motion with constant linear acceleration. ▍

TABLE 9.1 Comparison of Linear and Angular Motions with Constant Acceleration			
Straight-Line Motion with Constant Linear Acceleration		**Fixed-Axis Rotation with Constant Angular Acceleration**	
a_x = constant		α_z = constant	
$v_x = v_{0x} + a_x t$	(2.8)	$\omega_z = \omega_{0z} + \alpha_z t$	(9.7)
$x = x_0 + v_{0x}t + \tfrac{1}{2}a_x t^2$	(2.12)	$\theta = \theta_0 + \omega_{0z}t + \tfrac{1}{2}\alpha_z t^2$	(9.11)
$v_x^2 = v_{0x}^2 + 2a_x(x - x_0)$	(2.13)	$\omega_z^2 = \omega_{0z}^2 + 2\alpha_z(\theta - \theta_0)$	(9.12)
$x - x_0 = \tfrac{1}{2}(v_{0x} + v_x)t$	(2.14)	$\theta - \theta_0 = \tfrac{1}{2}(\omega_{0z} + \omega_z)t$	(9.10)

EXAMPLE 9.3 ROTATION WITH CONSTANT ANGULAR ACCELERATION

You have finished watching a movie on Blu-ray and the disc is slowing to a stop. The disc's angular velocity at $t = 0$ is 27.5 rad/s, and its angular acceleration is a constant -10.0 rad/s^2. A line PQ on the disc's surface lies along the $+x$-axis at $t = 0$ (**Fig. 9.8**). (a) What is the disc's angular velocity at $t = 0.300$ s? (b) What angle does the line PQ make with the $+x$-axis at this time?

9.8 A line PQ on a rotating Blu-ray disc at $t = 0$.

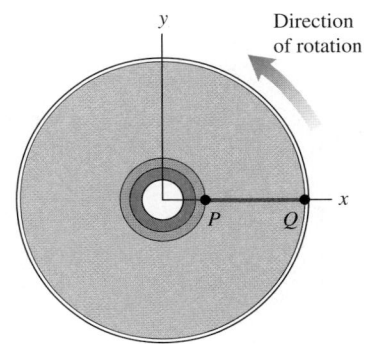

SOLUTION

IDENTIFY and SET UP: The angular acceleration of the disc is constant, so we can use any of the equations derived in this section (Table 9.1). Our target variables are the angular velocity ω_z and the angular displacement θ at $t = 0.300$ s. Given $\omega_{0z} = 27.5$ rad/s, $\theta_0 = 0$, and $\alpha_z = -10.0$ rad/s^2, it's easiest to use Eqs. (9.7) and (9.11) to find the target variables.

Continued

EXECUTE: (a) From Eq. (9.7), at $t = 0.300$ s we have

$$\omega_z = \omega_{0z} + \alpha_z t = 27.5 \text{ rad/s} + (-10.0 \text{ rad/s}^2)(0.300 \text{ s})$$
$$= 24.5 \text{ rad/s}$$

(b) From Eq. (9.11),

$$\theta = \theta_0 + \omega_{0z}t + \tfrac{1}{2}\alpha_z t^2$$
$$= 0 + (27.5 \text{ rad/s})(0.300 \text{ s}) + \tfrac{1}{2}(-10.0 \text{ rad/s}^2)(0.300 \text{ s})^2$$
$$= 7.80 \text{ rad} = 7.80 \text{ rad}\left(\frac{1 \text{ rev}}{2\pi \text{ rad}}\right) = 1.24 \text{ rev}$$

The disc has made one complete revolution plus an additional 0.24 revolution—that is, $360°$ plus $(0.24 \text{ rev})(360°/\text{rev}) = 87°$. Hence the line PQ makes an angle of $87°$ with the $+x$-axis.

EVALUATE: Our answer to part (a) tells us that the disc's angular velocity has decreased, as it should since $\alpha_z < 0$. We can use our result for ω_z from part (a) with Eq. (9.12) to check our result for θ from part (b). To do so, we solve Eq. (9.12) for θ:

$$\omega_z^2 = \omega_{0z}^2 + 2\alpha_z(\theta - \theta_0)$$
$$\theta = \theta_0 + \left(\frac{\omega_z^2 - \omega_{0z}^2}{2\alpha_z}\right)$$
$$= 0 + \frac{(24.5 \text{ rad/s})^2 - (27.5 \text{ rad/s})^2}{2(-10.0 \text{ rad/s}^2)} = 7.80 \text{ rad}$$

This agrees with our previous result from part (b).

TEST YOUR UNDERSTANDING OF SECTION 9.2 Suppose the disc in Example 9.3 was initially spinning at twice the rate (55.0 rad/s rather than 27.5 rad/s) and slowed down at twice the rate (-20.0 rad/s^2 rather than -10.0 rad/s^2). (a) Compared to the situation in Example 9.3, how long would it take the disc to come to a stop? (i) The same amount of time; (ii) twice as much time; (iii) 4 times as much time; (iv) $\tfrac{1}{2}$ as much time; (v) $\tfrac{1}{4}$ as much time. (b) Compared to the situation in Example 9.3, through how many revolutions would the disc rotate before coming to a stop? (i) The same number of revolutions; (ii) twice as many revolutions; (iii) 4 times as many revolutions; (iv) $\tfrac{1}{2}$ as many revolutions; (v) $\tfrac{1}{4}$ as many revolutions. ∎

9.3 RELATING LINEAR AND ANGULAR KINEMATICS

How do we find the linear speed and acceleration of a particular point in a rotating rigid body? We need to answer this question to proceed with our study of rotation. For example, to find the kinetic energy of a rotating body, we have to start from $K = \tfrac{1}{2}mv^2$ for a particle, and this requires that we know the speed v for each particle in the body. So it's worthwhile to develop general relationships between the *angular* speed and acceleration of a rigid body rotating about a fixed axis and the *linear* speed and acceleration of a specific point or particle in the body.

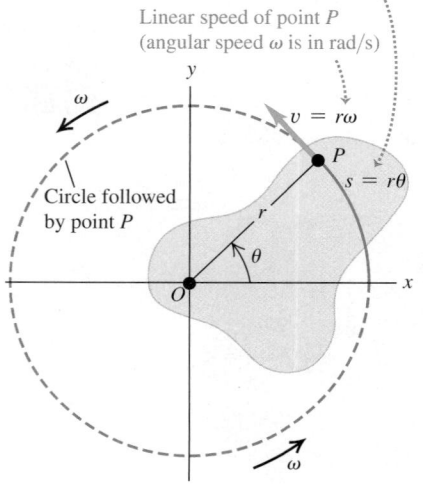

9.9 A rigid body rotating about a fixed axis through point O.

Distance through which point P on the body moves (angle θ is in radians)

Linear speed of point P (angular speed ω is in rad/s)

$v = r\omega$

$s = r\theta$

Circle followed by point P

Linear Speed in Rigid-Body Rotation

When a rigid body rotates about a fixed axis, every particle in the body moves in a circular path that lies in a plane perpendicular to the axis and is centered on the axis. A particle's speed is directly proportional to the body's angular velocity; the faster the rotation, the greater the speed of each particle. In **Fig. 9.9**, point P is a constant distance r from the axis, so it moves in a circle of radius r. At any time, Eq. (9.1) relates the angle θ (in radians) and the arc length s:

$$s = r\theta$$

We take the time derivative of this, noting that r is constant for any specific particle, and take the absolute value of both sides:

$$\left|\frac{ds}{dt}\right| = r\left|\frac{d\theta}{dt}\right|$$

Now $|ds/dt|$ is the absolute value of the rate of change of arc length, which is equal to the instantaneous *linear* speed v of the particle. The absolute value of the rate of change of the angle, $|d\theta/dt|$, is the instantaneous **angular speed** ω—that is, the magnitude of the instantaneous angular velocity in rad/s. Thus

$$\underset{\substack{\text{Linear speed of a point}\\ \text{on a rotating rigid body}}}{} v = r\omega \underset{\substack{\text{Angular speed of the}\\ \text{rotating rigid body}}}{} \qquad (9.13)$$
$$\underset{\text{Distance of that point from rotation axis}}{}$$

The farther a point is from the axis, the greater its linear speed. The *direction* of the linear velocity *vector* is tangent to its circular path at each point (Fig. 9.9).

CAUTION **Speed vs. velocity** Keep in mind the distinction between the linear and angular *speeds* v and ω, which appear in Eq. (9.13), and the linear and angular *velocities* v_x and ω_z. The quantities without subscripts, v and ω, are never negative; they are the magnitudes of the vectors $\vec{v}$ and $\vec{\omega}$, respectively, and their values tell you only how fast a particle is moving (v) or how fast a body is rotating (ω). The quantities with subscripts, v_x and ω_z, can be either positive or negative; their signs tell you the direction of the motion.

Linear Acceleration in Rigid-Body Rotation

We can represent the acceleration $\vec{a}$ of a particle moving in a circle in terms of its centripetal and tangential components, a_{rad} and a_{tan} (**Fig. 9.10**), as we did in Section 3.4. (You should review that section now.) We found that the **tangential component of acceleration** a_{tan}, the component parallel to the instantaneous velocity, acts to change the *magnitude* of the particle's velocity (i.e., the speed) and is equal to the rate of change of speed. Taking the derivative of Eq. (9.13), we find

$$\underset{\substack{\text{Tangential}\\ \text{acceleration of a}\\ \text{point on a rotating}\\ \text{rigid body}}}{} a_{tan} = \frac{dv}{dt} = r\frac{d\omega}{dt} = r\alpha \qquad (9.14)$$
$$\underset{\substack{\text{Rate of change of}\\ \text{linear speed of that point}}}{} \qquad \underset{\substack{\text{Rate of change of}\\ \text{angular speed of body}}}{}$$

This component of $\vec{a}$ is always tangent to the circular path of point P (Fig. 9.10).

The quantity $\alpha = d\omega/dt$ in Eq. (9.14) is the rate of change of the angular *speed*. It is not quite the same as $\alpha_z = d\omega_z/dt$, which is the rate of change of the angular *velocity*. For example, consider a body rotating so that its angular velocity vector points in the $-z$-direction (see Fig. 9.5b). If the body is gaining angular speed at a rate of 10 rad/s per second, then $\alpha = 10 \text{ rad/s}^2$. But ω_z is negative and becoming more negative as the rotation gains speed, so $\alpha_z = -10 \text{ rad/s}^2$. The rule for rotation about a fixed axis is that α is equal to α_z if ω_z is positive but equal to $-\alpha_z$ if ω_z is negative.

The component of $\vec{a}$ in Fig. 9.10 directed toward the rotation axis, the **centripetal component of acceleration** a_{rad}, is associated with the change of *direction* of the velocity of point P. In Section 3.4 we worked out the relationship $a_{rad} = v^2/r$. We can express this in terms of ω by using Eq. (9.13):

$$\underset{\substack{\text{Centripetal}\\ \text{acceleration of a}\\ \text{point on a rotating}\\ \text{rigid body}}}{} \overset{\text{Linear speed of that point}}{a_{rad} = \frac{v^2}{r}} = \omega^2 r \overset{\text{Angular speed of body}}{} \qquad (9.15)$$
$$\underset{\text{Distance of that point from rotation axis}}{}$$

This is true at each instant, *even when ω and v are not constant*. The centripetal component always points toward the axis of rotation.

CAUTION **Use angles in radians** Remember that Eq. (9.1), $s = r\theta$, is valid *only* when θ is measured in radians. The same is true of any equation derived from this, including Eqs. (9.13), (9.14), and (9.15). When you use these equations, you *must* express the angular quantities in radians, not revolutions or degrees (**Fig. 9.11**).

PhET: Ladybug Revolution

9.10 A rigid body whose rotation is speeding up. The acceleration of point P has a component a_{rad} toward the rotation axis (perpendicular to $\vec{v}$) and a component a_{tan} along the circle that point P follows (parallel to $\vec{v}$).

Radial and tangential acceleration components:
• $a_{rad} = \omega^2 r$ is point P's centripetal acceleration.
• $a_{tan} = r\alpha$ means that P's rotation is speeding up (the body has angular acceleration).

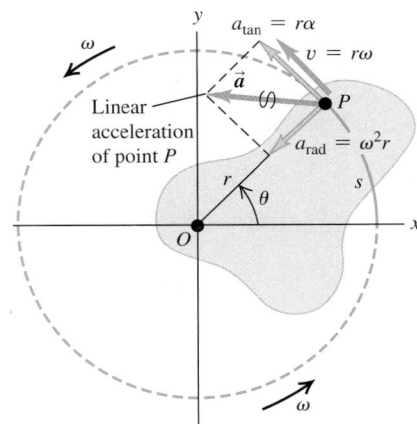

9.11 Always use radians when relating linear and angular quantities.

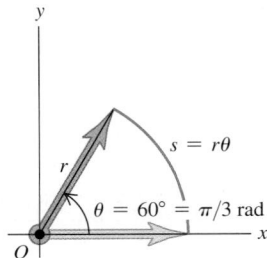

In any equation that relates linear quantities to angular quantities, the angles MUST be expressed in radians ...

RIGHT! ▶ $s = (\pi/3)r$

... never in degrees or revolutions.

WRONG ▶ $s = 60r$

Equations (9.1), (9.13), and (9.14) also apply to any particle that has the same tangential velocity as a point in a rotating rigid body. For example, when a rope wound around a circular cylinder unwraps without stretching or slipping, its speed and acceleration at any instant are equal to the speed and tangential acceleration of the point at which it is tangent to the cylinder. The same principle holds for situations such as bicycle chains and sprockets, belts and pulleys that turn without slipping, and so on. We will have several opportunities to use these relationships later in this chapter and in Chapter 10. Note that Eq. (9.15) for the centripetal component a_{rad} is applicable to the rope or chain *only* at points that are in contact with the cylinder or sprocket. Other points do not have the same acceleration toward the center of the circle that points on the cylinder or sprocket have.

EXAMPLE 9.4 THROWING A DISCUS

An athlete whirls a discus in a circle of radius 80.0 cm. At a certain instant, the athlete is rotating at 10.0 rad/s and the angular speed is increasing at 50.0 rad/s². For this instant, find the tangential and centripetal components of the acceleration of the discus and the magnitude of the acceleration.

SOLUTION

IDENTIFY and SET UP: We treat the discus as a particle traveling in a circular path (**Fig. 9.12a**), so we can use the ideas developed in this section. We are given $r = 0.800$ m, $\omega = 10.0$ rad/s, and $\alpha = 50.0$ rad/s² (Fig. 9.12b). We'll use Eqs. (9.14) and (9.15) to find the acceleration components a_{tan} and a_{rad}, respectively; we'll then find the magnitude a by using the Pythagorean theorem.

EXECUTE: From Eqs. (9.14) and (9.15),

$$a_{tan} = r\alpha = (0.800 \text{ m})(50.0 \text{ rad/s}^2) = 40.0 \text{ m/s}^2$$

$$a_{rad} = \omega^2 r = (10.0 \text{ rad/s})^2(0.800 \text{ m}) = 80.0 \text{ m/s}^2$$

Then

$$a = \sqrt{a_{tan}^2 + a_{rad}^2} = 89.4 \text{ m/s}^2$$

EVALUATE: Note that we dropped the unit "radian" from our results for a_{tan}, a_{rad}, and a. We can do this because "radian" is a dimensionless quantity. Can you show that if the angular speed doubles to 20.0 rad/s while α remains the same, the acceleration magnitude a increases to 322 m/s²?

9.12 (a) Whirling a discus in a circle. (b) Our sketch showing the acceleration components for the discus.

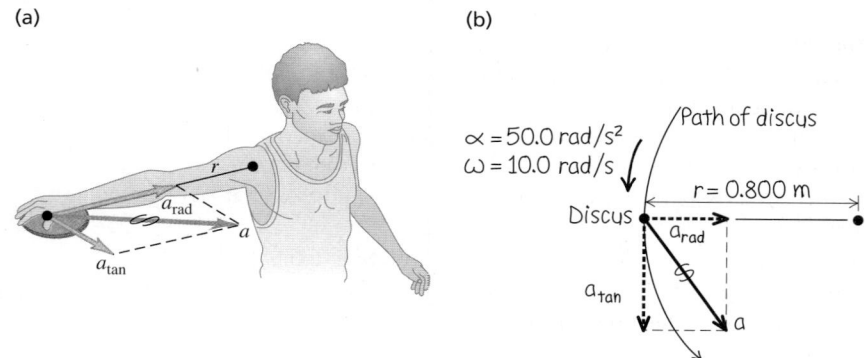

EXAMPLE 9.5 DESIGNING A PROPELLER

You are designing an airplane propeller that is to turn at 2400 rpm (**Fig. 9.13a**). The forward airspeed of the plane is to be 75.0 m/s, and the speed of the propeller tips through the air must not exceed 270 m/s. (This is about 80% of the speed of sound in air. If the propeller tips moved faster, they would produce a lot of noise.) (a) What is the maximum possible propeller radius? (b) With this radius, what is the acceleration of the propeller tip?

SOLUTION

IDENTIFY and SET UP: We consider a particle at the tip of the propeller; our target variables are the particle's distance from the axis and its acceleration. The speed of this particle through the air, which cannot exceed 270 m/s, is due to both the propeller's rotation and the forward motion of the airplane. Figure 9.13b shows

9.13 (a) A propeller-driven airplane in flight. (b) Our sketch showing the velocity components for the propeller tip.

(a)

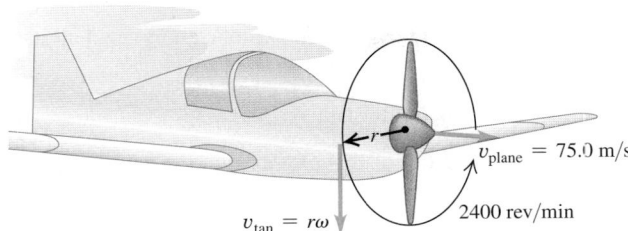

(b)

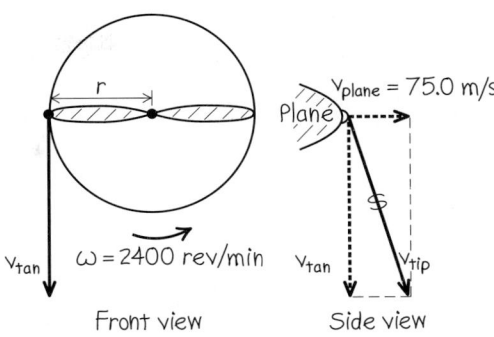

that the particle's velocity $\vec{v}_{tip}$ is the vector sum of its tangential velocity due to the propeller's rotation of magnitude $v_{tan} = \omega r$, given by Eq. (9.13), and the forward velocity of the airplane of magnitude $v_{plane} = 75.0$ m/s. The propeller rotates in a plane perpendicular to the direction of flight, so $\vec{v}_{tan}$ and $\vec{v}_{plane}$ are perpendicular to each other, and we can use the Pythagorean theorem to obtain an expression for v_{tip} from v_{tan} and v_{plane}. We will then set $v_{tip} = 270$ m/s and solve for the radius r. The angular speed of the propeller is constant, so the acceleration of the propeller tip has only a radial component; we'll find it by using Eq. (9.15).

EXECUTE: We first convert ω to rad/s (see Fig. 9.11):

$$\omega = 2400 \text{ rpm} = \left(2400 \frac{\text{rev}}{\text{min}}\right)\left(\frac{2\pi \text{ rad}}{1 \text{ rev}}\right)\left(\frac{1 \text{ min}}{60 \text{ s}}\right) = 251 \text{ rad/s}$$

(a) From Fig. 9.13b and Eq. (9.13),

$$v_{tip}^2 = v_{plane}^2 + v_{tan}^2 = v_{plane}^2 + r^2\omega^2 \quad \text{so}$$

$$r^2 = \frac{v_{tip}^2 - v_{plane}^2}{\omega^2} \quad \text{and} \quad r = \frac{\sqrt{v_{tip}^2 - v_{plane}^2}}{\omega}$$

If $v_{tip} = 270$ m/s, the maximum propeller radius is

$$r = \frac{\sqrt{(270 \text{ m/s})^2 - (75.0 \text{ m/s})^2}}{251 \text{ rad/s}} = 1.03 \text{ m}$$

(b) The centripetal acceleration of the particle is, from Eq. (9.15),

$$a_{rad} = \omega^2 r = (251 \text{ rad/s})^2(1.03 \text{ m})$$
$$= 6.5 \times 10^4 \text{ m/s}^2 = 6600g$$

The tangential acceleration a_{tan} is zero because ω is constant.

EVALUATE: From $\Sigma\vec{F} = m\vec{a}$, the propeller must exert a force of 6.5×10^4 N on each kilogram of material at its tip! This is why propellers are made out of tough material, usually aluminum alloy.

TEST YOUR UNDERSTANDING OF SECTION 9.3 Information is stored on a Blu-ray disc (see Fig. 9.8) in a coded pattern of tiny pits. The pits are arranged in a track that spirals outward toward the rim of the disc. As the disc spins inside a player, the track is scanned at a constant *linear* speed. How must the rotation speed ω of the disc change as the player's scanning head moves outward over the track? (i) ω must increase; (ii) ω must decrease; (iii) ω must stay the same. ▮

9.4 ENERGY IN ROTATIONAL MOTION

A rotating rigid body consists of mass in motion, so it has kinetic energy. As we will see, we can express this kinetic energy in terms of the body's angular speed and a new quantity, called *moment of inertia*, that depends on the body's mass and how the mass is distributed.

To begin, we think of a body as being made up of a large number of particles, with masses $m_1, m_2, \ldots$ at distances $r_1, r_2, \ldots$ from the axis of rotation. We label the particles with the index i: The mass of the ith particle is m_i and r_i is the *perpendicular* distance from the axis to the ith particle. (The particles need not all lie in the same plane.)

When a rigid body rotates about a fixed axis, the speed v_i of the ith particle is given by Eq. (9.13), $v_i = r_i\omega$, where ω is the body's angular speed. Different

DATA *SPEAKS*

Relating Linear and Angular Quantities

When students were given a problem about the motion of points on a rotating rigid body, more than 21% gave an incorrect response. Common errors:

- Confusing centripetal and tangential acceleration. Points on a rigid body have a centripetal (radial) acceleration a_{rad} whenever the body is rotating but have a tangential acceleration a_{tan} only if the angular speed is changing.

- Forgetting that the values of a_{rad} and a_{tan} at a point depend on the point's distance from the rotation axis.

particles have different values of r_i, but ω is the same for all (otherwise, the body wouldn't be rigid). The kinetic energy of the ith particle can be expressed as

$$\tfrac{1}{2}m_i v_i^2 = \tfrac{1}{2}m_i r_i^2 \omega^2$$

The body's *total* kinetic energy is the sum of the kinetic energies of all its particles:

$$K = \tfrac{1}{2}m_1 r_1^2 \omega^2 + \tfrac{1}{2}m_2 r_2^2 \omega^2 + \cdots = \sum_i \tfrac{1}{2}m_i r_i^2 \omega^2$$

Taking the common factor $\omega^2/2$ out of this expression, we get

$$K = \tfrac{1}{2}(m_1 r_1^2 + m_2 r_2^2 + \cdots)\omega^2 = \tfrac{1}{2}\Big(\sum_i m_i r_i^2\Big)\omega^2$$

DEMO

The quantity in parentheses, obtained by multiplying the mass of each particle by the square of its distance from the axis of rotation and adding these products, is called the **moment of inertia** I of the body for this rotation axis:

Moment of inertia of a body for a given rotation axis $\longrightarrow$ $I = m_1 r_1^2 + m_2 r_2^2 + \cdots = \sum_i m_i r_i^2$ $\quad$ (9.16)

Masses of the particles that make up the body

Perpendicular distances of the particles from rotation axis

"Moment" means that I depends on how the body's mass is distributed in space; it has nothing to do with a "moment" of time. For a body with a given rotation axis and a given total mass, the greater the distances from the axis to the particles that make up the body, the greater the moment of inertia I. In a rigid body, all distances r_i are constant and I is independent of how the body rotates around the given axis. The SI unit of I is the kilogram-meter^2 ($\text{kg} \cdot \text{m}^2$).

Using Eq. (9.16), we see that the **rotational kinetic energy** K of a rigid body is

Rotational kinetic energy of a rigid body rotating around an axis $\longrightarrow$ $K = \tfrac{1}{2}I\omega^2$ $\quad$ Moment of inertia of body for given rotation axis $\quad$ (9.17)

Angular speed of body

The kinetic energy given by Eq. (9.17) is *not* a new form of energy; it's simply the sum of the kinetic energies of the individual particles that make up the rotating rigid body. To use Eq. (9.17), ω *must* be measured in radians per second, not revolutions or degrees per second, to give K in joules. That's because we used $v_i = r_i \omega$ in our derivation.

Equation (9.17) gives a simple physical interpretation of moment of inertia: *The greater the moment of inertia, the greater the kinetic energy of a rigid body rotating with a given angular speed ω.* We learned in Chapter 6 that the kinetic energy of a body equals the amount of work done to accelerate that body from rest. So the greater a body's moment of inertia, the harder it is to start the body rotating if it's at rest and the harder it is to stop its rotation if it's already rotating (**Fig. 9.14**). For this reason, I is also called the *rotational inertia*.

9.14 An apparatus free to rotate around a vertical axis. To vary the moment of inertia, the two equal-mass cylinders can be locked into different positions on the horizontal shaft.

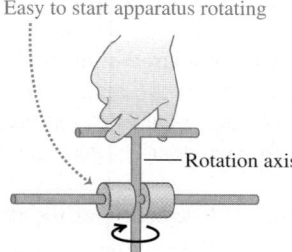

• Mass close to axis
• Small moment of inertia
• Easy to start apparatus rotating

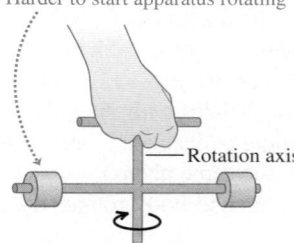

• Mass farther from axis
• Greater moment of inertia
• Harder to start apparatus rotating

EXAMPLE 9.6   MOMENTS OF INERTIA FOR DIFFERENT ROTATION AXES

A machine part (**Fig. 9.15**) consists of three disks linked by light-weight struts. (a) What is this body's moment of inertia about axis 1 through the center of disk A, perpendicular to the plane of the diagram? (b) What is its moment of inertia about axis 2 through the centers of disks B and C? (c) What is the body's kinetic energy if it rotates about axis 1 with angular speed $\omega = 4.0$ rad/s?

SOLUTION

IDENTIFY and SET UP: We'll consider the disks as massive particles located at the centers of the disks, and consider the struts as

9.15 An oddly shaped machine part.

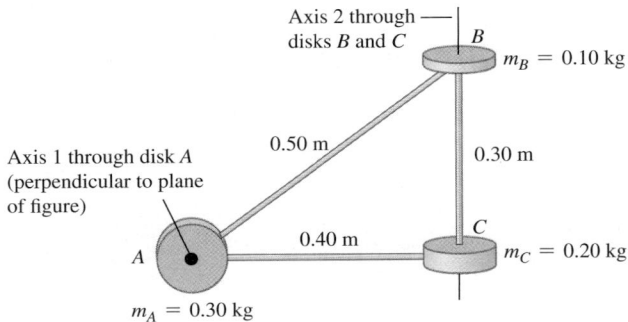

massless. In parts (a) and (b), we'll use Eq. (9.16) to find the moments of inertia. Given the moment of inertia about axis 1, we'll use Eq. (9.17) in part (c) to find the rotational kinetic energy.

EXECUTE: (a) The particle at point A lies *on* axis 1 through A, so its distance r from the axis is zero and it contributes nothing to the moment of inertia. Hence only B and C contribute in Eq. (9.16):

$$I_1 = \sum m_i r_i^2 = (0.10\text{ kg})(0.50\text{ m})^2 + (0.20\text{ kg})(0.40\text{ m})^2$$
$$= 0.057\text{ kg}\cdot\text{m}^2$$

(b) The particles at B and C both lie on axis 2, so neither particle contributes to the moment of inertia. Hence only A contributes:

$$I_2 = \sum m_i r_i^2 = (0.30\text{ kg})(0.40\text{ m})^2 = 0.048\text{ kg}\cdot\text{m}^2$$

(c) From Eq. (9.17),

$$K_1 = \tfrac{1}{2}I_1\omega^2 = \tfrac{1}{2}(0.057\text{ kg}\cdot\text{m}^2)(4.0\text{ rad/s})^2 = 0.46\text{ J}$$

EVALUATE: The moment of inertia about axis 2 is smaller than that about axis 1. Hence, of the two axes, it's easier to make the machine part rotate about axis 2.

CAUTION **Moment of inertia depends on the choice of axis** Example 9.6 shows that the moment of inertia of a body depends on the location and orientation of the axis. It's not enough to say, "The moment of inertia is 0.048 kg·m²." We have to be specific and say, "The moment of inertia *about the axis through B and C* is 0.048 kg·m²." ∎

In Example 9.6 we represented the body as several point masses, and we evaluated the sum in Eq. (9.16) directly. When the body is a *continuous* distribution of matter, such as a solid cylinder or plate, the sum becomes an integral, and we need to use calculus to calculate the moment of inertia. We will give several examples of such calculations in Section 9.6; meanwhile, **Table 9.2** (next page) gives moments of inertia for several familiar shapes in terms of their masses and dimensions. Each body shown in Table 9.2 is *uniform;* that is, the density has the same value at all points within the solid parts of the body. **?**

CAUTION **Computing moments of inertia** You may be tempted to try to compute the moment of inertia of a body by assuming that all the mass is concentrated at the center of mass and multiplying the total mass by the square of the distance from the center of mass to the axis. That doesn't work! For example, when a uniform thin rod of length L and mass M is pivoted about an axis through one end, perpendicular to the rod, the moment of inertia is $I = ML^2/3$ [case (b) in Table 9.2]. If we took the mass as concentrated at the center, a distance $L/2$ from the axis, we would obtain the *incorrect* result $I = M(L/2)^2 = ML^2/4$. ∎

Now that we know how to calculate the kinetic energy of a rotating rigid body, we can apply the energy principles of Chapter 7 to rotational motion. The Problem-Solving Strategy on the next page, along with the examples that follow, show how this is done.

BIO **Application Moment of Inertia of a Bird's Wing** When a bird flaps its wings, it rotates the wings up and down around the shoulder. A hummingbird has small wings with a small moment of inertia, so the bird can move its wings rapidly (up to 70 beats per second). By contrast, the Andean condor (*Vultur gryphus*) has immense wings that are hard to move due to their large moment of inertia. Condors flap their wings at about one beat per second on takeoff, but at most times prefer to soar while holding their wings steady.

| TABLE 9.2 | Moments of Inertia of Various Bodies |

(a) Slender rod, axis through center

$$I = \frac{1}{12}ML^2$$

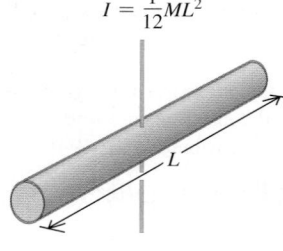

(b) Slender rod, axis through one end

$$I = \frac{1}{3}ML^2$$

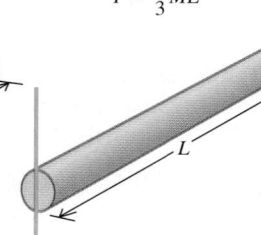

(c) Rectangular plate, axis through center

$$I = \frac{1}{12}M(a^2 + b^2)$$

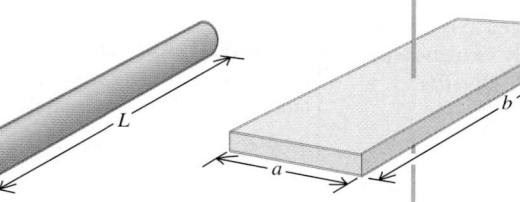

(d) Thin rectangular plate, axis along edge

$$I = \frac{1}{3}Ma^2$$

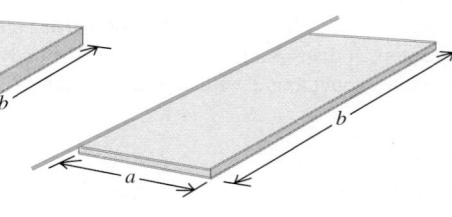

(e) Hollow cylinder

$$I = \frac{1}{2}M(R_1^2 + R_2^2)$$

(f) Solid cylinder

$$I = \frac{1}{2}MR^2$$

(g) Thin-walled hollow cylinder

$$I = MR^2$$

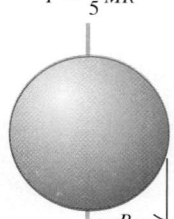
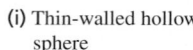
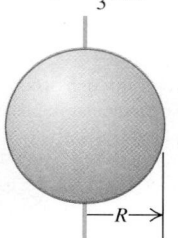

(h) Solid sphere

$$I = \frac{2}{5}MR^2$$

(i) Thin-walled hollow sphere

$$I = \frac{2}{3}MR^2$$

PROBLEM-SOLVING STRATEGY 9.1 ROTATIONAL ENERGY

IDENTIFY *the relevant concepts:* You can use work–energy relationships and conservation of energy to find relationships involving the position and motion of a rigid body rotating around a fixed axis. The energy method is usually not helpful for problems that involve elapsed time. In Chapter 10 we'll see how to approach rotational problems of this kind.

SET UP *the problem* using Problem-Solving Strategy 7.1 (Section 7.1), with the following additional steps:

5. You can use Eqs. (9.13) and (9.14) in problems involving a rope (or the like) wrapped around a rotating rigid body, if the rope doesn't slip. These equations relate the linear speed and tangential acceleration of a point on the body to the body's angular velocity and angular acceleration. (See Examples 9.7 and 9.8.)

6. Use Table 9.2 to find moments of inertia. Use the parallel-axis theorem, Eq. (9.19) (to be derived in Section 9.5), to find

moments of inertia for rotation about axes parallel to those shown in the table.

EXECUTE *the solution:* Write expressions for the initial and final kinetic and potential energies K_1, K_2, U_1, and U_2 and for the non-conservative work W_{other} (if any), where K_1 and K_2 must now include any rotational kinetic energy $K = \frac{1}{2}I\omega^2$. Substitute these expressions into Eq. (7.14), $K_1 + U_1 + W_{other} = K_2 + U_2$ (if non-conservative work is done), or Eq. (7.12), $K_1 + U_1 = K_2 + U_2$ (if only conservative work is done), and solve for the target variables. It's helpful to draw bar graphs showing the initial and final values of K, U, and $E = K + U$.

EVALUATE *your answer:* Check whether your answer makes physical sense.

EXAMPLE 9.7 AN UNWINDING CABLE I

We wrap a light, nonstretching cable around a solid cylinder, of mass 50 kg and diameter 0.120 m, that rotates in frictionless bearings about a stationary horizontal axis (**Fig. 9.16**). We pull the free end of the cable with a constant 9.0-N force for a distance of 2.0 m; it turns the cylinder as it unwinds without slipping. The cylinder is initially at rest. Find its final angular speed and the final speed of the cable.

SOLUTION

IDENTIFY: We'll solve this problem by using energy methods. We'll assume that the cable is massless, so only the cylinder has kinetic energy. There are no changes in gravitational potential energy. There is friction between the cable and the cylinder, but because the cable doesn't slip, there is no motion of the cable relative to

9.16 A cable unwinds from a cylinder (side view).

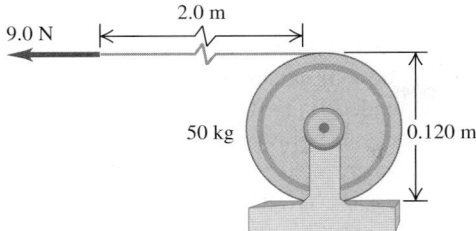

the cylinder and no mechanical energy is lost in frictional work. Because the cable is massless, the force that the cable exerts on the cylinder rim is equal to the applied force F.

SET UP: Point 1 is when the cable begins to move. The cylinder starts at rest, so $K_1 = 0$. Point 2 is when the cable has moved a distance $s = 2.0$ m and the cylinder has kinetic energy $K_2 = \frac{1}{2}I\omega^2$. One of our target variables is ω; the other is the speed of the cable at point 2, which is equal to the tangential speed v of the cylinder at that point. We'll use Eq. (9.13) to find v from ω.

EXECUTE: The work done on the cylinder is $W_{\text{other}} = Fs = (9.0 \text{ N})(2.0 \text{ m}) = 18$ J. From Table 9.2 the moment of inertia is

$$I = \tfrac{1}{2}mR^2 = \tfrac{1}{2}(50 \text{ kg})(0.060 \text{ m})^2 = 0.090 \text{ kg} \cdot \text{m}^2$$

(The radius R is half the diameter.) From Eq. (7.14), $K_1 + U_1 + W_{\text{other}} = K_2 + U_2$, so

$$0 + 0 + W_{\text{other}} = \tfrac{1}{2}I\omega^2 + 0$$

$$\omega = \sqrt{\frac{2W_{\text{other}}}{I}} = \sqrt{\frac{2(18 \text{ J})}{0.090 \text{ kg} \cdot \text{m}^2}} = 20 \text{ rad/s}$$

From Eq. (9.13), the final tangential speed of the cylinder, and hence the final speed of the cable, is

$$v = R\omega = (0.060 \text{ m})(20 \text{ rad/s}) = 1.2 \text{ m/s}$$

EVALUATE: If the cable mass is not negligible, some of the 18 J of work would go into the kinetic energy of the cable. Then the cylinder would have less kinetic energy and a lower angular speed than we calculated here.

EXAMPLE 9.8 **AN UNWINDING CABLE II**

We wrap a light, nonstretching cable around a solid cylinder with mass M and radius R. The cylinder rotates with negligible friction about a stationary horizontal axis. We tie the free end of the cable to a block of mass m and release the block from rest at a distance h above the floor. As the block falls, the cable unwinds without stretching or slipping. Find the speed of the falling block and the angular speed of the cylinder as the block strikes the floor.

SOLUTION

IDENTIFY: As in Example 9.7, the cable doesn't slip and so friction does no work. We assume that the cable is massless, so that the forces it exerts on the cylinder and the block have equal magnitudes. At its upper end the force and displacement are in the same direction, and at its lower end they are in opposite directions, so the cable does no *net* work and $W_{\text{other}} = 0$. Only gravity does work, and mechanical energy is conserved.

SET UP: Figure 9.17a shows the situation before the block begins to fall (point 1). The initial kinetic energy is $K_1 = 0$. We take the gravitational potential energy to be zero when the block is at floor level (point 2), so $U_1 = mgh$ and $U_2 = 0$. (We ignore the gravitational potential energy for the rotating cylinder, since its height doesn't change.) Just before the block hits the floor (Fig. 9.17b), both the block and the cylinder have kinetic energy, so

$$K_2 = \tfrac{1}{2}mv^2 + \tfrac{1}{2}I\omega^2$$

The moment of inertia of the cylinder is $I = \frac{1}{2}MR^2$. Also, $v = R\omega$ since the speed of the falling block must be equal to the tangential speed at the outer surface of the cylinder.

9.17 Our sketches for this problem.

(a) Initial (block at point 1)

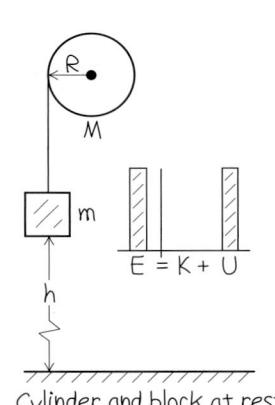

Cylinder and block at rest

(b) Final (block at point 2)

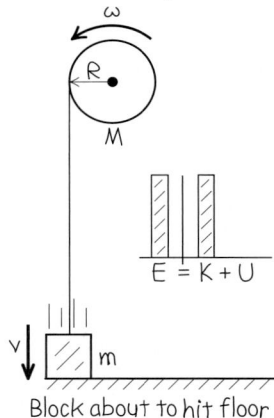

Block about to hit floor

EXECUTE: We use our expressions for K_1, U_1, K_2, and U_2 and the relationship $\omega = v/R$ in Eq. (7.4), $K_1 + U_1 = K_2 + U_2$, and solve for v:

$$0 + mgh = \tfrac{1}{2}mv^2 + \tfrac{1}{2}\left(\tfrac{1}{2}MR^2\right)\left(\frac{v}{R}\right)^2 + 0 = \tfrac{1}{2}\left(m + \tfrac{1}{2}M\right)v^2$$

$$v = \sqrt{\frac{2gh}{1 + M/2m}}$$

The final angular speed of the cylinder is $\omega = v/R$.

EVALUATE: When M is much larger than m, v is very small; when M is much smaller than m, v is nearly equal to $\sqrt{2gh}$, the speed of a body that falls freely from height h. Both of these results are as we would expect.

9.18 In a technique called the "Fosbury flop" after its innovator, this athlete arches her body as she passes over the bar in the high jump. As a result, her center of mass actually passes *under* the bar. This technique requires a smaller increase in gravitational potential energy [Eq. (9.18)] than the older method of straddling the bar.

Gravitational Potential Energy for an Extended Body

In Example 9.8 the cable was of negligible mass, so we could ignore its kinetic energy as well as the gravitational potential energy associated with it. If the mass is *not* negligible, we need to know how to calculate the *gravitational potential energy* associated with such an extended body. If the acceleration of gravity g is the same at all points on the body, the gravitational potential energy is the same as though all the mass were concentrated at the center of mass of the body. Suppose we take the y-axis vertically upward. Then for a body with total mass M, the gravitational potential energy U is simply

$$U = Mgy_{cm} \qquad \text{(gravitational potential energy for an extended body)} \qquad (9.18)$$

where y_{cm} is the y-coordinate of the center of mass. This expression applies to any extended body, whether it is rigid or not (**Fig. 9.18**).

To prove Eq. (9.18), we again represent the body as a collection of mass elements m_i. The potential energy for element m_i is $m_i g y_i$, so the total potential energy is

$$U = m_1 g y_1 + m_2 g y_2 + \cdots = (m_1 y_1 + m_2 y_2 + \cdots)g$$

But from Eq. (8.28), which defines the coordinates of the center of mass,

$$m_1 y_1 + m_2 y_2 + \cdots = (m_1 + m_2 + \cdots)y_{cm} = My_{cm}$$

where $M = m_1 + m_2 + \cdots$ is the total mass. Combining this with the above expression for U, we find $U = Mgy_{cm}$ in agreement with Eq. (9.18).

We leave the application of Eq. (9.18) to the problems. In Chapter 10 we'll use this equation to help us analyze rigid-body problems in which the axis of rotation moves.

TEST YOUR UNDERSTANDING OF SECTION 9.4 Suppose the cylinder and block in Example 9.8 have the same mass, so $m = M$. Just before the block strikes the floor, which statement is correct about the relationship between the kinetic energy of the falling block and the rotational kinetic energy of the cylinder? (i) The block has more kinetic energy than the cylinder. (ii) The block has less kinetic energy than the cylinder. (iii) The block and the cylinder have equal amounts of kinetic energy. ▌

9.5 PARALLEL-AXIS THEOREM

9.19 The parallel-axis theorem.

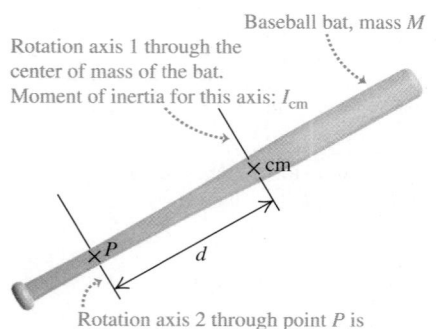

Baseball bat, mass M

Rotation axis 1 through the center of mass of the bat. Moment of inertia for this axis: I_{cm}

×cm

×P

d

Rotation axis 2 through point P is parallel to, and a distance d from, axis 1. Moment of inertia for this axis: I_P

Parallel-axis theorem: $I_P = I_{cm} + Md^2$

We pointed out in Section 9.4 that a body doesn't have just one moment of inertia. In fact, it has infinitely many, because there are infinitely many axes about which it might rotate. But there is a simple relationship, called the **parallel-axis theorem,** between the moment of inertia of a body about an axis through its center of mass and the moment of inertia about any other axis parallel to the original axis (**Fig. 9.19**):

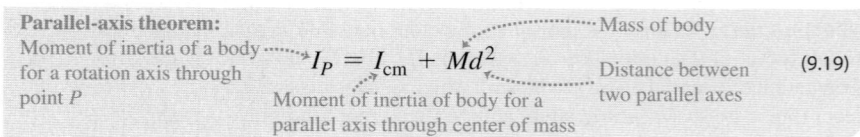

Parallel-axis theorem:
Moment of inertia of a body for a rotation axis through point P $\quad I_P = I_{cm} + Md^2 \quad$ Mass of body

Moment of inertia of body for a parallel axis through center of mass

Distance between two parallel axes   (9.19)

To prove this theorem, we consider two axes, both parallel to the z-axis: one through the center of mass and the other through a point P (**Fig. 9.20**). First we take a very thin slice of the body, parallel to the xy-plane and perpendicular to the z-axis. We take the origin of our coordinate system to be at the center of mass of the body; the coordinates of the center of mass are then $x_{cm} = y_{cm} = z_{cm} = 0$. The axis through the center of mass passes through this thin slice at point O, and the parallel axis passes through point P, whose x- and y-coordinates are (a, b). The distance of this axis from the axis through the center of mass is d, where $d^2 = a^2 + b^2$.

We can write an expression for the moment of inertia I_P about the axis through point P. Let m_i be a mass element in our slice, with coordinates (x_i, y_i, z_i). Then the moment of inertia I_{cm} of the slice about the axis through the center of mass (at O) is

$$I_{cm} = \sum_i m_i(x_i^2 + y_i^2)$$

The moment of inertia of the slice about the axis through P is

$$I_P = \sum_i m_i[(x_i - a)^2 + (y_i - b)^2]$$

These expressions don't involve the coordinates z_i measured perpendicular to the slices, so we can extend the sums to include *all* particles in *all* slices. Then I_P becomes the moment of inertia of the *entire* body for an axis through P. We then expand the squared terms and regroup, and obtain

$$I_P = \sum_i m_i(x_i^2 + y_i^2) - 2a\sum_i m_i x_i - 2b\sum_i m_i y_i + (a^2 + b^2)\sum_i m_i$$

The first sum is I_{cm}. From Eq. (8.28), the definition of the center of mass, the second and third sums are proportional to x_{cm} and y_{cm}; these are zero because we have taken our origin to be the center of mass. The final term is d^2 multiplied by the total mass, or Md^2. This completes our proof that $I_P = I_{cm} + Md^2$.

As Eq. (9.19) shows, a rigid body has a lower moment of inertia about an axis through its center of mass than about any other parallel axis. Thus it's easier to start a body rotating if the rotation axis passes through the center of mass. This suggests that it's somehow most natural for a rotating body to rotate about an axis through its center of mass; we'll make this idea more quantitative in Chapter 10.

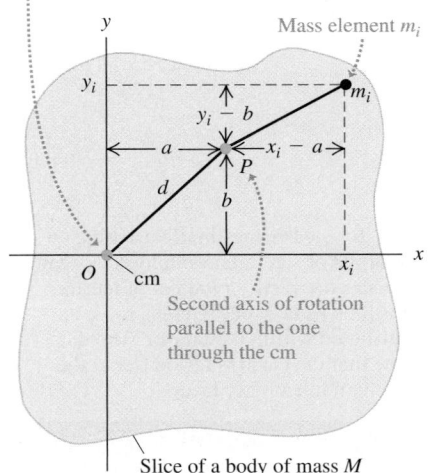

9.20 The mass element m_i has coordinates (x_i, y_i) with respect to an axis of rotation through the center of mass (cm) and coordinates $(x_i - a, y_i - b)$ with respect to the parallel axis through point P.

Axis of rotation passing through cm and perpendicular to the plane of the figure

Mass element m_i

Second axis of rotation parallel to the one through the cm

Slice of a body of mass M

EXAMPLE 9.9   USING THE PARALLEL-AXIS THEOREM

A part of a mechanical linkage (**Fig. 9.21**) has a mass of 3.6 kg. Its moment of inertia I_P about an axis 0.15 m from its center of mass is $I_P = 0.132$ kg·m². What is the moment of inertia I_{cm} about a parallel axis through the center of mass?

9.21 Calculating I_{cm} from a measurement of I_P.

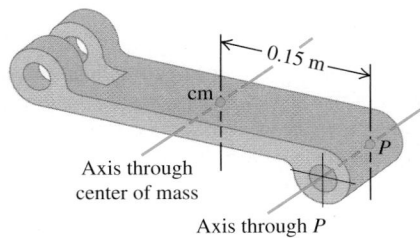

0.15 m

cm

P

Axis through center of mass

Axis through P

SOLUTION

IDENTIFY, SET UP, and EXECUTE: We'll determine the target variable I_{cm} by using the parallel-axis theorem, Eq. (9.19). Rearranging the equation, we obtain

$$I_{cm} = I_P - Md^2 = 0.132\ \text{kg·m}^2 - (3.6\ \text{kg})(0.15\ \text{m})^2$$
$$= 0.051\ \text{kg·m}^2$$

EVALUATE: As we expect, I_{cm} is less than I_P; the moment of inertia for an axis through the center of mass has a lower value than for any other parallel axis.

TEST YOUR UNDERSTANDING OF SECTION 9.5   A pool cue is a wooden rod of uniform composition and is tapered with a larger diameter at one end than at the other end. Use the parallel-axis theorem to decide whether a pool cue has a larger moment of inertia (i) for an axis through the thicker end and perpendicular to the length of the rod, or (ii) for an axis through the thinner end and perpendicular to the length of the rod. ▮

9.6 MOMENT-OF-INERTIA CALCULATIONS

If a rigid body is a continuous distribution of mass—like a solid cylinder or a solid sphere—it cannot be represented by a few point masses. In this case the *sum* of masses and distances that defines the moment of inertia [Eq. (9.16)] becomes an *integral*. Imagine dividing the body into elements of mass dm that are very small, so that all points in a particular element are at essentially the same perpendicular distance from the axis of rotation. We call this distance r, as before. Then the moment of inertia is

$$I = \int r^2 \, dm \tag{9.20}$$

To evaluate the integral, we have to represent r and dm in terms of the same integration variable. When the object is effectively one-dimensional, such as the slender rods (a) and (b) in Table 9.2, we can use a coordinate x along the length and relate dm to an increment dx. For a three-dimensional object it is usually easiest to express dm in terms of an element of volume dV and the *density* ρ of the body. Density is mass per unit volume, $\rho = dm/dV$, so we may write Eq. (9.20) as

$$I = \int r^2 \rho \, dV$$

This expression tells us that a body's moment of inertia depends on how its density varies within its volume (**Fig. 9.22**). If the body is uniform in density, then we may take ρ outside the integral:

$$I = \rho \int r^2 \, dV \tag{9.21}$$

To use this equation, we have to express the volume element dV in terms of the differentials of the integration variables, such as $dV = dx \, dy \, dz$. The element dV must always be chosen so that all points within it are at very nearly the same distance from the axis of rotation. The limits on the integral are determined by the shape and dimensions of the body. For regularly shaped bodies, this integration is often easy to do.

9.22 By measuring small variations in the orbits of satellites, geophysicists can measure the earth's moment of inertia. This tells us how our planet's mass is distributed within its interior. The data show that the earth is far denser at the core than in its outer layers.

EXAMPLE 9.10   **HOLLOW OR SOLID CYLINDER, ROTATING ABOUT AXIS OF SYMMETRY**

Figure 9.23 shows a hollow cylinder of uniform mass density ρ with length L, inner radius R_1, and outer radius R_2. (It might be a steel cylinder in a printing press.) Using integration, find its moment of inertia about its axis of symmetry.

9.23 Finding the moment of inertia of a hollow cylinder about its symmetry axis.

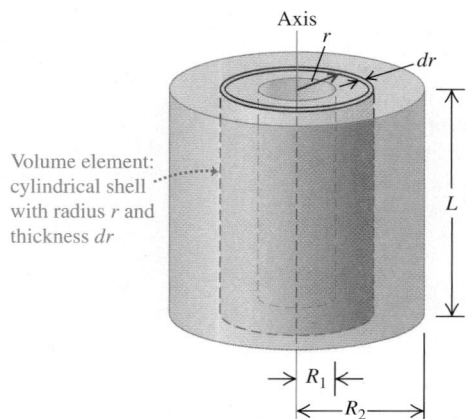

SOLUTION

IDENTIFY and SET UP: We choose as a volume element a thin cylindrical shell of radius r, thickness dr, and length L. All parts of this shell are at very nearly the same distance r from the axis. The volume of the shell is very nearly that of a flat sheet with thickness dr, length L, and width $2\pi r$ (the circumference of the shell). The mass of the shell is

$$dm = \rho \, dV = \rho(2\pi r L \, dr)$$

We'll use this expression in Eq. (9.20), integrating from $r = R_1$ to $r = R_2$.

EXECUTE: From Eq. (9.20), the moment of inertia is

$$I = \int r^2 \, dm = \int_{R_1}^{R_2} r^2 \rho(2\pi r L \, dr)$$

$$= 2\pi\rho L \int_{R_1}^{R_2} r^3 \, dr = \frac{2\pi\rho L}{4}(R_2^4 - R_1^4)$$

$$= \frac{\pi\rho L}{2}(R_2^2 - R_1^2)(R_2^2 + R_1^2)$$

(In the last step we used the identity $a^2 - b^2 = (a - b)(a + b)$.) Let's express this result in terms of the total mass M of the body, which is its density ρ multiplied by the total volume V. The cylinder's volume is

$$V = \pi L(R_2^2 - R_1^2)$$

so its total mass M is

$$M = \rho V = \pi L\rho(R_2^2 - R_1^2)$$

Comparing with the above expression for I, we see that

$$I = \tfrac{1}{2}M(R_1^2 + R_2^2)$$

EVALUATE: Our result agrees with Table 9.2, case (e). If the cylinder is solid, with outer radius $R_2 = R$ and inner radius $R_1 = 0$, its moment of inertia is

$$I = \tfrac{1}{2}MR^2$$

in agreement with case (f). If the cylinder wall is very thin, we have $R_1 \approx R_2 = R$ and the moment of inertia is

$$I = MR^2$$

in agreement with case (g). We could have predicted this last result without calculation; in a thin-walled cylinder, all the mass is at the same distance $r = R$ from the axis, so $I = \int r^2 \, dm = R^2 \int dm = MR^2$.

EXAMPLE 9.11 UNIFORM SPHERE WITH RADIUS R, AXIS THROUGH CENTER

Find the moment of inertia of a solid sphere of uniform mass density ρ (like a billiard ball) about an axis through its center.

SOLUTION

IDENTIFY and SET UP: We divide the sphere into thin, solid disks of thickness dx (**Fig. 9.24**), whose moment of inertia we know from Table 9.2, case (f). We'll integrate over these to find the total moment of inertia.

EXECUTE: The radius and hence the volume and mass of a disk depend on its distance x from the center of the sphere. The radius r of the disk shown in Fig. 9.24 is

$$r = \sqrt{R^2 - x^2}$$

Its volume is

$$dV = \pi r^2 \, dx = \pi(R^2 - x^2) \, dx$$

and so its mass is

$$dm = \rho \, dV = \pi\rho(R^2 - x^2) \, dx$$

From Table 9.2, case (f), the moment of inertia of a disk of radius r and mass dm is

$$dI = \tfrac{1}{2}r^2 \, dm = \tfrac{1}{2}(R^2 - x^2)\left[\pi\rho(R^2 - x^2)dx\right]$$

$$= \frac{\pi\rho}{2}(R^2 - x^2)^2 \, dx$$

Integrating this expression from $x = 0$ to $x = R$ gives the moment of inertia of the right hemisphere. The total I for the entire sphere, including both hemispheres, is just twice this:

$$I = (2)\frac{\pi\rho}{2}\int_0^R (R^2 - x^2)^2 \, dx$$

Carrying out the integration, we find

$$I = \frac{8\pi\rho R^5}{15}$$

The volume of the sphere is $V = 4\pi R^3/3$, so in terms of its mass M its density is

$$\rho = \frac{M}{V} = \frac{3M}{4\pi R^3}$$

Hence our expression for I becomes

$$I = \left(\frac{8\pi R^5}{15}\right)\left(\frac{3M}{4\pi R^3}\right) = \tfrac{2}{5}MR^2$$

EVALUATE: This is just as in Table 9.2, case (h). Note that the moment of inertia $I = \tfrac{2}{5}MR^2$ of a solid sphere of mass M and radius R is less than the moment of inertia $I = \tfrac{1}{2}MR^2$ of a solid *cylinder* of the same mass and radius, because more of the sphere's mass is located close to the axis.

9.24 Finding the moment of inertia of a sphere about an axis through its center.

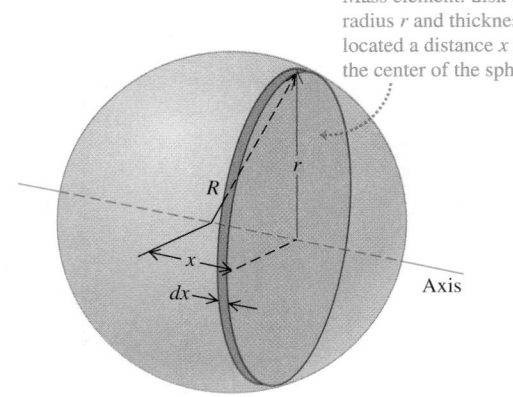

Mass element: disk of radius r and thickness dx located a distance x from the center of the sphere

R
r
x
dx
Axis

TEST YOUR UNDERSTANDING OF SECTION 9.6 Two hollow cylinders have the same inner and outer radii and the same mass, but they have different lengths. One is made of low-density wood and the other of high-density lead. Which cylinder has the greater moment of inertia around its axis of symmetry? (i) The wood cylinder; (ii) the lead cylinder; (iii) the two moments of inertia are equal. ∎

Rotational kinematics: When a rigid body rotates about a stationary axis (usually called the z-axis), the body's position is described by an angular coordinate θ. The angular velocity ω_z is the time derivative of θ, and the angular acceleration α_z is the time derivative of ω_z or the second derivative of θ. (See Examples 9.1 and 9.2.) If the angular acceleration is constant, then θ, ω_z, and α_z are related by simple kinematic equations analogous to those for straight-line motion with constant linear acceleration. (See Example 9.3.)

$$\omega_z = \lim_{\Delta t \to 0} \frac{\Delta \theta}{\Delta t} = \frac{d\theta}{dt} \quad (9.3)$$

$$\alpha_z = \lim_{\Delta t \to 0} \frac{\Delta \omega_z}{\Delta t} = \frac{d\omega_z}{dt} \quad (9.5)$$

Constant α_z only:

$$\theta = \theta_0 + \omega_{0z}t + \tfrac{1}{2}\alpha_z t^2 \quad (9.11)$$

$$\theta - \theta_0 = \tfrac{1}{2}(\omega_{0z} + \omega_z)t \quad (9.10)$$

$$\omega_z = \omega_{0z} + \alpha_z t \quad (9.7)$$

$$\omega_z^2 = \omega_{0z}^2 + 2\alpha_z(\theta - \theta_0) \quad (9.12)$$

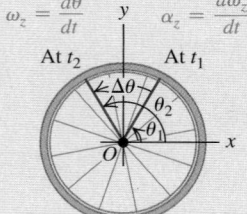

$$\omega_z = \frac{d\theta}{dt} \qquad \alpha_z = \frac{d\omega_z}{dt}$$

Relating linear and angular kinematics: The angular speed ω of a rigid body is the magnitude of the body's angular velocity. The rate of change of ω is $\alpha = d\omega/dt$. For a particle in the body a distance r from the rotation axis, the speed v and the components of the acceleration $\vec{a}$ are related to ω and α. (See Examples 9.4 and 9.5.)

$$v = r\omega \quad (9.13)$$

$$a_{\text{tan}} = \frac{dv}{dt} = r\frac{d\omega}{dt} = r\alpha \quad (9.14)$$

$$a_{\text{rad}} = \frac{v^2}{r} = \omega^2 r \quad (9.15)$$

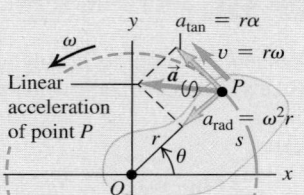

Moment of inertia and rotational kinetic energy: The moment of inertia I of a body about a given axis is a measure of its rotational inertia: The greater the value of I, the more difficult it is to change the state of the body's rotation. The moment of inertia can be expressed as a sum over the particles m_i that make up the body, each of which is at its own perpendicular distance r_i from the axis. The rotational kinetic energy of a rigid body rotating about a fixed axis depends on the angular speed ω and the moment of inertia I for that rotation axis. (See Examples 9.6–9.8.)

$$I = m_1 r_1^2 + m_2 r_2^2 + \cdots$$

$$= \sum_i m_i r_i^2 \quad (9.16)$$

$$K = \tfrac{1}{2}I\omega^2 \quad (9.17)$$

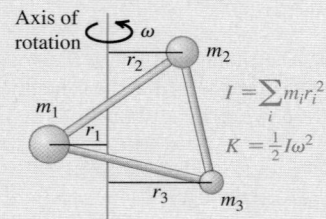

Calculating the moment of inertia: The parallel-axis theorem relates the moments of inertia of a rigid body of mass M about two parallel axes: an axis through the center of mass (moment of inertia I_{cm}) and a parallel axis a distance d from the first axis (moment of inertia I_P). (See Example 9.9.) If the body has a continuous mass distribution, the moment of inertia can be calculated by integration. (See Examples 9.10 and 9.11.)

$$I_P = I_{\text{cm}} + Md^2 \quad (9.19)$$

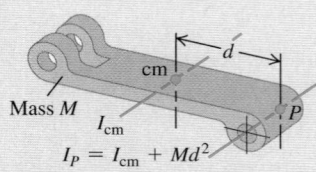

BRIDGING PROBLEM   A ROTATING, UNIFORM THIN ROD

Figure 9.25 shows a slender uniform rod with mass M and length L. It might be a baton held by a twirler in a marching band (without the rubber end caps). (a) Use integration to compute its moment of inertia about an axis through O, at an arbitrary distance h from one end. (b) Initially the rod is at rest. It is given a constant angular acceleration of magnitude α around the axis through O. Find how much work is done on the rod in a time t. (c) At time t, what is the *linear* acceleration of the point on the rod farthest from the axis?

SOLUTION GUIDE

IDENTIFY and SET UP
1. Make a list of the target variables for this problem.
2. To calculate the moment of inertia of the rod, you'll have to divide the rod into infinitesimal elements of mass. If an element

has length dx, what is the mass of the element? What are the limits of integration?
3. What is the angular speed of the rod at time t? How does the work required to accelerate the rod from rest to this angular speed compare to the rod's kinetic energy at time t?
4. At time t, does the point on the rod farthest from the axis have a centripetal acceleration? A tangential acceleration? Why or why not?

EXECUTE
5. Do the integration required to find the moment of inertia.
6. Use your result from step 5 to calculate the work done in time t to accelerate the rod from rest.
7. Find the linear acceleration components for the point in question at time t. Use these to find the magnitude of the acceleration.

EVALUATE
8. Check your results for the special cases $h = 0$ (the axis passes through one end of the rod) and $h = L/2$ (the axis passes through the middle of the rod). Are these limits consistent with Table 9.2? With the parallel-axis theorem?
9. Is the acceleration magnitude from step 7 constant? Would you expect it to be?

9.25 A thin rod with an axis through O.

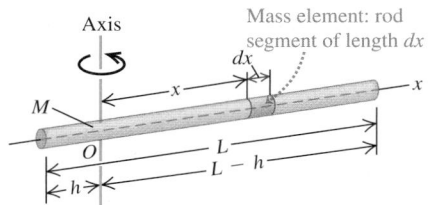

Problems

For assigned homework and other learning materials, go to MasteringPhysics®.   (MP)

•, ••, •••: Difficulty levels. **CP**: Cumulative problems incorporating material from earlier chapters. **CALC**: Problems requiring calculus. **DATA**: Problems involving real data, scientific evidence, experimental design, and/or statistical reasoning. **BIO**: Biosciences problems.

DISCUSSION QUESTIONS

Q9.1 Which of the following formulas is valid if the angular acceleration of an object is *not* constant? Explain your reasoning in each case. (a) $v = r\omega$; (b) $a_{\text{tan}} = r\alpha$; (c) $\omega = \omega_0 + \alpha t$; (d) $a_{\text{tan}} = r\omega^2$; (e) $K = \frac{1}{2}I\omega^2$.

Q9.2 A diatomic molecule can be modeled as two point masses, m_1 and m_2, slightly separated (**Fig. Q9.2**). If the molecule is oriented along the y-axis, it has kinetic energy K when it spins about the x-axis. What will its kinetic energy (in terms of K) be if it spins at the same angular speed about (a) the z-axis and (b) the y-axis?

Figure **Q9.2**

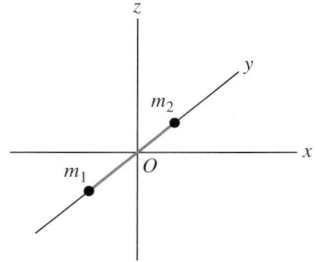

Q9.3 What is the difference between tangential and radial acceleration for a point on a rotating body?

Q9.4 In **Fig. Q9.4**, all points on the chain have the same linear speed. Is the magnitude of the linear acceleration also the same for all points on the chain? How are the angular accelerations of the two sprockets related? Explain.

Figure **Q9.4**

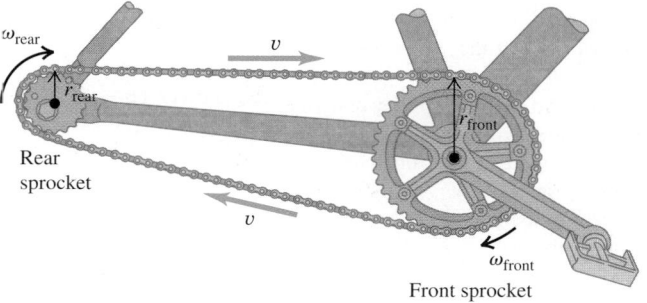

Front sprocket

Q9.5 In Fig. Q9.4, how are the radial accelerations of points at the teeth of the two sprockets related? Explain.
Q9.6 A flywheel rotates with constant angular velocity. Does a point on its rim have a tangential acceleration? A radial acceleration?

Are these accelerations constant in magnitude? In direction? In each case give your reasoning.

Q9.7 What is the purpose of the spin cycle of a washing machine? Explain in terms of acceleration components.

Q9.8 You are designing a flywheel to store kinetic energy. If all of the following uniform objects have the same mass and same angular velocity, which one will store the greatest amount of kinetic energy? Which will store the least? Explain. (a) A solid sphere of diameter D rotating about a diameter; (b) a solid cylinder of diameter D rotating about an axis perpendicular to each face through its center; (c) a thin-walled hollow cylinder of diameter D rotating about an axis perpendicular to the plane of the circular face at its center; (d) a solid, thin bar of length D rotating about an axis perpendicular to it at its center.

Q9.9 Can you think of a body that has the same moment of inertia for all possible axes? If so, give an example, and if not, explain why this is not possible. Can you think of a body that has the same moment of inertia for all axes passing through a certain point? If so, give an example and indicate where the point is located.

Q9.10 To maximize the moment of inertia of a flywheel while minimizing its weight, what shape and distribution of mass should it have? Explain.

Q9.11 How might you determine experimentally the moment of inertia of an irregularly shaped body about a given axis?

Q9.12 A cylindrical body has mass M and radius R. Can the mass be distributed within the body in such a way that its moment of inertia about its axis of symmetry is greater than MR^2? Explain.

Q9.13 Describe how you could use part (b) of Table 9.2 to derive the result in part (d).

Q9.14 A hollow spherical shell of radius R that is rotating about an axis through its center has rotational kinetic energy K. If you want to modify this sphere so that it has three times as much kinetic energy at the same angular speed while keeping the same mass, what should be its radius in terms of R?

Q9.15 For the equations for I given in parts (a) and (b) of Table 9.2 to be valid, must the rod have a circular cross section? Is there any restriction on the size of the cross section for these equations to apply? Explain.

Q9.16 In part (d) of Table 9.2, the thickness of the plate must be much less than a for the expression given for I to apply. But in part (c), the expression given for I applies no matter how thick the plate is. Explain.

Q9.17 Two identical balls, A and B, are each attached to very light string, and each string is wrapped around the rim of a frictionless pulley of mass M. The only difference is that the pulley for ball A is a solid disk, while the one for ball B is a hollow disk, like part (e) in Table 9.2. If both balls are released from rest and fall the same distance, which one will have more kinetic energy, or will they have the same kinetic energy? Explain your reasoning.

Q9.18 An elaborate pulley consists of four identical balls at the ends of spokes extending out from a rotating drum (**Fig. Q9.18**). A box is connected to a light, thin rope wound around the rim of the drum. When it is released from rest, the box acquires a speed V after having fallen a distance d. Now the four balls are moved inward closer to the drum, and the box is again released from rest. After it has fallen a distance d,

Figure **Q9.18**

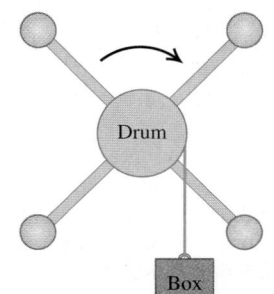

Drum

Box

will its speed be equal to V, greater than V, or less than V? Show or explain why.

Q9.19 You can use any angular measure—radians, degrees, or revolutions—in some of the equations in Chapter 9, but you can use only radian measure in others. Identify those for which using radians is necessary and those for which it is not, and in each case give your reasoning.

Q9.20 When calculating the moment of inertia of an object, can we treat all its mass as if it were concentrated at the center of mass of the object? Justify your answer.

Q9.21 A wheel is rotating about an axis perpendicular to the plane of the wheel and passing through the center of the wheel. The angular speed of the wheel is increasing at a constant rate. Point A is on the rim of the wheel and point B is midway between the rim and center of the wheel. For each of the following quantities, is its magnitude larger at point A or at point B, or is it the same at both points? (a) angular speed; (b) tangential speed; (c) angular acceleration; (d) tangential acceleration; (e) radial acceleration. Justify each answer.

Q9.22 Estimate your own moment of inertia about a vertical axis through the center of the top of your head when you are standing up straight with your arms outstretched. Make reasonable approximations and measure or estimate necessary quantities.

EXERCISES

Section 9.1 Angular Velocity and Acceleration

9.1 • (a) What angle in radians is subtended by an arc 1.50 m long on the circumference of a circle of radius 2.50 m? What is this angle in degrees? (b) An arc 14.0 cm long on the circumference of a circle subtends an angle of 128°. What is the radius of the circle? (c) The angle between two radii of a circle with radius 1.50 m is 0.700 rad. What length of arc is intercepted on the circumference of the circle by the two radii?

9.2 • An airplane propeller is rotating at 1900 rpm (rev/min). (a) Compute the propeller's angular velocity in rad/s. (b) How many seconds does it take for the propeller to turn through 35°?

9.3 • CP CALC The angular velocity of a flywheel obeys the equation $\omega_z(t) = A + Bt^2$, where t is in seconds and A and B are constants having numerical values 2.75 (for A) and 1.50 (for B). (a) What are the units of A and B if ω_z is in rad/s? (b) What is the angular acceleration of the wheel at (i) $t = 0$ and (ii) $t = 5.00$ s? (c) Through what angle does the flywheel turn during the first 2.00 s? (*Hint:* See Section 2.6.)

9.4 •• CALC A fan blade rotates with angular velocity given by $\omega_z(t) = \gamma - \beta t^2$, where $\gamma = 5.00$ rad/s and $\beta = 0.800$ rad/s³. (a) Calculate the angular acceleration as a function of time. (b) Calculate the instantaneous angular acceleration α_z at $t = 3.00$ s and the average angular acceleration $\alpha_{av\text{-}z}$ for the time interval $t = 0$ to $t = 3.00$ s. How do these two quantities compare? If they are different, why?

9.5 •• CALC A child is pushing a merry-go-round. The angle through which the merry-go-round has turned varies with time according to $\theta(t) = \gamma t + \beta t^3$, where $\gamma = 0.400$ rad/s and $\beta = 0.0120$ rad/s³. (a) Calculate the angular velocity of the merry-go-round as a function of time. (b) What is the initial value of the angular velocity? (c) Calculate the instantaneous value of the angular velocity ω_z at $t = 5.00$ s and the average angular velocity $\omega_{av\text{-}z}$ for the time interval $t = 0$ to $t = 5.00$ s. Show that $\omega_{av\text{-}z}$ is *not* equal to the average of the instantaneous angular velocities at $t = 0$ and $t = 5.00$ s, and explain.

9.6 • CALC At $t = 0$ the current to a dc electric motor is reversed, resulting in an angular displacement of the motor shaft given by $\theta(t) = (250 \text{ rad/s})t - (20.0 \text{ rad/s}^2)t^2 - (1.50 \text{ rad/s}^3)t^3$. (a) At what time is the angular velocity of the motor shaft zero? (b) Calculate the angular acceleration at the instant that the motor shaft has zero angular velocity. (c) How many revolutions does the motor shaft turn through between the time when the current is reversed and the instant when the angular velocity is zero? (d) How fast was the motor shaft rotating at $t = 0$, when the current was reversed? (e) Calculate the average angular velocity for the time period from $t = 0$ to the time calculated in part (a).

9.7 • CALC The angle θ through which a disk drive turns is given by $\theta(t) = a + bt - ct^3$, where a, b, and c are constants, t is in seconds, and θ is in radians. When $t = 0$, $\theta = \pi/4$ rad and the angular velocity is 2.00 rad/s. When $t = 1.50$ s, the angular acceleration is 1.25 rad/s². (a) Find a, b, and c, including their units. (b) What is the angular acceleration when $\theta = \pi/4$ rad? (c) What are θ and the angular velocity when the angular acceleration is 3.50 rad/s²?

9.8 • A wheel is rotating about an axis that is in the z-direction. The angular velocity ω_z is -6.00 rad/s at $t = 0$, increases linearly with time, and is $+4.00$ rad/s at $t = 7.00$ s. We have taken counterclockwise rotation to be positive. (a) Is the angular acceleration during this time interval positive or negative? (b) During what time interval is the speed of the wheel increasing? Decreasing? (c) What is the angular displacement of the wheel at $t = 7.00$ s?

Section 9.2 Rotation with Constant Angular Acceleration

9.9 • A bicycle wheel has an initial angular velocity of 1.50 rad/s. (a) If its angular acceleration is constant and equal to 0.200 rad/s², what is its angular velocity at $t = 2.50$ s? (b) Through what angle has the wheel turned between $t = 0$ and $t = 2.50$ s?

9.10 •• An electric fan is turned off, and its angular velocity decreases uniformly from 500 rev/min to 200 rev/min in 4.00 s. (a) Find the angular acceleration in rev/s² and the number of revolutions made by the motor in the 4.00-s interval. (b) How many more seconds are required for the fan to come to rest if the angular acceleration remains constant at the value calculated in part (a)?

9.11 •• The rotating blade of a blender turns with constant angular acceleration 1.50 rad/s². (a) How much time does it take to reach an angular velocity of 36.0 rad/s, starting from rest? (b) Through how many revolutions does the blade turn in this time interval?

9.12 • (a) Derive Eq. (9.12) by combining Eqs. (9.7) and (9.11) to eliminate t. (b) The angular velocity of an airplane propeller increases from 12.0 rad/s to 16.0 rad/s while turning through 7.00 rad. What is the angular acceleration in rad/s²?

9.13 •• A turntable rotates with a constant 2.25 rad/s² angular acceleration. After 4.00 s it has rotated through an angle of 30.0 rad. What was the angular velocity of the wheel at the beginning of the 4.00-s interval?

9.14 • A circular saw blade 0.200 m in diameter starts from rest. In 6.00 s it accelerates with constant angular acceleration to an angular velocity of 140 rad/s. Find the angular acceleration and the angle through which the blade has turned.

9.15 •• A high-speed flywheel in a motor is spinning at 500 rpm when a power failure suddenly occurs. The flywheel has mass 40.0 kg and diameter 75.0 cm. The power is off for 30.0 s, and during this time the flywheel slows due to friction in its axle bearings. During the time the power is off, the flywheel makes 200 complete revolutions. (a) At what rate is the flywheel spinning when the power comes back on? (b) How long after the beginning of the power failure would it have taken the flywheel to stop if the power had not come back on, and how many revolutions would the wheel have made during this time?

9.16 •• At $t = 0$ a grinding wheel has an angular velocity of 24.0 rad/s. It has a constant angular acceleration of 30.0 rad/s² until a circuit breaker trips at $t = 2.00$ s. From then on, it turns through 432 rad as it coasts to a stop at constant angular acceleration. (a) Through what total angle did the wheel turn between $t = 0$ and the time it stopped? (b) At what time did it stop? (c) What was its acceleration as it slowed down?

9.17 •• A safety device brings the blade of a power mower from an initial angular speed of ω_1 to rest in 1.00 revolution. At the same constant acceleration, how many revolutions would it take the blade to come to rest from an initial angular speed ω_3 that was three times as great, $\omega_3 = 3\omega_1$?

Section 9.3 Relating Linear and Angular Kinematics

9.18 • In a charming 19th-century hotel, an old-style elevator is connected to a counterweight by a cable that passes over a rotating disk 2.50 m in diameter (**Fig. E9.18**). The elevator is raised and lowered by turning the disk, and the cable does not slip on the rim of the disk but turns with it. (a) At how many rpm must the disk turn to raise the elevator at 25.0 cm/s? (b) To start the elevator moving, it must be accelerated at $\frac{1}{8}g$. What must be the angular acceleration of the disk, in rad/s²? (c) Through what angle (in radians and degrees) has the disk turned when it has raised the elevator 3.25 m between floors?

Figure **E9.18**

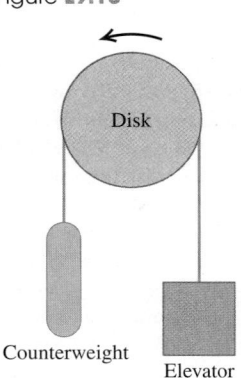

Disk

Counterweight

Elevator

9.19 • Using Appendix F, along with the fact that the earth spins on its axis once per day, calculate (a) the earth's orbital angular speed (in rad/s) due to its motion around the sun, (b) its angular speed (in rad/s) due to its axial spin, (c) the tangential speed of the earth around the sun (assuming a circular orbit), (d) the tangential speed of a point on the earth's equator due to the planet's axial spin, and (e) the radial and tangential acceleration components of the point in part (d).

9.20 • **Compact Disc.** A compact disc (CD) stores music in a coded pattern of tiny pits 10^{-7} m deep. The pits are arranged in a track that spirals outward toward the rim of the disc; the inner and outer radii of this spiral are 25.0 mm and 58.0 mm, respectively. As the disc spins inside a CD player, the track is scanned at a constant *linear* speed of 1.25 m/s. (a) What is the angular speed of the CD when the innermost part of the track is scanned? The outermost part of the track? (b) The maximum playing time of a CD is 74.0 min. What would be the length of the track on such a maximum-duration CD if it were stretched out in a straight line? (c) What is the average angular acceleration of a maximum-duration CD during its 74.0-min playing time? Take the direction of rotation of the disc to be positive.

9.21 •• A wheel of diameter 40.0 cm starts from rest and rotates with a constant angular acceleration of 3.00 rad/s². Compute the radial acceleration of a point on the rim for the instant the wheel completes its second revolution from the relationship (a) $a_{\text{rad}} = \omega^2 r$ and (b) $a_{\text{rad}} = v^2/r$.

9.22 •• You are to design a rotating cylindrical axle to lift 800-N buckets of cement from the ground to a rooftop 78.0 m above the ground. The buckets will be attached to a hook on the free end of a cable that wraps around the rim of the axle; as the axle turns, the buckets will rise. (a) What should the diameter of the axle be in order to raise the buckets at a steady 2.00 cm/s when it is turning at 7.5 rpm? (b) If instead the axle must give the buckets an upward acceleration of 0.400 m/s², what should the angular acceleration of the axle be?

9.23 • A flywheel with a radius of 0.300 m starts from rest and accelerates with a constant angular acceleration of 0.600 rad/s². Compute the magnitude of the tangential acceleration, the radial acceleration, and the resultant acceleration of a point on its rim (a) at the start; (b) after it has turned through 60.0°; (c) after it has turned through 120.0°.

9.24 •• An electric turntable 0.750 m in diameter is rotating about a fixed axis with an initial angular velocity of 0.250 rev/s and a constant angular acceleration of 0.900 rev/s². (a) Compute the angular velocity of the turntable after 0.200 s. (b) Through how many revolutions has the turntable spun in this time interval? (c) What is the tangential speed of a point on the rim of the turntable at $t = 0.200$ s? (d) What is the magnitude of the *resultant* acceleration of a point on the rim at $t = 0.200$ s?

9.25 •• **Centrifuge.** An advertisement claims that a centrifuge takes up only 0.127 m of bench space but can produce a radial acceleration of 3000g at 5000 rev/min. Calculate the required radius of the centrifuge. Is the claim realistic?

9.26 • At $t = 3.00$ s a point on the rim of a 0.200-m-radius wheel has a tangential speed of 50.0 m/s as the wheel slows down with a tangential acceleration of constant magnitude 10.0 m/s². (a) Calculate the wheel's constant angular acceleration. (b) Calculate the angular velocities at $t = 3.00$ s and $t = 0$. (c) Through what angle did the wheel turn between $t = 0$ and $t = 3.00$ s? (d) At what time will the radial acceleration equal g?

9.27 • **Electric Drill.** According to the shop manual, when drilling a 12.7-mm-diameter hole in wood, plastic, or aluminum, a drill should have a speed of 1250 rev/min. For a 12.7-mm-diameter drill bit turning at a constant 1250 rev/min, find (a) the maximum linear speed of any part of the bit and (b) the maximum radial acceleration of any part of the bit.

Section 9.4 Energy in Rotational Motion

9.28 • Four small spheres, each of which you can regard as a point of mass 0.200 kg, are arranged in a square 0.400 m on a side and connected by extremely light rods (**Fig. E9.28**). Find the moment of inertia of the system about an axis (a) through the center of the square, perpendicular to its plane

Figure **E9.28**

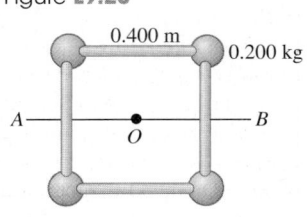

(an axis through point O in the figure); (b) bisecting two opposite sides of the square (an axis along the line AB in the figure); (c) that passes through the centers of the upper left and lower right spheres and through point O.

9.29 • Calculate the moment of inertia of each of the following uniform objects about the axes indicated. Consult Table 9.2 as needed. (a) A thin 2.50-kg rod of length 75.0 cm, about an axis perpendicular to it and passing through (i) one end and (ii) its center, and (iii) about an axis parallel to the rod and passing through it. (b) A 3.00-kg sphere 38.0 cm in diameter, about an axis through

its center, if the sphere is (i) solid and (ii) a thin-walled hollow shell. (c) An 8.00-kg cylinder, of length 19.5 cm and diameter 12.0 cm, about the central axis of the cylinder, if the cylinder is (i) thin-walled and hollow, and (ii) solid.

9.30 •• Small blocks, each with mass m, are clamped at the ends and at the center of a rod of length L and negligible mass. Compute the moment of inertia of the system about an axis perpendicular to the rod and passing through (a) the center of the rod and (b) a point one-fourth of the length from one end.

9.31 • A uniform bar has two small balls glued to its ends. The bar is 2.00 m long and has mass 4.00 kg, while the balls each have mass 0.300 kg and can be treated as point masses. Find the moment of inertia of this combination about an axis (a) perpendicular to the bar through its center; (b) perpendicular to the bar through one of the balls; (c) parallel to the bar through both balls; and (d) parallel to the bar and 0.500 m from it.

9.32 •• You are a project manager for a manufacturing company. One of the machine parts on the assembly line is a thin, uniform rod that is 60.0 cm long and has mass 0.400 kg. (a) What is the moment of inertia of this rod for an axis at its center, perpendicular to the rod? (b) One of your engineers has proposed to reduce the moment of inertia by bending the rod at its center into a V-shape, with a 60.0° angle at its vertex. What would be the moment of inertia of this bent rod about an axis perpendicular to the plane of the V at its vertex?

9.33 •• A wagon wheel is constructed as shown in **Fig. E9.33**. The radius of the wheel is 0.300 m, and the rim has mass 1.40 kg. Each of the eight spokes that lie along a diameter and are 0.300 m long has mass 0.280 kg. What is the moment of inertia of the wheel about an axis through its center and perpendicular to the plane of the wheel? (Use Table 9.2.)

Figure **E9.33**

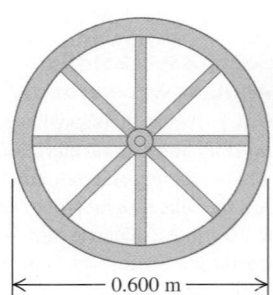

0.600 m

9.34 •• An airplane propeller is 2.08 m in length (from tip to tip) with mass 117 kg and is rotating at 2400 rpm (rev/min) about an axis through its center. You can model the propeller as a slender rod. (a) What is its rotational kinetic energy? (b) Suppose that, due to weight constraints, you had to reduce the propeller's mass to 75.0% of its original mass, but you still needed to keep the same size and kinetic energy. What would its angular speed have to be, in rpm?

9.35 •• A compound disk of outside diameter 140.0 cm is made up of a uniform solid disk of radius 50.0 cm and area density 3.00 g/cm² surrounded by a concentric ring of inner radius 50.0 cm, outer radius 70.0 cm, and area density 2.00 g/cm². Find the moment of inertia of this object about an axis perpendicular to the plane of the object and passing through its center.

9.36 • A wheel is turning about an axis through its center with constant angular acceleration. Starting from rest, at $t = 0$, the wheel turns through 8.20 revolutions in 12.0 s. At $t = 12.0$ s the kinetic energy of the wheel is 36.0 J. For an axis through its center, what is the moment of inertia of the wheel?

9.37 • A uniform sphere with mass 28.0 kg and radius 0.380 m is rotating at constant angular velocity about a stationary axis that lies along a diameter of the sphere. If the kinetic energy of the sphere is 236 J, what is the tangential velocity of a point on the rim of the sphere?

9.38 •• A hollow spherical shell has mass 8.20 kg and radius 0.220 m. It is initially at rest and then rotates about a stationary axis that lies along a diameter with a constant acceleration of 0.890 rad/s². What is the kinetic energy of the shell after it has turned through 6.00 rev?

9.39 •• The flywheel of a gasoline engine is required to give up 500 J of kinetic energy while its angular velocity decreases from 650 rev/min to 520 rev/min. What moment of inertia is required?

9.40 •• You need to design an industrial turntable that is 60.0 cm in diameter and has a kinetic energy of 0.250 J when turning at 45.0 rpm (rev/min). (a) What must be the moment of inertia of the turntable about the rotation axis? (b) If your workshop makes this turntable in the shape of a uniform solid disk, what must be its mass?

9.41 •• Energy is to be stored in a 70.0-kg flywheel in the shape of a uniform solid disk with radius $R = 1.20$ m. To prevent structural failure of the flywheel, the maximum allowed radial acceleration of a point on its rim is 3500 m/s². What is the maximum kinetic energy that can be stored in the flywheel?

9.42 • A light, flexible rope is wrapped several times around a *hollow* cylinder, with a weight of 40.0 N and a radius of 0.25 m, that rotates without friction about a fixed horizontal axis. The cylinder is attached to the axle by spokes of a negligible moment of inertia. The cylinder is initially at rest. The free end of the rope is pulled with a constant force P for a distance of 5.00 m, at which point the end of the rope is moving at 6.00 m/s. If the rope does not slip on the cylinder, what is P?

9.43 •• A frictionless pulley has the shape of a uniform solid disk of mass 2.50 kg and radius 20.0 cm. A 1.50-kg stone is attached to a very light wire that is wrapped around the rim of the pulley (**Fig. E9.43**), and the system is released from rest. (a) How far must the stone fall so that the pulley has 4.50 J of kinetic energy? (b) What percent of the total kinetic energy does the pulley have?

Figure **E9.43**

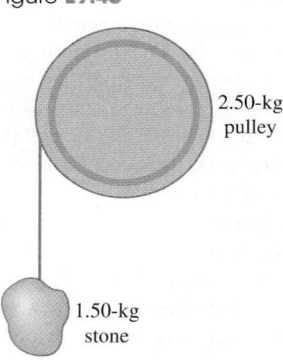

2.50-kg pulley

1.50-kg stone

9.44 •• A bucket of mass m is tied to a massless cable that is wrapped around the outer rim of a frictionless uniform pulley of radius R, similar to the system shown in Fig. E9.43. In terms of the stated variables, what must be the moment of inertia of the pulley so that it always has half as much kinetic energy as the bucket?

9.45 •• CP A thin, light wire is wrapped around the rim of a wheel (**Fig. E9.45**). The wheel rotates without friction about a stationary horizontal axis that passes through the center of the wheel. The wheel is a uniform disk with radius $R = 0.280$ m. An object of mass $m = 4.20$ kg is suspended from the free end of the wire. The system is released from rest and the suspended object descends with constant acceleration. If the suspended object moves downward a distance of 3.00 m in 2.00 s, what is the mass of the wheel?

Figure **E9.45**

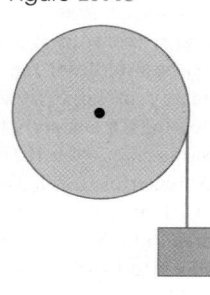

9.46 •• A uniform 2.00-m ladder of mass 9.00 kg is leaning against a vertical wall while making an angle of 53.0° with the floor. A worker pushes the ladder up against the wall until it is vertical. What is the increase in the gravitational potential energy of the ladder?

9.47 •• **How I Scales.** If we multiply all the design dimensions of an object by a scaling factor f, its volume and mass will be multiplied by f^3. (a) By what factor will its moment of inertia be multiplied? (b) If a $\frac{1}{48}$-scale model has a rotational kinetic energy of 2.5 J, what will be the kinetic energy for the full-scale object of the same material rotating at the same angular velocity?

Section 9.5 Parallel-Axis Theorem

9.48 •• Find the moment of inertia of a hoop (a thin-walled, hollow ring) with mass M and radius R about an axis perpendicular to the hoop's plane at an edge.

9.49 •• About what axis will a uniform, balsa-wood sphere have the same moment of inertia as does a thin-walled, hollow, lead sphere of the same mass and radius, with the axis along a diameter?

9.50 • (a) For the thin rectangular plate shown in part (d) of Table 9.2, find the moment of inertia about an axis that lies in the plane of the plate, passes through the center of the plate, and is parallel to the axis shown. (b) Find the moment of inertia of the plate for an axis that lies in the plane of the plate, passes through the center of the plate, and is perpendicular to the axis in part (a).

9.51 •• A thin, rectangular sheet of metal has mass M and sides of length a and b. Use the parallel-axis theorem to calculate the moment of inertia of the sheet for an axis that is perpendicular to the plane of the sheet and that passes through one corner of the sheet.

9.52 •• A thin uniform rod of mass M and length L is bent at its center so that the two segments are now perpendicular to each other. Find its moment of inertia about an axis perpendicular to its plane and passing through (a) the point where the two segments meet and (b) the midpoint of the line connecting its two ends.

Section 9.6 Moment-of-Inertia Calculations

9.53 •• CALC Use Eq. (9.20) to calculate the moment of inertia of a uniform, solid disk with mass M and radius R for an axis perpendicular to the plane of the disk and passing through its center.

9.54 • CALC Use Eq. (9.20) to calculate the moment of inertia of a slender, uniform rod with mass M and length L about an axis at one end, perpendicular to the rod.

9.55 •• CALC A slender rod with length L has a mass per unit length that varies with distance from the left end, where $x = 0$, according to $dm/dx = \gamma x$, where γ has units of kg/m². (a) Calculate the total mass of the rod in terms of γ and L. (b) Use Eq. (9.20) to calculate the moment of inertia of the rod for an axis at the left end, perpendicular to the rod. Use the expression you derived in part (a) to express I in terms of M and L. How does your result compare to that for a uniform rod? Explain. (c) Repeat part (b) for an axis at the right end of the rod. How do the results for parts (b) and (c) compare? Explain.

PROBLEMS

9.56 •• CALC A uniform disk with radius $R = 0.400$ m and mass 30.0 kg rotates in a horizontal plane on a frictionless vertical axle that passes through the center of the disk. The angle through which the disk has turned varies with time according to $\theta(t) = (1.10 \text{ rad/s})t + (6.30 \text{ rad/s}^2)t^2$. What is the resultant linear acceleration of a point on the rim of the disk at the instant when the disk has turned through 0.100 rev?

9.57 •• CP A circular saw blade with radius 0.120 m starts from rest and turns in a vertical plane with a constant angular acceleration of 2.00 rev/s². After the blade has turned through 155 rev, a small piece of the blade breaks loose from the top of the blade. After the piece breaks loose, it travels with a velocity that is initially horizontal and equal to the tangential velocity of the rim of the blade. The piece travels a vertical distance of 0.820 m to the floor. How far does the piece travel horizontally, from where it broke off the blade until it strikes the floor?

9.58 • CALC A roller in a printing press turns through an angle $\theta(t)$ given by $\theta(t) = \gamma t^2 - \beta t^3$, where $\gamma = 3.20$ rad/s² and $\beta = 0.500$ rad/s³. (a) Calculate the angular velocity of the roller as a function of time. (b) Calculate the angular acceleration of the roller as a function of time. (c) What is the maximum positive angular velocity, and at what value of t does it occur?

9.59 •• CP CALC A disk of radius 25.0 cm is free to turn about an axle perpendicular to it through its center. It has very thin but strong string wrapped around its rim, and the string is attached to a ball that is pulled tangentially away from the rim of the disk (**Fig. P9.59**). The pull increases in magnitude and produces an acceleration of the ball that obeys the equation $a(t) = At$, where t is in seconds and A is a constant. The cylinder starts from rest, and at the end of the third second, the ball's acceleration is 1.80 m/s². (a) Find A. (b) Express the angular acceleration of the disk as a function of time. (c) How much time after the disk has begun to turn does it reach an angular speed of 15.0 rad/s? (d) Through what angle has the disk turned just as it reaches 15.0 rad/s? (*Hint:* See Section 2.6.)

Figure **P9.59**

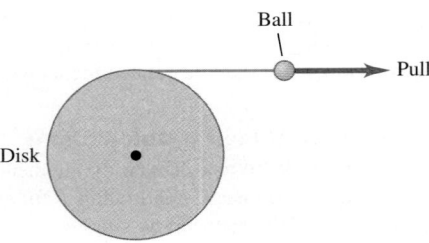

9.60 •• You are designing a rotating metal flywheel that will be used to store energy. The flywheel is to be a uniform disk with radius 25.0 cm. Starting from rest at $t = 0$, the flywheel rotates with constant angular acceleration 3.00 rad/s² about an axis perpendicular to the flywheel at its center. If the flywheel has a density (mass per unit volume) of 8600 kg/m³, what thickness must it have to store 800 J of kinetic energy at $t = 8.00$ s?

9.61 •• You must design a device for shooting a small marble vertically upward. The marble is in a small cup that is attached to the rim of a wheel of radius 0.260 m; the cup is covered by a lid. The wheel starts from rest and rotates about a horizontal axis that is perpendicular to the wheel at its center. After the wheel has turned through 20.0 rev, the cup is the same height as the center of the wheel. At this point in the motion, the lid opens and the marble travels vertically upward to a maximum height h above the center of the wheel. If the wheel rotates with a constant angular acceleration α, what value of α is required for the marble to reach a height of $h = 12.0$ m?

9.62 •• Engineers are designing a system by which a falling mass m imparts kinetic energy to a rotating uniform drum to which it is attached by thin, very light wire wrapped around the rim of the drum (**Fig. P9.62**). There is no appreciable friction in the axle of the drum, and everything starts from rest. This system is being tested on earth, but it is to be used on Mars, where the acceleration due to gravity is 3.71 m/s². In the earth tests, when m is set to 15.0 kg and allowed to fall through 5.00 m, it gives 250.0 J of kinetic energy to the drum. (a) If the system is operated on Mars, through what distance would the 15.0-kg mass have to fall to give the same amount of kinetic energy to the drum? (b) How fast would the 15.0-kg mass be moving on Mars just as the drum gained 250.0 J of kinetic energy?

Figure **P9.62**

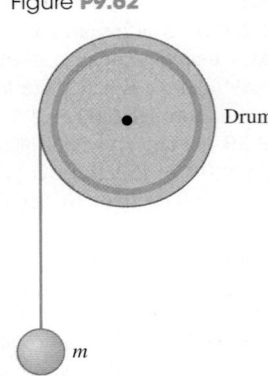
Drum

m

9.63 • A vacuum cleaner belt is looped over a shaft of radius 0.45 cm and a wheel of radius 1.80 cm. The arrangement of the belt, shaft, and wheel is similar to that of the chain and sprockets in Fig. Q9.4. The motor turns the shaft at 60.0 rev/s and the moving belt turns the wheel, which in turn is connected by another shaft to the roller that beats the dirt out of the rug being vacuumed. Assume that the belt doesn't slip on either the shaft or the wheel. (a) What is the speed of a point on the belt? (b) What is the angular velocity of the wheel, in rad/s?

9.64 •• The motor of a table saw is rotating at 3450 rev/min. A pulley attached to the motor shaft drives a second pulley of half the diameter by means of a V-belt. A circular saw blade of diameter 0.208 m is mounted on the same rotating shaft as the second pulley. (a) The operator is careless and the blade catches and throws back a small piece of wood. This piece of wood moves with linear speed equal to the tangential speed of the rim of the blade. What is this speed? (b) Calculate the radial acceleration of points on the outer edge of the blade to see why sawdust doesn't stick to its teeth.

9.65 ••• While riding a multispeed bicycle, the rider can select the radius of the rear sprocket that is fixed to the rear axle. The front sprocket of a bicycle has radius 12.0 cm. If the angular speed of the front sprocket is 0.600 rev/s, what is the radius of the rear sprocket for which the tangential speed of a point on the rim of the rear wheel will be 5.00 m/s? The rear wheel has radius 0.330 m.

9.66 ••• A computer disk drive is turned on starting from rest and has constant angular acceleration. If it took 0.0865 s for the drive to make its *second* complete revolution, (a) how long did it take to make the first complete revolution, and (b) what is its angular acceleration, in rad/s²?

9.67 ••• It has been argued that power plants should make use of off-peak hours (such as late at night) to generate mechanical energy and store it until it is needed during peak load times, such as the middle of the day. One suggestion has been to store the energy in large flywheels spinning on nearly frictionless ball bearings. Consider a flywheel made of iron (density 7800 kg/m³) in the shape of a 10.0-cm-thick uniform disk. (a) What would the diameter of such a disk need to be if it is to store 10.0 megajoules of kinetic energy when spinning at 90.0 rpm about an axis perpendicular to the disk at its center? (b) What would be the centripetal acceleration of a point on its rim when spinning at this rate?

9.68 •• A uniform disk has radius R_0 and mass M_0. Its moment of inertia for an axis perpendicular to the plane of the disk at the disk's center is $\frac{1}{2}M_0R_0^2$. You have been asked to halve the disk's moment of inertia by cutting out a circular piece at the center of the disk. In terms of R_0, what should be the radius of the circular piece that you remove?

9.69 •• **Measuring I.** As an intern at an engineering firm, you are asked to measure the moment of inertia of a large wheel for rotation about an axis perpendicular to the wheel at its center. You measure the diameter of the wheel to be 0.640 m. Then you mount the wheel on frictionless bearings on a horizontal frictionless axle at the center of the wheel. You wrap a light rope around the wheel and hang an 8.20-kg block of wood from the free end of the rope, as in Fig. E9.45. You release the system from rest and find that the block descends 12.0 m in 4.00 s. What is the moment of inertia of the wheel for this axis?

9.70 ••• A uniform, solid disk with mass m and radius R is pivoted about a horizontal axis through its center. A small object of the same mass m is glued to the rim of the disk. If the disk is released from rest with the small object at the end of a horizontal radius, find the angular speed when the small object is directly below the axis.

9.71 •• CP A meter stick with a mass of 0.180 kg is pivoted about one end so it can rotate without friction about a horizontal axis. The meter stick is held in a horizontal position and released. As it swings through the vertical, calculate (a) the change in gravitational potential energy that has occurred; (b) the angular speed of the stick; (c) the linear speed of the end of the stick opposite the axis. (d) Compare the answer in part (c) to the speed of a particle that has fallen 1.00 m, starting from rest.

9.72 •• A physics student of mass 43.0 kg is standing at the edge of the flat roof of a building, 12.0 m above the sidewalk. An unfriendly dog is running across the roof toward her. Next to her is a large wheel mounted on a horizontal axle at its center. The wheel, used to lift objects from the ground to the roof, has a light crank attached to it and a light rope wrapped around it; the free end of the rope hangs over the edge of the roof. The student grabs the end of the rope and steps off the roof. If the wheel has radius 0.300 m and a moment of inertia of 9.60 kg·m² for rotation about the axle, how long does it take her to reach the sidewalk, and how fast will she be moving just before she lands? Ignore friction.

9.73 ••• A slender rod is 80.0 cm long and has mass 0.120 kg. A small 0.0200-kg sphere is welded to one end of the rod, and a small 0.0500-kg sphere is welded to the other end. The rod, pivoting about a stationary, frictionless axis at its center, is held horizontal and released from rest. What is the linear speed of the 0.0500-kg sphere as it passes through its lowest point?

9.74 •• Exactly one turn of a flexible rope with mass m is wrapped around a uniform cylinder with mass M and radius R. The cylinder rotates without friction about a horizontal axle along the cylinder axis. One end of the rope is attached to the cylinder. The cylinder starts with angular speed ω_0. After one revolution of the cylinder the rope has unwrapped and, at this instant, hangs vertically down, tangent to the cylinder. Find the angular speed of the cylinder and the linear speed of the lower end of the rope at this time. Ignore the thickness of the rope. [*Hint:* Use Eq. (9.18).]

9.75 • The pulley in **Fig. P9.75** has radius R and a moment of inertia I. The rope does not slip over the pulley, and the pulley spins on a frictionless axle. The coefficient of kinetic friction between block A and the tabletop is μ_k. The system is released from rest, and block B descends. Block A has mass m_A and block B has mass m_B. Use energy methods to calculate the speed of block B as a function of the distance d that it has descended.

Figure **P9.75**

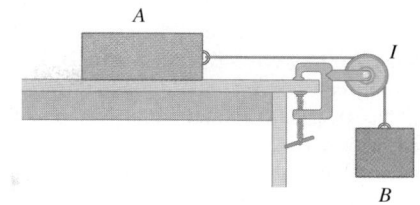

9.76 •• The pulley in **Fig. P9.76** has radius 0.160 m and moment of inertia 0.380 kg·m². The rope does not slip on the pulley rim. Use energy methods to calculate the speed of the 4.00-kg block just before it strikes the floor.

Figure **P9.76**

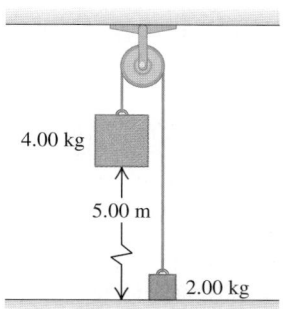

4.00 kg

5.00 m

2.00 kg

9.77 •• Two metal disks, one with radius $R_1 = 2.50$ cm and mass $M_1 = 0.80$ kg and the other with radius $R_2 = 5.00$ cm and mass $M_2 = 1.60$ kg, are welded together and mounted on a frictionless axis through their common center (**Fig. P9.77**). (a) What is the total moment of inertia of the two disks? (b) A light string is wrapped around the edge of the smaller disk, and a 1.50-kg block is suspended from the free end of the string. If the block is released from rest at a distance of 2.00 m above the floor, what is its speed just before it strikes the floor? (c) Repeat part (b), this time with the string wrapped around the edge of the larger disk. In which case is the final speed of the block greater? Explain.

Figure **P9.77**

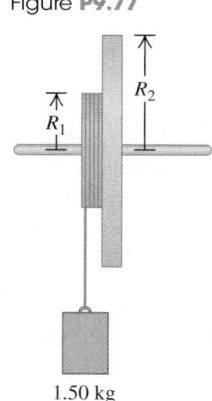

R_1
R_2

1.50 kg

9.78 •• A thin, light wire is wrapped around the rim of a wheel as shown in Fig. E9.45. The wheel rotates about a stationary horizontal axle that passes through the center of the wheel. The wheel has radius 0.180 m and moment of inertia for rotation about the axle of $I = 0.480$ kg·m². A small block with mass 0.340 kg is suspended from the free end of the wire. When the system is released from rest, the block descends with constant acceleration. The bearings in the wheel at the axle are rusty, so friction there does -9.00 J of work as the block descends 3.00 m. What is the magnitude of the angular velocity of the wheel after the block has descended 3.00 m?

9.79 ••• In the system shown in Fig. 9.17, a 12.0-kg mass is released from rest and falls, causing the uniform 10.0-kg cylinder of diameter 30.0 cm to turn about a frictionless axle through its center. How far will the mass have to descend to give the cylinder 480 J of kinetic energy?

9.80 • In **Fig. P9.80**, the cylinder and pulley turn without friction about stationary horizontal axles that pass through their centers. A light rope is wrapped around the cylinder, passes over the pulley, and has a 3.00-kg box suspended from its free end. There is no slipping between the rope and the pulley surface. The uniform cylinder has mass 5.00 kg and radius 40.0 cm. The pulley is a uniform disk with mass 2.00 kg and radius 20.0 cm. The box is released from rest and descends as the rope unwraps from the cylinder. Find the speed of the box when it has fallen 2.50 m.

Figure **P9.80**

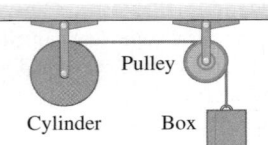

9.81 •• BIO **The Kinetic Energy of Walking.** If a person of mass M simply moved forward with speed V, his kinetic energy would be $\frac{1}{2}MV^2$. However, in addition to possessing a forward motion, various parts of his body (such as the arms and legs) undergo rotation. Therefore, his total kinetic energy is the sum of the energy from his forward motion plus the rotational kinetic energy of his arms and legs. The purpose of this problem is to see how much this rotational motion contributes to the person's kinetic energy. Biomedical measurements show that the arms and hands together typically make up 13% of a person's mass, while the legs and feet together account for 37%. For a rough (but reasonable) calculation, we can model the arms and legs as thin uniform bars pivoting about the shoulder and hip, respectively. In a brisk walk, the arms and legs each move through an angle of about $\pm 30°$ (a total of 60°) from the vertical in approximately 1 second. Assume that they are held straight, rather than being bent, which is not quite true. Consider a 75-kg person walking at 5.0 km/h, having arms 70 cm long and legs 90 cm long. (a) What is the average angular velocity of his arms and legs? (b) Using the average angular velocity from part (a), calculate the amount of rotational kinetic energy in this person's arms and legs as he walks. (c) What is the total kinetic energy due to both his forward motion and his rotation? (d) What percentage of his kinetic energy is due to the rotation of his legs and arms?

9.82 •• BIO **The Kinetic Energy of Running.** Using Problem 9.81 as a guide, apply it to a person running at 12 km/h, with his arms and legs each swinging through $\pm 30°$ in $\frac{1}{2}$ s. As before, assume that the arms and legs are kept straight.

9.83 •• BIO **Human Rotational Energy.** A dancer is spinning at 72 rpm about an axis through her center with her arms outstretched (**Fig. P9.83**). From biomedical measurements, the typical distribution of mass in a human body is as follows:

Figure **P9.83**

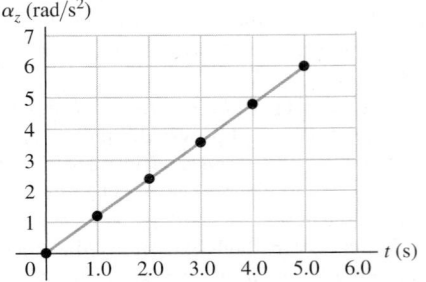

Head: 7.0%
Arms: 13% (for both)
Trunk and legs: 80.0%

Suppose you are this dancer. Using this information plus length measurements on your own body, calculate (a) your moment of inertia about your spin axis and (b) your rotational kinetic energy. Use Table 9.2 to model reasonable approximations for the pertinent parts of your body.

9.84 ••• A thin, uniform rod is bent into a square of side length a. If the total mass is M, find the moment of inertia about an axis through the center and perpendicular to the plane of the square. (*Hint:* Use the parallel-axis theorem.)

9.85 •• CALC A sphere with radius $R = 0.200$ m has density ρ that decreases with distance r from the center of the sphere according to $\rho = 3.00 \times 10^3 \text{ kg/m}^3 - (9.00 \times 10^3 \text{ kg/m}^4)r$. (a) Calculate the total mass of the sphere. (b) Calculate the moment of inertia of the sphere for an axis along a diameter.

9.86 •• CALC **Neutron Stars and Supernova Remnants.** The Crab Nebula is a cloud of glowing gas about 10 light-years across, located about 6500 light-years from the earth (**Fig. P9.86**). It is the remnant of a star that underwent a *supernova explosion,* seen on earth in 1054 A.D. Energy is released by the Crab Nebula at a rate of about 5×10^{31} W, about 10^5 times the rate at which the sun radiates energy. The Crab Nebula obtains its energy from the rotational kinetic energy of a rapidly spinning *neutron star* at its center. This object rotates once every 0.0331 s, and this period is increasing by 4.22×10^{-13} s for each second of time that elapses. (a) If the rate at which energy is lost by the neutron star is equal to the rate at which energy is released by the nebula, find the moment of inertia of the neutron star. (b) Theories of supernovae predict that the neutron star in the Crab Nebula has a mass about 1.4 times that of the sun. Modeling the neutron star as a solid uniform sphere, calculate its radius in kilometers. (c) What is the linear speed of a point on the equator of the neutron star? Compare to the speed of light. (d) Assume that the neutron star is uniform and calculate its density. Compare to the density of ordinary rock (3000 kg/m^3) and to the density of an atomic nucleus (about 10^{17} kg/m^3). Justify the statement that a neutron star is essentially a large atomic nucleus.

Figure **P9.86**

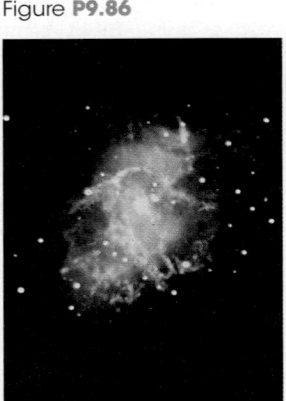

9.87 •• DATA A technician is testing a computer-controlled, variable-speed motor. She attaches a thin disk to the motor shaft, with the shaft at the center of the disk. The disk starts from rest, and sensors attached to the motor shaft measure the angular acceleration α_z of the shaft as a function of time. The results from one test run are shown in **Fig. P9.87**: (a) Through how many revolutions has the disk turned in the first 5.0 s? Can you use Eq. (9.11)? Explain. What is the angular velocity, in rad/s, of the disk (b) at $t = 5.0$ s; (c) when it has turned through 2.00 rev?

Figure **P9.87**

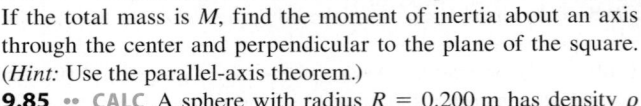

9.88 •• DATA You are analyzing the motion of a large flywheel that has radius 0.800 m. In one test run, the wheel starts from rest and turns with constant angular acceleration. An accelerometer on the rim of the flywheel measures the magnitude of the resultant acceleration a of a point on the rim of the flywheel as a function of the angle $\theta - \theta_0$ through which the wheel has turned. You collect these results:

$\theta - \theta_0$ (rad)	0.50	1.00	1.50	2.00	2.50	3.00	3.50	4.00
a (m/s^2)	0.678	1.07	1.52	1.98	2.45	2.92	3.39	3.87

Construct a graph of a^2 (in m^2/s^4) versus $(\theta - \theta_0)^2$ in (rad^2). (a) What are the slope and y-intercept of the straight line that gives the best fit to the data? (b) Use the slope from part (a) to find the angular acceleration of the flywheel. (c) What is the linear speed of a point on the rim of the flywheel when the wheel has turned through an angle of 135°? (d) When the flywheel has turned through an angle of 90.0°, what is the angle between the linear velocity of a point on its rim and the resultant acceleration of that point?

9.89 •• DATA You are rebuilding a 1965 Chevrolet. To decide whether to replace the flywheel with a newer, lighter-weight one, you want to determine the moment of inertia of the original, 35.6-cm-diameter flywheel. It is not a uniform disk, so you can't use $I = \frac{1}{2}MR^2$ to calculate the moment of inertia. You remove the flywheel from the car and use low-friction bearings to mount it on a horizontal, stationary rod that passes through the center of the flywheel, which can then rotate freely (about 2 m above the ground). After gluing one end of a long piece of flexible fishing line to the rim of the flywheel, you wrap the line a number of turns around the rim and suspend a 5.60-kg metal block from the free end of the line. When you release the block from rest, it descends as the flywheel rotates. With high-speed photography you measure the distance d the block has moved downward as a function of the time since it was released. The equation for the graph shown in **Fig. P9.89** that gives a good fit to the data points is $d = (165 \text{ cm/s}^2)t^2$. (a) Based on the graph, does the block fall with constant acceleration? Explain. (b) Use the graph to calculate the speed of the block when it has descended 1.50 m. (c) Apply conservation of mechanical energy to the system of flywheel and block to calculate the moment of inertia of the flywheel. (d) You are relieved that the fishing line doesn't break. Apply Newton's second law to the block to find the tension in the line as the block descended.

Figure **P9.89**

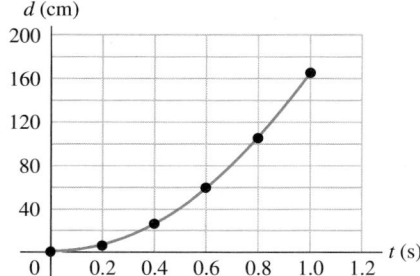

CHALLENGE PROBLEMS

9.90 ••• CALC Calculate the moment of inertia of a uniform solid cone about an axis through its center (**Fig. P9.90**). The cone has mass M and altitude h. The radius of its circular base is R.

Figure **P9.90**

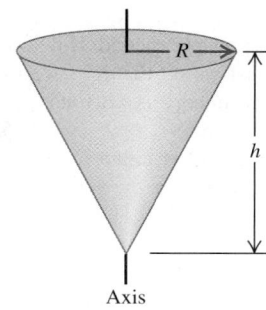

Axis

9.91 ••• CALC On a compact disc (CD), music is coded in a pattern of tiny pits arranged in a track that spirals outward toward the rim of the disc. As the disc spins inside a CD player, the track is scanned at a constant *linear* speed of $v = 1.25$m/s. Because the radius of the track varies as it spirals outward, the *angular* speed of the disc must change as the CD is played. (See Exercise 9.20.) Let's see what angular acceleration is required to keep v constant. The equation of a spiral is $r(\theta) = r_0 + \beta\theta$, where r_0 is the radius of the spiral at $\theta = 0$ and β is a constant. On a CD, r_0 is the inner radius of the spiral track. If we take the rotation direction of the CD to be positive, β must be positive so that r increases as the disc turns and θ increases. (a) When the disc rotates through a small angle $d\theta$, the distance scanned along the track is $ds = r\,d\theta$. Using the above expression for $r(\theta)$, integrate ds to find the total distance s scanned along the track as a function of the total angle θ through which the disc has rotated. (b) Since the track is scanned at a constant linear speed v, the distance s found in part (a) is equal to vt. Use this to find θ as a function of time. There will be two solutions for θ; choose the positive one, and explain why this is the solution to choose. (c) Use your expression for $\theta(t)$ to find the angular velocity ω_z and the angular acceleration α_z as functions of time. Is α_z constant? (d) On a CD, the inner radius of the track is 25.0 mm, the track radius increases by 1.55 μm per revolution, and the playing time is 74.0 min. Find r_0, β, and the total number of revolutions made during the playing time. (e) Using your results from parts (c) and (d), make graphs of ω_z (in rad/s) versus t and α_z (in rad/s^2) versus t between $t = 0$ and $t = 74.0$ min.

| PASSAGE PROBLEMS |

BIO **THE SPINNING EEL.** American eels (*Anguilla rostrata*) are freshwater fish with long, slender bodies that we can treat as uniform cylinders 1.0 m long and 10 cm in diameter. An eel compensates for its small jaw and teeth by holding onto prey with its mouth and then rapidly spinning its body around its long axis to tear off a piece of flesh. Eels have been recorded to spin at up to 14 revolutions per second when feeding in this way. Although this feeding method is costly in terms of energy, it allows the eel to feed on larger prey than it otherwise could.

9.92 A field researcher uses the slow-motion feature on her phone's camera to shoot a video of an eel spinning at its maximum rate. The camera records at 120 frames per second. Through what angle does the eel rotate from one frame to the next? (a) 1°; (b) 10°; (c) 22°; (d) 42°.

9.93 The eel is observed to spin at 14 spins per second clockwise, and 10 seconds later it is observed to spin at 8 spins per second counterclockwise. What is the magnitude of the eel's average angular acceleration during this time? (a) 6/10 rad/s^2; (b) 6π/10 rad/s^2; (c) 12π/10 rad/s^2; (d) 44π/10 rad/s^2.

9.94 The eel has a certain amount of rotational kinetic energy when spinning at 14 spins per second. If it swam in a straight line instead, about how fast would the eel have to swim to have the same amount of kinetic energy as when it is spinning? (a) 0.5 m/s; (b) 0.7 m/s; (c) 3 m/s; (d) 5 m/s.

9.95 A new species of eel is found to have the same mass but one-quarter the length and twice the diameter of the American eel.

How does its moment of inertia for spinning around its long axis compare to that of the American eel? The new species has (a) half the moment of inertia as the American eel; (b) the same moment of inertia as the American eel; (c) twice the moment of inertia as the American eel; (d) four times the moment of inertia as the American eel.

Answers

Chapter Opening Question ?

(ii) The rotational kinetic energy of a rigid body rotating around an axis is $K = \frac{1}{2}I\omega^2$, where I is the body's moment of inertia for that axis and ω is the rotational speed. Table 9.2 shows that the moment of inertia for a slender rod of mass M and length L with an axis through one end (like a wind turbine blade) is $I = \frac{1}{3}ML^2$. If we double L while M and ω stay the same, both the moment of inertia I and the kinetic energy K increase by a factor of $2^2 = 4$.

Test Your Understanding Questions

9.1 (a) (i) and (iii), (b) (ii) The rotation is speeding up when the angular velocity and angular acceleration have the same sign, and slowing down when they have opposite signs. Hence it is speeding up for $0 < t < 2$ s (both ω_z and α_z are positive) and for 4 s $< t < 6$ s (both ω_z and α_z are negative) but is slowing down for 2 s $< t < 4$ s (ω_z is positive and α_z is negative). Note that the body is rotating in one direction for $t < 4$ s (ω_z is positive) and in the opposite direction for $t > 4$ s (ω_z is negative).

9.2 (a) (i), (b) (ii) When the disc comes to rest, $\omega_z = 0$. From Eq. (9.7), the *time* when this occurs is $t = (\omega_z - \omega_{0z})/\alpha_z = -\omega_{0z}/\alpha_z$ (this is a positive time because α_z is negative). If we double the initial angular velocity ω_{0z} and also double the angular acceleration α_z, their ratio is unchanged and the rotation stops in the same amount of time. The *angle* through which the disc rotates is given by Eq. (9.10): $\theta - \theta_0 = \frac{1}{2}(\omega_{0z} + \omega_z)t = \frac{1}{2}\omega_{0z}t$ (since the final angular velocity is $\omega_z = 0$). The initial angular velocity ω_{0z} has been doubled but the time t is the same, so the

angular displacement $\theta - \theta_0$ (and hence the number of revolutions) has doubled. You can also come to the same conclusion by using Eq. (9.12).

9.3 (ii) From Eq. (9.13), $v = r\omega$. To maintain a constant linear speed v, the angular speed ω must decrease as the scanning head moves outward (greater r).

9.4 (i) The kinetic energy in the falling block is $\frac{1}{2}mv^2$, and the kinetic energy in the rotating cylinder is $\frac{1}{2}I\omega^2 = \frac{1}{2}\left(\frac{1}{2}mR^2\right)\left(v/R\right)^2 = \frac{1}{4}mv^2$. Hence the total kinetic energy of the system is $\frac{3}{4}mv^2$, of which two-thirds is in the block and one-third is in the cylinder.

9.5 (ii) More of the mass of the pool cue is concentrated at the thicker end, so the center of mass is closer to that end. The moment of inertia through a point P at either end is $I_P = I_{cm} + Md^2$; the thinner end is farther from the center of mass, so the distance d and the moment of inertia I_P are greater for the thinner end.

9.6 (iii) Our result from Example 9.10 does *not* depend on the cylinder length L. The moment of inertia depends on only the *radial* distribution of mass, not on its distribution along the axis.

Bridging Problem

(a) $I = \left[\dfrac{M}{L}\left(\dfrac{x^3}{3}\right)\right]_{-h}^{L-h} = \frac{1}{3}M(L^2 - 3Lh + 3h^2)$

(b) $W = \frac{1}{6}M(L^2 - 3Lh + 3h^2)\alpha^2 t^2$

(c) $a = (L - h)\alpha\sqrt{1 + \alpha^2 t^4}$

? These jugglers toss the pins so that they rotate in midair. Each pin is of uniform composition, so its weight is concentrated toward its thick end. If we ignore air resistance but not the effects of gravity, will the angular speed of a pin in flight (i) increase continuously; (ii) decrease continuously; (iii) alternately increase and decrease; or (iv) remain the same?

10 DYNAMICS OF ROTATIONAL MOTION

LEARNING GOALS

Looking forward at ...

10.1 What is meant by the torque produced by a force.

10.2 How the net torque on a body affects the body's rotational motion.

10.3 How to analyze the motion of a body that both rotates and moves as a whole through space.

10.4 How to solve problems that involve work and power for rotating bodies.

10.5 What is meant by the angular momentum of a particle or rigid body.

10.6 How the angular momentum of a body can remain constant even if the body changes shape.

10.7 Why a spinning gyroscope undergoes precession.

Looking back at ...

We learned in Chapters 4 and 5 that a net force applied to a body gives that body an acceleration. But what does it take to give a body an *angular* acceleration? That is, what does it take to start a stationary body rotating or to bring a spinning body to a halt? A force is required, but it must be applied in a way that gives a twisting or turning action.

In this chapter we'll define a new physical quantity, *torque,* that describes the twisting or turning effort of a force. We'll find that the net torque acting on a rigid body determines its angular acceleration, in the same way that the net force on a body determines its linear acceleration. We'll also look at work and power in rotational motion so as to understand, for example, how energy is transferred by an electric motor. Next we'll develop a new conservation principle, *conservation of angular momentum,* that is tremendously useful for understanding the rotational motion of both rigid and nonrigid bodies. We'll finish this chapter by studying *gyroscopes,* rotating devices that don't fall over when you might think they should— but that actually behave in accordance with the dynamics of rotational motion.

10.1 TORQUE

We know that forces acting on a body can affect its **translational motion**—that is, the motion of the body as a whole through space. Now we want to learn which aspects of a force determine how effective it is in causing or changing *rotational* motion. The magnitude and direction of the force are important, but so is the point on the body where the force is applied. In **Fig. 10.1** (next page) a wrench is being used to loosen a tight bolt. Force $\vec{F}_b$, applied near the end of the handle, is more effective than an equal force $\vec{F}_a$ applied near the bolt. Force $\vec{F}_c$ does no good; it's applied at the same point and has the same magnitude as $\vec{F}_b$, but it's directed along the length of the handle. The quantitative measure of the tendency of a force to cause or change a body's rotational motion is called *torque;* we say that $\vec{F}_a$

10.1 Which of these three equal-magnitude forces is most likely to loosen the tight bolt?

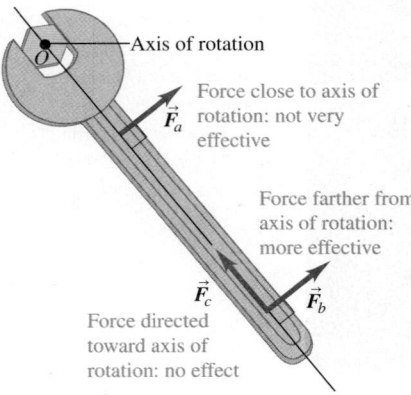

10.2 The torque of a force about a point is the product of the force magnitude and the lever arm of the force.

$\vec{F}_1$ tends to cause *counterclockwise* rotation about point O, so its torque is *positive*: $\tau_1 = +F_1 l_1$

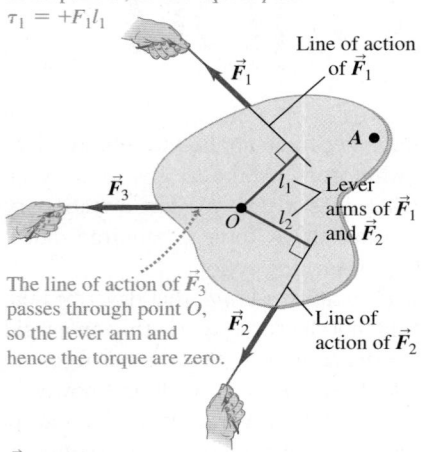

The line of action of $\vec{F}_3$ passes through point O, so the lever arm and hence the torque are zero.

$\vec{F}_2$ tends to cause *clockwise* rotation about point O, so its torque is *negative*: $\tau_2 = -F_2 l_2$

10.3 Three ways to calculate the torque of force $\vec{F}$ about point O. In this figure, $\vec{r}$ and $\vec{F}$ are in the plane of the page and the torque vector $\vec{\tau}$ points out of the page toward you.

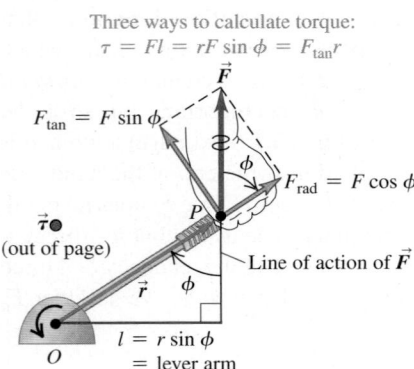

applies a torque about point O to the wrench in Fig. 10.1, $\vec{F}_b$ applies a greater torque about O, and $\vec{F}_c$ applies zero torque about O.

Figure 10.2 shows three examples of how to calculate torque. The body can rotate about an axis that is perpendicular to the plane of the figure and passes through point O. Three forces act on the body in the plane of the figure. The tendency of the first of these forces, $\vec{F}_1$, to cause a rotation about O depends on its magnitude F_1. It also depends on the *perpendicular* distance l_1 between point O and the **line of action** of the force (that is, the line along which the force vector lies). We call the distance l_1 the **lever arm** (or **moment arm**) of force $\vec{F}_1$ about O. The twisting effort is directly proportional to both F_1 and l_1, so we define the **torque** (or *moment*) of the force $\vec{F}_1$ with respect to O as the product $F_1 l_1$. We use the Greek letter τ (tau) for torque. If a force of magnitude F has a line of action that is a perpendicular distance l from O, the torque is

$$\tau = Fl \tag{10.1}$$

Physicists usually use the term "torque," while engineers usually use "moment" (unless they are talking about a rotating shaft).

The lever arm of $\vec{F}_1$ in Fig. 10.2 is the perpendicular distance l_1, and the lever arm of $\vec{F}_2$ is the perpendicular distance l_2. The line of action of $\vec{F}_3$ passes through point O, so the lever arm for $\vec{F}_3$ is zero and its torque with respect to O is zero. In the same way, force $\vec{F}_c$ in Fig. 10.1 has zero torque with respect to point O; $\vec{F}_b$ has a greater torque than $\vec{F}_a$ because its lever arm is greater.

CAUTION **Torque is always measured about a point** Torque is *always* defined with reference to a specific point. If we shift the position of this point, the torque of each force may change. For example, the torque of force $\vec{F}_3$ in Fig. 10.2 is zero with respect to point O but *not* with respect to point A. It's not enough to refer to "the torque of $\vec{F}$"; you must say "the torque of $\vec{F}$ with respect to point X" or "the torque of $\vec{F}$ about point X." ▮

Force $\vec{F}_1$ in Fig. 10.2 tends to cause *counterclockwise* rotation about O, while $\vec{F}_2$ tends to cause *clockwise* rotation. To distinguish between these two possibilities, we need to choose a positive sense of rotation. With the choice that *counterclockwise torques are positive and clockwise torques are negative,* the torques of $\vec{F}_1$ and $\vec{F}_2$ about O are

$$\tau_1 = +F_1 l_1 \qquad \tau_2 = -F_2 l_2$$

Figure 10.2 shows this choice for the sign of torque. We will often use the symbol ⊕ to indicate our choice of the positive sense of rotation.

The SI unit of torque is the newton-meter. In our discussion of work and energy we called this combination the joule. But torque is *not* work or energy, and torque should be expressed in newton-meters, *not* joules.

Figure 10.3 shows a force $\vec{F}$ applied at point P, located at position $\vec{r}$ with respect to point O. There are three ways to calculate the torque of $\vec{F}$:

1. Find the lever arm l and use $\tau = Fl$.

2. Determine the angle ϕ between the vectors $\vec{r}$ and $\vec{F}$; the lever arm is $r \sin \phi$, so $\tau = rF \sin \phi$.

3. Represent $\vec{F}$ in terms of a radial component F_{rad} along the direction of $\vec{r}$ and a tangential component F_{tan} at right angles, perpendicular to $\vec{r}$. (We call this component *tangential* because if the body rotates, the point where the force acts moves in a circle, and this component is tangent to that circle.) Then $F_{\text{tan}} = F \sin \phi$ and $\tau = r(F \sin \phi) = F_{\text{tan}} r$. The component F_{rad} produces *no* torque with respect to O because its lever arm with respect to that point is zero (compare to forces $\vec{F}_c$ in Fig. 10.1 and $\vec{F}_3$ in Fig. 10.2).

Summarizing these three expressions for torque, we have

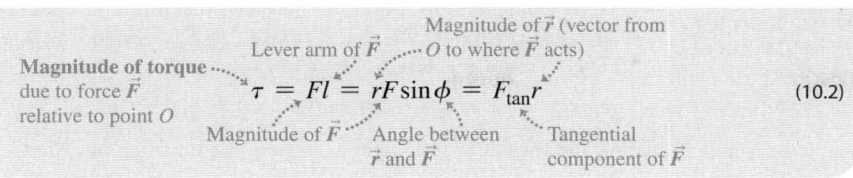

$$\tau = Fl = rF\sin\phi = F_{tan}r \qquad (10.2)$$

Magnitude of torque due to force $\vec{F}$ relative to point O ···

Lever arm of $\vec{F}$

Magnitude of $\vec{r}$ (vector from O to where $\vec{F}$ acts)

Magnitude of $\vec{F}$ ··· Angle between $\vec{r}$ and $\vec{F}$ ··· Tangential component of $\vec{F}$

DEMO

Torque As a Vector

We saw in Section 9.1 that angular velocity and angular acceleration can be represented as vectors; the same is true for torque. To see how to do this, note that the quantity $rF\sin\phi$ in Eq. (10.2) is the magnitude of the *vector product* $\vec{r} \times \vec{F}$ that we defined in Section 1.10. (Go back and review that definition.) We generalize the definition of torque as follows: When a force $\vec{F}$ acts at a point having a position vector $\vec{r}$ with respect to an origin O, as in Fig. 10.3, the torque $\vec{\tau}$ of the force with respect to O is the *vector* quantity

$$\vec{\tau} = \vec{r} \times \vec{F} \qquad (10.3)$$

Torque vector due to force $\vec{F}$ relative to point O ··· Vector from O to where $\vec{F}$ acts ··· Force $\vec{F}$

The torque as defined in Eq. (10.2) is the magnitude of the torque vector $\vec{r} \times \vec{F}$. The direction of $\vec{\tau}$ is perpendicular to both $\vec{r}$ and $\vec{F}$. In particular, if both $\vec{r}$ and $\vec{F}$ lie in a plane perpendicular to the axis of rotation, as in Fig. 10.3, then the torque vector $\vec{\tau} = \vec{r} \times \vec{F}$ is directed along the axis of rotation, with a sense given by the right-hand rule (see Fig. 1.30 and **Fig. 10.4**).

Because $\vec{\tau} = \vec{r} \times \vec{F}$ is perpendicular to the plane of the vectors $\vec{r}$ and $\vec{F}$, it's common to have diagrams like Fig. 10.4, in which one of the vectors is perpendicular to the page. We use a dot (•) to represent a vector that points out of the page and a cross (×) to represent a vector that points into the page (see Figs. 10.3 and 10.4).

In the following sections we will usually be concerned with rotation of a body about an axis oriented in a specified constant direction. In that case, only the component of torque along that axis will matter. We often call that component the torque with respect to the specified *axis*.

10.4 The torque vector $\vec{\tau} = \vec{r} \times \vec{F}$ is directed along the axis of the bolt, perpendicular to both $\vec{r}$ and $\vec{F}$. The fingers of the right hand curl in the direction of the rotation that the torque tends to cause.

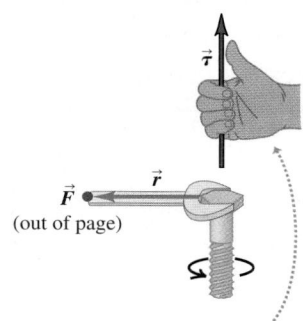

If you point the fingers of your right hand in the direction of $\vec{r}$ and then curl them in the direction of $\vec{F}$, your outstretched thumb points in the direction of $\vec{\tau}$.

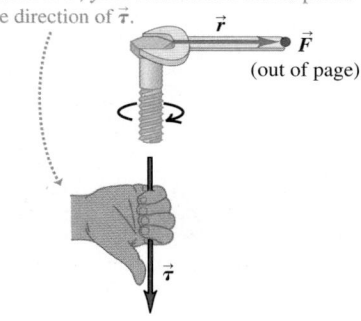

EXAMPLE 10.1 **APPLYING A TORQUE**

To loosen a pipe fitting, a plumber slips a piece of scrap pipe (a "cheater") over his wrench handle. He stands on the end of the cheater, applying his 900-N weight at a point 0.80 m from the center of the fitting (**Fig. 10.5a**). The wrench handle and cheater make an angle of 19° with the horizontal. Find the magnitude and direction of the torque he applies about the center of the fitting.

10.5 (a) Loosening a pipe fitting by standing on a "cheater." (b) Our vector diagram to find the torque about O.

(a) Diagram of situation

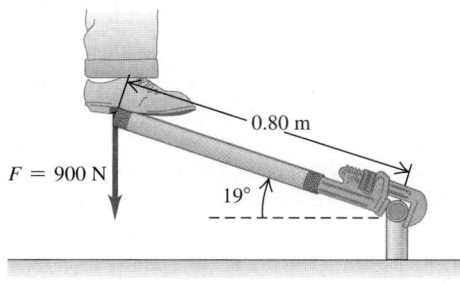

(b) Free-body diagram

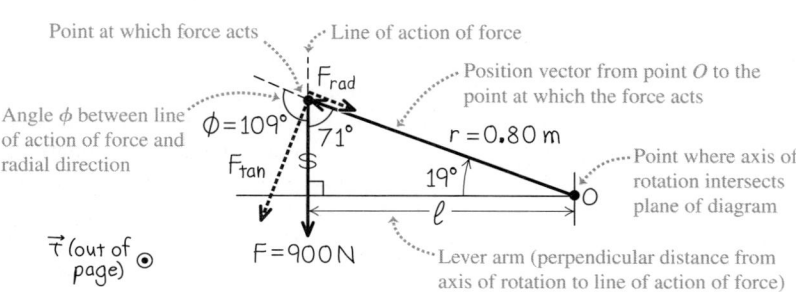

Point at which force acts

Line of action of force

Angle ϕ between line of action of force and radial direction

$\phi = 109°$

$71°$

F_{rad}

F_{tan}

Position vector from point O to the point at which the force acts

$r = 0.80$ m

$19°$

ℓ

$\vec{\tau}$ (out of page) ⊙

$F = 900$ N

Point where axis of rotation intersects plane of diagram

Lever arm (perpendicular distance from axis of rotation to line of action of force)

Continued

SOLUTION

IDENTIFY and SET UP: Figure 10.5b shows the vectors $\vec{r}$ and $\vec{F}$ and the angle between them ($\phi = 109°$). Equation (10.1) or (10.2) will tell us the magnitude of the torque. The right-hand rule with Eq. (10.3), $\vec{\tau} = \vec{r} \times \vec{F}$, will tell us the direction of the torque.

EXECUTE: To use Eq. (10.1), we first calculate the lever arm l. As Fig. 10.5b shows,

$$l = r \sin\phi = (0.80 \text{ m}) \sin 109° = 0.76 \text{ m}$$

Then Eq. (10.1) tells us that the magnitude of the torque is

$$\tau = Fl = (900 \text{ N})(0.76 \text{ m}) = 680 \text{ N} \cdot \text{m}$$

We get the same result from Eq. (10.2):

$$\tau = rF \sin\phi = (0.80 \text{ m})(900 \text{ N})(\sin 109°) = 680 \text{ N} \cdot \text{m}$$

Alternatively, we can find F_{tan}, the tangential component of $\vec{F}$ that acts perpendicular to $\vec{r}$. Figure 10.5b shows that this component is at an angle of $109° - 90° = 19°$ from $\vec{F}$, so $F_{\text{tan}} = F(\cos 19°) = (900 \text{ N})(\cos 19°) = 851 \text{ N}$. Then, from Eq. 10.2,

$$\tau = F_{\text{tan}}r = (851 \text{ N})(0.80 \text{ m}) = 680 \text{ N} \cdot \text{m}$$

Curl the fingers of your right hand from the direction of $\vec{r}$ (in the plane of Fig. 10.5b, to the left and up) into the direction of $\vec{F}$ (straight down). Then your right thumb points out of the plane of the figure: This is the direction of $\vec{\tau}$.

EVALUATE: To check the direction of $\vec{\tau}$, note that the force in Fig. 10.5 tends to produce a counterclockwise rotation about O. If you curl the fingers of your right hand in a counterclockwise direction, the thumb points out of the plane of Fig. 10.5, which is indeed the direction of the torque.

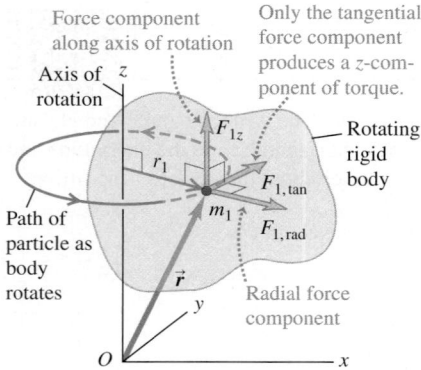

TEST YOUR UNDERSTANDING OF SECTION 10.1 The accompanying figure shows a force P being applied to one end of a lever of length L. What is the magnitude of the torque of this force about point A? (i) $PL\sin\theta$; (ii) $PL\cos\theta$; (iii) $PL\tan\theta$; (iv) $PL/\sin\theta$; (v) PL. ▮

10.2 TORQUE AND ANGULAR ACCELERATION FOR A RIGID BODY

We're now ready to develop the fundamental relationship for the rotational dynamics of a rigid body. We'll show that the angular acceleration of a rotating rigid body is directly proportional to the sum of the torque components along the axis of rotation. The proportionality factor is the moment of inertia.

To develop this relationship, let's begin as we did in Section 9.4 by envisioning the rigid body as being made up of a large number of particles. We choose the axis of rotation to be the z-axis; the first particle has mass m_1 and distance r_1 from this axis (**Fig. 10.6**). The *net force* $\vec{F}_1$ acting on this particle has a component $F_{1,\text{rad}}$ along the radial direction, a component $F_{1,\text{tan}}$ that is tangent to the circle of radius r_1 in which the particle moves as the body rotates, and a component F_{1z} along the axis of rotation. Newton's second law for the tangential component is

$$F_{1,\text{tan}} = m_1 a_{1,\text{tan}} \tag{10.4}$$

We can express the tangential acceleration of the first particle in terms of the angular acceleration α_z of the body by using Eq. (9.14): $a_{1,\text{tan}} = r_1\alpha_z$. Using this relationship and multiplying both sides of Eq. (10.4) by r_1, we obtain

$$F_{1,\text{tan}}r_1 = m_1 r_1^2 \alpha_z \tag{10.5}$$

From Eq. (10.2), $F_{1,\text{tan}}r_1$ is the *torque* of the net force with respect to the rotation axis, equal to the component τ_{1z} of the torque vector along the rotation axis. The subscript z is a reminder that the torque affects rotation around the z-axis, in the same way that the subscript on F_{1z} is a reminder that this force affects the motion of particle 1 along the z-axis.

Neither of the components $F_{1,\text{rad}}$ or F_{1z} contributes to the torque about the z-axis, since neither tends to change the particle's rotation about that axis. So $\tau_{1z} = F_{1,\text{tan}}r_1$ is the total torque acting on the particle with respect to the rotation axis. Also, $m_1 r_1^2$ is I_1, the moment of inertia of the particle about the rotation axis. Hence we can rewrite Eq. (10.5) as

$$\tau_{1z} = I_1\alpha_z = m_1 r_1^2 \alpha_z$$

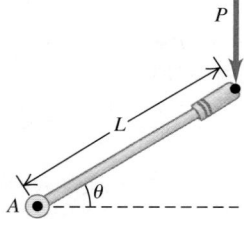

10.6 As a rigid body rotates around the z-axis, a net force $\vec{F}_1$ acts on one particle of the body. Only the force component $F_{1,\text{tan}}$ can affect the rotation, because only $F_{1,\text{tan}}$ exerts a torque about O with a z-component (along the rotation axis).

We write such an equation for every particle in the body, then add all these equations:

$$\tau_{1z} + \tau_{2z} + \cdots = I_1\alpha_z + I_2\alpha_z + \cdots$$
$$= m_1 r_1^2 \alpha_z + m_2 r_2^2 \alpha_z + \cdots$$

or

$$\sum \tau_{iz} = \left(\sum m_i r_i^2 \right) \alpha_z \tag{10.6}$$

The left side of Eq. (10.6) is the sum of all the torques about the rotation axis that act on all the particles. The right side is $I = \sum m_i r_i^2$, the total moment of inertia about the rotation axis, multiplied by the angular acceleration α_z. Note that α_z is the same for every particle because this is a *rigid* body. Thus Eq. (10.6) says that for the rigid body as a whole,

> **Rotational analog of Newton's second law for a rigid body:**
>
> Net torque on a rigid body about z-axis ⟶ $\sum \tau_z = I\alpha_z$ ⟵ Moment of inertia of rigid body about z-axis · · · Angular acceleration of rigid body about z-axis (10.7)

Just as Newton's second law says that a net *force* on a particle causes an *acceleration* in the direction of the net force, Eq. (10.7) says that a net *torque* on a rigid body about an axis causes an *angular acceleration* about that axis (**Fig. 10.7**).

Our derivation assumed that the angular acceleration α_z is the same for all particles in the body. So Eq. (10.7) is valid *only* for *rigid* bodies. Hence this equation doesn't apply to a rotating tank of water or a swirling tornado of air, different parts of which have different angular accelerations. Note that since our derivation used Eq. (9.14), $a_{\text{tan}} = r\alpha_z$, α_z must be measured in rad/s².

The torque on each particle is due to the net force on that particle, which is the vector sum of external and internal forces (see Section 8.2). According to Newton's third law, the *internal* forces that any pair of particles in the rigid body exert on each other are equal in magnitude and opposite in direction (**Fig. 10.8**). If these forces act along the line joining the two particles, their lever arms with respect to any axis are also equal. So the torques for each such pair are equal and opposite, and add to zero. Hence *all* the internal torques add to zero, so the sum $\sum \tau_z$ in Eq. (10.7) includes only the torques of the *external* forces.

Often, an important external force acting on a body is its *weight*. This force is not concentrated at a single point; it acts on every particle in the entire body. Nevertheless, if $\vec{g}$ has the same value at all points, we always get the correct torque (about any specified axis) if we assume that all the weight is concentrated at the *center of mass* of the body. We'll prove this statement in Chapter 11, but meanwhile we'll use it for some of the problems in this chapter.

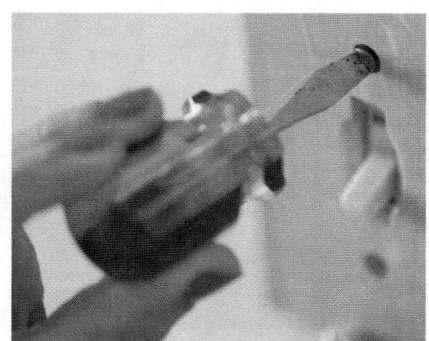

10.7 Loosening or tightening a screw requires giving it an angular acceleration and hence applying a torque. This is made easier by using a screwdriver with a large-radius handle, which provides a large lever arm for the force you apply with your hand.

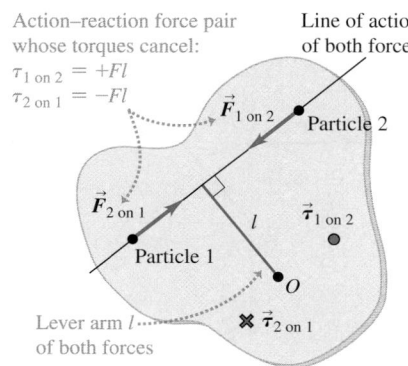

10.8 Why only *external* torques affect a rigid body's rotation: Any two particles in the body exert equal and opposite forces on each other. If the forces act along the line joining the particles, the lever arms of the forces with respect to an axis through O are the same and the torques due to the two forces are equal and opposite.

Action–reaction force pair whose torques cancel:
$\tau_{1 \text{ on } 2} = +Fl$
$\tau_{2 \text{ on } 1} = -Fl$

Line of action of both forces

$\vec{F}_{1 \text{ on } 2}$
Particle 2

$\vec{F}_{2 \text{ on } 1}$

l $\vec{\tau}_{1 \text{ on } 2}$

Particle 1

O

Lever arm l of both forces   ✕ $\vec{\tau}_{2 \text{ on } 1}$

PROBLEM-SOLVING STRATEGY 10.1 | ROTATIONAL DYNAMICS FOR RIGID BODIES

Our strategy for solving problems in rotational dynamics is very similar to Problem-Solving Strategy 5.2 for solving problems involving Newton's second law.

IDENTIFY *the relevant concepts:* Equation (10.7), $\sum \tau_z = I\alpha_z$, is useful whenever torques act on a rigid body. Sometimes you can use an energy approach instead, as we did in Section 9.4. However, if the target variable is a force, a torque, an acceleration, an angular acceleration, or an elapsed time, using $\sum \tau_z = I\alpha_z$ is almost always best.

SET UP *the problem* using the following steps:
1. Sketch the situation and identify the body or bodies to be analyzed. Indicate the rotation axis.
2. For each body, draw a free-body diagram that shows the body's *shape,* including all dimensions and angles. Label pertinent quantities with algebraic symbols.
3. Choose coordinate axes for each body and indicate a positive sense of rotation (clockwise or counterclockwise) for each rotating body. If you know the sense of α_z, pick that as the positive sense of rotation.

EXECUTE *the solution:*
1. For each body, decide whether it undergoes translational motion, rotational motion, or both. Then apply $\sum \vec{F} = m\vec{a}$ (as in Section 5.2), $\sum \tau_z = I\alpha_z$, or both to the body.
2. Express in algebraic form any *geometrical* relationships between the motions of two or more bodies. An example is a string that unwinds, without slipping, from a pulley or a wheel that rolls without slipping (discussed in Section 10.3). These relationships usually appear as relationships between linear and/or angular accelerations.
3. Ensure that you have as many independent equations as there are unknowns. Solve the equations to find the target variables.

EVALUATE *your answer:* Check that the algebraic signs of your results make sense. As an example, if you are unrolling thread from a spool, your answers should not tell you that the spool is turning in the direction that rolls the thread back onto the spool! Check that any algebraic results are correct for special cases or for extreme values of quantities.

EXAMPLE 10.2 | AN UNWINDING CABLE I

Figure 10.9a shows the situation that we analyzed in Example 9.7 using energy methods. What is the cable's acceleration?

SOLUTION

IDENTIFY and SET UP: We can't use the energy method of Section 9.4, which doesn't involve acceleration. Instead we'll apply rotational dynamics to find the angular acceleration of the cylinder (Fig. 10.9b). We'll then find a relationship between the motion of the cable and the motion of the cylinder rim, and use this to find the acceleration of the cable. The cylinder rotates counterclockwise when the cable is pulled, so we take counterclockwise rotation to be positive. The net force on the cylinder must be zero because its center of mass remains at rest. The force F exerted by the cable produces a torque about the rotation axis. The weight (magnitude Mg) and the normal force (magnitude n) exerted by the cylinder's bearings produce *no* torque about the rotation axis because both act along lines through that axis.

EXECUTE: The lever arm of F is equal to the radius $R = 0.060$ m of the cylinder, so the torque is $\tau_z = FR$. (This torque is positive,

as it tends to cause a counterclockwise rotation.) From Table 9.2, case (f), the moment of inertia of the cylinder about the rotation axis is $I = \frac{1}{2}MR^2$. Then Eq. (10.7) tells us that

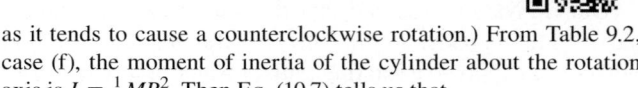

$$\alpha_z = \frac{\tau_z}{I} = \frac{FR}{MR^2/2} = \frac{2F}{MR} = \frac{2(9.0 \text{ N})}{(50 \text{ kg})(0.060 \text{ m})} = 6.0 \text{ rad/s}^2$$

(We can add "rad" to our result because radians are dimensionless.)

To get the linear acceleration of the cable, recall from Section 9.3 that the acceleration of a cable unwinding from a cylinder is the same as the tangential acceleration of a point on the surface of the cylinder where the cable is tangent to it. This tangential acceleration is given by Eq. (9.14):

$$a_{\text{tan}} = R\alpha_z = (0.060 \text{ m})(6.0 \text{ rad/s}^2) = 0.36 \text{ m/s}^2$$

EVALUATE: Can you use this result, together with an equation from Chapter 2, to determine the speed of the cable after it has been pulled 2.0 m? Does your result agree with that of Example 9.7?

10.9 (a) Cylinder and cable. (b) Our free-body diagram for the cylinder.

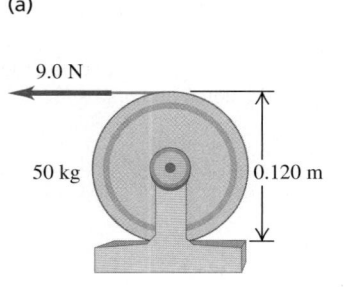

(a)

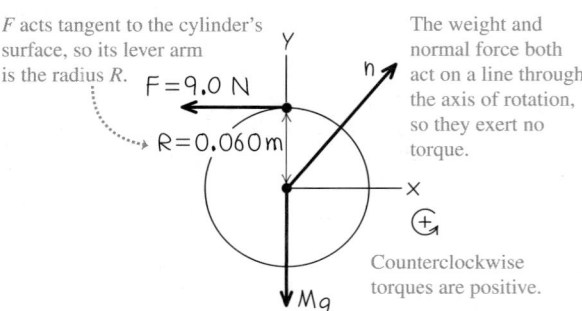

(b) *F* acts tangent to the cylinder's surface, so its lever arm is the radius *R*.

The weight and normal force both act on a line through the axis of rotation, so they exert no torque.

Counterclockwise torques are positive.

EXAMPLE 10.3   AN UNWINDING CABLE II

In Example 9.8 (Section 9.4), what are the acceleration of the falling block and the tension in the cable?

SOLUTION

IDENTIFY and SET UP: We'll apply translational dynamics to the block and rotational dynamics to the cylinder. As in Example 10.2, we'll relate the linear acceleration of the block (our target variable) to the angular acceleration of the cylinder. **Figure 10.10** shows our sketch of the situation and a free-body diagram for each body. We take the positive sense of rotation for the cylinder to be counterclockwise and the positive direction of the y-coordinate for the block to be downward.

10.10 (a) Our diagram of the situation. (b) Our free-body diagrams for the cylinder and the block. We assume the cable has negligible mass.

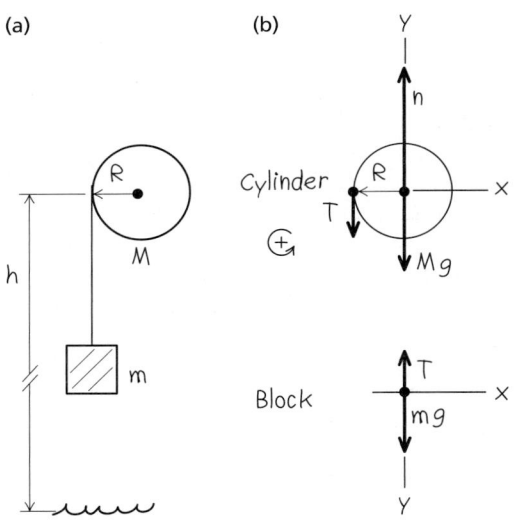

EXECUTE: For the block, Newton's second law gives

$$\Sigma F_y = mg + (-T) = ma_y$$

For the cylinder, the only torque about its axis is that due to the cable tension T. Hence Eq. (10.7) gives

$$\Sigma \tau_z = RT = I\alpha_z = \tfrac{1}{2}MR^2\alpha_z$$

As in Example 10.2, the acceleration of the cable is the same as the tangential acceleration of a point on the cylinder rim. From Eq. (9.14), this acceleration is $a_y = a_{tan} = R\alpha_z$. We use this to replace $R\alpha_z$ with a_y in the cylinder equation above, and divide by R. The result is $T = \tfrac{1}{2}Ma_y$. Now we substitute this expression for T into Newton's second law for the block and solve for the acceleration a_y:

$$mg - \tfrac{1}{2}Ma_y = ma_y$$

$$a_y = \frac{g}{1 + M/2m}$$

To find the cable tension T, we substitute our expression for a_y into the block equation:

$$T = mg - ma_y = mg - m\left(\frac{g}{1 + M/2m}\right) = \frac{mg}{1 + 2m/M}$$

EVALUATE: The acceleration is positive (in the downward direction) and less than g, as it should be, since the cable is holding back the block. The cable tension is *not* equal to the block's weight mg; if it were, the block could not accelerate.

Let's check some particular cases. When M is much larger than m, the tension is nearly equal to mg and the acceleration is correspondingly much less than g. When M is zero, $T = 0$ and $a_y = g$; the object falls freely. If the object starts from rest ($v_{0y} = 0$) a height h above the floor, its y-velocity when it strikes the floor is given by $v_y^2 = v_{0y}^2 + 2a_yh = 2a_yh$, so

$$v_y = \sqrt{2a_yh} = \sqrt{\frac{2gh}{1 + M/2m}}$$

We found this result from energy considerations in Example 9.8.

TEST YOUR UNDERSTANDING OF SECTION 10.2 The figure shows a glider of mass m_1 that can slide without friction on a horizontal air track. It is attached to an object of mass m_2 by a massless string. The pulley has radius R and moment of inertia I about its axis of rotation. When released, the hanging object accelerates downward, the glider accelerates to the right, and the string turns the pulley without slipping or stretching. Rank the magnitudes of the following forces that act during the motion, in order from largest to smallest magnitude. (i) The tension force (magnitude T_1) in the horizontal part of the string; (ii) the tension force (magnitude T_2) in the vertical part of the string; (iii) the weight m_2g of the hanging object. ∎

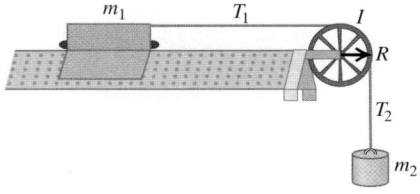

10.3 RIGID-BODY ROTATION ABOUT A MOVING AXIS

We can extend our analysis of the dynamics of rotational motion to some cases in which the axis of rotation moves. When that happens, the motion of the body is **combined translation and rotation.** The key to understanding such situations is this: Every possible motion of a rigid body can be represented as a

10.11 The motion of a rigid body is a combination of translational motion of the center of mass and rotation around the center of mass.

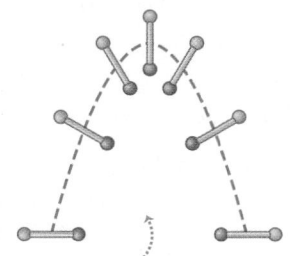

The motion of this tossed baton can be represented as a combination of ...

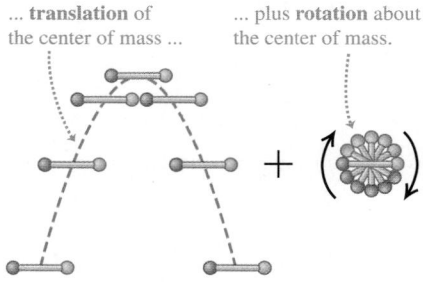

... **translation** of the center of mass ...   ... plus **rotation** about the center of mass.

combination of *translational motion of the center of mass* and *rotation about an axis through the center of mass.* This is true even when the center of mass accelerates, so it is not at rest in any inertial frame. **Figure 10.11** illustrates this for the motion of a tossed baton: The center of mass of the baton follows a parabolic curve, as though the baton were a particle located at the center of mass. A rolling ball is another example of combined translational and rotational motions.

Combined Translation and Rotation: Energy Relationships

It's beyond our scope to prove that rigid-body motion can always be divided into translation of the center of mass and rotation about the center of mass. But we *can* prove this for the kinetic energy K of a rigid body that has both translational and rotational motions. For such a body, K is the sum of two parts:

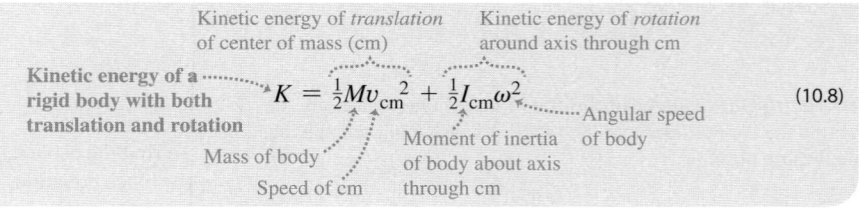

Kinetic energy of *translation* of center of mass (cm)   Kinetic energy of *rotation* around axis through cm

Kinetic energy of a rigid body with both translation and rotation

$$K = \tfrac{1}{2}Mv_{\text{cm}}^2 + \tfrac{1}{2}I_{\text{cm}}\omega^2 \qquad (10.8)$$

Mass of body   Angular speed of body
Speed of cm   Moment of inertia of body about axis through cm

To prove this relationship, we again imagine the rigid body to be made up of particles. For a typical particle with mass m_i (**Fig. 10.12**), the velocity $\vec{v}_i$ of this particle relative to an inertial frame is the vector sum of the velocity $\vec{v}_{\text{cm}}$ of the center of mass and the velocity $\vec{v}_i{}'$ of the particle *relative to* the center of mass:

$$\vec{v}_i = \vec{v}_{\text{cm}} + \vec{v}_i{}' \qquad (10.9)$$

10.12 A rigid body with both translational and rotational motions.

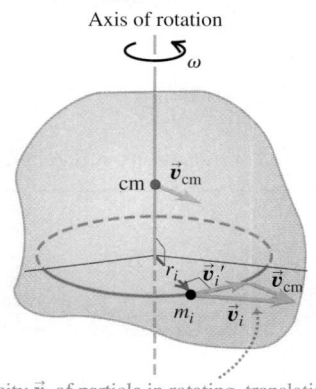

Axis of rotation

cm $\vec{v}_{\text{cm}}$

r_i $\vec{v}_i{}'$ $\vec{v}_{\text{cm}}$

m_i $\vec{v}_i$

Velocity $\vec{v}_i$ of particle in rotating, translating rigid body = (velocity $\vec{v}_{\text{cm}}$ of center of mass) + (particle's velocity $\vec{v}_i{}'$ relative to center of mass)

The kinetic energy K_i of this particle in the inertial frame is $\tfrac{1}{2}m_i v_i^2$, which we can also express as $\tfrac{1}{2}m_i(\vec{v}_i \cdot \vec{v}_i)$. Substituting Eq. (10.9) into this, we get

$$K_i = \tfrac{1}{2}m_i(\vec{v}_{\text{cm}} + \vec{v}_i{}') \cdot (\vec{v}_{\text{cm}} + \vec{v}_i{}')$$
$$= \tfrac{1}{2}m_i(\vec{v}_{\text{cm}} \cdot \vec{v}_{\text{cm}} + 2\vec{v}_{\text{cm}} \cdot \vec{v}_i{}' + \vec{v}_i{}' \cdot \vec{v}_i{}')$$
$$= \tfrac{1}{2}m_i(v_{\text{cm}}^2 + 2\vec{v}_{\text{cm}} \cdot \vec{v}_i{}' + v_i{}'^2)$$

The total kinetic energy is the sum $\sum K_i$ for all the particles making up the body. Expressing the three terms in this equation as separate sums, we get

$$K = \sum K_i = \sum \left(\tfrac{1}{2}m_i v_{\text{cm}}^2\right) + \sum (m_i \vec{v}_{\text{cm}} \cdot \vec{v}_i{}') + \sum \left(\tfrac{1}{2}m_i v_i{}'^2\right)$$

The first and second terms have common factors that we take outside the sum:

$$K = \tfrac{1}{2}\left(\sum m_i\right)v_{\text{cm}}^2 + \vec{v}_{\text{cm}} \cdot \left(\sum m_i \vec{v}_i{}'\right) + \sum \left(\tfrac{1}{2}m_i v_i{}'^2\right) \qquad (10.10)$$

Now comes the reward for our effort. In the first term, $\sum m_i$ is the total mass M. The second term is zero because $\sum m_i \vec{v}_i{}'$ is M times the velocity of the center of mass *relative to the center of mass,* and this is zero by definition. The last term is the sum of the kinetic energies of the particles computed by using their speeds with respect to the center of mass; this is just the kinetic energy of rotation around the center of mass. Using the same steps that led to Eq. (9.17) for the rotational kinetic energy of a rigid body, we can write this last term as $\tfrac{1}{2}I_{\text{cm}}\omega^2$, where I_{cm} is the moment of inertia with respect to the axis through the center of mass and ω is the angular speed. So Eq. (10.10) becomes Eq. (10.8):

$$K = \tfrac{1}{2}Mv_{\text{cm}}^2 + \tfrac{1}{2}I_{\text{cm}}\omega^2$$

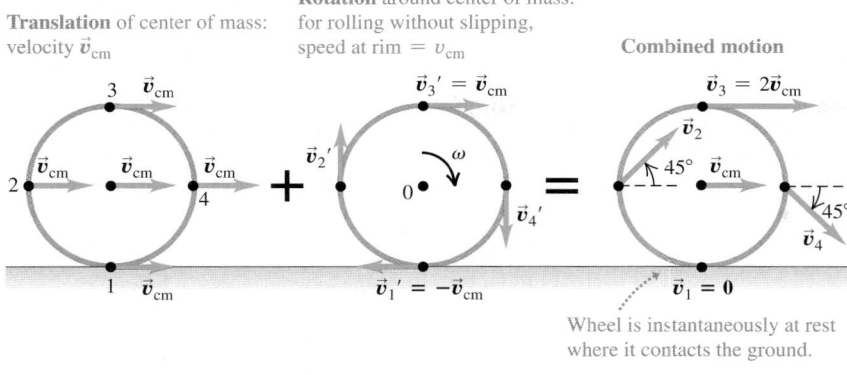

Translation of center of mass: velocity $\vec{v}_{cm}$

Rotation around center of mass: for rolling without slipping, speed at rim = v_{cm}

Combined motion

Wheel is instantaneously at rest where it contacts the ground.

10.13 The motion of a rolling wheel is the sum of the translational motion of the center of mass plus the rotational motion of the wheel around the center of mass.

Rolling Without Slipping

An important case of combined translation and rotation is **rolling without slipping.** The rolling wheel in **Fig. 10.13** is symmetrical, so its center of mass is at its geometric center. We view the motion in an inertial frame of reference in which the surface on which the wheel rolls is at rest. In this frame, the point on the wheel that contacts the surface must be instantaneously *at rest* so that it does not slip. Hence the velocity $\vec{v}_1{}'$ of the point of contact relative to the center of mass must have the same magnitude but opposite direction as the center-of-mass velocity $\vec{v}_{cm}$. If the wheel's radius is R and its angular speed about the center of mass is ω, then the magnitude of $\vec{v}_1{}'$ is $R\omega$; hence

Condition for rolling without slipping:

Speed of center of mass of rolling wheel ⋯⋯ $v_{cm} = R\omega$ ⋯⋯ Radius of wheel ⋯⋯ Angular speed of wheel (10.11)

As Fig. 10.13 shows, the velocity of a point on the wheel is the vector sum of the velocity of the center of mass and the velocity of the point relative to the center of mass. Thus while point 1, the point of contact, is instantaneously at rest, point 3 at the top of the wheel is moving forward *twice as fast* as the center of mass, and points 2 and 4 at the sides have velocities at 45° to the horizontal.

At any instant we can think of the wheel as rotating about an "instantaneous axis" of rotation that passes through the point of contact with the ground. The angular velocity ω is the same for this axis as for an axis through the center of mass; an observer at the center of mass sees the rim make the same number of revolutions per second as does an observer at the rim watching the center of mass spin around him. If we think of the motion of the rolling wheel in Fig. 10.13 in this way, the kinetic energy of the wheel is $K = \frac{1}{2}I_1\omega^2$, where I_1 is the moment of inertia of the wheel about an axis through point 1. But by the parallel-axis theorem, Eq. (9.19), $I_1 = I_{cm} + MR^2$, where M is the total mass of the wheel and I_{cm} is the moment of inertia with respect to an axis through the center of mass. Using Eq. (10.11), we find that the wheel's kinetic energy is as given by Eq. (10.8):

$$K = \tfrac{1}{2}I_1\omega^2 = \tfrac{1}{2}I_{cm}\omega^2 + \tfrac{1}{2}MR^2\omega^2 = \tfrac{1}{2}I_{cm}\omega^2 + \tfrac{1}{2}Mv_{cm}{}^2$$

CAUTION **Rolling without slipping** The relationship $v_{cm} = R\omega$ holds *only* if there is rolling without slipping. When a drag racer first starts to move, the rear tires are spinning very fast even though the racer is hardly moving, so $R\omega$ is greater than v_{cm} (**Fig. 10.14**). If a driver applies the brakes too heavily so that the car skids, the tires will spin hardly at all and $R\omega$ is less than v_{cm}. ▮

If a rigid body changes height as it moves, we must also consider gravitational potential energy. We saw in Section 9.4 that for any extended body of mass M, rigid or not, the gravitational potential energy U is the same as if we replaced the body by a particle of mass M located at the body's center of mass, so

$$U = Mgy_{cm}$$

BIO Application Combined Translation and Rotation A maple seed consists of a pod attached to a much lighter, flattened wing. Airflow around the wing slows the falling seed to about 1 m/s and causes the seed to rotate about its center of mass. The seed's slow fall means that a breeze can carry the seed some distance from the parent tree. In the absence of wind, the seed's center of mass falls straight down.

Maple seed

Maple seed falling

10.14 The smoke rising from this drag racer's rear tires shows that the tires are slipping on the road, so v_{cm} is *not* equal to $R\omega$.

SOLUTION

EXAMPLE 10.4   SPEED OF A PRIMITIVE YO-YO

A primitive yo-yo has a massless string wrapped around a solid cylinder with mass M and radius R (**Fig. 10.15**). You hold the free end of the string stationary and release the cylinder from rest. The string unwinds but does not slip or stretch as the cylinder descends and rotates. Using energy considerations, find the speed v_{cm} of the cylinder's center of mass after it has descended a distance h.

SOLUTION

IDENTIFY and SET UP: Since you hold the upper end of the string fixed, your hand does no work on the string–cylinder system. There is friction between the string and the cylinder, but the string doesn't slip so no mechanical energy is lost. Hence we can use

10.15 Calculating the speed of a primitive yo-yo.

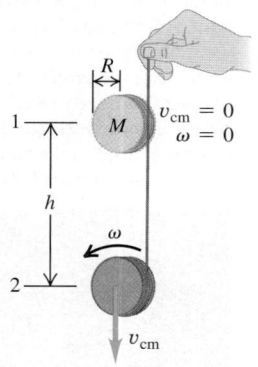

conservation of mechanical energy. The initial kinetic energy of the cylinder is $K_1 = 0$, and its final kinetic energy K_2 is given by Eq. (10.8); the massless string has no kinetic energy. The moment of inertia is $I_{cm} = \frac{1}{2}MR^2$, and by Eq. (9.13) $\omega = v_{cm}/R$ because the string doesn't slip. The potential energies are $U_1 = Mgh$ and $U_2 = 0$.

EXECUTE: From Eq. (10.8), the kinetic energy at point 2 is

$$K_2 = \tfrac{1}{2}Mv_{cm}^2 + \tfrac{1}{2}\left(\tfrac{1}{2}MR^2\right)\left(\frac{v_{cm}}{R}\right)^2 = \tfrac{3}{4}Mv_{cm}^2$$

The kinetic energy is $1\frac{1}{2}$ times what it would be if the yo-yo were falling at speed v_{cm} without rotating. Two-thirds of the total kinetic energy $\left(\frac{1}{2}Mv_{cm}^2\right)$ is translational and one-third $\left(\frac{1}{4}Mv_{cm}^2\right)$ is rotational. Using conservation of energy,

$$K_1 + U_1 = K_2 + U_2$$
$$0 + Mgh = \tfrac{3}{4}Mv_{cm}^2 + 0$$
$$v_{cm} = \sqrt{\tfrac{4}{3}gh}$$

EVALUATE: No mechanical energy was lost or gained, so from the energy standpoint the string is merely a way to convert some of the gravitational potential energy (which is released as the cylinder falls) into rotational kinetic energy rather than translational kinetic energy. Because not all of the released energy goes into translation, v_{cm} is less than the speed $\sqrt{2gh}$ of an object dropped from height h with no strings attached.

SOLUTION

EXAMPLE 10.5   RACE OF THE ROLLING BODIES

In a physics demonstration, an instructor "races" various bodies that roll without slipping from rest down an inclined plane (**Fig. 10.16**). What shape should a body have to reach the bottom of the incline first?

SOLUTION

IDENTIFY and SET UP: Kinetic friction does no work if the bodies roll without slipping. We can also ignore the effects of *rolling friction,* introduced in Section 5.3, if the bodies and the surface of the incline are rigid. (Later in this section we'll explain why this is so.) We can therefore use conservation of energy. Each body starts from rest at the top of an incline with height h, so $K_1 = 0$, $U_1 = Mgh$, and $U_2 = 0$. Equation (10.8) gives the kinetic energy at the bottom of the incline; since the bodies roll without slipping, $\omega = v_{cm}/R$. We can express the moments of inertia of the four round bodies in

10.16 Which body rolls down the incline fastest, and why?

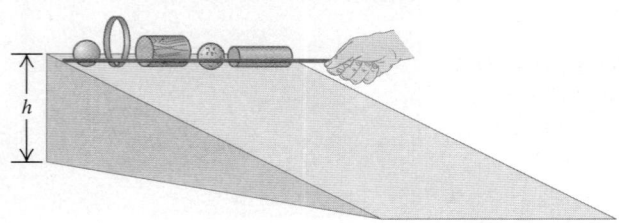

Table 9.2, cases (f)–(i), as $I_{cm} = cMR^2$, where c is a number less than or equal to 1 that depends on the shape of the body. Our goal is to find the value of c that gives the body the greatest speed v_{cm} after its center of mass has descended a vertical distance h.

EXECUTE: From conservation of energy,

$$K_1 + U_1 = K_2 + U_2$$
$$0 + Mgh = \tfrac{1}{2}Mv_{cm}^2 + \tfrac{1}{2}cMR^2\left(\frac{v_{cm}}{R}\right)^2 + 0$$
$$Mgh = \tfrac{1}{2}(1 + c)Mv_{cm}^2$$
$$v_{cm} = \sqrt{\frac{2gh}{1 + c}}$$

EVALUATE: For a given value of c, the speed v_{cm} after descending a distance h is *independent* of the body's mass M and radius R. Hence *all* uniform solid cylinders $\left(c = \frac{1}{2}\right)$ have the same speed at the bottom, regardless of their mass and radii. The values of c tell us that the order of finish for uniform bodies will be as follows: (1) any solid sphere $\left(c = \frac{2}{5}\right)$, (2) any solid cylinder $\left(c = \frac{1}{2}\right)$, (3) any thin-walled, hollow sphere $\left(c = \frac{2}{3}\right)$, and (4) any thin-walled, hollow cylinder $(c = 1)$. Small-c bodies always beat large-c bodies because less of their kinetic energy is tied up in rotation, so more is available for translation.

Combined Translation and Rotation: Dynamics

We can also analyze the combined translational and rotational motions of a rigid body from the standpoint of dynamics. We showed in Section 8.5 that for an extended body, the acceleration of the center of mass is the same as that of a particle of the same mass acted on by all the external forces on the actual body:

$$\sum \vec{F}_{ext} = M\vec{a}_{cm} \qquad (10.12)$$

Net external ····· force on a body ↘ ↙ ····· Total mass of body ····· Acceleration of center of mass

The rotational motion about the center of mass is described by the rotational analog of Newton's second law, Eq. (10.7):

$$\sum \tau_z = I_{cm}\alpha_z \qquad (10.13)$$

Net torque on a rigid ····· body about z-axis through center of mass ↘ ↙ ····· Moment of inertia of rigid body about z-axis ····· Angular acceleration of rigid body about z-axis

It's not immediately obvious that Eq. (10.13) should apply to the motion of a translating rigid body; after all, our derivation of $\sum \tau_z = I\alpha_z$ in Section 10.2 assumed that the axis of rotation was stationary. But Eq. (10.13) is valid *even when the axis of rotation moves,* provided the following two conditions are met:

1. The axis through the center of mass must be an axis of symmetry.
2. The axis must not change direction.

These conditions are satisfied for many types of rotation (**Fig. 10.17**). Note that in general this moving axis of rotation is *not* at rest in an inertial frame of reference.

We can now solve dynamics problems involving a rigid body that undergoes translational and rotational motions at the same time, provided that the rotation axis satisfies the two conditions just mentioned. Problem-Solving Strategy 10.1 (Section 10.2) is equally useful here, and you should review it now. Keep in mind that when a body undergoes translational and rotational motions at the same time, we need two separate equations of motion *for the same body:* Eq. (10.12) for the translation of the center of mass and Eq. (10.13) for rotation about an axis through the center of mass.

10.17 The axle of a bicycle wheel passes through the wheel's center of mass and is an axis of symmetry. Hence the rotation of the wheel is described by Eq. (10.13), provided the bicycle doesn't turn or tilt to one side (which would change the orientation of the axle).

EXAMPLE 10.6   ACCELERATION OF A PRIMITIVE YO-YO

SOLUTION

For the primitive yo-yo in Example 10.4 (**Fig. 10.18a**), find the downward acceleration of the cylinder and the tension in the string.

SOLUTION

IDENTIFY and SET UP: Figure 10.18b shows our free-body diagram for the yo-yo, including our choice of positive coordinate directions. Our target variables are $a_{cm\text{-}y}$ and T. We'll use Eq. (10.12) for the translational motion of the center of mass and Eq. (10.13) for the rotational motion around the center of mass. We'll also use Eq. (10.11), which says that the string unwinds without slipping. As in Example 10.4, the moment of inertia of the yo-yo for an axis through its center of mass is $I_{cm} = \frac{1}{2}MR^2$.

EXECUTE: From Eq. (10.12),

$$\sum F_y = Mg + (-T) = Ma_{cm\text{-}y} \qquad (10.14)$$

10.18 Dynamics of a primitive yo-yo (see Fig. 10.15).

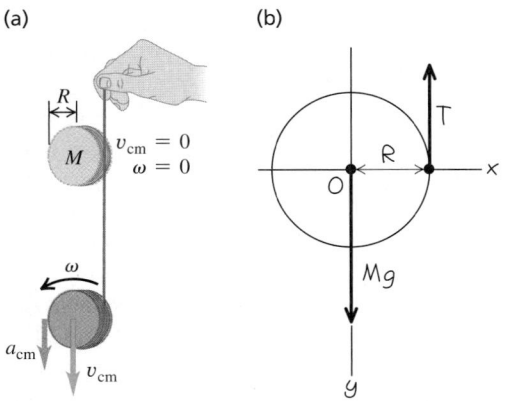

Continued

From Eq. (10.13),

$$\Sigma \tau_z = TR = I_{cm}\alpha_z = \tfrac{1}{2}MR^2\alpha_z \qquad (10.15)$$

From Eq. (10.11), $v_{cm-z} = R\omega_z$; the derivative of this expression with respect to time gives us

$$a_{cm-y} = R\alpha_z \qquad (10.16)$$

We now use Eq. (10.16) to eliminate α_z from Eq. (10.15) and then solve Eqs. (10.14) and (10.15) simultaneously for T and a_{cm-y}:

$$a_{cm-y} = \tfrac{2}{3}g \qquad T = \tfrac{1}{3}Mg$$

EVALUATE: The string slows the fall of the yo-yo, but not enough to stop it completely. Hence a_{cm-y} is less than the free-fall value g and T is less than the yo-yo weight Mg.

EXAMPLE 10.7 ACCELERATION OF A ROLLING SPHERE

A bowling ball rolls without slipping down a ramp that is inclined at an angle β to the horizontal (**Fig. 10.19a**). What are the ball's acceleration and the magnitude of the friction force on the ball? Treat the ball as a uniform solid sphere, ignoring the finger holes.

SOLUTION

IDENTIFY and SET UP: The free-body diagram (Fig. 10.19b) shows that only the friction force exerts a torque about the center of mass. Our target variables are the acceleration a_{cm-x} of the ball's center of mass and the magnitude f of the friction force. (Because the ball does not slip at the instantaneous point of contact with the ramp, this is a *static* friction force; it prevents slipping and gives the ball its angular acceleration.) We use Eqs. (10.12) and (10.13) as in Example 10.6.

10.19 A bowling ball rolling down a ramp.

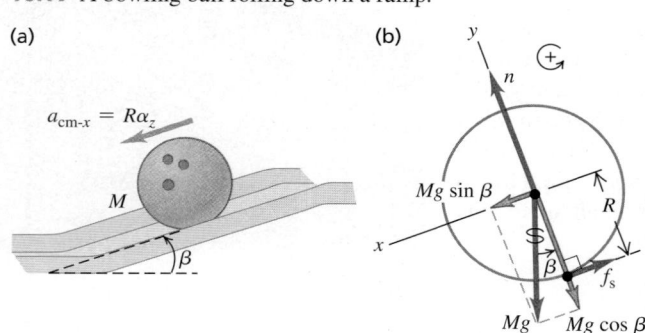

(a)

(b)

EXECUTE: The ball's moment of inertia is $I_{cm} = \tfrac{2}{5}MR^2$. The equations of motion are

$$\Sigma F_x = Mg\sin\beta + (-f) = Ma_{cm-x} \qquad (10.17)$$

$$\Sigma \tau_z = fR = I_{cm}\alpha_z = \left(\tfrac{2}{5}MR^2\right)\alpha_z \qquad (10.18)$$

The ball rolls without slipping, so as in Example 10.6 we use $a_{cm-x} = R\alpha_z$ to eliminate α_z from Eq. (10.18):

$$fR = \tfrac{2}{5}MRa_{cm-x}$$

This equation and Eq. (10.17) are two equations for the unknowns a_{cm-x} and f. We solve Eq. (10.17) for f, substitute that expression into the above equation to eliminate f, and solve for a_{cm-x}:

$$a_{cm-x} = \tfrac{5}{7}g\sin\beta$$

Finally, we substitute this acceleration into Eq. (10.17) and solve for f:

$$f = \tfrac{2}{7}Mg\sin\beta$$

EVALUATE: The ball's acceleration is just $\tfrac{5}{7}$ as large as that of an object *sliding* down the slope without friction. If the ball descends a vertical distance h as it rolls down the ramp, its displacement along the ramp is $h/\sin\beta$. You can show that the speed of the ball at the bottom of the ramp is $v_{cm} = \sqrt{\tfrac{10}{7}gh}$, the same as our result from Example 10.5 with $c = \tfrac{2}{5}$.

If the ball were rolling *uphill* without slipping, the force of friction would still be directed uphill as in Fig. 10.19b. Can you see why?

Rolling Friction

In Example 10.5 we said that we can ignore rolling friction if both the rolling body and the surface over which it rolls are perfectly rigid. In **Fig. 10.20a** a perfectly rigid sphere is rolling down a perfectly rigid incline. The line of action of the normal force passes through the center of the sphere, so its torque is zero; there is no sliding at the point of contact, so the friction force does no work. Figure 10.20b shows a more realistic situation, in which the surface "piles up" in front of the sphere and the sphere rides in a shallow trench. Because of these deformations, the contact forces on the sphere no longer act along a single point but over an area; the forces are concentrated on the front of the sphere as shown. As a result, the normal force now exerts a torque that opposes the rotation. In addition, there is some sliding of the sphere over the surface due to the deformation, causing mechanical energy to be lost. The combination of these two effects is the phenomenon of *rolling friction*. Rolling friction also occurs

(a) Perfectly rigid sphere rolling on a perfectly rigid surface

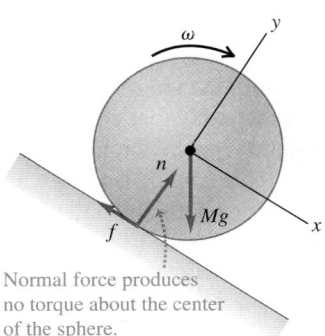

Normal force produces no torque about the center of the sphere.

(b) Rigid sphere rolling on a deformable surface

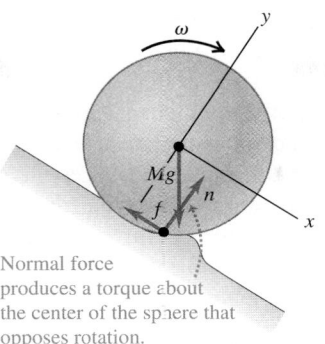

Normal force produces a torque about the center of the sphere that opposes rotation.

10.20 Rolling down **(a)** a perfectly rigid surface and **(b)** a deformable surface. In (b) the deformation is greatly exaggerated, and the force *n* is the component of the contact force that points normal to the plane of the surface before it is deformed.

if the rolling body is deformable, such as an automobile tire. Often the rolling body and the surface are rigid enough that rolling friction can be ignored, as we have assumed in all the examples in this section.

TEST YOUR UNDERSTANDING OF SECTION 10.3 Suppose the solid cylinder used as a yo-yo in Example 10.6 is replaced by a hollow cylinder of the same mass and radius. (a) Will the acceleration of the yo-yo (i) increase, (ii) decrease, or (iii) remain the same? (b) Will the string tension (i) increase, (ii) decrease, or (iii) remain the same? ❚

10.4 WORK AND POWER IN ROTATIONAL MOTION

When you pedal a bicycle, you apply forces to a rotating body and do work on it. Similar things happen in many other real-life situations, such as a rotating motor shaft driving a power tool or a car engine propelling the vehicle. Let's see how to apply our ideas about work from Chapter 6 to rotational motion.

Suppose a tangential force $\vec{F}_{\text{tan}}$ acts at the rim of a pivoted disk—for example, a child running while pushing on a playground merry-go-round (**Fig. 10.21a**). The disk rotates through an infinitesimal angle $d\theta$ about a fixed axis during an infinitesimal time interval dt (Fig. 10.21b). The work dW done by the force $\vec{F}_{\text{tan}}$ while a point on the rim moves a distance ds is $dW = F_{\text{tan}}\,ds$. If $d\theta$ is measured in radians, then $ds = R\,d\theta$ and

$$dW = F_{\text{tan}} R\,d\theta$$

Now $F_{\text{tan}}R$ is the *torque* τ_z due to the force $\vec{F}_{\text{tan}}$, so

$$dW = \tau_z\,d\theta \qquad (10.19)$$

As the disk rotates from θ_1 to θ_2, the total work done by the torque is

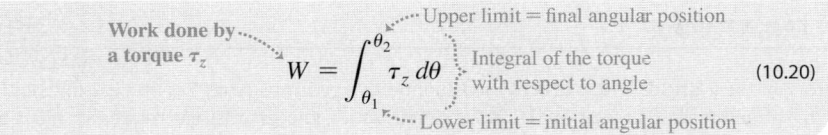

Work done by a torque τ_z ····· Upper limit = final angular position

$$W = \int_{\theta_1}^{\theta_2} \tau_z\,d\theta \qquad (10.20)$$

Integral of the torque with respect to angle ·· Lower limit = initial angular position

If the torque remains *constant* while the angle changes, then the work is the product of torque and angular displacement:

Work done by a constant torque τ_z ·····Torque·····

$$W = \tau_z(\theta_2 - \theta_1) = \tau_z\Delta\theta \qquad (10.21)$$

Final minus initial angular position = angular displacement

10.21 A tangential force applied to a rotating body does work.

(a)

Child applies tangential force.

$\vec{F}_{\text{tan}}$

(b) Overhead view of merry-go-round

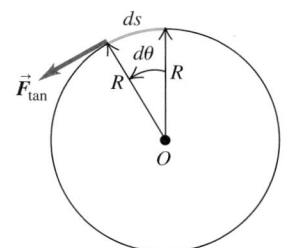

If torque is expressed in newton-meters $(\text{N} \cdot \text{m})$ and angular displacement in radians, the work is in joules. Equation (10.21) is the rotational analog of Eq. (6.1), $W = Fs$, and Eq. (10.20) is the analog of Eq. (6.7), $W = \int F_x \, dx$, for the work done by a force in a straight-line displacement.

If the force in Fig. 10.21 had an axial component (parallel to the rotation axis) or a radial component (directed toward or away from the axis), that component would do no work because the displacement of the point of application has only a tangential component. An axial or radial component of force would also make no contribution to the torque about the axis of rotation. So Eqs. (10.20) and (10.21) are correct for *any* force, no matter what its components.

When a torque does work on a rotating rigid body, the kinetic energy changes by an amount equal to the work done. We can prove this by using exactly the same procedure that we used in Eqs. (6.11) through (6.13) for the translational kinetic energy of a particle. Let τ_z represent the *net* torque on the body so that $\tau_z = I\alpha_z$ from Eq. (10.7), and assume that the body is rigid so that the moment of inertia I is constant. We then transform the integrand in Eq. (10.20) into an integrand with respect to ω_z as follows:

$$\tau_z \, d\theta = (I\alpha_z) \, d\theta = I\frac{d\omega_z}{dt} \, d\theta = I\frac{d\theta}{dt} \, d\omega_z = I\omega_z \, d\omega_z$$

Since τ_z is the net torque, the integral in Eq. (10.20) is the *total* work done on the rotating rigid body. This equation then becomes

Total work done on a rotating rigid body = work done by the net external torque

Final rotational kinetic energy

Initial rotational kinetic energy

$$W_{\text{tot}} = \int_{\omega_1}^{\omega_2} I\omega_z \, d\omega_z = \tfrac{1}{2}I\omega_2{}^2 - \tfrac{1}{2}I\omega_1{}^2 \qquad (10.22)$$

The change in the rotational kinetic energy of a *rigid* body equals the work done by forces exerted from outside the body (**Fig. 10.22**). This equation is analogous to Eq. (6.13), the work–energy theorem for a particle.

How does *power* relate to torque? When we divide both sides of Eq. (10.19) by the time interval dt during which the angular displacement occurs, we find

$$\frac{dW}{dt} = \tau_z \frac{d\theta}{dt}$$

But dW/dt is the rate of doing work, or *power P,* and $d\theta/dt$ is angular velocity ω_z:

Power due to a torque acting on a rigid body

Torque with respect to body's rotation axis

$$P = \tau_z\omega_z \qquad (10.23)$$

Angular velocity of body about axis

This is the analog of the relationship $P = \vec{F} \cdot \vec{v}$ that we developed in Section 6.4 for particle motion.

10.22 The rotational kinetic energy of a helicopter's main rotor is equal to the total work done to set it spinning. When it is spinning at a constant rate, positive work is done on the rotor by the engine and negative work is done on it by air resistance. Hence the net work being done is zero and the kinetic energy remains constant.

EXAMPLE 10.8 **CALCULATING POWER FROM TORQUE**

An electric motor exerts a constant $10\text{-N} \cdot \text{m}$ torque on a grindstone, which has a moment of inertia of $2.0 \text{ kg} \cdot \text{m}^2$ about its shaft. The system starts from rest. Find the work W done by the motor in 8.0 s and the grindstone's kinetic energy K at this time. What average power P_{av} is delivered by the motor?

SOLUTION

IDENTIFY and SET UP: The only torque acting is that due to the motor. Since this torque is constant, the grindstone's angular acceleration α_z is constant. We'll use Eq. (10.7) to find α_z, and then

use this in the kinematics equations from Section 9.2 to calculate the angle $\Delta\theta$ through which the grindstone rotates in 8.0 s and its final angular velocity ω_z. From these we'll calculate W, K, and P_{av}.

EXECUTE: We have $\sum \tau_z = 10 \text{ N} \cdot \text{m}$ and $I = 2.0 \text{ kg} \cdot \text{m}^2$, so $\sum \tau_z = I\alpha_z$ yields $\alpha_z = 5.0 \text{ rad/s}^2$. From Eqs. (9.11) and (10.21),

$$\Delta\theta = \tfrac{1}{2}\alpha_z t^2 = \tfrac{1}{2}(5.0 \text{ rad/s}^2)(8.0 \text{ s})^2 = 160 \text{ rad}$$

$$W = \tau_z\Delta\theta = (10 \text{ N} \cdot \text{m})(160 \text{ rad}) = 1600 \text{ J}$$

From Eqs. (9.7) and (9.17),

$$\omega_z = \alpha_z t = (5.0 \text{ rad/s}^2)(8.0 \text{ s}) = 40 \text{ rad/s}$$

$$K = \tfrac{1}{2}I\omega_z^2 = \tfrac{1}{2}(2.0 \text{ kg} \cdot \text{m}^2)(40 \text{ rad/s})^2 = 1600 \text{ J}$$

The average power is the work done divided by the time interval:

$$P_{av} = \frac{1600 \text{ J}}{8.0 \text{ s}} = 200 \text{ J/s} = 200 \text{ W}$$

EVALUATE: The initial kinetic energy was zero, so the work done W must equal the final kinetic energy K [Eq. (10.22)]. This is just as we calculated. We can check our result $P_{av} = 200$ W by considering the *instantaneous* power $P = \tau_z\omega_z$. Because ω_z increases continuously, P increases continuously as well; its value increases from zero at $t = 0$ to $(10 \text{ N} \cdot \text{m})(40 \text{ rad/s}) = 400$ W at $t = 8.0$ s. Both ω_z and P increase *uniformly* with time, so the *average* power is just half this maximum value, or 200 W.

TEST YOUR UNDERSTANDING OF SECTION 10.4 You apply equal torques to two different cylinders. Cylinder 1 has a moment of inertia twice as large as cylinder 2. Each cylinder is initially at rest. After one complete rotation, which cylinder has the greater kinetic energy? (i) Cylinder 1; (ii) cylinder 2; (iii) both cylinders have the same kinetic energy. ▌

10.5 ANGULAR MOMENTUM

Every rotational quantity that we have encountered in Chapters 9 and 10 is the analog of some quantity in the translational motion of a particle. The analog of *momentum* of a particle is **angular momentum,** a vector quantity denoted as $\vec{L}$. Its relationship to momentum $\vec{p}$ (which we will often call *linear momentum* for clarity) is exactly the same as the relationship of torque to force, $\vec{\tau} = \vec{r} \times \vec{F}$. For a particle with constant mass m and velocity $\vec{v}$, the angular momentum is

Angular momentum of····   Position vector of particle relative to O
a particle relative to
origin O of an inertial $$\vec{L} = \vec{r} \times \vec{p} = \vec{r} \times m\vec{v} \qquad (10.24)$$
frame of reference   Linear momentum of particle = mass times velocity

The value of $\vec{L}$ depends on the choice of origin O, since it involves the particle's position vector $\vec{r}$ relative to O. The units of angular momentum are kg $\cdot$ m^2/s.

In **Fig. 10.23** a particle moves in the xy-plane; its position vector $\vec{r}$ and momentum $\vec{p} = m\vec{v}$ are shown. The angular momentum vector $\vec{L}$ is perpendicular to the xy-plane. The right-hand rule for vector products shows that its direction is along the $+z$-axis, and its magnitude is

$$L = mvr\sin\phi = mvl \qquad (10.25)$$

where l is the perpendicular distance from the line of $\vec{v}$ to O. This distance plays the role of "lever arm" for the momentum vector.

When a net force $\vec{F}$ acts on a particle, its velocity and momentum change, so its angular momentum may also change. We can show that the *rate of change* of angular momentum is equal to the torque of the net force. We take the time derivative of Eq. (10.24), using the rule for the derivative of a product:

$$\frac{d\vec{L}}{dt} = \left(\frac{d\vec{r}}{dt} \times m\vec{v}\right) + \left(\vec{r} \times m\frac{d\vec{v}}{dt}\right) = (\vec{v} \times m\vec{v}) + (\vec{r} \times m\vec{a})$$

The first term is zero because it contains the vector product of the vector $\vec{v} = d\vec{r}/dt$ with itself. In the second term we replace $m\vec{a}$ with the net force $\vec{F}$:

$$\frac{d\vec{L}}{dt} = \vec{r} \times \vec{F} = \vec{\tau} \qquad \text{(for a particle acted on by net force } \vec{F}\text{)} \qquad (10.26)$$

The rate of change of angular momentum of a particle equals the torque of the net force acting on it. Compare this result to Eq. (8.4): The rate of change $d\vec{p}/dt$ of the *linear* momentum of a particle equals the net force that acts on it.

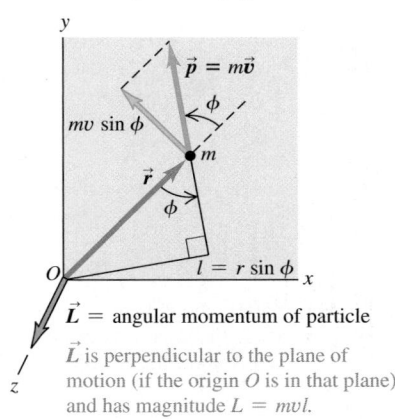

10.23 Calculating the angular momentum $\vec{L} = \vec{r} \times m\vec{v} = \vec{r} \times \vec{p}$ of a particle with mass m moving in the xy-plane.

$\vec{L}$ = angular momentum of particle

$\vec{L}$ is perpendicular to the plane of motion (if the origin O is in that plane) and has magnitude $L = mvl$.

10.24 Calculating the angular momentum of a particle of mass m_i in a rigid body rotating at angular speed ω. (Compare Fig. 10.23.)

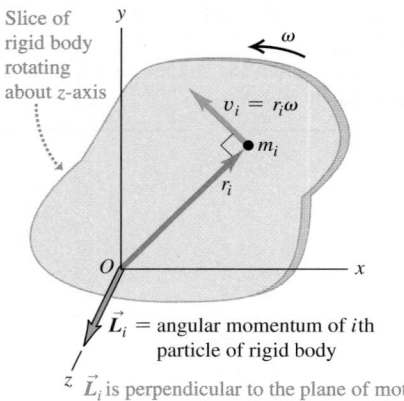

$\vec{L}_i$ = angular momentum of ith particle of rigid body

$\vec{L}_i$ is perpendicular to the plane of motion (if the origin O is in that plane) and has magnitude $L_i = m_i v_i r_i = m_i r_i^2 \omega$.

10.25 Two particles of the same mass located symmetrically on either side of the rotation axis of a rigid body. The angular momentum vectors $\vec{L}_1$ and $\vec{L}_2$ of the two particles do not lie along the rotation axis, but their vector sum $\vec{L}_1 + \vec{L}_2$ does.

Another slice of a rigid body rotating about the z-axis (viewed edge-on)

$\vec{L}_1 + \vec{L}_2$ is along the rotation axis.

$m_2 = m_1$

This particle of the body is moving toward you.

This particle of the body is moving away from you.

Angular Momentum of a Rigid Body

We can use Eq. (10.25) to find the total angular momentum of a *rigid body* rotating about the z-axis with angular speed ω. First consider a thin slice of the body lying in the xy-plane (**Fig. 10.24**). Each particle in the slice moves in a circle centered at the origin, and at each instant its velocity $\vec{v}_i$ is perpendicular to its position vector $\vec{r}_i$, as shown. Hence in Eq. (10.25), $\phi = 90°$ for every particle. A particle with mass m_i at a distance r_i from O has a speed v_i equal to $r_i\omega$. From Eq. (10.25) the magnitude L_i of its angular momentum is

$$L_i = m_i(r_i\omega)\,r_i = m_i r_i^2 \omega \qquad (10.27)$$

The direction of each particle's angular momentum, as given by the right-hand rule for the vector product, is along the $+z$-axis.

The *total* angular momentum of the slice of the body lying in the xy-plane is the sum ΣL_i of the angular momenta L_i of the particles. Summing Eq. (10.27), we have

$$L = \Sigma L_i = \left(\Sigma m_i r_i^2\right)\omega = I\omega$$

where I is the moment of inertia of the slice about the z-axis.

We can do this same calculation for the other slices of the body, all parallel to the xy-plane. For points that do not lie in the xy-plane, a complication arises because the $\vec{r}$ vectors have components in the z-direction as well as in the x- and y-directions; this gives the angular momentum of each particle a component perpendicular to the z-axis. But *if the z-axis is an axis of symmetry,* the perpendicular components for particles on opposite sides of this axis add up to zero (**Fig. 10.25**). So when a body rotates about an axis of symmetry, its angular momentum vector $\vec{L}$ lies along the symmetry axis, and its magnitude is $L = I\omega$.

The angular velocity vector $\vec{\omega}$ also lies along the rotation axis, as we saw in Section 9.1. Hence for a rigid body rotating around an axis of symmetry, $\vec{L}$ and $\vec{\omega}$ are in the same direction (**Fig. 10.26**). So we have the *vector* relationship

Angular momentum of a rigid body rotating around a symmetry axis ···· $\vec{L} = I\vec{\omega}$ ···· Moment of inertia of body about symmetry axis

Angular velocity vector of body $\qquad (10.28)$

From Eq. (10.26) the rate of change of angular momentum of a particle equals the torque of the net force acting on the particle. For any system of particles (including both rigid and nonrigid bodies), the rate of change of the *total* angular momentum equals the sum of the torques of all forces acting on all the particles. The torques of the *internal* forces add to zero if these forces act along the line from one particle to another, as in Fig. 10.8, and so the sum of the torques includes only the torques of the *external* forces. (We saw a similar

10.26 For rotation about an axis of symmetry, $\vec{\omega}$ and $\vec{L}$ are parallel and along the axis. The directions of both vectors are given by the right-hand rule (compare Fig. 9.5).

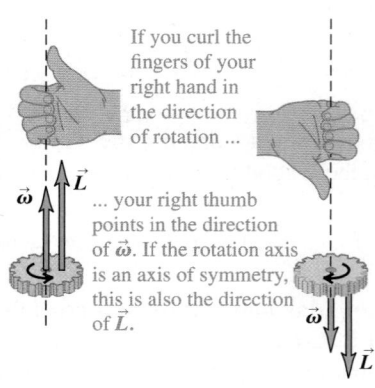

If you curl the fingers of your right hand in the direction of rotation ...

... your right thumb points in the direction of $\vec{\omega}$. If the rotation axis is an axis of symmetry, this is also the direction of $\vec{L}$.

cancellation in our discussion of center-of-mass motion in Section 8.5.) So we conclude that

> **For a system of particles:**
> Sum of external torques ·········▸ $\sum \vec{\tau} = \dfrac{d\vec{L}}{dt}$ ◂········· Rate of change of total angular momentum $\vec{L}$ of system   (10.29)
> on the system

Finally, if the system of particles is a rigid body rotating about a symmetry axis (the z-axis), then $L_z = I\omega_z$ and I is constant. If this axis has a fixed direction in space, then vectors $\vec{L}$ and $\vec{\omega}$ change only in magnitude, not in direction. In that case, $dL_z/dt = I \, d\omega_z/dt = I\alpha_z$, or

$$\sum \tau_z = I\alpha_z$$

which is again our basic relationship for the dynamics of rigid-body rotation. If the body is *not* rigid, I may change; in that case, L changes even when ω is constant. For a nonrigid body, Eq. (10.29) is still valid, even though Eq. (10.7) is not.

When the axis of rotation is *not* a symmetry axis, the angular momentum is in general *not* parallel to the axis (**Fig. 10.27**). As the body turns, the angular momentum vector $\vec{L}$ traces out a cone around the rotation axis. Because $\vec{L}$ changes, there must be a net external torque acting on the body even though the angular velocity magnitude ω may be constant. If the body is an unbalanced wheel on a car, this torque is provided by friction in the bearings, which causes the bearings to wear out. "Balancing" a wheel means distributing the mass so that the rotation axis is an axis of symmetry; then $\vec{L}$ points along the rotation axis, and no net torque is required to keep the wheel turning.

In fixed-axis rotation we often use the term "angular momentum of the body" to refer to only the *component* of $\vec{L}$ along the rotation axis of the body (the z-axis in Fig. 10.27), with a positive or negative sign to indicate the sense of rotation just as with angular velocity.

10.27 If the rotation axis of a rigid body is not a symmetry axis, $\vec{L}$ does not in general lie along the rotation axis. Even if $\vec{\omega}$ is constant, the direction of $\vec{L}$ changes and a net torque is required to maintain rotation.

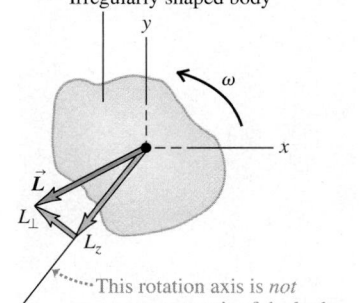

Irregularly shaped body

This rotation axis is *not* a symmetry axis of the body: $\vec{L}$ is not along the rotation axis.

EXAMPLE 10.9   ANGULAR MOMENTUM AND TORQUE

A turbine fan in a jet engine has a moment of inertia of 2.5 kg·m² about its axis of rotation. As the turbine starts up, its angular velocity is given by $\omega_z = (40 \text{ rad/s}^3)t^2$. (a) Find the fan's angular momentum as a function of time, and find its value at $t = 3.0$ s. (b) Find the net torque on the fan as a function of time, and find its value at $t = 3.0$ s.

SOLUTION

IDENTIFY and SET UP: The fan rotates about its axis of symmetry (the z-axis). Hence the angular momentum vector has only a z-component L_z, which we can determine from the angular velocity ω_z. Since the direction of angular momentum is constant, the net torque likewise has only a component τ_z along the rotation axis. We'll use Eq. (10.28) to find L_z from ω_z and then Eq. (10.29) to find τ_z.

EXECUTE: (a) From Eq. (10.28),

$$L_z = I\omega_z = (2.5 \text{ kg·m}^2)(40 \text{ rad/s}^3)t^2 = (100 \text{ kg·m}^2/\text{s}^3)t^2$$

(We dropped the dimensionless quantity "rad" from the final expression.) At $t = 3.0$ s, $L_z = 900$ kg·m²/s.

(b) From Eq. (10.29),

$$\tau_z = \frac{dL_z}{dt} = (100 \text{ kg·m}^2/\text{s}^3)(2t) = (200 \text{ kg·m}^2/\text{s}^3)t$$

At $t = 3.0$ s,

$$\tau_z = (200 \text{ kg·m}^2/\text{s}^3)(3.0 \text{ s}) = 600 \text{ kg·m}^2/\text{s}^2 = 600 \text{ N·m}$$

EVALUATE: As a check on our expression for τ_z, note that the angular acceleration of the turbine is $\alpha_z = d\omega_z/dt = (40 \text{ rad/s}^3)(2t) = (80 \text{ rad/s}^3)t$. Hence from Eq. (10.7), the torque on the fan is $\tau_z = I\alpha_z = (2.5 \text{ kg·m}^2)(80 \text{ rad/s}^3)t = (200 \text{ kg·m}^2/\text{s}^3)t$, just as we calculated.

TEST YOUR UNDERSTANDING OF SECTION 10.5 A ball is attached to one end of a piece of string. You hold the other end of the string and whirl the ball in a circle around your hand. (a) If the ball moves at a constant speed, is its linear momentum $\vec{p}$ constant? Why or why not? (b) Is its angular momentum $\vec{L}$ constant? Why or why not? ❙

DEMO   DEMO

PhET: Torque

10.28 A falling cat twists different parts of its body in different directions so that it lands feet first. At all times during this process the angular momentum of the cat as a whole remains zero.

10.6 CONSERVATION OF ANGULAR MOMENTUM

We have just seen that angular momentum can be used for an alternative statement of the basic dynamic principle for rotational motion. It also forms the basis for the **principle of conservation of angular momentum.** Like conservation of energy and of linear momentum, this principle is a universal conservation law, valid at all scales from atomic and nuclear systems to the motions of galaxies. This principle follows directly from Eq. (10.29): $\sum \vec{\tau} = d\vec{L}/dt$. If $\sum \vec{\tau} = 0$, then $d\vec{L}/dt = 0$, and $\vec{L}$ is constant.

> When the net external torque acting on a system is zero, the total angular momentum of the system is constant (conserved).

A circus acrobat, a diver, and an ice skater pirouetting on one skate all take advantage of this principle. Suppose an acrobat has just left a swing; she has her arms and legs extended and is rotating counterclockwise about her center of mass. When she pulls her arms and legs in, her moment of inertia I_{cm} with respect to her center of mass changes from a large value I_1 to a much smaller value I_2. The only external force acting on her is her weight, which has no torque with respect to an axis through her center of mass. So her angular momentum $L_z = I_{cm}\omega_z$ remains constant, and her angular velocity ω_z increases as I_{cm} decreases. That is,

$$I_1\omega_{1z} = I_2\omega_{2z} \qquad \text{(zero net external torque)} \qquad (10.30)$$

When a skater or ballerina spins with arms outstretched and then pulls her arms in, her angular velocity increases as her moment of inertia decreases. In each case there is conservation of angular momentum in a system in which the net external torque is zero.

When a system has several parts, the internal forces that the parts exert on one another cause changes in the angular momenta of the parts, but the *total* angular momentum doesn't change. Here's an example. Consider two bodies A and B that interact with each other but not with anything else, such as the astronauts we discussed in Section 8.2 (see Fig. 8.9). Suppose body A exerts a force $\vec{F}_{A \text{ on } B}$ on body B; the corresponding torque (with respect to whatever point we choose) is $\vec{\tau}_{A \text{ on } B}$. According to Eq. (10.29), this torque is equal to the rate of change of angular momentum of B:

$$\vec{\tau}_{A \text{ on } B} = \frac{d\vec{L}_B}{dt}$$

At the same time, body B exerts a force $\vec{F}_{B \text{ on } A}$ on body A, with a corresponding torque $\vec{\tau}_{B \text{ on } A}$, and

$$\vec{\tau}_{B \text{ on } A} = \frac{d\vec{L}_A}{dt}$$

From Newton's third law, $\vec{F}_{B \text{ on } A} = -\vec{F}_{A \text{ on } B}$. Furthermore, if the forces act along the same line, as in Fig. 10.8, their lever arms with respect to the chosen axis are equal. Thus the *torques* of these two forces are equal and opposite, and $\vec{\tau}_{B \text{ on } A} = -\vec{\tau}_{A \text{ on } B}$. So if we add the two preceding equations, we find

$$\frac{d\vec{L}_A}{dt} + \frac{d\vec{L}_B}{dt} = 0$$

or, because $\vec{L}_A + \vec{L}_B$ is the *total* angular momentum $\vec{L}$ of the system,

$$\frac{d\vec{L}}{dt} = 0 \qquad \text{(zero net external torque)} \qquad (10.31)$$

That is, the total angular momentum of the system is constant. The torques of the internal forces can transfer angular momentum from one body to the other, but they can't change the *total* angular momentum of the system (**Fig. 10.28**).

EXAMPLE 10.10   ANYONE CAN BE A BALLERINA

A physics professor stands at the center of a frictionless turntable with arms outstretched and a 5.0-kg dumbbell in each hand (**Fig. 10.29**). He is set rotating about the vertical axis, making one revolution in 2.0 s. Find his final angular velocity if he pulls the dumbbells inward to his stomach. His moment of inertia (without the dumbbells) is $3.0 \text{ kg} \cdot \text{m}^2$ with arms outstretched and $2.2 \text{ kg} \cdot \text{m}^2$ with his hands at his stomach. The dumbbells are 1.0 m from the axis initially and 0.20 m at the end.

SOLUTION

IDENTIFY, SET UP, and EXECUTE: No external torques act about the z-axis, so L_z is constant. We'll use Eq. (10.30) to find the final

10.29 Fun with conservation of angular momentum.

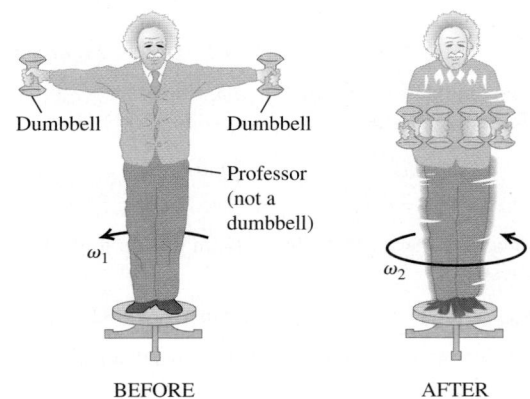

Dumbbell   Dumbbell

Professor
(not a
dumbbell)

ω_1

ω_2

BEFORE   AFTER

angular velocity ω_{2z}. The moment of inertia of the system is $I = I_{\text{prof}} + I_{\text{dumbbells}}$. We treat each dumbbell as a particle of mass m that contributes mr^2 to $I_{\text{dumbbells}}$, where r is the perpendicular distance from the axis to the dumbbell. Initially we have

$$I_1 = 3.0 \text{ kg} \cdot \text{m}^2 + 2(5.0 \text{ kg})(1.0 \text{ m})^2 = 13 \text{ kg} \cdot \text{m}^2$$

$$\omega_{1z} = \frac{1 \text{ rev}}{2.0 \text{ s}} = 0.50 \text{ rev/s}$$

The final moment of inertia is

$$I_2 = 2.2 \text{ kg} \cdot \text{m}^2 + 2(5.0 \text{ kg})(0.20 \text{ m})^2 = 2.6 \text{ kg} \cdot \text{m}^2$$

From Eq. (10.30), the final angular velocity is

$$\omega_{2z} = \frac{I_1}{I_2}\omega_{1z} = \frac{13 \text{ kg} \cdot \text{m}^2}{2.6 \text{ kg} \cdot \text{m}^2}(0.50 \text{ rev/s}) = 2.5 \text{ rev/s} = 5\omega_{1z}$$

Can you see why we didn't have to change "revolutions" to "radians" in this calculation?

EVALUATE: The angular momentum remained constant, but the angular velocity increased by a factor of 5, from $\omega_{1z} = (0.50 \text{ rev/s}) \times (2\pi \text{ rad/rev}) = 3.14 \text{ rad/s}$ to $\omega_{2z} = (2.5 \text{ rev/s})(2\pi \text{ rad/rev}) = 15.7 \text{ rad/s}$. The initial and final kinetic energies are then

$$K_1 = \tfrac{1}{2}I_1\omega_{1z}{}^2 = \tfrac{1}{2}(13 \text{ kg} \cdot \text{m}^2)(3.14 \text{ rad/s})^2 = 64 \text{ J}$$

$$K_2 = \tfrac{1}{2}I_2\omega_{2z}{}^2 = \tfrac{1}{2}(2.6 \text{ kg} \cdot \text{m}^2)(15.7 \text{ rad/s})^2 = 320 \text{ J}$$

The fivefold increase in kinetic energy came from the work that the professor did in pulling his arms and the dumbbells inward.

EXAMPLE 10.11   A ROTATIONAL "COLLISION"

Figure 10.30 shows two disks: an engine flywheel (A) and a clutch plate (B) attached to a transmission shaft. Their moments of inertia are I_A and I_B; initially, they are rotating with constant angular speeds ω_A and ω_B, respectively. We push the disks together with forces acting along the axis, so as not to apply any torque on either disk. The disks rub against each other and eventually reach a common angular speed ω. Derive an expression for ω.

SOLUTION

IDENTIFY, SET UP, and EXECUTE: There are no external torques, so the only torque acting on either disk is the torque applied by the other disk. Hence the total angular momentum of the system of two disks is conserved. At the end they rotate together as one body with total moment of inertia $I = I_A + I_B$ and angular speed ω. Figure 10.30 shows that all angular velocities are in the same direction, so we can regard ω_A, ω_B, and ω as components of angular velocity along the rotation axis. Conservation of angular momentum gives

$$I_A\omega_A + I_B\omega_B = (I_A + I_B)\omega$$

$$\omega = \frac{I_A\omega_A + I_B\omega_B}{I_A + I_B}$$

10.30 When the net external torque is zero, angular momentum is conserved.

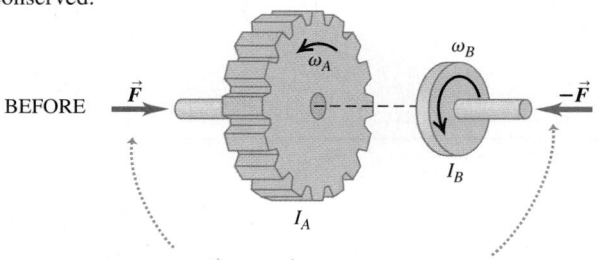

BEFORE $\vec{F}$ ω_B $-\vec{F}$

ω_A

I_B

I_A

Forces $\vec{F}$ and $-\vec{F}$ are along the axis of rotation, and thus exert no torque about this axis on either disk.

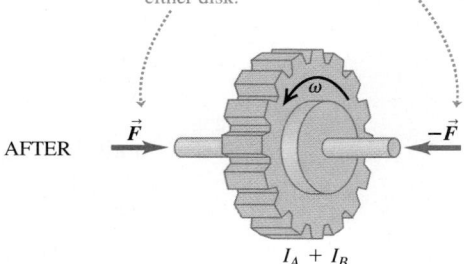

AFTER $\vec{F}$ ω $-\vec{F}$

$I_A + I_B$

Continued

EVALUATE: This "collision" is analogous to a completely inelastic collision (see Section 8.3). When two objects in translational motion along the same axis collide and stick, the linear momentum of the system is conserved. Here two objects in *rotational* motion around the same axis "collide" and stick, and the *angular* momentum of the system is conserved.

The kinetic energy of a system decreases in a completely inelastic collision. Here kinetic energy is lost because nonconservative (friction) internal forces act while the two disks rub together. Suppose flywheel *A* has a mass of 2.0 kg, a radius of 0.20 m, and an initial angular speed of 50 rad/s (about 500 rpm), and clutch plate *B* has a mass of 4.0 kg, a radius of 0.10 m, and an initial angular speed of 200 rad/s. Can you show that the final kinetic energy is only two-thirds of the initial kinetic energy?

EXAMPLE 10.12 ANGULAR MOMENTUM IN A CRIME BUST

A door 1.00 m wide, of mass 15 kg, can rotate freely about a vertical axis through its hinges. A bullet with a mass of 10 g and a speed of 400 m/s strikes the center of the door, in a direction perpendicular to the plane of the door, and embeds itself there. Find the door's angular speed. Is kinetic energy conserved?

SOLUTION

IDENTIFY and SET UP: We consider the door and bullet as a system. There is no external torque about the hinge axis, so angular momentum about this axis is conserved. **Figure 10.31** shows our sketch. The initial angular momentum is that of the bullet, as given by Eq. (10.25). The final angular momentum is that of a rigid body composed of the door and the embedded bullet. We'll equate these quantities and solve for the resulting angular speed ω of the door and bullet.

EXECUTE: From Eq. (10.25), the initial angular momentum of the bullet is

$$L = mvl = (0.010 \text{ kg})(400 \text{ m/s})(0.50 \text{ m}) = 2.0 \text{ kg} \cdot \text{m}^2/\text{s}$$

The final angular momentum is $I\omega$, where $I = I_{\text{door}} + I_{\text{bullet}}$. From Table 9.2, case (d), for a door of width $d = 1.00$ m,

$$I_{\text{door}} = \frac{Md^2}{3} = \frac{(15 \text{ kg})(1.00 \text{ m})^2}{3} = 5.0 \text{ kg} \cdot \text{m}^2$$

The moment of inertia of the bullet (with respect to the axis along the hinges) is

$$I_{\text{bullet}} = ml^2 = (0.010 \text{ kg})(0.50 \text{ m})^2 = 0.0025 \text{ kg} \cdot \text{m}^2$$

Conservation of angular momentum requires that $mvl = I\omega$, or

$$\omega = \frac{mvl}{I} = \frac{2.0 \text{ kg} \cdot \text{m}^2/\text{s}}{5.0 \text{ kg} \cdot \text{m}^2 + 0.0025 \text{ kg} \cdot \text{m}^2} = 0.40 \text{ rad/s}$$

The initial and final kinetic energies are

$$K_1 = \tfrac{1}{2}mv^2 = \tfrac{1}{2}(0.010 \text{ kg})(400 \text{ m/s})^2 = 800 \text{ J}$$

$$K_2 = \tfrac{1}{2}I\omega^2 = \tfrac{1}{2}(5.0025 \text{ kg} \cdot \text{m}^2)(0.40 \text{ rad/s})^2 = 0.40 \text{ J}$$

EVALUATE: The final kinetic energy is only $\frac{1}{2000}$ of the initial value! We did not expect kinetic energy to be conserved: The collision is inelastic because nonconservative friction forces act during the impact. The door's final angular speed is quite slow: At 0.40 rad/s, it takes 3.9 s to swing through 90° ($\pi/2$ radians).

10.31 The swinging door seen from above.

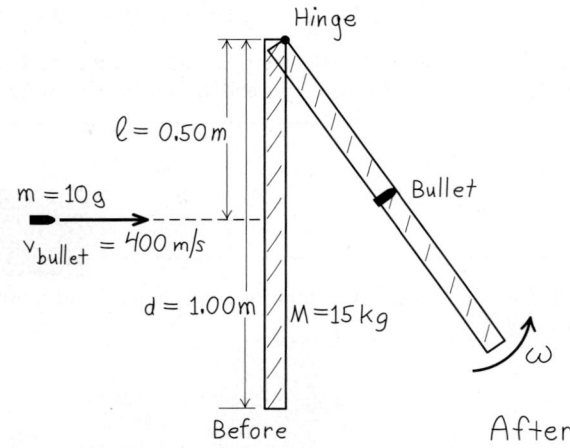

TEST YOUR UNDERSTANDING OF SECTION 10.6 If the polar ice caps were to melt completely due to global warming, the melted ice would redistribute itself over the earth. This change would cause the length of the day (the time needed for the earth to rotate once on its axis) to (i) increase; (ii) decrease; (iii) remain the same. (*Hint:* Use angular momentum ideas. Assume that the sun, moon, and planets exert negligibly small torques on the earth.) ▌

10.7 GYROSCOPES AND PRECESSION

In all the situations we've looked at so far in this chapter, the axis of rotation either has stayed fixed or has moved and kept the same direction (such as rolling without slipping). But a variety of new physical phenomena, some quite unexpected, can occur when the axis of rotation changes direction. For example, consider a toy gyroscope that's supported at one end (**Fig. 10.32**). If we hold it with the flywheel axis horizontal and let go, the free end of the axis simply drops owing to gravity—*if* the flywheel isn't spinning. But if the flywheel *is* spinning, what happens is quite different. One possible motion is a steady circular motion of the axis in a horizontal plane, combined with the spin motion of the flywheel about the axis. This surprising, nonintuitive motion of the axis is called **precession.** Precession is found in nature as well as in rotating machines such as gyroscopes. As you read these words, the earth itself is precessing; its spin axis (through the north and south poles) slowly changes direction, going through a complete cycle of precession every 26,000 years.

To study this strange phenomenon of precession, we must remember that angular velocity, angular momentum, and torque are all *vector* quantities. In particular, we need the general relationship between the net torque $\sum\vec{\tau}$ that acts on a body and the rate of change of the body's angular momentum $\vec{L}$, given by Eq. (10.29), $\sum\vec{\tau} = d\vec{L}/dt$. Let's first apply this equation to the case in which the flywheel is *not* spinning (**Fig. 10.33a**). We take the origin O at the pivot and assume that the flywheel is symmetrical, with mass M and moment of inertia I about the flywheel axis. The flywheel axis is initially along the x-axis. The only external forces on the gyroscope are the normal force $\vec{n}$ acting at the pivot (assumed to be friction-less) and the weight $\vec{w}$ of the flywheel that acts at its center of mass, a distance r from the pivot. The normal force has zero torque with respect to the pivot, and the weight has a torque $\vec{\tau}$ in the y-direction, as shown in Fig. 10.33a. Initially, there is no rotation, and the initial angular momentum $\vec{L}_i$ is zero. From Eq. (10.29) the *change* $d\vec{L}$ in angular momentum in a short time interval dt following this is

$$d\vec{L} = \vec{\tau}\,dt \qquad (10.32)$$

This change is in the y-direction because $\vec{\tau}$ is. As each additional time interval dt elapses, the angular momentum changes by additional increments $d\vec{L}$ in the y-direction because the direction of the torque is constant (Fig. 10.33b). The steadily increasing horizontal angular momentum means that the gyroscope rotates downward faster and faster around the y-axis until it hits either the stand or the table on which it sits.

Now let's see what happens if the flywheel *is* spinning initially, so the initial angular momentum $\vec{L}_i$ is not zero (**Fig. 10.34a**). Since the flywheel rotates around its symmetry axis, $\vec{L}_i$ lies along this axis. But each change in angular momentum $d\vec{L}$ is perpendicular to the flywheel axis because the torque $\vec{\tau} = \vec{r} \times \vec{w}$ is perpendicular to that axis (Fig. 10.34b). This causes the *direction* of $\vec{L}$ to change, but not its magnitude. The changes $d\vec{L}$ are always in the horizontal xy-plane, so the angular momentum vector and the flywheel axis with which it moves are always horizontal. That is, the axis doesn't fall—it precesses.

10.34 (a) The flywheel is spinning initially with angular momentum $\vec{L}_i$. The forces (not shown) are the same as those in Fig. 10.33a. (b) Because the initial angular momentum is not zero, each change $d\vec{L} = \vec{\tau}\,dt$ in angular momentum is perpendicular to $\vec{L}$. As a result, the magnitude of $\vec{L}$ remains the same but its direction changes continuously.

(a) Rotating flywheel

When the flywheel is rotating, the system starts with an angular momentum $\vec{L}_i$ parallel to the flywheel's axis of rotation.

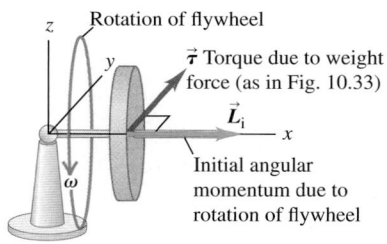

(b) View from above

Now the effect of the torque is to cause the angular momentum to precess around the pivot. The gyroscope circles around its pivot without falling.

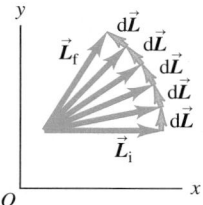

10.32 A gyroscope supported at one end. The horizontal circular motion of the flywheel and axis is called precession. The angular speed of precession is Ω.

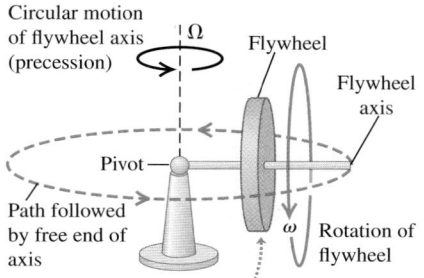

When the flywheel and its axis are stationary, they will fall to the table surface. When the flywheel spins, it and its axis "float" in the air while moving in a circle about the pivot.

10.33 (a) If the flywheel in Fig. 10.32 is initially not spinning, its initial angular momentum is zero. (b) In each successive time interval dt, the torque produces a change $d\vec{L} = \vec{\tau}\,dt$ in the angular momentum. The flywheel acquires an angular momentum $\vec{L}$ in the same direction as $\vec{\tau}$, and the flywheel axis falls.

(a) Nonrotating flywheel falls

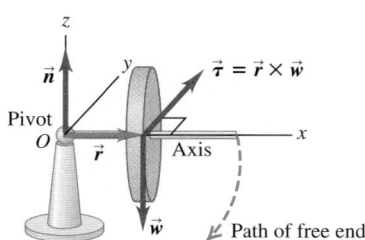

When the flywheel is not rotating, its weight creates a torque around the pivot, causing it to fall along a circular path until its axis rests on the table surface.

(b) View from above as flywheel falls

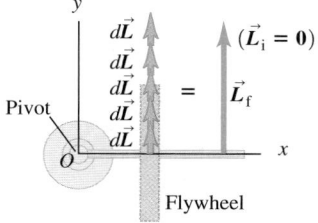

In falling, the flywheel rotates about the pivot and thus acquires an angular momentum $\vec{L}$. The *direction* of $\vec{L}$ stays constant.

If this still seems mystifying to you, think about a ball attached to a string. If the ball is initially at rest and you pull the string toward you, the ball moves toward you also. But if the ball is initially moving and you continuously pull the string in a direction perpendicular to the ball's motion, the ball moves in a circle around your hand; it does not approach your hand at all. In the first case the ball has zero linear momentum $\vec{p}$ to start with; when you apply a force $\vec{F}$ toward you for a time dt, the ball acquires a momentum $d\vec{p} = \vec{F}\,dt$, which is also toward you. But if the ball already has linear momentum $\vec{p}$, a change in momentum $d\vec{p}$ that's perpendicular to $\vec{p}$ changes the direction of motion, not the speed. Replace $\vec{p}$ with $\vec{L}$ and $\vec{F}$ with $\vec{\tau}$ in this argument, and you'll see that precession is simply the rotational analog of uniform circular motion.

At the instant shown in Fig. 10.34a, the gyroscope has angular momentum $\vec{L}$. A short time interval dt later, the angular momentum is $\vec{L} + d\vec{L}$; the infinitesimal change in angular momentum is $d\vec{L} = \vec{\tau}\,dt$, which is perpendicular to $\vec{L}$. As the vector diagram in **Fig. 10.35** shows, this means that the flywheel axis of the gyroscope has turned through a small angle $d\phi$ given by $d\phi = |d\vec{L}|/|\vec{L}|$. The rate at which the axis moves, $d\phi/dt$, is called the **precession angular speed;** denoting this quantity by Ω, we find

$$\Omega = \frac{d\phi}{dt} = \frac{|d\vec{L}|/|\vec{L}|}{dt} = \frac{\tau_z}{L_z} = \frac{wr}{I\omega} \qquad (10.33)$$

Thus the precession angular speed is *inversely* proportional to the angular speed of spin about the axis. A rapidly spinning gyroscope precesses slowly; if friction in its bearings causes the flywheel to slow down, the precession angular speed *increases*! The precession angular speed of the earth is very slow (1 rev/26,000 yr) because its spin angular momentum L_z is large and the torque τ_z, due to the gravitational influences of the moon and sun, is relatively small.

As a gyroscope precesses, its center of mass moves in a circle with radius r in a horizontal plane. Its vertical component of acceleration is zero, so the upward normal force $\vec{n}$ exerted by the pivot must be just equal in magnitude to the weight. The circular motion of the center of mass with angular speed Ω requires a force $\vec{F}$ directed toward the center of the circle, with magnitude $F = M\Omega^2 r$. This force must also be supplied by the pivot.

One key assumption that we made in our analysis of the gyroscope was that the angular momentum vector $\vec{L}$ is associated with only the spin of the flywheel and is purely horizontal. But there will also be a vertical component of angular momentum associated with the precessional motion of the gyroscope. By ignoring this, we've tacitly assumed that the precession is *slow*—that is, that the precession angular speed Ω is very much less than the spin angular speed ω. As Eq. (10.33) shows, a large value of ω automatically gives a small value of Ω, so this approximation is reasonable. When the precession is not slow, additional effects show up, including an up-and-down wobble or *nutation* of the flywheel axis that's superimposed on the precessional motion. You can see nutation occurring in a gyroscope as its spin slows down, so that Ω increases and the vertical component of $\vec{L}$ can no longer be ignored.

10.35 Detailed view of part of Fig. 10.34b.

In a time dt, the angular momentum vector and the flywheel axis (to which it is parallel) precess together through an angle $d\phi$.

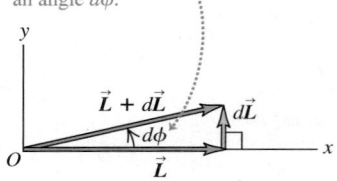

EXAMPLE 10.13   A PRECESSING GYROSCOPE

Figure 10.36a shows a top view of a spinning, cylindrical gyroscope wheel. The pivot is at O, and the mass of the axle is negligible. (a) As seen from above, is the precession clockwise or counterclockwise? (b) If the gyroscope takes 4.0 s for one revolution of precession, what is the angular speed of the wheel?

SOLUTION

IDENTIFY and SET UP: We'll determine the direction of precession by using the right-hand rule as in Fig. 10.34, which shows the same kind of gyroscope as Fig. 10.36. We'll use the relationship

10.36 In which direction and at what speed does this gyroscope precess?

(a) Top view

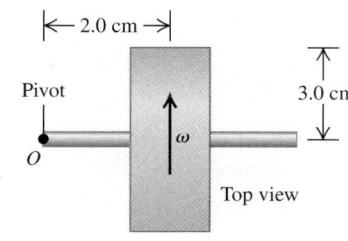

(b) Vector diagram

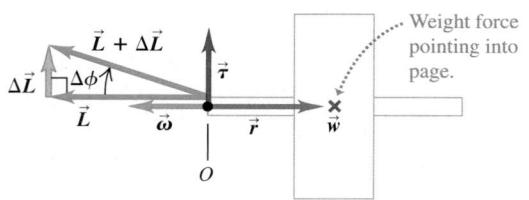

between precession angular speed Ω and spin angular speed ω, Eq. (10.33), to find ω.

EXECUTE: (a) The right-hand rule shows that $\vec{\omega}$ and $\vec{L}$ are to the left in Fig. 10.36b. The weight $\vec{w}$ points into the page in this top view and acts at the center of mass (denoted by $\times$ in the figure). The torque $\vec{\tau} = \vec{r} \times \vec{w}$ is toward the top of the page, so $d\vec{L}/dt$ is also toward the top of the page. Adding a small $d\vec{L}$ to the initial vector $\vec{L}$ changes the direction of $\vec{L}$ as shown, so the precession is clockwise as seen from above.

(b) Be careful not to confuse ω and Ω! The precession angular speed is $\Omega = (1\text{ rev})/(4.0\text{ s}) = (2\pi\text{ rad})/(4.0\text{ s}) = 1.57\text{ rad/s}$. The weight is mg, and if the wheel is a solid, uniform cylinder, its moment of inertia about its symmetry axis is $I = \frac{1}{2}mR^2$. From Eq. (10.33),

$$\omega = \frac{wr}{I\Omega} = \frac{mgr}{(mR^2/2)\Omega} = \frac{2gr}{R^2\Omega}$$

$$= \frac{2(9.8\text{ m/s}^2)(2.0 \times 10^{-2}\text{ m})}{(3.0 \times 10^{-2}\text{ m})^2(1.57\text{ rad/s})}$$

$$= 280\text{ rad/s} = 2600\text{ rev/min}$$

EVALUATE: The precession angular speed Ω is only about 0.6% of the spin angular speed ω, so this is an example of slow precession.

TEST YOUR UNDERSTANDING OF SECTION 10.7   Suppose the mass of the flywheel in Fig. 10.34 is doubled but all other dimensions and the spin angular speed remain the same. What effect would this change have on the precession angular speed Ω? (i) Ω would increase by a factor of 4; (ii) Ω would double; (iii) Ω would be unaffected; (iv) Ω would be one-half as much; (v) Ω would be one-quarter as much. ∎

Torque: When a force $\vec{F}$ acts on a body, the torque of that force with respect to a point O has a magnitude given by the product of the force magnitude F and the lever arm l. More generally, torque is a vector $\vec{\tau}$ equal to the vector product of $\vec{r}$ (the position vector of the point at which the force acts) and $\vec{F}$. (See Example 10.1.)

$$\tau = Fl = rF\sin\phi = F_{\text{tan}}r \quad (10.2)$$

$$\vec{\tau} = \vec{r} \times \vec{F} \quad (10.3)$$

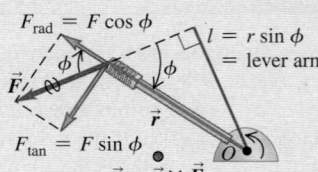

Rotational dynamics: The rotational analog of Newton's second law says that the net torque acting on a body equals the product of the body's moment of inertia and its angular acceleration. (See Examples 10.2 and 10.3.)

$$\sum \tau_z = I\alpha_z \quad (10.7)$$

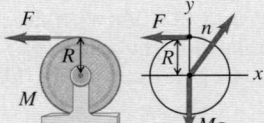

Combined translation and rotation: If a rigid body is both moving through space and rotating, its motion can be regarded as translational motion of the center of mass plus rotational motion about an axis through the center of mass. Thus the kinetic energy is a sum of translational and rotational kinetic energies. For dynamics, Newton's second law describes the motion of the center of mass, and the rotational equivalent of Newton's second law describes rotation about the center of mass. In the case of rolling without slipping, there is a special relationship between the motion of the center of mass and the rotational motion. (See Examples 10.4–10.7.)

$$K = \tfrac{1}{2}Mv_{\text{cm}}^2 + \tfrac{1}{2}I_{\text{cm}}\omega^2 \quad (10.8)$$

$$\sum \vec{F}_{\text{ext}} = M\vec{a}_{\text{cm}} \quad (10.12)$$

$$\sum \tau_z = I_{\text{cm}}\alpha_z \quad (10.13)$$

$$v_{\text{cm}} = R\omega \quad (10.11)$$
(rolling without slipping)

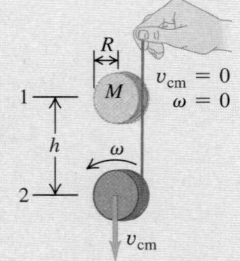

Work done by a torque: A torque that acts on a rigid body as it rotates does work on that body. The work can be expressed as an integral of the torque. The work–energy theorem says that the total rotational work done on a rigid body is equal to the change in rotational kinetic energy. The power, or rate at which the torque does work, is the product of the torque and the angular velocity (See Example 10.8.)

$$W = \int_{\theta_1}^{\theta_2} \tau_z \, d\theta \quad (10.20)$$

$$W = \tau_z(\theta_2 - \theta_1) = \tau_z \Delta\theta \quad (10.21)$$
(constant torque only)

$$W_{\text{tot}} = \tfrac{1}{2}I\omega_2^2 - \tfrac{1}{2}I\omega_1^2 \quad (10.22)$$

$$P = \tau_z\omega_z \quad (10.23)$$

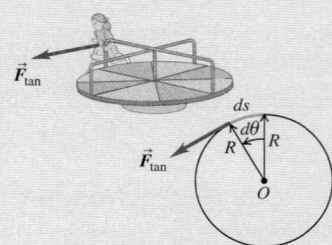

Angular momentum: The angular momentum of a particle with respect to point O is the vector product of the particle's position vector $\vec{r}$ relative to O and its momentum $\vec{p} = m\vec{v}$. When a symmetrical body rotates about a stationary axis of symmetry, its angular momentum is the product of its moment of inertia and its angular velocity vector $\vec{\omega}$. If the body is not symmetrical or the rotation (z) axis is not an axis of symmetry, the component of angular momentum along the rotation axis is $I\omega_z$. (See Example 10.9.)

$$\vec{L} = \vec{r} \times \vec{p} = \vec{r} \times m\vec{v} \quad (10.24)$$
(particle)

$$\vec{L} = I\vec{\omega} \quad (10.28)$$
(rigid body rotating about axis of symmetry)

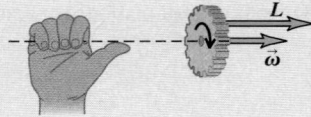

Rotational dynamics and angular momentum: The net external torque on a system is equal to the rate of change of its angular momentum. If the net external torque on a system is zero, the total angular momentum of the system is constant (conserved). (See Examples 10.10–10.13.)

$$\sum \vec{\tau} = \frac{d\vec{L}}{dt} \quad (10.29)$$

BRIDGING PROBLEM   BILLIARD PHYSICS

A cue ball (a uniform solid sphere of mass m and radius R) is at rest on a level pool table. Using a pool cue, you give the ball a sharp, horizontal hit of magnitude F at a height h above the center of the ball (**Fig. 10.37**). The force of the hit is much greater than the friction force f that the table surface exerts on the ball. The hit lasts for a short time Δt. (a) For what value of h will the ball roll without slipping? (b) If you hit the ball dead center ($h = 0$), the ball will slide across the table for a while, but eventually it will roll without slipping. What will the speed of its center of mass be then?

10.37

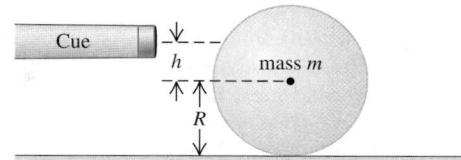

SOLUTION GUIDE

IDENTIFY and SET UP

1. Draw a free-body diagram for the ball for the situation in part (a), including your choice of coordinate axes. Note that the cue exerts both an impulsive force on the ball and an impulsive torque around the center of mass.
2. The cue force applied for a time Δt gives the ball's center of mass a speed v_{cm}, and the cue torque applied for that same time gives the ball an angular speed ω. How must v_{cm} and ω be related for the ball to roll without slipping?
3. Draw two free-body diagrams for the ball in part (b): one showing the forces during the hit and the other showing the forces after the hit but before the ball is rolling without slipping.
4. What is the angular speed of the ball in part (b) just after the hit? While the ball is sliding, does v_{cm} increase or decrease? Does ω increase or decrease? What is the relationship between v_{cm} and ω when the ball is finally rolling without slipping?

EXECUTE

5. In part (a), use the impulse–momentum theorem to find the speed of the ball's center of mass immediately after the hit.

Then use the rotational version of the impulse–momentum theorem to find the angular speed immediately after the hit. (*Hint:* To write the rotational version of the impulse–momentum theorem, remember that the relationship between torque and angular momentum is the same as that between force and linear momentum.)
6. Use your results from step 5 to find the value of h that will cause the ball to roll without slipping immediately after the hit.
7. In part (b), again find the ball's center-of-mass speed and angular speed immediately after the hit. Then write Newton's second law for the translational motion and rotational motion of the ball as it slides. Use these equations to write expressions for v_{cm} and ω as functions of the elapsed time t since the hit.
8. Using your results from step 7, find the time t when v_{cm} and ω have the correct relationship for rolling without slipping. Then find the value of v_{cm} at this time.

EVALUATE

9. If you have access to a pool table, test the results of parts (a) and (b) for yourself!
10. Can you show that if you used a hollow cylinder rather than a solid ball, you would have to hit the top of the cylinder to cause rolling without slipping as in part (a)?

Problems

For assigned homework and other learning materials, go to MasteringPhysics®. (MP)

°, °°, °°°: Difficulty levels. **CP**: Cumulative problems incorporating material from earlier chapters. **CALC**: Problems requiring calculus. **DATA**: Problems involving real data, scientific evidence, experimental design, and/or statistical reasoning. **BIO**: Biosciences problems.

DISCUSSION QUESTIONS

Q10.1 Can a single force applied to a body change both its translational and rotational motions? Explain.

Q10.2 Suppose you could use wheels of any type in the design of a soapbox-derby racer (an unpowered, four-wheel vehicle that coasts from rest down a hill). To conform to the rules on the total weight of the vehicle and rider, should you design with large massive wheels or small light wheels? Should you use solid wheels or wheels with most of the mass at the rim? Explain.

Q10.3 Serious bicyclists say that if you reduce the weight of a bike, it is more effective if you do so in the wheels rather than in the frame. Why would reducing weight in the wheels make it easier on the bicyclist than reducing the same amount in the frame?

Q10.4 The harder you hit the brakes while driving forward, the more the front end of your car will move down (and the rear end move up). Why? What happens when cars accelerate forward? Why do drag racers not use front-wheel drive only?

Q10.5 When an acrobat walks on a tightrope, she extends her arms straight out from her sides. She does this to make it easier for her to catch herself if she should tip to one side or the other. Explain how this works. [*Hint:* Think about Eq. (10.7).]

Q10.6 When you turn on an electric motor, it takes longer to come up to final speed if a grinding wheel is attached to the shaft. Why?

Q10.7 The work done by a force is the product of force and distance. The torque due to a force is the product of force and distance. Does this mean that torque and work are equivalent? Explain.

Q10.8 A valued client brings a treasured ball to your engineering firm, wanting to know whether the ball is solid or hollow. He has tried tapping on it, but that has given insufficient information. Design a simple, inexpensive experiment that you could perform quickly, without injuring the precious ball, to find out whether it is solid or hollow.

Q10.9 You make two versions of the same object out of the same material having uniform density. For one version, all the dimensions are exactly twice as great as for the other one. If the same torque acts on both versions, giving the smaller version angular acceleration α, what will be the angular acceleration of the larger version in terms of α?

Q10.10 Two identical masses are attached to frictionless pulleys by very light strings wrapped around the rim of the pulley and are released from rest. Both pulleys have the same mass and same diameter, but one is solid and the other is a hoop. As the masses fall, in which case is the tension in the string greater, or is it the same in both cases? Justify your answer.

Q10.11 The force of gravity acts on the baton in Fig. 10.11, and forces produce torques that cause a body's angular velocity to change. Why, then, is the angular velocity of the baton in the figure constant?

Q10.12 A certain solid uniform ball reaches a maximum height h_0 when it rolls up a hill without slipping. What maximum height (in terms of h_0) will it reach if you (a) double its diameter, (b) double its mass, (c) double both its diameter and mass, (d) double its angular speed at the bottom of the hill?

Q10.13 A wheel is rolling without slipping on a horizontal surface. In an inertial frame of reference in which the surface is at rest, is there any point on the wheel that has a velocity that is purely vertical? Is there any point that has a horizontal velocity component opposite to the velocity of the center of mass? Explain. Do your answers change if the wheel is slipping as it rolls? Why or why not?

Q10.14 A hoop, a uniform solid cylinder, a spherical shell, and a uniform solid sphere are released from rest at the top of an incline. What is the order in which they arrive at the bottom of the incline? Does it matter whether or not the masses and radii of the objects are all the same? Explain.

Q10.15 A ball is rolling along at speed v without slipping on a horizontal surface when it comes to a hill that rises at a constant angle above the horizontal. In which case will it go higher up the hill: if the hill has enough friction to prevent slipping or if the hill is perfectly smooth? Justify your answers in both cases in terms of energy conservation and in terms of Newton's second law.

Q10.16 You are standing at the center of a large horizontal turntable in a carnival funhouse. The turntable is set rotating on frictionless bearings, and it rotates freely (that is, there is no motor driving the turntable). As you walk toward the edge of the turntable, what happens to the combined angular momentum of you and the turntable? What happens to the rotation speed of the turntable? Explain.

Q10.17 Global Warming. If the earth's climate continues to warm, ice near the poles will melt, and the water will be added to the oceans. What effect will this have on the length of the day? Justify your answer. (*Hint:* Consult a map to see where the oceans lie.)

Q10.18 If two spinning objects have the same angular momentum, do they necessarily have the same rotational kinetic energy? If they have the same rotational kinetic energy, do they necessarily have the same angular momentum? Explain.

Q10.19 A student is sitting on a frictionless rotating stool with her arms outstretched as she holds equal heavy weights in each hand. If she suddenly lets go of the weights, will her angular speed increase, stay the same, or decrease? Explain.

Q10.20 A point particle travels in a straight line at constant speed, and the closest distance it comes to the origin of coordinates is a distance l. With respect to this origin, does the particle have nonzero angular momentum? As the particle moves along its straight-line path, does its angular momentum with respect to the origin change?

Q10.21 In Example 10.10 (Section 10.6) the angular speed ω changes, and this must mean that there is nonzero angular acceleration. But there is no torque about the rotation axis if the forces the professor applies to the weights are directly, radially inward. Then, by Eq. (10.7), α_z must be zero. Explain what is wrong with this reasoning that leads to this apparent contradiction.

Q10.22 In Example 10.10 (Section 10.6) the rotational kinetic energy of the professor and dumbbells increases. But since there are no external torques, no work is being done to change the rotational kinetic energy. Then, by Eq. (10.22), the kinetic energy must remain the same! Explain what is wrong with this reasoning, which leads to an apparent contradiction. Where *does* the extra kinetic energy come from?

Q10.23 As discussed in Section 10.6, the angular momentum of a circus acrobat is conserved as she tumbles through the air. Is her *linear* momentum conserved? Why or why not?

Q10.24 If you stop a spinning raw egg for the shortest possible instant and then release it, the egg will start spinning again. If you do the same to a hard-boiled egg, it will remain stopped. Try it. Explain it.

Q10.25 A helicopter has a large main rotor that rotates in a horizontal plane and provides lift. There is also a small rotor on the tail that rotates in a vertical plane. What is the purpose of the tail rotor? (*Hint:* If there were no tail rotor, what would happen when the pilot changed the angular speed of the main rotor?) Some helicopters have no tail rotor, but instead have two large main rotors that rotate in a horizontal plane. Why is it important that the two main rotors rotate in opposite directions?

Q10.26 In a common design for a gyroscope, the flywheel and flywheel axis are enclosed in a light, spherical frame with the flywheel at the center of the frame. The gyroscope is then balanced on top of a pivot so that the flywheel is directly above the pivot. Does the gyroscope precess if it is released while the flywheel is spinning? Explain.

Q10.27 A gyroscope is precessing about a vertical axis. What happens to the precession angular speed if the following changes are made, with all other variables remaining the same? (a) The angular speed of the spinning flywheel is doubled; (b) the total weight is doubled; (c) the moment of inertia about the axis of the spinning flywheel is doubled; (d) the distance from the pivot to the center of gravity is doubled. (e) What happens if all of the variables in parts (a) through (d) are doubled? In each case justify your answer.

Q10.28 A gyroscope takes 3.8 s to precess 1.0 revolution about a vertical axis. Two minutes later, it takes only 1.9 s to precess 1.0 revolution. No one has touched the gyroscope. Explain.

Q10.29 A gyroscope is precessing as in Fig. 10.32. What happens if you gently add some weight to the end of the flywheel axis farthest from the pivot?

Q10.30 A bullet spins on its axis as it emerges from a rifle. Explain how this prevents the bullet from tumbling and keeps the streamlined end pointed forward.

EXERCISES

Section 10.1 Torque

10.1 • Calculate the torque (magnitude and direction) about point O due to the force $\vec{F}$ in each of the cases sketched in **Fig. E10.1**. In each case, both the force $\vec{F}$ and the rod lie in the plane of the page, the rod has length 4.00 m, and the force has magnitude $F = 10.0$ N.

Figure **E10.1**

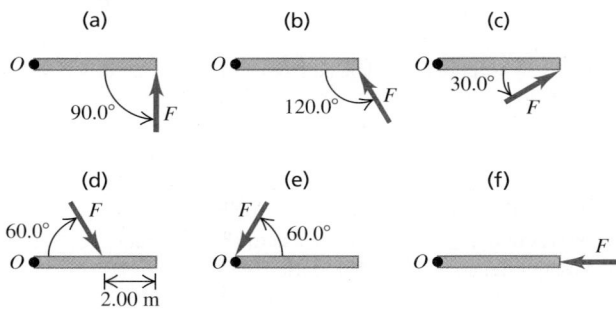

10.2 • Calculate the net torque about point O for the two forces applied as in **Fig. E10.2**. The rod and both forces are in the plane of the page.

Figure **E10.2**

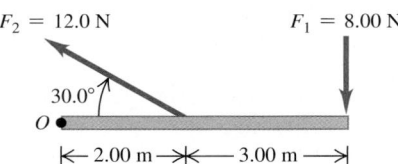

10.3 •• A square metal plate 0.180 m on each side is pivoted about an axis through point O at its center and perpendicular to the plate (**Fig. E10.3**). Calculate the net torque about this axis due to the three forces shown in the figure if the magnitudes of the forces are $F_1 = 18.0$ N, $F_2 = 26.0$ N, and $F_3 = 14.0$ N. The plate and all forces are in the plane of the page.

Figure **E10.3**

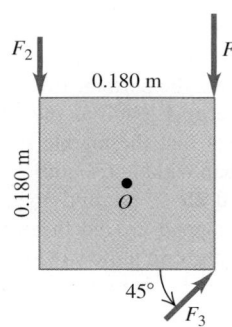

10.4 • Three forces are applied to a wheel of radius 0.350 m, as shown in **Fig. E10.4**. One force is perpendicular to the rim, one is tangent to it, and the other one makes a 40.0° angle with the radius. What is the net torque on the wheel due to these three forces for an axis perpendicular to the wheel and passing through its center?

Figure **E10.4**

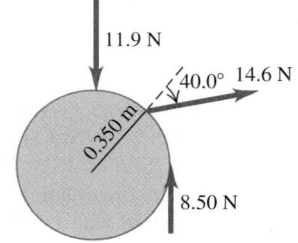

10.5 • One force acting on a machine part is $\vec{F} = (-5.00 \text{ N})\hat{\imath} + (4.00 \text{ N})\hat{\jmath}$. The vector from the origin to the point where the force is applied is $\vec{r} = (-0.450 \text{ m})\hat{\imath} + (0.150 \text{ m})\hat{\jmath}$. (a) In a sketch, show $\vec{r}$, $\vec{F}$, and the origin. (b) Use the right-hand rule to determine the direction of the torque. (c) Calculate the vector torque for an axis at the origin produced by this force. Verify that the direction of the torque is the same as you obtained in part (b).

10.6 • A metal bar is in the xy-plane with one end of the bar at the origin. A force $\vec{F} = (7.00 \text{ N})\hat{\imath} + (-3.00 \text{ N})\hat{\jmath}$ is applied to the bar at the point $x = 3.00$ m, $y = 4.00$ m. (a) In terms of unit vectors $\hat{\imath}$ and $\hat{\jmath}$, what is the position vector $\vec{r}$ for the point where the force is applied? (b) What are the magnitude and direction of the torque with respect to the origin produced by $\vec{F}$?

10.7 • A machinist is using a wrench to loosen a nut. The wrench is 25.0 cm long, and he exerts a 17.0-N force at the end of the handle at 37° with the handle (**Fig. E10.7**). (a) What torque does the machinist exert about the center of the nut? (b) What is the maximum torque he could exert with this force, and how should the force be oriented?

Figure **E10.7**

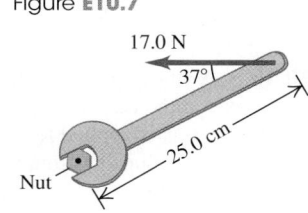

Section 10.2 Torque and Angular Acceleration for a Rigid Body

10.8 •• A uniform disk with mass 40.0 kg and radius 0.200 m is pivoted at its center about a horizontal, frictionless axle that is stationary. The disk is initially at rest, and then a constant force $F = 30.0$ N is applied tangent to the rim of the disk. (a) What is the magnitude v of the tangential velocity of a point on the rim of the disk after the disk has turned through 0.200 revolution? (b) What is the magnitude a of the resultant acceleration of a point on the rim of the disk after the disk has turned through 0.200 revolution?

10.9 •• The flywheel of an engine has moment of inertia $1.60 \text{ kg} \cdot \text{m}^2$ about its rotation axis. What constant torque is required to bring it up to an angular speed of 400 rev/min in 8.00 s, starting from rest?

10.10 • A cord is wrapped around the rim of a solid uniform wheel 0.250 m in radius and of mass 9.20 kg. A steady horizontal pull of 40.0 N to the right is exerted on the cord, pulling it off tangentially from the wheel. The wheel is mounted on frictionless bearings on a horizontal axle through its center. (a) Compute the angular acceleration of the wheel and the acceleration of the part of the cord that has already been pulled off the wheel. (b) Find the magnitude and direction of the force that the axle exerts on the wheel. (c) Which of the answers in parts (a) and (b) would change if the pull were upward instead of horizontal?

10.11 •• A machine part has the shape of a solid uniform sphere of mass 225 g and diameter 3.00 cm. It is spinning about a frictionless axle through its center, but at one point on its equator it is scraping against metal, resulting in a friction force of 0.0200 N at that point. (a) Find its angular acceleration. (b) How long will it take to decrease its rotational speed by 22.5 rad/s?

10.12 •• CP A stone is suspended from the free end of a wire that is wrapped around the outer rim of a pulley, similar to what is shown in Fig. 10.10. The pulley is a uniform disk with mass 10.0 kg and radius 30.0 cm and turns on frictionless bearings. You measure that the stone travels 12.6 m in the first 3.00 s starting from rest. Find (a) the mass of the stone and (b) the tension in the wire.

10.13 •• CP A 2.00-kg textbook rests on a frictionless, horizontal surface. A cord attached to the book passes over a pulley whose diameter is 0.150 m, to a hanging book with mass 3.00 kg. The system is released from rest, and the books are observed to move 1.20 m in 0.800 s. (a) What is the tension in each part of the cord? (b) What is the moment of inertia of the pulley about its rotation axis?

10.14 •• CP A 15.0-kg bucket of water is suspended by a very light rope wrapped around a solid uniform cylinder 0.300 m in diameter with mass 12.0 kg. The cylinder pivots on a frictionless axle through its center. The bucket is released from rest at the top of a well and falls 10.0 m to the water. (a) What is the tension in the rope while the bucket is falling? (b) With what speed does the bucket strike the water? (c) What is the time of fall? (d) While the bucket is falling, what is the force exerted on the cylinder by the axle?

10.15 • A wheel rotates without friction about a stationary horizontal axis at the center of the wheel. A constant tangential force equal to 80.0 N is applied to the rim of the wheel. The wheel has radius 0.120 m. Starting from rest, the wheel has an angular speed of 12.0 rev/s after 2.00 s. What is the moment of inertia of the wheel?

10.16 •• A 12.0-kg box resting on a horizontal, frictionless surface is attached to a 5.00-kg weight by a thin, light wire that passes over a frictionless pulley (**Fig. E10.16**). The pulley has the shape of a uniform solid disk of mass 2.00 kg and diameter 0.500 m. After the system is released, find (a) the tension in the wire on both sides of the pulley, (b) the acceleration of the box, and (c) the horizontal and vertical components of the force that the axle exerts on the pulley.

Figure **E10.16**

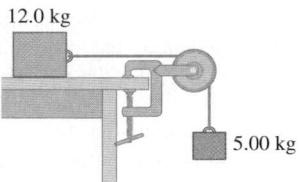

12.0 kg

5.00 kg

Section 10.3 Rigid-Body Rotation About a Moving Axis

10.17 • A 2.20-kg hoop 1.20 m in diameter is rolling to the right without slipping on a horizontal floor at a steady 2.60 rad/s. (a) How fast is its center moving? (b) What is the total kinetic energy of the hoop? (c) Find the velocity vector of each of the following points, as viewed by a person at rest on the ground: (i) the highest point on the hoop; (ii) the lowest point on the hoop; (iii) a point on the right side of the hoop, midway between the top and the bottom. (d) Find the velocity vector for each of the points in part (c), but this time as viewed by someone moving along with the same velocity as the hoop.

10.18 • BIO **Gymnastics.** We can roughly model a gymnastic tumbler as a uniform solid cylinder of mass 75 kg and diameter 1.0 m. If this tumbler rolls forward at 0.50 rev/s, (a) how much total kinetic energy does he have, and (b) what percent of his total kinetic energy is rotational?

10.19 • What fraction of the total kinetic energy is rotational for the following objects rolling without slipping on a horizontal surface? (a) A uniform solid cylinder; (b) a uniform sphere; (c) a thin-walled, hollow sphere; (d) a hollow cylinder with outer radius R and inner radius $R/2$.

10.20 •• A string is wrapped several times around the rim of a small hoop with radius 8.00 cm and mass 0.180 kg. The free end of the string is held in place and the hoop is released from rest (**Fig. E10.20**). After the hoop has descended 75.0 cm, calculate (a) the angular speed of the rotating hoop and (b) the speed of its center.

Figure **E10.20**

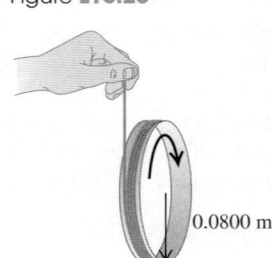

0.0800 m

10.21 •• A solid ball is released from rest and slides down a hillside that slopes downward at 65.0° from the horizontal. (a) What minimum value must the coefficient of static friction between the hill and ball surfaces have for no slipping to occur? (b) Would the coefficient of friction calculated in part (a) be sufficient to prevent a hollow ball (such as a soccer ball) from slipping? Justify your answer. (c) In part (a), why did we use the coefficient of static friction and not the coefficient of kinetic friction?

10.22 •• A hollow, spherical shell with mass 2.00 kg rolls without slipping down a 38.0° slope. (a) Find the acceleration, the friction force, and the minimum coefficient of friction needed to prevent slipping. (b) How would your answers to part (a) change if the mass were doubled to 4.00 kg?

10.23 •• A 392-N wheel comes off a moving truck and rolls without slipping along a highway. At the bottom of a hill it is rotating at 25.0 rad/s. The radius of the wheel is 0.600 m, and its moment of inertia about its rotation axis is $0.800MR^2$. Friction does work on the wheel as it rolls up the hill to a stop, a height h above the bottom of the hill; this work has absolute value 2600 J. Calculate h.

10.24 •• A uniform marble rolls down a symmetrical bowl, starting from rest at the top of the left side. The top of each side is a distance h above the bottom of the bowl. The left half of the bowl is rough enough to cause the marble to roll without slipping, but the right half has no friction because it is coated with oil. (a) How far up the smooth side will the marble go, measured vertically from the bottom? (b) How high would the marble go if both sides were as rough as the left side? (c) How do you account for the fact that the marble goes *higher* with friction on the right side than without friction?

10.25 •• A thin, light string is wrapped around the outer rim of a uniform hollow cylinder of mass 4.75 kg having inner and outer radii as shown in **Fig. E10.25**. The cylinder is then released from rest. (a) How far must the cylinder fall before its center is moving at 6.66 m/s? (b) If you just dropped this cylinder without any string, how fast would its center be moving when it had fallen the distance in part (a)? (c) Why do you get two different answers when the cylinder falls the same distance in both cases?

Figure **E10.25**

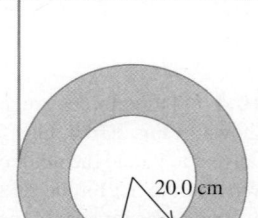

20.0 cm

35.0 cm

10.26 •• **A Ball Rolling Uphill.** A bowling ball rolls without slipping up a ramp that slopes upward at an angle β to the horizontal (see Example 10.7 in Section 10.3). Treat the ball as a uniform solid sphere, ignoring the finger holes. (a) Draw the free-body diagram for the ball. Explain why the friction force must be directed *uphill*. (b) What is the acceleration of the center of mass of the ball? (c) What minimum coefficient of static friction is needed to prevent slipping?

10.27 •• A size-5 soccer ball of diameter 22.6 cm and mass 426 g rolls up a hill without slipping, reaching a maximum height of 5.00 m above the base of the hill. We can model this ball as a thin-walled hollow sphere. (a) At what rate was it rotating at the base of the hill? (b) How much rotational kinetic energy did it have then?

10.28 •• A bicycle racer is going downhill at 11.0 m/s when, to his horror, one of his 2.25-kg wheels comes off as he is 75.0 m above the foot of the hill. We can model the wheel as a thin-walled cylinder 85.0 cm in diameter and ignore the small mass of the spokes. (a) How fast is the wheel moving when it reaches the foot of the hill if it rolled without slipping all the way down? (b) How much total kinetic energy does the wheel have when it reaches the bottom of the hill?

Section 10.4 Work and Power in Rotational Motion

10.29 • A playground merry-go-round has radius 2.40 m and moment of inertia 2100 kg · m² about a vertical axle through its center, and it turns with negligible friction. (a) A child applies an 18.0-N force tangentially to the edge of the merry-go-round for 15.0 s. If the merry-go-round is initially at rest, what is its angular speed after this 15.0-s interval? (b) How much work did the child do on the merry-go-round? (c) What is the average power supplied by the child?

10.30 • An engine delivers 175 hp to an aircraft propeller at 2400 rev/min. (a) How much torque does the aircraft engine provide? (b) How much work does the engine do in one revolution of the propeller?

10.31 • A 2.80-kg grinding wheel is in the form of a solid cylinder of radius 0.100 m. (a) What constant torque will bring it from rest to an angular speed of 1200 rev/min in 2.5 s? (b) Through what angle has it turned during that time? (c) Use Eq. (10.21) to calculate the work done by the torque. (d) What is the grinding wheel's kinetic energy when it is rotating at 1200 rev/min? Compare your answer to the result in part (c).

10.32 •• An electric motor consumes 9.00 kJ of electrical energy in 1.00 min. If one-third of this energy goes into heat and other forms of internal energy of the motor, with the rest going to the motor output, how much torque will this engine develop if you run it at 2500 rpm?

10.33 • (a) Compute the torque developed by an industrial motor whose output is 150 kW at an angular speed of 4000 rev/min. (b) A drum with negligible mass, 0.400 m in diameter, is attached to the motor shaft, and the power output of the motor is used to raise a weight hanging from a rope wrapped around the drum. How heavy a weight can the motor lift at constant speed? (c) At what constant speed will the weight rise?

10.34 •• An airplane propeller is 2.08 m in length (from tip to tip) and has a mass of 117 kg. When the airplane's engine is first started, it applies a constant torque of 1950 N · m to the propeller, which starts from rest. (a) What is the angular acceleration of the propeller? Model the propeller as a slender rod and see Table 9.2. (b) What is the propeller's angular speed after making 5.00 revolutions? (c) How much work is done by the engine during the first 5.00 revolutions? (d) What is the average power output of the engine during the first 5.00 revolutions? (e) What is the instantaneous power output of the motor at the instant that the propeller has turned through 5.00 revolutions?

Section 10.5 Angular Momentum

10.35 • A 2.00-kg rock has a horizontal velocity of magnitude 12.0 m/s when it is at point P in **Fig. E10.35**. (a) At this instant, what are the magnitude and direction of its angular momentum relative to point O? (b) If the only force acting on the rock is its weight, what is the rate of change (magnitude and direction) of its angular momentum at this instant?

Figure **E10.35**

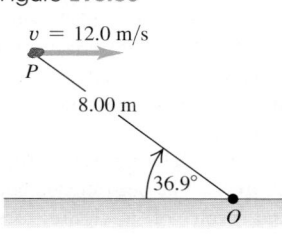

10.36 •• A woman with mass 50 kg is standing on the rim of a large disk that is rotating at 0.80 rev/s about an axis through its center. The disk has mass 110 kg and radius 4.0 m. Calculate the magnitude of the total angular momentum of the woman–disk system. (Assume that you can treat the woman as a point.)

10.37 •• Find the magnitude of the angular momentum of the second hand on a clock about an axis through the center of the clock face. The clock hand has a length of 15.0 cm and a mass of 6.00 g. Take the second hand to be a slender rod rotating with constant angular velocity about one end.

10.38 •• (a) Calculate the magnitude of the angular momentum of the earth in a circular orbit around the sun. Is it reasonable to model it as a particle? (b) Calculate the magnitude of the angular momentum of the earth due to its rotation around an axis through the north and south poles, modeling it as a uniform sphere. Consult Appendix E and the astronomical data in Appendix F.

10.39 •• CALC A hollow, thin-walled sphere of mass 12.0 kg and diameter 48.0 cm is rotating about an axle through its center. The angle (in radians) through which it turns as a function of time (in seconds) is given by $\theta(t) = At^2 + Bt^4$, where A has numerical value 1.50 and B has numerical value 1.10. (a) What are the units of the constants A and B? (b) At the time 3.00 s, find (i) the angular momentum of the sphere and (ii) the net torque on the sphere.

Section 10.6 Conservation of Angular Momentum

10.40 • CP A small block on a frictionless, horizontal surface has a mass of 0.0250 kg. It is attached to a massless cord passing through a hole in the surface (**Fig. E10.40**). The block is originally revolving at a distance of 0.300 m from the hole with an angular speed of 2.85 rad/s. The cord is then pulled from below, shortening the radius of the circle in which the block revolves to 0.150 m. Model the block as a particle. (a) Is the angular momentum of the block conserved? Why or why not? (b) What is the new angular speed? (c) Find the change in kinetic energy of the block. (d) How much work was done in pulling the cord?

Figure **E10.40**

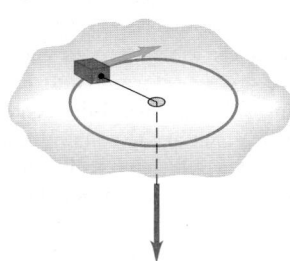

10.41 •• Under some circumstances, a star can collapse into an extremely dense object made mostly of neutrons and called a *neutron star*. The density of a neutron star is roughly 10^{14} times as great as that of ordinary solid matter. Suppose we represent the star as a uniform, solid, rigid sphere, both before and after the collapse. The star's initial radius was 7.0×10^5 km (comparable to our sun); its final radius is 16 km. If the original star rotated once in 30 days, find the angular speed of the neutron star.

10.42 •• A diver comes off a board with arms straight up and legs straight down, giving her a moment of inertia about her rotation axis of 18 kg·m². She then tucks into a small ball, decreasing this moment of inertia to 3.6 kg·m². While tucked, she makes two complete revolutions in 1.0 s. If she hadn't tucked at all, how many revolutions would she have made in the 1.5 s from board to water?

10.43 •• **The Spinning Figure Skater.** The outstretched hands and arms of a figure skater preparing for a spin can be considered a slender rod pivoting about an axis through its center (**Fig. E10.43**). When the skater's hands and arms are brought in and wrapped around his body to execute the spin, the hands and arms can be considered a thin-walled, hollow cylinder. His hands and arms have a combined mass of 8.0 kg. When outstretched, they span 1.8 m; when wrapped, they form a cylinder of radius 25 cm. The moment of inertia about the rotation axis of the remainder of his body is constant and equal to 0.40 kg·m². If his original angular speed is 0.40 rev/s, what is his final angular speed?

Figure **E10.43**

10.44 •• A solid wood door 1.00 m wide and 2.00 m high is hinged along one side and has a total mass of 40.0 kg. Initially open and at rest, the door is struck at its center by a handful of sticky mud with mass 0.500 kg, traveling perpendicular to the door at 12.0 m/s just before impact. Find the final angular speed of the door. Does the mud make a significant contribution to the moment of inertia?

10.45 •• A large wooden turntable in the shape of a flat uniform disk has a radius of 2.00 m and a total mass of 120 kg. The turntable is initially rotating at 3.00 rad/s about a vertical axis through its center. Suddenly, a 70.0-kg parachutist makes a soft landing on the turntable at a point near the outer edge. (a) Find the angular speed of the turntable after the parachutist lands. (Assume that you can treat the parachutist as a particle.) (b) Compute the kinetic energy of the system before and after the parachutist lands. Why are these kinetic energies not equal?

10.46 •• **Asteroid Collision!** Suppose that an asteroid traveling straight toward the center of the earth were to collide with our planet at the equator and bury itself just below the surface. What would have to be the mass of this asteroid, in terms of the earth's mass *M*, for the day to become 25.0% longer than it presently is as a result of the collision? Assume that the asteroid is very small compared to the earth and that the earth is uniform throughout.

10.47 •• A small 10.0-g bug stands at one end of a thin uniform bar that is initially at rest on a smooth horizontal table. The other end of the bar pivots about a nail driven into the table and can rotate freely, without friction. The bar has mass 50.0 g and is 100 cm in length. The bug jumps off in the horizontal direction, perpendicular to the bar, with a speed of 20.0 cm/s relative to the table. (a) What is the angular speed of the bar just after the frisky insect leaps? (b) What is the total kinetic energy of the system just after the bug leaps? (c) Where does this energy come from?

10.48 •• A thin uniform rod has a length of 0.500 m and is rotating in a circle on a frictionless table. The axis of rotation is perpendicular to the length of the rod at one end and is stationary. The rod has an angular velocity of 0.400 rad/s and a

moment of inertia about the axis of 3.00 × 10⁻³ kg·m². A bug initially standing on the rod at the axis of rotation decides to crawl out to the other end of the rod. When the bug has reached the end of the rod and sits there, its tangential speed is 0.160 m/s. The bug can be treated as a point mass. What is the mass of (a) the rod; (b) the bug?

10.49 •• A thin, uniform metal bar, 2.00 m long and weighing 90.0 N, is hanging vertically from the ceiling by a frictionless pivot. Suddenly it is struck 1.50 m below the ceiling by a small 3.00-kg ball, initially traveling horizontally at 10.0 m/s. The ball rebounds in the opposite direction with a speed of 6.00 m/s. (a) Find the angular speed of the bar just after the collision. (b) During the collision, why is the angular momentum conserved but not the linear momentum?

10.50 •• A uniform, 4.5-kg, square, solid wooden gate 1.5 m on each side hangs vertically from a frictionless pivot at the center of its upper edge. A 1.1-kg raven flying horizontally at 5.0 m/s flies into this door at its center and bounces back at 2.0 m/s in the opposite direction. (a) What is the angular speed of the gate just after it is struck by the unfortunate raven? (b) During the collision, why is the angular momentum conserved but not the linear momentum?

Section 10.7 Gyroscopes and Precession

10.51 •• The rotor (flywheel) of a toy gyroscope has mass 0.140 kg. Its moment of inertia about its axis is 1.20 × 10⁻⁴ kg·m². The mass of the frame is 0.0250 kg. The gyroscope is supported on a single pivot (**Fig. E10.51**) with its center of mass a horizontal distance of 4.00 cm from the pivot. The gyroscope is precessing in a horizontal plane at the rate of one revolution in 2.20 s. (a) Find the upward force exerted by the pivot. (b) Find the angular speed with which the rotor is spinning about its axis, expressed in rev/min. (c) Copy the diagram and draw vectors to show the angular momentum of the rotor and the torque acting on it.

Figure **E10.51**

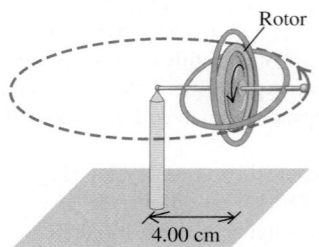

Rotor

4.00 cm

10.52 • **A Gyroscope on the Moon.** A certain gyroscope precesses at a rate of 0.50 rad/s when used on earth. If it were taken to a lunar base, where the acceleration due to gravity is 0.165*g*, what would be its precession rate?

10.53 • **Stabilization of the Hubble Space Telescope.** The Hubble Space Telescope is stabilized to within an angle of about 2-millionths of a degree by means of a series of gyroscopes that spin at 19,200 rpm. Although the structure of these gyroscopes is actually quite complex, we can model each of the gyroscopes as a thin-walled cylinder of mass 2.0 kg and diameter 5.0 cm, spinning about its central axis. How large a torque would it take to cause these gyroscopes to precess through an angle of 1.0 × 10⁻⁶ degree during a 5.0-hour exposure of a galaxy?

PROBLEMS

10.54 ·· A 50.0-kg grindstone is a solid disk 0.520 m in diameter. You press an ax down on the rim with a normal force of 160 N (**Fig. P10.54**). The coefficient of kinetic friction between the blade and the stone is 0.60, and there is a constant friction torque of 6.50 N·m between the axle of the stone and its bearings. (a) How much force must be applied tangentially at the end of a crank handle 0.500 m long to bring the stone from rest to 120 rev/min in 9.00 s? (b) After the grindstone attains an angular speed of 120 rev/min, what tangential force at the end of the handle is needed to maintain a constant angular speed of 120 rev/min? (c) How much time does it take the grindstone to come from 120 rev/min to rest if it is acted on by the axle friction alone?

Figure **P10.54**

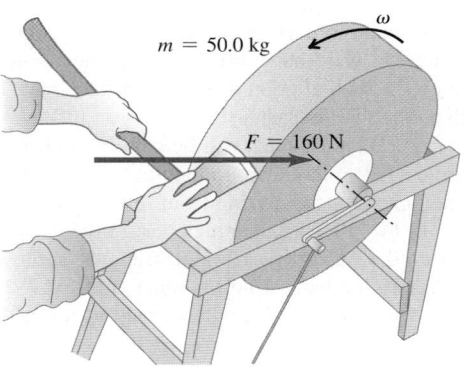

$m = 50.0$ kg
ω
$F = 160$ N

10.55 ··· A grindstone in the shape of a solid disk with diameter 0.520 m and a mass of 50.0 kg is rotating at 850 rev/min. You press an ax against the rim with a normal force of 160 N (Fig. P10.54), and the grindstone comes to rest in 7.50 s. Find the coefficient of friction between the ax and the grindstone. You can ignore friction in the bearings.

10.56 ··· A uniform, 8.40-kg, spherical shell 50.0 cm in diameter has four small 2.00-kg masses attached to its outer surface and equally spaced around it. This combination is spinning about an axis running through the center of the sphere and two of the small masses (**Fig. P10.56**). What friction torque is needed to reduce its angular speed from 75.0 rpm to 50.0 rpm in 30.0 s?

Figure **P10.56**

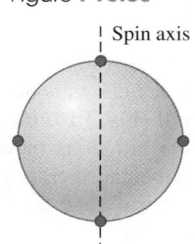

¦ Spin axis

10.57 ··· A thin, uniform, 3.80-kg bar, 80.0 cm long, has very small 2.50-kg balls glued on at either end (**Fig. P10.57**). It is supported horizontally by a thin, horizontal, frictionless axle passing through its center and perpendicular to the bar. Suddenly the right-hand ball becomes detached and falls off, but the other ball remains glued to the bar. (a) Find the angular acceleration of the bar just after the ball falls off. (b) Will the angular acceleration remain constant as the bar continues to swing? If not, will it increase or decrease? (c) Find the angular velocity of the bar just as it swings through its vertical position.

Figure **P10.57**

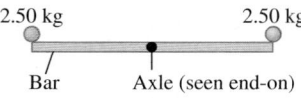

2.50 kg   2.50 kg

Bar   Axle (seen end-on)

10.58 ·· You are designing a simple elevator system for an old warehouse that is being converted to loft apartments. A 22,500-N elevator is to be accelerated upward by connecting it to a counterweight by means of a light (but strong!) cable passing over a solid uniform disk-shaped pulley. The cable does not slip where it is in contact with the surface of the pulley. There is no appreciable friction at the axle of the pulley, but its mass is 875 kg and it is 1.50 m in diameter. (a) What mass should the counterweight have so that it will accelerate the elevator upward through 6.75 m in the first 3.00 s, starting from rest? (b) What is the tension in the cable on each side of the pulley?

10.59 ·· The Atwood's Machine. **Figure P10.59** illustrates an Atwood's machine. Find the linear accelerations of blocks A and B, the angular acceleration of the wheel C, and the tension in each side of the cord if there is no slipping between the cord and the surface of the wheel. Let the masses of blocks A and B be 4.00 kg and 2.00 kg, respectively, the moment of inertia of the wheel about its axis be 0.220 kg·m², and the radius of the wheel be 0.120 m.

Figure **P10.59**

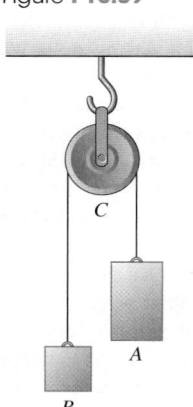

C
A
B

10.60 ··· The mechanism shown in **Fig. P10.60** is used to raise a crate of supplies from a ship's hold. The crate has total mass 50 kg. A rope is wrapped around a wooden cylinder that turns on a metal axle. The cylinder has radius 0.25 m and moment of inertia $I = 2.9$ kg·m² about the axle. The crate is suspended from the free end of the rope. One end of the axle pivots on frictionless bearings; a crank handle is attached to the other end. When the crank is turned, the end of the handle rotates about the axle in a vertical circle of radius 0.12 m, the cylinder turns, and the crate is raised. What magnitude of the force $\vec{F}$ applied tangentially to the rotating crank is required to raise the crate with an acceleration of 1.40 m/s²? (You can ignore the mass of the rope as well as the moments of inertia of the axle and the crank.)

Figure **P10.60**

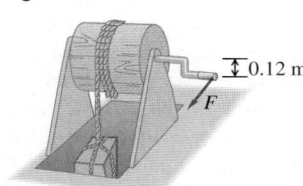

0.12 m
F

10.61 ·· A large 16.0-kg roll of paper with radius $R = 18.0$ cm rests against the wall and is held in place by a bracket attached to a rod through the center of the roll (**Fig. P10.61**). The rod turns without friction in the bracket, and the moment of inertia of the paper and rod about the axis is 0.260 kg·m². The other end of the bracket is attached by a frictionless hinge to the wall such that the bracket makes an angle of 30.0° with the wall. The weight of the bracket is negligible. The coefficient of kinetic friction between the paper and the wall is $\mu_k = 0.25$. A constant vertical force $F = 60.0$ N is applied to the paper, and the paper unrolls. What is the magnitude of (a) the force that the rod exerts on the paper as it unrolls; (b) the angular acceleration of the roll?

Figure **P10.61**

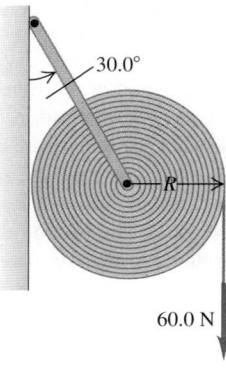

30.0°
R
60.0 N

10.62 ·· A block with mass $m = 5.00$ kg slides down a surface inclined 36.9° to the horizontal (**Fig. P10.62**). The coefficient of kinetic friction is 0.25. A string attached to the block is wrapped around a flywheel on a fixed axis at O. The flywheel has mass 25.0 kg and moment of inertia 0.500 kg · m² with respect to the axis of rotation.

Figure **P10.62**

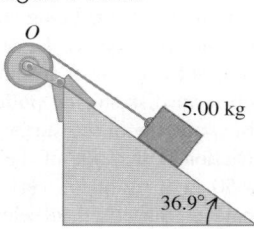

The string pulls without slipping at a perpendicular distance of 0.200 m from that axis. (a) What is the acceleration of the block down the plane? (b) What is the tension in the string?

10.63 ··· Two metal disks, one with radius $R_1 = 2.50$ cm and mass $M_1 = 0.80$ kg and the other with radius $R_2 = 5.00$ cm and mass $M_2 = 1.60$ kg, are welded together and mounted on a frictionless axis through their common center, as in Problem 9.77. (a) A light string is wrapped around the edge of the smaller disk, and a 1.50-kg block is suspended from the free end of the string. What is the magnitude of the downward acceleration of the block after it is released? (b) Repeat the calculation of part (a), this time with the string wrapped around the edge of the larger disk. In which case is the acceleration of the block greater? Does your answer make sense?

10.64 ·· A lawn roller in the form of a thin-walled, hollow cylinder with mass M is pulled horizontally with a constant horizontal force F applied by a handle attached to the axle. If it rolls without slipping, find the acceleration and the friction force.

10.65 · Two weights are connected by a very light, flexible cord that passes over an 80.0-N frictionless pulley of radius 0.300 m. The pulley is a solid uniform disk and is supported by a hook connected to the ceiling (**Fig. P10.65**). What force does the ceiling exert on the hook?

Figure **P10.65**

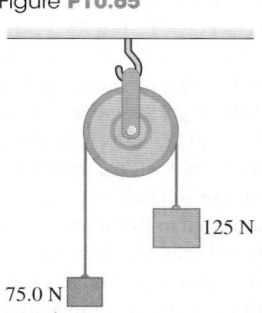

125 N

75.0 N

10.66 ·· You complain about fire safety to the landlord of your high-rise apartment building. He is willing to install an evacuation device if it is cheap and reliable, and he asks you to design it. Your proposal is to mount a large wheel (radius 0.400 m) on an axle at its center and wrap a long, light rope around the wheel, with the free end of the rope hanging just past the edge of the roof. Residents would evacuate to the roof and, one at a time, grasp the free end of the rope, step off the roof, and be lowered to the ground below. (Ignore friction at the axle.) You want a 90.0-kg person to descend with an acceleration of $g/4$. (a) If the wheel can be treated as a uniform disk, what mass must it have? (b) As the person descends, what is the tension in the rope?

10.67 · **The Yo-yo.** A yo-yo is made from two uniform disks, each with mass m and radius R, connected by a light axle of radius b. A light, thin string is wound several times around the axle and then held stationary while the yo-yo is released from rest, dropping as the string unwinds. Find the linear acceleration and angular acceleration of the yo-yo and the tension in the string.

10.68 ·· CP A thin-walled, hollow spherical shell of mass m and radius r starts from rest and rolls without slipping down a track (**Fig. P10.68**). Points A and B are on a circular part of the track

Figure **P10.68**

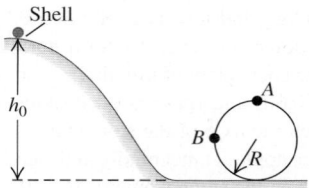

having radius R. The diameter of the shell is very small compared to h_0 and R, and the work done by rolling friction is negligible. (a) What is the minimum height h_0 for which this shell will make a complete loop-the-loop on the circular part of the track? (b) How hard does the track push on the shell at point B, which is at the same level as the center of the circle? (c) Suppose that the track had no friction and the shell was released from the same height h_0 you found in part (a). Would it make a complete loop-the-loop? How do you know? (d) In part (c), how hard does the track push on the shell at point A, the top of the circle? How hard did it push on the shell in part (a)?

10.69 ·· A basketball (which can be closely modeled as a hollow spherical shell) rolls down a mountainside into a valley and then up the opposite side, starting from rest at a height H_0 above the bottom. In **Fig. P10.69**, the rough part of the terrain prevents slipping while the smooth part has no friction. (a) How high, in terms of H_0, will the ball go up the other side? (b) Why doesn't the ball return to height H_0? Has it lost any of its original potential energy?

Figure **P10.69**

10.70 ·· CP A solid uniform ball rolls without slipping up a hill (**Fig. P10.70**). At the top of the hill, it is moving horizontally, and then it goes over the vertical cliff. (a) How far from the foot of the cliff does the

Figure **P10.70**

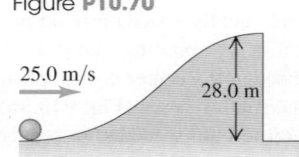

ball land, and how fast is it moving just before it lands? (b) Notice that when the balls lands, it has a greater translational speed than when it was at the bottom of the hill. Does this mean that the ball somehow gained energy? Explain!

10.71 ·· **Rolling Stones.** A solid, uniform, spherical boulder starts from rest and rolls down a 50.0-m-high hill, as shown in **Fig. P10.71**. The top half of the hill is rough enough to cause the boulder to roll without slipping, but the lower half is covered with

Figure **P10.71**

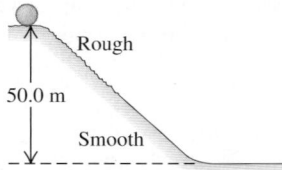

ice and there is no friction. What is the translational speed of the boulder when it reaches the bottom of the hill?

10.72 ••• You are designing a system for moving aluminum cylinders from the ground to a loading dock. You use a sturdy wooden ramp that is 6.00 m long and inclined at 37.0° above the horizontal. Each cylinder is fitted with a light, frictionless yoke through its center, and a light (but strong) rope is attached to the yoke. Each cylinder is uniform and has mass 460 kg and radius 0.300 m. The cylinders are pulled up the ramp by applying a constant force $\vec{F}$ to the free end of the rope. $\vec{F}$ is parallel to the surface of the ramp and exerts no torque on the cylinder. The coefficient of static friction between the ramp surface and the cylinder is 0.120. (a) What is the largest magnitude $\vec{F}$ can have so that the cylinder still rolls without slipping as it moves up the ramp? (b) If the cylinder starts from rest at the bottom of the ramp and rolls without slipping as it moves up the ramp, what is the shortest time it can take the cylinder to reach the top of the ramp?

10.73 •• A 42.0-cm-diameter wheel, consisting of a rim and six spokes, is constructed from a thin, rigid plastic material having a linear mass density of 25.0 g/cm. This wheel is released from rest at the top of a hill 58.0 m high. (a) How fast is it rolling when it reaches the bottom of the hill? (b) How would your answer change if the linear mass density and the diameter of the wheel were each doubled?

10.74 ••• A uniform, 0.0300-kg rod of length 0.400 m rotates in a horizontal plane about a fixed axis through its center and perpendicular to the rod. Two small rings, each with mass 0.0200 kg, are mounted so that they can slide along the rod. They are initially held by catches at positions 0.0500 m on each side of the center of the rod, and the system is rotating at 48.0 rev/min. With no other changes in the system, the catches are released, and the rings slide outward along the rod and fly off at the ends. What is the angular speed (a) of the system at the instant when the rings reach the ends of the rod; (b) of the rod after the rings leave it?

10.75 • A uniform solid cylinder with mass M and radius $2R$ rests on a horizontal tabletop. A string is attached by a yoke to a frictionless axle through the center of the cylinder so that the cylinder can rotate about the axle. The string runs over a disk-shaped pulley with mass M and radius R that is mounted on a frictionless axle through its center. A block of mass M is suspended from the free end of the string (**Fig. P10.75**). The string doesn't slip over the pulley surface, and the cylinder rolls without slipping on the tabletop. Find the magnitude of the acceleration of the block after the system is released from rest.

Figure **P10.75**

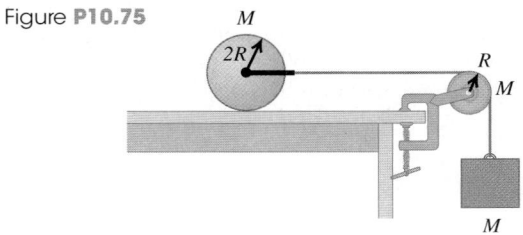

10.76 •• **Tarzan and Jane in the 21st Century.** Tarzan has foolishly gotten himself into another scrape with the animals and must be rescued once again by Jane. The 60.0-kg Jane starts from rest at a height of 5.00 m in the trees and swings down to the ground using a thin, but very rigid, 30.0-kg vine 8.00 m long.

She arrives just in time to snatch the 72.0-kg Tarzan from the jaws of an angry hippopotamus. What is Jane's (and the vine's) angular speed (a) just before she grabs Tarzan and (b) just after she grabs him? (c) How high will Tarzan and Jane go on their first swing after this daring rescue?

10.77 ••• A 5.00-kg ball is dropped from a height of 12.0 m above one end of a uniform bar that pivots at its center. The bar has mass 8.00 kg and is 4.00 m in length. At the other end of the bar sits another 5.00-kg ball, unattached to the bar. The dropped ball sticks to the bar after the collision. How high will the other ball go after the collision?

10.78 •• The solid wood door of a gymnasium is 1.00 m wide and 2.00 m high, has total mass 35.0 kg, and is hinged along one side. The door is open and at rest when a stray basketball hits the center of the door head-on, applying an average force of 1500 N to the door for 8.00 ms. Find the angular speed of the door after the impact. [*Hint:* Integrating Eq. (10.29) yields $\Delta L_z = \int_{t_1}^{t_2} (\sum \tau_z) dt = (\sum \tau_z)_{av} \Delta t$. The quantity $\int_{t_1}^{t_2} (\sum \tau_z) dt$ is called the angular impulse.]

10.79 •• A uniform rod of length L rests on a frictionless horizontal surface. The rod pivots about a fixed frictionless axis at one end. The rod is initially at rest. A bullet traveling parallel to the horizontal surface and perpendicular to the rod with speed v strikes the rod at its center and becomes embedded in it. The mass of the bullet is one-fourth the mass of the rod. (a) What is the final angular speed of the rod? (b) What is the ratio of the kinetic energy of the system after the collision to the kinetic energy of the bullet before the collision?

10.80 •• CP A large turntable with radius 6.00 m rotates about a fixed vertical axis, making one revolution in 8.00 s. The moment of inertia of the turntable about this axis is 1200 kg · m². You stand, barefooted, at the rim of the turntable and very slowly walk toward the center, along a radial line painted on the surface of the turntable. Your mass is 70.0 kg. Since the radius of the turntable is large, it is a good approximation to treat yourself as a point mass. Assume that you can maintain your balance by adjusting the positions of your feet. You find that you can reach a point 3.00 m from the center of the turntable before your feet begin to slip. What is the coefficient of static friction between the bottoms of your feet and the surface of the turntable?

10.81 •• In your job as a mechanical engineer you are designing a flywheel and clutch-plate system like the one in Example 10.11. Disk A is made of a lighter material than disk B, and the moment of inertia of disk A about the shaft is one-third that of disk B. The moment of inertia of the shaft is negligible. With the clutch disconnected, A is brought up to an angular speed ω_0; B is initially at rest. The accelerating torque is then removed from A, and A is coupled to B. (Ignore bearing friction.) The design specifications allow for a maximum of 2400 J of thermal energy to be developed when the connection is made. What can be the maximum value of the original kinetic energy of disk A so as not to exceed the maximum allowed value of the thermal energy?

10.82 •• A local ice hockey team has asked you to design an apparatus for measuring the speed of the hockey puck after a slap shot. Your design is a 2.00-m-long, uniform rod pivoted about one end so that it is free to rotate horizontally on the ice without friction. The 0.800-kg rod has a light basket at the other end to catch the 0.163-kg puck. The puck slides across the ice with velocity $\vec{v}$ (perpendicular to the rod), hits the basket, and is caught. After the collision, the rod rotates. If the rod makes one revolution every 0.736 s after the puck is caught, what was the puck's speed just before it hit the rod?

10.83 ••• You are designing a slide for a water park. In a sitting position, park guests slide a vertical distance h down the waterslide, which has negligible friction. When they reach the bottom of the slide, they grab a handle at the bottom end of a 6.00-m-long uniform pole. The pole hangs vertically, initially at rest. The upper end of the pole is pivoted about a stationary, frictionless axle. The pole with a person hanging on the end swings up through an angle of 72.0°, and then the person lets go of the pole and drops into a pool of water. Treat the person as a point mass. The pole's moment of inertia is given by $I = \frac{1}{3}ML^2$, where $L = 6.00$ m is the length of the pole and $M = 24.0$ kg is its mass. For a person of mass 70.0 kg, what must be the height h in order for the pole to have a maximum angle of swing of 72.0° after the collision?

10.84 •• **Neutron Star Glitches.** Occasionally, a rotating neutron star (see Exercise 10.41) undergoes a sudden and unexpected speedup called a *glitch*. One explanation is that a glitch occurs when the crust of the neutron star settles slightly, decreasing the moment of inertia about the rotation axis. A neutron star with angular speed $\omega_0 = 70.4$ rad/s underwent such a glitch in October 1975 that increased its angular speed to $\omega = \omega_0 + \Delta\omega$, where $\Delta\omega/\omega_0 = 2.01 \times 10^{-6}$. If the radius of the neutron star before the glitch was 11 km, by how much did its radius decrease in the starquake? Assume that the neutron star is a uniform sphere.

10.85 ••• A 500.0-g bird is flying horizontally at 2.25 m/s, not paying much attention, when it suddenly flies into a stationary vertical bar, hitting it 25.0 cm below the top (**Fig. P10.85**). The bar is uniform, 0.750 m long, has a mass of 1.50 kg, and is hinged at its base. The collision stuns the bird so that it just drops to the ground afterward (but soon recovers to fly happily away). What is the angular velocity of the bar (a) just after it is hit by the bird and (b) just as it reaches the ground?

Figure **P10.85**

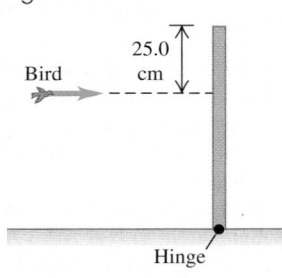

10.86 ••• **CP** A small block with mass 0.130 kg is attached to a string passing through a hole in a frictionless, horizontal surface (see Fig. E10.40). The block is originally revolving in a circle with a radius of 0.800 m about the hole with a tangential speed of 4.00 m/s. The string is then pulled slowly from below, shortening the radius of the circle in which the block revolves. The breaking strength of the string is 30.0 N. What is the radius of the circle when the string breaks?

10.87 • A 55-kg runner runs around the edge of a horizontal turntable mounted on a vertical, frictionless axis through its center. The runner's velocity relative to the earth has magnitude 2.8 m/s. The turntable is rotating in the opposite direction with an angular velocity of magnitude 0.20 rad/s relative to the earth. The radius of the turntable is 3.0 m, and its moment of inertia about the axis of rotation is 80 kg·m². Find the final angular velocity of the system if the runner comes to rest relative to the turntable. (You can model the runner as a particle.)

10.88 •• **DATA** The V6 engine in a 2014 Chevrolet Silverado 1500 pickup truck is reported to produce a maximum power of 285 hp at 5300 rpm and a maximum torque of 305 ft·lb at 3900 rpm. (a) Calculate the torque, in both ft·lb and N·m, at 5300 rpm. Is your answer in ft·lb smaller than the specified maximum value? (b) Calculate the power, in both horsepower and watts, at 3900 rpm. Is your answer in hp smaller than the specified maximum value? (c) The relationship between power in hp and torque in ft·lb

at a particular angular velocity in rpm is often written as hp = [torque (in ft·lb) × rpm]/c, where c is a constant. What is the numerical value of c? (d) The engine of a 2012 Chevrolet Camaro ZL1 is reported to produce 580 hp at 6000 rpm. What is the torque (in ft·lb) at 6000 rpm?

10.89 •• **DATA** You have one object of each of these shapes, all with mass 0.840 kg: a uniform solid cylinder, a thin-walled hollow cylinder, a uniform solid sphere, and a thin-walled hollow sphere. You release each object from rest at the same vertical height h above the bottom of a long wooden ramp that is inclined at 35.0° from the horizontal. Each object rolls without slipping down the ramp. You measure the time t that it takes each one to reach the bottom of the ramp; **Fig. P10.89** shows the results. (a) From the bar graphs, identify objects A through D by shape. (b) Which of objects A through D has the greatest total kinetic energy at the bottom of the ramp, or do all have the same kinetic energy? (c) Which of objects A through D has the greatest rotational kinetic energy $\frac{1}{2}I\omega^2$ at the bottom of the ramp, or do all have the same rotational kinetic energy? (d) What minimum coefficient of static friction is required for all four objects to roll without slipping?

Figure **P10.89**

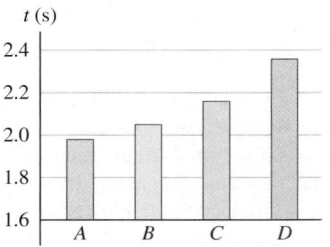

10.90 ••• **DATA** You are testing a small flywheel (radius 0.166 m) that will be used to store a small amount of energy. The flywheel is pivoted with low-friction bearings about a horizontal shaft through the flywheel's center. A thin, light cord is wrapped multiple times around the rim of the flywheel. Your lab has a device that can apply a specified horizontal force $\vec{F}$ to the free end of the cord. The device records both the magnitude of that force as a function of the horizontal distance the end of the cord has traveled and the time elapsed since the force was first applied. The flywheel is initially at rest. (a) You start with a test run to determine the flywheel's moment of inertia I. The magnitude F of the force is a constant 25.0 N, and the end of the rope moves 8.35 m in 2.00 s. What is I? (b) In a second test, the flywheel again starts from rest but the free end of the rope travels 6.00 m; **Fig. P10.90** shows the force magnitude F as a function of the distance d that the end of the rope has moved. What is the kinetic energy of the flywheel when $d = 6.00$ m? (c) What is the angular speed of the flywheel, in rev/min, when $d = 6.00$ m?

Figure **P10.90**

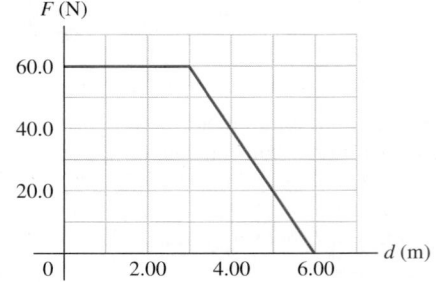

CHALLENGE PROBLEMS

10.91 ••• CP CALC A block with mass m is revolving with linear speed v_1 in a circle of radius r_1 on a frictionless horizontal surface (see Fig. E10.40). The string is slowly pulled from below until the radius of the circle in which the block is revolving is reduced to r_2. (a) Calculate the tension T in the string as a function of r, the distance of the block from the hole. Your answer will be in terms of the initial velocity v_1 and the radius r_1. (b) Use $W = \int_{r_1}^{r_2} \vec{T}(r) \cdot d\vec{r}$ to calculate the work done by $\vec{T}$ when r changes from r_1 to r_2. (c) Compare the results of part (b) to the change in the kinetic energy of the block.

10.92 ••• When an object is rolling without slipping, the rolling friction force is much less than the friction force when the object is sliding; a silver dollar will roll on its edge much farther than it will slide on its flat side (see Section 5.3). When an object is rolling without slipping on a horizontal surface, we can approximate the friction force to be zero, so that a_x and α_z are approximately zero and v_x and ω_z are approximately constant. Rolling without slipping means $v_x = r\omega_z$ and $a_x = r\alpha_z$. If an object is set in motion on a surface *without* these equalities, sliding (kinetic) friction will act on the object as it slips until rolling without slipping is established. A solid cylinder with mass M and radius R, rotating with angular speed ω_0 about an axis through its center, is set on a horizontal surface for which the kinetic friction coefficient is μ_k. (a) Draw a free-body diagram for the cylinder on the surface. Think carefully about the direction of the kinetic friction force on the cylinder. Calculate the accelerations a_x of the center of mass and α_z of rotation about the center of mass. (b) The cylinder is initially slipping completely, so initially $\omega_z = \omega_0$ but $v_x = 0$. Rolling without slipping sets in when $v_x = r\omega_z$. Calculate the *distance* the cylinder rolls before slipping stops. (c) Calculate the work done by the friction force on the cylinder as it moves from where it was set down to where it begins to roll without slipping.

10.93 ••• A demonstration gyroscope wheel is constructed by removing the tire from a bicycle wheel 0.650 m in diameter, wrapping lead wire around the rim, and taping it in place. The shaft projects 0.200 m at each side of the wheel, and a woman holds the ends of the shaft in her hands. The mass of the system is 8.00 kg; its entire mass may be assumed to be located at its rim. The shaft is horizontal, and the wheel is spinning about the shaft at 5.00 rev/s. Find the magnitude and direction of the force each hand exerts on the shaft (a) when the shaft is at rest; (b) when the shaft is rotating in a horizontal plane about its center at 0.050 rev/s; (c) when the shaft is rotating in a horizontal plane about its center at 0.300 rev/s. (d) At what rate must the shaft rotate in order that it may be supported at one end only?

BIO **HUMAN MOMENT OF INERTIA.** The moment of inertia of the human body about an axis through its center of mass is important in the application of biomechanics to sports such as diving and gymnastics. We can measure the body's moment of inertia in a particular position while a person remains in that position on a horizontal turntable, with the body's center of mass on the turntable's rotational axis. The turntable with the person on it is then accelerated from rest by a torque that is produced by using a rope wound around a pulley on the shaft of the turntable. From the measured tension in the rope and the angular acceleration, we can calculate the body's moment of inertia about an axis through its center of mass.

Overhead view of a female gymnast lying in somersault position atop a turntable

10.94 The moment of inertia of the empty turntable is 1.5 kg m². With a constant torque of 2.5 N·m, the turntable–person system takes 3.0 s to spin from rest to an angular speed of 1.0 rad/s. What is the person's moment of inertia about an axis through her center of mass? Ignore friction in the turntable axle. (a) 2.5 kg·m²; (b) 6.0 kg·m²; (c) 7.5 kg·m²; (d) 9.0 kg·m².

10.95 While the turntable is being accelerated, the person suddenly extends her legs. What happens to the turntable? (a) It suddenly speeds up; (b) it rotates with constant speed; (c) its acceleration decreases; (d) it suddenly stops rotating.

10.96 A doubling of the torque produces a greater angular acceleration. Which of the following would do this, assuming that the tension in the rope doesn't change? (a) Increasing the pulley diameter by a factor of $\sqrt{2}$; (b) increasing the pulley diameter by a factor of 2; (c) increasing the pulley diameter by a factor of 4; (d) decreasing the pulley diameter by a factor of $\sqrt{2}$.

10.97 If the body's center of mass were not placed on the rotational axis of the turntable, how would the person's measured moment of inertia compare to the moment of inertia for rotation about the center of mass? (a) The measured moment of inertia would be too large; (b) the measured moment of inertia would be too small; (c) the two moments of inertia would be the same; (d) it depends on where the body's center of mass is placed relative to the center of the turntable.

Answers

Chapter Opening Question ?

(iv) A tossed pin rotates around its center of mass (which is located toward its thick end). This is also the point at which the gravitational force acts on the pin, so this force exerts no torque on the pin. Hence the pin rotates with constant angular momentum, and its angular speed remains the same.

Test Your Understanding Questions

10.1 (ii) Force P acts along a vertical line, so the lever arm is the horizontal distance from A to the line of action. This is the horizontal component of distance L, which is $L\cos\theta$. Hence the magnitude of the torque is the product of the force magnitude P and the lever arm $L\cos\theta$, or $\tau = PL\cos\theta$.

10.2 (iii), (ii), (i) For the hanging object of mass m_2 to accelerate downward, the net force on it must be downward. Hence the magnitude $m_2 g$ of the downward weight force must be greater than the magnitude T_2 of the upward tension force. For the pulley to have a clockwise angular acceleration, the net torque on the pulley must be clockwise. Tension T_2 tends to rotate the pulley clockwise, while tension T_1 tends to rotate the pulley counterclockwise. Both tension forces have the same lever arm R, so there is a clockwise torque $T_2 R$ and a counterclockwise torque $T_1 R$. For the net torque to be clockwise, T_2 must be greater than T_1. Hence $m_2 g > T_2 > T_1$.

10.3 (a) (ii), (b) (i) If you redo the calculation of Example 10.6 with a hollow cylinder (moment of inertia $I_{cm} = MR^2$) instead of a solid cylinder (moment of inertia $I_{cm} = \frac{1}{2}MR^2$), you will find $a_{cm-y} = \frac{1}{2}g$ and $T = \frac{1}{2}Mg$ (instead of $a_{cm-y} = \frac{2}{3}g$ and $T = \frac{1}{3}Mg$ for a solid cylinder). Hence the acceleration is less but the tension is greater. You can come to the same conclusion without doing the calculation. The greater moment of inertia means that the hollow cylinder will rotate more slowly and hence will roll downward more slowly. To slow the downward motion, a greater upward tension force is needed to oppose the downward force of gravity.

10.4 (iii) You apply the same torque over the same angular displacement to both cylinders. Hence, by Eq. (10.21), you do the same amount of work to both cylinders and impart the same kinetic energy to both. (The one with the smaller moment of inertia ends up with a greater angular speed, but that isn't what we are asked. Compare Conceptual Example 6.5 in Section 6.2.)

10.5 (a) **no**, (b) **yes** As the ball goes around the circle, the magnitude of $\vec{p} = m\vec{v}$ remains the same (the speed is constant) but its direction changes, so the linear momentum vector isn't constant. But $\vec{L} = \vec{r} \times \vec{p}$ is constant: It has a constant magnitude (both the speed and the perpendicular distance from your hand to the ball are constant) and a constant direction (along the rotation axis, perpendicular to the plane of the ball's motion). The linear momentum changes because there is a net *force* $\vec{F}$ on the ball (toward the center of the circle). The angular momentum remains constant because there is no net *torque;* the vector $\vec{r}$ points from your hand to the ball and the force $\vec{F}$ on the ball is directed toward your hand, so the vector product $\vec{\tau} = \vec{r} \times \vec{F}$ is zero.

10.6 (i) In the absence of external torques, the earth's angular momentum $L_z = I\omega_z$ would remain constant. The melted ice would move from the poles toward the equator—that is, away from our planet's rotation axis—and the earth's moment of inertia I would increase slightly. Hence the angular velocity ω_z would decrease slightly and the day would be slightly longer.

10.7 (iii) Doubling the flywheel mass would double both its moment of inertia I and its weight w, so the ratio I/w would be unchanged. Equation (10.33) shows that the precession angular speed depends on this ratio, so there would be *no* effect on the value of Ω.

Bridging Problem

(a) $h = \dfrac{2R}{5}$ (b) $\frac{5}{7}$ of the speed it had just after the hit

? This Roman aqueduct uses the principle of the arch to sustain the weight of the structure and the water it carries. Are the blocks that make up the arch being (i) compressed, (ii) stretched, (iii) a combination of these, or (iv) neither compressed nor stretched?

11 EQUILIBRIUM AND ELASTICITY

W e've devoted a good deal of effort to understanding why and how bodies accelerate in response to the forces that act on them. But very often we're interested in making sure that bodies *don't* accelerate. Any building, from a multistory skyscraper to the humblest shed, must be designed so that it won't topple over. Similar concerns arise with a suspension bridge, a ladder leaning against a wall, or a crane hoisting a bucket full of concrete.

A body that can be modeled as a *particle* is in equilibrium whenever the vector sum of the forces acting on it is zero. But for the situations we've just described, that condition isn't enough. If forces act at different points on an extended body, an additional requirement must be satisfied to ensure that the body has no tendency to *rotate:* The sum of the *torques* about any point must be zero. This requirement is based on the principles of rotational dynamics developed in Chapter 10. We can compute the torque due to the weight of a body by using the concept of center of gravity, which we introduce in this chapter.

Idealized rigid bodies don't bend, stretch, or squash when forces act on them. But all real materials are *elastic* and do deform to some extent. Elastic properties of materials are tremendously important. You want the wings of an airplane to be able to bend a little, but you'd rather not have them break off. Tendons in your limbs need to stretch when you exercise, but they must return to their relaxed lengths when you stop. Many of the necessities of everyday life, from rubber bands to suspension bridges, depend on the elastic properties of materials. In this chapter we'll introduce the concepts of *stress*, *strain*, and *elastic modulus* and a simple principle called *Hooke's law*, which helps us predict what deformations will occur when forces are applied to a real (not perfectly rigid) body.

11.1 CONDITIONS FOR EQUILIBRIUM

We learned in Sections 4.2 and 5.1 that a particle is in *equilibrium*—that is, the particle does not accelerate—in an inertial frame of reference if the vector sum of all the forces acting on the particle is zero, $\sum \vec{F} = 0$. For an *extended* body, the equivalent statement is that the center of mass of the body has zero acceleration if the vector sum of all external forces acting on the body is zero, as discussed in Section 8.5. This is often called the **first condition for equilibrium:**

11.1 To be in static equilibrium, a body at rest must satisfy *both* conditions for equilibrium: It can have no tendency to accelerate as a whole or to start rotating.

(a) This body is in static equilibrium.

Equilibrium conditions:

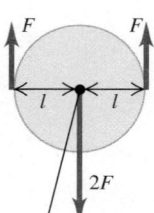

First condition satisfied: Net force = 0, so body at rest has no tendency to start moving as a whole.

Second condition satisfied: Net torque about the axis = 0, so body at rest has no tendency to start rotating.

Axis of rotation (perpendicular to figure)

(b) This body has no tendency to accelerate as a whole, but it has a tendency to start rotating.

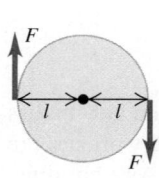

First condition satisfied: Net force = 0, so body at rest has no tendency to start moving as a whole.

Second condition NOT satisfied: There is a net clockwise torque about the axis, so body at rest will start rotating clockwise.

(c) This body has a tendency to accelerate as a whole but no tendency to start rotating.

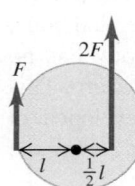

First condition NOT satisfied: There is a net upward force, so body at rest will start moving upward.

Second condition satisfied: Net torque about the axis = 0, so body at rest has no tendency to start rotating.

First condition for equilibrium: For the center of mass of a body at rest to remain at rest ...	$\sum \vec{F} = 0$	... net external force on the body must be *zero*.	(11.1)

A second condition for an extended body to be in equilibrium is that the body must have no tendency to *rotate*. A rigid body that, in an inertial frame, is not rotating about a certain point has zero angular momentum about that point. If it is not to start rotating about that point, the rate of change of angular momentum must *also* be zero. From the discussion in Section 10.5, particularly Eq. (10.29), this means that the sum of torques due to all the external forces acting on the body must be zero. A rigid body in equilibrium can't have any tendency to start rotating about *any* point, so the sum of external torques must be zero about any point. This is the **second condition for equilibrium:**

Second condition for equilibrium: For a nonrotating body to remain nonrotating ...	$\sum \vec{\tau} = 0$	... net external torque *around any point* on the body must be *zero*.	(11.2)

In this chapter we'll apply the first and second conditions for equilibrium to situations in which a rigid body is at rest (no translation or rotation). Such a body is said to be in **static equilibrium (Fig. 11.1)**. But the same conditions apply to a rigid body in uniform *translational* motion (without rotation), such as an airplane in flight with constant speed, direction, and altitude. Such a body is in equilibrium but is not static.

TEST YOUR UNDERSTANDING OF SECTION 11.1 Which situation satisfies both the first and second conditions for equilibrium? (i) A seagull gliding at a constant angle below the horizontal and at a constant speed; (ii) an automobile crankshaft turning at an increasing angular speed in the engine of a parked car; (iii) a thrown baseball that does not rotate as it sails through the air. ▮

11.2 CENTER OF GRAVITY

In most equilibrium problems, one of the forces acting on the body is its weight. We need to be able to calculate the *torque* of this force. The weight doesn't act at a single point; it is distributed over the entire body. But we can always calculate the torque due to the body's weight by assuming that the entire force of gravity (weight) is concentrated at a point called the **center of gravity** (abbreviated "cg"). The acceleration due to gravity decreases with altitude; but if we can ignore this variation over the vertical dimension of the body, then the body's center of gravity is identical to its *center of mass* (abbreviated "cm"), which we defined in Section 8.5. We stated this result without proof in Section 10.2, and now we'll prove it.

First let's review the definition of the center of mass. For a collection of particles with masses m_1, m_2, ... and coordinates (x_1, y_1, z_1), (x_2, y_2, z_2), ..., the coordinates x_{cm}, y_{cm}, and z_{cm} of the center of mass of the collection are

$$x_{cm} = \frac{m_1 x_1 + m_2 x_2 + m_3 x_3 + \cdots}{m_1 + m_2 + m_3 + \cdots} = \frac{\sum_i m_i x_i}{\sum_i m_i}$$

$$y_{cm} = \frac{m_1 y_1 + m_2 y_2 + m_3 y_3 + \cdots}{m_1 + m_2 + m_3 + \cdots} = \frac{\sum_i m_i y_i}{\sum_i m_i} \qquad \text{(center of mass)} \qquad (11.3)$$

$$z_{cm} = \frac{m_1 z_1 + m_2 z_2 + m_3 z_3 + \cdots}{m_1 + m_2 + m_3 + \cdots} = \frac{\sum_i m_i z_i}{\sum_i m_i}$$

Also, x_{cm}, y_{cm}, and z_{cm} are the components of the position vector $\vec{r}_{cm}$ of the center of mass, so Eqs. (11.3) are equivalent to the vector equation

Position vector of center of mass $\cdots\cdots\rightarrow$
of a system of particles

Position vectors of individual particles

$$\vec{r}_{cm} = \frac{m_1 \vec{r}_1 + m_2 \vec{r}_2 + m_3 \vec{r}_3 + \cdots}{m_1 + m_2 + m_3 + \cdots} = \frac{\sum_i m_i \vec{r}_i}{\sum_i m_i} \qquad (11.4)$$

Masses of individual particles

Now consider the gravitational torque on a body of arbitrary shape (**Fig. 11.2**). We assume that the acceleration due to gravity $\vec{g}$ is the same at every point in the body. Every particle in the body experiences a gravitational force, and the total weight of the body is the vector sum of a large number of parallel forces. A typical particle has mass m_i and weight $\vec{w}_i = m_i \vec{g}$. If $\vec{r}_i$ is the position vector of this particle with respect to an arbitrary origin O, then the torque vector $\vec{\tau}_i$ of the weight $\vec{w}_i$ with respect to O is, from Eq. (10.3),

$$\vec{\tau}_i = \vec{r}_i \times \vec{w}_i = \vec{r}_i \times m_i \vec{g}$$

The *total* torque due to the gravitational forces on all the particles is

$$\vec{\tau} = \sum_i \vec{\tau}_i = \vec{r}_1 \times m_1 \vec{g} + \vec{r}_2 \times m_2 \vec{g} + \cdots$$

$$= (m_1 \vec{r}_1 + m_2 \vec{r}_2 + \cdots) \times \vec{g}$$

$$= \left(\sum_i m_i \vec{r}_i \right) \times \vec{g}$$

When we multiply and divide this by the total mass of the body,

$$M = m_1 + m_2 + \cdots = \sum_i m_i$$

we get

$$\vec{\tau} = \frac{m_1 \vec{r}_1 + m_2 \vec{r}_2 + \cdots}{m_1 + m_2 + \cdots} \times M\vec{g} = \frac{\sum_i m_i \vec{r}_i}{\sum_i m_i} \times M\vec{g}$$

11.2 The center of gravity (cg) and center of mass (cm) of an extended body.

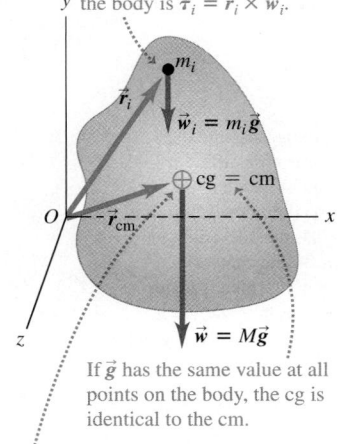

The gravitational torque about O on a particle of mass m_i within the body is $\vec{\tau}_i = \vec{r}_i \times \vec{w}_i$.

If $\vec{g}$ has the same value at all points on the body, the cg is identical to the cm.

The net gravitational torque about O on the entire body is the same as if all the weight acted at the cg: $\vec{\tau} = \vec{r}_{cm} \times \vec{w}$.

11.3 The acceleration due to gravity at the bottom of the 452-m-tall Petronas Towers in Malaysia is only 0.014% greater than at the top. The center of gravity of the towers is only about 2 cm below the center of mass.

11.4 Finding the center of gravity of an irregularly shaped body—in this case, a coffee mug.

Where is the center of gravity of this mug?

(1) Suspend the mug from any point. A vertical line extending down from the point of suspension passes through the center of gravity.

(2) Now suspend the mug from a different point. A vertical line extending down from this point intersects the first line at the center of gravity (which is inside the mug).

Center of gravity

The fraction in this equation is just the position vector $\vec{r}_{cm}$ of the center of mass, with components x_{cm}, y_{cm}, and z_{cm}, as given by Eq. (11.4), and $M\vec{g}$ is equal to the total weight $\vec{w}$ of the body. Thus

$$\vec{\tau} = \vec{r}_{cm} \times M\vec{g} = \vec{r}_{cm} \times \vec{w} \qquad (11.5)$$

The total gravitational torque, given by Eq. (11.5), is the same as though the total weight $\vec{w}$ were acting at the position $\vec{r}_{cm}$ of the center of mass, which we also call the *center of gravity*. **If $\vec{g}$ has the same value at all points on a body, its center of gravity is identical to its center of mass.** Note, however, that the center of mass is defined independently of any gravitational effect.

While the value of $\vec{g}$ varies somewhat with elevation, the variation is extremely slight (**Fig. 11.3**). We'll assume throughout this chapter that the center of gravity and center of mass are identical unless explicitly stated otherwise.

Finding and Using the Center of Gravity

We can often use symmetry considerations to locate the center of gravity of a body, just as we did for the center of mass. The center of gravity of a homogeneous sphere, cube, or rectangular plate is at its geometric center. The center of gravity of a right circular cylinder or cone is on its axis of symmetry.

For a body with a more complex shape, we can sometimes locate the center of gravity by thinking of the body as being made of symmetrical pieces. For example, we could approximate the human body as a collection of solid cylinders, with a sphere for the head. Then we can locate the center of gravity of the combination with Eqs. (11.3), letting m_1, m_2, ... be the masses of the individual pieces and (x_1, y_1, z_1), (x_2, y_2, z_2), ... be the coordinates of their centers of gravity.

When a body in rotational equilibrium and acted on by gravity is supported or suspended at a single point, the center of gravity is always at or directly above or below the point of suspension. If it were anywhere else, the weight would have a torque with respect to the point of suspension, and the body could not be in rotational equilibrium. **Figure 11.4** shows an application of this idea.

Using the same reasoning, we can see that a body supported at several points must have its center of gravity somewhere within the area bounded by the supports. This explains why a car can drive on a straight but slanted road if the slant angle is relatively small (**Fig. 11.5a**) but will tip over if the angle is too steep (Fig. 11.5b). The truck in Fig. 11.5c has a higher center of gravity than the car and will tip over on a shallower incline.

The lower the center of gravity and the larger the area of support, the harder it is to overturn a body. Four-legged animals such as deer and horses have a large area of support bounded by their legs; hence they are naturally stable and need only small feet or hooves. Animals that walk on two legs, such as humans and birds, need relatively large feet to give them a reasonable area of support. If a

11.5 In (a) the center of gravity is within the area bounded by the supports, and the car is in equilibrium. The car in (b) and the truck in (c) will tip over because their centers of gravity lie outside the area of support.

(a)   (b)   (c)

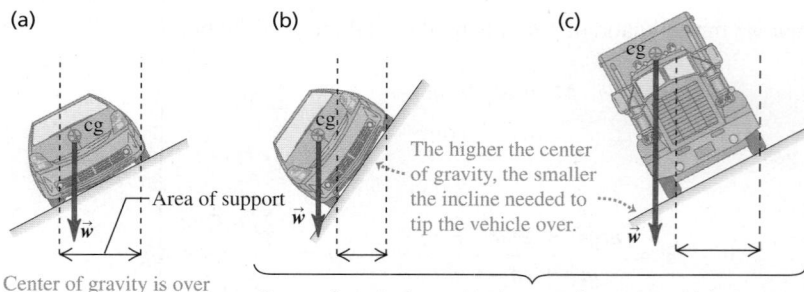

The higher the center of gravity, the smaller the incline needed to tip the vehicle over.

Area of support

Center of gravity is over the area of support: car is in equilibrium.

Center of gravity is outside the area of support: vehicle tips over.

two-legged animal holds its body approximately horizontal, like a chicken or the dinosaur *Tyrannosaurus rex,* it must perform a balancing act as it walks to keep its center of gravity over the foot that is on the ground. A chicken does this by moving its head; *T. rex* probably did it by moving its massive tail.

DEMO

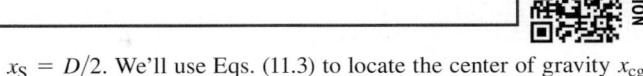

SOLUTION

EXAMPLE 11.1 WALKING THE PLANK

A uniform plank of length $L = 6.0$ m and mass $M = 90$ kg rests on sawhorses separated by $D = 1.5$ m and equidistant from the center of the plank. Cousin Throckmorton wants to stand on the right-hand end of the plank. If the plank is to remain at rest, how massive can Throckmorton be?

SOLUTION

IDENTIFY and SET UP: To just balance, Throckmorton's mass m must be such that the center of gravity of the plank–Throcky system is directly over the right-hand sawhorse (**Fig. 11.6**). We take the origin at C, the geometric center and center of gravity of the plank, and take the positive x-axis horizontally to the right. Then the centers of gravity of the plank and Throcky are at $x_P = 0$ and $x_T = L/2 = 3.0$ m, respectively, and the right-hand sawhorse is at

11.6 Our sketch for this problem.

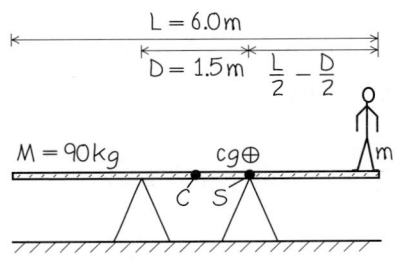

$x_S = D/2$. We'll use Eqs. (11.3) to locate the center of gravity x_{cg} of the plank–Throcky system.

EXECUTE: From the first of Eqs. (11.3),

$$x_{cg} = \frac{M(0) + m(L/2)}{M + m} = \frac{m}{M + m} \frac{L}{2}$$

We set $x_{cg} = x_S$ and solve for m:

$$\frac{m}{M + m} \frac{L}{2} = \frac{D}{2}$$

$$mL = (M + m)D$$

$$m = M\frac{D}{L - D} = (90 \text{ kg})\frac{1.5 \text{ m}}{6.0 \text{ m} - 1.5 \text{ m}} = 30 \text{ kg}$$

EVALUATE: As a check, let's repeat the calculation with the origin at the right-hand sawhorse. Now $x_S = 0$, $x_P = -D/2$, and $x_T = (L/2) - (D/2)$, and we require $x_{cg} = x_S = 0$:

$$x_{cg} = \frac{M(-D/2) + m[(L/2) - (D/2)]}{M + m} = 0$$

$$m = \frac{MD/2}{(L/2) - (D/2)} = M\frac{D}{L - D} = 30 \text{ kg}$$

The result doesn't depend on our choice of origin.

A 60-kg adult could stand only halfway between the right-hand sawhorse and the end of the plank. Can you see why?

TEST YOUR UNDERSTANDING OF SECTION 11.2 A rock is attached to the left end of a uniform meter stick that has the same mass as the rock. In order for the combination of rock and meter stick to balance atop the triangular object in **Fig. 11.7**, how far from the left end of the stick should the triangular object be placed? (i) Less than 0.25 m; (ii) 0.25 m; (iii) between 0.25 m and 0.50 m; (iv) 0.50 m; (v) more than 0.50 m. ▮

11.7 At what point will the meter stick with rock attached be in balance?

Rock, mass m Meter stick, mass m

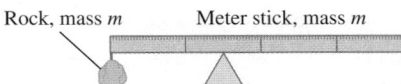

11.3 SOLVING RIGID-BODY EQUILIBRIUM PROBLEMS

There are just two key conditions for rigid-body equilibrium: The vector sum of the forces on the body must be zero, and the sum of the torques about any point must be zero. To keep things simple, we'll restrict our attention to situations in which we can treat all forces as acting in a single plane, which we'll call the xy-plane. Then we need consider only the x- and y-components of force in Eq. (11.1), and in Eq. (11.2) we need consider only the z-components of torque (perpendicular to the plane). The first and second conditions for equilibrium are then

$$\sum F_x = 0 \quad \text{and} \quad \sum F_y = 0 \quad \text{(first condition for equilibrium, forces in xy-plane)}$$

$$\sum \tau_z = 0 \quad \text{(second condition for equilibrium, forces in xy-plane)}$$

(11.6)

CAUTION **Choosing the reference point for calculating torques** In equilibrium problems, the choice of reference point for calculating torques in $\sum \tau_z$ is completely arbitrary. But once you make your choice, you must use the *same* point to calculate *all* the torques on a body. Choose the point so as to simplify the calculations as much as possible. ▮

The challenge is to apply these simple conditions to specific problems. Problem-Solving Strategy 11.1 is very similar to the suggestions given in Section 5.1 for the equilibrium of a particle. You should compare it with Problem-Solving Strategy 10.1 (Section 10.2) for rotational dynamics problems.

PROBLEM-SOLVING STRATEGY 11.1 | EQUILIBRIUM OF A RIGID BODY

IDENTIFY *the relevant concepts:* The first and second conditions for equilibrium ($\sum F_x = 0$, $\sum F_y = 0$, and $\sum \tau_z = 0$) are applicable to any rigid body that is not accelerating in space and not rotating.

SET UP *the problem* using the following steps:
1. Sketch the physical situation and identify the body in equilibrium to be analyzed. Sketch the body accurately; do *not* represent it as a point. Include dimensions.
2. Draw a free-body diagram showing all forces acting *on* the body. Show the point on the body at which each force acts.
3. Choose coordinate axes and specify their direction. Specify a positive direction of rotation for torques. Represent forces in terms of their components with respect to the chosen axes.
4. Choose a reference point about which to compute torques. Choose wisely; you can eliminate from your torque equation

any force whose line of action goes through the point you choose. The body doesn't actually have to be pivoted about an axis through the reference point.

EXECUTE *the solution* as follows:
1. Write equations expressing the equilibrium conditions. Remember that $\sum F_x = 0$, $\sum F_y = 0$, and $\sum \tau_z = 0$ are *separate* equations. You can compute the torque of a force by finding the torque of each of its components separately, each with its appropriate lever arm and sign, and adding the results.
2. To obtain as many equations as you have unknowns, you may need to compute torques with respect to two or more reference points; choose them wisely, too.

EVALUATE *your answer:* Check your results by writing $\sum \tau_z = 0$ with respect to a different reference point. You should get the same answers.

EXAMPLE 11.2 LOCATING YOUR CENTER OF GRAVITY WHILE YOU WORK OUT

The *plank* (**Fig. 11.8a**) is a great way to strengthen abdominal, back, and shoulder muscles. You can also use this exercise position to locate your center of gravity. Holding plank position with a scale under his toes and another under his forearms, one athlete measured that 66.0% of his weight was supported by his forearms and 34.0% by his toes. (That is, the total normal forces on his forearms and toes were $0.660w$ and $0.340w$, respectively, where w is the athlete's weight.) He is 1.80 m tall, and in plank position

the distance from his toes to the middle of his forearms is 1.53 m. How far from his toes is his center of gravity?

SOLUTION

IDENTIFY and SET UP: We can use the two conditions for equilibrium, Eqs. (11.6), for an athlete at rest. So both the net force and net torque on the athlete are zero. Figure 11.8b shows a free-body diagram, including x- and y-axes and our convention that counterclockwise torques are positive. The weight w acts at the center of gravity, which is between the two supports (as it must be; see Section 11.2). Our target variable is the distance L_{cg}, the lever arm of the weight with respect to the toes T, so it is wise to take torques with respect to T. The torque due to the weight is negative (it tends to cause a clockwise rotation around T), and the torque due to the upward normal force at the forearms F is positive (it tends to cause a counterclockwise rotation around T).

EXECUTE: The first condition for equilibrium is satisfied (Fig. 11.8b): $\sum F_x = 0$ because there are no x-components and $\sum F_y = 0$ because $0.340w + 0.660w + (-w) = 0$. We write the torque equation and solve for L_{cg}:

$$\sum \tau_R = 0.340w(0) - wL_{cg} + 0.660w(1.53 \text{ m}) = 0$$

$$L_{cg} = 1.01 \text{ m}$$

EVALUATE: The center of gravity is slightly below our athlete's navel (as it is for most people) and closer to his forearms than to his toes, which is why his forearms support most of his weight. You can check our result by writing the torque equation about the forearms F. You'll find that his center of gravity is 0.52 m from his forearms, or $(1.53 \text{ m}) - (0.52 \text{ m}) = 1.01 \text{ m}$ from his toes.

11.8 An athlete in plank position.

(a)

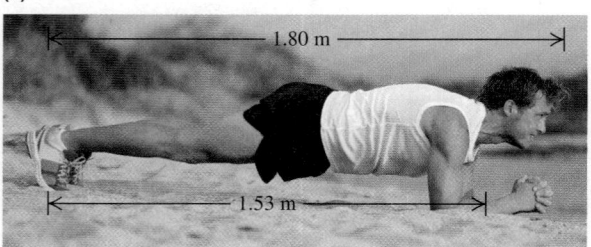

(b)

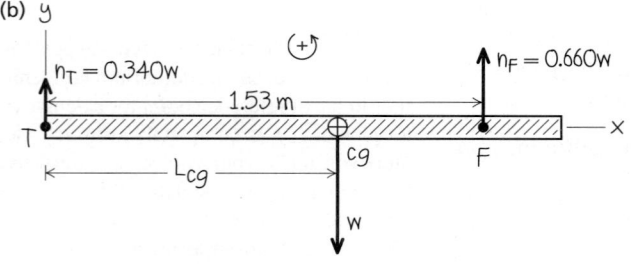

EXAMPLE 11.3 WILL THE LADDER SLIP?

Sir Lancelot, who weighs 800 N, is assaulting a castle by climbing a uniform ladder that is 5.0 m long and weighs 180 N (**Fig. 11.9a**). The bottom of the ladder rests on a ledge and leans across the moat in equilibrium against a frictionless, vertical castle wall. The ladder makes an angle of 53.1° with the horizontal. Lancelot pauses one-third of the way up the ladder. (a) Find the normal and friction forces on the base of the ladder. (b) Find the minimum coefficient of static friction needed to prevent slipping at the base. (c) Find the magnitude and direction of the contact force on the base of the ladder.

SOLUTION

IDENTIFY and SET UP: The ladder–Lancelot system is stationary, so we can use the two conditions for equilibrium to solve part (a). In part (b), we need the relationship among the static friction force, coefficient of static friction, and normal force (see Section 5.3). In part (c), the contact force is the vector sum of the normal and friction forces acting at the base of the ladder, found in part (a). Figure 11.9b shows the free-body diagram, with x- and y-directions as shown and with counterclockwise torques taken to be positive. The ladder's center of gravity is at its geometric center. Lancelot's 800-N weight acts at a point one-third of the way up the ladder.

The wall exerts only a normal force n_1 on the top of the ladder. The forces on the base are an upward normal force n_2 and a static friction force f_s, which must point to the right to prevent slipping. The magnitudes n_2 and f_s are the target variables in part (a). From Eq. (5.4), these magnitudes are related by $f_s \leq \mu_s n_2$; the coefficient of static friction μ_s is the target variable in part (b).

EXECUTE: (a) From Eqs. (11.6), the first condition for equilibrium gives

$$\Sigma F_x = f_s + (-n_1) = 0$$
$$\Sigma F_y = n_2 + (-800\ \text{N}) + (-180\ \text{N}) = 0$$

These are two equations for the three unknowns n_1, n_2, and f_s. The second equation gives $n_2 = 980$ N. To obtain a third equation, we use the second condition for equilibrium. We take torques about point B, about which n_2 and f_s have no torque. The 53.1° angle creates a 3-4-5 right triangle, so from Fig. 11.9b the lever arm for the ladder's weight is 1.5 m, the lever arm for Lancelot's

weight is 1.0 m, and the lever arm for n_1 is 4.0 m. The torque equation for point B is then

$$\Sigma \tau_B = n_1(4.0\ \text{m}) - (180\ \text{N})(1.5\ \text{m})$$
$$- (800\ \text{N})(1.0\ \text{m}) + n_2(0) + f_s(0) = 0$$

Solving for n_1, we get $n_1 = 268$ N. We substitute this into the $\Sigma F_x = 0$ equation and get $f_s = 268$ N.

(b) The static friction force f_s cannot exceed $\mu_s n_2$, so the *minimum* coefficient of static friction to prevent slipping is

$$(\mu_s)_{\text{min}} = \frac{f_s}{n_2} = \frac{268\ \text{N}}{980\ \text{N}} = 0.27$$

(c) The components of the contact force $\vec{F}_B$ at the base are the static friction force f_s and the normal force n_2, so

$$\vec{F}_B = f_s \hat{\imath} + n_2 \hat{\jmath} = (268\ \text{N})\hat{\imath} + (980\ \text{N})\hat{\jmath}$$

The magnitude and direction of $\vec{F}_B$ (Fig. 11.9c) are

$$F_B = \sqrt{(268\ \text{N})^2 + (980\ \text{N})^2} = 1020\ \text{N}$$
$$\theta = \arctan \frac{980\ \text{N}}{268\ \text{N}} = 75°$$

EVALUATE: As Fig. 11.9c shows, the contact force $\vec{F}_B$ is *not* directed along the length of the ladder. Can you show that if $\vec{F}_B$ were directed along the ladder, there would be a net counterclockwise torque with respect to the top of the ladder, and equilibrium would be impossible?

As Lancelot climbs higher on the ladder, the lever arm and torque of his weight about B increase. This increases the values of n_1, f_s, and the required friction coefficient $(\mu_s)_{\text{min}}$, so the ladder is more and more likely to slip as he climbs (see Exercise 11.10). A simple way to make slipping less likely is to use a larger ladder angle (say, 75° rather than 53.1°). This decreases the lever arms with respect to B of the weights of the ladder and Lancelot and increases the lever arm of n_1, all of which decrease the required friction force.

If we had assumed friction on the wall as well as on the floor, the problem would be impossible to solve by using the equilibrium conditions alone. (Try it!) The difficulty is that it's no longer adequate to treat the body as being perfectly rigid. Another problem of this kind is a four-legged table; there's no way to use the equilibrium conditions alone to find the force on each separate leg.

11.9 (a) Sir Lancelot pauses a third of the way up the ladder, fearing it will slip. (b) Free-body diagram for the system of Sir Lancelot and the ladder. (c) The contact force at B is the superposition of the normal force and the static friction force.

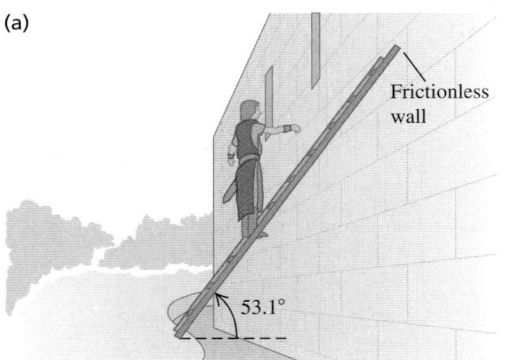

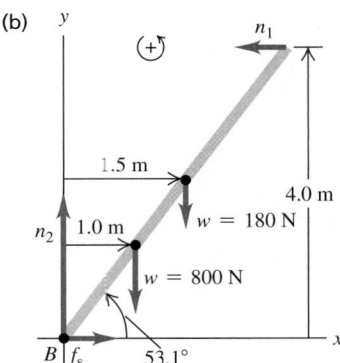

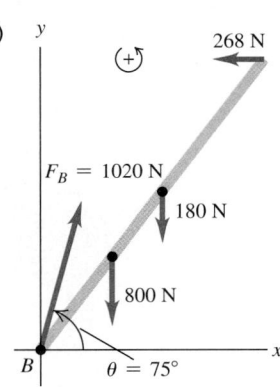

EXAMPLE 11.4 EQUILIBRIUM AND PUMPING IRON

Figure 11.10a shows a horizontal human arm lifting a dumbbell. The forearm is in equilibrium under the action of the weight $\vec{w}$ of the dumbbell, the tension $\vec{T}$ in the tendon connected to the biceps muscle, and the force $\vec{E}$ exerted on the forearm by the upper arm at the elbow joint. We ignore the weight of the forearm itself. (For clarity, in the drawing we've exaggerated the distance from the elbow to the point A where the tendon is attached.) Given the weight w and the angle θ between the tension force and the horizontal, find T and the two components of $\vec{E}$ (three unknown scalar quantities in all).

SOLUTION

IDENTIFY and SET UP: The system is at rest, so we use the conditions for equilibrium. We represent $\vec{T}$ and $\vec{E}$ in terms of their components (Fig. 11.10b). We guess that the directions of E_x and E_y are as shown; the signs of E_x and E_y as given by our solution will tell us the actual directions. Our target variables are T, E_x, and E_y.

EXECUTE: To find T, we take torques about the elbow joint so that the torque equation does not contain E_x, E_y, or T_x, then solve for T_y and hence T:

$$\sum \tau_{\text{elbow}} = Lw - DT_y = 0$$

$$T_y = \frac{Lw}{D} = T\sin\theta \quad \text{and} \quad T = \frac{Lw}{D\sin\theta}$$

To find E_x and E_y, we use the first conditions for equilibrium:

$$\sum F_x = T_x + (-E_x) = 0$$

$$E_x = T_x = T\cos\theta = \frac{Lw}{D\sin\theta}\cos\theta$$

$$= \frac{Lw}{D}\cot\theta = \frac{Lw}{D}\frac{D}{h} = \frac{Lw}{h}$$

$$\sum F_y = T_y + E_y + (-w) = 0$$

$$E_y = w - \frac{Lw}{D} = -\frac{(L-D)w}{D}$$

The negative sign for E_y tells us that it should actually point *down* in Fig. 11.10b.

EVALUATE: We can check our results for E_x and E_y by taking torques about points A and B, about both of which T has zero torque:

$$\sum \tau_A = (L-D)w + DE_y = 0 \quad \text{so} \quad E_y = -\frac{(L-D)w}{D}$$

$$\sum \tau_B = Lw - hE_x = 0 \quad \text{so} \quad E_x = \frac{Lw}{h}$$

As a realistic example, take $w = 200$ N, $D = 0.050$ m, $L = 0.30$ m, and $\theta = 80°$, so that $h = D\tan\theta = (0.050$ m$)(5.67) = 0.28$ m. Using our results for T, E_x, and E_y, we find

$$T = \frac{Lw}{D\sin\theta} = \frac{(0.30\text{ m})(200\text{ N})}{(0.050\text{ m})(0.98)} = 1220\text{ N}$$

$$E_y = -\frac{(L-D)w}{D} = -\frac{(0.30\text{ m} - 0.050\text{ m})(200\text{ N})}{0.050\text{ m}}$$

$$= -1000\text{ N}$$

$$E_x = \frac{Lw}{h} = \frac{(0.30\text{ m})(200\text{ N})}{0.28\text{ m}} = 210\text{ N}$$

The magnitude of the force at the elbow is

$$E = \sqrt{E_x^2 + E_y^2} = 1020\text{ N}$$

Note that T and E are *much* larger than the 200-N weight of the dumbbell. A forearm weighs only about 20 N, so it was reasonable to ignore its weight.

11.10 (a) The situation. (b) Our free-body diagram for the forearm. The weight of the forearm is ignored, and the distance D is greatly exaggerated for clarity.

(a)

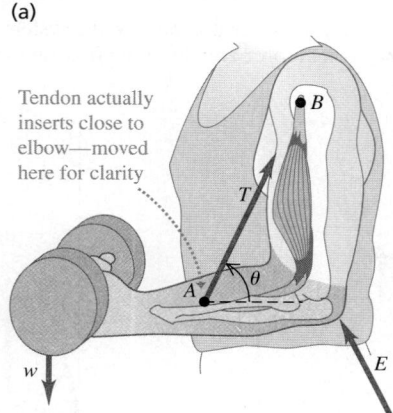

(b)

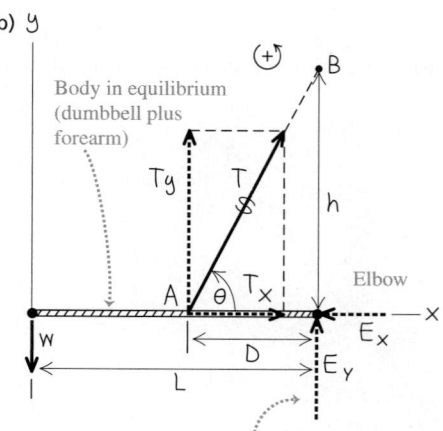

TEST YOUR UNDERSTANDING OF SECTION 11.3 A metal advertising sign (weight *w*) for a specialty shop is suspended from the end of a horizontal rod of length *L* and negligible mass (**Fig. 11.11**). The rod is supported by a cable at an angle θ from the horizontal and by a hinge at point *P*. Rank the following force magnitudes in order from greatest to smallest: (i) the weight *w* of the sign; (ii) the tension in the cable; (iii) the vertical component of force exerted on the rod by the hinge at *P*. ❚

11.11 What are the tension in the diagonal cable and the vertical component of force exerted by the hinge at *P*?

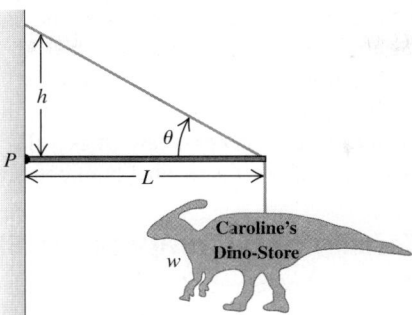

11.4 STRESS, STRAIN, AND ELASTIC MODULI

The rigid body is a useful idealized model, but the stretching, squeezing, and twisting of real bodies when forces are applied are often too important to ignore. **Figure 11.12** shows three examples. We want to study the relationship between the forces and deformations for each case.

You don't have to look far to find a deformable body; it's as plain as the nose on your face (**Fig. 11.13**). If you grasp the tip of your nose between your index finger and thumb, you'll find that the harder you pull your nose outward or push it inward, the more it stretches or compresses. Likewise, the harder you squeeze your index finger and thumb together, the more the tip of your nose compresses. If you try to twist the tip of your nose, you'll get a greater amount of twist if you apply stronger forces.

These observations illustrate a general rule. In each case you apply a **stress** to your nose; the amount of stress is a measure of the forces causing the deformation, on a "force per unit area" basis. And in each case the stress causes a deformation, or **strain.** More careful versions of the experiments with your nose suggest that for relatively small stresses, the resulting strain is proportional to the stress: The greater the deforming forces, the greater the resulting deformation. This proportionality is called **Hooke's law,** and the ratio of stress to strain is called the **elastic modulus:**

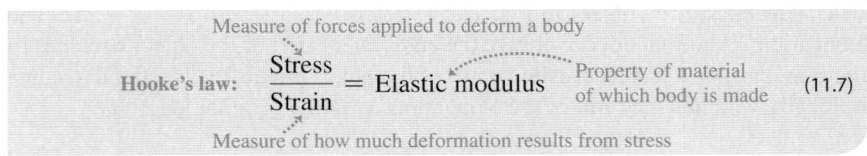

Measure of forces applied to deform a body

Hooke's law: $\dfrac{\text{Stress}}{\text{Strain}} = \text{Elastic modulus}$ Property of material of which body is made   (11.7)

Measure of how much deformation results from stress

11.12 Three types of stress. (a) Guitar strings under *tensile stress,* being stretched by forces acting at their ends. (b) A diver under *bulk stress,* being squeezed from all sides by forces due to water pressure. (c) A ribbon under *shear stress,* being deformed and eventually cut by forces exerted by the scissors.

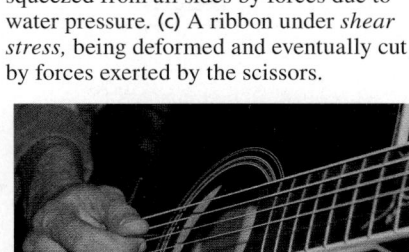

11.13 When you pinch your nose, the force per area that you apply to your nose is called *stress.* The fractional change in the size of your nose (the change in size divided by the initial size) is called *strain.* The deformation is *elastic* because your nose springs back to its initial size when you stop pinching.

The value of the elastic modulus depends on what the body is made of but not its shape or size. If a material returns to its original state after the stress is removed, it is called **elastic;** Hooke's law is a special case of elastic behavior. If a material instead remains deformed after the stress is removed, it is called **plastic.** Here we'll consider elastic behavior only; we'll return to plastic behavior in Section 11.5.

We used one form of Hooke's law in Section 6.3: The elongation of an ideal spring is proportional to the stretching force. Remember that Hooke's "law" is not really a general law; it is valid over only a limited range of stresses. In Section 11.5 we'll see what happens beyond that limited range.

Tensile and Compressive Stress and Strain

The simplest elastic behavior to understand is the stretching of a bar, rod, or wire when its ends are pulled (Fig. 11.12a). **Figure 11.14** shows an object that initially has uniform cross-sectional area A and length l_0. We then apply forces of equal magnitude $F_\perp$ but opposite directions at the ends (this ensures that the object has no tendency to move left or right). We say that the object is in **tension.** We've already talked a lot about tension in ropes and strings; it's the same concept here. The subscript $\perp$ is a reminder that the forces act perpendicular to the cross section.

We define the **tensile stress** at the cross section as the ratio of the force $F_\perp$ to the cross-sectional area A:

$$\text{Tensile stress} = \frac{F_\perp}{A} \tag{11.8}$$

This is a *scalar* quantity because $F_\perp$ is the *magnitude* of the force. The SI unit of stress is the **pascal** (abbreviated Pa and named for the 17th-century French scientist and philosopher Blaise Pascal). Equation (11.8) shows that 1 pascal equals 1 newton per square meter (N/m^2):

$$1 \text{ pascal} = 1 \text{ Pa} = 1 \text{ N/m}^2$$

In the British system the most common unit of stress is the pound per square inch $(\text{lb/in.}^2 \text{ or psi})$. The conversion factors are

$$1 \text{ psi} = 6895 \text{ Pa} \qquad \text{and} \qquad 1 \text{ Pa} = 1.450 \times 10^{-4} \text{ psi}$$

The units of stress are the same as those of *pressure,* which we will encounter often in later chapters.

Under tension the object in Fig. 11.14 stretches to a length $l = l_0 + \Delta l$. The elongation Δl does not occur only at the ends; every part of the object stretches in the same proportion. The **tensile strain** of the object equals the fractional change in length, which is the ratio of the elongation Δl to the original length l_0:

$$\text{Tensile strain} = \frac{l - l_0}{l_0} = \frac{\Delta l}{l_0} \tag{11.9}$$

Tensile strain is stretch per unit length. It is a ratio of two lengths, always measured in the same units, and so is a pure (dimensionless) number with no units.

Experiment shows that for a sufficiently small tensile stress, stress and strain are proportional, as in Eq. (11.7). The corresponding elastic modulus is called **Young's modulus,** denoted by Y:

11.14 An object in tension. The net force on the object is zero, but the object deforms. The tensile stress (the ratio of the force to the cross-sectional area) produces a tensile strain (the elongation divided by the initial length). The elongation Δl is exaggerated for clarity.

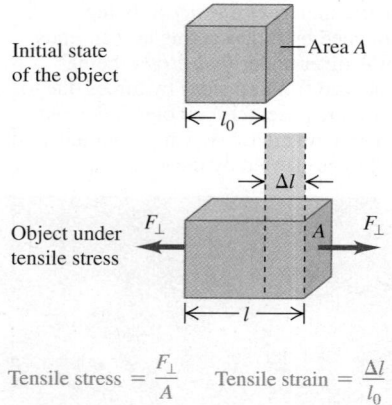

Initial state of the object — Area A

$\leftarrow l_0 \rightarrow$

$\rightarrow \Delta l \leftarrow$

Object under tensile stress — $F_\perp \quad A \quad F_\perp$

$\leftarrow l \rightarrow$

Tensile stress $= \dfrac{F_\perp}{A}$ Tensile strain $= \dfrac{\Delta l}{l_0}$

BIO Application Young's Modulus of a Tendon The anterior tibial tendon connects your foot to the large muscle that runs along the side of your shinbone. (You can feel this tendon at the front of your ankle.) Measurements show that this tendon has a Young's modulus of 1.2×10^9 Pa, much less than for the metals listed in Table 11.1. Hence this tendon stretches substantially (up to 2.5% of its length) in response to the stresses experienced in walking and running.

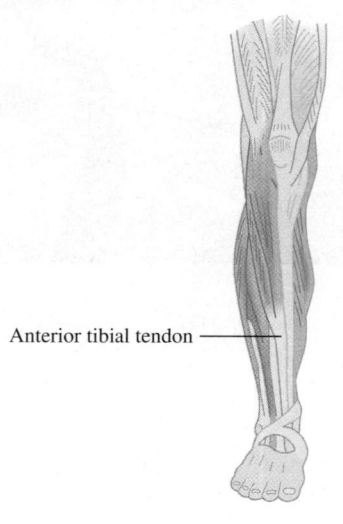

Anterior tibial tendon ——

$$Y = \frac{\text{Tensile stress}}{\text{Tensile strain}} = \frac{F_\perp/A}{\Delta l/l_0} = \frac{F_\perp}{A} \frac{l_0}{\Delta l} \tag{11.10}$$

Young's modulus for tension

Force applied perpendicular to cross section

Original length (see Fig. 11.14)

Cross-sectional area of object

Elongation (see Fig. 11.14)

TABLE 11.1	Approximate Elastic Moduli		
Material	**Young's Modulus, Y (Pa)**	**Bulk Modulus, B (Pa)**	**Shear Modulus, S (Pa)**
Aluminum	7.0×10^{10}	7.5×10^{10}	2.5×10^{10}
Brass	9.0×10^{10}	6.0×10^{10}	3.5×10^{10}
Copper	11×10^{10}	14×10^{10}	4.4×10^{10}
Iron	21×10^{10}	16×10^{10}	7.7×10^{10}
Lead	1.6×10^{10}	4.1×10^{10}	0.6×10^{10}
Nickel	21×10^{10}	17×10^{10}	7.8×10^{10}
Silicone rubber	0.001×10^{10}	0.2×10^{10}	0.0002×10^{10}
Steel	20×10^{10}	16×10^{10}	7.5×10^{10}
Tendon (typical)	0.12×10^{10}	—	—

Since strain is a pure number, the units of Young's modulus are the same as those of stress: force per unit area. **Table 11.1** lists some typical values. (This table also gives values of two other elastic moduli that we will discuss later in this chapter.) A material with a large value of Y is relatively unstretchable; a large stress is required for a given strain. For example, the value of Y for cast steel $(2 \times 10^{11}$ Pa$)$ is much larger than that for a tendon $(1.2 \times 10^9$ Pa$)$.

When the forces on the ends of a bar are pushes rather than pulls (**Fig. 11.15**), the bar is in **compression** and the stress is a **compressive stress.** The **compressive strain** of an object in compression is defined in the same way as the tensile strain, but Δl has the opposite direction. Hooke's law and Eq. (11.10) are valid for compression as well as tension if the compressive stress is not too great. For many materials, Young's modulus has the same value for both tensile and compressive stresses. Composite materials such as concrete and stone are an exception; they can withstand compressive stresses but fail under comparable tensile stresses. Stone was the primary building material used by ancient civilizations such as the Babylonians, Assyrians, and Romans, so their structures had to be designed to avoid tensile stresses. Hence they used arches in doorways and bridges, where the weight of the overlying material compresses the stones of the arch together and does not place them under tension.

In many situations, bodies can experience both tensile and compressive stresses at the same time. For example, a horizontal beam supported at each end sags under its own weight. As a result, the top of the beam is under compression while the bottom of the beam is under tension (**Fig. 11.16a**). To minimize the stress and hence the bending strain, the top and bottom of the beam are given a large cross-sectional area. There is neither compression nor tension along the centerline of the beam, so this part can have a small cross section; this helps keep the weight of the beam to a minimum and further helps reduce the stress. The result is an I-beam of the familiar shape used in building construction (Fig. 11.16b).

11.15 An object in compression. The compressive stress and compressive strain are defined in the same way as tensile stress and strain (see Fig. 11.14), except that Δl now denotes the distance that the object contracts.

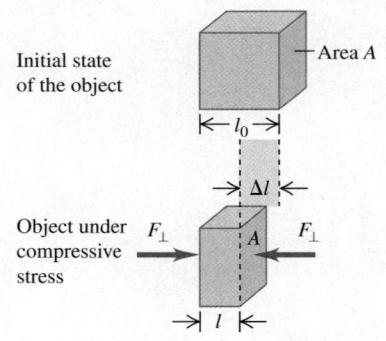

$$\text{Compressive stress} = \frac{F_\perp}{A} \qquad \text{Compressive strain} = \frac{\Delta l}{l_0}$$

(a)

Top of beam is under compression.

Beam's centerline is under neither tension nor compression.

Bottom of beam is under tension.

(b)

The top and bottom of an I-beam are broad to minimize the compressive and tensile stresses.

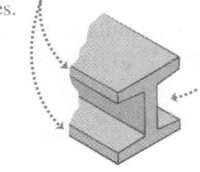

The beam can be narrow near its centerline, which is under neither compression nor tension.

11.16 (a) A beam supported at both ends is under both compression and tension. (b) The cross-sectional shape of an I-beam minimizes both stress and weight.

EXAMPLE 11.5 TENSILE STRESS AND STRAIN

A steel rod 2.0 m long has a cross-sectional area of 0.30 cm². It is hung by one end from a support, and a 550-kg milling machine is hung from its other end. Determine the stress on the rod and the resulting strain and elongation.

SOLUTION

IDENTIFY, SET UP, and EXECUTE: The rod is under tension, so we can use Eq. (11.8) to find the tensile stress; Eq. (11.9), with the value of Young's modulus Y for steel from Table 11.1, to find the corresponding strain; and Eq. (11.10) to find the elongation Δl:

$$\text{Tensile stress} = \frac{F_\perp}{A} = \frac{(550 \text{ kg})(9.8 \text{ m/s}^2)}{3.0 \times 10^{-5} \text{ m}^2} = 1.8 \times 10^8 \text{ Pa}$$

$$\text{Strain} = \frac{\Delta l}{l_0} = \frac{\text{Stress}}{Y} = \frac{1.8 \times 10^8 \text{ Pa}}{20 \times 10^{10} \text{ Pa}} = 9.0 \times 10^{-4}$$

$$\text{Elongation} = \Delta l = (\text{Strain}) \times l_0$$
$$= (9.0 \times 10^{-4})(2.0 \text{ m}) = 0.0018 \text{ m} = 1.8 \text{ mm}$$

EVALUATE: This small elongation, resulting from a load of over half a ton, is a testament to the stiffness of steel. (We've ignored the relatively small stress due to the weight of the rod itself.)

BIO Application Bulk Stress on an Anglerfish The anglerfish (*Melanocetus johnsoni*) is found in oceans throughout the world at depths as great as 1000 m, where the pressure (that is, the bulk stress) is about 100 atmospheres. Anglerfish are able to withstand such stress because they have no internal air spaces, unlike fish found in the upper ocean, where pressures are lower. The largest anglerfish are about 12 cm (5 in.) long.

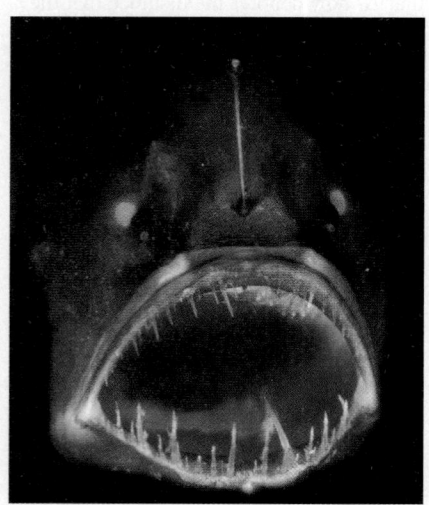

Bulk Stress and Strain

When a scuba diver plunges deep into the ocean, the water exerts nearly uniform pressure everywhere on his surface and squeezes him to a slightly smaller volume (see Fig. 11.12b). This is a different situation from the tensile and compressive stresses and strains we have discussed. The uniform pressure on all sides of the diver is a **bulk stress** (or **volume stress**), and the resulting deformation—a **bulk strain** (or **volume strain**)—is a change in his volume.

If an object is immersed in a fluid (liquid or gas) at rest, the fluid exerts a force on any part of the object's surface; this force is *perpendicular* to the surface. (If we tried to make the fluid exert a force parallel to the surface, the fluid would slip sideways to counteract the effort.) The force $F_\perp$ per unit area that the fluid exerts on an immersed object is called the **pressure** p in the fluid:

$$\text{Pressure in a fluid} \quad p = \frac{F_\perp}{A} \quad \begin{array}{l} \text{Force that fluid applies to} \\ \text{surface of an immersed object} \\ \text{Area over which force is exerted} \end{array} \quad (11.11)$$

Pressure has the same units as stress; commonly used units include 1 Pa ($= 1 \text{ N/m}^2$), 1 lb/in.² (1 psi), and 1 **atmosphere** (1 atm). One atmosphere is the approximate average pressure of the earth's atmosphere at sea level:

$$1 \text{ atmosphere} = 1 \text{ atm} = 1.013 \times 10^5 \text{ Pa} = 14.7 \text{ lb/in.}^2$$

CAUTION Pressure vs. force Unlike force, pressure has no intrinsic direction: The pressure on the surface of an immersed object is the same no matter how the surface is oriented. Hence pressure is a *scalar* quantity, not a vector quantity.

The pressure in a fluid increases with depth. For example, the pressure in the ocean increases by about 1 atm every 10 m. If an immersed object is relatively small, however, we can ignore these pressure differences for purposes of calculating bulk stress. We'll then treat the pressure as having the same value at all points on an immersed object's surface.

Pressure plays the role of stress in a volume deformation. The corresponding strain is the fractional change in volume (**Fig. 11.17**)—that is, the ratio of the volume change ΔV to the original volume V_0:

$$\text{Bulk (volume) strain} = \frac{\Delta V}{V_0} \quad (11.12)$$

Volume strain is the change in volume per unit volume. Like tensile or compressive strain, it is a pure number, without units.

When Hooke's law is obeyed, an increase in pressure (bulk stress) produces a *proportional* bulk strain (fractional change in volume). The corresponding elastic modulus (ratio of stress to strain) is called the **bulk modulus,** denoted by B. When the pressure on a body changes by a small amount Δp, from p_0 to $p_0 + \Delta p$, and the resulting bulk strain is $\Delta V/V_0$, Hooke's law takes the form

11.17 An object under bulk stress. Without the stress, the cube has volume V_0; when the stress is applied, the cube has a smaller volume V. The volume change ΔV is exaggerated for clarity.

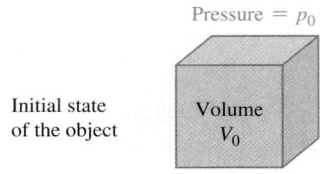

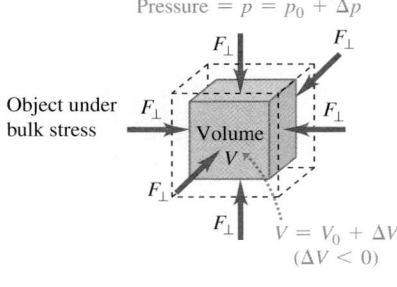

$$\underbrace{B}_{\substack{\text{Bulk modulus} \\ \text{for compression}}} = \frac{\text{Bulk stress}}{\text{Bulk strain}} = -\frac{\Delta p \overset{\text{Additional pressure on object}}{}}{\underset{\substack{\text{Original volume} \\ \text{(see Fig. 11.17)}}}{\Delta V/V_0}} \qquad (11.13)$$

$$\text{Change in volume (see Fig. 11.17)}$$

We include a minus sign in this equation because an *increase* of pressure always causes a *decrease* in volume. In other words, if Δp is positive, ΔV is negative. The bulk modulus B itself is a positive quantity.

For small pressure changes in a solid or a liquid, we consider B to be constant. The bulk modulus of a *gas*, however, depends on the initial pressure p_0. Table 11.1 includes values of B for several solid materials. Its units, force per unit area, are the same as those of pressure (and of tensile or compressive stress).

The reciprocal of the bulk modulus is called the **compressibility** and is denoted by k. From Eq. (11.13),

$$k = \frac{1}{B} = -\frac{\Delta V/V_0}{\Delta p} = -\frac{1}{V_0}\frac{\Delta V}{\Delta p} \qquad \text{(compressibility)} \qquad (11.14)$$

Compressibility is the fractional decrease in volume, $-\Delta V/V_0$, per unit increase Δp in pressure. The units of compressibility are those of *reciprocal pressure*, Pa^{-1} or atm^{-1}.

Table 11.2 lists the values of compressibility k for several liquids. For example, the compressibility of water is $46.4 \times 10^{-6} \text{ atm}^{-1}$, which means that the volume of water decreases by 46.4 parts per million for each 1-atmosphere increase in pressure. Materials with small bulk modulus B and large compressibility k are easiest to compress.

Compressibilities of Liquids

TABLE 11.2		
	Compressibility, k	
Liquid	Pa^{-1}	atm^{-1}
Carbon disulfide	93×10^{-11}	94×10^{-6}
Ethyl alcohol	110×10^{-11}	111×10^{-6}
Glycerine	21×10^{-11}	21×10^{-6}
Mercury	3.7×10^{-11}	3.8×10^{-6}
Water	45.8×10^{-11}	46.4×10^{-6}

EXAMPLE 11.6 BULK STRESS AND STRAIN

A hydraulic press contains 0.25 m^3 (250 L) of oil. Find the decrease in the volume of the oil when it is subjected to a pressure increase $\Delta p = 1.6 \times 10^7 \text{ Pa}$ (about 160 atm or 2300 psi). The bulk modulus of the oil is $B = 5.0 \times 10^9 \text{ Pa}$ (about $5.0 \times 10^4 \text{ atm}$), and its compressibility is $k = 1/B = 20 \times 10^{-6} \text{ atm}^{-1}$.

SOLUTION

IDENTIFY, SET UP, and EXECUTE: This example uses the ideas of bulk stress and strain. We are given both the bulk modulus and the compressibility, and our target variable is ΔV. Solving Eq. (11.13) for ΔV, we find

$$\Delta V = -\frac{V_0 \Delta p}{B} = -\frac{(0.25 \text{ m}^3)(1.6 \times 10^7 \text{ Pa})}{5.0 \times 10^9 \text{ Pa}}$$

$$= -8.0 \times 10^{-4} \text{ m}^3 = -0.80 \text{ L}$$

Alternatively, we can use Eq. (11.14) with the approximate unit conversions given above:

$$\Delta V = -kV_0\Delta p = -(20 \times 10^{-6} \text{ atm}^{-1})(0.25 \text{ m}^3)(160 \text{ atm})$$

$$= -8.0 \times 10^{-4} \text{ m}^3$$

EVALUATE: The negative value of ΔV means that the volume decreases when the pressure increases. The 160-atm pressure increase is large, but the *fractional* volume change is very small:

$$\frac{\Delta V}{V_0} = \frac{-8.0 \times 10^{-4} \text{ m}^3}{0.25 \text{ m}^3} = -0.0032 \qquad \text{or} \qquad -0.32\%$$

11.18 An object under shear stress. Forces are applied tangent to opposite surfaces of the object (in contrast to the situation in Fig. 11.14, in which the forces act perpendicular to the surfaces). The deformation x is exaggerated for clarity.

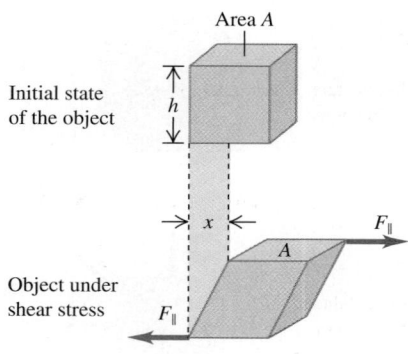

Initial state of the object

Object under shear stress

$$\text{Shear stress} = \frac{F_\parallel}{A} \qquad \text{Shear strain} = \frac{x}{h}$$

Shear Stress and Strain

The third kind of stress-strain situation is called *shear*. The ribbon in Fig. 11.12c is under **shear stress:** One part of the ribbon is being pushed up while an adjacent part is being pushed down, producing a deformation of the ribbon. **Figure 11.18** shows a body being deformed by a shear stress. In the figure, forces of equal magnitude but opposite direction act *tangent* to the surfaces of opposite ends of the object. We define the shear stress as the force $F_\parallel$ acting tangent to the surface divided by the area A on which it acts:

$$\text{Shear stress} = \frac{F_\parallel}{A} \tag{11.15}$$

Shear stress, like the other two types of stress, is a force per unit area.

Figure 11.18 shows that one face of the object under shear stress is displaced by a distance x relative to the opposite face. We define **shear strain** as the ratio of the displacement x to the transverse dimension h:

$$\text{Shear strain} = \frac{x}{h} \tag{11.16}$$

In real-life situations, x is typically much smaller than h. Like all strains, shear strain is a dimensionless number; it is a ratio of two lengths.

If the forces are small enough that Hooke's law is obeyed, the shear strain is *proportional* to the shear stress. The corresponding elastic modulus (ratio of shear stress to shear strain) is called the **shear modulus,** denoted by S:

Shear modulus for shear

Force applied tangent to surface of object ⋯⋯

Transverse dimension (see Fig. 11.18)

$$S = \frac{\text{Shear stress}}{\text{Shear strain}} = \frac{F_\parallel/A}{x/h} = \frac{F_\parallel}{A}\frac{h}{x} \tag{11.17}$$

Area over which force is exerted ⋯⋯

Deformation (see Fig. 11.18)

Table 11.1 gives several values of shear modulus. For a given material, S is usually one-third to one-half as large as Young's modulus Y for tensile stress. Keep in mind that the concepts of shear stress, shear strain, and shear modulus apply to *solid* materials only. The reason is that *shear* refers to deforming an object that has a definite shape (see Fig. 11.18). This concept doesn't apply to gases and liquids, which do not have definite shapes.

EXAMPLE 11.7   **SHEAR STRESS AND STRAIN**

Suppose the object in Fig. 11.18 is the brass base plate of an outdoor sculpture that experiences shear forces in an earthquake. The plate is 0.80 m square and 0.50 cm thick. What is the force exerted on each of its edges if the resulting displacement x is 0.16 mm?

SOLUTION

IDENTIFY and SET UP: This example uses the relationship among shear stress, shear strain, and shear modulus. Our target variable is the force $F_\parallel$ exerted parallel to each edge, as shown in Fig. 11.18. We'll find the shear strain from Eq. (11.16), the shear stress from Eq. (11.17), and $F_\parallel$ from Eq. (11.15). Table 11.1 gives the shear modulus of brass. In Fig. 11.18, h represents the 0.80-m length of each side of the plate. The area A in Eq. (11.15) is the product of the 0.80-m length and the 0.50-cm thickness.

EXECUTE: From Eq. (11.16),

$$\text{Shear strain} = \frac{x}{h} = \frac{1.6 \times 10^{-4}\,\text{m}}{0.80\,\text{m}} = 2.0 \times 10^{-4}$$

From Eq. (11.17),

$$\begin{aligned}\text{Shear stress} &= (\text{Shear strain}) \times S \\ &= (2.0 \times 10^{-4})(3.5 \times 10^{10}\,\text{Pa}) = 7.0 \times 10^6\,\text{Pa}\end{aligned}$$

Finally, from Eq. (11.15),

$$\begin{aligned}F_\parallel &= (\text{Shear stress}) \times A \\ &= (7.0 \times 10^6\,\text{Pa})(0.80\,\text{m})(0.0050\,\text{m}) = 2.8 \times 10^4\,\text{N}\end{aligned}$$

EVALUATE: The shear force supplied by the earthquake is more than 3 tons! The large shear modulus of brass makes it hard to deform. Further, the plate is relatively thick (0.50 cm), so the area A is relatively large and a substantial force $F_\parallel$ is needed to provide the necessary stress $F_\parallel/A$.

TEST YOUR UNDERSTANDING OF SECTION 11.4 A copper rod of cross-sectional area 0.500 cm² and length 1.00 m is elongated by 2.00×10^{-2} mm, and a steel rod of the same cross-sectional area but 0.100 m in length is elongated by 2.00×10^{-3} mm. (a) Which rod has greater tensile *strain*? (i) The copper rod; (ii) the steel rod; (iii) the strain is the same for both. (b) Which rod is under greater tensile *stress*? (i) The copper rod; (ii) the steel rod; (iii) the stress is the same for both. ▌

11.5 ELASTICITY AND PLASTICITY

Hooke's law—the proportionality of stress and strain in elastic deformations—has a limited range of validity. In the preceding section we used phrases such as "if the forces are small enough that Hooke's law is obeyed." Just what *are* the limitations of Hooke's law? What's more, if you pull, squeeze, or twist *anything* hard enough, it will bend or break. Can we be more precise than that?

To address these questions, let's look at a graph of tensile stress as a function of tensile strain. **Figure 11.19** shows a typical graph of this kind for a metal such as copper or soft iron. The strain is shown as the *percent* elongation; the horizontal scale is not uniform beyond the first portion of the curve, up to a strain of less than 1%. The first portion is a straight line, indicating Hooke's law behavior with stress directly proportional to strain. This straight-line portion ends at point *a*; the stress at this point is called the *proportional limit*.

From *a* to *b*, stress and strain are no longer proportional, and Hooke's law is *not* obeyed. However, from *a* to *b* (and *O* to *a*), the behavior of the material is *elastic:* If the load is gradually removed starting at any point between *O* and *b*, the curve is retraced until the material returns to its original length. This elastic deformation is *reversible*.

Point *b*, the end of the elastic region, is called the *yield point:* the stress at the yield point is called the *elastic limit*. When we increase the stress beyond point *b*, the strain continues to increase. But if we remove the load at a point like *c* beyond the elastic limit, the material does *not* return to its original length. Instead, it follows the red line in Fig. 11.19. The material has deformed *irreversibly* and acquired a *permanent set*. This is the *plastic* behavior mentioned in Section 11.4.

Once the material has become plastic, a small additional stress produces a relatively large increase in strain, until a point *d* is reached at which *fracture* takes place. That's what happens if a steel guitar string in Fig. 11.12a is tightened too much: The string breaks at the fracture point. Steel is *brittle* because it breaks soon after reaching its elastic limit; other materials, such as soft iron, are *ductile*—they can be given a large permanent stretch without breaking. (The material depicted in Fig. 11.19 is ductile, since it can stretch by more than 30% before breaking.)

Unlike uniform materials such as metals, stretchable biological materials such as tendons and ligaments have no true plastic region. That's because these materials are made of a collection of microscopic fibers; when stressed beyond the elastic limit, the fibers tear apart from each other. (A torn ligament or tendon is one that has fractured in this way.)

If a material is still within its elastic region, something very curious can happen when it is stretched and then allowed to relax. **Figure 11.20** is a stress-strain curve for vulcanized rubber that has been stretched by more than seven times its original length. The stress is not proportional to the strain, but the behavior is elastic because when the load is removed, the material returns to its original length. However, the material follows *different* curves for increasing and decreasing stress. This is called *elastic hysteresis*. The work done by the material when it returns to its original shape is less than the work required to deform it; that's due to internal friction. Rubber with large elastic hysteresis is very useful for absorbing vibrations, such as in engine mounts and shock-absorber bushings for cars. Tendons display similar behavior.

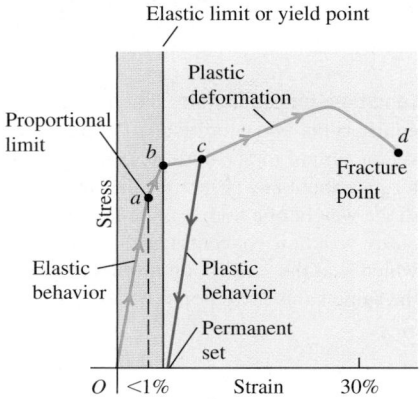

11.19 Typical stress-strain diagram for a ductile metal under tension.

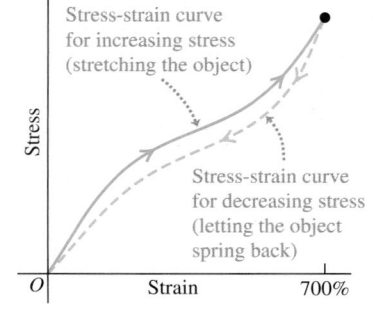

11.20 Typical stress-strain diagram for vulcanized rubber. The curves are different for increasing and decreasing stress, a phenomenon called elastic hysteresis.

Approximate Breaking
TABLE 11.3	Stresses

Material	Breaking Stress (Pa or N/m²)
Aluminum	2.2×10^8
Brass	4.7×10^8
Glass	10×10^8
Iron	3.0×10^8
Steel	$5-20 \times 10^8$
Tendon (typical)	1×10^8

The stress required to cause actual fracture of a material is called the *breaking stress,* the *ultimate strength,* or (for tensile stress) the *tensile strength.* Two materials, such as two types of steel, may have very similar elastic constants but vastly different breaking stresses. **Table 11.3** gives typical values of breaking stress for several materials in tension. Comparing Tables 11.1 and 11.3 shows that iron and steel are comparably *stiff* (they have almost the same value of Young's modulus), but steel is *stronger* (it has a larger breaking stress than does iron).

TEST YOUR UNDERSTANDING OF SECTION 11.5 While parking your car, you accidentally back into a steel post. You pull forward until the car no longer touches the post and then get out to inspect the damage. What does your rear bumper look like if the strain in the impact was (a) less than at the proportional limit; (b) greater than at the proportional limit but less than at the yield point; (c) greater than at the yield point but less than at the fracture point; and (d) greater than at the fracture point? ▮

CHAPTER 11 SUMMARY

 SOLUTIONS TO ALL EXAMPLES

Conditions for equilibrium: For a rigid body to be in equilibrium, two conditions must be satisfied. First, the vector sum of forces must be zero. Second, the sum of torques about any point must be zero. The torque due to the weight of a body can be found by assuming the entire weight is concentrated at the center of gravity, which is at the same point as the center of mass if $\vec{g}$ has the same value at all points. (See Examples 11.1–11.4.)

$$\sum \vec{F} = 0 \tag{11.1}$$
$$\sum \vec{\tau} = 0 \quad \text{about } any \text{ point} \tag{11.2}$$
$$\vec{r}_{cm} = \frac{m_1\vec{r}_1 + m_2\vec{r}_2 + m_3\vec{r}_3 + \cdots}{m_1 + m_2 + m_3 + \cdots} \tag{11.4}$$

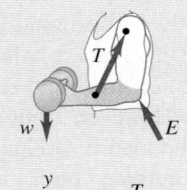

Stress, strain, and Hooke's law: Hooke's law states that in elastic deformations, stress (force per unit area) is proportional to strain (fractional deformation). The proportionality constant is called the elastic modulus.

$$\frac{\text{Stress}}{\text{Strain}} = \text{Elastic modulus} \tag{11.7}$$

Tensile and compressive stress: Tensile stress is tensile force per unit area, $F_\perp/A$. Tensile strain is fractional change in length, $\Delta l/l_0$. The elastic modulus for tension is called Young's modulus Y. Compressive stress and strain are defined in the same way. (See Example 11.5.)

$$Y = \frac{\text{Tensile stress}}{\text{Tensile strain}} = \frac{F_\perp/A}{\Delta l/l_0} = \frac{F_\perp}{A}\frac{l_0}{\Delta l} \tag{11.10}$$

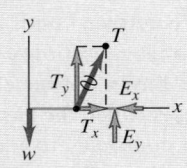

Bulk stress: Pressure in a fluid is force per unit area. Bulk stress is pressure change, Δp, and bulk strain is fractional volume change, $\Delta V/V_0$. The elastic modulus for compression is called the bulk modulus, B. Compressibility, k, is the reciprocal of bulk modulus: $k = 1/B$. (See Example 11.6.)

$$p = \frac{F_\perp}{A} \tag{11.11}$$
$$B = \frac{\text{Bulk stress}}{\text{Bulk strain}} = -\frac{\Delta p}{\Delta V/V_0} \tag{11.13}$$

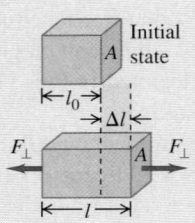
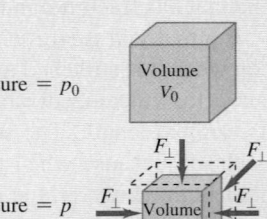
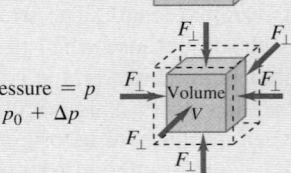

Shear stress: Shear stress is force per unit area, $F_\parallel/A$, for a force applied tangent to a surface. Shear strain is the displacement x of one side divided by the transverse dimension h. The elastic modulus for shear is called the shear modulus, S. (See Example 11.7.)

$$S = \frac{\text{Shear stress}}{\text{Shear strain}} = \frac{F_\parallel/A}{x/h} = \frac{F_\parallel \, h}{A \, x} \quad (11.17)$$

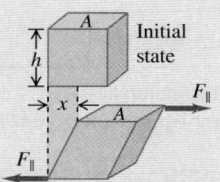

The limits of Hooke's law: The proportional limit is the maximum stress for which stress and strain are proportional. Beyond the proportional limit, Hooke's law is not valid. The elastic limit is the stress beyond which irreversible deformation occurs. The breaking stress, or ultimate strength, is the stress at which the material breaks.

BRIDGING PROBLEM   IN EQUILIBRIUM AND UNDER STRESS

A horizontal, uniform, solid copper rod has an original length l_0, cross-sectional area A, Young's modulus Y, bulk modulus B, shear modulus S, and mass m. It is supported by a frictionless pivot at its right end and by a cable a distance $l_0/4$ from its left end (**Fig. 11.21**). Both pivot and cable are attached so that they exert their forces uniformly over the rod's cross section. The cable makes an angle θ with the rod and compresses it. (a) Find the tension in the cable. (b) Find the magnitude and direction of the force exerted by the pivot on the right end of the rod. How does this magnitude compare to the cable tension? How does this angle compare to θ? (c) Find the change in length of the rod due to the stresses exerted by the cable and pivot on the rod. (The length change is small compared to the original length l_0.) (d) By what factor would your answer in part (c) increase if the solid copper rod were twice as long but had the same cross-sectional area?

11.21 What are the forces on the rod? What are the stress and strain?

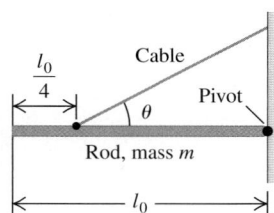

SOLUTION GUIDE

IDENTIFY and SET UP

1. Draw a free-body diagram for the rod. Be careful to place each force in the correct location.
2. List the unknown quantities, and decide which are the target variables.
3. What conditions must be met so that the rod remains at rest? What kind of stress (and resulting strain) is involved? Use your answers to select the appropriate equations.

EXECUTE

4. Use your equations to solve for the target variables. (*Hint:* You can make the solution easier by carefully choosing the point around which you calculate torques.)
5. Use trigonometry to decide whether the pivot force or the cable tension has the greater magnitude and whether the angle of the pivot force is greater than, less than, or equal to θ.

EVALUATE

6. Check whether your answers are reasonable. Which force, the cable tension or the pivot force, holds up more of the weight of the rod? Does this make sense?

Problems

For assigned homework and other learning materials, go to MasteringPhysics®.

•, ••, •••: Difficulty levels. **CP**: Cumulative problems incorporating material from earlier chapters. **CALC**: Problems requiring calculus. **DATA**: Problems involving real data, scientific evidence, experimental design, and/or statistical reasoning. **BIO**: Biosciences problems.

DISCUSSION QUESTIONS

Q11.1 Does a rigid object in uniform rotation about a fixed axis satisfy the first and second conditions for equilibrium? Why? Does it then follow that every particle in this object is in equilibrium? Explain.

Q11.2 (a) Is it possible for an object to be in translational equilibrium (the first condition) but *not* in rotational equilibrium (the second condition)? Illustrate your answer with a simple example. (b) Can an object be in rotational equilibrium yet *not* in translational equilibrium? Justify your answer with a simple example.

Q11.3 Car tires are sometimes "balanced" on a machine that pivots the tire and wheel about the center. Weights are placed around the wheel rim until it does not tip from the horizontal plane. Discuss this procedure in terms of the center of gravity.

Q11.4 Does the center of gravity of a solid body always lie within the material of the body? If not, give a counterexample.

Q11.5 In Section 11.2 we always assumed that the value of *g* was the same at all points on the body. This is *not* a good approximation if the dimensions of the body are great enough, because the value of *g* decreases with altitude. If this is taken into account, will the center of gravity of a long, vertical rod be above, below, or at its center of mass? Explain how this can be used to keep the long axis of an orbiting spacecraft pointed toward the earth. (This would be useful for a weather satellite that must always keep its camera lens trained on the earth.) The moon is not exactly spherical but is somewhat elongated. Explain why this same effect is responsible for keeping the same face of the moon pointed toward the earth at all times.

Q11.6 You are balancing a wrench by suspending it at a single point. Is the equilibrium stable, unstable, or neutral if the point is above, at, or below the wrench's center of gravity? In each case give the reasoning behind your answer. (For rotation, a rigid body is in *stable* equilibrium if a small rotation of the body produces a torque that tends to return the body to equilibrium; it is in *unstable* equilibrium if a small rotation produces a torque that tends to take the body farther from equilibrium; and it is in *neutral* equilibrium if a small rotation produces no torque.)

Q11.7 You can probably stand flatfooted on the floor and then rise up and balance on your tiptoes. Why are you unable do it if your toes are touching the wall of your room? (Try it!)

Q11.8 You freely pivot a horseshoe from a horizontal nail through one of its nail holes. You then hang a long string with a weight at its bottom from the same nail, so that the string hangs vertically in front of the horseshoe without touching it. How do you know that the horseshoe's center of gravity is along the line behind the string? How can you locate the center of gravity by repeating the process at another nail hole? Will the center of gravity be within the solid material of the horseshoe?

Q11.9 An object consists of a ball of weight *W* glued to the end of a uniform bar also of weight *W*. If you release it from rest, with the bar horizontal, what will its behavior be as it falls if air resistance is negligible? Will it (a) remain horizontal; (b) rotate about its center of gravity; (c) rotate about the ball; or (d) rotate so that the ball swings downward? Explain your reasoning.

Q11.10 Suppose that the object in Question 11.9 is released from rest with the bar tilted at 60° above the horizontal with the ball at

the upper end. As it is falling, will it (a) rotate about its center of gravity until it is horizontal; (b) rotate about its center of gravity until it is vertical with the ball at the bottom; (c) rotate about the ball until it is vertical with the ball at the bottom; or (d) remain at 60° above the horizontal?

Q11.11 Why must a water skier moving with constant velocity lean backward? What determines how far back she must lean? Draw a free-body diagram for the water skier to justify your answers.

Q11.12 In pioneer days, when a Conestoga wagon was stuck in the mud, people would grasp the wheel spokes and try to turn the wheels, rather than simply pushing the wagon. Why?

Q11.13 The mighty Zimbo claims to have leg muscles so strong that he can stand flat on his feet and lean forward to pick up an apple on the floor with his teeth. Should you pay to see him perform, or do you have any suspicions about his claim? Why?

Q11.14 Why is it easier to hold a 10-kg dumbbell in your hand at your side than it is to hold it with your arm extended horizontally?

Q11.15 Certain features of a person, such as height and mass, are fixed (at least over relatively long periods of time). Are the following features also fixed? (a) location of the center of gravity of the body; (b) moment of inertia of the body about an axis through the person's center of mass. Explain your reasoning.

Q11.16 During pregnancy, women often develop back pains from leaning backward while walking. Why do they have to walk this way?

Q11.17 Why is a tapered water glass with a narrow base easier to tip over than a glass with straight sides? Does it matter whether the glass is full or empty?

Q11.18 When a tall, heavy refrigerator is pushed across a rough floor, what factors determine whether it slides or tips?

Q11.19 A uniform beam is suspended horizontally and attached to a wall by a small hinge (**Fig. Q11.19**). What are the directions (upward or downward, and to the left or the right) of the components of the force that the hinge exerts *on the beam*? Explain.

Figure **Q11.19**

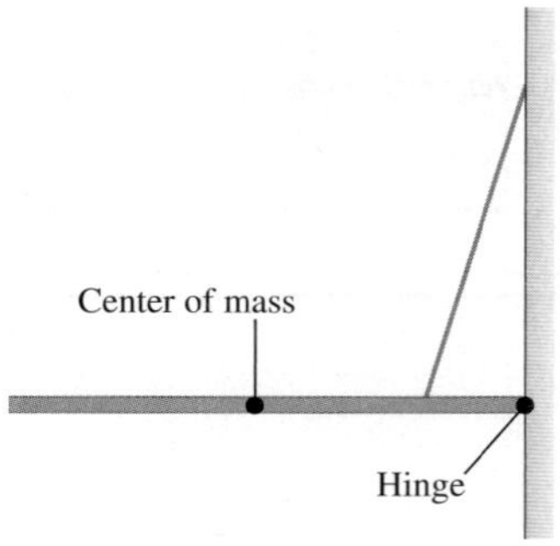

Center of mass

Hinge

Q11.20 If a metal wire has its length doubled and its diameter tripled, by what factor does its Young's modulus change?

Q11.21 A metal wire of diameter *D* stretches by 0.100 mm when supporting a weight *W*. If the same-length wire is used to support a weight three times as heavy, what would its diameter have to be (in terms of *D*) so it still stretches only 0.100 mm?

Q11.22 Compare the mechanical properties of a steel cable, made by twisting many thin wires together, with the properties of a solid steel rod of the same diameter. What advantages does each have?

Q11.23 The material in human bones and elephant bones is essentially the same, but an elephant has much thicker legs. Explain why, in terms of breaking stress.

Q11.24 There is a small but appreciable amount of elastic hysteresis in the large tendon at the back of a horse's leg. Explain how this can cause damage to the tendon if a horse runs too hard for too long a time.

Q11.25 When rubber mounting blocks are used to absorb machine vibrations through elastic hysteresis, as mentioned in Section 11.5, what becomes of the energy associated with the vibrations?

EXERCISES

Section 11.2 Center of Gravity

11.1 •• A 0.120-kg, 50.0-cm-long uniform bar has a small 0.055-kg mass glued to its left end and a small 0.110-kg mass glued to the other end. The two small masses can each be treated as point masses. You want to balance this system horizontally on a fulcrum placed just under its center of gravity. How far from the left end should the fulcrum be placed?

11.2 •• The center of gravity of a 5.00-kg irregular object is shown in **Fig. E11.2**. You need to move the center of gravity 2.20 cm to the left by gluing on a 1.50-kg mass, which will then be considered as part of the object. Where should the center of gravity of this additional mass be located?

Figure **E11.2**

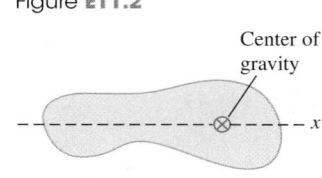

Center of gravity

11.3 • A uniform rod is 2.00 m long and has mass 1.80 kg. A 2.40-kg clamp is attached to the rod. How far should the center of gravity of the clamp be from the left-hand end of the rod in order for the center of gravity of the composite object to be 1.20 m from the left-hand end of the rod?

Section 11.3 Solving Rigid-Body Equilibrium Problems

11.4 • A uniform 300-N trapdoor in a floor is hinged at one side. Find the net upward force needed to begin to open it and the total force exerted on the door by the hinges (a) if the upward force is applied at the center and (b) if the upward force is applied at the center of the edge opposite the hinges.

11.5 •• **Raising a Ladder.** A ladder carried by a fire truck is 20.0 m long. The ladder weighs 3400 N and its center of gravity is at its center. The ladder is pivoted at one end (A) about a pin (**Fig. E11.5**); ignore the friction torque at the pin. The ladder is raised into position by a force applied by a hydraulic piston at C. Point C is 8.0 m from A, and the force $\vec{F}$ exerted by the piston makes an angle of 40° with the ladder. What magnitude must $\vec{F}$ have to just lift the ladder off the support bracket at B? Start with a free-body diagram of the ladder.

Figure **E11.5**

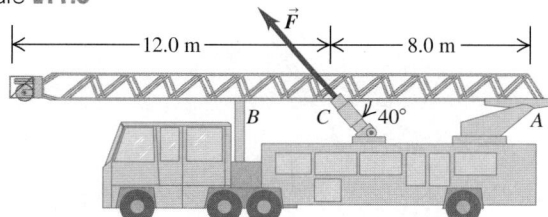

11.6 •• Two people are carrying a uniform wooden board that is 3.00 m long and weighs 160 N. If one person applies an upward force equal to 60 N at one end, at what point does the other person lift? Begin with a free-body diagram of the board.

11.7 •• Two people carry a heavy electric motor by placing it on a light board 2.00 m long. One person lifts at one end with a force of 400 N, and the other lifts the opposite end with a force of 600 N. (a) What is the weight of the motor, and where along the board is its center of gravity located? (b) Suppose the board is not light but weighs 200 N, with its center of gravity at its center, and the two people each exert the same forces as before. What is the weight of the motor in this case, and where is its center of gravity located?

11.8 •• A 60.0-cm, uniform, 50.0-N shelf is supported horizontally by two vertical wires attached to the sloping ceiling (**Fig. E11.8**). A very small 25.0-N tool is placed on the shelf midway between the points where the wires are attached to it. Find the tension in each wire. Begin by making a free-body diagram of the shelf.

Figure **E11.8**

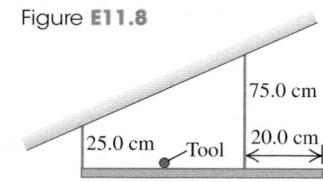

11.9 •• A 350-N, uniform, 1.50-m bar is suspended horizontally by two vertical cables at each end. Cable A can support a maximum tension of 500.0 N without breaking, and cable B can support up to 400.0 N. You want to place a small weight on this bar. (a) What is the heaviest weight you can put on without breaking either cable, and (b) where should you put this weight?

11.10 •• A uniform ladder 5.0 m long rests against a frictionless, vertical wall with its lower end 3.0 m from the wall. The ladder weighs 160 N. The coefficient of static friction between the foot of the ladder and the ground is 0.40. A man weighing 740 N climbs slowly up the ladder. Start by drawing a free-body diagram of the ladder. (a) What is the maximum friction force that the ground can exert on the ladder at its lower end? (b) What is the actual friction force when the man has climbed 1.0 m along the ladder? (c) How far along the ladder can the man climb before the ladder starts to slip?

11.11 • A diving board 3.00 m long is supported at a point 1.00 m from the end, and a diver weighing 500 N stands at the free end (**Fig. E11.11**). The diving board is of uniform cross section and weighs 280 N. Find (a) the force at the support point and (b) the force at the left-hand end.

Figure **E11.11**

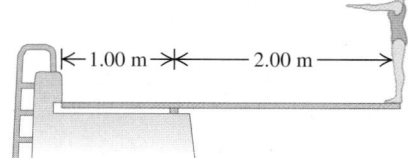

11.12 • A uniform aluminum beam 9.00 m long, weighing 300 N, rests symmetrically on two supports 5.00 m apart (**Fig. E11.12**). A boy weighing 600 N starts at point A and walks toward the right.

Figure **E11.12**

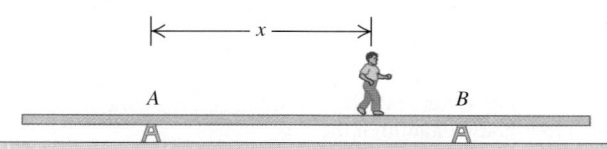

(a) In the same diagram construct two graphs showing the upward forces F_A and F_B exerted on the beam at points A and B, as functions of the coordinate x of the boy. Let 1 cm = 100 N vertically, and 1 cm = 1.00 m horizontally. (b) From your diagram, how far beyond point B can the boy walk before the beam tips? (c) How far from the right end of the beam should support B be placed so that the boy can walk just to the end of the beam without causing it to tip?

11.13 • Find the tension T in each cable and the magnitude and direction of the force exerted on the strut by the pivot in each of the arrangements in **Fig. E11.13**. In each case let w be the weight of the suspended crate full of priceless art objects. The strut is uniform and also has weight w. Start each case with a free-body diagram of the strut.

Figure **E11.13**

(a)   (b)

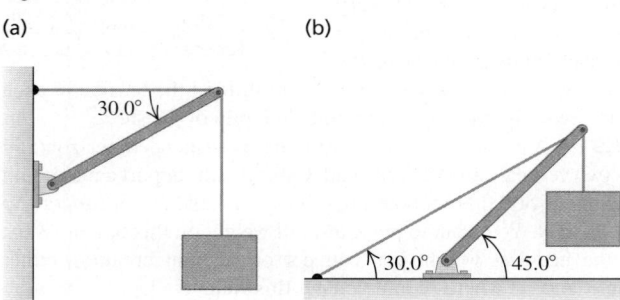

11.14 • The horizontal beam in **Fig. E11.14** weighs 190 N, and its center of gravity is at its center. Find (a) the tension in the cable and (b) the horizontal and vertical components of the force exerted on the beam at the wall.

Figure **E11.14**

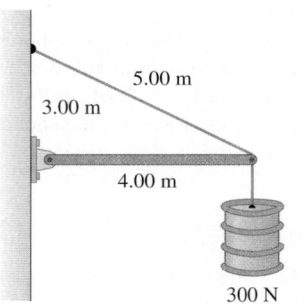

5.00 m
3.00 m
4.00 m
300 N

11.15 •• The boom shown in **Fig. E11.15** weighs 2600 N and is attached to a frictionless pivot at its lower end. It is not uniform; the distance of its center of gravity from the pivot is 35% of its length. Find (a) the tension in the guy wire and (b) the horizontal and vertical components of the force exerted on the boom at its lower end. Start with a free-body diagram of the boom.

Figure **E11.15**

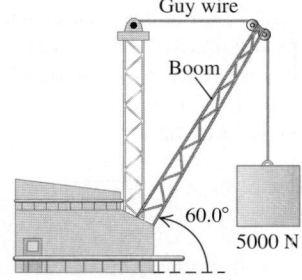

Guy wire
Boom
60.0°
5000 N

11.16 •• Suppose that you can lift no more than 650 N (around 150 lb) unaided. (a) How much can you lift using a 1.4-m-long wheelbarrow that weighs 80.0 N and whose center of gravity is 0.50 m from the center of the wheel (**Fig. E11.16**)? The center of gravity of the load carried in the wheelbarrow is also 0.50 m from the center of the wheel. (b) Where does the force come from to enable you to lift more than 650 N using the wheelbarrow?

Figure **E11.16**

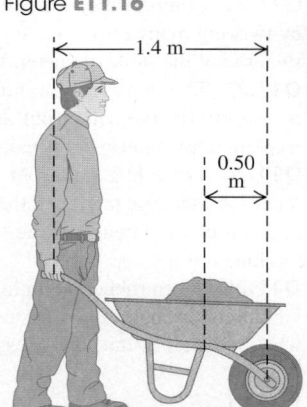

1.4 m
0.50 m

11.17 •• A 9.00-m-long uniform beam is hinged to a vertical wall and held horizontally by a 5.00-m-long cable attached to the wall 4.00 m above the hinge (**Fig. E11.17**). The metal of this cable has a test strength of 1.00 kN, which means that it will break if the tension in it exceeds that amount. (a) Draw a free-body diagram of the beam. (b) What is the heaviest beam that the cable can support in this configuration? (c) Find the horizontal and vertical components of the force the hinge exerts on the beam. Is the vertical component upward or downward?

Figure **E11.17**

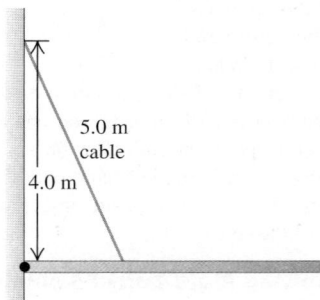

5.0 m
cable
4.0 m

11.18 •• A 15,000-N crane pivots around a friction-free axle at its base and is supported by a cable making a 25° angle with the crane (**Fig. E11.18**). The crane is 16 m long and is not uniform, its center of gravity being 7.0 m from the axle as measured along the crane. The cable is attached 3.0 m from the upper end of the crane. When the crane is raised to 55° above the horizontal holding an 11,000-N pallet of bricks by a 2.2-m, very light cord, find (a) the tension in the cable and (b) the horizontal and vertical components of the force that the axle exerts on the crane. Start with a free-body diagram of the crane.

Figure **E11.18**

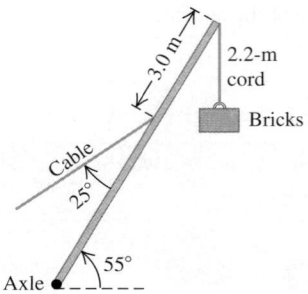

3.0 m
2.2-m cord
Bricks
Cable
25°
55°
Axle

11.19 •• A 3.00-m-long, 190-N, uniform rod at the zoo is held in a horizontal position by two ropes at its ends (**Fig. E11.19**). The left rope makes an angle of 150° with the rod, and the right rope makes an angle θ with the horizontal. A 90-N howler monkey (*Alouatta seniculus*) hangs motionless 0.50 m from the right end of the rod as he carefully studies you. Calculate the tensions in the two ropes and the angle θ. First make a free-body diagram of the rod.

Figure **E11.19**

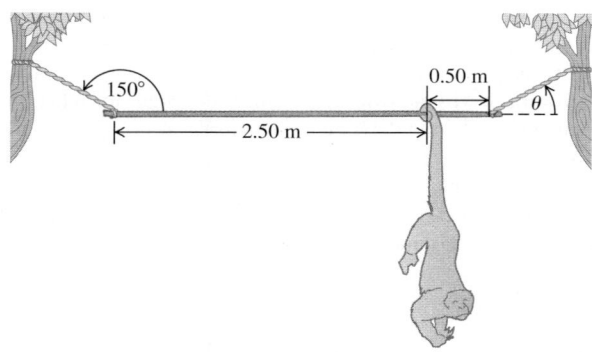

11.20 •• A nonuniform beam 4.50 m long and weighing 1.40 kN makes an angle of 25.0° below the horizontal. It is held in position by a frictionless pivot at its upper right end and by a cable 3.00 m farther down the beam and perpendicular to it (**Fig. E11.20**). The center of gravity of the beam is 2.00 m down the beam from the pivot. Lighting equipment exerts a 5.00-kN downward force on the lower left end of the beam. Find the tension T in the cable and the horizontal and vertical components of the force exerted on the beam by the pivot. Start by sketching a free-body diagram of the beam.

Figure **E11.20**

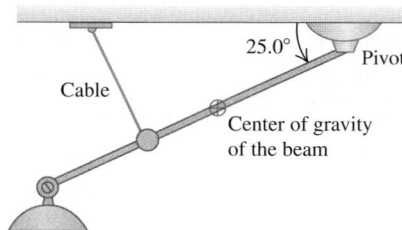

11.21 • **A Couple.** Two forces equal in magnitude and opposite in direction, acting on an object at two different points, form what is called a *couple*. Two antiparallel forces with equal magnitudes $F_1 = F_2 = 8.00$ N are applied to a rod as shown in **Fig. E11.21**. (a) What should the distance l between the forces be if they are to provide a net torque of 6.40 N·m about the left end of the rod? (b) Is the sense of this torque clockwise or counterclockwise? (c) Repeat parts (a) and (b) for a pivot at the point on the rod where $\vec{F}_2$ is applied.

Figure **E11.21**

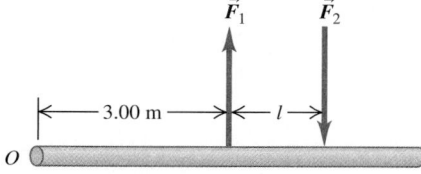

11.22 •• **BIO A Good Workout.** You are doing exercises on a Nautilus machine in a gym to strengthen your deltoid (shoulder) muscles. Your arms are raised vertically and can pivot around the shoulder joint, and you grasp the cable of the machine in your hand 64.0 cm from your shoulder joint. The deltoid muscle is attached to the humerus 15.0 cm from the shoulder joint and makes a 12.0° angle with that bone (**Fig. E11.22**). If you have set the tension in the cable of the machine to 36.0 N on each arm, what is the tension in each deltoid muscle if you simply hold your outstretched arms in place? (*Hint:* Start by making a clear free-body diagram of your arm.)

Figure **E11.22**

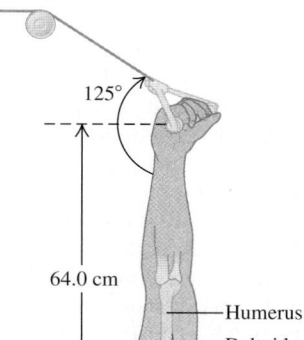

11.23 •• **BIO Neck Muscles.** A student bends her head at 40.0° from the vertical while intently reading her physics book, pivoting the head around the upper vertebra (point P in **Fig. E11.23**). Her head has a mass of 4.50 kg (which is typical), and its center of mass is 11.0 cm from the pivot point P. Her neck muscles are 1.50 cm from point P, as measured *perpendicular* to these muscles. The neck itself and the vertebrae are held vertical. (a) Draw a free-body diagram of the student's head. (b) Find the tension in her neck muscles.

Figure **E11.23**

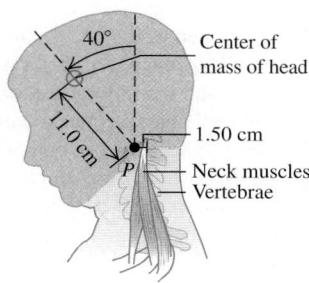

Section 11.4 Stress, Strain, and Elastic Moduli

11.24 • **BIO Biceps Muscle.** A relaxed biceps muscle requires a force of 25.0 N for an elongation of 3.0 cm; the same muscle under maximum tension requires a force of 500 N for the same elongation. Find Young's modulus for the muscle tissue under each of these conditions if the muscle is assumed to be a uniform cylinder with length 0.200 m and cross-sectional area 50.0 cm².

11.25 •• A circular steel wire 2.00 m long must stretch no more than 0.25 cm when a tensile force of 700 N is applied to each end of the wire. What minimum diameter is required for the wire?

11.26 •• Two circular rods, one steel and the other copper, are joined end to end. Each rod is 0.750 m long and 1.50 cm in diameter. The combination is subjected to a tensile force with magnitude 4000 N. For each rod, what are (a) the strain and (b) the elongation?

11.27 •• A metal rod that is 4.00 m long and 0.50 cm² in cross-sectional area is found to stretch 0.20 cm under a tension of 5000 N. What is Young's modulus for this metal?

11.28 •• **Stress on a Mountaineer's Rope.** A nylon rope used by mountaineers elongates 1.10 m under the weight of a 65.0-kg climber. If the rope is 45.0 m in length and 7.0 mm in diameter, what is Young's modulus for nylon?

11.29 •• In constructing a large mobile, an artist hangs an aluminum sphere of mass 6.0 kg from a vertical steel wire 0.50 m long and 2.5×10^{-3} cm^2 in cross-sectional area. On the bottom of the sphere he attaches a similar steel wire, from which he hangs a brass cube of mass 10.0 kg. For each wire, compute (a) the tensile strain and (b) the elongation.

11.30 •• A vertical, solid steel post 25 cm in diameter and 2.50 m long is required to support a load of 8000 kg. You can ignore the weight of the post. What are (a) the stress in the post; (b) the strain in the post; and (c) the change in the post's length when the load is applied?

11.31 •• **BIO Compression of Human Bone.** The bulk modulus for bone is 15 GPa. (a) If a diver-in-training is put into a pressurized suit, by how much would the pressure have to be raised (in atmospheres) above atmospheric pressure to compress her bones by 0.10% of their original volume? (b) Given that the pressure in the ocean increases by 1.0×10^4 Pa for every meter of depth below the surface, how deep would this diver have to go for her bones to compress by 0.10%? Does it seem that bone compression is a problem she needs to be concerned with when diving?

11.32 • A solid gold bar is pulled up from the hold of the sunken RMS *Titanic*. (a) What happens to its volume as it goes from the pressure at the ship to the lower pressure at the ocean's surface? (b) The pressure difference is proportional to the depth. How many times greater would the volume change have been had the ship been twice as deep? (c) The bulk modulus of lead is one-fourth that of gold. Find the ratio of the volume change of a solid lead bar to that of a gold bar of equal volume for the same pressure change.

11.33 • A specimen of oil having an initial volume of 600 cm^3 is subjected to a pressure increase of 3.6×10^6 Pa, and the volume is found to decrease by 0.45 cm^3. What is the bulk modulus of the material? The compressibility?

11.34 •• In the Challenger Deep of the Marianas Trench, the depth of seawater is 10.9 km and the pressure is 1.16×10^8 Pa (about 1.15×10^3 atm). (a) If a cubic meter of water is taken from the surface to this depth, what is the change in its volume? (Normal atmospheric pressure is about 1.0×10^5 Pa. Assume that k for seawater is the same as the freshwater value given in Table 11.2.) (b) What is the density of seawater at this depth? (At the surface, seawater has a density of 1.03×10^3 kg/m^3.)

11.35 •• A copper cube measures 6.00 cm on each side. The bottom face is held in place by very strong glue to a flat horizontal surface, while a horizontal force F is applied to the upper face parallel to one of the edges. (Consult Table 11.1.) (a) Show that the glue exerts a force F on the bottom face that is equal in magnitude but opposite to the force on the top face. (b) How large must F be to cause the cube to deform by 0.250 mm? (c) If the same experiment were performed on a lead cube of the same size as the copper one, by what distance would it deform for the same force as in part (b)?

11.36 •• A square steel plate is 10.0 cm on a side and 0.500 cm thick. (a) Find the shear strain that results if a force of magnitude 9.0×10^5 N is applied to each of the four sides, parallel to the side. (b) Find the displacement x in centimeters.

11.37 • In lab tests on a 9.25-cm cube of a certain material, a force of 1375 N directed at 8.50° to the cube (**Fig. E11.37**) causes the cube to deform through an angle of 1.24°. What is the shear modulus of the material?

Figure **E11.37**

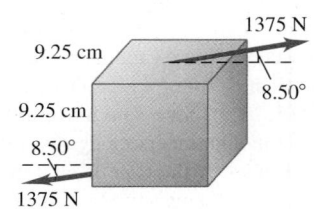

Section 11.5 Elasticity and Plasticity

11.38 •• A brass wire is to withstand a tensile force of 350 N without breaking. What minimum diameter must the wire have?

11.39 •• In a materials testing laboratory, a metal wire made from a new alloy is found to break when a tensile force of 90.8 N is applied perpendicular to each end. If the diameter of the wire is 1.84 mm, what is the breaking stress of the alloy?

11.40 • A 4.0-m-long steel wire has a cross-sectional area of 0.050 cm^2. Its proportional limit has a value of 0.0016 times its Young's modulus (see Table 11.1). Its breaking stress has a value of 0.0065 times its Young's modulus. The wire is fastened at its upper end and hangs vertically. (a) How great a weight can be hung from the wire without exceeding the proportional limit? (b) How much will the wire stretch under this load? (c) What is the maximum weight that the wire can support?

11.41 •• **CP** A steel cable with cross-sectional area 3.00 cm^2 has an elastic limit of 2.40×10^8 Pa. Find the maximum upward acceleration that can be given a 1200-kg elevator supported by the cable if the stress is not to exceed one-third of the elastic limit.

PROBLEMS

11.42 ••• A door 1.00 m wide and 2.00 m high weighs 330 N and is supported by two hinges, one 0.50 m from the top and the other 0.50 m from the bottom. Each hinge supports half the total weight of the door. Assuming that the door's center of gravity is at its center, find the horizontal components of force exerted on the door by each hinge.

11.43 ••• A box of negligible mass rests at the left end of a 2.00-m, 25.0-kg plank (**Fig. P11.43**). The width of the box is 75.0 cm, and sand is to be distributed uniformly throughout it. The center of gravity of the nonuniform plank is 50.0 cm from the right end. What mass of sand should be put into the box so that the plank balances horizontally on a fulcrum placed just below its midpoint?

Figure **P11.43**

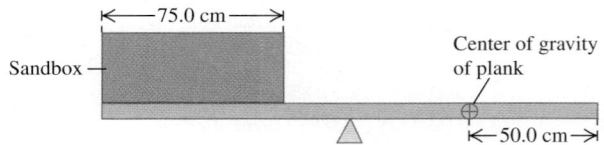

11.44 • Sir Lancelot rides slowly out of the castle at Camelot and onto the 12.0-m-long drawbridge that passes over the moat (**Fig. P11.44**). Unbeknownst to him, his enemies have partially severed the vertical cable holding up the front end of the bridge so that it will break under a tension of 5.80×10^3 N. The bridge has mass 200 kg and its center of gravity is at its center. Lancelot, his lance, his armor, and his horse together have a combined mass of 600 kg. Will the cable break before Lancelot reaches the end of the drawbridge? If so, how far from the castle end of the bridge will the center of gravity of the horse plus rider be when the cable breaks?

Figure **P11.44**

11.45 ••• **Mountain Climbing.** Mountaineers often use a rope to lower themselves down the face of a cliff (this is called *rappelling*). They do this with their body nearly horizontal and their feet pushing against the cliff (**Fig. P11.45**). Suppose that an 82.0-kg climber, who is 1.90 m tall and has a center of gravity 1.1 m from his feet, rappels down a vertical cliff with his body raised 35.0° above the horizontal. He holds the rope 1.40 m from his feet, and it makes a 25.0° angle with the cliff face. (a) What tension does his rope need to support? (b) Find the horizontal and vertical components of the force that the cliff face exerts on the climber's feet. (c) What minimum coefficient of static friction is needed to prevent the climber's feet from slipping on the cliff face if he has one foot at a time against the cliff?

Figure **P11.45**

11.46 •• A uniform, 8.0-m, 1150-kg beam is hinged to a wall and supported by a thin cable attached 2.0 m from the free end of the beam (**Fig. P11.46**). The beam is supported at an angle of 30.0° above the horizontal. (a) Draw a free-body diagram of the beam. (b) Find the tension in the cable. (c) How hard does the beam push inward on the wall?

Figure **P11.46**

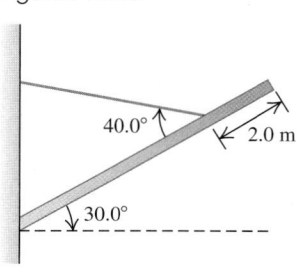

40.0°   2.0 m

30.0°

11.47 •• A uniform, 255-N rod that is 2.00 m long carries a 225-N weight at its right end and an unknown weight W toward the left end (**Fig. P11.47**). When W is placed 50.0 cm from the left end of the rod, the system just balances horizontally when the fulcrum is located 75.0 cm from the right end. (a) Find W. (b) If W is now moved 25.0 cm to the right, how far and in what direction must the fulcrum be moved to restore balance?

Figure **P11.47**

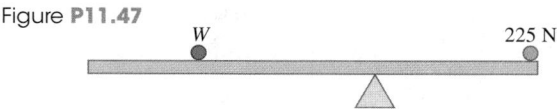

W 225 N

11.48 ••• A claw hammer is used to pull a nail out of a board (**Fig. P11.48**). The nail is at an angle of 60° to the board, and a force $\vec{F}_1$ of magnitude 400 N applied to the nail is required to pull it from the board. The hammer head contacts the board at point A, which is 0.080 m from where the nail enters the board. A horizontal force $\vec{F}_2$ is applied to the hammer handle at a distance of 0.300 m above the board. What magnitude of force $\vec{F}_2$ is required to apply the required 400-N force (F_1) to the nail? (Ignore the weight of the hammer.)

Figure **P11.48**

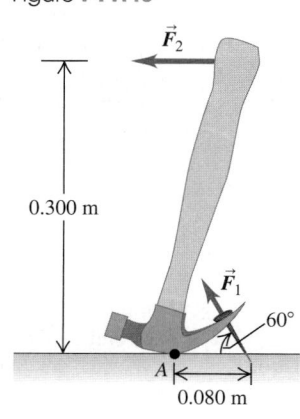

$\vec{F}_2$

0.300 m

$\vec{F}_1$

60°

A

0.080 m

11.49 •• You open a restaurant and hope to entice customers by hanging out a sign (**Fig. P11.49**). The uniform horizontal beam supporting the sign is 1.50 m long, has a mass of 16.0 kg, and is hinged to the wall. The sign itself is uniform with a mass of 28.0 kg and overall length of 1.20 m. The two wires supporting the sign are each 32.0 cm long, are 90.0 cm apart, and are equally spaced from the middle of the sign. The cable supporting the beam is 2.00 m long. (a) What minimum tension must your cable be able to support without having your sign come crashing down? (b) What minimum vertical force must the hinge be able to support without pulling out of the wall?

Figure **P11.49**

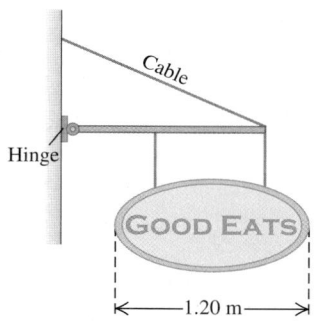

Cable

Hinge

GOOD EATS

⟵—1.20 m—⟶

11.50 • End A of the bar AB in **Fig. P11.50** rests on a frictionless horizontal surface, and end B is hinged. A horizontal force $\vec{F}$ of magnitude 220 N is exerted on end A. Ignore the weight of the bar. What are the horizontal and vertical components of the force exerted by the bar on the hinge at B?

Figure **P11.50**

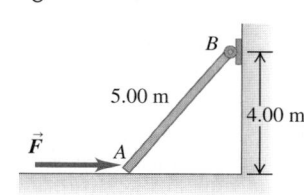

B

5.00 m

4.00 m

$\vec{F}$

A

11.51 •• BIO **Supporting a Broken Leg.** A therapist tells a 74-kg patient with a broken leg that he must have his leg in a cast suspended horizontally. For minimum discomfort, the leg should be supported by a vertical strap attached at the center of mass of the leg–cast system (**Fig. P11.51**). To comply with these instructions, the patient consults a table of typical mass distributions and finds that both upper legs (thighs) together typically account for 21.5% of body weight and the center of mass of each thigh is 18.0 cm from the hip joint. The patient also reads that the two lower legs (including the feet) are 14.0% of body weight, with a center of mass 69.0 cm from the hip joint. The cast has a mass of 5.50 kg, and its center of mass is 78.0 cm from the hip joint. How far from the hip joint should the supporting strap be attached to the cast?

Figure **P11.51**

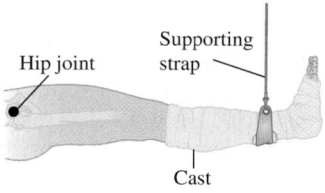

Supporting
strap

Hip joint

Cast

11.52 · A Truck on a Drawbridge. A loaded cement mixer drives onto an old drawbridge, where it stalls with its center of gravity three-quarters of the way across the span. The truck driver radios for help, sets the handbrake, and waits. Meanwhile, a boat approaches, so the drawbridge is raised by means of a cable attached to the end opposite the hinge (**Fig. P11.52**). The drawbridge is 40.0 m long and has a mass of 18,000 kg; its center of gravity is at its midpoint. The cement mixer, with driver, has mass 30,000 kg. When the drawbridge has been raised to an angle of 30° above the horizontal, the cable makes an angle of 70° with the surface of the bridge. (a) What is the tension T in the cable when the drawbridge is held in this position? (b) What are the horizontal and vertical components of the force the hinge exerts on the span?

Figure **P11.52**

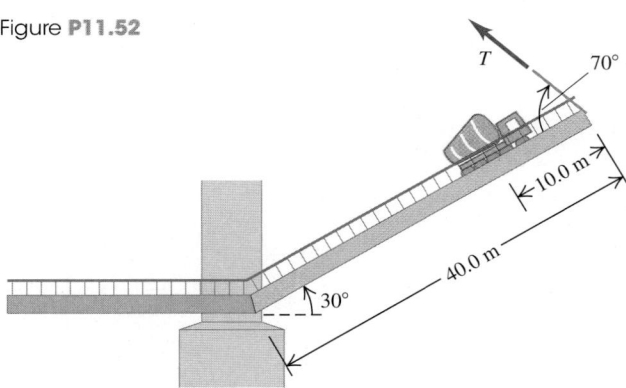

11.53 ·· BIO Leg Raises.
In a simplified version of the musculature action in leg raises, the abdominal muscles pull on the femur (thigh bone) to raise the leg by pivoting it about one end (**Fig. P11.53**). When you are lying horizontally, these muscles make an angle of approximately 5° with the femur, and if you raise your legs, the muscles remain approximately horizontal, so the angle θ increases. Assume for simplicity that these muscles attach to the femur in only one place, 10 cm from the hip joint (although, in reality, the situation is more complicated). For a certain 80-kg person having a leg 90 cm long, the mass of the leg is 15 kg and its center of mass is 44 cm from his hip joint as measured along the leg. If the person raises his leg to 60° above the horizontal, the angle between the abdominal muscles and his femur would also be about 60°. (a) With his leg raised to 60°, find the tension in the abdominal muscle on each leg. Draw a free-body diagram. (b) When is the tension in this muscle greater: when the leg is raised to 60° or when the person just starts to raise it off the ground? Why? (Try this yourself.) (c) If the abdominal muscles attached to the femur were perfectly horizontal when a person was lying down, could the person raise his leg? Why or why not?

11.54 ·· BIO Pumping Iron. A 72.0-kg weightlifter doing arm raises holds a 7.50-kg weight. Her arm pivots around the elbow joint, starting 40.0° below the horizontal (**Fig. P11.54**). Biometric measurements have shown that, together, the forearms and the hands account for 6.00% of a person's weight. Since the

Figure **P11.53**

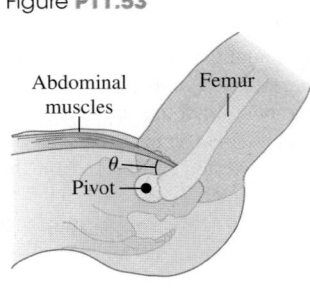

Figure **P11.54**

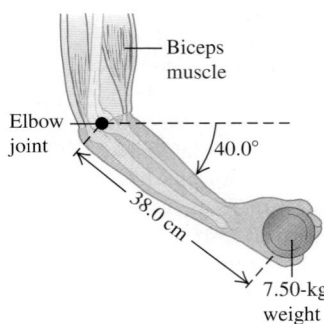

upper arm is held vertically, the biceps muscle always acts vertically and is attached to the bones of the forearm 5.50 cm from the elbow joint. The center of mass of this person's forearm–hand combination is 16.0 cm from the elbow joint, along the bones of the forearm, and she holds the weight 38.0 cm from her elbow joint. (a) Draw a free-body diagram of the forearm. (b) What force does the biceps muscle exert on the forearm? (c) Find the magnitude and direction of the force that the elbow joint exerts on the forearm. (d) As the weightlifter raises her arm toward a horizontal position, will the force in the biceps muscle increase, decrease, or stay the same? Why?

11.55 ·· BIO Back Pains During Pregnancy. Women often suffer from back pains during pregnancy. Model a woman (not including her fetus) as a uniform cylinder of diameter 30 cm and mass 60 kg. Model the fetus as a 10-kg sphere that is 25 cm in diameter and centered about 5 cm *outside* the front of the woman's body. (a) By how much does her pregnancy change the horizontal location of the woman's center of mass? (b) How does the change in part (a) affect the way the pregnant woman must stand and walk? In other words, what must she do to her posture to make up for her shifted center of mass? (c) Can you explain why she might have backaches?

11.56 · You are asked to design the decorative mobile shown in **Fig. P11.56**. The strings and rods have negligible weight, and the rods are to hang horizontally. (a) Draw a free-body diagram for each rod. (b) Find the weights of the balls A, B, and C. Find the tensions in the strings S_1, S_2, and S_3. (c) What can you say about the horizontal location of the mobile's center of gravity? Explain.

Figure **P11.56**

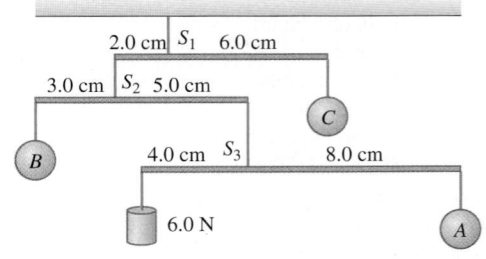

11.57 ·· A uniform, 7.5-m-long beam weighing 6490 N is hinged to a wall and supported by a thin cable attached 1.5 m from the free end of the beam. The cable runs between the beam and the wall and makes a 40° angle with the beam. What is the tension in the cable when the beam is at an angle of 30° above the horizontal?

11.58 •• CP A uniform drawbridge must be held at a 37° angle above the horizontal to allow ships to pass underneath. The drawbridge weighs 45,000 N and is 14.0 m long. A cable is connected 3.5 m from the hinge where the bridge pivots (measured along the bridge) and pulls horizontally on the bridge to hold it in place. (a) What is the tension in the cable? (b) Find the magnitude and direction of the force the hinge exerts on the bridge. (c) If the cable suddenly breaks, what is the magnitude of the angular acceleration of the drawbridge just after the cable breaks? (d) What is the angular speed of the drawbridge as it becomes horizontal?

11.59 •• BIO **Tendon-Stretching Exercises.** As part of an exercise program, a 75-kg person does toe raises in which he raises his entire body weight on the ball of one foot (**Fig. P11.59**). The Achilles tendon pulls straight upward on the heel bone of his foot. This tendon is 25 cm long and has a cross-sectional area of 78 mm² and a Young's modulus of 1470 MPa. (a) Draw a free-body diagram of the person's foot (everything below the ankle joint). Ignore the weight of the foot. (b) What force does the Achilles tendon exert on the heel during this exercise? Express your answer in newtons and in multiples of his weight. (c) By how many millimeters does the exercise stretch his Achilles tendon?

Figure **P11.59**

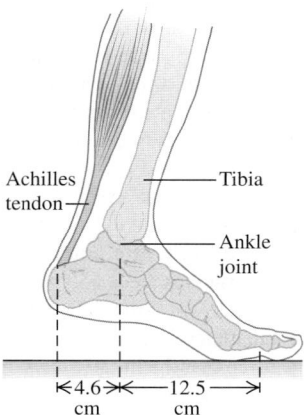

11.60 •• (a) In **Fig. P11.60** a 6.00-m-long, uniform beam is hanging from a point 1.00 m to the right of its center. The beam weighs 140 N and makes an angle of 30.0° with the vertical. At the right-hand end of the beam a 100.0-N weight is hung; an unknown weight *w* hangs at the left end. If the system is in equilibrium, what is *w*? You can ignore the thickness of the beam. (b) If the beam makes, instead, an angle of 45.0° with the vertical, what is *w*?

Figure **P11.60**

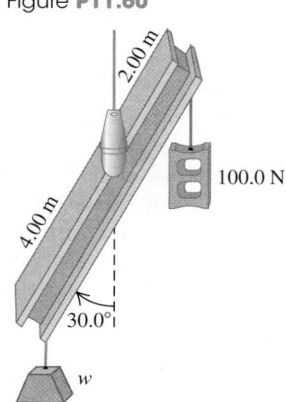

11.61 ••• A uniform, horizontal flagpole 5.00 m long with a weight of 200 N is hinged to a vertical wall at one end. A 600-N stuntwoman hangs from its other end. The flagpole is supported by a guy wire running from its outer end to a point on the wall directly above the pole. (a) If the tension in this wire is not to exceed 1000 N, what is the minimum height above the pole at

which it may be fastened to the wall? (b) If the flagpole remains horizontal, by how many newtons would the tension be increased if the wire were fastened 0.50 m below this point?

11.62 • A holiday decoration consists of two shiny glass spheres with masses 0.0240 kg and 0.0360 kg suspended from a uniform rod with mass 0.120 kg and length 1.00 m (**Fig. P11.62**). The rod is suspended from the ceiling by a vertical cord at each end, so that it is horizontal. Calculate the tension in each of the cords A through F.

Figure **P11.62**

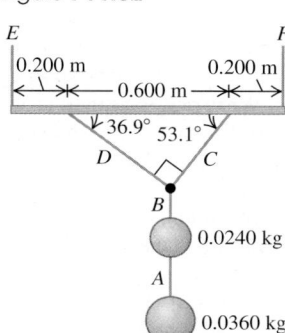

11.63 •• BIO **Downward-Facing Dog.** The yoga exercise "Downward-Facing Dog" requires stretching your hands straight out above your head and bending down to lean against the floor. This exercise is performed by a 750-N person as shown in **Fig. P11.63**. When he bends his body at the hip to a 90° angle between his legs and trunk, his legs, trunk, head, and arms have the dimensions indicated. Furthermore, his legs and feet weigh a total of 277 N, and their center of mass is 41 cm from his hip, measured along his legs. The person's trunk, head, and arms weigh 473 N, and their center of gravity is 65 cm from his hip, measured along the upper body. (a) Find the normal force that the floor exerts on each foot and on each hand, assuming that the person does not favor either hand or either foot. (b) Find the friction force on each foot and on each hand, assuming that it is the same on both feet and on both hands (but not necessarily the same on the feet as on the hands). [*Hint:* First treat his entire body as a system; then isolate his legs (or his upper body).]

Figure **P11.63**

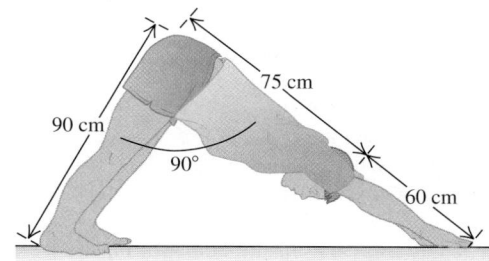

11.64 •• A uniform metal bar that is 8.00 m long and has mass 30.0 kg is attached at one end to the side of a building by a frictionless hinge. The bar is held at an angle of 64.0° above the horizontal by a thin, light cable that runs from the end of the bar opposite the hinge to a point on the wall that is above the hinge. The cable makes an angle of 37.0° with the bar. Your mass is 65.0 kg. You grab the bar near the hinge and hang beneath it, with your hands close together and your feet off the ground. To impress your friends, you intend to shift your hands slowly toward the top end of the bar. (a) If the cable breaks when its tension exceeds 455 N, how far from the upper end of the bar are you when the cable breaks? (b) Just before the cable breaks, what are the magnitude and direction of the resultant force that the hinge exerts on the bar?

11.65 • A worker wants to turn over a uniform, 1250-N, rectangular crate by pulling at 53.0° on one of its vertical sides (**Fig. P11.65**). The floor is rough enough to prevent the crate from slipping. (a) What pull is needed to just start the crate to tip? (b) How hard does the floor push upward on the crate? (c) Find the friction force on the crate. (d) What is the minimum coefficient of static friction needed to prevent the crate from slipping on the floor?

Figure **P11.65**

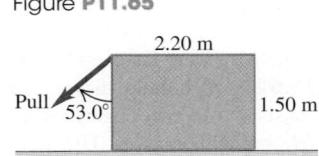

11.66 ••• One end of a uniform meter stick is placed against a vertical wall (**Fig. P11.66**). The other end is held by a lightweight cord that makes an angle θ with the stick. The coefficient of static friction between the end of the meter stick and the wall is 0.40. (a) What is the maximum value the angle θ can have if the stick is to remain in equilibrium? (b) Let the angle θ be 15°. A block of the same weight as the meter stick is suspended from the stick, as shown, at a distance x from the wall. What is the minimum value of x for which the stick will remain in equilibrium? (c) When $\theta = 15°$, how large must the coefficient of static friction be so that the block can be attached 10 cm from the left end of the stick without causing it to slip?

Figure **P11.66**

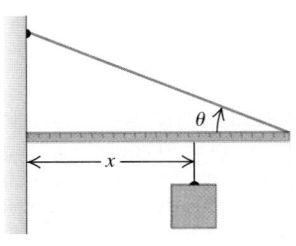

11.67 •• Two friends are carrying a 200-kg crate up a flight of stairs. The crate is 1.25 m long and 0.500 m high, and its center of gravity is at its center. The stairs make a 45.0° angle with respect to the floor. The crate also is carried at a 45.0° angle, so that its bottom side is parallel to the slope of the stairs (**Fig. P11.67**). If the force each person applies is vertical, what is the magnitude of each of these forces? Is it better to be the person above or below on the stairs?

Figure **P11.67**

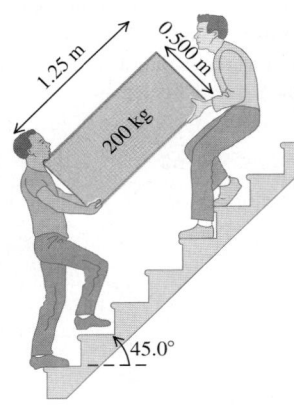

11.68 •• BIO **Forearm.** In the human arm, the forearm and hand pivot about the elbow joint. Consider a simplified model in which the biceps muscle is attached to the forearm 3.80 cm from the elbow joint. Assume that the person's hand and forearm together weigh 15.0 N and that their center of gravity is 15.0 cm from the elbow (not quite halfway to the hand). The forearm is held horizontally at a right angle to the upper arm, with the biceps muscle exerting its force perpendicular to the forearm. (a) Draw a free-body diagram for the forearm, and find the force exerted by the biceps when the hand is empty. (b) Now the person holds an 80.0-N weight in his hand, with the forearm still horizontal. Assume that the center of gravity of this weight is 33.0 cm from the elbow. Draw a free-body diagram for the forearm, and find the force now exerted by the biceps. Explain why the biceps muscle needs to be very strong. (c) Under the conditions of part (b), find the magnitude and direction of the force that the elbow joint exerts on the forearm. (d) While holding the 80.0-N weight, the person

raises his forearm until it is at an angle of 53.0° above the horizontal. If the biceps muscle continues to exert its force perpendicular to the forearm, what is this force now? Has the force increased or decreased from its value in part (b)? Explain why this is so, and test your answer by doing this with your own arm.

11.69 •• BIO CALC Refer to the discussion of holding a dumbbell in Example 11.4 (Section 11.3). The maximum weight that can be held in this way is limited by the maximum allowable tendon tension T (determined by the strength of the tendons) and by the distance D from the elbow to where the tendon attaches to the forearm. (a) Let T_{max} represent the maximum value of the tendon tension. Use the results of Example 11.4 to express w_{max} (the maximum weight that can be held) in terms of T_{max}, L, D, and h. Your expression should *not* include the angle θ. (b) The tendons of different primates are attached to the forearm at different values of D. Calculate the derivative of w_{max} with respect to D, and determine whether the derivative is positive or negative. (c) A chimpanzee tendon is attached to the forearm at a point farther from the elbow than for humans. Use this to explain why chimpanzees have stronger arms than humans. (The disadvantage is that chimpanzees have less flexible arms than do humans.)

11.70 ••• In a city park a nonuniform wooden beam 4.00 m long is suspended horizontally by a light steel cable at each end. The cable at the left-hand end makes an angle of 30.0° with the vertical and has tension 620 N. The cable at the right-hand end of the beam makes an angle of 50.0° with the vertical. As an employee of the Parks and Recreation Department, you are asked to find the weight of the beam and the location of its center of gravity.

11.71 •• You are a summer intern for an architectural firm. An 8.00-m-long uniform steel rod is to be attached to a wall by a frictionless hinge at one end. The rod is to be held at 22.0° below the horizontal by a light cable that is attached to the end of the rod opposite the hinge. The cable makes an angle of 30.0° with the rod and is attached to the wall at a point above the hinge. The cable will break if its tension exceeds 650 N. (a) For what mass of the rod will the cable break? (b) If the rod has a mass that is 10.0 kg less than the value calculated in part (a), what are the magnitude and direction of the force that the hinge exerts on the rod?

11.72 •• You are trying to raise a bicycle wheel of mass m and radius R up over a curb of height h. To do this, you apply a horizontal force $\vec{F}$ (**Fig. P11.72**). What is the smallest magnitude of the force $\vec{F}$ that will succeed in raising the wheel onto the curb when the force is applied (a) at the center of the wheel and (b) at the top of the wheel? (c) In which case is less force required?

Figure **P11.72**

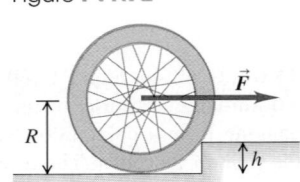

11.73 • **The Farmyard Gate.** A gate 4.00 m wide and 2.00 m high weighs 700 N. Its center of gravity is at its center, and it is hinged at A and B. To relieve the strain on the top hinge, a wire CD is connected as shown in **Fig. P11.73**. The tension in CD is increased until the horizontal force at hinge A is zero. What is (a) the tension in the wire CD; (b) the magnitude of the horizontal component of the force at hinge B; (c) the combined vertical force exerted by hinges A and B?

Figure **P11.73**

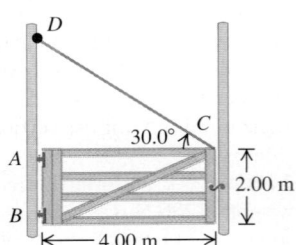

11.74 • If you put a uniform block at the edge of a table, the center of the block must be over the table for the block not to fall off. (a) If you stack two identical blocks at the table edge, the center of the top block must be over the bottom block, and the center of gravity of the two blocks together must be over the table. In terms of the length L of each block, what is the maximum overhang possible (**Fig. P11.74**)? (b) Repeat part (a) for three identical blocks and for four identical blocks. (c) Is it possible to make a stack of blocks such that the uppermost block is not directly over the table at all? How many blocks would it take to do this? (Try.)

Figure **P11.74**

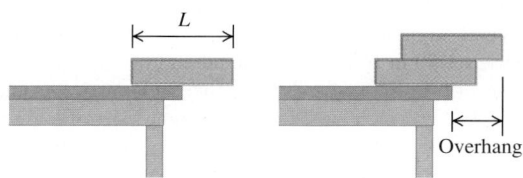

11.75 ••• Two uniform, 75.0-g marbles 2.00 cm in diameter are stacked as shown in **Fig. P11.75** in a container that is 3.00 cm wide. (a) Find the force that the container exerts on the marbles at the points of contact A, B, and C. (b) What force does each marble exert on the other?

Figure **P11.75**

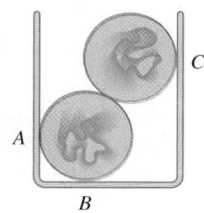

11.76 •• Two identical, uniform beams weighing 260 N each are connected at one end by a frictionless hinge. A light horizontal crossbar attached at the midpoints of the beams maintains an angle of 53.0° between the beams. The beams are suspended from the ceiling by vertical wires such that they form a "V" (**Fig. P11.76**). (a) What force does the crossbar exert on each beam? (b) Is the crossbar under tension or compression? (c) What force (magnitude and direction) does the hinge at point A exert on each beam?

Figure **P11.76**

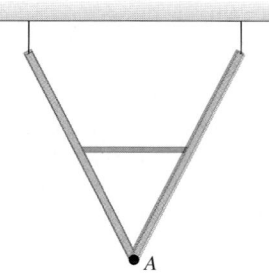

11.77 • An engineer is designing a conveyor system for loading hay bales into a wagon (**Fig. P11.77**). Each bale is 0.25 m wide, 0.50 m high, and 0.80 m long (the dimension perpendicular to the plane of the figure), with mass 30.0 kg. The center of gravity of each bale is at its geometrical center. The coefficient of static friction between a bale and the conveyor belt is 0.60, and the belt moves with constant speed. (a) The angle β of the conveyor is slowly increased. At some critical angle a bale will tip (if it doesn't slip first), and at some different critical angle it will slip (if it doesn't tip first). Find the two critical angles and determine which happens at the smaller angle. (b) Would the outcome of part (a) be different if the coefficient of friction were 0.40?

Figure **P11.77**

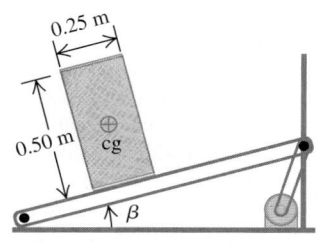

11.78 • A weight W is supported by attaching it to a vertical uniform metal pole by a thin cord passing over a pulley having negligible mass and friction. The cord is attached to the pole 40.0 cm below the top and pulls horizontally on it (**Fig. P11.78**). The pole is pivoted about a hinge at its base, is 1.75 m tall, and weighs 55.0 N. A thin wire connects the top of the pole to a vertical wall. The nail that holds this wire to the wall will pull out if an *outward* force greater than 22.0 N acts on it. (a) What is the greatest weight W that can be supported this way without pulling out the nail? (b) What is the *magnitude* of the force that the hinge exerts on the pole?

Figure **P11.78**

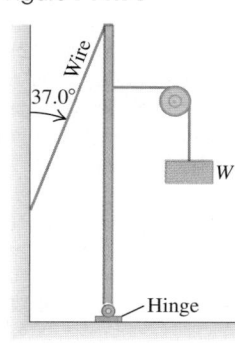

11.79 •• A garage door is mounted on an overhead rail (**Fig. P11.79**). The wheels at A and B have rusted so that they do not roll, but rather slide along the track. The coefficient of kinetic friction is 0.52. The distance between the wheels is 2.00 m, and each is 0.50 m from the vertical sides of the door. The door is uniform and weighs 950 N. It is pushed to the left at constant speed by a horizontal force $\vec{F}$. (a) If the distance h is 1.60 m, what is the vertical component of the force exerted on each wheel by the track? (b) Find the maximum value h can have without causing one wheel to leave the track.

Figure **P11.79**

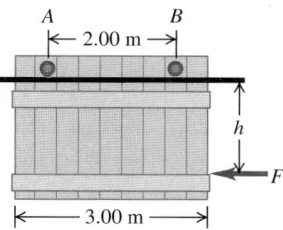

11.80 ••• **Pyramid Builders.** Ancient pyramid builders are balancing a uniform rectangular slab of stone tipped at an angle θ above the horizontal using a rope (**Fig. P11.80**). The rope is held by five workers who share the force equally. (a) If $\theta = 20.0°$, what force does each worker exert on the rope? (b) As θ increases, does each worker have to exert more or less force than in part (a), assuming they do not change the angle of the rope? Why? (c) At what angle do the workers need to exert *no force* to balance the slab? What happens if θ exceeds this value?

Figure **P11.80**

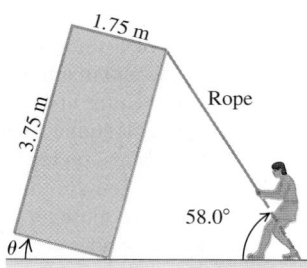

11.81 ••• CP A 12.0-kg mass, fastened to the end of an aluminum wire with an unstretched length of 0.70 m, is whirled in a vertical circle with a constant angular speed of 120 rev/min. The cross-sectional area of the wire is 0.014 cm². Calculate the elongation of the wire when the mass is (a) at the lowest point of the path and (b) at the highest point of its path.

11.82 •• **Hooke's Law for a Wire.** A wire of length l_0 and cross-sectional area A supports a hanging weight W. (a) Show that if the wire obeys Eq. (11.7), it behaves like a spring of force constant AY/l_0, where Y is Young's modulus for the wire material. (b) What would the force constant be for a 75.0-cm length of 16-gauge (diameter = 1.291 mm) copper wire? See Table 11.1. (c) What would W have to be to stretch the wire in part (b) by 1.25 mm?

11.83 ••• A 1.05-m-long rod of negligible weight is supported at its ends by wires A and B of equal length (**Fig. P11.83**). The cross-sectional area of A is 2.00 mm^2 and that of B is 4.00 mm^2. Young's modulus for wire A is 1.80×10^{11} Pa; that for B is 1.20×10^{11} Pa. At what point along the rod should a weight w be suspended to produce (a) equal stresses in A and B and (b) equal strains in A and B?

Figure **P11.83**

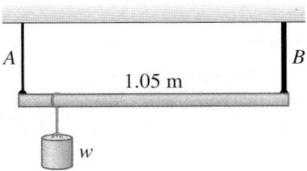

11.84 ••• CP An amusement park ride consists of airplane-shaped cars attached to steel rods (**Fig. P11.84**). Each rod has a length of 15.0 m and a cross-sectional area of 8.00 cm^2. (a) How much is each rod stretched when it is vertical and the ride is at rest? (Assume that each car plus two people seated in it has a total weight of 1900 N.) (b) When operating, the ride has a maximum angular speed of 12.0 rev/min. How much is the rod stretched then?

Figure **P11.84**

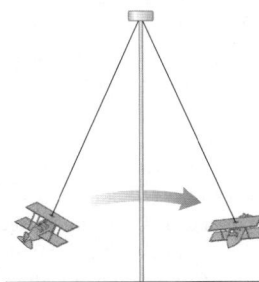

11.85 ••• CP BIO **Stress on the Shin Bone.** The compressive strength of our bones is important in everyday life. Young's modulus for bone is about 1.4×10^{10} Pa. Bone can take only about a 1.0% change in its length before fracturing. (a) What is the maximum force that can be applied to a bone whose minimum cross-sectional area is 3.0 cm^2? (This is approximately the cross-sectional area of a tibia, or shin bone, at its narrowest point.) (b) Estimate the maximum height from which a 70-kg man could jump and not fracture his tibia. Take the time between when he first touches the floor and when he has stopped to be 0.030 s, and assume that the stress on his two legs is distributed equally.

11.86 •• DATA You are to use a long, thin wire to build a pendulum in a science museum. The wire has an unstretched length of 22.0 m and a circular cross section of diameter 0.860 mm; it is made of an alloy that has a large breaking stress. One end of the wire will be attached to the ceiling, and a 9.50-kg metal sphere will be attached to the other end. As the pendulum swings back and forth, the wire's maximum angular displacement from the vertical will be 36.0°. You must determine the maximum amount the wire will stretch during this motion. So, before you attach the metal sphere, you suspend a test mass (mass m) from the wire's lower end. You then measure the increase in length Δl of the wire for several different test masses. **Figure P11.86**, a graph of Δl

Figure **P11.86**

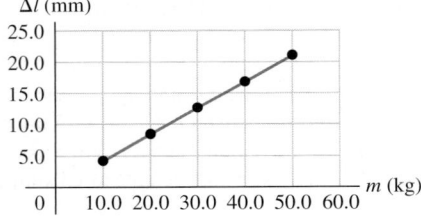

versus m, shows the results and the straight line that gives the best fit to the data. The equation for this line is $\Delta l = (0.422 \text{ mm/kg})m$. (a) Assume that $g = 9.80$ m/s^2, and use Fig. P11.86 to calculate Young's modulus Y for this wire. (b) You remove the test masses, attach the 9.50-kg sphere, and release the sphere from rest, with the wire displaced by 36.0°. Calculate the amount the wire will stretch as it swings through the vertical. Ignore air resistance.

11.87 •• DATA You need to measure the mass M of a 4.00-m-long bar. The bar has a square cross section but has some holes drilled along its length, so you suspect that its center of gravity isn't in the middle of the bar. The bar is too long for you to weigh on your scale. So, first you balance the bar on a knife-edge pivot and determine that the bar's center of gravity is 1.88 m from its left-hand end. You then place the bar on the pivot so that the point of support is 1.50 m from the left-hand end of the bar. Next you suspend a 2.00-kg mass (m_1) from the bar at a point 0.200 m from the left-hand end. Finally, you suspend a mass $m_2 = 1.00$ kg from the bar at a distance x from the left-hand end and adjust x so that the bar is balanced. You repeat this step for other values of m_2 and record each corresponding value of x. The table gives your results.

m_2 (kg)	1.00	1.50	2.00	2.50	3.00	4.00
x (m)	3.50	2.83	2.50	2.32	2.16	2.00

(a) Draw a free-body diagram for the bar when m_1 and m_2 are suspended from it. (b) Apply the static equilibrium equation $\Sigma \tau_z = 0$ with the axis at the location of the knife-edge pivot. Solve the equation for x as a function of m_2. (c) Plot x versus $1/m_2$. Use the slope of the best-fit straight line and the equation you derived in part (b) to calculate that bar's mass M. Use $g = 9.80$ m/s^2. (d) What is the y-intercept of the straight line that fits the data? Explain why it has this value.

11.88 ••• DATA You are a construction engineer working on the interior design of a retail store in a mall. A 2.00-m-long uniform bar of mass 8.50 kg is to be attached at one end to a wall, by means of a hinge that allows the bar to rotate freely with very little friction. The bar will be held in a horizontal position by a light cable from a point on the bar (a distance x from the hinge) to a point on the wall above the hinge. The cable makes an angle θ with the bar. The architect has proposed four possible ways to connect the cable and asked you to assess them:

Alternative	A	B	C	D
x (m)	2.00	1.50	0.75	0.50
θ (degrees)	30	60	37	75

(a) There is concern about the strength of the cable that will be required. Which set of x and θ values in the table produces the smallest tension in the cable? The greatest? (b) There is concern about the breaking strength of the sheetrock wall where the hinge will be attached. Which set of x and θ values produces the smallest horizontal component of the force the bar exerts on the hinge? The largest? (c) There is also concern about the required strength of the hinge and the strength of its attachment to the wall. Which set of x and θ values produces the smallest magnitude of the vertical component of the force the bar exerts on the hinge? The largest? (Hint: Does the direction of the vertical component of the force the hinge exerts on the bar depend on where along the bar the cable is attached?) (d) Is one of the alternatives given in the table preferable? Should any of the alternatives be avoided? Discuss.

CHALLENGE PROBLEMS

11.89 ••• Two ladders, 4.00 m and 3.00 m long, are hinged at point A and tied together by a horizontal rope 0.90 m above the floor (**Fig. P11.89**). The ladders weigh 480 N and 360 N, respectively, and the center of gravity of each is at its center. Assume that the floor is freshly waxed and frictionless. (a) Find the upward force at the bottom of each ladder. (b) Find the tension in the rope. (c) Find the magnitude of the force one ladder exerts on the other at point A. (d) If an 800-N painter stands at point A, find the tension in the horizontal rope.

Figure **P11.89**

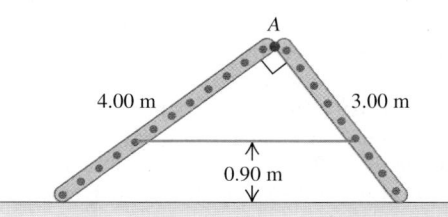

11.90 ••• **Knocking Over a Post.** One end of a post weighing 400 N and with height h rests on a rough horizontal surface with $\mu_s = 0.30$. The upper end is held by a rope fastened to the surface and making an angle of 36.9° with the post (**Fig. P11.90**). A horizontal force $\vec{F}$ is exerted on the post as shown. (a) If the force $\vec{F}$ is applied at the midpoint of the post, what is the largest value it can have without causing the post to slip? (b) How large can the force be without causing the post to slip if its point of application is $\frac{6}{10}$ of the way from the ground to the top of the post? (c) Show that if the point of application of the force is too high, the post cannot be made to slip, no matter how great the force. Find the critical height for the point of application.

Figure **P11.90**

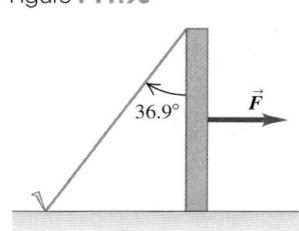

11.91 ••• CP An angler hangs a 4.50-kg fish from a vertical steel wire 1.50 m long and 5.00×10^{-3} cm^2 in cross-sectional area. The upper end of the wire is securely fastened to a support. (a) Calculate the amount the wire is stretched by the hanging fish. The angler now applies a varying force $\vec{F}$ at the lower end of the wire, pulling it very slowly downward by 0.500 mm from its equilibrium position. For this downward motion, calculate (b) the work done by gravity; (c) the work done by the force $\vec{F}$, (d) the work done by the force the wire exerts on the fish; and (e) the change in the elastic potential energy (the potential energy associated with the tensile stress in the wire). Compare the answers in parts (d) and (e).

PASSAGE PROBLEMS

BIO **TORQUES AND TUG-OF-WAR.** In a study of the biomechanics of the tug-of-war, a 2.0-m-tall, 80.0-kg competitor in the middle of the line is considered to be a rigid body leaning back at an angle of 30.0° to the vertical. The competitor is pulling on a rope that is held horizontal a distance of 1.5 m from his feet (as measured along the line of the body). At the moment shown in the figure, the man is stationary and the tension in the rope in front of him is $T_1 = 1160$ N. Since there is friction between the rope and his hands, the tension in the rope behind him, T_2, is not equal to T_1. His center of mass is halfway between his feet and the top of his head. The coefficient of static friction between his feet and the ground is 0.65.

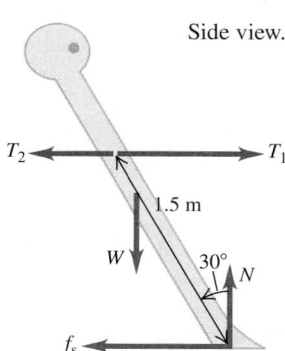

11.92 What is tension T_2 in the rope behind him? (a) 590 N; (b) 650 N; (c) 860 N; (d) 1100 N.

11.93 If he leans slightly farther back (increasing the angle between his body and the vertical) but remains stationary in this new position, which of the following statements is true? Assume that the rope remains horizontal. (a) The difference between T_1 and T_2 will increase, balancing the increased torque about his feet that his weight produces when he leans farther back; (b) the difference between T_1 and T_2 will decrease, balancing the increased torque about his feet that his weight produces when he leans farther back; (c) neither T_1 nor T_2 will change, because no other forces are changing; (d) both T_1 and T_2 will change, but the difference between them will remain the same.

11.94 His body is again leaning back at 30.0° to the vertical, but now the height at which the rope is held above—but still parallel to—the ground is varied. The tension in the rope in front of the competitor (T_1) is measured as a function of the shortest distance between the rope and the ground (the holding height). Tension T_1 is found to decrease as the holding height increases. What could explain this observation? As the holding height increases, (a) the moment arm of the rope about his feet decreases due to the angle that his body makes with the vertical; (b) the moment arm of the weight about his feet decreases due to the angle that his body makes with the vertical; (c) a smaller tension in the rope is needed to produce a torque sufficient to balance the torque of the weight about his feet; (d) his center of mass moves down to compensate, so less tension in the rope is required to maintain equilibrium.

11.95 His body is leaning back at 30.0° to the vertical, but the coefficient of static friction between his feet and the ground is suddenly reduced to 0.50. What will happen? (a) His entire body will accelerate forward; (b) his feet will slip forward; (c) his feet will slip backward; (d) his feet will not slip.

Answers

Chapter Opening Question ?

(i) Each stone in the arch is under compression, not tension. This is because the forces on the stones tend to push them inward toward the center of the arch and thus squeeze them together. Compared to a solid supporting wall, a wall with arches is just as strong yet much more economical to build.

Test Your Understanding Questions

11.1 (i) Situation (i) satisfies both equilibrium conditions because the seagull has zero acceleration (so $\Sigma \vec{F} = 0$) and no tendency to start rotating (so $\Sigma \vec{\tau} = 0$). Situation (ii) satisfies the first condition because the crankshaft as a whole does not accelerate through space, but it does not satisfy the second condition; the crankshaft has an angular acceleration, so $\Sigma \vec{\tau}$ is not zero. Situation (iii) satisfies the second condition (there is no tendency to rotate) but not the first one; the baseball accelerates in its flight (due to gravity and air resistance), so $\Sigma \vec{F}$ is not zero.

11.2 (ii) In equilibrium, the center of gravity must be at the point of support. Since the rock and meter stick have the same mass and hence the same weight, the center of gravity of the system is midway between their respective centers. The center of gravity of the meter stick alone is 0.50 m from the left end (that is, at the middle of the meter stick), so the center of gravity of the combination of rock and meter stick is 0.25 m from the left end.

11.3 (ii), (i), (iii) This is the same situation described in Example 11.4, with the rod replacing the forearm, the hinge replacing the elbow, and the cable replacing the tendon. The only difference is that the cable attachment point is at the end of the rod, so the distances D and L are identical. From Example 11.4, the tension is

$$T = \frac{Lw}{L\sin\theta} = \frac{w}{\sin\theta}$$

Since $\sin\theta$ is less than 1, the tension T is greater than the weight w. The vertical component of the force exerted by the hinge is

$$E_y = -\frac{(L-L)w}{L} = 0$$

In this situation, the hinge exerts *no* vertical force. To see this, calculate torques around the right end of the horizontal rod: The only force that exerts a torque around this point is the vertical component of the hinge force, so this force component must be zero.

11.4 (a) (iii), (b) (ii) In (a), the copper rod has 10 times the elongation Δl of the steel rod, but it also has 10 times the original length l_0. Hence the tensile strain $\Delta l / l_0$ is the same for both rods. In (b), the stress is equal to Young's modulus Y multiplied by the strain. From Table 11.1, steel has a larger value of Y, so a greater stress is required to produce the same strain.

11.5 In (a) and (b), the bumper will have sprung back to its original shape (although the paint may be scratched). In (c), the bumper will have a permanent dent or deformation. In (d), the bumper will be torn or broken.

Bridging Problem

(a) $T = \dfrac{2mg}{3\sin\theta}$

(b) $F = \dfrac{2mg}{3\sin\theta}\sqrt{\cos^2\theta + \tfrac{1}{4}\sin^2\theta}, \quad \phi = \arctan\left(\tfrac{1}{2}\tan\theta\right)$

(c) $\Delta l = \dfrac{2mgl_0}{3AY\tan\theta}$ **(d)** 4

12 FLUID MECHANICS

Fluids play a vital role in many aspects of everyday life. We drink them, breathe them, swim in them. They circulate through our bodies and control our weather. The physics of fluids is therefore crucial to our understanding of both nature and technology.

We begin our study with **fluid statics,** the study of fluids at rest in equilibrium situations. Like other equilibrium situations, it is based on Newton's first and third laws. We will explore the key concepts of density, pressure, and buoyancy. **Fluid dynamics,** the study of fluids in motion, is much more complex; indeed, it is one of the most complex branches of mechanics. Fortunately, we can analyze many important situations by using simple idealized models and familiar principles such as Newton's laws and conservation of energy. Even so, we will barely scratch the surface of this broad and interesting topic.

12.1 GASES, LIQUIDS, AND DENSITY

A **fluid** is any substance that can flow and change the shape of the volume that it occupies. (By contrast, a solid tends to maintain its shape.) We use the term "fluid" for both gases and liquids. The key difference between them is that a liquid has *cohesion,* while a gas does not. The molecules in a liquid are close to one another, so they can exert attractive forces on each other and thus tend to stay together (that is, to cohere). That's why a quantity of liquid maintains the same volume as it flows: If you pour 500 mL of water into a pan, the water will still occupy a volume of 500 mL. The molecules of a gas, by contrast, are separated on average by distances far larger than the size of a molecule. Hence the forces between molecules are weak, there is little or no cohesion, and a gas can easily change in volume. If you open the valve on a tank of compressed oxygen that has a volume of 500 mL, the oxygen will expand to a far greater volume.

An important property of *any* material, fluid or solid, is its **density,** defined as its mass per unit volume. A homogeneous material such as ice or iron has the

12.1 Two objects with different masses and different volumes but the same density.

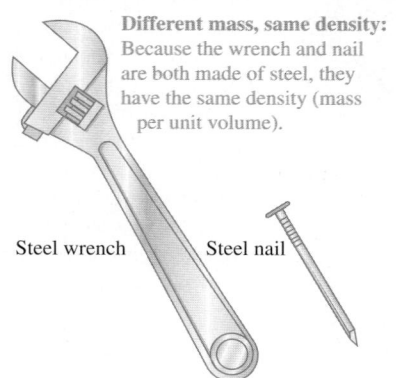

Different mass, same density: Because the wrench and nail are both made of steel, they have the same density (mass per unit volume).

Steel wrench   Steel nail

BIO Application Liquid Cohesion in Trees How do trees—some of which grow to heights greater than 100 m—supply water to their highest leaves? The answer lies in the strong cohesive forces between molecules of liquid water. Narrow pipes within the tree extend upward from the roots to the leaves. As water evaporates from the leaves, cohesive forces pull replacement water upward through these pipes.

same density throughout. We use ρ (the Greek letter rho) for density. For a homogeneous material,

$$\text{Density of a homogeneous material} \cdots\rightarrow \rho = \frac{m \,\cdots\cdots\; \text{Mass of material}}{V \,\cdots\cdots\; \text{Volume occupied by material}} \qquad (12.1)$$

Two objects made of the same material have the same density even though they may have different masses and different volumes. That's because the *ratio* of mass to volume is the same for both objects (**Fig. 12.1**).

The SI unit of density is the kilogram per cubic meter (1 kg/m^3). The cgs unit, the gram per cubic centimeter (1 g/cm^3), is also widely used:

$$1 \text{ g/cm}^3 = 1000 \text{ kg/m}^3$$

The densities of some common substances at ordinary temperatures are given in **Table 12.1.** Note the wide range of magnitudes. The densest material found on earth is the metal osmium ($\rho = 22{,}500 \text{ kg/m}^3$), but its density pales by comparison to the densities of exotic astronomical objects, such as white dwarf stars and neutron stars.

The **specific gravity** of a material is the ratio of its density to the density of water at $4.0°\text{C}$, 1000 kg/m^3; it is a pure number without units. For example, the specific gravity of aluminum is 2.7. "Specific gravity" is a poor term, since it has nothing to do with gravity; "relative density" would have been a better choice.

The density of some materials varies from point to point within the material. One example is the material of the human body, which includes low-density fat (about 940 kg/m^3) and high-density bone (from 1700 to 2500 kg/m^3). Two others are the earth's atmosphere (which is less dense at high altitudes) and oceans (which are denser at greater depths). For these materials, Eq. (12.1) describes the **average density.** In general, the density of a material depends on environmental factors such as temperature and pressure.

TABLE 12.1	Densities of Some Common Substances		
Material	**Density (kg/m^3)***	**Material**	**Density (kg/m^3)***
Air (1 atm, 20°C)	1.20	Iron, steel	7.8×10^3
Ethanol	0.81×10^3	Brass	8.6×10^3
Benzene	0.90×10^3	Copper	8.9×10^3
Ice	0.92×10^3	Silver	10.5×10^3
Water	1.00×10^3	Lead	11.3×10^3
Seawater	1.03×10^3	Mercury	13.6×10^3
Blood	1.06×10^3	Gold	19.3×10^3
Glycerine	1.26×10^3	Platinum	21.4×10^3
Concrete	2×10^3	White dwarf star	10^{10}
Aluminum	2.7×10^3	Neutron star	10^{18}

*To obtain the densities in grams per cubic centimeter, simply divide by 10^3.

EXAMPLE 12.1   **THE WEIGHT OF A ROOMFUL OF AIR**

Find the mass and weight of the air at 20°C in a living room with a 4.0 m × 5.0 m floor and a ceiling 3.0 m high, and the mass and weight of an equal volume of water.

SOLUTION

IDENTIFY and SET UP: We assume that the air density is the same throughout the room. (Air is less dense at high elevations than near sea level, but the density varies negligibly over the room's

3.0-m height; see Section 12.2.) We use Eq. (12.1) to relate the mass m_{air} to the room's volume V (which we'll calculate) and the air density ρ_{air} (given in Table 12.1).

EXECUTE: We have $V = (4.0 \text{ m})(5.0 \text{ m})(3.0 \text{ m}) = 60 \text{ m}^3$, so from Eq. (12.1),

$$m_{\text{air}} = \rho_{\text{air}}V = (1.20 \text{ kg/m}^3)(60 \text{ m}^3) = 72 \text{ kg}$$
$$w_{\text{air}} = m_{\text{air}}g = (72 \text{ kg})(9.8 \text{ m/s}^2) = 700 \text{ N} = 160 \text{ lb}$$

The mass and weight of an equal volume of water are

$$m_{\text{water}} = \rho_{\text{water}}V = (1000 \text{ kg/m}^3)(60 \text{ m}^3) = 6.0 \times 10^4 \text{ kg}$$

$$w_{\text{water}} = m_{\text{water}}g = (6.0 \times 10^4 \text{ kg})(9.8 \text{ m/s}^2)$$

$$= 5.9 \times 10^5 \text{ N} = 1.3 \times 10^5 \text{ lb} = 66 \text{ tons}$$

EVALUATE: A roomful of air weighs about the same as an average adult. Water is nearly a thousand times denser than air, so its mass and weight are larger by the same factor. The weight of a roomful of water would collapse the floor of an ordinary house.

TEST YOUR UNDERSTANDING OF SECTION 12.1   Rank the following objects in order from highest to lowest average density: (i) mass $m = 4.00$ kg, volume $V = 1.60 \times 10^{-3}$ m^3; (ii) $m = 8.00$ kg, $V = 1.60 \times 10^{-3}$ m^3; (iii) $m = 8.00$ kg, $V = 3.20 \times 10^{-3}$ m^3; (iv) $m = 2560$ kg, $V = 0.640$ m^3; (v) $m = 2560$ kg, $V = 1.28$ m^3. ∎

12.2 PRESSURE IN A FLUID

A fluid exerts a force perpendicular to any surface in contact with it, such as a container wall or a body immersed in the fluid. This is the force that you feel pressing on your legs when you dangle them in a swimming pool. Even when a fluid as a whole is at rest, the molecules that make up the fluid are in motion; the force exerted by the fluid is due to molecules colliding with their surroundings.

Imagine a surface *within* a fluid at rest. For this surface and the fluid to remain at rest, the fluid must exert forces of equal magnitude but opposite direction on the surface's two sides. Consider a small surface of area dA centered on a point in the fluid; the normal force exerted by the fluid on each side is $dF_\perp$ (**Fig. 12.2**). We define the **pressure** p at that point as the normal force per unit area—that is, the ratio of $dF_\perp$ to dA (**Fig. 12.3**):

$$p = \frac{dF_\perp}{dA} \qquad (12.2)$$

If the pressure is the same at all points of a finite plane surface with area A, then

$$p = \frac{F_\perp}{A} \qquad (12.3)$$

where $F_\perp$ is the net normal force on one side of the surface. The SI unit of pressure is the **pascal**, where

$$1 \text{ pascal} = 1 \text{ Pa} = 1 \text{ N/m}^2$$

We introduced the pascal in Chapter 11. Two related units, used principally in meteorology, are the *bar,* equal to 10^5 Pa, and the *millibar,* equal to 100 Pa.

Atmospheric pressure p_a is the pressure of the earth's atmosphere, the pressure at the bottom of this sea of air in which we live. This pressure varies with weather changes and with elevation. Normal atmospheric pressure at sea level (an average value) is 1 *atmosphere* (atm), defined to be exactly 101,325 Pa. To four significant figures,

$$(p_a)_{\text{av}} = 1 \text{ atm} = 1.013 \times 10^5 \text{ Pa}$$
$$= 1.013 \text{ bar} = 1013 \text{ millibar} = 14.70 \text{ lb/in.}^2$$

CAUTION **Don't confuse pressure and force** In everyday language "pressure" and "force" mean pretty much the same thing. In fluid mechanics, however, these words describe very different quantities. Pressure acts perpendicular to any surface in a fluid, no matter how that surface is oriented (Fig. 12.3). Hence pressure has no direction of its own; it's a scalar. By contrast, force is a vector with a definite direction. Remember, too, that pressure is force per unit area. As Fig. 12.3 shows, a surface with twice the area has twice as much force exerted on it by the fluid, so the pressure is the same. ∎

12.2 Forces acting on a small surface within a fluid at rest.

The surface does not accelerate, so the surrounding fluid exerts equal normal forces on both sides of it. (The fluid cannot exert any force parallel to the surface, since that would cause the surface to accelerate.)

12.3 Pressure is a scalar with units of newtons per square meter. By contrast, force is a vector with units of newtons.

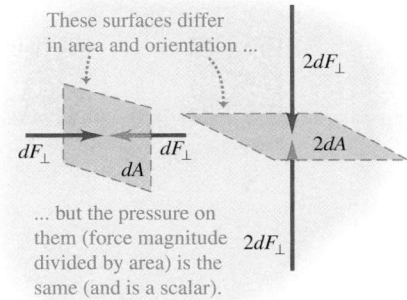

EXAMPLE 12.2 **THE FORCE OF AIR**

In the room described in Example 12.1, what is the total downward force on the floor due to an air pressure of 1.00 atm?

SOLUTION

IDENTIFY and SET UP: This example uses the relationship among the pressure p of a fluid (air), the area A subjected to that pressure, and the resulting normal force $F_\perp$ the fluid exerts. The pressure is uniform, so we use Eq. (12.3), $F_\perp = pA$, to determine $F_\perp$. The floor is horizontal, so $F_\perp$ is vertical (downward).

EXECUTE: We have $A = (4.0 \text{ m})(5.0 \text{ m}) = 20 \text{ m}^2$, so from Eq. (12.3),

$$F_\perp = pA = (1.013 \times 10^5 \text{ N/m}^2)(20 \text{ m}^2)$$
$$= 2.0 \times 10^6 \text{ N} = 4.6 \times 10^5 \text{ lb} = 230 \text{ tons}$$

EVALUATE: Unlike the water in Example 12.1, $F_\perp$ will not collapse the floor here, because there is an *upward* force of equal magnitude on the floor's underside. If the house has a basement, this upward force is exerted by the air underneath the floor. In this case, if we ignore the thickness of the floor, the *net* force due to air pressure is zero.

12.4 The forces on an element of fluid in equilibrium.

(a)

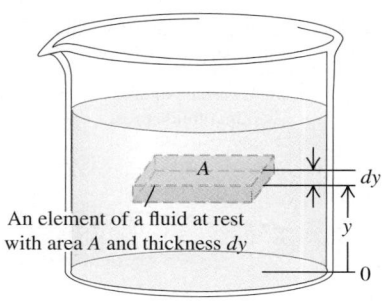

An element of a fluid at rest with area A and thickness dy

(b)

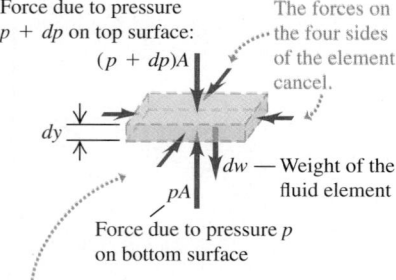

Force due to pressure $p + dp$ on top surface: $(p + dp)A$

The forces on the four sides of the element cancel.

dw — Weight of the fluid element

pA

Force due to pressure p on bottom surface

Because the fluid is in equilibrium, the vector sum of the vertical forces on the fluid element must be zero: $pA - (p + dp)A - dw = 0$.

DEMO

DEMO

Pressure, Depth, and Pascal's Law

If the weight of the fluid can be ignored, the pressure in a fluid is the same throughout its volume. We used that approximation in our discussion of bulk stress and strain in Section 11.4. But often the fluid's weight is *not* negligible, and pressure variations are important. Atmospheric pressure is less at high altitude than at sea level, which is why airliner cabins have to be pressurized. When you dive into deep water, you can feel the increased pressure on your ears.

We can derive a relationship between the pressure p at any point in a fluid at rest and the elevation y of the point. We'll assume that the density ρ has the same value throughout the fluid (that is, the density is *uniform*), as does the acceleration due to gravity g. If the fluid is in equilibrium, any thin element of the fluid with thickness dy is also in equilibrium (**Fig. 12.4a**). The bottom and top surfaces each have area A, and they are at elevations y and $y + dy$ above some reference level where $y = 0$. The fluid element has volume $dV = A \, dy$, mass $dm = \rho \, dV = \rho A \, dy$, and weight $dw = dm \, g = \rho g A \, dy$.

What are the other forces on this fluid element (Fig 12.4b)? Let's call the pressure at the bottom surface p; then the total y-component of upward force on this surface is pA. The pressure at the top surface is $p + dp$, and the total y-component of (downward) force on the top surface is $-(p + dp)A$. The fluid element is in equilibrium, so the total y-component of force, including the weight and the forces at the bottom and top surfaces, must be zero:

$$\sum F_y = 0 \qquad \text{so} \qquad pA - (p + dp)A - \rho g A \, dy = 0$$

When we divide out the area A and rearrange, we get

$$\frac{dp}{dy} = -\rho g \tag{12.4}$$

This equation shows that when y increases, p decreases; that is, as we move upward in the fluid, pressure decreases, as we expect. If p_1 and p_2 are the pressures at elevations y_1 and y_2, respectively, and if ρ and g are constant, then

Pressure difference between two points in a **fluid of uniform density**

Uniform density of fluid

$$p_2 - p_1 = -\rho g(y_2 - y_1) \tag{12.5}$$

Heights of the two points

Acceleration due to gravity ($g > 0$)

It's often convenient to express Eq. (12.5) in terms of the *depth* below the surface of a fluid (**Fig. 12.5**). Take point 1 at any level in the fluid and let p represent the pressure at this point. Take point 2 at the *surface* of the fluid, where the pressure is p_0 (subscript zero for zero depth). The depth of point 1 below the surface is $h = y_2 - y_1$, and Eq. (12.5) becomes

$$p_0 - p = -\rho g(y_2 - y_1) = -\rho g h \quad \text{or}$$

Pressure at depth h in a **fluid of uniform** density $\cdots\rightarrow p = p_0 + \rho g h \cdots$ Uniform density of fluid / Depth below surface (12.6)

Pressure at surface of fluid · Acceleration due to gravity ($g > 0$)

The pressure p at a depth h is greater than the pressure p_0 at the surface by an amount $\rho g h$. Note that the pressure is the same at any two points at the same level in the fluid. The *shape* of the container does not matter (**Fig. 12.6**).

Equation (12.6) shows that if we increase the pressure p_0 at the top surface, possibly by using a piston that fits tightly inside the container to push down on the fluid surface, the pressure p at any depth increases by exactly the same amount. This observation is called *Pascal's law*.

> **PASCAL'S LAW: Pressure applied to an enclosed fluid is transmitted undiminished to every portion of the fluid and the walls of the containing vessel.**

The hydraulic lift (**Fig. 12.7**) illustrates Pascal's law. A piston with small cross-sectional area A_1 exerts a force F_1 on the surface of a liquid such as oil. The applied pressure $p = F_1/A_1$ is transmitted through the connecting pipe to a larger piston of area A_2. The applied pressure is the same in both cylinders, so

$$p = \frac{F_1}{A_1} = \frac{F_2}{A_2} \quad \text{and} \quad F_2 = \frac{A_2}{A_1}F_1 \quad (12.7)$$

The hydraulic lift is a force-multiplying device with a multiplication factor equal to the ratio of the areas of the two pistons. Dentist's chairs, car lifts and jacks, many elevators, and hydraulic brakes all use this principle.

For gases the assumption that the density ρ is uniform is realistic over only short vertical distances. In a room with a ceiling height of 3.0 m filled with air of uniform density 1.2 kg/m³, the difference in pressure between floor and ceiling, given by Eq. (12.6), is

$$\rho g h = (1.2 \text{ kg/m}^3)(9.8 \text{ m/s}^2)(3.0 \text{ m}) = 35 \text{ Pa}$$

or about 0.00035 atm, a very small difference. But between sea level and the summit of Mount Everest (8882 m) the density of air changes by nearly a factor of 3, and in this case we cannot use Eq. (12.6). Liquids, by contrast, are nearly incompressible, and it is usually a very good approximation to regard their density as independent of pressure.

Absolute Pressure and Gauge Pressure

If the pressure inside a car tire is equal to atmospheric pressure, the tire is flat. The pressure has to be *greater* than atmospheric to support the car, so the significant quantity is the *difference* between the inside and outside pressures. When we say that the pressure in a car tire is "32 pounds" (actually 32 lb/in.², equal to 220 kPa or 2.2×10^5 Pa), we mean that it is *greater* than atmospheric pressure (14.7 lb/in.² or 1.01×10^5 Pa) by this amount. The *total* pressure in the tire is then 47 lb/in.² or 320 kPa. The excess pressure above atmospheric pressure is usually called **gauge pressure,** and the total pressure is called **absolute pressure.** Engineers use the abbreviations psig and psia for "pounds per square inch gauge" and "pounds per square inch absolute," respectively. If the pressure is *less* than atmospheric, as in a partial vacuum, the gauge pressure is negative.

12.5 How pressure varies with depth in a fluid with uniform density.

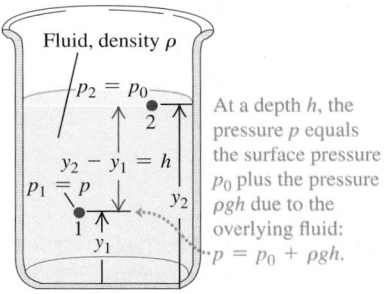

At a depth h, the pressure p equals the surface pressure p_0 plus the pressure $\rho g h$ due to the overlying fluid: $p = p_0 + \rho g h$.

Pressure difference between levels 1 and 2:
$$p_2 - p_1 = -\rho g(y_2 - y_1)$$
The pressure is greater at the lower level.

12.6 Each fluid column has the same height, no matter what its shape.

The pressure at the top of each liquid column is atmospheric pressure, p_0.

The pressure at the bottom of each liquid column has the same value p.

The difference between p and p_0 is $\rho g h$, where h is the distance from the top to the bottom of the liquid column. Hence all columns have the same height.

12.7 The hydraulic lift is an application of Pascal's law. The size of the fluid-filled container is exaggerated for clarity.

A small force is applied to a small piston.

Because the pressure p is the same at all points at a given height in the fluid ...

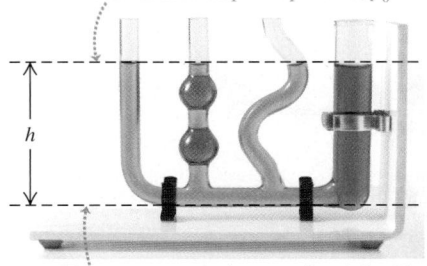

... a piston of larger area at the same height experiences a larger force.

EXAMPLE 12.3 FINDING ABSOLUTE AND GAUGE PRESSURES

Water stands 12.0 m deep in a storage tank whose top is open to the atmosphere. What are the absolute and gauge pressures at the bottom of the tank?

SOLUTION

IDENTIFY and SET UP: Table 11.2 indicates that water is nearly incompressible, so we can treat it as having uniform density. The level of the top of the tank corresponds to point 2 in Fig. 12.5, and the level of the bottom of the tank corresponds to point 1. Our target variable is p in Eq. (12.6). We have $h = 12.0$ m and $p_0 = 1$ atm $= 1.01 \times 10^5$ Pa.

EXECUTE: From Eq. (12.6), the pressures are

absolute:

$$p = p_0 + \rho g h$$
$$= (1.01 \times 10^5 \text{ Pa}) + (1000 \text{ kg/m}^3)(9.80 \text{ m/s}^2)(12.0 \text{ m})$$
$$= 2.19 \times 10^5 \text{ Pa} = 2.16 \text{ atm} = 31.8 \text{ lb/in.}^2$$

gauge: $\quad p - p_0 = (2.19 - 1.01) \times 10^5 \text{ Pa}$
$$= 1.18 \times 10^5 \text{ Pa} = 1.16 \text{ atm} = 17.1 \text{ lb/in.}^2$$

EVALUATE: A pressure gauge at the bottom of such a tank would probably be calibrated to read gauge pressure rather than absolute pressure.

Pressure Gauges

The simplest pressure gauge is the open-tube *manometer* (**Fig. 12.8a**). The U-shaped tube contains a liquid of density ρ, often mercury or water. The left end of the tube is connected to the container where the pressure p is to be measured, and the right end is open to the atmosphere at pressure $p_0 = p_{atm}$. The pressure at the bottom of the tube due to the fluid in the left column is $p + \rho g y_1$, and the pressure at the bottom due to the fluid in the right column is $p_{atm} + \rho g y_2$. These pressures are measured at the same level, so they must be equal:

$$p + \rho g y_1 = p_{atm} + \rho g y_2$$
$$p - p_{atm} = \rho g (y_2 - y_1) = \rho g h \tag{12.8}$$

In Eq. (12.8), p is the *absolute pressure,* and the difference $p - p_{atm}$ between absolute and atmospheric pressure is the gauge pressure. Thus the gauge pressure is proportional to the difference in height $h = y_2 - y_1$ of the liquid columns.

Another common pressure gauge is the **mercury barometer.** It consists of a long glass tube, closed at one end, that has been filled with mercury and then inverted in a dish of mercury (Fig. 12.8b). The space above the mercury column

12.8 Two types of pressure gauge.

(a) Open-tube manometer

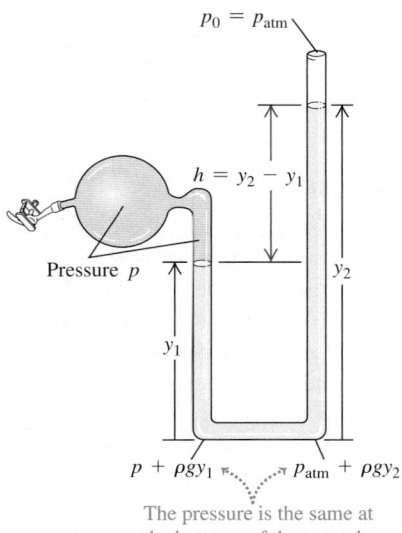

(b) Mercury barometer

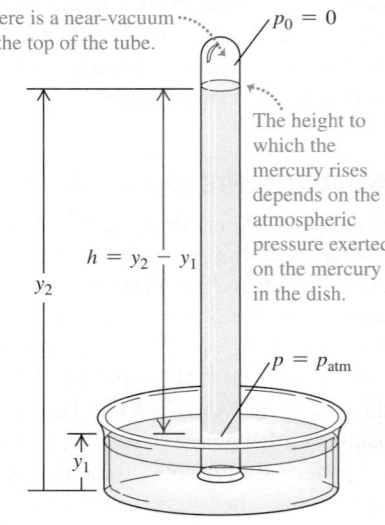

12.9 (a) A Bourdon pressure gauge. When the pressure inside the flexible tube increases, the tube straightens out a little, deflecting the attached pointer. (b) This Bourdon-type pressure gauge is connected to a high-pressure gas line. The gauge pressure shown is just over 5 bars (1 bar = 10^5 Pa).

(a)

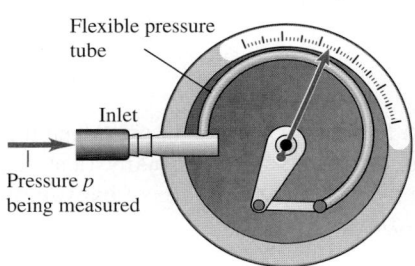

Changes in the inlet pressure cause the tube to coil or uncoil, which moves the pointer.

Flexible pressure tube

Inlet

Pressure p being measured

(b)

contains only mercury vapor; its pressure is negligibly small, so the pressure p_0 at the top of the mercury column is practically zero. From Eq. (12.6),

$$p_{atm} = p = 0 + \rho g(y_2 - y_1) = \rho g h \qquad (12.9)$$

So the height h of the mercury column indicates the atmospheric pressure p_{atm}.

Pressures are often described in terms of the height of the corresponding mercury column, as so many "inches of mercury" or "millimeters of mercury" (abbreviated mm Hg). A pressure of 1 mm Hg is called *1 torr,* after Evangelista Torricelli, inventor of the mercury barometer. But these units depend on the density of mercury, which varies with temperature, and on the value of g, which varies with location, so the pascal is the preferred unit of pressure.

Many types of pressure gauges use a flexible sealed tube (**Fig. 12.9**). A change in the pressure either inside or outside the tube causes a change in its dimensions. This change is detected optically, electrically, or mechanically.

BIO Application Gauge Pressure of Blood Blood-pressure readings, such as 130/80, give the maximum and minimum gauge pressures in the arteries, measured in mm Hg or torr. Blood pressure varies with vertical position within the body; the standard reference point is the upper arm, level with the heart.

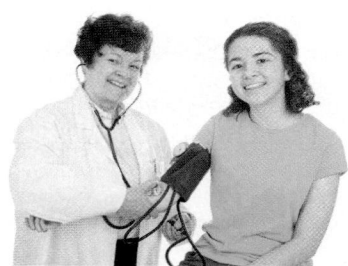

EXAMPLE 12.4   A TALE OF TWO FLUIDS

A manometer tube is partially filled with water. Oil (which does not mix with water) is poured into the left arm of the tube until the oil–water interface is at the midpoint of the tube as shown in **Fig. 12.10.** Both arms of the tube are open to the air. Find a relationship between the heights h_{oil} and h_{water}.

SOLUTION

IDENTIFY and SET UP: Figure 12.10 shows our sketch. The relationship between pressure and depth given by Eq. (12.6) applies to fluids of uniform density only; we have two fluids of different

12.10 Our sketch for this problem.

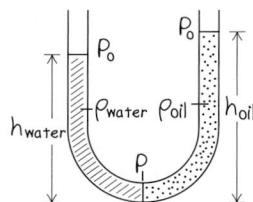

densities, so we must write a separate pressure–depth relationship for each. Both fluid columns have pressure p at the bottom (where they are in contact and in equilibrium), and both are at atmospheric pressure p_0 at the top (where both are in contact with and in equilibrium with the air).

EXECUTE: Writing Eq. (12.6) for each fluid gives

$$p = p_0 + \rho_{water} g h_{water}$$
$$p = p_0 + \rho_{oil} g h_{oil}$$

Since the pressure p at the bottom of the tube is the same for both fluids, we set these two expressions equal to each other and solve for h_{oil} in terms of h_{water}:

$$h_{oil} = \frac{\rho_{water}}{\rho_{oil}} h_{water}$$

EVALUATE: Water ($\rho_{water} = 1000 \text{ kg/m}^3$) is denser than oil ($\rho_{oil} \approx 850 \text{ kg/m}^3$), so h_{oil} is greater than h_{water} as Fig. 12.10 shows. It takes a greater height of low-density oil to produce the same pressure p at the bottom of the tube.

TEST YOUR UNDERSTANDING OF SECTION 12.2 Mercury is less dense at high temperatures than at low temperatures. Suppose you move a mercury barometer from the cold interior of a tightly sealed refrigerator to outdoors on a hot summer day. You find that the column of mercury remains at the same height in the tube. Compared to the air pressure inside the refrigerator, is the air pressure outdoors (i) higher, (ii) lower, or (iii) the same? (Ignore the very small change in the dimensions of the glass tube due to the temperature change.) ❙

PhET: Balloons & Buoyancy

DEMO

12.3 BUOYANCY

A body immersed in water seems to weigh less than when it is in air. When the body is less dense than the fluid, it floats. The human body usually floats in water, and a helium-filled balloon floats in air. These are examples of **buoyancy,** a phenomenon described by *Archimedes's principle:*

> **ARCHIMEDES'S PRINCIPLE: When a body is completely or partially immersed in a fluid, the fluid exerts an upward force on the body equal to the weight of the fluid displaced by the body.**

DATA SPEAKS

Buoyancy

When students were given a problem about buoyancy, more than 25% gave an incorrect response. Common errors:

- Forgetting that the buoyant force on an object depends on the density of the fluid and the submerged volume of the object but not on the density of the object.
- Forgetting that the buoyant force on an object equals the weight of displaced fluid—which need not be the same as the weight of the object.

To prove this principle, we consider an arbitrary element of fluid at rest. The dashed curve in **Fig. 12.11a** outlines such an element. The arrows labeled $dF_\perp$ represent the forces exerted on the element's surface by the surrounding fluid.

The entire fluid is in equilibrium, so the sum of all the *y*-components of force on this element of fluid is zero. Hence the sum of the *y*-components of the *surface* forces must be an upward force equal in magnitude to the weight mg of the fluid inside the surface. Also, the sum of the torques on the element of fluid must be zero, so the line of action of the resultant *y*-component of surface force must pass through the center of gravity of this element of fluid.

Now we replace the fluid inside the surface with a solid body having exactly the same shape (Fig. 12.11b). The pressure at every point is the same as before. So the total upward force exerted on the body by the fluid is also the same, again equal in magnitude to the weight mg of the fluid displaced to make way for the body. We call this upward force the **buoyant force** on the solid body. The line of action of the buoyant force again passes through the center of gravity of the displaced fluid (which doesn't necessarily coincide with the center of gravity of the body).

When a balloon floats in equilibrium in air, its weight (including the gas inside it) must be the same as the weight of the air displaced by the balloon. A fish's flesh is denser than water, yet many fish can float while submerged. These fish have a gas-filled cavity within their bodies, which makes the fish's

12.11 Archimedes's principle.

(a) Arbitrary element of fluid in equilibrium

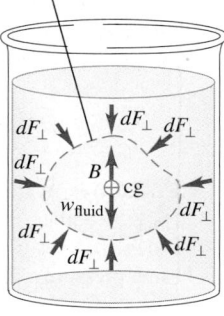

The forces on the fluid element due to pressure must sum to a buoyant force equal in magnitude to the element's weight.

(b) Fluid element replaced with solid body of the same size and shape

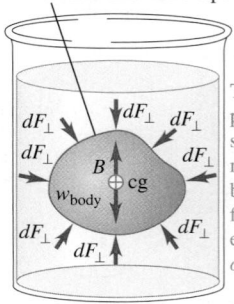

The forces due to pressure are the same, so the body must be acted upon by the same buoyant force as the fluid element, *regardless of the body's weight.*

average density the same as water's. So the net weight of the fish is the same as the weight of the water it displaces. A body whose average density is *less* than that of a liquid can float partially submerged at the free upper surface of the liquid. A ship made of steel (which is much denser than water) can float because the ship is hollow, with air occupying much of its interior volume, so its average density is less than that of water. The greater the density of the liquid, the less of the body is submerged. When you swim in seawater (density 1030 kg/m³), your body floats higher than in freshwater (1000 kg/m³).

A practical example of buoyancy is the hydrometer, used to measure the density of liquids (**Fig. 12.12a**). The calibrated float sinks into the fluid until the weight of the fluid it displaces is exactly equal to its own weight. The hydrometer floats *higher* in denser liquids than in less dense liquids, and a scale in the top stem permits direct density readings. Hydrometers like this are used in medical diagnosis to measure the density of urine (which depends on a patient's level of hydration). Figure 12.12b shows a type of hydrometer used to measure the density of battery acid or antifreeze. The bottom of the large tube is immersed in the liquid; the bulb is squeezed to expel air and is then released, like a giant medicine dropper. The liquid rises into the outer tube, and the hydrometer floats in this liquid.

12.12 Measuring the density of a fluid.

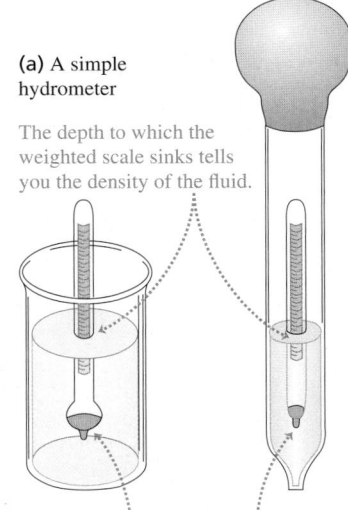

(b) Using a hydrometer to measure the density of battery acid or antifreeze

(a) A simple hydrometer

The depth to which the weighted scale sinks tells you the density of the fluid.

The weight at the bottom makes the scale float upright.

EXAMPLE 12.5   **BUOYANCY**

A 15.0-kg solid gold statue is raised from the sea bottom (**Fig. 12.13a**). What is the tension in the hoisting cable (assumed massless) when the statue is (a) at rest and completely underwater and (b) at rest and completely out of the water?

SOLUTION

IDENTIFY and SET UP: In both cases the statue is in equilibrium and experiences three forces: its weight, the cable tension, and a buoyant force equal in magnitude to the weight of the fluid displaced by the statue (seawater in part (a), air in part (b)). Figure 12.13b shows the free-body diagram for the statue. Our target variables are the values of the tension in seawater (T_{sw}) and in air (T_{air}). We are

12.13 What is the tension in the cable hoisting the statue?

(a) Immersed statue in equilibrium   **(b)** Free-body diagram of statue

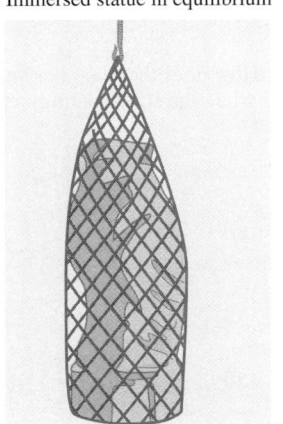
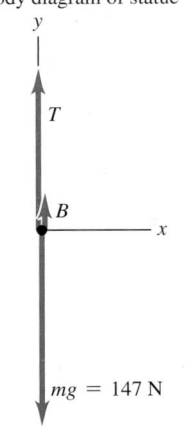

$mg = 147$ N

given the mass m_{statue}, and we can calculate the buoyant force in seawater (B_{sw}) and in air (B_{air}) by using Archimedes's principle.

EXECUTE: (a) To find B_{sw}, we first find the statue's volume V by using the density of gold from Table 12.1:

$$V = \frac{m_{statue}}{\rho_{gold}} = \frac{15.0 \text{ kg}}{19.3 \times 10^3 \text{ kg/m}^3} = 7.77 \times 10^{-4} \text{ m}^3$$

The buoyant force B_{sw} equals the weight of this same volume of seawater. Using Table 12.1 again:

$$B_{sw} = w_{sw} = m_{sw}g = \rho_{sw}Vg$$
$$= (1.03 \times 10^3 \text{ kg/m}^3)(7.77 \times 10^{-4} \text{ m}^3)(9.80 \text{ m/s}^2)$$
$$= 7.84 \text{ N}$$

The statue is at rest, so the net external force acting on it is zero. From Fig. 12.13b,

$$\Sigma F_y = B_{sw} + T_{sw} + (-m_{statue}g) = 0$$
$$T_{sw} = m_{statue}g - B_{sw} = (15.0 \text{ kg})(9.80 \text{ m/s}^2) - 7.84 \text{ N}$$
$$= 147 \text{ N} - 7.84 \text{ N} = 139 \text{ N}$$

A spring scale attached to the upper end of the cable will indicate a tension 7.84 N less than the statue's actual weight $m_{statue}g = 147$ N.

(b) The density of air is about 1.2 kg/m³, so the buoyant force of air on the statue is

$$B_{air} = \rho_{air}Vg$$
$$= (1.2 \text{ kg/m}^3)(7.77 \times 10^{-4} \text{ m}^3)(9.80 \text{ m/s}^2)$$
$$= 9.1 \times 10^{-3} \text{ N}$$

Continued

This is negligible compared to the statue's actual weight $m_{statue}g =$ 147 N. So within the precision of our data, the tension in the cable with the statue in air is $T_{air} = m_{statue}g = 147$ N.

EVALUATE: Note that the buoyant force is proportional to the density of the *fluid* in which the statue is immersed, *not* the density of the statue. The denser the fluid, the greater the buoyant force and the smaller the cable tension. If the fluid had the same density

as the statue, the buoyant force would be equal to the statue's weight and the tension would be zero (the cable would go slack). If the fluid were denser than the statue, the tension would be *negative:* The buoyant force would be greater than the statue's weight, and a downward force would be required to keep the statue from rising upward.

12.14 The surface of the water acts like a membrane under tension, allowing this water strider to "walk on water."

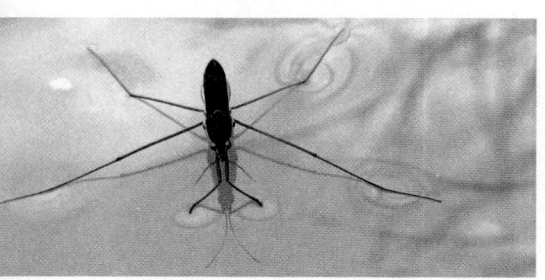

12.15 A molecule at the surface of a liquid is attracted into the bulk liquid, which tends to reduce the liquid's surface area.

Molecules in a liquid are attracted by neighboring molecules.

At the surface, the unbalanced attractions cause the surface to resist being stretched.

Water molecules

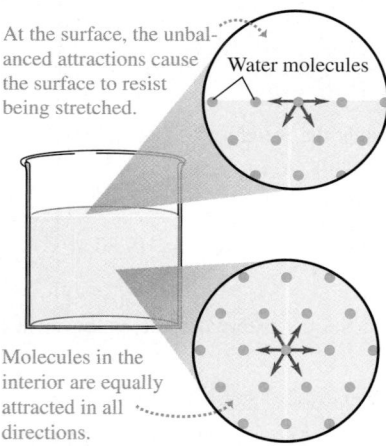

Molecules in the interior are equally attracted in all directions.

Surface Tension

We've seen that if an object is less dense than water, it will float partially submerged. But a paper clip can rest *atop* a water surface even though its density is several times that of water. This is an example of **surface tension:** The surface of the liquid behaves like a membrane under tension (**Fig. 12.14**). Surface tension arises because the molecules of the liquid exert attractive forces on each other. There is zero net force on a molecule within the interior of the liquid, but a surface molecule is drawn into the interior (**Fig. 12.15**). Thus the liquid tends to minimize its surface area, just as a stretched membrane does.

Surface tension explains why raindrops are spherical (*not* teardrop-shaped): A sphere has a smaller surface area for its volume than any other shape. It also explains why hot, soapy water is used for washing. To wash clothing thoroughly, water must be forced through the tiny spaces between the fibers (**Fig. 12.16**). This requires increasing the surface area of the water, which is difficult to achieve because of surface tension. The job is made easier by increasing the temperature of the water and adding soap, both of which decrease the surface tension.

Surface tension is important for a millimeter-sized water drop, which has a relatively large surface area for its volume. (A sphere of radius r has surface area $4\pi r^2$ and volume $(4\pi/3)r^3$. The ratio of surface area to volume is $3/r$, which increases with decreasing radius.) But for large quantities of liquid, the ratio of surface area to volume is relatively small, and surface tension is negligible compared to pressure forces. For the remainder of this chapter, we'll consider only fluids in bulk and ignore the effects of surface tension.

TEST YOUR UNDERSTANDING OF SECTION 12.3 You place a container of seawater on a scale and note the reading on the scale. You now suspend the statue of Example 12.5 in the water (**Fig. 12.17**). How does the scale reading change? (i) It increases by 7.84 N; (ii) it decreases by 7.84 N; (iii) it remains the same; (iv) none of these. ∎

12.16 Surface tension makes it difficult to force water through small crevices. The required water pressure p can be reduced by using hot, soapy water, which has less surface tension.

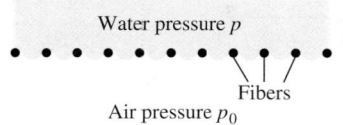

Water pressure p

Fibers

Air pressure p_0

12.17 How does the scale reading change when the statue is immersed in water?

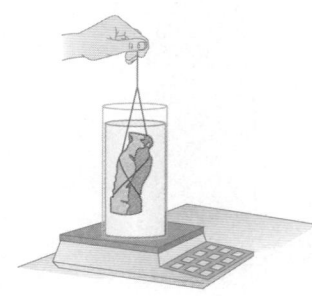

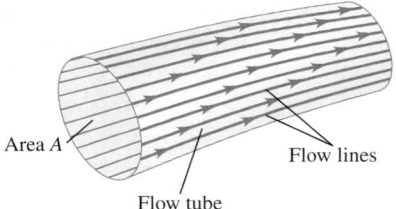

12.18 A flow tube bounded by flow lines. In steady flow, fluid cannot cross the walls of a flow tube.

Area A

Flow lines

Flow tube

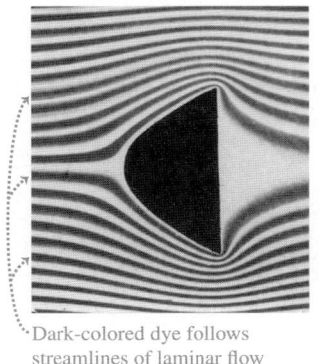

12.19 Laminar flow around an obstacle.

Dark-colored dye follows streamlines of laminar flow (flow is from left to right).

12.4 FLUID FLOW

We are now ready to consider *motion* of a fluid. Fluid flow can be extremely complex, as shown by the currents in river rapids or the swirling flames of a campfire. But we can represent some situations by relatively simple idealized models. An **ideal fluid** is a fluid that is *incompressible* (that is, its density cannot change) and has no internal friction (called **viscosity**). Liquids are approximately incompressible in most situations, and we may also treat a gas as incompressible if the pressure differences from one region to another are not too great. Internal friction in a fluid causes shear stresses when two adjacent layers of fluid move relative to each other, as when fluid flows inside a tube or around an obstacle. In some cases we can ignore these shear forces in comparison with forces arising from gravitation and pressure differences.

The path of an individual particle in a moving fluid is called a **flow line.** In **steady flow,** the overall flow pattern does not change with time, so every element passing through a given point follows the same flow line. In this case the "map" of the fluid velocities at various points in space remains constant, although the velocity of a particular particle may change in both magnitude and direction during its motion. A **streamline** is a curve whose tangent at any point is in the direction of the fluid velocity at that point. When the flow pattern changes with time, the streamlines do not coincide with the flow lines. We'll consider only steady-flow situations, for which flow lines and streamlines are identical.

The flow lines passing through the edge of an imaginary element of area, such as area A in **Fig. 12.18**, form a tube called a **flow tube.** From the definition of a flow line, in steady flow no fluid can cross the side walls of a given flow tube.

Figure 12.19 shows the pattern of fluid flow from left to right around an obstacle. The photograph was made by injecting dye into water flowing between two closely spaced glass plates. This pattern is typical of **laminar flow,** in which adjacent layers of fluid slide smoothly past each other and the flow is steady. (A *lamina* is a thin sheet.) At sufficiently high flow rates, or when boundary surfaces cause abrupt changes in velocity, the flow can become irregular and chaotic. This is called **turbulent flow (Fig. 12.20).** In turbulent flow there is no steady-state pattern; the flow pattern changes continuously.

The Continuity Equation

The mass of a moving fluid doesn't change as it flows. This leads to an important relationship called the **continuity equation.** Consider a portion of a flow tube between two stationary cross sections with areas A_1 and A_2 (**Fig. 12.21**). The fluid speeds at these sections are v_1 and v_2, respectively. As we mentioned above, no fluid flows in or out across the side walls of such a tube. During a small time interval dt, the fluid at A_1 moves a distance $ds_1 = v_1\,dt$, so a cylinder of fluid with height $v_1\,dt$ and volume $dV_1 = A_1 v_1\,dt$ flows into the tube across A_1. During this same interval, a cylinder of volume $dV_2 = A_2 v_2\,dt$ flows out of the tube across A_2.

Let's first consider the case of an incompressible fluid so that the density ρ has the same value at all points. The mass dm_1 flowing into the tube across A_1 in

12.20 The flow of smoke rising from this burnt match is laminar up to a certain point, and then becomes turbulent.

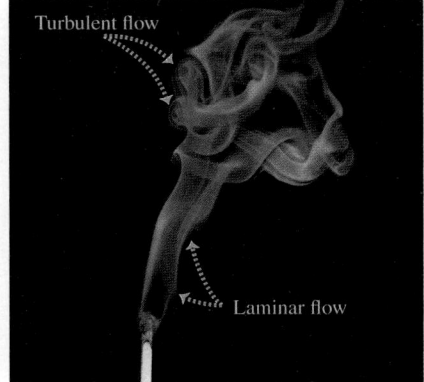

Turbulent flow

Laminar flow

12.21 A flow tube with changing cross-sectional area.

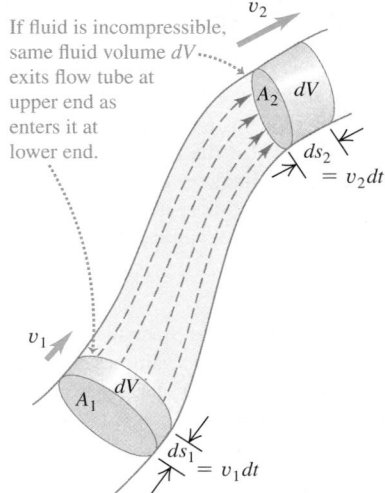

If fluid is incompressible, same fluid volume dV exits flow tube at upper end as enters it at lower end.

v_2

A_2 dV

$ds_2 = v_2 dt$

v_1

A_1 dV

$ds_1 = v_1 dt$

If fluid is incompressible, product Av (tube area times speed) has same value at all points along tube.

time dt is $dm_1 = \rho A_1 v_1\,dt$. Similarly, the mass dm_2 that flows out across A_2 in the same time is $dm_2 = \rho A_2 v_2\,dt$. In steady flow the total mass in the tube is constant, so $dm_1 = dm_2$ and

$$\rho A_1 v_1\,dt = \rho A_2 v_2\,dt \qquad \text{or}$$

Continuity equation for an **incompressible fluid**	$A_1 v_1 = A_2 v_2$	Cross-sectional area of flow tube at two points (see Fig. 12.21)	(12.10)
		Speed of flow at the two points	

The product Av is the *volume flow rate* dV/dt, the rate at which volume crosses a section of the tube:

Volume flow rate of a fluid	$\dfrac{dV}{dt} = Av$	Cross-sectional area of flow tube Speed of flow	(12.11)

The *mass flow rate* is the mass flow per unit time through a cross section. This is equal to the density ρ times the volume flow rate dV/dt.

Equation (12.10) shows that the volume flow rate has the same value at all points along any flow tube (**Fig. 12.22**). When the cross section of a flow tube decreases, the speed increases, and vice versa. A broad, deep part of a river has a larger cross section and slower current than a narrow, shallow part, but the volume flow rates are the same in both. This is the essence of the familiar maxim, "Still waters run deep." If a water pipe with 2-cm diameter is connected to a pipe with 1-cm diameter, the flow speed is four times as great in the 1-cm part as in the 2-cm part.

We can generalize Eq. (12.10) for the case in which the fluid is *not* incompressible. If ρ_1 and ρ_2 are the densities at sections 1 and 2, then

$$\rho_1 A_1 v_1 = \rho_2 A_2 v_2 \qquad \text{(continuity equation, compressible fluid)} \qquad (12.12)$$

If the fluid is denser at point 2 than at point 1 ($\rho_2 > \rho_1$), the volume flow rate at point 2 will be less than at point 1 ($A_2 v_2 < A_1 v_1$). We leave the details to you. If the fluid is incompressible so that ρ_1 and ρ_2 are always equal, Eq. (12.12) reduces to Eq. (12.10).

12.22 The continuity equation, Eq. (12.10), helps explain the shape of a stream of honey poured from a spoon.

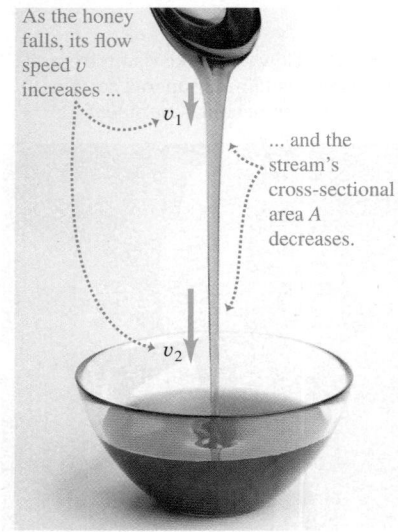

As the honey falls, its flow speed v increases ...

v_1

... and the stream's cross-sectional area A decreases.

v_2

The volume flow rate $dV/dt = Av$ remains constant.

EXAMPLE 12.6   **FLOW OF AN INCOMPRESSIBLE FLUID**

Incompressible oil of density $850\ \text{kg/m}^3$ is pumped through a cylindrical pipe at a rate of 9.5 liters per second. (a) The first section of the pipe has a diameter of 8.0 cm. What is the flow speed of the oil? What is the mass flow rate? (b) The second section of the pipe has a diameter of 4.0 cm. What are the flow speed and mass flow rate in that section?

SOLUTION

IDENTIFY and SET UP: Since the oil is incompressible, the volume flow rate has the *same* value (9.5 L/s) in both sections of pipe. The mass flow rate (the density times the volume flow rate) also has the same value in both sections. (This is just the statement that no fluid is lost or added anywhere along the pipe.) We use the volume flow rate equation, Eq. (12.11), to determine the speed v_1 in the 8.0-cm-diameter section and the continuity equation for incompressible flow, Eq. (12.10), to find the speed v_2 in the 4.0-cm-diameter section.

EXECUTE: (a) From Eq. (12.11) the volume flow rate in the first section is $dV/dt = A_1 v_1$, where A_1 is the cross-sectional area of the pipe of diameter 8.0 cm and radius 4.0 cm. Hence

$$v_1 = \frac{dV/dt}{A_1} = \frac{(9.5\ \text{L/s})(10^{-3}\ \text{m}^3/\text{L})}{\pi(4.0 \times 10^{-2}\ \text{m})^2} = 1.9\ \text{m/s}$$

The mass flow rate is $\rho\,dV/dt = (850\ \text{kg/m}^3)(9.5 \times 10^{-3}\ \text{m}^3/\text{s}) = 8.1\ \text{kg/s}$.

(b) From the continuity equation, Eq. (12.10),

$$v_2 = \frac{A_1}{A_2}v_1 = \frac{\pi(4.0 \times 10^{-2}\ \text{m})^2}{\pi(2.0 \times 10^{-2}\ \text{m})^2}(1.9\ \text{m/s}) = 7.6\ \text{m/s} = 4v_1$$

The volume and mass flow rates are the same as in part (a).

EVALUATE: The second section of pipe has one-half the diameter and one-fourth the cross-sectional area of the first section. Hence the speed must be four times greater in the second section, which is just what our result shows.

TEST YOUR UNDERSTANDING OF SECTION 12.4 A maintenance crew is working on a section of a three-lane highway, leaving only one lane open to traffic. The result is much slower traffic flow (a traffic jam). Do cars on a highway behave like the molecules of (i) an incompressible fluid or (ii) a compressible fluid? ❙

12.5 BERNOULLI'S EQUATION

According to the continuity equation, the speed of fluid flow can vary along the paths of the fluid. The pressure can also vary; it depends on height as in the static situation (see Section 12.2), and it also depends on the speed of flow. We can derive an important relationship called *Bernoulli's equation,* which relates the pressure, flow speed, and height for flow of an ideal, incompressible fluid. Bernoulli's equation is useful in analyzing many kinds of fluid flow.

The dependence of pressure on speed follows from the continuity equation, Eq. (12.10). When an incompressible fluid flows along a flow tube with varying cross section, its speed *must* change, and so an element of fluid must have an acceleration. If the tube is horizontal, the force that causes this acceleration has to be applied by the surrounding fluid. This means that the pressure *must* be different in regions of different cross section; if it were the same everywhere, the net force on every fluid element would be zero. When a horizontal flow tube narrows and a fluid element speeds up, it must be moving toward a region of lower pressure in order to have a net forward force to accelerate it. If the elevation also changes, this causes an additional pressure difference.

Deriving Bernoulli's Equation

To derive Bernoulli's equation, we apply the work–energy theorem to the fluid in a section of a flow tube. In **Fig. 12.23** we consider the element of fluid that at some initial time lies between the two cross sections a and c. The speeds at the lower and upper ends are v_1 and v_2. In a small time interval dt, the fluid that is initially at a moves to b, a distance $ds_1 = v_1\,dt$, and the fluid that is initially at c moves to d, a distance $ds_2 = v_2\,dt$. The cross-sectional areas at the two ends are A_1 and A_2, as shown. The fluid is incompressible; hence by the continuity equation, Eq. (12.10), the volume of fluid dV passing *any* cross section during time dt is the same. That is, $dV = A_1\,ds_1 = A_2\,ds_2$.

Let's compute the *work* done on this fluid element during dt. If there is negligible internal friction in the fluid (i.e., no viscosity), the only nongravitational forces that do work on the element are due to the pressure of the surrounding fluid. The pressures at the two ends are p_1 and p_2; the force on the cross section at a is p_1A_1, and the force at c is p_2A_2. The net work dW done on the element by the surrounding fluid during this displacement is therefore

$$dW = p_1A_1\,ds_1 - p_2A_2\,ds_2 = (p_1 - p_2)\,dV \qquad (12.13)$$

The term $p_2A_2\,ds_2$ has a negative sign because the force at c opposes the displacement of the fluid.

The work dW is due to forces other than the conservative force of gravity, so it equals the change in the total mechanical energy (kinetic energy plus gravitational potential energy) associated with the fluid element. The mechanical energy for the fluid between sections b and c does not change. At the beginning of dt the fluid between a and b has volume $A_1\,ds_1$, mass $\rho A_1\,ds_1$, and kinetic energy $\frac{1}{2}\rho(A_1\,ds_1)v_1^2$. At the end of dt the fluid between c and d has kinetic energy $\frac{1}{2}\rho(A_2\,ds_2)v_2^2$. The net change in kinetic energy dK during time dt is

$$dK = \tfrac{1}{2}\rho\,dV(v_2^2 - v_1^2) \qquad (12.14)$$

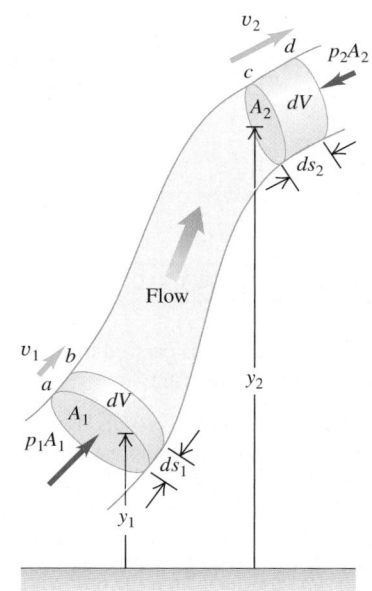

12.23 Deriving Bernoulli's equation. The net work done on a fluid element by the pressure of the surrounding fluid equals the change in the kinetic energy plus the change in the gravitational potential energy.

Bernoulli's equation suggests that as blood flows upward at roughly constant speed v from the heart to the brain, the pressure p will drop as the blood's height y increases. For blood to reach the brain with the required minimal pressure, the human heart provides a maximum (systolic) gauge pressure of about 120 mm Hg. The vertical distance from heart to brain is much larger for a giraffe, so its heart must produce a much greater maximum gauge pressure (about 280 mm Hg).

What about the change in gravitational potential energy? At the beginning of time interval dt, the potential energy for the mass between a and b is $dm\, gy_1 = \rho\, dV\, gy_1$. At the end of dt, the potential energy for the mass between c and d is $dm\, gy_2 = \rho\, dV\, gy_2$. The net change in potential energy dU during dt is

$$dU = \rho\, dV\, g(y_2 - y_1) \qquad (12.15)$$

Combining Eqs. (12.13), (12.14), and (12.15) in the energy equation $dW = dK + dU$, we obtain

$$(p_1 - p_2)dV = \tfrac{1}{2}\rho\, dV(v_2^2 - v_1^2) + \rho\, dV\, g(y_2 - y_1)$$
$$p_1 - p_2 = \tfrac{1}{2}\rho(v_2^2 - v_1^2) + \rho g(y_2 - y_1) \qquad (12.16)$$

This is **Bernoulli's equation.** It states that the work done on a unit volume of fluid by the surrounding fluid is equal to the sum of the changes in kinetic and potential energies per unit volume that occur during the flow. We may also interpret Eq. (12.16) in terms of pressures. The first term on the right is the pressure difference associated with the change of speed of the fluid. The second term on the right is the additional pressure difference caused by the weight of the fluid and the difference in elevation of the two ends.

We can also express Eq. (12.16) in a more convenient form as

$$p_1 + \rho gy_1 + \tfrac{1}{2}\rho v_1^2 = p_2 + \rho gy_2 + \tfrac{1}{2}\rho v_2^2 \qquad (12.17)$$

Subscripts 1 and 2 refer to *any* two points along the flow tube, so we can write

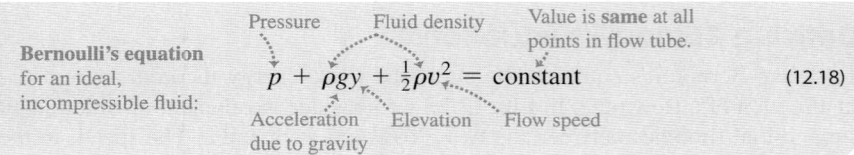

Bernoulli's equation for an ideal, incompressible fluid:

Pressure — Fluid density — Value is **same** at all points in flow tube.

$$p + \rho gy + \tfrac{1}{2}\rho v^2 = \text{constant} \qquad (12.18)$$

Acceleration due to gravity — Elevation — Flow speed

Note that when the fluid is *not* moving (so $v_1 = v_2 = 0$), Eq. (12.17) reduces to the pressure relationship we derived for a fluid at rest, Eq. (12.5).

DEMO

CAUTION **Bernoulli's equation applies in certain situations only** We stress again that Bernoulli's equation is valid for only incompressible, steady flow of a fluid with no internal friction (no viscosity). It's a simple equation, but don't be tempted to use it in situations in which it doesn't apply! ▮

PROBLEM-SOLVING STRATEGY 12.1 | BERNOULLI'S EQUATION

Bernoulli's equation is derived from the work–energy theorem, so much of Problem-Solving Strategy 7.1 (Section 7.1) applies here.

IDENTIFY *the relevant concepts:* Bernoulli's equation is applicable to steady flow of an incompressible fluid that has no internal friction (see Section 12.6). It is generally applicable to flows through large pipes and to flows within bulk fluids (e.g., air flowing around an airplane or water flowing around a fish).

SET UP *the problem* using the following steps:
1. Identify the points 1 and 2 referred to in Bernoulli's equation, Eq. (12.17).
2. Define your coordinate system, particularly the level at which $y = 0$. Take the positive y-direction to be upward.

3. List the unknown and known quantities in Eq. (12.17). Decide which unknowns are the target variables.

EXECUTE *the solution* as follows: Write Bernoulli's equation and solve for the unknowns. You may need the continuity equation, Eq. (12.10), to relate the two speeds in terms of cross-sectional areas of pipes or containers. You may also need Eq. (12.11) to find the volume flow rate.

EVALUATE *your answer:* Verify that the results make physical sense. Check that you have used consistent units: In SI units, pressure is in pascals, density in kilograms per cubic meter, and speed in meters per second. The pressures must be either *all* absolute pressures or *all* gauge pressures.

EXAMPLE 12.7 WATER PRESSURE IN THE HOME

Water enters a house (**Fig. 12.24**) through a pipe with an inside diameter of 2.0 cm at an absolute pressure of 4.0×10^5 Pa (about 4 atm). A 1.0-cm-diameter pipe leads to the second-floor bathroom 5.0 m above. When the flow speed at the inlet pipe is 1.5 m/s, find the flow speed, pressure, and volume flow rate in the bathroom.

SOLUTION

IDENTIFY and SET UP: We assume that the water flows at a steady rate. Water is effectively incompressible, so we can use the continuity equation. It's reasonable to ignore internal friction because

12.24 What is the water pressure in the second-story bathroom of this house?

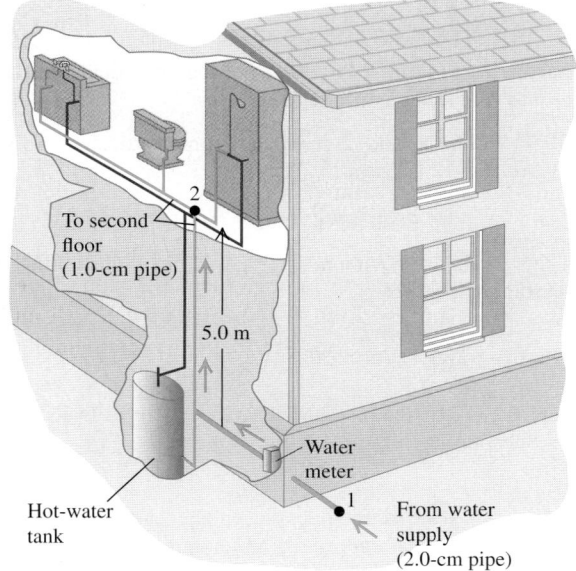

To second floor (1.0-cm pipe)

5.0 m

Hot-water tank

Water meter

From water supply (2.0-cm pipe)

the pipe has a relatively large diameter, so we can also use Bernoulli's equation. Let points 1 and 2 be at the inlet pipe and at the bathroom, respectively. We are given the pipe diameters at points 1 and 2, from which we calculate the areas A_1 and A_2, as well as the speed $v_1 = 1.5$ m/s and pressure $p_1 = 4.0 \times 10^5$ Pa at the inlet pipe. We take $y_1 = 0$ and $y_2 = 5.0$ m. We find the speed v_2 from the continuity equation and the pressure p_2 from Bernoulli's equation. Knowing v_2, we calculate the volume flow rate v_2A_2.

EXECUTE: From the continuity equation, Eq. (12.10),

$$v_2 = \frac{A_1}{A_2}v_1 = \frac{\pi(1.0 \text{ cm})^2}{\pi(0.50 \text{ cm})^2}(1.5 \text{ m/s}) = 6.0 \text{ m/s}$$

From Bernoulli's equation, Eq. (12.16),

$$p_2 = p_1 - \tfrac{1}{2}\rho(v_2^2 - v_1^2) - \rho g(y_2 - y_1)$$
$$= 4.0 \times 10^5 \text{ Pa}$$
$$\quad - \tfrac{1}{2}(1.0 \times 10^3 \text{ kg/m}^3)(36 \text{ m}^2/\text{s}^2 - 2.25 \text{ m}^2/\text{s}^2)$$
$$\quad - (1.0 \times 10^3 \text{ kg/m}^3)(9.8 \text{ m/s}^2)(5.0 \text{ m})$$
$$= 4.0 \times 10^5 \text{ Pa} - 0.17 \times 10^5 \text{ Pa} - 0.49 \times 10^5 \text{ Pa}$$
$$= 3.3 \times 10^5 \text{ Pa} = 3.3 \text{ atm} = 48 \text{ lb/in.}^2$$

The volume flow rate is

$$\frac{dV}{dt} = A_2v_2 = \pi(0.50 \times 10^{-2} \text{ m})^2(6.0 \text{ m/s})$$
$$= 4.7 \times 10^{-4} \text{ m}^3/\text{s} = 0.47 \text{ L/s}$$

EVALUATE: This is a reasonable flow rate for a bathroom faucet or shower. Note that if the water is turned off, both v_1 and v_2 are zero, the term $\tfrac{1}{2}\rho(v_2^2 - v_1^2)$ in Bernoulli's equation vanishes, and p_2 rises from 3.3×10^5 Pa to 3.5×10^5 Pa.

EXAMPLE 12.8 SPEED OF EFFLUX

Figure 12.25 shows a gasoline storage tank with cross-sectional area A_1, filled to a depth h. The space above the gasoline contains air at pressure p_0, and the gasoline flows out the bottom of the tank through a short pipe with cross-sectional area A_2. Derive expressions for the flow speed in the pipe and the volume flow rate.

SOLUTION

IDENTIFY and SET UP: We consider the entire volume of moving liquid as a single flow tube of an incompressible fluid with negligible internal friction. Hence, we can use Bernoulli's equation. Points 1 and 2 are at the surface of the gasoline and at the exit pipe, respectively. At point 1 the pressure is p_0, which we assume to be fixed; at point 2 it is atmospheric pressure p_{atm}. We take $y = 0$ at the exit pipe, so $y_1 = h$ and $y_2 = 0$. Because A_1 is very much larger than A_2, the upper surface of the gasoline will drop very slowly and so v_1 is essentially zero. We find v_2 from Eq. (12.17) and the volume flow rate from Eq. (12.11).

12.25 Calculating the speed of efflux for gasoline flowing out the bottom of a storage tank.

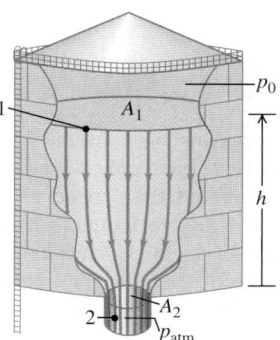

Continued

EXECUTE: We apply Bernoulli's equation to points 1 and 2:

$$p_0 + \tfrac{1}{2}\rho v_1^2 + \rho g h = p_{\text{atm}} + \tfrac{1}{2}\rho v_2^2 + \rho g(0)$$

$$v_2^2 = v_1^2 + 2\left(\frac{p_0 - p_{\text{atm}}}{\rho}\right) + 2gh$$

Using $v_1 = 0$, we find

$$v_2 = \sqrt{2\left(\frac{p_0 - p_{\text{atm}}}{\rho}\right) + 2gh}$$

From Eq. (12.11), the volume flow rate is $dV/dt = v_2 A_2$.

EVALUATE: The speed v_2, sometimes called the *speed of efflux,* depends on both the pressure difference $(p_0 - p_{\text{atm}})$ and the height h of the liquid level in the tank. If the top of the tank is vented to the atmosphere, $p_0 = p_{\text{atm}}$ and $p_0 - p_{\text{atm}} = 0$. Then

$$v_2 = \sqrt{2gh}$$

That is, the speed of efflux from an opening at a distance h below the top surface of the liquid is the *same* as the speed a body would acquire in falling freely through a height h. This result is called *Torricelli's theorem.* It is valid also for a hole in a side wall at a depth h below the surface. If $p_0 = p_{\text{atm}}$, the volume flow rate is

$$\frac{dV}{dt} = A_2\sqrt{2gh}$$

EXAMPLE 12.9   THE VENTURI METER

Figure 12.26 shows a *Venturi meter,* used to measure flow speed in a pipe. Derive an expression for the flow speed v_1 in terms of the cross-sectional areas A_1 and A_2 and the difference in height h of the liquid levels in the two vertical tubes.

SOLUTION

IDENTIFY and SET UP: The flow is steady, and we assume the fluid is incompressible and has negligible internal friction. Hence we can apply Bernoulli's equation to the wide part (point 1)

12.26 The Venturi meter.

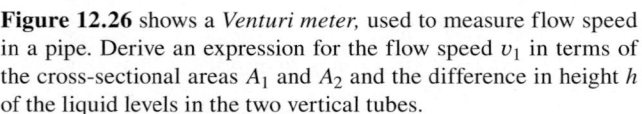

Difference in height results from reduced pressure in throat (point 2).

and narrow part (point 2, the *throat*) of the pipe. Equation (12.6) relates h to the pressure difference $p_1 - p_2$.

EXECUTE: Points 1 and 2 have the same vertical coordinate $y_1 = y_2$, so Eq. (12.17) says

$$p_1 + \tfrac{1}{2}\rho v_1^2 = p_2 + \tfrac{1}{2}\rho v_2^2$$

From the continuity equation, $v_2 = (A_1/A_2)v_1$. Substituting this and rearranging, we get

$$p_1 - p_2 = \tfrac{1}{2}\rho v_1^2\left[\left(\frac{A_1}{A_2}\right)^2 - 1\right]$$

From Eq. (12.6), the pressure difference $p_1 - p_2$ is also equal to $\rho g h$. Substituting this and solving for v_1, we get

$$v_1 = \sqrt{\frac{2gh}{(A_1/A_2)^2 - 1}}$$

EVALUATE: Because A_1 is greater than A_2, v_2 is greater than v_1 and the pressure p_2 in the throat is *less* than p_1. Those pressure differences produce a net force to the right that makes the fluid speed up as it enters the throat, and a net force to the left that slows it as it leaves.

CONCEPTUAL EXAMPLE 12.10   LIFT ON AN AIRPLANE WING

Figure 12.27a shows flow lines around a cross section of an airplane wing. The flow lines crowd together above the wing, corresponding to increased flow speed and reduced pressure, just as in the Venturi throat in Example 12.9. Hence the downward force of the air on the top side of the wing is less than the upward force of the air on the underside of the wing, and there is a net upward force or *lift.* Lift is not simply due to the impulse of air striking the underside of the wing; in fact, the reduced pressure on the upper wing surface makes the greatest contribution to the lift. (This simplified discussion ignores the formation of vortices.)

We can understand the lift force on the basis of momentum changes instead. The vector diagram in Fig. 12.27a shows that there is a net *downward* change in the vertical component of momentum of the air flowing past the wing, corresponding to the downward force the wing exerts on the air. The reaction force *on* the wing is *upward,* as we concluded above.

Similar flow patterns and lift forces are found in the vicinity of any humped object in a wind. A moderate wind makes an umbrella "float"; a strong wind can turn it inside out. At high speed, lift can reduce traction on a car's tires; a "spoiler" at the car's tail, shaped like an upside-down wing, provides a compensating downward force.

12.27 Flow around an airplane wing.

(a) Flow lines around an airplane wing

Flow lines are crowded together above the wing, so flow speed is higher there and pressure is lower.

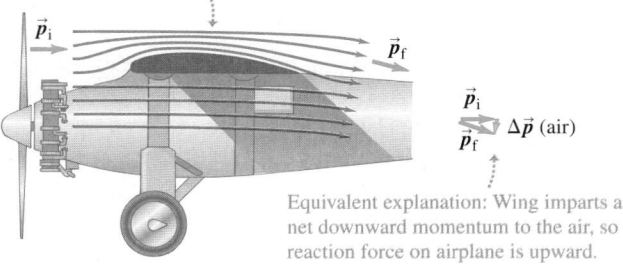

$\vec{p}_i$

$\vec{p}_f$

$\dfrac{\vec{p}_i}{\vec{p}_f}$ $\Delta\vec{p}$ (air)

Equivalent explanation: Wing imparts a net downward momentum to the air, so reaction force on airplane is upward.

(b) Computer simulation of air parcels flowing around a wing, showing that air moves much faster over the top than over the bottom.

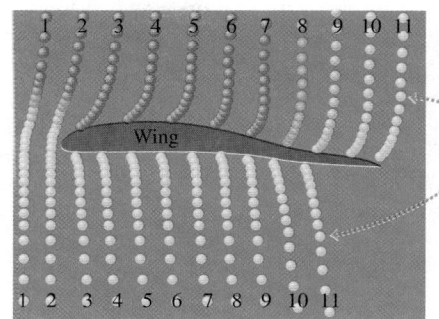

Wing

Notice that air particles that are together at the leading edge of the wing do *not* meet up at the trailing edge!

CAUTION **A misconception about wings** Some discussions of lift claim that air travels faster over the top of a wing because "it has farther to travel." This claim assumes that air molecules that part company at the front of the wing, one traveling over the wing and one under it, must meet again at the wing's trailing edge. Not so! Figure 12.27b shows a computer simulation of parcels of air flowing around an airplane wing. Parcels that are adjacent at the front of the wing do *not* meet at the trailing edge; the flow over the top of the wing is much faster than if the parcels had to meet. In accordance with Bernoulli's equation, this faster speed means that there is even lower pressure above the wing (and hence greater lift) than the "farther-to-travel" claim would suggest. ▌

TEST YOUR UNDERSTANDING OF SECTION 12.5 Which is the most accurate statement of Bernoulli's principle? (i) Fast-moving air causes lower pressure; (ii) lower pressure causes fast-moving air; (iii) both (i) and (ii) are equally accurate. ▌

12.6 VISCOSITY AND TURBULENCE

In our discussion of fluid flow we assumed that the fluid had no internal friction and that the flow was laminar. While these assumptions are often quite valid, in many important physical situations the effects of viscosity (internal friction) and turbulence (nonlaminar flow) are extremely important. Let's take a brief look at some of these situations.

Viscosity

Viscosity is internal friction in a fluid. Viscous forces oppose the motion of one portion of a fluid relative to another. Viscosity is the reason it takes effort to paddle a canoe through calm water, but it is also the reason the paddle works. Viscous effects are important in the flow of fluids in pipes, the flow of blood, the lubrication of engine parts, and many other situations.

Fluids that flow readily, such as water or gasoline, have smaller viscosities than do "thick" liquids such as honey or motor oil. Viscosities of all fluids are strongly temperature dependent, increasing for gases and decreasing for liquids as the temperature increases (**Fig. 12.28**). Oils for engine lubrication must flow equally well in cold and warm conditions, and so are designed to have as *little* temperature variation of viscosity as possible.

A viscous fluid always tends to cling to a solid surface in contact with it. There is always a thin *boundary layer* of fluid near the surface, in which the fluid is nearly at rest with respect to the surface. That's why dust particles can cling to a fan blade even when it is rotating rapidly, and why you can't get all the dirt off your car by just squirting a hose at it.

Viscosity has important effects on the flow of liquids through pipes, including the flow of blood in the circulatory system. First think about a fluid with zero viscosity so that we can apply Bernoulli's equation, Eq. (12.17). If the two ends of a long cylindrical pipe are at the same height ($y_1 = y_2$) and the flow speed is

12.28 Lava is an example of a viscous fluid. The viscosity decreases with increasing temperature: The hotter the lava, the more easily it can flow.

12.29 Velocity profile for a viscous fluid in a cylindrical pipe.

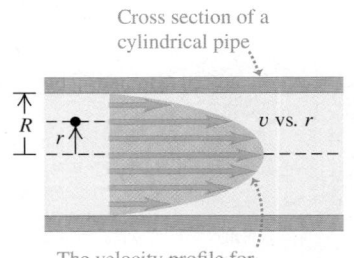

Cross section of a cylindrical pipe

v vs. r

The velocity profile for viscous fluid flowing in the pipe has a parabolic shape.

BIO Application Listening for Turbulent Flow Normal blood flow in the human aorta is laminar, but a small disturbance such as a heart pathology can cause the flow to become turbulent. Turbulence makes noise, which is why listening to blood flow with a stethoscope is a useful diagnostic technique.

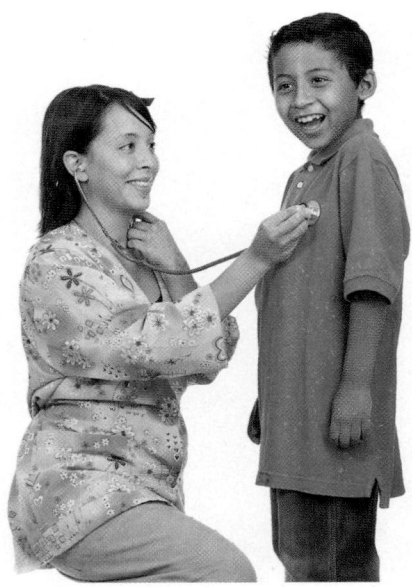

the same at both ends $(v_1 = v_2)$, Bernoulli's equation tells us that the pressure is the same at both ends of the pipe. But this isn't true if we account for viscosity. To see why, consider **Fig. 12.29**, which shows the flow-speed profile for laminar flow of a viscous fluid in a long cylindrical pipe. Due to viscosity, the speed is *zero* at the pipe walls (to which the fluid clings) and is greatest at the center of the pipe. The motion is like a lot of concentric tubes sliding relative to one another, with the central tube moving fastest and the outermost tube at rest. Viscous forces between the tubes oppose this sliding, so to keep the flow going we must apply a greater pressure at the back of the flow than at the front. That's why you have to keep squeezing a tube of toothpaste or a packet of ketchup (both viscous fluids) to keep the fluid coming out of its container. Your fingers provide a pressure at the back of the flow that is far greater than the atmospheric pressure at the front of the flow.

The pressure difference required to sustain a given volume flow rate through a cylindrical pipe of length L and radius R turns out to be proportional to L/R^4. If we decrease R by one-half, the required pressure increases by $2^4 = 16$; decreasing R by a factor of 0.90 (a 10% reduction) increases the required pressure difference by a factor of $(1/0.90)^4 = 1.52$ (a 52% increase). This simple relationship explains the connection between a high-cholesterol diet (which tends to narrow the arteries) and high blood pressure. Due to the R^4 dependence, even a small narrowing of the arteries can result in substantially elevated blood pressure and added strain on the heart muscle.

Turbulence

When the speed of a flowing fluid exceeds a certain critical value, the flow is no longer laminar. Instead, the flow pattern becomes extremely irregular and complex, and it changes continuously with time; there is no steady-state pattern. This irregular, chaotic flow is called **turbulence.** Figure 12.20 shows the contrast between laminar and turbulent flow for smoke rising in air. Bernoulli's equation is *not* applicable to regions where turbulence occurs because the flow is not steady.

Whether a flow is laminar or turbulent depends in part on the fluid's viscosity. The greater the viscosity, the greater the tendency for the fluid to flow in sheets (laminae) and the more likely the flow is to be laminar. (When we discussed Bernoulli's equation in Section 12.5, we assumed that the flow was laminar and that the fluid had zero viscosity. In fact, a *little* viscosity is needed to ensure that the flow is laminar.)

For a fluid of a given viscosity, flow speed is a determining factor for the onset of turbulence. A flow pattern that is stable at low speeds suddenly becomes unstable when a critical speed is reached. Irregularities in the flow pattern can be caused by roughness in the pipe wall, variations in the density of the fluid, and many other factors. At low flow speeds, these disturbances damp out; the flow pattern is *stable* and tends to maintain its laminar nature (**Fig. 12.30a**). When the critical speed is reached, however, the flow pattern becomes unstable. The disturbances no longer damp out but grow until they destroy the entire laminar-flow pattern (Fig. 12.30b).

12.30 The flow of water from a faucet can be (a) laminar or (b) turbulent.

(a) Low speed: laminar flow

(b) High speed: turbulent flow

CONCEPTUAL EXAMPLE 12.11 THE CURVE BALL

Does a curve ball *really* curve? Yes, it does, and the reason is turbulence. **Figure 12.31a** shows a nonspinning ball moving through the air from left to right. The flow lines show that to an observer moving with the ball, the air stream appears to move from right to left. Because of the high speeds involved (typically near 35 m/s, or 75 mi/h), there is a region of *turbulent* flow behind the ball.

Figure 12.31b shows a *spinning* ball with "top spin." Layers of air near the ball's surface are pulled around in the direction of the spin by friction between the ball and air and by the air's internal friction (viscosity). Hence air moves relative to the ball's surface more slowly at the top of the ball than at the bottom, and turbulence occurs farther forward on the top side than on the bottom. As a result, the average pressure at the top of the ball is now greater than that at the bottom, and the resulting net force deflects the ball downward (Fig. 12.31c). "Top spin" is used in tennis to keep a fast serve in the court (Fig. 12.31d).

In baseball, a curve ball spins about a nearly *vertical* axis and the resulting deflection is sideways. In that case, Fig. 12.31c is a *top* view of the situation. A curve ball thrown by a left-handed pitcher spins as shown in Fig. 12.31e and will curve *toward* a right-handed batter, making it harder to hit.

A similar effect occurs when golf balls acquire "backspin" from impact with the grooved, slanted club face. Figure 12.31f shows the backspin of a golf ball just after impact. The resulting pressure difference between the top and bottom of the ball causes a *lift* force that keeps the ball in the air longer than would be possible without spin. A well-hit drive appears, from the tee, to "float" or even curve *upward* during the initial portion of its flight. This is a real effect, not an illusion. The dimples on the golf ball play an essential role; the viscosity of air gives a dimpled ball a much longer trajectory than an undimpled one with the same initial velocity and spin.

12.31 (a)–(e) Analyzing the motion of a spinning ball through the air. (f) Stroboscopic photograph of a golf ball being struck by a club. The picture was taken at 1000 flashes per second. The ball rotates about once in eight flashes, corresponding to an angular speed of 125 rev/s, or 7500 rpm. Source: Harold Edgerton at MIT, copyright 2014. Courtesy of Palm Press, Inc.

(a) Motion of air relative to a nonspinning ball

(b) Motion of a spinning ball

(c) Force generated when a spinning ball moves through air

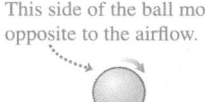

This side of the ball moves opposite to the airflow.

This side moves in the direction of the airflow.

A moving ball drags the adjacent air with it. So, when air moves past a spinning ball:

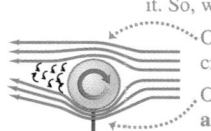

On one side, the ball **slows the air**, creating a region of **high pressure**.

On the other side, the ball **speeds the air**, creating a region of **low pressure**.

The resultant force points in the direction of the low-pressure side.

(d) Spin pushing a tennis ball downward

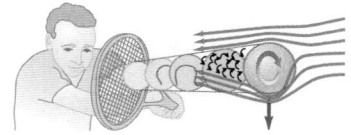

(e) Spin causing a curve ball to be deflected sideways

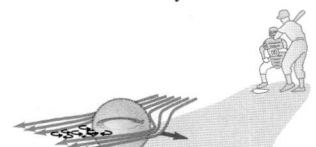

(f) Backspin of a golf ball

TEST YOUR UNDERSTANDING OF SECTION 12.6 How much more thumb pressure must a nurse use to administer an injection with a hypodermic needle of inside diameter 0.30 mm compared to one with inside diameter 0.60 mm? Assume that the two needles have the same length and that the volume flow rate is the same in both cases. (i) Twice as much; (ii) 4 times as much; (iii) 8 times as much; (iv) 16 times as much; (v) 32 times as much. ❚

Density and pressure: Density is mass per unit volume. If a mass m of homogeneous material has volume V, its density ρ is the ratio m/V. Specific gravity is the ratio of the density of a material to the density of water. (See Example 12.1.)

Pressure is normal force per unit area. Pascal's law states that pressure applied to an enclosed fluid is transmitted undiminished to every portion of the fluid. Absolute pressure is the total pressure in a fluid; gauge pressure is the difference between absolute pressure and atmospheric pressure. The SI unit of pressure is the pascal (Pa): $1 \text{ Pa} = 1 \text{ N/m}^2$. (See Example 12.2.)

$$\rho = \frac{m}{V} \quad (12.1)$$

$$p = \frac{dF_\perp}{dA} \quad (12.2)$$

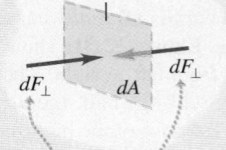

Small area dA within fluid at rest

Equal normal forces exerted on both sides by surrounding fluid

Pressures in a fluid at rest: The pressure difference between points 1 and 2 in a static fluid of uniform density ρ (an incompressible fluid) is proportional to the difference between the elevations y_1 and y_2. If the pressure at the surface of an incompressible liquid at rest is p_0, then the pressure at a depth h is greater by an amount $\rho g h$. (See Examples 12.3 and 12.4.)

$$p_2 - p_1 = -\rho g (y_2 - y_1)$$
(pressure in a fluid of uniform density) $\quad (12.5)$

$$p = p_0 + \rho g h$$
(pressure in a fluid of uniform density) $\quad (12.6)$

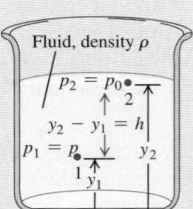

Fluid, density ρ

$p_2 = p_0$

$y_2 - y_1 = h$

$p_1 = p$

Buoyancy: Archimedes's principle states that when a body is immersed in a fluid, the fluid exerts an upward buoyant force on the body equal to the weight of the fluid that the body displaces. (See Example 12.5.)

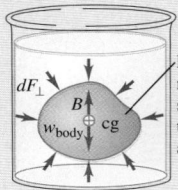

Fluid element replaced with solid body of the same size and shape

Fluid flow: An ideal fluid is incompressible and has no viscosity (no internal friction). A flow line is the path of a fluid particle; a streamline is a curve tangent at each point to the velocity vector at that point. A flow tube is a tube bounded at its sides by flow lines. In laminar flow, layers of fluid slide smoothly past each other. In turbulent flow, there is great disorder and a constantly changing flow pattern.

Conservation of mass in an incompressible fluid is expressed by the continuity equation, which relates the flow speeds v_1 and v_2 for two cross sections A_1 and A_2 in a flow tube. The product Av equals the volume flow rate, dV/dt, the rate at which volume crosses a section of the tube. (See Example 12.6.)

Bernoulli's equation states that a quantity involving the pressure p, flow speed v, and elevation y has the same value anywhere in a flow tube, assuming steady flow in an ideal fluid. This equation can be used to relate the properties of the flow at any two points. (See Examples 12.7–12.10.)

$$A_1 v_1 = A_2 v_2$$
(continuity equation, incompressible fluid) $\quad (12.10)$

$$\frac{dV}{dt} = Av \quad (12.11)$$
(volume flow rate)

$$p + \rho g y + \tfrac{1}{2}\rho v^2 = \text{constant}$$
(Bernoulli's equation) $\quad (12.18)$

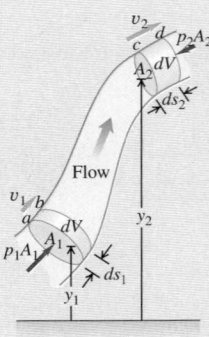

BRIDGING PROBLEM HOW LONG TO DRAIN?

A large cylindrical tank with diameter D is open to the air at the top. The tank contains water to a height H. A small circular hole with diameter d, where $d \ll D$, is then opened at the bottom of the tank (**Fig. 12.32**). Ignore any effects of viscosity. (a) Find y, the height of water in the tank a time t after the hole is opened, as a function of t. (b) How long does it take to drain the tank completely? (c) If you double height H, by what factor does the time to drain the tank increase?

SOLUTION GUIDE

IDENTIFY and SET UP

1. Draw a sketch of the situation that shows all of the relevant dimensions.
2. List the unknown quantities, and decide which of these are the target variables.
3. At what speed does water flow out of the bottom of the tank? How is this related to the volume flow rate of water out of the tank? How is the volume flow rate related to the rate of change of y?

EXECUTE

4. Use your results from step 3 to write an equation for dy/dt.
5. Your result from step 4 is a relatively simple differential equation. With your knowledge of calculus, you can integrate it to find y as a function of t. (*Hint:* Once you've done the integration, you'll still have to do a little algebra.)

12.32 A water tank that is open at the top and has a hole at the bottom.

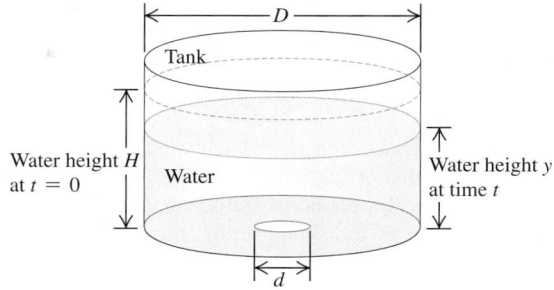

6. Use your result from step 5 to find the time when the tank is empty. How does your result depend on the initial height H?

EVALUATE

7. Check whether your answers are reasonable. A good check is to draw a graph of y versus t. According to your graph, what is the algebraic sign of dy/dt at different times? Does this make sense?

Problems For assigned homework and other learning materials, go to MasteringPhysics®. **MP**

•, ••, •••: Difficulty levels. **CP**: Cumulative problems incorporating material from earlier chapters. **CALC**: Problems requiring calculus. **DATA**: Problems involving real data, scientific evidence, experimental design, and/or statistical reasoning. **BIO**: Biosciences problems.

DISCUSSION QUESTIONS

Q12.1 A cube of oak wood with very smooth faces normally floats in water. Suppose you submerge it completely and press one face flat against the bottom of a tank so that no water is under that face. Will the block float to the surface? Is there a buoyant force on it? Explain.

Q12.2 A rubber hose is attached to a funnel, and the free end is bent around to point upward. When water is poured into the funnel, it rises in the hose to the same level as in the funnel, even though the funnel has a lot more water in it than the hose does. Why? What supports the extra weight of the water in the funnel?

Q12.3 Comparing Example 12.1 (Section 12.1) and Example 12.2 (Section 12.2), it seems that 700 N of air is exerting a downward force of 2.0×10^6 N on the floor. How is this possible?

Q12.4 Equation (12.7) shows that an area ratio of 100 to 1 can give 100 times more output force than input force. Doesn't this violate conservation of energy? Explain.

Q12.5 You have probably noticed that the lower the tire pressure, the larger the contact area between the tire and the road. Why?

Q12.6 In hot-air ballooning, a large balloon is filled with air heated by a gas burner at the bottom. Why must the air be heated? How does the balloonist control ascent and descent?

Q12.7 In describing the size of a large ship, one uses such expressions as "it displaces 20,000 tons." What does this mean? Can the weight of the ship be obtained from this information?

Q12.8 You drop a solid sphere of aluminum in a bucket of water that sits on the ground. The buoyant force equals the weight of water displaced; this is less than the weight of the sphere, so the sphere sinks to the bottom. If you take the bucket with you on an elevator that accelerates upward, the apparent weight of the water increases and the buoyant force on the sphere increases. Could the acceleration of the elevator be great enough to make the sphere pop up out of the water? Explain.

Q12.9 A rigid, lighter-than-air dirigible filled with helium cannot continue to rise indefinitely. Why? What determines the maximum height it can attain?

Q12.10 Which has a greater buoyant force on it: a 25-cm^3 piece of wood floating with part of its volume above water or a 25-cm^3 piece of submerged iron? Or, must you know their masses before you can answer? Explain.

Q12.11 The purity of gold can be tested by weighing it in air and in water. How? Do you think you could get away with making a fake gold brick by gold-plating some cheaper material?

Q12.12 During the Great Mississippi Flood of 1993, the levees in St. Louis tended to rupture first at the bottom. Why?

Q12.13 A cargo ship travels from the Atlantic Ocean (salt water) to Lake Ontario (freshwater) via the St. Lawrence River. The ship rides several centimeters lower in the water in Lake Ontario than it did in the ocean. Explain.

Q12.14 You push a piece of wood under the surface of a swimming pool. After it is completely submerged, you keep pushing it deeper and deeper. As you do this, what will happen to the buoyant force on it? Will the force keep increasing, stay the same, or decrease? Why?

Q12.15 An old question is "Which weighs more, a pound of feathers or a pound of lead?" If the weight in pounds is the gravitational force, will a pound of feathers balance a pound of lead on opposite pans of an equal-arm balance? Explain, taking into account buoyant forces.

Q12.16 Suppose the door of a room makes an airtight but frictionless fit in its frame. Do you think you could open the door if the air pressure on one side were standard atmospheric pressure and the air pressure on the other side differed from standard by 1%? Explain.

Q12.17 At a certain depth in an incompressible liquid, the absolute pressure is p. At twice this depth, will the absolute pressure be equal to $2p$, greater than $2p$, or less than $2p$? Justify your answer.

Q12.18 A piece of iron is glued to the top of a block of wood. When the block is placed in a bucket of water with the iron on top, the block floats. The block is now turned over so that the iron is submerged beneath the wood. Does the block float or sink? Does the water level in the bucket rise, drop, or stay the same? Explain.

Q12.19 You take an empty glass jar and push it into a tank of water with the open mouth of the jar downward, so that the air inside the jar is trapped and cannot get out. If you push the jar deeper into the water, does the buoyant force on the jar stay the same? If not, does it increase or decrease? Explain.

Q12.20 You are floating in a canoe in the middle of a swimming pool. Your friend is at the edge of the pool, carefully noting the level of the water on the side of the pool. You have a bowling ball with you in the canoe. If you carefully drop the bowling ball over the side of the canoe and it sinks to the bottom of the pool, does the water level in the pool rise or fall?

Q12.21 You are floating in a canoe in the middle of a swimming pool. A large bird flies up and lights on your shoulder. Does the water level in the pool rise or fall?

Q12.22 Two identical buckets are filled to the brim with water, but one of them has a piece of wood floating in it. Which bucket of water weighs more? Explain.

Q12.23 An ice cube floats in a glass of water. When the ice melts, will the water level in the glass rise, fall, or remain unchanged? Explain.

Q12.24 A helium-filled balloon is tied to a light string inside a car at rest. The other end of the string is attached to the floor of the car, so the balloon pulls the string vertical. The car now accelerates forward. Does the balloon move? If so, does it move forward or backward? Justify your reasoning with reference to buoyancy. (If you have a chance, try this experiment yourself—but with someone else driving!)

Q12.25 If the velocity at each point in space in steady-state fluid flow is constant, how can a fluid particle accelerate?

Q12.26 In a store-window vacuum cleaner display, a table-tennis ball is suspended in midair in a jet of air blown from the outlet hose of a tank-type vacuum cleaner. The ball bounces around a

little but always moves back toward the center of the jet, even if the jet is tilted from the vertical. How does this behavior illustrate Bernoulli's equation?

Q12.27 A tornado consists of a rapidly whirling air vortex. Why is the pressure always much lower in the center than at the outside? How does this condition account for the destructive power of a tornado?

Q12.28 Airports at high elevations have longer runways for takeoffs and landings than do airports at sea level. One reason is that aircraft engines develop less power in the thin air well above sea level. What is another reason?

Q12.29 When a smooth-flowing stream of water comes out of a faucet, it narrows as it falls. Explain.

Q12.30 Identical-size lead and aluminum cubes are suspended at different depths by two wires in a large vat of water (**Fig. Q12.30**). (a) Which cube experiences a greater buoyant force? (b) For which cube is the tension in the wire greater? (c) Which cube experiences a greater force on its lower face? (d) For which cube is the difference in pressure between the upper and lower faces greater?

Figure **Q12.30**

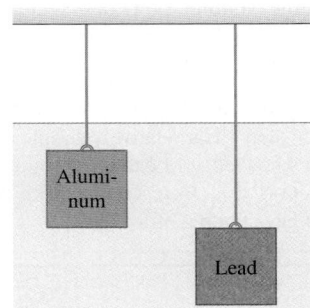

EXERCISES

Section 12.1 Gases, Liquids, and Density

12.1 •• On a part-time job, you are asked to bring a cylindrical iron rod of length 85.8 cm and diameter 2.85 cm from a storage room to a machinist. Will you need a cart? (To answer, calculate the weight of the rod.)

12.2 •• A cube 5.0 cm on each side is made of a metal alloy. After you drill a cylindrical hole 2.0 cm in diameter all the way through and perpendicular to one face, you find that the cube weighs 6.30 N. (a) What is the density of this metal? (b) What did the cube weigh before you drilled the hole in it?

12.3 • You purchase a rectangular piece of metal that has dimensions $5.0 \times 15.0 \times 30.0$ mm and mass 0.0158 kg. The seller tells you that the metal is gold. To check this, you compute the average density of the piece. What value do you get? Were you cheated?

12.4 •• **Gold Brick.** You win the lottery and decide to impress your friends by exhibiting a million-dollar cube of gold. At the time, gold is selling for $1282 per troy ounce, and 1.0000 troy ounce equals 31.1035 g. How tall would your million-dollar cube be?

12.5 •• A uniform lead sphere and a uniform aluminum sphere have the same mass. What is the ratio of the radius of the aluminum sphere to the radius of the lead sphere?

12.6 • (a) What is the average density of the sun? (b) What is the average density of a neutron star that has the same mass as the sun but a radius of only 20.0 km?

12.7 •• A hollow cylindrical copper pipe is 1.50 m long and has an outside diameter of 3.50 cm and an inside diameter of 2.50 cm. How much does it weigh?

Section 12.2 Pressure in a Fluid

12.8 •• **Black Smokers.** Black smokers are hot volcanic vents that emit smoke deep in the ocean floor. Many of them teem with exotic creatures, and some biologists think that life on earth may have begun around such vents. The vents range in depth from about 1500 m to 3200 m below the surface. What is the gauge pressure at a 3200-m deep vent, assuming that the density of water does not vary? Express your answer in pascals and atmospheres.

12.9 •• **Oceans on Mars.** Scientists have found evidence that Mars may once have had an ocean 0.500 km deep. The acceleration due to gravity on Mars is 3.71 m/s². (a) What would be the gauge pressure at the bottom of such an ocean, assuming it was freshwater? (b) To what depth would you need to go in the earth's ocean to experience the same gauge pressure?

12.10 •• **BIO** (a) Calculate the difference in blood pressure between the feet and top of the head for a person who is 1.65 m tall. (b) Consider a cylindrical segment of a blood vessel 2.00 cm long and 1.50 mm in diameter. What *additional* outward force would such a vessel need to withstand in the person's feet compared to a similar vessel in her head?

12.11 • **BIO** In intravenous feeding, a needle is inserted in a vein in the patient's arm and a tube leads from the needle to a reservoir of fluid (density 1050 kg/m³) located at height h above the arm. The top of the reservoir is open to the air. If the gauge pressure inside the vein is 5980 Pa, what is the minimum value of h that allows fluid to enter the vein? Assume the needle diameter is large enough that you can ignore the viscosity (see Section 12.6) of the fluid.

12.12 • A barrel contains a 0.120-m layer of oil floating on water that is 0.250 m deep. The density of the oil is 600 kg/m³. (a) What is the gauge pressure at the oil–water interface? (b) What is the gauge pressure at the bottom of the barrel?

12.13 • **BIO** **Standing on Your Head.** (a) What is the *difference* between the pressure of the blood in your brain when you stand on your head and the pressure when you stand on your feet? Assume that you are 1.85 m tall. The density of blood is 1060 kg/m³. (b) What effect does the increased pressure have on the blood vessels in your brain?

12.14 •• You are designing a diving bell to withstand the pressure of seawater at a depth of 250 m. (a) What is the gauge pressure at this depth? (You can ignore changes in the density of the water with depth.) (b) At this depth, what is the net force due to the water outside and the air inside the bell on a circular glass window 30.0 cm in diameter if the pressure inside the diving bell equals the pressure at the surface of the water? (Ignore the small variation of pressure over the surface of the window.)

12.15 •• **BIO** **Ear Damage from Diving.** If the force on the tympanic membrane (eardrum) increases by about 1.5 N above the force from atmospheric pressure, the membrane can be damaged. When you go scuba diving in the ocean, below what depth could damage to your eardrum start to occur? The eardrum is typically 8.2 mm in diameter. (Consult Table 12.1.)

12.16 •• The liquid in the open-tube manometer in Fig. 12.8a is mercury, $y_1 = 3.00$ cm, and $y_2 = 7.00$ cm. Atmospheric pressure is 980 millibars. What is (a) the absolute pressure at the bottom of the U-shaped tube; (b) the absolute pressure in the open tube at a depth of 4.00 cm below the free surface; (c) the absolute pressure of the gas in the container; (d) the gauge pressure of the gas in pascals?

12.17 • **BIO** There is a maximum depth at which a diver can breathe through a snorkel tube (**Fig. E12.17**) because as the depth increases, so does the pressure difference, which tends to collapse the diver's lungs. Since the snorkel connects the air in the lungs to the atmosphere at the surface, the pressure inside the lungs is atmospheric pressure. What is the external–internal pressure difference when the diver's lungs are at a depth of 6.1 m (about 20 ft)? Assume that the diver is in freshwater. (A scuba diver breathing from compressed air tanks can operate at greater depths than can a snorkeler, since the pressure of the air inside the scuba diver's lungs increases to match the external pressure of the water.)

Figure **E12.17**

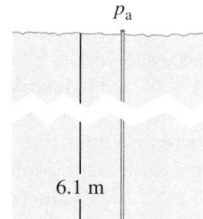

12.18 •• **BIO** The lower end of a long plastic straw is immersed below the surface of the water in a plastic cup. An average person sucking on the upper end of the straw can pull water into the straw to a vertical height of 1.1 m above the surface of the water in the cup. (a) What is the lowest gauge pressure that the average person can achieve inside his lungs? (b) Explain why your answer in part (a) is negative.

12.19 •• An electrical short cuts off all power to a submersible diving vehicle when it is 30 m below the surface of the ocean. The crew must push out a hatch of area 0.75 m² and weight 300 N on the bottom to escape. If the pressure inside is 1.0 atm, what downward force must the crew exert on the hatch to open it?

12.20 •• A tall cylinder with a cross-sectional area 12.0 cm² is partially filled with mercury; the surface of the mercury is 8.00 cm above the bottom of the cylinder. Water is slowly poured in on top of the mercury, and the two fluids don't mix. What volume of water must be added to double the gauge pressure at the bottom of the cylinder?

12.21 •• A cylindrical disk of wood weighing 45.0 N and having a diameter of 30.0 cm floats on a cylinder of oil of density 0.850 g/cm³ (**Fig. E12.21**). The cylinder of oil is 75.0 cm deep and has a diameter the same as that of the wood. (a) What is the gauge pressure at the top of the oil column? (b) Suppose now that someone puts a weight of 83.0 N on top of the wood, but no oil seeps around the edge of the wood. What is the *change* in pressure at (i) the bottom of the oil and (ii) halfway down in the oil?

Figure **E12.21**

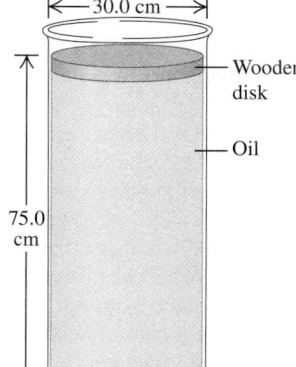

12.22 •• A closed container is partially filled with water. Initially, the air above the water is at atmospheric pressure (1.01×10^5 Pa) and the gauge pressure at the bottom of the water is 2500 Pa. Then additional air is pumped in, increasing the pressure of the air above the water by 1500 Pa. (a) What is the gauge pressure at the bottom of the water? (b) By how much must the water level in the container be reduced, by drawing some water out through a valve at the bottom of the container, to return the gauge pressure at the bottom of the water to its original value of 2500 Pa? The pressure of the air above the water is maintained at 1500 Pa above atmospheric pressure.

12.23 •• **Hydraulic Lift I.** For the hydraulic lift shown in Fig. 12.7, what must be the ratio of the diameter of the vessel at the car to the diameter of the vessel where the force F_1 is applied so that a 1520-kg car can be lifted with a force F_1 of just 125 N?

12.24 • **Hydraulic Lift II.** The piston of a hydraulic automobile lift is 0.30 m in diameter. What gauge pressure, in pascals, is required to lift a car with a mass of 1200 kg? Also express this pressure in atmospheres.

12.25 •• **Exploring Venus.** The surface pressure on Venus is 92 atm, and the acceleration due to gravity there is $0.894g$. In a future exploratory mission, an upright cylindrical tank of benzene is sealed at the top but still pressurized at 92 atm just above the benzene. The tank has a diameter of 1.72 m, and the benzene column is 11.50 m tall. Ignore any effects due to the very high temperature on Venus. (a) What total force is exerted on the inside surface of the bottom of the tank? (b) What force does the Venusian atmosphere exert on the outside surface of the bottom of the tank? (c) What total inward force does the atmosphere exert on the vertical walls of the tank?

Section 12.3 Buoyancy

12.26 •• A rock has mass 1.80 kg. When the rock is suspended from the lower end of a string and totally immersed in water, the tension in the string is 12.8 N. What is the smallest density of a liquid in which the rock will float?

12.27 • A 950-kg cylindrical can buoy floats vertically in seawater. The diameter of the buoy is 0.900 m. Calculate the additional distance the buoy will sink when an 80.0-kg man stands on top of it.

12.28 •• A slab of ice floats on a freshwater lake. What minimum volume must the slab have for a 65.0-kg woman to be able to stand on it without getting her feet wet?

12.29 •• An ore sample weighs 17.50 N in air. When the sample is suspended by a light cord and totally immersed in water, the tension in the cord is 11.20 N. Find the total volume and the density of the sample.

12.30 •• You are preparing some apparatus for a visit to a newly discovered planet Caasi having oceans of glycerine and a surface acceleration due to gravity of 5.40 m/s². If your apparatus floats in the oceans on earth with 25.0% of its volume submerged, what percentage will be submerged in the glycerine oceans of Caasi?

12.31 •• A rock with density 1200 kg/m³ is suspended from the lower end of a light string. When the rock is in air, the tension in the string is 28.0 N. What is the tension in the string when the rock is totally immersed in a liquid with density 750 kg/m³?

12.32 • A hollow plastic sphere is held below the surface of a freshwater lake by a cord anchored to the bottom of the lake. The sphere has a volume of 0.650 m³ and the tension in the cord is 1120 N. (a) Calculate the buoyant force exerted by the water on the sphere. (b) What is the mass of the sphere? (c) The cord breaks and the sphere rises to the surface. When the sphere comes to rest, what fraction of its volume will be submerged?

12.33 •• A cubical block of wood, 10.0 cm on a side, floats at the interface between oil and water with its lower surface 1.50 cm below the interface (**Fig. E12.33**). The density of the oil is 790 kg/m³. (a) What is the gauge pressure at

Figure **E12.33**

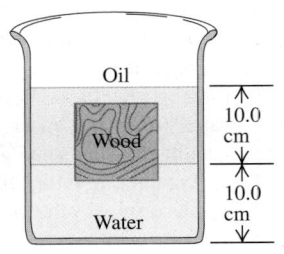

the upper face of the block? (b) What is the gauge pressure at the lower face of the block? (c) What are the mass and density of the block?

12.34 • A solid aluminum ingot weighs 89 N in air. (a) What is its volume? (b) The ingot is suspended from a rope and totally immersed in water. What is the tension in the rope (the *apparent weight* of the ingot in water)?

12.35 •• A rock is suspended by a light string. When the rock is in air, the tension in the string is 39.2 N. When the rock is totally immersed in water, the tension is 28.4 N. When the rock is totally immersed in an unknown liquid, the tension is 21.5 N. What is the density of the unknown liquid?

Section 12.4 Fluid Flow

12.36 •• Water runs into a fountain, filling all the pipes, at a steady rate of 0.750 m³/s. (a) How fast will it shoot out of a hole 4.50 cm in diameter? (b) At what speed will it shoot out if the diameter of the hole is three times as large?

12.37 •• A shower head has 20 circular openings, each with radius 1.0 mm. The shower head is connected to a pipe with radius 0.80 cm. If the speed of water in the pipe is 3.0 m/s, what is its speed as it exits the shower-head openings?

12.38 • Water is flowing in a pipe with a varying cross-sectional area, and at all points the water completely fills the pipe. At point 1 the cross-sectional area of the pipe is 0.070 m², and the magnitude of the fluid velocity is 3.50 m/s. (a) What is the fluid speed at points in the pipe where the cross-sectional area is (a) 0.105 m² and (b) 0.047 m²? (c) Calculate the volume of water discharged from the open end of the pipe in 1.00 hour.

12.39 • Water is flowing in a pipe with a circular cross section but with varying cross-sectional area, and at all points the water completely fills the pipe. (a) At one point in the pipe the radius is 0.150 m. What is the speed of the water at this point if water is flowing into this pipe at a steady rate of 1.20 m³/s? (b) At a second point in the pipe the water speed is 3.80 m/s. What is the radius of the pipe at this point?

12.40 • **Home Repair.** You need to extend a 2.50-inch-diameter pipe, but you have only a 1.00-inch-diameter pipe on hand. You make a fitting to connect these pipes end to end. If the water is flowing at 6.00 cm/s in the wide pipe, how fast will it be flowing through the narrow one?

Section 12.5 Bernoulli's Equation

12.41 •• A sealed tank containing seawater to a height of 11.0 m also contains air above the water at a gauge pressure of 3.00 atm. Water flows out from the bottom through a small hole. How fast is this water moving?

12.42 •• BIO **Artery Blockage.** A medical technician is trying to determine what percentage of a patient's artery is blocked by plaque. To do this, she measures the blood pressure just before the region of blockage and finds that it is 1.20×10^4 Pa, while in the region of blockage it is 1.15×10^4 Pa. Furthermore, she knows that blood flowing through the normal artery just before the point of blockage is traveling at 30.0 cm/s, and the specific gravity of this patient's blood is 1.06. What percentage of the cross-sectional area of the patient's artery is blocked by the plaque?

12.43 • What gauge pressure is required in the city water mains for a stream from a fire hose connected to the mains to reach a vertical height of 15.0 m? (Assume that the mains have a much larger diameter than the fire hose.)

12.44 • A small circular hole 6.00 mm in diameter is cut in the side of a large water tank, 14.0 m below the water level in the tank. The top of the tank is open to the air. Find (a) the speed of efflux of the water and (b) the volume discharged per second.

12.45 • At a certain point in a horizontal pipeline, the water's speed is 2.50 m/s and the gauge pressure is 1.80×10^4 Pa. Find the gauge pressure at a second point in the line if the cross-sectional area at the second point is twice that at the first.

12.46 •• At one point in a pipeline the water's speed is 3.00 m/s and the gauge pressure is 5.00×10^4 Pa. Find the gauge pressure at a second point in the line, 11.0 m lower than the first, if the pipe diameter at the second point is twice that at the first.

12.47 •• A golf course sprinkler system discharges water from a horizontal pipe at the rate of 7200 cm³/s. At one point in the pipe, where the radius is 4.00 cm, the water's absolute pressure is 2.40×10^5 Pa. At a second point in the pipe, the water passes through a constriction where the radius is 2.00 cm. What is the water's absolute pressure as it flows through this constriction?

12.48 • A soft drink (mostly water) flows in a pipe at a beverage plant with a mass flow rate that would fill 220 0.355-L cans per minute. At point 2 in the pipe, the gauge pressure is 152 kPa and the cross-sectional area is 8.00 cm². At point 1, 1.35 m above point 2, the cross-sectional area is 2.00 cm². Find the (a) mass flow rate; (b) volume flow rate; (c) flow speeds at points 1 and 2; (d) gauge pressure at point 1.

Section 12.6 Viscosity and Turbulence

12.49 •• BIO **Clogged Artery.** Viscous blood is flowing through an artery partially clogged by cholesterol. A surgeon wants to remove enough of the cholesterol to double the flow rate of blood through this artery. If the original diameter of the artery is D, what should be the new diameter (in terms of D) to accomplish this for the same pressure gradient?

12.50 • A pressure difference of 6.00×10^4 Pa is required to maintain a volume flow rate of 0.800 m³/s for a viscous fluid flowing through a section of cylindrical pipe that has radius 0.210 m. What pressure difference is required to maintain the same volume flow rate if the radius of the pipe is decreased to 0.0700 m?

PROBLEMS

12.51 ••• In a lecture demonstration, a professor pulls apart two hemispherical steel shells (diameter D) with ease using their attached handles. She then places them together, pumps out the air to an absolute pressure of p, and hands them to a bodybuilder in the back row to pull apart. (a) If atmospheric pressure is p_0, how much force must the bodybuilder exert on each shell? (b) Evaluate your answer for the case $p = 0.025$ atm, $D = 10.0$ cm.

12.52 •• CP The deepest point known in any of the earth's oceans is in the Marianas Trench, 10.92 km deep. (a) Assuming water is incompressible, what is the pressure at this depth? Use the density of seawater. (b) The actual pressure is 1.16×10^8 Pa; your calculated value will be less because the density actually varies with depth. Using the compressibility of water and the actual pressure, find the density of the water at the bottom of the Marianas Trench. What is the percent change in the density of the water?

12.53 ••• CALC A swimming pool is 5.0 m long, 4.0 m wide, and 3.0 m deep. Compute the force exerted by the water against (a) the bottom and (b) either end. (*Hint:* Calculate the force on a thin, horizontal strip at a depth h, and integrate this over the end of the pool.) Do not include the force due to air pressure.

12.54 •• BIO **Fish Navigation.** (a) As you can tell by watching them in an aquarium, fish are able to remain at any depth in water with no effort. What does this ability tell you about their density? (b) Fish are able to inflate themselves using a sac (called the *swim bladder*) located under their spinal column. These sacs can be filled with an oxygen–nitrogen mixture that comes from the blood. If a 2.75-kg fish in freshwater inflates itself and increases its volume by 10%, find the *net* force that the *water* exerts on it. (c) What is the net *external* force on it? Does the fish go up or down when it inflates itself?

12.55 ••• CP CALC The upper edge of a gate in a dam runs along the water surface. The gate is 2.00 m high and 4.00 m wide and is hinged along a horizontal line through its center (**Fig. P12.55**). Calculate the torque about the hinge arising from the force due to the water. (*Hint:* Use a procedure similar to that used in Problem 12.53; calculate the torque on a thin, horizontal strip at a depth h and integrate this over the gate.)

Figure **P12.55**

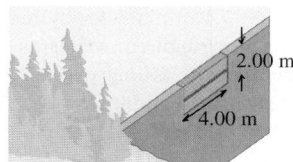

2.00 m

4.00 m

12.56 •• **Ballooning on Mars.** It has been proposed that we could explore Mars using inflated balloons to hover just above the surface. The buoyancy of the atmosphere would keep the balloon aloft. The density of the Martian atmosphere is 0.0154 kg/m³ (although this varies with temperature). Suppose we construct these balloons of a thin but tough plastic having a density such that each square meter has a mass of 5.00 g. We inflate them with a very light gas whose mass we can ignore. (a) What should be the radius and mass of these balloons so they just hover above the surface of Mars? (b) If we released one of the balloons from part (a) on earth, where the atmospheric density is 1.20 kg/m³, what would be its initial acceleration assuming it was the same size as on Mars? Would it go up or down? (c) If on Mars these balloons have five times the radius found in part (a), how heavy an instrument package could they carry?

12.57 •• A 0.180-kg cube of ice (frozen water) is floating in glycerine. The gylcerine is in a tall cylinder that has inside radius 3.50 cm. The level of the glycerine is well below the top of the cylinder. If the ice completely melts, by what distance does the height of liquid in the cylinder change? Does the level of liquid rise or fall? That is, is the surface of the water above or below the original level of the glycerine before the ice melted?

12.58 •• A narrow, U-shaped glass tube with open ends is filled with 25.0 cm of oil (of specific gravity 0.80) and 25.0 cm of water on opposite sides, with a barrier separating the liquids (**Fig. P12.58**). (a) Assume that the two liquids do not mix, and find the final heights of the columns of liquid in each side of the tube after the barrier is removed. (b) For the following cases, arrive at your answer by simple physical reasoning, not by calculations: (i) What would be the height on each side if the oil and water had equal densities? (ii) What would the heights be if the oil's density were much less than that of water?

Figure **P12.58**

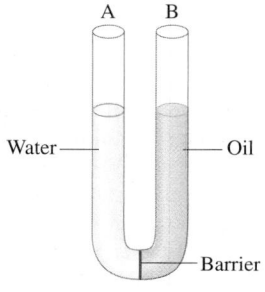

A   B

Water

Oil

Barrier

12.59 • A U-shaped tube open to the air at both ends contains some mercury. A quantity of water is carefully poured into the left arm of the U-shaped tube until the vertical height of the water column is 15.0 cm (**Fig. P12.59**). (a) What is the gauge pressure at the water–mercury interface? (b) Calculate the vertical distance h from the top of the mercury in the right-hand arm of the tube to the top of the water in the left-hand arm.

Figure **P12.59**

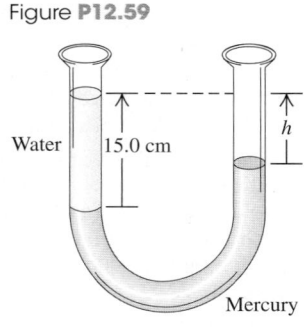

12.60 •• CALC **The Great Molasses Flood.** On the afternoon of January 15, 1919, an unusually warm day in Boston, a 17.7-m-high, 27.4-m-diameter cylindrical metal tank used for storing molasses ruptured. Molasses flooded into the streets in a 5-m-deep stream, killing pedestrians and horses and knocking down buildings. The molasses had a density of 1600 kg/m³. If the tank was full before the accident, what was the total outward force the molasses exerted on its sides? (*Hint:* Consider the outward force on a circular ring of the tank wall of width dy and at a depth y below the surface. Integrate to find the total outward force. Assume that before the tank ruptured, the pressure at the surface of the molasses was equal to the air pressure outside the tank.)

12.61 •• A large, 40.0-kg cubical block of wood with uniform density is floating in a freshwater lake with 20.0% of its volume above the surface of the water. You want to load bricks onto the floating block and then push it horizontally through the water to an island where you are building an outdoor grill. (a) What is the volume of the block? (b) What is the maximum mass of bricks that you can place on the block without causing it to sink below the water surface?

12.62 ••• A hot-air balloon has a volume of 2200 m³. The balloon fabric (the envelope) weighs 900 N. The basket with gear and full propane tanks weighs 1700 N. If the balloon can barely lift an additional 3200 N of passengers, breakfast, and champagne when the outside air density is 1.23 kg/m³, what is the average density of the heated gases in the envelope?

12.63 • An open barge has the dimensions shown in **Fig. P12.63**. If the barge is made out of 4.0-cm-thick steel plate on each of its four sides and its bottom, what mass of coal can the barge carry in freshwater without sinking? Is there enough room in the barge to hold this amount of coal? (The density of coal is about 1500 kg/m³.)

Figure **P12.63**

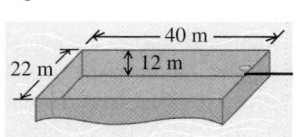

12.64 • A single ice cube with mass 16.4 g floats in a glass completely full of 420 cm³ of water. Ignore the water's surface tension and its variation in density with temperature (as long as it remains a liquid). (a) What volume of water does the ice cube displace? (b) When the ice cube has completely melted, has any water overflowed? If so, how much? If not, explain why this is so. (c) Suppose the water in the glass had been very salty water of density 1050 kg/m³. What volume of salt water would the 9.70-g ice cube displace? (d) Redo part (b) for the freshwater ice cube in the salty water.

12.65 •• Advertisements for a certain small car claim that it floats in water. (a) If the car's mass is 900 kg and its interior volume is 3.0 m³, what fraction of the car is immersed when it floats? Ignore the volume of steel and other materials. (b) Water gradually leaks in and displaces the air in the car. What fraction of the interior volume is filled with water when the car sinks?

12.66 ••• A piece of wood is 0.600 m long, 0.250 m wide, and 0.080 m thick. Its density is 700 kg/m³. What volume of lead must be fastened underneath it to sink the wood in calm water so that its top is just even with the water level? What is the mass of this volume of lead?

12.67 •• The densities of air, helium, and hydrogen (at $p = 1.0$ atm and $T = 20°C$) are 1.20 kg/m³, 0.166 kg/m³, and 0.0899 kg/m³, respectively. (a) What is the volume in cubic meters displaced by a hydrogen-filled airship that has a total "lift" of 90.0 kN? (The "lift" is the amount by which the buoyant force exceeds the weight of the gas that fills the airship.) (b) What would be the "lift" if helium were used instead of hydrogen? In view of your answer, why is helium used in modern airships like advertising blimps?

12.68 •• When an open-faced boat has a mass of 5750 kg, including its cargo and passengers, it floats with the water just up to the top of its gunwales (sides) on a freshwater lake. (a) What is the volume of this boat? (b) The captain decides that it is too dangerous to float with his boat on the verge of sinking, so he decides to throw some cargo overboard so that 20% of the boat's volume will be above water. How much mass should he throw out?

12.69 •• CP A firehose must be able to shoot water to the top of a building 28.0 m tall when aimed straight up. Water enters this hose at a steady rate of 0.500 m³/s and shoots out of a round nozzle. (a) What is the maximum diameter this nozzle can have? (b) If the only nozzle available has a diameter twice as great, what is the highest point the water can reach?

12.70 •• In seawater, a life preserver with a volume of 0.0400 m³ will support a 75.0-kg person (average density 980 kg/m³), with 20% of the person's volume above the water surface when the life preserver is fully submerged. What is the density of the material composing the life preserver?

12.71 ••• CALC A closed and elevated vertical cylindrical tank with diameter 2.00 m contains water to a depth of 0.800 m. A worker accidently pokes a circular hole with diameter 0.0200 m in the bottom of the tank. As the water drains from the tank, compressed air above the water in the tank maintains a gauge pressure of 5.00×10^3 Pa at the surface of the water. Ignore any effects of viscosity. (a) Just after the hole is made, what is the speed of the water as it emerges from the hole? What is the ratio of this speed to the efflux speed if the top of the tank is open to the air? (b) How much time does it take for all the water to drain from the tank? What is the ratio of this time to the time it takes for the tank to drain if the top of the tank is open to the air?

12.72 •• Block A in **Fig. P12.72** hangs by a cord from spring balance D and is submerged in a liquid C contained in beaker B. The mass of the beaker is 1.00 kg; the mass of the liquid is 1.80 kg. Balance D reads 3.50 kg, and balance E reads 7.50 kg. The volume of block A is 3.80×10^{-3} m³. (a) What is the density of the liquid? (b) What will each balance read if block A is pulled up out of the liquid?

Figure **P12.72**

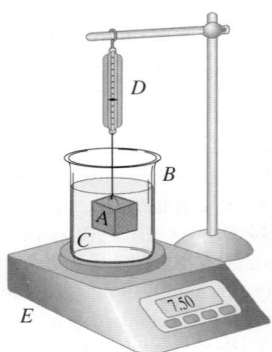

12.73 •• A plastic ball has radius 12.0 cm and floats in water with 24.0% of its volume submerged. (a) What force must you apply to the ball to hold it at rest totally below the surface of the water? (b) If you let go of the ball, what is its acceleration the instant you release it?

12.74 •• Assume that crude oil from a supertanker has density 750 kg/m³. The tanker runs aground on a sandbar. To refloat the tanker, its oil cargo is pumped out into steel barrels, each of which has a mass of 15.0 kg when empty and holds 0.120 m³ of oil. You can ignore the volume occupied by the steel from which the barrel is made. (a) If a salvage worker accidentally drops a filled, sealed barrel overboard, will it float or sink in the seawater? (b) If the barrel floats, what fraction of its volume will be above the water surface? If it sinks, what minimum tension would have to be exerted by a rope to haul the barrel up from the ocean floor? (c) Repeat parts (a) and (b) if the density of the oil is 910 kg/m³ and the mass of each empty barrel is 32.0 kg.

12.75 ••• A cubical block of density ρ_B and with sides of length L floats in a liquid of greater density ρ_L. (a) What fraction of the block's volume is above the surface of the liquid? (b) The liquid is denser than water (density ρ_W) and does not mix with it. If water is poured on the surface of that liquid, how deep must the water layer be so that the water surface just rises to the top of the block? Express your answer in terms of L, ρ_B, ρ_L, and ρ_W. (c) Find the depth of the water layer in part (b) if the liquid is mercury, the block is made of iron, and $L = 10.0$ cm.

12.76 •• A barge is in a rectangular lock on a freshwater river. The lock is 60.0 m long and 20.0 m wide, and the steel doors on each end are closed. With the barge floating in the lock, a 2.50×10^6 N load of scrap metal is put onto the barge. The metal has density 7200 kg/m³. (a) When the load of scrap metal, initially on the bank, is placed onto the barge, what vertical distance does the water in the lock rise? (b) The scrap metal is now pushed overboard into the water. Does the water level in the lock rise, fall, or remain the same? If it rises or falls, by what vertical distance does it change?

12.77 • CP Water stands at a depth H in a large, open tank whose side walls are vertical (**Fig. P12.77**). A hole is made in one of the walls at a depth h below the water surface. (a) At what distance R from the foot of the wall does the emerging stream strike the floor? (b) How far above the bottom of the tank could a second hole be cut so that the stream emerging from it could have the same range as for the first hole?

Figure **P12.77**

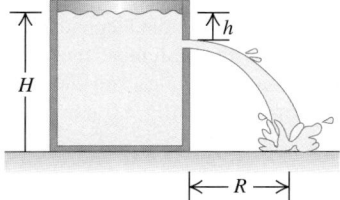

12.78 •• Your uncle is in the below-deck galley of his boat while you are spear fishing in the water nearby. An errant spear makes a small hole in the boat's hull, and water starts to leak into the galley. (a) If the hole is 0.900 m below the water surface and has area 1.20 cm², how long does it take 10.0 L of water to leak into the boat? (b) Do you need to take into consideration the fact that the boat sinks lower into the water as water leaks in?

12.79 •• CP You hold a hose at waist height and spray water horizontally with it. The hose nozzle has a diameter of 1.80 cm, and the water splashes on the ground a distance of 0.950 m horizontally from the nozzle. If you constrict the nozzle to a diameter of 0.750 cm, how far from the nozzle, horizontally, will the water travel before it hits the ground? (Ignore air resistance.)

12.80 ••• A cylindrical bucket, open at the top, is 25.0 cm high and 10.0 cm in diameter. A circular hole with a cross-sectional area 1.50 cm² is cut in the center of the bottom of the bucket. Water flows into the bucket from a tube above it at the rate of 2.40×10^{-4} m³/s. How high will the water in the bucket rise?

12.81 • Water flows steadily from an open tank as in **Fig. P12.81**. The elevation of point 1 is 10.0 m, and the elevation of points 2 and 3 is 2.00 m. The cross-sectional area at point 2 is 0.0480 m²; at point 3 it is 0.0160 m². The area of the tank is very large compared with the cross-sectional area of the pipe. Assuming that Bernoulli's equation applies, compute (a) the discharge rate in cubic meters per second and (b) the gauge pressure at point 2.

Figure **P12.81**

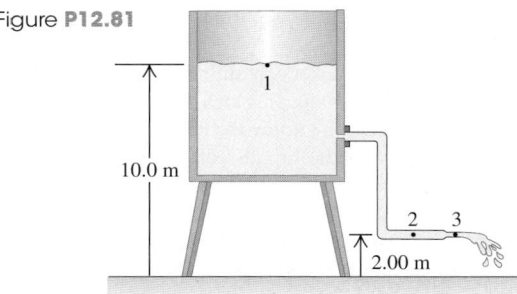

12.82 •• CP In 1993 the radius of Hurricane Emily was about 350 km. The wind speed near the center ("eye") of the hurricane, whose radius was about 30 km, reached about 200 km/h. As air swirled in from the rim of the hurricane toward the eye, its angular momentum remained roughly constant. Estimate (a) the wind speed at the rim of the hurricane; (b) the pressure difference at the earth's surface between the eye and the rim. (*Hint:* See Table 12.1.) Where is the pressure greater? (c) If the kinetic energy of the swirling air in the eye could be converted completely to gravitational potential energy, how high would the air go? (d) In fact, the air in the eye is lifted to heights of several kilometers. How can you reconcile this with your answer to part (c)?

12.83 •• Two very large open tanks A and F (**Fig. P12.83**) contain the same liquid. A horizontal pipe BCD, having a constriction at C and open to the air at D, leads out of the bottom of tank A, and a vertical pipe E opens into the constriction at C and dips into the liquid in tank F. Assume streamline flow and no viscosity. If the cross-sectional area at C is one-half the area at D and if D is a distance h_1 below the level of the liquid in A, to what height h_2 will liquid rise in pipe E? Express your answer in terms of h_1.

Figure **P12.83**

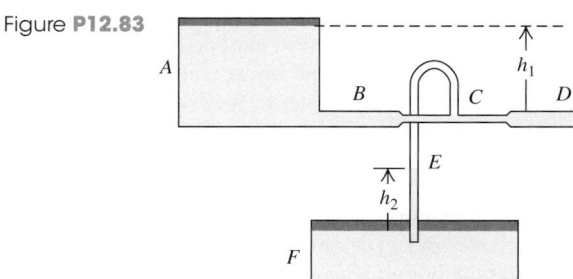

12.84 • A liquid flowing from a vertical pipe has a definite shape as it flows from the pipe. To get the equation for this shape, assume that the liquid is in free fall once it leaves the pipe. Just as it leaves the pipe, the liquid has speed v_0 and the radius of the stream of liquid is r_0. (a) Find an equation for the speed of the liquid as a function of the distance y it has fallen. Combining this with the equation of continuity, find an expression for the radius of the stream as a function of y. (b) If water flows out of a vertical pipe at a speed of 1.20 m/s, how far below the outlet will the radius be one-half the original radius of the stream?

12.85 •• DATA The density values in Table 12.1 are listed in increasing order. A chemistry student notices that the first four chemical elements that are included are also listed in order of increasing atomic mass. (a) See whether there is a simple relationship between density and atomic mass by plotting a graph of density (in g/cm^3) versus atomic mass for all eight elements in that table. (See Appendix D for their atomic masses in grams per mole.) (b) Can you draw a straight line or simple curve through the points to find a "simple" relationship? (c) Explain why "More massive atoms result in more dense solids" does not tell the whole story.

12.86 •• DATA You have a bucket containing an unknown liquid. You also have a cube-shaped wooden block that you measure to be 8.0 cm on a side, but you don't know the mass or density of the block. To find the density of the liquid, you perform an experiment. First you place the wooden block in the liquid and measure the height of the top of the floating block above the liquid surface. Then you stack various numbers of U.S. quarter-dollar coins onto the block and measure the new value of h. The straight line that gives the best fit to the data you have collected is shown in **Fig. P12.86**. Find the mass of one quarter (see www.usmint.gov for quarters dated 2012). Use this information and the slope and intercept of the straight-line fit to your data to calculate (a) the density of the liquid (in kg/m^3) and (b) the mass of the block (in kg).

Figure **P12.86**

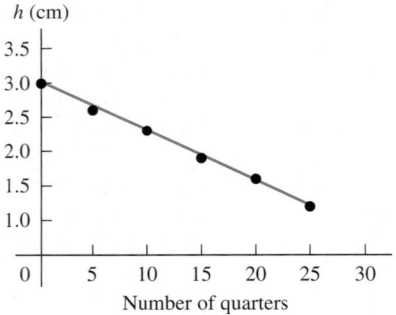

12.87 ••• DATA The Environmental Protection Agency is investigating an abandoned chemical plant. A large, closed cylindrical tank contains an unknown liquid. You must determine the liquid's density and the height of the liquid in the tank (the vertical distance from the surface of the liquid to the bottom of the tank). To maintain various values of the gauge pressure in the air that is above the liquid in the tank, you can use compressed air. You make a small hole at the bottom of the side of the tank, which is on a concrete platform—so the hole is 50.0 cm above the ground. The table gives your measurements of the horizontal distance R that the initially horizontal stream of liquid pouring out of the tank travels before it strikes the ground and the gauge pressure p_g of the air in the tank.

p_g (atm)	0.50	1.00	2.00	3.00	4.00
R (m)	5.4	6.5	8.2	9.7	10.9

(a) Graph R^2 as a function of p_g. Explain why the data points fall close to a straight line. Find the slope and intercept of that line. (b) Use the slope and intercept found in part (a) to calculate the height h (in meters) of the liquid in the tank and the density of the liquid (in kg/m^3). Use $g = 9.80$ m/s^2. Assume that the liquid is nonviscous and that the hole is small enough compared to the tank's diameter so that the change in h during the measurements is very small.

CHALLENGE PROBLEM

12.88 ••• A *siphon* (**Fig. P12.88**) is a convenient device for removing liquids from containers. To establish the flow, the tube must be initially filled with fluid. Let the fluid have density ρ, and let the atmospheric pressure be p_{atm}. Assume that the cross-sectional area of the tube is the same at all points along it. (a) If the lower end of the siphon is at a distance h below the surface of the liquid in the container, what is the speed of the fluid as it flows out the lower end of the siphon? (Assume that the container has a very large diameter, and ignore any effects of viscosity.) (b) A curious feature of a siphon is that the fluid initially flows "uphill." What is the greatest height H that the high point of the tube can have if flow is still to occur?

Figure **P12.88**

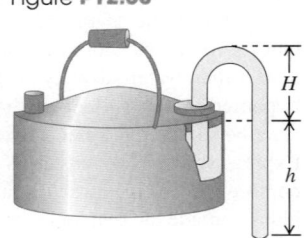

BIO **ELEPHANTS UNDER PRESSURE.** An elephant can swim or walk with its chest several meters underwater while the animal breathes through its trunk, which remains above the water surface and acts like a snorkel. The elephant's tissues are at an increased pressure due to the surrounding water, but the lungs are at atmospheric pressure because they are connected to the air through the trunk. The figure shows the gauge pressures in an elephant's lungs and abdomen when the elephant's chest is submerged to a particular depth in a lake. In this situation, the elephant's diaphragm, which separates the lungs from the abdomen, must sustain the difference in pressure between the lungs and the abdomen. The diaphragm of an elephant is typically 3.0 cm thick and 120 cm in

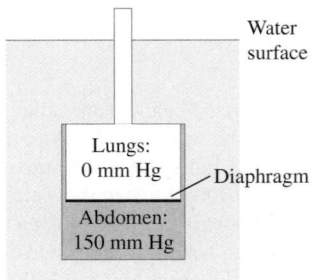

diameter. (See "Why Doesn't the Elephant Have a Pleural Space?" by John B. West, *Physiology,* Vol. 17:47–50, April 1, 2002.)

12.89 For the situation shown, the tissues in the elephant's abdomen are at a gauge pressure of 150 mm Hg. This pressure corresponds to what distance below the surface of a lake? (a) 1.5 m; (b) 2.0 m; (c) 3.0 m; (d) 15 m.

12.90 The maximum force the muscles of the diaphragm can exert is 24,000 N. What maximum pressure difference can the diaphragm withstand? (a) 160 mm Hg; (b) 760 mm Hg; (c) 920 mm Hg; (d) 5000 mm Hg.

12.91 How does the force the diaphragm experiences due to the difference in pressure between the lungs and abdomen depend on the abdomen's distance below the water surface? The force

(a) increases linearly with distance; (b) increases as distance squared; (c) increases as distance cubed; (d) increases exponentially with distance.

12.92 If the elephant were to snorkel in salt water, which is more dense than freshwater, would the maximum depth at which it could snorkel be different from that in freshwater? (a) Yes—that depth would increase, because the pressure would be lower at a given depth in salt water than in freshwater; (b) yes—that depth would decrease, because the pressure would be higher at a given depth in salt water than in freshwater; (c) no, because pressure differences within the submerged elephant depend on only the density of air, not the density of the water; (d) no, because the buoyant force on the elephant would be the same in both cases.

Answers

Chapter Opening Question **?**

(v) The ratio of mass to volume is density. The flesh of both the wrasse and the ray is denser than seawater, but a wrasse has a gas-filled body cavity called a swimbladder. Hence the *average* density of the wrasse's body is the same as that of seawater, and the fish neither sinks nor rises. Rays have no such cavity, so they must swim continuously to avoid sinking: Their fins provide lift, much like the wings of a bird or airplane (see Section 12.5).

Test Your Understanding Questions

12.1 (ii), (iv), (i) and (iii) (tie), (v) In each case the average density equals the mass divided by the volume:

(i) $\rho = (4.00\ \text{kg})/(1.60 \times 10^{-3}\ \text{m}^3) = 2.50 \times 10^3\ \text{kg/m}^3$;

(ii) $\rho = (8.00\ \text{kg})/(1.60 \times 10^{-3}\ \text{m}^3) = 5.00 \times 10^3\ \text{kg/m}^3$;

(iii) $\rho = (8.00\ \text{kg})/(3.20 \times 10^{-3}\ \text{m}^3) = 2.50 \times 10^3\ \text{kg/m}^3$;

(iv) $\rho = (2560\ \text{kg})/(0.640\ \text{m}^3) = 4.00 \times 10^3\ \text{kg/m}^3$;

(v) $\rho = (2560\ \text{kg})/(1.28\ \text{m}^3) = 2.00 \times 10^3\ \text{kg/m}^3$.

Note that compared to object (i), object (ii) has double the mass but the same volume and so has double the average density. Object (iii) has double the mass and double the volume of object (i), so (i) and (iii) have the same average density. Finally, object (v) has the same mass as object (iv) but double the volume, so (v) has half the average density of (iv).

12.2 (ii) From Eq. (12.9), the pressure outside the barometer is equal to the product ρgh. When the barometer is taken out of the refrigerator, the density ρ decreases while the height h of the mercury column remains the same. Hence the air pressure must be lower outdoors than inside the refrigerator.

12.3 (i) Consider the water, the statue, and the container together as a system; the total weight of the system does not depend on whether the statue is immersed. The total supporting force, including the tension T and the upward force F of the scale on the container (equal to the scale reading), is the same in both cases.

But we saw in Example 12.5 that T decreases by 7.84 N when the statue is immersed, so the scale reading F must *increase* by 7.84 N. An alternative viewpoint is that the water exerts an upward buoyant force of 7.84 N on the statue, so the statue must exert an equal downward force on the water, making the scale reading 7.84 N greater than the weight of water and container.

12.4 (ii) A highway that narrows from three lanes to one is like a pipe whose cross-sectional area narrows to one-third of its value. If cars behaved like the molecules of an incompressible fluid, then as the cars encountered the one-lane section, the spacing between cars (the "density") would stay the same but the cars would triple their speed. This would keep the "volume flow rate" (number of cars per second passing a point on the highway) the same. In real life cars behave like the molecules of a *compressible* fluid: They end up packed closer (the "density" increases) and fewer cars per second pass a point on the highway (the "volume flow rate" decreases).

12.5 (ii) Newton's second law tells us that a body accelerates (its velocity changes) in response to a net force. In fluid flow, a pressure difference between two points means that fluid particles moving between those two points experience a force, and this force causes the fluid particles to accelerate and change speed.

12.6 (iv) The required pressure is proportional to $1/R^4$, where R is the inside radius of the needle (half the inside diameter). With the smaller-diameter needle, the pressure is greater by a factor of $[(0.60\ \text{mm})/(0.30\ \text{mm})]^4 = 2^4 = 16$.

Bridging Problem

(a) $y = H - \left(\dfrac{d}{D}\right)^2 \sqrt{2gH}\, t + \left(\dfrac{d}{D}\right)^4 \dfrac{gt^2}{2}$

(b) $T = \sqrt{\dfrac{2H}{g}} \left(\dfrac{D}{d}\right)^2$ **(c)** $\sqrt{2}$

? The rings of Saturn are made of countless individual orbiting particles. Compared with a ring particle that orbits far from Saturn, does a ring particle close to Saturn orbit with (i) the same speed and greater acceleration; (ii) a faster speed and the same acceleration; (iii) a slower speed and the same acceleration; (iv) a faster speed and greater acceleration; or (v) none of these?

13 GRAVITATION

S ome of the earliest investigations in physical science started with questions that people asked about the night sky. Why doesn't the moon fall to earth? Why do the planets move across the sky? Why doesn't the earth fly off into space rather than remaining in orbit around the sun? The study of gravitation provides the answers to these and many related questions.

As we remarked in Chapter 5, gravitation is one of the four classes of interactions found in nature, and it was the earliest of the four to be studied extensively. Newton discovered in the 17th century that the same interaction that makes an apple fall out of a tree also keeps the planets in their orbits around the sun. This was the beginning of *celestial mechanics,* the study of the dynamics of objects in space. Today, our knowledge of celestial mechanics allows us to determine how to put a satellite into any desired orbit around the earth or to choose just the right trajectory to send a spacecraft to another planet.

In this chapter you will learn the basic law that governs gravitational interactions. This law is *universal:* Gravity acts in the same fundamental way between the earth and your body, between the sun and a planet, and between a planet and one of its moons. We'll apply the law of gravitation to phenomena such as the variation of weight with altitude, the orbits of satellites around the earth, and the orbits of planets around the sun.

13.1 NEWTON'S LAW OF GRAVITATION

The gravitational attraction that's most familiar to you is your *weight,* the force that attracts you toward the earth. By studying the motions of the moon and planets, Newton discovered a fundamental **law of gravitation** that describes the gravitational attraction between *any* two bodies. Newton published this law in 1687 along with his three laws of motion. In modern language, it says

> **Every particle of matter in the universe attracts every other particle with a force that is directly proportional to the product of the masses of the particles and inversely proportional to the square of the distance between them.**

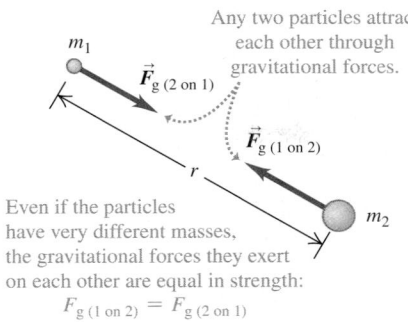

Any two particles attract each other through gravitational forces.

m_1

$\vec{F}_{g\,(2\,\text{on}\,1)}$

$\vec{F}_{g\,(1\,\text{on}\,2)}$

r

m_2

Even if the particles have very different masses, the gravitational forces they exert on each other are equal in strength:

$F_{g\,(1\,\text{on}\,2)} = F_{g\,(2\,\text{on}\,1)}$

13.1 The gravitational forces between two particles of masses m_1 and m_2.

Figure 13.1 depicts this law, which we can express as an equation:

Gravitational constant (same for any two particles)

Newton's **law of gravitation:**
Magnitude of attractive gravitational force between any two particles

$$F_g = \frac{Gm_1m_2}{r^2}$$

Masses of particles

Distance between particles

(13.1)

The **gravitational constant** G in Eq. (13.1) is a fundamental physical constant that has the same value for *any* two particles. We'll see shortly what the value of G is and how this value is measured.

Equation (13.1) tells us that the gravitational force between two particles decreases with increasing distance r: If the distance is doubled, the force is only one-fourth as great, and so on. Although many of the stars in the night sky are far more massive than the sun, they are so far away that their gravitational force on the earth is negligibly small.

CAUTION **Don't confuse g and G** The symbols g and G are similar, but they represent two very different gravitational quantities. Lowercase g is the acceleration due to gravity, which relates the weight w of a body to its mass m: $w = mg$. The value of g is different at different locations on the earth's surface and on the surfaces of other planets. By contrast, capital G relates the gravitational force between any two bodies to their masses and the distance between them. We call G a *universal* constant because it has the same value for any two bodies, no matter where in space they are located. We'll soon see how the values of g and G are related. ▮

Gravitational forces always act along the line joining the two particles and form an action–reaction pair. Even when the masses of the particles are different, the two interaction forces have equal magnitude (Fig. 13.1). The attractive force that your body exerts on the earth has the same magnitude as the force that the earth exerts on you. When you fall from a diving board into a swimming pool, the entire earth rises up to meet you! (You don't notice this because the earth's mass is greater than yours by a factor of about 10^{23}. Hence the earth's acceleration is only 10^{-23} as great as yours.)

Gravitation and Spherically Symmetric Bodies

We have stated the law of gravitation in terms of the interaction between two *particles*. It turns out that the gravitational interaction of any two bodies having *spherically symmetric* mass distributions (such as solid spheres or spherical shells) is the same as though we concentrated all the mass of each at its center, as in **Fig. 13.2**. Thus, if we model the earth as a spherically symmetric body with mass m_E, the force it exerts on a particle or on a spherically symmetric body with mass m, at a distance r between centers, is

$$F_g = \frac{Gm_Em}{r^2}$$

(13.2)

provided that the body lies outside the earth. A force of the same magnitude is exerted *on* the earth by the body. (We will prove these statements in Section 13.6.)

13.2 The gravitational effect *outside* any spherically symmetric mass distribution is the same as though all of the mass were concentrated at its center.

(a) The gravitational force between two spherically symmetric masses m_1 and m_2 ...

(b) ... is the same as if we concentrated all the mass of each sphere at the sphere's center.

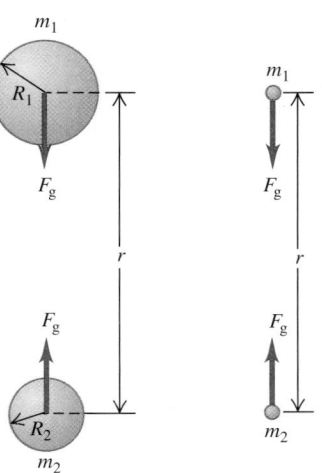

13.3 Spherical and nonspherical bodies: the planet Jupiter and one of Jupiter's small moons, Amalthea.

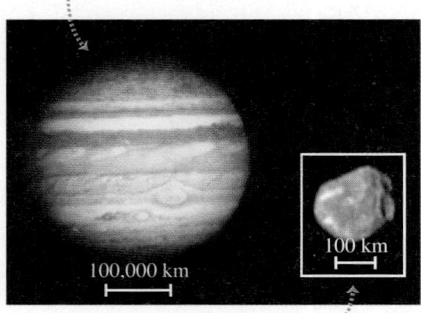

Jupiter's mass is very large (1.90×10^{27} kg), so the mutual gravitational attraction of its parts has pulled it into a nearly spherical shape.

100,000 km

100 km

Amalthea, one of Jupiter's small moons, has a relatively tiny mass (7.17×10^{18} kg, only about 3.8×10^{-9} the mass of Jupiter) and weak mutual gravitation, so it has an irregular shape.

At points *inside* the earth the situation is different. If we could drill a hole to the center of the earth and measure the gravitational force on a body at various depths, we would find that toward the center of the earth the force *decreases,* rather than increasing as $1/r^2$. As the body enters the interior of the earth (or other spherical body), some of the earth's mass is on the side of the body opposite from the center and pulls in the opposite direction. Exactly at the center, the earth's gravitational force on the body is zero.

Spherically symmetric bodies are an important case because moons, planets, and stars all tend to be spherical. Since all particles in a body gravitationally attract each other, the particles tend to move to minimize the distance between them. As a result, the body naturally tends to assume a spherical shape, just as a lump of clay forms into a sphere if you squeeze it with equal forces on all sides. This effect is greatly reduced in celestial bodies of low mass, since the gravitational attraction is less, and these bodies tend *not* to be spherical (**Fig. 13.3**).

Determining the Value of G

To determine the value of the gravitational constant G, we have to *measure* the gravitational force between two bodies of known masses m_1 and m_2 at a known distance r. The force is extremely small for bodies that are small enough to be brought into the laboratory, but it can be measured with an instrument called a *torsion balance,* which Sir Henry Cavendish used in 1798 to determine G.

Figure 13.4 shows a modern version of the Cavendish torsion balance. A light, rigid rod shaped like an inverted T is supported by a very thin, vertical quartz fiber. Two small spheres, each of mass m_1, are mounted at the ends of the horizontal arms of the T. When we bring two large spheres, each of mass m_2, to the positions shown, the attractive gravitational forces twist the T through a small angle. To measure this angle, we shine a beam of light on a mirror fastened to the T. The reflected beam strikes a scale, and as the T twists, the reflected beam moves along the scale.

After calibrating the Cavendish balance, we can measure gravitational forces and thus determine G. The presently accepted value is

$$G = 6.67384(80) \times 10^{-11} \, \text{N} \cdot \text{m}^2/\text{kg}^2$$

To three significant figures, $G = 6.67 \times 10^{-11} \, \text{N} \cdot \text{m}^2/\text{kg}^2$. Because $1 \, \text{N} = 1 \, \text{kg} \cdot \text{m}/\text{s}^2$, the units of G can also be expressed as $\text{m}^3/(\text{kg} \cdot \text{s}^2)$.

Gravitational forces combine vectorially. If each of two masses exerts a force on a third, the *total* force on the third mass is the vector sum of the individual forces of the first two. Example 13.3 makes use of this property, which is often called *superposition of forces* (see Section 4.1).

13.4 The principle of the Cavendish balance, used for determining the value of G. The angle of deflection has been exaggerated here for clarity.

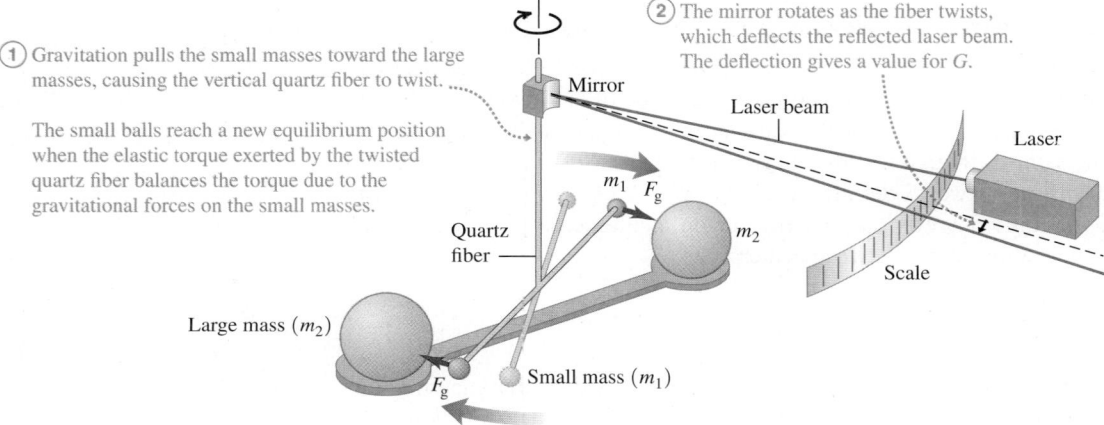

① Gravitation pulls the small masses toward the large masses, causing the vertical quartz fiber to twist.

The small balls reach a new equilibrium position when the elastic torque exerted by the twisted quartz fiber balances the torque due to the gravitational forces on the small masses.

② The mirror rotates as the fiber twists, which deflects the reflected laser beam. The deflection gives a value for G.

Mirror

Laser beam

Laser

m_1 F_g

m_2

Quartz fiber

Scale

Large mass (m_2)

F_g

Small mass (m_1)

EXAMPLE 13.1 CALCULATING GRAVITATIONAL FORCE

The mass m_1 of one of the small spheres of a Cavendish balance is 0.0100 kg, the mass m_2 of the nearest large sphere is 0.500 kg, and the center-to-center distance between them is 0.0500 m. Find the gravitational force F_g on each sphere due to the other.

SOLUTION

IDENTIFY, SET UP, and EXECUTE: Because the spheres are spherically symmetric, we can calculate F_g by treating them as *particles* separated by 0.0500 m, as in Fig. 13.2. Each sphere experiences the same magnitude of force from the other sphere. We use Newton's

law of gravitation, Eq. (13.1), to determine F_g:

$$F_g = \frac{(6.67 \times 10^{-11}\ \text{N} \cdot \text{m}^2/\text{kg}^2)(0.0100\ \text{kg})(0.500\ \text{kg})}{(0.0500\ \text{m})^2}$$

$$= 1.33 \times 10^{-10}\ \text{N}$$

EVALUATE: It's remarkable that such a small force could be measured—or even detected—more than 200 years ago. Only a very massive object such as the earth exerts a gravitational force we can feel.

EXAMPLE 13.2 ACCELERATION DUE TO GRAVITATIONAL ATTRACTION

Suppose the two spheres in Example 13.1 are placed with their centers 0.0500 m apart at a point in space far removed from all other bodies. What is the magnitude of the acceleration of each, relative to an inertial system?

SOLUTION

IDENTIFY, SET UP, and EXECUTE: Each sphere exerts on the other a gravitational force of the same magnitude F_g, which we found in Example 13.1. We can ignore any other forces. The *acceleration* magnitudes a_1 and a_2 are different because the masses are different.

To determine these we'll use Newton's second law:

$$a_1 = \frac{F_g}{m_1} = \frac{1.33 \times 10^{-10}\ \text{N}}{0.0100\ \text{kg}} = 1.33 \times 10^{-8}\ \text{m/s}^2$$

$$a_2 = \frac{F_g}{m_2} = \frac{1.33 \times 10^{-10}\ \text{N}}{0.500\ \text{kg}} = 2.66 \times 10^{-10}\ \text{m/s}^2$$

EVALUATE: The larger sphere has 50 times the mass of the smaller one and hence has $\frac{1}{50}$ the acceleration. These accelerations are *not* constant; the gravitational forces increase as the spheres move toward each other.

EXAMPLE 13.3 SUPERPOSITION OF GRAVITATIONAL FORCES

Many stars belong to *systems* of two or more stars held together by their mutual gravitational attraction. **Figure 13.5** shows a three-star system at an instant when the stars are at the vertices of a 45° right triangle. Find the total gravitational force exerted on the small star by the two large ones.

SOLUTION

IDENTIFY, SET UP, and EXECUTE: We use the principle of superposition: The total force $\vec{F}$ on the small star is the vector sum of the forces $\vec{F}_1$ and $\vec{F}_2$ due to each large star, as Fig. 13.5 shows. We assume that the stars are spheres as in Fig. 13.2. We first calculate the magnitudes F_1 and F_2 from Eq. (13.1) and then compute the vector sum by using components:

$$F_1 = \frac{\left[\begin{array}{c} (6.67 \times 10^{-11}\ \text{N} \cdot \text{m}^2/\text{kg}^2) \\ \times\ (8.00 \times 10^{30}\ \text{kg})(1.00 \times 10^{30}\ \text{kg}) \end{array} \right]}{(2.00 \times 10^{12}\ \text{m})^2 + (2.00 \times 10^{12}\ \text{m})^2}$$

$$= 6.67 \times 10^{25}\ \text{N}$$

$$F_2 = \frac{\left[\begin{array}{c} (6.67 \times 10^{-11}\ \text{N} \cdot \text{m}^2/\text{kg}^2) \\ \times\ (8.00 \times 10^{30}\ \text{kg})(1.00 \times 10^{30}\ \text{kg}) \end{array} \right]}{(2.00 \times 10^{12}\ \text{m})^2}$$

$$= 1.33 \times 10^{26}\ \text{N}$$

13.5 The total gravitational force on the small star (at O) is the vector sum of the forces exerted on it by the two larger stars. (For comparison, the mass of the sun—a rather ordinary star—is 1.99×10^{30} kg and the earth–sun distance is 1.50×10^{11} m.)

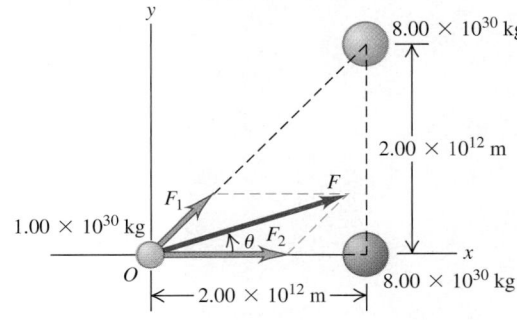

The x- and y-components of these forces are

$$F_{1x} = (6.67 \times 10^{25}\ \text{N})(\cos 45°) = 4.72 \times 10^{25}\ \text{N}$$

$$F_{1y} = (6.67 \times 10^{25}\ \text{N})(\sin 45°) = 4.72 \times 10^{25}\ \text{N}$$

$$F_{2x} = 1.33 \times 10^{26}\ \text{N}$$

$$F_{2y} = 0$$

Continued

The components of the total force $\vec{F}$ on the small star are

$$F_x = F_{1x} + F_{2x} = 1.81 \times 10^{26} \text{ N}$$
$$F_y = F_{1y} + F_{2y} = 4.72 \times 10^{25} \text{ N}$$

The magnitude of $\vec{F}$ and its angle θ (see Fig. 13.5) are

$$F = \sqrt{F_x^2 + F_y^2} = \sqrt{(1.81 \times 10^{26} \text{ N})^2 + (4.72 \times 10^{25} \text{ N})^2}$$

$$= 1.87 \times 10^{26} \text{ N}$$

$$\theta = \arctan\frac{F_y}{F_x} = \arctan\frac{4.72 \times 10^{25} \text{ N}}{1.81 \times 10^{26} \text{ N}} = 14.6°$$

EVALUATE: While the force magnitude F is tremendous, the magnitude of the resulting acceleration is not: $a = F/m = (1.87 \times 10^{26} \text{ N})/(1.00 \times 10^{30} \text{ kg}) = 1.87 \times 10^{-4} \text{ m/s}^2$. Furthermore, the force $\vec{F}$ is *not* directed toward the center of mass of the two large stars.

DATA *SPEAKS*

Gravitation

When students were given a problem about superposition of gravitational forces, more than 60% gave an incorrect response. Common errors:

- Assuming that equal-mass objects *A* and *B* must exert equally strong gravitational attraction on an object *C* (which is not true when *A* and *B* are different distances from *C*).
- Neglecting to account for the vector nature of force. (To add two forces that point in different directions, you can't just add the force magnitudes.)

13.6 Our solar system is part of a spiral galaxy like this one, which contains roughly 10^{11} stars as well as gas, dust, and other matter. The entire assemblage is held together by the mutual gravitational attraction of all the matter in the galaxy.

Why Gravitational Forces Are Important

Comparing Examples 13.1 and 13.3 shows that gravitational forces are negligible between ordinary household-sized objects but very substantial between objects that are the size of stars. Indeed, gravitation is *the* most important force on the scale of planets, stars, and galaxies (**Fig. 13.6**). It is responsible for holding our earth together and for keeping the planets in orbit about the sun. The mutual gravitational attraction between different parts of the sun compresses material at the sun's core to very high densities and temperatures, making it possible for nuclear reactions to take place there. These reactions generate the sun's energy output, which makes it possible for life to exist on earth and for you to read these words.

The gravitational force is so important on the cosmic scale because it acts *at a distance*, without any direct contact between bodies. Electric and magnetic forces have this same remarkable property, but they are less important on astronomical scales because large accumulations of matter are electrically neutral; that is, they contain equal amounts of positive and negative charge. As a result, the electric and magnetic forces between stars or planets are very small or zero. The strong and weak interactions that we discussed in Section 5.5 also act at a distance, but their influence is negligible at distances much greater than the diameter of an atomic nucleus (about 10^{-14} m).

A useful way to describe forces that act at a distance is in terms of a *field*. One body sets up a disturbance or field at all points in space, and the force that acts on a second body at a particular point is its response to the first body's field at that point. There is a field associated with each force that acts at a distance, and so we refer to gravitational fields, electric fields, magnetic fields, and so on. We won't need the field concept for our study of gravitation in this chapter, so we won't discuss it further here. But in later chapters we'll find that the field concept is an extraordinarily powerful tool for describing electric and magnetic interactions.

TEST YOUR UNDERSTANDING OF SECTION 13.1 The planet Saturn has about 100 times the mass of the earth and is about 10 times farther from the sun than the earth is. Compared to the acceleration of the earth caused by the sun's gravitational pull, how great is the acceleration of Saturn due to the sun's gravitation? (i) 100 times greater; (ii) 10 times greater; (iii) the same; (iv) $\frac{1}{10}$ as great; (v) $\frac{1}{100}$ as great. ∎

13.2 WEIGHT

We defined the *weight* of a body in Section 4.4 as the attractive gravitational force exerted on it by the earth. We can now broaden our definition and say that *the weight of a body is the total gravitational force exerted on the body by all other bodies in the universe.* When the body is near the surface of the earth, we can ignore all other gravitational forces and consider the weight as just the earth's gravitational attraction. At the surface of the *moon* we consider a body's weight to be the gravitational attraction of the moon, and so on.

PhET: Lunar Lander

If we again model the earth as a spherically symmetric body with radius R_E, the weight of a small body at the earth's surface (a distance R_E from its center) is

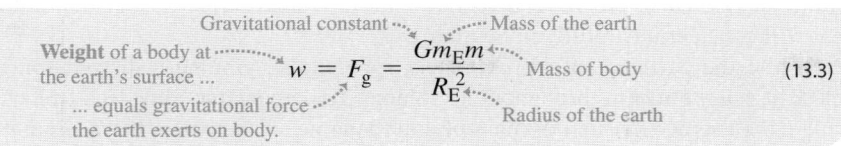

Weight of a body at the earth's surface ... equals gravitational force the earth exerts on body.

Gravitational constant ... Mass of the earth

$$w = F_g = \frac{Gm_E m}{R_E^2}$$ (13.3)

Mass of body
Radius of the earth

But we also know from Section 4.4 that the weight w of a body is the force that causes the acceleration g of free fall, so by Newton's second law, $w = mg$. Equating this with Eq. (13.3) and dividing by m, we find

Acceleration due to gravity at the earth's surface

Gravitational constant ... Mass of the earth

$$g = \frac{Gm_E}{R_E^2}$$ (13.4)

Radius of the earth

The acceleration due to gravity g is independent of the mass m of the body because m doesn't appear in this equation. We already knew that, but we can now see how it follows from the law of gravitation.

We can *measure* all the quantities in Eq. (13.4) except for m_E, so this relationship allows us to compute the mass of the earth. Solving Eq. (13.4) for m_E and using $R_E = 6370$ km $= 6.37 \times 10^6$ m and $g = 9.80$ m/s^2, we find

$$m_E = \frac{gR_E^2}{G} = 5.96 \times 10^{24} \text{ kg}$$

This is very close to the currently accepted value of 5.972×10^{24} kg. Once Cavendish had measured G, he computed the mass of the earth in just this way.

At a point above the earth's surface a distance r from the center of the earth (a distance $r - R_E$ above the surface), the weight of a body is given by Eq. (13.3) with R_E replaced by r:

$$w = F_g = \frac{Gm_E m}{r^2}$$ (13.5)

The weight of a body decreases inversely with the square of its distance from the earth's center (**Fig. 13.7**). **Figure 13.8** shows how the weight varies with height above the earth for an astronaut who weighs 700 N at the earth's surface.

13.8 An astronaut who weighs 700 N at the earth's surface experiences less gravitational attraction when above the surface. The relevant distance r is from the astronaut to the *center* of the earth (*not* from the astronaut to the earth's surface).

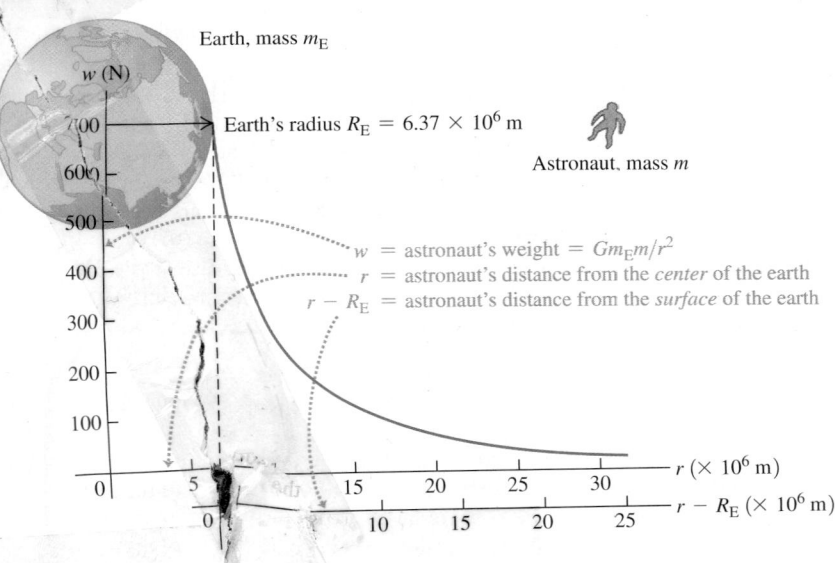

Earth, mass m_E

w (N)

Earth's radius $R_E = 6.37 \times 10^6$ m

Astronaut, mass m

w = astronaut's weight = $Gm_E m/r^2$
r = astronaut's distance from the *center* of the earth
$r - R_E$ = astronaut's distance from the *surface* of the earth

Application **Walking and Running on the Moon** You automatically transition from a walk to a run when the vertical force you exert on the ground—which, by Newton's third law, equals the vertical force the ground exerts on you—exceeds your weight. This transition from walking to running happens at much lower speeds on the moon, where objects weigh only 17% as much as on earth. Hence, the Apollo astronauts found themselves running even when moving relatively slowly during their moon "walks."

13.7 In an airliner at high altitude, you are farther from the center of the earth than when on the ground and hence weigh slightly less. Can you show that at an altitude of 10 km above the surface, you weigh 0.3% less than you do on the ground?

13.9 The density ρ of the earth decreases with increasing distance r from its center.

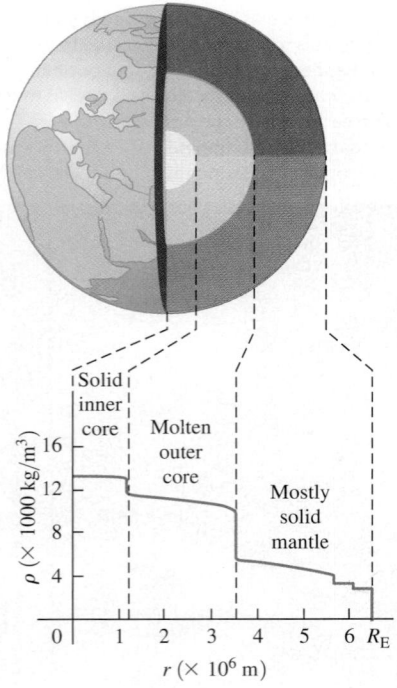

The *apparent* weight of a body on earth differs slightly from the earth's gravitational force because the earth rotates and is therefore not precisely an inertial frame of reference. We've ignored this relatively small effect in our discussion but will consider it carefully in Section 13.7.

While the earth is an approximately spherically symmetric distribution of mass, it is *not* uniform throughout its volume. To demonstrate this, let's first calculate the average *density,* or mass per unit volume, of the earth. If we assume a spherical earth, the volume is

$$V_E = \tfrac{4}{3}\pi R_E^3 = \tfrac{4}{3}\pi (6.37 \times 10^6 \text{ m})^3 = 1.08 \times 10^{21} \text{ m}^3$$

The average density ρ (the Greek letter rho) of the earth is the total mass divided by the total volume:

$$\rho = \frac{m_E}{V_E} = \frac{5.97 \times 10^{24} \text{ kg}}{1.08 \times 10^{21} \text{ m}^3}$$

$$= 5500 \text{ kg/m}^3 = 5.5 \text{ g/cm}^3$$

(Compare to the density of water, $1000 \text{ kg/m}^3 = 1.00 \text{ g/cm}^3$.) If the earth were uniform, rocks near the earth's surface would have this same density. In fact, the density of surface rocks is substantially lower, ranging from about 2000 kg/m^3 for sedimentary rocks to about 3300 kg/m^3 for basalt. So the earth *cannot* be uniform, and its interior must be much more dense than its surface in order that the *average* density be 5500 kg/m^3. According to geophysical models of the earth's interior, the maximum density at the center is about $13{,}000 \text{ kg/m}^3$. **Figure 13.9** is a graph of density as a function of distance from the center.

EXAMPLE 13.4   **GRAVITY ON MARS**

A robotic lander with an earth weight of 3430 N is sent to Mars, which has radius $R_M = 3.39 \times 10^6$ m and mass $m_M = 6.42 \times 10^{23}$ kg (see Appendix F). Find the weight F_g of the lander on the Martian surface and the acceleration there due to gravity, g_M.

SOLUTION

IDENTIFY and SET UP: To find F_g we use Eq. (13.3), replacing m_E and R_E with m_M and R_M. We determine the lander mass m from the lander's earth weight w and then find g_M from $F_g = mg_M$.

EXECUTE: The lander's earth weight is $w = mg$, so

$$m = \frac{w}{g} = \frac{3430 \text{ N}}{9.80 \text{ m/s}^2} = 350 \text{ kg}$$

The mass is the same no matter where the lander is. From Eq. (13.3), the lander's weight on Mars is

$$F_g = \frac{Gm_M m}{R_M^2}$$

$$= \frac{(6.67 \times 10^{-11} \text{ N} \cdot \text{m}^2/\text{kg}^2)(6.42 \times 10^{23} \text{ kg})(350 \text{ kg})}{(3.39 \times 10^6 \text{ m})^2}$$

$$= 1.30 \times 10^3 \text{ N}$$

The acceleration due to gravity on Mars is

$$g_M = \frac{F_g}{m} = \frac{1.30 \times 10^3 \text{ N}}{350 \text{ kg}} = 3.7 \text{ m/s}^2$$

EVALUATE: Even though Mars has just 11% of the earth's mass (6.42×10^{23} kg versus 5.97×10^{24} kg), the acceleration due to gravity g_M (and hence an object's weight F_g) is roughly 40% as large as on earth. That's because g_M is also inversely proportional to the square of the planet's radius, and Mars has only 53% the radius of earth (3.39×10^6 m versus 6.37×10^6 m).

You can check our result for g_M by using Eq. (13.4), with appropriate replacements. Do you get the same answer?

TEST YOUR UNDERSTANDING OF SECTION 13.2 Rank the following hypothetical planets in order from highest to lowest value of g at the surface: (i) mass = 2 times the mass of the earth, radius = 2 times the radius of the earth; (ii) mass = 4 times the mass of the earth, radius = 4 times the radius of the earth; (iii) mass = 4 times the mass of the earth, radius = 2 times the radius of the earth; (iv) mass = 2 times the mass of the earth, radius = 4 times the radius of the earth. ❚

13.3 GRAVITATIONAL POTENTIAL ENERGY

When we first introduced gravitational potential energy in Section 7.1, we assumed that the earth's gravitational force on a body of mass m doesn't depend on the body's height. This led to the expression $U = mgy$. But Eq. (13.2), $F_g = Gm_Em/r^2$, shows that the gravitational force exerted by the earth (mass m_E) *does* in general depend on the distance r from the body to the earth's center. For problems in which a body can be far from the earth's surface, we need a more general expression for gravitational potential energy.

To find this expression, we follow the same steps as in Section 7.1. We consider a body of mass m outside the earth, and first compute the work W_{grav} done by the gravitational force when the body moves directly away from or toward the center of the earth from $r = r_1$ to $r = r_2$, as in **Fig. 13.10**. This work is given by

$$W_{grav} = \int_{r_1}^{r_2} F_r \, dr \qquad (13.6)$$

where F_r is the radial component of the gravitational force $\vec{F}$—that is, the component in the direction *outward* from the center of the earth. Because $\vec{F}$ points directly *inward* toward the center of the earth, F_r is negative. It differs from Eq. (13.2), the magnitude of the gravitational force, by a minus sign:

$$F_r = -\frac{Gm_Em}{r^2} \qquad (13.7)$$

Substituting Eq. (13.7) into Eq. (13.6), we see that W_{grav} is given by

$$W_{grav} = -Gm_Em \int_{r_1}^{r_2} \frac{dr}{r^2} = \frac{Gm_Em}{r_2} - \frac{Gm_Em}{r_1} \qquad (13.8)$$

The path doesn't have to be a straight line; it could also be a curve like the one in Fig. 13.10. By an argument similar to that in Section 7.1, this work depends on only the initial and final values of r, not on the path taken. This also proves that the gravitational force is always *conservative*.

We now define the corresponding potential energy U so that $W_{grav} = U_1 - U_2$, as in Eq. (7.3). Comparing this with Eq. (13.8), we see that the appropriate definition for **gravitational potential energy** is

Gravitational constant ⋯⋯ ⋯⋯ Mass of the earth

Gravitational ⋯⋯⋯⋯ $U = -\dfrac{Gm_Em}{r}$ ⋯⋯ Mass of body
potential energy ⋯⋯⋯⋯⋯ Distance of body from $\qquad (13.9)$
(general expression) the earth's center

Figure 13.11 shows how the gravitational potential energy depends on the distance r between the body of mass m and the center of the earth. When the body moves away from the earth, r increases, the gravitational force does negative work, and U increases (i.e., becomes less negative). When the body "falls" toward earth, r decreases, the gravitational work is positive, and the potential energy decreases (i.e., becomes more negative).

You may be troubled by Eq. (13.9) because it states that gravitational potential energy is always negative. But in fact you've seen negative values of U before. In using the formula $U = mgy$ in Section 7.1, we found that U was negative whenever the body of mass m was at a value of y below the arbitrary height we chose to be $y = 0$—that is, whenever the body and the earth were closer together than some arbitrary distance. (See, for instance, Example 7.2 in Section 7.1.) In defining U by Eq. (13.9), we have chosen U to be zero when the body of mass m is infinitely far from the earth ($r = \infty$). As the body moves toward the earth, gravitational potential energy decreases and so becomes negative.

13.10 Calculating the work done on a body by the gravitational force as the body moves from radial coordinate r_1 to r_2.

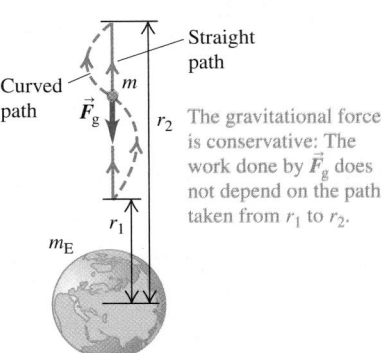

13.11 A graph of the gravitational potential energy U for the system of the earth (mass m_E) and an astronaut (mass m) versus the astronaut's distance r from the center of the earth.

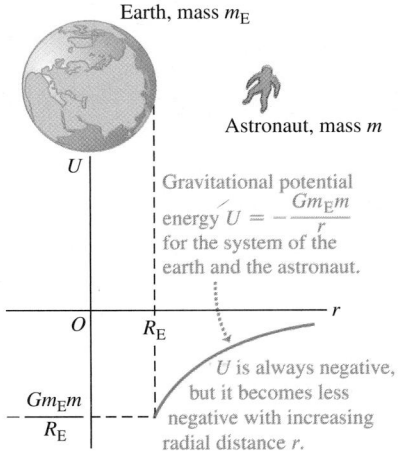

Gravitational force vs. gravitational potential energy Don't confuse the expressions for gravitational force, Eq. (13.7), and gravitational potential energy, Eq. (13.9). The force F_r is proportional to $1/r^2$, while potential energy U is proportional to $1/r$. ▮

If we wanted, we could make $U = 0$ at the earth's surface, where $r = R_E$, by adding the quantity Gm_Em/R_E to Eq. (13.9). This would make U positive when $r > R_E$. We won't do this for two reasons: One, it would complicate the expression for U; two, the added term would not affect the *difference* in potential energy between any two points, which is the only physically significant quantity.

If the earth's gravitational force on a body is the only force that does work, then the total mechanical energy of the system of the earth and body is constant, or *conserved*. In the following example we'll use this principle to calculate **escape speed,** the speed required for a body to escape completely from a planet.

EXAMPLE 13.5 "FROM THE EARTH TO THE MOON"

In Jules Verne's 1865 story with this title, three men went to the moon in a shell fired from a giant cannon sunk in the earth in Florida. (a) Find the minimum muzzle speed needed to shoot a shell straight up to a height above the earth equal to the earth's radius R_E. (b) Find the minimum muzzle speed that would allow a shell to escape from the earth completely (the *escape speed*). Neglect air resistance, the earth's rotation, and the gravitational pull of the moon. The earth's radius and mass are $R_E = 6.37 \times 10^6$ m and $m_E = 5.97 \times 10^{24}$ kg.

SOLUTION

IDENTIFY and SET UP: Once the shell leaves the cannon muzzle, only the (conservative) gravitational force does work. Hence we can use conservation of mechanical energy to find the speed at which the shell must leave the muzzle so as to come to a halt (a) at two earth radii from the earth's center and (b) at an infinite distance from earth. The energy-conservation equation is $K_1 + U_1 = K_2 + U_2$, with U given by Eq. (13.9).

In **Fig. 13.12** point 1 is at $r_1 = R_E$, where the shell leaves the cannon with speed v_1 (the target variable). Point 2 is where the shell reaches its maximum height; in part (a) $r_2 = 2R_E$ (Fig. 13.12a), and

13.12 Our sketches for this problem.

(a)

(b)

in part (b) $r_2 = \infty$ (Fig 13.12b). In both cases $v_2 = 0$ and $K_2 = 0$. Let m be the mass of the shell (with passengers).

EXECUTE: (a) We solve the energy-conservation equation for v_1:

$$K_1 + U_1 = K_2 + U_2$$

$$\tfrac{1}{2}mv_1^2 + \left(-\frac{Gm_Em}{R_E}\right) = 0 + \left(-\frac{Gm_Em}{2R_E}\right)$$

$$v_1 = \sqrt{\frac{Gm_E}{R_E}} = \sqrt{\frac{(6.67 \times 10^{-11}\ \text{N}\cdot\text{m}^2/\text{kg}^2)(5.97 \times 10^{24}\ \text{kg})}{6.37 \times 10^6\ \text{m}}}$$

$$= 7910\ \text{m/s}\ (= 28{,}500\ \text{km/h} = 17{,}700\ \text{mi/h})$$

(b) Now $r_2 = \infty$ so $U_2 = 0$ (see Fig. 13.11). Since $K_2 = 0$, the total mechanical energy $K_2 + U_2$ is zero in this case. Again we solve the energy-conservation equation for v_1:

$$\tfrac{1}{2}mv_1^2 + \left(-\frac{Gm_Em}{R_E}\right) = 0 + 0$$

$$v_1 = \sqrt{\frac{2Gm_E}{R_E}}$$

$$= \sqrt{\frac{2(6.67 \times 10^{-11}\ \text{N}\cdot\text{m}^2/\text{kg}^2)(5.97 \times 10^{24}\ \text{kg})}{6.37 \times 10^6\ \text{m}}}$$

$$= 1.12 \times 10^4\ \text{m/s}\ (= 40{,}200\ \text{km/h} = 25{,}000\ \text{mi/h})$$

EVALUATE: Our results don't depend on the mass of the shell or the direction of launch. A modern spacecraft launched from Florida must attain essentially the speed found in part (b) to escape the earth; however, before launch it's already moving at 410 m/s to the east because of the earth's rotation. Launching to the east takes advantage of this "free" contribution toward escape speed.

To generalize, the initial speed v_1 needed for a body to escape from the surface of a spherical body of mass M and radius R (ignoring air resistance) is $v_1 = \sqrt{2GM/R}$ (escape speed). This equation yields 5.03×10^3 m/s for Mars, 6.02×10^4 m/s for Jupiter, and 6.18×10^5 m/s for the sun.

More on Gravitational Potential Energy

As a final note, let's show that when we are close to the earth's surface, Eq. (13.9) reduces to the familiar $U = mgy$ from Chapter 7. We first rewrite Eq. (13.8) as

$$W_{\text{grav}} = Gm_Em\frac{r_1 - r_2}{r_1 r_2}$$

If the body stays close to the earth, then in the denominator we may replace r_1 and r_2 by R_E, the earth's radius, so

$$W_{grav} = Gm_E m \frac{r_1 - r_2}{R_E^2}$$

According to Eq. (13.4), $g = Gm_E/R_E^2$, so

$$W_{grav} = mg(r_1 - r_2)$$

If we replace the r's by y's, this is just Eq. (7.1) for the work done by a constant gravitational force. In Section 7.1 we used this equation to derive Eq. (7.2), $U = mgy$, so we may consider Eq. (7.2) for gravitational potential energy to be a special case of the more general Eq. (13.9).

TEST YOUR UNDERSTANDING OF SECTION 13.3   If a planet has the same surface gravity as the earth (that is, the same value of g at the surface), what is its escape speed? (i) The same as the earth's; (ii) less than the earth's; (iii) greater than the earth's; (iv) any of these are possible. ▌

13.4 THE MOTION OF SATELLITES

Artificial satellites orbiting the earth are a familiar part of technology (**Fig. 13.13**). But how do they stay in orbit, and what determines the properties of their orbits? We can use Newton's laws and the law of gravitation to provide the answers. In the next section we'll analyze the motion of planets in the same way.

To begin, think back to the discussion of projectile motion in Section 3.3. In Example 3.6 a motorcycle rider rides horizontally off the edge of a cliff, launching himself into a parabolic path that ends on the flat ground at the base of the cliff. If he survives and repeats the experiment with increased launch speed, he will land farther from the starting point. We can imagine him launching himself with great enough speed that the earth's curvature becomes significant. As he falls, the earth curves away beneath him. If he is going fast enough, and if his launch point is high enough that he clears the mountaintops, he may be able to go right on around the earth without ever landing.

Figure 13.14 shows a variation on this theme. We launch a projectile from point A in the direction AB, tangent to the earth's surface. Trajectories 1 through 7 show the effect of increasing the initial speed. In trajectories 3 through 5 the projectile misses the earth and becomes a satellite. If there is no retarding force such

13.13 With a mass of approximately 4.5×10^5 kg and a width of over 108 m, the International Space Station is the largest satellite ever placed in orbit.

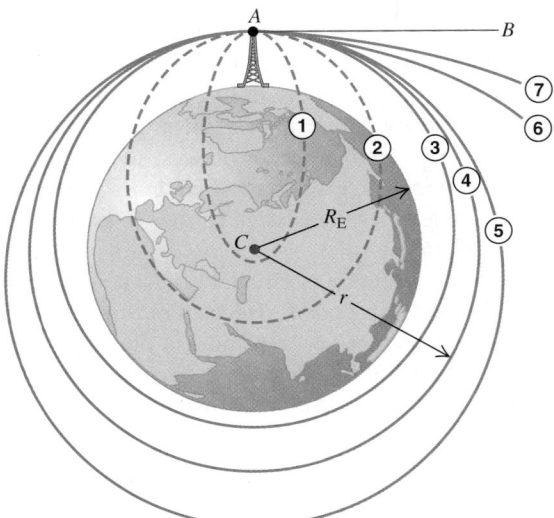

A projectile is launched from A toward B. Trajectories ① through ⑦ show the effect of increasing initial speed.

13.14 Trajectories of a projectile launched from a great height (ignoring air resistance). Orbits 1 and 2 would be completed as shown if the earth were a point mass at C. (This illustration is based on one in Isaac Newton's *Principia*.)

PhET: My Solar System

as air resistance, the projectile's speed when it returns to point A is the same as its initial speed and it repeats its motion indefinitely.

Trajectories 1 through 5 close on themselves and are called **closed orbits.** All closed orbits are ellipses or segments of ellipses; trajectory 4 is a circle, a special case of an ellipse. (We'll discuss the properties of an ellipse in Section 13.5.) Trajectories 6 and 7 are **open orbits.** For these paths the projectile never returns to its starting point but travels ever farther away from the earth.

Satellites: Circular Orbits

A *circular* orbit, like trajectory 4 in Fig. 13.14, is the simplest case. It is also an important case, since many artificial satellites have nearly circular orbits and the orbits of the planets around the sun are also fairly circular. The only force acting on a satellite in circular orbit around the earth is the earth's gravitational attraction, which is directed toward the center of the earth and hence toward the center of the orbit (**Fig. 13.15**). As we discussed in Section 5.4, this means that the satellite is in *uniform* circular motion and its speed is constant. The satellite isn't falling *toward* the earth; rather, it's constantly falling *around* the earth. In a circular orbit the speed is just right to keep the distance from the satellite to the center of the earth constant.

13.15 The force $\vec{F}_g$ due to the earth's gravitational attraction provides the centripetal acceleration that keeps a satellite in orbit. Compare to Fig. 5.28.

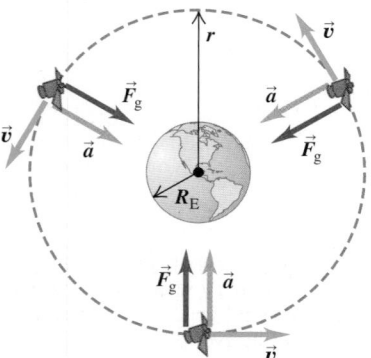

The satellite is in a circular orbit: Its acceleration $\vec{a}$ is always perpendicular to its velocity $\vec{v}$, so its speed v is constant.

Let's see how to find the constant speed v of a satellite in a circular orbit. The radius of the orbit is r, measured from the *center* of the earth; the acceleration of the satellite has magnitude $a_{rad} = v^2/r$ and is always directed toward the center of the circle. By the law of gravitation, the net force (gravitational force) on the satellite of mass m has magnitude $F_g = Gm_Em/r^2$ and is in the same direction as the acceleration. Newton's second law ($\sum \vec{F} = m\vec{a}$) then tells us that

$$\frac{Gm_Em}{r^2} = \frac{mv^2}{r}$$

Solving this for v, we find

Speed of satellite in a circular orbit around the earth
$$v = \sqrt{\frac{Gm_E}{r}}$$
Gravitational constant
Mass of the earth
Radius of orbit

(13.10)

This relationship shows that we can't choose the orbit radius r and the speed v independently; for a given radius r, the speed v for a circular orbit is determined.

The satellite's mass m doesn't appear in Eq. (13.10), which shows that the motion of a satellite does not depend on its mass. An astronaut on board an orbiting space station is herself a satellite of the earth, held by the earth's gravity in the same orbit as the station. The astronaut has the same velocity and acceleration as the station, so nothing is pushing her against the station's floor or walls. She is in a state of *apparent weightlessness,* as in a freely falling elevator; see the discussion following Example 5.9 in Section 5.2. (*True* weightlessness would occur only if the astronaut were infinitely far from any other masses, so that the gravitational force on her would be zero.) Indeed, every part of her body is apparently weightless; she feels nothing pushing her stomach against her intestines or her head against her shoulders (**Fig. 13.16**).

13.16 These astronauts are in a state of apparent weightlessness. Which are right side up and which are upside down?

Apparent weightlessness is not just a feature of circular orbits; it occurs whenever gravity is the only force acting on a spacecraft. Hence it occurs for orbits of any shape, including open orbits such as trajectories 6 and 7 in Fig. 13.14.

We can derive a relationship between the radius r of a circular orbit and the period T, the time for one revolution. The speed v is the distance $2\pi r$ traveled in one revolution, divided by the period:

$$v = \frac{2\pi r}{T}$$

(13.11)

We solve Eq. (13.11) for T and substitute v from Eq. (13.10):

Period of a circular orbit around the earth

$$T = \frac{2\pi r}{v} = 2\pi r \sqrt{\frac{r}{Gm_E}} = \frac{2\pi r^{3/2}}{\sqrt{Gm_E}} \qquad (13.12)$$

·····Radius of orbit·····
Orbital speed
Gravitational constant····· ·····Mass of the earth

Equations (13.10) and (13.12) show that larger orbits correspond to slower speeds and longer periods. As an example, the International Space Station (Fig. 13.13) orbits 6800 km from the center of the earth (400 km above the earth's surface) with an orbital speed of 7.7 km/s and an orbital period of 93 min. The moon orbits the earth in a much larger orbit of radius 384,000 km, and so has a much slower orbital speed (1.0 km/s) and a much longer orbital period (27.3 days).

It's interesting to compare Eq. (13.10) to the calculation of escape speed in Example 13.5. We see that the escape speed from a spherical body with radius R is $\sqrt{2}$ times greater than the speed of a satellite in a circular orbit at that radius. If our spacecraft is in circular orbit around *any* planet, we have to multiply our speed by a factor of $\sqrt{2}$ to escape to infinity, regardless of the planet's mass.

Since the speed v in a circular orbit is determined by Eq. (13.10) for a given orbit radius r, the total mechanical energy $E = K + U$ is determined as well. Using Eqs. (13.9) and (13.10), we have

$$E = K + U = \tfrac{1}{2}mv^2 + \left(-\frac{Gm_E m}{r} \right)$$

$$= \tfrac{1}{2}m\left(\frac{Gm_E}{r} \right) - \frac{Gm_E m}{r}$$

$$= -\frac{Gm_E m}{2r} \qquad \text{(circular orbit)} \qquad (13.13)$$

The total mechanical energy in a circular orbit is negative and equal to one-half the potential energy. Increasing the orbit radius r means increasing the mechanical energy (that is, making E less negative). If the satellite is in a relatively low orbit that encounters the outer fringes of earth's atmosphere, mechanical energy decreases due to negative work done by the force of air resistance; as a result, the orbit radius decreases until the satellite hits the ground or burns up in the atmosphere.

We have talked mostly about earth satellites, but we can apply the same analysis to the circular motion of *any* body under its gravitational attraction to a stationary body. **Figure 13.17** shows an example.

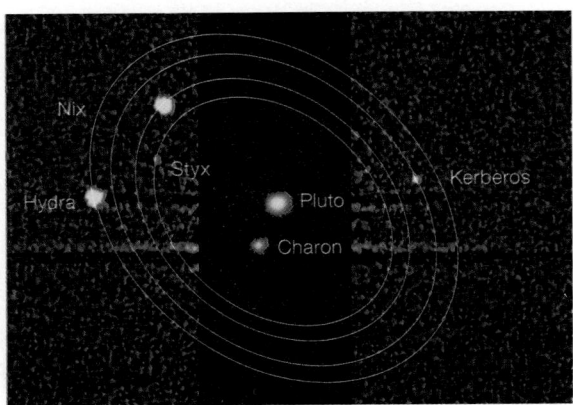

13.17 The dwarf planet Pluto is known to have at least five satellites. In accordance with Eqs. (13.10) and (13.12), the larger the satellite's orbit around Pluto, the slower the satellite's orbital speed and the longer its orbital period.

EXAMPLE 13.6 **A SATELLITE ORBIT**

You wish to put a 1000-kg satellite into a circular orbit 300 km above the earth's surface. (a) What speed, period, and radial acceleration will it have? (b) How much work must be done to the satellite to put it in orbit? (c) How much additional work would have to be done to make the satellite escape the earth? The earth's radius and mass are given in Example 13.5 (Section 13.3).

SOLUTION

IDENTIFY and SET UP: The satellite is in a circular orbit, so we can use the equations derived in this section. In part (a), we first find the radius r of the satellite's orbit from its altitude. We then calculate the speed v and period T from Eqs. (13.10) and (13.12) and the acceleration from $a_{\text{rad}} = v^2/r$. In parts (b) and (c), the work required is the difference between the initial and final mechanical energies, which for a circular orbit is given by Eq. (13.13).

EXECUTE: (a) The radius of the satellite's orbit is $r = 6370$ km + 300 km = 6670 km = 6.67×10^6 m. From Eq. (13.10), the orbital speed is

$$v = \sqrt{\frac{Gm_E}{r}} = \sqrt{\frac{(6.67 \times 10^{-11} \text{ N} \cdot \text{m}^2/\text{kg}^2)(5.97 \times 10^{24} \text{ kg})}{6.67 \times 10^6 \text{ m}}}$$

$$= 7730 \text{ m/s}$$

We find the orbital period from Eq. (13.12):

$$T = \frac{2\pi r}{v} = \frac{2\pi (6.67 \times 10^6 \text{ m})}{7730 \text{ m/s}} = 5420 \text{ s} = 90.4 \text{ min}$$

Finally, the radial acceleration is

$$a_{\text{rad}} = \frac{v^2}{r} = \frac{(7730 \text{ m/s})^2}{6.67 \times 10^6 \text{ m}} = 8.96 \text{ m/s}^2$$

This is the value of g at a height of 300 km above the earth's surface; it is about 10% less than the value of g at the surface.

(b) The work required is the difference between E_2, the total mechanical energy when the satellite is in orbit, and E_1, the total mechanical energy when the satellite was at rest on the launch pad. From Eq. (13.13), the energy in orbit is

$$E_2 = -\frac{Gm_E m}{2r}$$

$$= -\frac{(6.67 \times 10^{-11} \text{ N} \cdot \text{m}^2/\text{kg}^2)(5.97 \times 10^{24} \text{ kg})(1000 \text{ kg})}{2(6.67 \times 10^6 \text{ m})}$$

$$= -2.98 \times 10^{10} \text{ J}$$

The satellite's kinetic energy is zero on the launch pad ($r = R_E$), so

$$E_1 = K_1 + U_1 = 0 + \left(-\frac{Gm_E m}{R_E} \right)$$

$$= -\frac{(6.67 \times 10^{-11} \text{ N} \cdot \text{m}^2/\text{kg}^2)(5.97 \times 10^{24} \text{ kg})(1000 \text{ kg})}{6.37 \times 10^6 \text{ m}}$$

$$= -6.25 \times 10^{10} \text{ J}$$

Hence the work required is

$$W_{\text{required}} = E_2 - E_1 = (-2.98 \times 10^{10} \text{ J}) - (-6.25 \times 10^{10} \text{ J})$$

$$= 3.27 \times 10^{10} \text{ J}$$

(c) We saw in part (b) of Example 13.5 that the minimum total mechanical energy for a satellite to escape to infinity is zero. Here, the total mechanical energy in the circular orbit is $E_2 = -2.98 \times 10^{10}$ J; to increase this to zero, an amount of work equal to 2.98×10^{10} J would have to be done on the satellite, presumably by rocket engines attached to it.

EVALUATE: In part (b) we ignored the satellite's initial kinetic energy (while it was still on the launch pad) due to the rotation of the earth. How much difference does this make? (See Example 13.5 for useful data.)

TEST YOUR UNDERSTANDING OF SECTION 13.4 Your personal spacecraft is in a low-altitude circular orbit around the earth. Air resistance from the outer regions of the atmosphere does negative work on the spacecraft, causing the radius of the circular orbit to decrease slightly. Does the speed of the spacecraft (i) remain the same, (ii) increase, or (iii) decrease? ❙

13.5 KEPLER'S LAWS AND THE MOTION OF PLANETS

The name *planet* comes from a Greek word meaning "wanderer," and indeed the planets continuously change their positions in the sky relative to the background of stars. One of the great intellectual accomplishments of the 16th and 17th centuries was the threefold realization that the earth is also a planet, that all planets orbit the sun, and that the apparent motions of the planets as seen from the earth can be used to determine their orbits precisely.

The first and second of these ideas were published by Nicolaus Copernicus in Poland in 1543. The nature of planetary orbits was deduced between 1601 and 1619 by the German astronomer and mathematician Johannes Kepler, using precise data on apparent planetary motions compiled by his mentor, the Danish astronomer Tycho Brahe. By trial and error, Kepler discovered three empirical laws that accurately described the motions of the planets:

1. **Each planet moves in an elliptical orbit, with the sun at one focus of the ellipse.**
2. **A line from the sun to a given planet sweeps out equal areas in equal times.**
3. **The periods of the planets are proportional to the $\frac{3}{2}$ powers of the major axis lengths of their orbits.**

Kepler did not know *why* the planets moved in this way. Three generations later, when Newton turned his attention to the motion of the planets, he discovered that each of Kepler's laws can be *derived;* they are consequences of Newton's laws of motion and the law of gravitation. Let's see how each of Kepler's laws arises.

Kepler's First Law

First consider the elliptical orbits described in Kepler's first law. **Figure 13.18** shows the geometry of an ellipse. The longest dimension is the *major axis,* with half-length a; this half-length is called the **semi-major axis.** The sum of the distances from S to P and from S' to P is the same for all points on the curve. S and S' are the *foci* (plural of *focus*). The sun is at S (*not* at the center of the ellipse) and the planet is at P; we think of both as points because the size of each is very small in comparison to the distance between them. There is nothing at the other focus, S'.

The distance of each focus from the center of the ellipse is ea, where e is a dimensionless number between 0 and 1 called the **eccentricity.** If $e = 0$, the two foci coincide and the ellipse is a circle. The actual orbits of the planets are fairly circular; their eccentricities range from 0.007 for Venus to 0.206 for Mercury. (The earth's orbit has $e = 0.017$.) The point in the planet's orbit closest to the sun is the *perihelion,* and the point most distant is the *aphelion.*

Newton showed that for a body acted on by an attractive force proportional to $1/r^2$, the only possible closed orbits are a circle or an ellipse; he also showed that open orbits (trajectories 6 and 7 in Fig. 13.14) must be parabolas or hyperbolas. These results can be derived from Newton's laws and the law of gravitation, together with a lot more differential equations than we're ready for.

Kepler's Second Law

Figure 13.19 shows Kepler's second law. In a small time interval dt, the line from the sun S to the planet P turns through an angle $d\theta$. The area swept out is the colored triangle with height r, base length $r\,d\theta$, and area $dA = \frac{1}{2}r^2\,d\theta$ in Fig. 13.19b. The rate at which area is swept out, dA/dt, is called the *sector velocity:*

$$\frac{dA}{dt} = \frac{1}{2}r^2\frac{d\theta}{dt} \tag{13.14}$$

The essence of Kepler's second law is that the sector velocity has the same value at all points in the orbit. When the planet is close to the sun, r is small and $d\theta/dt$ is large; when the planet is far from the sun, r is large and $d\theta/dt$ is small.

To see how Kepler's second law follows from Newton's laws, we express dA/dt in terms of the velocity vector $\vec{v}$ of the planet P. The component of $\vec{v}$ perpendicular to the radial line is $v_\perp = v\sin\phi$. From Fig. 13.19b the displacement along the direction of $v_\perp$ during time dt is $r\,d\theta$, so we also have $v_\perp = r\,d\theta/dt$. Using this relationship in Eq. (13.14), we find

$$\frac{dA}{dt} = \frac{1}{2}rv\sin\phi \qquad \text{(sector velocity)} \tag{13.15}$$

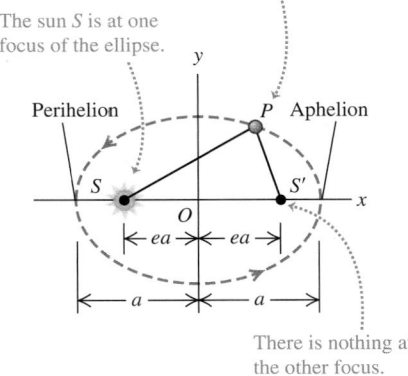

13.18 Geometry of an ellipse. The sum of the distances SP and $S'P$ is the same for every point on the curve. The sizes of the sun (S) and planet (P) are exaggerated for clarity.

A planet P follows an elliptical orbit.

The sun S is at one focus of the ellipse.

Perihelion P Aphelion

There is nothing at the other focus.

13.19 (a) The planet (P) moves about the sun (S) in an elliptical orbit. (b) In a time dt the line SP sweeps out an area $dA = \frac{1}{2}(r\,d\theta)r = \frac{1}{2}r^2\,d\theta$. (c) The planet's speed varies so that the line SP sweeps out the same area A in a given time t regardless of the planet's position in its orbit.

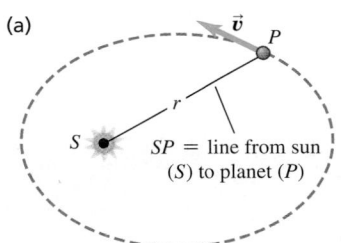

(a)

$SP =$ line from sun (S) to planet (P)

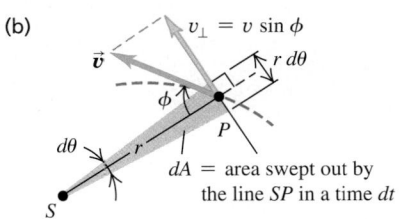

(b)

$v_\perp = v\sin\phi$

$dA =$ area swept out by the line SP in a time dt

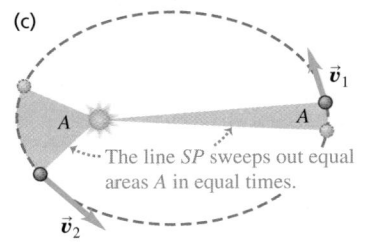

(c)

The line SP sweeps out equal areas A in equal times.

13.20 Because the gravitational force that the sun exerts on a planet produces zero torque around the sun, the planet's angular momentum around the sun remains constant.

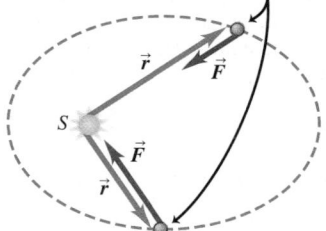

Same planet at two points in its orbit

- Gravitational force $\vec{F}$ on planet has different magnitudes at different points but is always opposite to vector $\vec{r}$ from sun S to planet.
- Hence $\vec{F}$ produces zero torque around sun.

BIO Application Biological Hazards of Interplanetary Travel A spacecraft sent from earth to another planet spends most of its journey coasting along an elliptical orbit with the sun at one focus. Rockets are used at only the start and end of the journey, and even the trip to a nearby planet like Mars takes several months. During its journey, the spacecraft is exposed to cosmic rays—radiation that emanates from elsewhere in our galaxy. (On earth we're shielded from this radiation by our planet's magnetic field, as we'll describe in Chapter 27.) This poses no problem for a robotic spacecraft but would be a severe medical hazard for astronauts undertaking such a voyage.

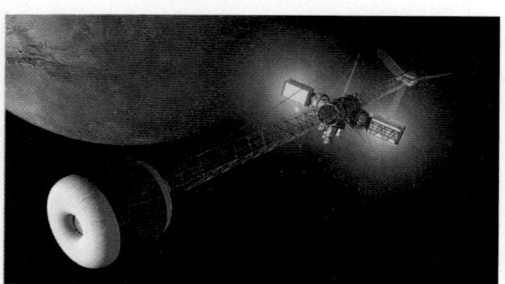

Now $rv\sin\phi$ is the magnitude of the vector product $\vec{r} \times \vec{v}$, which in turn is $1/m$ times the angular momentum $\vec{L} = \vec{r} \times m\vec{v}$ of the planet with respect to the sun. So we have

$$\frac{dA}{dt} = \frac{1}{2m}|\vec{r} \times m\vec{v}| = \frac{L}{2m} \tag{13.16}$$

Thus Kepler's second law—that sector velocity is constant—means that angular momentum is constant!

It is easy to see why the angular momentum of the planet *must* be constant. According to Eq. (10.26), the rate of change of $\vec{L}$ equals the torque of the gravitational force $\vec{F}$ acting on the planet:

$$\frac{d\vec{L}}{dt} = \vec{\tau} = \vec{r} \times \vec{F}$$

In our situation, $\vec{r}$ is the vector from the sun to the planet, and the force $\vec{F}$ is directed from the planet to the sun (**Fig. 13.20**). So these vectors always lie along the same line, and their vector product $\vec{r} \times \vec{F}$ is zero. Hence $d\vec{L}/dt = \mathbf{0}$. This conclusion does not depend on the $1/r^2$ behavior of the force; angular momentum is conserved for *any* force that acts always along the line joining the particle to a fixed point. Such a force is called a *central force*. (Kepler's first and third laws are valid for a $1/r^2$ force *only*.)

Conservation of angular momentum also explains why the orbit lies in a plane. The vector $\vec{L} = \vec{r} \times m\vec{v}$ is always perpendicular to the plane of the vectors $\vec{r}$ and $\vec{v}$; since $\vec{L}$ is constant in magnitude *and* direction, $\vec{r}$ and $\vec{v}$ always lie in the same plane, which is just the plane of the planet's orbit.

Kepler's Third Law

We have already derived Kepler's third law for the particular case of circular orbits. Equation (13.12) shows that the period of a satellite or planet in a circular orbit is proportional to the $\frac{3}{2}$ power of the orbit radius. Newton was able to show that this same relationship holds for an *elliptical* orbit, with the orbit radius r replaced by the semi-major axis a:

$$T = \frac{2\pi a^{3/2}}{\sqrt{Gm_S}} \qquad \text{(elliptical orbit around the sun)} \tag{13.17}$$

Since the planet orbits the sun, not the earth, we have replaced the earth's mass m_E in Eq. (13.12) with the sun's mass m_S. Note that the period does not depend on the eccentricity e. An asteroid in an elongated elliptical orbit with semi-major axis a will have the same orbital period as a planet in a circular orbit of radius a. The key difference is that the asteroid moves at different speeds at different points in its elliptical orbit (Fig. 13.19c), while the planet's speed is constant around its circular orbit.

CONCEPTUAL EXAMPLE 13.7 ORBITAL SPEEDS

At what point in an elliptical orbit (see Fig. 13.19) does a planet move the fastest? The slowest?

SOLUTION

Mechanical energy is conserved as a planet moves in its orbit. The planet's kinetic energy $K = \frac{1}{2}mv^2$ is maximum when the potential energy $U = -Gm_S m/r$ is minimum (that is, most negative; see Fig. 13.11), which occurs when the sun–planet distance r is a minimum. Hence the speed v is greatest at perihelion. Similarly, K is minimum when r is maximum, so the speed is slowest at aphelion.

Your intuition about falling bodies is helpful here. As the planet falls inward toward the sun, it picks up speed, and its speed is maximum when closest to the sun. The planet slows down as it moves away from the sun, and its speed is minimum at aphelion.

EXAMPLE 13.8   KEPLER'S THIRD LAW

The asteroid Pallas has an orbital period of 4.62 years and an orbital eccentricity of 0.233. Find the semi-major axis of its orbit.

SOLUTION

IDENTIFY and SET UP: We need Kepler's third law, which relates the period T and the semi-major axis a for an orbiting object (such as an asteroid). We use Eq. (13.17) to determine a; from Appendix F we have $m_S = 1.99 \times 10^{30}$ kg, and a conversion factor from Appendix E gives $T = (4.62 \text{ yr})(3.156 \times 10^7 \text{ s/yr}) = 1.46 \times 10^8$ s. Note that we don't need the value of the eccentricity.

EXECUTE: From Eq. (13.17), $a^{3/2} = [(Gm_S)^{1/2}T]/2\pi$. To solve for a, we raise both sides of this expression to the $\frac{2}{3}$ power and then substitute the values of G, m_S, and T:

$$a = \left(\frac{Gm_S T^2}{4\pi^2}\right)^{1/3} = 4.15 \times 10^{11} \text{ m}$$

(Plug in the numbers yourself to check.)

EVALUATE: Our result is intermediate between the semi-major axes of Mars and Jupiter (see Appendix F). Most known asteroids orbit in an "asteroid belt" between the orbits of these two planets.

EXAMPLE 13.9   COMET HALLEY

Comet Halley moves in an elongated elliptical orbit around the sun (**Fig. 13.21**). Its distances from the sun at perihelion and aphelion are 8.75×10^7 km and 5.26×10^9 km, respectively. Find the orbital semi-major axis, eccentricity, and period.

SOLUTION

IDENTIFY and SET UP: We are to find the semi-major axis a, eccentricity e, and orbital period T. We can use Fig. 13.18 to find a and e from the given perihelion and aphelion distances. Knowing a, we can find T from Kepler's third law, Eq. (13.17).

EXECUTE: From Fig. 13.18, the length $2a$ of the major axis equals the sum of the comet–sun distance at perihelion and the comet–sun distance at aphelion. Hence

$$a = \frac{(8.75 \times 10^7 \text{ km}) + (5.26 \times 10^9 \text{ km})}{2} = 2.67 \times 10^9 \text{ km}$$

Figure 13.18 also shows that the comet–sun distance at perihelion is $a - ea = a(1 - e)$. This distance is 8.75×10^7 km, so

$$e = 1 - \frac{8.75 \times 10^7 \text{ km}}{a} = 1 - \frac{8.75 \times 10^7 \text{ km}}{2.67 \times 10^9 \text{ km}} = 0.967$$

From Eq. (13.17), the period is

$$T = \frac{2\pi a^{3/2}}{\sqrt{Gm_S}} = \frac{2\pi(2.67 \times 10^{12} \text{ m})^{3/2}}{\sqrt{(6.67 \times 10^{-11} \text{ N} \cdot \text{m}^2/\text{kg}^2)(1.99 \times 10^{30} \text{ kg})}}$$

$$= 2.38 \times 10^9 \text{ s} = 75.5 \text{ years}$$

EVALUATE: The eccentricity is close to 1, so the orbit is very elongated (see Fig. 13.21a). Comet Halley was at perihelion in early 1986 (Fig. 13.21b); it will next reach perihelion one period later, in 2061.

13.21 (a) The orbit of Comet Halley. (b) Comet Halley as it appeared in 1986. At the heart of the comet is an icy body, called the nucleus, that is about 10 km across. When the comet's orbit carries it close to the sun, the heat of sunlight causes the nucleus to partially evaporate. The evaporated material forms the tail, which can be tens of millions of kilometers long.

(a)

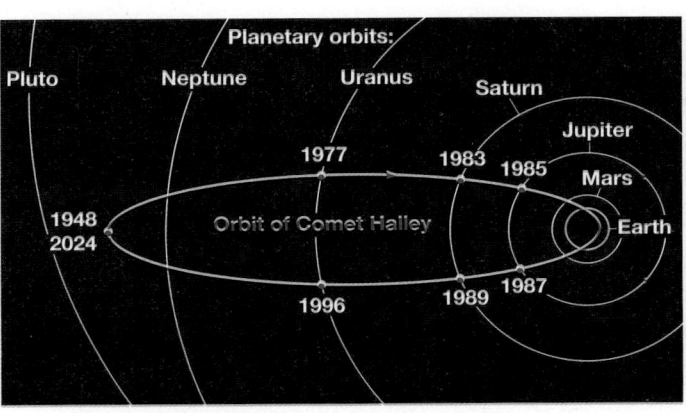

(b)

13.22 Both a star and its planet orbit about their common center of mass.

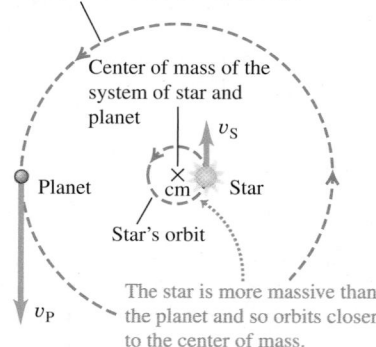

Planet's orbit around the center of mass

Center of mass of the system of star and planet

v_S

Planet

cm

Star

Star's orbit

v_P

The star is more massive than the planet and so orbits closer to the center of mass.

The planet and star are always on opposite sides of the center of mass.

13.23 Calculating the gravitational potential energy of interaction between a point mass m outside a spherical shell and a ring on the surface of the shell of mass M.

(a) Geometry of the situation

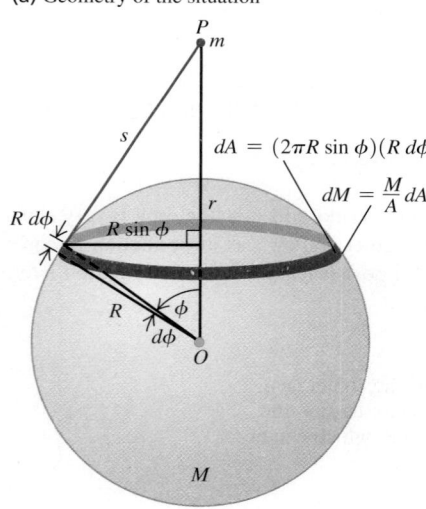

P
m

s

$dA = (2\pi R \sin \phi)(R\, d\phi)$

$dM = \dfrac{M}{A} dA$

$R\, d\phi$

$R \sin \phi$

r

R

ϕ

$d\phi$

O

M

(b) The distance s is the hypotenuse of a right triangle with sides $(r - R \cos \phi)$ and $R \sin \phi$.

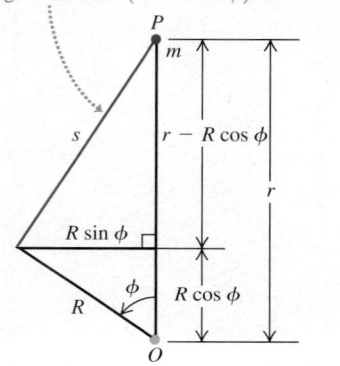

P
m

s

$r - R \cos \phi$

r

$R \sin \phi$

ϕ

$R \cos \phi$

R

O

Planetary Motions and the Center of Mass

We have assumed that as a planet or comet orbits the sun, the sun remains absolutely stationary. This can't be correct; because the sun exerts a gravitational force on the planet, the planet exerts a gravitational force on the sun of the same magnitude but opposite direction. In fact, *both* the sun and the planet orbit around their common center of mass (**Fig. 13.22**). We've made only a small error by ignoring this effect, however; the sun's mass is about 750 times the total mass of all the planets combined, so the center of mass of the solar system is not far from the center of the sun. Remarkably, astronomers have used this effect to detect the presence of planets orbiting other stars. Sensitive telescopes are able to detect the apparent "wobble" of a star as it orbits the common center of mass of the star and an unseen companion planet. (The planets are too faint to observe directly.) By analyzing these "wobbles," astronomers have discovered planets in orbit around hundreds of other stars.

The most remarkable result of Newton's analysis of planetary motion is that bodies in the heavens obey the *same* laws of motion as do bodies on the earth. This *Newtonian synthesis,* as it has come to be called, is one of the great unifying principles of science. It has had profound effects on the way that humanity looks at the universe—not as a realm of impenetrable mystery, but as a direct extension of our everyday world, subject to scientific study and calculation.

TEST YOUR UNDERSTANDING OF SECTION 13.5 The orbit of Comet X has a semi-major axis that is four times longer than the semi-major axis of Comet Y. What is the ratio of the orbital period of X to the orbital period of Y? (i) 2; (ii) 4; (iii) 8; (iv) 16; (v) 32; (vi) 64. ∎

13.6 SPHERICAL MASS DISTRIBUTIONS

We have stated without proof that the gravitational interaction between two spherically symmetric mass distributions is the same as though all the mass of each were concentrated at its center. Now we're ready to prove this statement. Newton searched for a proof for several years, and he delayed publication of the law of gravitation until he found one.

Rather than starting with two spherically symmetric masses, we'll tackle the simpler problem of a point mass m interacting with a thin spherical shell with total mass M. We'll show that when m is outside the sphere, the *potential energy* associated with this gravitational interaction is the same as though M were concentrated in a point at the center of the sphere. We learned in Section 7.4 that the force is the negative derivative of the potential energy, so the *force* on m is also the same as for a point mass M. Our result will also hold for *any* spherically symmetric mass distribution, which we can think of as being made of many concentric spherical shells.

A Point Mass Outside a Spherical Shell

We start by considering a ring on the surface of a shell (**Fig. 13.23a**), centered on the line from the center of the shell to m. We do this because all of the particles that make up the ring are the same distance s from the point mass m. From Eq. (13.9) the potential energy of interaction between the earth (mass m_E) and a point mass m, separated by a distance r, is $U = -Gm_E m/r$. From this expression, we see that the potential energy of interaction between the point mass m and a particle of mass m_i within the ring is

$$U_i = -\frac{Gmm_i}{s}$$

To find the potential energy dU of interaction between m and the entire ring of mass $dM = \sum_i m_i$, we sum this expression for U_i over all particles in the ring:

$$dU = \sum_i U_i = \sum_i \left(-\frac{Gmm_i}{s} \right) = -\frac{Gm}{s} \sum_i m_i = -\frac{Gm\, dM}{s} \qquad (13.18)$$

To proceed, we need to know the mass dM of the ring. We can find this with the aid of a little geometry. The radius of the shell is R, so in terms of the angle ϕ shown in the figure, the radius of the ring is $R\sin\phi$, and its circumference is $2\pi R\sin\phi$. The width of the ring is $R\, d\phi$, and its area dA is approximately equal to its width times its circumference:

$$dA = 2\pi R^2 \sin\phi\, d\phi$$

The ratio of the ring mass dM to the total mass M of the shell is equal to the ratio of the area dA of the ring to the total area $A = 4\pi R^2$ of the shell:

$$\frac{dM}{M} = \frac{2\pi R^2 \sin\phi\, d\phi}{4\pi R^2} = \tfrac{1}{2} \sin\phi\, d\phi \qquad (13.19)$$

Now we solve Eq. (13.19) for dM and substitute the result into Eq. (13.18) to find the potential energy of interaction between point mass m and the ring:

$$dU = -\frac{GMm \sin\phi\, d\phi}{2s} \qquad (13.20)$$

The total potential energy of interaction between the point mass and the *shell* is the integral of Eq. (13.20) over the whole sphere as ϕ varies from 0 to π (*not* 2π!) and s varies from $r - R$ to $r + R$. To carry out the integration, we have to express the integrand in terms of a single variable; we choose s. To express ϕ and $d\phi$ in terms of s, we have to do a little more geometry. Figure 13.23b shows that s is the hypotenuse of a right triangle with sides $(r - R\cos\phi)$ and $R\sin\phi$, so the Pythagorean theorem gives

$$s^2 = (r - R\cos\phi)^2 + (R\sin\phi)^2$$
$$= r^2 - 2rR\cos\phi + R^2 \qquad (13.21)$$

We take differentials of both sides:

$$2s\, ds = 2rR\sin\phi\, d\phi$$

Next we divide this by $2rR$ and substitute the result into Eq. (13.20):

$$dU = -\frac{GMm}{2s}\frac{s\, ds}{rR} = -\frac{GMm}{2rR}\, ds \qquad (13.22)$$

We can now integrate Eq. (13.22), recalling that s varies from $r - R$ to $r + R$:

$$U = -\frac{GMm}{2rR} \int_{r-R}^{r+R} ds = -\frac{GMm}{2rR} [(r + R) - (r - R)] \qquad (13.23)$$

Finally, we have

$$U = -\frac{GMm}{r} \qquad \text{(point mass } m \text{ outside spherical shell } M\text{)} \qquad (13.24)$$

This is equal to the potential energy of two point masses m and M at a distance r. So we have proved that the gravitational potential energy of spherical shell M and point mass m at any distance r is the same as though they were point masses. Because the force is given by $F_r = -dU/dr$, the force is also the same.

The Gravitational Force Between Spherical Mass Distributions

Any spherically symmetric mass distribution can be thought of as a combination of concentric spherical shells. Because of the principle of superposition of forces, what is true of one shell is also true of the combination. So we have proved half of what we set out to prove: that the gravitational interaction between any spherically symmetric mass distribution and a point mass is the same as though all the mass of the spherically symmetric distribution were concentrated at its center.

The other half is to prove that *two* spherically symmetric mass distributions interact as though both were points. That's easier. In Fig. 13.23a the forces the two bodies exert on each other are an action–reaction pair, and they obey Newton's third law. So we have also proved that the force that m exerts *on* sphere M is the same as though M were a point. But now if we replace m with a spherically symmetric mass distribution centered at m's location, the resulting gravitational force on any part of M is the same as before, and so is the total force. This completes our proof.

A Point Mass Inside a Spherical Shell

We assumed at the beginning that the point mass m was outside the spherical shell, so our proof is valid only when m is outside a spherically symmetric mass distribution. When m is *inside* a spherical shell, the geometry is as shown in **Fig. 13.24**. The entire analysis goes just as before; Eqs. (13.18) through (13.22) are still valid. But when we get to Eq. (13.23), the limits of integration have to be changed to $R - r$ and $R + r$. We then have

$$U = -\frac{GMm}{2rR} \int_{R-r}^{R+r} ds = -\frac{GMm}{2rR}[(R + r) - (R - r)] \qquad (13.25)$$

and the final result is

$$U = -\frac{GMm}{R} \qquad \text{(point mass } m \text{ inside spherical shell } M) \qquad (13.26)$$

Compare this result to Eq. (13.24): Instead of having r, the distance between m and the center of M, in the denominator, we have R, the radius of the shell. This means that U in Eq. (13.26) doesn't depend on r and thus has the same value everywhere inside the shell. When m moves around inside the shell, no work is done on it, so the force on m at any point inside the shell must be zero.

More generally, at any point in the interior of any spherically symmetric mass distribution (not necessarily a shell), at a distance r from its center, the gravitational force on a point mass m is the same as though we removed all the mass at points farther than r from the center and concentrated all the remaining mass at the center.

13.24 When a point mass m is *inside* a uniform spherical shell of mass M, the potential energy is the same no matter where inside the shell the point mass is located. The force from the masses' mutual gravitational interaction is zero.

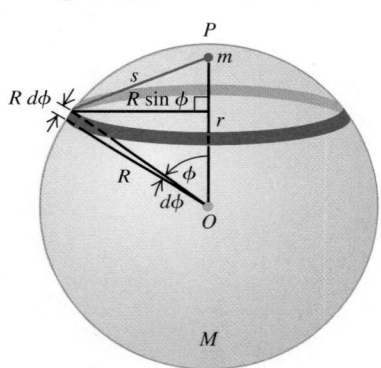

EXAMPLE 13.10   "JOURNEY TO THE CENTER OF THE EARTH"

Imagine that we drill a hole through the earth along a diameter and drop a mail pouch down the hole. Derive an expression for the gravitational force F_g on the pouch as a function of its distance from the earth's center. Assume that the earth's density is uniform (not a very realistic model; see Fig. 13.9).

SOLUTION

IDENTIFY and SET UP: From the discussion immediately above, the value of F_g at a distance r from the earth's center is determined by only the mass M within a spherical region of radius r

(**Fig. 13.25**). Hence F_g is the same as if all the mass within radius r were concentrated at the center of the earth. The mass of a uniform sphere is proportional to the volume of the sphere, which is $\frac{4}{3}\pi r^3$ for a sphere of arbitrary radius r and $\frac{4}{3}\pi R_E^3$ for the entire earth.

EXECUTE: The ratio of the mass M of the sphere of radius r to the mass m_E of the earth is

$$\frac{M}{m_E} = \frac{\frac{4}{3}\pi r^3}{\frac{4}{3}\pi R_E^3} = \frac{r^3}{R_E^3} \qquad \text{so} \qquad M = m_E \frac{r^3}{R_E^3}$$

13.25 A hole through the center of the earth (assumed to be uniform). When an object is a distance r from the center, only the mass inside a sphere of radius r exerts a net gravitational force on it.

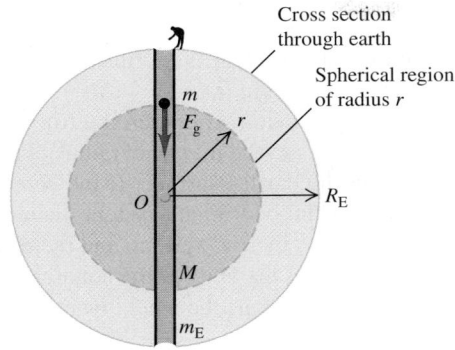

The magnitude of the gravitational force on m is then

$$F_{\text{g}} = \frac{GMm}{r^2} = \frac{Gm}{r^2}\left(m_{\text{E}}\frac{r^3}{R_{\text{E}}^3}\right) = \frac{Gm_{\text{E}}m}{R_{\text{E}}^3}r$$

EVALUATE: Inside this uniform-density sphere, F_{g} is *directly proportional* to the distance r from the center, rather than to $1/r^2$ as it is outside the sphere. At the surface $r = R_{\text{E}}$, we have $F_{\text{g}} = Gm_{\text{E}}m/R_{\text{E}}^2$, as we should. In the next chapter we'll learn how to compute the time it would take for the mail pouch to emerge on the other side of the earth.

TEST YOUR UNDERSTANDING OF SECTION 13.6   In the classic 1913 science-fiction novel *At the Earth's Core,* by Edgar Rice Burroughs, explorers discover that the earth is a hollow sphere and that an entire civilization lives on the inside of the sphere. Would it be possible to stand and walk on the inner surface of a hollow, nonrotating planet? ❙

13.7 APPARENT WEIGHT AND THE EARTH'S ROTATION

Because the earth rotates on its axis, it is not precisely an inertial frame of reference. For this reason the apparent weight of a body on earth is not precisely equal to the earth's gravitational attraction, which we will call the **true weight** $\vec{w}_0$ of the body. **Figure 13.26** is a cutaway view of the earth, showing three observers. Each one holds a spring scale with a body of mass m hanging from it. Each scale applies a tension force $\vec{F}$ to the body hanging from it, and the reading on each

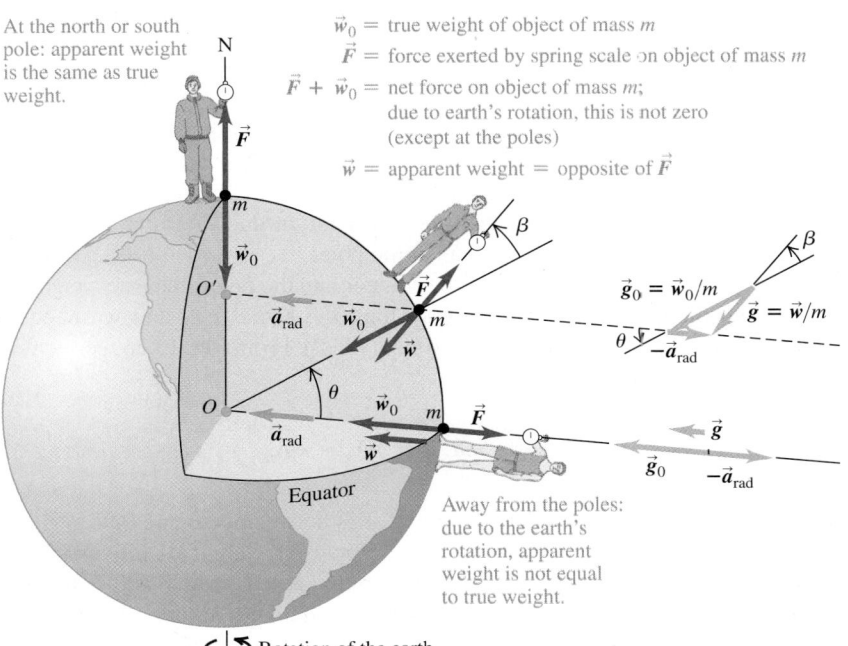

13.26 Except at the poles, the reading for an object being weighed on a scale (the *apparent weight*) is less than the gravitational force of attraction on the object (the *true weight*). The reason is that a net force is needed to provide a centripetal acceleration as the object rotates with the earth. For clarity, the illustration greatly exaggerates the angle β between the true and apparent weight vectors.

scale is the magnitude F of this force. If the observers are unaware of the earth's rotation, each one *thinks* that the scale reading equals the weight of the body because he thinks the body on his spring scale is in equilibrium. So each observer thinks that the tension $\vec{F}$ must be opposed by an equal and opposite force $\vec{w}$, which we call the **apparent weight.** But if the bodies are rotating with the earth, they are *not* precisely in equilibrium. Our problem is to find the relationship between the apparent weight $\vec{w}$ and the true weight $\vec{w}_0$.

If we assume that the earth is spherically symmetric, then the true weight $\vec{w}_0$ has magnitude Gm_Em/R_E^2, where m_E and R_E are the mass and radius of the earth. This value is the same for all points on the earth's surface. If the center of the earth can be taken as the origin of an inertial coordinate system, then the body at the north pole really *is* in equilibrium in an inertial system, and the reading on that observer's spring scale is equal to w_0. But the body at the equator is moving in a circle of radius R_E with speed v, and there must be a net inward force equal to the mass times the centripetal acceleration:

$$w_0 - F = \frac{mv^2}{R_E}$$

So the magnitude of the apparent weight (equal to the magnitude of F) is

$$w = w_0 - \frac{mv^2}{R_E} \qquad \text{(at the equator)} \qquad (13.27)$$

If the earth were not rotating, the body when released would have a free-fall acceleration $g_0 = w_0/m$. Since the earth *is* rotating, the falling body's actual acceleration relative to the observer at the equator is $g = w/m$. Dividing Eq. (13.27) by m and using these relationships, we find

$$g = g_0 - \frac{v^2}{R_E} \qquad \text{(at the equator)}$$

To evaluate v^2/R_E, we note that in 86,164 s a point on the equator moves a distance equal to the earth's circumference, $2\pi R_E = 2\pi(6.37 \times 10^6 \text{ m})$. (The solar day, 86,400 s, is $\frac{1}{365}$ longer than this because in one day the earth also completes $\frac{1}{365}$ of its orbit around the sun.) Thus we find

$$v = \frac{2\pi(6.37 \times 10^6 \text{ m})}{86,164 \text{ s}} = 465 \text{ m/s}$$

$$\frac{v^2}{R_E} = \frac{(465 \text{ m/s})^2}{6.37 \times 10^6 \text{ m}} = 0.0339 \text{ m/s}^2$$

So for a spherically symmetric earth the acceleration due to gravity should be about 0.03 m/s² less at the equator than at the poles.

At locations intermediate between the equator and the poles, the true weight $\vec{w}_0$ and the centripetal acceleration are not along the same line, and we need to write a vector equation corresponding to Eq. (13.27). From Fig. 13.26 we see that the appropriate equation is

$$\vec{w} = \vec{w}_0 - m\vec{a}_{rad} = m\vec{g}_0 - m\vec{a}_{rad} \qquad (13.28)$$

The difference in the magnitudes of g and g_0 lies between zero and 0.0339 m/s². As Fig. 13.26 shows, the *direction* of the apparent weight differs from the direction toward the center of the earth by a small angle β, which is 0.1° or less.

Table 13.1 gives the values of g at several locations. In addition to moderate variations with latitude, there are small variations due to elevation, differences in local density, and the earth's deviation from perfect spherical symmetry.

Variations of g with

TABLE 13.1 | Latitude and Elevation

Station	North Latitude	Elevation (m)	g (m/s²)
Canal Zone	09°	0	9.78243
Jamaica	18°	0	9.78591
Bermuda	32°	0	9.79806
Denver, CO	40°	1638	9.79609
Pittsburgh, PA	40.5°	235	9.80118
Cambridge, MA	42°	0	9.80398
Greenland	70°	0	9.82534

TEST YOUR UNDERSTANDING OF SECTION 13.7 Imagine a planet that has the same mass and radius as the earth but that makes 10 rotations during the time the earth makes one rotation. What would be the difference between the acceleration due to gravity at the planet's equator and the acceleration due to gravity at its poles? (i) 0.00339 m/s²; (ii) 0.0339 m/s²; (iii) 0.339 m/s²; (iv) 3.39 m/s². ❚

13.8 BLACK HOLES

In 1916 Albert Einstein presented his general theory of relativity, which included a new concept of the nature of gravitation. In his theory, a massive object actually changes the geometry of the space around it. Other objects sense this altered geometry and respond by being attracted to the first object. The general theory of relativity is beyond our scope in this chapter, but we can look at one of its most startling predictions: the existence of **black holes,** objects whose gravitational influence is so great that nothing—not even light—can escape them. We can understand the basic idea of a black hole by using Newtonian principles.

The Escape Speed from a Star

Think first about the properties of our own sun. Its mass $M = 1.99 \times 10^{30}$ kg and radius $R = 6.96 \times 10^8$ m are much larger than those of any planet, but compared to other stars, our sun is not exceptionally massive. You can find the sun's average density ρ in the same way we found the average density of the earth in Section 13.2:

$$\rho = \frac{M}{V} = \frac{M}{\frac{4}{3}\pi R^3} = \frac{1.99 \times 10^{30} \text{ kg}}{\frac{4}{3}\pi(6.96 \times 10^8 \text{ m})^3} = 1410 \text{ kg/m}^3$$

The sun's temperatures range from 5800 K (about 5500°C or 10,000°F) at the surface up to 1.5×10^7 K (about 2.7×10^{7}°F) in the interior, so it surely contains no solids or liquids. Yet gravitational attraction pulls the sun's gas atoms together until the sun is, on average, 41% denser than water and about 1200 times as dense as the air we breathe.

Now think about the escape speed for a body at the surface of the sun. In Example 13.5 (Section 13.3) we found that the escape speed from the surface of a spherical mass M with radius R is $v = \sqrt{2GM/R}$. Substituting $M = \rho V = \rho\left(\frac{4}{3}\pi R^3\right)$ into the expression for escape speed gives

$$v = \sqrt{\frac{2GM}{R}} = \sqrt{\frac{8\pi G\rho}{3}}R \qquad (13.29)$$

Using either form of this equation, you can show that the escape speed for a body at the surface of our sun is $v = 6.18 \times 10^5$ m/s (about 2.2 million km/h, or 1.4 million mi/h). This value, roughly $\frac{1}{500}$ the speed of light in vacuum, is independent of the mass of the escaping body; it depends on only the mass and radius (or average density and radius) of the sun.

Now consider various stars with the same average density ρ and different radii R. Equation (13.29) shows that for a given value of density ρ, the escape speed v is directly proportional to R. In 1783 the Rev. John Mitchell, an amateur astronomer, noted that if a body with the same average density as the sun had about 500 times the radius of the sun, its escape speed would be greater than the speed of light in vacuum, c. With his statement that "all light emitted from such a body would be made to return toward it," Mitchell became the first person to suggest the existence of what we now call a black hole.

Black Holes, the Schwarzschild Radius, and the Event Horizon

The first expression for escape speed in Eq. (13.29) suggests that a body of mass M will act as a black hole if its radius R is less than or equal to a certain critical radius. How can we determine this critical radius? You might think that you can find the answer by simply setting $v = c$ in Eq. (13.29). As a matter of fact, this does give the correct result, but only because of two compensating errors. The kinetic energy of light is *not* $mc^2/2$, and the gravitational potential energy near a black hole is *not* given by Eq. (13.9). In 1916, Karl Schwarzschild used Einstein's general theory of relativity to derive an expression for the critical radius R_S, now called the **Schwarzschild radius.** The result turns out to be the same as though we had set $v = c$ in Eq. (13.29), so

$$c = \sqrt{\frac{2GM}{R_S}}$$

Solving for the Schwarzschild radius R_S, we find

$$\text{Schwarzschild radius} \cdots \qquad \overset{\cdots \text{Gravitational constant}}{R_S = \frac{2GM}{c^2}} \overset{\cdots \text{Mass of black hole}}{} \qquad (13.30)$$
$$\text{of a black hole} \qquad \qquad \underset{\cdots \text{Speed of light in vacuum}}{}$$

If a spherical, nonrotating body with mass M has a radius less than R_S, then *nothing* (not even light) can escape from the surface of the body, and the body is a black hole (**Fig. 13.27**). In this case, any other body within a distance R_S of the center of the black hole is trapped by the gravitational attraction of the black hole and cannot escape from it.

The surface of the sphere with radius R_S surrounding a black hole is called the **event horizon:** Since light can't escape from within that sphere, we can't see events occurring inside. All that an observer outside the event horizon can know about a black hole is its mass (from its gravitational effects on other bodies), its electric charge (from the electric forces it exerts on other charged bodies), and its angular momentum (because a rotating black hole tends to drag space—and everything in that space—around with it). All other information about the body is irretrievably lost when it collapses inside its event horizon.

13.27 (a) A body with a radius R greater than the Schwarzschild radius R_S. (b) If the body collapses to a radius smaller than R_S, it is a black hole with an escape speed greater than the speed of light. The surface of the sphere of radius R_S is called the event horizon of the black hole.

(a) When the radius R of a body is greater than the Schwarzschild radius R_S, light can escape from the surface of the body.

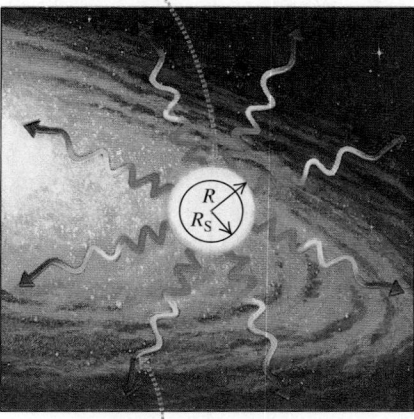

Gravity acting on the escaping light "red shifts" it to longer wavelengths.

(b) If all the mass of the body lies inside radius R_S, the body is a black hole: No light can escape from it.

EXAMPLE 13.11   BLACK HOLE CALCULATIONS

Astrophysical theory suggests that a burned-out star whose mass is at least three solar masses will collapse under its own gravity to form a black hole. If it does, what is the radius of its event horizon?

SOLUTION

IDENTIFY, SET UP, and EXECUTE: The radius in question is the Schwarzschild radius. We use Eq. (13.30) with a value of M equal to three solar masses, or $M = 3(1.99 \times 10^{30}\,\text{kg}) = 6.0 \times 10^{30}\,\text{kg}$:

$$R_S = \frac{2GM}{c^2} = \frac{2(6.67 \times 10^{-11}\,\text{N} \cdot \text{m}^2/\text{kg}^2)(6.0 \times 10^{30}\,\text{kg})}{(3.00 \times 10^8\,\text{m/s})^2}$$

$$= 8.9 \times 10^3\,\text{m} = 8.9\,\text{km} = 5.5\,\text{mi}$$

EVALUATE: The average density of such an object is

$$\rho = \frac{M}{\frac{4}{3}\pi R^3} = \frac{6.0 \times 10^{30}\,\text{kg}}{\frac{4}{3}\pi(8.9 \times 10^3\,\text{m})^3} = 2.0 \times 10^{18}\,\text{kg/m}^3$$

This is about 10^{15} times as great as the density of familiar matter on earth and is comparable to the densities of atomic nuclei. In fact, once the body collapses to a radius of R_S, nothing can prevent it from collapsing further. All of the mass ends up being crushed down to a single point called a *singularity* at the center of the event horizon. This point has zero volume and so has *infinite* density.

A Visit to a Black Hole

At points far from a black hole, its gravitational effects are the same as those of any normal body with the same mass. If the sun collapsed to form a black hole, the orbits of the planets would be unaffected. But things get dramatically different close to the black hole. If you decided to become a martyr for science and jump into a black hole, the friends you left behind would notice several odd effects as you moved toward the event horizon, most of them associated with effects of general relativity.

If you carried a radio transmitter to send back your comments on what was happening, your friends would have to retune their receiver continuously to lower and lower frequencies, an effect called the *gravitational red shift*. Consistent with this shift, they would observe that your clocks (electronic or biological) would appear to run more and more slowly, an effect called *time dilation*. In fact, during their lifetimes they would never see you make it to the event horizon.

In your frame of reference, you would make it to the event horizon in a rather short time but in a rather disquieting way. As you fell feet first into the black hole, the gravitational pull on your feet would be greater than that on your head, which would be slightly farther away from the black hole. The *differences* in gravitational force on different parts of your body would be great enough to stretch you along the direction toward the black hole and compress you perpendicular to it. These effects (called *tidal forces*) would rip you to atoms, and then rip your atoms apart, before you reached the event horizon.

Detecting Black Holes

If light cannot escape from a black hole and if black holes are as small as Example 13.11 suggests, how can we know that such things exist? The answer is that any gas or dust near the black hole tends to be pulled into an *accretion disk* that swirls around and into the black hole, rather like a whirlpool (**Fig. 13.28,** next page). Friction within the accretion disk's gas causes it to lose mechanical energy and spiral into the black hole; as it moves inward, it is compressed together. This causes heating of the gas, just as air compressed in a bicycle pump gets hotter. Temperatures in excess of 10^6 K can occur in the accretion disk, so hot that the disk emits not just visible light (as do bodies that are "red-hot" or "white-hot") but x rays. Astronomers look for these x rays (emitted by the gas material *before* it crosses the event horizon) to signal the presence of a black hole. Several promising candidates have been found, and astronomers now express considerable confidence in the existence of black holes.

13.28 In a *binary* star system, two stars orbit each other; in the special case shown here, one of the stars is a black hole. The black hole itself cannot be seen, but the x rays from its accretion disk can be detected.

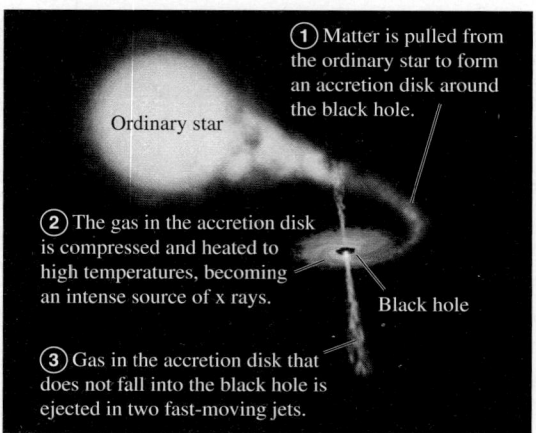

① Matter is pulled from the ordinary star to form an accretion disk around the black hole.

Ordinary star

② The gas in the accretion disk is compressed and heated to high temperatures, becoming an intense source of x rays.

Black hole

③ Gas in the accretion disk that does not fall into the black hole is ejected in two fast-moving jets.

13.29 This false-color image shows the motions of stars at the center of our galaxy over a 17-year period. Analysis of these orbits by using Kepler's third law indicates that the stars are moving about an unseen object that is some 4.1×10^6 times the mass of the sun. The scale bar indicates a length of 10^{14} m (670 times the distance from the earth to the sun).

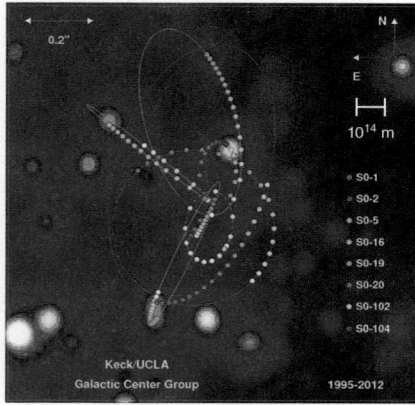

Black holes in binary star systems like the one depicted in Fig. 13.28 have masses a few times greater than the sun's mass. There is also mounting evidence for the existence of much larger *supermassive black holes*. One example lies at the center of our Milky Way galaxy, some 26,000 light-years from earth in the direction of the constellation Sagittarius. High-resolution images of the galactic center reveal stars moving at speeds greater than 1500 km/s about an unseen object that lies at the position of a source of radio waves called Sgr A* (**Fig. 13.29**). By analyzing these motions, astronomers can infer the period T and semi-major axis a of each star's orbit. The mass m_X of the unseen object can be calculated from Kepler's third law in the form given in Eq. (13.17), with the mass of the sun m_S replaced by m_X:

$$T = \frac{2\pi a^{3/2}}{\sqrt{Gm_X}} \qquad \text{so} \qquad m_X = \frac{4\pi^2 a^3}{GT^2}$$

The conclusion is that the mysterious dark object at the galactic center has a mass of 8.2×10^{36} kg, or 4.1 *million* times the mass of the sun. Yet observations with radio telescopes show that it has a radius no more than 4.4×10^{10} m, about one-third of the distance from the earth to the sun. These observations suggest that this massive, compact object is a black hole with a Schwarzschild radius of 1.1×10^{10} m. Astronomers hope to improve the resolution of their observations so that they can actually see the event horizon of this black hole.

Other lines of research suggest that even larger black holes, in excess of 10^9 times the mass of the sun, lie at the centers of other galaxies. Observational and theoretical studies of black holes of all sizes continue to be an exciting area of research in both physics and astronomy.

TEST YOUR UNDERSTANDING OF SECTION 13.8 If the sun somehow collapsed to form a black hole, what effect would this event have on the orbit of the earth? The orbit would (i) shrink; (ii) expand; (iii) remain the same size. ▌

Newton's law of gravitation: *Any* two particles with masses m_1 and m_2, a distance r apart, attract each other with forces inversely proportional to r^2. These forces form an action–reaction pair and obey Newton's third law. When two or more bodies exert gravitational forces on a particular body, the total gravitational force on that individual body is the vector sum of the forces exerted by the other bodies. The gravitational interaction between spherical mass distributions, such as planets or stars, is the same as if all the mass of each distribution were concentrated at the center. (See Examples 13.1–13.3 and 13.10.)

$$F_g = \frac{Gm_1m_2}{r^2} \qquad (13.1)$$

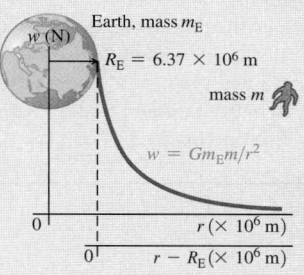

Gravitational force, weight, and gravitational potential energy: The weight w of a body is the total gravitational force exerted on it by all other bodies in the universe. Near the surface of the earth (mass m_E and radius R_E), the weight is essentially equal to the gravitational force of the earth alone. The gravitational potential energy U of two masses m and m_E separated by a distance r is inversely proportional to r. The potential energy is never positive; it is zero only when the two bodies are infinitely far apart. (See Examples 13.4 and 13.5.)

$$w = F_g = \frac{Gm_Em}{R_E^2} \qquad (13.3)$$
(weight at earth's surface)

$$g = \frac{Gm_E}{R_E^2} \qquad (13.4)$$
(acceleration due to gravity at earth's surface)

$$U = -\frac{Gm_Em}{r} \qquad (13.9)$$

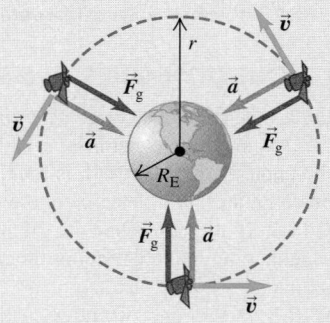

Orbits: When a satellite moves in a circular orbit, the centripetal acceleration is provided by the gravitational attraction of the earth. Kepler's three laws describe the more general case: an elliptical orbit of a planet around the sun or a satellite around a planet. (See Examples 13.6–13.9.)

$$v = \sqrt{\frac{Gm_E}{r}} \qquad (13.10)$$
(speed in circular orbit)

$$T = \frac{2\pi r}{v} = 2\pi r\sqrt{\frac{r}{Gm_E}} = \frac{2\pi r^{3/2}}{\sqrt{Gm_E}}$$
(period in circular orbit) $\qquad (13.12)$

Black holes: If a nonrotating spherical mass distribution with total mass M has a radius less than its Schwarzschild radius R_S, it is called a black hole. The gravitational interaction prevents anything, including light, from escaping from within a sphere with radius R_S. (See Example 13.11.)

$$R_S = \frac{2GM}{c^2} \qquad (13.30)$$
(Schwarzschild radius)

If all of the body is inside its Schwarzschild radius $R_S = 2GM/c^2$, the body is a black hole.

BRIDGING PROBLEM **SPEEDS IN AN ELLIPTICAL ORBIT**

A comet orbits the sun (mass m_S) in an elliptical orbit of semi-major axis a and eccentricity e. (a) Find expressions for the speeds of the comet at perihelion and aphelion. (b) Evaluate these expressions for Comet Halley (see Example 13.9), and find the kinetic energy, gravitational potential energy, and total mechanical energy for this comet at perihelion and aphelion. Take the mass of Comet Halley to be 2.2×10^{14} kg.

SOLUTION GUIDE

IDENTIFY and SET UP

1. Sketch the situation; show all relevant dimensions. Label the perihelion and aphelion. (See Figure 13.18.)
2. List the unknown quantities, and identify the target variables.
3. Just as for a satellite orbiting the earth, the mechanical energy is conserved for a comet orbiting the sun. (Why?) What other quantity is conserved as the comet moves in its orbit? (*Hint:* See Section 13.5.)

EXECUTE

4. You'll need at least two equations that involve the two unknown speeds, and you'll need expressions for the sun–comet distances at perihelion and aphelion. (*Hint:* See Fig. 13.18.)
5. Solve the equations for your target variables. Compare your expressions: Which speed is lower? Does this make sense?
6. Use your expressions from step 5 to find the perihelion and aphelion speeds for Comet Halley. (*Hint:* See Appendix F.)
7. Use your results from step 6 to find the kinetic energy K, gravitational potential energy U, and total mechanical energy E for Comet Halley at perihelion and aphelion.

EVALUATE

8. Check whether your results from part (a) make sense for the special case of a circular orbit ($e = 0$).
9. In part (b), how do your calculated values of E at perihelion and aphelion compare? Does this make sense? What does it mean that E is negative?

Problems

For assigned homework and other learning materials, go to MasteringPhysics®. **MP**

•, ••, •••: Difficulty levels. **CP**: Cumulative problems incorporating material from earlier chapters. **CALC**: Problems requiring calculus. **DATA**: Problems involving real data, scientific evidence, experimental design, and/or statistical reasoning. **BIO**: Biosciences problems.

DISCUSSION QUESTIONS

Q13.1 A student wrote: "The only reason an apple falls downward to meet the earth instead of the earth rising upward to meet the apple is that the earth is much more massive and so exerts a much greater pull." Please comment.

Q13.2 If all planets had the same average density, how would the acceleration due to gravity at the surface of a planet depend on its radius?

Q13.3 Is a pound of butter on the earth the same amount as a pound of butter on Mars? What about a kilogram of butter? Explain.

Q13.4 Example 13.2 (Section 13.1) shows that the acceleration of each sphere caused by the gravitational force is inversely proportional to the mass of that sphere. So why does the force of gravity give all masses the same acceleration when they are dropped near the surface of the earth?

Q13.5 When will you attract the sun more: today at noon, or tonight at midnight? Explain.

Q13.6 Since the moon is constantly attracted toward the earth by the gravitational interaction, why doesn't it crash into the earth?

Q13.7 A spaceship makes a circular orbit with period T around a star. If it were to orbit, at the same distance, a star with three times the mass of the original star, would the new period (in terms of T) be (a) $3T$, (b) $T\sqrt{3}$, (c) T, (d) $T/\sqrt{3}$, or (e) $T/3$?

Q13.8 A planet makes a circular orbit with period T around a star. If the planet were to orbit at the same distance around this star, but the planet had three times as much mass, what would the new period (in terms of T) be: (a) $3T$, (b) $T\sqrt{3}$, (c) T, (d) $T/\sqrt{3}$, or (e) $T/3$?

Q13.9 The sun pulls on the moon with a force that is more than twice the magnitude of the force with which the earth attracts the moon. Why, then, doesn't the sun take the moon away from the earth?

Q13.10 Which takes more fuel: a voyage from the earth to the moon or from the moon to the earth? Explain.

Q13.11 A planet is moving at constant speed in a circular orbit around a star. In one complete orbit, what is the net amount of work done on the planet by the star's gravitational force: positive, negative, or zero? What if the planet's orbit is an ellipse, so that the speed is not constant? Explain your answers.

Q13.12 Does the escape speed for an object at the earth's surface depend on the direction in which it is launched? Explain. Does your answer depend on whether or not you include the effects of air resistance?

Q13.13 If a projectile is fired straight up from the earth's surface, what would happen if the total mechanical energy (kinetic plus potential) is (a) less than zero, and (b) greater than zero? In each case, ignore air resistance and the gravitational effects of the sun, the moon, and the other planets.

Q13.14 Discuss whether this statement is correct: "In the absence of air resistance, the trajectory of a projectile thrown near the earth's surface is an *ellipse,* not a parabola."

Q13.15 The earth is closer to the sun in November than in May. In which of these months does it move faster in its orbit? Explain why.

Q13.16 A communications firm wants to place a satellite in orbit so that it is always directly above the earth's 45th parallel (latitude 45° north). This means that the plane of the orbit will not pass through the center of the earth. Is such an orbit possible? Why or why not?

Q13.17 At what point in an elliptical orbit is the acceleration maximum? At what point is it minimum? Justify your answers.

Q13.18 What would Kepler's third law be for circular orbits if an amendment to Newton's law of gravitation made the gravitational force inversely proportional to r^3? Would this change affect Kepler's other two laws? Explain.

Q13.19 In the elliptical orbit of Comet Halley shown in Fig. 13.21a, the sun's gravity is responsible for making the comet fall inward from aphelion to perihelion. But what is responsible for making the comet move from perihelion back outward to aphelion?

Q13.20 Many people believe that orbiting astronauts feel weightless because they are "beyond the pull of the earth's gravity." How far from the earth would a spacecraft have to travel to be truly beyond the earth's gravitational influence? If a spacecraft were really unaffected by the earth's gravity, would it remain in orbit? Explain. What is the real reason astronauts in orbit feel weightless?

Q13.21 As part of their training before going into orbit, astronauts ride in an airliner that is flown along the same parabolic trajectory as a freely falling projectile. Explain why this gives the same experience of apparent weightlessness as being in orbit.

EXERCISES

Section 13.1 Newton's Law of Gravitation

13.1 • What is the ratio of the gravitational pull of the sun on the moon to that of the earth on the moon? (Assume the distance of the moon from the sun can be approximated by the distance of the earth from the sun.) Use the data in Appendix F. Is it more accurate to say that the moon orbits the earth, or that the moon orbits the sun?

13.2 •• CP **Cavendish Experiment.** In the Cavendish balance apparatus shown in Fig. 13.4, suppose that $m_1 = 1.10$ kg, $m_2 = 25.0$ kg, and the rod connecting the m_1 pairs is 30.0 cm long. If, in each pair, m_1 and m_2 are 12.0 cm apart center to center, find (a) the net force and (b) the net torque (about the rotation axis) on the rotating part of the apparatus. (c) Does it seem that the torque in part (b) would be enough to easily rotate the rod? Suggest some ways to improve the sensitivity of this experiment.

13.3 • **Rendezvous in Space!** A couple of astronauts agree to rendezvous in space after hours. Their plan is to let gravity bring them together. One of them has a mass of 65 kg and the other a mass of 72 kg, and they start from rest 20.0 m apart. (a) Make a free-body diagram of each astronaut, and use it to find his or her initial acceleration. As a rough approximation, we can model the astronauts as uniform spheres. (b) If the astronauts' acceleration remained constant, how many days would they have to wait before reaching each other? (Careful! They *both* have acceleration toward each other.) (c) Would their acceleration, in fact, remain constant? If not, would it increase or decrease? Why?

13.4 •• Two uniform spheres, each with mass M and radius R, touch each other. What is the magnitude of their gravitational force of attraction?

13.5 • Two uniform spheres, each of mass 0.260 kg, are fixed at points A and B (**Fig. E13.5**). Find the magnitude and direction of the initial acceleration of a uniform sphere with mass 0.010 kg if released from rest at point P and acted on only by forces of gravitational attraction of the spheres at A and B.

Figure **E13.5**

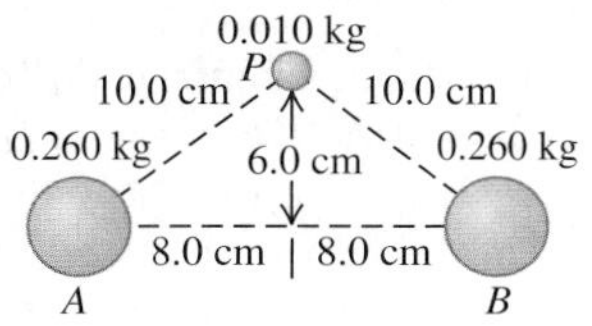

0.010 kg

10.0 cm P 10.0 cm

0.260 kg   6.0 cm   0.260 kg

8.0 cm | 8.0 cm

A B

13.6 •• Find the magnitude and direction of the net gravitational force on mass A due to masses B and C in **Fig. E13.6**. Each mass is 2.00 kg.

Figure **E13.6**

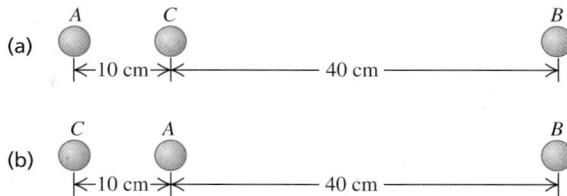

(a) A C B

10 cm   40 cm

(b) C A B

10 cm   40 cm

13.7 • A typical adult human has a mass of about 70 kg. (a) What force does a full moon exert on such a human when it is directly overhead with its center 378,000 km away? (b) Compare this force with the force exerted on the human by the earth.

13.8 •• An 8.00-kg point mass and a 12.0-kg point mass are held in place 50.0 cm apart. A particle of mass m is released from a point between the two masses 20.0 cm from the 8.00-kg mass along the line connecting the two fixed masses. Find the magnitude and direction of the acceleration of the particle.

13.9 •• A particle of mass $3m$ is located 1.00 m from a particle of mass m. (a) Where should you put a third mass M so that the net gravitational force on M due to the two masses is exactly zero? (b) Is the equilibrium of M at this point stable or unstable (i) for points along the line connecting m and $3m$, and (ii) for points along the line passing through M and perpendicular to the line connecting m and $3m$?

13.10 •• The point masses m and $2m$ lie along the x-axis, with m at the origin and $2m$ at $x = L$. A third point mass M is moved along the x-axis. (a) At what point is the net gravitational force on M due to the other two masses equal to zero? (b) Sketch the x-component of the net force on M due to m and $2m$, taking quantities to the right as positive. Include the regions $x < 0$, $0 < x < L$, and $x > L$. Be especially careful to show the behavior of the graph on either side of $x = 0$ and $x = L$.

Section 13.2 Weight

13.11 •• At what distance above the surface of the earth is the acceleration due to the earth's gravity 0.980 m/s² if the acceleration due to gravity at the surface has magnitude 9.80 m/s²?

13.12 • The mass of Venus is 81.5% that of the earth, and its radius is 94.9% that of the earth. (a) Compute the acceleration due to gravity on the surface of Venus from these data. (b) If a rock weighs 75.0 N on earth, what would it weigh at the surface of Venus?

13.13 • Titania, the largest moon of the planet Uranus, has $\frac{1}{8}$ the radius of the earth and $\frac{1}{1700}$ the mass of the earth. (a) What is the acceleration due to gravity at the surface of Titania? (b) What is the average density of Titania? (This is less than the density of rock, which is one piece of evidence that Titania is made primarily of ice.)

13.14 • Rhea, one of Saturn's moons, has a radius of 764 km and an acceleration due to gravity of 0.265 m/s² at its surface. Calculate its mass and average density.

13.15 •• Calculate the earth's gravity force on a 75-kg astronaut who is repairing the Hubble Space Telescope 600 km above the earth's surface, and then compare this value with his weight at the earth's surface. In view of your result, explain why it is said that astronauts are weightless when they orbit the earth in a satellite such as a space shuttle. Is it because the gravitational pull of the earth is negligibly small?

Section 13.3 Gravitational Potential Energy

13.16 •• **Volcanoes on Io.** Jupiter's moon Io has active volcanoes (in fact, it is the most volcanically active body in the solar system) that eject material as high as 500 km (or even higher) above the surface. Io has a mass of 8.93×10^{22} kg and a radius of 1821 km. For this calculation, ignore any variation in gravity over the 500-km range of the debris. How high would this material go on earth if it were ejected with the same speed as on Io?

13.17 • Use the results of Example 13.5 (Section 13.3) to calculate the escape speed for a spacecraft (a) from the surface of Mars and (b) from the surface of Jupiter. Use the data in Appendix F. (c) Why is the escape speed for a spacecraft independent of the spacecraft's mass?

13.18 •• Ten days after it was launched toward Mars in December 1998, the *Mars Climate Orbiter* spacecraft (mass 629 kg) was 2.87×10^6 km from the earth and traveling at 1.20×10^4 km/h relative to the earth. At this time, what were (a) the spacecraft's kinetic energy relative to the earth and (b) the potential energy of the earth–spacecraft system?

13.19 •• A planet orbiting a distant star has radius 3.24×10^6 m. The escape speed for an object launched from this planet's surface is 7.65×10^3 m/s. What is the acceleration due to gravity at the surface of the planet?

Section 13.4 The Motion of Satellites

13.20 • An earth satellite moves in a circular orbit with an orbital speed of 6200 m/s. Find (a) the time of one revolution of the satellite; (b) the radial acceleration of the satellite in its orbit.

13.21 • For a satellite to be in a circular orbit 890 km above the surface of the earth, (a) what orbital speed must it be given, and (b) what is the period of the orbit (in hours)?

13.22 •• **Aura Mission.** On July 15, 2004, NASA launched the *Aura* spacecraft to study the earth's climate and atmosphere. This satellite was injected into an orbit 705 km above the earth's surface. Assume a circular orbit. (a) How many hours does it take this satellite to make one orbit? (b) How fast (in km/s) is the *Aura* spacecraft moving?

13.23 •• Two satellites are in circular orbits around a planet that has radius 9.00×10^6 m. One satellite has mass 68.0 kg, orbital radius 7.00×10^7 m, and orbital speed 4800 m/s. The second satellite has mass 84.0 kg and orbital radius 3.00×10^7 m. What is the orbital speed of this second satellite?

13.24 •• **International Space Station.** In its orbit each day, the International Space Station makes 15.65 revolutions around the earth. Assuming a circular orbit, how high is this satellite above the surface of the earth?

13.25 • Deimos, a moon of Mars, is about 12 km in diameter with mass 1.5×10^{15} kg. Suppose you are stranded alone on Deimos and want to play a one-person game of baseball. You would be the pitcher, and you would be the batter! (a) With what speed would you have to throw a baseball so that it would go into a circular orbit just above the surface and return to you so you could hit it? Do you think you could actually throw it at this speed? (b) How long (in hours) after throwing the ball should you be ready to hit it? Would this be an action-packed baseball game?

Section 13.5 Kepler's Laws and the Motion of Planets

13.26 •• **Planet Vulcan.** Suppose that a planet were discovered between the sun and Mercury, with a circular orbit of radius equal to $\frac{2}{3}$ of the average orbit radius of Mercury. What would be the orbital period of such a planet? (Such a planet was once postulated, in part to explain the precession of Mercury's orbit. It was even given the name Vulcan, although we now have no evidence that it actually exists. Mercury's precession has been explained by general relativity.)

13.27 •• The star Rho[1] Cancri is 57 light-years from the earth and has a mass 0.85 times that of our sun. A planet has been detected in a circular orbit around Rho[1] Cancri with an orbital radius equal to 0.11 times the radius of the earth's orbit around the sun. What are (a) the orbital speed and (b) the orbital period of the planet of Rho[1] Cancri?

13.28 •• In March 2006, two small satellites were discovered orbiting Pluto, one at a distance of 48,000 km and the other at 64,000 km. Pluto already was known to have a large satellite Charon, orbiting at 19,600 km with an orbital period of 6.39 days. Assuming that the satellites do not affect each other, find the orbital periods of the two small satellites *without* using the mass of Pluto.

13.29 • The dwarf planet Pluto has an elliptical orbit with a semi-major axis of 5.91×10^{12} m and eccentricity 0.249. (a) Calculate Pluto's orbital period. Express your answer in seconds and in earth years. (b) During Pluto's orbit around the sun, what are its closest and farthest distances from the sun?

13.30 •• **Hot Jupiters.** In 2004 astronomers reported the discovery of a large Jupiter-sized planet orbiting very close to the star HD 179949 (hence the term "hot Jupiter"). The orbit was just $\frac{1}{9}$ the distance of Mercury from our sun, and it takes the planet only 3.09 days to make one orbit (assumed to be circular). (a) What is the mass of the star? Express your answer in kilograms and as a multiple of our sun's mass. (b) How fast (in km/s) is this planet moving?

13.31 •• **Planets Beyond the Solar System.** On October 15, 2001, a planet was discovered orbiting around the star HD 68988. Its orbital distance was measured to be 10.5 million kilometers from the center of the star, and its orbital period was estimated at 6.3 days. What is the mass of HD 68988? Express your answer in kilograms and in terms of our sun's mass. (Consult Appendix F.)

Section 13.6 Spherical Mass Distributions

13.32 • A uniform, spherical, 1000.0-kg shell has a radius of 5.00 m. (a) Find the gravitational force this shell exerts on a 2.00-kg point mass placed at the following distances from the center of the shell: (i) 5.01 m, (ii) 4.99 m, (iii) 2.72 m. (b) Sketch a qualitative graph of the magnitude of the gravitational force this sphere exerts on a point mass m as a function of the distance r of m from the center of the sphere. Include the region from $r = 0$ to $r \rightarrow \infty$.

13.33 •• A uniform, solid, 1000.0-kg sphere has a radius of 5.00 m. (a) Find the gravitational force this sphere exerts on a 2.00-kg point mass placed at the following distances from the center of the sphere: (i) 5.01 m, (ii) 2.50 m. (b) Sketch a qualitative graph of the magnitude of the gravitational force this sphere exerts on a point mass m as a function of the distance r of m from the center of the sphere. Include the region from $r = 0$ to $r \rightarrow \infty$.

13.34 • CALC A thin, uniform rod has length L and mass M. A small uniform sphere of mass m is placed a distance x from one end of the rod, along the axis of the rod (**Fig. E13.34**). (a) Calculate the gravitational potential energy of the rod–sphere system. Take the potential energy to be zero when the rod and

Figure **E13.34**

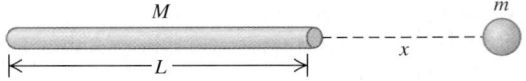

sphere are infinitely far apart. Show that your answer reduces to the expected result when x is much larger than L. (*Hint:* Use the power series expansion for $\ln(1 + x)$ given in Appendix B.) (b) Use $F_x = -dU/dx$ to find the magnitude and direction of the gravitational force exerted on the sphere by the rod (see Section 7.4). Show that your answer reduces to the expected result when x is much larger than L.

13.35 • **CALC** Consider the ring-shaped body of **Fig. E13.35**. A particle with mass m is placed a distance x from the center of the ring, along the line through the center of the ring and perpendicular to its plane. (a) Calculate the gravitational potential energy U of this system. Take the potential energy to be zero when the two objects are far apart. (b) Show that your answer to part (a) reduces to the expected result when x is much larger than the radius a of the ring. (c) Use $F_x = -dU/dx$ to find the magnitude and direction of the force on the particle (see Section 7.4). (d) Show that your answer to part (c) reduces to the expected result when x is much larger than a. (e) What are the values of U and F_x when $x = 0$? Explain why these results make sense.

Figure **E13.35**

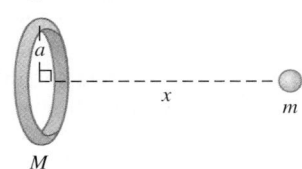

Section 13.7 Apparent Weight and the Earth's Rotation

13.36 •• **A Visit to Santa.** You decide to visit Santa Claus at the north pole to put in a good word about your splendid behavior throughout the year. While there, you notice that the elf Sneezy, when hanging from a rope, produces a tension of 395.0 N in the rope. If Sneezy hangs from a similar rope while delivering presents at the earth's equator, what will the tension in it be? (Recall that the earth is rotating about an axis through its north and south poles.) Consult Appendix F and start with a free-body diagram of Sneezy at the equator.

13.37 • The acceleration due to gravity at the north pole of Neptune is approximately 11.2 m/s². Neptune has mass 1.02×10^{26} kg and radius 2.46×10^4 km and rotates once around its axis in about 16 h. (a) What is the gravitational force on a 3.00-kg object at the north pole of Neptune? (b) What is the apparent weight of this same object at Neptune's equator? (Note that Neptune's "surface" is gaseous, not solid, so it is impossible to stand on it.)

Section 13.8 Black Holes

13.38 •• **Mini Black Holes.** Cosmologists have speculated that black holes the size of a proton could have formed during the early days of the Big Bang when the universe began. If we take the diameter of a proton to be 1.0×10^{-15} m, what would be the mass of a mini black hole?

13.39 •• **At the Galaxy's Core.** Astronomers have observed a small, massive object at the center of our Milky Way galaxy (see Section 13.8). A ring of material orbits this massive object; the ring has a diameter of about 15 light-years and an orbital speed of about 200 km/s. (a) Determine the mass of the object at the center of the Milky Way galaxy. Give your answer both in kilograms and in solar masses (one solar mass is the mass of the sun). (b) Observations of stars, as well as theories of the structure of stars, suggest that it is impossible for a single star to have a mass of more than about 50 solar masses. Can this massive object be a single, ordinary star? (c) Many astronomers believe that the massive object at the center of the Milky Way galaxy is a black hole. If so, what must the Schwarzschild radius of this black hole be? Would a black hole of this size fit inside the earth's orbit around the sun?

13.40 • In 2005 astronomers announced the discovery of a large black hole in the galaxy Markarian 766 having clumps of matter orbiting around once every 27 hours and moving at 30,000 km/s. (a) How far are these clumps from the center of the black hole? (b) What is the mass of this black hole, assuming circular orbits? Express your answer in kilograms and as a multiple of our sun's mass. (c) What is the radius of its event horizon?

PROBLEMS

13.41 ••• Neutron stars, such as the one at the center of the Crab Nebula, have about the same mass as our sun but have a *much* smaller diameter. If you weigh 675 N on the earth, what would you weigh at the surface of a neutron star that has the same mass as our sun and a diameter of 20 km?

13.42 ••• Four identical masses of 8.00 kg each are placed at the corners of a square whose side length is 2.00 m. What is the net gravitational force (magnitude and direction) on one of the masses, due to the other three?

13.43 • Three uniform spheres are fixed at the positions shown in **Fig. P13.43**. (a) What are the magnitude and direction of the force on a 0.0150-kg particle placed at P? (b) If the spheres are in deep outer space and a 0.0150-kg particle is released from rest 300 m from the origin along a line 45° below the $-x$-axis, what will the particle's speed be when it reaches the origin?

Figure **P13.43**

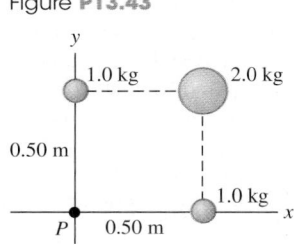

13.44 ••• **CP Exploring Europa.** There is strong evidence that Europa, a satellite of Jupiter, has a liquid ocean beneath its icy surface. Many scientists think we should land a vehicle there to search for life. Before launching it, we would want to test such a lander under the gravity conditions at the surface of Europa. One way to do this is to put the lander at the end of a rotating arm in an orbiting earth satellite. If the arm is 4.25 m long and pivots about one end, at what angular speed (in rpm) should it spin so that the acceleration of the lander is the same as the acceleration due to gravity at the surface of Europa? The mass of Europa is 4.80×10^{22} kg and its diameter is 3120 km.

13.45 •• A uniform sphere with mass 50.0 kg is held with its center at the origin, and a second uniform sphere with mass 80.0 kg is held with its center at the point $x = 0$, $y = 3.00$ m. (a) What are the magnitude and direction of the net gravitational force due to these objects on a third uniform sphere with mass 0.500 kg placed at the point $x = 4.00$ m, $y = 0$? (b) Where, other than infinitely far away, could the third sphere be placed such that the net gravitational force acting on it from the other two spheres is equal to zero?

13.46 •• **Mission to Titan.** On December 25, 2004, the *Huygens* probe separated from the *Cassini* spacecraft orbiting Saturn and began a 22-day journey to Saturn's giant moon Titan, on whose surface it landed. Besides the data in Appendix F, it is useful to know that Titan is 1.22×10^6 km from the center of Saturn and has a mass of 1.35×10^{23} kg and a diameter of 5150 km. At what distance from Titan should the gravitational pull of Titan just balance the gravitational pull of Saturn?

13.47 ••• CP An experiment is performed in deep space with two uniform spheres, one with mass 50.0 kg and the other with mass 100.0 kg. They have equal radii, $r = 0.20$ m. The spheres are released from rest with their centers 40.0 m apart. They accelerate toward each other because of their mutual gravitational attraction. You can ignore all gravitational forces other than that between the two spheres. (a) Explain why linear momentum is conserved. (b) When their centers are 20.0 m apart, find (i) the speed of each sphere and (ii) the magnitude of the relative velocity with which one sphere is approaching the other. (c) How far from the initial position of the center of the 50.0-kg sphere do the surfaces of the two spheres collide?

13.48 ••• At a certain instant, the earth, the moon, and a stationary 1250-kg spacecraft lie at the vertices of an equilateral triangle whose sides are 3.84×10^5 km in length. (a) Find the magnitude and direction of the net gravitational force exerted on the spacecraft by the earth and moon. State the direction as an angle measured from a line connecting the earth and the spacecraft. In a sketch, show the earth, the moon, the spacecraft, and the force vector. (b) What is the minimum amount of work that you would have to do to move the spacecraft to a point far from the earth and moon? Ignore any gravitational effects due to the other planets or the sun.

13.49 • **Geosynchronous Satellites.** Many satellites are moving in a circle in the earth's equatorial plane. They are at such a height above the earth's surface that they always remain above the same point. (a) Find the altitude of these satellites above the earth's surface. (Such an orbit is said to be *geosynchronous*.) (b) Explain, with a sketch, why the radio signals from these satellites cannot directly reach receivers on earth that are north of 81.3° N latitude.

13.50 •• CP **Submarines on Europa.** Some scientists are eager to send a remote-controlled submarine to Jupiter's moon Europa to search for life in its oceans below an icy crust. Europa's mass has been measured to be 4.80×10^{22} kg, its diameter is 3120 km, and it has no appreciable atmosphere. Assume that the layer of ice at the surface is not thick enough to exert substantial force on the water. If the windows of the submarine you are designing each have an area of 625 cm^2 and can stand a maximum inward force of 8750 N per window, what is the greatest depth to which this submarine can safely dive?

13.51 ••• What is the escape speed from a 300-km-diameter asteroid with a density of 2500 kg/m^3?

13.52 ••• A landing craft with mass 12,500 kg is in a circular orbit 5.75×10^5 m above the surface of a planet. The period of the orbit is 5800 s. The astronauts in the lander measure the diameter of the planet to be 9.60×10^6 m. The lander sets down at the north pole of the planet. What is the weight of an 85.6-kg astronaut as he steps out onto the planet's surface?

13.53 •• Planet X rotates in the same manner as the earth, around an axis through its north and south poles, and is perfectly spherical. An astronaut who weighs 943.0 N on the earth weighs 915.0 N at the north pole of Planet X and only 850.0 N at its equator. The distance from the north pole to the equator is 18,850 km, measured along the surface of Planet X. (a) How long is the day on Planet X? (b) If a 45,000-kg satellite is placed in a circular orbit 2000 km above the surface of Planet X, what will be its orbital period?

13.54 ••• (a) Suppose you are at the earth's equator and observe a satellite passing directly overhead and moving from west to east in the sky. Exactly 12.0 hours later, you again observe this satellite to be directly overhead. How far above the earth's surface is the satellite's orbit? (b) You observe another satellite directly overhead and traveling east to west. This satellite is again overhead in 12.0 hours. How far is this satellite's orbit above the surface of the earth?

13.55 •• CP An astronaut, whose mission is to go where no one has gone before, lands on a spherical planet in a distant galaxy. As she stands on the surface of the planet, she releases a small rock from rest and finds that it takes the rock 0.480 s to fall 1.90 m. If the radius of the planet is 8.60×10^7 m, what is the mass of the planet?

13.56 ••• CP Your starship, the *Aimless Wanderer,* lands on the mysterious planet Mongo. As chief scientist-engineer, you make the following measurements: A 2.50-kg stone thrown upward from the ground at 12.0 m/s returns to the ground in 4.80 s; the circumference of Mongo at the equator is 2.00×10^5 km; and there is no appreciable atmosphere on Mongo. The starship commander, Captain Confusion, asks for the following information: (a) What is the mass of Mongo? (b) If the *Aimless Wanderer* goes into a circular orbit 30,000 km above the surface of Mongo, how many hours will it take the ship to complete one orbit?

13.57 •• CP You are exploring a distant planet. When your spaceship is in a circular orbit at a distance of 630 km above the planet's surface, the ship's orbital speed is 4900 m/s. By observing the planet, you determine its radius to be 4.48×10^6 m. You then land on the surface and, at a place where the ground is level, launch a small projectile with initial speed 12.6 m/s at an angle of 30.8° above the horizontal. If resistance due to the planet's atmosphere is negligible, what is the horizontal range of the projectile?

13.58 •• The 0.100-kg sphere in **Fig. P13.58** is released from rest at the position shown in the sketch, with its center 0.400 m from the center of the 5.00-kg mass. Assume that the only forces on the 0.100-kg sphere are the gravitational forces exerted by the other two spheres and that the 5.00-kg and 10.0-kg spheres are held in place at their initial positions. What is the speed of the 0.100-kg sphere when it has moved 0.400 m to the right from its initial position?

Figure **P13.58**

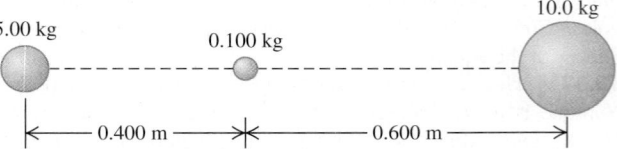

13.59 ••• An unmanned spacecraft is in a circular orbit around the moon, observing the lunar surface from an altitude of 50.0 km (see Appendix F). To the dismay of scientists on earth, an electrical fault causes an on-board thruster to fire, decreasing the speed of the spacecraft by 20.0 m/s. If nothing is done to correct its orbit, with what speed (in km/h) will the spacecraft crash into the lunar surface?

13.60 ••• **Mass of a Comet.** On July 4, 2005, the NASA spacecraft *Deep Impact* fired a projectile onto the surface of Comet Tempel 1. This comet is about 9.0 km across. Observations of surface debris released by the impact showed that dust with a speed as low as 1.0 m/s was able to escape the comet. (a) Assuming a spherical shape, what is the mass of this comet? (*Hint:* See Example 13.5 in Section 13.3.) (b) How far from the comet's center will this debris be when it has lost (i) 90.0% of its initial kinetic energy at the surface and (ii) all of its kinetic energy at the surface?

13.61 • **Falling Hammer.** A hammer with mass m is dropped from rest from a height h above the earth's surface. This height is not necessarily small compared with the radius R_E of the earth. Ignoring air resistance, derive an expression for the speed v of the hammer when it reaches the earth's surface. Your expression should involve h, R_E, and m_E (the earth's mass).

13.62 • (a) Calculate how much work is required to launch a spacecraft of mass m from the surface of the earth (mass m_E, radius R_E) and place it in a circular *low earth orbit*—that is, an orbit whose altitude above the earth's surface is much less than R_E. (As an example, the International Space Station is in low earth orbit at an altitude of about 400 km, much less than $R_E = 6370$ km.) Ignore the kinetic energy that the spacecraft has on the ground due to the earth's rotation. (b) Calculate the minimum amount of additional work required to move the spacecraft from low earth orbit to a very great distance from the earth. Ignore the gravitational effects of the sun, the moon, and the other planets. (c) Justify the statement "In terms of energy, low earth orbit is halfway to the edge of the universe."

13.63 • **Binary Star—Equal Masses.** Two identical stars with mass M orbit around their center of mass. Each orbit is circular and has radius R, so that the two stars are always on opposite sides of the circle. (a) Find the gravitational force of one star on the other. (b) Find the orbital speed of each star and the period of the orbit. (c) How much energy would be required to separate the two stars to infinity?

13.64 •• CP **Binary Star—Different Masses.** Two stars, with masses M_1 and M_2, are in circular orbits around their center of mass. The star with mass M_1 has an orbit of radius R_1; the star with mass M_2 has an orbit of radius R_2. (a) Show that the ratio of the orbital radii of the two stars equals the reciprocal of the ratio of their masses—that is, $R_1/R_2 = M_2/M_1$. (b) Explain why the two stars have the same orbital period, and show that the period T is given by $T = 2\pi(R_1 + R_2)^{3/2}/\sqrt{G(M_1 + M_2)}$. (c) The two stars in a certain binary star system move in circular orbits. The first star, Alpha, has an orbital speed of 36.0 km/s. The second star, Beta, has an orbital speed of 12.0 km/s. The orbital period is 137 d. What are the masses of each of the two stars? (d) One of the best candidates for a black hole is found in the binary system called A0620-0090. The two objects in the binary system are an orange star, V616 Monocerotis, and a compact object believed to be a black hole (see Fig. 13.28). The orbital period of A0620-0090 is 7.75 hours, the mass of V616 Monocerotis is estimated to be 0.67 times the mass of the sun, and the mass of the black hole is estimated to be 3.8 times the mass of the sun. Assuming that the orbits are circular, find the radius of each object's orbit and the orbital speed of each object. Compare these answers to the orbital radius and orbital speed of the earth in its orbit around the sun.

13.65 ••• Comets travel around the sun in elliptical orbits with large eccentricities. If a comet has speed 2.0×10^4 m/s when at a distance of 2.5×10^{11} m from the center of the sun, what is its speed when at a distance of 5.0×10^{10} m?

13.66 • The planet Uranus has a radius of 25,360 km and a surface acceleration due to gravity of 9.0 m/s^2 at its poles. Its moon Miranda (discovered by Kuiper in 1948) is in a circular orbit about Uranus at an altitude of 104,000 km above the planet's surface. Miranda has a mass of 6.6×10^{19} kg and a radius of 236 km. (a) Calculate the mass of Uranus from the given data. (b) Calculate the magnitude of Miranda's acceleration due to its orbital motion about Uranus. (c) Calculate the acceleration due to Miranda's gravity at the surface of Miranda. (d) Do the answers to parts (b) and (c) mean that an object released 1 m above Miranda's surface on the side toward Uranus will fall *up* relative to Miranda? Explain.

13.67 ••• CP Consider a spacecraft in an elliptical orbit around the earth. At the low point, or perigee, of its orbit, it is 400 km above the earth's surface; at the high point, or apogee, it is 4000 km above the earth's surface. (a) What is the period of the spacecraft's orbit? (b) Using conservation of angular momentum, find the

ratio of the spacecraft's speed at perigee to its speed at apogee. (c) Using conservation of energy, find the speed at perigee and the speed at apogee. (d) It is necessary to have the spacecraft escape from the earth completely. If the spacecraft's rockets are fired at perigee, by how much would the speed have to be increased to achieve this? What if the rockets were fired at apogee? Which point in the orbit is more efficient to use?

13.68 •• A rocket with mass 5.00×10^3 kg is in a circular orbit of radius 7.20×10^6 m around the earth. The rocket's engines fire for a period of time to increase that radius to 8.80×10^6 m, with the orbit again circular. (a) What is the change in the rocket's kinetic energy? Does the kinetic energy increase or decrease? (b) What is the change in the rocket's gravitational potential energy? Does the potential energy increase or decrease? (c) How much work is done by the rocket engines in changing the orbital radius?

13.69 ••• A 5000-kg spacecraft is in a circular orbit 2000 km above the surface of Mars. How much work must the spacecraft engines perform to move the spacecraft to a circular orbit that is 4000 km above the surface?

13.70 •• A satellite with mass 848 kg is in a circular orbit with an orbital speed of 9640 m/s around the earth. What is the new orbital speed after friction from the earth's upper atmosphere has done -7.50×10^9 J of work on the satellite? Does the speed increase or decrease?

13.71 ••• CALC Planets are not uniform inside. Normally, they are densest at the center and have decreasing density outward toward the surface. Model a spherically symmetric planet, with the same radius as the earth, as having a density that decreases linearly with distance from the center. Let the density be 15.0×10^3 kg/m^3 at the center and 2.0×10^3 kg/m^3 at the surface. What is the acceleration due to gravity at the surface of this planet?

13.72 •• One of the brightest comets of the 20th century was Comet Hyakutake, which passed close to the sun in early 1996. The orbital period of this comet is estimated to be about 30,000 years. Find the semi-major axis of this comet's orbit. Compare it to the average sun–Pluto distance and to the distance to Alpha Centauri, the nearest star to the sun, which is 4.3 light-years distant.

13.73 ••• CALC An object in the shape of a thin ring has radius a and mass M. A uniform sphere with mass m and radius R is placed with its center at a distance x to the right of the center of the ring, along a line through the center of the ring, and perpendicular to its plane (see Fig. E13.35). What is the gravitational force that the sphere exerts on the ring-shaped object? Show that your result reduces to the expected result when x is much larger than a.

13.74 •• CALC A uniform wire with mass M and length L is bent into a semicircle. Find the magnitude and direction of the gravitational force this wire exerts on a point with mass m placed at the center of curvature of the semicircle.

13.75 • CALC A shaft is drilled from the surface to the center of the earth (see Fig. 13.25). As in Example 13.10 (Section 13.6), make the unrealistic assumption that the density of the earth is uniform. With this approximation, the gravitational force on an object with mass m, that is inside the earth at a distance r from the center, has magnitude $F_g = Gm_E mr/R_E^3$ (as shown in Example 13.10) and points toward the center of the earth. (a) Derive an expression for the gravitational potential energy $U(r)$ of the object–earth system as a function of the object's distance from the center of the earth. Take the potential energy to be zero when the object is at the center of the earth. (b) If an object is released in the shaft at the earth's surface, what speed will it have when it reaches the center of the earth?

13.76 •• DATA For each of the eight planets Mercury to Neptune, the semi-major axis a of their orbit and their orbital period T are as follows:

Planet	Semi-major Axis (10^6 km)	Orbital Period (days)
Mercury	57.9	88.0
Venus	108.2	224.7
Earth	149.6	365.2
Mars	227.9	687.0
Jupiter	778.3	4331
Saturn	1426.7	10,747
Uranus	2870.7	30,589
Neptune	4498.4	59,800

(a) Explain why these values, when plotted as T^2 versus a^3, fall close to a straight line. Which of Kepler's laws is being tested? However, the values of T^2 and a^3 cover such a wide range that this plot is not a very practical way to graph the data. (Try it.) Instead, plot $\log(T)$ (with T in seconds) versus $\log(a)$ (with a in meters). Explain why the data should also fall close to a straight line in such a plot. (b) According to Kepler's laws, what should be the slope of your $\log(T)$ versus $\log(a)$ graph in part (a)? Does your graph have this slope? (c) Using $G = 6.674 \times 10^{-11} \ \text{N} \cdot \text{m}^2/\text{kg}^2$, calculate the mass of the sun from the y-intercept of your graph. How does your calculated value compare with the value given in Appendix F? (d) The only asteroid visible to the naked eye (and then only under ideal viewing conditions) is Vesta, which has an orbital period of 1325.4 days. What is the length of the semi-major axis of Vesta's orbit? Where does this place Vesta's orbit relative to the orbits of the eight major planets? Some scientists argue that Vesta should be called a minor planet rather than an asteroid.

13.77 •• DATA For a spherical planet with mass M, volume V, and radius R, derive an expression for the acceleration due to gravity at the planet's surface, g, in terms of the average density of the planet, $\rho = M/V$, and the planet's diameter, $D = 2R$. The table gives the values of D and g for the eight major planets:

Planet	D (km)	g (m/s^2)
Mercury	4879	3.7
Venus	12,104	8.9
Earth	12,756	9.8
Mars	6792	3.7
Jupiter	142,984	23.1
Saturn	120,536	9.0
Uranus	51,118	8.7
Neptune	49,528	11.0

(a) Treat the planets as spheres. Your equation for g as a function of ρ and D shows that if the average density of the planets is constant, a graph of g versus D will be well represented by a straight line. Graph g as a function of D for the eight major planets. What does the graph tell you about the variation in average density? (b) Calculate the average density for each major planet. List the planets in order of decreasing density, and give the calculated average density of each. (c) The earth is not a uniform sphere and has greater density near its center. It is reasonable to assume this might be true for the other planets. Discuss the effect this nonuniformity has on your analysis. (d) If Saturn had the same average density as the earth, what would be the value of g at Saturn's surface?

13.78 ••• DATA For a planet in our solar system, assume that the axis of orbit is at the sun and is circular. Then the angular momentum about that axis due to the planet's orbital motion is $L = MvR$. (a) Derive an expression for L in terms of the planet's mass M, orbital radius R, and period T of the orbit. (b) Using Appendix F, calculate the magnitude of the orbital angular momentum for each of the eight major planets. (Assume a circular orbit.) Add these values to obtain the total angular momentum of the major planets due to their orbital motion. (All the major planets orbit in the same direction in close to the same plane, so adding the magnitudes to get the total is a reasonable approximation.) (c) The rotational period of the sun is 24.6 days. Using Appendix F, calculate the angular momentum the sun has due to the rotation about its axis. (Assume that the sun is a uniform sphere.) (d) How does the rotational angular momentum of the sun compare with the total orbital angular momentum of the planets? How does the mass of the sun compare with the total mass of the planets? The fact that the sun has most of the mass of the solar system but only a small fraction of its total angular momentum must be accounted for in models of how the solar system formed. (e) The sun has a density that decreases with distance from its center. Does this mean that your calculation in part (c) overestimates or underestimates the rotational angular momentum of the sun? Or doesn't the nonuniform density have any effect?

CHALLENGE PROBLEMS

13.79 ••• **Interplanetary Navigation.** The most efficient way to send a spacecraft from the earth to another planet is to use a *Hohmann transfer orbit* (**Fig. P13.79**). If the orbits of the departure and destination planets are circular, the Hohmann transfer orbit is an elliptical orbit whose perihelion and aphelion are tangent to the orbits of the two planets. The rockets are fired briefly at the departure planet to put the spacecraft into the transfer orbit; the spacecraft then coasts until it reaches the destination planet. The rockets are then fired again to put the spacecraft into the same orbit about the sun as the destination planet. (a) For a flight from earth to Mars, in what direction must the rockets be fired at the earth and at Mars: in the direction of motion or opposite the direction of motion? What about for a flight from Mars to the earth? (b) How long does a one-way trip from the earth to Mars take, between the firings of the rockets? (c) To reach Mars from the earth, the launch must be timed so that Mars will be at the right spot when the spacecraft reaches Mars's orbit around the sun. At launch, what must the angle between a sun–Mars line and a sun–earth line be? Use Appendix F.

Figure **P13.79**

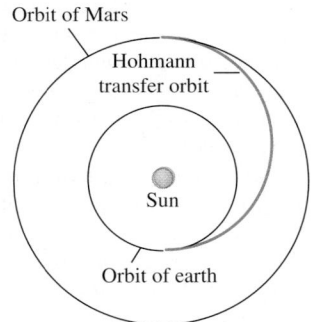

Orbit of Mars

Hohmann transfer orbit

Sun

Orbit of earth

13.80 ••• CP **Tidal Forces near a Black Hole.** An astronaut inside a spacecraft, which protects her from harmful radiation, is orbiting a black hole at a distance of 120 km from its center. The black hole is 5.00 times the mass of the sun and has a Schwarzschild radius of 15.0 km. The astronaut is positioned inside the spaceship such that one of her 0.030-kg ears is 6.0 cm farther from the black hole than the center of mass of the spacecraft and the other ear is 6.0 cm closer. (a) What is the tension between her ears? Would the astronaut find it difficult to keep from being torn apart by the gravitational forces? (Since her whole body orbits with the same angular velocity, one ear is moving too slowly for the radius of its orbit and the other is moving too fast. Hence her head must exert forces on her ears to keep them in their orbits.) (b) Is the center of gravity of her head at the same point as the center of mass? Explain.

13.81 ••• CALC Mass M is distributed uniformly over a disk of radius a. Find the gravitational force (magnitude and direction) between this disk-shaped mass and a particle with mass m located a distance x above the center of the disk (**Fig. P13.81**). Does your result reduce to the correct expression as x becomes very large? (*Hint:* Divide the disk into infinitesimally thin concentric rings, use the expression derived in Exercise 13.35 for the gravitational force due to each ring, and integrate to find the total force.)

Figure **P13.81**

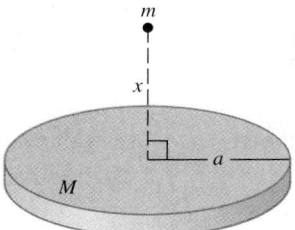

PASSAGE PROBLEMS

EXOPLANETS. As planets with a wide variety of properties are being discovered outside our solar system, astrobiologists are considering whether and how life could evolve on planets that might be very different from earth. One recently discovered extrasolar planet, or exoplanet, orbits a star whose mass is 0.70 times the mass of our sun. This planet has been found to have 2.3 times the earth's diameter and 7.9 times the earth's mass. For planets in this size range, computer models indicate a relationship between the planet's density and composition:

Density Compared with That of the Earth	Composition
2–3 times	Mostly iron
0.9–2 times	Iron core with a rock mantle
0.4–0.9 times	Iron core with a rock mantle and some lighter elements, such as (water) ice
< 0.4 times	Hydrogen and/or helium gas

Based on S. Seager et al., "Mass–Radius Relationships for Solid Exoplanets"; arXiv:0707.2895 [astro-ph].

13.82 Based on these data, what is the most likely composition of this planet? (a) Mostly iron; (b) iron and rock; (c) iron and rock with some lighter elements; (d) hydrogen and helium gases.

13.83 How many times the acceleration due to gravity g near the earth's surface is the acceleration due to gravity near the surface of this exoplanet? (a) About $0.29g$; (b) about $0.65g$; (c) about $1.5g$; (d) about $7.9g$.

13.84 Observations of this planet over time show that it is in a nearly circular orbit around its star and completes one orbit in only 9.5 days. How many times the orbital radius r of the earth around our sun is this exoplanet's orbital radius around its sun? Assume that the earth is also in a nearly circular orbit. (a) $0.026r$; (b) $0.078r$; (c) $0.70r$; (d) $2.3r$.

Answers

Chapter Opening Question ?

(iv) For a satellite a distance r from the center of its planet, the orbital speed is proportional to $\sqrt{1/r}$ and the acceleration due to gravity is proportional to $1/r^2$ (see Section 13.4). Hence a particle that orbits close to Saturn has a faster speed and a greater acceleration than one that orbits farther away.

Test Your Understanding Questions

13.1 (v) From Eq. (13.1), the gravitational force of the sun (mass m_1) on a planet (mass m_2) a distance r away has magnitude $F_g = Gm_1m_2/r^2$. Compared to the earth, Saturn has a value of r^2 that is $10^2 = 100$ times greater and a value of m_2 that is also 100 times greater. Hence the *force* that the sun exerts on Saturn has the same magnitude as the force that the sun exerts on earth. The *acceleration* of a planet equals the net force divided by the planet's mass: Since Saturn has 100 times more mass than the earth, its acceleration is $\frac{1}{100}$ as great as that of the earth.

13.2 (iii), (i), (ii), (iv) From Eq. (13.4), the acceleration due to gravity at the surface of a planet of mass m_P and radius R_P is $g_P = Gm_P/R_P^2$. That is, g_P is directly proportional to the planet's mass and inversely proportional to the square of its radius. It follows that compared to the value of g at the earth's surface, the value of g_P on each planet is (i) $2/2^2 = \frac{1}{2}$ as great; (ii) $4/4^2 = \frac{1}{4}$ as great; (iii) $4/2^2 = 1$ time as great—that is, the same as on earth; and (iv) $2/4^2 = \frac{1}{8}$ as great.

13.3 (iv) For a planet of mass m_P and radius R_P, the surface gravity is $g = Gm_P/R_P^2$ while the escape speed is $v_{esc} = \sqrt{2Gm_P/R_P}$. Comparing these two expressions, you get $v_{esc} = \sqrt{2gR_P}$. So even if a planet has the same value of g as the earth, its escape speed can be different, depending on how its radius R_P compares with the earth's radius. For the planet Saturn, for example, m_P is about 100 times the earth's mass and R_P is about 10 times the earth's radius. The value of g is different than on earth by a factor of $(100)/(10)^2 = 1$ (i.e., it is the same as on earth), while the escape speed is greater by a factor of $\sqrt{100}/10 = 3.2$.

13.4 (ii) Equation (13.10) shows that in a smaller-radius orbit, the spacecraft has a faster speed. The negative work done by air resistance decreases the *total* mechanical energy $E = K + U$; the kinetic energy K increases (becomes more positive), but the gravitational potential energy U decreases (becomes more negative) by a greater amount.

13.5 (iii) Equation (13.17) shows that the orbital period T is proportional to the $\frac{3}{2}$ power of the semi-major axis a. Hence the orbital period of Comet X is longer than that of Comet Y by a factor of $4^{3/2} = 8$.

13.6 No. Our analysis shows that there is *zero* gravitational force inside a hollow spherical shell. Hence visitors to the interior of a hollow planet would find themselves weightless, and they could not stand or walk on the planet's inner surface.

13.7 (iv) The discussion following Eq. (13.27) shows that the difference between the acceleration due to gravity at the equator and at the poles is v^2/R_E. Since this planet has the same radius and hence the same circumference as the earth, the speed v at its equator must be 10 times the speed at the earth's equator. Hence v^2/R_E is $10^2 = 100$ times greater than for the earth, or $100(0.0339 \text{ m/s}^2) = 3.39 \text{ m/s}^2$. The acceleration due to gravity at the poles is 9.80 m/s^2, while at the equator it is dramatically less, $9.80 \text{ m/s}^2 - 3.39 \text{ m/s}^2 = 6.41 \text{ m/s}^2$. You can show that if this planet were to rotate 17.0 times faster than the earth, the acceleration due to gravity at the equator would be *zero* and loose objects would fly off the equator's surface!

13.8 (iii) If the sun collapsed into a black hole (which, according to our understanding of stars, it cannot do), the sun would have a much smaller radius but the same mass. The sun's gravitational force on the earth doesn't depend on the sun's radius, so the earth's orbit would be unaffected.

Bridging Problem

(a) Perihelion: $v_P = \sqrt{\dfrac{Gm_S}{a} \dfrac{(1 + e)}{(1 - e)}}$

aphelion: $v_A = \sqrt{\dfrac{Gm_S}{a} \dfrac{(1 - e)}{(1 + e)}}$

(b) $v_P = 54.4 \text{ km/s}, v_A = 0.913 \text{ km/s}; K_P = 3.26 \times 10^{23} \text{ J},$
$U_P = -3.31 \times 10^{23} \text{ J}, E_P = -5.47 \times 10^{21} \text{ J};$
$K_A = 9.17 \times 10^{19} \text{ J}, \ U_A = -5.56 \times 10^{21} \text{ J},$
$E_A = -5.47 \times 10^{21} \text{ J}$

14 PERIODIC MOTION

LEARNING GOALS

Looking forward at...

14.1 How to describe oscillations in terms of amplitude, period, frequency, and angular frequency.

14.2 How to do calculations with simple harmonic motion, an important type of oscillation.

14.3 How to use energy concepts to analyze simple harmonic motion.

14.4 How to apply the ideas of simple harmonic motion to different physical situations.

14.5 How to analyze the motions of a simple pendulum.

14.6 What a physical pendulum is, and how to calculate the properties of its motion.

14.7 What determines how rapidly an oscillation dies out.

14.8 How a driving force applied to an oscillator at a particular frequency can cause a very large response, or resonance.

Looking back at ...

1.3 Time standards.

3.4 Uniform circular motion.

6.3 Hooke's law.

7.2, 7.4 Elastic potential energy; relating force and potential energy.

9.3 Relating angular motion and linear motion.

10.2 Newton's second law for rotational motion.

Many kinds of motion repeat themselves over and over: the vibration of a quartz crystal in a watch, the swinging pendulum of a grandfather clock, the sound vibrations produced by a clarinet or an organ pipe, and the back-and-forth motion of the pistons in a car engine. This kind of motion, called **periodic motion** or **oscillation,** is the subject of this chapter. Understanding periodic motion will be essential for our later study of waves, sound, alternating electric currents, and light.

A body that undergoes periodic motion always has a stable equilibrium position. When it is moved away from this position and released, a force or torque comes into play to pull it back toward equilibrium. But by the time it gets there, it has picked up some kinetic energy, so it overshoots, stopping somewhere on the other side, and is again pulled back toward equilibrium. Picture a ball rolling back and forth in a round bowl or a pendulum that swings back and forth past its straight-down position.

In this chapter we will concentrate on two simple examples of systems that can undergo periodic motions: spring-mass systems and pendulums. We will also study why oscillations often tend to die out with time and why some oscillations can build up to greater and greater displacements from equilibrium when periodically varying forces act.

14.1 DESCRIBING OSCILLATION

Figure 14.1 (next page) shows one of the simplest systems that can have periodic motion. A body with mass m rests on a frictionless horizontal guide system, such as a linear air track, so it can move along the x-axis only. The body is attached to a spring of negligible mass that can be either stretched or compressed. The left end of the spring is held fixed, and the right end is attached to the body. The spring force is the only horizontal force acting on the body; the vertical normal and gravitational forces always add to zero.

It's simplest to define our coordinate system so that the origin O is at the equilibrium position, where the spring is neither stretched nor compressed. Then x is

14.1 A system that can have periodic motion.

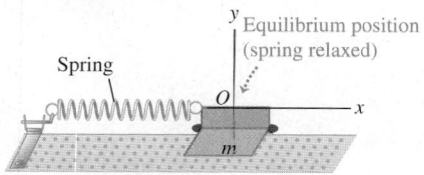

14.2 Model for periodic motion. When the body is displaced from its equilibrium position at $x = 0$, the spring exerts a restoring force back toward the equilibrium position.

(a)

$x > 0$: glider displaced to the right from the equilibrium position.

$F_x < 0$, so $a_x < 0$: stretched spring pulls glider toward equilibrium position.

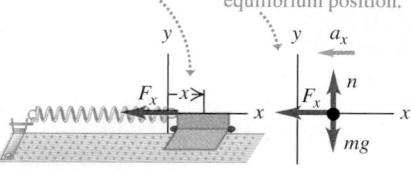

(b)

$x = 0$: The relaxed spring exerts no force on the glider, so the glider has zero acceleration.

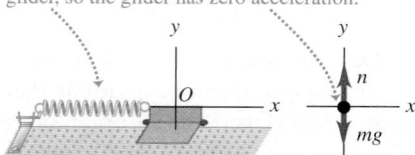

(c)

$x < 0$: glider displaced to the left from the equilibrium position.

$F_x > 0$, so $a_x > 0$: compressed spring pushes glider toward equilibrium position.

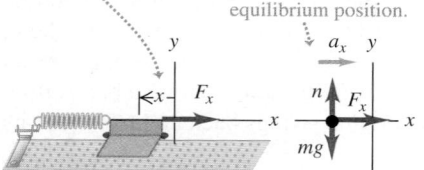

the x-component of the **displacement** of the body from equilibrium and is also the change in the length of the spring. The spring exerts a force on the body with x-component F_x, and the x-component of acceleration is $a_x = F_x/m$.

Figure 14.2 shows the body for three different displacements of the spring. Whenever the body is displaced from its equilibrium position, the spring force tends to restore it to the equilibrium position. We call a force with this character a **restoring force.** Oscillation can occur only when there is a restoring force tending to return the system to equilibrium.

Let's analyze how oscillation occurs in this system. If we displace the body to the right to $x = A$ and then let go, the net force and the acceleration are to the left (Fig. 14.2a). The speed increases as the body approaches the equilibrium position O. When the body is at O, the net force acting on it is zero (Fig. 14.2b), but because of its motion it *overshoots* the equilibrium position. On the other side of the equilibrium position the body is still moving to the left, but the net force and the acceleration are to the right (Fig. 14.2c); hence the speed decreases until the body comes to a stop. We will show later that with an ideal spring, the stopping point is at $x = -A$. The body then accelerates to the right, overshoots equilibrium again, and stops at the starting point $x = A$, ready to repeat the whole process. The body is oscillating! If there is no friction or other force to remove mechanical energy from the system, this motion repeats forever; the restoring force perpetually draws the body back toward the equilibrium position, only to have the body overshoot time after time.

In different situations the force may depend on the displacement x from equilibrium in different ways. But oscillation *always* occurs if the force is a *restoring* force that tends to return the system to equilibrium.

Amplitude, Period, Frequency, and Angular Frequency

Here are some terms that we'll use in discussing periodic motions of all kinds:

The **amplitude** of the motion, denoted by A, is the maximum magnitude of displacement from equilibrium—that is, the maximum value of $|x|$. It is always positive. If the spring in Fig. 14.2 is an ideal one, the total overall range of the motion is $2A$. The SI unit of A is the meter. A complete vibration, or **cycle,** is one complete round trip—say, from A to $-A$ and back to A, or from O to A, back through O to $-A$, and back to O. Note that motion from one side to the other (say, $-A$ to A) is a half-cycle, not a whole cycle.

The **period,** T, is the time to complete one cycle. It is always positive. The SI unit is the second, but it is sometimes expressed as "seconds per cycle."

The **frequency,** f, is the number of cycles in a unit of time. It is always positive. The SI unit of frequency is the *hertz,* named for the 19th-century German physicist Heinrich Hertz:

$$1 \text{ hertz} = 1 \text{ Hz} = 1 \text{ cycle/s} = 1 \text{ s}^{-1}$$

The **angular frequency,** ω, is 2π times the frequency:

$$\omega = 2\pi f$$

We'll learn shortly why ω is a useful quantity. It represents the rate of change of an angular quantity (not necessarily related to a rotational motion) that is always measured in radians, so its units are rad/s. Since f is in cycle/s, we may regard the number 2π as having units rad/cycle.

By definition, period and frequency are reciprocals of each other:

In periodic motion **frequency and period** are reciprocals of each other. ⋯⋯⋯→ Period

$$f = \frac{1}{T} \qquad T = \frac{1}{f} \qquad (14.1)$$

⋯⋯ Frequency

Also, from the definition of ω,

Angular frequency related to frequency and period ⋯⋯→ $\omega = 2\pi f = \dfrac{2\pi}{T}$ ⋯ Frequency

⋯ Period

(14.2)

EXAMPLE 14.1   PERIOD, FREQUENCY, AND ANGULAR FREQUENCY

An ultrasonic transducer used for medical diagnosis oscillates at 6.7 MHz = 6.7×10^6 Hz. How long does each oscillation take, and what is the angular frequency?

SOLUTION

IDENTIFY and SET UP: The target variables are the period T and the angular frequency ω. We can find these from the given frequency f in Eqs. (14.1) and (14.2).

EXECUTE: From Eqs. (14.1) and (14.2),

$$T = \frac{1}{f} = \frac{1}{6.7 \times 10^6 \text{ Hz}} = 1.5 \times 10^{-7} \text{ s} = 0.15 \ \mu\text{s}$$

$$\omega = 2\pi f = 2\pi(6.7 \times 10^6 \text{ Hz})$$
$$= (2\pi \text{ rad/cycle})(6.7 \times 10^6 \text{ cycle/s}) = 4.2 \times 10^7 \text{ rad/s}$$

EVALUATE: This is a very rapid vibration, with large f and ω and small T. A slow vibration has small f and ω and large T.

TEST YOUR UNDERSTANDING OF SECTION 14.1   A body like that shown in Fig. 14.2 oscillates back and forth. For each of the following values of the body's x-velocity v_x and x-acceleration a_x, state whether its displacement x is positive, negative, or zero. (a) $v_x > 0$ and $a_x > 0$; (b) $v_x > 0$ and $a_x < 0$; (c) $v_x < 0$ and $a_x > 0$; (d) $v_x < 0$ and $a_x < 0$; (e) $v_x = 0$ and $a_x < 0$; (f) $v_x > 0$ and $a_x = 0$. ∎

14.2 SIMPLE HARMONIC MOTION

The simplest kind of oscillation occurs when the restoring force F_x is *directly proportional* to the displacement from equilibrium x. This happens if the spring in Figs. 14.1 and 14.2 is an ideal one that obeys *Hooke's law* (see Section 6.3). The constant of proportionality between F_x and x is the force constant k. On either side of the equilibrium position, F_x and x always have opposite signs. In Section 6.3 we represented the force acting *on* a stretched ideal spring as $F_x = kx$. The x-component of force the spring exerts *on the body* is the negative of this, so

Restoring force exerted by an ideal spring ⋯⋯→ $F_x = -kx$ ← x-component of force

⋯⋯ Displacement

⋯ Force constant of spring

(14.3)

This equation gives the correct magnitude and sign of the force, whether x is positive, negative, or zero (**Fig. 14.3**). The force constant k is always positive and has units of N/m (a useful alternative set of units is kg/s²). We are assuming that there is no friction, so Eq. (14.3) gives the *net* force on the body.

14.3 An idealized spring exerts a restoring force that obeys Hooke's law, $F_x = -kx$. Oscillation with such a restoring force is called simple harmonic motion.

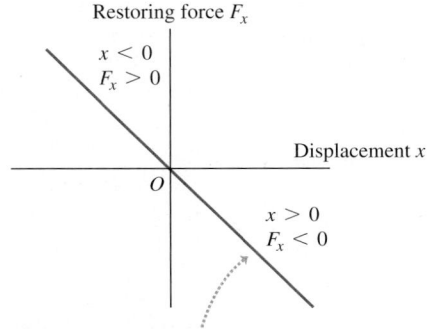

Restoring force F_x

$x < 0$
$F_x > 0$

Displacement x

O

$x > 0$
$F_x < 0$

The restoring force exerted by an idealized spring is directly proportional to the displacement (Hooke's law, $F_x = -kx$): the graph of F_x versus x is a straight line.

When the restoring force is directly proportional to the displacement from equilibrium, as given by Eq. (14.3), the oscillation is called **simple harmonic motion (SHM).** The acceleration $a_x = d^2x/dt^2 = F_x/m$ of a body in SHM is

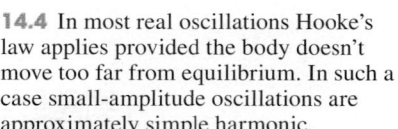

Equation for simple harmonic motion

x-component of acceleration ⋯⋯ Force constant of restoring force

$$a_x = \frac{d^2x}{dt^2} = -\frac{k}{m}x$$

Second derivative of displacement   Displacement
                                           Mass of object   (14.4)

The minus sign means that, in SHM, the acceleration and displacement always have opposite signs. This acceleration is *not* constant, so don't even think of using the constant-acceleration equations from Chapter 2. We'll see shortly how to solve this equation to find the displacement x as a function of time. A body that undergoes simple harmonic motion is called a **harmonic oscillator.**

Why is simple harmonic motion important? Not all periodic motions are simple harmonic; in periodic motion in general, the restoring force depends on displacement in a more complicated way than in Eq. (14.3). But in many systems the restoring force is *approximately* proportional to displacement if the displacement is sufficiently small (**Fig. 14.4**). That is, if the amplitude is small enough, the oscillations of such systems are approximately simple harmonic and therefore approximately described by Eq. (14.4). Thus we can use SHM as an approximate model for many different periodic motions, such as the vibration of a tuning fork, the electric current in an alternating-current circuit, and the oscillations of atoms in molecules and solids.

Circular Motion and the Equations of SHM

To explore the properties of simple harmonic motion, we must express the displacement x of the oscillating body as a function of time, $x(t)$. The second derivative of this function, d^2x/dt^2, must be equal to $(-k/m)$ times the function itself, as required by Eq. (14.4). As we mentioned, the formulas for constant acceleration from Section 2.4 are no help because the acceleration changes constantly as the displacement x changes. Instead, we'll find $x(t)$ by noting that SHM is related to *uniform circular motion,* which we studied in Section 3.4.

Figure 14.5a shows a top view of a horizontal disk of radius A with a ball attached to its rim at point Q. The disk rotates with constant angular speed ω (measured in rad/s), so the ball moves in uniform circular motion. A horizontal

14.4 In most real oscillations Hooke's law applies provided the body doesn't move too far from equilibrium. In such a case small-amplitude oscillations are approximately simple harmonic.

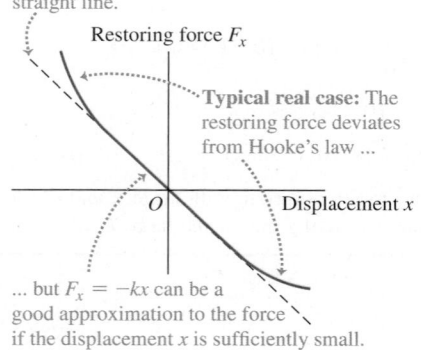

Ideal case: The restoring force obeys Hooke's law ($F_x = -kx$), so the graph of F_x versus x is a straight line.

Restoring force F_x

Typical real case: The restoring force deviates from Hooke's law ...

O Displacement x

... but $F_x = -kx$ can be a good approximation to the force if the displacement x is sufficiently small.

14.5 (a) Relating uniform circular motion and simple harmonic motion. (b) The ball's shadow moves exactly like a body oscillating on an ideal spring.

(a) Top view of apparatus for creating the reference circle

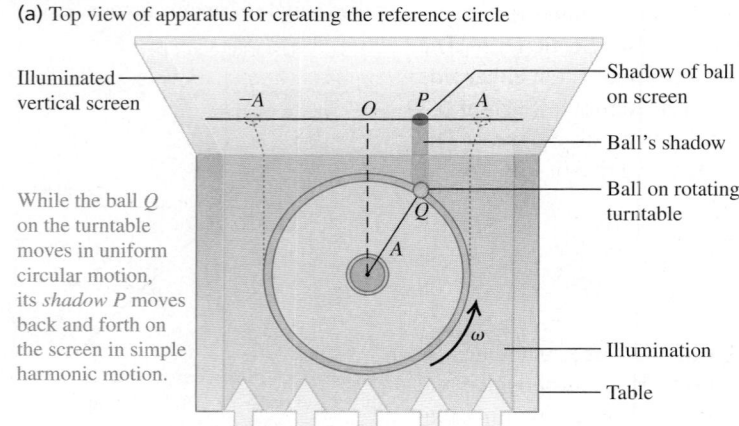

Illuminated vertical screen

Shadow of ball on screen

Ball's shadow

While the ball Q on the turntable moves in uniform circular motion, its *shadow P* moves back and forth on the screen in simple harmonic motion.

Ball on rotating turntable

ω

Illumination

Table

Light beam

(b) An abstract representation of the motion in **(a)**

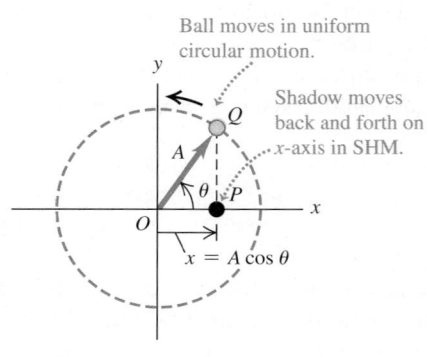

Ball moves in uniform circular motion.

Shadow moves back and forth on the x-axis in SHM.

$x = A \cos \theta$

light beam casts a shadow of the ball on a screen. The shadow at point P oscillates back and forth as the ball moves in a circle. We then arrange a body attached to an ideal spring, like the combination shown in Figs. 14.1 and 14.2, so that the body oscillates parallel to the shadow. We will prove that the motions of the body and of the ball's shadow are *identical* if the amplitude of the body's oscillation is equal to the disk radius A, and if the angular frequency $2\pi f$ of the oscillating body is equal to the angular speed ω of the rotating disk. That is, *simple harmonic motion is the projection of uniform circular motion onto a diameter.*

We can verify this remarkable statement by finding the acceleration of the shadow at P and comparing it to the acceleration of a body undergoing SHM, given by Eq. (14.4). The circle in which the ball moves so that its projection matches the motion of the oscillating body is called the **reference circle**; we will call the point Q the *reference point*. We take the reference circle to lie in the xy-plane, with the origin O at the center of the circle (Fig. 14.5b). At time t the vector OQ from the origin to reference point Q makes an angle θ with the positive x-axis. As point Q moves around the reference circle with constant angular speed ω, vector OQ rotates with the same angular speed. Such a rotating vector is called a **phasor**. (This term was in use long before the invention of the *Star Trek* stun gun with a similar name.) We'll use phasors again when we study alternating-current circuits in Chapter 31 and the interference of light in Chapters 35 and 36.

The x-component of the phasor at time t is just the x-coordinate of the point Q:

$$x = A\cos\theta \qquad (14.5)$$

This is also the x-coordinate of the shadow P, which is the *projection* of Q onto the x-axis. Hence the x-velocity of the shadow P along the x-axis is equal to the x-component of the velocity vector of point Q (**Fig. 14.6a**), and the x-acceleration of P is equal to the x-component of the acceleration vector of Q (Fig. 14.6b). Since point Q is in uniform circular motion, its acceleration vector $\vec{a}_Q$ is always directed toward O. Furthermore, the magnitude of $\vec{a}_Q$ is constant and given by the angular speed squared times the radius of the circle (see Section 9.3):

$$a_Q = \omega^2 A \qquad (14.6)$$

Figure 14.6b shows that the x-component of $\vec{a}_Q$ is $a_x = -a_Q\cos\theta$. Combining this with Eqs. (14.5) and (14.6), we get that the acceleration of point P is

$$a_x = -a_Q\cos\theta = -\omega^2 A\cos\theta \qquad \text{or} \qquad (14.7)$$

$$a_x = -\omega^2 x \qquad (14.8)$$

The acceleration of point P is directly proportional to the displacement x and always has the opposite sign. These are precisely the hallmarks of simple harmonic motion.

Equation (14.8) is *exactly* the same as Eq. (14.4) for the acceleration of a harmonic oscillator, provided that the angular speed ω of the reference point Q is related to the force constant k and mass m of the oscillating body by

$$\omega^2 = \frac{k}{m} \qquad \text{or} \qquad \omega = \sqrt{\frac{k}{m}} \qquad (14.9)$$

We have been using the same symbol ω for the angular *speed* of the reference point Q and the angular *frequency* of the oscillating point P. The reason is that these quantities are equal! If point Q makes one complete revolution in time T, then point P goes through one complete cycle of oscillation in the same time; hence T is the period of the oscillation. During time T the point Q moves through

14.6 The (a) x-velocity and (b) x-acceleration of the ball's shadow P (see Fig. 14.5) are the x-components of the velocity and acceleration vectors, respectively, of the ball Q.

(a) Using the reference circle to determine the x-velocity of point P

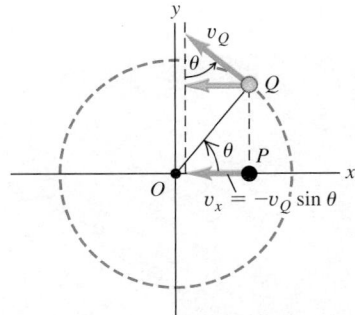

(b) Using the reference circle to determine the x-acceleration of point P

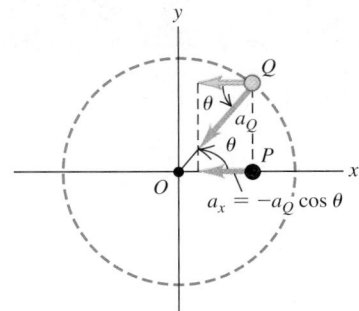

2π radians, so its angular speed is $\omega = 2\pi/T$. But this is the same as Eq. (14.2) for the angular frequency of the point P, which verifies our statement about the two interpretations of ω. This is why we introduced angular frequency in Section 14.1; this quantity makes the connection between oscillation and circular motion. So we reinterpret Eq. (14.9) as an expression for the angular frequency of simple harmonic motion:

$$\text{Angular frequency for simple harmonic motion} \rightarrow \omega = \sqrt{\frac{k \cdots\text{Force constant of restoring force}}{m \cdots\text{Mass of object}}} \qquad (14.10)$$

When you start a body oscillating in SHM, the value of ω is not yours to choose; it is predetermined by the values of k and m. The units of k are N/m or kg/s^2, so k/m is in $(\text{kg/s}^2)/\text{kg} = \text{s}^{-2}$. When we take the square root in Eq. (14.10), we get s^{-1}, or more properly rad/s because this is an *angular* frequency (recall that a radian is not a true unit).

According to Eqs. (14.1) and (14.2), the frequency f and period T are

$$\text{Frequency for simple harmonic motion} \rightarrow f = \frac{\overset{\text{Angular frequency}}{\omega}}{2\pi} = \frac{1}{2\pi}\sqrt{\frac{k \cdots\text{Force constant of restoring force}}{m \cdots\text{Mass of object}}} \qquad (14.11)$$

$$\text{Period for simple harmonic motion} \rightarrow T = \frac{1}{\underset{\text{Frequency}}{f}} = \frac{2\pi}{\underset{\text{Angular frequency}}{\omega}} = 2\pi\sqrt{\frac{m \cdots\text{Mass of object}}{k \cdots\text{Force constant of restoring force}}} \qquad (14.12)$$

We see from Eq. (14.12) that a larger mass m will have less acceleration and take a longer time for a complete cycle (**Fig. 14.7**). A stiffer spring (one with a larger force constant k) exerts a greater force at a given deformation x, causing greater acceleration and a shorter time T per cycle.

CAUTION **Don't confuse frequency and angular frequency** You can run into trouble if you don't make the distinction between frequency f and angular frequency $\omega = 2\pi f$. Frequency tells you how many cycles of oscillation occur per second, while angular frequency tells you how many radians per second this corresponds to on the reference circle. In solving problems, pay careful attention to whether the goal is to find f or ω. ▌

Period and Amplitude in SHM

Equations (14.11) and (14.12) show that the period and frequency of simple harmonic motion are completely determined by the mass m and the force constant k. *In simple harmonic motion the period and frequency do not depend on the amplitude A.* For given values of m and k, the time of one complete oscillation is the same whether the amplitude is large or small. Equation (14.3) shows why we should expect this. Larger A means that the body reaches larger values of $|x|$ and is subjected to larger restoring forces. This increases the average speed of the body over a complete cycle; this exactly compensates for having to travel a larger distance, so the same total time is involved.

The oscillations of a tuning fork are essentially simple harmonic motion, so it always vibrates with the same frequency, independent of amplitude. This is why a tuning fork can be used as a standard for musical pitch. If it were not for this characteristic of simple harmonic motion, it would be impossible to play most musical instruments in tune. If you encounter an oscillating body with a period that *does* depend on the amplitude, the oscillation is *not* simple harmonic motion.

14.7 The greater the mass m in a tuning fork's tines, the lower the frequency of oscillation $f = (1/2\pi)\sqrt{k/m}$ and the lower the pitch of the sound that the tuning fork produces.

Tines with large mass m:
low frequency $f = 128$ Hz

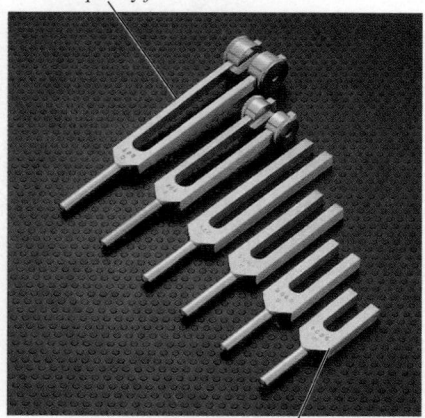

Tines with small mass m:
high frequency $f = 4096$ Hz

EXAMPLE 14.2 ANGULAR FREQUENCY, FREQUENCY, AND PERIOD IN SHM

A spring is mounted horizontally, with its left end fixed. A spring balance attached to the free end and pulled toward the right (**Fig. 14.8a**) indicates that the stretching force is proportional to the displacement, and a force of 6.0 N causes a displacement of 0.030 m. We replace the spring balance with a 0.50-kg glider, pull it 0.020 m to the right along a frictionless air track, and release it from rest (Fig. 14.8b). (a) Find the force constant k of the spring. (b) Find the angular frequency ω, frequency f, and period T of the resulting oscillation.

SOLUTION

IDENTIFY and SET UP: Because the spring force (equal in magnitude to the stretching force) is proportional to the displacement, the motion is simple harmonic. We find k from Hooke's law,

14.8 (a) The force exerted *on* the spring (shown by the vector F) has x-component $F_x = +6.0$ N. The force exerted *by* the spring has x-component $F_x = -6.0$ N. (b) A glider is attached to the same spring and allowed to oscillate.

(a)

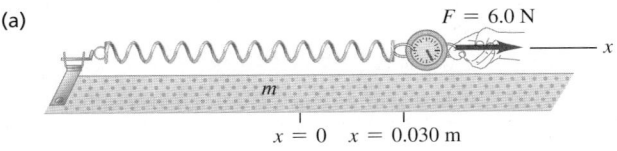

$$x = 0 \quad x = 0.030 \text{ m}$$

(b)

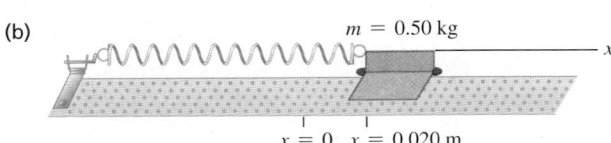

$$x = 0 \quad x = 0.020 \text{ m}$$

Eq. (14.3), and ω, f, and T from Eqs. (14.10), (14.11), and (14.12), respectively.

EXECUTE: (a) When $x = 0.030$ m, the force the spring exerts on the spring balance is $F_x = -6.0$ N. From Eq. (14.3),

$$k = -\frac{F_x}{x} = -\frac{-6.0 \text{ N}}{0.030 \text{ m}} = 200 \text{ N/m} = 200 \text{ kg/s}^2$$

(b) From Eq. (14.10), with $m = 0.50$ kg,

$$\omega = \sqrt{\frac{k}{m}} = \sqrt{\frac{200 \text{ kg/s}^2}{0.50 \text{ kg}}} = 20 \text{ rad/s}$$

$$f = \frac{\omega}{2\pi} = \frac{20 \text{ rad/s}}{2\pi \text{ rad/cycle}} = 3.2 \text{ cycle/s} = 3.2 \text{ Hz}$$

$$T = \frac{1}{f} = \frac{1}{3.2 \text{ cycle/s}} = 0.31 \text{ s}$$

EVALUATE: The amplitude of the oscillation is 0.020 m, the distance that we pulled the glider before releasing it. In SHM the angular frequency, frequency, and period are all independent of the amplitude. Note that a period is usually stated in "seconds" rather than "seconds per cycle."

Displacement, Velocity, and Acceleration in SHM

We still need to find the displacement x as a function of time for a harmonic oscillator. Equation (14.4) for a body in SHM along the x-axis is identical to Eq. (14.8) for the x-coordinate of the reference point in uniform circular motion with constant angular speed $\omega = \sqrt{k/m}$. Hence Eq. (14.5), $x = A\cos\theta$, describes the x-coordinate for both situations. If at $t = 0$ the phasor OQ makes an angle ϕ (the Greek letter phi) with the positive x-axis, then at any later time t this angle is $\theta = \omega t + \phi$. We substitute this into Eq. (14.5) to obtain

(**MP**)

PhET: Motion in 2D

Displacement in simple harmonic motion as a function of time ⋯ Amplitude ⋯ Time ⋯ Phase angle

$$x = A\cos(\omega t + \phi) \tag{14.13}$$

Angular frequency $= \sqrt{k/m}$

Figure 14.9 shows a graph of Eq. (14.13) for the particular case $\phi = 0$. We could also have written Eq. (14.13) in terms of a sine function rather than a cosine by using the identity $\cos\alpha = \sin(\alpha + \pi/2)$. *In simple harmonic motion the displacement is a periodic, sinusoidal function of time.* There are many other periodic functions, but none so simple as a sine or cosine function.

The value of the cosine function is always between -1 and 1, so in Eq. (14.13), x is always between $-A$ and A. This confirms that A is the amplitude of the motion.

14.9 Graph of x versus t [see Eq. (14.13)] for simple harmonic motion. The case shown has $\phi = 0$.

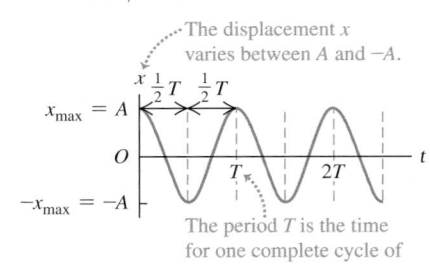

The displacement x varies between A and $-A$.

$x_{max} = A$

O

$-x_{max} = -A$

The period T is the time for one complete cycle of oscillation.

14.10 Variations of simple harmonic motion. All cases shown have $\phi = 0$ [see Eq. (14.13)].

(a) Increasing m; same A and k

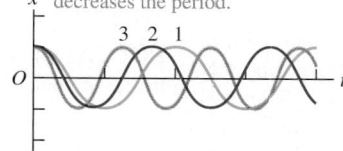

Mass m increases from curve 1 to 2 to 3. Increasing m alone increases the period.

(b) Increasing k; same A and m

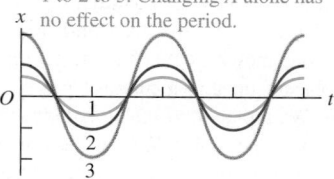

Force constant k increases from curve 1 to 2 to 3. Increasing k alone decreases the period.

(c) Increasing A; same k and m

Amplitude A increases from curve 1 to 2 to 3. Changing A alone has no effect on the period.

14.11 Variations of simple harmonic motion: same m, k, and A but different phase angles ϕ.

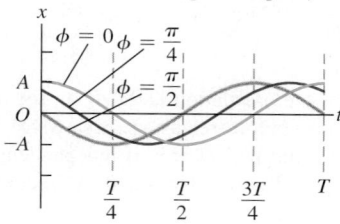

These three curves show SHM with the same period T and amplitude A but with different phase angles ϕ.

14.12 Graphs of **(a)** x versus t, **(b)** v_x versus t, and **(c)** a_x versus t for a body in SHM. For the motion depicted in these graphs, $\phi = \pi/3$.

(a) Displacement x as a function of time t

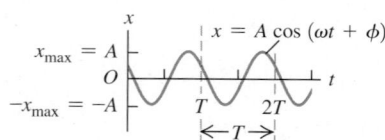

(b) Velocity v_x as a function of time t

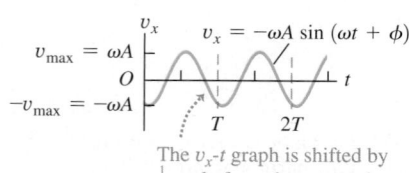

The v_x-t graph is shifted by $\frac{1}{4}$ cycle from the x-t graph.

(c) Acceleration a_x as a function of time t

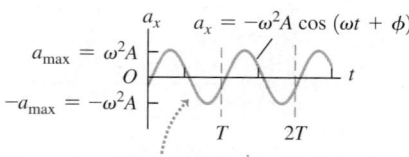

The a_x-t graph is shifted by $\frac{1}{4}$ cycle from the v_x-t graph and by $\frac{1}{2}$ cycle from the x-t graph.

The cosine function in Eq. (14.13) repeats itself whenever time t increases by one period T, or when $\omega t + \phi$ increases by 2π radians. Thus, if we start at time $t = 0$, the time T to complete one cycle is

$$\omega T = \sqrt{\frac{k}{m}}\,T = 2\pi \qquad \text{or} \qquad T = 2\pi\sqrt{\frac{m}{k}}$$

which is just Eq. (14.12). Changing either m or k changes the period T (**Figs. 14.10a** and 14.10b), but T does not depend on the amplitude A (Fig. 14.10c).

The constant ϕ in Eq. (14.13) is called the **phase angle.** It tells us at what point in the cycle the motion was at $t = 0$ (equivalent to where around the circle the point Q was at $t = 0$). We denote the displacement at $t = 0$ by x_0. Putting $t = 0$ and $x = x_0$ in Eq. (14.13), we get

$$x_0 = A\cos\phi \tag{14.14}$$

If $\phi = 0$, then $x_0 = A\cos 0 = A$, and the body starts at its maximum positive displacement. If $\phi = \pi$, then $x_0 = A\cos\pi = -A$, and the particle starts at its maximum *negative* displacement. If $\phi = \pi/2$, then $x_0 = A\cos(\pi/2) = 0$, and the particle is initially at the origin. **Figure 14.11** shows the displacement x versus time for three different phase angles.

We find the velocity v_x and acceleration a_x as functions of time for a harmonic oscillator by taking derivatives of Eq. (14.13) with respect to time:

$$v_x = \frac{dx}{dt} = -\omega A\sin(\omega t + \phi) \qquad \text{(velocity in SHM)} \tag{14.15}$$

$$a_x = \frac{dv_x}{dt} = \frac{d^2x}{dt^2} = -\omega^2 A\cos(\omega t + \phi) \qquad \text{(acceleration in SHM)} \tag{14.16}$$

The velocity v_x oscillates between $v_{\text{max}} = +\omega A$ and $-v_{\text{max}} = -\omega A$, and the acceleration a_x oscillates between $a_{\text{max}} = +\omega^2 A$ and $-a_{\text{max}} = -\omega^2 A$ (**Fig. 14.12**). Comparing Eq. (14.16) with Eq. (14.13) and recalling that $\omega^2 = k/m$ from Eq. (14.9), we see that

$$a_x = -\omega^2 x = -\frac{k}{m}x$$

which is just Eq. (14.4) for simple harmonic motion. This confirms that Eq. (14.13) for x as a function of time is correct.

We actually derived Eq. (14.16) earlier in a geometrical way by taking the x-component of the acceleration vector of the reference point Q. This was done in Fig. 14.6b and Eq. (14.7) (recall that $\theta = \omega t + \phi$). In the same way, we could have derived Eq. (14.15) by taking the x-component of the velocity vector of Q, as shown in Fig. 14.6b. We'll leave the details for you to work out.

Note that the sinusoidal graph of displacement versus time (Fig. 14.12a) is shifted by one-quarter period from the graph of velocity versus time (Fig. 14.12b) and by one-half period from the graph of acceleration versus time (Fig. 14.12c).

Figure 14.13 shows why this is so. When the body is passing through the equilibrium position so that $x = 0$, the velocity equals either v_{max} or $-v_{max}$ (depending on which way the body is moving) and the acceleration is zero. When the body is at either its most positive displacement, $x = +A$, or its most negative displacement, $x = -A$, the velocity is zero and the body is instantaneously at rest. At these points, the restoring force $F_x = -kx$ and the acceleration of the body have their maximum magnitudes. At $x = +A$ the acceleration is negative and equal to $-a_{max}$. At $x = -A$ the acceleration is positive: $a_x = +a_{max}$.

Here's how we can determine the amplitude A and phase angle ϕ for an oscillating body if we are given its initial displacement x_0 and initial velocity v_{0x}. The initial velocity v_{0x} is the velocity at time $t = 0$; putting $v_x = v_{0x}$ and $t = 0$ in Eq. (14.15), we find

$$v_{0x} = -\omega A \sin \phi \qquad (14.17)$$

To find ϕ, we divide Eq. (14.17) by Eq. (14.14). This eliminates A and gives an equation that we can solve for ϕ:

$$\frac{v_{0x}}{x_0} = \frac{-\omega A \sin \phi}{A \cos \phi} = -\omega \tan \phi$$

$$\phi = \arctan\left(-\frac{v_{0x}}{\omega x_0}\right) \qquad \text{(phase angle in SHM)} \qquad (14.18)$$

It is also easy to find the amplitude A if we are given x_0 and v_{0x}. We'll sketch the derivation, and you can fill in the details. Square Eq. (14.14); then divide Eq. (14.17) by ω, square it, and add to the square of Eq. (14.14). The right side will be $A^2(\sin^2 \phi + \cos^2 \phi)$, which is equal to A^2. The final result is

$$A = \sqrt{x_0^2 + \frac{v_{0x}^2}{\omega^2}} \qquad \text{(amplitude in SHM)} \qquad (14.19)$$

Note that when the body has both an initial displacement x_0 and a nonzero initial velocity v_{0x}, the amplitude A is *not* equal to the initial displacement. That's reasonable; if you start the body at a positive x_0 but give it a positive velocity v_{0x}, it will go *farther* than x_0 before it turns and comes back, and so $A > x_0$.

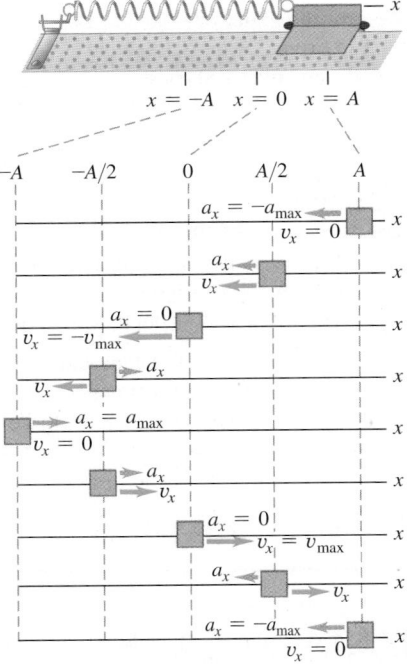

14.13 How x-velocity v_x and x-acceleration a_x vary during one cycle of SHM.

| PROBLEM-SOLVING STRATEGY 14.1 | SIMPLE HARMONIC MOTION I: DESCRIBING MOTION |

IDENTIFY *the relevant concepts:* An oscillating system undergoes simple harmonic motion (SHM) *only* if the restoring force is directly proportional to the displacement.

SET UP *the problem* using the following steps:
1. Identify the known and unknown quantities, and determine which are the target variables.
2. Distinguish between two kinds of quantities. *Properties of the system* include the mass m, the force constant k, and quantities derived from m and k, such as the period T, frequency f, and angular frequency ω. These are independent of *properties of the motion*, which describe how the system behaves when it is set into motion in a particular way; they include the amplitude A, maximum velocity v_{max}, and phase angle ϕ, and values of x, v_x, and a_x at particular times.
3. If necessary, define an x-axis as in Fig. 14.13, with the equilibrium position at $x = 0$.

EXECUTE *the solution* as follows:
1. Use the equations given in Sections 14.1 and 14.2 to solve for the target variables.
2. To find the values of x, v_x, and a_x at particular times, use Eqs. (14.13), (14.15), and (14.16), respectively. If both the initial displacement x_0 and initial velocity v_{0x} are given, determine ϕ and A from Eqs. (14.18) and (14.19). If the body has an initial positive displacement x_0 but zero initial velocity ($v_{0x} = 0$), then the amplitude is $A = x_0$ and the phase angle is $\phi = 0$. If it has an initial positive velocity v_{0x} but no initial displacement ($x_0 = 0$), the amplitude is $A = v_{0x}/\omega$ and the phase angle is $\phi = -\pi/2$. Express all phase angles in *radians*.

EVALUATE *your answer:* Make sure that your results are consistent. For example, suppose you used x_0 and v_{0x} to find general expressions for x and v_x at time t. If you substitute $t = 0$ into these expressions, you should get back the given values of x_0 and v_{0x}.

EXAMPLE 14.3 DESCRIBING SHM

We give the glider of Example 14.2 an initial displacement $x_0 = +0.015$ m and an initial velocity $v_{0x} = +0.40$ m/s. (a) Find the period, amplitude, and phase angle of the resulting motion. (b) Write equations for the displacement, velocity, and acceleration as functions of time.

SOLUTION

IDENTIFY and SET UP: As in Example 14.2, the oscillations are SHM. We use equations from this section and the given values $k = 200$ N/m, $m = 0.50$ kg, x_0, and v_{0x} to calculate the target variables A and ϕ and to obtain expressions for x, v_x, and a_x.

EXECUTE: (a) In SHM the period and angular frequency are *properties of the system* that depend on only k and m, not on the amplitude, and so are the same as in Example 14.2 ($T = 0.31$ s and $\omega = 20$ rad/s). From Eq. (14.19), the amplitude is

$$A = \sqrt{x_0^2 + \frac{v_{0x}^2}{\omega^2}} = \sqrt{(0.015 \text{ m})^2 + \frac{(0.40 \text{ m/s})^2}{(20 \text{ rad/s})^2}} = 0.025 \text{ m}$$

We use Eq. (14.18) to find the phase angle:

$$\phi = \arctan\left(-\frac{v_{0x}}{\omega x_0}\right)$$

$$= \arctan\left(-\frac{0.40 \text{ m/s}}{(20 \text{ rad/s})(0.015 \text{ m})}\right) = -53° = -0.93 \text{ rad}$$

(b) The displacement, velocity, and acceleration at any time are given by Eqs. (14.13), (14.15), and (14.16), respectively. We substitute the values of A, ω, and ϕ into these equations:

$$x = (0.025 \text{ m})\cos[(20 \text{ rad/s})t - 0.93 \text{ rad}]$$
$$v_x = -(0.50 \text{ m/s})\sin[(20 \text{ rad/s})t - 0.93 \text{ rad}]$$
$$a_x = -(10 \text{ m/s}^2)\cos[(20 \text{ rad/s})t - 0.93 \text{ rad}]$$

EVALUATE: You can check the expressions for x and v_x by confirming that if you substitute $t = 0$, they yield $x = x_0 = 0.015$ m and $v_x = v_{0x} = 0.40$ m/s.

DATA *SPEAKS*

Oscillations and SHM

When students were given a problem about oscillations and SHM, more than 26% gave an incorrect response. Common errors:

- Forgetting that the period *T* is the time for one complete cycle of motion, *not* the time to travel between $x = -A$ and $x = +A$.

- Not using Eq. (14.18) to determine the phase angle ϕ.

PhET: Masses & Springs

TEST YOUR UNDERSTANDING OF SECTION 14.2 A glider is attached to a spring as shown in Fig. 14.13. If the glider is moved to $x = 0.10$ m and released from rest at time $t = 0$, it will oscillate with amplitude $A = 0.10$ m and phase angle $\phi = 0$. (a) Suppose instead that at $t = 0$ the glider is at $x = 0.10$ m and is moving to the right in Fig. 14.13. In this situation is the amplitude greater than, less than, or equal to 0.10 m? Is the phase angle greater than, less than, or equal to zero? (b) Suppose instead that at $t = 0$ the glider is at $x = 0.10$ m and is moving to the left in Fig. 14.13. In this situation is the amplitude greater than, less than, or equal to 0.10 m? Is the phase angle greater than, less than, or equal to zero? ❙

14.3 ENERGY IN SIMPLE HARMONIC MOTION

We can learn even more about simple harmonic motion by using energy considerations. The only horizontal force on the body in SHM in Figs. 14.2 and 14.13 is the conservative force exerted by an ideal spring. The vertical forces do no work, so the total mechanical energy of the system is *conserved*. We also assume that the mass of the spring itself is negligible.

The kinetic energy of the body is $K = \frac{1}{2}mv^2$ and the potential energy of the spring is $U = \frac{1}{2}kx^2$, just as in Section 7.2. There are no nonconservative forces that do work, so the total mechanical energy $E = K + U$ is conserved:

$$E = \frac{1}{2}mv_x^2 + \frac{1}{2}kx^2 = \text{constant} \tag{14.20}$$

(Since the motion is one-dimensional, $v^2 = v_x^2$.)

The total mechanical energy E is also directly related to the amplitude A of the motion. When the body reaches the point $x = A$, its maximum displacement from equilibrium, it momentarily stops as it turns back toward the equilibrium position. That is, when $x = A$ (or $-A$), $v_x = 0$. At this point the energy is entirely potential, and $E = \frac{1}{2}kA^2$. Because E is constant, it is equal to $\frac{1}{2}kA^2$ at any other point. Combining this expression with Eq. (14.20), we get

| Total mechanical energy in simple harmonic motion | $E = \frac{1}{2}mv_x^2 + \frac{1}{2}kx^2 = \frac{1}{2}kA^2 = \text{constant}$ | (14.21) |

Mass   Force constant of restoring force

Velocity   Displacement   Amplitude

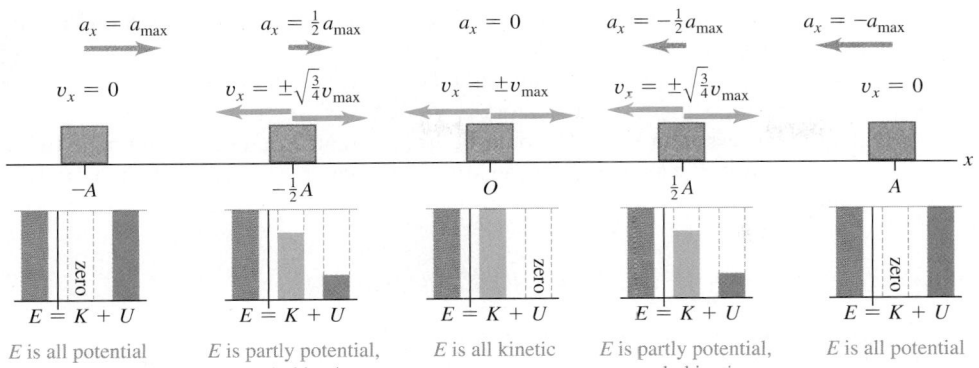

14.14 Graphs of E, K, and U versus displacement in SHM. The velocity of the body is *not* constant, so these images of the body at equally spaced positions are *not* equally spaced in time.

$E = K + U$ $E = K + U$ $E = K + U$ $E = K + U$ $E = K + U$

E is all potential energy. E is partly potential, partly kinetic energy. E is all kinetic energy. E is partly potential, partly kinetic energy. E is all potential energy.

We can verify this equation by substituting x and v_x from Eqs. (14.13) and (14.15) and using $\omega^2 = k/m$ from Eq. (14.9):

$$E = \tfrac{1}{2}mv_x^2 + \tfrac{1}{2}kx^2 = \tfrac{1}{2}m[-\omega A \sin(\omega t + \phi)]^2 + \tfrac{1}{2}k[A\cos(\omega t + \phi)]^2$$

$$= \tfrac{1}{2}kA^2 \sin^2(\omega t + \phi) + \tfrac{1}{2}kA^2 \cos^2(\omega t + \phi) = \tfrac{1}{2}kA^2$$

(Recall that $\sin^2\alpha + \cos^2\alpha = 1$.) Hence our expressions for displacement and velocity in SHM are consistent with energy conservation, as they must be.

We can use Eq. (14.21) to solve for the velocity v_x of the body at a given displacement x:

$$v_x = \pm\sqrt{\frac{k}{m}}\sqrt{A^2 - x^2} \tag{14.22}$$

The $\pm$ sign means that at a given value of x the body can be moving in either direction. For example, when $x = \pm A/2$,

$$v_x = \pm\sqrt{\frac{k}{m}}\sqrt{A^2 - \left(\pm\frac{A}{2}\right)^2} = \pm\sqrt{\frac{3}{4}}\sqrt{\frac{k}{m}}A$$

Equation (14.22) also shows that the *maximum* speed v_{max} occurs at $x = 0$. Using Eq. (14.10), $\omega = \sqrt{k/m}$, we find that

$$v_{max} = \sqrt{\frac{k}{m}}A = \omega A \tag{14.23}$$

This agrees with Eq. (14.15): v_x oscillates between $-\omega A$ and $+\omega A$.

Interpreting *E*, *K*, and *U* in SHM

Figure 14.14 shows the energy quantities E, K, and U at $x = 0$, $x = \pm A/2$, and $x = \pm A$. **Figure 14.15** is a graphical display of Eq. (14.21); energy (kinetic, potential, and total) is plotted vertically and the coordinate x is plotted horizontally. The parabolic curve in Fig. 14.15a represents the potential energy $U = \tfrac{1}{2}kx^2$. The horizontal line represents the total mechanical energy E, which is constant and does not vary with x. At any value of x between $-A$ and A, the vertical distance from the x-axis to the parabola is U; since $E = K + U$, the remaining vertical distance up to the horizontal line is K. Figure 14.15b shows both K and U as functions of x. The horizontal line for E intersects the potential-energy curve at $x = -A$ and $x = A$, so at these points the energy is entirely potential, the kinetic energy is zero, and the body comes momentarily to rest before reversing

14.15 Kinetic energy K, potential energy U, and total mechanical energy E as functions of displacement for SHM. At each value of x the sum of the values of K and U equals the constant value of E. Can you show that the energy is half kinetic and half potential at $x = \pm\sqrt{\tfrac{1}{2}}A$?

(a) The potential energy U and total mechanical energy E for a body in SHM as a function of displacement x

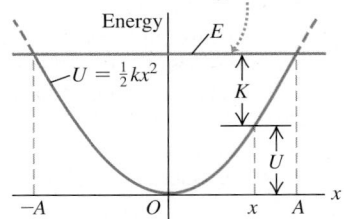

The total mechanical energy E is constant.

(b) The same graph as in **(a)**, showing kinetic energy K as well

At $x = \pm A$ the energy is all potential; $K = 0$.

At $x = 0$ the energy is all kinetic; $U = 0$.

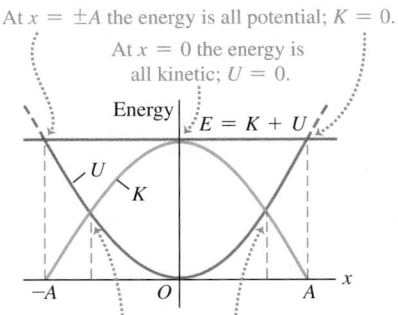

At these points the energy is half kinetic and half potential.

direction. As the body oscillates between $-A$ and A, the energy is continuously transformed from potential to kinetic and back again.

Figure 14.15a shows the connection between the amplitude A and the corresponding total mechanical energy $E = \frac{1}{2}kA^2$. If we tried to make x greater than A (or less than $-A$), U would be greater than E, and K would have to be negative. But K can never be negative, so x can't be greater than A or less than $-A$.

PROBLEM-SOLVING STRATEGY 14.2 | **SIMPLE HARMONIC MOTION II: ENERGY**

The SHM energy equation, Eq. (14.21), is a useful relationship among velocity, displacement, and total mechanical energy. If a problem requires you to relate displacement, velocity, and acceleration without reference to time, consider using Eq. (14.4) (from Newton's second law) or Eq. (14.21) (from energy conservation).

Because Eq. (14.21) involves x^2 and v_x^2, you must infer the *signs* of x and v_x from the situation. For instance, if the body is moving from the equilibrium position toward the point of greatest positive displacement, then x is positive and v_x is positive.

EXAMPLE 14.4 **VELOCITY, ACCELERATION, AND ENERGY IN SHM**

(a) Find the maximum and minimum velocities attained by the oscillating glider of Example 14.2. (b) Find the maximum and minimum accelerations. (c) Find the velocity v_x and acceleration a_x when the glider is halfway from its initial position to the equilibrium position $x = 0$. (d) Find the total energy, potential energy, and kinetic energy at this position.

SOLUTION

IDENTIFY and SET UP: The problem concerns properties of the motion at specified *positions*, not at specified *times*, so we can use the energy relationships of this section. Figure 14.13 shows our choice of x-axis. The maximum displacement from equilibrium is $A = 0.020$ m. We use Eqs. (14.22) and (14.4) to find v_x and a_x for a given x. We then use Eq. (14.21) for given x and v_x to find the total, potential, and kinetic energies E, U, and K.

EXECUTE: (a) From Eq. (14.22), the velocity v_x at any displacement x is

$$v_x = \pm\sqrt{\frac{k}{m}}\sqrt{A^2 - x^2}$$

The glider's maximum *speed* occurs when it is moving through $x = 0$:

$$v_{max} = \sqrt{\frac{k}{m}}\,A = \sqrt{\frac{200\text{ N/m}}{0.50\text{ kg}}}(0.020\text{ m}) = 0.40\text{ m/s}$$

Its maximum and minimum (most negative) *velocities* are $+0.40$ m/s and -0.40 m/s, which occur when it is moving through $x = 0$ to the right and left, respectively.

(b) From Eq. (14.4), $a_x = -(k/m)x$. The glider's maximum (most positive) acceleration occurs at the most negative value of x, $x = -A$:

$$a_{max} = -\frac{k}{m}(-A) = -\frac{200\text{ N/m}}{0.50\text{ kg}}(-0.020\text{ m}) = 8.0\text{ m/s}^2$$

The minimum (most negative) acceleration is $a_{min} = -8.0\text{ m/s}^2$, which occurs at $x = +A = +0.020$ m.

(c) The point halfway from $x = x_0 = A$ to $x = 0$ is $x = A/2 = 0.010$ m. From Eq. (14.22), at this point

$$v_x = -\sqrt{\frac{200\text{ N/m}}{0.50\text{ kg}}}\sqrt{(0.020\text{ m})^2 - (0.010\text{ m})^2}$$

$$= -0.35\text{ m/s}$$

We choose the negative square root because the glider is moving from $x = A$ toward $x = 0$. From Eq. (14.4),

$$a_x = -\frac{200\text{ N/m}}{0.50\text{ kg}}(0.010\text{ m}) = -4.0\text{ m/s}^2$$

Figure 14.14 shows the conditions at $x = 0$, $\pm A/2$, and $\pm A$.

(d) The energies are

$$E = \tfrac{1}{2}kA^2 = \tfrac{1}{2}(200\text{ N/m})(0.020\text{ m})^2 = 0.040\text{ J}$$

$$U = \tfrac{1}{2}kx^2 = \tfrac{1}{2}(200\text{ N/m})(0.010\text{ m})^2 = 0.010\text{ J}$$

$$K = \tfrac{1}{2}mv_x^2 = \tfrac{1}{2}(0.50\text{ kg})(-0.35\text{ m/s})^2 = 0.030\text{ J}$$

EVALUATE: At $x = A/2$, the total energy is one-fourth potential energy and three-fourths kinetic energy. You can confirm this by inspecting Fig. 14.15b.

EXAMPLE 14.5 ENERGY AND MOMENTUM IN SHM

A block of mass M attached to a horizontal spring with force constant k is moving in SHM with amplitude A_1. As the block passes through its equilibrium position, a lump of putty of mass m is dropped from a small height and sticks to it. (a) Find the new amplitude and period of the motion. (b) Repeat part (a) if the putty is dropped onto the block when it is at one end of its path.

SOLUTION

IDENTIFY and SET UP: The problem involves the motion at a given position, not a given time, so we can use energy methods. **Figure 14.16** shows our sketches. Before the putty falls, the mechanical energy of the block–spring system is constant. In part (a), the putty–block collision is completely inelastic: The horizontal component of momentum is conserved, kinetic energy decreases, and the amount of mass that's oscillating increases. After the collision, the mechanical energy remains constant at its new value. In part (b) the oscillating mass also increases, but the block isn't moving when the putty is added; there is effectively no collision at all, and no mechanical energy is lost. We find the amplitude A_2 after each collision from the final energy of the system by using Eq. (14.21) and conservation of momentum. The period T_2 after the collision is the same in both parts (a) and (b) because the final mass is the same; we find it by using Eq. (14.12).

EXECUTE: (a) Before the collision the total mechanical energy of the block and spring is $E_1 = \frac{1}{2}kA_1^2$. The block is at $x = 0$, so $U = 0$ and the energy is purely kinetic (Fig. 14.16a). If we let v_1 be the speed of the block at this point, then $E_1 = \frac{1}{2}kA_1^2 = \frac{1}{2}Mv_1^2$ and

$$v_1 = \sqrt{\frac{k}{M}}A_1$$

During the collision the x-component of momentum of the block–putty system is conserved. (Why?) Just before the collision this component is the sum of Mv_1 (for the block) and zero (for the putty). Just after the collision the block and putty move together with speed v_2, so their combined x-component of momentum is $(M + m)v_2$. From conservation of momentum,

$$Mv_1 + 0 = (M + m)v_2 \quad \text{so} \quad v_2 = \frac{M}{M + m}v_1$$

The collision lasts a very short time, so the block and putty are still at the equilibrium position just after the collision. The energy is still purely kinetic but is *less* than before the collision:

$$E_2 = \frac{1}{2}(M + m)v_2^2 = \frac{1}{2}\frac{M^2}{M + m}v_1^2$$

$$= \frac{M}{M + m}\left(\frac{1}{2}Mv_1^2\right) = \left(\frac{M}{M + m}\right)E_1$$

14.16 Our sketches for this problem.

(a)
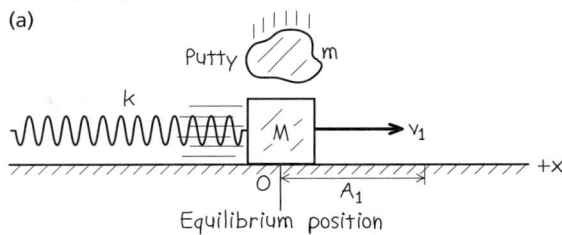

(b)
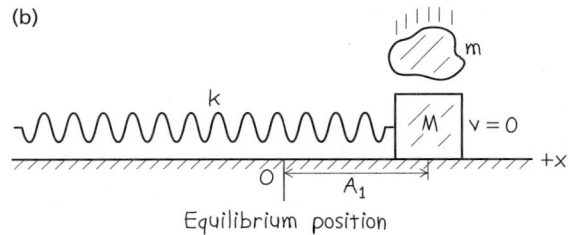

Since $E_2 = \frac{1}{2}kA_2^2$, where A_2 is the amplitude after the collision,

$$\frac{1}{2}kA_2^2 = \left(\frac{M}{M + m}\right)\frac{1}{2}kA_1^2$$

$$A_2 = A_1\sqrt{\frac{M}{M + m}}$$

From Eq. (14.12), the period of oscillation after the collision is

$$T_2 = 2\pi\sqrt{\frac{M + m}{k}}$$

(b) When the putty falls, the block is instantaneously at rest (Fig. 14.16b). The x-component of momentum is zero both before and after the collision. The block and putty have zero kinetic energy just before and just after the collision. The energy is all potential energy stored in the spring, so adding the putty has *no effect* on the mechanical energy. That is, $E_2 = E_1 = \frac{1}{2}kA_1^2$, and the amplitude is unchanged: $A_2 = A_1$. The period is again $T_2 = 2\pi\sqrt{(M + m)/k}$.

EVALUATE: Energy is lost in part (a) because the putty slides against the moving block during the collision, and energy is dissipated by kinetic friction. No energy is lost in part (b) because there is no sliding during the collision.

TEST YOUR UNDERSTANDING OF SECTION 14.3 (a) To double the total energy for a mass-spring system oscillating in SHM, by what factor must the amplitude increase? (i) 4; (ii) 2; (iii) $\sqrt{2} = 1.414$; (iv) $\sqrt[4]{2} = 1.189$. (b) By what factor will the frequency change due to this amplitude increase? (i) 4; (ii) 2; (iii) $\sqrt{2} = 1.414$; (iv) $\sqrt[4]{2} = 1.189$; (v) it does not change. ∎

14.17 A body attached to a hanging spring.

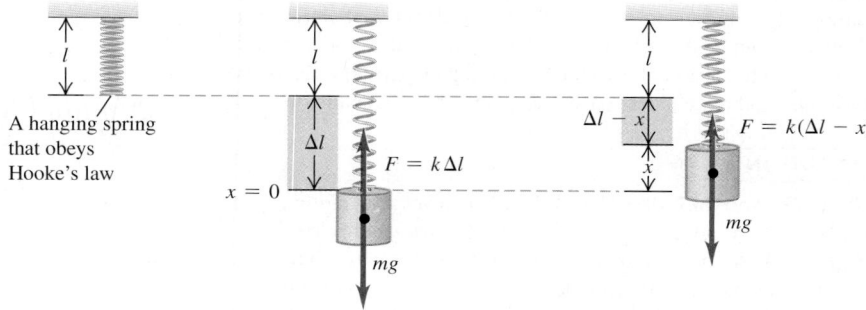

(a)

(b) A body is suspended from the spring. It is in equilibrium when the upward force exerted by the stretched spring equals the body's weight.

(c) If the body is displaced from equilibrium, the net force on the body is proportional to its displacement. The oscillations are SHM.

A hanging spring that obeys Hooke's law

$F = k\,\Delta l$

$x = 0$

$\Delta l - x$

$F = k(\Delta l - x)$

mg

mg

14.4 APPLICATIONS OF SIMPLE HARMONIC MOTION

So far, we've looked at a grand total of *one* situation in which simple harmonic motion (SHM) occurs: a body attached to an ideal horizontal spring. But SHM can occur in any system in which there is a restoring force that is directly proportional to the displacement from equilibrium, as given by Eq. (14.3), $F_x = -kx$. The restoring force originates in different ways in different situations, so we must find the force constant k for each case by examining the net force on the system. Once this is done, it's straightforward to find the angular frequency ω, frequency f, and period T; we just substitute the value of k into Eqs. (14.10), (14.11), and (14.12), respectively. Let's use these ideas to examine several examples of simple harmonic motion.

Vertical SHM

Suppose we hang a spring with force constant k (**Fig. 14.17a**) and suspend from it a body with mass m. Oscillations will now be vertical; will they still be SHM? In Fig. 14.17b the body hangs at rest, in equilibrium. In this position the spring is stretched an amount Δl just great enough that the spring's upward vertical force $k\,\Delta l$ on the body balances its weight mg:

$$k\,\Delta l = mg$$

14.18 If the weight mg compresses the spring a distance Δl, the force constant is $k = mg/\Delta l$ and the angular frequency for vertical SHM is $\omega = \sqrt{k/m}$—the same as if the body were suspended from the spring (see Fig. 14.17).

A body is placed atop the spring. It is in equilibrium when the upward force exerted by the compressed spring equals the body's weight.

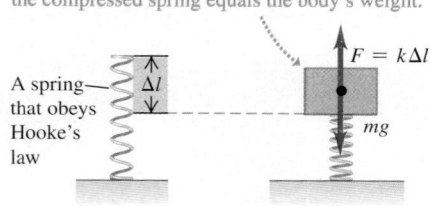

A spring that obeys Hooke's law

Δl

$F = k\,\Delta l$

mg

Take $x = 0$ to be this equilibrium position and take the positive x-direction to be upward. When the body is a distance x *above* its equilibrium position (Fig. 14.17c), the extension of the spring is $\Delta l - x$. The upward force it exerts on the body is then $k(\Delta l - x)$, and the net x-component of force on the body is

$$F_{net} = k(\Delta l - x) + (-mg) = -kx$$

that is, a net downward force of magnitude kx. Similarly, when the body is *below* the equilibrium position, there is a net upward force with magnitude kx. In either case there is a restoring force with magnitude kx. If the body is set in vertical motion, it oscillates in SHM with the same angular frequency as though it were horizontal, $\omega = \sqrt{k/m}$. So vertical SHM doesn't differ in any essential way from horizontal SHM. The only real change is that the equilibrium position $x = 0$ no longer corresponds to the point at which the spring is unstretched. The same ideas hold if a body with weight mg is placed atop a compressible spring (**Fig. 14.18**) and compresses it a distance Δl.

EXAMPLE 14.6   VERTICAL SHM IN AN OLD CAR

The shock absorbers in an old car with mass 1000 kg are completely worn out. When a 980-N person climbs slowly into the car at its center of gravity, the car sinks 2.8 cm. The car (with the person aboard) hits a bump, and the car starts oscillating up and down in SHM. Model the car and person as a single body on a single spring, and find the period and frequency of the oscillation.

SOLUTION

IDENTIFY and SET UP: The situation is like that shown in Fig. 14.18. The compression of the spring when the person's weight is added tells us the force constant, which we can use to find the period and frequency (the target variables).

EXECUTE: When the force increases by 980 N, the spring compresses an additional 0.028 m, and the x-coordinate of the car

changes by -0.028 m. Hence the effective force constant (including the effect of the entire suspension) is

$$k = -\frac{F_x}{x} = -\frac{980 \text{ N}}{-0.028 \text{ m}} = 3.5 \times 10^4 \text{ kg/s}^2$$

The person's mass is $w/g = (980 \text{ N})/(9.8 \text{ m/s}^2) = 100$ kg. The *total* oscillating mass is $m = 1000$ kg $+ 100$ kg $= 1100$ kg. The period T is

$$T = 2\pi\sqrt{\frac{m}{k}} = 2\pi\sqrt{\frac{1100 \text{ kg}}{3.5 \times 10^4 \text{ kg/s}^2}} = 1.11 \text{ s}$$

The frequency is $f = 1/T = 1/(1.11 \text{ s}) = 0.90$ Hz.

EVALUATE: A persistent oscillation with a period of about 1 second makes for a very unpleasant ride. The purpose of shock absorbers is to make such oscillations die out (see Section 14.7).

Angular SHM

A mechanical watch keeps time based on the oscillations of a balance wheel (**Fig. 14.19**). The wheel has a moment of inertia I about its axis. A coil spring exerts a restoring torque τ_z that is proportional to the angular displacement θ from the equilibrium position. We write $\tau_z = -\kappa\theta$, where κ (the Greek letter kappa) is a constant called the *torsion constant*. Using the rotational analog of Newton's second law for a rigid body, $\sum\tau_z = I\alpha_z = I\,d^2\theta/dt^2$, Eq. (10.7), we find

$$-\kappa\theta = I\alpha \qquad \text{or} \qquad \frac{d^2\theta}{dt^2} = -\frac{\kappa}{I}\theta$$

This equation is exactly the same as Eq. (14.4) for simple harmonic motion, with x replaced by θ and k/m replaced by κ/I. So we are dealing with a form of *angular* simple harmonic motion. The angular frequency ω and frequency f are given by Eqs. (14.10) and (14.11), respectively, with the same replacement:

14.19 The balance wheel of a mechanical watch. The spring exerts a restoring torque that is proportional to the angular displacement θ, so the motion is angular SHM.

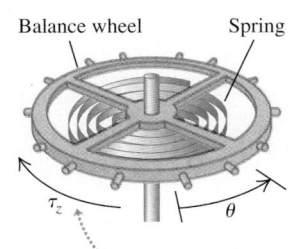

Balance wheel   Spring

τ_z θ

The spring torque τ_z opposes the angular displacement θ.

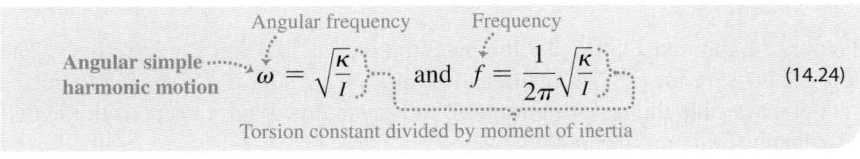

Angular simple harmonic motion

Angular frequency   Frequency

$$\omega = \sqrt{\frac{\kappa}{I}} \qquad \text{and} \qquad f = \frac{1}{2\pi}\sqrt{\frac{\kappa}{I}} \qquad (14.24)$$

Torsion constant divided by moment of inertia

The angular displacement θ as a function of time is given by

$$\theta = \Theta\cos(\omega t + \phi)$$

where Θ (the capital Greek letter theta) plays the role of an angular amplitude.

It's a good thing that the motion of a balance wheel *is* simple harmonic. If it weren't, the frequency might depend on the amplitude, and the watch would run too fast or too slow as the spring ran down.

Vibrations of Molecules

The following discussion of the vibrations of molecules uses the binomial theorem. If you aren't familiar with this theorem, you should read about it in the appropriate section of a math textbook.

14.20 (a) Two atoms with centers separated by r. (b) Potential energy U and (c) force F_r in the van der Waals interaction.

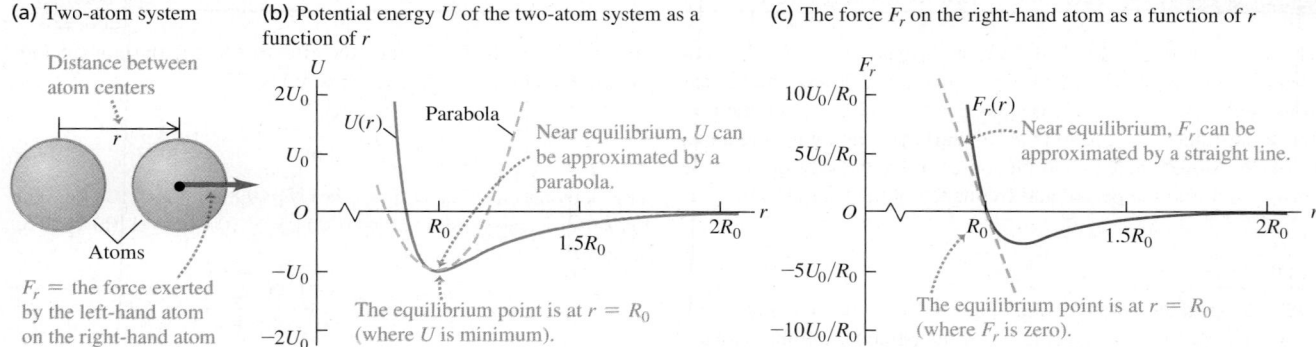

(a) Two-atom system

Distance between atom centers

r

Atoms

F_r = the force exerted by the left-hand atom on the right-hand atom

(b) Potential energy U of the two-atom system as a function of r

Parabola

$U(r)$

Near equilibrium, U can be approximated by a parabola.

The equilibrium point is at $r = R_0$ (where U is minimum).

(c) The force F_r on the right-hand atom as a function of r

$F_r(r)$

Near equilibrium, F_r can be approximated by a straight line.

The equilibrium point is at $r = R_0$ (where F_r is zero).

When two atoms are separated by a few atomic diameters, they can exert attractive forces on each other. But if the atoms are so close that their electron shells overlap, the atoms repel each other. Between these limits, there can be an equilibrium separation distance at which two atoms form a *molecule*. If these atoms are displaced slightly from equilibrium, they will oscillate.

Let's consider one type of interaction between atoms called the *van der Waals interaction*. Our immediate task here is to study oscillations, so we won't go into the details of how this interaction arises. Let the center of one atom be at the origin and let the center of the other atom be a distance r away (**Fig. 14.20a**); the equilibrium distance between centers is $r = R_0$. Experiment shows that the van der Waals interaction can be described by the potential-energy function

$$U = U_0 \left[\left(\frac{R_0}{r} \right)^{12} - 2 \left(\frac{R_0}{r} \right)^6 \right] \tag{14.25}$$

where U_0 is a positive constant with units of joules. When the two atoms are very far apart, $U = 0$; when they are separated by the equilibrium distance $r = R_0$, $U = -U_0$. From Section 7.4, the force on the second atom is the negative derivative of Eq. (14.25):

$$F_r = -\frac{dU}{dr} = U_0 \left[\frac{12R_0^{12}}{r^{13}} - 2 \frac{6R_0^6}{r^7} \right] = 12 \frac{U_0}{R_0} \left[\left(\frac{R_0}{r} \right)^{13} - \left(\frac{R_0}{r} \right)^7 \right] \tag{14.26}$$

Figures 14.20b and 14.20c plot the potential energy and force, respectively. The force is positive for $r < R_0$ and negative for $r > R_0$, so it is a *restoring* force.

Let's examine the restoring force F_r in Eq. (14.26). We let x represent the displacement from equilibrium:

$$x = r - R_0 \qquad \text{so} \qquad r = R_0 + x$$

In terms of x, the force F_r in Eq. (14.26) becomes

$$F_r = 12 \frac{U_0}{R_0} \left[\left(\frac{R_0}{R_0 + x} \right)^{13} - \left(\frac{R_0}{R_0 + x} \right)^7 \right]$$

$$= 12 \frac{U_0}{R_0} \left[\frac{1}{(1 + x/R_0)^{13}} - \frac{1}{(1 + x/R_0)^7} \right] \tag{14.27}$$

This looks nothing like Hooke's law, $F_x = -kx$, so we might be tempted to conclude that molecular oscillations cannot be SHM. But let us restrict ourselves to *small-amplitude* oscillations so that the absolute value of the displacement x is small in comparison to R_0 and the absolute value of the ratio x/R_0 is much less than 1. We can then simplify Eq. (14.27) by using the *binomial theorem:*

$$(1 + u)^n = 1 + nu + \frac{n(n - 1)}{2!}u^2 + \frac{n(n - 1)(n - 2)}{3!}u^3 + \cdots \quad (14.28)$$

If $|u|$ is much less than 1, each successive term in Eq. (14.28) is much smaller than the one it follows, and we can safely approximate $(1 + u)^n$ by just the first two terms. In Eq. (14.27), u is replaced by x/R_0 and n equals -13 or -7, so

$$\frac{1}{(1 + x/R_0)^{13}} = (1 + x/R_0)^{-13} \approx 1 + (-13)\frac{x}{R_0}$$

$$\frac{1}{(1 + x/R_0)^7} = (1 + x/R_0)^{-7} \approx 1 + (-7)\frac{x}{R_0}$$

$$F_r \approx 12\frac{U_0}{R_0}\left[\left(1 + (-13)\frac{x}{R_0}\right) - \left(1 + (-7)\frac{x}{R_0}\right)\right] = -\left(\frac{72U_0}{R_0^2}\right)x \quad (14.29)$$

This is just Hooke's law, with force constant $k = 72U_0/R_0^2$. (Note that k has the correct units, J/m^2 or N/m.) So oscillations of molecules bound by the van der Waals interaction can be simple harmonic motion, provided that the amplitude is small in comparison to R_0 so that the approximation $|x/R_0| \ll 1$ used in the derivation of Eq. (14.29) is valid.

You can also use the binomial theorem to show that the potential energy U in Eq. (14.25) can be written as $U \approx \frac{1}{2}kx^2 + C$, where $C = -U_0$ and k is again equal to $72U_0/R_0^2$. Adding a constant to the potential-energy function has no effect on the physics, so the system of two atoms is fundamentally no different from a mass attached to a horizontal spring for which $U = \frac{1}{2}kx^2$.

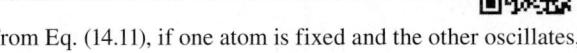

EXAMPLE 14.7 MOLECULAR VIBRATION

Two argon atoms form the molecule Ar_2 as a result of a van der Waals interaction with $U_0 = 1.68 \times 10^{-21}$ J and $R_0 = 3.82 \times 10^{-10}$ m. Find the frequency of small oscillations of one Ar atom about its equilibrium position.

SOLUTION

IDENTIFY and SET UP: This is the situation shown in Fig. 14.20. Because the oscillations are small, we can use Eq. (14.29) to find the force constant k and Eq. (14.11) to find the frequency f of SHM.

EXECUTE: From Eq. (14.29),

$$k = \frac{72U_0}{R_0^2} = \frac{72(1.68 \times 10^{-21} \text{ J})}{(3.82 \times 10^{-10} \text{ m})^2} = 0.829 \text{ J/m}^2 = 0.829 \text{ N/m}$$

(This force constant is comparable to that of a loose toy spring like a Slinky™.) From Appendix D, the average atomic mass of argon is $(39.948 \text{ u})(1.66 \times 10^{-27} \text{ kg/1 u}) = 6.63 \times 10^{-26} \text{ kg}$.

From Eq. (14.11), if one atom is fixed and the other oscillates,

$$f = \frac{1}{2\pi}\sqrt{\frac{k}{m}} = \frac{1}{2\pi}\sqrt{\frac{0.829 \text{ N/m}}{6.63 \times 10^{-26} \text{ kg}}} = 5.63 \times 10^{11} \text{ Hz}$$

EVALUATE: Our answer for f isn't quite right. If no net external force acts on the molecule, its center of mass (halfway between the atoms) doesn't accelerate, so *both* atoms must oscillate with the same amplitude in opposite directions. It turns out that we can account for this by replacing m with $m/2$ in our expression for f. This makes f larger by a factor of $\sqrt{2}$, so the correct frequency is $f = \sqrt{2}(5.63 \times 10^{11} \text{ Hz}) = 7.96 \times 10^{11} \text{ Hz}$. What's more, on the atomic scale we must use *quantum mechanics* rather than Newtonian mechanics to describe motion; happily, quantum mechanics also yields $f = 7.96 \times 10^{11} \text{ Hz}$.

TEST YOUR UNDERSTANDING OF SECTION 14.4 A block attached to a hanging ideal spring oscillates up and down with a period of 10 s on earth. If you take the block and spring to Mars, where the acceleration due to gravity is only about 40% as large as on earth, what will be the new period of oscillation? (i) 10 s; (ii) more than 10 s; (iii) less than 10 s. ∎

14.21 The dynamics of a simple pendulum.

(a) A real pendulum

(b) An idealized simple pendulum

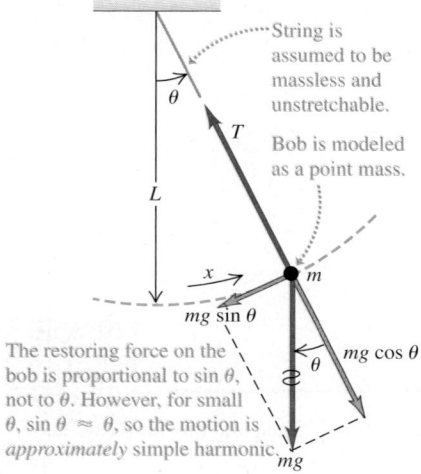

String is assumed to be massless and unstretchable.

Bob is modeled as a point mass.

The restoring force on the bob is proportional to sin θ, not to θ. However, for small θ, sin $\theta \approx \theta$, so the motion is *approximately* simple harmonic.

14.22 For small angular displacements θ, the restoring force $F_\theta = -mg\sin\theta$ on a simple pendulum is approximately equal to $-mg\theta$; that is, it is approximately proportional to the displacement θ. Hence for small angles the oscillations are simple harmonic.

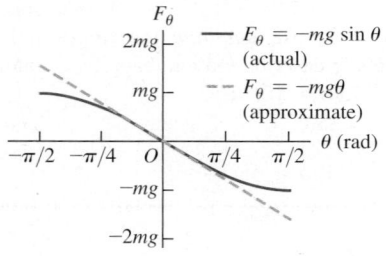

14.5 THE SIMPLE PENDULUM

A **simple pendulum** is an idealized model consisting of a point mass suspended by a massless, unstretchable string. When the point mass is pulled to one side of its straight-down equilibrium position and released, it oscillates about the equilibrium position. Familiar situations such as a wrecking ball on a crane's cable or a person on a swing (**Fig. 14.21a**) can be modeled as simple pendulums.

The path of the point mass (sometimes called a *pendulum bob*) is not a straight line but the arc of a circle with radius L equal to the length of the string (Fig. 14.21b). We use as our coordinate the distance x measured along the arc. If the motion is simple harmonic, the restoring force must be directly proportional to x or (because $x = L\theta$) to θ. Is it?

Figure 14.21b shows the radial and tangential components of the forces on the mass. The restoring force F_θ is the tangential component of the net force:

$$F_\theta = -mg\sin\theta \tag{14.30}$$

Gravity provides the restoring force F_θ; the tension T merely acts to make the point mass move in an arc. Since F_θ is proportional to $\sin\theta$, not to θ, the motion is *not* simple harmonic. However, if angle θ is *small,* $\sin\theta$ is very nearly equal to θ in radians (**Fig. 14.22**). (When $\theta = 0.1$ rad, about $6°$, $\sin\theta = 0.998$. That's only 0.2% different.) With this approximation, Eq. (14.30) becomes

$$F_\theta = -mg\theta = -mg\frac{x}{L} = -\frac{mg}{L}x \tag{14.31}$$

The restoring force is then proportional to the coordinate for small displacements, and the force constant is $k = mg/L$. From Eq. (14.10) the angular frequency ω of a simple pendulum with small amplitude is

Angular frequency of simple pendulum, small amplitude
$$\omega = \sqrt{\frac{k}{m}} = \sqrt{\frac{mg/L}{m}} = \sqrt{\frac{g}{L}} \tag{14.32}$$
Acceleration due to gravity — Pendulum length — Pendulum mass (cancels)

The corresponding frequency and period relationships are

Frequency of simple pendulum, small amplitude
$$f = \frac{\omega}{2\pi} = \frac{1}{2\pi}\sqrt{\frac{g}{L}} \tag{14.33}$$
Angular frequency — Acceleration due to gravity — Pendulum length

Period of simple pendulum, small amplitude
$$T = \frac{2\pi}{\omega} = \frac{1}{f} = 2\pi\sqrt{\frac{L}{g}} \tag{14.34}$$
Angular frequency — Frequency — Pendulum length — Acceleration due to gravity

These expressions don't involve the *mass* of the particle. That's because the gravitational restoring force is proportional to m, so the mass appears on *both* sides of $\sum\vec{F} = m\vec{a}$ and cancels out. (The same physics explains why bodies of different masses fall with the same acceleration in a vacuum.) For small oscillations, the period of a pendulum for a given value of g is determined entirely by its length.

Equations (14.32) through (14.34) tell us that a long pendulum (large L) has a longer period than a shorter one. Increasing g increases the restoring force, causing the frequency to increase and the period to decrease.

The motion of a pendulum is only *approximately* simple harmonic. When the maximum angular displacement Θ (amplitude) is not small, the departures from simple harmonic motion can be substantial. In general, the period T is given by

$$T = 2\pi\sqrt{\frac{L}{g}}\left(1 + \frac{1^2}{2^2}\sin^2\frac{\Theta}{2} + \frac{1^2 \cdot 3^2}{2^2 \cdot 4^2}\sin^4\frac{\Theta}{2} + \cdots\right) \tag{14.35}$$

We can compute T to any desired degree of precision by taking enough terms in the series. You can confirm that when $\Theta = 15°$, the true period is longer than that given by the approximate Eq. (14.34) by less than 0.5%.

A pendulum is a useful timekeeper because the period is *very nearly* independent of amplitude, provided that the amplitude is small. Thus, as a pendulum clock runs down and the amplitude of the swings decreases a little, the clock still keeps very nearly correct time.

EXAMPLE 14.8   A SIMPLE PENDULUM

Find the period and frequency of a simple pendulum 1.000 m long at a location where $g = 9.800$ m/s^2.

SOLUTION

IDENTIFY and SET UP: This is a simple pendulum, so we can use Eq. (14.34) to determine the pendulum's period T from its length and Eq. (14.1) to find the frequency f from T.

EXECUTE: From Eqs. (14.34) and (14.1),

$$T = 2\pi \sqrt{\frac{L}{g}} = 2\pi \sqrt{\frac{1.000 \text{ m}}{9.800 \text{ m/s}^2}} = 2.007 \text{ s}$$

and

$$f = \frac{1}{T} = \frac{1}{2.007 \text{ s}} = 0.4983 \text{ Hz}$$

EVALUATE: The period is almost exactly 2 s. When the metric system was established, the second was *defined* as half the period of a 1-m simple pendulum. This was a poor standard, however, because the value of g varies from place to place. We discussed more modern time standards in Section 1.3.

TEST YOUR UNDERSTANDING OF SECTION 14.5   When a body oscillating on a horizontal spring passes through its equilibrium position, its acceleration is zero (see Fig. 14.2b). When the bob of an oscillating simple pendulum passes from left to right through its equilibrium position, is its acceleration (i) zero; (ii) to the left; (iii) to the right; (iv) upward; or (v) downward?

14.6 THE PHYSICAL PENDULUM

A **physical pendulum** is any *real* pendulum that uses an extended body, as contrasted to the idealized *simple* pendulum with all of its mass concentrated at a point. **Figure 14.23** shows a body of irregular shape pivoted so that it can turn without friction about an axis through point O. In equilibrium the center of gravity (cg) is directly below the pivot; in the position shown, the body is displaced from equilibrium by an angle θ, which we use as a coordinate for the system. The distance from O to the center of gravity is d, the moment of inertia of the body about the axis of rotation through O is I, and the total mass is m. When the body is displaced as shown, the weight mg causes a restoring torque

$$\tau_z = -(mg)(d\sin\theta) \tag{14.36}$$

The negative sign shows that the restoring torque is clockwise when the displacement is counterclockwise, and vice versa.

When the body is released, it oscillates about its equilibrium position. The motion is not simple harmonic because the torque τ_z is proportional to $\sin\theta$ rather than to θ itself. However, if θ is small, we can approximate $\sin\theta$ by θ in radians, just as we did in analyzing the simple pendulum. Then the motion is *approximately* simple harmonic. With this approximation,

$$\tau_z = -(mgd)\theta$$

From Section 10.2, the equation of motion is $\Sigma\tau_z = I\alpha_z$, so

$$-(mgd)\theta = I\alpha_z = I\frac{d^2\theta}{dt^2}$$

$$\frac{d^2\theta}{dt^2} = -\frac{mgd}{I}\theta \tag{14.37}$$

14.23 Dynamics of a physical pendulum.

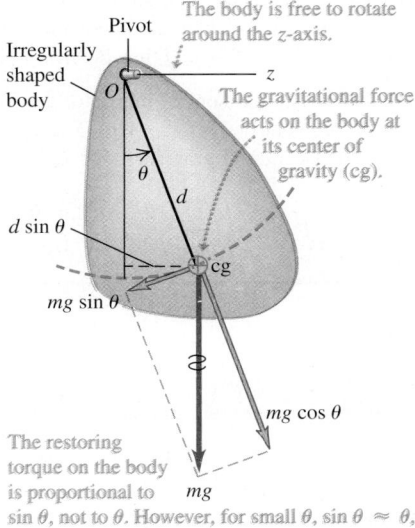

The body is free to rotate around the z-axis.

Pivot

Irregularly shaped body

The gravitational force acts on the body at its center of gravity (cg).

$d\sin\theta$

$mg\sin\theta$

$mg\cos\theta$

mg

The restoring torque on the body is proportional to $\sin\theta$, not to θ. However, for small θ, $\sin\theta \approx \theta$, so the motion is *approximately* simple harmonic.

Comparing this with Eq. (14.4), we see that the role of (k/m) for the spring-mass system is played here by the quantity (mgd/I). Thus the angular frequency is

Angular frequency of physical pendulum, small amplitude $\cdots\cdots\to \omega = \sqrt{\dfrac{mgd}{I}}$ (14.38)

Mass $\cdots$ ·· Acceleration due to gravity ···· Distance from rotation axis to center of gravity ·· Moment of inertia

The frequency f is $1/2\pi$ times this, and the period T is $1/f$:

Period of physical pendulum, small amplitude $\cdots\to T = 2\pi\sqrt{\dfrac{I}{mgd}}$ (14.39)

Moment of inertia ···· Distance from rotation axis to center of gravity ···· Mass ·· Acceleration due to gravity

Equation (14.39) is the basis of a common method for experimentally determining the moment of inertia of a body with a complicated shape. First locate the center of gravity by balancing the body. Then suspend the body so that it is free to oscillate about an axis, and measure the period T of small-amplitude oscillations. Finally, use Eq. (14.39) to calculate the moment of inertia I of the body about this axis from T, the body's mass m, and the distance d from the axis to the center of gravity (see Exercise 14.55). Biomechanics researchers use this method to find the moments of inertia of an animal's limbs. This information is important for analyzing how an animal walks, as we'll see in the second of the two following examples.

EXAMPLE 14.9 PHYSICAL PENDULUM VERSUS SIMPLE PENDULUM

If the body in Fig. 14.23 is a uniform rod with length L, pivoted at one end, what is the period of its motion as a pendulum?

SOLUTION

IDENTIFY and SET UP: Our target variable is the oscillation period T of a rod that acts as a physical pendulum. We find the rod's moment of inertia in Table 9.2, and then determine T from Eq. (14.39).

EXECUTE: The moment of inertia of a uniform rod about an axis through one end is $I = \frac{1}{3}ML^2$. The distance from the pivot to the rod's center of gravity is $d = L/2$. Then from Eq. (14.39),

$$T = 2\pi\sqrt{\frac{I}{mgd}} = 2\pi\sqrt{\frac{\frac{1}{3}ML^2}{MgL/2}} = 2\pi\sqrt{\frac{2L}{3g}}$$

EVALUATE: If the rod is a meter stick $(L = 1.00 \text{ m})$ and $g = 9.80 \text{ m/s}^2$, then

$$T = 2\pi\sqrt{\frac{2(1.00 \text{ m})}{3(9.80 \text{ m/s}^2)}} = 1.64 \text{ s}$$

The period is smaller by a factor of $\sqrt{\frac{2}{3}} = 0.816$ than that of a simple pendulum of the same length (see Example 14.8). The rod's moment of inertia around one end, $I = \frac{1}{3}ML^2$, is one-third that of the simple pendulum, and the rod's cg is half as far from the pivot as that of the simple pendulum. You can show that, taken together in Eq. (14.39), these two differences account for the factor $\sqrt{\frac{2}{3}}$ by which the periods differ.

EXAMPLE 14.10 *TYRANNOSAURUS REX* AND THE PHYSICAL PENDULUM

All walking animals, including humans, have a natural walking pace—a number of steps per minute that is more comfortable than a faster or slower pace. Suppose that this pace corresponds to the oscillation of the leg as a physical pendulum. (a) How does this pace depend on the length L of the leg from hip to foot? Treat the leg as a uniform rod pivoted at the hip joint. (b) Fossil evidence shows that *T. rex*, a two-legged dinosaur that lived about 65 million years ago, had a leg length $L = 3.1$ m and a stride length $S = 4.0$ m (the distance from one footprint to the next print of the same foot; see **Fig. 14.24**). Estimate the walking speed of *T. rex*.

SOLUTION

IDENTIFY and SET UP: Our target variables are (a) the relationship between walking pace and leg length L and (b) the walking speed of *T. rex*. We treat the leg as a physical pendulum, with a period of oscillation as found in Example 14.9. We can find the walking speed from the period and the stride length.

EXECUTE: (a) From Example 14.9 the period of oscillation of the leg is $T = 2\pi\sqrt{2L/3g}$, which is proportional to $\sqrt{L}$. Each step

14.24 The walking speed of *Tyrannosaurus rex* can be estimated from leg length L and stride length S.

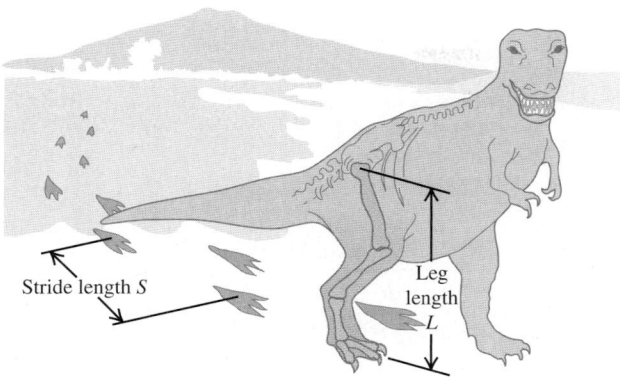

Stride length S

Leg length L

takes half of one period, so the walking pace (in steps per second) equals twice the oscillation frequency $f = 1/T$ and is proportional to $1/\sqrt{L}$. The longer the leg, the slower the pace.

(b) In our model, *T. rex* traveled one stride length S in a time

$$T = 2\pi\sqrt{\frac{2L}{3g}} = 2\pi\sqrt{\frac{2(3.1\ \text{m})}{3(9.8\ \text{m/s}^2)}} = 2.9\ \text{s}$$

so its walking speed was

$$v = \frac{S}{T} = \frac{4.0\ \text{m}}{2.9\ \text{s}} = 1.4\ \text{m/s} = 5.0\ \text{km/h} = 3.1\ \text{mi/h}$$

This is roughly the walking speed of an adult human.

EVALUATE: A uniform rod isn't a very good model for a leg. The legs of many animals, including both *T. rex* and humans, are tapered; there is more mass between hip and knee than between knee and foot. The center of mass is therefore less than $L/2$ from the hip; a reasonable guess would be about $L/4$. The moment of inertia is therefore *considerably* less than $ML^2/3$—say, $ML^2/15$. Use the analysis of Example 14.9 with these corrections; you'll get a shorter oscillation period and an even greater walking speed for *T. rex*.

TEST YOUR UNDERSTANDING OF SECTION 14.6 The center of gravity of a simple pendulum of mass m and length L is located at the pendulum bob, a distance L from the pivot point. The center of gravity of a uniform rod of the same mass m and length $2L$ pivoted at one end is also a distance L from the pivot point. Compared to the period of the simple pendulum, is the period of this uniform rod (i) longer; (ii) shorter; or (iii) the same? ▌

14.7 DAMPED OSCILLATIONS

The idealized oscillating systems we have discussed so far are frictionless. There are no nonconservative forces, the total mechanical energy is constant, and a system set into motion continues oscillating forever with no decrease in amplitude.

Real-world systems always have some dissipative forces, however, and oscillations die out with time unless we replace the dissipated mechanical energy (**Fig. 14.25**). A mechanical pendulum clock continues to run because potential energy stored in the spring or a hanging weight system replaces the mechanical energy lost due to friction in the pivot and the gears. But eventually the spring runs down or the weights reach the bottom of their travel. Then no more energy is available, and the pendulum swings decrease in amplitude and stop.

The decrease in amplitude caused by dissipative forces is called **damping** (*not* "dampening"), and the corresponding motion is called **damped oscillation.** The simplest case is a simple harmonic oscillator with a frictional damping force that is directly proportional to the *velocity* of the oscillating body. This behavior occurs in friction involving viscous fluid flow, such as in shock absorbers or sliding between oil-lubricated surfaces. We then have an additional force on the body due to friction, $F_x = -bv_x$, where $v_x = dx/dt$ is the velocity and b is a constant that describes the strength of the damping force. The negative sign shows that the force is always opposite in direction to the velocity. The *net* force on the body is then

$$\sum F_x = -kx - bv_x \qquad (14.40)$$

and Newton's second law for the system is

$$-kx - bv_x = ma_x \quad \text{or} \quad -kx - b\frac{dx}{dt} = m\frac{d^2x}{dt^2} \qquad (14.41)$$

Equation (14.41) is a differential equation for x; it's the same as Eq. (14.4), the equation for the acceleration in SHM, but with the added term $-b\,dx/dt$. We won't

14.25 A swinging bell left to itself will eventually stop oscillating due to damping forces (air resistance and friction at the point of suspension).

go into how to solve this equation; we'll just present the solution. If the damping force is relatively small, the motion is described by

Displacement of oscillator, little damping — Initial amplitude — Damping constant — Mass — Time

$$x = Ae^{-(b/2m)t}\cos(\omega't + \phi)$$ (14.42)

Angular frequency of damped oscillations — Phase angle

The angular frequency of these damped oscillations is given by

Angular frequency of oscillator, little damping — Force constant of restoring force

$$\omega' = \sqrt{\frac{k}{m} - \frac{b^2}{4m^2}}$$ — Damping constant — Mass (14.43)

You can verify that Eq. (14.42) is a solution of Eq. (14.41) by calculating the first and second derivatives of x, substituting them into Eq. (14.41), and checking whether the left and right sides are equal.

The motion described by Eq. (14.42) differs from the undamped case in two ways. First, the amplitude $Ae^{-(b/2m)t}$ is not constant but decreases with time because of the exponential factor $e^{-(b/2m)t}$. **Figure 14.26** is a graph of Eq. (14.42) for $\phi = 0$; the larger the value of b, the more quickly the amplitude decreases.

Second, the angular frequency ω', given by Eq. (14.43), is no longer equal to $\omega = \sqrt{k/m}$ but is somewhat smaller. It becomes zero when b becomes so large that

$$\frac{k}{m} - \frac{b^2}{4m^2} = 0 \qquad \text{or} \qquad b = 2\sqrt{km}$$ (14.44)

When Eq. (14.44) is satisfied, the condition is called **critical damping.** The system no longer oscillates but returns to its equilibrium position without oscillation when it is displaced and released.

If b is greater than $2\sqrt{km}$, the condition is called **overdamping.** Again there is no oscillation, but the system returns to equilibrium more slowly than with critical damping. For the overdamped case the solutions of Eq. (14.41) have the form

$$x = C_1 e^{-a_1 t} + C_2 e^{-a_2 t}$$

where C_1 and C_2 are constants that depend on the initial conditions and a_1 and a_2 are constants determined by m, k, and b.

When b is less than the critical value, as in Eq. (14.42), the condition is called **underdamping.** The system oscillates with steadily decreasing amplitude.

In a vibrating tuning fork or guitar string, it is usually desirable to have as little damping as possible. By contrast, damping plays a beneficial role in the oscillations of an automobile's suspension system. The shock absorbers provide a velocity-dependent damping force so that when the car goes over a bump, it doesn't continue bouncing forever (**Fig. 14.27**). For optimal passenger comfort, the system should be critically damped or slightly underdamped. Too much damping would be counterproductive; if the suspension is overdamped and the car hits a second bump just after the first one, the springs in the suspension will still be compressed somewhat from the first bump and will not be able to fully absorb the impact.

Energy in Damped Oscillations

In damped oscillations the damping force is nonconservative; the mechanical energy of the system is not constant but decreases continuously, approaching zero after a long time. To derive an expression for the rate of change of energy, we first write an expression for the total mechanical energy E at any instant:

$$E = \tfrac{1}{2}mv_x^2 + \tfrac{1}{2}kx^2$$

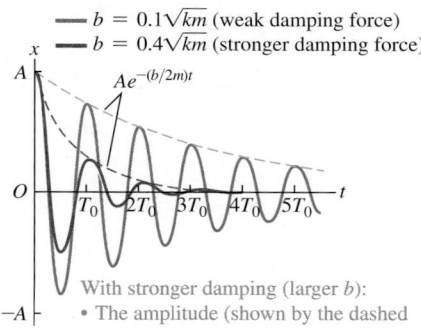

14.26 Graph of displacement versus time for an oscillator with little damping [see Eq. (14.42)] and with phase angle $\phi = 0$. The curves are for two values of the damping constant b.

— $b = 0.1\sqrt{km}$ (weak damping force)
— $b = 0.4\sqrt{km}$ (stronger damping force)

With stronger damping (larger b):
• The amplitude (shown by the dashed curves) decreases more rapidly.
• The period T increases (T_0 = period with zero damping).

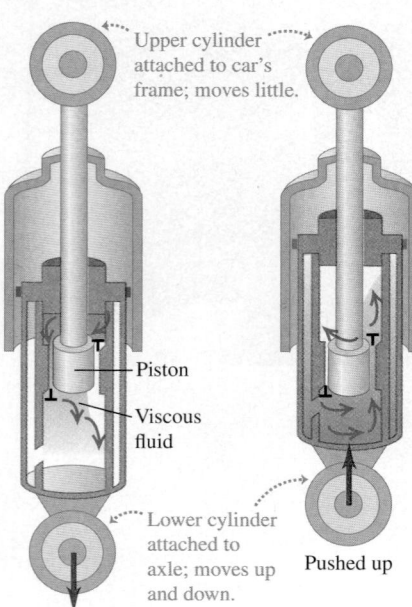

14.27 An automobile shock absorber. The viscous fluid causes a damping force that depends on the relative velocity of the two ends of the unit.

Upper cylinder attached to car's frame; moves little.

Piston

Viscous fluid

Lower cylinder attached to axle; moves up and down.

Pushed up

Pushed down

To find the rate of change of this quantity, we take its time derivative:

$$\frac{dE}{dt} = mv_x\frac{dv_x}{dt} + kx\frac{dx}{dt}$$

But $dv_x/dt = a_x$ and $dx/dt = v_x$, so

$$\frac{dE}{dt} = v_x(ma_x + kx)$$

From Eq. (14.41), $ma_x + kx = -bdx/dt = -bv_x$, so

$$\frac{dE}{dt} = v_x(-bv_x) = -bv_x^2 \qquad \text{(damped oscillations)} \qquad (14.45)$$

The right side of Eq. (14.45) is negative whenever the oscillating body is in motion, whether the x-velocity v_x is positive or negative. This shows that as the body moves, the energy decreases, though not at a uniform rate. The term $-bv_x^2 = (-bv_x)v_x$ (force times velocity) is the rate at which the damping force does (negative) work on the system (that is, the damping *power*). This equals the rate of change of the total mechanical energy of the system.

Similar behavior occurs in electric circuits containing inductance, capacitance, and resistance. There is a natural frequency of oscillation. and the resistance plays the role of the damping constant b. We will study these circuits in detail in Chapters 30 and 31.

TEST YOUR UNDERSTANDING OF SECTION 14.7 An airplane is flying in a straight line at a constant altitude. If a wind gust strikes and raises the nose of the airplane, the nose will bob up and down until the airplane eventually returns to its original attitude. Are these oscillations (i) undamped; (ii) underdamped; (iii) critically damped; or (iv) overdamped? ▌

14.8 FORCED OSCILLATIONS AND RESONANCE

A damped oscillator left to itself will eventually stop moving. But we can maintain a constant-amplitude oscillation by applying a force that varies with time in a periodic way. As an example, consider your cousin Throckmorton on a playground swing. You can keep him swinging with constant amplitude by giving him a push once each cycle. We call this additional force a **driving force.**

Damped Oscillation with a Periodic Driving Force

If we apply a periodic driving force with angular frequency ω_d to a damped harmonic oscillator, the motion that results is called a **forced oscillation** or a *driven oscillation*. It is different from the motion that occurs when the system is simply displaced from equilibrium and then left alone, in which case the system oscillates with a **natural angular frequency** ω' determined by m, k, and b, as in Eq. (14.43). In a forced oscillation, however, the angular frequency with which the mass oscillates is equal to the driving angular frequency ω_d. This does *not* have to be equal to the natural angular frequency ω'. If you grab the ropes of Throckmorton's swing, you can force the swing to oscillate with any frequency you like.

Suppose we force the oscillator to vibrate with an angular frequency ω_d that is nearly *equal* to the angular frequency ω' it would have with no driving force. What happens? The oscillator is naturally disposed to oscillate at $\omega = \omega'$, so we expect the amplitude of the resulting oscillation to be larger than when the two frequencies are very different. Detailed analysis and experiment show that this is just what happens. The easiest case to analyze is a *sinusoidally* varying force—say, $F(t) = F_{max}\cos\omega_d t$. If we vary the frequency ω_d of the driving force, the amplitude

BIO Application Forced Oscillations
This lady beetle (or "ladybug," family Coccinellidae) flies by means of a forced oscillation. Unlike the wings of birds, this insect's wings are extensions of its exoskeleton. Muscles attached to the inside of the exoskeleton apply a periodic driving force that deforms the exoskeleton rhythmically, causing the attached wings to beat up and down. The oscillation frequency of the wings and exoskeleton is the same as the frequency of the driving force.

14.28 Graph of the amplitude A of forced oscillation as a function of the angular frequency ω_d of the driving force. The horizontal axis shows the ratio of ω_d to the angular frequency $\omega = \sqrt{k/m}$ of an undamped oscillator. Each curve has a different value of the damping constant b.

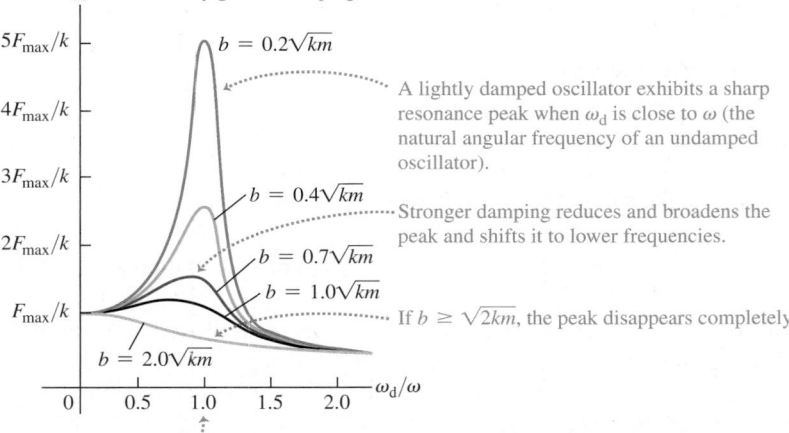

Each curve shows the amplitude A for an oscillator subjected to a driving force at various angular frequencies ω_d. Successive curves from blue to gold represent successively greater damping.

A lightly damped oscillator exhibits a sharp resonance peak when ω_d is close to ω (the natural angular frequency of an undamped oscillator).

Stronger damping reduces and broadens the peak and shifts it to lower frequencies.

If $b \geq \sqrt{2km}$, the peak disappears completely.

Driving frequency ω_d equals natural angular frequency ω of an undamped oscillator.

of the resulting forced oscillation varies in an interesting way (**Fig. 14.28**). When there is very little damping (small b), the amplitude goes through a sharp peak as the driving angular frequency ω_d nears the natural oscillation angular frequency ω'. When the damping is increased (larger b), the peak becomes broader and smaller in height and shifts toward lower frequencies.

Using more differential equations than we're ready for, we could find an expression for the amplitude A of the forced oscillation as a function of the driving angular frequency. Here is the result:

DEMO

Amplitude of a forced oscillator
Maximum value of driving force

$$A = \frac{F_{max}}{\sqrt{(k - m\omega_d^2)^2 + b^2\omega_d^2}}$$ (14.46)

Force constant of restoring force
Mass
Driving angular frequency
Damping constant

When $k - m\omega_d^2 = 0$, the first term under the radical is zero, so A has a maximum near $\omega_d = \sqrt{k/m}$. The height of the curve at this point is proportional to $1/b$; the less damping, the higher the peak. At the low-frequency extreme, when $\omega_d = 0$, we get $A = F_{max}/k$. This corresponds to a *constant* force F_{max} and a constant displacement $A = F_{max}/k$ from equilibrium, as we might expect.

Resonance and Its Consequences

The peaking of the amplitude at driving frequencies close to the natural frequency of the system is called **resonance.** Physics is full of examples of resonance; building up the oscillations of a child on a swing by pushing with a frequency equal to the swing's natural frequency is one. A vibrating rattle in a car that occurs only at a certain engine speed is another example. Inexpensive loudspeakers often have an annoying boom or buzz when a musical note coincides with the natural frequency of the speaker cone or housing. In Chapter 16 we will study examples of resonance that involve sound. Resonance also occurs in electric circuits, as we will see in Chapter 31; a tuned circuit in a radio receiver responds strongly to waves with frequencies near its natural frequency. This phenomenon lets us select one radio station and reject other stations.

Resonance in mechanical systems can be destructive. A company of soldiers once destroyed a bridge by marching across it in step; the frequency of their steps was close to a natural frequency of the bridge, and the resulting oscillation had large enough amplitude to tear the bridge apart. Ever since, marching soldiers have been ordered to break step before crossing a bridge. Some years ago,

BIO Application Canine Resonance Unlike humans, dogs have no sweat glands and so must pant in order to cool down. The frequency at which a dog pants is very close to the resonant frequency of its respiratory system. This causes the maximum amount of air inflow and outflow and so minimizes the effort that the dog must exert to cool itself.

vibrations of the engines of a particular type of airplane had just the right frequency to resonate with the natural frequencies of its wings. Large oscillations built up, and occasionally the wings fell off.

TEST YOUR UNDERSTANDING OF SECTION 14.8   When driven at a frequency near its natural frequency, an oscillator with very little damping has a much greater response than the same oscillator with more damping. When driven at a frequency that is much higher or lower than the natural frequency, which oscillator will have the greater response: (i) the one with very little damping or (ii) the one with more damping? ▮

CHAPTER 14 SUMMARY

SOLUTIONS TO ALL EXAMPLES

Periodic motion: Periodic motion is motion that repeats itself in a definite cycle. It occurs whenever a body has a stable equilibrium position and a restoring force that acts when the body is displaced from equilibrium. Period T is the time for one cycle. Frequency f is the number of cycles per unit time. Angular frequency ω is 2π times the frequency. (See Example 14.1.)

$$f = \frac{1}{T} \qquad T = \frac{1}{f} \tag{14.1}$$

$$\omega = 2\pi f = \frac{2\pi}{T} \tag{14.2}$$

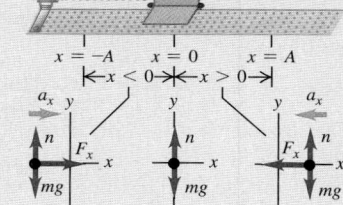

Simple harmonic motion: If the restoring force F_x in periodic motion is directly proportional to the displacement x, the motion is called simple harmonic motion (SHM). In many cases this condition is satisfied if the displacement from equilibrium is small. The angular frequency, frequency, and period in SHM do not depend on the amplitude but on only the mass m and force constant k. The displacement, velocity, and acceleration in SHM are sinusoidal functions of time; the amplitude A and phase angle ϕ of the oscillation are determined by the initial displacement and velocity of the body. (See Examples 14.2, 14.3, 14.6, and 14.7.)

$$F_x = -kx \tag{14.3}$$

$$a_x = \frac{F_x}{m} = -\frac{k}{m}x \tag{14.4}$$

$$\omega = \sqrt{\frac{k}{m}} \tag{14.10}$$

$$f = \frac{\omega}{2\pi} = \frac{1}{2\pi}\sqrt{\frac{k}{m}} \tag{14.11}$$

$$T = \frac{1}{f} = 2\pi\sqrt{\frac{m}{k}} \tag{14.12}$$

$$x = A\cos(\omega t + \phi) \tag{14.13}$$

Energy in simple harmonic motion: Energy is conserved in SHM. The total energy can be expressed in terms of the force constant k and amplitude A. (See Examples 14.4 and 14.5.)

$$E = \tfrac{1}{2}mv_x^2 + \tfrac{1}{2}kx^2 = \tfrac{1}{2}kA^2 = \text{constant} \tag{14.21}$$

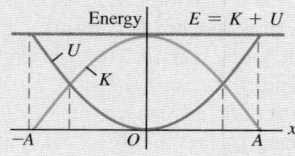

Angular simple harmonic motion: In angular SHM, the frequency and angular frequency are related to the moment of inertia I and the torsion constant κ.

$$\omega = \sqrt{\frac{\kappa}{I}} \quad \text{and}$$

$$f = \frac{1}{2\pi}\sqrt{\frac{\kappa}{I}} \tag{14.24}$$

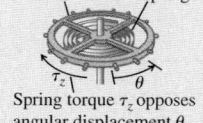

Balance wheel   Spring

Spring torque τ_z opposes angular displacement θ.

Simple pendulum: A simple pendulum consists of a point mass m at the end of a massless string of length L. Its motion is approximately simple harmonic for sufficiently small amplitude; the angular frequency, frequency, and period then depend on only g and L, not on the mass or amplitude. (See Example 14.8.)

$$\omega = \sqrt{\frac{g}{L}} \tag{14.32}$$

$$f = \frac{\omega}{2\pi} = \frac{1}{2\pi}\sqrt{\frac{g}{L}} \tag{14.33}$$

$$T = \frac{2\pi}{\omega} = \frac{1}{f} = 2\pi\sqrt{\frac{L}{g}} \tag{14.34}$$

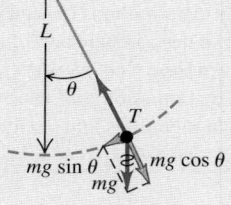

Physical pendulum: A physical pendulum is any body suspended from an axis of rotation. The angular frequency and period for small-amplitude oscillations are independent of amplitude but depend on the mass m, distance d from the axis of rotation to the center of gravity, and moment of inertia I about the axis. (See Examples 14.9 and 14.10.)

$$\omega = \sqrt{\frac{mgd}{I}} \quad (14.38)$$

$$T = 2\pi\sqrt{\frac{I}{mgd}} \quad (14.39)$$

Damped oscillations: When a force $F_x = -bv_x$ is added to a simple harmonic oscillator, the motion is called a damped oscillation. If $b < 2\sqrt{km}$ (called underdamping), the system oscillates with a decaying amplitude and an angular frequency ω' that is lower than it would be without damping. If $b = 2\sqrt{km}$ (called critical damping) or $b > 2\sqrt{km}$ (called overdamping), when the system is displaced it returns to equilibrium without oscillating.

$$x = Ae^{-(b/2m)t}\cos(\omega't + \phi) \quad (14.42)$$

$$\omega' = \sqrt{\frac{k}{m} - \frac{b^2}{4m^2}} \quad (14.43)$$

$$— \; b = 0.1\sqrt{km}$$
$$— \; b = 0.4\sqrt{km}$$

Forced oscillations and resonance: When a sinusoidally varying driving force is added to a damped harmonic oscillator, the resulting motion is called a forced oscillation or driven oscillation. The amplitude is a function of the driving frequency ω_d and reaches a peak at a driving frequency close to the natural frequency of the system. This behavior is called resonance.

$$A = \frac{F_{max}}{\sqrt{(k - m\omega_d^2)^2 + b^2\omega_d^2}} \quad (14.46)$$

$$b = 0.2\sqrt{km}$$
$$b = 0.4\sqrt{km}$$
$$b = 0.7\sqrt{km}$$
$$b = 1.0\sqrt{km}$$
$$b = 2.0\sqrt{km}$$

BRIDGING PROBLEM   OSCILLATING AND ROLLING

Two uniform, solid cylinders of radius R and total mass M are connected along their common axis by a short, light rod and rest on a horizontal tabletop (**Fig. 14.29**). A frictionless ring at the center of the rod is attached to a spring with force constant k; the other end of the spring is fixed. The cylinders are pulled to the left a distance x, stretching the spring, and then released from rest. Due to friction between the tabletop and the cylinders, the cylinders roll without slipping as they oscillate. Show that the motion of the center of mass of the cylinders is simple harmonic, and find its period.

14.29 Rolling cylinders attached to a spring.

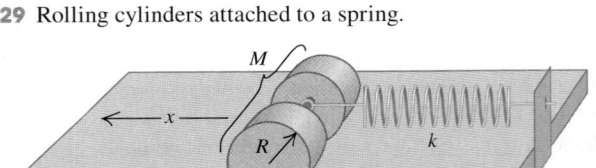

SOLUTION GUIDE

IDENTIFY and SET UP

1. What condition must be satisfied for the motion of the center of mass of the cylinders to be simple harmonic?
2. Which equations should you use to describe the translational and rotational motions of the cylinders? Which equation should you use to describe the condition that the cylinders roll without slipping? (*Hint:* See Section 10.3.)
3. Sketch the situation and choose a coordinate system. List the unknown quantities and decide which is the target variable.

EXECUTE

4. Draw a free-body diagram for the cylinders when they are displaced a distance x from equilibrium.
5. Solve the equations to find an expression for the acceleration of the center of mass of the cylinders. What does this expression tell you?
6. Use your result from step 5 to find the period of oscillation of the center of mass of the cylinders.

EVALUATE

7. What would be the period of oscillation if there were no friction and the cylinders didn't roll? Is this period larger or smaller than your result from step 6? Is this reasonable?

Problems

•, ••, •••: Difficulty levels. CP: Cumulative problems incorporating material from earlier chapters. CALC: Problems requiring calculus.
DATA: Problems involving real data, scientific evidence, experimental design, and/or statistical reasoning. BIO: Biosciences problems.

DISCUSSION QUESTIONS

Q14.1 An object is moving with SHM of amplitude A on the end of a spring. If the amplitude is doubled, what happens to the total distance the object travels in one period? What happens to the period? What happens to the maximum speed of the object? Discuss how these answers are related.

Q14.2 Think of several examples in everyday life of motions that are, at least approximately, simple harmonic. In what respects does each differ from SHM?

Q14.3 Does a tuning fork or similar tuning instrument undergo SHM? Why is this a crucial question for musicians?

Q14.4 A box containing a pebble is attached to an ideal horizontal spring and is oscillating on a friction-free air table. When the box has reached its maximum distance from the equilibrium point, the pebble is suddenly lifted out vertically without disturbing the box. Will the following characteristics of the motion increase, decrease, or remain the same in the subsequent motion of the box? Justify each answer. (a) Frequency; (b) period; (c) amplitude; (d) the maximum kinetic energy of the box; (e) the maximum speed of the box.

Q14.5 If a uniform spring is cut in half, what is the force constant of each half? Justify your answer. How would the frequency of SHM using a half-spring differ from the frequency using the same mass and the entire spring?

Q14.6 A glider is attached to a fixed ideal spring and oscillates on a horizontal, friction-free air track. A coin rests atop the glider and oscillates with it. At what points in the motion is the friction force on the coin greatest? The least? Justify your answers.

Q14.7 Two identical gliders on an air track are connected by an ideal spring. Could such a system undergo SHM? Explain. How would the period compare with that of a single glider attached to a spring whose other end is rigidly attached to a stationary object? Explain.

Q14.8 You are captured by Martians, taken into their ship, and put to sleep. You awake some time later and find yourself locked in a small room with no windows. All the Martians have left you with is your digital watch, your school ring, and your long silver-chain necklace. Explain how you can determine whether you are still on earth or have been transported to Mars.

Q14.9 The system shown in Fig. 14.17 is mounted in an elevator. What happens to the period of the motion (does it increase, decrease, or remain the same) if the elevator (a) accelerates upward at 5.0 m/s^2; (b) moves upward at a steady 5.0 m/s; (c) accelerates downward at 5.0 m/s^2? Justify your answers.

Q14.10 If a pendulum has a period of 2.5 s on earth, what would be its period in a space station orbiting the earth? If a mass hung from a vertical spring has a period of 5.0 s on earth, what would its period be in the space station? Justify your answers.

Q14.11 A simple pendulum is mounted in an elevator. What happens to the period of the pendulum (does it increase, decrease, or remain the same) if the elevator (a) accelerates upward at 5.0 m/s^2; (b) moves upward at a steady 5.0 m/s; (c) accelerates downward at 5.0 m/s^2; (d) accelerates downward at 9.8 m/s^2? Justify your answers.

Q14.12 What should you do to the length of the string of a simple pendulum to (a) double its frequency; (b) double its period; (c) double its angular frequency?

Q14.13 If a pendulum clock is taken to a mountaintop, does it gain or lose time, assuming it is correct at a lower elevation? Explain.

Q14.14 When the amplitude of a simple pendulum increases, should its period increase or decrease? Give a qualitative argument; do not rely on Eq. (14.35). Is your argument also valid for a physical pendulum?

Q14.15 Why do short dogs (like Chihuahuas) walk with quicker strides than do tall dogs (like Great Danes)?

Q14.16 At what point in the motion of a simple pendulum is the string tension greatest? Least? In each case give the reasoning behind your answer.

Q14.17 Could a standard of time be based on the period of a certain standard pendulum? What advantages and disadvantages would such a standard have compared to the actual present-day standard discussed in Section 1.3?

Q14.18 For a simple pendulum, clearly distinguish between ω (the angular speed) and ω (the angular frequency). Which is constant and which is variable?

Q14.19 In designing structures in an earthquake-prone region, how should the natural frequencies of oscillation of a structure relate to typical earthquake frequencies? Why? Should the structure have a large or small amount of damping?

EXERCISES

Section 14.1 Describing Oscillation

14.1 • BIO (a) **Music.** When a person sings, his or her vocal cords vibrate in a repetitive pattern that has the same frequency as the note that is sung. If someone sings the note B flat, which has a frequency of 466 Hz, how much time does it take the person's vocal cords to vibrate through one complete cycle, and what is the angular frequency of the cords? (b) **Hearing.** When sound waves strike the eardrum, this membrane vibrates with the same frequency as the sound. The highest pitch that young humans can hear has a period of $50.0 \ \mu\text{s}$. What are the frequency and angular frequency of the vibrating eardrum for this sound? (c) **Vision.** When light having vibrations with angular frequency ranging from 2.7×10^{15} rad/s to 4.7×10^{15} rad/s strikes the retina of the eye, it stimulates the receptor cells there and is perceived as visible light. What are the limits of the period and frequency of this light? (d) **Ultrasound.** High-frequency sound waves (ultrasound) are used to probe the interior of the body, much as x rays do. To detect small objects such as tumors, a frequency of around 5.0 MHz is used. What are the period and angular frequency of the molecular vibrations caused by this pulse of sound?

14.2 • If an object on a horizontal, frictionless surface is attached to a spring, displaced, and then released, it will oscillate. If it is displaced 0.120 m from its equilibrium position and released with zero initial speed, then after 0.800 s its displacement is found to be 0.120 m on the opposite side, and it has passed the equilibrium position once during this interval. Find (a) the amplitude; (b) the period; (c) the frequency.

14.3 • The tip of a tuning fork goes through 440 complete vibrations in 0.500 s. Find the angular frequency and the period of the motion.

14.4 • The displacement of an oscillating object as a function of time is shown in **Fig. E14.4**. What are (a) the frequency; (b) the amplitude; (c) the period; (d) the angular frequency of this motion?

Figure **E14.4**

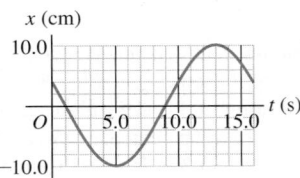

14.5 •• A machine part is undergoing SHM with a frequency of 4.00 Hz and amplitude 1.80 cm. How long does it take the part to go from $x = 0$ to $x = -1.80$ cm?

14.6 • BIO The wings of the blue-throated hummingbird (*Lampornis clemenciae*), which inhabits Mexico and the south-western United States, beat at a rate of up to 900 times per minute. Calculate (a) the period of vibration of this bird's wings, (b) the frequency of the wings' vibration, and (c) the angular frequency of the bird's wing beats.

Section 14.2 Simple Harmonic Motion

14.7 • A 2.40-kg ball is attached to an unknown spring and allowed to oscillate. **Figure E14.7** shows a graph of the ball's position x as a function of time t. What are the oscillation's (a) period, (b) frequency, (c) angular frequency, and (d) amplitude? (e) What is the force constant of the spring?

Figure **E14.7**

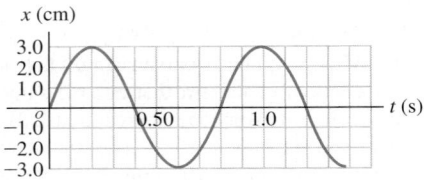

14.8 •• In a physics lab, you attach a 0.200-kg air-track glider to the end of an ideal spring of negligible mass and start it oscillating. The elapsed time from when the glider first moves through the equilibrium point to the second time it moves through that point is 2.60 s. Find the spring's force constant.

14.9 • When a body of unknown mass is attached to an ideal spring with force constant 120 N/m, it is found to vibrate with a frequency of 6.00 Hz. Find (a) the period of the motion; (b) the angular frequency; (c) the mass of the body.

14.10 • When a 0.750-kg mass oscillates on an ideal spring, the frequency is 1.75 Hz. What will the frequency be if 0.220 kg are (a) added to the original mass and (b) subtracted from the original mass? Try to solve this problem *without* finding the force constant of the spring.

14.11 •• An object is undergoing SHM with period 0.900 s and amplitude 0.320 m. At $t = 0$ the object is at $x = 0.320$ m and is instantaneously at rest. Calculate the time it takes the object to go (a) from $x = 0.320$ m to $x = 0.160$ m and (b) from $x = 0.160$ m to $x = 0$.

14.12 • A small block is attached to an ideal spring and is moving in SHM on a horizontal, frictionless surface. When the block is at $x = 0.280$ m, the acceleration of the block is -5.30 m/s^2. What is the frequency of the motion?

14.13 • A 2.00-kg, frictionless block is attached to an ideal spring with force constant 300 N/m. At $t = 0$ the spring is neither stretched nor compressed and the block is moving in the negative direction at 12.0 m/s. Find (a) the amplitude and (b) the phase angle. (c) Write an equation for the position as a function of time.

14.14 •• Repeat Exercise 14.13, but assume that at $t = 0$ the block has velocity -4.00 m/s and displacement $+0.200$ m.

14.15 • The point of the needle of a sewing machine moves in SHM along the x-axis with a frequency of 2.5 Hz. At $t = 0$ its position and velocity components are $+1.1$ cm and -15 cm/s, respectively. (a) Find the acceleration component of the needle at $t = 0$. (b) Write equations giving the position, velocity, and acceleration components of the point as a function of time.

14.16 •• A small block is attached to an ideal spring and is moving in SHM on a horizontal, frictionless surface. When the amplitude of the motion is 0.090 m, it takes the block 2.70 s to travel from $x = 0.090$ m to $x = -0.090$ m. If the amplitude is doubled, to 0.180 m, how long does it take the block to travel (a) from $x = 0.180$ m to $x = -0.180$ m and (b) from $x = 0.090$ m to $x = -0.090$ m?

14.17 • BIO **Weighing Astronauts.** This procedure has been used to "weigh" astronauts in space: A 42.5-kg chair is attached to a spring and allowed to oscillate. When it is empty, the chair takes 1.30 s to make one complete vibration. But with an astronaut sitting in it, with her feet off the floor, the chair takes 2.54 s for one cycle. What is the mass of the astronaut?

14.18 • A 0.400-kg object undergoing SHM has $a_x = -1.80$ m/s^2 when $x = 0.300$ m. What is the time for one oscillation?

14.19 • On a frictionless, horizontal air track, a glider oscillates at the end of an ideal spring of force constant 2.50 N/cm. The graph in **Fig. E14.19** shows the acceleration of the glider as a function of time. Find (a) the mass of the glider; (b) the maximum displacement of the glider from the equilibrium point; (c) the maximum force the spring exerts on the glider.

Figure **E14.19**

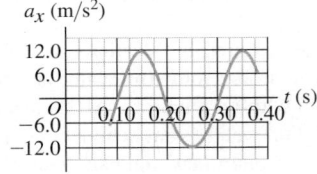

14.20 • A 0.500-kg mass on a spring has velocity as a function of time given by $v_x(t) = -(3.60 \text{ cm/s}) \sin[(4.71 \text{ rad/s})t - (\pi/2)]$. What are (a) the period; (b) the amplitude; (c) the maximum acceleration of the mass; (d) the force constant of the spring?

14.21 • A 1.50-kg mass on a spring has displacement as a function of time given by

$$x(t) = (7.40 \text{ cm}) \cos[(4.16 \text{ rad/s})t - 2.42]$$

Find (a) the time for one complete vibration; (b) the force constant of the spring; (c) the maximum speed of the mass; (d) the maximum force on the mass; (e) the position, speed, and acceleration of the mass at $t = 1.00$ s; (f) the force on the mass at that time.

14.22 • BIO **Weighing a Virus.** In February 2004, scientists at Purdue University used a highly sensitive technique to measure the mass of a vaccinia virus (the kind used in smallpox vaccine). The procedure involved measuring the frequency of oscillation of a tiny sliver of silicon (just 30 nm long) with a laser, first without the virus and then after the virus had attached itself to the silicon.

The difference in mass caused a change in the frequency. We can model such a process as a mass on a spring. (a) Show that the ratio of the frequency with the virus attached (f_{S+V}) to the frequency without the virus (f_S) is given by $f_{S+V}/f_S = 1/\sqrt{1 + (m_V/m_S)}$, where m_V is the mass of the virus and m_S is the mass of the silicon sliver. Notice that it is *not* necessary to know or measure the force constant of the spring. (b) In some data, the silicon sliver has a mass of 2.10×10^{-16} g and a frequency of 2.00×10^{15} Hz without the virus and 2.87×10^{14} Hz with the virus. What is the mass of the virus, in grams and in femtograms?

14.23 •• CALC **Jerk.** A guitar string vibrates at a frequency of 440 Hz. A point at its center moves in SHM with an amplitude of 3.0 mm and a phase angle of zero. (a) Write an equation for the position of the center of the string as a function of time. (b) What are the maximum values of the magnitudes of the velocity and acceleration of the center of the string? (c) The derivative of the acceleration with respect to time is a quantity called the *jerk*. Write an equation for the jerk of the center of the string as a function of time, and find the maximum value of the magnitude of the jerk.

Section 14.3 Energy in Simple Harmonic Motion

14.24 •• For the oscillating object in Fig. E14.4, what are (a) its maximum speed and (b) its maximum acceleration?

14.25 • A small block is attached to an ideal spring and is moving in SHM on a horizontal frictionless surface. The amplitude of the motion is 0.165 m. The maximum speed of the block is 3.90 m/s. What is the maximum magnitude of the acceleration of the block?

14.26 • A small block is attached to an ideal spring and is moving in SHM on a horizontal, frictionless surface. The amplitude of the motion is 0.250 m and the period is 3.20 s. What are the speed and acceleration of the block when $x = 0.160$ m?

14.27 • A 0.150-kg toy is undergoing SHM on the end of a horizontal spring with force constant $k = 300$ N/m. When the toy is 0.0120 m from its equilibrium position, it is observed to have a speed of 0.400 m/s. What are the toy's (a) total energy at any point of its motion; (b) amplitude of motion; (c) maximum speed during its motion?

14.28 •• A harmonic oscillator has angular frequency ω and amplitude A. (a) What are the magnitudes of the displacement and velocity when the elastic potential energy is equal to the kinetic energy? (Assume that $U = 0$ at equilibrium.) (b) How often does this occur in each cycle? What is the time between occurrences? (c) At an instant when the displacement is equal to $A/2$, what fraction of the total energy of the system is kinetic and what fraction is potential?

14.29 • A 0.500-kg glider, attached to the end of an ideal spring with force constant $k = 450$ N/m, undergoes SHM with an amplitude of 0.040 m. Compute (a) the maximum speed of the glider; (b) the speed of the glider when it is at $x = -0.015$ m; (c) the magnitude of the maximum acceleration of the glider; (d) the acceleration of the glider at $x = -0.015$ m; (e) the total mechanical energy of the glider at any point in its motion.

14.30 •• A cheerleader waves her pom-pom in SHM with an amplitude of 18.0 cm and a frequency of 0.850 Hz. Find (a) the maximum magnitude of the acceleration and of the velocity; (b) the acceleration and speed when the pom-pom's coordinate is $x = +9.0$ cm; (c) the time required to move from the equilibrium position directly to a point 12.0 cm away. (d) Which of the quantities asked for in parts (a), (b), and (c) can be found by using the energy approach used in Section 14.3, and which cannot? Explain.

14.31 • CP For the situation described in part (a) of Example 14.5, what should be the value of the putty mass m so that the amplitude after the collision is one-half the original amplitude? For this value of m, what fraction of the original mechanical energy is converted into thermal energy?

14.32 •• A block with mass $m = 0.300$ kg is attached to one end of an ideal spring and moves on a horizontal frictionless surface. The other end of the spring is attached to a wall. When the block is at $x = +0.240$ m, its acceleration is $a_x = -12.0$ m/s² and its velocity is $v_x = +4.00$ m/s. What are (a) the spring's force constant k; (b) the amplitude of the motion; (c) the maximum speed of the block during its motion; and (d) the maximum magnitude of the block's acceleration during its motion?

14.33 •• You are watching an object that is moving in SHM. When the object is displaced 0.600 m to the right of its equilibrium position, it has a velocity of 2.20 m/s to the right and an acceleration of 8.40 m/s² to the left. How much farther from this point will the object move before it stops momentarily and then starts to move back to the left?

14.34 • A 2.00-kg frictionless block is attached to an ideal spring with force constant 315 N/m. Initially the spring is neither stretched nor compressed, but the block is moving in the negative direction at 12.0 m/s. Find (a) the amplitude of the motion, (b) the block's maximum acceleration, and (c) the maximum force the spring exerts on the block.

14.35 • A 2.00-kg frictionless block attached to an ideal spring with force constant 315 N/m is undergoing simple harmonic motion. When the block has displacement +0.200 m, it is moving in the negative x-direction with a speed of 4.00 m/s. Find (a) the amplitude of the motion; (b) the block's maximum acceleration; and (c) the maximum force the spring exerts on the block.

14.36 •• A mass is oscillating with amplitude A at the end of a spring. How far (in terms of A) is this mass from the equilibrium position of the spring when the elastic potential energy equals the kinetic energy?

Section 14.4 Applications of Simple Harmonic Motion

14.37 • A 175-g glider on a horizontal, frictionless air track is attached to a fixed ideal spring with force constant 155 N/m. At the instant you make measurements on the glider, it is moving at 0.815 m/s and is 3.00 cm from its equilibrium point. Use *energy conservation* to find (a) the amplitude of the motion and (b) the maximum speed of the glider. (c) What is the angular frequency of the oscillations?

14.38 • A proud deep-sea fisherman hangs a 65.0-kg fish from an ideal spring having negligible mass. The fish stretches the spring 0.180 m. (a) Find the force constant of the spring. The fish is now pulled down 5.00 cm and released. (b) What is the period of oscillation of the fish? (c) What is the maximum speed it will reach?

14.39 • A thrill-seeking cat with mass 4.00 kg is attached by a harness to an ideal spring of negligible mass and oscillates vertically in SHM. The amplitude is 0.050 m, and at the highest point of the motion the spring has its natural unstretched length. Calculate the elastic potential energy of the spring (take it to be zero for the unstretched spring), the kinetic energy of the cat, the gravitational potential energy of the system relative to the lowest point of the motion, and the sum of these three energies when the cat is (a) at its highest point; (b) at its lowest point; (c) at its equilibrium position.

14.40 •• A uniform, solid metal disk of mass 6.50 kg and diameter 24.0 cm hangs in a horizontal plane, supported at its center by a vertical metal wire. You find that it requires a horizontal force of 4.23 N tangent to the rim of the disk to turn it by 3.34°, thus twisting the wire. You now remove this force and release the disk from rest. (a) What is the torsion constant for the metal wire? (b) What are the frequency and period of the torsional oscillations of the disk? (c) Write the equation of motion for $\theta(t)$ for the disk.

14.41 •• A certain alarm clock ticks four times each second, with each tick representing half a period. The balance wheel consists of a thin rim with radius 0.55 cm, connected to the balance shaft by thin spokes of negligible mass. The total mass of the balance wheel is 0.90 g. (a) What is the moment of inertia of the balance wheel about its shaft? (b) What is the torsion constant of the coil spring (Fig. 14.19)?

14.42 • A thin metal disk with mass 2.00×10^{-3} kg and radius 2.20 cm is attached at its center to a long fiber (**Fig. E14.42**). The disk, when twisted and released, oscillates with a period of 1.00 s. Find the torsion constant of the fiber.

Figure **E14.42**

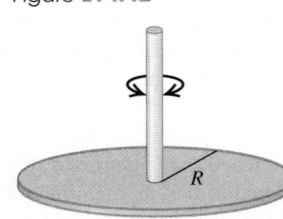

14.43 •• You want to find the moment of inertia of a complicated machine part about an axis through its center of mass. You suspend it from a wire along this axis. The wire has a torsion constant of 0.450 N · m/rad. You twist the part a small amount about this axis and let it go, timing 165 oscillations in 265 s. What is its moment of inertia?

14.44 •• CALC The balance wheel of a watch vibrates with an angular amplitude Θ, angular frequency ω, and phase angle $\phi = 0$. (a) Find expressions for the angular velocity $d\theta/dt$ and angular acceleration $d^2\theta/dt^2$ as functions of time. (b) Find the balance wheel's angular velocity and angular acceleration when its angular displacement is Θ, and when its angular displacement is $\Theta/2$ and θ is decreasing. (*Hint:* Sketch a graph of θ versus t.)

Section 14.5 The Simple Pendulum

14.45 •• You pull a simple pendulum 0.240 m long to the side through an angle of 3.50° and release it. (a) How much time does it take the pendulum bob to reach its highest speed? (b) How much time does it take if the pendulum is released at an angle of 1.75° instead of 3.50°?

14.46 • An 85.0-kg mountain climber plans to swing down, starting from rest, from a ledge using a light rope 6.50 m long. He holds one end of the rope, and the other end is tied higher up on a rock face. Since the ledge is not very far from the rock face, the rope makes a small angle with the vertical. At the lowest point of his swing, he plans to let go and drop a short distance to the ground. (a) How long after he begins his swing will the climber first reach his lowest point? (b) If he missed the first chance to drop off, how long after first beginning his swing will the climber reach his lowest point for the second time?

14.47 • A building in San Francisco has light fixtures consisting of small 2.35-kg bulbs with shades hanging from the ceiling at the end of light, thin cords 1.50 m long. If a minor earthquake occurs, how many swings per second will these fixtures make?

14.48 • **A Pendulum on Mars.** A certain simple pendulum has a period on the earth of 1.60 s. What is its period on the surface of Mars, where $g = 3.71 \text{ m/s}^2$?

14.49 • After landing on an unfamiliar planet, a space explorer constructs a simple pendulum of length 50.0 cm. She finds that the pendulum makes 100 complete swings in 136 s. What is the value of g on this planet?

14.50 •• In the laboratory, a student studies a pendulum by graphing the angle θ that the string makes with the vertical as a function of time t, obtaining the graph shown in **Fig. E14.50**. (a) What are the period, frequency, angular frequency, and amplitude of the pendulum's motion? (b) How long is the pendulum? (c) Is it possible to determine the mass of the bob?

Figure **E14.50**

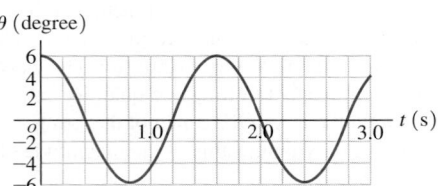

14.51 • A simple pendulum 2.00 m long swings through a maximum angle of 30.0° with the vertical. Calculate its period (a) assuming a small amplitude, and (b) using the first three terms of Eq. (14.35). (c) Which of the answers in parts (a) and (b) is more accurate? What is the percentage error of the less accurate answer compared with the more accurate one?

14.52 •• A small sphere with mass m is attached to a massless rod of length L that is pivoted at the top, forming a simple pendulum. The pendulum is pulled to one side so that the rod is at an angle θ from the vertical, and released from rest. (a) In a diagram, show the pendulum just after it is released. Draw vectors representing the *forces* acting on the small sphere and the *acceleration* of the sphere. Accuracy counts! At this point, what is the linear acceleration of the sphere? (b) Repeat part (a) for the instant when the pendulum rod is at an angle $\theta/2$ from the vertical. (c) Repeat part (a) for the instant when the pendulum rod is vertical. At this point, what is the linear speed of the sphere?

Section 14.6 The Physical Pendulum

14.53 • Two pendulums have the same dimensions (length L) and total mass (m). Pendulum A is a very small ball swinging at the end of a uniform massless bar. In pendulum B, half the mass is in the ball and half is in the uniform bar. Find the period of each pendulum for small oscillations. Which one takes longer for a swing?

14.54 •• We want to hang a thin hoop on a horizontal nail and have the hoop make one complete small-angle oscillation each 2.0 s. What must the hoop's radius be?

14.55 • A 1.80-kg connecting rod from a car engine is pivoted about a horizontal knife edge as shown in **Fig. E14.55**. The center of gravity of the rod was located by balancing and is 0.200 m from the pivot. When the rod is set into small-amplitude oscillation, it makes 100 complete swings in 120 s. Calculate the moment of inertia of the rod about the rotation axis through the pivot.

Figure **E14.55**

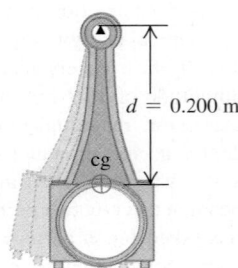

$d = 0.200$ m

cg

14.56 •• A 1.80-kg monkey wrench is pivoted 0.250 m from its center of mass and allowed to swing as a physical pendulum. The period for small-angle oscillations is 0.940 s. (a) What is the moment of inertia of the wrench about an axis through the pivot? (b) If the wrench is initially displaced 0.400 rad from its equilibrium position, what is the angular speed of the wrench as it passes through the equilibrium position?

14.57 •• The two pendulums shown in **Fig. E14.57** each consist of a uniform solid ball of mass M supported by a rigid massless rod, but the ball for pendulum A is very tiny while the ball for pendulum B is much larger. Find the period of each pendulum for small displacements. Which ball takes longer to complete a swing?

Figure **E14.57**

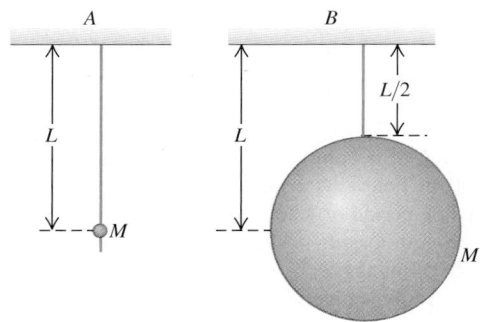

14.58 •• CP A holiday ornament in the shape of a hollow sphere with mass $M = 0.015$ kg and radius $R = 0.050$ m is hung from a tree limb by a small loop of wire attached to the surface of the sphere. If the ornament is displaced a small distance and released, it swings back and forth as a physical pendulum with negligible friction. Calculate its period. (*Hint:* Use the parallel-axis theorem to find the moment of inertia of the sphere about the pivot at the tree limb.)

Section 14.7 Damped Oscillations

14.59 • A 1.35-kg object is attached to a horizontal spring of force constant 2.5 N/cm. The object is started oscillating by pulling it 6.0 cm from its equilibrium position and releasing it so that it is free to oscillate on a frictionless horizontal air track. You observe that after eight cycles its maximum displacement from equilibrium is only 3.5 cm. (a) How much energy has this system lost to damping during these eight cycles? (b) Where did the "lost" energy go? Explain physically how the system could have lost energy.

14.60 •• A 50.0-g hard-boiled egg moves on the end of a spring with force constant $k = 25.0$ N/m. Its initial displacement is 0.300 m. A damping force $F_x = -bv_x$ acts on the egg, and the amplitude of the motion decreases to 0.100 m in 5.00 s. Calculate the magnitude of the damping constant b.

14.61 • An unhappy 0.300-kg rodent, moving on the end of a spring with force constant $k = 2.50$ N/m, is acted on by a damping force $F_x = -bv_x$. (a) If the constant b has the value 0.900 kg/s, what is the frequency of oscillation of the rodent? (b) For what value of the constant b will the motion be critically damped?

14.62 •• A mass is vibrating at the end of a spring of force constant 225 N/m. **Figure E14.62** shows a graph of its position x as a function of time t. (a) At what times is the mass not moving? (b) How much energy did this system originally contain? (c) How much energy did the system lose between $t = 1.0$ s and $t = 4.0$ s? Where did this energy go?

Figure **E14.62**

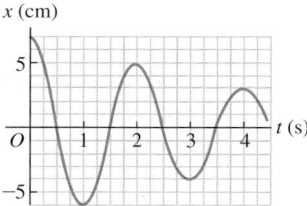

Section 14.8 Forced Oscillations and Resonance

14.63 • A sinusoidally varying driving force is applied to a damped harmonic oscillator of force constant k and mass m. If the damping constant has a value b_1, the amplitude is A_1 when the driving angular frequency equals $\sqrt{k/m}$. In terms of A_1, what is the amplitude for the same driving frequency and the same driving force amplitude F_{max}, if the damping constant is (a) $3b_1$ and (b) $b_1/2$?

PROBLEMS

14.64 ••• An object is undergoing SHM with period 0.300 s and amplitude 6.00 cm. At $t = 0$ the object is instantaneously at rest at $x = 6.00$ cm. Calculate the time it takes the object to go from $x = 6.00$ cm to $x = -1.50$ cm.

14.65 •• An object is undergoing SHM with period 1.200 s and amplitude 0.600 m. At $t = 0$ the object is at $x = 0$ and is moving in the negative x-direction. How far is the object from the equilibrium position when $t = 0.480$ s?

14.66 • Four passengers with combined mass 250 kg compress the springs of a car with worn-out shock absorbers by 4.00 cm when they get in. Model the car and passengers as a single body on a single ideal spring. If the loaded car has a period of vibration of 1.92 s, what is the period of vibration of the empty car?

14.67 •• At the end of a ride at a winter-theme amusement park, a sleigh with mass 250 kg (including two passengers) slides without friction along a horizontal, snow-covered surface. The sleigh hits one end of a light horizontal spring that obeys Hooke's law and has its other end attached to a wall. The sleigh latches onto the end of the spring and subsequently moves back and forth in SHM on the end of the spring until a braking mechanism is engaged, which brings the sleigh to rest. The frequency of the SHM is 0.225 Hz, and the amplitude is 0.950 m. (a) What was the speed of the sleigh just before it hit the end of the spring? (b) What is the maximum magnitude of the sleigh's acceleration during its SHM?

14.68 •• CP A block with mass M rests on a frictionless surface and is connected to a horizontal spring of force constant k. The other end of the spring is attached to a wall (**Fig. P14.68**). A second block with mass m rests on top of the first block. The coefficient of static friction between the blocks is μ_s. Find the *maximum* amplitude of oscillation such that the top block will not slip on the bottom block.

Figure **P14.68**

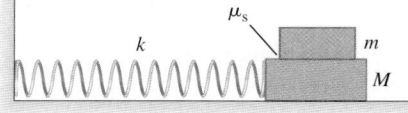

14.69 ••• A 1.50-kg, horizontal, uniform tray is attached to a vertical ideal spring of force constant 185 N/m and a 275-g metal ball is in the tray. The spring is below the tray, so it can oscillate up and down. The tray is then pushed down to point A, which is 15.0 cm below the equilibrium point, and released from rest. (a) How high above point A will the tray be when the metal ball leaves the tray? (*Hint:* This does *not* occur when the ball and tray reach their maximum speeds.) (b) How much time elapses between releasing the system at point A and the ball leaving the tray? (c) How fast is the ball moving just as it leaves the tray?

14.70 • CP A 10.0-kg mass is traveling to the right with a speed of 2.00 m/s on a smooth horizontal surface when it collides with and sticks to a second 10.0-kg mass that is initially at rest but is attached to a light spring with force constant 170.0 N/m. (a) Find the frequency, amplitude, and period of the subsequent oscillations. (b) How long does it take the system to return the first time to the position it had immediately after the collision?

14.71 ••• An apple weighs 1.00 N. When you hang it from the end of a long spring of force constant 1.50 N/m and negligible mass, it bounces up and down in SHM. If you stop the bouncing and let the apple swing from side to side through a small angle, the frequency of this simple pendulum is half the bounce frequency. (Because the angle is small, the back-and-forth swings do not cause any appreciable change in the length of the spring.) What is the unstretched length of the spring (with the apple removed)?

14.72 ••• CP **SHM of a Floating Object.** An object with height h, mass M, and a uniform cross-sectional area A floats upright in a liquid with density ρ. (a) Calculate the vertical distance from the surface of the liquid to the bottom of the floating object at equilibrium. (b) A downward force with magnitude F is applied to the top of the object. At the new equilibrium position, how much farther below the surface of the liquid is the bottom of the object than it was in part (a)? (Assume that some of the object remains above the surface of the liquid.) (c) Your result in part (b) shows that if the force is suddenly removed, the object will oscillate up and down in SHM. Calculate the period of this motion in terms of the density ρ of the liquid, the mass M, and the cross-sectional area A of the object. You can ignore the damping due to fluid friction (see Section 14.7).

14.73 •• CP A square object of mass m is constructed of four identical uniform thin sticks, each of length L, attached together. This object is hung on a hook at its upper corner (**Fig. P14.73**). If it is rotated slightly to the left and then released, at what frequency will it swing back and forth?

Figure **P14.73**

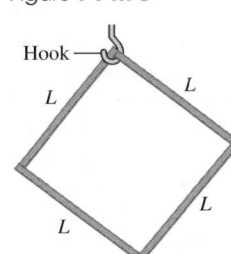

14.74 ••• An object with mass 0.200 kg is acted on by an elastic restoring force with force constant 10.0 N/m. (a) Graph elastic potential energy U as a function of displacement x over a range of x from -0.300 m to $+0.300$ m. On your graph, let 1 cm = 0.05 J vertically and 1 cm = 0.05 m horizontally. The object is set into oscillation with an initial potential energy of 0.140 J and an initial kinetic energy of 0.060 J. Answer the following questions by referring to the graph. (b) What is the amplitude of oscillation? (c) What is the potential energy when the displacement is one-half the amplitude? (d) At what displacement are the kinetic and potential energies equal? (e) What is the value of the phase angle ϕ if the initial velocity is positive and the initial displacement is negative?

14.75 • CALC A 2.00-kg bucket containing 10.0 kg of water is hanging from a vertical ideal spring of force constant 450 N/m and oscillating up and down with an amplitude of 3.00 cm. Suddenly the bucket springs a leak in the bottom such that water drops out at a steady rate of 2.00 g/s. When the bucket is half full, find (a) the period of oscillation and (b) the rate at which the period is changing with respect to time. Is the period getting longer or shorter? (c) What is the shortest period this system can have?

14.76 •• A uniform beam is suspended horizontally by two identical vertical springs that are attached between the ceiling and each end of the beam. The beam has mass 225 kg, and a 175-kg sack of gravel sits on the middle of it. The beam is oscillating in SHM with an amplitude of 40.0 cm and a frequency of 0.600 cycle/s. (a) The sack falls off the beam when the beam has its maximum upward displacement. What are the frequency and amplitude of the subsequent SHM of the beam? (b) If the sack instead falls off when the beam has its maximum speed, repeat part (a).

14.77 •• A 5.00-kg partridge is suspended from a pear tree by an ideal spring of negligible mass. When the partridge is pulled down 0.100 m below its equilibrium position and released, it vibrates with a period of 4.20 s. (a) What is its speed as it passes through the equilibrium position? (b) What is its acceleration when it is 0.050 m above the equilibrium position? (c) When it is moving upward, how much time is required for it to move from a point 0.050 m below its equilibrium position to a point 0.050 m above it? (d) The motion of the partridge is stopped, and then it is removed from the spring. How much does the spring shorten?

14.78 •• A 0.0200-kg bolt moves with SHM that has an amplitude of 0.240 m and a period of 1.500 s. The displacement of the bolt is $+0.240$ m when $t = 0$. Compute (a) the displacement of the bolt when $t = 0.500$ s; (b) the magnitude and direction of the force acting on the bolt when $t = 0.500$ s; (c) the minimum time required for the bolt to move from its initial position to the point where $x = -0.180$ m; (d) the speed of the bolt when $x = -0.180$ m.

14.79 •• CP **SHM of a Butcher's Scale.** A spring of negligible mass and force constant $k = 400$ N/m is hung vertically, and a 0.200-kg pan is suspended from its lower end. A butcher drops a 2.2-kg steak onto the pan from a height of 0.40 m. The steak makes a totally inelastic collision with the pan and sets the system into vertical SHM. What are (a) the speed of the pan and steak immediately after the collision; (b) the amplitude of the subsequent motion; (c) the period of that motion?

14.80 •• A 40.0-N force stretches a vertical spring 0.250 m. (a) What mass must be suspended from the spring so that the system will oscillate with a period of 1.00 s? (b) If the amplitude of the motion is 0.050 m and the period is that specified in part (a), where is the object and in what direction is it moving 0.35 s after it has passed the equilibrium position, moving downward? (c) What force (magnitude and direction) does the spring exert on the object when it is 0.030 m below the equilibrium position, moving upward?

14.81 •• **Don't Miss the Boat.** While on a visit to Minnesota ("Land of 10,000 Lakes"), you sign up to take an excursion around one of the larger lakes. When you go to the dock where the 1500-kg boat is tied, you find that the boat is bobbing up and down in the waves, executing simple harmonic motion with amplitude 20 cm. The boat takes 3.5 s to make one complete up-and-down cycle. When the boat is at its highest point, its deck is at the same height as the stationary dock. As you watch the boat bob up and down, you (mass 60 kg) begin to feel a bit woozy, due in part to the previous night's dinner of lutefisk. As a result, you refuse to board the boat unless the level of the boat's deck is within 10 cm of the dock level. How much time do you have to board the boat comfortably during each cycle of up-and-down motion?

14.82 • CP An interesting, though highly impractical example of oscillation is the motion of an object dropped down a hole that extends from one side of the earth, through its center, to the other side. With the assumption (not realistic) that the earth is a sphere of uniform density, prove that the motion is simple harmonic and find the period. [*Note:* The gravitational force on the object as a function of the object's distance r from the center of the earth was derived in Example 13.10 (Section 13.6). The motion is simple harmonic if the acceleration a_x and the displacement from equilibrium x are related by Eq. (14.8), and the period is then $T = 2\pi/\omega$.]

14.83 ••• CP A rifle bullet with mass 8.00 g and initial horizontal velocity 280 m/s strikes and embeds itself in a block with mass 0.992 kg that rests on a frictionless surface and is attached to one end of an ideal spring. The other end of the spring is attached to the wall. The impact compresses the spring a maximum distance of 15.0 cm. After the impact, the block moves in SHM. Calculate the period of this motion.

14.84 ••• CP Two uniform solid spheres, each with mass $M = 0.800$ kg and radius $R = 0.0800$ m, are connected by a short, light rod that is along a diameter of each sphere and are at rest on a horizontal tabletop. A spring with force constant $k = 160$ N/m has one end attached to the wall and the other end attached to a frictionless ring that passes over the rod at the center of mass of the spheres, which is midway between the centers of the two spheres. The spheres are each pulled the same distance from the wall, stretching the spring, and released. There is sufficient friction between the tabletop and the spheres for the spheres to roll without slipping as they move back and forth on the end of the spring. Show that the motion of the center of mass of the spheres is simple harmonic and calculate the period.

14.85 • CP In **Fig. P14.85** the upper ball is released from rest, collides with the stationary lower ball, and sticks to it. The strings are both 50.0 cm long. The upper ball has mass 2.00 kg, and it is initially 10.0 cm higher than the lower ball, which has mass 3.00 kg. Find the frequency and maximum angular displacement of the motion after the collision.

Figure **P14.85**

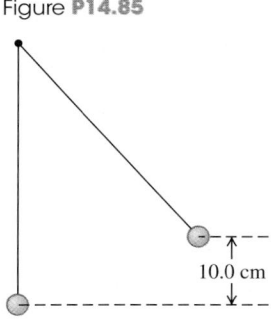

10.0 cm

14.86 •• **The Silently Ringing Bell.** A large, 34.0-kg bell is hung from a wooden beam so it can swing back and forth with negligible friction. The bell's center of mass is 0.60 m below the pivot. The bell's moment of inertia about an axis at the pivot is 18.0 kg · m². The clapper is a small, 1.8-kg mass attached to one end of a slender rod of length L and negligible mass. The other end of the rod is attached to the inside of the bell; the rod can swing freely about the same axis as the bell. What should be the length L of the clapper rod for the bell to ring silently—that is, for the period of oscillation for the bell to equal that of the clapper?

14.87 •• CALC A slender, uniform, metal rod with mass M is pivoted without friction about an axis through its midpoint and perpendicular to the rod. A horizontal spring with force constant k is attached to the lower end of the rod, with the other end of the spring attached to a rigid support. If the rod is displaced by a small angle Θ from the vertical (**Fig. P14.87**) and released, show that it moves in angular SHM and calculate the period. (*Hint:* Assume that the angle Θ is small enough for the approximations $\sin\Theta \approx \Theta$ and $\cos\Theta \approx 1$ to be valid. The motion is simple harmonic if $d^2\theta/dt^2 = -\omega^2\theta$, and the period is then $T = 2\pi/\omega$.)

Figure **P14.87**

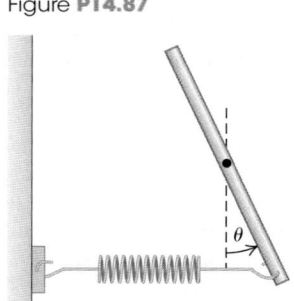

14.88 ••• Two identical thin rods, each with mass m and length L, are joined at right angles to form an L-shaped object. This object is balanced on top of a sharp edge (**Fig. P14.88**). If the L-shaped object is deflected slightly, it oscillates. Find the frequency of oscillation.

Figure **P14.88**

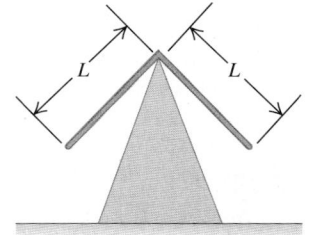

L L

14.89 •• DATA A mass m is attached to a spring of force constant 75 N/m and allowed to oscillate. **Figure P14.89** shows a graph of its velocity component v_x as a function of time t. Find (a) the period, (b) the frequency, and (c) the angular frequency of this motion. (d) What is the amplitude (in cm), and at what times does the mass reach this position? (e) Find the maximum acceleration magnitude of the mass and the times at which it occurs. (f) What is the value of m?

Figure **P14.89** v_x (cm/s)

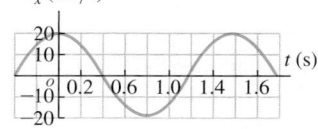

14.90 •• DATA You hang various masses m from the end of a vertical, 0.250-kg spring that obeys Hooke's law and is tapered, which means the diameter changes along the length of the spring. Since the mass of the spring is not negligible, you must replace m in the equation $T = 2\pi\sqrt{m/k}$ with $m + m_{eff}$, where m_{eff} is the effective mass of the oscillating spring. (See Challenge Problem 14.93.) You vary the mass m and measure the time for 10 complete oscillations, obtaining these data:

m (kg)	0.100	0.200	0.300	0.400	0.500
Time (s)	8.7	10.5	12.2	13.9	15.1

(a) Graph the square of the period T versus the mass suspended from the spring, and find the straight line of best fit. (b) From the slope of that line, determine the force constant of the spring. (c) From the vertical intercept of the line, determine the spring's effective mass. (d) What fraction is m_{eff} of the spring's mass? (e) If a 0.450-kg mass oscillates on the end of the spring, find its period, frequency, and angular frequency.

14.91 ••• DATA Experimenting with pendulums, you attach a light string to the ceiling and attach a small metal sphere to the lower end of the string. When you displace the sphere 2.00 m to the left, it nearly touches a vertical wall; with the string taut, you release the sphere from rest. The sphere swings back and forth as a simple pendulum, and you measure its period T. You repeat this act for strings of various lengths L, each time starting the motion with the sphere displaced 2.00 m to the left of the vertical position of the string. In each case the sphere's radius is very small compared with L. Your results are given in the table:

L (m)	12.00	10.00	8.00	6.00	5.00	4.00	3.00	2.50	2.30
T (s)	6.96	6.36	5.70	4.95	4.54	4.08	3.60	3.35	3.27

(a) For the five largest values of L, graph T^2 versus L. Explain why the data points fall close to a straight line. Does the slope of this line have the value you expected? (b) Add the remaining data to your graph. Explain why the data start to deviate from the straight-line fit as L decreases. To see this effect more clearly, plot T/T_0 versus L, where $T_0 = 2\pi\sqrt{L/g}$ and $g = 9.80$ m/s². (c) Use your graph of T/T_0 versus L to estimate the angular amplitude of the pendulum (in degrees) for which the equation $T = 2\pi\sqrt{L/g}$ is in error by 5%.

CHALLENGE PROBLEMS

14.92 ••• **The Effective Force Constant of Two Springs.** Two springs with the same unstretched length but different force constants k_1 and k_2 are attached to a block with mass m on a level, frictionless surface. Calculate the effective force constant k_{eff} in each of the three cases (a), (b), and (c) depicted in **Fig. P14.92**. (The effective force constant is defined by $\sum F_x = -k_{eff}x$.) (d) An object with mass m, suspended from a uniform spring with a force constant k, vibrates with a frequency f_1. When the spring is cut in half and the same object is suspended from one of the halves, the frequency is f_2. What is the ratio f_1/f_2?

Figure **P14.92**

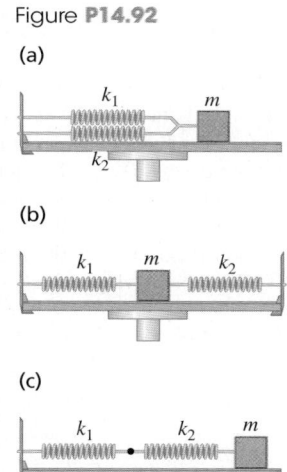

14.93 ••• CALC **A Spring with Mass.** The preceding problems in this chapter have assumed that the springs had negligible mass. But of course no spring is completely massless. To find the effect of the spring's mass, consider a spring with mass M, equilibrium length L_0, and spring constant k. When stretched or compressed to a length L, the potential energy is $\frac{1}{2}kx^2$, where $x = L - L_0$. (a) Consider a spring, as described above, that has one end fixed and the other end moving with speed v. Assume that the speed of points along the length of the spring varies linearly with distance l from the fixed end. Assume also that the

mass M of the spring is distributed uniformly along the length of the spring. Calculate the kinetic energy of the spring in terms of M and v. (*Hint:* Divide the spring into pieces of length dl; find the speed of each piece in terms of l, v, and L; find the mass of each piece in terms of dl, M, and L; and integrate from 0 to L. The result is *not* $\frac{1}{2}Mv^2$, since not all of the spring moves with the same speed.) (b) Take the time derivative of the conservation of energy equation, Eq. (14.21), for a mass m moving on the end of a *massless* spring. By comparing your results to Eq. (14.8), which defines ω, show that the angular frequency of oscillation is $\omega = \sqrt{k/m}$. (c) Apply the procedure of part (b) to obtain the angular frequency of oscillation ω of the spring considered in part (a). If the *effective mass* M' of the spring is defined by $\omega = \sqrt{k/M'}$, what is M' in terms of M?

BIO **"SEEING" SURFACES AT THE NANOSCALE.** One technique for making images of surfaces at the nanometer scale, including membranes and biomolecules, is dynamic atomic force microscopy. In this technique, a small tip is attached to a cantilever, which is a flexible, rectangular slab supported at one end, like a diving board. The cantilever vibrates, so the tip moves up and down in simple harmonic motion. In one operating mode, the resonant frequency for a cantilever with force constant $k = 1000$ N/m is 100 kHz. As the oscillating tip is brought within a few nanometers of the surface of a sample (as shown in the figure), it experiences an attractive force from the surface. For an oscillation with a small amplitude (typically, 0.050 nm), the force F that the sample surface exerts on the tip varies linearly with the displacement x of the tip, $|F| = k_{surf}x$, where k_{surf} is the effective force constant for this force. The net force on the tip is therefore $(k + k_{surf})x$, and the frequency of the oscillation changes slightly due to the interaction with the surface. Measurements of the frequency as the tip moves over different parts of the sample's surface can provide information about the sample.

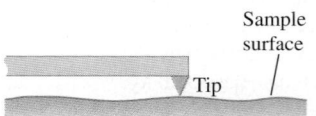

14.94 If we model the vibrating system as a mass on a spring, what is the mass necessary to achieve the desired resonant frequency when the tip is not interacting with the surface? (a) 25 ng; (b) 100 ng; (c) 2.5 μg; (d) 100 μg.

14.95 In the model of Problem 14.94, what is the mechanical energy of the vibration when the tip is not interacting with the surface? (a) 1.2×10^{-18} J; (b) 1.2×10^{-16} J; (c) 1.2×10^{-9} J; (d) 5.0×10^{-8} J.

14.96 By what percentage does the frequency of oscillation change if $k_{surf} = 5$ N/m? (a) 0.1%; (b) 0.2%; (c) 0.5%; (d) 1.0%.

Answers

Chapter Opening Question ❓

(i) The back-and-forth motion of a leg during walking is like a physical pendulum, for which the oscillation period is $T = 2\pi\sqrt{I/mgd}$ [see Eq. (14.39)]. In this expression I is the moment of inertia of the pendulum, m is its mass, and d is the distance from the rotation axis to the pendulum center of mass. I is proportional to m, so the mass cancels out of this expression for T. Hence only the dimensions of the leg matter. (See Examples 14.9 and 14.10.)

Test Your Understanding Questions

14.1 (a) $x < 0$, (b) $x > 0$, (c) $x < 0$, (d) $x > 0$, (e) $x > 0$, (f) $x = 0$ Figure 14.2 shows that both the net x-component of force F_x and the x-acceleration a_x are positive when $x < 0$ (so the body is displaced to the left and the spring is compressed), while both F_x and a_x are negative when $x > 0$ (so the body is displaced to the right and the spring is stretched). Hence x and a_x always have *opposite* signs. This is true whether the object is moving to the right $(v_x > 0)$, to the left $(v_x < 0)$, or not at all $(v_x = 0)$, since the force exerted by the spring depends on only whether it is compressed or stretched and by what distance. This explains the answers to (a) through (e). If the acceleration is zero as in (f), the net force must also be zero and so the spring must be relaxed; hence $x = 0$.

14.2 (a) $A > 0.10$ m, $\phi < 0$; (b) $A > 0.10$ m, $\phi > 0$ In both situations the initial $(t = 0)$ x-velocity v_{0x} is nonzero, so from Eq. (14.19) the amplitude $A = \sqrt{x_0^2 + (v_{0x}^2/\omega^2)}$ is greater than the initial x-coordinate $x_0 = 0.10$ m. From Eq. (14.18) the phase angle is $\phi = \arctan(-v_{0x}/\omega x_0)$, which is positive if the quantity $-v_{0x}/\omega x_0$ (the argument of the arctangent function) is positive and negative if $-v_{0x}/\omega x_0$ is negative. In part (a) both x_0 and v_{0x} are positive, so $-v_{0x}/\omega x_0 < 0$ and $\phi < 0$. In part (b) x_0 is positive and v_{0x} is negative, so $-v_{0x}/\omega x_0 > 0$ and $\phi > 0$.

14.3 (a) (iii), (b) (v) To increase the total energy $E = \frac{1}{2}kA^2$ by a factor of 2, the amplitude A must increase by a factor of $\sqrt{2}$. Because the motion is SHM, changing the amplitude has no effect on the frequency.

14.4 (i) The oscillation period of a body of mass m attached to a hanging spring of force constant k is given by $T = 2\pi\sqrt{m/k}$, the same expression as for a body attached to a horizontal spring. Neither m nor k changes when the apparatus is taken to Mars, so the period is unchanged. The only difference is that in equilibrium, the spring will stretch a shorter distance on Mars than on earth due to the weaker gravity.

14.5 (iv) Just as for an object oscillating on a spring, at the equilibrium position the *speed* of a pendulum bob is instantaneously not changing (this is where the speed is maximum, so its derivative at this time is zero). But the *direction* of motion is changing because the pendulum bob follows a circular path. Hence the bob must have a component of acceleration perpendicular to the path and toward the center of the circle (see Section 3.4). To cause this acceleration at the equilibrium position when the string is vertical, the upward tension force at this position must be greater than the weight of the bob. This causes a net upward force on the bob and an upward acceleration toward the center of the circular path.

14.6 (i) The period of a physical pendulum is given by Eq. (14.39), $T = 2\pi\sqrt{I/mgd}$. The distance $d = L$ from the pivot to the center of gravity is the same for both the rod and the simple pendulum, as is the mass m. Thus for any displacement angle θ the same restoring torque acts on both the rod and the simple pendulum. However, the rod has a greater moment of inertia: $I_{rod} = \frac{1}{3}m(2L)^2 = \frac{4}{3}mL^2$ and $I_{simple} = mL^2$ (all the mass of the pendulum is a distance L from the pivot). Hence the rod has a longer period.

14.7 (ii) The oscillations are underdamped with a decreasing amplitude on each cycle of oscillation, like those graphed in Fig. 14.26. If the oscillations were undamped, they would continue indefinitely with the same amplitude. If they were critically damped or overdamped, the nose would not bob up and down but would return smoothly to the original equilibrium attitude without overshooting.

14.8 (i) Figure 14.28 shows that the curve of amplitude versus driving frequency moves upward at *all* frequencies as the value of the damping constant b is decreased. Hence for fixed values of k and m, the oscillator with the least damping (smallest value of b) will have the greatest response at any driving frequency.

Bridging Problem

$T = 2\pi\sqrt{3M/2k}$

? When an earthquake strikes, the news of the event travels through the body of the earth in the form of seismic waves. Which aspects of a seismic wave determine how much power is carried by the wave: (i) the amplitude; (ii) the frequency; (iii) both the amplitude and the frequency; or (iv) neither the amplitude nor the frequency?

15 MECHANICAL WAVES

LEARNING GOALS

Looking forward at …

15.1 What is meant by a mechanical wave, and the different varieties of mechanical waves.

15.2 How to use the relationship among speed, frequency, and wavelength for a periodic wave.

15.3 How to interpret and use the mathematical expression for a sinusoidal periodic wave.

15.4 How to calculate the speed of waves on a rope or string.

15.5 How to calculate the rate at which a mechanical wave transports energy.

15.6 What happens when mechanical waves overlap and interfere.

15.7 The properties of standing waves on a string, and how to analyze these waves.

15.8 How stringed instruments produce sounds of specific frequencies.

Looking back at …

8.1 The impulse–momentum theorem.

14.1, 14.2 Periodic motion and simple harmonic motion.

Ripples on a pond, musical sounds, seismic tremors triggered by an earthquake—all these are *wave* phenomena. Waves can occur whenever a system is disturbed from equilibrium and when the disturbance can travel, or *propagate,* from one region of the system to another. As a wave propagates, it carries energy. The energy in light waves from the sun warms the surface of our planet; the energy in seismic waves can crack our planet's crust.

This chapter and the next are about mechanical waves—waves that travel within some material called a *medium.* (Chapter 16 is concerned with sound, an important type of mechanical wave.) We'll begin this chapter by deriving the basic equations for describing waves, including the important special case of *sinusoidal* waves in which the wave pattern is a repeating sine or cosine function. To help us understand waves in general, we'll look at the simple case of waves that travel on a stretched string or rope.

Waves on a string are important in music. When a musician strums a guitar or bows a violin, she makes waves that travel in opposite directions along the instrument's strings. What happens when these oppositely directed waves overlap is called *interference.* We'll discover that sinusoidal waves can occur on a guitar or violin string only for certain special frequencies, called *normal-mode frequencies,* determined by the properties of the string. These normal-mode frequencies determine the pitch of the musical sounds that a stringed instrument produces. (In the next chapter we'll find that interference also helps explain the pitches of *wind* instruments such as pipe organs.)

Not all waves are mechanical in nature. *Electromagnetic* waves—including light, radio waves, infrared and ultraviolet radiation, and x rays—can propagate even in empty space, where there is *no* medium. We'll explore these and other nonmechanical waves in later chapters.

15.1 TYPES OF MECHANICAL WAVES

A **mechanical wave** is a disturbance that travels through some material or substance called the **medium** for the wave. As the wave travels through the medium, the particles that make up the medium undergo displacements of various kinds, depending on the nature of the wave.

15.1 Three ways to make a wave that moves to the right. (a) The hand moves the string up and then returns, producing a transverse wave. (b) The piston moves to the right, compressing the gas or liquid, and then returns, producing a longitudinal wave. (c) The board moves to the right and then returns, producing a combination of longitudinal and transverse waves.

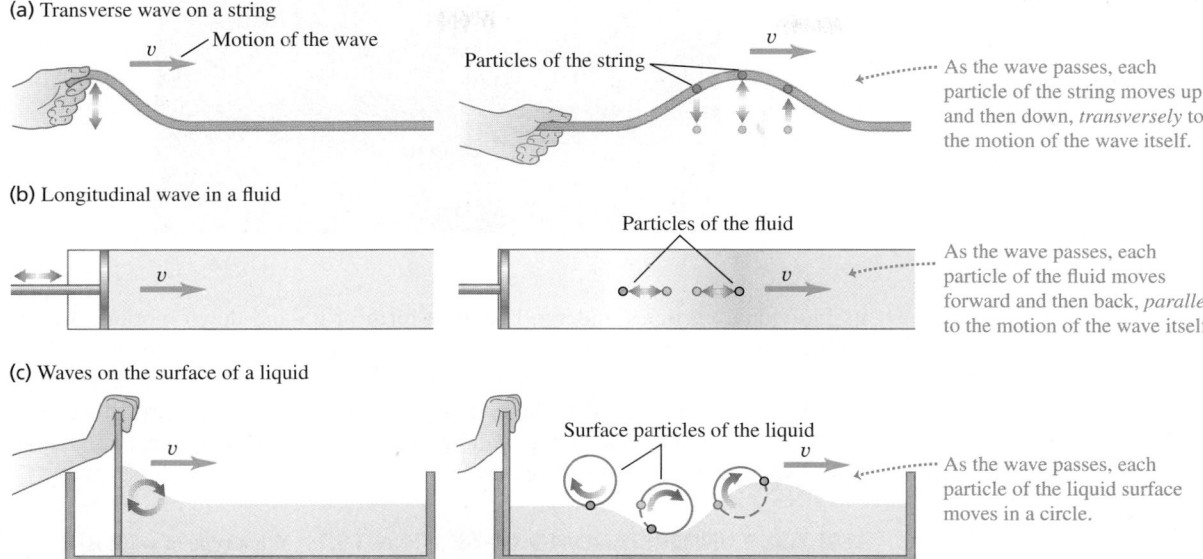

(a) Transverse wave on a string

As the wave passes, each particle of the string moves up and then down, *transversely* to the motion of the wave itself.

(b) Longitudinal wave in a fluid

As the wave passes, each particle of the fluid moves forward and then back, *parallel* to the motion of the wave itself.

(c) Waves on the surface of a liquid

As the wave passes, each particle of the liquid surface moves in a circle.

Figure 15.1 shows three varieties of mechanical waves. In Fig. 15.1a the medium is a string or rope under tension. If we give the left end a small upward shake or wiggle, the wiggle travels along the length of the string. Successive sections of string go through the same motion that we gave to the end, but at successively later times. Because the displacements of the medium are perpendicular or *transverse* to the direction of travel of the wave along the medium, this is called a **transverse wave.**

In Fig. 15.1b the medium is a liquid or gas in a tube with a rigid wall at the right end and a movable piston at the left end. If we give the piston a single back-and-forth motion, displacement and pressure fluctuations travel down the length of the medium. This time the motions of the particles of the medium are back and forth along the *same* direction that the wave travels. We call this a **longitudinal wave.**

In Fig. 15.1c the medium is a liquid in a channel, such as water in an irrigation ditch or canal. When we move the flat board at the left end forward and back once, a wave disturbance travels down the length of the channel. In this case the displacements of the water have *both* longitudinal and transverse components.

Each of these systems has an equilibrium state. For the stretched string it is the state in which the system is at rest, stretched out along a straight line. For the fluid in a tube it is a state in which the fluid is at rest with uniform pressure. And for the liquid in a trough it is a smooth, level water surface. In each case the wave motion is a disturbance from equilibrium that travels from one region of the medium to another. And in each case forces tend to restore the system to its equilibrium position when it is displaced, just as the force of gravity tends to pull a pendulum toward its straight-down equilibrium position when it is displaced.

These examples have three things in common. First, in each case the disturbance travels or *propagates* with a definite speed through the medium. This speed is called the speed of propagation, or simply the **wave speed.** Its value is determined in each case by the mechanical properties of the medium. We will use the symbol v for wave speed. (The wave speed is *not* the same as the speed with which particles move when they are disturbed by the wave. We'll return to this point in Section 15.3.) Second, the medium itself does not travel through space;

BIO Application Waves on a Snake's Body A snake moves itself along the ground by producing waves that travel backward along its body from its head to its tail. The waves remain stationary with respect to the ground as they push against the ground, so the snake moves forward.

15.2 "Doing the wave" at a sports stadium is an example of a mechanical wave: The disturbance propagates through the crowd, but there is no transport of matter (none of the spectators moves from one seat to another).

its individual particles undergo back-and-forth or up-and-down motions around their equilibrium positions. The overall pattern of the wave disturbance is what travels. Third, to set any of these systems into motion, we have to put in energy by doing mechanical work on the system. The wave motion transports this energy from one region of the medium to another. *Waves transport energy, but not matter, from one region to another* (**Fig. 15.2**).

TEST YOUR UNDERSTANDING OF SECTION 15.1 What type of wave is "the wave" shown in Fig. 15.2? (i) Transverse; (ii) longitudinal; (iii) a combination of transverse and longitudinal. ∎

15.2 PERIODIC WAVES

The transverse wave on a stretched string in Fig. 15.1a is an example of a *wave pulse*. The hand exerts a transverse force that shakes the string up and down just once, producing a single "wiggle," or pulse, that travels along the length of the string. The tension in the string restores its straight-line shape once the pulse has passed.

A more interesting situation develops when we give the free end of the string a repetitive, or *periodic,* motion. (You should review the discussion of periodic motion in Chapter 14 before going ahead.) Each particle in the string undergoes periodic motion as the wave propagates, and we have a **periodic wave.**

Periodic Transverse Waves

Suppose we move one end of the string up and down with *simple harmonic motion* (SHM) as in **Fig. 15.3**, with amplitude A, frequency f, angular frequency $\omega = 2\pi f$, and period $T = 1/f = 2\pi/\omega$. The wave that results is a symmetric sequence of *crests* and *troughs.* As we will see, periodic waves with SHM are

15.3 A block of mass m attached to a spring undergoes simple harmonic motion, producing a sinusoidal wave that travels to the right on the string. (In a real-life system a driving force would have to be applied to the block to replace the energy carried away by the wave.)

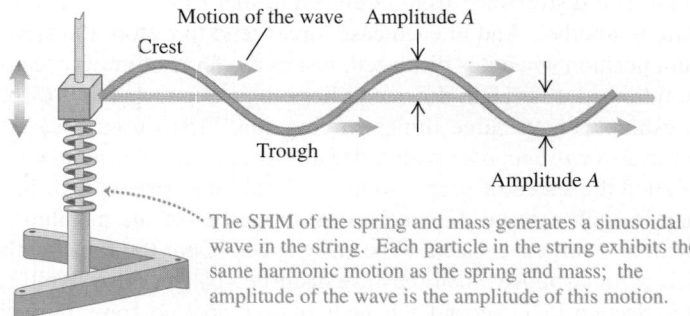

The SHM of the spring and mass generates a sinusoidal wave in the string. Each particle in the string exhibits the same harmonic motion as the spring and mass; the amplitude of the wave is the amplitude of this motion.

particularly easy to analyze; we call them **sinusoidal waves.** It turns out that *any* periodic wave can be represented as a combination of sinusoidal waves. So this kind of wave motion is worth special attention.

In Fig. 15.3 the wave is a *continuous succession* of transverse sinusoidal disturbances. **Figure 15.4** shows the shape of a part of the string near the left end at time intervals of $\frac{1}{8}$ of a period, for a total time of one period. The wave shape advances steadily toward the right, as indicated by the highlighted area. As the wave moves, any point on the string (any of the red dots, for example) oscillates up and down about its equilibrium position with simple harmonic motion. *When a sinusoidal wave passes through a medium, every particle in the medium undergoes simple harmonic motion with the same frequency.*

CAUTION **Wave motion vs. particle motion** Don't confuse the motion of the *transverse wave* along the string and the motion of a *particle* of the string. The wave moves with constant speed v *along* the length of the string, while the motion of the particle is simple harmonic and *transverse* (perpendicular) to the length of the string. ∎

For a periodic wave, the shape of the string at any instant is a repeating pattern. The **wavelength** λ (the Greek letter lambda) of the wave is the distance from one crest to the next, or from one trough to the next, or from any point to the corresponding point on the next repetition of the wave shape. The wave pattern travels with constant speed v and advances a distance of one wavelength λ in a time interval of one period T. So the wave speed is $v = \lambda/T$ or, because $f = 1/T$ from Eq. (14.1),

For a **periodic wave:**   Wave speed···→   ···Wavelength
$$v = \lambda f$$ ····· Frequency   (15.1)

The speed of propagation equals the product of wavelength and frequency. The frequency is a property of the *entire* periodic wave because all points on the string oscillate with the same frequency f.

Waves on a string propagate in just one dimension (in Fig. 15.4, along the x-axis). But the ideas of frequency, wavelength, and amplitude apply equally well to waves that propagate in two or three dimensions. **Figure 15.5** shows a wave propagating in two dimensions on the surface of a tank of water. As with waves on a string, the wavelength is the distance from one crest to the next, and the amplitude is the height of a crest above the equilibrium level.

In many important situations including waves on a string, the wave speed v is determined entirely by the mechanical properties of the medium. In this case, increasing f causes λ to decrease so the product $v = \lambda f$ remains the same, and waves of *all* frequencies propagate with the same wave speed. In this chapter we will consider *only* waves of this kind. (In later chapters we'll study the propagation of light waves in transparent materials where the wave speed depends on frequency; this turns out to be the reason raindrops create a rainbow.)

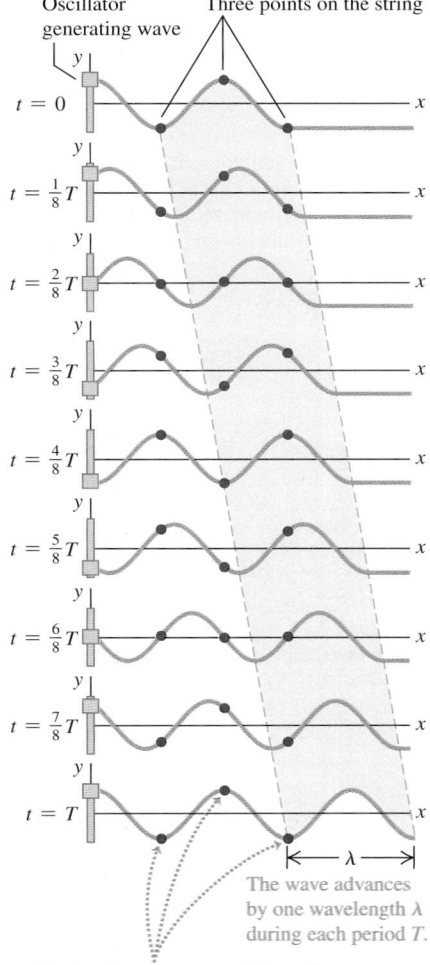

15.4 A sinusoidal transverse wave traveling to the right along a string. The vertical scale is exaggerated.

The string is shown at time intervals of $\frac{1}{8}$ period for a total of one period T. The highlighting shows the motion of one wavelength of the wave.

The wave advances by one wavelength λ during each period T.

Each point moves up and down in place. Particles one wavelength apart move in phase with each other.

15.5 A series of drops falling into water produces a periodic wave that spreads radially outward. The wave crests and troughs are concentric circles. The wavelength λ is the radial distance between adjacent crests or adjacent troughs.

15.6 Using an oscillating piston to make a sinusoidal longitudinal wave in a fluid.

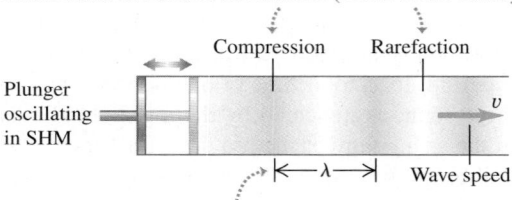

Forward motion of the plunger creates a compression (a zone of high density); backward motion creates a rarefaction (a zone of low density).

Wavelength λ is the distance between corresponding points on successive cycles.

15.7 A sinusoidal longitudinal wave traveling to the right in a fluid. The wave has the same amplitude A and period T as the oscillation of the piston.

Longitudinal waves are shown at intervals of $\frac{1}{8}T$ for one period T.

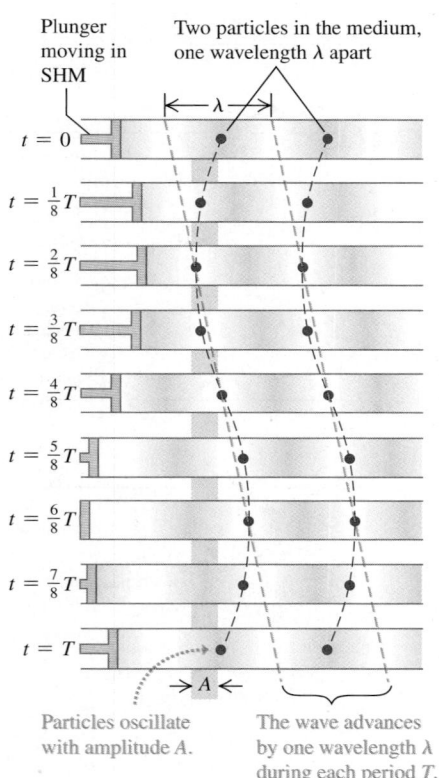

Particles oscillate with amplitude A.

The wave advances by one wavelength λ during each period T.

Periodic Longitudinal Waves

To understand the mechanics of a periodic *longitudinal* wave, consider a long tube filled with a fluid, with a piston at the left end as in Fig. 15.1b. If we push the piston in, we compress the fluid near the piston, increasing the pressure in this region. This region then pushes against the neighboring region of fluid, and so on, and a wave pulse moves along the tube.

Now suppose we move the piston back and forth in SHM along a line parallel to the axis of the tube (**Fig. 15.6**). This motion forms regions in the fluid where the pressure and density are greater or less than the equilibrium values. We call a region of increased density a *compression;* a region of reduced density is a *rarefaction.* Figure 15.6 shows compressions as darkly shaded areas and rarefactions as lightly shaded areas. The wavelength is the distance from one compression to the next or from one rarefaction to the next.

Figure 15.7 shows the wave propagating in the fluid-filled tube at time intervals of $\frac{1}{8}$ of a period, for a total time of one period. The pattern of compressions and rarefactions moves steadily to the right, just like the pattern of crests and troughs in a sinusoidal transverse wave (compare Fig. 15.4). Each particle in the fluid oscillates in SHM parallel to the direction of wave propagation (that is, left and right) with the same amplitude A and period T as the piston. The particles shown by the two red dots in Fig. 15.7 are one wavelength apart, and so oscillate in phase with each other.

Just like the sinusoidal transverse wave shown in Fig. 15.4, in one period T the longitudinal wave in Fig. 15.7 travels one wavelength λ to the right. Hence the fundamental equation $v = \lambda f$ holds for longitudinal waves as well as for transverse waves, and indeed for *all* types of periodic waves. Just as for transverse waves, in this chapter and the next we will consider only situations in which the speed of longitudinal waves does not depend on the frequency.

EXAMPLE 15.1   **WAVELENGTH OF A MUSICAL SOUND**

Sound waves are longitudinal waves in air. The speed of sound depends on temperature; at 20°C it is 344 m/s (1130 ft/s). What is the wavelength of a sound wave in air at 20°C if the frequency is 262 Hz (the approximate frequency of middle C on a piano)?

SOLUTION

IDENTIFY and SET UP: This problem involves Eq. (15.1), $v = \lambda f$, which relates wave speed v, wavelength λ, and frequency f for a periodic wave. The target variable is the wavelength λ. We are given $v = 344$ m/s and $f = 262$ Hz $= 262$ s^{-1}.

EXECUTE: We solve Eq. (15.1) for λ:

$$\lambda = \frac{v}{f} = \frac{344 \text{ m/s}}{262 \text{ Hz}} = \frac{344 \text{ m/s}}{262 \text{ s}^{-1}} = 1.31 \text{ m}$$

EVALUATE: The speed v of sound waves does *not* depend on the frequency. Hence $\lambda = v/f$ says that wavelength changes in inverse proportion to frequency. As an example, high (soprano) C is two octaves above middle C. Each octave corresponds to a factor of 2 in frequency, so the frequency of high C is four times that of middle C: $f = 4(262 \text{ Hz}) = 1048$ Hz. Hence the *wavelength* of high C is *one-fourth* as large: $\lambda = (1.31 \text{ m})/4 = 0.328$ m.

15.3 MATHEMATICAL DESCRIPTION OF A WAVE

Many characteristics of periodic waves can be described by using the concepts of wave speed, amplitude, period, frequency, and wavelength. Often, though, we need a more detailed description of the positions and motions of individual particles of the medium at particular times during wave propagation.

As a specific example, let's look at waves on a stretched string. If we ignore the sag of the string due to gravity, the equilibrium position of the string is along a straight line. We take this to be the x-axis of a coordinate system. Waves on a string are *transverse;* during wave motion a particle with equilibrium position x is displaced some distance y in the direction perpendicular to the x-axis. The value of y depends on which particle we are talking about (that is, y depends on x) and also on the time t when we look at it. Thus y is a *function* of both x and t; $y = y(x, t)$. We call $y(x, t)$ the **wave function** that describes the wave. If we know this function for a particular wave motion, we can use it to find the displacement (from equilibrium) of any particle at any time. From this we can find the velocity and acceleration of any particle, the shape of the string, and anything else we want to know about the behavior of the string at any time.

Wave Function for a Sinusoidal Wave

Let's see how to determine the form of the wave function for a sinusoidal wave. Suppose a sinusoidal wave travels from left to right (the direction of increasing x) along the string, as in **Fig. 15.8**. Every particle of the string oscillates in simple harmonic motion with the same amplitude and frequency. But the oscillations of particles at different points on the string are *not* all in step with each other. The particle at point B in Fig. 15.8 is at its maximum positive value of y at $t = 0$ and returns to $y = 0$ at $t = \frac{2}{8}T$; these same events occur for a particle at point A or point C at $t = \frac{4}{8}T$ and $t = \frac{6}{8}T$, exactly one half-period later. For any two particles of the string, the motion of the particle on the right (in terms of the wave, the "downstream" particle) lags behind the motion of the particle on the left by an amount proportional to the distance between the particles.

Hence the cyclic motions of various points on the string are out of step with each other by various fractions of a cycle. We call these differences *phase differences,* and we say that the *phase* of the motion is different for different points. For example, if one point has its maximum positive displacement at the same time that another has its maximum negative displacement, the two are a half-cycle out of phase. (This is the case for points A and B, or points B and C.)

Suppose that the displacement of a particle at the left end of the string ($x = 0$), where the wave originates, is given by

$$y(x = 0, t) = A\cos\omega t = A\cos 2\pi ft \qquad (15.2)$$

That is, the particle oscillates in SHM with amplitude A, frequency f, and angular frequency $\omega = 2\pi f$. The notation $y(x = 0, t)$ reminds us that the motion of this particle is a special case of the wave function $y(x, t)$ that describes the entire wave. At $t = 0$ the particle at $x = 0$ is at its maximum positive displacement ($y = A$) and is instantaneously at rest (because y is a maximum).

The wave disturbance travels from $x = 0$ to some point x to the right of the origin in an amount of time given by x/v, where v is the wave speed. So the motion of point x at time t is the same as the motion of point $x = 0$ at the earlier time $t - x/v$. Hence we can find the displacement of point x at time t

15.8 Tracking the oscillations of three points on a string as a sinusoidal wave propagates along it.

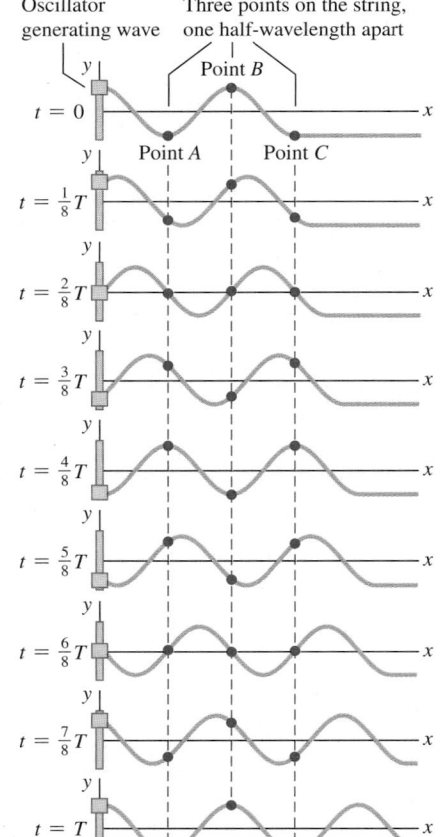

The string is shown at time intervals of $\frac{1}{8}$ period for a total of one period T.

by simply replacing t in Eq. (15.2) by $(t - x/v)$:

$$y(x, t) = A\cos\left[\omega\left(t - \frac{x}{v}\right)\right]$$

Because $\cos(-\theta) = \cos\theta$, we can rewrite the wave function as

Wave function for a sinusoidal wave propagating in +*x*-direction

Amplitude, Position, Time

$$y(x, t) = A\cos\left[\omega\left(\frac{x}{v} - t\right)\right] \qquad (15.3)$$

Angular frequency $= 2\pi f$ Wave speed

The displacement $y(x, t)$ is a function of both the location x of the point and the time t. We could make Eq. (15.3) more general by allowing for different values of the phase angle, as we did for SHM in Section 14.2, but for now we omit this.

We can rewrite the wave function given by Eq. (15.3) in several different but useful forms. We can express it in terms of the period $T = 1/f$ and the wavelength $\lambda = v/f = 2\pi v/\omega$:

Wave function for a sinusoidal wave propagating in +*x*-direction

Amplitude, Position, Time

$$y(x, t) = A\cos\left[2\pi\left(\frac{x}{\lambda} - \frac{t}{T}\right)\right] \qquad (15.4)$$

Wavelength   Period

It's convenient to define a quantity k, called the **wave number**:

$$k = \frac{2\pi}{\lambda} \qquad \text{(wave number)} \qquad (15.5)$$

Substituting $\lambda = 2\pi/k$ and $f = \omega/2\pi$ into Eq. (15.1), $v = \lambda f$, gives

$$\omega = vk \qquad \text{(periodic wave)} \qquad (15.6)$$

We can then rewrite Eq. (15.4) as

Wave function for a sinusoidal wave propagating in +*x*-direction

Amplitude, Position, Time

$$y(x, t) = A\cos(kx - \omega t) \qquad (15.7)$$

Wave number $= 2\pi/\lambda$ Angular frequency $= 2\pi f$

Which of these various forms for the wave function $y(x, t)$ we use in any specific problem is a matter of convenience. Note that ω has units rad/s, so for unit consistency in Eqs. (15.6) and (15.7) the wave number k must have the units rad/m. (*Warning:* Some textbooks define the wave number as $1/\lambda$ rather than $2\pi/\lambda$.)

Graphing the Wave Function

Figure 15.9a graphs the wave function $y(x, t)$ as a function of x for a specific time t. This graph gives the displacement y of a particle from its equilibrium position as a function of the coordinate x of the particle. If the wave is a transverse wave on a string, the graph in Fig. 15.9a represents the shape of the string at that instant, like a flash photograph of the string. In particular, at time $t = 0$,

$$y(x, t = 0) = A\cos kx = A\cos 2\pi\frac{x}{\lambda}$$

Figure 15.9b is a graph of the wave function versus time t for a specific coordinate x. This graph gives the displacement y of the particle at x as a function of time; that is, it describes the motion of that particle. At position $x = 0$,

$$y(x = 0, t) = A\cos(-\omega t) = A\cos\omega t = A\cos 2\pi\frac{t}{T}$$

15.9 Two graphs of the wave function $y(x, t)$ in Eq. (15.7). **(a)** Graph of displacement y versus coordinate x at time $t = 0$. **(b)** Graph of displacement y versus time t at coordinate $x = 0$. The vertical scale is exaggerated in both (a) and (b).

(a) If we use Eq. (15.7) to plot y as a function of x for time $t = 0$, the curve shows the *shape* of the string at $t = 0$.

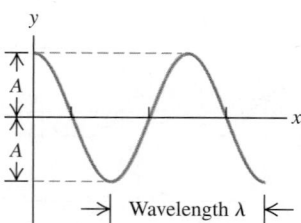

(b) If we use Eq. (15.7) to plot y as a function of t for position $x = 0$, the curve shows the *displacement* y of the particle at $x = 0$ as a function of time.

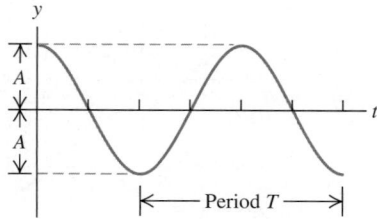

This is consistent with our original statement about the motion at $x = 0$, Eq. (15.2).

CAUTION **Wave graphs** Although they may look the same, Figs. 15.9a and 15.9b are *not* identical. Figure 15.9a is a picture of the shape of the string at $t = 0$, while Fig. 15.9b is a graph of the displacement y of a particle at $x = 0$ as a function of time. ❚

More on the Wave Function

We can modify Eqs. (15.3) through (15.7) to represent a wave traveling in the *negative* x-direction. In this case the displacement of point x at time t is the same as the motion of point $x = 0$ at the *later* time $(t + x/v)$, so in Eq. (15.2) we replace t by $(t + x/v)$. For a wave traveling in the negative x-direction,

$$y(x, t) = A\cos\left[\omega\left(\frac{x}{v} + t\right)\right] = A\cos\left[2\pi\left(\frac{x}{\lambda} + \frac{t}{T}\right)\right] = A\cos(kx + \omega t) \quad (15.8)$$

$$\text{(sinusoidal wave moving in } -x\text{-direction)}$$

In the expression $y(x, t) = A\cos(kx \pm \omega t)$ for a wave traveling in the $-x$- or $+x$-direction, the quantity $(kx \pm \omega t)$ is called the **phase.** It plays the role of an angular quantity (always measured in radians) in Eq. (15.7) or (15.8), and its value for any values of x and t determines what part of the sinusoidal cycle is occurring at a particular point and time. For a crest (where $y = A$ and the cosine function has the value 1), the phase could be 0, $\pm 2\pi$, $\pm 4\pi$, and so on; for a trough (where $y = -A$ and the cosine has the value -1), it could be $\pm \pi$, $\pm 3\pi$, $\pm 5\pi$, and so on.

The wave speed is the speed with which we have to move along with the wave to keep alongside a point of a given phase, such as a particular crest of a wave on a string. For a wave traveling in the $+x$-direction, that means $kx - \omega t = \text{constant}$. Taking the derivative with respect to t, we find $k\, dx/dt = \omega$, or

$$\frac{dx}{dt} = \frac{\omega}{k}$$

Comparing this with Eq. (15.6), we see that dx/dt is equal to the speed v of the wave. Because of this relationship, v is sometimes called the *phase velocity* of the wave. (*Phase speed* would be a better term.)

PROBLEM-SOLVING STRATEGY 15.1 | MECHANICAL WAVES

IDENTIFY *the relevant concepts:* As always, identify the target variables; these may include mathematical *expressions* (for example, the wave function for a given situation). Note that wave problems fall into two categories. *Kinematics* problems, concerned with describing wave motion, involve wave speed v, wavelength λ (or wave number k), frequency f (or angular frequency ω), and amplitude A. They may also involve the position, velocity, and acceleration of individual particles in the medium. *Dynamics* problems also use concepts from Newton's laws. Later in this chapter we'll encounter problems that involve the relationship of wave speed to the mechanical properties of the medium.

SET UP *the problem* using the following steps:
1. List the given quantities. Sketch graphs of y versus x (like Fig. 15.9a) and of y versus t (like Fig. 15.9b), and label them with known values.
2. Identify useful equations. These may include Eq. (15.1) $(v = \lambda f)$, Eq. (15.6) $(\omega = vk)$, and Eqs. (15.3), (15.4), and

(15.7), which express the wave function in various forms. From the wave function, you can find the value of y at any point (value of x) and at any time t.

3. If you need to determine the wave speed v and don't know both λ and f, you may be able to use a relationship between v and the mechanical properties of the system. (In the next section we'll develop this relationship for waves on a string.)

EXECUTE *the solution:* Solve for the unknown quantities using the equations you've identified. To determine the wave function from Eq. (15.3), (15.4), or (15.7), you must know A and any two of v, λ, and f (or v, k, and ω).

EVALUATE *your answer:* Confirm that the values of v, f, and λ (or v, ω, and k) agree with the relationships given in Eq. (15.1) or (15.6). If you've calculated the wave function, check one or more special cases for which you can predict the results.

EXAMPLE 15.2 WAVE ON A CLOTHESLINE

Cousin Throckmorton holds one end of the clothesline taut and wiggles it up and down sinusoidally with frequency 2.00 Hz and amplitude 0.075 m. The wave speed on the clothesline is $v = 12.0$ m/s. At $t = 0$ Throcky's end has maximum positive displacement and is instantaneously at rest. Assume that no wave bounces back from the far end. (a) Find the wave amplitude A, angular frequency ω, period T, wavelength λ, and wave number k. (b) Write a wave function describing the wave. (c) Write equations for the displacement, as a function of time, of Throcky's end of the clothesline and of a point 3.00 m from that end.

SOLUTION

IDENTIFY and SET UP: This is a kinematics problem about the clothesline's wave motion. Throcky produces a sinusoidal wave that propagates along the clothesline, so we can use all of the expressions of this section. In part (a) our target variables are A, ω, T, λ, and k. We use the relationships $\omega = 2\pi f$, $f = 1/T$, $v = \lambda f$, and $k = 2\pi/\lambda$. In parts (b) and (c) our target "variables" are expressions for displacement, which we'll obtain from an appropriate equation for the wave function. We take the positive x-direction to be the direction in which the wave propagates, so either Eq. (15.4) or (15.7) will yield the desired expression. A photograph of the clothesline at time $t = 0$ would look like Fig. 15.9a, with the maximum displacement at $x = 0$ (the end that Throcky holds).

EXECUTE: (a) The wave amplitude and frequency are the same as for the oscillations of Throcky's end of the clothesline, $A = 0.075$ m and $f = 2.00$ Hz. Hence

$$\omega = 2\pi f = \left(2\pi \frac{\text{rad}}{\text{cycle}}\right)\left(2.00\frac{\text{cycles}}{\text{s}}\right)$$

$$= 4.00\pi \text{ rad/s} = 12.6 \text{ rad/s}$$

The period is $T = 1/f = 0.500$ s, and from Eq. (15.1),

$$\lambda = \frac{v}{f} = \frac{12.0 \text{ m/s}}{2.00 \text{ s}^{-1}} = 6.00 \text{ m}$$

We find the wave number from Eq. (15.5) or (15.6):

$$k = \frac{2\pi}{\lambda} = \frac{2\pi \text{ rad}}{6.00 \text{ m}} = 1.05 \text{ rad/m}$$

or

$$k = \frac{\omega}{v} = \frac{4.00\pi \text{ rad/s}}{12.0 \text{ m/s}} = 1.05 \text{ rad/m}$$

(b) We write the wave function using Eq. (15.4) and the values of A, T, and λ from part (a):

$$y(x, t) = A\cos 2\pi\left(\frac{x}{\lambda} - \frac{t}{T}\right)$$

$$= (0.075 \text{ m})\cos 2\pi\left(\frac{x}{6.00 \text{ m}} - \frac{t}{0.500 \text{ s}}\right)$$

$$= (0.075 \text{ m})\cos[(1.05 \text{ rad/m})x - (12.6 \text{ rad/s})t]$$

We can also get this same expression from Eq. (15.7) by using the values of ω and k from part (a).

(c) We can find the displacement as a function of time at $x = 0$ and $x = +3.00$ m by substituting these values into the wave function from part (b):

$$y(x = 0, t) = (0.075 \text{ m})\cos 2\pi\left(\frac{0}{6.00 \text{ m}} - \frac{t}{0.500 \text{ s}}\right)$$

$$= (0.075 \text{ m})\cos(12.6 \text{ rad/s})t$$

$$y(x = +3.00 \text{ m}, t) = (0.075 \text{ m})\cos 2\pi\left(\frac{3.00 \text{ m}}{6.00 \text{ m}} - \frac{t}{0.500 \text{ s}}\right)$$

$$= (0.075 \text{ m})\cos[\pi - (12.6 \text{ rad/s})t]$$

$$= -(0.075 \text{ m})\cos(12.6 \text{ rad/s})t$$

EVALUATE: In part (b), the quantity $(1.05 \text{ rad/m})x - (12.6 \text{ rad/s})t$ is the *phase* of a point x on the string at time t. The two points in part (c) oscillate in SHM with the same frequency and amplitude, but their oscillations differ in phase by $(1.05 \text{ rad/m})(3.00 \text{ m}) = 3.15 \text{ rad} = \pi$ radians—that is, one half-cycle—because the points are separated by one half-wavelength: $\lambda/2 = (6.00 \text{ m})/2 = 3.00$ m. Thus, while a graph of y versus t for the point at $x = 0$ is a cosine curve (like Fig. 15.9b), a graph of y versus t for the point $x = 3.00$ m is a *negative* cosine curve (the same as a cosine curve shifted by one half-cycle).

Using the expression for $y(x = 0, t)$ in part (c), can you show that the end of the string at $x = 0$ is instantaneously at rest at $t = 0$, as stated at the beginning of this example? (*Hint:* Calculate the y-velocity at this point by taking the derivative of y with respect to t.)

Particle Velocity and Acceleration in a Sinusoidal Wave

From the wave function we can get an expression for the transverse velocity of any *particle* in a transverse wave. We call this v_y to distinguish it from the wave propagation speed v. To find the transverse velocity v_y at a particular point x, we take the derivative of the wave function $y(x, t)$ with respect to t, keeping x constant. If the wave function is

$$y(x, t) = A\cos(kx - \omega t)$$

then

$$v_y(x, t) = \frac{\partial y(x, t)}{\partial t} = \omega A \sin(kx - \omega t) \qquad (15.9)$$

The ∂ in this expression is a modified d, used to remind us that $y(x, t)$ is a function of *two* variables and that we are allowing only one (t) to vary. The other (x) is constant because we are looking at a particular point on the string. This derivative is called a *partial derivative*. If you haven't yet encountered partial derivatives in your study of calculus, don't fret; it's a simple idea.

Equation (15.9) shows that the transverse velocity of a particle varies with time, as we expect for simple harmonic motion. The maximum particle speed is ωA; this can be greater than, less than, or equal to the wave speed v, depending on the amplitude and frequency of the wave.

The *acceleration* of any particle is the *second* partial derivative of $y(x, t)$ with respect to t:

$$a_y(x, t) = \frac{\partial^2 y(x, t)}{\partial t^2} = -\omega^2 A \cos(kx - \omega t) \tag{15.10}$$

$$= -\omega^2 y(x, t)$$

The acceleration of a particle equals $-\omega^2$ times its displacement, which is the result we obtained in Section 14.2 for simple harmonic motion.

We can also compute partial derivatives of $y(x, t)$ with respect to x, holding t constant. The first partial derivative $\partial y(x, t)/\partial x$ is the *slope* of the string at point x and at time t. The second partial derivative with respect to x tells us the *curvature* of the string:

$$\frac{\partial^2 y(x, t)}{\partial x^2} = -k^2 A \cos(kx - \omega t) = -k^2 y(x, t) \tag{15.11}$$

From Eqs. (15.10) and (15.11) and the relationship $\omega = vk$ we see that

$$\frac{\partial^2 y(x, t)/\partial t^2}{\partial^2 y(x, t)/\partial x^2} = \frac{\omega^2}{k^2} = v^2 \qquad \text{and}$$

Wave equation involves second partial derivatives of wave function:

$$\underset{\text{Wave speed}}{\overset{\text{Second derivative with respect to } x}{\frac{\partial^2 y(x, t)}{\partial x^2} = \frac{1}{v^2} \frac{\partial^2 y(x, t)}{\partial t^2}}} \overset{\text{Second derivative}}{\text{with respect to } t} \tag{15.12}$$

We've derived Eq. (15.12) for a wave traveling in the positive x-direction. You can show that the wave function for a sinusoidal wave propagating in the *negative* x-direction, $y(x, t) = A \cos(kx + \omega t)$, also satisfies this equation.

Equation (15.12), called the **wave equation,** is one of the most important equations in all of physics. Whenever it occurs, we know that a disturbance can propagate as a wave along the x-axis with wave speed v. The disturbance need not be a sinusoidal wave; we'll see in the next section that *any* wave on a string obeys Eq. (15.12), whether the wave is periodic or not. In Chapter 32 we will find that electric and magnetic fields satisfy the wave equation; the wave speed will turn out to be the speed of light, which will lead us to the conclusion that light is an electromagnetic wave.

Figure 15.10a (next page) shows the transverse velocity v_y and transverse acceleration a_y, given by Eqs. (15.9) and (15.10), for several points on a string as a sinusoidal wave passes along it. At points where the string has an upward curvature ($\partial^2 y/\partial x^2 > 0$), the acceleration is positive ($a_y = \partial^2 y/\partial t^2 > 0$); this follows from the wave equation, Eq. (15.12). For the same reason the acceleration is negative ($a_y = \partial^2 y/\partial t^2 < 0$) at points where the string has a downward curvature ($\partial^2 y/\partial x^2 < 0$), and the acceleration is zero ($a_y = \partial^2 y/\partial t^2 = 0$) at points of inflection where the curvature is zero ($\partial^2 y/\partial x^2 = 0$). Remember that v_y and a_y

15.10 (a) Another view of the wave at $t = 0$ in Fig. 15.9a. The vectors show the transverse velocity v_y and transverse acceleration a_y at several points on the string. (b) From $t = 0$ to $t = 0.05T$, a particle at point 1 is displaced to point 1′, a particle at point 2 is displaced to point 2′, and so on.

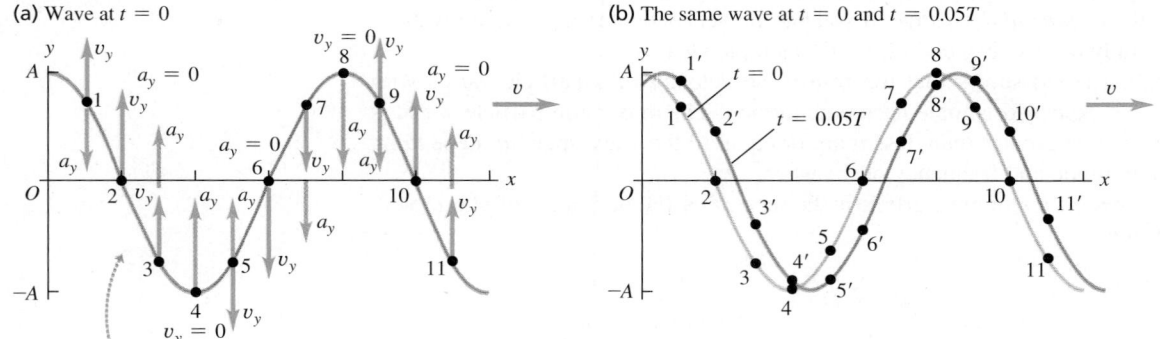

(a) Wave at $t = 0$

(b) The same wave at $t = 0$ and $t = 0.05T$

• Acceleration a_y at each point on the string is proportional to displacement y at that point.
• Acceleration is upward where string curves upward, downward where string curves downward.

are the *transverse* velocity and acceleration of points on the string; these points move along the y-direction, not along the propagation direction of the wave. Figure 15.10b shows these motions for several points on the string.

For *longitudinal* waves, the wave function $y(x, t)$ still measures the displacement of a particle of the medium from its equilibrium position. The difference is that for a longitudinal wave, this displacement is *parallel* to the x-axis instead of perpendicular to it. We'll discuss longitudinal waves in detail in Chapter 16.

TEST YOUR UNDERSTANDING OF SECTION 15.3 Figure 15.8 shows a sinusoidal wave of period T on a string at times $0, \frac{1}{8}T, \frac{2}{8}T, \frac{3}{8}T, \frac{4}{8}T, \frac{5}{8}T, \frac{6}{8}T, \frac{7}{8}T$, and T. (a) At which time is point A on the string moving upward with maximum speed? (b) At which time does point B on the string have the greatest upward acceleration? (c) At which time does point C on the string have a downward acceleration and a downward velocity? ▮

15.4 SPEED OF A TRANSVERSE WAVE

One of the key properties of any wave is the wave *speed*. Light waves in air have a much greater speed of propagation than do sound waves in air (3.00×10^8 m/s versus 344 m/s); that's why you see the flash from a bolt of lightning before you hear the clap of thunder. In this section we'll see what determines the speed of propagation of one particular kind of wave: transverse waves on a string. The speed of these waves is important to understand because it is an essential part of analyzing stringed musical instruments, as we'll discuss later in this chapter. Furthermore, the speeds of many kinds of mechanical waves turn out to have the same basic mathematical expression as does the speed of waves on a string.

What determines the speed of transverse waves on a string are the *tension* in the string and its *mass per unit length* (also called *linear mass density*). Increasing the tension also increases the restoring forces that tend to straighten the string when it is disturbed, thus increasing the wave speed. Increasing the mass per unit length makes the motion more sluggish, and so decreases the wave speed. We'll develop the exact relationship among wave speed, tension, and mass per unit length by two different methods. The first is simple in concept and considers a specific wave shape; the second is more general but also more formal.

Wave Speed on a String: First Method

We consider a perfectly flexible string (**Fig. 15.11**). In the equilibrium position the tension is F and the linear mass density (mass per unit length) is μ. (When portions of the string are displaced from equilibrium, the mass per unit length

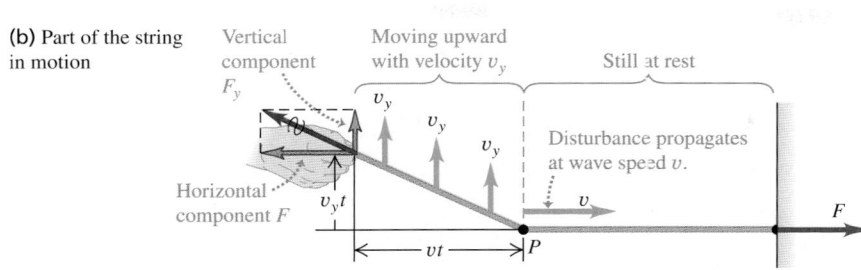

15.11 Propagation of a transverse wave on a string.

decreases a little, and the tension increases a little.) We ignore the weight of the string so that when the string is at rest in the equilibrium position, the string forms a perfectly straight line as in Fig. 15.11a.

Starting at time $t = 0$, we apply a constant upward force F_y at the left end of the string. We might expect that the end would move with constant acceleration; that would happen if the force were applied to a *point* mass. But here the effect of the force F_y is to set successively more and more mass in motion. The wave travels with constant speed v, so the division point P between moving and non-moving portions moves with the same constant speed v (Fig. 15.11b).

Figure 15.11b shows that all particles in the moving portion of the string move upward with constant *velocity* v_y, not constant acceleration. To see why this is so, we note that the *impulse* of the force F_y up to time t is $F_y t$. According to the impulse–momentum theorem (see Section 8.1), the impulse is equal to the change in the total transverse component of momentum of the moving part of the string. Because the system started with *zero* transverse momentum, this is equal to the total transverse momentum mv_y at time t:

$$\text{Transverse impulse} = \text{Transverse momentum}$$
$$F_y t = mv_y \tag{15.13}$$

The total momentum thus must increase proportionately with time. But since the division point P moves with constant speed, the length of string that is in motion and hence the total mass m in motion are also proportional to the time t that the force has been acting. So the *change* of momentum must be associated entirely with the increasing amount of mass in motion, not with an increasing velocity of an individual mass element. That is, mv_y changes because m, not v_y, changes.

At time t, the left end of the string has moved up a distance $v_y t$, and the boundary point P has advanced a distance vt. The total force at the left end of the string has components F and F_y. Why F? There is no motion in the direction along the length of the string, so there is no unbalanced horizontal force. Therefore F, the magnitude of the horizontal component, does not change when the string is displaced. In the displaced position the tension is $(F^2 + F_y^2)^{1/2}$; this is greater than F, so the string stretches somewhat.

To derive an expression for the wave speed v, we note that in Fig. 15.11b the right triangle whose vertex is at P, with sides $v_y t$ and vt, is similar to the right triangle whose vertex is at the position of the hand, with sides F_y and F. Hence

$$\frac{F_y}{F} = \frac{v_y t}{vt} \qquad F_y = F\frac{v_y}{v}$$

and

$$\text{Transverse impulse} = F_y t = F\frac{v_y}{v}t$$

15.12 These transmission cables have a relatively large amount of mass per unit length (μ) and a low tension (F). If the cables are disturbed—say, by a bird landing on them—transverse waves will travel along them at a slow speed $v = \sqrt{F/\mu}$.

The mass m of the moving portion of the string is the product of the mass per unit length μ and the length vt, or μvt. The transverse momentum is the product of this mass and the transverse velocity v_y:

$$\text{Transverse momentum} = mv_y = (\mu vt)v_y$$

Substituting these into Eq. (15.13), we obtain

$$F\frac{v_y}{v}t = \mu vtv_y$$

We solve this for the wave speed v:

$$\begin{array}{ll}\text{Speed of a} \\ \text{transverse wave} \cdots\!\!\rightarrow v = \sqrt{\dfrac{F}{\mu}} & \begin{array}{l}\leftarrow\cdots\text{Tension in string} \\ \leftarrow\cdots\text{Mass per unit length}\end{array}\end{array} \qquad (15.14)$$

Equation (15.14) confirms that the wave speed v increases when the tension F increases but decreases when the mass per unit length μ increases (**Fig. 15.12**).

Note that v_y does not appear in Eq. (15.14); thus the wave speed doesn't depend on v_y. Our calculation considered only a very special kind of pulse, but we can consider *any* shape of wave disturbance as a series of pulses with different values of v_y. So even though we derived Eq. (15.14) for a special case, it is valid for *any* transverse wave motion on a string, including the sinusoidal and other periodic waves we discussed in Section 15.3. Note also that the wave speed doesn't depend on the amplitude or frequency of the wave, in accordance with our assumptions in Section 15.3.

Wave Speed on a String: Second Method

Here is an alternative derivation of Eq. (15.14). If you aren't comfortable with partial derivatives, it can be omitted. We apply Newton's second law, $\Sigma \vec{F} = m\vec{a}$, to a small segment of string whose length in the equilibrium position is Δx (**Fig. 15.13**). The mass of the segment is $m = \mu \, \Delta x$. The x-components of the forces have equal magnitude F and add to zero because the motion is transverse and there is no component of acceleration in the x-direction. To obtain F_{1y} and F_{2y}, we note that the ratio F_{1y}/F is equal in magnitude to the *slope* of the string at point x and that F_{2y}/F is equal to the slope at point $x + \Delta x$. Taking proper account of signs, we find

$$\frac{F_{1y}}{F} = -\left(\frac{\partial y}{\partial x}\right)_x \qquad \frac{F_{2y}}{F} = \left(\frac{\partial y}{\partial x}\right)_{x+\Delta x} \qquad (15.15)$$

The notation reminds us that the derivatives are evaluated at points x and $x + \Delta x$, respectively. From Eq. (15.15) we find that the net y-component of force is

$$F_y = F_{1y} + F_{2y} = F\left[\left(\frac{\partial y}{\partial x}\right)_{x+\Delta x} - \left(\frac{\partial y}{\partial x}\right)_x\right] \qquad (15.16)$$

We now equate F_y from Eq. (15.16) to the mass $\mu \, \Delta x$ times the y-component of acceleration $\partial^2 y/\partial t^2$:

$$F\left[\left(\frac{\partial y}{\partial x}\right)_{x+\Delta x} - \left(\frac{\partial y}{\partial x}\right)_x\right] = \mu \, \Delta x \frac{\partial^2 y}{\partial t^2} \qquad (15.17)$$

or, dividing Eq. (15.17) by $F\Delta x$,

$$\frac{\left(\dfrac{\partial y}{\partial x}\right)_{x+\Delta x} - \left(\dfrac{\partial y}{\partial x}\right)_x}{\Delta x} = \frac{\mu}{F}\frac{\partial^2 y}{\partial t^2} \qquad (15.18)$$

We now take the limit as $\Delta x \to 0$. In this limit, the left side of Eq. (15.18) becomes the derivative of $\partial y/\partial x$ with respect to x (at constant t)—that is, the

15.13 Free-body diagram for a segment of string. The force at each end of the string is tangent to the string at the point of application.

The string to the right of the segment (not shown) exerts a force $\vec{F}_2$ on the segment.

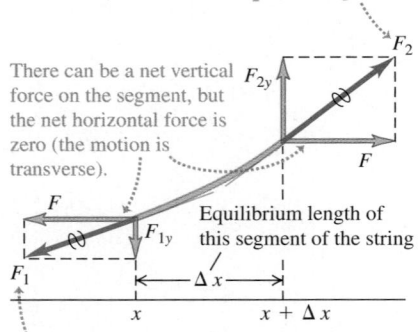

There can be a net vertical force on the segment, but the net horizontal force is zero (the motion is transverse).

Equilibrium length of this segment of the string

The string to the left of the segment (not shown) exerts a force $\vec{F}_1$ on the segment.

second (partial) derivative of y with respect to x:

$$\frac{\partial^2 y}{\partial x^2} = \frac{\mu}{F}\frac{\partial^2 y}{\partial t^2} \tag{15.19}$$

Now, Eq. (15.19) has exactly the same form as the *wave equation,* Eq. (15.12), that we derived at the end of Section 15.3. That equation and Eq. (15.19) describe the very same wave motion, so they must be identical. Comparing the two equations, we see that for this to be so, we must have

$$v = \sqrt{\frac{F}{\mu}}$$

which is the same expression as Eq. (15.14).

In going through this derivation, we didn't make any special assumptions about the shape of the wave. Since our derivation led us to rediscover Eq. (15.12), the wave equation, we conclude that the wave equation is valid for waves on a string that have *any* shape.

The Speed of Mechanical Waves

Equation (15.14) gives the wave speed for only the special case of mechanical waves on a stretched string or rope. Remarkably, it turns out that for many types of mechanical waves, including waves on a string, the expression for wave speed has the same general form:

$$v = \sqrt{\frac{\text{Restoring force returning the system to equilibrium}}{\text{Inertia resisting the return to equilibrium}}}$$

To interpret this expression, let's look at the now-familiar case of waves on a string. The tension F in the string plays the role of the restoring force; it tends to bring the string back to its undisturbed, equilibrium configuration. The mass of the string—or, more properly, the linear mass density μ—provides the inertia that prevents the string from returning instantaneously to equilibrium. Hence we have $v = \sqrt{F/\mu}$ for the speed of waves on a string.

In Chapter 16 we'll see a similar expression for the speed of sound waves in a gas. Roughly speaking, the gas pressure provides the force that tends to return the gas to its undisturbed state when a sound wave passes through. The inertia is provided by the density, or mass per unit volume, of the gas.

BIO Application Eating and Transverse Waves Swallowing food causes peristalsis, in which a transverse wave propagates down your esophagus. The wave is a radial contraction of the esophagus that pushes the bolus (the mass of swallowed food) toward the stomach. Unlike the speed of waves on a uniform string, the speed of this peristaltic wave is not constant: It averages about 3 cm/s in the upper esophagus, about 5 cm/s in the mid-esophagus, and about 2.5 cm/s in the lower esophagus.

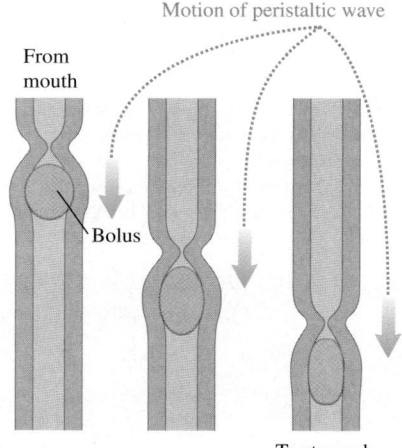

EXAMPLE 15.3 **CALCULATING WAVE SPEED**

One end of a 2.00-kg rope is tied to a support at the top of a mine shaft 80.0 m deep (**Fig. 15.14**). The rope is stretched taut by a 20.0-kg box of rocks attached at the bottom. (a) A geologist at the bottom of the shaft signals to a colleague at the top by jerking the rope sideways. What is the speed of a transverse wave on the rope? (b) If a point on the rope is in transverse SHM with $f = 2.00$ Hz, how many cycles of the wave are there in the rope's length?

15.14 Using transverse waves to send signals along a vertical rope.

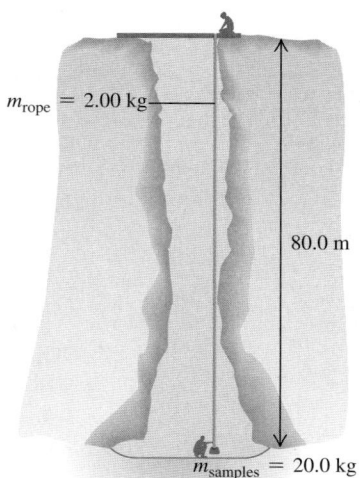

SOLUTION

IDENTIFY and SET UP: In part (a) we can find the wave speed (our target variable) by using the *dynamic* relationship $v = \sqrt{F/\mu}$ [Eq. (15.14)]. In part (b) we find the wavelength from the *kinematic* relationship $v = f\lambda$; from that we can find the target variable, the number of wavelengths that fit into the rope's 80.0-m length. We'll assume that the rope is massless (even though its weight is 10% that of the box), so that the box alone provides the tension in the rope.

Continued

EXECUTE: (a) The tension in the rope due to the box is

$$F = m_{box}g = (20.0 \text{ kg})(9.80 \text{ m/s}^2) = 196 \text{ N}$$

and the rope's linear mass density is

$$\mu = \frac{m_{rope}}{L} = \frac{2.00 \text{ kg}}{80.0 \text{ m}} = 0.0250 \text{ kg/m}$$

Hence, from Eq. (15.14), the wave speed is

$$v = \sqrt{\frac{F}{\mu}} = \sqrt{\frac{196 \text{ N}}{0.0250 \text{ kg/m}}} = 88.5 \text{ m/s}$$

(b) From Eq. (15.1), the wavelength is

$$\lambda = \frac{v}{f} = \frac{88.5 \text{ m/s}}{2.00 \text{ s}^{-1}} = 44.3 \text{ m}$$

There are $(80.0 \text{ m})/(44.3 \text{ m}) = 1.81$ wavelengths (that is, cycles of the wave) in the rope.

EVALUATE: Because of the rope's weight, its tension is greater at the top than at the bottom. Hence both the wave speed and the wavelength increase as a wave travels up the rope. If you take account of this, can you verify that the wave speed at the top of the rope is 92.9 m/s?

TEST YOUR UNDERSTANDING OF SECTION 15.4   The six strings of a guitar are the same length and under nearly the same tension, but they have different thicknesses. On which string do waves travel the fastest? (i) The thickest string; (ii) the thinnest string; (iii) the wave speed is the same on all strings. ∎

BIO Application Surface Waves and the Swimming Speed of Ducks
When a duck swims, it necessarily produces waves on the surface of the water. The faster the duck swims, the larger the wave amplitude and the more power the duck must supply to produce these waves. The maximum power available from their leg muscles limits the maximum swimming speed of ducks to only about 0.7 m/s (2.5 km/h = 1.6 mi/h).

15.5 ENERGY IN WAVE MOTION

Every wave motion has *energy* associated with it. The energy we receive from sunlight and the destructive effects of ocean surf and earthquakes bear this out. To produce any of the wave motions we have discussed in this chapter, we have to apply a force to a portion of the wave medium; the point where the force is applied moves, so we do *work* on the system. As the wave propagates, each portion of the medium exerts a force and does work on the adjoining portion. In this way a wave can transport energy from one region of space to another.

As an example, let's look again at transverse waves on a string. How is energy transferred from one portion of the string to another? Picture a wave traveling from left to right (the positive *x*-direction) past a point *a* on the string (**Fig. 15.15a**). The string to the left of point *a* exerts a force on the string to the right of it, and vice versa. In Fig. 15.15b we show the components F_x and F_y of the force that the string to the left of *a* exerts on the string to the right of *a*. As in Figs. 15.11 and 15.13, the magnitude of the horizontal component F_x equals the tension F in the undisturbed string. Note that F_y/F is equal to the negative of the *slope* of the string at *a*, and this slope is also given by $\partial y/\partial x$. Putting these together, we have

$$F_y(x, t) = -F\frac{\partial y(x, t)}{\partial x} \tag{15.20}$$

We need the negative sign because F_y is negative when the slope is positive (as in Fig. 15.15b). We write the vertical force as $F_y(x, t)$ as a reminder that its value may be different at different points along the string and at different times.

When point *a* moves in the *y*-direction, the force F_y does *work* on this point and therefore transfers energy into the part of the string to the right of *a*. The corresponding power P (rate of doing work) at the point *a* is the transverse force $F_y(x, t)$ at *a* times the transverse velocity $v_y(x, t) = \partial y(x, t)/\partial t$ of that point:

$$P(x, t) = F_y(x, t)v_y(x, t) = -F\frac{\partial y(x, t)}{\partial x}\frac{\partial y(x, t)}{\partial t} \tag{15.21}$$

This power is the *instantaneous* rate at which energy is transferred along the string at position *x* and time *t*. Note that energy is transferred only at points where the string has a nonzero slope ($\partial y/\partial x$ is nonzero), so that the tension force has a transverse component, and where the string has a nonzero transverse velocity ($\partial y/\partial t$ is nonzero) so that the transverse force can do work.

15.15 (a) Point *a* on a string carrying a wave from left to right. (b) The components of the force exerted on the part of the string to the right of point *a* by the part of the string to the left of point *a*.

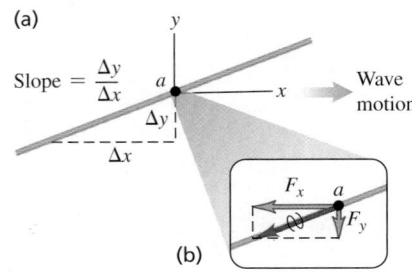

Equation (15.21) is valid for *any* wave on a string, sinusoidal or not. For a sinusoidal wave with wave function given by Eq. (15.7), we have

$$y(x, t) = A \cos(kx - \omega t)$$

$$\frac{\partial y(x, t)}{\partial x} = -kA \sin(kx - \omega t)$$

$$\frac{\partial y(x, t)}{\partial t} = \omega A \sin(kx - \omega t)$$

$$P(x, t) = Fk\omega A^2 \sin^2(kx - \omega t) \qquad (15.22)$$

By using the relationships $\omega = vk$ and $v^2 = F/\mu$, we can also express Eq. (15.22) in the alternative form

$$P(x, t) = \sqrt{\mu F}\, \omega^2 A^2 \sin^2(kx - \omega t) \qquad (15.23)$$

The $\sin^2$ function is never negative, so the instantaneous power in a sinusoidal wave is either positive (so that energy flows in the positive x-direction) or zero (at points where there is no energy transfer). Energy is never transferred in the direction opposite to the direction of wave propagation (**Fig. 15.16**).

The maximum value of the instantaneous power $P(x, t)$ occurs when the $\sin^2$ function has the value unity:

$$P_{max} = \sqrt{\mu F}\, \omega^2 A^2 \qquad (15.24)$$

The *average* value of the $\sin^2$ function, averaged over any whole number of cycles, is $\frac{1}{2}$. Hence we see from Eq. (15.23) that the *average* power P_{av} is just one-half the maximum instantaneous power P_{max} (Fig. 15.16):

Average power, sinusoidal wave on a string $\cdots\cdots$ Wave angular frequency $\cdots$

$$P_{av} = \tfrac{1}{2}\sqrt{\mu F}\, \omega^2 A^2 \qquad (15.25)$$

Mass per unit length $\cdots$ Tension in string $\cdots$ Wave amplitude

The average rate of energy transfer is proportional to the square of the amplitude and to the square of the frequency. This proportionality is a general result for mechanical waves of all types, including seismic waves (see the photo that opens this chapter). For a mechanical wave, the rate of energy transfer quadruples if the frequency is doubled (for the same amplitude) or if the amplitude is doubled (for the same frequency).

Electromagnetic waves turn out to be a bit different. While the average rate of energy transfer in an electromagnetic wave is proportional to the square of the amplitude, just as for mechanical waves, it is independent of the value of ω.

15.16 The instantaneous power $P(x, t)$ in a sinusoidal wave as given by Eq. (15.23), shown as a function of time at coordinate $x = 0$. The power is never negative, which means that energy never flows opposite to the direction of wave propagation.

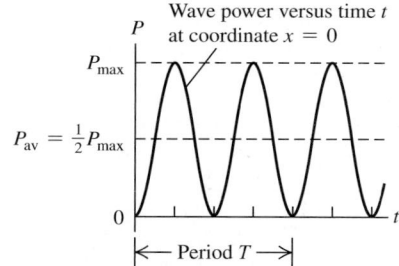

Wave power versus time t at coordinate $x = 0$

EXAMPLE 15.4   POWER IN A WAVE

(a) In Example 15.2 (Section 15.3), at what maximum rate does Throcky put energy into the clothesline? That is, what is his maximum instantaneous power? The linear mass density of the clothesline is $\mu = 0.250$ kg/m, and Throcky applies tension $F = 36.0$ N. (b) What is his average power? (c) As Throcky tires, the amplitude decreases. What is the average power when the amplitude is 7.50 mm?

SOLUTION

IDENTIFY and SET UP: In part (a) our target variable is the *maximum instantaneous* power P_{max}, while in parts (b) and (c) it

is the *average* power. For part (a) we'll use Eq. (15.24), and for parts (b) and (c) we'll use Eq. (15.25); Example 15.2 gives us all the needed quantities.

EXECUTE: (a) From Eq. (15.24),

$$P_{max} = \sqrt{\mu F}\, \omega^2 A^2$$

$$= \sqrt{(0.250 \text{ kg/m})(36.0 \text{ N})}\,(4.00\pi \text{ rad/s})^2 (0.075 \text{ m})^2$$

$$= 2.66 \text{ W}$$

Continued

SOLUTION

(b) From Eqs. (15.24) and (15.25), the average power is one-half of the maximum instantaneous power, so

$$P_{av} = \tfrac{1}{2}P_{max} = \tfrac{1}{2}(2.66 \text{ W}) = 1.33 \text{ W}$$

(c) The new amplitude is $\tfrac{1}{10}$ of the value we used in parts (a) and (b). From Eq. (15.25), the average power is proportional to A^2, so the new average power is

$$P_{av} = \left(\tfrac{1}{10}\right)^2(1.33 \text{ W}) = 0.0133 \text{ W} = 13.3 \text{ mW}$$

EVALUATE: Equation (15.23) shows that P_{max} occurs when $\sin^2(kx - \omega t) = 1$. At any given position x, this happens twice per period of the wave—once when the sine function is equal to $+1$, and once when it's equal to -1. The *minimum* instantaneous power is zero; this occurs when $\sin^2(kx - \omega t) = 0$, which also happens twice per period.

Can you confirm that the given values of μ and F give the wave speed mentioned in Example 15.2?

Wave Intensity

Waves on a string carry energy in one dimension (along the direction of the string). But other types of waves, including sound waves in air and seismic waves within the earth, carry energy across all three dimensions of space. For waves of this kind, we define the **intensity** (denoted by I) to be *the time average rate at which energy is transported by the wave, per unit area,* across a surface perpendicular to the direction of propagation. Intensity I is average power per unit area and is usually measured in watts per square meter (W/m^2).

If waves spread out equally in all directions from a source, the intensity at a distance r from the source is inversely proportional to r^2 (**Fig. 15.17**). This result, called the *inverse-square law for intensity,* follows directly from energy conservation. If the power output of the source is P, then the average intensity I_1 through a sphere with radius r_1 and surface area $4\pi r_1^2$ is

$$I_1 = \frac{P}{4\pi r_1^2}$$

A similar expression gives the average intensity I_2 through a sphere with a different radius r_2. If no energy is absorbed between the two spheres, the power P must be the same for both, and

$$4\pi r_1^2 I_1 = 4\pi r_2^2 I_2$$

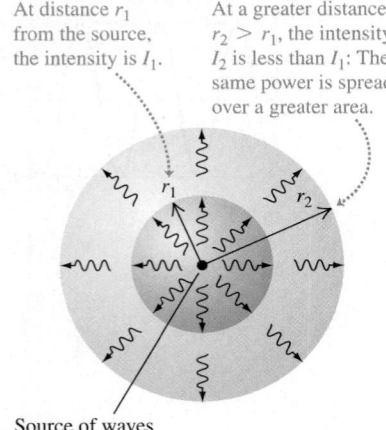

15.17 The greater the distance from a wave source, the greater the area over which the wave power is distributed and the smaller the wave intensity.

At distance r_1 from the source, the intensity is I_1.

At a greater distance $r_2 > r_1$, the intensity I_2 is less than I_1: The same power is spread over a greater area.

Source of waves

> **Inverse-square law for intensity:**
> Intensity is inversely proportional to the square of the distance from source.
>
> Intensity at point 1 ......→ $\dfrac{I_1}{I_2} = \dfrac{r_2^2}{r_1^2}$ ......Distance from source to point 2
> Intensity at point 2 ......→ ......Distance from source to point 1 (15.26)

EXAMPLE 15.5   THE INVERSE-SQUARE LAW

A siren on a tall pole radiates sound waves uniformly in all directions. At a distance of 15.0 m from the siren, the sound intensity is 0.250 W/m^2. At what distance is the intensity 0.010 W/m^2?

SOLUTION

IDENTIFY and SET UP: Because sound is radiated uniformly in all directions, we can use the inverse-square law, Eq. (15.26). At $r_1 = 15.0$ m the intensity is $I_1 = 0.250 \text{ W/m}^2$, and the target variable is the distance r_2 at which the intensity is $I_2 = 0.010 \text{ W/m}^2$.

EXECUTE: We solve Eq. (15.26) for r_2:

$$r_2 = r_1\sqrt{\frac{I_1}{I_2}} = (15.0 \text{ m})\sqrt{\frac{0.250 \text{ W/m}^2}{0.010 \text{ W/m}^2}} = 75.0 \text{ m}$$

EVALUATE: As a check on our answer, note that r_2 is five times greater than r_1. By the inverse-square law, the intensity I_2 should be $1/5^2 = 1/25$ as great as I_1, and indeed it is.

By using the inverse-square law, we've assumed that the sound waves travel in straight lines away from the siren. A more realistic solution, which is beyond our scope, would account for the reflection of sound waves from the ground.

15.6 WAVE INTERFERENCE, BOUNDARY CONDITIONS, AND SUPERPOSITION

Up to this point we've been discussing waves that propagate continuously in the same direction. But when a wave strikes the boundaries of its medium, all or part of the wave is *reflected*. When you yell at a building wall or a cliff face some distance away, the sound wave is reflected from the rigid surface and you hear an echo. When you flip the end of a rope whose far end is tied to a rigid support, a pulse travels the length of the rope and is reflected back to you. In both cases, the initial and reflected waves overlap in the same region of the medium. We use the term **interference** to refer to what happens when two or more waves pass through the same region at the same time.

As a simple example of wave reflections and the role of the boundary of a wave medium, let's look again at transverse waves on a stretched string. What happens when a wave pulse or a sinusoidal wave arrives at the *end* of the string?

If the end is fastened to a rigid support as in **Fig. 15.18**, it is a *fixed* end that cannot move. The arriving wave exerts a force on the support (drawing 4 in Fig. 15.18); the reaction to this force, exerted by the support on the string, "kicks back" on the string and sets up a reflected pulse or wave traveling in the reverse direction (drawing 7). The reflected pulse moves in the opposite direction from the initial, or *incident,* pulse, and its displacement is also opposite.

The opposite situation from an end that is held stationary is a *free* end, one that is perfectly free to move in the direction perpendicular to the length of the string. For example, the string might be tied to a light ring that slides on a frictionless rod perpendicular to the string, as in **Fig. 15.19**. The ring and rod maintain the tension but exert no transverse force. When a wave arrives at this free end, the ring slides along the rod. The ring reaches a maximum displacement, and both it and the string come momentarily to rest, as in drawing 4 in Fig. 15.19. But the string is now stretched, giving increased tension, so the free end of the string is pulled back down, and again a reflected pulse is produced (drawing 7). As for a fixed end, the reflected pulse moves in the opposite direction from the initial pulse, but now the direction of the displacement is the same as for the initial pulse. The conditions at the end of the string, such as a rigid support or the complete absence of transverse force, are called **boundary conditions.**

The formation of the reflected pulse is similar to the overlap of two pulses traveling in opposite directions. **Figure 15.20** (next page) shows two pulses with the same shape, one inverted with respect to the other, traveling in opposite directions. As the pulses overlap and pass each other, the total displacement of the string is the *algebraic sum* of the displacements at that point in the individual pulses. Because these two pulses have the same shape, the total displacement at point *O* in the middle of the figure is zero at all times. Thus the motion of the left half

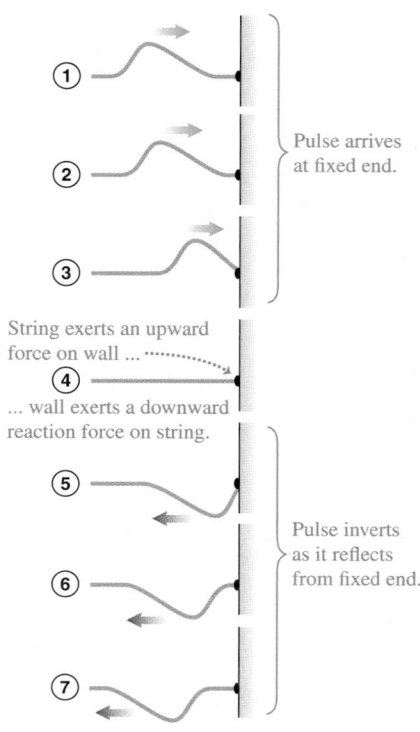

15.18 Reflection of a wave pulse at a fixed end of a string. Time increases from top to bottom.

String exerts an upward force on wall ...

... wall exerts a downward reaction force on string.

Pulse arrives at fixed end.

Pulse inverts as it reflects from fixed end.

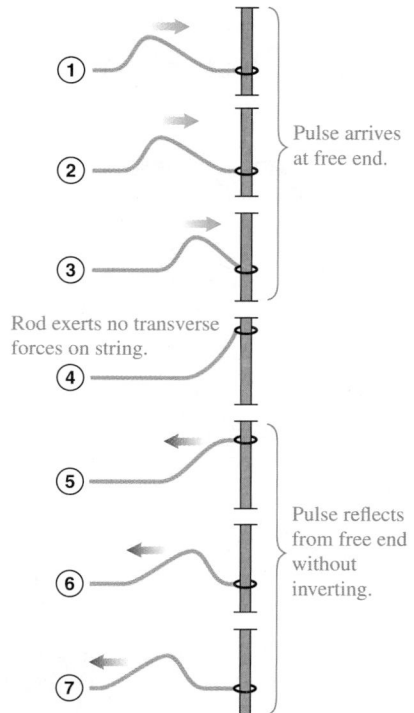

15.19 Reflection of a wave pulse at a free end of a string. Time increases from top to bottom. (Compare to Fig. 15.18.)

Pulse arrives at free end.

Rod exerts no transverse forces on string.

Pulse reflects from free end without inverting.

15.20 Overlap of two wave pulses—one right side up, one inverted—traveling in opposite directions. Time increases from top to bottom.

As the pulses overlap, the displacement of the string at any point is the algebraic sum of the displacements due to the individual pulses.

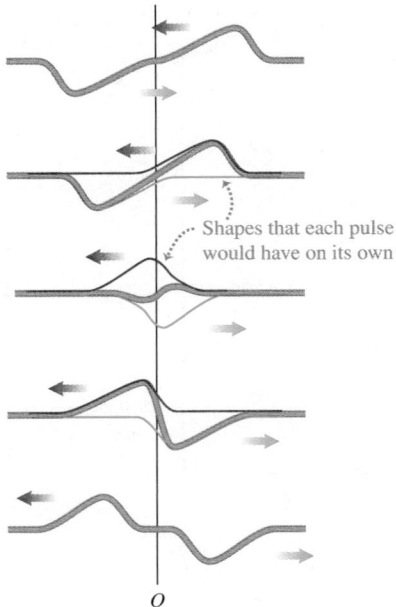

··· Shapes that each pulse would have on its own

O

15.21 Overlap of two wave pulses—both right side up—traveling in opposite directions. Time increases from top to bottom. Compare to Fig. 15.20.

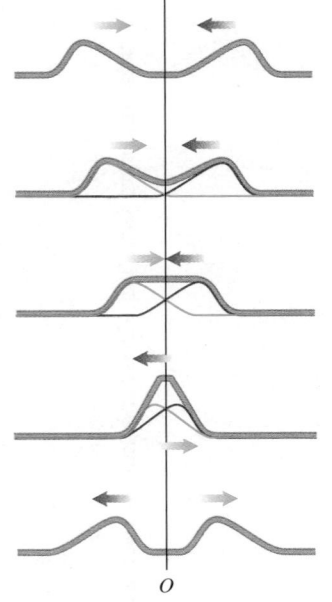

O

DEMO

of the string would be the same if we cut the string at point *O*, threw away the right side, and held the end at *O* fixed. The two pulses on the left side then correspond to the incident and reflected pulses, combining so that the total displacement at *O* is *always* zero. For this to occur, the reflected pulse must be inverted relative to the incident pulse, just as for reflection from the fixed end in Fig. 15.18.

Figure 15.21 shows two pulses with the same shape, traveling in opposite directions but *not* inverted relative to each other. The displacement at point *O* in the middle of the figure is not zero, but the slope of the string at this point is always zero. According to Eq. (15.20), this corresponds to the absence of any transverse force at this point. In this case the motion of the left half of the string would be the same as if we cut the string at point *O* and attached the end to a frictionless sliding ring (Fig. 15.19) that maintains tension without exerting any transverse force. In other words, this situation corresponds to reflection of a pulse at a free end of a string at point *O*. In this case the reflected pulse is *not* inverted.

The Principle of Superposition

Combining the displacements of the separate pulses at each point to obtain the actual displacement is an example of the **principle of superposition:** When two waves overlap, the actual displacement of any point on the string at any time is obtained by adding the displacement the point would have if only the first wave were present and the displacement it would have if only the second wave were present. In other words, the wave function $y(x, t)$ for the resulting motion is obtained by *adding* the two wave functions for the two separate waves:

Wave functions of two overlapping waves

Principle of superposition:
$$y(x, t) = y_1(x, t) + y_2(x, t) \tag{15.27}$$

Wave function of combined wave = sum of individual wave functions

Mathematically, this additive property of wave functions follows from the form of the wave equation, Eq. (15.12) or (15.19), which every physically possible wave function must satisfy. Specifically, the wave equation is *linear;* that is, it contains the function $y(x, t)$ only to the first power (there are no terms involving $y(x, t)^2$, $y(x, t)^{1/2}$, etc.). As a result, if any two functions $y_1(x, t)$ and $y_2(x, t)$ satisfy the wave equation separately, their sum $y_1(x, t) + y_2(x, t)$ also satisfies it and is therefore a physically possible motion. Because this principle depends on the linearity of the wave equation and the corresponding linear-combination property of its solutions, it is also called the *principle of linear superposition.* For some physical systems, such as a medium that does not obey Hooke's law, the wave equation is *not* linear; this principle does not hold for such systems.

The principle of superposition is of central importance in all types of waves. When a friend talks to you while you are listening to music, you can distinguish the speech and the music from each other. This is precisely because the total sound wave reaching your ears is the algebraic sum of the wave produced by your friend's voice and the wave produced by the speakers of your stereo. If two sound waves did *not* combine in this simple linear way, the sound you would hear in this situation would be a hopeless jumble. Superposition also applies to electromagnetic waves (such as light).

TEST YOUR UNDERSTANDING OF SECTION 15.6 **Figure 15.22** shows two wave pulses with different shapes traveling in different directions along a string. Make a series of sketches like Fig. 15.21 showing the shape of the string as the two pulses approach, overlap, and then pass each other. ❚

15.22 Two wave pulses with different shapes.

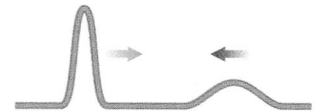

15.7 STANDING WAVES ON A STRING

We've looked at the reflection of a wave *pulse* on a string when it arrives at a boundary point (either a fixed end or a free end). Now let's consider what happens when a *sinusoidal* wave on a string is reflected by a fixed end. We'll again approach the problem by considering the superposition of two waves propagating through the string, one representing the incident wave and the other representing the wave reflected at the fixed end.

Figure 15.23 shows a string that is fixed at its left end. Its right end is moved up and down in simple harmonic motion to produce a wave that travels to the left; the wave reflected from the fixed end travels to the right. The resulting motion when the two waves combine no longer looks like two waves traveling in opposite directions. The string appears to be subdivided into segments, as in the time-exposure photographs of Figs. 15.23a, 15.23b, 15.23c, and 15.23d. Figure 15.23e shows two instantaneous shapes of the string in Fig. 15.23b. Let's compare this behavior with the waves we studied in Sections 15.1 through 15.5. In a wave that travels along the string, the amplitude is constant and the wave pattern moves with a speed equal to the wave speed. Here, instead, the wave pattern remains in the same position along the string and its amplitude fluctuates. There are particular points called **nodes** (labeled *N* in Fig. 15.23e) that never move at all. Midway between the nodes are points called **antinodes** (labeled *A* in Fig. 15.23e) where the amplitude of motion is greatest. Because the wave pattern doesn't appear to be moving in either direction along the string, it is called a **standing wave.** (To emphasize the difference, a wave that *does* move along the string is called a **traveling wave.**)

15.23 (a)–(d) Time exposures of standing waves in a stretched string. From (a) to (d), the frequency of oscillation of the right-hand end increases and the wavelength of the standing wave decreases. (e) The extremes of the motion of the standing wave in part (b), with nodes at the center and at the ends. The right-hand end of the string moves very little compared to the antinodes and so is essentially a node.

(a) String is one-half wavelength long.

(b) String is one wavelength long.

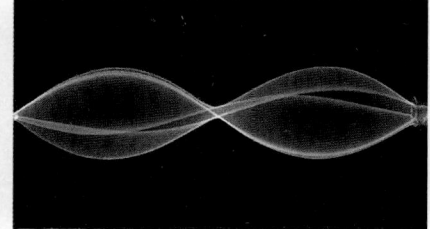

(c) String is one and a half wavelengths long.

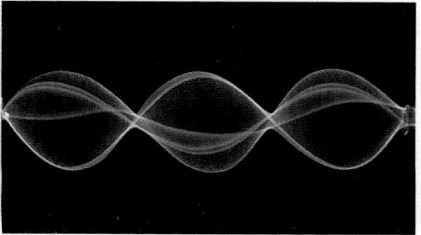

(d) String is two wavelengths long.

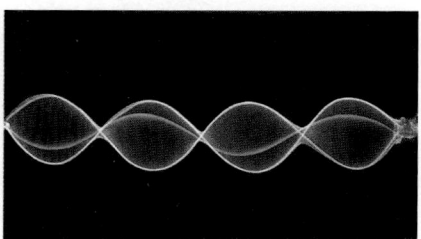

(e) The shape of the string in **(b)** at two different instants

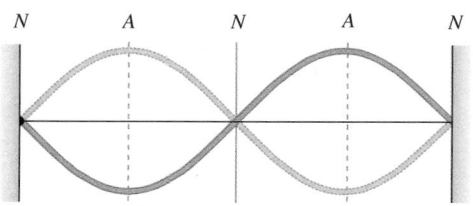

N = **nodes:** points at which the string never moves

A = **antinodes:** points at which the amplitude of string motion is greatest

The principle of superposition explains how the incident and reflected waves combine to form a standing wave. In **Fig. 15.24** the red curves show a wave traveling to the left. The blue curves show a wave traveling to the right with the same propagation speed, wavelength, and amplitude. The waves are shown at nine instants, $\frac{1}{16}$ of a period apart. At each point along the string, we add the displacements (the values of y) for the two separate waves; the result is the total wave on the string, shown in gold.

At certain instants, such as $t = \frac{1}{4}T$, the two wave patterns are exactly in phase with each other, and the shape of the string is a sine curve with twice the amplitude of either individual wave. At other instants, such as $t = \frac{1}{2}T$, the two waves are exactly out of phase with each other, and the total wave at that instant is zero. The resultant displacement is *always* zero at those places marked N at the bottom of Fig. 15.24. These are the *nodes*. At a node the displacements of the two waves in red and blue are always equal and opposite and cancel each other out. This cancellation is called **destructive interference.** Midway between the nodes are the points of *greatest* amplitude, or the *antinodes,* marked A. At the antinodes the displacements of the two waves in red and blue are always identical, giving a large resultant displacement; this phenomenon is called **constructive interference.** We can see from the figure that the distance between successive nodes or between successive antinodes is one half-wavelength, or $\lambda/2$.

15.24 Formation of a standing wave. A wave traveling to the left (red curves) combines with a wave traveling to the right (blue curves) to form a standing wave (gold curves).

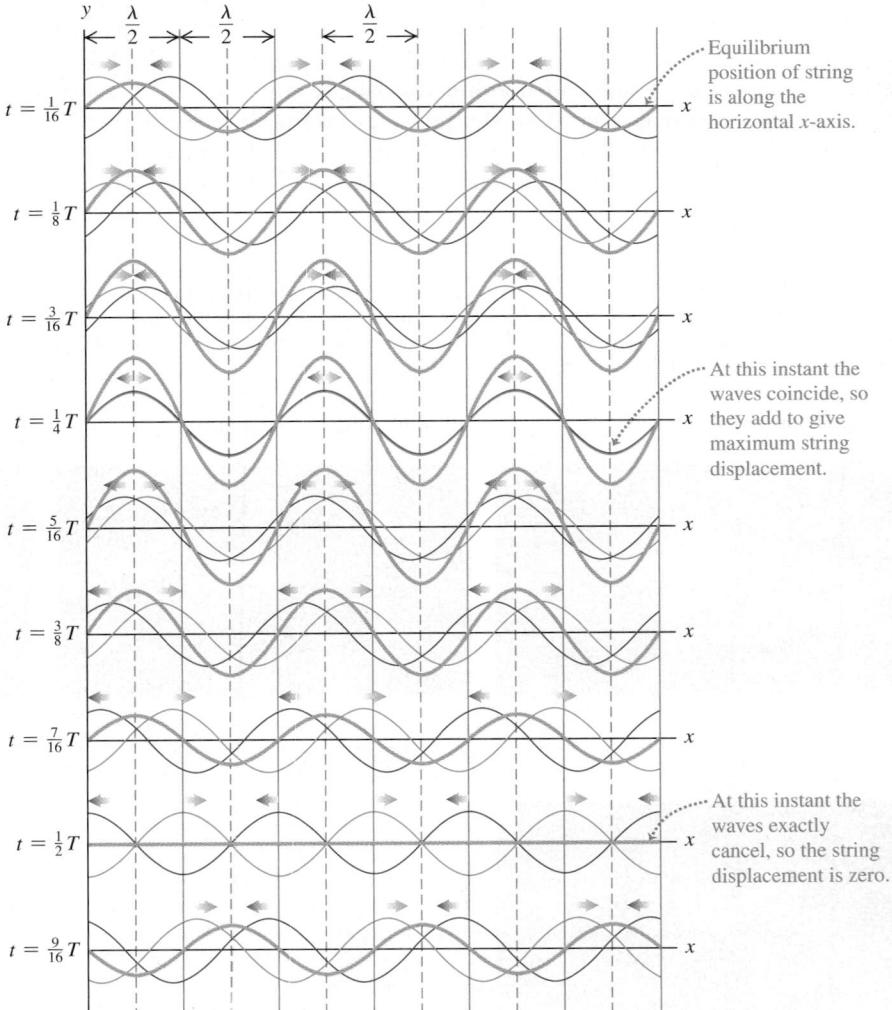

We can derive a wave function for the standing wave of Fig. 15.24 by adding the wave functions $y_1(x, t)$ and $y_2(x, t)$ for two waves with equal amplitude, period, and wavelength traveling in opposite directions. Here $y_1(x, t)$ (the red curves in Fig. 15.24) represents an incoming, or *incident,* wave traveling to the left along the $+x$-axis, arriving at the point $x = 0$ and being reflected; $y_2(x, t)$ (the blue curves in Fig. 15.24) represents the *reflected* wave traveling to the right from $x = 0$. We noted in Section 15.6 that the wave reflected from a fixed end of a string is inverted, so we give a negative sign to one of the waves:

$$y_1(x, t) = -A\cos(kx + \omega t) \qquad \text{(incident wave traveling to the left)}$$

$$y_2(x, t) = A\cos(kx - \omega t) \qquad \text{(reflected wave traveling to the right)}$$

The change in sign corresponds to a shift in *phase* of $180°$ or π radians. At $x = 0$ the motion from the reflected wave is $A\cos\omega t$ and the motion from the incident wave is $-A\cos\omega t$, which we can also write as $A\cos(\omega t + \pi)$. From Eq. (15.27), the wave function for the standing wave is the sum of the individual wave functions:

$$y(x, t) = y_1(x, t) + y_2(x, t) = A\left[-\cos(kx + \omega t) + \cos(kx - \omega t)\right]$$

We can rewrite each of the cosine terms by using the identities for the cosine of the sum and difference of two angles: $\cos(a \pm b) = \cos a\cos b \mp \sin a\sin b$. Applying these and combining terms, we obtain the wave function for the standing wave:

$$y(x, t) = y_1(x, t) + y_2(x, t) = (2A\sin kx)\sin\omega t \qquad \text{or}$$

Standing wave on a string, fixed end at $x = 0$:	Wave function $\qquad$ Standing-wave amplitude $$y(x, t) = (A_{SW}\sin kx)\sin\omega t \cdots\text{Time}$$ Wave number $\cdots$ Position $\quad$ Angular frequency	(15.28)

The standing-wave amplitude A_{SW} is twice the amplitude A of either of the original traveling waves: $A_{SW} = 2A$.

Equation (15.28) has two factors: a function of x and a function of t. The factor $A_{SW}\sin kx$ shows that at each instant the shape of the string is a sine curve. But unlike a wave traveling along a string, the wave shape stays in the same position, oscillating up and down as described by the $\sin\omega t$ factor. This behavior is shown by the gold curves in Fig. 15.24. Each point in the string still undergoes simple harmonic motion, but all the points between any successive pair of nodes oscillate *in phase.* This is in contrast to the phase differences between oscillations of adjacent points that we see with a traveling wave.

We can use Eq. (15.28) to find the positions of the nodes; these are the points for which $\sin kx = 0$, so the displacement is *always* zero. This occurs when $kx = 0, \pi, 2\pi, 3\pi, \ldots$, or, using $k = 2\pi/\lambda$,

$$x = 0, \frac{\pi}{k}, \frac{2\pi}{k}, \frac{3\pi}{k}, \ldots$$

$$= 0, \frac{\lambda}{2}, \frac{2\lambda}{2}, \frac{3\lambda}{2}, \ldots$$

(nodes of a standing wave on a string, fixed end at $x = 0$) $\qquad$ (15.29)

In particular, there is a node at $x = 0$, as there should be, since this point is a fixed end of the string.

A standing wave, unlike a traveling wave, *does not* transfer energy from one end to the other. The two waves that form it would individually carry equal amounts of power in opposite directions. There is a local flow of energy from each node to the adjacent antinodes and back, but the *average* rate of energy transfer is zero at every point. If you use the wave function of Eq. (15.28) to evaluate the wave power given by Eq. (15.21), you will find that the average power is zero.

PROBLEM-SOLVING STRATEGY 15.2 | STANDING WAVES

IDENTIFY *the relevant concepts:* Identify the target variables. Then determine whether the problem is purely *kinematic* (involving only such quantities as wave speed v, wavelength λ, and frequency f) or whether *dynamic* properties of the medium (such as F and μ for transverse waves on a string) are also involved.

SET UP *the problem* using the following steps:
1. Sketch the shape of the standing wave at a particular instant. This will help you visualize the nodes (label them N) and antinodes (A). The distance between adjacent nodes (or antinodes) is $\lambda/2$; the distance between a node and the adjacent antinode is $\lambda/4$.
2. Choose the equations you'll use. The wave function for the standing wave, like Eq. (15.28), is often useful.

3. You can determine the wave speed if you know λ and f (or, equivalently, $k = 2\pi/\lambda$ and $\omega = 2\pi f$) or if you know the relevant properties of the medium (for a string, F and μ).

EXECUTE *the solution:* Solve for the target variables. Once you've found the wave function, you can find the displacement y at any point x and at any time t. You can find the velocity and acceleration of a particle in the medium by taking the first and second partial derivatives of y with respect to time.

EVALUATE *your answer:* Compare your numerical answers with your sketch. Check that the wave function satisfies the boundary conditions (for example, the displacement should be zero at a fixed end).

EXAMPLE 15.6 | STANDING WAVES ON A GUITAR STRING

A guitar string lies along the x-axis when in equilibrium. The end of the string at $x = 0$ (the bridge of the guitar) is fixed. A sinusoidal wave with amplitude $A = 0.750$ mm $= 7.50 \times 10^{-4}$ m and frequency $f = 440$ Hz, corresponding to the red curves in Fig. 15.24, travels along the string in the $-x$-direction at 143 m/s. It is reflected from the fixed end, and the superposition of the incident and reflected waves forms a standing wave. (a) Find the equation giving the displacement of a point on the string as a function of position and time. (b) Locate the nodes. (c) Find the amplitude of the standing wave and the maximum transverse velocity and acceleration.

SOLUTION

IDENTIFY and SET UP: This is a *kinematics* problem (see Problem-Solving Strategy 15.1 in Section 15.3). The target variables are: in part (a), the wave function of the standing wave; in part (b), the locations of the nodes; and in part (c), the maximum displacement y, transverse velocity v_y, and transverse acceleration a_y. Since there is a fixed end at $x = 0$, we can use Eqs. (15.28) and (15.29) to describe this standing wave. We will need the relationships $\omega = 2\pi f$, $v = \omega/k$, and $v = \lambda f$.

EXECUTE: (a) The standing-wave amplitude is $A_{SW} = 2A = 1.50 \times 10^{-3}$ m (twice the amplitude of either the incident or reflected wave). The angular frequency and wave number are

$$\omega = 2\pi f = (2\pi \text{ rad})(440 \text{ s}^{-1}) = 2760 \text{ rad/s}$$

$$k = \frac{\omega}{v} = \frac{2760 \text{ rad/s}}{143 \text{ m/s}} = 19.3 \text{ rad/m}$$

Equation (15.28) then gives

$$y(x, t) = (A_{SW} \sin kx) \sin \omega t$$

$$= [(1.50 \times 10^{-3} \text{ m}) \sin(19.3 \text{ rad/m})x] \sin(2760 \text{ rad/s})t$$

(b) From Eq. (15.29), the positions of the nodes are $x = 0$, $\lambda/2$, λ, $3\lambda/2$, ... The wavelength is $\lambda = v/f = (143 \text{ m/s})/(440 \text{ Hz}) =$

0.325 m, so the nodes are at $x = 0$, 0.163 m, 0.325 m, 0.488 m, ...

(c) From the expression for $y(x, t)$ in part (a), the maximum displacement from equilibrium is $A_{SW} = 1.50 \times 10^{-3}$ m $=$ 1.50 mm. This occurs at the *antinodes,* which are midway between adjacent nodes (that is, at $x = 0.081$ m, 0.244 m, 0.406 m, ...).

For a particle on the string at any point x, the transverse (y-) velocity is

$$v_y(x, t) = \frac{\partial y(x, t)}{\partial t}$$

$$= [(1.50 \times 10^{-3} \text{ m}) \sin(19.3 \text{ rad/m})x]$$
$$\times [(2760 \text{ rad/s}) \cos(2760 \text{ rad/s})t]$$

$$= [(4.15 \text{ m/s}) \sin(19.3 \text{ rad/m})x] \cos(2760 \text{ rad/s})t$$

At an antinode, $\sin(19.3 \text{ rad/m})x = \pm 1$ and the transverse velocity varies between $+4.15$ m/s and -4.15 m/s. As is always the case in SHM, the maximum velocity occurs when the particle is passing through the equilibrium position ($y = 0$).

The transverse acceleration $a_y(x, t)$ is the *second* partial derivative of $y(x, t)$ with respect to time. You can show that

$$a_y(x, t) = \frac{\partial v_y(x, t)}{\partial t} = \frac{\partial^2 y(x, t)}{\partial t^2}$$

$$= [(-1.15 \times 10^4 \text{ m/s}^2) \sin(19.3 \text{ rad/m})x]$$
$$\times \sin(2760 \text{ rad/s})t$$

At the antinodes, the transverse acceleration varies between $+1.15 \times 10^4$ m/s^2 and -1.15×10^4 m/s^2.

EVALUATE: The maximum transverse velocity at an antinode is quite respectable (about 15 km/h, or 9.3 mi/h). But the maximum transverse acceleration is tremendous, 1170 times the acceleration due to gravity! Guitar strings are actually fixed at *both* ends; we'll see the consequences of this in the next section.

TEST YOUR UNDERSTANDING OF SECTION 15.7   Suppose the frequency of the standing wave in Example 15.6 were doubled from 440 Hz to 880 Hz. Would all of the nodes for $f = 440$ Hz also be nodes for $f = 880$ Hz? If so, would there be additional nodes for $f = 880$ Hz? If not, which nodes are absent for $f = 880$ Hz? ▮

15.8 NORMAL MODES OF A STRING

When we described standing waves on a string rigidly held at one end, as in Fig. 15.23, we made no assumptions about the length of the string or about what was happening at the other end. Let's now consider a string of a definite length L, rigidly held at *both* ends. Such strings are found in many musical instruments, including pianos, violins, and guitars. When a guitar string is plucked, a wave is produced in the string; this wave is reflected and re-reflected from the ends of the string, making a standing wave. This standing wave on the string in turn produces a sound wave in the air, with a frequency determined by the properties of the string. This is what makes stringed instruments so useful in making music.

To understand a standing wave on a string fixed at both ends, we first note that the standing wave must have a node at *both* ends of the string. We saw in the preceding section that adjacent nodes are one half-wavelength $(\lambda/2)$ apart, so the length of the string must be $\lambda/2$, or $2(\lambda/2)$, or $3(\lambda/2)$, or in general some integer number of half-wavelengths:

$$L = n\frac{\lambda}{2} \quad (n = 1, 2, 3, \ldots) \qquad \text{(string fixed at both ends)} \qquad (15.30)$$

That is, if a string with length L is fixed at both ends, a standing wave can exist only if its wavelength satisfies Eq. (15.30).

Solving this equation for λ and labeling the possible values of λ as λ_n, we find

$$\lambda_n = \frac{2L}{n} \quad (n = 1, 2, 3, \ldots) \qquad \text{(string fixed at both ends)} \qquad (15.31)$$

Waves can exist on the string if the wavelength is *not* equal to one of these values, but there cannot be a steady wave pattern with nodes and antinodes, and the total wave cannot be a standing wave. Equation (15.31) is illustrated by the standing waves shown in Figs. 15.23a, 15.23b, 15.23c, and 15.23d; these represent $n = 1, 2, 3,$ and 4, respectively.

Corresponding to the series of possible standing-wave wavelengths λ_n is a series of possible standing-wave frequencies f_n, each related to its corresponding wavelength by $f_n = v/\lambda_n$. The smallest frequency f_1 corresponds to the largest wavelength (the $n = 1$ case), $\lambda_1 = 2L$:

$$f_1 = \frac{v}{2L} \qquad \text{(string fixed at both ends)} \qquad (15.32)$$

This is called the **fundamental frequency.** The other standing-wave frequencies are $f_2 = 2v/2L$, $f_3 = 3v/2L$, and so on. These are all integer multiples of f_1, such as $2f_1, 3f_1, 4f_1,$ and so on. We can express *all* the frequencies as

Standing-wave frequencies, string fixed at both ends:
$$f_n = n\frac{v}{2L} = nf_1 \quad (n = 1, 2, 3, \ldots) \qquad (15.33)$$
Wave speed — v Fundamental frequency $= v/2L$ Length of string

These frequencies are called **harmonics,** and the series is called a **harmonic series.** Musicians sometimes call $f_2, f_3,$ and so on **overtones;** f_2 is the second harmonic or the first overtone, f_3 is the third harmonic or the second overtone, and so on. The first harmonic is the same as the fundamental frequency (**Fig. 15.25**).

PhET: Fourier: Making Waves
PhET: Waves on a String

15.25 Each string of a violin naturally oscillates at its harmonic frequencies, producing sound waves in the air with the same frequencies.

15.26 The first four normal modes of a string fixed at both ends. (Compare these to the photographs in Fig. 15.23.)

(a) $n = 1$: fundamental frequency, f_1

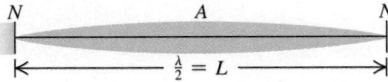

(b) $n = 2$: second harmonic, f_2 (first overtone)

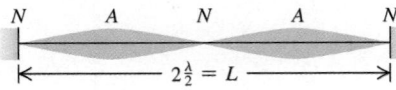

(c) $n = 3$: third harmonic, f_3 (second overtone)

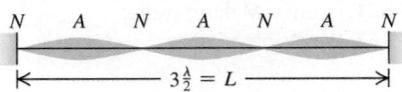

(d) $n = 4$: fourth harmonic, f_4 (third overtone)

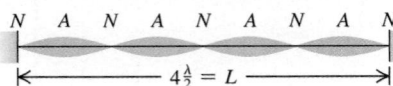

For a string with fixed ends at $x = 0$ and $x = L$, the wave function $y(x, t)$ of the nth standing wave is given by Eq. (15.28) (which satisfies the condition that there is a node at $x = 0$), with $\omega = \omega_n = 2\pi f_n$ and $k = k_n = 2\pi/\lambda_n$:

$$y_n(x, t) = A_{\text{SW}} \sin k_n x \sin \omega_n t \tag{15.34}$$

You can confirm that this wave function has nodes at both $x = 0$ and $x = L$.

A **normal mode** of an oscillating system is a motion in which all particles of the system move sinusoidally with the same frequency. For a system made up of a string of length L fixed at both ends, each of the frequencies given by Eq. (15.33) corresponds to a possible normal-mode pattern. **Figure 15.26** shows the first four normal-mode patterns and their associated frequencies and wavelengths; these correspond to Eq. (15.34) with $n = 1, 2, 3,$ and 4. By contrast, a harmonic oscillator, which has only one oscillating particle, has only one normal mode and one characteristic frequency. The string fixed at both ends has infinitely many normal modes ($n = 1, 2, 3, \dots$) because it is made up of a very large (effectively infinite) number of particles. More complicated oscillating systems also have infinite numbers of normal modes, though with more complex normal-mode patterns (**Fig. 15.27**).

Complex Standing Waves

If we could displace a string so that its shape is the same as one of the normal-mode patterns and then release it, it would vibrate with the frequency of that mode. Such a vibrating string would displace the surrounding air with the same frequency, producing a traveling sinusoidal sound wave that your ears would perceive as a pure tone. But when a string is struck (as in a piano) or plucked (as is done to guitar strings), the shape of the displaced string is *not* one of the patterns in Fig. 15.26. The motion is therefore a combination or *superposition* of many normal modes. Several simple harmonic motions of different frequencies are present simultaneously, and the displacement of any point on the string is the superposition of the displacements associated with the individual modes. The sound produced by the vibrating string is likewise a superposition of traveling sinusoidal sound waves, which you perceive as a rich, complex tone with the fundamental frequency f_1. The standing wave on the string and the traveling sound wave in the air have similar **harmonic content** (the extent to which frequencies higher than the fundamental are present). The harmonic content depends on how the string is initially set into motion. If you pluck the strings of an acoustic guitar

15.27 Astronomers have discovered that the sun oscillates in several different normal modes. This computer simulation shows one such mode.

Cross section of the sun's interior

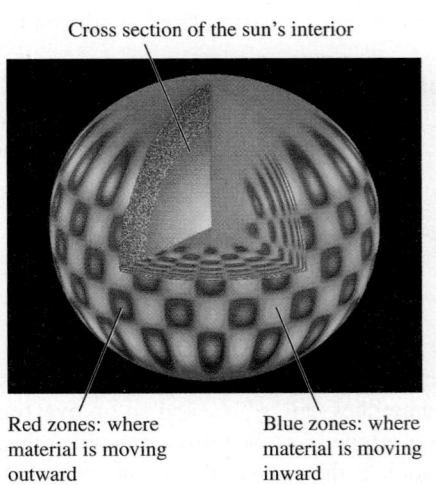

Red zones: where material is moving outward

Blue zones: where material is moving inward

in the normal location over the sound hole, the sound that you hear has a different harmonic content than if you pluck the strings next to the fixed end on the guitar body.

It is possible to represent every possible motion of the string as some superposition of normal-mode motions. Finding this representation for a given vibration pattern is called *harmonic analysis*. The sum of sinusoidal functions that represents a complex wave is called a *Fourier series*. **Figure 15.28** shows how a standing wave that is produced by plucking a guitar string of length L at a point $L/4$ from one end can be represented as a combination of sinusoidal functions.

Standing Waves and String Instruments

From Eq. (15.32), the fundamental frequency of a vibrating string is $f_1 = v/2L$. The speed v of waves on the string is determined by Eq. (15.14), $v = \sqrt{F/\mu}$. Combining these equations, we find

$$\text{Fundamental frequency,} \atop \text{string fixed at both ends} \cdots\!\!\rightarrow f_1 = \frac{1}{2L}\sqrt{\frac{F}{\mu}} \quad \begin{matrix} \leftarrow\cdots \text{Tension in string} \\ \\ \leftarrow\cdots \text{Mass per unit length} \end{matrix} \tag{15.35}$$
$$\underset{\text{Length of string}}{}$$

This is also the fundamental frequency of the sound wave created in the surrounding air by the vibrating string. The inverse dependence of frequency on length L is illustrated by the long strings of the bass (low-frequency) section of the piano or the bass viol compared with the shorter strings of the treble section of the piano or the violin (**Fig. 15.29**). The pitch of a violin or guitar is usually varied by pressing a string against the fingerboard with the fingers to change the length L of the vibrating portion of the string. Increasing the tension F increases the wave speed v and thus increases the frequency (and the pitch). All string instruments are "tuned" to the correct frequencies by varying the tension; you tighten the string to raise the pitch. Finally, increasing the mass per unit length μ decreases the wave speed and thus the frequency. The lower notes on a steel guitar are produced by thicker strings, and one reason for winding the bass strings of a piano with wire is to obtain the desired low frequency from a relatively short string.

Wind instruments such as saxophones and trombones also have normal modes. As for stringed instruments, the frequencies of these normal modes determine the pitch of the musical tones that these instruments produce. We'll discuss these instruments and many other aspects of sound in Chapter 16.

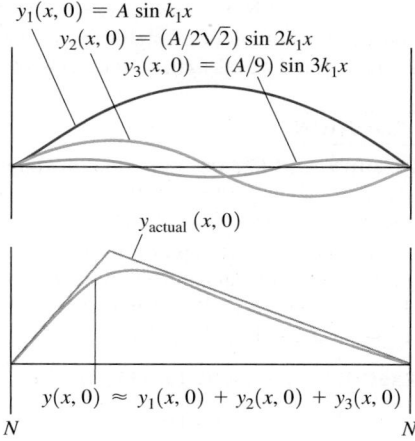

15.28 When a guitar string is plucked (pulled into a triangular shape) and released, a standing wave results. The standing wave is well represented (except at the sharp maximum point) by the sum of just three sinusoidal functions. Including additional sinusoidal functions further improves the representation.

$y_1(x, 0) = A \sin k_1 x$
$y_2(x, 0) = (A/2\sqrt{2}) \sin 2k_1 x$
$y_3(x, 0) = (A/9) \sin 3k_1 x$

$y_{\text{actual}}(x, 0)$

$y(x, 0) \approx y_1(x, 0) + y_2(x, 0) + y_3(x, 0)$

N N

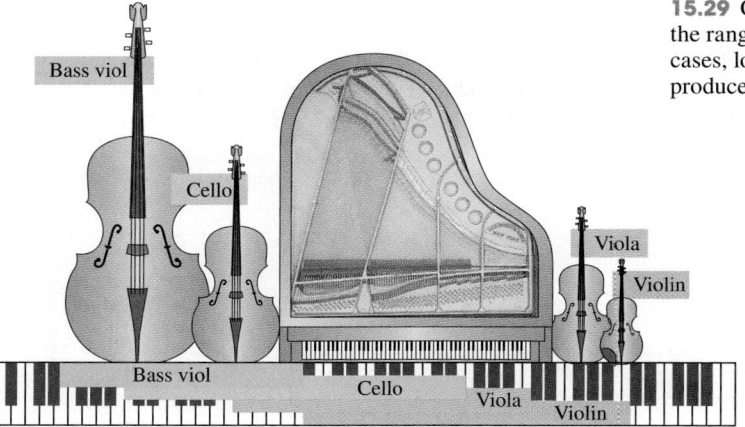

15.29 Comparing the range of a concert grand piano to the ranges of a bass viol, a cello, a viola, and a violin. In all cases, longer strings produce bass notes and shorter strings produce treble notes.

EXAMPLE 15.7 A GIANT BASS VIOL

In an attempt to get your name in *Guinness World Records,* you build a bass viol with strings of length 5.00 m between fixed points. One string, with linear mass density 40.0 g/m, is tuned to a 20.0-Hz fundamental frequency (the lowest frequency that the human ear can hear). Calculate (a) the tension of this string, (b) the frequency and wavelength on the string of the second harmonic, and (c) the frequency and wavelength on the string of the second overtone.

SOLUTION

IDENTIFY and SET UP: In part (a) the target variable is the string tension F; we'll use Eq. (15.35), which relates F to the known values $f_1 = 20.0$ Hz, $L = 5.00$ m, and $\mu = 40.0$ g/m. In parts (b) and (c) the target variables are the frequency and wavelength of a given harmonic and a given overtone. We determine these from the given length of the string and the fundamental frequency, using Eqs. (15.31) and (15.33).

EXECUTE: (a) We solve Eq. (15.35) for F:

$$F = 4\mu L^2 f_1^2 = 4(40.0 \times 10^{-3}\text{ kg/m})(5.00\text{ m})^2(20.0\text{ s}^{-1})^2$$

$$= 1600\text{ N} = 360\text{ lb}$$

(b) From Eqs. (15.33) and (15.31), the frequency and wavelength of the second harmonic ($n = 2$) are

$$f_2 = 2f_1 = 2(20.0\text{ Hz}) = 40.0\text{ Hz}$$

$$\lambda_2 = \frac{2L}{2} = \frac{2(5.00\text{ m})}{2} = 5.00\text{ m}$$

(c) The second overtone is the "second tone over" (above) the fundamental—that is, $n = 3$. Its frequency and wavelength are

$$f_3 = 3f_1 = 3(20.0\text{ Hz}) = 60.0\text{ Hz}$$

$$\lambda_3 = \frac{2L}{3} = \frac{2(5.00\text{ m})}{3} = 3.33\text{ m}$$

EVALUATE: The string tension in a real bass viol is typically a few hundred newtons; the tension in part (a) is a bit higher than that. The wavelengths in parts (b) and (c) are equal to the length of the string and two-thirds the length of the string, respectively, which agrees with the drawings of standing waves in Fig. 15.26.

EXAMPLE 15.8 FROM WAVES ON A STRING TO SOUND WAVES IN AIR

What are the frequency and wavelength of the sound waves produced in the air when the string in Example 15.7 is vibrating at its fundamental frequency? The speed of sound in air at 20°C is 344 m/s.

SOLUTION

IDENTIFY and SET UP: Our target variables are the frequency and wavelength for the *sound wave* produced by the bass viol string. The frequency of the sound wave is the same as the fundamental frequency f_1 of the standing wave, because the string forces the surrounding air to vibrate at the same frequency. The wavelength of the sound wave is $\lambda_{1(\text{sound})} = v_{\text{sound}}/f_1$.

EXECUTE: We have $f = f_1 = 20.0$ Hz, so

$$\lambda_{1(\text{sound})} = \frac{v_{\text{sound}}}{f_1} = \frac{344\text{ m/s}}{20.0\text{ Hz}} = 17.2\text{ m}$$

EVALUATE: In Example 15.7, the wavelength of the fundamental on the string was $\lambda_{1(\text{string})} = 2L = 2(5.00\text{ m}) = 10.0$ m. Here $\lambda_{1(\text{sound})} = 17.2$ m is greater than that by the factor of $17.2/10.0 = 1.72$. This is as it should be: Because the frequencies of the sound wave and the standing wave are equal, $\lambda = v/f$ says that the wavelengths in air and on the string are in the same ratio as the corresponding wave speeds; here $v_{\text{sound}} = 344$ m/s is greater than $v_{\text{string}} = (10.0\text{ m})(20.0\text{ Hz}) = 200$ m/s by just the factor 1.72.

TEST YOUR UNDERSTANDING OF SECTION 15.8 While a guitar string is vibrating, you gently touch the midpoint of the string to ensure that the string does not vibrate at that point. Which normal modes *cannot* be present on the string while you are touching it in this way? ▮

Waves and their properties: A wave is any disturbance that propagates from one region to another. A mechanical wave travels within some material called the medium. The wave speed v depends on the type of wave and the properties of the medium.

In a periodic wave, the motion of each point of the medium is periodic with frequency f and period T. The wavelength λ is the distance over which the wave pattern repeats, and the amplitude A is the maximum displacement of a particle in the medium. The product of λ and f equals the wave speed. A sinusoidal wave is a special periodic wave in which each point moves in simple harmonic motion. (See Example 15.1.)

$$v = \lambda f \qquad (15.1)$$

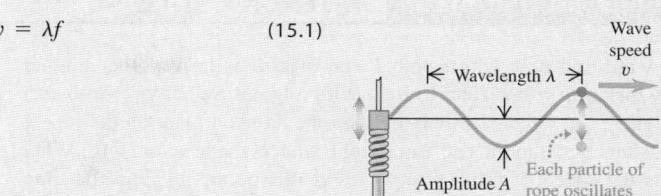

Wave functions and wave dynamics: The wave function $y(x, t)$ describes the displacements of individual particles in the medium. Equations (15.3), (15.4), and (15.7) give the wave equation for a sinusoidal wave traveling in the $+x$-direction. If the wave is moving in the $-x$-direction, the minus signs in the cosine functions are replaced by plus signs. (See Example 15.2.)

The wave function obeys a partial differential equation called the wave equation, Eq. (15.12).

The speed of transverse waves on a string depends on the tension F and mass per unit length μ. (See Example 15.3.)

$$y(x, t) = A \cos\left[\omega\left(\frac{x}{v} - t\right) \right] \qquad (15.3)$$

$$y(x, t) = A \cos 2\pi\left[\left(\frac{x}{\lambda} - \frac{t}{T}\right) \right] \qquad (15.4)$$

$$y(x, t) = A \cos(kx - \omega t) \qquad (15.7)$$

where $k = 2\pi/\lambda$ and $\omega = 2\pi f = vk$

$$\frac{\partial^2 y(x, t)}{\partial x^2} = \frac{1}{v^2}\frac{\partial^2 y(x, t)}{\partial t^2} \qquad (15.12)$$

$$v = \sqrt{\frac{F}{\mu}} \quad \text{(waves on a string)} \qquad (15.14)$$

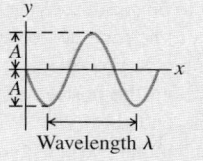

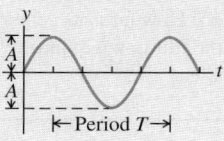

Wave power: Wave motion conveys energy from one region to another. For a sinusoidal mechanical wave, the average power P_{av} is proportional to the square of the wave amplitude and the square of the frequency. For waves that spread out in three dimensions, the wave intensity I is inversely proportional to the square of the distance from the source. (See Examples 15.4 and 15.5.)

$$P_{av} = \tfrac{1}{2}\sqrt{\mu F}\,\omega^2 A^2 \qquad (15.25)$$
(average power, sinusoidal wave)

$$\frac{I_1}{I_2} = \frac{r_2^2}{r_1^2} \qquad (15.26)$$
(inverse-square law for intensity)

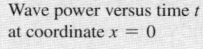

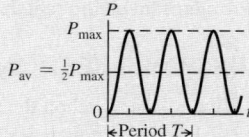

Wave superposition: A wave reflects when it reaches a boundary of its medium. At any point where two or more waves overlap, the total displacement is the sum of the displacements of the individual waves (principle of superposition).

$$y(x, t) = y_1(x, t) + y_2(x, t) \qquad (15.27)$$
(principle of superposition)

Standing waves on a string: When a sinusoidal wave is reflected from a fixed or free end of a stretched string, the incident and reflected waves combine to form a standing sinusoidal wave with nodes and antinodes. Adjacent nodes are spaced a distance $\lambda/2$ apart, as are adjacent antinodes. (See Example 15.6.)

When both ends of a string with length L are held fixed, standing waves can occur only when L is an integer multiple of $\lambda/2$. Each frequency with its associated vibration pattern is called a normal mode. (See Examples 15.7 and 15.8.)

$$y(x, t) = (A_{SW}\sin kx)\sin\omega t \qquad (15.28)$$
(standing wave on a string, fixed end at $x = 0$)

$$f_n = n\frac{v}{2L} = nf_1 \qquad (15.33)$$
$$(n = 1, 2, 3, \dots)$$

$$f_1 = \frac{1}{2L}\sqrt{\frac{F}{\mu}} \qquad (15.35)$$
(string fixed at both ends)

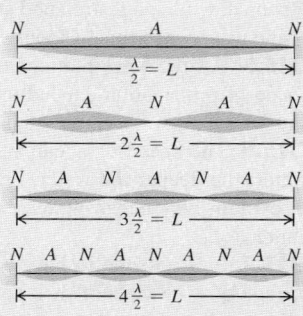

BRIDGING PROBLEM WAVES ON A ROTATING ROPE

A uniform rope with length L and mass m is held at one end and whirled in a horizontal circle with angular velocity ω. You can ignore the force of gravity on the rope. (a) At a point on the rope a distance r from the end that is held, what is the tension F? (b) What is the speed of transverse waves at this point? (c) Find the time required for a transverse wave to travel from one end of the rope to the other.

SOLUTION GUIDE

IDENTIFY and SET UP

1. Draw a sketch of the situation and label the distances r and L. The tension in the rope will be different at different values of r. Do you see why? Where on the rope do you expect the tension to be greatest? Where do you expect it will be least?

2. Where on the rope do you expect the wave speed to be greatest? Where do you expect it will be least?

3. Think about the portion of the rope that is farther out than r from the end that is held. What forces act on this portion? (Remember that you can ignore gravity.) What is the mass of this portion? How far is its center of mass from the rotation axis?

4. List the unknown quantities and decide which are your target variables.

EXECUTE

5. Draw a free-body diagram for the portion of the rope that is farther out than r from the end that is held.

6. Use your free-body diagram to help you determine the tension in the rope at distance r.

7. Use your result from step 6 to find the wave speed at distance r.

8. Use your result from step 7 to find the time for a wave to travel from one end to the other. (*Hint:* The wave speed is $v = dr/dt$, so the time for the wave to travel a distance dr along the rope is $dt = dr/v$. Integrate this to find the total time. See Appendix B.)

EVALUATE

9. Do your results for parts (a) and (b) agree with your expectations from steps 1 and 2? Are the units correct?

10. Check your result for part (a) by considering the net force on a small segment of the rope at distance r with length dr and mass $dm = (m/L)dr$. [*Hint:* The tension forces on this segment are $F(r)$ on one side and $F(r + dr)$ on the other side. You will get an equation for dF/dr that you can integrate to find F as a function of r.]

Problems

For assigned homework and other learning materials, go to MasteringPhysics®.

°, °°, °°°: Difficulty levels. **CP**: Cumulative problems incorporating material from earlier chapters. **CALC**: Problems requiring calculus. **DATA**: Problems involving real data, scientific evidence, experimental design, and/or statistical reasoning. **BIO**: Biosciences problems.

DISCUSSION QUESTIONS

Q15.1 Two waves travel on the same string. Is it possible for them to have (a) different frequencies; (b) different wavelengths; (c) different speeds; (d) different amplitudes; (e) the same frequency but different wavelengths? Explain your reasoning.

Q15.2 Under a tension F, it takes 2.00 s for a pulse to travel the length of a taut wire. What tension is required (in terms of F) for the pulse to take 6.00 s instead? Explain how you arrive at your answer.

Q15.3 What kinds of energy are associated with waves on a stretched string? How could you detect such energy experimentally?

Q15.4 The amplitude of a wave decreases gradually as the wave travels down a long, stretched string. What happens to the energy of the wave when this happens?

Q15.5 For the wave motions discussed in this chapter, does the speed of propagation depend on the amplitude? What makes you say this?

Q15.6 The speed of ocean waves depends on the depth of the water; the deeper the water, the faster the wave travels. Use this to explain why ocean waves crest and "break" as they near the shore.

Q15.7 Is it possible to have a longitudinal wave on a stretched string? Why or why not? Is it possible to have a transverse wave on a steel rod? Again, why or why not? If your answer is yes in either case, explain how you would create such a wave.

Q15.8 For transverse waves on a string, is the wave speed the same as the speed of any part of the string? Explain the difference between these two speeds. Which one is constant?

Q15.9 The four strings on a violin have different thicknesses, but are all under approximately the same tension. Do waves travel faster on the thick strings or the thin strings? Why? How does the fundamental vibration frequency compare for the thick versus the thin strings?

Q15.10 A sinusoidal wave can be described by a cosine function, which is negative just as often as positive. So why isn't the average power delivered by this wave zero?

Q15.11 Two strings of different mass per unit length μ_1 and μ_2 are tied together and stretched with a tension F. A wave travels along the string and passes the discontinuity in μ. Which of the following wave properties will be the same on both sides of the discontinuity, and which will change: speed of the wave; frequency; wavelength? Explain the physical reasoning behind each answer.

Q15.12 A long rope with mass m is suspended from the ceiling and hangs vertically. A wave pulse is produced at the lower end of the rope, and the pulse travels up the rope. Does the speed of the wave pulse change as it moves up the rope, and if so, does it increase or decrease? Explain.

Q15.13 In a transverse wave on a string, the motion of the string is perpendicular to the length of the string. How, then, is it possible for energy to move along the length of the string?

Q15.14 Energy can be transferred along a string by wave motion. However, in a standing wave on a string, no energy can ever be transferred past a node. Why not?

Q15.15 Can a standing wave be produced on a string by superposing two waves traveling in opposite directions with the same frequency but different amplitudes? Why or why not? Can a standing wave be produced by superposing two waves traveling in opposite directions with different frequencies but the same amplitude? Why or why not?

Q15.16 If you stretch a rubber band and pluck it, you hear a (somewhat) musical tone. How does the frequency of this tone change as you stretch the rubber band further? (Try it!) Does this agree with Eq. (15.35) for a string fixed at both ends? Explain.

Q15.17 A musical interval of an *octave* corresponds to a factor of 2 in frequency. By what factor must the tension in a guitar or violin string be increased to raise its pitch one octave? To raise it two octaves? Explain your reasoning. Is there any danger in attempting these changes in pitch?

Q15.18 By touching a string lightly at its center while bowing, a violinist can produce a note exactly one octave above the note to which the string is tuned—that is, a note with exactly twice the frequency. Why is this possible?

Q15.19 As we discussed in Section 15.1, water waves are a combination of longitudinal and transverse waves. Defend the following statement: "When water waves hit a vertical wall, the wall is a node of the longitudinal displacement but an antinode of the transverse displacement."

Q15.20 Violins are short instruments, while cellos and basses are long. In terms of the frequency of the waves they produce, explain why this is so.

Q15.21 What is the purpose of the frets on a guitar? In terms of the frequency of the vibration of the strings, explain their use.

EXERCISES

Section 15.2 Periodic Waves

15.1 • The speed of sound in air at 20°C is 344 m/s. (a) What is the wavelength of a sound wave with a frequency of 784 Hz, corresponding to the note G_5 on a piano, and how many milliseconds does each vibration take? (b) What is the wavelength of a sound wave one octave higher (twice the frequency) than the note in part (a)?

15.2 • BIO **Audible Sound.** Provided the amplitude is sufficiently great, the human ear can respond to longitudinal waves over a range of frequencies from about 20.0 Hz to about 20.0 kHz. (a) If you were to mark the beginning of each complete wave pattern with a red dot for the long-wavelength sound and a blue dot for the short-wavelength sound, how far apart would the red dots be, and how far apart would the blue dots be? (b) In reality would adjacent dots in each set be far enough apart for you to easily measure their separation with a meter stick? (c) Suppose you repeated part (a) in water, where sound travels at 1480 m/s. How far apart would the dots be in each set? Could you readily measure their separation with a meter stick?

15.3 • **Tsunami!** On December 26, 2004, a great earthquake occurred off the coast of Sumatra and triggered immense waves (tsunami) that killed some 200,000 people. Satellites observing these waves from space measured 800 km from one wave crest to the next and a period between waves of 1.0 hour. What was the speed of these waves in m/s and in km/h? Does your answer help you understand why the waves caused such devastation?

15.4 • BIO **Ultrasound Imaging.** Sound having frequencies above the range of human hearing (about 20,000 Hz) is called *ultrasound.* Waves above this frequency can be used to penetrate the body and to produce images by reflecting from surfaces. In a typical ultrasound scan, the waves travel through body tissue with a speed of 1500 m/s. For a good, detailed image, the wavelength should be no more than 1.0 mm. What frequency sound is required for a good scan?

15.5 • BIO (a) **Audible wavelengths.** The range of audible frequencies is from about 20 Hz to 20,000 Hz. What is the range of the wavelengths of audible sound in air? (b) **Visible light.** The range of visible light extends from 380 nm to 750 nm. What is the range of visible frequencies of light? (c) **Brain surgery.** Surgeons can remove brain tumors by using a cavitron ultrasonic surgical aspirator, which produces sound waves of frequency 23 kHz. What is the wavelength of these waves in air? (d) **Sound in the body.** What would be the wavelength of the sound in part (c) in bodily fluids in which the speed of sound is 1480 m/s but the frequency is unchanged?

15.6 •• A fisherman notices that his boat is moving up and down periodically, owing to waves on the surface of the water. It takes 2.5 s for the boat to travel from its highest point to its lowest, a total distance of 0.53 m. The fisherman sees that the wave crests are spaced 4.8 m apart. (a) How fast are the waves traveling? (b) What is the amplitude of each wave? (c) If the total vertical distance traveled by the boat were 0.30 m but the other data remained the same, how would the answers to parts (a) and (b) change?

Section 15.3 Mathematical Description of a Wave

15.7 • Transverse waves on a string have wave speed 8.00 m/s, amplitude 0.0700 m, and wavelength 0.320 m. The waves travel in the $-x$-direction, and at $t = 0$ the $x = 0$ end of the string has its maximum upward displacement. (a) Find the frequency, period, and wave number of these waves. (b) Write a wave function describing the wave. (c) Find the transverse displacement of a particle at $x = 0.360$ m at time $t = 0.150$ s. (d) How much time must elapse from the instant in part (c) until the particle at $x = 0.360$ m next has maximum upward displacement?

15.8 • A certain transverse wave is described by

$$y(x, t) = (6.50 \text{ mm}) \cos 2\pi \left(\frac{x}{28.0 \text{ cm}} - \frac{t}{0.0360 \text{ s}} \right)$$

Determine the wave's (a) amplitude; (b) wavelength; (c) frequency; (d) speed of propagation; (e) direction of propagation.

15.9 • CALC Which of the following wave functions satisfies the wave equation, Eq. (15.12)? (a) $y(x, t) = A \cos(kx + \omega t)$; (b) $y(x, t) = A \sin(kx + \omega t)$; (c) $y(x, t) = A(\cos kx + \cos \omega t)$. (d) For the wave of part (b), write the equations for the transverse velocity and transverse acceleration of a particle at point x.

15.10 • A water wave traveling in a straight line on a lake is described by the equation

$$y(x, t) = (2.75 \text{ cm}) \cos(0.410 \text{ rad/cm } x + 6.20 \text{ rad/s } t)$$

where y is the displacement perpendicular to the undisturbed surface of the lake. (a) How much time does it take for one complete wave pattern to go past a fisherman in a boat at anchor, and what horizontal distance does the wave crest travel in that time? (b) What are the wave number and the number of waves per second that pass the fisherman? (c) How fast does a wave crest travel past the fisherman, and what is the maximum speed of his cork floater as the wave causes it to bob up and down?

15.11 · A sinusoidal wave is propagating along a stretched string that lies along the x-axis. The displacement of the string as a function of time is graphed in **Fig. E15.11** for particles at $x = 0$ and at $x = 0.0900$ m.

Figure **E15.11**

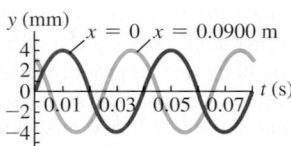

(a) What is the amplitude of the wave? (b) What is the period of the wave? (c) You are told that the two points $x = 0$ and $x = 0.0900$ m are within one wavelength of each other. If the wave is moving in the $+x$-direction, determine the wavelength and the wave speed. (d) If instead the wave is moving in the $-x$-direction, determine the wavelength and the wave speed. (e) Would it be possible to determine definitively the wavelengths in parts (c) and (d) if you were not told that the two points were within one wavelength of each other? Why or why not?

15.12 ·· CALC **Speed of Propagation vs. Particle Speed.** (a) Show that Eq. (15.3) may be written as

$$y(x, t) = A \cos\left[\frac{2\pi}{\lambda}(x - vt)\right]$$

(b) Use $y(x, t)$ to find an expression for the transverse velocity v_y of a particle in the string on which the wave travels. (c) Find the maximum speed of a particle of the string. Under what circumstances is this equal to the propagation speed v? Less than v? Greater than v?

15.13 ·· A transverse wave on a string has amplitude 0.300 cm, wavelength 12.0 cm, and speed 6.00 cm/s. It is represented by $y(x, t)$ as given in Exercise 15.12. (a) At time $t = 0$, compute y at 1.5-cm intervals of x (that is, at $x = 0, x = 1.5$ cm, $x = 3.0$ cm, and so on) from $x = 0$ to $x = 12.0$ cm. Graph the results. This is the shape of the string at time $t = 0$. (b) Repeat the calculations for the same values of x at times $t = 0.400$ s and $t = 0.800$ s. Graph the shape of the string at these instants. In what direction is the wave traveling?

15.14 · A wave on a string is described by $y(x, t) = A \cos(kx - \omega t)$. (a) Graph y, v_y, and a_y as functions of x for time $t = 0$. (b) Consider the following points on the string: (i) $x = 0$; (ii) $x = \pi/4k$; (iii) $x = \pi/2k$; (iv) $x = 3\pi/4k$; (v) $x = \pi/k$; (vi) $x = 5\pi/4k$; (vii) $x = 3\pi/2k$; (viii) $x = 7\pi/4k$. For a particle at each of these points at $t = 0$, describe in words whether the particle is moving and in what direction, and whether the particle is speeding up, slowing down, or instantaneously not accelerating.

Section 15.4 Speed of a Transverse Wave

15.15 · One end of a horizontal rope is attached to a prong of an electrically driven tuning fork that vibrates the rope transversely at 120 Hz. The other end passes over a pulley and supports a 1.50-kg mass. The linear mass density of the rope is 0.0480 kg/m. (a) What is the speed of a transverse wave on the rope? (b) What is the wavelength? (c) How would your answers to parts (a) and (b) change if the mass were increased to 3.00 kg?

15.16 · With what tension must a rope with length 2.50 m and mass 0.120 kg be stretched for transverse waves of frequency 40.0 Hz to have a wavelength of 0.750 m?

15.17 ·· The upper end of a 3.80-m-long steel wire is fastened to the ceiling, and a 54.0-kg object is suspended from the lower end of the wire. You observe that it takes a transverse pulse 0.0492 s to travel from the bottom to the top of the wire. What is the mass of the wire?

15.18 ·· A 1.50-m string of weight 0.0125 N is tied to the ceiling at its upper end, and the lower end supports a weight W. Ignore the very small variation in tension along the length of the string that is produced by the weight of the string. When you pluck the string slightly, the waves traveling up the string obey the equation

$$y(x, t) = (8.50 \text{ mm}) \cos(172 \text{ rad/m } x - 4830 \text{ rad/s } t)$$

Assume that the tension of the string is constant and equal to W. (a) How much time does it take a pulse to travel the full length of the string? (b) What is the weight W? (c) How many wavelengths are on the string at any instant of time? (d) What is the equation for waves traveling *down* the string?

15.19 · A thin, 75.0-cm wire has a mass of 16.5 g. One end is tied to a nail, and the other end is attached to a screw that can be adjusted to vary the tension in the wire. (a) To what tension (in newtons) must you adjust the screw so that a transverse wave of wavelength 3.33 cm makes 625 vibrations per second? (b) How fast would this wave travel?

15.20 ·· A heavy rope 6.00 m long and weighing 29.4 N is attached at one end to a ceiling and hangs vertically. A 0.500-kg mass is suspended from the lower end of the rope. What is the speed of transverse waves on the rope at the (a) bottom of the rope, (b) middle of the rope, and (c) top of the rope? (d) Is the tension in the middle of the rope the average of the tensions at the top and bottom? Is the wave speed at the middle of the rope the average of the wave speeds at the top and bottom? Explain.

15.21 · A simple harmonic oscillator at the point $x = 0$ generates a wave on a rope. The oscillator operates at a frequency of 40.0 Hz and with an amplitude of 3.00 cm. The rope has a linear mass density of 50.0 g/m and is stretched with a tension of 5.00 N. (a) Determine the speed of the wave. (b) Find the wavelength. (c) Write the wave function $y(x, t)$ for the wave. Assume that the oscillator has its maximum upward displacement at time $t = 0$. (d) Find the maximum transverse acceleration of points on the rope. (e) In the discussion of transverse waves in this chapter, the force of gravity was ignored. Is that a reasonable approximation for this wave? Explain.

Section 15.5 Energy in Wave Motion

15.22 ·· A piano wire with mass 3.00 g and length 80.0 cm is stretched with a tension of 25.0 N. A wave with frequency 120.0 Hz and amplitude 1.6 mm travels along the wire. (a) Calculate the average power carried by the wave. (b) What happens to the average power if the wave amplitude is halved?

15.23 · A horizontal wire is stretched with a tension of 94.0 N, and the speed of transverse waves for the wire is 406 m/s. What must the amplitude of a traveling wave of frequency 69.0 Hz be for the average power carried by the wave to be 0.365 W?

15.24 ·· A light wire is tightly stretched with tension F. Transverse traveling waves of amplitude A and wavelength λ_1 carry average power $P_{av,1} = 0.400$ W. If the wavelength of the waves is doubled, so $\lambda_2 = 2\lambda_1$, while the tension F and amplitude A are not altered, what then is the average power $P_{av,2}$ carried by the waves?

15.25 ·· A jet plane at takeoff can produce sound of intensity 10.0 W/m² at 30.0 m away. But you prefer the tranquil sound of normal conversation, which is 1.0 μW/m². Assume that the plane behaves like a point source of sound. (a) What is the closest distance you should live from the airport runway to preserve your peace of mind? (b) What intensity from the jet does your friend experience if she lives twice as far from the runway as you do? (c) What power of sound does the jet produce at takeoff?

15.26 •• **Threshold of Pain.** You are investigating the report of a UFO landing in an isolated portion of New Mexico, and you encounter a strange object that is radiating sound waves uniformly in all directions. Assume that the sound comes from a point source and that you can ignore reflections. You are slowly walking toward the source. When you are 7.5 m from it, you measure its intensity to be 0.11 W/m². An intensity of 1.0 W/m² is often used as the "threshold of pain." How much closer to the source can you move before the sound intensity reaches this threshold?

15.27 • **Energy Output.** By measurement you determine that sound waves are spreading out equally in all directions from a point source and that the intensity is 0.026 W/m² at a distance of 4.3 m from the source. (a) What is the intensity at a distance of 3.1 m from the source? (b) How much sound energy does the source emit in one hour if its power output remains constant?

15.28 • A fellow student with a mathematical bent tells you that the wave function of a traveling wave on a thin rope is $y(x, t) = 2.30 \text{ mm} \cos[(6.98 \text{ rad/m})x + (742 \text{ rad/s})t]$. Being more practical, you measure the rope to have a length of 1.35 m and a mass of 0.00338 kg. You are then asked to determine the following: (a) amplitude; (b) frequency; (c) wavelength; (d) wave speed; (e) direction the wave is traveling; (f) tension in the rope; (g) average power transmitted by the wave.

15.29 • At a distance of 7.00×10^{12} m from a star, the intensity of the radiation from the star is 15.4 W/m². Assuming that the star radiates uniformly in all directions, what is the total power output of the star?

Section 15.6 Wave Interference, Boundary Conditions, and Superposition

15.30 • **Reflection.** A wave pulse on a string has the dimensions shown in **Fig. E15.30** at $t = 0$. The wave speed is 40 cm/s. (a) If point O is a fixed end, draw the total wave on the string at $t = 15$ ms, 20 ms, 25 ms, 30 ms, 35 ms, 40 ms, and 45 ms. (b) Repeat part (a) for the case in which point O is a free end.

Figure **E15.30**

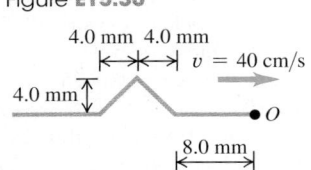

15.31 • **Reflection.** A wave pulse on a string has the dimensions shown in **Fig. E15.31** at $t = 0$. The wave speed is 5.0 m/s. (a) If point O is a fixed end, draw the total wave on the string at $t = 1.0$ ms, 2.0 ms, 3.0 ms, 4.0 ms, 5.0 ms, 6.0 ms, and 7.0 ms. (b) Repeat part (a) for the case in which point O is a free end.

Figure **E15.31**

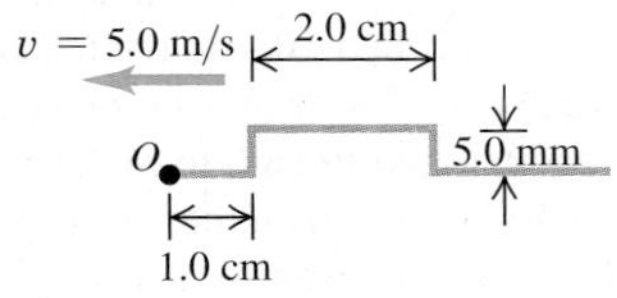

15.32 • **Interference of Triangular Pulses.** Two triangular wave pulses are traveling toward each other on a stretched string as shown in **Fig. E15.32**. Each pulse is identical to the other and travels at 2.00 cm/s. The leading edges of the pulses are 1.00 cm apart at $t = 0$. Sketch the shape of the string at $t = 0.250$ s, $t = 0.500$ s, $t = 0.750$ s, $t = 1.000$ s, and $t = 1.250$ s.

Figure **E15.32**

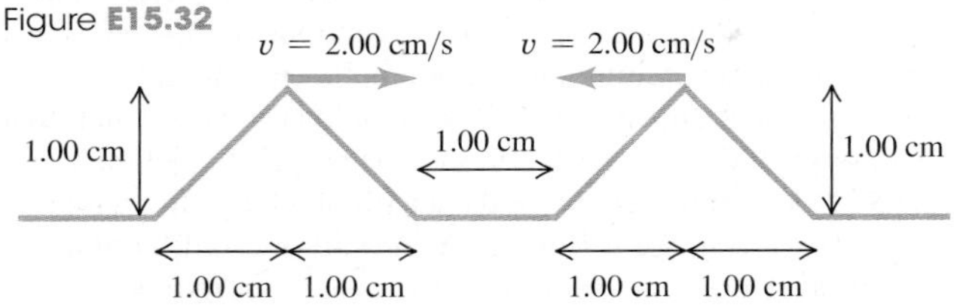

15.33 • Suppose that the left-traveling pulse in Exercise 15.32 is *below* the level of the unstretched string instead of above it. Make the same sketches that you did in that exercise.

15.34 •• Two pulses are moving in opposite directions at 1.0 cm/s on a taut string, as shown in **Fig. E15.34**. Each square is 1.0 cm. Sketch the shape of the string at the end of (a) 6.0 s; (b) 7.0 s; (c) 8.0 s.

Figure **E15.34**

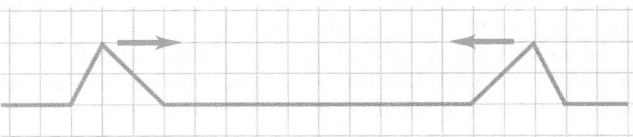

15.35 •• **Interference of Rectangular Pulses.** Figure E15.35 shows two rectangular wave pulses on a stretched string traveling toward each other. Each pulse is traveling with a speed of 1.00 mm/s and has the height and width shown in the figure. If the leading edges of the pulses are 8.00 mm apart at $t = 0$, sketch the shape of the string at $t = 4.00$ s, $t = 6.00$ s, and $t = 10.0$ s.

Figure **E15.35**

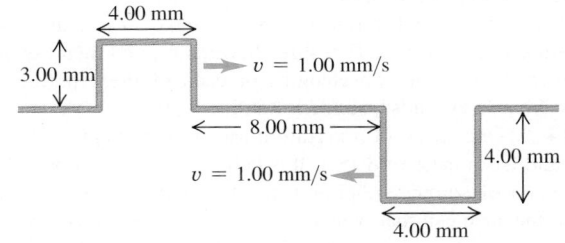

Section 15.7 Standing Waves on a String
Section 15.8 Normal Modes of a String

15.36 •• **CALC** Adjacent antinodes of a standing wave on a string are 15.0 cm apart. A particle at an antinode oscillates in simple harmonic motion with amplitude 0.850 cm and period 0.0750 s. The string lies along the $+x$-axis and is fixed at $x = 0$. (a) How far apart are the adjacent nodes? (b) What are the wavelength, amplitude, and speed of the two traveling waves that form this pattern? (c) Find the maximum and minimum transverse speeds of a point at an antinode. (d) What is the shortest distance along the string between a node and an antinode?

15.37 • Standing waves on a wire are described by Eq. (15.28), with $A_{SW} = 2.50$ mm, $\omega = 942$ rad/s, and $k = 0.750\pi$ rad/m. The left end of the wire is at $x = 0$. At what distances from the left end are (a) the nodes of the standing wave and (b) the antinodes of the standing wave?

15.38 • A 1.50-m-long rope is stretched between two supports with a tension that makes the speed of transverse waves 62.0 m/s. What are the wavelength and frequency of (a) the fundamental; (b) the second overtone; (c) the fourth harmonic?

15.39 • A wire with mass 40.0 g is stretched so that its ends are tied down at points 80.0 cm apart. The wire vibrates in its fundamental mode with frequency 60.0 Hz and with an amplitude at the antinodes of 0.300 cm. (a) What is the speed of propagation of transverse waves in the wire? (b) Compute the tension in the wire. (c) Find the maximum transverse velocity and acceleration of particles in the wire.

15.40 • A piano tuner stretches a steel piano wire with a tension of 800 N. The steel wire is 0.400 m long and has a mass of 3.00 g. (a) What is the frequency of its fundamental mode of vibration? (b) What is the number of the highest harmonic that could be heard by a person who is capable of hearing frequencies up to 10,000 Hz?

15.41 • CALC A thin, taut string tied at both ends and oscillating in its third harmonic has its shape described by the equation $y(x, t) = (5.60 \text{ cm}) \sin[(0.0340 \text{ rad/cm})x] \sin[(50.0 \text{ rad/s})t]$, where the origin is at the left end of the string, the x-axis is along the string, and the y-axis is perpendicular to the string. (a) Draw a sketch that shows the standing-wave pattern. (b) Find the amplitude of the two traveling waves that make up this standing wave. (c) What is the length of the string? (d) Find the wavelength, frequency, period, and speed of the traveling waves. (e) Find the maximum transverse speed of a point on the string. (f) What would be the equation $y(x,t)$ for this string if it were vibrating in its eighth harmonic?

15.42 • The wave function of a standing wave is $y(x, t) = 4.44 \text{ mm} \sin[(32.5 \text{ rad/m})x] \sin[(754 \text{ rad/s})t]$. For the two traveling waves that make up this standing wave, find the (a) amplitude; (b) wavelength; (c) frequency; (d) wave speed; (e) wave functions. (f) From the information given, can you determine which harmonic this is? Explain.

15.43 • **Waves on a Stick.** A flexible stick 2.0 m long is not fixed in any way and is free to vibrate. Make clear drawings of this stick vibrating in its first three harmonics, and then use your drawings to find the wavelengths of each of these harmonics. (*Hint:* Should the ends be nodes or antinodes?)

15.44 •• One string of a certain musical instrument is 75.0 cm long and has a mass of 8.75 g. It is being played in a room where the speed of sound is 344 m/s. (a) To what tension must you adjust the string so that, when vibrating in its second overtone, it produces sound of wavelength 0.765 m? (Assume that the breaking stress of the wire is very large and isn't exceeded.) (b) What frequency sound does this string produce in its fundamental mode of vibration?

15.45 • The portion of the string of a certain musical instrument between the bridge and upper end of the finger board (that part of the string that is free to vibrate) is 60.0 cm long, and this length of the string has mass 2.00 g. The string sounds an A_4 note (440 Hz) when played. (a) Where must the player put a finger (what distance x from the bridge) to play a D_5 note (587 Hz)? (See **Fig. E15.45**.)

Figure **E15.45**

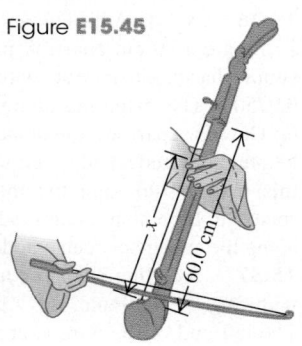

For both the A_4 and D_5 notes, the string vibrates in its fundamental mode. (b) Without retuning, is it possible to play a G_4 note (392 Hz) on this string? Why or why not?

15.46 • (a) A horizontal string tied at both ends is vibrating in its fundamental mode. The traveling waves have speed v, frequency f, amplitude A, and wavelength λ. Calculate the maximum transverse velocity and maximum transverse acceleration of points located at (i) $x = \lambda/2$, (ii) $x = \lambda/4$, and (iii) $x = \lambda/8$, from the left-hand end of the string. (b) At each of the points in part (a), what is the amplitude of the motion? (c) At each of the points in part (a), how much time does it take the string to go from its largest upward displacement to its largest downward displacement?

15.47 • **Guitar String.** One of the 63.5-cm-long strings of an ordinary guitar is tuned to produce the note B_3 (frequency 245 Hz) when vibrating in its fundamental mode. (a) Find the speed of transverse waves on this string. (b) If the tension in this string is increased by 1.0%, what will be the new fundamental frequency of the string? (c) If the speed of sound in the surrounding air is 344 m/s, find the frequency and wavelength of the sound wave produced in the air by the vibration of the B_3 string. How do these compare to the frequency and wavelength of the standing wave on the string?

PROBLEMS

15.48 • A transverse wave on a rope is given by

$$y(x, t) = (0.750 \text{ cm}) \cos \pi[(0.400 \text{ cm}^{-1})x + (250 \text{ s}^{-1})t]$$

(a) Find the amplitude, period, frequency, wavelength, and speed of propagation. (b) Sketch the shape of the rope at these values of t: 0, 0.0005 s, 0.0010 s. (c) Is the wave traveling in the $+x$- or $-x$-direction? (d) The mass per unit length of the rope is 0.0500 kg/m. Find the tension. (e) Find the average power of this wave.

15.49 • CALC A transverse sine wave with an amplitude of 2.50 mm and a wavelength of 1.80 m travels from left to right along a long, horizontal, stretched string with a speed of 36.0 m/s. Take the origin at the left end of the undisturbed string. At time $t = 0$ the left end of the string has its maximum upward displacement. (a) What are the frequency, angular frequency, and wave number of the wave? (b) What is the function $y(x, t)$ that describes the wave? (c) What is $y(t)$ for a particle at the left end of the string? (d) What is $y(t)$ for a particle 1.35 m to the right of the origin? (e) What is the maximum magnitude of transverse velocity of any particle of the string? (f) Find the transverse displacement and the transverse velocity of a particle 1.35 m to the right of the origin at time $t = 0.0625$ s.

15.50 •• CP A 1750-N irregular beam is hanging horizontally by its ends from the ceiling by two vertical wires (A and B), each 1.25 m long and weighing 0.290 N. The center of gravity of this beam is one-third of the way along the beam from the end where wire A is attached. If you pluck both strings at the same time at the beam, what is the time delay between the arrival of the two pulses at the ceiling? Which pulse arrives first? (Ignore the effect of the weight of the wires on the tension in the wires.)

15.51 •• Three pieces of string, each of length L, are joined together end to end, to make a combined string of length $3L$. The first piece of string has mass per unit length μ_1, the second piece has mass per unit length $\mu_2 = 4\mu_1$, and the third piece has mass per unit length $\mu_3 = \mu_1/4$. (a) If the combined string is under tension F, how much time does it take a transverse wave to travel the entire length $3L$? Give your answer in terms of L, F, and μ_1. (b) Does your answer to part (a) depend on the order in which the three pieces are joined together? Explain.

15.52 •• **Weightless Ant.** An ant with mass m is standing peacefully on top of a horizontal, stretched rope. The rope has mass per unit length μ and is under tension F. Without warning, Cousin Throckmorton starts a sinusoidal transverse wave of wavelength λ propagating along the rope. The motion of the rope is in a vertical plane. What minimum wave amplitude will make the ant become momentarily weightless? Assume that m is so small that the presence of the ant has no effect on the propagation of the wave.

15.53 •• You must determine the length of a long, thin wire that is suspended from the ceiling in the atrium of a tall building. A 2.00-cm-long piece of the wire is left over from its installation.

Using an analytical balance, you determine that the mass of the spare piece is 14.5 μg. You then hang a 0.400-kg mass from the lower end of the long, suspended wire. When a small-amplitude transverse wave pulse is sent up that wire, sensors at both ends measure that it takes the wave pulse 26.7 ms to travel the length of the wire. (a) Use these measurements to calculate the length of the wire. Assume that the weight of the wire has a negligible effect on the speed of the transverse waves. (b) Discuss the accuracy of the approximation made in part (a).

15.54 •• **Music.** You are designing a two-string instrument with metal strings 35.0 cm long, as shown in **Fig. P15.54**. Both strings are under the *same tension*. String S_1 has a mass of 8.00 g and produces the note middle C (frequency 262 Hz) in its fundamental mode. (a) What should be the tension in the string? (b) What should be the mass of string S_2 so that it will produce A-sharp (frequency 466 Hz) as its fundamental? (c) To extend the range of your instrument, you include a fret located just under the strings but not normally touching them. How far from the upper end should you put this fret so that when you press S_1 tightly against it, this string will produce C-sharp (frequency 277 Hz) in its fundamental? That is, what is x in the figure? (d) If you press S_2 against the fret, what frequency of sound will it produce in its fundamental?

Figure **P15.54**

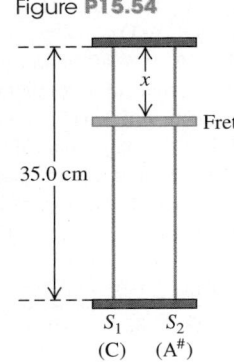

35.0 cm

Fret

x

S_1 S_2
(C)   (A#)

15.55 ••• CP   A 5.00-m, 0.732-kg wire is used to support two uniform 235-N posts of equal length (**Fig. P15.55**). Assume that the wire is essentially horizontal and that the speed of sound is 344 m/s. A strong wind is blowing, causing the wire to vibrate in its 5th overtone. What are the frequency and wavelength of the sound this wire produces?

Figure **P15.55**

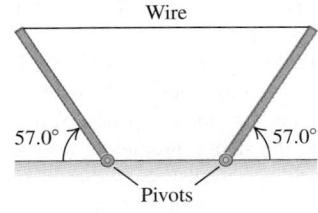

Wire

57.0°   57.0°

Pivots

15.56 ••• CP   You are exploring a newly discovered planet. The radius of the planet is 7.20×10^7 m. You suspend a lead weight from the lower end of a light string that is 4.00 m long and has mass 0.0280 kg. You measure that it takes 0.0685 s for a transverse pulse to travel from the lower end to the upper end of the string. On the earth, for the same string and lead weight, it takes 0.0390 s for a transverse pulse to travel the length of the string. The weight of the string is small enough that you ignore its effect on the tension in the string. Assuming that the mass of the planet is distributed with spherical symmetry, what is its mass?

15.57 •• For a string stretched between two supports, two successive standing-wave frequencies are 525 Hz and 630 Hz. There are other standing-wave frequencies lower than 525 Hz and higher than 630 Hz. If the speed of transverse waves on the string is 384 m/s, what is the length of the string? Assume that the mass of the wire is small enough for its effect on the tension in the wire to be ignored.

15.58 •• A 0.800-m-long string with linear mass density $\mu = 7.50$ g/m is stretched between two supports. The string has tension F and a standing-wave pattern (not the fundamental) of frequency 624 Hz. With the same tension, the next higher standing-wave frequency is 780 Hz. (a) What are the frequency

and wavelength of the fundamental standing wave for this string? (b) What is the value of F?

15.59 ••• CP   A 1.80-m-long uniform bar that weighs 638 N is suspended in a horizontal position by two vertical wires that are attached to the ceiling. One wire is aluminum and the other is copper. The aluminum wire is attached to the left-hand end of the bar, and the copper wire is attached 0.40 m to the left of the right-hand end. Each wire has length 0.600 m and a circular cross section with radius 0.280 mm. What is the fundamental frequency of transverse standing waves for each wire?

15.60 ••• A continuous succession of sinusoidal wave pulses are produced at one end of a very long string and travel along the length of the string. The wave has frequency 70.0 Hz, amplitude 5.00 mm, and wavelength 0.600 m. (a) How long does it take the wave to travel a distance of 8.00 m along the length of the string? (b) How long does it take a point on the string to travel a distance of 8.00 m, once the wave train has reached the point and set it into motion? (c) In parts (a) and (b), how does the time change if the amplitude is doubled?

15.61 •• A horizontal wire is tied to supports at each end and vibrates in its second-overtone standing wave. The tension in the wire is 5.00 N, and the node-to-node distance in the standing wave is 6.28 cm. (a) What is the length of the wire? (b) A point at an antinode of the standing wave on the wire travels from its maximum upward displacement to its maximum downward displacement in 8.40 ms. What is the wire's mass?

15.62 ••• CP   A vertical, 1.20-m length of 18-gauge (diameter of 1.024 mm) copper wire has a 100.0-N ball hanging from it. (a) What is the wavelength of the third harmonic for this wire? (b) A 500.0-N ball now *replaces* the original ball. What is the change in the wavelength of the third harmonic caused by replacing the light ball with the heavy one? (*Hint:* See Table 11.1 for Young's modulus.)

15.63 ••• A sinusoidal transverse wave travels on a string. The string has length 8.00 m and mass 6.00 g. The wave speed is 30.0 m/s, and the wavelength is 0.200 m. (a) If the wave is to have an average power of 50.0 W, what must be the amplitude of the wave? (b) For this same string, if the amplitude and wavelength are the same as in part (a), what is the average power for the wave if the tension is increased such that the wave speed is doubled?

15.64 •• A vibrating string 50.0 cm long is under a tension of 1.00 N. The results from five successive stroboscopic pictures are shown in **Fig. P15.64**. The strobe rate is set at 5000 flashes per minute, and observations reveal that the maximum displacement occurred at flashes 1 and 5 with no other maxima in between. (a) Find the period, frequency, and wavelength for the traveling waves on this string. (b) In what normal mode (harmonic) is the string vibrating? (c) What is the speed of the traveling waves on the string? (d) How fast is point P moving when the string is in (i) position 1 and (ii) position 3? (e) What is the mass of this string?

Figure **P15.64**

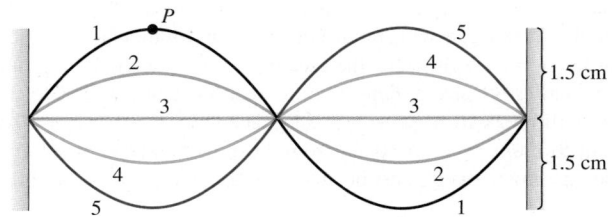

P

1   5

2   4

3   3

4   2

5   1

1.5 cm

1.5 cm

15.65 • **Clothesline Nodes.** Cousin Throckmorton is once again playing with the clothesline in Example 15.2 (Section 15.3). One end of the clothesline is attached to a vertical post. Throcky holds the other end loosely in his hand, so the speed of waves on the clothesline is a relatively slow 0.720 m/s. He finds several frequencies at which he can oscillate his end of the clothesline so that a light clothespin 45.0 cm from the post doesn't move. What are these frequencies?

15.66 •• A strong string of mass 3.00 g and length 2.20 m is tied to supports at each end and is vibrating in its fundamental mode. The maximum transverse speed of a point at the middle of the string is 9.00 m/s. The tension in the string is 330 N. (a) What is the amplitude of the standing wave at its antinode? (b) What is the magnitude of the maximum transverse acceleration of a point at the antinode?

15.67 •• A thin string 2.50 m in length is stretched with a tension of 90.0 N between two supports. When the string vibrates in its first overtone, a point at an antinode of the standing wave on the string has an amplitude of 3.50 cm and a maximum transverse speed of 28.0 m/s. (a) What is the string's mass? (b) What is the magnitude of the maximum transverse acceleration of this point on the string?

15.68 ••• CALC A guitar string is vibrating in its fundamental mode, with nodes at each end. The length of the segment of the string that is free to vibrate is 0.386 m. The maximum transverse acceleration of a point at the middle of the segment is 8.40×10^3 m/s^2 and the maximum transverse velocity is 3.80 m/s. (a) What is the amplitude of this standing wave? (b) What is the wave speed for the transverse traveling waves on this string?

15.69 ••• A uniform cylindrical steel wire, 55.0 cm long and 1.14 mm in diameter, is fixed at both ends. To what tension must it be adjusted so that, when vibrating in its first overtone, it produces the note D-sharp of frequency 311 Hz? Assume that it stretches an insignificant amount. (*Hint:* See Table 12.1.)

15.70 •• A string with both ends held fixed is vibrating in its third harmonic. The waves have a speed of 192 m/s and a frequency of 240 Hz. The amplitude of the standing wave at an antinode is 0.400 cm. (a) Calculate the amplitude at points on the string a distance of (i) 40.0 cm; (ii) 20.0 cm; and (iii) 10.0 cm from the left end of the string. (b) At each point in part (a), how much time does it take the string to go from its largest upward displacement to its largest downward displacement? (c) Calculate the maximum transverse velocity and the maximum transverse acceleration of the string at each of the points in part (a).

15.71 ••• CP A large rock that weighs 164.0 N is suspended from the lower end of a thin wire that is 3.00 m long. The density of the rock is 3200 kg/m^3. The mass of the wire is small enough that its effect on the tension in the wire can be ignored. The upper end of the wire is held fixed. When the rock is in air, the fundamental frequency for transverse standing waves on the wire is 42.0 Hz. When the rock is totally submerged in a liquid, with the top of the rock just below the surface, the fundamental frequency for the wire is 28.0 Hz. What is the density of the liquid?

15.72 • **Holding Up Under Stress.** A string or rope will break apart if it is placed under too much tensile stress [see Eq. (11.8)]. Thicker ropes can withstand more tension without breaking because the thicker the rope, the greater the cross-sectional area and the smaller the stress. One type of steel has density 7800 kg/m^3 and will break if the tensile stress exceeds 7.0×10^8 N/m^2. You want to make a guitar string from 4.0 g of this type of steel. In use, the guitar string must be able to withstand a tension of 900 N

without breaking. Your job is to determine (a) the maximum length and minimum radius the string can have; (b) the highest possible fundamental frequency of standing waves on this string, if the entire length of the string is free to vibrate.

15.73 •• **Tuning an Instrument.** A musician tunes the C-string of her instrument to a fundamental frequency of 65.4 Hz. The vibrating portion of the string is 0.600 m long and has a mass of 14.4 g. (a) With what tension must the musician stretch it? (b) What percent increase in tension is needed to increase the frequency from 65.4 Hz to 73.4 Hz, corresponding to a rise in pitch from C to D?

15.74 •• DATA *Scale length* is the length of the part of a guitar string that is free to vibrate. A standard value of scale length for an acoustic guitar is 25.5 in. The frequency of the fundamental standing wave on a string is determined by the string's scale length, tension, and linear mass density. The standard frequencies f to which the strings of a six-string guitar are tuned are given in the table:

String	E2	A2	D3	G3	B3	E4
f (Hz)	82.4	110.0	146.8	196.0	246.9	329.6

Assume that a typical value of the tension of a guitar string is 78.0 N (although tension varies somewhat for different strings). (a) Calculate the linear mass density μ (in g/cm) for the E2, G3, and E4 strings. (b) Just before your band is going to perform, your G3 string breaks. The only replacement string you have is an E2. If your strings have the linear mass densities calculated in part (a), what must be the tension in the replacement string to bring its fundamental frequency to the G3 value of 196.0 Hz?

15.75 •• DATA In your physics lab, an oscillator is attached to one end of a horizontal string. The other end of the string passes over a frictionless pulley. You suspend a mass M from the free end of the string, producing tension Mg in the string. The oscillator produces transverse waves of frequency f on the string. You don't vary this frequency during the experiment, but you try strings with three different linear mass densities μ. You also keep a fixed distance between the end of the string where the oscillator is attached and the point where the string is in contact with the pulley's rim. To produce standing waves on the string, you vary M; then you measure the node-to-node distance d for each standing-wave pattern and obtain the following data:

String	A	A	B	B	C
μ (g/cm)	0.0260	0.0260	0.0374	0.0374	0.0482
M (g)	559	249	365	207	262
d (cm)	48.1	31.9	32.0	24.2	23.8

(a) Explain why you obtain only certain values of d. (b) Graph μd^2 (in kg·m) versus M (in kg). Explain why the data plotted this way should fall close to a straight line. (c) Use the slope of the best straight-line fit to the data to determine the frequency f of the waves produced on the string by the oscillator. Take $g = 9.80$ m/s^2. (d) For string A ($\mu = 0.0260$ g/cm), what value of M (in grams) would be required to produce a standing wave with a node-to-node distance of 24.0 cm? Use the value of f that you calculated in part (c).

15.76 •• DATA You are measuring the frequency dependence of the average power P_{av} transmitted by traveling waves on a wire. In your experiment you use a wire with linear mass density 3.5 g/m. For a transverse wave on the wire with amplitude 4.0 mm, you measure P_{av} (in watts) as a function of the frequency f of the wave (in Hz). You have chosen to plot P_{av} as a function of f^2 (**Fig. P15.76**). (a) Explain why values of P_{av} plotted versus f^2 should be well fit by a straight line. (b) Use the slope of the straight-line fit to the data shown in Fig P15.76 to calculate the speed of the waves. (c) What angular frequency ω would result in $P_{av} = 10.0$ W?

Figure **P15.76**

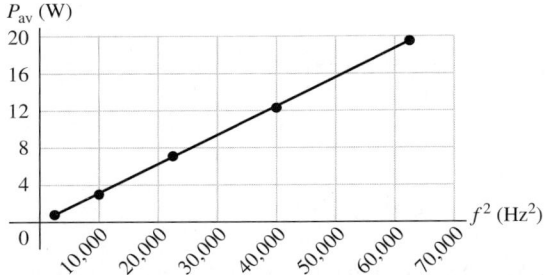

CHALLENGE PROBLEM

15.77 ••• **CP CALC** A deep-sea diver is suspended beneath the surface of Loch Ness by a 100-m-long cable that is attached to a boat on the surface (**Fig. P15.77**). The diver and his suit have a total mass of 120 kg and a volume of 0.0800 m³. The cable has a diameter of 2.00 cm and a linear mass density of $\mu = 1.10$ kg/m. The diver thinks he sees something moving in the murky depths and jerks the end of the cable back and forth to send transverse waves up the cable as a signal to his companions in the boat. (a) What is the tension in the cable at its lower end, where it is attached to the diver? Do not forget to include the buoyant force that the water (density 1000 kg/m³) exerts on him. (b) Calculate the tension in the cable a distance x above the diver. In your calculation, include the buoyant force on the cable. (c) The speed of transverse waves on the cable is given by $v = \sqrt{F/\mu}$ (Eq. 15.14). The speed therefore varies along the cable, since the tension is not constant. (This expression ignores the damping force that the water exerts on the moving cable.) Integrate to find the time required for the first signal to reach the surface.

Figure **P15.77**

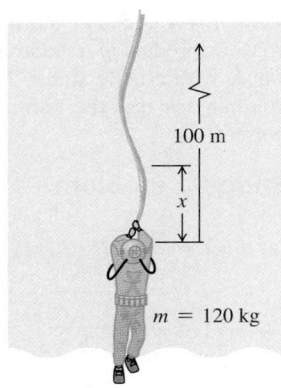

PASSAGE PROBLEMS

BIO WAVES ON VOCAL FOLDS. In the larynx, sound is produced by the vibration of the *vocal folds* (also called "vocal cords"). The accompanying figure is a cross section of the vocal tract at one instant in time. Air flows upward (in the +z-direction) through the vocal tract, causing a transverse wave to propagate vertically upward along the surface of the vocal folds. In a typical adult male, the thickness of the vocal folds in the direction of airflow is $d = 2.0$ mm. High-speed photography shows that for a frequency of vibration of $f = 125$ Hz, the wave along the surface of the vocal folds travels upward at a speed of $v = 375$ cm/s. Use t for time, z for displacement in the +z-direction, and λ for wavelength.

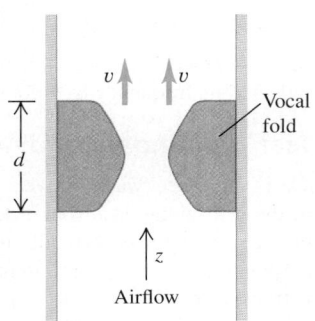

15.78 What is the wavelength of the wave that travels on the surface of the vocal folds when they are vibrating at frequency f? (a) 2.0 mm; (b) 3.3 mm; (c) 0.50 cm; (d) 3.0 cm.

15.79 Which of these is a possible mathematical description of the wave in Problem 15.78? (a) $A\sin[2\pi f(t + z/v)]$; (b) $A\sin[2\pi f(t - z/v)]$; (c) $A\sin(2\pi ft)\cos(2\pi z/\lambda)$; (d) $A\sin(2\pi ft)\sin(2\pi z/\lambda)$.

15.80 The wave speed is measured for different vibration frequencies. A graph of the wave speed as a function of frequency (**Fig. P15.80**) indicates that as the frequency increases, the wavelength (a) increases; (b) decreases; (c) doesn't change; (d) becomes undefined.

Figure **P15.80**

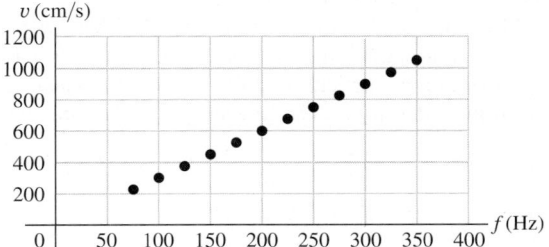

Answers

Chapter Opening Question ?

(iii) The power of a mechanical wave depends on both its amplitude and its frequency [see Eq. (15.25)].

Test Your Understanding Questions

15.1 (i) The "wave" travels horizontally from one spectator to the next along each row of the stadium, but the displacement of each spectator is vertically upward. Since the displacement is perpendicular to the direction in which the wave travels, the wave is transverse.

15.2 (iv) The speed of waves on a string, v, does not depend on the wavelength. We can rewrite the relationship $v = \lambda f$ as $f = v/\lambda$, which tells us that if the wavelength λ doubles, the frequency f becomes one-half as great.

15.3 (a) $\frac{2}{8}T$, **(b)** $\frac{4}{8}T$, **(c)** $\frac{5}{8}T$ Since the wave is sinusoidal, each point on the string oscillates in simple harmonic motion (SHM). Hence we can apply all of the ideas from Chapter 14 about SHM to the wave depicted in Fig. 15.8. (a) A particle in SHM has its maximum speed when it is passing through the equilibrium position ($y = 0$ in Fig. 15.8). The particle at point A is moving upward through this position at $t = \frac{2}{8}T$. (b) In vertical SHM the greatest *upward* acceleration occurs when a particle is at its maximum *downward* displacement. This occurs for the particle at point B at $t = \frac{4}{8}T$. (c) A particle in vertical SHM has a *downward* acceleration when its displacement is *upward*. The particle at C has an upward displacement and is moving downward at $t = \frac{5}{8}T$.

15.4 (ii) The relationship $v = \sqrt{F/\mu}$ [Eq. (15.14)] says that the wave speed is greatest on the string with the smallest linear mass density. This is the thinnest string, which has the smallest amount of mass m and hence the smallest linear mass density $\mu = m/L$ (all strings are the same length).

15.5 (iii), (iv), (ii), (i) Equation (15.25) says that the average power in a sinusoidal wave on a string is $P_{av} = \frac{1}{2}\sqrt{\mu F}\,\omega^2 A^2$. All four strings are identical, so all have the same mass, length, and linear mass density μ. The frequency f is the same for each wave, as is the angular frequency $\omega = 2\pi f$. Hence the average wave power for each string is proportional to the square root of the string tension F and the square of the amplitude A. Compared to string (i), the average power in each string is (ii) $\sqrt{4} = 2$ times greater; (iii) $4^2 = 16$ times greater; and (iv) $\sqrt{2}\,(2)^2 = 4\sqrt{2}$ times greater.

15.6

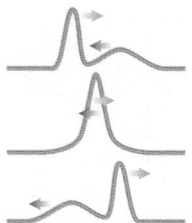

15.7 Yes, yes Doubling the frequency makes the wavelength half as large. Hence the spacing between nodes (equal to $\lambda/2$) is also half as large. There are nodes at all of the previous positions, but there is also a new node between every pair of old nodes.

15.8 $n = 1, 3, 5, \ldots$ When you touch the string at its center, you are producing a node at the center. Hence only standing waves with a node at $x = L/2$ are allowed. From Figure 15.26 you can see that the normal modes $n = 1, 3, 5, \ldots$ cannot be present.

Bridging Problem

(a) $F(r) = \dfrac{m\omega^2}{2L}(L^2 - r^2)$

(b) $v(r) = \omega\sqrt{\dfrac{L^2 - r^2}{2}}$

(c) $\dfrac{\pi}{\omega\sqrt{2}}$

? The sound from a horn travels more slowly on a cold winter day high in the mountains than on a warm summer day at sea level. This is because at high elevations in winter, the air has lower (i) pressure; (ii) density; (iii) humidity; (iv) temperature; (v) mass per mole.

16 SOUND AND HEARING

Of all the mechanical waves that occur in nature, the most important in our everyday lives are longitudinal waves in a medium—usually air—called *sound* waves. The reason is that the human ear is tremendously sensitive and can detect sound waves even of very low intensity. The ability to hear an unseen nocturnal predator was essential to the survival of our ancestors, so it is no exaggeration to say that we humans owe our existence to our highly evolved sense of hearing.

In Chapter 15 we described mechanical waves primarily in terms of displacement; however, because the ear is primarily sensitive to changes in pressure, it's often more appropriate to describe sound waves in terms of *pressure* fluctuations. We'll study the relationships among displacement, pressure fluctuation, and intensity and the connections between these quantities and human sound perception.

When a source of sound or a listener moves through the air, the listener may hear a frequency different from the one emitted by the source. This is the Doppler effect, which has important applications in medicine and technology.

16.1 SOUND WAVES

The most general definition of **sound** is a longitudinal wave in a medium. Our main concern is with sound waves in air, but sound can travel through any gas, liquid, or solid. You may be all too familiar with the propagation of sound through a solid if your neighbor's stereo speakers are right next to your wall.

The simplest sound waves are sinusoidal waves, which have definite frequency, amplitude, and wavelength. The human ear is sensitive to waves in the frequency range from about 20 to 20,000 Hz, called the **audible range,** but we also use the term "sound" for similar waves with frequencies above (**ultrasonic**) and below (**infrasonic**) the range of human hearing.

16.1 A sinusoidal longitudinal wave traveling to the right in a fluid. (Compare to Fig. 15.7.)

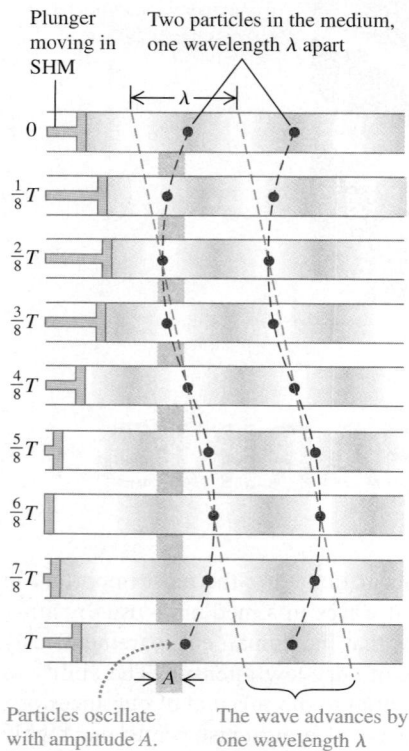

Longitudinal waves are shown at intervals of $\frac{1}{8}T$ for one period T.

Plunger moving in SHM

Two particles in the medium, one wavelength λ apart

Particles oscillate with amplitude A.

The wave advances by one wavelength λ during each period T.

Sound waves usually travel outward in all directions from the source of sound, with an amplitude that depends on the direction and distance from the source. We'll return to this point in the next section. For now, we concentrate on the idealized case of a sound wave that propagates in the positive x-direction only. As we discussed in Section 15.3, for such a wave, the wave function $y(x, t)$ gives the instantaneous displacement y of a particle in the medium at position x at time t. If the wave is sinusoidal, we can express it by using Eq. (15.7):

$$y(x, t) = A\cos(kx - \omega t) \qquad \text{(sound wave propagating in the +x-direction)} \qquad (16.1)$$

In a longitudinal wave the displacements are *parallel* to the direction of travel of the wave, so distances x and y are measured parallel to each other, not perpendicular as in a transverse wave. The amplitude A is the maximum displacement of a particle in the medium from its equilibrium position (**Fig. 16.1**). Hence A is also called the **displacement amplitude.**

Sound Waves As Pressure Fluctuations

We can also describe sound waves in terms of variations of *pressure* at various points. In a sinusoidal sound wave in air, the pressure fluctuates sinusoidally above and below atmospheric pressure p_a with the same frequency as the motions of the air particles. The human ear operates by sensing such pressure variations. A sound wave entering the ear canal exerts a fluctuating pressure on one side of the eardrum; the air on the other side of the eardrum, vented to the outside by the Eustachian tube, is at atmospheric pressure. The pressure difference on the two sides of the eardrum sets it into motion. Microphones and similar devices also usually sense pressure differences, not displacements.

Let $p(x, t)$ be the instantaneous pressure fluctuation in a sound wave at any point x at time t. That is, $p(x, t)$ is the amount by which the pressure *differs* from normal atmospheric pressure p_a. Think of $p(x, t)$ as the *gauge pressure* defined in Section 12.2; it can be either positive or negative. The *absolute* pressure at a point is then $p_a + p(x, t)$.

To see the connection between the pressure fluctuation $p(x, t)$ and the displacement $y(x, t)$ in a sound wave propagating in the +x-direction, consider an imaginary cylinder of a wave medium (gas, liquid, or solid) with cross-sectional area S and its axis along the direction of propagation (**Fig. 16.2**). When no sound wave is present, the cylinder has length Δx and volume $V = S\,\Delta x$, as shown by the shaded volume in Fig. 16.2. When a wave is present, at time t the end of the cylinder that is initially at x is displaced by $y_1 = y(x, t)$, and the end that is initially at $x + \Delta x$ is displaced by $y_2 = y(x + \Delta x, t)$; this is shown by the red lines. If $y_2 > y_1$, as shown in Fig. 16.2, the cylinder's volume has increased, which causes a decrease in pressure. If $y_2 < y_1$, the cylinder's volume has decreased and the pressure has increased. If $y_2 = y_1$, the cylinder is simply shifted to the left or right; there is no volume change and no pressure fluctuation. The pressure fluctuation depends on the *difference* between the displacements at neighboring points in the medium.

Quantitatively, the change in volume ΔV of the cylinder is

$$\Delta V = S(y_2 - y_1) = S[y(x + \Delta x, t) - y(x, t)]$$

In the limit as $\Delta x \to 0$, the fractional change in volume dV/V (volume change divided by original volume) is

16.2 As a sound wave propagates along the x-axis, the left and right ends undergo different displacements y_1 and y_2.

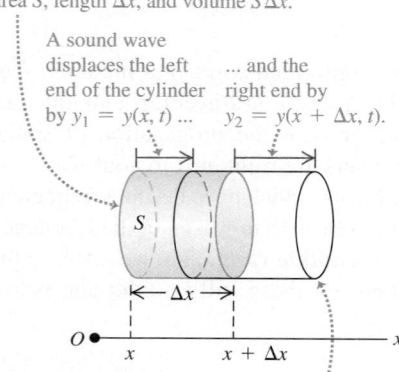

Undisturbed cylinder of fluid has cross-sectional area S, length Δx, and volume $S\Delta x$.

A sound wave displaces the left end of the cylinder by $y_1 = y(x, t)$ ...

... and the right end by $y_2 = y(x + \Delta x, t)$.

The change in volume of the disturbed cylinder of fluid is $S(y_2 - y_1)$.

$$\frac{dV}{V} = \lim_{\Delta x \to 0} \frac{S[y(x + \Delta x, t) - y(x, t)]}{S\,\Delta x} = \frac{\partial y(x, t)}{\partial x} \qquad (16.2)$$

16.3 Three ways to describe a sound wave.

(a) A graph of displacement y versus position x at $t = 0$

Where $y > 0$, particles are displaced to the right.

Where $y < 0$, particles are displaced to the left.

Undisplaced particles

(b) A cartoon showing the displacement of individual particles in the fluid at $t = 0$

Displaced particles

Rarefaction: particles pulled apart; pressure is most negative.

Compression: particles pile up; pressure is most positive.

(c) A graph of pressure fluctuation p versus position x at $t = 0$

The fractional volume change is related to the pressure fluctuation by the bulk modulus B, which by definition [Eq. (11.13)] is $B = -p(x, t)/(dV/V)$ (see Section 11.4). Solving for $p(x, t)$, we have

$$p(x, t) = -B\frac{\partial y(x, t)}{\partial x} \tag{16.3}$$

The negative sign arises because when $\partial y(x, t)/\partial x$ is positive, the displacement is greater at $x + \Delta x$ than at x, corresponding to an increase in volume, a decrease in pressure, and a negative pressure fluctuation.

When we evaluate $\partial y(x, t)/\partial x$ for the sinusoidal wave of Eq. (16.1), we find

$$p(x, t) = BkA \sin(kx - \omega t) \tag{16.4}$$

Figure 16.3 shows $y(x, t)$ and $p(x, t)$ for a sinusoidal sound wave at $t = 0$. It also shows how individual particles of the wave are displaced at this time. While $y(x, t)$ and $p(x, t)$ describe the same wave, these two functions are one-quarter cycle out of phase: At any time, the displacement is greatest where the pressure fluctuation is zero, and vice versa. In particular, note that the compressions (points of greatest pressure and density) and rarefactions (points of lowest pressure and density) are points of *zero* displacement.

Equation (16.4) shows that the quantity BkA represents the maximum pressure fluctuation. We call this the **pressure amplitude,** denoted by p_{max}:

Pressure amplitude, sinusoidal sound wave

Bulk modulus of medium

$$p_{max} = BkA$$

Displacement amplitude

Wave number $= 2\pi/\lambda$ $\tag{16.5}$

Waves of shorter wavelength λ (larger wave number $k = 2\pi/\lambda$) have greater pressure variations for a given displacement amplitude because the maxima and minima are squeezed closer together. A medium with a large value of bulk modulus B is less compressible and so requires a greater pressure amplitude for a given volume change (that is, a given displacement amplitude).

CAUTION **Graphs of a sound wave** The graphs in Fig. 16.3 show the wave at only *one* instant of time. Because the wave is propagating in the $+x$-direction, as time goes by the wave patterns described by the functions $y(x, t)$ and $p(x, t)$ move to the right at the wave speed $v = \omega/k$. The particles, by contrast, simply oscillate back and forth in simple harmonic motion as shown in Fig. 16.1. ∎

EXAMPLE 16.1 AMPLITUDE OF A SOUND WAVE IN AIR

In a sinusoidal sound wave of moderate loudness, the maximum pressure variations are about 3.0×10^{-2} Pa above and below atmospheric pressure. Find the corresponding maximum displacement if the frequency is 1000 Hz. In air at normal atmospheric pressure and density, the speed of sound is 344 m/s and the bulk modulus is 1.42×10^5 Pa.

SOLUTION

IDENTIFY and SET UP: This problem involves the relationship between two ways of describing a sound wave: in terms of displacement and in terms of pressure. The target variable is the displacement amplitude A. We are given the pressure amplitude p_{max}, wave speed v, frequency f, and bulk modulus B. Equation (16.5) relates the target variable A to p_{max}. We use $\omega = vk$

[Eq. (15.6)] to determine the wave number k from v and the angular frequency $\omega = 2\pi f$.

EXECUTE: From Eq. (15.6),

$$k = \frac{\omega}{v} = \frac{2\pi f}{v} = \frac{(2\pi \text{ rad})(1000 \text{ Hz})}{344 \text{ m/s}} = 18.3 \text{ rad/m}$$

Then from Eq. (16.5), the maximum displacement is

$$A = \frac{p_{max}}{Bk} = \frac{3.0 \times 10^{-2} \text{ Pa}}{(1.42 \times 10^5 \text{ Pa})(18.3 \text{ rad/m})} = 1.2 \times 10^{-8} \text{ m}$$

EVALUATE: This displacement amplitude is only about $\frac{1}{100}$ the size of a human cell. The ear actually senses pressure fluctuations; it detects these minuscule displacements only indirectly.

EXAMPLE 16.2 AMPLITUDE OF A SOUND WAVE IN THE INNER EAR

A sound wave that enters the human ear sets the eardrum into oscillation, which in turn causes oscillation of the *ossicles*, a chain of three tiny bones in the middle ear (**Fig. 16.4**). The ossicles transmit this oscillation to the fluid (mostly water) in the inner ear; there the fluid motion disturbs hair cells that send nerve impulses to the brain with information about the sound. The area of the moving part of the eardrum is about 43 mm², and that of the stapes (the smallest of the ossicles) where it connects to the inner ear is about 3.2 mm². For the sound in Example 16.1, determine (a) the pressure amplitude and (b) the displacement amplitude of the wave in the fluid of the inner ear, in which the speed of sound is 1500 m/s.

SOLUTION

IDENTIFY and SET UP: Although the sound wave here travels in liquid rather than air, the same principles and relationships among the properties of the wave apply. We can ignore the mass of the tiny ossicles (about 58 mg = 5.8×10^{-5} kg), so the force they exert on the inner-ear fluid is the same as that exerted on the eardrum and ossicles by the incident sound wave. (In Chapters 4 and 5 we used the same idea to say that the tension is the same at either end of a massless rope.) Hence the pressure amplitude in the inner ear, $p_{max \text{ (inner ear)}}$, is greater than in the outside air, $p_{max \text{ (air)}}$, because the same force is exerted on a smaller area (the area of the stapes versus the area of the eardrum). Given $p_{max \text{ (inner ear)}}$, we find the displacement amplitude $A_{inner \text{ ear}}$ from Eq. (16.5).

EXECUTE: (a) From the area of the eardrum and the pressure amplitude in air found in Example 16.1, the maximum force exerted by the sound wave in air on the eardrum is $F_{max} = p_{max \text{ (air)}} S_{eardrum}$. Then

$$p_{max \text{ (inner ear)}} = \frac{F_{max}}{S_{stapes}} = p_{max \text{ (air)}} \frac{S_{eardrum}}{S_{stapes}}$$

$$= (3.0 \times 10^{-2} \text{ Pa}) \frac{43 \text{ mm}^2}{3.2 \text{ mm}^2} = 0.40 \text{ Pa}$$

16.4 The anatomy of the human ear. The middle ear is the size of a small marble; the ossicles (incus, malleus, and stapes) are the smallest bones in the human body.

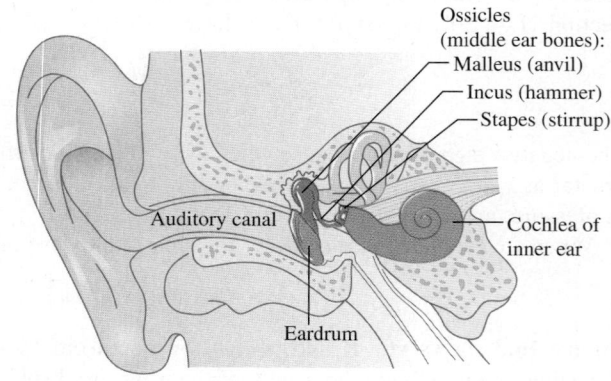

(b) To find the maximum displacement $A_{inner \text{ ear}}$, we use $A = p_{max}/Bk$ as in Example 16.1. The inner-ear fluid is mostly water, which has a much greater bulk modulus B than air. From Table 11.2 the compressibility of water (unfortunately also called k) is 45.8×10^{-11} Pa⁻¹, so $B_{fluid} = 1/(45.8 \times 10^{-11} \text{ Pa}^{-1}) = 2.18 \times 10^9$ Pa.

The wave in the inner ear has the same angular frequency ω as the wave in the air because the air, eardrum, ossicles, and inner-ear fluid all oscillate together (see Example 15.8 in Section 15.8). But because the wave speed v is greater in the inner ear than in the air (1500 m/s versus 344 m/s), the wave number $k = \omega/v$ is smaller. Using the value of ω from Example 16.1,

$$k_{inner \text{ ear}} = \frac{\omega}{v_{inner \text{ ear}}} = \frac{(2\pi \text{ rad})(1000 \text{ Hz})}{1500 \text{ m/s}} = 4.2 \text{ rad/m}$$

Putting everything together, we have

$$A_{\text{inner ear}} = \frac{p_{\max(\text{inner ear})}}{B_{\text{fluid}} k_{\text{inner ear}}} = \frac{0.40\ \text{Pa}}{(2.18 \times 10^9\ \text{Pa})(4.2\ \text{rad/m})}$$

$$= \frac{0.40\ \text{Pa}}{(2.18 \times 10^9\ \text{Pa})(4.2\ \text{rad/m})}$$

$$= 4.4 \times 10^{-11}\ \text{m}$$

EVALUATE: In part (a) we see that the ossicles increase the pressure amplitude by a factor of $(43\ \text{mm}^2)/(3.2\ \text{mm}^2) = 13$. This amplification helps give the human ear its great sensitivity.

The displacement amplitude in the inner ear is even smaller than in the air. But *pressure* variations within the inner-ear fluid are what set the hair cells into motion, so what matters is that the pressure amplitude is larger in the inner ear than in the air.

Perception of Sound Waves

The physical characteristics of a sound wave are directly related to the perception of that sound by a listener. For a given frequency, the greater the pressure amplitude of a sinusoidal sound wave, the greater the perceived **loudness.** The relationship between pressure amplitude and loudness is not a simple one, and it varies from one person to another. One important factor is that the ear is not equally sensitive to all frequencies in the audible range. A sound at one frequency may seem louder than one of equal pressure amplitude at a different frequency. At 1000 Hz the minimum pressure amplitude that can be perceived with normal hearing is about 3×10^{-5} Pa; to produce the same loudness at 200 Hz or 15,000 Hz requires about 3×10^{-4} Pa. Perceived loudness also depends on the health of the ear. Age usually brings a loss of sensitivity at high frequencies.

The frequency of a sound wave is the primary factor in determining the **pitch** of a sound, the quality that lets us classify the sound as "high" or "low." The higher the frequency of a sound (within the audible range), the higher the pitch that a listener will perceive. Pressure amplitude also plays a role in determining pitch. When a listener compares two sinusoidal sound waves with the same frequency but different pressure amplitudes, the one with the greater pressure amplitude is usually perceived as louder but also as slightly lower in pitch.

Musical sounds have wave functions that are more complicated than a simple sine function. **Figure 16.5a** shows the pressure fluctuation in the sound wave produced by a clarinet. The pattern is so complex because the column of air in a wind instrument like a clarinet vibrates at a fundamental frequency and at many harmonics at the same time. (In Section 15.8, we described this same behavior for a string that has been plucked, bowed, or struck. We'll examine the physics of wind instruments in Section 16.4.) The sound wave produced in the surrounding air has a similar amount of each harmonic—that is, a similar *harmonic content.* Figure 16.5b shows the harmonic content of the sound of a clarinet. The mathematical process of translating a pressure–time graph like Fig. 16.5a into a graph of harmonic content like Fig. 16.5b is called *Fourier analysis.*

Two tones produced by different instruments might have the same fundamental frequency (and thus the same pitch) but sound different because of different harmonic content. The difference in sound is called **timbre** and is often described in subjective terms such as reedy, mellow, and tinny. A tone that is rich in harmonics, like the clarinet tone in Figs. 16.5a and 16.5b, usually sounds thin and "reedy," while a tone containing mostly a fundamental, like the alto recorder tone in Figs. 16.5c and 16.5d, is more mellow and flutelike. The same principle applies to the human voice, which is another wind instrument; the vowels "a" and "e" sound different because of differences in harmonic content.

Another factor in determining tone quality is the behavior at the beginning (*attack*) and end (*decay*) of a tone. A piano tone begins with a thump and then dies away gradually. A harpsichord tone, in addition to having different harmonic content, begins much more quickly with a click, and the higher harmonics begin before the lower ones. When the key is released, the sound also dies away much more rapidly with a harpsichord than with a piano. Similar effects are present in other musical instruments.

16.5 Different representations of the sound of (a), (b) a clarinet and (c), (d) an alto recorder. (Graphs adapted from R.E. Berg and D.G. Stork, *The Physics of Sound,* Prentice-Hall, 1982.)

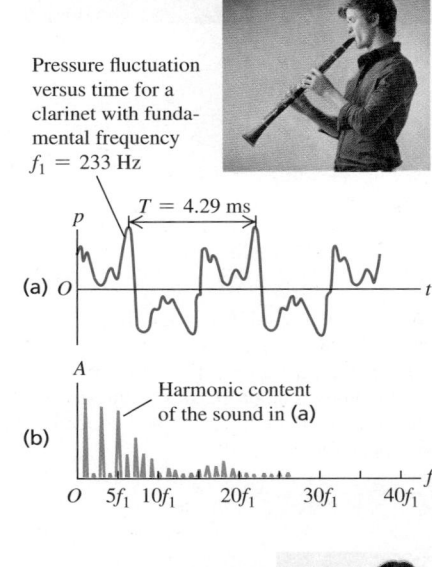

Pressure fluctuation versus time for a clarinet with fundamental frequency $f_1 = 233$ Hz

(a)

(b) Harmonic content of the sound in (a)

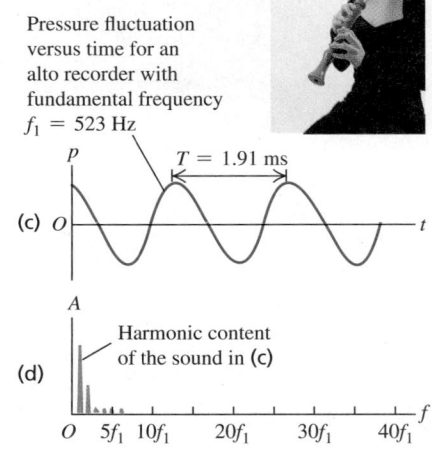

Pressure fluctuation versus time for an alto recorder with fundamental frequency $f_1 = 523$ Hz

(c)

(d) Harmonic content of the sound in (c)

Unlike the tones made by musical instruments, **noise** is a combination of *all* frequencies, not just frequencies that are integer multiples of a fundamental frequency. (An extreme case is "white noise," which contains equal amounts of all frequencies across the audible range.) Examples include the sound of the wind and the hissing sound you make in saying the consonant "s."

TEST YOUR UNDERSTANDING OF SECTION 16.1   You use an electronic signal generator to produce a sinusoidal sound wave in air. You then increase the frequency of the wave from 100 Hz to 400 Hz while keeping the pressure amplitude constant. What effect does this have on the displacement amplitude of the sound wave? (i) It becomes four times greater; (ii) it becomes twice as great; (iii) it is unchanged; (iv) it becomes $\frac{1}{2}$ as great; (v) it becomes $\frac{1}{4}$ as great. ∎

16.2 SPEED OF SOUND WAVES

We found in Section 15.4 that the speed v of a transverse wave on a string depends on the string tension F and the linear mass density μ: $v = \sqrt{F/\mu}$. What, we may ask, is the corresponding expression for the speed of sound waves in a gas or liquid? On what properties of the medium does the speed depend?

We can make an educated guess about these questions by remembering a claim that we made in Section 15.4: For mechanical waves in general, the expression for the wave speed is of the form

$$v = \sqrt{\frac{\text{Restoring force returning the system to equilibrium}}{\text{Inertia resisting the return to equilibrium}}}$$

A sound wave in a bulk fluid causes compressions and rarefactions of the fluid, so the restoring-force term in the above expression must be related to how difficult it is to compress the fluid. This is precisely what the bulk modulus B of the medium tells us. According to Newton's second law, inertia is related to mass. The "massiveness" of a bulk fluid is described by its density, or mass per unit volume, ρ. Hence we expect that the speed of sound waves should be of the form $v = \sqrt{B/\rho}$.

To check our guess, we'll derive the speed of sound waves in a fluid in a pipe. This is a situation of some importance, since all musical wind instruments are pipes in which a longitudinal wave (sound) propagates in a fluid (air) (**Fig. 16.6**). Human speech works on the same principle; sound waves propagate in your vocal tract, which is an air-filled pipe connected to the lungs at one end (your larynx) and to the outside air at the other end (your mouth). The steps in our derivation are completely parallel to those we used in Section 15.4 to find the speed of transverse waves.

16.6 When a wind instrument like this French horn is played, sound waves propagate through the air within the instrument's pipes. The properties of the sound that emerges from the large bell depend on the speed of these waves.

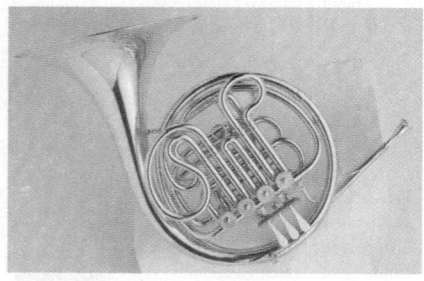

Speed of Sound in a Fluid

Figure 16.7 shows a fluid with density ρ in a pipe with cross-sectional area A. In equilibrium (Fig. 16.7a), the fluid is at rest and under a uniform pressure p. We take the x-axis along the length of the pipe. This is also the direction in which we make a longitudinal wave propagate, so the displacement y is also measured along the pipe, just as in Section 16.1 (see Fig. 16.2).

At time $t = 0$ we start the piston at the left end moving toward the right with constant speed v_y. This initiates a wave motion that travels to the right along the length of the pipe, in which successive sections of fluid begin to move and become compressed at successively later times.

Figure 16.7b shows the fluid at time t. All portions of fluid to the left of point P are moving to the right with speed v_y, and all portions to the right of P are still at rest. The boundary between the moving and stationary portions travels to the right with a speed equal to the speed of propagation or wave speed v. At time t the piston has moved a distance $v_y t$, and the boundary has advanced a distance vt. As with a transverse disturbance in a string, we can compute the speed of propagation from the impulse–momentum theorem.

The quantity of fluid set in motion in time t originally occupied a section of the cylinder with length vt, cross-sectional area A, volume vtA, and mass ρvtA. Its longitudinal momentum (that is, momentum along the length of the pipe) is

$$\text{Longitudinal momentum} = (\rho vtA)v_y$$

Next we compute the increase of pressure, Δp, in the moving fluid. The original volume of the moving fluid, Avt, has decreased by an amount Av_yt. From the definition of the bulk modulus B, Eq. (11.13) in Section 11.5.

$$B = \frac{-\text{Pressure change}}{\text{Fractional volume change}} = \frac{-\Delta p}{-Av_yt/Avt} \quad \text{and} \quad \Delta p = B\frac{v_y}{v}$$

The pressure in the moving fluid is $p + \Delta p$, and the force exerted on it by the piston is $(p + \Delta p)A$. The net force on the moving fluid (see Fig. 16.7b) is ΔpA, and the longitudinal impulse is

$$\text{Longitudinal impulse} = \Delta pAt = B\frac{v_y}{v}At$$

Because the fluid was at rest at time $t = 0$, the change in momentum up to time t is equal to the momentum at that time. Applying the impulse–momentum theorem (see Section 8.1), we find

$$B\frac{v_y}{v}At = \rho vtAv_y \tag{16.6}$$

When we solve this expression for v, we get

> Speed of a longitudinal wave in a fluid $\quad v = \sqrt{\dfrac{B}{\rho}} \quad$ ⋯→ Bulk modulus of fluid
> ⋯→ Density of fluid $\tag{16.7}$

which agrees with our educated guess.

While we derived Eq. (16.7) for waves in a pipe, it also applies to longitudinal waves in a bulk fluid, including sound waves traveling in air or water.

Speed of Sound in a Solid

When a longitudinal wave propagates in a *solid* rod or bar, the situation is somewhat different. The rod expands sideways slightly when it is compressed longitudinally, while a fluid in a pipe with constant cross section cannot move sideways. Using the same kind of reasoning that led us to Eq. (16.7), we can show that the speed of a longitudinal pulse in the rod is given by

> Speed of a longitudinal wave in a solid rod $\quad v = \sqrt{\dfrac{Y}{\rho}} \quad$ ⋯→ Young's modulus of rod material
> ⋯→ Density of rod material $\tag{16.8}$

We defined Young's modulus in Section 11.4.

CAUTION **Solid rods vs. bulk solids** Equation (16.8) applies to only rods whose sides are free to bulge and shrink a little as the wave travels. It does not apply to longitudinal waves in a *bulk* solid because sideways motion in any element of material is prevented by the surrounding material. The speed of longitudinal waves in a bulk solid depends on the density, the bulk modulus, and the *shear* modulus. ▮

Note that Eqs. (16.7) and (16.8) are valid for sinusoidal and other periodic waves, not just for the special case discussed here.

Table 16.1 lists the speed of sound in several bulk materials. Sound waves travel more slowly in lead than in aluminum or steel because lead has a lower bulk modulus and shear modulus and a higher density.

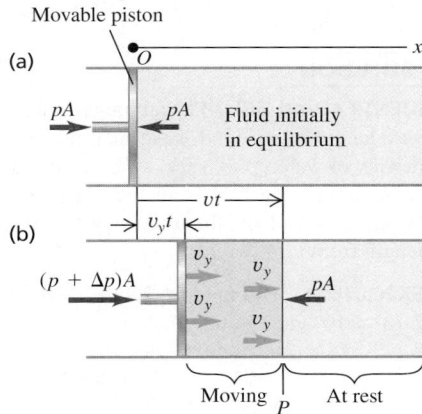

16.7 A sound wave propagating in a fluid confined to a tube. (a) Fluid in equilibrium. (b) A time t after the piston begins moving to the right at speed v_y, the fluid between the piston and point P is in motion. The speed of sound waves is v.

TABLE 16.1	Speed of Sound in Various Bulk Materials
Material	**Speed of Sound (m/s)**
Gases	
Air (20°C)	344
Helium (20°C)	999
Hydrogen (20°C)	1330
Liquids	
Liquid helium (4 K)	211
Mercury (20°C)	1451
Water (0°C)	1402
Water (20°C)	1482
Water (100°C)	1543
Solids	
Aluminum	6420
Lead	1960
Steel	5941

EXAMPLE 16.3   **WAVELENGTH OF SONAR WAVES**

A ship uses a sonar system (**Fig. 16.8**) to locate underwater objects. Find the speed of sound waves in water using Eq. (16.7), and find the wavelength of a 262-Hz wave.

SOLUTION

IDENTIFY and SET UP: Our target variables are the speed and wavelength of a sound wave in water. In Eq. (16.7), we use the density of water, $\rho = 1.00 \times 10^3 \text{ kg/m}^3$, and the bulk modulus of water, which we find from the compressibility (see Table 11.2). Given the speed and the frequency $f = 262$ Hz, we find the wavelength from $v = f\lambda$.

EXECUTE: In Example 16.2, we used Table 11.2 to find $B = 2.18 \times 10^9$ Pa. Then

$$v = \sqrt{\frac{B}{\rho}} = \sqrt{\frac{2.18 \times 10^9 \text{ Pa}}{1.00 \times 10^3 \text{ kg/m}^3}} = 1480 \text{ m/s}$$

and

$$\lambda = \frac{v}{f} = \frac{1480 \text{ m/s}}{262 \text{ s}^{-1}} = 5.65 \text{ m}$$

EVALUATE: The calculated value of v agrees well with the value in Table 16.1. Water is denser than air (ρ is larger) but is also

16.8 A sonar system uses underwater sound waves to detect and locate submerged objects.

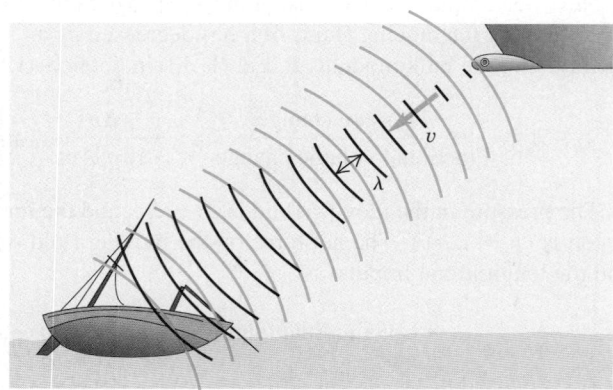

much more incompressible (B is much larger), and so the speed $v = \sqrt{B/\rho}$ is greater than the 344-m/s speed of sound in air at ordinary temperatures. The relationship $\lambda = v/f$ then says that a sound wave in water must have a longer wavelength than a wave of the same frequency in air. Indeed, we found in Example 15.1 (Section 15.2) that a 262-Hz sound wave in air has a wavelength of only 1.31 m.

16.9 This three-dimensional image of a fetus in the womb was made using a sequence of ultrasound scans. Each individual scan reveals a two-dimensional "slice" through the fetus; many such slices were then combined digitally. Ultrasound imaging is also used to study heart valve action and to detect tumors.

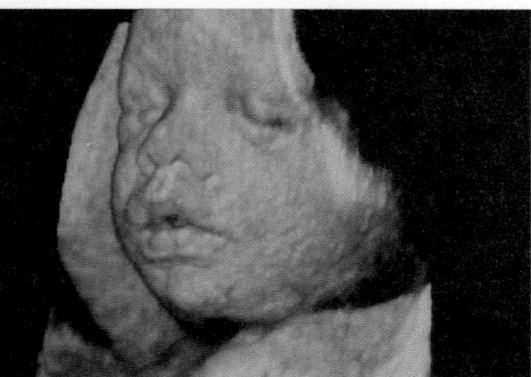

Dolphins emit high-frequency sound waves (typically 100,000 Hz) and use the echoes for guidance and for hunting. The corresponding wavelength in water is 1.48 cm. With this high-frequency "sonar" system they can sense objects that are roughly as small as the wavelength (but not much smaller). *Ultrasonic imaging* in medicine uses the same principle; sound waves of very high frequency and very short wavelength, called *ultrasound,* are scanned over the human body, and the "echoes" from interior organs are used to create an image. With ultrasound of frequency 5 MHz $= 5 \times 10^6$ Hz, the wavelength in water (the primary constituent of the body) is 0.3 mm, and features as small as this can be discerned in the image (**Fig. 16.9**). Ultrasound is more sensitive than x rays in distinguishing various kinds of tissues and does not have the radiation hazards associated with x rays.

Speed of Sound in a Gas

Most of the sound waves that we encounter propagate in air. To use Eq. (16.7) to find the speed of sound waves in air, we note that the bulk modulus of a gas depends on pressure: The greater the pressure applied to compress a gas, the more it resists further compression and hence the greater the bulk modulus. (That's why specific values of the bulk modulus for gases are not given in Table 11.1.) The expression for the bulk modulus of a gas for use in Eq. (16.7) is

$$B = \gamma p_0 \tag{16.9}$$

where p_0 is the equilibrium pressure of the gas. The quantity γ (the Greek letter gamma) is called the *ratio of heat capacities.* It is a dimensionless number that characterizes the thermal properties of the gas. (We'll learn more about this quantity in Chapter 19.) As an example, the ratio of heat capacities for air is $\gamma = 1.40$. At normal atmospheric pressure $p_0 = 1.013 \times 10^5$ Pa, so $B = (1.40)(1.013 \times 10^5 \text{ Pa}) = 1.42 \times 10^5$ Pa. This value is minuscule compared to the bulk modulus of a typical solid (see Table 11.1), which is approximately 10^{10} to 10^{11} Pa. This shouldn't be surprising: It's simply a statement that air is far easier to compress than steel.

The density ρ of a gas also depends on the pressure, which in turn depends on the temperature. It turns out that the ratio B/ρ for a given type of ideal gas does *not* depend on the pressure at all, only the temperature. From Eq. (16.7), this means that the speed of sound in a gas is fundamentally a function of temperature T:

Ratio of heat capacities ···. ···· Gas constant

Speed of sound ········· in an ideal gas $$v = \sqrt{\frac{\gamma RT}{M}}$$ ····· Absolute temperature ····· Molar mass (16.10)

This expression incorporates several quantities that we will study in Chapters 17, 18, and 19. The temperature T is the *absolute* temperature in kelvins (K), equal to the Celsius temperature plus 273.15; thus 20.00°C corresponds to $T = 293.15$ K. The quantity M is the *molar mass,* or mass per mole of the substance of which the gas is composed. The *gas constant R* has the same value for all gases. The current best numerical value of R is

$$R = 8.3144621(75) \text{ J/mol} \cdot \text{K}$$

which for practical calculations we can write as 8.314 J/mol·K.

For any particular gas, γ, R, and M are constants, and the wave speed is proportional to the square root of the absolute temperature. We will see in Chapter 18 that Eq. (16.10) is almost identical to the expression for the average speed of molecules in an ideal gas. This shows that sound speeds and molecular speeds are closely related.

EXAMPLE 16.4 **SPEED OF SOUND IN AIR**

Find the speed of sound in air at $T = 20°C$, and find the range of wavelengths in air to which the human ear (which can hear frequencies in the range of 20–20,000 Hz) is sensitive. The mean molar mass for air (a mixture of mostly nitrogen and oxygen) is $M = 28.8 \times 10^{-3}$ kg/mol and the ratio of heat capacities is $\gamma = 1.40$.

SOLUTION

IDENTIFY and SET UP: We use Eq. (16.10) to find the sound speed from γ, T, and M, and we use $v = f\lambda$ to find the wavelengths corresponding to the frequency limits. Note that in Eq. (16.10) temperature T *must* be expressed in kelvins, not Celsius degrees.

EXECUTE: At $T = 20°C = 293$ K, we find

$$v = \sqrt{\frac{\gamma RT}{M}} = \sqrt{\frac{(1.40)(8.314 \text{ J/mol} \cdot \text{K})(293 \text{ K})}{28.8 \times 10^{-3} \text{ kg/mol}}} = 344 \text{ m/s}$$

Using this value of v in $\lambda = v/f$, we find that at 20°C the frequency $f = 20$ Hz corresponds to $\lambda = 17$ m and $f = 20,000$ Hz to $\lambda = 1.7$ cm.

EVALUATE: Our calculated value of v agrees with the measured sound speed at $T = 20°C$.

A gas is actually composed of molecules in random motion, separated by distances that are large in comparison with their diameters. The vibrations that constitute a wave in a gas are superposed on the random thermal motion. At atmospheric pressure, a molecule travels an average distance of about 10^{-7} m between collisions, while the displacement amplitude of a faint sound may be only 10^{-9} m. We can think of a gas with a sound wave passing through as being comparable to a swarm of bees; the swarm as a whole oscillates slightly while individual insects move about within the swarm, apparently at random.

TEST YOUR UNDERSTANDING OF SECTION 16.2 Mercury is 13.6 times denser than water. Based on Table 16.1, at 20°C which of these liquids has the greater bulk modulus? (i) Mercury; (ii) water; (iii) both are about the same; (iv) not enough information is given to decide. ∎

16.3 SOUND INTENSITY

Traveling sound waves, like all other traveling waves, transfer energy from one region of space to another. In Section 15.5 we introduced the *wave intensity I,* equal to the time average rate at which wave energy is transported per unit area across a surface perpendicular to the direction of propagation. Let's see how to express the intensity of a sound wave in a fluid in terms of the displacement amplitude A or pressure amplitude p_{max}.

Let's consider a sound wave propagating in the $+x$-direction so that we can use our expressions from Section 16.1 for the displacement $y(x, t)$ [Eq. (16.1)] and pressure fluctuation $p(x, t)$ [Eq. (16.4)]. In Section 6.4 we saw that power equals the product of force and velocity [see Eq. (6.18)]. So the power per unit area in this sound wave equals the product of $p(x, t)$ (force per unit area) and the *particle* velocity $v_y(x, t)$, which is the velocity at time t of that portion of the wave medium at coordinate x. Using Eqs. (16.1) and (16.4), we find

$$v_y(x, t) = \frac{\partial y(x, t)}{\partial t} = \omega A \sin(kx - \omega t)$$

$$p(x, t)v_y(x, t) = [BkA \sin(kx - \omega t)][\omega A \sin(kx - \omega t)]$$

$$= B\omega kA^2 \sin^2(kx - \omega t)$$

CAUTION **Wave velocity vs. particle velocity** Remember that the velocity of the wave as a whole is *not* the same as the particle velocity. While the wave continues to move in the direction of propagation, individual particles in the wave medium merely slosh back and forth, as shown in Fig. 16.1. Furthermore, the maximum speed of a particle of the medium can be very different from the wave speed. ❚

The intensity is the time average value of the power per unit area $p(x, t)v_y(x, t)$. For any value of x the average value of the function $\sin^2(kx - \omega t)$ over one period $T = 2\pi/\omega$ is $\frac{1}{2}$, so

$$I = \frac{1}{2}B\omega kA^2 \tag{16.11}$$

Using the relationships $\omega = vk$ and $v = \sqrt{B/\rho}$, we can rewrite Eq. (16.11) as

Intensity of a sinusoidal sound wave in a fluid
$$I = \frac{1}{2}\sqrt{\rho B}\,\omega^2 A^2 \tag{16.12}$$

Angular frequency $= 2\pi f$
Displacement amplitude
Density of fluid
Bulk modulus of fluid

It is usually more useful to express I in terms of the pressure amplitude p_{max}. Using Eqs. (16.5) and (16.12) and the relationship $\omega = vk$, we find

$$I = \frac{\omega p_{max}^2}{2Bk} = \frac{v p_{max}^2}{2B} \tag{16.13}$$

By using the wave speed relationship $v = \sqrt{B/\rho}$, we can also write Eq. (16.13) in the alternative forms

Intensity of a sinusoidal sound wave in a fluid
$$I = \frac{p_{max}^2}{2\rho v} = \frac{p_{max}^2}{2\sqrt{\rho B}} \tag{16.14}$$

Pressure amplitude
Wave speed
Bulk modulus of fluid
Density of fluid

You should verify these expressions. Comparison of Eqs. (16.12) and (16.14) shows that sinusoidal sound waves of the same intensity but different frequency have different displacement amplitudes A but the *same* pressure amplitude p_{max}.

This is another reason it is usually more convenient to describe a sound wave in terms of pressure fluctuations, not displacement.

The *total* average power carried across a surface by a sound wave equals the product of the intensity at the surface and the surface area, if the intensity over the surface is uniform. The total average sound power emitted by a person speaking in an ordinary conversational tone is about 10^{-5} W, while a loud shout corresponds to about 3×10^{-2} W. If all the residents of New York City were to talk at the same time, the total sound power would be about 100 W, equivalent to the electric power requirement of a medium-sized light bulb. On the other hand, the power required to fill a large auditorium or stadium with loud sound is considerable (see Example 16.7.)

If the sound source emits waves in all directions equally, the intensity decreases with increasing distance r from the source according to the inverse-square law (Section 15.5): The intensity is proportional to $1/r^2$. The intensity can be increased by confining the sound waves to travel in the desired direction only (**Fig. 16.10**), although the $1/r^2$ law still applies.

The inverse-square relationship also does not apply indoors because sound energy can reach a listener by reflection from the walls and ceiling. Indeed, part of the architect's job in designing an auditorium is to tailor these reflections so that the intensity is as nearly uniform as possible over the entire auditorium.

16.10 By cupping your hands like this, you direct the sound waves emerging from your mouth so that they don't propagate to the sides. Hence you can be heard at greater distances.

PROBLEM-SOLVING STRATEGY 16.1 | SOUND INTENSITY

IDENTIFY *the relevant concepts:* The relationships between the intensity and amplitude of a sound wave are straightforward. Other quantities are involved in these relationships, however, so it's particularly important to decide which is your target variable.

SET UP *the problem* using the following steps:
1. Sort the physical quantities into categories. Wave properties include the displacement and pressure amplitudes A and p_{max}. The frequency f can be determined from the angular frequency ω, the wave number k, or the wavelength λ. These quantities are related through the wave speed v, which is determined by properties of the medium (B and ρ for a liquid, and γ, T, and M for a gas).

2. List the given quantities and identify the target variables. Find relationships that take you where you want to go.

EXECUTE *the solution*: Use your selected equations to solve for the target variables. Express the temperature in kelvins (Celsius temperature plus 273.15) to calculate the speed of sound in a gas.

EVALUATE *your answer:* If possible, use an alternative relationship to check your results.

EXAMPLE 16.5 | INTENSITY OF A SOUND WAVE IN AIR

Find the intensity of the sound wave in Example 16.1, with $p_{max} = 3.0 \times 10^{-2}$ Pa. Assume the temperature is 20°C so that the density of air is $\rho = 1.20$ kg/m³ and the speed of sound is $v = 344$ m/s.

SOLUTION

IDENTIFY and SET UP: Our target variable is the intensity I of the sound wave. We are given the pressure amplitude p_{max} of the wave as well as the density ρ and wave speed v for the medium. We can determine I from p_{max}, ρ, and v from Eq. (16.14).

EXECUTE: From Eq. (16.14),

$$I = \frac{p_{max}^2}{2\rho v} = \frac{(3.0 \times 10^{-2}\ \text{Pa})^2}{2(1.20\ \text{kg/m}^3)(344\ \text{m/s})}$$

$$= 1.1 \times 10^{-6}\ \text{J/(s} \cdot \text{m}^2) = 1.1 \times 10^{-6}\ \text{W/m}^2$$

EVALUATE: This seems like a very low intensity, but it is well within the range of sound intensities encountered on a daily basis. A very loud sound wave at the threshold of pain has a pressure amplitude of about 30 Pa and an intensity of about 1 W/m². The pressure amplitude of the faintest sound wave that can be heard is about 3×10^{-5} Pa, and the corresponding intensity is about 10^{-12} W/m². (Try these values of p_{max} in Eq. (16.14) to check that the corresponding intensities are as we have stated.)

EXAMPLE 16.6 SAME INTENSITY, DIFFERENT FREQUENCIES

What are the pressure and displacement amplitudes of a 20-Hz sound wave with the same intensity as the 1000-Hz sound wave of Examples 16.1 and 16.5?

SOLUTION

IDENTIFY and SET UP: In Examples 16.1 and 16.5 we found that for a 1000-Hz sound wave with $p_{max} = 3.0 \times 10^{-2}$ Pa, $A = 1.2 \times 10^{-8}$ m and $I = 1.1 \times 10^{-6}$ W/m^2. Our target variables are p_{max} and A for a 20-Hz sound wave of the same intensity I. We can find these using Eqs. (16.14) and (16.12), respectively.

EXECUTE: We can rearrange Eqs. (16.14) and (16.12) as $p_{max}^2 = 2I\sqrt{\rho B}$ and $\omega^2 A^2 = 2I/\sqrt{\rho B}$, respectively. These tell us that for a given sound intensity I in a given medium (constant ρ and B), the

quantities p_{max} and ωA (or, equivalently, fA) are *constants* that don't depend on frequency. From the first result we immediately have $p_{max} = 3.0 \times 10^{-2}$ Pa for $f = 20$ Hz, the same as for $f = 1000$ Hz. If we write the second result as $f_{20}A_{20} = f_{1000}A_{1000}$, we have

$$A_{20} = \left(\frac{f_{1000}}{f_{20}}\right)A_{1000}$$

$$= \left(\frac{1000 \text{ Hz}}{20 \text{ Hz}}\right)(1.2 \times 10^{-8} \text{ m}) = 6.0 \times 10^{-7} \text{ m} = 0.60 \text{ } \mu\text{m}$$

EVALUATE: Our result reinforces the idea that pressure amplitude is a more convenient description of a sound wave and its intensity than displacement amplitude.

EXAMPLE 16.7 "PLAY IT LOUD!"

For an outdoor concert we want the sound intensity to be 1 W/m^2 at a distance of 20 m from the speaker array. If the sound intensity is uniform in all directions, what is the required average acoustic power output of the array?

SOLUTION

IDENTIFY, SET UP, and EXECUTE: This example uses the definition of sound intensity as power per unit area. The total power is the target variable; the area in question is a hemisphere centered on the speaker array. We assume that the speakers are on the ground

and that none of the acoustic power is directed into the ground, so the acoustic power is uniform over a hemisphere 20 m in radius. The surface area of this hemisphere is $\left(\frac{1}{2}\right)(4\pi)(20 \text{ m})^2$, or about 2500 m^2. The required power is the product of this area and the intensity: $(1 \text{ W/m}^2)(2500 \text{ m}^2) = 2500 \text{ W} = 2.5 \text{ kW}$.

EVALUATE: The electrical power input to the speaker would need to be considerably greater than 2.5 kW, because speaker efficiency is not very high (typically a few percent for ordinary speakers, and up to 25% for horn-type speakers).

The Decibel Scale

Because the ear is sensitive over a broad range of intensities, a *logarithmic* measure of intensity called **sound intensity level** is often used:

$$\text{Sound intensity level} \cdots \beta = (10 \text{ dB}) \log \frac{I}{I_0} \cdots \overset{\text{Intensity of sound wave}}{\underset{\text{Logarithm to base 10}}{\cdots}} \overset{\text{Reference intensity}}{= 10^{-12} \text{ W/m}^2} \tag{16.15}$$

The chosen reference intensity I_0 in Eq. (16.15) is approximately the threshold of human hearing at 1000 Hz. Sound intensity levels are expressed in **decibels,** abbreviated dB. A decibel is $\frac{1}{10}$ of a *bel,* a unit named for Alexander Graham Bell (the inventor of the telephone). The bel is inconveniently large for most purposes, and the decibel is the usual unit of sound intensity level.

If the intensity of a sound wave equals I_0 or 10^{-12} W/m^2, its sound intensity level is $\beta = 0$ dB. An intensity of 1 W/m^2 corresponds to 120 dB. **Table 16.2** gives the sound intensity levels of some familiar sounds. You can use Eq. (16.15) to check the value of β given for each intensity in the table.

Because the ear is not equally sensitive to all frequencies in the audible range, some sound-level meters weight the various frequencies unequally. One such scheme leads to the so-called dBA scale; this scale deemphasizes the low and very high frequencies, where the ear is less sensitive.

TABLE 16.2 Sound Intensity Levels from Various Sources (Representative Values)

Source or Description of Sound	Sound Intensity Level, β (dB)	Intensity, I (W/m^2)
Military jet aircraft 30 m away	140	10^2
Threshold of pain	120	1
Riveter	95	3.2×10^{-3}
Elevated train	90	10^{-3}
Busy street traffic	70	10^{-5}
Ordinary conversation	65	3.2×10^{-6}
Quiet automobile	50	10^{-7}
Quiet radio in home	40	10^{-8}
Average whisper	20	10^{-10}
Rustle of leaves	10	10^{-11}
Threshold of hearing at 1000 Hz	0	10^{-12}

EXAMPLE 16.8 TEMPORARY—OR PERMANENT—HEARING LOSS

A 10-min exposure to 120-dB sound will temporarily shift your threshold of hearing at 1000 Hz from 0 dB up to 28 dB. Ten years of exposure to 92-dB sound will cause a *permanent* shift to 28 dB. What sound intensities correspond to 28 dB and 92 dB?

SOLUTION

IDENTIFY and SET UP: We are given two sound intensity levels β; our target variables are the corresponding intensities. We can solve Eq. (16.15) to find the intensity I that corresponds to each value of β.

EXECUTE: We solve Eq. (16.15) for I by dividing both sides by 10 dB and using the relationship $10^{\log x} = x$:

$$I = I_0 10^{\beta/(10\ \text{dB})}$$

For $\beta = 28$ dB and $\beta = 92$ dB, the exponents are $\beta/(10\ \text{dB}) = 2.8$ and 9.2, respectively, so that

$$I_{28\ \text{dB}} = (10^{-12}\ \text{W/m}^2)10^{2.8} = 6.3 \times 10^{-10}\ \text{W/m}^2$$
$$I_{92\ \text{dB}} = (10^{-12}\ \text{W/m}^2)10^{9.2} = 1.6 \times 10^{-3}\ \text{W/m}^2$$

EVALUATE: If your answers are a factor of 10 too large, you may have entered 10×10^{-12} in your calculator instead of 1×10^{-12}. Be careful!

EXAMPLE 16.9 A BIRD SINGS IN A MEADOW

Consider an idealized bird (treated as a point source) that emits constant sound power, with intensity obeying the inverse-square law (**Fig. 16.11**). If you move twice the distance from the bird, by how many decibels does the sound intensity level drop?

SOLUTION

IDENTIFY and SET UP: The decibel scale is logarithmic, so the *difference* between two sound intensity levels (the target variable) corresponds to the *ratio* of the corresponding intensities, which is determined by the inverse-square law. We label the two points P_1 and P_2 (Fig. 16.11). We use Eq. (16.15), the definition of sound intensity level, at each point. We use Eq. (15.26), the inverse-square law, to relate the intensities at the two points.

EXECUTE: The difference $\beta_2 - \beta_1$ between any two sound intensity levels is related to the corresponding intensities by

$$\beta_2 - \beta_1 = (10\ \text{dB})\left(\log\frac{I_2}{I_0} - \log\frac{I_1}{I_0}\right)$$
$$= (10\ \text{dB})\left[(\log I_2 - \log I_0) - (\log I_1 - \log I_0)\right]$$
$$= (10\ \text{dB})\log\frac{I_2}{I_1}$$

16.11 When you double your distance from a point source of sound, by how much does the sound intensity level decrease?

For this inverse-square-law source, Eq. (15.26) yields $I_2/I_1 = r_1^2/r_2^2 = \frac{1}{4}$, so

$$\beta_2 - \beta_1 = (10\ \text{dB})\log\frac{I_2}{I_1} = (10\ \text{dB})\log\frac{1}{4} = -6.0\ \text{dB}$$

Continued

EVALUATE: Our result is negative, which tells us (correctly) that the sound intensity level is less at P_2 than at P_1. The 6-dB difference doesn't depend on the sound intensity level at P_1; *any* doubling of the distance from an inverse-square-law source reduces the sound intensity level by 6 dB.

Note that the perceived *loudness* of a sound is not directly proportional to its intensity. For example, most people interpret an increase of 8 dB to 10 dB in sound intensity level (corresponding to increasing intensity by a factor of 6 to 10) as a doubling of loudness.

TEST YOUR UNDERSTANDING OF SECTION 16.3 You double the intensity of a sound wave in air while leaving the frequency unchanged. (The pressure, density, and temperature of the air remain unchanged as well.) What effect does this have on the displacement amplitude, pressure amplitude, bulk modulus, sound speed, and sound intensity level? ∎

16.4 STANDING SOUND WAVES AND NORMAL MODES

When longitudinal (sound) waves propagate in a fluid in a pipe, the waves are reflected from the ends in the same way that transverse waves on a string are reflected at its ends. The superposition of the waves traveling in opposite directions again forms a standing wave. Just as for transverse standing waves on a string (see Section 15.7), standing sound waves in a pipe can be used to create sound waves in the surrounding air. This is the principle of the human voice as well as many musical instruments, including woodwinds, brasses, and pipe organs.

Transverse waves on a string, including standing waves, are usually described only in terms of the displacement of the string. But, as we have seen, sound waves in a fluid may be described either in terms of the displacement of the fluid or in terms of the pressure variation in the fluid. To avoid confusion, we'll use the terms **displacement node** and **displacement antinode** to refer to points where particles of the fluid have zero displacement and maximum displacement, respectively.

We can demonstrate standing sound waves in a column of gas using an apparatus called a Kundt's tube (**Fig. 16.12**). A horizontal glass tube a meter or so long is closed at one end and has a flexible diaphragm at the other end that can transmit vibrations. A nearby loudspeaker is driven by an audio oscillator and amplifier; this produces sound waves that force the diaphragm to vibrate sinusoidally with a frequency that we can vary. The sound waves within the tube are reflected at the other, closed end of the tube. We spread a small amount of light powder uniformly along the bottom of the tube. As we vary the frequency of the sound, we pass through frequencies at which the amplitude of the standing waves becomes large enough for the powder to be swept along the tube at those points where the gas is in motion. The powder therefore collects at the displacement nodes (where the gas is not moving). Adjacent nodes are separated by a distance equal to $\lambda/2$.

16.12 Demonstrating standing sound waves using a Kundt's tube. The blue shading represents the density of the gas at an instant when the gas pressure at the displacement nodes is a maximum or a minimum.

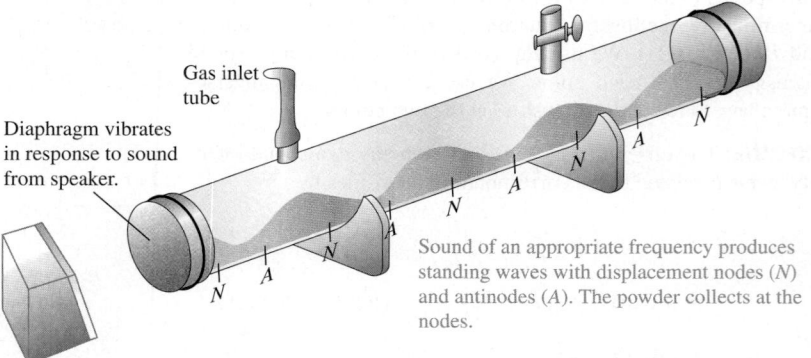

Gas inlet tube

Diaphragm vibrates in response to sound from speaker.

Sound of an appropriate frequency produces standing waves with displacement nodes (*N*) and antinodes (*A*). The powder collects at the nodes.

Speaker

Figure 16.13 shows the motions of nine different particles within a gas-filled tube in which there is a standing sound wave. A particle at a displacement node (N) does not move, while a particle at a displacement antinode (A) oscillates with maximum amplitude. Note that particles on opposite sides of a displacement node vibrate in opposite phase. When these particles approach each other, the gas between them is compressed and the pressure rises; when they recede from each other, there is an expansion and the pressure drops. Hence at a displacement *node* the gas undergoes the maximum amount of compression and expansion, and the variations in pressure and density above and below the average have their maximum value. By contrast, particles on opposite sides of a displacement *antinode* vibrate *in phase;* the distance between the particles is nearly constant, and there is *no* variation in pressure or density at a displacement antinode.

We use the term **pressure node** to describe a point in a standing sound wave at which the pressure and density do not vary and the term **pressure antinode** to describe a point at which the variations in pressure and density are greatest. Using these terms, we can summarize our observations as follows:

> **A pressure node is always a displacement antinode, and a pressure antinode is always a displacement node.**

Figure 16.12 depicts a standing sound wave at an instant at which the pressure variations are greatest; the blue shading shows that the density and pressure of the gas have their maximum and minimum values at the displacement nodes.

When reflection takes place at a *closed* end of a pipe (an end with a rigid barrier or plug), the displacement of the particles at this end must always be zero, analogous to a fixed end of a string. Thus a closed end of a pipe is a displacement node and a pressure antinode; the particles do not move, but the pressure variations are maximum. An *open* end of a pipe is a pressure node because it is open to the atmosphere, where the pressure is constant. Because of this, an open end is always a displacement *antinode,* in analogy to a free end of a string; the particles oscillate with maximum amplitude, but the pressure does not vary. (The pressure node actually occurs somewhat beyond an open end of a pipe. But if the diameter of the pipe is small in comparison to the wavelength, which is true for most musical instruments, this effect can safely be ignored.) Thus longitudinal sound waves are reflected at the closed and open ends of a pipe in the same way that transverse waves in a string are reflected at fixed and free ends, respectively.

16.13 In a standing sound wave, a displacement node N is a pressure antinode (a point where the pressure fluctuates the most) and a displacement antinode A is a pressure node (a point where the pressure does not fluctuate at all).

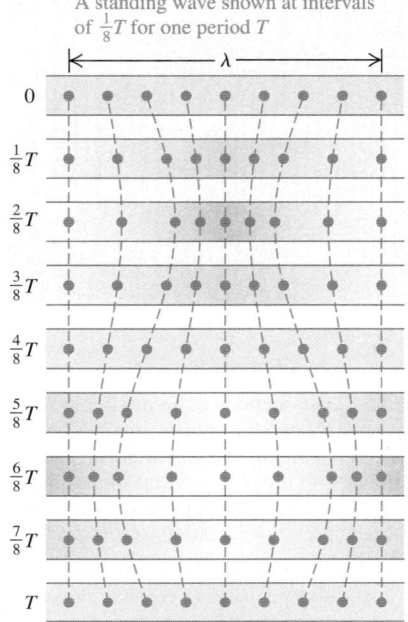

N = a displacement node = a pressure antinode
A = a displacement antinode = a pressure node

CONCEPTUAL EXAMPLE 16.10   THE SOUND OF SILENCE

A directional loudspeaker directs a sound wave of wavelength λ at a wall (**Fig. 16.14**). At what distances from the wall could you stand and hear no sound at all?

SOLUTION

Your ear detects pressure variations in the air; you will therefore hear no sound if your ear is at a *pressure node,* which is a displacement antinode. The wall is at a displacement node; the distance from any node to an adjacent antinode is $\lambda/4$, and the distance from one antinode to the next is $\lambda/2$ (Fig. 16.14). Hence the displacement antinodes (pressure nodes), at which no sound will be heard, are at distances $d = \lambda/4$, $d = \lambda/4 + \lambda/2 = 3\lambda/4$, $d = 3\lambda/4 + \lambda/2 = 5\lambda/4, \dots$ from the wall. If the loudspeaker is not highly directional, this effect is hard to notice because of reflections of sound waves from the floor, ceiling, and other walls.

16.14 When a sound wave is directed at a wall, it interferes with the reflected wave to create a standing wave. The N's and A's are *displacement* nodes and antinodes.

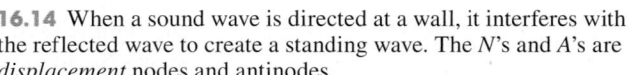

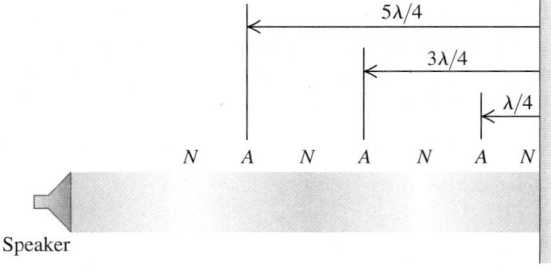

Speaker

16.15 Organ pipes of different sizes produce tones with different frequencies.

Organ Pipes and Wind Instruments

The most important application of standing sound waves is the production of musical tones. Organ pipes are one of the simplest examples (**Fig. 16.15**). Air is supplied by a blower to the bottom end of the pipe (**Fig. 16.16**). A stream of air emerges from the narrow opening at the edge of the horizontal surface and is directed against the top edge of the opening, which is called the *mouth* of the pipe. The column of air in the pipe is set into vibration, and there is a series of possible normal modes, just as with the stretched string. The mouth acts as an open end; it is a pressure node and a displacement antinode. The other end of the pipe (at the top in Fig. 16.16) may be either open or closed.

In **Fig. 16.17**, both ends of the pipe are open, so both ends are pressure nodes and displacement antinodes. An organ pipe that is open at both ends is called an *open pipe*. The fundamental frequency f_1 corresponds to a standing-wave pattern with a displacement antinode at each end and a displacement node in the middle (Fig. 16.17a). The distance between adjacent antinodes is always equal to one half-wavelength, and in this case that is equal to the length L of the pipe; $\lambda/2 = L$. The corresponding frequency, obtained from the relationship $f = v/\lambda$, is

16.16 Cross sections of an organ pipe at two instants one half-period apart. The N's and A's are *displacement* nodes and antinodes; as the blue shading shows, these are points of maximum pressure variation and zero pressure variation, respectively.

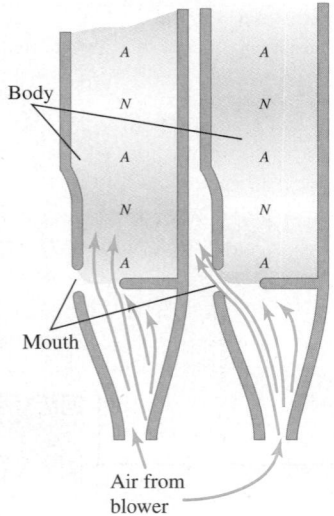

Vibrations from turbulent airflow set up standing waves in the pipe.

$$f_1 = \frac{v}{2L} \quad \text{(open pipe)} \tag{16.16}$$

Figures 16.17b and 16.17c show the second and third harmonics; their vibration patterns have two and three displacement nodes, respectively. For these, a half-wavelength is equal to $L/2$ and $L/3$, respectively, and the frequencies are twice and three times the fundamental, respectively: $f_2 = 2f_1$ and $f_3 = 3f_1$. For *every* normal mode of an open pipe the length L must be an integer number of half-wavelengths, and the possible wavelengths λ_n are given by

$$L = n\frac{\lambda_n}{2} \quad \text{or} \quad \lambda_n = \frac{2L}{n} \quad (n = 1, 2, 3, \dots) \quad \text{(open pipe)} \tag{16.17}$$

The corresponding frequencies f_n are given by $f_n = v/\lambda_n$, so all the normal-mode frequencies for a pipe that is open at both ends are given by

Standing waves, open pipe:	Frequency of nth harmonic ($n = 1, 2, 3, ...$) $$f_n = \frac{nv}{2L}$$ Speed of sound in pipe — Length of pipe	(16.18)

The value $n = 1$ corresponds to the fundamental frequency, $n = 2$ to the second harmonic (or first overtone), and so on. Alternatively, we can say

$$f_n = nf_1 \quad (n = 1, 2, 3, \dots) \quad \text{(open pipe)} \tag{16.19}$$

with f_1 given by Eq. (16.16).

Figure 16.18 shows a *stopped pipe:* It is open at the left end but closed at the right end. The left (open) end is a displacement antinode (pressure node), but the right (closed) end is a displacement node (pressure antinode). Figure 16.18a shows the lowest-frequency mode; the length of the pipe is the distance between

16.17 A cross section of an open pipe showing the first three normal modes. The shading indicates the pressure variations. The red curves are graphs of the displacement along the pipe axis at two instants separated in time by one half-period. The N's and A's are the *displacement* nodes and antinodes; interchange these to show the *pressure* nodes and antinodes.

(a) Fundamental: $f_1 = \dfrac{v}{2L}$

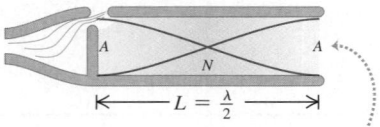

Open end is always a displacement antinode.

(b) Second harmonic: $f_2 = 2\dfrac{v}{2L} = 2f_1$

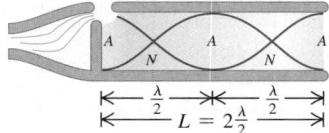

(c) Third harmonic: $f_3 = 3\dfrac{v}{2L} = 3f_1$

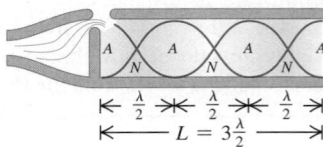

16.18 A cross section of a stopped pipe showing the first three normal modes as well as the *displacement* nodes and antinodes. Only odd harmonics are possible.

(a) Fundamental: $f_1 = \dfrac{v}{4L}$

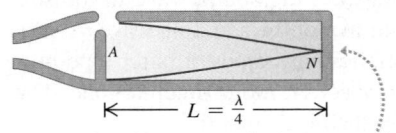

$\overleftarrow{\hspace{0.5cm}} L = \frac{\lambda}{4} \overrightarrow{\hspace{0.5cm}}$

Closed end is always a displacement node.

(b) Third harmonic: $f_3 = 3\dfrac{v}{4L} = 3f_1$

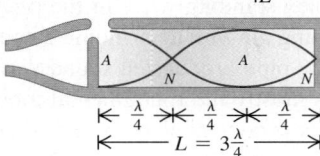

$\overleftarrow{\frac{\lambda}{4}} \overleftrightarrow{\frac{\lambda}{4}} \overleftrightarrow{\frac{\lambda}{4}}$
$\overleftarrow{\hspace{0.5cm}} L = 3\frac{\lambda}{4} \overrightarrow{\hspace{0.5cm}}$

(c) Fifth harmonic: $f_5 = 5\dfrac{v}{4L} = 5f_1$

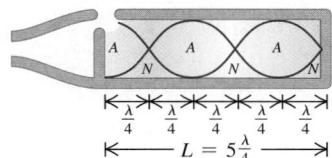

$\frac{\lambda}{4}\,\frac{\lambda}{4}\,\frac{\lambda}{4}\,\frac{\lambda}{4}\,\frac{\lambda}{4}$
$\overleftarrow{\hspace{0.5cm}} L = 5\frac{\lambda}{4} \overrightarrow{\hspace{0.5cm}}$

a node and the adjacent antinode, or a quarter-wavelength ($L = \lambda_1/4$). The fundamental frequency is $f_1 = v/\lambda_1$, or

$$f_1 = \frac{v}{4L} \qquad \text{(stopped pipe)} \qquad (16.20)$$

This is one-half the fundamental frequency for an *open* pipe of the same length. In musical language, the *pitch* of a closed pipe is one octave lower (a factor of 2 in frequency) than that of an open pipe of the same length. Figure 16.18b shows the next mode, for which the length of the pipe is *three-quarters* of a wavelength, corresponding to a frequency $3f_1$. For Fig. 16.18c, $L = 5\lambda/4$ and the frequency is $5f_1$. The possible wavelengths are given by

$$L = n\frac{\lambda_n}{4} \quad \text{or} \quad \lambda_n = \frac{4L}{n} \quad (n = 1, 3, 5, \ldots) \quad \text{(stopped pipe)} \quad (16.21)$$

The normal-mode frequencies are given by $f_n = v/\lambda_n$, or

Standing waves, stopped pipe:

Frequency of nth harmonic ($n = 1, 3, 5, \ldots$)

$$f_n = \frac{nv}{4L} \qquad \text{Speed of sound in pipe} \atop \text{Length of pipe} \qquad (16.22)$$

or

$$f_n = nf_1 \quad (n = 1, 3, 5, \ldots) \qquad \text{(stopped pipe)} \qquad (16.23)$$

with f_1 given by Eq. (16.20). We see that the second, fourth, and all *even* harmonics are missing. In a stopped pipe, the fundamental frequency is $f_1 = v/4L$, and only the odd harmonics in the series ($3f_1, 5f_1, \ldots$) are possible.

A final possibility is a pipe that is closed at *both* ends, with displacement nodes and pressure antinodes at both ends. This wouldn't be of much use as a musical instrument because the vibrations couldn't get out of the pipe.

EXAMPLE 16.11   A TALE OF TWO PIPES

On a day when the speed of sound is 345 m/s, the fundamental frequency of a particular stopped organ pipe is 220 Hz. (a) How long is this pipe? (b) The second *overtone* of this pipe has the same wavelength as the third *harmonic* of an *open* pipe. How long is the open pipe?

SOLUTION

IDENTIFY and SET UP: This problem uses the relationship between the length and normal-mode frequencies of open pipes (Fig. 16.17) and stopped pipes (Fig. 16.18). In part (a), we determine the length of the stopped pipe from Eq. (16.22). In part (b), we must determine the length of an open pipe, for which Eq. (16.18) gives the frequencies.

EXECUTE: (a) For a stopped pipe $f_1 = v/4L$, so

$$L_{\text{stopped}} = \frac{v}{4f_1} = \frac{345 \text{ m/s}}{4(220 \text{ s}^{-1})} = 0.392 \text{ m}$$

(b) The frequency of the second overtone of a stopped pipe (the *third* possible frequency) is $f_5 = 5f_1 = 5(220 \text{ Hz}) = 1100 \text{ Hz}$. If the wavelengths for the two pipes are the same, the frequencies are also the same. Hence the frequency of the third harmonic of the open pipe, which is at $3f_1 = 3(v/2L)$, equals 1100 Hz. Then

$$1100 \text{ Hz} = 3\left(\frac{345 \text{ m/s}}{2L_{\text{open}}}\right) \quad \text{and} \quad L_{\text{open}} = 0.470 \text{ m}$$

EVALUATE: The 0.392-m stopped pipe has a fundamental frequency of 220 Hz; the *longer* (0.470-m) open pipe has a *higher* fundamental frequency, $(1100 \text{ Hz})/3 = 367 \text{ Hz}$. This is not a contradiction, as you can see if you compare Figs. 16.17a and 16.18a.

In an organ pipe in actual use, several modes are always present at once; the motion of the air is a superposition of these modes. This situation is analogous to a string that is struck or plucked, as in Fig. 15.28. Just as for a vibrating string, a complex standing wave in the pipe produces a traveling sound wave in the surrounding air with a harmonic content similar to that of the standing wave. A very narrow pipe produces a sound wave rich in higher harmonics; a fatter pipe produces mostly the fundamental mode, heard as a softer, more flutelike tone. The harmonic content also depends on the shape of the pipe's mouth.

We have talked about organ pipes, but this discussion is also applicable to other wind instruments. The flute and the recorder are directly analogous. The most significant difference is that those instruments have holes along the pipe. Opening and closing the holes with the fingers changes the effective length L of the air column and thus changes the pitch. Any individual organ pipe, by comparison, can play only a single note. The flute and recorder behave as *open* pipes, while the clarinet acts as a *stopped* pipe (closed at the reed end, open at the bell).

Equations (16.18) and (16.22) show that the frequencies of any wind instrument are proportional to the speed of sound v in the air column inside the instrument. As Eq. (16.10) shows, v depends on temperature; it increases when temperature increases. Thus the pitch of all wind instruments rises with increasing temperature. An organ that has some of its pipes at one temperature and others at a different temperature is bound to sound out of tune.

TEST YOUR UNDERSTANDING OF SECTION 16.4 If you connect a hose to one end of a metal pipe and blow compressed air into it, the pipe produces a musical tone. If instead you blow compressed helium into the pipe at the same pressure and temperature, will the pipe produce (i) the same tone, (ii) a higher-pitch tone, or (iii) a lower-pitch tone? ∎

16.5 RESONANCE AND SOUND

Many mechanical systems have normal modes of oscillation. As we have seen, these include columns of air (as in an organ pipe) and stretched strings (as in a guitar; see Section 15.8). In each mode, every particle of the system oscillates with simple harmonic motion at the same frequency as the mode. Air columns and stretched strings have an infinite series of normal modes, but the basic concept is closely related to the simple harmonic oscillator, discussed in Chapter 14, which has only a single normal mode (that is, only one frequency at which it oscillates after being disturbed).

Suppose we apply a periodically varying force to a system that can oscillate. The system is then forced to oscillate with a frequency equal to the frequency of the applied force (called the *driving frequency*). This motion is called a *forced oscillation*. We talked about forced oscillations of the harmonic oscillator in Section 14.8, including the phenomenon of mechanical **resonance**. A simple example of resonance is pushing Cousin Throckmorton on a swing. The swing is a pendulum; it has only a single normal mode, with a frequency determined by its length. If we push the swing periodically with this frequency, we can build up the amplitude of the motion. But if we push with a very different frequency, the swing hardly moves at all.

Resonance also occurs when a periodically varying force is applied to a system with many normal modes. In **Fig. 16.19a** an open organ pipe is placed next to a loudspeaker that emits pure sinusoidal sound waves of frequency f, which can be varied by adjusting the amplifier. The air in the pipe is forced to vibrate with the same frequency f as the *driving force* provided by the loudspeaker. In general the amplitude of this motion is relatively small, and the air inside the pipe will not move in any of the normal-mode patterns shown in Fig. 16.17. But if the frequency f of the force is close to one of the normal-mode frequencies, the air in the pipe moves in the normal-mode pattern for that frequency, and

16.19 (a) The air in an open pipe is forced to oscillate at the same frequency as the sinusoidal sound waves coming from the loudspeaker. (b) The resonance curve of the open pipe graphs the amplitude of the standing sound wave in the pipe as a function of the driving frequency.

(a)

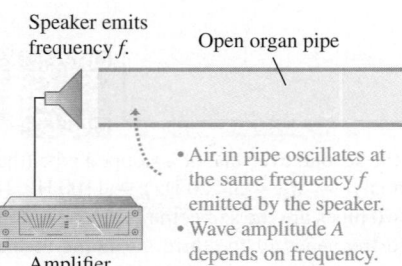

Speaker emits frequency f.

Open organ pipe

• Air in pipe oscillates at the same frequency f emitted by the speaker.
• Wave amplitude A depends on frequency.

Amplifier

(b) Resonance curve: graph of amplitude A versus driving frequency f. Peaks occur at normal-mode frequencies of the pipe: $f_1, f_2 = 2f_1, f_3 = 3f_1, \ldots$.

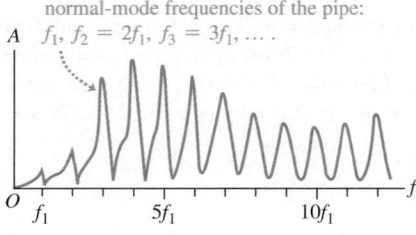

the amplitude can become quite large. Figure 16.19b shows the amplitude of oscillation of the air in the pipe as a function of the driving frequency f. This **resonance curve** of the pipe has peaks where f equals the normal-mode frequencies of the pipe. The detailed shape of the resonance curve depends on the geometry of the pipe.

If the frequency of the force is precisely *equal* to a normal-mode frequency, the system is in resonance, and the amplitude of the forced oscillation is maximum. If there were no friction or other energy-dissipating mechanism, a driving force at a normal-mode frequency would continue to add energy to the system, the amplitude would increase indefinitely, and the peaks in the resonance curve of Fig. 16.19b would be infinitely high. But in any real system there is always some dissipation of energy, or damping, as we discussed in Section 14.8; the amplitude of oscillation in resonance may be large, but it cannot be infinite.

The "sound of the ocean" you hear when you put your ear next to a large seashell is due to resonance. The noise of the outside air moving past the seashell is a mixture of sound waves of almost all audible frequencies, which forces the air inside the seashell to oscillate. The seashell behaves like an organ pipe, with a set of normal-mode frequencies; hence the inside air oscillates most strongly at those frequencies, producing the seashell's characteristic sound. To hear a similar phenomenon, uncap a full bottle of your favorite beverage and blow across the open top. The noise is provided by your breath blowing across the top, and the "organ pipe" is the column of air inside the bottle above the surface of the liquid. If you take a drink and repeat the experiment, you will hear a lower tone because the "pipe" is longer and the normal-mode frequencies are lower.

Resonance also occurs when a stretched string is forced to oscillate (see Section 15.8). Suppose that one end of a stretched string is held fixed while the other is given a transverse sinusoidal motion with small amplitude, setting up standing waves. If the frequency of the driving mechanism is *not* equal to one of the normal-mode frequencies of the string, the amplitude at the antinodes is fairly small. However, if the frequency is equal to any one of the normal-mode frequencies, the string is in resonance, and the amplitude at the antinodes is very much larger than that at the driven end. The driven end is not precisely a node, but it lies much closer to a node than to an antinode when the string is in resonance. The photographs of standing waves in Fig. 15.23 were made this way, with the left end of the string fixed and the right end oscillating vertically with small amplitude.

It is easy to demonstrate resonance with a piano. Push down the damper pedal (the right-hand pedal) so that the dampers are lifted and the strings are free to vibrate, and then sing a steady tone into the piano. When you stop singing, the piano seems to continue to sing the same note. The sound waves from your voice excite vibrations in the strings that have natural frequencies close to the frequencies (fundamental and harmonics) present in the note you sang.

A more spectacular example is a singer breaking a wine glass with her amplified voice. A good-quality wine glass has normal-mode frequencies that you can hear by tapping it. If the singer emits a loud note with a frequency corresponding exactly to one of these normal-mode frequencies, large-amplitude oscillations can build up and break the glass (**Fig. 16.20**).

BIO Application Resonance and the Sensitivity of the Ear The auditory canal of the human ear (see Fig. 16.4) is an air-filled pipe open at one end and closed at the other (eardrum) end. The canal is about $2.5\text{ cm} = 0.025\text{ m}$ long, so it has a resonance at its fundamental frequency $f_1 = v/4L = (344\text{ m/s})/[4(0.025\text{ m})] = 3440\text{ Hz}$. The resonance means that a sound at this frequency produces a strong oscillation of the eardrum. That's why your ear is most sensitive to sounds near 3440 Hz.

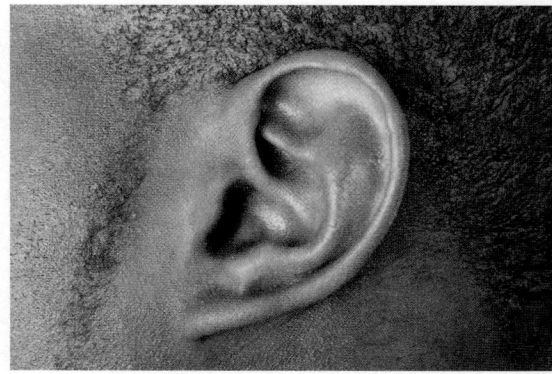

16.20 The frequency of the sound from this trumpet exactly matches one of the normal-mode frequencies of the goblet. The resonant vibrations of the goblet have such large amplitude that the goblet tears itself apart.

EXAMPLE 16.12 **AN ORGAN–GUITAR DUET**

A stopped organ pipe is sounded near a guitar, causing one of the strings to vibrate with large amplitude. We vary the string tension until we find the maximum amplitude. The string is 80% as long as the pipe. If both pipe and string vibrate at their fundamental frequency, calculate the ratio of the wave speed on the string to the speed of sound in air.

SOLUTION

IDENTIFY and SET UP: The large response of the string is an example of resonance. It occurs because the organ pipe and the guitar string have the same fundamental frequency. Letting the subscripts

Continued

a and s stand for the air in the pipe and the string, respectively, the condition for resonance is $f_{1a} = f_{1s}$. Equation (16.20) gives the fundamental frequency for a stopped pipe, and Eq. (15.32) gives the fundamental frequency for a guitar string held at both ends. These expressions involve the wave speed in air (v_a) and on the string (v_s) and the lengths of the pipe and string. We are given that $L_s = 0.80L_a$; our target variable is the ratio v_s/v_a.

EXECUTE: From Eqs. (16.20) and (15.32), $f_{1a} = v_a/4L_a$ and $f_{1s} = v_s/2L_s$. These frequencies are equal, so

$$\frac{v_a}{4L_a} = \frac{v_s}{2L_s}$$

Substituting $L_s = 0.80L_a$ and rearranging, we get $v_s/v_a = 0.40$.

EVALUATE: As an example, if the speed of sound in air is 345 m/s, the wave speed on the string is $(0.40)(345 \text{ m/s}) = 138 \text{ m/s}$. Note that while the standing waves in the pipe and on the string have the same frequency, they have different *wavelengths* $\lambda = v/f$ because the two media have different wave speeds v. Which standing wave has the greater wavelength?

TEST YOUR UNDERSTANDING OF SECTION 16.5 A stopped organ pipe of length L has a fundamental frequency of 220 Hz. For which of the following organ pipes will there be a resonance if a tuning fork of frequency 660 Hz is sounded next to the pipe? (There may be more than one correct answer.) (i) A stopped organ pipe of length L; (ii) a stopped organ pipe of length $2L$; (iii) an open organ pipe of length L; (iii) an open organ pipe of length $2L$. ∎

16.6 INTERFERENCE OF WAVES

Wave phenomena that occur when two or more waves overlap in the same region of space are grouped under the heading *interference*. As we have seen, standing waves are a simple example of an interference effect: Two waves traveling in opposite directions in a medium can combine to produce a standing-wave pattern with nodes and antinodes that do not move.

Figure 16.21 shows an example of another type of interference that involves waves that spread out in space. Two speakers, driven in phase by the same amplifier, emit identical sinusoidal sound waves with the same constant frequency. We place a microphone at point P in the figure, equidistant from the speakers. Wave crests emitted from the two speakers at the same time travel equal distances and arrive at point P at the same time; hence the waves arrive in phase, and there is constructive interference. The total wave amplitude that we measure at P is twice the amplitude from each individual wave.

Now let's move the microphone to point Q, where the distances from the two speakers to the microphone differ by a half-wavelength. Then the two waves arrive a half-cycle out of step, or *out of phase;* a positive crest from one speaker arrives at the same time as a negative crest from the other. Destructive interference takes place, and the amplitude measured by the microphone is much *smaller* than when only one speaker is present. If the amplitudes from the two speakers are equal, the two waves cancel each other out completely at point Q, and the total amplitude there is zero.

16.21 Two speakers driven by the same amplifier. Constructive interference occurs at point P, and destructive interference occurs at point Q.

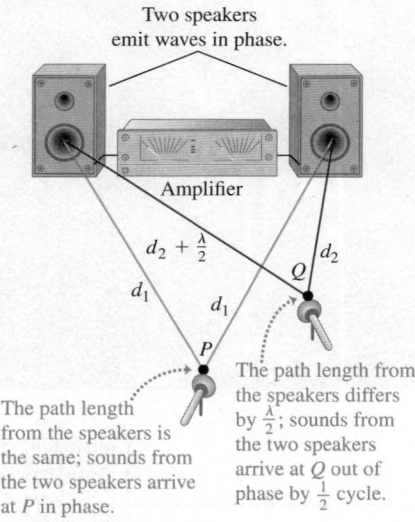

Two speakers emit waves in phase.

Amplifier

$d_2 + \frac{\lambda}{2}$ d_2

d_1 d_1 Q

P

The path length from the speakers is the same; sounds from the two speakers arrive at P in phase.

The path length from the speakers differs by $\frac{\lambda}{2}$; sounds from the two speakers arrive at Q out of phase by $\frac{1}{2}$ cycle.

CAUTION **Interference and traveling waves** The total wave in Fig. 16.21 is a *traveling* wave, not a standing wave. In a standing wave there is no net flow of energy in any direction; by contrast, in Fig. 16.21 there *is* an overall flow of energy from the speakers into the surrounding air, characteristic of a traveling wave. The interference between the waves from the two speakers simply causes the energy flow to be *channeled* into certain directions (for example, toward P) and away from other directions (for example, away from Q). You can see another difference between Fig. 16.21 and a standing wave by considering a point, such as Q, where destructive interference occurs. Such a point is *both* a displacement node *and* a pressure node because there is no wave at all at this point. In a standing wave, a pressure node is a displacement antinode, and vice versa. ∎

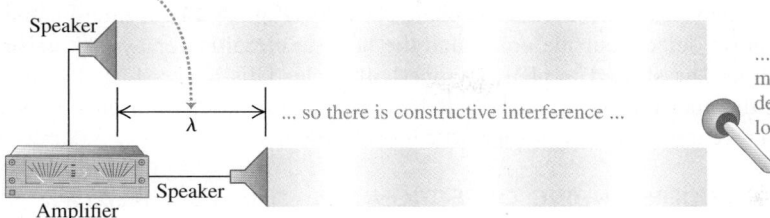

(a) The path lengths from the speakers to the microphone differ by λ ...

Speaker

λ

... so there is constructive interference ...

Speaker

Amplifier

... and the microphone detects a loud sound.

(b) The path lengths from the speakers to the microphone differ by $\frac{\lambda}{2}$ ...

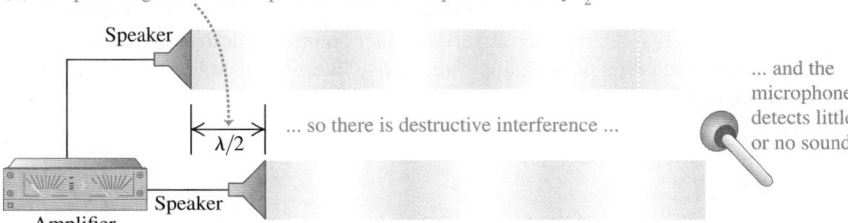

Speaker

λ/2

... so there is destructive interference ...

Speaker

Amplifier

... and the microphone detects little or no sound.

16.22 Two speakers driven by the same amplifier, emitting waves in phase. Only the waves directed toward the microphone are shown, and they are separated for clarity. **(a)** Constructive interference occurs when the path difference is 0, $\lambda, 2\lambda, 3\lambda, \ldots$. **(b)** Destructive interference occurs when the path difference is $\lambda/2, 3\lambda/2, 5\lambda/2, \ldots$.

Constructive interference occurs wherever the distances traveled by the two waves differ by a whole number of wavelengths, $0, \lambda, 2\lambda, 3\lambda, \ldots$; then the waves arrive at the microphone in phase (**Fig. 16.22a**). If the distances from the two speakers to the microphone differ by any half-integer number of wavelengths, $\lambda/2, 3\lambda/2, 5\lambda/2, \ldots$, the waves arrive at the microphone out of phase and there will be destructive interference (Fig. 16.22b). In this case, little or no sound energy flows toward the microphone. The energy instead flows in other directions, to where constructive interference occurs.

EXAMPLE 16.13 LOUDSPEAKER INTERFERENCE

Two small loudspeakers, A and B (**Fig. 16.23**), are driven by the same amplifier and emit pure sinusoidal waves in phase. (a) For what frequencies does constructive interference occur at point P? (b) For what frequencies does destructive interference occur? The speed of sound is 350 m/s.

SOLUTION

IDENTIFY and SET UP: The nature of the interference at P depends on the difference d in path lengths from point A to P and from point B to P. We calculate the path lengths using the Pythagorean theorem. Constructive interference occurs when d equals a whole number of wavelengths, while destructive interference occurs

16.23 What sort of interference occurs at P?

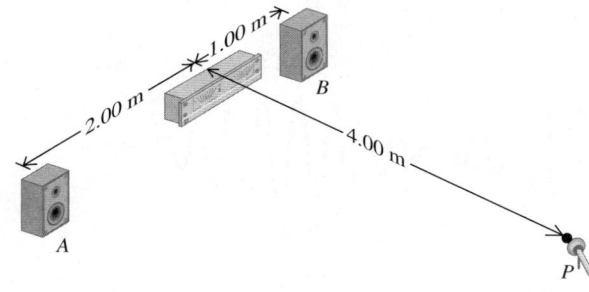

1.00 m

B

2.00 m

4.00 m

A

P

when d is a half-integer number of wavelengths. To find the corresponding frequencies, we use $v = f\lambda$.

EXECUTE: The A-to-P distance is $[(2.00 \text{ m})^2 + (4.00 \text{ m})^2]^{1/2} = 4.47$ m, and the B-to-P distance is $[(1.00 \text{ m})^2 + (4.00 \text{ m})^2]^{1/2} = 4.12$ m. The path difference is $d = 4.47 \text{ m} - 4.12 \text{ m} = 0.35$ m.

(a) Constructive interference occurs when $d = 0, \lambda, 2\lambda, \ldots$ or $d = 0, v/f, 2v/f, \ldots = nv/f$. So the possible frequencies are

$$f_n = \frac{nv}{d} = n\frac{350 \text{ m/s}}{0.35 \text{ m}} \qquad (n = 1, 2, 3, \ldots)$$

$$= 1000 \text{ Hz}, 2000 \text{ Hz}, 3000 \text{ Hz}, \ldots$$

(b) Destructive interference occurs when $d = \lambda/2, 3\lambda/2, 5\lambda/2, \ldots$ or $d = v/2f, 3v/2f, 5v/2f, \ldots$. The possible frequencies are

$$f_n = \frac{nv}{2d} = n\frac{350 \text{ m/s}}{2(0.35 \text{ m})} \qquad (n = 1, 3, 5, \ldots)$$

$$= 500 \text{ Hz}, 1500 \text{ Hz}, 2500 \text{ Hz}, \ldots$$

EVALUATE: As we increase the frequency, the sound at point P alternates between large and small (near zero) amplitudes, with maxima and minima at the frequencies given above. This effect may not be strong in an ordinary room because of reflections from the walls, floor, and ceiling.

16.24 This aviation headset uses destructive interference to minimize the amount of noise from wind and propellers that reaches the wearer's ears.

Interference is the principle behind active noise-reduction headsets, which are used in loud environments such as airplane cockpits (**Fig. 16.24**). A microphone on the headset detects outside noise, and the headset circuitry replays the noise inside the headset shifted in phase by one half-cycle. This phase-shifted sound interferes destructively with the sounds that enter the headset from outside, so the headset wearer experiences very little unwelcome noise.

TEST YOUR UNDERSTANDING OF SECTION 16.6 Suppose that speaker A in Fig. 16.23 emits a sinusoidal sound wave of frequency 500 Hz and speaker B emits a sinusoidal sound wave of frequency 1000 Hz. What sort of interference will there be between these two waves? (i) Constructive interference at various points, including point P, and destructive interference at various other points; (ii) destructive interference at various points, including point P, and constructive interference at various points; (iii) neither (i) nor (ii). ∎

16.7 BEATS

In Section 16.6 we talked about *interference* effects that occur when two different waves with the same frequency overlap in the same region of space. Now let's look at what happens when we have two waves with equal amplitude but slightly different frequencies. This occurs, for example, when two tuning forks with slightly different frequencies are sounded together, or when two organ pipes that are supposed to have exactly the same frequency are slightly "out of tune."

Consider a particular point in space where the two waves overlap. In **Fig. 16.25a** we plot the displacements of the individual waves at this point as functions of time. The total length of the time axis represents 1 second, and the frequencies are 16 Hz (blue graph) and 18 Hz (red graph). Applying the principle of superposition, we add the two displacement functions to find the total displacement function. The result is the graph of Fig. 16.25b. At certain times the two waves are in phase; their maxima coincide and their amplitudes add. But at certain times (like $t = 0.50$ s in Fig. 16.25) the two waves are exactly *out* of phase. The two waves then cancel each other, and the total amplitude is zero.

The resultant wave in Fig. 16.25b looks like a single sinusoidal wave with an amplitude that varies from a maximum to zero and back. In this example the amplitude goes through two maxima and two minima in 1 second, so the frequency of this amplitude variation is 2 Hz. The amplitude variation causes variations of loudness called **beats,** and the frequency with which the loudness varies is called the **beat frequency.** In this example the beat frequency is the *difference* of the two frequencies. If the beat frequency is a few hertz, we hear it as a waver or pulsation in the tone.

16.25 Beats are fluctuations in amplitude produced by two sound waves of slightly different frequency, here 16 Hz and 18 Hz. (a) Individual waves. (b) Resultant wave formed by superposition of the two waves. The beat frequency is 18 Hz − 16 Hz = 2 Hz.

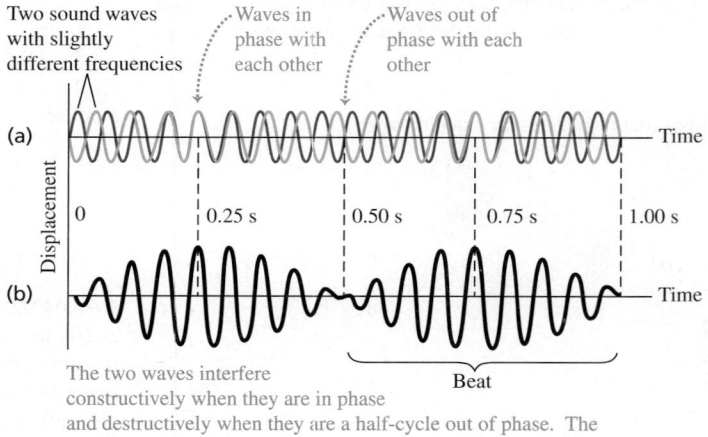

Two sound waves with slightly different frequencies

Waves in phase with each other

Waves out of phase with each other

(a)

Displacement

0   0.25 s   0.50 s   0.75 s   1.00 s

Time

(b)   Time

The two waves interfere constructively when they are in phase and destructively when they are a half-cycle out of phase. The resultant wave rises and falls in intensity, forming beats.

Beat

We can prove that the beat frequency is *always* the difference of the two frequencies f_a and f_b. Suppose f_a is larger than f_b; the corresponding periods are T_a and T_b, with $T_a < T_b$. If the two waves start out in phase at time $t = 0$, they are again in phase when the first wave has gone through exactly one more cycle than the second. This happens at a value of t equal to T_{beat}, the *period* of the beat. Let n be the number of cycles of the first wave in time T_{beat}; then the number of cycles of the second wave in the same time is $(n - 1)$, and we have the relationships

$$T_{\text{beat}} = nT_a \quad \text{and} \quad T_{\text{beat}} = (n - 1)T_b$$

Eliminating n between these two equations, we find

$$T_{\text{beat}} = \frac{T_a T_b}{T_b - T_a}$$

The reciprocal of the beat period is the beat *frequency*, $f_{\text{beat}} = 1/T_{\text{beat}}$, so

$$f_{\text{beat}} = \frac{T_b - T_a}{T_a T_b} = \frac{1}{T_a} - \frac{1}{T_b}$$

and finally

> Beat frequency for $\cdots$ — Frequency of wave a
> waves a and b $\quad f_{\text{beat}} = f_a - f_b \cdots$ Frequency of wave b $\qquad$ (16.24)
> (lower than f_a)

As claimed, the beat frequency is the difference of the two frequencies.

An alternative way to derive Eq. (16.24) is to write functions to describe the curves in Fig. 16.25a and then add them. Suppose that at a certain position the two waves are given by $y_a(t) = A \sin 2\pi f_a t$ and $y_b(t) = -A \sin 2\pi f_b t$. We use the trigonometric identity

$$\sin a - \sin b = 2 \sin\tfrac{1}{2}(a - b) \cos\tfrac{1}{2}(a + b)$$

We can then express the total wave $y(t) = y_a(t) + y_b(t)$ as

$$y_a(t) + y_b(t) = \left[2A \sin\tfrac{1}{2}(2\pi)(f_a - f_b)t \right] \cos\tfrac{1}{2}(2\pi)(f_a + f_b)t$$

The amplitude factor (the quantity in brackets) varies slowly with frequency $\tfrac{1}{2}(f_a - f_b)$. The cosine factor varies with a frequency equal to the *average* frequency $\tfrac{1}{2}(f_a + f_b)$. The *square* of the amplitude factor, which is proportional to the intensity that the ear hears, goes through two maxima and two minima per cycle. So the beat frequency f_{beat} that is heard is twice the quantity $\tfrac{1}{2}(f_a - f_b)$, or just $f_a - f_b$, in agreement with Eq. (16.24).

Beats between two tones can be heard up to a beat frequency of about 6 or 7 Hz. Two piano strings or two organ pipes differing in frequency by 2 or 3 Hz sound wavery and "out of tune," although some organ stops contain two sets of pipes deliberately tuned to beat frequencies of about 1 to 2 Hz for a gently undulating effect. Listening for beats is an important technique in tuning all musical instruments. *Avoiding* beats is part of the task of flying a multiengine propeller airplane (**Fig. 16.26**).

At frequency differences greater than about 6 or 7 Hz, we no longer hear individual beats, and the sensation merges into one of *consonance* or *dissonance*, depending on the frequency ratio of the two tones. In some cases the ear perceives a tone called a *difference tone*, with a pitch equal to the beat frequency of the two tones. For example, if you listen to a whistle that produces sounds at 1800 Hz and 1900 Hz when blown, you will hear not only these tones but also a much lower 100-Hz tone.

16.26 If the two propellers on this airplane are not precisely synchronized, the pilots, passengers, and listeners on the ground will hear beats as loud, annoying, throbbing sounds. On some airplanes the propellers are synched electronically; on others the pilot does it by ear, like tuning a piano.

TEST YOUR UNDERSTANDING OF SECTION 16.7   One tuning fork vibrates at 440 Hz, while a second tuning fork vibrates at an unknown frequency. When both tuning forks are sounded simultaneously, you hear a tone that rises and falls in intensity three times per second. What is the frequency of the second tuning fork? (i) 434 Hz; (ii) 437 Hz; (iii) 443 Hz; (iv) 446 Hz; (v) either 434 Hz or 446 Hz; (vi) either 437 Hz or 443 Hz. ∎

16.8 THE DOPPLER EFFECT

When a car approaches you with its horn sounding, the pitch seems to drop as the car passes. This phenomenon, first described by the 19th-century Austrian scientist Christian Doppler, is called the **Doppler effect.** When a source of sound and a listener are in motion relative to each other, the frequency of the sound heard by the listener is not the same as the source frequency. A similar effect occurs for light and radio waves; we'll return to this later in this section.

To analyze the Doppler effect for sound, we'll work out a relationship between the frequency shift and the velocities of source and listener relative to the medium (usually air) through which the sound waves propagate. To keep things simple, we consider only the special case in which the velocities of both source and listener lie along the line joining them. Let v_S and v_L be the velocity components along this line for the source and the listener, respectively, relative to the medium. We choose the positive direction for both v_S and v_L to be the direction from the listener L to the source S. The speed of sound relative to the medium, v, is always considered positive.

Moving Listener and Stationary Source

Let's think first about a listener L moving with velocity v_L toward a stationary source S (**Fig. 16.27**). The source emits a sound wave with frequency f_S and wavelength $\lambda = v/f_S$. The figure shows four wave crests, separated by equal distances λ. The wave crests approaching the moving listener have a speed of propagation *relative to the listener* of $(v + v_L)$. So the frequency f_L with which the crests arrive at the listener's position (that is, the frequency the listener hears) is

$$f_L = \frac{v + v_L}{\lambda} = \frac{v + v_L}{v/f_S} \tag{16.25}$$

or

$$f_L = \left(\frac{v + v_L}{v}\right)f_S = \left(1 + \frac{v_L}{v}\right)f_S \quad \text{(moving listener, stationary source)} \tag{16.26}$$

16.27 A listener moving toward a stationary source hears a frequency that is higher than the source frequency. This is because the relative speed of listener and wave is greater than the wave speed v.

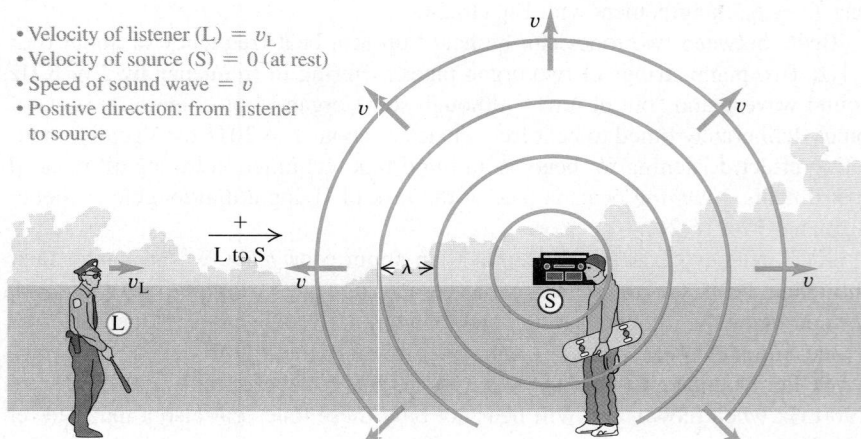

• Velocity of listener (L) = v_L
• Velocity of source (S) = 0 (at rest)
• Speed of sound wave = v
• Positive direction: from listener to source

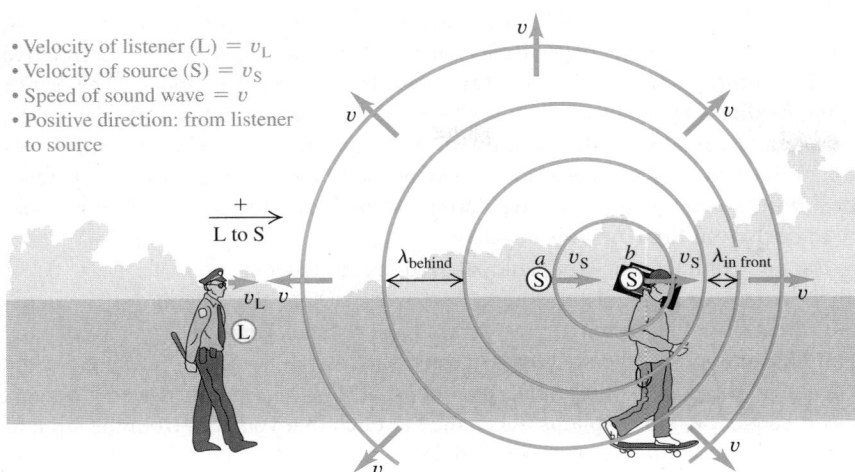

- Velocity of listener (L) = v_L
- Velocity of source (S) = v_S
- Speed of sound wave = v
- Positive direction: from listener to source

16.28 Wave crests emitted by a source moving from *a* to *b* are crowded together in front of the source (to the right of this source) and stretched out behind it (to the left of this source).

So a listener moving toward a source ($v_L > 0$), as in Fig. 16.27, hears a higher frequency (higher pitch) than does a stationary listener. A listener moving away from the source ($v_L < 0$) hears a lower frequency (lower pitch).

Moving Source and Moving Listener

Now suppose the source is also moving, with velocity v_S (**Fig. 16.28**). The wave speed relative to the wave medium (air) is still v; it is determined by the properties of the medium and is not changed by the motion of the source. But the wavelength is no longer equal to v/f_S. Here's why. The time for emission of one cycle of the wave is the period $T = 1/f_S$. During this time, the wave travels a distance $vT = v/f_S$ and the source moves a distance $v_S T = v_S/f_S$. The wavelength is the distance between successive wave crests, and this is determined by the *relative* displacement of source and wave. As Fig. 16.28 shows, this is different in front of and behind the source. In the region to the right of the source in Fig. 16.28 (that is, in front of the source), the wavelength is

$$\lambda_{\text{in front}} = \frac{v}{f_S} - \frac{v_S}{f_S} = \frac{v - v_S}{f_S} \qquad \begin{array}{l}\text{(wavelength in front} \\ \text{of a moving source)}\end{array} \qquad (16.27)$$

In the region to the left of the source (that is, behind the source), it is

$$\lambda_{\text{behind}} = \frac{v + v_S}{f_S} \qquad \begin{array}{l}\text{(wavelength behind} \\ \text{a moving source)}\end{array} \qquad (16.28)$$

The waves in front of and behind the source are compressed and stretched out, respectively, by the motion of the source.

To find the frequency heard by the listener behind the source, we substitute Eq. (16.28) into the first form of Eq. (16.25):

$$f_L = \frac{v + v_L}{\lambda_{\text{behind}}} = \frac{v + v_L}{(v + v_S)/f_S}$$

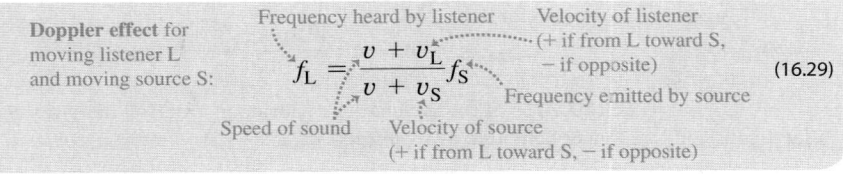

Doppler effect for moving listener L and moving source S:

Frequency heard by listener

Velocity of listener (+ if from L toward S, − if opposite)

$$f_L = \frac{v + v_L}{v + v_S} f_S \qquad (16.29)$$

Frequency emitted by source

Speed of sound · Velocity of source (+ if from L toward S, − if opposite)

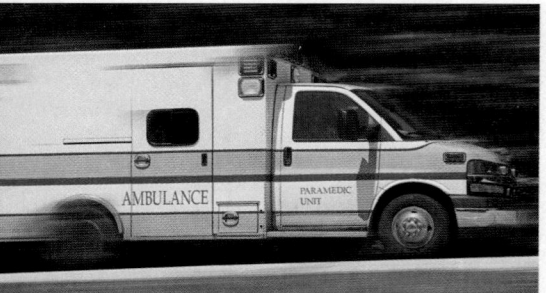

16.29 The Doppler effect explains why the siren on a fire engine or ambulance has a high pitch ($f_L > f_S$) when it is approaching you ($v_S < 0$) and a low pitch ($f_L < f_S$) when it is moving away ($v_S > 0$).

Although we derived it for the particular situation shown in Fig. 16.28, Eq. (16.29) includes *all* possibilities for motion of source and listener (relative to the medium) along the line joining them. If the listener happens to be at rest in the medium, v_L is zero. When both source and listener are at rest or have the same velocity relative to the medium, $v_L = v_S$ and $f_L = f_S$. Whenever the direction of the source or listener velocity is opposite to the direction from the listener toward the source (which we have defined as positive), the corresponding velocity to be used in Eq. (16.29) is negative.

As an example, the frequency heard by a listener at rest ($v_L = 0$) is $f_L = [v/(v + v_S)]f_S$. If the source is moving toward the listener (in the negative direction), then $v_S < 0, f_L > f_S$, and the listener hears a higher frequency than that emitted by the source. If instead the source is moving away from the listener (in the positive direction), then $v_S > 0$, $f_L < f_S$, and the listener hears a lower frequency. This explains the change in pitch that you hear from the siren of an ambulance as it passes you (**Fig. 16.29**).

PROBLEM-SOLVING STRATEGY 16.2 | **DOPPLER EFFECT**

IDENTIFY *the relevant concepts:* The Doppler effect occurs whenever the source of waves, the wave detector (listener), or both are in motion.

SET UP *the problem* using the following steps:
1. Establish a coordinate system, with the positive direction from the listener toward the source. Carefully determine the signs of all relevant velocities. A velocity in the direction from the listener toward the source is positive; a velocity in the opposite direction is negative. All velocities must be measured relative to the air in which the sound travels.
2. Use consistent subscripts to identify the various quantities: S for source and L for listener.
3. Identify which unknown quantities are the target variables.

EXECUTE *the solution* as follows:
1. Use Eq. (16.29) to relate the frequencies at the source and the listener, the sound speed, and the velocities of the source and

the listener according to the sign convention of step 1. If the source is moving, you can find the wavelength measured by the listener using Eq. (16.27) or (16.28).
2. When a wave is reflected from a stationary or moving surface, solve the problem in two steps. In the first, the surface is the "listener"; the frequency with which the wave crests arrive at the surface is f_L. In the second, the surface is the "source," emitting waves with this same frequency f_L. Finally, determine the frequency heard by a listener detecting this new wave.

EVALUATE *your answer:* Is the *direction* of the frequency shift reasonable? If the source and the listener are moving toward each other, $f_L > f_S$; if they are moving apart, $f_L < f_S$. If the source and the listener have no relative motion, $f_L = f_S$.

EXAMPLE 16.14 | **DOPPLER EFFECT I: WAVELENGTHS**

A police car's siren emits a sinusoidal wave with frequency $f_S = 300$ Hz. The speed of sound is 340 m/s and the air is still. (a) Find the wavelength of the waves if the siren is at rest. (b) Find the wavelengths of the waves in front of and behind the siren if it is moving at 30 m/s.

SOLUTION

IDENTIFY and SET UP: In part (a) there is no Doppler effect because neither source nor listener is moving with respect to the air; $v = \lambda f$ gives the wavelength. **Figure 16.30** shows the situation in part (b): The source is in motion, so we find the wavelengths using Eqs. (16.27) and (16.28) for the Doppler effect.

EXECUTE: (a) When the source is at rest,

$$\lambda = \frac{v}{f_S} = \frac{340 \text{ m/s}}{300 \text{ Hz}} = 1.13 \text{ m}$$

16.30 Our sketch for this problem.

Police car

$\lambda_{behind} = ?$ $v_S = 30$ m/s $\lambda_{in\ front} = ?$

(b) From Eq. (16.27), in front of the siren

$$\lambda_{in\ front} = \frac{v - v_S}{f_S} = \frac{340 \text{ m/s} - 30 \text{ m/s}}{300 \text{ Hz}} = 1.03 \text{ m}$$

From Eq. (16.28), behind the siren

$$\lambda_{behind} = \frac{v + v_S}{f_S} = \frac{340 \text{ m/s} + 30 \text{ m/s}}{300 \text{ Hz}} = 1.23 \text{ m}$$

EVALUATE: The wavelength is shorter in front of the siren and longer behind it, as we expect.

SOLUTION

EXAMPLE 16.15 DOPPLER EFFECT II: FREQUENCIES

If a listener L is at rest and the siren in Example 16.14 is moving away from L at 30 m/s, what frequency does the listener hear?

SOLUTION

IDENTIFY and SET UP: Our target variable is the frequency f_L heard by a listener behind the moving source. **Figure 16.31** shows the situation. We have $v_L = 0$ and $v_S = +30$ m/s (positive, since the velocity of the source is in the direction from listener to source).

EXECUTE: From Eq. (16.29),

$$f_L = \frac{v}{v + v_S}f_S = \frac{340 \text{ m/s}}{340 \text{ m/s} + 30 \text{ m/s}}(300 \text{ Hz}) = 276 \text{ Hz}$$

EVALUATE: The source and listener are moving apart, so $f_L < f_S$. Here's a check on our numerical result. From Example 16.14, the

16.31 Our sketch for this problem.

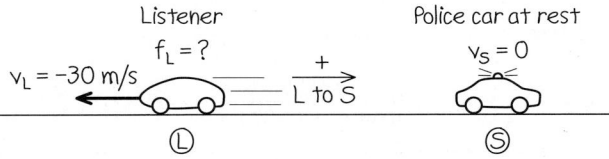

wavelength behind the source (where the listener in Fig. 16.31 is located) is 1.23 m. The wave speed relative to the stationary listener is $v = 340$ m/s even though the source is moving, so

$$f_L = \frac{v}{\lambda} = \frac{340 \text{ m/s}}{1.23 \text{ m}} = 276 \text{ Hz}$$

EXAMPLE 16.16 DOPPLER EFFECT III: A MOVING LISTENER

If the siren is at rest and the listener is moving away from it at 30 m/s, what frequency does the listener hear?

SOLUTION

IDENTIFY and SET UP: Again our target variable is f_L, but now L is in motion and S is at rest. **Figure 16.32** shows the situation. The velocity of the listener is $v_L = -30$ m/s (negative, since the motion is in the direction from source to listener).

EXECUTE: From Eq. (16.29),

$$f_L = \frac{v + v_L}{v}f_S = \frac{340 \text{ m/s} + (-30 \text{ m/s})}{340 \text{ m/s}}(300 \text{ Hz}) = 274 \text{ Hz}$$

16.32 Our sketch for this problem.

EVALUATE: Again the source and listener are moving apart, so $f_L < f_S$. Note that the *relative velocity* of source and listener is the same as in Example 16.15, but the Doppler shift is different because v_S and v_L are different.

EXAMPLE 16.17 DOPPLER EFFECT IV: MOVING SOURCE, MOVING LISTENER

The siren is moving away from the listener with a speed of 45 m/s relative to the air, and the listener is moving toward the siren with a speed of 15 m/s relative to the air. What frequency does the listener hear?

SOLUTION

IDENTIFY and SET UP: Now *both* L and S are in motion (**Fig. 16.33**). Again our target variable is f_L. Both the source velocity $v_S = +45$ m/s and the listener's velocity $v_L = +15$ m/s are positive because both velocities are in the direction from listener to source.

EXECUTE: From Eq. (16.29),

$$f_L = \frac{v + v_L}{v + v_S}f_S = \frac{340 \text{ m/s} + 15 \text{ m/s}}{340 \text{ m/s} + 45 \text{ m/s}}(300 \text{ Hz}) = 277 \text{ Hz}$$

16.33 Our sketch for this problem.

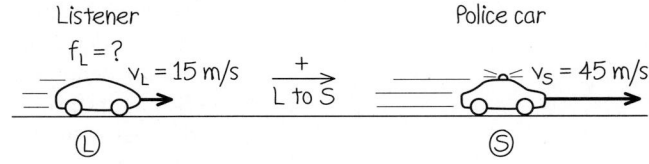

EVALUATE: As in Examples 16.15 and 16.16, the source and listener again move away from each other at 30 m/s, so again $f_L < f_S$. But f_L is different in all three cases because the Doppler effect for sound depends on how the source and listener are moving relative to the *air*, not simply on how they move relative to each other.

EXAMPLE 16.18 **DOPPLER EFFECT V: A DOUBLE DOPPLER SHIFT**

The police car is moving toward a warehouse at 30 m/s. What frequency does the driver hear reflected from the warehouse?

SOLUTION

IDENTIFY: This situation has *two* Doppler shifts (**Fig. 16.34**). In the first shift, the warehouse is the stationary "listener." The frequency of sound reaching the warehouse, which we call f_W, is greater than 300 Hz because the source is approaching. In the second shift, the warehouse acts as a source of sound with frequency f_W, and the listener is the driver of the police car; she hears a frequency greater than f_W because she is approaching the source.

SET UP: To determine f_W, we use Eq. (16.29) with f_L replaced by f_W. For this part of the problem, $v_L = v_W = 0$ (the warehouse is at rest) and $v_S = -30$ m/s (the siren is moving in the negative direction from source to listener).

To determine the frequency heard by the driver (our target variable), we again use Eq. (16.29) but now with f_S replaced by f_W. For this second part of the problem, $v_S = 0$ because the stationary warehouse is the source and the velocity of the listener (the driver) is $v_L = +30$ m/s. (The listener's velocity is positive because it is in the direction from listener to source.)

EXECUTE: The frequency reaching the warehouse is

$$f_W = \frac{v}{v + v_S}f_S = \frac{340 \text{ m/s}}{340 \text{ m/s} + (-30 \text{ m/s})}(300 \text{ Hz}) = 329 \text{ Hz}$$

16.34 Two stages of the sound wave's motion from the police car to the warehouse and back to the police car.

(a) Sound travels from police car's siren (source S) to warehouse ("listener" L).

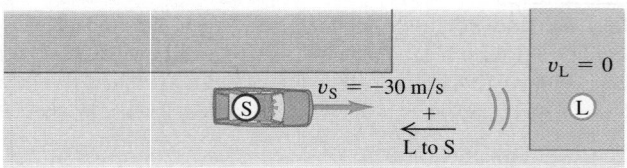

(b) Reflected sound travels from warehouse (source S) to police car (listener L).

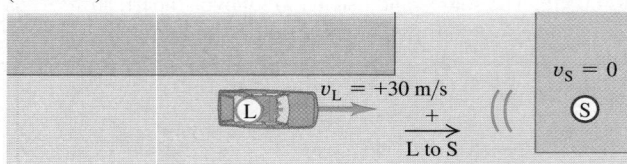

Then the frequency heard by the driver is

$$f_L = \frac{v + v_L}{v}f_W = \frac{340 \text{ m/s} + 30 \text{ m/s}}{340 \text{ m/s}}(329 \text{ Hz}) = 358 \text{ Hz}$$

EVALUATE: Because there are two Doppler shifts, the reflected sound heard by the driver has an even higher frequency than the sound heard by a stationary listener in the warehouse.

Doppler Effect for Electromagnetic Waves

In the Doppler effect for sound, the velocities v_L and v_S are always measured relative to the *air* or whatever medium we are considering. There is also a Doppler effect for *electromagnetic* waves in empty space, such as light waves or radio waves. In this case there is no medium that we can use as a reference to measure velocities, and all that matters is the *relative* velocity of source and receiver. (By contrast, the Doppler effect for sound does not depend simply on this relative velocity, as discussed in Example 16.17.)

To derive the expression for the Doppler frequency shift for light, we have to use the special theory of relativity. We will discuss this in Chapter 37, but for now we quote the result without derivation. The wave speed is the speed of light, usually denoted by c, and it is the same for both source and receiver. In the frame of reference in which the receiver is at rest, the source is moving away from the receiver with velocity v. (If the source is *approaching* the receiver, v is negative.) The source frequency is again f_S. The frequency f_R measured by the receiver R (the frequency of arrival of the waves at the receiver) is then

$$f_R = \sqrt{\frac{c - v}{c + v}}f_S \qquad \text{(Doppler effect for light)} \qquad (16.30)$$

When v is positive, the source is moving directly *away* from the receiver and f_R is always *less* than f_S; when v is negative, the source is moving directly *toward* the receiver and f_R is *greater* than f_S. The qualitative effect is the same as for sound, but the quantitative relationship is different.

A familiar application of the Doppler effect for radio waves is the radar device mounted on the side window of a police car to check other cars' speeds. The electromagnetic wave emitted by the device is reflected from a moving car, which acts as a moving source, and the wave reflected back to the device is Doppler-shifted in frequency. The transmitted and reflected signals are combined to produce beats, and the speed can be computed from the frequency of the beats. Similar techniques ("Doppler radar") are used to measure wind velocities in the atmosphere.

The Doppler effect is also used to track satellites and other space vehicles. In **Fig. 16.35** a satellite emits a radio signal with constant frequency f_S. As the satellite orbits past, it first approaches and then moves away from the receiver; the frequency f_R of the signal received on earth changes from a value greater than f_S to a value less than f_S as the satellite passes overhead.

16.35 Change of velocity component along the line of sight of a satellite passing a tracking station. The frequency received at the tracking station changes from high to low as the satellite passes overhead.

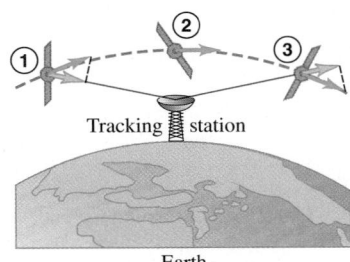

TEST YOUR UNDERSTANDING OF SECTION 16.8 You are at an outdoor concert with a wind blowing at 10 m/s from the performers toward you. Is the sound you hear Doppler-shifted? If so, is it shifted to lower or higher frequencies? ▌

16.9 SHOCK WAVES

You may have experienced "sonic booms" caused by an airplane flying overhead faster than the speed of sound. We can see qualitatively why this happens from **Fig. 16.36**. Let v_S denote the *speed* of the airplane relative to the air, so that it is always positive. The motion of the airplane through the air produces sound; if v_S is less than the speed of sound v, the waves in front of the airplane are crowded together with a wavelength given by Eq. (16.27):

$$\lambda_{\text{in front}} = \frac{v - v_S}{f_S}$$

As the speed v_S of the airplane approaches the speed of sound v, the wavelength approaches zero and the wave crests pile up on each other (Fig. 16.36a). The airplane must exert a large force to compress the air in front of it; by Newton's third law, the air exerts an equally large force back on the airplane. Hence there is a large increase in aerodynamic drag (air resistance) as the airplane approaches the speed of sound, a phenomenon known as the "sound barrier."

16.36 Wave crests around a sound source S moving **(a)** slightly slower than the speed of sound v and **(b)** faster than the sound speed v. **(c)** This photograph shows a T-38 jet airplane moving at 1.1 times the speed of sound. Separate shock waves are produced by the nose, wings, and tail. The angles of these waves vary because the air speeds up and slows down as it moves around the airplane, so the relative speed v_S of the airplane and air is different for shock waves produced at different points.

(a) Sound source S (airplane) moving at nearly the speed of sound

(b) Sound source moving faster than the speed of sound

(c) Shock waves around a supersonic airplane

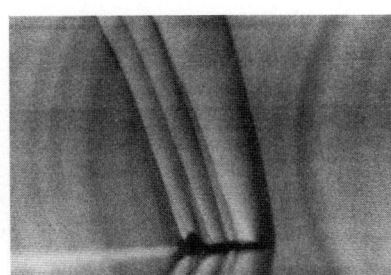

When v_S is greater in magnitude than v, the source of sound is **supersonic,** and Eqs. (16.27) and (16.29) for the Doppler effect no longer describe the sound wave in front of the source. Figure 16.36b shows a cross section of what happens. As the airplane moves, it displaces the surrounding air and produces sound. A series of wave crests is emitted from the nose of the airplane; each spreads out in a circle centered at the position of the airplane when it emitted the crest. After a time t the crest emitted from point S_1 has spread to a circle with radius vt, and the airplane has moved a greater distance v_St to position S_2. You can see that the circular crests interfere constructively at points along the blue line that makes an angle α with the direction of the airplane velocity, leading to a very-large-amplitude wave crest along this line. This large-amplitude crest is called a **shock wave** (Fig. 16.36c).

From the right triangle in Fig. 16.36b we can see that $\sin \alpha = vt/v_St$, or

16.37 The first supersonic airplane, the Bell X-1, was shaped much like a 50-caliber bullet—which was known to be able to travel faster than sound.

Shock wave produced by sound source moving faster than sound:

$$\sin\alpha = \frac{v}{v_S} \qquad \text{(16.31)}$$

Angle of shock wave ⋯ $\overset{\text{Speed of sound}}{\cdots}$ ⋯ Speed of source

The ratio v_S/v is called the **Mach number.** It is greater than unity for all supersonic speeds, and $\sin \alpha$ in Eq. (16.31) is the reciprocal of the Mach number. The first person to break the sound barrier was Capt. Chuck Yeager of the U.S. Air Force, flying the Bell X-1 at Mach 1.06 on October 14, 1947 (**Fig. 16.37**).

Shock waves are actually three-dimensional; a shock wave forms a *cone* around the direction of motion of the source. If the source (possibly a supersonic jet airplane or a rifle bullet) moves with constant velocity, the angle α is constant, and the shock-wave cone moves along with the source. It's the arrival of this shock wave that causes the sonic boom you hear after a supersonic airplane has passed by. In front of the shock-wave cone, there is no sound. Inside the cone a stationary listener hears the Doppler-shifted sound of the airplane moving away.

CAUTION Shock waves A shock wave is produced *continuously* by any object that moves through the air at supersonic speed, not only at the instant that it "breaks the sound barrier." The sound waves that combine to form the shock wave, as in Fig. 16.36b, are created by the motion of the object itself, not by any sound source that the object may carry. The cracking noises of a bullet and of the tip of a circus whip are due to their supersonic motion. A supersonic jet airplane may have very loud engines, but these do not cause the shock wave. If the pilot were to shut the engines off, the airplane would continue to produce a shock wave as long as its speed remained supersonic. ▮

Shock waves have applications outside of aviation. They are used to break up kidney stones and gallstones without invasive surgery, using a technique with the impressive name *extracorporeal shock-wave lithotripsy.* A shock wave produced outside the body is focused by a reflector or acoustic lens so that as much of it as possible converges on the stone. When the resulting stresses in the stone exceed its tensile strength, it breaks into small pieces and can be eliminated. This technique requires accurate determination of the location of the stone, which may be done using ultrasonic imaging techniques (see Fig. 16.9).

EXAMPLE 16.19 SONIC BOOM FROM A SUPERSONIC AIRPLANE

An airplane is flying at Mach 1.75 at an altitude of 8000 m, where the speed of sound is 320 m/s. How long after the plane passes directly overhead will you hear the sonic boom?

SOLUTION

IDENTIFY and SET UP: The shock wave forms a cone trailing backward from the airplane, so the problem is really asking for how much time elapses from when the airplane flies overhead to when the shock wave reaches you at point L (**Fig. 16.38**). During the time t (our target variable) since the airplane traveling at speed v_S passed overhead, it has traveled a distance v_St. Equation (16.31) gives the shock cone angle α; we use trigonometry to solve for t.

EXECUTE: From Eq. (16.31) the angle α of the shock cone is

$$\alpha = \arcsin\frac{1}{1.75} = 34.8°$$

16.38 You hear a sonic boom when the shock wave reaches you at L (*not* just when the plane breaks the sound barrier). A listener to the right of L has not yet heard the sonic boom but will shortly; a listener to the left of L has already heard the sonic boom.

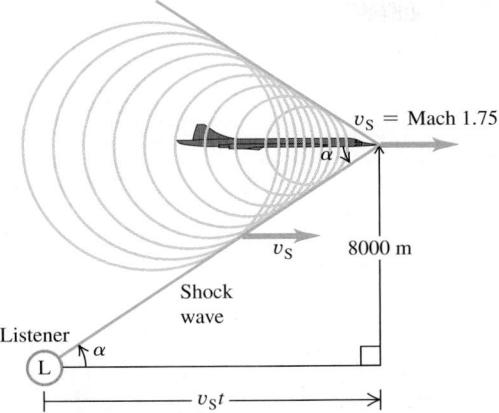

The speed of the plane is the speed of sound multiplied by the Mach number:

$$v_S = (1.75)(320 \text{ m/s}) = 560 \text{ m/s}$$

From Fig. 16.38 we have

$$\tan \alpha = \frac{8000 \text{ m}}{v_S t}$$

$$t = \frac{8000 \text{ m}}{(560 \text{ m/s})(\tan 34.8°)} = 20.5 \text{ s}$$

EVALUATE: You hear the boom 20.5 s after the airplane passes overhead, at which time it has traveled $(560 \text{ m/s})(20.5 \text{ s}) = 11.5$ km since it passed overhead. We have assumed that the speed of sound is the same at all altitudes, so that $\alpha = \arcsin v/v_S$ is constant and the shock wave forms a perfect cone. In fact, the speed of sound decreases with increasing altitude. How would this affect the value of t?

TEST YOUR UNDERSTANDING OF SECTION 16.9 What would you hear if you were directly behind (to the left of) the supersonic airplane in Fig. 16.38? (i) A sonic boom; (ii) the sound of the airplane, Doppler-shifted to higher frequencies; (iii) the sound of the airplane, Doppler-shifted to lower frequencies; (iv) nothing. ∎

CHAPTER 16 SUMMARY

SOLUTIONS TO ALL EXAMPLES

Sound waves: Sound consists of longitudinal waves in a medium. A sinusoidal sound wave is characterized by its frequency f and wavelength λ (or angular frequency ω and wave number k) and by its displacement amplitude A. The pressure amplitude p_{max} is directly proportional to the displacement amplitude, the wave number, and the bulk modulus B of the wave medium. (See Examples 16.1 and 16.2.)

The speed of a sound wave in a fluid depends on the bulk modulus B and density ρ. If the fluid is an ideal gas, the speed can be expressed in terms of the temperature T, molar mass M, and ratio of heat capacities γ of the gas. The speed of longitudinal waves in a solid rod depends on the density and Young's modulus Y. (See Examples 16.3 and 16.4.)

$$p_{max} = BkA \quad (16.5)$$
(sinusoidal sound wave)

$$v = \sqrt{\frac{B}{\rho}} \quad (16.7)$$
(longitudinal wave in a fluid)

$$v = \sqrt{\frac{\gamma RT}{M}} \quad (16.10)$$
(sound wave in an ideal gas)

$$v = \sqrt{\frac{Y}{\rho}} \quad (16.8)$$
(longitudinal wave in a solid rod)

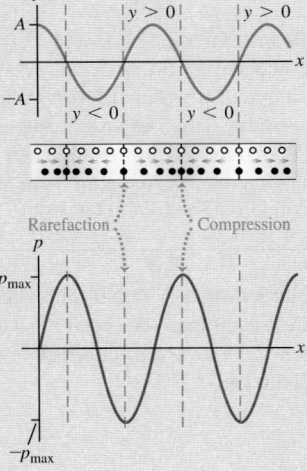

Intensity and sound intensity level: The intensity I of a sound wave is the time average rate at which energy is transported by the wave, per unit area. For a sinusoidal wave, the intensity can be expressed in terms of the displacement amplitude A or the pressure amplitude p_{max}. (See Examples 16.5–16.7.)

The sound intensity level β of a sound wave is a logarithmic measure of its intensity. It is measured relative to I_0, an arbitrary intensity defined to be 10^{-12} W/m². Sound intensity levels are expressed in decibels (dB). (See Examples 16.8 and 16.9.)

$$I = \frac{1}{2}\sqrt{\rho B}\,\omega^2 A^2 = \frac{p_{max}^2}{2\rho v}$$

$$= \frac{p_{max}^2}{2\sqrt{\rho B}} \quad (16.12), (16.14)$$
(intensity of a sinusoidal sound wave in a fluid)

$$\beta = (10 \text{ dB}) \log\frac{I}{I_0} \quad (16.15)$$
(definition of sound intensity level)

Standing sound waves: Standing sound waves can be set up in a pipe or tube. A closed end is a displacement node and a pressure antinode; an open end is a displacement antinode and a pressure node. For a pipe of length L open at both ends, the normal-mode frequencies are integer multiples of the sound speed divided by $2L$. For a stopped pipe (one that is open at only one end), the normal-mode frequencies are the odd multiples of the sound speed divided by $4L$. (See Examples 16.10 and 16.11.)

A pipe or other system with normal-mode frequencies can be driven to oscillate at any frequency. A maximum response, or resonance, occurs if the driving frequency is close to one of the normal-mode frequencies of the system. (See Example 16.12.)

$$f_n = \frac{nv}{2L} \quad (n = 1, 2, 3, \ldots) \quad (16.18)$$
(open pipe)

$$f_n = \frac{nv}{4L} \quad (n = 1, 3, 5, \ldots) \quad (16.22)$$
(stopped pipe)

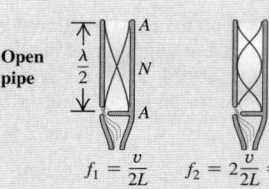

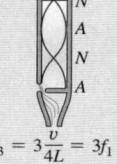

Interference: When two or more waves overlap in the same region of space, the resulting effects are called interference. The resulting amplitude can be either larger or smaller than the amplitude of each individual wave, depending on whether the waves are in phase (constructive interference) or out of phase (destructive interference). (See Example 16.13.)

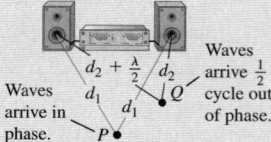

Beats: Beats are heard when two tones with slightly different frequencies f_a and f_b are sounded together. The beat frequency f_{beat} is the difference between f_a and f_b.

$$f_{\text{beat}} = f_a - f_b \quad (16.24)$$
(beat frequency)

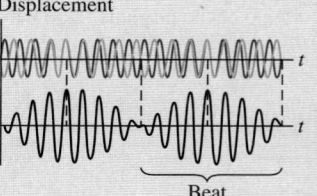

Doppler effect: The Doppler effect for sound is the frequency shift that occurs when there is motion of a source of sound, a listener, or both, relative to the medium. The source and listener frequencies f_S and f_L are related by the source and listener velocities v_S and v_L relative to the medium and to the speed of sound v. (See Examples 16.14–16.18.)

$$f_L = \frac{v + v_L}{v + v_S} f_S \quad (16.29)$$
(Doppler effect, moving source and moving listener)

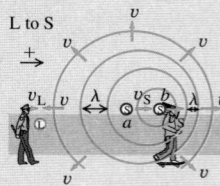

Shock waves: A sound source moving with a speed v_S greater than the speed of sound v creates a shock wave. The wave front is a cone with angle α. (See Example 16.19.)

$$\sin \alpha = \frac{v}{v_S} \quad \text{(shock wave)} \quad (16.31)$$

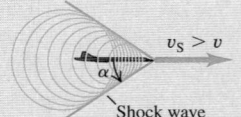

BRIDGING PROBLEM | **LOUDSPEAKER INTERFERENCE**

Loudspeakers *A* and *B* are 7.00 m apart and vibrate in phase at 172 Hz. They radiate sound uniformly in all directions. Their acoustic power outputs are 8.00×10^{-4} W and 6.00×10^{-5} W, respectively. The air temperature is 20°C. (a) Determine the difference in phase of the two signals at a point *C* along the line joining *A* and *B*, 3.00 m from *B* and 4.00 m from *A* (**Fig. 16.39**). (b) Determine the intensity and sound intensity level at *C* from speaker *A* alone (with *B* turned off) and from speaker *B* alone (with *A* turned off). (c) Determine the intensity and sound intensity level at *C* from both speakers together.

SOLUTION GUIDE

IDENTIFY and SET UP

1. Choose the equations that relate power, distance from the source, intensity, pressure amplitude, and sound intensity level.

16.39 The situation for this problem.

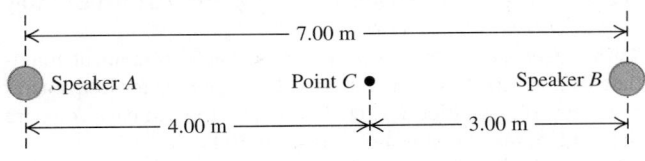

2. Decide how you will determine the phase difference in part (a). Once you have found the phase difference, how can you use it to find the amplitude of the combined wave at *C* due to both sources?

3. List the unknown quantities for each part of the problem and identify your target variables.

EXECUTE

4. Determine the phase difference at point *C*.
5. Find the intensity, sound intensity level, and pressure amplitude at *C* due to each speaker alone.
6. Use your results from steps 4 and 5 to find the pressure amplitude at *C* due to both loudspeakers together.
7. Use your result from step 6 to find the intensity and sound intensity level at *C* due to both loudspeakers together.

EVALUATE

8. How do your results from part (c) for intensity and sound intensity level at *C* compare to those from part (b)? Does this make sense?
9. What result would you have gotten in part (c) if you had (incorrectly) combined the *intensities* from *A* and *B* directly, rather than (correctly) combining the *pressure amplitudes* as you did in step 6?

Problems

For assigned homework and other learning materials, go to MasteringPhysics®. **MP**

•, ••, •••: Difficulty levels. **CP**: Cumulative problems incorporating material from earlier chapters. **CALC**: Problems requiring calculus.
DATA: Problems involving real data, scientific evidence, experimental design, and/or statistical reasoning. **BIO**: Biosciences problems.

DISCUSSION QUESTIONS

Q16.1 When sound travels from air into water, does the frequency of the wave change? The speed? The wavelength? Explain your reasoning.

Q16.2 The hero of a western movie listens for an oncoming train by putting his ear to the track. Why does this method give an earlier warning of the approach of a train than just listening in the usual way?

Q16.3 Would you expect the pitch (or frequency) of an organ pipe to increase or decrease with increasing temperature? Explain.

Q16.4 In most modern wind instruments the pitch is changed by using keys or valves to change the length of the vibrating air column. The bugle, however, has no valves or keys, yet it can play many notes. How might this be possible? Are there restrictions on what notes a bugle can play?

Q16.5 Symphonic musicians always "warm up" their wind instruments by blowing into them before a performance. What purpose does this serve?

Q16.6 In a popular and amusing science demonstration, a person inhales helium and then his voice becomes high and squeaky. Why does this happen? (*Warning:* Inhaling too much helium can cause unconsciousness or death.)

Q16.7 Lane dividers on highways sometimes have regularly spaced ridges or ripples. When the tires of a moving car roll along such a divider, a musical note is produced. Why? Explain how this phenomenon could be used to measure the car's speed.

Q16.8 (a) Does a sound level of 0 dB mean that there is no sound? (b) Is there any physical meaning to a sound having a negative intensity level? If so, what is it? (c) Does a sound intensity of zero mean that there is no sound? (d) Is there any physical meaning to a sound having a negative intensity? Why?

Q16.9 Which has a more direct influence on the loudness of a sound wave: the *displacement* amplitude or the *pressure* amplitude? Explain.

Q16.10 If the pressure amplitude of a sound wave is halved, by what factor does the intensity of the wave decrease? By what factor must the pressure amplitude of a sound wave be increased in order to increase the intensity by a factor of 16? Explain.

Q16.11 Does the sound intensity level β obey the inverse-square law? Why?

Q16.12 A small fraction of the energy in a sound wave is absorbed by the air through which the sound passes. How does this modify the inverse-square relationship between intensity and distance from the source? Explain.

Q16.13 A small metal band is slipped onto one of the tines of a tuning fork. As this band is moved closer and closer to the end of the tine, what effect does this have on the wavelength and frequency of the sound the tine produces? Why?

Q16.14 An organist in a cathedral plays a loud chord and then releases the keys. The sound persists for a few seconds and gradually dies away. Why does it persist? What happens to the sound energy when the sound dies away?

Q16.15 Two loudspeakers, A and B, are driven by the same amplifier and emit sinusoidal waves in phase. The frequency of the waves emitted by each speaker is 860 Hz. Point P is 12.0 m from A and 13.4 m from B. Is the interference at P constructive or destructive? Give the reasoning behind your answer.

Q16.16 Two vibrating tuning forks have identical frequencies, but one is stationary and the other is mounted at the rim of a rotating platform. What does a listener hear? Explain.

Q16.17 A large church has part of the organ in the front of the church and part in the back. A person walking rapidly down the aisle while both segments are playing at once reports that the two segments sound out of tune. Why?

Q16.18 A sound source and a listener are both at rest on the earth, but a strong wind is blowing from the source toward the listener. Is there a Doppler effect? Why or why not?

Q16.19 Can you think of circumstances in which a Doppler effect would be observed for surface waves in water? For elastic waves propagating in a body of water deep below the surface? If so, describe the circumstances and explain your reasoning. If not, explain why not.

Q16.20 Stars other than our sun normally appear featureless when viewed through telescopes. Yet astronomers can readily use the light from these stars to determine that they are rotating and even measure the speed of their surface. How do you think they can do this?

Q16.21 If you wait at a railroad crossing as a train approaches and passes, you hear a Doppler shift in its sound. But if you listen closely, you hear that the change in frequency is continuous; it does not suddenly go from one high frequency to another low frequency. Instead the frequency *smoothly* (but rather quickly) changes from high to low as the train passes. Why does this smooth change occur?

Q16.22 In case 1, a source of sound approaches a stationary observer at speed u. In case 2, the observer moves toward the stationary source at the same speed u. If the source is always producing the same frequency sound, will the observer hear the same frequency in both cases, since the relative speed is the same each time? Why or why not?

Q16.23 Does an aircraft make a sonic boom only at the instant its speed exceeds Mach 1? Explain.

Q16.24 If you are riding in a supersonic aircraft, what do you hear? Explain. In particular, do you hear a continuous sonic boom? Why or why not?

Q16.25 A jet airplane is flying at a constant altitude at a steady speed v_S greater than the speed of sound. Describe what observers at points A, B, and C hear at the instant shown in **Fig. Q16.25**, when the shock wave has just reached point B. Explain.

Figure **Q16.25**

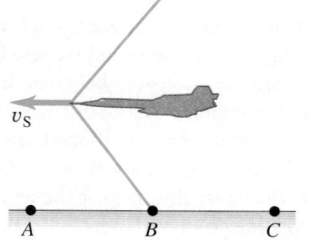

EXERCISES

Unless indicated otherwise, assume the speed of sound in air to be $v = 344$ m/s.

Section 16.1 Sound Waves

16.1 • Example 16.1 (Section 16.1) showed that for sound waves in air with frequency 1000 Hz, a displacement amplitude of 1.2×10^{-8} m produces a pressure amplitude of 3.0×10^{-2} Pa. (a) What is the wavelength of these waves? (b) For 1000-Hz waves in air, what displacement amplitude would be needed for the pressure amplitude to be at the pain threshold, which is 30 Pa? (c) For what wavelength and frequency will waves with a displacement amplitude of 1.2×10^{-8} m produce a pressure amplitude of 1.5×10^{-3} Pa?

16.2 • Example 16.1 (Section 16.1) showed that for sound waves in air with frequency 1000 Hz, a displacement amplitude of 1.2×10^{-8} m produces a pressure amplitude of 3.0×10^{-2} Pa. Water at 20°C has a bulk modulus of 2.2×10^9 Pa, and the speed of sound in water at this temperature is 1480 m/s. For 1000-Hz sound waves in 20°C water, what displacement amplitude is produced if the pressure amplitude is 3.0×10^{-2} Pa? Explain why your answer is much less than 1.2×10^{-8} m.

16.3 • Consider a sound wave in air that has displacement amplitude 0.0200 mm. Calculate the pressure amplitude for frequencies of (a) 150 Hz; (b) 1500 Hz; (c) 15,000 Hz. In each case compare the result to the pain threshold, which is 30 Pa.

16.4 • A loud factory machine produces sound having a displacement amplitude of 1.00 μm, but the frequency of this sound can be adjusted. In order to prevent ear damage to the workers, the maximum pressure amplitude of the sound waves is limited to 10.0 Pa. Under the conditions of this factory, the bulk modulus of air is 1.42×10^5 Pa. What is the highest-frequency sound to which this machine can be adjusted without exceeding the prescribed limit? Is this frequency audible to the workers?

16.5 • **BIO** **Ultrasound and Infrasound.** (a) **Whale communication.** Blue whales apparently communicate with each other using sound of frequency 17 Hz, which can be heard nearly 1000 km away in the ocean. What is the wavelength of such a sound in seawater, where the speed of sound is 1531 m/s? (b) **Dolphin clicks.** One type of sound that dolphins emit is a sharp click of wavelength 1.5 cm in the ocean. What is the frequency of such clicks? (c) **Dog whistles.** One brand of dog whistles claims a frequency of 25 kHz for its product. What is the wavelength of this sound? (d) **Bats.** While bats emit a wide variety of sounds, one type emits pulses of sound having a frequency between 39 kHz and 78 kHz. What is the range of wavelengths of this sound? (e) **Sonograms.** Ultrasound is used to view the interior of the body, much as x rays are utilized. For sharp imagery, the wavelength of the sound should be around one-fourth (or less) the size of the objects to be viewed. Approximately what frequency of sound is needed to produce a clear image of a tumor that is 1.0 mm across if the speed of sound in the tissue is 1550 m/s?

Section 16.2 Speed of Sound Waves

16.6 • (a) In a liquid with density 1300 kg/m³, longitudinal waves with frequency 400 Hz are found to have wavelength 8.00 m. Calculate the bulk modulus of the liquid. (b) A metal bar with a length of 1.50 m has density 6400 kg/m³. Longitudinal sound waves take 3.90×10^{-4} s to travel from one end of the bar to the other. What is Young's modulus for this metal?

16.7 • A submerged scuba diver hears the sound of a boat horn directly above her on the surface of the lake. At the same time, a friend on dry land 22.0 m from the boat also hears the horn (**Fig. E16.7**). The horn is 1.2 m above the surface of the water. What is the distance (labeled "?") from the horn to the diver? Both air and water are at 20°C.

Figure **E16.7**

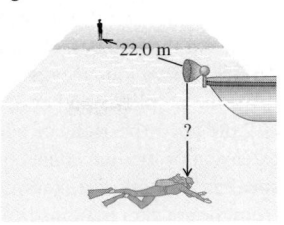
22.0 m
?

16.8 • At a temperature of 27.0°C, what is the speed of longitudinal waves in (a) hydrogen (molar mass 2.02 g/mol); (b) helium (molar mass 4.00 g/mol); (c) argon (molar mass 39.9 g/mol)? See Table 19.1 for values of γ. (d) Compare your answers for parts (a), (b), and (c) with the speed in air at the same temperature.

16.9 • An oscillator vibrating at 1250 Hz produces a sound wave that travels through an ideal gas at 325 m/s when the gas temperature is 22.0°C. For a certain experiment, you need to have the same oscillator produce sound of wavelength 28.5 cm in this gas. What should the gas temperature be to achieve this wavelength?

16.10 •• CALC (a) Show that the fractional change in the speed of sound (dv/v) due to a very small temperature change dT is given by $dv/v = \frac{1}{2}dT/T$. (*Hint:* Start with Eq. 16.10.) (b) The speed of sound in air at 20°C is found to be 344 m/s. Use the result in part (a) to find the change in the speed of sound for a 1.0°C change in air temperature.

16.11 •• A 60.0-m-long brass rod is struck at one end. A person at the other end hears two sounds as a result of two longitudinal waves, one traveling in the metal rod and the other traveling in air. What is the time interval between the two sounds? (The speed of sound in air is 344 m/s; see Tables 11.1 and 12.1 for relevant information about brass.)

16.12 •• What must be the stress (F/A) in a stretched wire of a material whose Young's modulus is Y for the speed of longitudinal waves to equal 30 times the speed of transverse waves?

Section 16.3 Sound Intensity

16.13 •• BIO **Energy Delivered to the Ear.** Sound is detected when a sound wave causes the tympanic membrane (the eardrum) to vibrate. Typically, the diameter of this membrane is about 8.4 mm in humans. (a) How much energy is delivered to the eardrum each second when someone whispers (20 dB) a secret in your ear? (b) To comprehend how sensitive the ear is to very small amounts of energy, calculate how fast a typical 2.0-mg mosquito would have to fly (in mm/s) to have this amount of kinetic energy.

16.14 • (a) By what factor must the sound intensity be increased to raise the sound intensity level by 13.0 dB? (b) Explain why you don't need to know the original sound intensity.

16.15 •• **Eavesdropping!** You are trying to overhear a juicy conversation, but from your distance of 15.0 m, it sounds like only an average whisper of 20.0 dB. How close should you move to the chatterboxes for the sound level to be 60.0 dB?

16.16 •• BIO **Human Hearing.** A fan at a rock concert is 30 m from the stage, and at this point the sound intensity level is 110 dB. (a) How much energy is transferred to her eardrums each second? (b) How fast would a 2.0-mg mosquito have to fly (in mm/s) to have this much kinetic energy? Compare the mosquito's speed with that found for the whisper in part (a) of Exercise 16.13.

16.17 • A sound wave in air at 20°C has a frequency of 320 Hz and a displacement amplitude of 5.00×10^{-3} mm. For this sound wave calculate the (a) pressure amplitude (in Pa); (b) intensity (in W/m²); (c) sound intensity level (in decibels).

16.18 •• You live on a busy street, but as a music lover, you want to reduce the traffic noise. (a) If you install special sound-reflecting windows that reduce the sound intensity level (in dB) by 30 dB, by what fraction have you lowered the sound intensity (in W/m²)? (b) If, instead, you reduce the intensity by half, what change (in dB) do you make in the sound intensity level?

16.19 • BIO For a person with normal hearing, the faintest sound that can be heard at a frequency of 400 Hz has a pressure amplitude of about 6.0×10^{-5} Pa. Calculate the (a) intensity; (b) sound intensity level; (c) displacement amplitude of this sound wave at 20°C.

16.20 •• The intensity due to a number of independent sound sources is the sum of the individual intensities. (a) When four quadruplets cry simultaneously, how many decibels greater is the sound intensity level than when a single one cries? (b) To increase the sound intensity level again by the same number of decibels as in part (a), how many more crying babies are required?

16.21 • CP A baby's mouth is 30 cm from her father's ear and 1.50 m from her mother's ear. What is the difference between the sound intensity levels heard by the father and by the mother?

16.22 •• The Sacramento City Council adopted a law to reduce the allowed sound intensity level of the much-despised leaf blowers from their current level of about 95 dB to 70 dB. With the new law, what is the ratio of the new allowed intensity to the previously allowed intensity?

16.23 •• CP At point A, 3.0 m from a small source of sound that is emitting uniformly in all directions, the sound intensity level is 53 dB. (a) What is the intensity of the sound at A? (b) How far from the source must you go so that the intensity is one-fourth of what it was at A? (c) How far must you go so that the sound intensity level is one-fourth of what it was at A? (d) Does intensity obey the inverse-square law? What about sound intensity level?

16.24 •• (a) If two sounds differ by 5.00 dB, find the ratio of the intensity of the louder sound to that of the softer one. (b) If one sound is 100 times as intense as another, by how much do they differ in sound intensity level (in decibels)? (c) If you increase the volume of your stereo so that the intensity doubles, by how much does the sound intensity level increase?

Section 16.4 Standing Sound Waves and Normal Modes

16.25 • Standing sound waves are produced in a pipe that is 1.20 m long. For the fundamental and first two overtones, determine the locations along the pipe (measured from the left end) of the displacement nodes and the pressure nodes if (a) the pipe is open at both ends and (b) the pipe is closed at the left end and open at the right end.

16.26 • The fundamental frequency of a pipe that is open at both ends is 524 Hz. (a) How long is this pipe? If one end is now closed, find (b) the wavelength and (c) the frequency of the new fundamental.

16.27 • BIO **The Human Voice.** The human vocal tract is a pipe that extends about 17 cm from the lips to the vocal folds (also called "vocal cords") near the middle of your throat. The vocal folds behave rather like the reed of a clarinet, and the vocal tract acts like a stopped pipe. Estimate the first three standing-wave frequencies of the vocal tract. Use $v = 344$ m/s. (The answers are only an estimate, since the position of lips and tongue affects the motion of air in the vocal tract.)

16.28 ⋅⋅ BIO **The Vocal Tract.** Many opera singers (and some pop singers) have a range of about $2\frac{1}{2}$ octaves or even greater. Suppose a soprano's range extends from A below middle C (frequency 220 Hz) up to E-flat above high C (frequency 1244 Hz). Although the vocal tract is quite complicated, we can model it as a resonating air column, like an organ pipe, that is open at the top and closed at the bottom. The column extends from the mouth down to the diaphragm in the chest cavity, and we can also assume that the lowest note is the fundamental. How long is this column of air if $v = 354$ m/s? Does your result seem reasonable, on the basis of observations of your own body?

16.29 ⋅ The longest pipe found in most medium-size pipe organs is 4.88 m (16 ft) long. What is the frequency of the note corresponding to the fundamental mode if the pipe is (a) open at both ends, (b) open at one end and closed at the other?

16.30 ⋅ **Singing in the Shower.** A pipe closed at both ends can have standing waves inside of it, but you normally don't hear them because little of the sound can get out. But you *can* hear them if you are *inside* the pipe, such as someone singing in the shower. (a) Show that the wavelengths of standing waves in a pipe of length L that is closed at both ends are $\lambda_n = 2L/n$ and the frequencies are given by $f_n = nv/2L = nf_1$, where $n = 1, 2, 3, \dots$. (b) Modeling it as a pipe, find the frequency of the fundamental and the first two overtones for a shower 2.50 m tall. Are these frequencies audible?

Section 16.5 Resonance and Sound

16.31 ⋅ You blow across the open mouth of an empty test tube and produce the fundamental standing wave of the air column inside the test tube. The speed of sound in air is 344 m/s and the test tube acts as a stopped pipe. (a) If the length of the air column in the test tube is 14.0 cm, what is the frequency of this standing wave? (b) What is the frequency of the fundamental standing wave in the air column if the test tube is half filled with water?

16.32 ⋅⋅ CP You have a stopped pipe of adjustable length close to a taut 62.0-cm, 7.25-g wire under a tension of 4110 N. You want to adjust the length of the pipe so that, when it produces sound at its fundamental frequency, this sound causes the wire to vibrate in its second *overtone* with very large amplitude. How long should the pipe be?

16.33 ⋅⋅ A 75.0-cm-long wire of mass 5.625 g is tied at both ends and adjusted to a tension of 35.0 N. When it is vibrating in its second overtone, find (a) the frequency and wavelength at which it is vibrating and (b) the frequency and wavelength of the sound waves it is producing.

Section 16.6 Interference of Waves

16.34 ⋅ Small speakers A and B are driven in phase at 725 Hz by the same audio oscillator. Both speakers start out 4.50 m from the listener, but speaker A is slowly moved away (**Fig. E16.34**). (a) At what distance d will the sound from the speakers first produce destructive interference at the listener's location? (b) If A is moved even farther away than in part (a), at what distance d will the speakers next produce destructive interference at the listener's

Figure **E16.34**

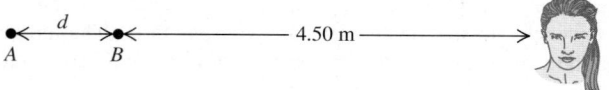

location? (c) After A starts moving away from its original spot, at what distance d will the speakers first produce constructive interference at the listener's location?

16.35 ⋅ Two loudspeakers, A and B (**Fig. E16.35**), are driven by the same amplifier and emit sinusoidal waves in phase. Speaker B is 2.00 m to the right of speaker A. Consider point Q along the extension of the line connecting the speakers, 1.00 m to the right of speaker B. Both speakers emit sound waves that travel directly from the speaker to point Q. What is the lowest frequency for which (a) *constructive* interference occurs at point Q; (b) *destructive* interference occurs at point Q?

Figure **E16.35**

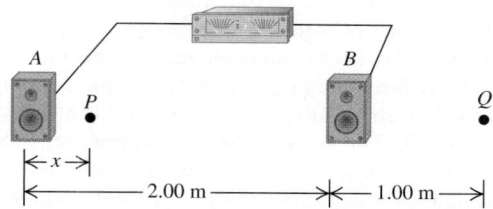

16.36 ⋅⋅ Two loudspeakers, A and B (see Fig. E16.35), are driven by the same amplifier and emit sinusoidal waves in phase. Speaker B is 2.00 m to the right of speaker A. The frequency of the sound waves produced by the loudspeakers is 206 Hz. Consider a point P between the speakers and along the line connecting them, a distance x to the right of A. Both speakers emit sound waves that travel directly from the speaker to point P. For what values of x will (a) *destructive* interference occur at P; (b) *constructive* interference occur at P? (c) Interference effects like those in parts (a) and (b) are almost never a factor in listening to home stereo equipment. Why not?

16.37 ⋅⋅ Two loudspeakers, A and B, are driven by the same amplifier and emit sinusoidal waves in phase. Speaker B is 12.0 m to the right of speaker A. The frequency of the waves emitted by each speaker is 688 Hz. You are standing between the speakers, along the line connecting them, and are at a point of constructive interference. How far must you walk toward speaker B to move to a point of destructive interference?

16.38 ⋅ Two loudspeakers, A and B, are driven by the same amplifier and emit sinusoidal waves in phase. The frequency of the waves emitted by each speaker is 172 Hz. You are 8.00 m from A. What is the closest you can be to B and be at a point of destructive interference?

16.39 ⋅⋅ Two small stereo speakers are driven in step by the same variable-frequency oscillator. Their sound is picked up by a microphone arranged as shown in **Fig. E16.39**. For what frequencies does their sound at the speakers produce (a) constructive interference and (b) destructive interference?

Figure **E16.39**

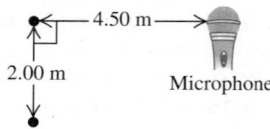

Section 16.7 Beats

16.40 ⋅⋅ Two guitarists attempt to play the same note of wavelength 64.8 cm at the same time, but one of the instruments is slightly out of tune and plays a note of wavelength 65.2 cm instead. What is the frequency of the beats these musicians hear when they play together?

16.41 •• **Tuning a Violin.** A violinist is tuning her instrument to concert A (440 Hz). She plays the note while listening to an electronically generated tone of exactly that frequency and hears a beat frequency of 3 Hz, which increases to 4 Hz when she tightens her violin string slightly. (a) What was the frequency of the note played by her violin when she heard the 3-Hz beats? (b) To get her violin perfectly tuned to concert A, should she tighten or loosen her string from what it was when she heard the 3-Hz beats?

16.42 •• **Adjusting Airplane Motors.** The motors that drive airplane propellers are, in some cases, tuned by using beats. The whirring motor produces a sound wave having the same frequency as the propeller. (a) If one single-bladed propeller is turning at 575 rpm and you hear 2.0-Hz beats when you run the second propeller, what are the two possible frequencies (in rpm) of the second propeller? (b) Suppose you increase the speed of the second propeller slightly and find that the beat frequency changes to 2.1 Hz. In part (a), which of the two answers was the correct one for the frequency of the second single-bladed propeller? How do you know?

16.43 •• Two organ pipes, open at one end but closed at the other, are each 1.14 m long. One is now lengthened by 2.00 cm. Find the beat frequency that they produce when playing together in their fundamentals.

Section 16.8 The Doppler Effect

16.44 •• In Example 16.18 (Section 16.8), suppose the police car is moving away from the warehouse at 20 m/s. What frequency does the driver of the police car hear reflected from the warehouse?

16.45 •• On the planet Arrakis a male ornithoid is flying toward his mate at 25.0 m/s while singing at a frequency of 1200 Hz. If the stationary female hears a tone of 1240 Hz, what is the speed of sound in the atmosphere of Arrakis?

16.46 • A railroad train is traveling at 25.0 m/s in still air. The frequency of the note emitted by the locomotive whistle is 400 Hz. What is the wavelength of the sound waves (a) in front of the locomotive and (b) behind the locomotive? What is the frequency of the sound heard by a stationary listener (c) in front of the locomotive and (d) behind the locomotive?

16.47 • Two train whistles, A and B, each have a frequency of 392 Hz. A is stationary and B is moving toward the right (away from A) at a speed of 35.0 m/s. A listener is between the two whistles and is moving toward the right with a speed of 15.0 m/s (**Fig. E16.47**). No wind is blowing. (a) What is the frequency from A as heard by the listener? (b) What is the frequency from B as heard by the listener? (c) What is the beat frequency detected by the listener?

Figure **E16.47**

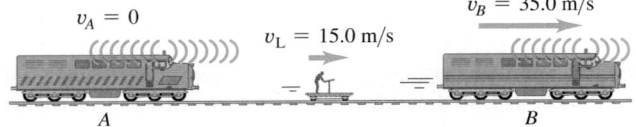

16.48 • **Moving Source vs. Moving Listener.** (a) A sound source producing 1.00-kHz waves moves toward a stationary listener at one-half the speed of sound. What frequency will the listener hear? (b) Suppose instead that the source is stationary and the listener moves toward the source at one-half the speed of sound. What frequency does the listener hear? How does your

answer compare to that in part (a)? Explain on physical grounds why the two answers differ.

16.49 • A swimming duck paddles the water with its feet once every 1.6 s, producing surface waves with this period. The duck is moving at constant speed in a pond where the speed of surface waves is 0.32 m/s, and the crests of the waves ahead of the duck are spaced 0.12 m apart. (a) What is the duck's speed? (b) How far apart are the crests behind the duck?

16.50 • A railroad train is traveling at 30.0 m/s in still air. The frequency of the note emitted by the train whistle is 352 Hz. What frequency is heard by a passenger on a train moving in the opposite direction to the first at 18.0 m/s and (a) approaching the first and (b) receding from the first?

16.51 • A car alarm is emitting sound waves of frequency 520 Hz. You are on a motorcycle, traveling directly away from the parked car. How fast must you be traveling if you detect a frequency of 490 Hz?

16.52 •• While sitting in your car by the side of a country road, you are approached by your friend, who happens to be in an identical car. You blow your car's horn, which has a frequency of 260 Hz. Your friend blows his car's horn, which is identical to yours, and you hear a beat frequency of 6.0 Hz. How fast is your friend approaching you?

16.53 • Two swift canaries fly toward each other, each moving at 15.0 m/s relative to the ground, each warbling a note of frequency 1750 Hz. (a) What frequency note does each bird hear from the other one? (b) What wavelength will each canary measure for the note from the other one?

16.54 •• The siren of a fire engine that is driving northward at 30.0 m/s emits a sound of frequency 2000 Hz. A truck in front of this fire engine is moving northward at 20.0 m/s. (a) What is the frequency of the siren's sound that the fire engine's driver hears reflected from the back of the truck? (b) What wavelength would this driver measure for these reflected sound waves?

16.55 •• A stationary police car emits a sound of frequency 1200 Hz that bounces off a car on the highway and returns with a frequency of 1250 Hz. The police car is right next to the highway, so the moving car is traveling directly toward or away from it. (a) How fast was the moving car going? Was it moving toward or away from the police car? (b) What frequency would the police car have received if it had been traveling toward the other car at 20.0 m/s?

16.56 •• How fast (as a percentage of light speed) would a star have to be moving so that the frequency of the light we receive from it is 10.0% higher than the frequency of the light it is emitting? Would it be moving away from us or toward us? (Assume it is moving either directly away from us or directly toward us.)

Section 16.9 Shock Waves

16.57 •• A jet plane flies overhead at Mach 1.70 and at a constant altitude of 1250 m. (a) What is the angle α of the shock-wave cone? (b) How much time after the plane passes directly overhead do you hear the sonic boom? Neglect the variation of the speed of sound with altitude.

16.58 • The shock-wave cone created by a space shuttle at one instant during its reentry into the atmosphere makes an angle of 58.0° with its direction of motion. The speed of sound at this altitude is 331 m/s. (a) What is the Mach number of the shuttle at this instant, and (b) how fast (in m/s and in mi/h) is it traveling relative to the atmosphere? (c) What would be its Mach number and the angle of its shock-wave cone if it flew at the same speed but at low altitude where the speed of sound is 344 m/s?

PROBLEMS

16.59 •• A soprano and a bass are singing a duet. While the soprano sings an A-sharp at 932 Hz, the bass sings an A-sharp but three octaves lower. In this concert hall, the density of air is 1.20 kg/m^3 and its bulk modulus is 1.42×10^5 Pa. In order for their notes to have the same sound intensity level, what must be (a) the ratio of the pressure amplitude of the bass to that of the soprano and (b) the ratio of the displacement amplitude of the bass to that of the soprano? (c) What displacement amplitude (in m and in nm) does the soprano produce to sing her A-sharp at 72.0 dB?

16.60 •• CP The sound from a trumpet radiates uniformly in all directions in 20°C air. At a distance of 5.00 m from the trumpet the sound intensity level is 52.0 dB. The frequency is 587 Hz. (a) What is the pressure amplitude at this distance? (b) What is the displacement amplitude? (c) At what distance is the sound intensity level 30.0 dB?

16.61 • CP A person is playing a small flute 10.75 cm long, open at one end and closed at the other, near a taut string having a fundamental frequency of 600.0 Hz. If the speed of sound is 344.0 m/s, for which harmonics of the flute will the string resonate? In each case, which harmonic of the string is in resonance?

16.62 •• CP A uniform 165-N bar is supported horizontally by two identical wires A and B (**Fig. P16.62**). A small 185-N cube of lead is placed three-fourths of the way from A to B. The wires are each 75.0 cm long and have a mass of 5.50 g. If both of them are simultaneously plucked at the center, what is the frequency of the beats that they will produce when vibrating in their fundamental?

Figure **P16.62**

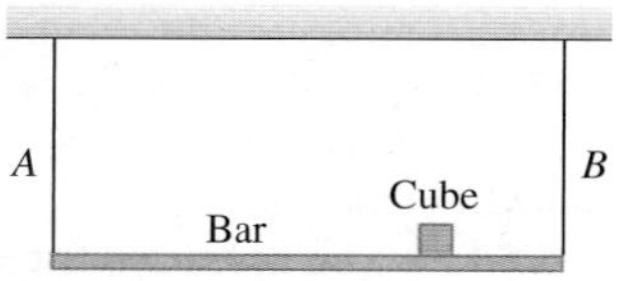

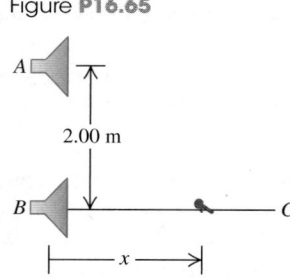

16.63 • An organ pipe has two successive harmonics with frequencies 1372 and 1764 Hz. (a) Is this an open or a stopped pipe? Explain. (b) What two harmonics are these? (c) What is the length of the pipe?

16.64 ••• The frequency of the note F_4 is 349 Hz. (a) If an organ pipe is open at one end and closed at the other, what length must it have for its fundamental mode to produce this note at 20.0°C? (b) At what air temperature will the frequency be 370 Hz, corresponding to a rise in pitch from F to F-sharp? (Ignore the change in length of the pipe due to the temperature change.)

16.65 •• Two identical loudspeakers are located at points A and B, 2.00 m apart. The loudspeakers are driven by the same amplifier and produce sound waves with a frequency of 784 Hz. Take the speed of sound in air to be 344 m/s. A small microphone is moved out from point B along a line perpendicular to the line connecting A and B (line BC in **Fig. P16.65**). (a) At what distances from B will there be *destructive* interference? (b) At what distances from B will there be *constructive* interference? (c) If the frequency is made low enough, there will be no positions along the line BC at which destructive interference occurs. How low must the frequency be for this to be the case?

Figure **P16.65**

16.66 •• A bat flies toward a wall, emitting a steady sound of frequency 1.70 kHz. This bat hears its own sound plus the sound reflected by the wall. How fast should the bat fly in order to hear a beat frequency of 8.00 Hz?

16.67 •• The sound source of a ship's sonar system operates at a frequency of 18.0 kHz. The speed of sound in water (assumed to be at a uniform 20°C) is 1482 m/s. (a) What is the wavelength of the waves emitted by the source? (b) What is the difference in frequency between the directly radiated waves and the waves reflected from a whale traveling directly toward the ship at 4.95 m/s? The ship is at rest in the water.

16.68 ••• BIO **Ultrasound in Medicine.** A 2.00-MHz sound wave travels through a pregnant woman's abdomen and is reflected from the fetal heart wall of her unborn baby. The heart wall is moving toward the sound receiver as the heart beats. The reflected sound is then mixed with the transmitted sound, and 72 beats per second are detected. The speed of sound in body tissue is 1500 m/s. Calculate the speed of the fetal heart wall at the instant this measurement is made.

16.69 ••• BIO Horseshoe bats (genus *Rhinolophus*) emit sounds from their nostrils and then listen to the frequency of the sound reflected from their prey to determine the prey's speed. (The "horseshoe" that gives the bat its name is a depression around the nostrils that acts like a focusing mirror, so that the bat emits sound in a narrow beam like a flashlight.) A *Rhinolophus* flying at speed v_{bat} emits sound of frequency f_{bat}; the sound it hears reflected from an insect flying toward it has a higher frequency f_{refl}. (a) Show that the speed of the insect is

$$v_{\text{insect}} = v \left[\frac{f_{\text{refl}}(v - v_{\text{bat}}) - f_{\text{bat}}(v + v_{\text{bat}})}{f_{\text{refl}}(v - v_{\text{bat}}) + f_{\text{bat}}(v + v_{\text{bat}})} \right]$$

where v is the speed of sound. (b) If $f_{\text{bat}} = 80.7$ kHz, $f_{\text{refl}} = 83.5$ kHz, and $v_{\text{bat}} = 3.9$ m/s, calculate the speed of the insect.

16.70 • CP A police siren of frequency f_{siren} is attached to a vibrating platform. The platform and siren oscillate up and down in simple harmonic motion with amplitude A_{p} and frequency f_{p}. (a) Find the maximum and minimum sound frequencies that you would hear at a position directly above the siren. (b) At what point in the motion of the platform is the maximum frequency heard? The minimum frequency? Explain.

16.71 •• CP A turntable 1.50 m in diameter rotates at 75 rpm. Two speakers, each giving off sound of wavelength 31.3 cm, are attached to the rim of the table at opposite ends of a diameter. A listener stands in front of the turntable. (a) What is the greatest beat frequency the listener will receive from this system? (b) Will the listener be able to distinguish individual beats?

16.72 •• DATA A long, closed cylindrical tank contains a diatomic gas that is maintained at a uniform temperature that can be varied. When you measure the speed of sound v in the gas as a function of the temperature T of the gas, you obtain these results:

T (°C)	−20.0	0.0	20.0	40.0	60.0	80.0
v (m/s)	324	337	349	361	372	383

(a) Explain how you can plot these results so that the graph will be well fit by a straight line. Construct this graph and verify that the plotted points do lie close to a straight line. (b) Because the gas is diatomic, $\gamma = 1.40$. Use the slope of the line in part (a) to calculate M, the molar mass of the gas. Express M in grams/mole. What type of gas is in the tank?

16.73 •• DATA A long tube contains air at a pressure of 1.00 atm and a temperature of 77.0°C. The tube is open at one end and closed at the other by a movable piston. A tuning fork that vibrates with a frequency of 500 Hz is placed near the open end. Resonance is produced when the piston is at distances 18.0 cm, 55.5 cm, and 93.0 cm from the open end. (a) From these values, what is the speed of sound in air at 77.0°C? (b) From the result of part (a), what is the value of γ? (c) These results show that a displacement antinode is slightly outside the open end of the tube. How far outside is it?

16.74 ••• DATA **Supernova!** (a) Equation (16.30) can be written as

$$f_R = f_S\left(1 - \frac{v}{c}\right)^{1/2}\left(1 + \frac{v}{c}\right)^{-1/2}$$

where c is the speed of light in vacuum, 3.00×10^8 m/s. Most objects move much slower than this (v/c is very small), so calculations made with Eq. (16.30) must be done carefully to avoid rounding errors. Use the binomial theorem to show that if $v \ll c$, Eq. (16.30) approximately reduces to $f_R = f_S[1 - (v/c)]$. (b) The gas cloud known as the Crab Nebula can be seen with even a small telescope. It is the remnant of a *supernova*, a cataclysmic explosion of a star. (The explosion was seen on the earth on July 4, 1054 C.E.) Its streamers glow with the characteristic red color of heated hydrogen gas. In a laboratory on the earth, heated hydrogen produces red light with frequency 4.568×10^{14} Hz; the red light received from streamers in the Crab Nebula that are pointed toward the earth has frequency 4.586×10^{14} Hz. Estimate the speed with which the outer edges of the Crab Nebula are expanding. Assume that the speed of the center of the nebula relative to the earth is negligible. (c) Assuming that the expansion speed of the Crab Nebula has been constant since the supernova that produced it, estimate the diameter of the Crab Nebula. Give your answer in meters and in light-years. (d) The angular diameter of the Crab Nebula as seen from the earth is about 5 arc-minutes $\left(\text{1 arc-minute} = \frac{1}{60} \text{ degree}\right)$. Estimate the distance (in light-years) to the Crab Nebula, and estimate the year in which the supernova actually took place.

CHALLENGE PROBLEMS

16.75 ••• CALC **Figure P16.75** shows the pressure fluctuation p of a nonsinusoidal sound wave as a function of x for $t = 0$. The wave is traveling in the $+x$-direction. (a) Graph the pressure fluctuation p as a function of t for $x = 0$. Show at least two cycles of oscillation. (b) Graph the displacement y in this sound wave as a function of x at $t = 0$. At $x = 0$, the displacement at $t = 0$ is zero. Show at least two wavelengths of the wave. (c) Graph the displacement y as a function of t for $x = 0$. Show at least two cycles of oscillation. (d) Calculate the maximum velocity and the maximum acceleration of an element of the air through which this sound wave is traveling. (e) Describe how the cone of a loudspeaker must move as a function of time to produce the sound wave in this problem.

Figure **P16.75**

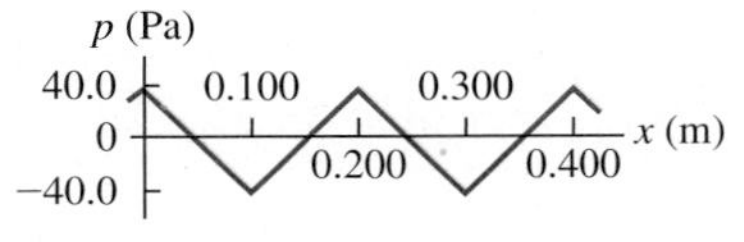

PASSAGE PROBLEMS

16.76 ••• CP **Longitudinal Waves on a Spring.** A long spring such as a Slinky™ is often used to demonstrate longitudinal waves. (a) Show that if a spring that obeys Hooke's law has mass m, length L, and force constant k', the speed of longitudinal waves on the spring is $v = L\sqrt{k'/m}$ (see Section 16.2). (b) Evaluate v for a spring with $m = 0.250$ kg, $L = 2.00$ m, and $k' = 1.50$ N/m.

BIO **ULTRASOUND IMAGING.** A typical ultrasound transducer used for medical diagnosis produces a beam of ultrasound with a frequency of 1.0 MHz. The beam travels from the transducer through tissue and partially reflects when it encounters different structures in the tissue. The same transducer that produces the ultrasound also detects the reflections. The transducer emits a short pulse of ultrasound and waits to receive the reflected echoes before emitting the next pulse. By measuring the time between the initial pulse and the arrival of the reflected signal, we can use the speed of ultrasound in tissue, 1540 m/s, to determine the distance from the transducer to the structure that produced the reflection.

As the ultrasound beam passes through tissue, the beam is attenuated through absorption. Thus deeper structures return weaker echoes. A typical attenuation in tissue is -100 dB/m · MHz; in bone it is -500 dB/m · MHz. In determining attenuation, we take the reference intensity to be the intensity produced by the transducer.

16.77 If the deepest structure you wish to image is 10.0 cm from the transducer, what is the maximum number of pulses per second that can be emitted? (a) 3850; (b) 7700; (c) 15,400; (d) 1,000,000.

16.78 After a beam passes through 10 cm of tissue, what is the beam's intensity as a fraction of its initial intensity from the transducer? (a) 1×10^{-11}; (b) 0.001; (c) 0.01; (d) 0.1.

16.79 Because the speed of ultrasound in bone is about twice the speed in soft tissue, the distance to a structure that lies beyond a bone can be measured incorrectly. If a beam passes through 4 cm of tissue, then 2 cm of bone, and then another 1 cm of tissue before echoing off a cyst and returning to the transducer, what is the difference between the true distance to the cyst and the distance that is measured by assuming the speed is always 1540 m/s? Compared with the measured distance, the structure is actually (a) 1 cm farther; (b) 2 cm farther; (c) 1 cm closer; (d) 2 cm closer.

16.80 In some applications of ultrasound, such as its use on cranial tissues, large reflections from the surrounding bones can produce standing waves. This is of concern because the large pressure amplitude in an antinode can damage tissues. For a frequency of 1.0 MHz, what is the distance between antinodes in tissue? (a) 0.38 mm; (b) 0.75 mm; (c) 1.5 mm; (d) 3.0 mm.

16.81 For cranial ultrasound, why is it advantageous to use frequencies in the kHZ range rather than the MHz range? (a) The antinodes of the standing waves will be closer together at the lower frequencies than at the higher frequencies; (b) there will be no standing waves at the lower frequencies; (c) cranial bones will attenuate the ultrasound more at the lower frequencies than at the higher frequencies; (d) cranial bones will attenuate the ultrasound less at the lower frequencies than at the higher frequencies.

Answers

Chapter Opening Question ?

(iv) Equation (16.10) in Section 16.2 says that the speed of sound in a gas depends on the temperature and on the kind of gas (through the ratio of heat capacities and the molar mass). Winter air in the mountains has a lower temperature than summer air at sea level, but they have essentially the same composition. Hence the lower temperature alone explains the slower speed of sound in winter in the mountains.

Test Your Understanding Questions

16.1 (v) From Eq. (16.5), the displacement amplitude is $A = p_{max}/Bk$. The pressure amplitude p_{max} and bulk modulus B remain the same, but the frequency f increases by a factor of 4. Hence the wave number $k = \omega/v = 2\pi f/v$ also increases by a factor of 4. Since A is inversely proportional to k, the displacement amplitude becomes $\frac{1}{4}$ as great. In other words, at higher frequency a smaller maximum displacement is required to produce the same maximum pressure fluctuation.

16.2 (i) From Eq. (16.7), the speed of longitudinal waves (sound) in a fluid is $v = \sqrt{B/\rho}$. We can rewrite this to give an expression for the bulk modulus B in terms of the fluid density ρ and the sound speed v: $B = \rho v^2$. At 20°C the speed of sound in mercury is slightly less than in water (1451 m/s versus 1482 m/s), but the density of mercury is greater than that of water by a large factor (13.6). Hence the bulk modulus of mercury is greater than that of water by a factor of $(13.6)(1451/1482)^2 = 13.0$.

16.3 A and p_{max} increase by a factor of $\sqrt{2}$, B and v are unchanged, β increases by 3.0 dB Equations (16.9) and (16.10) show that the bulk modulus B and sound speed v remain the same because the physical properties of the air are unchanged. From Eqs. (16.12) and (16.14), the intensity is proportional to the square of the displacement amplitude or the square of the pressure amplitude. Hence doubling the intensity means that A and p_{max} both increase by a factor of $\sqrt{2}$. Example 16.9 shows that *multiplying* the intensity by a factor of 2 ($I_2/I_1 = 2$) corresponds to *adding* to the sound intensity level by $(10 \text{ dB}) \log(I_2/I_1) = (10 \text{ dB}) \log 2 = 3.0$ dB.

16.4 (ii) Helium is less dense and has a lower molar mass than air, so sound travels faster in helium than in air. The normal-mode frequencies for a pipe are proportional to the sound speed v, so the frequency and hence the pitch increase when the air in the pipe is replaced with helium.

16.5 (i) and (iv) There will be a resonance if 660 Hz is one of the pipe's normal-mode frequencies. A stopped organ pipe has normal-mode frequencies that are odd multiples of its fundamental frequency [see Eq. (16.22) and Fig. 16.18]. Hence pipe (i), which has fundamental frequency 220 Hz, also has a normal-mode

frequency of 3(220 Hz) = 660 Hz. Pipe (ii) has twice the length of pipe (i); from Eq. (16.20), the fundamental frequency of a stopped pipe is inversely proportional to the length, so pipe (ii) has a fundamental frequency of $\left(\frac{1}{2}\right)(220 \text{ Hz}) = 110$ Hz. Its other normal-mode frequencies are 330 Hz, 550 Hz, 770 Hz, . . . , so a 660-Hz tuning fork will not cause resonance. Pipe (iii) is an open pipe of the same length as pipe (i), so its fundamental frequency is twice as great as for pipe (i) [compare Eqs. (16.16) and (16.20)], or 2(220 Hz) = 440 Hz. Its other normal-mode frequencies are integer multiples of the fundamental frequency [see Eq. (16.19)], or 880 Hz, 1320 Hz, . . . , none of which match the 660-Hz frequency of the tuning fork. Pipe (iv) is also an open pipe but with twice the length of pipe (iii) [see Eq. (16.18)], so its normal-mode frequencies are one-half those of pipe (iii): 220 Hz, 440 Hz, 660 Hz, . . . , so the third harmonic will resonate with the tuning fork.

16.6 (iii) Constructive and destructive interference between two waves can occur only if the two waves have the same frequency. In this case the frequencies are different, so there are no points where the two waves always reinforce each other (constructive interference) or always cancel each other (destructive interference).

16.7 (vi) The beat frequency is 3 Hz, so the difference between the two tuning fork frequencies is also 3 Hz. Hence the second tuning fork vibrates at a frequency of either 443 Hz or 437 Hz. You can distinguish between the two possibilities by comparing the pitches of the two tuning forks sounded one at a time: The frequency is 437 Hz if the second tuning fork has a lower pitch and 443 Hz if it has a higher pitch.

16.8 no The air (the medium for sound waves) is moving from the source toward the listener. Hence, relative to the air, both the source and the listener are moving in the direction from listener to source. So both velocities are positive and $v_S = v_L = +10$ m/s. The equality of these two velocities means that the numerator and the denominator in Eq. (16.29) are the same, so $f_L = f_S$ and there is *no* Doppler shift.

16.9 (iii) Figure 16.38 shows that there are sound waves inside the cone of the shock wave. Behind the airplane the wave crests are spread apart, just as they are behind the moving source in Fig. 16.28. Hence the waves that reach you have an increased wavelength and a lower frequency.

Bridging Problem

(a) $180° = \pi$ rad
(b) A alone: $I = 3.98 \times 10^{-6}$ W/m^2, $\beta = 66.0$ dB;
     B alone: $I = 5.31 \times 10^{-7}$ W/m^2, $\beta = 57.2$ dB
(c) $I = 1.60 \times 10^{-6}$ W/m^2, $\beta = 62.1$ dB

? At a steelworks, molten iron is heated to 1500° Celsius to remove impurities. It is most accurate to say that the molten iron contains a large amount of (i) temperature; (ii) heat; (iii) energy; (iv) two of these; (v) all three of these.

17 TEMPERATURE AND HEAT

Whether it's a sweltering summer day or a frozen midwinter night, your body needs to be kept at a nearly constant temperature. It has effective temperature-control mechanisms, but sometimes it needs help. On a hot day you wear less clothing to improve heat transfer from your body to the air and for better cooling by evaporation of perspiration. On a cold day you may sit by a roaring fire to absorb the energy that it radiates. The concepts in this chapter will help you understand the basic physics of keeping warm or cool.

The terms "temperature" and "heat" are often used interchangeably in everyday language. In physics, however, these two terms have very different meanings. In this chapter we'll define temperature in terms of how it's measured and see how temperature changes affect the dimensions of objects. We'll see that heat refers to energy transfer caused by temperature differences only and learn how to calculate and control such energy transfers.

Our emphasis in this chapter is on the concepts of temperature and heat as they relate to *macroscopic* objects such as cylinders of gas, ice cubes, and the human body. In Chapter 18 we'll look at these same concepts from a *microscopic* viewpoint in terms of the behavior of individual atoms and molecules. These two chapters lay the groundwork for the subject of **thermodynamics,** the study of energy transformations involving heat, mechanical work, and other aspects of energy and how these transformations relate to the properties of matter. Thermodynamics forms an indispensable part of the foundation of physics, chemistry, and the life sciences, and its applications turn up in such places as car engines, refrigerators, biochemical processes, and the structure of stars. We'll explore the key ideas of thermodynamics in Chapters 19 and 20.

17.1 TEMPERATURE AND THERMAL EQUILIBRIUM

The concept of **temperature** is rooted in qualitative ideas based on our sense of touch. A body that feels "hot" usually has a higher temperature than a similar body that feels "cold." That's pretty vague, and the senses can be deceived. But many properties of matter that we can *measure*—including the length of a metal

17.1 Two devices for measuring temperature.

(a) Changes in temperature cause the liquid's volume to change.

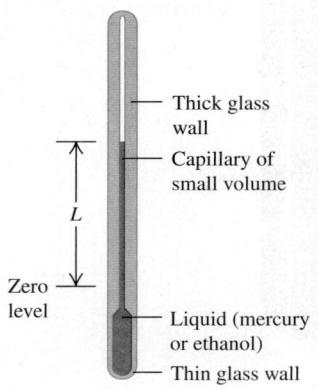

Thick glass wall

Capillary of small volume

L

Zero level

Liquid (mercury or ethanol)

Thin glass wall

(b) Changes in temperature cause the pressure of the gas to change.

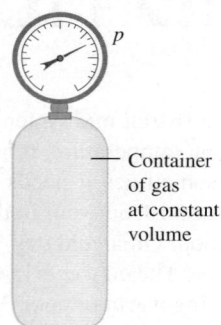

p

Container of gas at constant volume

rod, steam pressure in a boiler, the ability of a wire to conduct an electric current, and the color of a very hot glowing object—depend on temperature.

Temperature is also related to the kinetic energies of the molecules of a material. In general this relationship is fairly complex, so it's not a good place to start in *defining* temperature. In Chapter 18 we will look at the relationship between temperature and the energy of molecular motion for an ideal gas. However, we can define temperature and heat independently of any detailed molecular picture. In this section we'll develop a *macroscopic* definition of temperature.

To use temperature as a measure of hotness or coldness, we need to construct a temperature scale. To do this, we can use any measurable property of a system that varies with its "hotness" or "coldness." **Figure 17.1a** shows a familiar system that is used to measure temperature. When the system becomes hotter, the colored liquid (usually mercury or ethanol) expands and rises in the tube, and the value of L increases. Another simple system is a quantity of gas in a constant-volume container (Fig. 17.1b). The pressure p, measured by the gauge, increases or decreases as the gas becomes hotter or colder. A third example is the electrical resistance R of a conducting wire, which also varies when the wire becomes hotter or colder. Each of these properties gives us a number (L, p, or R) that varies with hotness and coldness, so each property can be used to make a **thermometer.**

To measure the temperature of a body, you place the thermometer in contact with the body. If you want to know the temperature of a cup of hot coffee, you stick the thermometer in the coffee; as the two interact, the thermometer becomes hotter and the coffee cools off a little. After the thermometer settles down to a steady value, you read the temperature. The system has reached an *equilibrium* condition, in which the interaction between the thermometer and the coffee causes no further change in the system. We call this a state of **thermal equilibrium.**

If two systems are separated by an insulating material or **insulator** such as wood, plastic foam, or fiberglass, they influence each other more slowly. Camping coolers are made with insulating materials to delay the cold food inside from warming up and attaining thermal equilibrium with the hot summer air outside. An *ideal insulator* is an idealized material that permits no interaction at all between the two systems. It prevents the systems from attaining thermal equilibrium if they aren't in thermal equilibrium at the start. Real insulators, like those in camping coolers, aren't ideal, so the contents of the cooler will warm up eventually. But an ideal insulator is nonetheless a useful idealization, like a massless rope or a frictionless incline.

The Zeroth Law of Thermodynamics

We can discover an important property of thermal equilibrium by considering three systems, A, B, and C, that initially are not in thermal equilibrium (**Fig. 17.2**). We surround them with an ideal insulating box so that they cannot interact with anything except each other. We separate systems A and B with an ideal insulating wall (the green slab in Fig. 17.2a), but we let system C interact with both systems

17.2 The zeroth law of thermodynamics.

(a) If systems A and B are each in thermal equilibrium with system C ...

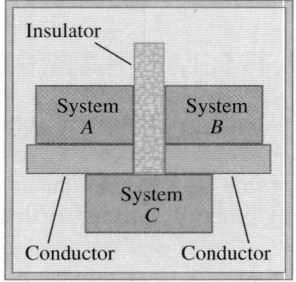

Insulator

System A

System B

System C

Conductor   Conductor

(b) ... then systems A and B are in thermal equilibrium with each other.

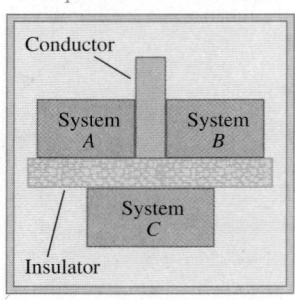

Conductor

System A

System B

System C

Insulator

A and *B*. We show this interaction in the figure by a yellow slab representing a thermal **conductor,** a material that *permits* thermal interactions through it. We wait until thermal equilibrium is attained; then *A* and *B* are each in thermal equilibrium with *C*. But are they in thermal equilibrium *with each other?*

To find out, we separate system *C* from systems *A* and *B* with an ideal insulating wall (Fig. 17.2b), then replace the insulating wall between *A* and *B* with a *conducting* wall that lets *A* and *B* interact. What happens? Experiment shows that *nothing* happens; there are no additional changes to *A* or *B*. This result is called the **zeroth law of thermodynamics:**

> If *C* is initially in thermal equilibrium with both *A* and *B*, then *A* and *B* are also in thermal equilibrium with each other.

(The importance of this law was recognized only after the first, second, and third laws of thermodynamics had been named. Since it is fundamental to all of them, the name "zeroth" seemed appropriate.)

Now suppose system *C* is a thermometer, such as the liquid-in-tube system of Fig. 17.1a. In Fig. 17.2a the thermometer *C* is in contact with both *A* and *B*. In thermal equilibrium, when the thermometer reading reaches a stable value, the thermometer measures the temperature of both *A* and *B*; hence both *A* and *B* have the *same* temperature. Experiment shows that thermal equilibrium isn't affected by adding or removing insulators, so the reading of thermometer *C* wouldn't change if it were in contact only with *A* or only with *B*. We conclude:

> Two systems are in thermal equilibrium if and only if they have the same temperature.

This is what makes a thermometer useful; a thermometer actually measures *its own* temperature, but when a thermometer is in thermal equilibrium with another body, the temperatures must be equal. When the temperatures of two systems are different, they *cannot* be in thermal equilibrium.

TEST YOUR UNDERSTANDING OF SECTION 17.1   You put a thermometer in a pot of hot water and record the reading. What temperature have you recorded? (i) The temperature of the water; (ii) the temperature of the thermometer; (iii) an equal average of the temperatures of the water and thermometer; (iv) a weighted average of the temperatures of the water and thermometer, with more emphasis on the temperature of the water; (v) a weighted average of the water and thermometer, with more emphasis on the temperature of the thermometer. ❚

17.2 THERMOMETERS AND TEMPERATURE SCALES

To make the liquid-in-tube device shown in Fig. 17.1a into a useful thermometer, we need to mark a scale on the tube wall with numbers on it. Suppose we label the thermometer's liquid level at the freezing temperature of pure water "zero" and the level at the boiling temperature "100," and divide the distance between these two points into 100 equal intervals called *degrees*. The result is the **Celsius temperature scale** (formerly called the *centigrade* scale in English-speaking countries). The Celsius temperature for a state colder than freezing water is a negative number. The Celsius scale is used, both in everyday life and in science and industry, almost everywhere in the world.

Another common type of thermometer uses a *bimetallic strip,* made by bonding strips of two different metals together (**Fig. 17.3a**). When the temperature of the composite strip increases, one metal expands more than the other and the strip bends (Fig. 17.3b). This strip is usually formed into a spiral, with the outer end anchored to the thermometer case and the inner end attached to a pointer (Fig. 17.3c). The pointer rotates in response to temperature changes.

17.3 Use of a bimetallic strip as a thermometer.

(a) A bimetallic strip

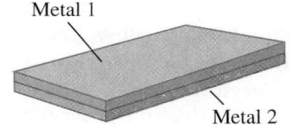

Metal 1

Metal 2

(b) The strip bends when its temperature is raised.

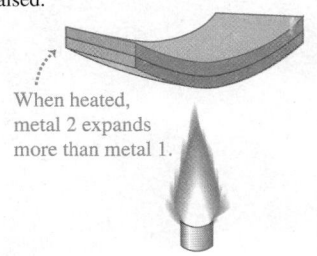

When heated, metal 2 expands more than metal 1.

(c) A bimetallic strip used in a thermometer

17.4 A temporal artery thermometer measures infrared radiation from the skin that overlies one of the important arteries in the head. Although the thermometer cover touches the skin, the infrared detector inside the cover does not.

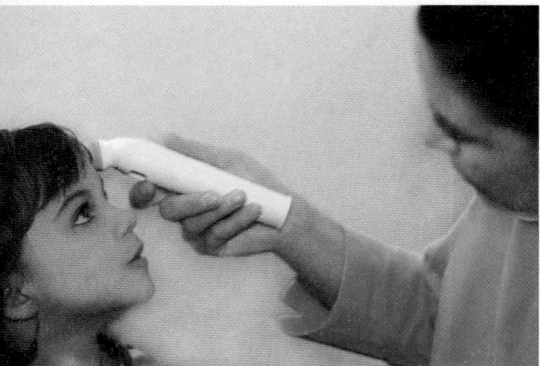

In a *resistance thermometer* the changing electrical resistance of a coil of fine wire, a carbon cylinder, or a germanium crystal is measured. Resistance thermometers are usually more precise than most other types.

Some thermometers detect the amount of infrared radiation emitted by an object. (We'll see in Section 17.7 that *all* objects emit electromagnetic radiation, including infrared, as a consequence of their temperature.) One example is a *temporal artery thermometer* (**Fig. 17.4**). A nurse runs this over a patient's forehead in the vicinity of the temporal artery, and an infrared sensor in the thermometer measures the radiation from the skin. This device gives more accurate values of body temperature than do oral or ear thermometers.

In the **Fahrenheit temperature scale,** still used in the United States, the freezing temperature of water is 32°F and the boiling temperature is 212°F, both at standard atmospheric pressure. There are 180 degrees between freezing and boiling, compared to 100 on the Celsius scale, so one Fahrenheit degree represents only $\frac{100}{180}$, or $\frac{5}{9}$, as great a temperature change as one Celsius degree.

To convert temperatures from Celsius to Fahrenheit, note that a Celsius temperature T_C is the number of Celsius degrees above freezing; the number of Fahrenheit degrees above freezing is $\frac{9}{5}$ of this. But freezing on the Fahrenheit scale is at 32°F, so to obtain the actual Fahrenheit temperature T_F, multiply the Celsius value by $\frac{9}{5}$ and then add 32°. Symbolically,

$$\underset{\substack{\text{Fahrenheit}\\\text{temperature}}}{} T_F = \tfrac{9}{5}T_C + 32° \quad \underset{\substack{\text{Celsius}\\\text{temperature}}}{} \tag{17.1}$$

To convert Fahrenheit to Celsius, solve this equation for T_C:

$$\underset{\substack{\text{Celsius}\\\text{temperature}}}{} T_C = \tfrac{5}{9}\left(T_F - 32°\right) \quad \underset{\substack{\text{Fahrenheit}\\\text{temperature}}}{} \tag{17.2}$$

In words, subtract 32° to get the number of Fahrenheit degrees above freezing, and then multiply by $\frac{5}{9}$ to obtain the number of Celsius degrees above freezing—that is, the Celsius temperature.

We don't recommend memorizing Eqs. (17.1) and (17.2). Instead, understand the reasoning that led to them so that you can derive them on the spot when you need them, checking your reasoning with the relationship 100°C = 212°F.

It is useful to distinguish between an actual temperature and a temperature *interval* (a difference or change in temperature). An actual temperature of 20° is stated as 20°C (twenty degrees Celsius), and a temperature *interval* of 10° is 10 C° (ten Celsius degrees). A beaker of water heated from 20°C to 30°C undergoes a temperature change of 10 C°.

BIO Application Mammalian Body Temperatures Most mammals maintain body temperatures in the range from 36°C to 40°C (309 K to 313 K). A high metabolic rate warms the animal from within, and insulation (such as fur and body fat) slows heat loss.

TEST YOUR UNDERSTANDING OF SECTION 17.2 Which of the following types of thermometers have to be in thermal equilibrium with the object being measured in order to give accurate readings? (i) A bimetallic strip; (ii) a resistance thermometer; (iii) a temporal artery thermometer; (iv) both (i) and (ii); (v) all of (i), (ii), and (iii). ❚

17.3 GAS THERMOMETERS AND THE KELVIN SCALE

When we calibrate two thermometers, such as a liquid-in-tube system and a resistance thermometer, so that they agree at 0°C and 100°C, they may not agree exactly at intermediate temperatures. Any temperature scale defined in this way always depends somewhat on the specific properties of the material used. Ideally, we would like to define a temperature scale that *doesn't* depend on the properties

17.5 **(a)** Using a constant-volume gas thermometer to measure temperature. **(b)** The greater the amount of gas in the thermometer, the higher the graph of pressure *p* versus temperature *T*.

(a) A constant-volume gas thermometer

(b) Graphs of pressure versus temperature at constant volume for three different types and quantities of gas

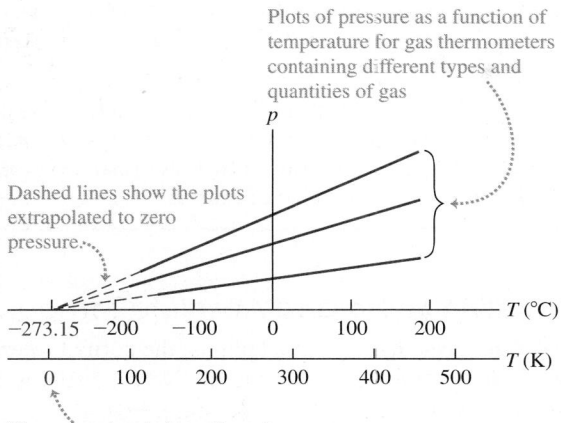

Plots of pressure as a function of temperature for gas thermometers containing different types and quantities of gas

Dashed lines show the plots extrapolated to zero pressure.

The extrapolated plots all reach zero pressure at the same temperature: −273.15°C.

of a particular material. To establish a truly material-independent scale, we first need to develop some principles of thermodynamics. We'll return to this fundamental problem in Chapter 20. Here we'll discuss a thermometer that comes close to the ideal, the *constant-volume gas thermometer*.

The principle of a constant-volume gas thermometer is that the pressure of a gas at constant volume increases with temperature. We place a quantity of gas in a constant-volume container (**Fig. 17.5a**) and measure its pressure by one of the devices described in Section 12.2. To calibrate this thermometer, we measure the pressure at two temperatures, say 0°C and 100°C, plot these points on a graph, and draw a straight line between them. Then we can read from the graph the temperature corresponding to any other pressure. Figure 17.5b shows the results of three such experiments, each using a different type and quantity of gas.

By extrapolating this graph, we see that there is a hypothetical temperature, −273.15°C, at which the absolute pressure of the gas would become zero. This temperature turns out to be the *same* for many different gases (at least in the limit of very low gas density). We can't actually observe this zero-pressure condition. Gases liquefy and solidify at very low temperatures, and the proportionality of pressure to temperature no longer holds.

We use this extrapolated zero-pressure temperature as the basis for a temperature scale with its zero at this temperature. This is the **Kelvin temperature scale,** named for the British physicist Lord Kelvin (1824–1907). The units are the same size as those on the Celsius scale, but the zero is shifted so that 0 K = −273.15°C and 273.15 K = 0°C (Fig. 17.5b); that is,

$$\text{Kelvin} \cdots \rightarrow T_{\text{K}} = T_{\text{C}} + 273.15 \leftarrow \cdots \text{Celsius} \tag{17.3}$$
$$\text{temperature} \qquad\qquad\qquad\qquad \text{temperature}$$

A common room temperature, 20°C (= 68°F), is 20 + 273.15, or about 293 K.

_____

CAUTION **Never say "degrees kelvin"** In SI nomenclature, the temperature mentioned above is read "293 kelvins," not "degrees kelvin" (**Fig. 17.6**). We capitalize Kelvin when it refers to the temperature scale; however, the *unit* of temperature is the *kelvin,* which is not capitalized (but is nonetheless abbreviated as a capital K). ▮

17.6 Correct and incorrect uses of the Kelvin scale.

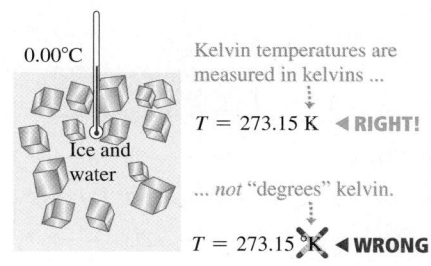

0.00°C

Ice and water

Kelvin temperatures are measured in kelvins ...

T = 273.15 K ◀ **RIGHT!**

... *not* "degrees" kelvin.

T = 273.15 °K ◀ **WRONG**

EXAMPLE 17.1 **BODY TEMPERATURE**

You place a small piece of ice in your mouth. Eventually, the water all converts from ice at $T_1 = 32.00°F$ to body temperature, $T_2 = 98.60°F$. Express these temperatures in both Celsius degrees and kelvins, and find $\Delta T = T_2 - T_1$ in both cases.

EXECUTE: From Eq. (17.2), $T_1 = 0.00°C$ and $T_2 = 37.00°C$; then $\Delta T = T_2 - T_1 = 37.00\ C°$. To get the Kelvin temperatures, just add 273.15 to each Celsius temperature: $T_1 = 273.15\ K$ and $T_2 = 310.15\ K$. The temperature difference is $\Delta T = T_2 - T_1 = 37.00\ K$.

SOLUTION

IDENTIFY and SET UP: Our target variables are stated above. We convert Fahrenheit temperatures to Celsius by using Eq. (17.2), and Celsius temperatures to Kelvin by using Eq. (17.3).

EVALUATE: The Celsius and Kelvin scales have different zero points but the same size degrees. Therefore *any* temperature difference ΔT is the *same* on the Celsius and Kelvin scales. However, ΔT is *not* the same on the Fahrenheit scale; here, for example, $\Delta T = 66.60\ F°$.

The Kelvin Scale and Absolute Temperature

The Celsius scale has two fixed points: the normal freezing and boiling temperatures of water. But we can define the Kelvin scale by using a gas thermometer with only a single reference temperature. Figure 17.5b shows that the pressure p in a gas thermometer is directly proportional to the Kelvin temperature. So we can define the ratio of any two Kelvin temperatures T_1 and T_2 as the ratio of the corresponding gas-thermometer pressures p_1 and p_2:

Definition of Kelvin scale:
Ratio of two **temperatures** in kelvins ...
$$\frac{T_2}{T_1} = \frac{p_2}{p_1}$$
... equals ratio of corresponding **pressures** in **constant-volume gas thermometer.** (17.4)

To complete the definition of T, we need only specify the Kelvin temperature of a single state. For reasons of precision and reproducibility, the state chosen is the *triple point* of water, the unique combination of temperature and pressure at which solid water (ice), liquid water, and water vapor can all coexist. It occurs at a temperature of 0.01°C and a water-vapor pressure of 610 Pa (about 0.006 atm). (This is the pressure of the *water*, not the gas pressure in the *thermometer*.) The triple-point temperature of water is *defined* to have the value $T_{\text{triple}} = 273.16\ K$, corresponding to 0.01°C. From Eq. (17.4), if p_{triple} is the pressure in a gas thermometer at temperature T_{triple} and p is the pressure at some other temperature T, then T is given on the Kelvin scale by

$$T = T_{\text{triple}} \frac{p}{p_{\text{triple}}} = (273.16\ K) \frac{p}{p_{\text{triple}}}$$ (17.5)

Gas thermometers are impractical for everyday use. They are bulky and very slow to come to thermal equilibrium. They are used principally to establish high-precision standards and to calibrate other thermometers.

Figure 17.7 shows the relationships among the three temperature scales we have discussed. The Kelvin scale is called an **absolute temperature scale,** and its zero point ($T = 0\ K = -273.15°C$, the temperature at which $p = 0$ in Eq. [17.5]) is called **absolute zero.** At absolute zero a system of molecules (such as a quantity of a gas, a liquid, or a solid) has its *minimum* possible total energy (kinetic plus potential); because of quantum effects, it is *not* correct to say that all molecular motion ceases at absolute zero. In Chapter 20 we'll define more completely what we mean by absolute zero through thermodynamic principles that we'll develop in the next several chapters.

17.7 Relationships among Kelvin (K), Celsius (C), and Fahrenheit (F) temperature scales. Temperatures have been rounded off to the nearest degree.

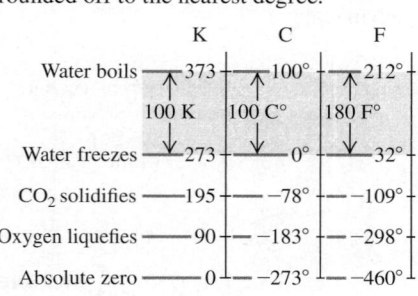

TEST YOUR UNDERSTANDING OF SECTION 17.3 Rank the following temperatures from highest to lowest: (i) 0.00°C; (ii) 0.00°F; (iii) 260.00 K; (iv) 77.00 K; (v) −180.00°C. ▮

17.4 THERMAL EXPANSION

Most materials expand when their temperatures increase. Rising temperatures make the liquid expand in a liquid-in-tube thermometer (Fig. 17.1a) and bend bimetallic strips (Fig. 17.3b). A completely filled and tightly capped bottle of water cracks when it is heated, but you can loosen a metal jar lid by running hot water over it. These are all examples of *thermal expansion*.

Linear Expansion

Suppose a solid rod has a length L_0 at some initial temperature T_0. When the temperature changes by ΔT, the length changes by ΔL. Experiments show that if ΔT is not too large (say, less than 100 C° or so), ΔL is *directly proportional* to ΔT (**Fig. 17.8a**). If two rods made of the same material have the same temperature change, but one is twice as long as the other, then the *change* in its length is also twice as great. Therefore ΔL must also be proportional to L_0 (Fig. 17.8b). We may express these relationships in an equation:

Linear thermal expansion:
Change in length ·········· Original length
$$\Delta L = \alpha L_0 \Delta T \qquad \text{Temperature change}$$
Coefficient of linear expansion
(17.6)

The constant α, which has different values for different materials, is called the **coefficient of linear expansion.** The units of α are K^{-1} or $(C°)^{-1}$. (Remember that a temperature interval is the same on the Kelvin and Celsius scales.) If a body has length L_0 at temperature T_0, then its length L at a temperature $T = T_0 + \Delta T$ is

$$L = L_0 + \Delta L = L_0 + \alpha L_0 \Delta T = L_0(1 + \alpha \Delta T) \qquad (17.7)$$

For many materials, every linear dimension changes according to Eq. (17.6) or (17.7). Thus L could be the thickness of a rod, the side length of a square sheet, or the diameter of a hole. Some materials, such as wood or single crystals, expand differently in different directions. We won't consider this complication.

We can understand thermal expansion qualitatively on a molecular basis. Picture the interatomic forces in a solid as springs, as in **Fig. 17.9a**. (We explored the analogy between spring forces and interatomic forces in Section 14.4.) Each atom vibrates about its equilibrium position. When the temperature increases, the energy and amplitude of the vibration also increase. The interatomic spring forces are not symmetrical about the equilibrium position; they usually behave like a spring that is easier to stretch than to compress. As a result, when the amplitude of vibration increases, the *average* distance between atoms also increases (Fig. 17.9b). As the atoms get farther apart, every dimension increases.

17.8 How the length of a rod changes with a change in temperature. (Length changes are exaggerated for clarity.)

(a) For moderate temperature changes, ΔL is directly proportional to ΔT.

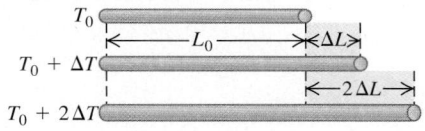

(b) ΔL is also directly proportional to L_0.

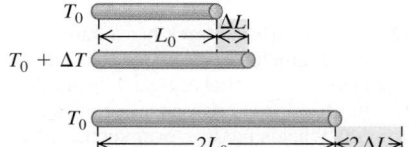

(a) A model of the forces between neighboring atoms in a solid

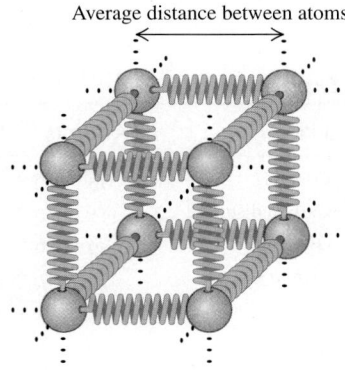

Average distance between atoms

(b) A graph of the "spring" potential energy $U(x)$

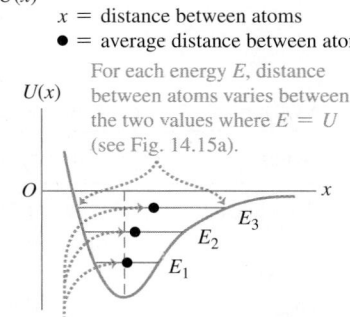

x = distance between atoms
● = average distance between atoms

For each energy E, distance between atoms varies between the two values where $E = U$ (see Fig. 14.15a).

Average distance between atoms is midway between two limits. As energy increases from E_1 to E_2 to E_3, average distance increases.

17.9 (a) We can model atoms in a solid as being held together by "springs" that are easier to stretch than to compress. (b) A graph of the "spring" potential energy $U(x)$ versus distance x between neighboring atoms is *not* symmetrical (compare Fig. 14.20b). As the energy increases and the atoms oscillate with greater amplitude, the average distance increases.

17.10 When an object undergoes thermal expansion, any holes in the object expand as well. (The expansion is exaggerated.)

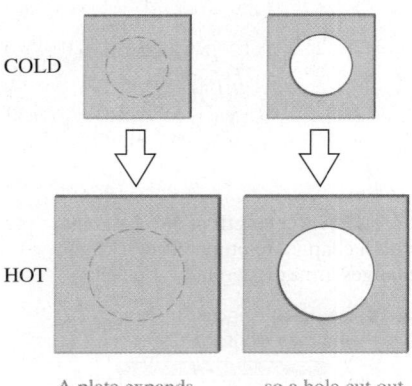

COLD

HOT

A plate expands when heated ...

... so a hole cut out of the plate must expand, too.

17.11 This railroad track has a gap between segments to allow for thermal expansion. (The "clickety-clack" sound familiar to railroad passengers comes from the wheels passing over such gaps.) On hot days, the segments expand and fill in the gap. If there were no gaps, the track could buckle under very hot conditions.

Gap

CAUTION **Heating an object with a hole** If a solid object has a hole in it, what happens to the size of the hole when the temperature of the object increases? A common misconception is that if the object expands, the hole will shrink because material expands into the hole. But, in fact, if the object expands, the hole will expand too (**Fig. 17.10**); *every* linear dimension of an object changes in the same way when the temperature changes. Think of the atoms in Fig. 17.9a as outlining a cubical hole. When the object expands, the atoms move apart and the hole increases in size. The only situation in which a "hole" will fill in due to thermal expansion is when two separate objects expand and close the gap between them (**Fig. 17.11**).

The direct proportionality in Eq. (17.6) is not exact; it is *approximately* correct only for sufficiently small temperature changes. For a given material, α varies somewhat with the initial temperature T_0 and the size of the temperature interval. We'll ignore this complication here, however. **Table 17.1** lists values of α for several materials. Within the precision of these values we don't need to worry whether T_0 is 0°C or 20°C or some other temperature. Typical values of α are very small; even for a temperature change of 100 C°, the fractional length change $\Delta L/L_0$ is only of the order of $\frac{1}{1000}$ for the metals in the table.

Volume Expansion

Increasing temperature usually causes increases in *volume* for both solids and liquids. Just as with linear expansion, experiments show that if the temperature change ΔT is less than 100 C° or so, the increase in volume ΔV is approximately proportional to both the temperature change ΔT and the initial volume V_0:

Volume thermal expansion: $\Delta V = \beta V_0 \Delta T$ — Change in volume — Original volume — Temperature change — Coefficient of volume expansion (17.8)

The constant β characterizes the volume expansion properties of a particular material; it is called the **coefficient of volume expansion.** The units of β are K^{-1} or $(C°)^{-1}$. As with linear expansion, β varies somewhat with temperature, and Eq. (17.8) is an approximate relationship that is valid only for small temperature changes. For many substances, β decreases at low temperatures. **Table 17.2** lists values of β for several materials near room temperature. Note that the values for liquids are generally much larger than those for solids.

For solid materials we can find a simple relationship between the volume expansion coefficient β and the linear expansion coefficient α. Consider a cube of material with side length L and volume $V = L^3$. At the initial temperature the values are L_0 and V_0. When the temperature increases by dT, the side length increases by dL and the volume increases by an amount dV:

$$dV = \frac{dV}{dL} dL = 3L^2\, dL$$

TABLE 17.1	Coefficients of Linear Expansion
Material	$\alpha\,[\mathbf{K^{-1}}$ or $\mathbf{(C°)^{-1}}]$
Aluminum	2.4×10^{-5}
Brass	2.0×10^{-5}
Copper	1.7×10^{-5}
Glass	$0.4–0.9 \times 10^{-5}$
Invar (nickel–iron alloy)	0.09×10^{-5}
Quartz (fused)	0.04×10^{-5}
Steel	1.2×10^{-5}

TABLE 17.2	Coefficients of Volume Expansion		
Solids	$\beta\,[\mathbf{K^{-1}}$ or $\mathbf{(C°)^{-1}}]$	**Liquids**	$\beta\,[\mathbf{K^{-1}}$ or $\mathbf{(C°)^{-1}}]$
Aluminum	7.2×10^{-5}	Ethanol	75×10^{-5}
Brass	6.0×10^{-5}	Carbon disulfide	115×10^{-5}
Copper	5.1×10^{-5}	Glycerin	49×10^{-5}
Glass	$1.2–2.7 \times 10^{-5}$	Mercury	18×10^{-5}
Invar	0.27×10^{-5}		
Quartz (fused)	0.12×10^{-5}		
Steel	3.6×10^{-5}		

Now we replace L and V by the initial values L_0 and V_0. From Eq. (17.6), dL is

$$dL = \alpha L_0\, dT$$

Since $V_0 = L_0^3$, this means that dV can also be expressed as

$$dV = 3L_0^2\alpha L_0\, dT = 3\alpha V_0\, dT$$

This is consistent with the infinitesimal form of Eq. (17.8), $dV = \beta V_0\, dT$, only if

$$\beta = 3\alpha \qquad\qquad (17.9)$$

(Check this relationship for some of the materials listed in Tables 17.1 and 17.2.)

PROBLEM-SOLVING STRATEGY 17.1 | THERMAL EXPANSION

IDENTIFY *the relevant concepts:* Decide whether the problem involves changes in length (linear thermal expansion) or in volume (volume thermal expansion).

SET UP *the problem* using the following steps:
1. List the known and unknown quantities and identify the target variables.
2. Choose Eq. (17.6) for linear expansion and Eq. (17.8) for volume expansion.

EXECUTE *the solution* as follows:
1. Solve for the target variables. If you are given an initial temperature T_0 and must find a final temperature T corresponding

to a given length or volume change, find ΔT and calculate $T = T_0 + \Delta T$. Remember that the size of a hole in a material varies with temperature just as any other linear dimension, and that the volume of a hole (such as the interior of a container) varies just as that of the corresponding solid shape.
2. Maintain unit consistency. Both L_0 and ΔL (or V_0 and ΔV) must have the same units. If you use a value of α or β in K^{-1} or $(\text{C}°)^{-1}$, then ΔT must be in either kelvins or Celsius degrees; from Example 17.1, the two scales are equivalent *for temperature differences.*

EVALUATE *your answer:* Check whether your results make sense.

EXAMPLE 17.2 | LENGTH CHANGE DUE TO TEMPERATURE CHANGE

A surveyor uses a steel measuring tape that is exactly 50.000 m long at a temperature of 20°C. The markings on the tape are calibrated for this temperature. (a) What is the length of the tape when the temperature is 35°C? (b) When it is 35°C, the surveyor uses the tape to measure a distance. The value that she reads off the tape is 35.794 m. What is the actual distance?

SOLUTION

IDENTIFY and SET UP: This problem concerns the linear expansion of a measuring tape. We are given the tape's initial length $L_0 = 50.000$ m at $T_0 = 20°C$. In part (a) we use Eq. (17.6) to find the change ΔL in the tape's length at $T = 35°C$, and use Eq. (17.7) to find L. (Table 17.1 gives the value of α for steel.) Since the tape expands, at 35°C the distance between two successive meter marks is greater than 1 m. Hence the actual distance in part (b) is *larger* than the distance read off the tape by a factor equal to the ratio of the tape's length L at 35°C to its length L_0 at 20°C.

EXECUTE: (a) The temperature change is $\Delta T = T - T_0 = 15\ \text{C}°$; from Eqs. (17.6) and (17.7),

$$\Delta L = \alpha L_0\, \Delta T = (1.2 \times 10^{-5}\ \text{K}^{-1})(50\ \text{m})(15\ \text{K})$$

$$= 9.0 \times 10^{-3}\ \text{m} = 9.0\ \text{mm}$$

$$L = L_0 + \Delta L = 50.000\ \text{m} + 0.009\ \text{m} = 50.009\ \text{m}$$

(b) Our result from part (a) shows that at 35°C, the slightly expanded tape reads a distance of 50.000 m when the true distance is 50.009 m. We can rewrite the algebra of part (a) as $L = L_0(1 + \alpha\,\Delta T)$; at 35°C, *any* true distance will be greater than the reading by the factor $50.009/50.000 = 1 + \alpha\,\Delta T = 1 + 1.8 \times 10^{-4}$. The true distance is therefore

$$(1 + 1.8 \times 10^{-4})(35.794\ \text{m}) = 35.800\ \text{m}$$

EVALUATE: In part (a) we needed only two of the five significant figures of L_0 to compute ΔL to the same number of decimal places as L_0. Our result shows that metals expand very little under moderate temperature changes. However, even the small difference $0.009\ \text{m} = 9\ \text{mm}$ found in part (b) between the scale reading and the true distance can be important in precision work.

EXAMPLE 17.3 VOLUME CHANGE DUE TO TEMPERATURE CHANGE

A 200-cm³ glass flask is filled to the brim with mercury at 20°C. How much mercury overflows when the temperature of the system is raised to 100°C? The coefficient of *linear* expansion of the glass is 0.40×10^{-5} K⁻¹.

SOLUTION

IDENTIFY and SET UP: This problem involves the volume expansion of the glass and of the mercury. The amount of overflow depends on the *difference* between the volume changes ΔV for these two materials, both given by Eq. (17.8). The mercury will overflow if its coefficient of volume expansion β (see Table 17.2) is greater than that of glass, which we find from Eq. (17.9) using the given value of α.

EXECUTE: From Table 17.2, $\beta_{\text{Hg}} = 18 \times 10^{-5}$ K⁻¹. That is indeed greater than $\beta_{\text{glass}} = 3\alpha_{\text{glass}} = 3(0.40 \times 10^{-5} \text{ K}^{-1}) = 1.2 \times 10^{-5}$ K⁻¹, from Eq. (17.9). The volume overflow is then

$$\Delta V_{\text{Hg}} - \Delta V_{\text{glass}} = \beta_{\text{Hg}} V_0 \Delta T - \beta_{\text{glass}} V_0 \Delta T$$
$$= V_0 \Delta T (\beta_{\text{Hg}} - \beta_{\text{glass}})$$
$$= (200 \text{ cm}^3)(80 \text{ C}°)(18 \times 10^{-5} - 1.2 \times 10^{-5})$$
$$= 2.7 \text{ cm}^3$$

EVALUATE: This is basically how a mercury-in-glass thermometer works; the column of mercury inside a sealed tube rises as T increases because mercury expands faster than glass.

As Tables 17.1 and 17.2 show, glass has smaller coefficients of expansion α and β than do most metals. This is why you can use hot water to loosen a metal lid on a glass jar; the metal expands more than the glass does.

17.12 The volume of 1 gram of water in the temperature range from 0°C to 100°C. By 100°C the volume has increased to 1.043 cm³. If the coefficient of volume expansion were constant, the curve would be a straight line.

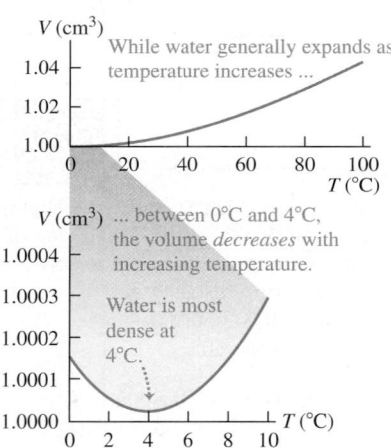

Thermal Expansion of Water

Water, in the temperature range from 0°C to 4°C, *decreases* in volume with increasing temperature. In this range its coefficient of volume expansion is *negative*. Above 4°C, water expands when heated (**Fig. 17.12**). Hence water has its greatest density at 4°C. Water also expands when it freezes, which is why ice humps up in the middle of the compartments in an ice cube tray. By contrast, most materials contract when they freeze.

This anomalous behavior of water has an important effect on plant and animal life in lakes. A lake cools from the surface down; above 4°C, the cooled water at the surface flows to the bottom because of its greater density. But when the surface temperature drops below 4°C, the water near the surface is less dense than the warmer water below. Hence the downward flow ceases, and the water near the surface remains colder than that at the bottom. As the surface freezes, the ice floats because it is less dense than water. The water at the bottom remains at 4°C until nearly the entire lake is frozen. If water behaved like most substances, contracting continuously on cooling and freezing, lakes would freeze from the bottom up. Circulation due to density differences would continuously carry warmer water to the surface for efficient cooling, and lakes would freeze solid much more easily. This would destroy all plant and animal life that cannot withstand freezing. If water did not have its special properties, the evolution of life would have taken a very different course.

17.13 Expansion joints on bridges are needed to accommodate changes in length that result from thermal expansion.

Thermal Stress

If we clamp the ends of a rod rigidly to prevent expansion or contraction and then change the temperature, **thermal stresses** develop. The rod would like to expand or contract, but the clamps won't let it. The resulting stresses may become large enough to strain the rod irreversibly or even break it. (You may want to review the discussion of stress and strain in Section 11.4).

Engineers must account for thermal stress when designing structures (see Fig. 17.11). Concrete highways and bridge decks usually have gaps between sections, filled with a flexible material or bridged by interlocking teeth (**Fig. 17.13**), to permit expansion and contraction of the concrete. Long steam pipes have expansion joints or U-shaped sections to prevent buckling or stretching with temperature changes. If one end of a steel bridge is rigidly fastened to its abutment, the other end usually rests on rollers.

To calculate the thermal stress in a clamped rod, we compute the amount the rod *would* expand (or contract) if not held and then find the stress needed to compress (or stretch) it back to its original length. Suppose that a rod with length L_0 and cross-sectional area A is held at constant length while the temperature is reduced (negative ΔT), causing a tensile stress. From Eq. (17.6), the fractional change in length if the rod were free to contract would be

$$\left(\frac{\Delta L}{L_0}\right)_{\text{thermal}} = \alpha\,\Delta T \tag{17.10}$$

Both ΔL and ΔT are negative. The tension must increase by an amount F that is just enough to produce an equal and opposite fractional change in length $(\Delta L/L_0)_{\text{tension}}$. From the definition of Young's modulus, Eq. (11.10),

$$Y = \frac{F/A}{\Delta L/L_0} \qquad \text{so} \qquad \left(\frac{\Delta L}{L_0}\right)_{\text{tension}} = \frac{F}{AY} \tag{17.11}$$

If the length is to be constant, the *total* fractional change in length must be zero. From Eqs. (17.10) and (17.11), this means that

$$\left(\frac{\Delta L}{L_0}\right)_{\text{thermal}} + \left(\frac{\Delta L}{L_0}\right)_{\text{tension}} = \alpha\,\Delta T + \frac{F}{AY} = 0$$

Solve for the tensile stress F/A required to keep the rod's length constant:

Thermal stress: ⋯⋯⋯⋯⋯Young's modulus

Force needed to ⋯⋯ $\dfrac{F}{A} = -Y\alpha\,\Delta T$ ⋯Temperature change (17.12)
keep length of rod
constant ⋯⋯⋯Coefficient of linear expansion
⋯⋯Cross-sectional area of rod

For a decrease in temperature, ΔT is negative, so F and F/A are positive; this means that a *tensile* force and stress are needed to maintain the length. If ΔT is positive, F and F/A are negative, and the required force and stress are *compressive*.

If there are temperature differences within a body, nonuniform expansion or contraction will result and thermal stresses can be induced. You can break a glass bowl by pouring very hot water into it; the thermal stress between the hot and cold parts of the bowl exceeds the breaking stress of the glass, causing cracks. The same phenomenon makes ice cubes crack when dropped into warm water.

EXAMPLE 17.4 THERMAL STRESS

An aluminum cylinder 10 cm long, with a cross-sectional area of 20 cm², is used as a spacer between two steel walls. At 17.2°C it just slips between the walls. Calculate the stress in the cylinder and the total force it exerts on each wall when it warms to 22.3°C, assuming that the walls are perfectly rigid and a constant distance apart.

SOLUTION

IDENTIFY and SET UP: See **Fig. 17.14**. Our target variables are the thermal stress F/A in the cylinder, whose cross-sectional area A is given, and the associated force F it exerts on the walls. We use Eq. (17.12) to relate F/A to the temperature change ΔT, and from

that calculate F. (The length of the cylinder is irrelevant.) We find Young's modulus Y_{Al} and the coefficient of linear expansion α_{Al} from Tables 11.1 and 17.1, respectively.

17.14 Our sketch for this problem.

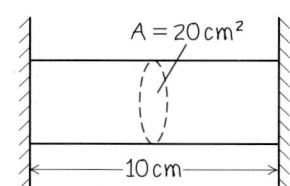

Continued

EXECUTE: We have $Y_{Al} = 7.0 \times 10^{10}$ Pa and $\alpha_{Al} = 2.4 \times 10^{-5}$ K^{-1}, and $\Delta T = 22.3°C - 17.2°C = 5.1$ C° = 5.1 K. From Eq. (17.12), the stress is

$$\frac{F}{A} = -Y_{Al}\alpha_{Al}\Delta T$$

$$= -(7.0 \times 10^{10} \text{ Pa})(2.4 \times 10^{-5} \text{ K}^{-1})(5.1 \text{ K})$$

$$= -8.6 \times 10^{6} \text{ Pa} = -1200 \text{ lb/in.}^{2}$$

The total force is the cross-sectional area times the stress:

$$F = A\left(\frac{F}{A}\right) = (20 \times 10^{-4} \text{ m}^{2})(-8.6 \times 10^{6} \text{ Pa})$$

$$= -1.7 \times 10^{4} \text{ N} = -1.9 \text{ tons}$$

EVALUATE: The stress on the cylinder and the force it exerts on each wall are immense. Such thermal stresses must be accounted for in engineering.

TEST YOUR UNDERSTANDING OF SECTION 17.4 In the bimetallic strip shown in Fig. 17.3a, metal 1 is copper. Which of the following materials could be used for metal 2? (There may be more than one correct answer). (i) Steel; (ii) brass; (iii) aluminum. ▌

17.5 QUANTITY OF HEAT

17.15 The same temperature change of the same system may be accomplished by (a) doing work on it or (b) adding heat to it.

(a) Raising the temperature of water by doing work on it

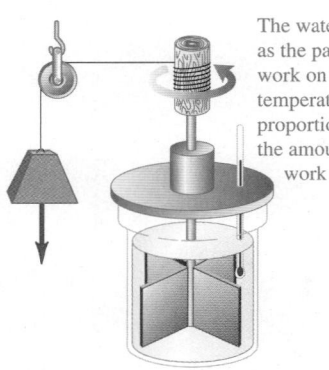

The water warms as the paddle does work on it; the temperature rise is proportional to the amount of work done.

(b) Raising the temperature of water by direct heating

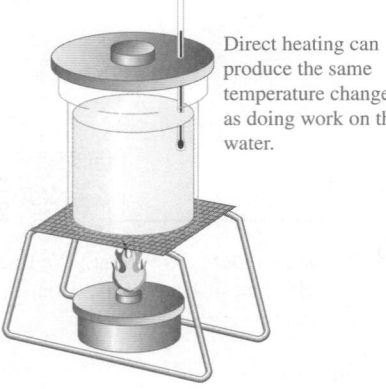

Direct heating can produce the same temperature change as doing work on the water.

When you put a cold spoon into a cup of hot coffee, the spoon warms up and the coffee cools down as they approach thermal equilibrium. What causes these temperature changes is a transfer of *energy* from one substance to another. Energy transfer that takes place solely because of a temperature difference is called *heat flow* or *heat transfer,* and energy transferred in this way is called **heat.**

An understanding of the relationship between heat and other forms of energy emerged during the 18th and 19th centuries. Sir James Joule (1818–1889) studied how water can be warmed by vigorous stirring with a paddle wheel (**Fig. 17.15a**). The paddle wheel adds energy to the water by doing *work* on it, and Joule found that *the temperature rise is directly proportional to the amount of work done.* The same temperature change can also be caused by putting the water in contact with some hotter body (Fig. 17.15b); hence this interaction must also involve an energy exchange. We'll explore the relationship between heat and mechanical energy in Chapters 19 and 20.

CAUTION **Temperature vs. heat** It is absolutely essential for you to distinguish between *temperature* and *heat*. Temperature depends on the physical state of a material and is a quantitative description of its hotness or coldness. In physics the term "heat" always refers to energy in transit from one body or system to another because of a temperature difference, never to the amount of energy contained within a particular system. We can change the temperature of a body by adding heat to it or taking heat away, or by adding or subtracting energy in other ways, such as mechanical work (Fig. 17.15a). If we cut a body in half, each half has the same temperature as the whole; but to raise the temperature of each half by a given interval, we add *half* as much heat as for the whole. ▌

We can define a *unit* of quantity of heat based on temperature changes of some specific material. The **calorie** (abbreviated cal) is *the amount of heat required to raise the temperature of 1 gram of water from 14.5°C to 15.5°C.* A food-value calorie is actually a kilocalorie (kcal), equal to 1000 cal. A corresponding unit of heat that uses Fahrenheit degrees and British units is the **British thermal unit,** or Btu. One Btu is the quantity of heat required to raise the temperature of 1 pound (weight) of water 1 F° from 63°F to 64°F.

Because heat is energy in transit, there must be a definite relationship between these units and the familiar mechanical energy units such as the joule (**Fig. 17.16**). Experiments similar in concept to Joule's have shown that

$$1 \text{ cal} = 4.186 \text{ J}$$

$$1 \text{ kcal} = 1000 \text{ cal} = 4186 \text{ J}$$

$$1 \text{ Btu} = 778 \text{ ft} \cdot \text{lb} = 252 \text{ cal} = 1055 \text{ J}$$

The calorie is not a fundamental SI unit. The International Committee on Weights and Measures recommends using the joule as the basic unit of energy in all forms, including heat. We will follow that recommendation in this book.

Specific Heat

We use the symbol Q for quantity of heat. When it is associated with an infinitesimal temperature change dT, we call it dQ. The quantity of heat Q required to increase the temperature of a mass m of a certain material from T_1 to T_2 is found to be approximately proportional to the temperature change $\Delta T = T_2 - T_1$. It is also proportional to the mass m of material. When you're heating water to make tea, you need twice as much heat for two cups as for one if the temperature change is the same. The quantity of heat needed also depends on the nature of the material; raising the temperature of 1 kilogram of water by 1 C° requires 4190 J of heat, but only 910 J is needed to raise the temperature of 1 kilogram of aluminum by 1 C°.

Putting all these relationships together, we have

Heat required to change temperature of a certain mass ⟶ Mass of material ⟶ Temperature change ⟶ Specific heat of material

$$Q = mc\,\Delta T \qquad (17.13)$$

The **specific heat** c has different values for different materials. For an infinitesimal temperature change dT and corresponding quantity of heat dQ,

$$dQ = mc\,dT \qquad (17.14)$$

$$c = \frac{1}{m}\frac{dQ}{dT} \qquad \text{(specific heat)} \qquad (17.15)$$

In Eqs. (17.13), (17.14), and (17.15), when Q (or dQ) and ΔT (or dT) are positive, heat enters the body and its temperature increases. When they are negative, heat leaves the body and its temperature decreases.

CAUTION **The definition of heat** Remember that dQ does not represent a change in the amount of heat *contained* in a body. Heat is always energy *in transit* as a result of a temperature difference. There is no such thing as "the amount of heat in a body." ▮

The specific heat of water is approximately

$$4190 \text{ J/kg} \cdot \text{K} \qquad 1 \text{ cal/g} \cdot \text{C}° \qquad \text{or} \qquad 1 \text{ Btu/lb} \cdot \text{F}°$$

The specific heat of a material always depends somewhat on the initial temperature and the temperature interval. **Figure 17.17** shows this dependence for water. In this chapter we will usually ignore this small variation.

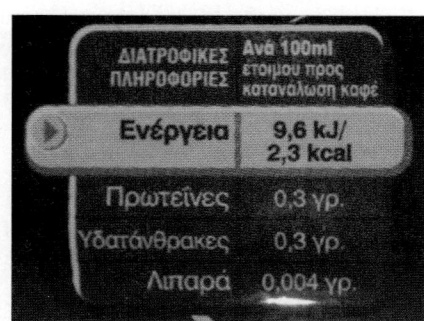

17.16 The word "energy" is of Greek origin. This label on a can of Greek coffee shows that 100 milliliters of prepared coffee have an energy content ($\varepsilon\nu\acute{\varepsilon}\rho\gamma\varepsilon\iota\alpha$) of 9.6 kilojoules or 2.3 kilocalories.

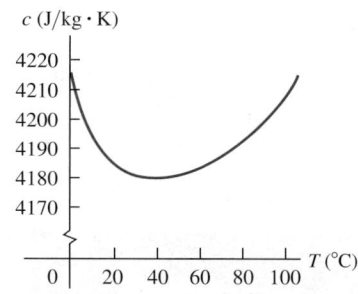

17.17 Specific heat of water as a function of temperature. The value of c varies by less than 1% between 0°C and 100°C.

EXAMPLE 17.5   FEED A COLD, STARVE A FEVER

During a bout with the flu an 80-kg man ran a fever of 39.0°C (102.2°F) instead of the normal body temperature of 37.0°C (98.6°F). Assuming that the human body is mostly water, how much heat is required to raise his temperature by that amount?

SOLUTION

IDENTIFY and SET UP: This problem uses the relationship among heat (the target variable), mass, specific heat, and temperature change. We use Eq. (17.13) to determine the required heat Q, with $m = 80$ kg, $c = 4190 \text{ J/kg} \cdot \text{K}$ (for water), and $\Delta T = 39.0°C - 37.0°C = 2.0 \text{ C}° = 2.0 \text{ K}$.

EXECUTE: From Eq. (17.13),

$$Q = mc\,\Delta T = (80 \text{ kg})(4190 \text{ J/kg} \cdot \text{K})(2.0 \text{ K}) = 6.7 \times 10^5 \text{ J}$$

EVALUATE: This corresponds to 160 kcal. In fact, the specific heat of the human body is about 3480 J/kg·K, 83% that of water, because protein, fat, and minerals have lower specific heats. Hence a more accurate answer is $Q = 5.6 \times 10^5 \text{ J} = 133$ kcal. Either result shows us that were it not for the body's temperature-regulating systems, taking in energy in the form of food would produce measurable changes in body temperature. (The elevated temperature of a person with the flu results from the body's extra activity in response to infection.)

EXAMPLE 17.6 OVERHEATING ELECTRONICS

You are designing an electronic circuit element made of 23 mg of silicon. The electric current through it adds energy at the rate of 7.4 mW = 7.4×10^{-3} J/s. If your design doesn't allow any heat transfer out of the element, at what rate does its temperature increase? The specific heat of silicon is 705 J/kg·K.

SOLUTION

IDENTIFY and SET UP: The energy added to the circuit element gives rise to a temperature increase, just as if heat were flowing into the element at the rate $dQ/dt = 7.4 \times 10^{-3}$ J/s. Our target variable is the rate of temperature change dT/dt. We can use Eq. (17.14),

which relates infinitesimal temperature changes dT to the corresponding heat dQ, to obtain an expression for dQ/dt in terms of dT/dt.

EXECUTE: We divide both sides of Eq. (17.14) by dt and rearrange:

$$\frac{dT}{dt} = \frac{dQ/dt}{mc} = \frac{7.4 \times 10^{-3} \text{ J/s}}{(23 \times 10^{-6} \text{ kg})(705 \text{ J/kg·K})} = 0.46 \text{ K/s}$$

EVALUATE: At this rate of temperature rise (27 K/min), the circuit element would soon self-destruct. Heat transfer is an important design consideration in electronic circuit elements.

Molar Heat Capacity

Sometimes it's more convenient to describe a quantity of substance in terms of the number of *moles n* rather than the *mass m* of material. Recall from your study of chemistry that a mole of any pure substance always contains the same number of molecules. (We will discuss this point in more detail in Chapter 18.) The *molar mass* of any substance, denoted by M, is the mass per mole. (The quantity M is sometimes called *molecular weight,* but *molar mass* is preferable; the quantity depends on the mass of a molecule, not its weight.) For example, the molar mass of water is 18.0 g/mol = 18.0×10^{-3} kg/mol; 1 mole of water has a mass of 18.0 g = 0.0180 kg. The total mass m of material is equal to the mass per mole M times the number of moles n:

$$m = nM \tag{17.16}$$

Replacing the mass m in Eq. (17.13) by the product nM, we find

$$Q = nMc\,\Delta T \tag{17.17}$$

17.18 Water has a much higher specific heat than the glass or metals used to make cookware. This helps explain why it takes several minutes to boil water on a stove, even though the pot or kettle reaches a high temperature very quickly.

The product Mc is called the **molar heat capacity** (or *molar specific heat*) and is denoted by C (capitalized). With this notation we rewrite Eq. (17.17) as

Heat required to change temperature of a certain number of moles $\cdots\!\!\rightarrow Q = nC\,\Delta T$ $\cdots\cdots$ Number of moles of material — Temperature change — Molar heat capacity of material $\tag{17.18}$

Comparing to Eq. (17.15), we can express the molar heat capacity C (heat per mole per temperature change) in terms of the specific heat c (heat per mass per temperature change) and the molar mass M (mass per mole):

$$C = \frac{1}{n}\frac{dQ}{dT} = Mc \qquad \text{(molar heat capacity)} \tag{17.19}$$

For example, the molar heat capacity of water is

$$C = Mc = (0.0180 \text{ kg/mol})(4190 \text{ J/kg·K}) = 75.4 \text{ J/mol·K}$$

Table 17.3 gives values of specific heat and molar heat capacity for several substances. Note the remarkably large specific heat for water (**Fig. 17.18**).

DEMO

DEMO

CAUTION **The meaning of "heat capacity"** The term "heat capacity" is unfortunate because it gives the erroneous impression that a body *contains* a certain amount of heat. Remember, heat is energy in transit to or from a body, not the energy residing in the body. ▌

TABLE 17.3 Approximate Specific Heats and Molar Heat Capacities (Constant Pressure)

Substance	Specific Heat, c (J/kg · K)	Molar Mass, M (kg/mol)	Molar Heat Capacity, C (J/mol · K)
Aluminum	910	0.0270	24.6
Beryllium	1970	0.00901	17.7
Copper	390	0.0635	24.8
Ethanol	2428	0.0461	111.9
Ethylene glycol	2386	0.0620	148.0
Ice (near 0°C)	2100	0.0180	37.8
Iron	470	0.0559	26.3
Lead	130	0.207	26.9
Marble ($CaCO_3$)	879	0.100	87.9
Mercury	138	0.201	27.7
Salt (NaCl)	879	0.0585	51.4
Silver	234	0.108	25.3
Water (liquid)	4190	0.0180	75.4

Measurements of specific heats and molar heat capacities for solid materials are usually made at constant atmospheric pressure; the corresponding values are called the *specific heat* and *molar heat capacity at constant pressure,* denoted by c_p and C_p. For a gas it is usually easier to keep the substance in a container with constant *volume;* the corresponding values are called the *specific heat* and *molar heat capacity at constant volume,* denoted by c_V and C_V. For a given substance, C_V and C_p are different. If the system can expand while heat is added, there is additional energy exchange through the performance of *work* by the system on its surroundings. If the volume is constant, the system does no work. For gases the difference between C_p and C_V is substantial. We will study heat capacities of gases in detail in Section 19.7.

The last column of Table 17.3 shows something interesting. The molar heat capacities for most elemental solids are about the same: about 25 J/mol · K. This correlation, named the *rule of Dulong and Petit* (for its discoverers), forms the basis for a very important idea. The number of atoms in 1 mole is the same for all elemental substances. This means that on a *per atom* basis, about the same amount of heat is required to raise the temperature of each of these elements by a given amount, even though the *masses* of the atoms are very different. The heat required for a given temperature increase depends only on *how many* atoms the sample contains, not on the mass of an individual atom. We will see the reason the rule of Dulong and Petit works so well when we study the molecular basis of heat capacities in greater detail in Chapter 18.

TEST YOUR UNDERSTANDING OF SECTION 17.5 You wish to raise the temperature of each of the following samples from 20°C to 21°C. Rank these in order of the amount of heat needed to do this, from highest to lowest. (i) 1 kilogram of mercury; (ii) 1 kilogram of ethanol; (iii) 1 mole of mercury; (iv) 1 mole of ethanol. ∎

17.6 CALORIMETRY AND PHASE CHANGES

Calorimetry means "measuring heat." We have discussed the energy transfer (heat) involved in temperature changes. Heat is also involved in *phase changes,* such as the melting of ice or boiling of water. Once we understand these additional heat relationships, we can analyze a variety of problems involving quantity of heat.

17.19 The surrounding air is at room temperature, but this ice–water mixture remains at 0°C until all of the ice has melted and the phase change is complete.

PhET: States of Matter

Phase Changes

We use the term **phase** to describe a specific state of matter, such as a solid, liquid, or gas. The compound H_2O exists in the *solid phase* as ice, in the *liquid phase* as water, and in the *gaseous phase* as steam. (These are also referred to as **states of matter:** the solid state, the liquid state, and the gaseous state.) A transition from one phase to another is called a **phase change** or *phase transition.* For any given pressure a phase change takes place at a definite temperature, usually accompanied by heat flowing in or out and a change of volume and density.

A familiar phase change is the melting of ice. When we add heat to ice at 0°C and normal atmospheric pressure, the temperature of the ice *does not* increase. Instead, some of it melts to form liquid water. If we add the heat slowly, to maintain the system very close to thermal equilibrium, the temperature remains at 0°C until all the ice is melted (**Fig. 17.19**). The effect of adding heat to this system is not to raise its temperature but to change its *phase* from solid to liquid.

To change 1 kg of ice at 0°C to 1 kg of liquid water at 0°C and normal atmospheric pressure requires 3.34×10^5 J of heat. The heat required per unit mass is called the **heat of fusion** (or sometimes *latent heat of fusion*), denoted by L_f. For water at normal atmospheric pressure the heat of fusion is

$$L_f = 3.34 \times 10^5 \text{ J/kg} = 79.6 \text{ cal/g} = 143 \text{ Btu/lb}$$

More generally, to melt a mass m of material that has a heat of fusion L_f requires a quantity of heat Q given by

$$Q = mL_f$$

This process is *reversible.* To freeze liquid water to ice at 0°C, we have to *remove* heat; the magnitude is the same, but in this case, Q is negative because heat is removed rather than added. To cover both possibilities and to include other kinds of phase changes, we write

Heat transfer in a phase change $\cdots\!\rightarrow Q = \pm mL \leftarrow\cdots$ Mass of material that changes phase / Latent heat for this phase change (17.20) / + if heat enters material, − if heat leaves

The plus sign (heat entering) is used when the material melts; the minus sign (heat leaving) is used when it freezes. The heat of fusion is different for different materials, and it also varies somewhat with pressure.

For any given material at any given pressure, the freezing temperature is the same as the melting temperature. At this unique temperature the liquid and solid phases can coexist in a condition called **phase equilibrium.**

We can go through this whole story again for *boiling* or *evaporation,* a phase transition between liquid and gaseous phases. The corresponding heat (per unit mass) is called the **heat of vaporization** L_v. At normal atmospheric pressure the heat of vaporization L_v for water is

$$L_v = 2.256 \times 10^6 \text{ J/kg} = 539 \text{ cal/g} = 970 \text{ Btu/lb}$$

That is, it takes 2.256×10^6 J to change 1 kg of liquid water at 100°C to 1 kg of water vapor at 100°C. By comparison, to raise the temperature of 1 kg of water from 0°C to 100°C requires $Q = mc\,\Delta T = (1.00 \text{ kg})(4190 \text{ J/kg} \cdot \text{C}°) \times (100 \text{ C}°) = 4.19 \times 10^5$ J, less than one-fifth as much heat as is required for vaporization at 100°C. This agrees with everyday kitchen experience; a pot of water may reach boiling temperature in a few minutes, but it takes a much longer time to completely evaporate all the water away.

Phase of water changes. During these periods, temperature stays constant and the phase change proceeds as heat is added: $Q = +mL$.

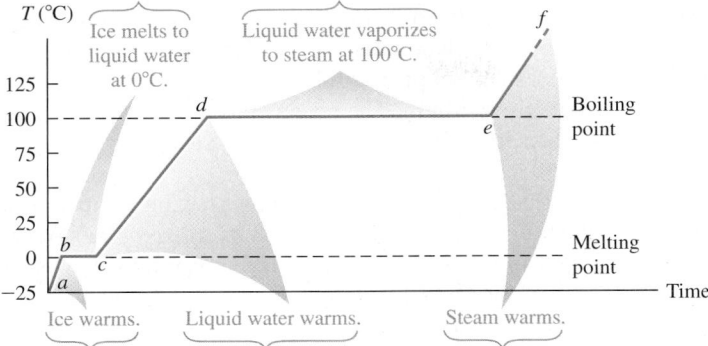

Temperature of water changes. During these periods, temperature rises as heat is added: $Q = mc\Delta T$.

17.20 Graph of temperature versus time for a specimen of water initially in the solid phase (ice). Heat is added to the specimen at a constant rate. The temperature remains constant during each change of phase, provided that the pressure remains constant.

$a \rightarrow b$: Ice initially at $-25°C$ is warmed to $0°C$.
$b \rightarrow c$: Temperature remains at $0°C$ until all ice melts.
$c \rightarrow d$: Water is warmed from $0°C$ to $100°C$.
$d \rightarrow e$: Temperature remains at $100°C$ until all water vaporizes.
$e \rightarrow f$: Steam is warmed to temperatures above $100°C$.

Like melting, boiling is a reversible transition. When heat is removed from a gas at the boiling temperature, the gas returns to the liquid phase, or *condenses,* giving up to its surroundings the same quantity of heat (heat of vaporization) that was needed to vaporize it. At a given pressure the boiling and condensation temperatures are always the same; at this temperature the liquid and gaseous phases can coexist in phase equilibrium.

Both L_v and the boiling temperature of a material depend on pressure. Water boils at a lower temperature (about $95°C$) in Denver than in Pittsburgh because Denver is at higher elevation and the average atmospheric pressure is lower. The heat of vaporization is somewhat greater at this lower pressure, about 2.27×10^6 J/kg.

Figure 17.20 summarizes these ideas about phase changes. **Table 17.4** lists heats of fusion and vaporization for some materials and their melting and boiling temperatures at normal atmospheric pressure. Very few *elements* have melting temperatures in the vicinity of ordinary room temperatures; one of the few is the metal gallium, shown in **Fig. 17.21**.

17.21 The metal gallium, shown here melting in a person's hand, is one of the few elements that melt in the vicinity of room temperature. Its melting temperature is $29.8°C$, and its heat of fusion is 8.04×10^4 J/kg.

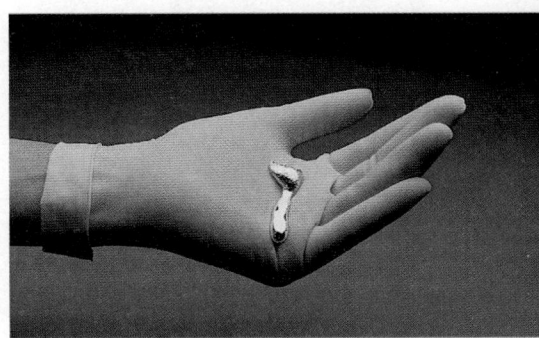

TABLE 17.4	**Heats of Fusion and Vaporization**					
	Normal Melting Point		**Heat of Fusion, L_f (J/kg)**	**Normal Boiling Point**		**Heat of Vaporization, L_v (J/kg)**
Substance	**K**	**°C**		**K**	**°C**	
Helium	*	*	*	4.216	−268.93	20.9×10^3
Hydrogen	13.84	−259.31	58.6×10^3	20.26	−252.89	452×10^3
Nitrogen	63.18	−209.97	25.5×10^3	77.34	−195.8	201×10^3
Oxygen	54.36	−218.79	13.8×10^3	90.18	−183.0	213×10^3
Ethanol	159	−114	104.2×10^3	351	78	854×10^3
Mercury	234	−39	11.8×10^3	630	357	272×10^3
Water	273.15	0.00	334×10^3	373.15	100.00	2256×10^3
Sulfur	392	119	38.1×10^3	717.75	444.60	326×10^3
Lead	600.5	327.3	24.5×10^3	2023	1750	871×10^3
Antimony	903.65	630.50	165×10^3	1713	1440	561×10^3
Silver	1233.95	960.80	88.3×10^3	2466	2193	2336×10^3
Gold	1336.15	1063.00	64.5×10^3	2933	2660	1578×10^3
Copper	1356	1083	134×10^3	1460	1187	5069×10^3

*A pressure in excess of 25 atmospheres is required to make helium solidify. At 1 atmosphere pressure, helium remains a liquid down to absolute zero.

17.22 When this airplane flew into a cloud at a temperature just below freezing, the plane struck supercooled water droplets in the cloud that rapidly crystallized and formed ice on the plane's nose (shown here) and wings. Such inflight icing can be extremely hazardous, which is why commercial airliners are equipped with devices to remove ice.

17.23 The water may be warm and it may be a hot day, but these children will feel cold when they first step out of the swimming pool. That's because as water evaporates from their skin, it removes the heat of vaporization from their bodies. To stay warm, they will need to dry off immediately.

A substance can sometimes change directly from the solid to the gaseous phase. This process is called *sublimation,* and the solid is said to *sublime.* The corresponding heat is called the *heat of sublimation, L_s.* Liquid carbon dioxide cannot exist at a pressure lower than about 5×10^5 Pa (about 5 atm), and "dry ice" (solid carbon dioxide) sublimes at atmospheric pressure. Sublimation of water from frozen food causes freezer burn. The reverse process, a phase change from gas to solid, occurs when frost forms on cold bodies such as refrigerator cooling coils.

Very pure water can be cooled several degrees below the freezing temperature without freezing; the resulting unstable state is described as *supercooled.* When a small ice crystal is dropped in or the water is agitated, it crystallizes within a second or less (**Fig. 17.22**). Supercooled water *vapor* condenses quickly into fog droplets when a disturbance, such as dust particles or ionizing radiation, is introduced. This principle is used in "seeding" clouds, which often contain supercooled water vapor, to cause condensation and rain.

A liquid can sometimes be *superheated* above its normal boiling temperature. Any small disturbance such as agitation causes local boiling with bubble formation.

Steam heating systems for buildings use a boiling–condensing process to transfer heat from the furnace to the radiators. Each kilogram of water that is turned to steam in the boiler absorbs over 2×10^6 J (the heat of vaporization L_v of water) from the boiler and gives it up when it condenses in the radiators. Boiling–condensing processes are also used in refrigerators, air conditioners, and heat pumps. We'll discuss these systems in Chapter 20.

The temperature-control mechanisms of many warm-blooded animals make use of heat of vaporization, removing heat from the body by using it to evaporate water from the tongue (panting) or from the skin (sweating). Such *evaporative cooling* enables humans to maintain normal body temperature in hot, dry desert climates where the air temperature may reach 55°C (about 130°F). The skin temperature may be as much as 30°C cooler than the surrounding air. Under these conditions a normal person may perspire several liters per day, and this lost water must be replaced. Evaporative cooling also explains why you feel cold when you first step out of a swimming pool (**Fig. 17.23**).

Evaporative cooling is also used to condense and recirculate "used" steam in coal-fired or nuclear-powered electric-generating plants. That's what goes on in the large, tapered concrete towers that you see at such plants.

Chemical reactions such as combustion are analogous to phase changes in that they involve definite quantities of heat. Complete combustion of 1 gram of gasoline produces about 46,000 J or about 11,000 cal, so the **heat of combustion** L_c of gasoline is

$$L_c = 46,000 \text{ J/g} = 4.6 \times 10^7 \text{ J/kg}$$

Energy values of foods are defined similarly. When we say that a gram of peanut butter "contains 6 calories," we mean that 6 kcal of heat (6000 cal or 25,000 J) is released when the carbon and hydrogen atoms in the peanut butter react with oxygen (with the help of enzymes) and are completely converted to CO_2 and H_2O. Not all of this energy is directly useful for mechanical work. We'll study the *efficiency* of energy utilization in Chapter 20.

Heat Calculations

Let's look at some examples of calorimetry calculations (calculations with heat). The basic principle is very simple: When heat flow occurs between two bodies that are isolated from their surroundings, the amount of heat lost by one body must equal the amount gained by the other. Heat is energy in transit, so this principle is really just conservation of energy. Calorimetry, dealing entirely with one conserved quantity, is in many ways the simplest of all physical theories!

PROBLEM-SOLVING STRATEGY 17.2 | CALORIMETRY PROBLEMS

IDENTIFY *the relevant concepts:* When heat flow occurs between two or more bodies that are isolated from their surroundings, the *algebraic sum* of the quantities of heat transferred to all the bodies is zero. We take a quantity of heat *added* to a body as *positive* and a quantity *leaving* a body as *negative*.

SET UP *the problem* using the following steps:
1. Identify the objects that exchange heat.
2. Each object may undergo a temperature change only, a phase change at constant temperature, or both. Use Eq. (17.13) for the heat transferred in a temperature change and Eq. (17.20) for the heat transferred in a phase change.
3. Consult Table 17.3 for values of specific heat or molar heat capacity and Table 17.4 for heats of fusion or vaporization.
4. List the known and unknown quantities and identify the target variables.

EXECUTE *the solution* as follows:
1. Use Eq. (17.13) and/or Eq. (17.20) and the energy-conservation relation $\Sigma Q = 0$ to solve for the target variables. Ensure that you use the correct algebraic signs for Q and ΔT terms, and that you correctly write $\Delta T = T_{\text{final}} - T_{\text{initial}}$ and not the reverse.
2. If a phase change occurs, you may not know in advance whether all, or only part, of the material undergoes a phase change. Make a reasonable guess; if that leads to an unreasonable result (such as a final temperature higher or lower than any initial temperature), the guess was wrong. Try again!

EVALUATE *your answer:* Double-check your calculations, and ensure that the results are physically sensible.

EXAMPLE 17.7   A TEMPERATURE CHANGE WITH NO PHASE CHANGE

A camper pours 0.300 kg of coffee, initially in a pot at 70.0°C, into a 0.120-kg aluminum cup initially at 20.0°C. What is the equilibrium temperature? Assume that coffee has the same specific heat as water and that no heat is exchanged with the surroundings.

SOLUTION

IDENTIFY and SET UP: The target variable is the common final temperature T of the cup and coffee. No phase changes occur, so we need only Eq. (17.13). With subscripts C for coffee, W for water, and Al for aluminum, we have $T_{0C} = 70.0°$ and $T_{0Al} = 20.0°$; Table 17.3 gives $c_W = 4190 \text{ J/kg} \cdot \text{K}$ and $c_{Al} = 910 \text{ J/kg} \cdot \text{K}$.

EXECUTE: The (negative) heat gained by the coffee is $Q_C = m_C c_W \Delta T_C$. The (positive) heat gained by the cup is $Q_{Al} = m_{Al} c_{Al} \Delta T_{Al}$. We set $Q_C + Q_{Al} = 0$ (see Problem-Solving Strategy 17.2) and substitute $\Delta T_C = T - T_{0C}$ and $\Delta T_{Al} = T - T_{0Al}$:

$$Q_C + Q_{Al} = m_C c_W \Delta T_C + m_{Al} c_{Al} \Delta T_{Al} = 0$$
$$m_C c_W (T - T_{0C}) + m_{Al} c_{Al} (T - T_{0Al}) = 0$$

Then we solve this expression for the final temperature T. A little algebra gives

$$T = \frac{m_C c_W T_{0C} + m_{Al} c_{Al} T_{0Al}}{m_C c_W + m_{Al} c_{Al}} = 66.0°C$$

EVALUATE: The final temperature is much closer to the initial temperature of the coffee than to that of the cup; water has a much higher specific heat than aluminum, and we have more than twice as much mass of water. We can also find the quantities of heat by substituting the value $T = 66.0°C$ back into the original equations. We find $Q_C = -5.0 \times 10^3$ J and $Q_{Al} = +5.0 \times 10^3$ J. As expected, Q_C is negative: The coffee loses heat to the cup.

EXAMPLE 17.8   CHANGES IN BOTH TEMPERATURE AND PHASE

A glass contains 0.25 kg of Omni-Cola (mostly water) initially at 25°C. How much ice, initially at −20°C, must you add to obtain a final temperature of 0°C with all the ice melted? Ignore the heat capacity of the glass.

SOLUTION

IDENTIFY and SET UP: The Omni-Cola and ice exchange heat. The cola undergoes a temperature change; the ice undergoes both a temperature change and a phase change from solid to liquid. We use subscripts C for cola, I for ice, and W for water. The target variable is the mass of ice, m_I. We use Eq. (17.13) to obtain an expression for the amount of heat involved in cooling the drink to $T = 0°C$ and warming the ice to $T = 0°C$, and Eq. (17.20) to obtain an expression for the heat required to melt the ice at 0°C. We have $T_{0C} = 25°C$ and $T_{0I} = -20°C$, Table 17.3 gives $c_W = 4190 \text{ J/kg} \cdot \text{K}$ and $c_I = 2100 \text{ J/kg} \cdot \text{K}$, and Table 17.4 gives $L_f = 3.34 \times 10^5 \text{ J/kg}$.

EXECUTE: From Eq. (17.13), the (negative) heat gained by the Omni-Cola is $Q_C = m_C c_W \Delta T_C$. The (positive) heat gained by the ice in warming is $Q_I = m_I c_I \Delta T_I$. The (positive) heat required to melt the ice is $Q_2 = m_I L_f$. We set $Q_C + Q_I + Q_2 = 0$, insert $\Delta T_C = T - T_{0C}$ and $\Delta T_I = T - T_{0I}$, and solve for m_I:

$$m_C c_W \Delta T_C + m_I c_I \Delta T_I + m_I L_f = 0$$
$$m_C c_W (T - T_{0C}) + m_I c_I (T - T_{0I}) + m_I L_f = 0$$
$$m_I [c_I (T - T_{0I}) + L_f] = -m_C c_W (T - T_{0C})$$
$$m_I = m_C \frac{c_W (T_{0C} - T)}{c_I (T - T_{0I}) + L_f}$$

Substituting numerical values, we find that $m_I = 0.070 \text{ kg} = 70$ g.

EVALUATE: Three or four medium-size ice cubes would make about 70 g, which seems reasonable given the 250 g of Omni-Cola to be cooled.

EXAMPLE 17.9 | WHAT'S COOKING?

A hot copper pot of mass 2.0 kg (including its copper lid) is at a temperature of 150°C. You pour 0.10 kg of cool water at 25°C into the pot, then quickly replace the lid so no steam can escape. Find the final temperature of the pot and its contents, and determine the phase of the water (liquid, gas, or a mixture). Assume that no heat is lost to the surroundings.

SOLUTION

IDENTIFY and SET UP: The water and the pot exchange heat. Three outcomes are possible: (1) No water boils, and the final temperature T is less than 100°C; (2) some water boils, giving a mixture of water and steam at 100°C; or (3) all the water boils, giving 0.10 kg of steam at 100°C or greater. We use Eq. (17.13) for the heat transferred in a temperature change and Eq. (17.20) for the heat transferred in a phase change.

EXECUTE: First consider case (1), which parallels Example 17.8 exactly. The equation that states that the heat flow into the water equals the heat flow out of the pot is

$$Q_W + Q_{Cu} = m_W c_W(T - T_{0W}) + m_{Cu} c_{Cu}(T - T_{0Cu}) = 0$$

Here we use subscripts W for water and Cu for copper, with $m_W = 0.10$ kg, $m_{Cu} = 2.0$ kg, $T_{0W} = 25$°C, and $T_{0Cu} = 150$°C. From Table 17.3, $c_W = 4190$ J/kg·K and $c_{Cu} = 390$ J/kg·K. Solving for the final temperature T and substituting these values, we get

$$T = \frac{m_W c_W T_{0W} + m_{Cu} c_{Cu} T_{0Cu}}{m_W c_W + m_{Cu} c_{Cu}} = 106°C$$

But this is above the boiling point of water, which contradicts our assumption that no water boils! So at least some of the water boils.

So consider case (2), in which the final temperature is $T = 100$°C and some unknown fraction x of the water boils, where (if this case is correct) x is greater than zero and less than or equal to 1. The (positive) amount of heat needed to vaporize this water is $xm_W L_v$. The energy-conservation condition $Q_W + Q_{Cu} = 0$ is then

$$m_W c_W(100°C - T_{0W}) + xm_W L_v + m_{Cu} c_{Cu}(100°C - T_{0Cu}) = 0$$

We solve for the target variable x:

$$x = \frac{-m_{Cu} c_{Cu}(100°C - T_{0Cu}) - m_W c_W(100°C - T_{0W})}{m_W L_v}$$

With $L_v = 2.256 \times 10^6$ J from Table 17.4, this yields $x = 0.034$. We conclude that the final temperature of the water and copper is 100°C and that $0.034(0.10$ kg$) = 0.0034$ kg $= 3.4$ g of the water is converted to steam at 100°C.

EVALUATE: Had x turned out to be greater than 1, case (3) would have held; all the water would have vaporized, and the final temperature would have been greater than 100°C. Can you show that this would have been the case if we had originally poured less than 15 g of 25°C water into the pot?

EXAMPLE 17.10 | COMBUSTION, TEMPERATURE CHANGE, AND PHASE CHANGE

In a particular camp stove, only 30% of the energy released in burning gasoline goes to heating the water in a pot on the stove. How much gasoline must we burn to heat 1.00 L (1.00 kg) of water from 20°C to 100°C and boil away 0.25 kg of it?

SOLUTION

IDENTIFY and SET UP: All of the water undergoes a temperature change and part of it undergoes a phase change, from liquid to gas. We determine the heat required to cause both of these changes, and then use the 30% combustion efficiency to determine the amount of gasoline that must be burned (the target variable). We use Eqs. (17.13) and (17.20) and the idea of heat of combustion.

EXECUTE: To raise the temperature of the water from 20°C to 100°C requires

$$Q_1 = mc\,\Delta T = (1.00\text{ kg})(4190\text{ J/kg·K})(80\text{ K})$$
$$= 3.35 \times 10^5 \text{ J}$$

To boil 0.25 kg of water at 100°C requires

$$Q_2 = mL_v = (0.25\text{ kg})(2.256 \times 10^6\text{ J/kg}) = 5.64 \times 10^5 \text{ J}$$

The total energy needed is $Q_1 + Q_2 = 8.99 \times 10^5$ J. This is 30% $= 0.30$ of the total heat of combustion, which is therefore $(8.99 \times 10^5$ J$)/0.30 = 3.00 \times 10^6$ J. As we mentioned earlier, the combustion of 1 g of gasoline releases 46,000 J, so the mass of gasoline required is $(3.00 \times 10^6$ J$)/(46,000$ J/g$) = 65$ g, or a volume of about 0.09 L of gasoline.

EVALUATE: This result suggests the tremendous amount of energy released in burning even a small quantity of gasoline. Another 123 g of gasoline would be required to boil away the remaining water; can you prove this?

TEST YOUR UNDERSTANDING OF SECTION 17.6 You take a block of ice at 0°C and add heat to it at a steady rate. It takes a time t to completely convert the block of ice to steam at 100°C. What do you have at time $t/2$? (i) All ice at 0°C; (ii) a mixture of ice and water at 0°C; (iii) water at a temperature between 0°C and 100°C; (iv) a mixture of water and steam at 100°C. ∎

17.7 MECHANISMS OF HEAT TRANSFER

We have talked about *conductors* and *insulators,* materials that permit or prevent heat transfer between bodies. Now let's look in more detail at *rates* of energy transfer. In the kitchen you use a metal or glass pot for good heat transfer from the stove to whatever you're cooking, but your refrigerator is insulated with a material that *prevents* heat from flowing into the food inside the refrigerator. How do we describe the difference between these two materials?

The three mechanisms of heat transfer are conduction, convection, and radiation. *Conduction* occurs within a body or between two bodies in contact. *Convection* depends on motion of mass from one region of space to another. *Radiation* is heat transfer by electromagnetic radiation, such as sunshine, with no need for matter to be present in the space between bodies.

Conduction

If you hold one end of a copper rod and place the other end in a flame, the end you are holding gets hotter and hotter, even though it is not in direct contact with the flame. Heat reaches the cooler end by **conduction** through the material. The atoms in the hotter regions have more kinetic energy, on the average, than their cooler neighbors. They jostle their neighbors, giving them some of their energy. The neighbors jostle *their* neighbors, and so on through the material. The atoms don't move from one region of material to another, but their energy does.

Most metals also conduct heat by another, more effective mechanism. Within the metal, some electrons can leave their parent atoms and wander through the metal. These "free" electrons can rapidly carry energy from hotter to cooler regions of the metal, so metals are generally good conductors of heat. A metal rod at 20°C feels colder than a piece of wood at 20°C because heat can flow more easily from your hand into the metal. The presence of "free" electrons also causes most metals to be good electrical conductors.

In conduction, the direction of heat flow is always from higher to lower temperature. **Figure 17.24a** shows a rod of conducting material with cross-sectional area A and length L. The left end of the rod is kept at a temperature T_H and the right end at a lower temperature T_C, so heat flows from left to right. The sides of the rod are covered by an ideal insulator, so no heat transfer occurs at the sides.

When a quantity of heat dQ is transferred through the rod in a time dt, the rate of heat flow is dQ/dt. We call this rate the **heat current,** denoted by H. That is, $H = dQ/dt$. Experiments show that the heat current is proportional to the cross-sectional area A of the rod (Fig. 17.24b) and to the temperature difference $(T_H - T_C)$ and is inversely proportional to the rod length L (Fig. 17.24c):

Rate of heat flow   Temperatures of hot and cold ends of rod

$$\text{Heat current in conduction} \quad H = \frac{dQ}{dt} = kA\frac{T_H - T_C}{L} \quad \text{Length of rod} \qquad (17.21)$$

Thermal conductivity of rod material   Cross-sectional area of rod

The quantity $(T_H - T_C)/L$ is the temperature difference *per unit length;* it is called the magnitude of the **temperature gradient.** The numerical value of the **thermal conductivity** k depends on the material of the rod. Materials with large k are good conductors of heat; materials with small k are poor conductors, or insulators. Equation (17.21) also gives the heat current through a slab or through *any* homogeneous body with uniform cross section A perpendicular to the direction of flow; L is the length of the heat-flow path.

The units of heat current H are units of energy per time, or power; the SI unit of heat current is the watt $(1 \text{ W} = 1 \text{ J/s})$. We can find the units of k by solving Eq. (17.21) for k; you can show that the SI units are $\text{W/m} \cdot \text{K}$. **Table 17.5** gives some numerical values of k.

17.24 Steady-state heat flow due to conduction in a uniform rod.

(a) Heat current H

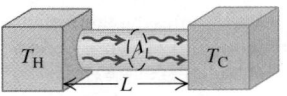

(b) Doubling the cross-sectional area of the conductor doubles the heat current (H is proportional to A).

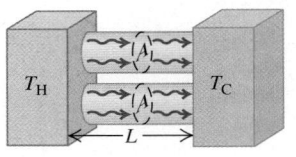

(c) Doubling the length of the conductor halves the heat current (H is inversely proportional to L).

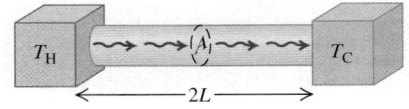

TABLE 17.5	Thermal Conductivities
Substance	**k (W/m · K)**
Metals	
Aluminum	205.0
Brass	109.0
Copper	385.0
Lead	34.7
Mercury	8.3
Silver	406.0
Steel	50.2
Solids (representative values)	
Brick, insulating	0.15
Brick, red	0.6
Concrete	0.8
Cork	0.04
Felt	0.04
Fiberglass	0.04
Glass	0.8
Ice	1.6
Rock wool	0.04
Styrofoam	0.027
Wood	0.12–0.04
Gases	
Air	0.024
Argon	0.016
Helium	0.14
Hydrogen	0.14
Oxygen	0.023

The thermal conductivity of "dead" (nonmoving) air is very small. A wool sweater keeps you warm because it traps air between the fibers. Many insulating materials such as Styrofoam and fiberglass are mostly dead air.

If the temperature varies in a nonuniform way along the length of the conducting rod, we introduce a coordinate x along the length and generalize the temperature gradient to be dT/dx. The corresponding generalization of Eq. (17.21) is

$$H = \frac{dQ}{dt} = -kA\frac{dT}{dx} \qquad (17.22)$$

The negative sign indicates that heat flows in the direction of *decreasing* temperature. If temperature increases with increasing x, then $dT/dx > 0$ and $H < 0$; the negative value of H in this case means that heat flows in the negative x-direction, from high to low temperature.

For thermal insulation in buildings, engineers use the concept of **thermal resistance,** denoted by R. The thermal resistance R of a slab of material with area A is defined so that the heat current H through the slab is

$$H = \frac{A(T_H - T_C)}{R} \qquad (17.23)$$

where T_H and T_C are the temperatures on the two sides of the slab. Comparing this with Eq. (17.21), we see that R is given by

$$R = \frac{L}{k} \qquad (17.24)$$

where L is the thickness of the slab. The SI unit of R is 1 m² · K/W. In the units used for commercial insulating materials in the United States, H is expressed in Btu/h, A is in ft², and $T_H - T_C$ in F°. (1 Btu/h = 0.293 W.) The units of R are then ft² · F° · h/Btu, though values of R are usually quoted without units; a 6-inch-thick layer of fiberglass has an R value of 19 (that is, $R = 19$ ft² · F° · h/Btu), a 2-inch-thick slab of polyurethane foam has an R value of 12, and so on. Doubling the thickness doubles the R value. Common practice in new construction in severe northern climates is to specify R values of around 30 for exterior walls and ceilings. When the insulating material is in layers, such as a plastered wall, fiberglass insulation, and wood exterior siding, the R values are additive. Do you see why?

PROBLEM-SOLVING STRATEGY 17.3 | HEAT CONDUCTION

IDENTIFY *the relevant concepts:* Heat conduction occurs whenever two objects at different temperatures are placed in contact.

SET UP *the problem* using the following steps:
1. Identify the direction of heat flow (from hot to cold). In Eq. (17.21), L is measured along this direction, and A is an area perpendicular to this direction. You can often approximate an irregular-shaped container with uniform wall thickness as a flat slab with the same thickness and total wall area.
2. List the known and unknown quantities and identify the target variable.

EXECUTE *the solution* as follows:
1. If heat flows through a single object, use Eq. (17.21) to solve for the target variable.
2. If the heat flows through two different materials in succession (in *series*), the temperature T at the interface between them is

intermediate between T_H and T_C, so that the temperature differences across the two materials are $(T_H - T)$ and $(T - T_C)$. In steady-state heat flow, the same heat must pass through both materials, so the heat current H must be the *same* in both materials.
3. If heat flows through two or more *parallel* paths, then the total heat current H is the sum of the currents H_1, H_2, ... for the separate paths. An example is heat flow from inside a room to outside, both through the glass in a window and through the surrounding wall. In parallel heat flow the temperature difference is the same for each path, but L, A, and k may be different for each path.
4. Be consistent with units. If k is expressed in W/m · K, for example, use distances in meters, heat in joules, and T in kelvins.

EVALUATE *your answer:* Are the results physically reasonable?

EXAMPLE 17.11 CONDUCTION INTO A PICNIC COOLER

A Styrofoam cooler (**Fig. 17.25a**) has total wall area (including the lid) of 0.80 m² and wall thickness 2.0 cm. It is filled with ice, water, and cans of Omni-Cola, all at 0°C. What is the rate of heat flow into the cooler if the temperature of the outside wall is 30°C? How much ice melts in 3 hours?

SOLUTION

IDENTIFY and SET UP: The target variables are the heat current H and the mass m of ice melted. We use Eq. (17.21) to determine H and Eq. (17.20) to determine m.

17.25 Conduction of heat across the walls of a Styrofoam cooler.

(a) A cooler at the beach

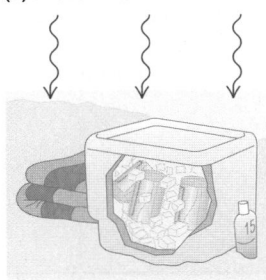

(b) Our sketch for this problem

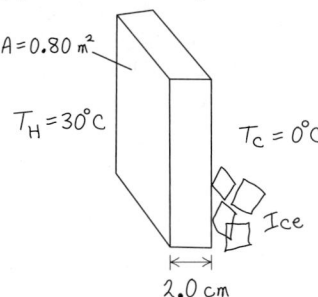

EXECUTE: We assume that the total heat flow is the same as it would be through a flat Styrofoam slab of area 0.80 m² and thickness 2.0 cm = 0.020 m (Fig. 17.25b). We find k from Table 17.5. From Eq. (17.21),

$$H = kA\frac{T_H - T_C}{L} = (0.027 \text{ W/m} \cdot \text{K})(0.80 \text{ m}^2)\frac{30°C - 0°C}{0.020 \text{ m}}$$

$$= 32.4 \text{ W} = 32.4 \text{ J/s}$$

The total heat flow is $Q = Ht$, with $t = 3 \text{ h} = 10{,}800 \text{ s}$. From Table 17.4, the heat of fusion of ice is $L_f = 3.34 \times 10^5 \text{ J/kg}$, so from Eq. (17.20) the mass of ice that melts is

$$m = \frac{Q}{L_f} = \frac{(32.4 \text{ J/s})(10{,}800 \text{ s})}{3.34 \times 10^5 \text{ J/kg}} = 1.0 \text{ kg}$$

EVALUATE: The low heat current is a result of the low thermal conductivity of Styrofoam.

EXAMPLE 17.12 CONDUCTION THROUGH TWO BARS I

A steel bar 10.0 cm long is welded end to end to a copper bar 20.0 cm long. Each bar has a square cross section, 2.00 cm on a side. The free end of the steel bar is kept at 100°C by placing it in contact with steam, and the free end of the copper bar is kept at 0°C by placing it in contact with ice. Both bars are perfectly insulated on their sides. Find the steady-state temperature at the junction of the two bars and the total rate of heat flow through the bars.

SOLUTION

IDENTIFY and SET UP: Figure 17.26 shows the situation. The heat currents in these end-to-end bars must be the same (see Problem-Solving Strategy 17.3). We are given "hot" and "cold" temperatures $T_H = 100°C$ and $T_C = 0°C$. With subscripts S for steel and Cu for copper, we write Eq. (17.21) separately for the heat currents H_S and H_{Cu} and set the resulting expressions equal to each other.

17.26 Our sketch for this problem.

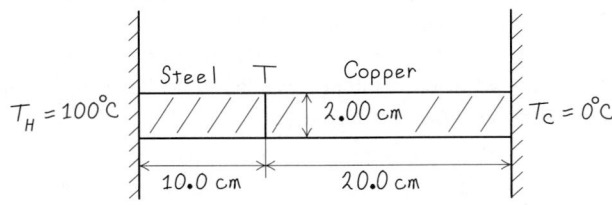

EXECUTE: Setting $H_S = H_{Cu}$, we have from Eq. (17.21)

$$H_S = k_S A\frac{T_H - T}{L_S} = H_{Cu} = k_{Cu} A\frac{T - T_C}{L_{Cu}}$$

We divide out the equal cross-sectional areas A and solve for T:

$$T = \frac{\dfrac{k_S}{L_S}T_H + \dfrac{k_{Cu}}{L_{Cu}}T_C}{\left(\dfrac{k_S}{L_S} + \dfrac{k_{Cu}}{L_{Cu}}\right)}$$

Substituting $L_S = 10.0$ cm and $L_{Cu} = 20.0$ cm, the given values of T_H and T_C, and the values of k_S and k_{Cu} from Table 17.5, we find $T = 20.7°C$.

We can find the total heat current by substituting this value of T into either the expression for H_S or the one for H_{Cu}:

$$H_S = (50.2 \text{ W/m} \cdot \text{K})(0.0200 \text{ m})^2\frac{100°C - 20.7°C}{0.100 \text{ m}}$$

$$= 15.9 \text{ W}$$

$$H_{Cu} = (385 \text{ W/m} \cdot \text{K})(0.0200 \text{ m})^2\frac{20.7°C}{0.200 \text{ m}} = 15.9 \text{ W}$$

EVALUATE: Even though the steel bar is shorter, the temperature drop across it is much greater (from 100°C to 20.7°C) than across the copper bar (from 20.7°C to 0°C). That's because steel is a much poorer conductor than copper.

EXAMPLE 17.13 **CONDUCTION THROUGH TWO BARS II**

Suppose the two bars of Example 17.12 are separated. One end of each bar is kept at 100°C and the other end of each bar is kept at 0°C. What is the *total* heat current in the two bars?

SOLUTION

IDENTIFY and SET UP: Figure 17.27 shows the situation. For each bar, $T_H - T_C = 100°C - 0°C = 100$ K. The total heat current is the sum of the currents in the two bars, $H_S + H_{Cu}$.

17.27 Our sketch for this problem.

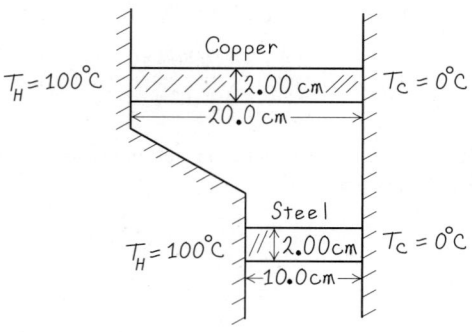

EXECUTE: We write the heat currents for the two rods individually, and then add them to get the total heat current:

$$H = H_S + H_{Cu} = k_S A \frac{T_H - T_C}{L_S} + k_{Cu} A \frac{T_H - T_C}{L_{Cu}}$$

$$= (50.2 \text{ W/m} \cdot \text{K})(0.0200 \text{ m})^2 \frac{100 \text{ K}}{0.100 \text{ m}}$$

$$+ (385 \text{ W/m} \cdot \text{K})(0.0200 \text{ m})^2 \frac{100 \text{ K}}{0.200 \text{ m}}$$

$$= 20.1 \text{ W} + 77.0 \text{ W} = 97.1 \text{ W}$$

EVALUATE: The heat flow in the copper bar is much greater than that in the steel bar, even though it is longer, because the thermal conductivity of copper is much larger. The total heat flow is greater than in Example 17.12 because the total cross section for heat flow is greater and because the full 100-K temperature difference appears across each bar.

Convection

Convection is the transfer of heat by mass motion of a fluid from one region of space to another. Familiar examples include hot-air and hot-water home heating systems, the cooling system of an automobile engine, and the flow of blood in the body. If the fluid is circulated by a blower or pump, the process is called *forced convection;* if the flow is caused by differences in density due to thermal expansion, such as hot air rising, the process is called *free convection* (**Fig. 17.28**).

Free convection in the atmosphere plays a dominant role in determining the daily weather, and convection in the oceans is an important global heat-transfer mechanism. On a smaller scale, soaring hawks and glider pilots make use of thermal updrafts from the warm earth. The most important mechanism for heat transfer within the human body (needed to maintain nearly constant temperature in various environments) is *forced* convection of blood, with the heart as the pump.

Convective heat transfer is a very complex process, and there is no simple equation to describe it. Here are a few experimental facts:

1. The heat current due to convection is directly proportional to the surface area. That's why radiators and cooling fins, which use convection to transfer heat, have large surface areas.

2. The viscosity of fluids slows natural convection near a stationary surface, giving a surface film that on a vertical surface typically has about the same insulating value as 1.3 cm of plywood (*R* value = 0.7). Forced convection decreases the thickness of this film, increasing the rate of heat transfer. This is the reason for the "wind-chill factor"; you get cold faster in a cold wind than in still air with the same temperature.

3. The heat current due to convection is found to be approximately proportional to the $\frac{5}{4}$ power of the temperature difference between the surface and the main body of fluid.

DEMO

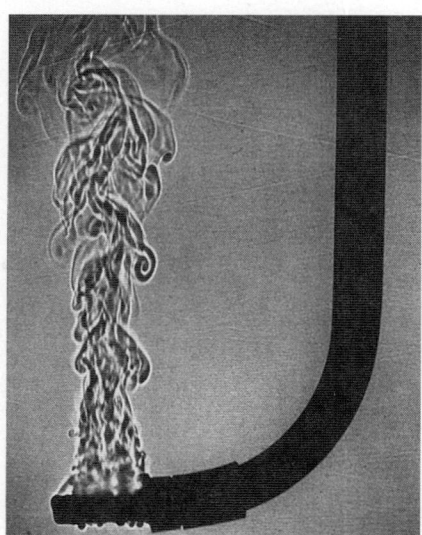

17.28 A heating element in the tip of this submerged tube warms the surrounding water, producing a complex pattern of free convection.

Radiation

Radiation is the transfer of heat by electromagnetic waves such as visible light, infrared, and ultraviolet radiation. Everyone has felt the warmth of the sun's radiation and the intense heat from a charcoal grill or the glowing coals in a fireplace. Most of the heat from these very hot bodies reaches you not by conduction or convection in the intervening air but by *radiation*. This heat transfer would occur even if there were nothing but vacuum between you and the source of heat.

Every body, even at ordinary temperatures, emits energy in the form of electromagnetic radiation. Around 20°C, nearly all the energy is carried by infrared waves with wavelengths much longer than those of visible light (see Fig. 17.4 and **Fig. 17.29**). As the temperature rises, the wavelengths shift to shorter values. At 800°C, a body emits enough visible radiation to appear "red-hot," although even at this temperature most of the energy is carried by infrared waves. At 3000°C, the temperature of an incandescent lamp filament, the radiation contains enough visible light that the body appears "white-hot."

The rate of energy radiation from a surface is proportional to the surface area A and to the fourth power of the absolute (Kelvin) temperature T. The rate also depends on the nature of the surface; we describe this dependence by a quantity e called **emissivity.** A dimensionless number between 0 and 1, e is the ratio of the rate of radiation from a particular surface to the rate of radiation from an equal area of an ideal radiating surface at the same temperature. Emissivity also depends somewhat on temperature. Thus we can express the heat current $H = dQ/dt$ due to radiation from a surface as

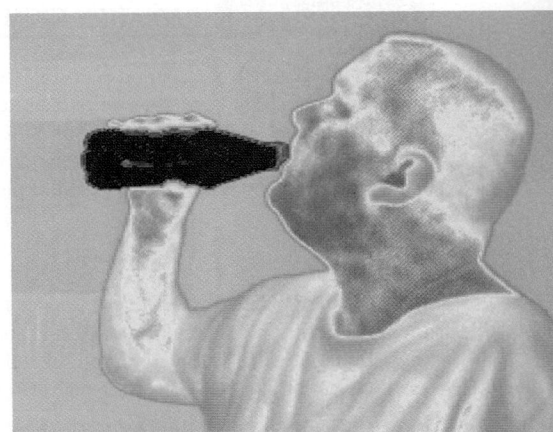

17.29 This false-color infrared photograph reveals radiation emitted by various parts of the man's body. The strongest emission (colored red) comes from the warmest areas, while there is very little emission from the bottle of cold beverage.

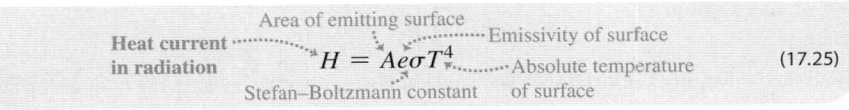

$$\underset{\text{Heat current in radiation}}{\qquad} \overset{\text{Area of emitting surface}}{\underset{\text{Stefan–Boltzmann constant}}{\overset{\qquad}{H = Ae\sigma T^4}}} \overset{\text{Emissivity of surface}}{\underset{\text{Absolute temperature of surface}}{\qquad}} \qquad (17.25)$$

This relationship is called the **Stefan–Boltzmann law** in honor of its late-19th-century discoverers. The **Stefan–Boltzmann constant** σ (Greek sigma) is a fundamental constant; its current best numerical value is

$$\sigma = 5.670373(21) \times 10^{-8} \text{ W/m}^2 \cdot \text{K}^4$$

You should check unit consistency in Eq. (17.25). Emissivity (e) is often larger for dark surfaces than for light ones. The emissivity of a smooth copper surface is about 0.3, but e for a dull black surface can be close to unity.

EXAMPLE 17.14 **HEAT TRANSFER BY RADIATION**

A thin, square steel plate, 10 cm on a side, is heated in a blacksmith's forge to 800°C. If the emissivity is 0.60, what is the total rate of radiation of energy from the plate?

SOLUTION

IDENTIFY and SET UP: The target variable is H, the rate of emission of energy from the plate's two surfaces. We use Eq. (17.25) to calculate H.

EXECUTE: The total surface area is $2(0.10 \text{ m})^2 = 0.020 \text{ m}^2$, and $T = 800°\text{C} = 1073 \text{ K}$. Then Eq. (17.25) gives

$$H = Ae\sigma T^4$$
$$= (0.020 \text{ m}^2)(0.60)(5.67 \times 10^{-8} \text{ W/m}^2 \cdot \text{K}^4)(1073 \text{ K})^4$$
$$= 900 \text{ W}$$

EVALUATE: The nearby blacksmith will easily feel the heat radiated from this plate.

Radiation and Absorption

While a body at absolute temperature T is radiating, its surroundings at temperature T_s are also radiating, and the body *absorbs* some of this radiation. If it is in thermal equilibrium with its surroundings, $T = T_s$ and the rates of radiation and absorption must be equal. For this to be true, the rate of absorption must be given in general by $H = Ae\sigma T_s^4$. Then the *net* rate of radiation from a body at temperature T with surroundings at temperature T_s is $Ae\sigma T^4 - Ae\sigma T_s^4$, or

Area of emitting surface .... Emissivity of surface

Net heat current ····· $H_{net} = Ae\sigma(T^4 - T_s^4)$ (17.26)
in radiation

Stefan–Boltzmann ····   Absolute temperatures of surface (T)
constant   and surroundings (T_s)

In Eq. (17.26) a positive value of H means a net heat flow *out of* the body. This will be the case if $T > T_s$.

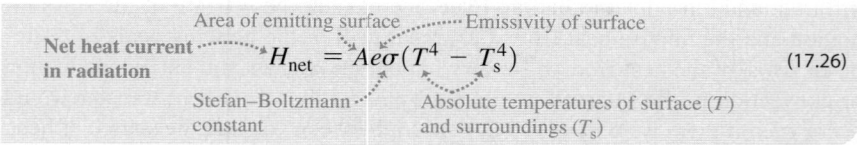

EXAMPLE 17.15   RADIATION FROM THE HUMAN BODY

What is the total rate of radiation of energy from a human body with surface area 1.20 m² and surface temperature 30°C = 303 K? If the surroundings are at a temperature of 20°C, what is the *net* rate of radiative heat loss from the body? The emissivity of the human body is very close to unity, irrespective of skin pigmentation.

SOLUTION

IDENTIFY and SET UP: We must consider both the radiation that the body emits and the radiation that it absorbs from its surroundings. Equation (17.25) gives the rate of radiation of energy from the body, and Eq. (17.26) gives the net rate of heat loss.

EXECUTE: Taking $e = 1$ in Eq. (17.25), we find that the body radiates at a rate

$$H = Ae\sigma T^4$$
$$= (1.20 \text{ m}^2)(1)(5.67 \times 10^{-8} \text{ W/m}^2 \cdot \text{K}^4)(303 \text{ K})^4 = 574 \text{ W}$$

This loss is partly offset by absorption of radiation, which depends on the temperature of the surroundings. From Eq. (17.26), the *net* rate of radiative energy transfer is

$$H_{net} = Ae\sigma(T^4 - T_s^4)$$
$$= (1.20 \text{ m}^2)(1)(5.67 \times 10^{-8} \text{ W/m}^2 \cdot \text{K}^4)$$
$$\times [(303 \text{ K}) - (293 \text{ K})^4]$$
$$= 72 \text{ W}$$

EVALUATE: The value of H_{net} is positive because the body is losing heat to its colder surroundings.

Applications of Radiation

Heat transfer by radiation is important in some surprising places. A premature baby in an incubator can be cooled dangerously by radiation if the walls of the incubator happen to be cold, even when the *air* in the incubator is warm. Some incubators regulate the air temperature by measuring the baby's skin temperature.

A body that is a good absorber must also be a good emitter. An ideal radiator, with emissivity $e = 1$, is also an ideal absorber, absorbing *all* of the radiation that strikes it. Such an ideal surface is called an ideal black body or simply a **blackbody.** Conversely, an ideal *reflector*, which absorbs *no* radiation at all, is also a very ineffective radiator.

This is the reason for the silver coatings on vacuum ("Thermos") bottles, invented by Sir James Dewar (1842–1923). A vacuum bottle has double glass walls. The air is pumped out of the spaces between the walls; this eliminates nearly all heat transfer by conduction and convection. The silver coating on the walls reflects most of the radiation from the contents back into the container, and the wall itself is a very poor emitter. Thus a vacuum bottle can keep coffee or soup hot for several hours. The Dewar flask, used to store very cold liquefied gases, is exactly the same in principle.

Radiation, Climate, and Climate Change

Our planet constantly absorbs radiation coming from the sun. In thermal equilibrium, the rate at which our planet absorbs solar radiation must equal the rate at which it emits radiation into space. The presence of an atmosphere on our planet has a significant effect on this equilibrium.

Most of the radiation emitted by the sun (which has a surface temperature of 5800 K) is in the visible part of the spectrum, to which our atmosphere is transparent. But the average surface temperature of the earth is only 287 K (14°C). Hence most of the radiation that our planet emits into space is infrared radiation, just like the radiation from the person shown in Fig. 17.29. However, our atmosphere is *not* completely transparent to infrared radiation. This is because our atmosphere contains carbon dioxide (CO_2), which is its fourth most abundant constituent (after nitrogen, oxygen, and argon). Molecules of CO_2 in the atmosphere *absorb* some of the infrared radiation coming upward from the surface. They then re-radiate the absorbed energy, but some of the re-radiated energy is directed back down toward the surface instead of escaping into space. In order to maintain thermal equilibrium, the earth's surface must compensate for this by increasing its temperature T and hence its total rate of radiating energy (which is proportional to T^4). This phenomenon, called the **greenhouse effect,** makes our planet's surface temperature about 33°C higher than it would be if there were no atmospheric CO_2. If CO_2 were absent, the earth's average surface temperature would be below the freezing point of water, and life as we know it would be impossible.

While atmospheric CO_2 has benefits, too much of it can have extremely negative consequences. Measurements of air trapped in ancient Antarctic ice show that over the past 650,000 years CO_2 has constituted less than 300 parts per million of our atmosphere. Since the beginning of the industrial age, however, the burning of fossil fuels such as coal and petroleum has elevated the atmospheric CO_2 concentration to unprecedented levels (**Fig. 17.30a**). As a consequence, since the 1950s the global average surface temperature has increased by 0.6°C and the earth has experienced the hottest years ever recorded (Fig. 17.30b). If we continue to consume fossil fuels at the same rate, by 2050 the atmospheric CO_2 concentration will reach 600 parts per million, well off the scale of Fig. 17.30a. The resulting temperature increase will have dramatic effects on global climate. In polar regions massive quantities of ice will melt and run from solid land to the sea, thus raising ocean levels worldwide and threatening the homes and lives of hundreds of millions of people who live near the coast. Coping with these threats is one of the greatest challenges facing 21st-century civilization.

PhET: The Greenhouse Effect

TEST YOUR UNDERSTANDING OF SECTION 17.7 A room has one wall made of concrete, one wall made of copper, and one wall made of steel. All of the walls are the same size and at the same temperature of 20°C. Which wall feels coldest to the touch? (i) The concrete wall; (ii) the copper wall; (iii) the steel wall; (iv) all three walls feel equally cold. ❚

17.30 (a) Due to humans burning fossil fuels, the concentration of carbon dioxide in the atmosphere is now more than 33% greater than in the pre-industrial era. (b) Due to the increased CO_2 concentration, during the past 50 years the global average temperature has increased at an average rate of approximately 0.13 C° per decade.

(a)

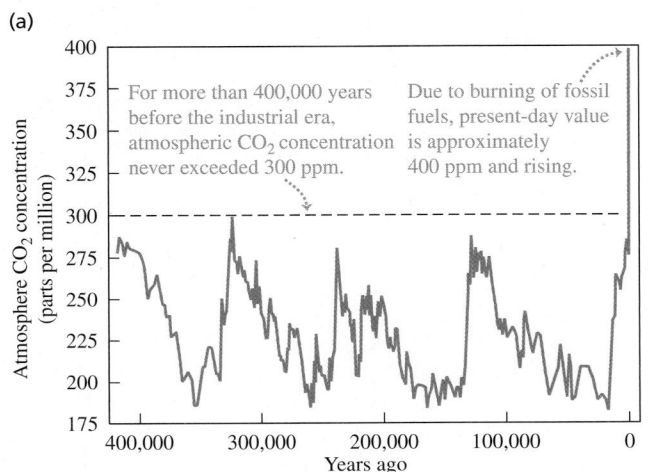

(b)

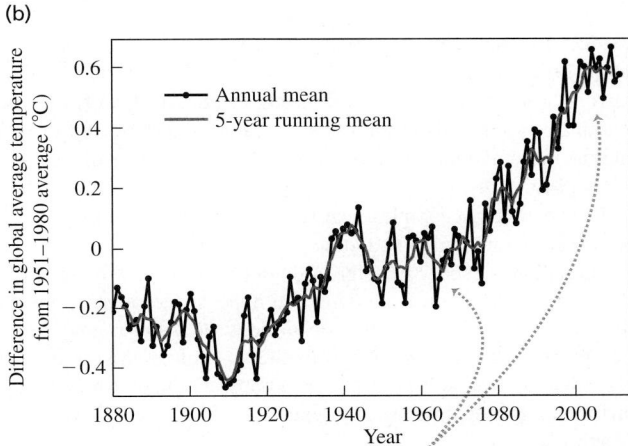

Increased atmospheric CO_2 due to burning of fossil fuels is the cause of this continuing increase in global average temperatures.

Temperature and temperature scales: Two bodies in thermal equilibrium must have the same temperature. A conducting material between two bodies permits them to interact and come to thermal equilibrium; an insulating material impedes this interaction.

The Celsius and Fahrenheit temperature scales are based on the freezing ($0°C = 32°F$) and boiling ($100°C = 212°F$) temperatures of water. One Celsius degree equals $\frac{9}{5}$ Fahrenheit degrees. (See Example 17.1.)

The Kelvin scale has its zero at the extrapolated zero-pressure temperature for a gas thermometer, $-273.15°C = 0$ K. In the gas-thermometer scale, the ratio of two temperatures T_1 and T_2 is defined to be equal to the ratio of the two corresponding gas-thermometer pressures p_1 and p_2.

$$T_F = \tfrac{9}{5}T_C + 32° \quad (17.1)$$

$$T_C = \tfrac{5}{9}(T_F - 32°) \quad (17.2)$$

$$T_K = T_C + 273.15 \quad (17.3)$$

$$\frac{T_2}{T_1} = \frac{p_2}{p_1} \quad (17.4)$$

If systems A and B are each in thermal equilibrium with system C ...

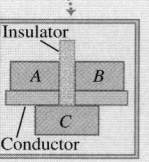 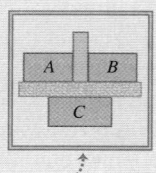

... then systems A and B are in thermal equilibrium with each other.

Thermal expansion and thermal stress: A temperature change ΔT causes a change in any linear dimension L_0 of a solid body. The change ΔL is approximately proportional to L_0 and ΔT. Similarly, a temperature change causes a change ΔV in the volume V_0 of any solid or liquid; ΔV is approximately proportional to V_0 and ΔT. The quantities α and β are the coefficients of linear expansion and volume expansion, respectively. For solids, $\beta = 3\alpha$. (See Examples 17.2 and 17.3.)

When a material is cooled or heated and held so it cannot contract or expand, it is under a tensile stress F/A. (See Example 17.4.)

$$\Delta L = \alpha L_0 \Delta T \quad (17.6)$$

$$\Delta V = \beta V_0 \Delta T \quad (17.8)$$

$$\frac{F}{A} = -Y\alpha \Delta T \quad (17.12)$$

$$L = L_0 + \Delta L$$
$$= L_0(1 + \alpha \Delta T)$$

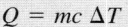

Heat, phase changes, and calorimetry: Heat is energy in transit from one body to another as a result of a temperature difference. Equations (17.13) and (17.18) give the quantity of heat Q required to cause a temperature change ΔT in a quantity of material with mass m and specific heat c (alternatively, with number of moles n and molar heat capacity $C = Mc$, where M is the molar mass and $m = nM$). When heat is added to a body, Q is positive; when it is removed, Q is negative. (See Examples 17.5 and 17.6.)

To change a mass m of a material to a different phase at the same temperature (such as liquid to vapor), a quantity of heat given by Eq. (17.20) must be added or subtracted. Here L is the heat of fusion, vaporization, or sublimation.

In an isolated system whose parts interact by heat exchange, the algebraic sum of the Q's for all parts of the system must be zero. (See Examples 17.7–17.10.)

$$Q = mc \, \Delta T \quad (17.13)$$

$$Q = nC \, \Delta T \quad (17.18)$$

$$Q = \pm mL \quad (17.20)$$

Phase changes, temperature is constant:
$Q = +mL$

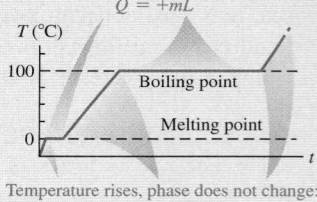

Temperature rises, phase does not change:
$Q = mc\,\Delta T$

Conduction, convection, and radiation: Conduction is the transfer of heat within materials without bulk motion of the materials. The heat current H depends on the area A through which the heat flows, the length L of the heat-flow path, the temperature difference $(T_H - T_C)$, and the thermal conductivity k of the material. (See Examples 17.11–17.13.)

Convection is a complex heat-transfer process that involves mass motion from one region to another.

Radiation is energy transfer through electromagnetic radiation. The radiation heat current H depends on the surface area A, the emissivity e of the surface (a pure number between 0 and 1), and the Kelvin temperature T. Here σ is the Stefan–Boltzmann constant. The *net* radiation heat current H_{net} from a body at temperature T to its surroundings at temperature T_s depends on both T and T_s. (See Examples 17.14 and 17.15.)

$$H = \frac{dQ}{dt} = kA\frac{T_H - T_C}{L} \quad (17.21)$$

$$H = Ae\sigma T^4 \quad (17.25)$$

$$H_{\text{net}} = Ae\sigma(T^4 - T_s^4) \quad (17.26)$$

Heat current H

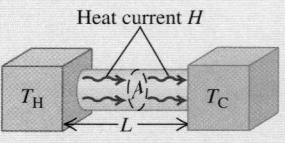

Heat current $H = kA\dfrac{T_H - T_C}{L}$

BRIDGING PROBLEM STEADY-STATE HEAT FLOW: RADIATION AND CONDUCTION

One end of a solid cylindrical copper rod 0.200 m long and 0.0250 m in radius is inserted into a large block of solid hydrogen at its melting temperature, 13.84 K. The other end is blackened and exposed to thermal radiation from surrounding walls at 500.0 K. (Some telescopes in space employ a similar setup. A solid refrigerant keeps the telescope very cold—required for proper operation—even though it is exposed to direct sunlight.) The sides of the rod are insulated, so no energy is lost or gained except at the ends of the rod. (a) When equilibrium is reached, what is the temperature of the blackened end? The thermal conductivity of copper at temperatures near 20 K is 1670 W/m·K. b) At what rate (in kg/h) does the solid hydrogen melt?

SOLUTION GUIDE

IDENTIFY and SET UP

1. Draw a sketch of the situation, showing all relevant dimensions.
2. List the known and unknown quantities, and identify the target variables.
3. In order for the rod to be in equilibrium, how must the radiation heat current from the walls into the blackened end of the rod compare to the conduction heat current from this end to the other end and into the solid hydrogen? Use your answers to select the appropriate equations for part (a).
4. How does the heat current from the rod into the hydrogen determine the rate at which the hydrogen melts? (*Hint:* See Table 17.4.) Use your answer to select the appropriate equations for part (b).

EXECUTE

5. Solve for the temperature of the blackened end of the rod. (*Hint:* Since copper is an excellent conductor of heat at low temperature, you can assume that the temperature of the blackened end is only slightly higher than 13.84 K.)
6. Use your result from step 5 to find the rate at which the hydrogen melts.

EVALUATE

7. Is your result from step 5 consistent with the hint in that step?
8. How would your results from steps 5 and 6 be affected if the rod had twice the radius?

Problems

For assigned homework and other learning materials, go to MasteringPhysics®.

•, ••, •••: Difficulty levels. **CP**: Cumulative problems incorporating material from earlier chapters. **CALC**: Problems requiring calculus. **DATA**: Problems involving real data, scientific evidence, experimental design, and/or statistical reasoning. **BIO**: Biosciences problems.

DISCUSSION QUESTIONS

Q17.1 Explain why it would not make sense to use a full-size glass thermometer to measure the temperature of a thimbleful of hot water.

Q17.2 If you heat the air inside a rigid, sealed container until its Kelvin temperature doubles, the air pressure in the container will also double. Is the same thing true if you double the Celsius temperature of the air in the container? Explain.

Q17.3 Many automobile engines have cast-iron cylinders and aluminum pistons. What kinds of problems could occur if the engine gets too hot? (The coefficient of volume expansion of cast iron is approximately the same as that of steel.)

Q17.4 Why do frozen water pipes burst? Would a mercury thermometer break if the temperature went below the freezing temperature of mercury? Why or why not?

Q17.5 Two bodies made of the same material have the same external dimensions and appearance, but one is solid and the other is hollow. When their temperature is increased, is the overall volume expansion the same or different? Why?

Q17.6 Why is it sometimes possible to loosen caps on screw-top bottles by dipping the capped bottle briefly into hot water?

Q17.7 The inside of an oven is at a temperature of 200°C (392°F). You can put your hand in the oven without injury as long as you don't touch anything. But since the air inside the oven is also at 200°C, why isn't your hand burned just the same?

Q17.8 A newspaper article about the weather states that "the temperature of a body measures how much heat the body contains." Is this description correct? Why or why not?

Q17.9 A student asserts that a suitable unit for specific heat is $1 \text{ m}^2/\text{s}^2 \cdot \text{C}°$. Is she correct? Why or why not?

Q17.10 In some household air conditioners used in dry climates, air is cooled by blowing it through a water-soaked filter, evaporating some of the water. How does this cool the air? Would such a system work well in a high-humidity climate? Why or why not?

Q17.11 The units of specific heat c are J/kg·K, but the units of heat of fusion L_f or heat of vaporization L_v are simply J/kg. Why do the units of L_f and L_v not include a factor of $(\text{K})^{-1}$ to account for a temperature change?

Q17.12 Why is a hot, humid day in the tropics generally more uncomfortable for human beings than a hot, dry day in the desert?

Q17.13 A piece of aluminum foil used to wrap a potato for baking in a hot oven can usually be handled safely within a few seconds after the potato is removed from the oven. The same is not true of the potato, however! Give two reasons for this difference.

Q17.14 Desert travelers sometimes keep water in a canvas bag. Some water seeps through the bag and evaporates. How does this cool the water inside the bag?

Q17.15 When you first step out of the shower, you feel cold. But as soon as you are dry you feel warmer, even though the room temperature does not change. Why?

Q17.16 The climate of regions adjacent to large bodies of water (like the Pacific and Atlantic coasts) usually features a narrower range of temperature than the climate of regions far from large bodies of water (like the prairies). Why?

Q17.17 When water is placed in ice-cube trays in a freezer, why doesn't the water freeze all at once when the temperature has reached 0°C? In fact, the water freezes first in a layer adjacent to the sides of the tray. Why?

Q17.18 Before giving you an injection, a physician swabs your arm with isopropyl alcohol at room temperature. Why does this make your arm feel cold? (*Hint:* The reason is *not* the fear of the injection! The boiling point of isopropyl alcohol is 82.4°C.)

Q17.19 A cold block of metal feels colder than a block of wood at the same temperature. Why? A *hot* block of metal feels hotter than a block of wood at the same temperature. Again, why? Is there any temperature at which the two blocks feel equally hot or cold? What temperature is this?

Q17.20 A person pours a cup of hot coffee, intending to drink it five minutes later. To keep the coffee as hot as possible, should she put cream in it now or wait until just before she drinks it? Explain.

Q17.21 When a freshly baked apple pie has just been removed from the oven, the crust and filling are both at the same temperature. Yet if you sample the pie, the filling will burn your tongue but the crust will not. Why is there a difference? (*Hint:* The filling is moist while the crust is dry.)

Q17.22 Old-time kitchen lore suggests that things cook better (evenly and without burning) in heavy cast-iron pots. What desirable characteristics do such pots have?

Q17.23 In coastal regions in the winter, the temperature over the land is generally colder than the temperature over the nearby ocean; in the summer, the reverse is usually true. Explain. (*Hint:* The specific heat of soil is only 0.2–0.8 times as great as that of water.)

Q17.24 It is well known that a potato bakes faster if a large nail is stuck through it. Why? Does an aluminum nail work better than a steel one? Why or why not? (*Note:* Don't try this in a microwave oven!) There is also a gadget on the market to hasten the roasting of meat; it consists of a hollow metal tube containing a wick and some water. This is claimed to work much better than a solid metal rod. How does it work?

Q17.25 Glider pilots in the Midwest know that thermal updrafts are likely to occur in the vicinity of freshly plowed fields. Why?

Q17.26 Some folks claim that ice cubes freeze faster if the trays are filled with hot water, because hot water cools off faster than cold water. What do you think?

Q17.27 We're lucky that the earth isn't in thermal equilibrium with the sun (which has a surface temperature of 5800 K). But why aren't the two bodies in thermal equilibrium?

Q17.28 When energy shortages occur, magazine articles sometimes urge us to keep our homes at a constant temperature day and night to conserve fuel. They argue that when we turn down the heat at night, the walls, ceilings, and other areas cool off and must be reheated in the morning. So if we keep the temperature constant, these parts of the house will not cool off and will not have to be reheated. Does this argument make sense? Would we really save energy by following this advice?

EXERCISES

Section 17.2 Thermometers and Temperature Scales

17.1 • Convert the following Celsius temperatures to Fahrenheit: (a) −62.8°C, the lowest temperature ever recorded in North America (February 3, 1947, Snag, Yukon); (b) 56.7°C, the highest temperature ever recorded in the United States (July 10, 1913, Death Valley, California); (c) 31.1°C, the world's highest average annual temperature (Lugh Ferrandi, Somalia).

17.2 • **BIO Temperatures in Biomedicine.** (a) **Normal body temperature.** The average normal body temperature measured in the mouth is 310 K. What would Celsius and Fahrenheit thermometers read for this temperature? (b) **Elevated body temperature.** During very vigorous exercise, the body's temperature can go as high as 40°C. What would Kelvin and Fahrenheit thermometers read for this temperature? (c) **Temperature difference in the body.** The surface temperature of the body is normally about 7 C° lower than the internal temperature. Express this temperature difference in kelvins and in Fahrenheit degrees. (d) **Blood storage.** Blood stored at 4.0°C lasts safely for about 3 weeks, whereas blood stored at −160°C lasts for 5 years. Express both temperatures on the Fahrenheit and Kelvin scales. (e) **Heat stroke.** If the body's temperature is above 105°F for a prolonged period, heat stroke can result. Express this temperature on the Celsius and Kelvin scales.

17.3 • (a) On January 22, 1943, the temperature in Spearfish, South Dakota, rose from −4.0°F to 45.0°F in just 2 minutes. What was the temperature change in Celsius degrees? (b) The temperature in Browning, Montana, was 44.0°F on January 23, 1916. The next day the temperature plummeted to −56°F. What was the temperature change in Celsius degrees?

Section 17.3 Gas Thermometers and the Kelvin Scale

17.4 • (a) Calculate the one temperature at which Fahrenheit and Celsius thermometers agree with each other. (b) Calculate the one temperature at which Fahrenheit and Kelvin thermometers agree with each other.

17.5 •• You put a bottle of soft drink in a refrigerator and leave it until its temperature has dropped 10.0 K. What is its temperature change in (a) F° and (b) C°?

17.6 • Convert the following Kelvin temperatures to the Celsius and Fahrenheit scales: (a) the midday temperature at the surface of the moon (400 K); (b) the temperature at the tops of the clouds in the atmosphere of Saturn (95 K); (c) the temperature at the center of the sun (1.55×10^7 K).

17.7 • The pressure of a gas at the triple point of water is 1.35 atm. If its volume remains unchanged, what will its pressure be at the temperature at which CO_2 solidifies?

17.8 •• A constant-volume gas thermometer registers an absolute pressure corresponding to 325 mm of mercury when in contact with water at the triple point. What pressure does it read when in contact with water at the normal boiling point?

17.9 •• **A Constant-Volume Gas Thermometer.** An experimenter using a gas thermometer found the pressure at the triple point of water (0.01°C) to be 4.80×10^4 Pa and the pressure at the normal boiling point (100°C) to be 6.50×10^4 Pa. (a) Assuming that the pressure varies linearly with temperature, use these two data points to find the Celsius temperature at which the gas pressure would be zero (that is, find the Celsius temperature of absolute zero). (b) Does the gas in this thermometer obey Eq. (17.4) precisely? If that equation were precisely obeyed and the pressure at 100°C were 6.50×10^4 Pa, what pressure would the experimenter have measured at 0.01°C? (As we will learn in Section 18.1, Eq. (17.4) is accurate only for gases at very low density.)

17.10 • Like the Kelvin scale, the *Rankine scale* is an absolute temperature scale: Absolute zero is zero degrees Rankine (0°R). However, the units of this scale are the same size as those of the Fahrenheit scale rather than the Celsius scale. What is the numerical value of the triple-point temperature of water on the Rankine scale?

Section 17.4 Thermal Expansion

17.11 • The Humber Bridge in England has the world's longest single span, 1410 m. Calculate the change in length of the steel deck of the span when the temperature increases from $-5.0°C$ to $18.0°C$.

17.12 • One of the tallest buildings in the world is the Taipei 101 in Taiwan, at a height of 1671 feet. Assume that this height was measured on a cool spring day when the temperature was $15.5°C$. You could use the building as a sort of giant thermometer on a hot summer day by carefully measuring its height. Suppose you do this and discover that the Taipei 101 is 0.471 foot taller than its official height. What is the temperature, assuming that the building is in thermal equilibrium with the air and that its entire frame is made of steel?

17.13 • A U.S. penny has a diameter of 1.9000 cm at $20.0°C$. The coin is made of a metal alloy (mostly zinc) for which the coefficient of linear expansion is $2.6 \times 10^{-5} \ K^{-1}$. What would its diameter be on a hot day in Death Valley ($48.0°C$)? On a cold night in the mountains of Greenland ($-53°C$)?

17.14 • **Ensuring a Tight Fit.** Aluminum rivets used in airplane construction are made slightly larger than the rivet holes and cooled by "dry ice" (solid CO_2) before being driven. If the diameter of a hole is 4.500 mm, what should be the diameter of a rivet at $23.0°C$ if its diameter is to equal that of the hole when the rivet is cooled to $-78.0°C$, the temperature of dry ice? Assume that the expansion coefficient remains constant at the value given in Table 17.1.

17.15 •• A copper cylinder is initially at $20.0°C$. At what temperature will its volume be 0.150% larger than it is at $20.0°C$?

17.16 •• A geodesic dome constructed with an aluminum framework is a nearly perfect hemisphere; its diameter measures 55.0 m on a winter day at a temperature of $-15°C$. How much more interior space does the dome have in the summer, when the temperature is $35°C$?

17.17 •• A glass flask whose volume is $1000.00 \ cm^3$ at $0.0°C$ is completely filled with mercury at this temperature. When flask and mercury are warmed to $55.0°C$, $8.95 \ cm^3$ of mercury overflow. If the coefficient of volume expansion of mercury is $18.0 \times 10^{-5} \ K^{-1}$, compute the coefficient of volume expansion of the glass.

17.18 •• A steel tank is completely filled with $1.90 \ m^3$ of ethanol when both the tank and the ethanol are at $32.0°C$. When the tank and its contents have cooled to $18.0°C$, what additional volume of ethanol can be put into the tank?

17.19 •• A machinist bores a hole of diameter 1.35 cm in a steel plate that is at $25.0°C$. What is the cross-sectional area of the hole (a) at $25.0°C$ and (b) when the temperature of the plate is increased to $175°C$? Assume that the coefficient of linear expansion remains constant over this temperature range.

17.20 •• As a new mechanical engineer for Engines Inc., you have been assigned to design brass pistons to slide inside steel cylinders. The engines in which these pistons will be used will operate between $20.0°C$ and $150.0°C$. Assume that the coefficients of expansion are constant over this temperature range. (a) If the piston just fits inside the chamber at $20.0°C$, will the engines be able to run at higher temperatures? Explain. (b) If the cylindrical pistons are 25.000 cm in diameter at $20.0°C$, what should be the minimum diameter of the cylinders at that temperature so the pistons will operate at $150.0°C$?

17.21 •• Steel train rails are laid in 12.0-m-long segments placed end to end. The rails are laid on a winter day when their temperature is $-9.0°C$. (a) How much space must be left between adjacent rails if they are just to touch on a summer day when their temperature is $33.0°C$? (b) If the rails are originally laid in contact, what is the stress in them on a summer day when their temperature is $33.0°C$?

17.22 •• A brass rod is 185 cm long and 1.60 cm in diameter. What force must be applied to each end of the rod to prevent it from contracting when it is cooled from $120.0°C$ to $10.0°C$?

Section 17.5 Quantity of Heat

17.23 •• An aluminum tea kettle with mass 1.10 kg and containing 1.80 kg of water is placed on a stove. If no heat is lost to the surroundings, how much heat must be added to raise the temperature from $20.0°C$ to $85.0°C$?

17.24 • In an effort to stay awake for an all-night study session, a student makes a cup of coffee by first placing a 200-W electric immersion heater in 0.320 kg of water. (a) How much heat must be added to the water to raise its temperature from $20.0°C$ to $80.0°C$? (b) How much time is required? Assume that all of the heater's power goes into heating the water.

17.25 • BIO While running, a 70-kg student generates thermal energy at a rate of 1200 W. For the runner to maintain a constant body temperature of $37°C$, this energy must be removed by perspiration or other mechanisms. If these mechanisms failed and the energy could not flow out of the student's body, for what amount of time could a student run before irreversible body damage occurred? (*Note:* Protein structures in the body are irreversibly damaged if body temperature rises to $44°C$ or higher. The specific heat of a typical human body is $3480 \ J/kg \cdot K$, slightly less than that of water. The difference is due to the presence of protein, fat, and minerals, which have lower specific heats.)

17.26 • BIO **Heat Loss During Breathing.** In very cold weather a significant mechanism for heat loss by the human body is energy expended in warming the air taken into the lungs with each breath. (a) On a cold winter day when the temperature is $-20°C$, what amount of heat is needed to warm to body temperature ($37°C$) the 0.50 L of air exchanged with each breath? Assume that the specific heat of air is $1020 \ J/kg \cdot K$ and that 1.0 L of air has mass 1.3×10^{-3} kg. (b) How much heat is lost per hour if the respiration rate is 20 breaths per minute?

17.27 • You are given a sample of metal and asked to determine its specific heat. You weigh the sample and find that its weight is 28.4 N. You carefully add 1.25×10^4 J of heat energy to the sample and find that its temperature rises $18.0 \ C°$. What is the sample's specific heat?

17.28 •• **On-Demand Water Heaters.** Conventional hot-water heaters consist of a tank of water maintained at a fixed temperature. The hot water is to be used when needed. The drawbacks are that energy is wasted because the tank loses heat when it is not in use and that you can run out of hot water if you use too much. Some utility companies are encouraging the use of *on-demand* water heaters (also known as *flash heaters*), which consist of heating units to heat the water as you use it. No water tank is involved, so no heat is wasted. A typical household shower flow rate is 2.5 gal/min (9.46 L/min) with the tap water being heated from $50°F$ ($10°C$) to $120°F$ ($49°C$) by the on-demand heater. What rate of heat input (either electrical or from gas) is required to operate such a unit, assuming that all the heat goes into the water?

17.29 • CP While painting the top of an antenna 225 m in height, a worker accidentally lets a 1.00-L water bottle fall from his lunchbox. The bottle lands in some bushes at ground level and does not break. If a quantity of heat equal to the magnitude of the change in mechanical energy of the water goes into the water, what is its increase in temperature?

17.30 • **CP** A 25,000-kg subway train initially traveling at 15.5 m/s slows to a stop in a station and then stays there long enough for its brakes to cool. The station's dimensions are 65.0 m long by 20.0 m wide by 12.0 m high. Assuming all the work done by the brakes in stopping the train is transferred as heat uniformly to all the air in the station, by how much does the air temperature in the station rise? Take the density of the air to be 1.20 kg/m^3 and its specific heat to be 1020 J/kg·K.

17.31 • **CP** A nail driven into a board increases in temperature. If we assume that 60% of the kinetic energy delivered by a 1.80-kg hammer with a speed of 7.80 m/s is transformed into heat that flows into the nail and does not flow out, what is the temperature increase of an 8.00-g aluminum nail after it is struck ten times?

17.32 • A technician measures the specific heat of an unidentified liquid by immersing an electrical resistor in it. Electrical energy is converted to heat transferred to the liquid for 120 s at a constant rate of 65.0 W. The mass of the liquid is 0.780 kg, and its temperature increases from 18.55°C to 22.54°C. (a) Find the average specific heat of the liquid in this temperature range. Assume that negligible heat is transferred to the container that holds the liquid and that no heat is lost to the surroundings. (b) Suppose that in this experiment heat transfer from the liquid to the container or surroundings cannot be ignored. Is the result calculated in part (a) an *overestimate* or an *underestimate* of the average specific heat? Explain.

17.33 •• **CP** A 15.0-g bullet traveling horizontally at 865 m/s passes through a tank containing 13.5 kg of water and emerges with a speed of 534 m/s. What is the maximum temperature increase that the water could have as a result of this event?

Section 17.6 Calorimetry and Phase Changes

17.34 • You have 750 g of water at 10.0°C in a large insulated beaker. How much boiling water at 100.0°C must you add to this beaker so that the final temperature of the mixture will be 75°C?

17.35 •• A 500.0-g chunk of an unknown metal, which has been in boiling water for several minutes, is quickly dropped into an insulating Styrofoam beaker containing 1.00 kg of water at room temperature (20.0°C). After waiting and gently stirring for 5.00 minutes, you observe that the water's temperature has reached a constant value of 22.0°C. (a) Assuming that the Styrofoam absorbs a negligibly small amount of heat and that no heat was lost to the surroundings, what is the specific heat of the metal? (b) Which is more useful for storing thermal energy: this metal or an equal weight of water? Explain. (c) If the heat absorbed by the Styrofoam actually is not negligible, how would the specific heat you calculated in part (a) be in error? Would it be too large, too small, or still correct? Explain.

17.36 • **BIO Treatment for a Stroke.** One suggested treatment for a person who has suffered a stroke is immersion in an ice-water bath at 0°C to lower the body temperature, which prevents damage to the brain. In one set of tests, patients were cooled until their internal temperature reached 32.0°C. To treat a 70.0-kg patient, what is the minimum amount of ice (at 0°C) you need in the bath so that its temperature remains at 0°C? The specific heat of the human body is 3480 J/kg·C°, and recall that normal body temperature is 37.0°C.

17.37 •• A blacksmith cools a 1.20-kg chunk of iron, initially at 650.0°C, by trickling 15.0°C water over it. All of the water boils away, and the iron ends up at 120.0°C. How much water did the blacksmith trickle over the iron?

17.38 •• A copper calorimeter can with mass 0.100 kg contains 0.160 kg of water and 0.0180 kg of ice in thermal equilibrium at atmospheric pressure. If 0.750 kg of lead at 255°C is dropped into the calorimeter can, what is the final temperature? Assume that no heat is lost to the surroundings.

17.39 •• A copper pot with a mass of 0.500 kg contains 0.170 kg of water, and both are at 20.0°C. A 0.250-kg block of iron at 85.0°C is dropped into the pot. Find the final temperature of the system, assuming no heat loss to the surroundings.

17.40 • In a container of negligible mass, 0.200 kg of ice at an initial temperature of −40.0°C is mixed with a mass m of water that has an initial temperature of 80.0°C. No heat is lost to the surroundings. If the final temperature of the system is 28.0°C, what is the mass m of the water that was initially at 80.0°C?

17.41 • A 6.00-kg piece of solid copper metal at an initial temperature T is placed with 2.00 kg of ice that is initially at −20.0°C. The ice is in an insulated container of negligible mass and no heat is exchanged with the surroundings. After thermal equilibrium is reached, there is 1.20 kg of ice and 0.80 kg of liquid water. What was the initial temperature of the piece of copper?

17.42 • **BIO** Before going in for his annual physical, a 70.0-kg man whose body temperature is 37.0°C consumes an entire 0.355-L can of a soft drink (mostly water) at 12.0°C. (a) What will his body temperature be after equilibrium is attained? Ignore any heating by the man's metabolism. The specific heat of the man's body is 3480 J/kg·K. (b) Is the change in his body temperature great enough to be measured by a medical thermometer?

17.43 •• **BIO Basal Metabolic Rate.** In the situation described in Exercise 17.42, the man's metabolism will eventually return the temperature of his body (and of the soft drink that he consumed) to 37.0°C. If his body releases energy at a rate of 7.00×10^3 kJ/day (the *basal metabolic rate*, or BMR), how long does this take? Assume that all of the released energy goes into raising the temperature.

17.44 •• An ice-cube tray of negligible mass contains 0.290 kg of water at 18.0°C. How much heat must be removed to cool the water to 0.00°C and freeze it? Express your answer in joules, calories, and Btu.

17.45 • How much heat is required to convert 18.0 g of ice at −10.0°C to steam at 100.0°C? Express your answer in joules, calories, and Btu.

17.46 •• An open container holds 0.550 kg of ice at −15.0°C. The mass of the container can be ignored. Heat is supplied to the container at the constant rate of 800.0 J/min for 500.0 min. (a) After how many minutes does the ice *start* to melt? (b) After how many minutes, from the time when the heating is first started, does the temperature begin to rise above 0.0°C? (c) Plot a curve showing the temperature as a function of the elapsed time.

17.47 • **CP** What must the initial speed of a lead bullet be at 25.0°C so that the heat developed when it is brought to rest will be just sufficient to melt it? Assume that all the initial mechanical energy of the bullet is converted to heat and that no heat flows from the bullet to its surroundings. (Typical rifles have muzzle speeds that exceed the speed of sound in air, which is 347 m/s at 25.0°C.)

17.48 •• **BIO Steam Burns Versus Water Burns.** What is the amount of heat input to your skin when it receives the heat released (a) by 25.0 g of steam initially at 100.0°C, when it is cooled to skin temperature (34.0°C)? (b) By 25.0 g of water initially at 100.0°C, when it is cooled to 34.0°C? (c) What does this tell you about the relative severity of burns from steam versus burns from hot water?

17.49 · BIO **"The Ship of the Desert."** Camels require very little water because they are able to tolerate relatively large changes in their body temperature. While humans keep their body temperatures constant to within one or two Celsius degrees, a dehydrated camel permits its body temperature to drop to 34.0°C overnight and rise to 40.0°C during the day. To see how effective this mechanism is for saving water, calculate how many liters of water a 400-kg camel would have to drink if it attempted to keep its body temperature at a constant 34.0°C by evaporation of sweat during the day (12 hours) instead of letting it rise to 40.0°C. (*Note:* The specific heat of a camel or other mammal is about the same as that of a typical human, 3480 J/kg · K. The heat of vaporization of water at 34°C is 2.42×10^6 J/kg.)

17.50 · BIO Evaporation of sweat is an important mechanism for temperature control in some warm-blooded animals. (a) What mass of water must evaporate from the skin of a 70.0-kg man to cool his body 1.00 C°? The heat of vaporization of water at body temperature (37°C) is 2.42×10^6 J/kg. The specific heat of a typical human body is 3480 J/kg · K (see Exercise 17.25). (b) What volume of water must the man drink to replenish the evaporated water? Compare to the volume of a soft-drink can (355 cm^3).

17.51 ·· CP An asteroid with a diameter of 10 km and a mass of 2.60×10^{15} kg impacts the earth at a speed of 32.0 km/s, landing in the Pacific Ocean. If 1.00% of the asteroid's kinetic energy goes to boiling the ocean water (assume an initial water temperature of 10.0°C), what mass of water will be boiled away by the collision? (For comparison, the mass of water contained in Lake Superior is about 2×10^{15} kg.)

17.52 · A laboratory technician drops a 0.0850-kg sample of unknown solid material, at 100.0°C, into a calorimeter. The calorimeter can, initially at 19.0°C, is made of 0.150 kg of copper and contains 0.200 kg of water. The final temperature of the calorimeter can and contents is 26.1°C. Compute the specific heat of the sample.

17.53 ·· An insulated beaker with negligible mass contains 0.250 kg of water at 75.0°C. How many kilograms of ice at −20.0°C must be dropped into the water to make the final temperature of the system 40.0°C?

17.54 · A 4.00-kg silver ingot is taken from a furnace, where its temperature is 750.0°C, and placed on a large block of ice at 0.0°C. Assuming that all the heat given up by the silver is used to melt the ice, how much ice is melted?

17.55 ·· A vessel whose walls are thermally insulated contains 2.40 kg of water and 0.450 kg of ice, all at 0.0°C. The outlet of a tube leading from a boiler in which water is boiling at atmospheric pressure is inserted into the water. How many grams of steam must condense inside the vessel (also at atmospheric pressure) to raise the temperature of the system to 28.0°C? You can ignore the heat transferred to the container.

Section 17.7 Mechanisms of Heat Transfer

17.56 ·· Two rods, one made of brass and the other made of copper, are joined end to end. The length of the brass section is 0.300 m and the length of the copper section is 0.800 m. Each segment has cross-sectional area 0.00500 m². The free end of the brass segment is in boiling water and the free end of the copper segment is in an ice–water mixture, in both cases under normal atmospheric pressure. The sides of the rods are insulated so there is no heat loss to the surroundings. (a) What is the temperature of the point where the brass and copper segments are joined? (b) What mass of ice is melted in 5.00 min by the heat conducted by the composite rod?

17.57 · Suppose that the rod in Fig. 17.24a is made of copper, is 45.0 cm long, and has a cross-sectional area of 1.25 cm². Let $T_H = 100.0°C$ and $T_C = 0.0°C$. (a) What is the final steady-state temperature gradient along the rod? (b) What is the heat current in the rod in the final steady state? (c) What is the final steady-state temperature at a point in the rod 12.0 cm from its left end?

17.58 ·· One end of an insulated metal rod is maintained at 100.0°C, and the other end is maintained at 0.00°C by an ice–water mixture. The rod is 60.0 cm long and has a cross-sectional area of 1.25 cm². The heat conducted by the rod melts 8.50 g of ice in 10.0 min. Find the thermal conductivity k of the metal.

17.59 ·· A carpenter builds an exterior house wall with a layer of wood 3.0 cm thick on the outside and a layer of Styrofoam insulation 2.2 cm thick on the inside wall surface. The wood has $k = 0.080$ W/m · K, and the Styrofoam has $k = 0.027$ W/m · K. The interior surface temperature is 19.0°C, and the exterior surface temperature is −10.0°C. (a) What is the temperature at the plane where the wood meets the Styrofoam? (b) What is the rate of heat flow per square meter through this wall?

17.60 · An electric kitchen range has a total wall area of 1.40 m² and is insulated with a layer of fiberglass 4.00 cm thick. The inside surface of the fiberglass has a temperature of 175°C, and its outside surface is at 35.0°C. The fiberglass has a thermal conductivity of 0.040 W/m · K. (a) What is the heat current through the insulation, assuming it may be treated as a flat slab with an area of 1.40 m²? (b) What electric-power input to the heating element is required to maintain this temperature?

17.61 · BIO **Conduction Through the Skin.** The blood plays an important role in removing heat from the body by bringing this energy directly to the surface where it can radiate away. Nevertheless, this heat must still travel through the skin before it can radiate away. Assume that the blood is brought to the bottom layer of skin at 37.0°C and that the outer surface of the skin is at 30.0°C. Skin varies in thickness from 0.50 mm to a few millimeters on the palms and soles, so assume an average thickness of 0.75 mm. A 165-lb, 6-ft-tall person has a surface area of about 2.0 m² and loses heat at a net rate of 75 W while resting. On the basis of our assumptions, what is the thermal conductivity of this person's skin?

17.62 · A long rod, insulated to prevent heat loss along its sides, is in perfect thermal contact with boiling water (at atmospheric pressure) at one end and with an ice–water mixture at the other (**Fig. E17.62**). The rod consists of a 1.00-m section of copper (one end in boiling water) joined end to end to a length L_2 of steel (one end in the ice–water mixture). Both sections of the rod have cross-sectional areas of 4.00 cm². The temperature of the copper–steel junction is 65.0°C after a steady state has been set up. (a) How much heat per second flows from the boiling water to the ice–water mixture? (b) What is the length L_2 of the steel section?

Figure **E17.62**

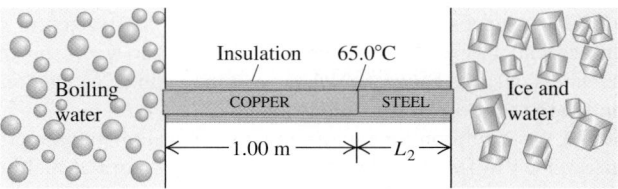

17.63 · A pot with a steel bottom 8.50 mm thick rests on a hot stove. The area of the bottom of the pot is 0.150 m². The water inside the pot is at 100.0°C, and 0.390 kg are evaporated every 3.00 min. Find the temperature of the lower surface of the pot, which is in contact with the stove.

17.64 •• You are asked to design a cylindrical steel rod 50.0 cm long, with a circular cross section, that will conduct 190.0 J/s from a furnace at 400.0°C to a container of boiling water under 1 atmosphere. What must the rod's diameter be?

17.65 •• A picture window has dimensions of 1.40 m × 2.50 m and is made of glass 5.20 mm thick. On a winter day, the temperature of the outside surface of the glass is −20.0°C, while the temperature of the inside surface is a comfortable 19.5°C. (a) At what rate is heat being lost through the window by conduction? (b) At what rate would heat be lost through the window if you covered it with a 0.750-mm-thick layer of paper (thermal conductivity 0.0500 W/m · K)?

17.66 • What is the rate of energy radiation per unit area of a blackbody at (a) 273 K and (b) 2730 K?

17.67 •• A spherical pot contains 0.75 L of hot coffee (essentially water) at an initial temperature of 95°C. The pot has an emissivity of 0.60, and the surroundings are at 20.0°C. Calculate the coffee's rate of heat loss by radiation.

17.68 •• The emissivity of tungsten is 0.350. A tungsten sphere with radius 1.50 cm is suspended within a large evacuated enclosure whose walls are at 290.0 K. What power input is required to maintain the sphere at 3000.0 K if heat conduction along the supports is ignored?

17.69 • **Size of a Light-Bulb Filament.** The operating temperature of a tungsten filament in an incandescent light bulb is 2450 K, and its emissivity is 0.350. Find the surface area of the filament of a 150-W bulb if all the electrical energy consumed by the bulb is radiated by the filament as electromagnetic waves. (Only a fraction of the radiation appears as visible light.)

17.70 • **The Sizes of Stars.** The hot glowing surfaces of stars emit energy in the form of electromagnetic radiation. It is a good approximation to assume $e = 1$ for these surfaces. Find the radii of the following stars (assumed to be spherical): (a) Rigel, the bright blue star in the constellation Orion, which radiates energy at a rate of 2.7×10^{32} W and has surface temperature 11,000 K; (b) Procyon B (visible only using a telescope), which radiates energy at a rate of 2.1×10^{23} W and has surface temperature 10,000 K. (c) Compare your answers to the radius of the earth, the radius of the sun, and the distance between the earth and the sun. (Rigel is an example of a *supergiant* star, and Procyon B is an example of a *white dwarf* star.)

PROBLEMS

17.71 •• CP A Foucault pendulum consists of a brass sphere with a diameter of 35.0 cm suspended from a steel cable 10.5 m long (both measurements made at 20.0°C). Due to a design oversight, the swinging sphere clears the floor by a distance of only 2.00 mm when the temperature is 20.0°C. At what temperature will the sphere begin to brush the floor?

17.72 •• Suppose that a steel hoop could be constructed to fit just around the earth's equator at 20.0°C. What would be the thickness of space between the hoop and the earth if the temperature of the hoop were increased by 0.500 C°?

17.73 ••• You propose a new temperature scale with temperatures given in °M. You define 0.0°M to be the normal melting point of mercury and 100.0°M to be the normal boiling point of mercury. (a) What is the normal boiling point of water in °M? (b) A temperature change of 10.0 M° corresponds to how many C°?

17.74 • CP CALC A 250-kg weight is hanging from the ceiling by a thin copper wire. In its fundamental mode, this wire vibrates

at the frequency of concert A (440 Hz). You then increase the temperature of the wire by 40 C°. (a) By how much will the fundamental frequency change? Will it increase or decrease? (b) By what percentage will the speed of a wave on the wire change? (c) By what percentage will the wavelength of the fundamental standing wave change? Will it increase or decrease?

17.75 ••• You are making pesto for your pasta and have a cylindrical measuring cup 10.0 cm high made of ordinary glass $[\beta = 2.7 \times 10^{-5}(C°)^{-1}]$ that is filled with olive oil $[\beta = 6.8 \times 10^{-4}(C°)^{-1}]$ to a height of 3.00 mm below the top of the cup. Initially, the cup and oil are at room temperature (22.0°C). You get a phone call and forget about the olive oil, which you inadvertently leave on the hot stove. The cup and oil heat up slowly and have a common temperature. At what temperature will the olive oil start to spill out of the cup?

17.76 •• A surveyor's 30.0-m steel tape is correct at 20.0°C. The distance between two points, as measured by this tape on a day when its temperature is 5.00°C, is 25.970 m. What is the true distance between the points?

17.77 •• A metal rod that is 30.0 cm long expands by 0.0650 cm when its temperature is raised from 0.0°C to 100.0°C. A rod of a different metal and of the same length expands by 0.0350 cm for the same rise in temperature. A third rod, also 30.0 cm long, is made up of pieces of each of the above metals placed end to end and expands 0.0580 cm between 0.0°C and 100.0°C. Find the length of each portion of the composite rod.

17.78 •• On a cool (4.0°C) Saturday morning, a pilot fills the fuel tanks of her Pitts S-2C (a two-seat aerobatic airplane) to their full capacity of 106.0 L. Before flying on Sunday morning, when the temperature is again 4.0°C, she checks the fuel level and finds only 103.4 L of gasoline in the aluminum tanks. She realizes that it was hot on Saturday afternoon and that thermal expansion of the gasoline caused the missing fuel to empty out of the tank's vent. (a) What was the maximum temperature (in °C) of the fuel and the tank on Saturday afternoon? The coefficient of volume expansion of gasoline is 9.5×10^{-4} K^{-1}. (b) To have the maximum amount of fuel available for flight, when should the pilot have filled the fuel tanks?

17.79 ••• (a) Equation (17.12) gives the stress required to keep the length of a rod constant as its temperature changes. Show that if the length is permitted to change by an amount ΔL when its temperature changes by ΔT, the stress is equal to

$$\frac{F}{A} = Y\left(\frac{\Delta L}{L_0} - \alpha \Delta T\right)$$

where F is the tension on the rod, L_0 is the original length of the rod, A its cross-sectional area, α its coefficient of linear expansion, and Y its Young's modulus. (b) A heavy brass bar has projections at its ends (**Fig. P17.79**). Two fine steel wires, fastened between the projections, are just taut (zero tension) when the whole system is at 20°C. What is the tensile stress in the steel wires when the temperature of the system is raised to 140°C? Make any simplifying assumptions you think are justified, but state them.

Figure **P17.79**

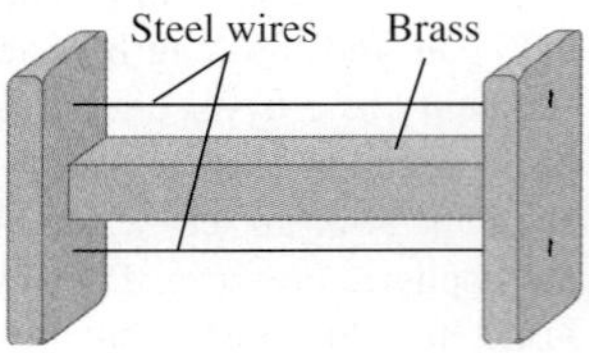

Steel wires   Brass

17.80 •• CP A metal wire, with density ρ and Young's modulus Y, is stretched between rigid supports. At temperature T, the speed of a transverse wave is found to be v_1. When the temperature is increased to $T + \Delta T$, the speed decreases to $v_2 < v_1$. Determine the coefficient of linear expansion of the wire.

17.81 •• A steel ring with a 2.5000-in. inside diameter at 20.0°C is to be warmed and slipped over a brass shaft with a 2.5020-in. outside diameter at 20.0°C. (a) To what temperature should the ring be warmed? (b) If the ring and the shaft together are cooled by some means such as liquid air, at what temperature will the ring just slip off the shaft?

17.82 • **BIO Doughnuts: Breakfast of Champions!** A typical doughnut contains 2.0 g of protein, 17.0 g of carbohydrates, and 7.0 g of fat. Average food energy values are 4.0 kcal/g for protein and carbohydrates and 9.0 kcal/g for fat. (a) During heavy exercise, an average person uses energy at a rate of 510 kcal/h. How long would you have to exercise to "work off" one doughnut? (b) If the energy in the doughnut could somehow be converted into the kinetic energy of your body as a whole, how fast could you move after eating the doughnut? Take your mass to be 60 kg, and express your answer in m/s and in km/h.

17.83 •• **BIO Shivering.** Shivering is your body's way of generating heat to restore its internal temperature to the normal 37°C, and it produces approximately 290 W of heat power per square meter of body area. A 68-kg, 1.78-m-tall woman has approximately 1.8 m² of surface area. How long would this woman have to shiver to raise her body temperature by 1.0 C°, assuming that the body loses none of this heat? The body's specific heat capacity is about 3500 J/kg · K.

17.84 •• You cool a 100.0-g slug of red-hot iron (temperature 745°C) by dropping it into an insulated cup of negligible mass containing 85.0 g of water at 20.0°C. Assuming no heat exchange with the surroundings, (a) what is the final temperature of the water and (b) what is the final mass of the iron and the remaining water?

17.85 •• **CALC Debye's T^3 Law.** At very low temperatures the molar heat capacity of rock salt varies with temperature according to Debye's T^3 law:

$$C = k\frac{T^3}{\theta^3}$$

where $k = 1940$ J/mol · K and $\theta = 281$ K. (a) How much heat is required to raise the temperature of 1.50 mol of rock salt from 10.0 K to 40.0 K? (*Hint:* Use Eq. (17.18) in the form $dQ = nC \, dT$ and integrate.) (b) What is the average molar heat capacity in this range? (c) What is the true molar heat capacity at 40.0 K?

17.86 •• **CP** A person of mass 70.0 kg is sitting in the bathtub. The bathtub is 190.0 cm by 80.0 cm; before the person got in, the water was 24.0 cm deep. The water is at 37.0°C. Suppose that the water were to cool down spontaneously to form ice at 0.0°C, and that all the energy released was used to launch the hapless bather vertically into the air. How high would the bather go? (As you will see in Chapter 20, this event is allowed by energy conservation but is prohibited by the second law of thermodynamics.)

17.87 • **Hot Air in a Physics Lecture.** (a) A typical student listening attentively to a physics lecture has a heat output of 100 W. How much heat energy does a class of 140 physics students release into a lecture hall over the course of a 50-min lecture? (b) Assume that all the heat energy in part (a) is transferred to the 3200 m³ of air in the room. The air has specific heat 1020 J/kg · K and density 1.20 kg/m³. If none of the heat escapes and the air conditioning system is off, how much will the temperature of the air in the room rise during the 50-min lecture? (c) If the class is taking an exam, the heat output per student rises to 280 W. What is the temperature rise during 50 min in this case?

17.88 ••• **CALC** The molar heat capacity of a certain substance varies with temperature according to the empirical equation

$$C = 29.5 \text{ J/mol} \cdot \text{K} + (8.20 \times 10^{-3} \text{ J/mol} \cdot \text{K}^2)T$$

How much heat is necessary to change the temperature of 3.00 mol of this substance from 27°C to 227°C? (*Hint:* Use Eq. (17.18) in the form $dQ = nC \, dT$ and integrate.)

17.89 •• **BIO Bicycling on a Warm Day.** If the air temperature is the same as the temperature of your skin (about 30°C), your body cannot get rid of heat by transferring it to the air. In that case, it gets rid of the heat by evaporating water (sweat). During bicycling, a typical 70-kg person's body produces energy at a rate of about 500 W due to metabolism, 80% of which is converted to heat. (a) How many kilograms of water must the person's body evaporate in an hour to get rid of this heat? The heat of vaporization of water at body temperature is 2.42×10^6 J/kg. (b) The evaporated water must, of course, be replenished, or the person will dehydrate. How many 750-mL bottles of water must the bicyclist drink per hour to replenish the lost water? (Recall that the mass of a liter of water is 1.0 kg.)

17.90 •• **BIO Overheating.** (a) By how much would the body temperature of the bicyclist in Problem 17.89 increase in an hour if he were unable to get rid of the excess heat? (b) Is this temperature increase large enough to be serious? To find out, how high a fever would it be equivalent to, in °F? (Recall that the normal internal body temperature is 98.6°F and the specific heat of the body is 3480 J/kg · C°.)

17.91 • **BIO A Thermodynamic Process in an Insect.** The African bombardier beetle (*Stenaptinus insignis*) can emit a jet of defensive spray from the movable tip of its abdomen (**Fig. P17.91**). The beetle's body has reservoirs containing two chemicals; when the beetle is disturbed, these chemicals combine in a reaction chamber, producing a compound that is warmed from 20°C to 100°C by the heat of reaction. The high pressure produced allows the compound to be sprayed out at speeds up to 19 m/s (68 km/h), scaring away predators of all kinds. (The beetle shown in Fig. P17.91 is 2 cm long.) Calculate the heat of reaction of the two chemicals (in J/kg). Assume that the specific heat of the chemicals and of the spray is the same as that of water, 4.19×10^3 J/kg · K, and that the initial temperature of the chemicals is 20°C.

Figure **P17.91**

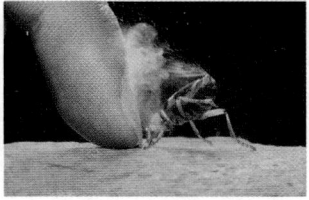

17.92 •• **Hot Water Versus Steam Heating.** In a household hot-water heating system, water is delivered to the radiators at 70.0°C (158.0°F) and leaves at 28.0°C (82.4°F). The system is to be replaced by a steam system in which steam at atmospheric pressure condenses in the radiators and the condensed steam leaves the radiators at 35.0°C (95.0°F). How many kilograms of steam will supply the same heat as was supplied by 1.00 kg of hot water in the first system?

17.93 •• You have 1.50 kg of water at 28.0°C in an insulated container of negligible mass. You add 0.600 kg of ice that is initially at −22.0°C. Assume that no heat exchanges with the surroundings. (a) After thermal equilibrium has been reached, has all of the ice melted? (b) If all of the ice has melted, what is the final temperature of the water in the container? If some ice remains, what is the final temperature of the water in the container, and how much ice remains?

17.94 •• A thirsty nurse cools a 2.00-L bottle of a soft drink (mostly water) by pouring it into a large aluminum mug of mass 0.257 kg and adding 0.120 kg of ice initially at −15.0°C. If the soft drink and mug are initially at 20.0°C, what is the final temperature of the system, assuming that no heat is lost?

17.95 ••• A copper calorimeter can with mass 0.446 kg contains 0.0950 kg of ice. The system is initially at 0.0°C. (a) If 0.0350 kg of steam at 100.0°C and 1.00 atm pressure is added to the can, what is the final temperature of the calorimeter can and its contents? (b) At the final temperature, how many kilograms are there of ice, how many of liquid water, and how many of steam?

17.96 • A Styrofoam bucket of negligible mass contains 1.75 kg of water and 0.450 kg of ice. More ice, from a refrigerator at −15.0°C, is added to the mixture in the bucket, and when thermal equilibrium has been reached, the total mass of ice in the bucket is 0.884 kg. Assuming no heat exchange with the surroundings, what mass of ice was added?

17.97 ••• In a container of negligible mass, 0.0400 kg of steam at 100°C and atmospheric pressure is added to 0.200 kg of water at 50.0°C. (a) If no heat is lost to the surroundings, what is the final temperature of the system? (b) At the final temperature, how many kilograms are there of steam and how many of liquid water?

17.98 •• BIO **Mammal Insulation.** Animals in cold climates often depend on *two* layers of insulation: a layer of body fat (of thermal conductivity 0.20 W/m·K) surrounded by a layer of air trapped inside fur or down. We can model a black bear (*Ursus americanus*) as a sphere 1.5 m in diameter having a layer of fat 4.0 cm thick. (Actually, the thickness varies with the season, but we are interested in hibernation, when the fat layer is thickest.) In studies of bear hibernation, it was found that the outer surface layer of the fur is at 2.7°C during hibernation, while the inner surface of the fat layer is at 31.0°C. (a) What is the temperature at the fat–inner fur boundary so that the bear loses heat at a rate of 50.0 W? (b) How thick should the air layer (contained within the fur) be?

17.99 •• **Effect of a Window in a Door.** A carpenter builds a solid wood door with dimensions 2.00 m × 0.95 m × 5.0 cm. Its thermal conductivity is $k = 0.120$ W/m·K. The air films on the inner and outer surfaces of the door have the same combined thermal resistance as an additional 1.8-cm thickness of solid wood. The inside air temperature is 20.0°C, and the outside air temperature is −8.0°C. (a) What is the rate of heat flow through the door? (b) By what factor is the heat flow increased if a window 0.500 m on a side is inserted in the door? The glass is 0.450 cm thick, and the glass has a thermal conductivity of 0.80 W/m·K. The air films on the two sides of the glass have a total thermal resistance that is the same as an additional 12.0 cm of glass.

17.100 •• One experimental method of measuring an insulating material's thermal conductivity is to construct a box of the material and measure the power input to an electric heater inside the box that maintains the interior at a measured temperature above the outside surface. Suppose that in such an apparatus a power input of 180 W is required to keep the interior surface of the box 65.0 C° (about 120 F°) above the temperature of the outer surface. The total area of the box is 2.18 m², and the wall thickness is 3.90 cm. Find the thermal conductivity of the material in SI units.

17.101 •• Compute the ratio of the rate of heat loss through a single-pane window with area 0.15 m² to that for a double-pane window with the same area. The glass of a single pane is 4.2 mm thick, and the air space between the two panes of the double-pane window is 7.0 mm thick. The glass has thermal conductivity 0.80 W/m·K. The air films on the room and outdoor surfaces of either window have a combined thermal resistance of 0.15 m²·K/W.

17.102 • Rods of copper, brass, and steel—each with cross-sectional area of 2.00 cm²—are welded together to form a Y-shaped figure. The free end of the copper rod is maintained at 100.0°C, and the free ends of the brass and steel rods at 0.0°C. Assume that there is no heat loss from the surfaces of the rods. The lengths of the rods are: copper, 13.0 cm; brass, 18.0 cm; steel, 24.0 cm. What is (a) the temperature of the junction point; (b) the heat current in each of the three rods?

17.103 ••• A brass rod 12.0 cm long, a copper rod 18.0 cm long, and an aluminum rod 24.0 cm long—each with cross-sectional area 2.30 cm³—are welded together end to end to form a rod 54.0 cm long, with copper as the middle section. The free end of the brass section is maintained at 100.0°C, and the free end of the aluminum section is maintained at 0.0°C. Assume that there is no heat loss from the curved surfaces and that the steady-state heat current has been established. What is (a) the temperature T_1 at the junction of the brass and copper sections; (b) the temperature T_2 at the junction of the copper and aluminum sections; (c) the heat current in the aluminum section?

17.104 •• BIO **Basal Metabolic Rate.** The *basal metabolic rate* is the rate at which energy is produced in the body when a person is at rest. A 75-kg (165-lb) person of height 1.83 m (6 ft) has a body surface area of approximately 2.0 m². (a) What is the net amount of heat this person could radiate per second into a room at 18°C (about 65°F) if his skin's surface temperature is 30°C? (At such temperatures, nearly all the heat is infrared radiation, for which the body's emissivity is 1.0, regardless of the amount of pigment.) (b) Normally, 80% of the energy produced by metabolism goes into heat, while the rest goes into things like pumping blood and repairing cells. Also normally, a person at rest can get rid of this excess heat just through radiation. Use your answer to part (a) to find this person's basal metabolic rate.

17.105 ••• CALC **Time Needed for a Lake to Freeze Over.** (a) When the air temperature is below 0°C, the water at the surface of a lake freezes to form an ice sheet. Why doesn't freezing occur throughout the entire volume of the lake? (b) Show that the thickness of the ice sheet formed on the surface of a lake is proportional to the square root of the time if the heat of fusion of the water freezing on the underside of the ice sheet is conducted through the sheet. (c) Assuming that the upper surface of the ice sheet is at −10°C and the bottom surface is at 0°C, calculate the time it will take to form an ice sheet 25 cm thick. (d) If the lake in part (c) is uniformly 40 m deep, how long would it take to freeze all the water in the lake? Is this likely to occur?

17.106 • The rate at which radiant energy from the sun reaches the earth's upper atmosphere is about 1.50 kW/m². The distance from the earth to the sun is 1.50×10^{11} m, and the radius of the sun is 6.96×10^8 m. (a) What is the rate of radiation of energy per unit area from the sun's surface? (b) If the sun radiates as an ideal blackbody, what is the temperature of its surface?

17.107 ••• **A Thermos for Liquid Helium.** A physicist uses a cylindrical metal can 0.250 m high and 0.090 m in diameter to store liquid helium at 4.22 K; at that temperature the heat of vaporization of helium is 2.09×10^4 J/kg. Completely surrounding the metal can are walls maintained at the temperature of liquid nitrogen, 77.3 K, with vacuum between the can and the surrounding walls. How much helium is lost per hour? The emissivity of the metal can is 0.200. The only heat transfer between the metal can and the surrounding walls is by radiation.

17.108 •• A metal sphere with radius 3.20 cm is suspended in a large metal box with interior walls that are maintained at 30.0°C. A small electric heater is embedded in the sphere. Heat energy must be supplied to the sphere at the rate of 0.660 J/s to maintain the sphere at a constant temperature of 41.0°C. (a) What is the emissivity of the metal sphere? (b) What power input to the sphere is required to maintain it at 82.0°C? What is the ratio of the power required for 82.0°C to the power required for 41.0°C? How does this ratio compare with 2^4? Explain.

17.109 •• **BIO Jogging in the Heat of the Day.** You have probably seen people jogging in extremely hot weather. There are good reasons not to do this! When jogging strenuously, an average runner of mass 68 kg and surface area 1.85 m^2 produces energy at a rate of up to 1300 W, 80% of which is converted to heat. The jogger radiates heat but actually absorbs more from the hot air than he radiates away. At such high levels of activity, the skin's temperature can be elevated to around 33°C instead of the usual 30°C. (Ignore conduction, which would bring even more heat into his body.) The only way for the body to get rid of this extra heat is by evaporating water (sweating). (a) How much heat per second is produced just by the act of jogging? (b) How much *net* heat per second does the runner gain just from radiation if the air temperature is 40.0°C (104°F)? (Remember: He radiates out, but the environment radiates back in.) (c) What is the *total* amount of excess heat this runner's body must get rid of per second? (d) How much water must his body evaporate every minute due to his activity? The heat of vaporization of water at body temperature is 2.42 × 10^6 J/kg. (e) How many 750-mL bottles of water must he drink after (or preferably before!) jogging for a half hour? Recall that a liter of water has a mass of 1.0 kg.

17.110 •• The icecaps of Greenland and Antarctica contain about 1.75% of the total water (by mass) on the earth's surface; the oceans contain about 97.5%, and the other 0.75% is mainly groundwater. Suppose the icecaps, currently at an average temperature of about −30°C, somehow slid into the ocean and melted. What would be the resulting temperature decrease of the ocean? Assume that the average temperature of ocean water is currently 5.00°C.

17.111 •• **DATA** As a physicist, you put heat into a 500.0-g solid sample at the rate of 10.0 kJ/min while recording its temperature as a function of time. You plot your data as shown in **Fig. P17.111**. (a) What is the latent heat of fusion for this solid? (b) What are the specific heats of the liquid and solid states of this material?

Figure **P17.111**

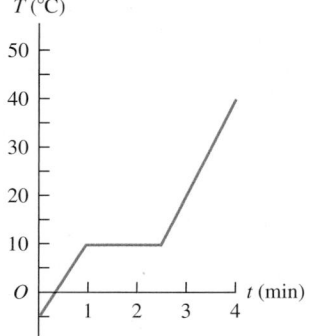

17.112 ••• **DATA** At a chemical plant where you are an engineer, a tank contains an unknown liquid. You must determine the liquid's specific heat capacity. You put 0.500 kg of the liquid into

an insulated metal cup of mass 0.200 kg. Initially the liquid and cup are at 20.0°C. You add 0.500 kg of water that has a temperature of 80.0°C. After thermal equilibrium has been reached, the final temperature of the two liquids and the cup is 58.1°C. You then empty the cup and repeat the experiment with the same initial temperatures, but this time with 1.00 kg of the unknown liquid. The final temperature is 49.3°C. Assume that the specific heat capacities are constant over the temperature range of the experiment and that no heat is lost to the surroundings. Calculate the specific heat capacity of the liquid and of the metal from which the cup is made.

17.113 •• **DATA** During your mechanical engineering internship, you are given two uniform metal bars A and B, which are made from different metals, to determine their thermal conductivities. Measuring the bars, you determine that both have length 40.0 cm and uniform cross-sectional area 2.50 cm^2. You place one end of bar A in thermal contact with a very large vat of boiling water at 100.0°C and the other end in thermal contact with an ice–water mixture at 0.0°C. To prevent heat loss along the bar's sides, you wrap insulation around the bar. You weigh the amount of ice initially and find it to be 300 g. After 45.0 min has elapsed, you weigh the ice again and find that 191 g of ice remains. The ice–water mixture is in an insulated container, so the only heat entering or leaving it is the heat conducted by the metal bar.

You are confident that your data will allow you to calculate the thermal conductivity k_A of bar A. But this measurement was tedious—you don't want to repeat it for bar B. Instead, you glue the bars together end to end, with adhesive that has very large thermal conductivity, to make a composite bar 80.0 m long. You place the free end of A in thermal contact with the boiling water and the free end of B in thermal contact with the ice–water mixture. As in the first measurement, the composite bar is thermally insulated. You go to lunch; when you return, you notice that ice remains in the ice–water mixture. Measuring the temperature at the junction of the two bars, you find that it is 62.4°C. After 10 minutes you repeat that measurement and get the same temperature, with ice remaining in the ice–water mixture. From your data, calculate the thermal conductivities of bar A and of bar B.

CHALLENGE PROBLEMS

17.114 ••• **BIO A Walk in the Sun.** Consider a poor lost soul walking at 5 km/h on a hot day in the desert, wearing only a bathing suit. This person's skin temperature tends to rise due to four mechanisms: (i) energy is generated by metabolic reactions in the body at a rate of 280 W, and almost all of this energy is converted to heat that flows to the skin; (ii) heat is delivered to the skin by convection from the outside air at a rate equal to $k'A_{skin}(T_{air} - T_{skin})$, where k' is 54 J/h·C°·m^2, the exposed skin area A_{skin} is 1.5 m^2, the air temperature T_{air} is 47°C, and the skin temperature T_{skin} is 36°C; (iii) the skin absorbs radiant energy from the sun at a rate of 1400 W/m^2; (iv) the skin absorbs radiant energy from the environment, which has temperature 47°C. (a) Calculate the net rate (in watts) at which the person's skin is heated by all four of these mechanisms. Assume that the emissivity of the skin is $e = 1$ and that the skin temperature is initially 36°C. Which mechanism is the most important? (b) At what rate (in L/h) must perspiration evaporate from this person's skin to maintain a constant skin temperature? (The heat of vaporization of water at 36°C is 2.42 × 10^6 J/kg.) (c) Suppose instead the person is protected by light-colored clothing ($e \approx 0$) so that the exposed skin area is only 0.45 m^2. What rate of perspiration is required now? Discuss the usefulness of the traditional clothing worn by desert peoples.

17.115 ••• A hollow cylinder has length L, inner radius a, and outer radius b, and the temperatures at the inner and outer surfaces are T_2 and T_1. (The cylinder could represent an insulated hot-water pipe.) The thermal conductivity of the material of which the cylinder is made is k. Derive an equation for (a) the total heat current through the walls of the cylinder; (b) the temperature variation inside the cylinder walls. (c) Show that the equation for the total heat current reduces to Eq. (17.21) for linear heat flow when the cylinder wall is very thin. (d) A steam pipe with a radius of 2.00 cm, carrying steam at 140°C, is surrounded by a cylindrical jacket with inner and outer radii 2.00 cm and 4.00 cm and made of a type of cork with thermal conductivity 4.00×10^{-2} W/m·K. This in turn is surrounded by a cylindrical jacket made of a brand of Styrofoam with thermal conductivity 2.70×10^{-2} W/m·K and having inner and outer radii 4.00 cm and 6.00 cm (**Fig. P17.115**). The outer surface of the Styrofoam has a temperature of 15°C. What is the temperature at a radius of 4.00 cm, where the two insulating layers meet? (e) What is the total rate of transfer of heat out of a 2.00-m length of pipe?

Figure **P17.115**

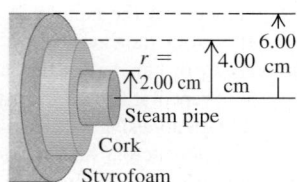

```
                          6.00
              4.00  cm
    r =       cm
   2.00 cm
        Steam pipe
        Cork
        Styrofoam
```

PASSAGE PROBLEMS

BIO **PRESERVING CELLS AT COLD TEMPERATURES.** In cryopreservation, biological materials are cooled to a very low temperature to slow down chemical reactions that might damage the cells or tissues. It is important to prevent the materials from forming ice crystals during freezing. One method for preventing ice formation is to place the material in a protective solution called a *cryoprotectant*. Stated values of the thermal properties of one cryoprotectant are listed here:

Melting point	−20°C
Latent heat of fusion	2.80×10^5 J/kg
Specific heat (liquid)	4.5×10^3 J/kg·K
Specific heat (solid)	2.0×10^3 J/kg·K
Thermal conductivity (liquid)	1.2 W/m·K
Thermal conductivity (solid)	2.5 W/m·K

17.116 You place 35 g of this cryoprotectant at 22°C in contact with a cold plate that is maintained at the boiling temperature of liquid nitrogen (77 K). The cryoprotectant is thermally insulated from everything but the cold plate. Use the values in the table to determine how much heat will be transferred from the cryoprotectant as it reaches thermal equilibrium with the cold plate. (a) 1.5×10^4 J; (b) 2.9×10^4 J; (c) 3.4×10^4 J; (d) 4.4×10^4 J.

17.117 Careful measurements show that the specific heat of the solid phase depends on temperature (**Fig. P17.117**). How will the actual time needed for this cryoprotectant to come to equilibrium with the cold plate compare with the time predicted by using the values in the table? Assume that all values other than the specific heat (solid) are correct. The actual time (a) will be shorter; (b) will be longer; (c) will be the same; (d) depends on the density of the cryoprotectant.

Figure **P17.117**

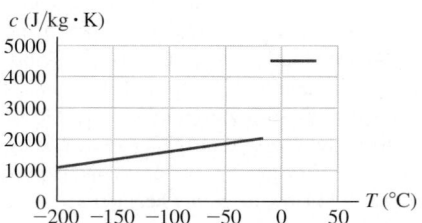

17.118 In another experiment, you place a layer of this cryoprotectant between one 10 cm × 10 cm cold plate maintained at −40°C and a second cold plate of the same size maintained at liquid nitrogen's boiling temperature (77 K). Then you measure the rate of heat transfer. Another lab wants to repeat the experiment but uses cold plates that are 20 cm × 20 cm, with one at −40°C and the other at 77 K. How thick does the layer of cryoprotectant have to be so that the rate of heat transfer by conduction is the same as that when you use the smaller plates? (a) One-quarter the thickness; (b) half the thickness; (c) twice the thickness; (d) four times the thickness.

17.119 To measure the specific heat in the liquid phase of a newly developed cryoprotectant, you place a sample of the new cryoprotectant in contact with a cold plate until the solution's temperature drops from room temperature to its freezing point. Then you measure the heat transferred to the cold plate. If the system isn't sufficiently isolated from its room-temperature surroundings, what will be the effect on the measurement of the specific heat? (a) The measured specific heat will be greater than the actual specific heat; (b) the measured specific heat will be less than the actual specific heat; (c) there will be no effect because the thermal conductivity of the cryoprotectant is so low; (d) there will be no effect on the specific heat, but the temperature of the freezing point will change.

Answers

Chapter Opening Question ?

(iii) The molten iron contains a large amount of energy. A body *has* a temperature but does not *contain* temperature. By "heat" we mean energy that is in transit from one body to another as a result of temperature difference between the bodies. Bodies do not *contain* heat.

Test Your Understanding Questions

17.1 (ii) A liquid-in-tube thermometer actually measures its own temperature. If the thermometer stays in the hot water long enough, it will come to thermal equilibrium with the water and its temperature will be the same as that of the water.

17.2 (iv) Both a bimetallic strip and a resistance thermometer measure their own temperature. For this to be equal to the temperature of the object being measured, the thermometer and object must be in thermal equilibrium. A temporal artery thermometer detects the infrared radiation from a person's skin; the detector and skin need not be at the same temperature.

17.3 (i), (iii), (ii), (v), (iv) To compare these temperatures, convert them all to the Kelvin scale. For (i), the Kelvin temperature is $T_K = T_C + 273.15 = 0.00 + 273.15 = 273.15$ K; for (ii), $T_C = \frac{5}{9}(T_F - 32°) = \frac{5}{9}(0.00° - 32°) = -17.78°C$ and $T_K = T_C + 273.15 = -17.78 + 273.15 = 255.37$ K; for (iii), $T_K = 260.00$ K; for (iv), $T_K = 77.00$ K; and for (v), $T_K = T_C + 273.15 = -180.00 + 273.15 = 93.15$ K.

17.4 (ii) and (iii) Metal 2 must expand more than metal 1 when heated and so must have a larger coefficient of linear expansion α. From Table 17.1, brass and aluminum have larger values of α than copper, but steel does not.

17.5 (ii), (i), (iv), (iii) For (i) and (ii), the relevant quantity is the specific heat c of the substance, which is the amount of heat required to raise the temperature of 1 *kilogram* of that substance by 1 K (1 C°). From Table 17.3, these values are (i) 138 J for mercury and (ii) 2428 J for ethanol. For (iii) and (iv) we need the molar heat capacity C, which is the amount of heat required to raise the temperature of 1 *mole* of that substance by 1 C°. Again from Table 17.3, these values are (iii) 27.7 J for mercury and (iv) 111.9 J for ethanol. (The ratio of molar heat capacities is different from the ratio of the specific heats because a mole of mercury and a mole of ethanol have different masses.)

17.6 (iv) In time t the system goes from point b to point e in Fig. 17.20. According to this figure, at time $t/2$ (halfway along the horizontal axis from b to e), the system is at 100°C and is still boiling; that is, it is a mixture of liquid and gas. This says that most of the heat added goes into boiling the water.

17.7 (ii) When you touch one of the walls, heat flows from your hand to the lower-temperature wall. The more rapidly heat flows from your hand, the colder you will feel. Equation (17.21) shows that the rate of heat flow is proportional to the thermal conductivity k. From Table 17.5, copper has a much higher thermal conductivity $(385.0 \text{ W/m} \cdot \text{K})$ than steel $(50.2 \text{ W/m} \cdot \text{K})$ or concrete $(0.8 \text{ W/m} \cdot \text{K})$, and so the copper wall feels the coldest.

Bridging Problem

(a) 14.26 K   **(b)** 0.427 kg/h

? The higher the temperature of a gas, the greater the average kinetic energy of its molecules. How much faster are molecules moving in the air above a frying pan (100°C) than in the surrounding kitchen air (25°C)? (i) 4 times faster; (ii) twice as fast; (iii) 1.25 times as fast; (iv) 1.12 times as fast; (v) 1.06 times as fast.

18 THERMAL PROPERTIES OF MATTER

LEARNING GOALS

Looking forward at ...

18.1 How to relate the pressure, volume, and temperature of a gas.

18.2 How the interactions between the molecules of a substance determine the properties of the substance.

18.3 How the pressure and temperature of a gas are related to the kinetic energy of its molecules.

18.4 How the heat capacities of a gas reveal whether its molecules are rotating or vibrating.

18.5 How the speeds of molecules are distributed in a gas.

18.6 What determines whether a substance is a gas, a liquid, or a solid.

Looking back at ...

7.4 Potential energy and force.

11.4 Bulk stress.

12.2 Fluids in equilibrium.

13.3 Escape speed.

14.4 Interatomic forces and oscillations.

17.1–17.6 Temperature, heat, thermal expansion, specific heat, molar heat capacity, phase changes.

The kitchen is a great place to learn about how the properties of matter depend on temperature. When you boil water in a tea kettle, the increase in temperature produces steam that whistles out of the spout at high pressure. If you forget to poke holes in a potato before baking it, the high-pressure steam produced inside the potato can cause it to explode messily. Water vapor in the air can condense into liquid on the sides of a glass of ice water; if the glass is just out of the freezer, water vapor will solidify and form frost on its sides.

These examples show the relationships among the large-scale or *macroscopic* properties of a substance, such as pressure, volume, temperature, and mass. But we can also describe a substance by using a *microscopic* perspective. This means investigating small-scale quantities such as the masses, speeds, kinetic energies, and momenta of the individual molecules that make up a substance.

The macroscopic and microscopic descriptions are intimately related. For example, the (microscopic) forces that occur when air molecules strike a solid surface (such as your skin) cause (macroscopic) atmospheric pressure. To produce standard atmospheric pressure of 1.01×10^5 Pa, 10^{32} molecules strike your skin every day with an average speed of over 1700 km/h (1000 mi/h)!

In this chapter we'll begin by looking at some macroscopic aspects of matter in general. We'll pay special attention to the *ideal gas*, one of the simplest types of matter to understand. We'll relate the macroscopic properties of an ideal gas to the microscopic behavior of its molecules. We'll also use microscopic ideas to understand the heat capacities of gases and solids. Finally, we'll look at the various phases of matter—gas, liquid, and solid—and the conditions under which each occurs.

18.1 EQUATIONS OF STATE

Quantities such as pressure, volume, temperature, and amount of substance describe the conditions, or *state,* in which a particular material exists. (For example, a tank of medical oxygen has a pressure gauge and a label stating the volume within the tank. We can add a thermometer and put the tank on a scale to measure the mass of oxygen.) These quantities are called **state variables**.

The volume V of a substance is usually determined by its pressure p, temperature T, and amount of substance, described by the mass m_{total} or number of moles n. (We are calling the total mass of a substance m_{total} because later in the chapter we will use m for the mass of one molecule.) Ordinarily, we can't change one of these variables without causing a change in another. When the tank of oxygen gets hotter, the pressure increases. If the tank gets too hot, it explodes.

In a few cases the relationship among p, V, T, and m_{total} (or n) is simple enough that we can express it as an equation called the **equation of state.** When it's too complicated for that, we can use graphs or numerical tables. Even then, the relationship among the variables still exists; we call it an equation of state even when we don't know the actual equation.

Here's a simple (though approximate) equation of state for a solid material. The temperature coefficient of volume expansion β (see Section 17.4) is the fractional volume change $\Delta V/V_0$ per unit temperature change, and the compressibility k (see Section 11.4) is the negative of the fractional volume change $\Delta V/V_0$ per unit pressure change. If a certain amount of material has volume V_0 when the pressure is p_0 and the temperature is T_0, the volume V at slightly differing pressure p and temperature T is approximately

$$V = V_0[1 + \beta(T - T_0) - k(p - p_0)] \tag{18.1}$$

(There is a negative sign in front of the term $k(p - p_0)$ because an *increase* in pressure causes a *decrease* in volume.)

The Ideal-Gas Equation

Another simple equation of state is the one for an *ideal gas.* **Figure 18.1** shows an experimental setup to study the behavior of a gas. The cylinder has a movable piston to vary the volume, the temperature can be varied by heating, and we can pump in any desired amount of gas. We then measure the pressure, volume, temperature, and amount of gas. Note that *pressure* refers both to the force per unit area exerted by the cylinder on the gas and to that exerted by the gas on the cylinder; by Newton's third law, these must be equal.

It is usually easiest to describe the amount of gas in terms of the number of moles n, rather than the mass. (We did this when we defined molar heat capacity in Section 17.5.) The **molar mass** M of a compound (sometimes confusingly called *molecular weight*) is the mass per mole:

Total mass of substance → $m_{total} = nM$ ← Number of moles of substance
Molar mass of substance $\tag{18.2}$

Hence if we know the number of moles of gas in the cylinder, we can determine the mass of gas from Eq. (18.2).

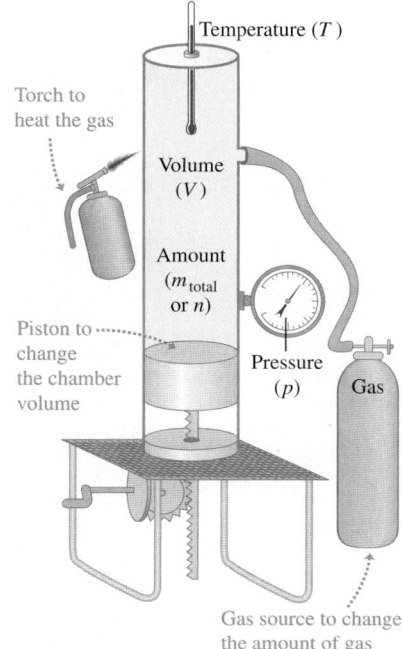

18.1 A hypothetical setup for studying the behavior of gases. By heating the gas, varying the volume with a movable piston, and adding more gas, we can control the gas pressure p, volume V, temperature T, and number of moles n.

Temperature (T)

Torch to heat the gas

Volume (V)

Amount (m_{total} or n)

Piston to change the chamber volume

Pressure (p)

Gas

Gas source to change the amount of gas

18.2 The ideal-gas equation $pV = nRT$ gives a good description of the air inside an inflated vehicle tire, where the pressure is about 3 atmospheres and the temperature is much too high for nitrogen or oxygen to liquefy. As the tire warms (T increases), the volume V changes only slightly but the pressure p increases.

BIO Application Respiration and the Ideal-Gas Equation To breathe, you rely on the ideal-gas equation $pV = nRT$. Contraction of the dome-shaped diaphragm muscle increases the volume V of the thoracic cavity (which encloses the lungs), decreasing its pressure p. The lowered pressure causes the lungs to expand and fill with air. (The temperature T is kept constant.) When you exhale, the diaphragm relaxes, allowing the lungs to contract and expel the air.

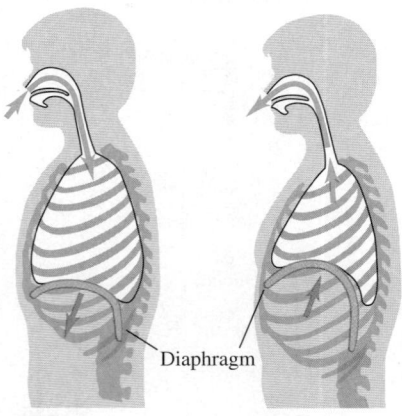

Diaphragm

Inhalation: Diaphragm contracts; lungs expand.

Exhalation: Diaphragm relaxes; lungs contract.

CAUTION **Density vs. pressure** When using Eq. (18.5), be certain that you distinguish between the Greek letter ρ (rho) for density and the letter p for pressure. ▮

Measurements of the behavior of various gases lead to three conclusions:

1. The volume V is proportional to the number of moles n. If we double n, keeping pressure and temperature constant, the volume doubles.

2. The volume varies *inversely* with the absolute pressure p. If we double p while holding the temperature T and number of moles n constant, the gas compresses to one-half of its initial volume. In other words, $pV = $ constant when n and T are constant.

3. The pressure is proportional to the *absolute* temperature T. If we double T, keeping the volume and number of moles constant, the pressure doubles. In other words, $p = $ (constant) $\times T$ when n and V are constant.

We can combine these three relationships into a single **ideal-gas equation:**

Ideal-gas equation: Gas pressure ⋯⋯ $pV = nRT$ ⋯⋯ Number of moles of gas / Absolute temperature of gas / Gas constant / Gas volume ⋯⋯ (18.3)

An **ideal gas** is one for which Eq. (18.3) holds precisely for *all* pressures and temperatures. This is an idealized model; it works best at very low pressures and high temperatures, when the gas molecules are far apart and in rapid motion. It is valid within a few percent at moderate pressures (such as a few atmospheres) and at temperatures well above those at which the gas liquefies (**Fig. 18.2**).

We might expect that the proportionality constant R in the ideal-gas equation would have different values for different gases, but it turns out to have the same value for *all* gases, at least at sufficiently high temperature and low pressure. It is called the **gas constant** (or *ideal-gas constant*). In SI units, in which the unit of p is Pa ($1 \text{ Pa} = 1 \text{ N/m}^2$) and the unit of V is m^3, the current best numerical value of R is

$$R = 8.3144621(75) \text{ J/mol} \cdot \text{K}$$

or $R = 8.314 \text{ J/mol} \cdot \text{K}$ to four significant figures. Note that the units of pressure times volume are the same as the units of work or energy (for example, N/m^2 times m^3); that's why R has units of energy per mole per unit of absolute temperature. In chemical calculations, volumes are often expressed in liters (L) and pressures in atmospheres (atm). In this system, to four significant figures,

$$R = 0.08206 \frac{\text{L} \cdot \text{atm}}{\text{mol} \cdot \text{K}}$$

We can express the ideal-gas equation, Eq. (18.3), in terms of the mass m_{total} of gas, using $m_{\text{total}} = nM$ from Eq. (18.2):

$$pV = \frac{m_{\text{total}}}{M} RT \qquad (18.4)$$

From this we can get an expression for the density $\rho = m_{\text{total}}/V$ of the gas:

$$\rho = \frac{pM}{RT} \qquad (18.5)$$

For a *constant mass* (or constant number of moles) of an ideal gas the product nR is constant, so the quantity pV/T is also constant. If the subscripts 1 and 2 refer to any two states of the same mass of a gas, then

$$\frac{p_1 V_1}{T_1} = \frac{p_2 V_2}{T_2} = \text{constant} \qquad \text{(ideal gas, constant mass)} \qquad (18.6)$$

Notice that you don't need the value of R to use this equation.

We used the proportionality of pressure to absolute temperature in Chapter 17 to define a temperature scale in terms of pressure in a constant-volume gas thermometer. That may make it seem that the pressure–temperature relationship in the ideal-gas equation, Eq. (18.3), is just a result of the way we define temperature. But the ideal-gas equation also tells us what happens when we change the volume or the amount of substance. Also, we'll see in Chapter 20 that the gas-thermometer scale corresponds closely to a temperature scale that does *not* depend on the properties of any particular material. For now, consider Eq. (18.6) as being based on this genuinely material-independent temperature scale.

DATA *SPEAKS*

The Ideal-Gas Equation

When students were given a problem using Eq. (18.3), more than 47% gave an incorrect response. Common errors:

- Forgetting that in Eq. (18.3) pressure p is *absolute*, not gauge, pressure (Section 12.2), and temperature T is *absolute* (Kelvin), not Celsius, temperature.
- Not correctly interpreting Eq. (18.3) to graph p versus V for constant T, p versus T for constant V, or V versus T for constant p.

PROBLEM-SOLVING STRATEGY 18.1 | IDEAL GASES

IDENTIFY *the relevant concepts:* Unless the problem states otherwise, you can use the ideal-gas equation to find quantities related to the state of a gas, such as pressure p, volume V, temperature T, and/or number of moles n.

SET UP *the problem* using the following steps:
1. List the known and unknown quantities. Identify the target variables.
2. If the problem concerns only one state of the system, use Eq. (18.3), $pV = nRT$ (or Eq. (18.5), $\rho = pM/RT$ if the problem involves the density ρ rather than n and V).
3. In problems that concern two states (call them 1 and 2) of the same amount of gas, if all but one of the six quantities p_1, p_2, V_1, V_2, T_1, and T_2 are known, use Eq. (18.6), $p_1V_1/T_1 = p_2V_2/T_2 = $ constant. Otherwise, use Eq. (18.3) or Eq. (18.5).

EXECUTE *the solution* as follows:
1. Use consistent units. (SI units are entirely consistent.) The problem statement may make one system of units more convenient than others. Make appropriate unit conversions, such as from atmospheres to pascals or from liters to cubic meters.
2. You may have to convert between mass m_{total} and number of moles n, using $m_{total} = Mn$, where M is the molar mass. If you

use Eq. (18.4), you *must* use the same mass units for m_{total} and M. So if M is in grams per mole (the usual units for molar mass), then m_{total} must also be in grams. To use m_{total} in kilograms, you must convert M to kg/mol. For example, the molar mass of oxygen is 32 g/mol or 32×10^{-3} kg/mol.
3. Remember that in the ideal-gas equations, T is always an *absolute* (Kelvin) temperature and p is always an absolute (not gauge) pressure.
4. Solve for the target variables.

EVALUATE *your answer:* Do your results make physical sense? Use benchmarks, such as the result of Example 18.1 below that a mole of an ideal gas at 1 atmosphere pressure occupies a volume of 22.4 liters.

EXAMPLE 18.1 VOLUME OF AN IDEAL GAS AT STP

What is the volume of a container that holds exactly 1 mole of an ideal gas at *standard temperature and pressure* (STP), defined as $T = 0°C = 273.15$ K and $p = 1$ atm $= 1.013 \times 10^5$ Pa?

SOLUTION

IDENTIFY and SET UP: This problem involves the properties of a single state of an ideal gas, so we use Eq. (18.3). We are given the pressure p, temperature T, and number of moles n; our target variable is the corresponding volume V.

EXECUTE: From Eq. (18.3), using R in J/mol $\cdot$ K, we get

$$V = \frac{nRT}{p} = \frac{(1 \text{ mol})(8.314 \text{ J/mol} \cdot \text{K})(273.15 \text{ K})}{1.013 \times 10^5 \text{ Pa}}$$

$$= 0.0224 \text{ m}^3 = 22.4 \text{ L}$$

EVALUATE: At STP, 1 mole of an ideal gas occupies 22.4 L. This is the volume of a cube 0.282 m (11.1 in.) on a side, or of a sphere 0.350 m (13.8 in.) in diameter.

EXAMPLE 18.2 COMPRESSING GAS IN AN AUTOMOBILE ENGINE

In an automobile engine, a mixture of air and vaporized gasoline is compressed in the cylinders before being ignited. A typical engine has a compression ratio of 9.00 to 1; that is, the gas in the cylinders is compressed to $\frac{1}{9.00}$ of its original volume (**Fig. 18.3**). The intake and exhaust valves are closed during the compression, so the quantity of gas is constant. What is the final temperature of the compressed gas if its initial temperature is 27°C and the initial and final pressures are 1.00 atm and 21.7 atm, respectively?

SOLUTION

IDENTIFY and SET UP: We must compare two states of the same quantity of ideal gas, so we use Eq. (18.6). In the uncompressed state 1, $p_1 = 1.00$ atm and $T_1 = 27°C = 300$ K. In the compressed state 2, $p_2 = 21.7$ atm. The cylinder volumes are not given, but we have $V_1 = 9.00V_2$. The temperature T_2 of the compressed gas is the target variable.

EXECUTE: We solve Eq. (18.6) for T_2:

$$T_2 = T_1 \frac{p_2 V_2}{p_1 V_1} = (300 \text{ K}) \frac{(21.7 \text{ atm})V_2}{(1.00 \text{ atm})(9.00V_2)} = 723 \text{ K} = 450°C$$

18.3 Cutaway of an automobile engine. While the air–gasoline mixture is being compressed prior to ignition, both the intake and exhaust valves are in the closed (up) position.

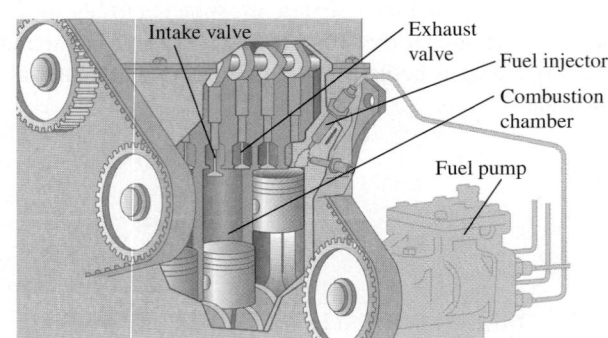

Intake valve
Exhaust valve
Fuel injector
Combustion chamber
Fuel pump

EVALUATE: This is the temperature of the air–gasoline mixture *before* the mixture is ignited; when burning starts, the temperature becomes higher still.

EXAMPLE 18.3 MASS OF AIR IN A SCUBA TANK

An "empty" aluminum scuba tank contains 11.0 L of air at 21°C and 1 atm. When the tank is filled rapidly from a compressor, the air temperature is 42°C and the gauge pressure is 2.10×10^7 Pa. What mass of air was added? (Air is about 78% nitrogen, 21% oxygen, and 1% miscellaneous; its average molar mass is 28.8 g/mol = 28.8×10^{-3} kg/mol.)

SOLUTION

IDENTIFY and SET UP: Our target variable is the difference $m_2 - m_1$ between the masses present at the end (state 2) and at the beginning (state 1). We are given the molar mass M of air, so we can use Eq. (18.2) to find the target variable if we know the number of moles present in states 1 and 2. We determine n_1 and n_2 by applying Eq. (18.3) to each state individually.

EXECUTE: We convert temperatures to the Kelvin scale by adding 273 and convert the pressure to absolute by adding 1.013×10^5 Pa.

The tank's volume is hardly affected by the increased temperature and pressure, so $V_2 = V_1$. From Eq. (18.3), the numbers of moles in the empty tank (n_1) and the full tank (n_2) are

$$n_1 = \frac{p_1 V_1}{RT_1} = \frac{(1.013 \times 10^5 \text{ Pa})(11.0 \times 10^{-3} \text{ m}^3)}{(8.314 \text{ J/mol} \cdot \text{K})(294 \text{ K})} = 0.46 \text{ mol}$$

$$n_2 = \frac{p_2 V_2}{RT_2} = \frac{(2.11 \times 10^7 \text{ Pa})(11.0 \times 10^{-3} \text{ m}^3)}{(8.314 \text{ J/mol} \cdot \text{K})(315 \text{ K})} = 88.6 \text{ mol}$$

We added $n_2 - n_1 = 88.6$ mol $- 0.46$ mol $= 88.1$ mol to the tank. From Eq. (18.2), the added mass is $M(n_2 - n_1) = (28.8 \times 10^{-3} \text{ kg/mol})(88.1 \text{ mol}) = 2.54$ kg.

EVALUATE: The added mass is not insubstantial: You could certainly use a scale to determine whether the tank was empty or full.

EXAMPLE 18.4 VARIATION OF ATMOSPHERIC PRESSURE WITH ELEVATION

Find the variation of atmospheric pressure with elevation in the earth's atmosphere. Assume that at all elevations, $T = 0°C$ and $g = 9.80$ m/s^2.

SOLUTION

IDENTIFY and SET UP: As the elevation y increases, both the atmospheric pressure p and the density ρ decrease. Hence we have *two* unknown functions of y; to solve for them, we need two independent equations. One is the ideal-gas equation, Eq. (18.5), which is expressed in terms of p and ρ. The other is Eq. (12.4), the rela-

tionship that we found in Section 12.2 among p, ρ, and y in a fluid in equilibrium: $dp/dy = -\rho g$. We are told to assume that g and T are the same at all elevations; we also assume that the atmosphere has the same chemical composition, and hence the same molar mass M, at all heights. We combine the two equations and solve for $p(y)$.

EXECUTE: We substitute $\rho = pM/RT$ into $dp/dy = -\rho g$, separate variables, and integrate, letting p_1 be the pressure at elevation y_1 and p_2 be the pressure at y_2:

$$\frac{dp}{dy} = -\frac{pM}{RT}g$$

$$\int_{p_1}^{p_2}\frac{dp}{p} = -\frac{Mg}{RT}\int_{y_1}^{y_2}dy$$

$$\ln\frac{p_2}{p_1} = -\frac{Mg}{RT}(y_2 - y_1)$$

$$\frac{p_2}{p_1} = e^{-Mg(y_2-y_1)/RT}$$

Now let $y_1 = 0$ be at sea level and let the pressure at that point be $p_0 = 1.013 \times 10^5$ Pa. Then the pressure p at any height y is

$$p = p_0 e^{-Mgy/RT}$$

EVALUATE: According to our calculation, the pressure decreases exponentially with elevation. The graph in **Fig. 18.4** shows that the slope dp/dy becomes less negative with greater elevation. That result makes sense, since $dp/dy = -\rho g$ and the density also decreases with elevation. At the summit of Mount Everest, where $y = 8848$ m,

$$\frac{Mgy}{RT} = \frac{(28.8 \times 10^{-3}\text{ kg/mol})(9.80\text{ m/s}^2)(8848\text{ m})}{(8.314\text{ J/mol}\cdot\text{K})(273\text{ K})} = 1.10$$

$$p = (1.013 \times 10^5\text{ Pa})e^{-1.10} = 0.337 \times 10^5\text{ Pa} = 0.33\text{ atm}$$

18.4 The variation of atmospheric pressure p with elevation y, assuming a constant temperature T.

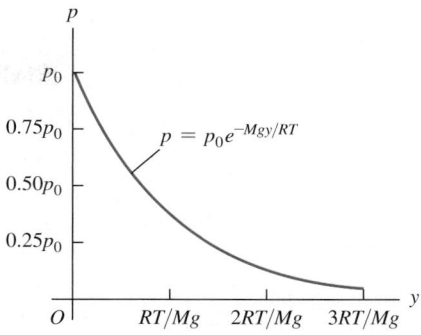

The assumption of constant temperature isn't realistic, and g decreases a little with increasing elevation (see Challenge Problem 18.84). Even so, this example shows why most mountaineers carry oxygen on Mount Everest. It also shows why jet airliners, which typically fly at altitudes of 8000 to 12,000 m, *must* have pressurized cabins for passenger comfort and health.

The van der Waals Equation

In Section 18.3 we'll obtain the ideal-gas equation, Eq. (18.3), from a simple molecular model that ignores the volumes of the molecules themselves and the attractive forces between them (**Fig. 18.5a**). Another equation of state, the **van der Waals equation,** makes approximate corrections for these two omissions (Fig. 18.5b). This equation was developed by the 19th-century Dutch physicistJ. D. van der Waals; the interaction between atoms that we discussed in Section 14.4 is named the *van der Waals interaction*. The van der Waals equation is

$$\left(p + \frac{an^2}{V^2}\right)(V - nb) = nRT \tag{18.7}$$

The constants a and b are different for different gases. Roughly speaking, b represents the volume of a mole of molecules; the total volume of the molecules is nb, and the volume remaining in which the molecules can move is $V - nb$. The constant a depends on the attractive intermolecular forces, which reduce the pressure of the gas by *pulling* the molecules together as they *push* on the walls of the container. The decrease in pressure is proportional to the number of molecules per unit volume in a layer near the wall (which are exerting the pressure on the wall) and is also proportional to the number per unit volume in the next layer beyond the wall (which are doing the attracting). Hence the decrease in pressure due to intermolecular forces is proportional to n^2/V^2.

When n/V is small (that is, when the gas is *dilute*), the average distance between molecules is large, the corrections in the van der Waals equation become insignificant, and Eq. (18.7) reduces to the ideal-gas equation. As an example, for carbon dioxide gas (CO_2) the constants in the van der Waals equation are $a = 0.364$ J$\cdot$m^3/mol^2 and $b = 4.27 \times 10^{-5}$ m^3/mol. We saw in Example 18.1

18.5 A gas as modeled by (a) the ideal-gas equation and (b) the van der Waals equation.

(a) An idealized model of a gas

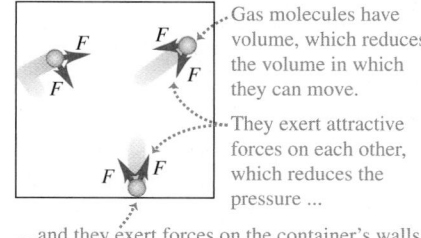

Gas molecules are infinitely small.

They exert forces on the walls of the container but not on each other.

(b) A more realistic model of a gas

Gas molecules have volume, which reduces the volume in which they can move.

They exert attractive forces on each other, which reduces the pressure ...

... and they exert forces on the container's walls.

that 1 mole of an ideal gas at $T = 0°C = 273.15$ K and $p = 1$ atm $= 1.013 \times 10^5$ Pa occupies a volume $V = 0.0224$ m^3; according to Eq. (18.7), 1 mole of CO_2 occupying this volume at this temperature would be at a pressure 532 Pa less than 1 atm, a difference of only 0.5% from the ideal-gas value.

pV-Diagrams

We could in principle represent the *p-V-T* relationship graphically as a *surface* in a three-dimensional space with coordinates *p*, *V*, and *T*. This representation is useful (see Section 18.6), but ordinary two-dimensional graphs are usually more convenient. One of the most useful of these is a set of graphs of pressure as a function of volume, each for a particular constant temperature. Such a diagram is called a *pV*-**diagram.** Each curve, representing behavior at a specific temperature, is called an **isotherm,** or a *pV-isotherm*.

Figure 18.6 shows *pV*-isotherms for a constant amount of an ideal gas. Since $p = nRT/V$ from Eq. (18.3), along an isotherm (constant *T*) the pressure *p* is inversely proportional to the volume *V* and the isotherms are hyperbolic curves.

Figure 18.7 shows a *pV*-diagram for a material that *does not* obey the ideal-gas equation. At temperatures below T_c the isotherms develop flat regions in which we can compress the material (that is, reduce the volume *V*) without increasing the pressure *p*. Observation shows that the gas is *condensing* from the vapor (gas) to the liquid phase. The flat parts of the isotherms in the shaded area of Fig. 18.7 represent conditions of liquid-vapor *phase equilibrium*. As the volume decreases, more and more material goes from vapor to liquid, but the pressure does not change. (To keep the temperature constant during condensation, we have to remove the heat of vaporization, discussed in Section 17.6.)

When we compress such a gas at a constant temperature T_2 in Fig. 18.7, it is vapor until point *a* is reached. Then it begins to liquefy; as the volume decreases further, more material liquefies, and *both* the pressure and the temperature remain constant. At point *b*, all the material is in the liquid state. After this, any further compression requires a very rapid rise of pressure, because liquids are in general much less compressible than gases. At a lower constant temperature T_1, similar behavior occurs, but the condensation begins at lower pressure and greater volume than at the constant temperature T_2. At temperatures greater than T_c, *no* phase transition occurs as the material is compressed; at the highest temperatures, such as T_4, the curves resemble the ideal-gas curves of Fig. 18.6. We call T_c the *critical temperature* for this material. In Section 18.6 we'll discuss what happens to the phase of the gas above the critical temperature.

We will use *pV*-diagrams often in the next two chapters. We will show that the *area* under a *pV*-curve (whether or not it is an isotherm) represents the *work* done by the system during a volume change. This work, in turn, is directly related to heat transfer and changes in the *internal energy* of the system.

18.6 Isotherms, or constant-temperature curves, for a constant amount of an ideal gas. The highest temperature is T_4; the lowest is T_1. This is a graphical representation of the ideal-gas equation of state.

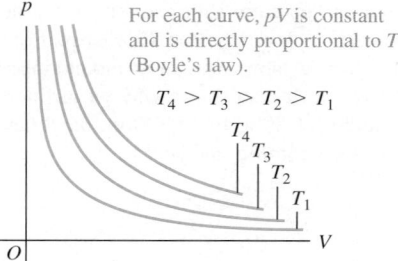

Each curve represents pressure as a function of volume for an ideal gas at a single temperature.

For each curve, *pV* is constant and is directly proportional to *T* (Boyle's law).

$$T_4 > T_3 > T_2 > T_1$$

18.7 A *pV*-diagram for a nonideal gas, showing isotherms for temperatures above and below the critical temperature T_c. The liquid–vapor equilibrium region is shown as a green shaded area. At still lower temperatures the material might undergo phase transitions from liquid to solid or from gas to solid; these are not shown here.

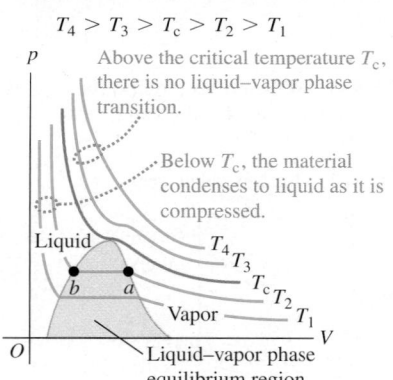

$$T_4 > T_3 > T_c > T_2 > T_1$$

Above the critical temperature T_c, there is no liquid–vapor phase transition.

Below T_c, the material condenses to liquid as it is compressed.

Liquid–vapor phase equilibrium region

TEST YOUR UNDERSTANDING OF SECTION 18.1 Rank the following ideal gases in order from highest to lowest number of moles: (i) Pressure 1 atm, volume 1 L, and temperature 300 K; (ii) pressure 2 atm, volume 1 L, and temperature 300 K; (iii) pressure 1 atm, volume 2 L, and temperature 300 K; (iv) pressure 1 atm, volume 1 L, and temperature 600 K; (v) pressure 2 atm, volume 1 L, and temperature 600 K. ▌

18.2 MOLECULAR PROPERTIES OF MATTER

We have studied several properties of matter in bulk, including elasticity, density, surface tension, heat capacities, and equations of state. Now we want to look in more detail at the relationship of bulk behavior to *molecular* structure. We begin with a general discussion of the molecular structure of matter. Then in the next two sections we develop the kinetic-molecular model of an ideal gas, obtaining from this molecular model the equation of state and an expression for heat capacity.

Molecules and Intermolecular Forces

Any specific chemical compound is made up of identical **molecules.** The smallest molecules contain one atom each and are of the order of 10^{-10} m in size; the largest contain many atoms and are at least 10,000 times larger. In gases the molecules move nearly independently; in liquids and solids they are held together by intermolecular forces. These forces arise from interactions among the electrically charged particles that make up the molecules. Gravitational forces between molecules are negligible in comparison with electric forces.

The interaction of two *point* electric charges is described by a force (repulsive for like charges, attractive for unlike charges) with a magnitude proportional to $1/r^2$, where r is the distance between the points. We will study this relationship, called *Coulomb's law,* in Chapter 21. Molecules are *not* point charges but complex structures containing both positive and negative charge, and their interactions are more complex. The force between molecules in a gas varies with the distance r between molecules somewhat as shown in **Fig. 18.8**, where a positive F_r corresponds to a repulsive force and a negative F_r to an attractive force. When molecules are far apart, the intermolecular forces are very small and usually attractive. As a gas is compressed and its molecules are brought closer together, the attractive forces increase. The intermolecular force becomes zero at an equilibrium spacing r_0, corresponding roughly to the spacing between molecules in the liquid and solid states. In liquids and solids, relatively large pressures are needed to compress the substance appreciably. This shows that at molecular distances slightly *less* than r_0, the forces become *repulsive* and relatively large.

Figure 18.8 also shows the potential energy as a function of r. This function has a *minimum* at r_0, where the force is zero. The two curves are related by $F_r(r) = -dU/dr$, as we showed in Section 7.4. Such a potential-energy function is often called a **potential well.** A molecule at rest at a distance r_0 from a second molecule would need an additional energy $|U_0|$, the "depth" of the potential well, to "escape" to an indefinitely large value of r.

Molecules are always in motion; their kinetic energies usually increase with temperature. At very low temperatures the average kinetic energy of a molecule may be much *less* than the depth of the potential well. The molecules then condense into the liquid or solid phase with average intermolecular spacings of about r_0. But at higher temperatures the average kinetic energy becomes larger than the depth $|U_0|$ of the potential well. Molecules can then escape the intermolecular force and become free to move independently, as in the gaseous phase of matter.

In *solids,* molecules vibrate about more or less fixed points. (See Section 17.4.) In a crystalline solid these points are arranged in a *crystal lattice.* **Figure 18.9** shows the cubic crystal structure of sodium chloride, and **Fig. 18.10** shows a scanning tunneling microscope image of individual silicon atoms on the surface of a crystal.

18.8 How the force between molecules and their potential energy of interaction depend on their separation r.

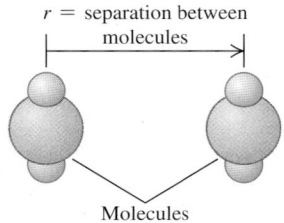

r = separation between molecules

Molecules

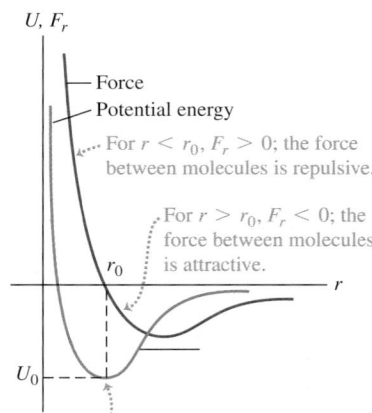

U, F_r

Force

Potential energy

For $r < r_0$, $F_r > 0$; the force between molecules is repulsive.

For $r > r_0$, $F_r < 0$; the force between molecules is attractive.

r_0

r

U_0

At a separation $r = r_0$, the potential energy of the two molecules is minimum and the force between the molecules is zero.

18.9 Schematic representation of the cubic crystal structure of sodium chloride (ordinary salt).

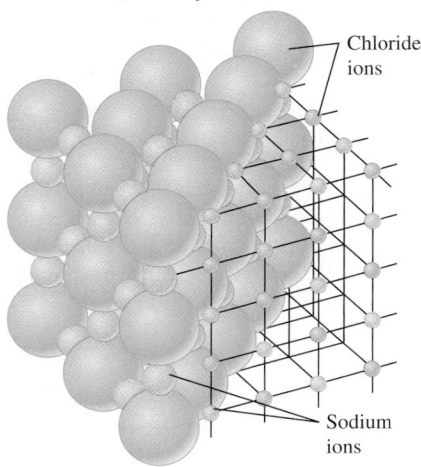

Chloride ions

Sodium ions

18.10 A scanning tunneling microscope image of the surface of a silicon crystal. The area shown is only 9.0 nm (9.0×10^{-9} m) across. Each blue "bead" is one silicon atom; these atoms are arranged in a (nearly) perfect array of hexagons.

In a *liquid,* the intermolecular distances are usually only slightly greater than in the solid phase of the same substance, but the molecules have much greater freedom of movement. Liquids show regularity of structure only in the immediate neighborhood of a few molecules.

The molecules of a *gas* are usually widely separated and so have only very small attractive forces. A gas molecule moves in a straight line until it collides with another molecule or with a wall of the container. In molecular terms, an *ideal gas* is a gas whose molecules exert *no* attractive forces on each other (see Fig. 18.5a) and therefore have no *potential* energy.

At low temperatures, most common substances are in the solid phase. As the temperature rises, a substance melts and then vaporizes. From a molecular point of view, these transitions are in the direction of increasing molecular kinetic energy. Thus temperature and molecular kinetic energy are closely related.

Moles and Avogadro's Number

We have used the mole as a measure of quantity of substance. One **mole** of any pure chemical element or compound contains a definite number of molecules, the same number for all elements and compounds. The official SI definition is:

> One mole is the amount of substance that contains as many elementary entities as there are atoms in 0.012 kilogram of carbon-12.

In our discussion, the "elementary entities" are molecules. (In a monatomic substance such as carbon or helium, each molecule is a single atom.) Atoms of a given element may occur in any of several isotopes, which are chemically identical but have different atomic masses; "carbon-12" is a specific isotope of carbon.

The number of molecules in a mole is called **Avogadro's number,** denoted by N_A. The current best numerical value of N_A is

$$N_A = 6.02214129(27) \times 10^{23} \text{ molecules/mol (Avogadro's number)}$$

The *molar mass M* of a compound is the mass of 1 mole. It is equal to the mass m of a single molecule multiplied by Avogadro's number:

$$\underset{\text{of a substance}}{\text{Molar mass}} \cdots\!\!\rightarrow M = N_A m \leftarrow\!\!\cdots \overset{\text{Avogadro's number}}{\underset{\text{Mass of a molecule of substance}}{}} \qquad (18.8)$$

When the molecule consists of a single atom, the term *atomic mass* is often used instead of molar mass.

EXAMPLE 18.5   ATOMIC AND MOLECULAR MASS

Find the mass of a single hydrogen atom and of a single oxygen molecule.

SOLUTION

IDENTIFY and SET UP: This problem involves the relationship between the mass of a molecule or atom (our target variable) and the corresponding molar mass M. We use Eq. (18.8) in the form $m = M/N_A$ and the values of the atomic masses from the periodic table of the elements (see Appendix D).

EXECUTE: For atomic hydrogen the atomic mass (molar mass) is $M_H = 1.008$ g/mol, so the mass m_H of a single hydrogen atom is

$$m_H = \frac{1.008 \text{ g/mol}}{6.022 \times 10^{23} \text{ atoms/mol}} = 1.674 \times 10^{-24} \text{ g/atom}$$

For oxygen the atomic mass is 16.0 g/mol, so for the diatomic (two-atom) oxygen molecule the molar mass is 32.0 g/mol. Then the mass of a single oxygen molecule is

$$m_{O_2} = \frac{32.0 \text{ g/mol}}{6.022 \times 10^{23} \text{ molecules/mol}} = 53.1 \times 10^{-24} \text{ g/molecule}$$

EVALUATE: We note that the values in Appendix D are for the *average* atomic masses of a natural sample of each element. Such a sample may contain several *isotopes* of the element, each with a different atomic mass. Natural samples of hydrogen and oxygen are almost entirely made up of just one isotope.

Suppose you could adjust the value of r_0 for the molecules of a certain chemical compound (Fig. 18.8) by turning a dial. If you doubled the value of r_0, the density of the solid form of this compound would become (i) twice as great; (ii) four times as great; (iii) eight times as great; (iv) $\frac{1}{2}$ as great; (v) $\frac{1}{4}$ as great; (vi) $\frac{1}{8}$ as great. ∎

18.3 KINETIC-MOLECULAR MODEL OF AN IDEAL GAS

PhET: Balloons & Buoyancy
PhET: Friction
PhET: Gas Properties

The goal of any molecular theory of matter is to understand the *macroscopic* properties of matter in terms of its atomic or molecular structure and behavior. Once we have this understanding, we can design materials to have specific desired properties. Theories have led to the development of high-strength steels, semiconductor materials for electronic devices, and countless other materials essential to contemporary technology.

Let's consider a simple molecular model of an ideal gas. This *kinetic-molecular model* represents the gas as a large number of particles bouncing around in a closed container. In this section we use the kinetic-molecular model to understand how the ideal-gas equation of state, Eq. (18.3), is related to Newton's laws. In the following section we use the kinetic-molecular model to predict the molar heat capacity of an ideal gas. We'll go on to elaborate the model to include "particles" that are not points but have a finite size.

Our discussion of the kinetic-molecular model has several steps, and you may need to go over them several times. Don't get discouraged!

Here are the assumptions of our model:

1. A container with volume V contains a very large number N of identical molecules, each with mass m.

2. The molecules behave as point particles that are small compared to the size of the container and to the average distance between molecules.

3. The molecules are in constant motion. Each molecule collides occasionally with a wall of the container. These collisions are perfectly elastic.

4. The container walls are rigid and infinitely massive and do not move.

CAUTION **Molecules vs. moles** Don't confuse N, the number of *molecules* in the gas, with n, the number of *moles*. The number of molecules is equal to the number of moles multiplied by Avogadro's number: $N = nN_\text{A}$. ∎

Collisions and Gas Pressure

During collisions the molecules exert *forces* on the walls of the container; this is the origin of the *pressure* that the gas exerts. In a typical collision (**Fig. 18.11**) the velocity component parallel to the wall is unchanged, and the component perpendicular to the wall reverses direction but does not change in magnitude.

We'll first determine the *number* of collisions that occur per unit time for a certain area A of wall. Then we find the total momentum change associated with these collisions and the force needed to cause this momentum change. From this we can determine the pressure (force per unit area) and compare to the ideal-gas equation. We'll find a direct connection between the temperature of the gas and the kinetic energy of its molecules.

To begin, we will assume that all molecules in the gas have the same *magnitude* of x-velocity, $|v_x|$. Later we'll see that our results don't depend on making this overly simplistic assumption.

As Fig. 18.11 shows, for each collision the x-component of velocity changes from $-|v_x|$ to $+|v_x|$. So the x-component of momentum p_x changes from $-m|v_x|$ to $+m|v_x|$, and the *change* in p_x is $m|v_x| - (-m|v_x|) = 2m|v_x|$.

18.11 Elastic collision of a molecule with an idealized container wall.

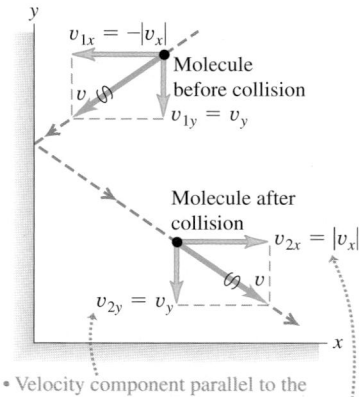

• Velocity component parallel to the wall (y-component) does not change.
• Velocity component perpendicular to the wall (x-component) reverses direction.
• Speed v does not change.

18.12 For a molecule to strike the wall in area A during a time interval dt, the molecule must be headed for the wall and be within the shaded cylinder of length $|v_x|\,dt$ at the beginning of the interval.

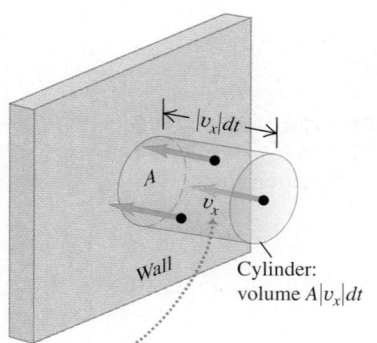

All molecules are assumed to have the same magnitude $|v_x|$ of x-velocity.

If a molecule is going to collide with a given wall area A during a small time interval dt, then at the beginning of dt it must be within a distance $|v_x|\,dt$ from the wall (**Fig. 18.12**) and it must be headed toward the wall. So the number of molecules that collide with A during dt is equal to the number of molecules within a cylinder with base area A and length $|v_x|\,dt$ that have their x-velocity aimed toward the wall. The volume of such a cylinder is $A|v_x|\,dt$. Assuming that the number of molecules per unit volume (N/V) is uniform, the *number* of molecules in this cylinder is $(N/V)(A|v_x|\,dt)$. On the average, half of these molecules are moving toward the wall and half are moving away from it. So the number of collisions with A during dt is

$$\tfrac{1}{2}\left(\frac{N}{V}\right)(A|v_x|\,dt)$$

For the system of all molecules in the gas, the total momentum change dP_x during dt is the *number* of collisions multiplied by $2m|v_x|$:

$$dP_x = \tfrac{1}{2}\left(\frac{N}{V}\right)(A|v_x|\,dt)(2m|v_x|) = \frac{NAmv_x^2\,dt}{V} \qquad (18.9)$$

(We are using capital P for total momentum and lowercase p for pressure. Be careful!) We wrote v_x^2 rather than $|v_x|^2$ in the final expression because the square of the absolute value of a number is equal to the square of that number. The *rate* of change of momentum component P_x is

$$\frac{dP_x}{dt} = \frac{NAmv_x^2}{V} \qquad (18.10)$$

According to Newton's second law, this rate of change of momentum equals the force exerted by the wall area A on the gas molecules. From Newton's *third* law this is equal and opposite to the force exerted *on* the wall *by* the molecules. Pressure p is the magnitude of the force exerted on the wall per unit area:

$$p = \frac{F}{A} = \frac{Nmv_x^2}{V} \qquad (18.11)$$

The pressure exerted by the gas depends on the number of molecules per volume (N/V), the mass m per molecule, and the speed of the molecules.

Pressure and Molecular Kinetic Energies

We mentioned that $|v_x|$ is really *not* the same for all the molecules. But we could have sorted the molecules into groups having the same $|v_x|$ within each group, then added up the resulting contributions to the pressure. The net effect of all this is just to replace v_x^2 in Eq. (18.11) by the *average* value of v_x^2, which we denote by $(v_x^2)_{\mathrm{av}}$. We can relate $(v_x^2)_{\mathrm{av}}$ to the *speeds* of the molecules. The speed v of a molecule is related to the velocity components v_x, v_y, and v_z by

$$v^2 = v_x^2 + v_y^2 + v_z^2$$

We can average this relation over all molecules:

$$(v^2)_{\mathrm{av}} = (v_x^2)_{\mathrm{av}} + (v_y^2)_{\mathrm{av}} + (v_z^2)_{\mathrm{av}}$$

But there is no real difference in our model between the x-, y-, and z-directions. (Molecular speeds are very fast in a typical gas, so the effects of gravity are negligibly small.) It follows that $(v_x^2)_{\mathrm{av}}$, $(v_y^2)_{\mathrm{av}}$, and $(v_z^2)_{\mathrm{av}}$ must all be *equal*. Hence $(v^2)_{\mathrm{av}}$ is equal to $3(v_x^2)_{\mathrm{av}}$ and

$$(v_x^2)_{\mathrm{av}} = \tfrac{1}{3}(v^2)_{\mathrm{av}}$$

so Eq. (18.11) becomes

$$pV = \tfrac{1}{3}Nm(v^2)_{\mathrm{av}} = \tfrac{2}{3}N\left[\tfrac{1}{2}m(v^2)_{\mathrm{av}}\right] \qquad (18.12)$$

We notice that $\frac{1}{2}m(v^2)_{av}$ is the average translational kinetic energy of a single molecule. The product of this and the total number of molecules N equals the total random kinetic energy K_{tr} of translational motion of all the molecules. (The notation K_{tr} reminds us that this is the energy of *translational* motion. There may also be energies associated with molecular rotation and vibration.) The product pV equals two-thirds of the total translational kinetic energy:

$$pV = \tfrac{2}{3}K_{tr} \tag{18.13}$$

Now compare Eq. (18.13) to the ideal-gas equation $pV = nRT$, Eq. (18.3), which is based on experimental studies of gas behavior. For the two equations to agree, we must have

Average translational ···· Number of moles of gas
kinetic energy of an $\qquad K_{tr} = \tfrac{3}{2}nRT$ ···· Absolute temperature of gas $\qquad$ (18.14)
ideal gas $\qquad\qquad\qquad$ ······· Gas constant

So K_{tr} is *directly proportional* to the absolute temperature T (**Fig. 18.13**).

The average translational kinetic energy of a single molecule is the total translational kinetic energy K_{tr} of all molecules divided by the number of molecules, N:

$$\frac{K_{tr}}{N} = \tfrac{1}{2}m(v^2)_{av} = \frac{3nRT}{2N}$$

Also, the total number of molecules N is the number of moles n multiplied by Avogadro's number N_A, so $N = nN_A$ and $n/N = 1/N_A$. Thus the above equation becomes

$$\frac{K_{tr}}{N} = \tfrac{1}{2}m(v^2)_{av} = \tfrac{3}{2}\left(\frac{R}{N_A}\right)T \tag{18.15}$$

The ratio R/N_A is called the **Boltzmann constant, k**:

$$k = \frac{R}{N_A} = \frac{8.314 \text{ J/mol} \cdot \text{K}}{6.022 \times 10^{23} \text{ molecules/mol}} = 1.381 \times 10^{-23} \text{ J/molecule} \cdot \text{K}$$

(The current best numerical value of k is $1.3806488(13) \times 10^{-23}$ J/molecule $\cdot$ K). In terms of k we can rewrite Eq. (18.15) as

Average translational ···· Mass of a molecule
kinetic energy of a $\qquad \tfrac{1}{2}m(v^2)_{av} = \tfrac{3}{2}kT$ ···· Absolute temperature $\qquad$ (18.16)
gas molecule $\qquad\qquad$ Average value of the $\qquad$ of gas
$\qquad\qquad\qquad$ square of molecular speeds $\qquad$ ···· Boltzmann constant

This shows that the average translational kinetic energy *per molecule* depends only on the temperature, not on the pressure, volume, or kind of molecule. We can obtain the average translational kinetic energy *per mole* by multiplying Eq. (18.16) by Avogadro's number and using the relation $M = N_A m$:

$$N_A\tfrac{1}{2}m(v^2)_{av} = \tfrac{1}{2}M(v^2)_{av} = \tfrac{3}{2}RT \qquad \text{(average translational kinetic energy per mole of gas)} \tag{18.17}$$

The translational kinetic energy of a mole of an ideal gas depends only on T.

Finally, it can be helpful to rewrite the ideal-gas equation on a "per-molecule" basis. We use $N = N_A n$ and $R = N_A k$ to obtain this alternative form:

$$pV = NkT \tag{18.18}$$

This shows that we can think of the Boltzmann constant k as a gas constant on a "per-molecule" basis instead of the usual "per-mole" basis for R.

18.13 Summer air (top) is warmer than winter air (bottom); that is, the average translational kinetic energy of air molecules is greater in summer.

Molecular Speeds

From Eqs. (18.16) and (18.17) we can obtain expressions for the square root of $(v^2)_{av}$, called the **root-mean-square speed** (or **rms speed**) v_{rms}:

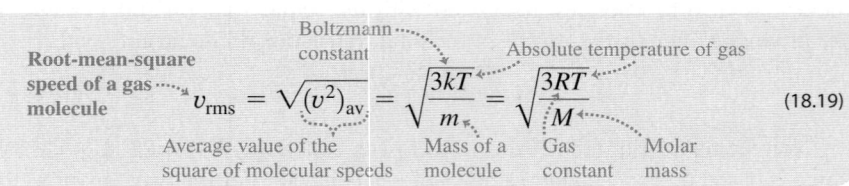

Root-mean-square
speed of a gas
molecule

$$v_{rms} = \sqrt{(v^2)_{av}} = \sqrt{\frac{3kT}{m}} = \sqrt{\frac{3RT}{M}} \qquad (18.19)$$

Boltzmann constant — Absolute temperature of gas

Average value of the square of molecular speeds — Mass of a molecule — Gas constant — Molar mass

It might seem more natural to give the *average* speed rather than v_{rms}, but v_{rms} follows more directly from Eqs. (18.16) and (18.17). To compute the rms speed, we square each molecular speed, add, divide by the number of molecules, and take the square root; v_{rms} is the *root* of the *mean* of the *squares*.

Equations (18.16) and (18.19) show that at a given temperature T, gas molecules of different mass m have the same average kinetic energy but different root-mean-square speeds. On average, the nitrogen molecules ($M = 28$ g/mol) in the air around you are moving faster than are the oxygen molecules ($M = 32$ g/mol). Hydrogen molecules ($M = 2$ g/mol) are fastest of all; this is why there is hardly any hydrogen in the earth's atmosphere, despite its being the most common element in the universe (**Fig. 18.14**). A sizable fraction of any H_2 molecules in the atmosphere would have speeds greater than the earth's escape speed of 1.12×10^4 m/s (calculated in Example 13.5 in Section 13.3) and would escape into space. The heavier, slower-moving gases cannot escape so easily, which is why they predominate in our atmosphere.

The assumption that individual molecules undergo perfectly elastic collisions with the container wall is a little too simple. In most cases, molecules actually adhere to the wall for a short time and then leave again with speeds that are characteristic of the temperature *of the wall*. However, the gas and the wall are ordinarily in thermal equilibrium and have the same temperature. So there is no net energy transfer between gas and wall, and our conclusions remain valid.

18.14 While hydrogen is a desirable fuel for vehicles, it is only a trace constituent of our atmosphere (0.00005% by volume). Hence hydrogen fuel has to be generated by electrolysis of water, which is itself an energy-intensive process.

PROBLEM-SOLVING STRATEGY 18.2 | **KINETIC-MOLECULAR THEORY**

IDENTIFY *the relevant concepts:* Use the results of the kinetic-molecular model to relate the macroscopic properties of a gas, such as temperature and pressure, to microscopic properties, such as molecular speeds.

SET UP *the problem* using the following steps:
1. List knowns and unknowns; identify the target variables.
2. Choose appropriate equation(s) from among Eqs. (18.14), (18.16), and (18.19).

EXECUTE *the solution* as follows: Maintain consistency in units.
1. The usual units for molar mass M are grams per mole; these units are often omitted in tables. In equations such as Eq. (18.19), when you use SI units you must express M in kilograms per mole. For example, for oxygen $M_{O_2} = 32$ g/mol $= 32 \times 10^{-3}$ kg/mol.

2. Are you working on a "per-molecule" basis (with m, N, and k) or a "per-mole" basis (with M, n, and R)? To check units, think of N as having units of "molecules"; then m has units of mass per molecule, and k has units of joules per molecule per kelvin. Similarly, n has units of moles; then M has units of mass per mole and R has units of joules per mole per kelvin.
3. Remember that T is always *absolute* (Kelvin) temperature.

EVALUATE *your answer:* Are your answers reasonable? Here's a benchmark: Typical molecular speeds at room temperature are several hundred meters per second.

EXAMPLE 18.6   **MOLECULAR KINETIC ENERGY AND v_{rms}**

(a) What is the average translational kinetic energy of an ideal-gas molecule at 27°C? (b) What is the total random translational kinetic energy of the molecules in 1 mole of this gas? (c) What is the rms speed of oxygen molecules at this temperature?

SOLUTION

IDENTIFY and SET UP: This problem involves the translational kinetic energy of an ideal gas on a per-molecule and per-mole basis, as well as the root-mean-square molecular speed v_{rms}. We are given $T = 27°C = 300$ K and $n = 1$ mol; we use the molecular mass m for oxygen. We use Eq. (18.16) to determine the average kinetic energy of a molecule, Eq. (18.14) to find the total molecular kinetic energy K_{tr} of 1 mole, and Eq. (18.19) to find v_{rms}.

EXECUTE: (a) From Eq. (18.16),

$$\tfrac{1}{2}m(v^2)_{av} = \tfrac{3}{2}kT = \tfrac{3}{2}(1.38 \times 10^{-23} \text{ J/K})(300 \text{ K})$$
$$= 6.21 \times 10^{-21} \text{ J}$$

(b) From Eq. (18.14), the kinetic energy of one mole is

$$K_{tr} = \tfrac{3}{2}nRT = \tfrac{3}{2}(1 \text{ mol})(8.314 \text{ J/mol} \cdot \text{K})(300 \text{ K}) = 3740 \text{ J}$$

(c) We found the mass per molecule m and molar mass M of molecular oxygen in Example 18.5. Using Eq. (18.19), we can calculate v_{rms} in two ways:

$$v_{rms} = \sqrt{\frac{3kT}{m}} = \sqrt{\frac{3(1.38 \times 10^{-23} \text{ J/K})(300 \text{ K})}{5.31 \times 10^{-26} \text{ kg}}}$$
$$= 484 \text{ m/s} = 1740 \text{ km/h} = 1080 \text{ mi/h}$$

$$v_{rms} = \sqrt{\frac{3RT}{M}} = \sqrt{\frac{3(8.314 \text{ J/mol} \cdot \text{K})(300 \text{ K})}{32.0 \times 10^{-3} \text{ kg/mol}}} = 484 \text{ m/s}$$

EVALUATE: The answer in part (a) does not depend on the mass of the molecule. We can check our result in part (b) by noting that the translational kinetic energy per mole must be equal to the product of the average translational kinetic energy per molecule from part (a) and Avogadro's number N_A: $K_{tr} = (6.21 \times 10^{-21} \text{ J/molecule}) \times (6.022 \times 10^{23} \text{ molecules}) = 3740 \text{ J}$.

EXAMPLE 18.7   **CALCULATING RMS AND AVERAGE SPEEDS**

Five gas molecules have speeds of 500, 600, 700, 800, and 900 m/s. What is the rms speed? What is the *average* speed?

SOLUTION

IDENTIFY and SET UP: We use the definitions of the root mean square and the average of a collection of numbers. To find v_{rms}, we square each speed, find the average (mean) of the squares, and take the square root of the result. We find v_{av} as usual.

EXECUTE: The average value of v^2 and the resulting v_{rms} for the five molecules are

$$(v^2)_{av} = \frac{500^2 + 600^2 + 700^2 + 800^2 + 900^2}{5} \text{ m}^2/\text{s}^2$$
$$= 5.10 \times 10^5 \text{ m}^2/\text{s}^2$$
$$v_{rms} = \sqrt{(v^2)_{av}} = 714 \text{ m/s}$$

The average speed v_{av} is

$$v_{av} = \frac{500 + 600 + 700 + 800 + 900}{5} \text{ m/s} = 700 \text{ m/s}$$

EVALUATE: In general v_{rms} and v_{av} are *not* the same. Roughly speaking, v_{rms} gives greater weight to the higher speeds than does v_{av}.

Collisions Between Molecules

We have ignored the possibility that two gas molecules might collide. If they are really points, they *never* collide. But consider a more realistic model in which the molecules are rigid spheres with radius r. How often do they collide with other molecules? How far do they travel, on average, between collisions? We can get approximate answers from the following rather primitive model.

Consider N spherical molecules with radius r in a volume V. Suppose only one molecule is moving. When it collides with another molecule, the distance between centers is $2r$. Suppose we draw a cylinder with radius $2r$, with its axis

18.15 In a time dt a molecule with radius r will collide with any other molecule within a cylindrical volume of radius $2r$ and length $v\,dt$.

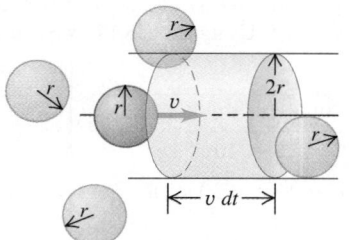

parallel to the velocity of the molecule (**Fig. 18.15**). The moving molecule collides with any other molecule whose center is inside this cylinder. In a short time dt a molecule with speed v travels a distance $v\,dt$; during this time it collides with any molecule that is in the cylindrical volume of radius $2r$ and length $v\,dt$. The volume of the cylinder is $4\pi r^2 v\,dt$. There are N/V molecules per unit volume, so the number dN with centers in this cylinder is

$$dN = 4\pi r^2 v\,dt\,N/V$$

Thus the number of collisions *per unit time* is

$$\frac{dN}{dt} = \frac{4\pi r^2 vN}{V}$$

This result assumes that only one molecule is moving. It turns out that collisions are more frequent when all the molecules move at once, and the above equation has to be multiplied by a factor of $\sqrt{2}$:

$$\frac{dN}{dt} = \frac{4\pi\sqrt{2}\,r^2 vN}{V}$$

The average time t_{mean} between collisions, called the *mean free time,* is the reciprocal of this expression:

$$t_{\text{mean}} = \frac{V}{4\pi\sqrt{2}\,r^2 vN} \tag{18.20}$$

The average distance traveled between collisions is called the **mean free path.** In our model, this is just the molecule's speed v multiplied by t_{mean}:

18.16 If you try to walk through a crowd, your mean free path—the distance you can travel on average without running into another person—depends on how large the people are and how closely they are spaced.

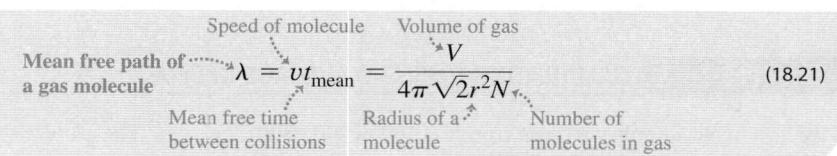

Mean free path of a gas molecule $\cdots\!\!\rightarrow \lambda = vt_{\text{mean}} = \dfrac{V}{4\pi\sqrt{2}r^2 N}$ $\tag{18.21}$

Speed of molecule · Volume of gas · Mean free time between collisions · Radius of a molecule · Number of molecules in gas

The mean free path λ (the Greek letter lambda) is inversely proportional to the number of molecules per unit volume (N/V) and inversely proportional to the cross-sectional area πr^2 of a molecule; the more molecules there are and the larger the molecule, the shorter the mean distance between collisions (**Fig. 18.16**). Note that the mean free path *does not* depend on the speed of the molecule.

We can express Eq. (18.21) in terms of macroscopic properties of the gas, using the ideal-gas equation in the form of Eq. (18.18), $pV = NkT$. We find

$$\lambda = \frac{kT}{4\pi\sqrt{2}\,r^2 p} \tag{18.22}$$

If the temperature is increased at constant pressure, the gas expands, the average distance between molecules increases, and λ increases. If the pressure is increased at constant temperature, the gas compresses and λ decreases.

EXAMPLE 18.8 CALCULATING MEAN FREE PATH

(a) Estimate the mean free path of a molecule of air at 27°C and 1 atm. Model the molecules as spheres with radius $r = 2.0 \times 10^{-10}$ m. (b) Estimate the mean free time of an oxygen molecule with $v = v_{\text{rms}}$ at 27°C and 1 atm.

SOLUTION

IDENTIFY and SET UP: This problem uses the concepts of mean free path and mean free time (our target variables). We use Eq. (18.22)

to determine the mean free path λ. We then use $\lambda = vt_{\text{mean}}$ in Eq. (18.21), with $v = v_{\text{rms}}$, to find the mean free time t_{mean}.

EXECUTE: (a) From Eq. (18.22),

$$\lambda = \frac{kT}{4\pi\sqrt{2}r^2 p} = \frac{(1.38 \times 10^{-23}\ \text{J/K})(300\ \text{K})}{4\pi\sqrt{2}(2.0 \times 10^{-10}\ \text{m})^2(1.01 \times 10^5\ \text{Pa})}$$

$$= 5.8 \times 10^{-8}\ \text{m}$$

(b) From Example 18.6, for oxygen at 27°C the root-mean-square speed is $v_{\text{rms}} = 484$ m/s, so the mean free time for a molecule with this speed is

$$t_{\text{mean}} = \frac{\lambda}{v} = \frac{5.8 \times 10^{-8}\ \text{m}}{484\ \text{m/s}} = 1.2 \times 10^{-10}\ \text{s}$$

This molecule undergoes about 10^{10} collisions per second!

EVALUATE: Note that from Eqs. (18.21) and (18.22) the mean free *path* doesn't depend on the molecule's speed, but the mean free *time* does. Slower molecules have a longer average time interval t_{mean} between collisions than do fast ones, but the average *distance* λ between collisions is the same no matter what the molecule's speed. Our answer to part (a) says that the molecule doesn't go far between collisions, but the mean free path is still several hundred times the molecular radius r.

TEST YOUR UNDERSTANDING OF SECTION 18.3   Rank the following gases in order from (a) highest to lowest rms speed of molecules and (b) highest to lowest average translational kinetic energy of a molecule: (i) oxygen ($M = 32.0$ g/mol) at 300 K; (ii) nitrogen ($M = 28.0$ g/mol) at 300 K; (iii) oxygen at 330 K; (iv) nitrogen at 330 K. ❚

18.4 HEAT CAPACITIES

When we introduced the concept of heat capacity in Section 17.5, we talked about ways to *measure* the specific heat or molar heat capacity of a particular material. Now we'll see how to *predict* these on theoretical grounds.

Heat Capacities of Gases

The basis of our analysis is that heat is *energy* in transit. When we add heat to a substance, we are increasing its molecular energy. We'll assume that the volume of the gas remains constant; if we were to let the gas expand, it would do work by pushing on the moving walls of its container, and this additional energy transfer would have to be included in our calculations. We'll return to this more general case in Chapter 19. For now we are concerned with C_V, the molar heat capacity *at constant volume*.

In the simple kinetic-molecular model of Section 18.3 the molecular energy consists of only the translational kinetic energy K_{tr} of the pointlike molecules. This energy is directly proportional to the absolute temperature T, as shown by Eq. (18.14), $K_{\text{tr}} = \frac{3}{2}nRT$. When the temperature changes by a small amount dT, the corresponding change in kinetic energy is

$$dK_{\text{tr}} = \tfrac{3}{2}nR\, dT \qquad (18.23)$$

From the definition of molar heat capacity at constant volume, C_V (see Section 17.5), we also have

$$dQ = nC_V\, dT \qquad (18.24)$$

where dQ is the heat input needed for a temperature change dT. Now if K_{tr} represents the total molecular energy, as we have assumed, then dQ and dK_{tr} must be *equal* (**Fig. 18.17**). From Eqs. (18.23) and (18.24), this says

$$nC_V\, dT = \tfrac{3}{2}nR\, dT$$

Molar heat capacity ⋯⋯
at constant volume, $C_V = \tfrac{3}{2}R$ ⋯⋯ Gas constant   (18.25)
ideal gas of point particles

This surprisingly simple result says that the molar heat capacity at constant volume is $3R/2$ for *any* gas whose molecules can be represented as points.

Does Eq. (18.25) agree with experiment? In SI units, Eq. (18.25) gives

$$C_V = \tfrac{3}{2}(8.314\ \text{J/mol} \cdot \text{K}) = 12.47\ \text{J/mol} \cdot \text{K}$$

18.17 (a) A fixed volume V of a monatomic ideal gas. (b) When an amount of heat dQ is added to the gas, the total translational kinetic energy increases by $dK_{\text{tr}} = dQ$ and the temperature increases by $dT = dQ/nC_V$.

(a)

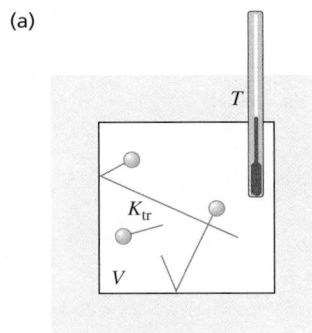

(b)

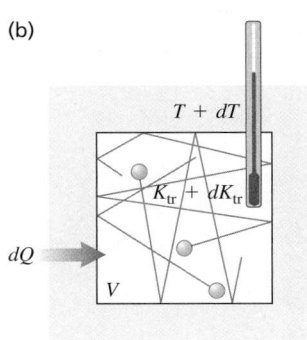

	Molar Heat Capacities of Gases	
TABLE 18.1		
Type of Gas	**Gas**	**C_V (J/mol · K)**
Monatomic	He	12.47
	Ar	12.47
Diatomic	H_2	20.42
	N_2	20.76
	O_2	20.85
	CO	20.85
Polyatomic	CO_2	28.46
	SO_2	31.39
	H_2S	25.95

18.18 Motions of a diatomic molecule.

(a) **Translational motion.** The molecule moves as a whole; its velocity may be described as the *x*-, *y*-, and *z*-velocity components of its center of mass.

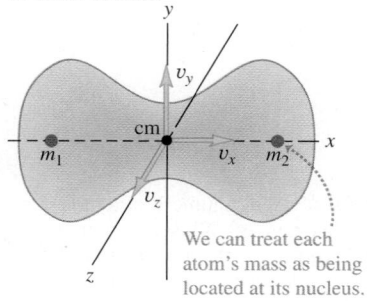

We can treat each atom's mass as being located at its nucleus.

(b) **Rotational motion.** The molecule rotates about its center of mass. This molecule has two independent axes of rotation.

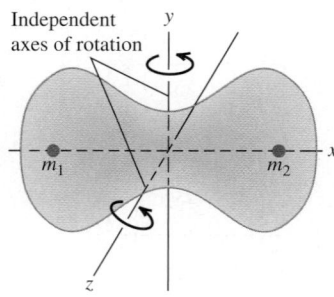

(c) **Vibrational motion.** The molecule oscillates as though the nuclei were connected by a spring.

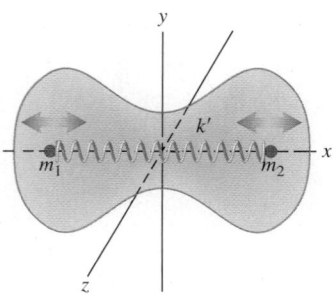

For comparison, **Table 18.1** gives measured values of C_V for several gases. We see that for *monatomic* gases our prediction is right on the money, but that it is way off for diatomic and polyatomic gases.

This comparison tells us that our point-molecule model is good enough for monatomic gases but that for diatomic and polyatomic molecules we need something more sophisticated. For example, we can picture a diatomic molecule as *two* point masses, like a little elastic dumbbell (see **Fig. 18.18**), with an interaction force between the atoms of the kind shown in Fig. 18.8. Such a molecule can have additional kinetic energy associated with *rotation* about axes through its center of mass. The atoms may also *vibrate* along the line joining them, with additional kinetic and potential energies.

When heat flows into a *monatomic* gas at constant volume, *all* of the added energy goes into an increase in random *translational* molecular kinetic energy. Equation (18.23) shows that this gives rise to an increase in temperature. But when the temperature is increased by the same amount in a *diatomic* or *polyatomic* gas, additional heat is needed to supply the increased rotational and vibrational energies. Thus polyatomic gases have *larger* molar heat capacities than monatomic gases, as Table 18.1 shows.

But how do we know how much energy is associated with each additional kind of motion of a complex molecule, compared to the translational kinetic energy? The new principle that we need is called the principle of **equipartition of energy.** It can be derived from sophisticated statistical-mechanics considerations; that derivation is beyond our scope, and we will treat the principle as an axiom.

The principle of equipartition of energy states that each velocity component (either linear or angular) has, on average, an associated kinetic energy per molecule of $\frac{1}{2}kT$, or one-half the product of the Boltzmann constant and the absolute temperature. The number of velocity components needed to describe the motion of a molecule completely is called the number of **degrees of freedom.** For a monatomic gas, there are three degrees of freedom (for the velocity components v_x, v_y, and v_z); this gives a total average kinetic energy per molecule of $3\left(\frac{1}{2}kT\right)$, consistent with Eq. (18.16).

For a *diatomic* molecule there are two possible axes of rotation, perpendicular to each other and to the molecule's axis. (We don't include rotation about the molecule's own axis because in ordinary collisions there is no way for this rotational motion to change.) If we add two rotational degrees of freedom for a diatomic molecule, the average total kinetic energy per molecule is $\frac{5}{2}kT$ instead of $\frac{3}{2}kT$. The total kinetic energy of n moles is $K_{\text{total}} = nN_A\left(\frac{5}{2}kT\right) = \frac{5}{2}n(kN_A)T = \frac{5}{2}nRT$, and the molar heat capacity (at constant volume) is

> Molar heat capacity at constant volume, ideal diatomic gas $\quad C_V = \frac{5}{2}R \quad$ Gas constant $\qquad$ (18.26)

In SI units,

$$C_V = \frac{5}{2}(8.314 \text{ J/mol} \cdot \text{K}) = 20.79 \text{ J/mol} \cdot \text{K}$$

This value is close to the measured values for diatomic gases in Table 18.1.

Vibrational motion can also contribute to the heat capacities of gases. Molecular bonds can stretch and bend, and the resulting vibrations lead to additional degrees of freedom and additional energies. For most diatomic gases, however, vibration does *not* contribute appreciably to heat capacity. The reason for this involves some concepts of quantum mechanics. Briefly, vibrational energy can change only in finite steps. If the energy change of the first step is much larger than the energy possessed by most molecules, then nearly all the molecules remain in the minimum-energy state of motion. Changing the temperature does not change their average vibrational energy appreciably, and the vibrational degrees of freedom are said to be "frozen out." In more complex molecules the

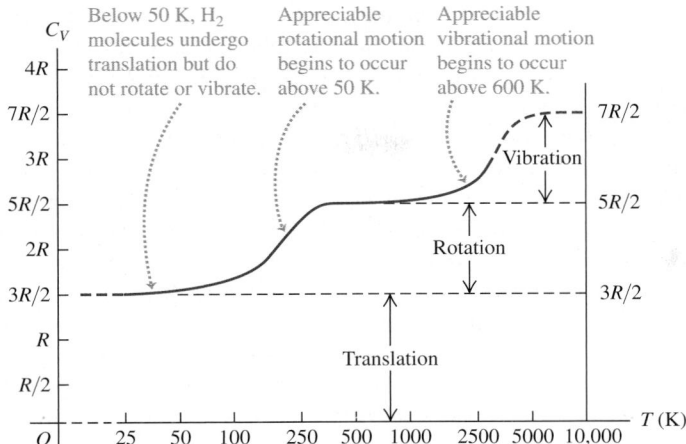

18.19 Experimental values of C_V, the molar heat capacity at constant volume, for hydrogen gas (H_2). The temperature is plotted on a logarithmic scale.

gaps between permitted energy levels can be much smaller, and then vibration *does* contribute to heat capacity. The rotational energy of a molecule also changes by finite steps, but they are usually much smaller; the "freezing out" of rotational degrees of freedom occurs only in rare instances.

In Table 18.1 the large values of C_V for polyatomic molecules show the effects of vibrational energy. In addition, a molecule with three or more atoms that are not in a straight line has *three* rotational degrees of freedom.

From this discussion we expect heat capacities to be temperature-dependent, generally increasing with increasing temperature. **Figure 18.19** is a graph of the temperature dependence of C_V for hydrogen gas (H_2), showing the temperatures at which the rotational and vibrational energies begin to contribute.

Heat Capacities of Solids

We can carry out a similar heat-capacity analysis for a crystalline solid. Consider a crystal consisting of N identical atoms (a *monatomic solid*). Each atom is bound to an equilibrium position by interatomic forces. Solid materials are elastic, so forces must permit stretching and bending of the bonds. We can think of a crystal as an array of atoms connected by little springs (**Fig. 18.20**).

Each atom can *vibrate* around its equilibrium position and has three degrees of freedom, corresponding to its three components of velocity. According to the equipartition principle, each atom has an average kinetic energy of $\frac{1}{2}kT$ for each degree of freedom. In addition, there is *potential* energy associated with the elastic deformation. For a simple harmonic oscillator (discussed in Chapter 14) it is not hard to show that the average kinetic energy is *equal* to the average potential energy. In our model of a crystal, each atom is a three-dimensional harmonic oscillator; it can be shown that the equality of average kinetic and potential energies also holds here, provided that the "spring" forces obey Hooke's law.

Thus we expect each atom to have an average kinetic energy $\frac{3}{2}kT$ and an average potential energy $\frac{3}{2}kT$, or an average total energy $3kT$ per atom. If the crystal contains N atoms or n moles, its total energy is

$$E_{\text{total}} = 3NkT = 3nRT \qquad (18.27)$$

From this we conclude that the molar heat capacity of a crystal should be

Molar heat capacity of an ideal monatomic solid (rule of Dulong and Petit) $C_V = 3R$ ········ Gas constant   (18.28)

18.20 To visualize the forces between neighboring atoms in a crystal, envision every atom as being attached to its neighbors by springs.

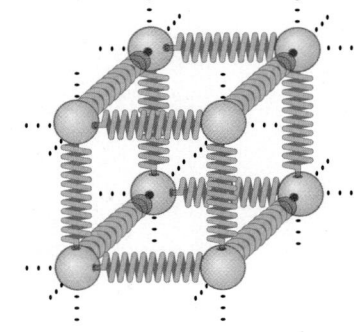

In SI units,

$$C_V = (3)(8.314 \text{ J/mol} \cdot \text{K}) = 24.9 \text{ J/mol} \cdot \text{K}$$

We have *derived* the **rule of Dulong and Petit,** which we encountered as an *empirical* finding in Section 17.5: Monatomic solids all have molar heat capacities of about 25 J/mol·K. The agreement is only approximate, but given the very simple nature of our model, it is quite significant.

At low temperatures, the heat capacities of most solids *decrease* with decreasing temperature (**Fig. 18.21**) for the same reason that vibrational degrees of freedom of molecules are frozen out at low temperatures. At very low temperatures the quantity kT is much *smaller* than the smallest energy step the vibrating atoms can take. Hence most of the atoms remain in their lowest energy states because the next higher energy level is out of reach. The average vibrational energy per atom is then *less* than $3kT$, and the heat capacity per molecule is *less* than $3k$. At higher temperatures when kT is *large* in comparison to the minimum energy step, the equipartition principle holds, and the total heat capacity is $3k$ per molecule or $3R$ per mole as the rule of Dulong and Petit predicts. Quantitative understanding of the temperature variation of heat capacities was one of the triumphs of quantum mechanics during its initial development in the 1920s.

18.21 Experimental values of C_V for lead, aluminum, silicon, and diamond. At high temperatures, C_V for each solid approaches about $3R$, in agreement with the rule of Dulong and Petit. At low temperatures, C_V is much less than $3R$.

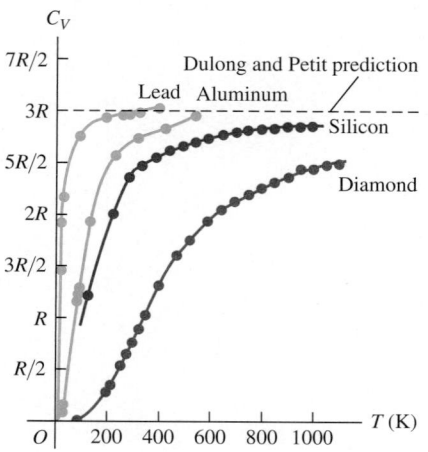

TEST YOUR UNDERSTANDING OF SECTION 18.4 A cylinder with a fixed volume contains hydrogen gas (H_2) at 25 K. You then add heat to the gas at a constant rate until its temperature reaches 500 K. Does the temperature of the gas increase at a constant rate? Why or why not? If not, does the temperature increase most rapidly near the beginning or near the end of this process? ∎

18.5 MOLECULAR SPEEDS

As we mentioned in Section 18.3, the molecules in a gas don't all have the same speed. **Figure 18.22** shows one experimental scheme for measuring the distribution of molecular speeds. A substance is vaporized in a hot oven; molecules of the vapor escape through an aperture in the oven wall and into a vacuum chamber. A series of slits blocks all molecules except those in a narrow beam, which is aimed at a pair of rotating disks. A molecule passing through the slit in the first disk is blocked by the second disk unless it arrives just as the slit in the second disk is lined up with the beam. The disks function as a speed selector that passes only molecules within a certain narrow speed range. This range can be varied by changing the disk rotation speed, and we can measure how many molecules lie within each of various speed ranges.

To describe the results of such measurements, we define a function $f(v)$ called a *distribution function*. If we observe a total of N molecules, the number dN having speeds in the range between v and $v + dv$ is given by

$$dN = Nf(v)\,dv \tag{18.29}$$

18.22 A molecule with a speed v passes through the slit in the first rotating disk. When the molecule reaches the second rotating disk, the disks have rotated through the offset angle θ. If $v = \omega x/\theta$, the molecule passes through the slit in the second rotating disk and reaches the detector.

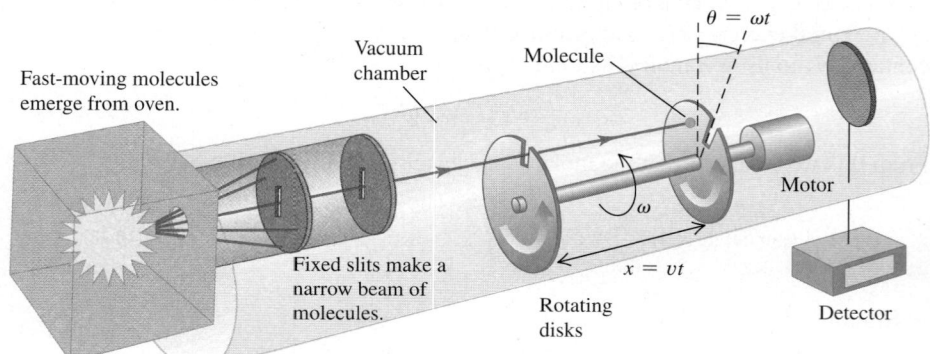

The *probability* that a randomly chosen molecule will have a speed in the interval v to $v + dv$ is $f(v)dv$. Hence $f(v)$ is the probability per unit speed *interval;* it is *not* the probability that a molecule has speed exactly equal to v. Since a probability is a pure number, $f(v)$ has units of reciprocal speed (s/m).

Figure 18.23a shows distribution functions for three different temperatures. At each temperature the height of the curve for any value of v is proportional to the number of molecules with speeds near v. The peak of the curve represents the *most probable speed* v_{mp} for the corresponding temperature. As the temperature increases, the average molecular kinetic energy increases, and so the peak of $f(v)$ shifts to higher and higher speeds.

Figure 18.23b shows that the area under a curve between any two values of v represents the fraction of all the molecules having speeds in that range. Every molecule must have *some* value of v, so the integral of $f(v)$ over all v must be unity for any T.

If we know $f(v)$, we can calculate the most probable speed v_{mp}, the average speed v_{av}, and the rms speed v_{rms}. To find v_{mp}, we simply find the point where $df/dv = 0$; this gives the value of the speed where the curve has its peak. To find v_{av}, we take the number $Nf(v)dv$ having speeds in each interval dv, multiply each number by the corresponding speed v, add all these products (by integrating over all v from zero to infinity), and finally divide by N. That is,

$$v_{av} = \int_0^\infty vf(v)\,dv \tag{18.30}$$

We can find the rms speed in a similar way; the average of v^2 is

$$(v^2)_{av} = \int_0^\infty v^2 f(v)\,dv \tag{18.31}$$

and v_{rms} is the square root of this.

The Maxwell–Boltzmann Distribution

The function $f(v)$ describing the actual distribution of molecular speeds is called the **Maxwell–Boltzmann distribution.** It can be derived from statistical-mechanics considerations, but that derivation is beyond our scope. Here is the result:

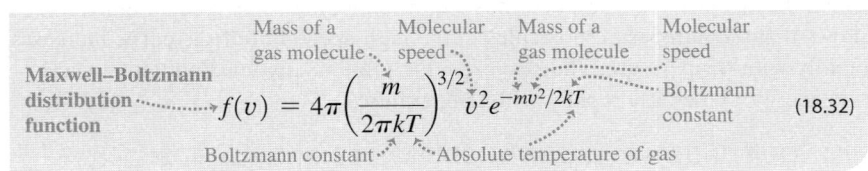

Maxwell–Boltzmann distribution function

$$f(v) = 4\pi \left(\frac{m}{2\pi kT} \right)^{3/2} v^2 e^{-mv^2/2kT} \tag{18.32}$$

Mass of a gas molecule · Molecular speed · Mass of a gas molecule · Molecular speed · Boltzmann constant · Boltzmann constant · Absolute temperature of gas

We can also express this function in terms of the translational kinetic energy of a molecule, which we denote by ϵ; that is, $\epsilon = \frac{1}{2}mv^2$. We invite you to verify that when this is substituted into Eq. (18.32), the result is

$$f(\epsilon) = \frac{8\pi}{m} \left(\frac{m}{2\pi kT} \right)^{3/2} \epsilon e^{-\epsilon/kT} \tag{18.33}$$

This form shows that the exponent in the Maxwell–Boltzmann distribution function is $-\epsilon/kT$; the shape of the curve is determined by the relative magnitude of ϵ and kT at any point. You can prove that the *peak* of each curve occurs where $\epsilon = kT$, corresponding to a most probable speed v_{mp}:

$$v_{mp} = \sqrt{\frac{2kT}{m}} \tag{18.34}$$

18.23 (a) Curves of the Maxwell–Boltzmann distribution function $f(v)$ for three temperatures. (b) The shaded areas under the curve represent the fractions of molecules within certain speed ranges. The most probable speed v_{mp} for a given temperature is at the peak of the curve.

(a)

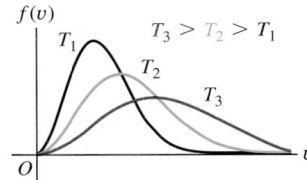

As temperature increases:
• the curve flattens.
• the maximum shifts to higher speeds.

(b)

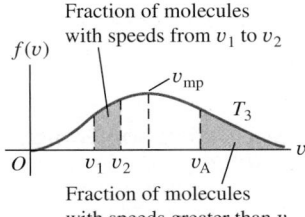

Fraction of molecules with speeds from v_1 to v_2

Fraction of molecules with speeds greater than v_A

TABLE 18.2	Fractions of Molecules in an Ideal Gas with Speeds Less Than Various Multiples of v_{rms}
v/v_{rms}	**Fraction**
0.20	0.011
0.40	0.077
0.60	0.218
0.80	0.411
1.00	0.608
1.20	0.771
1.40	0.882
1.60	0.947
1.80	0.979
2.00	0.993

BIO Application Activation Energy and Moth Activity This hawkmoth of genus *Manduca* cannot fly if the temperature of its muscles is below 29°C. The reason is that the enzyme-catalyzed reactions that power aerobic metabolism and enable muscle action require a minimum molecular energy (activation energy). Just like the molecules in an ideal gas, at low temperatures very few of the molecules involved in these reactions have high energy. As the temperature increases, more molecules have the required minimum energy and the reactions take place at a greater rate. Above 29°C, enough power is generated to allow the hawkmoth to fly.

To find the average speed, we substitute Eq. (18.32) into Eq. (18.30), make a change of variable $v^2 = x$, and integrate by parts. The result is

$$v_{av} = \sqrt{\frac{8kT}{\pi m}} \qquad (18.35)$$

Finally, to find the rms speed, we substitute Eq. (18.32) into Eq. (18.31). Evaluating the resulting integral takes some mathematical acrobatics, but we can find it in a table of integrals. The result is

$$v_{rms} = \sqrt{\frac{3kT}{m}} \qquad (18.36)$$

This result agrees with Eq. (18.19); it *must* agree if the Maxwell–Boltzmann distribution is to be consistent with our kinetic-theory calculations.

Table 18.2 shows the fraction of all the molecules in an ideal gas that have speeds *less than* various multiples of v_{rms}. These numbers were obtained by numerical integration; they are the same for all ideal gases.

The distribution of molecular speeds in liquids is similar, although not identical, to that for gases. We can understand evaporation and the vapor pressure of a liquid on this basis. Suppose a molecule must have a speed at least as great as v_A in Fig. 18.23b to escape from the surface of a liquid into the adjacent vapor. The number of such molecules, represented by the area under the "tail" of each curve (to the right of v_A), increases rapidly with temperature. Thus the rate at which molecules can escape is strongly temperature-dependent. This process is balanced by one in which molecules in the vapor phase collide inelastically with the surface and are trapped into the liquid phase. The number of molecules suffering this fate per unit time is proportional to the pressure in the vapor phase. Phase equilibrium between liquid and vapor occurs when these two competing processes proceed at the same rate. So if the molecular speed distributions are known for various temperatures, we can make a theoretical prediction of vapor pressure as a function of temperature. When liquid evaporates, it's the high-speed molecules that escape from the surface. The ones that are left have less energy on average; this gives us a molecular view of evaporative cooling.

Rates of chemical reactions are often strongly temperature-dependent, and the reason is contained in the Maxwell–Boltzmann distribution. When two reacting molecules collide, the reaction can occur only when the molecules are close enough for their electrons to interact strongly. This requires a minimum energy, called the *activation energy*, and thus a minimum molecular speed. Figure 18.23a shows that the number of molecules in the high-speed tail of the curve increases rapidly with temperature. Thus we expect the rate of any reaction with an activation energy to increase rapidly with temperature.

TEST YOUR UNDERSTANDING OF SECTION 18.5 A quantity of gas containing N molecules has a speed distribution function $f(v)$. How many molecules have speeds between v_1 and $v_2 > v_1$? (i) $\int_0^{v_2} f(v)\,dv - \int_0^{v_1} f(v)\,dv$; (ii) $N[\int_0^{v_2} f(v)\,dv - \int_0^{v_1} f(v)\,dv]$; (iii) $\int_0^{v_1} f(v)\,dv - \int_0^{v_2} f(v)\,dv$; (iv) $N[\int_0^{v_1} f(v)\,dv - \int_0^{v_2} f(v)\,dv]$; (v) none of these. ∎

18.6 PHASES OF MATTER

An ideal gas is the simplest system to analyze from a molecular viewpoint because we ignore the interactions between molecules. But those interactions are the very thing that makes matter condense into the liquid and solid phases under some conditions. So it's not surprising that theoretical analysis of liquid and solid structure and behavior is a lot more complicated than that for gases. We won't try to go far here with a microscopic picture, but we can talk in general about phases of matter, phase equilibrium, and phase transitions.

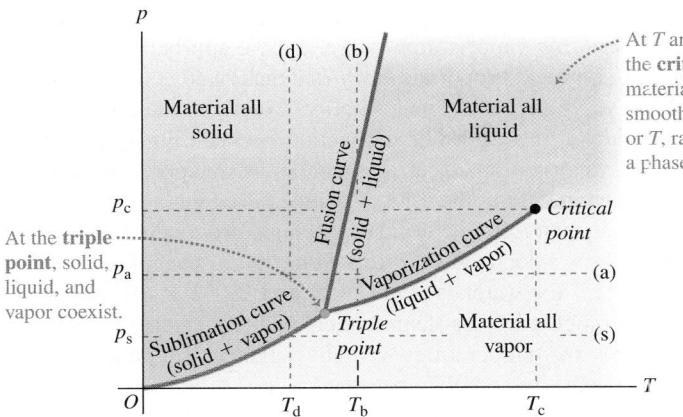

18.24 A typical pT phase diagram, showing regions of temperature and pressure at which the various phases exist and where phase changes occur.

In Section 17.6 we learned that each phase is stable in only certain ranges of temperature and pressure. A transition from one phase to another ordinarily requires **phase equilibrium** between the two phases, and for a given pressure this occurs at only one specific temperature. We can represent these conditions on a graph with axes p and T, called a **phase diagram; Fig. 18.24** shows an example. Each point on the diagram represents a pair of values of p and T.

Only a single phase can exist at each point in Fig. 18.24, except for points on the solid lines, where two phases can coexist in phase equilibrium. The fusion curve separates the solid and liquid areas and represents possible conditions of solid-liquid phase equilibrium. The vaporization curve separates the liquid and vapor areas, and the sublimation curve separates the solid and vapor areas. All three curves meet at the **triple point,** the only condition under which all three phases can coexist (**Fig. 18.25**). In Section 17.3 we used the triple-point temperature of water to define the Kelvin temperature scale. **Table 18.3** gives triple-point data for several substances.

If we heat a substance at a constant pressure p_a, it goes through a series of states represented by the horizontal line (a) in Fig. 18.24. The melting and boiling temperatures at this pressure are the temperatures at which the line intersects the fusion and vaporization curves, respectively. When the pressure is p_s, constant-pressure heating transforms a substance from solid directly to vapor. This process is called *sublimation;* the intersection of line (s) with the sublimation curve gives the temperature T_s at which it occurs for a pressure p_s. At any pressure less than the triple-point pressure, no liquid phase is possible. The triple-point pressure for carbon dioxide (CO_2) is 5.1 atm. At normal atmospheric pressure, solid CO_2 ("dry ice") undergoes sublimation; there is no liquid phase.

Line (b) in Fig. 18.24 represents compression at a constant temperature T_b. The material passes from vapor to liquid and then to solid at the points where line (b) crosses the vaporization curve and fusion curve, respectively. Line (d) shows constant-temperature compression at a lower temperature T_d; the material passes from vapor to solid at the point where line (d) crosses the sublimation curve.

We saw in the pV-diagram of Fig. 18.7 that a liquid-vapor phase transition occurs only when the temperature and pressure are less than those at the point at the top of the green shaded area labeled "Liquid-vapor phase equilibrium region." This point corresponds to the endpoint at the top of the vaporization curve in Fig. 18.24. It is called the **critical point,** and the corresponding values of p and T are called the critical pressure and temperature, p_c and T_c. A gas at a pressure *above* the critical pressure does not separate into two phases when it is cooled at constant pressure (along a horizontal line above the critical point in Fig. 18.24). Instead, its properties change gradually and continuously from those we ordinarily associate with a gas (low density, large compressibility) to those of a liquid (high density, small compressibility) *without a phase transition.*

18.25 Atmospheric pressure on earth is higher than the triple-point pressure of water (see line (a) in Fig. 18.24). Depending on the temperature, water can exist as a vapor (in the atmosphere), as a liquid (in the ocean), or as a solid (like the iceberg shown here).

TABLE 18.3	Triple-Point Data	
Substance	**Temperature (K)**	**Pressure (Pa)**
Hydrogen	13.80	0.0704×10^5
Deuterium	18.63	0.171×10^5
Neon	24.56	0.432×10^5
Nitrogen	63.18	0.125×10^5
Oxygen	54.36	0.00152×10^5
Ammonia	195.40	0.0607×10^5
Carbon dioxide	216.55	5.17×10^5
Sulfur dioxide	197.68	0.00167×10^5
Water	273.16	0.00610×10^5

You can understand this by thinking about liquid-phase transitions at successively higher points on the vaporization curve. As we approach the critical point, the *differences* in physical properties (such as density and compressibility) between the liquid and vapor phases become smaller. Exactly *at* the critical point they all become zero, and at this point the distinction between liquid and vapor disappears. The heat of vaporization also grows smaller as we approach the critical point, and it too becomes zero at the critical point.

For nearly all familiar materials the critical pressures are much greater than atmospheric pressure, so we don't observe this behavior in everyday life. For example, the critical point for water is at 647.4 K and 221.2×10^5 Pa (about 218 atm or 3210 psi). But high-pressure steam boilers in electric generating plants regularly run at pressures and temperatures well above the critical point.

Many substances can exist in more than one solid phase. A familiar example is carbon, which exists as noncrystalline soot and crystalline graphite and diamond. Water is another example; more than a dozen types of ice, differing in crystal structure and physical properties, have been observed at very high pressures.

pVT-Surfaces

We remarked in Section 18.1 that for any material, it can be useful to represent the equation of state as a surface in a three-dimensional space with coordinates p, V, and T. **Figure 18.26** shows a typical *pVT*-surface. The light lines represent *pV*-isotherms; projecting them onto the *pV*-plane gives a diagram similar to Fig. 18.7. The *pV*-isotherms represent contour lines on the *pVT*-surface, just as contour lines on a topographic map represent the elevation (the third dimension) at each point. The projections of the edges of the surface onto the *pT*-plane give the *pT* phase diagram of Fig. 18.24.

Line *abcdef* in Fig. 18.26 represents constant-pressure heating, with melting along *bc* and vaporization along *de*. Note the volume changes that occur as T increases along this line. Line *ghjklm* corresponds to an isothermal (constant temperature) compression, with liquefaction along *hj* and solidification along *kl*. Between these, segments *gh* and *jk* represent isothermal compression with increase in pressure; the pressure increases are much greater in the liquid region *jk* and the solid region *lm* than in the vapor region *gh*. Finally, line *nopq* represents isothermal solidification directly from vapor, as in the formation of snowflakes or frost.

18.26 A *pVT*-surface for a substance that expands on melting. Projections of the boundaries on the surface onto the *pT*- and *pV*-planes are also shown.

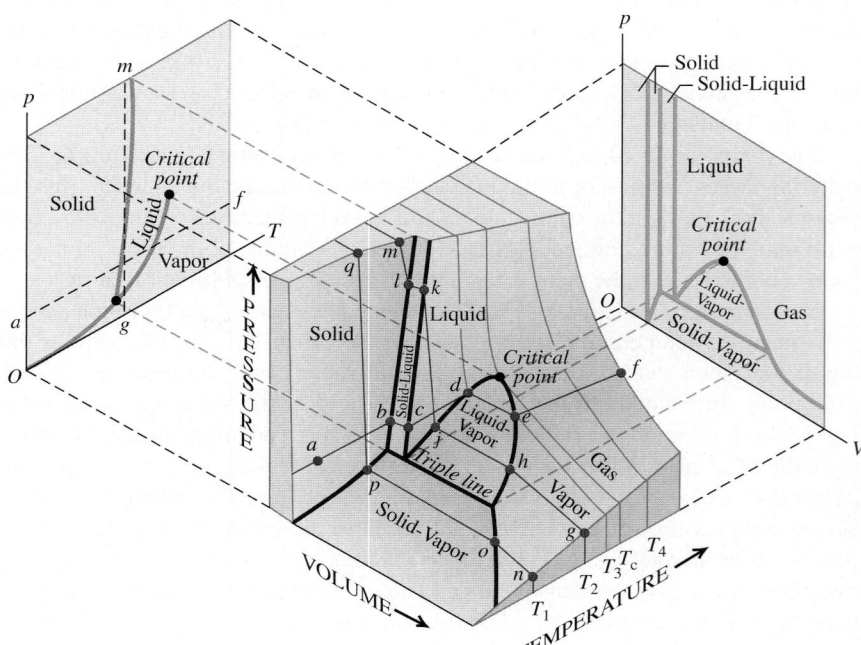

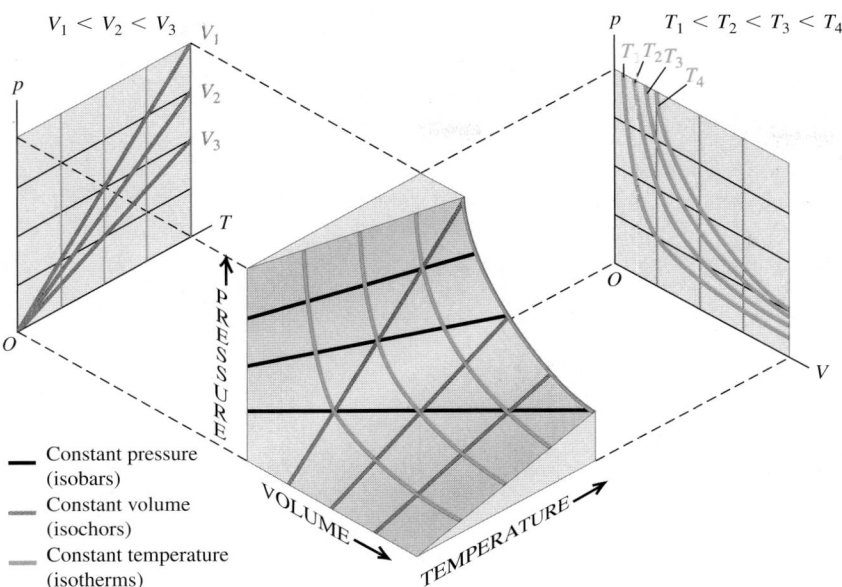

$V_1 < V_2 < V_3$ V_1
V_2
V_3
p
T
O

p $T_1 < T_2 < T_3 < T_4$
T_1 $T_2 T_3$
T_4
O
V

PRESSURE

VOLUME

TEMPERATURE

— Constant pressure
(isobars)
— Constant volume
(isochors)
— Constant temperature
(isotherms)

18.27 A pVT-surface for an ideal gas. At the left, each orange line corresponds to a certain constant volume; at the right, each green line corresponds to a certain constant temperature.

Figure 18.27 shows the much simpler pVT-surface for a substance that obeys the ideal-gas equation of state under all conditions. The projections of the constant-temperature curves onto the pV-plane correspond to the curves of Fig. 18.6, and the projections of the constant-volume curves onto the pT-plane show that pressure is directly proportional to absolute temperature. Figure 18.27 also shows the *isobars* (curves of constant pressure) and *isochors* (curves of constant volume) for an ideal gas.

TEST YOUR UNDERSTANDING OF SECTION 18.6 The average atmospheric pressure on Mars is 6.0×10^2 Pa. Could there be lakes or rivers of liquid water on Mars today? What about in the past, when the atmospheric pressure is thought to have been substantially greater than today? ❙

CHAPTER **18 SUMMARY**

SOLUTIONS TO ALL EXAMPLES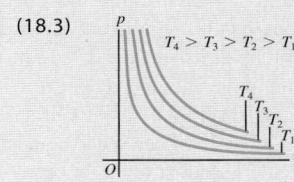

Equations of state: The pressure p, volume V, and absolute temperature T of a given quantity of a substance are related by an equation of state. This relationship applies only for equilibrium states, in which p and T are uniform throughout the system. The ideal-gas equation of state, Eq. (18.3), involves the number of moles n and a constant R that is the same for all gases. (See Examples 18.1–18.4.)

$$pV = nRT \qquad (18.3)$$

Molecular properties of matter: The molar mass M of a pure substance is the mass per mole. The mass m_{total} of a quantity of substance equals M multiplied by the number of moles n. Avogadro's number N_A is the number of molecules in a mole. The mass m of an individual molecule is M divided by N_A. (See Example 18.5.)

$$m_{total} = nM \qquad (18.2)$$
$$M = N_A m \qquad (18.3)$$

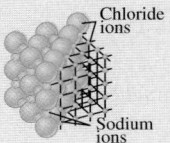

Kinetic-molecular model of an ideal gas: In an ideal gas, the total translational kinetic energy of the gas as a whole (K_{tr}) and the average translational kinetic energy per molecule [$\frac{1}{2}m(v^2)_{av}$] are proportional to the absolute temperature T, and the root-mean-square speed of molecules is proportional to the square root of T. These expressions involve the Boltzmann constant $k = R/N_A$. (See Examples 18.6 and 18.7.) The mean free path λ of molecules in an ideal gas depends on the number of molecules per volume (N/V) and the molecular radius r. (See Example 18.8.)

$$K_{tr} = \tfrac{3}{2}nRT \qquad (18.14)$$

$$\tfrac{1}{2}m(v^2)_{av} = \tfrac{3}{2}kT \qquad (18.16)$$

$$v_{rms} = \sqrt{(v^2)_{av}} = \sqrt{\frac{3kT}{m}} \qquad (18.19)$$

$$\lambda = vt_{mean} = \frac{V}{4\pi\sqrt{2}\,r^2 N} \qquad (18.21)$$

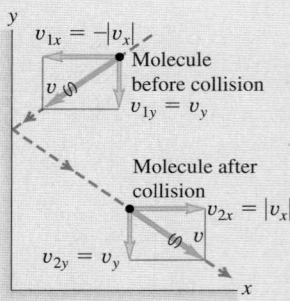

Heat capacities: The molar heat capacity at constant volume C_V is a simple multiple of the gas constant R for certain idealized cases: an ideal monatomic gas [Eq. (18.25)]; an ideal diatomic gas including rotational energy [Eq. (18.26)]; and an ideal monatomic solid [Eq. (18.28)]. Many real systems are approximated well by these idealizations.

$$C_V = \tfrac{3}{2}R \text{ (monatomic gas)} \qquad (18.25)$$

$$C_V = \tfrac{5}{2}R \text{ (diatomic gas)} \qquad (18.26)$$

$$C_V = 3R \text{ (monatomic solid)} \qquad (18.28)$$

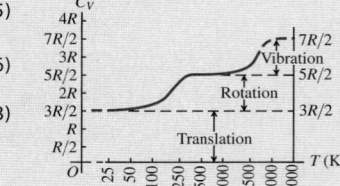

Molecular speeds: The speeds of molecules in an ideal gas are distributed according to the Maxwell–Boltzmann distribution $f(v)$. The quantity $f(v)\,dv$ describes what fraction of the molecules have speeds between v and $v + dv$.

$$f(v) = 4\pi\left(\frac{m}{2\pi kT}\right)^{3/2} v^2 e^{-mv^2/2kT} \qquad (18.32)$$

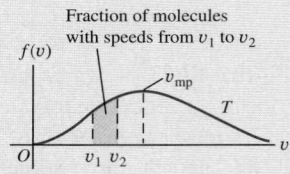

Phases of matter: Ordinary matter exists in the solid, liquid, and gas phases. A phase diagram shows conditions under which two phases can coexist in phase equilibrium. All three phases can coexist at the triple point. The vaporization curve ends at the critical point, above which the distinction between the liquid and gas phases disappears.

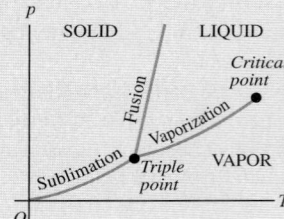

SOLUTION

BRIDGING PROBLEM GAS ON JUPITER'S MOON EUROPA

An astronaut visiting Jupiter's satellite Europa leaves a canister of 1.20 mol of nitrogen gas (28.0 g/mol) at 25.0°C on the satellite's surface. Europa has no significant atmosphere, and the acceleration due to gravity at its surface is 1.30 m/s². The canister springs a leak, allowing molecules to escape from a small hole. (a) What is the maximum height (in km) above Europa's surface that is reached by a nitrogen molecule whose speed equals the rms speed? Assume that the molecule is shot straight up out of the hole in the canister, and ignore the variation in g with altitude. (b) The escape speed from Europa is 2025 m/s. Can any of the nitrogen molecules escape from Europa and into space?

SOLUTION GUIDE

IDENTIFY AND SET UP

1. Sketch the situation, showing all relevant dimensions.
2. List the unknown quantities, and decide which are the target variables.

3. How will you find the rms speed of the nitrogen molecules? What principle will you use to find the maximum height that a molecule with this speed can reach?
4. Does the rms molecular speed in the gas represent the maximum molecular speed? If not, what is the maximum speed?

EXECUTE

5. Solve for the rms speed. Use this to calculate the maximum height that a molecule with this speed can reach.
6. Use your result from step 5 to answer the question in part (b).

EVALUATE

7. Do your results depend on the amount of gas in the container? Why or why not?
8. How would your results from steps 5 and 6 be affected if the gas cylinder were instead left on Jupiter's satellite Ganymede, which has higher surface gravity than Europa and a higher escape speed? Like Europa, Ganymede has no significant atmosphere.

Problems

•, ••, •••: Difficulty levels. CP: Cumulative problems incorporating material from earlier chapters. CALC: Problems requiring calculus. DATA: Problems involving real data, scientific evidence, experimental design, and/or statistical reasoning. BIO: Biosciences problems.

DISCUSSION QUESTIONS

Q18.1 Section 18.1 states that ordinarily, pressure, volume, and temperature cannot change individually without one affecting the others. Yet when a liquid evaporates, its volume changes, even though its pressure and temperature are constant. Is this inconsistent? Why or why not?

Q18.2 In the ideal-gas equation, could an equivalent Celsius temperature be used instead of the Kelvin one if an appropriate numerical value of the constant R is used? Why or why not?

Q18.3 When a car is driven some distance, the air pressure in the tires increases. Why? Should you let out some air to reduce the pressure? Why or why not?

Q18.4 The coolant in an automobile radiator is kept at a pressure higher than atmospheric pressure. Why is this desirable? The radiator cap will release coolant when the gauge pressure of the coolant reaches a certain value, typically 15 lb/in.² or so. Why not just seal the system completely?

Q18.5 Unwrapped food placed in a freezer experiences dehydration, known as "freezer burn." Why?

Q18.6 A group of students drove from their university (near sea level) up into the mountains for a skiing weekend. Upon arriving at the slopes, they discovered that the bags of potato chips they had brought for snacks had all burst open. What caused this to happen?

Q18.7 The derivation of the ideal-gas equation included the assumption that the number of molecules is very large, so that we could compute the average force due to many collisions. However, the ideal-gas equation holds accurately only at low pressures, where the molecules are few and far between. Is this inconsistent? Why or why not?

Q18.8 A rigid, perfectly insulated container has a membrane dividing its volume in half. One side contains a gas at an absolute temperature T_0 and pressure p_0, while the other half is completely empty. Suddenly a small hole develops in the membrane, allowing the gas to leak out into the other half until it eventually occupies twice its original volume. In terms of T_0 and p_0, what will be the new temperature and pressure of the gas when it is distributed equally in both halves of the container? Explain your reasoning.

Q18.9 (a) Which has more atoms: a kilogram of hydrogen or a kilogram of lead? Which has more mass? (b) Which has more atoms: a mole of hydrogen or a mole of lead? Which has more mass? Explain your reasoning.

Q18.10 Use the concepts of the kinetic-molecular model to explain: (a) why the pressure of a gas in a rigid container increases as heat is added to the gas and (b) why the pressure of a gas increases as we compress it, even if we do not change its temperature.

Q18.11 The proportions of various gases in the earth's atmosphere change somewhat with altitude. Would you expect the proportion of oxygen at high altitude to be greater or less than at sea level compared to the proportion of nitrogen? Why?

Q18.12 Comment on the following statement: *When two gases are mixed, if they are to be in thermal equilibrium, they must have the same average molecular speed.* Is the statement correct? Why or why not?

Q18.13 The kinetic-molecular model contains a hidden assumption about the temperature of the container walls. What is this assumption? What would happen if this assumption were not valid?

Q18.14 The temperature of an ideal gas is directly proportional to the average kinetic energy of its molecules. If a container of ideal gas is moving past you at 2000 m/s, is the temperature of the gas higher than if the container was at rest? Explain your reasoning.

Q18.15 If the pressure of an ideal monatomic gas is increased while the number of moles is kept constant, what happens to the average translational kinetic energy of one atom of the gas? Is it possible to change *both* the volume and the pressure of an ideal gas and keep the average translational kinetic energy of the atoms constant? Explain.

Q18.16 In deriving the ideal-gas equation from the kinetic-molecular model, we ignored potential energy due to the earth's gravity. Is this omission justified? Why or why not?

Q18.17 Imagine a special air filter placed in a window of a house. The tiny holes in the filter allow only air molecules moving faster than a certain speed to exit the house, and allow only air molecules moving slower than that speed to enter the house from outside. What effect would this filter have on the temperature inside the house? (It turns out that the second law of thermodynamics—which we will discuss in Chapter 20—tells us that such a wonderful air filter would be impossible to make.)

Q18.18 A gas storage tank has a small leak. The pressure in the tank drops more quickly if the gas is hydrogen or helium than if it is oxygen. Why?

Q18.19 Consider two specimens of ideal gas at the same temperature. Specimen A has the same total mass as specimen B, but the molecules in specimen A have greater molar mass than they do in specimen B. In which specimen is the total kinetic energy of the gas greater? Does your answer depend on the molecular structure of the gases? Why or why not?

Q18.20 The temperature of an ideal monatomic gas is increased from 25°C to 50°C. Does the average translational kinetic energy of each gas atom double? Explain. If your answer is no, what would the final temperature be if the average translational kinetic energy was doubled?

Q18.21 If the root-mean-square speed of the atoms of an ideal gas is to be doubled, by what factor must the Kelvin temperature of the gas be increased? Explain.

Q18.22 (a) If you apply the same amount of heat to 1.00 mol of an ideal monatomic gas and 1.00 mol of an ideal diatomic gas, which one (if any) will increase more in temperature? (b) Physically, *why* do diatomic gases have a greater molar heat capacity than monatomic gases?

Q18.23 The discussion in Section 18.4 concluded that all ideal monatomic gases have the same heat capacity C_V. Does this mean that it takes the same amount of heat to raise the temperature of 1.0 g of each one by 1.0 K? Explain your reasoning.

Q18.24 In a gas that contains N molecules, is it accurate to say that the number of molecules with speed v is equal to $f(v)$? Is it accurate to say that this number is given by $Nf(v)$? Explain your answers.

Q18.25 The atmosphere of the planet Mars is 95.3% carbon dioxide (CO_2) and about 0.03% water vapor. The atmospheric pressure is only about 600 Pa, and the surface temperature varies from -30°C to -100°C. The polar ice caps contain both CO_2 ice and water ice. Could there be *liquid* CO_2 on the surface of Mars? Could there be liquid water? Why or why not?

Q18.26 A beaker of water at room temperature is placed in an enclosure, and the air pressure in the enclosure is slowly reduced. When the air pressure is reduced sufficiently, the water begins to boil. The temperature of the water does not rise when it boils; in fact, the temperature *drops* slightly. Explain these phenomena.

Q18.27 Ice is slippery to walk on, and especially slippery if you wear ice skates. What does this tell you about how the melting temperature of ice depends on pressure? Explain.

Q18.28 Hydrothermal vents are openings in the ocean floor that discharge very hot water. The water emerging from one such vent off the Oregon coast, 2400 m below the surface, is at 279°C. Despite its high temperature, the water doesn't boil. Why not?

Q18.29 The dark areas on the moon's surface are called *maria*, Latin for "seas," and were once thought to be bodies of water. In fact, the maria are not "seas" at all, but plains of solidified lava. Given that there is no atmosphere on the moon, how can you explain the absence of liquid water on the moon's surface?

Q18.30 In addition to the normal cooking directions printed on the back of a box of rice, there are also "high-altitude directions." The only difference is that the "high-altitude directions" suggest increasing the cooking time and using a greater volume of boiling water in which to cook the rice. Why should the directions depend on the altitude in this way?

EXERCISES

Section 18.1 Equations of State

18.1 • A 20.0-L tank contains 4.86×10^{-4} kg of helium at 18.0°C. The molar mass of helium is 4.00 g/mol. (a) How many moles of helium are in the tank? (b) What is the pressure in the tank, in pascals and in atmospheres?

18.2 ••• Helium gas with a volume of 3.20 L, under a pressure of 0.180 atm and at 41.0°C, is warmed until both pressure and volume are doubled. (a) What is the final temperature? (b) How many grams of helium are there? The molar mass of helium is 4.00 g/mol.

18.3 • A cylindrical tank has a tight-fitting piston that allows the volume of the tank to be changed. The tank originally contains 0.110 m^3 of air at a pressure of 0.355 atm. The piston is slowly pulled out until the volume of the gas is increased to 0.390 m^3. If the temperature remains constant, what is the final value of the pressure?

18.4 • A 3.00-L tank contains air at 3.00 atm and 20.0°C. The tank is sealed and cooled until the pressure is 1.00 atm. (a) What is the temperature then in degrees Celsius? Assume that the volume of the tank is constant. (b) If the temperature is kept at the value found in part (a) and the gas is compressed, what is the volume when the pressure again becomes 3.00 atm?

18.5 • **Planetary Atmospheres.** (a) Calculate the density of the atmosphere at the surface of Mars (where the pressure is 650 Pa and the temperature is typically 253 K, with a CO_2 atmosphere), Venus (with an average temperature of 730 K and pressure of 92 atm, with a CO_2 atmosphere), and Saturn's moon Titan (where the pressure is 1.5 atm and the temperature is −178°C, with a N_2 atmosphere). (b) Compare each of these densities with that of the earth's atmosphere, which is 1.20 kg/m^3. Consult Appendix D to determine molar masses.

18.6 •• You have several identical balloons. You experimentally determine that a balloon will break if its volume exceeds 0.900 L. The pressure of the gas inside the balloon equals air pressure (1.00 atm). (a) If the air inside the balloon is at a constant 22.0°C and behaves as an ideal gas, what mass of air can you blow into one of the balloons before it bursts? (b) Repeat part (a) if the gas is helium rather than air.

18.7 •• A Jaguar XK8 convertible has an eight-cylinder engine. At the beginning of its compression stroke, one of the cylinders contains 499 cm^3 of air at atmospheric pressure (1.01×10^5 Pa) and a temperature of 27.0°C. At the end of the stroke, the air has been compressed to a volume of 46.2 cm^3 and the gauge pressure has increased to 2.72×10^6 Pa. Compute the final temperature.

18.8 •• A welder using a tank of volume 0.0750 m^3 fills it with oxygen (molar mass 32.0 g/mol) at a gauge pressure of 3.00×10^5 Pa and temperature of 37.0°C. The tank has a small leak, and in time some of the oxygen leaks out. On a day when the temperature is 22.0°C, the gauge pressure of the oxygen in the tank is 1.80×10^5 Pa. Find (a) the initial mass of oxygen and (b) the mass of oxygen that has leaked out.

18.9 •• A large cylindrical tank contains 0.750 m^3 of nitrogen gas at 27°C and 7.50×10^3 Pa (absolute pressure). The tank has a tight-fitting piston that allows the volume to be changed. What will be the pressure if the volume is decreased to 0.410 m^3 and the temperature is increased to 157°C?

18.10 • An empty cylindrical canister 1.50 m long and 90.0 cm in diameter is to be filled with pure oxygen at 22.0°C to store in a space station. To hold as much gas as possible, the absolute pressure of the oxygen will be 21.0 atm. The molar mass of oxygen is 32.0 g/mol. (a) How many moles of oxygen does this canister hold? (b) For someone lifting this canister, by how many kilograms does this gas increase the mass to be lifted?

18.11 • The gas inside a balloon will always have a pressure nearly equal to atmospheric pressure, since that is the pressure applied to the outside of the balloon. You fill a balloon with helium (a nearly ideal gas) to a volume of 0.600 L at 19.0°C. What is the volume of the balloon if you cool it to the boiling point of liquid nitrogen (77.3 K)?

18.12 • An ideal gas has a density of 1.33×10^{-6} g/cm^3 at 1.00×10^{-3} atm and 20.0°C. Identify the gas.

18.13 •• If a certain amount of ideal gas occupies a volume V at STP on earth, what would be its volume (in terms of V) on Venus, where the temperature is 1003°C and the pressure is 92 atm?

18.14 • A diver observes a bubble of air rising from the bottom of a lake (where the absolute pressure is 3.50 atm) to the surface (where the pressure is 1.00 atm). The temperature at the bottom is 4.0°C, and the temperature at the surface is 23.0°C. (a) What is the ratio of the volume of the bubble as it reaches the surface to its volume at the bottom? (b) Would it be safe for the diver to hold his breath while ascending from the bottom of the lake to the surface? Why or why not?

18.15 • A metal tank with volume 3.10 L will burst if the absolute pressure of the gas it contains exceeds 100 atm. (a) If 11.0 mol of an ideal gas is put into the tank at 23.0°C, to what temperature can the gas be warmed before the tank ruptures? Ignore the thermal expansion of the tank. (b) Based on your answer to part (a), is it reasonable to ignore the thermal expansion of the tank? Explain.

18.16 • Three moles of an ideal gas are in a rigid cubical box with sides of length 0.300 m. (a) What is the force that the gas exerts on each of the six sides of the box when the gas temperature is 20.0°C? (b) What is the force when the temperature of the gas is increased to 100.0°C?

18.17 • With the assumptions of Example 18.4 (Section 18.1), at what elevation above sea level is air pressure 90% of the pressure at sea level?

18.18 •• With the assumption that the air temperature is a uniform 0.0°C, what is the density of the air at an altitude of 1.00 km as a percentage of the density at the surface?

18.19 •• (a) Calculate the mass of nitrogen present in a volume of 3000 cm^3 if the gas is at 22.0°C and the absolute pressure of 2.00×10^{-13} atm is a partial vacuum easily obtained in laboratories. (b) What is the density (in kg/m^3) of the N_2?

18.20 • At an altitude of 11,000 m (a typical cruising altitude for a jet airliner), the air temperature is −56.5°C and the air density is 0.364 kg/m^3. What is the pressure of the atmosphere at that altitude? (*Note:* The temperature at this altitude is not the same as at the surface of the earth, so the calculation of Example 18.4 in Section 18.1 doesn't apply.)

Section 18.2 Molecular Properties of Matter

18.21 • How many moles are in a 1.00-kg bottle of water? How many molecules? The molar mass of water is 18.0 g/mol.

18.22 • A large organic molecule has a mass of 1.41×10^{-21} kg. What is the molar mass of this compound?

18.23 •• Modern vacuum pumps make it easy to attain pressures of the order of 10^{-13} atm in the laboratory. Consider a volume of air and treat the air as an ideal gas. (a) At a pressure of 9.00×10^{-14} atm and an ordinary temperature of 300.0 K, how many molecules are present in a volume of 1.00 cm^3? (b) How many molecules would be present at the same temperature but at 1.00 atm instead?

18.24 •• The Lagoon Nebula (**Fig. E18.24**) is a cloud of hydrogen gas located 3900 light-years from the earth. The cloud is about 45 light-years in diameter and glows because of its high temperature of 7500 K. (The gas is raised to this temperature by the stars that lie within the nebula.) The cloud is also very thin; there are only 80 molecules per cubic centimeter. (a) Find the gas pressure (in atmospheres) in the Lagoon Nebula. Compare it to the laboratory pressure referred to in Exercise 18.23. (b) Science-fiction films sometimes show starships being buffeted by turbulence as they fly through gas clouds such as the Lagoon Nebula. Does this seem realistic? Why or why not?

Figure **E18.24**

18.25 •• In a gas at standard conditions, what is the length of the side of a cube that contains a number of molecules equal to the population of the earth (about 7×10^9 people)?

18.26 •• **How Close Together Are Gas Molecules?** Consider an ideal gas at 27°C and 1.00 atm. To get some idea how close these molecules are to each other, on the average, imagine them to be uniformly spaced, with each molecule at the center of a small cube. (a) What is the length of an edge of each cube if adjacent cubes touch but do not overlap? (b) How does this distance compare with the diameter of a typical molecule? (c) How does their separation compare with the spacing of atoms in solids, which typically are about 0.3 nm apart?

Section 18.3 Kinetic-Molecular Model of an Ideal Gas

18.27 • (a) What is the total translational kinetic energy of the air in an empty room that has dimensions 8.00 m × 12.00 m × 4.00 m if the air is treated as an ideal gas at 1.00 atm? (b) What is the speed of a 2000-kg automobile if its kinetic energy equals the translational kinetic energy calculated in part (a)?

18.28 • A flask contains a mixture of neon (Ne), krypton (Kr), and radon (Rn) gases. Compare (a) the average kinetic energies of the three types of atoms and (b) the root-mean-square speeds. (*Hint:* Appendix D shows the molar mass (in g/mol) of each element under the chemical symbol for that element.)

18.29 • We have two equal-size boxes, *A* and *B*. Each box contains gas that behaves as an ideal gas. We insert a thermometer into each box and find that the gas in box *A* is at 50°C while the gas in box *B* is at 10°C. This is all we know about the gas in the boxes. Which of the following statements *must* be true? Which *could* be true? Explain your reasoning. (a) The pressure in *A* is higher than in *B*. (b) There are more molecules in *A* than in *B*. (c) *A* and *B* do not contain the same type of gas. (d) The molecules in *A* have more average kinetic energy per molecule than those in *B*. (e) The molecules in *A* are moving faster than those in *B*.

18.30 • A container with volume 1.64 L is initially evacuated. Then it is filled with 0.226 g of N_2. Assume that the pressure of the gas is low enough for the gas to obey the ideal-gas law to a high degree of accuracy. If the root-mean-square speed of the gas molecules is 182 m/s, what is the pressure of the gas?

18.31 •• (a) A deuteron, ^2_1H, is the nucleus of a hydrogen isotope and consists of one proton and one neutron. The plasma of deuterons in a nuclear fusion reactor must be heated to about 300 million K. What is the rms speed of the deuterons? Is this a significant fraction of the speed of light in vacuum ($c = 3.0 \times 10^8$ m/s)? (b) What would the temperature of the plasma be if the deuterons had an rms speed equal to 0.10*c*?

18.32 • **Martian Climate.** The atmosphere of Mars is mostly CO_2 (molar mass 44.0 g/mol) under a pressure of 650 Pa, which we shall assume remains constant. In many places the temperature varies from 0.0°C in summer to −100°C in winter. Over the course of a Martian year, what are the ranges of (a) the rms speeds of the CO_2 molecules and (b) the density (in mol/m^3) of the atmosphere?

18.33 •• Oxygen (O_2) has a molar mass of 32.0 g/mol. What is (a) the average translational kinetic energy of an oxygen molecule at a temperature of 300 K; (b) the average value of the square of its speed; (c) the root-mean-square speed; (d) the momentum of an oxygen molecule traveling at this speed? (e) Suppose an oxygen molecule traveling at this speed bounces back and forth between opposite sides of a cubical vessel 0.10 m on a side. What is the average force the molecule exerts on one of the walls of the container? (Assume that the molecule's velocity is perpendicular to the two sides that it strikes.) (f) What is the average force per unit area? (g) How many oxygen molecules traveling at this speed are necessary to produce an average pressure of 1 atm? (h) Compute the number of oxygen molecules that are contained in a vessel of this size at 300 K and atmospheric pressure. (i) Your answer for part (h) should be three times as large as the answer for part (g). Where does this discrepancy arise?

18.34 •• Calculate the mean free path of air molecules at 3.50×10^{-13} atm and 300 K. (This pressure is readily attainable in the laboratory; see Exercise 18.23.) As in Example 18.8, model the air molecules as spheres of radius 2.0×10^{-10} m.

18.35 •• At what temperature is the root-mean-square speed of nitrogen molecules equal to the root-mean-square speed of hydrogen molecules at 20.0°C? (*Hint:* Appendix D shows the molar mass (in g/mol) of each element under the chemical symbol for that element. The molar mass of H_2 is twice the molar mass of hydrogen atoms, and similarly for N_2.)

18.36 • Smoke particles in the air typically have masses of the order of 10^{-16} kg. The Brownian motion (rapid, irregular movement) of these particles, resulting from collisions with air molecules, can be observed with a microscope. (a) Find the root-mean-square speed of Brownian motion for a particle with a mass of 3.00×10^{-16} kg in air at 300 K. (b) Would the root-mean-square speed be different if the particle were in hydrogen gas at the same temperature? Explain.

Section 18.4 Heat Capacities

18.37 • How much heat does it take to increase the temperature of 1.80 mol of an ideal gas by 50.0 K near room temperature if the gas is held at constant volume and is (a) diatomic; (b) monatomic?

18.38 •• Perfectly rigid containers each hold n moles of ideal gas, one being hydrogen (H_2) and the other being neon (Ne). If it takes 300 J of heat to increase the temperature of the hydrogen by 2.50°C, by how many degrees will the same amount of heat raise the temperature of the neon?

18.39 •• (a) Compute the specific heat at constant volume of nitrogen (N_2) gas, and compare it with the specific heat of liquid water. The molar mass of N_2 is 28.0 g/mol. (b) You warm 1.00 kg of water at a constant volume of 1.00 L from 20.0°C to 30.0°C in a kettle. For the same amount of heat, how many kilograms of 20.0°C air would you be able to warm to 30.0°C? What volume (in liters) would this air occupy at 20.0°C and a pressure of 1.00 atm? Make the simplifying assumption that air is 100% N_2.

18.40 •• (a) Calculate the specific heat at constant volume of water vapor, assuming the nonlinear triatomic molecule has three translational and three rotational degrees of freedom and that vibrational motion does not contribute. The molar mass of water is 18.0 g/mol. (b) The actual specific heat of water vapor at low pressures is about 2000 J/kg·K. Compare this with your calculation and comment on the actual role of vibrational motion.

Section 18.5 Molecular Speeds

18.41 • For diatomic carbon dioxide gas (CO_2, molar mass 44.0 g/mol) at $T = 300$ K, calculate (a) the most probable speed v_{mp}; (b) the average speed v_{av}; (c) the root-mean-square speed v_{rms}.

18.42 • For a gas of nitrogen molecules (N_2), what must the temperature be if 94.7% of all the molecules have speeds less than (a) 1500 m/s; (b) 1000 m/s; (c) 500 m/s? Use Table 18.2. The molar mass of N_2 is 28.0 g/mol.

Section 18.6 Phases of Matter

18.43 • Solid water (ice) is slowly warmed from a very low temperature. (a) What minimum external pressure p_1 must be applied to the solid if a melting phase transition is to be observed? Describe the sequence of phase transitions that occur if the applied pressure p is such that $p < p_1$. (b) Above a certain maximum pressure p_2, no boiling transition is observed. What is this pressure? Describe the sequence of phase transitions that occur if $p_1 < p < p_2$.

18.44 • **Meteorology.** The *vapor pressure* is the pressure of the vapor phase of a substance when it is in equilibrium with the solid or liquid phase of the substance. The *relative humidity* is the partial pressure of water vapor in the air divided by the vapor pressure of water at that same temperature, expressed as a percentage. The air is saturated when the humidity is 100%. (a) The vapor pressure of water at 20.0°C is 2.34×10^3 Pa. If the air temperature is 20.0°C and the relative humidity is 60%, what is the partial pressure of water vapor in the atmosphere (that is, the pressure due to water vapor alone)? (b) Under the conditions of part (a), what is the mass of water in 1.00 m³ of air? (The molar mass of water is 18.0 g/mol. Assume that water vapor can be treated as an ideal gas.)

18.45 • Calculate the volume of 1.00 mol of liquid water at 20°C (at which its density is 998 kg/m³), and compare that with the volume occupied by 1.00 mol of water at the critical point, which is 56×10^{-6} m³. Water has a molar mass of 18.0 g/mol.

PROBLEMS

18.46 • A physics lecture room at 1.00 atm and 27.0°C has a volume of 216 m³. (a) Use the ideal-gas law to estimate the number of air molecules in the room. Assume that all of the air is N_2. Calculate (b) the particle density—that is, the number of N_2 molecules per cubic centimeter—and (c) the mass of the air in the room.

18.47 •• CP BIO **The Effect of Altitude on the Lungs.** (a) Calculate the *change* in air pressure you will experience if you climb a 1000-m mountain, assuming that the temperature and air density do not change over this distance and that they were 22°C and 1.2 kg/m³, respectively, at the bottom of the mountain. (*Note:* The result of Example 18.4 doesn't apply, since the expression derived in that example accounts for the variation of air density with altitude and we are told to ignore that here.) (b) If you took a 0.50-L breath at the foot of the mountain and managed to hold it until you reached the top, what would be the volume of this breath when you exhaled it there?

18.48 •• CP BIO **The Bends.** If deep-sea divers rise to the surface too quickly, nitrogen bubbles in their blood can expand and prove fatal. This phenomenon is known as the *bends*. If a scuba diver rises quickly from a depth of 25 m in Lake Michigan (which is fresh water), what will be the volume at the surface of an N_2 bubble that occupied 1.0 mm³ in his blood at the lower depth? Does it seem that this difference is large enough to be a problem? (Assume that the pressure difference is due to only the changing water pressure, not to any temperature difference. This assumption is reasonable, since we are warm-blooded creatures.)

18.49 ••• CP A hot-air balloon stays aloft because hot air at atmospheric pressure is less dense than cooler air at the same pressure. If the volume of the balloon is 500.0 m³ and the surrounding air is at 15.0°C, what must the temperature of the air in the balloon be for it to lift a total load of 290 kg (in addition to the mass of the hot air)? The density of air at 15.0°C and atmospheric pressure is 1.23 kg/m³.

18.50 •• In an evacuated enclosure, a vertical cylindrical tank of diameter D is sealed by a 3.00-kg circular disk that can move up and down without friction. Beneath the disk is a quantity of ideal gas at temperature T in the cylinder (**Fig. P18.50**). Initially the disk is at rest at a distance of $h = 4.00$ m above the bottom of the tank. When a lead brick of mass 9.00 kg is gently placed on the disk, the disk moves downward. If the temperature of the gas is kept constant and no gas escapes from the tank, what distance above the bottom of the tank is the disk when it again comes to rest?

Figure **P18.50**

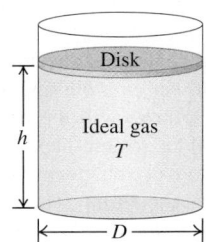

18.51 ••• A cylinder 1.00 m tall with inside diameter 0.120 m is used to hold propane gas (molar mass 44.1 g/mol) for use in a barbecue. It is initially filled with gas until the gauge pressure is 1.30×10^6 Pa at 22.0°C. The temperature of the gas remains constant as it is partially emptied out of the tank, until the gauge pressure is 3.40×10^5 Pa. Calculate the mass of propane that has been used.

18.52 • CP During a test dive in 1939, prior to being accepted by the U.S. Navy, the submarine *Squalus* sank at a point where the depth of water was 73.0 m. The temperature was 27.0°C at the surface and 7.0°C at the bottom. The density of seawater is 1030 kg/m³. (a) A diving bell was used to rescue 33 trapped crewmen from the *Squalus*. The diving bell was in the form of a circular cylinder 2.30 m high, open at the bottom and closed at the top. When the diving bell was lowered to the bottom of the sea, to what height did water rise within the diving bell? (*Hint:* Ignore the relatively small variation in water pressure between the bottom of the bell and the surface of the water within the bell.) (b) At what gauge pressure must compressed air have been supplied to the bell while on the bottom to expel all the water from it?

18.53 • **Atmosphere of Titan.** Titan, the largest satellite of Saturn, has a thick nitrogen atmosphere. At its surface, the pressure is 1.5 earth-atmospheres and the temperature is 94 K. (a) What is the surface temperature in °C? (b) Calculate the surface density in Titan's atmosphere in molecules per cubic meter. (c) Compare the density of Titan's surface atmosphere to the density of earth's atmosphere at 22°C. Which body has denser atmosphere?

18.54 • **Pressure on Venus.** At the surface of Venus the average temperature is a balmy 460°C due to the greenhouse effect (global warming!), the pressure is 92 earth-atmospheres, and the acceleration due to gravity is $0.894 g_{earth}$. The atmosphere is nearly all CO_2 (molar mass 44.0 g/mol), and the temperature remains remarkably constant. Assume that the temperature does not change with altitude. (a) What is the atmospheric pressure 1.00 km above the surface of Venus? Express your answer in Venus-atmospheres and earth-atmospheres. (b) What is the root-mean-square speed of the CO_2 molecules at the surface of Venus and at an altitude of 1.00 km?

18.55 •• An automobile tire has a volume of 0.0150 m³ on a cold day when the temperature of the air in the tire is 5.0°C and atmospheric pressure is 1.02 atm. Under these conditions the gauge pressure is measured to be 1.70 atm (about 25 lb/in.²).

After the car is driven on the highway for 30 min, the temperature of the air in the tires has risen to 45.0°C and the volume has risen to 0.0159 m³. What then is the gauge pressure?

18.56 •• A flask with a volume of 1.50 L, provided with a stopcock, contains ethane gas (C_2H_6) at 300 K and atmospheric pressure (1.013×10^5 Pa). The molar mass of ethane is 30.1 g/mol. The system is warmed to a temperature of 550 K, with the stopcock open to the atmosphere. The stopcock is then closed, and the flask is cooled to its original temperature. (a) What is the final pressure of the ethane in the flask? (b) How many grams of ethane remain in the flask?

18.57 •• CP A balloon of volume 750 m³ is to be filled with hydrogen at atmospheric pressure (1.01×10^5 Pa). (a) If the hydrogen is stored in cylinders with volumes of 1.90 m³ at a gauge pressure of 1.20×10^6 Pa, how many cylinders are required? Assume that the temperature of the hydrogen remains constant. (b) What is the total weight (in addition to the weight of the gas) that can be supported by the balloon if both the gas in the balloon and the surrounding air are at 15.0°C? The molar mass of hydrogen (H_2) is 2.02 g/mol. The density of air at 15.0°C and atmospheric pressure is 1.23 kg/m³. See Chapter 12 for a discussion of buoyancy. (c) What weight could be supported if the balloon were filled with helium (molar mass 4.00 g/mol) instead of hydrogen, again at 15.0°C?

18.58 •• A vertical cylindrical tank contains 1.80 mol of an ideal gas under a pressure of 0.300 atm at 20.0°C. The round part of the tank has a radius of 10.0 cm, and the gas is supporting a piston that can move up and down in the cylinder without friction. There is a vacuum above the piston. (a) What is the mass of this piston? (b) How tall is the column of gas that is supporting the piston?

18.59 •• CP A large tank of water has a hose connected to it (**Fig. P18.59**). The tank is sealed at the top and has compressed air between the water surface and the top. When the water height h has the value 3.50 m, the absolute pressure p of the compressed air is 4.20×10^5 Pa.

Figure **P18.59**

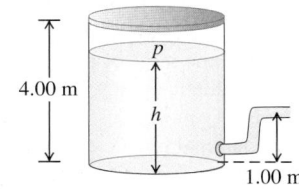

Assume that the air above the water expands at constant temperature, and take the atmospheric pressure to be 1.00×10^5 Pa. (a) What is the speed with which water flows out of the hose when $h = 3.50$ m? (b) As water flows out of the tank, h decreases. Calculate the speed of flow for $h = 3.00$ m and for $h = 2.00$ m. (c) At what value of h does the flow stop?

18.60 •• CP A light, plastic sphere with mass $m = 9.00$ g and density $\rho = 4.00$ kg/m³ is suspended in air by thread of negligible mass. (a) What is the tension T in the thread if the air is at 5.00°C and $p = 1.00$ atm? The molar mass of air is 28.8 g/mol. (b) How much does the tension in the thread change if the temperature of the gas is increased to 35.0°C? Ignore the change in volume of the plastic sphere when the temperature is changed.

18.61 •• BIO **How Many Atoms Are You?** Estimate the number of atoms in the body of a 50-kg physics student. Note that the human body is mostly water, which has molar mass 18.0 g/mol, and that each water molecule contains three atoms.

18.62 •• BIO A person at rest inhales 0.50 L of air with each breath at a pressure of 1.00 atm and a temperature of 20.0°C. The inhaled air is 21.0% oxygen. (a) How many oxygen molecules does this person inhale with each breath? (b) Suppose this person is now resting at an elevation of 2000 m but the temperature is still 20.0°C. Assuming that the oxygen percentage and volume per inhalation are the same as stated above, how many oxygen molecules does this person now inhale with each breath? (c) Given that the body still requires the same number of oxygen molecules per second as at sea level to maintain its functions, explain why some people report "shortness of breath" at high elevations.

18.63 •• You have two identical containers, one containing gas A and the other gas B. The masses of these molecules are $m_A = 3.34 \times 10^{-27}$ kg and $m_B = 5.34 \times 10^{-26}$ kg. Both gases are under the same pressure and are at 10.0°C. (a) Which molecules (A or B) have greater translational kinetic energy per molecule and rms speeds? (b) Now you want to raise the temperature of only one of these containers so that both gases will have the same rms speed. For which gas should you raise the temperature? (c) At what temperature will you accomplish your goal? (d) Once you have accomplished your goal, which molecules (A or B) now have greater average translational kinetic energy per molecule?

18.64 • The size of an oxygen molecule is about 2.0×10^{-10} m. Make a rough estimate of the pressure at which the finite volume of the molecules should cause noticeable deviations from ideal-gas behavior at ordinary temperatures ($T = 300$ K).

18.65 •• A sealed box contains a monatomic ideal gas. The number of gas atoms per unit volume is 5.00×10^{20} atoms/cm³, and the average translational kinetic energy of each atom is 1.80×10^{-23} J. (a) What is the gas pressure? (b) If the gas is neon (molar mass 20.18 g/mol), what is v_{rms} for the gas atoms?

18.66 •• Helium gas is in a cylinder that has rigid walls. If the pressure of the gas is 2.00 atm, then the root-mean-square speed of the helium atoms is $v_{rms} = 176$ m/s. By how much (in atmospheres) must the pressure be increased to increase the v_{rms} of the He atoms by 100 m/s? Ignore any change in the volume of the cylinder.

18.67 •• You blow up a spherical balloon to a diameter of 50.0 cm until the absolute pressure inside is 1.25 atm and the temperature is 22.0°C. Assume that all the gas is N_2, of molar mass 28.0 g/mol. (a) Find the mass of a single N_2 molecule. (b) How much translational kinetic energy does an average N_2 molecule have? (c) How many N_2 molecules are in this balloon? (d) What is the *total* translational kinetic energy of all the molecules in the balloon?

18.68 • CP (a) Compute the increase in gravitational potential energy for a nitrogen molecule (molar mass 28.0 g/mol) for an increase in elevation of 400 m near the earth's surface. (b) At what temperature is this equal to the average kinetic energy of a nitrogen molecule? (c) Is it possible that a nitrogen molecule near sea level where $T = 15.0$°C could rise to an altitude of 400 m? Is it likely that it could do so without hitting any other molecules along the way? Explain.

18.69 •• CP CALC **The Lennard-Jones Potential.** A commonly used potential-energy function for the interaction of two molecules (see Fig. 18.8) is the Lennard-Jones 6-12 potential:

$$U(r) = U_0\left[\left(\frac{R_0}{r}\right)^{12} - 2\left(\frac{R_0}{r}\right)^6\right]$$

where r is the distance between the centers of the molecules and U_0 and R_0 are positive constants. The corresponding force $F(r)$ is given in Eq. (14.26). (a) Graph $U(r)$ and $F(r)$ versus r. (b) Let r_1 be the value of r at which $U(r) = 0$, and let r_2 be the value of r at which $F(r) = 0$. Show the locations of r_1 and r_2 on your graphs of $U(r)$ and $F(r)$. Which of these values represents the equilibrium separation between the molecules? (c) Find the values of r_1 and r_2 in terms of R_0, and find the ratio r_1/r_2. (d) If the molecules are located a distance r_2 apart [as calculated in part (c)], how much work must be done to pull them apart so that $r \to \infty$?

18.70 • (a) What is the total random translational kinetic energy of 5.00 L of hydrogen gas (molar mass 2.016 g/mol) with pressure 1.01×10^5 Pa and temperature 300 K? (*Hint:* Use the procedure of Problem 18.67 as a guide.) (b) If the tank containing the gas is placed on a swift jet moving at 300.0 m/s, by what percentage is the *total* kinetic energy of the gas increased? (c) Since the kinetic energy of the gas molecules is greater when it is on the jet, does this mean that its temperature has gone up? Explain.

18.71 •• It is possible to make crystalline solids that are only one layer of atoms thick. Such "two-dimensional" crystals can be created by depositing atoms on a very flat surface. (a) If the atoms in such a two-dimensional crystal can move only within the plane of the crystal, what will be its molar heat capacity near room temperature? Give your answer as a multiple of R and in J/mol·K. (b) At very low temperatures, will the molar heat capacity of a two-dimensional crystal be greater than, less than, or equal to the result you found in part (a)? Explain why.

18.72 • **Hydrogen on the Sun.** The surface of the sun has a temperature of about 5800 K and consists largely of hydrogen atoms. (a) Find the rms speed of a hydrogen atom at this temperature. (The mass of a single hydrogen atom is 1.67×10^{-27} kg.) (b) The escape speed for a particle to leave the gravitational influence of the sun is given by $(2GM/R)^{1/2}$, where M is the sun's mass, R its radius, and G the gravitational constant (see Example 13.5 of Section 13.3). Use Appendix F to calculate this escape speed. (c) Can appreciable quantities of hydrogen escape from the sun? Can *any* hydrogen escape? Explain.

18.73 •• CP (a) Show that a projectile with mass m can "escape" from the surface of a planet if it is launched vertically upward with a kinetic energy greater than mgR_p, where g is the acceleration due to gravity at the planet's surface and R_p is the planet's radius. Ignore air resistance. (See Problem 18.72.) (b) If the planet in question is the earth, at what temperature does the average translational kinetic energy of a nitrogen molecule (molar mass 28.0 g/mol) equal that required to escape? What about a hydrogen molecule (molar mass 2.02 g/mol?) (c) Repeat part (b) for the moon, for which $g = 1.63$ m/s² and $R_p = 1740$ km. (d) While the earth and the moon have similar average surface temperatures, the moon has essentially no atmosphere. Use your results from parts (b) and (c) to explain why.

18.74 • **Planetary Atmospheres.** (a) The temperature near the top of Jupiter's multicolored cloud layer is about 140 K. The temperature at the top of the earth's troposphere, at an altitude of about 20 km, is about 220 K. Calculate the rms speed of hydrogen molecules in both these environments. Give your answers in m/s and as a fraction of the escape speed from the respective planet (see Problem 18.72). (b) Hydrogen gas (H_2) is a rare element in the earth's atmosphere. In the atmosphere of Jupiter, by contrast, 89% of all molecules are H_2. Explain why, using your results from part (a). (c) Suppose an astronomer claims to have discovered an oxygen (O_2) atmosphere on the asteroid Ceres. How likely is this? Ceres has a mass equal to 0.014 times the mass of the moon, a density of 2400 kg/m³, and a surface temperature of about 200 K.

18.75 •• CALC Calculate the integral in Eq. (18.31), $\int_0^\infty v^2 f(v)\, dv$, and compare this result to $(v^2)_{av}$ as given by Eq. (18.16). (*Hint:* You may use the tabulated integral

$$\int_0^\infty x^{2n} e^{-\alpha x^2}\, dx = \frac{1 \cdot 3 \cdot 5 \cdots (2n-1)}{2^{n+1}\alpha^n}\sqrt{\frac{\pi}{\alpha}}$$

where n is a positive integer and α is a positive constant.)

18.76 •• (a) Calculate the total *rotational* kinetic energy of the molecules in 1.00 mol of a diatomic gas at 300 K. (b) Calculate the moment of inertia of an oxygen molecule (O_2) for rotation about either the y- or z-axis shown in Fig. 18.18b. Treat the molecule as two massive points (representing the oxygen atoms) separated by a distance of 1.21×10^{-10} m. The molar mass of oxygen *atoms* is 16.0 g/mol. (c) Find the rms angular velocity of rotation of an oxygen molecule about either the y- or z-axis shown in Fig. 18.18b. How does your answer compare to the angular velocity of a typical piece of rapidly rotating machinery (10,000 rev/min)?

18.77 •• CALC (a) Explain why in a gas of N molecules, the number of molecules having speeds in the *finite* interval v to $v + \Delta v$ is $\Delta N = N \int_v^{v + \Delta v} f(v)\, dv$. (b) If Δv is small, then $f(v)$ is approximately constant over the interval and $\Delta N \approx Nf(v)\Delta v$. For oxygen gas ($O_2$, molar mass 32.0 g/mol) at $T = 300$ K, use this approximation to calculate the number of molecules with speeds within $\Delta v = 20$ m/s of v_{mp}. Express your answer as a multiple of N. (c) Repeat part (b) for speeds within $\Delta v = 20$ m/s of $7v_{mp}$. (d) Repeat parts (b) and (c) for a temperature of 600 K. (e) Repeat parts (b) and (c) for a temperature of 150 K. (f) What do your results tell you about the shape of the distribution as a function of temperature? Do your conclusions agree with what is shown in Fig. 18.23?

18.78 •• CALC Calculate the integral in Eq. (18.30), $\int_0^\infty vf(v)\,dv$, and compare this result to v_{av} as given by Eq. (18.35). (*Hint:* Make the change of variable $v^2 = x$ and use the tabulated integral

$$\int_0^\infty x^n e^{-\alpha x} dx = \frac{n!}{\alpha^{n+1}}$$

where n is a positive integer and α is a positive constant.)

18.79 ••• CP **Oscillations of a Piston.** A vertical cylinder of radius r contains an ideal gas and is fitted with a piston of mass m that is free to move (**Fig. P18.79**). The piston and the walls of the cylinder are frictionless, and the entire cylinder is placed in a constant-temperature bath. The outside air pressure is p_0. In equilibrium, the piston sits at a height h above the bottom of the cylinder. (a) Find the absolute pressure of the gas trapped below the piston when in equilibrium. (b) The piston is pulled up by a small distance and released. Find the net force acting on the piston when its base is a distance $h + y$ above the bottom of the cylinder, where $y \ll h$. (c) After the piston is displaced from equilibrium and released, it oscillates up and down. Find the frequency of these small oscillations. If the displacement is not small, are the oscillations simple harmonic? How can you tell?

Figure **P18.79**

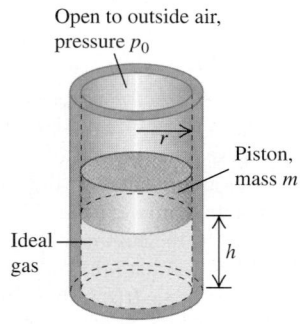

Open to outside air, pressure p_0

r

Piston, mass m

Ideal gas

h

18.80 •• DATA A steel cylinder with rigid walls is evacuated to a high degree of vacuum; you then put a small amount of helium into the cylinder. The cylinder has a pressure gauge that measures the pressure of the gas inside the cylinder. You place the cylinder in various temperature environments, wait for thermal equilibrium to be established, and then measure the pressure of the gas. You obtain these results:

	T (°C)	p (Pa)
Normal boiling point of nitrogen	−195.8	254
Ice–water mixture	0.0	890
Outdoors on a warm day	33.3	999
Normal boiling point of water	100.0	1214
Hot oven	232	1635

(a) Recall (Chapter 17) that absolute zero is the temperature at which the pressure of an ideal gas becomes zero. Use the data in the table to calculate the value of absolute zero in °C. Assume that the pressure of the gas is low enough for it to be treated as an ideal gas, and ignore the change in volume of the cylinder as its temperature is changed. (b) Use the coefficient of volume expansion for steel in Table 17.2 to calculate the percentage change in the volume of the cylinder between the lowest and highest temperatures in the table. Is it accurate to ignore the volume change of the cylinder as the temperature changes? Justify your answer.

18.81 ••• DATA **The Dew Point and Clouds.** The vapor pressure of water (see Exercise 18.44) decreases as the temperature decreases. The table lists the vapor pressure of water at various temperatures:

Temperature (°C)	Vapor Pressure (Pa)
10.0	1.23×10^3
12.0	1.40×10^3
14.0	1.60×10^3
16.0	1.81×10^3
18.0	2.06×10^3
20.0	2.34×10^3
22.0	2.65×10^3
24.0	2.99×10^3
26.0	3.36×10^3
28.0	3.78×10^3
30.0	4.25×10^3

If the amount of water vapor in the air is kept constant as the air is cooled, the *dew point* temperature is reached, at which the partial pressure and vapor pressure coincide and the vapor is saturated. If the air is cooled further, the vapor condenses to liquid until the partial pressure again equals the vapor pressure at that temperature. The temperature in a room is 30.0°C. (a) A meteorologist cools a metal can by gradually adding cold water. When the can's temperature reaches 16.0°C, water droplets form on its outside surface. What is the relative humidity of the 30.0°C air in the room? On a spring day in the midwestern United States, the air temperature at the surface is 28.0°C. Puffy cumulus clouds form at an altitude where the air temperature equals the dew point. If the air temperature decreases with altitude at a rate of 0.6 C°/100 m, at approximately what height above the ground will clouds form if the relative humidity at the surface is (b) 35%; (c) 80%?

18.82 •• DATA The statistical quantities "average value" and "root-mean-square value" can be applied to any distribution. **Figure P18.82** shows the scores of a class of 150 students on a 100-point quiz. (a) Find the average score for the class. (b) Find the rms score for the class. (c) Which is higher: the average score or the rms score? Why?

Figure **P18.82**

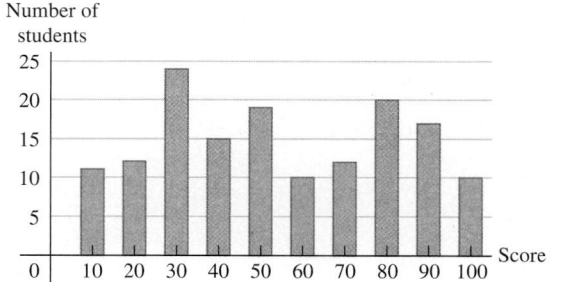

Number of students

CHALLENGE PROBLEMS

18.83 ••• **CP Dark Nebulae and the Interstellar Medium.** The dark area in **Fig. P18.83** that appears devoid of stars is a *dark nebula,* a cold gas cloud in interstellar space that contains enough material to block out light from the stars behind it. A typical dark nebula is about 20 light-years in diameter and contains about 50 hydrogen atoms per cubic centimeter (monatomic hydrogen, *not* H_2) at about 20 K. (A light-year is the distance light travels in vacuum in one year and is equal to 9.46×10^{15} m.) (a) Estimate the mean free path for a hydrogen atom in a dark nebula. The radius of a hydrogen atom is 5.0×10^{-11} m. (b) Estimate the rms speed of a hydrogen atom and the mean free time (the average time between collisions for a given atom). Based on this result, do you think that atomic collisions, such as those leading to H_2 molecule formation, are very important in determining the composition of the nebula? (c) Estimate the pressure inside a dark nebula. (d) Compare the rms speed of a hydrogen atom to the escape speed at the surface of the nebula (assumed spherical). If the space around the nebula were a vacuum, would such a cloud be stable or would it tend to evaporate? (e) The stability of dark nebulae is explained by the presence of the *interstellar medium* (ISM), an even thinner gas that permeates space and in which the dark nebulae are embedded. Show that for dark nebulae to be in equilibrium with the ISM, the numbers of atoms per volume (N/V) and the temperatures (T) of dark nebulae and the ISM must be related by

$$\frac{(N/V)_{\text{nebula}}}{(N/V)_{\text{ISM}}} = \frac{T_{\text{ISM}}}{T_{\text{nebula}}}$$

(f) In the vicinity of the sun, the ISM contains about 1 hydrogen atom per 200 cm³. Estimate the temperature of the ISM in the vicinity of the sun. Compare to the temperature of the sun's surface, about 5800 K. Would a spacecraft coasting through interstellar space burn up? Why or why not?

Figure **P18.83**

18.84 ••• **CALC Earth's Atmosphere.** In the *troposphere,* the part of the atmosphere that extends from earth's surface to an altitude of about 11 km, the temperature is not uniform but decreases with increasing elevation. (a) Show that if the temperature variation is approximated by the linear relationship

$$T = T_0 - \alpha y$$

where T_0 is the temperature at the earth's surface and T is the temperature at height y, the pressure p at height y is

$$\ln\left(\frac{p}{p_0}\right) = \frac{Mg}{R\alpha}\ln\left(\frac{T_0 - \alpha y}{T_0}\right)$$

where p_0 is the pressure at the earth's surface and M is the molar mass for air. The coefficient α is called the lapse rate of temperature. It varies with atmospheric conditions, but an average value is about 0.6 C°/100 m. (b) Show that the above result reduces to the result of Example 18.4 (Section 18.1) in the limit that $\alpha \to 0$. (c) With $\alpha = 0.6$ C°/100 m, calculate p for $y = 8863$ m and compare your answer to the result of Example 18.4. Take $T_0 = 288$ K and $p_0 = 1.00$ atm.

PASSAGE PROBLEMS

INSULATING WINDOWS. One way to improve insulation in windows is to fill a sealed space between two glass panes with a gas that has a lower thermal conductivity than that of air. The thermal conductivity k of a gas depends on its molar heat capacity C_V, molar mass M, and molecular radius r. The dependence on those quantities at a given temperature is approximated by $k \propto C_V/r^2\sqrt{M}$. The noble gases have properties that make them particularly good choices as insulating gases. Noble gases range from helium (molar mass 4.0 g/mol, molecular radius 0.13 nm) to xenon (molar mass 131 g/mol, molecular radius 0.22 nm). (The noble gas radon is heavier than xenon, but radon is radioactive and so is not suitable for this purpose.)

18.85 What is one reason the noble gases are *preferable* to air (which is mostly nitrogen and oxygen) as an insulating material? (a) Noble gases are monatomic, so no rotational modes contribute to their molar heat capacity; (b) noble gases are monatomic, so they have lower molecular masses than do nitrogen and oxygen; (c) molecular radii in noble gases are much larger than those of gases that consist of diatomic molecules; (d) because noble gases are monatomic, they have many more degrees of freedom than do diatomic molecules, and their molar heat capacity is reduced by the number of degrees of freedom.

18.86 Estimate the ratio of the thermal conductivity of Xe to that of He. (a) 0.015; (b) 0.061; (c) 0.10; (d) 0.17.

18.87 The rate of *effusion*—that is, leakage of a gas through tiny cracks—is proportional to v_{rms}. If tiny cracks exist in the material that's used to seal the space between two glass panes, how many times greater is the rate of He leakage out of the space between the panes than the rate of Xe leakage at the same temperature? (a) 370 times; (b) 19 times; (c) 6 times; (d) no greater—the He leakage rate is the same as for Xe.

Answers

Chapter Opening Question ?

(iv) From Eq. (18.19), the root-mean-square speed of a gas molecule is proportional to the square root of the absolute temperature T. The temperature range we're considering is $(25 + 273.15)$ K = 298 K to $(100 + 273.15)$ K = 373 K. Hence the speeds increase by a factor of $\sqrt{(373\text{ K})/(298\text{ K})} = 1.12$; that is, there is a 12% increase. While 100°C feels far warmer than 25°C, the difference in molecular speeds is relatively small.

Test Your Understanding Questions

18.1 (ii) and (iii) (tie), (i) and (v) (tie), (iv) We can rewrite the ideal-gas equation, Eq. (18.3), as $n = pV/RT$. This tells us that the number of moles n is proportional to the pressure and volume and inversely proportional to the absolute temperature. Hence, compared to (i), the number of moles in each case is (ii) $(2)(1)/(1) = 2$ times as much, (iii) $(1)(2)/(1) = 2$ times as much, (iv) $(1)(1)/(2) = \frac{1}{2}$ as much, and (v) $(2)(1)/(2) = 1$ time as much (that is, equal).

18.2 (vi) The value of r_0 determines the equilibrium separation of the molecules in the solid phase, so doubling r_0 means that the separation doubles as well. Hence a solid cube of this compound might grow from 1 cm on a side to 2 cm on a side. The volume would then be $2^3 = 8$ times larger, and the density (mass divided by volume) would be $\frac{1}{8}$ as great.

18.3 (a) (iv), (ii), (iii), (i); (b) (iii) and (iv) (tie), (i) and (ii) (tie) (a) Equation (18.19) tells us that $v_{\text{rms}} = \sqrt{3RT/M}$, so the rms speed is proportional to the square root of the ratio of absolute temperature T to molar mass M. Compared to (i) oxygen at 300 K, v_{rms} in the other cases is (ii) $\sqrt{(32.0\text{ g/mol})/(28.0\text{ g/mol})} = 1.07$ times faster, (iii) $\sqrt{(330\text{ K})/(300\text{ K})} = 1.05$ times faster, and (iv) $\sqrt{(330\text{ K})(32.0\text{ g/mol})/(300\text{ K})(28.0\text{ g/mol})} = 1.12$ times faster. (b) From Eq. (18.16), the average translational kinetic energy per molecule is $\frac{1}{2}m(v^2)_{\text{av}} = \frac{3}{2}kT$, which is directly proportional to T and independent of M. We have $T = 300$ K for cases (i) and (ii) and $T = 330$ K for cases (iii) and (iv), so $\frac{1}{2}m(v^2)_{\text{av}}$ has equal values for cases (iii) and (iv) and equal (but smaller) values for cases (i) and (ii).

18.4 no, near the beginning Adding a small amount of heat dQ to the gas changes the temperature by dT, where $dQ = nC_V dT$ from Eq. (18.24). Figure 18.19 shows that C_V for H_2 varies with temperature between 25 K and 500 K, so a given amount of heat gives rise to different amounts of temperature change during the process. Hence the temperature will *not* increase at a constant rate. The temperature change $dT = dQ/nC_V$ is inversely proportional to C_V, so the temperature increases most rapidly at the beginning of the process when the temperature is lowest and C_V is smallest (see Fig. 18.19).

18.5 (ii) Figure 18.23b shows that the *fraction* of molecules with speeds between v_1 and v_2 equals the area under the curve of $f(v)$ versus v from $v = v_1$ to $v = v_2$. This is equal to the integral $\int_{v_1}^{v_2} f(v)\,dv$, which in turn is equal to the difference between the integrals $\int_0^{v_2} f(v)\,dv$ (the fraction of molecules with speeds from 0 to v_2) and $\int_0^{v_1} f(v)\,dv$ (the fraction of molecules with speeds from 0 to the slower speed v_1). The *number* of molecules with speeds from v_1 to v_2 equals the fraction of molecules in this speed range multiplied by N, the total number of molecules.

18.6 no, yes The triple-point pressure of water from Table 18.3 is 6.10×10^2 Pa. The present-day pressure on Mars is just less than this value, corresponding to the line labeled p_s in Fig. 18.24. Hence liquid water cannot exist on the present-day Martian surface, and there are no rivers or lakes. Planetary scientists conclude that liquid water could have existed and almost certainly did exist on Mars in the past, when the atmosphere was thicker.

Bridging Problem

(a) 102 km   **(b)** yes

? A steam locomotive uses the first law of thermodynamics: Water is heated and boils, and the expanding steam does work to propel the locomotive. Would it be possible for the steam to propel the locomotive by doing work as it *condenses?* (i) Yes; (ii) no; (iii) answer depends on the details of how the steam condenses.

19

THE FIRST LAW OF THERMODYNAMICS

E very time you drive a gasoline-powered car, turn on an air conditioner, or cook a meal, you reap the benefits of *thermodynamics,* the study of relationships involving heat, mechanical work, and other aspects of energy and energy transfer. For example, in a car engine heat is generated by the chemical reaction of oxygen and vaporized gasoline in the engine's cylinders. The heated gas pushes on the pistons within the cylinders, doing mechanical work that is used to propel the car. This is an example of a *thermodynamic process.*

The first law of thermodynamics, central to the understanding of such processes, is an extension of the principle of conservation of energy. It broadens this principle to include energy exchange by both heat transfer and mechanical work and introduces the concept of the *internal energy* of a system. Conservation of energy plays a vital role in every area of physical science, and the first law has extremely broad usefulness. To state energy relationships precisely, we need the concept of a *thermodynamic system.* We'll discuss *heat* and *work* as two means of transferring energy into or out of such a system.

19.1 THERMODYNAMIC SYSTEMS

We have studied energy transfer through mechanical work (Chapter 6) and through heat transfer (Chapters 17 and 18). Now we are ready to combine and generalize these principles.

We always talk about energy transfer to or from some specific *system.* The system might be a mechanical device, a biological organism, or a specified quantity of material, such as the refrigerant in an air conditioner or steam expanding in a turbine. In general, a **thermodynamic system** is any collection of objects that is convenient to regard as a unit, and that may have the potential to exchange energy with its surroundings. A familiar example is a quantity of popcorn kernels in a pot with a lid. When the pot is placed on a stove, energy is added to the popcorn by conduction of heat. As the popcorn pops and expands, it does work as it exerts an upward force on the lid and moves it through a displacement (**Fig. 19.1**). The *state* of the popcorn—its volume, temperature, and pressure—changes as it pops. A process such as this one, in which there are changes in the state of a thermodynamic system, is called a **thermodynamic process.**

In mechanics we used the concept of *system* with free-body diagrams and with conservation of energy and momentum. For *thermodynamic* systems, as for all others, it is essential to define clearly at the start exactly what is and is not included in the system. Only then can we describe unambiguously the energy transfers into and out of that system. For instance, in our popcorn example we defined the system to include the popcorn but not the pot, lid, or stove.

Thermodynamics has its roots in many practical problems other than popping popcorn (**Fig. 19.2**). The gasoline engine in an automobile, the jet engines in an airplane, and the rocket engines in a launch vehicle use the heat of combustion of their fuel to perform mechanical work in propelling the vehicle. Muscle tissue in living organisms metabolizes chemical energy in food and performs mechanical work on the organism's surroundings. A steam engine or steam turbine uses the heat of combustion of coal or other fuel to perform mechanical work such as driving an electric generator or pulling a train.

19.1 The popcorn in the pot is a thermodynamic system. In the thermodynamic process shown here, heat is added to the system, and the system does work on its surroundings to lift the lid of the pot.

Signs for Heat and Work in Thermodynamics

We describe the energy relationships in any thermodynamic process in terms of the quantity of heat Q added *to* the system and the work W done *by* the system. Both Q and W may be positive, negative, or zero (**Fig. 19.3**). A positive value of Q represents heat flow *into* the system, with a corresponding input of energy to it; negative Q represents heat flow *out of* the system. A positive value of W represents work done *by* the system against its surroundings, such as work done by an expanding gas, and hence corresponds to energy *leaving* the system. Negative W, such as work done during compression of a gas in which work is done *on the gas* by its surroundings, represents energy *entering* the system. We will use these conventions consistently in the examples in this chapter and the next.

TEST YOUR UNDERSTANDING OF SECTION 19.1 In Example 17.7 (Section 17.6), what is the sign of Q for the coffee? For the aluminum cup? If a block slides along a horizontal surface with friction, what is the sign of W for the block? ▌

> **CAUTION** Be careful with the sign of work W Note that our sign rule for work is *opposite* to the one we used in mechanics, in which we always spoke of the work done by the forces acting *on* a body. In thermodynamics it is usually more convenient to call W the work done *by* the system so that when a system expands, the pressure, volume change, and work are all positive. Use the sign rules for work and heat consistently! ▌

19.2 (a) A rocket engine uses the heat of combustion of its fuel to do work propelling the launch vehicle. (b) Humans and other biological organisms are more complicated systems than we can analyze fully in this book, but the same basic principles of thermodynamics apply to them.

(a)　　　(b)

19.3 A thermodynamic system may exchange energy with its surroundings (environment) by means of heat, work, or both. Note the sign conventions for Q and W.

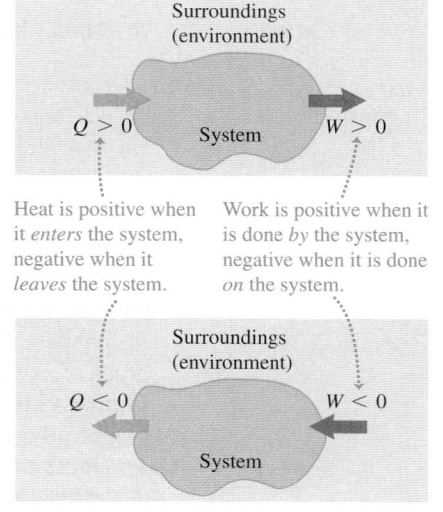

Surroundings (environment)

$Q > 0$　System　$W > 0$

Heat is positive when it *enters* the system, negative when it *leaves* the system.

Work is positive when it is done *by* the system, negative when it is done *on* the system.

Surroundings (environment)

$Q < 0$　　　$W < 0$

System

19.4 A molecule striking a piston (a) does positive work if the piston is moving away from the molecule and (b) does negative work if the piston is moving toward the molecule. Hence a gas does positive work when it expands as in (a) but does negative work when it compresses as in (b).

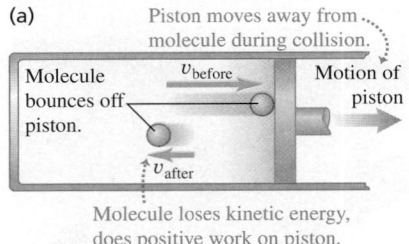

(a)

Piston moves away from molecule during collision.

Molecule bounces off piston.

v_{before}

Motion of piston

v_{after}

Molecule loses kinetic energy, does positive work on piston.

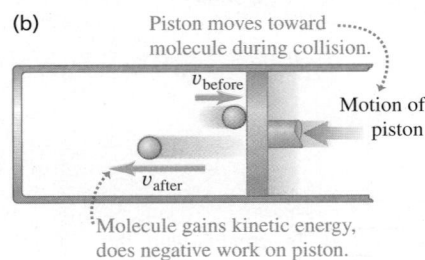

(b)

Piston moves toward molecule during collision.

v_{before}

Motion of piston

v_{after}

Molecule gains kinetic energy, does negative work on piston.

19.5 The infinitesimal work done by the system during the small expansion dx is $dW = pA\,dx$.

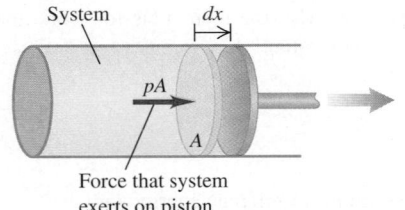

System

dx

pA

A

Force that system exerts on piston

19.2 WORK DONE DURING VOLUME CHANGES

A simple example of a thermodynamic system is a quantity of gas enclosed in a cylinder with a movable piston. Internal-combustion engines, steam engines, and compressors in refrigerators and air conditioners all use some version of such a system. In the next several sections we will use the gas-in-cylinder system to explore several kinds of thermodynamic processes.

We'll use a microscopic viewpoint, based on the kinetic and potential energies of individual molecules in a material, to develop intuition about thermodynamic quantities. But it is important to understand that the central principles of thermodynamics can be treated in a completely *macroscopic* way, without reference to microscopic models. Indeed, part of the great power and generality of thermodynamics is that it does *not* depend on details of the structure of matter.

First we consider the *work* done by the system during a volume change. When a gas expands, it pushes outward on its boundary surfaces as they move outward. Hence an expanding gas always does positive work. The same thing is true of any material that expands under pressure, such as the popcorn in Fig. 19.1.

We can understand the work done by a gas in a volume change by considering the molecules that make up the gas. When one such molecule ? collides with a stationary surface, it exerts a momentary force on the wall but does no work because the wall does not move. But if the surface is moving, like a piston in a gasoline engine, the molecule *does* do work on the surface during the collision. If the piston in **Fig. 19.4a** moves to the right, so the volume of the gas increases, the molecules that strike the piston exert a force through a distance and do *positive* work on the piston. If the piston moves toward the left as in Fig. 19.4b, so the volume of the gas decreases, positive work is done *on* the molecule during the collision. Hence the gas molecules do *negative* work on the piston.

Figure 19.5 shows a system whose volume can change (a gas, liquid, or solid) in a cylinder with a movable piston. Suppose that the cylinder has cross-sectional area A and that the pressure exerted by the system at the piston face is p. The total force F exerted by the system on the piston is $F = pA$. When the piston moves out an infinitesimal distance dx, the work dW done by this force is

$$dW = F\,dx = pA\,dx$$

But

$$A\,dx = dV$$

where dV is the infinitesimal change of volume of the system. Thus we can express the work done by the system in this infinitesimal volume change as

$$dW = p\,dV \qquad (19.1)$$

In a finite change of volume from V_1 to V_2,

Work done in a volume change

$$W = \int_{V_1}^{V_2} p\,dV \qquad (19.2)$$

Upper limit = final volume

Integral of the pressure with respect to volume

Lower limit = initial volume

In general, the pressure of the system may vary during the volume change. For example, this is the case in the cylinders of an automobile engine as the pistons move back and forth. To evaluate the integral in Eq. (19.2), we have to know how the pressure varies as a function of volume. We can represent this relationship as a graph of p as a function of V (a pV-diagram, described at the end of Section 18.1). **Figure 19.6** shows a simple example. In this figure, Eq. (19.2) is represented graphically as the *area* under the curve of p versus V between the limits V_1 and V_2. (In Section 6.3 we used a similar interpretation of the work done by a force F as the area under the curve of F versus x between the limits x_1 and x_2.)

19.6 The work done equals the area under the curve on a pV-diagram.

(a) pV-diagram for a system undergoing an expansion with varying pressure

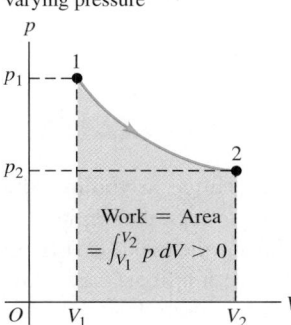

(b) pV-diagram for a system undergoing a compression with varying pressure

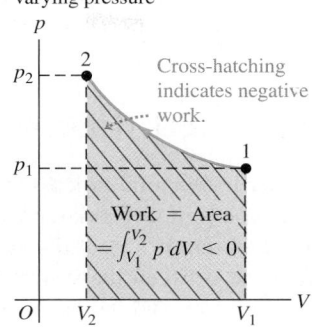

(c) pV-diagram for a system undergoing an expansion with constant pressure

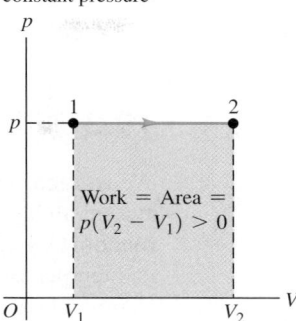

According to the rule we stated in Section 19.1, work is *positive* when a system *expands*. In an expansion from state 1 to state 2 in Fig. 19.6a, the area under the curve and the work are positive. A *compression* from 1 to 2 in Fig. 19.6b gives a *negative* area; when a system is compressed, its volume decreases and it does *negative* work on its surroundings (see also Fig. 19.4b).

If the pressure p remains constant while the volume changes from V_1 to V_2 (Fig. 19.6c), the work done by the system is

> **CAUTION** **Be careful with subscripts 1 and 2** In Eq. (19.2), V_1 is the *initial* volume and V_2 is the *final* volume. That's why labels 1 and 2 are reversed in Fig. 19.6b compared to Fig. 19.6a, even though both processes move between the same two thermodynamic states. ∎

Work done in a volume change at constant pressure

$$W = p(V_2 - V_1) \qquad (19.3)$$

Pressure — Final volume — Initial volume

If the volume is *constant,* there is no displacement and the system does no work.

EXAMPLE 19.1   **ISOTHERMAL EXPANSION OF AN IDEAL GAS**

As an ideal gas undergoes an *isothermal* (constant-temperature) expansion at temperature T, its volume changes from V_1 to V_2. How much work does the gas do?

SOLUTION

IDENTIFY and SET UP: The ideal-gas equation, Eq. (18.3), tells us that if the temperature T of n moles of an ideal gas is constant, the quantity $pV = nRT$ is also constant. If V changes, p changes as well, so we *cannot* use Eq. (19.3) to calculate the work done. Instead we must evaluate the integral in Eq. (19.2), so we must know p as a function of V; for this we use Eq. (18.3).

EXECUTE: From Eq. (18.3),

$$p = \frac{nRT}{V}$$

We substitute this into the integral of Eq. (19.2), take the constant factor nRT outside, and evaluate the integral:

$$W = \int_{V_1}^{V_2} p \, dV = nRT \int_{V_1}^{V_2} \frac{dV}{V}$$

$$= nRT \ln \frac{V_2}{V_1} \quad \text{(ideal gas, isothermal process)}$$

We can rewrite this expression for W in terms of p_1 and p_2. Because $pV = nRT$ is constant,

$$p_1 V_1 = p_2 V_2 \qquad \text{or} \qquad \frac{V_2}{V_1} = \frac{p_1}{p_2}$$

so

$$W = nRT \ln \frac{p_1}{p_2} \quad \text{(ideal gas, isothermal process)}$$

EVALUATE: We check our result by noting that in an expansion $V_2 > V_1$ and the ratio V_2/V_1 is greater than 1. The logarithm of a number greater than 1 is positive, so $W > 0$, as it should be. As an additional check, look at our second expression for W: In an isothermal expansion the volume increases and the pressure drops, so $p_2 < p_1$, the ratio $p_1/p_2 > 1$, and $W = nRT \ln(p_1/p_2)$ is again positive.

Our result for W also applies to an isothermal *compression* of a gas, for which $V_2 < V_1$ and $p_2 > p_1$.

19.3 PATHS BETWEEN THERMODYNAMIC STATES

We've seen that if a thermodynamic process involves a change in volume, the system undergoing the process does work (either positive or negative) on its surroundings. Heat also flows into or out of the system during the process if there is a temperature difference between the system and its surroundings. Let's now examine how the work done by and the heat added to the system during a thermodynamic process depend on the details of how the process takes place.

Work Done in a Thermodynamic Process

When a thermodynamic system changes from an initial state to a final state, it passes through a series of intermediate states. We call this series of states a **path.** There are always infinitely many possibilities for these intermediate states. When all are equilibrium states, the path can be plotted on a pV-diagram (**Fig. 19.7a**). Point 1 represents an initial state with pressure p_1 and volume V_1, and point 2 represents a final state with pressure p_2 and volume V_2. To pass from state 1 to state 2, we could keep the pressure constant at p_1 while the system expands to volume V_2 (point 3 in Fig. 19.7b), then reduce the pressure to p_2 (probably by decreasing the temperature) while keeping the volume constant at V_2 (to point 2). The work done by the system during this process is the area under the line $1 \rightarrow 3$; no work is done during the constant-volume process $3 \rightarrow 2$. Or the system might traverse the path $1 \rightarrow 4 \rightarrow 2$ (Fig. 19.7c); then the work is the area under the line $4 \rightarrow 2$, since no work is done during the constant-volume process $1 \rightarrow 4$. The smooth curve from 1 to 2 is another possibility (Fig. 19.7d), and the work for this path is different from that for either of the other paths.

We conclude that *the work done by the system depends not only on the initial and final states, but also on the intermediate states—that is, on the path.* Furthermore, we can take the system through a series of states forming a closed loop, such as $1 \rightarrow 3 \rightarrow 2 \rightarrow 4 \rightarrow 1$. In this case the final state is the same as the initial state, but the total work done by the system is *not* zero. (In fact, it is represented on the graph by the area enclosed by the loop; see Exercise 19.7.) So we can't talk about the amount of work *contained in* a system. In a particular state, a system may have definite values of the state coordinates p, V, and T, but it wouldn't make sense to say that it has a definite value of W.

Heat Added in a Thermodynamic Process

Like work, the *heat* added to a thermodynamic system when it undergoes a change of state depends on the path from the initial state to the final state. Here's an example. Suppose we want to change the volume of a certain quantity of an ideal gas from 2.0 L to 5.0 L while keeping the temperature constant at $T = 300$ K. **Figure 19.8** shows two different ways to do this. In Fig. 19.8a the gas is contained in a cylinder with a piston, with an initial volume of 2.0 L. We let the gas expand slowly, supplying heat from the electric heater to keep the temperature at 300 K until the gas reaches its final volume of 5.0 L. The gas absorbs a definite amount of heat in this isothermal process.

Figure 19.8b shows a different process leading to the same final state. The container is surrounded by insulating walls and is divided by a thin, breakable partition into two compartments. The lower part has volume 2.0 L and the upper part has volume 3.0 L. In the lower compartment we place the same amount of

19.7 The work done by a system during a transition between two states depends on the path chosen.

(a)
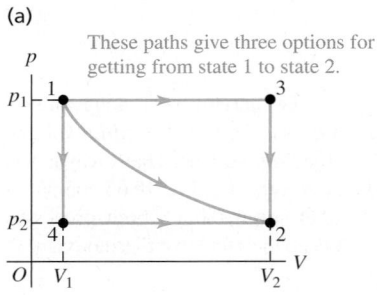
These paths give three options for getting from state 1 to state 2.

(b)
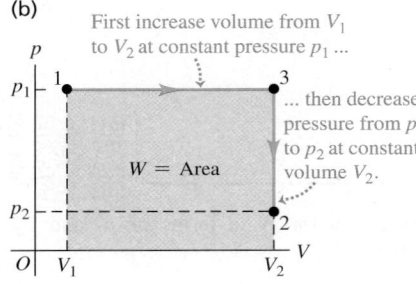
First increase volume from V_1 to V_2 at constant pressure p_1 ...
... then decrease pressure from p_1 to p_2 at constant volume V_2.
$W = $ Area

(c)
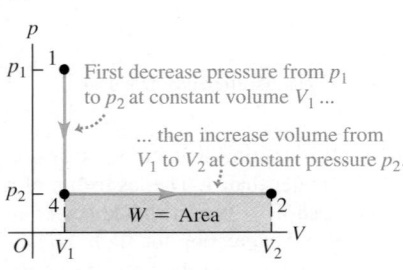
First decrease pressure from p_1 to p_2 at constant volume V_1 ...
... then increase volume from V_1 to V_2 at constant pressure p_2.
$W = $ Area

(d)
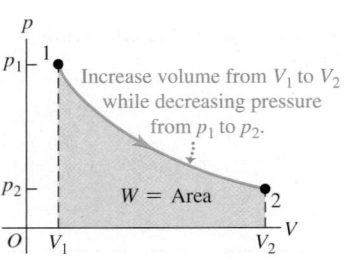
Increase volume from V_1 to V_2 while decreasing pressure from p_1 to p_2.
$W = $ Area

the same gas as in Fig. 19.8a, again at $T = 300$ K. The initial state is the same as before. Now we break the partition; the gas expands rapidly, with no heat passing through the insulating walls. The final volume is 5.0 L, the same as in Fig. 19.8a. The expanding gas does no work because it doesn't push against anything that moves. This uncontrolled expansion of a gas into vacuum is called a **free expansion;** we'll discuss it further in Section 19.6.

Experiments show that when an ideal gas undergoes a free expansion, there is no temperature change. So the final state of the gas is the same as in Fig. 19.8a. The intermediate pressures and volumes during the transition from state 1 to state 2 are entirely different in the two cases; Figs. 19.8a and 19.8b represent *two different paths* connecting the *same states* 1 and 2. For the path in Fig. 19.8b, *no* heat is transferred into the system, and the system does no work. Like work, *heat depends not only on the initial and final states but also on the path.*

Because of this path dependence, it would not make sense to say that a system "contains" a certain quantity of heat. To see this, suppose we assign an arbitrary value to the "heat in a body" in some reference state. Then presumably the "heat in the body" in some other state would equal the heat in the reference state plus the heat added when the body goes to the second state. But that's ambiguous, as we have just seen; the heat added depends on the *path* we take from the reference state to the second state. We are forced to conclude that there is *no* consistent way to define "heat in a body"; it is not a useful concept.

While it doesn't make sense to talk about "work in a body" or "heat in a body," it *does* make sense to speak of the amount of *internal energy* in a body. This important concept is our next topic.

TEST YOUR UNDERSTANDING OF SECTION 19.3   The system described in Fig. 19.7a undergoes four different thermodynamic processes. Each process is represented in a pV-diagram as a straight line from the initial state to the final state. (These processes are different from those shown in the pV-diagrams of Fig. 19.7.) Rank the processes in order of the amount of work done by the system, from the most positive to the most negative. (i) $1 \rightarrow 2$; (ii) $2 \rightarrow 1$; (iii) $3 \rightarrow 4$; (iv) $4 \rightarrow 3$. ❚

19.4 INTERNAL ENERGY AND THE FIRST LAW OF THERMODYNAMICS

Internal energy is one of the most important concepts in thermodynamics. In Section 7.3, when we discussed energy changes for a body sliding with friction, we stated that warming a body increased its internal energy and that cooling the body decreased its internal energy. But what *is* internal energy? We can look at it in various ways; let's start with one based on the ideas of mechanics. Matter consists of atoms and molecules, and these are made up of particles having kinetic and potential energies. We *tentatively* define the **internal energy** of a system as the sum of the kinetic energies of all of its constituent particles, plus the sum of all the potential energies of interaction among these particles.

⸻ CAUTION ⸻   **Is it internal?**   Internal energy does *not* include potential energy arising from the interaction between the system and its surroundings. If the system is a glass of water, placing it on a high shelf increases the gravitational potential energy arising from the interaction between the glass and the earth. But this has no effect on the interactions among the water molecules, and so the internal energy of the water does not change. ❚

We use the symbol U for internal energy. (We used this symbol in our study of mechanics to represent potential energy. However, U has a different meaning in thermodynamics.) During a change of state of the system, the internal energy may change from an initial value U_1 to a final value U_2. We denote the change in internal energy as $\Delta U = U_2 - U_1$.

19.8 (a) Slow, controlled isothermal expansion of a gas from an initial state 1 to a final state 2 with the same temperature but lower pressure. (b) Rapid, uncontrolled expansion of the same gas starting at the same state 1 and ending at the same state 2.

(a) System does work on piston; hot plate adds heat to system ($W > 0$ and $Q > 0$).

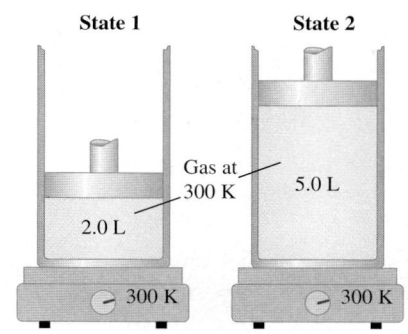

(b) System does no work; no heat enters or leaves system ($W = 0$ and $Q = 0$).

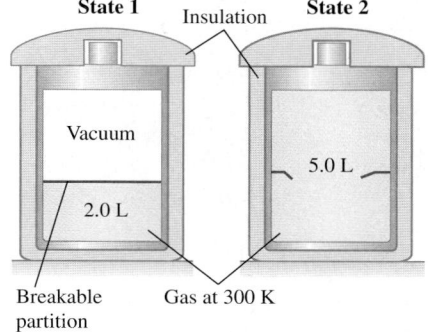

19.9 In a thermodynamic process, the internal energy U of a system may (a) increase ($\Delta U > 0$), (b) decrease ($\Delta U < 0$), or (c) remain the same ($\Delta U = 0$).

(a) More heat is added to system than system does work: Internal energy of system increases.

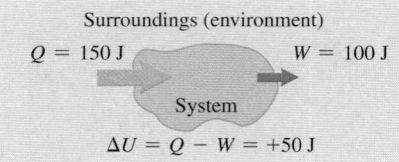

Surroundings (environment)

$Q = 150$ J $W = 100$ J

System

$\Delta U = Q - W = +50$ J

(b) More heat flows out of system than work is done: Internal energy of system decreases.

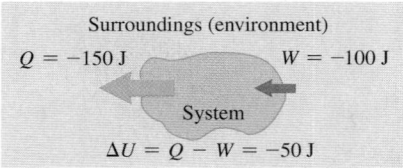

Surroundings (environment)

$Q = -150$ J $W = -100$ J

System

$\Delta U = Q - W = -50$ J

(c) Heat added to system equals work done by system: Internal energy of system unchanged.

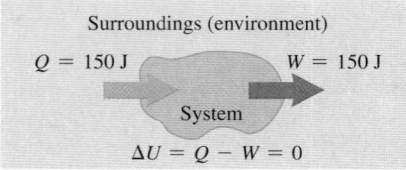

Surroundings (environment)

$Q = 150$ J $W = 150$ J

System

$\Delta U = Q - W = 0$

BIO Application The First Law of Exercise Thermodynamics Your body is a thermodynamic system. When you exercise, your body does work (such as the work done to lift your body as a whole in a push-up). Hence $W > 0$. Your body also warms up during exercise; by perspiration and other means the body rids itself of this heat, so $Q < 0$. Since Q is negative and W is positive, $\Delta U = Q - W < 0$ and the body's internal energy decreases. That's why exercise helps you lose weight: It uses up some of the internal energy stored in your body in the form of fat.

When we add a quantity of heat Q to a system and the system does no work during the process (so $W = 0$), the internal energy increases by an amount equal to Q; that is, $\Delta U = Q$. When a system does work W by expanding against its surroundings and no heat is added during the process, energy leaves the system and the internal energy decreases: W is positive, Q is zero, and $\Delta U = -W$. When *both* heat transfer and work occur, the *total* change in internal energy is

First law of thermodynamics:

Internal energy change of thermodynamic system

$$\Delta U = Q - W \qquad (19.4)$$

Heat added to system ···· ·····Work done by system

We can rearrange this to the form

$$Q = \Delta U + W \qquad (19.5)$$

The message of Eq. (19.5) is that when heat Q is added to a system, some of this added energy remains within the system, changing its internal energy by ΔU; the remainder leaves the system as the system does work W on its surroundings. Because W and Q may be positive, negative, or zero, ΔU can be positive, negative, or zero for different processes (**Fig. 19.9**).

Equation (19.4) or (19.5) is the **first law of thermodynamics.** It is a generalization of the principle of conservation of energy to include energy transfer through heat as well as mechanical work. As you will see in later chapters, this principle can be extended to ever-broader classes of phenomena by identifying additional forms of energy and energy transfer. In every situation in which it seems that the total energy in all known forms is not conserved, it has been possible to identify a new form of energy such that the total energy, including the new form, *is* conserved.

Understanding the First Law of Thermodynamics

At the beginning of this discussion we tentatively defined internal energy in terms of microscopic kinetic and potential energies. But actually *calculating* internal energy in this way for any real system would be hopelessly complicated. Furthermore, this definition isn't an *operational* one: It doesn't describe how to determine internal energy from physical quantities that we can measure.

So let's look at internal energy in another way. Starting over, we define the *change* in internal energy ΔU during any change of a system as the quantity given by Eq. (19.4), $\Delta U = Q - W$. This *is* an operational definition because we can measure Q and W. It does not define U itself, only ΔU. This is not a shortcoming because we can *define* the internal energy of a system to have a specified value in some reference state, and then use Eq. (19.4) to define the internal energy in any other state. This is analogous to our treatment of potential energy in Chapter 7, in which we arbitrarily defined the potential energy of a mechanical system to be zero at a certain position.

This new definition trades one difficulty for another. If we define ΔU by Eq. (19.4), then when the system goes from state 1 to state 2 by two different paths, how do we know that ΔU is the same for the two paths? We have already seen that Q and W are, in general, *not* the same for different paths. If ΔU, which equals $Q - W$, is also path dependent, then ΔU is ambiguous. If so, the concept of internal energy of a system is subject to the same criticism as the erroneous concept of quantity of heat in a system, as we discussed at the end of Section 19.3.

The only way to answer this question is through *experiment*. For various materials we measure Q and W for various changes of state and various paths to learn whether ΔU is or is not path dependent. The results of many such investigations are clear and unambiguous: While Q and W depend on the path, $\Delta U = Q - W$ *is independent of path. The change in internal energy of a system during any thermodynamic process depends only on the initial and final states, not on the path leading from one to the other.*

Experiment, then, is the ultimate justification for believing that a thermodynamic system in a specific state has a unique internal energy that depends only on that state. An equivalent statement is that the internal energy U of a system is a function of the state coordinates p, V, and T (actually, any two of these, since the three variables are related by the equation of state).

To say that the first law of thermodynamics, given by Eq. (19.4) or (19.5), represents conservation of energy for thermodynamic processes is correct, as far as it goes. But an important *additional* aspect of the first law is the fact that internal energy depends only on the state of a system (**Fig. 19.10**). In changes of state, the change in internal energy is independent of the path.

All this may seem a little abstract if you are satisfied to think of internal energy as microscopic mechanical energy. There's nothing wrong with that view, and we will make use of it at various times during our discussion. But as for heat, a precise *operational* definition of internal energy must be independent of the detailed microscopic structure of the material.

19.10 The internal energy of a cup of coffee depends on just its thermodynamic state—how much water and ground coffee it contains, and what its temperature is. It does not depend on the history of how the coffee was prepared—that is, the thermodynamic path that led to its current state.

Cyclic Processes and Isolated Systems

Two special cases of the first law of thermodynamics are worth mentioning. A process that eventually returns a system to its initial state is called a *cyclic process.* For such a process, the final state is the same as the initial state, and so the *total* internal energy change must be zero. Then

$$U_2 = U_1 \quad \text{and} \quad Q = W$$

If a net quantity of work W is done by the system during this process, an equal amount of energy must have flowed into the system as heat Q. But there is no reason either Q or W individually has to be zero (**Fig. 19.11**).

Another special case occurs in an *isolated system,* one that does no work on its surroundings and has no heat flow to or from its surroundings. For any process taking place in an isolated system,

$$W = Q = 0$$

and therefore

$$U_2 = U_1 = \Delta U = 0$$

In other words, *the internal energy of an isolated system is constant.*

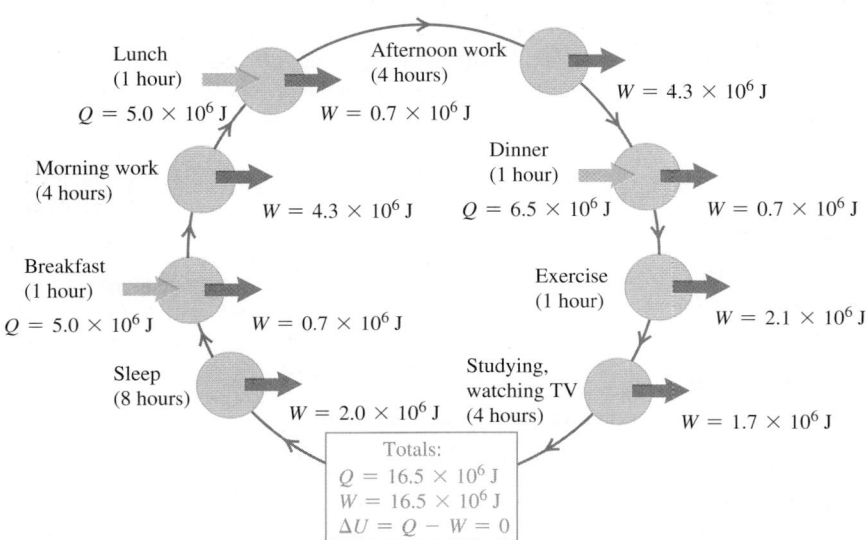

Lunch
(1 hour)
$Q = 5.0 \times 10^6$ J

Afternoon work
(4 hours)
$W = 0.7 \times 10^6$ J

$W = 4.3 \times 10^6$ J

Morning work
(4 hours)

$W = 4.3 \times 10^6$ J

Dinner
(1 hour)
$Q = 6.5 \times 10^6$ J

$W = 0.7 \times 10^6$ J

Breakfast
(1 hour)
$Q = 5.0 \times 10^6$ J

$W = 0.7 \times 10^6$ J

Exercise
(1 hour)

$W = 2.1 \times 10^6$ J

Sleep
(8 hours)
$W = 2.0 \times 10^6$ J

Studying,
watching TV
(4 hours)

$W = 1.7 \times 10^6$ J

Totals:
$Q = 16.5 \times 10^6$ J
$W = 16.5 \times 10^6$ J
$\Delta U = Q - W = 0$

19.11 Every day, your body (a thermodynamic system) goes through a cyclic thermodynamic process like this one. Heat Q is added by metabolizing food, and your body does work W in breathing, walking, and other activities. If you return to the same state at the end of the day, $Q = W$ and the net change in your internal energy is zero.

PROBLEM-SOLVING STRATEGY 19.1 | THE FIRST LAW OF THERMODYNAMICS

IDENTIFY *the relevant concepts:* The first law of thermodynamics is the statement of the law of conservation of energy in its most general form. You can apply it to *any* thermodynamic process in which the internal energy of a system changes, heat flows into or out of the system, and/or work is done by or on the system.

SET UP *the problem* using the following steps:
1. Define the thermodynamic system to be considered.
2. If the thermodynamic process has more than one step, identify the initial and final states for each step.
3. List the known and unknown quantities and identify the target variables.
4. Confirm that you have enough equations. You can apply the first law, $\Delta U = Q - W$, just once to each step in a thermodynamic process, so you will often need additional equations. These may include Eq. (19.2), $W = \int_{V_1}^{V_2} p\, dV$, which gives the work W done in a volume change, and the equation of state of the material that makes up the thermodynamic system (for an ideal gas, $pV = nRT$).

EXECUTE *the solution* as follows:
1. Be sure to use consistent units. If p is in Pa and V in m^3, then W is in joules. If a heat capacity is given in terms of calories,

convert it to joules. When you use $n = m_{\text{total}}/M$ to relate total mass m_{total} to number of moles n, remember that if m_{total} is in kilograms, M must be in *kilograms* per mole; M is usually tabulated in *grams* per mole.
2. The internal energy change ΔU in any thermodynamic process or series of processes is independent of the path, whether the substance is an ideal gas or not. If you can calculate ΔU for *any* path between given initial and final states, you know ΔU for *every possible path* between those states; you can then relate the various energy quantities for any of those other paths.
3. In a process comprising several steps, tabulate Q, W, and ΔU for each step, with one line per step and with the Q's, W's, and ΔU's forming columns (see Example 19.4). You can apply the first law to each line, and you can add each column and apply the first law to the sums. Do you see why?
4. Using steps 1–3, solve for the target variables.

EVALUATE *your answer:* Check your results for reasonableness. Ensure that each of your answers has the correct algebraic sign. A positive Q means that heat flows *into* the system; a negative Q means that heat flows *out of* the system. A positive W means that work is done *by* the system on its environment; a negative W means that work is done *on* the system by its environment.

EXAMPLE 19.2 WORKING OFF YOUR DESSERT

You propose to climb several flights of stairs to work off the energy you took in by eating a 900-calorie hot fudge sundae. How high must you climb? Assume that your mass is 60.0 kg.

SOLUTION

IDENTIFY and SET UP: The thermodynamic system is your body. You climb the stairs to make the final state of the system the same as the initial state (no fatter, no leaner). There is therefore no net change in internal energy: $\Delta U = 0$. Eating the hot fudge sundae corresponds to a heat flow into your body, and you do work climbing the stairs. We can relate these quantities by using the first law of thermodynamics. We are given that $Q = 900$ food calories (900 kcal) of heat flow into your body. The work you must do to raise your mass m a height h is $W = mgh$; our target variable is h.

EXECUTE: From the first law of thermodynamics, $\Delta U = 0 = Q - W$, so $W = mgh = Q$. Hence you must climb to height $h = Q/mg$. First convert units: $Q = (900 \text{ kcal})(4186 \text{ J}/1 \text{ kcal}) = 3.77 \times 10^6$ J. Then

$$h = \frac{Q}{mg} = \frac{3.77 \times 10^6 \text{ J}}{(60.0 \text{ kg})(9.80 \text{ m/s}^2)} = 6410 \text{ m}$$

EVALUATE: We have unrealistically assumed 100% efficiency in the conversion of food energy into mechanical work. The actual efficiency is roughly 25%, so the work W you do as you "burn off" the sundae is only about $(0.25)(900 \text{ kcal}) = 225$ kcal. (The remaining 75%, or 675 kcal, is transferred to your surroundings as heat.) Hence you actually must climb about $(0.25)(6410 \text{ m}) = 1600$ m, or one *mile!* Do you really want that sundae?

EXAMPLE 19.3 A CYCLIC PROCESS

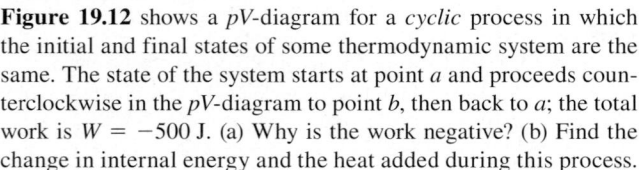

Figure 19.12 shows a *pV*-diagram for a *cyclic* process in which the initial and final states of some thermodynamic system are the same. The state of the system starts at point a and proceeds counterclockwise in the *pV*-diagram to point b, then back to a; the total work is $W = -500$ J. (a) Why is the work negative? (b) Find the change in internal energy and the heat added during this process.

SOLUTION

IDENTIFY and SET UP: We must relate the change in internal energy, the heat added, and the work done in a thermodynamic

process. Hence we can apply the first law of thermodynamics. The process is cyclic, and it has two steps: $a \rightarrow b$ via the lower curve in Fig. 19.12 and $b \rightarrow a$ via the upper curve. We are asked only about the *entire* cyclic process $a \rightarrow b \rightarrow a$.

EXECUTE: (a) The work done in any step equals the area under the curve in the *pV*-diagram, with the area taken as positive if $V_2 > V_1$ and negative if $V_2 < V_1$; this rule yields the signs that result from the actual integrations in Eq. (19.2), $W = \int_{V_1}^{V_2} p\, dV$. The area under the lower curve $a \rightarrow b$ is therefore positive, but it

19.12 The net work done by the system in process $a \rightarrow b \rightarrow a$ is -500 J. What would it have been if the process had proceeded clockwise in this pV-diagram?

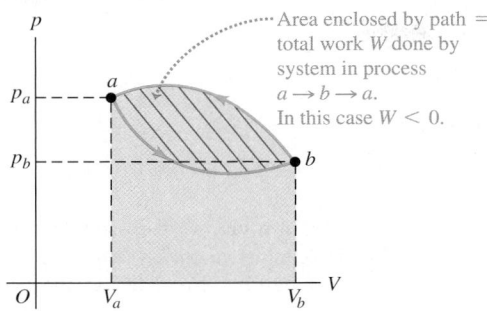

Area enclosed by path = total work W done by system in process $a \rightarrow b \rightarrow a$. In this case $W < 0$.

is smaller than the absolute value of the (negative) area under the upper curve $b \rightarrow a$. Therefore the net area (the area enclosed by the path, shown with red stripes) and the net work W are negative. In other words, 500 J more work is done *on* the system than *by* the system in the complete process.

(b) In any cyclic process, $\Delta U = 0$, so $Q = W$. Here, that means $Q = -500$ J; that is, 500 J of heat flows *out of* the system.

EVALUATE: In cyclic processes, the total work is positive if the process goes clockwise around the pV-diagram representing the cycle, and negative if the process goes counterclockwise (as here).

EXAMPLE 19.4 COMPARING THERMODYNAMIC PROCESSES

The pV-diagram of **Fig. 19.13** shows a series of thermodynamic processes. In process ab, 150 J of heat is added to the system; in process bd, 600 J of heat is added. Find (a) the internal energy change in process ab; (b) the internal energy change in process abd (shown in light blue); and (c) the total heat added in process acd (shown in dark blue).

SOLUTION

IDENTIFY and SET UP: In each process we use $\Delta U = Q - W$ to determine the desired quantity. We are given $Q_{ab} = +150$ J and $Q_{bd} = +600$ J (both values are positive because heat is *added* to the system). Our target variables are (a) ΔU_{ab}, (b) ΔU_{abd}, and (c) Q_{acd}.

EXECUTE: (a) No volume change occurs during process ab, so the system does no work: $W_{ab} = 0$ and so $\Delta U_{ab} = Q_{ab} = 150$ J.

(b) Process bd is an expansion at constant pressure, so from Eq. (19.3),

$$W_{bd} = p(V_2 - V_1)$$
$$= (8.0 \times 10^4 \text{ Pa})(5.0 \times 10^{-3} \text{ m}^3 - 2.0 \times 10^{-3} \text{ m}^3)$$
$$= 240 \text{ J}$$

19.13 A pV-diagram showing the various thermodynamic processes.

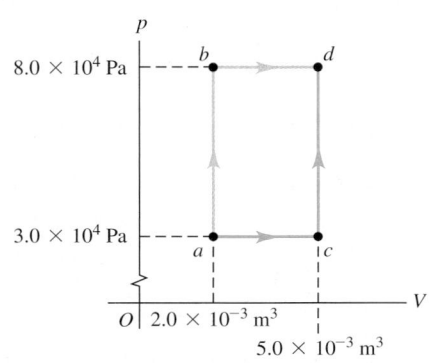

The total work for the two-step process abd is then

$$W_{abd} = W_{ab} + W_{bd} = 0 + 240 \text{ J} = 240 \text{ J}$$

and the total heat is

$$Q_{abd} = Q_{ab} + Q_{bd} = 150 \text{ J} + 600 \text{ J} = 750 \text{ J}$$

Applying Eq. (19.4) to abd, we then have

$$\Delta U_{abd} = Q_{abd} - W_{abd} = 750 \text{ J} - 240 \text{ J} = 510 \text{ J}$$

(c) Because ΔU is *independent of the path* from a to d, the internal energy change is the same for path acd as for path abd:

$$\Delta U_{acd} = \Delta U_{abd} = 510 \text{ J}$$

The total work for path acd is

$$W_{acd} = W_{ac} + W_{cd} = p(V_2 - V_1) + 0$$
$$= (3.0 \times 10^4 \text{ Pa})(5.0 \times 10^{-3} \text{ m}^3 - 2.0 \times 10^{-3} \text{ m}^3)$$
$$= 90 \text{ J}$$

Now we apply Eq. (19.5) to process acd:

$$Q_{acd} = \Delta U_{acd} + W_{acd} = 510 \text{ J} + 90 \text{ J} = 600 \text{ J}$$

We tabulate the quantities above:

Step	Q	W	$\Delta U = Q - W$	Step	Q	W	$\Delta U = Q - W$
ab	150 J	0 J	150 J	ac	?	90 J	?
bd	600 J	240 J	360 J	cd	?	0 J	?
abd	750 J	240 J	510 J	acd	600 J	90 J	510 J

EVALUATE: Be sure that you understand how each entry in the table above was determined. Although ΔU is the same (510 J) for abd and acd, W (240 J versus 90 J) and Q (750 J versus 600 J) are quite different. Although we couldn't find Q or ΔU for processes ac and cd, we could analyze the composite process acd by comparing it with process abd, which has the same initial and final states and for which we have more information.

EXAMPLE 19.5 **THERMODYNAMICS OF BOILING WATER**

One gram of water (1 cm^3) becomes 1671 cm^3 of steam when boiled at a constant pressure of 1 atm $(1.013 \times 10^5 \text{ Pa})$. The heat of vaporization at this pressure is $L_v = 2.256 \times 10^6 \text{ J/kg}$. Compute (a) the work done by the water when it vaporizes and (b) its increase in internal energy.

SOLUTION

IDENTIFY and SET UP: The heat added causes the system (water) to change phase from liquid to vapor. We can analyze this process by using the first law of thermodynamics. The water is boiled at constant pressure, so we can use Eq. (19.3) to calculate the work W done by the vaporizing water as it expands. We are given the mass of water and the heat of vaporization, so we can use Eq. (17.20), $Q = mL_v$, to calculate the heat Q added to the water. We can then find the internal energy change from Eq. (19.4), $\Delta U = Q - W$.

EXECUTE: (a) From Eq. (19.3), the water does work

$$W = p(V_2 - V_1)$$
$$= (1.013 \times 10^5 \text{ Pa})(1671 \times 10^{-6} \text{ m}^3 - 1 \times 10^{-6} \text{ m}^3)$$
$$= 169 \text{ J}$$

(b) From Eq. (17.20), the heat added to the water is

$$Q = mL_v = (10^{-3} \text{ kg})(2.256 \times 10^6 \text{ J/kg}) = 2256 \text{ J}$$

Then from Eq. (19.4),

$$\Delta U = Q - W = 2256 \text{ J} - 169 \text{ J} = 2087 \text{ J}$$

EVALUATE: To vaporize 1 g of water, we must add 2256 J of heat, most of which (2087 J) remains in the system as an increase in internal energy. The remaining 169 J leaves the system as the system expands from liquid to vapor and does work against the surroundings. (The increase in internal energy is associated mostly with the attractive intermolecular forces. The associated potential energies are greater after work has been done to pull apart the molecules in the liquid, forming the vapor state. It's like increasing gravitational potential energy by pulling an elevator farther from the center of the earth.)

Infinitesimal Changes of State

In the preceding examples the initial and final states differ by a finite amount. Later we will consider *infinitesimal* changes of state in which a small amount of heat dQ is added to the system, the system does a small amount of work dW, and its internal energy changes by an amount dU. For such a process,

> **First law of thermodynamics, infinitesimal process:**
>
> Infinitesimal internal energy change
> $$dU = dQ - dW \qquad (19.6)$$
> Infinitesimal heat added   Infinitesimal work done

For the systems we will discuss, the work dW is given by $dW = p \, dV$, so we can also state the first law as

$$dU = dQ - p \, dV \qquad (19.7)$$

TEST YOUR UNDERSTANDING OF SECTION 19.4 Rank the following thermodynamic processes according to the change in internal energy in each process, from most positive to most negative. (i) As you do 250 J of work on a system, it transfers 250 J of heat to its surroundings; (ii) as you do 250 J of work on a system, it absorbs 250 J of heat from its surroundings; (iii) as a system does 250 J of work on you, it transfers 250 J of heat to its surroundings; (iv) as a system does 250 J of work on you, it absorbs 250 J of heat from its surroundings. ▮

19.5 KINDS OF THERMODYNAMIC PROCESSES

In this section we describe four specific kinds of thermodynamic processes that occur often in practical situations. We can summarize these briefly as "no heat transfer" or *adiabatic,* "constant volume" or *isochoric,* "constant pressure" or *isobaric,* and "constant temperature" or *isothermal.* For some of these processes we can use a simplified form of the first law of thermodynamics.

Adiabatic Process

An **adiabatic process** (pronounced "ay-dee-ah-*bat*-ic") is defined as one with no heat transfer into or out of a system; $Q = 0$. We can prevent heat flow either by surrounding the system with thermally insulating material or by carrying out the process so quickly that there is not enough time for appreciable heat flow. From the first law we find that for every adiabatic process,

$$U_2 - U_1 = \Delta U = -W \qquad \text{(adiabatic process)} \qquad (19.8)$$

When a system expands adiabatically, W is positive (the system does work on its surroundings), so ΔU is negative and the internal energy decreases. When a system is *compressed* adiabatically, W is negative (work is done on the system by its surroundings) and U increases. In many (but not all) systems an increase of internal energy is accompanied by a rise in temperature, and a decrease in internal energy by a drop in temperature (**Fig. 19.14**).

The compression stroke in an internal-combustion engine is an approximately adiabatic process. The temperature rises as the air–fuel mixture in the cylinder is compressed. The expansion of the burned fuel during the power stroke is also an approximately adiabatic expansion with a drop in temperature. In Section 19.8 we'll consider adiabatic processes in an ideal gas.

Isochoric Process

An **isochoric process** (pronounced "eye-so-*kor*-ic") is a *constant-volume* process. When the volume of a thermodynamic system is constant, it does no work on its surroundings. Then $W = 0$ and

$$U_2 - U_1 = \Delta U = Q \qquad \text{(isochoric process)} \qquad (19.9)$$

In an isochoric process, all the energy added as heat remains in the system as an increase in internal energy. Heating a gas in a closed constant-volume container is an example of an isochoric process. The processes *ab* and *cd* in Example 19.4 are also examples of isochoric processes. (Note that there are types of work that do not involve a volume change. For example, we can do work on a fluid by stirring it. In some literature, "isochoric" is used to mean that no work of any kind is done.)

Isobaric Process

An **isobaric process** (pronounced "eye-so-*bear*-ic") is a *constant-pressure* process. In general, none of the three quantities ΔU, Q, and W is zero in an isobaric process, but calculating W is easy nonetheless. From Eq. (19.3),

$$W = p(V_2 - V_1) \qquad \text{(isobaric process)} \qquad (19.10)$$

Boiling water at constant pressure is an isobaric process (**Fig. 19.15**).

Isothermal Process

An **isothermal process** is a *constant-temperature* process. For a process to be isothermal, any heat flow into or out of the system must occur slowly enough that thermal equilibrium is maintained. In general, none of the quantities ΔU, Q, or W is zero in an isothermal process.

In some special cases the internal energy of a system depends *only* on its temperature, not on its pressure or volume. The most familiar system having this special property is an ideal gas, as we'll discuss in the next section. For such systems, if the temperature is constant, the internal energy is also constant;

19.14 When the cork is popped on a bottle of champagne, the pressurized gases inside the bottle expand rapidly and do work on the outside air ($W > 0$). There is little time for the gases to exchange heat with their surroundings, so the expansion is nearly adiabatic ($Q = 0$). Hence the internal energy of the expanding gases decreases ($\Delta U = -W < 0$) and their temperature drops. This makes water vapor condense and form a miniature cloud.

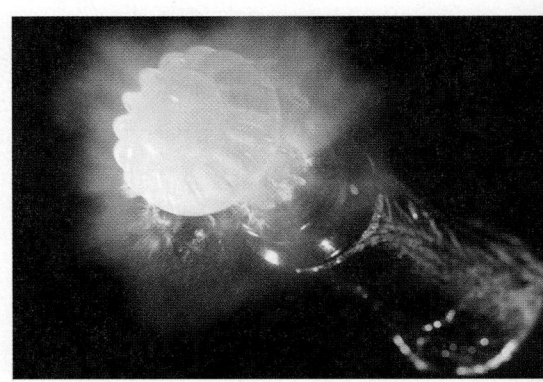

19.15 Most cooking involves isobaric processes. That's because the air pressure above a saucepan or frying pan, or inside a microwave oven, remains essentially constant while the food is being heated.

19.16 Four different processes for a constant amount of an ideal gas, all starting at state a. For the adiabatic process, $Q = 0$; for the isochoric process, $W = 0$; and for the isothermal process, $\Delta U = 0$. The temperature increases only during the isobaric expansion.

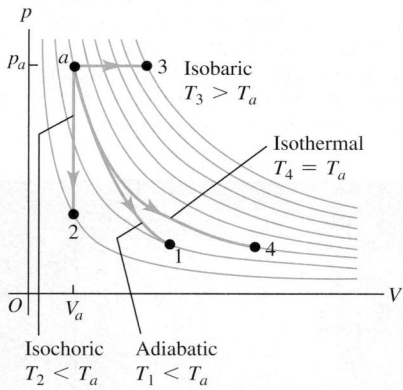

19.17 The partition is broken (or removed) to start the free expansion of gas into the vacuum region.

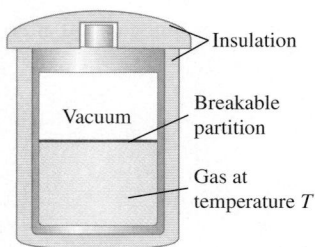

$\Delta U = 0$ and $Q = W$. That is, any energy entering the system as heat Q must leave it again as work W done by the system. Example 19.1, involving an ideal gas, is an example of an isothermal process in which U is also constant. For most systems other than ideal gases, the internal energy depends on pressure as well as temperature, so U may vary even when T is constant.

Figure 19.16 shows a pV-diagram for these four processes for a constant amount of an ideal gas. The path followed in an adiabatic process (a to 1) is called an **adiabat.** A vertical line (constant volume) is an **isochor,** a horizontal line (constant pressure) is an **isobar,** and a curve of constant temperature (shown as light blue lines in Fig. 19.16) is an **isotherm.**

TEST YOUR UNDERSTANDING OF SECTION 19.5 Which of the processes in Fig. 19.7 are isochoric? Which are isobaric? Is it possible to tell if any of the processes are isothermal or adiabatic? ∎

19.6 INTERNAL ENERGY OF AN IDEAL GAS

We now show that for an ideal gas, the internal energy U depends only on temperature, not on pressure or volume. Let's think again about the free-expansion experiment described in Section 19.3. A thermally insulated container with rigid walls is divided into two compartments by a partition (**Fig. 19.17**). One compartment has a quantity of an ideal gas and the other is evacuated.

When the partition is removed or broken, the gas expands to fill both parts of the container. There is no heat flow through the insulation, and the gas does no work on its surroundings because the walls of the container don't move. So both Q and W are zero and the internal energy U is constant.

Does the *temperature T* change during a free expansion? Suppose it *does* change, while the internal energy stays the same. In that case we have to conclude that the internal energy depends on both T and the volume V or on both T and the pressure p, but certainly not on T alone. But if T is constant during a free expansion, for which we know that U is constant even though both p and V change, then we have to conclude that U depends only on T, not on p or V.

Many experiments have shown that when a low-density gas (essentially an ideal gas) undergoes a free expansion, its temperature *does not* change. The conclusion is:

> The internal energy U of an ideal gas depends only on its temperature T, not on its pressure or volume.

This property, in addition to the ideal-gas equation of state, is part of the ideal-gas model. We'll make frequent use of this property.

For nonideal gases, some temperature change occurs during free expansions, even though the internal energy is constant. This shows that the internal energy cannot depend *only* on temperature; it must depend on pressure as well. From the microscopic viewpoint, in which internal energy U is the sum of the kinetic and potential energies for all the particles that make up the system, this is not surprising. Nonideal gases usually have attractive intermolecular forces, and when molecules move farther apart, the associated potential energies increase. If the total internal energy is constant, the kinetic energies must decrease. Temperature is directly related to molecular *kinetic* energy, and for such a gas a free expansion is usually accompanied by a *drop* in temperature.

TEST YOUR UNDERSTANDING OF SECTION 19.6 Is the internal energy of a solid likely to be independent of its volume, as is the case for an ideal gas? Explain your reasoning. (*Hint:* See Fig. 18.20.) ∎

19.7 HEAT CAPACITIES OF AN IDEAL GAS

We defined specific heat and molar heat capacity in Section 17.5. We also remarked at the end of that section that the specific heat or molar heat capacity of a substance depends on the conditions under which the heat is added. The heat capacity of a gas is usually measured in a closed container under constant-volume conditions. The corresponding heat capacity is the **molar heat capacity at constant volume,** denoted by C_V. Heat capacity measurements for solids and liquids are usually carried out under constant atmospheric pressure, and we call the corresponding heat capacity the **molar heat capacity at constant pressure, C_p.**

Let's consider C_V and C_p for an ideal gas. To measure C_V, we raise the temperature of an ideal gas in a rigid container with constant volume, ignoring its thermal expansion (**Fig. 19.18a**). To measure C_p, we let the gas expand just enough to keep the pressure constant as the temperature rises (Fig. 19.18b).

Why should these two molar heat capacities be different? The answer lies in the first law of thermodynamics. In a constant-volume temperature increase, the system does no work, and the change in internal energy ΔU equals the heat added Q. In a constant-pressure temperature increase, on the other hand, the volume *must* increase; otherwise, the pressure (given by the ideal-gas equation of state, $p = nRT/V$) could not remain constant. As the material expands, it does an amount of work W. According to the first law,

$$Q = \Delta U + W \qquad (19.11)$$

For a given temperature increase, the internal energy change ΔU of an ideal gas has the same value no matter what the process (remember that the internal energy of an ideal gas depends only on temperature, not on pressure or volume). Equation (19.11) then shows that the heat input for a constant-pressure process must be *greater* than that for a constant-volume process because additional energy must be supplied to account for the work W done during the expansion. So C_p is greater than C_V for an ideal gas. The pV-diagram in **Fig. 19.19** shows this relationship. For air, C_p is 40% greater than C_V.

For a very few substances (one of which is water between 0°C and 4°C) the volume *decreases* during heating. In this case, W is negative and the internal energy change ΔU is greater than the heat input Q.

Relating C_p and C_V for an Ideal Gas

We can derive a simple relationship between C_p and C_V for an ideal gas. First consider the constant-*volume* process. We place n moles of an ideal gas at temperature T in a constant-volume container. We place it in thermal contact with a hotter body; an infinitesimal quantity of heat dQ flows into the gas, and its temperature increases by an infinitesimal amount dT. By the definition of C_V, the molar heat capacity at constant volume,

$$dQ = nC_V \, dT \qquad (19.12)$$

The pressure increases during this process, but the gas does no work ($dW = 0$) because the volume is constant. The first law in differential form, Eq. (19.6), is $dQ = dU + dW$. Since $dW = 0$, $dQ = dU$ and Eq. (19.12) can also be written as

$$dU = nC_V \, dT \qquad (19.13)$$

Now consider a constant-*pressure* process with the same temperature change dT. We place the same gas in a cylinder with a piston that we allow to move just enough to maintain constant pressure (Fig. 19.18b). Again we bring the system into contact with a hotter body. As heat flows into the gas, it expands at constant pressure and does work. By the definition of C_p, the molar heat capacity at

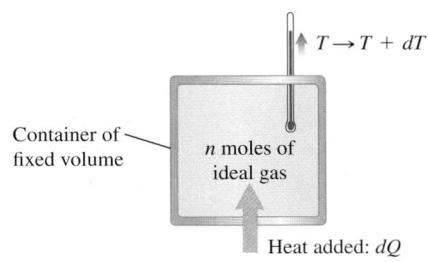

19.18 Measuring the molar heat capacity of an ideal gas (a) at constant volume and (b) at constant pressure.

(a) Constant volume: $dQ = nC_V \, dT$

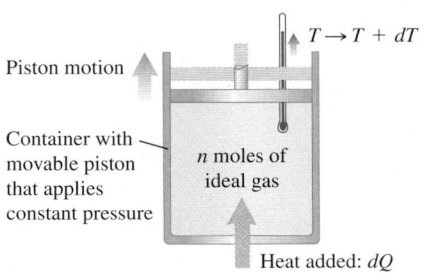

(b) Constant pressure: $dQ = nC_p \, dT$

19.19 Raising the temperature of an ideal gas from T_1 to T_2 by a constant-volume or a constant-pressure process. For an ideal gas, U depends only on T, so ΔU is the same for both processes. But for the constant-pressure process, more heat Q must be added to both increase U and do work W. Hence $C_p > C_V$.

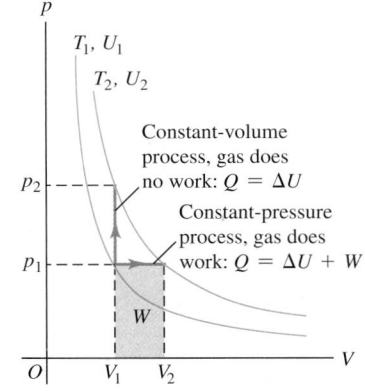

constant pressure, the amount of heat entering the gas is

$$dQ = nC_p\,dT \qquad (19.14)$$

The work dW done by the gas in this constant-pressure process is

$$dW = p\,dV$$

We can also express dW in terms of the temperature change dT by using the ideal-gas equation of state, $pV = nRT$. Because p is constant, the change in V is proportional to the change in T:

$$dW = p\,dV = nR\,dT \qquad (19.15)$$

Now substitute Eqs. (19.14) and (19.15) into the first law, $dQ = dU + dW$:

$$nC_p\,dT = dU + nR\,dT \qquad (19.16)$$

Here comes the crux of the calculation. The internal energy change dU for the constant-pressure process is again given by Eq. (19.13), $dU = nC_V\,dT$, *even though now the volume is not constant.* Why is this so? From Section 19.6, one of the special properties of an ideal gas is that its internal energy depends *only* on temperature. Thus the *change* in internal energy during any process must be determined only by the temperature change. If Eq. (19.13) is valid for an ideal gas for one particular kind of process, it must be valid for an ideal gas for *every* kind of process with the same dT. So we may replace dU in Eq. (19.16) by $nC_V\,dT$:

$$nC_p\,dT = nC_V\,dT + nR\,dT$$

When we divide each term by the common factor $n\,dT$, we get

For an ideal gas:
$$\underset{\text{Molar heat capacity at constant }volume}{\overset{\text{Molar heat capacity at constant }pressure}{C_p = C_V + R}} \qquad \text{Gas constant} \qquad (19.17)$$

As we predicted, the molar heat capacity of an ideal gas at constant pressure is *greater* than the molar heat capacity at constant volume; the difference is the gas constant R.

We have used the ideal-gas model to derive Eq. (19.17), but it turns out to be obeyed to within a few percent by many real gases at moderate pressures. **Table 19.1** gives measured values of C_p and C_V for several real gases at low pressures; the difference in most cases is approximately $R = 8.314\text{ J/mol}\cdot\text{K}$.

The table also shows that the molar heat capacity of a gas is related to its molecular structure, as we discussed in Section 18.4. In fact, the first two columns of Table 19.1 are the same as Table 18.1.

TABLE 19.1 | **Molar Heat Capacities of Gases at Low Pressure**

Type of Gas	Gas	C_V (J/mol · K)	C_p (J/mol · K)	$C_p - C_V$ (J/mol · K)	$\gamma = C_p/C_V$
Monatomic	He	12.47	20.78	8.31	1.67
	Ar	12.47	20.78	8.31	1.67
Diatomic	H_2	20.42	28.74	8.32	1.41
	N_2	20.76	29.07	8.31	1.40
	O_2	20.85	29.17	8.32	1.40
	CO	20.85	29.16	8.31	1.40
Polyatomic	CO_2	28.46	36.94	8.48	1.30
	SO_2	31.39	40.37	8.98	1.29
	H_2S	25.95	34.60	8.65	1.33

The Ratio of Heat Capacities

The last column of Table 19.1 lists the values of the dimensionless **ratio of heat capacities,** C_p/C_V, denoted by γ (the Greek letter gamma):

Ratio of heat capacities $\longrightarrow \gamma = \dfrac{C_p}{C_V}$ $\cdots\cdots$ Molar heat capacity at constant *pressure*

$\cdots\cdots$ Molar heat capacity at constant *volume* (19.18)

(This is sometimes called the "ratio of specific heats.") For gases, C_p is always greater than C_V and γ is always greater than unity. We'll see in the next section that γ plays an important role in adiabatic processes for an ideal gas.

We can use our kinetic-theory discussion of the molar heat capacity of an ideal gas (see Section 18.4) to predict values of γ. As an example, an ideal monatomic gas has $C_V = \frac{3}{2}R$. From Eq. (19.17),

$$C_p = C_V + R = \tfrac{3}{2}R + R = \tfrac{5}{2}R$$

so

$$\gamma = \frac{C_p}{C_V} = \frac{\frac{5}{2}R}{\frac{3}{2}R} = \tfrac{5}{3} = 1.67$$

As Table 19.1 shows, this agrees well with values of γ computed from measured heat capacities. For most diatomic gases near room temperature, $C_V = \frac{5}{2}R$, $C_p = C_V + R = \frac{7}{2}R$, and

$$\gamma = \frac{C_p}{C_V} = \frac{\frac{7}{2}R}{\frac{5}{2}R} = \tfrac{7}{5} = 1.40$$

also in good agreement with measured values.

Here's a final reminder: For an ideal gas, the internal energy change in *any* process is given by $\Delta U = nC_V\,\Delta T$, *whether the volume is constant or not.* This relationship holds for other substances *only* when the volume is constant.

SOLUTION

EXAMPLE 19.6 COOLING YOUR ROOM

A typical dorm room or bedroom contains about 2500 moles of air. Find the change in the internal energy of this much air when it is cooled from 35.0°C to 26.0°C at a constant pressure of 1.00 atm. Treat the air as an ideal gas with $\gamma = 1.400$.

SOLUTION

IDENTIFY and SET UP: Our target variable is the change in the internal energy ΔU of an ideal gas in a constant-pressure process. We are given the number of moles, the temperature change, and the value of γ for air. We use Eq. (19.13), $\Delta U = nC_V\,\Delta T$, which gives the internal energy change for an ideal gas in *any* process, *whether the volume is constant or not.* [See the discussion following Eq. (19.16).] We use Eqs. (19.17) and (19.18) to find C_V.

EXECUTE: From Eqs. (19.17) and (19.18),

$$\gamma = \frac{C_p}{C_V} = \frac{C_V + R}{C_V} = 1 + \frac{R}{C_V}$$

$$C_V = \frac{R}{\gamma - 1} = \frac{8.314\ \text{J/mol} \cdot \text{K}}{1.400 - 1} = 20.79\ \text{J/mol} \cdot \text{K}$$

Then from Eq. (19.13),

$$\Delta U = nC_V\,\Delta T$$
$$= (2500\ \text{mol})(20.79\ \text{J/mol} \cdot \text{K})(26.0°\text{C} - 35.0°\text{C})$$
$$= -4.68 \times 10^5\ \text{J}$$

EVALUATE: To cool 2500 moles of air from 35.0°C to 26.0°C, a room air conditioner must extract this much internal energy from the air and transfer it to the air outside. In Chapter 20 we'll discuss how this is done.

TEST YOUR UNDERSTANDING OF SECTION 19.7 You want to cool a storage cylinder containing 10 moles of compressed gas from 30°C to 20°C. For which kind of gas would this be easiest? (i) A monatomic gas; (ii) a diatomic gas; (iii) a polyatomic gas; (iv) it would be equally easy for all of these. ∎

19.8 ADIABATIC PROCESSES FOR AN IDEAL GAS

An adiabatic process, defined in Section 19.5, is a process in which no heat transfer takes place between a system and its surroundings. Zero heat transfer is an idealization, but a process is approximately adiabatic if the system is well insulated or if the process takes place so quickly that there is not enough time for appreciable heat flow to occur.

In an adiabatic process, $Q = 0$, so from the first law, $\Delta U = -W$. An adiabatic process for an ideal gas is shown in the pV-diagram of **Fig. 19.20**. As the gas expands from volume V_a to V_b, it does positive work, so its internal energy decreases and its temperature drops. If point a, representing the initial state, lies on an isotherm at temperature $T + dT$, then point b for the final state is on a different isotherm at a lower temperature T. For an adiabatic *compression* from V_b to V_a the situation is reversed and the temperature rises.

The air in the output hoses of air compressors used to inflate tires and to fill scuba tanks is always warmer than the air entering the compressor; this is because the compression is rapid and hence approximately adiabatic. Adiabatic *cooling* occurs when you open a bottle of your favorite carbonated beverage. The gas just above the beverage surface expands rapidly in a nearly adiabatic process; the gas temperature drops so much that water vapor in the gas condenses, forming a miniature cloud (see Fig. 19.14).

CAUTION **"Heating" and "cooling" without heat** When we talk about "adiabatic heating" and "adiabatic cooling," we really mean "raising the temperature" and "lowering the temperature," respectively. In an adiabatic process, the temperature change is due to work done by or on the system; there is *no* heat flow at all. ▌

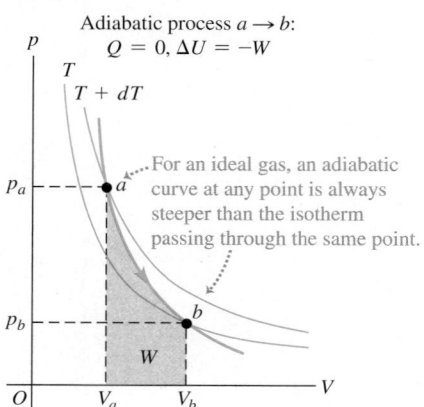

19.20 A pV-diagram of an adiabatic ($Q = 0$) process for an ideal gas. As the gas expands from V_a to V_b, it does positive work W on its environment, its internal energy decreases ($\Delta U = -W < 0$), and its temperature drops from $T + dT$ to T. (An adiabatic process is also shown in Fig. 19.16.)

Adiabatic process $a \rightarrow b$:
$Q = 0, \Delta U = -W$

For an ideal gas, an adiabatic curve at any point is always steeper than the isotherm passing through the same point.

Adiabatic Ideal Gas: Relating V, T, and p

We can derive a relationship between volume and temperature changes for an infinitesimal adiabatic process in an ideal gas. Equation (19.13) gives the internal energy change dU for *any* process for an ideal gas, adiabatic or not, so we have $dU = nC_V\, dT$. Also, the work done by the gas during the process is given by $dW = p\, dV$. Then, since $dU = -dW$ for an adiabatic process, we have

$$nC_V\, dT = -p\, dV \qquad (19.19)$$

To obtain a relationship containing only the volume V and temperature T, we eliminate p by using the ideal-gas equation in the form $p = nRT/V$. Substituting this into Eq. (19.19) and rearranging, we get

$$nC_V\, dT = -\frac{nRT}{V}dV$$

$$\frac{dT}{T} + \frac{R}{C_V}\frac{dV}{V} = 0$$

The coefficient R/C_V can be expressed in terms of $\gamma = C_p/C_V$. We have

$$\frac{R}{C_V} = \frac{C_p - C_V}{C_V} = \frac{C_p}{C_V} - 1 = \gamma - 1$$

$$\frac{dT}{T} + (\gamma - 1)\frac{dV}{V} = 0 \qquad (19.20)$$

Because γ is always greater than unity for a gas, $(\gamma - 1)$ is always positive. This means that in Eq. (19.20), dV and dT always have opposite signs. An adiabatic *expansion* of an ideal gas ($dV > 0$) always occurs with a *drop* in temperature ($dT < 0$), and an adiabatic *compression* ($dV < 0$) always occurs with a *rise* in temperature ($dT > 0$); this confirms our earlier prediction.

For finite changes in temperature and volume we integrate Eq. (19.20), obtaining

$$\ln T + (\gamma - 1)\ln V = \text{constant}$$

$$\ln T + \ln V^{\gamma-1} = \text{constant}$$

$$\ln(TV^{\gamma-1}) = \text{constant}$$

and finally,

$$TV^{\gamma-1} = \text{constant} \qquad (19.21)$$

Thus for an initial state (T_1, V_1) and a final state (T_2, V_2),

$$T_1 V_1^{\gamma-1} = T_2 V_2^{\gamma-1} \qquad \text{(adiabatic process, ideal gas)} \qquad (19.22)$$

Because we have used the ideal-gas equation in our derivation of Eqs. (19.21) and (19.22), the T's must always be *absolute* (Kelvin) temperatures.

We can also convert Eq. (19.21) into a relationship between pressure and volume by eliminating T, using the ideal-gas equation in the form $T = pV/nR$. Substituting this into Eq. (19.21), we find

$$\frac{pV}{nR}V^{\gamma-1} = \text{constant}$$

or, because n and R are constant,

$$pV^{\gamma} = \text{constant} \qquad (19.23)$$

For an initial state (p_1, V_1) and a final state (p_2, V_2), Eq. (19.23) becomes

$$p_1 V_1^{\gamma} = p_2 V_2^{\gamma} \qquad \text{(adiabatic process, ideal gas)} \qquad (19.24)$$

We can also calculate the *work* done by an ideal gas during an adiabatic process. We know that $Q = 0$ and $W = -\Delta U$ for *any* adiabatic process. For an ideal gas, $\Delta U = nC_V(T_2 - T_1)$. If the number of moles n and the initial and final temperatures T_1 and T_2 are known, we have simply

Work done by an ideal gas, adiabatic process · · · · Number of moles · · · · Initial temperature
$$W = nC_V(T_1 - T_2) \qquad (19.25)$$
Final temperature
Molar heat capacity at constant volume

We may also use $pV = nRT$ in this equation to obtain

Work done by an ideal gas, adiabatic process · · · Molar heat capacity at constant volume · · · Initial pressure, volume
$$W = \frac{C_V}{R}(p_1 V_1 - p_2 V_2) = \frac{1}{\gamma - 1}(p_1 V_1 - p_2 V_2) \qquad (19.26)$$
Gas constant · · · Ratio of heat capacities · · · Final pressure, volume

(We used the result $C_V = R/(\gamma - 1)$ from Example 19.6.) If the process is an expansion, the temperature drops, T_1 is greater than T_2, $p_1 V_1$ is greater than $p_2 V_2$, and the work is *positive*. If the process is a compression, the work is negative.

Throughout this analysis of adiabatic processes we have used the ideal-gas equation of state, which is valid only for *equilibrium* states. Strictly speaking, our results are valid only for a process that is fast enough to prevent appreciable heat exchange with the surroundings (so that $Q = 0$ and the process is adiabatic), yet slow enough that the system does not depart very much from thermal and mechanical equilibrium. Even when these conditions are not strictly satisfied, though, Eqs. (19.22), (19.24), and (19.26) give useful approximate results.

BIO **Application Exhaling Adiabatically** Put your hand a few centimeters in front of your mouth, open your mouth wide, and exhale. Your breath will feel warm on your hand, because the exhaled gases emerge at roughly the temperature of your body's interior. Now purse your lips as though you were going to whistle, and again blow on your hand. The exhaled gases will feel much cooler. In this case the gases undergo a rapid, essentially adiabatic expansion as they emerge from between your lips, so the temperature of the exhaled gases decreases.

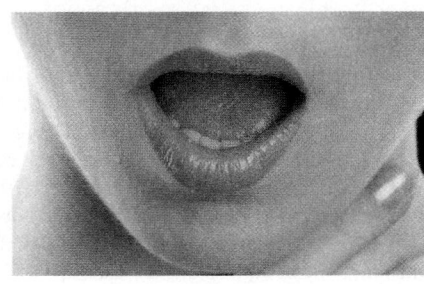

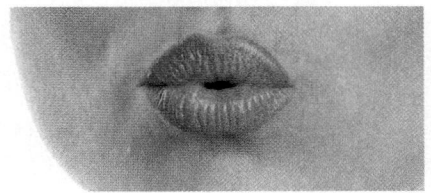

EXAMPLE 19.7 ADIABATIC COMPRESSION IN A DIESEL ENGINE

The compression ratio of a diesel engine is 15.0 to 1; that is, air in a cylinder is compressed to $\frac{1}{(15.0)}$ of its initial volume (**Fig. 19.21**). (a) If the initial pressure is 1.01×10^5 Pa and the initial temperature is 27°C (300 K), find the final pressure and the temperature after adiabatic compression. (b) How much work does the gas do during the compression if the initial volume of the cylinder is 1.00 L $= 1.00 \times 10^{-3}$ m³? Use the values $C_V = 20.8$ J/mol·K and $\gamma = 1.400$ for air.

SOLUTION

IDENTIFY and SET UP: This problem involves the adiabatic compression of an ideal gas, so we can use the ideas of this section. In part (a) we are given the initial pressure and temperature $p_1 = 1.01 \times 10^5$ Pa and $T_1 = 300$ K; the ratio of initial and final volumes is $V_1/V_2 = 15.0$. We use Eq. (19.22) to find the final temperature T_2 and Eq. (19.24) to find the final pressure p_2. In part (b) our target variable is W, the work done *by* the gas during the adiabatic compression. We use Eq. (19.26) to calculate W.

EXECUTE: (a) From Eqs. (19.22) and (19.24),

$$T_2 = T_1 \left(\frac{V_1}{V_2}\right)^{\gamma-1} = (300 \text{ K})(15.0)^{0.40} = 886 \text{ K} = 613°C$$

$$p_2 = p_1 \left(\frac{V_1}{V_2}\right)^{\gamma} = (1.01 \times 10^5 \text{ Pa})(15.0)^{1.40}$$

$$= 44.8 \times 10^5 \text{ Pa} = 44 \text{ atm}$$

(b) From Eq. (19.26), the work done is

$$W = \frac{1}{\gamma - 1}(p_1 V_1 - p_2 V_2)$$

Using $V_1/V_2 = 15.0$, we have

$$W = \frac{1}{1.400 - 1}\left[\begin{array}{l}(1.01 \times 10^5 \text{ Pa})(1.00 \times 10^{-3} \text{ m}^3) \\ - (44.8 \times 10^5 \text{ Pa})\left(\dfrac{1.00 \times 10^{-3} \text{ m}^3}{15.0}\right)\end{array}\right]$$

$$= -494 \text{ J}$$

19.21 Adiabatic compression of air in a cylinder of a diesel engine.

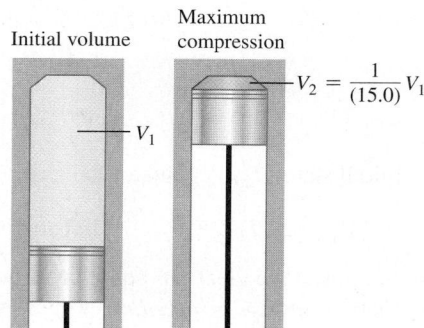

EVALUATE: If the compression had been isothermal, the final pressure would have been 15.0 atm. Because the temperature also increases during an adiabatic compression, the final pressure is much greater. When fuel is injected into the cylinders near the end of the compression stroke, the high temperature of the air attained during compression causes the fuel to ignite spontaneously without the need for spark plugs.

We can check our result in part (b) by using Eq. (19.25). The number of moles of gas in the cylinder is

$$n = \frac{p_1 V_1}{RT_1} = \frac{(1.01 \times 10^5 \text{ Pa})(1.00 \times 10^{-3} \text{ m}^3)}{(8.314 \text{ J/mol·K})(300 \text{ K})} = 0.0405 \text{ mol}$$

Then Eq. (19.25) gives

$$W = nC_V(T_1 - T_2)$$

$$= (0.0405 \text{ mol})(20.8 \text{ J/mol·K})(300 \text{ K} - 886 \text{ K})$$

$$= -494 \text{ J}$$

The work is negative because the gas is compressed.

TEST YOUR UNDERSTANDING OF SECTION 19.8 You have four samples of ideal gas, each of which contains the same number of moles of gas and has the same initial temperature, volume, and pressure. You compress each sample to one-half of its initial volume. Rank the four samples in order from highest to lowest value of the final pressure. (i) A monatomic gas compressed isothermally; (ii) a monatomic gas compressed adiabatically; (iii) a diatomic gas compressed isothermally; (iv) a diatomic gas compressed adiabatically. ∎

Heat and work in thermodynamic processes: A thermodynamic system has the potential to exchange energy with its surroundings by heat transfer or by mechanical work. When a system at pressure p changes volume from V_1 to V_2, it does an amount of work W given by the integral of p with respect to volume. If the pressure is constant, the work done is equal to p times the change in volume. A negative value of W means that work is done on the system. (See Example 19.1.)

In any thermodynamic process, the heat added to the system and the work done by the system depend not only on the initial and final states, but also on the path (the series of intermediate states through which the system passes).

$$W = \int_{V_1}^{V_2} p \, dV \tag{19.2}$$

$$W = p(V_2 - V_1) \tag{19.3}$$
(constant pressure only)

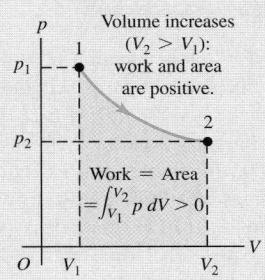

The first law of thermodynamics: The first law of thermodynamics states that when heat Q is added to a system while the system does work W, the internal energy U changes by an amount equal to $Q - W$. This law can also be expressed for an infinitesimal process. (See Examples 19.2, 19.3, and 19.5.)

The internal energy of any thermodynamic system depends only on its state. The change in internal energy in any process depends only on the initial and final states, not on the path. The internal energy of an isolated system is constant. (See Example 19.4.)

$$\Delta U = Q - W \tag{19.4}$$

$$dU = dQ - dW \tag{19.6}$$
(infinitesimal process)

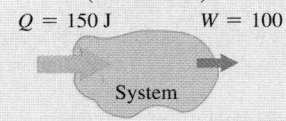

Important kinds of thermodynamic processes:

- Adiabatic process: No heat transfer into or out of a system; $Q = 0$.
- Isochoric process: Constant volume; $W = 0$.
- Isobaric process: Constant pressure; $W = p(V_2 - V_1)$.
- Isothermal process: Constant temperature.

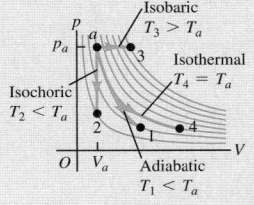

Thermodynamics of ideal gases: The internal energy of an ideal gas depends only on its temperature, not on its pressure or volume. For other substances the internal energy generally depends on both pressure and temperature.

The molar heat capacities C_V and C_p of an ideal gas differ by R, the ideal-gas constant. The dimensionless ratio of heat capacities, C_p/C_V, is denoted by γ. (See Example 19.6.)

$$C_p = C_V + R \tag{19.17}$$

$$\gamma = \frac{C_p}{C_V} \tag{19.18}$$

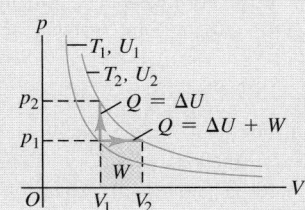

Adiabatic processes in ideal gases: For an adiabatic process for an ideal gas, the quantities $TV^{\gamma-1}$ and pV^γ are constant. The work done by an ideal gas during an adiabatic expansion can be expressed in terms of the initial and final values of temperature, or in terms of the initial and final values of pressure and volume. (See Example 19.7.)

$$W = nC_V(T_1 - T_2) \tag{19.25}$$
$$= \frac{C_V}{R}(p_1 V_1 - p_2 V_2)$$
$$= \frac{1}{\gamma - 1}(p_1 V_1 - p_2 V_2) \tag{19.26}$$

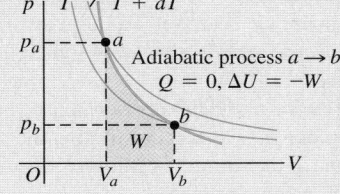

637

BRIDGING PROBLEM WORK DONE BY A VAN DER WAALS GAS

The van der Waals equation of state, an approximate representation of the behavior of gases at high pressure, is given by Eq. (18.7): $[p + (an^2/V^2)](V - nb) = nRT$, where a and b are constants having different values for different gases. (In the special case of $a = b = 0$, this is the ideal-gas equation.) (a) Calculate the work done by a gas with this equation of state in an isothermal expansion from V_1 to V_2. (b) For ethane gas (C_2H_6), $a = 0.554 \text{ J} \cdot \text{m}^3/\text{mol}^2$ and $b = 6.38 \times 10^{-5} \text{ m}^3/\text{mol}$. Calculate the work W done by 1.80 mol of ethane when it expands from $2.00 \times 10^{-3} \text{ m}^3$ to $4.00 \times 10^{-3} \text{ m}^3$ at a constant temperature of 300 K. Do the calculation by using (i) the van der Waals equation of state and (ii) the ideal-gas equation of state. (c) For which equation of state is W larger? Why should this be so?

SOLUTION GUIDE

IDENTIFY and SET UP

1. Review the discussion of the van der Waals equation of state in Section 18.1. What is the significance of the quantities a and b?
2. Decide how to find the work done by an expanding gas whose pressure p does not depend on V in the same way as for an ideal gas. (*Hint:* See Section 19.2.)
3. How will you find the work done by an expanding ideal gas?

EXECUTE

4. Find the general expression for the work done by a van der Waals gas as it expands from volume V_1 to volume V_2 (**Fig. 19.22**).

(*Hint:* If you set $a = b = 0$ in your result, it should reduce to the expression for the work done by an expanding ideal gas.)
5. Use your result from step 4 to solve part (b) for ethane treated as a van der Waals gas.
6. Use the formula you chose in step 3 to solve part (b) for ethane treated as an ideal gas.

EVALUATE

7. Is the difference between W for the two equations of state large enough to be significant?
8. Does the term with a in the van der Waals equation of state increase or decrease the amount of work done? What about the term with b? Which one is more important for the ethane in this problem?

19.22 A gas undergoes an isothermal expansion.

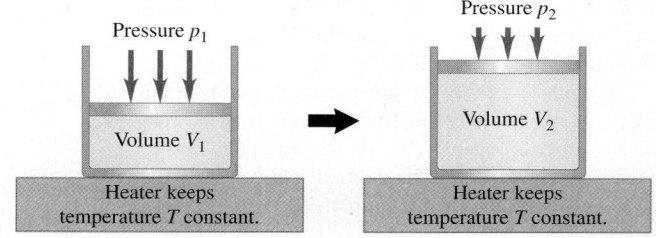

Problems

For assigned homework and other learning materials, go to MasteringPhysics®. MP

•, ••, •••: Difficulty levels. **CP**: Cumulative problems incorporating material from earlier chapters. **CALC**: Problems requiring calculus. **DATA**: Problems involving real data, scientific evidence, experimental design, and/or statistical reasoning. **BIO**: Biosciences problems.

DISCUSSION QUESTIONS

Q19.1 For the following processes, is the work done by the system (defined as the expanding or contracting gas) on the environment positive or negative? (a) expansion of the burned gasoline–air mixture in the cylinder of an automobile engine; (b) opening a bottle of champagne; (c) filling a scuba tank with compressed air; (d) partial crumpling of a sealed, empty water bottle as you drive from the mountains down to sea level.

Q19.2 It is not correct to say that a body contains a certain amount of heat, yet a body can transfer heat to another body. How can a body give away something it does not have in the first place?

Q19.3 In which situation must you do more work: inflating a balloon at sea level or inflating the same balloon to the same volume at the summit of Mt. McKinley? Explain in terms of pressure and volume change.

Q19.4 If you are told the initial and final states of a system and the associated change in internal energy, can you determine whether the internal energy change was due to work or to heat transfer? Explain.

Q19.5 Discuss the application of the first law of thermodynamics to a mountaineer who eats food, gets warm and perspires a lot during a climb, and does a lot of mechanical work in raising herself to the summit. The mountaineer also gets warm during the descent. Is the source of this energy the same as the source during the ascent?

Q19.6 When ice melts at 0°C, its volume decreases. Is the internal energy change greater than, less than, or equal to the heat added? How can you tell?

Q19.7 You hold an inflated balloon over a hot-air vent in your house and watch it slowly expand. You then remove it and let it cool back to room temperature. During the expansion, which was larger: the heat added to the balloon or the work done by the air inside it? Explain. (Assume that air is an ideal gas.) Once the balloon has returned to room temperature, how does the net heat gained or lost by the air inside it compare to the net work done on or by the surrounding air?

Q19.8 You bake chocolate chip cookies and put them, still warm, in a container with a loose (not airtight) lid. What kind of process does the air inside the container undergo as the cookies gradually cool to room temperature (isothermal, isochoric, adiabatic, isobaric, or some combination)? Explain.

Q19.9 Imagine a gas made up entirely of negatively charged electrons. Like charges repel, so the electrons exert repulsive forces on each other. Would you expect that the temperature of such a gas would rise, fall, or stay the same in a free expansion? Why?

Q19.10 In an adiabatic process for an ideal gas, the pressure decreases. In this process does the internal energy of the gas increase or decrease? Explain.

Q19.11 When you blow on the back of your hand with your mouth wide open, your breath feels warm. But if you partially close your mouth to form an "o" and then blow on your hand, your breath feels cool. Why?

Q19.12 An ideal gas expands while the pressure is kept constant. During this process, does heat flow into the gas or out of the gas? Justify your answer.

Q19.13 A liquid is irregularly stirred in a well-insulated container and thereby undergoes a rise in temperature. Regard the liquid as the system. Has heat been transferred? How can you tell? Has work been done? How can you tell? Why is it important that the stirring is irregular? What is the sign of ΔU? How can you tell?

Q19.14 When you use a hand pump to inflate the tires of your bicycle, the pump gets warm after a while. Why? What happens to the temperature of the air in the pump as you compress it? Why does this happen? When you raise the pump handle to draw outside air into the pump, what happens to the temperature of the air taken in? Again, why does this happen?

Q19.15 In the carburetor of an aircraft or automobile engine, air flows through a relatively small aperture and then expands. In cool, foggy weather, ice sometimes forms in this aperture even though the outside air temperature is above freezing. Why?

Q19.16 On a sunny day, large "bubbles" of air form on the sun-warmed earth, gradually expand, and finally break free to rise through the atmosphere. Soaring birds and glider pilots are fond of using these "thermals" to gain altitude easily. This expansion is essentially an adiabatic process. Why?

Q19.17 The prevailing winds on the Hawaiian island of Kauai blow from the northeast. The winds cool as they go up the slope of Mt. Waialeale (elevation 1523 m), causing water vapor to condense and rain to fall. There is much more precipitation at the summit than at the base of the mountain. In fact, Mt. Waialeale is the rainiest spot on earth, averaging 11.7 m of rainfall a year. But what makes the winds cool?

Q19.18 Applying the same considerations as in Question Q19.17, explain why the island of Niihau, a few kilometers to the southwest of Kauai, is almost a desert and farms there need to be irrigated.

Q19.19 In a constant-volume process, $dU = nC_V \, dT$. But in a constant-pressure process, it is *not* true that $dU = nC_p \, dT$. Why not?

Q19.20 When a gas surrounded by air is compressed adiabatically, its temperature rises even though there is no heat input to the gas. Where does the energy come from to raise the temperature?

Q19.21 When a gas expands adiabatically, it does work on its surroundings. But if there is no heat input to the gas, where does the energy come from to do the work?

Q19.22 The gas used in separating the two uranium isotopes ^{235}U and ^{238}U has the formula UF_6. If you added heat at equal rates to a mole of UF_6 gas and a mole of H_2 gas, which one's temperature would you expect to rise faster? Explain.

Q19.23 A system is taken from state a to state b along the three paths shown in **Fig. Q19.23**. (a) Along which path is the work done by the system the greatest? The least? (b) If $U_b > U_a$, along which path is the absolute value of the heat transfer, $|Q|$, the greatest? For this path, is heat absorbed or liberated by the system? Explain.

Figure **Q19.23**

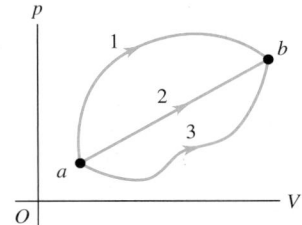

Q19.24 A thermodynamic system undergoes a cyclic process as shown in **Fig. Q19.24**. The cycle consists of two closed loops: I and II. (a) Over one complete cycle, does the system do positive or negative work? (b) In each loop, is the net work done by the system positive or negative? (c) Over one complete cycle, does heat flow into or out of the system? (d) In each loop, does heat flow into or out of the system? Explain.

Figure **Q19.24**

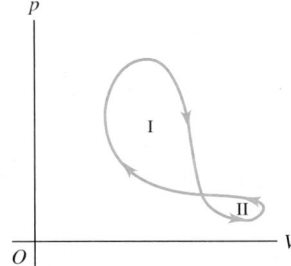

EXERCISES

Section 19.2 Work Done During Volume Changes and Section 19.3 Paths Between Thermodynamic States

19.1 •• Two moles of an ideal gas are heated at constant pressure from $T = 27°C$ to $T = 107°C$. (a) Draw a pV-diagram for this process. (b) Calculate the work done by the gas.

19.2 • Six moles of an ideal gas are in a cylinder fitted at one end with a movable piston. The initial temperature of the gas is $27.0°C$ and the pressure is constant. As part of a machine design project, calculate the final temperature of the gas after it has done 2.40×10^3 J of work.

19.3 •• CALC Two moles of an ideal gas are compressed in a cylinder at a constant temperature of $65.0°C$ until the original pressure has tripled. (a) Sketch a pV-diagram for this process. (b) Calculate the amount of work done.

19.4 •• **BIO Work Done by the Lungs.** The graph in **Fig. E19.4** shows a pV-diagram of the air in a human lung when a person is inhaling and then exhaling a deep breath. Such graphs, obtained in clinical practice, are normally somewhat curved, but we have modeled one as a set of straight lines of the same general shape. (*Important:* The pressure shown is the *gauge* pressure, *not* the absolute pressure.) (a) How many joules of *net* work does this person's lung do during one complete breath? (b) The process illustrated here is somewhat different from those we have been studying, because the pressure change is due to changes in the amount of gas in the lung, not to temperature changes. (Think of your own breathing. Your lungs do not expand because they've gotten hot.) If the temperature of the air in the lung remains a reasonable 20°C, what is the maximum number of moles in this person's lung during a breath?

Figure **E19.4**

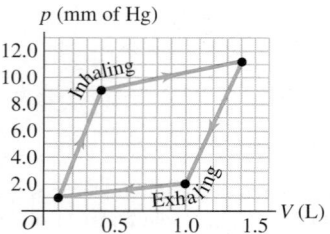

19.5 •• **CALC** During the time 0.305 mol of an ideal gas undergoes an isothermal compression at 22.0°C, 392 J of work is done on it by the surroundings. (a) If the final pressure is 1.76 atm, what was the initial pressure? (b) Sketch a pV-diagram for the process.

19.6 •• A gas undergoes two processes. In the first, the volume remains constant at $0.200 \, m^3$ and the pressure increases from 2.00×10^5 Pa to 5.00×10^5 Pa. The second process is a compression to a volume of $0.120 \, m^3$ at a constant pressure of 5.00×10^5 Pa. (a) In a pV-diagram, show both processes. (b) Find the total work done by the gas during both processes.

19.7 • **Work Done in a Cyclic Process.** (a) In Fig. 19.7a, consider the closed loop $1 \rightarrow 3 \rightarrow 2 \rightarrow 4 \rightarrow 1$. This is a *cyclic* process in which the initial and final states are the same. Find the total work done by the system in this cyclic process, and show that it is equal to the area enclosed by the loop. (b) How is the work done for the process in part (a) related to the work done if the loop is traversed in the opposite direction, $1 \rightarrow 4 \rightarrow 2 \rightarrow 3 \rightarrow 1$? Explain.

Section 19.4 Internal Energy and the First Law of Thermodynamics

19.8 •• **Figure E19.8** shows a pV-diagram for an ideal gas in which its absolute temperature at b is one-fourth of its absolute temperature at a. (a) What volume does this gas occupy at point b? (b) How many joules of work was done by or on the gas in this process? Was it done by or on the gas? (c) Did the internal energy of the gas increase or decrease from a to b? How do you know? (d) Did heat enter or leave the gas from a to b? How do you know?

Figure **E19.8**

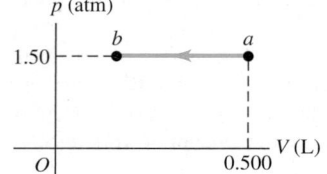

19.9 • A gas in a cylinder expands from a volume of $0.110 \, m^3$ to $0.320 \, m^3$. Heat flows into the gas just rapidly enough to keep the pressure constant at 1.65×10^5 Pa during the expansion. The total heat added is 1.15×10^5 J. (a) Find the work done by the gas. (b) Find the change in internal energy of the gas. (c) Does it matter whether the gas is ideal? Why or why not?

19.10 •• Five moles of an ideal monatomic gas with an initial temperature of 127°C expand and, in the process, absorb 1500 J of heat and do 2100 J of work. What is the final temperature of the gas?

19.11 •• The process abc shown in the pV-diagram in **Fig. E19.11** involves 0.0175 mol of an ideal gas. (a) What was the lowest temperature the gas reached in this process? Where did it occur? (b) How much work was done by or on the gas from a to b? From b to c? (c) If 215 J of heat was put into the gas during abc, how many of those joules went into internal energy?

Figure **E19.11**

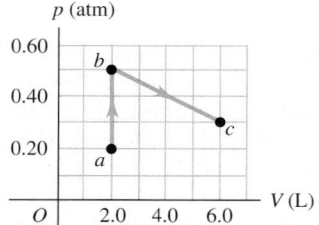

19.12 • A gas in a cylinder is held at a constant pressure of 1.80×10^5 Pa and is cooled and compressed from $1.70 \, m^3$ to $1.20 \, m^3$. The internal energy of the gas decreases by 1.40×10^5 J. (a) Find the work done by the gas. (b) Find the absolute value of the heat flow, $|Q|$, into or out of the gas, and state the direction of the heat flow. (c) Does it matter whether the gas is ideal? Why or why not?

19.13 •• The pV-diagram in **Fig. E19.13** shows a process abc involving 0.450 mol of an ideal gas. (a) What was the temperature of this gas at points a, b, and c? (b) How much work was done by or on the gas in this process? (c) How much heat had to be added during the process to increase the internal energy of the gas by 15,000 J?

Figure **E19.13**

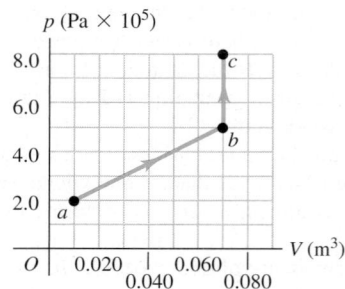

19.14 • **Boiling Water at High Pressure.** When water is boiled at a pressure of 2.00 atm, the heat of vaporization is 2.20×10^6 J/kg and the boiling point is 120°C. At this pressure, 1.00 kg of water has a volume of $1.00 \times 10^{-3} \, m^3$, and 1.00 kg of steam has a volume of $0.824 \, m^3$. (a) Compute the work done when 1.00 kg of steam is formed at this temperature. (b) Compute the increase in internal energy of the water.

19.15 • An ideal gas is taken from a to b on the pV-diagram shown in **Fig. E19.15**. During this process, 700 J of heat is added and the pressure doubles. (a) How much work is done by or on the gas? Explain. (b) How does the temperature of the gas at a compare to its temperature at b? Be specific. (c) How does the internal energy of the gas at a compare to the internal energy at b? Be specific and explain.

Figure **E19.15**

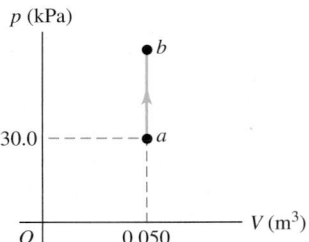

Section 19.5 Kinds of Thermodynamic Processes, Section 19.6 Internal Energy of an Ideal Gas, and Section 19.7 Heat Capacities of an Ideal Gas

19.16 • During an isothermal compression of an ideal gas, 410 J of heat must be removed from the gas to maintain constant temperature. How much work is done by the gas during the process?

19.17 • A cylinder contains 0.250 mol of carbon dioxide (CO_2) gas at a temperature of 27.0°C. The cylinder is provided with a frictionless piston, which maintains a constant pressure of 1.00 atm on the gas. The gas is heated until its temperature increases to 127.0°C. Assume that the CO_2 may be treated as an ideal gas. (a) Draw a pV-diagram for this process. (b) How much work is done by the gas in this process? (c) On what is this work done? (d) What is the change in internal energy of the gas? (e) How much heat was supplied to the gas? (f) How much work would have been done if the pressure had been 0.50 atm?

19.18 • A cylinder contains 0.0100 mol of helium at $T = 27.0$°C. (a) How much heat is needed to raise the temperature to 67.0°C while keeping the volume constant? Draw a pV-diagram for this process. (b) If instead the pressure of the helium is kept constant, how much heat is needed to raise the temperature from 27.0°C to 67.0°C? Draw a pV-diagram for this process. (c) What accounts for the difference between your answers to parts (a) and (b)? In which case is more heat required? What becomes of the additional heat? (d) If the gas is ideal, what is the change in its internal energy in part (a)? In part (b)? How do the two answers compare? Why?

19.19 • In an experiment to simulate conditions inside an automobile engine, 0.185 mol of air at 780 K and 3.00×10^6 Pa is contained in a cylinder of volume 40.0 cm³. Then 645 J of heat is transferred to the cylinder. (a) If the volume of the cylinder is constant while the heat is added, what is the final temperature of the air? Assume that the air is essentially nitrogen gas, and use the data in Table 19.1 even though the pressure is not low. Draw a pV-diagram for this process. (b) If instead the volume of the cylinder is allowed to increase while the pressure remains constant, repeat part (a).

19.20 •• When a quantity of monatomic ideal gas expands at a constant pressure of 4.00×10^4 Pa, the volume of the gas increases from 2.00×10^{-3} m³ to 8.00×10^{-3} m³. What is the change in the internal energy of the gas?

19.21 • Heat Q flows into a monatomic ideal gas, and the volume increases while the pressure is kept constant. What fraction of the heat energy is used to do the expansion work of the gas?

19.22 • Three moles of an ideal monatomic gas expands at a constant pressure of 2.50 atm; the volume of the gas changes from 3.20×10^{-2} m³ to 4.50×10^{-2} m³. Calculate (a) the initial and final temperatures of the gas; (b) the amount of work the gas does in expanding; (c) the amount of heat added to the gas; (d) the change in internal energy of the gas.

19.23 • An experimenter adds 970 J of heat to 1.75 mol of an ideal gas to heat it from 10.0°C to 25.0°C at constant pressure. The gas does +223 J of work during the expansion. (a) Calculate the change in internal energy of the gas. (b) Calculate γ for the gas.

19.24 • Propane gas (C_3H_8) behaves like an ideal gas with $\gamma = 1.127$. Determine the molar heat capacity at constant volume and the molar heat capacity at constant pressure.

19.25 • CALC The temperature of 0.150 mol of an ideal gas is held constant at 77.0°C while its volume is reduced to 25.0% of its initial volume. The initial pressure of the gas is 1.25 atm. (a) Determine the work done by the gas. (b) What is the change in its internal energy? (c) Does the gas exchange heat with its surroundings? If so, how much? Does the gas absorb or liberate heat?

Section 19.8 Adiabatic Processes for an Ideal Gas

19.26 •• Five moles of monatomic ideal gas have initial pressure 2.50×10^3 Pa and initial volume 2.10 m³. While undergoing an adiabatic expansion, the gas does 1480 J of work. What is the final pressure of the gas after the expansion?

19.27 • A monatomic ideal gas that is initially at 1.50×10^5 Pa and has a volume of 0.0800 m³ is compressed adiabatically to a volume of 0.0400 m³. (a) What is the final pressure? (b) How much work is done by the gas? (c) What is the ratio of the final temperature of the gas to its initial temperature? Is the gas heated or cooled by this compression?

19.28 • The engine of a Ferrari F355 F1 sports car takes in air at 20.0°C and 1.00 atm and compresses it adiabatically to 0.0900 times the original volume. The air may be treated as an ideal gas with $\gamma = 1.40$. (a) Draw a pV-diagram for this process. (b) Find the final temperature and pressure.

19.29 • During an adiabatic expansion the temperature of 0.450 mol of argon (Ar) drops from 66.0°C to 10.0°C. The argon may be treated as an ideal gas. (a) Draw a pV-diagram for this process. (b) How much work does the gas do? (c) What is the change in internal energy of the gas?

19.30 •• A player bounces a basketball on the floor, compressing it to 80.0% of its original volume. The air (assume it is essentially N_2 gas) inside the ball is originally at 20.0°C and 2.00 atm. The ball's inside diameter is 23.9 cm. (a) What temperature does the air in the ball reach at its maximum compression? Assume the compression is adiabatic and treat the gas as ideal. (b) By how much does the internal energy of the air change between the ball's original state and its maximum compression?

19.31 •• On a warm summer day, a large mass of air (atmospheric pressure 1.01×10^5 Pa) is heated by the ground to 26.0°C and then begins to rise through the cooler surrounding air. (This can be treated approximately as an adiabatic process; why?) Calculate the temperature of the air mass when it has risen to a level at which atmospheric pressure is only 0.850×10^5 Pa. Assume that air is an ideal gas, with $\gamma = 1.40$. (This rate of cooling for dry, rising air, corresponding to roughly 1 C° per 100 m of altitude, is called the *dry adiabatic lapse rate*.)

19.32 • A cylinder contains 0.100 mol of an ideal monatomic gas. Initially the gas is at 1.00×10^5 Pa and occupies a volume of 2.50×10^{-3} m³. (a) Find the initial temperature of the gas in kelvins. (b) If the gas is allowed to expand to twice the initial volume, find the final temperature (in kelvins) and pressure of the gas if the expansion is (i) isothermal; (ii) isobaric; (iii) adiabatic.

PROBLEMS

19.33 •• A quantity of air is taken from state a to state b along a path that is a straight line in the pV-diagram (**Fig. P19.33**). (a) In this process, does the temperature of the gas increase, decrease, or stay the same? Explain. (b) If $V_a = 0.0700$ m³, $V_b = 0.1100$ m³, $p_a = 1.00 \times 10^5$ Pa, and $p_b = 1.40 \times 10^5$ Pa, what is the work W done by the gas in this process? Assume that the gas may be treated as ideal.

Figure **P19.33**

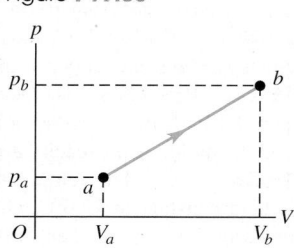

19.34 • One-half mole of an ideal gas is taken from state a to state c as shown in **Fig. P19.34**. (a) Calculate the final temperature of the gas. (b) Calculate the work done on (or by) the gas as it moves from state a to state c. (c) Does heat leave the system or enter the system during this process? How much heat? Explain.

Figure **P19.34**

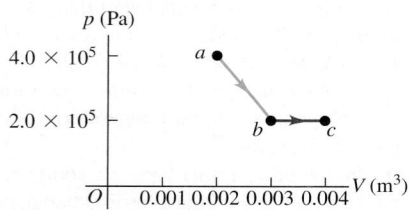

19.35 •• **Figure P19.35** shows the pV-diagram for a process in which the temperature of the ideal gas remains constant at 85°C. (a) How many moles of gas are involved? (b) What volume does this gas occupy at a? (c) How much work was done by or on the gas from a to b? (d) By how much did the internal energy of the gas change during this process?

Figure **P19.35**

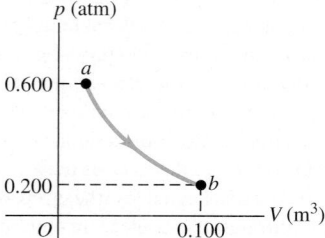

19.36 •• The graph in **Fig. P19.36** shows a pV-diagram for 3.25 mol of *ideal* helium (He) gas. Part ca of this process is isothermal. (a) Find the pressure of the He at point a. (b) Find the temperature

Figure **P19.36**

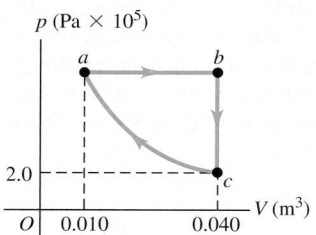

of the He at points a, b, and c. (c) How much heat entered or left the He during segments ab, bc, and ca? In each segment, did the heat enter or leave? (d) By how much did the internal energy of the He change from a to b, from b to c, and from c to a? Indicate whether this energy increased or decreased.

19.37 •• When a system is taken from state a to state b in **Fig. P19.37** along path acb, 90.0 J of heat flows into the system and 60.0 J of work is done by the system. (a) How much heat flows into the system along path adb if the work done by the system is 15.0 J? (b) When the system is returned from b to a along the curved path, the absolute value of the work done by the system is 35.0 J. Does the system absorb or liberate heat? How much heat? (c) If $U_a = 0$ and $U_d = 8.0$ J, find the heat absorbed in the processes ad and db.

Figure **P19.37**

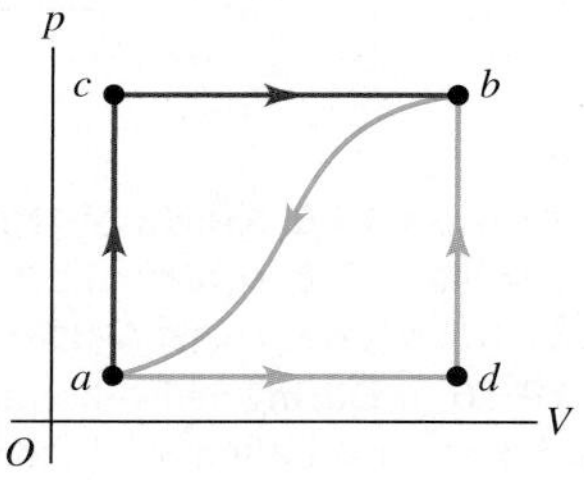

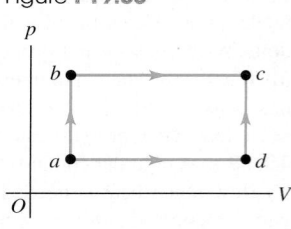

19.38 • A thermodynamic system is taken from state a to state c in **Fig. P19.38** along either path abc or path adc. Along path abc, the work W done by the system is 450 J. Along path adc, W is 120 J. The internal energies of each of the four states shown in the figure are $U_a = 150$ J, $U_b = 240$ J, $U_c = 680$ J, and $U_d = 330$ J. Calculate the heat flow Q for each of the four processes ab, bc, ad, and dc. In each process, does the system absorb or liberate heat?

Figure **P19.38**

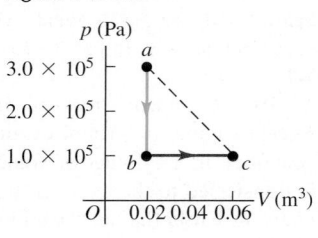

19.39 • A volume of air (assumed to be an ideal gas) is first cooled without changing its volume and then expanded without changing its pressure, as shown by path abc in **Fig. P19.39**. (a) How does the final temperature of the gas compare with its initial temperature? (b) How much heat does the air exchange with its surroundings during process abc? Does the air absorb heat or release heat during this process? Explain. (c) If the air instead expands from state a to state c by the straight-line path shown, how much heat does it exchange with its surroundings?

Figure **P19.39**

19.40 • Three moles of argon gas (assumed to be an ideal gas) originally at 1.50×10^4 Pa and a volume of 0.0280 m³ are first heated and expanded at constant pressure to a volume of 0.0435 m³, then heated at constant volume until the pressure reaches 3.50 × 10⁴ Pa, then cooled and compressed at constant pressure until the volume is again 0.0280 m³, and finally cooled at constant volume until the pressure drops to its original value of 1.50 × 10⁴ Pa. (a) Draw the pV-diagram for this cycle. (b) Calculate the total work done by (or on) the gas during the cycle. (c) Calculate the net heat exchanged with the surroundings. Does the gas gain or lose heat overall?

19.41 •• Two moles of an ideal monatomic gas go through the cycle abc. For the complete cycle, 800 J of heat flows out of the gas. Process ab is at constant pressure, and process bc is at constant volume. States a and b have temperatures $T_a = 200$ K and $T_b = 300$ K. (a) Sketch the pV-diagram for the cycle. (b) What is the work W for the process ca?

19.42 •• Three moles of an ideal gas are taken around cycle acb shown in **Fig. P19.42**. For this gas, $C_p = 29.1$ J/mol·K. Process ac is at constant pressure, process ba is at constant volume, and process cb is adiabatic. The temperatures of the gas in states a, c, and b are $T_a = 300$ K, $T_c = 492$ K, and $T_b = 600$ K. Calculate the total work W for the cycle.

Figure **P19.42**

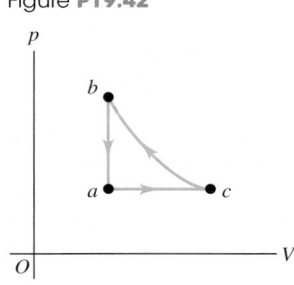

19.43 •• **Figure P19.43** shows a pV-diagram for 0.0040 mol of *ideal* H₂ gas. The temperature of the gas does not change during segment bc. (a) What volume does this gas occupy at point c? (b) Find the temperature of the gas at points a, b, and c. (c) How much heat went into or out of the gas during segments ab, ca, and bc? Indicate whether the heat has gone into or out of the gas. (d) Find the change in the internal energy of this hydrogen during segments ab, bc, and ca. Indicate whether the internal energy increased or decreased during each segment.

Figure **P19.43**

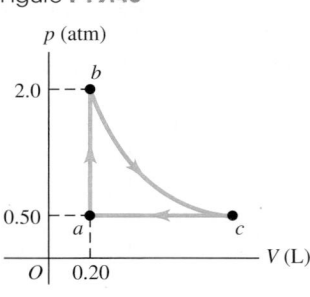

19.44 •• (a) One-third of a mole of He gas is taken along the path abc shown in **Fig. P19.44**. Assume that the gas may be treated as ideal. How much heat is transferred into or out of the gas? (b) If the gas instead went directly from state a to state c

Figure **P19.44**

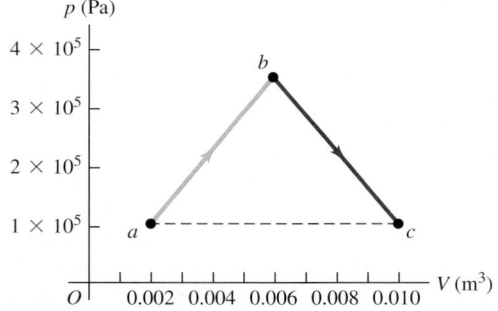

along the horizontal dashed line in Fig. P19.44, how much heat would be transferred into or out of the gas? (c) How does Q in part (b) compare with Q in part (a)? Explain.

19.45 ••• Starting with 2.50 mol of N₂ gas (assumed to be ideal) in a cylinder at 1.00 atm and 20.0°C, a chemist first heats the gas at constant volume, adding 1.36 × 10⁴ J of heat, then continues heating and allows the gas to expand at constant pressure to twice its original volume. Calculate (a) the final temperature of the gas; (b) the amount of work done by the gas; (c) the amount of heat added to the gas while it was expanding; (d) the change in internal energy of the gas for the whole process.

19.46 •• Nitrogen gas in an expandable container is cooled from 50.0°C to 10.0°C with the pressure held constant at 3.00 × 10⁵ Pa. The total heat liberated by the gas is 2.50 × 10⁴ J. Assume that the gas may be treated as ideal. Find (a) the number of moles of gas; (b) the change in internal energy of the gas; (c) the work done by the gas. (d) How much heat would be liberated by the gas for the same temperature change if the volume were constant?

19.47 • CALC A cylinder with a frictionless, movable piston like that shown in Fig. 19.5 contains a quantity of helium gas. Initially the gas is at 1.00 × 10⁵ Pa and 300 K and occupies a volume of 1.50 L. The gas then undergoes two processes. In the first, the gas is heated and the piston is allowed to move to keep the temperature at 300 K. This continues until the pressure reaches 2.50 × 10⁴ Pa. In the second process, the gas is compressed at constant pressure until it returns to its original volume of 1.50 L. Assume that the gas may be treated as ideal. (a) In a pV-diagram, show both processes. (b) Find the volume of the gas at the end of the first process, and the pressure and temperature at the end of the second process. (c) Find the total work done by the gas during both processes. (d) What would you have to do to the gas to return it to its original pressure and temperature?

19.48 • CP **A Thermodynamic Process in a Solid.** A cube of copper 2.00 cm on a side is suspended by a string. (The physical properties of copper are given in Tables 14.1, 17.2, and 17.3.) The cube is heated with a burner from 20.0°C to 90.0°C. The air surrounding the cube is at atmospheric pressure (1.01 × 10⁵ Pa). Find (a) the increase in volume of the cube; (b) the mechanical work done by the cube to expand against the pressure of the surrounding air; (c) the amount of heat added to the cube; (d) the change in internal energy of the cube. (e) Based on your results, explain whether there is any substantial difference between the specific heats c_p (at constant pressure) and c_V (at constant volume) for copper under these conditions.

19.49 ••• **Chinook.** During certain seasons strong winds called chinooks blow from the west across the eastern slopes of the Rockies and downhill into Denver and nearby areas. Although the mountains are cool, the wind in Denver is very hot; within a few minutes after the chinook wind arrives, the temperature can climb 20 C° ("chinook" refers to a Native American people of the Pacific Northwest). Similar winds occur in the Alps (called foehns) and in southern California (called Santa Anas). (a) Explain why the temperature of the chinook wind rises as it descends the slopes. Why is it important that the wind be fast moving? (b) Suppose a strong wind is blowing toward Denver (elevation 1630 m) from Grays Peak (80 km west of Denver, at an elevation of 4350 m), where the air pressure is 5.60 × 10⁴ Pa and the air temperature is −15.0°C. The temperature and pressure in Denver before the wind arrives are 2.0°C and 8.12 × 10⁴ Pa. By how many Celsius degrees will the temperature in Denver rise when the chinook arrives?

19.50 ••• High-Altitude Research. A large research balloon containing 2.00×10^3 m^3 of helium gas at 1.00 atm and a temperature of 15.0°C rises rapidly from ground level to an altitude at which the atmospheric pressure is only 0.900 atm (**Fig. P19.50**). Assume the helium behaves like an ideal gas and the balloon's ascent is too rapid to permit much heat exchange with the surrounding air. (a) Calculate the volume of the gas at the higher altitude. (b) Calculate the temperature of the gas at the higher altitude. (c) What is the change in internal energy of the helium as the balloon rises to the higher altitude?

Figure **P19.50**

19.51 ••• An air pump has a cylinder 0.250 m long with a movable piston. The pump is used to compress air from the atmosphere (at absolute pressure 1.01×10^5 Pa) into a very large tank at 3.80×10^5 Pa gauge pressure. (For air, $C_V = 20.8$ J/mol·K.) (a) The piston begins the compression stroke at the open end of the cylinder. How far down the length of the cylinder has the piston moved when air first begins to flow from the cylinder into the tank? Assume that the compression is adiabatic. (b) If the air is taken into the pump at 27.0°C, what is the temperature of the compressed air? (c) How much work does the pump do in putting 20.0 mol of air into the tank?

19.52 •• A certain ideal gas has molar heat capacity at constant volume C_V. A sample of this gas initially occupies a volume V_0 at pressure p_0 and absolute temperature T_0. The gas expands isobarically to a volume $2V_0$ and then expands further adiabatically to a final volume $4V_0$. (a) Draw a pV-diagram for this sequence of processes. (b) Compute the total work done by the gas for this sequence of processes. (c) Find the final temperature of the gas. (d) Find the absolute value of the total heat flow $|Q|$ into or out of the gas for this sequence of processes, and state the direction of heat flow.

19.53 • A monatomic ideal gas expands slowly to twice its original volume, doing 450 J of work in the process. Find the heat added to the gas and the change in internal energy of the gas if the process is (a) isothermal; (b) adiabatic; (c) isobaric.

19.54 •• CALC A cylinder with a piston contains 0.250 mol of oxygen at 2.40×10^5 Pa and 355 K. The oxygen may be treated as an ideal gas. The gas first expands isobarically to twice its original volume. It is then compressed isothermally back to its original volume, and finally it is cooled isochorically to its original pressure. (a) Show the series of processes on a pV-diagram. Compute (b) the temperature during the isothermal compression; (c) the maximum pressure; (d) the total work done by the piston on the gas during the series of processes.

19.55 • Use the conditions and processes of Problem 19.54 to compute (a) the work done by the gas, the heat added to it, and its internal energy change during the initial expansion; (b) the work done, the heat added, and the internal energy change during the final cooling; (c) the internal energy change during the isothermal compression.

19.56 •• CALC A cylinder with a piston contains 0.150 mol of nitrogen at 1.80×10^5 Pa and 300 K. The nitrogen may be treated as an ideal gas. The gas is first compressed isobarically to half its original volume. It then expands adiabatically back to its original volume, and finally it is heated isochorically to its original pressure. (a) Show the series of processes in a pV-diagram. (b) Compute the temperatures at the beginning and end of the adiabatic expansion. (c) Compute the minimum pressure.

19.57 • Use the conditions and processes of Problem 19.56 to compute (a) the work done by the gas, the heat added to it, and its internal energy change during the initial compression; (b) the work done by the gas, the heat added to it, and its internal energy change during the adiabatic expansion; (c) the work done, the heat added, and the internal energy change during the final heating.

19.58 • Comparing Thermodynamic Processes. In a cylinder, 1.20 mol of an ideal monatomic gas, initially at 3.60×10^5 Pa and 300 K, expands until its volume triples. Compute the work done by the gas if the expansion is (a) isothermal; (b) adiabatic; (c) isobaric. (d) Show each process in a pV-diagram. In which case is the absolute value of the work done by the gas greatest? Least? (e) In which case is the absolute value of the heat transfer greatest? Least? (f) In which case is the absolute value of the change in internal energy of the gas greatest? Least?

19.59 •• DATA You have recorded measurements of the heat flow Q into 0.300 mol of a gas that starts at $T_1 = 20.0$°C and ends at a temperature T_2. You measured Q for three processes: one isobaric, one isochoric, and one adiabatic. In each case, T_2 was the same. **Figure P19.59** summarizes your results. But you lost a page from your lab notebook and don't have a record of the value of T_2; you also don't know which process was isobaric, isochoric, or adiabatic. Each process was done at a sufficiently low pressure for the gas to be treated as ideal. (a) Identify each process a, b, or c as isobaric, isochoric, or adiabatic. (b) What is the value of T_2? (c) How much work is done by the gas in each process? (d) For which process is the magnitude of the volume change the greatest? (e) For each process, does the volume of the gas increase, decrease, or stay the same?

Figure **P19.59**

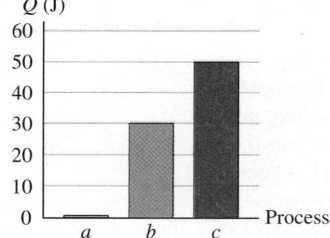

19.60 •• DATA You compress a gas in an insulated cylinder—no heat flows into or out of the gas. The gas pressure is fairly low, so treating the gas as ideal is a good approximation. When you measure the pressure as a function of the volume of the gas, you obtain these results:

V (L)	2.50	2.02	1.48	1.01	0.50
p (atm)	0.101	0.139	0.202	0.361	0.952

(a) Graph $\log(p)$ versus $\log(V)$, with p in Pa and V in m^3. Explain why the data points fall close to a straight line. (b) Use your graph to calculate γ for the gas. Is the gas monatomic, diatomic, or polyatomic? (c) When $p = 0.101$ atm and $V = 2.50$ L, the temperature is 22.0°C. Apply the ideal-gas equation and calculate the temperature for each of the other pairs of p and V values. In this compression, does the temperature of the gas increase, decrease, or stay constant?

19.61 •• DATA You place a quantity of gas into a metal cylinder that has a movable piston at one end. No gas leaks out of the cylinder as the piston moves. The external force applied to the piston can be varied to change the gas pressure as you move the piston to change the volume of the gas. A pressure gauge attached to the interior wall of the cylinder measures the gas pressure, and you can calculate the volume of the gas from a measurement of the piston's position in the cylinder.

You start with a pressure of 1.0 atm and a gas volume of 3.0 L. Holding the pressure constant, you increase the volume to 5.0 L. Then, keeping the volume constant at 5.0 L, you increase the pressure to 3.0 atm. Next you decrease the pressure linearly as a function of volume until the volume is 3.0 L and the pressure is 2.0 atm. Finally, you keep the volume constant at 3.0 L and decrease the pressure to 1.0 atm, returning the gas to its initial pressure and volume. The walls of the cylinder are good conductors of heat, and you provide the required heat sources and heat sinks so that the necessary heat flows can occur. At these relatively high pressures, you suspect that the ideal-gas equation will not apply with much accuracy. You don't know what gas is in the cylinder or whether it is monatomic, diatomic, or polyatomic. (a) Plot the cycle in the pV-plane. (b) What is the net heat flow for the gas during this cycle? Is there net heat flow into or out of the gas?

CHALLENGE PROBLEM

19.62 •• **Engine Turbochargers and Intercoolers.** The power output of an automobile engine is directly proportional to the mass of air that can be forced into the volume of the engine's cylinders to react chemically with gasoline. Many cars have a *turbocharger*, which compresses the air before it enters the engine, giving a greater mass of air per volume. This rapid, essentially adiabatic compression also heats the air. To compress it further, the air then passes through an *intercooler* in which the air exchanges heat with its surroundings at essentially constant pressure. The air is then drawn into the cylinders. In a typical installation, air is taken into the turbocharger at atmospheric pressure $(1.01 \times 10^5 \, \text{Pa})$, density $\rho = 1.23 \, \text{kg/m}^3$, and temperature 15.0°C. It is compressed adiabatically to $1.45 \times 10^5 \, \text{Pa}$. In the intercooler, the air is cooled to the original temperature of 15.0°C at a constant pressure of $1.45 \times 10^5 \, \text{Pa}$. (a) Draw a pV-diagram for this sequence of processes. (b) If the volume of one of the engine's cylinders is 575 cm³, what mass of air exiting from the intercooler will fill the cylinder at $1.45 \times 10^5 \, \text{Pa}$? Compared to the power output of an engine that takes in air at $1.01 \times 10^5 \, \text{Pa}$ at 15.0°C, what percentage increase in power is obtained by using the turbocharger and intercooler? (c) If the intercooler is not used, what mass of air exiting from the turbocharger will fill the cylinder at $1.45 \times 10^5 \, \text{Pa}$? Compared to the power output of an engine that takes in air at $1.01 \times 10^5 \, \text{Pa}$ at 15.0°C, what percentage increase in power is obtained by using the turbocharger alone?

PASSAGE PROBLEMS

BIO **ANESTHETIC GASES.** One type of gas mixture used in anesthesiology is a 50%/50% mixture (by volume) of nitrous oxide (N_2O) and oxygen (O_2), which can be premixed and kept in a cylinder for later use. Because these two gases don't react chemically at or below 2000 psi, at typical room temperatures they form a homogeneous single gas phase, which can be considered an ideal gas. If the temperature drops below −6°C, however, N_2O may begin to condense out of the gas phase. Then any gas removed from the cylinder will initially be nearly pure O_2; as the cylinder empties, the proportion of O_2 will decrease until the gas coming from the cylinder is nearly pure N_2O.

19.63 In a test of the effects of low temperatures on the gas mixture, a cylinder filled at 20.0°C to 2000 psi (gauge pressure) is cooled slowly and the pressure is monitored. What is the expected pressure at −5.00°C if the gas remains a homogeneous mixture? (a) 500 psi; (b) 1500 psi; (c) 1830 psi; (d) 1920 psi.

19.64 In another test, the valve of a 500-L cylinder full of the gas mixture at 2000 psi (gauge pressure) is opened wide so that the gas rushes out of the cylinder very rapidly. Why might some N_2O condense during this process? (a) This is an isochoric process in which the pressure decreases, so the temperature also decreases. (b) Because of the rapid expansion, heat is removed from the system, so the internal energy and temperature of the gas decrease. (c) This is an isobaric process, so as the volume increases, the temperature decreases proportionally. (d) With the rapid expansion, the expanding gas does work with no heat input, so the internal energy and temperature of the gas decrease.

19.65 You have a cylinder that contains 500 L of the gas mixture pressurized to 2000 psi (gauge pressure). A regulator sets the gas flow to deliver 8.2 L/min at atmospheric pressure. Assume that this flow is slow enough that the expansion is isothermal and the gases remain mixed. How much time will it take to empty the cylinder? (a) 1 h; (b) 33 h; (c) 57 h; (d) 140 h.

19.66 In a hospital, pure oxygen may be delivered at 50 psi (gauge pressure) and then mixed with N_2O. What volume of oxygen at 20°C and 50 psi (gauge pressure) should be mixed with 1.7 kg of N_2O to get a 50%/50% mixture by volume at 20°C? (a) 0.21 m³; (b) 0.27 m³; (c) 1.9 m³; (d) 100 m³.

Answers

Chapter Opening Question ?

(ii) The work done by a gas as its volume changes from V_1 to V_2 is equal to the integral $\int p\, dV$ between those two volume limits. If the volume of the gas contracts, the final volume V_2 is less than the initial volume V_1 and the gas does negative work. Propelling the locomotive requires that the gas do positive work, so the gas doesn't contribute to propulsion while contracting.

Test Your Understanding Questions

19.1 negative, positive, positive Heat flows out of the coffee, so $Q_{coffee} < 0$; heat flows into the aluminum cup, so $Q_{aluminum} > 0$. In mechanics, we would say that negative work is done *on* the block, since the surface exerts a force on the block that opposes the block's motion. But in thermodynamics we use the opposite convention and say that $W > 0$, which means that positive work is done *by* the block on the surface.

19.2 (ii) The work done in an expansion is represented by the area under the curve of pressure p versus volume V. In an isothermal expansion the pressure decreases as the volume increases, so the pV-diagram looks like Fig. 19.6a and the work done equals the shaded area under the blue curve from point 1 to point 2. If, however, the expansion is at constant pressure, the curve of p versus V would be the same as the dashed horizontal line at pressure p_2 in Fig. 19.6a. The area under this dashed line is smaller than the area under the blue curve for an isothermal expansion, so less work is done in the constant-pressure expansion than in the isothermal expansion.

19.3 (i) and (iv) (tie), (ii) and (iii) (tie) The accompanying figure shows the pV-diagrams for each of the four processes. The trapezoidal area under the curve, and hence the absolute value of the work, is the same in all four cases. In cases (i) and (iv) the volume increases, so the system does positive work as it expands against its surroundings. In cases (ii) and (iii) the volume decreases, so the system does negative work (shown by cross-hatching) as the surroundings push inward on it.

19.4 (ii), (i) and (iv) (tie), (iii) In the expression $\Delta U = Q - W$, Q is the heat *added* to the system and W is the work done *by* the system. If heat is transferred from the system to its surroundings, Q is negative; if work is done on the system, W is negative. Hence we have (i) $Q = -250$ J, $W = -250$ J, $\Delta U = -250$ J $- (-250$ J$) = 0$; (ii) $Q = 250$ J, $W = -250$ J, $\Delta U = 250$ J $- (-250$ J$) = 500$ J; (iii) $Q = -250$ J, $W = 250$ J, $\Delta U = -250$ J $- 250$ J $= -500$ J; and (iv) $Q = 250$ J, $W = 250$ J, $\Delta U = 250$ J $- 250$ J $= 0$.

19.5 $1 \to 4$ and $3 \to 2$ are isochoric; $1 \to 3$ and $4 \to 2$ are isobaric; no In a pV-diagram like those shown in Fig. 19.7, isochoric processes are represented by vertical lines (lines of constant volume) and isobaric processes are represented by horizontal lines (lines of constant pressure). The process $1 \to 2$ in Fig. 19.7 is shown as a curved line, which superficially resembles the adiabatic and isothermal processes for an ideal gas in Fig. 19.16. Without more information we can't tell whether process $1 \to 2$ is isothermal, adiabatic, or neither.

19.6 no Using the model of a solid in Fig. 18.20, we can see that the internal energy of a solid *does* depend on its volume. Compressing the solid means compressing the "springs" between the atoms, thereby increasing their stored potential energy and hence the internal energy of the solid.

19.7 (i) For a given number of moles n and a given temperature change ΔT, the amount of heat that must be transferred out of a fixed volume of air is $Q = nC_V\Delta T$. Hence the amount of heat transfer required is least for the gas with the smallest value of C_V. From Table 19.1, C_V is smallest for monatomic gases.

19.8 (ii), (iv), (i) and (iii) (tie) Samples (i) and (iii) are compressed isothermally, so $pV = $ constant. The volume of each sample decreases to one-half of its initial value, so the final pressure is twice the initial pressure. Samples (ii) and (iv) are compressed adiabatically, so $pV^\gamma = $ constant and the pressure increases by a factor of 2^γ. Sample (ii) is a monatomic gas for which $\gamma = \frac{5}{3}$, so its final pressure is $2^{5/3} = 3.17$ times greater than the initial pressure. Sample (iv) is a diatomic gas for which $\gamma = \frac{7}{5}$, so its final pressure is greater than the initial pressure by a factor of $2^{7/5} = 2.64$.

Bridging Problem

(a) $W = nRT \ln\left[\dfrac{V_2 - nb}{V_1 - nb}\right] + an^2\left[\dfrac{1}{V_2} - \dfrac{1}{V_1}\right]$

(b) (i) $W = 2.80 \times 10^3$ J, (ii) $W = 3.11 \times 10^3$ J

(c) Ideal gas, for which there is no attraction between molecules

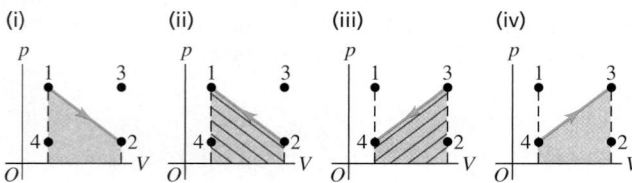

The second law of thermo-dynamics tells us that heat naturally flows from a hot body (such as molten lava, shown here flowing into the ocean in Hawaii) to a cold one (such as ocean water, which is heated to make steam). Is it *ever* possible for heat to flow from a cold body to a hot one? (i) Yes, no matter what the temperature difference; (ii) yes, but for only certain temperature differences; (iii) no; (iv) answer depends on the compositions of the two bodies.

20 THE SECOND LAW OF THERMODYNAMICS

LEARNING GOALS

Looking forward at ...

20.1 The difference between reversible and irreversible processes.

20.2 What a heat engine is, and how to calculate its efficiency.

20.3 The physics of internal-combustion engines.

20.4 How refrigerators and heat engines are related, and how to analyze the performance of a refrigerator.

20.5 How the second law of thermodynamics sets limits on the efficiency of engines and the performance of refrigerators.

20.6 How to do calculations involving the idealized Carnot cycle for engines and refrigerators.

20.7 What is meant by entropy, and how to use this concept to analyze thermodynamic processes.

20.8 How to use the concept of microscopic states to understand entropy.

Looking back at ...

17.3 The Kelvin scale.

18.3 The Boltzmann constant.

19.1–19.8 Thermodynamic processes; first law of thermodynamics; free expansion of a gas.

Many thermodynamic processes proceed naturally in one direction but not the opposite. For example, heat by itself always flows from a hot body to a cooler body, never the reverse. Heat flow from a cool body to a hot body would not violate the first law of thermodynamics; energy would be conserved. But it doesn't happen in nature. Why not? As another example, note that it is easy to convert mechanical energy completely into heat; this happens every time we use a car's brakes to stop it. In the reverse direction, there are plenty of devices that convert heat *partially* into mechanical energy. (An automobile engine is an example.) But no one has ever managed to build a machine that converts heat *completely* into mechanical energy. Again, why not?

The answer to both of these questions has to do with the *directions* of thermodynamic processes and is called the *second law of thermodynamics*. This law places fundamental limitations on the efficiency of an engine or a power plant. It also places limitations on the minimum energy input needed to operate a refrigerator. So the second law is directly relevant for many important practical problems.

We can also state the second law in terms of the concept of *entropy*, a quantitative measure of the degree of randomness of a system. The idea of entropy helps explain why ink mixed with water never spontaneously unmixes and why we never observe a host of other seemingly possible processes.

20.1 DIRECTIONS OF THERMODYNAMIC PROCESSES

Thermodynamic processes that occur in nature are all **irreversible processes.** These are processes that proceed spontaneously in one direction but not the other (**Fig. 20.1a**, next page). The flow of heat from a hot body to a cooler body is irreversible, as is the free expansion of a gas discussed in Sections 19.3 and 19.6. Sliding a book across a table converts mechanical energy into heat by friction; this process is

20.1 Reversible and irreversible processes.

(a) A block of ice melts *irreversibly* when we place it in a hot (70°C) metal box.

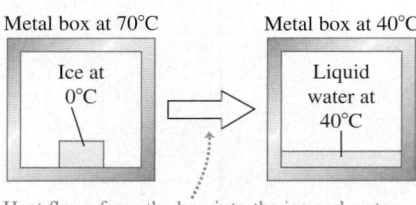

Heat flows from the box into the ice and water, never the reverse.

(b) A block of ice at 0°C can be melted *reversibly* if we put it in a 0°C metal box.

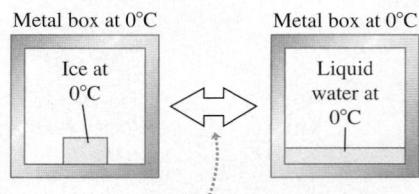

By infinitesimally raising or lowering the temperature of the box, we can make heat flow into the ice to melt it or make heat flow out of the water to refreeze it.

irreversible, for no one has ever observed the reverse process (in which a book initially at rest on the table would spontaneously start moving and the table and book would cool down). Our main topic for this chapter is the *second law of thermodynamics,* which determines the preferred direction for such processes.

Despite this preferred direction for every natural process, we can think of a class of idealized processes that *would* be reversible. A system that undergoes such an idealized **reversible process** is always very close to being in thermodynamic equilibrium within itself and with its surroundings. Any change of state that takes place can then be reversed by making only an infinitesimal change in the conditions of the system. For example, we can reverse heat flow between two bodies whose temperatures differ only infinitesimally by making only a very small change in one temperature or the other (Fig. 20.1b).

Reversible processes are thus **equilibrium processes,** with the system always in thermodynamic equilibrium. Of course, if a system were *truly* in thermodynamic equilibrium, no change of state would take place. Heat would not flow into or out of a system with truly uniform temperature throughout, and a system that is truly in mechanical equilibrium would not expand and do work against its surroundings. A reversible process is an idealization that can never be precisely attained in the real world. But by making the temperature gradients and the pressure differences in the substance very small, we can keep the system very close to equilibrium states and make the process nearly reversible.

By contrast, heat flow with finite temperature difference, free expansion of a gas, and conversion of work to heat by friction are all *irreversible* processes; no small change in conditions could make any of them go the other way. They are also all *nonequilibrium* processes, in that the system is not in thermodynamic equilibrium at any point until the end of the process.

Disorder and Thermodynamic Processes

There is a relationship between the direction of a process and the *disorder* or *randomness* of the resulting state. For example, imagine a thousand names written on file cards and arranged in alphabetical order. Throw the alphabetized stack of cards into the air, and they will likely come down in a random, disordered state. In the free expansion of a gas discussed in Sections 19.3 and 19.6, the air is more disordered after it has expanded into the entire box than when it was confined in one side, just as your clothes are more disordered when scattered all over your floor than when confined to your closet.

Similarly, macroscopic kinetic energy is energy associated with organized, coordinated motions of many molecules, but heat transfer involves changes in energy of random, disordered molecular motion. Therefore conversion of mechanical energy into heat involves an increase of randomness or disorder.

In the following sections we will introduce the second law of thermodynamics by considering two broad classes of devices: *heat engines,* which are partly successful in converting heat into work, and *refrigerators,* which are partly successful in transporting heat from cooler to hotter bodies.

Your left and right hands are normally at the same temperature, just like the metal box and ice in Fig. 20.1b. Is rubbing your hands together to warm them (i) a reversible process or (ii) an irreversible process? ❚

20.2 HEAT ENGINES

The essence of our technological society is the ability to use sources of energy other than muscle power. Sometimes, mechanical energy is directly available; water power and wind power are examples. But most of our energy comes from the burning of fossil fuels (coal, oil, and gas) and from nuclear reactions. They supply energy that is transferred as *heat*. This is directly useful for heating buildings, for cooking, and for chemical processing, but to operate a machine or propel a vehicle, we need *mechanical* energy.

Thus it's important to know how to take heat from a source and convert as much of it as possible into mechanical energy or work. This is what happens in gasoline engines in automobiles, jet engines in airplanes, steam turbines in electric power plants, and many other systems. Closely related processes occur in the animal kingdom; food energy is "burned" (that is, carbohydrates combine with oxygen to yield water, carbon dioxide, and energy) and partly converted to mechanical energy as an animal's muscles do work on its surroundings.

Any device that transforms heat partly into work or mechanical energy is called a **heat engine** (**Fig. 20.2**). Usually, a quantity of matter inside the engine undergoes inflow and outflow of heat, expansion and compression, and sometimes change of phase. We call this matter the **working substance** of the engine. In internal-combustion engines, such as those used in automobiles, the working substance is a mixture of air and fuel; in a steam turbine it is water.

The simplest kind of engine to analyze is one in which the working substance undergoes a **cyclic process**, a sequence of processes that eventually leaves the substance in the same state in which it started. In a steam turbine the water is recycled and used over and over. Internal-combustion engines do not use the same air over and over, but we can still analyze them in terms of cyclic processes that approximate their actual operation.

20.2 All motorized vehicles other than purely electric vehicles use heat engines for propulsion. (Hybrid vehicles use their internal-combustion engine to help charge the batteries for the electric motor.)

Hot and Cold Reservoirs

All heat engines *absorb* heat from a source at a relatively high temperature, perform some mechanical work, and *discard* or *reject* some heat at a lower temperature. As far as the engine is concerned, the discarded heat is wasted. In internal-combustion engines the waste heat is that discarded in the hot exhaust gases and the cooling system; in a steam turbine it is the heat that must flow out of the used steam to condense and recycle the water.

When a system is carried through a cyclic process, its initial and final internal energies are equal, so the first law of thermodynamics requires that

$$U_2 - U_1 = 0 = Q - W \qquad \text{so} \qquad Q = W$$

That is, the net heat flowing into the engine in a cyclic process equals the net work done by the engine.

When we analyze heat engines, it helps to think of two bodies with which the working substance of the engine can interact. One of these, called the *hot reservoir*, represents the heat source; it can give the working substance large amounts of heat at a constant temperature T_H without appreciably changing its own temperature. The other body, called the *cold reservoir*, can absorb large amounts of discarded heat from the engine at a constant lower temperature T_C.

In a steam-turbine system the flames and hot gases in the boiler are the hot reservoir, and the cold water and air used to condense and cool the used steam are the cold reservoir.

We denote the quantities of heat transferred from the hot and cold reservoirs as Q_H and Q_C, respectively. A quantity of heat Q is positive when heat is transferred *into* the working substance and is negative when heat leaves the working substance. Thus in a heat engine, Q_H is positive but Q_C is negative, representing heat *leaving* the working substance. This sign convention is consistent with the rules we stated in Section 19.1; we will continue to use those rules here. For clarity, we'll often state the relationships in terms of the absolute values of the Q's and W's because absolute values are always positive.

20.3 Schematic energy-flow diagram for a heat engine.

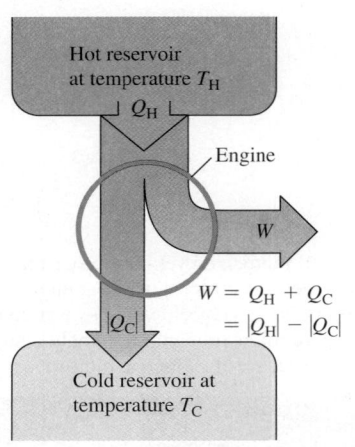

Energy-Flow Diagrams and Efficiency

We can represent the energy transformations in a heat engine by the *energy-flow diagram* of **Fig. 20.3**. The engine itself is represented by the circle. The amount of heat Q_H supplied to the engine by the hot reservoir is proportional to the width of the incoming "pipeline" at the top of the diagram. The width of the outgoing pipeline at the bottom is proportional to the magnitude $|Q_C|$ of the heat rejected in the exhaust. The branch line to the right represents the portion of the heat supplied that the engine converts to mechanical work, W.

When an engine repeats the same cycle over and over, Q_H and Q_C represent the quantities of heat absorbed and rejected by the engine *during one cycle*; Q_H is positive, and Q_C is negative. The *net* heat Q absorbed per cycle is

$$Q = Q_H + Q_C = |Q_H| - |Q_C| \tag{20.1}$$

The useful output of the engine is the net work W done by the working substance. From the first law,

$$W = Q = Q_H + Q_C = |Q_H| - |Q_C| \tag{20.2}$$

Ideally, we would like to convert *all* the heat Q_H into work; in that case we would have $Q_H = W$ and $Q_C = 0$. Experience shows that this is impossible; there is always some heat wasted, and Q_C *is never zero*. We define the **thermal efficiency** of an engine, denoted by e, as the quotient

$$e = \frac{W}{Q_H} \tag{20.3}$$

The thermal efficiency e represents the fraction of Q_H that *is* converted to work. To put it another way, e is what you get divided by what you pay for. This is always less than unity, an all-too-familiar experience! In terms of the flow diagram of Fig. 20.3, the most efficient engine is one for which the branch pipeline representing the work output is as wide as possible and the exhaust pipeline representing the heat thrown away is as narrow as possible.

When we substitute the two expressions for W given by Eq. (20.2) into Eq. (20.3), we get the following equivalent expressions for e:

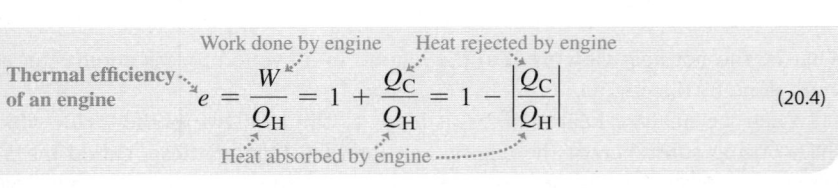

Work done by engine   Heat rejected by engine

$$\text{Thermal efficiency of an engine} \quad e = \frac{W}{Q_H} = 1 + \frac{Q_C}{Q_H} = 1 - \left|\frac{Q_C}{Q_H}\right| \tag{20.4}$$

Heat absorbed by engine

Note that e is a quotient of two energy quantities and thus is a pure number, without units. Of course, we must always express W, Q_H, and Q_C in the same units.

PROBLEM-SOLVING STRATEGY 20.1 | HEAT ENGINES

Problems involving heat engines are, fundamentally, problems in the first law of thermodynamics. You should review Problem-Solving Strategy 19.1 (Section 19.4).

IDENTIFY *the relevant concepts:* A heat engine is any device that converts heat partially to work, as shown schematically in Fig. 20.3. We will see in Section 20.4 that a refrigerator is essentially a heat engine running in reverse, so many of the same concepts apply.

SET UP *the problem* as suggested in Problem-Solving Strategy 19.1. Use Eq. (20.4) if the thermal efficiency of the engine is relevant. Sketch an energy-flow diagram like Fig. 20.3.

EXECUTE *the solution* as follows:
1. Be careful with the sign conventions for W and the various Q's. W is positive when the system expands and does work; W is

negative when the system is compressed and work is done on it. Each Q is positive if it represents heat entering the system and is negative if it represents heat leaving the system. When you know that a quantity is negative, such as Q_C in the above discussion, it sometimes helps to write it as $Q_C = -|Q_C|$.
2. Power is work per unit time ($P = W/t$), and rate of heat transfer (heat current) H is heat transfer per unit time ($H = Q/t$). In problems involving these concepts it helps to ask, "What is W or Q in one second (or one hour)?"
3. Keeping steps 1 and 2 in mind, solve for the target variables.

EVALUATE *your answer:* Use the first law of thermodynamics to check your results. Pay particular attention to algebraic signs.

SOLUTION

EXAMPLE 20.1   ANALYZING A HEAT ENGINE

A gasoline truck engine takes in 10,000 J of heat and delivers 2000 J of mechanical work per cycle. The heat is obtained by burning gasoline with heat of combustion $L_c = 5.0 \times 10^4$ J/g. (a) What is the thermal efficiency of this engine? (b) How much heat is discarded in each cycle? (c) If the engine goes through 25 cycles per second, what is its power output in watts? In horsepower? (d) How much gasoline is burned in each cycle? (e) How much gasoline is burned per second? Per hour?

SOLUTION

IDENTIFY and SET UP: This problem concerns a heat engine, so we can use the ideas of this section. **Figure 20.4** is our energy-flow diagram for one cycle. In each cycle the engine does $W = 2000$ J of work and takes in heat $Q_H = 10,000$ J. We use Eq. (20.4), in the form $e = W/Q_H$, to find the thermal efficiency. We use Eq. (20.2) to find the amount of heat Q_C rejected per cycle. The heat of combustion tells us how much gasoline must be burned per cycle and hence per unit time. The power output is the time rate at which the work W is done.

20.4 Our sketch for this problem.

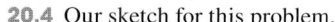

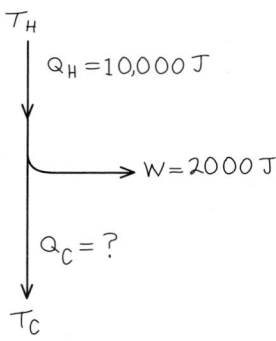

EXECUTE: (a) From Eq. (20.4), the thermal efficiency is

$$e = \frac{W}{Q_H} = \frac{2000 \text{ J}}{10,000 \text{ J}} = 0.20 = 20\%$$

(b) From Eq. (20.2), $W = Q_H + Q_C$, so

$$Q_C = W - Q_H = 2000 \text{ J} - 10,000 \text{ J} = -8000 \text{ J}$$

That is, 8000 J of heat leaves the engine during each cycle.

(c) The power P equals the work per cycle multiplied by the number of cycles per second:

$$P = (2000 \text{ J/cycle})(25 \text{ cycles/s}) = 50,000 \text{ W} = 50 \text{ kW}$$

$$= (50,000 \text{ W}) \frac{1 \text{ hp}}{746 \text{ W}} = 67 \text{ hp}$$

(d) Let m be the mass of gasoline burned during each cycle. Then $Q_H = mL_c$ and

$$m = \frac{Q_H}{L_c} = \frac{10,000 \text{ J}}{5.0 \times 10^4 \text{ J/g}} = 0.20 \text{ g}$$

(e) The mass of gasoline burned per second equals the mass per cycle multiplied by the number of cycles per second:

$$(0.20 \text{ g/cycle})(25 \text{ cycles/s}) = 5.0 \text{ g/s}$$

The mass burned per hour is

$$(5.0 \text{ g/s}) \frac{3600 \text{ s}}{1 \text{ h}} = 18,000 \text{ g/h} = 18 \text{ kg/h}$$

EVALUATE: An efficiency of 20% is fairly typical for cars and trucks if W includes only the work delivered to the wheels. We can check the mass burned per hour by expressing it in miles per gallon ("mileage"). The density of gasoline is about 0.70 g/cm³, so this is about 25,700 cm³, 25.7 L, or 6.8 gallons of gasoline per hour. If the truck is traveling at 55 mi/h (88 km/h), this represents fuel consumption of 8.1 miles/gallon (3.4 km/L). This is a fairly typical mileage for large trucks.

TEST YOUR UNDERSTANDING OF SECTION 20.2 Rank the following heat engines in order from highest to lowest thermal efficiency. (i) An engine that in one cycle absorbs 5000 J of heat and rejects 4500 J of heat; (ii) an engine that in one cycle absorbs 25,000 J of heat and does 2000 J of work; (iii) an engine that in one cycle does 400 J of work and rejects 2800 J of heat. ❙

20.3 INTERNAL-COMBUSTION ENGINES

The gasoline engine, used in automobiles and many other types of machinery, is a familiar example of a heat engine. Let's look at its thermal efficiency. **Figure 20.5** shows the operation of one type of gasoline engine. First a mixture of air and gasoline vapor flows into a cylinder through an open intake valve while the piston descends, increasing the volume of the cylinder from a minimum of V (when the piston is all the way up) to a maximum of rV (when it is all the way down). The quantity r is called the **compression ratio;** for present-day automobile engines its value is typically 8 to 10. At the end of this *intake stroke,* the intake valve closes and the mixture is compressed, approximately adiabatically, to volume V during the *compression stroke.* The mixture is then ignited by the spark plug, and the heated gas expands, approximately adiabatically, back to volume rV, pushing on the piston and doing work; this is the *power stroke.* Finally, the exhaust valve opens, and the combustion products are pushed out (during the *exhaust stroke*), leaving the cylinder ready for the next intake stroke.

The Otto Cycle

Figure 20.6 is a *pV*-diagram for an idealized model of the thermodynamic processes in a gasoline engine. This model is called the **Otto cycle.** At point a the gasoline–air mixture has entered the cylinder. The mixture is compressed adiabatically to point b and is then ignited. Heat Q_H is added to the system by the burning gasoline along line bc, and the power stroke is the adiabatic expansion to d. The gas is cooled to the temperature of the outside air along line da; during this process, heat $|Q_C|$ is rejected. This gas leaves the engine as exhaust and does not enter the engine again. But since an equivalent amount of gasoline and air enters, we may consider the process to be cyclic.

20.5 Cycle of a four-stroke internal-combustion engine.

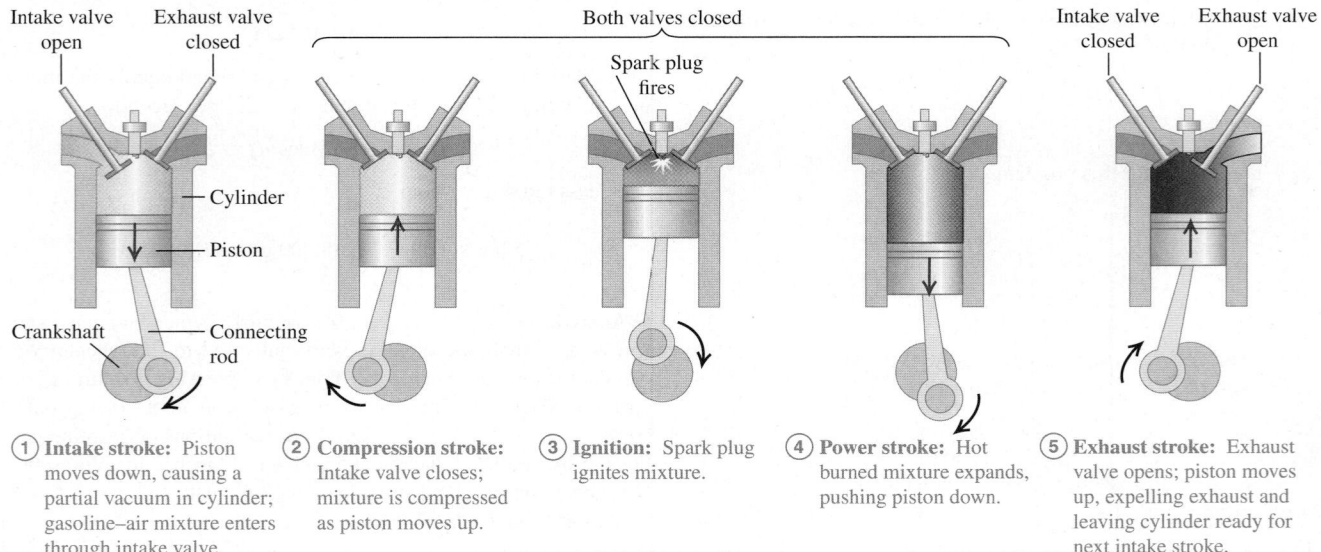

① **Intake stroke:** Piston moves down, causing a partial vacuum in cylinder; gasoline–air mixture enters through intake valve.

② **Compression stroke:** Intake valve closes; mixture is compressed as piston moves up.

③ **Ignition:** Spark plug ignites mixture.

④ **Power stroke:** Hot burned mixture expands, pushing piston down.

⑤ **Exhaust stroke:** Exhaust valve opens; piston moves up, expelling exhaust and leaving cylinder ready for next intake stroke.

We can calculate the efficiency of this idealized cycle. Processes bc and da are constant-volume, so the heats Q_H and Q_C are related simply to the temperatures at points a, b, c, and d:

$$Q_H = nC_V(T_c - T_b) > 0$$

$$Q_C = nC_V(T_a - T_d) < 0$$

The thermal efficiency is given by Eq. (20.4). Inserting the above expressions and cancelling out the common factor nC_V, we find

$$e = \frac{Q_H + Q_C}{Q_H} = \frac{T_c - T_b + T_a - T_d}{T_c - T_b} \tag{20.5}$$

To simplify this further, we use the temperature–volume relationship for adiabatic processes for an ideal gas, Eq. (19.22). For the two adiabatic processes ab and cd,

$$T_a(rV)^{\gamma-1} = T_b V^{\gamma-1} \qquad \text{and} \qquad T_d(rV)^{\gamma-1} = T_c V^{\gamma-1}$$

where γ is the ratio of heat capacities for the gas in the engine (see Section 19.7). We divide each of these equations by the common factor $V^{\gamma-1}$ and substitute the resulting expressions for T_b and T_c back into Eq. (20.5). The result is

$$e = \frac{T_d r^{\gamma-1} - T_a r^{\gamma-1} + T_a - T_d}{T_d r^{\gamma-1} - T_a r^{\gamma-1}} = \frac{(T_d - T_a)(r^{\gamma-1} - 1)}{(T_d - T_a)r^{\gamma-1}}$$

Dividing out the common factor $(T_d - T_a)$, we get

Thermal efficiency in Otto cycle
$$e = 1 - \frac{1}{r^{\gamma-1}} \tag{20.6}$$
Compression ratio ···→ ← ··· Ratio of heat capacities

The thermal efficiency given by Eq. (20.6) is always less than unity, even for this idealized model. With $r = 8$ and $\gamma = 1.4$ (the value for air) the theoretical efficiency is $e = 0.56$, or 56%. The efficiency can be increased by increasing r. However, this also increases the temperature at the end of the adiabatic compression of the air–fuel mixture. If the temperature is too high, the mixture explodes spontaneously during compression instead of burning evenly after the spark plug ignites it. This is called *pre-ignition* or *detonation;* it causes a knocking sound and can damage the engine. The octane rating of a gasoline is a measure of its antiknock qualities. The maximum practical compression ratio for high-octane, or "premium," gasoline is about 10 to 13.

The Otto cycle is a highly idealized model. It assumes that the mixture behaves as an ideal gas; it ignores friction, turbulence, loss of heat to cylinder walls, and many other effects that reduce the efficiency of an engine. Efficiencies of real gasoline engines are typically around 35%.

The Diesel Cycle

The Diesel engine is similar in operation to the gasoline engine. The most important difference is that there is no fuel in the cylinder at the beginning of the compression stroke. A little before the beginning of the power stroke, the injectors start to inject fuel directly into the cylinder, just fast enough to keep the pressure approximately constant during the first part of the power stroke. Because of the high temperature developed during the adiabatic compression, the fuel ignites spontaneously as it is injected; no spark plugs are needed.

Figure 20.7 shows the idealized **Diesel cycle.** Starting at point a, air is compressed adiabatically to point b, heated at constant pressure to point c, expanded adiabatically to point d, and cooled at constant volume to point a. Because there is no fuel in the cylinder during most of the compression stroke,

20.6 The pV-diagram for the Otto cycle, an idealized model of the thermodynamic processes in a gasoline engine.

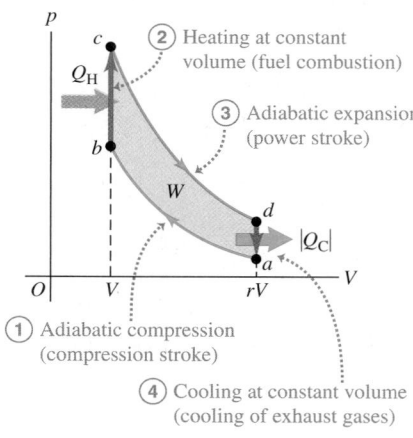

Otto cycle

② Heating at constant volume (fuel combustion)

③ Adiabatic expansion (power stroke)

① Adiabatic compression (compression stroke)

④ Cooling at constant volume (cooling of exhaust gases)

20.7 The pV-diagram for the idealized Diesel cycle.

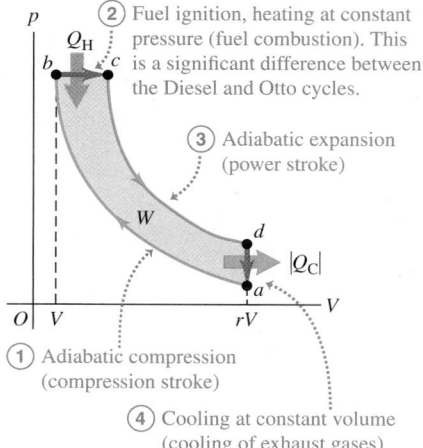

Diesel cycle

② Fuel ignition, heating at constant pressure (fuel combustion). This is a significant difference between the Diesel and Otto cycles.

③ Adiabatic expansion (power stroke)

① Adiabatic compression (compression stroke)

④ Cooling at constant volume (cooling of exhaust gases)

pre-ignition cannot occur, and the compression ratio r can be much higher than for a gasoline engine. This improves efficiency and ensures reliable ignition when the fuel is injected (because of the high temperature reached during the adiabatic compression). Values of r of 15 to 20 are typical; with these values and $\gamma = 1.4$, the theoretical efficiency of the idealized Diesel cycle is about 0.65 to 0.70. As with the Otto cycle, the efficiency of any actual engine is substantially less than this. While Diesel engines are very efficient, they must be built to much tighter tolerances than gasoline engines and the fuel-injection system requires careful maintenance.

TEST YOUR UNDERSTANDING OF SECTION 20.3   For an Otto-cycle engine with cylinders of a fixed size and a fixed compression ratio, which of the following aspects of the pV-diagram in Fig. 20.6 would change if you doubled the amount of fuel burned per cycle? (There may be more than one correct answer.) (i) The vertical distance between points b and c; (ii) the vertical distance between points a and d; (iii) the horizontal distance between points b and a. ∎

20.4 REFRIGERATORS

We can think of a **refrigerator** as a heat engine operating in reverse. A heat engine takes heat from a hot place and gives off heat to a colder place. A refrigerator does the opposite; it takes heat from a cold place (the inside of the refrigerator) and gives it off to a warmer place (usually the air in the room where the refrigerator is located). A heat engine has a net *output* of mechanical work; the refrigerator requires a net *input* of mechanical work. With the sign conventions from Section 20.2, for a refrigerator Q_C is positive but both W and Q_H are negative; hence $|W| = -W$ and $|Q_H| = -Q_H$.

Figure 20.8 shows an energy-flow diagram for a refrigerator. From the first law for a cyclic process,

$$Q_H + Q_C - W = 0 \qquad \text{or} \qquad -Q_H = Q_C - W$$

or, because both Q_H and W are negative,

$$|Q_H| = Q_C + |W| \tag{20.7}$$

Thus, as the diagram shows, the heat $|Q_H|$ leaving the working substance and given to the hot reservoir is always *greater* than the heat Q_C taken from the cold reservoir. Note that the absolute-value relationship

$$|Q_H| = |Q_C| + |W| \tag{20.8}$$

is valid for both heat engines and refrigerators.

From an economic point of view, the best refrigeration cycle is one that removes the greatest amount of heat $|Q_C|$ from the inside of the refrigerator for the least expenditure of mechanical work, $|W|$. The relevant ratio is therefore $|Q_C|/|W|$; the larger this ratio, the better the refrigerator. We call this ratio the **coefficient of performance, K**. From Eq. (20.8), $|W| = |Q_H| - |Q_C|$, so

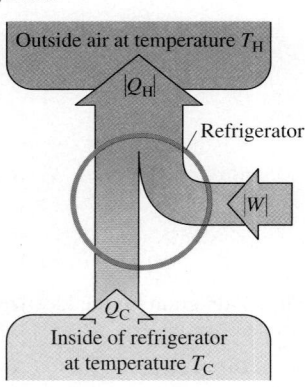

20.8 Schematic energy-flow diagram of a refrigerator.

Outside air at temperature T_H

$|Q_H|$

Refrigerator

$|W|$

Q_C

Inside of refrigerator at temperature T_C

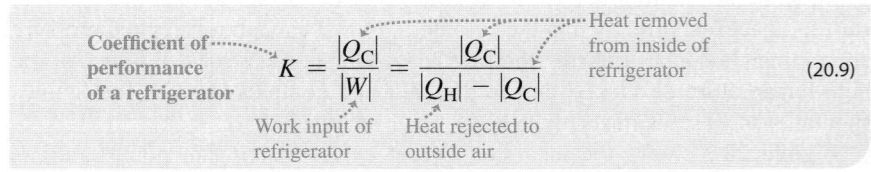

$$\begin{array}{c}\text{Coefficient of} \\ \text{performance} \\ \text{of a refrigerator}\end{array} \qquad K = \frac{|Q_C|}{|W|} = \frac{|Q_C|}{|Q_H| - |Q_C|} \tag{20.9}$$

Heat removed from inside of refrigerator

Work input of refrigerator   Heat rejected to outside air

As always, we measure Q_H, Q_C, and W all in the same energy units; K is then a dimensionless number.

20.9 (a) Principle of the mechanical refrigeration cycle. (b) How the key elements are arranged in a practical refrigerator.

(a)

(b)

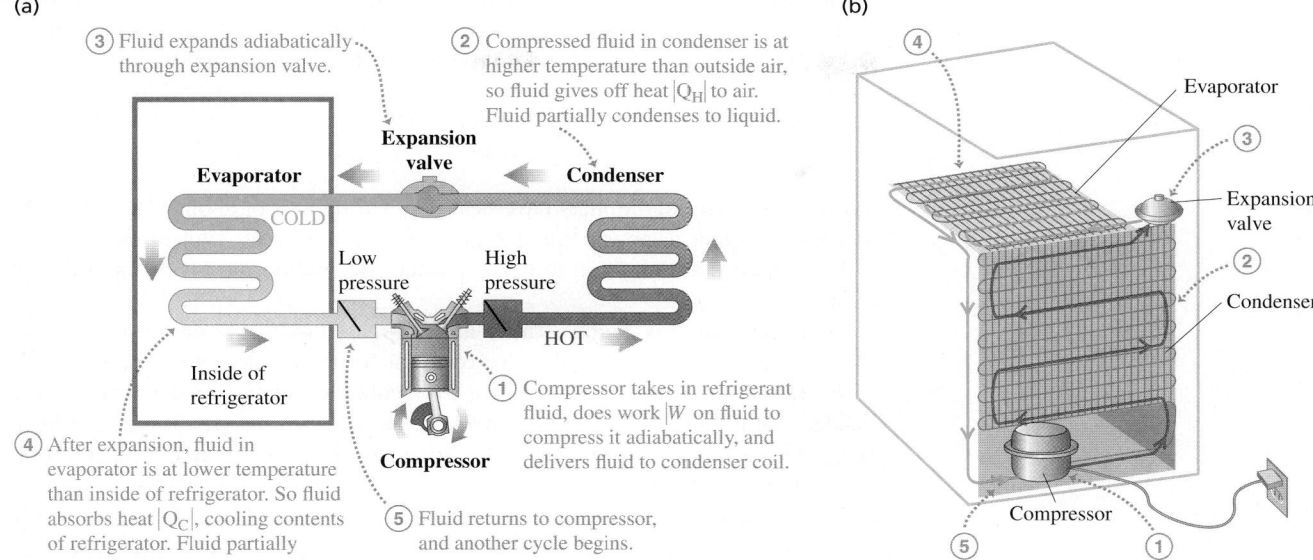

③ Fluid expands adiabatically through expansion valve.

② Compressed fluid in condenser is at higher temperature than outside air, so fluid gives off heat $|Q_H|$ to air. Fluid partially condenses to liquid.

Expansion valve

Evaporator

COLD

Condenser

Low pressure

High pressure

HOT

Inside of refrigerator

① Compressor takes in refrigerant fluid, does work $|W|$ on fluid to compress it adiabatically, and delivers fluid to condenser coil.

Compressor

④ After expansion, fluid in evaporator is at lower temperature than inside of refrigerator. So fluid absorbs heat $|Q_C|$, cooling contents of refrigerator. Fluid partially evaporates to vapor.

⑤ Fluid returns to compressor, and another cycle begins.

④ Evaporator

③ Expansion valve

② Condenser

⑤ Compressor ①

Practical Refrigerators

Figure 20.9a shows the principles of the cycle used in home refrigerators. The fluid "circuit" contains a refrigerant fluid (the working substance). The left side of the circuit (including the cooling coils inside the refrigerator) is at low temperature and low pressure; the right side (including the condenser coils outside the refrigerator) is at high temperature and high pressure. Ordinarily, both sides contain liquid and vapor in phase equilibrium. In each cycle the fluid absorbs heat $|Q_C|$ from the inside of the refrigerator on the left side and gives off heat $|Q_H|$ to the surrounding air on the right side. The compressor, usually driven by an electric motor (Fig. 20.9b), does work $|W|$ *on* the fluid during each cycle. So the compressor requires energy input, which is why refrigerators have to be plugged in.

An air conditioner operates on exactly the same principle. In this case the refrigerator box becomes a room or an entire building. The evaporator coils are inside, the condenser is outside, and fans circulate air through these (**Fig. 20.10**). In large installations the condenser coils are often cooled by water.

20.10 An air conditioner works on the same principle as a refrigerator.

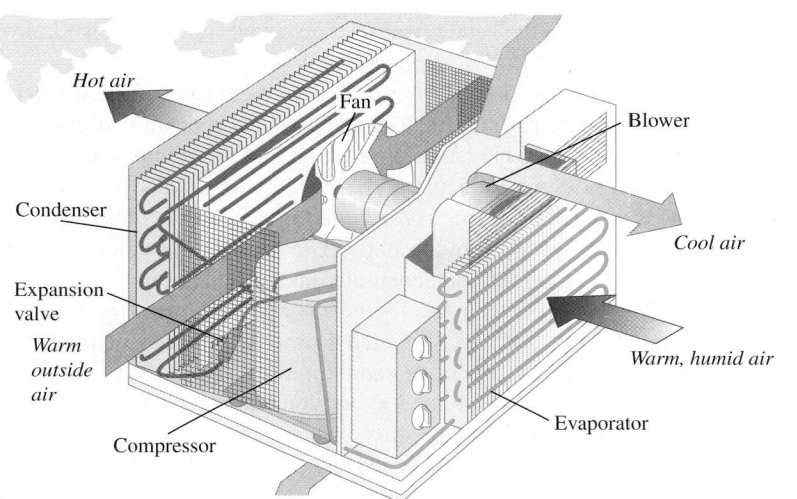

Hot air

Fan

Blower

Condenser

Cool air

Expansion valve

Warm outside air

Warm, humid air

Compressor

Evaporator

For air conditioners the quantities of greatest practical importance are the *rate* of heat removal (the heat current H from the region being cooled) and the *power* input $P = W/t$ to the compressor. If heat $|Q_C|$ is removed in time t, then $H = |Q_C|/t$. Then we can express the coefficient of performance as

$$K = \frac{|Q_C|}{|W|} = \frac{Ht}{Pt} = \frac{H}{P}$$

Typical room air conditioners have heat removal rates H of about 1500–3000 W and require electric power input of about 600 to 1200 W. Typical coefficients of performance are about 3; the actual values depend on the inside and outside temperatures.

A variation on this theme is the **heat pump,** used to heat buildings by cooling the outside air. It functions like a refrigerator turned inside out. The evaporator coils are outside, where they take heat from cold air, and the condenser coils are inside, where they give off heat to the warmer air. With proper design, the heat $|Q_H|$ delivered to the inside per cycle can be considerably greater than the work $|W|$ required to get it there.

Work is *always* needed to transfer heat from a colder to a hotter body. **?** Heat flows spontaneously from hotter to colder, and to reverse this flow requires the addition of work from the outside. Experience shows that it is impossible to make a refrigerator that transports heat from a colder body to a hotter body without the addition of work. If no work were needed, the coefficient of performance would be infinite. We call such a device a *workless refrigerator;* it is a mythical beast, like the unicorn and the free lunch.

TEST YOUR UNDERSTANDING OF SECTION 20.4   Can you cool your house by leaving the refrigerator door open? ∎

20.5 THE SECOND LAW OF THERMODYNAMICS

Experimental evidence suggests strongly that it is *impossible* to build a heat engine that converts heat completely to work—that is, an engine with 100% thermal efficiency. This impossibility is the basis of one statement of the **second law of thermodynamics,** as follows:

> **It is impossible for any system to undergo a process in which it absorbs heat from a reservoir at a single temperature and converts the heat completely into mechanical work, with the system ending in the same state in which it began.**

We will call this the "engine" statement of the second law. (It is also known to physicists as the *Kelvin–Planck statement* of this law.)

The basis of the second law of thermodynamics is the difference between the nature of internal energy and that of macroscopic mechanical energy. In a moving body the molecules have random motion, but superimposed on this is a coordinated motion of every molecule in the direction of the body's velocity. The kinetic energy associated with this *coordinated* macroscopic motion is what we call the kinetic energy of the moving body. The kinetic and potential energies associated with the *random* motion constitute the internal energy.

When a body sliding on a surface comes to rest as a result of friction, the organized motion of the body is converted to random motion of molecules in the body and in the surface. Since we cannot control the motions of individual molecules, we cannot convert this random motion completely back to organized motion. We can convert *part* of it, and this is what a heat engine does.

If the second law were *not* true, we could power an automobile or run a power plant by cooling the surrounding air. Neither of these impossibilities violates the *first* law of thermodynamics. The second law, therefore, is not a deduction from the first but stands by itself as a separate law of nature. The first law denies the possibility of creating or destroying energy; the second law limits the *availability* of energy and the ways in which it can be used and converted.

Restating the Second Law

Our analysis of refrigerators in Section 20.4 forms the basis for an alternative statement of the second law of thermodynamics. Heat flows spontaneously from hotter to colder bodies, never the reverse. A refrigerator does take heat from a colder to a hotter body, but its operation requires an input of mechanical energy or work. We can generalize this observation:

> **It is impossible for any process to have as its sole result the transfer of heat from a cooler to a hotter body.**

We'll call this the "refrigerator" statement of the second law. (It is also known as the *Clausius statement*.) It may not seem to be very closely related to the "engine" statement. In fact, though, the two statements are completely equivalent. For example, if we could build a workless refrigerator, violating the second or "refrigerator" statement of the second law, we could use it in conjunction with a heat engine, pumping the heat rejected by the engine back to the hot reservoir to be reused. This composite machine (**Fig. 20.11a**) would violate the "engine" statement of the second law because its net effect would be to take a net quantity of heat $Q_H - |Q_C|$ from the hot reservoir and convert it completely to work W.

20.11 Energy-flow diagrams showing that the two forms of the second law are equivalent.

(a) The "engine" statement of the second law of thermodynamics

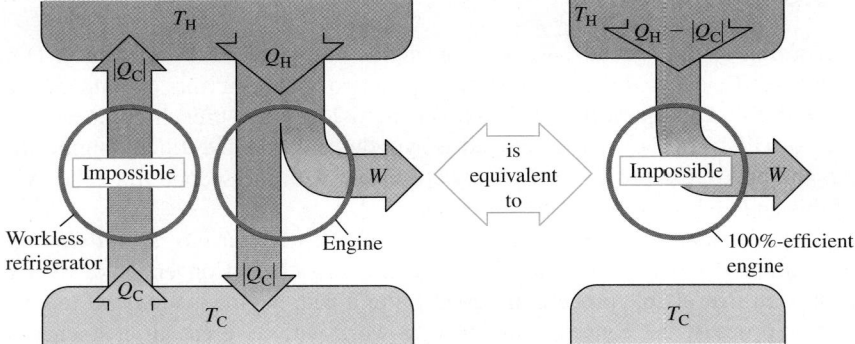

If a workless refrigerator were possible, it could be used in conjunction with an ordinary heat engine to form a 100%-efficient engine, converting heat $Q_H - |Q_C|$ completely to work.

(b) The "refrigerator" statement of the second law of thermodynamics

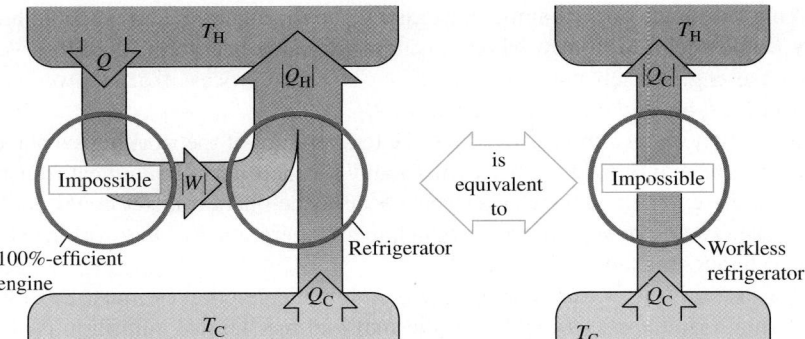

If a 100%-efficient engine were possible, it could be used in conjunction with an ordinary refrigerator to form a workless refrigerator, transferring heat Q_C from the cold to the hot reservoir with no input of work.

Alternatively, if we could make an engine with 100% thermal efficiency, in violation of the first statement, we could run it using heat from the hot reservoir and use the work output to drive a refrigerator that pumps heat from the cold reservoir to the hot (Fig. 20.11b). This composite device would violate the "refrigerator" statement because its net effect would be to take heat Q_C from the cold reservoir and deliver it to the hot reservoir without requiring any input of work. Thus any device that violates one form of the second law can be used to make a device that violates the other form. If violations of the first form are impossible, so are violations of the second!

The conversion of work to heat and the heat flow from hot to cold across a finite temperature gradient are *irreversible* processes. The "engine" and "refrigerator" statements of the second law state that these processes can be only partially reversed. We could cite other examples. Gases naturally flow from a region of high pressure to a region of low pressure; gases and miscible liquids left by themselves always tend to mix, not to unmix. The second law of thermodynamics is an expression of the inherent one-way aspect of these and many other irreversible processes. Energy conversion is an essential aspect of all plant and animal life and of human technology, so the second law of thermodynamics is of fundamental importance.

TEST YOUR UNDERSTANDING OF SECTION 20.5   Would a 100%-efficient engine (Fig. 20.11a) violate the *first* law of thermodynamics? What about a workless refrigerator (Fig. 20.11b)? ▮

20.6 THE CARNOT CYCLE

According to the second law, no heat engine can have 100% efficiency. How great an efficiency *can* an engine have, given two heat reservoirs at temperatures T_H and T_C? This question was answered in 1824 by the French engineer Sadi Carnot (1796–1832), who developed a hypothetical, idealized heat engine that has the maximum possible efficiency consistent with the second law. The cycle of this engine is called the **Carnot cycle.**

To understand the rationale of the Carnot cycle, we return to *reversibility* and its relationship to directions of thermodynamic processes. Conversion of work to heat is an irreversible process; the purpose of a heat engine is a *partial* reversal of this process, the conversion of heat to work with as great an efficiency as possible. For maximum heat-engine efficiency, therefore, *we must avoid all irreversible processes* (**Fig. 20.12**).

Heat flow through a finite temperature drop is an irreversible process. Therefore, during heat transfer in the Carnot cycle there must be *no* finite temperature difference. When the engine takes heat from the hot reservoir at temperature T_H, the working substance of the engine must also be at T_H; otherwise, irreversible heat flow would occur. Similarly, when the engine discards heat to the cold reservoir at T_C, the engine itself must be at T_C. That is, every process that involves heat transfer must be *isothermal* at either T_H or T_C.

Conversely, in any process in which the temperature of the working substance of the engine is intermediate between T_H and T_C, there must be *no* heat transfer between the engine and either reservoir because such heat transfer could not be reversible. Therefore any process in which the temperature T of the working substance changes must be *adiabatic.*

The bottom line is that every process in our idealized cycle must be either isothermal or adiabatic. In addition, thermal and mechanical equilibrium must be maintained at all times so that each process is completely reversible.

20.12 The temperature of the firebox of a steam engine is much higher than the temperature of water in the boiler, so heat flows irreversibly from firebox to water. Carnot's quest to understand the efficiency of steam engines led him to the idea that an ideal engine would involve only *reversible* processes.

20.13 The Carnot cycle for an ideal gas. The light blue lines in the pV-diagram are isotherms (curves of constant temperature) and the dark blue lines are adiabats (curves of zero heat flow).

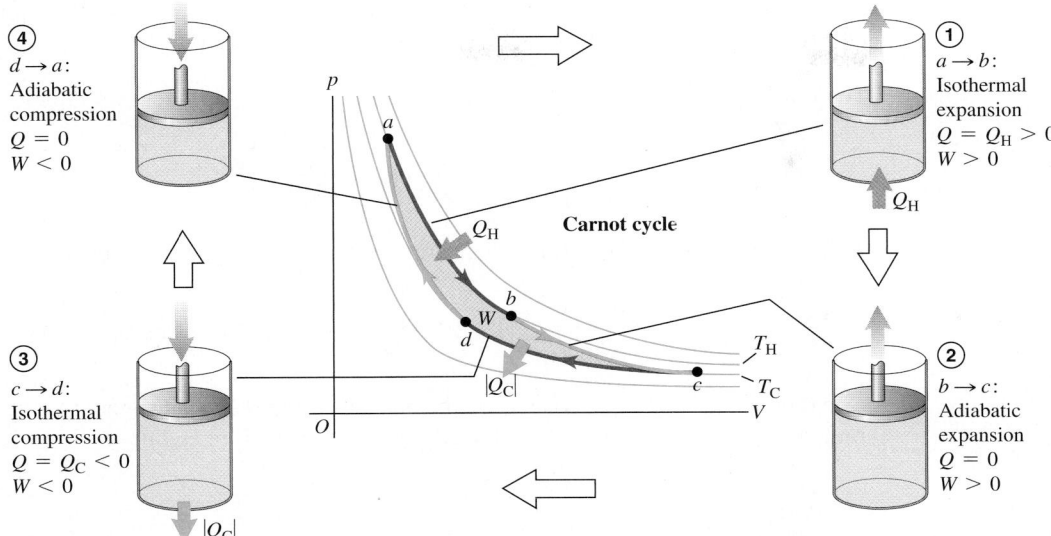

④ $d \rightarrow a$:
Adiabatic
compression
$Q = 0$
$W < 0$

① $a \rightarrow b$:
Isothermal
expansion
$Q = Q_H > 0$
$W > 0$

Q_H **Carnot cycle**

③ $c \rightarrow d$:
Isothermal
compression
$Q = Q_C < 0$
$W < 0$

② $b \rightarrow c$:
Adiabatic
expansion
$Q = 0$
$W > 0$

$|Q_C|$

T_H
T_C

Steps of the Carnot Cycle

The Carnot cycle consists of two reversible isothermal and two reversible adiabatic processes. **Figure 20.13** shows a Carnot cycle using as its working substance an ideal gas in a cylinder with a piston. It consists of the following steps:

1. The gas expands isothermally at temperature T_H, absorbing heat Q_H (ab).
2. It expands adiabatically until its temperature drops to T_C (bc).
3. It is compressed isothermally at T_C, rejecting heat $|Q_C|$ (cd).
4. It is compressed adiabatically back to its initial state at temperature T_H (da).

We can calculate the thermal efficiency e of a Carnot engine in the special case shown in Fig. 20.13 in which the working substance is an *ideal gas*. To do this, we'll first find the ratio Q_C/Q_H of the quantities of heat transferred in the two isothermal processes and then use Eq. (20.4) to find e.

For an ideal gas the internal energy U depends only on temperature and is thus constant in any isothermal process. For the isothermal expansion ab, $\Delta U_{ab} = 0$ and Q_H is equal to the work W_{ab} done by the gas during its isothermal expansion at temperature T_H. We calculated this work in Example 19.1 (Section 19.2); using that result, we have

$$Q_H = W_{ab} = nRT_H \ln \frac{V_b}{V_a} \qquad (20.10)$$

Similarly,

$$Q_C = W_{cd} = nRT_C \ln \frac{V_d}{V_c} = -nRT_C \ln \frac{V_c}{V_d} \qquad (20.11)$$

Because V_d is less than V_c, Q_C is negative ($Q_C = -|Q_C|$); heat flows out of the gas during the isothermal compression at temperature T_C.

The ratio of the two quantities of heat is thus

$$\frac{Q_C}{Q_H} = -\left(\frac{T_C}{T_H}\right) \frac{\ln(V_c/V_d)}{\ln(V_b/V_a)} \qquad (20.12)$$

This can be simplified further by use of the temperature–volume relationship for an adiabatic process, Eq. (19.22). We find for the two adiabatic processes:

$$T_H V_b^{\gamma-1} = T_C V_c^{\gamma-1} \qquad \text{and} \qquad T_H V_a^{\gamma-1} = T_C V_d^{\gamma-1}$$

Dividing the first of these by the second, we find

$$\frac{V_b^{\gamma-1}}{V_a^{\gamma-1}} = \frac{V_c^{\gamma-1}}{V_d^{\gamma-1}} \qquad \text{and} \qquad \frac{V_b}{V_a} = \frac{V_c}{V_d}$$

Thus the two logarithms in Eq. (20.12) are equal, and that equation reduces to

$$\frac{Q_C}{Q_H} = -\frac{T_C}{T_H} \quad \text{or} \quad \frac{|Q_C|}{|Q_H|} = \frac{T_C}{T_H} \qquad \text{(heat transfer in a Carnot engine)} \quad (20.13)$$

The ratio of the heat rejected at T_C to the heat absorbed at T_H is just equal to the ratio T_C/T_H. Then from Eq. (20.4) the efficiency of the Carnot engine is

Efficiency of a Carnot engine
$$e_{Carnot} = 1 - \frac{T_C}{T_H} = \frac{T_H - T_C}{T_H} \qquad (20.14)$$
Temperature of cold reservoir
Temperature of hot reservoir

CAUTION **Use Kelvin temperature in all Carnot calculations** In Carnot-cycle calculations, you must use *absolute* (Kelvin) temperatures only. That's because Eqs. (20.10) through (20.14) come from the ideal-gas equation $pV = nRT$, in which T is absolute temperature. ∎

This simple result says that the efficiency of a Carnot engine depends only on the temperatures of the two heat reservoirs. The efficiency is large when the temperature *difference* is large, and it is very small when the temperatures are nearly equal. The efficiency can never be exactly unity unless $T_C = 0$; we'll see later that this, too, is impossible.

EXAMPLE 20.2 ANALYZING A CARNOT ENGINE I

A Carnot engine takes 2000 J of heat from a reservoir at 500 K, does some work, and discards some heat to a reservoir at 350 K. How much work does it do, how much heat is discarded, and what is its efficiency?

SOLUTION

IDENTIFY and SET UP: This problem involves a Carnot engine, so we can use the ideas of this section and those of Section 20.2 (which apply to heat engines of all kinds). **Figure 20.14** shows the energy-flow diagram. We have $Q_H = 2000$ J, $T_H = 500$ K, and $T_C = 350$ K. We use Eq. (20.13) to find Q_C, and then use the first law of thermodynamics as given by Eq. (20.2) to find W. We find the efficiency e from T_C and T_H from Eq. (20.14).

EXECUTE: From Eq. (20.13),

$$Q_C = -Q_H \frac{T_C}{T_H} = -(2000 \text{ J})\frac{350 \text{ K}}{500 \text{ K}} = -1400 \text{ J}$$

Then from Eq. (20.2), the work done is

$$W = Q_H + Q_C = 2000 \text{ J} + (-1400 \text{ J}) = 600 \text{ J}$$

From Eq. (20.14), the thermal efficiency is

$$e = 1 - \frac{T_C}{T_H} = 1 - \frac{350 \text{ K}}{500 \text{ K}} = 0.30 = 30\%$$

EVALUATE: The negative sign of Q_C is correct: It shows that 1400 J of heat flows *out* of the engine and into the cold reservoir. We can check our result for e by using the basic definition of thermal efficiency, Eq. (20.3):

$$e = \frac{W}{Q_H} = \frac{600 \text{ J}}{2000 \text{ J}} = 0.30 = 30\%$$

20.14 Our sketch for this problem.

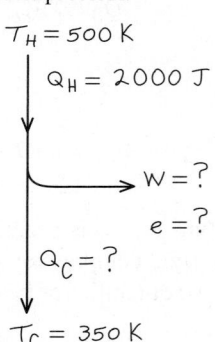

EXAMPLE 20.3   ANALYZING A CARNOT ENGINE II

Suppose 0.200 mol of an ideal diatomic gas ($\gamma = 1.40$) undergoes a Carnot cycle between 227°C and 27°C, starting at $p_a = 10.0 \times 10^5$ Pa at point a in the pV-diagram of Fig. 20.13. The volume doubles during the isothermal expansion step $a \to b$. (a) Find the pressure and volume at points a, b, c, and d. (b) Find Q, W, and ΔU for each step and for the entire cycle. (c) Find the efficiency directly from the results of part (b), and compare with the value calculated from Eq. (20.14).

SOLUTION

IDENTIFY and SET UP: This problem involves the properties of the Carnot cycle and those of an ideal gas. We are given the number of moles n and the pressure and temperature at point a (which is at the higher of the two reservoir temperatures); we can find the volume at a from the ideal-gas equation $pV = nRT$. We then find the pressure and volume at points b, c, and d from the known doubling of volume in step $a \to b$, from equations given in this section, and from $pV = nRT$. In each step we use Eqs. (20.10) and (20.11) to find the heat flow and work done and Eq. (19.13) to find the internal energy change.

EXECUTE: (a) With $T_H = (227 + 273.15)$ K $= 500$ K and $T_C = (27 + 273.15)$ K $= 300$ K, $pV = nRT$ yields

$$V_a = \frac{nRT_H}{p_a} = \frac{(0.200 \text{ mol})(8.314 \text{ J/mol} \cdot \text{K})(500 \text{ K})}{10.0 \times 10^5 \text{ Pa}}$$

$$= 8.31 \times 10^{-4} \text{ m}^3$$

The volume doubles during the isothermal expansion $a \to b$:

$$V_b = 2V_a = 2(8.31 \times 10^{-4} \text{ m}^3) = 16.6 \times 10^{-4} \text{ m}^3$$

Because the expansion $a \to b$ is isothermal, $p_a V_a = p_b V_b$, so

$$p_b = \frac{p_a V_a}{V_b} = 5.00 \times 10^5 \text{ Pa}$$

For the adiabatic expansion $b \to c$, we use the equation $T_H V_b^{\gamma-1} = T_C V_c^{\gamma-1}$ that follows Eq. (20.12) as well as the ideal-gas equation:

$$V_c = V_b \left(\frac{T_H}{T_C}\right)^{1/(\gamma-1)} = (16.6 \times 10^{-4} \text{ m}^3)\left(\frac{500 \text{ K}}{300 \text{ K}}\right)^{2.5}$$

$$= 59.6 \times 10^{-4} \text{ m}^3$$

$$p_c = \frac{nRT_C}{V_c} = \frac{(0.200 \text{ mol})(8.314 \text{ J/mol} \cdot \text{K})(300 \text{ K})}{59.6 \times 10^{-4} \text{ m}^3}$$

$$= 0.837 \times 10^5 \text{ Pa}$$

For the adiabatic compression $d \to a$, we have $T_C V_d^{\gamma-1} = T_H V_a^{\gamma-1}$ and so

$$V_d = V_a \left(\frac{T_H}{T_C}\right)^{1/(\gamma-1)} = (8.31 \times 10^{-4} \text{ m}^3)\left(\frac{500 \text{ K}}{300 \text{ K}}\right)^{2.5}$$

$$= 29.8 \times 10^{-4} \text{ m}^3$$

$$p_d = \frac{nRT_C}{V_d} = \frac{(0.200 \text{ mol})(8.314 \text{ J/mol} \cdot \text{K})(300 \text{ K})}{29.8 \times 10^{-4} \text{ m}^3}$$

$$= 1.67 \times 10^5 \text{ Pa}$$

(b) For the isothermal expansion $a \to b$, $\Delta U_{ab} = 0$. From Eq. (20.10),

$$W_{ab} = Q_H = nRT_H \ln\frac{V_b}{V_a}$$

$$= (0.200 \text{ mol})(8.314 \text{ J/mol} \cdot \text{K})(500 \text{ K})(\ln 2) = 576 \text{ J}$$

For the adiabatic expansion $b \to c$, $Q_{bc} = 0$. From the first law of thermodynamics, $\Delta U_{bc} = Q_{bc} - W_{bc} = -W_{bc}$; the work W_{bc} done by the gas in this adiabatic expansion equals the negative of the change in internal energy of the gas. From Eq. (19.13) we have $\Delta U = nC_V \Delta T$, where $\Delta T = T_C - T_H$. Using $C_V = 20.8$ J/mol $\cdot$ K for an ideal diatomic gas, we find

$$W_{bc} = -\Delta U_{bc} = -nC_V(T_C - T_H) = nC_V(T_H - T_C)$$

$$= (0.200 \text{ mol})(20.8 \text{ J/mol} \cdot \text{K})(500 \text{ K} - 300 \text{ K}) = 832 \text{ J}$$

For the isothermal compression $c \to d$, $\Delta U_{cd} = 0$; Eq. (20.11) gives

$$W_{cd} = Q_C = nRT_C \ln\frac{V_d}{V_c}$$

$$= (0.200 \text{ mol})(8.314 \text{ J/mol} \cdot \text{K})(300 \text{ K})\left(\ln\frac{29.8 \times 10^{-4} \text{ m}^3}{59.6 \times 10^{-4} \text{ m}^3}\right)$$

$$= -346 \text{ J}$$

For the adiabatic compression $d \to a$, $Q_{da} = 0$ and

$$W_{da} = -\Delta U_{da} = -nC_V(T_H - T_C) = nC_V(T_C - T_H)$$

$$= (0.200 \text{ mol})(20.8 \text{ J/mol} \cdot \text{K})(300 \text{ K} - 500 \text{ K}) = -832 \text{ J}$$

We can tabulate these results as follows:

Process	Q	W	ΔU
$a \to b$	576 J	576 J	0
$b \to c$	0	832 J	−832 J
$c \to d$	−346 J	−346 J	0
$d \to a$	0	−832 J	832 J
Total	230 J	230 J	0

(c) From the above table, $Q_H = 576$ J and the total work is 230 J. Thus

$$e = \frac{W}{Q_H} = \frac{230 \text{ J}}{576 \text{ J}} = 0.40 = 40\%$$

We can compare this to the result from Eq. (20.14),

$$e = \frac{T_H - T_C}{T_H} = \frac{500 \text{ K} - 300 \text{ K}}{500 \text{ K}} = 0.40 = 40\%$$

EVALUATE: The table in part (b) shows that for the entire cycle $Q = W$ and $\Delta U = 0$, just as we would expect: In a complete cycle, the *net* heat input is used to do work, and there is zero net change in the internal energy of the system. Note also that the quantities of work in the two adiabatic processes are negatives of each other. Can you show from the analysis leading to Eq. (20.13) that this must *always* be the case in a Carnot cycle?

The Carnot Refrigerator

Because each step in the Carnot cycle is reversible, the *entire cycle* may be reversed, converting the engine into a refrigerator. The coefficient of performance of the Carnot refrigerator is obtained by combining the general definition of K, Eq. (20.9), with Eq. (20.13) for the Carnot cycle. We first rewrite Eq. (20.9) as

$$K = \frac{|Q_C|}{|Q_H| - |Q_C|} = \frac{|Q_C|/|Q_H|}{1 - |Q_C|/|Q_H|}$$

Then we substitute Eq. (20.13), $|Q_C|/|Q_H| = T_C/T_H$, into this expression:

Coefficient of performance of a Carnot refrigerator $\longrightarrow K_{\text{Carnot}} = \dfrac{T_C \longleftarrow \text{Temperature of cold reservoir}}{T_H - T_C}$ (20.15)

Temperature of hot reservoir

When the temperature difference $T_H - T_C$ is small, K is much larger than unity; in this case a lot of heat can be "pumped" from the lower to the higher temperature with only a little expenditure of work. But the greater the temperature difference, the smaller the value of K and the more work is required to transfer a given quantity of heat.

EXAMPLE 20.4   ANALYZING A CARNOT REFRIGERATOR

If the cycle described in Example 20.3 is run backward as a refrigerator, what is its coefficient of performance?

SOLUTION

IDENTIFY and SET UP: This problem uses the ideas of Section 20.3 (for refrigerators in general) and the above discussion of Carnot refrigerators. Equation (20.9) gives the coefficient of performance K of *any* refrigerator in terms of the heat Q_C extracted from the cold reservoir per cycle and the work W that must be done per cycle.

EXECUTE: In Example 20.3 we found that in one cycle the Carnot engine rejects heat $Q_C = -346$ J to the cold reservoir and does work $W = 230$ J. When run in reverse as a refrigerator, the system

extracts heat $Q_C = +346$ J from the cold reservoir while requiring a work input of $W = -230$ J. From Eq. (20.9),

$$K = \frac{|Q_C|}{|W|} = \frac{346 \text{ J}}{230 \text{ J}} = 1.50$$

Because this is a Carnot cycle, we can also use Eq. (20.15):

$$K = \frac{T_C}{T_H - T_C} = \frac{300 \text{ K}}{500 \text{ K} - 300 \text{ K}} = 1.50$$

EVALUATE: Equations (20.14) and (20.15) show that e and K for a Carnot cycle depend only on T_H and T_C, and we don't need to calculate Q and W. For cycles containing irreversible processes, however, these two equations are not valid, and more detailed calculations are necessary.

The Carnot Cycle and the Second Law

We can prove that **no engine can be more efficient than a Carnot engine operating between the same two temperatures.** The key to the proof is the above observation that since each step in the Carnot cycle is reversible, the *entire cycle* may be reversed. Run backward, the engine becomes a refrigerator. Suppose we have an engine that is more efficient than a Carnot engine (**Fig. 20.15**). Let the Carnot engine, run backward as a refrigerator by negative work $-|W|$, take in heat Q_C from the cold reservoir and expel heat $|Q_H|$ to the hot reservoir. The superefficient engine expels heat $|Q_C|$, but to do this, it takes in a greater amount of heat $Q_H + \Delta$. Its work output is then $W + \Delta$, and the net effect of the two machines together is to take a quantity of heat Δ and convert it completely into work. This violates the "engine" statement of the second law. We could construct a similar argument that a superefficient engine could be used to violate the "refrigerator" statement of the second law. Note that we don't have to assume that the superefficient engine is reversible. In a similar way we can

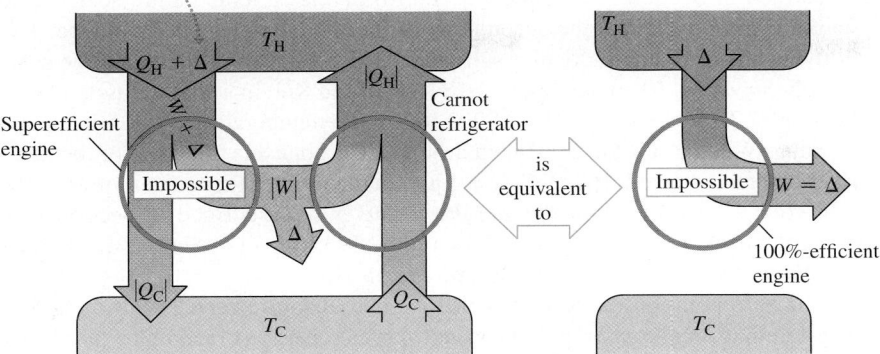

If a superefficient engine were possible, it could be used in conjunction with a Carnot refrigerator to convert the heat Δ completely to work, with no net transfer to the cold reservoir.

20.15 Proving that the Carnot engine has the highest possible efficiency. A "superefficient" engine (more efficient than a Carnot engine) combined with a Carnot refrigerator could convert heat completely into work with no net heat transfer to the cold reservoir. This would violate the second law of thermodynamics.

show that *no refrigerator can have a greater coefficient of performance than a Carnot refrigerator operating between the same two temperatures.*

Thus the statement that no engine can be more efficient than a Carnot engine is yet another equivalent statement of the second law of thermodynamics. It also follows directly that **all Carnot engines operating between the same two temperatures have the same efficiency, irrespective of the nature of the working substance.** Although we derived Eq. (20.14) for a Carnot engine by using an ideal gas as its working substance, it is in fact valid for *any* Carnot engine, no matter what its working substance.

Equation (20.14), the expression for the efficiency of a Carnot engine, sets an upper limit to the efficiency of a real engine such as a steam turbine. To maximize this upper limit and the actual efficiency of the real engine, the designer must make the intake temperature T_H as high as possible and the exhaust temperature T_C as low as possible (**Fig. 20.16**).

The exhaust temperature cannot be lower than the lowest temperature available for cooling the exhaust. For a steam turbine at an electric power plant, T_C may be the temperature of river or lake water; then we want the boiler temperature T_H to be as high as possible. The vapor pressures of all liquids increase rapidly with temperature, so we are limited by the mechanical strength of the boiler. At 500°C the vapor pressure of water is about 240×10^5 Pa (235 atm); this is about the maximum practical pressure in large present-day steam boilers.

20.16 To maximize efficiency, the temperatures inside a jet engine are made as high as possible. Exotic ceramic materials are used that can withstand temperatures in excess of 1000°C without melting or becoming soft.

The Kelvin Temperature Scale

In Chapter 17 we expressed the need for a temperature scale that doesn't depend on the properties of any particular material. We can now use the Carnot cycle to define such a scale. The thermal efficiency of a Carnot engine operating between two heat reservoirs at temperatures T_H and T_C is independent of the nature of the working substance and depends only on the temperatures. From Eq. (20.4), this thermal efficiency is

$$e = \frac{Q_H + Q_C}{Q_H} = 1 + \frac{Q_C}{Q_H}$$

Therefore the ratio Q_C/Q_H is the same for *all* Carnot engines operating between two given temperatures T_H and T_C.

Kelvin proposed that we *define* the ratio of the temperatures, T_C/T_H, to be equal to the magnitude of the ratio Q_C/Q_H of the quantities of heat absorbed and rejected:

$$\frac{T_C}{T_H} = \frac{|Q_C|}{|Q_H|} = -\frac{Q_C}{Q_H} \qquad \text{(definition of Kelvin temperature)} \qquad (20.16)$$

Equation (20.16) looks identical to Eq. (20.13), but there is a subtle and crucial difference. The temperatures in Eq. (20.13) are based on an ideal-gas thermometer, as defined in Section 17.3, while Eq. (20.16) *defines* a temperature scale based on the Carnot cycle and the second law of thermodynamics and is independent of the behavior of any particular substance. Thus the **Kelvin temperature scale** is truly *absolute*. To complete the definition of the Kelvin scale, we assign, as in Section 17.3, the arbitrary value of 273.16 K to the temperature of the triple point of water. When a substance is taken around a Carnot cycle, the ratio of the heats absorbed and rejected, $|Q_H|/|Q_C|$, is equal to the ratio of the temperatures of the reservoirs *as expressed on the gas-thermometer scale* defined in Section 17.3. Since the triple point of water is chosen to be 273.16 K in both scales, it follows that *the Kelvin and ideal-gas scales are identical.*

The zero point on the Kelvin scale is called **absolute zero.** At absolute zero a system has its *minimum* possible total internal energy (kinetic plus potential). Because of quantum effects, however, it is *not* true that at $T = 0$, all molecular motion ceases. There are theoretical reasons for believing that absolute zero cannot be attained experimentally, although temperatures below 10^{-7} K have been achieved. The more closely we approach absolute zero, the more difficult it is to get closer. One statement of the *third law of thermodynamics* is that it is impossible to reach absolute zero in a finite number of thermodynamic steps.

TEST YOUR UNDERSTANDING OF SECTION 20.6   An inventor looking for financial support comes to you with an idea for a gasoline engine that runs on a novel type of thermodynamic cycle. His design is made entirely of copper and is air-cooled. He claims that the engine will be 85% efficient. Should you invest in this marvelous new engine? (*Hint:* See Table 17.4.) ▮

20.7 ENTROPY

The second law of thermodynamics, as we have stated it, is not an equation or a quantitative relationship but rather a statement of *impossibility*. However, the second law *can* be stated as a quantitative relationship with the concept of *entropy*, the subject of this section.

We have talked about several processes that proceed naturally in the direction of increasing randomness. Irreversible heat flow increases randomness: The molecules are initially sorted into hotter and cooler regions, but this sorting is lost when the system comes to thermal equilibrium. Adding heat to a body also increases average molecular speeds; therefore, molecular motion becomes more random. In the free expansion of a gas, the molecules have greater randomness of position after the expansion than before. **Figure 20.17** shows another process in which randomness increases.

20.17 When firecrackers explode, randomness increases: The neatly packaged chemicals within each firecracker are dispersed in all directions, and the stored chemical energy is converted to random kinetic energy of the fragments.

Entropy and Randomness

Entropy provides a *quantitative* measure of randomness. To introduce this concept, let's consider an infinitesimal isothermal expansion of an ideal gas. We add heat dQ and let the gas expand just enough to keep the temperature constant. Because the internal energy of an ideal gas depends on only its temperature, the internal energy is also constant; thus from the first law, the work dW done by the gas is equal to the heat dQ added. That is,

$$dQ = dW = p\, dV = \frac{nRT}{V}dV \quad \text{so} \quad \frac{dV}{V} = \frac{dQ}{nRT}$$

The gas is more disordered after the expansion than before: The molecules are moving in a larger volume and have more randomness of position. Thus the fractional volume change dV/V is a measure of the increase in randomness, and the above equation shows that it is proportional to the quantity dQ/T. We introduce the symbol S for the entropy of the system, and we define the infinitesimal entropy change dS during an infinitesimal reversible process at absolute temperature T as

$$dS = \frac{dQ}{T} \qquad \text{(infinitesimal reversible process)} \qquad (20.17)$$

If a total amount of heat Q is added during a reversible isothermal process at absolute temperature T, the total entropy change $\Delta S = S_2 - S_1$ is given by

$$\Delta S = S_2 - S_1 = \frac{Q}{T} \qquad \text{(reversible isothermal process)} \qquad (20.18)$$

Entropy has units of energy divided by temperature; the SI unit of entropy is 1 J/K.

We can see how the quotient Q/T is related to the increase in randomness. Higher temperature means greater randomness of motion. If the substance is initially cold, with little molecular motion, adding heat Q causes a substantial fractional increase in molecular motion and randomness. But if the substance is already hot, the same quantity of heat adds relatively little to the greater molecular motion already present. So Q/T is an appropriate characterization of the increase in randomness when heat flows into a system.

EXAMPLE 20.5 **ENTROPY CHANGE IN MELTING**

What is the change of entropy of 1 kg of ice that is melted reversibly at 0°C and converted to water at 0°C? The heat of fusion of water is $L_f = 3.34 \times 10^5$ J/kg.

SOLUTION

IDENTIFY and SET UP: The melting occurs at a constant temperature $T = 0°C = 273$ K, so this is an *isothermal* reversible process. We can calculate the added heat Q required to melt the ice, then calculate the entropy change ΔS from Eq. (20.18).

EXECUTE: The heat needed to melt the ice is $Q = mL_f = 3.34 \times 10^5$ J. Then from Eq. (20.18),

$$\Delta S = S_2 - S_1 = \frac{Q}{T} = \frac{3.34 \times 10^5 \text{ J}}{273 \text{ K}} = 1.22 \times 10^3 \text{ J/K}$$

EVALUATE: This entropy increase corresponds to the increase in disorder when the water molecules go from the state of a crystalline solid to the much more randomly arranged state of a liquid. In *any* isothermal reversible process, the entropy change equals the heat transferred divided by the absolute temperature. When we refreeze the water, Q has the opposite sign, and the entropy change is $\Delta S = -1.22 \times 10^3$ J/K. The water molecules rearrange themselves into a crystal to form ice, so both randomness and entropy decrease.

Entropy in Reversible Processes

We can generalize the definition of entropy change to include *any* reversible process leading from one state to another, whether it is isothermal or not. We represent the process as a series of infinitesimal reversible steps. During a typical step, an infinitesimal quantity of heat dQ is added to the system at absolute temperature T. Then we sum (integrate) the quotients dQ/T for the entire process; that is,

Entropy change
in a reversible process $\cdots\cdots$ $\Delta S = \int_1^2 \dfrac{dQ}{T}$ $\quad$ Upper limit = final state $\quad$ Infinitesimal heat flow into system $\hspace{1cm}$ (20.19)

Lower limit = initial state $\qquad$ Absolute temperature

Because entropy is a measure of the randomness of a system in any specific state, it must depend only on the current state of the system, not on its past history. (We will verify this later.) When a system proceeds from an initial state with entropy S_1 to a final state with entropy S_2, the change in entropy $\Delta S = S_2 - S_1$ defined by Eq. (20.19) does not depend on the path leading from the initial to the final state but is the same for *all possible* processes leading from state 1 to state 2. Thus the entropy of a system must also have a definite value for any given state of the system. *Internal energy,* introduced in Chapter 19, also has this property, although entropy and internal energy are very different quantities.

Since entropy is a function only of the state of a system, we can also compute entropy changes in *irreversible* (nonequilibrium) processes for which Eqs. (20.17) and (20.19) are not applicable. We simply invent a path connecting the given initial and final states that *does* consist entirely of reversible equilibrium processes and compute the total entropy change for that path. It is not the actual path, but the entropy change must be the same as for the actual path.

As with internal energy, the above discussion does not tell us how to calculate entropy itself, but only the change in entropy in any given process. Just as with internal energy, we may arbitrarily assign a value to the entropy of a system in a specified reference state and then calculate the entropy of any other state with reference to this.

EXAMPLE 20.6 ENTROPY CHANGE IN A TEMPERATURE CHANGE

One kilogram of water at 0°C is heated to 100°C. Compute its change in entropy. Assume that the specific heat of water is constant at 4190 J/kg · K over this temperature range.

SOLUTION

IDENTIFY and SET UP: The entropy change of the water depends only on the initial and final states of the system, no matter whether the process is reversible or irreversible. We can imagine a reversible process in which the water temperature is increased in a sequence of infinitesimal steps dT. We can use Eq. (20.19) to integrate over all these steps and calculate the entropy change for such a reversible process. (Heating the water on a stove whose cooking surface is maintained at 100°C would be an irreversible process. The entropy change would be the same, however.)

EXECUTE: From Eq. (17.14) the heat required to carry out each infinitesimal step is $dQ = mc\,dT$. Substituting this into Eq. (20.19) and integrating, we find

$$\Delta S = S_2 - S_1 = \int_1^2 \frac{dQ}{T} = \int_{T_1}^{T_2} mc\,\frac{dT}{T} = mc\,\ln\frac{T_2}{T_1}$$

$$= (1.00\ \text{kg})(4190\ \text{J/kg}\cdot\text{K})\left(\ln\frac{373\ \text{K}}{273\ \text{K}}\right) = 1.31 \times 10^3\ \text{J/K}$$

EVALUATE: The entropy change is positive, as it must be for a process in which the system absorbs heat. Our assumption about the specific heat is a pretty good one, since c for water varies by less than 1% between 0°C and 100°C (see Fig. 17.17).

CAUTION **When $\Delta S = Q/T$ can (and cannot) be used** In solving this problem you might be tempted to avoid doing an integral by using the simpler expression in Eq. (20.18), $\Delta S = Q/T$. This would be incorrect, however, because Eq. (20.18) is applicable only to *isothermal* processes, and the initial and final temperatures in our example are *not* the same. The *only* correct way to find the entropy change in a process with different initial and final temperatures is to use Eq. (20.19). ▮

CONCEPTUAL EXAMPLE 20.7 ENTROPY CHANGE IN A REVERSIBLE ADIABATIC PROCESS

A gas expands adiabatically and reversibly. What is its change in entropy?

SOLUTION

In an adiabatic process, no heat enters or leaves the system. Hence $dQ = 0$ and there is *no* change in entropy in this reversible process:

$\Delta S = 0$. Every *reversible* adiabatic process is a constant-entropy process. (That's why such processes are also called *isentropic* processes.) The increase in randomness resulting from the gas occupying a greater volume is exactly balanced by the decrease in randomness associated with the lowered temperature and reduced molecular speeds.

EXAMPLE 20.8 ENTROPY CHANGE IN A FREE EXPANSION

A partition divides a thermally insulated box into two compartments, each of volume V (**Fig. 20.18**). Initially, one compartment contains n moles of an ideal gas at temperature T, and the other compartment is evacuated. We break the partition and the gas expands, filling both compartments. What is the entropy change in this free-expansion process?

SOLUTION

IDENTIFY and SET UP: For this process, $Q = 0$, $W = 0$, $\Delta U = 0$, and therefore (because the system is an ideal gas) $\Delta T = 0$. We might think that the entropy change is zero because there is no heat exchange. But Eq. (20.19) can be used to calculate entropy changes for *reversible* processes only; this free expansion is *not*

reversible, and there *is* an entropy change. As we mentioned at the beginning of this section, entropy increases in a free expansion because the positions of the molecules are more random than before the expansion. To calculate ΔS, we recall that the entropy change depends only on the initial and final states. We can devise a *reversible* process having the same endpoints as this free expansion, and in general we can then use Eq. (20.19) to calculate its entropy change, which will be the same as for the free expansion. An appropriate reversible process is an *isothermal* expansion from V to $2V$ at temperature T, which allows us to use the simpler Eq. (20.18) to calculate ΔS. The gas does work W during this expansion, so an equal amount of heat Q must be supplied to keep the internal energy constant.

EXECUTE: We saw in Example 19.1 that the work done by n moles of ideal gas in an isothermal expansion from V_1 to V_2 is $W = nRT \ln(V_2/V_1)$. With $V_1 = V$ and $V_2 = 2V$, we have

$$Q = W = nRT \ln\frac{2V}{V} = nRT \ln 2$$

From Eq. (20.18), the entropy change is

$$\Delta S = \frac{Q}{T} = nR \ln 2$$

20.18 (a, b) Free expansion of an insulated ideal gas. (c) The free-expansion process doesn't pass through equilibrium states from a to b. However, the entropy change $S_b - S_a$ can be calculated by using the isothermal path shown or *any* reversible path from a to b.

(a)

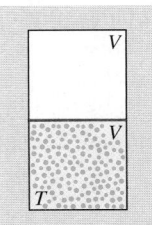

(b)

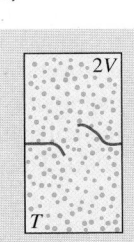

(c)

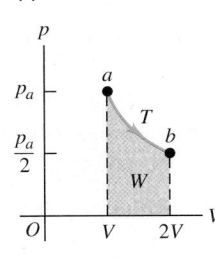

EVALUATE: For 1 mole, $\Delta S = (1 \text{ mol})(8.314 \text{ J/mol} \cdot \text{K})(\ln 2) = 5.76 \text{ J/K}$. The entropy change is positive, as we predicted. The factor $(\ln 2)$ in our answer is a result of the volume having increased by a factor of 2, from V to $2V$. Can you show that if the volume increases in a free expansion from V to xV, where x is an arbitrary number, the entropy change is $\Delta S = nR \ln x$?

EXAMPLE 20.9 ENTROPY AND THE CARNOT CYCLE

For the Carnot engine in Example 20.2 (Section 20.6), what is the total entropy change during one cycle?

SOLUTION

IDENTIFY and SET UP: All four steps in the Carnot cycle (see Fig. 20.13) are reversible, so we can use our expressions for the entropy change ΔS in a reversible process. We find ΔS for each step and add them to get ΔS for the complete cycle.

EXECUTE: There is no entropy change during the adiabatic expansion $b \rightarrow c$ or the adiabatic compression $d \rightarrow a$. During the isothermal expansion $a \rightarrow b$ at $T_H = 500$ K, the engine takes in 2000 J of heat, and from Eq. (20.18),

$$\Delta S_H = \frac{Q_H}{T_H} = \frac{2000 \text{ J}}{500 \text{ K}} = 4.0 \text{ J/K}$$

Continued

During the isothermal compression $c \rightarrow d$ at $T_C = 350$ K, the engine gives off 1400 J of heat, and

$$\Delta S_C = \frac{Q_C}{T_C} = \frac{-1400 \text{ J}}{350 \text{ K}} = -4.0 \text{ J/K}$$

The total entropy change in the engine during one cycle is $\Delta S_{\text{tot}} = \Delta S_H + \Delta S_C = 4.0 \text{ J/K} + (-4.0 \text{ J/K}) = 0$.

EVALUATE: The result $\Delta S_{\text{total}} = 0$ tells us that when the Carnot engine completes a cycle, it has the same entropy as it did at the beginning of the cycle. We'll explore this result in the next subsection.

What is the total entropy change of the engine's *environment* during this cycle? During the reversible isothermal expansion $a \rightarrow b$, the hot (500 K) reservoir gives off 2000 J of heat, so its entropy change is $(-2000 \text{ J})/(500 \text{ K}) = -4.0 \text{ J/K}$. During the reversible isothermal compression $c \rightarrow d$, the cold (350 K) reservoir absorbs 1400 J of heat, so its entropy change is $(+1400 \text{ J})/(350 \text{ K}) = +4.0 \text{ J/K}$. Thus the hot and cold reservoirs each have an entropy change, but the sum of these changes—that is, the total entropy change of the system's environment—is zero.

These results apply to the special case of the Carnot cycle, for which *all* of the processes are reversible. In this case the total entropy change of the system and the environment together is zero. We will see that if the cycle includes irreversible processes (as is the case for the Otto and Diesel cycles of Section 20.3), the total entropy change of the system and the environment *cannot* be zero, but rather must be positive.

Entropy in Cyclic Processes

Example 20.9 showed that the total entropy change for a cycle of a particular Carnot engine, which uses an ideal gas as its working substance, is zero. This result follows directly from Eq. (20.13), which we can rewrite as

$$\frac{Q_H}{T_H} + \frac{Q_C}{T_C} = 0 \tag{20.20}$$

The quotient Q_H/T_H equals ΔS_H, the entropy change of the engine that occurs at $T = T_H$. Likewise, Q_C/T_C equals ΔS_C, the (negative) entropy change of the engine that occurs at $T = T_C$. Hence Eq. (20.20) says that $\Delta S_H + \Delta S_C = 0$; that is, there is zero net entropy change in one cycle.

What about Carnot engines that use a different working substance? According to the second law, *any* Carnot engine operating between given temperatures T_H and T_C has the same efficiency $e = 1 - T_C/T_H$ [Eq. (20.14)]. Combining this expression for e with Eq. (20.4), $e = 1 + Q_C/Q_H$, just reproduces Eq. (20.20). So Eq. (20.20) is valid for any Carnot engine working between these temperatures, whether its working substance is an ideal gas or not. We conclude that *the total entropy change in one cycle of any Carnot engine is zero.*

This result can be generalized to show that the total entropy change during *any* reversible cyclic process is zero. A reversible cyclic process appears on a pV-diagram as a closed path (**Fig. 20.19a**). We can approximate such a path as closely as we like by a sequence of isothermal and adiabatic processes forming parts of many long, thin Carnot cycles (Fig. 20.19b). The total entropy change

20.19 (a) A reversible cyclic process for an ideal gas is shown as a red closed path on a pV-diagram. Several ideal-gas isotherms are shown in blue. (b) We can approximate the path in (a) by a series of long, thin Carnot cycles; one of these is highlighted in gold. The total entropy change is zero for each Carnot cycle and for the actual cyclic process. (c) The entropy change between points a and b is independent of the path.

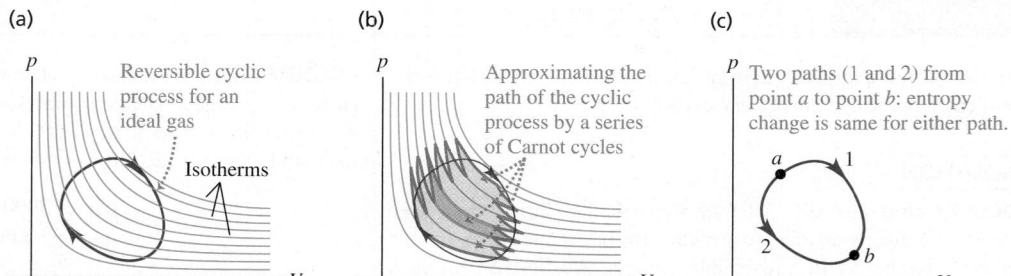

for the full cycle is the sum of the entropy changes for each small Carnot cycle, each of which is zero. So **the total entropy change during** *any* **reversible cycle is zero:**

$$\int \frac{dQ}{T} = 0 \qquad \text{(reversible cyclic process)} \qquad (20.21)$$

It follows that when a system undergoes a reversible process leading from any state *a* to any other state *b*, *the entropy change of the system is independent of the path* (Fig. 20.19c). If the entropy change for path 1 were different from the change for path 2, the system could be taken along path 1 and then backward along path 2 to the starting point, with a nonzero net change in entropy. This would violate the conclusion that the total entropy change in such a cyclic process must be zero. Because the entropy change in such processes is independent of path, we conclude that in any given state, the system has a definite value of entropy that depends only on the state, not on the processes that led to that state.

Entropy in Irreversible Processes

In an idealized, reversible process involving only equilibrium states, the total entropy change of the system and its surroundings is zero. But all *irreversible* processes involve an increase in entropy. Unlike energy, *entropy is not a conserved quantity*. The entropy of an isolated system *can* change, but as we shall see, it can never decrease. The free expansion of a gas (Example 20.8) is an irreversible process in an isolated system in which there is an entropy increase.

BIO Application Entropy Changes in a Living Organism When a puppy or other growing animal eats, it takes organized chemical energy from the food and uses it to make new cells that are even more highly organized. This process alone lowers entropy. But most of the energy in the food is either excreted in the animal's feces or used to generate heat, processes that lead to a large increase in entropy. So while the entropy of the animal alone decreases, the *total* entropy of animal plus food *increases*.

EXAMPLE 20.10 ENTROPY CHANGE IN AN IRREVERSIBLE PROCESS

Suppose 1.00 kg of water at 100°C is placed in thermal contact with 1.00 kg of water at 0°C. What is the total change in entropy? Assume that the specific heat of water is constant at 4190 J/kg·K over this temperature range.

SOLUTION

IDENTIFY and SET UP: This process involves irreversible heat flow because of the temperature differences. There are equal masses of 0°C water and 100°C water, so the final temperature is the average of these two temperatures: 50°C = 323 K. Although the processes are irreversible, we can calculate the entropy changes for the (initially) hot water and the (initially) cold water by assuming that the process occurs reversibly. As in Example 20.6, we must use Eq. (20.19) to calculate ΔS for each substance because the temperatures are not constant.

EXECUTE: The entropy changes of the hot water (subscript H) and the cold water (subscript C) are

$$\Delta S_H = mc \int_{T_1}^{T_2} \frac{dT}{T} = (1.00 \text{ kg})(4190 \text{ J/kg·K}) \int_{373 \text{ K}}^{323 \text{ K}} \frac{dT}{T}$$

$$= (4190 \text{ J/K})\left(\ln\frac{323 \text{ K}}{373 \text{ K}}\right) = -603 \text{ J/K}$$

$$\Delta S_C = (4190 \text{ J/K})\left(\ln\frac{323 \text{ K}}{273 \text{ K}}\right) = +705 \text{ J/K}$$

The *total* entropy change of the system is

$$\Delta S_{\text{tot}} = \Delta S_H + \Delta S_C = (-603 \text{ J/K}) + 705 \text{ J/K} = +102 \text{ J/K}$$

EVALUATE: An irreversible heat flow in an isolated system is accompanied by an increase in entropy. We could reach the same end state by mixing the hot and cold water, which is also an irreversible process; the total entropy change, which depends only on the initial and final states of the system, would again be 102 J/K.

Note that the entropy of the system increases *continuously* as the two quantities of water come to equilibrium. For example, the first 4190 J of heat transferred cools the hot water to 99°C and warms the cold water to 1°C. The net change in entropy for this step is approximately

$$\Delta S = \frac{-4190 \text{ J}}{373 \text{ K}} + \frac{4190 \text{ J}}{273 \text{ K}} = +4.1 \text{ J/K}$$

Can you show in a similar way that the net entropy change is positive for *any* one-degree temperature change leading to the equilibrium condition?

20.20 The mixing of colored ink and water starts from a state of low entropy in which each fluid is separate and distinct from the other. In the final state, both the ink and water molecules are spread randomly throughout the volume of liquid, so the entropy is greater. Spontaneous unmixing of the ink and water, a process in which there would be a net decrease in entropy, is never observed.

Entropy and the Second Law

The results of Example 20.10 about the flow of heat from a higher to a lower temperature are characteristic of *all* natural (that is, irreversible) processes. When we include the entropy changes of all the systems taking part in the process, the increases in entropy are always greater than the decreases. In the special case of a *reversible* process, the increases and decreases are equal. Hence we can state the general principle: **When all systems taking part in a process are included, the entropy either remains constant or increases.** In other words: **No process is possible in which the total entropy decreases, when all systems taking part in the process are included.** This is an alternative statement of the second law of thermodynamics in terms of entropy. Thus it is equivalent to the "engine" and "refrigerator" statements discussed earlier. **Figure 20.20** shows a specific example of this general principle.

The increase of entropy in every natural, irreversible process measures the increase of randomness in the universe associated with that process. Consider again the example of mixing hot and cold water (Example 20.10). We *might* have used the hot and cold water as the high- and low-temperature reservoirs of a heat engine. While removing heat from the hot water and giving heat to the cold water, we could have obtained some mechanical work. But once the hot and cold water have been mixed and have come to a uniform temperature, this opportunity to convert heat to mechanical work is lost irretrievably. The lukewarm water will never *unmix* itself and separate into hotter and colder portions. No decrease in *energy* occurs when the hot and cold water are mixed. What has been lost is the *opportunity* to convert part of the heat from the hot water into mechanical work. Hence when entropy increases, energy becomes less *available,* and the universe becomes more random or "run down."

TEST YOUR UNDERSTANDING OF SECTION 20.7   Suppose 2.00 kg of water at 50°C spontaneously changes temperature, so that half of the water cools to 0°C while the other half spontaneously warms to 100°C. (All of the water remains liquid, so it doesn't freeze or boil.) What would be the entropy change of the water? Is this process possible? (*Hint:* See Example 20.10.) ▐

20.8 MICROSCOPIC INTERPRETATION OF ENTROPY

We described in Section 19.4 how the internal energy of a system could be calculated, at least in principle, by adding up all the kinetic energies of its constituent particles and all the potential energies of interaction among the particles. This is called a *microscopic* calculation of the internal energy. We can also make a microscopic calculation of the entropy S of a system. Unlike energy, however, entropy is not something that belongs to each individual particle or pair of particles in the system. Rather, entropy is a measure of the randomness of the system as a whole. To see how to calculate entropy microscopically, we first have to introduce the idea of *macroscopic* and *microscopic states*.

Suppose you toss N identical coins on the floor, and half of them show heads and half show tails. This is a description of the large-scale or **macroscopic state** of the system of N coins. A description of the **microscopic state** of the system includes information about each individual coin: Coin 1 was heads, coin 2 was tails, coin 3 was tails, and so on. There can be many microscopic states that correspond to the same macroscopic description. For instance, with $N = 4$ coins there are six possible states in which half are heads and half are tails (**Fig. 20.21**). The number of microscopic states grows rapidly with increasing N; for $N = 100$ there are $2^{100} = 1.27 \times 10^{30}$ microscopic states, of which 1.01×10^{29} are half heads and half tails.

The least probable outcomes of the coin toss are the states that are either all heads or all tails. It is certainly possible that you could throw 100 heads in a row, but don't bet on it; the probability of doing this is only 1 in 1.27×10^{30}. The most probable outcome of tossing N coins is that half are heads and half are tails. The reason is that this *macroscopic* state has the greatest number of corresponding *microscopic* states, as Fig. 20.21 shows.

To make the connection to the concept of entropy, note that the macroscopic description "all heads" completely specifies the state of each one of the N coins. The same is true if the coins are all tails. But the macroscopic description "half heads, half tails" by itself tells you very little about the state (heads or tails) of each individual coin. Compared to the state "all heads" or "all tails," the state "half heads, half tails" has much greater *randomness* because the system could be in any of a much greater number of possible microscopic states. Hence the "half heads, half tails" state has much greater entropy (which is a quantitative measure of randomness).

Now instead of N coins, consider a mole of an ideal gas containing Avogadro's number of molecules. The macroscopic state of this gas is given by its pressure p, volume V, and temperature T; a description of the microscopic state involves stating the position and velocity for each molecule in the gas. At a given pressure, volume, and temperature, the gas may be in any one of an astronomically large number of microscopic states, depending on the positions and velocities of its 6.02×10^{23} molecules. If the gas undergoes a free expansion into a greater volume, the range of possible positions increases, as does the number of possible microscopic states. The system becomes more random, and the entropy increases as calculated in Example 20.8 (Section 20.7).

We can draw the following general conclusion: **For any thermodynamic system, the most probable macroscopic state is the one with the greatest number of corresponding microscopic states, which is also the macroscopic state with the greatest randomness and the greatest entropy.**

Calculating Entropy: Microscopic States

Let w represent the number of possible microscopic states for a given macroscopic state. (For the four coins shown in Fig. 20.21 the state of four heads has $w = 1$, the state of three heads and one tails has $w = 4$, and so on.) Then the entropy S of a macroscopic state can be shown to be given by

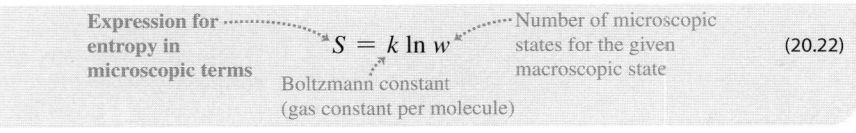

Expression for entropy in microscopic terms — $S = k \ln w$ — Boltzmann constant (gas constant per molecule) — Number of microscopic states for the given macroscopic state   (20.22)

(We introduced the Boltzmann constant in Section 18.3.) As Eq. (20.22) shows, increasing the number of possible microscopic states w increases the entropy S.

20.21 All possible microscopic states of four coins. There can be several possible microscopic states for each macroscopic state.

Macroscopic state	Corresponding microscopic states
Four heads	
Three heads, one tails	
Two heads, two tails	
One heads, three tails	
Four tails	

What matters in a thermodynamic process is not the absolute entropy S but the *difference* in entropy between the initial and final states. Hence an equally valid and useful definition would be $S = k \ln w + C$, where C is a constant, since C cancels in any calculation of an entropy difference between two states. But it's convenient to set this constant equal to zero and use Eq. (20.22). With this choice, since the smallest possible value of w is unity, the smallest possible value of S for any system is $k \ln 1 = 0$. Entropy can *never* be negative.

In practice, calculating w is a difficult task, so Eq. (20.22) is typically used only to calculate the absolute entropy S of certain special systems. But we can use this relationship to calculate *differences* in entropy between one state and another. Consider a system that undergoes a thermodynamic process that takes it from macroscopic state 1, for which there are w_1 possible microscopic states, to macroscopic state 2, with w_2 associated microscopic states. The change in entropy in this process is

$$\Delta S = S_2 - S_1 = k \ln w_2 - k \ln w_1 = k \ln \frac{w_2}{w_1} \qquad (20.23)$$

The *difference* in entropy between the two macroscopic states depends on the *ratio* of the numbers of possible microscopic states.

As the following example shows, using Eq. (20.23) to calculate a change in entropy from one macroscopic state to another gives the same results as considering a reversible process connecting those two states and using Eq. (20.19).

EXAMPLE 20.11 **A MICROSCOPIC CALCULATION OF ENTROPY CHANGE**

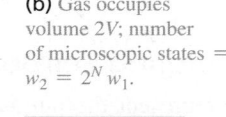

Use Eq. (20.23) to calculate the entropy change in the free expansion of n moles of gas at temperature T described in Example 20.8 (**Fig. 20.22**).

SOLUTION

IDENTIFY and SET UP: We are asked to calculate the entropy change by using the number of microscopic states in the initial and final macroscopic states (Figs. 20.22a and b). When the partition is broken, no work is done, so the velocities of the molecules are unaffected. But each molecule now has twice as much volume in which it can move and hence has twice the number of possible positions. This is all we need to calculate the entropy change using Eq. (20.23).

EXECUTE: Let w_1 be the number of microscopic states of the system as a whole when the gas occupies volume V (Fig. 20.22a). The number of molecules is $N = nN_A$, and each of these N molecules has twice as many possible states after the partition is broken. Hence the number w_2 of microscopic states when the gas occupies volume $2V$ (Fig. 20.22b) is greater by a factor of 2^N; that is, $w_2 = 2^N w_1$. The change in entropy in this process is

$$\Delta S = k \ln \frac{w_2}{w_1} = k \ln \frac{2^N w_1}{w_1}$$
$$= k \ln 2^N = Nk \ln 2$$

Since $N = nN_A$ and $k = R/N_A$, this becomes

$$\Delta S = (nN_A)(R/N_A) \ln 2$$
$$= nR \ln 2$$

EVALUATE: We found the same result as in Example 20.8, but without any reference to the thermodynamic path taken.

20.22 In a free expansion of N molecules in which the volume doubles, the number of possible microscopic states increases by a factor of 2^N.

(a) Gas occupies volume V; number of microscopic states $= w_1$.

(b) Gas occupies volume $2V$; number of microscopic states $= w_2 = 2^N w_1$.

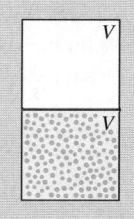

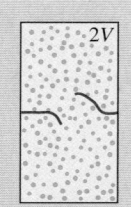

Microscopic States and the Second Law

The relationship between entropy and the number of microscopic states gives us new insight into the entropy statement of the second law of thermodynamics: that the entropy of a closed system can never decrease. From Eq. (20.22) this means that a closed system can never spontaneously undergo a process that decreases the number of possible microscopic states.

An example of such a forbidden process would be if all of the air in your room spontaneously moved to one half of the room, leaving a vacuum in the other half. Such a "free compression" would be the reverse of the free expansion of Examples 20.8 and 20.11. This would decrease the number of possible microscopic states by a factor of 2^N. Strictly speaking, this process is not impossible! The probability of finding a given molecule in one half of the room is $\frac{1}{2}$, so the probability of finding all of the molecules in one half of the room at once is $\left(\frac{1}{2}\right)^N$. (This is exactly the same as the probability of having a tossed coin come up heads N times in a row.) This probability is *not* zero. But lest you worry about suddenly finding yourself gasping for breath in the evacuated half of your room, consider that a typical room might hold 1000 moles of air, and so $N = 1000N_A = 6.02 \times 10^{26}$ molecules. The probability of all the molecules being in the same half of the room is therefore $\left(\frac{1}{2}\right)^{6.02 \times 10^{26}}$. Expressed as a decimal, this number has more than 10^{26} zeros to the right of the decimal point!

Because the probability of such a "free compression" taking place is so vanishingly small, it has almost certainly never occurred anywhere in the universe since the beginning of time. We conclude that for all practical purposes the second law of thermodynamics is never violated.

TEST YOUR UNDERSTANDING OF SECTION 20.8   A quantity of N molecules of an ideal gas initially occupies volume V. The gas then expands to volume $2V$. The number of microscopic states of the gas increases in this expansion. Under which of the following circumstances will this number increase the most? (i) If the expansion is reversible and isothermal; (ii) if the expansion is reversible and adiabatic; (iii) the number will change by the same amount for both circumstances. ❚

Application Polymers Coil in Solution
A molecule of polyethylene, the most common plastic, is a polymer—a long chain of monomer units (C_2H_4). In solution these molecules coil on themselves, and the entropy concept explains why. The polymer can coil in many ways (microscopic states), but there is only one microscopic state in which the polymer is fully stretched out. Thus the entropy of the coiled polymer is much greater than that of a stretched-out polymer. The second law of thermodynamics says that isolated systems always move toward greater entropy, so we expect a polymer chain in solution to be in a coiled state.

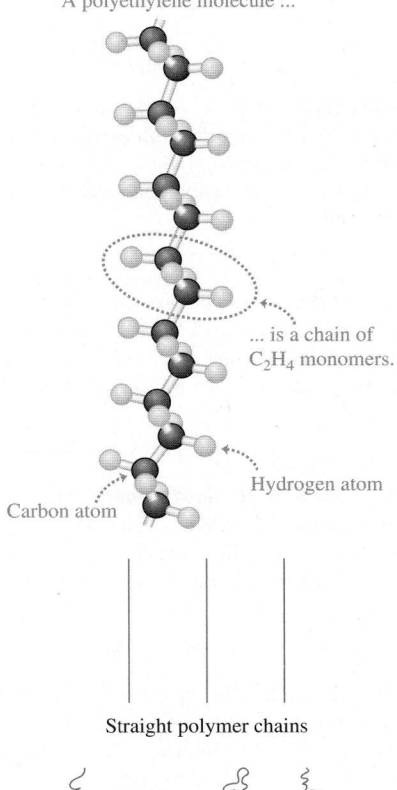

A polyethylene molecule ...

... is a chain of C_2H_4 monomers.

Hydrogen atom

Carbon atom

Straight polymer chains

Coiled polymer chains

Reversible and irreversible processes: A reversible process is one whose direction can be reversed by an infinitesimal change in the conditions of the process, and in which the system is always in or very close to thermal equilibrium. All other thermodynamic processes are irreversible.

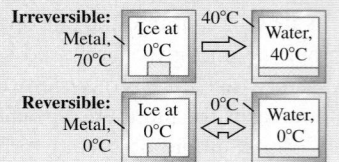

Heat engines: A heat engine takes heat Q_H from a source, converts part of it to work W, and discards the remainder $|Q_C|$ at a lower temperature. The thermal efficiency e of a heat engine measures how much of the absorbed heat is converted to work. (See Example 20.1.)

$$e = \frac{W}{Q_H} = 1 + \frac{Q_C}{Q_H} = 1 - \left|\frac{Q_C}{Q_H}\right| \quad (20.4)$$

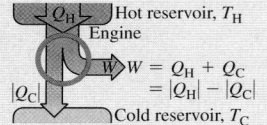

The Otto cycle: A gasoline engine operating on the Otto cycle has a theoretical maximum thermal efficiency e that depends on the compression ratio r and the ratio of heat capacities γ of the working substance.

$$e = 1 - \frac{1}{r^{\gamma-1}} \quad (20.6)$$

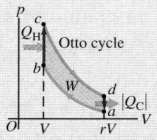

Refrigerators: A refrigerator takes heat Q_C from a colder place, has a work input $|W|$, and discards heat $|Q_H|$ at a warmer place. The effectiveness of the refrigerator is given by its coefficient of performance K.

$$K = \frac{|Q_C|}{|W|} = \frac{|Q_C|}{|Q_H| - |Q_C|} \quad (20.9)$$

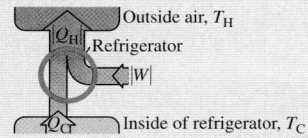

The second law of thermodynamics: The second law of thermodynamics describes the directionality of natural thermodynamic processes. It can be stated in several equivalent forms. The *engine* statement is that no cyclic process can convert heat completely into work. The *refrigerator* statement is that no cyclic process can transfer heat from a colder place to a hotter place with no input of mechanical work.

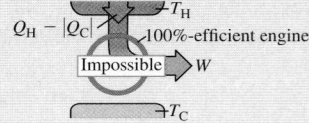

The Carnot cycle: The Carnot cycle operates between two heat reservoirs at temperatures T_H and T_C and uses only reversible processes. Its thermal efficiency depends only on T_H and T_C. An additional equivalent statement of the second law is that no engine operating between the same two temperatures can be more efficient than a Carnot engine. (See Examples 20.2 and 20.3.)

A Carnot engine run backward is a Carnot refrigerator. Its coefficient of performance depends on only T_H and T_C. Another form of the second law states that no refrigerator operating between the same two temperatures can have a larger coefficient of performance than a Carnot refrigerator. (See Example 20.4.)

$$e_{Carnot} = 1 - \frac{T_C}{T_H} = \frac{T_H - T_C}{T_H} \quad (20.14)$$

$$K_{Carnot} = \frac{T_C}{T_H - T_C} \quad (20.15)$$

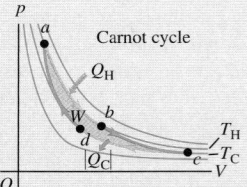

Entropy: Entropy is a quantitative measure of the randomness of a system. The entropy change in any reversible process depends on the amount of heat flow and the absolute temperature T. Entropy depends only on the state of the system, and the change in entropy between given initial and final states is the same for all processes leading from one state to the other. This fact can be used to find the entropy change in an irreversible process. (See Examples 20.5–20.10.)

$$\Delta S = \int_1^2 \frac{dQ}{T}$$
$$\text{(reversible process)}$$

(20.19)

An important statement of the second law of thermodynamics is that the entropy of an isolated system may increase but can never decrease. When a system interacts with its surroundings, the total entropy change of system and surroundings can never decrease. When the interaction involves only reversible processes, the total entropy is constant and $\Delta S = 0$; when there is any irreversible process, the total entropy increases and $\Delta S > 0$.

Entropy and microscopic states: When a system is in a particular macroscopic state, the particles that make up the system may be in any of w possible microscopic states. The greater the number w, the greater the entropy. (See Example 20.11.)

$$S = k \ln w$$

(20.22)

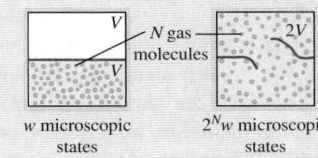

w microscopic states $2^N w$ microscopic states

BRIDGING PROBLEM   ENTROPY CHANGES: COLD ICE IN HOT WATER

An insulated container of negligible mass holds 0.600 kg of water at 45.0°C. You put a 0.0500-kg ice cube at −15.0°C in the water (**Fig. 20.23**). (a) Calculate the final temperature of the water once the ice has melted. (b) Calculate the change in entropy of the system.

SOLUTION GUIDE

IDENTIFY and SET UP

1. Make a list of the known and unknown quantities, and identify the target variables.
2. How will you find the final temperature of the ice–water mixture? How will you decide whether or not all the ice melts?
3. Once you find the final temperature of the mixture, how will you determine the changes in entropy of (i) the ice initially at −15.0°C and (ii) the water initially at 45.0°C?

EXECUTE

4. Use the methods of Chapter 17 to calculate the final temperature T. (*Hint:* First assume that all of the ice melts, then write an equation which says that the heat that flows into the ice equals the heat that flows out of the water. If your assumption is correct, the final temperature that you calculate will be greater than 0°C. If your assumption is incorrect, the final temperature will be 0°C or less, which means that some ice remains. You'll then need to redo the calculation to account for this.)

5. Use your result from step 4 to calculate the entropy changes of the ice and the water. (*Hint:* You must include the heat flow associated with temperature changes, as in Example 20.6, as well as the heat flow associated with the change of phase.)
6. Find the total change in entropy of the system.

EVALUATE

7. Do the signs of the entropy changes make sense? Why or why not?

20.23 What becomes of this ice–water mixture?

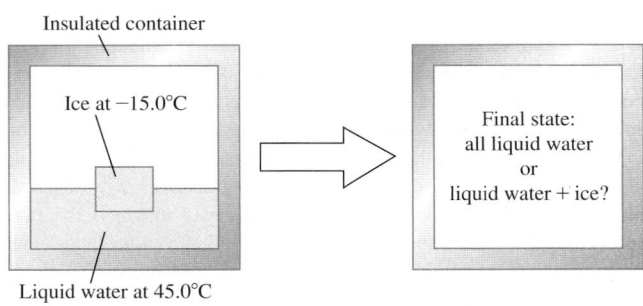

Problems

For assigned homework and other learning materials, go to MasteringPhysics®.   MP

•, ••, •••: Difficulty levels. **CP**: Cumulative problems incorporating material from earlier chapters. **CALC**: Problems requiring calculus. **DATA**: Problems involving real data, scientific evidence, experimental design, and/or statistical reasoning. **BIO**: Biosciences problems.

DISCUSSION QUESTIONS

Q20.1 A pot is half-filled with water, and a lid is placed on it, forming a tight seal so that no water vapor can escape. The pot is heated on a stove, forming water vapor inside the pot. The heat is then turned off and the water vapor condenses back to liquid. Is this cycle reversible or irreversible? Why?

Q20.2 Give two examples of reversible processes and two examples of irreversible processes in purely mechanical systems, such as blocks sliding on planes, springs, pulleys, and strings. Explain what makes each process reversible or irreversible.

Q20.3 Household refrigerators have arrays or coils of tubing on the outside, usually at the back or bottom. When the refrigerator is running, the tubing becomes quite hot. Where does the heat come from?

Q20.4 Suppose you try to cool the kitchen of your house by leaving the refrigerator door open. What happens? Why? Would the result be the same if you left open a picnic cooler full of ice? Explain the reason for any differences.

Q20.5 Why must a room air conditioner be placed in a window rather than just set on the floor and plugged in? Why can a refrigerator be set on the floor and plugged in?

Q20.6 Is it a violation of the second law of thermodynamics to convert mechanical energy completely into heat? To convert heat completely into work? Explain your answers.

Q20.7 Imagine a special air filter placed in a window of a house. The tiny holes in the filter allow only air molecules moving faster than a certain speed to exit the house, and allow only air molecules moving slower than that speed to enter the house from outside. Explain why such an air filter would cool the house, and why the second law of thermodynamics makes building such a filter an impossible task.

Q20.8 An electric motor has its shaft coupled to that of an electric generator. The motor drives the generator, and some current from the generator is used to run the motor. The excess current is used to light a home. What is wrong with this scheme?

Q20.9 When a wet cloth is hung up in a hot wind in the desert, it is cooled by evaporation to a temperature that may be 20 C° or so below that of the air. Discuss this process in light of the second law of thermodynamics.

Q20.10 Compare the pV-diagram for the Otto cycle in Fig. 20.6 with the diagram for the Carnot heat engine in Fig. 20.13. Explain some of the important differences between the two cycles.

Q20.11 The efficiency of heat engines is high when the temperature difference between the hot and cold reservoirs is large. Refrigerators, on the other hand, work better when the temperature difference is small. Thinking of the mechanical refrigeration cycle shown in Fig. 20.9, explain in physical terms why it takes less work to remove heat from the working substance if the two reservoirs (the inside of the refrigerator and the outside air) are at nearly the same temperature, than if the outside air is much warmer than the interior of the refrigerator.

Q20.12 What would be the efficiency of a Carnot engine operating with $T_H = T_C$? What would be the efficiency if $T_C = 0$ K and T_H were any temperature above 0 K? Interpret your answers.

Q20.13 Real heat engines, like the gasoline engine in a car, always have some friction between their moving parts, although lubricants keep the friction to a minimum. Would a heat engine with completely frictionless parts be 100% efficient? Why or why not? Does the answer depend on whether or not the engine runs on the Carnot cycle? Again, why or why not?

Q20.14 Does a refrigerator full of food consume more power if the room temperature is 20°C than if it is 15°C? Or is the power consumption the same? Explain your reasoning.

Q20.15 In Example 20.4, a Carnot refrigerator requires a work input of only 230 J to extract 346 J of heat from the cold reservoir. Doesn't this discrepancy imply a violation of the law of conservation of energy? Explain why or why not.

Q20.16 How can the thermal conduction of heat from a hot object to a cold object increase entropy when the same amount of heat that flows out of the hot object flows into the cold one?

Q20.17 Explain why each of the following processes is an example of increasing randomness: mixing hot and cold water; free expansion of a gas; irreversible heat flow; developing heat by mechanical friction. Are entropy increases involved in all of these? Why or why not?

Q20.18 The free expansion of an ideal gas is an adiabatic process and so no heat is transferred. No work is done, so the internal energy does not change. Thus, $Q/T = 0$, yet the randomness of the system and thus its entropy have increased after the expansion. Why does Eq. (20.19) not apply to this situation?

Q20.19 Are the earth and sun in thermal equilibrium? Are there entropy changes associated with the transmission of energy from the sun to the earth? Does radiation differ from other modes of heat transfer with respect to entropy changes? Explain your reasoning.

Q20.20 Suppose that you put a hot object in thermal contact with a cold object and observe (much to your surprise) that heat flows from the cold object to the hot object, making the cold one colder and the hot one hotter. Does this process necessarily violate the first law of thermodynamics? The second law of thermodynamics? Explain.

Q20.21 If you run a movie film backward, it is as if the direction of time were reversed. In the time-reversed movie, would you see processes that violate conservation of energy? Conservation of linear momentum? Would you see processes that violate the second law of thermodynamics? In each case, if law-breaking processes could occur, give some examples.

Q20.22 **BIO** Some critics of biological evolution claim that it violates the second law of thermodynamics, since evolution involves simple life forms developing into more complex and more highly ordered organisms. Explain why this is not a valid argument against evolution.

Q20.23 **BIO** A growing plant creates a highly complex and organized structure out of simple materials such as air, water, and trace minerals. Does this violate the second law of thermodynamics? Why or why not? What is the plant's ultimate source of energy? Explain.

EXERCISES

Section 20.2 Heat Engines

20.1 • A diesel engine performs 2200 J of mechanical work and discards 4300 J of heat each cycle. (a) How much heat must be supplied to the engine in each cycle? (b) What is the thermal efficiency of the engine?

20.2 • An aircraft engine takes in 9000 J of heat and discards 6400 J each cycle. (a) What is the mechanical work output of the engine during one cycle? (b) What is the thermal efficiency of the engine?

20.3 • **A Gasoline Engine.** A gasoline engine takes in 1.61×10^4 J of heat and delivers 3700 J of work per cycle. The heat is obtained by burning gasoline with a heat of combustion of 4.60×10^4 J/g. (a) What is the thermal efficiency? (b) How much heat is discarded in each cycle? (c) What mass of fuel is burned in each cycle? (d) If the engine goes through 60.0 cycles per second, what is its power output in kilowatts? In horsepower?

20.4 • A gasoline engine has a power output of 180 kW (about 241 hp). Its thermal efficiency is 28.0%. (a) How much heat must be supplied to the engine per second? (b) How much heat is discarded by the engine per second?

20.5 •• The pV-diagram in **Fig. E20.5** shows a cycle of a heat engine that uses 0.250 mol of an ideal gas with $\gamma = 1.40$. Process ab is adiabatic. (a) Find the pressure of the gas at point a. (b) How much heat enters this gas per cycle, and where does it happen? (c) How much heat leaves this gas in a cycle, and where does it occur? (d) How much work does this engine do in a cycle? (e) What is the thermal efficiency of the engine?

Figure **E20.5**

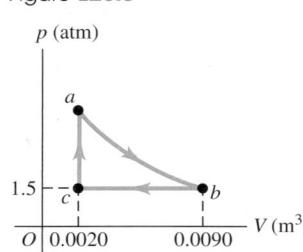

Section 20.3 Internal-Combustion Engines

20.6 • (a) Calculate the theoretical efficiency for an Otto-cycle engine with $\gamma = 1.40$ and $r = 9.50$. (b) If this engine takes in 10,000 J of heat from burning its fuel, how much heat does it discard to the outside air?

20.7 •• The Otto-cycle engine in a Mercedes-Benz SLK230 has a compression ratio of 8.8. (a) What is the ideal efficiency of the engine? Use $\gamma = 1.40$. (b) The engine in a Dodge Viper GT2 has a slightly higher compression ratio of 9.6. How much increase in the ideal efficiency results from this increase in the compression ratio?

Section 20.4 Refrigerators

20.8 • The coefficient of performance $K = H/P$ is a dimensionless quantity. Its value is independent of the units used for H and P, as long as the same units, such as watts, are used for both quantities. However, it is common practice to express H in Btu/h and P in watts. When these mixed units are used, the ratio H/P is called the *energy efficiency ratio* (*EER*). If a room air conditioner has $K = 3.0$, what is its EER?

20.9 • A refrigerator has a coefficient of performance of 2.10. In each cycle it absorbs 3.10×10^4 J of heat from the cold reservoir. (a) How much mechanical energy is required each cycle to operate the refrigerator? (b) During each cycle, how much heat is discarded to the high-temperature reservoir?

20.10 •• A freezer has a coefficient of performance of 2.40. The freezer is to convert 1.80 kg of water at 25.0°C to 1.80 kg of ice at −5.0°C in one hour. (a) What amount of heat must be removed from the water at 25.0°C to convert it to ice at −5.0°C? (b) How much electrical energy is consumed by the freezer during this hour? (c) How much wasted heat is delivered to the room in which the freezer sits?

20.11 •• A refrigerator has a coefficient of performance of 2.25, runs on an input of 135 W of electrical power, and keeps its inside compartment at 5°C. If you put a dozen 1.0-L plastic bottles of water at 31°C into this refrigerator, how long will it take for them to be cooled down to 5°C? (Ignore any heat that leaves the plastic.)

Section 20.6 The Carnot Cycle

20.12 • A Carnot engine is operated between two heat reservoirs at temperatures of 520 K and 300 K. (a) If the engine receives 6.45 kJ of heat energy from the reservoir at 520 K in each cycle, how many joules per cycle does it discard to the reservoir at 300 K? (b) How much mechanical work is performed by the engine during each cycle? (c) What is the thermal efficiency of the engine?

20.13 • A Carnot engine whose high-temperature reservoir is at 620 K takes in 550 J of heat at this temperature in each cycle and gives up 335 J to the low-temperature reservoir. (a) How much mechanical work does the engine perform during each cycle? What is (b) the temperature of the low-temperature reservoir; (c) the thermal efficiency of the cycle?

20.14 •• An ice-making machine operates in a Carnot cycle. It takes heat from water at 0.0°C and rejects heat to a room at 24.0°C. Suppose that 85.0 kg of water at 0.0°C are converted to ice at 0.0°C. (a) How much heat is discharged into the room? (b) How much energy must be supplied to the device?

20.15 • A Carnot engine has an efficiency of 66% and performs 2.5×10^4 J of work in each cycle. (a) How much heat does the engine extract from its heat source in each cycle? (b) Suppose the engine exhausts heat at room temperature (20.0°C). What is the temperature of its heat source?

20.16 •• A certain brand of freezer is advertised to use 730 kW · h of energy per year. (a) Assuming the freezer operates for 5 hours each day, how much power does it require while operating? (b) If the freezer keeps its interior at −5.0°C in a 20.0°C room, what is its theoretical maximum performance coefficient? (c) What is the theoretical maximum amount of ice this freezer could make in an hour, starting with water at 20.0°C?

20.17 • A Carnot refrigerator is operated between two heat reservoirs at temperatures of 320 K and 270 K. (a) If in each cycle the refrigerator receives 415 J of heat energy from the reservoir at 270 K, how many joules of heat energy does it deliver to the reservoir at 320 K? (b) If the refrigerator completes 165 cycles each minute, what power input is required to operate it? (c) What is the coefficient of performance of the refrigerator?

20.18 •• A Carnot heat engine uses a hot reservoir consisting of a large amount of boiling water and a cold reservoir consisting of a large tub of ice and water. In 5 minutes of operation, the heat rejected by the engine melts 0.0400 kg of ice. During this time, how much work W is performed by the engine?

20.19 •• You design an engine that takes in 1.50×10^4 J of heat at 650 K in each cycle and rejects heat at a temperature of 290 K. The engine completes 240 cycles in 1 minute. What is the theoretical maximum power output of your engine, in horsepower?

Section 20.7 Entropy

20.20 • A 4.50-kg block of ice at 0.00°C falls into the ocean and melts. The average temperature of the ocean is 3.50°C, including all the deep water. By how much does the change of this ice to water at 3.50°C alter the entropy of the world? Does the entropy increase or decrease? (*Hint:* Do you think that the ocean temperature will change appreciably as the ice melts?)

20.21 • A sophomore with nothing better to do adds heat to 0.350 kg of ice at 0.0°C until it is all melted. (a) What is the change in entropy of the water? (b) The source of heat is a very massive body at 25.0°C. What is the change in entropy of this body? (c) What is the total change in entropy of the water and the heat source?

20.22 • CALC You decide to take a nice hot bath but discover that your thoughtless roommate has used up most of the hot water. You fill the tub with 195 kg of 30.0°C water and attempt to warm it further by pouring in 5.00 kg of boiling water from the stove. (a) Is this a reversible or an irreversible process? Use physical reasoning to explain. (b) Calculate the final temperature of the bath water. (c) Calculate the net change in entropy of the system (bath water + boiling water), assuming no heat exchange with the air or the tub itself.

20.23 •• A 15.0-kg block of ice at 0.0°C melts to liquid water at 0.0°C inside a large room at 20.0°C. Treat the ice and the room as an isolated system, and assume that the room is large enough for its temperature change to be ignored. (a) Is the melting of the ice reversible or irreversible? Explain, using simple physical reasoning without resorting to any equations. (b) Calculate the net entropy change of the system during this process. Explain whether or not this result is consistent with your answer to part (a).

20.24 •• CALC You make tea with 0.250 kg of 85.0°C water and let it cool to room temperature (20.0°C). (a) Calculate the entropy change of the water while it cools. (b) The cooling process is essentially isothermal for the air in your kitchen. Calculate the change in entropy of the air while the tea cools, assuming that all of the heat lost by the water goes into the air. What is the total entropy change of the system tea + air?

20.25 • Three moles of an ideal gas undergo a reversible isothermal compression at 20.0°C. During this compression, 1850 J of work is done on the gas. What is the change of entropy of the gas?

20.26 •• What is the change in entropy of 0.130 kg of helium gas at the normal boiling point of helium when it all condenses isothermally to 1.00 L of liquid helium? (*Hint:* See Table 17.4 in Section 17.6.)

20.27 • (a) Calculate the change in entropy when 1.00 kg of water at 100°C is vaporized and converted to steam at 100°C (see Table 17.4). (b) Compare your answer to the change in entropy when 1.00 kg of ice is melted at 0°C, calculated in Example 20.5 (Section 20.7). Is the change in entropy greater for melting or for vaporization? Interpret your answer using the idea that entropy is a measure of the randomness of a system.

20.28 •• **Entropy Change Due to Driving.** Premium gasoline produces 1.23×10^8 J of heat per gallon when it is burned at approximately 400°C (although the amount can vary with the fuel mixture). If a car's engine is 25% efficient, three-fourths of that heat is expelled into the air, typically at 20°C. If your car gets 35 miles per gallon of gas, by how much does the car's engine change the entropy of the world when you drive 1.0 mi? Does it decrease or increase it?

Section 20.8 Microscopic Interpretation of Entropy

20.29 • CALC Two moles of an ideal gas occupy a volume V. The gas expands isothermally and reversibly to a volume $3V$. (a) Is the velocity distribution changed by the isothermal expansion? Explain. (b) Use Eq. (20.23) to calculate the change in entropy of the gas. (c) Use Eq. (20.18) to calculate the change in entropy of the gas. Compare this result to that obtained in part (b).

20.30 • A box is separated by a partition into two parts of equal volume. The left side of the box contains 500 molecules of nitrogen gas; the right side contains 100 molecules of oxygen gas. The two gases are at the same temperature. The partition is punctured, and equilibrium is eventually attained. Assume that the volume of the box is large enough for each gas to undergo a free expansion and not change temperature. (a) On average, how many molecules of each type will there be in either half of the box? (b) What is the change in entropy of the system when the partition is punctured? (c) What is the probability that the molecules will be found in the same distribution as they were before the partition was punctured—that is, 500 nitrogen molecules in the left half and 100 oxygen molecules in the right half?

20.31 • CALC A lonely party balloon with a volume of 2.40 L and containing 0.100 mol of air is left behind to drift in the temporarily uninhabited and depressurized International Space Station. Sunlight coming through a porthole heats and explodes the balloon, causing the air in it to undergo a free expansion into the empty station, whose total volume is 425 m³. Calculate the entropy change of the air during the expansion.

PROBLEMS

20.32 • You are designing a Carnot engine that has 2 mol of CO_2 as its working substance; the gas may be treated as ideal. The gas is to have a maximum temperature of 527°C and a maximum pressure of 5.00 atm. With a heat input of 400 J per cycle, you want 300 J of useful work. (a) Find the temperature of the cold reservoir. (b) For how many cycles must this engine run to melt completely a 10.0-kg block of ice originally at 0.0°C, using only the heat rejected by the engine?

20.33 •• CP An ideal Carnot engine operates between 500°C and 100°C with a heat input of 250 J per cycle. (a) How much heat is delivered to the cold reservoir in each cycle? (b) What minimum number of cycles is necessary for the engine to lift a 500-kg rock through a height of 100 m?

20.34 •• BIO **Entropy of Metabolism.** An average sleeping person metabolizes at a rate of about 80 W by digesting food or burning fat. Typically, 20% of this energy goes into bodily functions, such as cell repair, pumping blood, and other uses of mechanical energy, while the rest goes to heat. Most people get rid of all this excess heat by transferring it (by conduction and the flow of blood) to the surface of the body, where it is radiated away. The normal internal temperature of the body (where the metabolism takes place) is 37°C, and the skin is typically 7 C° cooler. By how much does the person's entropy change per second due to this heat transfer?

20.35 •• CP A certain heat engine operating on a Carnot cycle absorbs 410 J of heat per cycle at its hot reservoir at 135°C and has a thermal efficiency of 22.0%. (a) How much work does this engine do per cycle? (b) How much heat does the engine waste each cycle? (c) What is the temperature of the cold reservoir? (d) By how much does the engine change the entropy of the world each cycle? (e) What mass of water could this engine pump per cycle from a well 35.0 m deep?

20.36 • A heat engine takes 0.350 mol of a diatomic ideal gas around the cycle shown in the pV-diagram of **Fig. P20.36**. Process $1 \rightarrow 2$ is at constant volume, process $2 \rightarrow 3$ is

Figure **P20.36**

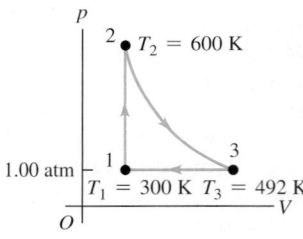

adiabatic, and process $3 \rightarrow 1$ is at a constant pressure of 1.00 atm. The value of γ for this gas is 1.40. (a) Find the pressure and volume at points 1, 2, and 3. (b) Calculate Q, W, and ΔU for each of the three processes. (c) Find the net work done by the gas in the cycle. (d) Find the net heat flow into the engine in one cycle. (e) What is the thermal efficiency of the engine? How does this compare to the efficiency of a Carnot-cycle engine operating between the same minimum and maximum temperatures T_1 and T_2?

20.37 •• BIO **Entropy Change from Digesting Fat.** Digesting fat produces 9.3 food calories per gram of fat, and typically 80% of this energy goes to heat when metabolized. (One food calorie is 1000 calories and therefore equals 4186 J.) The body then moves all this heat to the surface by a combination of thermal conductivity and motion of the blood. The internal temperature of the body (where digestion occurs) is normally 37°C, and the surface is usually about 30°C. By how much do the digestion and metabolism of a 2.50-g pat of butter change your body's entropy? Does it increase or decrease?

Figure **P20.38**

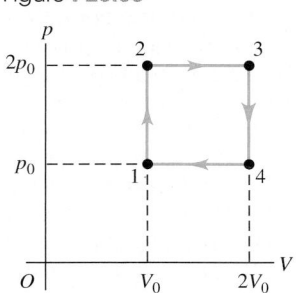

20.38 •• What is the thermal efficiency of an engine that operates by taking n moles of diatomic ideal gas through the cycle $1 \rightarrow 2 \rightarrow 3 \rightarrow 4 \rightarrow 1$ shown in **Fig. P20.38**?

20.39 •• CALC You build a heat engine that takes 1.00 mol of an ideal diatomic gas through the cycle shown in **Fig. P20.39**. (a) Show that process ab is an isothermal compression. (b) During which process(es) of the cycle is heat absorbed by the gas? During which process(es) is heat rejected? How do you know? Calculate (c) the temperature at points a, b, and c; (d) the net heat exchanged with the surroundings and net work done by the engine in one cycle; (e) the thermal efficiency of the engine.

Figure **P20.39**

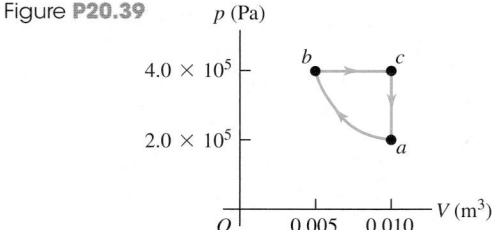

20.40 • CP As a budding mechanical engineer, you are called upon to design a Carnot engine that has 2.00 mol of a monatomic ideal gas as its working substance and operates from a high-temperature reservoir at 500°C. The engine is to lift a 15.0-kg weight 2.00 m per cycle, using 500 J of heat input. The gas in the engine chamber can have a minimum volume of 5.00 L during the cycle. (a) Draw a pV-diagram for this cycle. Show in your diagram where heat enters and leaves the gas. (b) What must be the temperature of the cold reservoir? (c) What is the thermal efficiency of the engine? (d) How much heat energy does this engine waste per cycle? (e) What is the maximum pressure that the gas chamber will have to withstand?

20.41 • CALC A heat engine operates using the cycle shown in **Fig. P20.41**. The working substance is 2.00 mol of helium gas, which reaches a maximum temperature of 327°C. Assume the helium can be treated as an ideal gas. Process bc is isothermal.

The pressure in states a and c is 1.00×10^5 Pa, and the pressure in state b is 3.00×10^5 Pa. (a) How much heat enters the gas and how much leaves the gas each cycle? (b) How much work does the engine do each cycle, and what is its efficiency? (c) Compare this engine's efficiency with the maximum possible efficiency attainable with the hot and cold reservoirs used by this cycle.

Figure **P20.41**

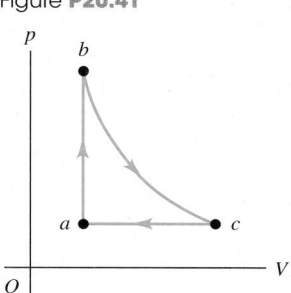

20.42 •• CP BIO **Human Entropy.** A person who has skin of surface area 1.85 m² and temperature 30.0°C is resting in an insulated room where the ambient air temperature is 20.0°C. In this state, a person gets rid of excess heat by radiation. By how much does the person change the entropy of the air in this room each second? (Recall that the room radiates back into the person and that the emissivity of the skin is 1.00.)

20.43 ••• An experimental power plant at the Natural Energy Laboratory of Hawaii generates electricity from the temperature gradient of the ocean. The surface and deep-water temperatures are 27°C and 6°C, respectively. (a) What is the maximum theoretical efficiency of this power plant? (b) If the power plant is to produce 210 kW of power, at what rate must heat be extracted from the warm water? At what rate must heat be absorbed by the cold water? Assume the maximum theoretical efficiency. (c) The cold water that enters the plant leaves it at a temperature of 10°C. What must be the flow rate of cold water through the system? Give your answer in kg/h and in L/h.

20.44 •• CP BIO **A Human Engine.** You decide to use your body as a Carnot heat engine. The operating gas is in a tube with one end in your mouth (where the temperature is 37.0°C) and the other end at the surface of your skin, at 30.0°C. (a) What is the maximum efficiency of such a heat engine? Would it be a very useful engine? (b) Suppose you want to use this human engine to lift a 2.50-kg box from the floor to a tabletop 1.20 m above the floor. How much must you increase the gravitational potential energy, and how much heat input is needed to accomplish this? (c) If your favorite candy bar has 350 food calories (1 food calorie = 4186 J) and 80% of the food energy goes into heat, how many of these candy bars must you eat to lift the box in this way?

20.45 • CALC A cylinder contains oxygen at a pressure of 2.00 atm. The volume is 4.00 L, and the temperature is 300 K. Assume that the oxygen may be treated as an ideal gas. The oxygen is carried through the following processes:

(i) Heated at constant pressure from the initial state (state 1) to state 2, which has $T = 450$ K.
(ii) Cooled at constant volume to 250 K (state 3).
(iii) Compressed at constant temperature to a volume of 4.00 L (state 4).
(iv) Heated at constant volume to 300 K, which takes the system back to state 1.

(a) Show these four processes in a pV-diagram, giving the numerical values of p and V in each of the four states. (b) Calculate Q and W for each of the four processes. (c) Calculate the net work done by the oxygen in the complete cycle. (d) What is the efficiency of this device as a heat engine? How does this compare to the efficiency of a Carnot-cycle engine operating between the same minimum and maximum temperatures of 250 K and 450 K?

20.46 • A monatomic ideal gas is taken around the cycle shown in **Fig. P20.46** in the direction shown in the figure. The path for process $c \rightarrow a$ is a straight line in the pV-diagram. (a) Calculate Q, W, and ΔU for each process $a \rightarrow b$, $b \rightarrow c$, and $c \rightarrow a$. (b) What are Q, W, and ΔU for one complete cycle? (c) What is the efficiency of the cycle?

Figure **P20.46**

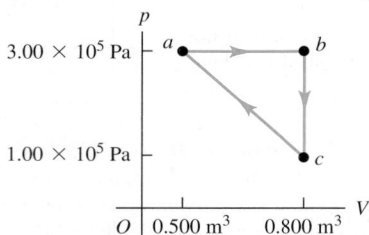

20.47 •• A Carnot engine operates between two heat reservoirs at temperatures T_H and T_C. An inventor proposes to increase the efficiency by running one engine between T_H and an intermediate temperature T' and a second engine between T' and T_C, using as input the heat expelled by the first engine. Compute the efficiency of this composite system, and compare it to that of the original engine.

20.48 ••• A typical coal-fired power plant generates 1000 MW of usable power at an overall thermal efficiency of 40%. (a) What is the rate of heat input to the plant? (b) The plant burns anthracite coal, which has a heat of combustion of 2.65×10^7 J/kg. How much coal does the plant use per day, if it operates continuously? (c) At what rate is heat ejected into the cool reservoir, which is the nearby river? (d) The river is at 18.0°C before it reaches the power plant and 18.5°C after it has received the plant's waste heat. Calculate the river's flow rate, in cubic meters per second. (e) By how much does the river's entropy increase each second?

20.49 • **Automotive Thermodynamics.** A Volkswagen Passat has a six-cylinder Otto-cycle engine with compression ratio $r = 10.6$. The diameter of each cylinder, called the *bore* of the engine, is 82.5 mm. The distance that the piston moves during the compression in Fig. 20.5, called the *stroke* of the engine, is 86.4 mm. The initial pressure of the air–fuel mixture (at point a in Fig. 20.6) is 8.50×10^4 Pa, and the initial temperature is 300 K (the same as the outside air). Assume that 200 J of heat is added to each cylinder in each cycle by the burning gasoline, and that the gas has $C_V = 20.5$ J/mol·K and $\gamma = 1.40$. (a) Calculate the total work done in one cycle in each cylinder of the engine, and the heat released when the gas is cooled to the temperature of the outside air. (b) Calculate the volume of the air–fuel mixture at point a in the cycle. (c) Calculate the pressure, volume, and temperature of the gas at points b, c, and d in the cycle. In a pV-diagram, show the numerical values of p, V, and T for each of the four states. (d) Compare the efficiency of this engine with the efficiency of a Carnot-cycle engine operating between the same maximum and minimum temperatures.

20.50 • An air conditioner operates on 800 W of power and has a performance coefficient of 2.80 with a room temperature of 21.0°C and an outside temperature of 35.0°C. (a) Calculate the rate of heat removal for this unit. (b) Calculate the rate at which heat is discharged to the outside air. (c) Calculate the total entropy change in the room if the air conditioner runs for 1 hour. Calculate the total entropy change in the outside air for the same time period. (d) What is the net change in entropy for the system (room + outside air)?

20.51 •• The pV-diagram in **Fig. P20.51** shows the cycle for a refrigerator operating on 0.850 mol of H_2. Assume that the gas can be treated as ideal. Process ab is isothermal. Find the coefficient of performance of this refrigerator.

Figure **P20.51**

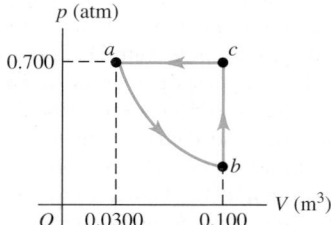

20.52 •• **BIO Human Entropy.** A person with skin of surface area 1.85 m^2 and temperature 30.0°C is resting in an insulated room where the ambient air temperature is 20.0°C. In this state, a person gets rid of excess heat by radiation. By how much does the person change the entropy of the air in this room each second? (Recall that the room radiates back into the person and that the emissivity of the skin is 1.00.)

20.53 •• **CALC** An object of mass m_1, specific heat c_1, and temperature T_1 is placed in contact with a second object of mass m_2, specific heat c_2, and temperature $T_2 > T_1$. As a result, the temperature of the first object increases to T and the temperature of the second object decreases to T'. (a) Show that the entropy increase of the system is

$$\Delta S = m_1 c_1 \ln \frac{T}{T_1} + m_2 c_2 \ln \frac{T'}{T_2}$$

and show that energy conservation requires that

$$m_1 c_1 (T - T_1) = m_2 c_2 (T_2 - T')$$

(b) Show that the entropy change ΔS, considered as a function of T, is a *maximum* if $T = T'$, which is just the condition of thermodynamic equilibrium. (c) Discuss the result of part (b) in terms of the idea of entropy as a measure of randomness.

20.54 •• **CALC** To heat 1 cup of water (250 cm^3) to make coffee, you place an electric heating element in the cup. As the water temperature increases from 20°C to 78°C, the temperature of the heating element remains at a constant 120°C. Calculate the change in entropy of (a) the water; (b) the heating element; (c) the system of water and heating element. (Make the same assumption about the specific heat of water as in Example 20.10 in Section 20.7, and ignore the heat that flows into the ceramic coffee cup itself.) (d) Is this process reversible or irreversible? Explain.

20.55 •• **DATA** In your summer job with a venture capital firm, you are given funding requests from four inventors of heat engines. The inventors claim the following data for their operating prototypes:

	Prototype			
	A	**B**	**C**	**D**
T_C (°C), low-temperature reservoir	47	17	−33	37
T_H (°C), high-temperature reservoir	177	197	247	137
Claimed efficiency e (%)	21	35	56	20

(a) Based on the T_C and T_H values for each prototype, find the maximum possible efficiency for each. (b) Are any of the claimed efficiencies impossible? Explain. (c) For all prototypes with an efficiency that is possible, rank the prototypes in decreasing order of the ratio of claimed efficiency to maximum possible efficiency.

20.56 •• DATA For a refrigerator or air conditioner, the coefficient of performance K (often denoted as COP) is, as in Eq. (20.9), the ratio of cooling output $|Q_C|$ to the required electrical energy input $|W|$, both in joules. The coefficient of performance is also expressed as a ratio of powers,

$$K = \frac{|Q_C|/t}{|W|/t}$$

where $|Q_C|/t$ is the cooling power and $|W|/t$ is the electrical power input to the device, both in watts. The energy efficiency ratio (EER) is the same quantity expressed in units of Btu for $|Q_C|$ and W·h for $|W|$. (a) Derive a general relationship that expresses EER in terms of K. (b) For a home air conditioner, EER is generally determined for a 95°F outside temperature and an 80°F return air temperature. Calculate EER for a Carnot device that operates between 95°F and 80°F. (c) You have an air conditioner with an EER of 10.9. Your home on average requires a total cooling output of $|Q_C| = 1.9 \times 10^{10}$ J per year. If electricity costs you 15.3 cents per kW·h, how much do you spend per year, on average, to operate your air conditioner? (Assume that the unit's EER accurately represents the operation of your air conditioner. A *seasonal energy efficiency ratio* (SEER) is often used. The SEER is calculated over a range of outside temperatures to get a more accurate seasonal average.) (d) You are considering replacing your air conditioner with a more efficient one with an EER of 14.6. Based on the EER, how much would that save you on electricity costs in an average year?

20.57 ••• DATA You are conducting experiments to study prototype heat engines. In one test, 4.00 mol of argon gas are taken around the cycle shown in **Fig. P20.57**. The pressure is low enough for the gas to be treated as ideal. You measure the gas temperature in states a, b, c, and d and find $T_a = 250.0$ K, $T_b = 300.0$ K, $T_c = 380.0$ K, and $T_d = 316.7$ K. (a) Calculate the efficiency e of the cycle. (b) Disappointed by the cycle's low efficiency, you consider doubling the number of moles of gas while keeping the pressure and volume the same. What would e be then? (c) You remember that the efficiency of a Carnot cycle increases if the temperature of the hot reservoir is increased. So, you return to using 4.00 mol of gas but double the volume in states c and d while keeping the pressures the same. The resulting temperatures in these states are $T_c = 760.0$ K and $T_d = 633.4$ K. T_a and T_b remain the same as in part (a). Calculate e for this cycle with the new T_c and T_d values. (d) Encouraged by the increase in efficiency, you raise T_c and T_d still further. But e doesn't increase very much; it seems to be approaching a limiting value. If $T_a = 250.0$ K and $T_b = 300.0$ K and you keep volumes V_a and V_b the same as in part (a), then $T_c/T_d = T_b/T_a$ and $T_c = 1.20T_d$. Derive an expression for e as a function of T_d for this cycle. What value does e approach as T_d becomes very large?

Figure **P20.57**

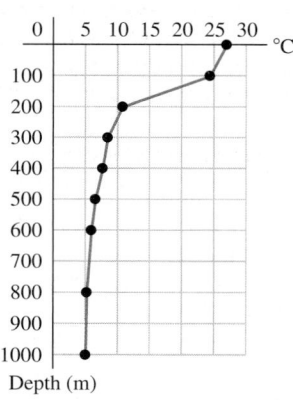

CHALLENGE PROBLEM

20.58 ••• Consider a Diesel cycle that starts (at point a in Fig. 20.7) with air at temperature T_a. The air may be treated as an ideal gas. (a) If the temperature at point c is T_c, derive an expression for the efficiency of the cycle in terms of the compression ratio r. (b) What is the efficiency if $T_a = 300$ K, $T_c = 950$ K, $\gamma = 1.40$, and $r = 21.0$?

PASSAGE PROBLEMS

POWER FROM THE SEA. *Ocean thermal energy conversion* is a process that uses the temperature difference between the warm surface water of tropical oceans and the cold deep-ocean water to run a heat engine. The graph shows a typical decrease of temperature with depth below the surface in tropical oceans. In the heat engine, the warmer surface water vaporizes a low-boiling-point fluid, such as ammonia. The heat of vaporization of ammonia is 260 cal/g at 27°C, the surface-water temperature. The vapor is used to turn a turbine and is then condensed back into a liquid by means of cold water brought from deep below the surface through a large intake pipe. A power plant producing 10 MW of useful power would require a cold seawater flow rate of about 30,000 kg/s.

20.59 If the power plant uses a Carnot cycle and the desired theoretical efficiency is 6.5%, from what depth must cold water be brought? (a) 100 m; (b) 400 m; (c) 800 m; (d) deeper than 1000 m.

20.60 What is the change in entropy of the ammonia vaporized per second in the 10-MW power plant, assuming an ideal Carnot efficiency of 6.5%? (a) $+6 \times 10^6$ J/K per second; (b) $+5 \times 10^5$ J/K per second; (c) $+1 \times 10^5$ J/K per second; (d) 0.

20.61 Compare the entropy change of the warmer water to that of the colder water during one cycle of the heat engine, assuming an ideal Carnot cycle. (a) The entropy does not change during one cycle in either case. (b) The entropy of both increases, but the entropy of the colder water increases by more because its initial temperature is lower. (c) The entropy of the warmer water decreases by more than the entropy of the colder water increases, because some of the heat removed from the warmer water goes to the work done by the engine. (d) The entropy of the warmer water decreases by the same amount that the entropy of the colder water increases.

20.62 If the proposed plant is built and produces 10 MW but the rate at which waste heat is exhausted to the cold water is 165 MW, what is the plant's actual efficiency? (a) 5.7%; (b) 6.1%; (c) 6.5%; (d) 16.5%.

Answers

Chapter Opening Question ?

(i) This is what a refrigerator does: It makes heat flow from the cold interior of the refrigerator to the warm outside. The second law of thermodynamics says that heat cannot *spontaneously* flow from a cold body to a hot one. A refrigerator has a motor that does work on the system to *force* the heat to flow in that way.

Test Your Understanding Questions

20.1 (ii) Like sliding a book across a table, rubbing your hands together uses friction to convert mechanical energy into heat. The (impossible) reverse process would involve your hands spontaneously getting colder, with the released energy forcing your hands to move rhythmically back and forth!

20.2 (iii), (i), (ii) From Eq. (20.4) the efficiency is $e = W/Q_H$, and from Eq. (20.2) $W = Q_H + Q_C = |Q_H| - |Q_C|$. For engine (i) $Q_H = 5000\,J$ and $Q_C = -4500\,J$, so $W = 5000\,J + (-4500\,J) = 500\,J$ and $e = (500\,J)/(5000\,J) = 0.100$. For engine (ii) $Q_H = 25{,}000\,J$ and $W = 2000\,J$, so $e = (2000\,J)/(25{,}000\,J) = 0.080$. For engine (iii) $W = 400\,J$ and $Q_C = -2800\,J$, so $Q_H = W - Q_C = 400\,J - (-2800\,J) = 3200\,J$ and $e = (400\,J)/(3200\,J) = 0.125$.

20.3 (i), (ii) Doubling the amount of fuel burned per cycle means that Q_H is doubled, so the resulting pressure increase from b to c in Fig. 20.6 is greater. The compression ratio and hence the efficiency remain the same, so $|Q_C|$ (the amount of heat rejected to the environment) must increase by the same factor as Q_H. Hence the pressure drop from d to a in Fig. 20.6 is also greater. The volume V and the compression ratio r don't change, so the horizontal dimensions of the pV-diagram don't change.

20.4 no A refrigerator uses an input of work to transfer heat from one system (the refrigerator's interior) to another system (its exterior, which includes the house in which the refrigerator is installed). If the door is open, these two systems are really the *same* system and will eventually come to the same temperature. By the first law of thermodynamics, all of the work input to the refrigerator motor will be converted into heat and the temperature in your house will actually *increase*. To cool the house you need a system that will transfer heat from it to the outside world, such as an air conditioner or heat pump.

20.5 no, no Both the 100%-efficient engine of Fig. 20.11a and the workless refrigerator of Fig. 20.11b return to the same state at the end of a cycle as at the beginning, so the net change in internal energy of each system is zero ($\Delta U = 0$). For the 100%-efficient engine, the net heat flow into the engine equals the net work done, so $Q = W$, $Q - W = 0$, and the first law ($\Delta U = Q - W$) is obeyed. For the workless refrigerator, no net work is done (so $W = 0$) and as much heat flows into it as out (so $Q = 0$), so again $Q - W = 0$ and $\Delta U = Q - W$ in accordance with the first law. It is the *second* law of thermodynamics that tells us that both the 100%-efficient engine and the workless refrigerator are impossible.

20.6 no The efficiency can be no better than that of a Carnot engine running between the same two temperature limits, $e_{Carnot} = 1 - (T_C/T_H)$ [Eq. (20.14)]. The temperature T_C of the cold reservoir for this air-cooled engine is about 300 K (ambient temperature), and the temperature T_H of the hot reservoir cannot exceed the melting point of copper, 1356 K (see Table 17.4). Hence the maximum possible Carnot efficiency is $e = 1 - (300\,K)/(1356\,K) = 0.78$, or 78%. The temperature of any real engine would be less than this, so it would be impossible for the inventor's engine to attain 85% efficiency. You should invest your money elsewhere.

20.7 $-102\,J/K$, no The process described is exactly the opposite of the process used in Example 20.10. The result violates the second law of thermodynamics, which states that the entropy of an isolated system cannot decrease.

20.8 (i) For case (i), we saw in Example 20.8 (Section 20.7) that for an ideal gas, the entropy change in a free expansion is the same as in an isothermal expansion. From Eq. (20.23), this implies that the ratio of the number of microscopic states after and before the expansion, w_2/w_1, is also the same for these two cases. From Example 20.11, $w_2/w_1 = 2^N$, so the number of microscopic states increases by a factor 2^N. For case (ii), in a reversible expansion the entropy change is $\Delta S = \int dQ/T = 0$; if the expansion is adiabatic there is no heat flow, so $\Delta S = 0$. From Eq. (20.23), $w_2/w_1 = 1$ and there is *no* change in the number of microscopic states. The difference is that in an adiabatic expansion the temperature drops and the molecules move more slowly, so they have fewer microscopic states available to them than in an isothermal expansion.

Bridging Problem

(a) 34.8°C **(b)** +10 J/K

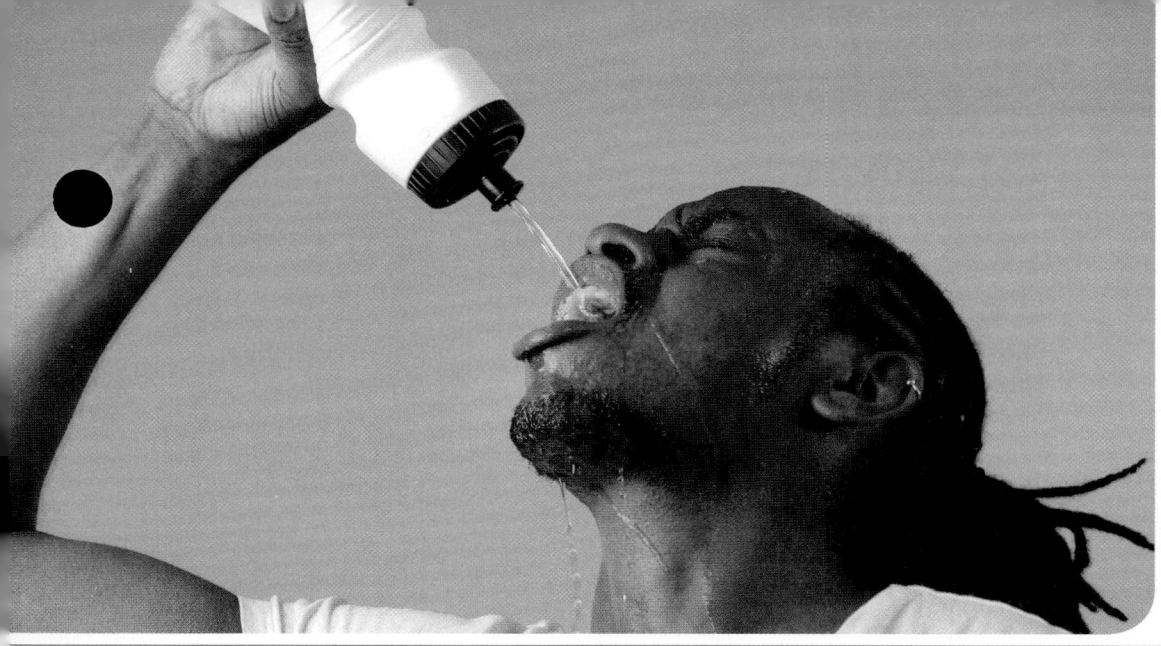

21 ELECTRIC CHARGE AND ELECTRIC FIELD

LEARNING GOALS

Looking forward at ...

21.1 The nature of electric charge, and how we know that electric charge is conserved.

21.2 How objects become electrically charged.

21.3 How to use Coulomb's law to calculate the electric force between charges.

21.4 The distinction between electric force and electric field.

21.5 How to calculate the electric field due to a collection of charges.

21.6 How to use the idea of electric field lines to visualize and interpret electric fields.

21.7 How to calculate the properties of electric dipoles.

Looking back at ...

1.7–1.10 Vector algebra, including the scalar (dot) product and the vector (cross) product.

4.3 Newton's second law.

7.5 Stable and unstable equilibria.

12.5 Streamlines in fluid flow.

I n Chapter 5 we mentioned the four kinds of fundamental forces. To this point the only one of these forces that we have examined in any detail is gravity. Now we are ready to examine the force of *electromagnetism,* which encompasses both electricity and magnetism.

Electromagnetic interactions involve particles that have *electric charge,* an attribute that is as fundamental as mass. Just as objects with mass are accelerated by gravitational forces, so electrically charged objects are accelerated by electric forces. The shock you feel when you scuff your shoes across a carpet and then reach for a metal doorknob is due to charged particles leaping between your finger and the doorknob. Electric currents are simply streams of charged particles flowing within wires in response to electric forces. Even the forces that hold atoms together to form solid matter, and that keep the atoms of solid objects from passing through each other, are fundamentally due to electric interactions between the charged particles within atoms.

We begin our study of electromagnetism in this chapter by examining the nature of electric charge. We'll find that charge is quantized and obeys a conservation principle. When charges are at rest in our frame of reference, they exert *electrostatic* forces on each other. These forces are of tremendous importance in chemistry and biology and have many technological applications. Electrostatic forces are governed by a simple relationship known as *Coulomb's law* and are most conveniently described by using the concept of *electric field.* In later chapters we'll expand our discussion to include electric charges in motion. This will lead us to an understanding of magnetism and, remarkably, of the nature of light.

While the key ideas of electromagnetism are conceptually simple, applying them to practical problems will make use of many of your mathematical skills, especially your knowledge of geometry and integral calculus. For this reason you may find this chapter and those that follow to be more mathematically demanding than earlier chapters. The reward for your extra effort will be a deeper understanding of principles that are at the heart of modern physics and technology.

21.1 ELECTRIC CHARGE

The ancient Greeks discovered as early as 600 B.C. that after they rubbed amber with wool, the amber could attract other objects. Today we say that the amber has acquired a net **electric charge,** or has become *charged*. The word "electric" is derived from the Greek word *elektron,* meaning amber. When you scuff your shoes across a nylon carpet, you become electrically charged, and you can charge a comb by passing it through dry hair.

Plastic rods and fur (real or fake) are particularly good for demonstrating **electrostatics,** the interactions between electric charges that are at rest (or nearly so). After we charge both plastic rods in **Fig. 21.1a** by rubbing them with the piece of fur, we find that the rods repel each other.

When we rub glass rods with silk, the glass rods also become charged and repel each other (Fig. 21.1b). But a charged plastic rod *attracts* a charged glass rod; furthermore, the plastic rod and the fur attract each other, and the glass rod and the silk attract each other (Fig. 21.1c).

These experiments and many others like them have shown that there are exactly two kinds of electric charge: the kind on the plastic rod rubbed with fur and the kind on the glass rod rubbed with silk. Benjamin Franklin (1706–1790) suggested calling these two kinds of charge *negative* and *positive,* respectively, and these names are still used. The plastic rod and the silk have negative charge; the glass rod and the fur have positive charge.

> **Two positive charges or two negative charges repel each other. A positive charge and a negative charge attract each other.**

CAUTION **Electric attraction and repulsion** The attraction and repulsion of two charged objects are sometimes summarized as "Like charges repel, and opposite charges attract." But "like charges" does *not* mean that the two charges are exactly identical, only that both charges have the same algebraic *sign* (both positive or both negative). "Opposite charges" means that both objects have an electric charge, and those charges have different signs (one positive and the other negative). ▌

21.1 Experiments in electrostatics. (a) Negatively charged objects repel each other. (b) Positively charged objects repel each other. (c) Positively charged objects and negatively charged objects attract each other.

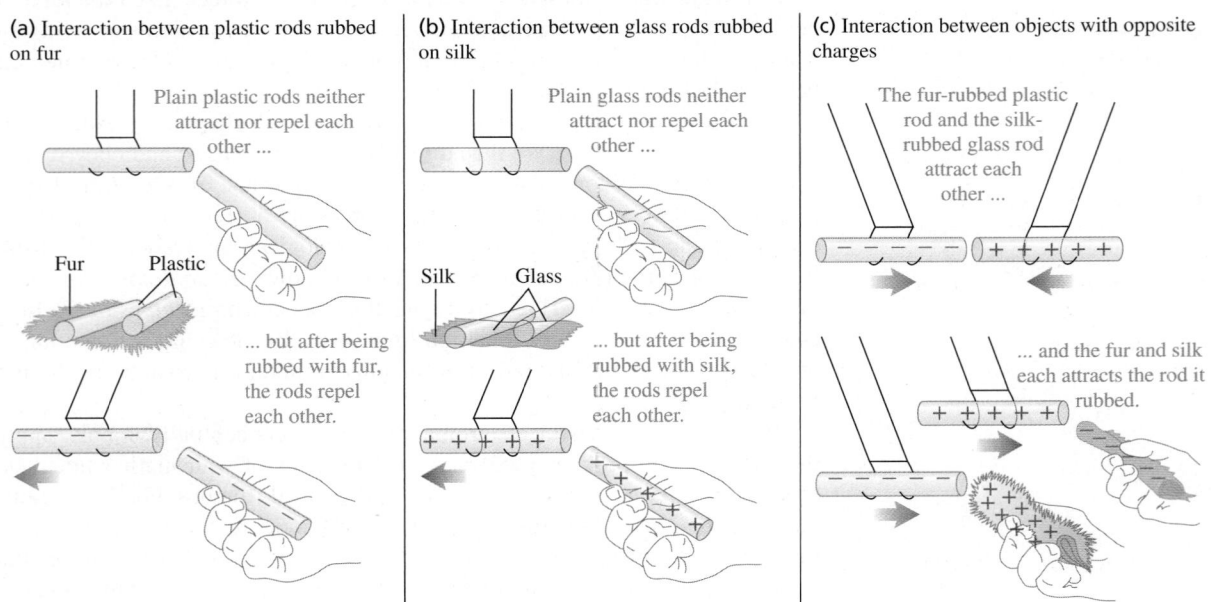

(a) Interaction between plastic rods rubbed on fur

Plain plastic rods neither attract nor repel each other ...

Fur   Plastic

... but after being rubbed with fur, the rods repel each other.

(b) Interaction between glass rods rubbed on silk

Plain glass rods neither attract nor repel each other ...

Silk   Glass

... but after being rubbed with silk, the rods repel each other.

(c) Interaction between objects with opposite charges

The fur-rubbed plastic rod and the silk-rubbed glass rod attract each other ...

... and the fur and silk each attracts the rod it rubbed.

21.2 Schematic diagram of the operation of a laser printer.

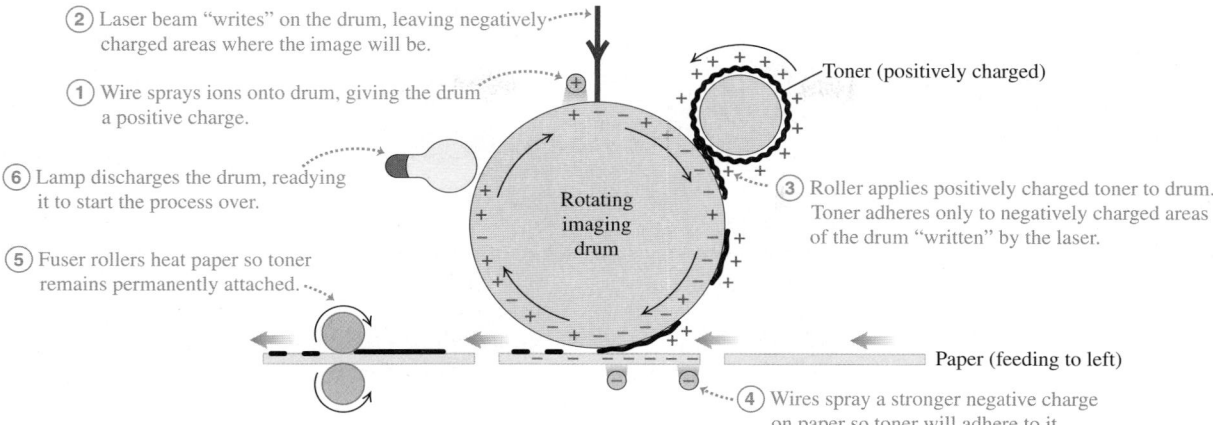

(2) Laser beam "writes" on the drum, leaving negatively charged areas where the image will be.

(1) Wire sprays ions onto drum, giving the drum a positive charge.

(6) Lamp discharges the drum, readying it to start the process over.

(5) Fuser rollers heat paper so toner remains permanently attached.

Toner (positively charged)

Rotating imaging drum

(3) Roller applies positively charged toner to drum. Toner adheres only to negatively charged areas of the drum "written" by the laser.

Paper (feeding to left)

(4) Wires spray a stronger negative charge on paper so toner will adhere to it.

A laser printer (**Fig. 21.2**) utilizes the forces between charged bodies. The printer's light-sensitive imaging drum is given a positive charge. As the drum rotates, a laser beam shines on selected areas of the drum, leaving those areas with a *negative* charge. Positively charged particles of toner adhere only to the areas of the drum "written" by the laser. When a piece of paper is placed in contact with the drum, the toner particles stick to the paper and form an image.

Electric Charge and the Structure of Matter

When you charge a rod by rubbing it with fur or silk as in Fig. 21.1, there is no visible change in the appearance of the rod. What, then, actually happens to the rod when you charge it? To answer this question, we must look more closely at the structure of atoms, the building blocks of ordinary matter.

The structure of atoms can be described in terms of three particles: the negatively charged **electron**, the positively charged **proton,** and the uncharged **neutron** (**Fig. 21.3**). The proton and neutron are combinations of other entities called *quarks,* which have charges of $\pm\frac{1}{3}$ and $\pm\frac{2}{3}$ times the electron charge. Isolated quarks have not been observed, and there are theoretical reasons to believe that it is impossible in principle to observe a quark in isolation.

The protons and neutrons in an atom make up a small, very dense core called the **nucleus,** with dimensions of the order of 10^{-15} m. Surrounding the nucleus are the electrons, extending out to distances of the order of 10^{-10} m from the nucleus. If an atom were a few kilometers across, its nucleus would be the size of a tennis ball. The negatively charged electrons are held within the atom by the attractive electric forces exerted on them by the positively charged nucleus. (The protons and neutrons are held within stable atomic nuclei by an attractive interaction, called the *strong nuclear force,* that overcomes the electric repulsion of the protons. The strong nuclear force has a short range, and its effects do not extend far beyond the nucleus.)

The masses of the individual particles, to the precision that they are presently known, are

$$\text{Mass of electron} = m_e = 9.10938291(40) \times 10^{-31} \text{ kg}$$

$$\text{Mass of proton} = m_p = 1.672621777(74) \times 10^{-27} \text{ kg}$$

$$\text{Mass of neutron} = m_n = 1.674927351(74) \times 10^{-27} \text{ kg}$$

The numbers in parentheses are the uncertainties in the last two digits. Note that the masses of the proton and neutron are nearly equal and are roughly 2000 times the mass of the electron. Over 99.9% of the mass of any atom is concentrated in its nucleus.

21.3 The structure of an atom. The particular atom depicted here is lithium (see Fig. 21.4a).

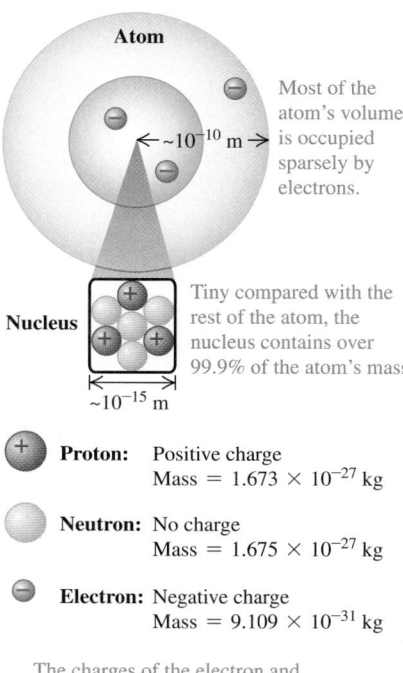

Atom

Most of the atom's volume is occupied sparsely by electrons.

$\leftarrow \sim 10^{-10}$ m $\rightarrow$

Nucleus

Tiny compared with the rest of the atom, the nucleus contains over 99.9% of the atom's mass.

$\sim 10^{-15}$ m

(+) **Proton:** Positive charge
Mass $= 1.673 \times 10^{-27}$ kg

○ **Neutron:** No charge
Mass $= 1.675 \times 10^{-27}$ kg

⊖ **Electron:** Negative charge
Mass $= 9.109 \times 10^{-31}$ kg

The charges of the electron and proton are equal in magnitude.

21.4 (a) A neutral atom has as many electrons as it does protons. (b) A positive ion has a deficit of electrons. (c) A negative ion has an excess of electrons. (The electron "shells" are a schematic representation of the actual electron distribution, a diffuse cloud many times larger than the nucleus.)

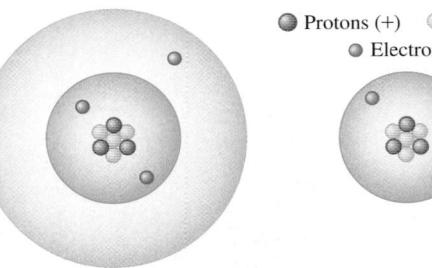

● Protons (+) ○ Neutrons
● Electrons (−)

(a) **Neutral lithium atom (Li):**

3 protons (3+)

4 neutrons

3 electrons (3−)

Electrons equal protons:
Zero net charge

(b) **Positive lithium ion (Li⁺):**

3 protons (3+)

4 neutrons

2 electrons (2−)

Fewer electrons than protons:
Positive net charge

(c) **Negative lithium ion (Li⁻):**

3 protons (3+)

4 neutrons

4 electrons (4−)

More electrons than protons:
Negative net charge

The negative charge of the electron has (within experimental error) *exactly* the same magnitude as the positive charge of the proton. In a neutral atom the number of electrons equals the number of protons in the nucleus, and the net electric charge (the algebraic sum of all the charges) is exactly zero (**Fig. 21.4a**). The number of protons or electrons in a neutral atom of an element is called the **atomic number** of the element. If one or more electrons are removed from an atom, what remains is called a **positive ion** (Fig. 21.4b). A **negative ion** is an atom that has *gained* one or more electrons (Fig. 21.4c). This gain or loss of electrons is called **ionization.**

When the total number of protons in a macroscopic body equals the total number of electrons, the total charge is zero and the body as a whole is electrically neutral. To give a body an excess negative charge, we may either *add negative* charges to a neutral body or *remove positive* charges from that body. Similarly, we can create an excess positive charge by either *adding positive* charge or *removing negative* charge. In most cases, negatively charged (and highly mobile) electrons are added or removed, and a "positively charged body" is one that has lost some of its normal complement of electrons. When we speak of the charge of a body, we always mean its *net* charge. The net charge is always a very small fraction (typically no more than 10^{-12}) of the total positive charge or negative charge in the body.

Electric Charge Is Conserved

Implicit in the foregoing discussion are two very important principles. First is the **principle of conservation of charge:**

> **The algebraic sum of all the electric charges in any closed system is constant.**

If we rub together a plastic rod and a piece of fur, both initially uncharged, the rod acquires a negative charge (since it takes electrons from the fur) and the fur acquires a positive charge of the *same* magnitude (since it has lost as many electrons as the rod has gained). Hence the total electric charge on the two bodies together does not change. In any charging process, charge is not created or destroyed; it is merely *transferred* from one body to another.

Conservation of charge is thought to be a *universal* conservation law. No experimental evidence for any violation of this principle has ever been observed. Even in high-energy interactions in which particles are created and destroyed, such as the creation of electron–positron pairs, the total charge of any closed system is exactly constant.

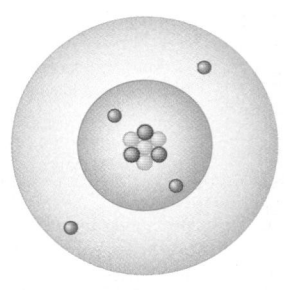

The second important principle is:

The magnitude of charge of the electron or proton is a natural unit of charge.

Every observable amount of electric charge is always an integer multiple of this basic unit. We say that charge is *quantized*. A familiar example of quantization is money. When you pay cash for an item in a store, you have to do it in one-cent increments. Cash can't be divided into amounts smaller than one cent, and electric charge can't be divided into amounts smaller than the charge of one electron or proton. (The quark charges, $\pm\frac{1}{3}$ and $\pm\frac{2}{3}$ of the electron charge, are probably not observable as isolated charges.) Thus the charge on any macroscopic body is always zero or an integer multiple (negative or positive) of the electron charge.

Understanding the electric nature of matter gives us insight into many aspects of the physical world (**Fig. 21.5**). The chemical bonds that hold atoms together to form molecules are due to electric interactions between the atoms. They include the strong ionic bonds that hold sodium and chlorine atoms together to make table salt and the relatively weak bonds between the strands of DNA that record your body's genetic code. When you stand, the normal force exerted on you by the floor arises from electric forces between charged particles in the atoms of your shoes and the atoms of the floor. The tension force in a stretched string and the adhesive force of glue are likewise due to electric interactions of atoms.

21.5 Most of the forces on this water skier are electric. Electric interactions between adjacent molecules give rise to the force of the water on the ski, the tension in the tow rope, and the resistance of the air on the skier's body. Electric interactions also hold the atoms of the skier's body together. Only one wholly nonelectric force acts on the skier: the force of gravity.

TEST YOUR UNDERSTANDING OF SECTION 21.1   Two charged objects repel each other through the electric force. The charges on the objects are (i) one positive and one negative; (ii) both positive; (iii) both negative; (iv) either (ii) or (iii); (v) any of (i), (ii), or (iii). ▌

21.2 CONDUCTORS, INSULATORS, AND INDUCED CHARGES

PhET: Balloons and Static Electricity
PhET: John Travoltage

Some materials permit electric charge to move easily from one region of the material to another, while others do not. For example, **Fig. 21.6a** (next page) shows a copper wire supported by a nylon thread. Suppose you touch one end of the wire to a charged plastic rod and attach the other end to a metal ball that is initially uncharged; you then remove the charged rod and the wire. When you bring another charged body up close to the ball (Figs. 21.6b and 21.6c), the ball is attracted or repelled, showing that the ball has become electrically charged. Electric charge has been transferred through the copper wire between the ball and the surface of the plastic rod.

The copper wire is called a **conductor** of electricity. If you repeat the experiment using a rubber band or nylon thread in place of the wire, you find that *no* charge is transferred to the ball. These materials are called **insulators.** Conductors permit the easy movement of charge through them, while insulators do not. (The supporting nylon threads shown in Fig. 21.6 are insulators, which prevents charge from leaving the metal ball and copper wire.)

As an example, carpet fibers on a dry day are good insulators. As you walk across a carpet, the rubbing of your shoes against the fibers causes charge to build up on you, and this charge remains on you because it can't flow through the insulating fibers. If you then touch a conducting object such as a doorknob, a rapid charge transfer takes place between your finger and the doorknob, and you feel a shock. One way to prevent this is to wind some of the carpet fibers around conducting cores so that any charge that builds up on you can be transferred harmlessly to the carpet. Another solution is to coat the carpet fibers with an antistatic layer that does not easily transfer electrons to or from your shoes; this prevents any charge from building up on you in the first place.

21.6 Copper is a good conductor of electricity; nylon is a good insulator. **(a)** The copper wire conducts charge between the metal ball and the charged plastic rod to charge the ball negatively. Afterward, the metal ball is **(b)** repelled by a negatively charged plastic rod and **(c)** attracted to a positively charged glass rod.

(a)

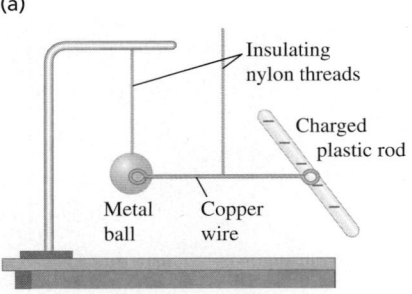

The wire conducts charge from the negatively charged plastic rod to the metal ball.

(b)

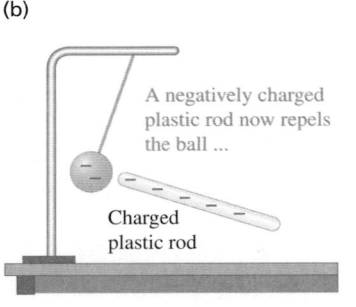

A negatively charged plastic rod now repels the ball ...

(c)

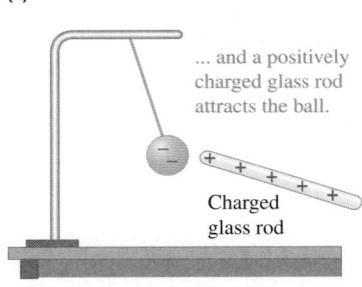

... and a positively charged glass rod attracts the ball.

Most metals are good conductors, while most nonmetals are insulators. Within a solid metal such as copper, one or more outer electrons in each atom become detached and can move freely throughout the material, just as the molecules of a gas can move through the spaces between the grains in a bucket of sand. The other electrons remain bound to the positively charged nuclei, which themselves are bound in nearly fixed positions within the material. In an insulator there are no, or very few, free electrons, and electric charge cannot move freely through the material. Some materials called *semiconductors* are intermediate in their properties between good conductors and good insulators.

Charging by Induction

We can charge a metal ball by using a copper wire and an electrically charged plastic rod, as in Fig. 21.6a. In this process, some of the excess electrons on the rod are transferred from it to the ball, leaving the rod with a smaller negative charge. But there is a different technique in which the plastic rod can give another body a charge of *opposite* sign without losing any of its own charge. This process is called charging by **induction.**

Figure 21.7 shows an example of charging by induction. An uncharged metal ball is supported on an insulating stand (Fig. 21.7a). When you bring a negatively charged rod near it, without actually touching it (Fig. 21.7b), the free electrons in the metal ball are repelled by the excess electrons on the rod, and they shift toward the right, away from the rod. They cannot escape from the ball because the supporting stand and the surrounding air are insulators. So we get excess negative charge at the right surface of the ball and a deficiency of negative charge (that is, a net positive charge) at the left surface. These excess charges are called **induced charges.**

Not all of the free electrons move to the right surface of the ball. As soon as any induced charge develops, it exerts forces toward the *left* on the other free electrons. These electrons are repelled by the negative induced charge on the right and attracted toward the positive induced charge on the left. The system reaches an equilibrium state in which the force toward the right on an electron, due to the charged rod, is just balanced by the force toward the left due to the induced charge. If we remove the charged rod, the free electrons shift back to the left, and the original neutral condition is restored.

What happens if, while the plastic rod is nearby, you touch one end of a conducting wire to the right surface of the ball and the other end to the earth (Fig. 21.7c)? The earth is a conductor, and it is so large that it can act as a practically infinite source of extra electrons or sink of unwanted electrons. Some of the negative charge flows through the wire to the earth. Now suppose you disconnect the wire (Fig. 21.7d) and then remove the rod (Fig. 21.7e); a net positive charge is left on the ball. The charge on the negatively charged rod has not changed during this process. The earth acquires a negative charge that is equal in magnitude to the induced positive charge remaining on the ball.

21.7 Charging a metal ball by induction.

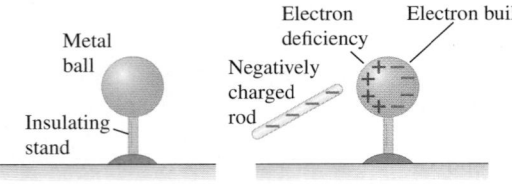

(a) Uncharged metal ball

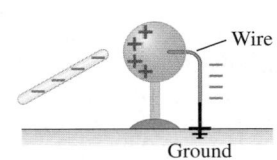

(b) Negative charge on rod repels electrons, creating zones of negative and positive **induced charge**.

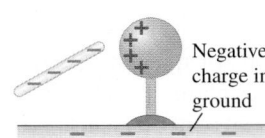

(c) Wire lets electron build-up (induced negative charge) flow into ground.

(d) Wire removed; ball now has only an electron-deficient region of positive charge.

(e) Rod removed; electrons rearrange themselves, ball has overall electron deficiency (net positive charge).

21.8 The charges within the molecules of an insulating material can shift slightly. As a result, a comb with either sign of charge attracts a neutral insulator. By Newton's third law the neutral insulator exerts an equal-magnitude attractive force on the comb.

(a) A charged comb picking up uncharged pieces of plastic

(b) How a negatively charged comb attracts an insulator

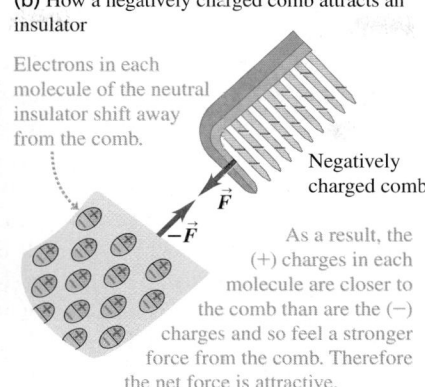

Electrons in each molecule of the neutral insulator shift away from the comb.

Negatively charged comb

$\vec{F}$

$-\vec{F}$

As a result, the (+) charges in each molecule are closer to the comb than are the (−) charges and so feel a stronger force from the comb. Therefore the net force is attractive.

(c) How a positively charged comb attracts an insulator

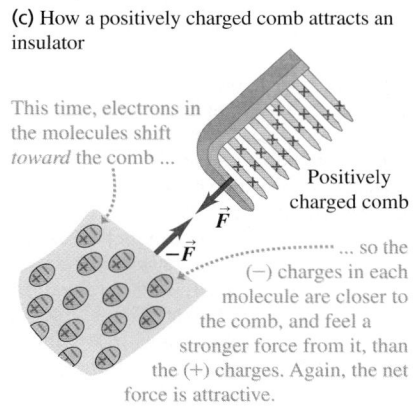

This time, electrons in the molecules shift *toward* the comb ...

Positively charged comb

$\vec{F}$

$-\vec{F}$

... so the (−) charges in each molecule are closer to the comb, and feel a stronger force from it, than the (+) charges. Again, the net force is attractive.

Electric Forces on Uncharged Objects

Finally, we note that a charged body can exert forces even on objects that are *not* charged themselves. If you rub a balloon on the rug and then hold the balloon against the ceiling, it sticks, even though the ceiling has no net electric charge. After you electrify a comb by running it through your hair, you can pick up uncharged bits of paper or plastic with it (**Fig. 21.8a**). How is this possible?

DEMO

This interaction is an induced-charge effect. Even in an insulator, electric charges can shift back and forth a little when there is charge nearby. This is shown in Fig. 21.8b; the negatively charged plastic comb causes a slight shifting of charge within the molecules of the neutral insulator, an effect called *polarization*. The positive and negative charges in the material are present in equal amounts, but the positive charges are closer to the plastic comb and so feel an attraction that is stronger than the repulsion felt by the negative charges, giving a net attractive force. (In Section 21.3 we will study how electric forces depend on distance.) Note that a neutral insulator is also attracted to a *positively* charged comb (Fig. 21.8c). Now the charges in the insulator shift in the opposite direction; the negative charges in the insulator are closer to the comb and feel an attractive force that is stronger than the repulsion felt by the positive charges in the insulator. Hence a charged object of *either* sign exerts an attractive force on an uncharged insulator. **Figure 21.9** shows an industrial application of this effect.

TEST YOUR UNDERSTANDING OF SECTION 21.2   You have two lightweight metal spheres, each hanging from an insulating nylon thread. One of the spheres has a net negative charge, while the other sphere has no net charge. (a) If the spheres are close together but do not touch, will they (i) attract each other, (ii) repel each other, or (iii) exert no force on each other? (b) You now allow the two spheres to touch. Once they have touched, will the two spheres (i) attract each other, (ii) repel each other, or (iii) exert no force on each other? ▌

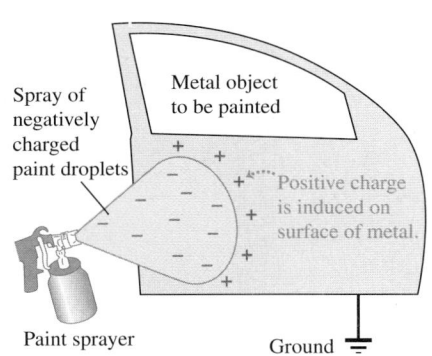

Spray of negatively charged paint droplets

Metal object to be painted

Positive charge is induced on surface of metal.

Paint sprayer

Ground

21.9 The electrostatic painting process (compare Figs. 21.7b and 21.7c). A metal object to be painted is connected to the earth ("ground"), and the paint droplets are given an electric charge as they exit the sprayer nozzle. Induced charges of the opposite sign appear in the object as the droplets approach, just as in Fig. 21.7b, and they attract the droplets to the surface. This process minimizes overspray from clouds of stray paint particles and gives a particularly smooth finish.

One way to test for the genetic disease cystic fibrosis (CF) is to measure the salt content of a person's sweat. Sweat is a mixture of water and ions, including the sodium (Na^+) and chloride (Cl^-) ions that make up ordinary salt (NaCl). When sweat is secreted by epithelial cells, some of the Cl^- ions flow from the sweat back into these cells (a process called reabsorption). The electric attraction between negative and positive charges pulls Na^+ ions along with the Cl^-. Water molecules cannot flow back into the epithelial cells, so sweat on the skin has a low salt content. However, in persons with CF the reabsorption of Cl^- ions is blocked. Hence the sweat of persons with CF is unusually salty, with up to four times the normal concentration of Cl^- and Na^+.

21.3 COULOMB'S LAW

Charles Augustin de Coulomb (1736–1806) studied the interaction forces of charged particles in detail in 1784. He used a torsion balance (**Fig. 21.10a**) similar to the one used 13 years later by Cavendish to study the much weaker gravitational interaction, as we discussed in Section 13.1. For **point charges,** charged bodies that are very small in comparison with the distance r between them, Coulomb found that the electric force is proportional to $1/r^2$. That is, when the distance r doubles, the force decreases to one-quarter of its initial value; when the distance is halved, the force increases to four times its initial value.

The electric force between two point charges also depends on the quantity of charge on each body, which we will denote by q or Q. To explore this dependence, Coulomb divided a charge into two equal parts by placing a small charged spherical conductor into contact with an identical but uncharged sphere; by symmetry, the charge is shared equally between the two spheres. (Note the essential role of the principle of conservation of charge in this procedure.) Thus he could obtain one-half, one-quarter, and so on, of any initial charge. He found that the forces that two point charges q_1 and q_2 exert on each other are proportional to each charge and therefore are proportional to the *product* q_1q_2 of the two charges.

Thus Coulomb established what we now call **Coulomb's law:**

> **The magnitude of the electric force between two point charges is directly proportional to the product of the charges and inversely proportional to the square of the distance between them.**

In mathematical terms, the magnitude F of the force that each of two point charges q_1 and q_2 a distance r apart exerts on the other can be expressed as

$$F = k\frac{|q_1q_2|}{r^2} \tag{21.1}$$

where k is a proportionality constant whose numerical value depends on the system of units used. The absolute value bars are used in Eq. (21.1) because the charges q_1 and q_2 can be either positive or negative, while the force magnitude F is always positive.

The directions of the forces the two charges exert on each other are always along the line joining them. When the charges q_1 and q_2 have the same sign, either both positive or both negative, the forces are repulsive; when the charges

21.10 (a) Measuring the electric force between point charges. (b) The electric forces between point charges obey Newton's third law: $\vec{F}_{1\text{ on }2} = -\vec{F}_{2\text{ on }1}$.

(a) A torsion balance of the type used by Coulomb to measure the electric force

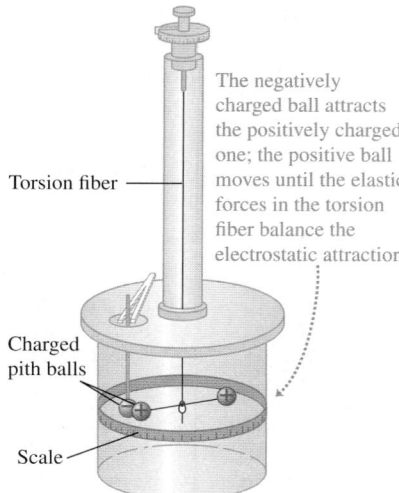

The negatively charged ball attracts the positively charged one; the positive ball moves until the elastic forces in the torsion fiber balance the electrostatic attraction.

Torsion fiber

Charged pith balls

Scale

(b) Interactions between point charges

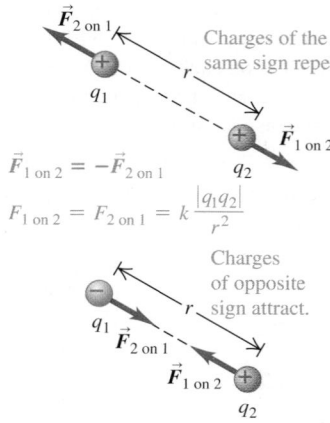

$\vec{F}_{2\text{ on }1}$

Charges of the same sign repel.

q_1

$\vec{F}_{1\text{ on }2}$

q_2

$\vec{F}_{1\text{ on }2} = -\vec{F}_{2\text{ on }1}$

$F_{1\text{ on }2} = F_{2\text{ on }1} = k\dfrac{|q_1q_2|}{r^2}$

Charges of opposite sign attract.

q_1

$\vec{F}_{2\text{ on }1}$

$\vec{F}_{1\text{ on }2}$

q_2

have opposite signs, the forces are attractive (Fig. 21.10b). The two forces obey Newton's third law; they are always equal in magnitude and opposite in direction, even when the charges are not equal in magnitude.

The proportionality of the electric force to $1/r^2$ has been verified with great precision. There is no reason to suspect that the exponent is different from precisely 2. Thus the form of Eq. (21.1) is the same as that of the law of gravitation. But electric and gravitational interactions are two distinct classes of phenomena. Electric interactions depend on electric charges and can be either attractive or repulsive, while gravitational interactions depend on mass and are always attractive (because there is no such thing as negative mass).

Fundamental Electric Constants

The value of the proportionality constant k in Coulomb's law depends on the system of units used. In our study of electricity and magnetism we will use SI units exclusively. The SI electric units include most of the familiar units such as the volt, the ampere, the ohm, and the watt. (There is *no* British system of electric units.) The SI unit of electric charge is called one **coulomb** (1 C). In SI units the constant k in Eq. (21.1) is

$$k = 8.987551787 \times 10^9 \, \text{N} \cdot \text{m}^2/\text{C}^2 \cong 8.988 \times 10^9 \, \text{N} \cdot \text{m}^2/\text{C}^2$$

The value of k is known to such a large number of significant figures because this value is closely related to the speed of light in vacuum. (We will show this in Chapter 32 when we study electromagnetic radiation.) As we discussed in Section 1.3, this speed is *defined* to be exactly $c = 2.99792458 \times 10^8$ m/s. The numerical value of k is defined in terms of c to be precisely

$$k = (10^{-7} \, \text{N} \cdot \text{s}^2/\text{C}^2)c^2$$

You should check this expression to confirm that k has the right units.

In principle we can measure the electric force F between two equal charges q at a measured distance r and use Coulomb's law to determine the charge. Thus we could regard the value of k as an operational definition of the coulomb. For reasons of experimental precision it is better to define the coulomb instead in terms of a unit of electric *current* (charge per unit time), the *ampere,* equal to 1 coulomb per second. We will return to this definition in Chapter 28.

In SI units we usually write the constant k in Eq. (21.1) as $1/4\pi\epsilon_0$, where ϵ_0 ("epsilon-nought" or "epsilon-zero") is called the **electric constant.** This shorthand simplifies many formulas that we will encounter in later chapters. From now on, we will usually write Coulomb's law as

Coulomb's law:
Magnitude of electric ⋯⋯⋯⋯➤ $$F = \frac{1}{4\pi\epsilon_0} \frac{|q_1 q_2|}{r^2}$$ ⋯⋯Values of the two charges
force between two
point charges   ⋯⋯Distance between the two charges
Electric constant ⋯⋯➤   (21.2)

The constants in Eq. (21.2) are approximately

$$\epsilon_0 = 8.854 \times 10^{-12} \, \text{C}^2/\text{N} \cdot \text{m}^2 \quad \text{and} \quad \frac{1}{4\pi\epsilon_0} = k = 8.988 \times 10^9 \, \text{N} \cdot \text{m}^2/\text{C}^2$$

In examples and problems we will often use the approximate value

$$\frac{1}{4\pi\epsilon_0} = 9.0 \times 10^9 \, \text{N} \cdot \text{m}^2/\text{C}^2$$

As we mentioned in Section 21.1, the most fundamental unit of charge is the magnitude of the charge of an electron or a proton, which is denoted by e. The most precise value available as of the writing of this book is

$$e = 1.602176565(35) \times 10^{-19} \, \text{C}$$

One coulomb represents the negative of the total charge of about 6×10^{18} electrons. For comparison, a copper cube 1 cm on a side contains about 2.4×10^{24} electrons. About 10^{19} electrons pass through the glowing filament of a flashlight bulb every second.

In electrostatics problems (problems that involve charges at rest), it's very unusual to encounter charges as large as 1 coulomb. Two 1-C charges separated by 1 m would exert forces on each other of magnitude 9×10^9 N (about 1 million tons)! The total charge of all the electrons in a copper one-cent coin is even greater, about 1.4×10^5 C, which shows that we can't disturb electric neutrality very much without using enormous forces. More typical values of charge range from about a microcoulomb ($1 \ \mu C = 10^{-6}$ C) to about a nanocoulomb ($1 \ nC = 10^{-9}$ C).

EXAMPLE 21.1 ELECTRIC FORCE VERSUS GRAVITATIONAL FORCE

An α particle (the nucleus of a helium atom) has mass $m = 6.64 \times 10^{-27}$ kg and charge $q = +2e = 3.2 \times 10^{-19}$ C. Compare the magnitude of the electric repulsion between two α ("alpha") particles with that of the gravitational attraction between them.

21.11 Our sketch for this problem.

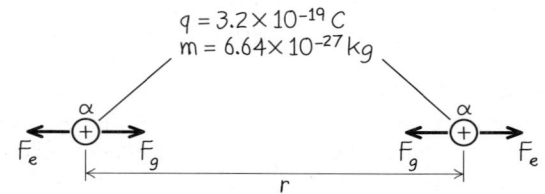

SOLUTION

IDENTIFY and SET UP: This problem involves Newton's law for the gravitational force F_g between particles (see Section 13.1) and Coulomb's law for the electric force F_e between point charges. To compare these forces, we make our target variable the ratio F_e/F_g. We use Eq. (21.2) for F_e and Eq. (13.1) for F_g.

EXECUTE: Figure 21.11 shows our sketch. From Eqs. (21.2) and (13.1),

$$F_e = \frac{1}{4\pi\epsilon_0} \frac{q^2}{r^2} \qquad F_g = G\frac{m^2}{r^2}$$

These are both inverse-square forces, so the r^2 factors cancel when we take the ratio:

$$\frac{F_e}{F_g} = \frac{1}{4\pi\epsilon_0 G} \frac{q^2}{m^2}$$

$$= \frac{9.0 \times 10^9 \ \text{N} \cdot \text{m}^2/\text{C}^2}{6.67 \times 10^{-11} \ \text{N} \cdot \text{m}^2/\text{kg}^2} \frac{(3.2 \times 10^{-19} \ \text{C})^2}{(6.64 \times 10^{-27} \ \text{kg})^2} = 3.1 \times 10^{35}$$

EVALUATE: This astonishingly large number shows that the gravitational force in this situation is completely negligible in comparison to the electric force. This is always true for interactions of atomic and subnuclear particles. But within objects the size of a person or a planet, the positive and negative charges are nearly equal in magnitude, and the net electric force is usually much *smaller* than the gravitational force.

Superposition of Forces

Coulomb's law as we have stated it describes only the interaction of two *point* charges. Experiments show that when two charges exert forces simultaneously on a third charge, the total force acting on that charge is the *vector sum* of the forces that the two charges would exert individually. This important property, called the **principle of superposition of forces,** holds for any number of charges. By using this principle, we can apply Coulomb's law to *any* collection of charges. Two of the examples at the end of this section use the superposition principle.

Strictly speaking, Coulomb's law as we have stated it should be used only for point charges *in a vacuum.* If matter is present in the space between the charges, the net force acting on each charge is altered because charges are induced in the molecules of the intervening material. We will describe this effect later. As a practical matter, though, we can use Coulomb's law unaltered for point charges in air. At normal atmospheric pressure, the presence of air changes the electric force from its vacuum value by only about one part in 2000.

PROBLEM-SOLVING STRATEGY 21.1 **COULOMB'S LAW**

IDENTIFY *the relevant concepts:* Coulomb's law describes the electric force between charged particles.

SET UP *the problem* using the following steps:
1. Sketch the locations of the charged particles and label each particle with its charge.
2. If the charges do not all lie on a single line, set up an *xy*-coordinate system.
3. The problem will ask you to find the electric force on one or more particles. Identify which these are.

EXECUTE *the solution* as follows:
1. For each particle that exerts an electric force on a given particle of interest, use Eq. (21.2) to calculate the magnitude of that force.
2. Using those magnitudes, sketch a free-body diagram showing the electric-force vectors acting on each particle of interest. The force exerted by particle 1 on particle 2 points from particle 2 toward particle 1 if the charges have opposite signs, but points from particle 2 directly away from particle 1 if the charges have the same sign.
3. Use the principle of superposition to calculate the total electric force—a *vector* sum—on each particle of interest. (Review the vector algebra in Sections 1.7 through 1.9. The method of components is often helpful.)

4. Use consistent units; SI units are completely consistent. With $1/4\pi\epsilon_0 = 9.0 \times 10^9 \text{ N} \cdot \text{m}^2/\text{C}^2$, distances must be in meters, charges in coulombs, and forces in newtons.
5. Some examples and problems in this and later chapters involve *continuous* distributions of charge along a line, over a surface, or throughout a volume. In these cases the vector sum in step 3 becomes a vector *integral*. We divide the charge distribution into infinitesimal pieces, use Coulomb's law for each piece, and integrate to find the vector sum. Sometimes this can be done without actual integration.
6. Exploit any symmetries in the charge distribution to simplify your problem solving. For example, two identical charges q exert zero net electric force on a charge Q midway between them, because the forces on Q have equal magnitude and opposite direction.

EVALUATE *your answer:* Check whether your numerical results are reasonable. Confirm that the direction of the net electric force agrees with the principle that charges of the same sign repel and charges of opposite sign attract.

SOLUTION

EXAMPLE 21.2 **FORCE BETWEEN TWO POINT CHARGES**

Two point charges, $q_1 = +25$ nC and $q_2 = -75$ nC, are separated by a distance $r = 3.0$ cm (**Fig. 21.12a**). Find the magnitude and direction of the electric force (a) that q_1 exerts on q_2 and (b) that q_2 exerts on q_1.

SOLUTION

IDENTIFY and SET UP: This problem asks for the electric forces that two charges exert on each other. We use Coulomb's law, Eq. (21.2), to calculate the magnitudes of the forces. The signs of the charges will determine the directions of the forces.

EXECUTE: (a) After converting the units of r to meters and the units of q_1 and q_2 to coulombs, Eq. (21.2) gives us

$$F_{1 \text{ on } 2} = \frac{1}{4\pi\epsilon_0} \frac{|q_1 q_2|}{r^2}$$

$$= (9.0 \times 10^9 \text{ N} \cdot \text{m}^2/\text{C}^2) \frac{|(+25 \times 10^{-9} \text{ C})(-75 \times 10^{-9} \text{ C})|}{(0.030 \text{ m})^2}$$

$$= 0.019 \text{ N}$$

The charges have opposite signs, so the force is attractive (to the left in Fig. 21.12b); that is, the force that acts on q_2 is directed toward q_1 along the line joining the two charges.

21.12 What force does q_1 exert on q_2, and what force does q_2 exert on q_1? Gravitational forces are negligible.

(a) The two charges

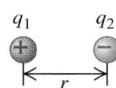
q_1 q_2

(b) Free-body diagram for charge q_2

$\vec{F}_{1 \text{ on } 2}$ q_2

(c) Free-body diagram for charge q_1

q_1 $\vec{F}_{2 \text{ on } 1}$

(b) Proceeding as in part (a), we have

$$F_{2 \text{ on } 1} = \frac{1}{4\pi\epsilon_0} \frac{|q_2 q_1|}{r^2} = F_{1 \text{ on } 2} = 0.019 \text{ N}$$

The attractive force that acts on q_1 is to the right, toward q_2 (Fig. 21.12c).

EVALUATE: Newton's third law applies to the electric force. Even though the charges have different magnitudes, the magnitude of the force that q_2 exerts on q_1 is the same as the magnitude of the force that q_1 exerts on q_2, and these two forces are in opposite directions.

EXAMPLE 21.3   VECTOR ADDITION OF ELECTRIC FORCES ON A LINE

Two point charges are located on the *x*-axis of a coordinate system: $q_1 = 1.0$ nC is at $x = +2.0$ cm, and $q_2 = -3.0$ nC is at $x = +4.0$ cm. What is the total electric force exerted by q_1 and q_2 on a charge $q_3 = 5.0$ nC at $x = 0$?

SOLUTION

IDENTIFY and SET UP: Figure 21.13a shows the situation. To find the total force on q_3, our target variable, we find the vector sum of the two electric forces on it.

EXECUTE: Figure 21.13b is a free-body diagram for q_3, which is repelled by q_1 (which has the same sign) and attracted to q_2 (which has the opposite sign): $\vec{F}_{1 \text{ on } 3}$ is in the $-x$-direction and $\vec{F}_{2 \text{ on } 3}$ is in the $+x$-direction. After unit conversions, we have from Eq. (21.2)

$$F_{1 \text{ on } 3} = \frac{1}{4\pi\epsilon_0} \frac{|q_1 q_3|}{r_{13}^2}$$

$$= (9.0 \times 10^9 \text{ N} \cdot \text{m}^2/\text{C}^2) \frac{(1.0 \times 10^{-9} \text{ C})(5.0 \times 10^{-9} \text{ C})}{(0.020 \text{ m})^2}$$

$$= 1.12 \times 10^{-4} \text{ N} = 112 \ \mu\text{N}$$

21.13 Our sketches for this problem.

(a) Our diagram of the situation

(b) Free-body diagram for q_3

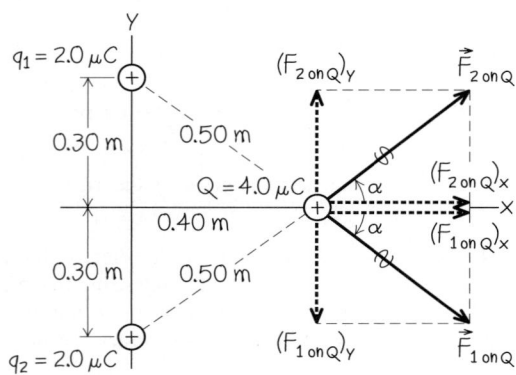

In the same way you can show that $F_{2 \text{ on } 3} = 84 \ \mu\text{N}$. We thus have $\vec{F}_{1 \text{ on } 3} = (-112 \ \mu\text{N})\hat{\imath}$ and $\vec{F}_{2 \text{ on } 3} = (84 \ \mu\text{N})\hat{\imath}$. The net force on q_3 is

$$\vec{F}_3 = \vec{F}_{1 \text{ on } 3} + \vec{F}_{2 \text{ on } 3} = (-112 \ \mu\text{N})\hat{\imath} + (84 \ \mu\text{N})\hat{\imath} = (-28 \ \mu\text{N})\hat{\imath}$$

EVALUATE: As a check, note that the magnitude of q_2 is three times that of q_1, but q_2 is twice as far from q_3 as q_1. Equation (21.2) then says that $F_{2 \text{ on } 3}$ must be $3/2^2 = 3/4 = 0.75$ as large as $F_{1 \text{ on } 3}$. This agrees with our calculated values: $F_{2 \text{ on } 3}/F_{1 \text{ on } 3} = (84 \ \mu\text{N})/(112 \ \mu\text{N}) = 0.75$. Because $F_{2 \text{ on } 3}$ is the weaker force, the direction of the net force is that of $\vec{F}_{1 \text{ on } 3}$—that is, in the negative *x*-direction.

EXAMPLE 21.4   VECTOR ADDITION OF ELECTRIC FORCES IN A PLANE

Two equal positive charges $q_1 = q_2 = 2.0 \ \mu\text{C}$ are located at $x = 0, y = 0.30$ m and $x = 0, y = -0.30$ m, respectively. What are the magnitude and direction of the total electric force that q_1 and q_2 exert on a third charge $Q = 4.0 \ \mu\text{C}$ at $x = 0.40$ m, $y = 0$?

SOLUTION

IDENTIFY and SET UP: As in Example 21.3, we must compute the force that each charge exerts on Q and then find the vector sum of those forces. **Figure 21.14** shows the situation. Since the three charges do not all lie on a line, the best way to calculate the forces is to use components.

21.14 Our sketch for this problem.

EXECUTE: Figure 21.14 shows the forces $\vec{F}_{1 \text{ on } Q}$ and $\vec{F}_{2 \text{ on } Q}$ due to the identical charges q_1 and q_2, which are at equal distances from Q. From Coulomb's law, *both* forces have magnitude

$$F_{1 \text{ or } 2 \text{ on } Q} = (9.0 \times 10^9 \text{ N} \cdot \text{m}^2/\text{C}^2)$$

$$\times \frac{(4.0 \times 10^{-6} \text{ C})(2.0 \times 10^{-6} \text{ C})}{(0.50 \text{ m})^2} = 0.29 \text{ N}$$

The *x*-components of the two forces are equal:

$$(F_{1 \text{ or } 2 \text{ on } Q})_x = (F_{1 \text{ or } 2 \text{ on } Q})\cos\alpha = (0.29 \text{ N})\frac{0.40 \text{ m}}{0.50 \text{ m}} = 0.23 \text{ N}$$

From symmetry we see that the *y*-components of the two forces are equal and opposite. Hence their sum is zero and the total force $\vec{F}$ on Q has only an *x*-component $F_x = 0.23$ N $+ 0.23$ N $= 0.46$ N. The total force on Q is in the $+x$-direction, with magnitude 0.46 N.

EVALUATE: The total force on Q points neither directly away from q_1 nor directly away from q_2. Rather, this direction is a compromise that points away from the *system* of charges q_1 and q_2. Can you see that the total force would *not* be in the $+x$-direction if q_1 and q_2 were not equal or if the geometrical arrangement of the changes were not so symmetric?

TEST YOUR UNDERSTANDING OF SECTION 21.3   Suppose that charge q_2 in Example 21.4 were $-2.0 \; \mu C$. In this case, the total electric force on Q would be (i) in the positive x-direction; (ii) in the negative x-direction; (iii) in the positive y-direction; (iv) in the negative y-direction; (v) zero; (vi) none of these. ∎

21.4 ELECTRIC FIELD AND ELECTRIC FORCES

When two electrically charged particles in empty space interact, how does each one know the other is there? We can begin to answer this question, and at the same time reformulate Coulomb's law in a very useful way, by using the concept of *electric field*.

Electric Field

To introduce this concept, let's look at the mutual repulsion of two positively charged bodies A and B (**Fig. 21.15a**). Suppose B has charge q_0, and let $\vec{F}_0$ be the electric force of A on B. One way to think about this force is as an "action-at-a-distance" force—that is, as a force that acts across empty space without needing physical contact between A and B. (Gravity can also be thought of as an "action-at-a-distance" force.) But a more fruitful way to visualize the repulsion between A and B is as a two-stage process. We first envision that body A, as a result of the charge that it carries, somehow *modifies the properties of the space around it*. Then body B, as a result of the charge that *it* carries, senses how space has been modified at its position. The response of body B is to experience the force $\vec{F}_0$.

To clarify how this two-stage process occurs, we first consider body A by itself: We remove body B and label its former position as point P (Fig. 21.15b). We say that the charged body A produces or causes an **electric field** at point P (and at all other points in the neighborhood). This electric field is present at P even if there is no charge at P; it is a consequence of the charge on body A only. If a point charge q_0 is then placed at point P, it experiences the force $\vec{F}_0$. We take the point of view that this force is exerted on q_0 *by the field* at P (Fig. 21.15c). Thus the electric field is the intermediary through which A communicates its presence to q_0. Because the point charge q_0 would experience a force at *any* point in the neighborhood of A, the electric field that A produces exists at all points in the region around A.

We can likewise say that the point charge q_0 produces an electric field in the space around it and that this electric field exerts the force $-\vec{F}_0$ on body A. For each force (the force of A on q_0 and the force of q_0 on A), one charge sets up an electric field that exerts a force on the second charge. We emphasize that this is an *interaction* between *two* charged bodies. A single charge produces an electric field in the surrounding space, but this electric field cannot exert a net force on the charge that created it; as we discussed in Section 4.3, a body cannot exert a net force on itself. (If this wasn't true, you would be able to lift yourself to the ceiling by pulling up on your belt!)

> The electric force on a charged body is exerted by the electric field created by *other* charged bodies.

To find out experimentally whether there is an electric field at a particular point, we place a small charged body, which we call a **test charge,** at the point (Fig. 21.15c). If the test charge experiences an electric force, then there is an electric field at that point. This field is produced by charges other than q_0.

Force is a vector quantity, so electric field is also a vector quantity. (Note the use of vector signs as well as boldface letters and plus, minus, and equals signs in the following discussion.) We define the *electric field* $\vec{E}$ at a point as the electric

21.15 A charged body creates an electric field in the space around it.

(a) *A* and *B* exert electric forces on each other.

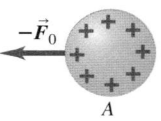
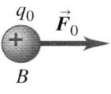

(b) Remove body *B* ...

... and label its former position as *P*.

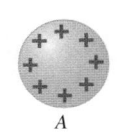
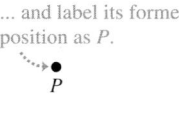

(c) Body *A* sets up an electric field $\vec{E}$ at point *P*.

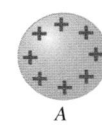
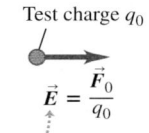

Test charge q_0

$$\vec{E} = \frac{\vec{F}_0}{q_0}$$

$\vec{E}$ is the force per unit charge exerted by A on a test charge at P.

BIO Application Sharks and the "Sixth Sense" Sharks have the ability to locate prey (such as flounder and other bottom-dwelling fish) that are completely hidden beneath the sand at the bottom of the ocean. They do this by sensing the weak electric fields produced by muscle contractions in their prey. Sharks derive their sensitivity to electric fields (a "sixth sense") from jelly-filled canals in their bodies. These canals end in pores on the shark's skin (shown in this photograph). An electric field as weak as 5×10^{-7} N/C causes charge flow within the canals and triggers a signal in the shark's nervous system. Because the shark has canals with different orientations, it can measure different components of the electric-field vector and hence determine the direction of the field.

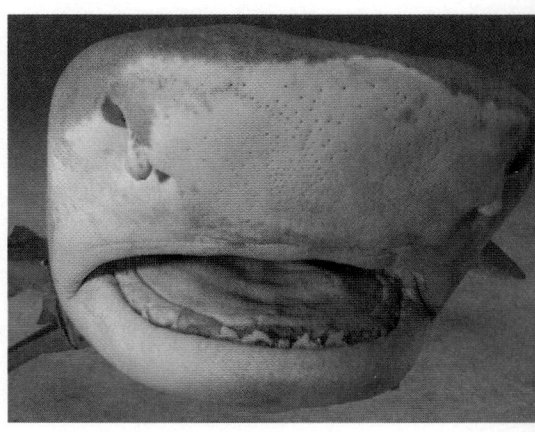

force $\vec{F}_0$ experienced by a test charge q_0 at the point, divided by the charge q_0. That is, the electric field at a certain point is equal to the *electric force per unit charge* experienced by a charge at that point:

$$
\underbrace{\text{Electric field} =}_{\substack{\text{electric force} \\ \text{per unit charge}}}\quad \vec{E} = \frac{\overbrace{\vec{F}_0}^{\substack{\text{Electric force} \\ \text{on a test charge } q_0 \\ \text{due to other charges}}}}{q_0 \underset{\text{Value of test charge}}{}} \tag{21.3}
$$

In SI units, in which the unit of force is 1 N and the unit of charge is 1 C, the unit of electric-field magnitude is 1 newton per coulomb (1 N/C).

If the field $\vec{E}$ at a certain point is known, rearranging Eq. (21.3) gives the force $\vec{F}_0$ experienced by a point charge q_0 placed at that point. This force is just equal to the electric field $\vec{E}$ produced at that point by charges other than q_0, multiplied by the charge q_0:

$$
\vec{F}_0 = q_0\vec{E} \qquad \begin{array}{l}\text{(force exerted on a point charge } q_0 \\ \text{by an electric field } \vec{E}) \end{array} \tag{21.4}
$$

The charge q_0 can be either positive or negative. If q_0 is *positive,* the force $\vec{F}_0$ experienced by the charge is in the same direction as $\vec{E}$; if q_0 is *negative,* $\vec{F}_0$ and $\vec{E}$ are in opposite directions (**Fig. 21.16**).

While the electric field concept may be new to you, the basic idea—that one body sets up a field in the space around it and a second body responds to that field—is one that you've actually used before. Compare Eq. (21.4) to the familiar expression for the gravitational force $\vec{F}_g$ that the earth exerts on a mass m_0:

$$
\vec{F}_g = m_0\vec{g} \tag{21.5}
$$

In this expression, $\vec{g}$ is the acceleration due to gravity. If we divide both sides of Eq. (21.5) by the mass m_0, we obtain

$$
\vec{g} = \frac{\vec{F}_g}{m_0}
$$

Thus $\vec{g}$ can be regarded as the gravitational force per unit mass. By analogy to Eq. (21.3), we can interpret $\vec{g}$ as the *gravitational field.* Thus we treat the gravitational interaction between the earth and the mass m_0 as a two-stage process: The earth sets up a gravitational field $\vec{g}$ in the space around it, and this gravitational field exerts a force given by Eq. (21.5) on the mass m_0 (which we can regard as a *test mass*). The gravitational field $\vec{g}$, or gravitational force per unit mass, is a useful concept because it does not depend on the mass of the body on which the gravitational force is exerted; likewise, the electric field $\vec{E}$, or electric force per unit charge, is useful because it does not depend on the charge of the body on which the electric force is exerted.

> CAUTION $\vec{F}_0 = q_0\vec{E}$ **is for *point* test charges only** The electric force experienced by a test charge q_0 can vary from point to point, so the electric field can also be different at different points. For this reason, use Eq. (21.4) to find the electric force on a *point* charge only. If a charged body is large enough in size, the electric field $\vec{E}$ may be noticeably different in magnitude and direction at different points on the body, and calculating the net electric force on it can be complicated. ∎

Electric Field of a Point Charge

If the source distribution is a point charge q, it is easy to find the electric field that it produces. We call the location of the charge the **source point,** and we call the point P where we are determining the field the **field point.** It is also useful to introduce a *unit vector* $\hat{r}$ that points along the line from source point

21.16 The force $\vec{F}_0 = q_0\vec{E}$ exerted on a point charge q_0 placed in an electric field $\vec{E}$.

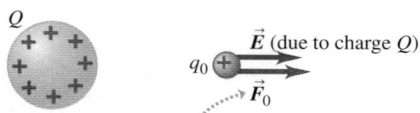

The force on a positive test charge q_0 points in the direction of the electric field.

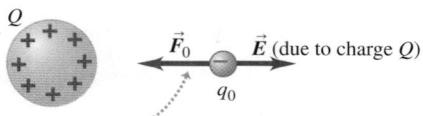

The force on a negative test charge q_0 points opposite to the electric field.

21.17 The electric field $\vec{E}$ produced at point P by an isolated point charge q at S. Note that in both (b) and (c), $\vec{E}$ is *produced* by q [see Eq. (21.7)] but *acts* on the charge q_0 at point P [see Eq. (21.4)].

(a)

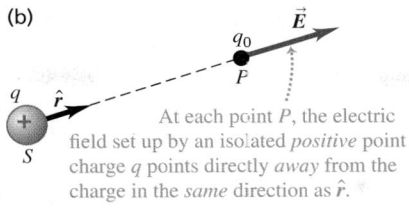

Unit vector $\hat{r}$ points from source point S to field point P.

(b)

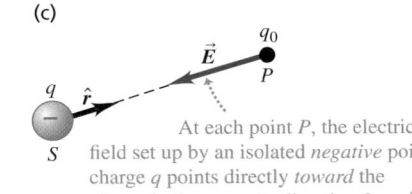

At each point P, the electric field set up by an isolated *positive* point charge q points directly *away* from the charge in the *same* direction as $\hat{r}$.

(c)

At each point P, the electric field set up by an isolated *negative* point charge q points directly *toward* the charge in the *opposite* direction from $\hat{r}$.

to field point (**Fig. 21.17a**). This unit vector is equal to the displacement vector $\vec{r}$ from the source point to the field point, divided by the distance $r = |\vec{r}|$ between these two points; that is, $\hat{r} = \vec{r}/r$. If we place a small test charge q_0 at the field point P, at a distance r from the source point, the magnitude F_0 of the force is given by Coulomb's law, Eq. (21.2):

$$ F_0 = \frac{1}{4\pi\epsilon_0} \frac{|qq_0|}{r^2} $$

From Eq. (21.3) the magnitude E of the electric field at P is

$$ E = \frac{1}{4\pi\epsilon_0} \frac{|q|}{r^2} \qquad \text{(magnitude of electric field of a point charge)} \qquad (21.6) $$

Using the unit vector $\hat{r}$, we can write a *vector* equation that gives both the magnitude and direction of the electric field $\vec{E}$:

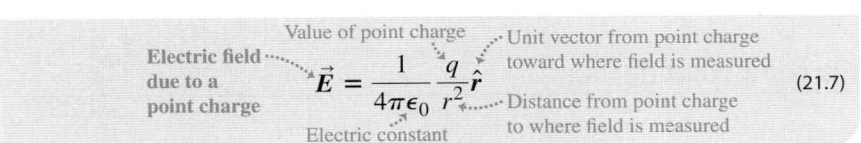

Electric field due to a point charge $\cdots\to \vec{E} = \dfrac{1}{4\pi\epsilon_0}\dfrac{q}{r^2}\hat{r}$ (21.7)

Value of point charge · Unit vector from point charge toward where field is measured · Distance from point charge to where field is measured · Electric constant

By definition, the electric field of a point charge always points *away from* a positive charge (that is, in the same direction as $\hat{r}$; see Fig. 21.17b) but *toward* a negative charge (that is, in the direction opposite $\hat{r}$; see Fig. 21.17c).

We have emphasized calculating the electric field $\vec{E}$ at a certain point. But since $\vec{E}$ can vary from point to point, it is not a single vector quantity but rather an *infinite* set of vector quantities, one associated with each point in space. This is an example of a **vector field. Figure 21.18** shows a number of the field vectors produced by a positive or negative point charge. If we use a rectangular (x, y, z) coordinate system, each component of $\vec{E}$ at any point is in general a function of the coordinates (x, y, z) of the point. We can represent the functions as $E_x(x, y, z)$, $E_y(x, y, z)$, and $E_z(x, y, z)$. Another example of a vector field is the velocity $\vec{v}$ of wind currents; the magnitude and direction of $\vec{v}$, and hence its vector components, vary from point to point in the atmosphere.

In some situations the magnitude and direction of the field (and hence its vector components) have the same values everywhere throughout a certain region; we then say that the field is *uniform* in this region. An important example of this is the electric field inside a *conductor*. If there is an electric field within a conductor, the field exerts a force on every charge in the conductor, giving the free charges a net motion. By definition an electrostatic situation is one in which the charges have *no* net motion. We conclude that *in electrostatics the electric field at every point within the material of a conductor must be zero.* (Note that we are not saying that the field is necessarily zero in a *hole* inside a conductor.)

21.18 A point charge q produces an electric field $\vec{E}$ at *all* points in space. The field strength decreases with increasing distance.

(a) The field produced by a positive point charge points *away from* the charge.

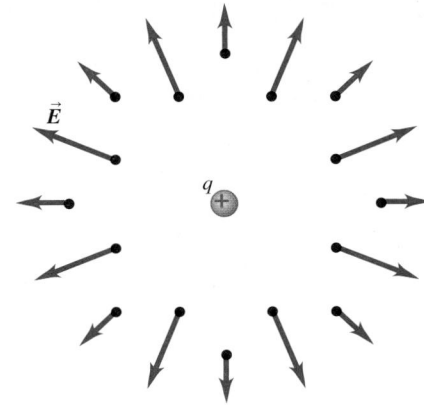

(b) The field produced by a negative point charge points *toward* the charge.

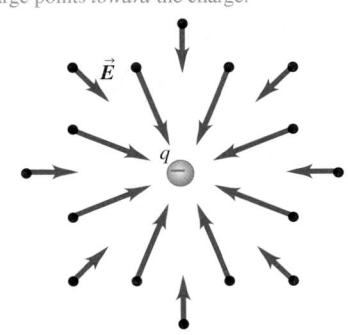

In summary, our description of electric interactions has two parts. First, a given charge distribution acts as a source of electric field. Second, the electric field exerts a force on any charge that is present in the field. Our analysis often has two corresponding steps: first, calculating the field caused by a source charge distribution; second, looking at the effect of the field in terms of force and motion. The second step often involves Newton's laws as well as the principles of electric interactions. In the next section we show how to calculate fields caused by various source distributions, but first here are three examples of calculating the field due to a point charge and of finding the force on a charge due to a given field $\vec{E}$.

EXAMPLE 21.5   ELECTRIC-FIELD MAGNITUDE FOR A POINT CHARGE

What is the magnitude of the electric field $\vec{E}$ at a field point 2.0 m from a point charge $q = 4.0$ nC?

SOLUTION

IDENTIFY and SET UP: This problem concerns the electric field due to a point charge. We are given the magnitude of the charge and the distance from the charge to the field point, so we use Eq. (21.6) to calculate the field magnitude E.

EXECUTE: From Eq. (21.6),

$$E = \frac{1}{4\pi\epsilon_0} \frac{|q|}{r^2} = (9.0 \times 10^9 \text{ N} \cdot \text{m}^2/\text{C}^2) \frac{4.0 \times 10^{-9} \text{ C}}{(2.0 \text{ m})^2}$$

$$= 9.0 \text{ N/C}$$

EVALUATE: Our result $E = 9.0$ N/C means that if we placed a 1.0-C charge at a point 2.0 m from q, it would experience a 9.0-N force. The force on a 2.0-C charge at that point would be $(2.0 \text{ C})(9.0 \text{ N/C}) = 18$ N, and so on.

EXAMPLE 21.6   ELECTRIC-FIELD VECTOR FOR A POINT CHARGE

A point charge $q = -8.0$ nC is located at the origin. Find the electric-field vector at the field point $x = 1.2$ m, $y = -1.6$ m.

SOLUTION

IDENTIFY and SET UP: We must find the electric-field *vector* $\vec{E}$ due to a point charge. **Figure 21.19** shows the situation. We use Eq. (21.7); to do this, we must find the distance r from the source point S (the position of the charge q, which in this example is at the origin O) to the field point P, and we must obtain an expression for the unit vector $\hat{r} = \vec{r}/r$ that points from S to P.

21.19 Our sketch for this problem.

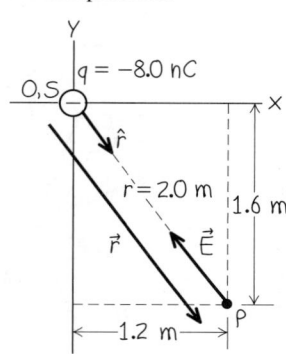

EXECUTE: The distance from S to P is

$$r = \sqrt{x^2 + y^2} = \sqrt{(1.2 \text{ m})^2 + (-1.6 \text{ m})^2} = 2.0 \text{ m}$$

The unit vector $\hat{r}$ is then

$$\hat{r} = \frac{\vec{r}}{r} = \frac{x\hat{i} + y\hat{j}}{r}$$

$$= \frac{(1.2 \text{ m})\hat{i} + (-1.6 \text{ m})\hat{j}}{2.0 \text{ m}} = 0.60\hat{i} - 0.80\hat{j}$$

Then, from Eq. (21.7),

$$\vec{E} = \frac{1}{4\pi\epsilon_0} \frac{q}{r^2} \hat{r}$$

$$= (9.0 \times 10^9 \text{ N} \cdot \text{m}^2/\text{C}^2) \frac{(-8.0 \times 10^{-9} \text{ C})}{(2.0 \text{ m})^2} (0.60\hat{i} - 0.80\hat{j})$$

$$= (-11 \text{ N/C})\hat{i} + (14 \text{ N/C})\hat{j}$$

EVALUATE: Since q is negative, $\vec{E}$ points from the field point to the charge (the source point), in the direction opposite to $\hat{r}$ (compare Fig. 21.17c). We leave the calculation of the magnitude and direction of $\vec{E}$ to you (see Exercise 21.30).

EXAMPLE 21.7 | ELECTRON IN A UNIFORM FIELD

When the terminals of a battery are connected to two parallel conducting plates with a small gap between them, the resulting charges on the plates produce a nearly uniform electric field $\vec{E}$ between the plates. (In the next section we'll see why this is.) If the plates are 1.0 cm apart and are connected to a 100-volt battery as shown in **Fig. 21.20**, the field is vertically upward and has magnitude $E = 1.00 \times 10^4$ N/C. (a) If an electron (charge $-e = -1.60 \times 10^{-9}$ C, mass $m = 9.11 \times 10^{-31}$ kg) is released from rest at the upper plate, what is its acceleration? (b) What speed and kinetic energy does it acquire while traveling 1.0 cm to the lower plate? (c) How long does it take to travel this distance?

SOLUTION

IDENTIFY and SET UP: This example involves the relationship between electric field and electric force. It also involves the relationship between force and acceleration, the definition of kinetic energy, and the kinematic relationships among acceleration, distance, velocity, and time. Figure 21.20 shows our coordinate system. We are given the electric field, so we use Eq. (21.4) to find the force on the electron and Newton's second law to find its acceleration. Because the field is uniform, the force is constant and we can use the constant-acceleration formulas from Chapter 2 to find

the electron's velocity and travel time. We find the kinetic energy from $K = \frac{1}{2}mv^2$.

EXECUTE: (a) Although $\vec{E}$ is upward (in the $+y$-direction), $\vec{F}$ is downward (because the electron's charge is negative) and so F_y is negative. Because F_y is constant, the electron's acceleration is constant:

$$a_y = \frac{F_y}{m} = \frac{-eE}{m} = \frac{(-1.60 \times 10^{-19}\text{ C})(1.00 \times 10^4\text{ N/C})}{9.11 \times 10^{-31}\text{ kg}}$$

$$= -1.76 \times 10^{15}\text{ m/s}^2$$

(b) The electron starts from rest, so its motion is in the y-direction only (the direction of the acceleration). We can find the electron's speed at any position y from the constant-acceleration Eq. (2.13), $v_y^2 = v_{0y}^2 + 2a_y(y - y_0)$. We have $v_{0y} = 0$ and $y_0 = 0$, so at $y = -1.0$ cm $= -1.0 \times 10^{-2}$ m we have

$$|v_y| = \sqrt{2a_y y} = \sqrt{2(-1.76 \times 10^{15}\text{ m/s}^2)(-1.0 \times 10^{-2}\text{ m})}$$

$$= 5.9 \times 10^6\text{ m/s}$$

The velocity is downward, so $v_y = -5.9 \times 10^6$ m/s. The electron's kinetic energy is

$$K = \frac{1}{2}mv^2 = \frac{1}{2}(9.11 \times 10^{-31}\text{ kg})(5.9 \times 10^6\text{ m/s})^2$$

$$= 1.6 \times 10^{-17}\text{ J}$$

(c) From Eq. (2.8) for constant acceleration, $v_y = v_{0y} + a_y t$,

$$t = \frac{v_y - v_{0y}}{a_y} = \frac{(-5.9 \times 10^6\text{ m/s}) - (0\text{ m/s})}{-1.76 \times 10^{15}\text{ m/s}^2}$$

$$= 3.4 \times 10^{-9}\text{ s}$$

EVALUATE: Our results show that in problems concerning subatomic particles such as electrons, many quantities—including acceleration, speed, kinetic energy, and time—will have *very* different values from those typical of everyday objects such as baseballs and automobiles.

21.20 A uniform electric field between two parallel conducting plates connected to a 100-volt battery. (The separation of the plates is exaggerated in this figure relative to the dimensions of the plates.)

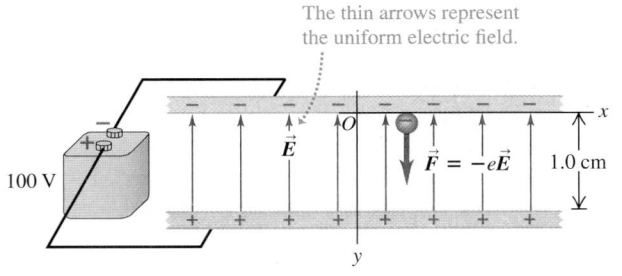

The thin arrows represent the uniform electric field.

$\vec{E}$ O $\vec{F} = -e\vec{E}$ 1.0 cm x

100 V

y

TEST YOUR UNDERSTANDING OF SECTION 21.4 (a) A negative point charge moves along a straight-line path directly toward a stationary positive point charge. Which aspect(s) of the electric force on the negative point charge will remain constant as it moves? (i) Magnitude; (ii) direction; (iii) both magnitude and direction; (iv) neither magnitude nor direction. (b) A negative point charge moves along a circular orbit around a positive point charge. Which aspect(s) of the electric force on the negative point charge will remain constant as it moves? (i) Magnitude; (ii) direction; (iii) both magnitude and direction; (iv) neither magnitude nor direction. ∎

21.5 ELECTRIC-FIELD CALCULATIONS

Equation (21.7) gives the electric field caused by a single point charge. But in most realistic situations that involve electric fields and forces, we encounter charge that is *distributed* over space. The charged plastic and glass rods in Fig. 21.1 have electric charge distributed over their surfaces, as does the imaging drum of a laser printer (Fig. 21.2). In this section we'll learn to calculate electric fields caused by various distributions of electric charge. Calculations of this kind are of tremendous importance for technological applications of electric forces.

DATA *SPEAKS*

To determine the trajectories of atomic nuclei in an accelerator for cancer radiotherapy or of charged particles in a semiconductor electronic device, you have to know the detailed nature of the electric field acting on the charges.

The Superposition of Electric Fields

To find the field caused by a charge distribution, we imagine the distribution to be made up of many point charges $q_1, q_2, q_3, \ldots$. (This is actually quite a realistic description, since we have seen that charge is carried by electrons and protons that are so small as to be almost pointlike.) At any given point P, each point charge produces its own electric field $\vec{E}_1, \vec{E}_2, \vec{E}_3, \ldots$, so a test charge q_0 placed at P experiences a force $\vec{F}_1 = q_0\vec{E}_1$ from charge q_1, a force $\vec{F}_2 = q_0\vec{E}_2$ from charge q_2, and so on. From the principle of superposition of forces discussed in Section 21.3, the *total* force $\vec{F}_0$ that the charge distribution exerts on q_0 is the vector sum of these individual forces:

$$\vec{F}_0 = \vec{F}_1 + \vec{F}_2 + \vec{F}_3 + \cdots = q_0\vec{E}_1 + q_0\vec{E}_2 + q_0\vec{E}_3 + \cdots$$

The combined effect of all the charges in the distribution is described by the *total* electric field $\vec{E}$ at point P. From the definition of electric field, Eq. (21.3), this is

$$\vec{E} = \frac{\vec{F}_0}{q_0} = \vec{E}_1 + \vec{E}_2 + \vec{E}_3 + \cdots$$

The total electric field at P is the vector sum of the fields at P due to each point charge in the charge distribution (**Fig. 21.21**). This statement is the **principle of superposition of electric fields.**

When charge is distributed along a line, over a surface, or through a volume, a few additional terms are useful. For a line charge distribution (such as a long, thin, charged plastic rod), we use λ (the Greek letter lambda) to represent the **linear charge density** (charge per unit length, measured in C/m). When charge is distributed over a surface (such as the surface of the imaging drum of a laser printer), we use σ (sigma) to represent the **surface charge density** (charge per unit area, measured in C/m^2). And when charge is distributed through a volume, we use ρ (rho) to represent the **volume charge density** (charge per unit volume, C/m^3).

Some of the calculations in the following examples may look complex. After you've worked through the examples one step at a time, the process will seem less formidable. We will use many of the calculational techniques in these examples in Chapter 28 to calculate the *magnetic* fields caused by charges in motion.

21.21 Illustrating the principle of superposition of electric fields.

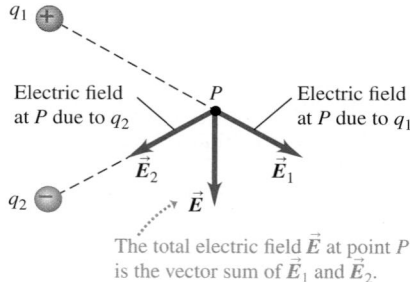

The total electric field $\vec{E}$ at point P is the vector sum of $\vec{E}_1$ and $\vec{E}_2$.

PROBLEM-SOLVING STRATEGY 21.2 | ELECTRIC-FIELD CALCULATIONS

IDENTIFY *the relevant concepts:* Use the principle of superposition to calculate the electric field due to a discrete or continuous charge distribution.

SET UP *the problem* using the following steps:
1. Make a drawing showing the locations of the charges and your choice of coordinate axes.
2. On your drawing, indicate the position of the *field point P* (the point at which you want to calculate the electric field $\vec{E}$).

EXECUTE *the solution* as follows:
1. Use consistent units. Distances must be in meters and charge must be in coulombs. If you are given centimeters or nanocoulombs, don't forget to convert.
2. Distinguish between the source point S and the field point P. The field produced by a point charge always points from S to P if the charge is positive, and from P to S if the charge is negative.

3. Use *vector* addition when applying the principle of superposition; review the treatment of vector addition in Chapter 1 if necessary.
4. Simplify your calculations by exploiting any symmetries in the charge distribution.
5. If the charge distribution is continuous, define a small element of charge that can be considered as a point, find its electric field at P, and find a way to add the fields of all the charge elements by doing an integral. Usually it is easiest to do this for each component of $\vec{E}$ separately, so you may need to evaluate more than one integral. Ensure that the limits on your integrals are correct; especially when the situation has symmetry, don't count a charge twice.

EVALUATE *your answer:* Check that the direction of $\vec{E}$ is reasonable. If your result for the electric-field magnitude E is a function of position (say, the coordinate x), check your result in any limits for which you know what the magnitude should be. When possible, check your answer by calculating it in a different way.

EXAMPLE 21.8 FIELD OF AN ELECTRIC DIPOLE

Point charges $q_1 = +12$ nC and $q_2 = -12$ nC are 0.100 m apart (**Fig. 21.22**). (Such pairs of point charges with equal magnitude and opposite sign are called *electric dipoles*.) Compute the electric field caused by q_1, the field caused by q_2, and the total field (a) at point a; (b) at point b; and (c) at point c.

SOLUTION

IDENTIFY and SET UP: We must find the total electric field at various points due to two point charges. We use the principle of superposition: $\vec{E} = \vec{E}_1 + \vec{E}_2$. Figure 21.22 shows the coordinate system and the locations of the field points a, b, and c.

EXECUTE: At each field point, $\vec{E}$ depends on $\vec{E}_1$ and $\vec{E}_2$ there; we first calculate the magnitudes E_1 and E_2 at each field point. At a the magnitude of the field $\vec{E}_{1a}$ caused by q_1 is

$$E_{1a} = \frac{1}{4\pi\epsilon_0} \frac{|q_1|}{r^2}$$

$$= (9.0 \times 10^9 \text{ N} \cdot \text{m}^2/\text{C}^2) \frac{12 \times 10^{-9} \text{ C}}{(0.060 \text{ m})^2}$$

$$= 3.0 \times 10^4 \text{ N/C}$$

We calculate the other field magnitudes in a similar way. The results are

$$E_{1a} = 3.0 \times 10^4 \text{ N/C}$$

$$E_{1b} = 6.8 \times 10^4 \text{ N/C}$$

$$E_{1c} = 6.39 \times 10^3 \text{ N/C}$$

$$E_{2a} = 6.8 \times 10^4 \text{ N/C}$$

$$E_{2b} = 0.55 \times 10^4 \text{ N/C}$$

$$E_{2c} = E_{1c} = 6.39 \times 10^3 \text{ N/C}$$

The *directions* of the corresponding fields are in all cases *away* from the positive charge q_1 and *toward* the negative charge q_2.

(a) At a, $\vec{E}_{1a}$ and $\vec{E}_{2a}$ are both directed to the right, so

$$\vec{E}_a = E_{1a}\hat{\imath} + E_{2a}\hat{\imath} = (9.8 \times 10^4 \text{ N/C})\hat{\imath}$$

(b) At b, $\vec{E}_{1b}$ is directed to the left and $\vec{E}_{2b}$ is directed to the right, so

$$\vec{E}_b = -E_{1b}\hat{\imath} + E_{2b}\hat{\imath} = (-6.2 \times 10^4 \text{ N/C})\hat{\imath}$$

(c) Figure 21.22 shows the directions of $\vec{E}_1$ and $\vec{E}_2$ at c. Both vectors have the same x-component:

$$E_{1cx} = E_{2cx} = E_{1c}\cos\alpha = (6.39 \times 10^3 \text{ N/C})\left(\tfrac{5}{13}\right)$$

$$= 2.46 \times 10^3 \text{ N/C}$$

21.22 Electric field at three points, a, b, and c, set up by charges q_1 and q_2, which form an electric dipole.

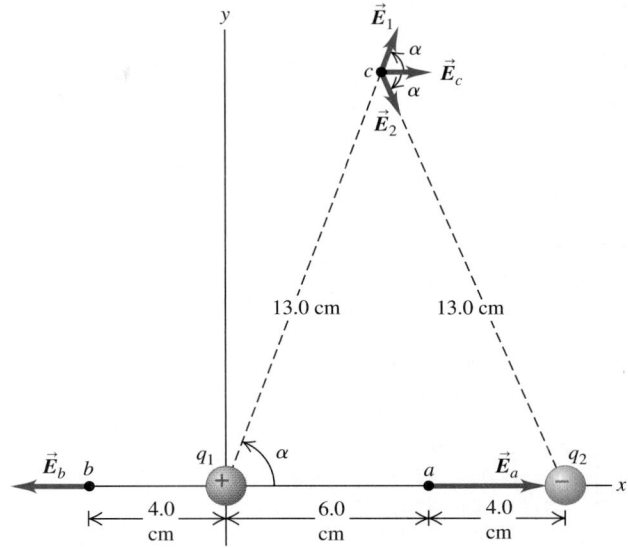

From symmetry, E_{1y} and E_{2y} are equal and opposite, so their sum is zero. Hence

$$\vec{E}_c = 2(2.46 \times 10^3 \text{ N/C})\hat{\imath} = (4.9 \times 10^3 \text{ N/C})\hat{\imath}$$

EVALUATE: We can also find $\vec{E}_c$ by using Eq. (21.7) for the field of a point charge. The displacement vector $\vec{r}_1$ from q_1 to point c is $\vec{r}_1 = r\cos\alpha\,\hat{\imath} + r\sin\alpha\,\hat{\jmath}$. Hence the unit vector that points from q_1 to point c is $\hat{r}_1 = \vec{r}_1/r = \cos\alpha\,\hat{\imath} + \sin\alpha\,\hat{\jmath}$. By symmetry, the unit vector that points from q_2 to point c has the opposite x-component but the same y-component: $\hat{r}_2 = -\cos\alpha\,\hat{\imath} + \sin\alpha\,\hat{\jmath}$. We can now use Eq. (21.7) to write the fields $\vec{E}_{1c}$ and $\vec{E}_{2c}$ at c in vector form, then find their sum. Since $q_2 = -q_1$ and the distance r to c is the same for both charges,

$$\vec{E}_c = \vec{E}_{1c} + \vec{E}_{2c} = \frac{1}{4\pi\epsilon_0}\frac{q_1}{r^2}\hat{r}_1 + \frac{1}{4\pi\epsilon_0}\frac{q_2}{r^2}\hat{r}_2$$

$$= \frac{1}{4\pi\epsilon_0 r^2}(q_1\hat{r}_1 + q_2\hat{r}_2)$$

$$= \frac{q_1}{4\pi\epsilon_0 r^2}(\hat{r}_1 - \hat{r}_2)$$

$$= \frac{1}{4\pi\epsilon_0}\frac{q_1}{r^2}(2\cos\alpha\,\hat{\imath})$$

$$= 2(9.0 \times 10^9 \text{ N}\cdot\text{m}^2/\text{C}^2)\frac{12 \times 10^{-9} \text{ C}}{(0.13 \text{ m})^2}\left(\tfrac{5}{13}\right)\hat{\imath}$$

$$= (4.9 \times 10^3 \text{ N/C})\hat{\imath}$$

This is the same as we calculated in part (c).

EXAMPLE 21.9 FIELD OF A RING OF CHARGE

Charge Q is uniformly distributed around a conducting ring of radius a (**Fig. 21.23**). Find the electric field at a point P on the ring axis at a distance x from its center.

SOLUTION

IDENTIFY and SET UP: This is a problem in the superposition of electric fields. Each bit of charge around the ring produces an electric field at an arbitrary point on the x-axis; our target variable is the total field at this point due to all such bits of charge.

EXECUTE: We divide the ring into infinitesimal segments ds as shown in Fig. 21.23. In terms of the linear charge density $\lambda = Q/2\pi a$, the charge in a segment of length ds is $dQ = \lambda\,ds$. Consider two identical segments, one as shown in the figure at $y = a$ and another halfway around the ring at $y = -a$. From Example 21.4, we see that the net force $d\vec{F}$ they exert on a point test charge at P, and thus their net field $d\vec{E}$, are directed along the x-axis. The same is true for any such pair of segments around the ring, so the *net* field at P is along the x-axis: $\vec{E} = E_x\hat{\imath}$.

To calculate E_x, note that the square of the distance r from a single ring segment to the point P is $r^2 = x^2 + a^2$. Hence the magnitude of this segment's contribution $d\vec{E}$ to the electric field at P is

$$dE = \frac{1}{4\pi\epsilon_0}\frac{dQ}{x^2 + a^2}$$

The x-component of this field is $dE_x = dE\cos\alpha$. We know $dQ = \lambda\,ds$ and Fig. 21.23 shows that $\cos\alpha = x/r = x/(x^2 + a^2)^{1/2}$, so

$$dE_x = dE\cos\alpha = \frac{1}{4\pi\epsilon_0}\frac{dQ}{x^2 + a^2}\frac{x}{\sqrt{x^2 + a^2}}$$

$$= \frac{1}{4\pi\epsilon_0}\frac{\lambda x}{(x^2 + a^2)^{3/2}}ds$$

To find E_x we integrate this expression over the entire ring—that is, for s from 0 to $2\pi a$ (the circumference of the ring). The integrand

21.23 Calculating the electric field on the axis of a ring of charge. In this figure, the charge is assumed to be positive.

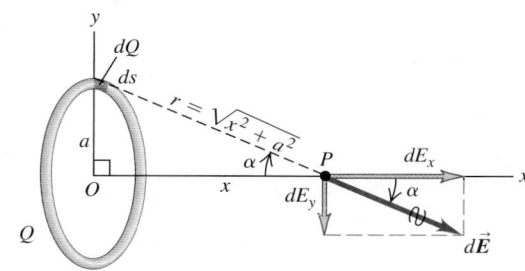

has the same value for all points on the ring, so it can be taken outside the integral. Hence we get

$$E_x = \int dE_x = \frac{1}{4\pi\epsilon_0}\frac{\lambda x}{(x^2 + a^2)^{3/2}}\int_0^{2\pi a} ds$$

$$= \frac{1}{4\pi\epsilon_0}\frac{\lambda x}{(x^2 + a^2)^{3/2}}(2\pi a)$$

$$\vec{E} = E_x\hat{\imath} = \frac{1}{4\pi\epsilon_0}\frac{Qx}{(x^2 + a^2)^{3/2}}\hat{\imath} \qquad (21.8)$$

EVALUATE: Equation (21.8) shows that $\vec{E} = \mathbf{0}$ at the center of the ring ($x = 0$). This makes sense; charges on opposite sides of the ring push in opposite directions on a test charge at the center, and the vector sum of each such pair of forces is zero. When the field point P is much farther from the ring than the ring's radius, we have $x \gg a$ and the denominator in Eq. (21.8) becomes approximately equal to x^3. In this limit the electric field at P is

$$\vec{E} = \frac{1}{4\pi\epsilon_0}\frac{Q}{x^2}\hat{\imath}$$

That is, when the ring is so far away that its radius is negligible in comparison to the distance x, its field is the same as that of a point charge.

EXAMPLE 21.10 FIELD OF A CHARGED LINE SEGMENT

Positive charge Q is distributed uniformly along the y-axis between $y = -a$ and $y = +a$. Find the electric field at point P on the x-axis at a distance x from the origin.

SOLUTION

IDENTIFY and SET UP: Figure 21.24 shows the situation. As in Example 21.9, we must find the electric field due to a continuous distribution of charge. Our target variable is an expression for the electric field at P as a function of x. The x-axis is a perpendicular bisector of the segment, so we can use a symmetry argument.

EXECUTE: We divide the line charge of length $2a$ into infinitesimal segments of length dy. The linear charge density is $\lambda = Q/2a$, and the charge in a segment is $dQ = \lambda\,dy = (Q/2a)dy$. The distance r

from a segment at height y to the field point P is $r = (x^2 + y^2)^{1/2}$, so the magnitude of the field at P due to the segment at height y is

$$dE = \frac{1}{4\pi\epsilon_0}\frac{dQ}{r^2} = \frac{1}{4\pi\epsilon_0}\frac{Q}{2a}\frac{dy}{(x^2 + y^2)}$$

Figure 21.24 shows that the x- and y-components of this field are $dE_x = dE\cos\alpha$ and $dE_y = -dE\sin\alpha$, where $\cos\alpha = x/r$ and $\sin\alpha = y/r$. Hence

$$dE_x = \frac{1}{4\pi\epsilon_0}\frac{Q}{2a}\frac{x\,dy}{(x^2 + y^2)^{3/2}}$$

$$dE_y = -\frac{1}{4\pi\epsilon_0}\frac{Q}{2a}\frac{y\,dy}{(x^2 + y^2)^{3/2}}$$

21.24 Our sketch for this problem.

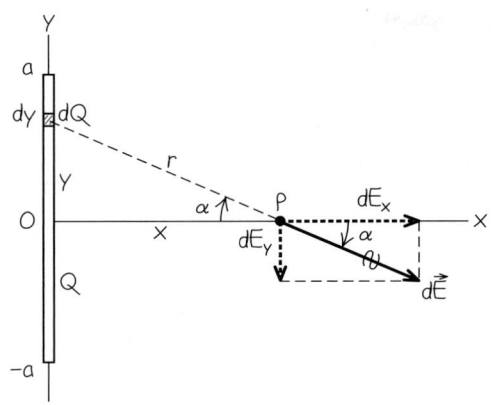

To find the total field at P, we must sum the fields from all segments along the line—that is, we must integrate from $y = -a$ to $y = +a$. You should work out the details of the integration (a table of integrals will help). The results are

$$E_x = \frac{1}{4\pi\epsilon_0}\frac{Q}{2a}\int_{-a}^{+a}\frac{x\,dy}{(x^2+y^2)^{3/2}} = \frac{Q}{4\pi\epsilon_0}\frac{1}{x\sqrt{x^2+a^2}}$$

$$E_y = -\frac{1}{4\pi\epsilon_0}\frac{Q}{2a}\int_{-a}^{+a}\frac{y\,dy}{(x^2+y^2)^{3/2}} = 0$$

or, in vector form,

$$\vec{E} = \frac{1}{4\pi\epsilon_0}\frac{Q}{x\sqrt{x^2+a^2}}\hat{\imath} \qquad (21.9)$$

$\vec{E}$ points away from the line of charge if λ is positive and toward the line of charge if λ is negative.

EVALUATE: Using a symmetry argument as in Example 21.9, we could have guessed that E_y would be zero; if we place a positive test charge at P, the upper half of the line of charge pushes

downward on it, and the lower half pushes up with equal magnitude. Symmetry also tells us that the upper and lower halves of the segment contribute equally to the total field at P.

If the segment is very *short* (or the field point is very far from the segment) so that $x \gg a$, we can ignore a in the denominator of Eq. (21.9). Then the field becomes that of a point charge, just as in Example 21.9:

$$\vec{E} = \frac{1}{4\pi\epsilon_0}\frac{Q}{x^2}\hat{\imath}$$

To see what happens if the segment is very *long* (or the field point is very close to it) so that $a \gg x$, we first rewrite Eq. (21.9) slightly:

$$\vec{E} = \frac{1}{2\pi\epsilon_0}\frac{\lambda}{x\sqrt{(x^2/a^2)+1}}\hat{\imath} \qquad (21.10)$$

In the limit $a \gg x$ we can ignore x^2/a^2 in the denominator of Eq. (21.10), so

$$\vec{E} = \frac{\lambda}{2\pi\epsilon_0 x}\hat{\imath}$$

This is the field of an *infinitely long* line of charge. At any point P at a perpendicular distance r from the line in *any* direction, $\vec{E}$ has magnitude

$$E = \frac{\lambda}{2\pi\epsilon_0 r} \qquad \text{(infinite line of charge)}$$

Note that this field is proportional to $1/r$ rather than to $1/r^2$ as for a point charge.

There's really no such thing in nature as an infinite line of charge. But when the field point is close enough to the line, there's very little difference between the result for an infinite line and the real-life finite case. For example, if the distance r of the field point from the center of the line is 1% of the length of the line, the value of E differs from the infinite-length value by less than 0.02%.

EXAMPLE 21.11 **FIELD OF A UNIFORMLY CHARGED DISK**

A nonconducting disk of radius R has a uniform positive surface charge density σ. Find the electric field at a point along the axis of the disk a distance x from its center. Assume that x is positive.

SOLUTION

IDENTIFY and SET UP: Figure 21.25 shows the situation. We represent the charge distribution as a collection of concentric rings of charge dQ. In Example 21.9 we obtained Eq. (21.8) for the field on the axis of a single uniformly charged ring, so all we need do here is integrate the contributions of our rings.

EXECUTE: A typical ring has charge dQ, inner radius r, and outer radius $r + dr$. Its area is approximately equal to its width dr times its circumference $2\pi r$, or $dA = 2\pi r\,dr$. The charge per unit area is $\sigma = dQ/dA$, so the charge of the ring is $dQ = \sigma\,dA = 2\pi\sigma r\,dr$. We use dQ in place of Q in Eq. (21.8), the expression for the field due to a ring that we found in Example 21.9, and replace the ring

21.25 Our sketch for this problem.

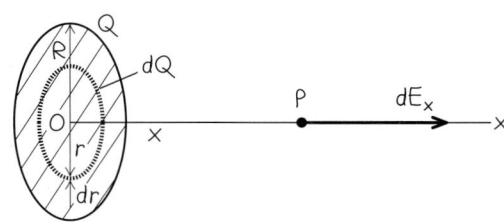

radius a with r. Then the field component dE_x at point P due to this ring is

$$dE_x = \frac{1}{4\pi\epsilon_0}\frac{2\pi\sigma rx\,dr}{(x^2+r^2)^{3/2}}$$

Continued

To find the total field due to all the rings, we integrate dE_x over r from $r = 0$ to $r = R$ (*not* from $-R$ to R):

$$E_x = \int_0^R \frac{1}{4\pi\epsilon_0} \frac{(2\pi\sigma r\, dr)x}{(x^2 + r^2)^{3/2}} = \frac{\sigma x}{4\epsilon_0} \int_0^R \frac{2r\, dr}{(x^2 + r^2)^{3/2}}$$

You can evaluate this integral by making the substitution $t = x^2 + r^2$ (which yields $dt = 2r\, dr$); you can work out the details. The result is

$$E_x = \frac{\sigma x}{2\epsilon_0}\left[-\frac{1}{\sqrt{x^2 + R^2}} + \frac{1}{x}\right]$$

$$= \frac{\sigma}{2\epsilon_0}\left[1 - \frac{1}{\sqrt{(R^2/x^2) + 1}}\right] \qquad (21.11)$$

EVALUATE: If the disk is very large (or if we are very close to it), so that $R \gg x$, the term $1/\sqrt{(R^2/x^2) + 1}$ in Eq. (21.11) is very much less than 1. Then Eq. (21.11) becomes

$$E = \frac{\sigma}{2\epsilon_0} \qquad (21.12)$$

Our final result does not contain the distance x from the plane. Hence the electric field produced by an *infinite* plane sheet of charge is *independent of the distance from the sheet*. The field direction is everywhere perpendicular to the sheet, away from it. There is no such thing as an infinite sheet of charge, but if the dimensions of the sheet are much larger than the distance x of the field point P from the sheet, the field is very nearly given by Eq. (21.12).

If P is to the *left* of the plane ($x < 0$), the result is the same except that the direction of $\vec{E}$ is to the left instead of the right. If the surface charge density is negative, the directions of the fields on both sides of the plane are toward it rather than away from it.

EXAMPLE 21.12 FIELD OF TWO OPPOSITELY CHARGED INFINITE SHEETS

Two infinite plane sheets with uniform surface charge densities $+\sigma$ and $-\sigma$ are placed parallel to each other with separation d (**Fig. 21.26**). Find the electric field between the sheets, above the upper sheet, and below the lower sheet.

SOLUTION

IDENTIFY and SET UP: Equation (21.12) gives the electric field due to a single infinite plane sheet of charge. To find the field due to *two* such sheets, we combine the fields by using the principle of superposition (Fig. 21.26).

EXECUTE: From Eq. (21.12), both $\vec{E}_1$ and $\vec{E}_2$ have the same magnitude at all points, independent of distance from either sheet:

$$E_1 = E_2 = \frac{\sigma}{2\epsilon_0}$$

From Example 21.11, $\vec{E}_1$ is everywhere directed away from sheet 1, and $\vec{E}_2$ is everywhere directed toward sheet 2.

Between the sheets, $\vec{E}_1$ and $\vec{E}_2$ reinforce each other; above the upper sheet and below the lower sheet, they cancel each other. Thus the total field is

$$\vec{E} = \vec{E}_1 + \vec{E}_2 = \begin{cases} \mathbf{0} & \text{above the upper sheet} \\ \dfrac{\sigma}{\epsilon_0}\hat{\jmath} & \text{between the sheets} \\ \mathbf{0} & \text{below the lower sheet} \end{cases}$$

21.26 Finding the electric field due to two oppositely charged infinite sheets. The sheets are seen edge-on; only a portion of the infinite sheets can be shown!

EVALUATE: Because we considered the sheets to be infinite, our result does not depend on the separation d. Our result shows that the field between oppositely charged plates is essentially uniform if the plate separation is much smaller than the dimensions of the plates. We actually used this result in Example 21.7 (Section 21.4).

CAUTION **Electric fields are not "flows"** You may have thought that the field $\vec{E}_1$ of sheet 1 would be unable to "penetrate" sheet 2, and that field $\vec{E}_2$ caused by sheet 2 would be unable to "penetrate" sheet 1. You might conclude this if you think of the electric field as some kind of physical substance that "flows" into or out of charges. But there is no such substance, and the electric fields $\vec{E}_1$ and $\vec{E}_2$ depend on only the individual charge distributions that create them. The *total* field at every point is just the vector sum of $\vec{E}_1$ and $\vec{E}_2$.

TEST YOUR UNDERSTANDING OF SECTION 21.5 Suppose that the line of charge in Fig. 21.24 (Example 21.10) had charge $+Q$ distributed uniformly between $y = 0$ and $y = +a$ and had charge $-Q$ distributed uniformly between $y = 0$ and $y = -a$. In this situation, the electric field at P would be (i) in the positive x-direction; (ii) in the negative x-direction; (iii) in the positive y-direction; (iv) in the negative y-direction; (v) zero; (vi) none of these.

21.6 ELECTRIC FIELD LINES

The concept of an electric field can be a little elusive because you can't see an electric field directly. Electric field *lines* can be a big help for visualizing electric fields and making them seem more real. An **electric field line** is an imaginary line or curve drawn through a region of space so that its tangent at any point is in the direction of the electric-field vector at that point. **Figure 21.27** shows the basic idea. (We used a similar concept in our discussion of fluid flow in Section 12.5. A *streamline* is a line or curve whose tangent at any point is in the direction of the velocity of the fluid at that point. However, the similarity between electric field lines and fluid streamlines is a mathematical one only; there is nothing "flowing" in an electric field.) The English scientist Michael Faraday (1791–1867) first introduced the concept of field lines. He called them "lines of force," but the term "field lines" is preferable.

Electric field lines show the direction of $\vec{E}$ at each point, and their spacing gives a general idea of the *magnitude* of $\vec{E}$ at each point. Where $\vec{E}$ is strong, we draw lines close together; where $\vec{E}$ is weaker, they are farther apart. At any particular point, the electric field has a unique direction, so only one field line can pass through each point of the field. In other words, *field lines never intersect.*

Figure 21.28 shows some of the electric field lines in a plane containing (a) a single positive charge; (b) two equal-magnitude charges, one positive and one negative (a dipole); and (c) two equal positive charges. Such diagrams are called *field maps;* they are cross sections of the actual three-dimensional patterns. The direction of the total electric field at every point in each diagram is along the tangent to the electric field line passing through the point. Arrowheads indicate the direction of the $\vec{E}$-field vector along each field line. The actual field vectors have been drawn at several points in each pattern. Notice that in general, the magnitude of the electric field is different at different points on a given field line; a field line is *not* a curve of constant electric-field magnitude!

Figure 21.28 shows that field lines are directed *away* from positive charges (since close to a positive point charge, $\vec{E}$ points away from the charge) and *toward* negative charges (since close to a negative point charge, $\vec{E}$ points toward the charge). In regions where the field magnitude is large, such as between the positive and negative charges in Fig. 21.28b, the field lines are drawn close together. In regions where the field magnitude is small, such as between the two positive charges in Fig. 21.28c, the lines are widely separated. In a *uniform* field, the field lines are straight, parallel, and uniformly spaced, as in Fig. 21.20.

21.27 The direction of the electric field at any point is tangent to the field line through that point.

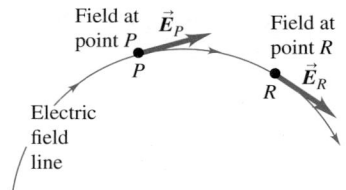

PhET: Charges and Fields
PhET: Electric Field of Dreams
PhET: Electric Field Hockey

21.28 Electric field lines for three different charge distributions. In general, the magnitude of $\vec{E}$ is different at different points along a given field line.

(a) A single positive charge   **(b)** Two equal and opposite charges (a dipole)   **(c)** Two equal positive charges

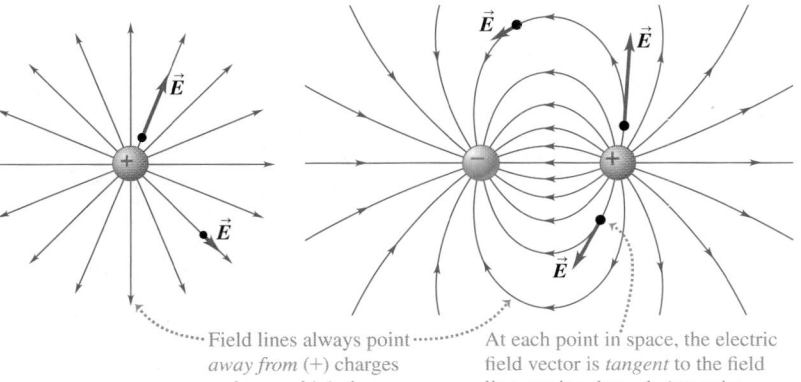
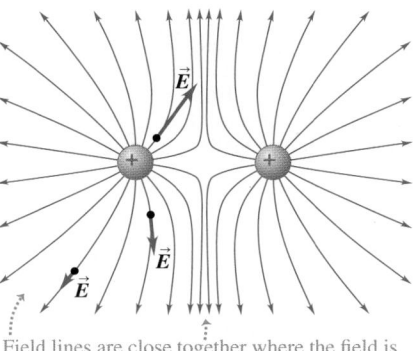

Field lines always point *away from* (+) charges and *toward* (−) charges.   At each point in space, the electric field vector is *tangent* to the field line passing through that point.   Field lines are close together where the field is strong, farther apart where it is weaker.

21.29 (a) Electric field lines produced by two opposite charges. The pattern is formed by grass seeds floating in mineral oil; the charges are on two wires whose tips are inserted into the oil. Compare this pattern with Fig. 21.28b. (b) The electric field causes polarization of the grass seeds, which in turn causes the seeds to align with the field.

(a)

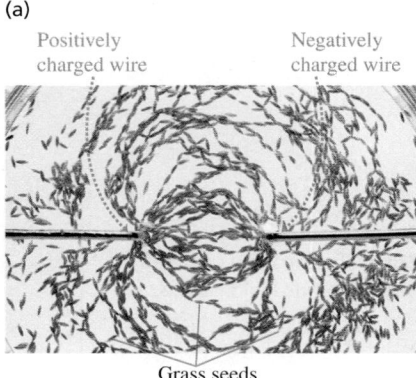

Positively charged wire

Negatively charged wire

Grass seeds

(b)

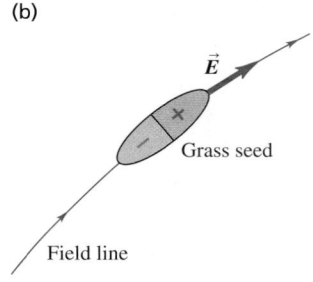

$\vec{E}$

Grass seed

Field line

Figure 21.29a is a view from above of a demonstration setup for visualizing electric field lines. In the arrangement shown here, the tips of two positively charged wires are inserted in a container of insulating liquid, and some grass seeds are floated on the liquid. The grass seeds are electrically neutral insulators, but the electric field of the two charged wires causes *polarization* of the grass seeds; there is a slight shifting of the positive and negative charges within the molecules of each seed, like that shown in Fig. 21.8. The positively charged end of each grass seed is pulled in the direction of $\vec{E}$ and the negatively charged end is pulled opposite $\vec{E}$. Hence the long axis of each grass seed tends to orient parallel to the electric field, in the direction of the field line that passes through the position of the seed (Fig. 21.29b).

_____

CAUTION **Electric field lines are not trajectories** It's a common misconception that if a particle of charge q is in motion where there is an electric field, the particle must move along an electric field line. Because $\vec{E}$ at any point is tangent to the field line that passes through that point, it is true that the *force* $\vec{F} = q\vec{E}$ on the particle, and hence the particle's acceleration, are tangent to the field line. But we learned in Chapter 3 that when a particle moves on a curved path, its acceleration *cannot* be tangent to the path. In general, the trajectory of a charged particle is *not* the same as a field line. ▌

TEST YOUR UNDERSTANDING OF SECTION 21.6 Suppose the electric field lines in a region of space are straight lines. If a charged particle is released from rest in that region, will the trajectory of the particle be along a field line? ▌

21.7 ELECTRIC DIPOLES

An **electric dipole** is a pair of point charges with equal magnitude and opposite sign (a positive charge q and a negative charge $-q$) separated by a distance d. We introduced electric dipoles in Example 21.8 (Section 21.5); the concept is worth exploring further because many physical systems, from molecules to TV antennas, can be described as electric dipoles. We will also use this concept extensively in our discussion of dielectrics in Chapter 24.

Figure 21.30a shows a molecule of water (H_2O), which in many ways behaves like an electric dipole. The water molecule as a whole is electrically neutral, but the chemical bonds within the molecule cause a displacement of charge; the result is a net negative charge on the oxygen end of the molecule and a net positive charge on the hydrogen end, forming an electric dipole. The effect is equivalent to shifting one electron only about 4×10^{-11} m (about the radius of a hydrogen atom), but the consequences of this shift are profound. Water is an excellent solvent for ionic substances such as table salt (sodium chloride, NaCl) precisely because the water molecule is an electric dipole (Fig. 21.30b).

21.30 (a) A water molecule is an example of an electric dipole. (b) Each test tube contains a solution of a different substance in water. The large electric dipole moment of water makes it an excellent solvent.

(a) A water molecule, showing positive charge as red and negative charge as blue

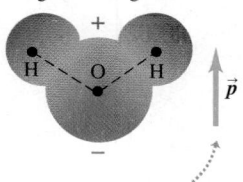

+

H   O   H

$\vec{p}$

−

The electric dipole moment $\vec{p}$ is directed from the negative end to the positive end of the molecule.

(b) Various substances dissolved in water

When dissolved in water, salt dissociates into a positive sodium ion (Na^+) and a negative chlorine ion (Cl^-), which tend to be attracted to the negative and positive ends, respectively, of water molecules; this holds the ions in solution. If water molecules were not electric dipoles, water would be a poor solvent, and almost all of the chemistry that occurs in aqueous solutions would be impossible. This includes all of the biochemical reactions that occur in all of the life on earth. In a very real sense, your existence as a living being depends on electric dipoles!

We examine two questions about electric dipoles. First, what forces and torques does an electric dipole experience when placed in an external electric field (that is, a field set up by charges outside the dipole)? Second, what electric field does an electric dipole itself produce?

Force and Torque on an Electric Dipole

To start with the first question, let's place an electric dipole in a *uniform* external electric field $\vec{E}$, as shown in **Fig. 21.31**. Both forces $\vec{F}_+$ and $\vec{F}_-$ on the two charges have magnitude qE, but their directions are opposite, and they add to zero. *The net force on an electric dipole in a uniform external electric field is zero.*

However, the two forces don't act along the same line, so their *torques* don't add to zero. We calculate torques with respect to the center of the dipole. Let the angle between the electric field $\vec{E}$ and the dipole axis be ϕ; then the lever arm for both $\vec{F}_+$ and $\vec{F}_-$ is $(d/2)\sin\phi$. The torque of $\vec{F}_+$ and the torque of $\vec{F}_-$ both have the same magnitude of $(qE)(d/2)\sin\phi$, and both torques tend to rotate the dipole clockwise (that is, $\vec{\tau}$ is directed into the page in Fig. 21.31). Hence the magnitude of the net torque is twice the magnitude of either individual torque:

$$\tau = (qE)(d\sin\phi) \tag{21.13}$$

where $d\sin\phi$ is the perpendicular distance between the lines of action of the two forces.

The product of the charge q and the separation d is the magnitude of a quantity called the **electric dipole moment,** denoted by p:

$$p = qd \quad \text{(magnitude of electric dipole moment)} \tag{21.14}$$

The units of p are charge times distance ($C \cdot m$). For example, the magnitude of the electric dipole moment of a water molecule is $p = 6.13 \times 10^{-30}\, C \cdot m$.

CAUTION **The symbol p has multiple meanings** Do not confuse dipole moment with momentum or pressure. There aren't as many letters in the alphabet as there are physical quantities, so some letters are used several times. The context usually makes it clear what we mean, but be careful. ∎

We further define the electric dipole moment to be a *vector* quantity $\vec{p}$. The magnitude of $\vec{p}$ is given by Eq. (21.14), and its direction is along the dipole axis from the negative charge to the positive charge as shown in Fig. 21.31.

In terms of p, Eq. (21.13) for the magnitude τ of the torque exerted by the field becomes

> Magnitude of torque ┈┈┈┈┈┈►   ⋯Magnitude of electric field $\vec{E}$
> on an electric dipole $\tau = pE\sin\phi$ ◄┈┈Angle between $\vec{p}$ and $\vec{E}$ (21.15)
>   ⋰Magnitude of electric dipole moment $\vec{p}$

Since the angle ϕ in Fig. 21.31 is the angle between the directions of the vectors $\vec{p}$ and $\vec{E}$, this is reminiscent of the expression for the magnitude of the *vector product*

21.31 The net force on this electric dipole is zero, but there is a torque directed into the page that tends to rotate the dipole clockwise.

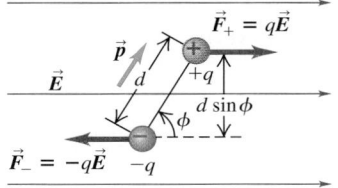

discussed in Section 1.10. (You may want to review that discussion.) Hence we can write the torque on the dipole in vector form as

$$\text{Vector torque on an electric dipole} \qquad \vec{\tau} = \vec{p} \times \vec{E} \qquad\qquad \text{(21.16)}$$

Vector torque on ⋯⋯ an electric dipole — Electric dipole moment ⋯ Electric field

You can use the right-hand rule for the vector product to verify that in the situation shown in Fig. 21.31, $\vec{\tau}$ is directed into the page. The torque is greatest when $\vec{p}$ and $\vec{E}$ are perpendicular and is zero when they are parallel or antiparallel. The torque always tends to turn $\vec{p}$ to line it up with $\vec{E}$. The position $\phi = 0$, with $\vec{p}$ parallel to $\vec{E}$, is a position of stable equilibrium, and the position $\phi = \pi$, with $\vec{p}$ and $\vec{E}$ antiparallel, is a position of unstable equilibrium. The polarization of a grass seed in the apparatus of Fig. 21.29b gives it an electric dipole moment; the torque exerted by $\vec{E}$ then causes the seed to align with $\vec{E}$ and hence with the field lines.

Potential Energy of an Electric Dipole

When a dipole changes direction in an electric field, the electric-field torque does *work* on it, with a corresponding change in potential energy. The work dW done by a torque τ during an infinitesimal displacement $d\phi$ is given by Eq. (10.19): $dW = \tau \, d\phi$. Because the torque is in the direction of decreasing ϕ, we must write the torque as $\tau = -pE\sin\phi$, and

$$dW = \tau \, d\phi = -pE\sin\phi \, d\phi$$

In a finite displacement from ϕ_1 to ϕ_2 the total work done on the dipole is

$$W = \int_{\phi_1}^{\phi_2} (-pE\sin\phi) \, d\phi$$

$$= pE\cos\phi_2 - pE\cos\phi_1$$

The work is the negative of the change of potential energy, just as in Chapter 7: $W = U_1 - U_2$. So a suitable definition of potential energy U for this system is

$$U(\phi) = -pE\cos\phi \qquad\qquad \text{(21.17)}$$

In this expression we recognize the *scalar product* $\vec{p} \cdot \vec{E} = pE\cos\phi$, so we can also write

$$\text{Potential energy for an electric dipole in an electric field} \qquad U = -\vec{p} \cdot \vec{E} \qquad\qquad \text{(21.18)}$$

Potential energy ⋯ for an electric dipole in an electric field — Electric field ⋯ Electric dipole moment

The potential energy has its minimum (most negative) value $U = -pE$ at the stable equilibrium position, where $\phi = 0$ and $\vec{p}$ is parallel to $\vec{E}$. The potential energy is maximum when $\phi = \pi$ and $\vec{p}$ is antiparallel to $\vec{E}$; then $U = +pE$. At $\phi = \pi/2$, where $\vec{p}$ is perpendicular to $\vec{E}$, U is zero. We could define U differently so that it is zero at some other orientation of $\vec{p}$, but our definition is simplest.

Equation (21.18) gives us another way to look at the effect illustrated in Fig. 21.29. The electric field $\vec{E}$ gives each grass seed an electric dipole moment, and the grass seed then aligns itself with $\vec{E}$ to minimize the potential energy.

PhET: Microwaves

EXAMPLE 21.13   FORCE AND TORQUE ON AN ELECTRIC DIPOLE

Figure 21.32a shows an electric dipole in a uniform electric field of magnitude 5.0×10^5 N/C that is directed parallel to the plane of the figure. The charges are $\pm 1.6 \times 10^{-19}$ C; both lie in the plane and are separated by 0.125 nm $= 0.125 \times 10^{-9}$ m. Find (a) the net force exerted by the field on the dipole; (b) the magnitude and direction of the electric dipole moment; (c) the magnitude and direction of the torque; (d) the potential energy of the system in the position shown.

SOLUTION

IDENTIFY and SET UP: This problem uses the ideas of this section about an electric dipole placed in an electric field. We use the relationship $\vec{F} = q\vec{E}$ for each point charge to find the force on the dipole as a whole. Equation (21.14) gives the dipole moment, Eq. (21.16) gives the torque on the dipole, and Eq. (21.18) gives the potential energy of the system.

21.32 (a) An electric dipole. (b) Directions of the electric dipole moment, electric field, and torque ($\vec{\tau}$ points out of the page).

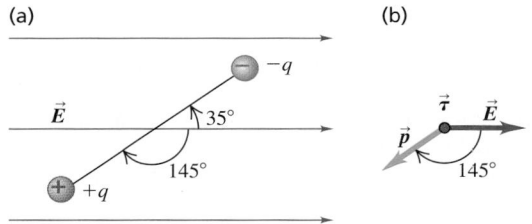

EXECUTE: (a) The field is uniform, so the forces on the two charges are equal and opposite. Hence the total force on the dipole is zero.

(b) The magnitude p of the electric dipole moment $\vec{p}$ is

$$p = qd = (1.6 \times 10^{-19}\,\text{C})(0.125 \times 10^{-9}\,\text{m})$$
$$= 2.0 \times 10^{-29}\,\text{C}\cdot\text{m}$$

The direction of $\vec{p}$ is from the negative to the positive charge, 145° clockwise from the electric-field direction (Fig. 21.32b).

(c) The magnitude of the torque is

$$\tau = pE\sin\phi = (2.0 \times 10^{-29}\,\text{C}\cdot\text{m})(5.0 \times 10^5\,\text{N/C})(\sin 145°)$$
$$= 5.7 \times 10^{-24}\,\text{N}\cdot\text{m}$$

From the right-hand rule for vector products (see Section 1.10), the direction of the torque $\vec{\tau} = \vec{p} \times \vec{E}$ is out of the page. This corresponds to a counterclockwise torque that tends to align $\vec{p}$ with $\vec{E}$.

(d) The potential energy

$$U = -pE\cos\phi$$
$$= -(2.0 \times 10^{-29}\,\text{C}\cdot\text{m})(5.0 \times 10^5\,\text{N/C})(\cos 145°)$$
$$= 8.2 \times 10^{-24}\,\text{J}$$

EVALUATE: The charge magnitude, the distance between the charges, the dipole moment, and the potential energy are all very small, but are all typical of molecules.

In this discussion we have assumed that $\vec{E}$ is uniform, so there is no net force on the dipole. If $\vec{E}$ is not uniform, the forces at the ends may not cancel completely, and the net force may not be zero. Thus a body with zero net charge but an electric dipole moment can experience a net force in a nonuniform electric field. As we mentioned in Section 21.1, an uncharged body can be polarized by an electric field, giving rise to a separation of charge and an electric dipole moment. This is how uncharged bodies can experience electrostatic forces (see Fig. 21.8).

Field of an Electric Dipole

Now let's think of an electric dipole as a *source* of electric field. Figure 21.28b shows the general shape of the field due to a dipole. At each point in the pattern the total $\vec{E}$ field is the vector sum of the fields from the two individual charges, as in Example 21.8 (Section 21.5). Try drawing diagrams showing this vector sum for several points.

To get quantitative information about the field of an electric dipole, we have to do some calculating, as illustrated in the next example. Notice the use of the principle of superposition of electric fields to add up the contributions to the field of the individual charges. Also notice that we need to use approximation techniques even for the relatively simple case of a field due to two charges. Field calculations often become very complicated, and computer analysis is typically used to determine the field due to an arbitrary charge distribution.

BIO Application A Fish with an Electric Dipole Moment Unlike the tiger shark (Section 21.4), which senses the electric fields produced by its prey, the African knifefish *Gymnarchus niloticus*—which is nocturnal and has poor vision—hunts other fish by generating its own electric field. It can make its tail negatively charged relative to its head, producing a dipole-like field similar to that in Fig. 21.28b. When a smaller fish ventures into the field, its body alters the field pattern and alerts *G. niloticus* that a meal is present.

EXAMPLE 21.14 FIELD OF AN ELECTRIC DIPOLE, REVISITED

An electric dipole is centered at the origin, with $\vec{p}$ in the direction of the $+y$-axis (**Fig. 21.33**). Derive an approximate expression for the electric field at a point P on the y-axis for which y is much larger than d. To do this, use the binomial expansion $(1 + x)^n \cong 1 + nx + n(n - 1)x^2/2 + \cdots$ (valid for the case $|x| < 1$).

SOLUTION

IDENTIFY and SET UP: We use the principle of superposition: The total electric field is the vector sum of the field produced by the positive charge and the field produced by the negative charge. At the field point P shown in Fig. 21.33, the field $\vec{E}_+$ of the positive charge has a positive (upward) y-component and the field $\vec{E}_-$

21.33 Finding the electric field of an electric dipole at a point on its axis.

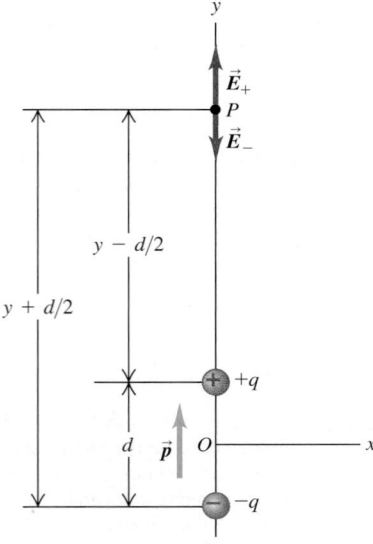

of the negative charge has a negative (downward) y-component. We add these components to find the total field and then apply the approximation that y is much greater than d.

EXECUTE: The total y-component E_y of electric field from the two charges is

$$E_y = \frac{q}{4\pi\epsilon_0}\left[\frac{1}{(y - d/2)^2} - \frac{1}{(y + d/2)^2}\right]$$
$$= \frac{q}{4\pi\epsilon_0 y^2}\left[\left(1 - \frac{d}{2y}\right)^{-2} - \left(1 + \frac{d}{2y}\right)^{-2}\right]$$

We used this same approach in Example 21.8 (Section 21.5). Now the approximation: When we are far from the dipole compared to its size, so $y \gg d$, we have $d/2y \ll 1$. With $n = -2$ and with $d/2y$ replacing x in the binomial expansion, we keep only the first two terms (the terms we discard are much smaller). We then have

$$\left(1 - \frac{d}{2y}\right)^{-2} \cong 1 + \frac{d}{y} \quad \text{and} \quad \left(1 + \frac{d}{2y}\right)^{-2} \cong 1 - \frac{d}{y}$$

Hence E_y is given approximately by

$$E_y \cong \frac{q}{4\pi\epsilon_0 y^2}\left[1 + \frac{d}{y} - \left(1 - \frac{d}{y}\right)\right] = \frac{qd}{2\pi\epsilon_0 y^3} = \frac{p}{2\pi\epsilon_0 y^3}$$

EVALUATE: An alternative route to this result is to put the fractions in the first expression for E_y over a common denominator, add, and then approximate the denominator $(y - d/2)^2(y + d/2)^2$ as y^4. We leave the details to you (see Exercise 21.58).

For points P off the coordinate axes, the expressions are more complicated, but at *all* points far away from the dipole (in any direction) the field drops off as $1/r^3$. We can compare this with the $1/r^2$ behavior of a point charge, the $1/r$ behavior of a long line charge, and the independence of r for a large sheet of charge. There are charge distributions for which the field drops off even more quickly. At large distances, the field of an *electric quadrupole*, which consists of two equal dipoles with opposite orientation, separated by a small distance, drops off as $1/r^4$.

TEST YOUR UNDERSTANDING OF SECTION 21.7 An electric dipole is placed in a region of uniform electric field $\vec{E}$, with the electric dipole moment $\vec{p}$ pointing in the direction opposite to $\vec{E}$. Is the dipole (i) in stable equilibrium, (ii) in unstable equilibrium, or (iii) neither? (*Hint:* You many want to review Section 7.5.) ∎

Electric charge, conductors, and insulators: The fundamental quantity in electrostatics is electric charge. There are two kinds of charge, positive and negative. Charges of the same sign repel each other; charges of opposite sign attract. Charge is conserved; the total charge in an isolated system is constant.

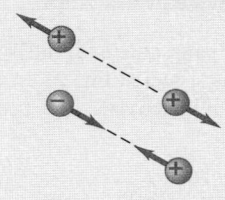

All ordinary matter is made of protons, neutrons, and electrons. The positive protons and electrically neutral neutrons in the nucleus of an atom are bound together by the nuclear force; the negative electrons surround the nucleus at distances much greater than the nuclear size. Electric interactions are chiefly responsible for the structure of atoms, molecules, and solids.

Conductors are materials in which charge moves easily; in insulators, charge does not move easily. Most metals are good conductors; most nonmetals are insulators.

Coulomb's law: For charges q_1 and q_2 separated by a distance r, the magnitude of the electric force on either charge is proportional to the product $q_1 q_2$ and inversely proportional to r^2. The force on each charge is along the line joining the two charges—repulsive if q_1 and q_2 have the same sign, attractive if they have opposite signs. In SI units the unit of electric charge is the coulomb, abbreviated C. (See Examples 21.1 and 21.2.)

When two or more charges each exert a force on a charge, the total force on that charge is the vector sum of the forces exerted by the individual charges. (See Examples 21.3 and 21.4.)

$$F = \frac{1}{4\pi\epsilon_0} \frac{|q_1 q_2|}{r^2} \quad (21.2)$$

$$\frac{1}{4\pi\epsilon_0} = 8.988 \times 10^9 \ \text{N} \cdot \text{m}^2/\text{C}^2$$

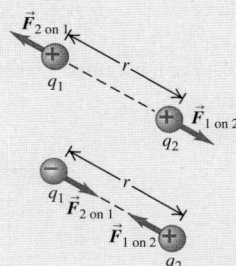

Electric field: Electric field $\vec{E}$, a vector quantity, is the force per unit charge exerted on a test charge at any point. The electric field produced by a point charge is directed radially away from or toward the charge. (See Examples 21.5–21.7.)

$$\vec{E} = \frac{\vec{F}_0}{q_0} \quad (21.3)$$

$$\vec{E} = \frac{1}{4\pi\epsilon_0} \frac{q}{r^2}\hat{r} \quad (21.7)$$

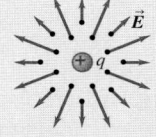

Superposition of electric fields: The electric field $\vec{E}$ of any combination of charges is the vector sum of the fields caused by the individual charges. To calculate the electric field caused by a continuous distribution of charge, divide the distribution into small elements, calculate the field caused by each element, and then carry out the vector sum, usually by integrating. Charge distributions are described by linear charge density λ, surface charge density σ, and volume charge density ρ. (See Examples 21.8–21.12.)

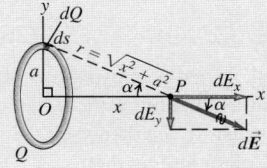

Electric field lines: Field lines provide a graphical representation of electric fields. At any point on a field line, the tangent to the line is in the direction of $\vec{E}$ at that point. The number of lines per unit area (perpendicular to their direction) is proportional to the magnitude of $\vec{E}$ at the point.

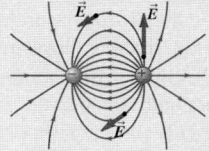

Electric dipoles: An electric dipole is a pair of electric charges of equal magnitude q but opposite sign, separated by a distance d. The electric dipole moment $\vec{p}$ has magnitude $p = qd$. The direction of $\vec{p}$ is from negative toward positive charge. An electric dipole in an electric field $\vec{E}$ experiences a torque $\vec{\tau}$ equal to the vector product of $\vec{p}$ and $\vec{E}$. The magnitude of the torque depends on the angle ϕ between $\vec{p}$ and $\vec{E}$. The potential energy U for an electric dipole in an electric field also depends on the relative orientation of $\vec{p}$ and $\vec{E}$. (See Examples 21.13 and 21.14.)

$$\tau = pE \sin\phi \quad (21.15)$$

$$\vec{\tau} = \vec{p} \times \vec{E} \quad (21.16)$$

$$U = -\vec{p} \cdot \vec{E} \quad (21.18)$$

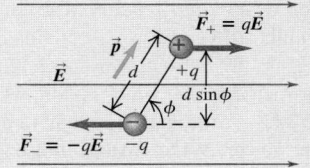

711

BRIDGING PROBLEM CALCULATING ELECTRIC FIELD: HALF A RING OF CHARGE

Positive charge Q is uniformly distributed around a semicircle of radius a as shown in **Fig. 21.34**. Find the magnitude and direction of the resulting electric field at point P, the center of curvature of the semicircle.

SOLUTION GUIDE

IDENTIFY and SET UP

1. The target variables are the components of the electric field at P.
2. Divide the semicircle into infinitesimal segments, each of which is a short circular arc of radius a and angle $d\theta$. What is the length of such a segment? How much charge is on a segment?

21.34 Charge uniformly distributed around a semicircle.

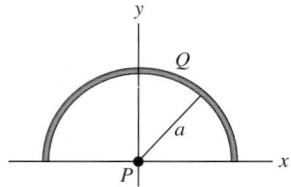

3. Consider an infinitesimal segment located at an angular position θ on the semicircle, measured from the lower right corner of the semicircle at $x = a$, $y = 0$. (Thus $\theta = \pi/2$ at $x = 0$, $y = a$ and $\theta = \pi$ at $x = -a$, $y = 0$.) What are the x- and y-components of the electric field at P (dE_x and dE_y) produced by just this segment?

EXECUTE

4. Integrate your expressions for dE_x and dE_y from $\theta = 0$ to $\theta = \pi$. The results will be the x-component and y-component of the electric field at P.
5. Use your results from step 4 to find the magnitude and direction of the field at P.

EVALUATE

6. Does your result for the electric-field magnitude have the correct units?
7. Explain how you could have found the x-component of the electric field by using a symmetry argument.
8. What would be the electric field at P if the semicircle were extended to a full circle centered at P?

Problems

For assigned homework and other learning materials, go to MasteringPhysics®. (**MP**)

•, ••, •••: Difficulty levels. **CP**: Cumulative problems incorporating material from earlier chapters. **CALC**: Problems requiring calculus.
DATA: Problems involving real data, scientific evidence, experimental design, and/or statistical reasoning. **BIO**: Biosciences problems.

DISCUSSION QUESTIONS

Q21.1 If you peel two strips of transparent tape off the same roll and immediately let them hang near each other, they will repel each other. If you then stick the sticky side of one to the shiny side of the other and rip them apart, they will attract each other. Give a plausible explanation, involving transfer of electrons between the strips of tape, for this sequence of events.

Q21.2 Two metal spheres are hanging from nylon threads. When you bring the spheres close to each other, they tend to attract. Based on this information alone, discuss all the possible ways that the spheres could be charged. Is it possible that after the spheres touch, they will cling together? Explain.

Q21.3 The electric force between two charged particles becomes weaker with increasing distance. Suppose instead that the electric force were *independent* of distance. In this case, would a charged comb still cause a neutral insulator to become polarized as in Fig. 21.8? Why or why not? Would the neutral insulator still be attracted to the comb? Again, why or why not?

Q21.4 Your clothing tends to cling together after going through the dryer. Why? Would you expect more or less clinging if all your clothing were made of the same material (say, cotton) than if you dried different kinds of clothing together? Again, why? (You may want to experiment with your next load of laundry.)

Q21.5 An uncharged metal sphere hangs from a nylon thread. When a positively charged glass rod is brought close to the metal

sphere, the sphere is drawn toward the rod. But if the sphere touches the rod, it suddenly flies away from the rod. Explain why the sphere is first attracted and then repelled.

Q21.6 **BIO** Estimate how many electrons there are in your body. Make any assumptions you feel are necessary, but clearly state what they are. (*Hint:* Most of the atoms in your body have equal numbers of electrons, protons, and neutrons.) What is the combined charge of all these electrons?

Q21.7 Figure Q21.7 shows some of the electric field lines due to three point charges arranged along the vertical axis. All three charges have the same magnitude. (a) What are the signs of the three charges? Explain your reasoning. (b) At what point(s) is the magnitude of the electric field the smallest? Explain your reasoning. Explain how the fields produced by each individual point charge combine to give a small net field at this point or points.

Figure **Q21.7**

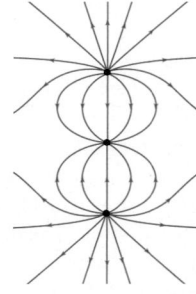

Q21.8 Good conductors of electricity, such as metals, are typically good conductors of heat; insulators, such as wood, are typically poor conductors of heat. Explain why there is a relationship between conduction of electricity and conduction of heat in these materials.

Q21.9 Suppose that the charge shown in Fig. 21.28a is fixed in position. A small, positively charged particle is then placed at some location and released. Will the trajectory of the particle follow an electric field line? Why or why not? Suppose instead that the particle is placed at some point in Fig. 21.28b and released (the positive and negative charges shown are fixed in position). Will its trajectory follow an electric field line? Again, why or why not? Explain any differences between your answers for the two situations.

Q21.10 Two identical metal objects are mounted on insulating stands. Describe how you could place charges of opposite sign but exactly equal magnitude on the two objects.

Q21.11 Because the charges on the electron and proton have the same absolute value, atoms are electrically neutral. Suppose that this is not precisely true, and the absolute value of the charge of the electron is less than the charge of the proton by 0.00100%. Estimate what the net charge of this textbook would be under these circumstances. Make any assumptions you feel are justified, but state clearly what they are. (*Hint:* Most of the atoms in this textbook have equal numbers of electrons, protons, and neutrons.) What would be the magnitude of the electric force between two textbooks placed 5.0 m apart? Would this force be attractive or repulsive? Discuss how the fact that ordinary matter is stable shows that the absolute values of the charges on the electron and proton must be identical to a *very* high level of accuracy.

Q21.12 If you walk across a nylon rug and then touch a large metal object such as a doorknob, you may get a spark and a shock. Why does this tend to happen more on dry days than on humid days? (*Hint:* See Fig. 21.30.) Why are you less likely to get a shock if you touch a *small* metal object, such as a paper clip?

Q21.13 You have a negatively charged object. How can you use it to place a net negative charge on an insulated metal sphere? To place a net positive charge on the sphere?

Q21.14 When two point charges of equal mass and charge are released on a frictionless table, each has an initial acceleration (magnitude) a_0. If instead you keep one fixed and release the other one, what will be its initial acceleration: a_0, $2a_0$, or $a_0/2$? Explain.

Q21.15 A point charge of mass m and charge Q and another point charge of mass m but charge $2Q$ are released on a frictionless table. If the charge Q has an initial acceleration a_0, what will be the acceleration of $2Q$: a_0, $2a_0$, $4a_0$, $a_0/2$, or $a_0/4$? Explain.

Q21.16 A proton is placed in a uniform electric field and then released. Then an electron is placed at this same point and released. Do these two particles experience the same force? The same acceleration? Do they move in the same direction when released?

Q21.17 In Example 21.1 (Section 21.3) we saw that the electric force between two α particles is of the order of 10^{35} times as strong as the gravitational force. So why do we readily feel the gravity of the earth but no electric force from it?

Q21.18 What similarities do electric forces have with gravitational forces? What are the most significant differences?

Q21.19 Two irregular objects A and B carry charges of opposite sign. **Figure Q21.19** shows the electric field lines near each of these objects. (a) Which object is positive, A or B? How do you know? (b) Where is the electric field stronger, close to A or close to B? How do you know?

Q21.20 Atomic nuclei are made of protons and neutrons. This shows that there must be another kind of interaction in addition to gravitational and electric forces. Explain.

Q21.21 Sufficiently strong electric fields can cause atoms to become positively ionized—that is, to lose one or more electrons. Explain how this can happen. What determines how strong the field must be to make this happen?

Q21.22 The electric fields at point P due to the positive charges q_1 and q_2 are shown in **Fig. Q21.22**. Does the fact that they cross each other violate the statement in Section 21.6 that electric field lines never cross? Explain.

Figure **Q21.22**

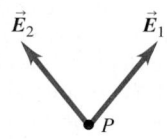

Q21.23 The air temperature and the velocity of the air have different values at different places in the earth's atmosphere. Is the air velocity a vector field? Why or why not? Is the air temperature a vector field? Again, why or why not?

EXERCISES

Section 21.3 Coulomb's Law

21.1 •• Excess electrons are placed on a small lead sphere with mass 8.00 g so that its net charge is -3.20×10^{-9} C. (a) Find the number of excess electrons on the sphere. (b) How many excess electrons are there per lead atom? The atomic number of lead is 82, and its atomic mass is 207 g/mol.

21.2 • Lightning occurs when there is a flow of electric charge (principally electrons) between the ground and a thundercloud. The maximum rate of charge flow in a lightning bolt is about 20,000 C/s; this lasts for 100 μs or less. How much charge flows between the ground and the cloud in this time? How many electrons flow during this time?

21.3 •• If a proton and an electron are released when they are 2.0×10^{-10} m apart (a typical atomic distance), find the initial acceleration of each particle.

21.4 • **Particles in a Gold Ring.** You have a pure (24-karat) gold ring of mass 10.8 g. Gold has an atomic mass of 197 g/mol and an atomic number of 79. (a) How many protons are in the ring, and what is their total positive charge? (b) If the ring carries no net charge, how many electrons are in it?

21.5 • **BIO Signal Propagation in Neurons.** *Neurons* are components of the nervous system of the body that transmit signals as electric impulses travel along their length. These impulses propagate when charge suddenly rushes into and then out of a part of the neuron called an *axon*. Measurements have shown that, during the inflow part of this cycle, approximately 5.6×10^{11} Na$^+$ (sodium ions) per meter, each with charge $+e$, enter the axon. How many coulombs of charge enter a 1.5-cm length of the axon during this process?

21.6 • Two small spheres spaced 20.0 cm apart have equal charge. How many excess electrons must be present on each sphere if the magnitude of the force of repulsion between them is 3.33×10^{-21} N?

21.7 •• An average human weighs about 650 N. If each of two average humans could carry 1.0 C of excess charge, one positive and one negative, how far apart would they have to be for the electric attraction between them to equal their 650-N weight?

Figure **Q21.19**

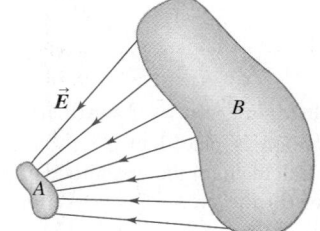

21.8 •• Two small aluminum spheres, each having mass 0.0250 kg, are separated by 80.0 cm. (a) How many electrons does each sphere contain? (The atomic mass of aluminum is 26.982 g/mol, and its atomic number is 13.) (b) How many electrons would have to be removed from one sphere and added to the other to cause an attractive force between the spheres of magnitude 1.00×10^4 N (roughly 1 ton)? Assume that the spheres may be treated as point charges. (c) What fraction of all the electrons in each sphere does this represent?

21.9 •• Two small plastic spheres are given positive electric charges. When they are 15.0 cm apart, the repulsive force between them has magnitude 0.220 N. What is the charge on each sphere (a) if the two charges are equal and (b) if one sphere has four times the charge of the other?

21.10 •• **Just How Strong Is the Electric Force?** Suppose you had two small boxes, each containing 1.0 g of protons. (a) If one were placed on the moon by an astronaut and the other were left on the earth, and if they were connected by a very light (and very long!) string, what would be the tension in the string? Express your answer in newtons and in pounds. Do you need to take into account the gravitational forces of the earth and moon on the protons? Why? (b) What gravitational force would each box of protons exert on the other box?

21.11 • In an experiment in space, one proton is held fixed and another proton is released from rest a distance of 2.50 mm away. (a) What is the initial acceleration of the proton after it is released? (b) Sketch qualitative (no numbers!) acceleration–time and velocity–time graphs of the released proton's motion.

21.12 • A negative charge of $-0.550\ \mu$C exerts an upward 0.600-N force on an unknown charge that is located 0.300 m directly below the first charge. What are (a) the value of the unknown charge (magnitude and sign); (b) the magnitude and direction of the force that the unknown charge exerts on the -0.550-μC charge?

21.13 •• Three point charges are arranged on a line. Charge $q_3 = +5.00$ nC and is at the origin. Charge $q_2 = -3.00$ nC and is at $x = +4.00$ cm. Charge q_1 is at $x = +2.00$ cm. What is q_1 (magnitude and sign) if the net force on q_3 is zero?

21.14 •• In Example 21.4, suppose the point charge on the y-axis at $y = -0.30$ m has negative charge $-2.0\ \mu$C, and the other charges remain the same. Find the magnitude and direction of the net force on Q. How does your answer differ from that in Example 21.4? Explain the differences.

21.15 •• In Example 21.3, calculate the net force on charge q_1.

21.16 •• In Example 21.4, what is the net force (magnitude and direction) on charge q_1 exerted by the other two charges?

21.17 •• Three point charges are arranged along the x-axis. Charge $q_1 = +3.00\ \mu$C is at the origin, and charge $q_2 = -5.00\ \mu$C is at $x = 0.200$ m. Charge $q_3 = -8.00\ \mu$C. Where is q_3 located if the net force on q_1 is 7.00 N in the $-x$-direction?

21.18 •• Repeat Exercise 21.17 for $q_3 = +8.00\ \mu$C.

21.19 •• Two point charges are located on the y-axis as follows: charge $q_1 = -1.50$ nC at $y = -0.600$ m, and charge $q_2 = +3.20$ nC at the origin $(y = 0)$. What is the total force (magnitude and direction) exerted by these two charges on a third charge $q_3 = +5.00$ nC located at $y = -0.400$ m?

21.20 •• Two point charges are placed on the x-axis as follows: Charge $q_1 = +4.00$ nC is located at $x = 0.200$ m, and charge $q_2 = +5.00$ nC is at $x = -0.300$ m. What are the magnitude and direction of the total force exerted by these two charges on a negative point charge $q_3 = -6.00$ nC that is placed at the origin?

21.21 •• **BIO** **Base Pairing in DNA, I.** The two sides of the DNA double helix are connected by pairs of bases (adenine, thymine, cytosine, and guanine). Because of the geometric shape of these molecules, adenine bonds with thymine and cytosine bonds with guanine. **Figure E21.21** shows the bonding of thymine and adenine. Each charge shown is $\pm e$, and the H—N distance is 0.110 nm. (a) Calculate the *net* force that thymine exerts on adenine. Is it attractive or repulsive? To keep the calculations fairly simple, yet reasonable, consider only the forces due to the O—H—N and the N—H—N combinations, assuming that these two combinations are parallel to each other. Remember, however, that in the O—H—N set, the O⁻ exerts a force on both the H⁺ and the N⁻, and likewise along the N—H—N set. (b) Calculate the force on the electron in the hydrogen atom, which is 0.0529 nm from the proton. Then compare the strength of the bonding force of the electron in hydrogen with the bonding force of the adenine–thymine molecules.

Figure **E21.21**

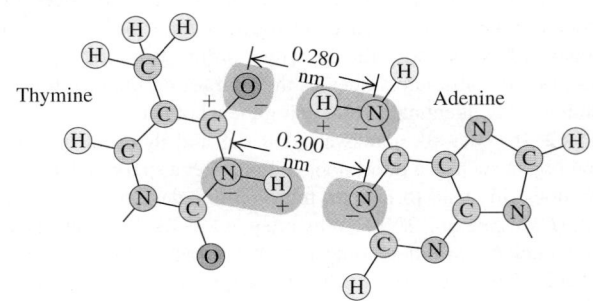

21.22 •• **BIO** **Base Pairing in DNA, II.** Refer to Exercise 21.21. **Figure E21.22** shows the bonding of cytosine and guanine. The O—H and H—N distances are each 0.110 nm. In this case, assume that the bonding is due only to the forces along the O—H—O, N—H—N, and O—H—N combinations, and assume also that these three combinations are parallel to each other. Calculate the *net* force that cytosine exerts on guanine due to the preceding three combinations. Is this force attractive or repulsive?

Figure **E21.22**

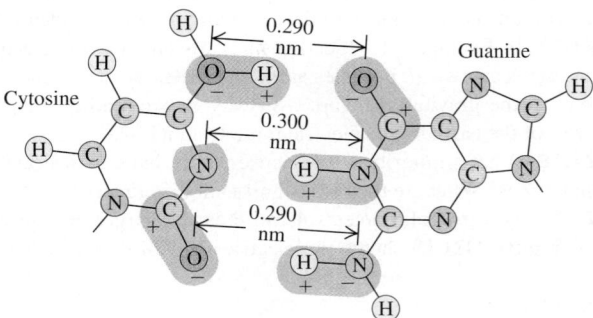

Section 21.4 Electric Field and Electric Forces

21.23 • **CP** A proton is placed in a uniform electric field of 2.75×10^3 N/C. Calculate (a) the magnitude of the electric force felt by the proton; (b) the proton's acceleration; (c) the proton's speed after 1.00 μs in the field, assuming it starts from rest.

21.24 • A particle has charge -5.00 nC. (a) Find the magnitude and direction of the electric field due to this particle at a point 0.250 m directly above it. (b) At what distance from this particle does its electric field have a magnitude of 12.0 N/C?

21.25 • CP A proton is traveling horizontally to the right at 4.50×10^6 m/s. (a) Find the magnitude and direction of the weakest electric field that can bring the proton uniformly to rest over a distance of 3.20 cm. (b) How much time does it take the proton to stop after entering the field? (c) What minimum field (magnitude and direction) would be needed to stop an electron under the conditions of part (a)?

21.26 • CP An electron is released from rest in a uniform electric field. The electron accelerates vertically upward, traveling 4.50 m in the first 3.00 μs after it is released. (a) What are the magnitude and direction of the electric field? (b) Are we justified in ignoring the effects of gravity? Justify your answer quantitatively.

21.27 •• (a) What must the charge (sign and magnitude) of a 1.45-g particle be for it to remain stationary when placed in a downward-directed electric field of magnitude 650 N/C? (b) What is the magnitude of an electric field in which the electric force on a proton is equal in magnitude to its weight?

21.28 •• **Electric Field of the Earth.** The earth has a net electric charge that causes a field at points near its surface equal to 150 N/C and directed in toward the center of the earth. (a) What magnitude and sign of charge would a 60-kg human have to acquire to overcome his or her weight by the force exerted by the earth's electric field? (b) What would be the force of repulsion between two people each with the charge calculated in part (a) and separated by a distance of 100 m? Is use of the earth's electric field a feasible means of flight? Why or why not?

21.29 •• CP An electron is projected with an initial speed $v_0 = 1.60 \times 10^6$ m/s into the uniform field between two parallel plates (**Fig. E21.29**). Assume that the field between the plates is uniform and directed vertically downward and

Figure **E21.29**

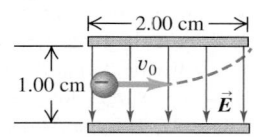

that the field outside the plates is zero. The electron enters the field at a point midway between the plates. (a) If the electron just misses the upper plate as it emerges from the field, find the magnitude of the electric field. (b) Suppose that the electron in Fig. E21.29 is replaced by a proton with the same initial speed v_0. Would the proton hit one of the plates? If not, what would be the magnitude and direction of its vertical displacement as it exits the region between the plates? (c) Compare the paths traveled by the electron and the proton, and explain the differences. (d) Discuss whether it is reasonable to ignore the effects of gravity for each particle.

21.30 • (a) Calculate the magnitude and direction (relative to the $+x$-axis) of the electric field in Example 21.6. (b) A -2.5-nC point charge is placed at point P in Fig. 21.19. Find the magnitude and direction of (i) the force that the -8.0-nC charge at the origin exerts on this charge and (ii) the force that this charge exerts on the -8.0-nC charge at the origin.

21.31 •• CP In Exercise 21.29, what is the speed of the electron as it emerges from the field?

21.32 •• CP A uniform electric field exists in the region between two oppositely charged plane parallel plates. A proton is released from rest at the surface of the positively charged plate and strikes the surface of the opposite plate, 1.60 cm distant from the first, in a time interval of 3.20×10^{-6} s. (a) Find the magnitude of the electric field. (b) Find the speed of the proton when it strikes the negatively charged plate.

21.33 • A point charge is at the origin. With this point charge as the source point, what is the unit vector $\hat{r}$ in the direction of the field point (a) at $x = 0$, $y = -1.35$ m; (b) at $x = 12.0$ cm, $y = 12.0$ cm; (c) at $x = -1.10$ m, $y = 2.60$ m? Express your results in terms of the unit vectors $\hat{i}$ and $\hat{j}$.

21.34 •• A $+8.75$-μC point charge is glued down on a horizontal frictionless table. It is tied to a -6.50-μC point charge by a light, nonconducting 2.50-cm wire. A uniform electric field of magnitude 1.85×10^8 N/C is directed parallel to the wire, as shown in **Fig. E21.34**. (a) Find the tension in the wire. (b) What would the tension be if both charges were negative?

Figure **E21.34**

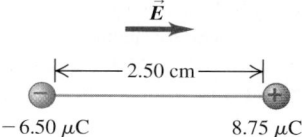

21.35 •• (a) An electron is moving east in a uniform electric field of 1.50 N/C directed to the west. At point A, the velocity of the electron is 4.50×10^5 m/s toward the east. What is the speed of the electron when it reaches point B, 0.375 m east of point A? (b) A proton is moving in the uniform electric field of part (a). At point A, the velocity of the proton is 1.90×10^4 m/s, east. What is the speed of the proton at point B?

Section 21.5 Electric-Field Calculations

21.36 • Two point charges Q and $+q$ (where q is positive) produce the net electric field shown at point P in **Fig. E21.36**. The field points parallel to the line connecting the two charges. (a) What can you conclude about the sign and magnitude of Q? Explain your reasoning. (b) If the lower charge were negative instead, would it be possible for the field to have the direction shown in the figure? Explain your reasoning.

Figure **E21.36**

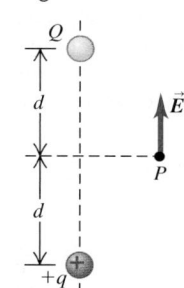

21.37 •• Two positive point charges q are placed on the x-axis, one at $x = a$ and one at $x = -a$. (a) Find the magnitude and direction of the electric field at $x = 0$. (b) Derive an expression for the electric field at points on the x-axis. Use your result to graph the x-component of the electric field as a function of x, for values of x between $-4a$ and $+4a$.

21.38 • The two charges q_1 and q_2 shown in **Fig. E21.38** have equal magnitudes. What is the direction of the net electric field due to these two charges at points A (midway between the charges), B, and C if (a) both charges are negative, (b) both charges are positive, (c) q_1 is positive and q_2 is negative.

Figure **E21.38**

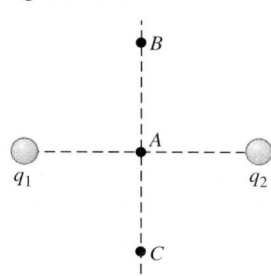

21.39 • A $+2.00$-nC point charge is at the origin, and a second -5.00-nC point charge is on the x-axis at $x = 0.800$ m. (a) Find the electric field (magnitude and direction) at each of the following points on the x-axis: (i) $x = 0.200$ m; (ii) $x = 1.20$ m; (iii) $x = -0.200$ m. (b) Find the net electric force that the two charges would exert on an electron placed at each point in part (a).

21.40 • Repeat Exercise 21.39, but now let the charge at the origin be -4.00 nC.

21.41 • Three negative point charges lie along a line as shown in **Fig. E21.41**. Find the magnitude and direction of the electric field this combination of charges produces at point P, which lies 6.00 cm from the -2.00-μC charge measured perpendicular to the line connecting the three charges.

Figure **E21.41**

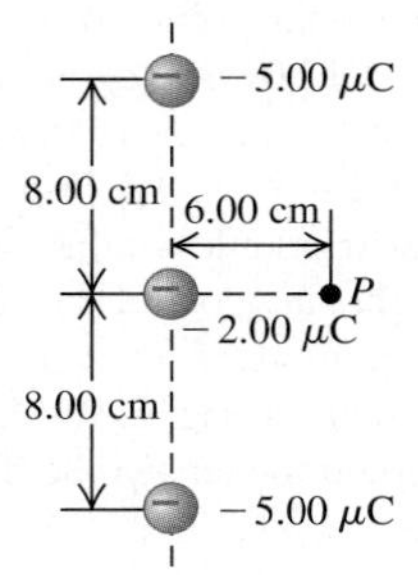

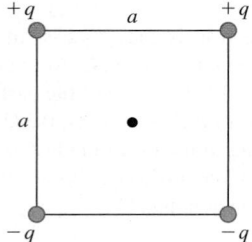

21.42 •• A point charge is placed at each corner of a square with side length a. All charges have magnitude q. Two of the charges are positive and two are negative (**Fig. E21.42**). What is the direction of the net electric field at the center of the square due to the four charges, and what is its magnitude in terms of q and a?

Figure **E21.42**

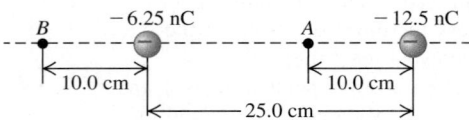

21.43 • Two point charges are separated by 25.0 cm (**Fig. E21.43**). Find the net electric field these charges produce at (a) point A and (b) point B. (c) What would be the magnitude and direction of the electric force this combination of charges would produce on a proton at A?

Figure **E21.43**

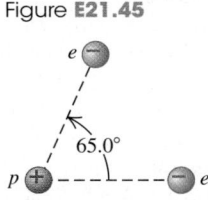

21.44 •• Point charge $q_1 = -5.00$ nC is at the origin and point charge $q_2 = +3.00$ nC is on the x-axis at $x = 3.00$ cm. Point P is on the y-axis at $y = 4.00$ cm. (a) Calculate the electric fields $\vec{E}_1$ and $\vec{E}_2$ at point P due to the charges q_1 and q_2. Express your results in terms of unit vectors (see Example 21.6). (b) Use the results of part (a) to obtain the resultant field at P, expressed in unit vector form.

21.45 •• If two electrons are each 1.50×10^{-10} m from a proton (**Fig. E21.45**), find the magnitude and direction of the net electric force they will exert on the proton.

Figure **E21.45**

21.46 •• **BIO Electric Field of Axons.** A nerve signal is transmitted through a neuron when an excess of Na^+ ions suddenly enters the axon, a long cylindrical part of the neuron. Axons are approximately 10.0 μm in diameter, and measurements show that about 5.6×10^{11} Na^+ ions per meter (each of charge $+e$) enter during this process. Although the axon is a long cylinder, the charge does not all enter everywhere at the same time. A plausible model would be a series of point charges moving along the axon. Consider a 0.10-mm length of the axon and model it as a point charge. (a) If the charge that enters each meter of the axon gets distributed uniformly along it, how many coulombs of charge enter a 0.10-mm length of the axon? (b) What electric field (magnitude and direction) does the sudden influx of charge produce at the surface of the body if the axon is 5.00 cm below the skin? (c) Certain sharks can respond to electric fields as weak as 1.0 μN/C. How far from this segment of axon could a shark be and still detect its electric field?

21.47 • In a rectangular coordinate system a positive point charge $q = 6.00 \times 10^{-9}$ C is placed at the point $x = +0.150$ m, $y = 0$, and an identical point charge is placed at $x = -0.150$ m, $y = 0$. Find the x- and y-components, the magnitude, and the direction of the electric field at the following points: (a) the origin; (b) $x = 0.300$ m, $y = 0$; (c) $x = 0.150$ m, $y = -0.400$ m; (d) $x = 0$, $y = 0.200$ m.

21.48 •• A point charge $q_1 = -4.00$ nC is at the point $x = 0.600$ m, $y = 0.800$ m, and a second point charge $q_2 = +6.00$ nC is at the point $x = 0.600$ m, $y = 0$. Calculate the magnitude and direction of the net electric field at the origin due to these two point charges.

21.49 •• A charge of -6.50 nC is spread uniformly over the surface of one face of a nonconducting disk of radius 1.25 cm. (a) Find the magnitude and direction of the electric field this disk produces at a point P on the axis of the disk a distance of 2.00 cm from its center. (b) Suppose that the charge were all pushed away from the center and distributed uniformly on the outer rim of the disk. Find the magnitude and direction of the electric field at point P. (c) If the charge is all brought to the center of the disk, find the magnitude and direction of the electric field at point P. (d) Why is the field in part (a) stronger than the field in part (b)? Why is the field in part (c) the strongest of the three fields?

21.50 •• A very long, straight wire has charge per unit length 3.20×10^{-10} C/m. At what distance from the wire is the electric-field magnitude equal to 2.50 N/C?

21.51 • A ring-shaped conductor with radius $a = 2.50$ cm has a total positive charge $Q = +0.125$ nC uniformly distributed around it (see Fig. 21.23). The center of the ring is at the origin of coordinates O. (a) What is the electric field (magnitude and direction) at point P, which is on the x-axis at $x = 40.0$ cm? (b) A point charge $q = -2.50$ μC is placed at P. What are the magnitude and direction of the force exerted by the charge q on the ring?

21.52 •• A straight, nonconducting plastic wire 8.50 cm long carries a charge density of $+175$ nC/m distributed uniformly along its length. It is lying on a horizontal tabletop. (a) Find the magnitude and direction of the electric field this wire produces at a point 6.00 cm directly above its midpoint. (b) If the wire is now bent into a circle lying flat on the table, find the magnitude and direction of the electric field it produces at a point 6.00 cm directly above its center.

Section 21.7 Electric Dipoles

21.53 • Point charges $q_1 = -4.5$ nC and $q_2 = +4.5$ nC are separated by 3.1 mm, forming an electric dipole. (a) Find the electric dipole moment (magnitude and direction). (b) The charges are in a uniform electric field whose direction makes an angle of 36.9° with the line connecting the charges. What is the magnitude of this field if the torque exerted on the dipole has magnitude 7.2×10^{-9} N·m?

21.54 • The ammonia molecule (NH_3) has a dipole moment of 5.0×10^{-30} C·m. Ammonia molecules in the gas phase are placed in a uniform electric field $\vec{E}$ with magnitude 1.6×10^6 N/C. (a) What is the change in electric potential energy when the dipole moment of a molecule changes its orientation with respect to $\vec{E}$ from parallel to perpendicular? (b) At what absolute temperature T is the average translational kinetic energy $\frac{3}{2}kT$ of a molecule equal to the change in potential energy calculated in part (a)? (*Note:* Above this temperature, thermal agitation prevents the dipoles from aligning with the electric field.)

21.55 • **Torque on a Dipole.** An electric dipole with dipole moment $\vec{p}$ is in a uniform external electric field $\vec{E}$. (a) Find the orientations of the dipole for which the torque on the dipole is zero. (b) Which of the orientations in part (a) is stable, and which is unstable? (*Hint:* Consider a small rotation away from the equilibrium position and see what happens.) (c) Show that for the stable orientation in part (b), the dipole's own electric field tends to oppose the external field.

21.56 • The dipole moment of the water molecule (H_2O) is 6.17×10^{-30} C·m. Consider a water molecule located at the origin whose dipole moment $\vec{p}$ points in the $+x$-direction. A chlorine ion (Cl^-), of charge -1.60×10^{-19} C, is located at $x = 3.00 \times 10^{-9}$ m. Find the magnitude and direction of the electric force that the water molecule exerts on the chlorine ion. Is this force attractive or repulsive? Assume that x is much larger than the separation d between the charges in the dipole, so that the approximate expression for the electric field along the dipole axis derived in Example 21.14 can be used.

21.57 • Three charges are at the corners of an isosceles triangle as shown in **Fig. E21.57**. The ± 5.00-μC charges form a dipole. (a) Find the force (magnitude and direction) the -10.00-μC charge exerts on the dipole. (b) For an axis perpendicular to the line connecting the ± 5.00-μC charges at the midpoint of this line, find the torque (magnitude and direction) exerted on the dipole by the -10.00-μC charge.

Figure **E21.57**

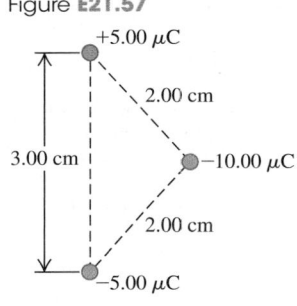

21.58 •• Consider the electric dipole of Example 21.14. (a) Derive an expression for the magnitude of the electric field produced by the dipole at a point on the x-axis in Fig. 21.33. What is the direction of this electric field? (b) How does the electric field at points on the x-axis depend on x when x is very large?

PROBLEMS

21.59 ••• Four identical charges Q are placed at the corners of a square of side L. (a) In a free-body diagram, show all of the forces that act on one of the charges. (b) Find the magnitude and direction of the total force exerted on one charge by the other three charges.

21.60 ••• Two charges are placed on the x-axis: one, of $2.50 \mu C$, at the origin and the other, of $-3.50 \mu C$, at $x = 0.600$ m (**Fig. P21.60**). Find the position on the x-axis where the net force on a small charge $+q$ would be zero.

Figure **P21.60**

$+2.50 \mu C$ $-3.50 \mu C$

—⊕————⊖—— x (m)

0   0.600 m

21.61 •• A charge $q_1 = +5.00$ nC is placed at the origin of an xy-coordinate system, and a charge $q_2 = -2.00$ nC is placed on the positive x-axis at $x = 4.00$ cm. (a) If a third charge $q_3 = +6.00$ nC is now placed at the point $x = 4.00$ cm, $y = 3.00$ cm, find the x- and y-components of the total force exerted on this charge by the other two. (b) Find the magnitude and direction of this force.

21.62 •• CP Two identical spheres with mass m are hung from silk threads of length L (**Fig. P21.62**). The spheres have the same charge, so $q_1 = q_2 = q$. The radius of each sphere is very small compared to the distance between the spheres, so they may be treated as point charges. Show that if the angle θ is small, the equilibrium separation d between the spheres is $d = (q^2 L/2\pi\epsilon_0 mg)^{1/3}$. (*Hint:* If θ is small, then $\tan\theta \cong \sin\theta$.)

Figure **P21.62**

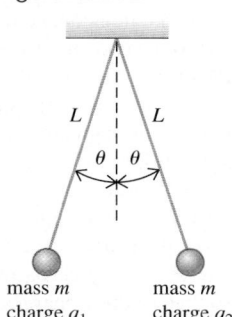

mass m mass m
charge q_1 charge q_2

21.63 ••• CP Two small spheres with mass $m = 15.0$ g are hung by silk threads of length $L = 1.20$ m from a common point (Fig. P21.62). When the spheres are given equal quantities of negative charge, so that $q_1 = q_2 = q$, each thread hangs at $\theta = 25.0°$ from the vertical. (a) Draw a diagram showing the forces on each sphere. Treat the spheres as point charges. (b) Find the magnitude of q. (c) Both threads are now shortened to length $L = 0.600$ m, while the charges q_1 and q_2 remain unchanged. What new angle will each thread make with the vertical? (*Hint:* This part of the problem can be solved numerically by using trial values for θ and adjusting the values of θ until a self-consistent answer is obtained.)

21.64 •• CP Two identical spheres are each attached to silk threads of length $L = 0.500$ m and hung from a common point (Fig. P21.62). Each sphere has mass $m = 8.00$ g. The radius of each sphere is very small compared to the distance between the spheres, so they may be treated as point charges. One sphere is given positive charge q_1, and the other a different positive charge q_2; this causes the spheres to separate so that when the spheres are in equilibrium, each thread makes an angle $\theta = 20.0°$ with the vertical. (a) Draw a free-body diagram for each sphere when in equilibrium, and label all the forces that act on each sphere. (b) Determine the magnitude of the electrostatic force that acts on each sphere, and determine the tension in each thread. (c) Based on the given information, what can you say about the magnitudes of q_1 and q_2? Explain. (d) A small wire is now connected between the spheres, allowing charge to be transferred from one sphere to the other until the two spheres have equal charges; the wire is then removed. Each thread now makes an angle of $30.0°$ with the vertical. Determine the original charges. (*Hint:* The total charge on the pair of spheres is conserved.)

21.65 •• CP A small 12.3-g plastic ball is tied to a very light 28.6-cm string that is attached to the vertical wall of a room (**Fig. P21.65**). A uniform horizontal electric field exists in this room. When the ball has been given an excess charge of $-1.11 \mu C$, you observe that it remains suspended, with the string making an angle of $17.4°$ with the wall. Find the magnitude and direction of the electric field in the room.

Figure **P21.65**

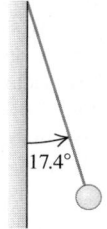

17.4°

21.66 ·· Point charge $q_1 = -6.00 \times 10^{-6}$ C is on the x-axis at $x = -0.200$ m. Point charge q_2 is on the x-axis at $x = +0.400$ m. Point charge $q_3 = +3.00 \times 10^{-6}$ C is at the origin. What is q_2 (magnitude and sign) (a) if the net force on q_3 is 6.00 N in the $+x$-direction; (b) if the net force on q_3 is 6.00 N in the $-x$-direction?

21.67 ·· Two particles having charges $q_1 = 0.500$ nC and $q_2 = 8.00$ nC are separated by a distance of 1.20 m. At what point along the line connecting the two charges is the total electric field due to the two charges equal to zero?

21.68 ·· A -3.00-nC point charge is on the x-axis at $x = 1.20$ m. A second point charge, Q, is on the x-axis at -0.600 m. What must be the sign and magnitude of Q for the resultant electric field at the origin to be (a) 45.0 N/C in the $+x$-direction, (b) 45.0 N/C in the $-x$-direction?

21.69 ··· A charge $+Q$ is located at the origin, and a charge $+4Q$ is at distance d away on the x-axis. Where should a third charge, q, be placed, and what should be its sign and magnitude, so that all three charges will be in equilibrium?

21.70 ·· A charge of -3.00 nC is placed at the origin of an xy-coordinate system, and a charge of 2.00 nC is placed on the y-axis at $y = 4.00$ cm. (a) If a third charge, of 5.00 nC, is now placed at the point $x = 3.00$ cm, $y = 4.00$ cm, find the x- and y-components of the total force exerted on this charge by the other two charges. (b) Find the magnitude and direction of this force.

21.71 · Three identical point charges q are placed at each of three corners of a square of side L. Find the magnitude and direction of the net force on a point charge $-3q$ placed (a) at the center of the square and (b) at the vacant corner of the square. In each case, draw a free-body diagram showing the forces exerted on the $-3q$ charge by each of the other three charges.

21.72 ··· Two point charges q_1 and q_2 are held in place 4.50 cm apart. Another point charge $Q = -1.75 \mu$C, of mass 5.00 g, is initially located 3.00 cm from both of these charges **(Fig. P21.72)** and released from rest. You observe that the initial acceleration of Q is 324 m/s² upward, parallel to the line connecting the two point charges. Find q_1 and q_2.

Figure **P21.72**

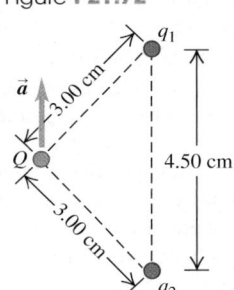

21.73 ·· **CP Strength of the Electric Force.** Imagine two 1.0-g bags of protons, one at the earth's north pole and the other at the south pole. (a) How many protons are in each bag? (b) Calculate the gravitational attraction and the electric repulsion that each bag exerts on the other. (c) Are the forces in part (b) large enough for you to feel if you were holding one of the bags?

21.74 ··· **CP** Two tiny spheres of mass 6.80 mg carry charges of equal magnitude, 72.0 nC, but opposite sign. They are tied to the same ceiling hook by light strings of length 0.530 m. When a horizontal uniform electric field E that is directed to the left is turned on, the spheres hang at rest with the angle θ between the strings equal to 58.0° **(Fig. P21.74)**. (a) Which ball (the one on the right or the one on the left) has positive charge? (b) What is the magnitude E of the field?

Figure **P21.74**

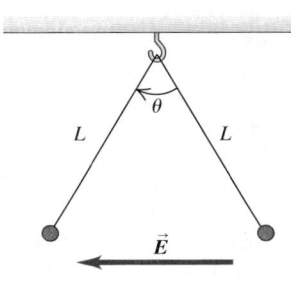

21.75 ·· **CP** Consider a model of a hydrogen atom in which an electron is in a circular orbit of radius $r = 5.29 \times 10^{-11}$ m around a stationary proton. What is the speed of the electron in its orbit?

21.76 ·· The earth has a downward-directed electric field near its surface of about 150 N/C. If a raindrop with a diameter of 0.020 mm is suspended, motionless, in this field, how many excess electrons must it have on its surface?

21.77 ·· **CP** A proton is projected into a uniform electric field that points vertically upward and has magnitude E. The initial velocity of the proton has a magnitude v_0 and is directed at an angle α below the horizontal. (a) Find the maximum distance h_{max} that the proton descends vertically below its initial elevation. Ignore gravitational forces. (b) After what horizontal distance d does the proton return to its original elevation? (c) Sketch the trajectory of the proton. (d) Find the numerical values of h_{max} and d if $E = 500$ N/C, $v_0 = 4.00 \times 10^5$ m/s, and $\alpha = 30.0°$.

21.78 ··· A small object with mass m, charge q, and initial speed $v_0 = 5.00 \times 10^3$ m/s is projected into a uniform electric field between two parallel metal plates of length 26.0 cm **(Fig. P21.78)**. The electric field between the plates is directed downward and has magnitude $E = 800$ N/C. Assume that the field is zero outside the region between the plates. The separation between the plates is large enough for the object to pass between the plates without hitting the lower plate. After passing through the field region, the object is deflected downward a vertical distance $d = 1.25$ cm from its original direction of motion and reaches a collecting plate that is 56.0 cm from the edge of the parallel plates. Ignore gravity and air resistance. Calculate the object's charge-to-mass ratio, q/m.

Figure **P21.78**

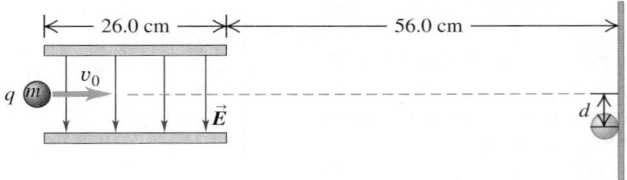

21.79 ·· **CALC** Positive charge Q is distributed uniformly along the x-axis from $x = 0$ to $x = a$. A positive point charge q is located on the positive x-axis at $x = a + r$, a distance r to the right of the end of Q **(Fig. P21.79)**. (a) Calculate the x- and y-components of the electric field produced by the charge distribution Q at points on the positive x-axis where $x > a$. (b) Calculate the force (magnitude and direction) that the charge distribution Q exerts on q. (c) Show that if $r \gg a$, the magnitude of the force in part (b) is approximately $Qq/4\pi\epsilon_0 r^2$. Explain why this result is obtained.

Figure **P21.79**

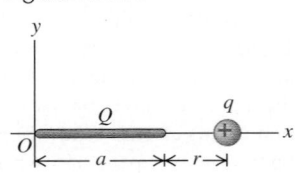

21.80 ·· In a region where there is a uniform electric field that is upward and has magnitude 3.60×10^4 N/C, a small object is projected upward with an initial speed of 1.92 m/s. The object travels upward a distance of 6.98 cm in 0.200 s. What is the object's charge-to-mass ratio q/m? Assume $g = 9.80$ m/s², and ignore air resistance.

21.81 · A negative point charge $q_1 = -4.00$ nC is on the x-axis at $x = 0.60$ m. A second point charge q_2 is on the x-axis at $x = -1.20$ m. What must the sign and magnitude of q_2 be for the net electric field at the origin to be (a) 50.0 N/C in the $+x$-direction and (b) 50.0 N/C in the $-x$-direction?

21.82 •• **CALC** Positive charge Q is distributed uniformly along the positive y-axis between $y = 0$ and $y = a$. A negative point charge $-q$ lies on the positive x-axis, a distance x from the origin (**Fig. P21.82**). (a) Calculate the x- and y-components of the electric field produced by the charge distribution Q at points on the positive x-axis. (b) Calculate the x- and y-components of the force that the charge distribution Q exerts on q. (c) Show that if $x \gg a$, $F_x \cong -Qq/4\pi\epsilon_0 x^2$ and $F_y \cong +Qqa/8\pi\epsilon_0 x^3$. Explain why this result is obtained.

Figure **P21.82**

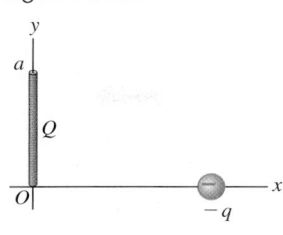

21.83 ••• A uniformly charged disk like the disk in Fig. 21.25 has radius 2.50 cm and carries a total charge of 7.0×10^{-12} C. (a) Find the electric field (magnitude and direction) on the x-axis at $x = 20.0$ cm. (b) Show that for $x \gg R$, Eq. (21.11) becomes $E = Q/4\pi\epsilon_0 x^2$, where Q is the total charge on the disk. (c) Is the magnitude of the electric field you calculated in part (a) larger or smaller than the electric field 20.0 cm from a point charge that has the same total charge as this disk? In terms of the approximation used in part (b) to derive $E = Q/4\pi\epsilon_0 x^2$ for a point charge from Eq. (21.11), explain why this is so. (d) What is the percent difference between the electric fields produced by the finite disk and by a point charge with the same charge at $x = 20.0$ cm and at $x = 10.0$ cm?

21.84 •• **CP** A small sphere with mass m carries a positive charge q and is attached to one end of a silk fiber of length L. The other end of the fiber is attached to a large vertical insulating sheet that has a positive surface charge density σ. Show that when the sphere is in equilibrium, the fiber makes an angle equal to $\arctan(q\sigma/2mg\epsilon_0)$ with the vertical sheet.

21.85 •• **CALC** Negative charge $-Q$ is distributed uniformly around a quarter-circle of radius a that lies in the first quadrant, with the center of curvature at the origin. Find the x- and y-components of the net electric field at the origin.

21.86 •• **CALC** A semicircle of radius a is in the first and second quadrants, with the center of curvature at the origin. Positive charge $+Q$ is distributed uniformly around the left half of the semicircle, and negative charge $-Q$ is distributed uniformly around the right half of the semicircle (**Fig. P21.86**). What are the magnitude and direction of the net electric field at the origin produced by this distribution of charge?

Figure **P21.86**

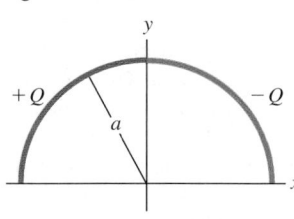

21.87 •• Two 1.20-m non-conducting rods meet at a right angle. One rod carries $+2.50\ \mu C$ of charge distributed uniformly along its length, and the other carries $-2.50\ \mu C$ distributed uniformly along it (**Fig. P21.87**). (a) Find the magnitude and direction of the electric field these rods produce at point P, which is 60.0 cm from each rod. (b) If an electron is released at P, what are the magnitude and direction of the net force that these rods exert on it?

Figure **P21.87**

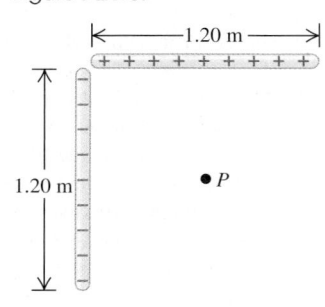

21.88 • Two very large parallel sheets are 5.00 cm apart. Sheet A carries a uniform surface charge density of $-8.80\ \mu C/m^2$, and sheet B, which is to the right of A, carries a uniform charge density of $-11.6\ \mu C/m^2$. Assume that the sheets are large enough to be treated as infinite. Find the magnitude and direction of the net electric field these sheets produce at a point (a) 4.00 cm to the right of sheet A; (b) 4.00 cm to the left of sheet A; (c) 4.00 cm to the right of sheet B.

21.89 • Repeat Problem 21.88 for the case where sheet B is positive.

21.90 • Two very large horizontal sheets are 4.25 cm apart and carry equal but opposite uniform surface charge densities of magnitude σ. You want to use these sheets to hold stationary in the region between them an oil droplet of mass 486 μg that carries an excess of five electrons. Assuming that the drop is in vacuum, (a) which way should the electric field between the plates point, and (b) what should σ be?

21.91 •• **CP** A thin disk with a circular hole at its center, called an *annulus*, has inner radius R_1 and outer radius R_2 (**Fig. P21.91**). The disk has a uniform positive surface charge density σ on its surface. (a) Determine the total electric charge on the annulus. (b) The annulus lies in the yz-plane, with its center at the origin. For an arbitrary point on the x-axis (the axis of the annulus), find the magnitude and direction of the electric field $\vec{E}$. Consider points both above and below the annulus. (c) Show that at points on the x-axis that are sufficiently close to the origin, the magnitude of the electric field is approximately proportional to the distance between the center of the annulus and the point. How close is "sufficiently close"? (d) A point particle with mass m and negative charge $-q$ is free to move along the x-axis (but cannot move off the axis). The particle is originally placed at rest at $x = 0.01R_1$ and released. Find the frequency of oscillation of the particle. (*Hint:* Review Section 14.2. The annulus is held stationary.)

Figure **P21.91**

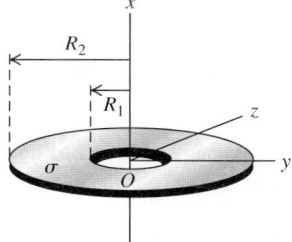

21.92 ••• **DATA CP** **Design of an Inkjet Printer.** Inkjet printers can be described as either continuous or drop-on-demand. In a continuous inkjet printer, letters are built up by squirting drops of ink at the paper from a rapidly moving nozzle. You are part of an engineering group working on the design of such a printer. Each ink drop will have a mass of 1.4×10^{-8} g. The drops will leave the nozzle and travel toward the paper at 50 m/s, passing through a charging unit that gives each drop a positive charge q by removing some electrons from it. The drops will then pass between parallel deflecting plates, 2.0 cm long, where there is a uniform vertical electric field with magnitude 8.0×10^4 N/C. Your team is working on the design of the charging unit that places the charge on the drops. (a) If a drop is to be deflected 0.30 mm by the time it reaches the end of the deflection plates, what magnitude of charge must be given to the drop? How many electrons must be removed from the drop to give it this charge? (b) If the unit that produces the stream of drops is redesigned so that it produces drops with a speed of 25 m/s, what q value is needed to achieve the same 0.30-mm deflection?

21.93 •• DATA Two small spheres, each carrying a net positive charge, are separated by 0.400 m. You have been asked to perform measurements that will allow you to determine the charge on each sphere. You set up a coordinate system with one sphere (charge q_1) at the origin and the other sphere (charge q_2) at $x = +0.400$ m. Available to you is a third sphere with net charge $q_3 = 4.00 \times 10^{-6}$ C and an apparatus that can accurately measure the location of this sphere and the net force on it. First you place the third sphere on the x-axis at $x = 0.200$ m; you measure the net force on it to be 4.50 N in the +x-direction. Then you move the third sphere to $x = +0.600$ m and measure the net force on it now to be 3.50 N in the +x-direction. (a) Calculate q_1 and q_2. (b) What is the net force (magnitude and direction) on q_3 if it is placed on the x-axis at $x = -0.200$ m? (c) At what value of x (other than $x = \pm\infty$) could q_3 be placed so that the net force on it is zero?

21.94 ••• DATA Positive charge Q is distributed uniformly around a very thin conducting ring of radius a, as in Fig. 21.23. You measure the electric field E at points on the ring axis, at a distance x from the center of the ring, over a wide range of values of x. (a) Your results for the larger values of x are plotted in **Fig. P21.94a** as Ex^2 versus x. Explain why the quantity Ex^2 approaches a constant value as x increases. Use Fig. P21.94a to calculate the net charge Q on the ring. (b) Your results for smaller values of x are plotted in Fig. P21.94b as E/x versus x. Explain why E/x approaches a constant value as x approaches zero. Use Fig. P21.94b to calculate a.

Figure **P21.94**

(a)

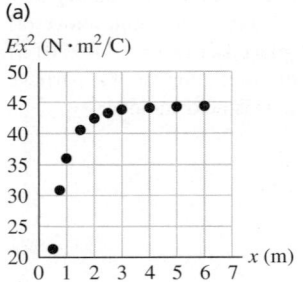

(b)

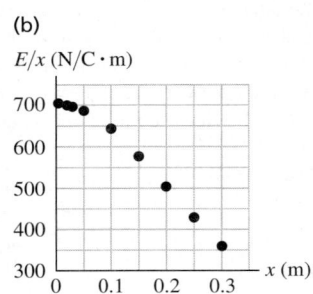

CHALLENGE PROBLEMS

21.95 ••• Three charges are placed as shown in **Fig. P21.95**. The magnitude of q_1 is 2.00 μC, but its sign and the value of the charge q_2 are not known. Charge q_3 is +4.00 μC, and the net force $\vec{F}$ on q_3 is entirely in the negative x-direction. (a) Considering the different possible signs of q_1, there are four possible force diagrams representing the forces $\vec{F}_1$ and $\vec{F}_2$ that q_1 and q_2 exert on q_3. Sketch these four possible force configurations. (b) Using the sketches from part (a) and the direction of $\vec{F}$, deduce the signs of the charges q_1 and q_2. (c) Calculate the magnitude of q_2. (d) Determine F, the magnitude of the net force on q_3.

Figure **P21.95**

$\vec{F}$ ← q_3
4.00 cm   3.00 cm
q_1 5.00 cm q_2

21.96 ••• Two charges are placed as shown in **Fig. P21.96**. The magnitude of q_1 is 3.00 μC, but its sign and the value of the charge q_2 are not known. The direction of the net electric field $\vec{E}$ at point P is entirely in the negative y-direction. (a) Considering the different possible signs of q_1 and q_2, four possible diagrams could represent the electric fields $\vec{E}_1$ and $\vec{E}_2$ produced by q_1 and q_2.

Figure **P21.96**

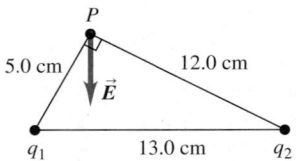

Sketch the four possible electric-field configurations. (b) Using the sketches from part (a) and the direction of $\vec{E}$, deduce the signs of q_1 and q_2. (c) Determine the magnitude of $\vec{E}$.

21.97 ••• CALC Two thin rods of length L lie along the x-axis, one between $x = \frac{1}{2}a$ and $x = \frac{1}{2}a + L$ and the other between $x = -\frac{1}{2}a$ and $x = -\frac{1}{2}a - L$. Each rod has positive charge Q distributed uniformly along its length. (a) Calculate the electric field produced by the second rod at points along the positive x-axis. (b) Show that the magnitude of the force that one rod exerts on the other is

$$F = \frac{Q^2}{4\pi\epsilon_0 L^2} \ln\left[\frac{(a+L)^2}{a(a+2L)}\right]$$

(c) Show that if $a \gg L$, the magnitude of this force reduces to $F = Q^2/4\pi\epsilon_0 a^2$. (Hint: Use the expansion $\ln(1+z) = z - \frac{1}{2}z^2 + \frac{1}{3}z^3 - \cdots$, valid for $|z| \ll 1$. Carry all expansions to at least order L^2/a^2.) Interpret this result.

PASSAGE PROBLEMS

BIO **ELECTRIC BEES.** Flying insects such as bees may accumulate a small positive electric charge as they fly. In one experiment, the mean electric charge of 50 bees was measured to be $+(30 \pm 5)$ pC per bee. Researchers also observed the electrical properties of a plant consisting of a flower atop a long stem. The charge on the stem was measured as a positively charged bee approached, landed, and flew away. Plants are normally electrically neutral, so the measured net electric charge on the stem was zero when the bee was very far away. As the bee approached the flower, a small net positive charge was detected in the stem, even before the bee landed. Once the bee landed, the whole plant became positively charged, and this positive charge remained on the plant after the bee flew away. By creating artificial flowers with various charge values, experimenters found that bees can distinguish between charged and uncharged flowers and may use the positive electric charge left by a previous bee as a cue indicating whether a plant has already been visited (in which case, little pollen may remain).

21.98 Consider a bee with the mean electric charge found in the experiment. This charge represents roughly how many missing electrons? (a) 1.9×10^8; (b) 3.0×10^8; (c) 1.9×10^{18}; (d) 3.0×10^{18}.

21.99 What is the best explanation for the observation that the electric charge on the stem became positive as the charged bee approached (before it landed)? (a) Because air is a good conductor, the positive charge on the bee's surface flowed through the air from bee to plant. (b) Because the earth is a reservoir of large amounts of charge, positive ions were drawn up the stem from the ground toward the charged bee. (c) The plant became electrically polarized as the charged bee approached. (d) Bees that had visited the plant earlier deposited a positive charge on the stem.

21.100 After one bee left a flower with a positive charge, that bee flew away and another bee with the same amount of positive charge flew close to the plant. Which diagram in **Fig. P21.100** best represents the electric field lines between the bee and the flower?

Figure **P21.100**

(a)   (b)   (c)   (d)

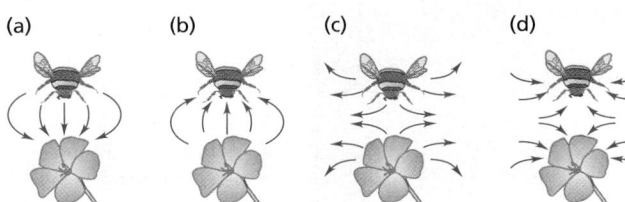

21.101 In a follow-up experiment, a charge of +40 pC was placed at the center of an artificial flower at the end of a 30-cm-long stem. Bees were observed to approach no closer than 15 cm from the center of this flower before they flew away. This observation suggests that the smallest external electric field to which bees may be sensitive is closest to which of these values? (a) 2.4 N/C; (b) 16 N/C; (c) 2.7×10^{-10} N/C; (d) 4.8×10^{-10} N/C.

Answers

Chapter Opening Question ?

(ii) Water molecules have a permanent electric dipole moment: One end of the molecule has a positive charge and the other end has a negative charge. These ends attract negative and positive ions, respectively, holding the ions apart in solution. Water is less effective as a solvent for materials whose molecules do not ionize (called *nonionic* substances), such as oils.

Test Your Understanding Questions

21.1 (iv) Two charged objects repel if their charges are of the same sign (either both positive or both negative).

21.2 (a) (i), (b) (ii) Before the two spheres touch, the negatively charged sphere exerts a repulsive force on the electrons in the other sphere, causing zones of positive and negative induced charge (see Fig. 21.7b). The positive zone is closer to the negatively charged sphere than the negative zone, so there is a net force of attraction that pulls the spheres together, like the comb and insulator in Fig. 21.8b. Once the two metal spheres touch, some of the excess electrons on the negatively charged sphere will flow onto the other sphere (because metals are conductors). Then both spheres will have a net negative charge and will repel each other.

21.3 (iv) The force exerted by q_1 on Q is still as in Example 21.4. The magnitude of the force exerted by q_2 on Q is still equal to $F_{1\ \text{on}\ Q}$, but the direction of the force is now *toward* q_2 at an angle α below the x-axis. Hence the x-components of the two forces cancel while the (negative) y-components add together, and the total electric force is in the negative y-direction.

21.4 (a) (ii), (b) (i) The electric field $\vec{E}$ produced by a positive point charge points directly away from the charge (see Fig. 21.18a) and has a magnitude that depends on the distance r from the charge to the field point. Hence a second, negative point charge $q < 0$ will feel a force $\vec{F} = q\vec{E}$ that points directly toward the positive charge and has a magnitude that depends on the distance r between the two charges. If the negative charge moves directly toward the positive charge, the direction of the force remains the same but the force magnitude increases as the distance r decreases. If the negative charge moves in a circle around the positive charge, the force magnitude stays the same (because the distance r is constant) but the force direction changes.

21.5 (iv) Think of a pair of segments of length dy, one at coordinate $y > 0$ and the other at coordinate $-y < 0$. The upper segment has a positive charge and produces an electric field $d\vec{E}$ at P that points away from the segment, so this $d\vec{E}$ has a positive x-component and a negative y-component, like the vector $d\vec{E}$ in Fig. 21.24. The lower segment has the same amount of negative charge. It produces a $d\vec{E}$ that has the same magnitude but points *toward* the lower segment, so it has a negative x-component and a negative y-component. By symmetry, the two x-components are equal but opposite, so they cancel. Thus the total electric field has only a negative y-component.

21.6 yes If the field lines are straight, $\vec{E}$ must point in the same direction throughout the region. Hence the force $\vec{F} = q\vec{E}$ on a particle of charge q is always in the same direction. A particle released from rest accelerates in a straight line in the direction of $\vec{F}$, and so its trajectory is a straight line along a field line.

21.7 (ii) Equations (21.17) and (21.18) tell us that the potential energy for a dipole in an electric field is $U = -\vec{p} \cdot \vec{E} = -pE\cos\phi$, where ϕ is the angle between the directions of $\vec{p}$ and $\vec{E}$. If $\vec{p}$ and $\vec{E}$ point in opposite directions, so that $\phi = 180°$, we have $\cos\phi = -1$ and $U = +pE$. This is the maximum value that U can have. From our discussion of energy diagrams in Section 7.5, it follows that this is a situation of unstable equilibrium.

Bridging Problem

$E = 2kQ/\pi a^2$ in the $-y$-direction

22 GAUSS'S LAW

LEARNING GOALS

Looking forward at …

22.1 How you can determine the amount of charge within a closed surface by examining the electric field on the surface.

22.2 What is meant by electric flux, and how to calculate it.

22.3 How Gauss's law relates the electric flux through a closed surface to the charge enclosed by the surface.

22.4 How to use Gauss's law to calculate the electric field due to a symmetric charge distribution.

22.5 Where the charge is located on a charged conductor.

Looking back at …

21.4–21.6 Electric fields and their properties.

The discussion of Gauss's law in this section is based on and inspired by the innovative ideas of Ruth W. Chabay and Bruce A. Sherwood in *Electric and Magnetic Interactions* (John Wiley & Sons, 1994).

I n physics, an important tool for simplifying problems is the *symmetry properties* of systems. Many physical systems have symmetry; for example, a cylindrical body doesn't look any different after you've rotated it around its axis, and a charged metal sphere looks just the same after you've turned it about any axis through its center.

In this chapter we'll use symmetry ideas along with a new principle, called *Gauss's law,* to simplify electric-field calculations. For example, the field of a straight-line or plane-sheet charge distribution, which we derived in Section 21.5 by using some fairly strenuous integrations, can be obtained in a few steps with the help of Gauss's law. But Gauss's law is more than just a way to make certain calculations easier. Indeed, it is a fundamental statement about the relationship between electric charges and electric fields. Among other things, Gauss's law can help us understand how electric charge distributes itself over conducting bodies.

Here's what Gauss's law is all about. Given any general distribution of charge, we surround it with an imaginary surface that encloses the charge. Then we look at the electric field at various points on this imaginary surface. Gauss's law is a relationship between the field at *all* the points on the surface and the total charge enclosed within the surface. This may sound like a rather indirect way of expressing things, but it turns out to be a tremendously useful relationship. In the next several chapters, we'll make frequent use of the insights that Gauss's law provides into the character of electric fields.

22.1 CHARGE AND ELECTRIC FLUX

In Chapter 21 we asked the question, "Given a charge distribution, what is the electric field produced by that distribution at a point *P*?" We saw that the answer could be found by representing the distribution as an assembly of point charges, each of which produces an electric field $\vec{E}$ given by Eq. (21.7). The total field at *P* is then the vector sum of the fields due to all the point charges.

But there is an alternative relationship between charge distributions and electric fields. To discover this relationship, let's stand the question of Chapter 21 on its head and ask, "If the electric-field pattern is known in a given region, what can we determine about the charge distribution in that region?"

Here's an example. Consider the box shown in **Fig. 22.1a**, which may or may not contain electric charge. We'll imagine that the box is made of a material that has no effect on any electric fields; it's of the same breed as the massless rope and the frictionless incline. Better still, let the box represent an *imaginary* surface that may or may not enclose some charge. We'll refer to the box as a **closed surface** because it completely encloses a volume. How can you determine how much (if any) electric charge lies within the box?

Knowing that a charge distribution produces an electric field and that an electric field exerts a force on a test charge, you move a test charge q_0 around the vicinity of the box. By measuring the force $\vec{F}$ experienced by the test charge at different positions, you make a three-dimensional map of the electric field $\vec{E} = \vec{F}/q_0$ outside the box. In the case shown in Fig. 22.1b, the map turns out to be the same as that of the electric field produced by a positive point charge (Fig. 21.28a). From the details of the map, you can find the exact value of the point charge inside the box.

To determine the contents of the box, we actually need to measure $\vec{E}$ on only the *surface* of the box. In **Fig. 22.2a** there is a single *positive* point charge inside the box, and in Fig. 22.2b there are two such charges. The field patterns on the surfaces of the boxes are different in detail, but in each case the electric field points *out* of the box. Figures 22.2c and 22.2d show cases with one and two *negative* point charges, respectively, inside the box. Again, the details of $\vec{E}$ are different for the two cases, but the electric field points *into* each box.

Electric Flux and Enclosed Charge

In Section 21.4 we mentioned the analogy between electric-field vectors and the velocity vectors of a fluid in motion. This analogy can be helpful, even though an electric field does not actually "flow." Using this analogy, in Figs. 22.2a and 22.2b, in which the electric-field vectors point out of the surface, we say that there is an *outward* **electric flux.** (The word "flux" comes from a Latin word meaning "flow.") In Figs. 22.2c and 22.2d the $\vec{E}$ vectors point into the surface, and the electric flux is *inward*.

Figure 22.2 suggests a simple relationship: Positive charge inside the box goes with an outward electric flux through the box's surface, and negative charge inside goes with an inward electric flux. What happens if there is *zero*

22.1 How can you measure the charge inside a box without opening it?

(a) A box containing an unknown amount of charge

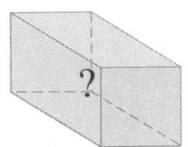

(b) Using a test charge outside the box to probe the amount of charge inside the box

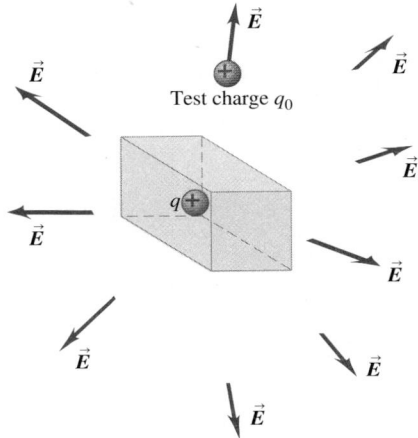

Test charge q_0

22.2 The electric field on the surface of boxes containing (a) a single positive point charge, (b) two positive point charges, (c) a single negative point charge, or (d) two negative point charges.

(a) Positive charge inside box, outward flux

(b) Positive charges inside box, outward flux

(c) Negative charge inside box, inward flux

(d) Negative charges inside box, inward flux

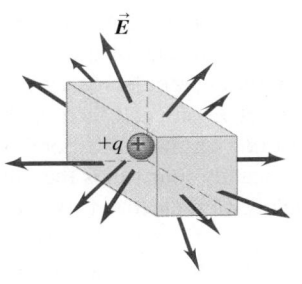

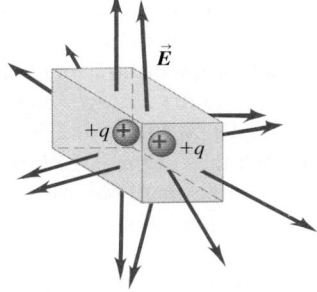

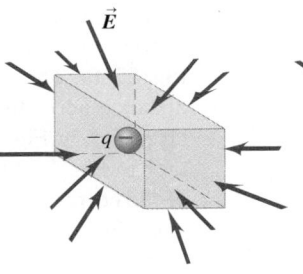

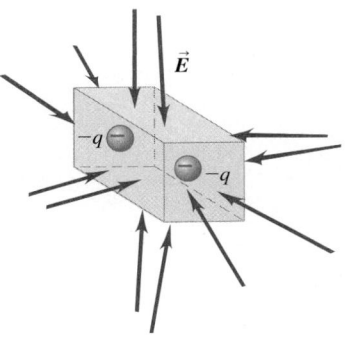

22.3 Three cases in which there is zero *net* charge inside a box and no net electric flux through the surface of the box. **(a)** An empty box with $\vec{E} = 0$. **(b)** A box containing one positive and one equal-magnitude negative point charge. **(c)** An empty box immersed in a uniform electric field.

(a) No charge inside box, zero flux

(b) Zero *net* charge inside box, inward flux cancels outward flux.

(c) No charge inside box, inward flux cancels outward flux.

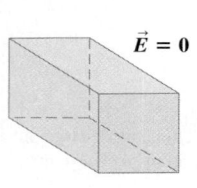

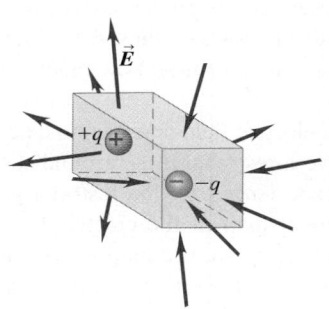

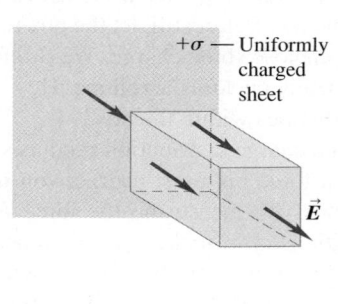

charge inside the box? In **Fig. 22.3a** the box is empty and $\vec{E} = 0$ everywhere, so there is no electric flux into or out of the box. In Fig. 22.3b, one positive and one negative point charge of equal magnitude are enclosed within the box, so the *net* charge inside the box is zero. There is an electric field, but it "flows into" the box on half of its surface and "flows out of" the box on the other half. Hence there is no *net* electric flux into or out of the box.

The box is again empty in Fig. 22.3c. However, there is charge present *outside* the box; the box has been placed with one end parallel to a uniformly charged infinite sheet, which produces a uniform electric field perpendicular to the sheet (see Example 21.11 of Section 21.5). On one end of the box, $\vec{E}$ points into the box; on the opposite end, $\vec{E}$ points out of the box; and on the sides, $\vec{E}$ is parallel to the surface and so points neither into nor out of the box. As in Fig. 22.3b, the inward electric flux on one part of the box exactly compensates for the outward electric flux on the other part. So in all of the cases shown in Fig. 22.3, there is no *net* electric flux through the surface of the box, and no *net* charge is enclosed in the box.

Figures 22.2 and 22.3 demonstrate a connection between the *sign* (positive, negative, or zero) of the *net* charge enclosed by a closed surface and the sense (outward, inward, or none) of the net electric flux through the surface. There is also a connection between the *magnitude* of the net charge inside the closed surface and the *strength* of the net "flow" of $\vec{E}$ over the surface. In both **Figs. 22.4a** and 22.4b there is a single point charge inside the box, but in Fig. 22.4b the magnitude of the charge is twice as great, and so $\vec{E}$ is everywhere twice as great in magnitude as in Fig. 22.4a. If we keep in mind the fluid-flow analogy, this means that the net outward electric flux is also twice as great in Fig. 22.4b as in Fig. 22.4a. This suggests that the net electric flux through the surface of the box is *directly proportional* to the magnitude of the net charge enclosed by the box.

This conclusion is independent of the size of the box. In Fig. 22.4c the point charge $+q$ is enclosed by a box with twice the linear dimensions of the box in Fig. 22.4a. The magnitude of the electric field of a point charge decreases with distance according to $1/r^2$, so the average magnitude of $\vec{E}$ on each face of the large box in Fig. 22.4c is just $\frac{1}{4}$ of the average magnitude on the corresponding face in Fig. 22.4a. But each face of the large box has exactly four times the area of the corresponding face of the small box. Hence the outward electric flux is the *same* for the two boxes if we *define* electric flux as follows: For each face of the box, take the product of the average perpendicular component of $\vec{E}$ and the area of that face; then add up the results from all faces of the box. With this definition the net electric flux due to a single point charge inside the box is independent of the size of the box and depends only on the net charge inside the box.

To summarize, for the special cases of a closed surface in the shape of a rectangular box and charge distributions made up of point charges or infinite charged sheets, we have found:

1. Whether there is a net outward or inward electric flux through a closed surface depends on the sign of the enclosed charge.

2. Charges *outside* the surface do not give a net electric flux through the surface.

3. The net electric flux is directly proportional to the net amount of charge enclosed within the surface but is otherwise independent of the size of the closed surface.

These observations are a qualitative statement of *Gauss's law.*

Do these observations hold true for other kinds of charge distributions and for closed surfaces of arbitrary shape? The answer to these questions will prove to be yes. But to explain why this is so, we need a precise mathematical statement of what we mean by electric flux. We develop this in the next section.

TEST YOUR UNDERSTANDING OF SECTION 22.1   If all of the dimensions of the box in Fig. 22.2a are increased by a factor of 3, how will the electric flux through the box change? (i) The flux will be $3^2 = 9$ times greater; (ii) the flux will be 3 times greater; (iii) the flux will be unchanged; (iv) the flux will be $\frac{1}{3}$ as great; (v) the flux will be $\left(\frac{1}{3}\right)^2 = \frac{1}{9}$ as great; (vi) not enough information is given to decide. ∎

22.2 CALCULATING ELECTRIC FLUX

In the preceding section we introduced the concept of *electric flux*. We used this to give a rough qualitative statement of Gauss's law: The net electric flux through a closed surface is directly proportional to the net charge inside that surface. To be able to make full use of this law, we need to know how to *calculate* electric flux. To do this, let's again make use of the analogy between an electric field $\vec{E}$ and the field of velocity vectors $\vec{v}$ in a flowing fluid. (Again, keep in mind that this is only an analogy; an electric field is *not* a flow.)

Flux: Fluid-Flow Analogy

Figure 22.5 shows a fluid flowing steadily from left to right. Let's examine the volume flow rate dV/dt (in, say, cubic meters per second) through the wire rectangle with area A. When the area is perpendicular to the flow velocity $\vec{v}$ (Fig. 22.5a) and the flow velocity is the same at all points in the fluid, the volume flow rate dV/dt is the area A multiplied by the flow speed v:

$$\frac{dV}{dt} = vA$$

When the rectangle is tilted at an angle ϕ (Fig. 22.5b) so that its face is not perpendicular to $\vec{v}$, the area that counts is the silhouette area that we see when we look in the direction of $\vec{v}$. This area, which is outlined in red and labeled $A_\perp$ in Fig. 22.5b, is the *projection* of the area A onto a surface perpendicular to $\vec{v}$. Two sides of the projected rectangle have the same length as the original one, but the

22.4 Three boxes, each of which encloses a positive point charge.

(a) A box containing a positive point charge $+q$

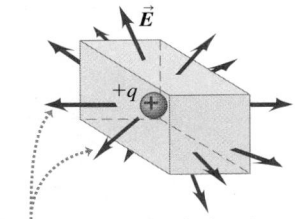

There is outward electric flux through the surface.

(b) The same box as in **(a)**, but containing a positive point charge $+2q$

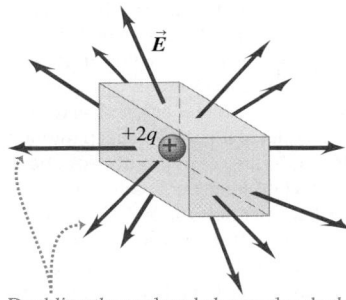

Doubling the enclosed charge also doubles the magnitude of the electric field on the surface, so the electric flux through the surface is twice as great as in **(a)**.

(c) The same positive point charge $+q$, but enclosed by a box with twice the dimensions

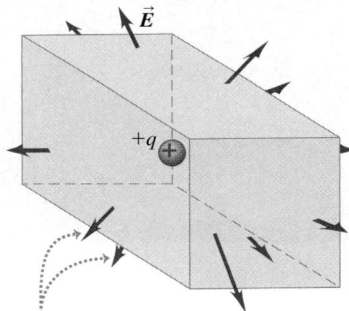

The electric flux is the same as in **(a)**: The magnitude of the electric field on the surface is $\frac{1}{4}$ as great as in **(a)**, but the area through which the field "flows" is 4 times greater.

(a) A wire rectangle in a fluid

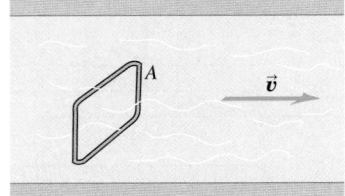

(b) The wire rectangle tilted by an angle ϕ

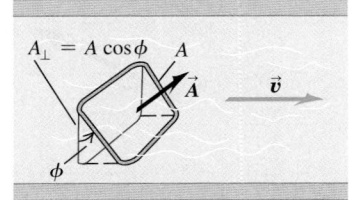

22.5 The volume flow rate of fluid through the wire rectangle **(a)** is vA when the area of the rectangle is perpendicular to $\vec{v}$ and **(b)** is $vA\cos\phi$ when the rectangle is tilted at an angle ϕ.

other two are foreshortened by a factor of $\cos\phi$, so the projected area $A_\perp$ is equal to $A\cos\phi$. Then the volume flow rate through A is

$$\frac{dV}{dt} = vA\cos\phi$$

If $\phi = 90°$, $dV/dt = 0$; the wire rectangle is edge-on to the flow, and no fluid passes through the rectangle.

Also, $v\cos\phi$ is the component of the vector $\vec{v}$ perpendicular to the plane of the area A. Calling this component $v_\perp$, we can rewrite the volume flow rate as

$$\frac{dV}{dt} = v_\perp A$$

We can express the volume flow rate more compactly by using the concept of *vector area $\vec{A}$*, a vector quantity with magnitude A and a direction perpendicular to the plane of the area we are describing. The vector area $\vec{A}$ describes both the size of an area and its orientation in space. In terms of $\vec{A}$, we can write the volume flow rate of fluid through the rectangle in Fig. 22.5b as a scalar (dot) product:

$$\frac{dV}{dt} = \vec{v} \cdot \vec{A}$$

Flux of a Uniform Electric Field

Using the analogy between electric field and fluid flow, we now define electric flux in the same way as we have just defined the volume flow rate of a fluid; we simply replace the fluid velocity $\vec{v}$ by the electric field $\vec{E}$. The symbol that we use for electric flux is Φ_E (the capital Greek letter phi; the subscript E is a reminder that this is *electric* flux). Consider first a flat area A perpendicular to a uniform electric field $\vec{E}$ (**Fig. 22.6a**). We define the electric flux through this area to be the product of the field magnitude E and the area A:

$$\Phi_E = EA$$

Roughly speaking, we can picture Φ_E in terms of the field lines passing through A. Increasing the area means that more lines of $\vec{E}$ pass through the area, increasing the flux; a stronger field means more closely spaced lines of $\vec{E}$ and therefore more lines per unit area, so again the flux increases.

If the area A is flat but not perpendicular to the field $\vec{E}$, then fewer field lines pass through it. In this case the area that counts is the silhouette area that we see

BIO Application Flux Through a Basking Shark's Mouth Unlike aggressive carnivorous sharks such as great whites, a basking shark feeds passively on plankton in the water that passes through the shark's gills as it swims. To survive on these tiny organisms requires a huge flux of water through a basking shark's immense mouth, which can be up to a meter across. The water flux—the product of the shark's speed through the water and the area of its mouth—can be up to 0.5 m^3/s (500 liters per second, or almost 5×10^5 gallons per hour). In a similar way, the flux of electric field through a surface depends on the magnitude of the field and the area of the surface (as well as the relative orientation of the field and surface).

22.6 A flat surface in a uniform electric field. The electric flux Φ_E through the surface equals the scalar product of the electric field $\vec{E}$ and the area vector $\vec{A}$.

(a) Surface is face-on to electric field:
- $\vec{E}$ and $\vec{A}$ are parallel (the angle between $\vec{E}$ and $\vec{A}$ is $\phi = 0$).
- The flux $\Phi_E = \vec{E} \cdot \vec{A} = EA$.

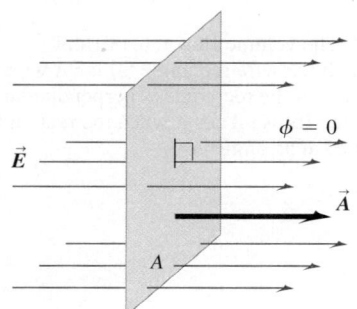

(b) Surface is tilted from a face-on orientation by an angle ϕ:
- The angle between $\vec{E}$ and $\vec{A}$ is ϕ.
- The flux $\Phi_E = \vec{E} \cdot \vec{A} = EA\cos\phi$.

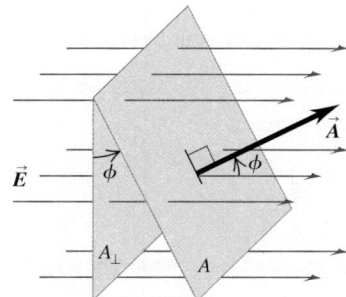

(c) Surface is edge-on to electric field:
- $\vec{E}$ and $\vec{A}$ are perpendicular (the angle between $\vec{E}$ and $\vec{A}$ is $\phi = 90°$).
- The flux $\Phi_E = \vec{E} \cdot \vec{A} = EA\cos 90° = 0$.

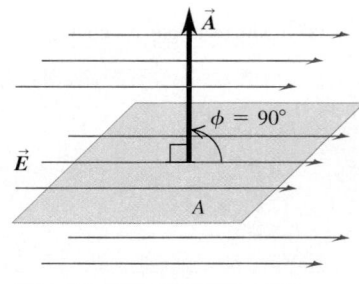

when looking in the direction of $\vec{E}$. This is the area $A_\perp$ in Fig. 22.6b and is equal to $A\cos\phi$ (compare to Fig. 22.5b). We generalize our definition of electric flux for a uniform electric field to

$$\Phi_E = EA\cos\phi \qquad \text{(electric flux for uniform } \vec{E}, \text{ flat surface)} \qquad (22.1)$$

Since $E\cos\phi$ is the component of $\vec{E}$ perpendicular to the area, we can rewrite Eq. (22.1) as

$$\Phi_E = E_\perp A \qquad \text{(electric flux for uniform } \vec{E}, \text{ flat surface)} \qquad (22.2)$$

In terms of the vector area $\vec{A}$ perpendicular to the area, we can write the electric flux as the scalar product of $\vec{E}$ and $\vec{A}$:

$$\Phi_E = \vec{E} \cdot \vec{A} \qquad \text{(electric flux for uniform } \vec{E}, \text{ flat surface)} \qquad (22.3)$$

Equations (22.1), (22.2), and (22.3) express the electric flux for a *flat* surface and a *uniform* electric field in different but equivalent ways. The SI unit for electric flux is $1 \text{ N} \cdot \text{m}^2/\text{C}$. Note that if the area is edge-on to the field, $\vec{E}$ and $\vec{A}$ are perpendicular and the flux is zero (Fig. 22.6c).

We can represent the direction of a vector area $\vec{A}$ by using a *unit vector* $\hat{n}$ perpendicular to the area; $\hat{n}$ stands for "normal." Then

$$\vec{A} = A\hat{n} \qquad (22.4)$$

A surface has two sides, so there are two possible directions for $\hat{n}$ and $\vec{A}$. We must always specify which direction we choose. In Section 22.1 we related the charge inside a *closed* surface to the electric flux through the surface. With a closed surface we will always choose the direction of $\hat{n}$ to be *outward*, and we will speak of the flux *out of* a closed surface. Thus what we called "outward electric flux" in Section 22.1 corresponds to a *positive* value of Φ_E, and what we called "inward electric flux" corresponds to a *negative* value of Φ_E.

Flux of a Nonuniform Electric Field

What happens if the electric field $\vec{E}$ isn't uniform but varies from point to point over the area A? Or what if A is part of a curved surface? Then we divide A into many small elements dA, each of which has a unit vector $\hat{n}$ perpendicular to it and a vector area $d\vec{A} = \hat{n}\,dA$. We calculate the electric flux through each element and integrate the results to obtain the total flux:

Electric flux through a surface —
Magnitude of electric field $\vec{E}$ —
Component of $\vec{E}$ perpendicular to surface —

$$\Phi_E = \int E\cos\phi\, dA = \int E_\perp\, dA = \int \vec{E} \cdot d\vec{A} \qquad (22.5)$$

Angle between $\vec{E}$ and normal to surface —
Element of surface area —
Vector element of surface area —

We call this integral the **surface integral** of the component $E_\perp$ over the area, or the surface integral of $\vec{E} \cdot d\vec{A}$. In specific problems, one form of the integral is sometimes more convenient than another. Example 22.3 at the end of this section illustrates the use of Eq. (22.5).

In Eq. (22.5) the electric flux $\int E_\perp\, dA$ is equal to the *average* value of the perpendicular component of the electric field, multiplied by the area of the surface. This is the same definition of electric flux that we were led to in Section 22.1, now expressed more mathematically. In the next section we will see the connection between the total electric flux through *any* closed surface, no matter what its shape, and the amount of charge enclosed within that surface.

EXAMPLE 22.1 ELECTRIC FLUX THROUGH A DISK

A disk of radius 0.10 m is oriented with its normal unit vector $\hat{n}$ at 30° to a uniform electric field $\vec{E}$ of magnitude 2.0×10^3 N/C (**Fig. 22.7**). (Since this isn't a closed surface, it has no "inside" or "outside." That's why we have to specify the direction of $\hat{n}$ in the figure.) (a) What is the electric flux through the disk? (b) What is the flux through the disk if it is turned so that $\hat{n}$ is perpendicular to $\vec{E}$? (c) What is the flux through the disk if $\hat{n}$ is parallel to $\vec{E}$?

22.7 The electric flux Φ_E through a disk depends on the angle between its normal $\hat{n}$ and the electric field $\vec{E}$.

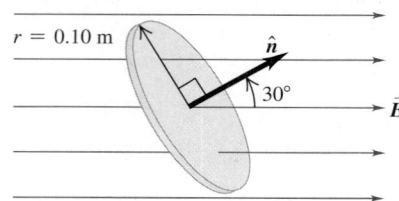

SOLUTION

IDENTIFY and SET UP: This problem is about a flat surface in a uniform electric field, so we can apply the ideas of this section. We calculate the electric flux from Eq. (22.1).

EXECUTE: (a) The area is $A = \pi(0.10 \text{ m})^2 = 0.0314 \text{ m}^2$ and the angle between $\vec{E}$ and $\vec{A} = A\hat{n}$ is $\phi = 30°$, so from Eq. (22.1),

$$\Phi_E = EA\cos\phi = (2.0 \times 10^3 \text{ N/C})(0.0314 \text{ m}^2)(\cos 30°)$$

$$= 54 \text{ N} \cdot \text{m}^2/\text{C}$$

(b) The normal to the disk is now perpendicular to $\vec{E}$, so $\phi = 90°$, $\cos\phi = 0$, and $\Phi_E = 0$.

(c) The normal to the disk is parallel to $\vec{E}$, so $\phi = 0$ and $\cos\phi = 1$:

$$\Phi_E = EA\cos\phi = (2.0 \times 10^3 \text{ N/C})(0.0314 \text{ m}^2)(1)$$

$$= 63 \text{ N} \cdot \text{m}^2/\text{C}$$

EVALUATE: Our answer to part (b) is smaller than that to part (a), which is in turn smaller than that to part (c). Is all this as it should be?

EXAMPLE 22.2 ELECTRIC FLUX THROUGH A CUBE

An imaginary cubical surface of side L is in a region of uniform electric field $\vec{E}$. Find the electric flux through each face of the cube and the total flux through the cube when (a) it is oriented with two of its faces perpendicular to $\vec{E}$ (**Fig. 22.8a**) and (b) the cube is turned by an angle θ about a vertical axis (Fig. 22.8b).

SOLUTION

IDENTIFY and SET UP: Since $\vec{E}$ is uniform and each of the six faces of the cube is flat, we find the flux Φ_{Ei} through each face from Eqs. (22.3) and (22.4). The total flux through the cube is the sum of the six individual fluxes.

EXECUTE: (a) Figure 22.8a shows the unit vectors $\hat{n}_1$ through $\hat{n}_6$ for each face; each unit vector points *outward* from the cube's closed surface. The angle between $\vec{E}$ and $\hat{n}_1$ is 180°, the angle between $\vec{E}$ and $\hat{n}_2$ is 0°, and the angle between $\vec{E}$ and each of the other four

22.8 Electric flux of a uniform field $\vec{E}$ through a cubical box of side L in two orientations.

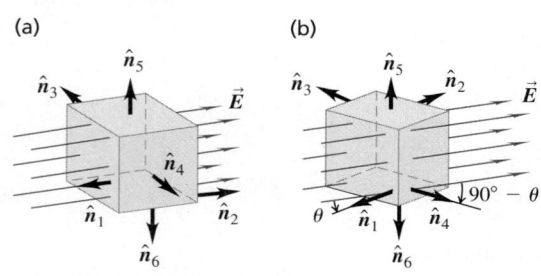

unit vectors is 90°. Each face of the cube has area L^2, so the fluxes through the faces are

$$\Phi_{E1} = \vec{E} \cdot \hat{n}_1 A = EL^2 \cos 180° = -EL^2$$

$$\Phi_{E2} = \vec{E} \cdot \hat{n}_2 A = EL^2 \cos 0° = +EL^2$$

$$\Phi_{E3} = \Phi_{E4} = \Phi_{E5} = \Phi_{E6} = EL^2 \cos 90° = 0$$

The flux is negative on face 1, where $\vec{E}$ is directed into the cube, and positive on face 2, where $\vec{E}$ is directed out of the cube. The total flux through the cube is

$$\Phi_E = \Phi_{E1} + \Phi_{E2} + \Phi_{E3} + \Phi_{E4} + \Phi_{E5} + \Phi_{E6}$$

$$= -EL^2 + EL^2 + 0 + 0 + 0 + 0 = 0$$

(b) The field $\vec{E}$ is directed into faces 1 and 3, so the fluxes through them are negative; $\vec{E}$ is directed out of faces 2 and 4, so the fluxes through them are positive. We find

$$\Phi_{E1} = \vec{E} \cdot \hat{n}_1 A = EL^2 \cos(180° - \theta) = -EL^2 \cos\theta$$

$$\Phi_{E2} = \vec{E} \cdot \hat{n}_2 A = +EL^2 \cos\theta$$

$$\Phi_{E3} = \vec{E} \cdot \hat{n}_3 A = EL^2 \cos(90° + \theta) = -EL^2 \sin\theta$$

$$\Phi_{E4} = \vec{E} \cdot \hat{n}_4 A = EL^2 \cos(90° - \theta) = +EL^2 \sin\theta$$

$$\Phi_{E5} = \Phi_{E6} = EL^2 \cos 90° = 0$$

The total flux $\Phi_E = \Phi_{E1} + \Phi_{E2} + \Phi_{E3} + \Phi_{E4} + \Phi_{E5} + \Phi_{E6}$ through the surface of the cube is again zero.

EVALUATE: We came to the same conclusion in our discussion of Fig. 22.3c: There is zero net flux of a uniform electric field through a closed surface that contains no electric charge.

EXAMPLE 22.3 **ELECTRIC FLUX THROUGH A SPHERE**

A point charge $q = +3.0 \, \mu\text{C}$ is surrounded by an imaginary sphere of radius $r = 0.20$ m centered on the charge (**Fig. 22.9**). Find the resulting electric flux through the sphere.

___
SOLUTION

IDENTIFY and SET UP: The surface is not flat and the electric field is not uniform, so to calculate the electric flux (our target variable) we must use the general definition, Eq. (22.5). Because the sphere is centered on the point charge, at any point on the spherical surface, $\vec{E}$ is directed out of the sphere perpendicular to the surface. The positive direction for both $\hat{n}$ and $E_\perp$ is outward, so $E_\perp = E$

22.9 Electric flux through a sphere centered on a point charge.

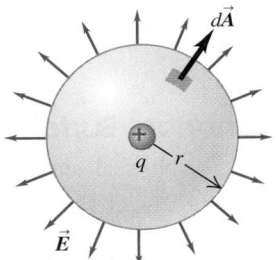

and the flux through a surface element dA is $\vec{E} \cdot d\vec{A} = E \, dA$. This greatly simplifies the integral in Eq. (22.5).

EXECUTE: We must evaluate the integral of Eq. (22.5), $\Phi_E = \int E \, dA$. At any point on the sphere of radius r the electric field has the same magnitude $E = q/4\pi\epsilon_0 r^2$. Hence E can be taken outside the integral, which becomes $\Phi_E = E \int dA = EA$, where A is the area of the spherical surface: $A = 4\pi r^2$. Hence the total flux through the sphere is

$$\Phi_E = EA = \frac{q}{4\pi\epsilon_0 r^2} 4\pi r^2 = \frac{q}{\epsilon_0}$$

$$= \frac{3.0 \times 10^{-6} \, \text{C}}{8.85 \times 10^{-12} \, \text{C}^2/\text{N} \cdot \text{m}^2} = 3.4 \times 10^5 \, \text{N} \cdot \text{m}^2/\text{C}$$

EVALUATE: The radius r of the sphere cancels out of the result for Φ_E. We would have obtained the same flux with a sphere of radius 2.0 m or 200 m. We came to essentially the same conclusion in our discussion of Fig. 22.4 in Section 22.1, where we considered rectangular closed surfaces of two different sizes enclosing a point charge. There we found that the flux of $\vec{E}$ was independent of the size of the surface; the same result holds true for a spherical surface. Indeed, the flux through *any* surface enclosing a single point charge is independent of the shape or size of the surface, as we'll soon see.

TEST YOUR UNDERSTANDING OF SECTION 22.2 Rank the following surfaces in order from most positive to most negative electric flux: (i) a flat rectangular surface with vector area $\vec{A} = (6.0 \, \text{m}^2)\hat{\imath}$ in a uniform electric field $\vec{E} = (4.0 \, \text{N/C})\hat{\jmath}$; (ii) a flat circular surface with vector area $\vec{A} = (3.0 \, \text{m}^2)\hat{\jmath}$ in a uniform electric field $\vec{E} = (4.0 \, \text{N/C})\hat{\imath} + (2.0 \, \text{N/C})\hat{\jmath}$; (iii) a flat square surface with vector area $\vec{A} = (3.0 \, \text{m}^2)\hat{\imath} + (7.0 \, \text{m}^2)\hat{\jmath}$ in a uniform electric field $\vec{E} = (4.0 \, \text{N/C})\hat{\imath} - (2.0 \, \text{N/C})\hat{\jmath}$; (iv) a flat oval surface with vector area $\vec{A} = (3.0 \, \text{m}^2)\hat{\imath} - (7.0 \, \text{m}^2)\hat{\jmath}$ in a uniform electric field $\vec{E} = (4.0 \, \text{N/C})\hat{\imath} - (2.0 \, \text{N/C})\hat{\jmath}$. ∎

22.3 GAUSS'S LAW

Gauss's law is an alternative to Coulomb's law. While completely equivalent to Coulomb's law, Gauss's law provides a different way to express the relationship between electric charge and electric field. It was formulated by Carl Friedrich Gauss (1777–1855), one of the greatest mathematicians of all time (**Fig. 22.10**).

Point Charge Inside a Spherical Surface

Gauss's law states that the total electric flux through any closed surface (a surface enclosing a definite volume) is proportional to the total (net) electric charge inside the surface. In Section 22.1 we observed this relationship qualitatively; now we'll develop it more rigorously. We'll start with the field of a single positive point charge q. The field lines radiate out equally in all directions. We place this charge at the center of an imaginary spherical surface with radius R. The magnitude E of the electric field at every point on the surface is given by

$$E = \frac{1}{4\pi\epsilon_0} \frac{q}{R^2}$$

At each point on the surface, $\vec{E}$ is perpendicular to the surface, and its magnitude is the same at every point, as in Example 22.3 (Section 22.2). The total electric

22.10 Carl Friedrich Gauss helped develop several branches of mathematics, including differential geometry, real analysis, and number theory. The "bell curve" of statistics is one of his inventions. Gauss also made state-of-the-art investigations of the earth's magnetism and calculated the orbit of the first asteroid to be discovered.

22.11 Projection of an element of area dA of a sphere of radius R onto a concentric sphere of radius $2R$. The projection multiplies each linear dimension by 2, so the area element on the larger sphere is 4 dA.

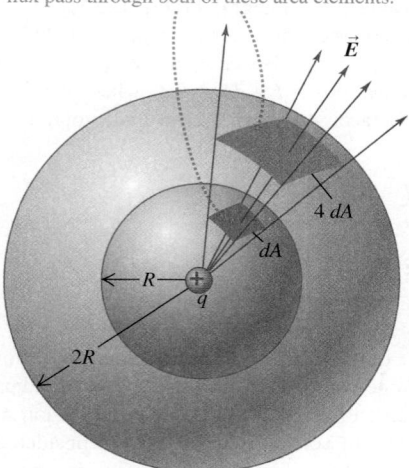

The same number of field lines and the same flux pass through both of these area elements.

flux is the product of the field magnitude E and the total area $A = 4\pi R^2$ of the sphere:

$$\Phi_E = EA = \frac{1}{4\pi\epsilon_0}\frac{q}{R^2}(4\pi R^2) = \frac{q}{\epsilon_0} \tag{22.6}$$

The flux is independent of the radius R of the sphere. It depends on only the charge q enclosed by the sphere.

We can also interpret this result in terms of field lines. **Figure 22.11** shows two spheres with radii R and $2R$ centered on the point charge q. Every field line that passes through the smaller sphere also passes through the larger sphere, so the total flux through each sphere is the same.

What is true of the entire sphere is also true of any portion of its surface. In Fig. 22.11 an area dA is outlined on the sphere of radius R and projected onto the sphere of radius $2R$ by drawing lines from the center through points on the boundary of dA. The area projected on the larger sphere is clearly 4 dA. But since the electric field due to a point charge is inversely proportional to r^2, the field magnitude is $\frac{1}{4}$ as great on the sphere of radius $2R$ as on the sphere of radius R. Hence the electric flux is the same for both areas and is independent of the radius of the sphere.

Point Charge Inside a Nonspherical Surface

We can extend this projection technique to nonspherical surfaces. Instead of a second sphere, let us surround the sphere of radius R by a surface of irregular shape, as in **Fig. 22.12a**. Consider a small element of area dA on the irregular surface; we note that this area is *larger* than the corresponding element on a spherical surface at the same distance from q. If a normal to dA makes an angle ϕ with a radial line from q, two sides of the area projected onto the spherical surface are foreshortened by a factor $\cos\phi$ (Fig. 22.12b). The other two sides are unchanged. Thus the electric flux through the spherical surface element is equal to the flux $E\,dA\cos\phi$ through the corresponding irregular surface element.

We can divide the entire irregular surface into elements dA, compute the electric flux $E\,dA\cos\phi$ for each, and sum the results by integrating, as in Eq. (22.5). Each of the area elements projects onto a corresponding spherical surface element. Thus the *total* electric flux through the irregular surface, given by any of the forms of Eq. (22.5), must be the same as the total flux through a sphere, which Eq. (22.6) shows is equal to q/ϵ_0. Thus, for the irregular surface,

$$\Phi_E = \oint \vec{E} \cdot d\vec{A} = \frac{q}{\epsilon_0} \tag{22.7}$$

22.12 Calculating the electric flux through a nonspherical surface.

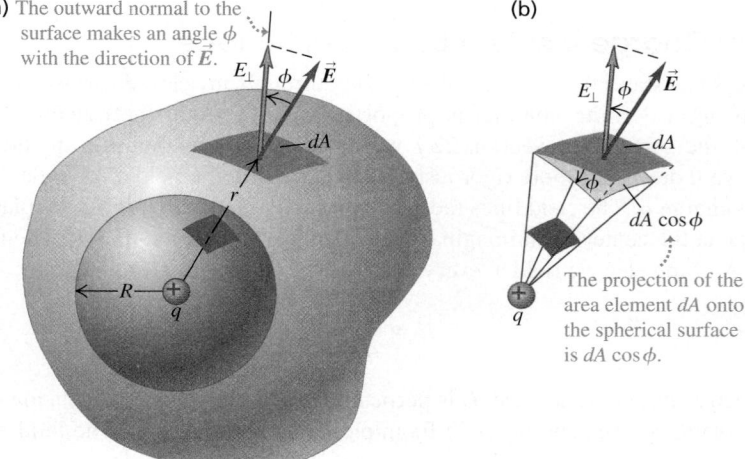

(a) The outward normal to the surface makes an angle ϕ with the direction of $\vec{E}$.

$E_\perp$ ϕ $\vec{E}$

dA

r

R

q

(b)

$E_\perp$ ϕ $\vec{E}$

dA

ϕ

$dA\cos\phi$

q

The projection of the area element dA onto the spherical surface is $dA\cos\phi$.

Equation (22.7) holds for a surface of *any* shape or size, provided only that it is a *closed* surface enclosing the charge q. The circle on the integral sign reminds us that the integral is always taken over a *closed* surface.

The area elements $d\vec{A}$ and the corresponding unit vectors $\hat{n}$ always point *out of* the volume enclosed by the surface. The electric flux is then positive in areas where the electric field points out of the surface and negative where it points inward. Also, $E_\perp$ is positive at points where $\vec{E}$ points out of the surface and negative at points where $\vec{E}$ points into the surface.

If the point charge in Fig. 22.12 is negative, the $\vec{E}$ field is directed radially *inward;* the angle ϕ is then greater than 90°, its cosine is negative, and the integral in Eq. (22.7) is negative. But since q is also negative, Eq. (22.7) holds.

For a closed surface enclosing *no* charge,

$$\Phi_E = \oint \vec{E} \cdot d\vec{A} = 0$$

This is the mathematical statement that when a region contains no charge, any field lines caused by charges *outside* the region that enter on one side must leave again on the other side. (In Section 22.1 we came to the same conclusion by considering the special case of a rectangular box in a uniform field.) **Figure 22.13** illustrates this point. *Electric field lines can begin or end inside a region of space only when there is charge in that region.*

General Form of Gauss's Law

Now comes the final step in obtaining the general form of Gauss's law. Suppose the surface encloses not just one point charge q but several charges q_1, q_2, q_3, ... The total (resultant) electric field $\vec{E}$ at any point is the vector sum of the $\vec{E}$ fields of the individual charges. Let Q_{encl} be the *total* charge enclosed by the surface: $Q_{encl} = q_1 + q_2 + q_3 + \cdots$. Also let $\vec{E}$ be the *total* field at the position of the surface area element $d\vec{A}$, and let $E_\perp$ be its component perpendicular to the plane of that element (that is, parallel to $d\vec{A}$). Then we can write an equation like Eq. (22.7) for each charge and its corresponding field and add the results. When we do, we obtain the general statement of Gauss's law:

Gauss's law:
$$\Phi_E = \oint \vec{E} \cdot d\vec{A} = \frac{Q_{encl}}{\epsilon_0} \qquad (22.8)$$

Total charge enclosed by surface

Electric flux through a closed surface of area A = surface integral of $\vec{E}$

Electric constant

The total electric flux through a closed surface is equal to the total (net) electric charge inside the surface, divided by ϵ_0.

Using the definition of Q_{encl} and the various ways to express electric flux given in Eq. (22.5), we can express Gauss's law in the following equivalent forms:

Various forms of Gauss's law:	Magnitude of electric field E	Component of $\vec{E}$ perpendicular to surface	Total charge enclosed by surface	

$$\Phi_E = \oint E\cos\phi\, dA = \oint E_\perp\, dA = \oint \vec{E} \cdot d\vec{A} = \frac{Q_{encl}}{\epsilon_0} \qquad (22.9)$$

Electric flux through a closed surface

Angle between $\vec{E}$ and normal to surface

Element of surface area

Vector element of surface area

Electric constant

As in Eq. (22.5), the various forms of the integral all express the same thing, the total electric flux through the Gaussian surface, in different terms. One form is sometimes more convenient than another.

22.13 A point charge *outside* a closed surface that encloses no charge. If an electric field line from the external charge enters the surface at one point, it must leave at another.

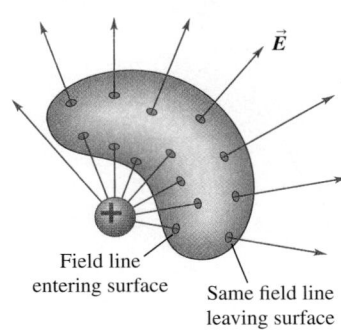

Field line entering surface

Same field line leaving surface

CAUTION Gaussian surfaces are imaginary Remember that the closed surface in Gauss's law is *imaginary;* there need not be any material object at the position of the surface. We often refer to a closed surface used in Gauss's law as a **Gaussian surface.** ∎

22.14 Spherical Gaussian surfaces around (a) a positive point charge and (b) a negative point charge.

(a) Gaussian surface around positive charge: positive (outward) flux

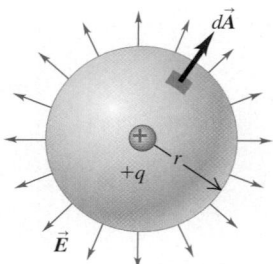

(b) Gaussian surface around negative charge: negative (inward) flux

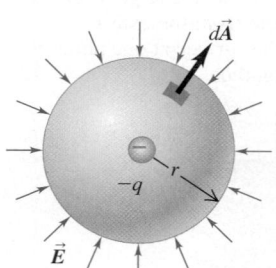

As an example, **Fig. 22.14a** shows a spherical Gaussian surface of radius r around a positive point charge $+q$. The electric field points out of the Gaussian surface, so at every point on the surface $\vec{E}$ is in the same direction as $d\vec{A}$, $\phi = 0$, and $E_\perp$ is equal to the field magnitude $E = q/4\pi\epsilon_0 r^2$. Since E is the same at all points on the surface, we can take it outside the integral in Eq. (22.9). Then the remaining integral is $\int dA = A = 4\pi r^2$, the area of the sphere. Hence Eq. (22.9) becomes

$$\Phi_E = \oint E_\perp \, dA = \oint \left(\frac{q}{4\pi\epsilon_0 r^2}\right) dA = \frac{q}{4\pi\epsilon_0 r^2} \oint dA = \frac{q}{4\pi\epsilon_0 r^2} 4\pi r^2 = \frac{q}{\epsilon_0}$$

The enclosed charge Q_{encl} is just the charge $+q$, so this agrees with Gauss's law. If the Gaussian surface encloses a *negative* point charge as in Fig. 22.14b, then $\vec{E}$ points *into* the surface at each point in the direction opposite $d\vec{A}$. Then $\phi = 180°$ and $E_\perp$ is equal to the negative of the field magnitude: $E_\perp = -E = -|-q|/4\pi\epsilon_0 r^2 = -q/4\pi\epsilon_0 r^2$. Equation (22.9) then becomes

$$\Phi_E = \oint E_\perp \, dA = \oint \left(\frac{-q}{4\pi\epsilon_0 r^2}\right) dA = \frac{-q}{4\pi\epsilon_0 r^2} \oint dA = \frac{-q}{4\pi\epsilon_0 r^2} 4\pi r^2 = \frac{-q}{\epsilon_0}$$

This again agrees with Gauss's law because the enclosed charge in Fig. 22.14b is $Q_{\text{encl}} = -q$.

In Eqs. (22.8) and (22.9), Q_{encl} is always the algebraic sum of all the positive and negative charges enclosed by the Gaussian surface, and $\vec{E}$ is the *total* field at each point on the surface. Also note that in general, this field is caused partly by charges inside the surface and partly by charges outside. But as Fig. 22.13 shows, the outside charges do *not* contribute to the total (net) flux through the surface. So Eqs. (22.8) and (22.9) are correct even when there are charges outside the surface that contribute to the electric field at the surface. When $Q_{\text{encl}} = 0$, the total flux through the Gaussian surface must be zero, even though some areas may have positive flux and others may have negative flux (see Fig. 22.3b).

Gauss's law is the definitive answer to the question we posed at the beginning of Section 22.1: "If the electric-field pattern is known in a given region, what can we determine about the charge distribution in that region?" It provides a relationship between the electric field on a closed surface and the charge distribution within that surface. But in some cases we can use Gauss's law to answer the reverse question: "If the charge distribution is known, what can we determine about the electric field that the charge distribution produces?" Gauss's law may seem like an unappealing way to address this question, since it may look as though evaluating the integral in Eq. (22.8) is a hopeless task. Sometimes it is, but other times it is surprisingly easy. Here's an example in which *no* integration is involved at all; we'll work out several more examples in the next section.

CONCEPTUAL EXAMPLE 22.4 | **ELECTRIC FLUX AND ENCLOSED CHARGE**

Figure 22.15 shows the field produced by two point charges $+q$ and $-q$ (an electric dipole). Find the electric flux through each of the closed surfaces A, B, C, and D.

SOLUTION

Gauss's law, Eq. (22.8), says that the total electric flux through a closed surface is equal to the total enclosed charge divided by ϵ_0.

In Fig. 22.15, surface A (shown in red) encloses the positive charge, so $Q_{\text{encl}} = +q$; surface B (in blue) encloses the negative charge, so $Q_{\text{encl}} = -q$; surface C (in purple) encloses *both* charges, so $Q_{\text{encl}} = +q + (-q) = 0$; and surface D (in yellow) encloses no charges, so $Q_{\text{encl}} = 0$. Hence, without having to do any integration, we have $\Phi_{EA} = +q/\epsilon_0$, $\Phi_{EB} = -q/\epsilon_0$, and $\Phi_{EC} = \Phi_{ED} = 0$. These results depend only on the charges enclosed within each Gaussian surface, not on the precise shapes of the surfaces.

22.15 The net number of field lines leaving a closed surface is proportional to the total charge enclosed by that surface.

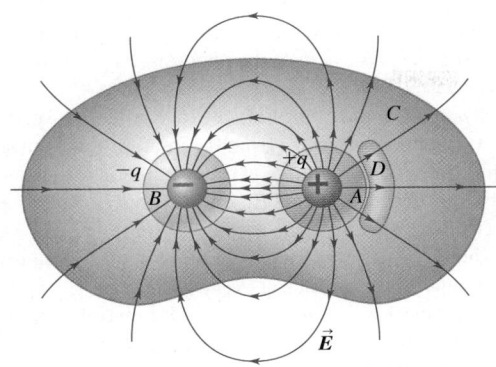

We can draw similar conclusions by examining the electric field lines. All the field lines that cross surface A are directed out of the surface, so the flux through A must be positive. Similarly, the flux through B must be negative since all of the field lines that cross that surface point inward. For both surface C and surface D, there are as many field lines pointing into the surface as there are field lines pointing outward, so the flux through each of these surfaces is zero.

TEST YOUR UNDERSTANDING OF SECTION 22.3 **Figure 22.16** shows six point charges that all lie in the same plane. Five Gaussian surfaces—S_1, S_2, S_3, S_4, and S_5—each enclose part of this plane, and Fig. 22.16 shows the intersection of each surface with the plane. Rank these five surfaces in order of the electric flux through them, from most positive to most negative. ▌

22.16 Five Gaussian surfaces and six point charges.

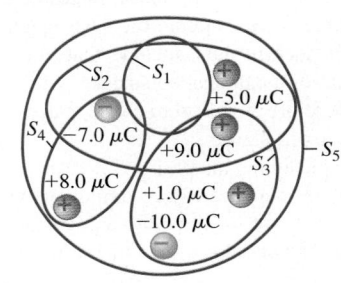

22.4 APPLICATIONS OF GAUSS'S LAW

Gauss's law is valid for *any* distribution of charges and for *any* closed surface. Gauss's law can be used in two ways. If we know the charge distribution, and if it has enough symmetry to let us evaluate the integral in Gauss's law, we can find the field. Or if we know the field, we can use Gauss's law to find the charge distribution, such as charges on conducting surfaces.

In this section we present examples of both kinds of applications. As you study them, watch for the role played by the symmetry properties of each system. We will use Gauss's law to calculate the electric fields caused by several simple charge distributions; the results are collected in a table in the chapter summary.

In practical problems we often encounter situations in which we want to know the electric field caused by a charge distribution on a conductor. These calculations are aided by the following remarkable fact: *When excess charge is placed on a solid conductor and is at rest, it resides entirely on the surface, not in the interior of the material.* (By *excess* we mean charges other than the ions and free electrons that make up the neutral conductor.) Here's the proof. We know from Section 21.4 that in an electrostatic situation (with all charges at rest) the electric field $\vec{E}$ at every point in the interior of a conducting material is zero. If $\vec{E}$ were *not* zero, the excess charges would move. Suppose we construct a Gaussian surface inside the conductor, such as surface A in **Fig. 22.17.** Because $\vec{E} = 0$ everywhere on this surface, Gauss's law requires that the net charge inside the surface is zero. Now imagine shrinking the surface like a collapsing balloon until it encloses a region so small that we may consider it as a point P; then the charge at that point must be zero. We can do this anywhere inside the conductor, so *there can be no excess charge at any point within a solid conductor; any excess charge must reside on the conductor's surface.* (This result is for a *solid* conductor. In the next section we'll discuss what can happen if the conductor has cavities in its interior.) We will make use of this fact frequently in the examples that follow.

22.17 Under electrostatic conditions (charges not in motion), any excess charge on a solid conductor resides entirely on the conductor's surface.

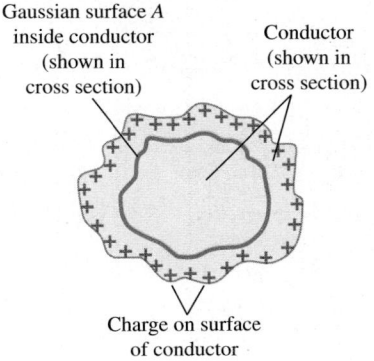

PROBLEM-SOLVING STRATEGY 22.1 | GAUSS'S LAW

IDENTIFY *the relevant concepts:* Gauss's law is most useful when the charge distribution has spherical, cylindrical, or planar symmetry. In these cases the symmetry determines the direction of $\vec{E}$. Then Gauss's law yields the magnitude of $\vec{E}$ if we are given the charge distribution, and vice versa. In either case, begin the analysis by asking the question: What is the symmetry?

SET UP *the problem* using the following steps:
1. List the known and unknown quantities and identify the target variable.
2. Select the appropriate closed, imaginary Gaussian surface. For spherical symmetry, use a concentric spherical surface. For cylindrical symmetry, use a coaxial cylindrical surface with flat ends perpendicular to the axis of symmetry (like a soup can). For planar symmetry, use a cylindrical surface (like a tuna can) with its flat ends parallel to the plane.

EXECUTE *the solution* as follows:
1. Determine the appropriate size and placement of your Gaussian surface. To evaluate the field magnitude at a particular point, the surface must include that point. It may help to place one end of a can-shaped surface within a conductor, where $\vec{E}$ and therefore Φ are zero, or to place its ends equidistant from a charged plane.
2. Evaluate the integral $\oint E_\perp \, dA$ in Eq. (22.9). In this equation $E_\perp$ is the perpendicular component of the *total* electric field at each point on the Gaussian surface. A well-chosen Gaussian surface should make integration trivial or unnecessary. If the surface comprises several separate surfaces, such as the sides

and ends of a cylinder, the integral $\oint E_\perp \, dA$ over the entire closed surface is the sum of the integrals $\int E_\perp \, dA$ over the separate surfaces. Consider points 3–6 as you work.
3. If $\vec{E}$ is *perpendicular* (normal) at every point to a surface with area A, if it points *outward* from the interior of the surface, and if it has the same *magnitude* at every point on the surface, then $E_\perp = E = $ constant, and $\int E_\perp \, dA$ over that surface is equal to EA. (If $\vec{E}$ is inward, then $E_\perp = -E$ and $\int E_\perp \, dA = -EA$.) This should be the case for part or all of your Gaussian surface. If $\vec{E}$ is tangent to a surface at every point, then $E_\perp = 0$ and the integral over that surface is zero. This may be the case for parts of a cylindrical Gaussian surface. If $\vec{E} = \mathbf{0}$ at every point on a surface, the integral is zero.
4. Even when there is *no* charge within a Gaussian surface, the field at any given point on the surface is not necessarily zero. In that case, however, the total electric flux through the surface is always zero.
5. The flux integral $\oint E_\perp \, dA$ can be approximated as the difference between the numbers of electric lines of force leaving and entering the Gaussian surface. In this sense the flux gives the sign of the enclosed charge, but is only proportional to it; zero flux corresponds to zero enclosed charge.
6. Once you have evaluated $\oint E_\perp \, dA$, use Eq. (22.9) to solve for your target variable.

EVALUATE *your answer:* If your result is a *function* that describes how the magnitude of the electric field varies with position, ensure that it makes sense.

EXAMPLE 22.5 FIELD OF A CHARGED CONDUCTING SPHERE

We place a total positive charge q on a solid conducting sphere with radius R (**Fig. 22.18**). Find $\vec{E}$ at any point inside or outside the sphere.

22.18 Calculating the electric field of a conducting sphere with positive charge q. Outside the sphere, the field is the same as if all of the charge were concentrated at the center of the sphere.

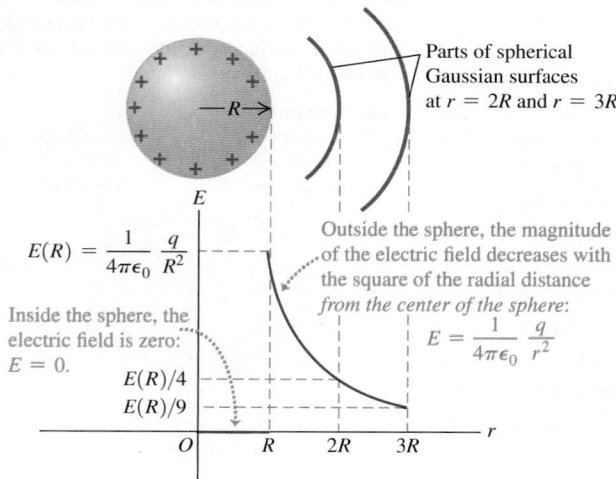

SOLUTION

IDENTIFY and SET UP: As we discussed earlier in this section, all of the charge must be on the surface of the sphere. The charge is free to move on the conductor, and there is no preferred position on the surface; the charge is therefore distributed *uniformly* over the surface, and the system is spherically symmetric. To exploit this symmetry, we take as our Gaussian surface a sphere of radius r centered on the conductor. We can calculate the field inside or outside the conductor by taking $r < R$ or $r > R$, respectively. In either case, the point at which we want to calculate $\vec{E}$ lies on the Gaussian surface.

EXECUTE: The spherical symmetry means that the direction of the electric field must be radial; that's because there is no preferred direction parallel to the surface, so $\vec{E}$ can have no component parallel to the surface. There is also no preferred orientation of the sphere, so the field magnitude E can depend only on the distance r from the center and must have the same value at all points on the Gaussian surface.

For $r > R$ the entire conductor is within the Gaussian surface, so the enclosed charge is q. The area of the Gaussian surface is $4\pi r^2$, and $\vec{E}$ is uniform over the surface and perpendicular to it at each point. The flux integral $\oint E_\perp \, dA$ is then just $E(4\pi r^2)$, and Eq. (22.8) gives

$$E(4\pi r^2) = \frac{q}{\epsilon_0} \quad \text{and}$$

$$E = \frac{1}{4\pi\epsilon_0} \frac{q}{r^2} \quad \begin{array}{l} \text{(outside a charged} \\ \text{conducting sphere)} \end{array}$$

This expression is the same as that for a point charge; outside the charged sphere, its field is the same as though the entire charge were concentrated at its center. Just outside the surface of the sphere, where $r = R$,

$$E = \frac{1}{4\pi\epsilon_0} \frac{q}{R^2} \quad \text{(at the surface of a charged conducting sphere)}$$

> CAUTION **Flux can be positive or negative** Remember that we have chosen the charge q to be *positive*. If the charge is negative, the electric field is radially *inward* instead of radially outward, and the electric flux through the Gaussian surface is negative. The electric-field magnitudes outside and at the surface of the sphere are given by the same expressions as above, except that q denotes the *magnitude* (absolute value) of the charge. ∎

For $r < R$ we again have $E(4\pi r^2) = Q_{\text{encl}}/\epsilon_0$. But now our Gaussian surface (which lies entirely within the conductor) encloses *no* charge, so $Q_{\text{encl}} = 0$. The electric field inside the conductor is therefore zero.

EVALUATE: We already knew that $\vec{E} = 0$ inside a solid conductor (whether spherical or not) when the charges are at rest. Figure 22.18 shows E as a function of the distance r from the center of the sphere. Note that in the limit as $R \to 0$, the sphere becomes a point charge; there is then only an "outside," and the field is everywhere given by $E = q/4\pi\epsilon_0 r^2$. Thus we have deduced Coulomb's law from Gauss's law. (In Section 22.3 we deduced Gauss's law from Coulomb's law; the two laws are equivalent.)

We can also use this method for a conducting spherical *shell* (a spherical conductor with a concentric spherical hole inside) if there is no charge inside the hole. We use a spherical Gaussian surface with radius r less than the radius of the hole. If there *were* a field inside the hole, it would have to be radial and spherically symmetric as before, so $E = Q_{\text{encl}}/4\pi\epsilon_0 r^2$. But now there is no enclosed charge, so $Q_{\text{encl}} = 0$ and $E = 0$ inside the hole.

Can you use this same technique to find the electric field in the region between a charged sphere and a concentric hollow conducting sphere that surrounds it?

EXAMPLE 22.6 **FIELD OF A UNIFORM LINE CHARGE**

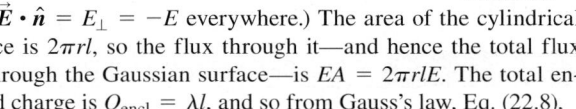

Electric charge is distributed uniformly along an infinitely long, thin wire. The charge per unit length is λ (assumed positive). Find the electric field by using Gauss's law.

SOLUTION

IDENTIFY and SET UP: We found in Example 21.10 (Section 21.5) that the field $\vec{E}$ of a uniformly charged, infinite wire is radially outward if λ is positive and radially inward if λ is negative, and that the field magnitude E depends on only the radial distance from the wire. This suggests that we use a *cylindrical* Gaussian surface, of radius r and arbitrary length l, coaxial with the wire and with its ends perpendicular to the wire (**Fig. 22.19**).

EXECUTE: The flux through the flat ends of our Gaussian surface is zero because the radial electric field is parallel to these ends, and so $\vec{E} \cdot \hat{n} = 0$. On the cylindrical part of our surface we have $\vec{E} \cdot \hat{n} = E_\perp = E$ everywhere. (If λ were negative, we would

have $\vec{E} \cdot \hat{n} = E_\perp = -E$ everywhere.) The area of the cylindrical surface is $2\pi r l$, so the flux through it—and hence the total flux Φ_E through the Gaussian surface—is $EA = 2\pi r l E$. The total enclosed charge is $Q_{\text{encl}} = \lambda l$, and so from Gauss's law, Eq. (22.8),

$$\Phi_E = 2\pi r l E = \frac{\lambda l}{\epsilon_0} \quad \text{and}$$

$$E = \frac{1}{2\pi\epsilon_0} \frac{\lambda}{r} \quad \text{(field of an infinite line of charge)}$$

We found this same result in Example 21.10 with *much* more effort.

If λ is *negative*, $\vec{E}$ is directed radially inward, and in the above expression for E we must interpret λ as the absolute value of the charge per unit length.

EVALUATE: We saw in Example 21.10 that the *entire* charge on the wire contributes to the field at any point, and yet we consider only that part of the charge $Q_{\text{encl}} = \lambda l$ within the Gaussian surface when we apply Gauss's law. There's nothing inconsistent here; it takes the entire charge to give the field the properties that allow us to calculate Φ_E so easily, and Gauss's law always applies to the enclosed charge only. If the wire is short, the symmetry of the infinite wire is lost, and E is not uniform over a coaxial, cylindrical Gaussian surface. Gauss's law then *cannot* be used to find Φ; we must solve the problem the hard way, as in Example 21.10.

We can use the Gaussian surface in Fig. 22.19 to show that the field outside a long, uniformly charged cylinder is the same as though all the charge were concentrated on a line along its axis (see Problem 22.41). We can also calculate the electric field in the space between a charged cylinder and a coaxial hollow conducting cylinder surrounding it (see Problem 22.39).

22.19 A coaxial cylindrical Gaussian surface is used to find the electric field outside an infinitely long, charged wire.

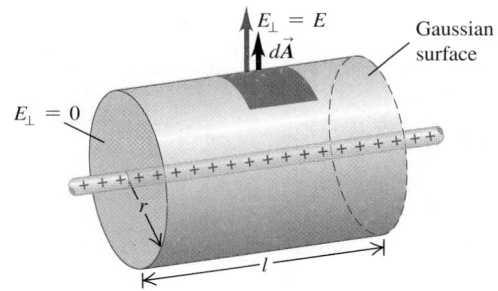

EXAMPLE 22.7 FIELD OF AN INFINITE PLANE SHEET OF CHARGE

Use Gauss's law to find the electric field caused by a thin, flat, infinite sheet with a uniform positive surface charge density σ.

SOLUTION

IDENTIFY and SET UP: In Example 21.11 (Section 21.5) we found that the field $\vec{E}$ of a uniformly charged infinite sheet is normal to the sheet, and that its magnitude is independent of the distance from the sheet. To take advantage of these symmetry properties, we use a cylindrical Gaussian surface with ends of area A and with its axis perpendicular to the sheet of charge (**Fig. 22.20**).

EXECUTE: The flux through the cylindrical part of our Gaussian surface is zero because $\vec{E} \cdot \hat{n} = 0$ everywhere. The flux through each flat end of the surface is $+EA$ because $\vec{E} \cdot \hat{n} = E_\perp = E$ everywhere, so the total flux through both ends—and hence the total flux Φ_E through the Gaussian surface—is $+2EA$. The total enclosed charge is $Q_{\text{encl}} = \sigma A$, and so from Gauss's law,

$$2EA = \frac{\sigma A}{\epsilon_0} \quad \text{and}$$

$$E = \frac{\sigma}{2\epsilon_0} \quad \text{(field of an infinite sheet of charge)}$$

22.20 A cylindrical Gaussian surface is used to find the field of an infinite plane sheet of charge.

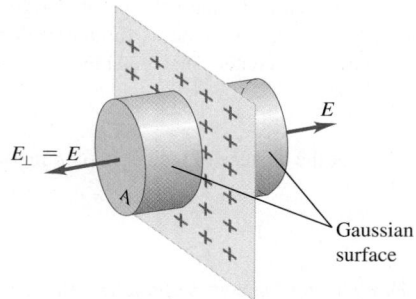

If σ is negative, $\vec{E}$ is directed *toward* the sheet, the flux through the Gaussian surface in Fig. 22.20 is negative, and σ in the expression $E = \sigma/2\epsilon_0$ denotes the magnitude (absolute value) of the charge density.

EVALUATE: We got the same result for the field of an infinite sheet of charge in Example 21.11 (Section 21.5). That calculation was much more complex and involved a fairly challenging integral. Thanks to the favorable symmetry, Gauss's law makes it much easier to solve this problem.

EXAMPLE 22.8 FIELD BETWEEN OPPOSITELY CHARGED PARALLEL CONDUCTING PLATES

Two large plane parallel conducting plates are given charges of equal magnitude and opposite sign; the surface charge densities are $+\sigma$ and $-\sigma$. Find the electric field in the region between the plates.

SOLUTION

IDENTIFY and SET UP: Figure 22.21a shows the field. Because opposite charges attract, most of the charge accumulates at the opposing faces of the plates. A small amount of charge resides on the *outer* surfaces of the plates, and there is some spreading or

"fringing" of the field at the edges. But if the plates are very large in comparison to the distance between them, the amount of charge on the outer surfaces is negligibly small, and the fringing can be ignored except near the edges. In this case we can assume that the field is uniform in the interior region between the plates, as in Fig. 22.21b, and that the charges are distributed uniformly over the opposing surfaces. To exploit this symmetry, we can use the shaded Gaussian surfaces S_1, S_2, S_3, and S_4. These surfaces are cylinders with flat ends of area A; one end of each surface lies *within* a plate.

22.21 Electric field between oppositely charged parallel plates.

(a) Realistic drawing

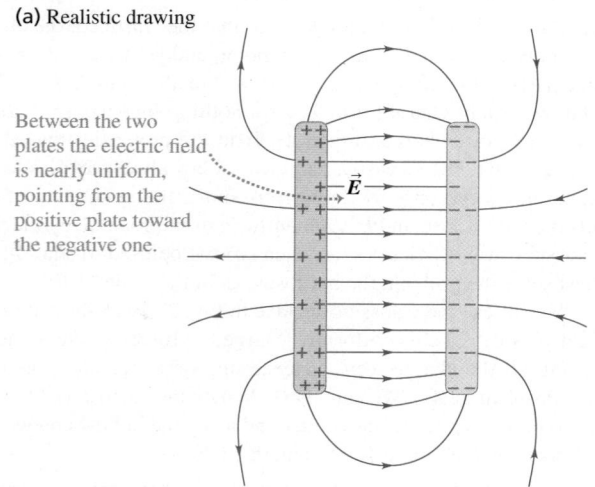

Between the two plates the electric field is nearly uniform, pointing from the positive plate toward the negative one.

(b) Idealized model

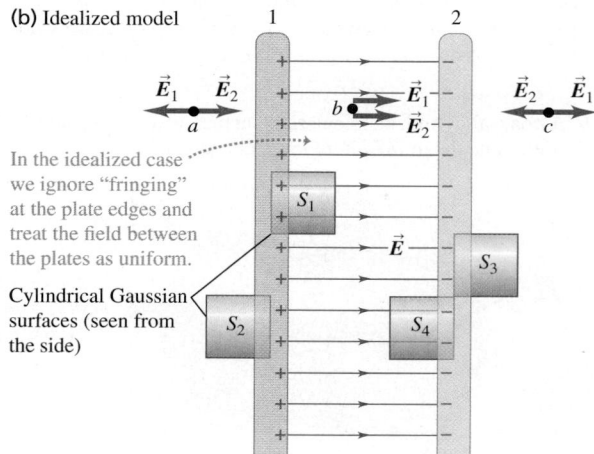

In the idealized case we ignore "fringing" at the plate edges and treat the field between the plates as uniform.

Cylindrical Gaussian surfaces (seen from the side)

EXECUTE: The left-hand end of surface S_1 is within the positive plate 1. Since the field is zero within the volume of any solid conductor under electrostatic conditions, there is no electric flux through this end. The electric field between the plates is perpendicular to the right-hand end, so on that end, $E_\perp$ is equal to E and the flux is EA; this is positive, since $\vec{E}$ is directed out of the Gaussian surface. There is no flux through the side walls of the cylinder, since these walls are parallel to $\vec{E}$. So the total flux integral in Gauss's law is EA. The net charge enclosed by the cylinder is σA, so Eq. (22.8) yields $EA = \sigma A/\epsilon_0$; we then have

$$E = \frac{\sigma}{\epsilon_0} \text{ (field between oppositely charged conducting plates)}$$

The field is uniform and perpendicular to the plates, and its magnitude is independent of the distance from either plate. The Gaussian

surface S_4 yields the same result. Surfaces S_2 and S_3 yield $E = 0$ to the left of plate 1 and to the right of plate 2, respectively. We leave these calculations to you (see Exercise 22.27).

EVALUATE: We obtained the same results in Example 21.12 by using the principle of superposition of electric fields. The fields due to the two sheets of charge (one on each plate) are $\vec{E}_1$ and $\vec{E}_2$; from Example 22.7, both of these have magnitude $\sigma/2\epsilon_0$. The total electric field at any point is the vector sum $\vec{E} = \vec{E}_1 + \vec{E}_2$. At points a and c in Fig. 22.21b, $\vec{E}_1$ and $\vec{E}_2$ point in opposite directions, and their sum is zero. At point b, $\vec{E}_1$ and $\vec{E}_2$ are in the same direction; their sum has magnitude $E = \sigma/\epsilon_0$, just as we found by using Gauss's law.

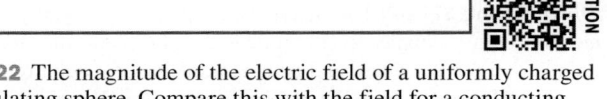

EXAMPLE 22.9 FIELD OF A UNIFORMLY CHARGED SPHERE

Positive electric charge Q is distributed uniformly *throughout the volume* of an *insulating* sphere with radius R. Find the magnitude of the electric field at a point P a distance r from the center of the sphere.

SOLUTION

IDENTIFY and SET UP: As in Example 22.5, the system is spherically symmetric. Hence we can use the conclusions of that example about the direction and magnitude of $\vec{E}$. To make use of the spherical symmetry, we choose as our Gaussian surface a sphere with radius r, concentric with the charge distribution.

EXECUTE: From symmetry, the direction of $\vec{E}$ is radial at every point on the Gaussian surface, so $E_\perp = E$ and the field magnitude E is the same at every point on the surface. Hence the total electric flux through the Gaussian surface is the product of E and the total area of the surface $A = 4\pi r^2$—that is, $\Phi_E = 4\pi r^2 E$.

The amount of charge enclosed within the Gaussian surface depends on r. To find E *inside* the sphere, we choose $r < R$. The volume charge density ρ is the charge Q divided by the volume of the entire charged sphere of radius R:

$$\rho = \frac{Q}{4\pi R^3/3}$$

The volume V_{encl} enclosed by the Gaussian surface is $\frac{4}{3}\pi r^3$, so the total charge Q_{encl} enclosed by that surface is

$$Q_{encl} = \rho V_{encl} = \left(\frac{Q}{4\pi R^3/3}\right)\left(\tfrac{4}{3}\pi r^3\right) = Q\frac{r^3}{R^3}$$

Then Gauss's law, Eq. (22.8), becomes

$$4\pi r^2 E = \frac{Q}{\epsilon_0}\frac{r^3}{R^3} \quad \text{or}$$

$$E = \frac{1}{4\pi\epsilon_0}\frac{Qr}{R^3} \quad \text{(field inside a uniformly charged sphere)}$$

The field magnitude is proportional to the distance r of the field point from the center of the sphere (see the graph of E versus r in **Fig. 22.22**).

22.22 The magnitude of the electric field of a uniformly charged insulating sphere. Compare this with the field for a conducting sphere (see Fig. 22.18).

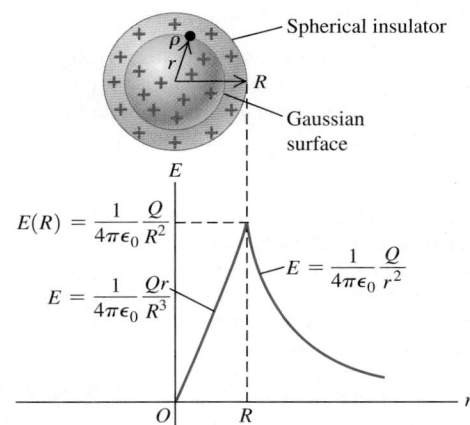

To find E *outside* the sphere, we take $r > R$. This surface encloses the entire charged sphere, so $Q_{encl} = Q$, and Gauss's law gives

$$4\pi r^2 E = \frac{Q}{\epsilon_0} \quad \text{or}$$

$$E = \frac{1}{4\pi\epsilon_0}\frac{Q}{r^2} \quad \text{(field outside a uniformly charged sphere)}$$

The field outside *any* spherically symmetric charged body varies as $1/r^2$, as though the entire charge were concentrated at the center. This is graphed in Fig. 22.22.

If the charge is *negative*, $\vec{E}$ is radially *inward* and in the expressions for E we interpret Q as the absolute value of the charge.

EVALUATE: Notice that if we set $r = R$ in either expression for E, we get the same result $E = Q/4\pi\epsilon_0 R^2$ for the magnitude of the field at the surface of the sphere. This is because the magnitude E is a *continuous* function of r. By contrast, for the charged conducting sphere of Example 22.5 the electric-field magnitude is *discontinuous* at $r = R$ (it jumps from $E = 0$ just inside the

Continued

sphere to $E = Q/4\pi\epsilon_0 R^2$ just outside the sphere). In general, the electric field $\vec{E}$ is discontinuous in magnitude, direction, or both wherever there is a *sheet* of charge, such as at the surface of a charged conducting sphere (Example 22.5), at the surface of an infinite charged sheet (Example 22.7), or at the surface of a charged conducting plate (Example 22.8).

The approach used here can be applied to *any* spherically symmetric distribution of charge, even if it is not radially uniform, as it was here. Such charge distributions occur within many atoms and atomic nuclei, so Gauss's law is useful in atomic and nuclear physics.

EXAMPLE 22.10 | CHARGE ON A HOLLOW SPHERE

A thin-walled, hollow sphere of radius 0.250 m has an unknown charge distributed uniformly over its surface. At a distance of 0.300 m from the center of the sphere, the electric field points radially inward and has magnitude 1.80×10^2 N/C. How much charge is on the sphere?

SOLUTION

IDENTIFY and SET UP: The charge distribution is spherically symmetric. As in Examples 22.5 and 22.9, it follows that the electric field is radial everywhere and its magnitude is a function of only the radial distance r from the center of the sphere. We use a spherical Gaussian surface that is concentric with the charge distribution and has radius $r = 0.300$ m. Our target variable is $Q_{encl} = q$.

EXECUTE: The charge distribution is the same as if the charge were on the surface of a 0.250-m-radius conducting sphere. Hence we can borrow the results of Example 22.5. We note that the electric

field here is directed toward the sphere, so that q must be *negative*. Furthermore, the electric field is directed into the Gaussian surface, so that $E_\perp = -E$ and $\Phi_E = \oint E_\perp \, dA = -E(4\pi r^2)$.

By Gauss's law, the flux is equal to the charge q on the sphere (all of which is enclosed by the Gaussian surface) divided by ϵ_0. Solving for q, we find

$$q = -E(4\pi\epsilon_0 r^2) = -(1.80 \times 10^2 \text{ N/C})(4\pi)$$
$$\times (8.854 \times 10^{-12} \text{ C}^2/\text{N} \cdot \text{m}^2)(0.300 \text{ m})^2$$
$$= -1.80 \times 10^{-9} \text{ C} = -1.80 \text{ nC}$$

EVALUATE: To determine the charge, we had to know the electric field at *all* points on the Gaussian surface so that we could calculate the flux integral. This was possible here because the charge distribution is highly symmetric. If the charge distribution is irregular or lacks symmetry, Gauss's law is not very useful for calculating the charge distribution from the field, or vice versa.

BIO Application Charge Distribution Inside a Nerve Cell The interior of a human nerve cell contains both positive potassium ions (K^+) and negatively charged protein molecules (Pr^-). Potassium ions can flow out of the cell through the cell membrane, but the much larger protein molecules cannot. The result is that the interior of the cell has a net negative charge. (The fluid outside the cell has a positive charge that balances this.) The fluid within the cell is a good conductor, so the Pr^- molecules distribute themselves on the outer surface of the fluid—that is, on the inner surface of the cell membrane, which is an insulator. This is true no matter what the shape of the cell.

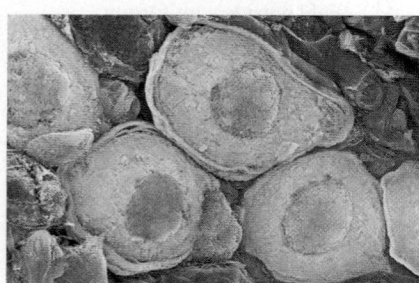

DEMO

TEST YOUR UNDERSTANDING OF SECTION 22.4 You place a known amount of charge Q on the irregularly shaped conductor shown in Fig. 22.17. If you know the size and shape of the conductor, can you use Gauss's law to calculate the electric field at an arbitrary position outside the conductor? ∎

22.5 CHARGES ON CONDUCTORS

We have learned that in an electrostatic situation (in which there is no net motion of charge) the electric field at every point within a conductor is zero and any excess charge on a solid conductor is located entirely on its surface (**Fig. 22.23a**). But what if there is a *cavity* inside the conductor (Fig. 22.23b)? If there is no charge within the cavity, we can use a Gaussian surface such as A (which lies completely within the material of the conductor) to show that the *net* charge on the *surface of the cavity* must be zero, because $\vec{E} = 0$ everywhere on the Gaussian surface. In fact, we can prove in this situation that there can't be any charge *anywhere* on the cavity surface. We will postpone detailed proof of this statement until Chapter 23.

Suppose we place a small body with a charge q inside a cavity within a conductor (Fig. 22.23c). The conductor is uncharged and is insulated from the charge q. Again $\vec{E} = 0$ everywhere on surface A, so according to Gauss's law the *total* charge inside this surface must be zero. Therefore there must be a charge $-q$ distributed on the surface of the cavity, drawn there by the charge q inside the cavity. The *total* charge on the conductor must remain zero, so a charge $+q$ must appear either on its outer surface or inside the material. But we showed that in an electrostatic situation there can't be any excess charge within the material of a conductor. So we conclude that charge $+q$ must appear on the outer surface. By the same reasoning, if the conductor originally had a charge q_C, then the total charge on the outer surface must be $q_C + q$ after charge q is inserted into the cavity.

22.23 Finding the electric field within a charged conductor.

(a) Solid conductor with charge q_C

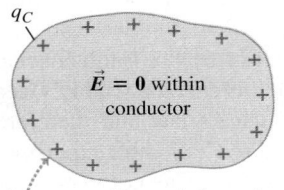

The charge q_C resides entirely on the surface of the conductor. The situation is electrostatic, so $\vec{E} = 0$ within the conductor.

(b) The same conductor with an internal cavity

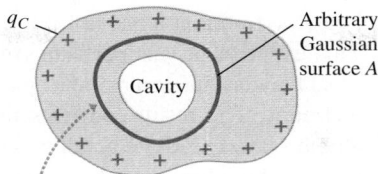

Because $\vec{E} = 0$ at all points within the conductor, the electric field at all points on the Gaussian surface must be zero.

(c) An isolated charge q placed in the cavity

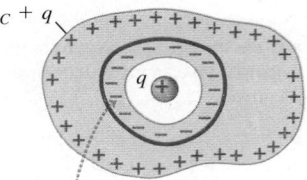

For $\vec{E}$ to be zero at all points on the Gaussian surface, the surface of the cavity must have a total charge $-q$.

CONCEPTUAL EXAMPLE 22.11   A CONDUCTOR WITH A CAVITY

A conductor with a cavity carries a total charge of $+7$ nC. Within the cavity, insulated from the conductor, is a point charge of -5 nC. How much charge is on each surface (inner and outer) of the conductor?

SOLUTION

Figure 22.24 shows the situation. If the charge in the cavity is $q = -5$ nC, the charge on the inner cavity surface must be $-q = -(-5$ nC$) = +5$ nC. The conductor carries a *total* charge of $+7$ nC, none of which is in the interior of the material. If $+5$ nC is on the inner surface of the cavity, then there must be $(+7$ nC$) - (+5$ nC$) = +2$ nC on the outer surface of the conductor.

22.24 Our sketch for this problem. There is zero electric field inside the bulk conductor and hence zero flux through the Gaussian surface shown, so the charge on the cavity wall must be the opposite of the point charge.

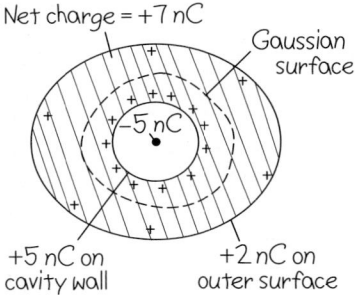

Testing Gauss's Law Experimentally

We can now consider a historic experiment, shown in **Fig. 22.25**. We mount a conducting container on an insulating stand. The container is initially uncharged. Then we hang a charged metal ball from an insulating thread (Fig. 22.25a), lower it into the container, and put the lid on (Fig. 22.25b). Charges are induced on the walls of the container, as shown. But now we let the ball *touch* the inner wall (Fig. 22.25c).

22.25 (a) A charged conducting ball suspended by an insulating thread outside a conducting container on an insulating stand. (b) The ball is lowered into the container, and the lid is put on. (c) The ball is touched to the inner surface of the container.

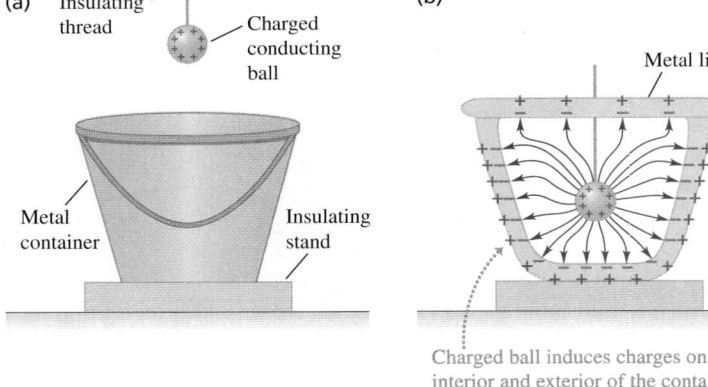

Charged ball induces charges on the interior and exterior of the container.

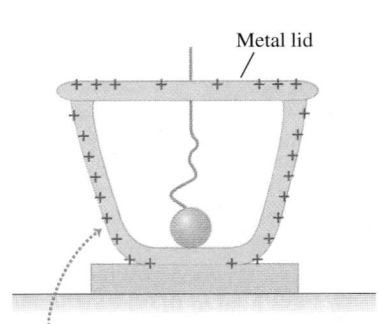

Once the ball touches the container, it is part of the interior surface; all the charge moves to the container's exterior.

DATA *SPEAKS*

The surface of the ball becomes part of the cavity surface. The situation is now the same as Fig. 22.23b; if Gauss's law is correct, the net charge on the cavity surface must be zero. Thus the ball must lose all its charge. Finally, we pull the ball out; we find that it has indeed lost all its charge.

This experiment was performed in the 19th century by the English scientist Michael Faraday, using a metal icepail with a lid, and it is called **Faraday's icepail experiment.** The result confirms the validity of Gauss's law and therefore of Coulomb's law. Faraday's result was significant because Coulomb's experimental method, using a torsion balance and dividing of charges, was not very precise; it is very difficult to confirm the $1/r^2$ dependence of the electrostatic force by direct force measurements. By contrast, experiments like Faraday's test the validity of Gauss's law, and therefore of Coulomb's law, with much greater precision. Modern versions of this experiment have shown that the exponent 2 in the $1/r^2$ of Coulomb's law does not differ from precisely 2 by more than 10^{-16}. So there is no reason to believe it is anything other than exactly 2.

The same principle behind Faraday's icepail experiment is used in a *Van de Graaff electrostatic generator* (**Fig. 22.26**). A charged belt continuously produces a buildup of charge on the inside of a conducting shell. By Gauss's law, there can never be any charge on the inner surface of this shell, so the charge is immediately carried away to the outside surface of the shell. As a result, the charge on the shell and the electric field around it can become very large very rapidly. The Van de Graaff generator is used as an accelerator of charged particles and for physics demonstrations.

This principle also forms the basis for *electrostatic shielding.* Suppose we have a very sensitive electronic instrument that we want to protect from stray electric fields that might cause erroneous measurements. We surround the instrument with a conducting box, or we line the walls, floor, and ceiling of the room with a conducting material such as sheet copper. The external electric field redistributes the free electrons in the conductor, leaving a net positive

22.26 Cutaway view of the essential parts of a Van de Graaff electrostatic generator. The electron sink at the bottom draws electrons from the belt, giving it a positive charge; at the top the belt attracts electrons away from the conducting shell, giving the shell a positive charge.

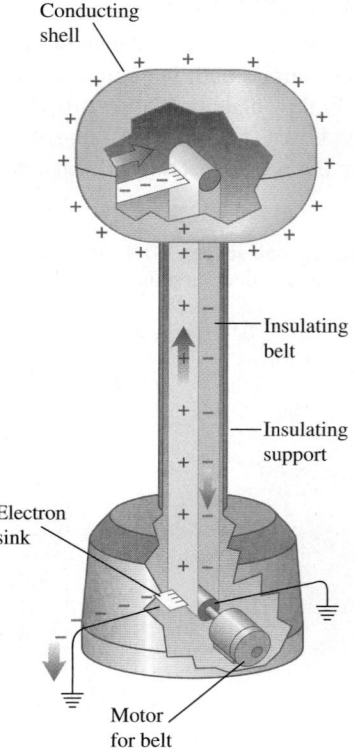

22.27 (a) A conducting box (a Faraday cage) immersed in a uniform electric field. The field of the induced charges on the box combines with the uniform field to give zero total field inside the box. (b) This person is inside a Faraday cage, and so is protected from the powerful electric discharge.

(a)

Field pushes electrons toward left side.   Net positive charge remains on right side.

$\vec{E}$ $\vec{E} = 0$ $\vec{E}$

Field perpendicular to conductor surface

(b)

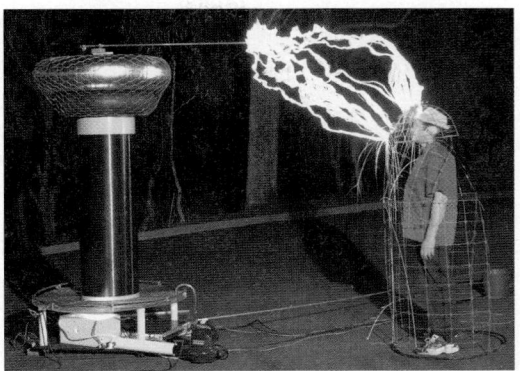

charge on the outer surface in some regions and a net negative charge in others (**Fig. 22.27**). This charge distribution causes an additional electric field such that the *total* field at every point inside the box is zero, as Gauss's law says it must be. The charge distribution on the box also alters the shapes of the field lines near the box, as the figure shows. Such a setup is often called a *Faraday cage*. The same physics tells you that one of the safest places to be in a lightning storm is inside a car; if the car is struck by lightning, the charge tends to remain on the metal skin of the vehicle, and little or no electric field is produced inside the passenger compartment.

Field at the Surface of a Conductor

Finally, we note that there is a direct relationship between the $\vec{E}$ field at a point just outside any conductor and the surface charge density σ at that point. In general, σ varies from point to point on the surface. We will show in Chapter 23 that at any such point, the direction of $\vec{E}$ is always *perpendicular* to the surface. (You can see this effect in Fig. 22.27a.)

To find a relationship between σ at any point on the surface and the perpendicular component of the electric field at that point, we construct a Gaussian surface in the form of a small cylinder (**Fig. 22.28**). One end face, with area A, lies within the conductor and the other lies just outside. The electric field is zero at all points within the conductor. Outside the conductor the component of $\vec{E}$ perpendicular to the side walls of the cylinder is zero, and over the end face the perpendicular component is equal to $E_\perp$. (If σ is positive, the electric field points out of the conductor and $E_\perp$ is positive; if σ is negative, the field points inward and $E_\perp$ is negative.) Hence the total flux through the surface is $E_\perp A$. The charge enclosed within the Gaussian surface is σA, so from Gauss's law, $E_\perp A = (\sigma A)/\epsilon_0$ and

Electric field at surface of a conductor, $\vec{E}$ perpendicular to surface
$$E_\perp = \frac{\sigma}{\epsilon_0}$$
Surface charge density ⋯ Electric constant   (22.10)

This agrees with our result for the field at the surface of a charged conducting plate (Example 22.8); soon we'll verify it for a charged conducting sphere.

We showed in Example 22.8 that the field magnitude between two infinite flat oppositely charged conducting plates also equals σ/ϵ_0. In this case the field magnitude E is the same at *all* distances from the plates, but in all other cases E decreases with increasing distance from the surface.

22.28 The field just outside a charged conductor is perpendicular to the surface, and its perpendicular component $E_\perp$ is equal to σ/ϵ_0.

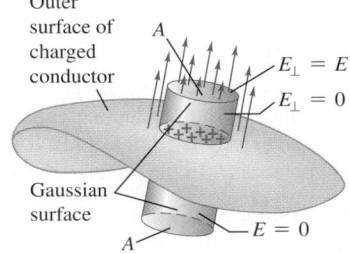

Outer surface of charged conductor $E_\perp = E$ $E_\perp = 0$

Gaussian surface $E = 0$

Application **Why Lightning Bolts Are Vertical** Our planet is a good conductor, and its surface has a negative charge. Hence, the electric field in the atmosphere above the surface points generally downward, toward the negative charge and perpendicular to the surface (see Example 22.13). The negative charge is balanced by positive charges in the atmosphere. In a lightning storm, the vertical electric field becomes great enough to cause charges to flow vertically through the air. The air is excited and ionized by the passage of charge through it, producing a visible lightning bolt.

CONCEPTUAL EXAMPLE 22.12 FIELD AT THE SURFACE OF A CONDUCTING SPHERE

Verify Eq. (22.10) for a conducting sphere with radius R and total charge q.

SOLUTION

In Example 22.5 (Section 22.4) we showed that the electric field just outside the surface is

$$E = \frac{1}{4\pi\epsilon_0}\frac{q}{R^2}$$

The surface charge density is uniform and equal to q divided by the surface area of the sphere:

$$\sigma = \frac{q}{4\pi R^2}$$

Comparing these two expressions, we see that $E = \sigma/\epsilon_0$, which verifies Eq. (22.10).

EXAMPLE 22.13 ELECTRIC FIELD OF THE EARTH

The earth (a conductor) has a net electric charge. The resulting electric field near the surface has an average value of about 150 N/C, directed toward the center of the earth. (a) What is the corresponding surface charge density? (b) What is the *total* surface charge of the earth?

SOLUTION

IDENTIFY and SET UP: We are given the electric-field magnitude at the surface of the conducting earth. We can calculate the surface charge density σ from Eq. (22.10). The total charge Q on the earth's surface is then the product of σ and the earth's surface area.

EXECUTE: (a) The direction of the field means that σ is negative (corresponding to $\vec{E}$ being directed *into* the surface, so $E_\perp$ is negative). From Eq. (22.10),

$$\sigma = \epsilon_0 E_\perp = (8.85 \times 10^{-12} \text{ C}^2/\text{N}\cdot\text{m}^2)(-150 \text{ N/C})$$

$$= -1.33 \times 10^{-9} \text{ C/m}^2 = -1.33 \text{ nC/m}^2$$

(b) The earth's surface area is $4\pi R_E^2$, where $R_E = 6.38 \times 10^6$ m is the radius of the earth (see Appendix F). The total charge Q is the product $4\pi R_E^2 \sigma$, or

$$Q = 4\pi(6.38 \times 10^6 \text{ m})^2(-1.33 \times 10^{-9} \text{ C/m}^2)$$

$$= -6.8 \times 10^5 \text{ C} = -680 \text{ kC}$$

EVALUATE: You can check our result in part (b) by using the result of Example 22.5. Solving for Q, we find

$$Q = 4\pi\epsilon_0 R^2 E_\perp$$

$$= \frac{1}{9.0 \times 10^9 \text{ N}\cdot\text{m}^2/\text{C}^2}(6.38 \times 10^6 \text{ m})^2 (-150 \text{ N/C})$$

$$= -6.8 \times 10^5 \text{ C}$$

One electron has a charge of -1.60×10^{-19} C. Hence this much excess negative electric charge corresponds to there being $(-6.8 \times 10^5 \text{ C})/(-1.60 \times 10^{-19} \text{ C}) = 4.2 \times 10^{24}$ excess electrons on the earth, or about 7 moles of excess electrons. This is compensated by an equal *deficiency* of electrons in the earth's upper atmosphere, so the combination of the earth and its atmosphere is electrically neutral.

TEST YOUR UNDERSTANDING OF SECTION 22.5 A hollow conducting sphere has no net charge. There is a positive point charge q at the center of the spherical cavity within the sphere. You connect a conducting wire from the outside of the sphere to ground. Will you measure an electric field outside the sphere? ∎

Electric flux: Electric flux is a measure of the "flow" of electric field through a surface. It is equal to the product of an area element and the perpendicular component of $\vec{E}$, integrated over a surface. (See Examples 22.1–22.3.)

$$\Phi_E = \int E \cos \phi \, dA$$

$$= \int E_\perp \, dA = \int \vec{E} \cdot d\vec{A} \qquad (22.5)$$

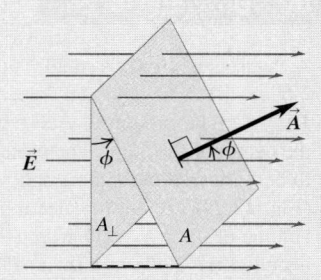

Gauss's law: Gauss's law states that the total electric flux through a closed surface, which can be written as the surface integral of the component of $\vec{E}$ normal to the surface, equals a constant times the total charge Q_{encl} enclosed by the surface. Gauss's law is logically equivalent to Coulomb's law, but its use greatly simplifies problems with a high degree of symmetry. (See Examples 22.4–22.10.)

When excess charge is placed on a solid conductor and is at rest, it resides entirely on the surface, and $\vec{E} = \mathbf{0}$ everywhere in the material of the conductor. (See Examples 22.11–22.13.)

$$\Phi_E = \oint E \cos \phi \, dA$$

$$= \oint E_\perp \, dA = \oint \vec{E} \cdot d\vec{A}$$

$$= \frac{Q_{\text{encl}}}{\epsilon_0} \qquad (22.8), (22.9)$$

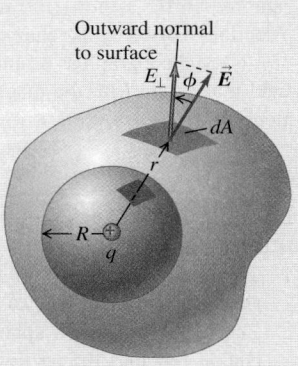

Electric field of various symmetric charge distributions: The following table lists electric fields caused by several symmetric charge distributions. In the table, q, Q, λ, and σ refer to the *magnitudes* of the quantities.

Charge Distribution	Point in Electric Field	Electric Field Magnitude
Single point charge q	Distance r from q	$E = \dfrac{1}{4\pi\epsilon_0} \dfrac{q}{r^2}$
Charge q on surface of conducting sphere with radius R	Outside sphere, $r > R$	$E = \dfrac{1}{4\pi\epsilon_0} \dfrac{q}{r^2}$
	Inside sphere, $r < R$	$E = 0$
Infinite wire, charge per unit length λ	Distance r from wire	$E = \dfrac{1}{2\pi\epsilon_0} \dfrac{\lambda}{r}$
Infinite conducting cylinder with radius R, charge per unit length λ	Outside cylinder, $r > R$	$E = \dfrac{1}{2\pi\epsilon_0} \dfrac{\lambda}{r}$
	Inside cylinder, $r < R$	$E = 0$
Solid insulating sphere with radius R, charge Q distributed uniformly throughout volume	Outside sphere, $r > R$	$E = \dfrac{1}{4\pi\epsilon_0} \dfrac{Q}{r^2}$
	Inside sphere, $r < R$	$E = \dfrac{1}{4\pi\epsilon_0} \dfrac{Qr}{R^3}$
Infinite sheet of charge with uniform charge per unit area σ	Any point	$E = \dfrac{\sigma}{2\epsilon_0}$
Two oppositely charged conducting plates with surface charge densities $+\sigma$ and $-\sigma$	Any point between plates	$E = \dfrac{\sigma}{\epsilon_0}$
Charged conductor	Just outside the conductor	$E = \dfrac{\sigma}{\epsilon_0}$

BRIDGING PROBLEM ELECTRIC FIELD INSIDE A HYDROGEN ATOM

A hydrogen atom is made up of a proton of charge $+Q = 1.60 \times 10^{-19}$ C and an electron of charge $-Q = -1.60 \times 10^{-19}$ C. The proton may be regarded as a point charge at $r = 0$, the center of the atom. The motion of the electron causes its charge to be "smeared out" into a spherical distribution around the proton (**Fig. 22.29**), so that the electron is equivalent to a charge per unit volume of $\rho(r) = -(Q/\pi a_0^3)e^{-2r/a_0}$, where $a_0 = 5.29 \times 10^{-11}$ m is called the *Bohr radius*. (a) Find the total amount of the hydrogen atom's charge that is enclosed within a sphere with radius r centered on the proton. (b) Find the electric field (magnitude and direction) caused by the charge of the hydrogen atom as a function of r. (c) Make a graph as a function of r of the ratio of the electric-field magnitude E to the magnitude of the field due to the proton alone.

22.29 The charge distribution in a hydrogen atom.

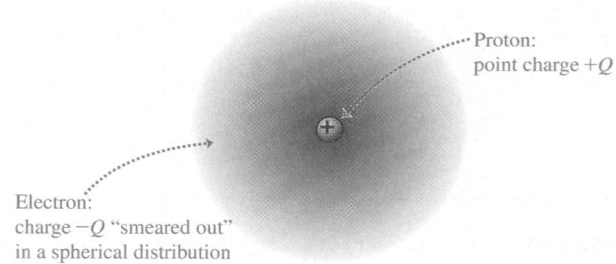

Proton:
point charge $+Q$

Electron:
charge $-Q$ "smeared out"
in a spherical distribution

--- SOLUTION GUIDE ---

IDENTIFY and SET UP

1. The charge distribution in this problem is spherically symmetric, as in Example 22.9, so you can solve it with Gauss's law.
2. The charge within a sphere of radius r includes the proton charge $+Q$ plus the portion of the electron charge distribution that lies within the sphere. The difference from Example 22.9 is that the electron charge distribution is *not* uniform, so the charge enclosed within a sphere of radius r is *not* simply the charge density multiplied by the volume $4\pi r^3/3$ of the sphere. Instead, you'll have to do an integral.
3. Consider a thin spherical shell centered on the proton, with radius r' and infinitesimal thickness dr'. Since the shell is so thin, every point within the shell is at essentially the same radius from the proton. Hence the amount of electron charge within this shell *is* equal to the electron charge density $\rho(r')$ at this radius multiplied by the volume dV of the shell. What is dV in terms of r'?

4. The total electron charge within a radius r equals the integral of $\rho(r')dV$ from $r' = 0$ to $r' = r$. Set up this integral (but don't solve it yet), and use it to write an expression for the total charge (including the proton) within a sphere of radius r.

EXECUTE

5. Integrate your expression from step 4 to find the charge within radius r. (*Hint:* Integrate by substitution: Change the integration variable from r' to $x = 2r'/a_0$. You can use integration by parts to calculate the integral $\int x^2 e^{-x}\, dx$, or you can look it up in a table of integrals or on the Web.)
6. Use Gauss's law and your results from step 5 to find the electric field at a distance r from the proton.
7. Find the ratio referred to in part (c) and graph it versus r. (You'll actually find it simplest to graph this function versus the quantity r/a_0.)

EVALUATE

8. How do your results for the enclosed charge and the electric-field magnitude behave in the limit $r \to 0$? In the limit $r \to \infty$? Explain your results.

Problems

For assigned homework and other learning materials, go to MasteringPhysics®. **MP**

•, ••, •••: Difficulty levels. **CP**: Cumulative problems incorporating material from earlier chapters. **CALC**: Problems requiring calculus.
DATA: Problems involving real data, scientific evidence, experimental design, and/or statistical reasoning. **BIO**: Biosciences problems.

DISCUSSION QUESTIONS

Q22.1 A rubber balloon has a single point charge in its interior. Does the electric flux through the balloon depend on whether or not it is fully inflated? Explain your reasoning.

Q22.2 Suppose that in Fig. 22.15 both charges were positive. What would be the fluxes through each of the four surfaces in the example?

Q22.3 In Fig. 22.15, suppose a third point charge were placed outside the purple Gaussian surface C. Would this affect the electric flux through any of the surfaces A, B, C, or D in the figure? Why or why not?

Q22.4 A certain region of space bounded by an imaginary closed surface contains no charge. Is the electric field always zero everywhere on the surface? If not, under what circumstances is it zero on the surface?

Q22.5 A spherical Gaussian surface encloses a point charge q. If the point charge is moved from the center of the sphere to a point away from the center, does the electric field at a point on the surface change? Does the total flux through the Gaussian surface change? Explain.

Q22.6 You find a sealed box on your doorstep. You suspect that the box contains several charged metal spheres packed in insulating material. How can you determine the total net charge inside the box without opening the box? Or isn't this possible?

Q22.7 A solid copper sphere has a net positive charge. The charge is distributed uniformly over the surface of the sphere, and the electric field inside the sphere is zero. Then a negative point charge outside the sphere is brought close to the surface of the sphere. Is all the net charge on the sphere still on its surface? If so, is this charge still distributed uniformly over the surface? If it is not uniform, how is it distributed? Is the electric field inside the sphere still zero? In each case justify your answers.

Q22.8 If the electric field of a point charge were proportional to $1/r^3$ instead of $1/r^2$, would Gauss's law still be valid? Explain your reasoning. (*Hint:* Consider a spherical Gaussian surface centered on a single point charge.)

Q22.9 In a conductor, one or more electrons from each atom are free to roam throughout the volume of the conductor. Does this contradict the statement that any excess charge on a solid conductor must reside on its surface? Why or why not?

Q22.10 You charge up the Van de Graaff generator shown in Fig. 22.26, and then bring an identical but uncharged hollow conducting sphere near it, without letting the two spheres touch. Sketch the distribution of charges on the second sphere. What is the net flux through the second sphere? What is the electric field inside the second sphere?

Q22.11 A lightning rod is a rounded copper rod mounted on top of a building and welded to a heavy copper cable running down into the ground. Lightning rods are used to protect houses and barns from lightning; the lightning current runs through the copper rather than through the building. Why? Why should the end of the rod be rounded?

Q22.12 A solid conductor has a cavity in its interior. Would the presence of a point charge inside the cavity affect the electric field outside the conductor? Why or why not? Would the presence of a point charge outside the conductor affect the electric field inside the cavity? Again, why or why not?

Q22.13 Explain this statement: "In a static situation, the electric field at the surface of a conductor can have no component parallel to the surface because this would violate the condition that the charges on the surface are at rest." Would this statement be valid for the electric field at the surface of an *insulator*? Explain your answer and the reason for any differences between the cases of a conductor and an insulator.

Q22.14 In a certain region of space, the electric field $\vec{E}$ is uniform. (a) Use Gauss's law to prove that this region of space must be electrically neutral; that is, the volume charge density ρ must be zero. (b) Is the converse true? That is, in a region of space where there is no charge, must $\vec{E}$ be uniform? Explain.

Q22.15 (a) In a certain region of space, the volume charge density ρ has a uniform positive value. Can $\vec{E}$ be uniform in this region? Explain. (b) Suppose that in this region of uniform positive ρ there is a "bubble" within which $\rho = 0$. Can $\vec{E}$ be uniform within this bubble? Explain.

Q22.16 A negative charge $-Q$ is placed inside the cavity of a hollow metal solid. The outside of the solid is grounded by connecting a conducting wire between it and the earth. Is any excess charge induced on the inner surface of the metal? Is there any excess charge on the outside surface of the metal? Why or why not? Would someone outside the solid measure an electric field due to the charge $-Q$? Is it reasonable to say that the grounded conductor has *shielded* the region outside the conductor from the effects of the charge $-Q$? In principle, could the same thing be done for gravity? Why or why not?

EXERCISES

Section 22.2 Calculating Electric Flux

22.1 • A flat sheet of paper of area 0.250 m² is oriented so that the normal to the sheet is at an angle of 60° to a uniform electric field of magnitude 14 N/C. (a) Find the magnitude of the electric flux through the sheet. (b) Does the answer to part (a) depend on the shape of the sheet? Why or why not? (c) For what angle ϕ between the normal to the sheet and the electric field is the magnitude of the flux through the sheet (i) largest and (ii) smallest? Explain your answers.

22.2 •• A flat sheet is in the shape of a rectangle with sides of lengths 0.400 m and 0.600 m. The sheet is immersed in a uniform electric field of magnitude 90.0 N/C that is directed at 20° from the plane of the sheet (**Fig. E22.2**). Find the magnitude of the electric flux through the sheet.

Figure **E22.2**

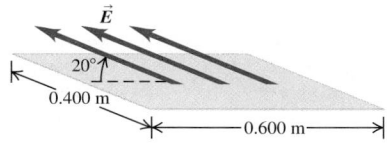

22.3 • You measure an electric field of 1.25×10^6 N/C at a distance of 0.150 m from a point charge. There is no other source of electric field in the region other than this point charge. (a) What is the electric flux through the surface of a sphere that has this charge at its center and that has radius 0.150 m? (b) What is the magnitude of this charge?

22.4 • It was shown in Example 21.10 (Section 21.5) that the electric field due to an infinite line of charge is perpendicular to the line and has magnitude $E = \lambda/2\pi\epsilon_0 r$. Consider an imaginary cylinder with radius $r = 0.250$ m and length $l = 0.400$ m that has an infinite line of positive charge running along its axis. The charge per unit length on the line is $\lambda = 3.00 \ \mu$C/m. (a) What is the electric flux through the cylinder due to this infinite line of charge? (b) What is the flux through the cylinder if its radius is increased to $r = 0.500$ m? (c) What is the flux through the cylinder if its length is increased to $l = 0.800$ m?

22.5 •• A hemispherical surface with radius r in a region of uniform electric field $\vec{E}$ has its axis aligned parallel to the direction of the field. Calculate the flux through the surface.

22.6 • The cube in **Fig. E22.6** has sides of length $L = 10.0$ cm. The electric field is uniform, has magnitude $E = 4.00 \times 10^3$ N/C, and is parallel to the *xy*-plane at an angle of 53.1° measured from the +*x*-axis toward the +*y*-axis. (a) What is the electric flux through each of the six cube faces S_1, S_2, S_3, S_4, S_5, and S_6? (b) What is the total electric flux through all faces of the cube?

Figure **E22.6**

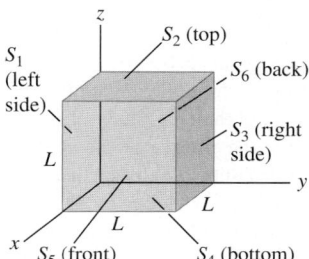

Section 22.3 Gauss's Law

22.7 • BIO As discussed in Section 22.5, human nerve cells have a net negative charge and the material in the interior of the cell is a good conductor. If a cell has a net charge of -8.65 pC, what are the magnitude and direction (inward or outward) of the net flux through the cell boundary?

22.8 • The three small spheres shown in **Fig. E22.8** carry charges $q_1 = 4.00$ nC, $q_2 = -7.80$ nC, and $q_3 = 2.40$ nC. Find the net electric flux through each of the following closed surfaces shown in cross section in the figure: (a) S_1; (b) S_2; (c) S_3; (d) S_4; (e) S_5. (f) Do your answers to parts (a)–(e) depend on how the charge is distributed over each small sphere? Why or why not?

Figure **E22.8**

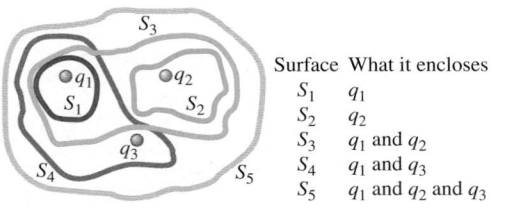

Surface	What it encloses
S_1	q_1
S_2	q_2
S_3	q_1 and q_2
S_4	q_1 and q_3
S_5	q_1 and q_2 and q_3

22.9 •• A charged paint is spread in a very thin uniform layer over the surface of a plastic sphere of diameter 12.0 cm, giving it a charge of $-49.0\ \mu$C. Find the electric field (a) just inside the paint layer; (b) just outside the paint layer; (c) 5.00 cm outside the surface of the paint layer.

22.10 • A point charge $q_1 = 4.00$ nC is located on the x-axis at $x = 2.00$ m, and a second point charge $q_2 = -6.00$ nC is on the y-axis at $y = 1.00$ m. What is the total electric flux due to these two point charges through a spherical surface centered at the origin and with radius (a) 0.500 m, (b) 1.50 m, (c) 2.50 m?

22.11 • A 6.20-μC point charge is at the center of a cube with sides of length 0.500 m. (a) What is the electric flux through one of the six faces of the cube? (b) How would your answer to part (a) change if the sides were 0.250 m long? Explain.

22.12 • **Electric Fields in an Atom.** The nuclei of large atoms, such as uranium, with 92 protons, can be modeled as spherically symmetric spheres of charge. The radius of the uranium nucleus is approximately 7.4×10^{-15} m. (a) What is the electric field this nucleus produces just outside its surface? (b) What magnitude of electric field does it produce at the distance of the electrons, which is about 1.0×10^{-10} m? (c) The electrons can be modeled as forming a uniform shell of negative charge. What net electric field do they produce at the location of the nucleus?

Section 22.4 Applications of Gauss's Law and Section 22.5 Charges on Conductors

22.13 •• Two very long uniform lines of charge are parallel and are separated by 0.300 m. Each line of charge has charge per unit length $+5.20\ \mu$C/m. What magnitude of force does one line of charge exert on a 0.0500-m section of the other line of charge?

22.14 •• A solid metal sphere with radius 0.450 m carries a net charge of 0.250 nC. Find the magnitude of the electric field (a) at a point 0.100 m outside the surface of the sphere and (b) at a point inside the sphere, 0.100 m below the surface.

22.15 •• How many excess electrons must be added to an isolated spherical conductor 26.0 cm in diameter to produce an electric field of magnitude 1150 N/C just outside the surface?

22.16 • Some planetary scientists have suggested that the planet Mars has an electric field somewhat similar to that of the earth, producing a net electric flux of -3.63×10^{16} N·m^2/C at the planet's surface. Calculate: (a) the total electric charge on the planet; (b) the electric field at the planet's surface (refer to the astronomical data inside the back cover); (c) the charge density on Mars, assuming all the charge is uniformly distributed over the planet's surface.

22.17 •• A very long uniform line of charge has charge per unit length 4.80 μC/m and lies along the x-axis. A second long uniform line of charge has charge per unit length $-2.40\ \mu$C/m and is parallel to the x-axis at $y = 0.400$ m. What is the net electric field (magnitude and direction) at the following points on the y-axis: (a) $y = 0.200$ m and (b) $y = 0.600$ m?

22.18 •• The electric field 0.400 m from a very long uniform line of charge is 840 N/C. How much charge is contained in a 2.00-cm section of the line?

22.19 •• A hollow, conducting sphere with an outer radius of 0.250 m and an inner radius of 0.200 m has a uniform surface charge density of $+6.37 \times 10^{-6}$ C/m^2. A charge of $-0.500\ \mu$C is now introduced at the center of the cavity inside the sphere. (a) What is the new charge density on the outside of the sphere? (b) Calculate the strength of the electric field just outside the sphere. (c) What is the electric flux through a spherical surface just inside the inner surface of the sphere?

22.20 • (a) At a distance of 0.200 cm from the center of a charged conducting sphere with radius 0.100 cm, the electric field is 480 N/C. What is the electric field 0.600 cm from the center of the sphere? (b) At a distance of 0.200 cm from the axis of a very long charged conducting cylinder with radius 0.100 cm, the electric field is 480 N/C. What is the electric field 0.600 cm from the axis of the cylinder? (c) At a distance of 0.200 cm from a large uniform sheet of charge, the electric field is 480 N/C. What is the electric field 1.20 cm from the sheet?

22.21 •• The electric field at a distance of 0.145 m from the surface of a solid insulating sphere with radius 0.355 m is 1750 N/C. (a) Assuming the sphere's charge is uniformly distributed, what is the charge density inside it? (b) Calculate the electric field inside the sphere at a distance of 0.200 m from the center.

22.22 •• A point charge of $-3.00\ \mu$C is located in the center of a spherical cavity of radius 6.50 cm that, in turn, is at the center of an insulating charged solid sphere. The charge density in the solid is $\rho = 7.35 \times 10^{-4}$ C/m^3. Calculate the electric field inside the solid at a distance of 9.50 cm from the center of the cavity.

22.23 •• CP An electron is released from rest at a distance of 0.300 m from a large insulating sheet of charge that has uniform surface charge density $+2.90 \times 10^{-12}$ C/m^2. (a) How much work is done on the electron by the electric field of the sheet as the electron moves from its initial position to a point 0.050 m from the sheet? (b) What is the speed of the electron when it is 0.050 m from the sheet?

22.24 •• Charge Q is distributed uniformly throughout the volume of an insulating sphere of radius $R = 4.00$ cm. At a distance of $r = 8.00$ cm from the center of the sphere, the electric field due to the charge distribution has magnitude $E = 940$ N/C. What are (a) the volume charge density for the sphere and (b) the electric field at a distance of 2.00 cm from the sphere's center?

22.25 • A conductor with an inner cavity, like that shown in Fig. 22.23c, carries a total charge of $+5.00$ nC. The charge within the cavity, insulated from the conductor, is -6.00 nC. How much charge is on (a) the inner surface of the conductor and (b) the outer surface of the conductor?

22.26 •• A very large, horizontal, nonconducting sheet of charge has uniform charge per unit area $\sigma = 5.00 \times 10^{-6} \, \text{C/m}^2$. (a) A small sphere of mass $m = 8.00 \times 10^{-6} \, \text{kg}$ and charge q is placed 3.00 cm above the sheet of charge and then released from rest. (a) If the sphere is to remain motionless when it is released, what must be the value of q? (b) What is q if the sphere is released 1.50 cm above the sheet?

22.27 • Apply Gauss's law to the Gaussian surfaces S_2, S_3, and S_4 in Fig. 22.21b to calculate the electric field between and outside the plates.

22.28 • A square insulating sheet 80.0 cm on a side is held horizontally. The sheet has 4.50 nC of charge spread uniformly over its area. (a) Calculate the electric field at a point 0.100 mm above the center of the sheet. (b) Estimate the electric field at a point 100 m above the center of the sheet. (c) Would the answers to parts (a) and (b) be different if the sheet were made of a conducting material? Why or why not?

22.29 • An infinitely long cylindrical conductor has radius R and uniform surface charge density σ. (a) In terms of σ and R, what is the charge per unit length λ for the cylinder? (b) In terms of σ, what is the magnitude of the electric field produced by the charged cylinder at a distance $r > R$ from its axis? (c) Express the result of part (b) in terms of λ and show that the electric field outside the cylinder is the same as if all the charge were on the axis. Compare your result to the result for a line of charge in Example 22.6 (Section 22.4).

22.30 • Two very large, nonconducting plastic sheets, each 10.0 cm thick, carry uniform charge densities σ_1, σ_2, σ_3, and σ_4 on their surfaces (**Fig. E22.30**). These surface charge densities have the values $\sigma_1 = -6.00 \, \mu\text{C/m}^2$, $\sigma_2 = +5.00 \, \mu\text{C/m}^2$, $\sigma_3 = +2.00 \, \mu\text{C/m}^2$, and $\sigma_4 = +4.00 \, \mu\text{C/m}^2$. Use Gauss's law to find the magnitude and direction of the electric field at the following points, far from the edges of these sheets: (a) point A, 5.00 cm from the left face of the left-hand sheet; (b) point B, 1.25 cm from the inner surface of the right-hand sheet; (c) point C, in the middle of the right-hand sheet.

Figure **E22.30**

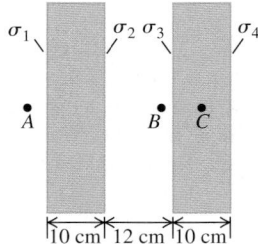

PROBLEMS

22.31 •• CP At time $t = 0$ a proton is a distance of 0.360 m from a very large insulating sheet of charge and is moving parallel to the sheet with speed $9.70 \times 10^2 \, \text{m/s}$. The sheet has uniform surface charge density $2.34 \times 10^{-9} \, \text{C/m}^2$. What is the speed of the proton at $t = 5.00 \times 10^{-8} \, \text{s}$?

22.32 •• CP A very small object with mass $8.20 \times 10^{-9} \, \text{kg}$ and positive charge $6.50 \times 10^{-9} \, \text{C}$ is projected directly toward a very large insulating sheet of positive charge that has uniform surface charge density $5.90 \times 10^{-8} \, \text{C/m}^2$. The object is initially 0.400 m from the sheet. What initial speed must the object have in order for its closest distance of approach to the sheet to be 0.100 m?

22.33 •• CP A small sphere with mass $4.00 \times 10^{-6} \, \text{kg}$ and charge $5.00 \times 10^{-8} \, \text{C}$ hangs from a thread near a very large, charged insulating sheet (**Fig. P22.33**). The charge density on the surface of the sheet is uniform and equal to $-2.50 \times 10^{-9} \, \text{C/m}^2$. Find the angle of the thread.

Figure **P22.33**

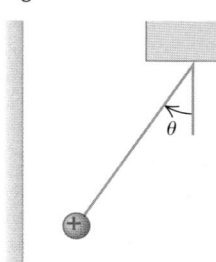

22.34 •• A cube has sides of length $L = 0.300 \, \text{m}$. One corner is at the origin (Fig. E22.6). The nonuniform electric field is given by $\vec{E} = (-5.00 \, \text{N/C} \cdot \text{m})x\hat{\imath} + (3.00 \, \text{N/C} \cdot \text{m})z\hat{k}$. (a) Find the electric flux through each of the six cube faces S_1, S_2, S_3, S_4, S_5, and S_6. (b) Find the total electric charge inside the cube.

22.35 • The electric field $\vec{E}$ in Fig. P22.35 is everywhere parallel to the x-axis, so the components E_y and E_z are zero. The x-component of the field E_x depends on x but not on y or z. At points in the yz-plane (where $x = 0$), $E_x = 125 \, \text{N/C}$. (a) What is the electric flux through surface I in Fig. P22.35? (b) What is the electric flux through surface II? (c) The volume shown is a small section of a very large insulating slab 1.0 m thick. If there is a total charge of $-24.0 \, \text{nC}$ within the volume shown, what are the magnitude and direction of $\vec{E}$ at the face opposite surface I? (d) Is the electric field produced by charges only within the slab, or is the field also due to charges outside the slab? How can you tell?

Figure **P22.35**

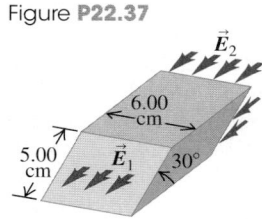

22.36 •• CALC In a region of space there is an electric field $\vec{E}$ that is in the z-direction and that has magnitude $E = [964 \, \text{N}/(\text{C} \cdot \text{m})]x$. Find the flux for this field through a square in the xy-plane at $z = 0$ and with side length 0.350 m. One side of the square is along the $+x$-axis and another side is along the $+y$-axis.

22.37 •• The electric field $\vec{E}_1$ at one face of a parallelepiped is uniform over the entire face and is directed out of the face. At the opposite face, the electric field $\vec{E}_2$ is also uniform over the entire face and is directed into that face (**Fig. P22.37**). The two faces in question are inclined at 30.0° from the horizontal, while both $\vec{E}_1$ and $\vec{E}_2$ are horizontal; $\vec{E}_1$ has a magnitude of $2.50 \times 10^4 \, \text{N/C}$, and $\vec{E}_2$ has a magnitude of $7.00 \times 10^4 \, \text{N/C}$. (a) Assuming that no other electric field lines cross the surfaces of the parallelepiped, determine the net charge contained within. (b) Is the electric field produced by the charges only within the parallelepiped, or is the field also due to charges outside the parallelepiped? How can you tell?

22.38 • A long line carrying a uniform linear charge density $+50.0 \, \mu\text{C/m}$ runs parallel to and 10.0 cm from the surface of a large, flat plastic sheet that has a uniform surface charge density of $-100 \, \mu\text{C/m}^2$ on one side. Find the location of all points where an α particle would feel no force due to this arrangement of charged objects.

22.39 • **The Coaxial Cable.** A long coaxial cable consists of an inner cylindrical conductor with radius a and an outer coaxial cylinder with inner radius b and outer radius c. The outer cylinder is mounted on insulating supports and has no net charge. The inner cylinder has a uniform positive charge per unit length λ. Calculate the electric field (a) at any point between the cylinders a distance r from the axis and (b) at any point outside the outer cylinder. (c) Graph the magnitude of the electric field as a function of the distance r from the axis of the cable, from $r = 0$ to $r = 2c$. (d) Find the charge per unit length on the inner surface and on the outer surface of the outer cylinder.

22.40 • A very long conducting tube (hollow cylinder) has inner radius a and outer radius b. It carries charge per unit length $+\alpha$, where α is a positive constant with units of C/m. A line of charge lies along the axis of the tube. The line of charge has charge per unit length $+\alpha$. (a) Calculate the electric field in terms of α and the distance r from the axis of the tube for (i) $r < a$; (ii) $a < r < b$; (iii) $r > b$. Show your results in a graph of E as a function of r. (b) What is the charge per unit length on (i) the inner surface of the tube and (ii) the outer surface of the tube?

22.41 • A very long, solid cylinder with radius R has positive charge uniformly distributed throughout it, with charge per unit volume ρ. (a) Derive the expression for the electric field inside the volume at a distance r from the axis of the cylinder in terms of the charge density ρ. (b) What is the electric field at a point outside the volume in terms of the charge per unit length λ in the cylinder? (c) Compare the answers to parts (a) and (b) for $r = R$. (d) Graph the electric-field magnitude as a function of r from $r = 0$ to $r = 3R$.

22.42 • **A Sphere in a Sphere.** A solid conducting sphere carrying charge q has radius a. It is inside a concentric hollow conducting sphere with inner radius b and outer radius c. The hollow sphere has no net charge. (a) Derive expressions for the electric-field magnitude in terms of the distance r from the center for the regions $r < a$, $a < r < b$, $b < r < c$, and $r > c$. (b) Graph the magnitude of the electric field as a function of r from $r = 0$ to $r = 2c$. (c) What is the charge on the inner surface of the hollow sphere? (d) On the outer surface? (e) Represent the charge of the small sphere by four plus signs. Sketch the field lines of the system within a spherical volume of radius $2c$.

22.43 • A solid conducting sphere with radius R that carries positive charge Q is concentric with a very thin insulating shell of radius $2R$ that also carries charge Q. The charge Q is distributed uniformly over the insulating shell. (a) Find the electric field (magnitude and direction) in each of the regions $0 < r < R$, $R < r < 2R$, and $r > 2R$. (b) Graph the electric-field magnitude as a function of r.

22.44 • A conducting spherical shell with inner radius a and outer radius b has a positive point charge Q located at its center. The total charge on the shell is $-3Q$, and it is insulated from its surroundings (**Fig. P22.44**). (a) Derive expressions for the electric-field magnitude E in terms of the distance r from the center for the regions $r < a$, $a < r < b$, and $r > b$. What is the surface charge density (b) on the inner surface of the conducting shell; (c) on the outer surface of the conducting shell? (d) Sketch the electric field lines and the location of all charges. (e) Graph E as a function of r.

Figure **P22.44**

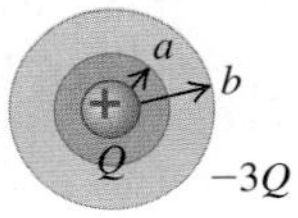

22.45 • **Concentric Spherical Shells.** A small conducting spherical shell with inner radius a and outer radius b is concentric with a larger conducting spherical shell with inner radius c and outer radius d (**Fig. P22.45**). The inner shell has total charge

$+2q$, and the outer shell has charge $+4q$. (a) Calculate the electric field $\vec{E}$ (magnitude and direction) in terms of q and the distance r from the common center of the two shells for (i) $r < a$; (ii) $a < r < b$; (iii) $b < r < c$; (iv) $c < r < d$; (v) $r > d$. Graph the radial component of $\vec{E}$ as a function of r. (b) What is the total charge on the (i) inner surface of the small shell; (ii) outer surface of the small shell; (iii) inner surface of the large shell; (iv) outer surface of the large shell?

Figure **P22.45**

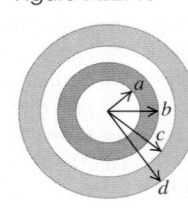

22.46 • Repeat Problem 22.45, but now let the outer shell have charge $-2q$. The inner shell still has charge $+2q$.

22.47 • Negative charge $-Q$ is distributed uniformly over the surface of a thin spherical insulating shell with radius R. Calculate the force (magnitude and direction) that the shell exerts on a positive point charge q located a distance (a) $r > R$ from the center of the shell (outside the shell); (b) $r < R$ from the center of the shell (inside the shell).

22.48 • A solid conducting sphere with radius R carries a positive total charge Q. The sphere is surrounded by an insulating shell with inner radius R and outer radius $2R$. The insulating shell has a uniform charge density ρ. (a) Find the value of ρ so that the net charge of the entire system is zero. (b) If ρ has the value found in part (a), find the electric field $\vec{E}$ (magnitude and direction) in each of the regions $0 < r < R$, $R < r < 2R$, and $r > 2R$. Graph the radial component of $\vec{E}$ as a function of r. (c) As a general rule, the electric field is discontinuous only at locations where there is a thin sheet of charge. Explain how your results in part (b) agree with this rule.

22.49 ••• **CALC** An insulating hollow sphere has inner radius a and outer radius b. Within the insulating material the volume charge density is given by $\rho(r) = \alpha/r$, where α is a positive constant. (a) In terms of α and a, what is the magnitude of the electric field at a distance r from the center of the shell, where $a < r < b$? (b) A point charge q is placed at the center of the hollow space, at $r = 0$. In terms of α and a, what value must q have (sign and magnitude) in order for the electric field to be constant in the region $a < r < b$, and what then is the value of the constant field in this region?

22.50 •• **CP** **Thomson's Model of the Atom.** Early in the 20th century, a leading model of the structure of the atom was that of English physicist J. J. Thomson (the discoverer of the electron). In Thomson's model, an atom consisted of a sphere of positively charged material in which were embedded negatively charged electrons, like chocolate chips in a ball of cookie dough. Consider such an atom consisting of one electron with mass m and charge $-e$, which may be regarded as a point charge, and a uniformly charged sphere of charge $+e$ and radius R. (a) Explain why the electron's equilibrium position is at the center of the nucleus. (b) In Thomson's model, it was assumed that the positive material provided little or no resistance to the electron's motion. If the electron is displaced from equilibrium by a distance less than R, show that the resulting motion of the electron will be simple harmonic, and calculate the frequency of oscillation. (*Hint:* Review the definition of SHM in Section 14.2. If it can be shown that the net force on the electron is of this form, then it follows that the motion is simple harmonic. Conversely, if the net force on the electron does not follow this form, the motion is not simple harmonic.) (c) By Thomson's time, it was known that excited atoms emit light waves of only certain frequencies. In his model, the frequency of emitted light is the same as the oscillation frequency of the electron(s) in the atom. What radius would a Thomson-model atom need for it to produce red light of frequency 4.57×10^{14} Hz? Compare your answer to the radii of real atoms, which are of the order of 10^{-10} m (see Appendix F). (d) If the electron were displaced from equilibrium by a distance greater than R, would the electron oscillate? Would its motion be simple harmonic? Explain your reasoning. (*Historical note:* In 1910, the atomic nucleus was discovered, proving the Thomson model to be incorrect. An atom's positive charge is not spread over its volume, as Thomson supposed, but is concentrated in the tiny nucleus of radius 10^{-14} to 10^{-15} m.)

22.51 • **Thomson's Model of the Atom, Continued.** Using Thomson's (outdated) model of the atom described in Problem 22.50, consider an atom consisting of two electrons, each of charge $-e$, embedded in a sphere of charge $+2e$ and radius R. In equilibrium, each electron is a distance d from the center of the atom (**Fig. P22.51**). Find the distance d in terms of the other properties of the atom.

Figure **P22.51**

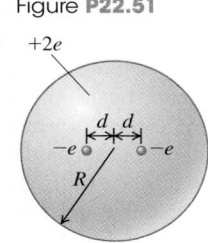

22.52 •• (a) How many excess electrons must be distributed uniformly within the volume of an isolated plastic sphere 30.0 cm in diameter to produce an electric field of magnitude 1390 N/C just outside the surface of the sphere? (b) What is the electric field at a point 10.0 cm outside the surface of the sphere?

22.53 •• CALC A nonuniform, but spherically symmetric, distribution of charge has a charge density $\rho(r)$ given as follows:

$$\rho(r) = \rho_0\left(1 - \frac{r}{R}\right) \qquad \text{for } r \leq R$$

$$\rho(r) = 0 \qquad \text{for } r \geq R$$

where $\rho_0 = 3Q/\pi R^3$ is a positive constant. (a) Show that the total charge contained in the charge distribution is Q. (b) Show that the electric field in the region $r \geq R$ is identical to that produced by a point charge Q at $r = 0$. (c) Obtain an expression for the electric field in the region $r \leq R$. (d) Graph the electric-field magnitude E as a function of r. (e) Find the value of r at which the electric field is maximum, and find the value of that maximum field.

22.54 • **A Uniformly Charged Slab.** A slab of insulating material has thickness $2d$ and is oriented so that its faces are parallel to the yz-plane and given by the planes $x = d$ and $x = -d$. The y- and z-dimensions of the slab are very large compared to d; treat them as essentially infinite. The slab has a uniform positive charge density ρ. (a) Explain why the electric field due to the slab is zero at the center of the slab ($x = 0$). (b) Using Gauss's law, find the electric field due to the slab (magnitude and direction) at all points in space.

22.55 • CALC **A Nonuniformly Charged Slab.** Repeat Problem 22.54, but now let the charge density of the slab be given by $\rho(x) = \rho_0(x/d)^2$, where ρ_0 is a positive constant.

22.56 • CALC A nonuniform, but spherically symmetric, distribution of charge has a charge density $\rho(r)$ given as follows:

$$\rho(r) = \rho_0\left(1 - \frac{4r}{3R}\right) \qquad \text{for } r \leq R$$

$$\rho(r) = 0 \qquad \text{for } r \geq R$$

where ρ_0 is a positive constant. (a) Find the total charge contained in the charge distribution. Obtain an expression for the electric field in the region (b) $r \geq R$; (c) $r \leq R$. (d) Graph the electric-field magnitude E as a function of r. (e) Find the value of r at which the electric field is maximum, and find the value of that maximum field.

22.57 • (a) An insulating sphere with radius a has a uniform charge density ρ. The sphere is not centered at the origin but at $\vec{r} = \vec{b}$. Show that the electric field inside the sphere is given by $\vec{E} = \rho(\vec{r} - \vec{b})/3\epsilon_0$. (b) An insulating sphere of radius R has a spherical hole of radius a located within its volume and centered a distance b from the center of the sphere, where $a < b < R$ (a cross section of the sphere is shown in **Fig. P22.57**). The solid part of the sphere has a uniform volume charge density ρ. Find the magnitude and direction of the electric field $\vec{E}$ inside the hole, and show that $\vec{E}$ is uniform over the entire hole. [*Hint:* Use the principle of superposition and the result of part (a).]

Figure **P22.57**

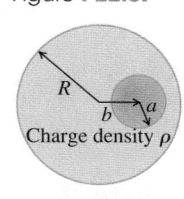

Charge density ρ

22.58 • A very long, solid insulating cylinder has radius R; bored along its entire length is a cylindrical hole with radius a. The axis of the hole is a distance b from the axis of the cylinder, where $a < b < R$ (**Fig. P22.58**). The solid material of the cylinder has a uniform volume charge density ρ. Find the magnitude and direction of the electric field $\vec{E}$ inside the hole, and show that $\vec{E}$ is uniform over the entire hole. (*Hint:* See Problem 22.57.)

Figure **P22.58**

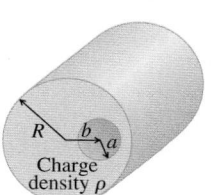

Charge density ρ

22.59 •• DATA In one experiment the electric field is measured for points at distances r from a uniform line of charge that has charge per unit length λ and length l, where $l \gg r$. In a second experiment the electric field is measured for points at distances r from the center of a uniformly charged insulating sphere that has volume charge density ρ and radius $R = 8.00$ mm, where $r > R$. The results of the two measurements are listed in the table, but you aren't told which set of data applies to which experiment:

r (cm)	1.00	1.50	2.00	2.50	3.00	3.50	4.00
Measurement A							
E (10^5 N/C)	2.72	1.79	1.34	1.07	0.902	0.770	0.677
Measurement B							
E (10^5 N/C)	5.45	2.42	1.34	0.861	0.605	0.443	0.335

For each set of data, draw two graphs: one for Er^2 versus r and one for Er versus r. (a) Use these graphs to determine which data set, A or B, is for the uniform line of charge and which set is for the uniformly charged sphere. Explain your reasoning. (b) Use the graphs in part (a) to calculate λ for the uniform line of charge and ρ for the uniformly charged sphere.

22.60 •• **DATA** The electric field is measured for points at distances r from the center of a uniformly charged insulating sphere that has volume charge density ρ and radius R, where $r < R$ (**Fig. P22.60**). Calculate ρ.

Figure **P22.60**

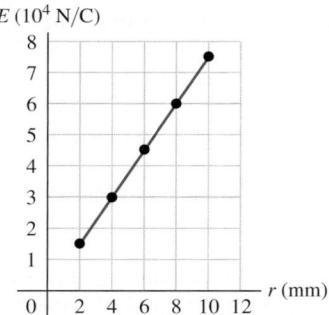

22.61 •• **DATA** The volume charge density ρ for a spherical charge distribution of radius $R = 6.00$ mm is not uniform. **Figure P22.61** shows ρ as a function of the distance r from the center of the distribution. Calculate the electric field at these values of r: (i) 1.00 mm; (ii) 3.00 mm; (iii) 5.00 mm; (iv) 7.00 mm.

Figure **P22.61**

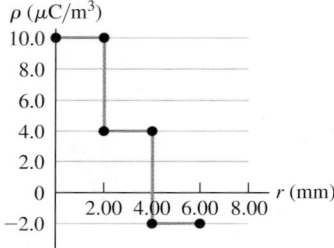

CHALLENGE PROBLEM

22.62 ••• **CP CALC** A region in space contains a total positive charge Q that is distributed spherically such that the volume charge density $\rho(r)$ is given by

$$\rho(r) = 3\alpha r/2R \qquad \text{for } r \leq R/2$$
$$\rho(r) = \alpha[1 - (r/R)^2] \qquad \text{for } R/2 \leq r \leq R$$
$$\rho(r) = 0 \qquad \text{for } r \geq R$$

Here α is a positive constant having units of C/m^3. (a) Determine α in terms of Q and R. (b) Using Gauss's law, derive an expression for the magnitude of the electric field as a function of r. Do this separately for all three regions. Express your answers in terms of Q. (c) What fraction of the total charge is contained within the region $R/2 \leq r \leq R$? (d) What is the magnitude of $\vec{E}$ at $r = R/2$? (e) If an electron with charge $q' = -e$ is released from rest at any point in any of the three regions, the resulting motion will be oscillatory but not simple harmonic. Why?

PASSAGE PROBLEMS

SPACE RADIATION SHIELDING. One of the hazards facing humans in space is space radiation: high-energy charged particles emitted by the sun. During a solar flare, the intensity of this radiation can reach lethal levels. One proposed method of protection for astronauts on the surface of the moon or Mars is an array of large, electrically charged spheres placed high above areas where people live and work. The spheres would produce a strong electric field $\vec{E}$ to deflect the charged particles that make up space radiation. The spheres would be similar in construction to a Mylar balloon, with a thin, electrically conducting layer on the outside surface on which a net positive or negative charge would be placed. A typical sphere might be 5 m in diameter.

22.63 Suppose that to repel electrons in the radiation from a solar flare, each sphere must produce an electric field $\vec{E}$ of magnitude 1×10^6 N/C at 25 m from the center of the sphere. What net charge on each sphere is needed? (a) -0.07 C; (b) -8 mC; (c) -80 μC; (d) -1×10^{-20} C.

22.64 What is the magnitude of $\vec{E}$ just outside the surface of such a sphere? (a) 0; (b) 10^6 N/C; (c) 10^7 N/C; (d) 10^8 N/C.

22.65 What is the direction of $\vec{E}$ just outside the surface of such a sphere? (a) Tangent to the surface of the sphere; (b) perpendicular to the surface, pointing toward the sphere; (c) perpendicular to the surface, pointing away from the sphere; (d) there is no electric field just outside the surface.

22.66 Which statement is true about $\vec{E}$ inside a negatively charged sphere as described here? (a) It points from the center of the sphere to the surface and is largest at the center. (b) It points from the surface to the center of the sphere and is largest at the surface. (c) It is zero. (d) It is constant but not zero.

Answers

Chapter Opening Question ?

(iii) The electric field inside a cavity within a conductor is zero, so there would be no electric effect on the child. (See Section 22.5.)

Test Your Understanding Questions

22.1 (iii) Each part of the surface of the box will be three times farther from the charge $+q$, so the electric field will be $\left(\frac{1}{3}\right)^2 = \frac{1}{9}$ as strong. But the area of the box will increase by a factor of $3^2 = 9$. Hence the electric flux will be multiplied by a factor of $\left(\frac{1}{9}\right)(9) = 1$. In other words, the flux will be unchanged.

22.2 (iv), (ii), (i), (iii) In each case the electric field is uniform, so the flux is $\Phi_E = \vec{E} \cdot \vec{A}$. We use the relationships for the scalar products of unit vectors: $\hat{\imath} \cdot \hat{\imath} = \hat{\jmath} \cdot \hat{\jmath} = 1$, $\hat{\imath} \cdot \hat{\jmath} = 0$. In case (i) we have $\Phi_E = (4.0 \text{ N/C})(6.0 \text{ m}^2)\hat{\imath} \cdot \hat{\jmath} = 0$ (the electric field and vector area are perpendicular, so there is zero flux). In case (ii) we have $\Phi_E = [(4.0 \text{ N/C})\hat{\imath} + (2.0 \text{ N/C})\hat{\jmath}] \cdot (3.0 \text{ m}^2)\hat{\jmath} = (2.0 \text{ N/C}) \cdot (3.0 \text{ m}^2) = 6.0 \text{ N} \cdot \text{m}^2/\text{C}$. Similarly, in case (iii) we have $\Phi_E = [(4.0 \text{ N/C})\hat{\imath} - (2.0 \text{ N/C})\hat{\jmath}] \cdot [(3.0 \text{ m}^2)\hat{\imath} + (7.0 \text{ m}^2)\hat{\jmath}] = (4.0 \text{ N/C})(3.0 \text{ m}^2) - (2.0 \text{ N/C})(7.0 \text{ m}^2) = -2 \text{ N} \cdot \text{m}^2/\text{C}$, and in case (iv) we have $\Phi_E = [(4.0 \text{ N/C})\hat{\imath} - (2.0 \text{ N/C})\hat{\jmath}] \cdot [(3.0 \text{ m}^2)\hat{\imath} - (7.0 \text{ m}^2)\hat{\jmath}] = (4.0 \text{ N/C})(3.0 \text{ m}^2) + (2.0 \text{ N/C}) \cdot (7.0 \text{ m}^2) = 26 \text{ N} \cdot \text{m}^2/\text{C}$.

22.3 S_2, S_5, S_4, S_1 and S_3 (tie) Gauss's law tells us that the flux through a closed surface is proportional to the amount of charge enclosed within that surface. So an ordering of these surfaces by their fluxes is the same as an ordering by the amount of enclosed charge. Surface S_1 encloses no charge, surface S_2 encloses $9.0 \ \mu\text{C} + 5.0 \ \mu\text{C} + (-7.0 \ \mu\text{C}) = 7.0 \ \mu\text{C}$, surface S_3 encloses $9.0 \ \mu\text{C} + 1.0 \ \mu\text{C} + (-10.0 \ \mu\text{C}) = 0$, surface S_4 encloses $8.0 \ \mu\text{C} + (-7.0 \ \mu\text{C}) = 1.0 \ \mu\text{C}$, and surface S_5 encloses $8.0 \ \mu\text{C} + (-7.0 \ \mu\text{C}) + (-10.0 \ \mu\text{C}) + (1.0 \ \mu\text{C}) + (9.0 \ \mu\text{C}) + (5.0 \ \mu\text{C}) = 6.0 \ \mu\text{C}$.

22.4 no You might be tempted to draw a Gaussian surface that is an enlarged version of the conductor, with the same shape and placed so that it completely encloses the conductor. While you know the flux through this Gaussian surface (by Gauss's law, it's $\Phi_E = Q/\epsilon_0$), the direction of the electric field need not be perpendicular to the surface and the magnitude of the field need not be the same at all points on the surface. It's not possible to do the flux integral $\oint E_\perp \, dA$, and we can't calculate the electric field. Gauss's law is useful for calculating the electric field only when the charge distribution is *highly* symmetric.

22.5 no Before you connect the wire to the sphere, the presence of the point charge will induce a charge $-q$ on the inner surface of the hollow sphere and a charge q on the outer surface (the net charge on the sphere is zero). There will be an electric field outside the sphere due to the charge on the outer surface. Once you touch the conducting wire to the sphere, however, electrons will flow from ground to the outer surface of the sphere to neutralize the charge there (see Fig. 21.7c). As a result the sphere will have no charge on its outer surface and no electric field outside.

Bridging Problem

(a) $Q(r) = Qe^{-2r/a_0}[2(r/a_0)^2 + 2(r/a_0) + 1]$

(b) $E = \dfrac{kQe^{-2r/a_0}}{r^2}[2(r/a_0)^2 + 2(r/a_0) + 1]$

(c)

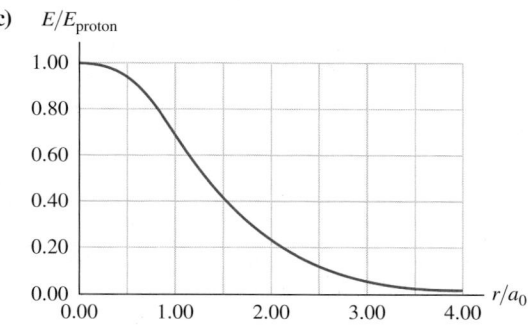

? In one type of welding, electric charge flows between the welding tool and the metal pieces that are to be joined. This produces a glowing arc whose high temperature fuses the pieces together. Why must the tool be held close to the metal pieces? (i) To maximize the potential difference between tool and pieces; (ii) to minimize this potential difference; (iii) to maximize the electric field between tool and pieces; (iv) to minimize this electric field; (v) more than one of these.

23 ELECTRIC POTENTIAL

This chapter is about energy associated with electrical interactions. Every time you turn on a light, use a mobile phone, or make toast in a toaster, you are using electrical energy, an indispensable ingredient of our technological society. In Chapters 6 and 7 we introduced the concepts of *work* and *energy* in the context of mechanics; now we'll combine these concepts with what we've learned about electric charge, electric forces, and electric fields. Just as we found for many problems in mechanics, using energy ideas makes it easier to solve a variety of problems in electricity.

When a charged particle moves in an electric field, the field exerts a force that can do *work* on the particle. This work can be expressed in terms of electric potential energy. Just as gravitational potential energy depends on the height of a mass above the earth's surface, electric potential energy depends on the position of the charged particle in the electric field. We'll use a new concept called *electric potential*, or simply *potential*, to describe electric potential energy. In circuits, a difference in potential from one point to another is often called *voltage*. The concepts of potential and voltage are crucial to understanding how electric circuits work and have equally important applications to electron beams used in cancer radiotherapy, high-energy particle accelerators, and many other devices.

23.1 ELECTRIC POTENTIAL ENERGY

The concepts of work, potential energy, and conservation of energy proved to be extremely useful in our study of mechanics. In this section we'll show that these concepts are just as useful for understanding and analyzing electrical interactions.

Let's begin by reviewing three essential points from Chapters 6 and 7. First, when a force $\vec{F}$ acts on a particle that moves from point a to point b, the work $W_{a \to b}$ done by the force is given by a *line integral*:

$$W_{a \to b} = \int_a^b \vec{F} \cdot d\vec{l} = \int_a^b F \cos \phi \, dl \quad \text{(work done by a force)} \quad (23.1)$$

where $d\vec{l}$ is an infinitesimal displacement along the particle's path and ϕ is the angle between $\vec{F}$ and $d\vec{l}$ at each point along the path.

Second, if the force $\vec{F}$ is *conservative*, as we defined the term in Section 7.3, the work done by $\vec{F}$ can always be expressed in terms of a **potential energy** U. When the particle moves from a point where the potential energy is U_a to a point where it is U_b, the change in potential energy is $\Delta U = U_b - U_a$ and

Work done ⋯⋯ | Potential energy at initial position
$$W_{a \to b} = U_a - U_b = -(U_b - U_a) = -\Delta U. \qquad (23.2)$$
by a conservative force | Potential energy at final position | Negative of change in potential energy

When $W_{a \to b}$ is positive, U_a is greater than U_b, ΔU is negative, and the potential energy *decreases*. That's what happens when a baseball falls from a high point (*a*) to a lower point (*b*) under the influence of the earth's gravity; the force of gravity does positive work, and the gravitational potential energy decreases (**Fig. 23.1**). When a tossed ball is moving upward, the gravitational force does negative work during the ascent, and the potential energy increases.

Third, the work–energy theorem says that the change in kinetic energy $\Delta K = K_b - K_a$ during a displacement equals the *total* work done on the particle. If only conservative forces do work, then Eq. (23.2) gives the total work, and $K_b - K_a = -(U_b - U_a)$. We usually write this as

$$K_a + U_a = K_b + U_b \qquad (23.3)$$

That is, the total mechanical energy (kinetic plus potential) is *conserved* under these circumstances.

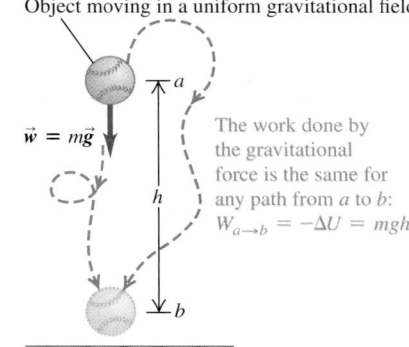

23.1 The work done on a baseball moving in a uniform gravitational field.

Object moving in a uniform gravitational field

The work done by the gravitational force is the same for any path from *a* to *b*:
$W_{a \to b} = -\Delta U = mgh$

Electric Potential Energy in a Uniform Field

Let's look at an electrical example of these concepts. In **Fig. 23.2** a pair of charged parallel metal plates sets up a uniform, downward electric field with magnitude E. The field exerts a downward force with magnitude $F = q_0 E$ on a positive test charge q_0. As the charge moves downward a distance d from point a to point b, the force on the test charge is constant and independent of its location. So the work done by the electric field is the product of the force magnitude and the component of displacement in the (downward) direction of the force:

$$W_{a \to b} = Fd = q_0 Ed \qquad (23.4)$$

This work is positive, since the force is in the same direction as the net displacement of the test charge.

The y-component of the electric force, $F_y = -q_0 E$, is constant, and there is no x- or z-component. This is exactly analogous to the gravitational force on a mass m near the earth's surface; for this force, there is a constant y-component $F_y = -mg$ and the x- and z-components are zero. Because of this analogy, we can conclude that the force exerted on q_0 by the uniform electric field in Fig. 23.2 is *conservative*, just as is the gravitational force. This means that the work $W_{a \to b}$ done by the field is independent of the path the particle takes from a to b. We can represent this work with a *potential-energy* function U, just as we did for gravitational potential energy in Section 7.1. The potential energy for the gravitational force $F_y = -mg$ was $U = mgy$; hence the potential energy for the electric force $F_y = -q_0 E$ is

$$U = q_0 Ey \qquad (23.5)$$

When the test charge moves from height y_a to height y_b, the work done on the charge by the field is given by

$$W_{a \to b} = -\Delta U = -(U_b - U_a) = -(q_0 Ey_b - q_0 Ey_a) = q_0 E(y_a - y_b) \quad (23.6)$$

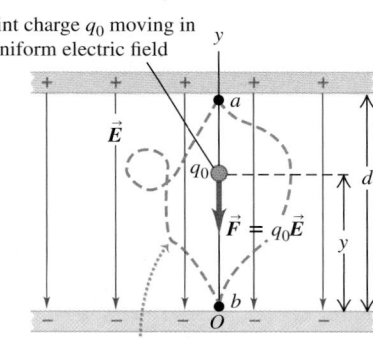

23.2 The work done on a point charge moving in a uniform electric field. Compare with Fig. 23.1.

Point charge q_0 moving in a uniform electric field

The work done by the electric force is the same for any path from *a* to *b*:
$W_{a \to b} = -\Delta U = q_0 Ed$

23.3 A positive charge moving (a) in the direction of the electric field $\vec{E}$ and (b) in the direction opposite $\vec{E}$.

(a) Positive charge q_0 moves in the direction of $\vec{E}$:
• Field does *positive* work on charge.
• U *decreases*.

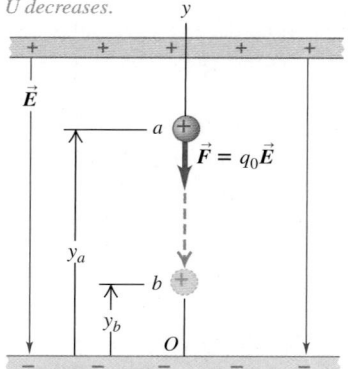

(b) Positive charge q_0 moves opposite $\vec{E}$:
• Field does *negative* work on charge.
• U *increases*.

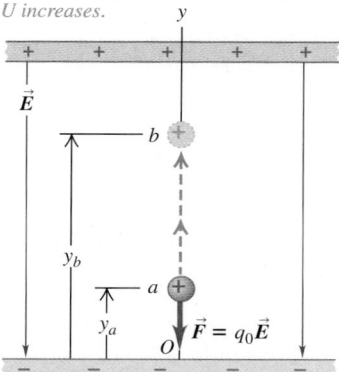

23.5 Test charge q_0 moves along a straight line extending radially from charge q. As it moves from a to b, the distance varies from r_a to r_b.

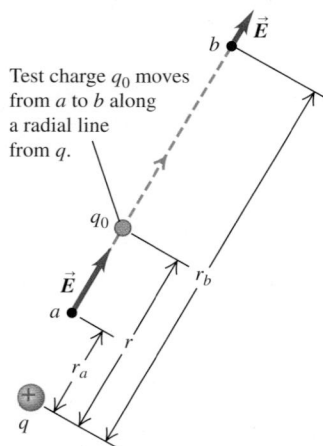

Test charge q_0 moves from a to b along a radial line from q.

23.4 A negative charge moving (a) in the direction of the electric field $\vec{E}$ and (b) in the direction opposite $\vec{E}$. Compare with Fig. 23.3.

(a) Negative charge q_0 moves in the direction of $\vec{E}$:
• Field does *negative* work on charge.
• U *increases*.

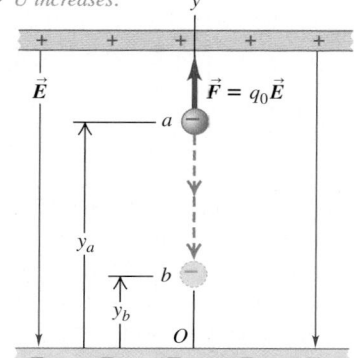

(b) Negative charge q_0 moves opposite $\vec{E}$:
• Field does *positive* work on charge.
• U *decreases*.

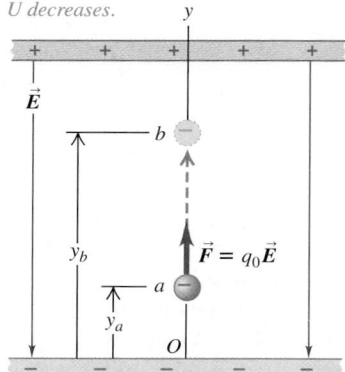

When y_a is greater than y_b (**Fig. 23.3a**), the positive test charge q_0 moves downward, in the same direction as $\vec{E}$; the displacement is in the same direction as the force $\vec{F} = q_0\vec{E}$, so the field does positive work and U decreases. [In particular, if $y_a - y_b = d$ as in Fig. 23.2, Eq. (23.6) gives $W_{a \to b} = q_0 E d$, in agreement with Eq. (23.4).] When y_a is less than y_b (Fig. 23.3b), the positive test charge q_0 moves upward, in the opposite direction to $\vec{E}$; the displacement is opposite the force, the field does negative work, and U increases.

If the test charge q_0 is *negative*, the potential energy increases when it moves with the field and decreases when it moves against the field (**Fig. 23.4**).

Whether the test charge is positive or negative, the following general rules apply: *U increases* if the test charge q_0 moves in the direction *opposite* the electric force $\vec{F} = q_0\vec{E}$ (Figs. 23.3b and Fig. 23.4a); *U decreases* if q_0 moves in the *same* direction as $\vec{F} = q_0\vec{E}$ (Figs. 23.3a and Fig. 23.4b). This is the same behavior as for gravitational potential energy, which increases if a mass m moves upward (opposite the direction of the gravitational force) and decreases if m moves downward (in the same direction as the gravitational force).

> **CAUTION** **Electric potential energy** The relationship between electric potential energy change and motion in an electric field is an important one that we'll use often, but that takes some effort to understand. Studying the preceding paragraph as well as Figs. 23.3 and Fig. 23.4 carefully now will help you tremendously later! ▮

Electric Potential Energy of Two Point Charges

The idea of electric potential energy isn't restricted to the special case of a uniform electric field. Indeed, we can apply this concept to a point charge in *any* electric field caused by a static charge distribution. Recall from Chapter 21 that we can represent any charge distribution as a collection of point charges. Therefore it's useful to calculate the work done on a test charge q_0 moving in the electric field caused by a single, stationary point charge q.

We'll consider first a displacement along the *radial* line in **Fig. 23.5**. The force on q_0 is given by Coulomb's law, and its radial component is

$$F_r = \frac{1}{4\pi\epsilon_0} \frac{qq_0}{r^2} \qquad (23.7)$$

If q and q_0 have the same sign $(+ \text{ or } -)$, the force is repulsive and F_r is positive; if the two charges have opposite signs, the force is attractive and F_r is negative. The force is *not* constant during the displacement, and we must integrate to calculate the work $W_{a\to b}$ done on q_0 by this force as q_0 moves from a to b:

$$W_{a\to b} = \int_{r_a}^{r_b} F_r\, dr = \int_{r_a}^{r_b} \frac{1}{4\pi\epsilon_0} \frac{qq_0}{r^2}\, dr = \frac{qq_0}{4\pi\epsilon_0}\left(\frac{1}{r_a} - \frac{1}{r_b}\right) \tag{23.8}$$

The work done by the electric force for this path depends on only the endpoints.

Now let's consider a more general displacement (**Fig. 23.6**) in which a and b do *not* lie on the same radial line. From Eq. (23.1) the work done on q_0 during this displacement is given by

$$W_{a\to b} = \int_{r_a}^{r_b} F\cos\phi\, dl = \int_{r_a}^{r_b} \frac{1}{4\pi\epsilon_0} \frac{qq_0}{r^2}\cos\phi\, dl$$

But Fig. 23.6 shows that $\cos\phi\, dl = dr$. That is, the work done during a small displacement $d\vec{l}$ depends only on the change dr in the distance r between the charges, which is the *radial component* of the displacement. Thus Eq. (23.8) is valid even for this more general displacement; the work done on q_0 by the electric field $\vec{E}$ produced by q depends only on r_a and r_b, not on the details of the path. Also, if q_0 returns to its starting point a by a different path, the total work done in the round-trip displacement is zero (the integral in Eq. (23.8) is from r_a back to r_a). These are the needed characteristics for a conservative force, as we defined it in Section 7.3. Thus the force on q_0 is a *conservative* force.

We see that Eqs. (23.2) and (23.8) are consistent if we define the potential energy to be $U_a = qq_0/4\pi\epsilon_0 r_a$ when q_0 is a distance r_a from q, and to be $U_b = qq_0/4\pi\epsilon_0 r_b$ when q_0 is a distance r_b from q. Thus

> **Electric potential energy** ········· $U = \dfrac{1}{4\pi\epsilon_0}\dfrac{qq_0}{r}$ ·······•Values of two charges
> of two point charges
> Electric constant ·······•Distance between two charges $\tag{23.9}$

Equation (23.9) is valid no matter what the signs of the charges q and q_0. The potential energy is positive if the charges q and q_0 have the same sign (**Fig. 23.7a**) and negative if they have opposite signs (Fig. 23.7b).

Potential energy is always defined relative to some reference point where $U = 0$. In Eq. (23.9), U is zero when q and q_0 are infinitely far apart and $r = \infty$. Therefore U represents the work that would be done on the test charge q_0 by the field of q if q_0 moved from an initial distance r to infinity. If q and q_0 have the same sign, the interaction is repulsive, this work is positive, and U is positive at any finite separation (Fig. 23.7a). If the charges have opposite signs, the interaction is attractive, the work done is negative, and U is negative (Fig. 23.7b).

We emphasize that the potential energy U given by Eq. (23.9) is a *shared* property of the two charges. If the distance between q and q_0 is changed from r_a to r_b, the change in potential energy is the same whether q is held fixed and q_0 is moved or q_0 is held fixed and q is moved. For this reason, we never use the phrase "the electric potential energy *of* a point charge." (Likewise, if a mass m is at a height h above the earth's surface, the gravitational potential energy is a shared property of the mass m and the earth. We emphasized this in Sections 7.1 and 13.3.)

Equation (23.9) also holds if the charge q_0 is outside a spherically symmetric charge *distribution* with total charge q; the distance r is from q_0 to the center of the distribution. That's because Gauss's law tells us that the electric field outside such a distribution is the same as if all of its charge q were concentrated at its center (see Example 22.9 in Section 22.4).

23.6 The work done on charge q_0 by the electric field of charge q does not depend on the path taken, but only on the distances r_a and r_b.

Test charge q_0 moves from a to b along an arbitrary path.

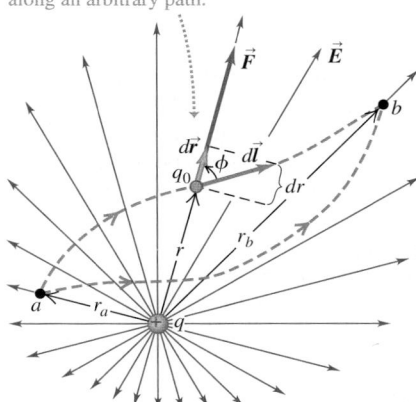

CAUTION Electric potential energy vs. electric force Don't confuse Eq. (23.9) for the potential energy of two point charges with Eq. (23.7) for the radial component of the electric force that one charge exerts on the other. Potential energy U is proportional to $1/r$, while the force component F_r is proportional to $1/r^2$. ▌

23.7 Graphs of the potential energy U of two point charges q and q_0 versus their separation r.

(a) q and q_0 have the same sign.

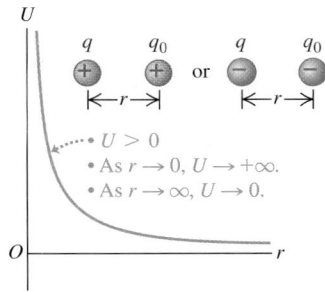

(b) q and q_0 have opposite signs.

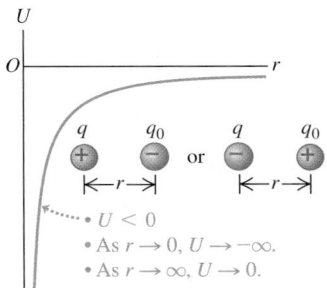

EXAMPLE 23.1 CONSERVATION OF ENERGY WITH ELECTRIC FORCES

A positron (the electron's antiparticle) has mass 9.11×10^{-31} kg and charge $q_0 = +e = +1.60 \times 10^{-19}$ C. Suppose a positron moves in the vicinity of an α (alpha) particle, which has charge $q = +2e = 3.20 \times 10^{-19}$ C and mass 6.64×10^{-27} kg. The α particle's mass is more than 7000 times that of the positron, so we assume that the α particle remains at rest. When the positron is 1.00×10^{-10} m from the α particle, it is moving directly away from the α particle at 3.00×10^6 m/s. (a) What is the positron's speed when the particles are 2.00×10^{-10} m apart? (b) What is the positron's speed when it is very far from the α particle? (c) Suppose the initial conditions are the same but the moving particle is an electron (with the same mass as the positron but charge $q_0 = -e$). Describe the subsequent motion.

SOLUTION

IDENTIFY and SET UP: The electric force between a positron (or an electron) and an α particle is conservative, so mechanical energy (kinetic plus potential) is conserved. Equation (23.9) gives the potential energy U at any separation r: The potential-energy function for parts (a) and (b) looks like that of Fig. 23.7a, and the function for part (c) looks like that of Fig. 23.7b. We are given the positron speed $v_a = 3.00 \times 10^6$ m/s when the separation between the particles is $r_a = 1.00 \times 10^{-10}$ m. In parts (a) and (b) we use Eqs. (23.3) and (23.9) to find the speed for $r = r_b = 2.00 \times 10^{-10}$ m and $r = r_c \to \infty$, respectively. In part (c) we replace the positron with an electron and reconsider the problem.

EXECUTE: (a) Both particles have positive charge, so the positron speeds up as it moves away from the α particle. From the energy-conservation equation, Eq. (23.3), the final kinetic energy is

$$K_b = \tfrac{1}{2}mv_b^2 = K_a + U_a - U_b$$

In this expression,

$$K_a = \tfrac{1}{2}mv_a^2 = \tfrac{1}{2}(9.11 \times 10^{-31} \text{ kg})(3.00 \times 10^6 \text{ m/s})^2$$
$$= 4.10 \times 10^{-18} \text{ J}$$

$$U_a = \frac{1}{4\pi\epsilon_0} \frac{qq_0}{r_a} = (9.0 \times 10^9 \text{ N} \cdot \text{m}^2/\text{C}^2)$$
$$\times \frac{(3.20 \times 10^{-19} \text{ C})(1.60 \times 10^{-19} \text{ C})}{1.00 \times 10^{-10} \text{ m}}$$
$$= 4.61 \times 10^{-18} \text{ J}$$

$$U_b = \frac{1}{4\pi\epsilon_0} \frac{qq_0}{r_b} = 2.30 \times 10^{-18} \text{ J}$$

Hence the positron kinetic energy and speed at $r = r_b = 2.00 \times 10^{-10}$ m are

$$K_b = \tfrac{1}{2}mv_b^2 = 4.10 \times 10^{-18} \text{ J} + 4.61 \times 10^{-18} \text{ J} - 2.30 \times 10^{-18} \text{ J}$$
$$= 6.41 \times 10^{-18} \text{ J}$$

$$v_b = \sqrt{\frac{2K_b}{m}} = \sqrt{\frac{2(6.41 \times 10^{-18} \text{ J})}{9.11 \times 10^{-31} \text{ kg}}} = 3.8 \times 10^6 \text{ m/s}$$

(b) When the positron and α particle are very far apart so that $r = r_c \to \infty$, the final potential energy U_c approaches zero. Again from energy conservation, the final kinetic energy and speed of the positron in this case are

$$K_c = K_a + U_a - U_c = 4.10 \times 10^{-18} \text{ J} + 4.61 \times 10^{-18} \text{ J} - 0$$
$$= 8.71 \times 10^{-18} \text{ J}$$

$$v_c = \sqrt{\frac{2K_c}{m}} = \sqrt{\frac{2(8.71 \times 10^{-18} \text{ J})}{9.11 \times 10^{-31} \text{ kg}}} = 4.4 \times 10^6 \text{ m/s}$$

(c) The electron and α particle have opposite charges, so the force is attractive and the electron slows down as it moves away. Changing the moving particle's sign from $+e$ to $-e$ means that the initial potential energy is now $U_a = -4.61 \times 10^{-18}$ J, which makes the total mechanical energy *negative*:

$$K_a + U_a = (4.10 \times 10^{-18} \text{ J}) - (4.61 \times 10^{-18} \text{ J})$$
$$= -0.51 \times 10^{-18} \text{ J}$$

The total mechanical energy would have to be positive for the electron to move infinitely far away from the α particle. Like a rock thrown upward at low speed from the earth's surface, it will reach a maximum separation $r = r_d$ from the α particle before reversing direction. At this point its speed and its kinetic energy K_d are zero, so at separation r_d we have

$$U_d = K_a + U_a - K_d = (-0.51 \times 10^{-18} \text{ J}) - 0$$
$$U_d = \frac{1}{4\pi\epsilon_0} \frac{qq_0}{r_d} = -0.51 \times 10^{-18} \text{ J}$$

$$r_d = \frac{1}{U_d} \frac{qq_0}{4\pi\epsilon_0}$$
$$= \frac{(9.0 \times 10^9 \text{ N} \cdot \text{m}^2/\text{C}^2)}{-0.51 \times 10^{-18} \text{ J}} (3.20 \times 10^{-19} \text{ C})(-1.60 \times 10^{-19} \text{ C})$$
$$= 9.0 \times 10^{-10} \text{ m}$$

For $r_b = 2.00 \times 10^{-10}$ m we have $U_b = -2.30 \times 10^{-18}$ J, so the electron kinetic energy and speed at this point are

$$K_b = \tfrac{1}{2}mv_b^2$$
$$= 4.10 \times 10^{-18} \text{ J} + (-4.61 \times 10^{-18} \text{ J}) - (-2.30 \times 10^{-18} \text{ J})$$
$$= 1.79 \times 10^{-18} \text{ J}$$

$$v_b = \sqrt{\frac{2K_b}{m}} = \sqrt{\frac{2(1.79 \times 10^{-18} \text{ J})}{9.11 \times 10^{-31} \text{ kg}}} = 2.0 \times 10^6 \text{ m/s}$$

EVALUATE: Both particles behave as expected as they move away from the α particle: The positron speeds up, and the electron slows down and eventually turns around. How fast would an electron have to be moving at $r_a = 1.00 \times 10^{-10}$ m to travel infinitely far from the α particle? (*Hint:* See Example 13.5 in Section 13.3.)

Electric Potential Energy with Several Point Charges

Suppose the electric field $\vec{E}$ in which charge q_0 moves is caused by *several* point charges $q_1, q_2, q_3, \ldots$ at distances $r_1, r_2, r_3, \ldots$ from q_0, as in **Fig. 23.8**. For example, q_0 could be a positive ion moving in the presence of other ions (**Fig. 23.9**). The total electric field at each point is the *vector sum* of the fields due to the individual charges, and the total work done on q_0 during any displacement is the sum of the contributions from the individual charges. From Eq. (23.9) we conclude that the potential energy associated with the test charge q_0 at point a in Fig. 23.8 is the *algebraic* sum (*not* a vector sum):

23.8 The potential energy associated with a charge q_0 at point a depends on the other charges q_1, q_2, and q_3 and on their distances r_1, r_2, and r_3 from point a.

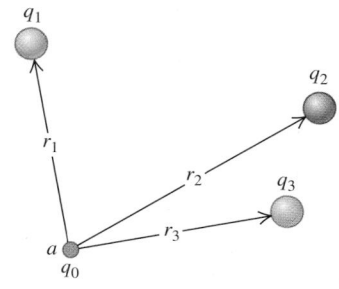

Electric potential energy of point charge q_0 and collection of charges $q_1, q_2, q_3, \ldots$

$$U = \frac{q_0}{4\pi\epsilon_0}\left(\frac{q_1}{r_1} + \frac{q_2}{r_2} + \frac{q_3}{r_3} + \cdots\right) = \frac{q_0}{4\pi\epsilon_0}\sum_i \frac{q_i}{r_i} \qquad (23.10)$$

Electric constant   Distances from q_0 to $q_1, q_2, q_3, \ldots$

When q_0 is at a different point b, the potential energy is given by the same expression, but $r_1, r_2, \ldots$ are the distances from $q_1, q_2, \ldots$ to point b. The work done on charge q_0 when it moves from a to b along any path is equal to the difference $U_a - U_b$ between the potential energies when q_0 is at a and at b.

We can represent *any* charge distribution as a collection of point charges, so Eq. (23.10) shows that we can always find a potential-energy function for *any* static electric field. It follows that **for every electric field due to a static charge distribution, the force exerted by that field is conservative.**

Equations (23.9) and (23.10) define U to be zero when distances $r_1, r_2, \ldots$ are infinite—that is, when the test charge q_0 is very far away from all the charges that produce the field. As with any potential-energy function, the point where $U = 0$ is arbitrary; we can always add a constant to make U equal zero at any point we choose. In electrostatics problems it's usually simplest to choose this point to be at infinity. When we analyze electric circuits in Chapters 25 and 26, other choices will be more convenient.

Equation (23.10) gives the potential energy associated with the presence of the test charge q_0 in the $\vec{E}$ field produced by $q_1, q_2, q_3, \ldots$. But there is also potential energy involved in assembling these charges. If we start with charges $q_1, q_2, q_3, \ldots$ all separated from each other by infinite distances and then bring them together so that the distance between q_i and q_j is r_{ij}, the *total* potential energy U is the sum of the potential energies of interaction for each pair of charges. We can write this as

$$U = \frac{1}{4\pi\epsilon_0}\sum_{i<j} \frac{q_i q_j}{r_{ij}} \qquad (23.11)$$

23.9 This ion engine for spacecraft uses electric forces to eject a stream of positive xenon ions (Xe^+) at speeds in excess of 30 km/s. The thrust produced is very low (about 0.09 newton) but can be maintained continuously for days, in contrast to chemical rockets, which produce a large thrust for a short time (see Fig. 8.34). Such ion engines have been used for maneuvering interplanetary spacecraft.

This sum extends over all *pairs* of charges; we don't let $i = j$ (because that would be an interaction of a charge with itself), and we include only terms with $i < j$ to make sure that we count each pair only once. Thus, to account for the interaction between q_3 and q_4, we include a term with $i = 3$ and $j = 4$ but not a term with $i = 4$ and $j = 3$.

Interpreting Electric Potential Energy

As a final comment, here are two viewpoints on electric potential energy. We have defined it in terms of the work done *by the electric field* on a charged particle moving in the field, just as in Chapter 7 we defined potential energy in terms of the work done by gravity or by a spring. When a particle moves from point a to point b, the work done on it by the electric field is $W_{a\rightarrow b} = U_a - U_b$. Thus the potential-energy difference $U_a - U_b$ equals *the work that is done by the electric force when the particle moves from a to b.* When U_a is greater than U_b, the field does positive work on the particle as it "falls" from a point of higher potential energy (a) to a point of lower potential energy (b).

An alternative but equivalent viewpoint is to consider how much work we would have to do to "raise" a particle from a point b where the potential energy is U_b to a point a where it has a greater value U_a (pushing two positive charges closer together, for example). To move the particle slowly (so as not to give it any kinetic energy), we need to exert an additional external force $\vec{F}_{\text{ext}}$ that is equal and opposite to the electric-field force and does positive work. The potential-energy difference $U_a - U_b$ is then defined as *the work that must be done by an external force to move the particle slowly from b to a against the electric force.* Because $\vec{F}_{\text{ext}}$ is the negative of the electric-field force and the displacement is in the opposite direction, this definition of the potential difference $U_a - U_b$ is equivalent to that given above. This alternative viewpoint also works if U_a is less than U_b, corresponding to "lowering" the particle; an example is moving two positive charges away from each other. In this case, $U_a - U_b$ is again equal to the work done by the external force, but now this work is negative.

We will use both of these viewpoints in the next section to interpret what is meant by electric *potential,* or potential energy per unit charge.

EXAMPLE 23.2 A SYSTEM OF POINT CHARGES

Two point charges are located on the x-axis, $q_1 = -e$ at $x = 0$ and $q_2 = +e$ at $x = a$. (a) Find the work that must be done by an external force to bring a third point charge $q_3 = +e$ from infinity to $x = 2a$. (b) Find the total potential energy of the system of three charges.

23.10 Our sketch of the situation after the third charge has been brought in from infinity.

$$q_1 = -e \qquad q_2 = +e \qquad q_3 = +e$$

$$x = 0 \qquad\quad x = a \qquad\quad x = 2a$$

SOLUTION

IDENTIFY and SET UP: Figure 23.10 shows the final arrangement of the three charges. In part (a) we need to find the work W that must be done on q_3 by an external force $\vec{F}_{\text{ext}}$ to bring q_3 in from infinity to $x = 2a$. We do this by using Eq. (23.10) to find the potential energy associated with q_3 in the presence of q_1 and q_2. In part (b) we use Eq. (23.11), the expression for the potential energy of a collection of point charges, to find the total potential energy of the system.

EXECUTE: (a) The work W equals the difference between (i) the potential energy U associated with q_3 when it is at $x = 2a$ and (ii) the potential energy when it is infinitely far away. The second of these is zero, so the work required is equal to U. The distances between the charges are $r_{13} = 2a$ and $r_{23} = a$, so from Eq. (23.10),

$$W = U = \frac{q_3}{4\pi\epsilon_0}\left(\frac{q_1}{r_{13}} + \frac{q_2}{r_{23}}\right) = \frac{+e}{4\pi\epsilon_0}\left(\frac{-e}{2a} + \frac{+e}{a}\right) = \frac{+e^2}{8\pi\epsilon_0 a}$$

This is positive, just as we should expect. If we bring q_3 in from infinity along the $+x$-axis, it is attracted by q_1 but is repelled more strongly by q_2. Hence we must do positive work to push q_3 to the position at $x = 2a$.

(b) From Eq. (23.11), the total potential energy of the three-charge system is

$$U = \frac{1}{4\pi\epsilon_0}\sum_{i<j}\frac{q_iq_j}{r_{ij}} = \frac{1}{4\pi\epsilon_0}\left(\frac{q_1q_2}{r_{12}} + \frac{q_1q_3}{r_{13}} + \frac{q_2q_3}{r_{23}}\right)$$

$$= \frac{1}{4\pi\epsilon_0}\left[\frac{(-e)(e)}{a} + \frac{(-e)(e)}{2a} + \frac{(e)(e)}{a}\right] = \frac{-e^2}{8\pi\epsilon_0 a}$$

EVALUATE: Our negative result in part (b) means that the system has lower potential energy than it would if the three charges were infinitely far apart. An external force would have to do *negative* work to bring the three charges from infinity to assemble this entire arrangement and would have to do *positive* work to move the three charges back to infinity.

TEST YOUR UNDERSTANDING OF SECTION 23.1 Consider the system of three point charges in Example 21.4 (Section 21.3) and shown in Fig. 21.14. (a) What is the sign of the total potential energy of this system? (i) Positive; (ii) negative; (iii) zero. (b) What is the sign of the total amount of work you would have to do to move these charges infinitely far from each other? (i) Positive; (ii) negative; (iii) zero. ∎

23.2 ELECTRIC POTENTIAL

PhET: Charges and Fields

In Section 23.1 we looked at the potential energy U associated with a test charge q_0 in an electric field. Now we want to describe this potential energy on a "per unit charge" basis, just as electric field describes the force per unit charge on a charged particle in the field. This leads us to the concept of *electric potential, often called simply *potential*. This concept is very useful in calculations involving energies of charged particles. It also facilitates many electric-field calculations because electric potential is closely related to the electric field $\vec{E}$. When we need to determine an electric field, it is often easier to determine the potential first and then find the field from it.

Potential is *potential energy per unit charge*. We define the potential V at any point in an electric field as the potential energy U *per unit charge* associated with a test charge q_0 at that point:

$$V = \frac{U}{q_0} \quad \text{or} \quad U = q_0 V \qquad (23.12)$$

Potential energy and charge are both scalars, so potential is a scalar. From Eq. (23.12) its units are the units of energy divided by those of charge. The SI unit of potential, called one **volt** (1 V) in honor of the Italian electrical experimenter Alessandro Volta (1745–1827), equals 1 joule per coulomb:

$$1 \text{ V} = 1 \text{ volt} = 1 \text{ J/C} = 1 \text{ joule/coulomb}$$

Let's put Eq. (23.2), which equates the work done by the electric force during a displacement from a to b to the quantity $-\Delta U = -(U_b - U_a)$, on a "work per unit charge" basis. We divide this equation by q_0, obtaining

$$\frac{W_{a \to b}}{q_0} = -\frac{\Delta U}{q_0} = -\left(\frac{U_b}{q_0} - \frac{U_a}{q_0} \right) = -(V_b - V_a) = V_a - V_b \qquad (23.13)$$

where $V_a = U_a/q_0$ is the potential energy per unit charge at point a and similarly for V_b. We call V_a and V_b the *potential at point a* and *potential at point b*, respectively. Thus the work done per unit charge by the electric force when a charged body moves from a to b is equal to the potential at a minus the potential at b.

The difference $V_a - V_b$ is called the *potential of a with respect to b*; we sometimes abbreviate this difference as $V_{ab} = V_a - V_b$ (note the order of the subscripts). This is often called the potential difference between a and b, but that's ambiguous unless we specify which is the reference point. In electric circuits, which we will analyze in later chapters, the potential difference between two points is often called **voltage** (**Fig. 23.11**). Equation (23.13) then states: V_{ab}, **the potential (in V) of a with respect to b, equals the work (in J) done by the electric force when a UNIT (1-C) charge moves from a to b.**

Another way to interpret the potential difference V_{ab} in Eq. (23.13) is to use the alternative viewpoint mentioned at the end of Section 23.1. In that viewpoint, $U_a - U_b$ is the amount of work that must be done by an *external* force to move a particle of charge q_0 slowly from b to a against the electric force. The work that must be done *per unit charge* by the external force is then $(U_a - U_b)/q_0 = V_a - V_b = V_{ab}$. In other words: V_{ab}, **the potential (in V) of a with respect to b, equals the work (in J) that must be done to move a UNIT (1-C) charge slowly from b to a against the electric force.**

An instrument that measures the difference of potential between two points is called a *voltmeter*. (In Chapter 26 we'll discuss how these devices work.) Voltmeters that can measure a potential difference of 1 μV are common, and sensitivities down to 10^{-12} V can be attained.

23.11 The voltage of this battery equals the difference in potential $V_{ab} = V_a - V_b$ between its positive terminal (point a) and its negative terminal (point b).

Point a (positive terminal)

$V_{ab} = 1.5$ volts

Point b (negative terminal)

Calculating Electric Potential

To find the potential V due to a single point charge q, we divide Eq. (23.9) by q_0:

$$\text{Electric potential due to a point charge} \quad V = \frac{1}{4\pi\epsilon_0}\frac{q}{r} \quad \begin{array}{l}\text{Value of point charge}\\ \text{Distance from point charge to where potential is measured}\end{array} \quad (23.14)$$

Electric constant

If q is positive, the potential that it produces is positive at all points; if q is negative, it produces a potential that is negative everywhere. In either case, V is equal to zero at $r = \infty$, an infinite distance from the point charge. Note that potential, like electric field, is independent of the test charge q_0 that we use to define it.

Similarly, we divide Eq. (23.10) by q_0 to find the potential due to a collection of point charges:

$$\text{Electric potential due to a collection of point charges} \quad V = \frac{1}{4\pi\epsilon_0}\sum_i \frac{q_i}{r_i} \quad \begin{array}{l}\text{Value of }i\text{th point charge}\\ \text{Distance from }i\text{th point charge to where potential is measured}\end{array} \quad (23.15)$$

Electric constant

Just as the electric field due to a collection of point charges is the *vector* sum of the fields produced by each charge, the electric potential due to a collection of point charges is the *scalar* sum of the potentials due to each charge. When we have a continuous distribution of charge along a line, over a surface, or through a volume, we divide the charge into elements dq, and the sum in Eq. (23.15) becomes an integral:

$$\text{Electric potential due to a continuous distribution of charge} \quad V = \frac{1}{4\pi\epsilon_0}\int \frac{dq}{r} \quad \begin{array}{l}\text{Integral over charge distribution}\\ \text{Charge element}\\ \text{Distance from charge element to where potential is measured}\end{array} \quad (23.16)$$

Electric constant

We'll work out several examples of such cases. The potential defined by Eqs. (23.15) and (23.16) is zero at points that are infinitely far away from *all* the charges. Later we'll encounter cases in which the charge distribution itself extends to infinity. We'll find that in such cases we cannot set $V = 0$ at infinity, and we'll need to exercise care in using and interpreting Eqs. (23.15) and (23.16).

CAUTION **What is electric potential?** Before getting too involved in the details of how to calculate electric potential, remind yourself what potential is. The electric *potential* at a certain point is the potential energy per *unit* charge placed at that point. That's why potential is measured in joules per coulomb, or volts. Keep in mind, too, that there doesn't have to be a charge at a given point for a potential V to exist at that point. (In the same way, an electric field can exist at a given point even if there's no charge there to respond to it.) ❚

Finding Electric Potential from Electric Field

When we are given a collection of point charges, Eq. (23.15) is usually the easiest way to calculate the potential V. But in some problems in which the electric field is known or can be found easily, it is easier to determine V from $\vec{E}$. The force $\vec{F}$ on a test charge q_0 can be written as $\vec{F} = q_0\vec{E}$, so from Eq. (23.1) the work done by the electric force as the test charge moves from a to b is given by

$$W_{a\rightarrow b} = \int_a^b \vec{F}\cdot d\vec{l} = \int_a^b q_0\vec{E}\cdot d\vec{l}$$

If we divide this by q_0 and compare the result with Eq. (23.13), we find

Integral along path from a to b

Electric
potential ····→
difference
$$V_a - V_b = \int_a^b \vec{E} \cdot d\vec{l} = \int_a^b E \cos\phi \, dl \qquad (23.17)$$

Scalar product of electric field and displacement vector

Electric-field magnitude

Displacement

Angle between $\vec{E}$ and $d\vec{l}$

The value of $V_a - V_b$ is independent of the path taken from a to b, just as the value of $W_{a \to b}$ is independent of the path. To interpret Eq. (23.17), remember that $\vec{E}$ is the electric force per unit charge on a test charge. If the line integral $\int_a^b \vec{E} \cdot d\vec{l}$ is positive, the electric field does positive work on a positive test charge as it moves from a to b. In this case the electric potential energy decreases as the test charge moves, so the potential energy per unit charge decreases as well; hence V_b is less than V_a and $V_a - V_b$ is positive.

As an illustration, consider a positive point charge (**Fig. 23.12a**). The electric field is directed away from the charge, and $V = q/4\pi\epsilon_0 r$ is positive at any finite distance from the charge. If you move away from the charge, in the direction of $\vec{E}$, you move toward lower values of V; if you move toward the charge, in the direction opposite $\vec{E}$, you move toward greater values of V. For the negative point charge in Fig. 23.12b, $\vec{E}$ is directed toward the charge and $V = q/4\pi\epsilon_0 r$ is negative at any finite distance from the charge. In this case, if you move toward the charge, you are moving in the direction of $\vec{E}$ and in the direction of decreasing (more negative) V. Moving away from the charge, in the direction opposite $\vec{E}$, moves you toward increasing (less negative) values of V. The general rule, valid for *any* electric field, is: Moving *with* the direction of $\vec{E}$ means moving in the direction of *decreasing V*, and moving *against* the direction of $\vec{E}$ means moving in the direction of *increasing V*.

Also, a positive test charge q_0 experiences an electric force in the direction of $\vec{E}$, toward lower values of V; a negative test charge experiences a force opposite $\vec{E}$, toward higher values of V. Thus a positive charge tends to "fall" from a high-potential region to a lower-potential region. The opposite is true for a negative charge.

Notice that Eq. (23.17) can be rewritten as

$$V_a - V_b = -\int_b^a \vec{E} \cdot d\vec{l} \qquad (23.18)$$

This has a negative sign compared to the integral in Eq. (23.17), and the limits are reversed; hence Eqs. (23.17) and (23.18) are equivalent. But Eq. (23.18) has a slightly different interpretation. To move a unit charge slowly against the electric force, we must apply an *external* force per unit charge equal to $-\vec{E}$, equal and opposite to the electric force per unit charge $\vec{E}$. Equation (23.18) says that $V_a - V_b = V_{ab}$, the potential of a with respect to b, equals the work done per unit charge by this external force to move a unit charge from b to a. This is the same alternative interpretation we discussed under Eq. (23.13).

Equations (23.17) and (23.18) show that the unit of potential difference (1 V) is equal to the unit of electric field (1 N/C) multiplied by the unit of distance (1 m). Hence the unit of electric field can be expressed as 1 *volt per meter* (1 V/m), as well as 1 N/C:

$$1 \text{ V/m} = 1 \text{ volt/meter} = 1 \text{ N/C} = 1 \text{ newton/coulomb}$$

In practice, the volt per meter is the usual unit of electric-field magnitude.

23.12 If you move in the direction of $\vec{E}$, electric potential V decreases; if you move in the direction opposite $\vec{E}$, V increases.

(a) A positive point charge

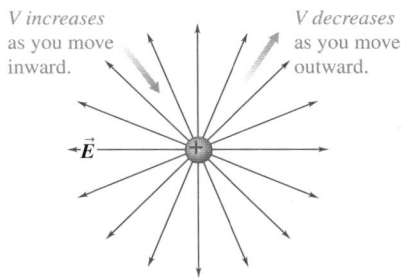

V increases as you move inward.

V decreases as you move outward.

(b) A negative point charge

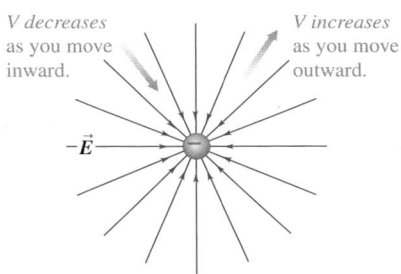

V decreases as you move inward.

V increases as you move outward.

DATA *SPEAKS*

Electric Potential and Electric Potential Energy

When students were given a problem involving electric potential and electric potential energy, more than 20% gave an incorrect response. Common errors:

- Confusing potential energy and potential. Electric potential energy is measured in joules; electric potential is potential energy *per unit charge* and is measured in volts.

- Forgetting the relationships among electric potential V, electric field $\vec{E}$, and electric force. $\vec{E}$ always points from regions of high V toward regions of low V; the direction of the electric force on a point charge q is in the direction of $\vec{E}$ if $q > 0$ but opposite $\vec{E}$ if $q < 0$.

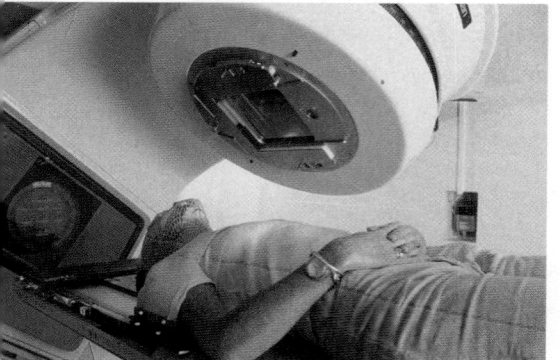

Electron Volts

The magnitude e of the electron charge can be used to define a unit of energy that is useful in many calculations with atomic and nuclear systems. When a particle with charge q moves from a point where the potential is V_b to a point where it is V_a, the change in the potential energy U is

$$U_a - U_b = q(V_a - V_b) = qV_{ab}$$

If charge q equals the magnitude e of the electron charge, 1.602×10^{-19} C, and the potential difference is $V_{ab} = 1$ V, the change in energy is

$$U_a - U_b = (1.602 \times 10^{-19}\,\text{C})(1\,\text{V}) = 1.602 \times 10^{-19}\,\text{J}$$

This quantity of energy is defined to be 1 **electron volt** (1 eV):

$$1\,\text{eV} = 1.602 \times 10^{-19}\,\text{J}$$

The multiples meV, keV, MeV, GeV, and TeV are often used.

CAUTION **Electron volts vs. volts** Remember that the electron volt is a unit of energy, *not* a unit of potential or potential difference! ∎

When a particle with charge e moves through a potential difference of 1 volt, the change in potential *energy* is 1 eV. If the charge is some multiple of e—say, Ne—the change in potential energy in electron volts is N times the potential difference in volts. For example, when an alpha particle, which has charge $2e$, moves between two points with a potential difference of 1000 V, the change in potential energy is $2(1000\,\text{eV}) = 2000$ eV. To confirm this, we write

$$U_a - U_b = qV_{ab} = (2e)(1000\,\text{V}) = (2)(1.602 \times 10^{-19}\,\text{C})(1000\,\text{V})$$
$$= 3.204 \times 10^{-16}\,\text{J} = 2000\,\text{eV}$$

Although we defined the electron volt in terms of *potential* energy, we can use it for *any* form of energy, such as the kinetic energy of a moving particle. When we speak of a "one-million-electron-volt proton," we mean a proton with a kinetic energy of one million electron volts (1 MeV), equal to $(10^6)(1.602 \times 10^{-19}\,\text{J}) = 1.602 \times 10^{-13}$ J. The Large Hadron Collider near Geneva, Switzerland, is designed to accelerate protons to a kinetic energy of 7 TeV (7×10^{12} eV).

EXAMPLE 23.3 ELECTRIC FORCE AND ELECTRIC POTENTIAL

A proton (charge $+e = 1.602 \times 10^{-19}$ C) moves a distance $d = 0.50$ m in a straight line between points a and b in a linear accelerator. The electric field is uniform along this line, with magnitude $E = 1.5 \times 10^7$ V/m $= 1.5 \times 10^7$ N/C in the direction from a to b. Determine (a) the force on the proton; (b) the work done on it by the field; (c) the potential difference $V_a - V_b$.

SOLUTION

IDENTIFY and SET UP: This problem uses the relationship between electric field and electric force. It also uses the relationship among force, work, and potential-energy difference. We are given the electric field, so it is straightforward to find the electric force on the proton. Calculating the work is also straightforward because $\vec{E}$ is uniform, so the force on the proton is constant. Once the work is known, we find $V_a - V_b$ from Eq. (23.13).

EXECUTE: (a) The force on the proton is in the same direction as the electric field, and its magnitude is

$$F = qE = (1.602 \times 10^{-19}\,\text{C})(1.5 \times 10^7\,\text{N/C})$$
$$= 2.4 \times 10^{-12}\,\text{N}$$

(b) The force is constant and in the same direction as the displacement, so the work done on the proton is

$$W_{a \to b} = Fd = (2.4 \times 10^{-12}\,\text{N})(0.50\,\text{m})$$
$$= 1.2 \times 10^{-12}\,\text{J}$$
$$= (1.2 \times 10^{-12}\,\text{J})\frac{1\,\text{eV}}{1.602 \times 10^{-19}\,\text{J}}$$
$$= 7.5 \times 10^6\,\text{eV} = 7.5\,\text{MeV}$$

(c) From Eq. (23.13) the potential difference is the work per unit charge, which is

$$V_a - V_b = \frac{W_{a\rightarrow b}}{q} = \frac{1.2 \times 10^{-12}\,\text{J}}{1.602 \times 10^{-19}\,\text{C}}$$

$$= 7.5 \times 10^6\,\text{J/C} = 7.5 \times 10^6\,\text{V} = 7.5\,\text{MV}$$

We can get this same result even more easily by remembering that 1 electron volt equals 1 volt multiplied by the charge e. The work done is 7.5×10^6 eV and the charge is e, so the potential difference is $(7.5 \times 10^6\,\text{eV})/e = 7.5 \times 10^6$ V.

EVALUATE: We can check our result in part (c) by using Eq. (23.17) or Eq. (23.18). The angle ϕ between the constant field $\vec{E}$ and the displacement is zero, so Eq. (23.17) becomes

$$V_a - V_b = \int_a^b E\cos\phi\,dl = \int_a^b E\,dl = E\int_a^b dl$$

The integral of dl from a to b is just the distance d, so we again find

$$V_a - V_b = Ed = (1.5 \times 10^7\,\text{V/m})(0.50\,\text{m}) = 7.5 \times 10^6\,\text{V}$$

EXAMPLE 23.4 POTENTIAL DUE TO TWO POINT CHARGES

An electric dipole consists of point charges $q_1 = +12$ nC and $q_2 = -12$ nC placed 10.0 cm apart (**Fig. 23.13**). Compute the electric potentials at points a, b, and c.

SOLUTION

IDENTIFY and SET UP: This is the same arrangement as in Example 21.8, in which we calculated the electric *field* at each point by doing a *vector* sum. Here our target variable is the electric *potential* V at three points, which we find by doing the *algebraic* sum in Eq. (23.15).

EXECUTE: At point a we have $r_1 = 0.060$ m and $r_2 = 0.040$ m, so Eq. (23.15) becomes

$$V_a = \frac{1}{4\pi\epsilon_0}\sum_i \frac{q_i}{r_i} = \frac{1}{4\pi\epsilon_0}\frac{q_1}{r_1} + \frac{1}{4\pi\epsilon_0}\frac{q_2}{r_2}$$

$$= (9.0 \times 10^9\,\text{N}\cdot\text{m}^2/\text{C}^2)\frac{12 \times 10^{-9}\,\text{C}}{0.060\,\text{m}}$$

$$+ (9.0 \times 10^9\,\text{N}\cdot\text{m}^2/\text{C}^2)\frac{(-12 \times 10^{-9}\,\text{C})}{0.040\,\text{m}}$$

$$= 1800\,\text{N}\cdot\text{m/C} + (-2700\,\text{N}\cdot\text{m/C})$$

$$= 1800\,\text{V} + (-2700\,\text{V}) = -900\,\text{V}$$

In a similar way you can show that the potential at point b (where $r_1 = 0.040$ m and $r_2 = 0.140$ m) is $V_b = 1930$ V and that the potential at point c (where $r_1 = r_2 = 0.130$ m) is $V_c = 0$.

23.13 What are the potentials at points a, b, and c due to this electric dipole?

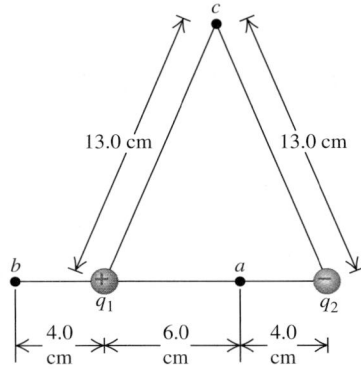

EVALUATE: Let's confirm that these results make sense. Point a is closer to the -12-nC charge than to the $+12$-nC charge, so the potential at a is negative. The potential is positive at point b, which is closer to the $+12$-nC charge than the -12-nC charge. Finally, point c is equidistant from the $+12$-nC charge and the -12-nC charge, so the potential there is zero. (The potential is also equal to zero at a point infinitely far from both charges.)

Comparing this example with Example 21.8 shows that it's much easier to calculate electric potential (a scalar) than electric field (a vector). We'll take advantage of this simplification whenever possible.

EXAMPLE 23.5 POTENTIAL AND POTENTIAL ENERGY

Compute the potential energy associated with a $+4.0$-nC point charge if it is placed at points a, b, and c in Fig. 23.13.

SOLUTION

IDENTIFY and SET UP: The potential energy U associated with a point charge q at a location where the electric potential is V is $U = qV$. We use the values of V from Example 23.4.

EXECUTE: At the three points we find

$$U_a = qV_a = (4.0 \times 10^{-9}\,\text{C})(-900\,\text{J/C}) = -3.6 \times 10^{-6}\,\text{J}$$

$$U_b = qV_b = (4.0 \times 10^{-9}\,\text{C})(1930\,\text{J/C}) = 7.7 \times 10^{-6}\,\text{J}$$

$$U_c = qV_c = 0$$

All of these values correspond to U and V being zero at infinity.

EVALUATE: Note that *zero* net work is done on the 4.0-nC charge if it moves from point c to infinity *by any path*. In particular, let the path be along the perpendicular bisector of the line joining the other two charges q_1 and q_2 in Fig. 23.13. As shown in Example 21.8 (Section 21.5), at points on the bisector, the direction of $\vec{E}$ is perpendicular to the bisector. Hence the force on the 4.0-nC charge is perpendicular to the path, and no work is done in any displacement along it.

EXAMPLE 23.6 FINDING POTENTIAL BY INTEGRATION

By integrating the electric field as in Eq. (23.17), find the potential at a distance r from a point charge q.

23.14 Calculating the potential by integrating $\vec{E}$ for a single point charge.

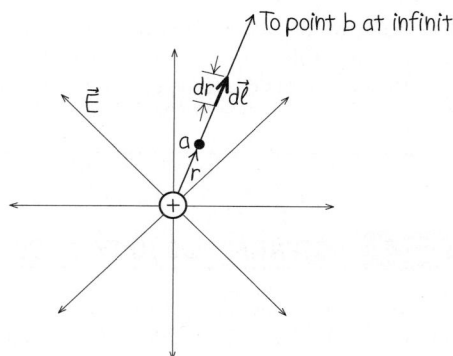

SOLUTION

IDENTIFY and SET UP: We let point a in Eq. (23.17) be at distance r and let point b be at infinity (**Fig. 23.14**). As usual, we choose the potential to be zero at an infinite distance from the charge q.

EXECUTE: To carry out the integral, we can choose any path we like between points a and b. The most convenient path is a radial line as shown in Fig. 23.14, so that $d\vec{l}$ is in the radial direction and has magnitude dr. Writing $d\vec{l} = \hat{r}\,dr$, we have from Eq. (23.17)

$$V - 0 = V = \int_r^\infty \vec{E} \cdot d\vec{l}$$

$$= \int_r^\infty \frac{1}{4\pi\epsilon_0} \frac{q}{r^2}\hat{r} \cdot \hat{r}\,dr = \int_r^\infty \frac{q}{4\pi\epsilon_0 r^2}\,dr$$

$$= -\frac{q}{4\pi\epsilon_0 r}\bigg|_r^\infty = 0 - \left(-\frac{q}{4\pi\epsilon_0 r}\right) = \frac{q}{4\pi\epsilon_0 r}$$

EVALUATE: Our result agrees with Eq. (23.14) and is correct for positive or negative q.

EXAMPLE 23.7 MOVING THROUGH A POTENTIAL DIFFERENCE

In **Fig. 23.15** a dust particle with mass $m = 5.0 \times 10^{-9}$ kg = 5.0 μg and charge $q_0 = 2.0$ nC starts from rest and moves in a straight line from point a to point b. What is its speed v at point b?

SOLUTION

IDENTIFY and SET UP: Only the conservative electric force acts on the particle, so mechanical energy is conserved: $K_a + U_a = K_b + U_b$. We get the potential energies U from the corresponding potentials V from Eq. (23.12): $U_a = q_0 V_a$ and $U_b = q_0 V_b$.

EXECUTE: We have $K_a = 0$ and $K_b = \frac{1}{2}mv^2$. We substitute these and our expressions for U_a and U_b into the energy-conservation equation, then solve for v. We find

$$0 + q_0 V_a = \tfrac{1}{2}mv^2 + q_0 V_b$$

$$v = \sqrt{\frac{2q_0(V_a - V_b)}{m}}$$

We calculate the potentials from Eq. (23.15), $V = q/4\pi\epsilon_0 r$:

$$V_a = (9.0 \times 10^9 \text{ N} \cdot \text{m}^2/\text{C}^2) \times$$

$$\left(\frac{3.0 \times 10^{-9} \text{ C}}{0.010 \text{ m}} + \frac{(-3.0 \times 10^{-9} \text{ C})}{0.020 \text{ m}}\right)$$

$$= 1350 \text{ V}$$

$$V_b = (9.0 \times 10^9 \text{ N} \cdot \text{m}^2/\text{C}^2) \times$$

$$\left(\frac{3.0 \times 10^{-9} \text{ C}}{0.020 \text{ m}} + \frac{(-3.0 \times 10^{-9} \text{ C})}{0.010 \text{ m}}\right)$$

$$= -1350 \text{ V}$$

$$V_a - V_b = (1350 \text{ V}) - (-1350 \text{ V}) = 2700 \text{ V}$$

Finally,

$$v = \sqrt{\frac{2(2.0 \times 10^{-9} \text{ C})(2700 \text{ V})}{5.0 \times 10^{-9} \text{ kg}}} = 46 \text{ m/s}$$

EVALUATE: Our result makes sense: The positive dust particle speeds up as it moves away from the $+3.0$-nC charge and toward the -3.0-nC charge. To check unit consistency in the final line of the calculation, note that 1 V = 1 J/C, so the numerator under the radical has units of J or kg · m²/s².

23.15 The particle moves from point a to point b; its acceleration is not constant.

Particle
3.0 nC   −3.0 nC
+ a b −
        |← 1.0 →|← 1.0 →|← 1.0 →|
           cm   cm   cm

TEST YOUR UNDERSTANDING OF SECTION 23.2   If the electric *potential* at a certain point is zero, does the electric *field* at that point have to be zero? (*Hint:* Consider point *c* in Example 23.4 and Example 21.8.) ▌

23.3 CALCULATING ELECTRIC POTENTIAL

When calculating the potential due to a charge distribution, we usually follow one of two routes. If we know the charge distribution, we can use Eq. (23.15) or (23.16). Or if we know how the electric field depends on position, we can use Eq. (23.17), defining the potential to be zero at some convenient place. Some problems require a combination of these approaches.

As you read through these examples, compare them with the related examples of calculating electric *field* in Section 21.5. You'll see how much easier it is to calculate scalar electric potentials than vector electric fields. The moral is clear: Whenever possible, solve problems by means of an energy approach (using electric potential and electric potential energy) rather than a dynamics approach (using electric fields and electric forces).

PROBLEM-SOLVING STRATEGY 23.1 ⟩ **CALCULATING ELECTRIC POTENTIAL**

IDENTIFY *the relevant concepts:* Remember that electric potential is *potential energy per unit charge.*

SET UP *the problem* using the following steps:
1. Make a drawing showing the locations and values of the charges (which may be point charges or a continuous distribution of charge) and your choice of coordinate axes.
2. Indicate on your drawing the position of the point at which you want to calculate the electric potential *V*. Sometimes this position will be an arbitrary one (say, a point a distance *r* from the center of a charged sphere).

EXECUTE *the solution* as follows:
1. To find the potential due to a collection of point charges, use Eq. (23.15). If you are given a continuous charge distribution, devise a way to divide it into infinitesimal elements and use Eq. (23.16). Carry out the integration, using appropriate limits to include the entire charge distribution.
2. If you are given the electric field, or if you can find it from any of the methods presented in Chapters 21 or 22, it may be

easier to find the potential difference between points *a* and *b* from Eq. (23.17) or (23.18). When appropriate, make use of your freedom to define *V* to be zero at some convenient place, and choose this place to be point *b*. (For point charges, this will usually be at infinity. For other distributions of charge—especially those that themselves extend to infinity—it may be necessary to define V_b to be zero at some finite distance from the charge distribution.) Then the potential at any other point, say *a*, can by found from Eq. (23.17) or (23.18) with $V_b = 0$.
3. Although potential *V* is a *scalar* quantity, you may have to use components of the vectors $\vec{E}$ and $d\vec{l}$ when you use Eq. (23.17) or (23.18) to calculate *V*.

EVALUATE *your answer:* Check whether your answer agrees with your intuition. If your result gives *V* as a function of position, graph the function to see whether it makes sense. If you know the electric field, you can make a rough check of your result for *V* by verifying that *V* decreases if you move in the direction of $\vec{E}$.

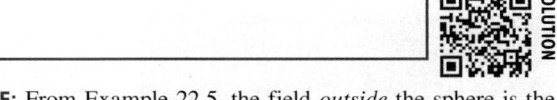

EXAMPLE 23.8   **A CHARGED CONDUCTING SPHERE**

A solid conducting sphere of radius *R* has a total charge *q*. Find the electric potential everywhere, both outside and inside the sphere.

SOLUTION

IDENTIFY and SET UP: In Example 22.5 (Section 22.4) we used Gauss's law to find the electric *field* at all points for this charge distribution. We can use that result to determine the potential.

EXECUTE: From Example 22.5, the field *outside* the sphere is the same as if the sphere were removed and replaced by a point charge *q*. We take *V* = 0 at infinity, as we did for a point charge. Then the potential at a point outside the sphere at a distance *r* from its center is the same as that due to a point charge *q* at the center:

$$V = \frac{1}{4\pi\epsilon_0}\frac{q}{r}$$

The potential at the surface of the sphere is $V_{\text{surface}} = q/4\pi\epsilon_0 R$.

Continued

Inside the sphere, $\vec{E}$ is zero everywhere. Hence no work is done on a test charge that moves from any point to any other point inside the sphere. Thus the potential is the same at every point inside the sphere and is equal to its value $q/4\pi\epsilon_0 R$ at the surface.

EVALUATE: Figure 23.16 shows the field and potential for a positive charge q. In this case the electric field points radially away from the sphere. As you move away from the sphere, in the direction of $\vec{E}$, V decreases (as it should).

23.16 Electric-field magnitude E and potential V at points inside and outside a positively charged spherical conductor.

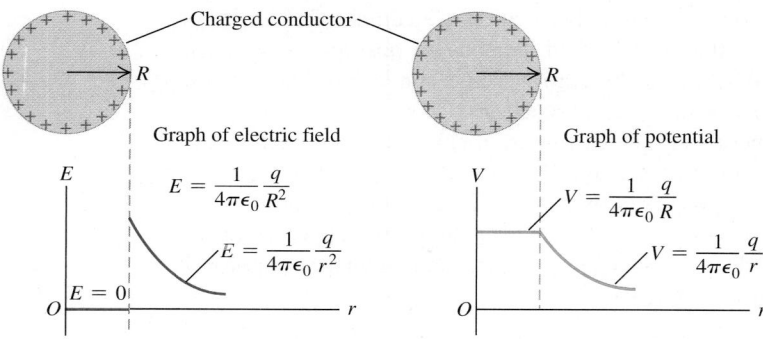

Ionization and Corona Discharge

The results of Example 23.8 have numerous practical consequences. One consequence relates to the maximum potential to which a conductor in air can be raised. This potential is limited because air molecules become *ionized*, and air becomes a conductor, at an electric-field magnitude of about 3×10^6 V/m. Assume for the moment that q is positive. When we compare the expressions in Example 23.8 for the potential $V_{surface}$ and field magnitude $E_{surface}$ at the surface of a charged conducting sphere, we note that $V_{surface} = E_{surface}R$. Thus, if E_m represents the electric-field magnitude at which air becomes conductive (known as the *dielectric strength* of air), then the maximum potential V_m to which a spherical conductor can be raised is

$$V_m = RE_m$$

For a conducting sphere 1 cm in radius in air, $V_m = (10^{-2} \text{ m})(3 \times 10^6 \text{ V/m}) = 30{,}000$ V. No amount of "charging" could raise the potential of a conducting sphere of this size in air higher than about 30,000 V; attempting to raise the potential further by adding extra charge would cause the surrounding air to become ionized and conductive, and the extra added charge would leak into the air.

To attain even higher potentials, high-voltage machines such as Van de Graaff generators use spherical terminals with very large radii (see Fig. 22.26 and the photograph that opens Chapter 22). For example, a terminal of radius $R = 2$ m has a maximum potential $V_m = (2 \text{ m})(3 \times 10^6 \text{ V/m}) = 6 \times 10^6 \text{ V} = 6 \text{ MV}$.

Our result in Example 23.8 also explains what happens with a charged conductor with a very *small* radius of curvature, such as a sharp point or thin wire. Because the maximum potential is proportional to the radius, even relatively small potentials applied to sharp points in air produce sufficiently high fields just outside the point to ionize the surrounding air, making it become a conductor. The resulting current and its associated glow (visible in a dark room) are called *corona discharge*. Laser printers and photocopying machines use corona discharge from fine wires to spray charge on the imaging drum (see Fig. 21.2).

A large-radius conductor is used in situations where it's important to *prevent* corona discharge. An example is the blunt end of a metal lightning rod (**Fig. 23.17**). If there is an excess charge in the atmosphere, as happens during thunderstorms, a substantial charge of the opposite sign can build up on this blunt end. As a result,

23.17 The metal mast at the top of the Empire State Building acts as a lightning rod. It is struck by lightning as many as 500 times each year.

when the atmospheric charge is discharged through a lightning bolt, it tends to be attracted to the charged lightning rod rather than to other structures that could be damaged. (A conducting wire connecting the lightning rod to the ground then allows the acquired charge to dissipate harmlessly.) A lightning rod with a sharp end would allow less charge buildup and hence be less effective.

EXAMPLE 23.9   **OPPOSITELY CHARGED PARALLEL PLATES**

Find the potential at any height y between the two oppositely charged parallel plates discussed in Section 23.1 (**Fig. 23.18**).

SOLUTION

IDENTIFY and SET UP: We discussed this situation in Section 23.1. From Eq. (23.5), we know the electric *potential energy U* for a test charge q_0 is $U = q_0Ey$. (We set $y = 0$ and $U = 0$ at the bottom plate.) We use Eq. (23.12), $U = q_0V$, to find the electric *potential V* as a function of y.

EXECUTE: The potential $V(y)$ at coordinate y is the potential energy per unit charge:

$$V(y) = \frac{U(y)}{q_0} = \frac{q_0Ey}{q_0} = Ey$$

The potential decreases as we move in the direction of $\vec{E}$ from the upper to the lower plate. At point a, where $y = d$ and $V(y) = V_a$,

$$V_a - V_b = Ed \quad \text{and} \quad E = \frac{V_a - V_b}{d} = \frac{V_{ab}}{d}$$

where V_{ab} is the potential of the positive plate with respect to the negative plate. That is, the electric field equals the potential difference between the plates divided by the distance between them. For a given potential difference V_{ab}, the smaller the distance d between the two plates, the greater the magnitude E of the electric field. (This relationship between E and V_{ab} holds *only* for the planar geometry we have described. It does *not* work for situations

23.18 The charged parallel plates from Fig. 23.2.

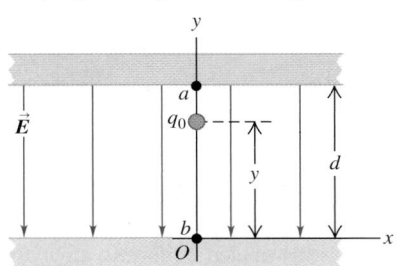

such as concentric cylinders or spheres in which the electric field is not uniform.)

EVALUATE: Our result shows that $V = 0$ at the bottom plate (at $y = 0$). This is consistent with our choice that $U = q_0V = 0$ for a test charge placed at the bottom plate.

CAUTION   **"Zero potential" is arbitrary** You might think that if a conducting body has zero potential, it must also have zero net charge. But that just isn't so! As an example, the plate at $y = 0$ in Fig. 23.18 has zero potential ($V = 0$) but has a nonzero charge per unit area $-\sigma$. There's nothing special about the place where potential is zero; we *define* this place to be wherever we want it to be. ∎

EXAMPLE 23.10   **AN INFINITE LINE CHARGE OR CHARGED CONDUCTING CYLINDER**

Find the potential at a distance r from a very long line of charge with linear charge density (charge per unit length) λ.

SOLUTION

IDENTIFY and SET UP: In both Example 21.10 (Section 21.5) and Example 22.6 (Section 22.4) we found that the electric field at a radial distance r from a long straight-line charge (**Fig. 23.19a**) has only a radial component $E_r = \lambda/2\pi\epsilon_0 r$. We use this expression to find the potential by integrating $\vec{E}$ as in Eq. (23.17).

EXECUTE: Since the field has only a radial component, we have $\vec{E} \cdot d\vec{l} = E_r dr$. Hence from Eq. (23.17) the potential of any point a with respect to any other point b, at radial distances r_a and r_b from the line of charge, is

$$V_a - V_b = \int_a^b \vec{E} \cdot d\vec{l} = \int_a^b E_r dr = \frac{\lambda}{2\pi\epsilon_0} \int_{r_a}^{r_b} \frac{dr}{r} = \frac{\lambda}{2\pi\epsilon_0} \ln\frac{r_b}{r_a}$$

23.19 Electric field outside **(a)** a long, positively charged wire and **(b)** a long, positively charged cylinder.

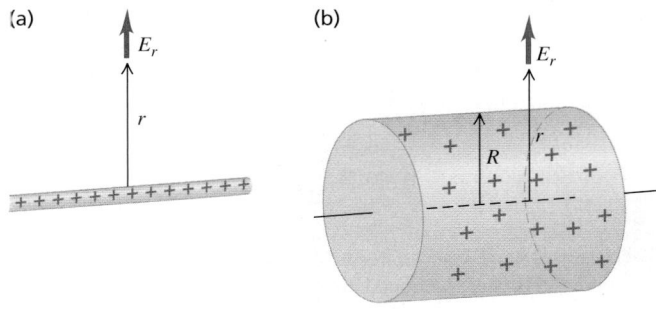

If we take point b at infinity and set $V_b = 0$, we find that V_a is *infinite* for any finite distance r_a from the line charge: $V_a = (\lambda/2\pi\epsilon_0) \ln(\infty/r_a) = \infty$. This is *not* a useful way to define

V for this problem! The difficulty is that the charge distribution itself extends to infinity.

Instead, as recommended in Problem-Solving Strategy 23.1, we set $V_b = 0$ at point *b* at an arbitrary but *finite* radial distance r_0. Then the potential $V = V_a$ at point *a* at a radial distance *r* is given by $V - 0 = (\lambda/2\pi\epsilon_0) \ln(r_0/r)$, or

$$V = \frac{\lambda}{2\pi\epsilon_0} \ln \frac{r_0}{r}$$

EVALUATE: According to our result, if λ is positive, then *V* decreases as *r* increases. This is as it should be: *V* decreases as we move in the direction of $\vec{E}$.

From Example 22.6, the expression for E_r with which we started also applies outside a long, charged conducting cylinder with charge per unit length λ (Fig. 23.19b). Hence our result also gives the potential for such a cylinder, but only for values of *r* (the distance from the cylinder axis) equal to or greater than the radius *R* of the cylinder. If we choose r_0 to be the radius *R*, so that $V = 0$ when $r = R$, then at any point for which $r > R$,

$$V = \frac{\lambda}{2\pi\epsilon_0} \ln \frac{R}{r}$$

Inside the cylinder, $\vec{E} = 0$, and *V* has the same value (zero) as on the cylinder's surface.

EXAMPLE 23.11   A RING OF CHARGE

Electric charge *Q* is distributed uniformly around a thin ring of radius *a* (**Fig. 23.20**). Find the potential at a point *P* on the ring axis at a distance *x* from the center of the ring.

SOLUTION

IDENTIFY and SET UP: We divide the ring into infinitesimal segments and use Eq. (23.16) to find *V*. All parts of the ring (and therefore all elements of the charge distribution) are at the same distance from *P*.

EXECUTE: Figure 23.20 shows that the distance from each charge element *dq* to *P* is $r = \sqrt{x^2 + a^2}$. Hence we can take the factor $1/r$ outside the integral in Eq. (23.16), and

$$V = \frac{1}{4\pi\epsilon_0} \int \frac{dq}{r} = \frac{1}{4\pi\epsilon_0} \frac{1}{\sqrt{x^2 + a^2}} \int dq = \frac{1}{4\pi\epsilon_0} \frac{Q}{\sqrt{x^2 + a^2}}$$

EVALUATE: When *x* is much larger than *a*, our expression for *V* becomes approximately $V = Q/4\pi\epsilon_0 x$, which is the potential at a

23.20 All the charge in a ring of charge *Q* is the same distance *r* from a point *P* on the ring axis.

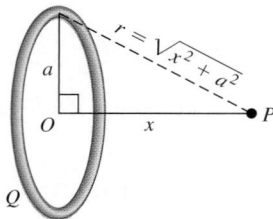

distance *x* from a point charge *Q*. Very far from a charged ring, its electric potential looks like that of a point charge. We drew a similar conclusion about the electric *field* of a ring in Example 21.9 (Section 21.5).

We know the electric field at all points along the *x*-axis from Example 21.9 (Section 21.5), so we can also find *V* along this axis by integrating $\vec{E} \cdot d\vec{l}$ as in Eq. (23.17).

EXAMPLE 23.12   POTENTIAL OF A LINE OF CHARGE

Positive electric charge *Q* is distributed uniformly along a line of length 2*a* lying along the *y*-axis between $y = -a$ and $y = +a$ (**Fig. 23.21**). Find the electric potential at a point *P* on the *x*-axis at a distance *x* from the origin.

23.21 Our sketch for this problem.

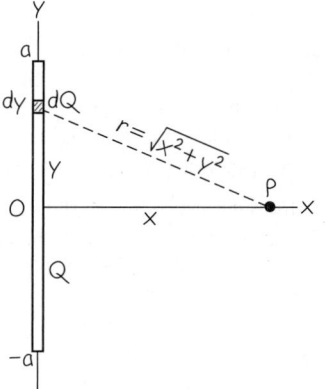

SOLUTION

IDENTIFY and SET UP: This is the situation of Example 21.10 (Section 21.5), where we found an expression for the electric field $\vec{E}$ at an arbitrary point on the *x*-axis. We can find *V* at point *P* by using Eq. (23.16) to integrate over the charge distribution. Unlike the situation in Example 23.11, each charge element *dQ* is a *different* distance from point *P*, so the integration will take a little more effort.

Continued

EXECUTE: As in Example 21.10, the element of charge dQ corresponding to an element of length dy on the rod is $dQ = (Q/2a)dy$. The distance from dQ to P is $\sqrt{x^2 + y^2}$, so the contribution dV that the charge element makes to the potential at P is

$$dV = \frac{1}{4\pi\epsilon_0} \frac{Q}{2a} \frac{dy}{\sqrt{x^2 + y^2}}$$

To find the potential at P due to the entire rod, we integrate dV over the length of the rod from $y = -a$ to $y = a$:

$$V = \frac{1}{4\pi\epsilon_0} \frac{Q}{2a} \int_{-a}^{a} \frac{dy}{\sqrt{x^2 + y^2}}$$

You can look up the integral in a table. The final result is

$$V = \frac{1}{4\pi\epsilon_0} \frac{Q}{2a} \ln\left(\frac{\sqrt{a^2 + x^2} + a}{\sqrt{a^2 + x^2} - a}\right)$$

EVALUATE: We can check our result by letting x approach infinity. In this limit the point P is infinitely far from all of the charge, so we expect V to approach zero; you can verify that it does.

We know the electric field at all points along the x-axis from Example 21.10. We invite you to use this information to find V along this axis by integrating $\vec{E}$ as in Eq. (23.17).

TEST YOUR UNDERSTANDING OF SECTION 23.3 If the electric *field* at a certain point is zero, does the electric *potential* at that point have to be zero? (*Hint:* Consider the center of the ring in Example 23.11 and Example 21.9.) ▌

23.4 EQUIPOTENTIAL SURFACES

Field lines (see Section 21.6) help us visualize electric fields. In a similar way, the potential at various points in an electric field can be represented graphically by *equipotential surfaces*. These use the same fundamental idea as topographic maps like those used by hikers and mountain climbers (**Fig. 23.22**). On a topographic map, contour lines are drawn through points that are all at the same elevation. Any number of these could be drawn, but typically only a few contour lines are shown at equal spacings of elevation. If a mass m is moved over the terrain along such a contour line, the gravitational potential energy mgy does not change because the elevation y is constant. Thus contour lines on a topographic map are really curves of constant gravitational potential energy. Contour lines are close together where the terrain is steep and there are large changes in elevation over a small horizontal distance; the contour lines are farther apart where the terrain is gently sloping. A ball allowed to roll downhill will experience the greatest downhill gravitational force where contour lines are closest together.

By analogy to contour lines on a topographic map, an **equipotential surface** is a three-dimensional surface on which the *electric potential V* is the same at every point. If a test charge q_0 is moved from point to point on such a surface, the *electric* potential energy q_0V remains constant. In a region where an electric field is present, we can construct an equipotential surface through any point. In diagrams we usually show only a few representative equipotentials, often with equal potential differences between adjacent surfaces. No point can be at two different potentials, so equipotential surfaces for different potentials can never touch or intersect.

23.22 Contour lines on a topographic map are curves of constant elevation and hence of constant gravitational potential energy.

Equipotential Surfaces and Field Lines

Because potential energy does not change as a test charge moves over an equipotential surface, the electric field can do no work on such a charge. It follows that $\vec{E}$ must be perpendicular to the surface at every point so that the electric force $q_0\vec{E}$ is always perpendicular to the displacement of a charge moving on the surface. **Field lines and equipotential surfaces are always mutually perpendicular.** In general, field lines are curves, and equipotentials are curved surfaces. For the special case of a *uniform* field, in which the field lines are straight, parallel, and equally spaced, the equipotentials are parallel *planes* perpendicular to the field lines.

23.23 Cross sections of equipotential surfaces (blue lines) and electric field lines (red lines) for assemblies of point charges. There are equal potential differences between adjacent surfaces. Compare these diagrams to those in Fig. 21.28, which showed only the electric field lines.

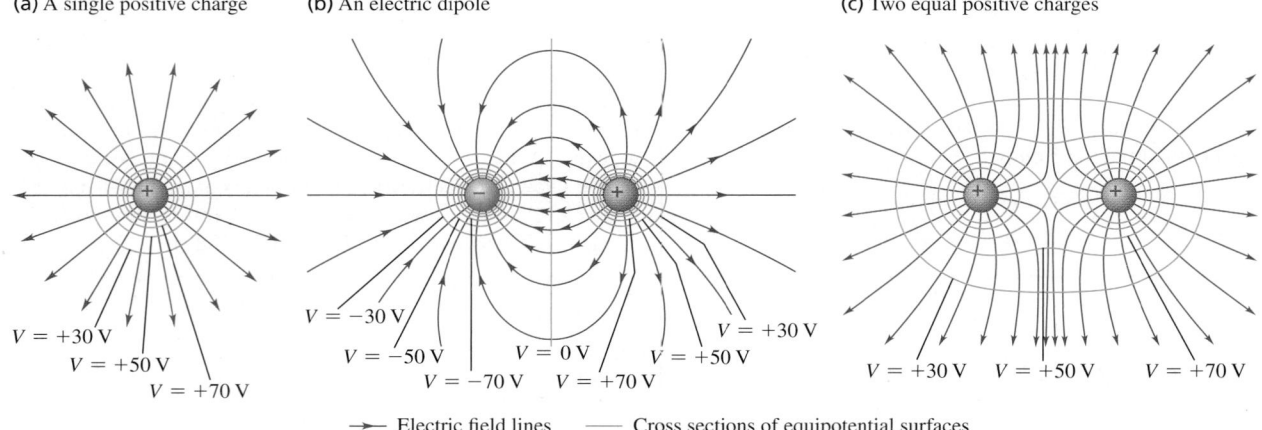

(a) A single positive charge

(b) An electric dipole

(c) Two equal positive charges

$V = +30$ V
$V = +50$ V
$V = +70$ V

$V = -30$ V
$V = -50$ V
$V = -70$ V
$V = 0$ V
$V = +70$ V
$V = +30$ V
$V = +50$ V

$V = +30$ V
$V = +50$ V
$V = +70$ V

→ Electric field lines — Cross sections of equipotential surfaces

Figure **23.23** shows three arrangements of charges. The field lines in the plane of the charges are represented by red lines, and the intersections of the equipotential surfaces with this plane (that is, cross sections of these surfaces) are shown as blue lines. The actual equipotential surfaces are three-dimensional. At each crossing of an equipotential and a field line, the two are perpendicular.

In Fig. 23.23 we have drawn equipotentials so that there are equal potential differences between adjacent surfaces. In regions where the magnitude of $\vec{E}$ is large, the equipotential surfaces are close together because the field does a relatively large amount of work on a test charge in a relatively small displacement. This is the case near the point charge in Fig. 23.23a or between the two point charges in Fig. 23.23b; note that in these regions the field lines are also closer together. This is directly analogous to the downhill force of gravity being greatest in regions on a topographic map where contour lines are close together. Conversely, in regions where the field is weaker, the equipotential surfaces are farther apart; this happens at larger radii in Fig. 23.23a, to the left of the negative charge or the right of the positive charge in Fig. 23.23b, and at greater distances from both charges in Fig. 23.23c. (It may appear that two equipotential surfaces intersect at the center of Fig. 23.23c, in violation of the rule that this can never happen. In fact this is a single figure-8–shaped equipotential surface.)

23.24 When charges are at rest, a conducting surface is always an equipotential surface. Field lines are perpendicular to a conducting surface.

> **CAUTION** *E* need not be constant over an equipotential surface On a given equipotential surface, the potential *V* has the same value at every point. In general, however, the electric-field magnitude *E* is *not* the same at all points on an equipotential surface. For instance, on equipotential surface "$V = -30$ V" in Fig. 23.23b, *E* is less to the left of the negative charge than it is between the two charges. On the figure-8–shaped equipotential surface in Fig. 23.23c, $E = 0$ at the middle point halfway between the two charges; at any other point on this surface, *E* is nonzero.

Equipotentials and Conductors

Here's an important statement about equipotential surfaces: **When all charges are at rest, the surface of a conductor is always an equipotential surface.** Since the electric field $\vec{E}$ is always perpendicular to an equipotential surface, we can prove this statement by proving that **when all charges are at rest, the electric field just outside a conductor must be perpendicular to the surface at every point (Fig. 23.24).** We know that $\vec{E} = 0$ everywhere inside the conductor; otherwise, charges would move. In particular, at any point just inside the surface the component of $\vec{E}$ tangent to the surface is zero. It follows that the tangential component of $\vec{E}$ is also zero just *outside* the surface. If it were not, a charge could

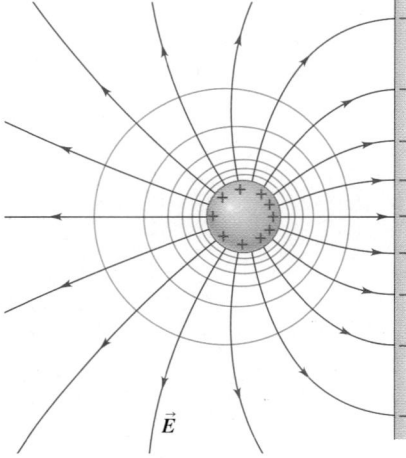

$\vec{E}$

— Cross sections of equipotential surfaces
→ Electric field lines

move around a rectangular path partly inside and partly outside (**Fig. 23.25**) and return to its starting point with a net amount of work having been done on it. This would violate the conservative nature of electrostatic fields, so the tangential component of $\vec{E}$ just outside the surface must be zero at every point on the surface. Thus $\vec{E}$ is perpendicular to the surface at each point, proving our statement.

It also follows that **when all charges are at rest, the entire solid volume of a conductor is at the same potential.** Equation (23.17) states that the potential difference between two points a and b within the conductor's solid volume, $V_a - V_b$, is equal to the line integral $\int_a^b \vec{E} \cdot d\vec{l}$ of the electric field from a to b. Since $\vec{E} = 0$ everywhere inside the conductor, the integral is guaranteed to be zero for any two such points a and b. Hence the potential is the same for any two points within the solid volume of the conductor. We describe this by saying that the solid volume of the conductor is an *equipotential volume*.

We can now prove a theorem that we quoted without proof in Section 22.5. The theorem is as follows: In an electrostatic situation, if a conductor contains a cavity and if no charge is present inside the cavity, then there can be no net charge *anywhere* on the surface of the cavity. This means that if you're inside a charged conducting box, you can safely touch any point on the inside walls of the box without being shocked. To prove this theorem, we first prove that *every point in the cavity is at the same potential*. In **Fig. 23.26** the conducting surface A of the cavity is an equipotential surface, as we have just proved. Suppose point P in the cavity is at a different potential; then we can construct a different equipotential surface B including point P.

Now consider a Gaussian surface, shown in Fig. 23.26, between the two equipotential surfaces. Because of the relationship between $\vec{E}$ and the equipotentials, we know that the field at every point between the equipotentials is from A toward B, or else at every point it is from B toward A, depending on which equipotential surface is at higher potential. In either case the flux through this Gaussian surface is certainly not zero. But then Gauss's law says that the charge enclosed by the Gaussian surface cannot be zero. This contradicts our initial assumption that there is *no* charge in the cavity. So the potential at P *cannot* be different from that at the cavity wall.

The entire region of the cavity must therefore be at the same potential. But for this to be true, *the electric field inside the cavity must be zero everywhere.* Finally, Gauss's law shows that the electric field at any point on the surface of a conductor is proportional to the surface charge density σ at that point. We conclude that *the surface charge density on the wall of the cavity is zero at every point.* This chain of reasoning may seem tortuous, but it is worth careful study.

CAUTION Equipotential surfaces vs. Gaussian surfaces Don't confuse equipotential surfaces with the Gaussian surfaces we encountered in Chapter 22. Gaussian surfaces have relevance only when we are using Gauss's law, and we can choose *any* Gaussian surface that's convenient. We *cannot* choose equipotential surfaces; the shape is determined by the charge distribution. ▌

TEST YOUR UNDERSTANDING OF SECTION 23.4 Would the shapes of the equipotential surfaces in Fig. 23.23 change if the sign of each charge were reversed? ▌

23.5 POTENTIAL GRADIENT

Electric field and potential are closely related. Equation (23.17), restated here, expresses one aspect of that relationship:

$$V_a - V_b = \int_a^b \vec{E} \cdot d\vec{l}$$

If we know $\vec{E}$ at various points, we can use this equation to calculate potential differences. In this section we show how to turn this around; if we know the

DEMO

23.25 At all points on the surface of a conductor, the electric field must be perpendicular to the surface. If $\vec{E}$ had a tangential component, a net amount of work would be done on a test charge by moving it around a loop as shown here—which is impossible because the electric force is conservative.

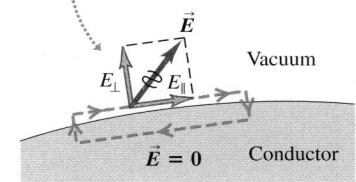

An impossible electric field
If the electric field just outside a conductor had a tangential component $E_\parallel$, a charge could move in a loop with net work done.

23.26 A cavity in a conductor. If the cavity contains no charge, every point in the cavity is at the same potential, the electric field is zero everywhere in the cavity, and there is no charge anywhere on the surface of the cavity.

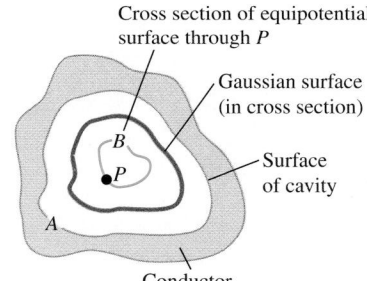

Cross section of equipotential surface through P

Gaussian surface (in cross section)

Surface of cavity

Conductor

potential V at various points, we can use it to determine $\vec{E}$. Regarding V as a function of the coordinates (x, y, z) of a point in space, we will show that the components of $\vec{E}$ are related to the *partial derivatives* of V with respect to x, y, and z.

In Eq. (23.17), $V_a - V_b$ is the potential of a with respect to b—that is, the change of potential encountered on a trip from b to a. We can write this as

$$V_a - V_b = \int_b^a dV = -\int_a^b dV$$

where dV is the infinitesimal change of potential accompanying an infinitesimal element $d\vec{l}$ of the path from b to a. Comparing to Eq. (23.17), we have

$$-\int_a^b dV = \int_a^b \vec{E} \cdot d\vec{l}$$

These two integrals must be equal for *any* pair of limits a and b, and for this to be true the *integrands* must be equal. Thus, for *any* infinitesimal displacement $d\vec{l}$,

$$-dV = \vec{E} \cdot d\vec{l}$$

To interpret this expression, we write $\vec{E}$ and $d\vec{l}$ in terms of their components: $\vec{E} = \hat{\imath} E_x + \hat{\jmath} E_y + \hat{k} E_z$ and $d\vec{l} = \hat{\imath} dx + \hat{\jmath} dy + \hat{k} dz$. Then

$$-dV = E_x dx + E_y dy + E_z dz$$

Suppose the displacement is parallel to the x-axis, so $dy = dz = 0$. Then $-dV = E_x dx$ or $E_x = -(dV/dx)_{y,\,z\text{ constant}}$, where the subscript reminds us that only x varies in the derivative; recall that V is in general a function of x, y, and z. But this is just what is meant by the partial derivative $\partial V/\partial x$. The y- and z-components of $\vec{E}$ are related to the corresponding derivatives of V in the same way, so

Electric field
components found
from potential:
···· Each electric field component ...

$$E_x = -\frac{\partial V}{\partial x} \qquad E_y = -\frac{\partial V}{\partial y} \qquad E_z = -\frac{\partial V}{\partial z} \qquad (23.19)$$

... equals the negative of the corresponding
partial derivative of electric potential function V.

This is consistent with the units of electric field being V/m. In terms of unit vectors we can write

Electric field
vector found
from potential:
··· Electric field

$$\vec{E} = -\left(\hat{\imath}\frac{\partial V}{\partial x} + \hat{\jmath}\frac{\partial V}{\partial y} + \hat{k}\frac{\partial V}{\partial z} \right) \qquad (23.20)$$

Partial derivatives of electric potential function V

The following operation is called the **gradient** of the function f:

$$\vec{\nabla} f = \left(\hat{\imath}\frac{\partial}{\partial x} + \hat{\jmath}\frac{\partial}{\partial y} + \hat{k}\frac{\partial}{\partial z} \right) f \qquad (23.21)$$

The operator denoted by $\vec{\nabla}$ is called "grad" or "del." Thus in vector notation,

$$\vec{E} = -\vec{\nabla} V \qquad (23.22)$$

This is read "$\vec{E}$ is the negative of the gradient of V" or "$\vec{E}$ equals negative grad V." The quantity $\vec{\nabla} V$ is called the *potential gradient*.

At each point, the potential gradient $\vec{\nabla}V$ points in the direction in which V *increases* most rapidly with a change in position. Hence at each point the direction of $\vec{E} = -\vec{\nabla}V$ is the direction in which V *decreases* most rapidly and is always perpendicular to the equipotential surface through the point. This agrees with our observation in Section 23.2 that moving in the direction of the electric field means moving in the direction of decreasing potential.

Equation (23.22) doesn't depend on the particular choice of the zero point for V. If we were to change the zero point, the effect would be to change V at every point by the same amount; the derivatives of V would be the same.

If $\vec{E}$ has a radial component E_r with respect to a point or an axis and r is the distance from the point or axis, the relationship corresponding to Eqs. (23.19) is

$$E_r = -\frac{\partial V}{\partial r} \qquad \text{(radial electric field)} \qquad (23.23)$$

Often we can compute the electric field caused by a charge distribution in either of two ways: directly, by adding the $\vec{E}$ fields of point charges, or by first calculating the potential and then taking its gradient to find the field. The second method is often easier because potential is a *scalar* quantity, requiring at worst the integration of a scalar function. Electric field is a *vector* quantity, requiring computation of components for each element of charge and a separate integration for each component. Thus, quite apart from its fundamental significance, potential offers a very useful computational technique in field calculations. In the next two examples, a knowledge of V is used to find the electric field.

We stress once more that if we know $\vec{E}$ as a function of position, we can calculate V from Eq. (23.17) or (23.18), and if we know V as a function of position, we can calculate $\vec{E}$ from Eq. (23.19), (23.20), or (23.23). Deriving V from $\vec{E}$ requires integration, and deriving $\vec{E}$ from V requires differentiation.

BIO Application Potential Gradient Across a Cell Membrane The interior of a human cell is at a lower electric potential V than the exterior. (The potential difference when the cell is inactive is about -70 mV in neurons and about -95 mV in skeletal muscle cells.) Hence there is a potential gradient $\vec{\nabla}V$ that points from the *interior* to the *exterior* of the cell membrane, and an electric field $\vec{E} = -\vec{\nabla}V$ that points from the *exterior* to the *interior*. This field affects how ions flow into or out of the cell through special channels in the membrane.

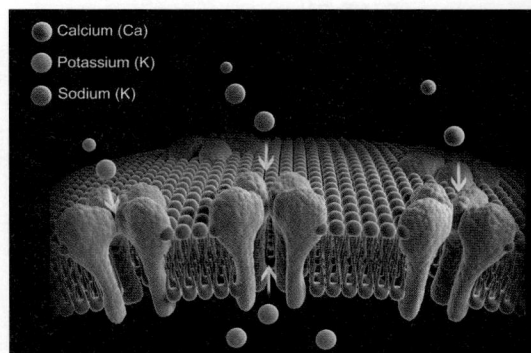

- Calcium (Ca)
- Potassium (K)
- Sodium (K)

EXAMPLE 23.13 POTENTIAL AND FIELD OF A POINT CHARGE

From Eq. (23.14) the potential at a radial distance r from a point charge q is $V = q/4\pi\epsilon_0 r$. Find the vector electric field from this expression for V.

SOLUTION

IDENTIFY and SET UP: This problem uses the general relationship between the electric potential as a function of position and the electric-field vector. By symmetry, the electric field here has only a radial component E_r. We use Eq. (23.23) to find this component.

EXECUTE: From Eq. (23.23),

$$E_r = -\frac{\partial V}{\partial r} = -\frac{\partial}{\partial r}\left(\frac{1}{4\pi\epsilon_0}\frac{q}{r}\right) = \frac{1}{4\pi\epsilon_0}\frac{q}{r^2}$$

so the vector electric field is

$$\vec{E} = \hat{r}E_r = \frac{1}{4\pi\epsilon_0}\frac{q}{r^2}\hat{r}$$

EVALUATE: Our result agrees with Eq. (21.7), as it must.

An alternative approach is to ignore the radial symmetry, write the radial distance as $r = \sqrt{x^2 + y^2 + z^2}$, and take the derivatives of V with respect to x, y, and z as in Eq. (23.20). We find

$$\frac{\partial V}{\partial x} = \frac{\partial}{\partial x}\left(\frac{1}{4\pi\epsilon_0}\frac{q}{\sqrt{x^2 + y^2 + z^2}}\right) = -\frac{1}{4\pi\epsilon_0}\frac{qx}{(x^2 + y^2 + z^2)^{3/2}}$$

$$= -\frac{qx}{4\pi\epsilon_0 r^3}$$

and similarly

$$\frac{\partial V}{\partial y} = -\frac{qy}{4\pi\epsilon_0 r^3} \qquad \frac{\partial V}{\partial z} = -\frac{qz}{4\pi\epsilon_0 r^3}$$

Then from Eq. (23.20),

$$\vec{E} = -\left[\hat{\imath}\left(-\frac{qx}{4\pi\epsilon_0 r^3}\right) + \hat{\jmath}\left(-\frac{qy}{4\pi\epsilon_0 r^3}\right) + \hat{k}\left(-\frac{qz}{4\pi\epsilon_0 r^3}\right)\right]$$

$$= \frac{1}{4\pi\epsilon_0}\frac{q}{r^2}\left(\frac{x\hat{\imath} + y\hat{\jmath} + z\hat{k}}{r}\right) = \frac{1}{4\pi\epsilon_0}\frac{q}{r^2}\hat{r}$$

This approach gives us the same answer, but with more effort. Clearly it's best to exploit the symmetry of the charge distribution whenever possible.

EXAMPLE 23.14 POTENTIAL AND FIELD OF A RING OF CHARGE

In Example 23.11 (Section 23.3) we found that for a ring of charge with radius a and total charge Q, the potential at a point P on the ring's symmetry axis a distance x from the center is

$$V = \frac{1}{4\pi\epsilon_0} \frac{Q}{\sqrt{x^2 + a^2}}$$

Find the electric field at P.

SOLUTION

IDENTIFY and SET UP: Figure 23.20 shows the situation. We are given V as a function of x along the x-axis, and we wish to find the electric field at a point on this axis. From the symmetry of the charge distribution, the electric field along the symmetry (x-) axis of the ring can have only an x-component. We find it by using the first of Eqs. (23.19).

EXECUTE: The x-component of the electric field is

$$E_x = -\frac{\partial V}{\partial x} = \frac{1}{4\pi\epsilon_0} \frac{Qx}{(x^2 + a^2)^{3/2}}$$

EVALUATE: This agrees with our result in Example 21.9.

CAUTION **Don't use expressions where they don't apply** In this example, V is not a function of y or z on the ring axis, so $\partial V/\partial y = \partial V/\partial z = 0$ and $E_y = E_z = 0$. But that does not mean that it's true *everywhere*; our expressions for V and E_x are valid *on the ring axis only*. If we had an expression for V valid at *all* points in space, we could use it to find the components of $\vec{E}$ at any point by using Eqs. (23.19). ∎

TEST YOUR UNDERSTANDING OF SECTION 23.5 In a certain region of space the potential is given by $V = A + Bx + Cy^3 + Dxy$, where A, B, C, and D are positive constants. Which of these statements about the electric field $\vec{E}$ in this region of space is correct? (There may be more than one correct answer.) (i) Increasing the value of A will increase the value of $\vec{E}$ at all points; (ii) increasing the value of A will decrease the value of $\vec{E}$ at all points; (iii) $\vec{E}$ has no z-component; (iv) the electric field is zero at the origin $(x = 0, y = 0, z = 0)$. ∎

Electric potential energy: The electric force caused by any collection of charges at rest is a conservative force. The work W done by the electric force on a charged particle moving in an electric field can be represented by the change in a potential-energy function U.

The electric potential energy for two point charges q and q_0 depends on their separation r. The electric potential energy for a charge q_0 in the presence of a collection of charges q_1, q_2, q_3 depends on the distance from q_0 to each of these other charges. (See Examples 23.1 and 23.2.)

$$W_{a \to b} = U_a - U_b \qquad (23.2)$$

$$U = \frac{1}{4\pi\epsilon_0} \frac{qq_0}{r} \qquad (23.9)$$

(two point charges)

$$U = \frac{q_0}{4\pi\epsilon_0}\left(\frac{q_1}{r_1} + \frac{q_2}{r_2} + \frac{q_3}{r_3} + \cdots\right)$$

$$= \frac{q_0}{4\pi\epsilon_0}\sum_i \frac{q_i}{r_i} \qquad (23.10)$$

(q_0 in presence of other point charges)

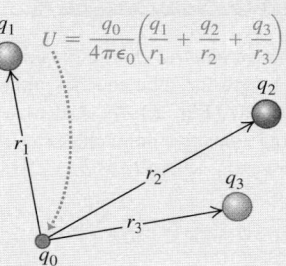

Electric potential: Potential, denoted by V, is potential energy per unit charge. The potential difference between two points equals the amount of work per charge that would be required to move a positive test charge between those points. The potential V due to a quantity of charge can be calculated by summing (if the charge is a collection of point charges) or by integrating (if the charge is a distribution). (See Examples 23.3, 23.4, 23.5, 23.7, 23.11, and 23.12.)

The potential difference between two points a and b, also called the potential of a with respect to b, is given by the line integral of $\vec{E}$. The potential at a given point can be found by first finding $\vec{E}$ and then carrying out this integral. (See Examples 23.6, 23.8, 23.9, and 23.10.)

$$V = \frac{1}{4\pi\epsilon_0}\frac{q}{r} \qquad (23.14)$$

(due to a point charge)

$$V = \frac{1}{4\pi\epsilon_0}\sum_i \frac{q_i}{r_i} \qquad (23.15)$$

(due to a collection of point charges)

$$V = \frac{1}{4\pi\epsilon_0}\int \frac{dq}{r} \qquad (23.16)$$

(due to a charge distribution)

$$V_a - V_b = \int_a^b \vec{E} \cdot d\vec{l} \qquad (23.17)$$

$$= \int_a^b E\cos\phi\, dl$$

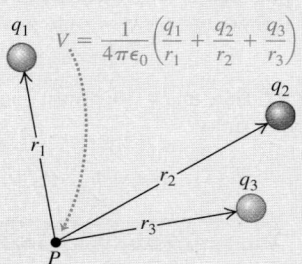

Equipotential surfaces: An equipotential surface is a surface on which the potential has the same value at every point. At a point where a field line crosses an equipotential surface, the two are perpendicular. When all charges are at rest, the surface of a conductor is always an equipotential surface and all points in the interior of a conductor are at the same potential. When a cavity within a conductor contains no charge, the entire cavity is an equipotential region and there is no surface charge anywhere on the surface of the cavity.

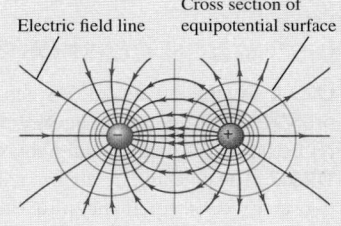

Finding electric field from electric potential: If the potential V is known as a function of the coordinates x, y, and z, the components of electric field $\vec{E}$ at any point are given by partial derivatives of V. (See Examples 23.13 and 23.14.)

$$E_x = -\frac{\partial V}{\partial x} \quad E_y = -\frac{\partial V}{\partial y} \quad E_z = -\frac{\partial V}{\partial z} \qquad (23.19)$$

$$\vec{E} = -\left(\hat{i}\frac{\partial V}{\partial x} + \hat{j}\frac{\partial V}{\partial y} + \hat{k}\frac{\partial V}{\partial z}\right) \qquad (23.20)$$

(vector form)

BRIDGING PROBLEM   A POINT CHARGE AND A LINE OF CHARGE

Positive electric charge Q is distributed uniformly along a thin rod of length $2a$. The rod lies along the x-axis between $x = -a$ and $x = +a$ (**Fig. 23.27**). Calculate how much work you must do to bring a positive point charge q from infinity to the point $x = +L$ on the x-axis, where $L > a$.

SOLUTION GUIDE

IDENTIFY and SET UP

1. In this problem you must first calculate the potential V at $x = +L$ due to the charged rod. You can then find the change in potential energy involved in bringing the point charge q from infinity (where $V = 0$) to $x = +L$.

23.27 How much work must you do to bring point charge q in from infinity?

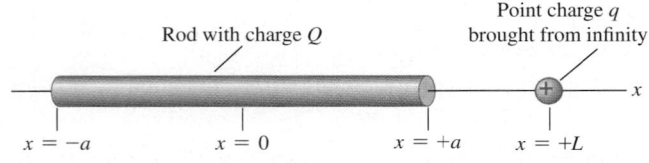

Rod with charge Q

Point charge q brought from infinity

$x = -a$ $x = 0$ $x = +a$ $x = +L$

2. To find V, divide the rod into infinitesimal segments of length dx'. How much charge is on such a segment? Consider one such segment located at $x = x'$, where $-a \le x' \le a$. What is the potential dV at $x = +L$ due to this segment?
3. The total potential at $x = +L$ is the integral of dV, including contributions from all of the segments for x' from $-a$ to $+a$. Set up this integral.

EXECUTE

4. Integrate your expression from step 3 to find the potential V at $x = +L$. A simple, standard substitution will do the trick; use a table of integrals only as a last resort.
5. Use your result from step 4 to find the potential energy for a point charge q placed at $x = +L$.
6. Use your result from step 5 to find the work you must do to bring the point charge from infinity to $x = +L$.

EVALUATE

7. What does your result from step 5 become in the limit $a \to 0$? Does this make sense?
8. Suppose the point charge q were negative rather than positive. How would this affect your result in step 4? In step 5?

Problems

For assigned homework and other learning materials, go to MasteringPhysics®.   (MP)

•, ••, •••: Difficulty levels. **CP**: Cumulative problems incorporating material from earlier chapters. **CALC**: Problems requiring calculus. **DATA**: Problems involving real data, scientific evidence, experimental design, and/or statistical reasoning. **BIO**: Biosciences problems.

DISCUSSION QUESTIONS

Q23.1 A student asked, "Since electrical potential is always proportional to potential energy, why bother with the concept of potential at all?" How would you respond?

Q23.2 The potential (relative to a point at infinity) midway between two charges of equal magnitude and opposite sign is zero. Is it possible to bring a test charge from infinity to this midpoint in such a way that no work is done in any part of the displacement? If so, describe how it can be done. If it is not possible, explain why.

Q23.3 Is it possible to have an arrangement of two point charges separated by a finite distance such that the electric potential energy of the arrangement is the same as if the two charges were infinitely far apart? Why or why not? What if there are three charges? Explain.

Q23.4 Since potential can have any value you want depending on the choice of the reference level of zero potential, how does a voltmeter know what to read when you connect it between two points?

Q23.5 If $\vec{E}$ is zero everywhere along a certain path that leads from point A to point B, what is the potential difference between those two points? Does this mean that $\vec{E}$ is zero everywhere along *any* path from A to B? Explain.

Q23.6 If $\vec{E}$ is zero throughout a certain region of space, is the potential necessarily also zero in this region? Why or why not? If not, what *can* be said about the potential?

Q23.7 Which way do electric field lines point, from high to low potential or from low to high? Explain.

Q23.8 (a) If the potential (relative to infinity) is zero at a point, is the electric field necessarily zero at that point? (b) If the electric field is zero at a point, is the potential (relative to infinity) necessarily zero there? Prove your answers, using simple examples.

Q23.9 If you carry out the integral of the electric field $\int \vec{E} \cdot d\vec{l}$ for a *closed* path like that shown in **Fig. Q23.9**, the integral will *always* be equal to zero, independent of the shape of the path and independent of where charges may be located relative to the path. Explain why.

Figure **Q23.9**

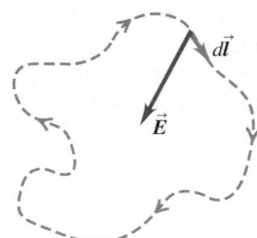

$d\vec{l}$

$\vec{E}$

Q23.10 The potential difference between the two terminals of an AA battery (used in flashlights and portable stereos) is 1.5 V. If two AA batteries are placed end to end with the positive terminal of one battery touching the negative terminal of the other, what is the potential difference between the terminals at the exposed ends of the combination? What if the two positive terminals are touching each other? Explain your reasoning.

Q23.11 It is easy to produce a potential difference of several thousand volts between your body and the floor by scuffing your shoes across a nylon carpet. When you touch a metal doorknob, you get a mild shock. Yet contact with a power line of comparable voltage would probably be fatal. Why is there a difference?

Q23.12 If the electric potential at a single point is known, can $\vec{E}$ at that point be determined? If so, how? If not, why not?

Q23.13 Because electric field lines and equipotential surfaces are always perpendicular, two equipotential surfaces can never cross; if they did, the direction of $\vec{E}$ would be ambiguous at the crossing points. Yet two equipotential surfaces appear to cross at the center of Fig. 23.23c. Explain why there is no ambiguity about the direction of $\vec{E}$ in this particular case.

Q23.14 A uniform electric field is directed due east. Point B is 2.00 m west of point A, point C is 2.00 m east of point A, and point D is 2.00 m south of A. For each point, B, C, and D, is the potential at that point larger, smaller, or the same as at point A? Give the reasoning behind your answers.

Q23.15 We often say that if point A is at a higher potential than point B, A is at positive potential and B is at negative potential. Does it necessarily follow that a point at positive potential is positively charged, or that a point at negative potential is negatively charged? Illustrate your answers with clear, simple examples.

Q23.16 A conducting sphere is to be charged by bringing in positive charge a little at a time until the total charge is Q. The total work required for this process is alleged to be proportional to Q^2. Is this correct? Why or why not?

Q23.17 In electronics it is customary to define the potential of ground (thinking of the earth as a large conductor) as zero. Is this consistent with the fact that the earth has a net electric charge that is not zero? (Refer to Exercise 21.28.)

Q23.18 A conducting sphere is placed between two charged parallel plates such as those shown in Fig. 23.2. Does the electric field inside the sphere depend on precisely where between the plates the sphere is placed? What about the electric potential inside the sphere? Do the answers to these questions depend on whether or not there is a net charge on the sphere? Explain your reasoning.

Q23.19 A conductor that carries a net charge Q has a hollow, empty cavity in its interior. Does the potential vary from point to point within the material of the conductor? What about within the cavity? How does the potential inside the cavity compare to the potential within the material of the conductor?

Q23.20 A high-voltage dc power line falls on a car, so the entire metal body of the car is at a potential of 10,000 V with respect to the ground. What happens to the occupants (a) when they are sitting in the car and (b) when they step out of the car? Explain your reasoning.

Q23.21 When a thunderstorm is approaching, sailors at sea sometimes observe a phenomenon called "St. Elmo's fire," a bluish flickering light at the tips of masts. What causes this? Why does it occur at the tips of masts? Why is the effect most pronounced when the masts are wet? (*Hint:* Seawater is a good conductor of electricity.)

Q23.22 A positive point charge is placed near a very large conducting plane. A professor of physics asserted that the field caused by this configuration is the same as would be obtained by removing the plane and placing a negative point charge of equal magnitude in the mirror-image position behind the initial position of the plane. Is this correct? Why or why not? (*Hint:* Inspect Fig. 23.23b.)

EXERCISES

Section 23.1 Electric Potential Energy

23.1 •• A point charge $q_1 = +2.40\ \mu C$ is held stationary at the origin. A second point charge $q_2 = -4.30\ \mu C$ moves from the point $x = 0.150$ m, $y = 0$ to the point $x = 0.250$ m, $y = 0.250$ m. How much work is done by the electric force on q_2?

23.2 • A point charge q_1 is held stationary at the origin. A second charge q_2 is placed at point a, and the electric potential energy of the pair of charges is $+5.4 \times 10^{-8}$ J. When the second charge is moved to point b, the electric force on the charge does -1.9×10^{-8} J of work. What is the electric potential energy of the pair of charges when the second charge is at point b?

23.3 •• **Energy of the Nucleus.** How much work is needed to assemble an atomic nucleus containing three protons (such as Li) if we model it as an equilateral triangle of side 2.00×10^{-15} m with a proton at each vertex? Assume the protons started from very far away.

23.4 •• (a) How much work would it take to push two protons very slowly from a separation of 2.00×10^{-10} m (a typical atomic distance) to 3.00×10^{-15} m (a typical nuclear distance)? (b) If the protons are both released from rest at the closer distance in part (a), how fast are they moving when they reach their original separation?

23.5 •• A small metal sphere, carrying a net charge of $q_1 = -2.80\ \mu C$, is held in a stationary position by insulating supports. A second small metal sphere, with a net charge of $q_2 = -7.80\ \mu C$ and mass 1.50 g, is projected toward q_1. When the two spheres are 0.800 m apart, q_2, is moving toward q_1 with speed 22.0 m/s (**Fig. E23.5**). Assume that the two spheres can be treated as point charges. You can ignore the force of gravity. (a) What is the speed of q_2 when the spheres are 0.400 m apart? (b) How close does q_2 get to q_1?

Figure **E23.5**

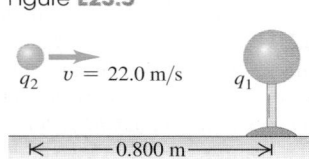

$q_2\ \ v = 22.0$ m/s $\quad q_1$

|← 0.800 m →|

23.6 •• **BIO** **Energy of DNA Base Pairing.** (See Exercise 21.21.) (a) Calculate the electric potential energy of the adenine–thymine bond, using the same combinations of molecules (O–H–N and N–H–N) as in Exercise 21.21. (b) Compare this energy with the potential energy of the proton–electron pair in the hydrogen atom.

23.7 •• Two protons, starting several meters apart, are aimed directly at each other with speeds of 2.00×10^5 m/s, measured relative to the earth. Find the maximum electric force that these protons will exert on each other.

23.8 •• Three equal 1.20-μC point charges are placed at the corners of an equilateral triangle with sides 0.400 m long. What is the potential energy of the system? (Take as zero the potential energy of the three charges when they are infinitely far apart.)

23.9 •• Two protons are released from rest when they are 0.750 nm apart. (a) What is the maximum speed they will reach? When does this speed occur? (b) What is the maximum acceleration they will achieve? When does this acceleration occur?

23.10 •• Four electrons are located at the corners of a square 10.0 nm on a side, with an alpha particle at its midpoint. How much work is needed to move the alpha particle to the midpoint of one of the sides of the square?

23.11 •• Three point charges, which initially are infinitely far apart, are placed at the corners of an equilateral triangle with sides d. Two of the point charges are identical and have charge q. If zero net work is required to place the three charges at the corners of the triangle, what must the value of the third charge be?

Section 23.2 Electric Potential

23.12 • An object with charge $q = -6.00 \times 10^{-9}$ C is placed in a region of uniform electric field and is released from rest at point A. After the charge has moved to point B, 0.500 m to the right, it has kinetic energy 3.00×10^{-7} J. (a) If the electric potential at point A is $+30.0$ V, what is the electric potential at point B? (b) What are the magnitude and direction of the electric field?

23.13 • A small particle has charge $-5.00 \ \mu$C and mass 2.00×10^{-4} kg. It moves from point A, where the electric potential is $V_A = +200$ V, to point B, where the electric potential is $V_B = +800$ V. The electric force is the only force acting on the particle. The particle has speed 5.00 m/s at point A. What is its speed at point B? Is it moving faster or slower at B than at A? Explain.

23.14 • A particle with charge $+4.20$ nC is in a uniform electric field $\vec{E}$ directed to the left. The charge is released from rest and moves to the left; after it has moved 6.00 cm, its kinetic energy is $+2.20 \times 10^{-6}$ J. What are (a) the work done by the electric force, (b) the potential of the starting point with respect to the end point, and (c) the magnitude of $\vec{E}$?

23.15 • A charge of 28.0 nC is placed in a uniform electric field that is directed vertically upward and has a magnitude of 4.00×10^4 V/m. What work is done by the electric force when the charge moves (a) 0.450 m to the right; (b) 0.670 m upward; (c) 2.60 m at an angle of $45.0°$ downward from the horizontal?

23.16 • Two stationary point charges $+3.00$ nC and $+2.00$ nC are separated by a distance of 50.0 cm. An electron is released from rest at a point midway between the two charges and moves along the line connecting the two charges. What is the speed of the electron when it is 10.0 cm from the $+3.00$-nC charge?

23.17 •• Point charges $q_1 = +2.00 \ \mu$C and $q_2 = -2.00 \ \mu$C are placed at adjacent corners of a square for which the length of each side is 3.00 cm. Point a is at the center of the square, and point b is at the empty corner closest to q_2. Take the electric potential to be zero at a distance far from both charges. (a) What is the electric potential at point a due to q_1 and q_2? (b) What is the electric potential at point b? (c) A point charge $q_3 = -5.00 \ \mu$C moves from point a to point b. How much work is done on q_3 by the electric forces exerted by q_1 and q_2? Is this work positive or negative?

23.18 • Two point charges of equal magnitude Q are held a distance d apart. Consider only points on the line passing through both charges. (a) If the two charges have the same sign, find the location of all points (if there are any) at which (i) the potential (relative to infinity) is zero (is the electric field zero at these points?), and (ii) the electric field is zero (is the potential zero at these points?). (b) Repeat part (a) for two point charges having opposite signs.

23.19 • Two point charges $q_1 = +2.40$ nC and $q_2 = -6.50$ nC are 0.100 m apart. Point A is midway between them; point B is 0.080 m from q_1 and 0.060 m from q_2 (**Fig. E23.19**). Take the electric potential to be zero at infinity. Find (a) the potential at point A; (b) the potential at point B; (c) the work done by the electric field on a charge of 2.50 nC that travels from point B to point A.

Figure **E23.19**

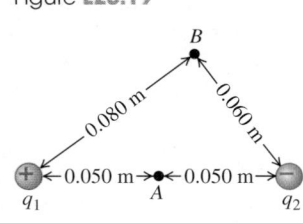

23.20 •• (a) An electron is to be accelerated from 3.00×10^6 m/s to 8.00×10^6 m/s. Through what potential difference must the electron pass to accomplish this? (b) Through what potential difference must the electron pass if it is to be slowed from 8.00×10^6 m/s to a halt?

23.21 •• A positive charge q is fixed at the point $x = 0$, $y = 0$, and a negative charge $-2q$ is fixed at the point $x = a$, $y = 0$. (a) Show the positions of the charges in a diagram. (b) Derive an expression for the potential V at points on the x-axis as a function of the coordinate x. Take V to be zero at an infinite distance from the charges. (c) At which positions on the x-axis is $V = 0$? (d) Graph V at points on the x-axis as a function of x in the range from $x = -2a$ to $x = +2a$. (e) What does the answer to part (b) become when $x \gg a$? Explain why this result is obtained.

23.22 • At a certain distance from a point charge, the potential and electric-field magnitude due to that charge are 4.98 V and 16.2 V/m, respectively. (Take $V = 0$ at infinity.) (a) What is the distance to the point charge? (b) What is the magnitude of the charge? (c) Is the electric field directed toward or away from the point charge?

23.23 • A uniform electric field has magnitude E and is directed in the negative x-direction. The potential difference between point a (at $x = 0.60$ m) and point b (at $x = 0.90$ m) is 240 V. (a) Which point, a or b, is at the higher potential? (b) Calculate the value of E. (c) A negative point charge $q = -0.200 \ \mu$C is moved from b to a. Calculate the work done on the point charge by the electric field.

23.24 • For each of the following arrangements of two point charges, find all the points along the line passing through both charges for which the electric potential V is zero (take $V = 0$ infinitely far from the charges) and for which the electric field E is zero: (a) charges $+Q$ and $+2Q$ separated by a distance d, and (b) charges $-Q$ and $+2Q$ separated by a distance d. (c) Are both V and E zero at the same places? Explain.

Section 23.3 Calculating Electric Potential

23.25 •• A thin spherical shell with radius $R_1 = 3.00$ cm is concentric with a larger thin spherical shell with radius $R_2 = 5.00$ cm. Both shells are made of insulating material. The smaller shell has charge $q_1 = +6.00$ nC distributed uniformly over its surface, and the larger shell has charge $q_2 = -9.00$ nC distributed uniformly over its surface. Take the electric potential to be zero at an infinite distance from both shells. (a) What is the electric potential due to the two shells at the following distance from their common center: (i) $r = 0$; (ii) $r = 4.00$ cm; (iii) $r = 6.00$ cm? (b) What is the magnitude of the potential difference between the surfaces of the two shells? Which shell is at higher potential: the inner shell or the outer shell?

23.26 • A total electric charge of 3.50 nC is distributed uniformly over the surface of a metal sphere with a radius of 24.0 cm. If the potential is zero at a point at infinity, find the value of the potential at the following distances from the center of the sphere: (a) 48.0 cm; (b) 24.0 cm; (c) 12.0 cm.

23.27 •• A uniformly charged, thin ring has radius 15.0 cm and total charge $+24.0$ nC. An electron is placed on the ring's axis a distance 30.0 cm from the center of the ring and is constrained to stay on the axis of the ring. The electron is then released from rest. (a) Describe the subsequent motion of the electron. (b) Find the speed of the electron when it reaches the center of the ring.

23.28 • A solid conducting sphere has net positive charge and radius $R = 0.400$ m. At a point 1.20 m from the center of the sphere, the electric potential due to the charge on the sphere is 24.0 V. Assume that $V = 0$ at an infinite distance from the sphere. What is the electric potential at the center of the sphere?

23.29 • Charge $Q = 5.00 \, \mu C$ is distributed uniformly over the volume of an insulating sphere that has radius $R = 12.0$ cm. A small sphere with charge $q = +3.00 \, \mu C$ and mass 6.00×10^{-5} kg is projected toward the center of the large sphere from an initial large distance. The large sphere is held at a fixed position and the small sphere can be treated as a point charge. What minimum speed must the small sphere have in order to come within 8.00 cm of the surface of the large sphere?

23.30 •• An infinitely long line of charge has linear charge density 5.00×10^{-12} C/m. A proton (mass 1.67×10^{-27} kg, charge $+1.60 \times 10^{-19}$ C) is 18.0 cm from the line and moving directly toward the line at 3.50×10^3 m/s. (a) Calculate the proton's initial kinetic energy. (b) How close does the proton get to the line of charge?

23.31 • A very long wire carries a uniform linear charge density λ. Using a voltmeter to measure potential difference, you find that when one probe of the meter is placed 2.50 cm from the wire and the other probe is 1.00 cm farther from the wire, the meter reads 575 V. (a) What is λ? (b) If you now place one probe at 3.50 cm from the wire and the other probe 1.00 cm farther away, will the voltmeter read 575 V? If not, will it read more or less than 575 V? Why? (c) If you place both probes 3.50 cm from the wire but 17.0 cm from each other, what will the voltmeter read?

23.32 •• A very long insulating cylinder of charge of radius 2.50 cm carries a uniform linear density of 15.0 nC/m. If you put one probe of a voltmeter at the surface, how far from the surface must the other probe be placed so that the voltmeter reads 175 V?

23.33 •• A very long insulating cylindrical shell of radius 6.00 cm carries charge of linear density $8.50 \, \mu C/m$ spread uniformly over its outer surface. What would a voltmeter read if it were connected between (a) the surface of the cylinder and a point 4.00 cm above the surface, and (b) the surface and a point 1.00 cm from the central axis of the cylinder?

23.34 • A ring of diameter 8.00 cm is fixed in place and carries a charge of $+5.00 \, \mu C$ uniformly spread over its circumference. (a) How much work does it take to move a tiny $+3.00$-μC charged ball of mass 1.50 g from very far away to the center of the ring? (b) Is it necessary to take a path along the axis of the ring? Why? (c) If the ball is slightly displaced from the center of the ring, what will it do and what is the maximum speed it will reach?

23.35 •• A very small sphere with positive charge $q = +8.00 \, \mu C$ is released from rest at a point 1.50 cm from a very long line of uniform linear charge density $\lambda = +3.00 \, \mu C/m$. What is the kinetic energy of the sphere when it is 4.50 cm from the line of charge if the only force on it is the force exerted by the line of charge?

23.36 • CP Two large, parallel conducting plates carrying opposite charges of equal magnitude are separated by 2.20 cm. (a) If the surface charge density for each plate has magnitude 47.0 nC/m^2, what is the magnitude of $\vec{E}$ in the region between the plates? (b) What is the potential difference between the two plates? (c) If the separation between the plates is doubled while the surface charge density is kept constant at the value in part (a), what happens to the magnitude of the electric field and to the potential difference?

23.37 • Two large, parallel, metal plates carry opposite charges of equal magnitude. They are separated by 45.0 mm, and the potential difference between them is 360 V. (a) What is the magnitude of the electric field (assumed to be uniform) in the region between the plates? (b) What is the magnitude of the force this field exerts on a particle with charge $+2.40$ nC? (c) Use the results of part (b) to compute the work done by the field on the particle as it moves from the higher-potential plate to the lower. (d) Compare

the result of part (c) to the change of potential energy of the same charge, computed from the electric potential.

23.38 • BIO **Electrical Sensitivity of Sharks.** Certain sharks can detect an electric field as weak as $1.0 \, \mu V/m$. To grasp how weak this field is, if you wanted to produce it between two parallel metal plates by connecting an ordinary 1.5-V AA battery across these plates, how far apart would the plates have to be?

23.39 • The electric field at the surface of a charged, solid, copper sphere with radius 0.200 m is 3800 N/C, directed toward the center of the sphere. What is the potential at the center of the sphere, if we take the potential to be zero infinitely far from the sphere?

23.40 •• (a) How much excess charge must be placed on a copper sphere 25.0 cm in diameter so that the potential of its center, relative to infinity, is 3.75 kV? (b) What is the potential of the sphere's surface relative to infinity?

Section 23.4 Equipotential Surfaces and
Section 23.5 Potential Gradient

23.41 •• CALC A metal sphere with radius r_a is supported on an insulating stand at the center of a hollow, metal, spherical shell with radius r_b. There is charge $+q$ on the inner sphere and charge $-q$ on the outer spherical shell. (a) Calculate the potential $V(r)$ for (i) $r < r_a$; (ii) $r_a < r < r_b$; (iii) $r > r_b$. (*Hint:* The net potential is the sum of the potentials due to the individual spheres.) Take V to be zero when r is infinite. (b) Show that the potential of the inner sphere with respect to the outer is

$$V_{ab} = \frac{q}{4\pi\epsilon_0}\left(\frac{1}{r_a} - \frac{1}{r_b}\right)$$

(c) Use Eq. (23.23) and the result from part (a) to show that the electric field at any point between the spheres has magnitude

$$E(r) = \frac{V_{ab}}{(1/r_a - 1/r_b)}\frac{1}{r^2}$$

(d) Use Eq. (23.23) and the result from part (a) to find the electric field at a point outside the larger sphere at a distance r from the center, where $r > r_b$. (e) Suppose the charge on the outer sphere is not $-q$ but a negative charge of different magnitude, say $-Q$. Show that the answers for parts (b) and (c) are the same as before but the answer for part (d) is different.

23.42 • A very large plastic sheet carries a uniform charge density of -6.00 nC/m^2 on one face. (a) As you move away from the sheet along a line perpendicular to it, does the potential increase or decrease? How do you know, without doing any calculations? Does your answer depend on where you choose the reference point for potential? (b) Find the spacing between equipotential surfaces that differ from each other by 1.00 V. What type of surfaces are these?

23.43 • CALC In a certain region of space, the electric potential is $V(x, y, z) = Axy - Bx^2 + Cy$, where A, B, and C are positive constants. (a) Calculate the x-, y-, and z-components of the electric field. (b) At which points is the electric field equal to zero?

23.44 • CALC In a certain region of space the electric potential is given by $V = +Ax^2y - Bxy^2$, where $A = 5.00$ V/m^3 and $B = 8.00$ V/m^3. Calculate the magnitude and direction of the electric field at the point in the region that has coordinates $x = 2.00$ m, $y = 0.400$ m, and $z = 0$.

23.45 • A metal sphere with radius $r_a = 1.20$ cm is supported on an insulating stand at the center of a hollow, metal, spherical shell with radius $r_b = 9.60$ cm. Charge $+q$ is put on the inner sphere and charge $-q$ on the outer spherical shell. The magnitude of q is chosen to make the potential difference between the spheres 500 V, with the inner sphere at higher potential. (a) Use the result of Exercise 23.41(b) to calculate q. (b) With the help of the result of Exercise 23.41(a), sketch the equipotential surfaces that correspond to 500, 400, 300, 200, 100, and 0 V. (c) In your sketch, show the electric field lines. Are the electric field lines and equipotential surfaces mutually perpendicular? Are the equipotential surfaces closer together when the magnitude of $\vec{E}$ is largest?

PROBLEMS

23.46 • CP A point charge $q_1 = +5.00\ \mu C$ is held fixed in space. From a horizontal distance of 6.00 cm, a small sphere with mass 4.00×10^{-3} kg and charge $q_2 = +2.00\ \mu C$ is fired toward the fixed charge with an initial speed of 40.0 m/s. Gravity can be neglected. What is the acceleration of the sphere at the instant when its speed is 25.0 m/s?

23.47 ••• A point charge $q_1 = 4.00$ nC is placed at the origin, and a second point charge $q_2 = -3.00$ nC is placed on the x-axis at $x = +20.0$ cm. A third point charge $q_3 = 2.00$ nC is to be placed on the x-axis between q_1 and q_2. (Take as zero the potential energy of the three charges when they are infinitely far apart.) (a) What is the potential energy of the system of the three charges if q_3 is placed at $x = +10.0$ cm? (b) Where should q_3 be placed to make the potential energy of the system equal to zero?

23.48 •• A positive point charge $q_1 = +5.00 \times 10^{-4}$ C is held at a fixed position. A small object with mass 4.00×10^{-3} kg and charge $q_2 = -3.00 \times 10^{-4}$ C is projected directly at q_1. Ignore gravity. When q_2 is 0.400 m away, its speed is 800 m/s. What is its speed when it is 0.200 m from q_1?

23.49 •• A gold nucleus has a radius of 7.3×10^{-15} m and a charge of $+79e$. Through what voltage must an alpha particle, with charge $+2e$, be accelerated so that it has just enough energy to reach a distance of 2.0×10^{-14} m from the surface of a gold nucleus? (Assume that the gold nucleus remains stationary and can be treated as a point charge.)

23.50 ••• A small sphere with mass 5.00×10^{-7} kg and charge $+7.00\ \mu C$ is released from rest a distance of 0.400 m above a large horizontal insulating sheet of charge that has uniform surface charge density $\sigma = +8.00$ pC/m². Using energy methods, calculate the speed of the sphere when it is 0.100 m above the sheet.

23.51 •• **Determining the Size of the Nucleus.** When radium-226 decays radioactively, it emits an alpha particle (the nucleus of helium), and the end product is radon-222. We can model this decay by thinking of the radium-226 as consisting of an alpha particle emitted from the surface of the spherically symmetric radon-222 nucleus, and we can treat the alpha particle as a point charge. The energy of the alpha particle has been measured in the laboratory and has been found to be 4.79 MeV when the alpha particle is essentially infinitely far from the nucleus. Since radon is much heavier than the alpha particle, we can assume that there is no appreciable recoil of the radon after the decay. The radon nucleus contains 86 protons, while the alpha particle has 2 protons and the radium nucleus has 88 protons. (a) What was the electric potential energy of the alpha–radon combination just before the decay, in MeV and in joules? (b) Use your result from part (a) to calculate the radius of the radon nucleus.

23.52 •• CP A proton and an alpha particle are released from rest when they are 0.225 nm apart. The alpha particle (a helium nucleus) has essentially four times the mass and two times the charge of a proton. Find the maximum *speed* and maximum *acceleration* of each of these particles. When do these maxima occur: just following the release of the particles or after a very long time?

23.53 • A particle with charge $+7.60$ nC is in a uniform electric field directed to the left. Another force, in addition to the electric force, acts on the particle so that when it is released from rest, it moves to the right. After it has moved 8.00 cm, the additional force has done 6.50×10^{-5} J of work and the particle has 4.35×10^{-5} J of kinetic energy. (a) What work was done by the electric force? (b) What is the potential of the starting point with respect to the end point? (c) What is the magnitude of the electric field?

23.54 •• Identical charges $q = +5.00\ \mu C$ are placed at opposite corners of a square that has sides of length 8.00 cm. Point A is at one of the empty corners, and point B is at the center of the square. A charge $q_0 = -3.00\ \mu C$ is placed at point A and moves along the diagonal of the square to point B. (a) What is the magnitude of the net electric force on q_0 when it is at point A? Sketch the placement of the charges and the direction of the net force. (b) What is the magnitude of the net electric force on q_0 when it is at point B? (c) How much work does the electric force do on q_0 during its motion from A to B? Is this work positive or negative? When it goes from A to B, does q_0 move to higher potential or to lower potential?

23.55 •• CALC A vacuum tube diode consists of concentric cylindrical electrodes, the negative cathode and the positive anode. Because of the accumulation of charge near the cathode, the electric potential between the electrodes is given by

$$V(x) = Cx^{4/3}$$

where x is the distance from the cathode and C is a constant, characteristic of a particular diode and operating conditions. Assume that the distance between the cathode and anode is 13.0 mm and the potential difference between electrodes is 240 V. (a) Determine the value of C. (b) Obtain a formula for the electric field between the electrodes as a function of x. (c) Determine the force on an electron when the electron is halfway between the electrodes.

23.56 •• Two oppositely charged, identical insulating spheres, each 50.0 cm in diameter and carrying a uniformly distributed charge of magnitude $250\ \mu C$, are placed 1.00 m apart center to center

Figure **P23.56**

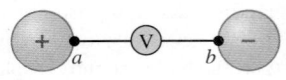

(**Fig. P23.56**). (a) If a voltmeter is connected between the nearest points (a and b) on their surfaces, what will it read? (b) Which point, a or b, is at the higher potential? How can you know this without any calculations?

23.57 •• **An Ionic Crystal.** **Figure P23.57** shows eight point charges arranged at the corners of a cube with sides of length d. The values of the charges are $+q$ and $-q$, as shown. This is a model of one cell of a cubic ionic crystal. In sodium chloride (NaCl), for instance, the positive ions are Na$^+$ and the negative ions are Cl$^-$. (a) Calculate the potential energy U of this arrangement. (Take as zero the potential energy of the eight charges when they are infinitely far apart.) (b) In part (a), you should have found that $U < 0$. Explain the relationship between this result and the observation that such ionic crystals exist in nature.

Figure **P23.57**

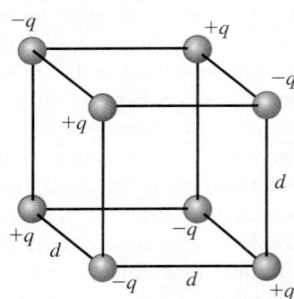

23.58 • (a) Calculate the potential energy of a system of two small spheres, one carrying a charge of $2.00\,\mu C$ and the other a charge of $-3.50\,\mu C$, with their centers separated by a distance of 0.180 m. Assume that $U = 0$ when the charges are infinitely separated. (b) Suppose that one sphere is held in place; the other sphere, with mass 1.50 g, is shot away from it. What minimum initial speed would the moving sphere need to escape completely from the attraction of the fixed sphere? (To escape, the moving sphere would have to reach a velocity of zero when it is infinitely far from the fixed sphere.)

23.59 •• CP A small sphere with mass 1.50 g hangs by a thread between two very large parallel vertical plates 5.00 cm apart (**Fig. P23.59**). The plates are insulating and have uniform surface charge densities $+\sigma$ and $-\sigma$. The charge on the sphere is $q = 8.90 \times 10^{-6}$ C. What potential difference between the plates will cause the thread to assume an angle of 30.0° with the vertical?

Figure **P23.59**

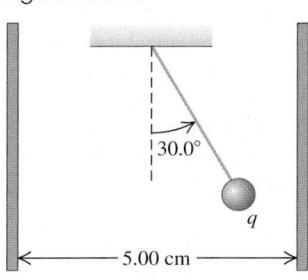

23.60 •• Two spherical shells have a common center. The inner shell has radius $R_1 = 5.00$ cm and charge $q_1 = +3.00 \times 10^{-6}$ C; the outer shell has radius $R_2 = 15.0$ cm and charge $q_2 = -5.00 \times 10^{-6}$ C. Both charges are spread uniformly over the shell surface. What is the electric potential due to the two shells at the following distances from their common center: (a) $r = 2.50$ cm; (b) $r = 10.0$ cm; (c) $r = 20.0$ cm? Take $V = 0$ at a large distance from the shells.

23.61 • CALC **Coaxial Cylinders.** A long metal cylinder with radius a is supported on an insulating stand on the axis of a long, hollow, metal tube with radius b. The positive charge per unit length on the inner cylinder is λ, and there is an equal negative charge per unit length on the outer cylinder. (a) Calculate the potential $V(r)$ for (i) $r < a$; (ii) $a < r < b$; (iii) $r > b$. (*Hint:* The net potential is the sum of the potentials due to the individual conductors.) Take $V = 0$ at $r = b$. (b) Show that the potential of the inner cylinder with respect to the outer is

$$V_{ab} = \frac{\lambda}{2\pi\epsilon_0}\ln\frac{b}{a}$$

(c) Use Eq. (23.23) and the result from part (a) to show that the electric field at any point between the cylinders has magnitude

$$E(r) = \frac{V_{ab}}{\ln(b/a)}\frac{1}{r}$$

(d) What is the potential difference between the two cylinders if the outer cylinder has no net charge?

23.62 •• A *Geiger counter* detects radiation such as alpha particles by using the fact that the radiation ionizes the air along its path. A thin wire lies on the axis of a hollow metal cylinder and is insulated from it (**Fig. P23.62**). A large potential difference is established between the wire and the outer cylinder, with the wire at higher potential; this sets up a strong electric field directed radially outward. When ionizing radiation enters the device, it ionizes a few air molecules. The free electrons produced are accelerated by the electric field toward the wire and, on the way there, ionize many more air molecules. Thus a current pulse is produced that can be detected by appropriate electronic circuitry and converted

Figure **P23.62**

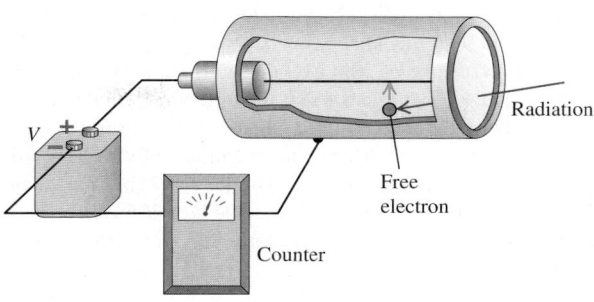

to an audible "click." Suppose the radius of the central wire is 145 μm and the radius of the hollow cylinder is 1.80 cm. What potential difference between the wire and the cylinder produces an electric field of 2.00×10^4 V/m at a distance of 1.20 cm from the axis of the wire? (The wire and cylinder are both very long in comparison to their radii, so the results of Problem 23.61 apply.)

23.63 • CP **Deflection in a CRT.** Cathode-ray tubes (CRTs) were often found in oscilloscopes and computer monitors. In **Fig. P23.63** an electron with an initial speed of 6.50×10^6 m/s is projected along the axis midway between the deflection plates of a cathode-ray tube. The potential difference between the two plates is 22.0 V and the lower plate is the one at higher potential. (a) What is the force (magnitude and direction) on the electron when it is between the plates? (b) What is the acceleration of the electron (magnitude and direction) when acted on by the force in part (a)? (c) How far below the axis has the electron moved when it reaches the end of the plates? (d) At what angle with the axis is it moving as it leaves the plates? (e) How far below the axis will it strike the fluorescent screen S?

Figure **P23.63**

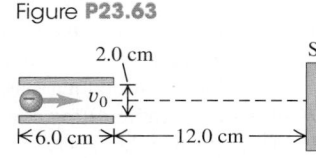

23.64 •• CP **Deflecting Plates of an Oscilloscope.** The vertical deflecting plates of a typical classroom oscilloscope are a pair of parallel square metal plates carrying equal but opposite charges. Typical dimensions are about 3.0 cm on a side, with a separation of about 5.0 mm. The potential difference between the plates is 25.0 V. The plates are close enough that we can ignore fringing at the ends. Under these conditions: (a) how much charge is on each plate, and (b) how strong is the electric field between the plates? (c) If an electron is ejected at rest from the negative plate, how fast is it moving when it reaches the positive plate?

23.65 •• *Electrostatic precipitators* use electric forces to remove pollutant particles from smoke, in particular in the smokestacks of coal-burning power plants. One form of precipitator consists of a vertical, hollow, metal cylinder with a thin wire, insulated from the cylinder, running along its axis (**Fig. P23.65**). A large potential difference is established between the wire and the outer cylinder, with the wire at lower potential. This sets up a strong radial electric field directed inward. The field produces a region of ionized air near the wire. Smoke enters the precipitator at the bottom, ash and dust in it pick up

Figure **P23.65**

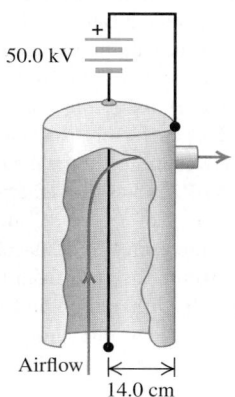

electrons, and the charged pollutants are accelerated toward the outer cylinder wall by the electric field. Suppose the radius of the central wire is 90.0 μm, the radius of the cylinder is 14.0 cm, and a potential difference of 50.0 kV is established between the wire and the cylinder. Also assume that the wire and cylinder are both very long in comparison to the cylinder radius, so the results of Problem 23.61 apply. (a) What is the magnitude of the electric field midway between the wire and the cylinder wall? (b) What magnitude of charge must a 30.0-μg ash particle have if the electric field computed in part (a) is to exert a force ten times the weight of the particle?

23.66 •• **CALC** A disk with radius R has uniform surface charge density σ. (a) By regarding the disk as a series of thin concentric rings, calculate the electric potential V at a point on the disk's axis a distance x from the center of the disk. Assume that the potential is zero at infinity. (*Hint:* Use the result of Example 23.11 in Section 23.3.) (b) Calculate $-\partial V/\partial x$. Show that the result agrees with the expression for E_x calculated in Example 21.11 (Section 21.5).

23.67 ••• **CALC Self-Energy of a Sphere of Charge.** A solid sphere of radius R contains a total charge Q distributed uniformly throughout its volume. Find the energy needed to assemble this charge by bringing infinitesimal charges from far away. This energy is called the "self-energy" of the charge distribution. (*Hint:* After you have assembled a charge q in a sphere of radius r, how much energy would it take to add a spherical shell of thickness dr having charge dq? Then integrate to get the total energy.)

23.68 • **CALC** A thin insulating rod is bent into a semicircular arc of radius a, and a total electric charge Q is distributed uniformly along the rod. Calculate the potential at the center of curvature of the arc if the potential is assumed to be zero at infinity.

23.69 •• Charge $Q = +4.00\,\mu$C is distributed uniformly over the volume of an insulating sphere that has radius $R = 5.00$ cm. What is the potential difference between the center of the sphere and the surface of the sphere?

23.70 • An insulating spherical shell with inner radius 25.0 cm and outer radius 60.0 cm carries a charge of $+150.0\,\mu$C uniformly distributed over its outer surface. Point a is at the center of the shell, point b is on the inner surface, and point c is on the outer surface. (a) What will a voltmeter read if it is connected between the following points: (i) a and b; (ii) b and c; (iii) c and infinity; (iv) a and c? (b) Which is at higher potential: (i) a or b; (ii) b or c; (iii) a or c? (c) Which, if any, of the answers would change sign if the charge were $-150\,\mu$C?

23.71 •• **CP** Two plastic spheres, each carrying charge uniformly distributed throughout its interior, are initially placed in contact and then released. One sphere is 60.0 cm in diameter, has mass 50.0 g, and contains $-10.0\,\mu$C of charge. The other sphere is 40.0 cm in diameter, has mass 150.0 g, and contains $-30.0\,\mu$C of charge. Find the maximum acceleration and the maximum speed achieved by each sphere (relative to the fixed point of their initial location in space), assuming that no other forces are acting on them. (*Hint:* The uniformly distributed charges behave as though they were concentrated at the centers of the two spheres.)

23.72 • (a) If a spherical raindrop of radius 0.650 mm carries a charge of -3.60 pC uniformly distributed over its volume, what is the potential at its surface? (Take the potential to be zero at an infinite distance from the raindrop.) (b) Two identical raindrops, each with radius and charge specified in part (a), collide and merge into one larger raindrop. What is the radius of this larger drop, and what is the potential at its surface, if its charge is uniformly distributed over its volume?

23.73 • **CALC** Electric charge is distributed uniformly along a thin rod of length a, with total charge Q. Take the potential to be zero at infinity. Find the potential at the following points (**Fig. P23.73**):

Figure **P23.73**

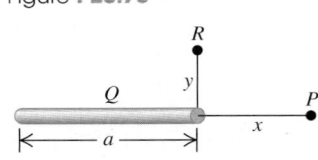

(a) point P, a distance x to the right of the rod, and (b) point R, a distance y above the right-hand end of the rod. (c) In parts (a) and (b), what does your result reduce to as x or y becomes much larger than a?

23.74 • An alpha particle with kinetic energy 9.50 MeV (when far away) collides head-on with a lead nucleus at rest. What is the distance of closest approach of the two particles? (Assume that the lead nucleus remains stationary and may be treated as a point charge. The atomic number of lead is 82. The alpha particle is a helium nucleus, with atomic number 2.)

23.75 •• Two metal spheres of different sizes are charged such that the electric potential is the same at the surface of each. Sphere A has a radius three times that of sphere B. Let Q_A and Q_B be the charges on the two spheres, and let E_A and E_B be the electric-field magnitudes at the surfaces of the two spheres. What are (a) the ratio Q_B/Q_A and (b) the ratio E_B/E_A?

23.76 • A metal sphere with radius R_1 has a charge Q_1. Take the electric potential to be zero at an infinite distance from the sphere. (a) What are the electric field and electric potential at the surface of the sphere? This sphere is now connected by a long, thin conducting wire to another sphere of radius R_2 that is several meters from the first sphere. Before the connection is made, this second sphere is uncharged. After electrostatic equilibrium has been reached, what are (b) the total charge on each sphere; (c) the electric potential at the surface of each sphere; (d) the electric field at the surface of each sphere? Assume that the amount of charge on the wire is much less than the charge on each sphere.

23.77 •• **CP Nuclear Fusion in the Sun.** The source of the sun's energy is a sequence of nuclear reactions that occur in its core. The first of these reactions involves the collision of two protons, which fuse together to form a heavier nucleus and release energy. For this process, called *nuclear fusion,* to occur, the two protons must first approach until their surfaces are essentially in contact. (a) Assume both protons are moving with the same speed and they collide head-on. If the radius of the proton is 1.2×10^{-15} m, what is the minimum speed that will allow fusion to occur? The charge distribution within a proton is spherically symmetric, so the electric field and potential outside a proton are the same as if it were a point charge. The mass of the proton is 1.67×10^{-27} kg. (b) Another nuclear fusion reaction that occurs in the sun's core involves a collision between two helium nuclei, each of which has 2.99 times the mass of the proton, charge $+2e$, and radius 1.7×10^{-15} m. Assuming the same collision geometry as in part (a), what minimum speed is required for this fusion reaction to take place if the nuclei must approach a center-to-center distance of about 3.5×10^{-15} m? As for the proton, the charge of the helium nucleus is uniformly distributed throughout its volume. (c) In Section 18.3 it was shown that the average translational kinetic energy of a particle with mass m in a gas at absolute temperature T is $\frac{3}{2}kT$, where k is the Boltzmann constant (given in Appendix F). For two protons with kinetic energy equal to this average value to be able to undergo the process described in part (a), what absolute temperature is required? What absolute temperature is required for two average helium nuclei to be able to undergo the process described in part (b)? (At these temperatures, atoms are completely ionized, so nuclei and electrons move separately.) (d) The temperature in the sun's core is about 1.5×10^7 K. How does this compare to the temperatures calculated in part (c)? How can the reactions described in parts (a) and (b) occur at all in the interior of the sun? (*Hint:* See the discussion of the distribution of molecular speeds in Section 18.5.)

23.78 • CALC The electric potential V in a region of space is given by

$$V(x, y, z) = A(x^2 - 3y^2 + z^2)$$

where A is a constant. (a) Derive an expression for the electric field $\vec{E}$ at any point in this region. (b) The work done by the field when a 1.50-μC test charge moves from the point $(x, y, z) = (0, 0, 0.250$ m$)$ to the origin is measured to be 6.00×10^{-5} J. Determine A. (c) Determine the electric field at the point $(0, 0, 0.250$ m$)$. (d) Show that in every plane parallel to the xz-plane the equipotential contours are circles. (e) What is the radius of the equipotential contour corresponding to $V = 1280$ V and $y = 2.00$ m?

23.79 •• DATA The electric potential in a region that is within 2.00 m of the origin of a rectangular coordinate system is given by $V = Ax^l + By^m + Cz^n + D$, where $A, B, C, D, l, m,$ and n are constants. The units of $A, B, C,$ and D are such that if $x, y,$ and z are in meters, then V is in volts. You measure V and each component of the electric field at four points and obtain these results:

Point	(x, y, z) (m)	V (V)	E_x (V/m)	E_y (V/m)	E_z (V/m)
1	(0, 0, 0)	10.0	0	0	0
2	(1.00, 0, 0)	4.0	12.0	0	0
3	(0, 1.00, 0)	6.0	0	12.0	0
4	(0, 0, 1.00)	8.0	0	0	12.0

(a) Use the data in the table to calculate $A, B, C, D, l, m,$ and n. (b) What are V and the magnitude of E at the points $(0, 0, 0)$, $(0.50$ m, 0.50 m, 0.50 m$)$, and $(1.00$ m, 1.00 m, 1.00 m$)$?

23.80 •• DATA A small, stationary sphere carries a net charge Q. You perform the following experiment to measure Q: From a large distance you fire a small particle with mass $m = 4.00 \times 10^{-4}$ kg and charge $q = 5.00 \times 10^{-8}$ C directly at the center of the sphere. The apparatus you are using measures the particle's speed v as a function of the distance x from the sphere. The sphere's mass is much greater than the mass of the projectile particle, so you assume that the sphere remains at rest. All of the measured values of x are much larger than the radius of either object, so you treat both objects as point particles. You plot your data on a graph of v^2 versus $(1/x)$ (**Fig. P23.80**). The straight line $v^2 = 400$ m^2/s^2 $- [(15.75$ m^3/s$^2)/x]$ gives a good fit to the data points. (a) Explain why the graph is a straight line. (b) What is the initial speed v_0 of the particle when it is very far from the sphere? (c) What is Q? (d) How close does the particle get to the sphere? Assume that this

Figure **P23.80**

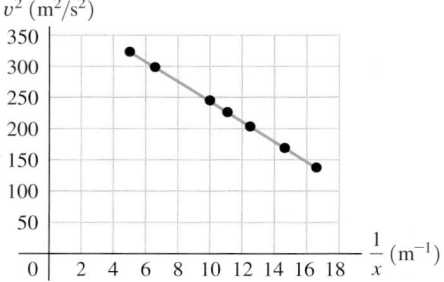

distance is much larger than the radii of the particle and sphere, so continue to treat them as point particles and to assume that the sphere remains at rest.

23.81 ••• DATA **The Millikan Oil-Drop Experiment.** The charge of an electron was first measured by the American physicist Robert Millikan during 1909–1913. In his experiment, oil was sprayed in very fine drops (about 10^{-4} mm in diameter) into the space between two parallel horizontal plates separated by a distance d. A potential difference V_{AB} was maintained between the plates, causing a downward electric field between them. Some of the oil drops acquired a negative charge because of frictional effects or because of ionization of the surrounding air by x rays or radioactivity. The drops were observed through a microscope. (a) Show that an oil drop of radius r at rest between the plates remained at rest if the magnitude of its charge was

$$q = \frac{4\pi}{3} \frac{\rho r^3 g d}{V_{AB}}$$

where ρ is oil's density. (Ignore the buoyant force of the air.) By adjusting V_{AB} to keep a given drop at rest, Millikan determined the charge on that drop, provided its radius r was known. (b) Millikan's oil drops were much too small to measure their radii directly. Instead, Millikan determined r by cutting off the electric field and measuring the *terminal speed* v_t of the drop as it fell. (We discussed terminal speed in Section 5.3.) The viscous force F on a sphere of radius r moving at speed v through a fluid with viscosity η is given by Stokes's law: $F = 6\pi\eta r v$. When a drop fell at v_t, the viscous force just balanced the drop's weight $w = mg$. Show that the magnitude of the charge on the drop was

$$q = 18\pi \frac{d}{V_{AB}} \sqrt{\frac{\eta^3 v_t^3}{2\rho g}}$$

(c) You repeat the Millikan oil-drop experiment. Four of your measured values of V_{AB} and v_t are listed in the table:

Drop	1	2	3	4
V_{AB} (V)	9.16	4.57	12.32	6.28
v_t (10^{-5} m/s)	2.54	0.767	4.39	1.52

In your apparatus, the separation d between the horizontal plates is 1.00 mm. The density of the oil you use is 824 kg/m^3. For the viscosity η of air, use the value 1.81×10^{-5} N·s/m^2. Assume that $g = 9.80$ m/s^2. Calculate the charge q of each drop. (d) If electric charge is *quantized* (that is, exists in multiples of the magnitude of the charge of an electron), then the charge on each drop is $-ne$, where n is the number of excess electrons on each drop. (All four drops in your table have negative charge.) Drop 2 has the smallest magnitude of charge observed in the experiment, for all 300 drops on which measurements were made, so assume that its charge is due to an excess charge of one electron. Determine the number of excess electrons n for each of the other three drops. (e) Use $q = -ne$ to calculate e from the data for each of the four drops, and average these four values to get your best experimental value of e.

CHALLENGE PROBLEMS

23.82 ••• CALC A hollow, thin-walled insulating cylinder of radius R and length L (like the cardboard tube in a roll of toilet paper) has charge Q uniformly distributed over its surface. (a) Calculate the electric potential at all points along the axis of the tube. Take the origin to be at the center of the tube, and take the potential to be zero at infinity. (b) Show that if $L \ll R$, the result of part (a) reduces to the potential on the axis of a ring of charge of radius R. (See Example 23.11 in Section 23.3.) (c) Use the result of part (a) to find the electric field at all points along the axis of the tube.

23.83 ••• **CP** In experiments in which atomic nuclei collide, head-on collisions like that described in Problem 23.74 do happen, but "near misses" are more common. Suppose the alpha particle in that problem is not "aimed" at the center of the lead nucleus but has an initial nonzero angular momentum (with respect to the stationary lead nucleus) of magnitude $L = p_0b$, where p_0 is the magnitude of the particle's initial momentum and $b = 1.00 \times 10^{-12}$ m. What is the distance of closest approach? Repeat for $b = 1.00 \times 10^{-13}$ m and $b = 1.00 \times 10^{-14}$ m.

PASSAGE PROBLEMS

MATERIALS ANALYSIS WITH IONS. *Rutherford backscattering spectrometry* (RBS) is a technique used to determine the structure and composition of materials. A beam of ions (typically helium ions) is accelerated to high energy and aimed at a sample. By analyzing the distribution and energy of the ions that are scattered from (that is, deflected by collisions with) the atoms in the sample, researchers can determine the sample's composition. To accelerate the ions to high energies, a *tandem electrostatic accelerator* may be used. In this device, negative ions (He^-) start at

a potential $V = 0$ and are accelerated by a high positive voltage at the midpoint of the accelerator. The high voltage produces a constant electric field in the acceleration tube through which the ions move. When accelerated ions reach the midpoint, the electrons are stripped off, turning the negative ions into doubly positively charged ions (He^{++}). These positive ions are then repelled from the midpoint by the high positive voltage there and continue to accelerate to the far end of the accelerator, where again $V = 0$.

23.84 For a particular experiment, helium ions are to be given a kinetic energy of 3.0 MeV. What should the voltage at the center of the accelerator be, assuming that the ions start essentially at rest? (a) -3.0 MV; (b) $+3.0$ MV; (c) $+1.5$ MV; (d) $+1.0$ MV.

23.85 A helium ion (He^{++}) that comes within about 10 fm of the center of the nucleus of an atom in the sample may induce a nuclear reaction instead of simply scattering. Imagine a helium ion with a kinetic energy of 3.0 MeV heading straight toward an atom at rest in the sample. Assume that the atom stays fixed. What minimum charge can the nucleus of the atom have such that the helium ion gets no closer than 10 fm from the center of the atomic nucleus? (1 fm $= 1 \times 10^{-15}$ m, and e is the magnitude of the charge of an electron or a proton.) (a) $2e$; (b) $11e$; (c) $20e$; (d) $22e$.

23.86 The maximum voltage at the center of a typical tandem electrostatic accelerator is 6.0 MV. If the distance from one end of the acceleration tube to the midpoint is 12 m, what is the magnitude of the average electric field in the tube under these conditions? (a) 41,000 V/m; (b) 250,000 V/m; (c) 500,000 V/m; (d) 6,000,000 V/m.

Answers

Chapter Opening Question **?**

(iii) A large, constant potential difference V_{ab} is maintained between the welding tool (a) and the metal pieces to be welded (b). For a given potential difference between two conductors a and b, the smaller the distance d separating the conductors, the greater is the magnitude E of the field between them. Hence d must be small in order for E to be large enough to ionize the gas between the conductors (see Section 23.3) and produce an arc through this gas.

Test Your Understanding Questions

23.1 (a) (i), (b) (ii) The three charges q_1, q_2, and q_3 are all positive, so all three of the terms in the sum in Eq. (23.11)—q_1q_2/r_{12}, q_1q_3/r_{13}, and q_2q_3/r_{23}—are positive. Hence the total electric potential energy U is positive. This means that it would take positive work to bring the three charges from infinity to the positions shown in Fig. 21.14, and hence *negative* work to move the three charges from these positions back to infinity.

23.2 no If $V = 0$ at a certain point, $\vec{E}$ does *not* have to be zero at that point. An example is point c in Figs. 21.23 and 23.13, for which there is an electric field in the $+x$-direction (see Example 21.9 in Section 21.5) even though $V = 0$ (see Example 23.4). This isn't a surprising result because V and $\vec{E}$ are quite different quantities: V is the net amount of work required to bring a unit charge from infinity to the point in question, whereas $\vec{E}$ is the electric force that acts on a unit charge when it arrives at that point.

23.3 no If $\vec{E} = 0$ at a certain point, V does *not* have to be zero at that point. An example is point O at the center of the charged ring in Figs. 21.23 and Fig. 23.21. From Example 21.9 (Section 21.5),

the electric field is zero at O because the electric-field contributions from different parts of the ring completely cancel. From Example 23.11, however, the potential at O is *not* zero: This point corresponds to $x = 0$, so $V = (1/4\pi\epsilon_0)(Q/a)$. This value of V corresponds to the work that would have to be done to move a unit positive test charge along a path from infinity to point O; it is nonzero because the charged ring repels the test charge, so positive work must be done to move the test charge toward the ring.

23.4 no If the positive charges in Fig. 23.23 were replaced by negative charges, and vice versa, the equipotential surfaces would be the same but the sign of the potential would be reversed. For example, the surfaces in Fig. 23.23b with potential $V = +30$ V and $V = -50$ V would have potential $V = -30$ V and $V = +50$ V, respectively.

23.5 (iii) From Eqs. (23.19), the components of the electric field are $E_x = -\partial V/\partial x = -(B + Dy)$, $E_y = -\partial V/\partial y = -(3Cy^2 + Dx)$, and $E_z = -\partial V/\partial z = 0$. The value of A has no effect, which means that we can add a constant to the electric potential at all points without changing $\vec{E}$ or the potential difference between two points. The potential does not depend on z, so the z-component of $\vec{E}$ is zero. Note that at the origin the electric field is not zero because it has a nonzero x-component: $E_x = -B$, $E_y = 0$, $E_z = 0$.

Bridging Problem

$$\frac{qQ}{8\pi\epsilon_0 a} \ln\left(\frac{L + a}{L - a}\right)$$

? In flash photography, the energy used to make the flash is stored in a capacitor, which consists of two closely spaced conductors that carry opposite charges. If the amount of charge on the conductors is doubled, by what factor does the stored energy increase? (i) $\sqrt{2}$; (ii) 2; (iii) $2\sqrt{2}$; (iv) 4; (v) 8.

24 CAPACITANCE AND DIELECTRICS

LEARNING GOALS

Looking forward at …

24.1 The nature of capacitors, and how to calculate a quantity that measures their ability to store charge.

24.2 How to analyze capacitors connected in a network.

24.3 How to calculate the amount of energy stored in a capacitor.

24.4 What dielectrics are, and how they make capacitors more effective.

24.5 How a dielectric inside a charged capacitor becomes polarized.

24.6 How to use Gauss's laws when dielectrics are present.

Looking back at …

21.2, 21.5, 21.7 Polarization; field of charged conductors; electric dipoles.

22.3–22.5 Gauss's law.

23.3, 23.4 Potential for charged conductors; potential due to a cylindrical charge distribution.

When you stretch the rubber band of a slingshot or pull back the string of an archer's bow, you are storing mechanical energy as elastic potential energy. A capacitor is a device that stores *electric* potential energy and electric charge. To make a capacitor, just insulate two conductors from each other. To store energy in this device, transfer charge from one conductor to the other so that one has a negative charge and the other has an equal amount of positive charge. Work must be done to move the charges through the resulting potential difference between the conductors, and the work done is stored as electric potential energy.

Capacitors have a tremendous number of practical applications in devices such as electronic flash units for photography, pulsed lasers, air bag sensors for cars, and radio and television receivers. We'll encounter many of these applications in later chapters (particularly Chapter 31, in which we'll see the crucial role played by capacitors in the alternating-current circuits that pervade our technological society). In this chapter, however, our emphasis is on the fundamental properties of capacitors. For a particular capacitor, the ratio of the charge on each conductor to the potential difference between the conductors is a constant, called the *capacitance*. The capacitance depends on the sizes and shapes of the conductors and on the insulating material (if any) between them. Compared to the case in which there is only vacuum between the conductors, the capacitance increases when an insulating material (a *dielectric*) is present. This happens because a redistribution of charge, called *polarization*, takes place within the insulating material. Studying polarization will give us added insight into the electrical properties of matter.

Capacitors also give us a new way to think about electric potential energy. The energy stored in a charged capacitor is related to the electric field in the space between the conductors. We will see that electric potential energy can be regarded as being stored *in the field itself*. The idea that the electric field is itself a storehouse of energy is at the heart of the theory of electromagnetic waves and our modern understanding of the nature of light, to be discussed in Chapter 32.

24.1 Any two conductors a and b insulated from each other form a capacitor.

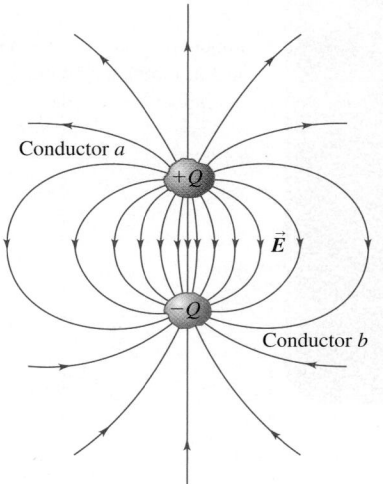

Conductor a

$+Q$

$\vec{E}$

$-Q$

Conductor b

24.1 CAPACITORS AND CAPACITANCE

Any two conductors separated by an insulator (or a vacuum) form a **capacitor** (**Fig. 24.1**). In most practical applications, each conductor initially has zero net charge and electrons are transferred from one conductor to the other; this is called *charging* the capacitor. Then the two conductors have charges with equal magnitude and opposite sign, and the *net* charge on the capacitor as a whole remains zero. We will assume throughout this chapter that this is the case. When we say that a capacitor has charge Q, or that a charge Q is *stored* on the capacitor, we mean that the conductor at higher potential has charge $+Q$ and the conductor at lower potential has charge $-Q$ (assuming that Q is positive). Keep this in mind in the following discussion and examples.

In circuit diagrams a capacitor is represented by either of these symbols:

The vertical lines (straight or curved) represent the conductors, and the horizontal lines represent wires connected to either conductor. One common way to charge a capacitor is to connect these two wires to opposite terminals of a battery. Once the charges Q and $-Q$ are established on the conductors, the battery is disconnected. This gives a fixed *potential difference* V_{ab} between the conductors (that is, the potential of the positively charged conductor a with respect to the negatively charged conductor b) that is just equal to the voltage of the battery.

The electric field at any point in the region between the conductors is proportional to the magnitude Q of charge on each conductor. It follows that the potential difference V_{ab} between the conductors is also proportional to Q. If we double the magnitude of charge on each conductor, the charge density at each point doubles, the electric field at each point doubles, and the potential difference between conductors doubles; however, the *ratio* of charge to potential difference does not change. This ratio is called the **capacitance** C of the capacitor:

$$\text{Capacitance} \cdots\!\!\rightarrow C = \frac{Q}{V_{ab}} \;\xleftarrow{\cdots}\; \begin{matrix} \cdots\text{Magnitude of charge on each conductor} \\ \cdots\text{Potential difference between} \\ \text{conductors (}a\text{ has charge }+Q, \\ b\text{ has charge }-Q\text{)} \end{matrix} \quad (24.1)$$

The SI unit of capacitance is called one **farad** (1 F), in honor of the 19th-century English physicist Michael Faraday. From Eq. (24.1), one farad is equal to one *coulomb per volt* (1 C/V):

$$1\text{ F} = 1\text{ farad} = 1\text{ C/V} = 1\text{ coulomb/volt}$$

The greater the capacitance C of a capacitor, the greater the magnitude Q of charge on either conductor for a given potential difference V_{ab} and hence the greater the amount of stored energy. (Remember that potential is potential energy per unit charge.) Thus *capacitance is a measure of the ability of a capacitor to store energy*. We will see that the capacitance depends only on the shapes, sizes, and relative positions of the conductors and on the nature of the insulator between them. (For special types of insulating materials, the capacitance *does* depend on Q and V_{ab}. We won't discuss these materials, however.)

CAUTION Capacitance vs. coulombs Don't confuse the symbol C for capacitance (which is always in italics) with the abbreviation C for coulombs (which is never italicized). ∎

Calculating Capacitance: Capacitors in Vacuum

We can calculate the capacitance C of a given capacitor by finding the potential difference V_{ab} between the conductors for a given magnitude of charge Q and then using Eq. (24.1). For now we'll consider only *capacitors in vacuum;* that is, empty space separates the conductors that make up the capacitor.

The simplest form of capacitor consists of two parallel conducting plates, each with area A, separated by a distance d that is small in comparison with their dimensions (**Fig. 24.2a**). When the plates are charged, the electric field is almost completely localized in the region between the plates (Fig. 24.2b). As we discussed in Example 22.8 (Section 22.4), the field between such plates is essentially *uniform,* and the charges on the plates are uniformly distributed over their opposing surfaces. We call this arrangement a **parallel-plate capacitor.**

We found the electric-field magnitude E for this arrangement in Example 21.12 (Section 21.5) by using the principle of superposition of electric fields and again in Example 22.8 (Section 22.4) by using Gauss's law. It would be a good idea to review those examples. We found that $E = \sigma/\epsilon_0$, where σ is the magnitude (absolute value) of the surface charge density on each plate. This is equal to the magnitude of the total charge Q on each plate divided by the area A of the plate, or $\sigma = Q/A$, so the field magnitude E can be expressed as

$$E = \frac{\sigma}{\epsilon_0} = \frac{Q}{\epsilon_0 A}$$

The field is uniform and the distance between the plates is d, so the potential difference (voltage) between the two plates is

$$V_{ab} = Ed = \frac{1}{\epsilon_0}\frac{Qd}{A}$$

Thus

Capacitance of a parallel-plate capacitor in vacuum

······ Magnitude of charge on each plate

$$C = \frac{Q}{V_{ab}} = \epsilon_0\frac{A}{d} \qquad (24.2)$$

Area of each plate ·······A
Distance between plates ·······d
Potential difference between plates ·······V_{ab}
Electric constant ·······ϵ_0

The capacitance depends on only the geometry of the capacitor; it is directly proportional to the area A of each plate and inversely proportional to their separation d. The quantities A and d are constants for a given capacitor, and ϵ_0 is a universal constant. Thus in vacuum the capacitance C is a constant independent of the charge on the capacitor or the potential difference between the plates. If one of the capacitor plates is flexible, C changes as the plate separation d changes. This is the operating principle of a condenser microphone (**Fig. 24.3**).

When matter is present between the plates, its properties affect the capacitance. We will return to this topic in Section 24.4. Meanwhile, we remark that if the space contains air at atmospheric pressure instead of vacuum, the capacitance differs from the prediction of Eq. (24.2) by less than 0.06%.

In Eq. (24.2), if A is in square meters and d in meters, then C is in farads. The units of the electric constant ϵ_0 are $C^2/N \cdot m^2$, so

$$1\ F = 1\ C^2/N \cdot m = 1\ C^2/J$$

Because $1\ V = 1\ J/C$ (energy per unit charge), this equivalence agrees with our definition $1\ F = 1\ C/V$. Finally, we can express the units of ϵ_0 as $1\ C^2/N \cdot m^2 = 1\ F/m$, so

$$\epsilon_0 = 8.85 \times 10^{-12}\ F/m$$

This relationship is useful in capacitance calculations, and it also helps us to verify that Eq. (24.2) is dimensionally consistent.

24.2 A charged parallel-plate capacitor.

(a) Arrangement of the capacitor plates

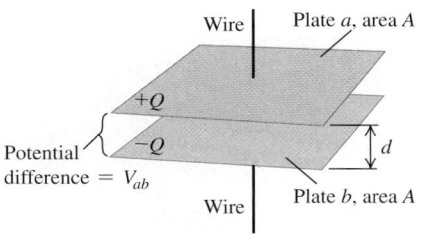

Wire | Plate a, area A
$+Q$
$-Q$
Potential difference $= V_{ab}$
d
Wire | Plate b, area A

(b) Side view of the electric field $\vec{E}$

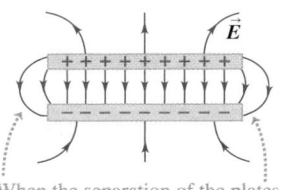

$\vec{E}$
$+ + + + + + + +$
$- - - - - - - -$

When the separation of the plates is small compared to their size, the fringing of the field is slight.

24.3 Inside a condenser microphone is a capacitor with one rigid plate and one flexible plate. The two plates are kept at a constant potential difference V_{ab}. Sound waves cause the flexible plate to move back and forth, varying the capacitance C and causing charge to flow to and from the capacitor in accordance with the relationship $C = Q/V_{ab}$. Thus a sound wave is converted to a charge flow that can be amplified and recorded digitally.

24.4 A commercial capacitor is labeled with the value of its capacitance. For these capacitors, $C = 2200 \ \mu\text{F}$, $1000 \ \mu\text{F}$, and $470 \ \mu\text{F}$.

One farad is a very large capacitance, as the following example shows. In many applications the most convenient units of capacitance are the *microfarad* $(1 \ \mu\text{F} = 10^{-6} \ \text{F})$ and the *picofarad* $(1 \ \text{pF} = 10^{-12} \ \text{F})$. For example, the flash unit in a point-and-shoot camera uses a capacitor of a few hundred microfarads (**Fig. 24.4**), while capacitances in a radio tuning circuit are typically from 10 to 100 picofarads.

For *any* capacitor in vacuum, the capacitance C depends only on the shapes, dimensions, and separation of the conductors that make up the capacitor. If the conductor shapes are more complex than those of the parallel-plate capacitor, the expression for capacitance is more complicated than in Eq. (24.2). In the following examples we show how to calculate C for two other conductor geometries.

EXAMPLE 24.1 SIZE OF A 1-F CAPACITOR

The parallel plates of a 1.0-F capacitor are 1.0 mm apart. What is their area?

SOLUTION

IDENTIFY and SET UP: This problem uses the relationship among the capacitance C, plate separation d, and plate area A (our target variable) for a parallel-plate capacitor. We solve Eq. (24.2) for A.

EXECUTE: From Eq. (24.2),

$$A = \frac{Cd}{\epsilon_0} = \frac{(1.0 \ \text{F})(1.0 \times 10^{-3} \ \text{m})}{8.85 \times 10^{-12} \ \text{F/m}} = 1.1 \times 10^8 \ \text{m}^2$$

EVALUATE: This corresponds to a square about 10 km (about 6 miles) on a side! The volume of such a capacitor would be at least $Ad = 1.1 \times 10^5 \ \text{m}^3$, equivalent to that of a cube about 50 m on a side. In fact, it's possible to make 1-F capacitors a few *centimeters* on a side. The trick is to have an appropriate substance between the plates rather than a vacuum, so that (among other things) the plate separation d can be greatly reduced. We'll explore this further in Section 24.4.

EXAMPLE 24.2 PROPERTIES OF A PARALLEL-PLATE CAPACITOR

The plates of a parallel-plate capacitor in vacuum are 5.00 mm apart and 2.00 m² in area. A 10.0-kV potential difference is applied across the capacitor. Compute (a) the capacitance; (b) the charge on each plate; and (c) the magnitude of the electric field between the plates.

SOLUTION

IDENTIFY and SET UP: We are given the plate area A, the plate spacing d, and the potential difference $V_{ab} = 1.00 \times 10^4 \ \text{V}$ for this parallel-plate capacitor. Our target variables are the capacitance C, the charge Q on each plate, and the electric-field magnitude E. We use Eq. (24.2) to calculate C and then use Eq. (24.1) and V_{ab} to find Q. We use $E = Q/\epsilon_0 A$ to find E.

EXECUTE: (a) From Eq. (24.2),

$$C = \epsilon_0 \frac{A}{d} = (8.85 \times 10^{-12} \ \text{F/m}) \frac{(2.00 \ \text{m}^2)}{5.00 \times 10^{-3} \ \text{m}}$$
$$= 3.54 \times 10^{-9} \ \text{F} = 0.00354 \ \mu\text{F}$$

(b) The charge on the capacitor is

$$Q = CV_{ab} = (3.54 \times 10^{-9} \ \text{C/V})(1.00 \times 10^4 \ \text{V})$$
$$= 3.54 \times 10^{-5} \ \text{C} = 35.4 \ \mu\text{C}$$

The plate at higher potential has charge $+35.4 \ \mu\text{C}$, and the other plate has charge $-35.4 \ \mu\text{C}$.

(c) The electric-field magnitude is

$$E = \frac{\sigma}{\epsilon_0} = \frac{Q}{\epsilon_0 A} = \frac{3.54 \times 10^{-5} \ \text{C}}{(8.85 \times 10^{-12} \ \text{C}^2/\text{N} \cdot \text{m}^2)(2.00 \ \text{m}^2)}$$
$$= 2.00 \times 10^6 \ \text{N/C}$$

EVALUATE: We can also find E by recalling that the electric field is equal in magnitude to the potential gradient [Eq. (23.22)]. The field between the plates is uniform, so

$$E = \frac{V_{ab}}{d} = \frac{1.00 \times 10^4 \ \text{V}}{5.00 \times 10^{-3} \ \text{m}} = 2.00 \times 10^6 \ \text{V/m}$$

(Remember that 1 N/C = 1 V/m.)

EXAMPLE 24.3 A SPHERICAL CAPACITOR

Two concentric spherical conducting shells are separated by vacuum (**Fig. 24.5**). The inner shell has total charge $+Q$ and outer radius r_a, and the outer shell has charge $-Q$ and inner radius r_b. Find the capacitance of this spherical capacitor.

SOLUTION

IDENTIFY and SET UP: By definition, the capacitance C is the magnitude Q of the charge on either sphere divided by the potential difference V_{ab} between the spheres. We first find V_{ab}, and then use Eq. (24.1) to find the capacitance $C = Q/V_{ab}$.

EXECUTE: Using a Gaussian surface such as that shown in Fig. 24.5, we found in Example 22.5 (Section 22.4) that the charge on a conducting sphere produces zero field *inside* the sphere, so the outer sphere makes no contribution to the field between the spheres. Therefore the electric field *and* the electric potential between the shells are the same as those outside a charged conducting sphere

24.5 A spherical capacitor.

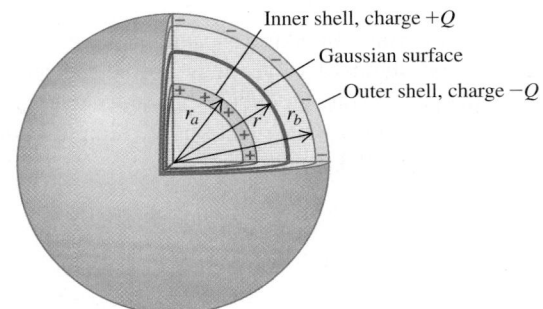

Inner shell, charge $+Q$

Gaussian surface

Outer shell, charge $-Q$

with charge $+Q$. We considered that problem in Example 23.8 (Section 23.3), so the same result applies here: The potential at any point between the spheres is $V = Q/4\pi\epsilon_0 r$. Hence the potential of the inner (positive) conductor at $r = r_a$ with respect to that of the outer (negative) conductor at $r = r_b$ is

$$V_{ab} = V_a - V_b = \frac{Q}{4\pi\epsilon_0 r_a} - \frac{Q}{4\pi\epsilon_0 r_b}$$

$$= \frac{Q}{4\pi\epsilon_0}\left(\frac{1}{r_a} - \frac{1}{r_b}\right) = \frac{Q}{4\pi\epsilon_0}\frac{r_b - r_a}{r_a r_b}$$

The capacitance is then

$$C = \frac{Q}{V_{ab}} = 4\pi\epsilon_0 \frac{r_a r_b}{r_b - r_a}$$

As an example, if $r_a = 9.5$ cm and $r_b = 10.5$ cm,

$$C = 4\pi(8.85 \times 10^{-12} \text{ F/m})\frac{(0.095 \text{ m})(0.105 \text{ m})}{0.010 \text{ m}}$$

$$= 1.1 \times 10^{-10} \text{ F} = 110 \text{ pF}$$

EVALUATE: We can relate our expression for C to that for a parallel-plate capacitor. The quantity $4\pi r_a r_b$ is intermediate between the areas $4\pi r_a^2$ and $4\pi r_b^2$ of the two spheres; in fact, it's the *geometric mean* of these two areas, which we can denote by A_{gm}. The distance between spheres is $d = r_b - r_a$, so we can write $C = 4\pi\epsilon_0 r_a r_b/(r_b - r_a) = \epsilon_0 A_{gm}/d$. This has the same form as for parallel plates: $C = \epsilon_0 A/d$. If the distance between spheres is very small in comparison to their radii, their capacitance is the same as that of parallel plates with the same area and spacing.

EXAMPLE 24.4 A CYLINDRICAL CAPACITOR

Two long, coaxial cylindrical conductors are separated by vacuum (**Fig. 24.6**). The inner cylinder has radius r_a and linear charge density $+\lambda$. The outer cylinder has inner radius r_b and linear charge density $-\lambda$. Find the capacitance per unit length for this capacitor.

SOLUTION

IDENTIFY and SET UP: As in Example 24.3, we use the definition of capacitance, $C = Q/V_{ab}$. We use the result of Example 23.10 (Section 23.3) to find the potential difference V_{ab} between the cylinders, and find the charge Q on a length L of the cylinders from the linear charge density. We then find the corresponding capacitance C from Eq. (24.1). Our target variable is this capacitance divided by L.

EXECUTE: As in Example 24.3, the potential V between the cylinders is not affected by the presence of the charged outer cylinder. Hence our result in Example 23.10 for the potential outside a charged conducting cylinder also holds in this example for potential in the space between the cylinders:

$$V = \frac{\lambda}{2\pi\epsilon_0}\ln\frac{r_0}{r}$$

24.6 A long cylindrical capacitor. The linear charge density λ is assumed to be positive in this figure. The magnitude of charge in a length L of either cylinder is λL.

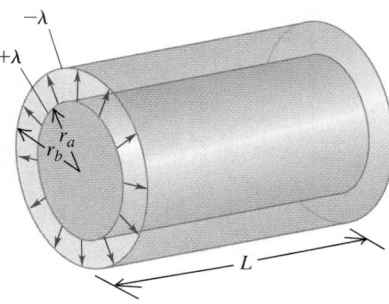

$-\lambda$

$+\lambda$

r_a

r_b

L

Here r_0 is the arbitrary, *finite* radius at which $V = 0$. We take $r_0 = r_b$, the radius of the inner surface of the outer cylinder. Then the potential at the outer surface of the inner cylinder (at which $r = r_a$) is just the potential V_{ab} of the inner (positive) cylinder a with respect to the outer (negative) cylinder b:

$$V_{ab} = \frac{\lambda}{2\pi\epsilon_0}\ln\frac{r_b}{r_a}$$

Continued

If λ is positive as in Fig. 24.6, then V_{ab} is positive as well: The inner cylinder is at higher potential than the outer.

The total charge Q in a length L is $Q = \lambda L$, so from Eq. (24.1) the capacitance C of a length L is

$$C = \frac{Q}{V_{ab}} = \frac{\lambda L}{\dfrac{\lambda}{2\pi\epsilon_0}\ln\dfrac{r_b}{r_a}} = \frac{2\pi\epsilon_0 L}{\ln(r_b/r_a)}$$

The capacitance per unit length is

$$\frac{C}{L} = \frac{2\pi\epsilon_0}{\ln(r_b/r_a)}$$

Substituting $\epsilon_0 = 8.85 \times 10^{-12}$ F/m $= 8.85$ pF/m, we get

$$\frac{C}{L} = \frac{55.6 \text{ pF/m}}{\ln(r_b/r_a)}$$

EVALUATE: The capacitance of coaxial cylinders is determined entirely by their dimensions, just as for parallel-plate and spherical capacitors. Ordinary coaxial cables are made like this but with an insulating material instead of vacuum between the conductors. A typical cable used for connecting a television to a cable TV feed has a capacitance per unit length of 69 pF/m.

24.7 An assortment of commercially available capacitors.

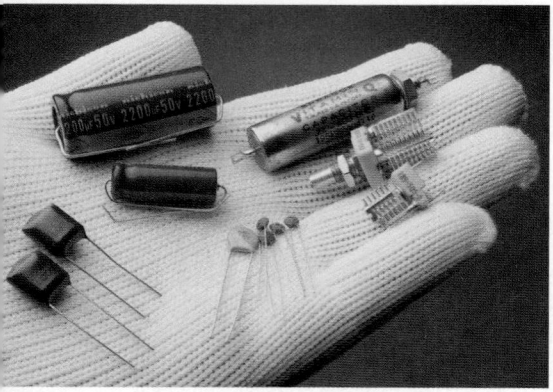

TEST YOUR UNDERSTANDING OF SECTION 24.1 A capacitor has vacuum in the space between the conductors. If you double the amount of charge on each conductor, what happens to the capacitance? (i) It increases; (ii) it decreases; (iii) it remains the same; (iv) the answer depends on the size or shape of the conductors. ∎

24.2 CAPACITORS IN SERIES AND PARALLEL

Capacitors are manufactured with certain standard capacitances and working voltages (**Fig. 24.7**). However, these standard values may not be the ones you actually need in a particular application. You can obtain the values you need by combining capacitors; many combinations are possible, but the simplest combinations are a series connection and a parallel connection.

Capacitors in Series

Figure 24.8a is a schematic diagram of a **series connection.** Two capacitors are connected in series (one after the other) by conducting wires between points a and b. Both capacitors are initially uncharged. When a constant positive potential difference V_{ab} is applied between points a and b, the capacitors become charged; the figure shows that the charge on *all* conducting plates has the same magnitude. To see why, note first that the top plate of C_1 acquires a positive charge Q. The electric field of this positive charge pulls negative charge up to the bottom plate of C_1 until all of the field lines that begin on the top plate end on the bottom plate. This requires that the bottom plate have charge $-Q$. These negative charges had to come from the top plate of C_2, which becomes positively charged with charge $+Q$. This positive charge then pulls negative charge $-Q$ from the connection at point b onto the bottom plate of C_2. The total charge on the lower plate of C_1 and the upper plate of C_2 together must always be zero because these plates aren't connected to anything except each other. Thus *in a series connection the magnitude of charge on all plates is the same.*

Referring to Fig. 24.8a, we can write the potential differences between points a and c, c and b, and a and b as

$$V_{ac} = V_1 = \frac{Q}{C_1}, \qquad V_{cb} = V_2 = \frac{Q}{C_2}, \qquad V_{ab} = V = V_1 + V_2 = Q\left(\frac{1}{C_1} + \frac{1}{C_2}\right)$$

and so

$$\frac{V}{Q} = \frac{1}{C_1} + \frac{1}{C_2} \tag{24.3}$$

Following a common convention, we use the symbols V_1, V_2, and V to denote the potential *differences* V_{ac} (across the first capacitor), V_{cb} (across the second capacitor), and V_{ab} (across the entire combination of capacitors), respectively.

24.8 A series connection of two capacitors.

(a) Two capacitors in series

Capacitors in series:
• The capacitors have the same charge Q.
• Their potential differences add:
$\quad V_{ac} + V_{cb} = V_{ab}.$

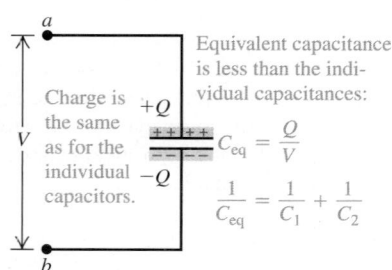

(b) The equivalent single capacitor

Charge is the same as for the individual capacitors.

Equivalent capacitance is less than the individual capacitances:

$$C_{eq} = \frac{Q}{V}$$

$$\frac{1}{C_{eq}} = \frac{1}{C_1} + \frac{1}{C_2}$$

The **equivalent capacitance** C_{eq} of the series combination is defined as the capacitance of a *single* capacitor for which the charge Q is the same as for the combination, when the potential difference V is the same. In other words, the combination can be replaced by an *equivalent capacitor* of capacitance C_{eq}. For such a capacitor, shown in Fig. 24.8b,

$$C_{eq} = \frac{Q}{V} \quad \text{or} \quad \frac{1}{C_{eq}} = \frac{V}{Q} \tag{24.4}$$

Combining Eqs. (24.3) and (24.4), we find

$$\frac{1}{C_{eq}} = \frac{1}{C_1} + \frac{1}{C_2}$$

We can extend this analysis to any number of capacitors in series. We find the following result for the *reciprocal* of the equivalent capacitance:

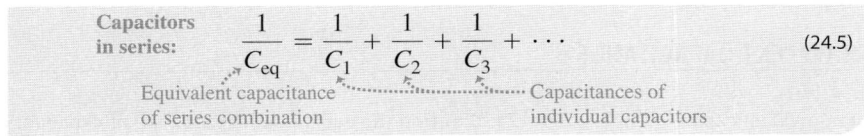

Capacitors in series:
$$\frac{1}{C_{eq}} = \frac{1}{C_1} + \frac{1}{C_2} + \frac{1}{C_3} + \cdots \tag{24.5}$$
Equivalent capacitance ⋯⋯ Capacitances of
of series combination   individual capacitors

The reciprocal of the equivalent capacitance of a series combination equals the sum of the reciprocals of the individual capacitances. In a series connection the equivalent capacitance is always *less than* any individual capacitance.

CAUTION **Capacitors in series** The magnitude of charge is the same on all plates of all the capacitors in a series combination; however, the potential differences of the individual capacitors are *not* the same unless their individual capacitances are the same. The potential differences of the individual capacitors add to give the total potential difference across the series combination: $V_{total} = V_1 + V_2 + V_3 + \cdots$. ▮

Capacitors in Parallel

The arrangement shown in **Fig. 24.9a** is called a **parallel connection**. Two capacitors are connected in parallel between points a and b. In this case the upper plates of the two capacitors are connected by conducting wires to form an equipotential surface, and the lower plates form another. Hence *in a parallel connection the potential difference for all individual capacitors is the same* and is equal to $V_{ab} = V$. The charges Q_1 and Q_2 are not necessarily equal, however, since charges can reach each capacitor independently from the source (such as a battery) of the voltage V_{ab}. The charges are

$$Q_1 = C_1 V \quad \text{and} \quad Q_2 = C_2 V$$

The *total* charge Q of the combination, and thus the total charge on the equivalent capacitor, is

$$Q = Q_1 + Q_2 = (C_1 + C_2)V$$

so

$$\frac{Q}{V} = C_1 + C_2 \tag{24.6}$$

The parallel combination is equivalent to a single capacitor with the same total charge $Q = Q_1 + Q_2$ and potential difference V as the combination (Fig. 24.9b). The equivalent capacitance of the combination, C_{eq}, is the same as the capacitance Q/V of this single equivalent capacitor. So from Eq. (24.6),

$$C_{eq} = C_1 + C_2$$

Application Touch Screens and Capacitance The touch screen on a mobile phone, an MP3 player, or (as shown here) a medical device uses the physics of capacitors. Behind the screen are two parallel layers, one behind the other, of thin strips of a transparent conductor such as indium tin oxide. A voltage is maintained between the two layers. The strips in one layer are oriented perpendicular to those in the other layer; the points where two strips overlap act as a grid of capacitors. When you bring your finger (a conductor) up to a point on the screen, your finger and the front conducting layer act like a second capacitor in series at that point. The circuitry attached to the conducting layers detects the location of the capacitance change, and so detects where you touched the screen.

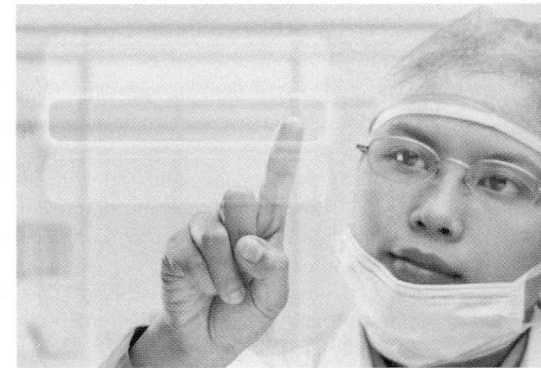

24.9 A parallel connection of two capacitors.

(a) Two capacitors in parallel

Capacitors in parallel:
• The capacitors have the same potential V.
• The charge on each capacitor depends on its capacitance: $Q_1 = C_1 V$, $Q_2 = C_2 V$.

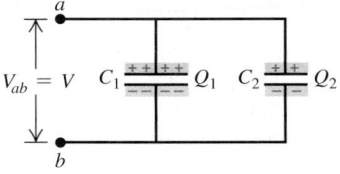

(b) The equivalent single capacitor

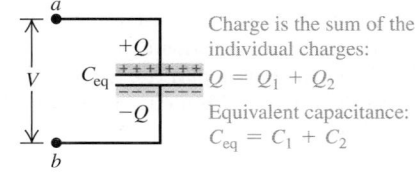

Charge is the sum of the individual charges:
$Q = Q_1 + Q_2$
Equivalent capacitance:
$C_{eq} = C_1 + C_2$

In the same way we can show that for any number of capacitors in parallel,

> **Capacitors in parallel:** $\underbrace{C_{eq}}_{\substack{\text{Equivalent capacitance} \\ \text{of parallel combination}}} = \underbrace{C_1 + C_2 + C_3 + \cdots}_{\substack{\text{Capacitances of} \\ \text{individual capacitors}}}$ (24.7)

The equivalent capacitance of a parallel combination equals the *sum* of the individual capacitances. In a parallel connection the equivalent capacitance is always *greater than* any individual capacitance.

DEMO

CAUTION **Capacitors in parallel** The potential differences are the same for all capacitors in a parallel combination; however, the charges on individual capacitors are *not* the same unless their individual capacitances are the same. The charges on the individual capacitors add to give the total charge on the parallel combination: $Q_{total} = Q_1 + Q_2 + Q_3 + \cdots$. [Compare these statements to those in the "Caution" paragraph following Eq. (24.5).]

PROBLEM-SOLVING STRATEGY 24.1 EQUIVALENT CAPACITANCE

IDENTIFY *the relevant concepts:* The concept of equivalent capacitance is useful whenever two or more capacitors are connected.

SET UP *the problem* using the following steps:
1. Make a drawing of the capacitor arrangement.
2. Identify all groups of capacitors that are connected in series or in parallel.
3. Keep in mind that when we say a capacitor "has charge Q," we mean that the plate at higher potential has charge $+Q$ and the other plate has charge $-Q$.

EXECUTE *the solution* as follows:
1. Use Eq. (24.5) to find the equivalent capacitance of capacitors connected in series, as in Fig. 24.8. Such capacitors each have the *same charge* if they were uncharged before they were connected; that charge is the same as that on the equivalent capacitor. The potential difference across the combination is the sum of the potential differences across the individual capacitors.

2. Use Eq. (24.7) to find the equivalent capacitance of capacitors connected in parallel, as in Fig. 24.9. Such capacitors all have the *same potential difference* across them; that potential difference is the same as that across the equivalent capacitor. The total charge on the combination is the sum of the charges on the individual capacitors.
3. After replacing all the series or parallel groups you initially identified, you may find that more such groups reveal themselves. Replace those groups by using the same procedure as above until no more replacements are possible. If you then need to find the charge or potential difference for an individual original capacitor, you may have to retrace your steps.

EVALUATE *your answer:* Check whether your result makes sense. If the capacitors are connected in series, the equivalent capacitance C_{eq} must be *smaller* than any of the individual capacitances. If the capacitors are connected in parallel, C_{eq} must be *greater* than any of the individual capacitances.

EXAMPLE 24.5 CAPACITORS IN SERIES AND IN PARALLEL

In Figs. 24.8 and 24.9, let $C_1 = 6.0\ \mu F$, $C_2 = 3.0\ \mu F$, and $V_{ab} = 18$ V. Find the equivalent capacitance and the charge and potential difference for each capacitor when the capacitors are connected (a) in series (see Fig. 24.8) and (b) in parallel (see Fig. 24.9).

SOLUTION

IDENTIFY and SET UP: In both parts of this example a target variable is the equivalent capacitance C_{eq}, which is given by Eq. (24.5) for the series combination in part (a) and by Eq. (24.7) for the parallel combination in part (b). In each part we find the charge and potential difference from the definition of capacitance, Eq. (24.1), and the rules outlined in Problem-Solving Strategy 24.1.

EXECUTE: (a) From Eq. (24.5) for a series combination,

$$\frac{1}{C_{eq}} = \frac{1}{C_1} + \frac{1}{C_2} = \frac{1}{6.0\ \mu F} + \frac{1}{3.0\ \mu F} \qquad C_{eq} = 2.0\ \mu F$$

The charge Q on each capacitor in series is the same as that on the equivalent capacitor:

$$Q = C_{eq}V = (2.0\ \mu F)(18\ V) = 36\ \mu C$$

The potential difference across each capacitor is inversely proportional to its capacitance:

$$V_{ac} = V_1 = \frac{Q}{C_1} = \frac{36\ \mu C}{6.0\ \mu F} = 6.0\ V$$

$$V_{cb} = V_2 = \frac{Q}{C_2} = \frac{36\ \mu C}{3.0\ \mu F} = 12.0\ V$$

(b) From Eq. (24.7) for a parallel combination,

$$C_{eq} = C_1 + C_2 = 6.0\ \mu F + 3.0\ \mu F$$
$$= 9.0\ \mu F$$

The potential difference across each of the capacitors is the same as that across the equivalent capacitor, 18 V. The charge on each capacitor is directly proportional to its capacitance:

$$Q_1 = C_1 V = (6.0 \ \mu\text{F})(18 \ \text{V}) = 108 \ \mu\text{C}$$

$$Q_2 = C_2 V = (3.0 \ \mu\text{F})(18 \ \text{V}) = 54 \ \mu\text{C}$$

EVALUATE: As expected, the equivalent capacitance C_{eq} for the series combination in part (a) is less than either C_1 or C_2, while that for the parallel combination in part (b) is greater than either C_1 or C_2. For two capacitors in series, as in part (a), the charge is the same on either capacitor and the *larger* potential difference appears across the capacitor with the *smaller* capacitance. Furthermore, the sum of the potential differences across the individual capacitors in series equals the potential difference across the equivalent capacitor: $V_{ac} + V_{cb} = V_{ab} = 18 \ \text{V}$. By contrast, for two capacitors in parallel, as in part (b), each capacitor has the same potential difference and the *larger* charge appears on the capacitor with the *larger* capacitance. Can you show that the total charge $Q_1 + Q_2$ on the parallel combination is equal to the charge $Q = C_{eq} V$ on the equivalent capacitor?

EXAMPLE 24.6   A CAPACITOR NETWORK

Find the equivalent capacitance of the five-capacitor network shown in **Fig. 24.10a**.

SOLUTION

IDENTIFY and SET UP: These capacitors are neither all in series nor all in parallel. We can, however, identify portions of the arrangement that *are* either in series or parallel. We combine these as described in Problem-Solving Strategy 24.1 to find the net equivalent capacitance, using Eq. (24.5) for series connections and Eq. (24.7) for parallel connections.

EXECUTE: The caption of Fig. 24.10 outlines our procedure. We first use Eq. (24.5) to replace the 12-μF and 6-μF series combination by its equivalent capacitance C':

$$\frac{1}{C'} = \frac{1}{12 \ \mu\text{F}} + \frac{1}{6 \ \mu\text{F}} \qquad C' = 4 \ \mu\text{F}$$

This gives us the equivalent combination of Fig. 24.10b. Now we see three capacitors in parallel, and we use Eq. (24.7) to replace them with their equivalent capacitance C'':

$$C'' = 3 \ \mu\text{F} + 11 \ \mu\text{F} + 4 \ \mu\text{F} = 18 \ \mu\text{F}$$

This gives us the equivalent combination of Fig. 24.10c, which has two capacitors in series. We use Eq. (24.5) to replace them with their equivalent capacitance C_{eq}, which is our target variable (Fig. 24.10d):

$$\frac{1}{C_{eq}} = \frac{1}{18 \ \mu\text{F}} + \frac{1}{9 \ \mu\text{F}} \qquad C_{eq} = 6 \ \mu\text{F}$$

EVALUATE: If the potential difference across the entire network in Fig. 24.10a is $V_{ab} = 9.0 \ \text{V}$, the net charge on the network is $Q = C_{eq} V_{ab} = (6 \ \mu\text{F})(9.0 \ \text{V}) = 54 \ \mu\text{C}$. Can you find the charge on, and the voltage across, each of the five individual capacitors?

24.10 (a) A capacitor network between points a and b. (b) The 12-μF and 6-μF capacitors in series in (a) are replaced by an equivalent 4-μF capacitor. (c) The 3-μF, 11-μF, and 4-μF capacitors in parallel in (b) are replaced by an equivalent 18-μF capacitor. (d) Finally, the 18-μF and 9-μF capacitors in series in (c) are replaced by an equivalent 6-μF capacitor.

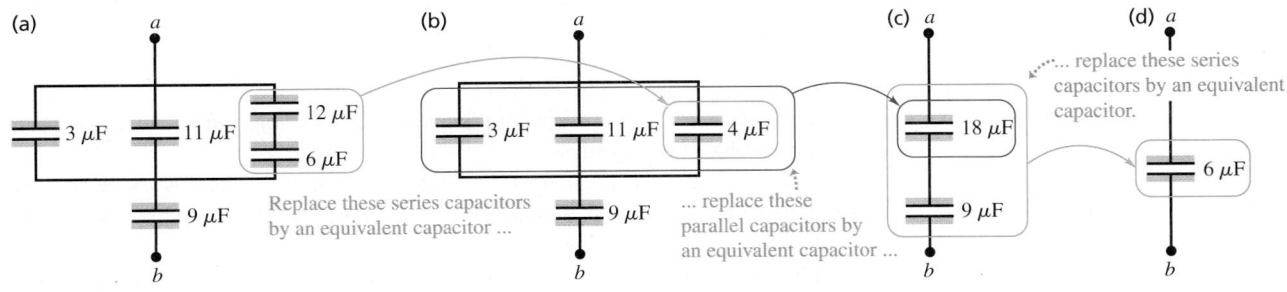

TEST YOUR UNDERSTANDING OF SECTION 24.2   You want to connect a 4-μF capacitor and an 8-μF capacitor. (a) With which type of connection will the 4-μF capacitor have a greater potential difference across it than the 8-μF capacitor? (i) Series; (ii) parallel; (iii) either series or parallel; (iv) neither series nor parallel. (b) With which type of connection will the 4-μF capacitor have a greater charge than the 8-μF capacitor? (i) Series; (ii) parallel; (iii) either series or parallel; (iv) neither series nor parallel. ∎

24.3 ENERGY STORAGE IN CAPACITORS AND ELECTRIC-FIELD ENERGY

Many of the most important applications of capacitors depend on their ability to store energy. The electric potential energy stored in a charged capacitor is just equal to the amount of work required to charge it—that is, to separate opposite charges and place them on different conductors. When the capacitor is discharged, this stored energy is recovered as work done by electrical forces.

We can calculate the potential energy U of a charged capacitor by calculating the work W required to charge it. Suppose that when we are done charging the capacitor, the final charge is Q and the final potential difference is V. From Eq. (24.1) these quantities are related by

$$V = \frac{Q}{C}$$

Let q and v be the charge and potential difference, respectively, at an intermediate stage during the charging process; then $v = q/C$. At this stage the work dW required to transfer an additional element of charge dq is

$$dW = v\,dq = \frac{q\,dq}{C}$$

The total work W needed to increase the capacitor charge q from zero to Q is

$$W = \int_0^W dW = \frac{1}{C}\int_0^Q q\,dq = \frac{Q^2}{2C} \quad \text{(work to charge a capacitor)} \quad (24.8)$$

This is also the total work done by the electric field on the charge when the capacitor discharges. Then q *decreases* from an initial value Q to zero as the elements of charge dq "fall" through potential differences v that vary from V down to zero.

If we define the potential energy of an *uncharged* capacitor to be zero, then W in Eq. (24.8) is equal to the potential energy U of the charged capacitor. The final stored charge is $Q = CV$, so we can express U (which is equal to W) as

Potential energy stored in a capacitor $\cdots\cdots\rightarrow U = \dfrac{Q^2}{2C} = \dfrac{1}{2}CV^2 = \dfrac{1}{2}QV$ $\qquad (24.9)$

Magnitude of charge on each plate

Capacitance $\cdots\cdots$ Potential difference between plates

When Q is in coulombs, C in farads (coulombs per volt), and V in volts (joules per coulomb), U is in joules.

The last form of Eq. (24.9), $U = \frac{1}{2}QV$, shows that the total work W required to charge the capacitor is equal to the total charge Q multiplied by the *average* potential difference $\frac{1}{2}V$ during the charging process.

The expression $U = \frac{1}{2}(Q^2/C)$ in Eq. (24.9) shows that a charged capacitor is the electrical analog of a stretched spring with elastic potential energy $U = \frac{1}{2}kx^2$. The charge Q is analogous to the elongation x, and the *reciprocal* of the capacitance, $1/C$, is analogous to the force constant k. The energy supplied to a capacitor in the charging process is analogous to the work we do on a spring when we stretch it.

Equations (24.8) and (24.9) tell us that capacitance measures the ability of a capacitor to store both energy and charge. If a capacitor is charged by connecting it to a battery or other source that provides a fixed potential difference V, then increasing the value of C gives a greater charge $Q = CV$ and a greater amount of stored energy $U = \frac{1}{2}CV^2$. If instead the goal is to transfer a given quantity of charge Q from one conductor to another, Eq. (24.8) shows that the work W required is inversely proportional to C; the greater the capacitance, the easier it is to give a capacitor a fixed amount of charge.

Applications of Capacitors: Energy Storage

Most practical applications of capacitors take advantage of their ability to store and release energy. In an electronic flash unit used in photography, the energy stored in a capacitor (see Fig. 24.4) is released when the button is pressed to take a photograph. This provides a conducting path from one capacitor plate to the other through the flash tube. Once this path is established, the stored energy is rapidly converted into a brief but intense flash of light. An extreme example of the same principle is the Z machine at Sandia National Laboratories in New Mexico, which is used in experiments in controlled nuclear fusion (**Fig. 24.11**). A bank of charged capacitors releases more than a million joules of energy in just a few billionths of a second. For that brief space of time, the power output of the Z machine is 2.9×10^{14} W, or about 80 times the power output of all the electric power plants on earth combined!

In other applications, the energy is released more slowly. Springs in the suspension of an automobile help smooth out the ride by absorbing the energy from sudden jolts and releasing that energy gradually; in an analogous way, a capacitor in an electronic circuit can smooth out unwanted variations in voltage due to power surges. We'll discuss these circuits in detail in Chapter 26.

24.11 The Z machine uses a large number of capacitors in parallel to give a tremendous equivalent capacitance C (see Section 24.2). Hence a large amount of energy $U = \frac{1}{2}CV^2$ can be stored with even a modest potential difference V. The arcs shown here are produced when the capacitors discharge their energy into a target, which is no larger than a spool of thread. This heats the target to a temperature higher than 2×10^9 K.

Electric-Field Energy

We can charge a capacitor by moving electrons directly from one plate to another. This requires doing work against the electric field between the plates. Thus we can think of the energy as being stored *in the field* in the region between the plates. To see this, let's find the energy *per unit volume* in the space between the plates of a parallel-plate capacitor with plate area A and separation d. We call this the **energy density,** denoted by u. From Eq. (24.9) the total stored potential energy is $\frac{1}{2}CV^2$ and the volume between the plates is Ad; hence

$$u = \text{Energy density} = \frac{\frac{1}{2}CV^2}{Ad} \qquad (24.10)$$

From Eq. (24.2) the capacitance C is given by $C = \epsilon_0 A/d$. The potential difference V is related to the electric-field magnitude E by $V = Ed$. If we use these expressions in Eq. (24.10), the geometric factors A and d cancel, and we find

Electric energy density ⋯ Magnitude of electric field
in a vacuum $\qquad u = \frac{1}{2}\epsilon_0 E^2 \qquad\qquad (24.11)$
⋯ Electric constant

Although we have derived this relationship for a parallel-plate capacitor only, it turns out to be valid for any capacitor in vacuum and indeed *for any electric field configuration in vacuum*. This result has an interesting implication. We think of vacuum as space with no matter in it, but vacuum can nevertheless have electric fields and therefore energy. Thus "empty" space need not be truly empty after all. We will use this idea and Eq. (24.11) in Chapter 32 in connection with the energy transported by electromagnetic waves.

CAUTION **Electric-field energy is electric potential energy** It's a common misconception that electric-field energy is a new kind of energy, different from the electric potential energy described before. This is *not* the case; it is simply a different way of interpreting electric potential energy. We can regard the energy of a given system of charges as being a shared property of all the charges, or we can think of the energy as being a property of the electric field that the charges create. Either interpretation leads to the same value of the potential energy. ▮

EXAMPLE 24.7 TRANSFERRING CHARGE AND ENERGY BETWEEN CAPACITORS

We connect a capacitor $C_1 = 8.0\ \mu\text{F}$ to a power supply, charge it to a potential difference $V_0 = 120\ \text{V}$, and disconnect the power supply (**Fig. 24.12**). Switch S is open. (a) What is the charge Q_0 on C_1? (b) What is the energy stored in C_1? (c) Capacitor $C_2 = 4.0\ \mu\text{F}$ is initially uncharged. We close switch S. After charge no longer flows, what is the potential difference across each capacitor, and what is the charge on each capacitor? (d) What is the final energy of the system?

SOLUTION

IDENTIFY and SET UP: In parts (a) and (b) we find the charge Q_0 and stored energy U_{initial} for the single charged capacitor C_1 from Eqs. (24.1) and (24.9), respectively. After we close switch S, one wire connects the upper plates of the two capacitors and another wire connects the lower plates; the capacitors are now connected in parallel. In part (c) we use the character of the parallel connection to determine how Q_0 is shared between the two capacitors. In part (d) we again use Eq. (24.9) to find the energy stored in capacitors C_1 and C_2; the energy of the system is the sum of these values.

EXECUTE: (a) The initial charge Q_0 on C_1 is

$$Q_0 = C_1 V_0 = (8.0\ \mu\text{F})(120\ \text{V}) = 960\ \mu\text{C}$$

(b) The energy initially stored in C_1 is

$$U_{\text{initial}} = \tfrac{1}{2}Q_0 V_0 = \tfrac{1}{2}(960 \times 10^{-6}\ \text{C})(120\ \text{V}) = 0.058\ \text{J}$$

(c) When we close the switch, the positive charge Q_0 is distributed over the upper plates of both capacitors and the negative charge $-Q_0$ is distributed over the lower plates. Let Q_1 and Q_2 be the magnitudes of the final charges on the capacitors. Conservation of charge requires that $Q_1 + Q_2 = Q_0$. The potential difference V

24.12 When the switch S is closed, the charged capacitor C_1 is connected to an uncharged capacitor C_2. The center part of the switch is an insulating handle; charge can flow only between the two upper terminals and between the two lower terminals.

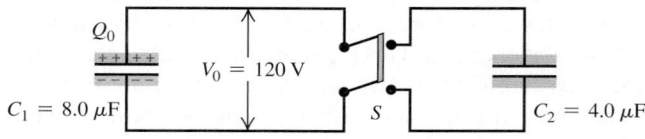

between the plates is the same for both capacitors because they are connected in parallel, so the charges are $Q_1 = C_1 V$ and $Q_2 = C_2 V$. We now have three independent equations relating the three unknowns Q_1, Q_2, and V. Solving these, we find

$$V = \frac{Q_0}{C_1 + C_2} = \frac{960\ \mu\text{C}}{8.0\ \mu\text{F} + 4.0\ \mu\text{F}} = 80\ \text{V}$$

$$Q_1 = 640\ \mu\text{C} \qquad Q_2 = 320\ \mu\text{C}$$

(d) The final energy of the system is

$$U_{\text{final}} = \tfrac{1}{2}Q_1 V + \tfrac{1}{2}Q_2 V = \tfrac{1}{2}Q_0 V$$

$$= \tfrac{1}{2}(960 \times 10^{-6}\ \text{C})(80\ \text{V}) = 0.038\ \text{J}$$

EVALUATE: The final energy is less than the initial energy; the difference was converted to energy of some other form. The conductors become a little warmer because of their resistance, and some energy is radiated as electromagnetic waves. We'll study the behavior of capacitors in more detail in Chapters 26 and Chapter 31.

EXAMPLE 24.8 ELECTRIC-FIELD ENERGY

(a) What is the magnitude of the electric field required to store 1.00 J of electric potential energy in a volume of 1.00 m^3 in vacuum? (b) If the field magnitude is 10 times larger than that, how much energy is stored per cubic meter?

SOLUTION

IDENTIFY and SET UP: We use the relationship between the electric-field magnitude E and the energy density u. In part (a) we use the given information to find u; then we use Eq. (24.11) to find the corresponding value of E. In part (b), Eq. (24.11) tells us how u varies with E.

EXECUTE: (a) The desired energy density is $u = 1.00\ \text{J/m}^3$. Then from Eq. (24.11),

$$E = \sqrt{\frac{2u}{\epsilon_0}} = \sqrt{\frac{2(1.00\ \text{J/m}^3)}{8.85 \times 10^{-12}\ \text{C}^2/\text{N}\cdot\text{m}^2}}$$

$$= 4.75 \times 10^5\ \text{N/C} = 4.75 \times 10^5\ \text{V/m}$$

(b) Equation (24.11) shows that u is proportional to E^2. If E increases by a factor of 10, u increases by a factor of $10^2 = 100$, so the energy density becomes $u = 100\ \text{J/m}^3$.

EVALUATE: Dry air can sustain an electric field of about $3 \times 10^6\ \text{V/m}$ without experiencing *dielectric breakdown,* which we will discuss in Section 24.4. There we will see that field magnitudes in practical insulators can be even larger than this.

EXAMPLE 24.9 TWO WAYS TO CALCULATE ENERGY STORED IN A CAPACITOR

The spherical capacitor described in Example 24.3 (Section 24.1) has charges $+Q$ and $-Q$ on its inner and outer conductors. Find the electric potential energy stored in the capacitor (a) by using the capacitance C found in Example 24.3 and (b) by integrating the electric-field energy density u.

SOLUTION

IDENTIFY and SET UP: We can determine the energy U stored in a capacitor in two ways: in terms of the work done to put the charges on the two conductors, and in terms of the energy in the electric field between the conductors. The descriptions are equivalent, so they must give us the same result. In Example 24.3 we found the capacitance C and the field magnitude E in the space between the conductors. (The electric field is zero inside the inner sphere and is also zero outside the inner surface of the outer sphere, because a Gaussian surface with radius $r < r_a$ or $r > r_b$ encloses zero net charge. Hence the energy density is nonzero only in the space between the spheres, $r_a < r < r_b$.) In part (a) we use Eq. (24.9) to find U. In part (b) we use Eq. (24.11) to find u, which we integrate over the volume between the spheres to find U.

EXECUTE: (a) From Example 24.3, the spherical capacitor has capacitance

$$C = 4\pi\epsilon_0 \frac{r_a r_b}{r_b - r_a}$$

where r_a and r_b are the radii of the inner and outer conducting spheres, respectively. From Eq. (24.9) the energy stored in this capacitor is

$$U = \frac{Q^2}{2C} = \frac{Q^2}{8\pi\epsilon_0} \frac{r_b - r_a}{r_a r_b}$$

(b) The electric field in the region $r_a < r < r_b$ between the two conducting spheres has magnitude $E = Q/4\pi\epsilon_0 r^2$. The energy density in this region is

$$u = \tfrac{1}{2}\epsilon_0 E^2 = \tfrac{1}{2}\epsilon_0 \left(\frac{Q}{4\pi\epsilon_0 r^2} \right)^2 = \frac{Q^2}{32\pi^2\epsilon_0 r^4}$$

The energy density is *not* uniform; it decreases rapidly with increasing distance from the center of the capacitor. To find the total electric-field energy, we integrate u (the energy per unit volume) over the region $r_a < r < r_b$. We divide this region into spherical shells of radius r, surface area $4\pi r^2$, thickness dr, and volume $dV = 4\pi r^2\, dr$. Then

$$U = \int u\, dV = \int_{r_a}^{r_b} \left(\frac{Q^2}{32\pi^2\epsilon_0 r^4} \right) 4\pi r^2\, dr$$

$$= \frac{Q^2}{8\pi\epsilon_0} \int_{r_a}^{r_b} \frac{dr}{r^2} = \frac{Q^2}{8\pi\epsilon_0} \left(-\frac{1}{r_b} + \frac{1}{r_a} \right)$$

$$= \frac{Q^2}{8\pi\epsilon_0} \frac{r_b - r_a}{r_a r_b}$$

EVALUATE: Electric potential energy can be associated with either the *charges,* as in part (a), or the *field,* as in part (b); the calculated amount of stored energy is the same in either case.

TEST YOUR UNDERSTANDING OF SECTION 24.3 You want to connect a 4-μF capacitor and an 8-μF capacitor. With which type of connection will the 4-μF capacitor have a greater amount of *stored energy* than the 8-μF capacitor? (i) Series; (ii) parallel; (iii) either series or parallel; (iv) neither series nor parallel. ▮

24.4 DIELECTRICS

Most capacitors have a nonconducting material, or **dielectric,** between their conducting plates. A common type of capacitor uses long strips of metal foil for the plates, separated by strips of plastic sheet such as Mylar. A sandwich of these materials is rolled up, forming a unit that can provide a capacitance of several microfarads in a compact package (**Fig. 24.13**).

Placing a solid dielectric between the plates of a capacitor serves three functions. First, it solves the mechanical problem of maintaining two large metal sheets at a very small separation without actual contact.

Second, using a dielectric increases the maximum possible potential difference between the capacitor plates. Any insulating material, when subjected to a sufficiently large electric field, experiences a partial ionization that permits conduction through it (Section 23.3). This phenomenon is called **dielectric breakdown.** Many dielectric materials can tolerate stronger electric fields without breakdown

24.13 A common type of capacitor uses dielectric sheets to separate the conductors.

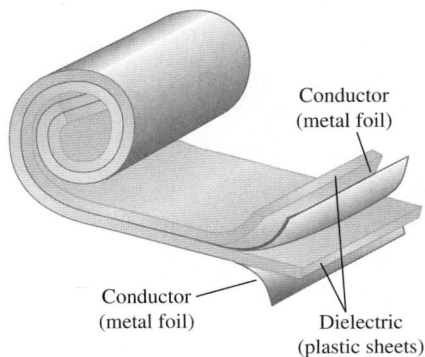

Conductor
(metal foil)

Conductor
(metal foil)

Dielectric
(plastic sheets)

24.14 Effect of a dielectric between the plates of a parallel-plate capacitor. (a) With a given charge, the potential difference is V_0. (b) With the same charge but with a dielectric between the plates, the potential difference V is smaller than V_0.

(a)

Vacuum

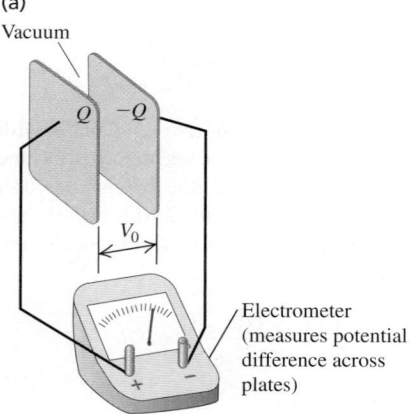

Electrometer (measures potential difference across plates)

(b)

Dielectric

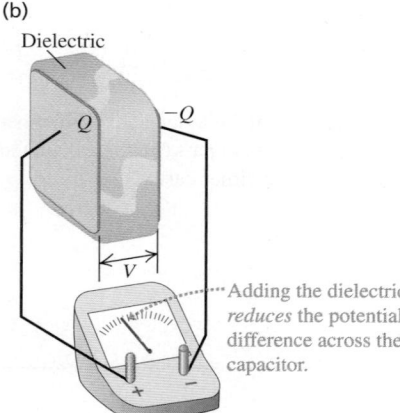

Adding the dielectric *reduces* the potential difference across the capacitor.

than can air. Thus using a dielectric allows a capacitor to sustain a higher potential difference V and so store greater amounts of charge and energy.

Third, the capacitance of a capacitor of given dimensions is *greater* when there is a dielectric material between the plates than when there is vacuum. We can demonstrate this effect with the aid of a sensitive *electrometer,* a device that measures the potential difference between two conductors without letting any appreciable charge flow from one to the other. **Figure 24.14a** shows an electrometer connected across a charged capacitor, with magnitude of charge Q on each plate and potential difference V_0. When we insert an uncharged sheet of dielectric, such as glass, paraffin, or polystyrene, between the plates, experiment shows that the potential difference *decreases* to a smaller value V (Fig. 24.14b). When we remove the dielectric, the potential difference returns to its original value V_0, showing that the original charges on the plates have not changed.

The original capacitance C_0 is given by $C_0 = Q/V_0$, and the capacitance C with the dielectric present is $C = Q/V$. The charge Q is the same in both cases, and V is less than V_0, so we conclude that the capacitance C with the dielectric present is *greater* than C_0. When the space between plates is completely filled by the dielectric, the ratio of C to C_0 (equal to the ratio of V_0 to V) is called the **dielectric constant** of the material, K:

$$K = \frac{C}{C_0} \quad \text{(definition of dielectric constant)} \quad (24.12)$$

When the charge is constant, $Q = C_0 V_0 = CV$ and $C/C_0 = V_0/V$. In this case,

$$V = \frac{V_0}{K} \quad \text{(when } Q \text{ is constant)} \quad (24.13)$$

With the dielectric present, the potential difference for a given charge Q is *reduced* by a factor K.

The dielectric constant K is a pure number. Because C is always greater than C_0, K is always greater than unity. **Table 24.1** gives some representative values of K. For vacuum, $K = 1$ by definition. For air at ordinary temperatures and pressures, K is about 1.0006; this is so nearly equal to 1 that for most purposes an air capacitor is equivalent to one in vacuum. Note that while water has a very large value of K, it is usually not a very practical dielectric for use in capacitors. The reason is that while pure water is a very poor conductor, it is also an excellent ionic solvent. Any ions that are dissolved in the water will cause charge to flow between the capacitor plates, so the capacitor discharges.

CAUTION **Dielectric constant vs. electric constant** Don't confuse the *dielectric* constant K with the *electric* constant ϵ_0. The value of K is a pure number with no units and is different for different materials (see Table 24.1). By contrast, ϵ_0 is a universal constant with units $C^2/N \cdot m^2$ or F/m. ▮

TABLE 24.1	Values of Dielectric Constant K at 20°C		
Material	**K**	**Material**	**K**
Vacuum	1	Polyvinyl chloride	3.18
Air (1 atm)	1.00059	Plexiglas®	3.40
Air (100 atm)	1.0548	Glass	5–10
Teflon	2.1	Neoprene	6.70
Polyethylene	2.25	Germanium	16
Benzene	2.28	Glycerin	42.5
Mica	3–6	Water	80.4
Mylar	3.1	Strontium titanate	310

No real dielectric is a perfect insulator. Hence there is always some *leakage current* between the charged plates of a capacitor with a dielectric. We tacitly ignored this effect in Section 24.2 when we derived expressions for the equivalent capacitances of capacitors in series, Eq. (24.5), and in parallel, Eq. (24.7). But if a leakage current flows for a long enough time to substantially change the charges from the values we used to derive Eqs. (24.5) and (24.7), those equations may no longer be accurate.

Induced Charge and Polarization

When a dielectric material is inserted between the plates while the charge is kept constant, the potential difference between the plates decreases by a factor K. Therefore the electric field between the plates must decrease by the same factor. If E_0 is the vacuum value and E is the value with the dielectric, then

$$E = \frac{E_0}{K} \quad \text{(when } Q \text{ is constant)} \tag{24.14}$$

Since the electric-field magnitude is smaller when the dielectric is present, the surface charge density (which causes the field) must be smaller as well. The surface charge on the conducting plates does not change, but an *induced* charge of the opposite sign appears on each surface of the dielectric (**Fig. 24.15**). The dielectric was originally electrically neutral and is still neutral; the induced surface charges arise as a result of *redistribution* of positive and negative charge within the dielectric material, a phenomenon called **polarization.** We first encountered polarization in Section 21.2, and we suggest that you reread the discussion of Fig. 21.8. We will assume that the induced surface charge is *directly proportional* to the electric-field magnitude E in the material; this is indeed the case for many common dielectrics. (This direct proportionality is analogous to Hooke's law for a spring.) In that case, K is a constant for any particular material. When the electric field is very strong or if the dielectric is made of certain crystalline materials, the relationship between induced charge and the electric field can be more complex; we won't consider such cases here.

We can derive a relationship between this induced surface charge and the charge on the plates. Let σ_i be the magnitude of the charge per unit area induced on the surfaces of the dielectric (the induced surface charge density). The magnitude of the surface charge density on the capacitor plates is σ, as usual. Then the *net* surface charge on each side of the capacitor has magnitude $(\sigma - \sigma_i)$; see Fig. 24.15b. As we found in Example 21.12 (Section 21.5) and in Example 22.8 (Section 22.4), the field between the plates is related to the net surface charge density by $E = \sigma_{net}/\epsilon_0$. Without and with the dielectric, respectively,

$$E_0 = \frac{\sigma}{\epsilon_0} \qquad E = \frac{\sigma - \sigma_i}{\epsilon_0} \tag{24.15}$$

Using these expressions in Eq. (24.14) and rearranging the result, we find

$$\sigma_i = \sigma\left(1 - \frac{1}{K}\right) \quad \text{(induced surface charge density)} \tag{24.16}$$

This equation shows that when K is very large, σ_i is nearly as large as σ. In this case, σ_i nearly cancels σ, and the field and potential difference are much smaller than their values in vacuum.

The product $K\epsilon_0$ is called the **permittivity** of the dielectric, denoted by ϵ:

$$\epsilon = K\epsilon_0 \quad \text{(definition of permittivity)} \tag{24.17}$$

In terms of ϵ we can express the electric field within the dielectric as

$$E = \frac{\sigma}{\epsilon} \tag{24.18}$$

24.15 Electric field lines with (a) vacuum between the plates and (b) dielectric between the plates.

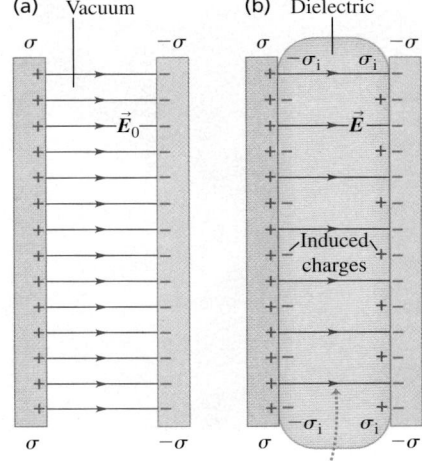

For a given charge density σ, the induced charges on the dielectric's surfaces reduce the electric field between the plates.

DATA *SPEAKS*

Capacitors and Capacitance

When students were given a problem involving capacitors and capacitance, more than 25% gave an incorrect response. Common errors:

- Forgetting that the capacitance C of a capacitor depends on only the capacitor geometry (the size, shape, and position of its conductors) and the presence or absence of a dielectric. C does not depend on the amount of charge Q placed on the conductors.

- Not understanding what happens if capacitance changes (for example, by inserting or removing a dielectric). If the capacitor is isolated, Q remains constant but the potential difference V_{ab} changes if C changes. If V_{ab} is held constant, Q changes if C changes.

Application Capacitors in the Toolbox Several practical devices rely on the way a capacitor responds to a change in dielectric constant. One example is an electric stud finder, used to locate metal fasteners hidden behind a wall's surface. It consists of a metal plate with associated circuitry. The plate acts as one half of a capacitor; the wall acts as the other half. If the stud finder moves over a metal fastener, the effective dielectric constant for the capacitor changes, which changes the capacitance and triggers a signal.

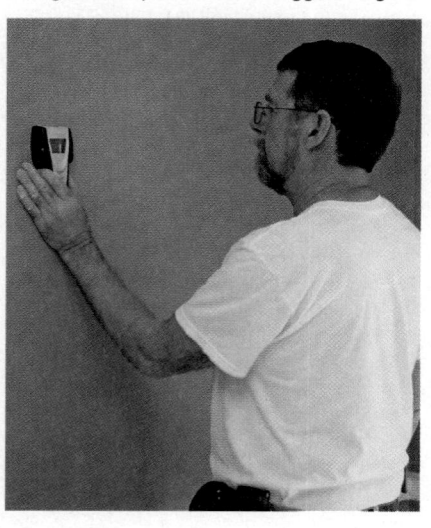

Then

Capacitance of a parallel-plate capacitor, dielectric between plates

$$C = KC_0 = K\epsilon_0 \frac{A}{d} = \epsilon \frac{A}{d} \quad (24.19)$$

Dielectric constant · Area of each plate · Permittivity = $K\epsilon_0$. Capacitance without dielectric · Electric constant · Distance between plates

We can repeat the derivation of Eq. (24.11) for the energy density u in an electric field for the case in which a dielectric is present. The result is

Electric energy density in a dielectric

$$u = \tfrac{1}{2}K\epsilon_0 E^2 = \tfrac{1}{2}\epsilon E^2 \quad (24.20)$$

Dielectric constant · Permittivity = $K\epsilon_0$. Electric constant · Magnitude of electric field

In empty space, where $K = 1$, $\epsilon = \epsilon_0$ and Eqs. (24.19) and (24.20) reduce to Eqs. (24.2) and (24.11), respectively, for a parallel-plate capacitor in vacuum. For this reason, ϵ_0 is sometimes called the "permittivity of free space" or the "permittivity of vacuum." Because K is a pure number, ϵ and ϵ_0 have the same units, $C^2/N \cdot m^2$ or F/m.

Equation (24.19) shows that extremely high capacitances can be obtained with plates that have a large surface area A and are separated by a small distance d by a dielectric with a large value of K. In an *electrolytic double-layer capacitor*, tiny carbon granules adhere to each plate: A is the combined surface area of the granules, which can be tremendous. The plates with granules attached are separated by a very thin dielectric sheet. Such a capacitor can have a capacitance of 5000 F yet fit in the palm of your hand (compare Example 24.1 in Section 24.1).

PROBLEM-SOLVING STRATEGY 24.2 | DIELECTRICS

IDENTIFY *the relevant concepts:* The relationships in this section are useful whenever there is an electric field in a dielectric, such as a dielectric between charged capacitor plates. Typically you must relate the potential difference V_{ab} between the plates, the electric field magnitude E in the capacitor, the charge density σ on the capacitor plates, and the induced charge density σ_i on the surfaces of the capacitor.

SET UP *the problem* using the following steps:
1. Make a drawing of the situation.
2. Identify the target variables, and choose which equations from this section will help you solve for those variables.

EXECUTE *the solution* as follows:
1. In problems such as the next example, it is easy to get lost in a blizzard of formulas. Ask yourself at each step what kind of

quantity each symbol represents. For example, distinguish clearly between charges and charge densities, and between electric fields and electric potential differences.
2. Check for consistency of units. Distances must be in meters. A microfarad is 10^{-6} farad, and so on. Don't confuse the numerical value of ϵ_0 with the value of $1/4\pi\epsilon_0$. Electric-field magnitude can be expressed in both N/C and V/m. The units of ϵ_0 are $C^2/N \cdot m^2$ or F/m.

EVALUATE *your answer:* With a dielectric present, (a) the capacitance is greater than without a dielectric; (b) for a given charge on the capacitor, the electric field and potential difference are less than without a dielectric; and (c) the magnitude of the induced surface charge density σ_i on the dielectric is less than that of the charge density σ on the capacitor plates.

EXAMPLE 24.10 **A CAPACITOR WITH AND WITHOUT A DIELECTRIC**

Suppose the parallel plates in Fig. 24.15 each have an area of 2000 cm^2 (2.00×10^{-1} m^2) and are 1.00 cm (1.00×10^{-2} m) apart. We connect the capacitor to a power supply, charge it to a potential difference $V_0 = 3.00$ kV, and disconnect the power supply. We then insert a sheet of insulating plastic material between the plates, completely filling the space between them. We find that the potential difference decreases to 1.00 kV while the charge on

each capacitor plate remains constant. Find (a) the original capacitance C_0; (b) the magnitude of charge Q on each plate; (c) the capacitance C after the dielectric is inserted; (d) the dielectric constant K of the dielectric; (e) the permittivity ϵ of the dielectric; (f) the magnitude of the induced charge Q_i on each face of the dielectric; (g) the original electric field E_0 between the plates; and (h) the electric field E after the dielectric is inserted.

SOLUTION

IDENTIFY and SET UP: This problem uses most of the relationships we have discussed for capacitors and dielectrics. (Energy relationships are treated in Example 24.11.) Most of the target variables can be obtained in several ways. The methods used below are a sample; we encourage you to think of others and compare your results.

EXECUTE: (a) With vacuum between the plates, we use Eq. (24.19) with $K = 1$:

$$C_0 = \epsilon_0 \frac{A}{d} = (8.85 \times 10^{-12}\,\text{F/m})\frac{2.00 \times 10^{-1}\,\text{m}^2}{1.00 \times 10^{-2}\,\text{m}}$$

$$= 1.77 \times 10^{-10}\,\text{F} = 177\,\text{pF}$$

(b) From the definition of capacitance, Eq. (24.1),

$$Q = C_0 V_0 = (1.77 \times 10^{-10}\,\text{F})(3.00 \times 10^3\,\text{V})$$

$$= 5.31 \times 10^{-7}\,\text{C} = 0.531\,\mu\text{C}$$

(c) When the dielectric is inserted, Q is unchanged but the potential difference decreases to $V = 1.00\,\text{kV}$. Hence from Eq. (24.1), the new capacitance is

$$C = \frac{Q}{V} = \frac{5.31 \times 10^{-7}\,\text{C}}{1.00 \times 10^3\,\text{V}} = 5.31 \times 10^{-10}\,\text{F}$$

$$= 531\,\text{pF}$$

(d) From Eq. (24.12), the dielectric constant is

$$K = \frac{C}{C_0} = \frac{5.31 \times 10^{-10}\,\text{F}}{1.77 \times 10^{-10}\,\text{F}} = \frac{531\,\text{pF}}{177\,\text{pF}}$$

$$= 3.00$$

Alternatively, from Eq. (24.13),

$$K = \frac{V_0}{V} = \frac{3000\,\text{V}}{1000\,\text{V}} = 3.00$$

(e) With K from part (d) in Eq. (24.17), the permittivity is

$$\epsilon = K\epsilon_0 = (3.00)(8.85 \times 10^{-12}\,\text{C}^2/\text{N}\cdot\text{m}^2)$$

$$= 2.66 \times 10^{-11}\,\text{C}^2/\text{N}\cdot\text{m}^2$$

(f) Multiplying both sides of Eq. (24.16) by the plate area A gives the induced charge $Q_i = \sigma_i A$ in terms of the charge $Q = \sigma A$ on each plate:

$$Q_i = Q\left(1 - \frac{1}{K}\right) = (5.31 \times 10^{-7}\,\text{C})\left(1 - \frac{1}{3.00}\right)$$

$$= 3.54 \times 10^{-7}\,\text{C}$$

(g) Since the electric field between the plates is uniform, its magnitude is the potential difference divided by the plate separation:

$$E_0 = \frac{V_0}{d} = \frac{3000\,\text{V}}{1.00 \times 10^{-2}\,\text{m}} = 3.00 \times 10^5\,\text{V/m}$$

(h) After the dielectric is inserted,

$$E = \frac{V}{d} = \frac{1000\,\text{V}}{1.00 \times 10^{-2}\,\text{m}} = 1.00 \times 10^5\,\text{V/m}$$

or, from Eq. (24.18),

$$E = \frac{\sigma}{\epsilon} = \frac{Q}{\epsilon A} = \frac{5.31 \times 10^{-7}\,\text{C}}{(2.66 \times 10^{-11}\,\text{C}^2/\text{N}\cdot\text{m}^2)(2.00 \times 10^{-1}\,\text{m}^2)}$$

$$= 1.00 \times 10^5\,\text{V/m}$$

or, from Eq. (24.15),

$$E = \frac{\sigma - \sigma_i}{\epsilon_0} = \frac{Q - Q_i}{\epsilon_0 A}$$

$$= \frac{(5.31 - 3.54) \times 10^{-7}\,\text{C}}{(8.85 \times 10^{-12}\,\text{C}^2/\text{N}\cdot\text{m}^2)(2.00 \times 10^{-1}\,\text{m}^2)}$$

$$= 1.00 \times 10^5\,\text{V/m}$$

or, from Eq. (24.14),

$$E = \frac{E_0}{K} = \frac{3.00 \times 10^5\,\text{V/m}}{3.00} = 1.00 \times 10^5\,\text{V/m}$$

EVALUATE: Inserting the dielectric increased the capacitance by a factor of $K = 3.00$ and reduced the electric field between the plates by a factor of $1/K = 1/3.00$. It did so by developing induced charges on the faces of the dielectric of magnitude $Q(1 - 1/K) = Q(1 - 1/3.00) = 0.667Q$.

EXAMPLE 24.11 ENERGY STORAGE WITH AND WITHOUT A DIELECTRIC

Find the energy stored in the electric field of the capacitor in Example 24.10 and the energy density, both before and after the dielectric sheet is inserted.

SOLUTION

IDENTIFY and SET UP: We consider the ideas of energy stored in a capacitor and of electric-field energy density. We use Eq. (24.9) to find the stored energy and Eq. (24.20) to find the energy density.

EXECUTE: From Eq. (24.9), the stored energies U_0 and U without and with the dielectric in place are

$$U_0 = \tfrac{1}{2}C_0 V_0^2 = \tfrac{1}{2}(1.77 \times 10^{-10}\,\text{F})(3000\,\text{V})^2 = 7.97 \times 10^{-4}\,\text{J}$$

$$U = \tfrac{1}{2}CV^2 = \tfrac{1}{2}(5.31 \times 10^{-10}\,\text{F})(1000\,\text{V})^2 = 2.66 \times 10^{-4}\,\text{J}$$

The final energy is one-third of the original energy.

Equation (24.20) gives the energy densities without and with the dielectric:

$$u_0 = \tfrac{1}{2}\epsilon_0 E_0^2 = \tfrac{1}{2}(8.85 \times 10^{-12}\,\text{C}^2/\text{N}\cdot\text{m}^2)(3.00 \times 10^5\,\text{N/C})^2$$

$$= 0.398\,\text{J/m}^3$$

$$u = \tfrac{1}{2}\epsilon E^2 = \tfrac{1}{2}(2.66 \times 10^{-11}\,\text{C}^2/\text{N}\cdot\text{m}^2)(1.00 \times 10^5\,\text{N/C})^2$$

$$= 0.133\,\text{J/m}^3$$

The energy density with the dielectric is one-third of the original energy density.

Continued

EVALUATE: We can check our answer for u_0 by noting that the volume between the plates is $V = (0.200 \text{ m}^2)(0.0100 \text{ m}) = 0.00200 \text{ m}^3$. Since the electric field between the plates is uniform, u_0 is uniform as well and the energy density is just the stored energy divided by the volume:

$$u_0 = \frac{U_0}{V} = \frac{7.97 \times 10^{-4} \text{ J}}{0.00200 \text{ m}^3} = 0.398 \text{ J/m}^3$$

This agrees with our earlier answer. You can use the same approach to check our result for u.

In general, when a dielectric is inserted into a capacitor while the charge on each plate remains the same, the permittivity ϵ increases by a factor of K (the dielectric constant), and the electric field E and the energy density $u = \frac{1}{2}\epsilon E^2$ decrease by a factor of $1/K$. Where does the energy go? The answer lies in the fringing field at the edges of a real parallel-plate capacitor. As **Fig. 24.16** shows, that field tends to pull the dielectric into the space between

24.16 The fringing field at the edges of the capacitor exerts forces $\vec{F}_{-i}$ and $\vec{F}_{+i}$ on the negative and positive induced surface charges of a dielectric, pulling the dielectric into the capacitor.

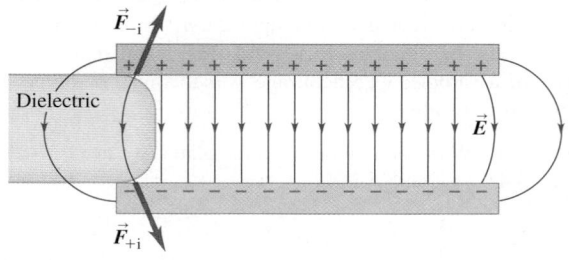

the plates, doing work on it as it does so. We could attach a spring to the left end of the dielectric in Fig. 24.16 and use this force to stretch the spring. Because work is done by the field, the field energy density decreases.

Dielectric Breakdown

We mentioned earlier that when a dielectric is subjected to a sufficiently strong electric field, *dielectric breakdown* takes place and the dielectric becomes a conductor. This occurs when the electric field is so strong that electrons are ripped loose from their molecules and crash into other molecules, liberating even more electrons. This avalanche of moving charge forms a spark or arc discharge. Lightning is a dramatic example of dielectric breakdown in air.

Because of dielectric breakdown, capacitors always have maximum voltage ratings. When a capacitor is subjected to excessive voltage, an arc may form through a layer of dielectric, burning or melting a hole in it. This arc creates a conducting path (a short circuit) between the conductors. If a conducting path remains after the arc is extinguished, the device is rendered permanently useless as a capacitor.

The maximum electric-field magnitude that a material can withstand without the occurrence of breakdown is called its **dielectric strength.** This quantity is affected significantly by temperature, trace impurities, small irregularities in the metal electrodes, and other factors that are difficult to control. For this reason we can give only approximate figures for dielectric strengths. The dielectric strength of dry air is about 3×10^6 V/m. **Table 24.2** lists the dielectric strengths of a few common insulating materials. All of the values are substantially greater than the value for air. For example, a layer of polycarbonate 0.01 mm thick (about the smallest practical thickness) has 10 times the dielectric strength of air and can withstand a maximum voltage of about $(3 \times 10^7 \text{ V/m})(1 \times 10^{-5} \text{ m}) = 300$ V.

BIO Application Dielectric Cell Membrane The membrane of a living cell behaves like a dielectric between the plates of a capacitor. The membrane is made of two sheets of lipid molecules, with their water-insoluble ends in the middle and their water-soluble ends (shown in red) on the outer surfaces. Conductive fluids on either side of the membrane (water with negative ions inside the cell, water with positive ions outside) act as charged capacitor plates, and the nonconducting membrane acts as a dielectric with K of about 10. The potential difference V across the membrane is about 0.07 V and the membrane thickness d is about 7×10^{-9} m, so the electric field $E = V/d$ in the membrane is about 10^7 V/m—close to the dielectric strength of the membrane. If the membrane were made of air, V and E would be larger by a factor of $K \approx 10$ and dielectric breakdown would occur.

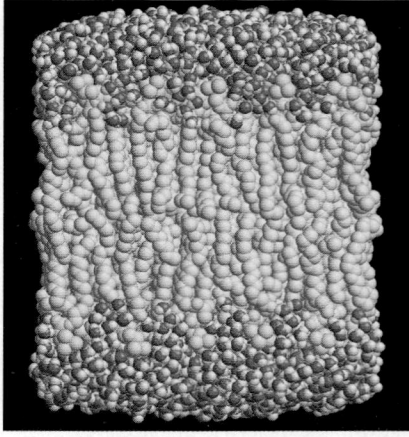

Dielectric Constant and Dielectric Strength of Some Insulating Materials

TABLE 24.2

Material	Dielectric Constant, K	Dielectric Strength, E_m (V/m)
Polycarbonate	2.8	3×10^7
Polyester	3.3	6×10^7
Polypropylene	2.2	7×10^7
Polystyrene	2.6	2×10^7
Pyrex glass	4.7	1×10^7

TEST YOUR UNDERSTANDING OF SECTION 24.4 The space between the plates of an isolated parallel-plate capacitor is filled by a slab of dielectric with dielectric constant K. The two plates of the capacitor have charges Q and $-Q$. You pull out the dielectric slab. If the charges do not change, how does the energy in the capacitor change when you remove the slab? (i) It increases; (ii) it decreases; (iii) it remains the same. ▮

24.5 MOLECULAR MODEL OF INDUCED CHARGE

In Section 24.4 we discussed induced surface charges on a dielectric in an electric field. Now let's look at how these surface charges can arise. If the material were a *conductor,* the answer would be simple. Conductors contain charge that is free to move, and when an electric field is present, some of the charge redistributes itself on the surface so that there is no electric field inside the conductor. But an ideal dielectric has *no* charges that are free to move, so how can a surface charge occur?

To understand this, we have to look again at rearrangement of charge at the *molecular* level. Some molecules, such as H_2O and N_2O, have equal amounts of positive and negative charges but a lopsided distribution, with excess positive charge concentrated on one side of the molecule and negative charge on the other. As we described in Section 21.7, such an arrangement is called an *electric dipole,* and the molecule is called a *polar molecule.* When no electric field is present in a gas or liquid with polar molecules, the molecules are oriented randomly (**Fig. 24.17a**). In an electric field, however, they tend to orient themselves as in Fig. 24.17b, as a result of the electric-field torques described in Section 21.7. Because of thermal agitation, the alignment of the molecules with $\vec{E}$ is not perfect.

Even a molecule that is *not* ordinarily polar *becomes* a dipole when it is placed in an electric field because the field pushes the positive charges in the molecules in the direction of the field and pushes the negative charges in the opposite direction. This causes a redistribution of charge within the molecule (**Fig. 24.18**). Such dipoles are called *induced* dipoles.

With either polar or nonpolar molecules, the redistribution of charge caused by the field leads to the formation of a layer of charge on each surface of the dielectric material (**Fig. 24.19**). These layers are the surface charges described in Section 24.4; their surface charge density is denoted by σ_i. The charges are *not* free to move indefinitely, as they would be in a conductor, because each charge is bound to a molecule. They are in fact called **bound charges** to distinguish them

24.19 Polarization of a dielectric in an electric field $\vec{E}$ gives rise to thin layers of bound charges on the surfaces, creating surface charge densities σ_i and $-\sigma_i$. The sizes of the molecules are greatly exaggerated for clarity.

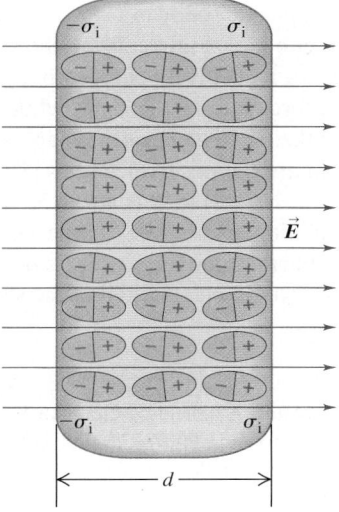

24.17 Polar molecules (a) without and (b) with an applied electric field $\vec{E}$.

(a)

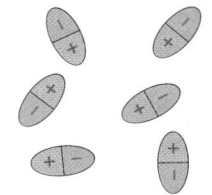

In the absence of an electric field, polar molecules orient randomly.

(b)

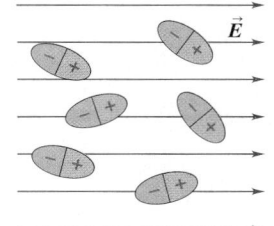

When an electric field is applied, the molecules tend to align with it.

(MP)

PhET: Molecular Motors
PhET: Optical Tweezers and Applications
PhET: Stretching DNA

24.18 Nonpolar molecules (a) without and (b) with an applied electric field $\vec{E}$.

(a)

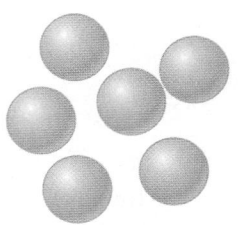

In the absence of an electric field, nonpolar molecules are not electric dipoles.

(b)

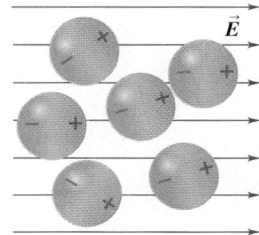

An electric field causes the molecules' positive and negative charges to separate slightly, making the molecule effectively polar.

24.20 (a) Electric field of magnitude E_0 between two charged plates. (b) Introduction of a dielectric of dielectric constant K. (c) The induced surface charges and their field. (d) Resultant field of magnitude E_0/K.

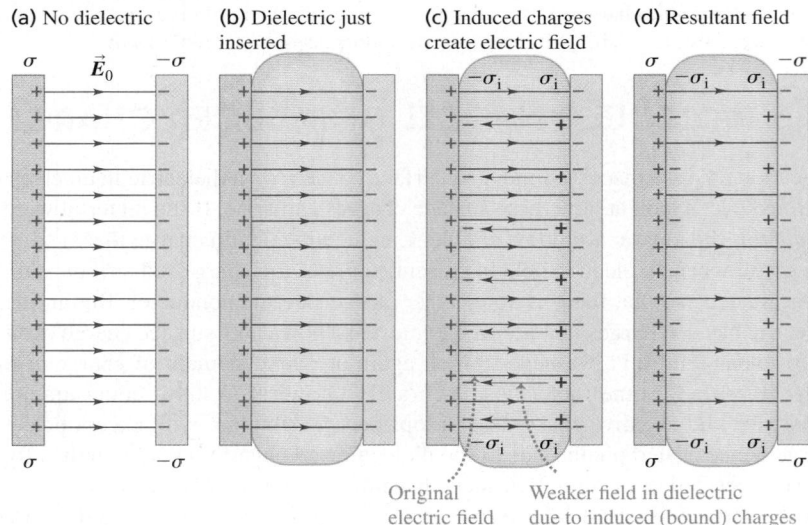

from the **free charges** that are added to and removed from the conducting capacitor plates. In the interior of the material the net charge per unit volume remains zero. As we have seen, this redistribution of charge is called *polarization,* and we say that the material is *polarized.*

The four parts of **Fig. 24.20** show the behavior of a slab of dielectric when it is inserted in the field between a pair of oppositely charged capacitor plates. Figure 24.20a shows the original field. Figure 24.20b is the situation after the dielectric has been inserted but before any rearrangement of charges has occurred. Figure 24.20c shows by thinner arrows the additional field set up in the dielectric by its induced surface charges. This field is *opposite* to the original field, but it is not great enough to cancel the original field completely because the charges in the dielectric are not free to move indefinitely. The resultant field in the dielectric, shown in Fig. 24.20d, is therefore decreased in magnitude. In the field-line representation, some of the field lines leaving the positive plate go through the dielectric, while others terminate on the induced charges on the faces of the dielectric.

As we discussed in Section 21.2, polarization is also the reason a charged body, such as an electrified plastic rod, can exert a force on an *uncharged* body such as a bit of paper or a pith ball. **Figure 24.21** shows an uncharged dielectric sphere B in the radial field of a positively charged body A. The induced positive charges on B experience a force toward the right, while the force on the induced negative charges is toward the left. The negative charges are closer to A, and thus are in a stronger field, than are the positive charges. The force toward the left is stronger than that toward the right, and B is attracted toward A, even though its net charge is zero. The attraction occurs whether the sign of A's charge is positive or negative (see Fig. 21.8). Furthermore, the effect is not limited to dielectrics; an uncharged conducting body would be attracted in the same way.

24.21 A neutral sphere B in the radial electric field of a positively charged sphere A is attracted to the charge because of polarization.

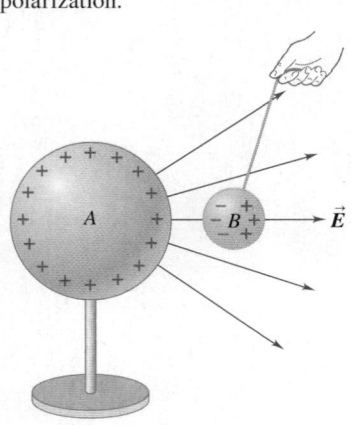

TEST YOUR UNDERSTANDING OF SECTION 24.5 A parallel-plate capacitor has charges Q and $-Q$ on its two plates. A dielectric slab with $K = 3$ is then inserted into the space between the plates as shown in Fig. 24.20. Rank the following electric-field magnitudes in order from largest to smallest. (i) The field before the slab is inserted; (ii) the resultant field after the slab is inserted; (iii) the field due to the bound charges. ❙

24.6 GAUSS'S LAW IN DIELECTRICS

We can extend the analysis of Section 24.4 to reformulate Gauss's law in a form that is particularly useful for dielectrics. **Figure 24.22** is a close-up view of the left capacitor plate and left surface of the dielectric in Fig. 24.15b. Let's apply Gauss's law to the rectangular box shown in cross section by the purple line; the surface area of the left and right sides is A. The left side is embedded in the conductor that forms the left capacitor plate, and so the electric field everywhere on that surface is zero. The right side is embedded in the dielectric, where the electric field has magnitude E, and $E_\perp = 0$ everywhere on the other four sides. The total charge enclosed, including both the charge on the capacitor plate and the induced charge on the dielectric surface, is $Q_{encl} = (\sigma - \sigma_i)A$, so Gauss's law gives

$$EA = \frac{(\sigma - \sigma_i)A}{\epsilon_0} \tag{24.21}$$

This equation is not very illuminating as it stands because it relates two unknown quantities: E inside the dielectric and the induced surface charge density σ_i. But now we can use Eq. (24.16), developed for this same situation, to simplify this equation by eliminating σ_i. Equation (24.16) is

$$\sigma_i = \sigma\left(1 - \frac{1}{K}\right)$$

or

$$\sigma - \sigma_i = \frac{\sigma}{K}$$

Combining this with Eq. (24.21), we get

$$EA = \frac{\sigma A}{K\epsilon_0}$$

or

$$KEA = \frac{\sigma A}{\epsilon_0} \tag{24.22}$$

Equation (24.22) says that the flux of $K\vec{E}$, not $\vec{E}$, through the Gaussian surface in Fig. 24.22 is equal to the enclosed *free* charge σA divided by ϵ_0. It turns out that for *any* Gaussian surface, whenever the induced charge is proportional to the electric field in the material, we can rewrite Gauss's law as

Gauss's law in a dielectric:

Dielectric constant ⋯⋯⋯⋯⋯⋯ Total free charge
$$\oint K\vec{E} \cdot d\vec{A} = \frac{Q_{encl\text{-}free}}{\epsilon_0} \qquad \text{enclosed by surface} \tag{24.23}$$
Surface integral of $K\vec{E}$ ⋯⋯⋯ Electric constant
over a closed surface

where $Q_{encl\text{-}free}$ is the total *free* charge (not bound charge) enclosed by the Gaussian surface. The significance of these results is that the right sides contain only the *free* charge on the conductor, not the bound (induced) charge. In fact, although we have not proved it, Eq. (24.23) remains valid even when different parts of the Gaussian surface are embedded in dielectrics having different values of K, provided that the value of K in each dielectric is independent of the electric field (usually the case for electric fields that are not too strong) and that we use the appropriate value of K for each point on the Gaussian surface.

24.22 Gauss's law with a dielectric. This figure shows a close-up of the left-hand capacitor plate in Fig. 24.15b. The Gaussian surface is a rectangular box that lies half in the conductor and half in the dielectric.

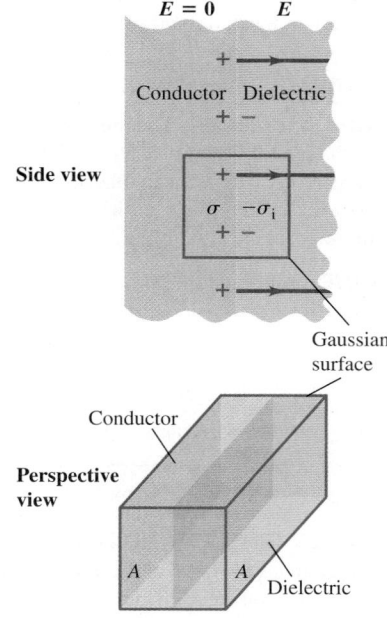

EXAMPLE 24.12 | A SPHERICAL CAPACITOR WITH DIELECTRIC

Use Gauss's law to find the capacitance of the spherical capacitor of Example 24.3 (Section 24.1) if the volume between the shells is filled with an insulating oil with dielectric constant K.

SOLUTION

IDENTIFY and SET UP: The spherical symmetry of the problem is not changed by the presence of the dielectric, so as in Example 24.3, we use a concentric spherical Gaussian surface of radius r between the shells. Since a dielectric is present, we use Gauss's law in the form of Eq. (24.23).

EXECUTE: From Eq. (24.23),

$$\oint K\vec{E} \cdot d\vec{A} = \oint KE\, dA = KE \oint dA = (KE)(4\pi r^2) = \frac{Q}{\epsilon_0}$$

$$E = \frac{Q}{4\pi K\epsilon_0 r^2} = \frac{Q}{4\pi\epsilon r^2}$$

where $\epsilon = K\epsilon_0$. Compared to the case in which there is vacuum between the shells, the electric field is reduced by a factor of $1/K$. The potential difference V_{ab} between the shells is reduced by the same factor, and so the capacitance $C = Q/V_{ab}$ is *increased* by a factor of K, just as for a parallel-plate capacitor when a dielectric is inserted. Using the result of Example 24.3, we find that the capacitance with the dielectric is

$$C = \frac{4\pi K\epsilon_0 r_a r_b}{r_b - r_a} = \frac{4\pi\epsilon r_a r_b}{r_b - r_a}$$

EVALUATE: If the dielectric fills the volume between the two conductors, the capacitance is just K times the value with no dielectric. The result is more complicated if the dielectric only partially fills this volume.

TEST YOUR UNDERSTANDING OF SECTION 24.6 A single point charge q is embedded in a very large block of dielectric of dielectric constant K. At a point inside the dielectric a distance r from the point charge, what is the magnitude of the electric field? (i) $q/4\pi\epsilon_0 r^2$; (ii) $Kq/4\pi\epsilon_0 r^2$; (iii) $q/4\pi K\epsilon_0 r^2$; (iv) none of these. ∎

CHAPTER 24 SUMMARY

Capacitors and capacitance: A capacitor is any pair of conductors separated by an insulating material. When the capacitor is charged, there are charges of equal magnitude Q and opposite sign on the two conductors, and the potential V_{ab} of the positively charged conductor with respect to the negatively charged conductor is proportional to Q. The capacitance C is defined as the ratio of Q to V_{ab}. The SI unit of capacitance is the farad (F): $1\text{ F} = 1\text{ C/V}$.

A parallel-plate capacitor consists of two parallel conducting plates, each with area A, separated by a distance d. If they are separated by vacuum, the capacitance depends on only A and d. For other geometries, the capacitance can be found by using the definition $C = Q/V_{ab}$. (See Examples 24.1–24.4.)

$$C = \frac{Q}{V_{ab}} \quad (24.1)$$

$$C = \frac{Q}{V_{ab}} = \epsilon_0 \frac{A}{d} \quad (24.2)$$

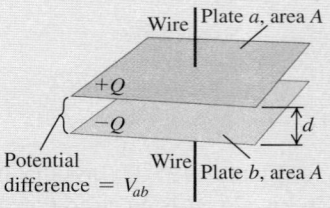

Capacitors in series and parallel: When capacitors with capacitances $C_1, C_2, C_3, \ldots$ are connected in series, the reciprocal of the equivalent capacitance C_{eq} equals the sum of the reciprocals of the individual capacitances. When capacitors are connected in parallel, the equivalent capacitance C_{eq} equals the sum of the individual capacitances. (See Examples 24.5 and 24.6.)

$$\frac{1}{C_{eq}} = \frac{1}{C_1} + \frac{1}{C_2} + \frac{1}{C_3} + \cdots \quad (24.5)$$
(capacitors in series)

$$C_{eq} = C_1 + C_2 + C_3 + \cdots \quad (24.7)$$
(capacitors in parallel)

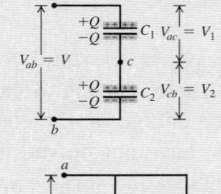

Energy in a capacitor: The energy U required to charge a capacitor C to a potential difference V and a charge Q is equal to the energy stored in the capacitor. This energy can be thought of as residing in the electric field between the conductors; the energy density u (energy per unit volume) is proportional to the square of the electric-field magnitude. (See Examples 24.7–24.9.)

$$U = \frac{Q^2}{2C} = \tfrac{1}{2}CV^2 = \tfrac{1}{2}QV \qquad (24.9)$$

$$u = \tfrac{1}{2}\epsilon_0 E^2 \qquad (24.11)$$

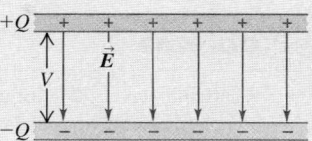

Dielectrics: When the space between the conductors is filled with a dielectric material, the capacitance increases by a factor K, the dielectric constant of the material. The quantity $\epsilon = K\epsilon_0$ is the permittivity of the dielectric. For a fixed amount of charge on the capacitor plates, induced charges on the surface of the dielectric decrease the electric field and potential difference between the plates by the same factor K. The surface charge results from polarization, a microscopic rearrangement of charge in the dielectric. (See Example 24.10.)

Under sufficiently strong electric fields, dielectrics become conductors, a situation called dielectric breakdown. The maximum field that a material can withstand without breakdown is called its dielectric strength.

In a dielectric, the expression for the energy density is the same as in vacuum but with ϵ_0 replaced by $\epsilon = K\epsilon_0$. (See Example 24.11.)

Gauss's law in a dielectric has almost the same form as in vacuum, with two key differences: $\vec{E}$ is replaced by $K\vec{E}$ and Q_{encl} is replaced by $Q_{\text{encl-free}}$, which includes only the free charge (not bound charge) enclosed by the Gaussian surface. (See Example 24.12.)

$$C = KC_0 = K\epsilon_0\frac{A}{d} = \epsilon\frac{A}{d} \qquad (24.19)$$
(parallel-plate capacitor filled with dielectric)

$$u = \tfrac{1}{2}K\epsilon_0 E^2 = \tfrac{1}{2}\epsilon E^2 \qquad (24.20)$$

$$\oint K\vec{E} \cdot d\vec{A} = \frac{Q_{\text{encl-free}}}{\epsilon_0} \qquad (24.23)$$

Dielectric between plates

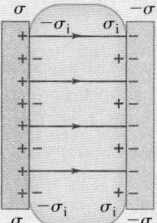

BRIDGING PROBLEM | **ELECTRIC-FIELD ENERGY AND CAPACITANCE OF A CONDUCTING SPHERE**

A solid conducting sphere of radius R carries a charge Q. Calculate the electric-field energy density at a point a distance r from the center of the sphere for (a) $r < R$ and (b) $r > R$. (c) Calculate the total electric-field energy associated with the charged sphere. (d) How much work is required to assemble the charge Q on the sphere? (e) Use the result of part (c) to find the capacitance of the sphere. (You can think of the second conductor as a hollow conducting shell of infinite radius.)

SOLUTION GUIDE

IDENTIFY and SET UP

1. You know the electric field for this situation at all values of r from Example 22.5 (Section 22.4). You'll use this to find the electric-field energy density u and *total* electric-field energy U. You can then find the capacitance from the relationship $U = Q^2/2C$.

2. To find U, consider a spherical shell of radius r and thickness dr that has volume $dV = 4\pi r^2 dr$. (It will help to make a drawing of such a shell concentric with the conducting sphere.) The energy stored in this volume is $u\,dV$, and the total energy is the integral of $u\,dV$ from $r = 0$ to $r \rightarrow \infty$. Set up this integral.

EXECUTE

3. Find u for $r < R$ and for $r > R$. (*Hint:* What is the field inside a solid conductor?)

4. Substitute your results from step 3 into the expression from step 2. Then calculate the integral to find the total electric-field energy U.

5. Use your understanding of the energy stored in a charge distribution to find the work required to assemble the charge Q.

6. Find the capacitance of the sphere.

EVALUATE

7. Where is the electric-field energy density greatest? Where is it least?

8. How would the results be affected if the solid sphere were replaced by a hollow conducting sphere of the same radius R?

9. You can find the potential difference between the sphere and infinity from $C = Q/V$. Does this agree with the result of Example 23.8 (Section 23.3)?

Problems

For assigned homework and other learning materials, go to MasteringPhysics®.

•, ••, •••: Difficulty levels. **CP**: Cumulative problems incorporating material from earlier chapters. **CALC**: Problems requiring calculus. **DATA**: Problems involving real data, scientific evidence, experimental design, and/or statistical reasoning. **BIO**: Biosciences problems.

DISCUSSION QUESTIONS

Q24.1 Equation (24.2) shows that the capacitance of a parallel-plate capacitor becomes larger as the plate separation d decreases. However, there is a practical limit to how small d can be made, which places limits on how large C can be. Explain what sets the limit on d. (*Hint:* What happens to the magnitude of the electric field as $d \rightarrow 0$?)

Q24.2 Suppose several different parallel-plate capacitors are charged up by a constant-voltage source. Thinking of the actual movement and position of the charges on an atomic level, why does it make sense that the capacitances are proportional to the surface areas of the plates? Why does it make sense that the capacitances are *inversely* proportional to the distance between the plates?

Q24.3 Suppose the two plates of a capacitor have different areas. When the capacitor is charged by connecting it to a battery, do the charges on the two plates have equal magnitude, or may they be different? Explain your reasoning.

Q24.4 To store the maximum amount of energy in a parallel-plate capacitor with a given battery (voltage source), would it be better to have the plates far apart or close together?

Q24.5 In the parallel-plate capacitor of Fig. 24.2, suppose the plates are pulled apart so that the separation d is much larger than the size of the plates. (a) Is it still accurate to say that the electric field between the plates is uniform? Why or why not? (b) In the situation shown in Fig. 24.2, the potential difference between the plates is $V_{ab} = Qd/\epsilon_0 A$. If the plates are pulled apart as described above, is V_{ab} more or less than this formula would indicate? Explain your reasoning. (c) With the plates pulled apart as described above, is the capacitance more than, less than, or the same as that given by Eq. (24.2)? Explain your reasoning.

Q24.6 A parallel-plate capacitor is charged by being connected to a battery and is kept connected to the battery. The separation between the plates is then doubled. How does the electric field change? The charge on the plates? The total energy? Explain.

Q24.7 A parallel-plate capacitor is charged by being connected to a battery and is then disconnected from the battery. The separation between the plates is then doubled. How does the electric field change? The potential difference? The total energy? Explain.

Q24.8 Two parallel-plate capacitors, identical except that one has twice the plate separation of the other, are charged by the same voltage source. Which capacitor has a stronger electric field between the plates? Which capacitor has a greater charge? Which has greater energy density? Explain your reasoning.

Q24.9 The charged plates of a capacitor attract each other, so to pull the plates farther apart requires work by some external force. What becomes of the energy added by this work? Explain.

Q24.10 You have two capacitors and want to connect them across a voltage source (battery) to store the maximum amount of energy. Should they be connected in series or in parallel?

Q24.11 As shown in Table 24.1, water has a very large dielectric constant $K = 80.4$. Why do you think water is not commonly used as a dielectric in capacitors?

Q24.12 Is dielectric strength the same thing as dielectric constant? Explain any differences between the two quantities. Is there a simple relationship between dielectric strength and dielectric constant (see Table 24.2)?

Q24.13 A capacitor made of aluminum foil strips separated by Mylar film was subjected to excessive voltage, and the resulting dielectric breakdown melted holes in the Mylar. After this, the capacitance was found to be about the same as before, but the breakdown voltage was much less. Why?

Q24.14 Suppose you bring a slab of dielectric close to the gap between the plates of a charged capacitor, preparing to slide it between the plates. What force will you feel? What does this force tell you about the energy stored between the plates once the dielectric is in place, compared to before the dielectric is in place?

Q24.15 The freshness of fish can be measured by placing a fish between the plates of a capacitor and measuring the capacitance. How does this work? (*Hint:* As time passes, the fish dries out. See Table 24.1.)

Q24.16 *Electrolytic* capacitors use as their dielectric an extremely thin layer of nonconducting oxide between a metal plate and a conducting solution. Discuss the advantage of such a capacitor over one constructed using a solid dielectric between the metal plates.

Q24.17 In terms of the dielectric constant K, what happens to the electric flux through the Gaussian surface shown in Fig. 24.22 when the dielectric is inserted into the previously empty space between the plates? Explain.

Q24.18 A parallel-plate capacitor is connected to a power supply that maintains a fixed potential difference between the plates. (a) If a sheet of dielectric is then slid between the plates, what happens to (i) the electric field between the plates, (ii) the magnitude of charge on each plate, and (iii) the energy stored in the capacitor? (b) Now suppose that before the dielectric is inserted, the charged capacitor is disconnected from the power supply. In this case, what happens to (i) the electric field between the plates, (ii) the magnitude of charge on each plate, and (iii) the energy stored in the capacitor? Explain any differences between the two situations.

Q24.19 Liquid dielectrics that have polar molecules (such as water) always have dielectric constants that decrease with increasing temperature. Why?

Q24.20 A conductor is an extreme case of a dielectric, since if an electric field is applied to a conductor, charges are free to move within the conductor to set up "induced charges." What is the dielectric constant of a perfect conductor? Is it $K = 0$, $K \rightarrow \infty$, or something in between? Explain your reasoning.

Q24.21 The two plates of a capacitor are given charges $\pm Q$. The capacitor is then disconnected from the charging device so that the charges on the plates can't change, and the capacitor is immersed in a tank of oil. Does the electric field between the plates increase, decrease, or stay the same? Explain your reasoning. How can this field be measured?

EXERCISES

Section 24.1 Capacitors and Capacitance

24.1 • The plates of a parallel-plate capacitor are 2.50 mm apart, and each carries a charge of magnitude 80.0 nC. The plates are in vacuum. The electric field between the plates has a magnitude of 4.00×10^6 V/m. What is (a) the potential difference between the plates; (b) the area of each plate; (c) the capacitance?

24.2 • The plates of a parallel-plate capacitor are 3.28 mm apart, and each has an area of 9.82 cm². Each plate carries a charge of magnitude 4.35×10^{-8} C. The plates are in vacuum. What is (a) the capacitance; (b) the potential difference between the plates; (c) the magnitude of the electric field between the plates?

24.3 • A parallel-plate air capacitor of capacitance 245 pF has a charge of magnitude 0.148 μC on each plate. The plates are 0.328 mm apart. (a) What is the potential difference between the plates? (b) What is the area of each plate? (c) What is the electric-field magnitude between the plates? (d) What is the surface charge density on each plate?

24.4 •• Cathode-ray-tube oscilloscopes have parallel metal plates inside them to deflect the electron beam. These plates are called the *deflecting plates*. Typically, they are squares 3.0 cm on a side and separated by 5.0 mm, with vacuum in between. What is the capacitance of these deflecting plates and hence of the oscilloscope? (*Note:* This capacitance can sometimes have an effect on the circuit you are trying to study and must be taken into consideration in your calculations.)

24.5 • A 10.0-μF parallel-plate capacitor with circular plates is connected to a 12.0-V battery. (a) What is the charge on each plate? (b) How much charge would be on the plates if their separation were doubled while the capacitor remained connected to the battery? (c) How much charge would be on the plates if the capacitor were connected to the 12.0-V battery after the radius of each plate was doubled without changing their separation?

24.6 • A 5.00-μF parallel-plate capacitor is connected to a 12.0-V battery. After the capacitor is fully charged, the battery is disconnected without loss of any of the charge on the plates. (a) A voltmeter is connected across the two plates without discharging them. What does it read? (b) What would the voltmeter read if (i) the plate separation were doubled; (ii) the radius of each plate were doubled but their separation was unchanged?

24.7 • A parallel-plate air capacitor is to store charge of magnitude 240.0 pC on each plate when the potential difference between the plates is 42.0 V. (a) If the area of each plate is 6.80 cm², what is the separation between the plates? (b) If the separation between the two plates is double the value calculated in part (a), what potential difference is required for the capacitor to store charge of magnitude 240.0 pC on each plate?

24.8 • A 5.00-pF, parallel-plate, air-filled capacitor with circular plates is to be used in a circuit in which it will be subjected to potentials of up to 1.00×10^2 V. The electric field between the plates is to be no greater than 1.00×10^4 N/C. As a budding electrical engineer for Live-Wire Electronics, your tasks are to (a) design the capacitor by finding what its physical dimensions and separation must be; (b) find the maximum charge these plates can hold.

24.9 • A capacitor is made from two hollow, coaxial, iron cylinders, one inside the other. The inner cylinder is negatively charged and the outer is positively charged; the magnitude of the charge on each is 10.0 pC. The inner cylinder has radius 0.50 mm, the outer one has radius 5.00 mm, and the length of each cylinder is 18.0 cm. (a) What is the capacitance? (b) What applied potential difference is necessary to produce these charges on the cylinders?

24.10 • A cylindrical capacitor consists of a solid inner conducting core with radius 0.250 cm, surrounded by an outer hollow conducting tube. The two conductors are separated by air, and the length of the cylinder is 12.0 cm. The capacitance is 36.7 pF. (a) Calculate the inner radius of the hollow tube. (b) When the capacitor is charged to 125 V, what is the charge per unit length λ on the capacitor?

24.11 •• A spherical capacitor contains a charge of 3.30 nC when connected to a potential difference of 220 V. If its plates are separated by vacuum and the inner radius of the outer shell is 4.00 cm, calculate: (a) the capacitance; (b) the radius of the inner sphere; (c) the electric field just outside the surface of the inner sphere.

24.12 •• A cylindrical capacitor has an inner conductor of radius 2.2 mm and an outer conductor of radius 3.5 mm. The two conductors are separated by vacuum, and the entire capacitor is 2.8 m long. (a) What is the capacitance per unit length? (b) The potential of the inner conductor is 350 mV higher than that of the outer conductor. Find the charge (magnitude and sign) on both conductors.

24.13 • A spherical capacitor is formed from two concentric, spherical, conducting shells separated by vacuum. The inner sphere has radius 15.0 cm and the capacitance is 116 pF. (a) What is the radius of the outer sphere? (b) If the potential difference between the two spheres is 220 V, what is the magnitude of charge on each sphere?

Section 24.2 Capacitors in Series and Parallel

24.14 • **Figure E24.14** shows a system of four capacitors, where the potential difference across *ab* is 50.0 V. (a) Find the equivalent capacitance of this system between *a* and *b*. (b) How much charge is stored by this combination of capacitors? (c) How much charge is stored in each of the 10.0-μF and the 9.0-μF capacitors?

Figure **E24.14**

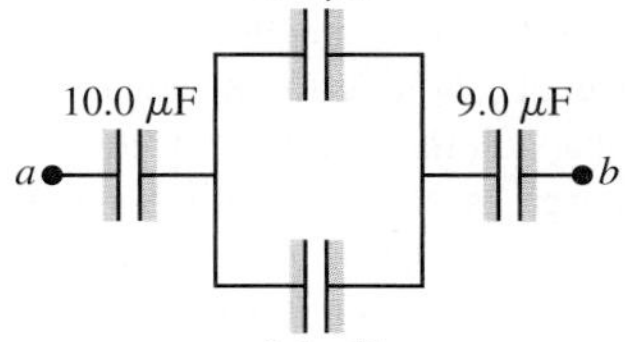

5.0 μF

10.0 μF   9.0 μF

a   *b*

8.0 μF

24.15 • **BIO** **Electric Eels.** Electric eels and electric fish generate large potential differences that are used to stun enemies and prey. These potentials are produced by cells that each can generate 0.10 V. We can plausibly model such cells as charged capacitors. (a) How should these cells be connected (in series or in parallel) to produce a total potential of more than 0.10 V? (b) Using the connection in part (a), how many cells must be connected together to produce the 500-V surge of the electric eel?

24.16 • For the system of capacitors shown in **Fig. E24.16**, find the equivalent capacitance (a) between *b* and *c*, and (b) between *a* and *c*.

Figure **E24.16**

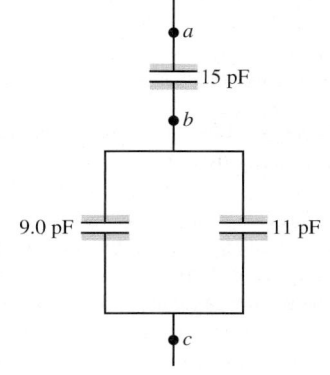

a

15 pF

b

9.0 pF   11 pF

c

24.17 • In **Fig. E24.17**, each capacitor has $C = 4.00\ \mu F$ and $V_{ab} = +28.0$ V. Calculate (a) the charge on each capacitor; (b) the potential difference across each capacitor; (c) the potential difference between points a and d.

Figure **E24.17**

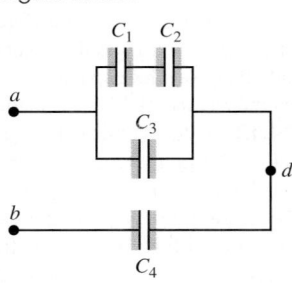

24.18 • In Fig. 24.8a, let $C_1 = 3.00\ \mu F$, $C_2 = 5.00\ \mu F$, and $V_{ab} = +64.0$ V. Calculate (a) the charge on each capacitor and (b) the potential difference across each capacitor.

24.19 • In Fig. 24.9a, let $C_1 = 3.00\ \mu F$, $C_2 = 5.00\ \mu F$, and $V_{ab} = +52.0$ V. Calculate (a) the charge on each capacitor and (b) the potential difference across each capacitor.

24.20 • In **Fig. E24.20**, $C_1 = 6.00\ \mu F$, $C_2 = 3.00\ \mu F$, and $C_3 = 5.00\ \mu F$. The capacitor network is connected to an applied potential V_{ab}. After the charges on the capacitors have reached their final values, the charge on C_2 is $30.0\ \mu C$. (a) What are the charges on capacitors C_1 and C_3? (b) What is the applied voltage V_{ab}?

Figure **E24.20**

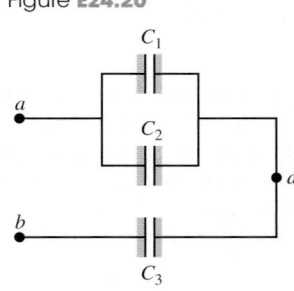

24.21 •• For the system of capacitors shown in **Fig. E24.21**, a potential difference of 25 V is maintained across ab. (a) What is the equivalent capacitance of this system between a and b? (b) How much charge is stored by this system? (c) How much charge does the 6.5-nF capacitor store? (d) What is the potential difference across the 7.5-nF capacitor?

Figure **E24.21**

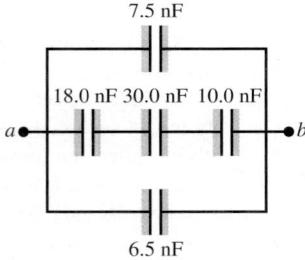

24.22 •• Suppose the 3-μF capacitor in Fig. 24.10a were removed and replaced by a different one, and that this changed the equivalent capacitance between points a and b to $8\ \mu F$. What would be the capacitance of the replacement capacitor?

Section 24.3 Energy Storage in Capacitors and Electric-Field Energy

24.23 • A 5.80-μF, parallel-plate, air capacitor has a plate separation of 5.00 mm and is charged to a potential difference of 400 V. Calculate the energy density in the region between the plates, in units of J/m^3.

24.24 • A parallel-plate air capacitor has a capacitance of 920 pF. The charge on each plate is $3.90\ \mu C$. (a) What is the potential difference between the plates? (b) If the charge is kept constant, what will be the potential difference if the plate separation is doubled? (c) How much work is required to double the separation?

24.25 • An air capacitor is made from two flat parallel plates 1.50 mm apart. The magnitude of charge on each plate is $0.0180\ \mu C$ when the potential difference is 200 V. (a) What is the capacitance? (b) What is the area of each plate? (c) What maximum voltage can be applied without dielectric breakdown? (Dielectric breakdown for air occurs at an electric-field strength of 3.0×10^6 V/m.) (d) When the charge is $0.0180\ \mu C$, what total energy is stored?

24.26 •• A parallel-plate vacuum capacitor has 8.38 J of energy stored in it. The separation between the plates is 2.30 mm. If the separation is decreased to 1.15 mm, what is the energy stored (a) if the capacitor is disconnected from the potential source so the charge on the plates remains constant, and (b) if the capacitor remains connected to the potential source so the potential difference between the plates remains constant?

24.27 • You have two identical capacitors and an external potential source. (a) Compare the total energy stored in the capacitors when they are connected to the applied potential in series and in parallel. (b) Compare the maximum amount of charge stored in each case. (c) Energy storage in a capacitor can be limited by the maximum electric field between the plates. What is the ratio of the electric field for the series and parallel combinations?

24.28 • For the capacitor network shown in **Fig. E24.28**, the potential difference across ab is 48 V. Find (a) the total charge stored in this network; (b) the charge on each capacitor; (c) the total energy stored in the network; (d) the energy stored in each capacitor; (e) the potential differences across each capacitor.

Figure **E24.28**

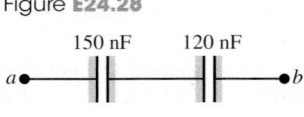

24.29 • For the capacitor network shown in **Fig. E24.29**, the potential difference across ab is 220 V. Find (a) the total charge stored in this network; (b) the charge on each capacitor; (c) the total energy stored in the network; (d) the energy stored in each capacitor; (e) the potential difference across each capacitor.

Figure **E24.29**

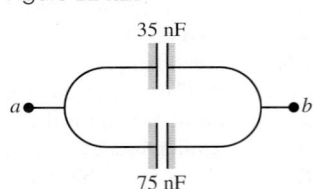

24.30 • A 0.350-m-long cylindrical capacitor consists of a solid conducting core with a radius of 1.20 mm and an outer hollow conducting tube with an inner radius of 2.00 mm. The two conductors are separated by air and charged to a potential difference of 6.00 V. Calculate (a) the charge per length for the capacitor; (b) the total charge on the capacitor; (c) the capacitance; (d) the energy stored in the capacitor when fully charged.

24.31 • A cylindrical air capacitor of length 15.0 m stores 3.20×10^{-9} J of energy when the potential difference between the two conductors is 4.00 V. (a) Calculate the magnitude of the charge on each conductor. (b) Calculate the ratio of the radii of the inner and outer conductors.

24.32 •• A capacitor is formed from two concentric spherical conducting shells separated by vacuum. The inner sphere has radius 12.5 cm, and the outer sphere has radius 14.8 cm. A potential difference of 120 V is applied to the capacitor. (a) What is the energy density at $r = 12.6$ cm, just outside the inner sphere? (b) What is the energy density at $r = 14.7$ cm, just inside the outer sphere? (c) For a parallel-plate capacitor the energy density is uniform in the region between the plates, except near the edges of the plates. Is this also true for a spherical capacitor?

Section 24.4 Dielectrics

24.33 • A 12.5-μF capacitor is connected to a power supply that keeps a constant potential difference of 24.0 V across the plates. A piece of material having a dielectric constant of 3.75 is placed between the plates, completely filling the space between them. (a) How much energy is stored in the capacitor before and after the dielectric is inserted? (b) By how much did the energy change during the insertion? Did it increase or decrease?

24.34 • A parallel-plate capacitor has capacitance $C_0 = 8.00$ pF when there is air between the plates. The separation between the plates is 1.50 mm. (a) What is the maximum magnitude of charge Q that can be placed on each plate if the electric field in the region between the plates is not to exceed 3.00×10^4 V/m? (b) A dielectric with $K = 2.70$ is inserted between the plates of the capacitor, completely filling the volume between the plates. Now what is the maximum magnitude of charge on each plate if the electric field between the plates is not to exceed 3.00×10^4 V/m?

24.35 • Two parallel plates have equal and opposite charges. When the space between the plates is evacuated, the electric field is $E = 3.20 \times 10^5$ V/m. When the space is filled with dielectric, the electric field is $E = 2.50 \times 10^5$ V/m. (a) What is the charge density on each surface of the dielectric? (b) What is the dielectric constant?

24.36 • A budding electronics hobbyist wants to make a simple 1.0-nF capacitor for tuning her crystal radio, using two sheets of aluminum foil as plates, with a few sheets of paper between them as a dielectric. The paper has a dielectric constant of 3.0, and the thickness of one sheet of it is 0.20 mm. (a) If the sheets of paper measure 22 × 28 cm and she cuts the aluminum foil to the same dimensions, how many sheets of paper should she use between her plates to get the proper capacitance? (b) Suppose for convenience she wants to use a single sheet of posterboard, with the same dielectric constant but a thickness of 12.0 mm, instead of the paper. What area of aluminum foil will she need for her plates to get her 1.0 nF of capacitance? (c) Suppose she goes high-tech and finds a sheet of Teflon of the same thickness as the posterboard to use as a dielectric. Will she need a larger or smaller area of Teflon than of posterboard? Explain.

24.37 • The dielectric to be used in a parallel-plate capacitor has a dielectric constant of 3.60 and a dielectric strength of 1.60×10^7 V/m. The capacitor is to have a capacitance of 1.25×10^{-9} F and must be able to withstand a maximum potential difference of 5500 V. What is the minimum area the plates of the capacitor may have?

24.38 •• **BIO Potential in Human Cells.** Some cell walls in the human body have a layer of negative charge on the inside surface and a layer of positive charge of equal magnitude on the outside surface. Suppose that the charge density on either surface is $\pm 0.50 \times 10^{-3}$ C/m^2, the cell wall is 5.0 nm thick, and the cell-wall material is air. (a) Find the magnitude of $\vec{E}$ in the wall between the two layers of charge. (b) Find the potential difference between the inside and the outside of the cell. Which is at the higher potential? (c) A typical cell in the human body has a volume of 10^{-16} m^3. Estimate the total electric-field energy stored in the wall of a cell of this size. (*Hint:* Assume that the cell is spherical, and calculate the volume of the cell wall.) (d) In reality, the cell wall is made up, not of air, but of tissue with a dielectric constant of 5.4. Repeat parts (a) and (b) in this case.

24.39 • A constant potential difference of 12 V is maintained between the terminals of a 0.25-μF, parallel-plate, air capacitor. (a) A sheet of Mylar is inserted between the plates of the capacitor, completely filling the space between the plates. When this is done, how much additional charge flows onto the positive plate of the capacitor (see Table 24.1)? (b) What is the total induced charge on either face of the Mylar sheet? (c) What effect does the Mylar sheet have on the electric field between the plates? Explain how you can reconcile this with the increase in charge on the plates, which acts to *increase* the electric field.

24.40 •• Polystyrene has dielectric constant 2.6 and dielectric strength 2.0×10^7 V/m. A piece of polystyrene is used as a dielectric in a parallel-plate capacitor, filling the volume between the plates. (a) When the electric field between the plates is 80% of the dielectric strength, what is the energy density of the stored energy? (b) When the capacitor is connected to a battery with voltage 500.0 V, the electric field between the plates is 80% of the dielectric strength. What is the area of each plate if the capacitor stores 0.200 mJ of energy under these conditions?

24.41 • When a 360-nF air capacitor (1 nF = 10^{-9} F) is connected to a power supply, the energy stored in the capacitor is 1.85×10^{-5} J. While the capacitor is kept connected to the power supply, a slab of dielectric is inserted that completely fills the space between the plates. This increases the stored energy by 2.32×10^{-5} J. (a) What is the potential difference between the capacitor plates? (b) What is the dielectric constant of the slab?

24.42 • A parallel-plate capacitor has capacitance $C = 12.5$ pF when the volume between the plates is filled with air. The plates are circular, with radius 3.00 cm. The capacitor is connected to a battery, and a charge of magnitude 25.0 pC goes onto each plate. With the capacitor still connected to the battery, a slab of dielectric is inserted between the plates, completely filling the space between the plates. After the dielectric has been inserted, the charge on each plate has magnitude 45.0 pC. (a) What is the dielectric constant K of the dielectric? (b) What is the potential difference between the plates before and after the dielectric has been inserted? (c) What is the electric field at a point midway between the plates before and after the dielectric has been inserted?

Section 24.6 Gauss's Law in Dielectrics

24.43 • A parallel-plate capacitor has the volume between its plates filled with plastic with dielectric constant K. The magnitude of the charge on each plate is Q. Each plate has area A, and the distance between the plates is d. (a) Use Gauss's law as stated in Eq. (24.23) to calculate the magnitude of the electric field in the dielectric. (b) Use the electric field determined in part (a) to calculate the potential difference between the two plates. (c) Use the result of part (b) to determine the capacitance of the capacitor. Compare your result to Eq. (24.12).

24.44 • A parallel-plate capacitor has plates with area 0.0225 m^2 separated by 1.00 mm of Teflon. (a) Calculate the charge on the plates when they are charged to a potential difference of 12.0 V. (b) Use Gauss's law (Eq. 24.23) to calculate the electric field inside the Teflon. (c) Use Gauss's law to calculate the electric field if the voltage source is disconnected and the Teflon is removed.

PROBLEMS

24.45 • Electronic flash units for cameras contain a capacitor for storing the energy used to produce the flash. In one such unit, the flash lasts for $\frac{1}{675}$ s with an average light power output of 2.70×10^5 W. (a) If the conversion of electrical energy to light is 95% efficient (the rest of the energy goes to thermal energy), how much energy must be stored in the capacitor for one flash? (b) The capacitor has a potential difference between its plates of 125 V when the stored energy equals the value calculated in part (a). What is the capacitance?

24.46 • A parallel-plate air capacitor is made by using two plates 12 cm square, spaced 3.7 mm apart. It is connected to a 12-V battery. (a) What is the capacitance? (b) What is the charge on each plate? (c) What is the electric field between the plates? (d) What is the energy stored in the capacitor? (e) If the battery is disconnected and then the plates are pulled apart to a separation of 7.4 mm, what are the answers to parts (a)–(d)?

24.47 ••• In one type of computer keyboard, each key holds a small metal plate that serves as one plate of a parallel-plate, air-filled capacitor. When the key is depressed, the plate separation decreases and the capacitance increases. Electronic circuitry detects the change in capacitance and thus detects that the key has been pressed. In one particular keyboard, the area of each metal plate is 42.0 mm², and the separation between the plates is 0.700 mm before the key is depressed. (a) Calculate the capacitance before the key is depressed. (b) If the circuitry can detect a change in capacitance of 0.250 pF, how far must the key be depressed before the circuitry detects its depression?

24.48 ••• BIO **Cell Membranes.** Cell membranes (the walled enclosure around a cell) are typically about 7.5 nm thick. They are partially permeable to allow charged material to pass in and out, as needed. Equal but opposite charge densities build up on the inside and outside faces of such a membrane, and these charges prevent additional charges from passing through the cell wall. We can model a cell membrane as a parallel-plate capacitor, with the membrane itself containing proteins embedded in an organic material to give the membrane a dielectric constant of about 10. (See **Fig. P24.48.**) (a) What is the capacitance per square centimeter of such a cell wall? (b) In its normal resting state, a cell has a potential difference of 85 mV across its membrane. What is the electric field inside this membrane?

Figure **P24.48**

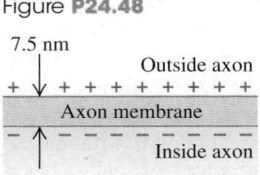

24.49 •• A 20.0-μF capacitor is charged to a potential difference of 800 V. The terminals of the charged capacitor are then connected to those of an uncharged 10.0-μF capacitor. Compute (a) the original charge of the system, (b) the final potential difference across each capacitor, (c) the final energy of the system, and (d) the decrease in energy when the capacitors are connected.

24.50 •• In Fig. 24.9a, let $C_1 = 9.0\ \mu F$, $C_2 = 4.0\ \mu F$, and $V_{ab} = 64$ V. Suppose the charged capacitors are disconnected from the source and from each other, and then reconnected to each other with plates of *opposite* sign together. By how much does the energy of the system decrease?

24.51 • For the capacitor network shown in **Fig. P24.51**, the potential difference across ab is 12.0 V. Find (a) the total energy stored in this network and (b) the energy stored in the 4.80-μF capacitor.

Figure **P24.51**

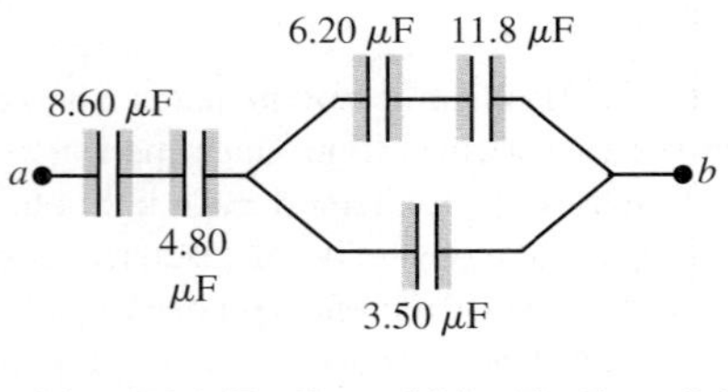

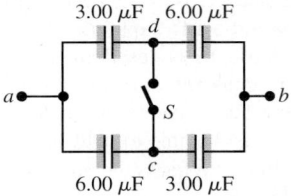

24.52 •• In Fig. E24.17, $C_1 = 6.00\ \mu F$, $C_2 = 3.00\ \mu F$, $C_3 = 4.00\ \mu F$, and $C_4 = 8.00\ \mu F$. The capacitor network is connected to an applied potential difference V_{ab}. After the charges on the capacitors have reached their final values, the voltage across C_3 is 40.0 V. What are (a) the voltages across C_1 and C_2, (b) the voltage across C_4, and (c) the voltage V_{ab} applied to the network?

24.53 • In **Fig. P24.53**, $C_1 = C_5 = 8.4\ \mu F$ and $C_2 = C_3 = C_4 = 4.2\ \mu F$. The applied potential is $V_{ab} = 220$ V. (a) What is the equivalent capacitance of the network between points a and b? (b) Calculate the charge on each capacitor and the potential difference across each capacitor.

Figure **P24.53**

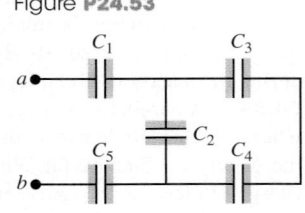

24.54 •• Current materials-science technology allows engineers to construct capacitors with much higher values of C than were previously possible. A capacitor has $C = 3000$ F and is rated to withstand a maximum potential difference of 2.7 V. The cylindrical capacitor has diameter 6.0 cm and length 13.5 cm. (a) Find the maximum electric potential energy that can be stored in this capacitor. (b) Does your value in part (a) agree with the 3.0-Wh value printed on the capacitor? (c) What is the maximum attainable energy density in this capacitor? (d) Compare this maximum energy density to the maximum possible energy density for polyester (see Table 24.2).

24.55 •• In Fig. E24.20, $C_1 = 3.00\ \mu F$ and $V_{ab} = 150$ V. The charge on capacitor C_1 is 150 μC and the charge on C_3 is 450 μC. What are the values of the capacitances of C_2 and C_3?

24.56 • The capacitors in **Fig. P24.56** are initially uncharged and are connected, as in the diagram, with switch S open. The applied potential difference is $V_{ab} = +210$ V. (a) What is the potential difference V_{cd}? (b) What is the potential difference across each capacitor after switch S is closed? (c) How much charge flowed through the switch when it was closed?

Figure **P24.56**

24.57 •• Three capacitors having capacitances of 8.4, 8.4, and 4.2 μF are connected in series across a 36-V potential difference. (a) What is the charge on the 4.2-μF capacitor? (b) What is the total energy stored in all three capacitors? (c) The capacitors are disconnected from the potential difference without allowing them to discharge. They are then reconnected in parallel with each other, with the positively charged plates connected together. What is the voltage across each capacitor in the parallel combination? (d) What is the total energy now stored in the capacitors?

24.58 • **Capacitance of a Thundercloud.** The charge center of a thundercloud, drifting 3.0 km above the earth's surface, contains 20 C of negative charge. Assuming the charge center has a radius of 1.0 km, and modeling the charge center and the earth's surface as parallel plates, calculate: (a) the capacitance of the system; (b) the potential difference between charge center and ground; (c) the average strength of the electric field between cloud and ground; (d) the electrical energy stored in the system.

24.59 •• In **Fig. P24.59**, each capacitance C_1 is 6.9 μF, and each capacitance C_2 is 4.6 μF. (a) Compute the equivalent capacitance of the network between points a and b. (b) Compute the charge on each of the three capacitors nearest a and b when $V_{ab} = 420$ V. (c) With 420 V across a and b, compute V_{cd}.

Figure **P24.59**

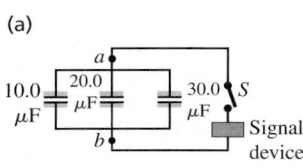

24.60 • Each combination of capacitors between points a and b in **Fig. P24.60** is first connected across a 120-V battery, charging the combination to 120 V. These combinations are then connected to make the circuits shown. When the switch S is thrown, a surge of charge for the discharging capacitors flows to trigger the signal device. How much charge flows through the signal device in each case?

Figure **P24.60**

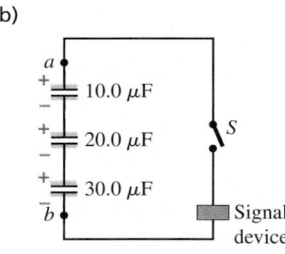

24.61 • A parallel-plate capacitor with only air between the plates is charged by connecting it to a battery. The capacitor is then disconnected from the battery, without any of the charge leaving the plates. (a) A voltmeter reads 45.0 V when placed across the capacitor. When a dielectric is inserted between the plates, completely filling the space, the voltmeter reads 11.5 V. What is the dielectric constant of this material? (b) What will the voltmeter read if the dielectric is now pulled partway out so it fills only one-third of the space between the plates?

24.62 •• An air capacitor is made by using two flat plates, each with area A, separated by a distance d. Then a metal slab having thickness a (less than d) and the same shape and size as the plates is inserted between them, parallel to the plates and not touching either plate (**Fig. P24.62**). (a) What is the capacitance of this arrangement? (b) Express the capacitance as a multiple of the capacitance C_0 when the metal slab is not present. (c) Discuss what happens to the capacitance in the limits $a \to 0$ and $a \to d$.

Figure **P24.62**

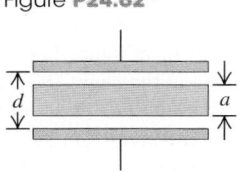

24.63 •• A potential difference $V_{ab} = 48.0$ V is applied across the capacitor network of Fig. E24.17. If $C_1 = C_2 = 4.00$ μF and $C_4 = 8.00$ μF, what must the capacitance C_3 be if the network is to store 2.90×10^{-3} J of electrical energy?

24.64 • CALC The inner cylinder of a long, cylindrical capacitor has radius r_a and linear charge density $+\lambda$. It is surrounded by a coaxial cylindrical conducting shell with inner radius r_b and linear charge density $-\lambda$ (see Fig. 24.6). (a) What is the energy density in the region between the conductors at a distance r from the axis? (b) Integrate the energy density calculated in part (a) over the volume between the conductors in a length L of the capacitor to obtain the total electric-field energy per unit length. (c) Use Eq. (24.9) and the capacitance per unit length calculated in Example 24.4 (Section 24.1) to calculate U/L. Does your result agree with that obtained in part (b)?

24.65 •• A parallel-plate capacitor has square plates that are 8.00 cm on each side and 3.80 mm apart. The space between the plates is completely filled with two square slabs of dielectric, each 8.00 cm on a side and 1.90 mm thick. One slab is Pyrex glass and the other is polystyrene. If the potential difference between the plates is 86.0 V, how much electrical energy is stored in the capacitor?

24.66 •• A parallel-plate capacitor is made from two plates 12.0 cm on each side and 4.50 mm apart. Half of the space between these plates contains only air, but the other half is filled with Plexiglas® of dielectric constant 3.40 (**Fig. P24.66**). An 18.0-V battery is connected across the plates. (a) What is the capacitance of this combination? (*Hint:* Can you think of this capacitor as equivalent to two capacitors in parallel?) (b) How much energy is stored in the capacitor? (c) If we remove the Plexiglas® but change nothing else, how much energy will be stored in the capacitor?

Figure **P24.66**

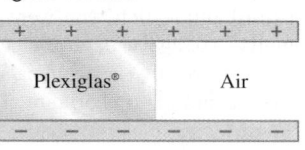

24.67 •• Three square metal plates A, B, and C, each 12.0 cm on a side and 1.50 mm thick, are arranged as in **Fig. P24.67**. The plates are separated by sheets of paper 0.45 mm thick and with dielectric constant 4.2. The outer plates are connected together and connected to point b. The inner plate is connected to point a. (a) Copy the diagram and show by plus and minus signs the charge distribution on the plates when point a is maintained at a positive potential relative to point b. (b) What is the capacitance between points a and b?

Figure **P24.67**

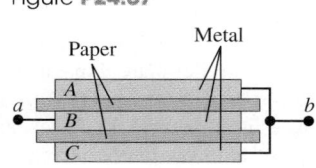

24.68 •• A fuel gauge uses a capacitor to determine the height of the fuel in a tank. The effective dielectric constant K_{eff} changes from a value of 1 when the tank is empty to a value of K, the dielectric constant of the fuel, when the tank is full. The appropriate electronic circuitry can determine the effective dielectric constant of the combined air and fuel between the capacitor plates. Each of the two rectangular plates has a width w and a length L (**Fig. P24.68**). The height of the fuel between the plates is h. You can ignore any fringing effects. (a) Derive an expression for K_{eff} as a function of h. (b) What is the effective dielectric constant for a tank $\frac{1}{4}$ full, $\frac{1}{2}$ full, and $\frac{3}{4}$ full if the fuel is gasoline ($K = 1.95$)? (c) Repeat part (b) for methanol ($K = 33.0$). (d) For which fuel is this fuel gauge more practical?

Figure **P24.68**

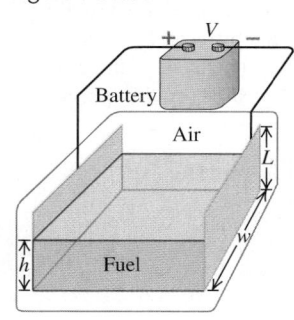

24.69 •• DATA Your electronics company has several identical capacitors with capacitance C_1 and several others with capacitance C_2. You must determine the values of C_1 and C_2 but don't have access to C_1 and C_2 individually. Instead, you have a network with C_1 and C_2 connected in series and a network with C_1 and C_2 connected in parallel. You have a 200.0-V battery and instrumentation that measures the total energy supplied by the battery when it is connected to the network. When the parallel combination is connected to the battery, 0.180 J of energy is stored in the network. When the series combination is connected, 0.0400 J of energy is stored. You are told that C_1 is greater than C_2. (a) Calculate C_1 and C_2. (b) For the series combination, does C_1 or C_2 store more charge, or are the values equal? Does C_1 or C_2 store more energy, or are the values equal? (c) Repeat part (b) for the parallel combination.

24.70 •• DATA You are designing capacitors for various applications. For one application, you want the maximum possible stored energy. For another, you want the maximum stored charge. For a third application, you want the capacitor to withstand a large applied voltage without dielectric breakdown. You start with an air-filled parallel-plate capacitor that has $C_0 = 6.00$ pF and a plate separation of 2.50 mm. You then consider the use of each of the dielectric materials listed in Table 24.2. In each application, the dielectric will fill the volume between the plates, and the electric field between the plates will be 50% of the dielectric strength given in the table. (a) For each of the five materials given in the table, calculate the energy stored in the capacitor. Which dielectric allows the maximum stored energy? (b) For each material, what is the charge Q stored on each plate of the capacitor? (c) For each material, what is the voltage applied across the capacitor? (d) Is one dielectric material in the table your best choice for all three applications?

24.71 •• DATA You are conducting experiments with an air-filled parallel-plate capacitor. You connect the capacitor to a battery with voltage 24.0 V. Initially the separation d between the plates is 0.0500 cm. In one experiment, you leave the battery connected to the capacitor, increase the separation between the plates, and measure the energy stored in the capacitor for each value of d. In a second experiment, you make the same measurements but disconnect the battery before you change the plate separation. One set of your data is given in **Fig. P24.71**, where you have plotted the stored energy U versus $1/d$. (a) For which experiment does this data set apply: the first (battery remains connected) or the second (battery disconnected before d is changed)? Explain. (b) Use the data plotted in Fig. P24.71 to calculate the area A of each plate. (c) For which case, the battery connected or the battery disconnected, is there more energy stored in the capacitor when $d = 0.400$ cm? Explain.

Figure **P24.71**

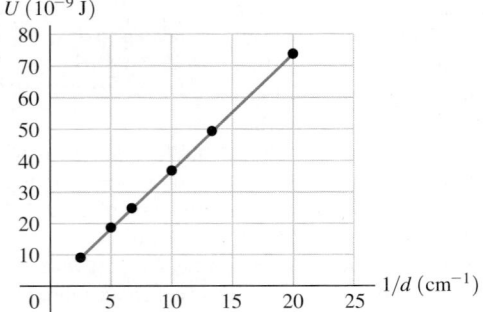

CHALLENGE PROBLEM

24.72 ••• Two square conducting plates with sides of length L are separated by a distance D. A dielectric slab with constant K with dimensions $L \times L \times D$ is inserted a distance x into the space between the plates, as shown in **Fig. P24.72**. (a) Find the capacitance C of this system. (b) Suppose that the capacitor is

connected to a battery that maintains a constant potential difference V between the plates. If the dielectric slab is inserted an additional distance dx into the space between the plates, show that the change in stored energy is

$$dU = +\frac{(K - 1)\epsilon_0 V^2 L}{2D} dx$$

(c) Suppose that before the slab is moved by dx, the plates are disconnected from the battery, so that the charges on the plates remain constant. Determine the magnitude of the charge on each plate, and then show that when the slab is moved dx farther into the space between the plates, the stored energy changes by an amount that is the *negative* of the expression for dU given in part (b). (d) If F is the force exerted on the slab by the charges on the plates, then dU should equal the work done *against* this force to move the slab a distance dx. Thus $dU = -F\,dx$. Show that applying this expression to the result of part (b) suggests that the electric force on the slab pushes it *out* of the capacitor, while the result of part (c) suggests that the force pulls the slab *into* the capacitor. (e) Figure 24.16 shows that the force in fact pulls the slab into the capacitor. Explain why the result of part (b) gives an incorrect answer for the direction of this force, and calculate the magnitude of the force. (This method does not require knowledge of the nature of the fringing field.)

Figure **P24.72**

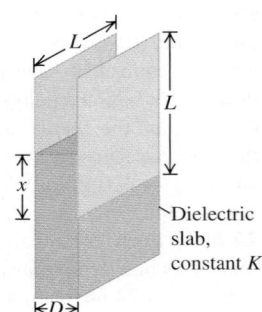

PASSAGE PROBLEMS

BIO **THE ELECTRIC EGG.** Upon fertilization, the eggs of many species undergo a rapid change in potential difference across their outer membrane. This change affects the physiological development of the eggs. The potential difference across the membrane is called the *membrane potential*, V_m, which is the potential inside the membrane minus the potential outside it. The membrane potential arises when enzymes use the energy available in ATP to expel three sodium ions (Na^+) actively and accumulate two potassium ions (K^+) inside the membrane—making the interior less positively charged than the exterior. For a sea urchin egg, V_m is about -70 mV; that is, the potential inside is 70 mV less than that outside. The egg membrane behaves as a capacitor with a capacitance of about 1 $\mu F/cm^2$. The membrane of the unfertilized egg is *selectively permeable* to K^+; that is, K^+ can readily pass through certain channels in the membrane, but other ions cannot. When a sea urchin egg is fertilized, Na^+ channels in the membrane open, Na^+ enters the egg, and V_m rapidly increases to $+30$ mV, where it remains for several minutes. The concentration of Na^+ is about 30 mmol/L in the egg's interior but 450 mmol/L in the surrounding seawater. The K^+ concentration is about 200 mmol/L inside but 10 mmol/L outside. A useful constant that connects electrical and chemical units is the *Faraday number*, which has a value of approximately 10^5 C/mol; that is, Avogadro's number (a mole) of monovalent ions, such as Na^+ or K^+, carries a charge of 10^5 C.

24.73 How many moles of Na^+ must m ove per unit area of membrane to change V_m from -70 mV to $+30$ mV, if we assume that the membrane behaves purely as a capacitor? (a) 10^{-4} mol/cm^2; (b) 10^{-9} mol/cm^2; (c) 10^{-12} mol/cm^2; (d) 10^{-14} mol/cm^2.

24.74 Suppose that the egg has a diameter of 200 μm. What fractional change in the internal Na$^+$ concentration results from the fertilization-induced change in V_m? Assume that Na$^+$ ions are distributed throughout the cell volume. The concentration increases by (a) 1 part in 10^4; (b) 1 part in 10^5; (c) 1 part in 10^6; (d) 1 part in 10^7.

24.75 Suppose that the change in V_m was caused by the entry of Ca^{2+} instead of Na$^+$. How many Ca^{2+} ions would have to enter the

cell per unit membrane to produce the change? (a) Half as many as for Na$^+$; (b) the same as for Na$^+$; (c) twice as many as for Na$^+$; (d) cannot say without knowing the inside and outside concentrations of Ca^{2+}.

24.76 What is the minimum amount of work that must be done by the cell to restore V_m to -70 mV? (a) 3 mJ; (b) 3 μJ; (c) 3 nJ; (d) 3 pJ.

Answers

Chapter Opening Question ?

(iv) Equation (24.9) shows that the energy stored in a capacitor with capacitance C and charge Q is $U = Q^2/2C$. If Q is doubled, the stored energy increases by a factor of $2^2 = 4$. Note that if the value of Q is too great, the electric-field magnitude inside the capacitor will exceed the dielectric strength of the material between the plates and dielectric breakdown will occur (see Section 24.4). This puts a practical limit on the amount of energy that can be stored.

Test Your Understanding Questions

24.1 (iii) The capacitance does not depend on the value of the charge Q. Doubling Q causes the potential difference V_{ab} to double, so the capacitance $C = Q/V_{ab}$ remains the same. These statements are true no matter what the geometry of the capacitor.

24.2 (a) (i), (b) (iv) In a series connection the two capacitors carry the same charge Q but have different potential differences $V_{ab} = Q/C$; the capacitor with the smaller capacitance C has the greater potential difference. In a parallel connection the two capacitors have the same potential difference V_{ab} but carry different charges $Q = CV_{ab}$; the capacitor with the larger capacitance C has the greater charge. Hence a 4-μF capacitor will have a greater potential difference than an 8-μF capacitor if the two are connected in series. The 4-μF capacitor cannot carry more charge than the 8-μF capacitor no matter how they are connected: In a series connection they will carry the same charge, and in a parallel connection the 8-μF capacitor will carry more charge.

24.3 (i) Capacitors connected in series carry the same charge Q. To compare the amount of energy stored, we use the expression $U = Q^2/2C$ from Eq. (24.9); it shows that the capacitor with the smaller capacitance ($C = 4\ \mu$F) has more stored energy in a series combination. By contrast, capacitors in parallel have the same potential difference V, so to compare them we use $U = \frac{1}{2}CV^2$ from

Eq. (24.9). It shows that in a parallel combination, the capacitor with the larger capacitance ($C = 8\ \mu$F) has more stored energy. (If we had instead used $U = \frac{1}{2}CV^2$ to analyze the series combination, we would have to account for the different potential differences across the two capacitors. Likewise, using $U = Q^2/2C$ to study the parallel combination would require us to account for the different charges on the capacitors.)

24.4 (i) Here Q remains the same, so we use $U = Q^2/2C$ from Eq. (24.9) for the stored energy. Removing the dielectric lowers the capacitance by a factor of $1/K$; since U is inversely proportional to C, the stored energy *increases* by a factor of K. It takes work to pull the dielectric slab out of the capacitor because the fringing field tries to pull the slab back in (Fig. 24.16). The work that you do goes into the energy stored in the capacitor.

24.5 (i), (iii), (ii) Equation (24.14) says that if E_0 is the initial electric-field magnitude (before the dielectric slab is inserted), then the resultant field magnitude after the slab is inserted is $E_0/K = E_0/3$. The magnitude of the resultant field equals the difference between the initial field magnitude and the magnitude E_i of the field due to the bound charges (see Fig. 24.20). Hence $E_0 - E_i = E_0/3$ and $E_i = 2E_0/3$.

24.6 (iii) Equation (24.23) shows that this situation is the same as an isolated point charge in vacuum but with $\vec{E}$ replaced by $K\vec{E}$. Hence KE at the point of interest is equal to $q/4\pi\epsilon_0 r^2$, and so $E = q/4\pi K\epsilon_0 r^2$. As in Example 24.12, filling the space with a dielectric reduces the electric field by a factor of $1/K$.

Bridging Problem

(a) 0 **(b)** $Q^2/32\pi^2\epsilon_0 r^4$ **(c)** $Q^2/8\pi\epsilon_0 R$
(d) $Q^2/8\pi\epsilon_0 R$ **(e)** $C = 4\pi\epsilon_0 R$

? In a flashlight, how does the amount of current that flows out of the bulb compare to the amount that flows into the bulb? (i) Current out is less than current in; (ii) current out is greater than current in; (iii) current out equals current in; (iv) the answer depends on the brightness of the bulb.

25 CURRENT, RESISTANCE, AND ELECTROMOTIVE FORCE

LEARNING GOALS

Looking forward at ...

25.1 The meaning of electric current, and how charges move in a conductor.

25.2 What is meant by the resistivity and conductivity of a substance.

25.3 How to calculate the resistance of a conductor from its dimensions and its resistivity.

25.4 How an electromotive force (emf) makes it possible for current to flow in a circuit.

25.5 How to do calculations involving energy and power in circuits.

25.6 How to use a simple model to understand the flow of current in metals.

Looking back at ...

17.7 Thermal conductivity.

23.2 Voltmeters, electric field, and electric potential.

24.4 Dielectric breakdown in insulators.

I n the past four chapters we studied the interactions of electric charges *at rest;* now we're ready to study charges *in motion.* An *electric current* consists of charges in motion from one region to another. If the charges follow a conducting path that forms a closed loop, the path is called an *electric circuit.*

Fundamentally, electric circuits are a means for conveying *energy* from one place to another. As charged particles move within a circuit, electric potential energy is transferred from a source (such as a battery or generator) to a device in which that energy is either stored or converted to another form: into sound in a stereo system or into heat and light in a toaster or light bulb. Electric circuits are useful because they allow energy to be transported without any moving parts (other than the moving charged particles themselves). They are at the heart of computers, television transmitters and receivers, and household and industrial power distribution systems. Your nervous system is a specialized electric circuit that carries vital signals from one part of your body to another.

In Chapter 26 we will see how to analyze electric circuits and will examine some practical applications of circuits. To prepare you for that, in this chapter we'll examine the basic properties of electric currents. We'll begin by describing the nature of electric conductors and considering how they are affected by temperature. We'll learn why a short, fat, cold copper wire is a better conductor than a long, skinny, hot steel wire. We'll study the properties of batteries and see how they cause current and energy transfer in a circuit. In this analysis we will use the concepts of current, potential difference (or voltage), resistance, and electromotive force. Finally, we'll look at electric current in a material from a microscopic viewpoint.

25.1 CURRENT

A **current** is any motion of charge from one region to another. In this section we'll discuss currents in conducting materials. The vast majority of technological applications of charges in motion involve currents of this kind.

In electrostatic situations (discussed in Chapters 21 through 24) the electric field is zero everywhere within the conductor, and there is *no* current. However, this does not mean that all charges within the conductor are at rest. In an ordinary metal such as copper or aluminum, some of the electrons are free to move within the conducting material. These free electrons move randomly in all directions, somewhat like the molecules of a gas but with much greater speeds, of the order of 10^6 m/s. The electrons nonetheless do not escape from the conducting material, because they are attracted to the positive ions of the material. The motion of the electrons is random, so there is no *net* flow of charge in any direction and hence no current.

Now consider what happens if a constant, steady electric field $\vec{E}$ is established inside a conductor. (We'll see later how this can be done.) A charged particle (such as a free electron) inside the conducting material is then subjected to a steady force $\vec{F} = q\vec{E}$. If the charged particle were moving in *vacuum,* this steady force would cause a steady acceleration in the direction of $\vec{F}$, and after a time the charged particle would be moving in that direction at high speed. But a charged particle moving in a *conductor* undergoes frequent collisions with the massive, nearly stationary ions of the material. In each such collision the particle's direction of motion undergoes a random change. The net effect of the electric field $\vec{E}$ is that in addition to the random motion of the charged particles within the conductor, there is also a very slow net motion or *drift* of the moving charged particles as a group in the direction of the electric force $\vec{F} = q\vec{E}$ (**Fig. 25.1**). This motion is described in terms of the **drift velocity** $\vec{v}_\mathrm{d}$ of the particles. As a result, there is a net current in the conductor.

While the random motion of the electrons has a very fast average speed of about 10^6 m/s, the drift speed is very slow, often on the order of 10^{-4} m/s. Given that the electrons move so slowly, you may wonder why the light comes on immediately when you turn on the switch of a flashlight. The reason is that the electric field is set up in the wire with a speed approaching the speed of light, and electrons start to move all along the wire at very nearly the same time. The time that it takes any individual electron to get from the switch to the light bulb isn't really relevant. A good analogy is a group of soldiers standing at attention when the sergeant orders them to start marching; the order reaches the soldiers' ears at the speed of sound, which is much faster than their marching speed, so all the soldiers start to march essentially in unison.

The Direction of Current Flow

The drift of moving charges through a conductor can be interpreted in terms of work and energy. The electric field $\vec{E}$ does work on the moving charges. The resulting kinetic energy is transferred to the material of the conductor by means of collisions with the ions, which vibrate about their equilibrium positions in the crystalline structure of the conductor. This energy transfer increases the average vibrational energy of the ions and therefore the temperature of the material. Thus much of the work done by the electric field goes into heating the conductor, *not* into making the moving charges move ever faster and faster. This heating is sometimes useful, as in an electric toaster, but in many situations is simply an unavoidable by-product of current flow.

In different current-carrying materials, the charges of the moving particles may be positive or negative. In metals the moving charges are always (negative) electrons, while in an ionized gas (plasma) or an ionic solution the moving charges may include both electrons and positively charged ions. In a semiconductor

25.1 If there is no electric field inside a conductor, an electron moves randomly from point P_1 to point P_2 in a time Δt. If an electric field $\vec{E}$ is present, the electric force $\vec{F} = q\vec{E}$ imposes a small drift (greatly exaggerated here) that takes the electron to point P'_2, a distance $v_\mathrm{d}\Delta t$ from P_2 in the direction of the force.

Conductor without internal $\vec{E}$ field

Path of electron without $\vec{E}$ field. Electron moves randomly.

Path of electron with $\vec{E}$ field. The motion is mostly random, but ...

P_1

P_2 P_2'

$v_\mathrm{d}\Delta t$

... the $\vec{E}$ field results in a net displacement along the wire.

Conductor with internal $\vec{E}$ field

$\vec{E}$ $\vec{F} = q\vec{E}$ $\vec{E}$

An electron has a negative charge q, so the force on it due to the $\vec{E}$ field is in the direction opposite to $\vec{E}$.

25.2 The same current is produced by (a) positive charges moving in the direction of the electric field $\vec{E}$ or (b) the same number of negative charges moving at the same speed in the direction opposite to $\vec{E}$.

(a)

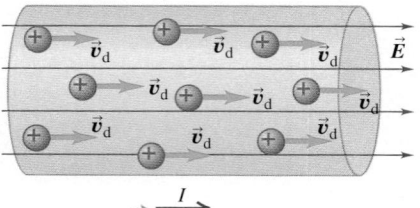

A **conventional current** is treated as a flow of positive charges, regardless of whether the free charges in the conductor are positive, negative, or both.

(b)

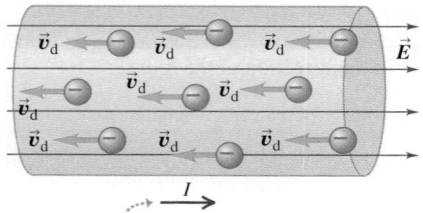

In a metallic conductor, the moving charges are electrons — but the *current* still points in the direction positive charges would flow.

25.3 The current I is the time rate of charge transfer through the cross-sectional area A. The random component of each moving charged particle's motion averages to zero, and the current is in the same direction as $\vec{E}$ whether the moving charges are positive (as shown here) or negative (see Fig. 25.2b).

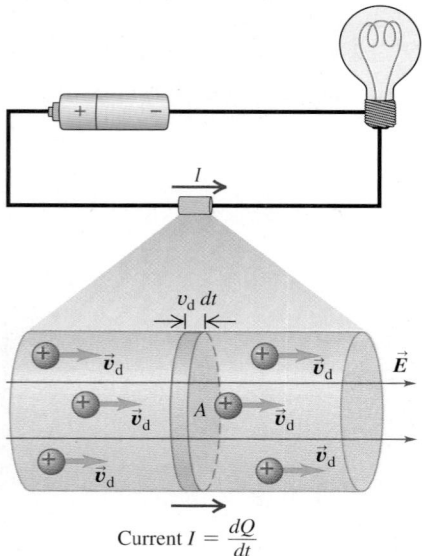

Current $I = \dfrac{dQ}{dt}$

material such as germanium or silicon, conduction is partly by electrons and partly by motion of *vacancies,* also known as *holes;* these are sites of missing electrons and act like positive charges.

Figure 25.2 shows segments of two different current-carrying materials. In Fig. 25.2a the moving charges are positive, the electric force is in the same direction as $\vec{E}$, and the drift velocity $\vec{v}_{\mathrm{d}}$ is from left to right. In Fig. 25.2b the charges are negative, the electric force is opposite to $\vec{E}$, and the drift velocity $\vec{v}_{\mathrm{d}}$ is from right to left. In both cases there is a net flow of positive charge from left to right, and positive charges end up to the right of negative ones. We *define* the current, denoted by I, to be in the direction in which there is a flow of *positive* charge. Thus we describe currents as though they consisted entirely of positive charge flow, even in cases in which we know that the actual current is due to electrons. Hence the current is to the right in both Figs. 25.2a and 25.2b. This choice or convention for the direction of current flow is called **conventional current.** While the direction of the conventional current is *not* necessarily the same as the direction in which charged particles are actually moving, we'll find that the sign of the moving charges is of little importance in analyzing electric circuits.

Figure 25.3 shows a segment of a conductor in which a current is flowing. We consider the moving charges to be *positive,* so they are moving in the same direction as the current. We define the current through the cross-sectional area A to be *the net charge flowing through the area per unit time.* Thus, if a net charge dQ flows through an area in a time dt, the current I through the area is

$$I = \frac{dQ}{dt} \qquad \text{(definition of current)} \qquad (25.1)$$

⬚ CAUTION **Current is not a vector** Although we refer to the *direction* of a current, current as defined by Eq. (25.1) is *not* a vector quantity. In a current-carrying wire, the current is always along the length of the wire, regardless of whether the wire is straight or curved. No single vector could describe motion along a curved path. We'll usually describe the direction of current either in words (as in "the current flows clockwise around the circuit") or by choosing a current to be positive if it flows in one direction along a conductor and negative if it flows in the other direction. ▌

The SI unit of current is the **ampere;** one ampere is defined to be *one coulomb per second* (1 A = 1 C/s). This unit is named in honor of the French scientist André Marie Ampère (1775–1836). When an ordinary flashlight (D-cell size) is turned on, the current in the flashlight is about 0.5 to 1 A; the current in the wires of a car engine's starter motor is around 200 A. Currents in radio and television circuits are usually expressed in *milliamperes* (1 mA = 10^{-3} A) or *microamperes* (1 μA = 10^{-6} A), and currents in computer circuits are expressed in *nanoamperes* (1 nA = 10^{-9} A) or *picoamperes* (1 pA = 10^{-12} A).

Current, Drift Velocity, and Current Density

We can express current in terms of the drift velocity of the moving charges. Let's reconsider the situation of Fig. 25.3 of a conductor with cross-sectional area A and an electric field $\vec{E}$ directed from left to right. To begin with, we'll assume that the free charges in the conductor are positive; then the drift velocity is in the same direction as the field.

Suppose there are n moving charged particles per unit volume. We call n the **concentration** of particles; its SI unit is m^{-3}. Assume that all the particles move with the same drift velocity with magnitude v_{d}. In a time interval dt, each particle moves a distance $v_{\mathrm{d}}\, dt$. The particles that flow out of the right end of the shaded cylinder with length $v_{\mathrm{d}}\, dt$ during dt are the particles that were within this cylinder at the beginning of the interval dt. The volume of the cylinder is $Av_{\mathrm{d}}\, dt$, and the number of particles within it is $nAv_d\,dt$. If each particle has a charge q, the charge dQ that flows out of the end of the cylinder during time dt is

$$dQ = q(nAv_d\,dt) = nqv_dA\,dt$$

and the current is

$$I = \frac{dQ}{dt} = nqv_dA$$

The current *per unit cross-sectional area* is called the **current density** J:

$$J = \frac{I}{A} = nqv_d$$

The units of current density are amperes per square meter (A/m^2).

If the moving charges are negative rather than positive, as in Fig. 25.2b, the drift velocity is opposite to $\vec{E}$. But the *current* is still in the same direction as $\vec{E}$ at each point in the conductor. Hence current I and current density J don't depend on the sign of the charge, and so we replace the charge q by its absolute value $|q|$:

$$\underbrace{I}_{\substack{\text{Current through} \\ \text{an area}}} = \underbrace{\frac{dQ}{dt}}_{\substack{\text{Rate at which charge flows through area}}} = n\underbrace{|q|}_{\substack{\text{Charge per particle}}}\underbrace{v_d}_{\substack{\text{Drift speed}}}\underbrace{A}_{\substack{\text{Cross-sectional area}}} \qquad (25.2)$$

Concentration of moving charged particles

$$J = \frac{I}{A} = n|q|v_d \qquad \text{(current density)} \qquad (25.3)$$

The current in a conductor is the product of the concentration of moving charged particles, the magnitude of charge of each such particle, the magnitude of the drift velocity, and the cross-sectional area of the conductor.

We can also define a *vector* current density $\vec{J}$ that includes the direction of the drift velocity:

$$\underbrace{\vec{J}}_{\substack{\text{Vector current density}}} = n\underbrace{q}_{\substack{\text{Charge per particle}}}\underbrace{\vec{v}_d}_{\substack{\text{Drift velocity}}} \qquad (25.4)$$

Concentration of moving charged particles

There are *no* absolute value signs in Eq. (25.4). If q is positive, $\vec{v}_d$ is in the same direction as $\vec{E}$; if q is negative, $\vec{v}_d$ is opposite to $\vec{E}$. In either case, $\vec{J}$ is in the same direction as $\vec{E}$. Equation (25.3) gives the *magnitude* J of the vector current density $\vec{J}$.

CAUTION **Current density vs. current** Current density $\vec{J}$ is a vector, but current I is not. The difference is that the current density $\vec{J}$ describes how charges flow at a certain point, and the vector's direction tells you about the direction of the flow at that point. By contrast, the current I describes how charges flow through an extended object such as a wire. For example, I has the same value at all points in the circuit of Fig. 25.3, but $\vec{J}$ does not: $\vec{J}$ is directed downward in the left-hand side of the loop and upward in the right-hand side. The magnitude of $\vec{J}$ can also vary around a circuit. In Fig. 25.3 $J = I/A$ is less in the battery (which has a large cross-sectional area A) than in the wires (which have a small A).

In general, a conductor may contain several different kinds of moving charged particles having charges $q_1, q_2, \ldots$, concentrations $n_1, n_2, \ldots$, and drift velocities with magnitudes $v_{d1}, v_{d2}, \ldots$. An example is current flow in an ionic solution (**Fig. 25.4**). In a sodium chloride solution, current can be carried by both positive sodium ions and negative chlorine ions; the total current I is found by adding up the currents due to each kind of charged particle, from Eq. (25.2). Likewise, the total vector current density $\vec{J}$ is found by using Eq. (25.4) for each kind of charged particle and adding the results.

25.4 Part of the electric circuit that includes this light bulb passes through a beaker with a solution of sodium chloride. The current in the solution is carried by both positive charges (Na^+ ions) and negative charges (Cl^- ions).

We will see in Section 25.4 that it is possible to have a current that is *steady* (constant in time) only if the conducting material forms a closed loop, called a *complete circuit*. In such a steady situation, the total charge in every segment of the conductor is constant. Hence the rate of flow of charge *out* at one end of a segment at any instant equals the rate of flow of charge *in* at the other end of the segment, and *the current is the same at all cross sections of the circuit.* We'll use this observation when we analyze electric circuits later in this chapter.

In many simple circuits, such as flashlights or cordless electric drills, the direction of the current is always the same; this is called *direct current*. But home appliances such as toasters, refrigerators, and televisions use *alternating current,* in which the current continuously changes direction. In this chapter we'll consider direct current only. Alternating current has many special features worthy of detailed study, which we'll examine in Chapter 31.

EXAMPLE 25.1 CURRENT DENSITY AND DRIFT VELOCITY IN A WIRE

An 18-gauge copper wire (the size usually used for lamp cords), with a diameter of 1.02 mm, carries a constant current of 1.67 A to a 200-W lamp. The free-electron density in the wire is 8.5×10^{28} per cubic meter. Find (a) the current density and (b) the drift speed.

SOLUTION

IDENTIFY and SET UP: This problem uses the relationships among current I, current density J, and drift speed v_d. We are given I and the wire diameter d, so we use Eq. (25.3) to find J. We use Eq. (25.3) again to find v_d from J and the known electron density n.

EXECUTE: (a) The cross-sectional area is

$$A = \frac{\pi d^2}{4} = \frac{\pi(1.02 \times 10^{-3}\text{ m})^2}{4} = 8.17 \times 10^{-7}\text{ m}^2$$

The magnitude of the current density is then

$$J = \frac{I}{A} = \frac{1.67\text{ A}}{8.17 \times 10^{-7}\text{ m}^2} = 2.04 \times 10^6\text{ A/m}^2$$

(b) From Eq. (25.3) for the drift velocity magnitude v_d, we find

$$v_d = \frac{J}{n|q|} = \frac{2.04 \times 10^6\text{ A/m}^2}{(8.5 \times 10^{28}\text{ m}^{-3})|-1.60 \times 10^{-19}\text{ C}|}$$

$$= 1.5 \times 10^{-4}\text{ m/s} = 0.15\text{ mm/s}$$

EVALUATE: At this speed an electron would require 6700 s (almost 2 h) to travel 1 m along this wire. The speeds of random motion of the electrons are roughly 10^6 m/s, around 10^{10} times the drift speed. Picture the electrons as bouncing around frantically, with a very slow drift!

TEST YOUR UNDERSTANDING OF SECTION 25.1 Suppose we replaced the wire in Example 25.1 with 12-gauge copper wire, which has twice the diameter of 18-gauge wire. If the current remains the same, what effect would this have on the magnitude of the drift velocity v_d? (i) None—v_d would be unchanged; (ii) v_d would be twice as great; (iii) v_d would be four times greater; (iv) v_d would be half as great; (v) v_d would be one-fourth as great. ▌

25.2 RESISTIVITY

The current density $\vec{J}$ in a conductor depends on the electric field $\vec{E}$ and on the properties of the material. In general, this dependence can be quite complex. But for some materials, especially metals, at a given temperature, $\vec{J}$ is nearly *directly proportional* to $\vec{E}$, and the ratio of the magnitudes of E and J is constant. This relationship, called Ohm's law, was discovered in 1826 by the German physicist Georg Simon Ohm (1787–1854). The word "law" should actually be in quotation marks, since **Ohm's law,** like the ideal-gas equation and Hooke's law, is an *idealized* model that describes the behavior of some materials quite well but is not a general description of *all* matter. In the following discussion we'll assume that Ohm's law is valid, even though there are many situations in which it is not.

TABLE 25.1	Resistivities at Room Temperature (20°C)			
	Substance	$\rho\ (\Omega \cdot m)$	Substance	$\rho\ (\Omega \cdot m)$
Conductors			**Semiconductors**	
Metals	Silver	1.47×10^{-8}	Pure carbon (graphite)	3.5×10^{-5}
	Copper	1.72×10^{-8}	Pure germanium	0.60
	Gold	2.44×10^{-8}	Pure silicon	2300
	Aluminum	2.75×10^{-8}	**Insulators**	
	Tungsten	5.25×10^{-8}	Amber	5×10^{14}
	Steel	20×10^{-8}	Glass	$10^{10} - 10^{14}$
	Lead	22×10^{-8}	Lucite	$>10^{13}$
	Mercury	95×10^{-8}	Mica	$10^{11} - 10^{15}$
Alloys	Manganin (Cu 84%, Mn 12%, Ni 4%)	44×10^{-8}	Quartz (fused)	75×10^{16}
	Constantan (Cu 60%, Ni 40%)	49×10^{-8}	Sulfur	10^{15}
	Nichrome	100×10^{-8}	Teflon	$>10^{13}$
			Wood	$10^{8} - 10^{11}$

We define the **resistivity** ρ of a material as

$$\text{Resistivity of a material} \quad \rho = \frac{E \quad \cdots\text{Magnitude of electric field in material}}{J \quad \cdots\text{Magnitude of current density caused by electric field}} \tag{25.5}$$

The greater the resistivity, the greater the field needed to cause a given current density, or the smaller the current density caused by a given field. From Eq. (25.5) the units of ρ are $(V/m)/(A/m^2) = V \cdot m/A$. As we will discuss in Section 25.3, 1 V/A is called one *ohm* (1 Ω; the Greek letter Ω, omega, is alliterative with "ohm"). So the SI units for ρ are $\Omega \cdot m$ (ohm-meters). **Table 25.1** lists some representative values of resistivity. A perfect conductor would have zero resistivity, and a perfect insulator would have an infinite resistivity. Metals and alloys have the smallest resistivities and are the best conductors. The resistivities of insulators are greater than those of the metals by an enormous factor, on the order of 10^{22}.

The reciprocal of resistivity is **conductivity.** Its units are $(\Omega \cdot m)^{-1}$. Good conductors of electricity have larger conductivity than insulators. Conductivity is the direct electrical analog of thermal conductivity. Comparing Table 25.1 with Table 17.5 (Thermal Conductivities), we note that good electrical conductors, such as metals, are usually also good conductors of heat. Poor electrical conductors, such as ceramic and plastic materials, are also poor thermal conductors. In a metal the free electrons that carry charge in electrical conduction also provide the principal mechanism for heat conduction, so we should expect a correlation between electrical and thermal conductivity. Because of the enormous difference in conductivity between electrical conductors and insulators, it is easy to confine electric currents to well-defined paths or circuits (**Fig. 25.5**). The variation in *thermal* conductivity is much less, only a factor of 10^3 or so, and it is usually impossible to confine heat currents to that extent.

Semiconductors have resistivities intermediate between those of metals and those of insulators. These materials are important because of the way their resistivities are affected by temperature and by small amounts of impurities.

A material that obeys Ohm's law reasonably well is called an *ohmic* conductor or a *linear* conductor. For such materials, at a given temperature, ρ is a *constant* that does not depend on the value of E. Many materials show substantial departures from Ohm's-law behavior; they are *nonohmic*, or *nonlinear*. In these materials, J depends on E in a more complicated manner.

Analogies with fluid flow can be a big help in developing intuition about electric current and circuits. For example, in the making of wine or maple syrup, the product is sometimes filtered to remove sediments. A pump forces the fluid

25.5 The copper "wires," or traces, on this circuit board are printed directly onto the surface of the dark-colored insulating board. Even though the traces are very close to each other (only about a millimeter apart), the board has such a high resistivity (and low conductivity) that essentially no current can flow between the traces.

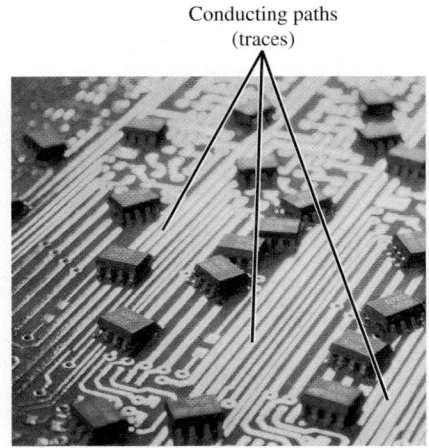

Conducting paths (traces)

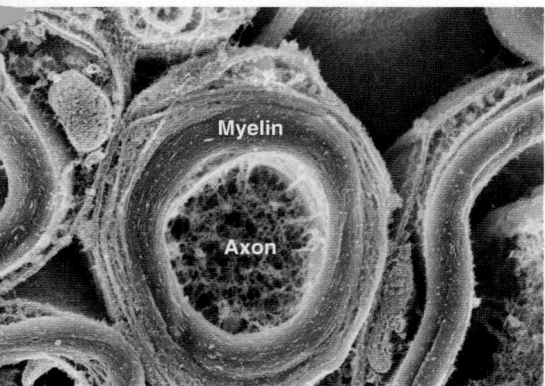

through the filter under pressure; if the flow rate (analogous to J) is proportional to the pressure difference between the upstream and downstream sides (analogous to E), the behavior is analogous to Ohm's law.

Resistivity and Temperature

The resistivity of a *metallic* conductor nearly always increases with increasing temperature, as shown in **Fig. 25.6a**. As temperature increases, the ions of the conductor vibrate with greater amplitude, making it more likely that a moving electron will collide with an ion as in Fig. 25.1; this impedes the drift of electrons through the conductor and hence reduces the current. Over a small temperature range (up to 100 C° or so), the resistivity of a metal can be represented approximately by the equation

Temperature dependence of resistivity:

Resistivity at temperature T ·········· Temperature coefficient of resistivity

$$\rho(T) = \rho_0[1 + \alpha(T - T_0)] \qquad (25.6)$$

Resistivity at reference temperature T_0

The reference temperature T_0 is often taken as 0°C or 20°C; the temperature T may be higher or lower than T_0. The factor α is called the **temperature coefficient of resistivity.** Some representative values are given in **Table 25.2**. The resistivity of the alloy manganin is practically independent of temperature.

The resistivity of graphite (a nonmetal) *decreases* with increasing temperature, since at higher temperatures, more electrons "shake loose" from the atoms and become mobile; hence the temperature coefficient of resistivity of graphite is negative. The same behavior occurs for semiconductors (Fig. 25.6b). Measuring the resistivity of a small semiconductor crystal is therefore a sensitive measure of temperature; this is the principle of a type of thermometer called a *thermistor*.

Some materials, including several metallic alloys and oxides, show a phenomenon called *superconductivity*. As the temperature decreases, the resistivity at first decreases smoothly, like that of any metal. But then at a certain critical temperature T_c a phase transition occurs and the resistivity suddenly drops to zero, as shown in Fig. 25.6c. Once a current has been established in a superconducting ring, it continues indefinitely without the presence of any driving field.

Superconductivity was discovered in 1911 by the Dutch physicist Heike Kamerlingh Onnes (1853–1926). He discovered that at very low temperatures, below 4.2 K, the resistivity of mercury suddenly dropped to zero. For the next 75 years, the highest T_c attained was about 20 K. This meant that superconductivity occurred only when the material was cooled by using expensive liquid helium, with a boiling-point temperature of 4.2 K, or explosive liquid hydrogen, with a boiling point of 20.3 K. But in 1986 Karl Müller and Johannes Bednorz discovered an oxide of barium, lanthanum, and copper with a T_c of nearly 40 K, and the race was on to develop "high-temperature" superconducting materials.

By 1987 a complex oxide of yttrium, copper, and barium had been found that has a value of T_c well above the 77 K boiling temperature of liquid nitrogen, a refrigerant that is both inexpensive and safe. The current (2014) record for T_c

25.6 Variation of resistivity ρ with absolute temperature T for **(a)** a normal metal, **(b)** a semiconductor, and **(c)** a superconductor. In (a) the linear approximation to ρ as a function of T is shown as a green line; the approximation agrees exactly at $T = T_0$, where $\rho = \rho_0$.

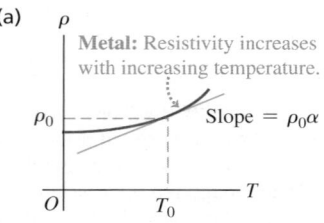

(a)

Metal: Resistivity increases with increasing temperature.

Slope $= \rho_0\alpha$

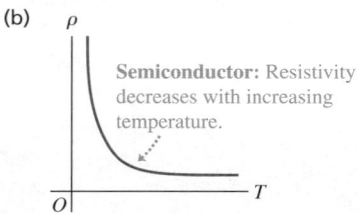

(b)

Semiconductor: Resistivity decreases with increasing temperature.

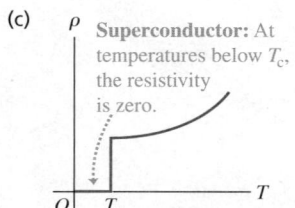

(c)

Superconductor: At temperatures below T_c, the resistivity is zero.

Temperature Coefficients of Resistivity

TABLE 25.2	**(Approximate Values Near Room Temperature)**		
Material	$\alpha\,[(°C)^{-1}]$	**Material**	$\alpha\,[(°C)^{-1}]$
Aluminum	0.0039	Lead	0.0043
Brass	0.0020	Manganin	0.00000
Carbon (graphite)	−0.0005	Mercury	0.00088
Constantan	0.00001	Nichrome	0.0004
Copper	0.00393	Silver	0.0038
Iron	0.0050	Tungsten	0.0045

at atmospheric pressure is 138 K, and materials that are superconductors at room temperature may become a reality. The implications of these discoveries for power-distribution systems, computer design, and transportation are enormous. Meanwhile, superconducting electromagnets cooled by liquid helium are used in particle accelerators and some experimental magnetic-levitation railroads. Superconductors have other exotic properties that require an understanding of magnetism to explore; we will discuss these further in Chapter 29.

TEST YOUR UNDERSTANDING OF SECTION 25.2   You maintain a constant electric field inside a piece of semiconductor while lowering the semiconductor's temperature. What happens to the current density in the semiconductor? (i) It increases; (ii) it decreases; (iii) it remains the same. ▌

25.3 RESISTANCE

For a conductor with resistivity ρ, the current density $\vec{J}$ at a point where the electric field is $\vec{E}$ is given by Eq. (25.5), which we can write as

PhET: Resistance in a Wire

$$\vec{E} = \rho\vec{J} \qquad (25.7)$$

When Ohm's law is obeyed, ρ is constant and independent of the magnitude of the electric field, so $\vec{E}$ is directly proportional to $\vec{J}$. Often, however, we are more interested in the total current I in a conductor than in $\vec{J}$ and more interested in the potential difference V between the ends of the conductor than in $\vec{E}$. This is so largely because I and V are much easier to measure than are $\vec{J}$ and $\vec{E}$.

Suppose our conductor is a wire with uniform cross-sectional area A and length L, as shown in **Fig. 25.7**. Let V be the potential difference between the higher-potential and lower-potential ends of the conductor, so that V is positive. (Another name for V is the *voltage across* the conductor.) The *direction* of the current is always from the higher-potential end to the lower-potential end. That's because current in a conductor flows in the direction of $\vec{E}$, no matter what the sign of the moving charges (Fig. 25.2), and because $\vec{E}$ points in the direction of *decreasing* electric potential (see Section 23.2). As the current flows through the potential difference, electric potential energy is lost; this energy is transferred to the ions of the conducting material during collisions.

25.7 A conductor with uniform cross section. The current density is uniform over any cross section, and the electric field is constant along the length.

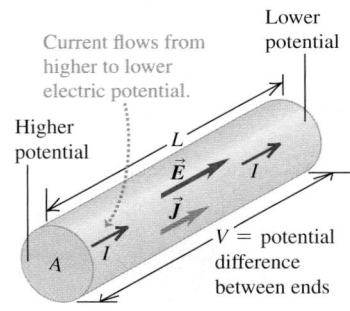

We can also relate the *value* of the current I to the potential difference between the ends of the conductor. If the magnitudes of the current density $\vec{J}$ and the electric field $\vec{E}$ are uniform throughout the conductor, the total current I is $I = JA$, and the potential difference V between the ends is $V = EL$. We solve these equations for J and E, respectively, and substitute the results into Eq. (25.7):

$$\frac{V}{L} = \frac{\rho I}{A} \quad \text{or} \quad V = \frac{\rho L}{A}I \qquad (25.8)$$

This shows that when ρ is constant, the total current I is proportional to the potential difference V.

The ratio of V to I for a particular conductor is called its **resistance** R:

$$R = \frac{V}{I} \qquad (25.9)$$

Comparing this definition of R to Eq. (25.8), we see that

Resistance of a conductor	$R = \dfrac{\rho L}{A}$	(25.10)

Resistivity of conductor material ···· Length of conductor ···· Cross-sectional area of conductor

If ρ is constant, as is the case for ohmic materials, then so is R.

The following equation is often called Ohm's law:

| Relationship among voltage, current, and resistance: | $V = IR$ | (25.11) |

Voltage between ends of conductor
Resistance of conductor
Current in conductor

However, it's important to understand that the real content of Ohm's law is the direct proportionality (for some materials) of V to I or of J to E. Equation (25.9) or (25.11) *defines* resistance R for *any* conductor, but only when R is constant can we correctly call this relationship Ohm's law.

Interpreting Resistance

Equation (25.10) shows that the resistance of a wire or other conductor of uniform cross section is directly proportional to its length and inversely proportional to its cross-sectional area. It is also proportional to the resistivity of the material of which the conductor is made.

The flowing-fluid analogy is again useful. In analogy to Eq. (25.10), a narrow water hose offers more resistance to flow than a fat one, and a long hose has more resistance than a short one (**Fig. 25.8**). We can increase the resistance to flow by stuffing the hose with cotton or sand; this corresponds to increasing the resistivity. The flow rate is approximately proportional to the pressure difference between the ends. Flow rate is analogous to current, and pressure difference is analogous to potential difference (voltage). Let's not stretch this analogy too far, though; the water flow rate in a pipe is usually *not* proportional to its cross-sectional area (see Section 14.6).

The SI unit of resistance is the **ohm,** equal to one volt per ampere ($1 \, \Omega = 1 \, \text{V/A}$). The *kilohm* ($1 \, \text{k}\Omega = 10^3 \, \Omega$) and the *megohm* ($1 \, \text{M}\Omega = 10^6 \, \Omega$) are also in common use. A 100-m length of 12-gauge copper wire, the size usually used in household wiring, has a resistance at room temperature of about $0.5 \, \Omega$. A 100-W, 120-V incandescent light bulb has a resistance (at operating temperature) of $140 \, \Omega$. If the same current I flows in both the copper wire and the light bulb, the potential difference $V = IR$ is much greater across the light bulb, and much more potential energy is lost per charge in the light bulb. This lost energy is converted by the light bulb filament into light and heat. You don't want your household wiring to glow white-hot, so its resistance is kept low by using wire of low resistivity and large cross-sectional area.

Because the resistivity of a material varies with temperature, the resistance of a specific conductor also varies with temperature. For temperature ranges that are not too great, this variation is approximately linear, analogous to Eq. (25.6):

$$R(T) = R_0[1 + \alpha(T - T_0)] \tag{25.12}$$

In this equation, $R(T)$ is the resistance at temperature T and R_0 is the resistance at temperature T_0, often taken to be 0°C or 20°C. The *temperature coefficient of resistance* α is the same constant that appears in Eq. (25.6) if the dimensions L and A in Eq. (25.10) do not change appreciably with temperature; this is indeed the case for most conducting materials. Within the limits of validity of Eq. (25.12), the *change* in resistance resulting from a temperature change $T - T_0$ is given by $R_0\alpha(T - T_0)$.

A circuit device made to have a specific value of resistance between its ends is called a **resistor**. Resistors in the range 0.01 to $10^7 \, \Omega$ can be bought off the shelf. Individual resistors used in electronic circuitry are often cylindrical, a few millimeters in diameter and length, with wires coming out of the ends. The resistance may be marked with a standard code that uses three or four color bands near one end (**Fig. 25.9**), according to the scheme in **Table 25.3**. The first two bands (starting with the band nearest an end) are digits, and the third is a power-of-10 multiplier. For example, green–violet–red means $57 \times 10^2 \, \Omega$, or $5.7 \, \text{k}\Omega$. The fourth band, if present, indicates the accuracy (tolerance) of the value;

25.8 A long fire hose offers substantial resistance to water flow. To make water pass through the hose rapidly, the upstream end of the hose must be at much higher pressure than the end where the water emerges. In an analogous way, there must be a large potential difference between the ends of a long wire in order to cause a substantial electric current through the wire.

25.9 This resistor has a resistance of $5.7 \, \text{k}\Omega$ with an accuracy (tolerance) of $\pm 10\%$.

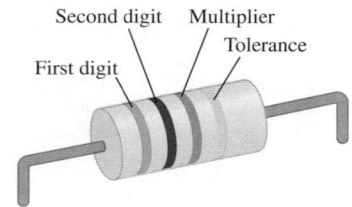

Second digit   Multiplier

First digit   Tolerance

Color Codes for Resistors

TABLE 25.3		
Color	**Value as Digit**	**Value as Multiplier**
Black	0	1
Brown	1	10
Red	2	10^2
Orange	3	10^3
Yellow	4	10^4
Green	5	10^5
Blue	6	10^6
Violet	7	10^7
Gray	8	10^8
White	9	10^9

(a)

Ohmic resistor (e.g., typical metal wire): At a given temperature, current is proportional to voltage.

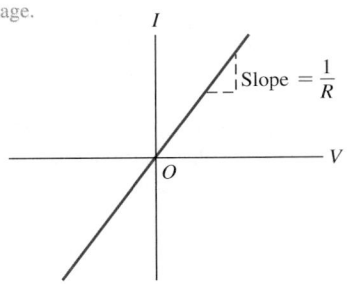

(b)

Semiconductor diode: a nonohmic resistor

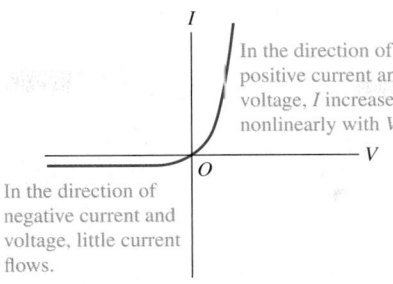

In the direction of positive current and voltage, I increases nonlinearly with V.

In the direction of negative current and voltage, little current flows.

25.10 Current–voltage relationships for two devices. Only for a resistor that obeys Ohm's law as in (a) is current I proportional to voltage V.

no band means $\pm 20\%$, a silver band $\pm 10\%$, and a gold band $\pm 5\%$. Another important characteristic of a resistor is the maximum *power* it can dissipate without damage. We'll return to this point in Section 25.5.

For a resistor that obeys Ohm's law, a graph of current as a function of potential difference (voltage) is a straight line (**Fig. 25.10a**). The slope of the line is $1/R$. If the sign of the potential difference changes, so does the sign of the current produced; in Fig. 25.7 this corresponds to interchanging the higher- and lower-potential ends of the conductor, so the electric field, current density, and current all reverse direction.

In devices that do not obey Ohm's law, the relationship of voltage to current may not be a direct proportion, and it may be different for the two directions of current. Figure 25.10b shows the behavior of a semiconductor *diode*, a device used to convert alternating current to direct current and to perform a wide variety of logic functions in computer circuitry. For positive potentials V of the anode (one of two terminals of the diode) with respect to the cathode (the other terminal), I increases exponentially with increasing V; for negative potentials the current is extremely small. Thus a positive V causes a current to flow in the positive direction, but a potential difference of the other sign causes little or no current. Hence a diode acts like a one-way valve in a circuit.

EXAMPLE 25.2 **ELECTRIC FIELD, POTENTIAL DIFFERENCE, AND RESISTANCE IN A WIRE**

The 18-gauge copper wire of Example 25.1 has a cross-sectional area of 8.20×10^{-7} m^2. It carries a current of 1.67 A. Find (a) the electric-field magnitude in the wire; (b) the potential difference between two points in the wire 50.0 m apart; (c) the resistance of a 50.0-m length of this wire.

SOLUTION

IDENTIFY and SET UP: We are given the cross-sectional area A and current I. Our target variables are the electric-field magnitude E, potential difference V, and resistance R. The current density is $J = I/A$. We find E from Eq. (25.5), $E = \rho J$ (Table 25.1 gives the resistivity ρ for copper). The potential difference is then the product of E and the length of the wire. We can use either Eq. (25.10) or Eq. (25.11) to find R.

EXECUTE: (a) From Table 25.1, $\rho = 1.72 \times 10^{-8}$ Ω · m. Hence, from Eq. (25.5),

$$E = \rho J = \frac{\rho I}{A} = \frac{(1.72 \times 10^{-8}\ \Omega \cdot \text{m})(1.67\ \text{A})}{8.20 \times 10^{-7}\ \text{m}^2} = 0.0350\ \text{V/m}$$

(b) The potential difference is

$$V = EL = (0.0350\ \text{V/m})(50.0\ \text{m}) = 1.75\ \text{V}$$

(c) From Eq. (25.10) the resistance of 50.0 m of this wire is

$$R = \frac{\rho L}{A} = \frac{(1.72 \times 10^{-8}\ \Omega \cdot \text{m})(50.0\ \text{m})}{8.20 \times 10^{-7}\ \text{m}^2} = 1.05\ \Omega$$

Alternatively, we can find R from Eq. (25.11):

$$R = \frac{V}{I} = \frac{1.75\ \text{V}}{1.67\ \text{A}} = 1.05\ \Omega$$

EVALUATE: We emphasize that the resistance of the wire is *defined* to be the ratio of voltage to current. If the wire is made of nonohmic material, then R is different for different values of V but is always given by $R = V/I$. Resistance is also always given by $R = \rho L/A$; if the material is nonohmic, ρ is not constant but depends on E (or, equivalently, on $V = EL$).

SOLUTION

EXAMPLE 25.3 **TEMPERATURE DEPENDENCE OF RESISTANCE**

Suppose the resistance of a copper wire is 1.05 Ω at 20°C. Find the resistance at 0°C and 100°C.

SOLUTION

IDENTIFY and SET UP: We are given the resistance $R_0 = 1.05$ Ω at a reference temperature $T_0 = 20$°C. We use Eq. (25.12) to find the resistances at $T = 0$°C and $T = 100$°C (our target variables), taking the temperature coefficient of resistivity from Table 25.2.

EXECUTE: From Table 25.2, $\alpha = 0.00393$ (C°)$^{-1}$ for copper. Then from Eq. (25.12),

$$R = R_0[1 + \alpha(T - T_0)]$$
$$= (1.05 \ \Omega)\{1 + [0.00393 \ (\text{C°})^{-1}][0°\text{C} - 20°\text{C}]\}$$
$$= 0.97 \ \Omega \text{ at } T = 0°\text{C}$$
$$R = (1.05 \ \Omega)\{1 + [0.00393 \ (\text{C°})^{-1}][100°\text{C} - 20°\text{C}]\}$$
$$= 1.38 \ \Omega \text{ at } T = 100°\text{C}$$

EVALUATE: The resistance at 100°C is greater than that at 0°C by a factor of $(1.38 \ \Omega)/(0.97 \ \Omega) = 1.42$: Raising the temperature of copper wire from 0°C to 100°C increases its resistance by 42%. From Eq. (25.11), $V = IR$, this means that 42% more voltage is required to produce the same current at 100°C than at 0°C. Designers of electric circuits that must operate over a wide temperature range must take this substantial effect into account.

TEST YOUR UNDERSTANDING OF SECTION 25.3 Suppose you increase the voltage across the copper wire in Examples 25.2 and 25.3. The increased voltage causes more current to flow, which makes the temperature of the wire increase. (The same thing happens to the coils of an electric oven or a toaster when a voltage is applied to them. We'll explore this issue in more depth in Section 25.5.) If you double the voltage across the wire, the current in the wire increases. By what factor does it increase? (i) 2; (ii) greater than 2; (iii) less than 2. ▮

25.11 If an electric field is produced inside a conductor that is *not* part of a complete circuit, current flows for only a very short time.

(a) An electric field $\vec{E}_1$ produced inside an isolated conductor causes a current.

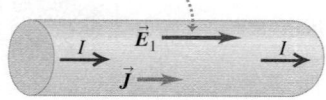

(b) The current causes charge to build up at the ends.

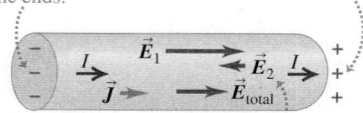

The charge buildup produces an opposing field $\vec{E}_2$, thus reducing the current.

(c) After a very short time $\vec{E}_2$ has the same magnitude as $\vec{E}_1$; then the total field is $\vec{E}_{\text{total}} = 0$ and the current stops completely.

25.4 ELECTROMOTIVE FORCE AND CIRCUITS

For a conductor to have a steady current, it must be part of a path that forms a closed loop or **complete circuit**. Here's why. If you establish an electric field $\vec{E}_1$ inside an isolated conductor with resistivity ρ that is *not* part of a complete circuit, a current begins to flow with current density $\vec{J} = \vec{E}_1/\rho$ (**Fig. 25.11a**). As a result a net positive charge quickly accumulates at one end of the conductor and a net negative charge accumulates at the other end (Fig. 25.11b). These charges themselves produce an electric field $\vec{E}_2$ in the direction opposite to $\vec{E}_1$, causing the total electric field and hence the current to decrease. Within a very small fraction of a second, enough charge builds up on the conductor ends that the total electric field $\vec{E} = \vec{E}_1 + \vec{E}_2 = 0$ inside the conductor. Then $\vec{J} = 0$ as well, and the current stops altogether (Fig. 25.11c). So there can be no steady motion of charge in such an *incomplete* circuit.

To see how to maintain a steady current in a *complete* circuit, we recall a basic fact about electric potential energy: If a charge q goes around a complete circuit and returns to its starting point, the potential energy must be the same at the end of the round trip as at the beginning. As described in Section 25.3, there is always a *decrease* in potential energy when charges move through an ordinary conducting material with resistance. So there must be some part of the circuit in which the potential energy *increases*.

The problem is analogous to an ornamental water fountain that recycles its water. The water pours out of openings at the top, cascades down over the terraces and spouts (moving in the direction of decreasing gravitational potential energy), and collects in a basin in the bottom. A pump then lifts it back to the top (increasing the potential energy) for another trip. Without the pump, the water would just fall to the bottom and stay there.

Electromotive Force

In an electric circuit there must be a device somewhere in the loop that acts like the water pump in a water fountain (**Fig. 25.12**). In this device a charge travels "uphill," from lower to higher potential energy, even though the electrostatic force is trying to push it from higher to lower potential energy. The direction of current in such a device is from lower to higher potential, just the opposite of what happens in an ordinary conductor.

The influence that makes current flow from lower to higher potential is called **electromotive force** (abbreviated **emf** and pronounced "ee-em-eff"), and a circuit device that provides emf is called a **source of emf.** Note that "electromotive force" is a poor term because emf is *not* a force but an energy-per-unit-charge quantity, like potential. The SI unit of emf is the same as that for potential, the volt ($1 \text{ V} = 1 \text{ J/C}$). A typical flashlight battery has an emf of 1.5 V; this means that the battery does 1.5 J of work on every coulomb of charge that passes through it. We'll use the symbol $\mathcal{E}$ (a script capital E) for emf.

Every complete circuit with a steady current must include a source of emf. Batteries, electric generators, solar cells, thermocouples, and fuel cells are all examples of sources of emf. All such devices convert energy of some form (mechanical, chemical, thermal, and so on) into electric potential energy and transfer it into the circuit to which the device is connected. An *ideal* source of emf maintains a constant potential difference between its terminals, independent of the current through it. We define electromotive force quantitatively as the magnitude of this potential difference. As we will see, such an ideal source is a mythical beast, like the frictionless plane and the massless rope. We will discuss later how real-life sources of emf differ in their behavior from this idealized model.

Figure 25.13 is a schematic diagram of an ideal source of emf that maintains a potential difference between conductors a and b, called the *terminals* of the device. Terminal a, marked $+$, is maintained at *higher* potential than terminal b, marked $-$. Associated with this potential difference is an electric field $\vec{E}$ in the region around the terminals, both inside and outside the source. The electric field inside the device is directed from a to b, as shown. A charge q within the source experiences an electric force $\vec{F}_e = q\vec{E}$. But the source also provides an additional influence, which we represent as a nonelectrostatic force $\vec{F}_n$. This force, operating inside the device, pushes charge from b to a in an "uphill" direction against the electric force $\vec{F}_e$. Thus $\vec{F}_n$ maintains the potential difference between the terminals. If $\vec{F}_n$ were not present, charge would flow between the terminals until the potential difference was zero. The origin of the additional influence $\vec{F}_n$ depends on the kind of source. In a generator it results from magnetic-field forces on moving charges. In a battery or fuel cell it is associated with diffusion processes and varying electrolyte concentrations resulting from chemical reactions. In an electrostatic machine such as a Van de Graaff generator (see Fig. 22.26), an actual mechanical force is applied by a moving belt or wheel.

If a positive charge q is moved from b to a inside the source, the nonelectrostatic force $\vec{F}_n$ does a positive amount of work $W_n = q\mathcal{E}$ on the charge. This displacement is *opposite* to the electrostatic force $\vec{F}_e$, so the potential energy associated with the charge *increases* by an amount equal to qV_{ab}, where $V_{ab} = V_a - V_b$ is the (positive) potential of point a with respect to point b. For the ideal source of emf that we've described, $\vec{F}_e$ and $\vec{F}_n$ are equal in magnitude but opposite in direction, so the total work done on the charge q is zero; there is an increase in potential energy but *no* change in the kinetic energy of the charge.

25.12 Just as a water fountain requires a pump, an electric circuit requires a source of electromotive force to sustain a steady current.

25.13 Schematic diagram of a source of emf in an "open-circuit" situation. The electric-field force $\vec{F}_e = q\vec{E}$ and the nonelectrostatic force $\vec{F}_n$ are shown for a positive charge q.

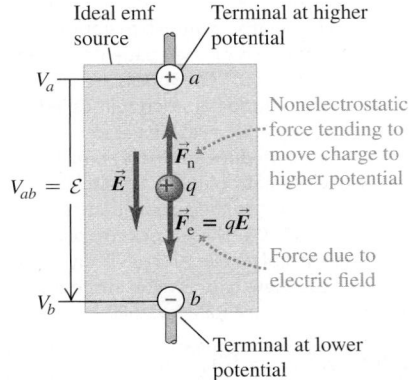

Ideal emf source

Terminal at higher potential

V_a

Nonelectrostatic force tending to move charge to higher potential

$V_{ab} = \mathcal{E}$ $\vec{E}$

$\vec{F}_n$

q

$\vec{F}_e = q\vec{E}$

Force due to electric field

V_b

Terminal at lower potential

When the emf source is not part of a closed circuit, $F_n = F_e$ and there is no net motion of charge between the terminals.

PhET: Battery Voltage
PhET: Signal Circuit

It's like lifting a book from the floor to a high shelf at constant speed. The increase in potential energy is just equal to the nonelectrostatic work W_n, so $q\mathcal{E} = qV_{ab}$, or

$$V_{ab} = \mathcal{E} \qquad \text{(ideal source of emf)} \qquad (25.13)$$

Now let's make a complete circuit by connecting a wire with resistance R to the terminals of a source (**Fig. 25.14**). The potential difference between terminals a and b sets up an electric field within the wire; this causes current to flow around the loop from a toward b, from higher to lower potential. Where the wire bends, equal amounts of positive and negative charge persist on the "inside" and "outside" of the bend. These charges exert the forces that cause the current to follow the bends in the wire.

From Eq. (25.11) the potential difference between the ends of the wire in Fig. 25.14 is given by $V_{ab} = IR$. Combining with Eq. (25.13), we have

$$\mathcal{E} = V_{ab} = IR \qquad \text{(ideal source of emf)} \qquad (25.14)$$

That is, when a positive charge q flows around the circuit, the potential *rise* $\mathcal{E}$ as it passes through the ideal source is numerically equal to the potential *drop* $V_{ab} = IR$ as it passes through the remainder of the circuit. Once $\mathcal{E}$ and R are known, this relationship determines the current in the circuit.

> CAUTION **Current is not "used up" in a circuit** It's a common misconception that in a closed circuit, current squirts out of the positive terminal of a battery and is consumed or "used up" by the time it reaches the negative terminal. In fact the current is the *same* at every point in a simple loop circuit like that in Fig. 25.14, even if the wire thickness is not constant throughout the circuit. This happens because charge is conserved (it can be neither created nor destroyed) and because charge cannot accumulate in the circuit devices we have described. It's like the flow of water in an ornamental fountain; water flows out of the top at the same rate at which it reaches the bottom, no matter what the dimensions of the fountain. None of the water is "used up" along the way! ∎

Internal Resistance

Real sources of emf in a circuit don't behave in exactly the way we have described; the potential difference across a real source in a circuit is *not* equal to the emf as in Eq. (25.14). The reason is that charge moving through the material of any real source encounters *resistance*. We call this the **internal resistance** of the source, denoted by r. If this resistance behaves according to Ohm's law, r is constant and independent of the current I. As the current moves through r, it experiences an associated drop in potential equal to Ir. Thus, when a current is flowing through a source from the negative terminal b to the positive terminal a, the potential difference V_{ab} between the terminals is

Terminal voltage, ····· emf of source ····· Current through source
source with $\quad$ $V_{ab} = \mathcal{E} - Ir$ ····· Internal resistance
internal resistance $\qquad\qquad\qquad\qquad\qquad$ of source $\qquad (25.15)$

The potential V_{ab}, called the **terminal voltage,** is less than the emf $\mathcal{E}$ because of the term Ir representing the potential drop across the internal resistance r. Hence the increase in potential energy qV_{ab} as a charge q moves from b to a within the source is less than the work $q\mathcal{E}$ done by the nonelectrostatic force $\vec{F}_n$, since some potential energy is lost in traversing the internal resistance.

A 1.5-V battery has an emf of 1.5 V, but the terminal voltage V_{ab} of the battery is equal to 1.5 V only if no current is flowing through it so that $I = 0$ in Eq. (25.15). If the battery is part of a complete circuit through which current

25.14 Schematic diagram of an ideal source of emf in a complete circuit. The electric-field force $\vec{F}_e = q\vec{E}$ and the nonelectrostatic force $\vec{F}_n$ are shown for a positive charge q. The current is in the direction from a to b in the external circuit and from b to a within the source.

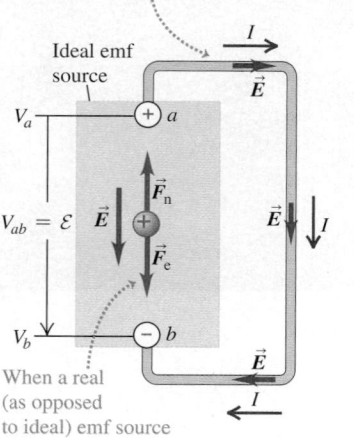

Potential across terminals creates electric field in circuit, causing charges to move.

When a real (as opposed to ideal) emf source is connected to a circuit, V_{ab} and thus F_e fall, so $F_n > F_e$ and $\vec{F}_n$ does work on the charges.

BIO Application Danger: Electric Ray! Electric rays deliver electric shocks to stun their prey and to discourage predators. (In ancient Rome, physicians practiced a primitive form of electroconvulsive therapy by placing electric rays on their patients to cure headaches and gout.) The shocks are produced by specialized flattened cells called electroplaques. Such a cell moves ions across membranes to produce an emf of about 0.05 V. Thousands of electroplaques are stacked on top of each other, so their emfs add to a total of as much as 200 V. These stacks make up more than half of an electric ray's body mass. A ray can use these to deliver an impressive current of up to 30 A for a few milliseconds.

is flowing, the terminal voltage will be less than 1.5 V. *For a real source of emf, the terminal voltage equals the emf only if no current is flowing through the source* (**Fig. 25.15**). Thus we can describe the behavior of a source in terms of two properties: an emf $\mathcal{E}$, which supplies a constant potential difference independent of current, in series with an internal resistance r.

The current in the external circuit connected to the source terminals a and b is still determined by $V_{ab} = IR$. Combining this with Eq. (25.15), we find

$$\mathcal{E} - Ir = IR \quad \text{or} \quad I = \frac{\mathcal{E}}{R + r} \qquad \begin{array}{l}\text{(current, source with}\\ \text{internal resistance)}\end{array} \qquad (25.16)$$

That is, the current equals the source emf divided by the *total* circuit resistance $(R + r)$.

CAUTION **A battery is not a "current source"** You might have thought that a battery or other source of emf always produces the same current, no matter what circuit it's used in. Equation (25.16) shows that this isn't so! The greater the resistance R of the external circuit, the less current the source will produce. ▌

Symbols for Circuit Diagrams

An important part of analyzing any electric circuit is drawing a schematic *circuit diagram*. **Table 25.4** shows the usual symbols used in circuit diagrams. We will use these symbols extensively in this chapter and the next. We usually assume that the wires that connect the various elements of the circuit have negligible resistance; from Eq. (25.11), $V = IR$, the potential difference between the ends of such a wire is zero.

Table 25.4 includes two *meters* that are used to measure the properties of circuits. Idealized meters do not disturb the circuit in which they are connected. A **voltmeter,** introduced in Section 23.2, measures the potential difference between its terminals; an idealized voltmeter has infinitely large resistance and measures potential difference without having any current diverted through it. An ammeter measures the current passing through it; an idealized **ammeter** has zero resistance and has no potential difference between its terminals. The following examples illustrate how to analyze circuits that include meters.

25.15 The emf of this battery—that is, the terminal voltage when it's not connected to anything—is 12 V. But because the battery has internal resistance, the terminal voltage of the battery is less than 12 V when it is supplying current to a light bulb.

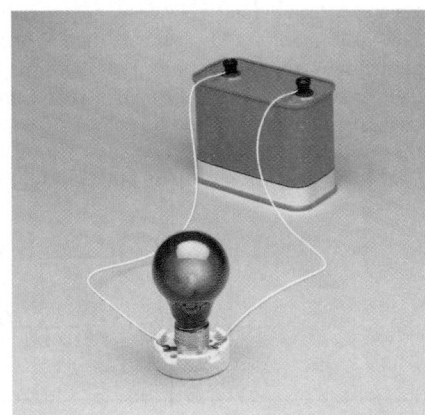

DATA *SPEAKS*

Circuits, emf, and Current

When students were given a problem involving a source of emf, more than 25% gave an incorrect response. Common errors:

- Forgetting that the internal resistance r affects the potential difference V_{ab} between the terminals of the source of emf $\mathcal{E}$. If the current I inside the source is from the negative terminal b to the positive terminal a, then $V_{ab} < \mathcal{E}$ by an amount Ir; if the current flows the other way, $V_{ab} > \mathcal{E}$ by an amount Ir.

- Forgetting that the internal resistance is an intrinsic part of a source of emf. Although we draw the emf and internal resistance as adjacent parts of the circuit, both are parts of the source and cannot be separated.

TABLE 25.4	Symbols for Circuit Diagrams	
		Conductor with negligible resistance
	R	Resistor
	$+ \| \| \mathcal{E}$	Source of emf (longer vertical line always represents the positive terminal, usually the terminal with higher potential)
	$\mathcal{E} | + $ or $+ \| \mathcal{E}$	Source of emf with internal resistance r (r can be placed on either side)
	V	Voltmeter (measures potential difference between its terminals)
	A	Ammeter (measures current through it)

CONCEPTUAL EXAMPLE 25.4   A SOURCE IN AN OPEN CIRCUIT

Figure 25.16 shows a source (a battery) with emf $\mathcal{E} = 12$ V and internal resistance $r = 2\ \Omega$. (For comparison, the internal resistance of a commercial 12-V lead storage battery is only a few thousandths of an ohm.) The wires to the left of a and to the right of the ammeter A are not connected to anything. Determine the respective readings V_{ab} and I of the idealized voltmeter V and the idealized ammeter A.

25.16 A source of emf in an open circuit.

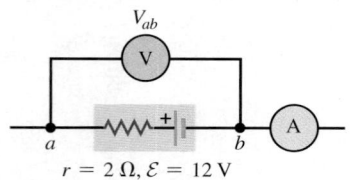

SOLUTION

There is *zero* current because there is no complete circuit. (Our idealized voltmeter has an infinitely large resistance, so no current flows through it.) Hence the ammeter reads $I = 0$. Because there is no current through the battery, there is no potential difference across its internal resistance. From Eq. (25.15) with $I = 0$, the potential difference V_{ab} across the battery terminals is equal to the emf. So the voltmeter reads $V_{ab} = \mathcal{E} = 12$ V. The terminal voltage of a real, nonideal source equals the emf *only* if there is no current flowing through the source, as in this example.

EXAMPLE 25.5   A SOURCE IN A COMPLETE CIRCUIT

We add a 4-Ω resistor to the battery in Conceptual Example 25.4, forming a complete circuit (**Fig. 25.17**). What are the voltmeter and ammeter readings V_{ab} and I now?

SOLUTION

IDENTIFY and SET UP: Our target variables are the current I through the circuit $aa'b'b$ and the potential difference V_{ab}. We first find I from Eq. (25.16). To find V_{ab}, we can use either Eq. (25.11) or Eq. (25.15).

25.17 A source of emf in a complete circuit.

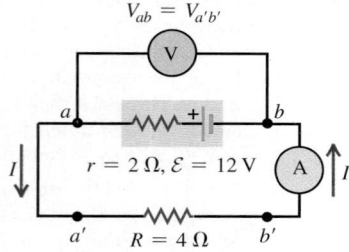

EXECUTE: The ideal ammeter has zero resistance, so the total resistance external to the source is $R = 4\ \Omega$. From Eq. (25.16), the current through the circuit $aa'b'b$ is then

$$I = \frac{\mathcal{E}}{R + r} = \frac{12\ \text{V}}{4\ \Omega + 2\ \Omega} = 2\ \text{A}$$

Our idealized conducting wires and the idealized ammeter have zero resistance, so there is no potential difference between points a and a' or between points b and b'; that is, $V_{ab} = V_{a'b'}$. We find V_{ab} by considering a and b as the terminals of the resistor: From Ohm's law, Eq. (25.11), we then have

$$V_{a'b'} = IR = (2\ \text{A})(4\ \Omega) = 8\ \text{V}$$

Alternatively, we can consider a and b as the terminals of the source. Then, from Eq. (25.15),

$$V_{ab} = \mathcal{E} - Ir = 12\ \text{V} - (2\ \text{A})(2\ \Omega) = 8\ \text{V}$$

Either way, we see that the voltmeter reading is 8 V.

EVALUATE: With this current flowing through the source, the terminal voltage V_{ab} is less than the emf $\mathcal{E}$. The smaller the internal resistance r, the less the difference between V_{ab} and $\mathcal{E}$.

CONCEPTUAL EXAMPLE 25.6   USING VOLTMETERS AND AMMETERS

We move the voltmeter and ammeter in Example 25.5 to different positions in the circuit. What are the readings of the ideal voltmeter and ammeter in the situations shown in (a) **Fig. 25.18a** and (b) Fig. 25.18b?

SOLUTION

(a) The voltmeter now measures the potential difference between points a' and b'. As in Example 25.5, $V_{ab} = V_{a'b'}$, so the voltmeter reads the same as in Example 25.5: $V_{a'b'} = 8$ V.

CAUTION **Current in a simple loop** As charges move through a resistor, there is a decrease in electric potential energy, but there is *no* change in the current. *The current in a simple loop is the same at every point;* it is not "used up" as it moves through a resistor. Hence the ammeter in Fig. 25.17 ("downstream" of the 4-Ω resistor) and the ammeter in Fig. 25.18b ("upstream" of the resistor) both read $I = 2$ A.

(b) There is no current through the ideal voltmeter because it has infinitely large resistance. Since the voltmeter is now part of the circuit, there is no current at all in the circuit, and the ammeter reads $I = 0$.

The voltmeter measures the potential difference $V_{bb'}$ between points b and b'. Since $I = 0$, the potential difference across the resistor is $V_{a'b'} = IR = 0$, and the potential difference between the ends a and a' of the idealized ammeter is also zero. So $V_{bb'}$ is equal to V_{ab}, the terminal voltage of the source. As in Conceptual Example 25.4, there is no current, so the terminal voltage equals the emf, and the voltmeter reading is $V_{ab} = \mathcal{E} = 12$ V.

This example shows that ammeters and voltmeters are circuit elements, too. Moving the voltmeter from the position in Fig. 25.18a to that in Fig. 25.18b makes large changes in the current and potential differences in the circuit. If you want to measure the potential difference between two points in a circuit without disturbing the circuit, use a voltmeter as in Fig. 25.17 or 25.18a, *not* as in Fig. 25.18b.

25.18 Different placements of a voltmeter and an ammeter in a complete circuit.

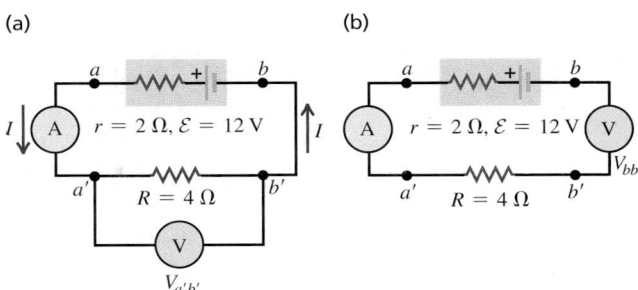

(a)

(b)

EXAMPLE 25.7   A SOURCE WITH A SHORT CIRCUIT

In the circuit of Example 25.5 we replace the 4-Ω resistor with a zero-resistance conductor. What are the meter readings now?

SOLUTION

IDENTIFY and SET UP: Figure 25.19 shows the new circuit. Our target variables are again I and V_{ab}. There is now a zero-resistance path between points a and b, through the lower loop, so the potential difference between these points must be zero.

EXECUTE: We must have $V_{ab} = IR = I(0) = 0$, no matter what the current. We can therefore find the current I from Eq. (25.15):

$$V_{ab} = \mathcal{E} - Ir = 0$$

$$I = \frac{\mathcal{E}}{r} = \frac{12 \text{ V}}{2 \text{ }\Omega} = 6 \text{ A}$$

EVALUATE: The current in this circuit has a different value than in Example 25.5, even though the same battery is used; the current depends on both the internal resistance r and the resistance of the external circuit.

25.19 Our sketch for this problem.

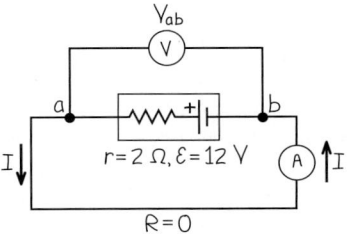

The situation here is called a *short circuit*. The external-circuit resistance is zero, because terminals of the battery are connected directly to each other. The short-circuit current is equal to the emf $\mathcal{E}$ divided by the internal resistance r. *Warning:* Short circuits can be dangerous! An automobile battery or a household power line has very small internal resistance (much less than in these examples), and the short-circuit current can be great enough to melt a small wire or cause a storage battery to explode.

Potential Changes around a Circuit

The net change in potential energy for a charge q making a round trip around a complete circuit must be zero. Hence the net change in *potential* around the circuit must also be zero; in other words, the algebraic sum of the potential differences and emfs around the loop is zero. We can see this by rewriting Eq. (25.16) in the form

$$\mathcal{E} - Ir - IR = 0$$

A potential gain of $\mathcal{E}$ is associated with the emf, and potential drops of Ir and IR are associated with the internal resistance of the source and the external circuit, respectively. **Figure 25.20** shows how the potential varies as we go around the complete circuit of Fig. 25.17. The horizontal axis doesn't necessarily represent actual distances, but rather various points in the loop. If we take the potential to be zero at the negative terminal of the battery, then we have a rise $\mathcal{E}$ and a drop Ir in the battery and an additional drop IR in the external resistor, and as we finish our trip around the loop, the potential is back where it started.

25.20 Potential rises and drops in a circuit.

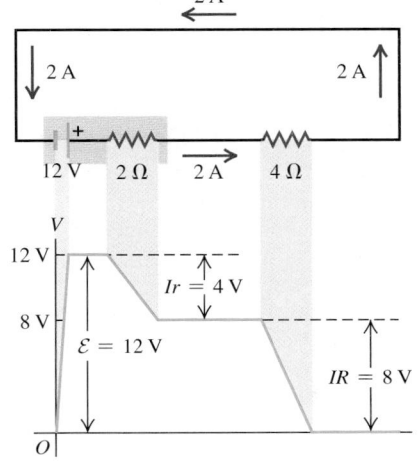

In this section we have considered only situations in which the resistances are ohmic. If the circuit includes a nonlinear device such as a diode (see Fig. 25.10b), Eq. (25.16) is still valid but cannot be solved algebraically because R is not a constant. In such a situation, I can be found by using numerical techniques.

Finally, we remark that Eq. (25.15) is not always an adequate representation of the behavior of a source. The emf may not be constant, and what we have described as an internal resistance may actually be a more complex voltage–current relationship that doesn't obey Ohm's law. Nevertheless, the concept of internal resistance frequently provides an adequate description of batteries, generators, and other energy converters. The principal difference between a fresh flashlight battery and an old one is not in the emf, which decreases only slightly with use, but in the internal resistance, which may increase from less than an ohm when the battery is fresh to as much as 1000 Ω or more after long use. Similarly, a car battery can deliver less current to the starter motor on a cold morning than when the battery is warm, not because the emf is appreciably less but because the internal resistance increases with decreasing temperature.

TEST YOUR UNDERSTANDING OF SECTION 25.4 Rank the following circuits in order from highest to lowest current: (i) A 1.4-Ω resistor connected to a 1.5-V battery that has an internal resistance of 0.10 Ω; (ii) a 1.8-Ω resistor connected to a 4.0-V battery that has a terminal voltage of 3.6 V but an unknown internal resistance; (iii) an unknown resistor connected to a 12.0-V battery that has an internal resistance of 0.20 Ω and a terminal voltage of 11.0 V. ∎

25.5 ENERGY AND POWER IN ELECTRIC CIRCUITS

Let's now look at some energy and power relationships in electric circuits. The box in **Fig. 25.21** represents a circuit element with potential difference $V_a - V_b = V_{ab}$ between its terminals and current I passing through it in the direction from a toward b. This element might be a resistor, a battery, or something else; the details don't matter. As charge passes through the circuit element, the electric field does work on the charge. In a source of emf, additional work is done by the force $\vec{F}_n$ that we mentioned in Section 25.4.

As an amount of charge q passes through the circuit element, there is a change in potential energy equal to qV_{ab}. For example, if $q > 0$ and $V_{ab} = V_a - V_b$ is positive, potential energy decreases as the charge "falls" from potential V_a to lower potential V_b. The moving charges don't gain *kinetic* energy, because the current (the rate of charge flow) out of the circuit element must be the same as the current into the element. Instead, the quantity qV_{ab} represents energy transferred into the circuit element. This situation occurs in the coils of a toaster or electric oven, in which electrical energy is converted to thermal energy.

If the potential at a is lower than at b, then V_{ab} is negative and there is a net transfer of energy *out* of the circuit element. The element then acts as a source, delivering electrical energy into the circuit to which it is attached. This is the usual situation for a battery, which converts chemical energy into electrical energy and delivers it to the external circuit. Thus qV_{ab} can denote a quantity of energy that is either delivered to a circuit element or extracted from that element.

In electric circuits we are most often interested in the *rate* at which energy is either delivered to or extracted from a circuit element. If the current through the element is I, then in a time interval dt an amount of charge $dQ = I\,dt$ passes through the element. The potential energy change for this amount of charge is $V_{ab}\,dQ = V_{ab}I\,dt$. Dividing this expression by dt, we obtain the *rate* at which energy is transferred either into or out of the circuit element. The time rate of energy transfer is *power*, denoted by P, so we write

25.21 The power input to the circuit element between a and b is $P = (V_a - V_b)I = V_{ab}I$.

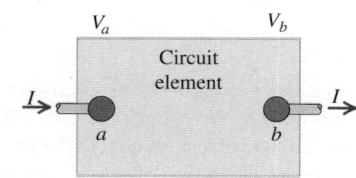

PhET: Battery-Resistor Circuit
PhET: Circuit Construction Kit (AC+DC)
PhET: Circuit Construction Kit (DC Only)
PhET: Ohm's Law

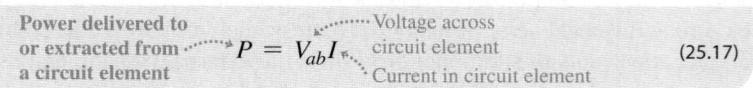

Power delivered to or extracted from a circuit element ⋯▸ $P = V_{ab}I$ ◂⋯ Voltage across circuit element · Current in circuit element   (25.17)

The unit of V_{ab} is one volt, or one joule per coulomb, and the unit of I is one ampere, or one coulomb per second. Hence the unit of $P = V_{ab}I$ is one watt:

$$(1 \text{ J/C})(1 \text{ C/s}) = 1 \text{ J/s} = 1 \text{ W}$$

Let's consider a few special cases.

Power Input to a Pure Resistance

If the circuit element in Fig. 25.21 is a resistor, the potential difference is $V_{ab} = IR$. From Eq. (25.17) the power delivered to the resistor by the circuit is

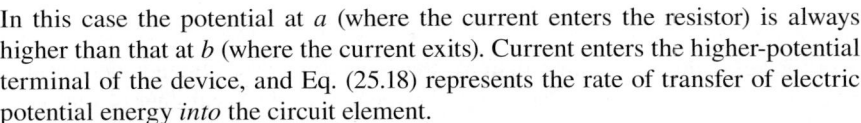

$$\text{Power delivered to a resistor} \quad P = V_{ab}I = I^2R = \frac{V_{ab}^2}{R} \quad (25.18)$$

Voltage across resistor
Current in resistor   Resistance of resistor

DEMO

In this case the potential at a (where the current enters the resistor) is always higher than that at b (where the current exits). Current enters the higher-potential terminal of the device, and Eq. (25.18) represents the rate of transfer of electric potential energy *into* the circuit element.

What becomes of this energy? The moving charges collide with atoms in the resistor and transfer some of their energy to these atoms, increasing the internal energy of the material. Either the temperature of the resistor increases or there is a flow of heat out of it, or both. In any of these cases we say that energy is *dissipated* in the resistor at a rate I^2R. Every resistor has a *power rating*, the maximum power the device can dissipate without becoming overheated and damaged. Some devices, such as electric heaters, are designed to get hot and transfer heat to their surroundings. But if the power rating is exceeded, even such a device may melt or even explode.

Power Output of a Source

The upper rectangle in **Fig. 25.22a** represents a source with emf $\mathcal{E}$ and internal resistance r, connected by ideal (resistanceless) conductors to an external circuit represented by the lower box. This could describe a car battery connected to one of the car's headlights (Fig. 25.22b). Point a is at higher potential than point b, so $V_a > V_b$ and V_{ab} is positive. Note that the current I is *leaving* the source at the higher-potential terminal (rather than entering there). Energy is being delivered to the external circuit, at a rate given by Eq. (25.17):

$$P = V_{ab}I$$

For a source that can be described by an emf $\mathcal{E}$ and an internal resistance r, we may use Eq. (25.15):

$$V_{ab} = \mathcal{E} - Ir$$

Multiplying this equation by I, we find

$$P = V_{ab}I = \mathcal{E}I - I^2r \quad (25.19)$$

What do the terms $\mathcal{E}I$ and I^2r mean? In Section 25.4 we defined the emf $\mathcal{E}$ as the work per unit charge performed on the charges by the nonelectrostatic force as the charges are pushed "uphill" from b to a in the source. In a time dt, a charge $dQ = I\,dt$ flows through the source; the work done on it by this nonelectrostatic force is $\mathcal{E}\,dQ = \mathcal{E}I\,dt$. Thus $\mathcal{E}I$ is the *rate* at which work is done on the circulating charges by whatever agency causes the nonelectrostatic force in the source. This term represents the rate of conversion of nonelectrical energy to electrical energy within the source. The term I^2r is the rate at which electrical energy is

25.22 Energy conversion in a simple circuit.

(a) Diagrammatic circuit

- The emf source converts nonelectrical to electrical energy at a rate $\mathcal{E}I$.
- Its internal resistance *dissipates* energy at a rate I^2r.
- The difference $\mathcal{E}I - I^2r$ is its power output.

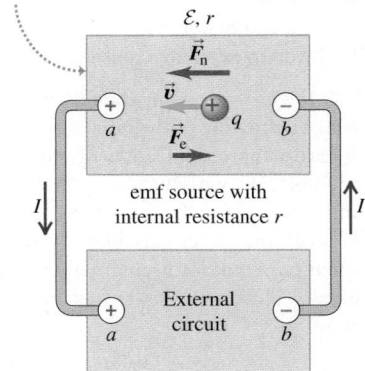

(b) A real circuit of the type shown in **(a)**

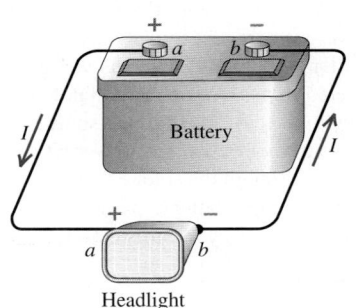

dissipated in the internal resistance of the source. The difference $\mathcal{E}I - I^2r$ is the *net* electrical power output of the source—that is, the rate at which the source delivers electrical energy to the remainder of the circuit.

Power Input to a Source

25.23 When two sources are connected in a simple loop circuit, the source with the larger emf delivers energy to the other source.

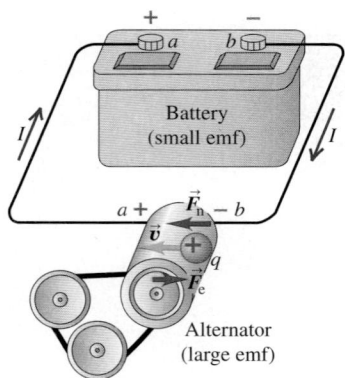

Battery (small emf)

Alternator (large emf)

Suppose that the lower rectangle in Fig. 25.22a is itself a source, with an emf *larger* than that of the upper source and opposite to that of the upper source. **Figure 25.23** shows a practical example, an automobile battery (the upper circuit element) being charged by the car's alternator (the lower element). The current I in the circuit is then *opposite* to that shown in Fig. 25.22; the lower source is pushing current backward through the upper source. Because of this reversal of current, instead of Eq. (25.15), we have for the upper source

$$V_{ab} = \mathcal{E} + Ir$$

and instead of Eq. (25.19), we have

$$P = V_{ab}I = \mathcal{E}I + I^2r \qquad (25.20)$$

Work is being done *on,* rather than *by,* the agent that causes the nonelectrostatic force in the upper source. There is a conversion of electrical energy into nonelectrical energy in the upper source at a rate $\mathcal{E}I$. The term I^2r in Eq. (25.20) is again the rate of dissipation of energy in the internal resistance of the upper source, and the sum $\mathcal{E}I + I^2r$ is the total electrical power *input* to the upper source. This is what happens when a rechargeable battery (a storage battery) is connected to a charger. The charger supplies electrical energy to the battery; part of it is converted to chemical energy, to be reconverted later, and the remainder is dissipated (wasted) in the battery's internal resistance, warming the battery and causing a heat flow out of it. If you have a power tool or laptop computer with a rechargeable battery, you may have noticed that it gets warm while it is charging.

PROBLEM-SOLVING STRATEGY 25.1 POWER AND ENERGY IN CIRCUITS

IDENTIFY *the relevant concepts:* The ideas of electrical power input and output can be applied to any electric circuit. Many problems will ask you to explicitly consider power or energy.

SET UP *the problem* using the following steps:
1. Make a drawing of the circuit.
2. Identify the circuit elements, including sources of emf and resistors. We will introduce other circuit elements later, including capacitors (Chapter 26) and inductors (Chapter 30).
3. Identify the target variables. Typically they will be the power input or output for each circuit element, or the total amount of energy put into or taken out of a circuit element in a given time.

EXECUTE *the solution* as follows:
1. A source of emf $\mathcal{E}$ delivers power $\mathcal{E}I$ into a circuit when current I flows through the source in the direction from $-$ to $+$. (For example, energy is converted from chemical energy in a battery, or from mechanical energy in a generator.) In this case there is a *positive* power output to the circuit or, equivalently, a *negative* power input to the source.
2. A source of emf takes power $\mathcal{E}I$ from a circuit when current passes through the source from $+$ to $-$. (This occurs in charging a storage battery, when electrical energy is converted to chemical energy.) In this case there is a *negative* power output

to the circuit or, equivalently, a *positive* power input to the source.
3. There is always a *positive* power input to a resistor through which current flows, irrespective of the direction of current flow. This process removes energy from the circuit, converting it to heat at the rate $VI = I^2R = V^2/R$, where V is the potential difference across the resistor.
4. Just as in item 3, there always is a positive power input to the internal resistance r of a source through which current flows, irrespective of the direction of current flow. This process likewise removes energy from the circuit, converting it into heat at the rate I^2r.
5. If the power into or out of a circuit element is constant, the energy delivered to or extracted from that element is the product of power and elapsed time. (In Chapter 26 we will encounter situations in which the power is not constant. In such cases, calculating the total energy requires an integral over the relevant time interval.)

EVALUATE *your answer:* Check your results; in particular, check that energy is conserved. This conservation can be expressed in either of two forms: "net power input = net power output" or "the algebraic sum of the power inputs to the circuit elements is zero."

EXAMPLE 25.8 POWER INPUT AND OUTPUT IN A COMPLETE CIRCUIT

For the circuit that we analyzed in Example 25.5, find the rates of energy conversion (chemical to electrical) and energy dissipation in the battery, the rate of energy dissipation in the 4-Ω resistor, and the battery's net power output.

SOLUTION

IDENTIFY and SET UP: Figure 25.24 shows the circuit, gives values of quantities known from Example 25.5, and indicates how we find the target variables. We use Eq. (25.19) to find the battery's net power output, the rate of chemical-to-electrical energy conversion, and the rate of energy dissipation in the battery's internal

25.24 Our sketch for this problem.

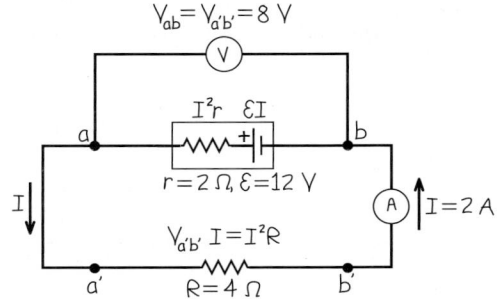

resistance. We use Eq. (25.18) to find the power delivered to (and dissipated in) the 4-Ω resistor.

EXECUTE: From the first term in Eq. (25.19), the rate of energy conversion in the battery is

$$\mathcal{E}I = (12\ \text{V})(2\ \text{A}) = 24\ \text{W}$$

From the second term in Eq. (25.19), the rate of dissipation of energy in the battery is

$$I^2r = (2\ \text{A})^2(2\ \Omega) = 8\ \text{W}$$

The *net* electrical power output of the battery is the difference between these: $\mathcal{E}I - I^2r = 16\ \text{W}$. From Eq. (25.18), the electrical power input to, and the equal rate of dissipation of electrical energy in, the 4-Ω resistor are

$$V_{a'b'}I = (8\ \text{V})(2\ \text{A}) = 16\ \text{W} \quad \text{and}$$

$$I^2R = (2\ \text{A})^2(4\ \Omega) = 16\ \text{W}$$

EVALUATE: The rate $V_{a'b'}I$ at which energy is supplied to the 4-Ω resistor equals the rate I^2R at which energy is dissipated there. This is also equal to the battery's net power output: $P = V_{ab}I = (8\ \text{V})(2\ \text{A}) = 16\ \text{W}$. In summary, the rate at which the source of emf supplies energy is $\mathcal{E}I = 24\ \text{W}$, of which $I^2r = 8\ \text{W}$ is dissipated in the battery's internal resistor and $I^2R = 16\ \text{W}$ is dissipated in the external resistor.

EXAMPLE 25.9 INCREASING THE RESISTANCE

Suppose we replace the external 4-Ω resistor in Fig. 25.24 with an 8-Ω resistor. How does this affect the electrical power dissipated in this resistor?

SOLUTION

IDENTIFY and SET UP: Our target variable is the power dissipated in the resistor to which the battery is connected. The situation is the same as in Example 25.8, but with a higher external resistance R.

EXECUTE: According to Eq. (25.18), the power dissipated in the resistor is $P = I^2R$. You might conclude that making the resistance R twice as great as in Example 25.8 should make the power twice as great, or $2(16\ \text{W}) = 32\ \text{W}$. If instead you used the formula $P = V_{ab}^2/R$, you might conclude that the power should be one-half as great as in the preceding example, or $(16\ \text{W})/2 = 8\ \text{W}$. Which answer is correct?

In fact, *both* of these answers are *incorrect*. The first is wrong because changing the resistance R also changes the current in the circuit (remember, a source of emf does *not* generate the same current in all situations). The second answer is wrong because the potential difference V_{ab} across the resistor changes when the current changes. To get the correct answer, we first find the current just as we did in Example 25.5:

$$I = \frac{\mathcal{E}}{R + r} = \frac{12\ \text{V}}{8\ \Omega + 2\ \Omega} = 1.2\ \text{A}$$

The greater resistance causes the current to decrease. The potential difference across the resistor is

$$V_{ab} = IR = (1.2\ \text{A})(8\ \Omega) = 9.6\ \text{V}$$

which is greater than that with the 4-Ω resistor. We can then find the power dissipated in the resistor in either of two ways:

$$P = I^2R = (1.2\ \text{A})^2(8\ \Omega) = 12\ \text{W} \quad \text{or}$$

$$P = \frac{V_{ab}^2}{R} = \frac{(9.6\ \text{V})^2}{8\ \Omega} = 12\ \text{W}$$

EVALUATE: Increasing the resistance R causes a *reduction* in the power input to the resistor. In the expression $P = I^2R$ the decrease in current is more important than the increase in resistance; in the expression $P = V_{ab}^2/R$ the increase in resistance is more important than the increase in V_{ab}. This same principle applies to ordinary light bulbs; a 50-W light bulb has a greater resistance than does a 100-W light bulb.

Can you show that replacing the 4-Ω resistor with an 8-Ω resistor decreases both the rate of energy conversion (chemical to electrical) in the battery and the rate of energy dissipation in the battery?

EXAMPLE 25.10 POWER IN A SHORT CIRCUIT

For the short-circuit situation of Example 25.7, find the rates of energy conversion and energy dissipation in the battery and the net power output of the battery.

SOLUTION

IDENTIFY and SET UP: Our target variables are again the power inputs and outputs associated with the battery. **Figure 25.25** shows the circuit. This is the same situation as in Example 25.8, but now the external resistance R is zero.

EXECUTE: We found in Example 25.7 that the current in this situation is $I = 6$ A. From Eq. (25.19), the rate of energy conversion (chemical to electrical) in the battery is then

$$\mathcal{E}I = (12 \text{ V})(6 \text{ A}) = 72 \text{ W}$$

and the rate of dissipation of energy in the battery is

$$I^2r = (6 \text{ A})^2(2 \text{ } \Omega) = 72 \text{ W}$$

The net power output of the source is $\mathcal{E}I - I^2r = 0$. We get this same result from the expression $P = V_{ab}I$, because the terminal voltage V_{ab} of the source is zero.

25.25 Our sketch for this problem.

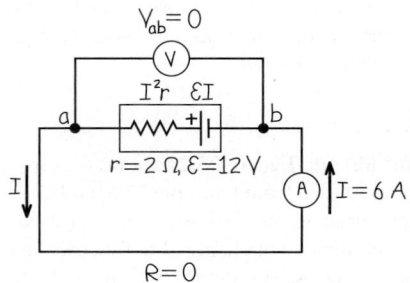

EVALUATE: With ideal wires and an ideal ammeter, so that $R = 0$, *all* of the converted energy from the source is dissipated within the source. This is why a short-circuited battery is quickly ruined and may explode.

TEST YOUR UNDERSTANDING OF SECTION 25.5 Rank the following circuits in order from highest to lowest values of the net power output of the battery. (i) A 1.4-Ω resistor connected to a 1.5-V battery that has an internal resistance of 0.10 Ω; (ii) a 1.8-Ω resistor connected to a 4.0-V battery that has a terminal voltage of 3.6 V but an unknown internal resistance; (iii) an unknown resistor connected to a 12.0-V battery that has an internal resistance of 0.20 Ω and a terminal voltage of 11.0 V. ❚

25.26 Random motions of an electron in a metallic crystal **(a)** with zero electric field and **(b)** with an electric field that causes drift. The curvatures of the paths are greatly exaggerated.

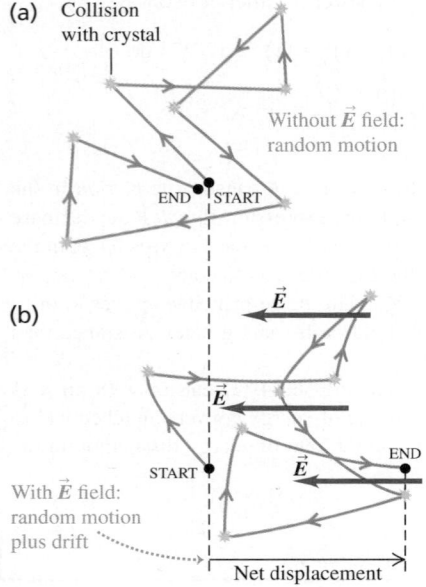

25.6 THEORY OF METALLIC CONDUCTION

We can gain additional insight into electrical conduction by looking at the microscopic origin of conductivity. We'll consider a very simple model that treats the electrons as classical particles and ignores their quantum-mechanical behavior in solids. Using this model, we'll derive an expression for the resistivity of a metal. Even though this model is not entirely correct, it will still help you to develop an intuitive idea of the microscopic basis of conduction.

In the simplest microscopic model of conduction in a metal, each atom in the metallic crystal gives up one or more of its outer electrons. These electrons are then free to move through the crystal, colliding at intervals with the stationary positive ions. The motion of the electrons is analogous to the motion of molecules of a gas moving through a porous bed of sand.

If there is no electric field, the electrons move in straight lines between collisions, the directions of their velocities are random, and on average they never get anywhere (**Fig. 25.26a**). But if an electric field is present, the paths curve slightly because of the acceleration caused by electric-field forces. Figure 25.26b shows a few paths of an electron in an electric field directed from right to left. As we mentioned in Section 25.1, the average speed of random motion is of the order of 10^6 m/s, while the average drift speed is *much* slower, of the order of 10^{-4} m/s.

The average time between collisions is called the **mean free time,** denoted by τ. **Figure 25.27** shows a mechanical analog of this electron motion.

We would like to derive from this model an expression for the resistivity ρ of a material, defined by Eq. (25.5):

$$\rho = \frac{E}{J} \tag{25.21}$$

where E and J are the magnitudes of electric field and current density, respectively. The current density $\vec{J}$ is in turn given by Eq. (25.4):

$$\vec{J} = nq\vec{v}_{\mathrm{d}} \tag{25.22}$$

where n is the number of free electrons per unit volume (the electron concentration), $q = -e$ is the charge of each, and $\vec{v}_{\mathrm{d}}$ is their average drift velocity.

We need to relate the drift velocity $\vec{v}_{\mathrm{d}}$ to the electric field $\vec{E}$. The value of $\vec{v}_{\mathrm{d}}$ is determined by a steady-state condition in which, on average, the velocity *gains* of the charges due to the force of the $\vec{E}$ field are just balanced by the velocity *losses* due to collisions. To clarify this process, let's imagine turning on the two effects one at a time. Suppose that before time $t = 0$ there is no field. The electron motion is then completely random. A typical electron has velocity $\vec{v}_0$ at time $t = 0$, and the value of $\vec{v}_0$ averaged over many electrons (that is, the initial velocity of an average electron) is zero: $(\vec{v}_0)_{\mathrm{av}} = \mathbf{0}$. Then at time $t = 0$ we turn on a constant electric field $\vec{E}$. The field exerts a force $\vec{F} = q\vec{E}$ on each charge, and this causes an acceleration $\vec{a}$ in the direction of the force, given by

$$\vec{a} = \frac{\vec{F}}{m} = \frac{q\vec{E}}{m}$$

where m is the electron mass. *Every* electron has this acceleration.

After a time τ, the average time between collisions, we "turn on" the collisions. At time $t = \tau$ an electron that has velocity $\vec{v}_0$ at time $t = 0$ has a velocity

$$\vec{v} = \vec{v}_0 + \vec{a}\tau$$

The velocity $\vec{v}_{\mathrm{av}}$ of an *average* electron at this time is the sum of the averages of the two terms on the right. As we have pointed out, the initial velocity $\vec{v}_0$ is zero for an average electron, so

$$\vec{v}_{\mathrm{av}} = \vec{a}\tau = \frac{q\tau}{m}\vec{E} \tag{25.23}$$

After time $t = \tau$, the tendency of the collisions to decrease the velocity of an average electron (by means of randomizing collisions) just balances the tendency of the $\vec{E}$ field to increase this velocity. Thus the velocity of an average electron, given by Eq. (25.23), is maintained over time and is equal to the drift velocity $\vec{v}_{\mathrm{d}}$:

$$\vec{v}_{\mathrm{d}} = \frac{q\tau}{m}\vec{E}$$

Now we substitute this equation for the drift velocity $\vec{v}_{\mathrm{d}}$ into Eq. (25.22):

$$\vec{J} = nq\vec{v}_{\mathrm{d}} = \frac{nq^2\tau}{m}\vec{E}$$

Comparing this with Eq. (25.21), which we can rewrite as $\vec{J} = \vec{E}/\rho$, and substituting $q = -e$ for an electron, we see that

Resistivity of a metal ⋯⋯⋯⋯ Electron mass ⋯⋯⋯⋯

$$\rho = \frac{m}{ne^2\tau} \tag{25.24}$$

Number of free electrons per unit volume ⋯➤ ⋯⋯ Average time between collisions

⋯⋯ Magnitude of electron charge

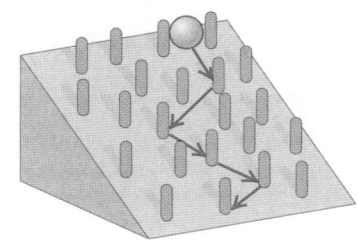

25.27 The motion of a ball rolling down an inclined plane and bouncing off pegs in its path is analogous to the motion of an electron in a metallic conductor with an electric field present.

PhET: Conductivity

If n and τ are independent of $\vec{E}$, then the resistivity is independent of $\vec{E}$ and the conducting material obeys Ohm's law.

Turning the interactions on one at a time may seem artificial. But the derivation would come out the same if each electron had its own clock and the $t = 0$ times were different for different electrons. If τ is the average time between collisions, then $\vec{v}_d$ is still the average electron drift velocity, even though the motions of the various electrons aren't actually correlated in the way we postulated.

What about the temperature dependence of resistivity? In a perfect crystal with no atoms out of place, a correct quantum-mechanical analysis would let the free electrons move through the crystal with no collisions at all. But the atoms vibrate about their equilibrium positions. As the temperature increases, the amplitudes of these vibrations increase, collisions become more frequent, and the mean free time τ decreases. So this theory predicts that the resistivity of a metal increases with temperature. In a superconductor, roughly speaking, there are no inelastic collisions, τ is infinite, and the resistivity ρ is zero.

In a pure semiconductor such as silicon or germanium, the number of charge carriers per unit volume, n, is not constant but increases very rapidly with increasing temperature. This increase in n far outweighs the decrease in the mean free time, and in a semiconductor the resistivity always decreases rapidly with increasing temperature. At low temperatures, n is very small, and the resistivity becomes so large that the material can be considered an insulator.

Electrons gain energy between collisions through the work done on them by the electric field. During collisions they transfer some of this energy to the atoms of the material of the conductor. This leads to an increase in the material's internal energy and temperature; that's why wires carrying current get warm. If the electric field in the material is large enough, an electron can gain enough energy between collisions to knock off electrons that are normally bound to atoms in the material. These can then knock off more electrons, and so on, leading to an avalanche of current. This is the basis of dielectric breakdown in insulators (see Section 24.4).

EXAMPLE 25.11 **MEAN FREE TIME IN COPPER**

Calculate the mean free time between collisions in copper at room temperature.

SOLUTION

IDENTIFY and SET UP: We can obtain an expression for mean free time τ in terms of n, ρ, e, and m by rearranging Eq. (25.24). From Example 25.1 and Table 25.1, for copper $n = 8.5 \times 10^{28}\text{ m}^{-3}$ and $\rho = 1.72 \times 10^{-8}\ \Omega \cdot \text{m}$. In addition, $e = 1.60 \times 10^{-19}$ C and $m = 9.11 \times 10^{-31}$ kg for electrons.

EXECUTE: From Eq. (25.24), we get

$$\tau = \frac{m}{ne^2\rho}$$

$$= \frac{9.11 \times 10^{-31}\text{ kg}}{(8.5 \times 10^{28}\text{ m}^{-3})(1.60 \times 10^{-19}\text{ C})^2(1.72 \times 10^{-8}\ \Omega \cdot \text{m})}$$

$$= 2.4 \times 10^{-14}\text{ s}$$

EVALUATE: The mean free time is the average time between collisions for a given electron. Taking the reciprocal, we find that each electron averages $1/\tau = 4.2 \times 10^{13}$ collisions per second!

TEST YOUR UNDERSTANDING OF SECTION 25.6 Which of the following factors will, if increased, make it more difficult to produce a certain amount of current in a conductor? (There may be more than one correct answer.) (i) The mass of the moving charged particles in the conductor; (ii) the number of moving charged particles per cubic meter; (iii) the amount of charge on each moving particle; (iv) the average time between collisions for a typical moving charged particle. ∎

Current and current density: Current is the amount of charge flowing through a specified area, per unit time. The SI unit of current is the ampere (1 A = 1 C/s). The current I through an area A depends on the concentration n and charge q of the charge carriers, as well as on the magnitude of their drift velocity $\vec{v}_d$. The current density is current per unit cross-sectional area. Current is usually described in terms of a flow of positive charge, even when the charges are actually negative or of both signs. (See Example 25.1.)

$$I = \frac{dQ}{dt} = n|q|v_d A \qquad (25.2)$$

$$\vec{J} = nq\vec{v}_d \qquad (25.4)$$

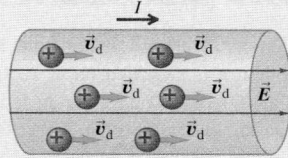

Resistivity: The resistivity ρ of a material is the ratio of the magnitudes of electric field and current density. Good conductors have small resistivity; good insulators have large resistivity. Ohm's law, obeyed approximately by many materials, states that ρ is a constant independent of the value of E. Resistivity usually increases with temperature; for small temperature changes this variation is represented approximately by Eq. (25.6), where α is the temperature coefficient of resistivity.

$$\rho = \frac{E}{J} \qquad (25.5)$$

$$\rho(T) = \rho_0[1 + \alpha(T - T_0)] \qquad (25.6)$$

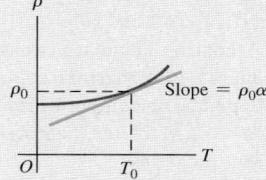

Metal: ρ increases with increasing T.

Resistors: The potential difference V across a sample of material that obeys Ohm's law is proportional to the current I through the sample. The ratio $V/I = R$ is the resistance of the sample. The SI unit of resistance is the ohm (1 Ω = 1 V/A). The resistance of a cylindrical conductor is related to its resistivity ρ, length L, and cross-sectional area A. (See Examples 25.2 and 25.3.)

$$V = IR \qquad (25.11)$$

$$R = \frac{\rho L}{A} \qquad (25.10)$$

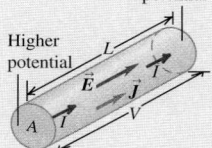

Circuits and emf: A complete circuit has a continuous current-carrying path. A complete circuit carrying a steady current must contain a source of electromotive force (emf) $\mathcal{E}$. The SI unit of electromotive force is the volt (V). Every real source of emf has some internal resistance r, so its terminal potential difference V_{ab} depends on current. (See Examples 25.4–25.7.)

$$V_{ab} = \mathcal{E} - Ir \qquad (25.15)$$
(source with internal resistance)

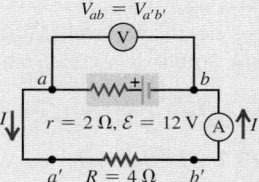

Energy and power in circuits: A circuit element puts energy into a circuit if the current direction is from lower to higher potential in the device, and it takes energy out of the circuit if the current is opposite. The power P equals the product of the potential difference $V_a - V_b = V_{ab}$ and the current I. A resistor always takes electrical energy out of a circuit. (See Examples 25.8–25.10.)

$$P = V_{ab}I \qquad (25.17)$$
(general circuit element)

$$P = V_{ab}I = I^2 R = \frac{V_{ab}^2}{R} \qquad (25.18)$$
(power delivered to a resistor)

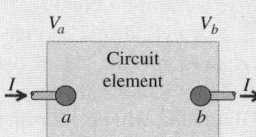

Conduction in metals: In a metal, current is due to the motion of electrons. They move freely through the metallic crystal but collide with positive ions. In a crude classical model of this motion, the resistivity of the material can be related to the electron mass, charge, speed of random motion, density, and mean free time between collisions. (See Example 25.11.)

$$\rho = \frac{m}{ne^2\tau} \qquad (25.24)$$

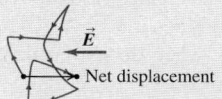

BRIDGING PROBLEM | RESISTIVITY, TEMPERATURE, AND POWER

A toaster using a Nichrome heating element operates on 120 V. When it is switched on at 20°C, the heating element carries an initial current of 1.35 A. A few seconds later the current reaches the steady value of 1.23 A. (a) What is the final temperature of the element? The average value of the temperature coefficient of resistivity for Nichrome over the relevant temperature range is 4.5×10^{-4} $(\text{C}°)^{-1}$. (b) What is the power dissipated in the heating element initially and when the current reaches 1.23 A?

SOLUTION GUIDE

IDENTIFY and SET UP

1. A heating element acts as a resistor that converts electrical energy into thermal energy. The resistivity ρ of Nichrome depends on temperature, and hence so does the resistance $R = \rho L/A$ of the heating element and the current $I = V/R$ that it carries.
2. We are given $V = 120$ V and the initial and final values of I. Select an equation that will allow you to find the initial and final values of resistance, and an equation that relates resistance to temperature [the target variable in part (a)].

3. The power P dissipated in the heating element depends on I and V. Select an equation that will allow you to calculate the initial and final values of P.

EXECUTE

4. Combine your equations from step 2 to give a relationship between the initial and final values of I and the initial and final temperatures (20°C and T_{final}).
5. Solve your expression from step 4 for T_{final}.
6. Use your equation from step 3 to find the initial and final powers.

EVALUATE

7. Is the final temperature greater than or less than 20°C? Does this make sense?
8. Is the final resistance greater than or less than the initial resistance? Again, does this make sense?
9. Is the final power greater than or less than the initial power? Does this agree with your observations in step 8?

Problems

For assigned homework and other learning materials, go to MasteringPhysics®. **MP**

•, ••, •••: Difficulty levels. **CP**: Cumulative problems incorporating material from earlier chapters. **CALC**: Problems requiring calculus. **DATA**: Problems involving real data, scientific evidence, experimental design, and/or statistical reasoning. **BIO**: Biosciences problems.

DISCUSSION QUESTIONS

Q25.1 The definition of resistivity $(\rho = E/J)$ implies that an electric field exists inside a conductor. Yet we saw in Chapter 21 that there can be no electrostatic electric field inside a conductor. Is there a contradiction here? Explain.

Q25.2 A cylindrical rod has resistance R. If we triple its length and diameter, what is its resistance, in terms of R?

Q25.3 A cylindrical rod has resistivity ρ. If we triple its length and diameter, what is its resistivity, in terms of ρ?

Q25.4 Two copper wires with different diameters are joined end to end. If a current flows in the wire combination, what happens to electrons when they move from the larger-diameter wire into the smaller-diameter wire? Does their drift speed increase, decrease, or stay the same? If the drift speed changes, what is the force that causes the change? Explain your reasoning.

Q25.5 When is a 1.5-V AAA battery *not* actually a 1.5-V battery? That is, when do its terminals provide a potential difference of less than 1.5 V?

Q25.6 Can the potential difference between the terminals of a battery ever be opposite in direction to the emf? If it can, give an example. If it cannot, explain why not.

Q25.7 A rule of thumb used to determine the internal resistance of a source is that it is the open-circuit voltage divided by the short-circuit current. Is this correct? Why or why not?

Q25.8 Batteries are always labeled with their emf; for instance, an AA flashlight battery is labeled "1.5 volts." Would it also be appropriate to put a label on batteries stating how much current they provide? Why or why not?

Q25.9 We have seen that a coulomb is an enormous amount of charge; it is virtually impossible to place a charge of 1 C on an object. Yet, a current of 10 A, 10 C/s, is quite reasonable. Explain this apparent discrepancy.

Q25.10 Electrons in an electric circuit pass through a resistor. The wire on either side of the resistor has the same diameter. (a) How does the drift speed of the electrons before entering the resistor compare to the speed after leaving the resistor? Explain your reasoning. (b) How does the potential energy for an electron before entering the resistor compare to the potential energy after leaving the resistor? Explain your reasoning.

Q25.11 Temperature coefficients of resistivity are given in Table 25.2. (a) If a copper heating element is connected to a source of constant voltage, does the electrical power consumed by the heating element increase or decrease as its temperature increases? Explain. (b) A resistor in the form of a carbon cylinder is connected to the voltage source. As the temperature of the cylinder increases, does the electrical power it consumes increase or decrease? Explain.

Q25.12 Which of the graphs in **Fig. Q25.12** best illustrates the current I in a real resistor as a function of the potential difference V across it? Explain.

Figure **Q25.12**

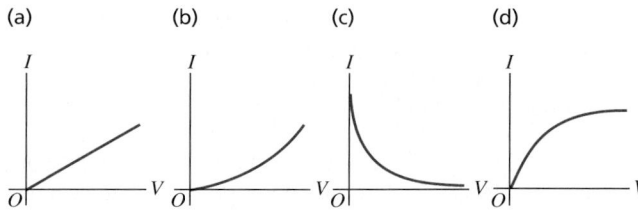

(a) (b) (c) (d)

Q25.13 Why does an electric light bulb nearly always burn out just as you turn on the light, almost never while the light is shining?

Q25.14 A light bulb glows because it has resistance. The brightness of a light bulb increases with the electrical power dissipated in the bulb. (a) In the circuit shown in **Fig. Q25.14a**, the two bulbs A and B are identical. Compared to bulb A, does bulb B glow more brightly, just as brightly, or less brightly? Explain your reasoning. (b) Bulb B is removed from the circuit and the circuit is completed as shown in Fig. Q25.14b. Compared to the brightness of bulb A in Fig. Q25.14a, does bulb A now glow more brightly, just as brightly, or less brightly? Explain your reasoning.

Figure **Q25.14**

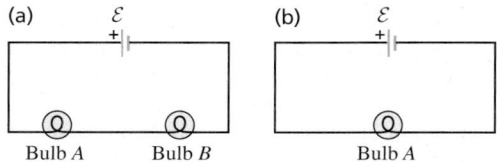

Bulb A Bulb B Bulb A

Q25.15 (See Discussion Question Q25.14.) An ideal ammeter A is placed in a circuit with a battery and a light bulb as shown in **Fig. Q25.15a**, and the ammeter reading is noted. The circuit is then reconnected as in Fig. Q25.15b, so that the positions of the ammeter and light bulb are reversed. (a) How does the ammeter reading in the situation shown in Fig. Q25.15a compare to the reading in the situation shown in Fig. Q25.15b? Explain your reasoning. (b) In which situation does the light bulb glow more brightly? Explain your reasoning.

Figure **Q25.15**

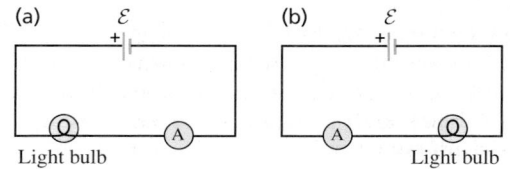

Light bulb Light bulb

Q25.16 (See Discussion Question Q25.14.) Will a light bulb glow more brightly when it is connected to a battery as shown in **Fig. Q25.16a**, in which an ideal ammeter A is placed in the circuit, or when it is connected as shown in Fig. 25.16b, in which an ideal voltmeter V is placed in the circuit? Explain your reasoning.

Figure **Q25.16**

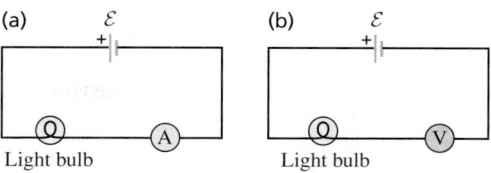

(a) (b)

Light bulb Light bulb

Q25.17 The energy that can be extracted from a storage battery is always less than the energy that goes into it while it is being charged. Why?

Q25.18 Eight flashlight batteries in series have an emf of about 12 V, similar to that of a car battery. Could they be used to start a car with a dead battery? Why or why not?

Q25.19 Small aircraft often have 24-V electrical systems rather than the 12-V systems in automobiles, even though the electrical power requirements are roughly the same in both applications. The explanation given by aircraft designers is that a 24-V system weighs less than a 12-V system because thinner wires can be used. Explain why this is so.

Q25.20 Long-distance, electric-power, transmission lines always operate at very high voltage, sometimes as much as 750 kV. What are the advantages of such high voltages? What are the disadvantages?

Q25.21 Ordinary household electric lines in North America usually operate at 120 V. Why is this a desirable voltage, rather than a value considerably larger or smaller? On the other hand, automobiles usually have 12-V electrical systems. Why is this a desirable voltage?

Q25.22 A fuse is a device designed to break a circuit, usually by melting when the current exceeds a certain value. What characteristics should the material of the fuse have?

Q25.23 High-voltage power supplies are sometimes designed intentionally to have rather large internal resistance as a safety precaution. Why is such a power supply with a large internal resistance safer than a supply with the same voltage but lower internal resistance?

Q25.24 The text states that good thermal conductors are also good electrical conductors. If so, why don't the cords used to connect toasters, irons, and similar heat-producing appliances get hot by conduction of heat from the heating element?

EXERCISES

Section 25.1 Current

25.1 • **Lightning Strikes.** During lightning strikes from a cloud to the ground, currents as high as 25,000 A can occur and last for about 40 μs. How much charge is transferred from the cloud to the earth during such a strike?

25.2 • A silver wire 2.6 mm in diameter transfers a charge of 420 C in 80 min. Silver contains 5.8×10^{28} free electrons per cubic meter. (a) What is the current in the wire? (b) What is the magnitude of the drift velocity of the electrons in the wire?

25.3 • A 5.00-A current runs through a 12-gauge copper wire (diameter 2.05 mm) and through a light bulb. Copper has 8.5×10^{28} free electrons per cubic meter. (a) How many electrons pass through the light bulb each second? (b) What is the current density in the wire? (c) At what speed does a typical electron pass by any given point in the wire? (d) If you were to use wire of twice the diameter, which of the above answers would change? Would they increase or decrease?

25.4 • An 18-gauge copper wire (diameter 1.02 mm) carries a current with a current density of 3.20×10^6 A/m². The density of free electrons for copper is 8.5×10^{28} electrons per cubic meter. Calculate (a) the current in the wire and (b) the drift velocity of electrons in the wire.

25.5 •• Copper has 8.5×10^{28} free electrons per cubic meter. A 71.0-cm length of 12-gauge copper wire that is 2.05 mm in diameter carries 4.85 A of current. (a) How much time does it take for an electron to travel the length of the wire? (b) Repeat part (a) for 6-gauge copper wire (diameter 4.12 mm) of the same length that carries the same current. (c) Generally speaking, how does changing the diameter of a wire that carries a given amount of current affect the drift velocity of the electrons in the wire?

25.6 •• You want to produce three 1.00-mm-diameter cylindrical wires, each with a resistance of 1.00 Ω at room temperature. One wire is gold, one is copper, and one is aluminum. Refer to Table 25.1 for the resistivity values. (a) What will be the length of each wire? (b) Gold has a density of 1.93×10^4 kg/m³. What will be the mass of the gold wire? If you consider the current price of gold, is this wire very expensive?

25.7 • CALC The current in a wire varies with time according to the relationship $I = 55$ A $- (0.65$ A/s²$)t^2$. (a) How many coulombs of charge pass a cross section of the wire in the time interval between $t = 0$ and $t = 8.0$ s? (b) What constant current would transport the same charge in the same time interval?

25.8 • Current passes through a solution of sodium chloride. In 1.00 s, 2.68×10^{16} Na$^+$ ions arrive at the negative electrode and 3.92×10^{16} Cl$^-$ ions arrive at the positive electrode. (a) What is the current passing between the electrodes? (b) What is the direction of the current?

25.9 • BIO **Transmission of Nerve Impulses.** Nerve cells transmit electric signals through their long tubular axons. These signals propagate due to a sudden rush of Na$^+$ ions, each with charge $+e$, into the axon. Measurements have revealed that typically about 5.6×10^{11} Na$^+$ ions enter each meter of the axon during a time of 10 ms. What is the current during this inflow of charge in a meter of axon?

Section 25.2 Resistivity and Section 25.3 Resistance

25.10 • (a) At room temperature, what is the strength of the electric field in a 12-gauge copper wire (diameter 2.05 mm) that is needed to cause a 4.50-A current to flow? (b) What field would be needed if the wire were made of silver instead?

25.11 •• A 1.50-m cylindrical rod of diameter 0.500 cm is connected to a power supply that maintains a constant potential difference of 15.0 V across its ends, while an ammeter measures the current through it. You observe that at room temperature (20.0°C) the ammeter reads 18.5 A, while at 92.0°C it reads 17.2 A. You can ignore any thermal expansion of the rod. Find (a) the resistivity at 20.0°C and (b) the temperature coefficient of resistivity at 20°C for the material of the rod.

25.12 • A copper wire has a square cross section 2.3 mm on a side. The wire is 4.0 m long and carries a current of 3.6 A. The density of free electrons is 8.5×10^{28}/m³. Find the magnitudes of (a) the current density in the wire and (b) the electric field in the wire. (c) How much time is required for an electron to travel the length of the wire?

25.13 • A 14-gauge copper wire of diameter 1.628 mm carries a current of 12.5 mA. (a) What is the potential difference across a 2.00-m length of the wire? (b) What would the potential difference in part (a) be if the wire were silver instead of copper, but all else were the same?

25.14 •• A wire 6.50 m long with diameter of 2.05 mm has a resistance of 0.0290 Ω. What material is the wire most likely made of?

25.15 •• A cylindrical tungsten filament 15.0 cm long with a diameter of 1.00 mm is to be used in a machine for which the temperature will range from room temperature (20°C) up to 120°C. It will carry a current of 12.5 A at all temperatures (consult Tables 25.1 and 25.2). (a) What will be the maximum electric field in this filament, and (b) what will be its resistance with that field? (c) What will be the maximum potential drop over the full length of the filament?

25.16 •• A ductile metal wire has resistance R. What will be the resistance of this wire in terms of R if it is stretched to three times its original length, assuming that the density and resistivity of the material do not change when the wire is stretched? (*Hint:* The amount of metal does not change, so stretching out the wire will affect its cross-sectional area.)

25.17 • In household wiring, copper wire 2.05 mm in diameter is often used. Find the resistance of a 24.0-m length of this wire.

25.18 •• What diameter must a copper wire have if its resistance is to be the same as that of an equal length of aluminum wire with diameter 2.14 mm?

25.19 • A strand of wire has resistance 5.60 $\mu\Omega$. Find the net resistance of 120 such strands if they are (a) placed side by side to form a cable of the same length as a single strand, and (b) connected end to end to form a wire 120 times as long as a single strand.

25.20 • You apply a potential difference of 4.50 V between the ends of a wire that is 2.50 m in length and 0.654 mm in radius. The resulting current through the wire is 17.6 A. What is the resistivity of the wire?

25.21 • A current-carrying gold wire has diameter 0.84 mm. The electric field in the wire is 0.49 V/m. What are (a) the current carried by the wire; (b) the potential difference between two points in the wire 6.4 m apart; (c) the resistance of a 6.4-m length of this wire?

25.22 • A hollow aluminum cylinder is 2.50 m long and has an inner radius of 2.75 cm and an outer radius of 4.60 cm. Treat each surface (inner, outer, and the two end faces) as an equipotential surface. At room temperature, what will an ohmmeter read if it is connected between (a) the opposite faces and (b) the inner and outer surfaces?

25.23 • (a) What is the resistance of a Nichrome wire at 0.0°C if its resistance is 100.00 Ω at 11.5°C? (b) What is the resistance of a carbon rod at 25.8°C if its resistance is 0.0160 Ω at 0.0°C?

25.24 • A carbon resistor is to be used as a thermometer. On a winter day when the temperature is 4.0°C, the resistance of the carbon resistor is 217.3 Ω. What is the temperature on a spring day when the resistance is 215.8 Ω? (Take the reference temperature T_0 to be 4.0°C.)

Section 25.4 Electromotive Force and Circuits

25.25 • A copper transmission cable 100 km long and 10.0 cm in diameter carries a current of 125 A. (a) What is the potential drop across the cable? (b) How much electrical energy is dissipated as thermal energy every hour?

25.26 • Consider the circuit shown in **Fig. E25.26**. The terminal voltage of the 24.0-V battery is 21.2 V. What are (a) the internal resistance r of the battery and (b) the resistance R of the circuit resistor?

Figure **E25.26**

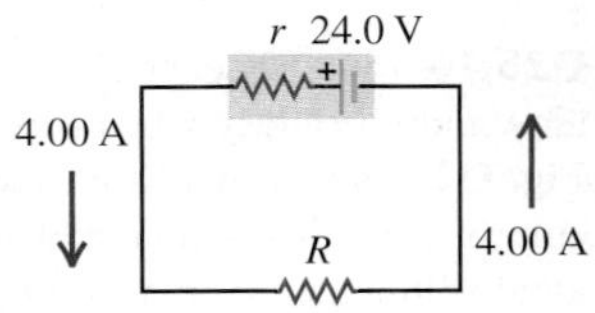

25.27 • An ideal voltmeter V is connected to a 2.0-Ω resistor and a battery with emf 5.0 V and internal resistance 0.5 Ω as shown in **Fig. E25.27**. (a) What is the current in the 2.0-Ω resistor? (b) What is the terminal voltage of the battery? (c) What is the reading on the voltmeter? Explain your answers.

Figure **E25.27**

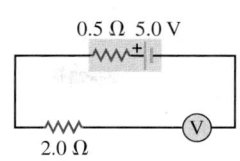

0.5 Ω 5.0 V

2.0 Ω

25.28 • An idealized ammeter is connected to a battery as shown in **Fig. E25.28**. Find (a) the reading of the ammeter, (b) the current through the 4.00-Ω resistor, (c) the terminal voltage of the battery.

Figure **E25.28**

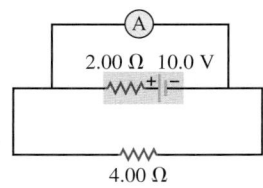

2.00 Ω 10.0 V

4.00 Ω

25.29 • When switch S in **Fig. E25.29** is open, the voltmeter V reads 3.08 V. When the switch is closed, the voltmeter reading drops to 2.97 V, and the ammeter A reads 1.65 A. Find the emf, the internal resistance of the battery, and the circuit resistance R. Assume that the two meters are ideal, so they don't affect the circuit.

Figure **E25.29**

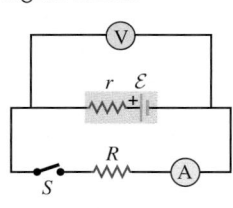

r $\mathcal{E}$

R

S

25.30 • The circuit shown in **Fig. E25.30** contains two batteries, each with an emf and an internal resistance, and two resistors. Find (a) the current in the circuit (magnitude *and* direction); (b) the terminal voltage V_{ab} of the 16.0-V battery; (c) the potential difference V_{ac} of point a with respect to point c. (d) Using Fig. 25.20 as a model, graph the potential rises and drops in this circuit.

Figure **E25.30**

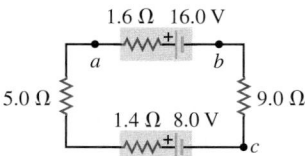

1.6 Ω 16.0 V

a b

5.0 Ω 9.0 Ω

1.4 Ω 8.0 V

c

25.31 • In the circuit shown in Fig. E25.30, the 16.0-V battery is removed and reinserted with the opposite polarity, so that its negative terminal is now next to point a. Find (a) the current in the circuit (magnitude *and* direction); (b) the terminal voltage V_{ba} of the 16.0-V battery; (c) the potential difference V_{ac} of point a with respect to point c. (d) Graph the potential rises and drops in this circuit (see Fig. 25.20).

25.32 • In the circuit of Fig. E25.30, the 5.0-Ω resistor is removed and replaced by a resistor of unknown resistance R. When this is done, an ideal voltmeter connected across the points b and c reads 1.9 V. Find (a) the current in the circuit and (b) the resistance R. (c) Graph the potential rises and drops in this circuit (see Fig. 25.20).

25.33 •• The circuit shown in **Fig. E25.33** contains two batteries, each with an emf and an internal resistance, and two resistors. Find (a) the current in the circuit (magnitude and direction) and (b) the terminal voltage V_{ab} of the 16.0-V battery.

Figure **E25.33**

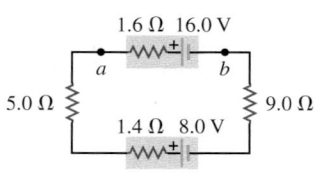

1.6 Ω 16.0 V

a b

5.0 Ω 9.0 Ω

1.4 Ω 8.0 V

Section 25.5 Energy and Power in Electric Circuits

25.34 •• When a resistor with resistance R is connected to a 1.50-V flashlight battery, the resistor consumes 0.0625 W of electrical power. (Throughout, assume that each battery has negligible internal resistance.) (a) What power does the resistor consume if it is connected to a 12.6-V car battery? Assume that R remains constant when the power consumption changes. (b) The resistor is connected to a battery and consumes 5.00 W. What is the voltage of this battery?

25.35 • **Light Bulbs.** The power rating of a light bulb (such as a 100-W bulb) is the power it dissipates when connected across a 120-V potential difference. What is the resistance of (a) a 100-W bulb and (b) a 60-W bulb? (c) How much current does each bulb draw in normal use?

25.36 • If a "75-W" bulb (see Problem 25.35) is connected across a 220-V potential difference (as is used in Europe), how much power does it dissipate? Ignore the temperature dependence of the bulb's resistance.

25.37 • **European Light Bulb.** In Europe the standard voltage in homes is 220 V instead of the 120 V used in the United States. Therefore a "100-W" European bulb would be intended for use with a 220-V potential difference (see Problem 25.36). (a) If you bring a "100-W" European bulb home to the United States, what should be its U.S. power rating? (b) How much current will the 100-W European bulb draw in normal use in the United States?

25.38 • A battery-powered global positioning system (GPS) receiver operating on 9.0 V draws a current of 0.13 A. How much electrical energy does it consume during 30 minutes?

25.39 • Consider the circuit of Fig. E25.30. (a) What is the total rate at which electrical energy is dissipated in the 5.0-Ω and 9.0-Ω resistors? (b) What is the power output of the 16.0-V battery? (c) At what rate is electrical energy being converted to other forms in the 8.0-V battery? (d) Show that the power output of the 16.0-V battery equals the overall rate of consumption of electrical energy in the rest of the circuit.

25.40 • **BIO Electric Eels.** Electric eels generate electric pulses along their skin that can be used to stun an enemy when they come into contact with it. Tests have shown that these pulses can be up to 500 V and produce currents of 80 mA (or even larger). A typical pulse lasts for 10 ms. What power and how much energy are delivered to the unfortunate enemy with a single pulse, assuming a steady current?

25.41 • **BIO Treatment of Heart Failure.** A heart defibrillator is used to enable the heart to start beating if it has stopped. This is done by passing a large current of 12 A through the body at 25 V for a very short time, usually about 3.0 ms. (a) What power does the defibrillator deliver to the body, and (b) how much energy is transferred?

25.42 •• The battery for a certain cell phone is rated at 3.70 V. According to the manufacturer it can produce 3.15×10^4 J of electrical energy, enough for 5.25 h of operation, before needing to be recharged. Find the average current that this cell phone draws when turned on.

25.43 •• The capacity of a storage battery, such as those used in automobile electrical systems, is rated in ampere-hours (A·h). A 50-A·h battery can supply a current of 50 A for 1.0 h, or 25 A for 2.0 h, and so on. (a) What total energy can be supplied by a 12-V, 60-A·h battery if its internal resistance is negligible? (b) What volume (in liters) of gasoline has a total heat of combustion equal to the energy obtained in part (a)? (See Section 17.6; the density of gasoline is 900 kg/m^3.) (c) If a generator with an average electrical power output of 0.45 kW is connected to the battery, how much time will be required for it to charge the battery fully?

25.44 • An idealized voltmeter is connected across the terminals of a 15.0-V battery, and a 75.0-Ω appliance is also connected across its terminals. If the voltmeter reads 11.9 V, (a) how much power is being dissipated by the appliance, and (b) what is the internal resistance of the battery?

25.45 •• A 25.0-Ω bulb is connected across the terminals of a 12.0-V battery having 3.50 Ω of internal resistance. What percentage of the power of the battery is dissipated across the internal resistance and hence is not available to the bulb?

25.46 •• A typical small flashlight contains two batteries, each having an emf of 1.5 V, connected in series with a bulb having resistance 17 Ω. (a) If the internal resistance of the batteries is negligible, what power is delivered to the bulb? (b) If the batteries last for 5.0 h, what is the total energy delivered to the bulb? (c) The resistance of real batteries increases as they run down. If the initial internal resistance is negligible, what is the combined internal resistance of both batteries when the power to the bulb has decreased to half its initial value? (Assume that the resistance of the bulb is constant. Actually, it will change somewhat when the current through the filament changes, because this changes the temperature of the filament and hence the resistivity of the filament wire.)

25.47 • In the circuit in **Fig. E25.47**, find (a) the rate of conversion of internal (chemical) energy to electrical energy within the battery; (b) the rate of dissipation of electrical energy in the battery; (c) the rate of dissipation of electrical energy in the external resistor.

Figure **E25.47**

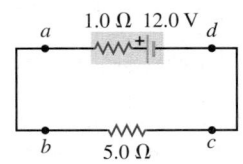

25.48 • A "540-W" electric heater is designed to operate from 120-V lines. (a) What is its operating resistance? (b) What current does it draw? (c) If the line voltage drops to 110 V, what power does the heater take? (Assume that the resistance is constant. Actually, it will change because of the change in temperature.) (d) The heater coils are metallic, so that the resistance of the heater decreases with decreasing temperature. If the change of resistance with temperature is taken into account, will the electrical power consumed by the heater be larger or smaller than what you calculated in part (c)? Explain.

Section 25.6 Theory of Metallic Conduction

25.49 •• Pure silicon at room temperature contains approximately 1.0×10^{16} free electrons per cubic meter. (a) Referring to Table 25.1, calculate the mean free time τ for silicon at room temperature. (b) Your answer in part (a) is much greater than the mean free time for copper given in Example 25.11. Why, then, does pure silicon have such a high resistivity compared to copper?

PROBLEMS

25.50 •• In an ionic solution, a current consists of Ca^{2+} ions (of charge $+2e$) and Cl^- ions (of charge $-e$) traveling in opposite directions. If 5.11×10^{18} Cl^- ions go from A to B every 0.50 min, while 3.24×10^{18} Ca^{2+} ions move from B to A, what is the current (in mA) through this solution, and in which direction (from A to B or from B to A) is it going?

25.51 • An electrical conductor designed to carry large currents has a circular cross section 2.50 mm in diameter and is 14.0 m long. The resistance between its ends is 0.104 Ω. (a) What is the resistivity of the material? (b) If the electric-field magnitude in the conductor is 1.28 V/m, what is the total current? (c) If the material has 8.5×10^{28} free electrons per cubic meter, find the average drift speed under the conditions of part (b).

25.52 •• An overhead transmission cable for electrical power is 2000 m long and consists of two parallel copper wires, each encased in insulating material. A short circuit has developed somewhere along the length of the cable where the insulation has worn thin and the two wires are in contact. As a power-company employee, you must locate the short so that repair crews can be sent to that location. Both ends of the cable have been disconnected from the power grid. At one end of the cable (point A), you connect the ends of the two wires to a 9.00-V battery that has negligible internal resistance and measure that 2.86 A of current flows through the battery. At the other end of the cable (point B), you attach those two wires to the battery and measure that 1.65 A of current flows through the battery. How far is the short from point A?

25.53 •• On your first day at work as an electrical technician, you are asked to determine the resistance per meter of a long piece of wire. The company you work for is poorly equipped. You find a battery, a voltmeter, and an ammeter, but no meter for directly measuring resistance (an ohmmeter). You put the leads from the voltmeter across the terminals of the battery, and the meter reads 12.6 V. You cut off a 20.0-m length of wire and connect it to the battery, with an ammeter in series with it to measure the current in the wire. The ammeter reads 7.00 A. You then cut off a 40.0-m length of wire and connect it to the battery, again with the ammeter in series to measure the current. The ammeter reads 4.20 A. Even though the equipment you have available to you is limited, your boss assures you of its high quality: The ammeter has very small resistance, and the voltmeter has very large resistance. What is the resistance of 1 meter of wire?

25.54 • A 2.0-m length of wire is made by welding the end of a 120-cm-long silver wire to the end of an 80-cm-long copper wire. Each piece of wire is 0.60 mm in diameter. The wire is at room temperature, so the resistivities are as given in Table 25.1. A potential difference of 9.0 V is maintained between the ends of the 2.0-m composite wire. What is (a) the current in the copper section; (b) the current in the silver section; (c) the magnitude of $\vec{E}$ in the copper; (d) the magnitude of $\vec{E}$ in the silver; (e) the potential difference between the ends of the silver section of wire?

25.55 • A 3.00-m length of copper wire at 20° C has a 1.20-m-long section with diameter 1.60 mm and a 1.80-m-long section with diameter 0.80 mm. There is a current of 2.5 mA in the 1.60-mm-diameter section. (a) What is the current in the 0.80-mm-diameter section? (b) What is the magnitude of $\vec{E}$ in the 1.60-mm-diameter section? (c) What is the magnitude of $\vec{E}$ in the 0.80-mm-diameter section? (d) What is the potential difference between the ends of the 3.00-m length of wire?

25.56 •• A heating element made of tungsten wire is connected to a large battery that has negligible internal resistance. When the heating element reaches 80.0° C, it consumes electrical energy at a rate of 480 W. What is its power consumption when its temperature is 150.0° C? Assume that the temperature coefficient of resistivity has the value given in Table 25.2 and that it is constant over the temperature range in this problem. In Eq. (25.12) take T_0 to be 20.0° C.

25.57 •• CP BIO **Struck by Lightning.** Lightning strikes can involve currents as high as 25,000 A that last for about 40 μs. If a person is struck by a bolt of lightning with these properties, the current will pass through his body. We shall assume that his mass is 75 kg, that he is wet (after all, he is in a rainstorm) and therefore has a resistance of 1.0 kΩ, and that his body is all water (which is reasonable for a rough, but plausible, approximation). (a) By how many degrees Celsius would this lightning bolt increase the temperature of 75 kg of water? (b) Given that the internal body temperature is about 37°C, would the person's temperature actually increase that much? Why not? What would happen first?

25.58 •• A resistor with resistance R is connected to a battery that has emf 12.0 V and internal resistance $r = 0.40\ \Omega$. For what two values of R will the power dissipated in the resistor be 80.0 W?

25.59 • CALC A material of resistivity ρ is formed into a solid, truncated cone of height h and radii r_1 and r_2 at either end (**Fig. P25.59**). (a) Calculate the resistance of the cone between the two flat end faces. (*Hint:* Imagine slicing the cone into very many thin disks, and calculate the resistance of one such disk.) (b) Show that your result agrees with Eq. (25.10) when $r_1 = r_2$.

Figure **P25.59**

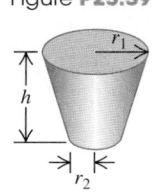

25.60 • CALC The region between two concentric conducting spheres with radii a and b is filled with a conducting material with resistivity ρ. (a) Show that the resistance between the spheres is given by

$$R = \frac{\rho}{4\pi}\left(\frac{1}{a} - \frac{1}{b}\right)$$

(b) Derive an expression for the current density as a function of radius, in terms of the potential difference V_{ab} between the spheres. (c) Show that the result in part (a) reduces to Eq. (25.10) when the separation $L = b - a$ between the spheres is small.

25.61 • The potential difference across the terminals of a battery is 8.40 V when there is a current of 1.50 A in the battery from the negative to the positive terminal. When the current is 3.50 A in the reverse direction, the potential difference becomes 10.20 V. (a) What is the internal resistance of the battery? (b) What is the emf of the battery?

25.62 • (a) What is the potential difference V_{ad} in the circuit of **Fig. P25.62**? (b) What is the terminal voltage of the 4.00-V battery? (c) A battery with emf 10.30 V and internal resistance 0.50 Ω is inserted in the circuit at d, with its negative terminal connected to the negative terminal of the 8.00-V battery. What is the difference of potential V_{bc} between the terminals of the 4.00-V battery now?

Figure **P25.62**

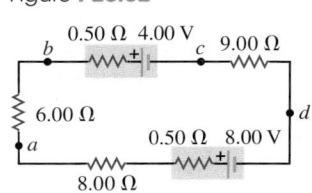

25.63 • BIO The average bulk resistivity of the human body (apart from surface resistance of the skin) is about 5.0 $\Omega \cdot$m. The conducting path between the hands can be represented approximately as a cylinder 1.6 m long and 0.10 m in diameter. The skin resistance can be made negligible by soaking the hands in salt water. (a) What is the resistance between the hands if the skin resistance is negligible? (b) What potential difference between the hands is needed for a lethal shock current of 100 mA? (Note that your result shows that small potential differences produce dangerous currents when the skin is damp.) (c) With the current in part (b), what power is dissipated in the body?

25.64 •• BIO A person with body resistance between his hands of 10 kΩ accidentally grasps the terminals of a 14-kV power supply. (a) If the internal resistance of the power supply is 2000 Ω, what is the current through the person's body? (b) What is the power dissipated in his body? (c) If the power supply is to be made safe by increasing its internal resistance, what should the internal resistance be for the maximum current in the above situation to be 1.00 mA or less?

25.65 • A typical cost for electrical power is $0.120 per kilowatt-hour. (a) Some people leave their porch light on all the time. What is the yearly cost to keep a 75-W bulb burning day and night? (b) Suppose your refrigerator uses 400 W of power when it's running, and it runs 8 hours a day. What is the yearly cost of operating your refrigerator?

25.66 •• In the circuit shown in Fig. P25.66, R is a *variable resistor* whose value ranges from 0 to ∞, and a and b are the terminals of a battery that has an emf $\mathcal{E} = 15.0$ V and an internal resistance of 4.00 Ω. The ammeter and voltmeter are idealized meters. As R varies over its full range of values, what will be the largest and smallest readings of (a) the voltmeter and (b) the ammeter? (c) Sketch qualitative graphs of the readings of both meters as functions of R.

Figure **P25.66**

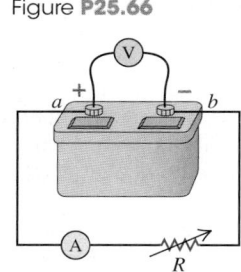

25.67 •• **A Nonideal Ammeter.** Unlike the idealized ammeter described in Section 25.4, any real ammeter has a nonzero resistance. (a) An ammeter with resistance R_A is connected in series with a resistor R and a battery of emf $\mathcal{E}$ and internal resistance r. The current measured by the ammeter is I_A. Find the current through the circuit if the ammeter is removed so that the battery and the resistor form a complete circuit. Express your answer in terms of I_A, r, R_A, and R. The more "ideal" the ammeter, the smaller the difference between this current and the current I_A. (b) If $R = 3.80\ \Omega$, $\mathcal{E} = 7.50$ V, and $r = 0.45\ \Omega$, find the maximum value of the ammeter resistance R_A so that I_A is within 1.0% of the current in the circuit when the ammeter is absent. (c) Explain why your answer in part (b) represents a *maximum* value.

25.68 •• A cylindrical copper cable 1.50 km long is connected across a 220.0-V potential difference. (a) What should be its diameter so that it produces heat at a rate of 90.0 W? (b) What is the electric field inside the cable under these conditions?

25.69 • CALC A 1.50-m cylinder of radius 1.10 cm is made of a complicated mixture of materials. Its resistivity depends on the distance x from the left end and obeys the formula $\rho(x) = a + bx^2$, where a and b are constants. At the left end, the resistivity is $2.25 \times 10^{-8}\ \Omega \cdot$m, while at the right end it is $8.50 \times 10^{-8}\ \Omega \cdot$m. (a) What is the resistance of this rod? (b) What is the electric field at its midpoint if it carries a 1.75-A current? (c) If we cut the rod into two 75.0-cm halves, what is the resistance of each half?

25.70 •• **Compact Fluorescent Bulbs.** Compact fluorescent bulbs are much more efficient at producing light than are ordinary incandescent bulbs. They initially cost much more, but they last far longer and use much less electricity. According to one study of these bulbs, a compact bulb that produces as much light as a 100-W incandescent bulb uses only 23 W of power. The compact bulb lasts 10,000 hours, on the average, and costs $11.00, whereas the incandescent bulb costs only $0.75, but lasts just 750 hours. The study assumed that electricity costs $0.080 per kilowatt-hour and that the bulbs are on for 4.0 h per day. (a) What is the total cost (including the price of the bulbs) to run each bulb for 3.0 years? (b) How much do you save over 3.0 years if you use a compact fluorescent bulb instead of an incandescent bulb? (c) What is the resistance of a "100-W" fluorescent bulb? (Remember, it actually uses only 23 W of power and operates across 120 V.)

25.71 • A lightning bolt strikes one end of a steel lightning rod, producing a 15,000-A current burst that lasts for 65 μs. The rod is 2.0 m long and 1.8 cm in diameter, and its other end is connected to the ground by 35 m of 8.0-mm-diameter copper wire. (a) Find the potential difference between the top of the steel rod and the lower end of the copper wire during the current burst. (b) Find the total energy deposited in the rod and wire by the current burst.

25.72 •• **CP** Consider the circuit shown in **Fig. P25.72**. The battery has emf 72.0 V and negligible internal resistance. $R_2 = 2.00\ \Omega$, $C_1 = 3.00\ \mu$F, and $C_2 = 6.00\ \mu$F. After the capacitors have attained their final charges, the charge on C_1 is $Q_1 = 18.0\ \mu$C. What is (a) the final charge on C_2; (b) the resistance R_1?

Figure **P25.72**

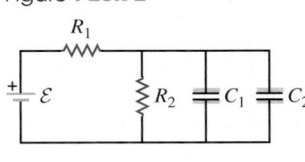

25.73 •• **CP** Consider the circuit shown in **Fig. P25.73**. The emf source has negligible internal resistance. The resistors have resistances $R_1 = 6.00\ \Omega$ and $R_2 = 4.00\ \Omega$. The capacitor has capacitance $C = 9.00\ \mu$F. When the capacitor is fully charged, the magnitude of the charge on its plates is $Q = 36.0\ \mu$C. Calculate the emf $\mathcal{E}$.

Figure **P25.73**

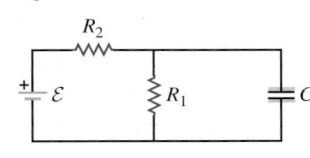

25.74 •• **DATA** An external resistor R is connected between the terminals of a battery. The value of R varies. For each R value, the current I in the circuit and the terminal voltage V_{ab} of the battery are measured. The results are plotted in **Fig. P25.74**, a graph of V_{ab} versus I that shows the best straight-line fit to the data. (a) Use the graph in Fig. P25.74 to calculate the battery's emf and internal resistance. (b) For what value of R is V_{ab} equal to 80.0% of the battery emf?

Figure **P25.74**

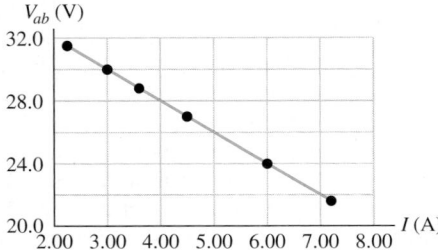

25.75 •• **DATA** The voltage drop V_{ab} across each of resistors A and B was measured as a function of the current I in the resistor. The results are shown in the table:

Resistor A

I (A)	0.50	1.00	2.00	4.00
V_{ab} (V)	2.55	3.11	3.77	4.58

Resistor B

I (A)	0.50	1.00	2.00	4.00
V_{ab} (V)	1.94	3.88	7.76	15.52

(a) For each resistor, graph V_{ab} as a function of I and graph the resistance $R = V_{ab}/I$ as a function of I. (b) Does resistor A obey Ohm's law? Explain. (c) Does resistor B obey Ohm's law? Explain. (d) What is the power dissipated in A if it is connected to a 4.00-V battery that has negligible internal resistance? (e) What is the power dissipated in B if it is connected to the battery?

25.76 •• **DATA** According to the U.S. National Electrical Code, copper wire used for interior wiring of houses, hotels, office buildings, and industrial plants is permitted to carry no more than a specified maximum amount of current. The table shows values of the maximum current I_{max} for several common sizes of wire with varnished cambric insulation. The "wire gauge" is a standard used to describe the diameter of wires. Note that the larger the diameter of the wire, the *smaller* the wire gauge.

Wire gauge	Diameter (cm)	I_{max} (A)
14	0.163	18
12	0.205	25
10	0.259	30
8	0.326	40
6	0.412	60
5	0.462	65
4	0.519	85

(a) What considerations determine the maximum current-carrying capacity of household wiring? (b) A total of 4200 W of power is to be supplied through the wires of a house to the household electrical appliances. If the potential difference across the group of appliances is 120 V, determine the gauge of the thinnest permissible wire that can be used. (c) Suppose the wire used in this house is of the gauge found in part (b) and has total length 42.0 m. At what rate is energy dissipated in the wires? (d) The house is built in a community where the consumer cost of electrical energy is $0.11 per kilowatt-hour. If the house were built with wire of the next larger diameter than that found in part (b), what would be the savings in electricity costs in one year? Assume that the appliances are kept on for an average of 12 hours a day.

CHALLENGE PROBLEMS

25.77 ••• **CALC** The resistivity of a semiconductor can be modified by adding different amounts of impurities. A rod of semiconducting material of length L and cross-sectional area A lies along the x-axis between $x = 0$ and $x = L$. The material obeys Ohm's law, and its resistivity varies along the rod according to $\rho(x) = \rho_0 \exp(-x/L)$. The end of the rod at $x = 0$ is at a potential V_0 greater than the end at $x = L$. (a) Find the total resistance of the rod and the current in the rod. (b) Find the electric-field magnitude $E(x)$ in the rod as a function of x. (c) Find the electric potential $V(x)$ in the rod as a function of x. (d) Graph the functions $\rho(x)$, $E(x)$, and $V(x)$ for values of x between $x = 0$ and $x = L$.

25.78 ••• An external resistor with resistance R is connected to a battery that has emf $\mathcal{E}$ and internal resistance r. Let P be the electrical power output of the source. By conservation of energy, P is equal to the power consumed by R. What is the value of P in the limit that R is (a) very small; (b) very large? (c) Show that the power output of the battery is a maximum when $R = r$. What is this maximum P in terms of $\mathcal{E}$ and r? (d) A battery has $\mathcal{E} = 64.0$ V and $r = 4.00\ \Omega$. What is the power output of this battery when it is connected to a resistor R, for $R = 2.00\ \Omega$, $R = 4.00\ \Omega$, and $R = 6.00\ \Omega$? Are your results consistent with the general result that you derived in part (b)?

BIO **SPIDERWEB CONDUCTIVITY.** Some types of spiders build webs that consist of threads made of dry silk coated with a solution of a variety of compounds. This coating leaves the threads, which are used to capture prey, *hygroscopic*—that is, they attract water from the atmosphere. It has been hypothesized that this aqueous coating makes the threads good electrical conductors. To test the electrical properties of coated thread, researchers placed a 5-mm length of thread between two electrical contacts.* The researchers stretched the thread in 1-mm increments to more than twice its original length, and then allowed it to return to its original length, again in 1-mm increments. Some of the resistance measurements are shown in the table:

Resistance of thread ($10^9\ \Omega$)	9	19	41	63	102	76	50	24
Length of thread (mm)	5	7	9	11	13	9	7	5

*Based on F. Vollrath and D. Edmonds, "Consequences of electrical conductivity in an orb spider's capture web," *Naturwissenschaften* (100:12, December 2013, pp. 1163–69).

25.79 What is the best explanation for the behavior exhibited in the data? (a) Longer threads can carry more current than shorter threads do and so make better electrical conductors. (b) The thread stops being a conductor when it is stretched to 13 mm, due to breaks that occur in the thin coating. (c) As the thread is stretched, the coating thins and its resistance increases; as the thread is relaxed, the coating returns nearly to its original state. (d) The resistance of the thread increases with distance from the end of the thread.

25.80 If the conductivity of the thread results from the aqueous coating only, how does the cross-sectional area A of the coating compare when the thread is 13 mm long versus the starting length of 5 mm? Assume that the resistivity of the coating remains constant and the coating is uniform along the thread. $A_{13\ mm}$ is about (a) $\frac{1}{10}A_{5\ mm}$; (b) $\frac{1}{4}A_{5\ mm}$; (c) $\frac{2}{5}A_{5\ mm}$; (d) the same as $A_{5\ mm}$.

25.81 What is the maximum current that flows in the thread during this experiment if the voltage source is a 9-V battery? (a) about 1 A; (b) about 0.1 A; (c) about 1 μA; (d) about 1 nA.

25.82 In another experiment, a piece of the web is suspended so that it can move freely. When either a positively charged object or a negatively charged object is brought near the web, the thread is observed to move toward the charged object. What is the best interpretation of this observation? The web is (a) a negatively charged conductor; (b) a positively charged conductor; (c) either a positively or negatively charged conductor; (d) an electrically neutral conductor.

Answers

Chapter Opening Question ?

(iii) The current out equals the current in. In other words, charge must enter the bulb at the same rate as it exits the bulb. It is not "used up" or consumed as it flows through the bulb.

Test Your Understanding Questions

25.1 (v) Doubling the diameter increases the cross-sectional area A by a factor of 4. Hence the current-density magnitude $J = I/A$ is reduced to $\frac{1}{4}$ of the value in Example 25.1, and the magnitude of the drift velocity $v_d = J/n|q|$ is reduced by the same factor. The new magnitude is $v_d = (0.15\ \text{mm/s})/4 = 0.038\ \text{mm/s}$. This behavior is the same as that of an incompressible fluid, which slows down when it moves from a narrow pipe to a broader one (see Section 12.4).

25.2 (ii) Figure 25.6b shows that the resistivity ρ of a semiconductor increases as the temperature decreases. From Eq. (25.5), the magnitude of the current density is $J = E/\rho$, so the current density decreases as the temperature drops and the resistivity increases.

25.3 (iii) Solving Eq. (25.11) for the current shows that $I = V/R$. If the resistance R of the wire remained the same, doubling the voltage V would make the current I double as well. However, we saw in Example 25.3 that the resistance is *not* constant: As the current increases and the temperature increases, R increases as well. Thus doubling the voltage produces a current that is *less* than double the original current. An ohmic conductor is one for which $R = V/I$ has the same value no matter what the voltage, so the wire is *nonohmic*. (In many practical problems the temperature change of the wire is so small that it can be ignored, so we can safely regard the wire as being ohmic. We do so in almost all examples in this book.)

25.4 (iii), (ii), (i) For circuit (i), we find the current from Eq. (25.16): $I = \mathcal{E}/(R + r) = (1.5\ \text{V})/(1.4\ \Omega + 0.10\ \Omega) = 1.0\ \text{A}$. For circuit (ii), we note that the terminal voltage $v_{ab} = 3.6\ \text{V}$ equals the voltage IR across the 1.8-Ω resistor: $V_{ab} = IR$, so $I = V_{ab}/R = (3.6\ \text{V})/(1.8\ \Omega) = 2.0\ \text{A}$. For circuit (iii), we use Eq. (25.15) for the terminal voltage: $V_{ab} = \mathcal{E} - Ir$, so $I = (\mathcal{E} - V_{ab})/r = (12.0\ \text{V} - 11.0\ \text{V})/(0.20\ \Omega) = 5.0\ \text{A}$.

25.5 (iii), (ii), (i) These are the same circuits that we analyzed in Test Your Understanding of Section 25.4. In each case the net power output of the battery is $P = V_{ab}I$, where V_{ab} is the battery terminal voltage. For circuit (i), we found that $I = 1.0\ \text{A}$, so $V_{ab} = \mathcal{E} - Ir = 1.5\ \text{V} - (1.0\ \text{A})(0.10\ \Omega) = 1.4\ \text{V}$, so $P = (1.4\ \text{V})(1.0\ \text{A}) = 1.4\ \text{W}$. For circuit (ii), we have $V_{ab} = 3.6\ \text{V}$ and found that $I = 2.0\ \text{A}$, so $P = (3.6\ \text{V})(2.0\ \text{A}) = 7.2\ \text{W}$. For circuit (iii), we have $V_{ab} = 11.0\ \text{V}$ and found that $I = 5.0\ \text{A}$, so $P = (11.0\ \text{V})(5.0\ \text{A}) = 55\ \text{W}$.

25.6 (i) The difficulty of producing a certain amount of current increases as the resistivity ρ increases. From Eq. (25.24), $\rho = m/ne^2\tau$, so increasing the mass m will increase the resistivity. That's because a more massive charged particle will respond more sluggishly to an applied electric field and hence drift more slowly. To produce the same current, a greater electric field would be needed. (Increasing n, e, or τ would decrease the resistivity and make it easier to produce a given current.)

Bridging Problem

(a) $237°C$ **(b)** 162 W initially, 148 W at 1.23 A

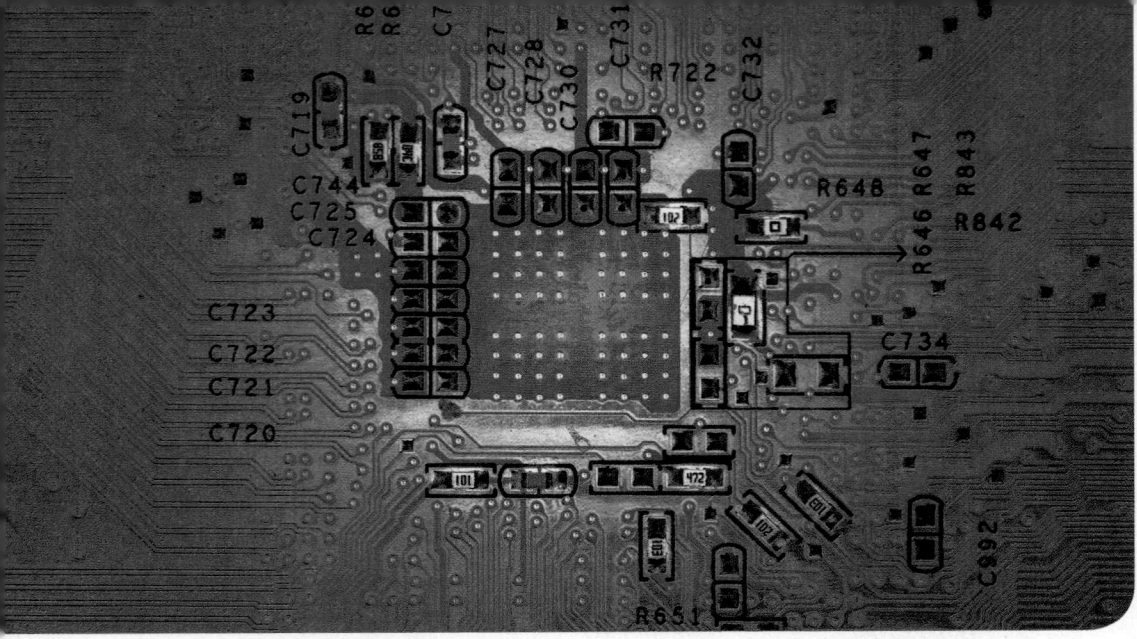

? In a complex circuit like the one on this circuit board, is it possible to connect several resistors with different resistances so that all of them have the same potential difference? (i) Yes, and the current will be the same through all of the resistors; (ii) yes, but the current may be different through different resistors; (iii) no; (iv) the answer depends on the value of the potential difference.

26 DIRECT-CURRENT CIRCUITS

I f you look inside a mobile phone, a computer, or under the hood of a car, you will find circuits of much greater complexity than the simple circuits we studied in Chapter 25. Whether connected by wires or integrated in a semiconductor chip, these circuits often include several sources, resistors, and other circuit elements interconnected in a *network*.

In this chapter we study general methods for analyzing such networks, including how to find voltages and currents of circuit elements. We'll learn how to determine the equivalent resistance for several resistors connected in series or in parallel. For more general networks we need two rules called *Kirchhoff's rules*. One is based on the principle of conservation of charge applied to a junction; the other is derived from energy conservation for a charge moving around a closed loop. We'll discuss instruments for measuring various electrical quantities. We'll also look at a circuit containing resistance and capacitance, in which the current varies with time.

Our principal concern in this chapter is with **direct-current** (dc) circuits, in which the direction of the current does not change with time. Flashlights and automobile wiring systems are examples of direct-current circuits. Household electrical power is supplied in the form of **alternating current** (ac), in which the current oscillates back and forth. The same principles for analyzing networks apply to both kinds of circuits, and we conclude this chapter with a look at household wiring systems. We'll discuss alternating-current circuits in detail in Chapter 31.

26.1 RESISTORS IN SERIES AND PARALLEL

Resistors turn up in all kinds of circuits, ranging from hair dryers and space heaters to circuits that limit or divide current or reduce or divide a voltage. Such circuits often contain several resistors, so it's appropriate to consider *combinations* of resistors. A simple example is a string of light bulbs used for holiday decorations; each bulb acts as a resistor, and from a circuit-analysis perspective the string of bulbs is simply a combination of resistors.

Suppose we have three resistors with resistances R_1, R_2, and R_3. **Figure 26.1** shows four different ways in which they might be connected between points a and b. When several circuit elements such as resistors, batteries, and motors are connected in sequence as in Fig. 26.1a, with only a single current path between the points, we say that they are connected in **series**. We studied *capacitors* in series in Section 24.2; we found that, because of conservation of charge, capacitors in series all have the same charge if they are initially uncharged. In circuits we're often more interested in the *current*, which is charge flow per unit time.

The resistors in Fig. 26.1b are said to be connected in **parallel** between points a and b. Each resistor provides an alternative path between the points. For circuit elements that are connected in parallel, the *potential difference* is the same across each element. We studied capacitors in parallel in Section 24.2.

In Fig. 26.1c, resistors R_2 and R_3 are in parallel, and this combination is in series with R_1. In Fig. 26.1d, R_2 and R_3 are in series, and this combination is in parallel with R_1.

For any combination of resistors we can always find a *single* resistor that could replace the combination and result in the same total current and potential difference. For example, a string of holiday light bulbs could be replaced by a single, appropriately chosen light bulb that would draw the same current and have the same potential difference between its terminals as the original string of bulbs. The resistance of this single resistor is called the **equivalent resistance** of the combination. If any one of the networks in Fig. 26.1 were replaced by its equivalent resistance R_{eq}, we could write

$$V_{ab} = IR_{eq} \quad \text{or} \quad R_{eq} = \frac{V_{ab}}{I}$$

where V_{ab} is the potential difference between terminals a and b of the network and I is the current at point a or b. To compute an equivalent resistance, we assume a potential difference V_{ab} across the actual network, compute the corresponding current I, and take the ratio V_{ab}/I.

Resistors in Series

We can derive general equations for the equivalent resistance of a series or parallel combination of resistors. In Fig. 26.1a, the three resistors are in *series,* so the current I is the same in all of them. (Recall from Section 25.4 that current is *not* "used up" as it passes through a circuit.) Applying $V = IR$ to each resistor, we have

$$V_{ax} = IR_1 \qquad V_{xy} = IR_2 \qquad V_{yb} = IR_3$$

The potential differences across each resistor need not be the same (except for the special case in which all three resistances are equal). The potential difference V_{ab} across the entire combination is the sum of these individual potential differences:

$$V_{ab} = V_{ax} + V_{xy} + V_{yb} = I(R_1 + R_2 + R_3)$$

and so

$$\frac{V_{ab}}{I} = R_1 + R_2 + R_3$$

The ratio V_{ab}/I is, by definition, the equivalent resistance R_{eq}. Therefore

$$R_{eq} = R_1 + R_2 + R_3$$

It is easy to generalize this to *any* number of resistors:

| Resistors in series: | $R_{eq} = R_1 + R_2 + R_3 + \cdots$ | (26.1) |

Equivalent resistance of series combination ⟵ Resistances of individual resistors

26.1 Four different ways of connecting three resistors.

(a) R_1, R_2, and R_3 in series

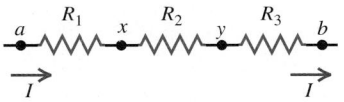

(b) R_1, R_2, and R_3 in parallel

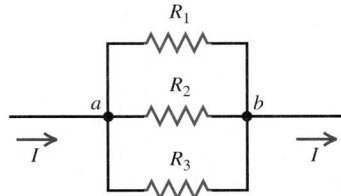

(c) R_1 in series with parallel combination of R_2 and R_3

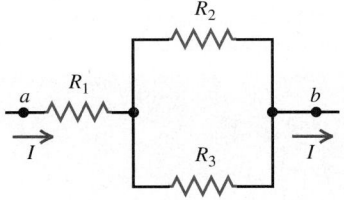

(d) R_1 in parallel with series combination of R_2 and R_3

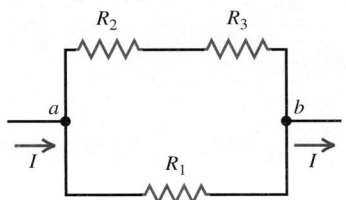

The equivalent resistance of a series combination equals the sum of the individual resistances. The equivalent resistance is *greater than* any individual resistance.

CAUTION **Resistors vs. capacitors in series** Don't confuse *resistors* in series with *capacitors* in series. Resistors in series add *directly* [Eq. (26.1)] because the voltage across each is directly proportional to its resistance and to the common current. Capacitors in series [Eq. (24.5)] add *reciprocally;* the voltage across each is directly proportional to the common charge but *inversely* proportional to the individual capacitance. ▮

Resistors in Parallel

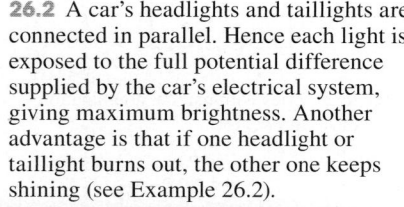

26.2 A car's headlights and taillights are connected in parallel. Hence each light is exposed to the full potential difference supplied by the car's electrical system, giving maximum brightness. Another advantage is that if one headlight or taillight burns out, the other one keeps shining (see Example 26.2).

If the resistors are in *parallel,* as in Fig. 26.1b, the current through each resistor need not be the same. But the potential difference between the terminals of each resistor must be the same and equal to V_{ab} (**Fig. 26.2**). (Remember that the potential difference between any two points does not depend on the path taken between the points.) Let's call the currents in the three resistors I_1, I_2, and I_3. Then from $I = V/R$,

$$I_1 = \frac{V_{ab}}{R_1} \qquad I_2 = \frac{V_{ab}}{R_2} \qquad I_3 = \frac{V_{ab}}{R_3}$$

In general, the current is different through each resistor. Because charge neither accumulates at nor drains out of point a, the total current I must equal the sum of the three currents in the resistors:

$$I = I_1 + I_2 + I_3 = V_{ab}\left(\frac{1}{R_1} + \frac{1}{R_2} + \frac{1}{R_3}\right) \quad \text{or}$$

$$\frac{I}{V_{ab}} = \frac{1}{R_1} + \frac{1}{R_2} + \frac{1}{R_3}$$

But by the definition of the equivalent resistance R_{eq}, $I/V_{ab} = 1/R_{eq}$, so

$$\frac{1}{R_{eq}} = \frac{1}{R_1} + \frac{1}{R_2} + \frac{1}{R_3}$$

Again it is easy to generalize to *any* number of resistors in parallel:

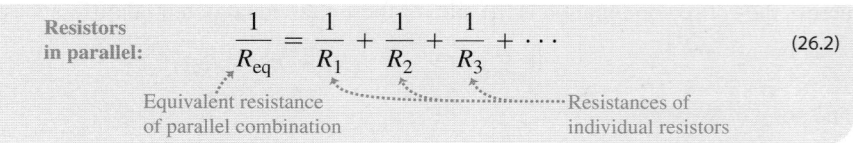

Resistors in parallel:
$$\frac{1}{R_{eq}} = \frac{1}{R_1} + \frac{1}{R_2} + \frac{1}{R_3} + \cdots \qquad (26.2)$$

Equivalent resistance of parallel combination   Resistances of individual resistors

DEMO

The reciprocal of the equivalent resistance of a parallel combination equals the sum of the reciprocals of the individual resistances. The equivalent resistance is always *less than* any individual resistance.

CAUTION **Resistors vs. capacitors in parallel** Note the differences between *resistors* in parallel and *capacitors* in parallel. Resistors in parallel add *reciprocally* [Eq. (26.2)] because the current in each is proportional to the common voltage across them and *inversely* proportional to the resistance of each. Capacitors in parallel add *directly* [Eq. (24.7)] because the charge on each is proportional to the common voltage across them and *directly* proportional to the capacitance of each. ▮

For the special case of *two* resistors in parallel,

$$\frac{1}{R_{eq}} = \frac{1}{R_1} + \frac{1}{R_2} = \frac{R_1 + R_2}{R_1 R_2} \quad \text{and}$$

$$R_{eq} = \frac{R_1 R_2}{R_1 + R_2} \qquad \text{(two resistors in parallel)} \qquad (26.3)$$

Because $V_{ab} = I_1 R_1 = I_2 R_2$, it follows that

$$\frac{I_1}{I_2} = \frac{R_2}{R_1} \qquad \text{(two resistors in parallel)} \qquad (26.4)$$

Thus the currents carried by two resistors in parallel are *inversely proportional* to their resistances. More current goes through the path of least resistance.

PROBLEM-SOLVING STRATEGY 26.1 | RESISTORS IN SERIES AND PARALLEL

IDENTIFY *the relevant concepts:* As in Fig. 26.1, many resistor networks are made up of resistors in series, in parallel, or a combination thereof. Such networks can be replaced by a single equivalent resistor. The logic is similar to that of Problem-Solving Strategy 24.1 for networks of capacitors.

SET UP *the problem* using the following steps:
1. Make a drawing of the resistor network.
2. Identify groups of resistors connected in series or parallel.
3. Identify the target variables. They could include the equivalent resistance of the network, the potential difference across each resistor, or the current through each resistor.

EXECUTE *the solution* as follows:
1. Use Eq. (26.1) or (26.2), respectively, to find the equivalent resistance for series or parallel combinations.
2. If the network is more complex, try reducing it to series and parallel combinations. For example, in Fig. 26.1c we first replace the parallel combination of R_2 and R_3 with its equivalent

resistance; this then forms a series combination with R_1. In Fig. 26.1d, the combination of R_2 and R_3 in series forms a parallel combination with R_1.
3. Keep in mind that the total potential difference across resistors connected in series is the sum of the individual potential differences. The potential difference across resistors connected in parallel is the same for every resistor and equals the potential difference across the combination.
4. The current through resistors connected in series is the same through every resistor and equals the current through the combination. The total current through resistors connected in parallel is the sum of the currents through the individual resistors.

EVALUATE *your answer:* Check whether your results are consistent. The equivalent resistance of resistors connected in series should be greater than that of any individual resistor; that of resistors in parallel should be less than that of any individual resistor.

EXAMPLE 26.1 EQUIVALENT RESISTANCE

Find the equivalent resistance of the network in **Fig. 26.3a** (next page) and the current in each resistor. The source of emf has negligible internal resistance.

SOLUTION

IDENTIFY and SET UP: This network of three resistors is a *combination* of series and parallel resistances, as in Fig. 26.1c. We determine the equivalent resistance of the parallel 6-Ω and 3-Ω resistors, and then that of their series combination with the 4-Ω resistor: This is the equivalent resistance R_{eq} of the network as a whole. We then find the current in the emf, which is the same as that in the 4-Ω resistor. The potential difference is the same across each of the parallel 6-Ω and 3-Ω resistors; we use this to determine how the current is divided between these.

EXECUTE: Figures 26.3b and 26.3c show successive steps in reducing the network to a single equivalent resistance R_{eq}. From Eq. (26.2), the 6-Ω and 3-Ω resistors in parallel in Fig. 26.3a are equivalent to the single 2-Ω resistor in Fig. 26.3b:

$$\frac{1}{R_{6\,\Omega+3\,\Omega}} = \frac{1}{6\ \Omega} + \frac{1}{3\ \Omega} = \frac{1}{2\ \Omega}$$

[Equation (26.3) gives the same result.] From Eq. (26.1) the series combination of this 2-Ω resistor with the 4-Ω resistor is equivalent to the single 6-Ω resistor in Fig. 26.3c.

We reverse these steps to find the current in each resistor of the original network. In the circuit shown in Fig. 26.3d (identical to Fig. 26.3c), the current is $I = V_{ab}/R = (18\text{ V})/(6\ \Omega) = 3$ A. So the current in the 4-Ω and 2-Ω resistors in Fig. 26.3e (identical

Continued

26.3 Steps in reducing a combination of resistors to a single equivalent resistor and finding the current in each resistor.

(a)

$\mathcal{E} = 18\ \text{V},\ r = 0$

(b)   (c)   (d)   (e)   (f)

to Fig. 26.3b) is also 3 A. The potential difference V_{cb} across the 2-Ω resistor is therefore $V_{cb} = IR = (3\ \text{A})(2\ \Omega) = 6\ \text{V}$. This potential difference must also be 6 V in Fig. 26.3f (identical to Fig. 26.3a). From $I = V_{cb}/R$, the currents in the 6-Ω and 3-Ω resistors in Fig. 26.3f are, respectively, $(6\ \text{V})/(6\ \Omega) = 1\ \text{A}$ and $(6\ \text{V})/(3\ \Omega) = 2\ \text{A}$.

EVALUATE: Note that for the two resistors in parallel between points c and b in Fig. 26.3f, there is twice as much current through the 3-Ω resistor as through the 6-Ω resistor; more current goes through the path of least resistance, in accordance with Eq. (26.4). Note also that the total current through these two resistors is 3 A, the same as it is through the 4-Ω resistor between points a and c.

EXAMPLE 26.2   SERIES VERSUS PARALLEL COMBINATIONS

Two identical incandescent light bulbs, each with resistance $R = 2\ \Omega$, are connected to a source with $\mathcal{E} = 8\ \text{V}$ and negligible internal resistance. Find the current through each bulb, the potential difference across each bulb, and the power delivered to each bulb and to the entire network if the bulbs are connected (a) in series and (b) in parallel. (c) Suppose one of the bulbs burns out; that is, its filament breaks and current can no longer flow through it. What happens to the other bulb in the series case? In the parallel case?

SOLUTION

IDENTIFY and SET UP: The light bulbs are just resistors in simple series and parallel connections (**Figs. 26.4a** and 26.4b). Once we find the current I through each bulb, we can find the power delivered to each bulb by using Eq. (25.18), $P = I^2R = V^2/R$.

EXECUTE: (a) From Eq. (26.1) the equivalent resistance of the two bulbs between points a and c in Fig. 26.4a is $R_{eq} = 2R = 2(2\ \Omega) = 4\ \Omega$. In series, the current is the same through each bulb:

$$I = \frac{V_{ac}}{R_{eq}} = \frac{8\ \text{V}}{4\ \Omega} = 2\ \text{A}$$

Since the bulbs have the same resistance, the potential difference is the same across each bulb:

$$V_{ab} = V_{bc} = IR = (2\ \text{A})(2\ \Omega) = 4\ \text{V}$$

26.4 Our sketches for this problem.

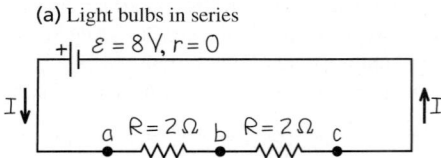

(a) Light bulbs in series

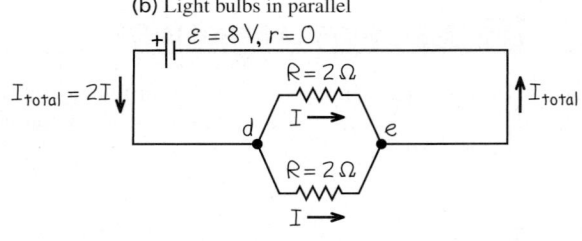

(b) Light bulbs in parallel

From Eq. (25.18), the power delivered to each bulb is

$$P = I^2R = (2\ \text{A})^2(2\ \Omega) = 8\ \text{W} \qquad \text{or}$$

$$P = \frac{V_{ab}^2}{R} = \frac{V_{bc}^2}{R} = \frac{(4\ \text{V})^2}{2\ \Omega} = 8\ \text{W}$$

The total power delivered to both bulbs is $P_{tot} = 2P = 16\ \text{W}$.

(b) If the bulbs are in parallel, as in Fig. 26.4b, the potential difference V_{de} across each bulb is the same and equal to 8 V, the terminal voltage of the source. Hence the current through each light bulb is

$$I = \frac{V_{de}}{R} = \frac{8 \text{ V}}{2 \text{ } \Omega} = 4 \text{ A}$$

and the power delivered to each bulb is

$$P = I^2 R = (4 \text{ A})^2 (2 \text{ } \Omega) = 32 \text{ W} \qquad \text{or}$$

$$P = \frac{V_{de}^2}{R} = \frac{(8 \text{ V})^2}{2 \text{ } \Omega} = 32 \text{ W}$$

Both the potential difference across each bulb and the current through each bulb are twice as great as in the series case. Hence the power delivered to each bulb is *four* times greater, and each bulb is brighter.

The total power delivered to the parallel network is $P_{\text{total}} = 2P = 64 \text{ W}$, four times greater than in the series case. The increased power compared to the series case isn't obtained "for free"; energy is extracted from the source four times more rapidly in the parallel case than in the series case. If the source is a battery, it will be used up four times as fast.

(c) In the series case the same current flows through both bulbs. If one bulb burns out, there will be no current in the circuit, and neither bulb will glow.

In the parallel case the potential difference across either bulb is unchanged if a bulb burns out. The current through the functional bulb and the power delivered to it are unchanged.

EVALUATE: Our calculation isn't completely accurate, because the resistance $R = V/I$ of real light bulbs depends on the potential difference V across the bulb. That's because the filament resistance increases with increasing operating temperature and therefore with increasing V. But bulbs connected in series across a source do in fact glow less brightly than when connected in parallel across the same source (**Fig. 26.5**).

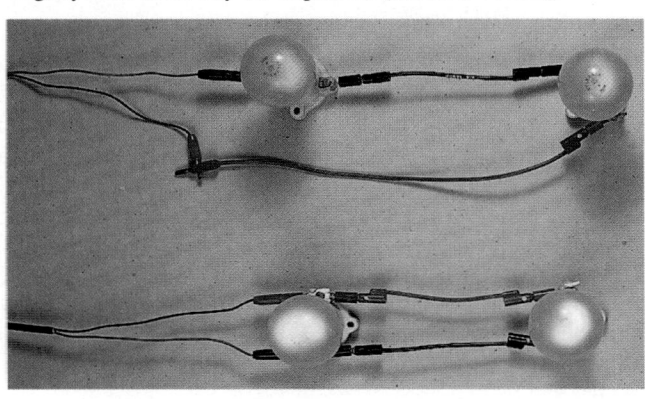

26.5 When connected to the same source, two incandescent light bulbs in series (shown at top) draw less power and glow less brightly than when they are in parallel (shown at bottom).

TEST YOUR UNDERSTANDING OF SECTION 26.1 Suppose all three of the resistors shown in Fig. 26.1 have the same resistance, so $R_1 = R_2 = R_3 = R$. Rank the four arrangements shown in parts (a)–(d) of Fig. 26.1 in order of their equivalent resistance, from highest to lowest. ∎

26.2 KIRCHHOFF'S RULES

Many practical resistor networks cannot be reduced to simple series-parallel combinations. **Figure 26.6a** shows a dc power supply with emf $\mathcal{E}_1$ charging a battery with a smaller emf $\mathcal{E}_2$ and feeding current to a light bulb with resistance R. Figure 26.6b is a "bridge" circuit, used in many different types of measurement and control systems. (Problem 26.74 describes one important application of a "bridge" circuit.) To analyze these networks, we'll use the techniques developed by the German physicist Gustav Robert Kirchhoff (1824–1887).

First, here are two terms that we will use often. A **junction** in a circuit is a point where three or more conductors meet. A **loop** is any closed conducting path. In Fig. 26.6a points a and b are junctions, but points c and d are not; in Fig. 26.6b points a, b, c, and d are junctions, but points e and f are not. The blue lines in Figs. 26.6a and 26.6b show some possible loops in these circuits.

Kirchhoff's rules are the following two statements:

26.6 Two networks that cannot be reduced to simple series-parallel combinations of resistors.

(a)

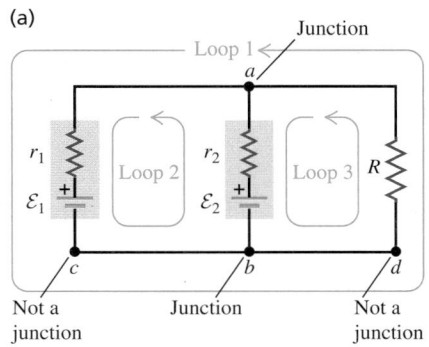

(b)

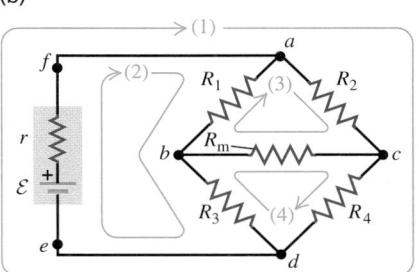

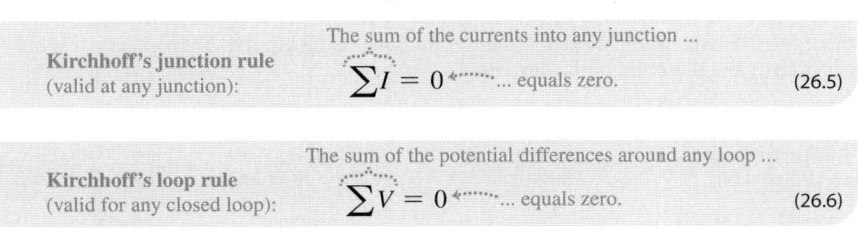

| **Kirchhoff's junction rule** (valid at any junction): | The sum of the currents into any junction ... $\sum I = 0$ ... equals zero. | (26.5) |

| **Kirchhoff's loop rule** (valid for any closed loop): | The sum of the potential differences around any loop ... $\sum V = 0$ ... equals zero. | (26.6) |

Note that the potential differences V in Eq. (26.6) include those associated with all circuit elements in the loop, including emfs and resistors.

26.7 Kirchhoff's junction rule states that as much current flows into a junction as flows out of it.

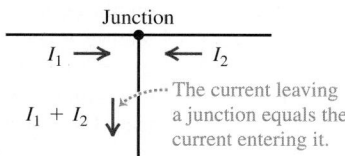

(a) Kirchhoff's junction rule

Junction

$I_1 \rightarrow$ $\leftarrow I_2$

$I_1 + I_2 \downarrow$

The current leaving a junction equals the current entering it.

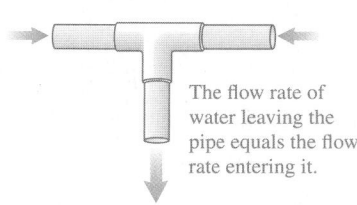

(b) Water-pipe analogy

The flow rate of water leaving the pipe equals the flow rate entering it.

The junction rule is based on *conservation of electric charge*. No charge can accumulate at a junction, so the total charge entering the junction per unit time must equal the total charge leaving per unit time (**Fig. 26.7a**). Charge per unit time is current, so if we consider the currents entering a junction to be positive and those leaving to be negative, the algebraic sum of currents into a junction must be zero. It's like a T branch in a water pipe (Fig. 26.7b); if you have a total of 1 liter per minute coming in the two pipes, you can't have 3 liters per minute going out the third pipe. We used the junction rule (without saying so) in Section 26.1 in the derivation of Eq. (26.2) for resistors in parallel.

The loop rule is a statement that the electrostatic force is *conservative*. Suppose we go around a loop, measuring potential differences across successive circuit elements as we go. When we return to the starting point, we must find that the *algebraic sum* of these differences is zero; otherwise, we could not say that the potential at this point has a definite value.

Sign Conventions for the Loop Rule

In applying the loop rule, we need some sign conventions. Problem-Solving Strategy 26.2 describes in detail how to use these, but here's a quick overview. We first assume a direction for the current in each branch of the circuit and mark it on a diagram of the circuit. Then, starting at any point in the circuit, we imagine traveling around a loop, adding emfs and *IR* terms as we come to them. When we travel through a source in the direction from − to +, the emf is considered to be *positive*; when we travel from + to −, the emf is considered to be *negative* (**Fig. 26.8a**). When we travel through a resistor in the *same* direction as the assumed current, the *IR* term is *negative* because the current goes in the direction of decreasing potential. When we travel through a resistor in the direction *opposite* to the assumed current, the *IR* term is *positive* because this represents a rise of potential (Fig. 26.8b).

Kirchhoff's two rules are all we need to solve a wide variety of network problems. Usually, some of the emfs, currents, and resistances are known, and others are unknown. We must always obtain from Kirchhoff's rules a number of independent equations equal to the number of unknowns so that we can solve the equations simultaneously. Often the hardest part of the solution is keeping track of algebraic signs!

26.8 Use these sign conventions when you apply Kirchhoff's loop rule. In each part of the figure "Travel" is the direction that we imagine going around the loop, which is not necessarily the direction of the current.

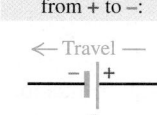

(a) Sign conventions for emfs

$+\mathcal{E}$: Travel direction from − to +:	$-\mathcal{E}$: Travel direction from + to −:
— Travel →	← Travel —
$-\|+$	$-\|+$
$\mathcal{E}$	$\mathcal{E}$

(b) Sign conventions for resistors

$-IR$: Travel *opposite* to current direction:	$-IR$: Travel *in* current direction:
— Travel →	← Travel —
$I \leftarrow$	$I \leftarrow$
$-\text{WWW}+$	$-\text{WWW}+$
R	R

PROBLEM-SOLVING STRATEGY 26.2) KIRCHHOFF'S RULES

IDENTIFY *the relevant concepts:* Kirchhoff's rules are useful for analyzing any electric circuit.

SET UP *the problem* using the following steps:
1. Draw a circuit diagram, leaving room to label all quantities, known and unknown. Indicate an assumed direction for each unknown current and emf. (Kirchhoff's rules will yield the magnitudes *and directions* of unknown currents and emfs. If the actual direction of a quantity is opposite to your assumption, the resulting quantity will have a negative sign.)
2. As you label currents, it's helpful to use Kirchhoff's junction rule, as in **Fig. 26.9**, so as to express the currents in terms of as few quantities as possible.
3. Identify the target variables.

EXECUTE *the solution* as follows:
1. Choose any loop in the network and choose a direction (clockwise or counterclockwise) to travel around the loop as you apply Kirchhoff's loop rule. The direction need not be the same as any assumed current direction.

2. Travel around the loop in the chosen direction, adding potential differences algebraically as you cross them. Use the sign conventions of Fig. 26.8.
3. Equate the sum obtained in step 2 to zero in accordance with the loop rule.
4. If you need more independent equations, choose another loop and repeat steps 1–3; continue until you have as many independent equations as unknowns or until every circuit element has been included in at least one loop.
5. Solve the equations simultaneously to determine the unknowns.
6. You can use the loop-rule bookkeeping system to find the potential V_{ab} of any point a with respect to any other point b. Start at b and add the potential changes you encounter in going from b to a; use the same sign rules as in step 2. The algebraic sum of these changes is $V_{ab} = V_a - V_b$.

EVALUATE *your answer:* Check all the steps in your algebra. Apply steps 1 and 2 to a loop you have not yet considered; if the sum of potential drops isn't zero, you've made an error somewhere.

26.9 Applying the junction rule to point a reduces the number of unknown currents from three to two.

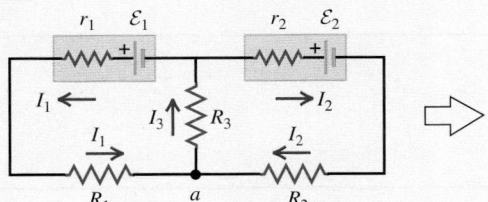

(a) Three unknown currents: I_1, I_2, I_3

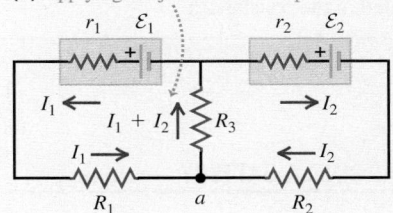

(b) Applying the junction rule to point a eliminates I_3.

EXAMPLE 26.3 A SINGLE-LOOP CIRCUIT

The circuit shown in **Fig. 26.10a** (next page) contains two batteries, each with an emf and an internal resistance, and two resistors. Find (a) the current in the circuit, (b) the potential difference V_{ab}, and (c) the power output of the emf of each battery.

SOLUTION

IDENTIFY and SET UP: There are no junctions in this single-loop circuit, so we don't need Kirchhoff's junction rule. To apply Kirchhoff's loop rule, we first assume a direction for the current; let's assume a counterclockwise direction as shown in Fig. 26.10a.

EXECUTE: (a) Starting at a and traveling counterclockwise with the current, we add potential increases and decreases and equate the sum to zero as in Eq. (26.6):

$$-I(4\,\Omega) - 4\,V - I(7\,\Omega) + 12\,V - I(2\,\Omega) - I(3\,\Omega) = 0$$

Collecting like terms and solving for I, we find

$$8\,V = I(16\,\Omega) \quad \text{and} \quad I = 0.5\,A$$

The positive result for I shows that our assumed current direction is correct.

(b) To find V_{ab}, the potential at a with respect to b, we start at b and add potential changes as we go toward a. There are two paths from b to a; taking the lower one, we find

$$V_{ab} = (0.5\,A)(7\,\Omega) + 4\,V + (0.5\,A)(4\,\Omega) = 9.5\,V$$

Point a is at 9.5 V higher potential than b. All the terms in this sum, including the IR terms, are positive because each represents an *increase* in potential as we go from b to a. For the upper path,

$$V_{ab} = 12\,V - (0.5\,A)(2\,\Omega) - (0.5\,A)(3\,\Omega) = 9.5\,V$$

Here the IR terms are negative because our path goes in the direction of the current, with potential decreases through the resistors. The results for V_{ab} are the same for both paths, as they must be in order for the total potential change around the loop to be zero.

(c) The power outputs of the emf of the 12-V and 4-V batteries are

$$P_{12V} = \mathcal{E}I = (12\,V)(0.5\,A) = 6\,W$$

$$P_{4V} = \mathcal{E}I = (-4\,V)(0.5\,A) = -2\,W$$

Continued

26.10 (a) In this example we travel around the loop in the same direction as the assumed current, so all the *IR* terms are negative. The potential decreases as we travel from + to − through the bottom emf but increases as we travel from − to + through the top emf. (b) A real-life example of a circuit of this kind.

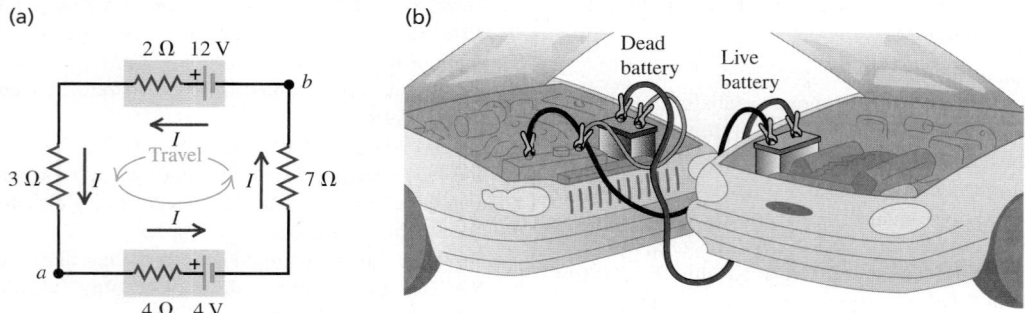

(a)

(b)

The negative sign in $\mathcal{E}$ for the 4-V battery appears because the current actually runs from the higher-potential side of the battery to the lower-potential side. The negative value of *P* means that we are *storing* energy in that battery; the 12-V battery is *recharging* it (if it is in fact rechargeable; otherwise, we're destroying it).

EVALUATE: By applying the expression $P = I^2R$ to each of the four resistors in Fig. 26.10a, you can show that the total power dissipated in all four resistors is 4 W. Of the 6 W provided by the emf of the 12-V battery, 2 W goes into storing energy in the 4-V battery and 4 W is dissipated in the resistances.

The circuit shown in Fig. 26.10a is much like that used when a fully charged 12-V storage battery (in a car with its engine running) "jump-starts" a car with a run-down battery (Fig. 26.10b). The run-down battery is slightly recharged in the process. The 3-Ω and 7-Ω resistors in Fig. 26.10a represent the resistances of the jumper cables and of the conducting path through the car with the run-down battery. (The values of the resistances in actual automobiles and jumper cables are considerably lower, and the emf of a run-down car battery isn't much less than 12 V.)

EXAMPLE 26.4 CHARGING A BATTERY

In the circuit shown in **Fig. 26.11**, a 12-V power supply with unknown internal resistance *r* is connected to a run-down rechargeable battery with unknown emf $\mathcal{E}$ and internal resistance 1 Ω and to an indicator light bulb of resistance 3 Ω carrying a current of 2 A. The current through the run-down battery is 1 A in the direction shown. Find *r*, $\mathcal{E}$, and the current *I* through the power supply.

SOLUTION

IDENTIFY and SET UP: This circuit has more than one loop, so we must apply both the junction and loop rules. We assume the direction of the current through the 12-V power supply, and the polarity of the run-down battery, to be as shown in Fig. 26.11. There are three target variables, so we need three equations.

26.11 In this circuit a power supply charges a run-down battery and lights a bulb. An assumption has been made about the polarity of the emf $\mathcal{E}$ of the battery. Is this assumption correct?

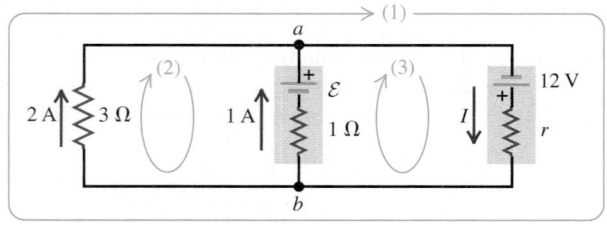

EXECUTE: We apply the junction rule, Eq. (26.5), to point *a*:

$$-I + 1\,\text{A} + 2\,\text{A} = 0 \qquad \text{so} \qquad I = 3\,\text{A}$$

To determine *r*, we apply the loop rule, Eq. (26.6), to the large, outer loop (1):

$$12\,\text{V} - (3\,\text{A})r - (2\,\text{A})(3\,\Omega) = 0 \qquad \text{so} \qquad r = 2\,\Omega$$

To determine $\mathcal{E}$, we apply the loop rule to the left-hand loop (2):

$$-\mathcal{E} + (1\,\text{A})(1\,\Omega) - (2\,\text{A})(3\,\Omega) = 0 \qquad \text{so} \qquad \mathcal{E} = -5\,\text{V}$$

The negative value for $\mathcal{E}$ shows that the actual polarity of this emf is opposite to that shown in Fig. 26.11. As in Example 26.3, the battery is being recharged.

EVALUATE: Try applying the junction rule at point *b* instead of point *a*, and try applying the loop rule counterclockwise rather than clockwise around loop (1). You'll get the same results for *I* and *r*. We can check our result for $\mathcal{E}$ by using loop (3):

$$12\,\text{V} - (3\,\text{A})(2\,\Omega) - (1\,\text{A})(1\,\Omega) + \mathcal{E} = 0$$

which again gives us $\mathcal{E} = -5$ V.

As an additional check, we note that $V_{ba} = V_b - V_a$ equals the voltage across the 3-Ω resistance, which is $(2\,\text{A})(3\,\Omega) = 6$ V. Going from *a* to *b* by the right-hand branch, we encounter potential differences $+12\,\text{V} - (3\,\text{A})(2\,\Omega) = +6$ V, and going by the middle branch, we find $-(-5\,\text{V}) + (1\,\text{A})(1\,\Omega) = +6$ V. The three ways of getting V_{ba} give the same results.

EXAMPLE 26.5 POWER IN A BATTERY-CHARGING CIRCUIT

In the circuit of Example 26.4 (shown in Fig. 26.11), find the power delivered by the 12-V power supply and by the battery being recharged, and find the power dissipated in each resistor.

SOLUTION

IDENTIFY and SET UP: We use the results of Section 25.5, in which we found that the power delivered *from* an emf to a circuit is $\mathcal{E}I$ and the power delivered *to* a resistor from a circuit is $V_{ab}I = I^2R$. We know the values of all relevant quantities from Example 26.4.

EXECUTE: The power output from the emf of the power supply is

$$P_{\text{supply}} = \mathcal{E}_{\text{supply}}I_{\text{supply}} = (12\text{ V})(3\text{ A}) = 36\text{ W}$$

The power dissipated in the power supply's internal resistance r is

$$P_{r-\text{supply}} = I_{\text{supply}}^2 r_{\text{supply}} = (3\text{ A})^2(2\text{ }\Omega) = 18\text{ W}$$

so the power supply's *net* power output is $P_{\text{net}} = 36\text{ W} - 18\text{ W} = 18\text{ W}$. Alternatively, from Example 26.4 the terminal voltage of the battery is $V_{ba} = 6$ V, so the net power output is

$$P_{\text{net}} = V_{ba}I_{\text{supply}} = (6\text{ V})(3\text{ A}) = 18\text{ W}$$

The power output of the emf $\mathcal{E}$ of the battery being charged is

$$P_{\text{emf}} = \mathcal{E}I_{\text{battery}} = (-5\text{ V})(1\text{ A}) = -5\text{ W}$$

This is negative because the 1-A current runs through the battery from the higher-potential side to the lower-potential side. (As we mentioned in Example 26.4, the polarity assumed for this battery in Fig. 26.11 was wrong.) We are storing energy in the battery as we charge it. Additional power is dissipated in the battery's internal resistance; this power is

$$P_{r-\text{battery}} = I_{\text{battery}}^2 r_{\text{battery}} = (1\text{ A})^2(1\text{ }\Omega) = 1\text{ W}$$

The total power input to the battery is thus $1\text{ W} + |-5\text{ W}| = 6\text{ W}$. Of this, 5 W represents useful energy stored in the battery; the remainder is wasted in its internal resistance.

The power dissipated in the light bulb is

$$P_{\text{bulb}} = I_{\text{bulb}}^2 R_{\text{bulb}} = (2\text{ A})^2(3\text{ }\Omega) = 12\text{ W}$$

EVALUATE: As a check, note that all of the power from the supply is accounted for. Of the 18 W of net power from the power supply, 5 W goes to recharge the battery, 1 W is dissipated in the battery's internal resistance, and 12 W is dissipated in the light bulb.

EXAMPLE 26.6 A COMPLEX NETWORK

Figure 26.12 shows a "bridge" circuit of the type described at the beginning of this section (see Fig. 26.6b). Find the current in each resistor and the equivalent resistance of the network of five resistors.

SOLUTION

IDENTIFY and SET UP: This network is neither a series combination nor a parallel combination. Hence we must use Kirchhoff's rules to find the values of the target variables. There are five unknown currents, but by applying the junction rule to junctions a and b, we can represent them in terms of three unknown currents I_1, I_2, and I_3, as shown in Fig. 26.12.

26.12 A network circuit with several resistors.

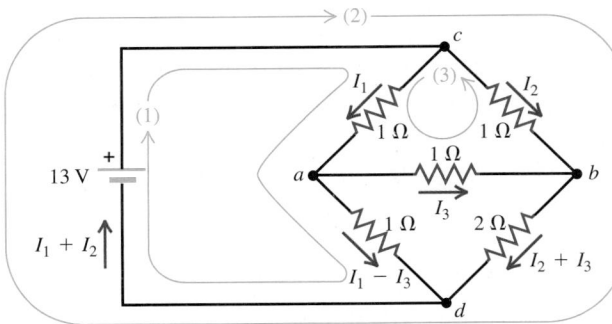

EXECUTE: We apply the loop rule to the three loops shown:

$$13\text{ V} - I_1(1\text{ }\Omega) - (I_1 - I_3)(1\text{ }\Omega) = 0 \quad (1)$$

$$-I_2(1\text{ }\Omega) - (I_2 + I_3)(2\text{ }\Omega) + 13\text{ V} = 0 \quad (2)$$

$$-I_1(1\text{ }\Omega) - I_3(1\text{ }\Omega) + I_2(1\text{ }\Omega) = 0 \quad (3)$$

One way to solve these simultaneous equations is to solve Eq. (3) for I_2, obtaining $I_2 = I_1 + I_3$, and then substitute this expression into Eq. (2) to eliminate I_2. We then have

$$13\text{ V} = I_1(2\text{ }\Omega) - I_3(1\text{ }\Omega) \quad (1')$$

$$13\text{ V} = I_1(3\text{ }\Omega) + I_3(5\text{ }\Omega) \quad (2')$$

Now we can eliminate I_3 by multiplying Eq. (1') by 5 and adding the two equations. We obtain

$$78\text{ V} = I_1(13\text{ }\Omega) \qquad I_1 = 6\text{ A}$$

We substitute this result into Eq. (1') to obtain $I_3 = -1$ A, and from Eq. (3) we find $I_2 = 5$ A. The negative value of I_3 tells us that its direction is opposite to the direction we assumed.

The total current through the network is $I_1 + I_2 = 11$ A, and the potential drop across it is equal to the battery emf, 13 V. The equivalent resistance of the network is therefore

$$R_{\text{eq}} = \frac{13\text{ V}}{11\text{ A}} = 1.2\text{ }\Omega$$

EVALUATE: You can check our results for I_1, I_2, and I_3 by substituting them back into Eqs. (1)–(3). What do you find?

EXAMPLE 26.7 **A POTENTIAL DIFFERENCE IN A COMPLEX NETWORK**

In the circuit of Example 26.6 (Fig. 26.12), find the potential difference V_{ab}.

SOLUTION

IDENTIFY and SET UP: Our target variable $V_{ab} = V_a - V_b$ is the potential at point a with respect to point b. To find it, we start at point b and follow a path to point a, adding potential rises and drops as we go. We can follow any of several paths from b to a; the result must be the same for all such paths, which gives us a way to check our result.

EXECUTE: The simplest path is through the center 1-Ω resistor. In Example 26.6 we found $I_3 = -1$ A, showing that the actual current direction through this resistor is from right to left. Thus, as we go from b to a, there is a *drop* of potential with magnitude $|I_3|R = (1 \text{ A})(1 \text{ } \Omega) = 1$ V. Hence $V_{ab} = -1$ V, and the potential at a is 1 V less than at point b.

EVALUATE: To check our result, let's try a path from b to a that goes through the lower two resistors. The currents through these are

$$I_2 + I_3 = 5 \text{ A} + (-1 \text{ A}) = 4 \text{ A} \quad \text{and}$$
$$I_1 - I_3 = 6 \text{ A} - (-1 \text{ A}) = 7 \text{ A}$$

and so

$$V_{ab} = -(4 \text{ A})(2 \text{ } \Omega) + (7 \text{ A})(1 \text{ } \Omega) = -1 \text{ V}$$

You can confirm this result by using some other paths from b to a.

26.13 Both this ammeter (top) and voltmeter (bottom) are d'Arsonval galvanometers. The difference has to do with their internal connections (see Fig. 26.15).

TEST YOUR UNDERSTANDING OF SECTION 26.2 Subtract Eq. (1) from Eq. (2) in Example 26.6. To which loop in Fig. 26.12 does this equation correspond? Would this equation have simplified the solution of Example 26.6? ▮

26.3 ELECTRICAL MEASURING INSTRUMENTS

We've been talking about potential difference, current, and resistance for two chapters, so it's about time we said something about how to *measure* these quantities. Many common devices, including car instrument panels, battery chargers, and inexpensive electrical instruments, measure potential difference (voltage), current, or resistance with a **d'Arsonval galvanometer** (**Fig. 26.13**). In the following discussion we'll often call it just a *meter*. A pivoted coil of fine wire is placed in the magnetic field of a permanent magnet (**Fig. 26.14**). Attached to the coil is a spring, similar to the hairspring on the balance wheel of a watch. In the equilibrium position, with no current in the coil, the pointer is at zero. When there is a current in the coil, the magnetic field exerts a torque on the coil that is proportional to the current. (We'll discuss this magnetic interaction in detail in Chapter 27.) As the coil turns, the spring exerts a restoring torque that is proportional to the angular displacement.

26.14 A d'Arsonval galvanometer, showing a pivoted coil with attached pointer, a permanent magnet supplying a magnetic field that is uniform in magnitude, and a spring to provide restoring torque, which opposes magnetic-field torque.

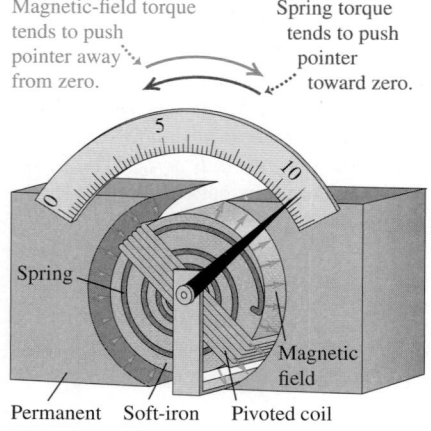

Thus the angular deflection of the coil and pointer is directly proportional to the coil current, and the device can be calibrated to measure current. The maximum deflection, typically 90° or so, is called *full-scale deflection*. The essential electrical characteristics of the meter are the current I_{fs} required for full-scale deflection (typically on the order of 10 μA to 10 mA) and the resistance R_c of the coil (typically on the order of 10 Ω to 1000 Ω).

The meter deflection is proportional to the *current* in the coil. If the coil obeys Ohm's law, the current is proportional to the *potential difference* between the terminals of the coil, and the deflection is also proportional to this potential difference. For example, consider a meter whose coil has a resistance $R_c = 20.0$ Ω and that deflects full scale when the current in its coil is $I_{fs} = 1.00$ mA. The corresponding potential difference for full-scale deflection is

$$V = I_{fs}R_c = (1.00 \times 10^{-3} \text{ A})(20.0 \text{ } \Omega) = 0.0200 \text{ V}$$

Ammeters

A current-measuring instrument is usually called an **ammeter** (or milliammeter, microammeter, and so forth, depending on the range). *An ammeter always measures the current passing through it.* An *ideal* ammeter, discussed in Section 25.4, would have *zero* resistance, so including it in a branch of a circuit would not affect the current in that branch. Real ammeters always have a finite resistance, but it is always desirable for an ammeter to have as little resistance as possible.

We can adapt any meter to measure currents that are larger than its full-scale reading by connecting a resistor in parallel with it (**Fig. 26.15a**) so that some of the current bypasses the meter coil. The parallel resistor is called a **shunt resistor** or simply a *shunt,* denoted as R_{sh}.

Suppose we want to make a meter with full-scale current I_{fs} and coil resistance R_c into an ammeter with full-scale reading I_a. To determine the shunt resistance R_{sh} needed, note that at full-scale deflection the total current through the parallel combination is I_a, the current through the coil of the meter is I_{fs}, and the current through the shunt is the difference $I_a - I_{fs}$. The potential difference V_{ab} is the same for both paths, so

$$I_{fs}R_c = (I_a - I_{fs})R_{sh} \qquad \text{(for an ammeter)} \qquad (26.7)$$

26.15 Using the same meter to measure (a) current and (b) voltage.

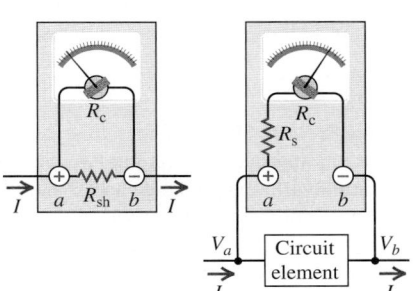

(a) Moving-coil ammeter (b) Moving-coil voltmeter

EXAMPLE 26.8 DESIGNING AN AMMETER

What shunt resistance is required to make the 1.00-mA, 20.0-Ω meter described above into an ammeter with a range of 0 to 50.0 mA?

SOLUTION

IDENTIFY and SET UP: Since the meter is being used as an ammeter, its internal connections are as shown in Fig. 26.15a. Our target variable is the shunt resistance R_{sh}, which we will find from Eq. (26.7). The ammeter must handle a maximum current $I_a = 50.0 \times 10^{-3}$ A. The coil resistance is $R_c = 20.0\ \Omega$, and the meter shows full-scale deflection when the current through the coil is $I_{fs} = 1.00 \times 10^{-3}$ A.

EXECUTE: Solving Eq. (26.7) for R_{sh}, we find

$$R_{sh} = \frac{I_{fs}R_c}{I_a - I_{fs}} = \frac{(1.00 \times 10^{-3}\ \text{A})(20.0\ \Omega)}{50.0 \times 10^{-3}\ \text{A} - 1.00 \times 10^{-3}\ \text{A}}$$

$$= 0.408\ \Omega$$

EVALUATE: It's useful to consider the equivalent resistance R_{eq} of the ammeter as a whole. From Eq. (26.2),

$$R_{eq} = \left(\frac{1}{R_c} + \frac{1}{R_{sh}}\right)^{-1} = \left(\frac{1}{20.0\ \Omega} + \frac{1}{0.408\ \Omega}\right)^{-1}$$

$$= 0.400\ \Omega$$

The shunt resistance is so small in comparison to the coil resistance that the equivalent resistance is very nearly equal to the shunt resistance. The result is an ammeter with a low equivalent resistance and the desired 0–50.0-mA range. At full-scale deflection, $I = I_a = 50.0$ mA, the current through the galvanometer is 1.00 mA, the current through the shunt resistor is 49.0 mA, and $V_{ab} = 0.0200$ V. If the current I is *less* than 50.0 mA, the coil current and the deflection are proportionally less.

Voltmeters

This same basic meter may also be used to measure potential difference or *voltage.* A voltage-measuring device is called a **voltmeter.** A voltmeter always measures the potential difference between two points, and its terminals must be connected to these points. (Example 25.6 in Section 25.4 described what can happen if a voltmeter is connected incorrectly.) As we discussed in Section 25.4, an ideal voltmeter would have *infinite* resistance, so connecting it between two points in a circuit would not alter any of the currents. Real voltmeters always have finite resistance, but a voltmeter should have large enough resistance that connecting it in a circuit does not change the other currents appreciably.

For the meter of Example 26.8, the voltage across the meter coil at full-scale deflection is only $I_{fs}R_c = (1.00 \times 10^{-3}\ \text{A})(20.0\ \Omega) = 0.0200$ V. We can extend this range by connecting a resistor R_s in *series* with the coil (Fig. 26.15b). Then only a fraction of the total potential difference appears across the coil itself, and the remainder appears across R_s. For a voltmeter with full-scale reading V_V, we need a series resistor R_s in Fig. 26.15b such that

$$V_V = I_{fs}(R_c + R_s) \qquad \text{(for a voltmeter)} \qquad (26.8)$$

BIO Application Electromyography
A fine needle containing two electrodes is being inserted into a muscle in this patient's hand. By using a sensitive voltmeter to measure the potential difference between these electrodes, a physician can probe the muscle's electrical activity. This is an important technique for diagnosing neurological and neuromuscular diseases.

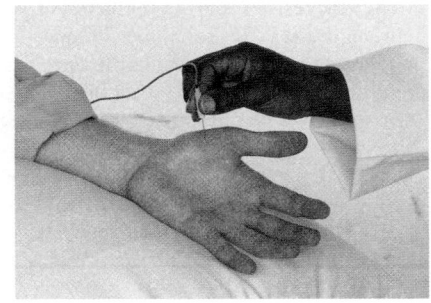

EXAMPLE 26.9 DESIGNING A VOLTMETER

What series resistance is required to make the 1.00-mA, 20.0-Ω meter described above into a voltmeter with a range of 0 to 10.0 V?

SOLUTION

IDENTIFY and SET UP: Since this meter is being used as a voltmeter, its internal connections are as shown in Fig. 26.15b. The maximum allowable voltage across the voltmeter is $V_V = 10.0$ V. We want this to occur when the current through the coil is $I_{fs} = 1.00 \times 10^{-3}$ A. Our target variable is the series resistance R_s, which we find from Eq. (26.8).

EXECUTE: From Eq. (26.8),

$$R_s = \frac{V_V}{I_{fs}} - R_c = \frac{10.0 \text{ V}}{0.00100 \text{ A}} - 20.0 \ \Omega = 9980 \ \Omega$$

EVALUATE: At full-scale deflection, $V_{ab} = 10.0$ V, the voltage across the meter is 0.0200 V, the voltage across R_s is 9.98 V, and the current through the voltmeter is 0.00100 A. Most of the voltage appears across the series resistor. The meter's equivalent resistance is a desirably high $R_{eq} = 20.0 \ \Omega + 9980 \ \Omega = 10,000 \ \Omega$. Such a meter is called a "1000 ohms-per-volt" meter, referring to the ratio of resistance to full-scale deflection. In normal operation the current through the circuit element being measured (I in Fig. 26.15b) is much greater than 0.00100 A, and the resistance between points a and b in the circuit is much less than 10,000 Ω. The voltmeter draws off only a small fraction of the current and thus disturbs the circuit being measured only slightly.

26.16 Ammeter–voltmeter method for measuring resistance.

(a)

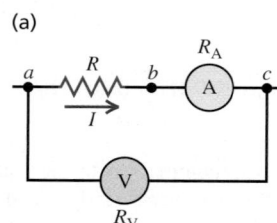

(b)

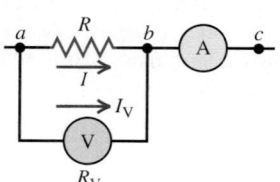

Ammeters and Voltmeters in Combination

A voltmeter and an ammeter can be used together to measure *resistance* and *power*. The resistance R of a resistor equals the potential difference V_{ab} between its terminals divided by the current I; that is, $R = V_{ab}/I$. The power input P to any circuit element is the product of the potential difference across it and the current through it: $P = V_{ab}I$. In principle, the most straightforward way to measure R or P is to measure V_{ab} and I simultaneously.

With practical ammeters and voltmeters this isn't quite as simple as it seems. In **Fig. 26.16a**, ammeter A reads the current I in the resistor R. Voltmeter V, however, reads the *sum* of the potential difference V_{ab} across the resistor and the potential difference V_{bc} across the ammeter. If we transfer the voltmeter terminal from c to b, as in Fig. 26.16b, then the voltmeter reads the potential difference V_{ab} correctly, but the ammeter now reads the *sum* of the current I in the resistor and the current I_V in the voltmeter. Either way, we have to correct the reading of one instrument or the other unless the corrections are small enough to be negligible.

EXAMPLE 26.10 MEASURING RESISTANCE I

The voltmeter in the circuit of Fig. 26.16a reads 12.0 V and the ammeter reads 0.100 A. The meter resistances are $R_V = 10,000 \ \Omega$ (for the voltmeter) and $R_A = 2.00 \ \Omega$ (for the ammeter). What are the resistance R and the power dissipated in the resistor?

SOLUTION

IDENTIFY and SET UP: The ammeter reads the current $I = 0.100$ A through the resistor, and the voltmeter reads the potential difference between a and c. If the ammeter were *ideal* (that is, if $R_A = 0$), there would be zero potential difference between b and c, the voltmeter reading $V = 12.0$ V would be equal to the potential difference V_{ab} across the resistor, and the resistance would be equal to $R = V/I = (12.0 \text{ V})/(0.100 \text{ A}) = 120 \ \Omega$. The ammeter is *not* ideal, however (its resistance is $R_A = 2.00 \ \Omega$), so the voltmeter reading V is actually the sum of the potential differences V_{bc} (across the ammeter) and V_{ab} (across the resistor). We use Ohm's

law to find the voltage V_{bc} from the known current and ammeter resistance. Then we solve for V_{ab} and R. Given these, we are able to calculate the power P into the resistor.

EXECUTE: From Ohm's law, $V_{bc} = IR_A = (0.100 \text{ A})(2.00 \ \Omega) = 0.200$ V and $V_{ab} = IR$. The sum of these is $V = 12.0$ V, so the potential difference across the resistor is $V_{ab} = V - V_{bc} = (12.0 \text{ V}) - (0.200 \text{ V}) = 11.8$ V. Hence the resistance is

$$R = \frac{V_{ab}}{I} = \frac{11.8 \text{ V}}{0.100 \text{ A}} = 118 \ \Omega$$

The power dissipated in this resistor is

$$P = V_{ab}I = (11.8 \text{ V})(0.100 \text{ A}) = 1.18 \text{ W}$$

EVALUATE: You can confirm this result for the power by using the alternative formula $P = I^2R$. Do you get the same answer?

EXAMPLE 26.11   MEASURING RESISTANCE II

Suppose the meters of Example 26.10 are connected to a different resistor as shown in Fig. 26.16b, and the readings obtained on the meters are the same as in Example 26.10. What is the value of this new resistance R, and what is the power dissipated in the resistor?

SOLUTION

IDENTIFY and SET UP: In Example 26.10 the ammeter read the actual current through the resistor, but the voltmeter reading was not the same as the potential difference across the resistor. Now the situation is reversed: The voltmeter reading $V = 12.0$ V shows the actual potential difference V_{ab} across the resistor, but the ammeter reading $I_A = 0.100$ A is *not* equal to the current I through the resistor. Applying the junction rule at b in Fig. 26.16b shows that $I_A = I + I_V$, where I_V is the current through the voltmeter. We find I_V from the given values of V and the voltmeter resistance R_V, and we use this value to find the resistor current I. We then determine the resistance R from I and the voltmeter reading, and calculate the power as in Example 26.10.

EXECUTE: We have $I_V = V/R_V = (12.0\text{ V})/(10{,}000\ \Omega) = 1.20$ mA. The actual current I in the resistor is $I = I_A - I_V = 0.100\text{ A} - 0.0012\text{ A} = 0.0988$ A, and the resistance is

$$R = \frac{V_{ab}}{I} = \frac{12.0\text{ V}}{0.0988\text{ A}} = 121\ \Omega$$

The power dissipated in the resistor is

$$P = V_{ab}I = (12.0\text{ V})(0.0988\text{ A}) = 1.19\text{ W}$$

EVALUATE: Had the meters been ideal, our results would have been $R = 12.0\text{ V}/0.100\text{ A} = 120\ \Omega$ and $P = VI = (12.0\text{ V}) \times (0.100\text{ A}) = 1.2$ W both here and in Example 26.10. The actual (correct) results are not too different in either case. That's because the ammeter and voltmeter are nearly ideal: Compared with the resistance R under test, the ammeter resistance R_A is very small and the voltmeter resistance R_V is very large. Under these conditions, treating the meters as ideal yields pretty good results; accurate work requires calculations as in these two examples.

Ohmmeters

An alternative method for measuring resistance is to use a d'Arsonval meter in an arrangement called an **ohmmeter.** It consists of a meter, a resistor, and a source (often a flashlight battery) connected in series (**Fig. 26.17**). The resistance R to be measured is connected between terminals x and y.

The series resistance R_s is variable; it is adjusted so that when terminals x and y are short-circuited (that is, when $R = 0$), the meter deflects full scale. When nothing is connected to terminals x and y, so that the circuit between x and y is *open* (that is, when $R \to \infty$), there is no current and hence no deflection. For any intermediate value of R the meter deflection depends on the value of R, and the meter scale can be calibrated to read the resistance R directly. Larger currents correspond to smaller resistances, so this scale reads backward compared to the scale showing the current.

In situations in which high precision is required, instruments containing d'Arsonval meters have been supplanted by electronic instruments with direct digital readouts. Digital voltmeters can be made with extremely high internal resistance, of the order of 100 MΩ. **Figure 26.18** shows a digital *multimeter,* an instrument that can measure voltage, current, or resistance over a wide range.

26.17 Ohmmeter circuit. The resistor R_s has a variable resistance, as is indicated by the arrow through the resistor symbol. To use the ohmmeter, first connect x directly to y and adjust R_s until the meter reads zero. Then connect x and y across the resistor R and read the scale.

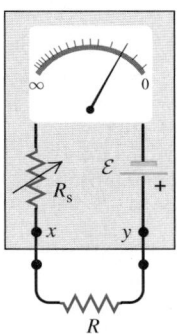

26.18 This digital multimeter can be used as a voltmeter (red arc), ammeter (yellow arc), or ohmmeter (green arc).

26.19 A potentiometer.

(a) Potentiometer circuit

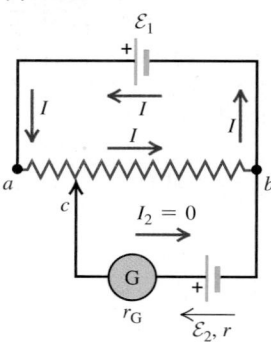

$I_2 = 0$

(b) Circuit symbol for potentiometer (variable resistor)

The Potentiometer

The *potentiometer* is an instrument that can be used to measure the emf of a source without drawing any current from the source; it also has a number of other useful applications. Essentially, it balances an unknown potential difference against an adjustable, measurable potential difference.

The principle of the potentiometer is shown schematically in **Fig. 26.19a**. A resistance wire ab of total resistance R_{ab} is permanently connected to the terminals of a source of known emf $\mathcal{E}_1$. A sliding contact c is connected through the galvanometer G to a second source whose emf $\mathcal{E}_2$ is to be measured. As contact c is moved along the resistance wire, the resistance R_{cb} between points c and b varies; if the resistance wire is uniform, R_{cb} is proportional to the length of wire between c and b. To determine the value of $\mathcal{E}_2$, contact c is moved until a position is found at which the galvanometer shows no deflection; this corresponds to zero current passing through $\mathcal{E}_2$. With $I_2 = 0$, Kirchhoff's loop rule gives

$$\mathcal{E}_2 = IR_{cb}$$

With $I_2 = 0$, the current I produced by the emf $\mathcal{E}_1$ has the same value no matter what the value of the emf $\mathcal{E}_2$. We calibrate the device by replacing $\mathcal{E}_2$ by a source of known emf; then to find any unknown emf $\mathcal{E}_2$, we measure the length of wire cb for which $I_2 = 0$. Note: For this to work, V_{ab} must be greater than $\mathcal{E}_2$.

The term *potentiometer* is also used for any variable resistor, usually having a circular resistance element and a sliding contact controlled by a rotating shaft and knob. The circuit symbol for a potentiometer is shown in Fig. 26.19b.

TEST YOUR UNDERSTANDING OF SECTION 26.3 You want to measure the current through and the potential difference across the 2-Ω resistor shown in Fig. 26.12 (Example 26.6 in Section 26.2). (a) How should you connect an ammeter and a voltmeter to do this? (i) Both ammeter and voltmeter in series with the 2-Ω resistor; (ii) ammeter in series with the 2-Ω resistor and voltmeter connected between points b and d; (iii) ammeter connected between points b and d and voltmeter in series with the 2-Ω resistor; (iv) both ammeter and voltmeter connected between points b and d. (b) What resistances should these meters have? (i) Both ammeter and voltmeter resistances should be much greater than 2 Ω; (ii) ammeter resistance should be much greater than 2 Ω and voltmeter resistance should be much less than 2 Ω; (iii) ammeter resistance should be much less than 2 Ω and voltmeter resistance should be much greater than 2 Ω; (iv) both ammeter and voltmeter resistances should be much less than 2 Ω. ▮

26.4 *R-C* CIRCUITS

In the circuits we have analyzed up to this point, we have assumed that all the emfs and resistances are *constant* (time independent) so that all the potentials, currents, and powers are also independent of time. But in the simple act of charging or discharging a capacitor we find a situation in which the currents, voltages, and powers *do* change with time.

Many devices incorporate circuits in which a capacitor is alternately charged and discharged. These include flashing traffic lights, automobile turn signals, and electronic flash units. Understanding what happens in such circuits is thus of great practical importance.

Charging a Capacitor

Figure 26.20 shows a simple circuit for charging a capacitor. A circuit such as this that has a resistor and a capacitor in series is called an ***R-C* circuit.** We idealize the battery (or power supply) to have a constant emf $\mathcal{E}$ and zero internal resistance ($r = 0$), and we ignore the resistance of all the connecting conductors.

We begin with the capacitor initially uncharged (Fig. 26.20a); then at some initial time $t = 0$ we close the switch, completing the circuit and permitting current around the loop to begin charging the capacitor (Fig. 26.20b). For all practical purposes, the current begins at the same instant in every conducting part of the circuit, and at each instant the current is the same in every part.

26.20 Charging a capacitor. (a) Just before the switch is closed, the charge q is zero. (b) When the switch closes (at $t = 0$), the current jumps from zero to $\mathcal{E}/R$. As time passes, q approaches Q_f and the current i approaches zero.

(a) Capacitor initially uncharged

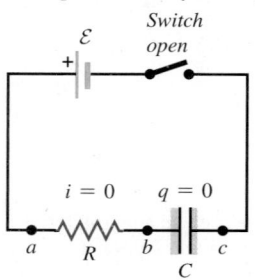

(b) Charging the capacitor

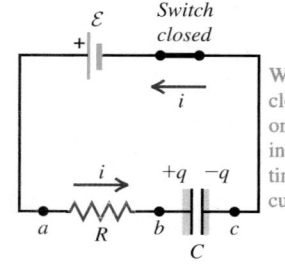

When the switch is closed, the charge on the capacitor increases over time while the current decreases.

Because the capacitor in Fig. 26.20 is initially uncharged, the potential difference v_{bc} across it is zero at $t = 0$. At this time, from Kirchhoff's loop law, the voltage v_{ab} across the resistor R is equal to the battery emf $\mathcal{E}$. The initial $(t = 0)$ current through the resistor, which we will call I_0, is given by Ohm's law: $I_0 = v_{ab}/R = \mathcal{E}/R$.

As the capacitor charges, its voltage v_{bc} increases and the potential difference v_{ab} across the resistor decreases, corresponding to a decrease in current. The sum of these two voltages is constant and equal to $\mathcal{E}$. After a long time the capacitor is fully charged, the current decreases to zero, and v_{ab} across the resistor becomes zero. Then the entire battery emf $\mathcal{E}$ appears across the capacitor and $v_{bc} = \mathcal{E}$.

Let q represent the charge on the capacitor and i the current in the circuit at some time t after the switch has been closed. We choose the positive direction for the current to correspond to positive charge flowing onto the left-hand capacitor plate, as in Fig. 26.20b. The instantaneous potential differences v_{ab} and v_{bc} are

$$v_{ab} = iR \qquad v_{bc} = \frac{q}{C}$$

Using these in Kirchhoff's loop rule, we find

$$\mathcal{E} - iR - \frac{q}{C} = 0 \qquad (26.9)$$

The potential drops by an amount iR as we travel from a to b and by q/C as we travel from b to c. Solving Eq. (26.9) for i, we find

$$i = \frac{\mathcal{E}}{R} - \frac{q}{RC} \qquad (26.10)$$

At time $t = 0$, when the switch is first closed, the capacitor is uncharged, and so $q = 0$. Substituting $q = 0$ into Eq. (26.10), we find that the *initial* current I_0 is given by $I_0 = \mathcal{E}/R$, as we have already noted. If the capacitor were not in the circuit, the last term in Eq. (26.10) would not be present; then the current would be *constant* and equal to $\mathcal{E}/R$.

As the charge q increases, the term q/RC becomes larger and the capacitor charge approaches its final value, which we will call Q_f. The current decreases and eventually becomes zero. When $i = 0$, Eq. (26.10) gives

$$\frac{\mathcal{E}}{R} = \frac{Q_f}{RC} \qquad Q_f = C\mathcal{E} \qquad (26.11)$$

Note that the final charge Q_f does not depend on R.

Figure 26.21 shows the current and capacitor charge as functions of time. At the instant the switch is closed $(t = 0)$, the current jumps from zero to its initial value $I_0 = \mathcal{E}/R$; after that, it gradually approaches zero. The capacitor charge starts at zero and gradually approaches the final value given by Eq. (26.11), $Q_f = C\mathcal{E}$.

CAUTION Lowercase means time-varying Up to this point we have been working with constant potential differences (voltages), currents, and charges, and we have used *capital* letters V, I, and Q, respectively, to denote these quantities. To distinguish between quantities that vary with time and those that are constant, we will use *lowercase* letters v, i, and q for time-varying voltages, currents, and charges, respectively. We suggest that you follow this same convention in your own work.

(a) Graph of current versus time for a charging capacitor

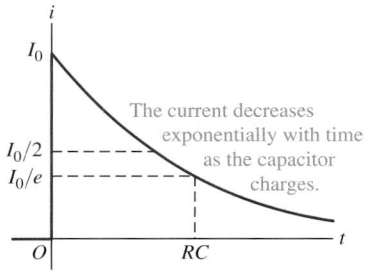

The current decreases exponentially with time as the capacitor charges.

(b) Graph of capacitor charge versus time for a charging capacitor

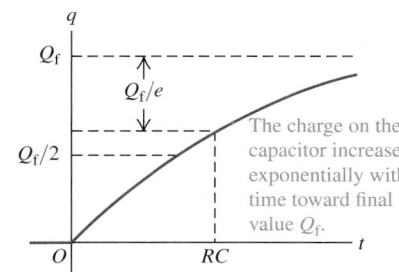

The charge on the capacitor increases exponentially with time toward final value Q_f.

26.21 Current i and capacitor charge q as functions of time for the circuit of Fig. 26.20. The initial current is I_0 and the initial capacitor charge is zero. The current asymptotically approaches zero, and the capacitor charge asymptotically approaches a final value of Q_f.

We can derive general expressions for charge q and current i as functions of time. With our choice of the positive direction for current (Fig. 26.20b), i equals the rate at which positive charge arrives at the left-hand (positive) plate of the capacitor, so $i = dq/dt$. Making this substitution in Eq. (26.10), we have

$$\frac{dq}{dt} = \frac{\mathcal{E}}{R} - \frac{q}{RC} = -\frac{1}{RC}(q - C\mathcal{E})$$

We can rearrange this to

$$\frac{dq}{q - C\mathcal{E}} = -\frac{dt}{RC}$$

and then integrate both sides. We change the integration variables to q' and t' so that we can use q and t for the upper limits. The lower limits are $q' = 0$ and $t' = 0$:

$$\int_0^q \frac{dq'}{q' - C\mathcal{E}} = -\int_0^t \frac{dt'}{RC}$$

When we carry out the integration, we get

$$\ln\left(\frac{q - C\mathcal{E}}{-C\mathcal{E}}\right) = -\frac{t}{RC}$$

Exponentiating both sides (that is, taking the inverse logarithm) and solving for q, we find

$$\frac{q - C\mathcal{E}}{-C\mathcal{E}} = e^{-t/RC}$$

PhET: Circuit Construction Kit (AC+DC)
PhET: Circuit Construction Kit (DC Only)

BIO Application Pacemakers and Capacitors This x-ray image shows a pacemaker implanted in a patient with a malfunctioning sinoatrial node, the part of the heart that generates the electrical signal to trigger heartbeats. The pacemaker circuit contains a battery, a capacitor, and a computer-controlled switch. To maintain regular beating, once per second the switch discharges the capacitor and sends an electrical pulse along the lead to the heart. The switch then flips to allow the capacitor to recharge for the next pulse.

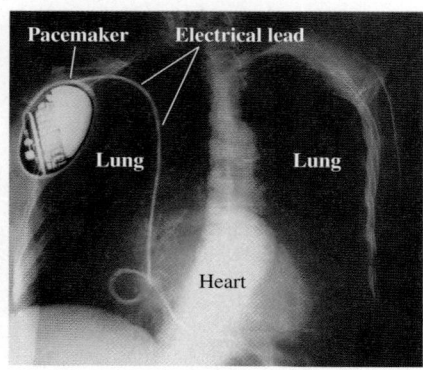

R-C circuit, charging capacitor:
Capacitor charge · · · Capacitance · · · Final capacitor charge $= C\mathcal{E}$

$$q = C\mathcal{E}(1 - e^{-t/RC}) = Q_f(1 - e^{-t/RC}) \qquad (26.12)$$

Battery emf · · Time since switch closed · · · Resistance

The instantaneous current i is just the time derivative of Eq. (26.12):

R-C circuit, charging capacitor:
Current · · · Battery emf · · · Time since switch closed

$$i = \frac{dq}{dt} = \frac{\mathcal{E}}{R}e^{-t/RC} = I_0 e^{-t/RC} \overset{\text{Initial current}}{\underset{= \mathcal{E}/R}{}} \qquad (26.13)$$

Rate of change of · · · Resistance · · · Capacitance
capacitor charge

The charge and current are both *exponential* functions of time. Figure 26.21a is a graph of Eq. (26.13) and Fig. 26.21b is a graph of Eq. (26.12).

Time Constant

After a time equal to RC, the current in the R-C circuit has decreased to $1/e$ (about 0.368) of its initial value. At this time, the capacitor charge has reached $(1 - 1/e) = 0.632$ of its final value $Q_f = C\mathcal{E}$. The product RC is therefore a measure of how quickly the capacitor charges. We call RC the **time constant,** or the **relaxation time**, of the circuit, denoted by τ:

$$\tau = RC \qquad \text{(time constant for } R\text{-}C \text{ circuit)} \qquad (26.14)$$

When τ is small, the capacitor charges quickly; when it is larger, the charging takes more time. If the resistance is small, it's easier for current to flow, and the capacitor charges more quickly. If R is in ohms and C in farads, τ is in seconds.

In Fig. 26.21a the horizontal axis is an *asymptote* for the curve. Strictly speaking, i never becomes exactly zero. But the longer we wait, the closer it gets. After a time equal to $10RC$, the current has decreased to 0.000045 of its initial value. Similarly, the curve in Fig. 26.21b approaches the horizontal dashed line labeled Q_f as an asymptote. The charge q never attains exactly this value, but after a time equal to $10RC$, the difference between q and Q_f is only 0.000045 of Q_f. We invite you to verify that the product RC has units of time.

Discharging a Capacitor

Now suppose that after the capacitor in Fig. 26.21b has acquired a charge Q_0, we remove the battery from our *R-C* circuit and connect points a and c to an open switch (**Fig. 26.22a**). We then close the switch and at the same instant reset our stopwatch to $t = 0$; at that time, $q = Q_0$. The capacitor then *discharges* through the resistor, and its charge eventually decreases to zero.

Again let i and q represent the time-varying current and charge at some instant after the connection is made. In Fig. 26.22b we make the same choice of the positive direction for current as in Fig. 26.20b. Then Kirchhoff's loop rule gives Eq. (26.10) but with $\mathcal{E} = 0$; that is,

$$i = \frac{dq}{dt} = -\frac{q}{RC} \tag{26.15}$$

The current i is now negative; this is because positive charge q is leaving the left-hand capacitor plate in Fig. 26.22b, so the current is in the direction opposite to that shown. At time $t = 0$, when $q = Q_0$, the initial current is $I_0 = -Q_0/RC$.

To find q as a function of time, we rearrange Eq. (26.15), again change the variables to q' and t', and integrate. This time the limits for q' are Q_0 to q:

$$\int_{Q_0}^{q} \frac{dq'}{q'} = -\frac{1}{RC} \int_{0}^{t} dt'$$

$$\ln \frac{q}{Q_0} = -\frac{t}{RC}$$

R-C circuit, discharging capacitor:	Capacitor charge ⋯⋯⋯⋯⋯ Initial capacitor charge $q = Q_0 e^{-t/RC}$ ⋯ Capacitance Time since switch closed ⋯⋯ Resistance	(26.16)

The instantaneous current i is the derivative of this with respect to time:

R-C circuit, discharging capacitor:	Current Initial capacitor charge ⋯ Capacitance $i = \frac{dq}{dt} = -\frac{Q_0}{RC} e^{-t/RC} = I_0 e^{-t/RC}$ Time since switch closed Rate of change of Resistance Initial current $= -Q_0/RC$ capacitor charge	(26.17)

We graph the current and the charge in **Fig. 26.23**; both quantities approach zero exponentially with time. Comparing these results with Eqs. (26.12) and (26.13), we note that the expressions for the current are identical, apart from the sign of I_0. The capacitor charge approaches zero asymptotically in Eq. (26.16), while the *difference* between q and Q approaches zero asymptotically in Eq. (26.12).

Energy considerations give us additional insight into the behavior of an *R-C* circuit. While the capacitor is charging, the instantaneous rate at which the battery delivers energy to the circuit is $P = \mathcal{E}i$. The instantaneous rate at which

26.22 Discharging a capacitor.
(a) Before the switch is closed at time $t = 0$, the capacitor charge is Q_0 and the current is zero. **(b)** At time t after the switch is closed, the capacitor charge is q and the current is i. The actual current direction is opposite to the direction shown; i is negative. After a long time, q and i both approach zero.

(a) Capacitor initially charged

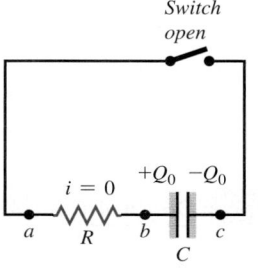

(b) Discharging the capacitor

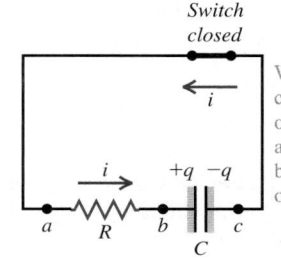

When the switch is closed, the charge on the capacitor and the current both decrease over time.

26.23 Current i and capacitor charge q as functions of time for the circuit of Fig. 26.22. The initial current is I_0 and the initial capacitor charge is Q_0. Both i and q asymptotically approach zero.

(a) Graph of current versus time for a discharging capacitor

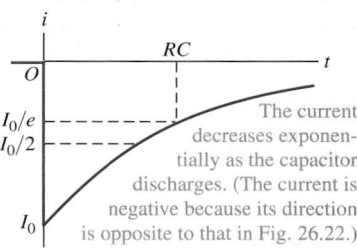

The current decreases exponentially as the capacitor discharges. (The current is negative because its direction is opposite to that in Fig. 26.22.)

(b) Graph of capacitor charge versus time for a discharging capacitor

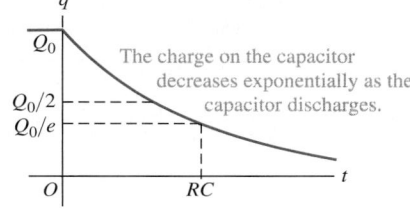

The charge on the capacitor decreases exponentially as the capacitor discharges.

electrical energy is dissipated in the resistor is i^2R, and the rate at which energy is stored in the capacitor is $iv_{bc} = iq/C$. Multiplying Eq. (26.9) by i, we find

$$\mathcal{E}i = i^2R + \frac{iq}{C} \tag{26.18}$$

This means that of the power $\mathcal{E}i$ supplied by the battery, part (i^2R) is dissipated in the resistor and part (iq/C) is stored in the capacitor.

The *total* energy supplied by the battery during charging of the capacitor equals the battery emf $\mathcal{E}$ multiplied by the total charge Q_f, or $\mathcal{E}Q_f$. The total energy stored in the capacitor, from Eq. (24.9), is $Q_f\mathcal{E}/2$. Thus, of the energy supplied by the battery, *exactly half* is stored in the capacitor, and the other half is dissipated in the resistor. This half-and-half division of energy doesn't depend on C, R, or $\mathcal{E}$. You can verify this result by taking the integral over time of each of the power quantities in Eq. (26.18).

EXAMPLE 26.12 CHARGING A CAPACITOR

A 10-MΩ resistor is connected in series with a 1.0-μF capacitor and a battery with emf 12.0 V. Before the switch is closed at time $t = 0$, the capacitor is uncharged. (a) What is the time constant? (b) What fraction of the final charge Q_f is on the capacitor at $t = 46$ s? (c) What fraction of the initial current I_0 is still flowing at $t = 46$ s?

SOLUTION

IDENTIFY and SET UP: This is the situation shown in Fig. 26.20, with $R = 10$ MΩ, $C = 1.0$ μF, and $\mathcal{E} = 12.0$ V. The charge q and current i vary with time as shown in Fig. 26.21. Our target variables are (a) the time constant τ, (b) the ratio q/Q_f at $t = 46$ s, and (c) the ratio i/I_0 at $t = 46$ s. Equation (26.14) gives τ. For a capacitor being charged, Eq. (26.12) gives q and Eq. (26.13) gives i.

EXECUTE: (a) From Eq. (26.14),

$$\tau = RC = (10 \times 10^6 \, \Omega)(1.0 \times 10^{-6} \, \text{F}) = 10 \text{ s}$$

(b) From Eq. (26.12),

$$\frac{q}{Q_f} = 1 - e^{-t/RC} = 1 - e^{-(46\,\text{s})/(10\,\text{s})} = 0.99$$

(c) From Eq. (26.13),

$$\frac{i}{I_0} = e^{-t/RC} = e^{-(46\,\text{s})/(10\,\text{s})} = 0.010$$

EVALUATE: After 4.6 time constants the capacitor is 99% charged and the charging current has decreased to 1.0% of its initial value. The circuit would charge more rapidly if we reduced the time constant by using a smaller resistance.

EXAMPLE 26.13 DISCHARGING A CAPACITOR

The resistor and capacitor of Example 26.12 are reconnected as shown in Fig. 26.22. The capacitor has an initial charge of 5.0 μC and is discharged by closing the switch at $t = 0$. (a) At what time will the charge be 0.50 μC? (b) What is the current at this time?

SOLUTION

IDENTIFY and SET UP: Now the capacitor is being discharged, so q and i vary with time as in Fig. 26.23, with $Q_0 = 5.0 \times 10^{-6}$ C. Again we have $RC = \tau = 10$ s. Our target variables are (a) the value of t at which $q = 0.50$ μC and (b) the value of i at this time. We first solve Eq. (26.16) for t, and then solve Eq. (26.17) for i.

EXECUTE: (a) Solving Eq. (26.16) for the time t gives

$$t = -RC \ln\frac{q}{Q_0} = -(10\,\text{s}) \ln\frac{0.50\,\mu\text{C}}{5.0\,\mu\text{C}} = 23 \text{ s} = 2.3\tau$$

(b) From Eq. (26.17), with $Q_0 = 5.0$ μC $= 5.0 \times 10^{-6}$ C,

$$i = -\frac{Q_0}{RC}e^{-t/RC} = -\frac{5.0 \times 10^{-6}\,\text{C}}{10\,\text{s}}e^{-2.3} = -5.0 \times 10^{-8} \text{ A}$$

EVALUATE: The current in part (b) is negative because i has the opposite sign when the capacitor is discharging than when it is charging. Note that we could have avoided evaluating $e^{-t/RC}$ by noticing that at the time in question, $q = 0.10Q_0$; from Eq. (26.16) this means that $e^{-t/RC} = 0.10$.

TEST YOUR UNDERSTANDING OF SECTION 26.4 The energy stored in a capacitor is equal to $q^2/2C$. When a capacitor is discharged, what fraction of the initial energy remains after an elapsed time of one time constant? (i) $1/e$; (ii) $1/e^2$; (iii) $1 - 1/e$; (iv) $(1 - 1/e)^2$; (v) answer depends on how much energy was stored initially. ∎

26.5 POWER DISTRIBUTION SYSTEMS

We conclude this chapter with a brief discussion of practical household and automotive electric-power distribution systems. Automobiles use direct-current (dc) systems, while nearly all household, commercial, and industrial systems use alternating current (ac) because of the ease of stepping voltage up and down with transformers. Most of the same basic wiring concepts apply to both. We'll talk about alternating-current circuits in greater detail in Chapter 31.

The various lamps, motors, and other appliances to be operated are always connected in *parallel* to the power source (the wires from the power company for houses, or from the battery and alternator for a car). If appliances were connected in series, shutting one appliance off would shut them all off (see Example 26.2 in Section 26.1). **Figure 26.24** shows the basic idea of house wiring. One side of the "line," as the pair of conductors is called, is called the *neutral* side; it is always connected to "ground" at the entrance panel. For houses, *ground* is an actual electrode driven into the earth (which is usually a good conductor) or sometimes connected to the household water pipes. Electricians speak of the "hot" side and the "neutral" side of the line. Most modern house wiring systems have *two* hot lines with opposite polarity with respect to the neutral. We'll return to this detail later.

Household voltage is nominally 120 V in the United States and Canada, and often 240 V in Europe. (For alternating current, which varies sinusoidally with time, these numbers represent the *root-mean-square* voltage, which is $1/\sqrt{2}$ times the peak voltage. We'll discuss this further in Section 31.1.) The amount of current I drawn by a given device is determined by its power input P, given by Eq. (25.17): $P = VI$. Hence $I = P/V$. For example, the current in a 100-W light bulb is

$$I = \frac{P}{V} = \frac{100 \text{ W}}{120 \text{ V}} = 0.83 \text{ A}$$

The power input to this bulb is actually determined by its resistance R. Using Eq. (25.18), which states that $P = VI = I^2R = V^2/R$ for a resistor, we get the resistance of this bulb at operating temperature:

$$R = \frac{V}{I} = \frac{120 \text{ V}}{0.83 \text{ A}} = 144 \ \Omega \quad \text{or} \quad R = \frac{V^2}{P} = \frac{(120 \text{ V})^2}{100 \text{ W}} = 144 \ \Omega$$

Similarly, a 1500-W waffle iron draws a current of $(1500 \text{ W})/(120 \text{ V}) = 12.5 \text{ A}$ and has a resistance, at operating temperature, of 9.6 Ω. Because of the temperature dependence of resistivity, the resistances of these devices are considerably less when they are cold. If you measure the resistance of a 100-W light bulb with

26.24 Schematic diagram of part of a house wiring system. Only two branch circuits are shown; an actual system might have four to thirty branch circuits. Lamps and appliances may be plugged into the outlets. The grounding wires, which normally carry no current, are not shown.

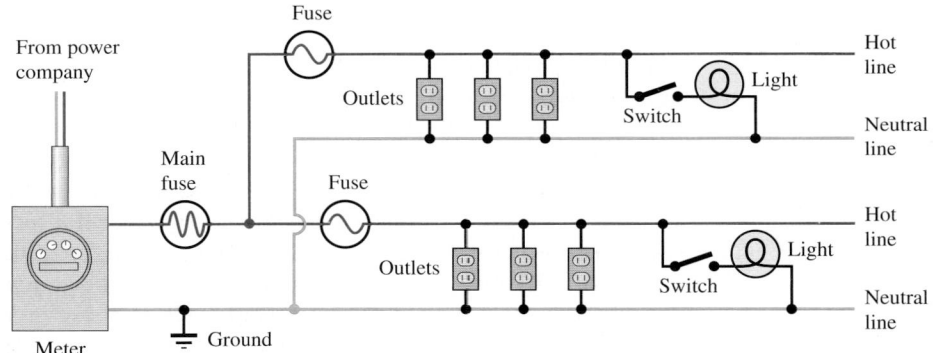

an ohmmeter (whose small current causes very little temperature rise), you will probably get a value of about 10 Ω. When a light bulb is turned on, this low resistance causes an initial surge of current until the filament heats up. That's why a light bulb that's ready to burn out nearly always does so just when you turn it on.

Circuit Overloads and Short Circuits

The maximum current available from an individual circuit is limited by the resistance of the wires. As we discussed in Section 25.5, the I^2R power loss in the wires causes them to become hot, and in extreme cases this can cause a fire or melt the wires. Ordinary lighting and outlet wiring in houses usually uses 12-gauge wire. This has a diameter of 2.05 mm and can carry a maximum current of 20 A safely (without overheating). Larger-diameter wires of the same length have lower resistance [see Eq. (25.10)]. Hence 8-gauge (3.26 mm) or 6-gauge (4.11 mm) is used for high-current appliances such as clothes dryers, and 2-gauge (6.54 mm) or larger is used for the main power lines entering a house.

Protection against overloading and overheating of circuits is provided by fuses or circuit breakers. A *fuse* contains a link of lead–tin alloy with a very low melting temperature; the link melts and breaks the circuit when its rated current is exceeded (**Fig. 26.25a**). A *circuit breaker* is an electromechanical device that performs the same function, using an electromagnet or a bimetallic strip to "trip" the breaker and interrupt the circuit when the current exceeds a specified value (Fig. 26.25b). Circuit breakers have the advantage that they can be reset after they are tripped, while a blown fuse must be replaced.

26.25 (a) Excess current will melt the thin wire of lead–tin alloy that runs along the length of a fuse, inside the transparent housing. (b) The switch on this circuit breaker will flip if the maximum allowable current is exceeded.

(a)

(b)

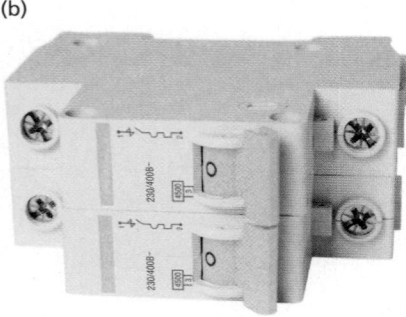

CAUTION **Fuses** If your system has fuses and you plug too many high-current appliances into the same outlet, the fuse blows. Do not replace the fuse with one that has a higher rating; if you do, you risk overheating the wires and starting a fire. The only safe solution is to distribute the appliances among several circuits. Modern kitchens often have three or four separate 20-A circuits. ▮

Contact between the hot and neutral sides of the line causes a *short circuit*. Such a situation, which can be caused by faulty insulation or by a variety of mechanical malfunctions, provides a very low-resistance current path, permitting a very large current that would quickly melt the wires and ignite their insulation if the current were not interrupted by a fuse or circuit breaker (see Example 25.10 in Section 25.5). An equally dangerous situation is a broken wire that interrupts the current path, creating an *open circuit*. This is hazardous because of the sparking that can occur at the point of intermittent contact.

In approved wiring practice, a fuse or breaker is placed *only* in the hot side of the line, never in the neutral side. Otherwise, if a short circuit should develop because of faulty insulation or other malfunction, the ground-side fuse could blow. The hot side would still be live and would pose a shock hazard if you touched the live conductor and a grounded object such as a water pipe. For similar reasons the wall switch for a light fixture is always in the hot side of the line, never the neutral side.

Further protection against shock hazard is provided by a third conductor called the *grounding wire,* included in all present-day wiring. This conductor corresponds to the long round or U-shaped prong of the three-prong connector plug on an appliance or power tool. It is connected to the neutral side of the line at the entrance panel. The grounding wire normally carries no current, but it connects the metal case or frame of the device to ground. If a conductor on the hot side of the line accidentally contacts the frame or case, the grounding conductor provides a current path, and the fuse blows. Without the ground wire, the frame could become "live"—that is, at a potential 120 V above ground. Then if you touched it and a water pipe (or even a damp floor) at the same time, you could

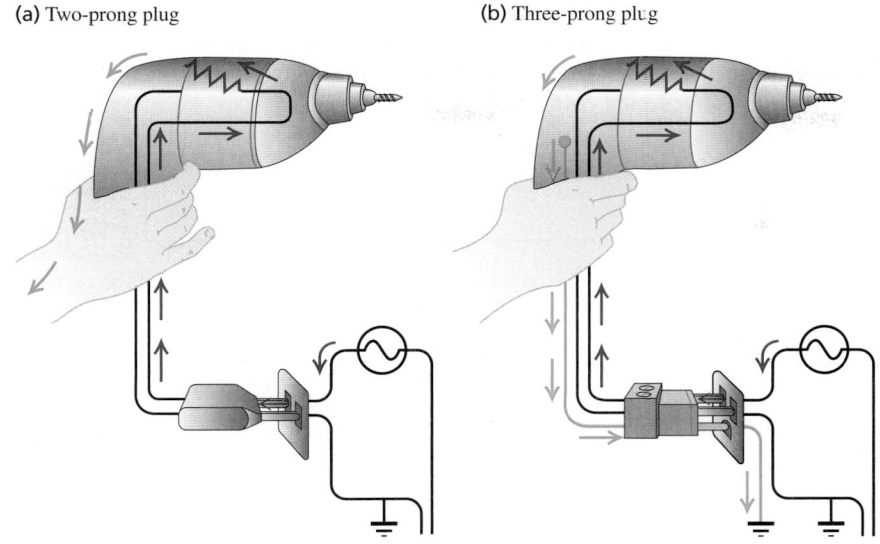

(a) Two-prong plug   (b) Three-prong plug

26.26 (a) If a malfunctioning electric drill is connected to a wall socket via a two-prong plug, a person may receive a shock. (b) When the drill malfunctions when connected via a three-prong plug, a person touching it receives no shock, because electric charge flows through the ground wire (shown in green) to the third prong and into the ground rather than into the person's body. If the ground current is appreciable, the fuse blows.

get a dangerous shock (**Fig. 26.26**). In some situations, especially outlets located outdoors or near a sink or other water pipes, a special kind of circuit breaker called a *ground-fault interrupter* (GFI or GFCI) is used. This device senses the difference in current between the hot and neutral conductors (which is normally zero) and trips when some very small value, typically 5 mA, is exceeded.

Household and Automotive Wiring

Most modern household wiring systems actually use a slight elaboration of the system described above. The power company provides *three* conductors. One is neutral; the other two are both at 120 V with respect to the neutral but with opposite polarity, giving a voltage between them of 240 V. The power company calls this a *three-wire line,* in contrast to the 120-V two-wire (plus ground wire) line described above. With a three-wire line, 120-V lamps and appliances can be connected between neutral and either hot conductor, and high-power devices requiring 240 V, such as electric ranges and clothes dryers, are connected between the two hot lines.

All of the above discussion can be applied directly to automobile wiring. The voltage is about 13 V (direct current); the power is supplied by the battery and by the alternator, which charges the battery when the engine is running. The neutral side of each circuit is connected to the body and frame of the vehicle. For this low voltage a separate grounding conductor is not required for safety. The fuse or circuit breaker arrangement is the same in principle as in household wiring. Because of the lower voltage (less energy per charge), more current (a greater number of charges per second) is required for the same power; a 100-W headlight bulb requires a current of about $(100 \text{ W})/(13 \text{ V}) = 8$ A.

Although we spoke of *power* in the above discussion, what we buy from the power company is *energy*. Power is energy transferred per unit time, so energy is average power multiplied by time. The usual unit of energy sold by the power company is the kilowatt-hour (1 kW·h):

$$1 \text{ kW·h} = (10^3 \text{ W})(3600 \text{ s}) = 3.6 \times 10^6 \text{ W·s} = 3.6 \times 10^6 \text{ J}$$

In the United States, one kilowatt-hour typically costs 8 to 27 cents, depending on the location and quantity of energy purchased. To operate a 1500-W (1.5-kW) waffle iron continuously for 1 hour requires 1.5 kW·h of energy; at 10 cents per kilowatt-hour, the energy cost is 15 cents. The cost of operating any lamp or appliance for a specified time can be calculated in the same way if the power rating is known. However, many electric cooking utensils (including waffle irons) cycle on and off to maintain a constant temperature, so the average power may be less than the power rating marked on the device.

EXAMPLE 26.14 A KITCHEN CIRCUIT

An 1800-W toaster, a 1.3-kW electric frying pan, and a 100-W lamp are plugged into the same 20-A, 120-V circuit. (a) What current is drawn by each device, and what is the resistance of each device? (b) Will this combination trip the circuit breaker?

SOLUTION

IDENTIFY and SET UP: When plugged into the same circuit, the three devices are connected in parallel, so the voltage across each appliance is $V = 120$ V. We find the current I drawn by each device from the relationship $P = VI$, where P is the power input of the device. To find the resistance R of each device we use the relationship $P = V^2/R$.

EXECUTE: (a) To simplify the calculation of current and resistance, we note that $I = P/V$ and $R = V^2/P$. Hence

$$I_{\text{toaster}} = \frac{1800 \text{ W}}{120 \text{ V}} = 15 \text{ A} \qquad R_{\text{toaster}} = \frac{(120 \text{ V})^2}{1800 \text{ W}} = 8 \text{ }\Omega$$

$$I_{\text{frying pan}} = \frac{1300 \text{ W}}{120 \text{ V}} = 11 \text{ A} \qquad R_{\text{frying pan}} = \frac{(120 \text{ V})^2}{1300 \text{ W}} = 11 \text{ }\Omega$$

$$I_{\text{lamp}} = \frac{100 \text{ W}}{120 \text{ V}} = 0.83 \text{ A} \qquad R_{\text{lamp}} = \frac{(120 \text{ V})^2}{100 \text{ W}} = 144 \text{ }\Omega$$

For constant voltage the device with the *least* resistance (in this case the toaster) draws the most current and receives the most power.

(b) The total current through the line is the sum of the currents drawn by the three devices:

$$I = I_{\text{toaster}} + I_{\text{frying pan}} + I_{\text{lamp}}$$

$$= 15 \text{ A} + 11 \text{ A} + 0.83 \text{ A} = 27 \text{ A}$$

This exceeds the 20-A rating of the line, and the circuit breaker will indeed trip.

EVALUATE: We could also find the total current by using $I = P/V$ and the total power P delivered to all three devices:

$$I = \frac{P_{\text{toaster}} + P_{\text{frying pan}} + P_{\text{lamp}}}{V}$$

$$= \frac{1800 \text{ W} + 1300 \text{ W} + 100 \text{ W}}{120 \text{ V}} = 27 \text{ A}$$

A third way to determine I is to use $I = V/R_{\text{eq}}$, where R_{eq} is the equivalent resistance of the three devices in parallel:

$$I = \frac{V}{R_{\text{eq}}} = (120 \text{ V})\left(\frac{1}{8 \text{ }\Omega} + \frac{1}{11 \text{ }\Omega} + \frac{1}{144 \text{ }\Omega}\right) = 27 \text{ A}$$

Appliances with such current demands are common, so modern kitchens have more than one 20-A circuit. To keep currents safely below 20 A, the toaster and frying pan should be plugged into different circuits.

TEST YOUR UNDERSTANDING OF SECTION 26.5 To prevent the circuit breaker in Example 26.14 from blowing, a home electrician replaces the circuit breaker with one rated at 40 A. Is this a reasonable thing to do? ❚

Resistors in series and parallel: When several resistors $R_1, R_2, R_3, \ldots$ are connected in series, the equivalent resistance R_{eq} is the sum of the individual resistances. The same *current* flows through all the resistors in a series connection. When several resistors are connected in parallel, the reciprocal of equivalent resistance R_{eq} is the sum of the reciprocals of the individual resistances. All resistors in a parallel connection have the same *potential difference* between their terminals. (See Examples 26.1 and 26.2.)

$$R_{eq} = R_1 + R_2 + R_3 + \cdots \quad (26.1)$$
(resistors in series)

$$\frac{1}{R_{eq}} = \frac{1}{R_1} + \frac{1}{R_2} + \frac{1}{R_3} + \cdots \quad (26.2)$$
(resistors in parallel)

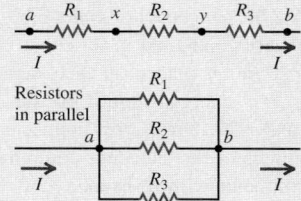

Kirchhoff's rules: Kirchhoff's junction rule is based on conservation of charge. It states that the algebraic sum of the currents into any junction must be zero. Kirchhoff's loop rule is based on conservation of energy and the conservative nature of electrostatic fields. It states that the algebraic sum of potential differences around any loop must be zero. Careful use of consistent sign rules is essential in applying Kirchhoff's rules. (See Examples 26.3–26.7.)

$$\sum I = 0 \quad \text{(junction rule)} \quad (26.5)$$
$$\sum V = 0 \quad \text{(loop rule)} \quad (26.6)$$

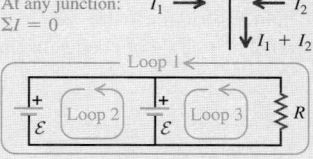

Electrical measuring instruments: In a d'Arsonval galvanometer, the deflection is proportional to the current in the coil. For a larger current range, a shunt resistor is added, so some of the current bypasses the meter coil. Such an instrument is called an ammeter. If the coil and any additional series resistance included obey Ohm's law, the meter can also be calibrated to read potential difference or voltage. The instrument is then called a voltmeter. A good ammeter has very low resistance; a good voltmeter has very high resistance. (See Examples 26.8–26.11.)

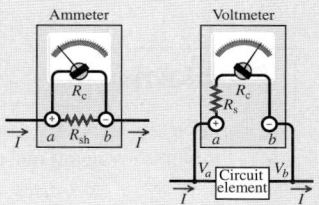

R-C circuits: When a capacitor is charged by a battery in series with a resistor, the current and capacitor charge are not constant. The charge approaches its final value asymptotically and the current approaches zero asymptotically. The charge and current in the circuit are given by Eqs. (26.12) and (26.13). After a time $\tau = RC$, the charge has approached within $1/e$ of its final value. This time is called the time constant or relaxation time of the circuit. When the capacitor discharges, the charge and current are given as functions of time by Eqs. (26.16) and (26.17). The time constant is the same for charging and discharging. (See Examples 26.12 and 26.13.)

Capacitor charging:
$$q = C\mathcal{E}\left(1 - e^{-t/RC}\right)$$
$$= Q_f\left(1 - e^{-t/RC}\right) \quad (26.12)$$

$$i = \frac{dq}{dt} = \frac{\mathcal{E}}{R}e^{-t/RC}$$
$$= I_0 e^{-t/RC} \quad (26.13)$$

Capacitor discharging:
$$q = Q_0 e^{-t/RC} \quad (26.16)$$

$$i = \frac{dq}{dt} = -\frac{Q_0}{RC}e^{-t/RC} \quad (26.17)$$
$$= I_0 e^{-t/RC}$$

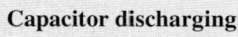
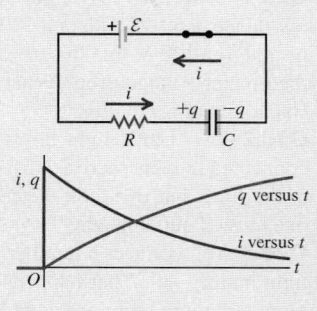

Household wiring: In household wiring systems, the various electrical devices are connected in parallel across the power line, which consists of a pair of conductors, one "hot" and the other "neutral." An additional "ground" wire is included for safety. The maximum permissible current in a circuit is determined by the size of the wires and the maximum temperature they can tolerate. Protection against excessive current and the resulting fire hazard is provided by fuses or circuit breakers. (See Example 26.14.)

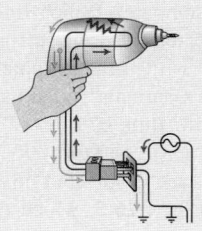

BRIDGING PROBLEM TWO CAPACITORS AND TWO RESISTORS

A 2.40-μF capacitor and a 3.60-μF capacitor are connected in series. (a) A charge of 5.20 mC is placed on each capacitor. What is the energy stored in the capacitors? (b) A 655-Ω resistor is connected to the terminals of the capacitor combination, and a voltmeter with resistance $4.58 \times 10^4 \ \Omega$ is connected across the resistor (**Fig. 26.27**). What is the rate of change of the energy stored in the capacitors just after the connection is made? (c) How long after the connection is made has the energy stored in the capacitors decreased to $1/e$ of its initial value? (d) At the instant calculated in part (c), what is the rate of change of the energy stored in the capacitors?

26.27 When the connection is made, the charged capacitors discharge.

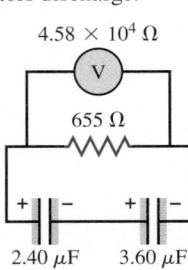

equivalent resistor. Select equations that will allow you to calculate the values of these equivalent circuit elements.
2. In part (a) you will need to use Eq. (24.9), which gives the energy stored in a capacitor.
3. For parts (b), (c), and (d), you will need to use Eq. (24.9) as well as Eqs. (26.16) and (26.17), which give the capacitor charge and current as functions of time. (*Hint:* The rate at which energy is lost by the capacitors equals the rate at which energy is dissipated in the resistances.)

EXECUTE
4. Find the stored energy at $t = 0$.
5. Find the rate of change of the stored energy at $t = 0$.
6. Find the value of t at which the stored energy has $1/e$ of the value you found in step 4.
7. Find the rate of change of the stored energy at the time you found in step 6.

EVALUATE
8. Check your results from steps 5 and 7 by calculating the rate of change in a different way. (*Hint:* The rate of change of the stored energy U is dU/dt.)

SOLUTION GUIDE

IDENTIFY AND SET UP
1. The two capacitors act as a single equivalent capacitor (see Section 24.2), and the resistor and voltmeter act as a single

---

Problems   For assigned homework and other learning materials, go to MasteringPhysics®. (MP)

•, ••, •••: Difficulty levels. **CP**: Cumulative problems incorporating material from earlier chapters. **CALC**: Problems requiring calculus.
DATA: Problems involving real data, scientific evidence, experimental design, and/or statistical reasoning. **BIO**: Biosciences problems.

DISCUSSION QUESTIONS

Q26.1 In which 120-V light bulb does the filament have greater resistance: a 60-W bulb or a 120-W bulb? If the two bulbs are connected to a 120-V line in series, through which bulb will there be the greater voltage drop? What if they are connected in parallel? Explain your reasoning.

Q26.2 Two 120-V light bulbs, one 25-W and one 200-W, were connected in series across a 240-V line. It seemed like a good idea at the time, but one bulb burned out almost immediately. Which one burned out, and why?

Q26.3 You connect a number of identical light bulbs to a flashlight battery. (a) What happens to the brightness of each bulb as more and more bulbs are added to the circuit if you connect them (i) in series and (ii) in parallel? (b) Will the battery last longer if the bulbs are in series or in parallel? Explain your reasoning.

Q26.4 In the circuit shown in **Fig. Q26.4**, three identical light bulbs are connected to a flashlight battery. How do the brightnesses of the bulbs compare? Which light bulb has the greatest current passing through it? Which light bulb has the greatest potential difference between its terminals? What happens if bulb A is unscrewed? Bulb B? Bulb C? Explain your reasoning.

Q26.5 If two resistors R_1 and R_2 ($R_2 > R_1$) are connected in series as shown in **Fig. Q26.5**, which of the following must be true? In each case justify

Figure **Q26.4**

Figure **Q26.5**

your answer. (a) $I_1 = I_2 = I_3$. (b) The current is greater in R_1 than in R_2. (c) The electrical power consumption is the same for both resistors. (d) The electrical power consumption is greater in R_2 than in R_1. (e) The potential drop is the same across both resistors. (f) The potential at point a is the same as at point c. (g) The potential at point b is lower than at point c. (h) The potential at point c is lower than at point b.

Q26.6 If two resistors R_1 and R_2 ($R_2 > R_1$) are connected in parallel as shown in **Fig. Q26.6**, which of the following must be true? In each case justify your answer. (a) $I_1 = I_2$. (b) $I_3 = I_4$. (c) The current is greater in R_1 than in R_2. (d) The rate of electrical energy consumption is the same for both resistors. (e) The rate of electrical energy consumption is greater in R_2 than in R_1. (f) $V_{cd} = V_{ef} = V_{ab}$. (g) Point c is at higher potential than point d. (h) Point f is at higher potential than point e. (i) Point c is at higher potential than point e.

Q26.7 A battery with no internal resistance is connected across identical light bulbs as shown in **Fig. Q26.7**. When you close the switch S, will the brightness of bulbs B_1 and B_2 change? If so, how will it change? Explain.

Figure **Q26.6**

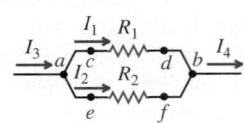

Figure **Q26.7**

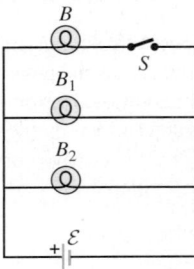

Q26.8 A resistor consists of three identical metal strips connected as shown in **Fig. Q26.8**. If one of the strips is cut out, does the ammeter reading increase, decrease, or stay the same? Why?

Figure **Q26.8**

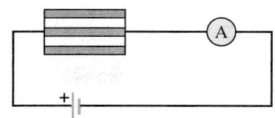

Q26.9 A light bulb is connected in the circuit shown in **Fig. Q26.9**. If we close the switch S, does the bulb's brightness increase, decrease, or remain the same? Explain why.

Figure **Q26.9**

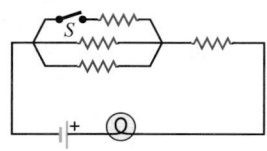

Q26.10 A real battery, having nonnegligible internal resistance, is connected across a light bulb as shown in **Fig. Q26.10**. When the switch S is closed, what happens to the brightness of the bulb? Why?

Figure **Q26.10**

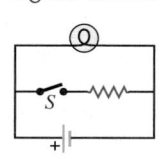

Q26.11 If the battery in Discussion Question Q26.10 is ideal with no internal resistance, what will happen to the brightness of the bulb when S is closed? Why?

Q26.12 Consider the circuit shown in **Fig. Q26.12**. What happens to the brightnesses of the bulbs when the switch S is closed if the battery (a) has no internal resistance and (b) has nonnegligible internal resistance? Explain why.

Figure **Q26.12**

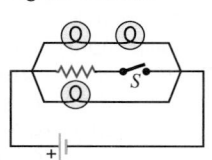

Q26.13 Is it possible to connect resistors together in a way that cannot be reduced to some combination of series and parallel combinations? If so, give examples. If not, state why not.

Q26.14 The battery in the circuit shown in **Fig. Q26.14** has no internal resistance. After you close the switch S, will the brightness of bulb B_1 increase, decrease, or stay the same?

Figure **Q26.14**

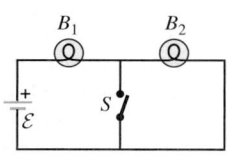

Q26.15 In a two-cell flashlight, the batteries are usually connected in series. Why not connect them in parallel? What possible advantage could there be in connecting several identical batteries in parallel?

Q26.16 Identical light bulbs A, B, and C are connected as shown in **Fig. Q26.16**. When the switch S is closed, bulb C goes out. Explain why. What happens to the brightness of bulbs A and B? Explain.

Figure **Q26.16**

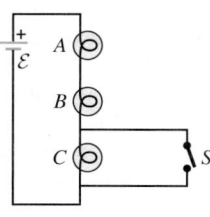

Q26.17 The emf of a flashlight battery is roughly constant with time, but its internal resistance increases with age and use. What sort of meter should be used to test the freshness of a battery?

Q26.18 Will the capacitors in the circuits shown in **Fig. Q26.18** charge at the same rate when the switch S is closed? If not, in which circuit will the capacitors charge more rapidly? Explain.

Q26.19 Verify that the time constant RC has units of time.

Figure **Q26.18**

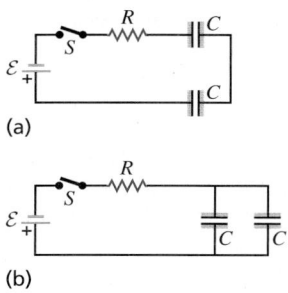

Q26.20 For very large resistances it is easy to construct R-C circuits that have time constants of several seconds or minutes. How might this fact be used to measure very large resistances, those that are too large to measure by more conventional means?

Q26.21 When a capacitor, battery, and resistor are connected in series, does the resistor affect the maximum charge stored on the capacitor? Why or why not? What purpose does the resistor serve?

EXERCISES

Section 26.1 Resistors in Series and Parallel

26.1 •• A uniform wire of resistance R is cut into three equal lengths. One of these is formed into a circle and connected between the other two (**Fig. E26.1**). What is the resistance between the opposite ends a and b?

Figure **E26.1**

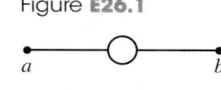

26.2 •• A machine part has a resistor X protruding from an opening in the side. This resistor is connected to three other resistors, as shown in **Fig. E26.2**. An ohmmeter connected across a and b reads 2.00 Ω. What is the resistance of X?

Figure **E26.2**

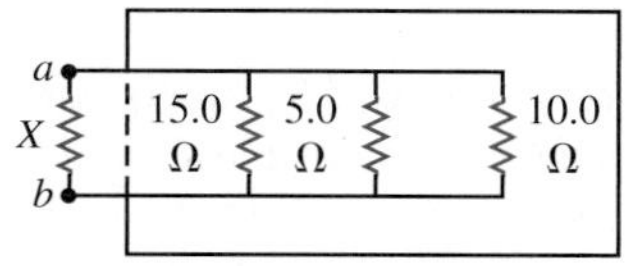

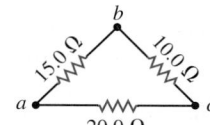

26.3 • A resistor with $R_1 = 25.0$ Ω is connected to a battery that has negligible internal resistance and electrical energy is dissipated by R_1 at a rate of 36.0 W. If a second resistor with $R_2 = 15.0$ Ω is connected in series with R_1, what is the total rate at which electrical energy is dissipated by the two resistors?

26.4 • A 42-Ω resistor and a 20-Ω resistor are connected in parallel, and the combination is connected across a 240-V dc line. (a) What is the resistance of the parallel combination? (b) What is the total current through the parallel combination? (c) What is the current through each resistor?

26.5 • A triangular array of resistors is shown in **Fig. E26.5**. What current will this array draw from a 35.0-V battery having negligible internal resistance if we connect it across (a) ab; (b) bc; (c) ac? (d) If the battery has an internal resistance of 3.00 Ω, what current will the array draw if the battery is connected across bc?

Figure **E26.5**

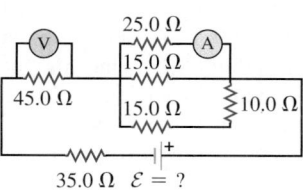

26.6 •• For the circuit shown in **Fig. E26.6** both meters are idealized, the battery has no appreciable internal resistance, and the ammeter reads 1.25 A. (a) What does the voltmeter read? (b) What is the emf $\mathcal{E}$ of the battery?

Figure **E26.6**

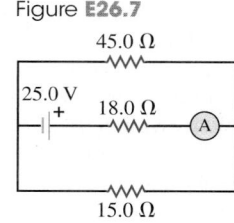

26.7 •• For the circuit shown in **Fig. E26.7** find the reading of the idealized ammeter if the battery has an internal resistance of 3.26 Ω.

Figure **E26.7**

26.8 • Three resistors having resistances of 1.60 Ω, 2.40 Ω, and 4.80 Ω are connected in parallel to a 28.0-V battery that has negligible internal resistance. Find (a) the equivalent resistance of the combination; (b) the current in each resistor; (c) the total current through the battery; (d) the voltage across each resistor; (e) the power dissipated in each resistor.

(f) Which resistor dissipates the most power: the one with the greatest resistance or the least resistance? Explain why this should be.

26.9 • Now the three resistors of Exercise 26.8 are connected in series to the same battery. Answer the same questions for this situation.

26.10 •• **Power Rating of a Resistor.** The *power rating* of a resistor is the maximum power the resistor can safely dissipate without too great a rise in temperature and hence damage to the resistor. (a) If the power rating of a 15-k Ω resistor is 5.0 W, what is the maximum allowable potential difference across the terminals of the resistor? (b) A 9.0-kΩ resistor is to be connected across a 120-V potential difference. What power rating is required? (c) A 100.0-Ω and a 150.0-Ω resistor, both rated at 2.00 W, are connected in series across a variable potential difference. What is the greatest this potential difference can be without overheating either resistor, and what is the rate of heat generated in each resistor under these conditions?

26.11 • In **Fig. E26.11**, $R_1 =$ 3.00 Ω, $R_2 = 6.00$ Ω, and $R_3 = 5.00$ Ω. The battery has negligible internal resistance. The current I_2 through R_2 is 4.00 A. (a) What are the currents I_1 and I_3? (b) What is the emf of the battery?

Figure **E26.11**

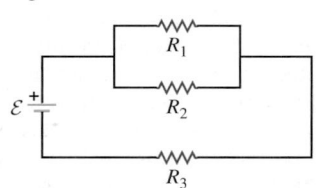

26.12 •• In Fig. E26.11 the battery has emf 35.0 V and negligible internal resistance. $R_1 = 5.00$ Ω. The current through R_1 is 1.50 A, and the current through $R_3 = 4.50$ A. What are the resistances R_2 and R_3?

26.13 • Compute the equivalent resistance of the network in **Fig. E26.13**, and find the current in each resistor. The battery has negligible internal resistance.

Figure **E26.13**

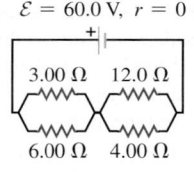

$\mathcal{E} = 60.0$ V, $r = 0$

3.00 Ω 12.0 Ω

6.00 Ω 4.00 Ω

26.14 • Compute the equivalent resistance of the network in **Fig. E26.14**, and find the current in each resistor. The battery has negligible internal resistance.

Figure **E26.14**

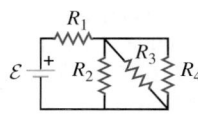

$\mathcal{E} = 48.0$ V, $r = 0$

1.00 Ω 3.00 Ω

7.00 Ω 5.00 Ω

26.15 • In the circuit of **Fig. E26.15**, each resistor represents a light bulb. Let $R_1 = R_2 = R_3 = R_4 = 4.50$ Ω and $\mathcal{E} = 9.00$ V. (a) Find the current in each bulb. (b) Find the power dissipated in each bulb. Which bulb or bulbs glow the brightest? (c) Bulb R_4 is now removed from the circuit, leaving a break in the wire at its position. Now what is the current in each of the remaining bulbs R_1, R_2, and R_3? (d) With bulb R_4 removed, what is the power dissipated in each of the remaining bulbs? (e) Which light bulb(s) glow brighter as a result of removing R_4? Which bulb(s) glow less brightly? Discuss why there are different effects on different bulbs.

Figure **E26.15**

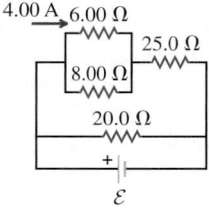

26.16 • Consider the circuit shown in **Fig. E26.16**. The current through the 6.00-Ω resistor is 4.00 A, in the direction

Figure **E26.16**

4.00 A 6.00 Ω

25.0 Ω

8.00 Ω

20.0 Ω

$\mathcal{E}$

shown. What are the currents through the 25.0-Ω and 20.0-Ω resistors?

26.17 • In the circuit shown in **Fig. E26.17**, the voltage across the 2.00-Ω resistor is 12.0 V. What are the emf of the battery and the current through the 6.00-Ω resistor?

Figure **E26.17**

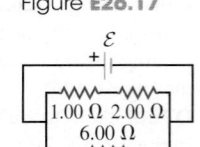

$\mathcal{E}$

1.00 Ω 2.00 Ω
6.00 Ω

26.18 •• In the circuit shown in **Fig. E26.18**, $\mathcal{E} = 36.0$ V, $R_1 = 4.00$ Ω, $R_2 = 6.00$ Ω, and $R_3 = 3.00$ Ω. (a) What is the potential difference V_{ab} between points a and b when the switch S is open and when S is closed? (b) For each resistor, calculate the current through the resistor with S open and with S closed. For each resistor, does the current increase or decrease when S is closed?

Figure **E26.18**

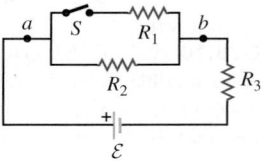

26.19 •• CP In the circuit in **Fig. E26.19**, a 20.0-Ω resistor is inside 100 g of pure water that is surrounded by insulating styrofoam. If the water is initially at 10.0°C, how long will it take for its temperature to rise to 58.0°C?

Figure **E26.19**

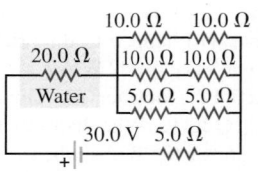

26.20 • In the circuit shown in **Fig. E26.20**, the rate at which R_1 is dissipating electrical energy is 15.0 W. (a) Find R_1 and R_2. (b) What is the emf of the battery? (c) Find the current through both R_2 and the 10.0-Ω resistor. (d) Calculate the total electrical power consumption in all the resistors and the electrical power delivered by the battery. Show that your results are consistent with conservation of energy.

Figure **E26.20**

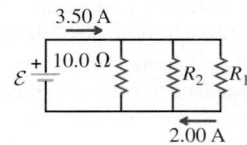

3.50 A

$\mathcal{E}$ 10.0 Ω R_2 R_1

2.00 A

26.21 • **Light Bulbs in Series and in Parallel.** Two light bulbs have constant resistances of 400 Ω and 800 Ω. If the two light bulbs are connected in series across a 120-V line, find (a) the current through each bulb; (b) the power dissipated in each bulb; (c) the total power dissipated in both bulbs. The two light bulbs are now connected in parallel across the 120-V line. Find (d) the current through each bulb; (e) the power dissipated in each bulb; (f) the total power dissipated in both bulbs. (g) In each situation, which of the two bulbs glows the brightest? (h) In which situation is there a greater total light output from both bulbs combined?

26.22 • **Light Bulbs in Series.** A 60-W, 120-V light bulb and a 200-W, 120-V light bulb are connected in series across a 240-V line. Assume that the resistance of each bulb does not vary with current. (*Note:* This description of a light bulb gives the power it dissipates when connected to the stated potential difference; that is, a 25-W, 120-V light bulb dissipates 25 W when connected to a 120-V line.) (a) Find the current through the bulbs. (b) Find the power dissipated in each bulb. (c) One bulb burns out very quickly. Which one? Why?

Section 26.2 Kirchhoff's Rules

26.23 •• In the circuit shown in **Fig. E26.23**, ammeter A_1 reads 10.0 A and the batteries have no appreciable internal resistance. (a) What is the resistance of R? (b) Find the readings in the other ammeters.

Figure E26.23

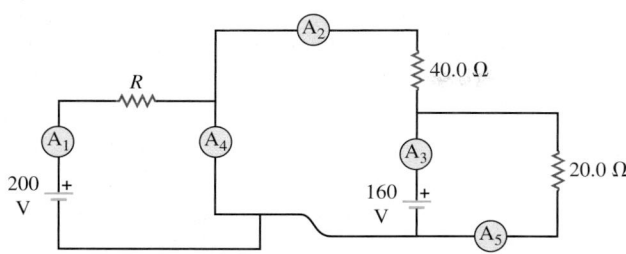

26.24 •• The batteries shown in the circuit in **Fig. E26.24** have negligibly small internal resistances. Find the current through (a) the 30.0-Ω resistor; (b) the 20.0-Ω resistor; (c) the 10.0-V battery.

Figure E26.24

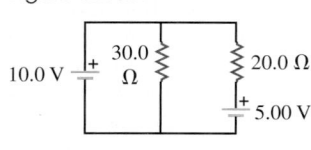

26.25 • In the circuit shown in **Fig. E26.25** find (a) the current in resistor R; (b) the resistance R; (c) the unknown emf $\mathcal{E}$. (d) If the circuit is broken at point x, what is the current in resistor R?

26.26 • Find the emfs $\mathcal{E}_1$ and $\mathcal{E}_2$ in the circuit of **Fig. E26.26**, and find the potential difference of point b relative to point a.

Figure E26.25

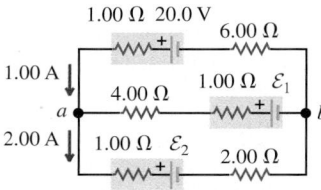

Figure E26.26

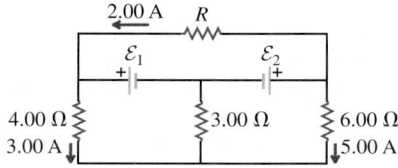

26.27 • In the circuit shown in **Fig. E26.27**, find (a) the current in the 3.00-Ω resistor; (b) the unknown emfs $\mathcal{E}_1$ and $\mathcal{E}_2$; (c) the resistance R. Note that three currents are given.

Figure E26.27

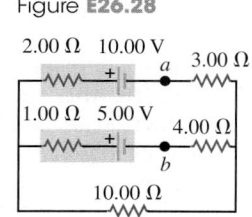

26.28 •• In the circuit shown in **Fig. E26.28**, find (a) the current in each branch and (b) the potential difference V_{ab} of point a relative to point b.

26.29 • The 10.00-V battery in Fig. E26.28 is removed from the circuit and reinserted with the opposite polarity, so that its positive terminal is now next to point a. The rest of the circuit is as shown in the figure. Find (a) the current in each branch and (b) the potential difference V_{ab} of point a relative to point b.

Figure E26.28

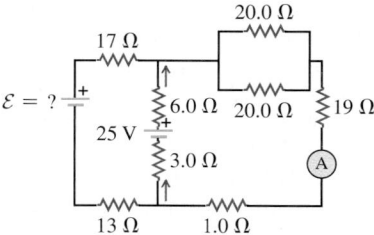

26.30 • The 5.00-V battery in Fig. E26.28 is removed from the circuit and replaced by a 15.00-V battery, with its negative terminal next to point b. The rest of the circuit is as shown in the figure. Find (a) the current in each branch and (b) the potential difference V_{ab} of point a relative to point b.

26.31 •• In the circuit shown in **Fig. E26.31** the batteries have negligible internal resistance and the meters are both idealized. With the switch S open, the voltmeter reads 15.0 V. (a) Find the emf $\mathcal{E}$ of the battery. (b) What will the ammeter read when the switch is closed?

Figure E26.31

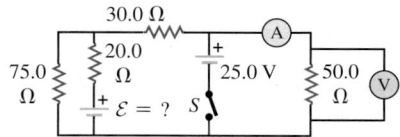

26.32 •• In the circuit shown in **Fig. E26.32** both batteries have insignificant internal resistance and the idealized ammeter reads 1.50 A in the direction shown. Find the emf $\mathcal{E}$ of the battery. Is the polarity shown correct?

Figure E26.32

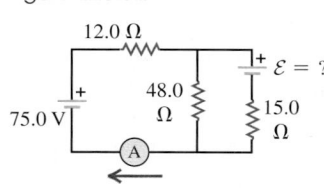

26.33 • In the circuit shown in **Fig. E26.33** all meters are idealized and the batteries have no appreciable internal resistance. (a) Find the reading of the voltmeter with the switch S open. Which point is at a higher potential: a or b? (b) With S closed, find the reading of the voltmeter and the ammeter. Which way (up or down) does the current flow through the switch?

Figure E26.33

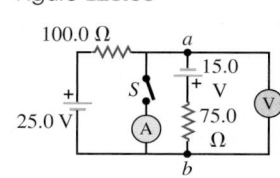

26.34 •• In the circuit shown in **Fig. E26.34**, the 6.0-Ω resistor is consuming energy at a rate of 24 J/s when the current through it flows as shown. (a) Find the current through the ammeter A. (b) What are the polarity and emf $\mathcal{E}$ of the unknown battery, assuming it has negligible internal resistance?

Figure E26.34

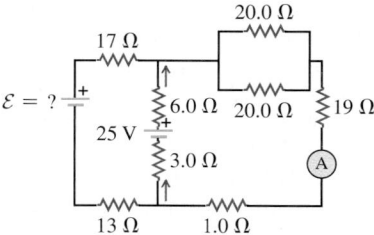

Section 26.3 Electrical Measuring Instruments

26.35 • The resistance of a galvanometer coil is 25.0 Ω, and the current required for full-scale deflection is 500 μA. (a) Show in a diagram how to convert the galvanometer to an ammeter reading 20.0 mA full scale, and compute the shunt resistance. (b) Show how to convert the galvanometer to a voltmeter reading 500 mV full scale, and compute the series resistance.

26.36 • The resistance of the coil of a pivoted-coil galvanometer is 9.36 Ω, and a current of 0.0224 A causes it to deflect full scale. We want to convert this galvanometer to an ammeter reading 20.0 A full scale. The only shunt available has a resistance of 0.0250 Ω. What resistance R must be connected in series with the coil (**Fig. E26.36**)?

Figure **E26.36**

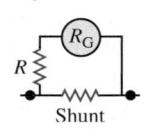

26.37 • A circuit consists of a series combination of 6.00-kΩ and 5.00-kΩ resistors connected across a 50.0-V battery having negligible internal resistance. You want to measure the true potential difference (that is, the potential difference without the meter present) across the 5.00-kΩ resistor using a voltmeter having an internal resistance of 10.0 kΩ. (a) What potential difference does the voltmeter measure across the 5.00-kΩ resistor? (b) What is the *true* potential difference across this resistor when the meter is not present? (c) By what percentage is the voltmeter reading in error from the true potential difference?

26.38 • A galvanometer having a resistance of 25.0 Ω has a 1.00-Ω shunt resistance installed to convert it to an ammeter. It is then used to measure the current in a circuit consisting of a 15.0-Ω resistor connected across the terminals of a 25.0-V battery having no appreciable internal resistance. (a) What current does the ammeter measure? (b) What should be the *true* current in the circuit (that is, the current without the ammeter present)? (c) By what percentage is the ammeter reading in error from the *true* current?

Section 26.4 *R-C* Circuits

26.39 • A capacitor is charged to a potential of 12.0 V and is then connected to a voltmeter having an internal resistance of 3.40 MΩ. After a time of 4.00 s the voltmeter reads 3.0 V. What are (a) the capacitance and (b) the time constant of the circuit?

26.40 •• You connect a battery, resistor, and capacitor as in Fig. 26.20a, where $\mathcal{E} = 36.0$ V, $C = 5.00\ \mu$F, and $R = 120\ \Omega$. The switch S is closed at $t = 0$. (a) When the voltage across the capacitor is 8.00 V, what is the magnitude of the current in the circuit? (b) At what time t after the switch is closed is the voltage across the capacitor 8.00 V? (c) When the voltage across the capacitor is 8.00 V, at what rate is energy being stored in the capacitor?

26.41 • A 4.60-μF capacitor that is initially uncharged is connected in series with a 7.50-kΩ resistor and an emf source with $\mathcal{E} = 245$ V and negligible internal resistance. Just after the circuit is completed, what are (a) the voltage drop across the capacitor; (b) the voltage drop across the resistor; (c) the charge on the capacitor; (d) the current through the resistor? (e) A long time after the circuit is completed (after many time constants) what are the values of the quantities in parts (a)–(d)?

26.42 •• You connect a battery, resistor, and capacitor as in Fig. 26.20a, where $R = 12.0\ \Omega$ and $C = 5.00 \times 10^{-6}$ F. The switch S is closed at $t = 0$. When the current in the circuit has magnitude 3.00 A, the charge on the capacitor is 40.0×10^{-6} C. (a) What is the emf of the battery? (b) At what time t after the switch is closed is the charge on the capacitor equal to 40.0×10^{-6} C? (c) When the current has magnitude 3.00 A, at what rate is energy being (i) stored in the capacitor, (ii) supplied by the battery?

26.43 •• CP In the circuit shown in **Fig. E26.43** both capacitors are initially charged to 45.0 V. (a) How long after closing the switch S will the potential across each capacitor be reduced to 10.0 V, and (b) what will be the current at that time?

Figure **E26.43**

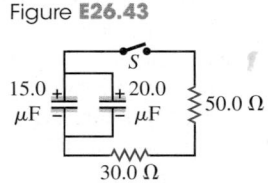

26.44 • A 12.4-μF capacitor is connected through a 0.895-MΩ resistor to a constant potential difference of 60.0 V. (a) Compute the charge on the capacitor at the following times after the connections are made: 0, 5.0 s, 10.0 s, 20.0 s, and 100.0 s. (b) Compute the charging currents at the same instants. (c) Graph the results of parts (a) and (b) for t between 0 and 20 s.

26.45 • An emf source with $\mathcal{E} = 120$ V, a resistor with $R = 80.0\ \Omega$, and a capacitor with $C = 4.00\ \mu$F are connected in series. As the capacitor charges, when the current in the resistor is 0.900 A, what is the magnitude of the charge on each plate of the capacitor?

26.46 • A resistor and a capacitor are connected in series to an emf source. The time constant for the circuit is 0.780 s. (a) A second capacitor, identical to the first, is added in series. What is the time constant for this new circuit? (b) In the original circuit a second capacitor, identical to the first, is connected in parallel with the first capacitor. What is the time constant for this new circuit?

26.47 •• CP In the circuit shown in **Fig. E26.47** each capacitor initially has a charge of magnitude 3.50 nC on its plates. After the switch S is closed, what will be the current in the circuit at the instant that the capacitors have lost 80.0% of their initial stored energy?

Figure **E26.47**

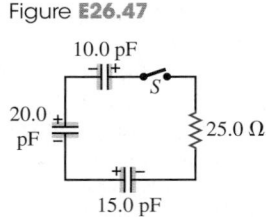

26.48 • A 1.50-μF capacitor is charging through a 12.0-Ω resistor using a 10.0-V battery. What will be the current when the capacitor has acquired $\frac{1}{4}$ of its maximum charge? Will it be $\frac{1}{4}$ of the maximum current?

26.49 • In the circuit in **Fig. E26.49** the capacitors are initially uncharged, the battery has no internal resistance, and the ammeter is idealized. Find the ammeter reading (a) just after the switch S is closed and (b) after S has been closed for a very long time.

Figure **E26.49**

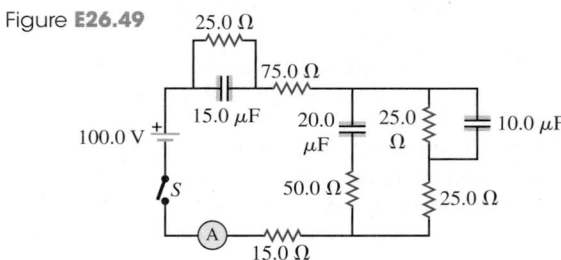

26.50 • A 12.0-μF capacitor is charged to a potential of 50.0 V and then discharged through a 225-Ω resistor. How long does it take the capacitor to lose (a) half of its charge and (b) half of its stored energy?

26.51 • In the circuit shown in **Fig. E26.51**, $C = 5.90\ \mu$F, $\mathcal{E} = 28.0$ V, and the emf has negligible resistance. Initially the capacitor is uncharged and the switch S is in position 1. The switch is then moved to position 2, so that the capacitor begins to charge. (a) What will be the charge on the capacitor a long time after S is moved to position 2? (b) After S has been in position 2 for 3.00 ms, the charge on the capacitor is measured to be 110 μC. What is the value of the resistance R? (c) How long after S is moved to position 2 will the charge on the capacitor be equal to 99.0% of the final value found in part (a)?

Figure **E26.51**

Switch S in position 1   Switch S in position 2

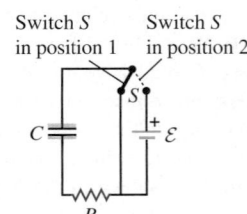

Section 26.5 Power Distribution Systems

26.52 • The heating element of an electric dryer is rated at 4.1 kW when connected to a 240-V line. (a) What is the current in the heating element? Is 12-gauge wire large enough to supply this current? (b) What is the resistance of the dryer's heating element at its operating temperature? (c) At 11 cents per kWh, how much does it cost per hour to operate the dryer?

26.53 • A 1500-W electric heater is plugged into the outlet of a 120-V circuit that has a 20-A circuit breaker. You plug an electric hair dryer into the same outlet. The hair dryer has power settings of 600 W, 900 W, 1200 W, and 1500 W. You start with the hair dryer on the 600-W setting and increase the power setting until the circuit breaker trips. What power setting caused the breaker to trip?

PROBLEMS

26.54 •• In **Fig. P26.54**, the battery has negligible internal resistance and $\mathcal{E} = 48.0$ V. $R_1 = R_2 = 4.00$ Ω and $R_4 = 3.00$ Ω. What must the resistance R_3 be for the resistor network to dissipate electrical energy at a rate of 295 W?

Figure **P26.54**

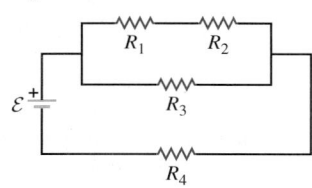

26.55 • The two identical light bulbs in Example 26.2 (Section 26.1) are connected in parallel to a different source, one with $\mathcal{E} = 8.0$ V and internal resistance 0.8 Ω. Each light bulb has a resistance $R = 2.0$ Ω (assumed independent of the current through the bulb). (a) Find the current through each bulb, the potential difference across each bulb, and the power delivered to each bulb. (b) Suppose one of the bulbs burns out, so that its filament breaks and current no longer flows through it. Find the power delivered to the remaining bulb. Does the remaining bulb glow more or less brightly after the other bulb burns out than before?

26.56 •• Each of the three resistors in **Fig. P26.56** has a resistance of 2.4 Ω and can dissipate a maximum of 48 W without becoming excessively heated. What is the maximum power the circuit can dissipate?

Figure **P26.56**

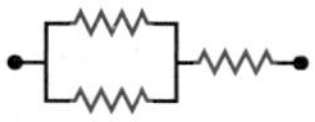

26.57 •• (a) Find the potential of point a with respect to point b in **Fig. P26.57**. (b) If points a and b are connected by a wire with negligible resistance, find the current in the 12.0-V battery.

Figure **P26.57**

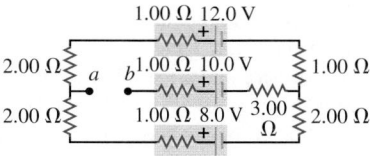

26.58 •• CP For the circuit shown in **Fig. P26.58** a 20.0-Ω resistor is embedded in a large block of ice at 0.00°C, and the battery has negligible internal resistance. At what rate (in g/s) is this circuit melting the ice? (The latent heat of fusion for ice is 3.34×10^5 J/kg.)

Figure **P26.58**

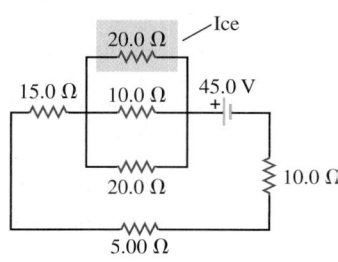

26.59 • Calculate the three currents I_1, I_2, and I_3 indicated in the circuit diagram shown in **Fig. P26.59**.

Figure **P26.59**

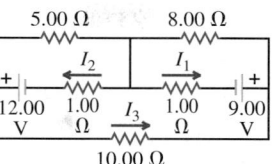

Figure **P26.60**

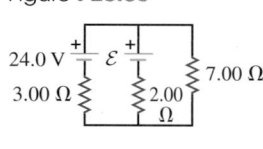

26.60 ••• What must the emf $\mathcal{E}$ in **Fig. P26.60** be in order for the current through the 7.00-Ω resistor to be 1.80 A? Each emf source has negligible internal resistance.

26.61 • Find the current through each of the three resistors of the circuit shown in **Fig. P26.61**. The emf sources have negligible internal resistance.

Figure **P26.61**

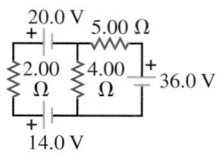

Figure **P26.62**

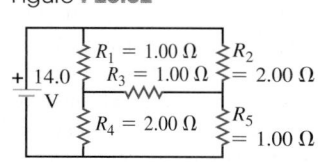

26.62 • (a) Find the current through the battery and each resistor in the circuit shown in **Fig. P26.62**. (b) What is the equivalent resistance of the resistor network?

26.63 •• Consider the circuit shown in **Fig. P26.63**. (a) What must the emf $\mathcal{E}$ of the battery be in order for a current of 2.00 A to flow through the 5.00-V battery as shown? Is the polarity of the battery correct as shown? (b) How long does it take for 60.0 J of thermal energy to be produced in the 10.0-Ω resistor?

Figure **P26.63**

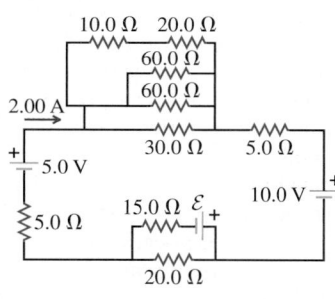

26.64 •• In the circuit shown in **Fig. P26.64**, $\mathcal{E} = 24.0$ V, $R_1 = 6.00$ Ω, $R_3 = 12.0$ Ω, and R_2 can vary between 3.00 Ω and 24.0 Ω. For what value of R_2 is the power dissipated by heating element R_1 the greatest? Calculate the magnitude of the greatest power.

Figure **P26.64**

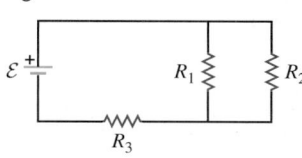

26.65 • In the circuit shown in **Fig. P26.65**, the current in the 20.0-V battery is 5.00 A in the direction shown and the voltage across the 8.00-Ω resistor is 16.0 V, with the lower end of the resistor at higher potential. Find (a) the emf (including its polarity) of the battery X; (b) the current I through the 200.0-V battery (including its direction); (c) the resistance R.

Figure **P26.65**

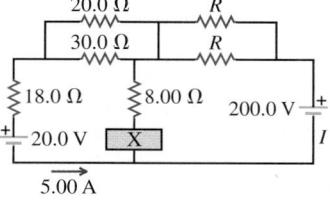

26.66 ••• In the circuit shown in **Fig. P26.66** all the resistors are rated at a maximum power of 2.00 W. What is the maximum emf $\mathcal{E}$ that the battery can have without burning up any of the resistors?

Figure **P26.66**

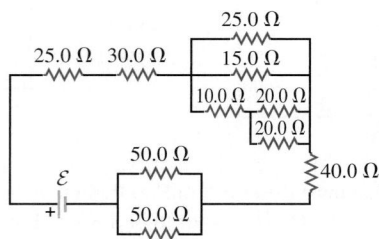

26.67 • **Figure P26.67** employs a convention often used in circuit diagrams. The battery (or other power supply) is not shown explicitly. It is understood that the point at the top, labeled "36.0 V," is connected to the positive terminal of a 36.0-V battery having negligible internal resistance, and that the *ground* symbol at the bottom is connected to the negative terminal of the battery. The circuit is completed through the battery, even though it is not shown. (a) What is the potential difference V_{ab}, the potential of point a relative to point b, when the switch S is open? (b) What is the current through S when it is closed? (c) What is the equivalent resistance when S is closed?

Figure **P26.67**

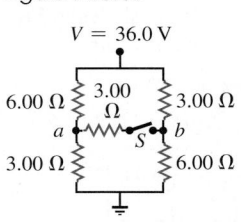

26.68 •• Three identical resistors are connected in series. When a certain potential difference is applied across the combination, the total power dissipated is 45.0 W. What power would be dissipated if the three resistors were connected in parallel across the same potential difference?

26.69 • A resistor R_1 consumes electrical power P_1 when connected to an emf $\mathcal{E}$. When resistor R_2 is connected to the same emf, it consumes electrical power P_2. In terms of P_1 and P_2, what is the total electrical power consumed when they are both connected to this emf source (a) in parallel and (b) in series?

26.70 • The capacitor in **Fig. P26.70** is initially uncharged. The switch S is closed at $t = 0$. (a) Immediately after the switch is closed, what is the current through each resistor? (b) What is the final charge on the capacitor?

Figure **P26.70**

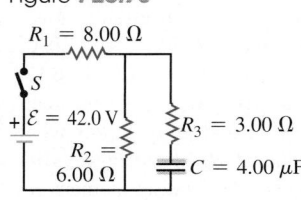

26.71 •• A 2.00-μF capacitor that is initially uncharged is connected in series with a 6.00-kΩ resistor and an emf source with $\mathcal{E} = 90.0$ V and negligible internal resistance. The circuit is completed at $t = 0$. (a) Just after the circuit is completed, what is the rate at which electrical energy is being dissipated in the resistor? (b) At what value of t is the rate at which electrical energy is being dissipated in the resistor equal to the rate at which electrical energy is being stored in the capacitor? (c) At the time calculated in part (b), what is the rate at which electrical energy is being dissipated in the resistor?

26.72 •• A 6.00-μF capacitor that is initially uncharged is connected in series with a 5.00-Ω resistor and an emf source with $\mathcal{E} = 50.0$ V and negligible internal resistance. At the instant when the resistor is dissipating electrical energy at a rate of 300 W, how much energy has been stored in the capacitor?

26.73 • Point a in **Fig. P26.73** is maintained at a constant potential of 400 V above ground. (See Problem 26.67.) (a) What is the reading of a voltmeter with the proper range and with resistance 5.00×10^4 Ω when connected between point b and ground? (b) What is the reading of a voltmeter with resistance 5.00×10^6 Ω? (c) What is the reading of a voltmeter with infinite resistance?

Figure **P26.73**

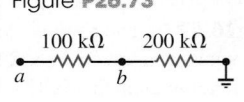

26.74 •• **The Wheatstone Bridge.** The circuit shown in **Fig. P26.74**, called a *Wheatstone bridge*, is used to determine the value of an unknown resistor X by comparison with three resistors M, N, and P whose resistances can be varied. For each setting, the resistance of each resistor is precisely known. With switches S_1 and S_2 closed, these resistors are varied until the current in the galvanometer G is zero; the bridge is then said to be *balanced*. (a) Show that under this condition the unknown resistance is given by $X = MP/N$. (This method permits very high precision in comparing resistors.) (b) If galvanometer G shows zero deflection when $M = 850.0$ Ω, $N = 15.00$ Ω, and $P = 33.48$ Ω, what is the unknown resistance X?

Figure **P26.74**

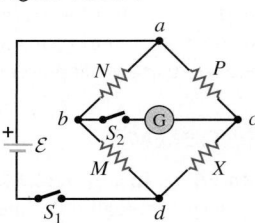

26.75 • (See Problem 26.67.) (a) What is the potential of point a with respect to point b in **Fig. P26.75** when the switch S is open? (b) Which point, a or b, is at the higher potential? (c) What is the final potential of point b with respect to ground when S is closed? (d) How much does the charge on each capacitor change when S is closed?

Figure **P26.75**

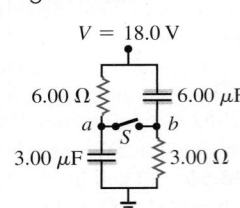

26.76 • A 2.36-μF capacitor that is initially uncharged is connected in series with a 5.86-Ω resistor and an emf source with $\mathcal{E} = 120$ V and negligible internal resistance. (a) Just after the connection is made, what are (i) the rate at which electrical energy is being dissipated in the resistor; (ii) the rate at which the electrical energy stored in the capacitor is increasing; (iii) the electrical power output of the source? How do the answers to parts (i), (ii), and (iii) compare? (b) Answer the same questions as in part (a) at a long time after the connection is made. (c) Answer the same questions as in part (a) at the instant when the charge on the capacitor is one-half its final value.

26.77 • A 224-Ω resistor and a 589-Ω resistor are connected in series across a 90.0-V line. (a) What is the voltage across each resistor? (b) A voltmeter connected across the 224-Ω resistor reads 23.8 V. Find the voltmeter resistance. (c) Find the reading of the same voltmeter if it is connected across the 589-Ω resistor. (d) The readings on this voltmeter are lower than the "true" voltages (that is, without the voltmeter present). Would it be possible to design a voltmeter that gave readings *higher* than the "true" voltages? Explain.

26.78 • A resistor with $R = 850$ Ω is connected to the plates of a charged capacitor with capacitance $C = 4.62$ μF. Just before the connection is made, the charge on the capacitor is 6.90 mC. (a) What is the energy initially stored in the capacitor? (b) What is the electrical power dissipated in the resistor just after the connection is made? (c) What is the electrical power dissipated in the resistor at the instant when the energy stored in the capacitor has decreased to half the value calculated in part (a)?

26.79 • A capacitor that is initially uncharged is connected in series with a resistor and an emf source with $\mathcal{E} = 110$ V and negligible internal resistance. Just after the circuit is completed, the current through the resistor is 6.5×10^{-5} A. The time constant for the circuit is 5.2 s. What are the resistance of the resistor and the capacitance of the capacitor?

26.80 •• DATA You set up the circuit shown in Fig. 26.22a, where $R = 196 \ \Omega$. You close the switch at time $t = 0$ and measure the magnitude i of the current in the resistor R as a function of time t since the switch was closed. Your results are shown in **Fig. P26.80**, where you have chosen to plot $\ln i$ as a function of t. (a) Explain why your data points lie close to a straight line. (b) Use the graph in Fig. P26.80 to calculate the capacitance C and the initial charge Q_0 on the capacitor. (c) When $i = 0.0500$ A, what is the charge on the capacitor? (d) When $q = 0.500 \times 10^{-4}$ C, what is the current in the resistor?

Figure **P26.80**

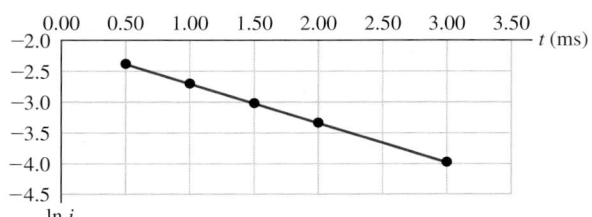

26.81 •• DATA You set up the circuit shown in Fig. 26.20, where $C = 5.00 \times 10^{-6}$ F. At time $t = 0$, you close the switch and then measure the charge q on the capacitor as a function of the current i in the resistor. Your results are given in the table:

i (mA)	56.0	48.0	40.0	32.0	24.0
q (μC)	10.1	19.8	30.2	40.0	49.9

(a) Graph q as a function of i. Explain why the data points, when plotted this way, fall close to a straight line. Find the slope and y-intercept of the straight line that gives the best fit to the data. (b) Use your results from part (a) to calculate the resistance R of the resistor and the emf $\mathcal{E}$ of the battery. (c) At what time t after the switch is closed is the voltage across the capacitor equal to 10.0 V? (d) When the voltage across the capacitor is 4.00 V, what is the voltage across the resistor?

26.82 •• DATA The electronics supply company where you work has two different resistors, R_1 and R_2, in its inventory, and you must measure the values of their resistances. Unfortunately, stock is low, and all you have are R_1 and R_2 in parallel and in series—and you can't separate these two resistor combinations. You separately connect each resistor network to a battery with emf 48.0 V and negligible internal resistance and measure the power P supplied by the battery in both cases. For the series combination, $P = 48.0$ W; for the parallel combination, $P = 256$ W. You are told that $R_1 > R_2$. (a) Calculate R_1 and R_2. (b) For the series combination, which resistor consumes more power, or do they consume the same power? Explain. (c) For the parallel combination, which resistor consumes more power, or do they consume the same power?

CHALLENGE PROBLEMS

26.83 ••• **An Infinite Network.** As shown in **Fig. P26.83**, a network of resistors of resistances R_1 and R_2 extends to infinity toward the right. Prove that the total resistance R_T of the infinite network is equal to

$$R_T = R_1 + \sqrt{R_1^2 + 2R_1R_2}$$

(*Hint:* Since the network is infinite, the resistance of the network to the right of points c and d is also equal to R_T.)

Figure **P26.83**

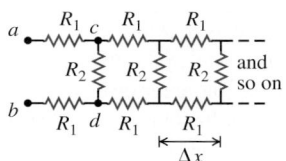

26.84 ••• Suppose a resistor R lies along each edge of a cube (12 resistors in all) with connections at the corners. Find the equivalent resistance between two diagonally opposite corners of the cube (points a and b in **Fig. P26.84**).

Figure **P26.84**

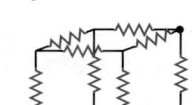

26.85 ••• BIO **Attenuator Chains and Axons.** The infinite network of resistors shown in Fig. P26.83 is known as an *attenuator chain,* since this chain of resistors causes the potential difference between the upper and lower wires to decrease, or attenuate, along the length of the chain. (a) Show that if the potential difference between the points a and b in Fig. 26.83 is V_{ab}, then the potential difference between points c and d is $V_{cd} = V_{ab}/(1 + \beta)$, where $\beta = 2R_1(R_T + R_2)/R_TR_2$ and R_T, the total resistance of the network, is given in Challenge Problem 26.83. (See the hint given in that problem.) (b) If the potential difference between terminals a and b at the left end of the infinite network is V_0, show that the potential difference between the upper and lower wires n segments from the left end is $V_n = V_0/(1 + \beta)^n$. If $R_1 = R_2$, how many segments are needed to decrease the potential difference V_n to less than 1.0% of V_0? (c) An infinite attenuator chain provides a model of the propagation of a voltage pulse along a nerve fiber, or axon. Each segment of the network in Fig. P26.83 represents a short segment of the axon of length Δx. The resistors R_1 represent the resistance of the fluid inside and outside the membrane wall of the axon. The resistance of the membrane to current flowing through the wall is represented by R_2. For an axon segment of length $\Delta x = 1.0 \ \mu$m, $R_1 = 6.4 \times 10^3 \ \Omega$ and $R_2 = 8.0 \times 10^8 \ \Omega$ (the membrane wall is a good insulator). Calculate the total resistance R_T and β for an infinitely long axon. (This is a good approximation, since the length of an axon is much greater than its width; the largest axons in the human nervous system are longer than 1 m but only about 10^{-7} m in radius.) (d) By what fraction does the potential difference between the inside and outside of the axon decrease over a distance of 2.0 mm? (e) The attenuation of the potential difference calculated in part (d) shows that the axon cannot simply be a passive, current-carrying electrical cable; the potential difference must periodically be reinforced along the axon's length. This reinforcement mechanism is slow, so a signal propagates along the axon at only about 30 m/s. In situations where faster response is required, axons are covered with a segmented sheath of fatty myelin. The segments are about 2 mm long, separated by gaps called the *nodes of Ranvier*. The myelin increases the resistance of a 1.0-μm-long segment of the membrane to $R_2 = 3.3 \times 10^{12}\ \Omega$. For such a myelinated axon, by what fraction does the potential difference between the inside and outside of the axon decrease over the distance from one node of Ranvier to the next? This smaller attenuation means the propagation speed is increased.

PASSAGE PROBLEMS

BIO **NERVE CELLS AND R-C CIRCUITS.** The portion of a nerve cell that conducts signals is called an *axon*. Many of the electrical properties of axons are governed by ion channels, which are protein molecules that span the axon's cell membrane. When open, each ion channel has a pore that is filled with fluid of low resistivity and connects the interior of the cell electrically to the medium outside the cell. In contrast, the lipid-rich cell membrane in which ion channels reside has very high resistivity.

26.86 Assume that a typical open ion channel spanning an axon's membrane has a resistance of $1 \times 10^{11}\ \Omega$. We can model this ion channel, with its pore, as a 12-nm-long cylinder of radius 0.3 nm. What is the resistivity of the fluid in the pore? (a) 10 $\Omega \cdot$ m; (b) 6 $\Omega \cdot$ m; (c) 2 $\Omega \cdot$ m; (d) 1 $\Omega \cdot$ m.

26.87 In a simple model of an axon conducting a nerve signal, ions move across the cell membrane through open ion channels, which act as purely resistive elements. If a typical current density (current per unit cross-sectional area) in the cell membrane is 5 mA/cm^2 when the voltage across the membrane (the *action potential*) is 50 mV, what is the number density of open ion channels in the membrane? (a) 1/cm^2; (b) 10/cm^2; (c) 10/mm^2; (d) 100/μm^2.

26.88 Cell membranes across a wide variety of organisms have a capacitance per unit area of 1 μF/cm^2. For the electrical signal in a nerve to propagate down the axon, the charge on the membrane "capacitor" must change. What time constant is required when the ion channels are open? (a) 1 μs; (b) 10 μs; (c) 100 μs; (d) 1 ms.

Answers

Chapter Opening Question ?

(ii) The potential difference V is the same across resistors connected in parallel. However, there is a different current I through each resistor if the resistances R are different: $I = V/R$.

Test Your Understanding Questions

26.1 (a), (c), (d), (b) Here's why: The three resistors in Fig. 26.1a are in series, so $R_{eq} = R + R + R = 3R$. In Fig. 26.1b the three resistors are in parallel, so $1/R_{eq} = 1/R + 1/R + 1/R = 3/R$ and $R_{eq} = R/3$. In Fig. 26.1c the second and third resistors are in parallel, so their equivalent resistance R_{23} is given by $1/R_{23} = 1/R + 1/R = 2/R$; hence $R_{23} = R/2$. This combination is in series with the first resistor, so the three resistors together have equivalent resistance $R_{eq} = R + R/2 = 3R/2$. In Fig. 26.1d the second and third resistors are in series, so their equivalent resistance is $R_{23} = R + R = 2R$. This combination is in parallel with the first resistor, so the equivalent resistance of the three-resistor combination is given by $1/R_{eq} = 1/R + 1/2R = 3/2R$. Hence $R_{eq} = 2R/3$.

26.2 loop cbdac, no Equation (2) minus Eq. (1) gives $-I_2(1\ \Omega) - (I_2 + I_3)(2\ \Omega) + (I_1 - I_3)(1\ \Omega) + I_1(1\ \Omega) = 0$. We can obtain this equation by applying the loop rule around the path from c to b to d to a to c in Fig. 26.12. This isn't an independent equation, so it would not have helped with the solution of Example 26.6.

26.3 (a) (ii), (b) (iii) An ammeter must always be placed in series with the circuit element of interest, and a voltmeter must always be placed in parallel. Ideally the ammeter would have zero resistance and the voltmeter would have infinite resistance so that their presence would have no effect on either the resistor current or the voltage. Neither of these idealizations is possible, but the ammeter resistance should be much less than 2 Ω and the voltmeter resistance should be much greater than 2 Ω.

26.4 (ii) After one time constant, $t = RC$ and the initial charge Q_0 has decreased to $Q_0 e^{-t/RC} = Q_0 e^{-RC/RC} = Q_0 e^{-1} = Q_0/e$. Hence the stored energy has decreased from $Q_0^2/2C$ to $(Q_0/e)^2/2C = Q_0^2/2Ce^2$, a fraction $1/e^2 = 0.135$ of its initial value. This result doesn't depend on the initial value of the energy.

26.5 no This is a very dangerous thing to do. The circuit breaker will allow currents up to 40 A, double the rated value of the wiring. The amount of power $P = I^2R$ dissipated in a section of wire can therefore be up to four times the rated value, so the wires could get very warm and start a fire. (This assumes the resistance R remains unchanged. In fact, R increases with temperature, so the dissipated power can be even greater, and more dangerous, than we have estimated.)

Bridging Problem

(a) 9.39 J   **(b)** 2.02×10^4 W   **(c)** 4.65×10^{-4} s
(d) 7.43×10^3 W

27

MAGNETIC FIELD AND MAGNETIC FORCES

LEARNING GOALS

Looking forward at …

27.1 The properties of magnets, and how magnets interact with each other.

27.2 The nature of the force that a moving charged particle experiences in a magnetic field.

27.3 How magnetic field lines are different from electric field lines.

27.4 How to analyze the motion of a charged particle in a magnetic field.

27.5 Some practical applications of magnetic fields in chemistry and physics.

27.6 How to analyze magnetic forces on current-carrying conductors.

27.7 How current loops behave when placed in a magnetic field.

27.8 How direct-current motors work.

27.9 How magnetic forces give rise to the Hall effect.

Looking back at …

1.10 Vector product of two vectors.

3.4, 5.4 Uniform circular motion.

10.1 Torque.

21.6, 21.7 Electric field lines and electric dipole moment.

22.2, 22.3 Electric flux and Gauss's law.

25.1 Electric current.

26.3 Galvanometers.

E verybody uses magnetic forces. They are at the heart of electric motors, microwave ovens, loudspeakers, computer printers, and disk drives. The most familiar examples of magnetism are permanent magnets, which attract unmagnetized iron objects and can also attract or repel other magnets. A compass needle aligning itself with the earth's magnetism is an example of this interaction. But the *fundamental* nature of magnetism is the interaction of moving electric charges. Unlike electric forces, which act on electric charges whether they are moving or not, magnetic forces act only on *moving* charges.

We saw in Chapter 21 that the electric force arises in two stages: (1) a charge produces an electric field in the space around it, and (2) a second charge responds to this field. Magnetic forces also arise in two stages. First, a *moving* charge or a collection of moving charges (that is, an electric current) produces a *magnetic* field. Next, a second current or moving charge responds to this magnetic field, and so experiences a magnetic force.

In this chapter we study the second stage in the magnetic interaction—that is, how moving charges and currents *respond* to magnetic fields. In particular, we will see how to calculate magnetic forces and torques, and we will discover why magnets can pick up iron objects like paper clips. In Chapter 28 we will complete our picture of the magnetic interaction by examining how moving charges and currents *produce* magnetic fields.

27.1 MAGNETISM

Magnetic phenomena were first observed at least 2500 years ago in fragments of magnetized iron ore found near the ancient city of Magnesia (now Manisa, in western Turkey). These fragments were what are now called **permanent magnets;** you probably have several permanent magnets on your refrigerator door at home. Permanent magnets were found to exert forces on each other as well as on pieces of

27.1 (a) Two bar magnets attract when opposite poles (N and S, or S and N) are next to each other. (b) The bar magnets repel when like poles (N and N, or S and S) are next to each other.

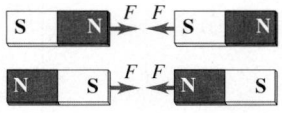

(a) Opposite poles attract.

(b) Like poles repel.

27.2 (a) Either pole of a bar magnet attracts an unmagnetized object that contains iron, such as a nail. (b) A real-life example of this effect.

(a)

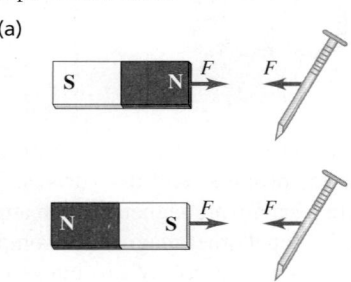

(b)

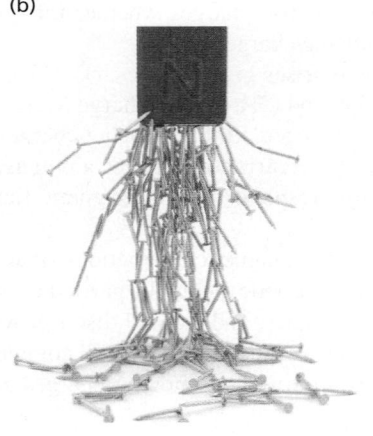

iron that were not magnetized. It was discovered that when an iron rod is brought in contact with a natural magnet, the rod also becomes magnetized. When such a rod is floated on water or suspended by a string from its center, it tends to line itself up in a north-south direction. The needle of an ordinary compass is just such a piece of magnetized iron.

Before the relationship of magnetic interactions to moving charges was understood, the interactions of permanent magnets and compass needles were described in terms of *magnetic poles.* If a bar-shaped permanent magnet, or *bar magnet,* is free to rotate, one end points north. This end is called a *north pole* or *N pole;* the other end is a *south pole* or *S pole.* Opposite poles attract each other, and like poles repel each other (**Fig. 27.1**). An object that contains iron but is not itself magnetized (that is, it shows no tendency to point north or south) is attracted by *either* pole of a permanent magnet (**Fig. 27.2**). This is the attraction that acts between a magnet and the unmagnetized steel door of a refrigerator. By analogy to electric interactions, we describe the interactions in Figs. 27.1 and 27.2 by saying that a bar magnet sets up a *magnetic field* in the space around it and a second body responds to that field. A compass needle tends to align with the magnetic field at the needle's position.

The earth itself is a magnet. Its north geographic pole is close to a magnetic *south* pole, which is why the north pole of a compass needle points north. The earth's magnetic axis is not quite parallel to its geographic axis (the axis of rotation), so a compass reading deviates somewhat from geographic north. This deviation, which varies with location, is called *magnetic declination* or *magnetic variation.* Also, the magnetic field is not horizontal at most points on the earth's surface; its angle up or down is called *magnetic inclination.* At the magnetic poles the magnetic field is vertical.

Figure 27.3 is a sketch of the earth's magnetic field. The lines, called *magnetic field lines,* show the direction that a compass would point at each location; they are discussed in detail in Section 27.3. The direction of the field at any point can be defined as the direction of the force that the field would exert on a magnetic north pole. In Section 27.2 we'll describe a more fundamental way to define the direction and magnitude of a magnetic field.

27.3 A sketch of the earth's magnetic field. The field, which is caused by currents in the earth's molten core, changes with time; geologic evidence shows that it reverses direction entirely at irregular intervals of 10^4 to 10^6 years.

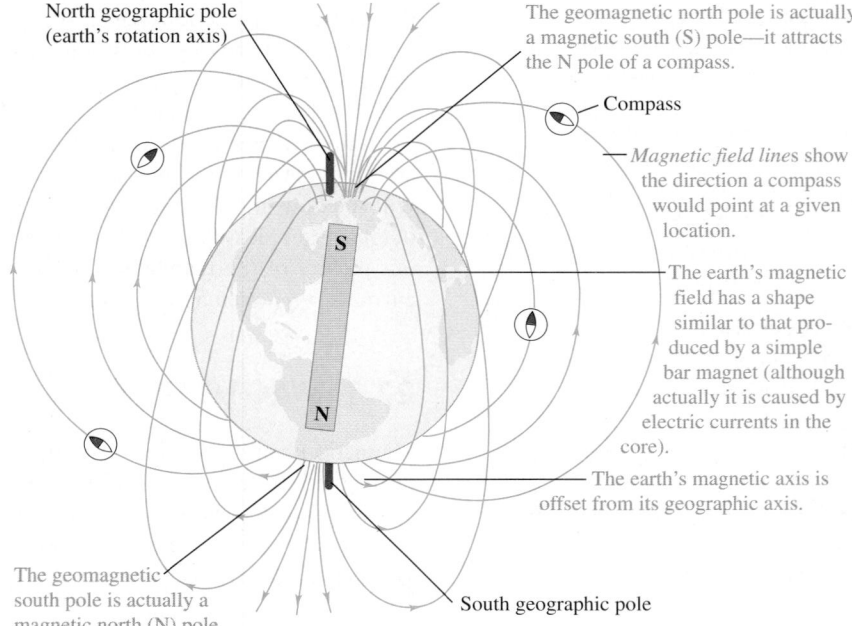

North geographic pole (earth's rotation axis)

The geomagnetic north pole is actually a magnetic south (S) pole—it attracts the N pole of a compass.

Compass

Magnetic field lines show the direction a compass would point at a given location.

The earth's magnetic field has a shape similar to that produced by a simple bar magnet (although actually it is caused by electric currents in the core).

The earth's magnetic axis is offset from its geographic axis.

The geomagnetic south pole is actually a magnetic north (N) pole.

South geographic pole

Magnetic Poles Versus Electric Charge

The concept of magnetic poles may appear similar to that of electric charge, and north and south poles may seem analogous to positive and negative charges. But the analogy can be misleading. While isolated positive and negative charges exist, there is *no* experimental evidence that one isolated magnetic pole exists; poles always appear in pairs. If a bar magnet is broken in two, each broken end becomes a pole (**Fig. 27.4**). The existence of an isolated magnetic pole, or **magnetic monopole,** would have sweeping implications for theoretical physics. Extensive searches for magnetic monopoles have been carried out, but so far without success.

The first evidence of the relationship of magnetism to moving charges was discovered in 1820 by the Danish scientist Hans Christian Oersted. He found that a compass needle was deflected by a current-carrying wire (**Fig. 27.5**). Similar investigations were carried out in France by André Ampère. A few years later, Michael Faraday in England and Joseph Henry in the United States discovered that moving a magnet near a conducting loop can cause a current in the loop. We now know that the magnetic forces between two bodies shown in Figs. 27.1 and 27.2 are fundamentally due to interactions between moving electrons in the atoms of the bodies. (There are also *electric* interactions between the two bodies, but these are far weaker than the magnetic interactions because the bodies are electrically neutral.) Inside a magnetized body such as a permanent magnet, the motion of certain of the atomic electrons is *coordinated;* in an unmagnetized body these motions are not coordinated. (We'll describe these motions further in Section 27.7 and see how the interactions shown in Figs. 27.1 and 27.2 come about.)

Electric and magnetic interactions prove to be intimately connected. Over the next several chapters we will develop the unifying principles of electromagnetism, culminating in the expression of these principles in *Maxwell's equations.* These equations represent the synthesis of electromagnetism, just as Newton's laws of motion are the synthesis of mechanics, and like Newton's laws they represent a towering achievement of the human intellect.

TEST YOUR UNDERSTANDING OF SECTION 27.1   Suppose you cut off the part of the compass needle shown in Fig. 27.5a that is painted gray. You discard this part, drill a hole in the remaining red part, and place the red part on the pivot at the center of the compass. Will the red part still swing when a current is applied as in Fig. 27.5b? ▮

27.2 MAGNETIC FIELD

To introduce the concept of magnetic field properly, let's review our formulation of *electric* interactions in Chapter 21, where we introduced the concept of *electric* field. We represented electric interactions in two steps:

1. A distribution of electric charge creates an electric field $\vec{E}$ in the surrounding space.
2. The electric field exerts a force $\vec{F} = q\vec{E}$ on any other charge q that is present in the field.

We can describe magnetic interactions in a similar way:

1. A moving charge or a current creates a **magnetic field** in the surrounding space (in addition to its *electric* field).
2. The magnetic field exerts a force $\vec{F}$ on any other moving charge or current that is present in the field.

27.4 Breaking a bar magnet. Each piece has a north and south pole, even if the pieces are different sizes. (The smaller the piece, the weaker its magnetism.)

In contrast to electric charges, magnetic poles always come in pairs and can't be isolated.

Breaking a magnet in two ...

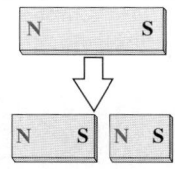

... yields two magnets, not two isolated poles.

27.5 In Oersted's experiment, a compass is placed directly over a horizontal wire (here viewed from above).

(a)

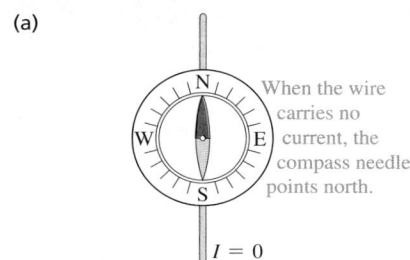

When the wire carries no current, the compass needle points north.

$I = 0$

(b)

When the wire carries a current, the compass needle deflects. The direction of deflection depends on the direction of the current.

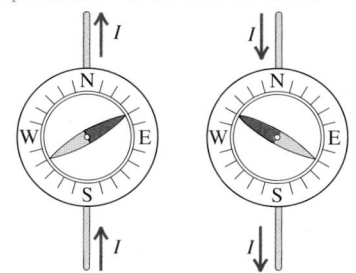

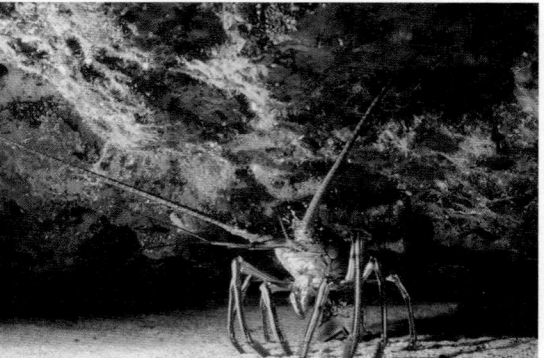

27.6 The magnetic force $\vec{F}$ acting on a positive charge q moving with velocity $\vec{v}$ is perpendicular to both $\vec{v}$ and the magnetic field $\vec{B}$. For given values of speed v and magnetic field strength B, the force is greatest when $\vec{v}$ and $\vec{B}$ are perpendicular.

(a)

A charge moving **parallel** to a magnetic field experiences **zero magnetic force.**

(b)

A charge moving at an angle ϕ to a magnetic field experiences a magnetic force with magnitude $F = |q|v_\perp B = |q|vB \sin \phi$.

$\vec{F}$ is perpendicular to the plane containing $\vec{v}$ and $\vec{B}$.

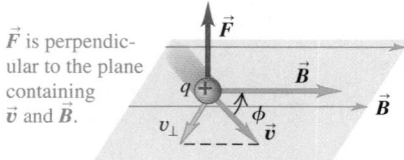

(c)

A charge moving **perpendicular** to a magnetic field experiences a maximal magnetic force with magnitude $F_{max} = qvB.$

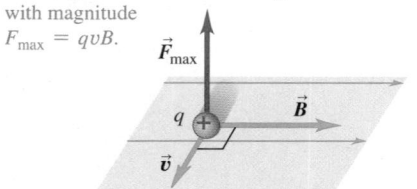

In this chapter we'll concentrate on the *second* aspect of the interaction: Given the presence of a magnetic field, what force does it exert on a moving charge or a current? In Chapter 28 we will come back to the problem of how magnetic fields are *created* by moving charges and currents.

Like electric field, magnetic field is a *vector field*—that is, a vector quantity associated with each point in space. We will use the symbol $\vec{B}$ for magnetic field. At any position the direction of $\vec{B}$ is defined as the direction in which the north pole of a compass needle tends to point. The arrows in Fig. 27.3 suggest the direction of the earth's magnetic field; for any magnet, $\vec{B}$ points out of its north pole and into its south pole.

Magnetic Forces on Moving Charges

There are four key characteristics of the magnetic force on a moving charge. First, its magnitude is proportional to the magnitude of the charge. If a 1-μC charge and a 2-μC charge move through a given magnetic field with the same velocity, experiments show that the force on the 2-μC charge is twice as great as the force on the 1-μC charge. Second, the magnitude of the force is also proportional to the magnitude, or "strength," of the field; if we double the magnitude of the field (for example, by using two identical bar magnets instead of one) without changing the charge or its velocity, the force doubles.

A third characteristic is that the magnetic force depends on the particle's velocity. This is quite different from the electric-field force, which is the same whether the charge is moving or not. A charged particle at rest experiences *no* magnetic force. And fourth, we find by experiment that the magnetic force $\vec{F}$ *does not* have the same direction as the magnetic field $\vec{B}$ but instead is always *perpendicular* to both $\vec{B}$ and the velocity $\vec{v}$. The magnitude F of the force is proportional to the component of $\vec{v}$ perpendicular to the field; when that component is zero (that is, when $\vec{v}$ and $\vec{B}$ are parallel or antiparallel), the force is zero.

Figure 27.6 shows these relationships. The direction of $\vec{F}$ is always perpendicular to the plane containing $\vec{v}$ and $\vec{B}$. Its magnitude is given by

$$F = |q|v_\perp B = |q|vB \sin \phi \qquad (27.1)$$

where $|q|$ is the magnitude of the charge and ϕ is the angle measured from the direction of $\vec{v}$ to the direction of $\vec{B}$, as shown in the figure.

This description does not specify the direction of $\vec{F}$ completely; there are always two directions, opposite to each other, that are both perpendicular to the plane of $\vec{v}$ and $\vec{B}$. To complete the description, we use the same right-hand rule that we used to define the vector product in Section 1.10. (It would be a good idea to review that section before you go on.) Draw the vectors $\vec{v}$ and $\vec{B}$ with their tails together, as in **Fig. 27.7a**. Imagine turning $\vec{v}$ until it points in the direction of $\vec{B}$ (turning through the smaller of the two possible angles). Wrap the fingers of your right hand around the line perpendicular to the plane of $\vec{v}$ and $\vec{B}$ so that they curl around with the sense of rotation from $\vec{v}$ to $\vec{B}$. Your thumb then points in the direction of the force $\vec{F}$ on a *positive* charge.

This discussion shows that the force on a charge q moving with velocity $\vec{v}$ in a magnetic field $\vec{B}$ is given, both in magnitude and in direction, by

Magnetic force on a moving charged particle $\cdots\rightarrow \vec{F} = q\vec{v} \times \vec{B} \leftarrow\cdots$ Magnetic field	(27.2)

Particle's charge
Particle's velocity

This is the first of several vector products we will encounter in our study of magnetic-field relationships. It's important to note that Eq. (27.2) was *not* deduced theoretically; it is an observation based on *experiment*.

27.7 Finding the direction of the magnetic force on a moving charged particle.

(a)

Right-hand rule for the direction of magnetic force on a **positive** charge moving in a magnetic field:

① Place the $\vec{v}$ and $\vec{B}$ vectors tail to tail.

② Imagine turning $\vec{v}$ toward $\vec{B}$ in the $\vec{v}$-$\vec{B}$ plane (through the smaller angle).

③ The force acts along a line perpendicular to the $\vec{v}$-$\vec{B}$ plane. Curl the fingers of your *right hand* around this line in the same direction you rotated $\vec{v}$. Your thumb now points in the direction the force acts.

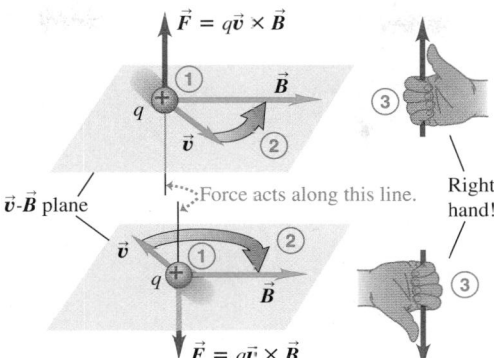

(b)

If the charge is negative, the direction of the force is *opposite* to that given by the right-hand rule.

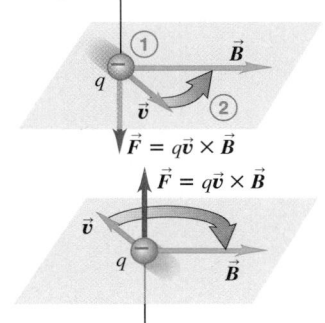

Equation (27.2) is valid for both positive and negative charges. When q is negative, the direction of the force $\vec{F}$ is opposite to that of $\vec{v} \times \vec{B}$ (Fig. 27.7b). If two charges with equal magnitude and opposite sign move in the same $\vec{B}$ field with the same velocity (**Fig. 27.8**), the forces have equal magnitude and opposite direction. Figures 27.6, 27.7, and 27.8 show several examples of the relationships of the directions of $\vec{F}$, $\vec{v}$, and $\vec{B}$ for both positive and negative charges. Be sure you understand the relationships shown in these figures.

Equation (27.1) gives the magnitude of the magnetic force $\vec{F}$ in Eq. (27.2). Since ϕ is the angle between the directions of vectors $\vec{v}$ and $\vec{B}$, we may interpret $B \sin \phi$ as the component of $\vec{B}$ perpendicular to $\vec{v}$—that is, $B_\perp$. With this notation the force magnitude is

$$F = |q| v B_\perp \qquad (27.3)$$

This form may be more convenient, especially in problems involving *currents* rather than individual particles. We'll discuss forces on currents later in this chapter.

From Eq. (27.1) the *units* of B must be the same as the units of F/qv. Therefore the SI unit of B is equivalent to $1 \text{ N} \cdot \text{s}/\text{C} \cdot \text{m}$, or, since one ampere is one coulomb per second ($1 \text{ A} = 1 \text{ C/s}$), $1 \text{ N/A} \cdot \text{m}$. This unit is called the **tesla** (abbreviated T), in honor of Nikola Tesla (1856–1943), the prominent Serbian-American scientist and inventor:

$$1 \text{ tesla} = 1 \text{ T} = 1 \text{ N/A} \cdot \text{m}$$

Another unit of B, the **gauss** ($1 \text{ G} = 10^{-4} \text{ T}$), is also in common use.

The magnetic field of the earth is of the order of 10^{-4} T or 1 G. Magnetic fields of the order of 10 T occur in the interior of atoms and are important in the analysis of atomic spectra. The largest steady magnetic field that can be produced at present in the laboratory is about 45 T. Some pulsed-current electromagnets can produce fields of the order of 120 T for millisecond time intervals.

Measuring Magnetic Fields with Test Charges

To explore an unknown magnetic field, we can measure the magnitude and direction of the force on a *moving* test charge and then use Eq. (27.2) to determine $\vec{B}$. The electron beam in a cathode-ray tube, such as that in an older television set (not a flat-screen TV), is a convenient device for this. The electron gun shoots out a narrow beam of electrons at a known speed. If there is no force to deflect the beam, it strikes the center of the screen.

27.8 Two charges of the same magnitude but opposite sign moving with the same velocity in the same magnetic field. The magnetic forces on the charges are equal in magnitude but opposite in direction.

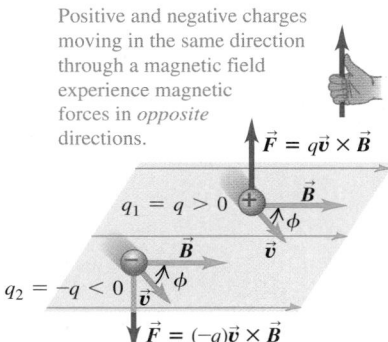

BIO Application Magnetic Fields of the Body All living cells are electrically active, and the feeble electric currents within your body produce weak but measurable magnetic fields. The fields produced by skeletal muscles have magnitudes less than 10^{-10} T, about one-millionth as strong as the earth's magnetic field. Your brain produces magnetic fields that are far weaker, only about 10^{-12} T.

27.9 Determining the direction of a magnetic field by using a cathode-ray tube. Because electrons have a negative charge, the magnetic force $\vec{F} = q\vec{v} \times \vec{B}$ in part (b) points opposite to the direction given by the right-hand rule (see Fig. 27.7b).

(a) If the tube axis is parallel to the y-axis, the beam is undeflected, so $\vec{B}$ is in either the +y- or the −y-direction.

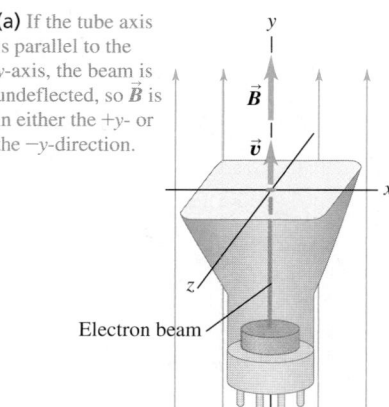

Electron beam

(b) If the tube axis is parallel to the x-axis, the beam is deflected in the −z-direction, so $\vec{B}$ is in the +y-direction.

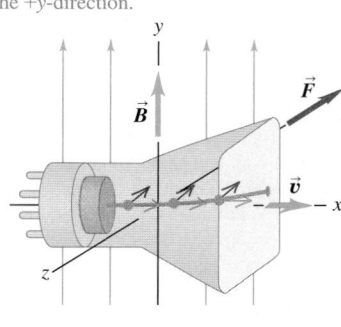

If a magnetic field is present, in general the electron beam is deflected. But if the beam is parallel or antiparallel to the field, then $\phi = 0$ or π in Eq. (27.1) and $F = 0$; there is no force and hence no deflection. If we find that the electron beam is not deflected when its direction is parallel to a certain axis as in **Fig. 27.9a**, the $\vec{B}$ vector must point either up or down along that axis.

If we then turn the tube 90° (Fig. 27.9b), $\phi = \pi/2$ in Eq. (27.1) and the magnetic force is maximum; the beam is deflected in a direction perpendicular to the plane of $\vec{B}$ and $\vec{v}$. The direction and magnitude of the deflection determine the direction and magnitude of $\vec{B}$. We can perform additional experiments in which the angle between $\vec{B}$ and $\vec{v}$ is between 0° and 90° to confirm Eq. (27.1). We note that the electron has a negative charge; the force in Fig. 27.9b is opposite in direction to the force on a positive charge.

When a charged particle moves through a region of space where *both* electric and magnetic fields are present, both fields exert forces on the particle. The total force $\vec{F}$ is the vector sum of the electric and magnetic forces:

$$\vec{F} = q(\vec{E} + \vec{v} \times \vec{B})$$ (27.4)

PROBLEM-SOLVING STRATEGY 27.1 | MAGNETIC FORCES

IDENTIFY *the relevant concepts:* The equation $\vec{F} = q\vec{v} \times \vec{B}$ allows you to determine the magnetic force on a moving charged particle.

SET UP *the problem* using the following steps:
1. Draw the velocity $\vec{v}$ and magnetic field $\vec{B}$ with their tails together so that you can visualize the plane that contains them.
2. Determine the angle ϕ between $\vec{v}$ and $\vec{B}$.
3. Identify the target variables.

EXECUTE *the solution* as follows:
1. Use Eq. (27.2), $\vec{F} = q\vec{v} \times \vec{B}$, to express the magnetic force. Equation (27.1) gives the magnitude of the force, $F = qvB\sin\phi$.

2. Remember that $\vec{F}$ is perpendicular to the plane containing $\vec{v}$ and $\vec{B}$. The right-hand rule (see Fig. 27.7) gives the direction of $\vec{v} \times \vec{B}$. If q is negative, $\vec{F}$ is *opposite* to $\vec{v} \times \vec{B}$.

EVALUATE *your answer:* Whenever possible, solve the problem in two ways to confirm that the results agree. Do it directly from the geometric definition of the vector product. Then find the components of the vectors in some convenient coordinate system and calculate the vector product from the components. Verify that the results agree.

EXAMPLE 27.1 MAGNETIC FORCE ON A PROTON

A beam of protons ($q = 1.6 \times 10^{-19}$ C) moves at 3.0×10^5 m/s through a uniform 2.0-T magnetic field directed along the positive z-axis (**Fig. 27.10**). The velocity of each proton lies in the xz-plane and is directed at 30° to the +z-axis. Find the force on a proton.

SOLUTION

IDENTIFY and SET UP: This problem uses the expression $\vec{F} = q\vec{v} \times \vec{B}$ for the magnetic force $\vec{F}$ on a moving charged particle. The target variable is $\vec{F}$.

EXECUTE: The charge is positive, so the force is in the same direction as the vector product $\vec{v} \times \vec{B}$. From the right-hand rule, this direction is along the negative y-axis. The magnitude of the force, from Eq. (27.1), is

$$F = qvB\sin\phi$$
$$= (1.6 \times 10^{-19}\,\text{C})(3.0 \times 10^5\,\text{m/s})(2.0\,\text{T})(\sin 30°)$$
$$= 4.8 \times 10^{-14}\,\text{N}$$

EVALUATE: To check our result, we evaluate the force by using vector language and Eq. (27.2). We have

$$\vec{v} = (3.0 \times 10^5\,\text{m/s})(\sin 30°)\hat{\imath} + (3.0 \times 10^5\,\text{m/s})(\cos 30°)\hat{k}$$
$$\vec{B} = (2.0\,\text{T})\hat{k}$$

27.10 Directions of $\vec{v}$ and $\vec{B}$ for a proton in a magnetic field.

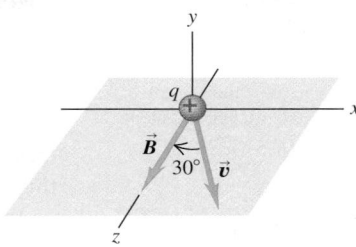

$$\vec{F} = q\vec{v} \times \vec{B}$$
$$= (1.6 \times 10^{-19}\,\text{C})(3.0 \times 10^5\,\text{m/s})(2.0\,\text{T})$$
$$\times (\sin 30°\hat{\imath} + \cos 30°\hat{k}) \times \hat{k}$$
$$= (-4.8 \times 10^{-14}\,\text{N})\hat{\jmath}$$

(Recall that $\hat{\imath} \times \hat{k} = -\hat{\jmath}$ and $\hat{k} \times \hat{k} = \mathbf{0}$.) We again find that the force is in the negative y-direction with magnitude 4.8×10^{-14} N.

If the beam consists of *electrons* rather than protons, the charge is negative ($q = -1.6 \times 10^{-19}$ C) and the direction of the force is reversed. The force is now directed along the *positive* y-axis, but the magnitude is the same as before, $F = 4.8 \times 10^{-14}$ N.

TEST YOUR UNDERSTANDING OF SECTION 27.2 The accompanying figure shows a uniform magnetic field $\vec{B}$ directed into the plane of the paper (shown by the blue ×'s). A particle with a negative charge moves in the plane. Which path—1, 2, or 3—does the particle follow? ▮

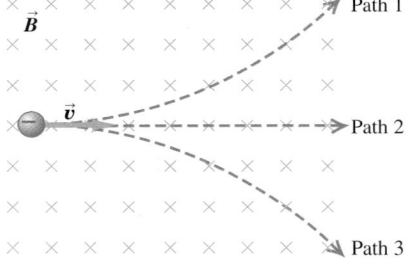

27.3 MAGNETIC FIELD LINES AND MAGNETIC FLUX

We can represent any magnetic field by **magnetic field lines,** just as we did for the earth's magnetic field in Fig. 27.3. The idea is the same as for the electric field lines we introduced in Section 21.6. We draw the lines so that the line through any point is tangent to the magnetic field vector $\vec{B}$ at that point (**Fig. 27.11**). Just as with electric field lines, we draw only a few representative lines; otherwise, the lines would fill up all of space. Where adjacent field lines are close together, the field magnitude is large; where these field lines are far apart, the field magnitude is small. Also, because the direction of $\vec{B}$ at each point is unique, field lines never intersect.

CAUTION **Magnetic field lines are not "lines of force"** Unlike electric field lines, magnetic field lines *do not* point in the direction of the force on a charge (**Fig. 27.12**, next page). Equation (27.2) shows that the force on a moving charged particle is always perpendicular to the magnetic field and hence to the magnetic field line that passes through the particle's position. The direction of the force depends on the particle's velocity and the sign of its charge, so just looking at magnetic field lines cannot tell you the direction of the force on an arbitrary moving charged particle. Magnetic field lines *do* have the direction that a compass needle would point at each location; this may help you visualize them. ▮

27.11 Magnetic field lines of a permanent magnet. Note that the field lines pass through the interior of the magnet.

At each point, the field line is tangent to the magnetic-field vector $\vec{B}$.

The more densely the field lines are packed, the stronger the field is at that point.

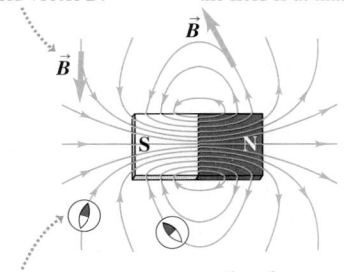

At each point, the field lines point in the same direction a compass would ...

... therefore, magnetic field lines point *away from* N poles and *toward* S poles.

27.12 Magnetic field lines are *not* "lines of force."

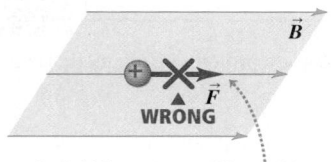

Magnetic field lines are *not* "lines of force." The force on a charged particle is not along the direction of a field line.

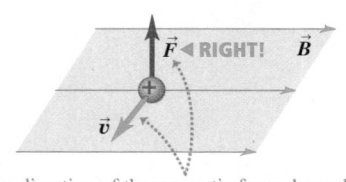

The direction of the magnetic force depends on the velocity $\vec{v}$, as expressed by the magnetic force law $\vec{F} = q\vec{v} \times \vec{B}$.

Figures 27.11 and 27.13 show magnetic field lines produced by several common sources of magnetic field. In the gap between the poles of the magnet shown in **Fig. 27.13a**, the field lines are approximately straight, parallel, and equally spaced, showing that the magnetic field in this region is approximately *uniform* (that is, constant in magnitude and direction).

Because magnetic-field patterns are three-dimensional, it's often necessary to draw magnetic field lines that point into or out of the plane of a drawing. To do this we use a dot (·) to represent a vector directed out of the plane and a cross (×) to represent a vector directed into the plane (Fig. 27.13b). To remember these, think of a dot as the head of an arrow coming directly toward you, and think of a cross as the feathers of an arrow flying directly away from you.

Iron filings, like compass needles, tend to align with magnetic field lines. Hence they provide an easy way to visualize field lines (**Fig. 27.14**).

Magnetic Flux and Gauss's Law for Magnetism

We define the **magnetic flux** Φ_B through a surface just as we defined electric flux in connection with Gauss's law in Section 22.2. We can divide any surface into elements of area dA (**Fig. 27.15**). For each element we determine $B_\perp$, the component of $\vec{B}$ normal to the surface at the position of that element, as shown.

27.13 Magnetic field lines produced by some common sources of magnetic field.

(a) Magnetic field of a C-shaped magnet

(b) Magnetic field of a straight current-carrying wire

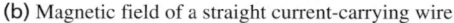

Between flat, parallel magnetic poles, the magnetic field is nearly uniform.

To represent a field coming out of or going into the plane of the paper, we use dots and crosses, respectively.

$\vec{B}$ directed out of plane

$\vec{B}$ directed into plane

Perspective view

Wire in plane of paper

(c) Magnetic fields of a current-carrying loop and a current-carrying coil (solenoid)

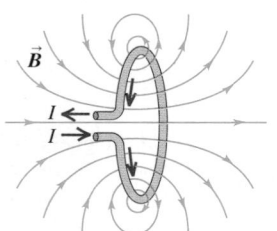

Notice that the field of the loop and, especially, that of the coil look like the field of a bar magnet (see Fig. 27.11).

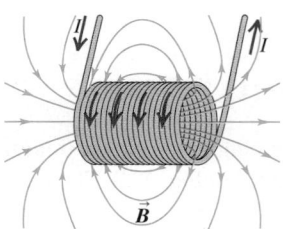

27.14 (a) Like little compass needles, iron filings line up tangent to magnetic field lines. **(b)** Drawing of field lines for the situation shown in (a).

(a)

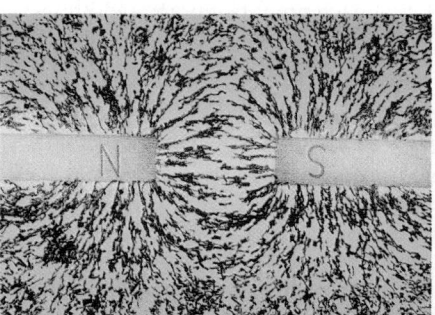

(b)

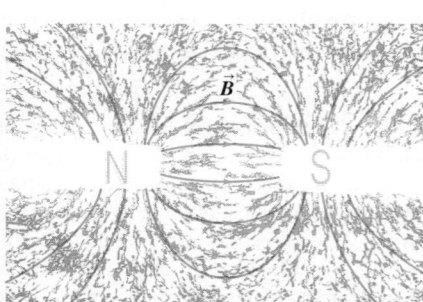

From the figure, $B_\perp = B\cos\phi$, where ϕ is the angle between the direction of $\vec{B}$ and a line perpendicular to the surface. (Be careful not to confuse ϕ with Φ_B.) In general, this component varies from point to point on the surface. We define the magnetic flux $d\Phi_B$ through this area as

$$d\Phi_B = B_\perp \, dA = B\cos\phi \, dA = \vec{B} \cdot d\vec{A} \qquad (27.5)$$

The *total* magnetic flux through the surface is the sum of the contributions from the individual area elements:

Magnetic flux through a surface

$$\Phi_B = \int B\cos\phi \, dA = \int B_\perp \, dA = \int \vec{B} \cdot d\vec{A} \qquad (27.6)$$

Magnitude of magnetic field $\vec{B}$ — Component of $\vec{B}$ perpendicular to surface — Angle between $\vec{B}$ and normal to surface — Element of surface area — Vector element of surface area

(Review the concepts of vector area and surface integral in Section 22.2.)

Magnetic flux is a *scalar* quantity. If $\vec{B}$ is uniform over a plane surface with total area A, then $B_\perp$ and ϕ are the same at all points on the surface, and

$$\Phi_B = B_\perp A = BA\cos\phi \qquad (27.7)$$

If $\vec{B}$ is also perpendicular to the surface (parallel to the area vector), then $\cos\phi = 1$ and Eq. (27.7) reduces to $\Phi_B = BA$. We will use the concept of magnetic flux extensively during our study of electromagnetic induction in Chapter 29.

The SI unit of magnetic flux is equal to the unit of magnetic field (1 T) times the unit of area (1 m^2). This unit is called the **weber** (1 Wb), in honor of the German physicist Wilhelm Weber (1804–1891):

$$1 \text{ Wb} = 1 \text{ T} \cdot \text{m}^2$$

Also, $1 \text{ T} = 1 \text{ N/A} \cdot \text{m}$, so

$$1 \text{ Wb} = 1 \text{ T} \cdot \text{m}^2 = 1 \text{ N} \cdot \text{m/A}$$

In Gauss's law the total *electric* flux through a closed surface is proportional to the total electric charge enclosed by the surface. For example, if the closed surface encloses an electric dipole, the total electric flux is zero because the total charge is zero. (You may want to review Section 22.3 on Gauss's law.) By analogy, if there were such a thing as a single magnetic charge (magnetic monopole), the total *magnetic* flux through a closed surface would be proportional to the total magnetic charge enclosed. But we have mentioned that no magnetic monopole has ever been observed, despite intensive searches. This leads us to **Gauss's law for magnetism:**

Gauss's law for magnetism:

The total magnetic flux through *any* **closed surface** ...

$$\oint \vec{B} \cdot d\vec{A} = 0 \qquad \text{... equals zero.} \qquad (27.8)$$

You can verify Gauss's law for magnetism by examining Figs. 27.11 and 27.13; if you draw a closed surface anywhere in any of the field maps shown in those figures, you will see that every field line that enters the surface also exits from it; the net flux through the surface is zero. It also follows from Eq. (27.8) that magnetic field lines always form closed loops.

CAUTION **Magnetic field lines have no ends** Unlike electric field lines, which begin and end on electric charges, magnetic field lines *never* have endpoints; such a point would indicate the presence of a monopole. You might be tempted to draw magnetic field lines that begin at the north pole of a magnet and end at a south pole. But as Fig. 27.11 shows, a magnet's field lines continue through the interior of the magnet. Like all other magnetic field lines, they form closed loops. ▮

For Gauss's law, which always deals with *closed* surfaces, the vector area element $d\vec{A}$ in Eq. (27.6) always points *out of* the surface. However, some

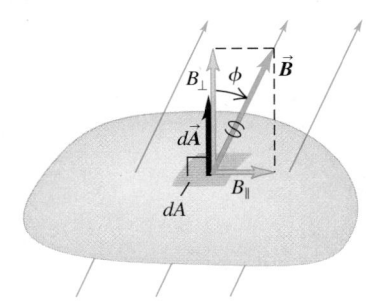

27.15 The magnetic flux through an area element dA is defined to be $d\Phi_B = B_\perp \, dA$.

PhET: Magnet and Compass
PhET: Magnets and Electromagnets

DATA *SPEAKS*

Magnetic Forces and Magnetic Field Lines

When students were given a problem involving magnetic forces and field lines, more than 16% gave an incorrect response. Common errors:

- Forgetting that only the component of the magnetic field that is perpendicular to the velocity of a charged particle causes a force on the particle. If there is no perpendicular component, there is no magnetic force.

- Forgetting that the magnetic force on a moving charged particle is not directed along a magnetic field line. That force is perpendicular to the magnetic field as well as to the particle's velocity.

applications of *magnetic* flux involve an *open* surface with a boundary line; there is then an ambiguity of sign in Eq. (27.6) because of the two possible choices of direction for $d\vec{A}$. In these cases we choose one of the two sides of the surface to be the "positive" side and use that choice consistently.

If the element of area dA in Eq. (27.5) is at right angles to the field lines, then $B_\perp = B$; calling the area $dA_\perp$, we have

$$B = \frac{d\Phi_B}{dA_\perp} \tag{27.9}$$

That is, the magnitude of magnetic field is equal to *flux per unit area* across an area at right angles to the magnetic field. For this reason, magnetic field $\vec{B}$ is sometimes called **magnetic flux density.**

EXAMPLE 27.2 MAGNETIC FLUX CALCULATIONS

Figure 27.16a is a perspective view of a flat surface with area 3.0 cm² in a uniform magnetic field $\vec{B}$. The magnetic flux through this surface is $+0.90$ mWb. Find the magnitude of the magnetic field and the direction of the area vector $\vec{A}$.

27.16 (a) A flat area A in a uniform magnetic field $\vec{B}$. (b) The area vector $\vec{A}$ makes a 60° angle with $\vec{B}$. (If we had chosen $\vec{A}$ to point in the opposite direction, ϕ would have been 120° and the magnetic flux Φ_B would have been negative.)

(a) Perspective view

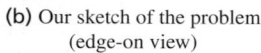

(b) Our sketch of the problem (edge-on view)

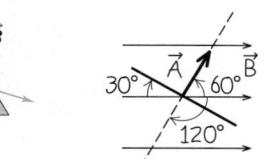

SOLUTION

IDENTIFY and SET UP: Our target variables are the field magnitude B and the direction of the area vector. Because $\vec{B}$ is uniform, B and ϕ are the same at all points on the surface. Hence we can use Eq. (27.7), $\Phi_B = BA\cos\phi$.

EXECUTE: The area A is 3.0×10^{-4} m²; the direction of $\vec{A}$ is perpendicular to the surface, so ϕ could be either 60° or 120°. But Φ_B, B, and A are all positive, so $\cos\phi$ must also be positive. This rules out 120°, so $\phi = 60°$ (Fig. 27.16b). Hence we find

$$B = \frac{\Phi_B}{A\cos\phi} = \frac{0.90 \times 10^{-3}\text{ Wb}}{(3.0 \times 10^{-4}\text{ m}^2)(\cos 60°)} = 6.0\text{ T}$$

EVALUATE: In many problems we are asked to calculate the flux of a given magnetic field through a given area. This example is somewhat different: It tests your understanding of the definition of magnetic flux.

TEST YOUR UNDERSTANDING OF SECTION 27.3 Imagine moving along the axis of the current-carrying loop in Fig. 27.13c, starting at a point well to the left of the loop and ending at a point well to the right of the loop. (a) How would the magnetic field strength vary as you moved along this path? (i) It would be the same at all points along the path; (ii) it would increase and then decrease; (iii) it would decrease and then increase. (b) Would the magnetic field direction vary as you moved along the path? ∎

27.4 MOTION OF CHARGED PARTICLES IN A MAGNETIC FIELD

When a charged particle moves in a magnetic field, it is acted on by the magnetic force given by Eq. (27.2), and the motion is determined by Newton's laws. **Figure 27.17a** shows a simple example. A particle with positive charge q is at point O, moving with velocity $\vec{v}$ in a uniform magnetic field $\vec{B}$ directed into the plane of the figure. The vectors $\vec{v}$ and $\vec{B}$ are perpendicular, so the magnetic force $\vec{F} = q\vec{v} \times \vec{B}$ has magnitude $F = qvB$ and a direction as shown in the figure. The force is *always* perpendicular to $\vec{v}$, so it cannot change the *magnitude* of the velocity, only its direction. To put it differently, the magnetic force never has a component parallel to the particle's motion, so the magnetic force can never do *work* on the particle. This is true even if the magnetic field is not uniform.

Motion of a charged particle under the action of a magnetic field alone is always motion with constant speed.

Using this principle, we see that in the situation shown in Fig. 27.17a the magnitudes of both $\vec{F}$ and $\vec{v}$ are constant. As the particle of mass m moves from O to P to S, the directions of force and velocity change but their magnitudes stay the same. The particle therefore moves under the influence of a constant-magnitude force that is always at right angles to the velocity of the particle. Comparing the discussion of circular motion in Sections 3.4 and 5.4, we see that the particle's path is a *circle*, traced out with constant speed v. The centripetal acceleration is v^2/R and only the magnetic force acts, so from Newton's second law,

$$F = |q|vB = m\frac{v^2}{R} \qquad (27.10)$$

We solve Eq. (27.10) for R:

Radius of a circular orbit in a magnetic field ⋯⋯ $R = \dfrac{mv}{|q|B}$ ← Particle's mass / Particle's speed / Magnetic-field magnitude / Particle's charge $\qquad (27.11)$

If the charge q is negative, the particle moves *clockwise* around the orbit in Fig. 27.17a.

The angular speed ω of the particle can be found from Eq. (9.13), $v = R\omega$. Combining this with Eq. (27.11), we get

$$\omega = \frac{v}{R} = v\frac{|q|B}{mv} = \frac{|q|B}{m} \qquad (27.12)$$

The number of revolutions per unit time is $f = \omega/2\pi$. This frequency f is independent of the radius R of the path. It is called the **cyclotron frequency;** in a particle accelerator called a *cyclotron*, particles moving in nearly circular paths are given a boost twice each revolution, increasing their energy and their orbital radii but not their angular speed or frequency. Similarly, one type of *magnetron*, a common source of microwave radiation for microwave ovens and radar systems, emits radiation with a frequency equal to the frequency of circular motion of electrons in a vacuum chamber between the poles of a magnet.

If the direction of the initial velocity is *not* perpendicular to the field, the velocity *component* parallel to the field is constant because there is no force parallel to the field. Then the particle moves in a helix (**Fig. 27.18**). The radius of the helix is given by Eq. (27.11), where v is now the component of velocity perpendicular to the $\vec{B}$ field.

Motion of a charged particle in a nonuniform magnetic field is more complex. **Figure 27.19** shows a field produced by two circular coils separated by some distance. Particles near either coil experience a magnetic force toward the center

27.17 A charged particle moves in a plane perpendicular to a uniform magnetic field $\vec{B}$.

(a) The orbit of a charged particle in a uniform magnetic field

A charge moving at right angles to a uniform $\vec{B}$ field moves in a circle at constant speed because $\vec{F}$ and $\vec{v}$ are always perpendicular to each other.

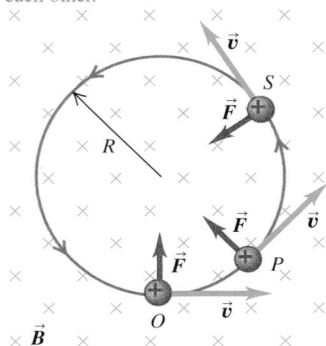

(b) An electron beam (seen as a white arc) curving in a magnetic field

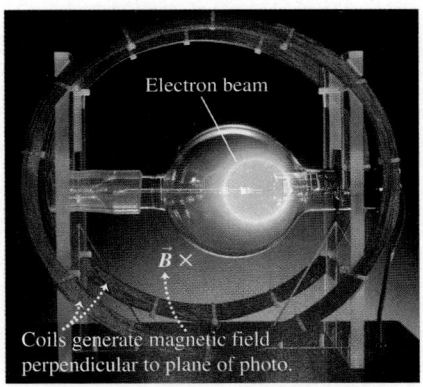

Electron beam

$\vec{B}$ ×

Coils generate magnetic field perpendicular to plane of photo.

DEMO

27.18 The general case of a charged particle moving in a uniform magnetic field $\vec{B}$. The magnetic field does no work on the particle, so its speed and kinetic energy remain constant.

This particle's motion has components both parallel ($v_\parallel$) and perpendicular ($v_\perp$) to the magnetic field, so it moves in a helical path.

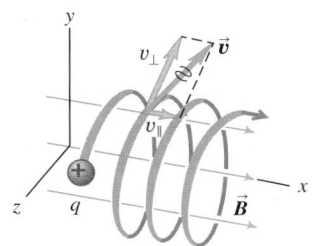

27.19 A magnetic bottle. Particles near either end of the region experience a magnetic force toward the center of the region. This is one way of containing an ionized gas that has a temperature of the order of 10^6 K, which would vaporize any material container.

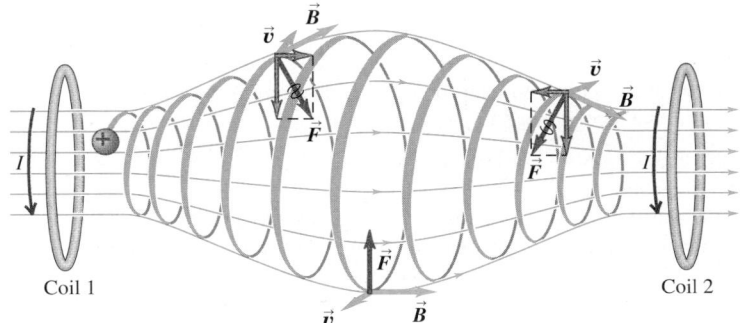

Coil 1

Coil 2

27.20 (a) The Van Allen radiation belts around the earth. Near the poles, charged particles from these belts can enter the atmosphere, producing the aurora borealis ("northern lights") and aurora australis ("southern lights"). (b) A photograph of the aurora borealis.

(a)

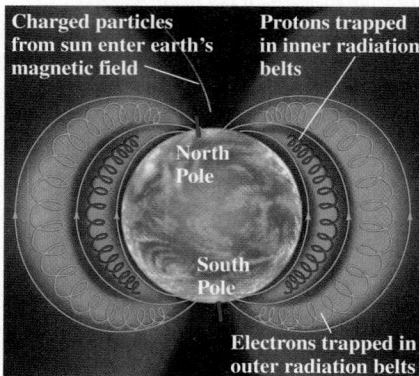

(b)

27.21 This bubble chamber image shows the result of a high-energy gamma ray (which does not leave a track) that collides with an electron in a hydrogen atom. This electron flies off to the right at high speed. Some of the energy in the collision is transformed into a second electron and a positron (a positively charged electron). A magnetic field is directed into the plane of the image, which makes the positive and negative particles curve off in different directions.

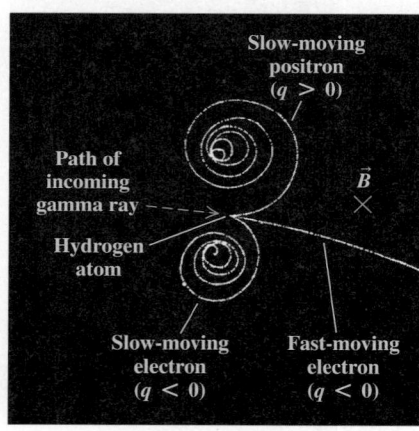

of the region; particles with appropriate speeds spiral repeatedly from one end of the region to the other and back. Because charged particles can be trapped in such a magnetic field, it is called a *magnetic bottle*. This technique is used to confine very hot plasmas with temperatures of the order of 10^6 K. In a similar way the earth's nonuniform magnetic field traps charged particles coming from the sun in doughnut-shaped regions around the earth, as shown in **Fig. 27.20**. These regions, called the *Van Allen radiation belts,* were discovered in 1958 from data obtained by instruments aboard the Explorer I satellite.

Magnetic forces on charged particles play an important role in studies of elementary particles. **Figure 27.21** shows a chamber filled with liquid hydrogen and with a magnetic field directed into the plane of the photograph. A high-energy gamma ray dislodges an electron from a hydrogen atom, sending it off at high speed and creating a visible track in the liquid hydrogen. The track shows the electron curving downward due to the magnetic force. The energy of the collision also produces another electron and a *positron* (a positively charged electron). Because of their opposite charges, the trajectories of the electron and the positron curve in opposite directions. As these particles plow through the liquid hydrogen, they collide with other charged particles, losing energy and speed. As a result, the radius of curvature decreases as suggested by Eq. (27.11). (The electron's speed is comparable to the speed of light, so Eq. (27.11) isn't directly applicable here.) Similar experiments allow physicists to determine the mass and charge of newly discovered particles.

PROBLEM-SOLVING STRATEGY 27.2) **MOTION IN MAGNETIC FIELDS**

IDENTIFY *the relevant concepts:* In analyzing the motion of a charged particle in electric and magnetic fields, you will apply Newton's second law of motion, $\sum \vec{F} = m\vec{a}$, with the net force given by $\sum \vec{F} = q(\vec{E} + \vec{v} \times \vec{B})$. Often other forces such as gravity can be ignored. Many of the problems are similar to the trajectory and circular-motion problems in Sections 3.3, 3.4, and 5.4; it would be a good idea to review those sections.

SET UP *the problem* using the following steps:
1. Determine the target variable(s).
2. Often the use of components is the most efficient approach. Choose a coordinate system and then express all vector quantities in terms of their components in this system.

EXECUTE *the solution* as follows:
1. If the particle moves perpendicular to a uniform magnetic field, the trajectory is a circle with a radius and angular speed given by Eqs. (27.11) and (27.12), respectively.
2. If your calculation involves a more complex trajectory, use $\sum \vec{F} = m\vec{a}$ in component form: $\sum F_x = ma_x$, and so forth. This approach is particularly useful when both electric and magnetic fields are present.

EVALUATE *your answer:* Check whether your results are reasonable.

EXAMPLE 27.3   ELECTRON MOTION IN A MAGNETRON

A magnetron in a microwave oven emits electromagnetic waves with frequency $f = 2450$ MHz. What magnetic field strength is required for electrons to move in circular paths with this frequency?

SOLUTION

IDENTIFY and SET UP: The problem refers to circular motion as shown in Fig. 27.17a. We use Eq. (27.12) to solve for the field magnitude B.

EXECUTE: The angular speed that corresponds to the frequency f is $\omega = 2\pi f = (2\pi)(2450 \times 10^6 \text{ s}^{-1}) = 1.54 \times 10^{10} \text{ s}^{-1}$. Then from Eq. (27.12),

$$B = \frac{m\omega}{|q|} = \frac{(9.11 \times 10^{-31} \text{ kg})(1.54 \times 10^{10} \text{ s}^{-1})}{1.60 \times 10^{-19} \text{ C}} = 0.0877 \text{ T}$$

EVALUATE: This is a moderate field strength, easily produced with a permanent magnet. Incidentally, 2450-MHz electromagnetic waves are useful for heating and cooking food because they are strongly absorbed by water molecules.

EXAMPLE 27.4   HELICAL PARTICLE MOTION IN A MAGNETIC FIELD

In a situation like that shown in Fig. 27.18, the charged particle is a proton ($q = 1.60 \times 10^{-19}$ C, $m = 1.67 \times 10^{-27}$ kg) and the uniform, 0.500-T magnetic field is directed along the x-axis. At $t = 0$ the proton has velocity components $v_x = 1.50 \times 10^5$ m/s, $v_y = 0$, and $v_z = 2.00 \times 10^5$ m/s. Only the magnetic force acts on the proton. (a) At $t = 0$, find the force on the proton and its acceleration. (b) Find the radius of the resulting helical path, the angular speed of the proton, and the *pitch* of the helix (the distance traveled along the helix axis per revolution).

SOLUTION

IDENTIFY and SET UP: The magnetic force is $\vec{F} = q\vec{v} \times \vec{B}$; Newton's second law gives the resulting acceleration. Because $\vec{F}$ is perpendicular to $\vec{v}$, the proton's speed does not change. Hence Eq. (27.11) gives the radius of the helical path if we replace v with the velocity component perpendicular to $\vec{B}$. Equation (27.12) gives the angular speed ω, which yields the time T for one revolution (the *period*). Given the velocity component parallel to the magnetic field, we can then determine the pitch.

EXECUTE: (a) With $\vec{B} = B\hat{\imath}$ and $\vec{v} = v_x\hat{\imath} + v_z\hat{k}$, Eq. (27.2) yields

$$\vec{F} = q\vec{v} \times \vec{B} = q(v_x\hat{\imath} + v_z\hat{k}) \times B\hat{\imath} = qv_zB\hat{\jmath}$$

$$= (1.60 \times 10^{-19} \text{ C})(2.00 \times 10^5 \text{ m/s})(0.500 \text{ T})\hat{\jmath}$$

$$= (1.60 \times 10^{-14} \text{ N})\hat{\jmath}$$

(Recall: $\hat{\imath} \times \hat{\imath} = 0$ and $\hat{k} \times \hat{\imath} = \hat{\jmath}$.) The resulting acceleration is

$$\vec{a} = \frac{\vec{F}}{m} = \frac{1.60 \times 10^{-14} \text{ N}}{1.67 \times 10^{-27} \text{ kg}}\hat{\jmath} = (9.58 \times 10^{12} \text{ m/s}^2)\hat{\jmath}$$

(b) Since $v_y = 0$, the component of velocity perpendicular to $\vec{B}$ is v_z; then from Eq. (27.11),

$$R = \frac{mv_z}{|q|B} = \frac{(1.67 \times 10^{-27} \text{ kg})(2.00 \times 10^5 \text{ m/s})}{(1.60 \times 10^{-19} \text{ C})(0.500 \text{ T})}$$

$$= 4.18 \times 10^{-3} \text{ m} = 4.18 \text{ mm}$$

From Eq. (27.12) the angular speed is

$$\omega = \frac{|q|B}{m} = \frac{(1.60 \times 10^{-19} \text{ C})(0.500 \text{ T})}{1.67 \times 10^{-27} \text{ kg}} = 4.79 \times 10^7 \text{ rad/s}$$

The period is $T = 2\pi/\omega = 2\pi/(4.79 \times 10^7 \text{ s}^{-1}) = 1.31 \times 10^{-7}$ s. The pitch is the distance traveled along the x-axis in this time, or

$$v_xT = (1.50 \times 10^5 \text{ m/s})(1.31 \times 10^{-7} \text{ s})$$

$$= 0.0197 \text{ m} = 19.7 \text{ mm}$$

EVALUATE: Although the magnetic force has a tiny magnitude, it produces an immense acceleration because the proton mass is so small. Note that the pitch of the helix is almost five times greater than the radius R, so this helix is much more "stretched out" than that shown in Fig. 27.18.

TEST YOUR UNDERSTANDING OF SECTION 27.4   (a) If you double the speed of the charged particle in Fig. 27.17a while keeping the magnetic field the same (as well as the charge and the mass), how does this affect the radius of the trajectory? (i) The radius is unchanged; (ii) the radius is twice as large; (iii) the radius is four times as large; (iv) the radius is $\frac{1}{2}$ as large; (v) the radius is $\frac{1}{4}$ as large. (b) How does this affect the time required for one complete circular orbit? (i) The time is unchanged; (ii) the time is twice as long; (iii) the time is four times as long; (iv) the time is $\frac{1}{2}$ as long; (v) the time is $\frac{1}{4}$ as long. ▮

27.5 APPLICATIONS OF MOTION OF CHARGED PARTICLES

This section describes several applications of the principles introduced in this chapter. Study them carefully, watching for applications of Problem-Solving Strategy 27.2 (Section 27.4).

Velocity Selector

27.22 (a) A velocity selector for charged particles uses perpendicular $\vec{E}$ and $\vec{B}$ fields. Only charged particles with $v = E/B$ move through undeflected. (b) The electric and magnetic forces on a positive charge. The forces are reversed if the charge is negative.

(a) Schematic diagram of velocity selector

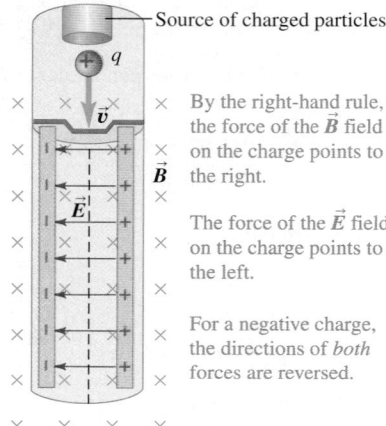

- Source of charged particles

By the right-hand rule, the force of the $\vec{B}$ field on the charge points to the right.

The force of the $\vec{E}$ field on the charge points to the left.

For a negative charge, the directions of *both* forces are reversed.

(b) Free-body diagram for a positive particle

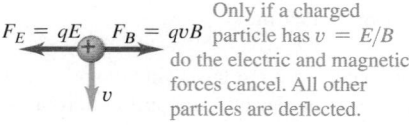

$F_E = qE \qquad F_B = qvB$

Only if a charged particle has $v = E/B$ do the electric and magnetic forces cancel. All other particles are deflected.

In a beam of charged particles produced by a heated cathode or a radioactive material, not all particles move with the same speed. Many applications, however, require a beam in which all the particle speeds are the same. Particles of a specific speed can be selected from the beam by using an arrangement of electric and magnetic fields called a *velocity selector*. In **Fig. 27.22a** a charged particle with mass m, charge q, and speed v enters a region of space where the electric and magnetic fields are perpendicular to the particle's velocity and to each other. The electric field $\vec{E}$ is to the left, and the magnetic field $\vec{B}$ is into the plane of the figure. If q is positive, the electric force is to the left, with magnitude qE, and the magnetic force is to the right, with magnitude qvB. For given field magnitudes E and B, for a particular value of v the electric and magnetic forces will be equal in magnitude; the total force is then zero, and the particle travels in a straight line with constant velocity. This will be the case if $qE = qvB$ (Fig. 27.22b), so the speed v for which there is no deflection is

$$v = \frac{E}{B} \qquad (27.13)$$

Only particles with speeds equal to E/B can pass through without being deflected by the fields. By adjusting E and B appropriately, we can select particles having a particular speed for use in other experiments. Because q divides out in Eq. (27.13), a velocity selector for positively charged particles also works for electrons or other negatively charged particles.

Thomson's *e/m* Experiment

In one of the landmark experiments in physics at the end of the 19th century, J. J. Thomson (1856–1940) used the idea just described to measure the ratio of charge to mass for the electron. For this experiment, carried out in 1897 at the Cavendish Laboratory in Cambridge, England, Thomson used the apparatus shown in **Fig. 27.23**. In a highly evacuated glass container, electrons from the hot cathode are accelerated and formed into a beam by a potential difference V between the two anodes A and A'. The speed v of the electrons is determined

27.23 Thomson's apparatus for measuring the ratio e/m for the electron.

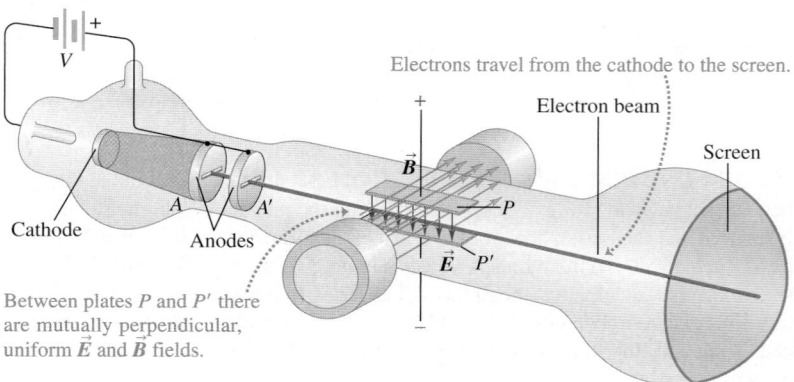

Electrons travel from the cathode to the screen.

Electron beam

Screen

Cathode

Anodes

Between plates P and P' there are mutually perpendicular, uniform $\vec{E}$ and $\vec{B}$ fields.

by the accelerating potential V. The gained kinetic energy $\frac{1}{2}mv^2$ equals the lost electric potential energy eV, where e is the magnitude of the electron charge:

$$\frac{1}{2}mv^2 = eV \qquad \text{or} \qquad v = \sqrt{\frac{2eV}{m}} \qquad (27.14)$$

The electrons pass between the plates P and P' and strike the screen at the end of the tube, which is coated with a material that fluoresces (glows) at the point of impact. The electrons pass straight through the plates when Eq. (27.13) is satisfied; combining this with Eq. (27.14), we get

$$\frac{E}{B} = \sqrt{\frac{2eV}{m}} \qquad \text{so} \qquad \frac{e}{m} = \frac{E^2}{2VB^2} \qquad (27.15)$$

All the quantities on the right side can be measured, so the ratio e/m of charge to mass can be determined. It is *not* possible to measure e or m separately by this method, only their ratio.

The most significant aspect of Thomson's e/m measurements was that he found a *single value* for this quantity. It did not depend on the cathode material, the residual gas in the tube, or anything else about the experiment. This independence showed that the particles in the beam, which we now call electrons, are a common constituent of all matter. Thus Thomson is credited with the first discovery of a subatomic particle, the electron.

The most precise value of e/m available as of this writing is

$$e/m = 1.758820088(39) \times 10^{11} \text{ C/kg}$$

In this expression, (39) indicates the likely uncertainty in the last two digits, 88.

Fifteen years after Thomson's experiments, the American physicist Robert Millikan succeeded in measuring the charge of the electron precisely (see Problem 23.81). This value, together with the value of e/m, enables us to determine the *mass* of the electron. The most precise value available at present is

$$m = 9.10938291(40) \times 10^{-31} \text{ kg}$$

Mass Spectrometers

Techniques similar to Thomson's e/m experiment can be used to measure masses of ions and thus measure atomic and molecular masses. In 1919, Francis Aston (1877–1945), a student of Thomson's, built the first of a family of instruments called **mass spectrometers.** A variation built by Bainbridge is shown in **Fig. 27.24**. Positive ions from a source pass through the slits S_1 and S_2, forming a narrow beam. Then the ions pass through a velocity selector with crossed $\vec{E}$ and $\vec{B}$ fields, as we have described, to block all ions except those with speeds v equal to E/B. Finally, the ions pass into a region with a magnetic field $\vec{B}'$ perpendicular to the figure, where they move in circular arcs with radius R determined by Eq. (27.11): $R = mv/qB'$. Ions with different masses strike the detector at different points, and the values of R can be measured. We assume that each ion has lost one electron, so the net charge of each ion is just $+e$. With everything known in this equation except m, we can compute the mass m of the ion.

One of the earliest results from this work was the discovery that neon has two species of atoms, with atomic masses 20 and 22 g/mol. We now call these species **isotopes** of the element. Later experiments have shown that many elements have several isotopes—atoms with identical chemical behaviors but different masses due to differing numbers of neutrons in their nuclei. This is just one of the many applications of mass spectrometers in chemistry and physics.

27.24 Bainbridge's mass spectrometer utilizes a velocity selector to produce particles with uniform speed v. In the region of magnetic field B', particles with greater mass $(m_2 > m_1)$ travel in paths with larger radius $(R_2 > R_1)$.

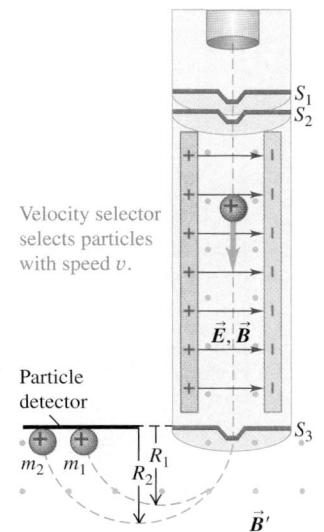

Velocity selector selects particles with speed v.

Particle detector

Magnetic field separates particles by mass; the greater a particle's mass, the larger is the radius of its path.

EXAMPLE 27.5 AN *e/m* DEMONSTRATION EXPERIMENT

You set out to reproduce Thomson's *e/m* experiment with an accelerating potential of 150 V and a deflecting electric field of magnitude 6.0×10^6 N/C. (a) How fast do the electrons move? (b) What magnetic-field magnitude will yield zero beam deflection? (c) With this magnetic field, how will the electron beam behave if you increase the accelerating potential above 150 V?

SOLUTION

IDENTIFY and SET UP: This is the situation shown in Fig. 27.23. We use Eq. (27.14) to determine the electron speed and Eq. (27.13) to determine the required magnetic field B.

EXECUTE: (a) From Eq. (27.14), the electron speed v is

$$v = \sqrt{2(e/m)V} = \sqrt{2(1.76 \times 10^{11} \text{ C/kg})(150 \text{ V})}$$
$$= 7.27 \times 10^6 \text{ m/s} = 0.024c$$

(b) From Eq. (27.13), the required field strength is

$$B = \frac{E}{v} = \frac{6.0 \times 10^6 \text{ N/C}}{7.27 \times 10^6 \text{ m/s}} = 0.83 \text{ T}$$

(c) Increasing the accelerating potential V increases the electron speed v. In Fig. 27.23 this doesn't change the upward electric force eE, but it increases the downward magnetic force evB. Therefore the electron beam will turn *downward* and will hit the end of the tube below the undeflected position.

EVALUATE: The required magnetic field is relatively large because the electrons are moving fairly rapidly (2.4% of the speed of light). If the maximum available magnetic field is less than 0.83 T, the electric field strength E would have to be reduced to maintain the desired ratio E/B in Eq. (27.15).

EXAMPLE 27.6 FINDING LEAKS IN A VACUUM SYSTEM

There is almost no helium in ordinary air, so helium sprayed near a leak in a vacuum system will quickly show up in the output of a vacuum pump connected to such a system. You are designing a leak detector that uses a mass spectrometer to detect He^+ ions (charge $+e = +1.60 \times 10^{-19}$ C, mass 6.65×10^{-27} kg). Ions emerge from the velocity selector with a speed of 1.00×10^5 m/s. They are curved in a semicircular path by a magnetic field B' and are detected at a distance of 10.16 cm from the slit S_3 in Fig. 27.24. Calculate the magnitude of the magnetic field B'.

SOLUTION

IDENTIFY and SET UP: After it passes through the slit, the ion follows a circular path as described in Section 27.4 (see Fig. 27.17). We solve Eq. (27.11) for B'.

EXECUTE: The distance given is the *diameter* of the semicircular path shown in Fig. 27.24, so the radius is $R = \frac{1}{2}(10.16 \times 10^{-2} \text{ m})$. From Eq. (27.11), $R = mv/qB'$, we get

$$B' = \frac{mv}{qR} = \frac{(6.65 \times 10^{-27} \text{ kg})(1.00 \times 10^5 \text{ m/s})}{(1.60 \times 10^{-19} \text{ C})(5.08 \times 10^{-2} \text{ m})}$$
$$= 0.0818 \text{ T}$$

EVALUATE: Helium leak detectors are widely used with high-vacuum systems. Our result shows that only a small magnetic field is required, so leak detectors can be relatively compact.

TEST YOUR UNDERSTANDING OF SECTION 27.5 In Example 27.6 He^+ ions with charge $+e$ move at 1.00×10^5 m/s in a straight line through a velocity selector. Suppose the He^+ ions were replaced with He^{2+} ions, in which both electrons have been removed from the helium atom and the ion charge is $+2e$. At what speed must the He^{2+} ions travel through the same velocity selector in order to move in a straight line? (i) 4.00×10^5 m/s; (ii) 2.00×10^5 m/s; (iii) 1.00×10^5 m/s; (iv) 0.50×10^5 m/s; (v) 0.25×10^5 m/s. ∎

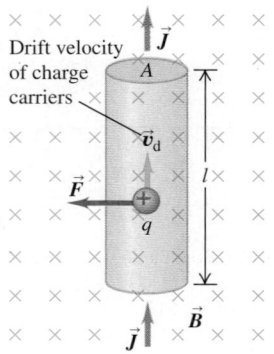

27.25 Forces on a moving positive charge in a current-carrying conductor.

27.6 MAGNETIC FORCE ON A CURRENT-CARRYING CONDUCTOR

What makes an electric motor work? Within the motor are conductors that carry currents (that is, whose charges are in motion), as well as magnets that exert forces on the moving charges. Hence there is a magnetic force on each current-carrying conductor, and these forces make the motor turn. The d'Arsonval galvanometer (Section 26.3) also uses magnetic forces on conductors.

We can compute the force on a current-carrying conductor starting with the magnetic force $\vec{F} = q\vec{v} \times \vec{B}$ on a single moving charge. **Figure 27.25** shows a straight segment of a conducting wire, with length l and cross-sectional area A;

the current is from bottom to top. The wire is in a uniform magnetic field $\vec{B}$, perpendicular to the plane of the diagram and directed *into* the plane. Let's assume first that the moving charges are positive. Later we'll see what happens when they are negative.

DEMO

The drift velocity $\vec{v}_d$ is upward, perpendicular to $\vec{B}$. The average force on each charge is $\vec{F} = q\vec{v}_d \times \vec{B}$, directed to the left as shown in the figure; since $\vec{v}_d$ and $\vec{B}$ are perpendicular, the magnitude of the force is $F = qv_dB$.

We can derive an expression for the *total* force on all the moving charges in a length l of conductor with cross-sectional area A by using the same language we used in Eqs. (25.2) and (25.3) of Section 25.1. The number of charges per unit volume, or charge concentration, is n; a segment of conductor with length l has volume Al and contains a number of charges equal to nAl. The total force $\vec{F}$ on *all* the moving charges in this segment has magnitude

$$F = (nAl)(qv_dB) = (nqv_dA)(lB) \qquad (27.16)$$

From Eq. (25.3) the current density is $J = nqv_d$. The product JA is the total current I, so we can rewrite Eq. (27.16) as

$$F = IlB \qquad (27.17)$$

If the $\vec{B}$ field is not perpendicular to the wire but makes an angle ϕ with it, as in **Fig. 27.26**, we handle the situation the same way we did in Section 27.2 for a single charge. Only the component of $\vec{B}$ perpendicular to the wire (and to the drift velocities of the charges) exerts a force; this component is $B_\perp = B\sin\phi$. The magnetic force on the wire segment is then

$$F = IlB_\perp = IlB\sin\phi \qquad (27.18)$$

The force is always perpendicular to both the conductor and the field, with the direction determined by the same right-hand rule we used for a moving positive charge (Fig. 27.26). Hence this force can be expressed as a vector product, like the force on a single moving charge. We represent the segment of wire with a vector $\vec{l}$ along the wire in the direction of the current; then the force $\vec{F}$ on this segment is

| Magnetic force on a straight wire segment | $\cdots\blacktriangleright \vec{F} = I\vec{l} \times \vec{B} \blacktriangleleft\cdots$ Magnetic field | (27.19) |

Current

Vector length of segment (points in current direction)

Figure 27.27 illustrates the directions of $\vec{B}$, $\vec{l}$, and $\vec{F}$ for several cases.

If the conductor is not straight, we can divide it into infinitesimal segments $d\vec{l}$. The force $d\vec{F}$ on each segment is

| Magnetic force on an infinitesimal wire segment | $\cdots\blacktriangleright d\vec{F} = I\,d\vec{l} \times \vec{B} \blacktriangleleft\cdots$ Magnetic field | (27.20) |

Current

Vector length of segment (points in current direction)

Then we can integrate this expression along the wire to find the total force on a conductor of any shape. The integral is a *line integral*, the same mathematical operation we have used to define work (Section 6.3) and electric potential (Section 23.2).

CAUTION **Current is not a vector** Recall from Section 25.1 that the current I is not a vector. The direction of current flow is described by $d\vec{l}$, not I. If the conductor is curved, I is the same at all points along its length, but $d\vec{l}$ changes direction—it is always tangent to the conductor. ∎

27.26 A straight wire segment of length $\vec{l}$ carries a current I in the direction of $\vec{l}$. The magnetic force on this segment is perpendicular to both $\vec{l}$ and the magnetic field $\vec{B}$.

Force $\vec{F}$ on a straight wire carrying a positive current and oriented at an angle ϕ to a magnetic field $\vec{B}$:

• Magnitude is $F = IlB_\perp = IlB\sin\phi$.
• Direction of $\vec{F}$ is given by the right-hand rule.

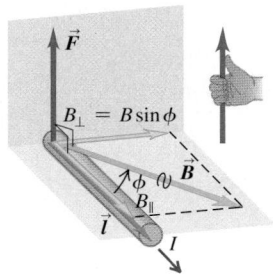

27.27 Magnetic field $\vec{B}$, length $\vec{l}$, and force $\vec{F}$ vectors for a straight wire carrying a current I.

(a)

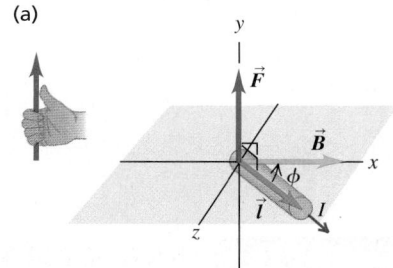

(b)

Reversing $\vec{B}$ reverses the force direction.

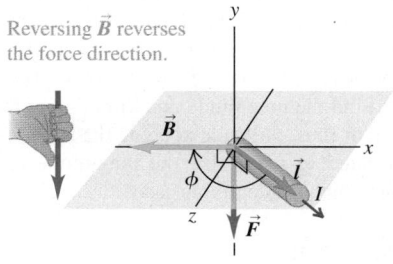

(c)

Reversing the current [relative to (b)] reverses the force direction.

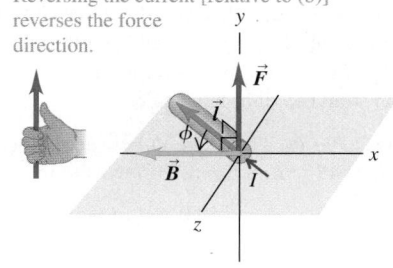

27.28 (a) Components of a loudspeaker. (b) The permanent magnet creates a magnetic field that exerts forces on the current in the voice coil; for a current I in the direction shown, the force is to the right. If the electric current in the voice coil oscillates, the speaker cone attached to the voice coil oscillates at the same frequency.

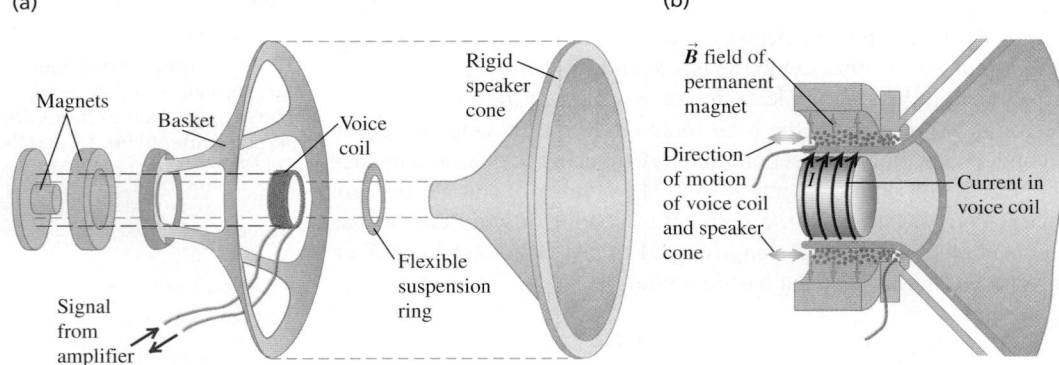

(a)

(b)

Finally, what happens when the moving charges are negative, such as electrons in a metal? Then in Fig. 27.25 an upward current corresponds to a downward drift velocity. But because q is now negative, the direction of the force $\vec{F}$ is the same as before. Thus Eqs. (27.17) through (27.20) are valid for *both* positive and negative charges and even when *both* signs of charge are present at once. This happens in some semiconductor materials and in ionic solutions.

A common application of the magnetic forces on a current-carrying wire is found in loudspeakers (**Fig. 27.28**). The radial magnetic field created by the permanent magnet exerts a force on the voice coil that is proportional to the current in the coil; the direction of the force is either to the left or to the right, depending on the direction of the current. The signal from the amplifier causes the current to oscillate in direction and magnitude. The coil and the speaker cone to which it is attached respond by oscillating with an amplitude proportional to the amplitude of the current in the coil. Turning up the volume knob on the amplifier increases the current amplitude and hence the amplitudes of the cone's oscillation and of the sound wave produced by the moving cone.

EXAMPLE 27.7 **MAGNETIC FORCE ON A STRAIGHT CONDUCTOR**

A straight horizontal copper rod carries a current of 50.0 A from west to east in a region between the poles of a large electromagnet. In this region there is a horizontal magnetic field toward the northeast (that is, 45° north of east) with magnitude 1.20 T. (a) Find the magnitude and direction of the force on a 1.00-m section of rod. (b) While keeping the rod horizontal, how should it be oriented to maximize the magnitude of the force? What is the force magnitude in this case?

SOLUTION

IDENTIFY and SET UP: Figure 27.29 shows the situation. This is a straight wire segment in a uniform magnetic field, as in Fig. 27.26. Our target variables are the force $\vec{F}$ on the segment and the angle ϕ for which the force magnitude F is greatest. We find the magnitude of the magnetic force from Eq. (27.18) and the direction from the right-hand rule.

EXECUTE: (a) The angle ϕ between the directions of current and field is 45°. From Eq. (27.18) we obtain

$$F = IlB\sin\phi = (50.0\text{ A})(1.00\text{ m})(1.20\text{ T})(\sin 45°) = 42.4\text{ N}$$

27.29 Our sketch of the copper rod as seen from overhead.

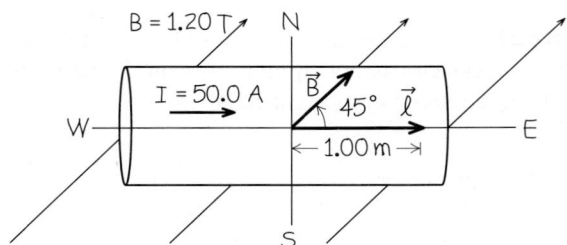

The direction of the force is perpendicular to the plane of the current and the field, both of which lie in the horizontal plane. Thus the force must be vertical; the right-hand rule shows that it is vertically *upward* (out of the plane of the figure).

(b) From $F = IlB\sin\phi$, F is maximum for $\phi = 90°$, so that $\vec{l}$ and $\vec{B}$ are perpendicular. To keep $\vec{F} = I\vec{l} \times \vec{B}$ upward, we rotate the rod *clockwise* by 45° from its orientation in Fig. 27.29, so that the current runs toward the southeast. Then $F = IlB = (50.0\text{ A})(1.00\text{ m})(1.20\text{ T}) = 60.0\text{ N}$.

EVALUATE: We check the result in part (a) by using Eq. (27.19) to calculate the force vector. If we use a coordinate system with the x-axis pointing east, the y-axis north, and the z-axis upward, we have $\vec{l} = (1.00 \text{ m})\hat{\imath}$, $\vec{B} = (1.20 \text{ T})[(\cos 45°)\hat{\imath} + (\sin 45°)\hat{\jmath}]$, and

$$\vec{F} = I\vec{l} \times \vec{B}$$
$$= (50.0 \text{ A})(1.00 \text{ m})\hat{\imath} \times (1.20 \text{ T})[(\cos 45°)\hat{\imath} + (\sin 45°)\hat{\jmath}]$$
$$= (42.4 \text{ N})\hat{k}$$

Note that the maximum upward force of 60.0 N can hold the conductor in midair against the force of gravity—that is, *magnetically levitate* the conductor—if its weight is 60.0 N and its mass is $m = w/g = (60.0 \text{ N})/(9.8 \text{ m/s}^2) = 6.12 \text{ kg}$. Magnetic levitation is used in some high-speed trains to suspend the train over the tracks. Eliminating rolling friction in this way allows the train to achieve speeds of over 400 km/h.

EXAMPLE 27.8 | MAGNETIC FORCE ON A CURVED CONDUCTOR

In **Fig. 27.30** the magnetic field $\vec{B}$ is uniform and perpendicular to the plane of the figure, pointing out of the page. The conductor, carrying current I to the left, has three segments: (1) a straight segment with length L perpendicular to the plane of the figure, (2) a semicircle with radius R, and (3) another straight segment with length L parallel to the x-axis. Find the total magnetic force on this conductor.

SOLUTION

IDENTIFY and SET UP: The magnetic field $\vec{B} = B\hat{k}$ is uniform, so we find the forces $\vec{F}_1$ and $\vec{F}_3$ on the straight segments (1) and (3) from Eq. (27.19). We divide the curved segment (2) into infinitesimal straight segments and find the corresponding force $d\vec{F}_2$ on each straight segment from Eq. (27.20). We then integrate to find $\vec{F}_2$. The total magnetic force on the conductor is then $\vec{F} = \vec{F}_1 + \vec{F}_2 + \vec{F}_3$.

27.30 What is the total magnetic force on the conductor?

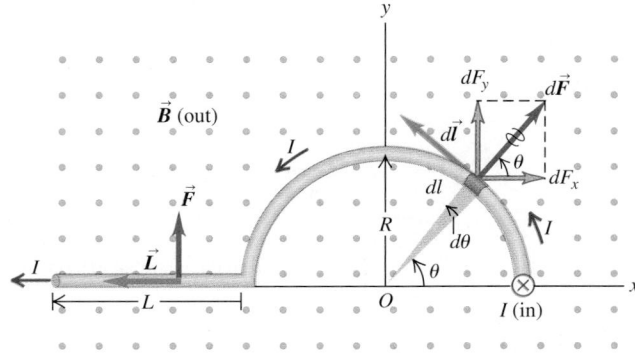

EXECUTE: For segment (1), $\vec{L} = -L\hat{k}$. Hence from Eq. (27.19), $\vec{F}_1 = I\vec{L} \times \vec{B} = 0$. For segment (3), $\vec{L} = -L\hat{\imath}$, so $\vec{F}_3 = I\vec{L} \times \vec{B} = I(-L\hat{\imath}) \times (B\hat{k}) = ILB\hat{\jmath}$.

For the curved segment (2), Fig. 27.30 shows a segment $d\vec{l}$ with length $dl = R\, d\theta$, at angle θ. The right-hand rule shows that the direction of $d\vec{l} \times \vec{B}$ is radially outward from the center; make sure you can verify this. Because $d\vec{l}$ and $\vec{B}$ are perpendicular, the magnitude dF_2 of the force on the segment $d\vec{l}$ is $dF_2 = I\, dl\, B = I(R\, d\theta)B$. The components of the force on this segment are

$$dF_{2x} = IR\, d\theta\, B \cos\theta \qquad dF_{2y} = IR\, d\theta\, B \sin\theta$$

To find the components of the total force, we integrate these expressions with respect to θ from $\theta = 0$ to $\theta = \pi$ to take in the whole semicircle. The results are

$$F_{2x} = IRB \int_0^\pi \cos\theta\, d\theta = 0$$

$$F_{2y} = IRB \int_0^\pi \sin\theta\, d\theta = 2IRB$$

Hence $\vec{F}_2 = 2IRB\hat{\jmath}$. Finally, adding the forces on all three segments, we find that the total force is in the positive y-direction:

$$\vec{F} = \vec{F}_1 + \vec{F}_2 + \vec{F}_3 = 0 + 2IRB\hat{\jmath} + ILB\hat{\jmath} = IB(2R + L)\hat{\jmath}$$

EVALUATE: We could have predicted from symmetry that the x-component of $\vec{F}_2$ would be zero: On the right half of the semicircle the x-component of the force is positive (to the right) and on the left half it is negative (to the left); the positive and negative contributions to the integral cancel. The result is that $\vec{F}_2$ is the force that would be exerted if we replaced the semicircle with a *straight* segment of length $2R$ along the x-axis. Do you see why?

TEST YOUR UNDERSTANDING OF SECTION 27.6 The accompanying figure shows a top view of two conducting rails on which a conducting bar can slide. A uniform magnetic field is directed perpendicular to the plane of the figure as shown. A battery is to be connected to the two rails so that when the switch is closed, current will flow through the bar and cause a magnetic force to push the bar to the right. In which orientation, A or B, should the battery be placed in the circuit? ∎

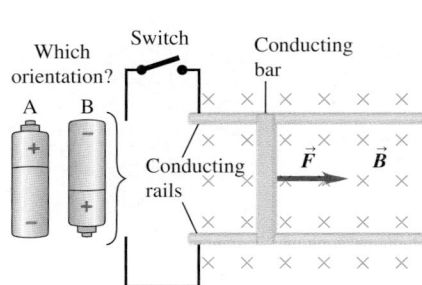

27.7 FORCE AND TORQUE ON A CURRENT LOOP

Current-carrying conductors usually form closed loops, so it is worthwhile to use the results of Section 27.6 to find the *total* magnetic force and torque on a conductor in the form of a loop. Many practical devices make use of the magnetic force or torque on a conducting loop, including loudspeakers (see Fig. 27.28) and galvanometers (see Section 26.3). Hence the results of this section are of substantial practical importance. These results will also help us understand the behavior of bar magnets described in Section 27.1.

As an example, let's look at a rectangular current loop in a uniform magnetic field. We can represent the loop as a series of straight line segments. We will find that the total *force* on the loop is zero but that there can be a net *torque* acting on the loop, with some interesting properties.

Figure 27.31a shows a rectangular loop of wire with side lengths a and b. A line perpendicular to the plane of the loop (i.e., a *normal* to the plane) makes an angle ϕ with the direction of the magnetic field $\vec{B}$, and the loop carries a current I. The wires leading the current into and out of the loop and the source of emf are omitted to keep the diagram simple.

The force $\vec{F}$ on the right side of the loop (length a) is to the right, in the $+x$-direction as shown. On this side, $\vec{B}$ is perpendicular to the current direction, and the force on this side has magnitude

$$F = IaB \tag{27.21}$$

A force $-\vec{F}$ with the same magnitude but opposite direction acts on the opposite side of the loop, as shown in the figure.

The sides with length b make an angle $(90° - \phi)$ with the direction of $\vec{B}$. The forces on these sides are the vectors $\vec{F}'$ and $-\vec{F}'$; their magnitude F' is given by

$$F' = IbB\sin(90° - \phi) = IbB\cos\phi$$

The lines of action of both forces lie along the y-axis.

27.31 Finding the torque on a current-carrying loop in a uniform magnetic field.

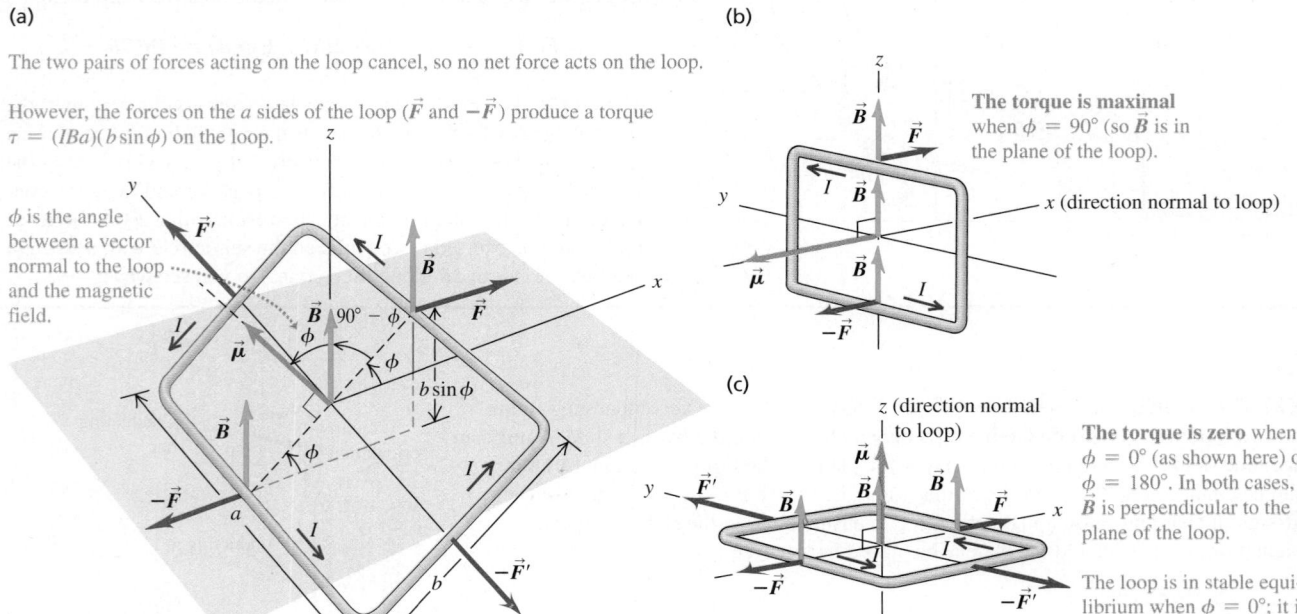

(a)

The two pairs of forces acting on the loop cancel, so no net force acts on the loop.

However, the forces on the a sides of the loop ($\vec{F}$ and $-\vec{F}$) produce a torque $\tau = (IBa)(b\sin\phi)$ on the loop.

ϕ is the angle between a vector normal to the loop and the magnetic field.

(b)

The torque is maximal when $\phi = 90°$ (so $\vec{B}$ is in the plane of the loop).

x (direction normal to loop)

(c)

z (direction normal to loop)

The torque is zero when $\phi = 0°$ (as shown here) or $\phi = 180°$. In both cases, $\vec{B}$ is perpendicular to the plane of the loop.

The loop is in stable equilibrium when $\phi = 0°$; it is in unstable equilibrium when $\phi = 180°$.

The *total* force on the loop is zero because the forces on opposite sides cancel out in pairs.

> **The net force on a current loop in a uniform magnetic field is zero. However, the net torque is not in general equal to zero.**

(You may find it helpful to review the discussion of torque in Section 10.1.) The two forces $\vec{F}'$ and $-\vec{F}'$ in Fig. 27.31a lie along the same line and so give rise to zero net torque with respect to any point. The two forces $\vec{F}$ and $-\vec{F}$ lie along different lines, and each gives rise to a torque about the y-axis. According to the right-hand rule for determining the direction of torques, the vector torques due to $\vec{F}$ and $-\vec{F}$ are both in the $+y$-direction; hence the net vector torque $\vec{\tau}$ is in the $+y$-direction as well. The moment arm for each of these forces (equal to the perpendicular distance from the rotation axis to the line of action of the force) is $(b/2)\sin\phi$, so the torque due to each force has magnitude $F(b/2)\sin\phi$. If we use Eq. (27.21) for F, the magnitude of the net torque is

$$\tau = 2F(b/2)\sin\phi = (IBa)(b\sin\phi) \tag{27.22}$$

The torque is greatest when $\phi = 90°$, $\vec{B}$ is in the plane of the loop, and the normal to this plane is perpendicular to $\vec{B}$ (Fig. 27.31b). The torque is zero when ϕ is $0°$ or $180°$ and the normal to the loop is parallel or antiparallel to the field (Fig. 27.31c). The value $\phi = 0°$ is a stable equilibrium position because the torque is zero there, and when the loop is rotated slightly from this position, the resulting torque tends to rotate it back toward $\phi = 0°$. The position $\phi = 180°$ is an *unstable* equilibrium position; if displaced slightly from this position, the loop tends to move farther away from $\phi = 180°$. Figure 27.31 shows rotation about the y-axis, but because the net force on the loop is zero, Eq. (27.22) for the torque is valid for *any* choice of axis. The torque always tends to rotate the loop in the direction of *decreasing* ϕ—that is, toward the stable equilibrium position $\phi = 0°$.

The area A of the loop is equal to ab, so we can rewrite Eq. (27.22) as

> **Magnitude of magnetic torque on a current loop** $\cdots\blacktriangleright \tau = IBA\sin\phi$
> Current $\cdots$ Magnetic-field magnitude
> Angle between normal to loop plane and field direction
> Area of loop
> $\tag{27.23}$

The product IA is called the **magnetic dipole moment** or **magnetic moment** of the loop, for which we use the symbol μ (the Greek letter mu):

$$\mu = IA \tag{27.24}$$

It is analogous to the electric dipole moment introduced in Section 21.7. In terms of μ, the magnitude of the torque on a current loop is

$$\tau = \mu B\sin\phi \tag{27.25}$$

where ϕ is the angle between the normal to the loop (the direction of the vector area $\vec{A}$) and $\vec{B}$. A current loop, or any other body that experiences a magnetic torque given by Eq. (27.25), is also called a **magnetic dipole.**

Magnetic Torque: Vector Form

We can also define a vector magnetic moment $\vec{\mu}$ with magnitude IA: This is shown in Fig. 27.31. The direction of $\vec{\mu}$ is defined to be perpendicular to the plane of the loop, with a sense determined by a right-hand rule, as shown in **Fig. 27.32**. Wrap the fingers of your right hand around the perimeter of the loop in the direction of the current. Then extend your thumb so that it is perpendicular to the plane of the loop; its direction is the direction of $\vec{\mu}$ (and of the vector area $\vec{A}$ of the loop). The torque is greatest when $\vec{\mu}$ and $\vec{B}$ are perpendicular and is zero when they are parallel or antiparallel. In the stable equilibrium position, $\vec{\mu}$ and $\vec{B}$ are parallel.

BIO Application Magnetic Resonance Imaging In magnetic resonance imaging (MRI), a patient is placed in a strong magnetic field. Each hydrogen nucleus in the patient acts like a miniature current loop with a magnetic dipole moment that tends to align with the applied field. Radio waves of just the right frequency then flip these magnetic moments out of alignment. The extent to which the radio waves are absorbed is proportional to the amount of hydrogen present. This makes it possible to image details in hydrogen-rich soft tissue that cannot be seen in x-ray images. (X rays are superior to MRI for imaging bone, which is hydrogen deficient.)

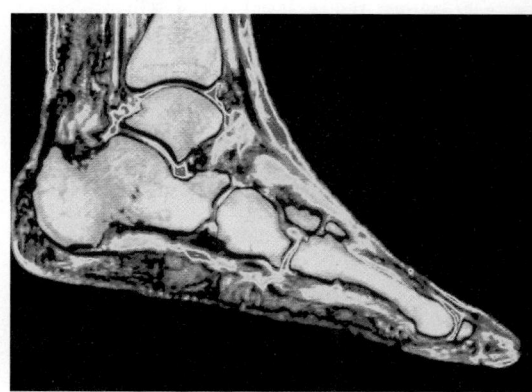

27.32 The right-hand rule determines the direction of the magnetic moment of a current-carrying loop. This is also the direction of the loop's area vector $\vec{A}$; $\vec{\mu} = I\vec{A}$ is a vector equation.

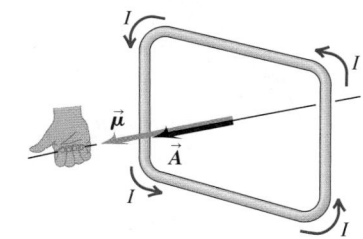

Finally, we can express this interaction in terms of the torque vector $\vec{\tau}$, which we used for *electric*-dipole interactions in Section 21.7. From Eq. (27.25) the magnitude of $\vec{\tau}$ is equal to the magnitude of $\vec{\mu} \times \vec{B}$, and reference to Fig. 27.31 shows that the directions are also the same. So we have

Vector magnetic torque on a current loop ········ Magnetic dipole moment

$$\vec{\tau} = \vec{\mu} \times \vec{B} \quad \text{Magnetic field} \qquad (27.26)$$

This result is directly analogous to the result we found in Section 21.7 for the torque exerted by an *electric* field $\vec{E}$ on an *electric* dipole with dipole moment $\vec{p}$.

Potential Energy for a Magnetic Dipole

When a magnetic dipole changes orientation in a magnetic field, the field does work on it. In an infinitesimal angular displacement $d\phi$, the work dW is given by $\tau\, d\phi$, and there is a corresponding change in potential energy. As the above discussion suggests, the potential energy U is least when $\vec{\mu}$ and $\vec{B}$ are parallel and greatest when they are antiparallel. To find an expression for U as a function of orientation, note that the torque on an *electric* dipole in an *electric* field is $\vec{\tau} = \vec{p} \times \vec{E}$; we found in Section 21.7 that the corresponding potential energy is $U = -\vec{p} \cdot \vec{E}$. The torque on a *magnetic* dipole in a *magnetic* field is $\vec{\tau} = \vec{\mu} \times \vec{B}$, so we can conclude immediately that

Potential energy for a magnetic dipole in a magnetic field ········ Magnetic dipole moment

$$U = -\vec{\mu} \cdot \vec{B} = -\mu B \cos\phi \quad \text{Angle between } \vec{\mu} \text{ and } \vec{B} \qquad (27.27)$$

Magnetic field

With this definition, U is zero when the magnetic dipole moment is perpendicular to the magnetic field ($\phi = 90°$); then $U = -\vec{\mu} \cdot \vec{B} = -\mu B \cos 90° = 0$.

Magnetic Torque: Loops and Coils

Although we have derived Eqs. (27.21) through (27.27) for a rectangular current loop, all these relationships are valid for a plane loop of any shape at all. Any planar loop may be approximated as closely as we wish by a very large number of rectangular loops, as shown in **Fig. 27.33**. If these loops all carry equal currents in the same clockwise sense, then the forces and torques on the sides of two loops adjacent to each other cancel, and the only forces and torques that do not cancel are due to currents around the boundary. Thus all the above relationships are valid for a plane current loop of any shape, with the magnetic moment $\vec{\mu} = I\vec{A}$.

We can also generalize this whole formulation to a coil consisting of N planar loops close together; the effect is simply to multiply each force, the magnetic moment, the torque, and the potential energy by a factor of N.

An arrangement of particular interest is the **solenoid,** a helical winding of wire, such as a coil wound on a circular cylinder (**Fig. 27.34**). If the windings are closely spaced, the solenoid can be approximated by a number of circular loops lying in planes at right angles to its long axis. The total torque on a solenoid in a magnetic field is simply the sum of the torques on the individual turns. For a solenoid with N turns in a uniform field B, the magnetic moment is $\mu = NIA$ and

$$\tau = NIAB \sin\phi \qquad (27.28)$$

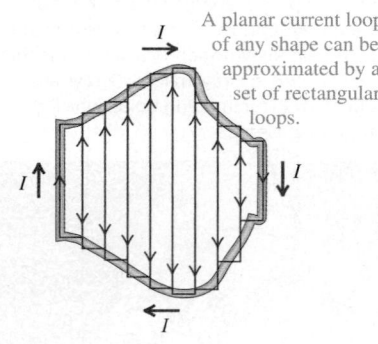

27.33 The collection of rectangles exactly matches the irregular plane loop in the limit as the number of rectangles approaches infinity and the width of each rectangle approaches zero.

A planar current loop of any shape can be approximated by a set of rectangular loops.

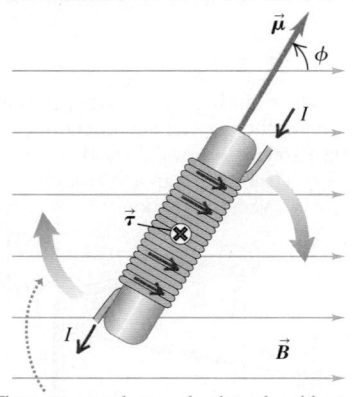

27.34 The torque $\vec{\tau} = \vec{\mu} \times \vec{B}$ on this solenoid in a uniform magnetic field is directed straight into the page. An actual solenoid has many more turns, wrapped closely together.

The torque tends to make the solenoid rotate clockwise in the plane of the page, aligning magnetic moment $\vec{\mu}$ with field $\vec{B}$.

where ϕ is the angle between the axis of the solenoid and the direction of the field. The magnetic moment vector $\vec{\mu}$ is along the solenoid axis. The torque is greatest when the solenoid axis is perpendicular to the magnetic field and zero when they are parallel. The effect of this torque is to tend to rotate the solenoid into a position where its axis is parallel to the field. Solenoids are also useful as *sources* of magnetic field, as we'll discuss in Chapter 28.

The d'Arsonval galvanometer, described in Section 26.3, makes use of a magnetic torque on a coil carrying a current. As Fig. 26.14 shows, the magnetic field is not uniform but is *radial*, so the side thrusts on the coil are always perpendicular to its plane. Thus the angle ϕ in Eq. (27.28) is always 90°, and the magnetic torque is directly proportional to the current, no matter what the orientation of the coil. A restoring torque proportional to the angular displacement of the coil is provided by two hairsprings, which also serve as current leads to the coil. When current is supplied to the coil, it rotates along with its attached pointer until the restoring spring torque just balances the magnetic torque. Thus the pointer deflection is proportional to the current.

EXAMPLE 27.9   MAGNETIC TORQUE ON A CIRCULAR COIL

A circular coil 0.0500 m in radius, with 30 turns of wire, lies in a horizontal plane. It carries a counterclockwise (as viewed from above) current of 5.00 A. The coil is in a uniform 1.20-T magnetic field directed toward the right. Find the magnitudes of the magnetic moment and the torque on the coil.

SOLUTION

IDENTIFY and SET UP: This problem uses the definition of magnetic moment and the expression for the torque on a magnetic dipole in a magnetic field. **Figure 27.35** shows the situation.

27.35 Our sketch for this problem.

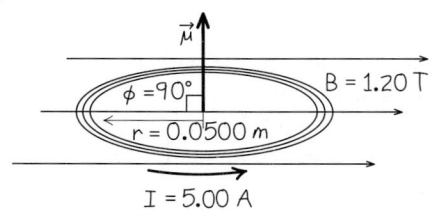

Equation (27.24) gives the magnitude μ of the magnetic moment of a single turn of wire; for N turns, the magnetic moment is N times greater. Equation (27.25) gives the magnitude τ of the torque.

EXECUTE: The area of the coil is $A = \pi r^2$. From Eq. (27.24), the total magnetic moment of all 30 turns is

$$\mu_{\text{total}} = NIA = 30(5.00\ \text{A})\pi(0.0500\ \text{m})^2 = 1.18\ \text{A} \cdot \text{m}^2$$

The angle ϕ between the direction of $\vec{B}$ and the direction of $\vec{\mu}$ (which is along the normal to the plane of the coil) is 90°. From Eq. (27.25), the torque on the coil is

$$\tau = \mu_{\text{total}}B \sin\phi = (1.18\ \text{A} \cdot \text{m}^2)(1.20\ \text{T})(\sin 90°)$$
$$= 1.41\ \text{N} \cdot \text{m}$$

EVALUATE: The torque tends to rotate the right side of the coil down and the left side up, into a position where the normal to its plane is parallel to $\vec{B}$.

EXAMPLE 27.10   POTENTIAL ENERGY FOR A COIL IN A MAGNETIC FIELD

If the coil in Example 27.9 rotates from its initial orientation to one in which its magnetic moment $\vec{\mu}$ is parallel to $\vec{B}$, what is the change in potential energy?

SOLUTION

IDENTIFY and SET UP: Equation (27.27) gives the potential energy for each orientation. The initial position is as shown in Fig. 27.35, with $\phi_1 = 90°$. In the final orientation, the coil has been rotated 90° clockwise so that $\vec{\mu}$ and $\vec{B}$ are parallel, so the angle between these vectors is $\phi_2 = 0$.

EXECUTE: From Eq. (27.27), the potential energy change is

$$\Delta U = U_2 - U_1 = -\mu B \cos\phi_2 - (-\mu B \cos\phi_1)$$
$$= -\mu B(\cos\phi_2 - \cos\phi_1)$$
$$= -(1.18\ \text{A} \cdot \text{m}^2)(1.20\ \text{T})(\cos 0° - \cos 90°) = -1.41\ \text{J}$$

EVALUATE: The potential energy decreases because the rotation is in the direction of the magnetic torque that we found in Example 27.9.

27.36 Forces on current loops in a nonuniform $\vec{B}$ field. In each case the axis of the bar magnet is perpendicular to the plane of the loop and passes through the center of the loop.

(a) Net force on this coil is away from north pole of magnet.

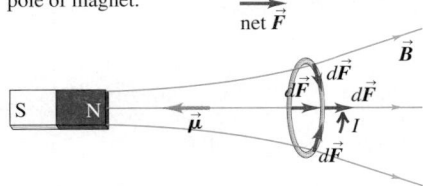

(b) Net force on same coil is toward south pole of magnet.

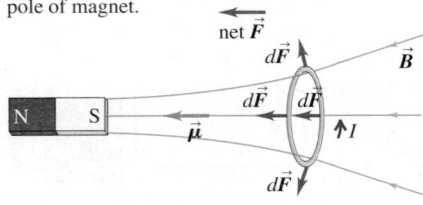

27.37 (a) An unmagnetized piece of iron. (Only a few representative atomic moments are shown.) (b) A magnetized piece of iron (bar magnet). The net magnetic moment of the bar magnet points from its south pole to its north pole. (c) A bar magnet in a magnetic field.

(a) Unmagnetized iron: magnetic moments are oriented randomly.

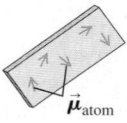

(b) In a bar magnet, the magnetic moments are aligned.

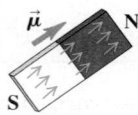

(c) A magnetic field creates a torque on the bar magnet that tends to align its dipole moment with the $\vec{B}$ field.

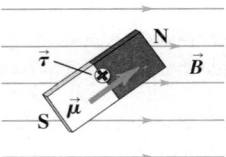

Magnetic Dipole in a Nonuniform Magnetic Field

We have seen that a current loop (that is, a magnetic dipole) experiences zero net force in a uniform magnetic field. **Figure 27.36** shows two current loops in the *nonuniform $\vec{B}$* field of a bar magnet; in both cases the net force on the loop is *not* zero. In Fig. 27.36a the magnetic moment $\vec{\mu}$ is in the direction opposite to the field, and the force $d\vec{F} = I\,d\vec{l} \times \vec{B}$ on a segment of the loop has both a radial component and a component to the right. When these forces are summed to find the net force $\vec{F}$ on the loop, the radial components cancel so that the net force is to the right, away from the magnet. Note that in this case the force is toward the region where the field lines are farther apart and the field magnitude B is less. The polarity of the bar magnet is reversed in Fig. 27.36b, so $\vec{\mu}$ and $\vec{B}$ are parallel; now the net force on the loop is to the left, toward the region of greater field magnitude near the magnet. Later in this section we'll use these observations to explain why bar magnets can pick up unmagnetized iron objects.

Magnetic Dipoles and How Magnets Work

The behavior of a solenoid in a magnetic field (see Fig. 27.34) resembles that of a bar magnet or compass needle; if free to turn, both the solenoid and the magnet orient themselves with their axes parallel to the $\vec{B}$ field. In both cases this is due to the interaction of moving electric charges with a magnetic field; the difference is that in a bar magnet the motion of charge occurs on the microscopic scale of the atom.

Think of an electron as being like a spinning ball of charge. In this analogy the circulation of charge around the spin axis is like a current loop, and so the electron has a net magnetic moment. (This analogy, while helpful, is inexact; an electron isn't really a spinning sphere. A full explanation of the origin of an electron's magnetic moment involves quantum mechanics, which is beyond our scope here.) In an iron atom a substantial fraction of the electron magnetic moments align with each other, and the atom has a nonzero magnetic moment. (By contrast, the atoms of most elements have little or no net magnetic moment.) In an unmagnetized piece of iron there is no overall alignment of the magnetic moments of the atoms; their vector sum is zero, and the net magnetic moment is zero (**Fig. 27.37a**). But in an iron bar magnet the magnetic moments of many of the atoms are parallel, and there is a substantial net magnetic moment $\vec{\mu}$ (Fig. 27.37b). If the magnet is placed in a magnetic field $\vec{B}$, the field exerts a torque given by Eq. (27.26) that tends to align $\vec{\mu}$ with $\vec{B}$ (Fig. 27.37c). A bar magnet tends to align with a $\vec{B}$ field so that a line from the south pole to the north pole of the magnet is in the direction of $\vec{B}$; hence the real significance of a magnet's north and south poles is that they represent the head and tail, respectively, of the magnet's dipole moment $\vec{\mu}$.

The torque experienced by a current loop in a magnetic field also explains how an unmagnetized iron object like that in Fig. 27.37a becomes magnetized. If an unmagnetized iron paper clip is placed next to a powerful magnet, the magnetic moments of the paper clip's atoms tend to align with the $\vec{B}$ field of the magnet. When the paper clip is removed, its atomic dipoles tend to remain aligned, and the paper clip has a net magnetic moment. The paper clip can be demagnetized by being dropped on the floor or heated; the added internal energy jostles and re-randomizes the atomic dipoles.

The magnetic-dipole picture of a bar magnet explains the attractive and repulsive forces between bar magnets shown in Fig. 27.1. The magnetic moment $\vec{\mu}$ of a bar magnet points from its south pole to its north pole, so the current loops in Figs. 27.36a and 27.36b are both equivalent to a magnet with its north pole on the left. Hence the situation in Fig. 27.36a is equivalent to two bar magnets with their north poles next to each other; the resultant force is repulsive, as in Fig. 27.1b. In Fig. 27.36b we again have the equivalent of two bar magnets end to end, but with the south pole of the left-hand magnet next to the north pole of the right-hand magnet. The resultant force is attractive, as in Fig. 27.1a.

Finally, we can explain how a magnet can attract an unmagnetized iron object (see Fig. 27.2). It's a two-step process. First, the atomic magnetic moments of the iron tend to align with the $\vec{B}$ field of the magnet, so the iron acquires a net magnetic dipole moment $\vec{\mu}$ parallel to the field. Second, the nonuniform field of the magnet attracts the magnetic dipole. **Figure 27.38a** shows an example. The north pole of the magnet is closer to the nail (which contains iron), and the magnetic dipole produced in the nail is equivalent to a loop with a current that circulates in a direction opposite to that shown in Fig. 27.36a. Hence the net magnetic force on the nail is opposite to the force on the loop in Fig. 27.36a, and the nail is attracted toward the magnet. Changing the polarity of the magnet, as in Fig. 27.38b, reverses the directions of both $\vec{B}$ and $\vec{\mu}$. The situation is now equivalent to that shown in Fig. 27.36b; like the loop in that figure, the nail is attracted toward the magnet. Hence a previously unmagnetized object containing iron is attracted to *either* pole of a magnet. By contrast, objects made of brass, aluminum, or wood hardly respond at all to a magnet; the atomic magnetic dipoles of these materials, if present at all, have less tendency to align with an external field.

Our discussion of how magnets and pieces of iron interact has just scratched the surface of a diverse subject known as *magnetic properties of materials*. We'll discuss these properties in more depth in Section 28.8.

TEST YOUR UNDERSTANDING OF SECTION 27.7 Figure 27.13c depicts the magnetic field lines due to a circular current-carrying loop. (a) What is the direction of the magnetic moment of this loop? (b) Which side of the loop is equivalent to the north pole of a magnet, and which side is equivalent to the south pole? ▮

27.8 THE DIRECT-CURRENT MOTOR

Electric motors play an important role in contemporary society. In a motor a magnetic torque acts on a current-carrying conductor, and electric energy is converted to mechanical energy. As an example, let's look at a simple type of direct-current (dc) motor, shown in **Fig. 27.39**.

The moving part of the motor is the *rotor,* a length of wire formed into an open-ended loop and free to rotate about an axis. The ends of the rotor wires are attached to circular conducting segments that form a *commutator.* In Fig. 27.39a, each of the two commutator segments makes contact with one of the terminals,

27.38 A bar magnet attracts an unmagnetized iron nail in two steps. First, the $\vec{B}$ field of the bar magnet gives rise to a net magnetic moment in the nail. Second, because the field of the bar magnet is not uniform, this magnetic dipole is attracted toward the magnet. The attraction is the same whether the nail is closer to (a) the magnet's north pole or (b) the magnet's south pole.

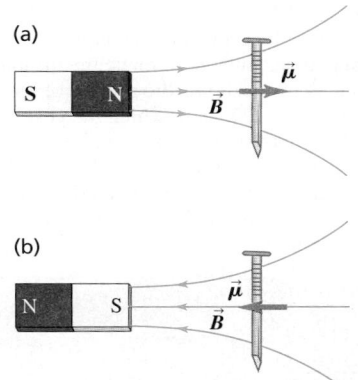

27.39 Schematic diagram of a simple dc motor. The rotor is a wire loop that is free to rotate about an axis; the rotor ends are attached to the two curved conductors that form the commutator. (The rotor halves are colored red and blue for clarity.) The commutator segments are insulated from one another.

(a) Brushes are aligned with commutator segments.

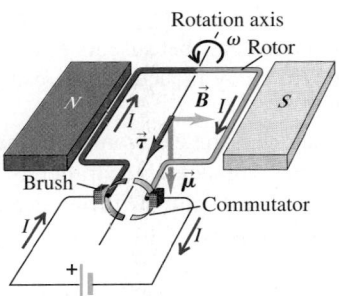

- Current flows into the red side of the rotor and out of the blue side.
- Therefore the magnetic torque causes the rotor to spin counterclockwise.

(b) Rotor has turned 90°.

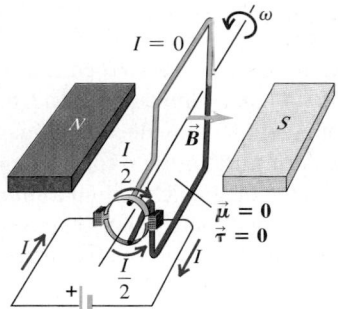

- Each brush is in contact with both commutator segments, so the current bypasses the rotor altogether.
- No magnetic torque acts on the rotor.

(c) Rotor has turned 180°.

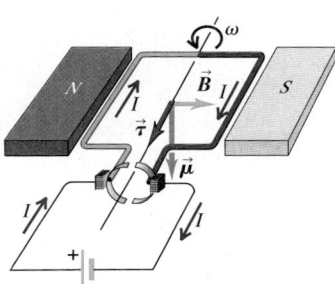

- The brushes are again aligned with commutator segments. This time the current flows into the blue side of the rotor and out of the red side.
- Therefore the magnetic torque again causes the rotor to spin counterclockwise.

or *brushes,* of an external circuit that includes a source of emf. This causes a current to flow into the rotor on one side, shown in red, and out of the rotor on the other side, shown in blue. Hence the rotor is a current loop with a magnetic moment $\vec{\mu}$. The rotor lies between opposing poles of a permanent magnet, so there is a magnetic field $\vec{B}$ that exerts a torque $\vec{\tau} = \vec{\mu} \times \vec{B}$ on the rotor. For the rotor orientation shown in Fig. 27.39a the torque causes the rotor to turn counterclockwise, in the direction that will align $\vec{\mu}$ with $\vec{B}$.

In Fig. 27.39b the rotor has rotated by $90°$ from its orientation in Fig. 27.39a. If the current through the rotor were constant, the rotor would now be in its equilibrium orientation; it would simply oscillate around this orientation. But here's where the commutator comes into play; each brush is now in contact with *both* segments of the commutator. There is no potential difference between the commutators, so at this instant no current flows through the rotor, and the magnetic moment is zero. The rotor continues to rotate counterclockwise because of its inertia, and current again flows through the rotor as in Fig. 27.39c. But now current enters on the *blue* side of the rotor and exits on the *red* side, just the opposite of the situation in Fig. 27.39a. While the direction of the current has reversed with respect to the rotor, the rotor itself has rotated $180°$ and the magnetic moment $\vec{\mu}$ is in the same direction with respect to the magnetic field. Hence the magnetic torque $\vec{\tau}$ is in the same direction in Fig. 27.39c as in Fig. 27.39a. Thanks to the commutator, the current reverses after every $180°$ of rotation, so the torque is always in the direction to rotate the rotor counterclockwise. When the motor has come "up to speed," the average magnetic torque is just balanced by an opposing torque due to air resistance, friction in the rotor bearings, and friction between the commutator and brushes.

The simple motor shown in Fig. 27.39 has only a single turn of wire in its rotor. In practical motors, the rotor has many turns; this increases the magnetic moment and the torque so that the motor can spin larger loads. The torque can also be increased by using a stronger magnetic field, which is why many motor designs use electromagnets instead of a permanent magnet. Another drawback of the simple design in Fig. 27.39 is that the magnitude of the torque rises and falls as the rotor spins. This can be remedied by having the rotor include several independent coils of wire oriented at different angles (**Fig. 27.40**).

27.40 This motor from a computer disk drive has 12 current-carrying coils. They interact with permanent magnets on the turntable (not shown) to make the turntable rotate. (This design is the reverse of the design in Fig. 27.39, in which the permanent magnets are stationary and the coil rotates.) Because there are multiple coils, the magnetic torque is very nearly constant and the turntable spins at a very constant rate.

Coils

Power for Electric Motors

Because a motor converts electric energy to mechanical energy or work, it requires electric energy input. If the potential difference between its terminals is V_{ab} and the current is I, then the power input is $P = V_{ab}I$. Even if the motor coils have negligible resistance, there must be a potential difference between the terminals if P is to be different from zero. This potential difference results principally from magnetic forces exerted on the currents in the conductors of the rotor as they rotate through the magnetic field. The associated electromotive force $\mathcal{E}$ is called an *induced* emf; it is also called a *back* emf because its sense is opposite to that of the current. In Chapter 29 we will study induced emfs resulting from motion of conductors in magnetic fields.

In a *series* motor the rotor is connected in series with the electromagnet that produces the magnetic field; in a *shunt* motor they are connected in parallel. In a series motor with internal resistance r, V_{ab} is greater than $\mathcal{E}$, and the difference is the potential drop Ir across the internal resistance. That is,

$$V_{ab} = \mathcal{E} + Ir \qquad (27.29)$$

Because the magnetic force is proportional to velocity, $\mathcal{E}$ is *not* constant but is proportional to the speed of rotation of the rotor.

EXAMPLE 27.11 A SERIES DC MOTOR

A dc motor with its rotor and field coils connected in series has an internal resistance of 2.00 Ω. When running at full load on a 120-V line, it draws a 4.00-A current. (a) What is the emf in the rotor? (b) What is the power delivered to the motor? (c) What is the rate of dissipation of energy in the internal resistance? (d) What is the mechanical power developed? (e) What is the motor's efficiency? (f) What happens if the machine being driven by the motor jams, so that the rotor suddenly stops turning?

SOLUTION

IDENTIFY and SET UP: This problem uses the ideas of power and potential drop in a series dc motor. We are given the internal resistance $r = 2.00 \ \Omega$, the voltage $V_{ab} = 120$ V across the motor, and the current $I = 4.00$ A through the motor. We use Eq. (27.29) to determine the emf $\mathcal{E}$ from these quantities. The power delivered to the motor is $V_{ab}I$, the rate of energy dissipation is I^2r, and the power output by the motor is the difference between the power input and the power dissipated. The efficiency e is the ratio of mechanical power output to electric power input.

EXECUTE: (a) From Eq. (27.29), $V_{ab} = \mathcal{E} + Ir$, we have

$$120 \text{ V} = \mathcal{E} + (4.00 \text{ A})(2.00 \ \Omega) \quad \text{and so} \quad \mathcal{E} = 112 \text{ V}$$

(b) The power delivered to the motor from the source is

$$P_{\text{input}} = V_{ab}I = (120 \text{ V})(4.00 \text{ A}) = 480 \text{ W}$$

(c) The power dissipated in the resistance r is

$$P_{\text{dissipated}} = I^2r = (4.00 \text{ A})^2(2.00 \ \Omega) = 32 \text{ W}$$

(d) The mechanical power output is the electric power input minus the rate of dissipation of energy in the motor's resistance (assuming that there are no other power losses):

$$P_{\text{output}} = P_{\text{input}} - P_{\text{dissipated}} = 480 \text{ W} - 32 \text{ W} = 448 \text{ W}$$

(e) The efficiency e is the ratio of mechanical power output to electric power input:

$$e = \frac{P_{\text{output}}}{P_{\text{input}}} = \frac{448 \text{ W}}{480 \text{ W}} = 0.93 = 93\%$$

(f) With the rotor stalled, the back emf $\mathcal{E}$ (which is proportional to rotor speed) goes to zero. From Eq. (27.29) the current becomes

$$I = \frac{V_{ab}}{r} = \frac{120 \text{ V}}{2.00 \ \Omega} = 60 \text{ A}$$

and the power dissipated in the resistance r becomes

$$P_{\text{dissipated}} = I^2r = (60 \text{ A})^2(2.00 \ \Omega) = 7200 \text{ W}$$

EVALUATE: If this massive overload doesn't blow a fuse or trip a circuit breaker, the coils will quickly melt. When the motor is first turned on, there's a momentary surge of current until the motor picks up speed. This surge causes greater-than-usual voltage drops ($V = IR$) in the power lines supplying the current. Similar effects are responsible for the momentary dimming of lights in a house when an air conditioner or dishwasher motor starts.

TEST YOUR UNDERSTANDING OF SECTION 27.8 In the circuit shown in Fig. 27.39, you add a switch in series with the source of emf so that the current can be turned on and off. When you close the switch and allow current to flow, will the rotor begin to turn no matter what its original orientation? ▮

27.9 THE HALL EFFECT

The reality of the forces acting on the moving charges in a conductor in a magnetic field is strikingly demonstrated by the *Hall effect,* an effect analogous to the transverse deflection of an electron beam in a magnetic field in vacuum. (The effect was discovered by the American physicist Edwin Hall in 1879 while he was still a graduate student.) To describe this effect, let's consider a conductor in the form of a flat strip, as shown in **Fig. 27.41** (next page). The current is in the direction of the $+x$-axis and there is a uniform magnetic field $\vec{B}$ perpendicular to the plane of the strip, in the $+y$-direction. The drift velocity of the moving charges (charge magnitude $|q|$) has magnitude v_d. Figure 27.41a shows the case of negative charges, such as electrons in a metal, and Fig. 27.41b shows positive charges. In both cases the magnetic force is upward, just as the magnetic force on a conductor is the same whether the moving charges are positive or negative. In either case a moving charge is driven toward the *upper* edge of the strip by the magnetic force $F_z = |q|v_dB$.

27.41 Forces on charge carriers in a conductor in a magnetic field.

(a) Negative charge carriers (electrons)

The charge carriers are pushed toward the top of the strip ...

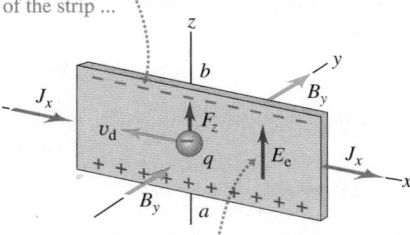

... so point a is at a higher potential than point b.

(b) Positive charge carriers

The charge carriers are again pushed toward the top of the strip ...

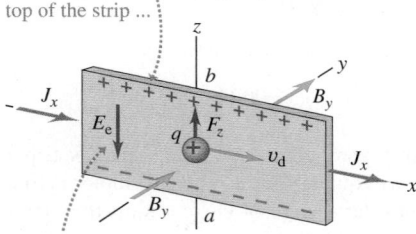

... so the polarity of the potential difference is opposite to that for negative charge carriers.

If the charge carriers are electrons, as in Fig. 27.41a, an excess negative charge accumulates at the upper edge of the strip, leaving an excess positive charge at its lower edge. This accumulation continues until the resulting transverse electrostatic field $\vec{E}_e$ becomes large enough to cause a force (magnitude $|q|E_e$) that is equal and opposite to the magnetic force (magnitude $|q|v_d B$). After that, there is no longer any net transverse force to deflect the moving charges. This electric field causes a transverse potential difference between opposite edges of the strip, called the *Hall voltage* or the *Hall emf.* The polarity depends on whether the moving charges are positive or negative. Experiments show that for metals the upper edge of the strip in Fig. 27.41a *does* become negatively charged, showing that the charge carriers in a metal are indeed negative electrons.

However, if the charge carriers are *positive,* as in Fig. 27.41b, then *positive* charge accumulates at the upper edge, and the potential difference is *opposite* to the situation with negative charges. Soon after the discovery of the Hall effect in 1879, it was observed that some materials, particularly some *semiconductors,* show a Hall emf opposite to that of the metals, as if their charge carriers were positively charged. We now know that these materials conduct by a process known as *hole conduction.* Within such a material there are locations, called *holes,* that would normally be occupied by an electron but are actually empty. A missing negative charge is equivalent to a positive charge. When an electron moves in one direction to fill a hole, it leaves another hole behind it. The hole migrates in the direction opposite to that of the electron.

In terms of the coordinate axes in Fig. 27.41b, the electrostatic field $\vec{E}_e$ for the positive-q case is in the $-z$-direction; its z-component E_z is negative. The magnetic field is in the $+y$-direction, and we write it as B_y. The magnetic force (in the $+z$-direction) is $qv_d B_y$. The current density J_x is in the $+x$-direction. In the steady state, when the forces qE_z and $qv_d B_y$ sum to zero,

$$qE_z + qv_d B_y = 0 \quad \text{or} \quad E_z = -v_d B_y$$

This confirms that when q is positive, E_z is negative. From Eq. (25.4),

$$J_x = nqv_d$$

Eliminating v_d between these equations, we find

Hall effect: Concentration of mobile charge carriers   Current density / Magnetic field

$$nq = \frac{-J_x B_y}{E_z}$$

Charge per carrier   Electrostatic field in conductor

(27.30)

Note that this result (as well as the entire derivation) is valid for both positive and negative q. When q is negative, E_z is positive, and conversely.

We can measure J_x, B_y, and E_z, so we can compute the product nq. In both metals and semiconductors, q is equal in magnitude to the electron charge, so the Hall effect permits a direct measurement of n, the concentration of current-carrying charges in the material. The *sign* of the charges is determined by the polarity of the Hall emf, as we have described.

The Hall effect can also be used for a direct measurement of electron drift speed v_d in metals. As we saw in Chapter 25, these speeds are very small, often of the order of 1 mm/s or less. If we move the entire conductor in the opposite direction to the current with a speed equal to the drift speed, then the electrons are at rest with respect to the magnetic field, and the Hall emf disappears. Thus the conductor speed needed to make the Hall emf vanish is equal to the drift speed.

EXAMPLE 27.12   A HALL-EFFECT MEASUREMENT

You place a strip of copper, 2.0 mm thick and 1.50 cm wide, in a uniform 0.40-T magnetic field as shown in Fig. 27.41a. When you run a 75-A current in the $+x$-direction, you find that the potential at the bottom of the slab is 0.81 μV higher than at the top. From this measurement, determine the concentration of mobile electrons in copper.

SOLUTION

IDENTIFY and SET UP: This problem describes a Hall-effect experiment. We use Eq. (27.30) to determine the mobile electron concentration n.

EXECUTE: First we find the current density J_x and the electric field E_z:

$$J_x = \frac{I}{A} = \frac{75 \text{ A}}{(2.0 \times 10^{-3} \text{ m})(1.50 \times 10^{-2} \text{ m})} = 2.5 \times 10^6 \text{ A/m}^2$$

$$E_z = \frac{V}{d} = \frac{0.81 \times 10^{-6} \text{ V}}{1.5 \times 10^{-2} \text{ m}} = 5.4 \times 10^{-5} \text{ V/m}$$

Then, from Eq. (27.30),

$$n = \frac{-J_x B_y}{q E_z} = \frac{-(2.5 \times 10^6 \text{ A/m}^2)(0.40 \text{ T})}{(-1.60 \times 10^{-19} \text{ C})(5.4 \times 10^{-5} \text{ V/m})}$$

$$= 11.6 \times 10^{28} \text{ m}^{-3}$$

EVALUATE: The actual value of n for copper is $8.5 \times 10^{28} \text{ m}^{-3}$. The difference shows that our simple model of the Hall effect, which ignores quantum effects and electron interactions with the ions, must be used with caution. This example also shows that with good conductors, the Hall emf is very small even with large current densities. In practice, Hall-effect devices for magnetic-field measurements use semiconductor materials, for which moderate current densities give much larger Hall emfs.

TEST YOUR UNDERSTANDING OF SECTION 27.9   A copper wire of square cross section is oriented vertically. The four sides of the wire face north, south, east, and west. There is a uniform magnetic field directed from east to west, and the wire carries current downward. Which side of the wire is at the highest electric potential? (i) North side; (ii) south side; (iii) east side; (iv) west side. ▮

CHAPTER 27 SUMMARY

SOLUTIONS TO ALL EXAMPLES

Magnetic forces: Magnetic interactions are fundamentally interactions between moving charged particles. These interactions are described by the vector magnetic field, denoted by $\vec{B}$. A particle with charge q moving with velocity $\vec{v}$ in a magnetic field $\vec{B}$ experiences a force $\vec{F}$ that is perpendicular to both $\vec{v}$ and $\vec{B}$. The SI unit of magnetic field is the tesla ($1 \text{ T} = 1 \text{ N/A} \cdot \text{m}$). (See Example 27.1.)

$$\vec{F} = q\vec{v} \times \vec{B} \qquad (27.2)$$

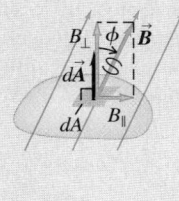

Magnetic field lines and flux: A magnetic field can be represented graphically by magnetic field lines. At each point a magnetic field line is tangent to the direction of $\vec{B}$ at that point. Where field lines are close together, the field magnitude is large, and vice versa. Magnetic flux Φ_B through an area is defined in an analogous way to electric flux. The SI unit of magnetic flux is the weber ($1 \text{ Wb} = 1 \text{ T} \cdot \text{m}^2$). The net magnetic flux through any closed surface is zero (Gauss's law for magnetism). As a result, magnetic field lines always close on themselves. (See Example 27.2.)

$$\Phi_B = \int B \cos \phi \, dA$$

$$= \int B_\perp \, dA \qquad (27.6)$$

$$= \int \vec{B} \cdot d\vec{A}$$

$$\oint \vec{B} \cdot d\vec{A} = 0 \quad \text{(closed surface)} \qquad (27.8)$$

Motion in a magnetic field: The magnetic force is always perpendicular to $\vec{v}$; a particle moving under the action of a magnetic field alone moves with constant speed. In a uniform field, a particle with initial velocity perpendicular to the field moves in a circle with radius R that depends on the magnetic field strength B and the particle mass m, speed v, and charge q. (See Examples 27.3 and 27.4.)

Crossed electric and magnetic fields can be used as a velocity selector. The electric and magnetic forces exactly cancel when $v = E/B$. (See Examples 27.5 and 27.6.)

$$R = \frac{mv}{|q|B} \qquad (27.11)$$

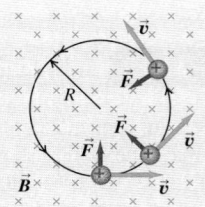

Magnetic force on a conductor: A straight segment of a conductor carrying current I in a uniform magnetic field $\vec{B}$ experiences a force $\vec{F}$ that is perpendicular to both $\vec{B}$ and the vector $\vec{l}$, which points in the direction of the current and has magnitude equal to the length of the segment. A similar relationship gives the force $d\vec{F}$ on an infinitesimal current-carrying segment $d\vec{l}$. (See Examples 27.7 and 27.8.)

$$\vec{F} = I\vec{l} \times \vec{B} \qquad (27.19)$$

$$d\vec{F} = I\,d\vec{l} \times \vec{B} \qquad (27.20)$$

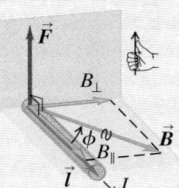

Magnetic torque: A current loop with area A and current I in a uniform magnetic field $\vec{B}$ experiences no net magnetic force, but does experience a magnetic torque of magnitude τ. The vector torque $\vec{\tau}$ can be expressed in terms of the magnetic moment $\vec{\mu} = I\vec{A}$ of the loop, as can the potential energy U of a magnetic moment in a magnetic field $\vec{B}$. The magnetic moment of a loop depends only on the current and the area; it is independent of the shape of the loop. (See Examples 27.9 and 27.10.)

$$\tau = IBA \sin\phi \qquad (27.23)$$

$$\vec{\tau} = \vec{\mu} \times \vec{B} \qquad (27.26)$$

$$U = -\vec{\mu} \cdot \vec{B} = -\mu B \cos\phi \qquad (27.27)$$

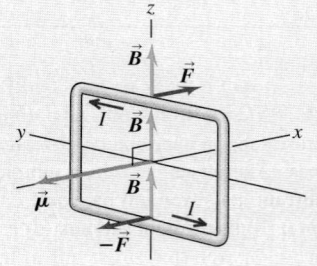

Electric motors: In a dc motor a magnetic field exerts a torque on a current in the rotor. Motion of the rotor through the magnetic field causes an induced emf called a back emf. For a series motor, in which the rotor coil is in series with coils that produce the magnetic field, the terminal voltage is the sum of the back emf and the drop Ir across the internal resistance. (See Example 27.11.)

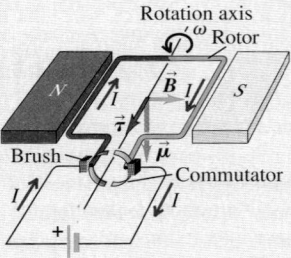

The Hall effect: The Hall effect is a potential difference perpendicular to the direction of current in a conductor, when the conductor is placed in a magnetic field. The Hall potential is determined by the requirement that the associated electric field must just balance the magnetic force on a moving charge. Hall-effect measurements can be used to determine the sign of charge carriers and their concentration n. (See Example 27.12.)

$$nq = \frac{-J_x B_y}{E_z} \qquad (27.30)$$

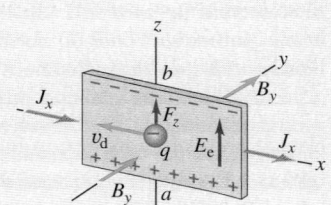

SOLUTION

BRIDGING PROBLEM MAGNETIC TORQUE ON A CURRENT-CARRYING RING

A circular ring with area 4.45 cm^2 is carrying a current of 12.5 A. The ring, initially at rest, is immersed in a region of uniform magnetic field given by $\vec{B} = (1.15 \times 10^{-2}\,\text{T})(12\hat{\imath} + 3\hat{\jmath} - 4\hat{k})$. The ring is positioned initially such that its magnetic moment is given by $\vec{\mu}_\text{i} = \mu(-0.800\hat{\imath} + 0.600\hat{\jmath})$, where μ is the (positive) magnitude of the magnetic moment. (a) Find the initial magnetic torque on the ring. (b) The ring (which is free to rotate around one diameter) is released and turns through an angle of 90.0°, at which point its magnetic moment is given by $\vec{\mu}_\text{f} = -\mu\hat{k}$. Determine the decrease in potential energy. (c) If the moment of inertia of the ring about a diameter is $8.50 \times 10^{-7}\,\text{kg}\cdot\text{m}^2$, determine the angular speed of the ring as it passes through the second position.

SOLUTION GUIDE

IDENTIFY and SET UP

1. The current-carrying ring acts as a magnetic dipole, so you can use the equations for a magnetic dipole in a uniform magnetic field.

2. There are no nonconservative forces acting on the ring as it rotates, so the sum of its rotational kinetic energy (discussed in Section 9.4) and the potential energy is conserved.

EXECUTE

3. Use the vector expression for the torque on a magnetic dipole to find the answer to part (a). (*Hint:* Review Section 1.10.)
4. Find the change in potential energy from the first orientation of the ring to the second orientation.
5. Use your result from step 4 to find the rotational kinetic energy of the ring when it is in the second orientation.
6. Use your result from step 5 to find the ring's angular speed when it is in the second orientation.

EVALUATE

7. If the ring were free to rotate around *any* diameter, in what direction would the magnetic moment point when the ring is in a state of stable equilibrium?

Problems

For assigned homework and other learning materials, go to MasteringPhysics®. **MP**

•, ••, •••: Difficulty levels. **CP**: Cumulative problems incorporating material from earlier chapters. **CALC**: Problems requiring calculus. **DATA**: Problems involving real data, scientific evidence, experimental design, and/or statistical reasoning. **BIO**: Biosciences problems.

DISCUSSION QUESTIONS

Q27.1 Can a charged particle move through a magnetic field without experiencing any force? If so, how? If not, why not?

Q27.2 At any point in space, the electric field $\vec{E}$ is defined to be in the direction of the electric force on a positively charged particle at that point. Why don't we similarly define the magnetic field $\vec{B}$ to be in the direction of the magnetic force on a moving, positively charged particle?

Q27.3 Section 27.2 describes a procedure for finding the direction of the magnetic force using your right hand. If you use the same procedure, but with your left hand, will you get the correct direction for the force? Explain.

Q27.4 The magnetic force on a moving charged particle is always perpendicular to the magnetic field $\vec{B}$. Is the trajectory of a moving charged particle always perpendicular to the magnetic field lines? Explain your reasoning.

Q27.5 A charged particle is fired into a cubical region of space where there is a uniform magnetic field. Outside this region, there is no magnetic field. Is it possible that the particle will remain inside the cubical region? Why or why not?

Q27.6 If the magnetic force does no work on a charged particle, how can it have any effect on the particle's motion? Are there other examples of forces that do no work but have a significant effect on a particle's motion?

Q27.7 A charged particle moves through a region of space with constant velocity (magnitude and direction). If the external

magnetic field is zero in this region, can you conclude that the external electric field in the region is also zero? Explain. (By "external" we mean fields other than those produced by the charged particle.) If the external electric field is zero in the region, can you conclude that the external magnetic field in the region is also zero?

Q27.8 How might a loop of wire carrying a current be used as a compass? Could such a compass distinguish between north and south? Why or why not?

Q27.9 How could the direction of a magnetic field be determined by making only *qualitative* observations of the magnetic force on a straight wire carrying a current?

Q27.10 A loose, floppy loop of wire is carrying current *I*. The loop of wire is placed on a horizontal table in a uniform magnetic field $\vec{B}$ perpendicular to the plane of the table. This causes the loop of wire to expand into a circular shape while still lying on the table. In a diagram, show all possible orientations of the current *I* and magnetic field $\vec{B}$ that could cause this to occur. Explain your reasoning.

Q27.11 Several charges enter a uniform magnetic field directed into the page. (a) What path would a positive charge *q* moving with a velocity of magnitude *v* follow through the field? (b) What path would a positive charge *q* moving with a velocity of magnitude 2*v* follow through the field? (c) What path would a negative charge −*q* moving with a velocity of magnitude *v* follow through the field? (d) What path would a neutral particle follow through the field?

Q27.12 Each of the lettered points at the corners of the cube in **Fig. Q27.12** represents a positive charge q moving with a velocity of magnitude v in the direction indicated. The region in the figure is in a uniform magnetic field $\vec{B}$, parallel to the x-axis and directed toward the right. Which charges experience a force due to $\vec{B}$? What is the direction of the force on each charge?

Figure **Q27.12**

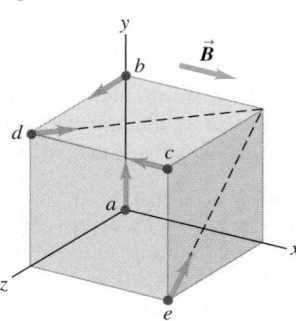

Q27.13 A student claims that if lightning strikes a metal flagpole, the force exerted by the earth's magnetic field on the current in the pole can be large enough to bend it. Typical lightning currents are of the order of 10^4 to 10^5 A. Is the student's opinion justified? Explain your reasoning.

Q27.14 Could an accelerator be built in which *all* the forces on the particles, for steering and for increasing speed, are magnetic forces? Why or why not?

Q27.15 The magnetic force acting on a charged particle can never do work because at every instant the force is perpendicular to the velocity. The torque exerted by a magnetic field can do work on a current loop when the loop rotates. Explain how these seemingly contradictory statements can be reconciled.

Q27.16 When the polarity of the voltage applied to a dc motor is reversed, the direction of motion does *not* reverse. Why not? How *could* the direction of motion be reversed?

Q27.17 In a Hall-effect experiment, is it possible that *no* transverse potential difference will be observed? Under what circumstances might this happen?

Q27.18 Hall-effect voltages are much greater for relatively poor conductors (such as germanium) than for good conductors (such as copper), for comparable currents, fields, and dimensions. Why?

EXERCISES

Section 27.2 Magnetic Field

27.1 • A particle with a charge of -1.24×10^{-8} C is moving with instantaneous velocity $\vec{v} = (4.19 \times 10^4 \text{ m/s})\hat{\imath} + (-3.85 \times 10^4 \text{ m/s})\hat{\jmath}$. What is the force exerted on this particle by a magnetic field (a) $\vec{B} = (1.40 \text{ T})\hat{\imath}$ and (b) $\vec{B} = (1.40 \text{ T})\hat{k}$?

27.2 • A particle of mass 0.195 g carries a charge of -2.50×10^{-8} C. The particle is given an initial horizontal velocity that is due north and has magnitude 4.00×10^4 m/s. What are the magnitude and direction of the minimum magnetic field that will keep the particle moving in the earth's gravitational field in the same horizontal, northward direction?

27.3 • In a 1.25-T magnetic field directed vertically upward, a particle having a charge of magnitude 8.50 μC and initially moving northward at 4.75 km/s is deflected toward the east. (a) What is the sign of the charge of this particle? Make a sketch to illustrate how you found your answer. (b) Find the magnetic force on the particle.

27.4 • A particle with mass 1.81×10^{-3} kg and a charge of 1.22×10^{-8} C has, at a given instant, a velocity $\vec{v} = (3.00 \times 10^4 \text{ m/s})\hat{\jmath}$. What are the magnitude and direction of the particle's acceleration produced by a uniform magnetic field $\vec{B} = (1.63 \text{ T})\hat{\imath} + (0.980 \text{ T})\hat{\jmath}$?

27.5 • An electron experiences a magnetic force of magnitude 4.60×10^{-15} N when moving at an angle of 60.0° with respect to a magnetic field of magnitude 3.50×10^{-3} T. Find the speed of the electron.

27.6 • An electron moves at 1.40×10^6 m/s through a region in which there is a magnetic field of unspecified direction and magnitude 7.40×10^{-2} T. (a) What are the largest and smallest possible magnitudes of the acceleration of the electron due to the magnetic field? (b) If the actual acceleration of the electron is one-fourth of the largest magnitude in part (a), what is the angle between the electron velocity and the magnetic field?

27.7 •• CP A particle with charge 7.80 μC is moving with velocity $\vec{v} = -(3.80 \times 10^3 \text{m/s})\hat{\jmath}$. The magnetic force on the particle is measured to be $\vec{F} = +(7.60 \times 10^{-3} \text{ N})\hat{\imath} - (5.20 \times 10^{-3} \text{ N})\hat{k}$. (a) Calculate all the components of the magnetic field you can from this information. (b) Are there components of the magnetic field that are not determined by the measurement of the force? Explain. (c) Calculate the scalar product $\vec{B} \cdot \vec{F}$. What is the angle between $\vec{B}$ and $\vec{F}$?

27.8 •• CP A particle with charge -5.60 nC is moving in a uniform magnetic field $\vec{B} = -(1.25 \text{ T})\hat{k}$. The magnetic force on the particle is measured to be $\vec{F} = -(3.40 \times 10^{-7}\text{N})\hat{\imath} + (7.40 \times 10^{-7} \text{ N})\hat{\jmath}$. (a) Calculate all the components of the velocity of the particle that you can from this information. (b) Are there components of the velocity that are not determined by the measurement of the force? Explain. (c) Calculate the scalar product $\vec{v} \cdot \vec{F}$. What is the angle between $\vec{v}$ and $\vec{F}$?

27.9 •• A group of particles is traveling in a magnetic field of unknown magnitude and direction. You observe that a proton moving at 1.50 km/s in the $+x$-direction experiences a force of 2.25×10^{-16} N in the $+y$-direction, and an electron moving at 4.75 km/s in the $-z$-direction experiences a force of 8.50×10^{-16} N in the $+y$-direction. (a) What are the magnitude and direction of the magnetic field? (b) What are the magnitude and direction of the magnetic force on an electron moving in the $-y$-direction at 3.20 km/s?

Section 27.3 Magnetic Field Lines and Magnetic Flux

27.10 • A flat, square surface with side length 3.40 cm is in the xy-plane at $z = 0$. Calculate the magnitude of the flux through this surface produced by a magnetic field $\vec{B} = (0.200 \text{ T})\hat{\imath} + (0.300 \text{ T})\hat{\jmath} - (0.500 \text{ T})\hat{k}$.

27.11 • A circular area with a radius of 6.50 cm lies in the xy-plane. What is the magnitude of the magnetic flux through this circle due to a uniform magnetic field $B = 0.230$ T (a) in the $+z$-direction; (b) at an angle of 53.1° from the $+z$-direction; (c) in the $+y$-direction?

27.12 • A horizontal rectangular surface has dimensions 2.80 cm by 3.20 cm and is in a uniform magnetic field that is directed at an angle of 30.0° above the horizontal. What must the magnitude of the magnetic field be to produce a flux of 3.10×10^{-4} Wb through the surface?

27.13 •• An open plastic soda bottle with an opening diameter of 2.5 cm is placed on a table. A uniform 1.75-T magnetic field directed upward and oriented 25° from vertical encompasses the bottle. What is the total magnetic flux through the plastic of the soda bottle?

27.14 •• The magnetic field $\vec{B}$ in a certain region is 0.128 T, and its direction is that of the $+z$-axis in **Fig. E27.14**. (a) What is the

Figure **E27.14**

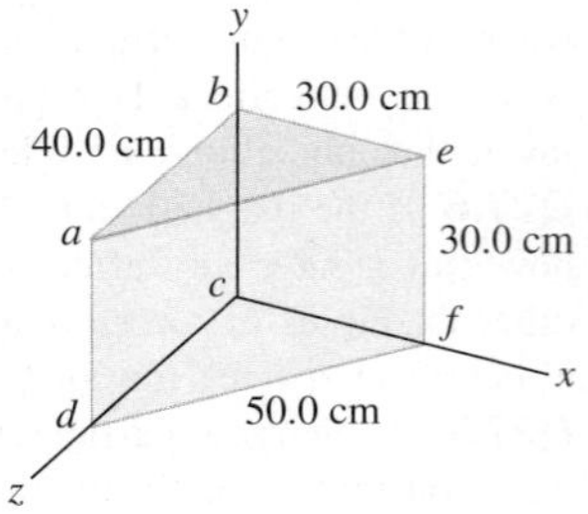

magnetic flux across the surface *abcd* in the figure? (b) What is the magnetic flux across the surface *befc*? (c) What is the magnetic flux across the surface *aefd*? (d) What is the net flux through all five surfaces that enclose the shaded volume?

Section 27.4 Motion of Charged Particles in a Magnetic Field

27.15 •• An electron at point *A* in **Fig. E27.15** has a speed v_0 of 1.41×10^6 m/s. Find (a) the magnitude and direction of the magnetic field that will cause the electron to follow the semicircular path from *A* to *B*, and (b) the time required for the electron to move from *A* to *B*.

Figure **E27.15**

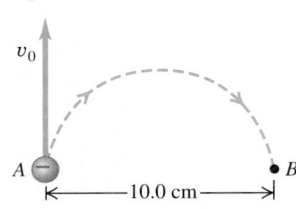

27.16 •• Repeat Exercise 27.15 for the case in which the particle is a proton rather than an electron.

27.17 • CP A 150-g ball containing 4.00×10^8 excess electrons is dropped into a 125-m vertical shaft. At the bottom of the shaft, the ball suddenly enters a uniform horizontal magnetic field that has magnitude 0.250 T and direction from east to west. If air resistance is negligibly small, find the magnitude and direction of the force that this magnetic field exerts on the ball just as it enters the field.

27.18 • An alpha particle (a He nucleus, containing two protons and two neutrons and having a mass of 6.64×10^{-27} kg) traveling horizontally at 35.6 km/s enters a uniform, vertical, 1.80-T magnetic field. (a) What is the diameter of the path followed by this alpha particle? (b) What effect does the magnetic field have on the speed of the particle? (c) What are the magnitude and direction of the acceleration of the alpha particle while it is in the magnetic field? (d) Explain why the speed of the particle does not change even though an unbalanced external force acts on it.

27.19 • In an experiment with cosmic rays, a vertical beam of particles that have charge of magnitude $3e$ and mass 12 times the proton mass enters a uniform horizontal magnetic field of 0.250 T and is bent in a semicircle of diameter 95.0 cm, as shown in **Fig. E27.19**. (a) Find the speed of the particles and the sign of their charge. (b) Is it reasonable to ignore the gravity force on the particles? (c) How does the speed of the particles as they enter the field compare to their speed as they exit the field?

Figure **E27.19**

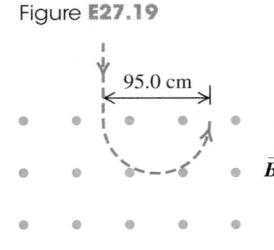

27.20 • BIO Cyclotrons are widely used in nuclear medicine for producing short-lived radioactive isotopes. These cyclotrons typically accelerate H⁻ (the *hydride* ion, which has one proton and two electrons) to an energy of 5 MeV to 20 MeV. This ion has a mass very close to that of a proton because the electron mass is negligible—about $\frac{1}{2000}$ of the proton's mass. A typical magnetic field in such cyclotrons is 1.9 T. (a) What is the speed of a 5.0-MeV H⁻? (b) If the H⁻ has energy 5.0 MeV and $B = 1.9$ T, what is the radius of this ion's circular orbit?

27.21 • A deuteron (the nucleus of an isotope of hydrogen) has a mass of 3.34×10^{-27} kg and a charge of $+e$. The deuteron travels in a circular path with a radius of 6.96 mm in a magnetic field with magnitude 2.50 T. (a) Find the speed of the deuteron. (b) Find the time required for it to make half a revolution. (c) Through what

potential difference would the deuteron have to be accelerated to acquire this speed?

27.22 •• In a cyclotron, the orbital radius of protons with energy 300 keV is 16.0 cm. You are redesigning the cyclotron to be used instead for alpha particles with energy 300 keV. An alpha particle has charge $q = +2e$ and mass $m = 6.64 \times 10^{-27}$ kg. If the magnetic field isn't changed, what will be the orbital radius of the alpha particles?

27.23 • An electron in the beam of a cathode-ray tube is accelerated by a potential difference of 2.00 kV. Then it passes through a region of transverse magnetic field, where it moves in a circular arc with radius 0.180 m. What is the magnitude of the field?

27.24 •• A beam of protons traveling at 1.20 km/s enters a uniform magnetic field, traveling perpendicular to the field. The beam exits the magnetic field, leaving the field in a direction perpendicular to its original direction (**Fig. E27.24**). The beam travels a distance of 1.18 cm *while in the field*. What is the magnitude of the magnetic field?

Figure **E27.24**

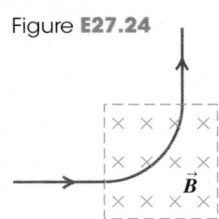

27.25 •• A proton ($q = 1.60 \times 10^{-19}$ C, $m = 1.67 \times 10^{-27}$ kg) moves in a uniform magnetic field $\vec{B} = (0.500 \text{ T})\hat{\imath}$. At $t = 0$ the proton has velocity components $v_x = 1.50 \times 10^5$ m/s, $v_y = 0$, and $v_z = 2.00 \times 10^5$ m/s (see Example 27.4). (a) What are the magnitude and direction of the magnetic force acting on the proton? In addition to the magnetic field there is a uniform electric field in the $+x$-direction, $\vec{E} = (+2.00 \times 10^4 \text{ V/m})\hat{\imath}$. (b) Will the proton have a component of acceleration in the direction of the electric field? (c) Describe the path of the proton. Does the electric field affect the radius of the helix? Explain. (d) At $t = T/2$, where T is the period of the circular motion of the proton, what is the x-component of the displacement of the proton from its position at $t = 0$?

27.26 • A singly charged ion of ⁷Li (an isotope of lithium) has a mass of 1.16×10^{-26} kg. It is accelerated through a potential difference of 220 V and then enters a magnetic field with magnitude 0.874 T perpendicular to the path of the ion. What is the radius of the ion's path in the magnetic field?

Section 27.5 Applications of Motion of Charged Particles

27.27 • **Crossed $\vec{E}$ and $\vec{B}$ Fields.** A particle with initial velocity $\vec{v}_0 = (5.85 \times 10^3 \text{m/s})\hat{\jmath}$ enters a region of uniform electric and magnetic fields. The magnetic field in the region is $\vec{B} = -(1.35 \text{ T})\hat{k}$. Calculate the magnitude and direction of the electric field in the region if the particle is to pass through undeflected, for a particle of charge (a) $+0.640$ nC and (b) -0.320 nC. You can ignore the weight of the particle.

27.28 • (a) What is the speed of a beam of electrons when the simultaneous influence of an electric field of 1.56×10^4 V/m and a magnetic field of 4.62×10^{-3} T, with both fields normal to the beam and to each other, produces no deflection of the electrons? (b) In a diagram, show the relative orientation of the vectors $\vec{v}$, $\vec{E}$, and $\vec{B}$. (c) When the electric field is removed, what is the radius of the electron orbit? What is the period of the orbit?

27.29 •• A 150-V battery is connected across two parallel metal plates of area 28.5 cm² and separation 8.20 mm. A beam of alpha particles (charge $+2e$, mass 6.64×10^{-27} kg) is accelerated from rest through a potential difference of 1.75 kV and enters the region between the plates perpendicular to the electric field, as shown in

Fig. E27.29. What magnitude and direction of magnetic field are needed so that the alpha particles emerge undeflected from between the plates?

Figure **E27.29**

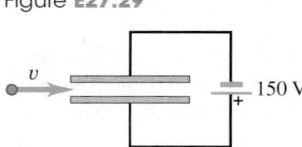

27.30 • A singly ionized (one electron removed) ^{40}K atom passes through a velocity selector consisting of uniform perpendicular electric and magnetic fields. The selector is adjusted to allow ions having a speed of 4.50 km/s to pass through undeflected when the magnetic field is 0.0250 T. The ions next enter a second uniform magnetic field (B') oriented at right angles to their velocity. ^{40}K contains 19 protons and 21 neutrons and has a mass of 6.64×10^{-26} kg. (a) What is the magnitude of the electric field in the velocity selector? (b) What must be the magnitude of B' so that the ions will be bent into a semicircle of radius 12.5 cm?

27.31 • Singly ionized (one electron removed) atoms are accelerated and then passed through a velocity selector consisting of perpendicular electric and magnetic fields. The electric field is 155 V/m and the magnetic field is 0.0315 T. The ions next enter a uniform magnetic field of magnitude 0.0175 T that is oriented perpendicular to their velocity. (a) How fast are the ions moving when they emerge from the velocity selector? (b) If the radius of the path of the ions in the second magnetic field is 17.5 cm, what is their mass?

27.32 • In the Bainbridge mass spectrometer (see Fig. 27.24), the magnetic-field magnitude in the velocity selector is 0.510 T, and ions having a speed of 1.82×10^{6} m/s pass through undeflected. (a) What is the electric-field magnitude in the velocity selector? (b) If the separation of the plates is 5.20 mm, what is the potential difference between the plates?

27.33 •• BIO **Ancient Meat Eating.** The amount of meat in prehistoric diets can be determined by measuring the ratio of the isotopes ^{15}N to ^{14}N in bone from human remains. Carnivores concentrate ^{15}N, so this ratio tells archaeologists how much meat was consumed. For a mass spectrometer that has a path radius of 12.5 cm for ^{12}C ions (mass 1.99×10^{-26} kg), find the separation of the ^{14}N (mass 2.32×10^{-26} kg) and ^{15}N (mass 2.49×10^{-26} kg) isotopes at the detector.

Section 27.6 Magnetic Force on a Current-Carrying Conductor

27.34 • A straight, 2.5-m wire carries a typical household current of 1.5 A (in one direction) at a location where the earth's magnetic field is 0.55 gauss from south to north. Find the magnitude and direction of the force that our planet's magnetic field exerts on this wire if it is oriented so that the current in it is running (a) from west to east, (b) vertically upward, (c) from north to south. (d) Is the magnetic force ever large enough to cause significant effects under normal household conditions?

27.35 •• A long wire carrying 4.50 A of current makes two 90° bends, as shown in **Fig. E27.35.** The bent part of the wire passes through a uniform 0.240-T magnetic field directed as shown in the figure and confined to a limited region of space. Find the magnitude and direction of the force that the magnetic field exerts on the wire.

Figure **E27.35**

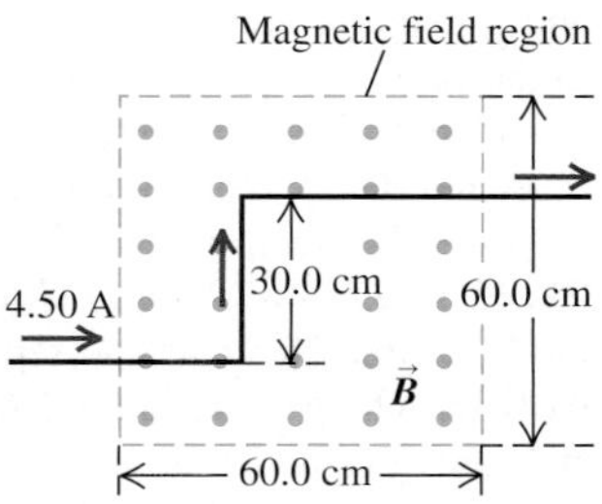

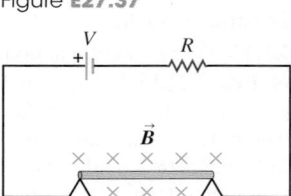

Magnetic field region

4.50 A
30.0 cm
60.0 cm
$\vec{B}$
60.0 cm

27.36 •• An electromagnet produces a magnetic field of 0.550 T in a cylindrical region of radius 2.50 cm between its poles. A straight wire carrying a current of 10.8 A passes through the center of this region and is perpendicular to both the axis of the cylindrical region and the magnetic field. What magnitude of force does this field exert on the wire?

27.37 • A thin, 50.0-cm-long metal bar with mass 750 g rests on, but is not attached to, two metallic supports in a uniform 0.450-T magnetic field, as shown in **Fig. E27.37.** A battery and a 25.0-Ω resistor in series are connected to the supports. (a) What is the highest voltage the battery can have without breaking the circuit at the supports? (b) The battery voltage has the maximum value calculated in part (a). If the resistor suddenly gets partially short-circuited, decreasing its resistance to 2.0 Ω, find the initial acceleration of the bar.

Figure **E27.37**

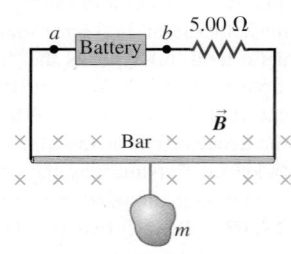

27.38 • A straight, vertical wire carries a current of 2.60 A downward in a region between the poles of a large superconducting electromagnet, where the magnetic field has magnitude $B = 0.588$ T and is horizontal. What are the magnitude and direction of the magnetic force on a 1.00-cm section of the wire that is in this uniform magnetic field, if the magnetic field direction is (a) east; (b) south; (c) 30.0° south of west?

27.39 • **Magnetic Balance.** The circuit shown in **Fig. E27.39** is used to make a magnetic balance to weigh objects. The mass m to be measured is hung from the center of the bar that is in a uniform magnetic field of 1.50 T, directed into the plane of the figure. The battery voltage can be adjusted to vary the current in the circuit. The horizontal bar is 60.0 cm long and is made of extremely light-weight material. It is connected to the battery by thin vertical wires that can support no appreciable tension; all the weight of the suspended mass m is supported by the magnetic force on the bar. A resistor with $R = 5.00$ Ω is in series with the bar; the resistance of the rest of the circuit is much less than this. (a) Which point, a or b, should be the positive terminal of the battery? (b) If the maximum terminal voltage of the battery is 175 V, what is the greatest mass m that this instrument can measure?

Figure **E27.39**

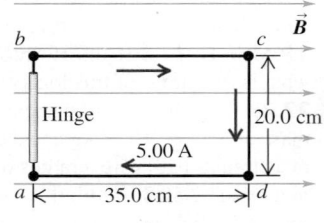

Section 27.7 Force and Torque on a Current Loop

27.40 •• The plane of a 5.0 cm × 8.0 cm rectangular loop of wire is parallel to a 0.19-T magnetic field. The loop carries a current of 6.2 A. (a) What torque acts on the loop? (b) What is the magnetic moment of the loop? (c) What is the maximum torque that can be obtained with the same total length of wire carrying the same current in this magnetic field?

27.41 • The 20.0 cm × 35.0 cm rectangular circuit shown in **Fig. E27.41** is hinged along side ab. It carries a clockwise

Figure **E27.41**

$\vec{B}$
b
c
Hinge
20.0 cm
5.00 A
a
35.0 cm
d

5.00-A current and is located in a uniform 1.20-T magnetic field oriented perpendicular to two of its sides, as shown. (a) Draw a clear diagram showing the direction of the force that the magnetic field exerts on each segment of the circuit (*ab*, *bc*, etc.). (b) Of the four forces you drew in part (a), decide which ones exert a torque about the hinge *ab*. Then calculate only those forces that exert this torque. (c) Use your results from part (b) to calculate the torque that the magnetic field exerts on the circuit about the hinge axis *ab*.

27.42 • A rectangular coil of wire, 22.0 cm by 35.0 cm and carrying a current of 1.95 A, is oriented with the plane of its loop perpendicular to a uniform 1.50-T magnetic field (**Fig. E27.42**). (a) Calculate the net force and torque that the magnetic field exerts on the coil. (b) The coil is rotated through a 30.0° angle about the axis shown, with the left side coming out of the plane of the figure and the right side going into the plane. Calculate the net force and torque that the magnetic field now exerts on the coil. (*Hint:* To visualize this three-dimensional problem, make a careful drawing of the coil as viewed along the rotation axis.)

Figure **E27.42**

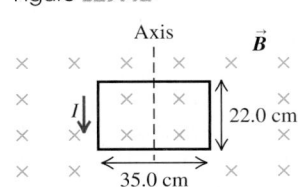

27.43 • **CP** A uniform rectangular coil of total mass 212 g and dimensions 0.500 m × 1.00 m is oriented with its plane parallel to a uniform 3.00-T magnetic field (**Fig. E27.43**). A current of 2.00 A is suddenly started in the coil. (a) About which axis (A_1 or A_2) will the coil begin to rotate? Why? (b) Find the initial angular acceleration of the coil just after the current is started.

Figure **E27.43**

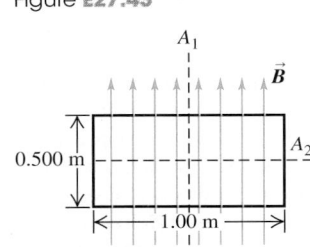

27.44 • Both circular coils A and B (**Fig. E27.44**) have area A and N turns. They are free to rotate about a diameter that coincides with the *x*-axis. Current I circulates in each coil in the direction shown. There is a uniform magnetic field $\vec{B}$ in the +*z*-direction. (a) What is the direction of the magnetic moment $\vec{\mu}$ for each coil? (b) Explain why the torque on both coils due to the magnetic field is zero, so the coil is in rotational equilibrium. (c) Use Eq. (27.27) to calculate the potential energy for each coil. (d) For each coil, is the equilibrium stable or unstable? Explain.

Figure **E27.44**

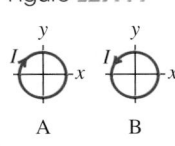

27.45 • A circular coil with area A and N turns is free to rotate about a diameter that coincides with the *x*-axis. Current I is circulating in the coil. There is a uniform magnetic field $\vec{B}$ in the positive *y*-direction. Calculate the magnitude and direction of the torque $\vec{\tau}$ and the value of the potential energy U, as given in Eq. (27.27), when the coil is oriented as shown in parts (a) through (d) of **Fig. E27.45**.

Figure **E27.45**

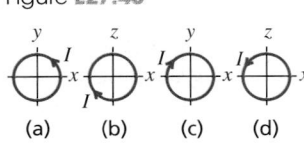

27.46 •• A coil with magnetic moment 1.45 A·m² is oriented initially with its magnetic moment antiparallel to a uniform 0.835-T magnetic field. What is the change in potential energy of the coil when it is rotated 180° so that its magnetic moment is parallel to the field?

Section 27.8 The Direct-Current Motor

27.47 •• In a shunt-wound dc motor with the field coils and rotor connected in parallel (**Fig. E27.47**), the resistance R_f of the field coils is 106 Ω, and the resistance R_r of the rotor is 5.9 Ω. When a potential difference of 120 V is applied to the brushes and the motor is running at full speed delivering mechanical power, the current supplied to it is 4.82 A. (a) What is the current in the field coils? (b) What is the current in the rotor? (c) What is the induced emf developed by the motor? (d) How much mechanical power is developed by this motor?

Figure **E27.47**

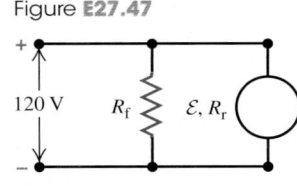

27.48 • A dc motor with its rotor and field coils connected in series has an internal resistance of 3.2 Ω. When the motor is running at full load on a 120-V line, the emf in the rotor is 105 V. (a) What is the current drawn by the motor from the line? (b) What is the power delivered to the motor? (c) What is the mechanical power developed by the motor?

Section 27.9 The Hall Effect

27.49 • **Figure E27.49** shows a portion of a silver ribbon with $z_1 = 11.8$ mm and $y_1 = 0.23$ mm, carrying a current of 120 A in the +*x*-direction. The ribbon lies in a uniform magnetic field, in the *y*-direction, with magnitude 0.95 T. Apply the simplified model of the Hall effect presented in Section 27.9. If there are 5.85×10^{28} free electrons per cubic meter, find (a) the magnitude of the drift velocity of the electrons in the *x*-direction; (b) the magnitude and direction of the electric field in the *z*-direction due to the Hall effect; (c) the Hall emf.

Figure **E27.49**

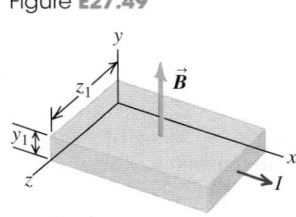

27.50 • Let Fig. E27.49 represent a strip of an unknown metal of the same dimensions as those of the silver ribbon in Exercise 27.49. When the magnetic field is 2.29 T and the current is 78.0 A, the Hall emf is found to be 131 μV. What does the simplified model of the Hall effect presented in Section 27.9 give for the density of free electrons in the unknown metal?

PROBLEMS

27.51 • When a particle of charge $q > 0$ moves with a velocity of $\vec{v}_1$ at 45.0° from the +*x*-axis in the *xy*-plane, a uniform magnetic field exerts a force $\vec{F}_1$ along the −*z*-axis (**Fig. P27.51**). When the same particle moves with a velocity $\vec{v}_2$ with the same magnitude as $\vec{v}_1$ but along the +*z*-axis, a force $\vec{F}_2$ of magnitude F_2 is exerted on it along the +*x*-axis. (a) What are the magnitude (in terms of q, v_1, and F_2) and direction of the magnetic field? (b) What is the magnitude of $\vec{F}_1$ in terms of F_2?

Figure **P27.51**

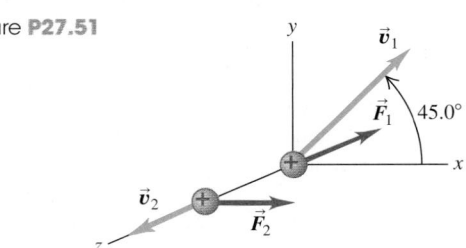

27.52 • A particle with charge 7.26×10^{-8} C is moving in a region where there is a uniform 0.650-T magnetic field in the $+x$-direction. At a particular instant, the velocity of the particle has components $v_x = -1.68 \times 10^4$ m/s, $v_y = -3.11 \times 10^4$ m/s, and $v_z = 5.85 \times 10^4$ m/s. What are the components of the force on the particle at this time?

27.53 ••• CP **Fusion Reactor.** If two deuterium nuclei (charge $+e$, mass 3.34×10^{-27} kg) get close enough together, the attraction of the strong nuclear force will fuse them to make an isotope of helium, releasing vast amounts of energy. The range of this force is about 10^{-15} m. This is the principle behind the fusion reactor. The deuterium nuclei are moving much too fast to be contained by physical walls, so they are confined magnetically. (a) How fast would two nuclei have to move so that in a head-on collision they would get close enough to fuse? (Assume their speeds are equal. Treat the nuclei as point charges, and assume that a separation of 1.0×10^{-15} is required for fusion.) (b) What strength magnetic field is needed to make deuterium nuclei with this speed travel in a circle of diameter 2.50 m?

27.54 •• **Magnetic Moment of the Hydrogen Atom.** In the Bohr model of the hydrogen atom (see Section 39.3), in the lowest energy state the electron orbits the proton at a speed of 2.2×10^6 m/s in a circular orbit of radius 5.3×10^{-11} m. (a) What is the orbital period of the electron? (b) If the orbiting electron is considered to be a current loop, what is the current I? (c) What is the magnetic moment of the atom due to the motion of the electron?

27.55 •• You wish to hit a target from several meters away with a charged coin having a mass of 4.25 g and a charge of $+2500 \, \mu$C. The coin is given an initial velocity of 12.8 m/s, and a downward, uniform electric field with field strength 27.5 N/C exists throughout the region. If you aim directly at the target and fire the coin horizontally, what magnitude and direction of uniform magnetic field are needed in the region for the coin to hit the target?

27.56 • The magnetic poles of a small cyclotron produce a magnetic field with magnitude 0.85 T. The poles have a radius of 0.40 m, which is the maximum radius of the orbits of the accelerated particles. (a) What is the maximum energy to which protons ($q = 1.60 \times 10^{-19}$ C, $m = 1.67 \times 10^{-27}$ kg) can be accelerated by this cyclotron? Give your answer in electron volts and in joules. (b) What is the time for one revolution of a proton orbiting at this maximum radius? (c) What would the magnetic-field magnitude have to be for the maximum energy to which a proton can be accelerated to be twice that calculated in part (a)? (d) For $B = 0.85$ T, what is the maximum energy to which alpha particles ($q = 3.20 \times 10^{-19}$ C, $m = 6.64 \times 10^{-27}$ kg) can be accelerated by this cyclotron? How does this compare to the maximum energy for protons?

27.57 •• A particle with negative charge q and mass $m = 2.58 \times 10^{-15}$ kg is traveling through a region containing a uniform magnetic field $\vec{B} = -(0.120 \text{ T})\hat{k}$. At a particular instant of time the velocity of the particle is $\vec{v} = (1.05 \times 10^6 \text{ m/s})(-3\hat{i} + 4\hat{j} + 12\hat{k})$ and the force $\vec{F}$ on the particle has a magnitude of 2.45 N. (a) Determine the charge q. (b) Determine the acceleration $\vec{a}$ of the particle. (c) Explain why the path of the particle is a helix, and determine the radius of curvature R of the circular component of the helical path. (d) Determine the cyclotron frequency of the particle. (e) Although helical motion is not periodic in the full sense of the word, the x- and y-coordinates do vary in a periodic way. If the coordinates of the particle at $t = 0$ are $(x, y, z) = (R, 0, 0)$, determine its coordinates at a time $t = 2T$, where T is the period of the motion in the xy-plane.

27.58 •• A particle of charge $q > 0$ is moving at speed v in the $+z$-direction through a region of uniform magnetic field $\vec{B}$. The magnetic force on the particle is $\vec{F} = F_0(3\hat{i} + 4\hat{j})$, where F_0 is a positive constant. (a) Determine the components B_x, B_y, and B_z, or at least as many of the three components as is possible from the information given. (b) If it is given in addition that the magnetic field has magnitude $6F_0/qv$, determine as much as you can about the remaining components of $\vec{B}$.

27.59 •• Suppose the electric field between the plates in Fig. 27.24 is 1.88×10^4 V/m and the magnetic field in both regions is 0.682 T. If the source contains the three isotopes of krypton, ^{82}Kr, ^{84}Kr, and ^{86}Kr, and the ions are singly charged, find the distance between the lines formed by the three isotopes on the particle detector. Assume the atomic masses of the isotopes (in atomic mass units) are equal to their mass numbers, 82, 84, and 86. (One atomic mass unit = 1 u = 1.66×10^{-27} kg.)

27.60 •• **Mass Spectrograph.** A mass spectrograph is used to measure the masses of ions, or to separate ions of different masses (see Section 27.5). In one design for such an instrument, ions with mass m and charge q are accelerated through a potential difference V. They then enter a uniform magnetic field that is perpendicular to their velocity, and they are deflected in a semicircular path of radius R. A detector measures where the ions complete the semicircle and from this it is easy to calculate R. (a) Derive the equation for calculating the mass of the ion from measurements of B, V, R, and q. (b) What potential difference V is needed so that singly ionized ^{12}C atoms will have $R = 50.0$ cm in a 0.150-T magnetic field? (c) Suppose the beam consists of a mixture of ^{12}C and ^{14}C ions. If v and B have the same values as in part (b), calculate the separation of these two isotopes at the detector. Do you think that this beam separation is sufficient for the two ions to be distinguished? (Make the assumption described in Problem 27.59 for the masses of the ions.)

27.61 •• A straight piece of conducting wire with mass M and length L is placed on a frictionless incline tilted at an angle θ from the horizontal (**Fig. P27.61**). There is a uniform, vertical magnetic field $\vec{B}$ at all points (produced by an arrangement of magnets not shown in the figure). To keep the wire from sliding down the incline, a voltage source is attached to the ends of the wire. When just the right amount of current flows through the wire, the wire remains at rest. Determine the magnitude and direction of the current in the wire that will cause the wire to remain at rest. Copy the figure and draw the direction of the current on your copy. In addition, show in a free-body diagram all the forces that act on the wire.

Figure **P27.61**

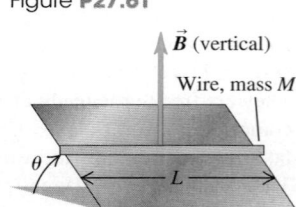

$\vec{B}$ (vertical)

Wire, mass M

θ

L

27.62 •• CP A 2.60-N metal bar, 0.850 m long and having a resistance of 10.0 Ω, rests horizontally on conducting wires connecting it to the circuit shown in **Fig. P27.62**. The bar is in a uniform, horizontal, 1.60-T magnetic field and is not attached to the wires in the circuit. What is the acceleration of the bar just after the switch S is closed?

Figure **P27.62**

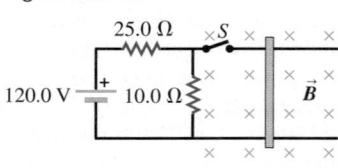

25.0 Ω

S

120.0 V

10.0 Ω

$\vec{B}$

27.63 •• BIO **Determining Diet.** One method for determining the amount of corn in early Native American diets is the *stable* *isotope ratio analysis* (SIRA) technique. As corn photosynthesizes, it concentrates the isotope carbon-13, whereas most other plants concentrate carbon-12. Overreliance on corn consumption can then be correlated with certain diseases, because corn lacks the essential amino acid lysine. Archaeologists use a mass spectrometer to separate the ^{12}C and ^{13}C isotopes in samples of human remains. Suppose you use a velocity selector to obtain singly ionized (missing one electron) atoms of speed 8.50 km/s, and you want to bend them within a uniform magnetic field in a semicircle of diameter 25.0 cm for the ^{12}C. The measured masses of these isotopes are 1.99×10^{-26} kg (^{12}C) and 2.16×10^{-26} kg (^{13}C). (a) What strength of magnetic field is required? (b) What is the diameter of the ^{13}C semicircle? (c) What is the separation of the ^{12}C and ^{13}C ions at the detector at the end of the semicircle? Is this distance large enough to be easily observed?

27.64 •• CP A plastic circular loop has radius R, and a positive charge q is distributed uniformly around the circumference of the loop. The loop is then rotated around its central axis, perpendicular to the plane of the loop, with angular speed ω. If the loop is in a region where there is a uniform magnetic field $\vec{B}$ directed parallel to the plane of the loop, calculate the magnitude of the magnetic torque on the loop.

27.65 •• CP **An Electromagnetic Rail Gun.** A conducting bar with mass m and length L slides over horizontal rails that are connected to a voltage source. The voltage source maintains a constant current I in the rails and bar, and a constant, uniform, vertical magnetic field $\vec{B}$ fills the region between the rails (**Fig. P27.65**). (a) Find the magnitude and direction of the net force on the conducting bar. Ignore friction, air resistance, and electrical resistance. (b) If the bar has mass m, find the distance d that the bar must move along the rails from rest to attain speed v. (c) It has been suggested that rail guns based on this principle could accelerate payloads into earth orbit or beyond. Find the distance the bar must travel along the rails if it is to reach the escape speed for the earth (11.2 km/s). Let $B = 0.80$ T, $I = 2.0 \times 10^3$ A, $m = 25$ kg, and $L = 50$ cm. For simplicity assume the net force on the object is equal to the magnetic force, as in parts (a) and (b), even though gravity plays an important role in an actual launch in space.

Figure **P27.65**

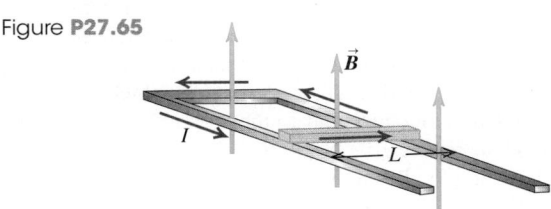

27.66 • A wire 25.0 cm long lies along the z-axis and carries a current of 7.40 A in the $+z$-direction. The magnetic field is uniform and has components $B_x = -0.242$ T, $B_y = -0.985$ T, and $B_z = -0.336$ T. (a) Find the components of the magnetic force on the wire. (b) What is the magnitude of the net magnetic force on the wire?

27.67 •• A long wire carrying 6.50 A of current makes two bends, as shown in **Fig. P27.67**. The bent part of the wire passes through a uniform 0.280-T magnetic

Figure **P27.67**

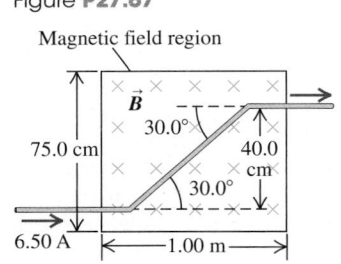

Magnetic field region

field directed as shown and confined to a limited region of space. Find the magnitude and direction of the force that the magnetic field exerts on the wire.

27.68 •• The rectangular loop shown in **Fig. P27.68** is pivoted about the y-axis and carries a current of 15.0 A in the direction indicated. (a) If the loop is in a uniform magnetic field with magnitude 0.48 T in the $+x$-direction, find the magnitude and direction of the torque required to hold the loop in the position shown. (b) Repeat part (a) for the case in which the field is in the $-z$-direction. (c) For each of the above magnetic fields, what torque would be required if the loop were pivoted about an axis through its center, parallel to the y-axis?

Figure **P27.68**

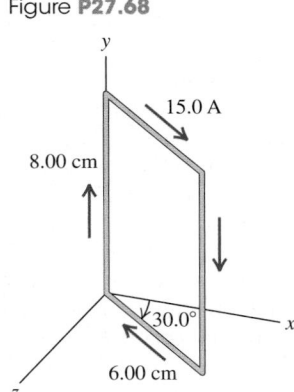

27.69 •• CP The rectangular loop of wire shown in **Fig. P27.69** has a mass of 0.15 g per centimeter of length and is pivoted about side ab on a frictionless axis. The current in the wire is 8.2 A in the direction shown. Find the magnitude and direction of the magnetic field parallel to the y-axis that will cause the loop to swing up until its plane makes an angle of 30.0° with the yz-plane.

Figure **P27.69**

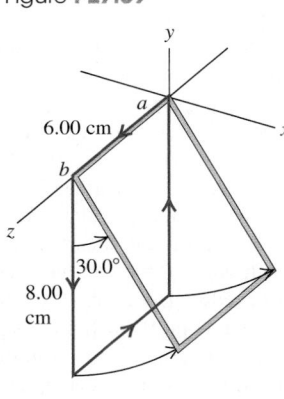

27.70 •• CALC A uniform bar of length L carries a current I in the direction from point a to point b (**Fig. P27.70**). The bar is in a uniform magnetic field that is directed into the page. Consider the torque about an axis perpendicular to the bar at point a that is due to the force that the magnetic field exerts on the bar. (a) Suppose that an infinitesimal section of the bar has length dx and is located a distance x from point a. Calculate the torque $d\tau$ about point a due to the magnetic force on this infinitesimal section. (b) Use $\tau = \int_a^b d\tau$ to calculate the total torque τ on the bar. (c) Show that τ is the same as though all of the magnetic force acted at the midpoint of the bar.

Figure **P27.70**

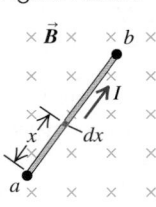

27.71 •• The loop of wire shown in **Fig. P27.71** forms a right triangle and carries a current $I = 5.00$ A in the direction shown. The loop is in a uniform magnetic field that has magnitude $B = 3.00$ T and the same direction as the current in side PQ of the loop. (a) Find the force exerted by the magnetic field on each side of the triangle. If the force is not zero, specify its direction. (b) What is the net force on the loop? (c) The loop is pivoted about an axis that lies along side PR. Use the forces calculated in part (a) to calculate the torque on each side of the loop

Figure **P27.71**

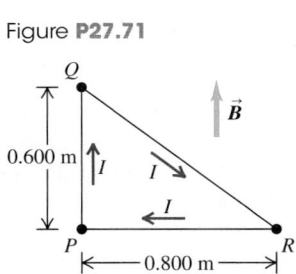

(see Problem 27.70). (d) What is the magnitude of the net torque on the loop? Calculate the net torque from the torques calculated in part (c) and also from Eq. (27.28). Do these two results agree? (e) Is the net torque directed to rotate point Q into the plane of the figure or out of the plane of the figure?

27.72 •• CP A uniform bar has mass 0.0120 kg and is 30.0 cm long. It pivots without friction about an axis perpendicular to the bar at point a (**Fig. P27.72**). The gravitational force on the bar acts in the $-y$-direction. The bar is in a uniform magnetic field that is directed into the page and has magnitude $B = 0.150$ T. (a) What must be the current I in the bar for the bar to be in rotational equilibrium when it is at an angle $\theta = 30.0°$ above the horizontal? Use your result from Problem 27.70. (b) For the bar to be in rotational equilibrium, should I be in the direction from a to b or b to a?

Figure **P27.72**

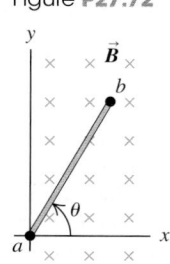

27.73 •• CALC A Voice Coil. It was shown in Section 27.7 that the net force on a current loop in a *uniform* magnetic field is zero. The magnetic force on the voice coil of a loudspeaker (see Fig. 27.28) is nonzero because the magnetic field at the coil is not uniform. A voice coil in a loudspeaker has 50 turns of wire and a diameter of 1.56 cm, and the current in the coil is 0.950 A. Assume that the magnetic field at each point of the coil has a constant magnitude of 0.220 T and is directed at an angle of 60.0° outward from the normal to the plane of the coil (**Fig. P27.73**). Let the axis of the coil be in the y-direction. The current in the coil is in the direction shown (counterclockwise as viewed from a point above the coil on the y-axis). Calculate the magnitude and direction of the net magnetic force on the coil.

Figure **P27.73**

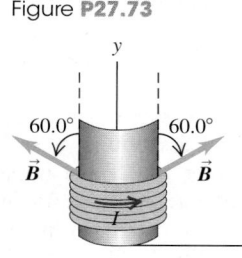

27.74 •• CP The lower end of the thin uniform rod in **Fig. P27.74** is attached to the floor by a frictionless hinge at point P. The rod has mass 0.0840 kg and length 18.0 cm and is in a uniform magnetic field $B = 0.120$ T that is directed into the page. The rod is held at an angle $\theta = 53.0°$ above the horizontal by a horizontal string that connects the top of the rod to the wall. The rod carries a current $I = 12.0$ A in the direction toward P. Calculate the tension in the string. Use your result from Problem 27.70 to calculate the torque due to the magnetic-field force.

Figure **P27.74**

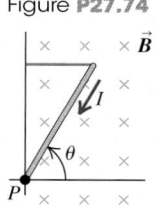

27.75 •• CALC Force on a Current Loop in a Nonuniform Magnetic Field. It was shown in Section 27.7 that the net force on a current loop in a *uniform* magnetic field is zero. But what if $\vec{B}$ is *not* uniform? **Figure P27.75** shows a square loop of wire that lies in the xy-plane. The loop has corners at $(0, 0)$, $(0, L)$, $(L, 0)$, and (L, L) and carries a constant current I in the clockwise direction. The magnetic field has no x-component but has both y- and z-components: $\vec{B} = (B_0 z/L)\hat{j} + (B_0 y/L)\hat{k}$, where B_0 is a positive constant. (a) Sketch the magnetic field lines in the yz-plane. (b) Find the magnitude and direction of the magnetic force exerted on each of the sides of the loop by integrating

Figure **P27.75**

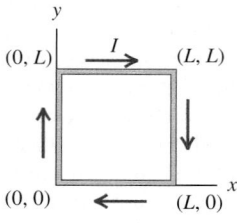

Eq. (27.20). (c) Find the magnitude and direction of the net magnetic force on the loop.

27.76 •• Quark Model of the Neutron. The neutron is a particle with zero charge. Nonetheless, it has a nonzero magnetic moment with z-component 9.66×10^{-27} A·m². This can be explained by the internal structure of the neutron. A substantial body of evidence indicates that a neutron is composed of three fundamental particles called *quarks:* an "up" (u) quark, of charge $+2e/3$, and two "down" (d) quarks, each of charge $-e/3$. The combination of the three quarks produces a net charge of $\frac{2}{3}e - \frac{1}{3}e - \frac{1}{3}e = 0$. If the quarks are in motion, they can produce a nonzero magnetic moment. As a very simple model, suppose the u quark moves in a counterclockwise circular path and the d quarks move in a clockwise circular path, all of radius r and all with the same speed v (**Fig. P27.76**). (a) Determine the current due to the circulation of the u quark. (b) Determine the magnitude of the magnetic moment due to the circulating u quark. (c) Determine the magnitude of the magnetic moment of the three-quark system. (Be careful to use the correct magnetic moment directions.) (d) With what speed v must the quarks move if this model is to reproduce the magnetic moment of the neutron? Use $r = 1.20 \times 10^{-15}$ m (the radius of the neutron) for the radius of the orbits.

Figure **P27.76**

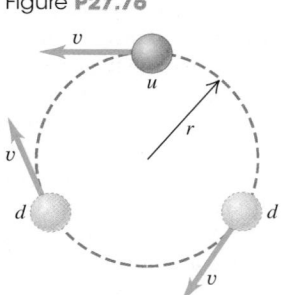

27.77 •• A circular loop of wire with area A lies in the xy-plane. As viewed along the z-axis looking in the $-z$-direction toward the origin, a current I is circulating clockwise around the loop. The torque produced by an external magnetic field $\vec{B}$ is given by $\vec{\tau} = D(4\hat{i} - 3\hat{j})$, where D is a positive constant, and for this orientation of the loop the magnetic potential energy $U = -\vec{\mu} \cdot \vec{B}$ is negative. The magnitude of the magnetic field is $B_0 = 13D/IA$. (a) Determine the vector magnetic moment of the current loop. (b) Determine the components B_x, B_y, and B_z of $\vec{B}$.

27.78 •• DATA You are using a type of mass spectrometer to measure charge-to-mass ratios of atomic ions. In the device, atoms are ionized with a beam of electrons to produce positive ions, which are then accelerated through a potential difference V. (The final speed of the ions is great enough that you can ignore their initial speed.) The ions then enter a region in which a uniform magnetic field $\vec{B}$ is perpendicular to the velocity of the ions and has magnitude $B = 0.250$ T. In this $\vec{B}$ region, the ions move in a semicircular path of radius R. You measure R as a function of the accelerating voltage V for one particular atomic ion:

V (kV)	10.0	12.0	14.0	16.0	18.0
R (cm)	19.9	21.8	23.6	25.2	26.8

(a) How can you plot the data points so that they will fall close to a straight line? Explain. (b) Construct the graph described in part (a). Use the slope of the best-fit straight line to calculate the charge-to-mass ratio (q/m) for the ion. (c) For $V = 20.0$ kV, what is the speed of the ions as they enter the $\vec{B}$ region? (d) If ions that have $R = 21.2$ cm for $V = 12.0$ kV are singly ionized, what is R when $V = 12.0$ kV for ions that are doubly ionized?

27.79 •• DATA You are a research scientist working at a high-energy particle accelerator. Using a modern version of the Thomson e/m apparatus, you want to measure the mass of a muon (a fundamental particle that has the same charge as an electron but greater mass). The magnetic field between the two charged plates is 0.340 T. You measure the electric field for zero particle deflection as a function of the accelerating potential V. This potential is large enough that you can assume the initial speed of the muons to be zero. **Figure P27.79** is an E^2-versus-V graph of your data. (a) Explain why the data points fall close to a straight line. (b) Use the graph in Fig. P27.79 to calculate the mass m of a muon. (c) If the two charged plates are separated by 6.00 mm, what must be the voltage between the plates in order for the electric field between the plates to be 2.00×10^5 V/m? Assume that the dimensions of the plates are much larger than the separation between them. (d) When $V = 400$ V, what is the speed of the muons as they enter the region between the plates?

Figure **P27.79**

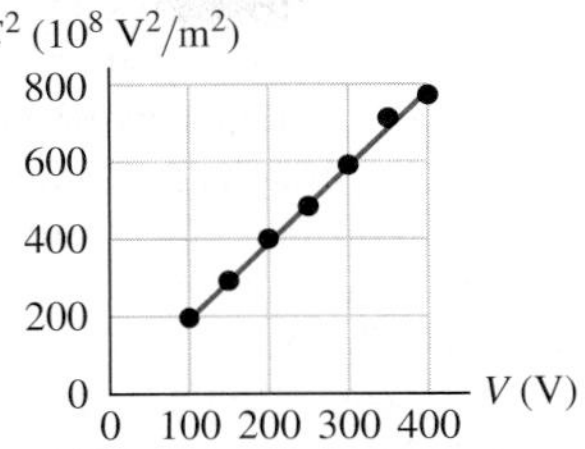

E²(10⁸ V²/m²)

(graph with vertical axis E^2 (10^8 V²/m²) marked 200, 400, 600, 800 and horizontal axis V (V) marked 0, 100, 200, 300, 400)

27.80 •• **DATA** You are a technician testing the operation of a cyclotron. An alpha particle in the device moves in a circular path in a magnetic field $\vec{B}$ that is directed perpendicular to the path of the alpha particle. You measure the number of revolutions per second (the frequency f) of the alpha particle as a function of the magnetic field strength B. **Figure P27.80** shows your results and the best straight-line fit to your data. (a) Use the graph in Fig. P27.80 to calculate the charge-to-mass ratio of the alpha particle, which has charge $+2e$. On the basis of your data, what is the mass of an alpha particle? (b) With $B = 0.300$ T, what are the cyclotron frequencies f of a proton and of an electron? How do these f values compare to the frequency of an alpha particle? (c) With $B = 0.300$ T, what speed and kinetic energy does an alpha particle have if the radius of its path is 12.0 cm?

Figure **P27.80**

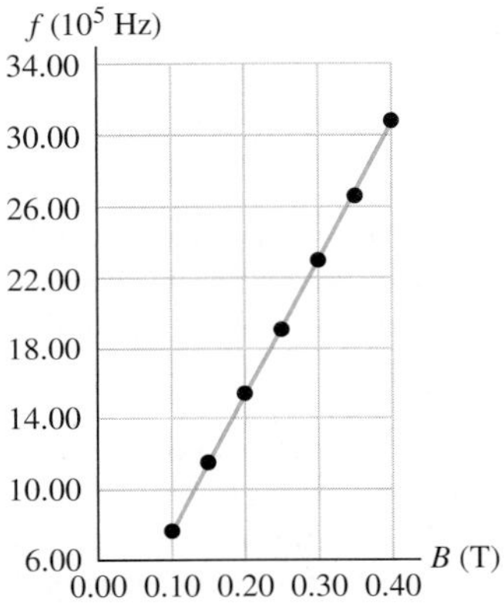

f (10⁵ Hz)

(graph with vertical axis f (10^5 Hz) marked 6.00, 10.00, 14.00, 18.00, 22.00, 26.00, 30.00, 34.00 and horizontal axis B (T) marked 0.00, 0.10, 0.20, 0.30, 0.40)

CHALLENGE PROBLEMS

27.81 ••• A particle with charge 2.15 μC and mass 3.20×10^{-11} kg is initially traveling in the $+y$-direction with a speed $v_0 = 1.45 \times 10^5$ m/s. It then enters a region containing a uniform magnetic field that is directed into, and perpendicular to, the page in **Fig. P27.81**. The magnitude of the field is 0.420 T. The region extends a distance of 25.0 cm along the initial direction of travel; 75.0 cm from the point of entry into the magnetic field region is a wall. The length of the field-free region is thus 50.0 cm. When the charged particle enters the magnetic field, it follows a curved path whose radius of curvature is R. It then leaves the magnetic field after a time t_1, having been deflected a distance Δx_1. The particle then travels in the field-free region and strikes the wall after undergoing a total deflection Δx. (a) Determine the radius R of the curved part of the path. (b) Determine t_1, the time the particle spends in the magnetic field. (c) Determine Δx_1, the horizontal deflection at the point of exit from the field. (d) Determine Δx, the total horizontal deflection.

Figure **P27.81**

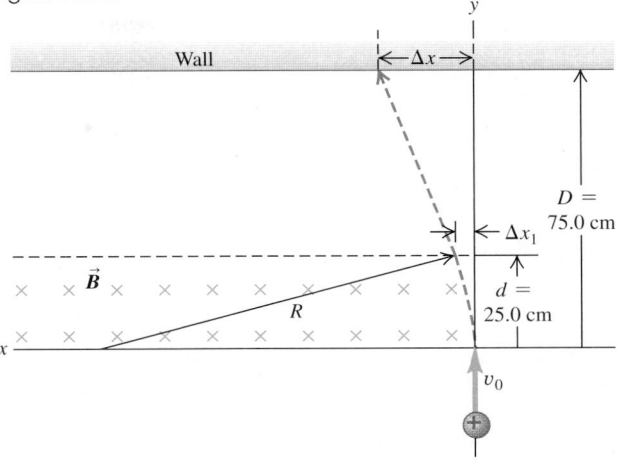

27.82 ••• **CP A Cycloidal Path.** A particle with mass m and positive charge q starts from rest at the origin shown in **Fig. P27.82**. There is a uniform electric field $\vec{E}$ in the $+y$-direction and a uniform magnetic field $\vec{B}$ directed out of the page. It is shown in more advanced books that the path is a cycloid whose radius of curvature at the top points is twice the y-coordinate at that level. (a) Explain why the path has this general shape and why it is repetitive. (b) Prove that the speed at any point is equal to $\sqrt{2qEy/m}$. (*Hint:* Use energy conservation.) (c) Applying Newton's second law at the top point and taking as given that the radius of curvature here equals $2y$, prove that the speed at this point is $2E/B$.

Figure **P27.82**

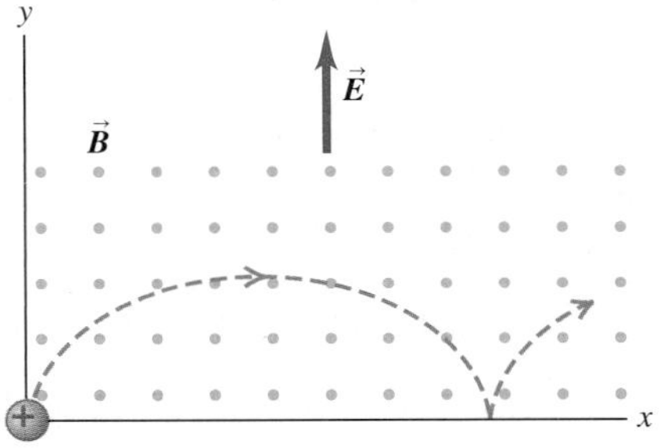

(diagram showing uniform field $\vec{E}$ in +y direction, $\vec{B}$ out of page, and cycloidal dashed path)

PASSAGE PROBLEMS

BIO MAGNETIC FIELDS AND MRI. *Magnetic resonance imaging* (MRI) is a powerful imaging method that, unlike x-ray imaging, allows sharp images of soft tissue to be made without exposing the patient to potentially damaging radiation. A rudimentary understanding of this method can be achieved by the relatively simple application of the classical (that is, non-quantum) physics of magnetism. The starting point for MRI is *nuclear magnetic resonance* (NMR), a technique that depends on the fact that protons in the atomic nucleus have a magnetic field $\vec{B}$. The origin of the proton's magnetic field is the spin of the proton. Being charged, the spinning proton constitutes an electric current analogous to a wire loop through which current flows. Like the wire loop, the proton has a magnetic moment $\vec{\mu}$; thus it will experience a torque when it is subjected to an external magnetic field $\vec{B}_0$. The magnitude of $\vec{\mu}$ is about 1.4×10^{-26} J/T. The proton can be thought of as being in one of two states, with $\vec{\mu}$ oriented parallel or antiparallel to the applied magnetic field, and work must be done to flip the proton from the low-energy state to the high-energy state, as the accompanying figure (next page) shows.

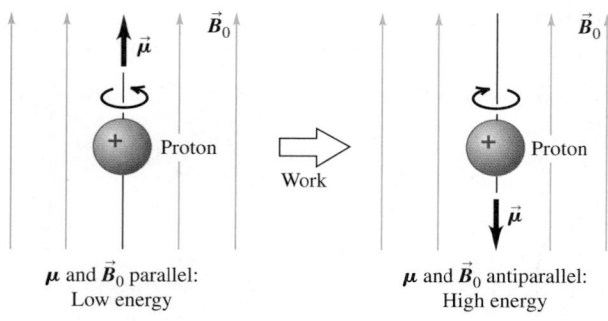

$\vec{\mu}$ and $\vec{B}_0$ parallel:
Low energy

$\vec{\mu}$ and $\vec{B}_0$ antiparallel:
High energy

An important consideration is that the net magnetic field of any nucleus, except for that of hydrogen (which has a proton only), consists of contributions from both protons and neutrons. If a nucleus has an even number of protons and neutrons, they will pair in such a way that half of the protons have spins in one orientation and half have spins in the other orientation. Thus the net magnetic

moment of the nucleus is zero. Only nuclei with a net magnetic moment are candidates for MRI. Hydrogen is the atom that is most commonly imaged.

27.83 If a proton is exposed to an external magnetic field of 2 T that has a direction perpendicular to the axis of the proton's spin, what will be the torque on the proton? (a) 0; (b) 1.4×10^{-26} N·m; (c) 2.8×10^{-26} N·m; (d) 0.7×10^{-26} N·m.

27.84 Which of following elements is a candidate for MRI? (a) $^{12}C_6$; (b) $^{16}O_8$; (c) $^{40}Ca_{20}$; (d) $^{31}P_{15}$.

27.85 The large magnetic fields used in MRI can produce forces on electric currents within the human body. This effect has been proposed as a possible method for imaging "biocurrents" flowing in the body, such as the current that flows in individual nerves. For a magnetic field strength of 2 T, estimate the magnitude of the maximum force on a 1-mm-long segment of a single cylindrical nerve that has a diameter of 1.5 mm. Assume that the entire nerve carries a current due to an applied voltage of 100 mV (that of a typical action potential). The resistivity of the nerve is 0.6 Ω·m. (a) 6×10^{-7} N; (b) 1×10^{-6} N; (c) 3×10^{-4} N; (d) 0.3 N.

Answers

Chapter Opening Question ?

(ii) A magnetized compass needle has a magnetic dipole moment along its length, and the earth's magnetic field (which points generally northward) exerts a torque that tends to align that dipole moment with the field. See Section 27.7 for details.

Test Your Understanding Questions

27.1 yes When a magnet is cut apart, each part has a north and south pole (see Fig. 27.4). Hence the small red part behaves much like the original, full-sized compass needle.

27.2 path 3 Applying the right-hand rule to the vectors $\vec{v}$ (which points to the right) and $\vec{B}$ (which points into the plane of the figure) says that the force $\vec{F} = q\vec{v} \times \vec{B}$ on a *positive* charge would point *upward*. Since the charge is *negative*, the force points *downward* and the particle follows a trajectory that curves downward.

27.3 (a) (ii), (b) no The magnitude of $\vec{B}$ would increase as you moved to the right, reaching a maximum as you passed through the plane of the loop. As you moved beyond the plane of the loop, the field magnitude would decrease. You can tell this from the spacing of the field lines: The closer the field lines, the stronger the field. The direction of the field would be to the right at all points along the path, since the path is along a field line and the direction of $\vec{B}$ at any point is tangent to the field line through that point.

27.4 (a) (ii), (b) (i) The radius of the orbit as given by Eq. (27.11) is directly proportional to the speed, so doubling the particle speed causes the radius to double as well. The particle has twice as far to travel to complete one orbit but is traveling at double the speed, so the time for one orbit is unchanged. This result also follows from Eq. (27.12), which states that the angular speed ω is independent of the linear speed v. Hence the time per orbit, $T = 2\pi/\omega$, likewise does not depend on v.

27.5 (iii) From Eq. (27.13), the speed $v = E/B$ at which particles travel straight through the velocity selector does not depend on the magnitude or sign of the charge or the mass of the particle. All that is required is that the particles (in this case, ions) have a nonzero charge.

27.6 A This orientation will cause current to flow clockwise around the circuit and hence through the conducting bar in the direction from the top to the bottom of the figure. From the right-hand rule, the magnetic force $\vec{F} = I\vec{l} \times \vec{B}$ on the bar will then point to the right.

27.7 (a) to the right; (b) north pole on the right, south pole on the left If you wrap the fingers of your right hand around the coil in the direction of the current, your right thumb points to the right (perpendicular to the plane of the coil). This is the direction of the magnetic moment $\vec{\mu}$. The magnetic moment points from the south pole to the north pole, so the right side of the loop is equivalent to a north pole and the left side is equivalent to a south pole.

27.8 no The rotor will not begin to turn when the switch is closed if the rotor is initially oriented as shown in Fig. 27.39b. In this case there is no current through the rotor and hence no magnetic torque. This situation can be remedied by using multiple rotor coils oriented at different angles around the rotation axis. With this arrangement, there is always a magnetic torque no matter what the orientation.

27.9 (ii) The mobile charge carriers in copper are negatively charged electrons, which move upward through the wire to give a downward current. From the right-hand rule, the force on a positively charged particle moving upward in a westward-pointing magnetic field would be to the south; hence the force on a negatively charged particle is to the north. The result is an excess of negative charge on the north side of the wire, leaving an excess of positive charge—and hence a higher electric potential—on the south side.

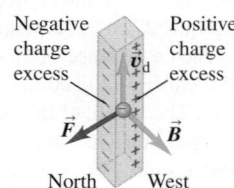

Bridging Problem

(a) $\tau_x = -1.54 \times 10^{-4}$ N·m,
$\tau_y = -2.05 \times 10^{-4}$ N·m,
$\tau_z = -6.14 \times 10^{-4}$ N·m
(b) -7.55×10^{-4} J   **(c)** 42.1 rad/s

28 SOURCES OF MAGNETIC FIELD

I n Chapter 27 we studied the forces exerted on moving charges and on current-carrying conductors in a magnetic field. We didn't worry about how the magnetic field got there; we simply took its existence as a given fact. But how are magnetic fields *created*? We know that both permanent magnets and electric currents in electromagnets create magnetic fields. In this chapter we will study these sources of magnetic field in detail.

We've learned that a charge creates an electric field and that an electric field exerts a force on a charge. But a *magnetic* field exerts a force on only a *moving* charge. Similarly, we'll see that only *moving* charges *create* magnetic fields. We'll begin our analysis with the magnetic field created by a single moving point charge. We can use this analysis to determine the field created by a small segment of a current-carrying conductor. Once we can do that, we can in principle find the magnetic field produced by *any* shape of conductor.

Then we will introduce Ampere's law, which plays a role in magnetism analogous to the role of Gauss's law in electrostatics. Ampere's law lets us exploit symmetry properties in relating magnetic fields to their sources.

Moving charged particles within atoms respond to magnetic fields and can also act as sources of magnetic field. We'll use these ideas to understand how certain magnetic materials can be used to intensify magnetic fields as well as why some materials such as iron act as permanent magnets.

28.1 MAGNETIC FIELD OF A MOVING CHARGE

Let's start with the basics, the magnetic field of a single point charge q moving with a constant velocity $\vec{v}$. In practical applications, such as the solenoid shown in the photo that opens this chapter, magnetic fields are produced by tremendous numbers of charged particles moving together in a current. But once we understand how to calculate the magnetic field due to a single point charge, it's a small leap to calculate the field due to a current-carrying wire or collection of wires.

28.1 (a) Magnetic-field vectors due to a moving positive point charge q. At each point, $\vec{B}$ is perpendicular to the plane of $\vec{r}$ and $\vec{v}$, and its magnitude is proportional to the sine of the angle between them. (b) Magnetic field lines in a plane containing a moving positive charge.

(a) Perspective view

Right-hand rule for the magnetic field due to a positive charge moving at constant velocity: Point the thumb of your right hand in the direction of the velocity. Your fingers now curl around the charge in the direction of the magnetic field lines. (If the charge is negative, the field lines are in the opposite direction.)

For these field points, $\vec{r}$ and $\vec{v}$ both lie in the beige plane, and $\vec{B}$ is perpendicular to this plane.

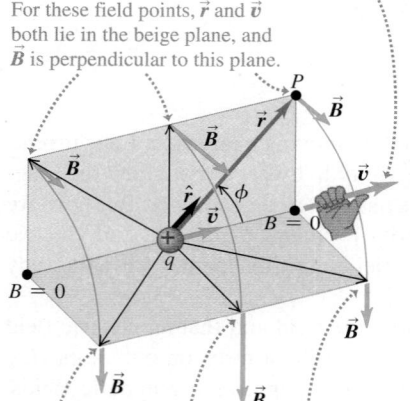

For these field points, $\vec{r}$ and $\vec{v}$ both lie in the gold plane, and $\vec{B}$ is perpendicular to this plane.

(b) View from behind the charge

The $\times$ symbol indicates that the charge is moving into the plane of the page (away from you).

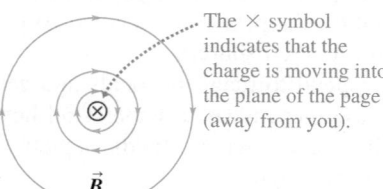

As we did for electric fields, we call the location of the moving charge at a given instant the **source point** and the point P where we want to find the field the **field point.** In Section 21.4 we found that at a field point a distance r from a point charge q, the magnitude of the *electric* field $\vec{E}$ caused by the charge is proportional to the charge magnitude $|q|$ and to $1/r^2$, and the direction of $\vec{E}$ (for positive q) is along the line from source point to field point. The corresponding relationship for the *magnetic* field $\vec{B}$ of a point charge q moving with constant velocity has some similarities and some interesting differences.

Experiments show that the magnitude of $\vec{B}$ is also proportional to $|q|$ and to $1/r^2$. But the *direction* of $\vec{B}$ is *not* along the line from source point to field point. Instead, $\vec{B}$ is perpendicular to the plane containing this line and the particle's velocity vector $\vec{v}$, as shown in **Fig. 28.1.** Furthermore, the field *magnitude* B is also proportional to the particle's speed v and to the sine of the angle ϕ. Thus the magnetic-field magnitude at point P is

$$B = \frac{\mu_0}{4\pi} \frac{|q|v\sin\phi}{r^2} \tag{28.1}$$

The quantity μ_0 (read as "mu-nought" or "mu-sub-zero") is called the **magnetic constant.** The reason for including the factor of 4π will emerge shortly. We did something similar with Coulomb's law in Section 21.3.

Moving Charge: Vector Magnetic Field

We can incorporate both the magnitude and direction of $\vec{B}$ into a single vector equation by using the vector product. To avoid having to say "the direction from the source q to the field point P" over and over, we introduce a *unit* vector $\hat{r}$ ("r-hat") that points from the source point to the field point. (We used $\hat{r}$ for the same purpose in Section 21.4.) This unit vector is equal to the vector $\vec{r}$ from the source to the field point divided by its magnitude: $\hat{r} = \vec{r}/r$. Then

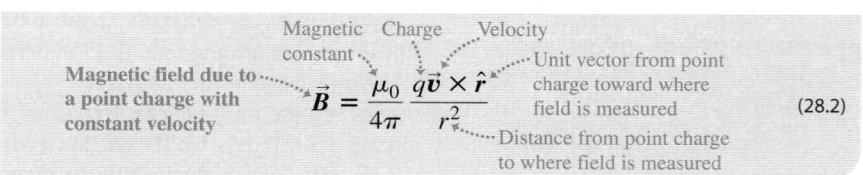

Magnetic field due to a point charge with constant velocity · $\vec{B} = \dfrac{\mu_0}{4\pi} \dfrac{q\vec{v} \times \hat{r}}{r^2}$ (28.2)

Magnetic constant · Charge · Velocity · Unit vector from point charge toward where field is measured · Distance from point charge to where field is measured

Figure 28.1 shows the relationship of $\hat{r}$ to P and shows the magnetic field $\vec{B}$ at several points in the vicinity of the charge. At all points along a line through the charge parallel to the velocity $\vec{v}$, the field is zero because $\sin\phi = 0$ at all such points. At any distance r from q, $\vec{B}$ has its greatest magnitude at points lying in the plane perpendicular to $\vec{v}$, because there $\phi = 90°$ and $\sin\phi = 1$. If q is negative, the directions of $\vec{B}$ are opposite to those shown in Fig. 28.1.

Moving Charge: Magnetic Field Lines

A point charge in motion also produces an *electric* field, with field lines that radiate outward from a positive charge. The *magnetic* field lines are completely different. For a point charge moving with velocity $\vec{v}$, the magnetic field lines are *circles* centered on the line of $\vec{v}$ and lying in planes perpendicular to this line. The field-line directions for a positive charge are given by the following *right-hand rule,* one of several that we will encounter in this chapter: Grasp the velocity vector $\vec{v}$ with your right hand so that your right thumb points in the direction of $\vec{v}$; your fingers then curl around the line of $\vec{v}$ in the same sense as the magnetic field lines, assuming q is positive. Figure 28.1a shows parts of a few field lines; Fig. 28.1b shows some field lines in a plane through q, perpendicular to $\vec{v}$. If the point charge is negative, the directions of the field and field lines are the opposite of those shown in Fig. 28.1.

Equations (28.1) and (28.2) describe the $\vec{B}$ field of a point charge moving with *constant* velocity. If the charge *accelerates,* the field can be much more complicated. We won't need these more complicated results for our purposes. (The moving charged particles that make up a current in a wire accelerate at points where the wire bends and the direction of $\vec{v}$ changes. But because the magnitude v_d of the drift velocity in a conductor is typically very small, the centripetal acceleration v_d^2/r is so small that we can ignore its effects.)

As we discussed in Section 27.2, the unit of B is one tesla (1 T):

$$1\,T = 1\,N\cdot s/C\cdot m = 1\,N/A\cdot m$$

Using this with Eq. (28.1) or (28.2), we find that the units of the constant μ_0 are

$$1\,N\cdot s^2/C^2 = 1\,N/A^2 = 1\,Wb/A\cdot m = 1\,T\cdot m/A$$

In SI units the numerical value of μ_0 is exactly $4\pi \times 10^{-7}$. Thus

$$\mu_0 = 4\pi \times 10^{-7}\,N\cdot s^2/C^2 = 4\pi \times 10^{-7}\,Wb/A\cdot m$$
$$= 4\pi \times 10^{-7}\,T\cdot m/A \qquad (28.3)$$

It may seem incredible that μ_0 has *exactly* this numerical value! In fact this is a *defined* value that arises from the definition of the ampere, as we'll discuss in Section 28.4.

We mentioned in Section 21.3 that the constant $1/4\pi\epsilon_0$ in Coulomb's law is related to the speed of light c:

$$k = \frac{1}{4\pi\epsilon_0} = (10^{-7}\,N\cdot s^2/C^2)c^2$$

When we study electromagnetic waves in Chapter 32, we will find that their speed of propagation in vacuum, which is equal to the speed of light c, is given by

$$c^2 = \frac{1}{\epsilon_0\mu_0} \qquad (28.4)$$

If we solve the equation $k = 1/4\pi\epsilon_0$ for ϵ_0, substitute the resulting expression into Eq. (28.4), and solve for μ_0, we indeed get the value of μ_0 stated above. This discussion is a little premature, but it may give you a hint that electric and magnetic fields are intimately related to the nature of light.

EXAMPLE 28.1   FORCES BETWEEN TWO MOVING PROTONS

Two protons move parallel to the x-axis in opposite directions (**Fig. 28.2**) at the same speed v (small compared to the speed of light c). At the instant shown, find the electric and magnetic forces on the upper proton and compare their magnitudes.

SOLUTION

IDENTIFY and SET UP: Coulomb's law [Eq. (21.2)] gives the electric force F_E on the upper proton. The magnetic force law [Eq. (27.2)] gives the magnetic force on the upper proton; to use it, we must first use Eq. (28.2) to find the magnetic field that the lower proton produces at the position of the upper proton. The unit vector from the lower proton (the source) to the position of the upper proton is $\hat{r} = \hat{\jmath}$.

EXECUTE: From Coulomb's law, the magnitude of the electric force on the upper proton is

$$F_E = \frac{1}{4\pi\epsilon_0}\frac{q^2}{r^2}$$

The forces are repulsive, and the force on the upper proton is vertically upward (in the $+y$-direction).

The velocity of the lower proton is $\vec{v} = v\hat{\imath}$. From the right-hand rule for the cross product $\vec{v} \times \hat{r}$ in Eq. (28.2), the $\vec{B}$ field

28.2 Electric and magnetic forces between two moving protons.

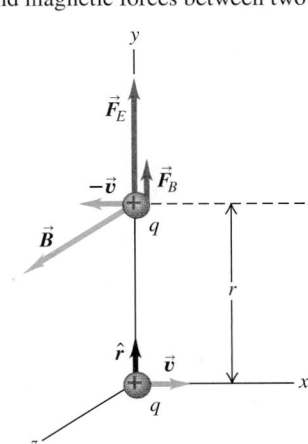

due to the lower proton at the position of the upper proton is in the $+z$-direction (see Fig. 28.2). From Eq. (28.2), the field is

$$\vec{B} = \frac{\mu_0}{4\pi}\frac{q(v\hat{\imath}) \times \hat{\jmath}}{r^2} = \frac{\mu_0}{4\pi}\frac{qv}{r^2}\hat{k} \qquad \textit{Continued}$$

The velocity of the upper proton is $-\vec{v} = -v\hat{\imath}$, so the magnetic force on it is

$$\vec{F}_B = q(-\vec{v}) \times \vec{B} = q(-v\hat{\imath}) \times \frac{\mu_0}{4\pi} \frac{qv}{r^2}\hat{k} = \frac{\mu_0}{4\pi} \frac{q^2v^2}{r^2}\hat{\jmath}$$

The magnetic interaction in this situation is also repulsive. The ratio of the force magnitudes is

$$\frac{F_B}{F_E} = \frac{\mu_0 q^2 v^2 / 4\pi r^2}{q^2/4\pi\epsilon_0 r^2} = \frac{\mu_0 v^2}{1/\epsilon_0} = \epsilon_0 \mu_0 v^2$$

With the relationship $\epsilon_0\mu_0 = 1/c^2$, Eq. (28.4), this becomes

$$\frac{F_B}{F_E} = \frac{v^2}{c^2}$$

When v is small in comparison to the speed of light, the magnetic force is much smaller than the electric force.

EVALUATE: We have described the velocities, fields, and forces as they are measured by an observer who is stationary in the coordinate system of Fig. 28.2. In a coordinate system that moves with one of the charges, one of the velocities would be zero, so there would be *no* magnetic force. The explanation of this apparent paradox provided one of the paths that led to the special theory of relativity.

TEST YOUR UNDERSTANDING OF SECTION 28.1 (a) If two protons are traveling parallel to each other in the *same* direction and at the same speed, is the magnetic force between them (i) attractive or (ii) repulsive? (b) Is the net force between them (i) attractive, (ii) repulsive, or (iii) zero? (Assume that the protons' speed is much slower than the speed of light.) ❙

28.3 (a) Magnetic-field vectors due to a current element $d\vec{l}$. (b) Magnetic field lines in a plane containing the current element $d\vec{l}$. Compare this figure to Fig. 28.1 for the field of a moving point charge.

(a) Perspective view

Right-hand rule for the magnetic field due to a current element: Point the thumb of your right hand in the direction of the current. Your fingers now curl around the current element in the direction of the magnetic field lines.

For these field points, $\vec{r}$ and $d\vec{l}$ both lie in the beige plane, and $d\vec{B}$ is perpendicular to this plane.

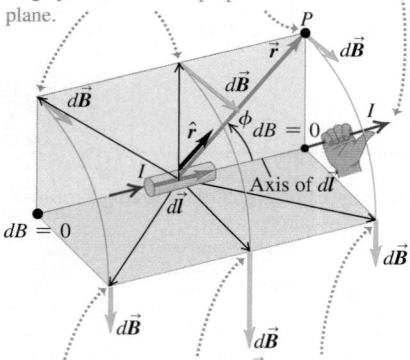

For these field points, $\vec{r}$ and $d\vec{l}$ both lie in the gold plane, and $d\vec{B}$ is perpendicular to this plane.

(b) View along the axis of the current element

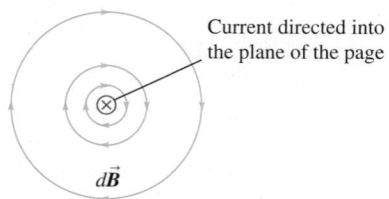

Current directed into the plane of the page

28.2 MAGNETIC FIELD OF A CURRENT ELEMENT

As for electric fields, there is a **principle of superposition of magnetic fields:**

> **The total magnetic field caused by several moving charges is the vector sum of the fields caused by the individual charges.**

We can use this principle with the results of Section 28.1 to find the magnetic field produced by a current in a conductor.

We begin by calculating the magnetic field caused by a short segment $d\vec{l}$ of a current-carrying conductor, as shown in **Fig. 28.3a**. The volume of the segment is $A\,dl$, where A is the cross-sectional area of the conductor. If there are n moving charged particles per unit volume, each of charge q, the total moving charge dQ in the segment is

$$dQ = nqA\,dl$$

The moving charges in this segment are equivalent to a single charge dQ, traveling with a velocity equal to the *drift* velocity $\vec{v}_d$. (Magnetic fields due to the *random* motions of the charges will, on average, cancel out at every point.) From Eq. (28.1) the magnitude of the resulting field $d\vec{B}$ at any field point P is

$$dB = \frac{\mu_0}{4\pi} \frac{|dQ|v_d \sin\phi}{r^2} = \frac{\mu_0}{4\pi} \frac{n|q|v_d A\,dl\sin\phi}{r^2}$$

But from Eq. (25.2), $n|q|v_d A$ equals the current I in the element. So

$$dB = \frac{\mu_0}{4\pi} \frac{I\,dl\sin\phi}{r^2} \tag{28.5}$$

Current Element: Vector Magnetic Field

In vector form, using the unit vector $\hat{r}$ as in Section 28.1, we have

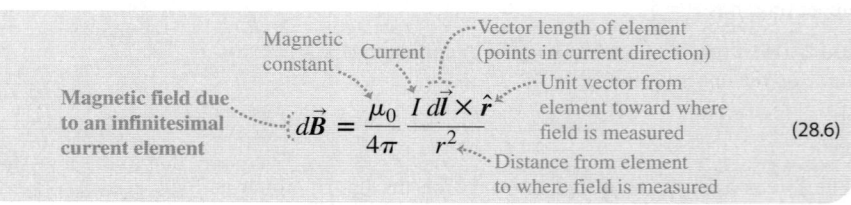

$$d\vec{B} = \frac{\mu_0}{4\pi} \frac{I\,d\vec{l} \times \hat{r}}{r^2} \tag{28.6}$$

Magnetic field due to an infinitesimal current element

Magnetic constant · Current · Vector length of element (points in current direction) · Unit vector from element toward where field is measured · Distance from element to where field is measured

where $d\vec{l}$ is a vector with length dl, in the same direction as the current.

Equations (28.5) and (28.6) are called the **law of Biot and Savart** (pronounced "Bee-oh" and "Suh-var"). We can use this law to find the total magnetic field $\vec{B}$ at any point in space due to the current in a complete circuit. To do this, we integrate Eq. (28.6) over all segments $d\vec{l}$ that carry current; symbolically,

$$\vec{B} = \frac{\mu_0}{4\pi} \int \frac{I\, d\vec{l} \times \hat{r}}{r^2} \qquad (28.7)$$

In the following sections we will carry out this vector integration for several examples.

Current Element: Magnetic Field Lines

As Fig. 28.3 shows, the field vectors $d\vec{B}$ and the magnetic field lines of a current element are exactly like those set up by a positive charge dQ moving in the direction of the drift velocity $\vec{v}_d$. The field lines are circles in planes perpendicular to $d\vec{l}$ and centered on the line of $d\vec{l}$. Their directions are given by the same right-hand rule that we introduced for point charges in Section 28.1.

We can't verify Eq. (28.5) or (28.6) directly because we can never experiment with an isolated segment of a current-carrying circuit. What we measure experimentally is the *total* $\vec{B}$ for a complete circuit. But we can still verify these equations indirectly by calculating $\vec{B}$ for various current configurations with Eq. (28.7) and comparing the results with experimental measurements.

If matter is present in the space around a current-carrying conductor, the field at a field point P in its vicinity will have an additional contribution resulting from the *magnetization* of the material. We'll return to this point in Section 28.8. However, unless the material is iron or some other ferromagnetic material, the additional field is small and is usually negligible. Additional complications arise if time-varying electric or magnetic fields are present or if the material is a superconductor; we'll return to these topics later.

Application Currents and Planetary Magnetism The earth's magnetic field is caused by currents circulating within its molten, conducting interior. These currents are stirred by our planet's relatively rapid spin (one rotation per 24 hours). The moon's internal currents are much weaker; it is much smaller than the earth, has a predominantly solid interior, and spins slowly (one rotation per 27.3 days). Hence the moon's magnetic field is only about 10^{-4} as strong as that of the earth.

PROBLEM-SOLVING STRATEGY 28.1 | **MAGNETIC-FIELD CALCULATIONS**

IDENTIFY *the relevant concepts:* The Biot–Savart law [Eqs. (28.5) and (28.6)] allows you to calculate the magnetic field at a field point P due to a current-carrying wire of any shape. The idea is to calculate the field element $d\vec{B}$ at P due to a representative current element in the wire and integrate all such field elements to find the field $\vec{B}$ at P.

SET UP *the problem* using the following steps:
1. Make a diagram showing a representative current element and the field point P.
2. Draw the current element $d\vec{l}$, being careful that it points in the direction of the current.
3. Draw the unit vector $\hat{r}$ directed *from* the current element (the source point) to P.
4. Identify the target variable (usually $\vec{B}$).

EXECUTE *the solution* as follows:
1. Use Eq. (28.5) or (28.6) to express the magnetic field $d\vec{B}$ at P from the representative current element.
2. Add up all the $d\vec{B}$'s to find the total field at point P. In some situations the $d\vec{B}$'s at point P have the same direction for all the current elements; then the magnitude of the total $\vec{B}$ field is the sum of the magnitudes of the $d\vec{B}$'s. But often the $d\vec{B}$'s have different directions for different current elements. Then you have to set up a coordinate system and represent each $d\vec{B}$ in terms of its components. The integral for the total $\vec{B}$ is then expressed in terms of an integral for each component.
3. Sometimes you can use the symmetry of the situation to prove that one component of $\vec{B}$ must vanish. Always be alert for ways to use symmetry to simplify the problem.
4. Look for ways to use the principle of superposition of magnetic fields. Later in this chapter we'll determine the fields produced by certain simple conductor shapes; if you encounter a conductor of a complex shape that can be represented as a combination of these simple shapes, you can use superposition to find the field of the complex shape. Examples include a rectangular loop and a semicircle with straight line segments on both sides.

EVALUATE *your answer:* Often your answer will be a mathematical expression for $\vec{B}$ as a function of the position of the field point. Check the answer by examining its behavior in as many limits as you can.

EXAMPLE 28.2   MAGNETIC FIELD OF A CURRENT SEGMENT

A copper wire carries a steady 125-A current to an electroplating tank (**Fig. 28.4**). Find the magnetic field due to a 1.0-cm segment of this wire at a point 1.2 m away from it, if the point is (a) point P_1, straight out to the side of the segment, and (b) point P_2, in the xy-plane and on a line at 30° to the segment.

SOLUTION

IDENTIFY and SET UP: Although Eqs. (28.5) and (28.6) apply only to infinitesimal current elements, we may use either of them here because the segment length is much less than the distance to the field point. The current element is shown in red in Fig. 28.4 and points in the $-x$-direction (the direction of the current), so $d\vec{l} = dl(-\hat{\imath})$. The unit vector $\hat{r}$ for each field point is directed from the current element toward that point: $\hat{r}$ is in the $+y$-direction for point P_1 and at an angle of 30° above the $-x$-direction for point P_2.

28.4 Finding the magnetic field at two points due to a 1.0-cm segment of current-carrying wire (not shown to scale).

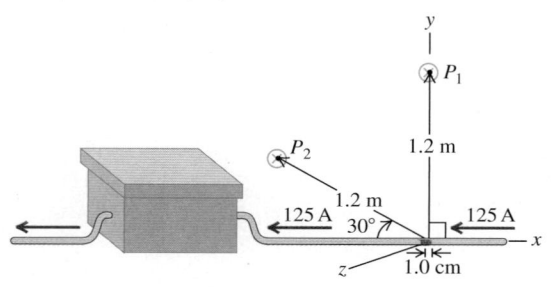

EXECUTE: (a) At point P_1, $\hat{r} = \hat{\jmath}$, so

$$\vec{B} = \frac{\mu_0}{4\pi}\frac{I\,d\vec{l}\times\hat{r}}{r^2} = \frac{\mu_0}{4\pi}\frac{I\,dl(-\hat{\imath})\times\hat{\jmath}}{r^2} = -\frac{\mu_0}{4\pi}\frac{I\,dl}{r^2}\hat{k}$$

$$= -(10^{-7}\,\text{T}\cdot\text{m/A})\frac{(125\,\text{A})(1.0\times10^{-2}\,\text{m})}{(1.2\,\text{m})^2}\hat{k}$$

$$= -(8.7\times10^{-8}\,\text{T})\hat{k}$$

The direction of $\vec{B}$ at P_1 is into the xy-plane of Fig. 28.4.

(b) At P_2, the unit vector is $\hat{r} = (-\cos30°)\hat{\imath} + (\sin30°)\hat{\jmath}$. From Eq. (28.6),

$$\vec{B} = \frac{\mu_0}{4\pi}\frac{I\,d\vec{l}\times\hat{r}}{r^2} = \frac{\mu_0}{4\pi}\frac{I\,dl(-\hat{\imath})\times(-\cos30°\hat{\imath}+\sin30°\hat{\jmath})}{r^2}$$

$$= -\frac{\mu_0 I}{4\pi}\frac{dl\sin30°}{r^2}\hat{k}$$

$$= -(10^{-7}\,\text{T}\cdot\text{m/A})\frac{(125\,\text{A})(1.0\times10^{-2}\,\text{m})(\sin30°)}{(1.2\,\text{m})^2}\hat{k}$$

$$= -(4.3\times10^{-8}\,\text{T})\hat{k}$$

The direction of $\vec{B}$ at P_2 is also into the xy-plane of Fig. 28.4.

EVALUATE: We can check our results for the direction of $\vec{B}$ by comparing them with Fig. 28.3. The xy-plane in Fig. 28.4 corresponds to the beige plane in Fig. 28.3, but here the direction of the current and hence of $d\vec{l}$ is the reverse of that shown in Fig. 28.3. Hence the direction of the magnetic field is reversed as well. Hence the field at points in the xy-plane in Fig. 28.4 must point *into,* not out of, that plane. This is just what we concluded above.

TEST YOUR UNDERSTANDING OF SECTION 28.2   An infinitesimal current element located at the origin ($x = y = z = 0$) carries current I in the positive y-direction. Rank the following locations in order of the strength of the magnetic field that the current element produces at that location, from largest to smallest value. (i) $x = L$, $y = 0$, $z = 0$; (ii) $x = 0$, $y = L$, $z = 0$; (iii) $x = 0$, $y = 0$, $z = L$; (iv) $x = L/\sqrt{2}$, $y = L/\sqrt{2}$, $z = 0$. ∎

28.5 Magnetic field produced by a straight current-carrying conductor of length $2a$.

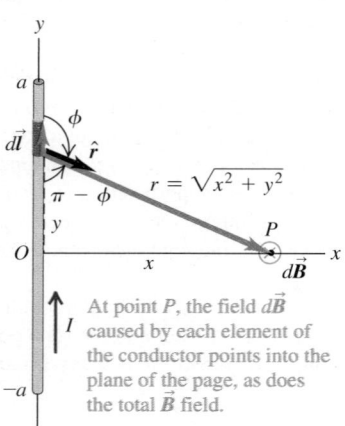

At point P, the field $d\vec{B}$ caused by each element of the conductor points into the plane of the page, as does the total $\vec{B}$ field.

28.3 MAGNETIC FIELD OF A STRAIGHT CURRENT-CARRYING CONDUCTOR

Let's use the law of Biot and Savart to find the magnetic field produced by a straight current-carrying conductor. This result is useful because straight conducting wires are found in essentially all electric and electronic devices. **Figure 28.5** shows such a conductor with length $2a$ carrying a current I. We will find $\vec{B}$ at a point a distance x from the conductor on its perpendicular bisector.

We first use the law of Biot and Savart, Eq. (28.5), to find the field $d\vec{B}$ caused by the element of conductor of length $dl = dy$ shown in Fig. 28.5. From the figure, $r = \sqrt{x^2 + y^2}$ and $\sin\phi = \sin(\pi - \phi) = x/\sqrt{x^2 + y^2}$. The right-hand rule for the vector product $d\vec{l}\times\hat{r}$ shows that the *direction* of $d\vec{B}$ is into the plane of the figure, perpendicular to the plane; furthermore, the directions of the $d\vec{B}$'s from *all* elements of the conductor are the same. Thus in integrating Eq. (28.7), we can just add the *magnitudes* of the $d\vec{B}$'s, a significant simplification.

Putting the pieces together, we find that the magnitude of the total $\vec{B}$ field is

$$B = \frac{\mu_0 I}{4\pi} \int_{-a}^{a} \frac{x \, dy}{(x^2 + y^2)^{3/2}}$$

We can integrate this by trigonometric substitution or by using an integral table:

$$B = \frac{\mu_0 I}{4\pi} \frac{2a}{x\sqrt{x^2 + a^2}} \tag{28.8}$$

When the length $2a$ of the conductor is much greater than its distance x from point P, we can consider it to be infinitely long. When a is much larger than x, $\sqrt{x^2 + a^2}$ is approximately equal to a; hence in the limit $a \to \infty$, Eq. (28.8) becomes

$$B = \frac{\mu_0 I}{2\pi x}$$

The physical situation has axial symmetry about the y-axis. Hence $\vec{B}$ must have the same *magnitude* at all points on a circle centered on the conductor and lying in a plane perpendicular to it, and the *direction* of $\vec{B}$ must be everywhere tangent to such a circle (**Fig. 28.6**). Thus, at all points on a circle of radius r around the conductor, the magnitude B is

Magnetic field near a long, straight, current-carrying conductor

Magnetic constant ⋯⋯

$$B = \frac{\mu_0 I}{2\pi r} \tag{28.9}$$

Current ⋯⋯

Distance from conductor ⋯⋯

The geometry in this case is similar to that of Example 21.10 (Section 21.5), in which we solved the problem of the *electric* field caused by an infinite line of charge. The same integral appears in both problems, and the field magnitudes in both problems are proportional to $1/r$. But the lines of $\vec{B}$ in the magnetic problem have completely different shapes than the lines of $\vec{E}$ in the analogous electrical problem. Electric field lines radiate outward from a positive line charge distribution (inward for negative charges). By contrast, magnetic field lines *encircle* the current that acts as their source. Electric field lines due to charges begin and end at those charges, but magnetic field lines always form closed loops and *never* have endpoints, irrespective of the shape of the current-carrying conductor that sets up the field. As we discussed in Section 27.3, this is a consequence of Gauss's law for magnetism, which states that the total magnetic flux through *any* closed surface is always zero:

$$\oint \vec{B} \cdot d\vec{A} = 0 \qquad \text{(magnetic flux through any closed surface)} \tag{28.10}$$

Any magnetic field line that enters a closed surface must emerge from that surface.

28.6 Magnetic field around a long, straight, current-carrying conductor. The field lines are circles, with directions determined by the right-hand rule.

Right-hand rule for the magnetic field around a current-carrying wire: Point the thumb of your right hand in the direction of the current. Your fingers now curl around the wire in the direction of the magnetic field lines.

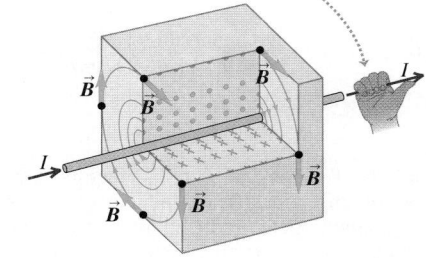

EXAMPLE 28.3 **MAGNETIC FIELD OF A SINGLE WIRE**

A long, straight conductor carries a 1.0-A current. At what distance from the axis of the conductor does the resulting magnetic field have magnitude $B = 0.5 \times 10^{-4}$ T (about that of the earth's magnetic field in Pittsburgh)?

SOLUTION

IDENTIFY and SET UP: The length of a "long" conductor is much greater than the distance from the conductor to the field point. Hence we can use the ideas of this section. The geometry is the same as that of Fig. 28.6, so we use Eq. (28.9). All of the quantities in this equation are known except the target variable, the distance r.

EXECUTE: We solve Eq. (28.9) for r:

$$r = \frac{\mu_0 I}{2\pi B} = \frac{(4\pi \times 10^{-7} \text{ T} \cdot \text{m/A})(1.0 \text{ A})}{(2\pi)(0.5 \times 10^{-4} \text{ T})}$$

$$= 4 \times 10^{-3} \text{ m} = 4 \text{ mm}$$

EVALUATE: As we saw in Example 26.14, currents of an ampere or more are typical of those found in the wiring of home appliances. This example shows that the magnetic fields produced by these appliances are very weak even very close to the wire; the fields are proportional to $1/r$, so they become even weaker at greater distances.

EXAMPLE 28.4 **MAGNETIC FIELD OF TWO WIRES**

Figure 28.7a is an end-on view of two long, straight, parallel wires perpendicular to the xy-plane, each carrying a current I but in opposite directions. (a) Find $\vec{B}$ at points P_1, P_2, and P_3. (b) Find an expression for $\vec{B}$ at any point on the x-axis to the right of wire 2.

SOLUTION

IDENTIFY and SET UP: We can find the magnetic fields $\vec{B}_1$ and $\vec{B}_2$ due to wires 1 and 2 at each point by using the ideas of this section. By the superposition principle, the magnetic field at each point is then $\vec{B} = \vec{B}_1 + \vec{B}_2$. We use Eq. (28.9) to find the magnitudes B_1 and B_2 of these fields and the right-hand rule to find the corresponding directions. Figure 28.7a shows $\vec{B}_1$, $\vec{B}_2$, and $\vec{B} = \vec{B}_{\text{total}}$ at each point; you should confirm that the directions and relative magnitudes shown are correct. Figure 28.7b shows some of the magnetic field lines due to this two-wire system.

EXECUTE: (a) Since point P_1 is a distance $2d$ from wire 1 and a distance $4d$ from wire 2, $B_1 = \mu_0 I/2\pi(2d) = \mu_0 I/4\pi d$ and $B_2 = \mu_0 I/2\pi(4d) = \mu_0 I/8\pi d$. The right-hand rule shows that $\vec{B}_1$ is in the negative y-direction and $\vec{B}_2$ is in the positive y-direction, so

$$\vec{B}_{\text{total}} = \vec{B}_1 + \vec{B}_2 = -\frac{\mu_0 I}{4\pi d}\hat{j} + \frac{\mu_0 I}{8\pi d}\hat{j} = -\frac{\mu_0 I}{8\pi d}\hat{j} \quad \text{(point } P_1\text{)}$$

28.7 (a) Two long, straight conductors carrying equal currents in opposite directions. The conductors are seen end-on. (b) Map of the magnetic field produced by the two conductors. The field lines are closest together between the conductors, where the field is strongest.

(a)

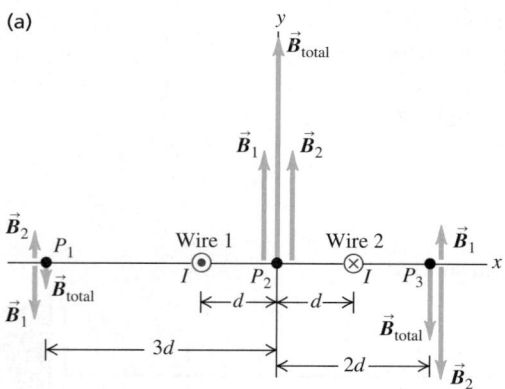

(b)

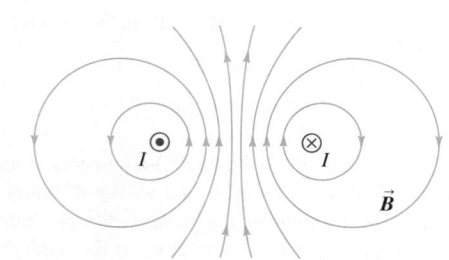

At point P_2, a distance d from both wires, $\vec{B}_1$ and $\vec{B}_2$ are both in the positive y-direction, and both have the same magnitude $B_1 = B_2 = \mu_0 I/2\pi d$. Hence

$$\vec{B}_{\text{total}} = \vec{B}_1 + \vec{B}_2 = \frac{\mu_0 I}{2\pi d}\hat{j} + \frac{\mu_0 I}{2\pi d}\hat{j} = \frac{\mu_0 I}{\pi d}\hat{j} \quad \text{(point } P_2\text{)}$$

Finally, at point P_3 the right-hand rule shows that $\vec{B}_1$ is in the positive y-direction and $\vec{B}_2$ is in the negative y-direction. This point is a distance $3d$ from wire 1 and a distance d from wire 2, so $B_1 = \mu_0 I/2\pi(3d) = \mu_0 I/6\pi d$ and $B_2 = \mu_0 I/2\pi d$. The total field at P_3 is

$$\vec{B}_{\text{total}} = \vec{B}_1 + \vec{B}_2 = \frac{\mu_0 I}{6\pi d}\hat{j} - \frac{\mu_0 I}{2\pi d}\hat{j} = -\frac{\mu_0 I}{3\pi d}\hat{j} \quad \text{(point } P_3\text{)}$$

The same technique can be used to find $\vec{B}_{\text{total}}$ at any point; for points off the x-axis, caution must be taken in vector addition, since $\vec{B}_1$ and $\vec{B}_2$ need no longer be simply parallel or antiparallel.

(b) At any point on the x-axis to the right of wire 2 (that is, for $x > d$), $\vec{B}_1$ and $\vec{B}_2$ are in the same directions as at P_3. Such a point is a distance $x + d$ from wire 1 and a distance $x - d$ from wire 2, so the total field is

$$\vec{B}_{\text{total}} = \vec{B}_1 + \vec{B}_2 = \frac{\mu_0 I}{2\pi(x + d)}\hat{j} - \frac{\mu_0 I}{2\pi(x - d)}\hat{j}$$

$$= -\frac{\mu_0 I d}{\pi(x^2 - d^2)}\hat{j}$$

where we used a common denominator to combine the two terms.

EVALUATE: Consider our result from part (b) at a point very far from the wires, so that x is much larger than d. Then the d^2 term in the denominator can be ignored, and the magnitude of the total field is approximately $B_{\text{total}} = \mu_0 I d/\pi x^2$. For one wire, Eq. (28.9) shows that the magnetic field decreases with distance in proportion to $1/x$; for two wires carrying opposite currents, $\vec{B}_1$ and $\vec{B}_2$ partially cancel each other, and so B_{total} decreases more rapidly, in proportion to $1/x^2$. This effect is used in communication systems such as telephone or computer networks. The wiring is arranged so that a conductor carrying a signal in one direction and the conductor carrying the return signal are side by side, as in Fig. 28.7a, or twisted around each other (**Fig. 28.8**). As a result, the magnetic field due to these signals *outside* the conductors is very small, making it less likely to exert unwanted forces on other information-carrying currents.

28.8 Computer cables, or cables for audio-video equipment, create little or no magnetic field. This is because within each cable, closely spaced wires carry current in both directions along the length of the cable. The magnetic fields from these opposing currents cancel each other.

TEST YOUR UNDERSTANDING OF SECTION 28.3 The accompanying figure shows a circuit that lies on a horizontal table. A compass is placed on top of the circuit as shown. A battery is to be connected to the circuit so that when the switch is closed, the compass needle deflects counterclockwise. In which orientation, A or B, should the battery be placed in the circuit? ▮

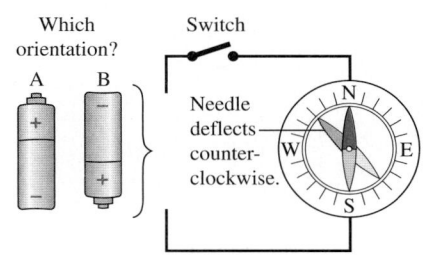

28.4 FORCE BETWEEN PARALLEL CONDUCTORS

Now that we know how to calculate the magnetic field produced by a long, current-carrying conductor, we can find the *magnetic force* that one such conductor exerts on another. This force plays a role in many practical situations in which current-carrying wires are close to each other. **Figure 28.9** shows segments of two long, straight, parallel conductors separated by a distance r and carrying currents I and I' in the same direction. Each conductor lies in the magnetic field set up by the other, so each experiences a force. The figure shows some of the field lines set up by the current in the lower conductor.

From Eq. (28.9) the lower conductor produces a $\vec{B}$ field that, at the position of the upper conductor, has magnitude

$$B = \frac{\mu_0 I}{2\pi r}$$

From Eq. (27.19) the force that this field exerts on a length L of the upper conductor is $\vec{F} = I'\vec{L} \times \vec{B}$, where the vector $\vec{L}$ is in the direction of the current I' and has magnitude L. Since $\vec{B}$ is perpendicular to the length of the conductor and hence to $\vec{L}$, the magnitude of this force is

$$F = I'LB = \frac{\mu_0 I I' L}{2\pi r}$$

and the force *per unit length* F/L is

Magnetic force per unit length between two long, parallel, current-carrying conductors	$\dfrac{F}{L} = \dfrac{\mu_0 I I'}{2\pi r}$	(28.11)

Magnetic constant ····· Current in first conductor
····· Current in second conductor
····· Distance between conductors

Applying the right-hand rule to $\vec{F} = I'\vec{L} \times \vec{B}$ shows that the force on the upper conductor is directed *downward*.

The current in the *upper* conductor also sets up a $\vec{B}$ field at the position of the *lower* conductor. Two successive applications of the right-hand rule for vector products (one to find the direction of the $\vec{B}$ field due to the upper conductor, as in Section 28.2, and one to find the direction of the force that this field exerts on the lower conductor, as in Section 27.6) show that the force on the lower conductor is *upward*. Thus *two parallel conductors carrying current in the same direction attract each other*. If the direction of either current is reversed, the forces also reverse. *Parallel conductors carrying currents in opposite directions repel each other.*

28.9 Parallel conductors carrying currents in the same direction attract each other. The diagrams show how the magnetic field $\vec{B}$ caused by the current in the lower conductor exerts a force $\vec{F}$ on the upper conductor.

The magnetic field of the lower wire exerts an attractive force on the upper wire. By the same token, the upper wire attracts the lower one.

If the wires had currents in *opposite* directions, they would *repel* each other.

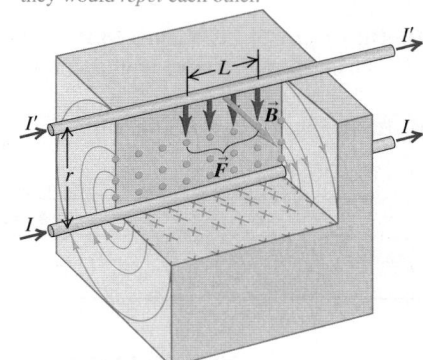

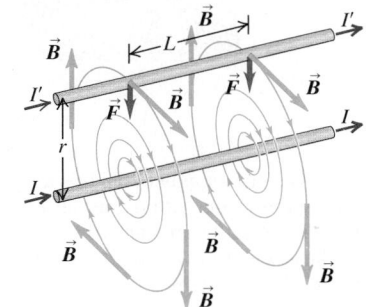

Magnetic Forces and Defining the Ampere

The attraction or repulsion between two straight, parallel, current-carrying conductors is the basis of the official SI definition of the **ampere**:

> *One ampere* is that unvarying current that, if present in each of two parallel conductors of infinite length and one meter apart in empty space, causes each conductor to experience a force of exactly 2×10^{-7} newtons per meter of length.

From Eq. (28.11) you can see that this definition of the ampere is what leads us to choose the value of $4\pi \times 10^{-7}$ T·m/A for μ_0. The SI definition of the coulomb is the amount of charge transferred in one second by a current of one ampere.

This is an *operational definition;* it gives us an actual experimental procedure for measuring current and defining a unit of current. For high-precision standardization of the ampere, coils of wire are used instead of straight wires, and their separation is only a few centimeters. Even more precise measurements of the standardized ampere are possible with a version of the Hall effect (see Section 27.9).

Mutual forces of attraction exist not only between *wires* carrying currents in the same direction, but also between the elements of a single current-carrying conductor. If the conductor is a liquid or an ionized gas (a plasma), these forces result in a constriction of the conductor called the *pinch effect*. The pinch effect in a plasma has been used in one technique to bring about nuclear fusion.

EXAMPLE 28.5 FORCES BETWEEN PARALLEL WIRES

Two straight, parallel, superconducting wires 4.5 mm apart carry equal currents of 15,000 A in opposite directions. What force, per unit length, does each wire exert on the other?

SOLUTION

IDENTIFY and SET UP: Figure 28.10 shows the situation. We find F/L, the magnetic force per unit length of wire, from Eq. (28.11).

28.10 Our sketch for this problem.

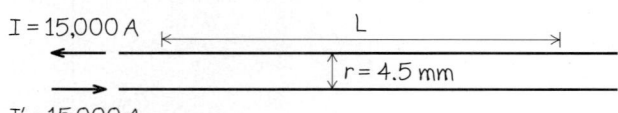

$I = 15,000\,A$
L
$r = 4.5\ mm$
$I' = 15,000\,A$

EXECUTE: The conductors *repel* each other because the currents are in opposite directions. From Eq. (28.11) the force per unit length is

$$\frac{F}{L} = \frac{\mu_0 I I'}{2\pi r} = \frac{(4\pi \times 10^{-7}\ \text{T·m/A})(15{,}000\ \text{A})^2}{(2\pi)(4.5 \times 10^{-3}\ \text{m})}$$

$$= 1.0 \times 10^4\ \text{N/m}$$

EVALUATE: This is a large force, more than one ton per meter. Currents and separations of this magnitude are used in superconducting electromagnets in particle accelerators, and mechanical stress analysis is a crucial part of the design process.

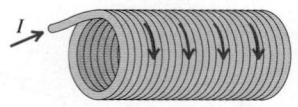

TEST YOUR UNDERSTANDING OF SECTION 28.4 A solenoid is a wire wound into a helical coil. The accompanying figure shows a solenoid that carries a current I. (a) Is the *magnetic* force that one turn of the coil exerts on an adjacent turn (i) attractive, (ii) repulsive, or (iii) zero? (b) Is the *electric* force that one turn of the coil exerts on an adjacent turn (i) attractive, (ii) repulsive, or (iii) zero? (c) Is the *magnetic* force between opposite sides of the same turn of the coil (i) attractive, (ii) repulsive, or (iii) zero? (d) Is the *electric* force between opposite sides of the same turn of the coil (i) attractive, (ii) repulsive, or (iii) zero? ▮

28.11 This electromagnet contains a current-carrying coil with numerous turns of wire. The resulting magnetic field can pick up large quantities of steel bars and other iron-bearing items.

28.5 MAGNETIC FIELD OF A CIRCULAR CURRENT LOOP

If you look inside a doorbell, a transformer, an electric motor, or an electromagnet (**Fig. 28.11**), you will find coils of wire with a large number of turns, spaced so closely that each turn is very nearly a planar circular loop. A current in such a coil is used to establish a magnetic field. In Section 27.7 we considered the force and torque on such a current loop placed in an external magnetic field produced by other currents; we are now about to find the magnetic field produced by such a loop or by a collection of closely spaced loops forming a coil.

Figure 28.12 shows a circular conductor with radius a. A current I is led into and out of the loop through two long, straight wires side by side; the currents in these straight wires are in opposite directions, and their magnetic fields very nearly cancel each other (see Example 28.4 in Section 28.3).

We can use the law of Biot and Savart, Eq. (28.5) or (28.6), to find the magnetic field at a point P on the axis of the loop, at a distance x from the center. As the figure shows, $d\vec{l}$ and $\hat{r}$ are perpendicular, and the direction of the field $d\vec{B}$ caused by this particular element $d\vec{l}$ lies in the xy-plane. Since $r^2 = x^2 + a^2$, the magnitude dB of the field due to element $d\vec{l}$ is

$$dB = \frac{\mu_0 I}{4\pi} \frac{dl}{(x^2 + a^2)} \qquad (28.12)$$

The components of the vector $d\vec{B}$ are

$$dB_x = dB\cos\theta = \frac{\mu_0 I}{4\pi} \frac{dl}{(x^2 + a^2)} \frac{a}{(x^2 + a^2)^{1/2}} \qquad (28.13)$$

$$dB_y = dB\sin\theta = \frac{\mu_0 I}{4\pi} \frac{dl}{(x^2 + a^2)} \frac{x}{(x^2 + a^2)^{1/2}} \qquad (28.14)$$

The *total* field $\vec{B}$ at P has only an x-component (it is perpendicular to the plane of the loop). Here's why: For every element $d\vec{l}$ there is a corresponding element on the opposite side of the loop, with opposite direction. These two elements give equal contributions to the x-component of $d\vec{B}$, given by Eq. (28.13), but *opposite* components perpendicular to the x-axis. Thus all the perpendicular components cancel and only the x-components survive.

To obtain the x-component of the total field $\vec{B}$, we integrate Eq. (28.13), including all the $d\vec{l}$'s around the loop. Everything in this expression except dl is constant and can be taken outside the integral, and we have

$$B_x = \int \frac{\mu_0 I}{4\pi} \frac{a \, dl}{(x^2 + a^2)^{3/2}} = \frac{\mu_0 I a}{4\pi(x^2 + a^2)^{3/2}} \int dl$$

The integral of dl is just the circumference of the circle, $\int dl = 2\pi a$, and so

Magnetic field on axis of a circular current-carrying loop ···· Magnetic constant ···· Current

$$B_x = \frac{\mu_0 I a^2}{2(x^2 + a^2)^{3/2}} \qquad (28.15)$$

Radius of loop

Distance along axis from center of loop to field point

The *direction* of this magnetic field is given by a right-hand rule. If you curl the fingers of your right hand around the loop in the direction of the current, your right thumb points in the direction of the field (**Fig. 28.13**).

Magnetic Field on the Axis of a Coil

Now suppose that instead of the single loop in Fig. 28.12 we have a coil consisting of N loops, all with the same radius. The loops are closely spaced so that the plane of each loop is essentially the same distance x from the field point P. Then the total field is N times the field of a single loop:

$$B_x = \frac{\mu_0 N I a^2}{2(x^2 + a^2)^{3/2}} \qquad \text{(on the axis of } N \text{ circular loops)} \qquad (28.16)$$

The factor N in Eq. (28.16) is the reason coils of wire, not single loops, are used to produce strong magnetic fields; for a desired field strength, using a single loop might require a current I so great as to exceed the rating of the loop's wire.

Figure 28.14 shows a graph of B_x as a function of x. The maximum value of the field is at $x = 0$, the center of the loop or coil:

Magnetic field at center of N circular current-carrying loops ···· Magnetic constant ···· Number of loops ···· Current

$$B_x = \frac{\mu_0 N I}{2a} \qquad (28.17)$$

Radius of loop

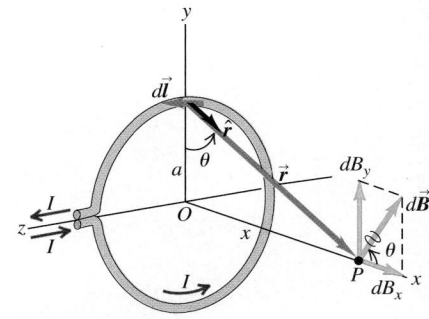

28.12 Magnetic field on the axis of a circular loop. The current in the segment $d\vec{l}$ causes the field $d\vec{B}$, which lies in the xy-plane. The currents in other $d\vec{l}$'s cause $d\vec{B}$'s with different components perpendicular to the x-axis; these components add to zero. The x-components of the $d\vec{B}$'s combine to give the total $\vec{B}$ field at point P.

MP

PhET: Faraday's Electromagnetic Lab
PhET: Magnets and Electromagnets

28.13 The right-hand rule for the direction of the magnetic field produced on the axis of a current-carrying coil.

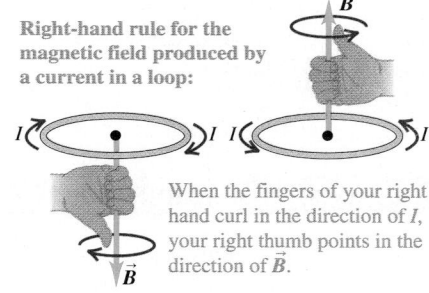

Right-hand rule for the magnetic field produced by a current in a loop:

When the fingers of your right hand curl in the direction of I, your right thumb points in the direction of $\vec{B}$.

28.14 Graph of the magnetic field along the axis of a circular coil with N turns. When x is much larger than a, the field magnitude decreases approximately as $1/x^3$.

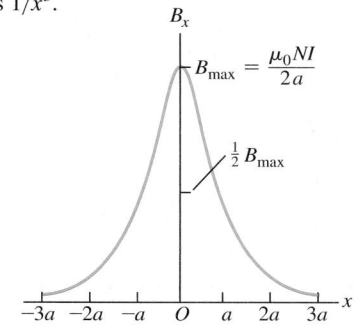

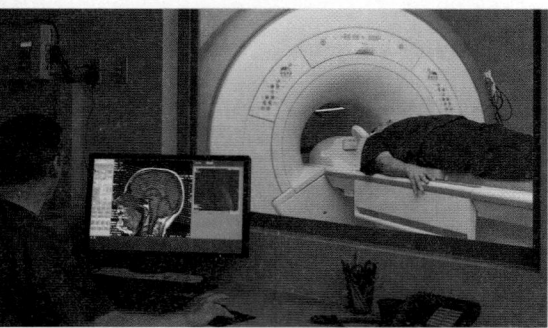

In Section 27.7 we defined the *magnetic dipole moment* μ (or *magnetic moment*) of a current-carrying loop to be equal to IA, where A is the cross-sectional area of the loop. If there are N loops, the total magnetic moment is NIA. The circular loop in Fig. 28.12 has area $A = \pi a^2$, so the magnetic moment of a single loop is $\mu = I\pi a^2$; for N loops, $\mu = NI\pi a^2$. Substituting these results into Eqs. (28.15) and (28.16), we find

$$B_x = \frac{\mu_0 \mu}{2\pi(x^2 + a^2)^{3/2}} \qquad \text{(on the axis of any number of circular loops)} \qquad (28.18)$$

We described a magnetic dipole in Section 27.7 in terms of its response to a magnetic field produced by currents outside the dipole. But a magnetic dipole is also a *source* of magnetic field; Eq. (28.18) describes the magnetic field *produced* by a magnetic dipole for points along the dipole axis. This field is directly proportional to the magnetic dipole moment μ. Note that the field at all points on the x-axis is in the same direction as the vector magnetic moment $\vec{\mu}$.

28.15 Magnetic field lines produced by the current in a circular loop. At points on the axis the $\vec{B}$ field has the same direction as the magnetic moment of the loop.

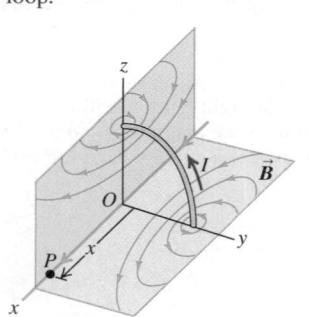

CAUTION **Magnetic field of a coil** Equations (28.15), (28.16), and (28.18) are valid only on the *axis* of a loop or coil. Don't attempt to apply these equations at other points! ▌

Figure 28.15 shows some of the magnetic field lines surrounding a circular current loop (magnetic dipole) in planes through the axis. The directions of the field lines are given by the same right-hand rule as for a long, straight conductor. Grab the conductor with your right hand, with your thumb in the direction of the current; your fingers curl around in the same direction as the field lines. The field lines for the circular current loop are closed curves that encircle the conductor; they are *not* circles, however.

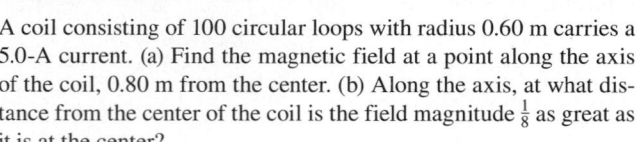

EXAMPLE 28.6   MAGNETIC FIELD OF A COIL

A coil consisting of 100 circular loops with radius 0.60 m carries a 5.0-A current. (a) Find the magnetic field at a point along the axis of the coil, 0.80 m from the center. (b) Along the axis, at what distance from the center of the coil is the field magnitude $\frac{1}{8}$ as great as it is at the center?

SOLUTION

IDENTIFY and SET UP: This problem concerns the magnetic field magnitude B along the axis of a current-carrying coil, so we can use Eq. (28.16). We are given $N = 100$, $I = 5.0$ A, and $a = 0.60$ m. In part (a) our target variable is B_x at a given value of x. In part (b) the target variable is the value of x at which the field has $\frac{1}{8}$ of the magnitude that it has at the origin.

EXECUTE: (a) Using $x = 0.80$ m, from Eq. (28.16) we have

$$B_x = \frac{(4\pi \times 10^{-7}\,\text{T}\cdot\text{m/A})(100)(5.0\,\text{A})(0.60\,\text{m})^2}{2[(0.80\,\text{m})^2 + (0.60\,\text{m})^2]^{3/2}}$$

$$= 1.1 \times 10^{-4}\,\text{T}$$

(b) Considering Eq. (28.16), we seek a value of x such that

$$\frac{1}{(x^2 + a^2)^{3/2}} = \frac{1}{8}\frac{1}{(0^2 + a^2)^{3/2}}$$

To solve this for x, we take the reciprocal of the whole thing and then take the 2/3 power of both sides; the result is

$$x = \pm\sqrt{3}a = \pm 1.04\,\text{m}$$

EVALUATE: We check our answer in part (a) by finding the coil's magnetic moment and substituting the result into Eq. (28.18):

$$\mu = NI\pi a^2 = (100)(5.0\,\text{A})\pi(0.60\,\text{m})^2 = 5.7 \times 10^2\,\text{A}\cdot\text{m}^2$$

$$B_x = \frac{(4\pi \times 10^{-7}\,\text{T}\cdot\text{m/A})(5.7 \times 10^2\,\text{A}\cdot\text{m}^2)}{2\pi[(0.80\,\text{m})^2 + (0.60\,\text{m})^2]^{3/2}} = 1.1 \times 10^{-4}\,\text{T}$$

The magnetic moment μ is relatively large, yet it produces a rather small field, comparable to that of the earth. This illustrates how difficult it is to produce strong fields of 1 T or more.

TEST YOUR UNDERSTANDING OF SECTION 28.5   Figure 28.12 shows the magnetic field $d\vec{B}$ produced at point P by a segment $d\vec{l}$ that lies on the positive y-axis (at the top of the loop). This field has components $dB_x > 0$, $dB_y > 0$, $dB_z = 0$. (a) What are the signs of the components of the field $d\vec{B}$ produced at P by a segment $d\vec{l}$ on the negative y-axis (at the bottom of the loop)? (i) $dB_x > 0$, $dB_y > 0$, $dB_z = 0$; (ii) $dB_x > 0$, $dB_y < 0$, $dB_z = 0$; (iii) $dB_x < 0$, $dB_y > 0$, $dB_z = 0$; (iv) $dB_x < 0$, $dB_y < 0$, $dB_z = 0$; (v) none of these. (b) What are the signs of the components of the field $d\vec{B}$ produced at P by a segment $d\vec{l}$ on the negative z-axis (at the right-hand side of the loop)? (i) $dB_x > 0$, $dB_y > 0$, $dB_z = 0$; (ii) $dB_x > 0$, $dB_y < 0$, $dB_z = 0$; (iii) $dB_x < 0$, $dB_y > 0$, $dB_z = 0$; (iv) $dB_x < 0$, $dB_y < 0$, $dB_z = 0$; (v) none of these. ∎

28.6 AMPERE'S LAW

So far our calculations of the magnetic field due to a current have involved finding the infinitesimal field $d\vec{B}$ due to a current element and then summing all the $d\vec{B}$'s to find the total field. This approach is directly analogous to our *electric-field* calculations in Chapter 21.

For the electric-field problem we found that in situations with a highly symmetric charge distribution, it was often easier to use Gauss's law to find $\vec{E}$. There is likewise a law that allows us to more easily find the *magnetic* fields caused by highly symmetric *current* distributions. But the law that allows us to do this, called *Ampere's law,* is rather different in character from Gauss's law.

Gauss's law for electric fields (Chapter 22) involves the flux of $\vec{E}$ through a closed surface; it states that this flux is equal to the total charge enclosed within the surface, divided by the constant ϵ_0. Thus this law relates electric fields and charge distributions. By contrast, Gauss's law for *magnetic* fields, Eq. (28.10), is *not* a relationship between magnetic fields and current distributions; it states that the flux of $\vec{B}$ through *any* closed surface is always zero, whether or not there are currents within the surface. So Gauss's law for $\vec{B}$ can't be used to determine the magnetic field produced by a particular current distribution.

Ampere's law is formulated not in terms of magnetic flux, but rather in terms of the *line integral* of $\vec{B}$ around a closed path, denoted by

$$\oint \vec{B} \cdot d\vec{l}$$

We used line integrals to define work in Chapter 6 and to calculate electric potential in Chapter 23. To evaluate this integral, we divide the path into infinitesimal segments $d\vec{l}$, calculate the scalar product of $\vec{B} \cdot d\vec{l}$ for each segment, and sum these products. In general, $\vec{B}$ varies from point to point, and we must use the value of $\vec{B}$ at the location of each $d\vec{l}$. An alternative notation is $\oint B_{\parallel} dl$, where $B_{\parallel}$ is the component of $\vec{B}$ parallel to $d\vec{l}$ at each point. The circle on the integral sign indicates that this integral is always computed for a *closed* path, one whose beginning and end points are the same.

Ampere's Law for a Long, Straight Conductor

To introduce the basic idea of Ampere's law, let's consider again the magnetic field caused by a long, straight conductor carrying a current I. We found in Section 28.3 that the field at a distance r from the conductor has magnitude

$$B = \frac{\mu_0 I}{2\pi r}$$

28.16 Three integration paths for the line integral of $\vec{B}$ in the vicinity of a long, straight conductor carrying current I *out* of the plane of the page (as indicated by the circle with a dot). The conductor is seen end-on.

(a) Integration path is a circle centered on the conductor; integration goes around the circle counterclockwise.

Result: $\oint \vec{B} \cdot d\vec{l} = \mu_0 I$

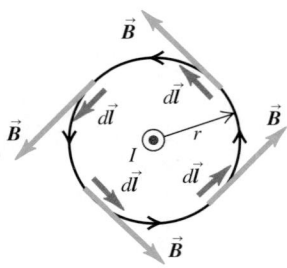

(b) Same integration path as in (a), but integration goes around the circle clockwise.

Result: $\oint \vec{B} \cdot d\vec{l} = -\mu_0 I$

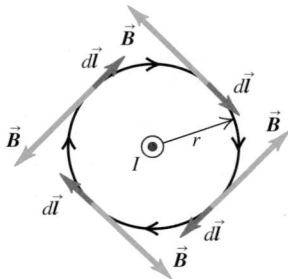

(c) An integration path that does not enclose the conductor

Result: $\oint \vec{B} \cdot d\vec{l} = 0$

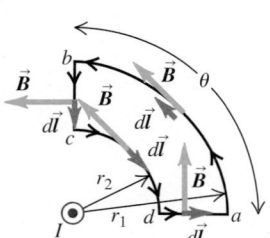

The magnetic field lines are circles centered on the conductor. Let's take the line integral of $\vec{B}$ around a circle with radius r, as in **Fig. 28.16a**. At every point on the circle, $\vec{B}$ and $d\vec{l}$ are parallel, and so $\vec{B} \cdot d\vec{l} = B\,dl$; since r is constant around the circle, B is constant as well. Alternatively, we can say that $B_{\parallel}$ is constant and equal to B at every point on the circle. Hence we can take B outside of the integral. The remaining integral is just the circumference of the circle, so

$$\oint \vec{B} \cdot d\vec{l} = \oint B_{\parallel}\,dl = B \oint dl = \frac{\mu_0 I}{2\pi r}(2\pi r) = \mu_0 I$$

The line integral is thus independent of the radius of the circle and is equal to μ_0 multiplied by the current passing through the area bounded by the circle.

In Fig. 28.16b the situation is the same, but the integration path now goes around the circle in the opposite direction. Now $\vec{B}$ and $d\vec{l}$ are antiparallel, so $\vec{B} \cdot d\vec{l} = -B\,dl$ and the line integral equals $-\mu_0 I$. We get the same result if the integration path is the same as in Fig. 28.16a, but the direction of the current is reversed. Thus $\oint \vec{B} \cdot d\vec{l}$ equals μ_0 multiplied by the current passing through the area bounded by the integration path, with a positive or negative sign depending on the direction of the current relative to the direction of integration.

There's a simple rule for the sign of the current; you won't be surprised to learn that it uses your right hand. Curl the fingers of your right hand around the integration path so that they curl in the direction of integration (that is, the direction that you use to evaluate $\oint \vec{B} \cdot d\vec{l}$). Then your right thumb indicates the positive current direction. Currents that pass through the integration path in this direction are positive; those in the opposite direction are negative. Using this rule, convince yourself that the current is positive in Fig. 28.16a and negative in Fig. 28.16b. Here's another way to say the same thing: Looking at the surface bounded by the integration path, integrate counterclockwise around the path as in Fig. 28.16a. Currents moving toward you through the surface are positive, and those going away from you are negative.

An integration path that does *not* enclose the conductor is used in Fig. 28.16c. Along the circular arc ab of radius r_1, $\vec{B}$ and $d\vec{l}$ are parallel, and $B_{\parallel} = B_1 = \mu_0 I/2\pi r_1$; along the circular arc cd of radius r_2, $\vec{B}$ and $d\vec{l}$ are antiparallel and $B_{\parallel} = -B_2 = -\mu_0 I/2\pi r_2$. The $\vec{B}$ field is perpendicular to $d\vec{l}$ at each point on the straight sections bc and da, so $B_{\parallel} = 0$ and these sections contribute zero to the line integral. The total line integral is then

$$\oint \vec{B} \cdot d\vec{l} = \oint B_{\parallel}\,dl = B_1 \int_a^b dl + (0) \int_b^c dl + (-B_2) \int_c^d dl + (0) \int_d^a dl$$

$$= \frac{\mu_0 I}{2\pi r_1}(r_1 \theta) + 0 - \frac{\mu_0 I}{2\pi r_2}(r_2 \theta) + 0 = 0$$

The magnitude of $\vec{B}$ is greater on arc cd than on arc ab, but the arc length is less, so the contributions from the two arcs exactly cancel. Even though there is a magnetic field everywhere along the integration path, the line integral $\oint \vec{B} \cdot d\vec{l}$ is zero if there is no current passing through the area bounded by the path.

We can also derive these results for more general integration paths, such as the one in **Fig. 28.17a**. At the position of the line element $d\vec{l}$, the angle between $d\vec{l}$ and $\vec{B}$ is ϕ, and

$$\vec{B} \cdot d\vec{l} = B\,dl\cos\phi$$

From the figure, $dl\cos\phi = r\,d\theta$, where $d\theta$ is the angle subtended by $d\vec{l}$ at the position of the conductor and r is the distance of $d\vec{l}$ from the conductor. Thus

$$\oint \vec{B} \cdot d\vec{l} = \oint \frac{\mu_0 I}{2\pi r}(r\,d\theta) = \frac{\mu_0 I}{2\pi} \oint d\theta$$

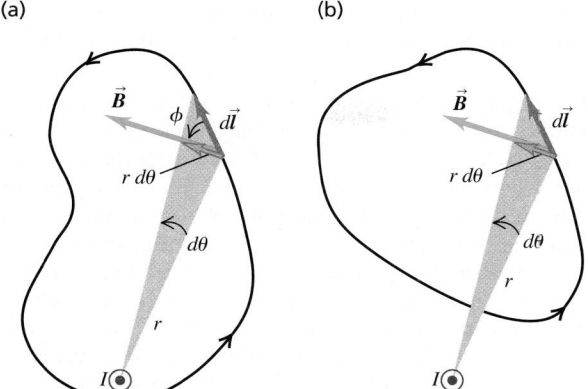

(a)   (b)

28.17 (a) A more general integration path for the line integral of $\vec{B}$ around a long, straight conductor carrying current I *out* of the plane of the page. The conductor is seen end-on. (b) A more general integration path that does not enclose the conductor.

But $\oint d\theta$ is just equal to 2π, the total angle swept out by the radial line from the conductor to $d\vec{l}$ during a complete trip around the path. So we get

$$\oint \vec{B} \cdot d\vec{l} = \mu_0 I \qquad (28.19)$$

This result doesn't depend on the shape of the path or on the position of the wire inside it. If the current in the wire is opposite to that shown, the integral has the opposite sign. But if the path doesn't enclose the wire (Fig. 28.17b), then the net change in θ during the trip around the integration path is zero; $\oint d\theta$ is zero instead of 2π and the line integral is zero.

Ampere's Law: General Statement

We can generalize Ampere's law even further. Suppose *several* long, straight conductors pass through the surface bounded by the integration path. The total magnetic field $\vec{B}$ at any point on the path is the vector sum of the fields produced by the individual conductors. Thus the line integral of the total $\vec{B}$ equals μ_0 times the *algebraic sum* of the currents. In calculating this sum, we use the sign rule for currents described above. If the integration path does not enclose a particular wire, the line integral of the $\vec{B}$ field of that wire is zero, because the angle θ for that wire sweeps through a net change of zero rather than 2π during the integration. Any conductors present that are not enclosed by a particular path may still contribute to the value of $\vec{B}$ at every point, but the *line integrals* of their fields around the path are zero.

Thus we can replace I in Eq. (28.19) with I_{encl}, the algebraic sum of the currents *enclosed* or *linked* by the integration path, with the sum evaluated by using the sign rule just described (**Fig. 28.18**). Then **Ampere's law** says

28.18 Ampere's law.

Perspective view

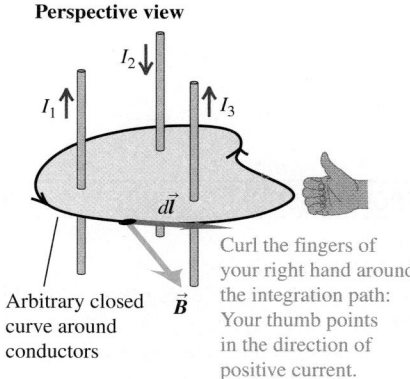

Arbitrary closed curve around conductors

Curl the fingers of your right hand around the integration path: Your thumb points in the direction of positive current.

Ampere's law:

$$\underbrace{\oint \vec{B} \cdot d\vec{l}}_{\text{Line integral around a closed path}} = \underbrace{\mu_0}_{\text{Magnetic constant}} \underbrace{I_{encl}}_{\substack{\text{Net current} \\ \text{enclosed by path}}} \qquad (28.20)$$

Scalar product of magnetic field and vector segment of path

While we have derived Ampere's law only for the special case of the field of several long, straight, parallel conductors, Eq. (28.20) is in fact valid for conductors and paths of *any* shape. The general derivation is no different in principle from what we have presented, but the geometry is more complicated.

If $\oint \vec{B} \cdot d\vec{l} = 0$, it *does not* necessarily mean that $\vec{B} = 0$ everywhere along the path, only that the total current through an area bounded by the path is zero. In Figs. 28.16c and 28.17b, the integration paths enclose no current at all; in

Top view

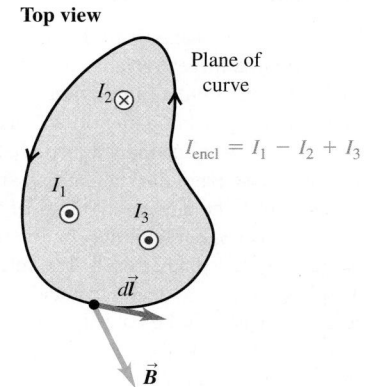

Plane of curve

$$I_{encl} = I_1 - I_2 + I_3$$

Ampere's law: If we calculate the line integral of the magnetic field around a closed curve, the result equals μ_0 times the total enclosed current: $\oint \vec{B} \cdot d\vec{l} = \mu_0 I_{encl}$.

28.19 Two long, straight conductors carrying equal currents in opposite directions. The conductors are seen end-on, and the integration path is counterclockwise. The line integral $\oint \vec{B} \cdot d\vec{l}$ gets zero contribution from the upper and lower segments, a positive contribution from the left segment, and a negative contribution from the right segment; the net integral is zero.

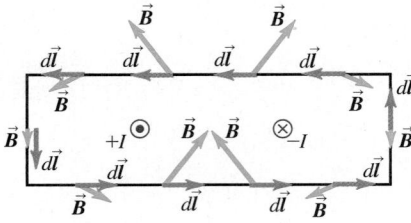

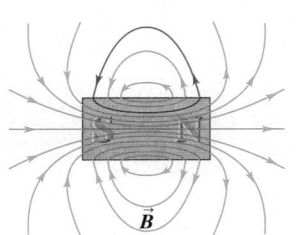

Fig. 28.19 there are positive and negative currents of equal magnitude through the area enclosed by the path. In both cases, $I_{encl} = 0$ and the line integral is zero.

> **CAUTION** **Line integrals of electric and magnetic fields** In Chapter 23 we saw that the line integral of the electrostatic field $\vec{E}$ around any closed path is equal to zero; this is a statement that the electrostatic force $\vec{F} = q\vec{E}$ on a point charge q is conservative, so this force does zero work on a charge that moves around a closed path that returns to the starting point. The value of the line integral $\oint \vec{B} \cdot d\vec{l}$ is not similarly related to the question of whether the *magnetic* force is conservative. Remember that the magnetic force $\vec{F} = q\vec{v} \times \vec{B}$ on a moving charged particle is always *perpendicular* to $\vec{B}$, so $\oint \vec{B} \cdot d\vec{l}$ is *not* related to the work done by the magnetic force; as stated in Ampere's law, this integral is related only to the total current through a surface bounded by the integration path. In fact, the magnetic force on a moving charged particle is *not* conservative. A conservative force depends on only the position of the body on which the force is exerted, but the magnetic force on a moving charged particle also depends on the *velocity* of the particle. ∎

Equation (28.20) turns out to be valid only if the currents are steady and if no magnetic materials or time-varying electric fields are present. In Chapter 29 we will see how to generalize Ampere's law for time-varying fields.

TEST YOUR UNDERSTANDING OF SECTION 28.6 The accompanying figure shows magnetic field lines through the center of a permanent magnet. The magnet is not connected to a source of emf. One of the field lines is colored red. What can you conclude about the currents inside the permanent magnet within the region enclosed by this field line? (i) There are no currents inside the magnet; (ii) there are currents directed out of the plane of the page; (iii) there are currents directed into the plane of the page; (iv) not enough information is given to decide. ∎

28.7 APPLICATIONS OF AMPERE'S LAW

Ampere's law is useful when we can exploit symmetry to evaluate the line integral of $\vec{B}$. Several examples are given below. Problem-Solving Strategy 28.2 is directly analogous to Problem-Solving Strategy 22.1 (Section 22.4) for applications of Gauss's law; we suggest you review that strategy now and compare the two methods.

PROBLEM-SOLVING STRATEGY 28.2 | **AMPERE'S LAW**

IDENTIFY *the relevant concepts:* Like Gauss's law, Ampere's law is most useful when the magnetic field is highly symmetric. In the form $\oint \vec{B} \cdot d\vec{l} = \mu_0 I_{encl}$, it can yield the magnitude of $\vec{B}$ as a function of position if we are given the magnitude and direction of the field-generating electric current.

SET UP *the problem* using the following steps:
1. Determine the target variable(s). Usually one will be the magnitude of the $\vec{B}$ field as a function of position.
2. Select the integration path you will use with Ampere's law. If you want to determine the magnetic field at a certain point, then the path must pass through that point. The integration path doesn't have to be any actual physical boundary. Usually it is a purely geometric curve; it may be in empty space, embedded in a solid body, or some of each. The integration path has to have enough *symmetry* to make evaluation of the integral possible. Ideally the path will be tangent to $\vec{B}$ in regions of interest; elsewhere the path should be perpendicular to $\vec{B}$ or should run through regions in which $\vec{B} = 0$.

EXECUTE *the solution* as follows:
1. Carry out the integral $\oint \vec{B} \cdot d\vec{l}$ along the chosen path. If $\vec{B}$ is tangent to all or some portion of the path and has the same magnitude B at every point, then its line integral is the product of B and the length of that portion of the path. If $\vec{B}$ is perpendicular to some portion of the path, or if $\vec{B} = 0$, that portion makes no contribution to the integral.
2. In the integral $\oint \vec{B} \cdot d\vec{l}$, $\vec{B}$ is the *total* magnetic field at each point on the path; it can be caused by currents enclosed *or not enclosed* by the path. If *no* net current is enclosed by the path, the field at points on the path need not be zero, but the integral $\oint \vec{B} \cdot d\vec{l}$ is always zero.
3. Determine the current I_{encl} enclosed by the integration path. A right-hand rule gives the sign of this current: If you curl the fingers of your right hand so that they follow the path in the direction of integration, then your right thumb points in the direction of positive current. If $\vec{B}$ is tangent to the path everywhere and I_{encl} is positive, the direction of $\vec{B}$ is the same as the direction of integration. If instead I_{encl} is negative, $\vec{B}$ is in the direction opposite to that of the integration.
4. Use Ampere's law $\oint \vec{B} \cdot d\vec{l} = \mu_0 I$ to solve for the target variable.

EVALUATE *your answer:* If your result is an expression for the field magnitude as a function of position, check it by examining how the expression behaves in different limits.

EXAMPLE 28.7   FIELD OF A LONG, STRAIGHT, CURRENT-CARRYING CONDUCTOR

In Section 28.6 we derived Ampere's law from Eq. (28.9) for the field $\vec{B}$ of a long, straight, current-carrying conductor. Reverse this process, and use Ampere's law to find $\vec{B}$ for this situation.

SOLUTION

IDENTIFY and SET UP: The situation has cylindrical symmetry, so in Ampere's law we take our integration path to be a circle with radius r centered on the conductor and lying in a plane perpendicular to it, as in Fig. 28.16a. The field $\vec{B}$ is everywhere tangent to this circle and has the same magnitude B everywhere on the circle.

EXECUTE: With our choice of the integration path, Ampere's law [Eq. (28.20)] becomes

$$\oint \vec{B} \cdot d\vec{l} = \oint B_{\parallel} \, dl = B(2\pi r) = \mu_0 I$$

Equation (28.9), $B = \mu_0 I / 2\pi r$, follows immediately.

Ampere's law determines the direction of $\vec{B}$ as well as its magnitude. Since we chose to go counterclockwise around the integration path, the positive direction for current is out of the plane of Fig. 28.16a; this is the same as the actual current direction in the figure, so I is positive and the integral $\oint \vec{B} \cdot d\vec{l}$ is also positive. Since the $d\vec{l}$'s run counterclockwise, the direction of $\vec{B}$ must be counterclockwise as well, as shown in Fig. 28.16a.

EVALUATE: Our results are consistent with those in Section 28.6.

EXAMPLE 28.8   FIELD OF A LONG CYLINDRICAL CONDUCTOR

A cylindrical conductor with radius R carries a current I (**Fig. 28.20**). The current is uniformly distributed over the cross-sectional area of the conductor. Find the magnetic field as a function of the distance r from the conductor axis for points both inside ($r < R$) and outside ($r > R$) the conductor.

SOLUTION

IDENTIFY and SET UP: As in Example 28.7, the current distribution has cylindrical symmetry, and the magnetic field lines must be circles concentric with the conductor axis. To find the magnetic field inside and outside the conductor, we choose circular integration paths with radii $r < R$ and $r > R$, respectively (see Fig. 28.20).

EXECUTE: In either case the field $\vec{B}$ has the same magnitude at every point on the circular integration path and is tangent to the path. Thus the magnitude of the line integral is simply $B(2\pi r)$. To find the current I_{encl} enclosed by a circular integration path inside the conductor ($r < R$), note that the current density (current per unit area) is $J = I/\pi R^2$ so $I_{encl} = J(\pi r^2) = Ir^2/R^2$. Hence Ampere's law gives $B(2\pi r) = \mu_0 Ir^2/R^2$, or

$$B = \frac{\mu_0 I}{2\pi} \frac{r}{R^2} \quad \text{(inside the conductor,} \atop r < R) \qquad (28.21)$$

A circular integration path outside the conductor encloses the total current in the conductor, so $I_{encl} = I$. Applying Ampere's law gives the same equation as in Example 28.7, with the same result for B:

$$B = \frac{\mu_0 I}{2\pi r} \quad \text{(outside the conductor,} \atop r > R) \qquad (28.22)$$

Outside the conductor, the magnetic field is the same as that of a long, straight conductor carrying current I, independent of the radius R over which the current is distributed. Indeed, the magnetic field outside *any* cylindrically symmetric current distribution is the same as if the entire current were concentrated along the axis of the distribution. This is analogous to the results of Examples 22.5 and 22.9 (Section 22.4), in which we found that the *electric* field outside a spherically symmetric *charged* body is the same as though the entire charge were concentrated at the center.

EVALUATE: At the surface of the conductor ($r = R$), Eqs. (28.21) and (28.22) agree, as they must. **Figure 28.21** shows a graph of B as a function of r.

28.20 To find the magnetic field at radius $r < R$, we apply Ampere's law to the circle enclosing the gray area. The current through the gray area is $(r^2/R^2)I$. To find the magnetic field at radius $r > R$, we apply Ampere's law to the circle enclosing the entire conductor.

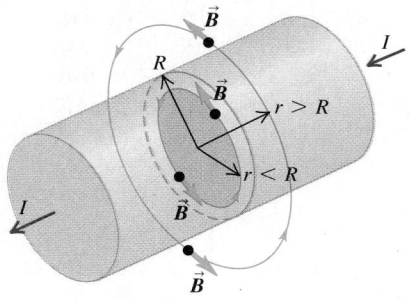

28.21 Magnitude of the magnetic field inside and outside a long, straight cylindrical conductor with radius R carrying a current I.

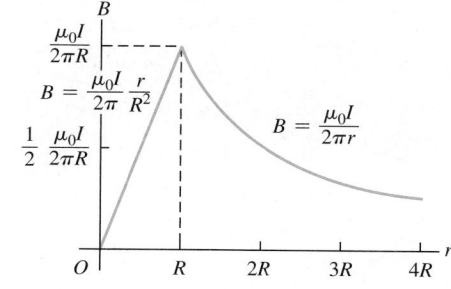

EXAMPLE 28.9 FIELD OF A SOLENOID

A solenoid consists of a helical winding of wire on a cylinder, usually circular in cross section. There can be thousands of closely spaced turns (often in several layers), each of which can be regarded as a circular loop. For simplicity, **Fig. 28.22** shows a solenoid with only a few turns. All turns carry the same current I, and the total $\vec{B}$ field at every point is the vector sum of the fields caused by the individual turns. The figure shows field lines in the xy- and xz-planes. We draw field lines that are uniformly spaced at the center of the solenoid. Exact calculations show that for a long, closely wound solenoid, half of these field lines emerge from the ends and half "leak out" through the windings between the center and the end, as the figure suggests.

If the solenoid is long in comparison with its cross-sectional diameter and the coils are tightly wound, the field inside the solenoid near its midpoint is very nearly uniform over the cross section and parallel to the axis; the *external* field near the midpoint is very small.

Use Ampere's law to find the field at or near the center of such a solenoid if it has n turns per unit length and carries current I.

SOLUTION

IDENTIFY and SET UP: We assume that $\vec{B}$ is uniform inside the solenoid and zero outside. **Figure 28.23** shows the situation and our chosen integration path, rectangle $abcd$. Side ab, with length L, is parallel to the axis of the solenoid. Sides bc and da are taken to be very long so that side cd is far from the solenoid; then the field at side cd is negligibly small.

EXECUTE: Along side ab, $\vec{B}$ is parallel to the path and is constant. Our Ampere's-law integration takes us along side ab in the same direction as $\vec{B}$, so here $B_\parallel = +B$ and

$$\int_a^b \vec{B} \cdot d\vec{l} = BL$$

Along sides bc and da, $\vec{B}$ is perpendicular to the path, and so $B_\parallel = 0$; along side cd, $\vec{B} = 0$ and so $B_\parallel = 0$. Around the entire closed path, then, we have $\oint \vec{B} \cdot d\vec{l} = BL$.

28.23 Our sketch for this problem.

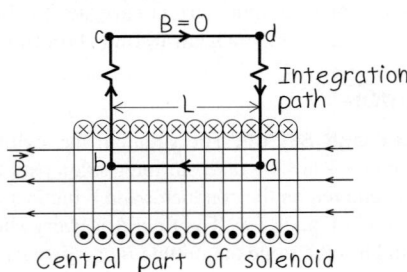

In a length L there are nL turns, each of which passes once through $abcd$ carrying current I. Hence the total current enclosed by the rectangle is $I_{encl} = nLI$. The integral $\oint \vec{B} \cdot d\vec{l}$ is positive, so from Ampere's law I_{encl} must be positive as well. This means that the current passing through the surface bounded by the integration path must be in the direction shown in Fig. 28.23. Ampere's law then gives $BL = \mu_0 nLI$, or

$$B = \mu_0 nI \quad \text{(solenoid)} \quad (28.23)$$

Side ab need not lie on the axis of the solenoid, so this result demonstrates that the field is uniform over the entire cross section at the center of the solenoid's length.

EVALUATE: Note that the *direction* of $\vec{B}$ inside the solenoid is in the same direction as the solenoid's vector magnetic moment $\vec{\mu}$, as we found in Section 28.5 for a single current-carrying loop.

For points along the axis, the field is strongest at the center of the solenoid and drops off near the ends. For a solenoid very long in comparison to its diameter, the field magnitude at each end is exactly half that at the center. This is approximately the case even for a relatively short solenoid, as **Fig. 28.24** shows.

28.22 Magnetic field lines produced by the current in a solenoid. For clarity, only a few turns are shown.

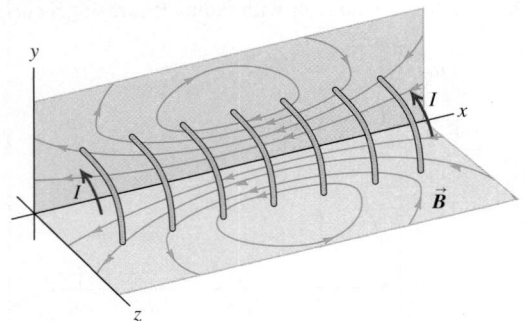

28.24 Magnitude of the magnetic field at points along the axis of a solenoid with length $4a$, equal to four times its radius a. The field magnitude at each end is about half its value at the center. (Compare with Fig. 28.14 for the field of N circular loops.)

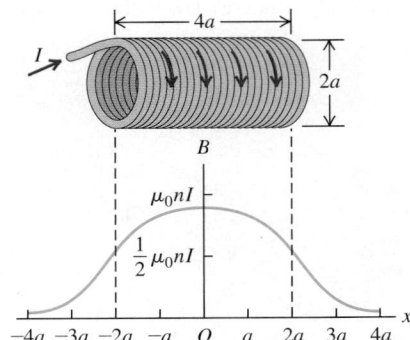

EXAMPLE 28.10 FIELD OF A TOROIDAL SOLENOID

Figure 28.25a shows a doughnut-shaped **toroidal solenoid,** tightly wound with N turns of wire carrying a current I. (In a practical solenoid the turns would be much more closely spaced than they are in the figure.) Find the magnetic field at all points.

SOLUTION

IDENTIFY and SET UP: Ignoring the slight pitch of the helical windings, we can consider each turn of a tightly wound toroidal solenoid as a loop lying in a plane perpendicular to the large, circular axis of the toroid. The symmetry of the situation then tells us that the magnetic field lines must be circles concentric with the toroid axis. Therefore we choose circular integration paths (of which Fig. 28.25b shows three) for use with Ampere's law, so that the field $\vec{B}$ (if any) is tangent to each path at all points along the path.

EXECUTE: Along each path, $\oint \vec{B} \cdot d\vec{l}$ equals the product of B and the path circumference $l = 2\pi r$. The total current enclosed by path 1 is zero, so from Ampere's law the field $\vec{B} = 0$ everywhere on this path.

Each turn of the winding passes *twice* through the area bounded by path 3, carrying equal currents in opposite directions. The *net* current enclosed is therefore zero, and hence $\vec{B} = 0$ at all points on this path as well. We conclude that *the field of an ideal toroidal solenoid is confined to the space enclosed by the windings.* We can think of such a solenoid as a tightly wound, straight solenoid that has been bent into a circle.

For path 2, we have $\oint \vec{B} \cdot d\vec{l} = 2\pi r B$. Each turn of the winding passes *once* through the area bounded by this path, so $I_{encl} = NI$. We note that I_{encl} is positive for the clockwise direction of integration in Fig. 28.25b, so $\vec{B}$ is in the direction shown. Ampere's law then says that $2\pi r B = \mu_0 NI$, so

$$B = \frac{\mu_0 NI}{2\pi r} \qquad \text{(toroidal solenoid)} \qquad (28.24)$$

EVALUATE: Equation (28.24) indicates that B is *not* uniform over the interior of the core, because different points in the interior

28.25 (a) A toroidal solenoid. For clarity, only a few turns of the winding are shown. (b) Integration paths (black circles) used to compute the magnetic field $\vec{B}$ set up by the current (shown as dots and crosses).

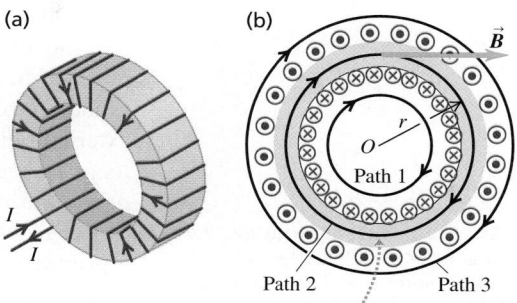

The magnetic field is confined almost entirely to the space enclosed by the windings (in blue).

are different distances r from the toroid axis. However, if the radial extent of the core is small in comparison to r, the variation is slight. In that case, considering that $2\pi r$ is the circumferential length of the toroid and that $N/2\pi r$ is the number of turns per unit length n, the field may be written as $B = \mu_0 nI$, just as it is at the center of a long, *straight* solenoid.

In a real toroidal solenoid the turns are not precisely circular loops but rather segments of a bent helix. As a result, the external field is not exactly zero. To estimate its magnitude, we imagine Fig. 28.25a as being *very* roughly equivalent, for points outside the torus, to a *single-turn* circular loop with radius r. At the center of such a loop, Eq. (28.17) gives $B = \mu_0 I/2r$; this is smaller than the field inside the solenoid by the factor N/π.

The equations we have derived for the field in a closely wound straight or toroidal solenoid are strictly correct only for windings in *vacuum*. For most practical purposes, however, they can be used for windings in air or on a core of any nonmagnetic, nonsuperconducting material. In the next section we will show how these equations are modified if the core is a magnetic material.

TEST YOUR UNDERSTANDING OF SECTION 28.7 Consider a conducting wire that runs along the central axis of a hollow conducting cylinder. Such an arrangement, called a *coaxial cable*, has many applications in telecommunications. (The cable that connects a television set to a local cable provider is an example of a coaxial cable.) In such a cable a current I runs in one direction along the hollow conducting cylinder and is spread uniformly over the cylinder's cross-sectional area. An equal current runs in the opposite direction along the central wire. How does the magnitude B of the magnetic field outside such a cable depend on the distance r from the central axis of the cable? (i) B is proportional to $1/r$; (ii) B is proportional to $1/r^2$; (iii) B is zero at all points outside the cable. ∎

Hollow conducting cylinder   Insulator   Central wire

28.8 MAGNETIC MATERIALS

In discussing how currents cause magnetic fields, we have assumed that the conductors are surrounded by vacuum. But the coils in transformers, motors, generators, and electromagnets nearly always have iron cores to increase the magnetic field and confine it to desired regions. Permanent magnets, magnetic recording tapes, and computer disks depend directly on the magnetic properties of materials; when you store information on a computer disk, you are actually setting up an array of microscopic permanent magnets on the disk. So it is worthwhile to examine some aspects of the magnetic properties of materials. After describing the atomic origins of magnetic properties, we will discuss three broad classes of magnetic behavior that occur in materials; these are called *paramagnetism,* *diamagnetism,* and *ferromagnetism.*

The Bohr Magneton

As we discussed briefly in Section 27.7, the atoms that make up all matter contain moving electrons, and these electrons form microscopic current loops that produce magnetic fields of their own. In many materials these currents are randomly oriented and cause no net magnetic field. But in some materials an external field (a field produced by currents outside the material) can cause these loops to become oriented preferentially with the field, so their magnetic fields *add* to the external field. We then say that the material is *magnetized.*

Let's look at how these microscopic currents come about. **Figure 28.26** shows a primitive model of an electron in an atom. We picture the electron (mass m, charge $-e$) as moving in a circular orbit with radius r and speed v. This moving charge is equivalent to a current loop. In Section 27.7 we found that a current loop with area A and current I has a magnetic dipole moment μ given by $\mu = IA$; for the orbiting electron the area of the loop is $A = \pi r^2$. To find the current associated with the electron, we note that the orbital period T (the time for the electron to make one complete orbit) is the orbit circumference divided by the electron speed: $T = 2\pi r/v$. The equivalent current I is the total charge passing any point on the orbit per unit time, which is just the magnitude e of the electron charge divided by the orbital period T:

$$I = \frac{e}{T} = \frac{ev}{2\pi r}$$

The magnetic moment $\mu = IA$ is then

$$\mu = \frac{ev}{2\pi r}(\pi r^2) = \frac{evr}{2} \tag{28.25}$$

It is useful to express μ in terms of the *angular momentum L* of the electron. For a particle moving in a circular path, the magnitude of angular momentum equals the magnitude of momentum mv multiplied by the radius r—that is, $L = mvr$ (see Section 10.5). Comparing this with Eq. (28.25), we can write

$$\mu = \frac{e}{2m}L \tag{28.26}$$

Equation (28.26) is useful in this discussion because atomic angular momentum is *quantized;* its component in a particular direction is always an integer multiple of $h/2\pi$, where h is a fundamental physical constant called *Planck's constant.* The numerical value of h is

$$h = 6.626 \times 10^{-34}\,\text{J} \cdot \text{s}$$

The quantity $h/2\pi$ thus represents a fundamental unit of angular momentum in atomic systems, just as e is a fundamental unit of charge. Associated with the quantization of $\vec{L}$ is a fundamental uncertainty in the *direction* of $\vec{L}$ and therefore of $\vec{\mu}$. In the following discussion, when we speak of the magnitude of a magnetic moment, a more precise statement would be "maximum component in a given direction." Thus, to say that a magnetic moment $\vec{\mu}$ is aligned with a magnetic field $\vec{B}$ really means that $\vec{\mu}$ has its maximum possible component in the direction of $\vec{B}$; such components are always quantized.

Equation (28.26) shows that associated with the fundamental unit of angular momentum is a corresponding fundamental unit of magnetic moment. If $L = h/2\pi$, then

$$\mu = \frac{e}{2m}\left(\frac{h}{2\pi}\right) = \frac{eh}{4\pi m} \tag{28.27}$$

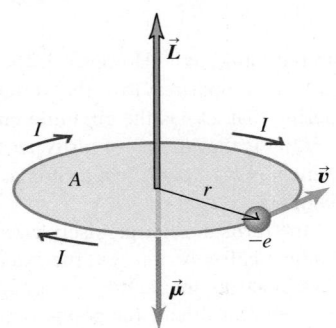

28.26 An electron moving with speed v in a circular orbit of radius r has an angular momentum $\vec{L}$ and an oppositely directed orbital magnetic dipole moment $\vec{\mu}$. It also has a spin angular momentum and an oppositely directed spin magnetic dipole moment.

This quantity is called the **Bohr magneton,** denoted by μ_B. Its numerical value is

$$\mu_B = 9.274 \times 10^{-24}\,\text{A} \cdot \text{m}^2 = 9.274 \times 10^{-24}\,\text{J/T}$$

You should verify that these two sets of units are consistent. The second set is useful when we compute the potential energy $U = -\vec{\mu} \cdot \vec{B}$ for a magnetic moment in a magnetic field.

Electrons also have an intrinsic angular momentum, called *spin,* that is not related to orbital motion but that can be pictured in a classical model as spinning on an axis. This angular momentum also has an associated magnetic moment, and its magnitude turns out to be almost exactly one Bohr magneton. (Effects having to do with quantization of the electromagnetic field cause the spin magnetic moment to be about $1.001\,\mu_B$.)

Paramagnetism

In an atom, most of the various orbital and spin magnetic moments of the electrons add up to zero. However, in some cases the atom has a net magnetic moment that is of the order of μ_B. When such a material is placed in a magnetic field, the field exerts a torque on each magnetic moment, as given by Eq. (27.26): $\vec{\tau} = \vec{\mu} \times \vec{B}$. These torques tend to align the magnetic moments with the field, as we discussed in Section 27.7. In this position, the directions of the current loops are such as to *add* to the externally applied magnetic field.

We saw in Section 28.5 that the $\vec{B}$ field produced by a current loop is proportional to the loop's magnetic dipole moment. In the same way, the additional $\vec{B}$ field produced by microscopic electron current loops is proportional to the total magnetic moment $\vec{\mu}_{\text{total}}$ per unit volume V in the material. We call this vector quantity the **magnetization** of the material, denoted by $\vec{M}$:

$$\vec{M} = \frac{\vec{\mu}_{\text{total}}}{V} \tag{28.28}$$

The additional magnetic field due to magnetization of the material turns out to be equal simply to $\mu_0\vec{M}$, where μ_0 is the same constant that appears in the law of Biot and Savart and Ampere's law. When such a material completely surrounds a current-carrying conductor, the total magnetic field $\vec{B}$ in the material is

$$\vec{B} = \vec{B}_0 + \mu_0\vec{M} \tag{28.29}$$

where $\vec{B}_0$ is the field caused by the current in the conductor.

To check that the units in Eq. (28.29) are consistent, note that magnetization $\vec{M}$ is the magnetic moment per unit volume. The units of magnetic moment are current times area $(\text{A} \cdot \text{m}^2)$, so the units of magnetization are $(\text{A} \cdot \text{m}^2)/\text{m}^3 = \text{A/m}$. From Section 28.1, the units of the constant μ_0 are $\text{T} \cdot \text{m/A}$. So the units of $\mu_0\vec{M}$ are the same as the units of $\vec{B}$: $(\text{T} \cdot \text{m/A})(\text{A/m}) = \text{T}$.

A material showing the behavior just described is said to be **paramagnetic.** The result is that the magnetic field at any point in such a material is greater by a dimensionless factor K_m, called the **relative permeability** of the material, than it would be if the material were replaced by vacuum. The value of K_m is different for different materials; for common paramagnetic solids and liquids at room temperature, K_m typically ranges from 1.00001 to 1.003.

All of the equations in this chapter that relate magnetic fields to their sources can be adapted to the situation in which the current-carrying conductor is embedded in a paramagnetic material. All that need be done is to replace μ_0 by $K_m\mu_0$. This product is usually denoted as μ and is called the **permeability** of the material:

$$\mu = K_m\mu_0 \tag{28.30}$$

CAUTION Two meanings of the symbol μ Equation (28.30) involves dangerous notation because we use μ for magnetic dipole moment as well as for permeability, as is customary. But beware: From now on, every time you see a μ, make sure you know whether it is permeability or magnetic moment. You can usually tell from the context. ∎

Material	$\chi_m = K_m - 1$ $(\times 10^{-5})$
Paramagnetic	
Iron ammonium alum	66
Uranium	40
Platinum	26
Aluminum	2.2
Sodium	0.72
Oxygen gas	0.19
Diamagnetic	
Bismuth	−16.6
Mercury	−2.9
Silver	−2.6
Carbon (diamond)	−2.1
Lead	−1.8
Sodium chloride	−1.4
Copper	−1.0

Magnetic Susceptibilities of Paramagnetic and Diamagnetic Materials

TABLE 28.1 at $T = 20°C$

The amount by which the relative permeability differs from unity is called the **magnetic susceptibility,** denoted by χ_m:

$$\chi_m = K_m - 1 \qquad (28.31)$$

Both K_m and χ_m are dimensionless quantities. **Table 28.1** lists values of magnetic susceptibility for several materials. For example, for aluminum, $\chi_m = 2.2 \times 10^{-5}$ and $K_m = 1.000022$. The first group in the table consists of paramagnetic materials; we'll soon discuss the second group, which contains *diamagnetic* materials.

The tendency of atomic magnetic moments to align themselves parallel to the magnetic field (where the potential energy is minimum) is opposed by random thermal motion, which tends to randomize their orientations. For this reason, paramagnetic susceptibility always decreases with increasing temperature. In many cases it is inversely proportional to the absolute temperature T, and the magnetization M can be expressed as

$$M = C\frac{B}{T} \qquad (28.32)$$

This relationship is called *Curie's law,* after its discoverer, Pierre Curie (1859–1906). The quantity C is a constant, different for different materials, called the *Curie constant.*

As we described in Section 27.7, a body with atomic magnetic dipoles is attracted to the poles of a magnet. In most paramagnetic substances this attraction is very weak due to thermal randomization of the atomic magnetic moments. But at very low temperatures the thermal effects are reduced, the magnetization increases in accordance with Curie's law, and the attractive forces are greater.

EXAMPLE 28.11 MAGNETIC DIPOLES IN A PARAMAGNETIC MATERIAL

Nitric oxide (NO) is a paramagnetic compound. The magnetic moment of each NO molecule has a maximum component in any direction of about one Bohr magneton. Compare the interaction energy of such magnetic moments in a 1.5-T magnetic field with the average translational kinetic energy of molecules at 300 K.

SOLUTION

IDENTIFY and SET UP: This problem involves the energy of a magnetic moment in a magnetic field and the average thermal kinetic energy. We have Eqs. (27.27), $U = -\vec{\mu} \cdot \vec{B}$, for the interaction energy of a magnetic moment $\vec{\mu}$ with a $\vec{B}$ field, and (18.16), $K = \frac{3}{2}kT$, for the average translational kinetic energy of a molecule at temperature T.

EXECUTE: We can write $U = -\mu_\parallel B$, where $\mu_\parallel$ is the component of the magnetic moment $\vec{\mu}$ in the direction of the $\vec{B}$ field. Here the maximum value of $\mu_\parallel$ is about μ_B, so

$$|U|_{max} \approx \mu_B B = (9.27 \times 10^{-24} \text{ J/T})(1.5 \text{ T})$$
$$= 1.4 \times 10^{-23} \text{ J} = 8.7 \times 10^{-5} \text{ eV}$$

The average translational kinetic energy K is

$$K = \tfrac{3}{2}kT = \tfrac{3}{2}(1.38 \times 10^{-23} \text{ J/K})(300 \text{ K})$$
$$= 6.2 \times 10^{-21} \text{ J} = 0.039 \text{ eV}$$

EVALUATE: At 300 K the magnetic interaction energy is only about 0.2% of the thermal kinetic energy, so we expect only a slight degree of alignment. This is why paramagnetic susceptibilities at ordinary temperature are usually very small.

Diamagnetism

In some materials the total magnetic moment of all the atomic current loops is zero when no magnetic field is present. But even these materials have magnetic effects because an external field alters electron motions within the atoms, causing additional current loops and induced magnetic dipoles comparable to the induced *electric* dipoles we studied in Section 28.5. In this case the additional field caused by these current loops is always *opposite* in direction to that of the external field. (This behavior is explained by Faraday's law of induction, which we will study in Chapter 29. An induced current always tends to cancel the field change that caused it.)

Such materials are said to be **diamagnetic.** They always have negative susceptibility, as shown in Table 28.1, and relative permeability K_m slightly *less* than unity, typically of the order of 0.99990 to 0.99999 for solids and liquids. Diamagnetic susceptibilities are very nearly temperature independent.

Ferromagnetism

There is a third class of materials, called **ferromagnetic** materials, that includes iron, nickel, cobalt, and many alloys containing these elements. In these materials, strong interactions between atomic magnetic moments cause them to line up parallel to each other in regions called **magnetic domains,** even when no external field is present. **Figure 28.27** shows an example of magnetic domain structure. Within each domain, nearly all of the atomic magnetic moments are parallel.

When there is no externally applied field, the domain magnetizations are randomly oriented. But when a field $\vec{B}_0$ (caused by external currents) is present, the domains tend to orient themselves parallel to the field. The domain boundaries also shift; the domains that are magnetized in the field direction grow, and those that are magnetized in other directions shrink. Because the total magnetic moment of a domain may be many thousands of Bohr magnetons, the torques that tend to align the domains with an external field are much stronger than occur with paramagnetic materials. The relative permeability K_m is *much* larger than unity, typically of the order of 1000 to 100,000. As a result, an object made of a ferromagnetic material such as iron is strongly magnetized by the field from a permanent magnet and is attracted to the magnet (see Fig. 27.38). A paramagnetic material such as aluminum is also attracted to a permanent magnet, but K_m for paramagnetic materials is so much smaller for such a material than for ferromagnetic materials that the attraction is very weak. Thus a magnet can pick up iron nails, but not aluminum cans.

As the external field is increased, a point is eventually reached at which nearly *all* the magnetic moments in the ferromagnetic material are aligned parallel to the external field. This condition is called *saturation magnetization;* after it is reached, further increase in the external field causes no increase in magnetization or in the additional field caused by the magnetization.

Figure 28.28 shows a "magnetization curve," a graph of magnetization M as a function of external magnetic field B_0, for soft iron. An alternative description of this behavior is that K_m is not constant but decreases as B_0 increases. (Paramagnetic materials also show saturation at sufficiently strong fields. But the magnetic fields required are so large that departures from a linear relationship between M and B_0 in these materials can be observed only at very low temperatures, 1 K or so.)

For many ferromagnetic materials the relationship of magnetization to external magnetic field is different when the external field is increasing from when it is decreasing. **Figure 28.29a** shows this relationship for such a material. When the material is magnetized to saturation and then the external field is reduced to

28.27 In this drawing adapted from a magnified photo, the arrows show the directions of magnetization in the domains of a single crystal of nickel. Domains that are magnetized in the direction of an applied magnetic field grow larger.

(a) No field

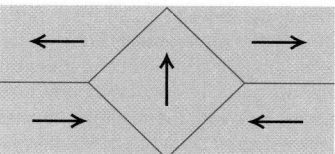

(b) Weak field

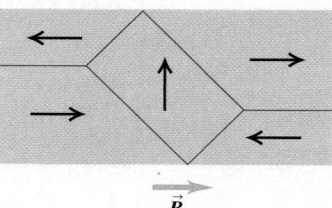

(c) Stronger field

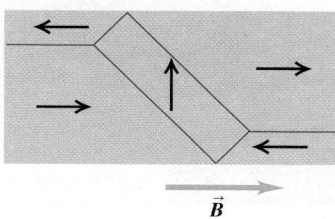

28.28 A magnetization curve for a ferromagnetic material. The magnetization M approaches its saturation value M_{sat} as the magnetic field B_0 (caused by external currents) becomes large.

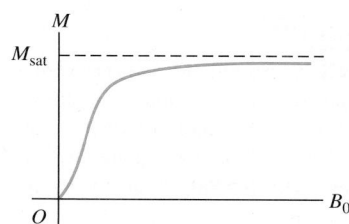

28.29 Hysteresis loops. The materials of both (a) and (b) remain strongly magnetized when B_0 is reduced to zero. Since (a) is also hard to demagnetize, it would be good for permanent magnets. Since (b) magnetizes and demagnetizes more easily, it could be used as a computer memory material. The material of (c) would be useful for transformers and other alternating-current devices where zero hysteresis would be optimal.

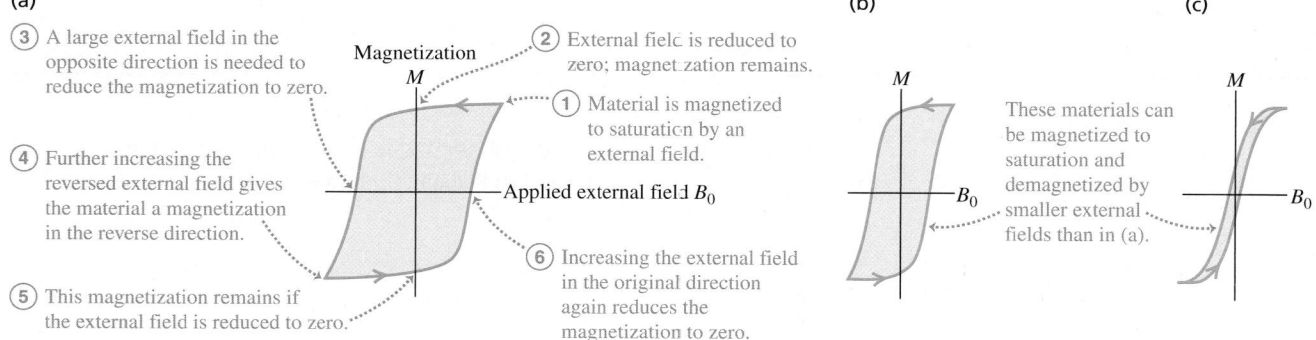

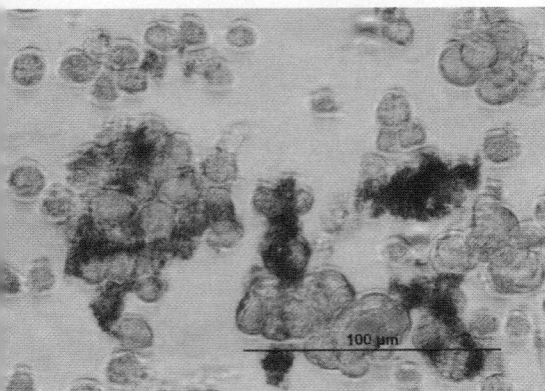

BIO Application Magnetic Nanoparticles for Cancer Therapy The violet blobs in this microscope image are cancer cells that have broken away from a tumor and threaten to spread throughout a patient's body. An experimental technique for fighting these cells uses particles of a magnetic material (shown in brown) injected into the body. These particles are coated with a chemical that preferentially attaches to cancer cells. A magnet outside the patient then "steers" the particles out of the body, taking the cancer cells with them. (Photo courtesy of cancer researcher Dr. Kenneth Scarberry.)

zero, some magnetization remains. This behavior is characteristic of permanent magnets, which retain most of their saturation magnetization when the magnetizing field is removed. To reduce the magnetization to zero requires a magnetic field in the reverse direction.

This behavior is called **hysteresis,** and the curves in Fig. 28.29 are called *hysteresis loops.* Magnetizing and demagnetizing a material that has hysteresis involve the dissipation of energy, and the temperature of the material increases during such a process.

Ferromagnetic materials are widely used in electromagnets, transformer cores, and motors and generators, in which it is desirable to have as large a magnetic field as possible for a given current. Because hysteresis dissipates energy, materials that are used in these applications should usually have as narrow a hysteresis loop as possible. Soft iron is often used; it has high permeability without appreciable hysteresis. For permanent magnets a broad hysteresis loop is usually desirable, with large zero-field magnetization and large reverse field needed to demagnetize. Many kinds of steel and many alloys, such as Alnico, are used for permanent magnets. The remaining magnetic field in such a material, after it has been magnetized to near saturation, is typically of the order of 1 T, corresponding to a remaining magnetization $M = B/\mu_0$ of about 800,000 A/m.

EXAMPLE 28.12   A FERROMAGNETIC MATERIAL

A cube-shaped permanent magnet is made of a ferromagnetic material with a magnetization M of about 8×10^5 A/m. The side length is 2 cm. (a) Find the magnetic dipole moment of the magnet. (b) Estimate the magnetic field due to the magnet at a point 10 cm from the magnet along its axis.

SOLUTION

IDENTIFY and SET UP: This problem uses the relationship between magnetization M and magnetic dipole moment μ_{total} and the idea that a magnetic dipole produces a magnetic field. We find μ_{total} from Eq. (28.28). To estimate the field, we approximate the magnet as a current loop with this same magnetic moment and use Eq. (28.18).

EXECUTE: (a) From Eq. (28.28),

$$\mu_{total} = MV = (8 \times 10^5 \text{ A/m})(2 \times 10^{-2} \text{ m})^3 = 6 \text{ A} \cdot \text{m}^2$$

(b) From Eq. (28.18), the magnetic field on the axis of a current loop with magnetic moment μ_{total} is

$$B = \frac{\mu_0 \mu_{total}}{2\pi(x^2 + a^2)^{3/2}}$$

where x is the distance from the loop and a is its radius. We can use this expression here if we take a to refer to the size of the permanent magnet. Strictly speaking, there are complications because our magnet does not have the same geometry as a circular current loop. But because $x = 10$ cm is fairly large in comparison to the 2-cm size of the magnet, the term a^2 is negligible in comparison to x^2 and can be ignored. So

$$B \approx \frac{\mu_0 \mu_{total}}{2\pi x^3} = \frac{(4\pi \times 10^{-7} \text{ T} \cdot \text{m/A})(6 \text{ A} \cdot \text{m}^2)}{2\pi(0.1 \text{ m})^3}$$

$$= 1 \times 10^{-3} \text{ T} = 10 \text{ G}$$

which is about ten times stronger than the earth's magnetic field.

EVALUATE: We calculated B at a point *outside* the magnetic material and therefore used μ_0, not the permeability μ of the magnetic material, in our calculation. You would substitute permeability μ for μ_0 if you were calculating B *inside* a material with relative permeability K_m, for which $\mu = K_m \mu_0$.

TEST YOUR UNDERSTANDING OF SECTION 28.8 Which of the following materials are attracted to a magnet? (i) Sodium; (ii) bismuth; (iii) lead; (iv) uranium. ❚

Magnetic field of a moving charge: The magnetic field $\vec{B}$ created by a charge q moving with velocity $\vec{v}$ depends on the distance r from the source point (the location of q) to the field point (where $\vec{B}$ is measured). The $\vec{B}$ field is perpendicular to $\vec{v}$ and to $\hat{r}$, the unit vector directed from the source point to the field point. The principle of superposition of magnetic fields states that the total $\vec{B}$ field produced by several moving charges is the vector sum of the fields produced by the individual charges. (See Example 28.1.)

$$\vec{B} = \frac{\mu_0}{4\pi}\frac{q\vec{v}\times\hat{r}}{r^2} \tag{28.2}$$

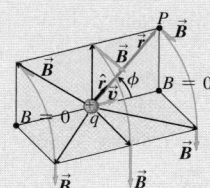

Magnetic field of a current-carrying conductor: The law of Biot and Savart gives the magnetic field $d\vec{B}$ created by an element $d\vec{l}$ of a conductor carrying current I. The field $d\vec{B}$ is perpendicular to both $d\vec{l}$ and $\hat{r}$, the unit vector from the element to the field point. The $\vec{B}$ field created by a finite current-carrying conductor is the integral of $d\vec{B}$ over the length of the conductor. (See Example 28.2.)

$$d\vec{B} = \frac{\mu_0}{4\pi}\frac{I\,d\vec{l}\times\hat{r}}{r^2} \tag{28.6}$$

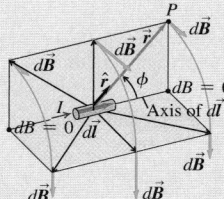

Magnetic field of a long, straight, current-carrying conductor: The magnetic field $\vec{B}$ at a distance r from a long, straight conductor carrying a current I has a magnitude that is inversely proportional to r. The magnetic field lines are circles coaxial with the wire, with directions given by the right-hand rule. (See Examples 28.3 and 28.4.)

$$B = \frac{\mu_0 I}{2\pi r} \tag{28.9}$$

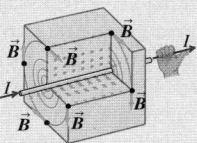

Magnetic force between current-carrying conductors: Two long, parallel, current-carrying conductors attract if the currents are in the same direction and repel if the currents are in opposite directions. The magnetic force per unit length between the conductors depends on their currents I and I' and separation r. The definition of the ampere is based on this relationship. (See Example 28.5.)

$$\frac{F}{L} = \frac{\mu_0 II'}{2\pi r} \tag{28.11}$$

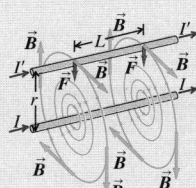

Magnetic field of a current loop: The law of Biot and Savart allows us to calculate the magnetic field produced along the axis of a circular conducting loop of radius a carrying current I. The field depends on the distance x along the axis from the center of the loop to the field point. If there are N loops, the field is multiplied by N. At the center of the loop, $x = 0$. (See Example 28.6.)

$$B_x = \frac{\mu_0 Ia^2}{2(x^2 + a^2)^{3/2}} \tag{28.15}$$
(circular loop)

$$B_x = \frac{\mu_0 NI}{2a} \tag{28.17}$$
(center of N circular loops)

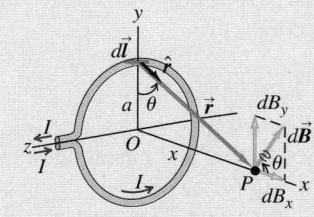

Ampere's law: Ampere's law states that the line integral of $\vec{B}$ around any closed path equals μ_0 times the net current through the area enclosed by the path. The positive sense of current is determined by a right-hand rule. (See Examples 28.7–28.10.)

$$\oint \vec{B} \cdot d\vec{l} = \mu_0 I_{\text{encl}} \tag{28.20}$$

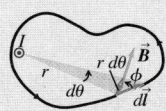

Magnetic fields due to current distributions: The table lists magnetic fields caused by several current distributions. In each case the conductor is carrying current I.

Current Distribution	Point in Magnetic Field	Magnetic-Field Magnitude
Long, straight conductor	Distance r from conductor	$B = \dfrac{\mu_0 I}{2\pi r}$
Circular loop of radius a	On axis of loop	$B = \dfrac{\mu_0 I a^2}{2(x^2 + a^2)^{3/2}}$
	At center of loop	$B = \dfrac{\mu_0 I}{2a}$ (for N loops, multiply these expressions by N)
Long cylindrical conductor of radius R	Inside conductor, $r < R$	$B = \dfrac{\mu_0 I}{2\pi} \dfrac{r}{R^2}$
	Outside conductor, $r > R$	$B = \dfrac{\mu_0 I}{2\pi r}$
Long, closely wound solenoid with n turns per unit length, near its midpoint	Inside solenoid, near center	$B = \mu_0 n I$
	Outside solenoid	$B \approx 0$
Tightly wound toroidal solenoid (toroid) with N turns	Within the space enclosed by the windings, distance r from symmetry axis	$B = \dfrac{\mu_0 N I}{2\pi r}$
	Outside the space enclosed by the windings	$B \approx 0$

Magnetic materials: When magnetic materials are present, the magnetization of the material causes an additional contribution to $\vec{B}$. For paramagnetic and diamagnetic materials, μ_0 is replaced in magnetic-field expressions by $\mu = K_m \mu_0$, where μ is the permeability of the material and K_m is its relative permeability. The magnetic susceptibility χ_m is defined as $\chi_m = K_m - 1$. Magnetic susceptibilities for paramagnetic materials are small positive quantities; those for diamagnetic materials are small negative quantities. For ferromagnetic materials, K_m is much larger than unity and is not constant. Some ferromagnetic materials are permanent magnets, retaining their magnetization even after the external magnetic field is removed. (See Examples 28.11 and 28.12.)

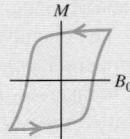

BRIDGING PROBLEM   MAGNETIC FIELD OF A CHARGED, ROTATING DIELECTRIC DISK

A thin dielectric disk with radius a has a total charge $+Q$ distributed uniformly over its surface (**Fig. 28.30**). It rotates n times per second about an axis perpendicular to the surface of the disk and passing through its center. Find the magnetic field at the center of the disk.

SOLUTION GUIDE

IDENTIFY and SET UP

1. Think of the rotating disk as a series of concentric rotating rings. Each ring acts as a circular current loop that produces a magnetic field at the center of the disk.
2. Use the results of Section 28.5 to find the magnetic field due to a single ring. Then integrate over all rings to find the total field.

EXECUTE

3. Find the charge on a ring with inner radius r and outer radius $r + dr$ (Fig. 28.30).
4. How long does it take the charge found in step 3 to make a complete trip around the rotating ring? Use this to find the current of the rotating ring.
5. Use a result from Section 28.5 to determine the magnetic field that this ring produces at the center of the disk.

28.30 Finding the $\vec{B}$ field at the center of a uniformly charged, rotating disk.

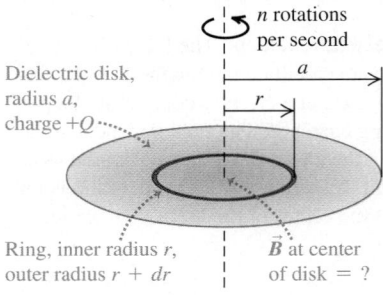

6. Integrate your result from step 5 to find the total magnetic field from all rings with radii from $r = 0$ to $r = a$.

EVALUATE

7. Does your answer have the correct units?
8. Suppose all of the charge were concentrated at the rim of the disk (at $r = a$). Would this increase or decrease the field at the center of the disk?

Problems

For assigned homework and other learning materials, go to MasteringPhysics®. **MP**

•, ••, •••: Difficulty levels. CP: Cumulative problems incorporating material from earlier chapters. CALC: Problems requiring calculus. DATA: Problems involving real data, scientific evidence, experimental design, and/or statistical reasoning. BIO: Biosciences problems.

DISCUSSION QUESTIONS

Q28.1 A topic of current interest in physics research is the search (thus far unsuccessful) for an isolated magnetic pole, or magnetic *monopole*. If such an entity were found, how could it be recognized? What would its properties be?

Q28.2 Streams of charged particles emitted from the sun during periods of solar activity create a disturbance in the earth's magnetic field. How does this happen?

Q28.3 The text discussed the magnetic field of an infinitely long, straight conductor carrying a current. Of course, there is no such thing as an infinitely long *anything*. How do you decide whether a particular wire is long enough to be considered infinite?

Q28.4 Two parallel conductors carrying current in the same direction attract each other. If they are permitted to move toward each other, the forces of attraction do work. From where does the energy come? Does this contradict the assertion in Chapter 27 that magnetic forces on moving charges do no work? Explain.

Q28.5 Pairs of conductors carrying current into or out of the power-supply components of electronic equipment are sometimes twisted together to reduce magnetic-field effects. Why does this help?

Q28.6 Suppose you have three long, parallel wires arranged so that in cross section they are at the corners of an equilateral triangle. Is there any way to arrange the currents so that all three wires attract each other? So that all three wires repel each other? Explain.

Q28.7 In deriving the force on one of the long, current-carrying conductors in Section 28.4, why did we use the magnetic field due to only one of the conductors? That is, why didn't we use the *total* magnetic field due to *both* conductors?

Q28.8 Two concentric, coplanar, circular loops of wire of different diameter carry currents in the same direction. Describe the nature of the force exerted on the inner loop by the outer loop and on the outer loop by the inner loop.

Q28.9 A current was sent through a helical coil spring. The spring contracted, as though it had been compressed. Why?

Q28.10 What are the relative advantages and disadvantages of Ampere's law and the law of Biot and Savart for practical calculations of magnetic fields?

Q28.11 Magnetic field lines never have a beginning or an end. Use this to explain why it is reasonable for the field of an ideal toroidal solenoid to be confined entirely to its interior, while a straight solenoid *must* have some field outside.

Q28.12 Two very long, parallel wires carry equal currents in opposite directions. (a) Is there any place that their magnetic fields completely cancel? If so, where? If not, why not? (b) How would the answer to part (a) change if the currents were in the same direction?

Q28.13 In the circuit shown in **Fig. Q28.13**, when switch S is suddenly closed, the wire L is pulled toward the lower wire carrying current I. Which (a or b) is the positive terminal of the battery? How do you know?

Figure Q28.13

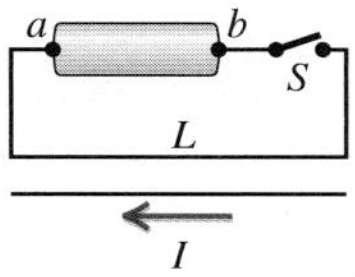

Q28.14 A metal ring carries a current that causes a magnetic field B_0 at the center of the ring and a field B at point P a distance x from the center along the axis of the ring. If the radius of the ring is doubled, find the magnetic field at the center. Will the field at point P change by the same factor? Why?

Q28.15 Show that the units $A \cdot m^2$ and J/T for the Bohr magneton are equivalent.

Q28.16 Why should the permeability of a paramagnetic material be expected to decrease with increasing temperature?

Q28.17 If a magnet is suspended over a container of liquid air, it attracts droplets to its poles. The droplets contain only liquid oxygen; even though nitrogen is the primary constituent of air, it is not attracted to the magnet. Explain what this tells you about the magnetic susceptibilities of oxygen and nitrogen, and explain why a magnet in ordinary, room-temperature air doesn't attract molecules of oxygen *gas* to its poles.

Q28.18 What features of atomic structure determine whether an element is diamagnetic or paramagnetic? Explain.

Q28.19 The magnetic susceptibility of paramagnetic materials is quite strongly temperature dependent, but that of diamagnetic materials is nearly independent of temperature. Why the difference?

Q28.20 A cylinder of iron is placed so that it is free to rotate around its axis. Initially the cylinder is at rest, and a magnetic field is applied to the cylinder so that it is magnetized in a direction parallel to its axis. If the direction of the *external* field is suddenly reversed, the direction of magnetization will also reverse and the cylinder will begin rotating around its axis. (This is called the *Einstein–de Haas effect.*) Explain why the cylinder begins to rotate.

EXERCISES

Section 28.1 Magnetic Field of a Moving Charge

28.1 •• A $+6.00$-μC point charge is moving at a constant 8.00×10^6 m/s in the $+y$-direction, relative to a reference frame. At the instant when the point charge is at the origin of this reference frame, what is the magnetic-field vector $\vec{B}$ it produces at the following points: (a) $x = 0.500$ m, $y = 0$, $z = 0$; (b) $x = 0$, $y = -0.500$ m, $z = 0$; (c) $x = 0$, $y = 0$, $z = +0.500$ m; (d) $x = 0$, $y = -0.500$ m, $z = +0.500$ m?

28.2 • **Fields Within the Atom.** In the Bohr model of the hydrogen atom, the electron moves in a circular orbit of radius 5.3×10^{-11} m with a speed of 2.2×10^6 m/s. If we are viewing the atom in such a way that the electron's orbit is in the plane of the paper with the electron moving clockwise, find the magnitude and direction of the electric and magnetic fields that the electron produces at the location of the nucleus (treated as a point).

28.3 • An electron moves at $0.100c$ as shown in **Fig. E28.3**. Find the magnitude and direction of the magnetic field this electron produces at the following points, each 2.00 μm from the electron: (a) points A and B; (b) point C; (c) point D.

Figure E28.3

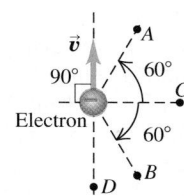

28.4 •• An alpha particle (charge $+2e$) and an electron move in opposite directions from the same point, each with the speed of 2.50×10^5 m/s (**Fig. E28.4**). Find the magnitude and direction of the total magnetic field these charges produce at point P, which is 8.65 nm from each charge.

Figure E28.4

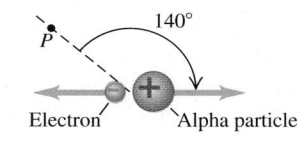

28.5 • A -4.80-μC charge is moving at a constant speed of 6.80×10^5 m/s in the $+x$-direction relative to a reference frame. At the instant when the point charge is at the origin, what is the magnetic-field vector it produces at the following points: (a) $x = 0.500$ m, $y = 0$, $z = 0$; (b) $x = 0$, $y = 0.500$ m, $z = 0$; (c) $x = 0.500$ m, $y = 0.500$ m, $z = 0$; (d) $x = 0$, $y = 0$, $z = 0.500$ m?

28.6 • Positive point charges $q = +8.00$ μC and $q' = +3.00$ μC are moving relative to an observer at point P, as shown in **Fig. E28.6**. The distance d is 0.120 m, $v = 4.50 \times 10^6$ m/s, and $v' = 9.00 \times 10^6$ m/s (a) When the two charges are at the locations shown in the figure, what are the magnitude and direction of the net magnetic field they produce at point P? (b) What are the magnitude and direction of the electric and magnetic forces that each charge exerts on the other, and what is the ratio of the magnitude of the electric force to the magnitude of the magnetic force? (c) If the direction of $\vec{v}'$ is reversed, so both charges are moving in the same direction, what are the magnitude and direction of the magnetic forces that the two charges exert on each other?

Figure **E28.6**

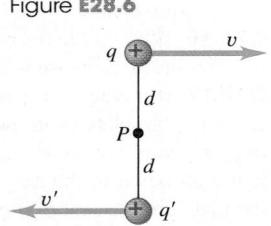

28.7 • A negative charge $q = -3.60 \times 10^{-6}$ C is located at the origin and has velocity $\vec{v} = (7.50 \times 10^4 \text{m/s})\hat{\imath} + (-4.90 \times 10^4 \text{ m/s})\hat{\jmath}$. At this instant what are the magnitude and direction of the magnetic field produced by this charge at the point $x = 0.200$ m, $y = -0.300$ m, $z = 0$?

28.8 •• An electron and a proton are each moving at 735 km/s in perpendicular paths as shown in **Fig. E28.8**. At the instant when they are at the positions shown, find the magnitude and direction of (a) the total magnetic field they produce at the origin; (b) the magnetic field the electron produces at the location of the proton; (c) the total electric force and the total magnetic force that the electron exerts on the proton.

Figure **E28.8**

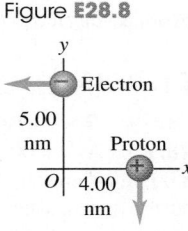

Section 28.2 Magnetic Field of a Current Element

28.9 • A straight wire carries a 10.0-A current (**Fig. E28.9**). $ABCD$ is a rectangle with point D in the middle of a 1.10-mm segment of the wire and point C in the wire. Find the magnitude and direction of the magnetic field due to this segment at (a) point A; (b) point B; (c) point C.

Figure **E28.9**

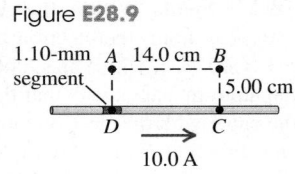

28.10 • A short current element $d\vec{l} = (0.500 \text{ mm})\hat{\jmath}$ carries a current of 5.40 A in the same direction as $d\vec{l}$. Point P is located at $\vec{r} = (-0.730 \text{ m})\hat{\imath} + (0.390 \text{ m})\hat{k}$. Use unit vectors to express the magnetic field at P produced by this current element.

28.11 •• A long, straight wire lies along the z-axis and carries a 4.00-A current in the $+z$-direction. Find the magnetic field (magnitude and direction) produced at the following points by a 0.500-mm segment of the wire centered at the origin: (a) $x = 2.00$ m, $y = 0$, $z = 0$; (b) $x = 0$, $y = 2.00$ m, $z = 0$; (c) $x = 2.00$ m, $y = 2.00$ m, $z = 0$; (d) $x = 0$, $y = 0$, $z = 2.00$ m,

28.12 •• Two parallel wires are 5.00 cm apart and carry currents in opposite directions, as shown in **Fig. E28.12**. Find the magnitude and direction of the magnetic field at point P due to two 1.50-mm segments of wire that are opposite each other and each 8.00 cm from P.

Figure **E28.12**

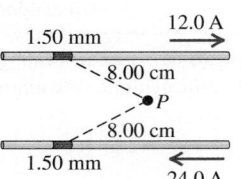

Figure **E28.13**

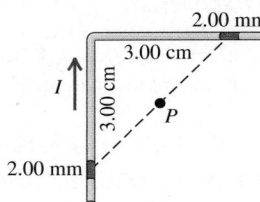

28.13 • A wire carrying a 28.0-A current bends through a right angle. Consider two 2.00-mm segments of wire, each 3.00 cm from the bend (**Fig. E28.13**). Find the magnitude and direction of the magnetic field these two segments produce at point P, which is midway between them.

28.14 •• A square wire loop 10.0 cm on each side carries a clockwise current of 8.00 A. Find the magnitude and direction of the magnetic field at its center due to the four 1.20-mm wire segments at the midpoint of each side.

Section 28.3 Magnetic Field of a Straight Current-Carrying Conductor

28.15 • **The Magnetic Field from a Lightning Bolt.** Lightning bolts can carry currents up to approximately 20 kA. We can model such a current as the equivalent of a very long, straight wire. (a) If you were unfortunate enough to be 5.0 m away from such a lightning bolt, how large a magnetic field would you experience? (b) How does this field compare to one you would experience by being 5.0 cm from a long, straight household current of 10 A?

28.16 • A very long, straight horizontal wire carries a current such that 8.20×10^{18} electrons per second pass any given point going from west to east. What are the magnitude and direction of the magnetic field this wire produces at a point 4.00 cm directly above it?

28.17 • **BIO Currents in the Heart.** The body contains many small currents caused by the motion of ions in the organs and cells. Measurements of the magnetic field around the chest due to currents in the heart give values of about 10 μG. Although the actual currents are rather complicated, we can gain a rough understanding of their magnitude if we model them as a long, straight wire. If the surface of the chest is 5.0 cm from this current, how large is the current in the heart?

28.18 • **BIO Bacteria Navigation.** Certain bacteria (such as *Aquaspirillum magnetotacticum*) tend to swim toward the earth's geographic north pole because they contain tiny particles, called magnetosomes, that are sensitive to a magnetic field. If a transmission line carrying 100 A is laid underwater, at what range of distances would the magnetic field from this line be great enough to interfere with the migration of these bacteria? (Assume that a field less than 5% of the earth's field would have little effect on the bacteria. Take the earth's field to be 5.0×10^{-5} T, and ignore the effects of the seawater.)

28.19 • (a) How large a current would a very long, straight wire have to carry so that the magnetic field 2.00 cm from the wire is equal to 1.00 G (comparable to the earth's northward-pointing magnetic field)? (b) If the wire is horizontal with the current running from east to west, at what locations would the magnetic field of the wire point in the same direction as the horizontal component of the earth's magnetic field? (c) Repeat part (b) except the wire is vertical with the current going upward.

28.20 • Two long, straight wires, one above the other, are separated by a distance $2a$ and are parallel to the x-axis. Let the $+y$-axis be in the plane of the wires in the direction from the lower wire to the upper wire. Each wire carries current I in the $+x$-direction. What are the magnitude and direction of the net magnetic field of the two wires at a point in the plane of the wires (a) midway between them; (b) at a distance a above the upper wire; (c) at a distance a below the lower wire?

28.21 •• A long, straight wire lies along the y-axis and carries a current $I = 8.00$ A in the $-y$-direction (**Fig. E28.21**). In addition to the magnetic field due to the current in the wire, a uniform magnetic field $\vec{B}_0$ with magnitude 1.50×10^{-6} T is in the $+x$-direction. What is the total field (magnitude and direction) at the following points in the xz-plane: (a) $x = 0$, $z = 1.00$ m; (b) $x = 1.00$ m, $z = 0$; (c) $x = 0$, $z = -0.25$ m?

Figure **E28.21**

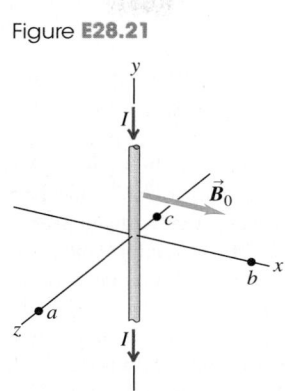

28.22 •• **BIO Transmission Lines and Health.** Currents in dc transmission lines can be 100 A or higher. Some people are concerned that the electromagnetic fields from such lines near their homes could pose health dangers. For a line that has current 150 A and a height of 8.0 m above the ground, what magnetic field does the line produce at ground level? Express your answer in teslas and as a percentage of the earth's magnetic field, which is 0.50 G. Is this value cause for worry?

28.23 • Two long, straight, parallel wires, 10.0 cm apart, carry equal 4.00-A currents in the same direction, as shown in **Fig. E28.23**. Find the magnitude and direction of the magnetic field at

Figure **E28.23**

$$I \odot \qquad\qquad \odot I$$
$$\longleftarrow\text{—10.0 cm —}\longrightarrow$$

(a) point P_1, midway between the wires; (b) point P_2, 25.0 cm to the right of P_1; (c) point P_3, 20.0 cm directly above P_1.

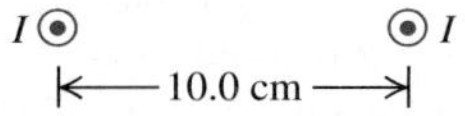

28.24 •• A rectangular loop with dimensions 4.20 cm by 9.50 cm carries current I. The current in the loop produces a magnetic field at the center of the loop that has magnitude 5.50×10^{-5} T and direction away from you as you view the plane of the loop. What are the magnitude and direction (clockwise or counterclockwise) of the current in the loop?

28.25 • Four, long, parallel power lines each carry 100-A currents. A cross-sectional diagram of these lines is a square, 20.0 cm on each side. For each of the three cases shown in **Fig. E28.25**, calculate the magnetic field at the center of the square.

Figure **E28.25**

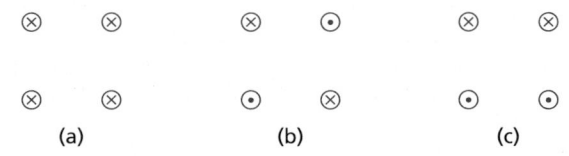

(a)   (b)   (c)

Figure **E28.26**

28.26 • Four very long, current-carrying wires in the same plane intersect to form a square 40.0 cm on each side, as shown in **Fig. E28.26**. Find the magnitude and direction of the current I so that the magnetic field at the center of the square is zero.

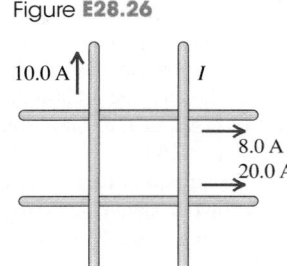

28.27 •• Two very long insulated wires perpendicular to each other in the same plane carry currents as shown in **Fig. E28.27**. Find the magnitude of the *net* magnetic field these wires produce at points P and Q if the 10.0-A current is (a) to the right or (b) to the left.

Figure **E28.27**

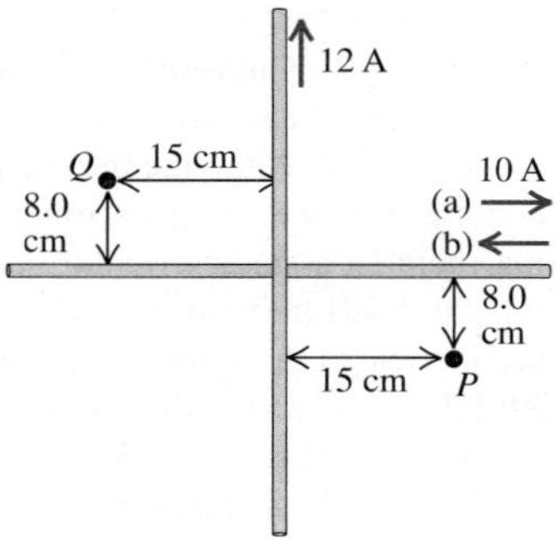

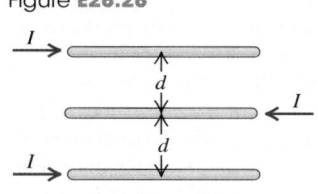

Section 28.4 Force Between Parallel Conductors

28.28 • Three very long parallel wires each carry current I in the directions shown in **Fig. E28.28**. If the separation between adjacent wires is d, calculate the magnitude and direction of the net magnetic force per unit length on each wire.

Figure **E28.28**

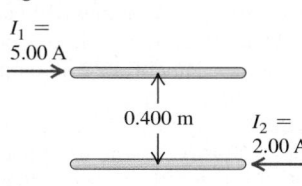

28.29 • Two long, parallel wires are separated by a distance of 0.400 m (**Fig. E28.29**). The currents I_1 and I_2 have the directions shown. (a) Calculate the magnitude of the force exerted by each wire on a 1.20-m length of the other. Is the force attractive or repulsive? (b) Each current is doubled, so that I_1 becomes 10.0 A and I_2 becomes 4.00 A. Now what is the magnitude of the force that each wire exerts on a 1.20-m length of the other?

Figure **E28.29**

$I_1 = 5.00$ A

0.400 m

$I_2 = 2.00$ A

28.30 • Two long, parallel wires are separated by a distance of 2.50 cm. The force per unit length that each wire exerts on the other is 4.00×10^{-5} N/m, and the wires repel each other. The current in one wire is 0.600 A. (a) What is the current in the second wire? (b) Are the two currents in the same direction or in opposite directions?

28.31 • **Lamp Cord Wires.** The wires in a household lamp cord are typically 3.0 mm apart center to center and carry equal currents in opposite directions. If the cord carries direct current to a 100-W light bulb connected across a 120-V potential difference, what force per meter does each wire of the cord exert on the other? Is the force attractive or repulsive? Is this force large enough so it should be considered in the design of the lamp cord? (Model the lamp cord as a very long straight wire.)

28.32 • A long, horizontal wire AB rests on the surface of a table and carries a current I. Horizontal wire CD is vertically above wire AB and is free to slide up and down on the two vertical metal guides C and D (**Fig. E28.32**). Wire CD is connected through the sliding contacts to another wire that also carries a current I, opposite in direction to the current in wire AB. The mass per unit length of the wire CD is λ. To what equilibrium height h will the wire CD rise, assuming that the magnetic force on it is due entirely to the current in the wire AB?

Figure **E28.32**

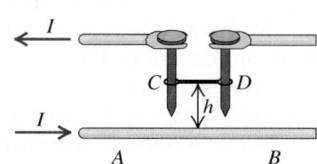

Section 28.5 Magnetic Field of a Circular Current Loop

28.33 • BIO **Currents in the Brain.** The magnetic field around the head has been measured to be approximately 3.0×10^{-8} G. Although the currents that cause this field are quite complicated, we can get a rough estimate of their size by modeling them as a single circular current loop 16 cm (the width of a typical head) in diameter. What is the current needed to produce such a field at the center of the loop?

28.34 • Calculate the magnitude and direction of the magnetic field at point P due to the current in the semicircular section of wire shown in **Fig. E28.34**. (*Hint: Does the current in the long, straight section of the wire produce any field at P?*)

Figure **E28.34**

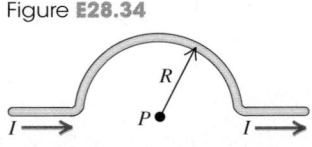

28.35 •• Calculate the magnitude of the magnetic field at point P of **Fig. E28.35** in terms of R, I_1, and I_2. What does your expression give when $I_1 = I_2$?

Figure **E28.35**

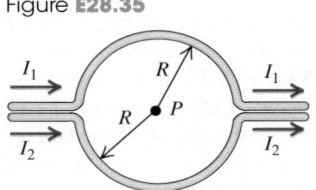

28.36 •• A closely wound, circular coil with radius 2.40 cm has 800 turns. (a) What must the current in the coil be if the magnetic field at the center of the coil is 0.0770 T? (b) At what distance x from the center of the coil, on the axis of the coil, is the magnetic field half its value at the center?

28.37 •• A single circular current loop 10.0 cm in diameter carries a 2.00-A current. (a) What is the magnetic field at the center of this loop? (b) Suppose that we now connect 1000 of these loops in series within a 500-cm length to make a solenoid 500 cm long. What is the magnetic field at the center of this solenoid? Is it 1000 times the field at the center of the loop in part (a)? Why or why not?

28.38 •• A closely wound coil has a radius of 6.00 cm and carries a current of 2.50 A. How many turns must it have if, at a point on the coil axis 6.00 cm from the center of the coil, the magnetic field is 6.39×10^{-4} T?

28.39 •• Two concentric circular loops of wire lie on a tabletop, one inside the other. The inner wire has a diameter of 20.0 cm and carries a clockwise current of 12.0 A, as viewed from above, and the outer wire has a diameter of 30.0 cm. What must be the magnitude and direction (as viewed from above) of the current in the outer wire so that the net magnetic field due to this combination of wires is zero at the common center of the wires?

Section 28.6 Ampere's Law

28.40 • **Figure E28.40** shows, in cross section, several conductors that carry currents through the plane of the figure. The currents have the magnitudes $I_1 = 4.0$ A, $I_2 = 6.0$ A, and $I_3 = 2.0$ A, and the directions shown. Four paths, labeled a through d, are shown. What is the line integral $\oint \vec{B} \cdot d\vec{l}$ for each path? Each integral involves going around the path in the counterclockwise direction. Explain your answers.

Figure **E28.40**

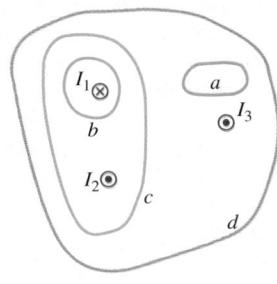

28.41 • A closed curve encircles several conductors. The line integral $\oint \vec{B} \cdot d\vec{l}$ around this curve is 3.83×10^{-4} T·m. (a) What is the net current in the conductors? (b) If you were to integrate around the curve in the opposite direction, what would be the value of the line integral? Explain.

Section 28.7 Applications of Ampere's Law

28.42 •• As a new electrical technician, you are designing a large solenoid to produce a uniform 0.150-T magnetic field near the center of the solenoid. You have enough wire for 4000 circular turns. This solenoid must be 55.0 cm long and 2.80 cm in diameter. What current will you need to produce the necessary field?

28.43 • **Coaxial Cable.** A solid conductor with radius a is supported by insulating disks on the axis of a conducting tube with inner radius b and outer radius c (**Fig. E28.43**). The central conductor and tube carry equal currents I in opposite directions. The currents are distributed uniformly over the cross sections of each conductor. Derive an expression for the magnitude of the magnetic field (a) at points outside the central, solid conductor but inside the tube ($a < r < b$) and (b) at points outside the tube ($r > c$).

Figure **E28.43**

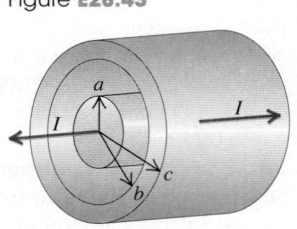

28.44 • Repeat Exercise 28.43 for the case in which the current in the central, solid conductor is I_1, the current in the tube is I_2, and these currents are in the same direction rather than in opposite directions.

28.45 •• A solenoid that is 35 cm long and contains 450 circular coils 2.0 cm in diameter carries a 1.75-A current. (a) What is the magnetic field at the center of the solenoid, 1.0 cm from the coils? (b) Suppose we now stretch out the coils to make a very long wire carrying the same current as before. What is the magnetic field 1.0 cm from the wire's center? Is it the same as that in part (a)? Why or why not?

28.46 •• A 15.0-cm-long solenoid with radius 0.750 cm is closely wound with 600 turns of wire. The current in the windings is 8.00 A. Compute the magnetic field at a point near the center of the solenoid.

28.47 •• A solenoid is designed to produce a magnetic field of 0.0270 T at its center. It has radius 1.40 cm and length 40.0 cm, and the wire can carry a maximum current of 12.0 A. (a) What minimum number of turns per unit length must the solenoid have? (b) What total length of wire is required?

28.48 • A toroidal solenoid has an inner radius of 12.0 cm and an outer radius of 15.0 cm. It carries a current of 1.50 A. How many equally spaced turns must it have so that it will produce a magnetic field of 3.75 mT at points within the coils 14.0 cm from its center?

28.49 • A magnetic field of 37.2 T has been achieved at the MIT Francis Bitter National Magnetic Laboratory. Find the current needed to achieve such a field (a) 2.00 cm from a long, straight wire; (b) at the center of a circular coil of radius 42.0 cm that has 100 turns; (c) near the center of a solenoid with radius 2.40 cm, length 32.0 cm, and 40,000 turns.

28.50 • An ideal toroidal solenoid (see Example 28.10) has inner radius $r_1 = 15.0$ cm and outer radius $r_2 = 18.0$ cm. The solenoid has 250 turns and carries a current of 8.50 A. What is the magnitude of the magnetic field at the following distances from the center of the torus: (a) 12.0 cm; (b) 16.0 cm; (c) 20.0 cm?

28.51 •• A wooden ring whose mean diameter is 14.0 cm is wound with a closely spaced toroidal winding of 600 turns. Compute the magnitude of the magnetic field at the center of the cross section of the windings when the current in the windings is 0.650 A.

Section 28.8 Magnetic Materials

28.52 •• A toroidal solenoid with 400 turns of wire and a mean radius of 6.0 cm carries a current of 0.25 A. The relative permeability of the core is 80. (a) What is the magnetic field in the core? (b) What part of the magnetic field is due to atomic currents?

28.53 • A long solenoid with 60 turns of wire per centimeter carries a current of 0.15 A. The wire that makes up the solenoid is wrapped around a solid core of silicon steel ($K_m = 5200$). (The wire of the solenoid is jacketed with an insulator so that none of the current flows into the core.) (a) For a point inside the core, find the magnitudes of (i) the magnetic field $\vec{B}_0$ due to the solenoid current; (ii) the magnetization $\vec{M}$; (iii) the total magnetic field $\vec{B}$. (b) In a sketch of the solenoid and core, show the directions of the vectors $\vec{B}$, $\vec{B}_0$, and $\vec{M}$ inside the core.

28.54 • The current in the windings of a toroidal solenoid is 2.400 A. There are 500 turns, and the mean radius is 25.00 cm. The toroidal solenoid is filled with a magnetic material. The magnetic field inside the windings is found to be 1.940 T. Calculate (a) the relative permeability and (b) the magnetic susceptibility of the material that fills the toroid.

PROBLEMS

28.55 •• A pair of point charges, $q = +8.00\ \mu$C and $q' = -5.00\ \mu$C, are moving as shown in **Fig. P28.55** with speeds $v = 9.00 \times 10^4$ m/s and $v' = 6.50 \times 10^4$ m/s. When the charges are at the locations shown in the figure, what are the magnitude and direction of (a) the magnetic field produced at the origin and (b) the magnetic force that q' exerts on q?

Figure **P28.55**

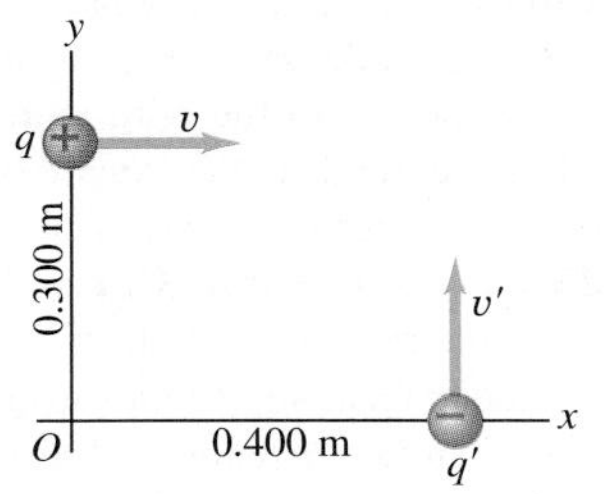

28.56 •• At a particular instant, charge $q_1 = +4.80 \times 10^{-6}$ C is at the point (0, 0.250 m, 0) and has velocity $\vec{v}_1 = (9.20 \times 10^5\ \text{m/s})\hat{\imath}$. Charge $q_2 = -2.90 \times 10^{-6}$ C is at the point (0.150 m, 0, 0) and has velocity $\vec{v}_2 = (-5.30 \times 10^5\ \text{m/s})\hat{\jmath}$. At this instant, what are the magnitude and direction of the magnetic force that q_1 exerts on q_2?

28.57 ••• Two long, parallel transmission lines, 40.0 cm apart, carry 25.0-A and 75.0-A currents. Find all locations where the net magnetic field of the two wires is zero if these currents are in (a) the same direction and (b) the opposite direction.

28.58 • A long, straight wire carries a current of 8.60 A. An electron is traveling in the vicinity of the wire. At the instant when the electron is 4.50 cm from the wire and traveling at a speed of 6.00×10^4 m/s directly toward the wire, what are the magnitude and direction (relative to the direction of the current) of the force that the magnetic field of the current exerts on the electron?

28.59 • CP A long, straight wire carries a 13.0-A current. An electron is fired parallel to this wire with a velocity of 250 km/s in the same direction as the current, 2.00 cm from the wire. (a) Find the magnitude and direction of the electron's initial acceleration. (b) What should be the magnitude and direction of a uniform electric field that will allow the electron to continue to travel parallel to the wire? (c) Is it necessary to include the effects of gravity? Justify your answer.

28.60 •• An electron is moving in the vicinity of a long, straight wire that lies along the x-axis. The wire has a constant current of 9.00 A in the $-x$-direction. At an instant when the electron is at point (0, 0.200 m, 0) and the electron's velocity is $\vec{v} = (5.00 \times 10^4\ \text{m/s})\hat{\imath} - (3.00 \times 10^4\ \text{m/s})\hat{\jmath}$, what is the force that the wire exerts on the electron? Express the force in terms of unit vectors, and calculate its magnitude.

28.61 •• An electric bus operates by drawing direct current from two parallel overhead cables, at a potential difference of 600 V, and spaced 55 cm apart. When the power input to the bus's motor is at its maximum power of 65 hp, (a) what current does it draw and (b) what is the attractive force per unit length between the cables?

28.62 •• **Figure P28.62** shows an end view of two long, parallel wires perpendicular to the xy-plane, each carrying a current I but in opposite directions. (a) Copy the diagram, and draw vectors to show the $\vec{B}$ field of each wire and the net $\vec{B}$ field at point P. (b) Derive the expression for the magnitude of $\vec{B}$ at any point on the x-axis in terms of the x-coordinate of the point. What is the direction of $\vec{B}$? (c) Graph the magnitude of $\vec{B}$ at points on the x-axis. (d) At what value of x is the magnitude of $\vec{B}$ a maximum? (e) What is the magnitude of $\vec{B}$ when $x \gg a$?

Figure **P28.62**

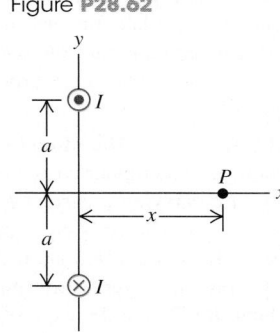

28.63 • Two long, straight, parallel wires are 1.00 m apart (**Fig. P28.63**). The wire on the left carries a current I_1 of 6.00 A into the plane of the paper. (a) What must the magnitude and direction of the current I_2 be for the net field at point P to be zero? (b) Then what are the magnitude and direction of the net field at Q? (c) Then what is the magnitude of the net field at S?

Figure **P28.63**

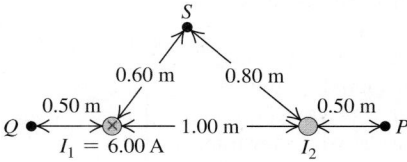

28.64 • The long, straight wire AB shown in **Fig. P28.64** carries a current of 14.0 A. The rectangular loop whose long edges are parallel to the wire carries a current of 5.00 A. Find the magnitude and direction of the net force exerted on the loop by the magnetic field of the wire.

Figure **P28.64**

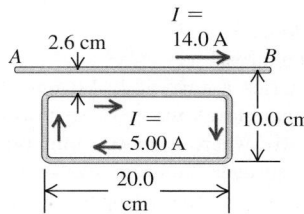

28.65 ••• CP Two long, parallel wires hang by 4.00-cm-long cords from a common axis (**Fig. P28.65**). The wires have a mass per unit length of 0.0125 kg/m and carry the same current in opposite directions. What is the current in each wire if the cords hang at an angle of 6.00° with the vertical?

Figure **P28.65**

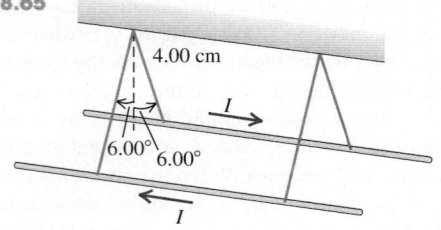

28.66 • The wire semicircles shown in **Fig. P28.66** have radii a and b. Calculate the net magnetic field (magnitude and direction) that the current in the wires produces at point P.

Figure **P28.66**

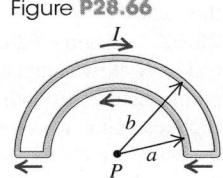

28.67 • CALC **Helmholtz Coils.** **Figure P28.67** is a sectional view of two circular coils with radius a, each wound with N turns of wire carrying a current I, circulating in the same direction in both coils. The coils are separated by a distance a equal to their radii. In this configuration the coils are called Helmholtz coils; they produce a very uniform magnetic field in the region between them. (a) Derive the expression for the magnitude B of the magnetic field at a point on the axis a distance x to the right of point P, which is midway between the coils. (b) Graph B versus x for $x = 0$ to $x = a/2$. Compare this graph to one for the magnetic field due to the right-hand coil alone. (c) From part (a), obtain an expression for the magnitude of the magnetic field at point P. (d) Calculate the magnitude of the magnetic field at P if $N = 300$ turns, $I = 6.00$ A, and $a = 8.00$ cm. (e) Calculate dB/dx and $d^2 B/dx^2$ at $P(x = 0)$. Discuss how your results show that the field is very uniform in the vicinity of P.

Figure **P28.67**

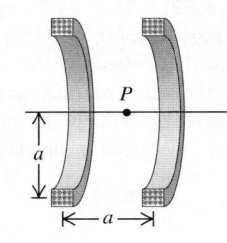

28.68 •• Calculate the magnetic field (magnitude and direction) at a point P due to a current $I = 12.0$ A in the wire shown in **Fig. P28.68**. Segment BC is an arc of a circle with radius 30.0 cm, and point P is at the center of curvature of the arc. Segment DA is an arc of a circle with radius 20.0 cm, and point P is at its center of curvature. Segments CD and AB are straight lines of length 10.0 cm each.

Figure **P28.68**

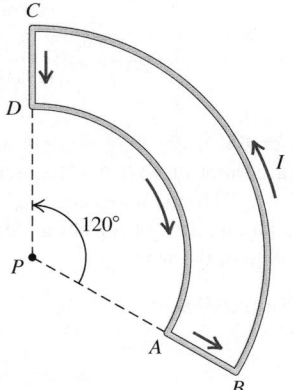

28.69 • CALC A long, straight wire with a circular cross section of radius R carries a current I. Assume that the current density is not constant across the cross section of the wire, but rather varies as $J = \alpha r$, where α is a constant. (a) By the requirement that J integrated over the cross section of the wire gives the total current I, calculate the constant α in terms of I and R. (b) Use Ampere's law to calculate the magnetic field $B(r)$ for (i) $r \leq R$ and (ii) $r \geq R$. Express your answers in terms of I.

28.70 • CALC The wire shown in **Fig. P28.70** is infinitely long and carries a current I. Calculate the magnitude and direction of the magnetic field that this current produces at point P.

Figure **P28.70**

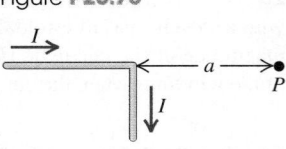

28.71 • CALC A long, straight, solid cylinder, oriented with its axis in the z-direction, carries a current whose current density is $\vec{J}$. The current density, although symmetric about the cylinder axis, is not constant but varies according to the relationship

$$\vec{J} = \frac{2I_0}{\pi a^2}\left[1 - \left(\frac{r}{a}\right)^2\right]\hat{k} \quad \text{for } r \leq a$$
$$= 0 \quad \text{for } r \geq a$$

where a is the radius of the cylinder, r is the radial distance from the cylinder axis, and I_0 is a constant having units of amperes. (a) Show that I_0 is the total current passing through the entire cross section of the wire. (b) Using Ampere's law, derive an expression for the magnitude of the magnetic field $\vec{B}$ in the region $r \geq a$. (c) Obtain an expression for the current I contained in a circular cross section of radius $r \leq a$ and centered at the cylinder axis. (d) Using Ampere's law, derive an expression for the magnitude of the magnetic field $\vec{B}$ in the region $r \leq a$. How do your results in parts (b) and (d) compare for $r = a$?

28.72 • A circular loop has radius R and carries current I_2 in a clockwise direction (**Fig. P28.72**). The center of the loop is a distance D above a long, straight wire. What are the magnitude and direction of the current I_1 in the wire if the magnetic field at the center of the loop is zero?

Figure **P28.72**

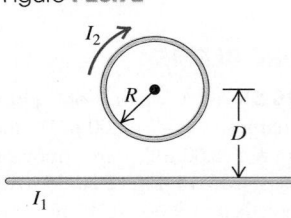

28.73 • **An Infinite Current Sheet.** Long, straight conductors with square cross sections and each carrying current I are laid side by side to form an infinite current sheet (**Fig. P28.73**). The conductors lie in the xy-plane, are parallel to the y-axis, and carry current in the $+y$-direction. There are n conductors per unit length measured along the x-axis. (a) What are the magnitude and direction of the magnetic field a distance a below the current sheet? (b) What are the magnitude and direction of the magnetic field a distance a above the current sheet?

Figure **P28.73**

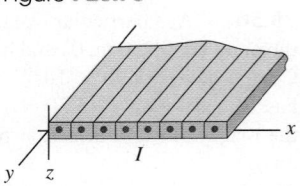

28.74 • Long, straight conductors with square cross section, each carrying current I, are laid side by side to form an infinite current sheet with current directed out of the plane of the page (**Fig. P28.74**). A second infinite current sheet is a distance d below the first and is parallel to it. The second sheet carries current into the plane of the page. Each sheet has n conductors per unit length. (Refer to Problem 28.73.) Calculate the magnitude and direction of the net magnetic field at (a) point P (above the upper sheet); (b) point R (midway between the two sheets); (c) point S (below the lower sheet).

Figure **P28.74**

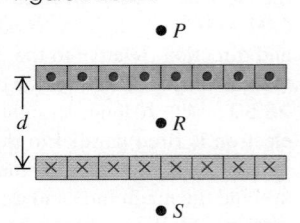

28.75 • A long, straight, solid cylinder, oriented with its axis in the z-direction, carries a current whose current density is $\vec{J}$. The current density, although symmetric about the cylinder axis, is not constant and varies according to the relationship

$$\vec{J} = \left(\frac{b}{r}\right)e^{(r-a)/\delta}\,\hat{k} \quad \text{for } r \le a$$

$$= 0 \quad \text{for } r \ge a$$

where the radius of the cylinder is $a = 5.00$ cm, r is the radial distance from the cylinder axis, b is a constant equal to 600 A/m, and δ is a constant equal to 2.50 cm. (a) Let I_0 be the total current passing through the entire cross section of the wire. Obtain an expression for I_0 in terms of b, δ, and a. Evaluate your expression to obtain a numerical value for I_0. (b) Using Ampere's law, derive an expression for the magnetic field $\vec{B}$ in the region $r \ge a$. Express your answer in terms of I_0 rather than b. (c) Obtain an expression for the current I contained in a circular cross section of radius $r \le a$ and centered at the cylinder axis. Express your answer in terms of I_0 rather than b. (d) Using Ampere's law, derive an expression for the magnetic field $\vec{B}$ in the region $r \le a$. (e) Evaluate the magnitude of the magnetic field at $r = \delta$, $r = a$, and $r = 2a$.

28.76 •• DATA As a summer intern at a research lab, you are given a long solenoid that has two separate windings that are wound close together, in the same direction, on the same hollow cylindrical form. You must determine the number of turns in each winding. The solenoid has length $L = 40.0$ cm and diameter 2.80 cm. You let a 2.00-mA current flow in winding 1 and vary the current I in winding 2; both currents flow in the same direction. Then you measure the magnetic-field magnitude B at the center of the solenoid as a function of I. You plot your results as BL/μ_0 versus I. The graph in **Fig. P28.76** shows the best-fit straight line to your data. (a) Explain why the data plotted in this way should fall close to a straight line. (b) Use Fig. P28.76 to calculate N_1 and N_2, the number of turns in windings 1 and 2. (c) If the current in winding 1 remains 2.00 mA in its original direction and winding 2 has $I = 5.00$ mA in the opposite direction, what is B at the center of the solenoid?

Figure **P28.76**

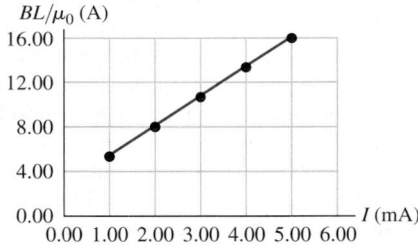

28.77 •• DATA You use a teslameter (a Hall-effect device) to measure the magnitude of the magnetic field at various distances from a long, straight, thick cylindrical copper cable that is carrying a large constant current. To exclude the earth's magnetic field from the measurement, you first set the meter to zero. You then measure the magnetic field B at distances x from the surface of the cable and obtain these data:

x (cm)	2.0	4.0	6.0	8.0	10.0
B (mT)	0.406	0.250	0.181	0.141	0.116

(a) You think you remember from your physics course that the magnetic field of a wire is inversely proportional to the distance from the wire. Therefore, you expect that the quantity Bx from your data will be constant. Calculate Bx for each data point in the table. Is Bx constant for this set of measurements? Explain. (b) Graph the data as x versus $1/B$. Explain why such a plot lies

close to a straight line. (c) Use the graph in part (b) to calculate the current I in the cable and the radius R of the cable.

28.78 ••• DATA A pair of long, rigid metal rods, each of length 0.50 m, lie parallel to each other on a frictionless table. Their ends are connected by identical, very light-weight conducting springs with unstretched length l_0 and force constant k (**Fig. P28.78**). When a current I runs through the circuit consisting of the rods and springs, the springs stretch. You measure the distance x each spring stretches for certain values of I. When $I = 8.05$ A, you measure that $x = 0.40$ cm. When $I = 13.1$ A, you find $x = 0.80$ cm. In both cases the rods are much longer than the stretched springs, so it is accurate to use Eq. (28.11) for two infinitely long, parallel conductors. (a) From these two measurements, calculate l_0 and k. (b) If $I = 12.0$ A, what distance x will each spring stretch? (c) What current is required for each spring to stretch 1.00 cm?

Figure **P28.78**

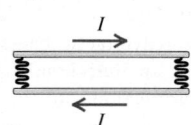

CHALLENGE PROBLEMS

28.79 ••• CP Two long, straight conducting wires with linear mass density λ are suspended from cords so that they are each horizontal, parallel to each other, and a distance d apart. The back ends of the wires are connected to each other by a slack, low-resistance connecting wire. A charged capacitor (capacitance C) is now added to the system; the positive plate of the capacitor (initial charge $+Q_0$) is connected to the front end of one of the wires, and the negative plate of the capacitor (initial charge $-Q_0$) is connected to the front end of the other wire (**Fig. P28.79**). Both of these connections are also made by slack, low-resistance wires. When the connection is made, the wires are pushed aside by the repulsive force between the wires, and each wire has an initial horizontal velocity of magnitude v_0. Assume that the time constant for the capacitor to discharge is negligible compared to the time it takes for any appreciable displacement in the position of the wires to occur. (a) Show that the initial speed v_0 of either wire is given by

Figure **P28.79**

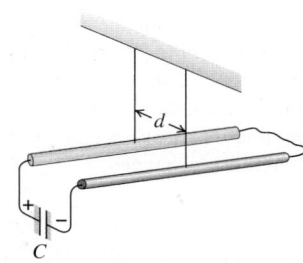

$$v_0 = \frac{\mu_0 Q_0^2}{4\pi \lambda R C d}$$

where R is the total resistance of the circuit. (b) To what height h will each wire rise as a result of the circuit connection?

28.80 ••• A wide, long, insulating belt has a uniform positive charge per unit area σ on its upper surface. Rollers at each end move the belt to the right at a constant speed v. Calculate the magnitude and direction of the magnetic field produced by the moving belt at a point just above its surface. (*Hint:* At points near the surface and far from its edges or ends, the moving belt can be considered to be an infinite current sheet like that in Problem 28.73.)

PASSAGE PROBLEMS

BIO **STUDYING MAGNETIC BACTERIA.** Some types of bacteria contain chains of ferromagnetic particles parallel to their long axis. The chains act like small bar magnets that align these *magnetotactic* bacteria with the earth's magnetic field. In one experiment to study the response of such bacteria to magnetic fields, a solenoid is constructed with copper wire 1.0 mm in diameter, evenly wound in a single layer to form a helical coil of length

40 cm and diameter 12 cm. The wire has a very thin layer of insulation, and the coil is wound so that adjacent turns are just touching. The solenoid, which generates a magnetic field, is in an enclosure that shields it from other magnetic fields. A sample of magnetotactic bacteria is placed inside the solenoid. The torque on an individual bacterium in the solenoid's magnetic field is proportional to the magnitude of the magnetic field and to the sine of the angle between the long axis of the bacterium and the magnetic-field direction.

28.81 What current is needed in the wire so that the magnetic field experienced by the bacteria has a magnitude of 150 μT? (a) 0.095 A; (b) 0.12 A; (c) 0.30 A; (d) 14 A.

28.82 To use a larger sample, the experimenters construct a solenoid that has the same length, type of wire, and loop spacing but twice the diameter of the original. How does the maximum

possible magnetic torque on a bacterium in this new solenoid compare with the torque the bacterium would have experienced in the original solenoid? Assume that the currents in the solenoids are the same. The maximum torque in the new solenoid is (a) twice that in the original one; (b) half that in the original one; (c) the same as that in the original one; (d) one-quarter that in the original one.

28.83 The solenoid is removed from the enclosure and then used in a location where the earth's magnetic field is 50 μT and points horizontally. A sample of bacteria is placed in the center of the solenoid, and the same current is applied that produced a magnetic field of 150 μT in the lab. Describe the field experienced by the bacteria: The field (a) is still 150 μT; (b) is now 200 μT; (c) is between 100 and 200 μT, depending on how the solenoid is oriented; (d) is between 50 and 150 μT, depending on how the solenoid is oriented.

Answers

Chapter Opening Question ?

(iv) There would be *no* change in the magnetic field strength. From Example 28.9 (Section 28.7), the field inside a solenoid has magnitude $B = \mu_0 nI$, where n is the number of turns of wire per unit length. Joining two solenoids end to end doubles both the number of turns and the length, so the number of turns per unit length is unchanged.

Test Your Understanding Questions

28.1 (a) (i), (b) (ii) The situation is the same as shown in Fig. 28.2 except that the upper proton has velocity $\vec{v}$ rather than $-\vec{v}$. The magnetic field due to the lower proton is the same as shown in Fig. 28.2, but the direction of the magnetic force $\vec{F} = q\vec{v} \times \vec{B}$ on the upper proton is reversed. Hence the magnetic force is attractive. Since the speed v is small compared to c, the magnetic force is much smaller in magnitude than the repulsive electric force and the net force is still repulsive.

28.2 (i) and (iii) (tie), (iv), (ii) From Eq. (28.5), the magnitude of the field dB due to a current element of length dl carrying current I is $dB = (\mu_0/4\pi)(I\,dl\sin\phi/r^2)$. In this expression r is the distance from the element to the field point, and ϕ is the angle between the direction of the current and a vector from the current element to the field point. All four points are the same distance $r = L$ from the current element, so the value of dB is proportional to the value of $\sin\phi$. For the four points the angle is (i) $\phi = 90°$, (ii) $\phi = 0$, (iii) $\phi = 90°$, and (iv) $\phi = 45°$, so the values of $\sin\phi$ are (i) 1, (ii) 0, (iii) 1, and (iv) $1/\sqrt{2}$.

28.3 A This orientation will cause current to flow clockwise around the circuit. Hence current will flow south through the wire that lies under the compass. From the right-hand rule for the magnetic field produced by a long, straight, current-carrying conductor, this will produce a magnetic field that points to the left at the position of the compass (which lies atop the wire). The combination of the northward magnetic field of the earth and the westward field produced by the current gives a net magnetic field to the northwest, so the compass needle will swing counterclockwise to align with this field.

28.4 (a) (i), (b) (iii), (c) (ii), (d) (iii) Current flows in the same direction in adjacent turns of the coil, so the magnetic forces between these turns are attractive. Current flows in opposite directions on opposite sides of the same turn, so the magnetic forces between these sides are repulsive. Thus the magnetic forces on the solenoid turns squeeze them together in the direction along

its axis but push them apart radially. The *electric* forces are zero because the wire is electrically neutral, with as much positive charge as there is negative charge.

28.5 (a) (ii), (b) (v) The vector $d\vec{B}$ is in the direction of $d\vec{l} \times \vec{r}$. For a segment on the negative y-axis, $d\vec{l} = -\hat{k}\,dl$ points in the negative z-direction and $\vec{r} = x\hat{\imath} + a\hat{\jmath}$. Hence $d\vec{l} \times \vec{r} = (a\,dl)\hat{\imath} - (x\,dl)\hat{\jmath}$, which has a positive x-component, a negative y-component, and zero z-component. For a segment on the negative z-axis, $d\vec{l} = \hat{\jmath}\,dl$ points in the positive y-direction and $\vec{r} = x\hat{\imath} + a\hat{k}$. Hence $d\vec{l} \times \vec{r} = 1(a\,dl)\hat{\imath} - 1(x\,dl)\hat{k}$, which has a positive x-component, zero y-component, and a negative z-component.

28.6 (ii) Imagine carrying out the integral $\oint \vec{B} \cdot d\vec{l}$ along an integration path that goes counterclockwise around the red magnetic field line. At each point along the path the magnetic field $\vec{B}$ and the infinitesimal segment $d\vec{l}$ are both tangent to the path, so $\vec{B} \cdot d\vec{l}$ is positive at each point and the integral $\oint \vec{B} \cdot d\vec{l}$ is likewise positive. It follows from Ampere's law $\oint \vec{B} \cdot d\vec{l} = \mu_0 I_{encl}$ and the right-hand rule that the integration path encloses a current directed out of the plane of the page. There are no currents in the empty space outside the magnet, so there must be currents inside the magnet (see Section 28.8).

28.7 (iii) By symmetry, any $\vec{B}$ field outside the cable must circulate around the cable, with circular field lines like those surrounding the solid cylindrical conductor in Fig. 28.20. Choose an integration path like the one shown in Fig. 28.20 with radius $r > R$, so that the path completely encloses the cable. As in Example 28.8, the integral $\oint \vec{B} \cdot d\vec{l}$ for this path has magnitude $B(2\pi r)$. From Ampere's law this is equal to $\mu_0 I_{encl}$. The net enclosed current I_{encl} is zero because it includes two currents of equal magnitude but opposite direction: one in the central wire and one in the hollow cylinder. Hence $B(2\pi r) = 0$, and so $B = 0$ for any value of r outside the cable. (The field is nonzero *inside* the cable; see Exercise 28.43.)

28.8 (i), (iv) Sodium and uranium are paramagnetic materials and hence are attracted to a magnet, while bismuth and lead are diamagnetic materials that are repelled by a magnet. (See Table 28.1.)

Bridging Problem

$$B = \frac{\mu_0 nQ}{a}$$

? The card reader at your bank's cash machine scans the information that is coded in a magnetic pattern on the back of your card. Why must you remove the card quickly rather than hold it motionless in the card reader's slot? (i) To maximize the magnetic force on the card; (ii) to maximize the magnetic force on the mobile charges in the card reader; (iii) to generate an electric force on the card; (iv) to generate an electric force on the mobile charges in the card reader.

29 ELECTROMAGNETIC INDUCTION

Almost every modern device or machine, from a computer to a washing machine to a power drill, has electric circuits at its heart. We learned in Chapter 25 that an electromotive force (emf) is required for a current to flow in a circuit; in Chapters 25 and 26 we almost always took the source of emf to be a battery. But for most devices that you plug into a wall socket, the source of emf is *not* a battery but an electric generating station. Such a station produces electrical energy by converting other forms of energy: gravitational potential energy at a hydroelectric plant, chemical energy in a coal-, gas-, or oil-fired plant, nuclear energy at a nuclear plant. But how is this energy conversion done?

The answer is a phenomenon known as *electromagnetic induction:* If the magnetic flux through a circuit changes, an emf and a current are induced in the circuit. In a power-generating station, magnets move relative to coils of wire to produce a changing magnetic flux in the coils and hence an emf.

The central principle of electromagnetic induction is *Faraday's law.* This law relates induced emf to changing magnetic flux in any loop, including a closed circuit. We also discuss Lenz's law, which helps us to predict the directions of induced emfs and currents. These principles will allow us to understand electrical energy-conversion devices such as motors, generators, and transformers.

Electromagnetic induction tells us that a time-varying magnetic field can act as a source of electric field. We will also see how a time-varying *electric* field can act as a source of *magnetic* field. These remarkable results form part of a neat package of formulas, called *Maxwell's equations,* that describe the behavior of electric and magnetic fields in general. Maxwell's equations pave the way toward an understanding of electromagnetic waves, the topic of Chapter 32.

29.1 INDUCTION EXPERIMENTS

During the 1830s, several pioneering experiments with magnetically induced emf were carried out in England by Michael Faraday and in the United States by Joseph Henry (1797–1878). **Figure 29.1** shows several examples. In Fig. 29.1a, a coil of wire is connected to a galvanometer. When the nearby magnet is stationary, the meter shows no current. This isn't surprising; there is no source of emf in the circuit. But when we *move* the magnet either toward or away from the coil, the meter shows current in the circuit, but *only* while the magnet is moving (Fig. 29.1b). If we keep the magnet stationary and move the coil, we again detect a current during the motion. We call this an **induced current,** and the corresponding emf required to cause this current is called an **induced emf.**

In Fig. 29.1c we replace the magnet with a second coil connected to a battery. When the second coil is stationary, there is no current in the first coil. However, when we move the second coil toward or away from the first or move the first toward or away from the second, there is current in the first coil, but again *only* while one coil is moving relative to the other.

Finally, using the two-coil setup in Fig. 29.1d, we keep both coils stationary and vary the current in the second coil by opening and closing the switch. As we open or close the switch, there is a momentary current pulse in the first coil. The induced current in the first coil is present only while the current in the second coil is changing.

To explore further the common elements in these observations, let's consider a more detailed series of experiments (**Fig. 29.2**). We connect a coil of wire to a galvanometer and then place the coil between the poles of an electromagnet whose magnetic field we can vary. Here's what we observe:

1. When there is no current in the electromagnet, so that $\vec{B} = 0$, the galvanometer shows no current.

2. When the electromagnet is turned on, there is a momentary current through the meter as $\vec{B}$ increases.

3. When $\vec{B}$ levels off at a steady value, the current drops to zero.

4. With the coil in a horizontal plane, we squeeze it so as to decrease the cross-sectional area of the coil. The meter detects current only *during* the deformation, not before or after. When we increase the area to return the coil to its original shape, there is current in the opposite direction, but only while the area of the coil is changing.

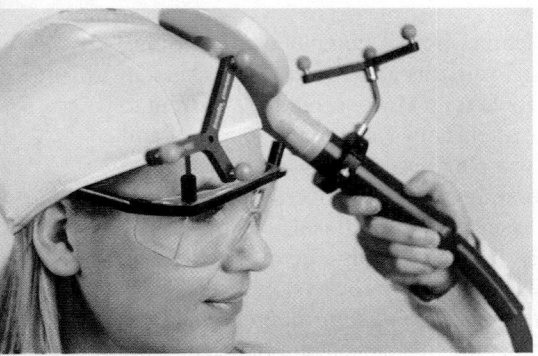

29.1 Demonstrating the phenomenon of induced current.

(a) A stationary magnet does NOT induce a current in a coil.

All these actions DO induce a current in the coil. What do they have in common?*

(b) Moving the magnet toward or away from the coil

(c) Moving a second, current-carrying coil toward or away from the coil

(d) Varying the current in the second coil (by closing or opening a switch)

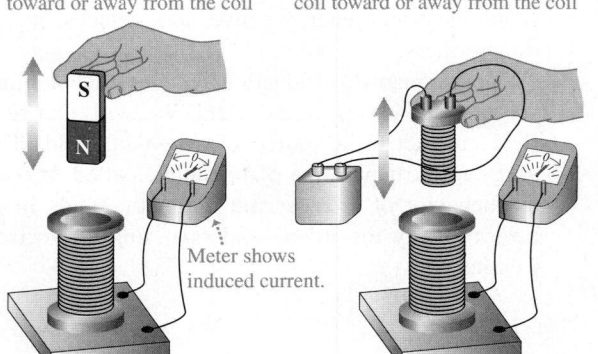

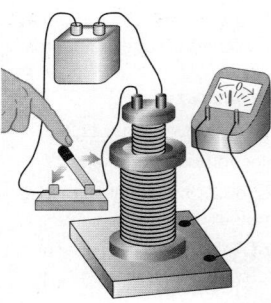

Meter shows zero current.

Meter shows induced current.

*They cause the magnetic field through the coil to *change.*

5. If we rotate the coil a few degrees about a horizontal axis, the meter detects current during the rotation, in the same direction as when we decreased the area. When we rotate the coil back, there is a current in the opposite direction during this rotation.

6. If we jerk the coil out of the magnetic field, there is a current during the motion, in the same direction as when we decreased the area.

7. If we decrease the number of turns in the coil by unwinding one or more turns, there is a current during the unwinding, in the same direction as when we decreased the area. If we wind more turns onto the coil, there is a current in the opposite direction during the winding.

8. When the magnet is turned off, there is a momentary current in the direction opposite to the current when it was turned on.

9. The faster we carry out any of these changes, the greater the current.

10. If all these experiments are repeated with a coil that has the same shape but different material and different resistance, the current in each case is inversely proportional to the total circuit resistance. This shows that the induced emfs that are causing the current do not depend on the material of the coil but only on its shape and the magnetic field.

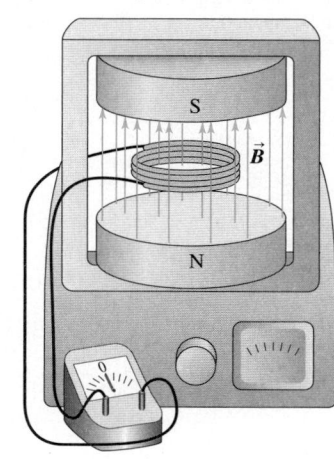

29.2 A coil in a magnetic field. When the $\vec{B}$ field is constant and the shape, location, and orientation of the coil do not change, no current is induced in the coil. A current is induced when any of these factors change.

The common element in these experiments is changing *magnetic flux* Φ_B through the coil connected to the galvanometer. In each case the flux changes either because the magnetic field changes with time or because the coil is moving through a nonuniform magnetic field. What's more, in each case the induced emf is proportional to the *rate of change* of magnetic flux Φ_B through the coil. The *direction* of the induced emf depends on whether the flux is increasing or decreasing. If the flux is constant, there is no induced emf.

Induced emfs have a tremendous number of practical applications. If you are reading these words indoors, you are making use of induced emfs right now! At the power plant that supplies your neighborhood, an electric generator produces an emf by varying the magnetic flux through coils of wire. (In the next section we'll see how this is done.) This emf supplies the voltage between the terminals of the wall sockets in your home, and this voltage supplies the power to your reading lamp.

Magnetically induced emfs, just like the emfs discussed in Section 25.4, are the result of *nonelectrostatic* forces. We have to distinguish carefully between the electrostatic electric fields produced by charges (according to Coulomb's law) and the nonelectrostatic electric fields produced by changing magnetic fields. We'll return to this distinction later in this chapter and the next.

29.2 FARADAY'S LAW

The common element in all induction effects is changing magnetic flux through a circuit. Before stating the simple physical law that summarizes all of the kinds of experiments described in Section 29.1, let's first review the concept of magnetic flux Φ_B (which we introduced in Section 27.3). For an infinitesimal-area element $d\vec{A}$ in a magnetic field $\vec{B}$ (**Fig. 29.3**), the magnetic flux $d\Phi_B$ through the area is

$$d\Phi_B = \vec{B} \cdot d\vec{A} = B_\perp \, dA = B \, dA \cos\phi$$

where $B_\perp$ is the component of $\vec{B}$ perpendicular to the surface of the area element and ϕ is the angle between $\vec{B}$ and $d\vec{A}$. (As in Chapter 27, be careful to distinguish between two quantities named "phi," ϕ and Φ_B.) The total magnetic flux Φ_B through a finite area is the integral of this expression over the area:

$$\Phi_B = \int \vec{B} \cdot d\vec{A} = \int B \, dA \cos\phi \qquad (29.1)$$

29.3 Calculating the magnetic flux through an area element.

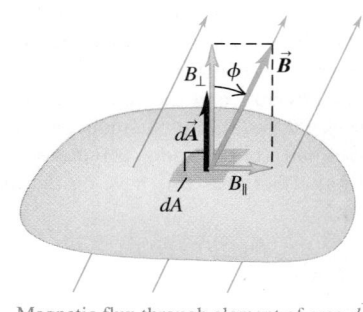

Magnetic flux through element of area $d\vec{A}$:
$d\Phi_B = \vec{B} \cdot d\vec{A} = B_\perp \, dA = B \, dA \cos\phi$

29.4 Calculating the flux of a uniform magnetic field through a flat area. (Compare to Fig. 22.6, which shows the rules for calculating the flux of a uniform *electric* field.)

Surface is face-on to magnetic field:
- $\vec{B}$ and $\vec{A}$ are parallel (the angle between $\vec{B}$ and $\vec{A}$ is $\phi = 0$).
- The magnetic flux $\Phi_B = \vec{B} \cdot \vec{A} = BA$.

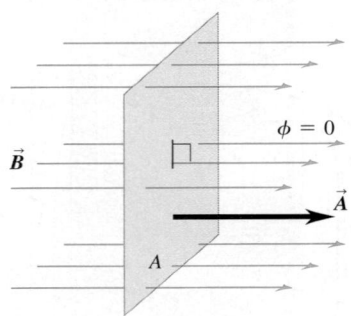

Surface is tilted from a face-on orientation by an angle ϕ:
- The angle between $\vec{B}$ and $\vec{A}$ is ϕ.
- The magnetic flux $\Phi_B = \vec{B} \cdot \vec{A} = BA\cos\phi$.

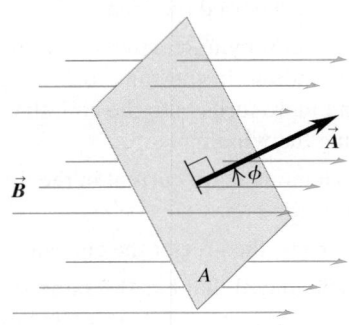

Surface is edge-on to magnetic field:
- $\vec{B}$ and $\vec{A}$ are perpendicular (the angle between $\vec{B}$ and $\vec{A}$ is $\phi = 90°$).
- The magnetic flux $\Phi_B = \vec{B} \cdot \vec{A} = BA\cos 90° = 0$.

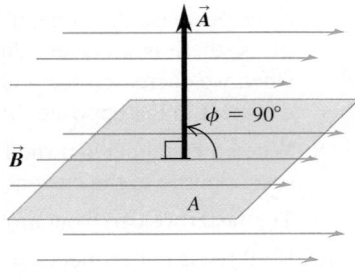

If $\vec{B}$ is uniform over a flat area $\vec{A}$, then

$$\Phi_B = \vec{B} \cdot \vec{A} = BA\cos\phi \tag{29.2}$$

Figure 29.4 reviews the rules for using Eq. (29.2).

> **CAUTION** **Choosing the direction of $d\vec{A}$ or $\vec{A}$** In Eqs. (29.1) and (29.2) we must define the direction of the vector area $d\vec{A}$ or $\vec{A}$ unambiguously. There are always two directions perpendicular to any given area, and the sign of the magnetic flux through the area depends on which one we choose. For example, in Fig. 29.3 we chose $d\vec{A}$ to point upward, so ϕ is less than 90° and $\vec{B} \cdot d\vec{A}$ is positive. We could have chosen $d\vec{A}$ to point downward, in which case ϕ would have been greater than 90° and $\vec{B} \cdot d\vec{A}$ would have been negative. Both choices are equally good, but once we make a choice we must stick with it. ▮

Faraday's law of induction states:

Faraday's law:
The induced emf $\cdots\cdots\rightarrow$ $\mathcal{E} = -\dfrac{d\Phi_B}{dt}$ $\cdots$equals the negative of the time rate of change of magnetic flux through the loop.
in a closed loop ...

$$\tag{29.3}$$

To understand the negative sign, we have to introduce a sign convention for the induced emf $\mathcal{E}$. But first let's look at a simple example of this law in action.

EXAMPLE 29.1 **EMF AND CURRENT INDUCED IN A LOOP**

The magnetic field between the poles of the electromagnet in **Fig. 29.5** is uniform at any time, but its magnitude is increasing at the rate of 0.020 T/s. The area of the conducting loop in the field is 120 cm², and the total circuit resistance, including the meter, is 5.0 Ω. (a) Find the induced emf and the induced current in the circuit. (b) If the loop is replaced by one made of an insulator, what effect does this have on the induced emf and induced current?

29.5 A stationary conducting loop in an increasing magnetic field.

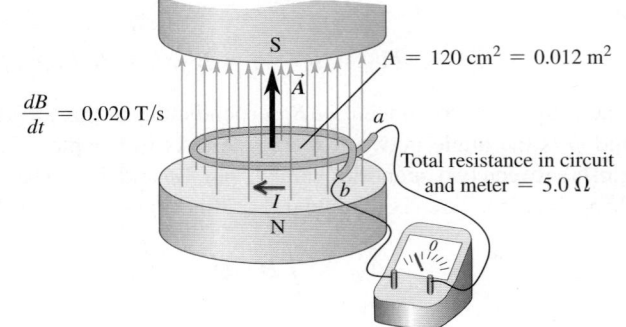

$\dfrac{dB}{dt} = 0.020$ T/s

$A = 120$ cm² $= 0.012$ m²

Total resistance in circuit and meter $= 5.0$ Ω

SOLUTION

IDENTIFY and SET UP: The magnetic flux Φ_B through the loop changes as the magnetic field changes. Hence there will be an induced emf $\mathcal{E}$ and an induced current I in the loop. We calculate

Φ_B from Eq. (29.2), then find $\mathcal{E}$ by using Faraday's law. Finally, we calculate I from $\mathcal{E} = IR$, where R is the total resistance of the circuit that includes the loop.

EXECUTE: (a) The area vector $\vec{A}$ for the loop is perpendicular to the plane of the loop; we take $\vec{A}$ to be vertically upward. Then $\vec{A}$ and $\vec{B}$ are parallel, and because $\vec{B}$ is uniform the magnetic flux through the loop is $\Phi_B = \vec{B} \cdot \vec{A} = BA\cos 0 = BA$. The area $A = 0.012 \text{ m}^2$ is constant, so the rate of change of magnetic flux is

$$\frac{d\Phi_B}{dt} = \frac{d(BA)}{dt} = \frac{dB}{dt}A = (0.020 \text{ T/s})(0.012 \text{ m}^2)$$

$$= 2.4 \times 10^{-4} \text{ V} = 0.24 \text{ mV}$$

This, apart from a sign that we haven't discussed yet, is the induced emf $\mathcal{E}$. The corresponding induced current is

$$I = \frac{\mathcal{E}}{R} = \frac{2.4 \times 10^{-4} \text{ V}}{5.0 \ \Omega} = 4.8 \times 10^{-5} \text{ A} = 0.048 \text{ mA}$$

(b) By changing to an insulating loop, we've made the resistance of the loop very high. Faraday's law, Eq. (29.3), does not involve the resistance of the circuit in any way, so the induced *emf* does not change. But the *current* will be smaller, as given by the equation $I = \mathcal{E}/R$. If the loop is made of a perfect insulator with infinite resistance, the induced current is zero. This situation is analogous to an isolated battery whose terminals aren't connected to anything: An emf is present, but no current flows.

EVALUATE: We can verify unit consistency in this calculation by noting that the magnetic-force relationship $\vec{F} = q\vec{v} \times \vec{B}$ implies that the units of $\vec{B}$ are the units of force divided by the units of (charge times velocity): $1 \text{ T} = (1 \text{ N})/(1 \text{ C} \cdot \text{m/s})$. The units of magnetic flux are then $(1 \text{ T})(1 \text{ m}^2) = 1 \text{ N} \cdot \text{s} \cdot \text{m/C}$, and the rate of change of magnetic flux is $1 \text{ N} \cdot \text{m/C} = 1 \text{ J/C} = 1 \text{ V}$. Thus the unit of $d\Phi_B/dt$ is the volt, as required by Eq. (29.3). Also recall that the unit of magnetic flux is the weber (Wb): $1 \text{ T} \cdot \text{m}^2 = 1 \text{ Wb}$, so $1 \text{ V} = 1 \text{ Wb/s}$.

Direction of Induced emf

We can find the direction of an induced emf or current by using Eq. (29.3) together with some simple sign rules. Here's the procedure:

1. Define a positive direction for the vector area $\vec{A}$.

2. From the directions of $\vec{A}$ and the magnetic field $\vec{B}$, determine the sign of the magnetic flux Φ_B and its rate of change $d\Phi_B/dt$. **Figure 29.6** shows several examples.

3. Determine the sign of the induced emf or current. If the flux is increasing, so $d\Phi_B/dt$ is positive, then the induced emf or current is negative; if the flux is decreasing, $d\Phi_B/dt$ is negative and the induced emf or current is positive.

4. Finally, use your right hand to determine the direction of the induced emf or current. Curl the fingers of your right hand around the $\vec{A}$ vector, with your right thumb in the direction of $\vec{A}$. If the induced emf or current in the circuit is *positive*, it is in the same direction as your curled fingers; if the induced emf or current is *negative*, it is in the opposite direction.

(a)

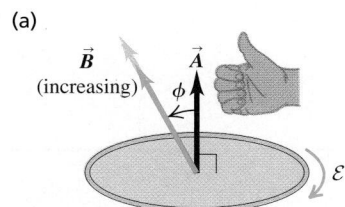

- Flux is positive ($\Phi_B > 0$) ...
- ... and becoming more positive ($d\Phi_B/dt > 0$).
- Induced emf is negative ($\mathcal{E} < 0$).

(b)

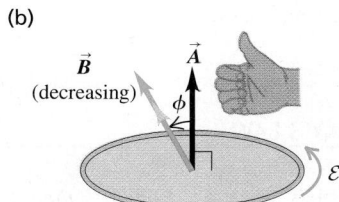

- Flux is positive ($\Phi_B > 0$) ...
- ... and becoming less positive ($d\Phi_B/dt < 0$).
- Induced emf is positive ($\mathcal{E} > 0$).

29.6 The magnetic flux is becoming (a) more positive, (b) less positive, (c) more negative, and (d) less negative. Therefore Φ_B is increasing in (a) and (d) and decreasing in (b) and (c). In (a) and (d) the emfs are negative (they are opposite to the direction of the curled fingers of your right hand when your right thumb points along $\vec{A}$). In (b) and (c) the emfs are positive (in the same direction as the curled fingers).

(c)

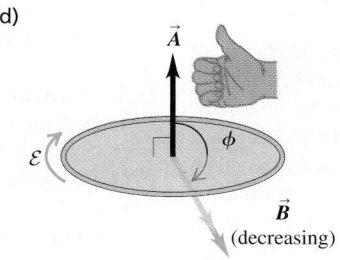

(d)

- Flux is negative ($\Phi_B < 0$) ...
- ... and becoming more negative ($d\Phi_B/dt < 0$).
- Induced emf is positive ($\mathcal{E} > 0$).

- Flux is negative ($\Phi_B < 0$) ...
- ... and becoming less negative ($d\Phi_B/dt > 0$).
- Induced emf is negative ($\mathcal{E} < 0$).

DATA *SPEAKS*

Magnetic Induction

When students were given a problem involving induced emf and induced currents, more than 68% gave an incorrect response. Common errors:

- Forgetting that *any* change in the magnetic flux through a loop induces an emf in the loop. This can include rotating the loop in a magnetic field or changing the loop's shape.

- Being careless with the sign of magnetic flux. Once you choose the direction of the area vector $\vec{A}$ for a loop, you must use it consistently in flux calculations.

In Example 29.1, in which $\vec{A}$ is upward, a positive $\mathcal{E}$ would be directed counterclockwise around the loop, as seen from above. Both $\vec{A}$ and $\vec{B}$ are upward in this example, so Φ_B is positive; the magnitude B is increasing, so $d\Phi_B/dt$ is positive. Hence by Eq. (29.3), $\mathcal{E}$ in Example 29.1 is *negative*. Its actual direction is thus *clockwise* around the loop, as seen from above.

If the loop in Fig. 29.5 is a conductor, the clockwise induced emf causes a clockwise induced current. This induced current produces an additional magnetic field through the loop, and the right-hand rule described in Section 28.5 shows that this field is *opposite* in direction to the increasing field produced by the electromagnet. This is an example of a general rule called *Lenz's law,* which says that any induction effect tends to oppose the change that caused it; in this case the change is the increase in the flux of the electromagnet's field through the loop. (We'll study Lenz's law in detail in the next section.)

You should check out the signs of the induced emfs and currents for the list of experiments in Section 29.1. For example, when the loop in Fig. 29.2 is in a constant field and we tilt it or squeeze it to *decrease* the flux through it, the induced emf and current are counterclockwise, as seen from above.

CAUTION **Induced emfs are caused by changes in flux** Since magnetic flux plays a central role in Faraday's law, it's tempting to think that *flux* is the cause of induced emf and that an induced emf will appear in a circuit whenever there is a magnetic field in the region bordered by the circuit. But Eq. (29.3) shows that only a *change* in flux through a circuit, not flux itself, can induce an emf in a circuit. If the flux through a circuit has a constant value, whether positive, negative, or zero, there is no induced emf. ∎

If a coil has N identical turns and if the flux varies at the same rate through each turn, the *total* rate of change through all turns is N times that for a single turn. If Φ_B is the flux through each turn, the total emf in a coil with N turns is

$$\mathcal{E} = -N\frac{d\Phi_B}{dt} \tag{29.4}$$

PhET: Faraday's Electromagnetic Lab
PhET: Faraday's Law
PhET: Generator

As we discussed in this chapter's introduction, induced emfs play an essential role in the generation of electrical power for commercial use. Several of the following examples explore different methods of producing emfs by changing the flux through a circuit.

PROBLEM-SOLVING STRATEGY 29.1 | FARADAY'S LAW

IDENTIFY *the relevant concepts:* Faraday's law applies when a magnetic flux is changing. To use the law, identify an area through which there is a flux of magnetic field. This will usually be the area enclosed by a loop made of a conducting material (though not always—see part (b) of Example 29.1). Identify the target variables.

SET UP *the problem* using the following steps:
1. Faraday's law relates the induced emf to the rate of change of magnetic flux. To calculate this rate of change, you first have to understand what is making the flux change. Is the conductor moving or changing orientation? Is the magnetic field changing?
2. The area vector $\vec{A}$ (or $d\vec{A}$) must be perpendicular to the plane of the area. You always have two choices of its direction; for example, if the area is in a horizontal plane, $\vec{A}$ could point up or down. Choose a direction and use it throughout the problem.

EXECUTE *the solution* as follows:
1. Calculate the magnetic flux from Eq. (29.2) if $\vec{B}$ is uniform over the area of the loop or Eq. (29.1) if it isn't uniform. Remember the direction you chose for the area vector.
2. Calculate the induced emf from Eq. (29.3) or (if your conductor has N turns in a coil) Eq. (29.4). Apply the sign rule (described just after Example 29.1) to determine the positive direction of emf.
3. If the circuit resistance is known, you can calculate the magnitude of the induced current I by using $\mathcal{E} = IR$.

EVALUATE *your answer:* Check your results for the proper units, and double-check that you have properly implemented the sign rules for magnetic flux and induced emf.

EXAMPLE 29.2 MAGNITUDE AND DIRECTION OF AN INDUCED EMF

A 500-loop circular wire coil with radius 4.00 cm is placed between the poles of a large electromagnet. The magnetic field is uniform and makes an angle of 60° with the plane of the coil; it decreases at 0.200 T/s. What are the magnitude and direction of the induced emf?

SOLUTION

IDENTIFY and SET UP: Our target variable is the emf induced by a varying magnetic flux through the coil. The flux varies because the magnetic field decreases in amplitude. We choose the area vector $\vec{A}$ to be in the direction shown in **Fig. 29.7**. With this choice, the geometry is similar to that of Fig. 29.6b. That figure will help us determine the direction of the induced emf.

29.7 Our sketch for this problem.

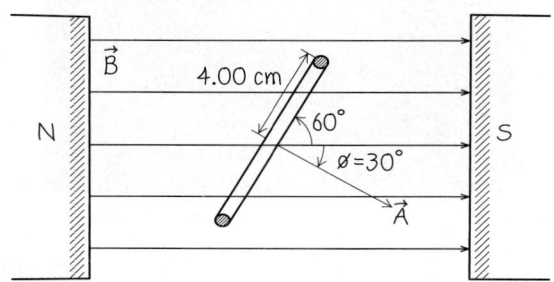

EXECUTE: The magnetic field is uniform over the loop, so we can calculate the flux from Eq. (29.2): $\Phi_B = BA\cos\phi$, where $\phi = 30°$. In this expression, the only quantity that changes with time is the magnitude B of the field, so $d\Phi_B/dt = (dB/dt)A\cos\phi$.

CAUTION **Remember how ϕ is defined** You may have been tempted to say that $\phi = 60°$ in this problem. If so, remember that ϕ is the angle between $\vec{A}$ and $\vec{B}$, *not* the angle between $\vec{B}$ and the plane of the loop.

From Eq. (29.4), the induced emf in the coil of $N = 500$ turns is

$$\mathcal{E} = -N\frac{d\Phi_B}{dt} = -N\frac{dB}{dt}A\cos\phi$$
$$= -500(-0.200\text{ T/s})\pi(0.0400\text{ m})^2(\cos 30°) = 0.435\text{ V}$$

The positive answer means that when you point your right thumb in the direction of area vector $\vec{A}$ (30° below field $\vec{B}$ in Fig. 29.7), the positive direction for $\mathcal{E}$ is in the direction of the curled fingers of your right hand. If you viewed the coil from the left in Fig. 29.7 and looked in the direction of $\vec{A}$, the emf would be clockwise.

EVALUATE: If the ends of the wire are connected, the direction of current in the coil is in the same direction as the emf—that is, clockwise as seen from the left side of the coil. A clockwise current increases the magnetic flux through the coil, and therefore tends to oppose the decrease in total flux. This is an example of Lenz's law, which we'll discuss in Section 29.3.

EXAMPLE 29.3 GENERATOR I: A SIMPLE ALTERNATOR

Figure 29.8a shows a simple *alternator,* a device that generates an emf. A rectangular loop is rotated with constant angular speed ω about the axis shown. The magnetic field $\vec{B}$ is uniform and constant. At time $t = 0$, $\phi = 0$. Determine the induced emf.

SOLUTION

IDENTIFY and SET UP: The magnetic field $\vec{B}$ and the loop area A are constant, but the flux through the loop varies because the loop rotates and so the angle ϕ between $\vec{B}$ and the area vector $\vec{A}$

29.8 (a) Schematic diagram of an alternator. A conducting loop rotates in a magnetic field, producing an emf. Connections from each end of the loop to the external circuit are made by means of that end's slip ring. The system is shown at the time when the angle $\phi = \omega t = 90°$. (b) Graph of the flux through the loop and the resulting emf between terminals a and b, along with the corresponding positions of the loop during one complete rotation.

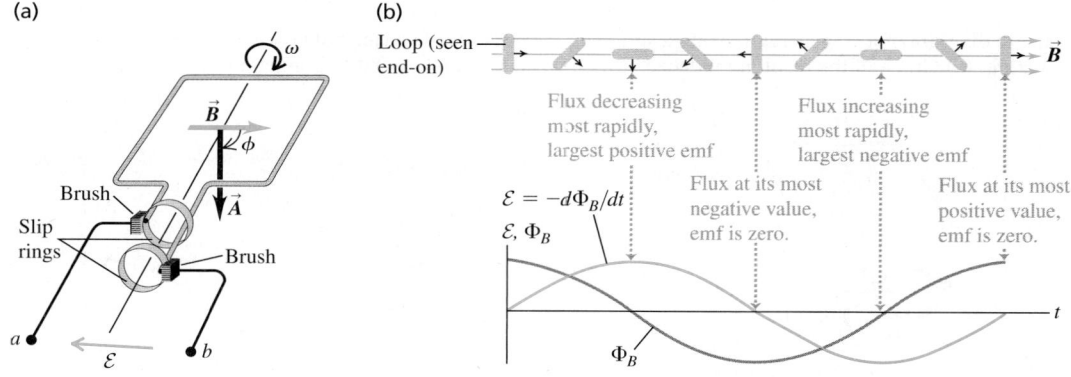

Continued

changes (Fig. 29.8a). Because the angular speed is constant and $\phi = 0$ at $t = 0$, the angle as a function of time is $\phi = \omega t$.

EXECUTE: The magnetic field is uniform over the loop, so the magnetic flux is $\Phi_B = BA\cos\phi = BA\cos\omega t$. Hence, by Faraday's law [Eq. (29.3)] the induced emf is

$$\mathcal{E} = -\frac{d\Phi_B}{dt} = -\frac{d}{dt}(BA\cos\omega t) = \omega BA\sin\omega t$$

EVALUATE: The induced emf $\mathcal{E}$ varies sinusoidally with time (see Fig. 29.8b). When the plane of the loop is perpendicular to $\vec{B}$ ($\phi = 0$ or $180°$), Φ_B reaches its maximum and minimum values. At these times, its instantaneous rate of change is zero and $\mathcal{E}$ is zero. Conversely, $\mathcal{E}$ reaches its maximum and minimum values when the plane of the loop is parallel to $\vec{B}$ ($\phi = 90°$ or $270°$) and Φ_B is changing most rapidly. We note that the induced emf does not depend on the *shape* of the loop, but only on its area.

We can use the alternator as a source of emf in an external circuit by using two *slip rings* that rotate with the loop, as shown in Fig. 29.8a. The rings slide against stationary contacts called *brushes,* which are connected to the output terminals a and b. Since the emf varies sinusoidally, the current that results in the circuit is an *alternating* current that also varies sinusoidally in magnitude and direction. The amplitude of the emf can be increased by increasing the rotation speed, the field magnitude, or the loop area or by using N loops instead of one, as in Eq. (29.4).

Alternators are used in automobiles to generate the currents in the ignition, the lights, and the entertainment system. The arrangement is a little different than in this example; rather than having a rotating loop in a magnetic field, the loop stays fixed and an electromagnet rotates. (The rotation is provided by a mechanical connection between the alternator and the engine.) But the result is the same; the flux through the loop varies sinusoidally, producing a sinusoidally varying emf. Larger alternators of this same type are used in electric power plants (**Fig. 29.9**).

29.9 A commercial alternator uses many loops of wire wound around a barrel-like structure called an armature. The armature and wire remain stationary while electromagnets rotate on a shaft (not shown) through the center of the armature. The resulting induced emf is far larger than would be possible with a single loop of wire.

EXAMPLE 29.4 GENERATOR II: A DC GENERATOR AND BACK EMF IN A MOTOR

The alternator in Example 29.3 produces a sinusoidally varying emf and hence an alternating current. **Figure 29.10a** shows a *direct-current* (dc) *generator* that produces an emf that always has the same sign. The arrangement of split rings, called a *commutator,* reverses the connections to the external circuit at angular positions at which the emf reverses. Figure 29.10b shows the resulting emf. Commercial dc generators have a large number of coils and commutator segments, smoothing out the bumps in the emf so that the terminal voltage is not only one-directional but also practically constant. This brush-and-commutator arrangement is the same as that in the direct-current motor discussed in Section 27.8. The motor's *back emf* is just the emf induced by the changing magnetic flux through its rotating coil. Consider a motor with a square, 500-turn coil 10.0 cm on a side. If the magnetic field has magnitude 0.200 T, at what rotation speed is the *average* back emf of the motor equal to 112 V?

SOLUTION

IDENTIFY and SET UP: As far as the rotating loop is concerned, the situation is the same as in Example 29.3 except that we now have N turns of wire. Without the commutator, the emf would alternate between positive and negative values and have an average value of zero (see Fig. 29.8b). With the commutator, the emf is never negative and its average value is positive (Fig. 29.10b). We'll use our result from Example 29.3 to obtain an expression for this average value and solve it for the rotational speed ω.

29.10 (a) Schematic diagram of a dc generator, using a split-ring commutator. The ring halves are attached to the loop and rotate with it. (b) Graph of the resulting induced emf between terminals a and b. Compare to Fig. 29.8b.

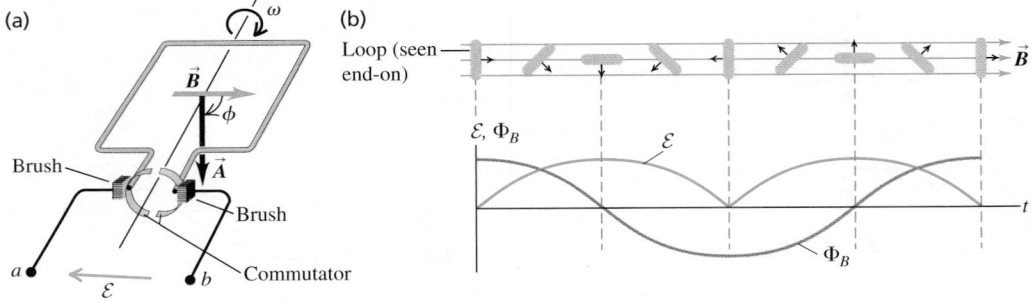

EXECUTE: Comparison of Figs. 29.8b and 29.10b shows that the back emf of the motor is just N times the absolute value of the emf found for an alternator in Example 29.3, as in Eq. (29.4): $|\mathcal{E}| = N\omega BA |\sin\omega t|$. To find the *average* back emf, we must replace $|\sin\omega t|$ by its average value. We find this by integrating $|\sin\omega t|$ over half a cycle, from $t = 0$ to $t = T/2 = \pi/\omega$, and dividing by the elapsed time π/ω. During this half cycle, the sine function is positive, so $|\sin\omega t| = \sin\omega t$, and we find

$$(|\sin\omega t|)_{av} = \frac{\int_0^{\pi/\omega} \sin\omega t \, dt}{\pi/\omega} = \frac{2}{\pi}$$

The average back emf is then

$$\mathcal{E}_{av} = \frac{2N\omega BA}{\pi}$$

Solving for ω, we obtain

$$\omega = \frac{\pi \mathcal{E}_{av}}{2NBA}$$

$$= \frac{\pi(112 \text{ V})}{2(500)(0.200 \text{ T})(0.100 \text{ m})^2} = 176 \text{ rad/s}$$

(Recall from Example 29.1 that $1 \text{ V} = 1 \text{ Wb/s} = 1 \text{ T} \cdot \text{m}^2/\text{s}$.)

EVALUATE: The average back emf is directly proportional to ω. Hence the slower the rotation speed, the less the back emf and the greater the possibility of burning out the motor, as we described in Example 27.11 (Section 27.8).

EXAMPLE 29.5 GENERATOR III: THE SLIDEWIRE GENERATOR

Figure 29.11 shows a U-shaped conductor in a uniform magnetic field $\vec{B}$ perpendicular to the plane of the figure and directed *into* the page. We lay a metal rod (the "slidewire") with length L across the two arms of the conductor, forming a circuit, and move it to the right with constant velocity $\vec{v}$. This induces an emf and a current, which is why this device is called a *slidewire generator*. Find the magnitude and direction of the resulting induced emf.

SOLUTION

IDENTIFY and SET UP: The magnetic flux changes because the area of the loop—bounded on the right by the moving rod—is increasing. Our target variable is the emf $\mathcal{E}$ induced in this expanding loop. The magnetic field is uniform over the area of the loop,

29.11 A slidewire generator. The magnetic field $\vec{B}$ and the vector area $\vec{A}$ are both directed into the figure. The increase in magnetic flux (caused by an increase in area) induces the emf and current.

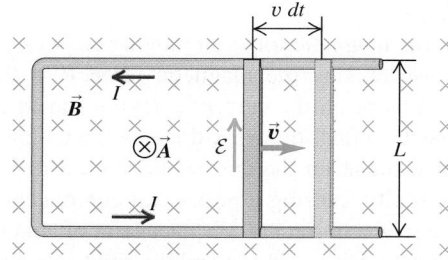

so we can find the flux from $\Phi_B = BA\cos\phi$. We choose the area vector $\vec{A}$ to point straight into the page, in the same direction as $\vec{B}$. With this choice a positive emf will be one that is directed clockwise around the loop. (You can check this with the right-hand rule: Using your right hand, point your thumb into the page and curl your fingers as in Fig. 29.6.)

EXECUTE: Since $\vec{B}$ and $\vec{A}$ point in the same direction, the angle $\phi = 0$ and $\Phi_B = BA$. The magnetic field magnitude B is constant, so the induced emf is

$$\mathcal{E} = -\frac{d\Phi_B}{dt} = -B\frac{dA}{dt}$$

To calculate dA/dt, note that in a time dt the sliding rod moves a distance $v \, dt$ (Fig. 29.11) and the loop area increases by an amount $dA = Lv \, dt$. Hence the induced emf is

$$\mathcal{E} = -B\frac{Lv \, dt}{dt} = -BLv$$

The minus sign tells us that the emf is directed *counterclockwise* around the loop. The induced current is also counterclockwise, as shown in the figure.

EVALUATE: The emf of a slidewire generator is constant if $\vec{v}$ is constant. Hence the slidewire generator is a *direct-current* generator. It's not a very practical device because the rod eventually moves beyond the U-shaped conductor and loses contact, after which the current stops.

EXAMPLE 29.6 WORK AND POWER IN THE SLIDEWIRE GENERATOR

In the slidewire generator of Example 29.5, energy is dissipated in the circuit owing to its resistance. Let the resistance of the circuit (made up of the moving slidewire and the U-shaped conductor that connects the ends of the slidewire) at a given point in the slidewire's motion be R. Find the rate at which energy is dissipated in the circuit and the rate at which work must be done to move the rod through the magnetic field.

SOLUTION

IDENTIFY and SET UP: Our target variables are the *rates* at which energy is dissipated and at which work is done. Energy is dissipated in the circuit at the rate $P_{dissipated} = I^2 R$. The current I in

Continued

the circuit equals $|\mathcal{E}|/R$; we found an expression for the induced emf $\mathcal{E}$ in this circuit in Example 29.5. There is a magnetic force $\vec{F} = I\vec{L} \times \vec{B}$ on the rod, where $\vec{L}$ points along the rod in the direction of the current. **Figure 29.12** shows that this force is opposite to the rod velocity $\vec{v}$; to maintain the motion, whoever is pushing the rod must apply a force of equal magnitude in the direction of $\vec{v}$. This force does work at the rate $P_{\text{applied}} = Fv$.

EXECUTE: First we'll calculate $P_{\text{dissipated}}$. From Example 29.5, $\mathcal{E} = -BLv$, so the current in the rod is $I = |\mathcal{E}|/R = Blv/R$. Hence

$$P_{\text{dissipated}} = I^2R = \left(\frac{BLv}{R}\right)^2 R = \frac{B^2L^2v^2}{R}$$

29.12 The magnetic force $\vec{F} = I\vec{L} \times \vec{B}$ that acts on the rod due to the induced current is to the left, opposite to $\vec{v}$.

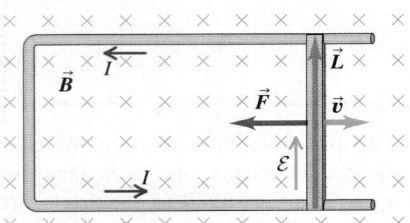

To calculate P_{applied}, we first calculate the magnitude of $\vec{F} = I\vec{L} \times \vec{B}$. Since $\vec{L}$ and $\vec{B}$ are perpendicular, this magnitude is

$$F = ILB = \frac{BLv}{R}LB = \frac{B^2L^2v}{R}$$

The applied force has the same magnitude and does work at the rate

$$P_{\text{applied}} = Fv = \frac{B^2L^2v^2}{R}$$

EVALUATE: The rate at which work is done is exactly *equal* to the rate at which energy is dissipated in the resistance.

CAUTION **You can't violate energy conservation** You might think that reversing the direction of $\vec{B}$ or of $\vec{v}$ would allow the magnetic force $\vec{F} = I\vec{L} \times \vec{B}$ to be in the *same* direction as $\vec{v}$. This would be a neat trick. Once the rod was moving, the changing magnetic flux would induce an emf and a current, and the magnetic force on the rod would make it move even faster, increasing the emf and current until the rod was moving at tremendous speed and producing electrical power at a prodigious rate. If this seems too good to be true and a violation of energy conservation, that's because it is. Reversing $\vec{B}$ also reverses the sign of the induced emf and current and hence the direction of $\vec{L}$, so the magnetic force still opposes the motion of the rod; a similar result holds true if we reverse $\vec{v}$. ∎

Generators as Energy Converters

Example 29.6 shows that the slidewire generator doesn't produce electrical energy out of nowhere; the energy is supplied by whatever body exerts the force that keeps the rod moving. All that the generator does is *convert* that energy into a different form. The equality between the rate at which *mechanical* energy is supplied to a generator and the rate at which *electrical* energy is generated holds for all types of generators, including the alternator described in Example 29.3. (We are ignoring the effects of friction in the bearings of an alternator or between the rod and the U-shaped conductor of a slidewire generator. The energy lost to friction is not available for conversion to electrical energy, so in real generators the friction is kept to a minimum.)

In Chapter 27 we stated that the magnetic force on moving charges can never do work. You might think, however, that the magnetic force $\vec{F} = I\vec{L} \times \vec{B}$ in Example 29.6 *is* doing (negative) work on the current-carrying rod as it moves, contradicting our earlier statement. In fact, the work done by the magnetic force is zero. The moving charges that make up the current in the rod in Fig. 29.12 have a vertical component of velocity, causing a horizontal component of force on these charges. As a result, there is a horizontal displacement of charge within the rod, the left side acquiring a net positive charge and the right side a net negative charge. The result is a horizontal component of electric field, perpendicular to the length of the rod (analogous to the Hall effect, described in Section 27.9). It is this field, in the direction of motion of the rod, that does work on the mobile charges in the rod and hence indirectly on the atoms making up the rod.

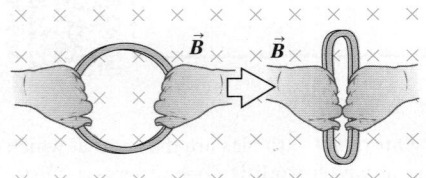

Circular wire coil   Coil squeezed
                             into oval

TEST YOUR UNDERSTANDING OF SECTION 29.2 The accompanying figure shows a wire coil being squeezed in a uniform magnetic field. (a) While the coil is being squeezed, is the induced emf in the coil (i) clockwise, (ii) counterclockwise, or (iii) zero? (b) Once the coil has reached its final squeezed shape, is the induced emf in the coil (i) clockwise, (ii) counterclockwise, or (iii) zero? ∎

29.3 LENZ'S LAW

Lenz's law is a convenient alternative method for determining the direction of an induced current or emf. Lenz's law, named for the Russian physicist H. F. E. Lenz (1804–1865), is not an independent principle; it can be derived from Faraday's law. It always gives the same results as the sign rules we introduced in connection with Faraday's law, but it is often easier to use. Lenz's law also helps us gain intuitive understanding of various induction effects and of the role of energy conservation. **Lenz's law** states:

> **The direction of any magnetic induction effect is such as to oppose the cause of the effect.**

The "cause" may be changing flux through a stationary circuit due to a varying magnetic field, changing flux due to motion of the conductors that make up the circuit, or any combination. If the flux in a stationary circuit changes, as in Examples 29.1 and 29.2, the induced current sets up a magnetic field of its own. Within the area bounded by the circuit, this field is *opposite* to the original field if the original field is *increasing* but is in the *same* direction as the original field if the latter is *decreasing*. That is, the induced current opposes the *change in flux* through the circuit (*not* the flux itself).

If the flux change is due to motion of the conductors, as in Examples 29.3 through 29.6, the direction of the induced current in the moving conductor is such that the direction of the magnetic-field force on the conductor is opposite in direction to its motion. Thus the motion of the conductor, which caused the induced current, is opposed. We saw this explicitly for the slidewire generator in Example 29.6. In all these cases the induced current tries to preserve the *status quo* by opposing motion or a change of flux.

Lenz's law is also directly related to energy conservation. If the induced current in Example 29.6 were in the direction opposite to that given by Lenz's law, the magnetic force on the rod would accelerate it to ever-increasing speed with no external energy source, even though electrical energy is being dissipated in the circuit. This would be a clear violation of energy conservation and doesn't happen in nature.

CONCEPTUAL EXAMPLE 29.7   LENZ'S LAW AND THE SLIDEWIRE GENERATOR

In Fig. 29.11, the induced current in the loop causes an additional magnetic field in the area bounded by the loop. The direction of the induced current is counterclockwise, so from the discussion of Section 28.5, this additional magnetic field is directed *out of* the plane of the figure. That direction is opposite that of the original magnetic field, so it tends to cancel the effect of that field. This is just what Lenz's law predicts.

CONCEPTUAL EXAMPLE 29.8   LENZ'S LAW AND THE DIRECTION OF INDUCED CURRENT

In **Fig. 29.13** there is a uniform magnetic field $\vec{B}$ through the coil. The magnitude of the field is increasing, so there is an induced emf. Use Lenz's law to determine the direction of the resulting induced current.

29.13 The induced current due to the change in $\vec{B}$ is clockwise, as seen from above the loop. The added field $\vec{B}_{\text{induced}}$ that it causes is downward, opposing the change in the upward field $\vec{B}$.

SOLUTION

This situation is the same as in Example 29.1 (Section 29.2). By Lenz's law the induced current must produce a magnetic field $\vec{B}_{\text{induced}}$ inside the coil that is downward, opposing the change in flux. From the right-hand rule we described in Section 28.5 for the direction of the magnetic field produced by a circular loop,

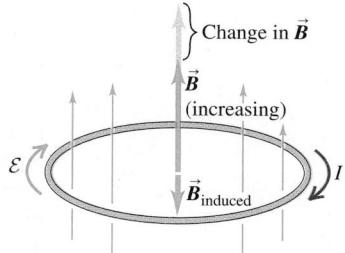

Continued

$\vec{B}_{induced}$ will be in the desired direction if the induced current flows as shown in Fig. 29.13.

Figure 29.14 shows several applications of Lenz's law to the similar situation of a magnet moving near a conducting loop. In each case, the induced current produces a magnetic field whose direction opposes the change in flux through the loop due to the magnet's motion.

29.14 Directions of induced currents as a bar magnet moves along the axis of a conducting loop. If the bar magnet is stationary, there is no induced current.

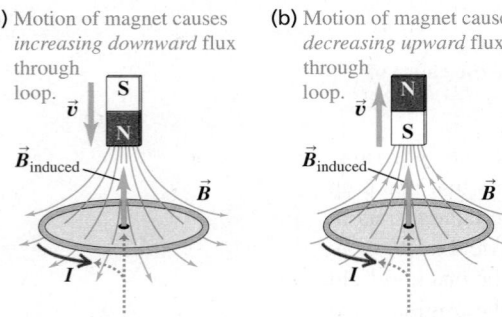

(a) Motion of magnet causes *increasing downward* flux through loop.

(b) Motion of magnet causes *decreasing upward* flux through loop.

The induced magnetic field is *upward* to oppose the flux change. To produce this induced field, the induced current must be *counterclockwise* as seen from above the loop.

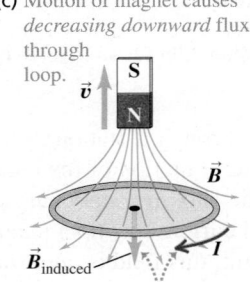

(c) Motion of magnet causes *decreasing downward* flux through loop.

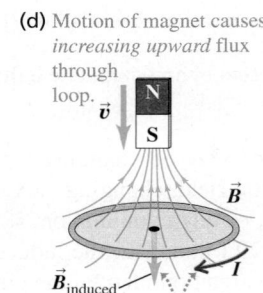

(d) Motion of magnet causes *increasing upward* flux through loop.

The induced magnetic field is *downward* to oppose the flux change. To produce this induced field, the induced current must be *clockwise* as seen from above the loop.

Lenz's Law and the Response to Flux Changes

Since an induced current always opposes any change in magnetic flux through a circuit, how is it possible for the flux to change at all? The answer is that Lenz's law gives only the *direction* of an induced current; the *magnitude* of the current depends on the resistance of the circuit. The greater the circuit resistance, the less the induced current that appears to oppose any change in flux and the easier it is for a flux change to take effect. If the loop in Fig. 29.14 were made out of wood (an insulator), there would be almost no induced current in response to changes in the flux through the loop.

Conversely, the less the circuit resistance, the greater the induced current and the more difficult it is to change the flux through the circuit. If the loop in Fig. 29.14 is a good conductor, an induced current flows as long as the magnet moves relative to the loop. Once the magnet and loop are no longer in relative motion, the induced current very quickly decreases to zero because of the non-zero resistance in the loop.

The extreme case occurs when the resistance of the circuit is *zero*. Then the induced current in Fig. 29.14 will continue to flow even after the induced emf has disappeared—that is, even after the magnet has stopped moving relative to the loop. Thanks to this *persistent current,* it turns out that the flux through the loop is exactly the same as it was before the magnet started to move, so the flux through a loop of zero resistance *never* changes. Exotic materials called *superconductors* do indeed have zero resistance; we discuss these further in Section 29.8.

TEST YOUR UNDERSTANDING OF SECTION 29.3 (a) Suppose the magnet in Fig. 29.14a were stationary and the loop of wire moved upward. Would the induced current in the loop be (i) in the same direction as shown in Fig. 29.14a, (ii) in the direction opposite to that shown in Fig. 29.14a, or (iii) zero? (b) Suppose the magnet and loop of wire in Fig. 29.14a both moved downward at the same velocity. Would the induced current in the loop be (i) in the same direction as shown in Fig. 29.14a, (ii) in the direction opposite to that shown in Fig. 29.14a, or (iii) zero? ▌

29.4 MOTIONAL ELECTROMOTIVE FORCE

We've seen several situations in which a conductor moves in a magnetic field, as in the generators discussed in Examples 29.3 through 29.6. We can gain additional insight into the origin of the induced emf in these situations by considering the magnetic forces on mobile charges in the conductor. **Figure 29.15a** shows the same moving rod that we discussed in Example 29.5, separated for the moment from the U-shaped conductor. The magnetic field $\vec{B}$ is uniform and directed into the page, and we move the rod to the right at a constant velocity $\vec{v}$. A charged particle q in the rod then experiences a magnetic force $\vec{F} = q\vec{v} \times \vec{B}$ with magnitude $F = |q|vB$. We'll assume in the following discussion that q is positive; in that case the direction of this force is upward along the rod, from b toward a.

This magnetic force causes the free charges in the rod to move, creating an excess of positive charge at the upper end a and negative charge at the lower end b. This in turn creates an electric field $\vec{E}$ within the rod, in the direction from a toward b (opposite to the magnetic force). Charge continues to accumulate at the ends of the rod until $\vec{E}$ becomes large enough for the downward electric force (with magnitude qE) to cancel exactly the *upward* magnetic force (with magnitude qvB). Then $qE = qvB$ and the charges are in equilibrium.

The magnitude of the potential difference $V_{ab} = V_a - V_b$ is equal to the electric-field magnitude E multiplied by the length L of the rod. From the above discussion, $E = vB$, so

$$V_{ab} = EL = vBL \tag{29.5}$$

with point a at higher potential than point b.

Now suppose the moving rod slides along a stationary U-shaped conductor, forming a complete circuit (Fig. 29.15b). No *magnetic* force acts on the charges in the stationary U-shaped conductor, but the charge that was near points a and b redistributes itself along the stationary conductor, creating an *electric* field within it. This field establishes a current in the direction shown. The moving rod has become a source of electromotive force; within it, charge moves from lower to higher potential, and in the remainder of the circuit, charge moves from higher to lower potential. We call this emf a **motional electromotive force,** denoted by $\mathcal{E}$. From the above discussion, the magnitude of this emf is

> **Motional emf,**
> **conductor length and velocity** ⋯⋯▶ $\mathcal{E} = vBL$ ◀⋯⋯ Conductor speed
> **perpendicular to uniform $\vec{B}$** ⋯⋯ Conductor length (29.6)
> Magnitude of uniform magnetic field

This corresponds to a force per unit charge of magnitude vB acting for a distance L along the moving rod. If the total circuit resistance of the U-shaped conductor and the sliding rod is R, the induced current I in the circuit is given by $vBL = IR$. This is the same result we obtained in Section 29.2 by using Faraday's law, and indeed motional emf is a particular case of Faraday's law. Verify that if we express v in meters per second, B in teslas, and L in meters, then $\mathcal{E}$ is in volts. (Recall that $1 \text{ V} = 1 \text{ J/C} = 1 \text{ T} \cdot \text{m}^2/\text{s}$.)

The emf associated with the moving rod in Fig. 29.15b is analogous to that of a battery with its positive terminal at a and its negative terminal at b, although the origins of the two emfs are quite different. In each case a nonelectrostatic force acts on the charges in the device, in the direction from b to a, and the emf is the work per unit charge done by this force when a charge moves from b to a in the device. When the device is connected to an external circuit, the direction of current is from b to a in the device and from a to b in the external circuit. Note that a motional emf is also present in the isolated moving rod in Fig. 29.15a, in the same way that a battery has an emf even when it's not part of a circuit.

29.15 A conducting rod moving in a uniform magnetic field. (a) The rod, the velocity, and the field are mutually perpendicular. (b) Direction of induced current in the circuit.

(a) Isolated moving rod

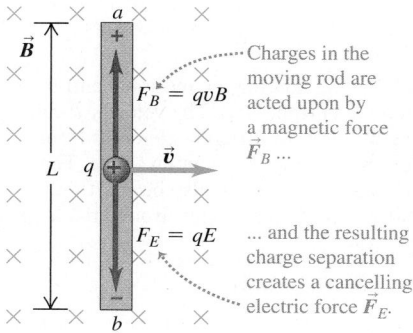

Charges in the moving rod are acted upon by a magnetic force $\vec{F}_B$ ...

... and the resulting charge separation creates a cancelling electric force $\vec{F}_E$.

(b) Rod connected to stationary conductor

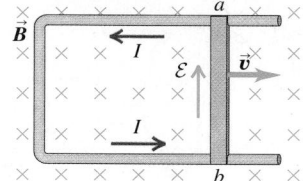

The motional emf $\mathcal{E}$ in the moving rod creates an electric field in the stationary conductor.

You can determine the direction of the induced emf in Fig. 29.15 by using Lenz's law, even if (as in Fig. 29.15a) the conductor does not form a complete circuit. In this case we can mentally complete the circuit between the ends of the conductor and use Lenz's law to determine the direction of the current. From this we can deduce the polarity of the ends of the open-circuit conductor. The direction from the $-$ end to the $+$ end within the conductor is the direction the current would have if the circuit were complete.

Motional emf: General Form

29.16 Calculating the motional emf for a moving current loop. The velocity $\vec{v}$ can be different for different elements if the loop is rotating or changing shape. The magnetic field $\vec{B}$ can also have different values at different points around the loop.

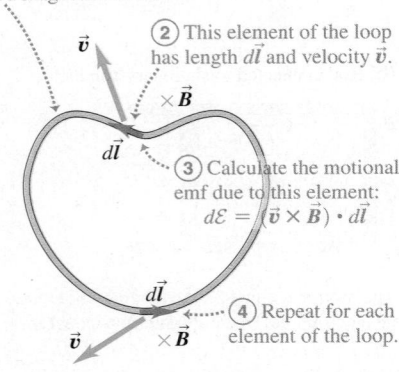

① A conducting loop moves in a magnetic field $\vec{B}$.

② This element of the loop has length $d\vec{l}$ and velocity $\vec{v}$.

③ Calculate the motional emf due to this element: $d\mathcal{E} = (\vec{v} \times \vec{B}) \cdot d\vec{l}$

④ Repeat for each element of the loop.

⑤ The total motional emf in the loop is the integral of the contributions from all elements:
$$\mathcal{E} = \oint (\vec{v} \times \vec{B}) \cdot d\vec{l}$$

We can generalize the concept of motional emf for a conductor with *any* shape, moving in any magnetic field, uniform or not, if we assume that the magnetic field at each point does not vary with time (**Fig. 29.16**). For an element $d\vec{l}$ of the conductor, the contribution $d\mathcal{E}$ to the emf is the magnitude dl multiplied by the component of $\vec{v} \times \vec{B}$ (the magnetic force per unit charge) parallel to $d\vec{l}$; that is,

$$d\mathcal{E} = (\vec{v} \times \vec{B}) \cdot d\vec{l}$$

For any closed conducting loop, the total emf is

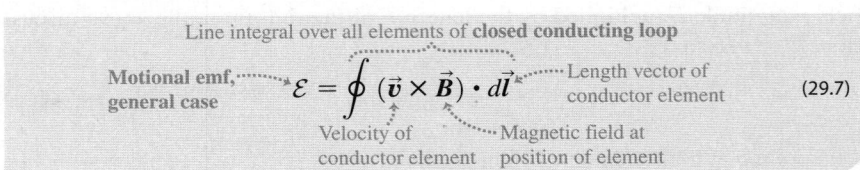

Line integral over all elements of **closed conducting loop**

Motional emf, general case $\qquad \mathcal{E} = \oint (\vec{v} \times \vec{B}) \cdot d\vec{l} \qquad$ (29.7)

Length vector of conductor element

Velocity of conductor element

Magnetic field at position of element

This expression looks very different from our original statement of Faraday's law, $\mathcal{E} = -d\Phi_B/dt$ [Eq. (29.3)]. In fact, though, the two statements are equivalent. It can be shown that the rate of change of magnetic flux through a moving conducting loop is always given by the negative of the expression in Eq. (29.7). Thus this equation gives us an alternative formulation of Faraday's law that is often convenient in problems with *moving* conductors. But when we have *stationary* conductors in changing magnetic fields, Eq. (29.7) *cannot* be used; in this case, $\mathcal{E} = -d\Phi_B/dt$ is the only correct way to express Faraday's law.

EXAMPLE 29.9 MOTIONAL EMF IN THE SLIDEWIRE GENERATOR

Suppose the moving rod in Fig. 29.15b is 0.10 m long, the velocity v is 2.5 m/s, the total resistance of the loop is 0.030 Ω, and B is 0.60 T. Find the motional emf, the induced current, and the force acting on the rod.

SOLUTION

IDENTIFY and SET UP: We'll find the motional emf $\mathcal{E}$ from Eq. (29.6) and the current from the values of $\mathcal{E}$ and the resistance R. The force on the rod is a *magnetic* force, exerted by $\vec{B}$ on the current in the rod; we'll find this force by using $\vec{F} = I\vec{L} \times \vec{B}$.

EXECUTE: From Eq. (29.6) the motional emf is

$$\mathcal{E} = vBL = (2.5 \text{ m/s})(0.60 \text{ T})(0.10 \text{ m}) = 0.15 \text{ V}$$

The induced current in the loop is

$$I = \frac{\mathcal{E}}{R} = \frac{0.15 \text{ V}}{0.030 \text{ }\Omega} = 5.0 \text{ A}$$

In the expression for the magnetic force, $\vec{F} = I\vec{L} \times \vec{B}$, the vector $\vec{L}$ points in the same direction as the induced current in the rod (from b to a in Fig. 29.15). The right-hand rule for vector products shows that this force is directed *opposite* to the rod's motion. Since $\vec{L}$ and $\vec{B}$ are perpendicular, the force has magnitude

$$F = ILB = (5.0 \text{ A})(0.10 \text{ m})(0.60 \text{ T}) = 0.30 \text{ N}$$

EVALUATE: We can check our answer for the direction of $\vec{F}$ by using Lenz's law. If we take the area vector $\vec{A}$ to point into the plane of the loop, the magnetic flux is positive and increasing as the rod moves to the right and increases the area of the loop. Lenz's law tells us that a force appears to oppose this increase in flux. Hence the force on the rod is to the left, opposite its motion.

EXAMPLE 29.10 THE FARADAY DISK DYNAMO

Figure 29.17 shows a conducting disk with radius R that lies in the xy-plane and rotates with constant angular velocity ω about the z-axis. The disk is in a uniform, constant $\vec{B}$ field in the z-direction. Find the induced emf between the center and the rim of the disk.

SOLUTION

IDENTIFY and SET UP: A motional emf arises because the conducting disk moves relative to $\vec{B}$. The complication is that different parts of the disk move at different speeds v, depending on their

29.17 A conducting disk with radius R rotating at an angular speed ω in a magnetic field $\vec{B}$. The emf is induced along radial lines of the disk and is applied to an external circuit through the two sliding contacts labeled b.

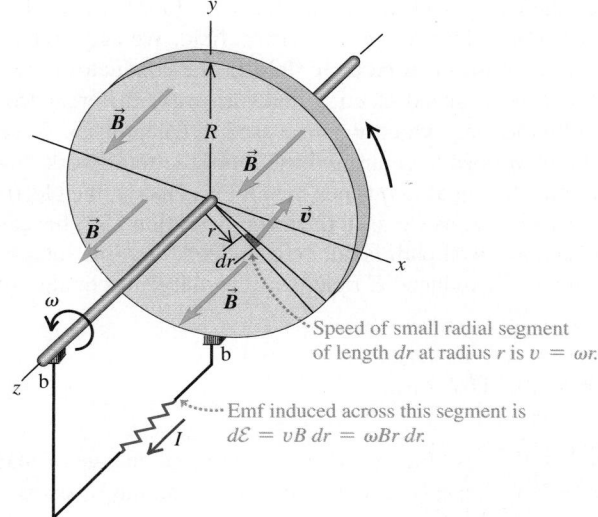

Speed of small radial segment of length dr at radius r is $v = \omega r$.

Emf induced across this segment is $d\mathcal{E} = vB\,dr = \omega Br\,dr$.

distance from the rotation axis. We'll address this by considering small segments of the disk and integrating their contributions to determine our target variable, the emf between the center and the rim. Consider the small segment of the disk shown in red in Fig. 29.17 and labeled by its velocity vector $\vec{v}$. The magnetic force per unit charge on this segment is $\vec{v} \times \vec{B}$, which points radially outward from the center of the disk. Hence the induced emf tends to make a current flow radially outward, which tells us that the moving conducting path to think about here is a straight line from the center to the rim. We can find the emf from each small disk segment along this line by using $d\mathcal{E} = (\vec{v} \times \vec{B}) \cdot d\vec{l}$ and then integrate to find the total emf.

EXECUTE: The length vector $d\vec{l}$ (of length dr) associated with the segment points radially outward, in the same direction as $\vec{v} \times \vec{B}$. The vectors $\vec{v}$ and $\vec{B}$ are perpendicular, and the magnitude of $\vec{v}$ is $v = \omega r$. The emf from the segment is then $d\mathcal{E} = \omega Br\,dr$. The total emf is the integral of $d\mathcal{E}$ from the center ($r = 0$) to the rim ($r = R$):

$$\mathcal{E} = \int_0^R \omega Br\,dr = \tfrac{1}{2}\omega BR^2$$

EVALUATE: We can use this device as a source of emf in a circuit by completing the circuit through two stationary brushes (labeled b in the figure) that contact the disk and its conducting shaft as shown. Such a disk is called a *Faraday disk dynamo* or a *homopolar generator*. Unlike the alternator in Example 29.3, the Faraday disk dynamo is a direct-current generator; it produces an emf that is constant in time. Can you use Lenz's law to show that for the direction of rotation in Fig. 29.17, the current in the external circuit must be in the direction shown?

TEST YOUR UNDERSTANDING OF SECTION 29.4 The earth's magnetic field points toward (magnetic) north. For simplicity, assume that the field has no vertical component (as is the case near the earth's equator). (a) If you hold a metal rod in your hand and walk toward the east, how should you orient the rod to get the maximum motional emf between its ends? (i) East-west; (ii) north-south; (iii) up-down; (iv) you get the same motional emf with all of these orientations. (b) How should you hold it to get *zero* emf as you walk toward the east? (i) East-west; (ii) north-south; (iii) up-down; (iv) none of these. (c) In which direction should you travel so that the motional emf across the rod is zero no matter how the rod is oriented? (i) West; (ii) north; (iii) south; (iv) straight up; (v) straight down. ❚

29.5 INDUCED ELECTRIC FIELDS

When a conductor moves in a magnetic field, we can understand the induced emf on the basis of magnetic forces on charges in the conductor, as described in Section 29.4. But an induced emf also occurs when there is a changing flux through a stationary conductor. What is it that pushes the charges around the circuit in this type of situation?

29.18 (a) The windings of a long solenoid carry a current I that is increasing at a rate dI/dt. The magnetic flux in the solenoid is increasing at a rate $d\Phi_B/dt$, and this changing flux passes through a wire loop. An emf $\mathcal{E} = -d\Phi_B/dt$ is induced in the loop, inducing a current I' that is measured by the galvanometer G. (b) Cross-sectional view.

(a)

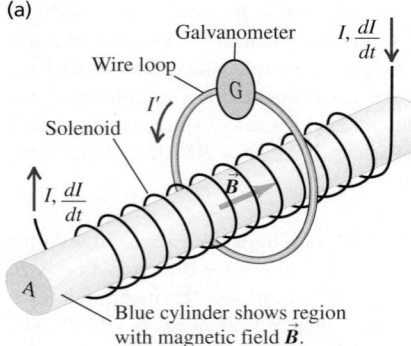

(b)

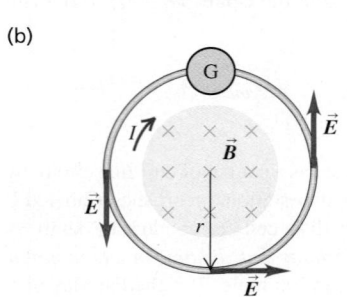

Let's consider the situation shown in **Fig. 29.18**. A long, thin solenoid with cross-sectional area A and n turns per unit length is encircled at its center by a circular conducting loop. The galvanometer G measures the current in the loop. A current I in the winding of the solenoid sets up a magnetic field $\vec{B}$ along the solenoid axis, as shown, with magnitude B as calculated in Example 28.9 (Section 28.7): $B = \mu_0 nI$, where n is the number of turns per unit length. If we ignore the small field outside the solenoid and take the area vector $\vec{A}$ to point in the same direction as $\vec{B}$, then the magnetic flux Φ_B through the loop is

$$\Phi_B = BA = \mu_0 nIA$$

When the solenoid current I changes with time, the magnetic flux Φ_B also changes, and according to Faraday's law the induced emf in the loop is given by

$$\mathcal{E} = -\frac{d\Phi_B}{dt} = -\mu_0 nA\frac{dI}{dt} \tag{29.8}$$

If the total resistance of the loop is R, the induced current in the loop, which we may call I', is $I' = \mathcal{E}/R$.

But what *force* makes the charges move around the wire loop? It can't be a magnetic force because the loop isn't even *in* a magnetic field. We are forced to conclude that there has to be an **induced electric field** in the conductor *caused by the changing magnetic flux*. Induced electric fields are *very* different from the electric fields caused by charges, which we discussed in Chapter 23. To see this, note that when a charge q goes once around the loop, the total work done on it by the electric field must be equal to q times the emf $\mathcal{E}$. That is, the electric field in the loop *is not conservative,* as we used the term in Section 23.1, because the line integral of $\vec{E}$ around a closed path is not zero. Indeed, this line integral, representing the work done by the induced $\vec{E}$ field per unit charge, is equal to the induced emf $\mathcal{E}$:

$$\oint \vec{E} \cdot d\vec{l} = \mathcal{E} \tag{29.9}$$

From Faraday's law the emf $\mathcal{E}$ is also the negative of the rate of change of magnetic flux through the loop. Thus for this case we can restate Faraday's law as

Faraday's law for a stationary integration path:	Line integral of electric field around path $$\oint \vec{E} \cdot d\vec{l} = -\frac{d\Phi_B}{dt}$$ Negative of the time rate of change of magnetic flux through path	(29.10)

Note that Faraday's law is *always* true in the form $\mathcal{E} = -d\Phi_B/dt$; the form given in Eq. (29.10) is valid *only* if the path around which we integrate is *stationary*.

Let's apply Eq. (29.10) to the stationary circular loop in Fig. 29.18b, which we take to have radius r. Because of cylindrical symmetry, $\vec{E}$ has the same magnitude at every point on the circle and is tangent to it at each point. (Symmetry would also permit the field to be *radial*, but then Gauss's law would require the presence of a net charge inside the circle, and there is none.) The line integral in Eq. (29.10) becomes simply the magnitude E times the circumference $2\pi r$ of the loop, $\oint \vec{E} \cdot d\vec{l} = 2\pi rE$, and Eq. (29.10) gives

$$E = \frac{1}{2\pi r}\left|\frac{d\Phi_B}{dt}\right| \tag{29.11}$$

The directions of $\vec{E}$ at points on the loop are shown in Fig. 29.18b. We know that $\vec{E}$ has to have the direction shown when $\vec{B}$ in the solenoid is increasing, because

$\oint \vec{E} \cdot d\vec{l}$ has to be negative when $d\Phi_B/dt$ is positive. The same approach can be used to find the induced electric field *inside* the solenoid when the solenoid $\vec{B}$ field is changing; we leave the details to you (see Exercise 29.37).

Nonelectrostatic Electric Fields

We've learned that Faraday's law, Eq. (29.3), is valid for two rather different situations. In one, an emf is induced by magnetic forces on charges when a conductor moves through a magnetic field. In the other, a time-varying magnetic field induces an electric field and hence an emf; the $\vec{E}$ field is induced even when no conductor is present. This $\vec{E}$ field differs from an electro*static* field in an important way. It is *nonconservative;* the line integral $\oint \vec{E} \cdot d\vec{l}$ around a closed path is not zero, and when a charge moves around a closed path, the field does a nonzero amount of work on it. It follows that for such a field the concept of *potential* has no meaning. We call such a field a **nonelectrostatic field.** In contrast, an *electrostatic* field is *always* conservative, as we discussed in Section 23.1, and always has an associated potential function. Despite this difference, the fundamental effect of *any* electric field is to exert a force $\vec{F} = q\vec{E}$ on a charge q. This relationship is valid whether $\vec{E}$ is conservative and produced by charges or nonconservative and produced by changing magnetic flux.

So a changing magnetic field acts as a source of electric field of a sort that we *cannot* produce with any static charge distribution. What's more, we'll see in Section 29.7 that a changing *electric* field acts as a source of *magnetic* field. We'll explore this symmetry between the two fields in our study of electromagnetic waves in Chapter 32.

If any doubt remains in your mind about the reality of magnetically induced electric fields, consider a few of the many practical applications (**Fig. 29.19**). Pickups in electric guitars use currents induced in stationary pickup coils by the vibration of nearby ferromagnetic strings. Alternators in most cars use rotating magnets to induce currents in stationary coils. Whether we realize it or not, magnetically induced electric fields play an important role in everyday life.

29.19 Applications of induced electric fields. (a) This hybrid automobile has both a gasoline engine and an electric motor. As the car comes to a halt, the spinning wheels run the motor backward so that it acts as a generator. The resulting induced current is used to recharge the car's batteries. (b) The rotating crankshaft of a piston-engine airplane spins a magnet, inducing an emf in an adjacent coil and generating the spark that ignites fuel in the engine cylinders. This keeps the engine running even if the airplane's other electrical systems fail.

(a)

(b)

EXAMPLE 29.11 **INDUCED ELECTRIC FIELDS**

Suppose the long solenoid in Fig. 29.18a has 500 turns per meter and cross-sectional area 4.0 cm^2. The current in its windings is increasing at 100 A/s. (a) Find the magnitude of the induced emf in the wire loop outside the solenoid. (b) Find the magnitude of the induced electric field within the loop if its radius is 2.0 cm.

SOLUTION

IDENTIFY and SET UP: As in Fig. 29.18b, the increasing magnetic field inside the solenoid causes a change in the magnetic flux through the wire loop and hence induces an electric field $\vec{E}$ around the loop. Our target variables are the induced emf $\mathcal{E}$ and the electric-field magnitude E. We use Eq. (29.8) to determine the emf. The loop and the solenoid share the same central axis. Hence, by symmetry, the electric field is tangent to the loop and has the same magnitude E all the way around its circumference. We can therefore use Eq. (29.9) to find E.

EXECUTE: (a) From Eq. (29.8), the induced emf is

$$\mathcal{E} = -\frac{d\Phi_B}{dt} = -\mu_0 nA\frac{dI}{dt}$$
$$= -(4\pi \times 10^{-7}\text{ Wb/A}\cdot\text{m})(500\text{ turns/m})$$
$$\times (4.0 \times 10^{-4}\text{ m}^2)(100\text{ A/s})$$
$$= -25 \times 10^{-6}\text{ Wb/s} = -25 \times 10^{-6}\text{ V} = -25\ \mu\text{V}$$

(b) By symmetry the line integral $\oint \vec{E} \cdot d\vec{l}$ has absolute value $2\pi rE$ no matter which direction we integrate around the loop. This is equal to the absolute value of the emf, so

$$E = \frac{|\mathcal{E}|}{2\pi r} = \frac{25 \times 10^{-6}\text{ V}}{2\pi(2.0 \times 10^{-2}\text{ m})} = 2.0 \times 10^{-4}\text{ V/m}$$

EVALUATE: In Fig. 29.18b the magnetic flux *into* the plane of the figure is increasing. According to the right-hand rule for induced emf (Fig. 29.6), a positive emf would be clockwise around the loop; the negative sign of $\mathcal{E}$ shows that the emf is in the counterclockwise direction. Can you also show this by using Lenz's law?

29.6 EDDY CURRENTS

29.20 Eddy currents induced in a rotating metal disk.

(a) Metal disk rotating through a magnetic field

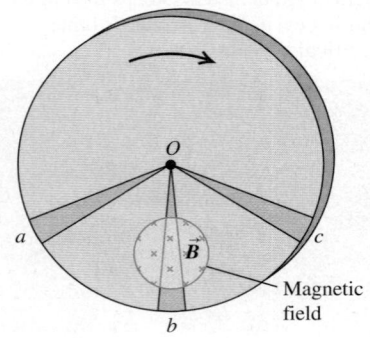

(b) Resulting eddy currents and braking force

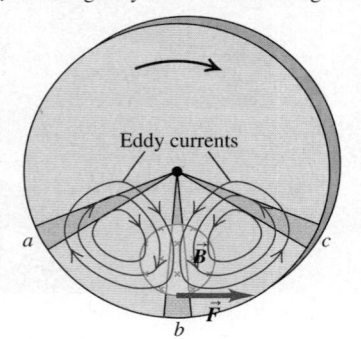

DEMO

In the examples of induction effects that we have studied, the induced currents have been confined to well-defined paths in conductors and other components forming a circuit. However, many pieces of electrical equipment contain masses of metal moving in magnetic fields or located in changing magnetic fields. In situations like these we can have induced currents that circulate throughout the volume of a material. Because their flow patterns resemble swirling eddies in a river, we call these **eddy currents.**

As an example, consider a metallic disk rotating in a magnetic field perpendicular to the plane of the disk but confined to a limited portion of the disk's area, as shown in **Fig. 29.20a**. Sector Ob is moving across the field and has an emf induced in it. Sectors Oa and Oc are not in the field, but they provide return conducting paths for charges displaced along Ob to return from b to O. The result is a circulation of eddy currents in the disk, somewhat as sketched in Fig. 29.20b.

We can use Lenz's law to decide on the direction of the induced current in the neighborhood of sector Ob. This current must experience a magnetic force $\vec{F} = I\vec{L} \times \vec{B}$ that *opposes* the rotation of the disk, and so this force must be to the right in Fig. 29.20b. Since $\vec{B}$ is directed into the plane of the disk, the current and hence $\vec{L}$ have downward components. The return currents lie outside the field, so they do not experience magnetic forces. The interaction between the eddy currents and the field causes a braking action on the disk. Such effects can be used to stop the rotation of a circular saw quickly when the power is turned off. Eddy current braking is used on some electrically powered rapid-transit vehicles. Electromagnets mounted in the cars induce eddy currents in the rails; the resulting magnetic fields cause braking forces on the electromagnets and thus on the cars.

Eddy currents have many other practical uses. In induction furnaces, eddy currents are used to heat materials in completely sealed containers for processes in which it is essential to avoid the slightest contamination of the materials. The metal detectors used at airport security checkpoints (**Fig. 29.21a**) operate by detecting eddy currents induced in metallic objects. Similar devices (Fig. 29.21b) are used to find buried treasure such as bottlecaps and lost pennies.

Eddy currents also have undesirable effects. In an alternating-current transformer, coils wrapped around an iron core carry a sinusoidally varying current. The resulting eddy currents in the core waste energy through I^2R heating and set up an unwanted opposing emf in the coils. To minimize these effects, the core

29.21 (a) A metal detector at an airport security checkpoint generates an alternating magnetic field $\vec{B}_0$. This induces eddy currents in a conducting object carried through the detector. The eddy currents in turn produce an alternating magnetic field $\vec{B}'$, which induces a current in the detector's receiver coil. (b) Portable metal detectors work on the same principle.

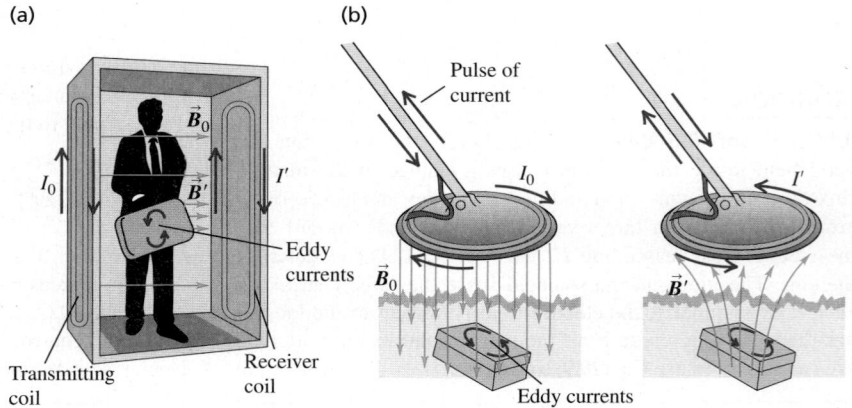

is designed so that the paths for eddy currents are as narrow as possible. We'll describe how this is done when we discuss transformers in Section 31.6.

TEST YOUR UNDERSTANDING OF SECTION 29.6   Suppose that the magnetic field in Fig. 29.20 were directed out of the plane of the figure and the disk were rotating counterclockwise. Compared to the directions of the force $\vec{F}$ and the eddy currents shown in Fig. 29.20b, what would the new directions be? (i) The force $\vec{F}$ and the eddy currents would both be in the same direction; (ii) the force $\vec{F}$ would be in the same direction, but the eddy currents would be in the opposite direction; (iii) the force $\vec{F}$ would be in the opposite direction, but the eddy currents would be in the same direction; (iv) the force $\vec{F}$ and the eddy currents would be in the opposite directions. ❙

29.7 DISPLACEMENT CURRENT AND MAXWELL'S EQUATIONS

We have seen that a varying magnetic field gives rise to an induced electric field. In one of the more remarkable examples of the symmetry of nature, it turns out that a varying *electric* field gives rise to a *magnetic* field. This effect is of tremendous importance, for it turns out to explain the existence of radio waves, gamma rays, and visible light, as well as all other forms of electromagnetic waves.

Generalizing Ampere's Law

To see the origin of the relationship between varying electric fields and magnetic fields, let's return to Ampere's law as given in Section 28.6, Eq. (28.20):

$$\oint \vec{B} \cdot d\vec{l} = \mu_0 I_{\text{encl}}$$

The problem with Ampere's law in this form is that it is *incomplete*. To see why, let's consider the process of charging a capacitor (**Fig. 29.22**). Conducting wires lead current i_C into one plate and out of the other; the charge Q increases, and the electric field $\vec{E}$ between the plates increases. The notation i_C indicates *conduction* current to distinguish it from another kind of current we are about to encounter, called *displacement* current i_D. We use lowercase i's and v's to denote instantaneous values of currents and potential differences, respectively, that may vary with time.

Let's apply Ampere's law to the circular path shown. The integral $\oint \vec{B} \cdot d\vec{l}$ around this path equals $\mu_0 I_{\text{encl}}$. For the plane circular area bounded by the circle, I_{encl} is just the current i_C in the left conductor. But the surface that bulges out to the right is bounded by the same circle, and the current through that surface is zero. So $\oint \vec{B} \cdot d\vec{l}$ is equal to $\mu_0 i_C$, and at the same time it is equal to zero! This is a clear contradiction.

However, something else is happening on the bulged-out surface. As the capacitor charges, the electric field $\vec{E}$ and the electric *flux* Φ_E through the surface are increasing. We can determine their rates of change in terms of the charge and current. The instantaneous charge is $q = Cv$, where C is the capacitance and v is the instantaneous potential difference. For a parallel-plate capacitor, $C = \epsilon_0 A/d$, where A is the plate area and d is the spacing. The potential difference v between plates is $v = Ed$, where E is the electric-field magnitude between plates. (We ignore fringing and assume that $\vec{E}$ is uniform in the region between the plates.) If this region is filled with a material with permittivity ϵ, we replace ϵ_0 by ϵ everywhere; we'll use ϵ in the following discussion.

Substituting these expressions for C and v into $q = Cv$, we can express the capacitor charge q in terms of the electric flux $\Phi_E = EA$ through the surface:

$$q = Cv = \frac{\epsilon A}{d}(Ed) = \epsilon EA = \epsilon \Phi_E \qquad (29.12)$$

Application Eddy Currents Help Power Io's Volcanoes Jupiter's moon Io is slightly larger than the earth's moon. It moves at more than 60,000 km/h through Jupiter's intense magnetic field (about ten times stronger than the earth's field), which sets up strong eddy currents within Io that dissipate energy at a rate of 10^{12} W. This dissipated energy helps to heat Io's interior and causes volcanic eruptions on its surface, as shown in the lower close-up image. (Gravitational effects from Jupiter cause even more heating.)

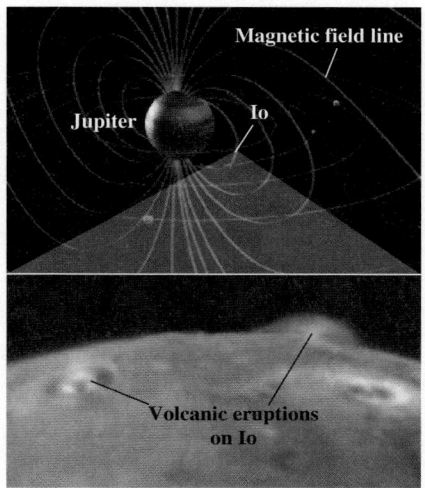

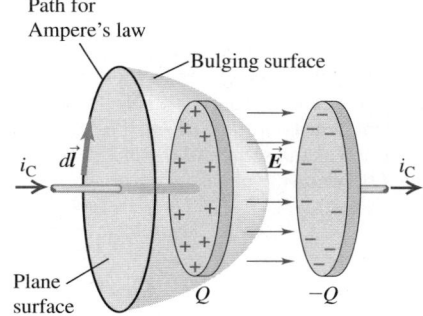

29.22 Parallel-plate capacitor being charged. The conduction current through the plane surface is i_C, but there is no conduction current through the surface that bulges out to pass between the plates. The two surfaces have a common boundary, so this difference in I_{encl} leads to an apparent contradiction in applying Ampere's law.

As the capacitor charges, the rate of change of q is the conduction current, $i_C = dq/dt$. Taking the derivative of Eq. (29.12) with respect to time, we get

$$i_C = \frac{dq}{dt} = \epsilon \frac{d\Phi_E}{dt} \tag{29.13}$$

Stretching our imagination a bit, we invent a fictitious **displacement current** i_D in the region between the plates, defined as

$$\underset{\substack{\text{Displacement current} \\ \text{through an area}}}{} i_D = \epsilon \frac{d\Phi_E}{dt} \underset{\substack{\text{Time rate of change of} \\ \text{electric flux through area}}}{} \tag{29.14}$$

Permittivity of material in area

That is, we imagine that the changing flux through the curved (bulged-out) surface in Fig. 29.22 is equivalent, in Ampere's law, to a conduction current through that surface. We include this fictitious current, along with the real conduction current i_C, in Ampere's law:

$$\oint \vec{B} \cdot d\vec{l} = \mu_0 (i_C + i_D)_{\text{encl}} \qquad \text{(generalized Ampere's law)} \tag{29.15}$$

Ampere's law in this form is obeyed no matter which surface we use in Fig. 29.22. For the flat surface, i_D is zero; for the curved surface, i_C is zero; and i_C for the flat surface equals i_D for the curved surface. Equation (29.15) remains valid in a magnetic material, provided that the magnetization is proportional to the external field and we replace μ_0 by μ.

The fictitious displacement current i_D was invented in 1865 by the Scottish physicist James Clerk Maxwell. There is a corresponding *displacement current density* $j_D = i_D/A$; using $\Phi_E = EA$ and dividing Eq. (29.14) by A, we find

$$j_D = \epsilon \frac{dE}{dt} \tag{29.16}$$

We have pulled the concept out of thin air, as Maxwell did, but we see that it enables us to save Ampere's law in situations such as that in Fig. 29.22.

Another benefit of displacement current is that it lets us generalize Kirchhoff's junction rule, discussed in Section 26.2. Considering the left plate of the capacitor, we have conduction current into it but none out of it. But when we include the displacement current, we have conduction current coming in one side and an equal displacement current coming out the other side. With this generalized meaning of the term "current," we can speak of current going *through* the capacitor.

The Reality of Displacement Current

You might well ask at this point whether displacement current has any real physical significance or whether it is just a ruse to satisfy Ampere's law and Kirchhoff's junction rule. Here's a fundamental experiment that helps to answer that question. We take a plane circular area between the capacitor plates (**Fig. 29.23**). If displacement current really plays the role in Ampere's law that we have claimed, then there ought to be a magnetic field in the region between the plates while the capacitor is charging. We can use our generalized Ampere's law, including displacement current, to predict what this field should be.

To be specific, let's picture round capacitor plates with radius R. To find the magnetic field at a point in the region between the plates at a distance r from the axis, we apply Ampere's law to a circle of radius r passing through the point, with $r < R$. This circle passes through points a and b in Fig. 29.23. The total current enclosed by the circle is j_D times its area, or $(i_D/\pi R^2)(\pi r^2)$. The integral

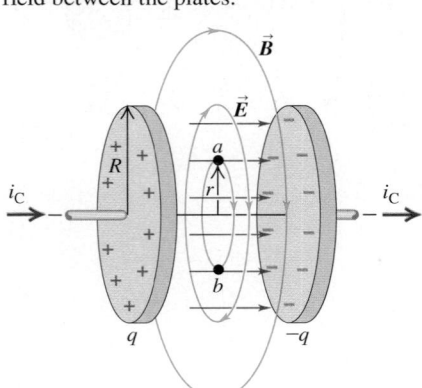

29.23 A capacitor being charged by a current i_C has a displacement current equal to i_C between the plates, with displacement-current density $j_D = \epsilon \, dE/dt$. This can be regarded as the source of the magnetic field between the plates.

$\oint \vec{B} \cdot d\vec{l}$ in Ampere's law is just B times the circumference $2\pi r$ of the circle, and because $i_\text{D} = i_\text{C}$ for the charging capacitor, Ampere's law becomes

$$\oint \vec{B} \cdot d\vec{l} = 2\pi r B = \mu_0 \frac{r^2}{R^2} i_\text{C} \quad \text{or}$$

$$B = \frac{\mu_0}{2\pi} \frac{r}{R^2} i_\text{C} \tag{29.17}$$

This result predicts that in the region between the plates $\vec{B}$ is zero at the axis and increases linearly with distance from the axis. A similar calculation shows that *outside* the region between the plates (that is, for $r > R$), $\vec{B}$ is the same as though the wire were continuous and the plates not present at all.

When we *measure* the magnetic field in this region, we find that it really is there and that it behaves just as Eq. (29.17) predicts. This confirms directly the role of displacement current as a source of magnetic field. It is now established beyond reasonable doubt that Maxwell's displacement current, far from being just an artifice, is a fundamental fact of nature.

Maxwell's Equations of Electromagnetism

We are now in a position to wrap up in a single package *all* of the relationships between electric and magnetic fields and their sources. This package consists of four equations, called **Maxwell's equations.** Maxwell did not discover all of these equations single-handedly (though he did develop the concept of displacement current). But he did put them together and recognized their significance, particularly in predicting the existence of electromagnetic waves.

For now we'll state Maxwell's equations in their simplest form, for the case in which we have charges and currents in otherwise empty space. In Chapter 32 we'll discuss how to modify these equations if a dielectric or a magnetic material is present.

Two of Maxwell's equations involve an integral of $\vec{E}$ or $\vec{B}$ over a closed surface. The first is simply Gauss's law for electric fields, Eq. (22.8):

Flux of electric field through a closed surface

Gauss's law for $\vec{E}$:
$$\oint \vec{E} \cdot d\vec{A} = \frac{Q_\text{encl}}{\epsilon_0} \tag{29.18}$$
Charge enclosed by surface
Electric constant

The second is the analogous relationship for *magnetic* fields, Eq. (27.8):

Flux of magnetic field through any closed surface ...

Gauss's law for $\vec{B}$:
$$\oint \vec{B} \cdot d\vec{A} = 0 \tag{29.19}$$
... equals zero.

This statement means, among other things, that there are no magnetic monopoles (single magnetic charges) to act as sources of magnetic field.

The third and fourth equations involve a line integral of $\vec{E}$ or $\vec{B}$ around a closed path. Faraday's law states that a changing magnetic flux acts as a source of electric field:

Line integral of electric field around path

Faraday's law for a stationary integration path:
$$\oint \vec{E} \cdot d\vec{l} = -\frac{d\Phi_B}{dt} \tag{29.20}$$
Negative of the time rate of change of magnetic flux through path

If there is a changing magnetic field, the line integral in Eq. (29.20)—which must be carried out over a *stationary* closed path—is not zero. Thus the $\vec{E}$ field produced by a changing $\vec{B}$ is not conservative.

The fourth and final equation is Ampere's law including displacement current. It states that both a conduction current and a changing electric flux act as sources of magnetic field:

Line integral of magnetic field around path · Electric constant · Time rate of change of electric flux through path

Ampere's law for a stationary integration path:

$$\oint \vec{B} \cdot d\vec{l} = \mu_0 \left(i_C + \epsilon_0 \frac{d\Phi_E}{dt} \right)_{encl} \tag{29.21}$$

Magnetic constant · Conduction current through path · Displacement current through path

It's worthwhile to look more carefully at the electric field $\vec{E}$ and its role in Maxwell's equations. In general, the total $\vec{E}$ field at a point in space can be the superposition of an electrostatic field $\vec{E}_c$ caused by a distribution of charges at rest and a magnetically induced, nonelectrostatic field $\vec{E}_n$. That is,

$$\vec{E} = \vec{E}_c + \vec{E}_n$$

The electrostatic part $\vec{E}_c$ is *always* conservative, so $\oint \vec{E}_c \cdot d\vec{l} = 0$. This conservative part of the field does not contribute to the integral in Faraday's law, so we can take $\vec{E}$ in Eq. (29.20) to be the *total* electric field $\vec{E}$, including both the part $\vec{E}_c$ due to charges and the magnetically induced part $\vec{E}_n$. Similarly, the nonconservative part $\vec{E}_n$ of the $\vec{E}$ field does not contribute to the integral in Gauss's law, because this part of the field is not caused by static charges. Hence $\oint \vec{E}_n \cdot d\vec{A}$ is always zero. We conclude that in all the Maxwell equations, $\vec{E}$ is the total electric field; these equations don't distinguish between conservative and nonconservative fields.

Symmetry in Maxwell's Equations

There is a remarkable symmetry in Maxwell's four equations. In empty space where there is no charge, the first two equations (Eqs. (29.18) and (29.19)) are identical in form, one containing $\vec{E}$ and the other containing $\vec{B}$ (**Fig. 29.24**). When we compare the second two equations, Eq. (29.20) says that a changing magnetic flux creates an electric field, and Eq. (29.21) says that a changing electric flux creates a magnetic field. In empty space, where there is no conduction current, $i_C = 0$ and the two equations have the same form, apart from a numerical constant and a negative sign, with the roles of $\vec{E}$ and $\vec{B}$ exchanged.

We can rewrite Eqs. (29.20) and (29.21) in a different but equivalent form by introducing the definitions of magnetic flux, $\Phi_B = \int \vec{B} \cdot d\vec{A}$, and electric flux, $\Phi_E = \int \vec{E} \cdot d\vec{A}$, respectively. In empty space, where there is no charge or conduction current, $i_C = 0$ and $Q_{encl} = 0$, and we have

$$\oint \vec{E} \cdot d\vec{l} = -\frac{d}{dt} \int \vec{B} \cdot d\vec{A} \tag{29.22}$$

$$\oint \vec{B} \cdot d\vec{l} = \mu_0 \epsilon_0 \frac{d}{dt} \int \vec{E} \cdot d\vec{A} \tag{29.23}$$

Again we notice the symmetry between the roles of $\vec{E}$ and $\vec{B}$ in these expressions.

The most remarkable feature of these equations is that a time-varying field of *either* kind induces a field of the other kind in neighboring regions of space. Maxwell recognized that these relationships predict the existence of electromagnetic disturbances consisting of time-varying electric and magnetic fields that travel or *propagate* from one region of space to another, even if no matter is present in the intervening space. Such disturbances, called *electromagnetic waves,* provide the physical basis for light, radio and television waves, infrared, ultraviolet, and x rays. We will return to this vitally important topic in Chapter 32.

29.24 Maxwell's equations in empty space are highly symmetric.

In empty space there are no charges, so the fluxes of $\vec{E}$ and $\vec{B}$ through any closed surface are equal to zero.

$$\oint \vec{E} \cdot d\vec{A} = 0$$

$$\oint \vec{B} \cdot d\vec{A} = 0$$

$$\oint \vec{E} \cdot d\vec{l} = -\frac{d\Phi_B}{dt}$$

$$\oint \vec{B} \cdot d\vec{l} = \mu_0 \epsilon_0 \frac{d\Phi_E}{dt}$$

In empty space there are no conduction currents, so the line integrals of $\vec{E}$ and $\vec{B}$ around any closed path are related to the rate of change of flux of the other field.

Although it may not be obvious, *all* the basic relationships between fields and their sources are contained in Maxwell's equations. We can derive Coulomb's law from Gauss's law, we can derive the law of Biot and Savart from Ampere's law, and so on. When we add the equation that defines the $\vec{E}$ and $\vec{B}$ fields in terms of the forces that they exert on a charge q, namely,

$$\vec{F} = q(\vec{E} + \vec{v} \times \vec{B}) \tag{29.24}$$

we have *all* the fundamental relationships of electromagnetism!

Maxwell's equations would have even greater symmetry between the $\vec{E}$ and $\vec{B}$ fields if single magnetic charges (magnetic monopoles) existed. The right side of Eq. (29.19) would contain the total *magnetic* charge enclosed by the surface, and the right side of Eq. (29.20) would include a magnetic monopole current term. However, no magnetic monopoles have yet been found.

In conciseness and generality, Maxwell's equations are in the same league with Newton's laws of motion and the laws of thermodynamics. Indeed, a major goal of science is learning how to express very broad and general relationships in a concise and compact form. Maxwell's synthesis of electromagnetism stands as a towering intellectual achievement, comparable to the Newtonian synthesis we described at the end of Section 13.5 and to the development of relativity and quantum mechanics in the 20th century.

TEST YOUR UNDERSTANDING OF SECTION 29.7 (a) Which of Maxwell's equations explains how a credit card reader works? (b) Which one describes how a wire carrying a steady current generates a magnetic field? ❚

29.8 SUPERCONDUCTIVITY

The most familiar property of a superconductor is the sudden disappearance of all electrical resistance when the material is cooled below a temperature called the *critical temperature,* denoted by T_c. We discussed this behavior and the circumstances of its discovery in Section 25.2. But superconductivity is far more than just the absence of measurable resistance. As we'll see in this section, superconductors also have extraordinary *magnetic* properties.

The first hint of unusual magnetic properties was the discovery that for any superconducting material the critical temperature T_c changes when the material is placed in an externally produced magnetic field $\vec{B}_0$. **Figure 29.25** shows this dependence for mercury, the first element in which superconductivity was observed. As the external field magnitude B_0 increases, the superconducting transition occurs at lower and lower temperature. When B_0 is greater than 0.0412 T, *no* superconducting transition occurs. The minimum magnitude of magnetic field that is needed to eliminate superconductivity at a temperature below T_c is called the *critical field,* denoted by B_c.

The Meissner Effect

Another aspect of the magnetic behavior of superconductors appears if we place a homogeneous sphere of a superconducting material in a uniform applied magnetic field $\vec{B}_0$ at a temperature T greater than T_c. The material is then in the normal phase, not the superconducting phase (**Fig. 29.26a**). Now we lower the temperature until the superconducting transition occurs. (We assume that the magnitude of $\vec{B}_0$ is not large enough to prevent the phase transition.) What happens to the field?

Measurements of the field outside the sphere show that the field lines become distorted as in Fig. 29.26b. There is no longer any field inside the material, except possibly in a very thin surface layer a hundred or so atoms thick. If a coil is wrapped

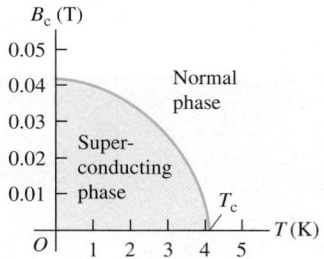

29.25 Phase diagram for pure mercury, showing the critical magnetic field B_c and its dependence on temperature. Superconductivity is impossible above the critical temperature T_c. The curves for other superconducting materials are similar but with different numerical values.

29.26 A superconducting material (a) above the critical temperature and (b), (c) below the critical temperature.

(a) Superconducting material in an external magnetic field $\vec{B}_0$ at $T > T_c$

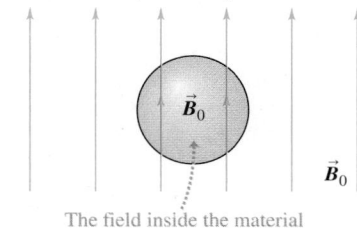

The field inside the material is very nearly equal to $\vec{B}_0$.

(b) The temperature is lowered to $T < T_c$, so the material becomes superconducting.

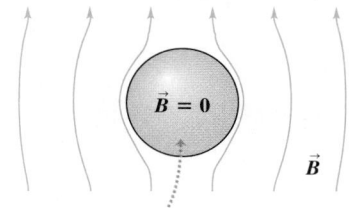

Magnetic flux is expelled from the material, and the field inside it is zero (Meissner effect).

(c) When the external field is turned off at $T < T_c$, the field is zero everywhere.

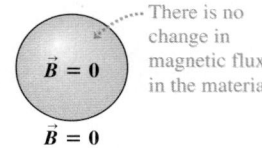

There is no change in magnetic flux in the material.

$\vec{B} = 0$

around the sphere, the emf induced in the coil shows that during the superconducting transition the magnetic flux through the coil decreases from its initial value to zero; this is consistent with the absence of field inside the material. Finally, if the field is now turned off while the material is still in its superconducting phase, no emf is induced in the coil, and measurements show no field outside the sphere (Fig. 29.26c).

We conclude that during a superconducting transition in the presence of the field $\vec{B}_0$, all of the magnetic flux is expelled from the bulk of the sphere, and the magnetic flux Φ_B through the coil becomes zero. This expulsion of magnetic flux is called the *Meissner effect*. As Fig. 29.26b shows, this expulsion crowds the magnetic field lines closer together to the side of the sphere, increasing $\vec{B}$ there.

Superconductor Levitation and Other Applications

The diamagnetic nature of a superconductor has some interesting *mechanical* consequences. A paramagnetic or ferromagnetic material is attracted by a permanent magnet because the magnetic dipoles in the material align with the nonuniform magnetic field of the permanent magnet. (We discussed this in Section 27.7.) For a diamagnetic material the magnetization is in the opposite sense, and a diamagnetic material is *repelled* by a permanent magnet. By Newton's third law the magnet is also repelled by the diamagnetic material. **Figure 29.27** shows the repulsion between a specimen of a high-temperature superconductor and a magnet; the magnet is supported ("levitated") by this repulsive magnetic force.

The behavior we have described is characteristic of what are called *type-I superconductors*. There is another class of superconducting materials called *type-II superconductors*. When such a material in the superconducting phase is placed in a magnetic field, the bulk of the material remains superconducting, but thin filaments of material, running parallel to the field, may return to the normal phase. Currents circulate around the boundaries of these filaments, and there *is* magnetic flux inside them. Type-II superconductors are used for electromagnets because they usually have much larger values of B_c than do type-I materials, permitting much larger magnetic fields without destroying the superconducting state. Type-II superconductors have *two* critical magnetic fields: The first, B_{c1}, is the field at which magnetic flux begins to enter the material, forming the filaments just described, and the second, B_{c2}, is the field at which the material becomes normal.

Superconducting electromagnets are in everyday use not only in research laboratories but also in medical MRI (magnetic resonance imaging) scanners. As we described in Section 27.7, scanning a patient through MRI requires a strong magnetic field to align the magnetic dipoles of the patient's atomic nuclei. A steady field of 1.5 T or more is needed, which is very difficult to produce with a conventional electromagnet, since this would require very high currents and hence very high energy losses due to resistance in the electromagnet coils. But with a superconducting electromagnet there is no resistive energy loss, and fields up to 10 T can routinely be attained.

Very sensitive measurements of magnetic fields can be made with superconducting quantum interference devices (SQUIDs), which can detect changes in magnetic flux of less than 10^{-14} Wb; these devices have applications in medicine, geology, and other fields. The number of potential uses for superconductors has increased greatly since the discovery in 1987 of high-temperature superconductors. These materials have critical temperatures that are above the temperature of liquid nitrogen (about 77 K) and so are comparatively easy to attain. Development of practical applications of superconductor science promises to be an exciting chapter in contemporary technology.

29.27 A superconductor exerts a repulsive force on a magnet, supporting the magnet in midair.

Magnet

Superconductor

Faraday's law: Faraday's law states that the induced emf in a closed loop equals the negative of the time rate of change of magnetic flux through the loop. This relationship is valid whether the flux change is caused by a changing magnetic field, motion of the loop, or both. (See Examples 29.1–29.6.)

$$\mathcal{E} = -\frac{d\Phi_B}{dt} \qquad (29.3)$$

The magnet's motion causes a *changing* magnetic field through the coil, inducing a current in the coil.

Lenz's law: Lenz's law states that an induced current or emf always tends to oppose or cancel out the change that caused it. Lenz's law can be derived from Faraday's law and is often easier to use. (See Examples 29.7 and 29.8.)

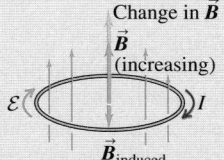

Motional emf: If a conductor moves in a magnetic field, a motional emf is induced. (See Examples 29.9 and 29.10.)

$$\mathcal{E} = vBL \qquad (29.6)$$
(conductor with length L moves in uniform $\vec{B}$ field, $\vec{L}$ and $\vec{v}$ both perpendicular to $\vec{B}$ and to each other)

$$\mathcal{E} = \oint (\vec{v} \times \vec{B}) \cdot d\vec{l} \qquad (29.7)$$

(all or part of a closed loop moves in a $\vec{B}$ field)

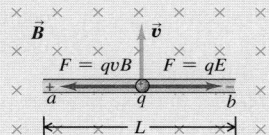
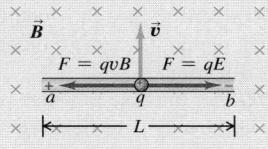

Induced electric fields: When an emf is induced by a changing magnetic flux through a stationary conductor, there is an induced electric field $\vec{E}$ of nonelectrostatic origin. This field is nonconservative and cannot be associated with a potential. (See Example 29.11.)

$$\oint \vec{E} \cdot d\vec{l} = -\frac{d\Phi_B}{dt} \qquad (29.10)$$

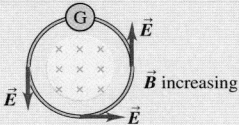

Displacement current and Maxwell's equations: A time-varying electric field generates displacement current i_D, which acts as a source of magnetic field in exactly the same way as conduction current. The relationships between electric and magnetic fields and their sources can be stated compactly in four equations, called Maxwell's equations. Together they form a complete basis for the relationship of $\vec{E}$ and $\vec{B}$ fields to their sources.

$$i_D = \epsilon \frac{d\Phi_E}{dt} \qquad (29.14)$$
(displacement current)

$$\oint \vec{E} \cdot d\vec{A} = \frac{Q_{encl}}{\epsilon_0} \qquad (29.18)$$
(Gauss's law for $\vec{E}$ fields)

$$\oint \vec{B} \cdot d\vec{A} = 0 \qquad (29.19)$$
(Gauss's law for $\vec{B}$ fields)

$$\oint \vec{E} \cdot d\vec{l} = -\frac{d\Phi_B}{dt} \qquad (29.20)$$
(Faraday's law)

$$\oint \vec{B} \cdot d\vec{l} = \mu_0 \left(i_C + \epsilon_0 \frac{d\Phi_E}{dt} \right)_{encl} \qquad (29.21)$$

(Ampere's law including displacement current)

BRIDGING PROBLEM A FALLING SQUARE LOOP

A vertically oriented square loop of copper wire falls from rest in a region in which the field $\vec{B}$ is horizontal, uniform, and perpendicular to the plane of the loop, into a field-free region (**Fig. 29.28**). The side length of the loop is s and the wire diameter is d. The resistivity of copper is ρ_R and the density of copper is ρ_m. If the loop reaches its terminal speed while its upper segment is still in the magnetic-field region, find an expression for the terminal speed.

29.28 A wire loop falling in a horizontal magnetic field $\vec{B}$. The plane of the loop is perpendicular to $\vec{B}$.

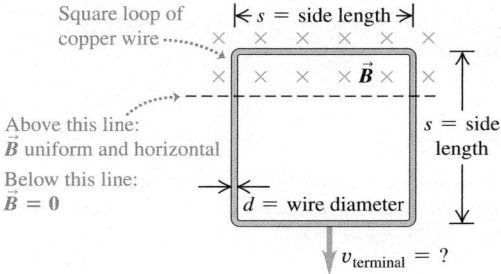

SOLUTION GUIDE

IDENTIFY and SET UP

1. The motion of the loop through the magnetic field induces an emf and a current in the loop. The field then gives rise to a magnetic force on this current that opposes the downward force of gravity. The loop reaches terminal speed (it no longer accelerates) when the upward magnetic force balances the downward force of gravity.

2. Consider the case in which the entire loop is in the magnetic-field region. Is there an induced emf in this case? If so, what is its direction?

3. Consider the case in which only the upper segment of the loop is in the magnetic-field region. Is there an induced emf in this case? If so, what is its direction?

4. For the case in which there is an induced emf and hence an induced current, what is the direction of the magnetic force on each of the four sides of the loop? What is the direction of the *net* magnetic force on the loop?

EXECUTE

5. For the case in which the loop is falling at speed v and there is an induced emf, find (i) the emf, (ii) the induced current, and (iii) the magnetic force on the loop in terms of its resistance R.

6. Find R and the mass of the loop in terms of the given information about the loop.

7. Use your results from steps 5 and 6 to find an expression for the terminal speed.

EVALUATE

8. How does the terminal speed depend on the magnetic-field magnitude B? Explain why this makes sense.

Problems For assigned homework and other learning materials, go to MasteringPhysics®.

•, ••, •••: Difficulty levels. **CP:** Cumulative problems incorporating material from earlier chapters. **CALC:** Problems requiring calculus.
DATA: Problems involving real data, scientific evidence, experimental design, and/or statistical reasoning. **BIO:** Biosciences problems.

DISCUSSION QUESTIONS

Q29.1 A sheet of copper is placed between the poles of an electromagnet with the magnetic field perpendicular to the sheet. When the sheet is pulled out, a considerable force is required, and the force required increases with speed. Explain. Is a force required also when the sheet is inserted between the poles? Explain.

Q29.2 In Fig. 29.8, if the angular speed ω of the loop is doubled, then the frequency with which the induced current changes direction doubles, and the maximum emf also doubles. Why? Does the torque required to turn the loop change? Explain.

Q29.3 Two circular loops lie side by side in the same plane. One is connected to a source that supplies an increasing current; the other is a simple closed ring. Is the induced current in the ring in the same direction as the current in the loop connected to the source, or opposite? What if the current in the first loop is decreasing? Explain.

Q29.4 For Eq. (29.6), show that if v is in meters per second, B in teslas, and L in meters, then the units of the right-hand side of the equation are joules per coulomb or volts (the correct SI units for $\mathcal{E}$).

Q29.5 A long, straight conductor passes through the center of a metal ring, perpendicular to its plane. If the current in the conductor increases, is a current induced in the ring? Explain.

Q29.6 A student asserted that if a permanent magnet is dropped down a vertical copper pipe, it eventually reaches a terminal velocity even if there is no air resistance. Why should this be? Or should it?

Q29.7 An airplane is in level flight over Antarctica, where the magnetic field of the earth is mostly directed upward away from the ground. As viewed by a passenger facing toward the front of the plane, is the left or the right wingtip at higher potential? Does your answer depend on the direction the plane is flying?

Q29.8 Consider the situation in Exercise 29.21. In part (a), find the direction of the force that the large circuit exerts on the small one. Explain how this result is consistent with Lenz's law.

Q29.9 A metal rectangle is close to a long, straight, current-carrying wire, with two of its sides parallel to the wire. If the current in the long wire is decreasing, is the rectangle repelled by or attracted to the wire? Explain why this result is consistent with Lenz's law.

Q29.10 A square conducting loop is in a region of uniform, constant magnetic field. Can the loop be rotated about an axis along one side and no emf be induced in the loop? Discuss, in terms of the orientation of the rotation axis relative to the magnetic-field direction.

Q29.11 Example 29.6 discusses the external force that must be applied to the slidewire to move it at constant speed. If there were a break in the left-hand end of the U-shaped conductor, how much force would be needed to move the slidewire at constant speed? As in the example, you can ignore friction.

Q29.12 In the situation shown in Fig. 29.18, would it be appropriate to ask how much *energy* an electron gains during a complete trip around the wire loop with current I'? Would it be appropriate to ask what *potential difference* the electron moves through during such a complete trip? Explain your answers.

Q29.13 A metal ring is oriented with the plane of its area perpendicular to a spatially uniform magnetic field that increases at a steady rate. If the radius of the ring is doubled, by what factor do (a) the emf induced in the ring and (b) the electric field induced in the ring change?

Q29.14 Small one-cylinder gasoline engines sometimes use a device called a *magneto* to supply current to the spark plug. A permanent magnet is attached to the flywheel, and a stationary coil is mounted adjacent to it. Explain how this device is able to generate current. What happens when the magnet passes the coil?

Q29.15 Does Lenz's law say that the induced current in a metal loop always flows to oppose the magnetic flux through that loop? Explain.

Q29.16 Does Faraday's law say that a large magnetic flux induces a large emf in a coil? Explain.

Q29.17 Can one have a displacement current as well as a conduction current within a conductor? Explain.

Q29.18 Your physics study partner asks you to consider a parallel-plate capacitor that has a dielectric completely filling the volume between the plates. He then claims that Eqs. (29.13) and (29.14) show that the conduction current in the dielectric equals the displacement current in the dielectric. Do you agree? Explain.

Q29.19 Match the mathematical statements of Maxwell's equations as given in Section 29.7 to these verbal statements. (a) Closed electric field lines are evidently produced only by changing magnetic flux. (b) Closed magnetic field lines are produced both by the motion of electric charge and by changing electric flux. (c) Electric field lines can start on positive charges and end on negative charges. (d) Evidently there are no magnetic monopoles on which to start and end magnetic field lines.

Q29.20 If magnetic monopoles existed, the right-hand side of Eq. (29.20) would include a term proportional to the current of magnetic monopoles. Suppose a steady monopole current is moving in a long straight wire. Sketch the *electric* field lines that such a current would produce.

Q29.21 A type-II superconductor in an external field between B_{c1} and B_{c2} has regions that contain magnetic flux and have resistance, and also has superconducting regions. What is the resistance of a long, thin cylinder of such material?

EXERCISES

Section 29.2 Faraday's Law

29.1 • A single loop of wire with an area of 0.0900 m^2 is in a uniform magnetic field that has an initial value of 3.80 T, is perpendicular to the plane of the loop, and is decreasing at a constant rate of 0.190 T/s. (a) What emf is induced in this loop? (b) If the loop has a resistance of 0.600 Ω, find the current induced in the loop.

29.2 •• In a physics laboratory experiment, a coil with 200 turns enclosing an area of 12 cm^2 is rotated in 0.040 s from a position where its plane is perpendicular to the earth's magnetic field to a position where its plane is parallel to the field. The earth's magnetic field at the lab location is 6.0×10^{-5} T. (a) What is the total magnetic flux through the coil before it is rotated? After it is rotated? (b) What is the average emf induced in the coil?

29.3 •• **Search Coils and Credit Cards.** One practical way to measure magnetic field strength uses a small, closely wound coil called a *search coil*. The coil is initially held with its plane perpendicular to a magnetic field. The coil is then either quickly rotated a quarter-turn about a diameter or quickly pulled out of the field. (a) Derive the equation relating the total charge Q that flows through a search coil to the magnetic-field magnitude B. The search coil has N turns, each with area A, and the flux through the coil is decreased from its initial maximum value to zero in a time Δt. The resistance of the coil is R, and the total charge is $Q = I\Delta t$, where I is the average current induced by the change in flux. (b) In a credit card reader, the magnetic strip on the back of a credit card is rapidly "swiped" past a coil within the reader. Explain, using the same ideas that underlie the operation of a search coil, how the reader can decode the information stored in the pattern of magnetization on the strip. (c) Is it necessary that the credit card be "swiped" through the reader at exactly the right speed? Why or why not?

29.4 • A closely wound search coil (see Exercise 29.3) has an area of 3.20 cm^2, 120 turns, and a resistance of 60.0 Ω. It is connected to a charge-measuring instrument whose resistance is 45.0 Ω. When the coil is rotated quickly from a position parallel to a uniform magnetic field to a position perpendicular to the field, the instrument indicates a charge of 3.56×10^{-5} C. What is the magnitude of the field?

29.5 • A circular loop of wire with a radius of 12.0 cm and oriented in the horizontal xy-plane is located in a region of uniform magnetic field. A field of 1.5 T is directed along the positive z-direction, which is upward. (a) If the loop is removed from the field region in a time interval of 2.0 ms, find the average emf that will be induced in the wire loop during the extraction process. (b) If the coil is viewed looking down on it from above, is the induced current in the loop clockwise or counterclockwise?

29.6 • CALC A coil 4.00 cm in radius, containing 500 turns, is placed in a uniform magnetic field that varies with time according to $B = (0.0120 \text{ T/s})t + (3.00 \times 10^{-5} \text{ T/s}^4)t^4$. The coil is connected to a 600-Ω resistor, and its plane is perpendicular to the magnetic field. You can ignore the resistance of the coil. (a) Find the magnitude of the induced emf in the coil as a function of time. (b) What is the current in the resistor at time $t = 5.00$ s?

29.7 • CALC The current in the long, straight wire AB shown in **Fig. E29.7** is upward and is increasing steadily at a rate di/dt. (a) At an instant when the current is i, what are the magnitude and direction of the field $\vec{B}$ at a distance r to the right of the wire? (b) What is the flux $d\Phi_B$ through the narrow, shaded strip? (c) What is the total flux through the loop? (d) What is the induced emf in the loop? (e) Evaluate the numerical value of the induced emf if $a = 12.0$ cm, $b = 36.0$ cm, $L = 24.0$ cm, and $di/dt = 9.60$ A/s.

Figure **E29.7**

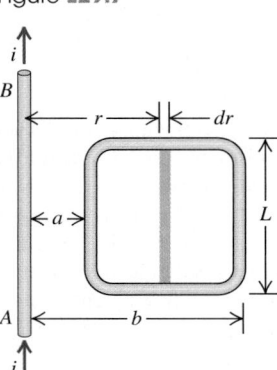

29.8 • CALC A flat, circular, steel loop of radius 75 cm is at rest in a uniform magnetic field, as shown in an edge-on view in **Fig. E29.8**. The field is changing with time, according to $B(t) = (1.4 \text{ T})e^{-(0.057 \text{ s}^{-1})t}$. (a) Find the emf induced in the loop as a function of time. (b) When is the induced emf equal to $\frac{1}{10}$ of its initial value? (c) Find the direction of the current induced in the loop, as viewed from above the loop.

Figure **E29.8**

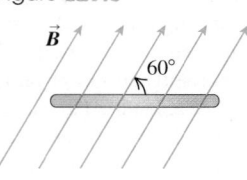

29.9 • **Shrinking Loop.** A circular loop of flexible iron wire has an initial circumference of 165.0 cm, but its circumference is decreasing at a constant rate of 12.0 cm/s due to a tangential pull on the wire. The loop is in a constant, uniform magnetic field oriented perpendicular to the plane of the loop and with magnitude 0.500 T. (a) Find the emf induced in the loop at the instant when 9.0 s have passed. (b) Find the direction of the induced current in the loop as viewed looking along the direction of the magnetic field.

29.10 • A closely wound rectangular coil of 80 turns has dimensions of 25.0 cm by 40.0 cm. The plane of the coil is rotated from a position where it makes an angle of 37.0° with a magnetic field of 1.70 T to a position perpendicular to the field. The rotation takes 0.0600 s. What is the average emf induced in the coil?

29.11 • CALC In a region of space, a magnetic field points in the $+x$-direction (toward the right). Its magnitude varies with position according to the formula $B_x = B_0 + bx$, where B_0 and b are positive constants, for $x \geq 0$. A flat coil of area A moves with uniform speed v from right to left with the plane of its area always perpendicular to this field. (a) What is the emf induced in this coil while it is to the right of the origin? (b) As viewed from the origin, what is the direction (clockwise or counterclockwise) of the current induced in the coil? (c) If instead the coil moved from left to right, what would be the answers to parts (a) and (b)?

29.12 • In many magnetic resonance imaging (MRI) systems, the magnetic field is produced by a superconducting magnet that must be kept cooled below the superconducting transition temperature. If the cryogenic cooling system fails, the magnet coils may lose their superconductivity and the strength of the magnetic field will rapidly decrease, or *quench*. The dissipation of energy as heat in the now-nonsuperconducting magnet coils can cause a rapid boil-off of the cryogenic liquid (usually liquid helium) that is used for cooling. Consider a superconducting MRI magnet for which the magnetic field decreases from 8.0 T to nearly 0 in 20 s. What is the average emf induced in a circular wedding ring of diameter 2.2 cm if the ring is at the center of the MRI magnet coils and the original magnetic field is perpendicular to the plane that is encircled by the ring?

29.13 •• The armature of a small generator consists of a flat, square coil with 120 turns and sides with a length of 1.60 cm. The coil rotates in a magnetic field of 0.0750 T. What is the angular speed of the coil if the maximum emf produced is 24.0 mV?

29.14 • A flat, rectangular coil of dimensions l and w is pulled with uniform speed v through a uniform magnetic field B with the plane of its area perpendicular to the field (**Fig. E29.14**). (a) Find the emf induced in this coil. (b) If the speed and magnetic field are both tripled, what is the induced emf?

Figure **E29.14**

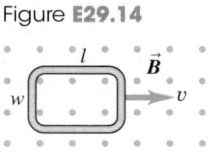

Section 29.3 Lenz's Law

29.15 • A circular loop of wire is in a region of spatially uniform magnetic field, as shown in **Fig. E29.15**. The magnetic field is directed into the plane of the figure. Determine the direction (clockwise or counterclockwise) of the induced current in the loop when (a) B is increasing; (b) B is decreasing; (c) B is constant with value B_0. Explain your reasoning.

Figure **E29.15**

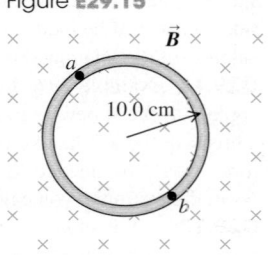

29.16 • The current I in a long, straight wire is constant and is directed toward the right as in **Fig. E29.16**. Conducting loops A, B, C, and D are moving, in the directions shown, near the wire. (a) For each loop, is the direction of the induced current clockwise or counterclockwise, or is the induced current zero? (b) For each loop, what is the direction of the net force that the wire exerts on the loop? Give your reasoning for each answer.

Figure **E29.16**

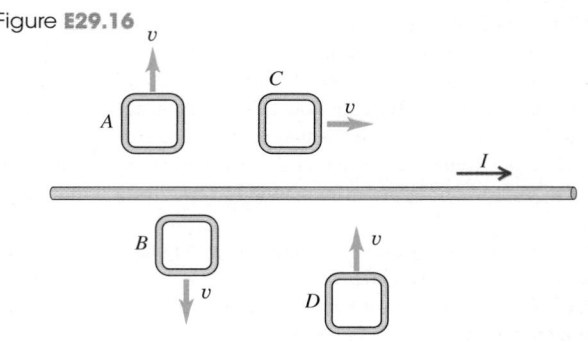

29.17 • Two closed loops A and C are close to a long wire carrying a current I (**Fig. E29.17**). (a) Find the direction (clockwise or counterclockwise) of the current induced in each loop if I is steadily decreasing. (b) While I is decreasing, what is the direction of the net force that the wire exerts on each loop? Explain how you obtain your answer.

Figure **E29.17**

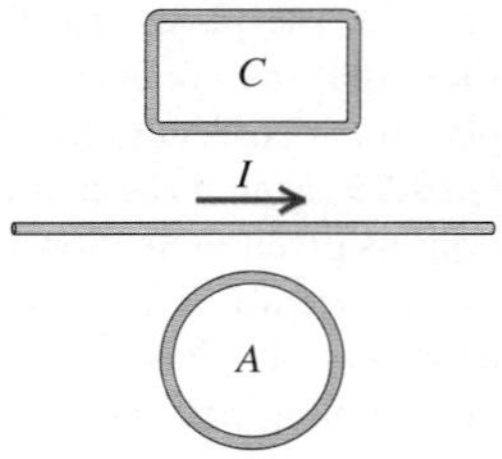

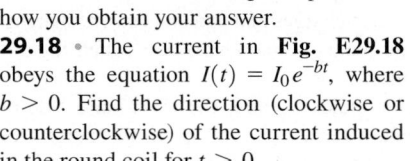

29.18 • The current in **Fig. E29.18** obeys the equation $I(t) = I_0 e^{-bt}$, where $b > 0$. Find the direction (clockwise or counterclockwise) of the current induced in the round coil for $t > 0$.

Figure **E29.18**

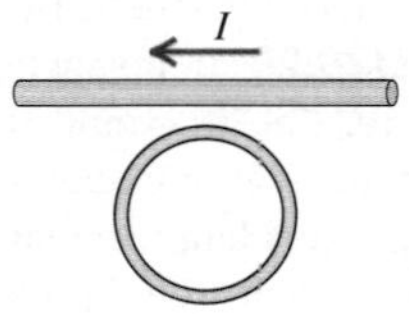

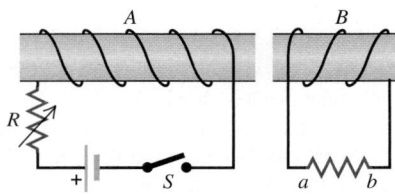

29.19 • Using Lenz's law, determine the direction of the current in resistor ab of **Fig. E29.19** when (a) switch S is opened after having been closed for several minutes; (b) coil B is brought closer to coil A with the switch closed; (c) the resistance of R is decreased while the switch remains closed.

Figure **E29.19**

29.20 • A cardboard tube is wrapped with two windings of insulated wire wound in opposite directions, as shown in **Fig. E29.20**. Terminals *a* and *b* of winding *A* may be connected to a battery through a reversing switch. State whether the induced current in the resistor *R* is from left to right or from right to left in the following circumstances: (a) the current in winding *A* is from *a* to *b* and is increasing; (b) the current in winding *A* is from *b* to *a* and is decreasing; (c) the current in winding *A* is from *b* to *a* and is increasing.

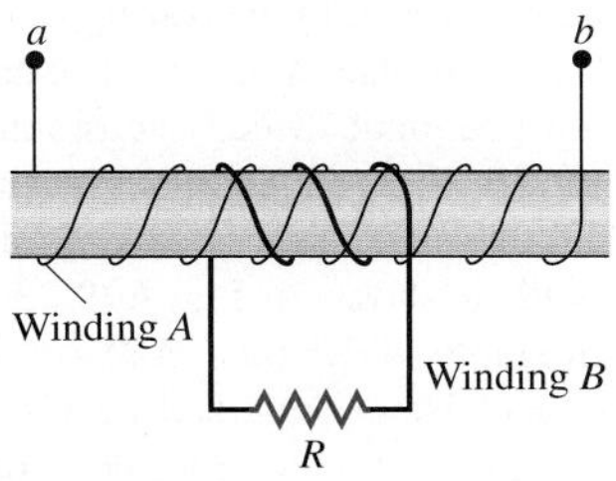

Figure **E29.20**

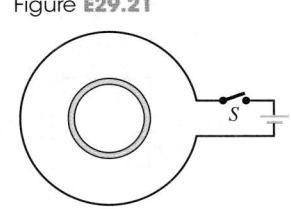

29.21 • A small, circular ring is inside a larger loop that is connected to a battery and a switch (**Fig. E29.21**). Use Lenz's law to find the direction of the current induced in the small ring (a) just after switch *S* is closed; (b) after *S* has been closed a long time; (c) just after *S* has been reopened after it was closed for a long time.

Figure **E29.21**

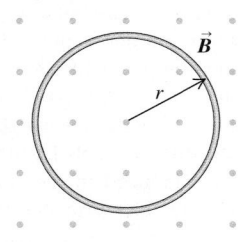

29.22 • A circular loop of wire with radius *r* = 0.0480 m and resistance *R* = 0.160 Ω is in a region of spatially uniform magnetic field, as shown in **Fig. E29.22**. The magnetic field is directed out of the plane of the figure. The magnetic field has an initial value of 8.00 T and is decreasing at a rate of $dB/dt = -0.680$ T/s. (a) Is the induced current in the loop clockwise or counterclockwise? (b) What is the rate at which electrical energy is being dissipated by the resistance of the loop?

Figure **E29.22**

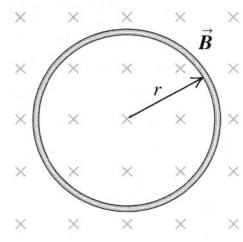

29.23 • **CALC** A circular loop of wire with radius *r* = 0.0250 m and resistance *R* = 0.390 Ω is in a region of spatially uniform magnetic field, as shown in **Fig. E29.23**. The magnetic field is directed into the plane of the figure. At *t* = 0, *B* = 0. The magnetic field then begins increasing, with $B(t) = (0.380 \text{ T/s}^3)t^3$. What is the current in the loop (magnitude and direction) at the instant when *B* = 1.33 T?

Figure **E29.23**

Section 29.4 Motional Electromotive Force

29.24 • A rectangular loop of wire with dimensions 1.50 cm by 8.00 cm and resistance *R* = 0.600 Ω is being pulled to the right out of a region of uniform magnetic field. The magnetic field has magnitude *B* = 2.40 T and is directed into the plane of **Fig. E29.24**. At the instant when the speed of the loop is 3.00 m/s and it is still

Figure **E29.24**

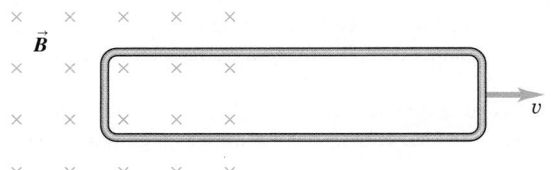

partially in the field region, what force (magnitude and direction) does the magnetic field exert on the loop?

29.25 • In **Fig. E29.25** a conducting rod of length *L* = 30.0 cm moves in a magnetic field $\vec{B}$ of magnitude 0.450 T directed into the plane of the figure. The rod moves with speed *v* = 5.00 m/s in the direction shown. (a) What is the potential difference between the ends of the rod? (b) Which point, *a* or *b*, is at higher potential? (c) When the charges in the rod are in equilibrium, what are the magnitude and direction of the electric field within the rod? (d) When the charges in the rod are in equilibrium, which point, *a* or *b*, has an excess of positive charge? (e) What is the potential difference across the rod if it moves (i) parallel to *ab* and (ii) directly out of the page?

Figure **E29.25**

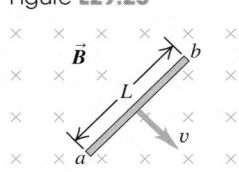

29.26 • A rectangle measuring 30.0 cm by 40.0 cm is located inside a region of a spatially uniform magnetic field of 1.25 T, with the field perpendicular to the plane of the coil (**Fig. E29.26**). The coil is pulled out at a steady rate of 2.00 cm/s traveling perpendicular to the field lines. The region of the field ends abruptly as shown. Find the emf induced in this coil when it is (a) all inside the field; (b) partly inside the field; (c) all outside the field.

Figure **E29.26**

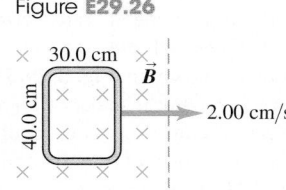

29.27 • **Are Motional emfs a Practical Source of Electricity?** How fast (in m/s and mph) would a 5.00-cm copper bar have to move at right angles to a 0.650-T magnetic field to generate 1.50 V (the same as a AA battery) across its ends? Does this seem like a practical way to generate electricity?

29.28 • **Motional emfs in Transportation.** Airplanes and trains move through the earth's magnetic field at rather high speeds, so it is reasonable to wonder whether this field can have a substantial effect on them. We shall use a typical value of 0.50 G for the earth's field. (a) The French TGV train and the Japanese "bullet train" reach speeds of up to 180 mph moving on tracks about 1.5 m apart. At top speed moving perpendicular to the earth's magnetic field, what potential difference is induced across the tracks as the wheels roll? Does this seem large enough to produce noticeable effects? (b) The Boeing 747-400 aircraft has a wingspan of 64.4 m and a cruising speed of 565 mph. If there is no wind blowing (so that this is also their speed relative to the ground), what is the maximum potential difference that could be induced between the opposite tips of the wings? Does this seem large enough to cause problems with the plane?

29.29 • The conducting rod *ab* shown in **Fig. E29.29** makes contact with metal rails *ca* and *db*. The apparatus is in a uniform magnetic field of 0.800 T, perpendicular to the plane of the figure. (a) Find the magnitude of the emf induced in the rod when it is moving toward the right with a speed 7.50 m/s. (b) In what direction does the current flow in the rod? (c) If the resistance of the circuit *abdc* is 1.50 Ω (assumed to be constant), find the force (magnitude and direction) required to keep the rod moving to the right with a constant speed of 7.50 m/s. You can ignore friction. (d) Compare the rate at which mechanical work is done by the force (Fv) with the rate at which thermal energy is developed in the circuit (I^2R).

Figure **E29.29**

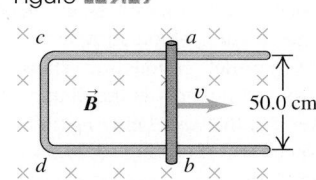

29.30 • A 0.650-m-long metal bar is pulled to the right at a steady 5.0 m/s perpendicular to a uniform, 0.750-T magnetic field. The bar rides on parallel metal rails connected through a 25.0-Ω resistor (**Fig. E29.30**), so the apparatus

Figure **E29.30**

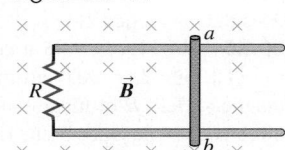

makes a complete circuit. Ignore the resistance of the bar and the rails. (a) Calculate the magnitude of the emf induced in the circuit. (b) Find the direction of the current induced in the circuit by using (i) the magnetic force on the charges in the moving bar; (ii) Faraday's law; (iii) Lenz's law. (c) Calculate the current through the resistor.

29.31 • A 0.360-m-long metal bar is pulled to the left by an applied force *F*. The bar rides on parallel metal rails connected through a 45.0-Ω resistor, as shown in **Fig. E29.31**, so the apparatus makes a complete circuit. You can ignore the resistance of the bar and

Figure **E29.31**

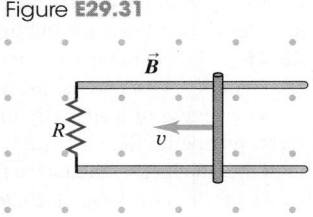

rails. The circuit is in a uniform 0.650-T magnetic field that is directed out of the plane of the figure. At the instant when the bar is moving to the left at 5.90 m/s, (a) is the induced current in the circuit clockwise or counterclockwise and (b) what is the rate at which the applied force is doing work on the bar?

29.32 • Consider the circuit shown in Fig. E29.31, but with the bar moving to the right with speed *v*. As in Exercise 29.31, the bar has length 0.360 m, *R* = 45.0 Ω, and *B* = 0.650 T. (a) Is the induced current in the circuit clockwise or counterclockwise? (b) At an instant when the 45.0-Ω resistor is dissipating electrical energy at a rate of 0.840 J/s, what is the speed of the bar?

29.33 • A 0.250-m-long bar moves on parallel rails that are connected through a 6.00-Ω resistor, as shown in **Fig. E29.33**, so the apparatus makes a complete circuit. You can ignore the resistance of the bar and rails. The cir-

Figure **E29.33**

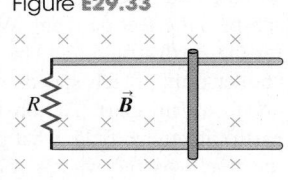

cuit is in a uniform magnetic field *B* = 1.20 T that is directed into the plane of the figure. At an instant when the induced current in the circuit is counterclockwise and equal to 1.75 A, what is the velocity of the bar (magnitude and direction)?

29.34 •• **BIO Measuring Blood Flow.** Blood contains positive and negative ions and thus is a conductor. A blood vessel, therefore, can be viewed as an electrical wire. We can even picture the flowing blood as a series of parallel conducting slabs whose thickness is the diameter *d* of the vessel moving with speed *v*. (See **Fig. E29.34**.) (a) If the blood vessel is placed in a magnetic field *B* perpendicular to the vessel, as in the figure, show that the motional potential difference induced

Figure **E29.34**

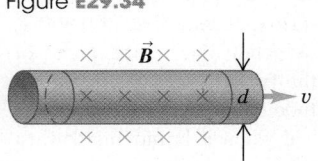

across it is $\mathcal{E} = vBd$. (b) If you expect that the blood will be flowing at 15 cm/s for a vessel 5.0 mm in diameter, what strength of magnetic field will you need to produce a potential difference of 1.0 mV? (c) Show that the volume rate of flow (*R*) of the blood

is equal to $R = \pi\mathcal{E}d/4B$. (*Note:* Although the method developed here is useful in measuring the rate of blood flow in a vessel, it is limited to use in surgery because measurement of the potential $\mathcal{E}$ must be made directly across the vessel.)

29.35 •• A rectangular circuit is moved at a constant velocity of 3.0 m/s into, through, and then out of a uniform 1.25-T magnetic field, as shown in **Fig. E29.35**. The magnetic-field region is considerably wider than 50.0 cm. Find the magnitude and direction (clockwise or counterclockwise) of the current induced in the circuit as it is (a) going into the magnetic field; (b) totally within the magnetic field, but still moving; and (c) moving out of the field. (d) Sketch a graph of the current in this circuit as a function of time, including the preceding three cases.

Figure **E29.35**

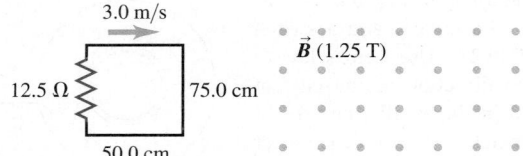

Section 29.5 Induced Electric Fields

29.36 • A metal ring 4.50 cm in diameter is placed between the north and south poles of large magnets with the plane of its area perpendicular to the magnetic field. These magnets produce an initial uniform field of 1.12 T between them but are gradually pulled apart, causing this field to remain uniform but decrease steadily at 0.250 T/s. (a) What is the magnitude of the electric field induced in the ring? (b) In which direction (clockwise or counterclockwise) does the current flow as viewed by someone on the south pole of the magnet?

29.37 • The magnetic field within a long, straight solenoid with a circular cross section and radius *R* is increasing at a rate of *dB/dt*. (a) What is the rate of change of flux through a circle with radius r_1 inside the solenoid, normal to the axis of the solenoid, and with center on the solenoid axis? (b) Find the magnitude of the induced electric field inside the solenoid, at a distance r_1 from its axis. Show the direction of this field in a diagram. (c) What is the magnitude of the induced electric field *outside* the solenoid, at a distance r_2 from the axis? (d) Graph the magnitude of the induced electric field as a function of the distance *r* from the axis from *r* = 0 to *r* = 2*R*. (e) What is the magnitude of the induced emf in a circular turn of radius *R*/2 that has its center on the solenoid axis? (f) What is the magnitude of the induced emf if the radius in part (e) is *R*? (g) What is the induced emf if the radius in part (e) is 2*R*?

29.38 •• A long, thin solenoid has 900 turns per meter and radius 2.50 cm. The current in the solenoid is increasing at a uniform rate of 36.0 A/s. What is the magnitude of the induced electric field at a point near the center of the solenoid and (a) 0.500 cm from the axis of the solenoid; (b) 1.00 cm from the axis of the solenoid?

29.39 •• A long, thin solenoid has 400 turns per meter and radius 1.10 cm. The current in the solenoid is increasing at a uniform rate *di/dt*. The induced electric field at a point near the center of the solenoid and 3.50 cm from its axis is 8.00×10^{-6} V/m. Calculate *di/dt*.

29.40 • The magnetic field $\vec{B}$ at all points within the colored circle shown in Fig. E29.15 has an initial magnitude of 0.750 T.

(The circle could represent approximately the space inside a long, thin solenoid.) The magnetic field is directed into the plane of the diagram and is decreasing at the rate of -0.0350 T/s. (a) What is the shape of the field lines of the induced electric field shown in Fig. E29.15, within the colored circle? (b) What are the magnitude and direction of this field at any point on the circular conducting ring with radius 0.100 m? (c) What is the current in the ring if its resistance is 4.00 Ω? (d) What is the emf between points a and b on the ring? (e) If the ring is cut at some point and the ends are separated slightly, what will be the emf between the ends?

29.41 • A long, straight solenoid with a cross-sectional area of 8.00 cm^2 is wound with 90 turns of wire per centimeter, and the windings carry a current of 0.350 A. A second winding of 12 turns encircles the solenoid at its center. The current in the solenoid is turned off such that the magnetic field of the solenoid becomes zero in 0.0400 s. What is the average induced emf in the second winding?

Section 29.7 Displacement Current and Maxwell's Equations

29.42 • A parallel-plate, air-filled capacitor is being charged as in Fig. 29.23. The circular plates have radius 4.00 cm, and at a particular instant the conduction current in the wires is 0.520 A. (a) What is the displacement current density j_D in the air space between the plates? (b) What is the rate at which the electric field between the plates is changing? (c) What is the induced magnetic field between the plates at a distance of 2.00 cm from the axis? (d) At 1.00 cm from the axis?

29.43 • **Displacement Current in a Dielectric.** Suppose that the parallel plates in Fig. 29.23 have an area of 3.00 cm^2 and are separated by a 2.50-mm-thick sheet of dielectric that completely fills the volume between the plates. The dielectric has dielectric constant 4.70. (You can ignore fringing effects.) At a certain instant, the potential difference between the plates is 120 V and the conduction current i_C equals 6.00 mA. At this instant, what are (a) the charge q on each plate; (b) the rate of change of charge on the plates; (c) the displacement current in the dielectric?

29.44 • **CALC** In Fig. 29.23 the capacitor plates have area 5.00 cm^2 and separation 2.00 mm. The plates are in vacuum. The charging current i_C has a *constant* value of 1.80 mA. At $t = 0$ the charge on the plates is zero. (a) Calculate the charge on the plates, the electric field between the plates, and the potential difference between the plates when $t = 0.500$ μs. (b) Calculate dE/dt, the time rate of change of the electric field between the plates. Does dE/dt vary in time? (c) Calculate the displacement current density j_D between the plates, and from this the total displacement current i_D. How do i_C and i_D compare?

Section 29.8 Superconductivity

29.45 • At temperatures near absolute zero, B_c approaches 0.142 T for vanadium, a type-I superconductor. The normal phase of vanadium has a magnetic susceptibility close to zero. Consider a long, thin vanadium cylinder with its axis parallel to an external magnetic field $\vec{B}_0$ in the $+x$-direction. At points far from the ends of the cylinder, by symmetry, all the magnetic vectors are parallel to the x-axis. At temperatures near absolute zero, what are the resultant magnetic field $\vec{B}$ and the magnetization $\vec{M}$ inside and outside the cylinder (far from the ends) for (a) $\vec{B}_0 = (0.130$ T$)\hat{i}$ and (b) $\vec{B}_0 = (0.260$ T$)\hat{i}$?

PROBLEMS

29.46 •• A very long, rectangular loop of wire can slide without friction on a horizontal surface. Initially the loop has part of its area in a region of uniform magnetic field that has magnitude $B = 2.90$ T and is perpendicular to the plane of the loop. The loop has dimensions 4.00 cm by 60.0 cm, mass 24.0 g, and resistance $R = 5.00 \times 10^{-3}$ Ω. The loop is initially at rest; then a constant force $F_{ext} = 0.180$ N is applied to the loop to pull it out of the field (**Fig. P29.46**). (a) What is the acceleration of the loop when $v = 3.00$ cm/s? (b) What are the loop's terminal speed and acceleration when the loop is moving at that terminal speed? (c) What is the acceleration of the loop when it is completely out of the magnetic field?

Figure **P29.46**

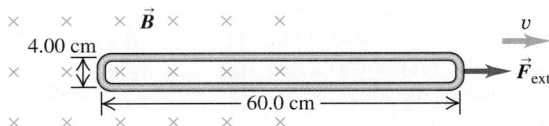

29.47 •• **CP CALC** In the circuit shown in **Fig. P29.47**, the capacitor has capacitance $C = 20$ μF and is initially charged to 100 V with the polarity shown. The resistor R_0 has resistance 10 Ω. At time $t = 0$ the switch S is closed. The small circuit is not connected in any way to the large one. The wire of the small circuit has a resistance of 1.0 Ω/m and contains 25 loops. The large circuit is a rectangle 2.0 m by 4.0 m, while the small one has dimensions $a = 10.0$ cm and $b = 20.0$ cm. The distance c is 5.0 cm. (The figure is not drawn

Figure **P29.47**

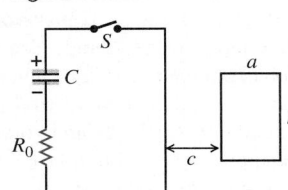

to scale.) Both circuits are held stationary. Assume that only the wire nearest the small circuit produces an appreciable magnetic field through it. (a) Find the current in the large circuit 200 μs after S is closed. (b) Find the current in the small circuit 200 μs after S is closed. (*Hint:* See Exercise 29.7.) (c) Find the direction of the current in the small circuit. (d) Justify why we can ignore the magnetic field from all the wires of the large circuit except for the wire closest to the small circuit.

29.48 •• **CP CALC** In the circuit in Fig. P29.47, an emf of 90.0 V is added in series with the capacitor and the resistor, and the capacitor is initially uncharged. The emf is placed between the capacitor and switch S, with the positive terminal of the emf adjacent to the capacitor. Otherwise, the two circuits are the same as in Problem 29.47. The switch is closed at $t = 0$. When the current in the large circuit is 5.00 A, what are the magnitude and direction of the induced current in the small circuit?

29.49 •• **CALC** A very long, straight solenoid with a cross-sectional area of 2.00 cm^2 is wound with 90.0 turns of wire per centimeter. Starting at $t = 0$, the current in the solenoid is increasing according to $i(t) = (0.160$ A/s$^2)t^2$. A secondary winding of 5 turns encircles the solenoid at its center, such that the secondary winding has the same cross-sectional area as the solenoid. What is the magnitude of the emf induced in the secondary winding at the instant that the current in the solenoid is 3.20 A?

29.50 • Suppose the loop in **Fig. P29.50** is (a) rotated about the y-axis; (b) rotated about the x-axis; (c) rotated about an edge parallel to the z-axis. What is the maximum induced emf in each case if $A = 600\ \text{cm}^2$, $\omega = 35.0\ \text{rad/s}$, and $B = 0.320\ \text{T}$?

Figure **P29.50**

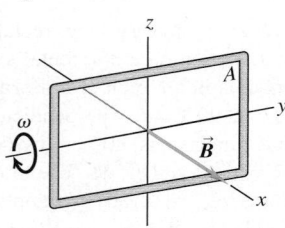

29.51 • In **Fig. P29.51** the loop is being pulled to the right at constant speed v. A constant current I flows in the long wire, in the direction shown. (a) Calculate the magnitude of the net emf $\mathcal{E}$ induced in the loop. Do this two ways: (i) by using Faraday's law of induction (*Hint:* See Exercise 29.7) and (ii) by looking at the emf induced in each segment of the loop due to its motion. (b) Find the direction (clockwise or counterclockwise) of the current induced in the loop. Do this two ways: (i) using Lenz's law and (ii) using the magnetic force on charges in the loop. (c) Check your answer for the emf in part (a) in the following special cases to see whether it is physically reasonable: (i) The loop is stationary; (ii) the loop is very thin, so $a \rightarrow 0$; (iii) the loop gets very far from the wire.

Figure **P29.51**

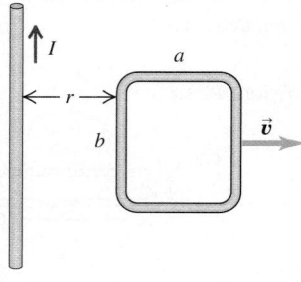

29.52 • Make a Generator? You are shipwrecked on a deserted tropical island. You have some electrical devices that you could operate using a generator but you have no magnets. The earth's magnetic field at your location is horizontal and has magnitude $8.0 \times 10^{-5}\ \text{T}$, and you decide to try to use this field for a generator by rotating a large circular coil of wire at a high rate. You need to produce a peak emf of 9.0 V and estimate that you can rotate the coil at 30 rpm by turning a crank handle. You also decide that to have an acceptable coil resistance, the maximum number of turns the coil can have is 2000. (a) What area must the coil have? (b) If the coil is circular, what is the maximum translational speed of a point on the coil as it rotates? Do you think this device is feasible? Explain.

29.53 • A flexible circular loop 6.50 cm in diameter lies in a magnetic field with magnitude 1.35 T, directed into the plane of the page as shown in **Fig. P29.53**. The loop is pulled at the points indicated by the arrows, forming a loop of zero area in 0.250 s. (a) Find the average induced emf in the circuit. (b) What is the direction of the current in R: from a to b or from b to a? Explain your reasoning.

Figure **P29.53**

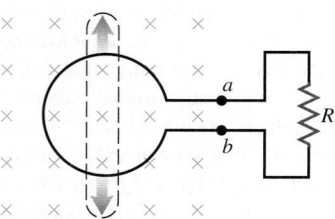

29.54 ••• CALC A conducting rod with length $L = 0.200\ \text{m}$, mass $m = 0.120\ \text{kg}$, and resistance $R = 80.0\ \Omega$ moves without friction on metal rails as shown in Fig. 29.11. A uniform magnetic

field with magnitude $B = 1.50\ \text{T}$ is directed into the plane of the figure. The rod is initially at rest, and then a constant force with magnitude $F = 1.90\ \text{N}$ and directed to the right is applied to the rod. How many seconds after the force is applied does the rod reach a speed of 25.0 m/s?

29.55 •• CALC A very long, cylindrical wire of radius R carries a current I_0 uniformly distributed across the cross section of the wire. Calculate the magnetic flux through a rectangle that has one side of length W running down the center of the wire and another side of length R, as shown in **Fig. P29.55** (see Exercise 29.7).

Figure **P29.55**

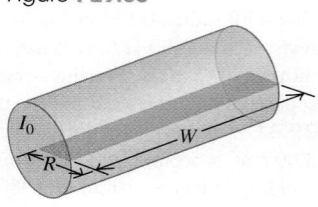

29.56 •• CP CALC Terminal Speed. A bar of length $L = 0.36\ \text{m}$ is free to slide without friction on horizontal rails as shown in **Fig. P29.56**. A uniform magnetic field $B = 2.4\ \text{T}$ is directed into the plane of the figure. At one end of the rails there is a battery with emf $\mathcal{E} = 12\ \text{V}$ and a switch S. The bar has mass 0.90 kg and resistance $5.0\ \Omega$; ignore all other resistance in the circuit. The switch is closed at time $t = 0$.

Figure **P29.56**

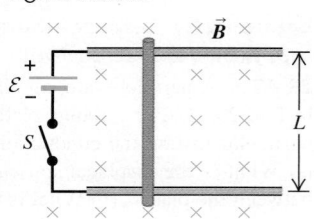

(a) Sketch the bar's speed as a function of time. (b) Just after the switch is closed, what is the acceleration of the bar? (c) What is the acceleration of the bar when its speed is 2.0 m/s? (d) What is the bar's terminal speed?

29.57 • CALC The long, straight wire shown in **Fig. P29.57a** carries constant current I. A metal bar with length L is moving at constant velocity $\vec{v}$, as shown in the figure. Point a is a distance d from the wire. (a) Calculate the emf induced in the bar. (b) Which point, a or b, is at higher potential? (c) If the bar is replaced by a rectangular wire loop of resistance R (Fig. P29.57b), what is the magnitude of the current induced in the loop?

Figure **P29.57**

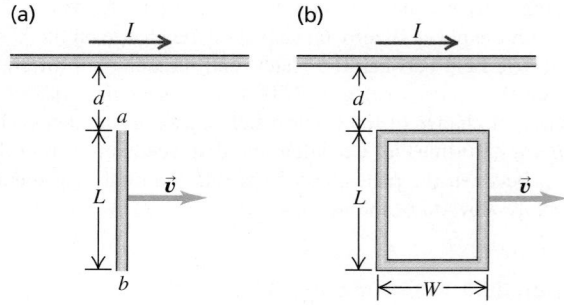

29.58 • CALC A circular conducting ring with radius $r_0 = 0.0420\ \text{m}$ lies in the xy-plane in a region of uniform magnetic field $\vec{B} = B_0[1 - 3(t/t_0)^2 + 2(t/t_0)^3]\hat{k}$. In this expression, $t_0 = 0.0100\ \text{s}$ and is constant, t is time, $\hat{k}$ is the unit vector in the $+z$-direction, and $B_0 = 0.0800\ \text{T}$ and is constant. At points a and b (**Fig. P29.58**) there is a small gap in the ring with wires leading to an external circuit of resistance $R = 12.0\ \Omega$. There is no magnetic field at the location of the external circuit. (a) Derive an expression, as a function of time, for the total magnetic flux Φ_B through the ring. (b) Determine the emf induced in the ring at time

$t = 5.00 \times 10^{-3}$ s. What is the polarity of the emf? (c) Because of the internal resistance of the ring, the current through R at the time given in part (b) is only 3.00 mA. Determine the internal resistance of the ring. (d) Determine the emf in the ring at a time $t = 1.21 \times 10^{-2}$ s. What is the polarity of the emf? (e) Determine the time at which the current through R reverses its direction.

Figure **P29.58**

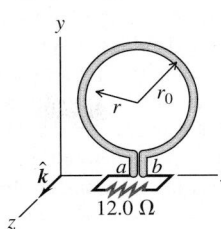

29.59 • CALC A slender rod, 0.240 m long, rotates with an angular speed of 8.80 rad/s about an axis through one end and perpendicular to the rod. The plane of rotation of the rod is perpendicular to a uniform magnetic field with a magnitude of 0.650 T. (a) What is the induced emf in the rod? (b) What is the potential difference between its ends? (c) Suppose instead the rod rotates at 8.80 rad/s about an axis through its center and perpendicular to the rod. In this case, what is the potential difference between the ends of the rod? Between the center of the rod and one end?

29.60 •• A 25.0-cm-long metal rod lies in the xy-plane and makes an angle of 36.9° with the positive x-axis and an angle of 53.1° with the positive y-axis. The rod is moving in the $+x$-direction with a speed of 6.80 m/s. The rod is in a uniform magnetic field $\vec{B} = (0.120\,\text{T})\hat{\imath} - (0.220\,\text{T})\hat{\jmath} - (0.0900\,\text{T})\hat{k}$. (a) What is the magnitude of the emf induced in the rod? (b) Indicate in a sketch which end of the rod is at higher potential.

29.61 •• CP CALC A rectangular loop with width L and a slide wire with mass m are as shown in **Fig. P29.61**. A uniform magnetic field $\vec{B}$ is directed perpendicular to the plane of the loop into the plane of the figure. The slide wire is given an initial speed of v_0 and then released. There is no friction between the slide wire and the loop, and the resistance of the loop is negligible in comparison to the resistance R of the slide wire. (a) Obtain an expression for F, the magnitude of the force exerted on the wire while it is moving at speed v. (b) Show that the distance x that the wire moves before coming to rest is $x = mv_0R/L^2B^2$.

Figure **P29.61**

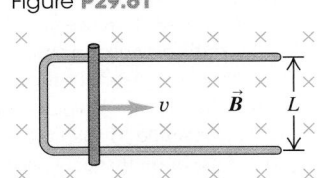

29.62 • CALC An airplane propeller of total length L rotates around its center with angular speed ω in a magnetic field that is perpendicular to the plane of rotation. Modeling the propeller as a thin, uniform bar, find the potential difference between (a) the center and either end of the propeller and (b) the two ends. (c) If the field is the earth's field of 0.50 G and the propeller turns at 220 rpm and is 2.0 m long, what is the potential difference between the middle and either end? It this large enough to be concerned about?

29.63 • The magnetic field $\vec{B}$, at all points within a circular region of radius R, is uniform in space and directed into the plane of the page as shown in **Fig. P29.63**. (The region could be a cross section inside the windings of a long, straight solenoid.) If the magnetic field is increasing at a rate dB/dt, what are the magnitude and direction of the force on a stationary positive point charge q located at points a, b, and c? (Point a is a distance r above the center of the region, point b is a distance r to the right of the center, and point c is at the center of the region.)

Figure **P29.63**

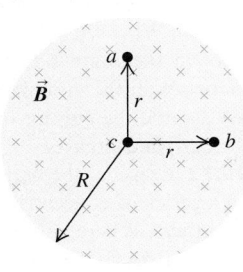

29.64 •• CP CALC A capacitor has two parallel plates with area A separated by a distance d. The space between plates is filled with a material having dielectric constant K. The material is not a perfect insulator but has resistivity ρ. The capacitor is initially charged with charge of magnitude Q_0 on each plate that gradually discharges by conduction through the dielectric. (a) Calculate the conduction current density $j_C(t)$ in the dielectric. (b) Show that at any instant the displacement current density in the dielectric is equal in magnitude to the conduction current density but opposite in direction, so the total current density is zero at every instant.

29.65 ••• CALC A dielectric of permittivity 3.5×10^{-11} F/m completely fills the volume between two capacitor plates. For $t > 0$ the electric flux through the dielectric is $(8.0 \times 10^3 \,\text{V}\cdot\text{m/s}^3)t^3$. The dielectric is ideal and nonmagnetic; the conduction current in the dielectric is zero. At what time does the displacement current in the dielectric equal 21 μA?

29.66 •• DATA You are evaluating the performance of a large electromagnet. The magnetic field of the electromagnet is zero at $t = 0$ and increases as the current through the windings of the electromagnet is increased. You determine the magnetic field as a function of time by measuring the time dependence of the current induced in a small coil that you insert between the poles of the electromagnet, with the plane of the coil parallel to the pole faces as in Fig. 29.5. The coil has 4 turns, a radius of 0.800 cm, and a resistance of 0.250 Ω. You measure the current i in the coil as a function of time t. Your results are shown in **Fig. P29.66**. Throughout your measurements, the current induced in the coil remains in the same direction. Calculate the magnetic field at the location of the coil for (a) $t = 2.00$ s, (b) $t = 5.00$ s, and (c) $t = 6.00$ s.

Figure **P29.66**

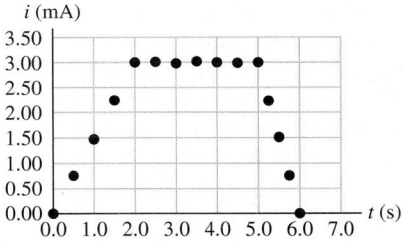

29.67 •• DATA You are conducting an experiment in which a metal bar of length 6.00 cm and mass 0.200 kg slides without friction on two parallel metal rails (**Fig. P29.67**). A resistor with resistance $R = 0.800\ \Omega$ is connected across one end of the rails so that the bar, rails, and resistor form a complete conducting path. The resistances of the rails and of the bar are much less than R and can be ignored. The entire apparatus is in a uniform magnetic field $\vec{B}$ that is directed into the plane of the figure. You give the bar an initial velocity $v = 20.0$ cm/s to the right and then release it, so that the only force on the bar then is the force exerted by the magnetic field. Using high-speed photography, you measure the magnitude of the acceleration of the bar as a function of its speed. Your results are given in the table:

Figure **P29.67**

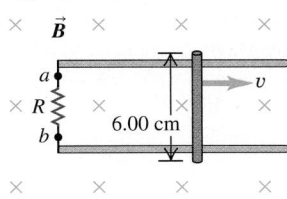

v (cm/s)	20.0	16.0	14.0	12.0	10.0	8.0
a (cm/s^2)	6.2	4.9	4.3	3.7	3.1	2.5

(a) Plot the data as a graph of a versus v. Explain why the data points plotted this way lie close to a straight line, and determine the slope of the best-fit straight line for the data. (b) Use your graph from part (a) to calculate the magnitude B of the magnetic field. (c) While the bar is moving, which end of the resistor, a or b, is at higher potential? (d) How many seconds does it take the speed of the bar to decrease from 20.0 cm/s to 10.0 cm/s?

29.68 ••• **DATA** You measure the magnitude of the external force $\vec{F}$ that must be applied to a rectangular conducting loop to pull it at constant speed v out of a region of uniform magnetic field $\vec{B}$ that is directed out of the plane of **Fig. P29.68**. The loop has dimensions 14.0 cm by 8.00 cm and resistance 4.00×10^{-3} Ω; it does not change shape as it moves. The measurements you collect are listed in the table.

Figure **P29.68**

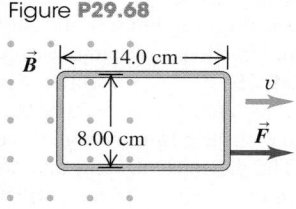

F (N)	0.10	0.21	0.31	0.41	0.52
v (cm/s)	2.0	4.0	6.0	8.0	10.0

(a) Plot the data as a graph of F versus v. Explain why the data points plotted this way lie close to a straight line, and determine the slope of the best-fit straight line for the data. (b) Use your graph from part (a) to calculate the magnitude B of the uniform magnetic field. (c) In Fig. P29.68, is the current induced in the loop clockwise or counterclockwise? (d) At what rate is electrical energy being dissipated in the loop when the speed of the loop is 5.00 cm/s?

CHALLENGE PROBLEMS

29.69 ••• A metal bar with length L, mass m, and resistance R is placed on frictionless metal rails that are inclined at an angle ϕ above the horizontal. The rails have negligible resistance. A uniform magnetic field of magnitude B is directed downward as shown in **Fig. P29.69**. The bar is released from rest and slides down the rails. (a) Is the direction of the current induced in the bar from a to b or from b to a? (b) What is the terminal speed of the bar? (c) What is the induced current in the bar when the terminal speed has been reached? (d) After the terminal speed has been reached, at what rate is electrical energy being converted to thermal energy in the resistance of the bar? (e) After the terminal speed has been reached, at what rate is work being done on the bar by gravity? Compare your answer to that in part (d).

Figure **P29.69**

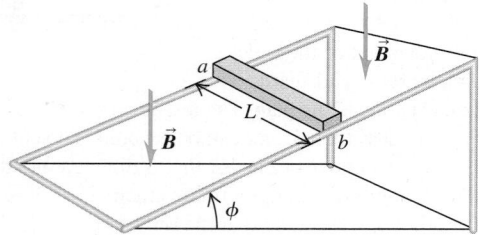

29.70 ••• **CP CALC** A square, conducting, wire loop of side L, total mass m, and total resistance R initially lies in the horizontal xy-plane, with corners at $(x, y, z) = (0, 0, 0)$, $(0, L, 0)$, $(L, 0, 0)$, and $(L, L, 0)$. There is a uniform, upward magnetic field $\vec{B} = B\hat{k}$ in the space within and around the loop. The side of the loop that

extends from $(0, 0, 0)$ to $(L, 0, 0)$ is held in place on the x-axis; the rest of the loop is free to pivot around this axis. When the loop is released, it begins to rotate due to the gravitational torque. (a) Find the *net* torque (magnitude and direction) that acts on the loop when it has rotated through an angle ϕ from its original orientation and is rotating downward at an angular speed ω. (b) Find the angular acceleration of the loop at the instant described in part (a). (c) Compared to the case with zero magnetic field, does it take the loop a longer or shorter time to rotate through 90°? Explain. (d) Is mechanical energy conserved as the loop rotates downward? Explain.

PASSAGE PROBLEMS

BIO STIMULATING THE BRAIN. Communication in the nervous system is based on the propagation of electrical signals called *action potentials* along axons, which are extensions of nerve cells (see the Passage Problems in Chapter 26). Action potentials are generated when the electric potential difference across the membrane of the nerve cell changes: Specifically, the inside of the cell becomes more positive. Researchers in clinical medicine and neurobiology cannot stimulate nerves (even noninvasively) at specific locations in conscious human subjects. Using electrodes to apply current to the skin is painful and requires large currents, which could be dangerous.

Anthony Barker and colleagues at the University of Sheffield in England developed a technique called *transcranial magnetic stimulation* (TMS). In this widely used procedure, a coil positioned near the skull produces a time-varying magnetic field that induces in the conductive tissue of the brain (see part (a) of the figure) electric currents that are sufficient to cause action potentials in nerve cells. For example, if the coil is placed near the motor cortex (the region of the brain that controls voluntary movement), scientists can monitor muscle contraction and assess the connections between brain and the muscles. Part (b) of the figure is a graph of the typical dependence on time t of the magnetic field B produced by the coil.

(a)

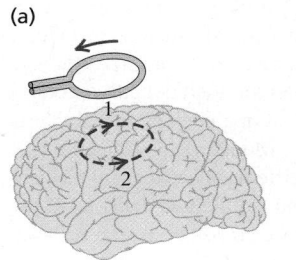

(b)

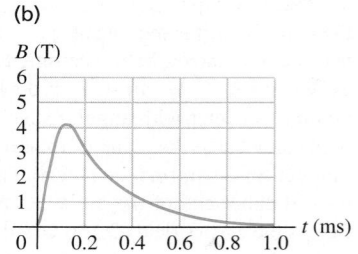

29.71 In part (a) of the figure, a current pulse increases to a peak and then decreases to zero in the direction shown in the stimulating coil. What will be the direction of the induced current (dashed line) in the brain tissue? (a) 1; (b) 2; (c) 1 while the current increases in the stimulating coil, 2 while the current decreases; (d) 2 while the current increases in the stimulating coil, 1 while the current decreases.

29.72 Consider the brain tissue at the level of the dashed line to be a series of concentric circles, each behaving independently of the others. Where will the induced emf be the greatest? (a) At the center of the dashed line; (b) at the periphery of the dashed line; (c) nowhere—it will be the same in all concentric circles; (d) at the center while the stimulating current increases, at the periphery while the current decreases.

29.73 It may be desirable to increase the maximum induced current in the brain tissue. In **Fig. P29.73**, which time-dependent graph of the magnetic field B in the coil achieves that goal? Assume that everything else remains constant. (a) A; (b) B; (c) either A or B; (d) neither A nor B.

Figure **P29.73**

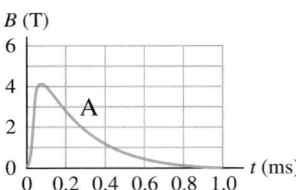 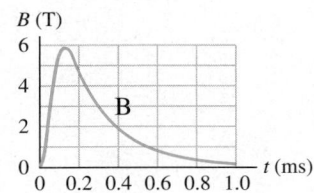

29.74 Which graph in **Fig. P29.74** best represents the time t dependence of the current i induced in the brain tissue, assuming that this tissue can be modeled as a resistive circuit? (The units of i are arbitrary.) (a) A; (b) B; (c) C; (d) D.

Figure **P29.74**

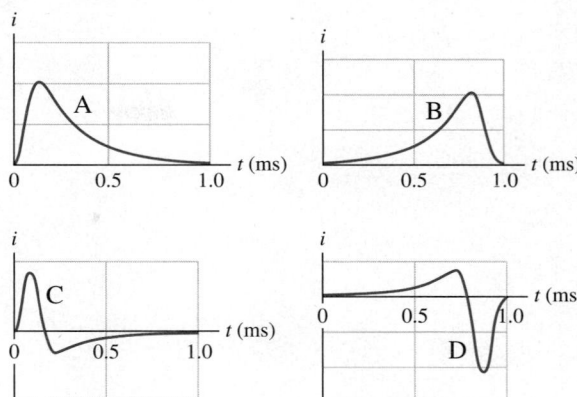

Answers

Chapter Opening Question ?

iv As the magnetic stripe moves through the card reader, the coded pattern of magnetization in the stripe causes a varying magnetic flux. An electric field is induced, which causes a current in the reader's circuits. If the card does not move, there is no induced current and none of the credit card's information is read.

Test Your Understanding Questions

29.2 (a) (i), (b) (iii) In (a), initially there is magnetic flux into the plane of the page, which we call positive. While the loop is being squeezed, the flux is becoming less positive ($d\Phi_B/dt < 0$) and so the induced emf is positive as in Fig. 29.6b ($\mathcal{E} = -d\Phi_B/dt > 0$). If you point the thumb of your right hand into the page, your fingers curl clockwise, so this is the direction of positive induced emf. In (b), since the coil's shape is no longer changing, the magnetic flux is not changing and there is no induced emf.

29.3 (a) (i), (b) (iii) In (a), as in the original situation, the magnet and loop are approaching each other and the downward flux through the loop is increasing. Hence the induced emf and induced current are the same. In (b), since the magnet and loop are moving together, the flux through the loop is not changing and no emf is induced.

29.4 (a) (iii); (b) (i) or (ii); (c) (ii) or (iii) You will get the maximum motional emf if you hold the rod vertically, so that its length is perpendicular to both the magnetic field and the direction of motion. With this orientation, $\vec{L}$ is parallel to $\vec{v} \times \vec{B}$. If you hold the rod in any horizontal orientation, $\vec{L}$ will be perpendicular to $\vec{v} \times \vec{B}$ and no emf will be induced. If you walk due north or south, $\vec{v} \times \vec{B} = 0$ and no emf will be induced for any orientation of the rod.

29.5 yes, no The magnetic field at a fixed position changes as you move the magnet, which induces an electric field. Such induced electric fields are *not* conservative.

29.6 (iii) By Lenz's law, the force must oppose the motion of the disk through the magnetic field. Since the disk material is now moving to the right through the field region, the force $\vec{F}$ is to the left—that is, in the opposite direction to that shown in Fig. 29.20b. To produce a leftward magnetic force $\vec{F} = I\vec{L} \times \vec{B}$ on currents moving through a magnetic field $\vec{B}$ directed out of the plane of the figure, the eddy currents must be moving downward in the figure—that is, in the same direction shown in Fig. 29.20b.

29.7 (a) Faraday's law, (b) Ampere's law A credit card reader works by inducing currents in the reader's coils as the card's magnetized stripe is swiped (see the answer to the chapter opening question). Ampere's law describes how currents of all kinds (both conduction currents and displacement currents) give rise to magnetic fields.

Bridging Problem

$v_{\text{terminal}} = 16\rho_m \rho_R g / B^2$

? Many traffic lights change when a car rolls up to the intersection. This process works because the car contains (i) conducting material; (ii) insulating material that carries a net electric charge; (iii) ferromagnetic material; (iv) ferromagnetic material that is already magnetized.

30 INDUCTANCE

Take a length of copper wire and wrap it around a pencil to form a coil. If you put this coil in a circuit, the coil behaves quite differently than a straight piece of wire. In an ordinary gasoline-powered car, a coil of this kind makes it possible for the 12-volt car battery to provide thousands of volts to the spark plugs in order for the plugs to fire and make the engine run. Other coils are used to keep fluorescent light fixtures shining. Larger coils placed under city streets are used to control the operation of traffic signals. All of these applications, and many others, involve the *induction* effects that we studied in Chapter 29.

A changing current in a coil induces an emf in an adjacent coil. The coupling between the coils is described by their *mutual inductance*. A changing current in a coil also induces an emf in that same coil. Such a coil is called an *inductor,* and the relationship of current to emf is described by the *inductance* (also called *self-inductance*) of the coil. If a coil is initially carrying a current, energy is released when the current decreases; this principle is used in automotive ignition systems. We'll find that this released energy was stored in the magnetic field caused by the current that was initially in the coil, and we'll look at some of the practical applications of magnetic-field energy.

We'll also take a first look at what happens when an inductor is part of a circuit. In Chapter 31 we'll go on to study how inductors behave in alternating-current circuits, and we'll learn why inductors play an essential role in modern electronics.

30.1 MUTUAL INDUCTANCE

In Section 28.4 we considered the magnetic interaction between two wires carrying *steady* currents; the current in one wire causes a magnetic field, which exerts a force on the current in the second wire. But an additional interaction arises between two circuits when there is a *changing* current in one of the circuits.

Consider two neighboring coils of wire, as in **Fig. 30.1**. A current flowing in coil 1 produces a magnetic field $\vec{B}$ and hence a magnetic flux through coil 2. If the current in coil 1 changes, the flux through coil 2 changes as well; according to Faraday's law (Section 29.2), this induces an emf in coil 2. In this way, a change in the current in one circuit can induce a current in a second circuit.

Let's analyze the situation shown in Fig. 30.1 in more detail. We'll use lower-case letters to represent quantities that vary with time; for example, a time-varying current is i, often with a subscript to identify the circuit. In Fig. 30.1 a current i_1 in coil 1 sets up a magnetic field $\vec{B}$, and some of the (blue) field lines pass through coil 2. We denote the magnetic flux through *each* turn of coil 2, caused by the current i_1 in coil 1, as Φ_{B2}. (If the flux is different through different turns of the coil, then Φ_{B2} denotes the *average* flux.) The magnetic field is proportional to i_1, so Φ_{B2} is also proportional to i_1. When i_1 changes, Φ_{B2} changes; this changing flux induces an emf $\mathcal{E}_2$ in coil 2, given by

$$\mathcal{E}_2 = -N_2 \frac{d\Phi_{B2}}{dt} \tag{30.1}$$

We could represent the proportionality of Φ_{B2} and i_1 in the form $\Phi_{B2} = $ (constant)i_1, but instead it is more convenient to include the number of turns N_2 in the relationship. Introducing a proportionality constant M_{21}, called the **mutual inductance** of the two coils, we write

$$N_2\Phi_{B2} = M_{21}i_1 \tag{30.2}$$

where Φ_{B2} is the flux through a *single* turn of coil 2. From this,

$$N_2\frac{d\Phi_{B2}}{dt} = M_{21}\frac{di_1}{dt}$$

and we can rewrite Eq. (30.1) as

$$\mathcal{E}_2 = -M_{21}\frac{di_1}{dt} \tag{30.3}$$

That is, a change in the current i_1 in coil 1 induces an emf in coil 2 that is directly proportional to the rate of change of i_1 (**Fig. 30.2**).

We may also write the definition of mutual inductance, Eq. (30.2), as

$$M_{21} = \frac{N_2\Phi_{B2}}{i_1}$$

If the coils are in vacuum, the flux Φ_{B2} through each turn of coil 2 is directly proportional to the current i_1. Then the mutual inductance M_{21} is a constant that depends only on the geometry of the two coils (the size, shape, number of turns, and orientation of each coil and the separation between the coils). If a magnetic material is present, M_{21} also depends on the magnetic properties of the material. If the material has nonlinear magnetic properties—that is, if the relative permeability K_m (defined in Section 28.8) is not constant and magnetization is not proportional to magnetic field—then Φ_{B2} is no longer directly proportional to i_1. In that case the mutual inductance also depends on the value of i_1. In this discussion we will assume that any magnetic material present has constant K_m so that flux *is* directly proportional to current and M_{21} depends on geometry only.

We can repeat our discussion for the opposite case in which a changing current i_2 in coil 2 causes a changing flux Φ_{B1} and an emf $\mathcal{E}_1$ in coil 1. It turns out that the corresponding constant M_{12} is *always* equal to M_{21}, even though in general the two coils are not identical and the flux through them is not the same.

30.1 A current i_1 in coil 1 gives rise to a magnetic flux through coil 2.

Mutual inductance: If the current in coil 1 is changing, the changing flux through coil 2 induces an emf in coil 2.

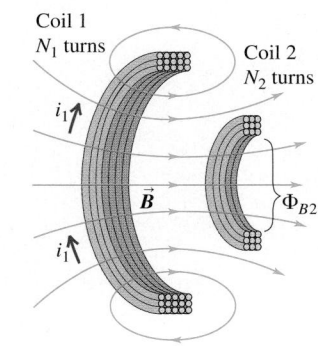

30.2 This electric toothbrush makes use of mutual inductance. The base contains a coil that is supplied with alternating current from a wall socket. Even though there is no direct electrical contact between the base and the toothbrush, this varying current induces an emf in a coil within the toothbrush itself, recharging the toothbrush battery.

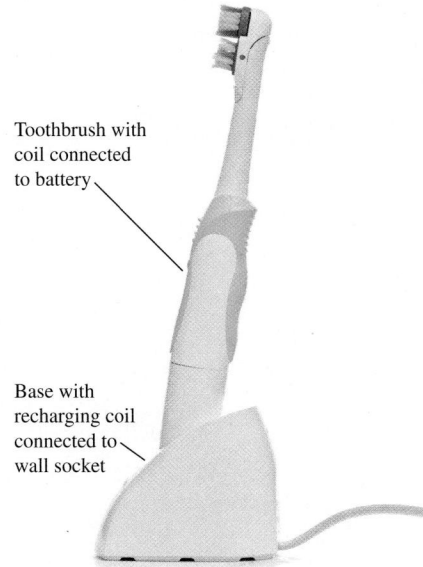

Toothbrush with coil connected to battery

Base with recharging coil connected to wall socket

We call this common value simply the mutual inductance, denoted by the symbol M without subscripts; it characterizes completely the induced-emf interaction of two coils. Then

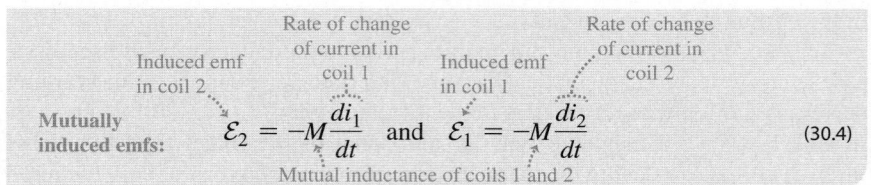

Mutually induced emfs:
$$\mathcal{E}_2 = -M\frac{di_1}{dt} \quad \text{and} \quad \mathcal{E}_1 = -M\frac{di_2}{dt} \qquad (30.4)$$

The negative signs in Eqs. (30.4) reflect Lenz's law (Section 29.3). The first equation says that a current change in coil 1 causes a flux change through coil 2, inducing an emf in coil 2 that opposes the flux change; in the second equation, the roles of the two coils are interchanged. The mutual inductance M is

Mutual inductance of coils 1 and 2
$$M = \frac{N_2\Phi_{B2}}{i_1} = \frac{N_1\Phi_{B1}}{i_2} \qquad (30.5)$$

CAUTION **Only a time-varying current induces an emf** Only a *time-varying* current in a coil can induce an emf and hence a current in a second coil. Equations (30.4) show that the induced emf in each coil is directly proportional to the *rate of change* of the current in the other coil, not to the *value* of the current. A steady current in one coil, no matter how strong, cannot induce a current in a neighboring coil.

The SI unit of mutual inductance is called the **henry** (1 H), in honor of the American physicist Joseph Henry (1797–1878), one of the discoverers of electromagnetic induction. From Eq. (30.5), one henry is equal to one weber per ampere. Other equivalent units, obtained by using Eqs. (30.4), are

$$1\ \text{H} = 1\ \text{Wb/A} = 1\ \text{V} \cdot \text{s/A} = 1\ \Omega \cdot \text{s} = 1\ \text{J/A}^2$$

Just as the farad is a rather large unit of capacitance (see Section 24.1), the henry is a rather large unit of mutual inductance. Typical values of mutual inductance can be in the millihenry (mH) or microhenry (μH) range.

Drawbacks and Uses of Mutual Inductance

Mutual inductance can be a nuisance in electric circuits, since variations in current in one circuit can induce unwanted emfs in other nearby circuits. To minimize these effects, multiple-circuit systems must be designed so that M is as small as possible; for example, two coils would be placed far apart.

Happily, mutual inductance also has many useful applications. A *transformer,* used in alternating-current circuits to raise or lower voltages, is fundamentally no different from the two coils shown in Fig. 30.1. A time-varying alternating current in one coil of the transformer produces an alternating emf in the other coil; the value of M, which depends on the geometry of the coils, determines the amplitude of the induced emf in the second coil and hence the amplitude of the output voltage. (We'll describe transformers in more detail in Chapter 31.)

EXAMPLE 30.1 CALCULATING MUTUAL INDUCTANCE

In one form of Tesla coil (a high-voltage generator popular in science museums), a long solenoid with length l and cross-sectional area A is closely wound with N_1 turns of wire. A coil with N_2 turns surrounds it at its center (**Fig. 30.3**). Find the mutual inductance M.

SOLUTION

IDENTIFY and SET UP: Mutual inductance occurs here because a current in either coil sets up a magnetic field that causes a flux through the other coil. From Example 28.9 (Section 28.7) we have an expression [Eq. (28.23)] for the field magnitude B_1 at the center of the solenoid (coil 1) in terms of the solenoid current i_1. This allows us to determine the flux through a cross section of the solenoid. Since there is almost no magnetic field outside a very

long solenoid, this is also equal to the flux Φ_{B2} through each turn of the *outer* coil (2). We then use Eq. (30.5), in the form $M = N_2\Phi_{B2}/i_1$, to determine M.

EXECUTE: Equation (28.23) is expressed in terms of the number of turns per unit length, which for solenoid (1) is $n_1 = N_1/L$. So

$$B_1 = \mu_0 n_1 i_1 = \frac{\mu_0 N_1 i_1}{l}$$

The flux through a cross section of the solenoid equals $B_1 A$. As we mentioned above, this also equals the flux Φ_{B2} through each turn of the outer coil, independent of its cross-sectional area. From Eq. (30.5), the mutual inductance M is then

$$M = \frac{N_2\Phi_{B2}}{i_1} = \frac{N_2 B_1 A}{i_1} = \frac{N_2}{i_1}\frac{\mu_0 N_1 i_1}{l}A = \frac{\mu_0 A N_1 N_2}{l}$$

EVALUATE: The mutual inductance M of any two coils is proportional to the product $N_1 N_2$ of their numbers of turns. Notice that M depends only on the geometry of the two coils, not on the current.

Here's a numerical example to give you an idea of magnitudes. Suppose $l = 0.50$ m, $A = 10$ cm^2 $= 1.0 \times 10^{-3}$ m^2, $N_1 = 1000$ turns, and $N_2 = 10$ turns. Then

$$M = \frac{(4\pi \times 10^{-7}\ \text{Wb/A}\cdot\text{m})(1.0 \times 10^{-3}\ \text{m}^2)(1000)(10)}{0.50\ \text{m}}$$

$$= 25 \times 10^{-6}\ \text{Wb/A} = 25 \times 10^{-6}\ \text{H} = 25\ \mu\text{H}$$

30.3 A long solenoid with cross-sectional area A and N_1 turns is surrounded at its center by a coil with N_2 turns.

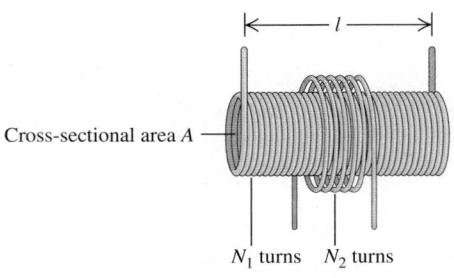

Cross-sectional area A

N_1 turns N_2 turns

EXAMPLE 30.2 EMF DUE TO MUTUAL INDUCTANCE

In Example 30.1, suppose the current i_2 in the outer coil is given by $i_2 = (2.0 \times 10^6\ \text{A/s})t$. (Currents in wires can indeed increase this rapidly for brief periods.) (a) At $t = 3.0\ \mu\text{s}$, what is the average magnetic flux through each turn of the solenoid (coil 1) due to the current in the outer coil? (b) What is the induced emf in the solenoid?

SOLUTION

IDENTIFY and SET UP: In Example 30.1 we found the mutual inductance by relating the current in the solenoid to the flux produced in the outer coil; to do that, we used Eq. (30.5) in the form $M = N_2\Phi_{B2}/i_1$. Here we are given the current i_2 in the outer coil and want to find the resulting flux Φ_1 in the solenoid. The mutual inductance is the *same* in either case, and we have $M = 25\ \mu\text{H}$ from Example 30.1. We use Eq. (30.5) in the form $M = N_1\Phi_{B1}/i_2$ to determine the average flux Φ_{B1} through each turn of the solenoid caused by current i_2 in the outer coil. We then use Eqs. (30.4) to find the emf induced in the solenoid by the time variation of i_2.

EXECUTE: (a) At $t = 3.0\ \mu\text{s} = 3.0 \times 10^{-6}$ s, the current in the outer coil is $i_2 = (2.0 \times 10^6\ \text{A/s})(3.0 \times 10^{-6}\ \text{s}) = 6.0$ A. We solve Eq. (30.5) for the flux Φ_{B1} through each turn of coil 1:

$$\Phi_{B1} = \frac{Mi_2}{N_1} = \frac{(25 \times 10^{-6}\ \text{H})(6.0\ \text{A})}{1000} = 1.5 \times 10^{-7}\ \text{Wb}$$

We emphasize that this is an *average* value; the flux can vary considerably between the center and the ends of the solenoid.

(b) We are given $i_2 = (2.0 \times 10^6\ \text{A/s})t$, so $di_2/dt = 2.0 \times 10^6$ A/s; then, from Eqs. (30.4), the induced emf in the solenoid is

$$\mathcal{E}_1 = -M\frac{di_2}{dt} = -(25 \times 10^{-6}\ \text{H})(2.0 \times 10^6\ \text{A/s}) = -50\ \text{V}$$

EVALUATE: This is a substantial induced emf in response to a very rapid current change. In an operating Tesla coil, there is a high-frequency alternating current rather than a continuously increasing current as in this example; both di_2/dt and $\mathcal{E}_1$ alternate as well, with amplitudes that can be thousands of times larger than in this example.

TEST YOUR UNDERSTANDING OF SECTION 30.1 Consider the Tesla coil described in Example 30.1. If you make the solenoid out of twice as much wire, so that it has twice as many turns and is twice as long, how much larger is the mutual inductance? (i) M is four times greater; (ii) M is twice as great; (iii) M is unchanged; (iv) M is $\frac{1}{2}$ as great; (v) M is $\frac{1}{4}$ as great. ∎

30.2 SELF-INDUCTANCE AND INDUCTORS

In our discussion of mutual inductance we considered two separate, independent circuits: A current in one circuit creates a magnetic field that gives rise to a flux through the second circuit. If the current in the first circuit changes, the flux through the second circuit changes and an emf is induced in the second circuit.

An important related effect occurs in a *single* isolated circuit. A current in a circuit sets up a magnetic field that causes a magnetic flux through the *same* circuit; this flux changes when the current changes. Thus any circuit that carries a varying current has an emf induced in it by the variation in *its own* magnetic field. Such an emf is called a **self-induced emf.** By Lenz's law, a self-induced emf opposes the change in the current that caused the emf and so tends to make it more difficult for variations in current to occur. Hence self-induced emfs can be of great importance whenever there is a varying current.

Self-induced emfs can occur in *any* circuit, since there is always some magnetic flux through the closed loop of a current-carrying circuit. But the effect is greatly enhanced if the circuit includes a coil with N turns of wire (**Fig. 30.4**). As a result of the current i, there is an average magnetic flux Φ_B through each turn of the coil. In analogy to Eq. (30.5) we define the **self-inductance** L of the circuit as

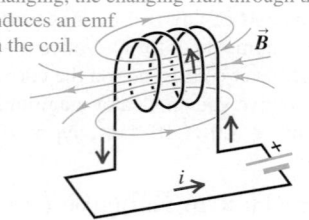

30.4 The current i in the circuit causes a magnetic field $\vec{B}$ in the coil and hence a flux through the coil.

Self-inductance: If the current i in the coil is changing, the changing flux through the coil induces an emf in the coil.

$$L = \frac{N\Phi_B}{i} \qquad (30.6)$$

Self-inductance (or **inductance**) of a coil — Number of turns in coil — Flux due to current through each turn of coil — Current in coil

When there is no danger of confusion with mutual inductance, the self-inductance is called simply the **inductance.** Comparing Eqs. (30.5) and (30.6), we see that the units of self-inductance are the same as those of mutual inductance; the SI unit of self-inductance is the henry.

If the current i changes, so does the flux Φ_B; from rearranging Eq. (30.6) and differentiating with respect to time, the rates of change are related by

$$N\frac{d\Phi_B}{dt} = L\frac{di}{dt}$$

From Faraday's law for a coil with N turns, Eq. (29.4), the self-induced emf is $\mathcal{E} = -N\, d\Phi_B/dt$, so it follows that

$$\mathcal{E} = -L\frac{di}{dt} \qquad (30.7)$$

Self-induced emf in a circuit — Inductance of circuit — Rate of change of current in circuit

The minus sign in Eq. (30.7) is a reflection of Lenz's law; it says that the self-induced emf in a circuit opposes any change in the current in that circuit.

Equation (30.7) states that the self-inductance of a circuit is the magnitude of the self-induced emf per unit rate of change of current. This relationship makes it possible to measure an unknown self-inductance: Change the current at a known rate di/dt, measure the induced emf, and take the ratio to determine L.

Inductors As Circuit Elements

A circuit device that is designed to have a particular inductance is called an **inductor,** or a *choke*. The usual circuit symbol for an inductor is

⟨⟨00000⟩⟩

Like resistors and capacitors, inductors are among the indispensable circuit elements of modern electronics. Their purpose is to oppose any variations in

Application Inductors, Power Transmission, and Lightning Strikes If lightning strikes part of an electrical power transmission system, it causes a sudden spike in voltage that can damage the components of the system as well as anything connected to that system (for example, home appliances). To minimize these effects, large inductors are incorporated into the transmission system. These use the principle that an inductor opposes and suppresses any rapid changes in the current.

the current through the circuit. An inductor in a direct-current circuit helps to maintain a steady current despite any fluctuations in the applied emf; in an alternating-current circuit, an inductor tends to suppress variations of the current that are more rapid than desired.

To understand the behavior of circuits containing inductors, we need to develop a general principle analogous to Kirchhoff's loop rule (discussed in Section 26.2). To apply that rule, we go around a conducting loop, measuring potential differences across successive circuit elements as we go. The algebraic sum of these differences around any closed loop must be zero because the electric field produced by charges distributed around the circuit is *conservative*. In Section 29.7 we denoted such a conservative field as $\vec{E}_c$.

When an inductor is included in the circuit, the situation changes. The magnetically induced electric field within the coils of the inductor is *not* conservative; as in Section 29.7, we'll denote it by $\vec{E}_n$. We need to think very carefully about the roles of the various fields. Let's assume we are dealing with an inductor whose coils have negligible resistance. Then a negligibly small electric field is required to make charge move through the coils, so the *total* electric field $\vec{E}_c + \vec{E}_n$ within the coils must be zero, even though neither field is individually zero. Because $\vec{E}_c$ is nonzero, there have to be accumulations of charge on the terminals of the inductor and the surfaces of its conductors to produce this field.

Consider the circuit shown in **Fig. 30.5**; the box contains some combination of batteries and variable resistors that enables us to control the current i in the circuit. According to Faraday's law, Eq. (29.10), the line integral of $\vec{E}_n$ around the circuit is the negative of the rate of change of flux through the circuit, which in turn is given by Eq. (30.7). Combining these two relationships, we get

$$\oint \vec{E}_n \cdot d\vec{l} = -L\frac{di}{dt}$$

where we integrate clockwise around the loop (the direction of the assumed current). But $\vec{E}_n$ is different from zero only within the inductor. Therefore the integral of $\vec{E}_n$ around the whole loop can be replaced by its integral only from a to b through the inductor; that is,

$$\int_a^b \vec{E}_n \cdot d\vec{l} = -L\frac{di}{dt}$$

Next, because $\vec{E}_c + \vec{E}_n = 0$ at each point within the inductor coils, $\vec{E}_n = -\vec{E}_c$. So we can rewrite the above equation as

$$\int_a^b \vec{E}_c \cdot d\vec{l} = L\frac{di}{dt}$$

But this integral is just the potential V_{ab} of point a with respect to point b, so

$$V_{ab} = V_a - V_b = L\frac{di}{dt} \tag{30.8}$$

We conclude that there is a genuine potential difference between the terminals of the inductor, associated with conservative, electrostatic forces, even though the electric field associated with magnetic induction is nonconservative. Thus we are justified in using Kirchhoff's loop rule to analyze circuits that include inductors. Equation (30.8) gives the potential difference across an inductor in a circuit.

_____

CAUTION **Self-induced emf opposes changes in current** Note that the self-induced emf does not oppose the current i itself; rather, it opposes any *change* (di/dt) in the current. Thus the circuit behavior of an inductor is quite different from that of a resistor. **Figure 30.6** compares the behaviors of a resistor and an inductor and summarizes the sign relationships. ▌

30.5 A circuit containing a source of emf and an inductor. The source is variable, so the current i and its rate of change di/dt can be varied.

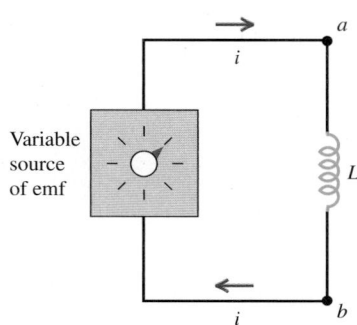

30.6 (a) The potential difference across a resistor depends on the current, whereas (b), (c), (d) the potential difference across an inductor depends on the rate of change of the current.

(a) Resistor with current i flowing from a to b: potential drops from a to b.

$$V_{ab} = iR > 0$$

(b) Inductor with *constant* current i flowing from a to b: no potential difference.

i constant: $di/dt = 0$

$\mathcal{E} = 0$

$$V_{ab} = L\frac{di}{dt} = 0$$

(c) Inductor with *increasing* current i flowing from a to b: potential drops from a to b.

i increasing: $di/dt > 0$

$\mathcal{E}$

$$V_{ab} = L\frac{di}{dt} > 0$$

(d) Inductor with *decreasing* current i flowing from a to b: potential increases from a to b.

i decreasing: $di/dt < 0$

$\mathcal{E}$

$$V_{ab} = L\frac{di}{dt} < 0$$

30.7 These fluorescent light tubes are wired in series with an inductor, or ballast, that helps to sustain the current flowing through the tubes.

Applications of Inductors

Because an inductor opposes changes in current, it plays an important role in fluorescent light fixtures (**Fig. 30.7**). In such fixtures, current flows from the wiring into the gas that fills the tube, ionizing the gas and causing it to glow. However, an ionized gas or *plasma* is a highly nonohmic conductor: The greater the current, the more highly ionized the plasma becomes and the lower its resistance. If a sufficiently large voltage is applied to the plasma, the current can grow so much that it damages the circuitry outside the fluorescent tube. To prevent this problem, an inductor or *magnetic ballast* is put in series with the fluorescent tube to keep the current from growing out of bounds.

The ballast also makes it possible for the fluorescent tube to work with the alternating voltage provided by household wiring. This voltage oscillates sinusoidally with a frequency of 60 Hz, so that it goes momentarily to zero 120 times per second. If there were no ballast, the plasma in the fluorescent tube would rapidly deionize when the voltage went to zero and the tube would shut off. With a ballast present, a self-induced emf sustains the current and keeps the tube lit. Magnetic ballasts are also used for this purpose in streetlights (which obtain their light from a glowing mercury or sodium vapor) and in neon lights. (In compact fluorescent lamps, the magnetic ballast is replaced by a more complicated scheme that utilizes transistors, discussed in Chapter 42.)

The self-inductance of a circuit depends on its size, shape, and number of turns. For N turns close together, it is always proportional to N^2. It also depends on the magnetic properties of the material enclosed by the circuit. In the following examples we will assume that the circuit encloses only vacuum (or air, which from the standpoint of magnetism is essentially vacuum). If, however, the flux is concentrated in a region containing a magnetic material with permeability μ, then in the expression for B we must replace μ_0 (the permeability of vacuum) by $\mu = K_m\mu_0$, as discussed in Section 28.8. If the material is diamagnetic or paramagnetic, this replacement makes very little difference, since K_m is very close to 1. If the material is *ferromagnetic,* however, the difference is of crucial importance. A solenoid wound on a soft iron core having $K_m = 5000$ can have an inductance approximately 5000 times as great as that of the same solenoid with an air core. Ferromagnetic-core inductors are very widely used in a variety of electronic and electric-power applications.

With ferromagnetic materials, the magnetization is in general not a linear function of magnetizing current, especially as saturation is approached. As a result, the inductance is not constant but can depend on current in a fairly complicated way. In our discussion we will ignore this complication and assume always that the inductance is constant. This is a reasonable assumption even for a ferromagnetic material if the magnetization remains well below the saturation level.

Because automobiles contain steel, a ferromagnetic material, driving an automobile over a coil causes an appreciable increase in the coil's inductance. This effect is used in traffic light sensors, which use a large, current-carrying coil embedded under the road surface near an intersection. The circuitry connected to the coil detects the inductance change as a car drives over. When a preprogrammed number of cars have passed over the coil, the light changes to green to allow the cars through the intersection.

EXAMPLE 30.3 **CALCULATING SELF-INDUCTANCE**

Determine the self-inductance of a toroidal solenoid with cross-sectional area A and mean radius r, closely wound with N turns of wire on a nonmagnetic core (**Fig. 30.8**). Assume that B is uniform across a cross section (that is, neglect the variation of B with distance from the toroid axis).

SOLUTION

IDENTIFY and SET UP: Our target variable is the self-inductance L of the toroidal solenoid. We can find L by using Eq. (30.6), which requires knowing the flux Φ_B through each turn and the

30.8 Determining the self-inductance of a closely wound toroidal solenoid. For clarity, only a few turns of the winding are shown. Part of the toroid has been cut away to show the cross-sectional area A and radius r.

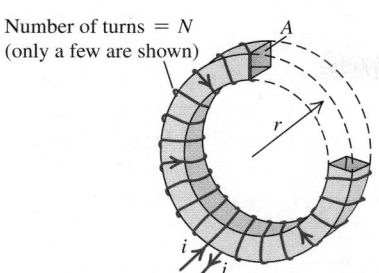

Number of turns $= N$
(only a few are shown)

current i in the coil. For this, we use the results of Example 28.10 (Section 28.7), in which we found the magnetic field in the interior of a toroidal solenoid as a function of the current.

EXECUTE: From Eq. (30.6), the self-inductance is $L = N\Phi_B/i$. From Example 28.10, the field magnitude at a distance r from the toroid axis is $B = \mu_0 Ni/2\pi r$. If we assume that the field has this magnitude over the entire cross-sectional area A, then

$$\Phi_B = BA = \frac{\mu_0 NiA}{2\pi r}$$

The flux Φ_B is the same through each turn, and

$$L = \frac{N\Phi_B}{i} = \frac{\mu_0 N^2 A}{2\pi r} \quad \text{(self-inductance of a toroidal solenoid)}$$

EVALUATE: Suppose $N = 200$ turns, $A = 5.0 \text{ cm}^2 = 5.0 \times 10^{-4} \text{ m}^2$, and $r = 0.10$ m; then

$$L = \frac{(4\pi \times 10^{-7} \text{ Wb/A} \cdot \text{m})(200)^2(5.0 \times 10^{-4} \text{ m}^2)}{2\pi(0.10 \text{ m})}$$

$$= 40 \times 10^{-6} \text{ H} = 40 \,\mu\text{H}$$

EXAMPLE 30.4   CALCULATING SELF-INDUCED EMF

If the current in the toroidal solenoid in Example 30.3 increases uniformly from 0 to 6.0 A in 3.0 μs, find the magnitude and direction of the self-induced emf.

SOLUTION

IDENTIFY and SET UP: We are given L, the self-inductance, and di/dt, the rate of change of the solenoid current. We find the magnitude of the self-induced emf $\mathcal{E}$ by using Eq. (30.7) and its direction by using Lenz's law.

EXECUTE: We have $di/dt = (6.0 \text{ A})/(3.0 \times 10^{-6} \text{ s}) = 2.0 \times 10^6$ A/s. From Eq. (30.7), the magnitude of the induced emf is

$$|\mathcal{E}| = L\left|\frac{di}{dt}\right| = (40 \times 10^{-6} \text{ H})(2.0 \times 10^6 \text{ A/s}) = 80 \text{ V}$$

The current is increasing, so according to Lenz's law the direction of the emf is opposite to that of the current. This corresponds to the situation in Fig. 30.6c; the emf is in the direction from b to a, like a battery with a as the $+$ terminal and b the $-$ terminal, tending to oppose the current increase from the external circuit.

EVALUATE: This example shows that even a small inductance L can give rise to a substantial induced emf if the current changes rapidly.

TEST YOUR UNDERSTANDING OF SECTION 30.2 Rank the following inductors in order of the potential difference V_{ab}, from most positive to most negative. In each case the inductor has zero resistance and the current flows from point a through the inductor to point b. (i) The current through a 2.0-μH inductor increases from 1.0 A to 2.0 A in 0.50 s; (ii) the current through a 4.0-μH inductor decreases from 3.0 A to 0 in 2.0 s; (iii) the current through a 1.0-μH inductor remains constant at 4.0 A; (iv) the current through a 1.0-μH inductor increases from 0 to 4.0 A in 0.25 s. ∎

30.3 MAGNETIC-FIELD ENERGY

Establishing a current in an inductor requires an input of energy, and an inductor carrying a current has energy stored in it. Let's see how this comes about. In Fig. 30.5, an increasing current i in the inductor causes an emf $\mathcal{E}$ between its terminals and a corresponding potential difference V_{ab} between the terminals of the source, with point a at higher potential than point b. Thus the source must be adding energy to the inductor, and the instantaneous power P (rate of transfer of energy into the inductor) is $P = V_{ab}i$.

Energy Stored in an Inductor

We can calculate the total energy input U needed to establish a final current I in an inductor with inductance L if the initial current is zero. We assume that the inductor has zero resistance, so no energy is dissipated within the inductor. Let the rate of change of the current i at some instant be di/dt; the current is increasing, so $di/dt > 0$. The voltage between the terminals a and b of the inductor at this instant is $V_{ab} = L\,di/dt$, and the rate P at which energy is being delivered to the inductor (equal to the instantaneous power supplied by the external source) is

$$P = V_{ab}i = Li\frac{di}{dt}$$

The energy dU supplied to the inductor during an infinitesimal time interval dt is $dU = P\,dt$, so

$$dU = Li\,di$$

The total energy U supplied while the current increases from zero to a final value I is

$$\text{Energy stored} \atop \text{in an inductor} \quad U = L\int_0^I i\,di = \tfrac{1}{2}LI^2 \qquad (30.9)$$

Inductance · · · · Final current

Integral from initial (zero) value of instantaneous current to final value

After the current has reached its final steady value I, $di/dt = 0$ and no more energy is input to the inductor. When there is no current, the stored energy U is zero; when the current is I, the energy is $\tfrac{1}{2}LI^2$.

When the current decreases from I to zero, the inductor acts as a source that supplies a total amount of energy $\tfrac{1}{2}LI^2$ to the external circuit. If we interrupt the circuit suddenly by opening a switch, the current decreases very rapidly, the induced emf is very large, and the energy may be dissipated in an arc across the switch contacts.

CAUTION **Energy, resistors, and inductors** Don't confuse the behavior of resistors and inductors where energy is concerned (**Fig. 30.9**). Energy flows into a resistor whenever a current passes through it, whether the current is steady or varying; this energy is dissipated in the form of heat. By contrast, energy flows into an ideal, zero-resistance inductor only when the current in the inductor *increases*. This energy is not dissipated; it is stored in the inductor and released when the current *decreases*. When a steady current flows through an inductor, there is no energy flow in or out. ▮

Magnetic Energy Density

The energy in an inductor is actually stored in the magnetic field of the coil, just as the energy of a capacitor is stored in the electric field between its plates. We can develop relationships for magnetic-field energy analogous to those we obtained for electric-field energy in Section 24.3 [Eqs. (24.9) and (24.11)]. We will concentrate on one simple case, the ideal toroidal solenoid. This system has the advantage that its magnetic field is confined completely to a finite region of space within its core. As in Example 30.3, we assume that the cross-sectional area A is small enough that we can pretend that the magnetic field is uniform over the area. The volume V enclosed by the toroidal solenoid is approximately equal to the circumference $2\pi r$ multiplied by the area A: $V = 2\pi rA$. From Example 30.3, the self-inductance of the toroidal solenoid with vacuum within its coils is

$$L = \frac{\mu_0 N^2 A}{2\pi r}$$

30.9 A resistor is a device in which energy is irrecoverably dissipated. By contrast, energy stored in a current-carrying inductor can be recovered when the current decreases to zero.

Resistor with current i: energy is *dissipated*.

Inductor with current i: energy is *stored*.

From Eq. (30.9), the energy U stored in the toroidal solenoid when the current is I is

$$U = \tfrac{1}{2}LI^2 = \tfrac{1}{2}\frac{\mu_0 N^2 A}{2\pi r}I^2$$

The magnetic field and therefore this energy are localized in the volume $V = 2\pi r A$ enclosed by the windings. The energy *per unit volume*, or *magnetic energy density*, is $u = U/V$:

$$u = \frac{U}{2\pi r A} = \tfrac{1}{2}\mu_0 \frac{N^2 I^2}{(2\pi r)^2}$$

We can express this in terms of the magnitude B of the magnetic field inside the toroidal solenoid. From Eq. (28.24) in Example 28.10 (Section 28.7), this is

$$B = \frac{\mu_0 N I}{2\pi r}$$

and so

$$\frac{N^2 I^2}{(2\pi r)^2} = \frac{B^2}{\mu_0^2}$$

When we substitute this into the above equation for u, we finally find the expression for **magnetic energy density** in vacuum:

$$\text{Magnetic energy density in vacuum} \quad u = \frac{B^2 \cdots \text{Magnetic-field magnitude}}{2\mu_0 \cdots \text{Magnetic constant}} \qquad (30.10)$$

This is the magnetic analog of the energy per unit volume in an *electric* field in vacuum, $u = \tfrac{1}{2}\epsilon_0 E^2$, which we derived in Section 24.3. As an example, the energy density in the 1.5-T magnetic field of an MRI scanner (see Section 27.7) is $u = B^2/2\mu_0 = (1.5\text{ T})^2/(2 \times 4\pi \times 10^{-7}\text{ T}\cdot\text{m/A}) = 9.0 \times 10^5\text{ J/m}^3$.

When the material inside the toroid is not vacuum but a material with (constant) magnetic permeability $\mu = K_m\mu_0$, we replace μ_0 by μ in Eq. (30.10):

$$\text{Magnetic energy density in a material} \quad u = \frac{B^2 \cdots \text{Magnetic-field magnitude}}{2\mu \cdots \text{Permeability of material}} \qquad (30.11)$$

Although we have derived Eq. (30.11) for only one special situation, it turns out to be the correct expression for the energy per unit volume associated with *any* magnetic-field configuration in a material with constant permeability. For vacuum, Eq. (30.11) reduces to Eq. (30.10). We will use the expressions for electric-field and magnetic-field energy in Chapter 32 when we study the energy associated with electromagnetic waves.

Magnetic-field energy plays an important role in the ignition systems of gasoline-powered automobiles. A primary coil of about 250 turns is connected to the car's battery and produces a strong magnetic field. This coil is surrounded by a secondary coil with some 25,000 turns of very fine wire. When it is time for a spark plug to fire (see Fig. 20.5 in Section 20.3), the current to the primary coil is interrupted, the magnetic field quickly drops to zero, and an emf of tens of thousands of volts is induced in the secondary coil. The energy stored in the magnetic field thus goes into a powerful pulse of current that travels through the secondary coil to the spark plug, generating the spark that ignites the fuel–air mixture in the engine's cylinders (**Fig. 30.10**).

Application A Magnetic Eruption on the Sun This composite of two images of the sun shows a coronal mass ejection, a dramatic event in which about 10^{12} kg (a billion tons) of material from the sun's outer atmosphere is ejected into space at speeds of 500 km/s or faster. Such ejections happen at intervals of a few hours to a few days. These immense eruptions are powered by the energy stored in the sun's magnetic field. Unlike the earth's relatively steady magnetic field, the sun's field is constantly changing, and regions of unusually strong field (and hence unusually high magnetic energy density) frequently form. A coronal mass ejection occurs when the energy stored in such a region is suddenly released.

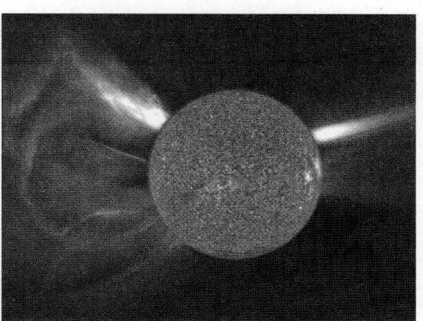

30.10 The energy required to fire an automobile spark plug is derived from magnetic-field energy stored in the ignition coil.

EXAMPLE 30.5 STORING ENERGY IN AN INDUCTOR

The electric-power industry would like to find efficient ways to store electrical energy generated during low-demand hours to help meet customer requirements during high-demand hours. Could a large inductor be used? What inductance would be needed to store 1.00 kW·h of energy in a coil carrying a 200-A current?

SOLUTION

IDENTIFY and SET UP: We are given the required amount of stored energy U and the current $I = 200$ A. We use Eq. (30.9) to find the self-inductance L.

EXECUTE: Here we have $I = 200$ A and $U = 1.00$ kW·h $= (1.00 \times 10^3$ W$)(3600$ s$) = 3.60 \times 10^6$ J. Solving Eq. (30.9) for L, we find

$$L = \frac{2U}{I^2} = \frac{2(3.60 \times 10^6 \text{ J})}{(200 \text{ A})^2} = 180 \text{ H}$$

EVALUATE: The required inductance is more than a million times greater than the self-inductance of the toroidal solenoid of Example 30.3. Conventional wires that are to carry 200 A would have to be of large diameter to keep the resistance low and avoid unacceptable energy losses due to I^2R heating. As a result, a 180-H inductor using conventional wire would be very large (room-size). A superconducting inductor could be much smaller, since the resistance of a superconductor is zero and much thinner wires could be used; but the wires would have to be kept at low temperature to remain superconducting, and maintaining this temperature would itself require energy. This scheme is impractical with present technology.

TEST YOUR UNDERSTANDING OF SECTION 30.3 The current in a solenoid is reversed in direction while keeping the same magnitude. (a) Does this change the magnetic field within the solenoid? (b) Does this change the magnetic energy density in the solenoid? ▮

30.4 THE *R-L* CIRCUIT

Let's look at some examples of the circuit behavior of an inductor. One thing is clear already; an inductor in a circuit makes it difficult for rapid changes in current to occur, thanks to the effects of self-induced emf. Equation (30.7) shows that the greater the rate of change of current di/dt, the greater the self-induced emf and the greater the potential difference between the inductor terminals. This equation, together with Kirchhoff's rules (see Section 26.2), gives us the principles we need to analyze circuits containing inductors.

PROBLEM-SOLVING STRATEGY 30.1 INDUCTORS IN CIRCUITS

IDENTIFY *the relevant concepts:* An inductor is just another circuit element, like a source of emf, a resistor, or a capacitor. One key difference is that when an inductor is included in a circuit, all the voltages, currents, and capacitor charges are in general functions of time, not constants as they have been in most of our previous circuit analysis. But even when the voltages and currents vary with time, Kirchhoff's rules (see Section 26.2) hold at each instant of time.

SET UP *the problem* using the following steps:
1. Follow the procedure given in Problem-Solving Strategy 26.2 (Section 26.2). Draw a circuit diagram and label all quantities, known and unknown. Apply the junction rule immediately to express the currents in terms of as few quantities as possible.
2. Determine which quantities are the target variables.

EXECUTE *the solution* as follows:
1. As in Problem-Solving Strategy 26.2, apply Kirchhoff's loop rule to each loop in the circuit.

2. Review the sign rules given in Problem-Solving Strategy 26.2. To get the correct sign for the potential difference between the terminals of an inductor, apply Lenz's law and the sign rule described in Section 30.2 in connection with Eq. (30.7) and Fig. 30.6. In Kirchhoff's loop rule, when we go through an inductor in the *same* direction as the assumed current, we encounter a voltage *drop* equal to $L\,di/dt$, so the corresponding term in the loop equation is $-L\,di/dt$. When we go through an inductor in the *opposite* direction from the assumed current, the potential difference is reversed and the term to use in the loop equation is $+L\,di/dt$.
3. Solve for the target variables.

EVALUATE *your answer:* Check whether your answer is consistent with the behavior of inductors. By Lenz's law, if the current through an inductor is changing, your result should indicate that the potential difference across the inductor opposes the change.

Current Growth in an *R-L* Circuit

We can learn a great deal about inductor behavior by analyzing the circuit of **Fig. 30.11**. A circuit that includes both a resistor and an inductor, and possibly a source of emf, is called an ***R-L* circuit.** The inductor helps to prevent rapid changes in current, which can be useful if a steady current is required but the source has a fluctuating emf. The resistor R may be a separate circuit element, or it may be the resistance of the inductor windings; every real-life inductor has some resistance unless it is made of superconducting wire. By closing switch S_1, we can connect the *R-L* combination to a source with constant emf $\mathcal{E}$. (We assume that the source has zero internal resistance, so the terminal voltage equals $\mathcal{E}$.)

Suppose both switches are open to begin with, and then at some initial time $t = 0$ we close switch S_1. The current cannot change suddenly from zero to some final value, since di/dt and the induced emf in the inductor would both be infinite. Instead, the current begins to grow at a rate that depends on the value of L in the circuit.

Let i be the current at some time t after switch S_1 is closed, and let di/dt be its rate of change at that time. The potential differences v_{ab} (across the resistor) and v_{bc} (across the inductor) are

$$v_{ab} = iR \quad \text{and} \quad v_{bc} = L\frac{di}{dt}$$

Note that if the current is in the direction shown in Fig. 30.11 and is increasing, then both v_{ab} and v_{bc} are positive; a is at a higher potential than b, which in turn is at a higher potential than c. (Compare to Figs. 30.6a and c.) We apply Kirchhoff's loop rule, starting at the negative terminal and proceeding counter-clockwise around the loop:

$$\mathcal{E} - iR - L\frac{di}{dt} = 0 \tag{30.12}$$

Solving this for di/dt, we find that the rate of increase of current is

$$\frac{di}{dt} = \frac{\mathcal{E} - iR}{L} = \frac{\mathcal{E}}{L} - \frac{R}{L}i \tag{30.13}$$

At the instant that switch S_1 is first closed, $i = 0$ and the potential drop across R is zero. The initial rate of change of current is

$$\left(\frac{di}{dt}\right)_{\text{initial}} = \frac{\mathcal{E}}{L}$$

The greater the inductance L, the more slowly the current increases.

As the current increases, the term $(R/L)i$ in Eq. (30.13) also increases, and the *rate* of increase of current given by Eq. (30.13) becomes smaller and smaller. This means that the current is approaching a final, steady-state value I. When the current reaches this value, its rate of increase is zero. Then Eq. (30.13) becomes

$$\left(\frac{di}{dt}\right)_{\text{final}} = 0 = \frac{\mathcal{E}}{L} - \frac{R}{L}I \quad \text{and}$$

$$I = \frac{\mathcal{E}}{R}$$

The *final* current I does not depend on the inductance L; it is the same as it would be if the resistance R alone were connected to the source with emf $\mathcal{E}$.

Figure 30.12 shows the behavior of the current as a function of time. To derive the equation for this curve (that is, an expression for current as a function of time), we proceed just as we did for the charging capacitor in Section 26.4. First we rearrange Eq. (30.13) to the form

$$\frac{di}{i - (\mathcal{E}/R)} = -\frac{R}{L}dt$$

30.11 An *R-L* circuit.

Closing switch S_1 connects the *R-L* combination in series with a source of emf $\mathcal{E}$.

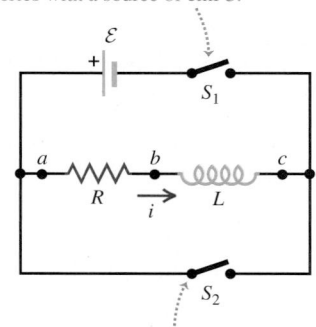

Closing switch S_2 while opening switch S_1 disconnnects the combination from the source.

30.12 Graph of i versus t for growth of current in an *R-L* circuit with an emf in series. The final current is $I = \mathcal{E}/R$; after one time constant τ, the current is $1 - 1/e$ of this value.

Switch S_1 is closed at $t = 0$.

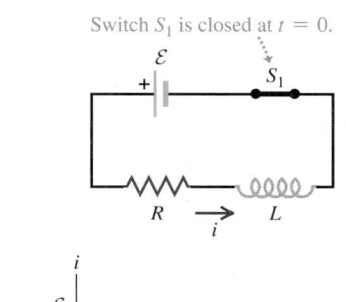

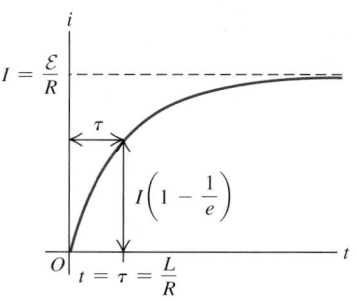

This separates the variables, with i on the left side and t on the right. Then we integrate both sides, renaming the integration variables i' and t' so that we can use i and t as the upper limits. (The lower limit for each integral is zero, corresponding to zero current at the initial time $t = 0$.) We get

$$\int_0^i \frac{di'}{i' - (\mathcal{E}/R)} = -\int_0^t \frac{R}{L} dt'$$

$$\ln\left(\frac{i - (\mathcal{E}/R)}{-\mathcal{E}/R}\right) = -\frac{R}{L}t$$

Now we take exponentials of both sides and solve for i. We leave the details for you to work out; the final result is the equation of the curve in Fig. 30.12:

$$i = \frac{\mathcal{E}}{R}(1 - e^{-(R/L)t}) \qquad \text{(current in an } R\text{-}L \text{ circuit with emf)} \qquad (30.14)$$

Taking the derivative of Eq. (30.14), we find

$$\frac{di}{dt} = \frac{\mathcal{E}}{L}e^{-(R/L)t} \qquad (30.15)$$

At time $t = 0$, $i = 0$ and $di/dt = \mathcal{E}/L$. As $t \to \infty$, $i \to \mathcal{E}/R$ and $di/dt \to 0$, as we predicted.

As Fig. 30.12 shows, the instantaneous current i first rises rapidly, then increases more slowly and approaches the final value $I = \mathcal{E}/R$ asymptotically. At a time equal to L/R, the current has risen to $(1 - 1/e)$, or about 63%, of its final value. The quantity L/R is therefore a measure of how quickly the current builds toward its final value; this quantity is called the **time constant** for the circuit, denoted by τ:

$$\text{Time constant for an } R\text{-}L \text{ circuit} \qquad \tau = \frac{L \leftarrow \text{Inductance}}{R \leftarrow \text{Resistance}} \qquad (30.16)$$

In a time equal to 2τ, the current reaches 86% of its final value; in 5τ, 99.3%; and in 10τ, 99.995%. (Compare the discussion in Section 26.4 of charging a capacitor of capacitance C that was in series with a resistor of resistance R; the time constant for that situation was the product RC.)

The graphs of i versus t have the same general shape for all values of L. For a given value of R, the time constant τ is greater for greater values of L. When L is small, the current rises rapidly to its final value; when L is large, it rises more slowly. For example, if $R = 100 \ \Omega$ and $L = 10 \ \text{H}$,

$$\tau = \frac{L}{R} = \frac{10 \ \text{H}}{100 \ \Omega} = 0.10 \ \text{s}$$

and the current increases to about 63% of its final value in 0.10 s. (Recall that $1 \ \text{H} = 1 \ \Omega \cdot \text{s}$.) But if $L = 0.010 \ \text{H}$, $\tau = 1.0 \times 10^{-4} \ \text{s} = 0.10 \ \text{ms}$, and the rise is much more rapid.

Energy considerations offer us additional insight into the behavior of an R-L circuit. The instantaneous rate at which the source delivers energy to the circuit is $P = \mathcal{E}i$. The instantaneous rate at which energy is dissipated in the resistor is i^2R, and the rate at which energy is stored in the inductor is $iv_{bc} = Li \, di/dt$ [or, equivalently, $(d/dt)\left(\frac{1}{2}Li^2\right) = Li \, di/dt$]. When we multiply Eq. (30.12) by i and rearrange, we find

$$\mathcal{E}i = i^2R + Li\frac{di}{dt} \qquad (30.17)$$

Of the power $\mathcal{E}i$ supplied by the source, part (i^2R) is dissipated in the resistor and part $(Li \, di/dt)$ goes to store energy in the inductor. This discussion is analogous to our power analysis for a charging capacitor, given at the end of Section 26.4.

EXAMPLE 30.6 ANALYZING AN *R-L* CIRCUIT

A sensitive electronic device of resistance $R = 175\ \Omega$ is to be connected to a source of emf (of negligible internal resistance) by a switch. The device is designed to operate with a 36-mA current, but to avoid damage to the device, the current can rise to no more than 4.9 mA in the first 58 μs after the switch is closed. An inductor is therefore connected in series with the device, as in Fig. 30.11; the switch in question is S_1. (a) What is the required source emf $\mathcal{E}$? (b) What is the required inductance L? (c) What is the R-L time constant τ?

SOLUTION

IDENTIFY and SET UP: This problem concerns current and current growth in an *R-L* circuit, so we can use the ideas of this section. Figure 30.12 shows the current i versus the time t that has elapsed since closing S_1. The graph shows that the final current is $I = \mathcal{E}/R$; we are given $R = 175\ \Omega$, so the emf is determined by the requirement that the final current be $I = 36$ mA. The other requirement is that the current be no more than $i = 4.9$ mA at $t = 58$ μs; to satisfy this, we use Eq. (30.14) for the current as a function of time and solve for the inductance, which is the only unknown quantity. Equation (30.16) then tells us the time constant.

EXECUTE: (a) We solve $I = \mathcal{E}/R$ for $\mathcal{E}$:

$$\mathcal{E} = IR = (0.036\ \text{A})(175\ \Omega) = 6.3\ \text{V}$$

(b) To find the required inductance, we solve Eq. (30.14) for L. First we multiply through by $(-R/\mathcal{E})$ and add 1 to both sides:

$$1 - \frac{iR}{\mathcal{E}} = e^{-(R/L)t}$$

Then we take natural logs of both sides, solve for L, and substitute:

$$L = \frac{-Rt}{\ln(1 - iR/\mathcal{E})}$$

$$= \frac{-(175\ \Omega)(58 \times 10^{-6}\ \text{s})}{\ln[1 - (4.9 \times 10^{-3}\ \text{A})(175\ \Omega)/(6.3\ \text{V})]} = 69\ \text{mH}$$

(c) From Eq. (30.16),

$$\tau = \frac{L}{R} = \frac{69 \times 10^{-3}\ \text{H}}{175\ \Omega} = 3.9 \times 10^{-4}\ \text{s} = 390\ \mu\text{s}$$

EVALUATE: Note that 58 μs is much less than the time constant. In 58 μs the current builds up from zero to 4.9 mA, a small fraction of its final value of 36 mA; after 390 μs the current equals $(1 - 1/e)$ of its final value, or about $(0.63)(36\ \text{mA}) = 23$ mA.

Current Decay in an *R-L* Circuit

Now suppose switch S_1 in the circuit of Fig. 30.11 has been closed for a while and the current has reached the value I_0. Resetting our stopwatch to redefine the initial time, we close switch S_2 at time $t = 0$, bypassing the battery. (At the same time we should open S_1 to protect the battery.) The current through R and L does not instantaneously go to zero but decays smoothly, as shown in **Fig. 30.13**. The Kirchhoff's-rule loop equation is obtained from Eq. (30.12) by omitting the $\mathcal{E}$ term. We challenge you to retrace the steps in the above analysis and show that the current i varies with time according to

$$i = I_0 e^{-(R/L)t} \tag{30.18}$$

where I_0 is the initial current at time $t = 0$. The time constant, $\tau = L/R$, is the time for current to decrease to $1/e$, or about 37%, of its original value. In time 2τ it has dropped to 13.5%, in time 5τ to 0.67%, and in 10τ to 0.0045%.

The energy that is needed to maintain the current during this decay is provided by the energy stored in the magnetic field of the inductor. The detailed energy analysis is simpler this time. In place of Eq. (30.17) we have

$$0 = i^2 R + Li\frac{di}{dt} \tag{30.19}$$

Now $Li\, di/dt$ is negative; Eq. (30.19) shows that the energy stored in the inductor *decreases* at the same rate i^2R at which energy is dissipated in the resistor.

This entire discussion should look familiar; the situation is very similar to that of a charging and discharging capacitor, analyzed in Section 26.4. It would be a good idea to compare that section with our discussion of the *R-L* circuit.

30.13 Graph of i versus t for decay of current in an *R-L* circuit. After one time constant τ, the current is $1/e$ of its initial value.

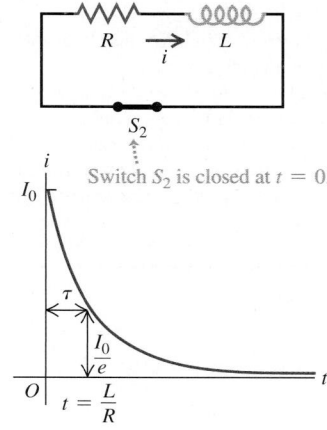

EXAMPLE 30.7   ENERGY IN AN *R-L* CIRCUIT

When the current in an *R-L* circuit is decaying, what fraction of the original energy stored in the inductor has been dissipated after 2.3 time constants?

SOLUTION

IDENTIFY and SET UP: This problem concerns current decay in an *R-L* circuit as well as the relationship between the current in an inductor and the amount of stored energy. The current i at any time t is given by Eq. (30.18); the stored energy associated with this current is given by Eq. (30.9), $U = \frac{1}{2}Li^2$.

EXECUTE: From Eq. (30.18), the current i at any time t is

$$i = I_0 e^{-(R/L)t}$$

We substitute this into $U = \frac{1}{2}Li^2$ to obtain an expression for the stored energy at any time:

$$U = \frac{1}{2}LI_0^2 e^{-2(R/L)t} = U_0 e^{-2(R/L)t}$$

where $U_0 = \frac{1}{2}LI_0^2$ is the energy at the initial time $t = 0$. When $t = 2.3\tau = 2.3L/R$, we have

$$U = U_0 e^{-2(2.3)} = U_0 e^{-4.6} = 0.010 U_0$$

That is, only 0.010 or 1.0% of the energy initially stored in the inductor remains, so 99.0% has been dissipated in the resistor.

EVALUATE: To get a sense of what this result means, consider the *R-L* circuit we analyzed in Example 30.6, for which $\tau = 390\ \mu$s. With $L = 69$ mH and $I_0 = 36$ mA, we have $U_0 = \frac{1}{2}LI_0^2 = \frac{1}{2}(0.069\ \text{H})(0.036\ \text{A})^2 = 4.5 \times 10^{-5}$ J. Of this, 99.0% or 4.4×10^{-5} J is dissipated in $2.3(390\ \mu\text{s}) = 9.0 \times 10^{-4}$ s $= 0.90$ ms. In other words, this circuit can be almost completely powered off (or powered on) in 0.90 ms, so the minimum time for a complete on–off cycle is 1.8 ms. Even shorter cycle times are required for many purposes, such as in fast switching networks for telecommunications. In such cases a smaller time constant $\tau = L/R$ is needed.

DATA *SPEAKS*

Inductors in Circuits

When students were given a problem involving an *R-L* circuit, more than 23% gave an incorrect response. Common errors:

- Confusion about current and its rate of change. The current i cannot change abruptly in a circuit with an inductor, so i must be a continuous function of time t. However, di/dt can change abruptly (say, when the emf in Fig. 30.11 is connected to the circuit).

- Confusion about initial and final values. When an emf is connected to an *R-L* circuit as in Fig. 30.12, the inductor opposes current change and so $i = 0$ just after the switch is closed. Long after the switch is closed and the current has stabilized, the inductor acts like a simple wire and has no effect.

TEST YOUR UNDERSTANDING OF SECTION 30.4   (a) In Fig. 30.11, what are the algebraic signs of the potential differences v_{ab} and v_{bc} when switch S_1 is closed and switch S_2 is open? (i) $v_{ab} > 0$, $v_{bc} > 0$; (ii) $v_{ab} > 0$, $v_{bc} < 0$; (iii) $v_{ab} < 0$, $v_{bc} > 0$; (iv) $v_{ab} < 0$, $v_{bc} < 0$. (b) What are the signs of v_{ab} and v_{bc} when S_1 is open, S_2 is closed, and current is flowing in the direction shown? (i) $v_{ab} > 0$, $v_{bc} > 0$; (ii) $v_{ab} > 0$, $v_{bc} < 0$; (iii) $v_{ab} < 0$, $v_{bc} > 0$; (iv) $v_{ab} < 0$, $v_{bc} < 0$. ▌

30.5 THE *L-C* CIRCUIT

A circuit containing an inductor and a capacitor shows an entirely new mode of behavior, characterized by *oscillating* current and charge. This is in sharp contrast to the *exponential* approach to a steady-state situation that we have seen with both *R-C* and *R-L* circuits. In the **L-C circuit** in **Fig. 30.14a** we charge the capacitor to a potential difference V_m and initial charge $Q_m = CV_m$ on its left-hand plate and then close the switch. What happens?

The capacitor begins to discharge through the inductor. Because of the induced emf in the inductor, the current cannot change instantaneously; it starts at zero and eventually builds up to a maximum value I_m. During this buildup the capacitor is discharging. At each instant the capacitor potential equals the induced emf, so as the capacitor discharges, the *rate of change* of current decreases. When the capacitor potential becomes zero, the induced emf is also zero, and the current has leveled off at its maximum value I_m. Figure 30.14b shows this situation; the capacitor has completely discharged. The potential difference between its terminals (and those of the inductor) has decreased to zero, and the current has reached its maximum value I_m.

During the discharge of the capacitor, the increasing current in the inductor has established a magnetic field in the space around it, and the energy that was initially stored in the capacitor's electric field is now stored in the inductor's magnetic field.

Although the capacitor is completely discharged in Fig. 30.14b, the current persists (it cannot change instantaneously), and the capacitor begins to charge with polarity opposite to that in the initial state. As the current decreases, the magnetic field also decreases, inducing an emf in the inductor in the *same* direction as the

30.14 In an oscillating *L-C* circuit, the charge on the capacitor and the current through the inductor both vary sinusoidally with time. Energy is transferred between magnetic energy in the inductor (U_B) and electrical energy in the capacitor (U_E). As in simple harmonic motion, the total energy E remains constant. (Compare Fig. 14.14 in Section 14.3.)

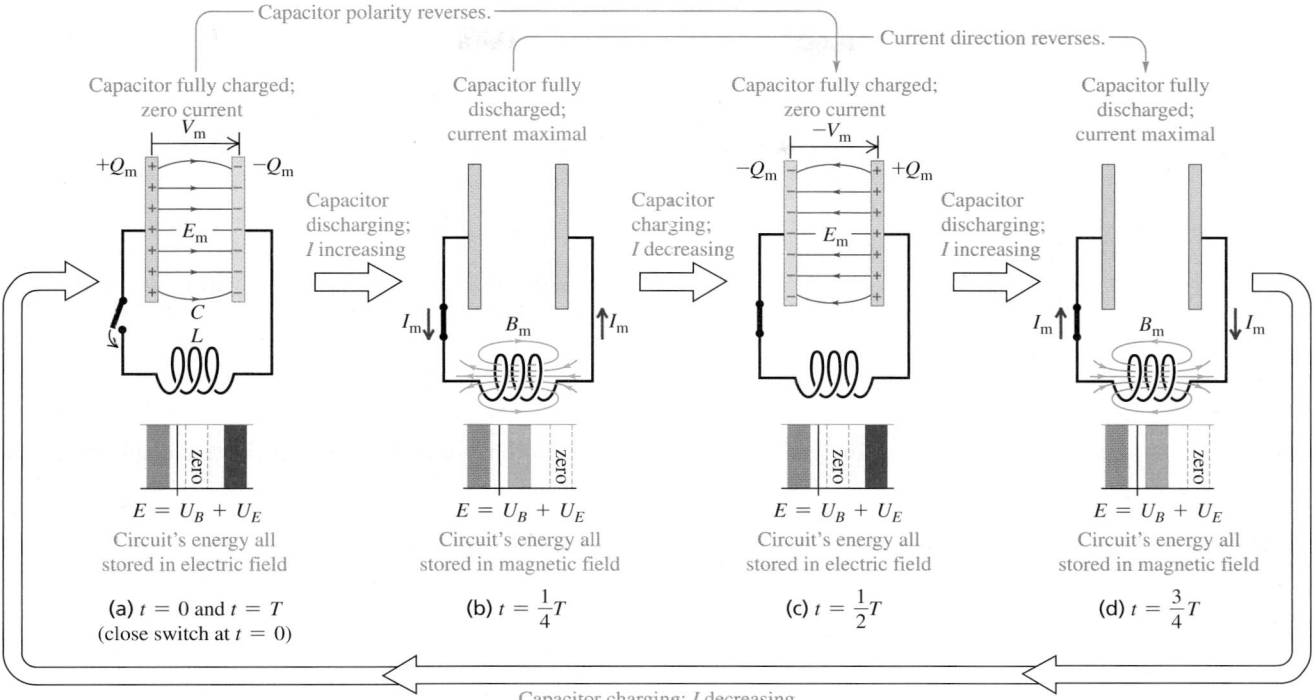

Capacitor polarity reverses.

Current direction reverses.

Capacitor fully charged; zero current

V_m

$+Q_m$ $-Q_m$

E_m

C
L

Capacitor discharging; *I* increasing

Capacitor fully discharged; current maximal

I_m I_m

B_m

Capacitor charging; *I* decreasing

Capacitor fully charged; zero current

$-V_m$

$-Q_m$ $+Q_m$

E_m

Capacitor discharging; *I* increasing

Capacitor fully discharged; current maximal

I_m B_m I_m

zero

$E = U_B + U_E$
Circuit's energy all stored in electric field

(a) $t = 0$ and $t = T$
(close switch at $t = 0$)

zero

$E = U_B + U_E$
Circuit's energy all stored in magnetic field

(b) $t = \frac{1}{4}T$

zero

$E = U_B + U_E$
Circuit's energy all stored in electric field

(c) $t = \frac{1}{2}T$

zero

$E = U_B + U_E$
Circuit's energy all stored in magnetic field

(d) $t = \frac{3}{4}T$

Capacitor charging; *I* decreasing

current; this slows down the decrease of the current. Eventually, the current and the magnetic field reach zero, and the capacitor has been charged in the sense *opposite* to its initial polarity (Fig. 30.14c), with potential difference $-V_m$ and charge $-Q_m$ on its left-hand plate.

The process now repeats in the reverse direction; a little later, the capacitor has again discharged, and there is a current in the inductor in the opposite direction (Fig. 30.14d). Still later, the capacitor charge returns to its original value (Fig. 30.14a), and the whole process repeats. If there are no energy losses, the charges on the capacitor continue to oscillate back and forth indefinitely. This process is called an **electrical oscillation.** (Before you read further, review the analogous case of *mechanical* oscillation in Sections 14.2 and 14.3.)

From an energy standpoint the oscillations of an electric circuit transfer energy from the capacitor's electric field to the inductor's magnetic field and back. The *total* energy associated with the circuit is constant. This is analogous to the transfer of energy in an oscillating mechanical system from potential energy to kinetic energy and back, with constant total energy (Section 14.3). As we will see, this analogy goes much further.

Electrical Oscillations in an *L-C* Circuit

To study the flow of charge in detail, we proceed just as we did for the *R-L* circuit. **Figure 30.15** shows our definitions of q and i.

CAUTION **Positive current in an *L-C* circuit** After you have examined Fig. 30.14, the positive direction for current in Fig. 30.15 may seem backward. In fact, we chose this direction to simplify the relationship between current and capacitor charge. We define the current at each instant to be $i = dq/dt$, the rate of change of the charge on the left-hand capacitor plate. If the capacitor is initially charged and begins to discharge as in Figs. 30.14a and 30.14b, then $dq/dt < 0$ and the initial current i is negative; the current direction is opposite to the (positive) direction shown in Fig. 30.15. ▌

30.15 Applying Kirchhoff's loop rule to the *L-C* circuit. The direction of travel around the loop in the loop equation is shown. Just after the circuit is completed and the capacitor first begins to discharge, as in Fig. 30.14a, the current is negative (opposite to the direction shown).

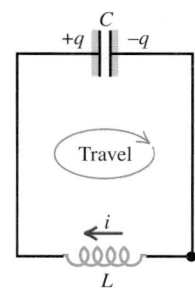

C
$+q$ $-q$

Travel

i

L

We apply Kirchhoff's loop rule to the circuit in Fig. 30.15. Starting at the lower-right corner of the circuit and adding voltages as we go clockwise around the loop, we obtain

$$-L\frac{di}{dt} - \frac{q}{C} = 0$$

Since $i = dq/dt$, it follows that $di/dt = d^2q/dt^2$. We substitute this expression into the above equation and divide by $-L$ to obtain

$$\frac{d^2q}{dt^2} + \frac{1}{LC}q = 0 \qquad (L\text{-}C \text{ circuit}) \qquad (30.20)$$

Equation (30.20) has exactly the same form as the equation we derived for simple harmonic motion in Section 14.2, Eq. (14.4): $d^2x/dt^2 = -(k/m)x$, or

$$\frac{d^2x}{dt^2} + \frac{k}{m}x = 0$$

In an L-C circuit the capacitor charge q plays the role of the displacement x, and the current $i = dq/dt$ is analogous to the particle's velocity $v_x = dx/dt$. The inductance L is analogous to the mass m, and the reciprocal of the capacitance, $1/C$, is analogous to the force constant k.

Pursuing this analogy, we recall that the angular frequency $\omega = 2\pi f$ of the harmonic oscillator is equal to $(k/m)^{1/2}$ [Eq. (14.10)], and the position is given as a function of time by Eq. (14.13),

$$x = A\cos(\omega t + \phi)$$

where the amplitude A and the phase angle ϕ depend on the initial conditions. In the analogous electrical situation, the capacitor charge q is given by

$$q = Q\cos(\omega t + \phi) \qquad (30.21)$$

and the angular frequency ω of oscillation is given by

$$\text{Angular frequency of} \cdots\cdots \omega = \sqrt{\frac{1}{LC}} \cdots \text{Capacitance} \qquad (30.22)$$
$$\text{oscillation in an } L\text{-}C \text{ circuit} \qquad\qquad\qquad \text{Inductance}$$

Verify that Eq. (30.21) satisfies the loop equation, Eq. (30.20), when ω has the value given by Eq. (30.22). In doing this, you will find that the instantaneous current $i = dq/dt$ is

$$i = -\omega Q\sin(\omega t + \phi) \qquad (30.23)$$

Thus the charge and current in an L-C circuit oscillate sinusoidally with time, with an angular frequency determined by the values of L and C. The ordinary frequency f, the number of cycles per second, is equal to $\omega/2\pi$. The constants Q and ϕ in Eqs. (30.21) and (30.23) are determined by the initial conditions. If at time $t = 0$ the left-hand capacitor plate in Fig. 30.15 has its maximum charge Q and the current i is zero, then $\phi = 0$. If $q = 0$ at $t = 0$, then $\phi = \pm\pi/2$ rad.

Energy in an *L-C* Circuit

We can also analyze the L-C circuit by using an energy approach. The analogy to simple harmonic motion is equally useful here. In the mechanical problem a body with mass m is attached to a spring with force constant k. Suppose we displace the body a distance A from its equilibrium position and release it from rest at time $t = 0$. The kinetic energy of the system at any later time is $\frac{1}{2}mv_x^2$, and its elastic potential energy is $\frac{1}{2}kx^2$. Because the system is conservative, the sum of these energies equals the initial energy of the system, $\frac{1}{2}kA^2$. We find the velocity v_x at any position x just as we did in Section 14.3, Eq. (14.22):

$$v_x = \pm\sqrt{\frac{k}{m}}\sqrt{A^2 - x^2} \qquad (30.24)$$

The *L-C* circuit is also a conservative system. Again let Q be the maximum capacitor charge. The magnetic-field energy $\frac{1}{2}Li^2$ in the inductor at any time corresponds to the kinetic energy $\frac{1}{2}mv^2$ of the oscillating body, and the electric-field energy $q^2/2C$ in the capacitor corresponds to the elastic potential energy $\frac{1}{2}kx^2$ of the spring. The sum of these energies is the total energy $Q^2/2C$ of the system:

$$\frac{1}{2}Li^2 + \frac{q^2}{2C} = \frac{Q^2}{2C} \qquad (30.25)$$

The total energy in the *L-C* circuit is *constant;* it oscillates between the magnetic and the electric forms, just as the constant total mechanical energy in simple harmonic motion is constant and oscillates between the kinetic and potential forms.

Solving Eq. (30.25) for i, we find that when the charge on the capacitor is q, the current i is

$$i = \pm\sqrt{\frac{1}{LC}}\sqrt{Q^2 - q^2} \qquad (30.26)$$

Verify this equation by substituting q from Eq. (30.21) and i from Eq. (30.23). Comparing Eqs. (30.24) and (30.26), we see that current $i = dq/dt$ and charge q are related in the same way as are velocity $v_x = dx/dt$ and position x in the mechanical problem.

Table 30.1 summarizes the analogies between simple harmonic motion and *L-C* circuit oscillations. The striking parallels shown there are so close that we can solve complicated mechanical problems by setting up analogous electric circuits and measuring the currents and voltages that correspond to the mechanical quantities to be determined. This is the basic principle of many analog computers. This analogy can be extended to *damped* oscillations, which we consider in the next section. In Chapter 31 we will extend the analogy further to include *forced* electrical oscillations, which occur in all alternating-current circuits.

Oscillation of a Mass-Spring System Compared with Electrical Oscillation in an *L-C* Circuit

TABLE 30.1

Mass-Spring System

Kinetic energy $= \frac{1}{2}mv_x^2$

Potential energy $= \frac{1}{2}kx^2$

$\frac{1}{2}mv_x^2 + \frac{1}{2}kx^2 = \frac{1}{2}kA^2$

$v_x = \pm\sqrt{k/m}\sqrt{A^2 - x^2}$

$v_x = dx/dt$

$\omega = \sqrt{\dfrac{k}{m}}$

$x = A\cos(\omega t + \phi)$

Inductor-Capacitor Circuit

Magnetic energy $= \frac{1}{2}Li^2$

Electrical energy $= q^2/2C$

$\frac{1}{2}Li^2 + q^2/2C = Q^2/2C$

$i = \pm\sqrt{1/LC}\sqrt{Q^2 - q^2}$

$i = dq/dt$

$\omega = \sqrt{\dfrac{1}{LC}}$

$q = Q\cos(\omega t + \phi)$

EXAMPLE 30.8   **AN OSCILLATING CIRCUIT**

A 300-V dc power supply is used to charge a 25-μF capacitor. After the capacitor is fully charged, it is disconnected from the power supply and connected across a 10-mH inductor. The resistance in the circuit is negligible. (a) Find the frequency and period of oscillation of the circuit. (b) Find the capacitor charge and the circuit current 1.2 ms after the inductor and capacitor are connected.

SOLUTION

IDENTIFY and SET UP: Our target variables are the oscillation frequency f and period T, as well as the charge q and current i at a particular time t. We are given the capacitance C and the inductance L, with which we can calculate the frequency and period from Eq. (30.22). We find the charge and current from Eqs. (30.21) and (30.23). Initially the capacitor is fully charged and the current is zero, as in Fig. 30.14a, so the phase angle is $\phi = 0$ [see the discussion that follows Eq. (30.23)].

EXECUTE: (a) The natural angular frequency is

$$\omega = \sqrt{\frac{1}{LC}} = \sqrt{\frac{1}{(10 \times 10^{-3}\text{ H})(25 \times 10^{-6}\text{ F})}}$$

$$= 2.0 \times 10^3 \text{ rad/s}$$

The frequency f and period T are then

$$f = \frac{\omega}{2\pi} = \frac{2.0 \times 10^3 \text{ rad/s}}{2\pi \text{ rad/cycle}} = 320 \text{ Hz}$$

$$T = \frac{1}{f} = \frac{1}{320 \text{ Hz}}$$

$$= 3.1 \times 10^{-3} \text{ s} = 3.1 \text{ ms}$$

Continued

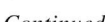

(b) Since the period of the oscillation is $T = 3.1$ ms, $t = 1.2$ ms equals $0.38T$; this corresponds to a situation intermediate between Fig. 30.14b ($t = T/4$) and Fig. 30.14c ($t = T/2$). Comparing those figures with Fig. 30.15, we expect the capacitor charge q to be negative (that is, there will be negative charge on the left-hand plate of the capacitor) and the current i to be negative as well (that is, current will flow counterclockwise).

To find q, we use Eq. (30.21), $q = Q \cos(\omega t + \phi)$. The charge is maximum at $t = 0$, so $\phi = 0$ and $Q = C\mathcal{E} = (25 \times 10^{-6}\,\text{F}) \times (300\,\text{V}) = 7.5 \times 10^{-3}\,\text{C}$. Hence Eq. (30.21) becomes

$$q = (7.5 \times 10^{-3}\,\text{C}) \cos \omega t$$

At time $t = 1.2 \times 10^{-3}$ s,

$$\omega t = (2.0 \times 10^3\,\text{rad/s})(1.2 \times 10^{-3}\,\text{s}) = 2.4\,\text{rad}$$

$$q = (7.5 \times 10^{-3}\,\text{C}) \cos(2.4\,\text{rad}) = -5.5 \times 10^{-3}\,\text{C}$$

From Eq. (30.23), the current i at any time is $i = -\omega Q \sin \omega t$. At $t = 1.2 \times 10^{-3}$ s,

$$i = -(2.0 \times 10^3\,\text{rad/s})(7.5 \times 10^{-3}\,\text{C}) \sin(2.4\,\text{rad}) = -10\,\text{A}$$

EVALUATE: The signs of both q and i are negative, as predicted.

EXAMPLE 30.9   ENERGY IN AN OSCILLATING CIRCUIT

For the L-C circuit of Example 30.8, find the magnetic and electrical energies (a) at $t = 0$ and (b) at $t = 1.2$ ms.

SOLUTION

IDENTIFY and SET UP: We must calculate the magnetic energy U_B (stored in the inductor) and the electrical energy U_E (stored in the capacitor) at two times during the L-C circuit oscillation. From Example 30.8 we know the values of the capacitor charge q and circuit current i for both times. We use them to calculate $U_B = \frac{1}{2}Li^2$ and $U_E = q^2/2C$.

EXECUTE: (a) At $t = 0$ there is no current and $q = Q$. Hence there is no magnetic energy, and all the energy in the circuit is in the form of electrical energy in the capacitor:

$$U_B = \tfrac{1}{2}Li^2 = 0 \qquad U_E = \frac{Q^2}{2C} = \frac{(7.5 \times 10^{-3}\,\text{C})^2}{2(25 \times 10^{-6}\,\text{F})} = 1.1\,\text{J}$$

(b) From Example 30.8, at $t = 1.2$ ms we have $i = -10$ A and $q = -5.5 \times 10^{-3}$ C. Hence

$$U_B = \tfrac{1}{2}Li^2 = \tfrac{1}{2}(10 \times 10^{-3}\,\text{H})(-10\,\text{A})^2 = 0.5\,\text{J}$$

$$U_E = \frac{q^2}{2C} = \frac{(-5.5 \times 10^{-3}\,\text{C})^2}{2(25 \times 10^{-6}\,\text{F})} = 0.6\,\text{J}$$

EVALUATE: The magnetic and electrical energies are the same at $t = 3T/8 = 0.375T$, halfway between the situations in Figs. 30.14b and 30.14c. We saw in Example 30.8 that the time considered in part (b), $t = 1.2$ ms, equals $0.38T$; this is slightly later than $0.375T$, so U_B is slightly less than U_E. At *all* times the *total* energy $E = U_B + U_E$ has the same value, 1.1 J. An L-C circuit without resistance is a conservative system; no energy is dissipated.

TEST YOUR UNDERSTANDING OF SECTION 30.5   One way to think about the energy stored in an L-C circuit is to say that the circuit elements do positive or negative work on the charges that move back and forth through the circuit. (a) Between stages (a) and (b) in Fig. 30.14, does the capacitor do positive or negative work on the charges? (b) What kind of force (electric or magnetic) does the capacitor exert on the charges to do this work? (c) During this process, does the inductor do positive or negative work on the charges? (d) What kind of force (electric or magnetic) does the inductor exert on the charges? ∎

30.6 THE *L-R-C* SERIES CIRCUIT

In our discussion of the L-C circuit we assumed that there was no *resistance* in the circuit. This is an idealization, of course; every real inductor has resistance in its windings, and there may also be resistance in the connecting wires. Because of resistance, the electromagnetic energy in the circuit is dissipated and converted to other forms, such as internal energy of the circuit materials. Resistance in an electric circuit is analogous to friction in a mechanical system.

Suppose an inductor with inductance L and a resistor of resistance R are connected in series across the terminals of a charged capacitor, forming an **L-R-C series circuit.** As before, the capacitor starts to discharge as soon as the circuit is completed. But due to i^2R losses in the resistor, the magnetic-field energy that the inductor acquires when the capacitor is completely discharged is *less* than the original electric-field energy of the capacitor. In the same way, the energy of the capacitor when the magnetic field has decreased to zero is still less and so on.

If the resistance R of the resistor is relatively small, the circuit still oscillates, but with **damped harmonic motion (Fig. 30.16a)**, and we say that the circuit is **underdamped.** If we increase R, the oscillations die out more rapidly. When R reaches a certain value, the circuit no longer oscillates; it is **critically damped** (Fig. 30.16b). For still larger values of R, the circuit is **overdamped** (Fig. 30.16c), and the capacitor charge approaches zero even more slowly. We used these same terms to describe the behavior of the analogous mechanical system, the damped harmonic oscillator, in Section 14.7.

Analyzing an *L-R-C* Series Circuit

To analyze *L-R-C* series circuit behavior in detail, consider the circuit shown in **Fig. 30.17**. It is like the *L-C* circuit of Fig. 30.15 except for the added resistor R; we also show the source that charges the capacitor initially. The labeling of the positive senses of q and i is the same as for the *L-C* circuit.

First we close the switch in the upward position, connecting the capacitor to a source of emf $\mathcal{E}$ for a long enough time to ensure that the capacitor acquires its final charge $Q = C\mathcal{E}$ and any initial oscillations have died out. Then at time $t = 0$ we flip the switch to the downward position, removing the source from the circuit and placing the capacitor in series with the resistor and inductor. Note that the initial current is negative, opposite to the direction of i shown in Fig. 30.17.

To find how q and i vary with time, we apply Kirchhoff's loop rule. Starting at point a and going around the loop in the direction $abcda$, we obtain

$$-iR - L\frac{di}{dt} - \frac{q}{C} = 0$$

Replacing i with dq/dt and rearranging, we get

$$\frac{d^2q}{dt^2} + \frac{R}{L}\frac{dq}{dt} + \frac{1}{LC}q = 0 \tag{30.27}$$

Note that when $R = 0$, this reduces to Eq. (30.20) for an *L-C* circuit.

There are general methods for obtaining solutions of Eq. (30.27). The form of the solution is different for the underdamped (small R) and overdamped (large R) cases. When R^2 is less than $4L/C$, the solution has the form

$$q = Ae^{-(R/2L)t}\cos\left(\sqrt{\frac{1}{LC} - \frac{R^2}{4L^2}}\,t + \phi\right) \tag{30.28}$$

where A and ϕ are constants. You can take the first and second derivatives of this function and show by direct substitution that it does satisfy Eq. (30.27).

This solution corresponds to the *underdamped* behavior shown in Fig. 30.16a; the function represents a sinusoidal oscillation with an exponentially decaying amplitude. (Note that the exponential factor $e^{-(R/2L)t}$ is *not* the same as the factor $e^{-(R/L)t}$ that we encountered in describing the *R-L* circuit in Section 30.4.) When $R = 0$, Eq. (30.28) reduces to Eq. (30.21) for the oscillations in an *L-C* circuit. If R is not zero, the angular frequency of the oscillation is *less* than $1/(LC)^{1/2}$ because of the term containing R. The angular frequency ω' of the damped oscillations is given by

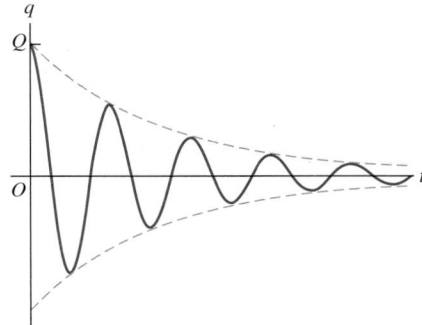

30.16 Graphs of capacitor charge as a function of time in an *L-R-C* series circuit with initial charge Q.

(a) Underdamped circuit (small resistance R)

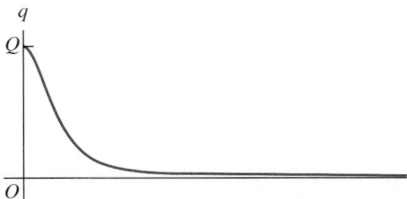

(b) Critically damped circuit (larger resistance R)

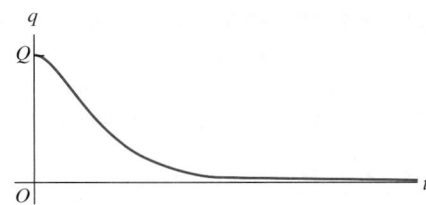

(c) Overdamped circuit (very large resistance R)

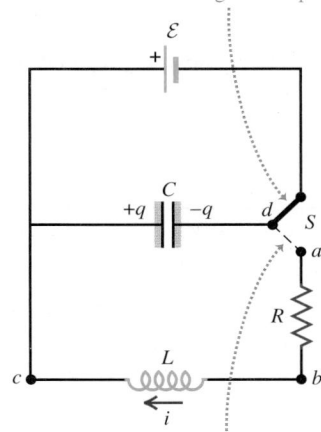

30.17 An *L-R-C* series circuit.

When switch S is in this position, the emf charges the capacitor.

When switch S is moved to this position, the capacitor discharges through the resistor and inductor.

$$\begin{array}{c}\text{Angular frequency of} \\ \text{underdamped oscillations} \\ \text{in an } L\text{-}R\text{-}C \text{ series circuit}\end{array} \cdots\!\!\rightarrow \omega' = \sqrt{\frac{1}{LC} - \frac{R^2}{4L^2}} \begin{array}{l}\cdots\text{ Resistance} \\[6pt] \cdots\text{ Inductance}\end{array} \tag{30.29}$$

Inductance Capacitance

When $R = 0$, this reduces to Eq. (30.22), $\omega = (1/LC)^{1/2}$. As R increases, ω' becomes smaller and smaller. When $R^2 = 4L/C$, the quantity under the radical becomes zero; the system no longer oscillates, and the case of *critical damping* (Fig. 30.16b) has been reached. For still larger values of R the system behaves as in Fig. 30.16c. In this case the circuit is *overdamped,* and q is given as a function of time by the sum of two decreasing exponential functions.

In the *underdamped* case the phase constant ϕ in the cosine function of Eq. (30.28) provides for the possibility of both an initial charge and an initial current at time $t = 0$, analogous to an underdamped harmonic oscillator given both an initial displacement and an initial velocity (see Exercise 30.41).

We emphasize once more that the behavior of the *L-R-C* series circuit is completely analogous to that of the damped harmonic oscillator, Section 14.7. We invite you to verify, for example, that if you start with Eq. (14.41) and substitute q for x, L for m, $1/C$ for k, and R for the damping constant b, the result is Eq. (30.27). Similarly, the cross-over point between underdamping and overdamping occurs at $b^2 = 4km$ for the mechanical system and at $R^2 = 4L/C$ for the electrical one. Can you find still other aspects of this analogy?

The practical applications of the *L-R-C* series circuit emerge when we include a sinusoidally varying source of emf in the circuit. This is analogous to the *forced oscillations* that we discussed in Section 14.7, and there are analogous *resonance* effects. Such a circuit is called an *alternating-current (ac) circuit.* The analysis of ac circuits is the principal topic of the next chapter.

EXAMPLE 30.10 AN UNDERDAMPED *L-R-C* SERIES CIRCUIT

What resistance R is required (in terms of L and C) to give an *L-R-C* series circuit a frequency that is one-half the undamped frequency?

SOLUTION

IDENTIFY and SET UP: This problem concerns an underdamped *L-R-C* series circuit (Fig. 30.16a). We want just enough resistance to reduce the oscillation frequency to one-half of the undamped value. Equation (30.29) gives the angular frequency ω' of an underdamped *L-R-C* series circuit; Eq. (30.22) gives the angular frequency ω of an undamped *L-C* circuit. We use these two equations to solve for R.

EXECUTE: From Eqs. (30.29) and (30.22), the requirement $\omega' = \omega/2$ yields

$$\sqrt{\frac{1}{LC} - \frac{R^2}{4L^2}} = \tfrac{1}{2}\sqrt{\frac{1}{LC}}$$

We square both sides and solve for R:

$$R = \sqrt{\frac{3L}{C}}$$

For example, adding $35\ \Omega$ to the circuit of Example 30.8 ($L = 10$ mH, $C = 25\ \mu$F) would reduce the frequency from 320 Hz to 160 Hz.

EVALUATE: The circuit becomes critically damped with no oscillations when $R = \sqrt{4L/C}$. Our result for R is smaller than that, as it should be; we want the circuit to be *under*damped.

TEST YOUR UNDERSTANDING OF SECTION 30.6 An *L-R-C* series circuit includes a 2.0-Ω resistor. At $t = 0$ the capacitor charge is 2.0 μC. For which of the following values of the inductance and capacitance will the charge on the capacitor *not* oscillate? (i) $L = 3.0\ \mu$H, $C = 6.0\ \mu$F; (ii) $L = 6.0\ \mu$H, $C = 3.0\ \mu$F; (iii) $L = 3.0\ \mu$H, $C = 3.0\ \mu$F. ▌

Mutual inductance: When a changing current i_1 in one circuit causes a changing magnetic flux in a second circuit, an emf $\mathcal{E}_2$ is induced in the second circuit. Likewise, a changing current i_2 in the second circuit induces an emf $\mathcal{E}_1$ in the first circuit. If the circuits are coils of wire with N_1 and N_2 turns, the mutual inductance M can be expressed in terms of the average flux Φ_{B2} through each turn of coil 2 caused by the current i_1 in coil 1, or in terms of the average flux Φ_{B1} through each turn of coil 1 caused by the current i_2 in coil 2. The SI unit of mutual inductance is the henry, abbreviated H. (See Examples 30.1 and 30.2.)

$$\mathcal{E}_2 = -M\frac{di_1}{dt} \quad \text{and}$$

$$\mathcal{E}_1 = -M\frac{di_2}{dt} \quad (30.4)$$

$$M = \frac{N_2\Phi_{B2}}{i_1} = \frac{N_1\Phi_{B1}}{i_2} \quad (30.5)$$

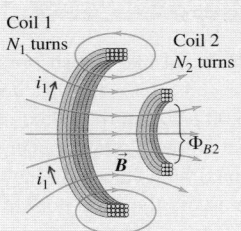

Self-inductance: A changing current i in any circuit causes a self-induced emf $\mathcal{E}$. The inductance (or self-inductance) L depends on the geometry of the circuit and the material surrounding it. The inductance of a coil of N turns is related to the average flux Φ_B through each turn caused by the current i in the coil. An inductor is a circuit device, usually including a coil of wire, intended to have a substantial inductance. (See Examples 30.3 and 30.4.)

$$\mathcal{E} = -L\frac{di}{dt} \quad (30.7)$$

$$L = \frac{N\Phi_B}{i} \quad (30.6)$$

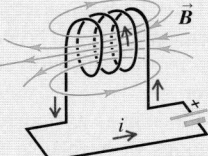

Magnetic-field energy: An inductor with inductance L carrying current I has energy U associated with the inductor's magnetic field. The magnetic energy density u (energy per unit volume) is proportional to the square of the magnetic-field magnitude. (See Example 30.5.)

$$U = \tfrac{1}{2}LI^2 \quad (30.9)$$

$$u = \frac{B^2}{2\mu_0} \quad \text{(in vacuum)} \quad (30.10)$$

$$u = \frac{B^2}{2\mu} \quad \begin{array}{l}\text{(in a material}\\ \text{with magnetic}\\ \text{permeability } \mu)\end{array} \quad (30.11)$$

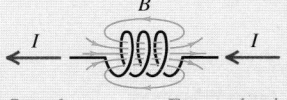

Stored energy $U = \frac{1}{2}LI^2$ Energy density $u = B^2/2\mu_0$

R-L circuits: In a circuit containing a resistor R, an inductor L, and a source of emf, the growth and decay of current are exponential. The time constant τ is the time required for the current to approach within a fraction $1/e$ of its final value. (See Examples 30.6 and 30.7.)

$$\tau = \frac{L}{R} \quad (30.16)$$

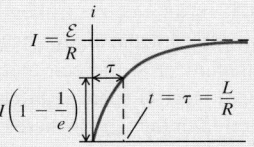

L-C circuits: A circuit that contains inductance L and capacitance C undergoes electrical oscillations with an angular frequency ω that depends on L and C. This is analogous to a mechanical harmonic oscillator, with inductance L analogous to mass m, the reciprocal of capacitance $1/C$ to force constant k, charge q to displacement x, and current i to velocity v_x. (See Examples 30.8 and 30.9.)

$$\omega = \sqrt{\frac{1}{LC}} \quad (30.22)$$

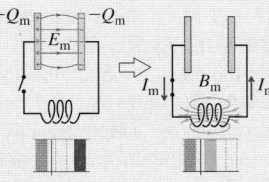

L-R-C series circuits: A circuit that contains inductance, resistance, and capacitance undergoes damped oscillations for sufficiently small resistance. The frequency ω' of damped oscillations depends on the values of L, R, and C. As R increases, the damping increases; if R is greater than a certain value, the behavior becomes over-damped and no longer oscillates. (See Example 30.10.)

$$\omega' = \sqrt{\frac{1}{LC} - \frac{R^2}{4L^2}} \quad (30.29)$$

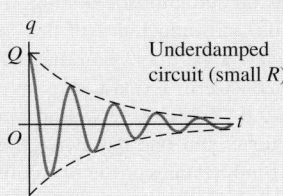

Underdamped circuit (small R)

BRIDGING PROBLEM   ANALYZING AN *L-C* CIRCUIT

An *L-C* circuit like that shown in Fig. 30.14 consists of a 60.0-mH inductor and a 250-μF capacitor. The initial charge on the capacitor is 6.00 μC, and the initial current in the inductor is 0.400 mA. (a) What is the maximum energy stored in the inductor? (b) What is the maximum current in the inductor? (c) What is the maximum voltage across the capacitor? (d) When the current in the inductor has half its maximum value, what are the energy stored in the inductor and the voltage across the capacitor?

SOLUTION GUIDE

IDENTIFY and SET UP

1. An *L-C* circuit is a conservative system—there is no resistance to dissipate energy. The energy oscillates between electrical energy in the capacitor and magnetic energy stored in the inductor.

2. Oscillations in an *L-C* circuit are analogous to the mechanical oscillations of a particle at the end of an ideal spring (see Table 30.1). Compare this problem to the analogous mechanical problem (see Example 14.3 in Section 14.2 and Example 14.4 in Section 14.3).

3. Which key equations are needed to describe the capacitor? To describe the inductor?

EXECUTE

4. Find the initial total energy in the circuit. Use it to determine the maximum energy stored in the inductor during the oscillation.

5. Use the result of step 4 to find the maximum current in the inductor.

6. Use the result of step 4 to find the maximum energy stored in the capacitor during the oscillation. Then use this to find the maximum capacitor voltage.

7. Find the energy in the inductor and the capacitor charge when the current has half the value that you found in step 5.

EVALUATE

8. Initially, what fraction of the total energy is in the inductor? Is it possible to tell whether this is initially increasing or decreasing?

Problems   For assigned homework and other learning materials, go to MasteringPhysics®.   (MP)

•, ••, •••: Difficulty levels. **CP**: Cumulative problems incorporating material from earlier chapters. **CALC**: Problems requiring calculus. **DATA**: Problems involving real data, scientific evidence, experimental design, and/or statistical reasoning. **BIO**: Biosciences problems.

DISCUSSION QUESTIONS

Q30.1 In an electric trolley or bus system, the vehicle's motor draws current from an overhead wire by means of a long arm with an attachment at the end that slides along the overhead wire. A brilliant electric spark is often seen when the attachment crosses a junction in the wires where contact is momentarily lost. Explain this phenomenon.

Q30.2 From Eq. (30.5) 1 H = 1 Wb/A, and from Eqs. (30.4) 1 H = 1 $\Omega \cdot$ s. Show that these two definitions are equivalent.

Q30.3 In Fig. 30.1, if coil 2 is turned 90° so that its axis is vertical, does the mutual inductance increase or decrease? Explain.

Q30.4 The tightly wound toroidal solenoid is one of the few configurations for which it is easy to calculate self-inductance. What features of the toroidal solenoid give it this simplicity?

Q30.5 Two identical, closely wound, circular coils, each having self-inductance *L*, are placed next to each other, so that they are coaxial and almost touching. If they are connected in series, what is the self-inductance of the combination? What if they are connected in parallel? Can they be connected so that the total inductance is zero? Explain.

Q30.6 Two closely wound circular coils have the same number of turns, but one has twice the radius of the other. How are the self-inductances of the two coils related? Explain your reasoning.

Q30.7 You are to make a resistor by winding a wire around a cylindrical form. To make the inductance as small as possible, it is proposed that you wind half the wire in one direction and the other half in the opposite direction. Would this achieve the desired result? Why or why not?

Q30.8 For the same magnetic field strength *B*, is the energy density greater in vacuum or in a magnetic material? Explain. Does Eq. (30.11) imply that for a long solenoid in which the current is *I* the energy stored is proportional to $1/\mu$? And does this mean that for the same current less energy is stored when the solenoid is filled with a ferromagnetic material rather than with air? Explain.

Q30.9 In an *R-C* circuit, a resistor, an uncharged capacitor, a dc battery, and an open switch are in series. In an *R-L* circuit, a resistor, an inductor, a dc battery, and an open switch are in series. Compare the behavior of the current in these circuits (a) just after the switch is closed and (b) long after the switch has been closed. In other words, compare the way in which a capacitor and an inductor affect a circuit.

Q30.10 A Differentiating Circuit. The current in a resistanceless inductor is caused to vary with time as shown in the graph of **Fig. Q30.10**. (a) Sketch the pattern that would be observed on the screen of an oscilloscope connected to the terminals of the inductor. (The oscilloscope spot sweeps horizontally across the screen at a constant speed, and its vertical deflection is proportional to the potential difference between the inductor terminals.) (b) Explain why a circuit with an inductor can be described as a "differentiating circuit."

Figure **Q30.10**

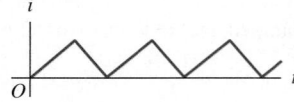

Q30.11 In Section 30.5 Kirchhoff's loop rule is applied to an L-C circuit where the capacitor is initially fully charged and the equation $-L\,(di/dt) - (q/C) = 0$ is derived. But as the capacitor starts to discharge, the current increases from zero. The equation says $L\,di/dt = -q/C$, so it says $L\,di/dt$ is negative. Explain how $L\,di/dt$ can be negative when the current is increasing.

Q30.12 In Section 30.5 the relationship $i = dq/dt$ is used in deriving Eq. (30.20). But a flow of current corresponds to a decrease in the charge on the capacitor. Explain, therefore, why this is the correct equation to use in the derivation, rather than $i = -dq/dt$.

Q30.13 In the R-L circuit shown in Fig. 30.11, when switch S_1 is closed, the potential v_{ac} changes suddenly and discontinuously, but the current does not. Explain why the voltage can change suddenly but the current can't.

Q30.14 In the R-L circuit shown in Fig. 30.11, is the current in the resistor always the same as the current in the inductor? How do you know?

Q30.15 Suppose there is a steady current in an inductor. If you attempt to reduce the current to zero instantaneously by quickly opening a switch, an arc can appear at the switch contacts. Why? Is it physically possible to stop the current instantaneously? Explain.

Q30.16 In an L-R-C series circuit, what criteria could be used to decide whether the system is overdamped or underdamped? For example, could we compare the maximum energy stored during one cycle to the energy dissipated during one cycle? Explain.

EXERCISES

Section 30.1 Mutual Inductance

30.1 • Two coils have mutual inductance $M = 3.25 \times 10^{-4}$ H. The current i_1 in the first coil increases at a uniform rate of 830 A/s. (a) What is the magnitude of the induced emf in the second coil? Is it constant? (b) Suppose that the current described is in the second coil rather than the first. What is the magnitude of the induced emf in the first coil?

30.2 • Two coils are wound around the same cylindrical form, like the coils in Example 30.1. When the current in the first coil is decreasing at a rate of -0.242 A/s, the induced emf in the second coil has magnitude 1.65×10^{-3} V. (a) What is the mutual inductance of the pair of coils? (b) If the second coil has 25 turns, what is the flux through each turn when the current in the first coil equals 1.20 A? (c) If the current in the second coil increases at a rate of 0.360 A/s, what is the magnitude of the induced emf in the first coil?

30.3 • A 10.0-cm-long solenoid of diameter 0.400 cm is wound uniformly with 800 turns. A second coil with 50 turns is wound around the solenoid at its center. What is the mutual inductance of the combination of the two coils?

30.4 • A solenoidal coil with 25 turns of wire is wound tightly around another coil with 300 turns (see Example 30.1). The inner solenoid is 25.0 cm long and has a diameter of 2.00 cm. At a certain time, the current in the inner solenoid is 0.120 A and is increasing at a rate of 1.75×10^3 A/s. For this time, calculate: (a) the average magnetic flux through each turn of the inner solenoid; (b) the mutual inductance of the two solenoids; (c) the emf induced in the outer solenoid by the changing current in the inner solenoid.

30.5 • Two toroidal solenoids are wound around the same form so that the magnetic field of one passes through the turns of the other. Solenoid 1 has 700 turns, and solenoid 2 has 400 turns.

When the current in solenoid 1 is 6.52 A, the average flux through each turn of solenoid 2 is 0.0320 Wb. (a) What is the mutual inductance of the pair of solenoids? (b) When the current in solenoid 2 is 2.54 A, what is the average flux through each turn of solenoid 1?

30.6 •• A toroidal solenoid with mean radius r and cross-sectional area A is wound uniformly with N_1 turns. A second toroidal solenoid with N_2 turns is wound uniformly on top of the first, so that the two solenoids have the same cross-sectional area and mean radius. (a) What is the mutual inductance of the two solenoids? Assume that the magnetic field of the first solenoid is uniform across the cross section of the two solenoids. (b) If $N_1 = 500$ turns, $N_2 = 300$ turns, $r = 10.0$ cm, and $A = 0.800$ cm^2, what is the value of the mutual inductance?

Section 30.2 Self-Inductance and Inductors

30.7 • A 2.50-mH toroidal solenoid has an average radius of 6.00 cm and a cross-sectional area of 2.00 cm^2. (a) How many coils does it have? (Make the same assumption as in Example 30.3.) (b) At what rate must the current through it change so that a potential difference of 2.00 V is developed across its ends?

30.8 • A toroidal solenoid has 500 turns, cross-sectional area 6.25 cm^2, and mean radius 4.00 cm. (a) Calculate the coil's self-inductance. (b) If the current decreases uniformly from 5.00 A to 2.00 A in 3.00 ms, calculate the self-induced emf in the coil. (c) The current is directed from terminal a of the coil to terminal b. Is the direction of the induced emf from a to b or from b to a?

30.9 • At the instant when the current in an inductor is increasing at a rate of 0.0640 A/s, the magnitude of the self-induced emf is 0.0160 V. (a) What is the inductance of the inductor? (b) If the inductor is a solenoid with 400 turns, what is the average magnetic flux through each turn when the current is 0.720 A?

30.10 •• When the current in a toroidal solenoid is changing at a rate of 0.0260 A/s, the magnitude of the induced emf is 12.6 mV. When the current equals 1.40 A, the average flux through each turn of the solenoid is 0.00285 Wb. How many turns does the solenoid have?

30.11 • The inductor in **Fig. E30.11** has inductance 0.260 H and carries a current in the direction shown that is decreasing at a uniform rate, $di/dt = -0.0180$ A/s. (a) Find the self-induced emf. (b) Which end of the inductor, a or b, is at a higher potential?

Figure **E30.11**

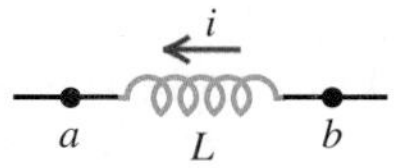

30.12 • The inductor shown in Fig. E30.11 has inductance 0.260 H and carries a current in the direction shown. The current is changing at a constant rate. (a) The potential between points a and b is $V_{ab} = 1.04$ V, with point a at higher potential. Is the current increasing or decreasing? (b) If the current at $t = 0$ is 12.0 A, what is the current at $t = 2.00$ s?

30.13 •• A toroidal solenoid has mean radius 12.0 cm and cross-sectional area 0.600 cm^2. (a) How many turns does the solenoid have if its inductance is 0.100 mH? (b) What is the resistance of the solenoid if the wire from which it is wound has a resistance per unit length of 0.0760 Ω/m?

30.14 • A long, straight solenoid has 800 turns. When the current in the solenoid is 2.90 A, the average flux through each turn of the solenoid is 3.25×10^{-3} Wb. What must be the magnitude of the rate of change of the current in order for the self-induced emf to equal 6.20 mV?

30.15 •• **Inductance of a Solenoid.** (a) A long, straight solenoid has N turns, uniform cross-sectional area A, and length l. Show that the inductance of this solenoid is given by the equation $L = \mu_0 A N^2 / l$. Assume that the magnetic field is uniform inside the solenoid and zero outside. (Your answer is approximate because B is actually smaller at the ends than at the center. For this reason, your answer is actually an upper limit on the inductance.) (b) A metallic laboratory spring is typically 5.00 cm long and 0.150 cm in diameter and has 50 coils. If you connect such a spring in an electric circuit, how much self-inductance must you include for it if you model it as an ideal solenoid?

Section 30.3 Magnetic-Field Energy

30.16 • An inductor used in a dc power supply has an inductance of 12.0 H and a resistance of 180 Ω. It carries a current of 0.500 A. (a) What is the energy stored in the magnetic field? (b) At what rate is thermal energy developed in the inductor? (c) Does your answer to part (b) mean that the magnetic-field energy is decreasing with time? Explain.

30.17 • An air-filled toroidal solenoid has a mean radius of 15.0 cm and a cross-sectional area of 5.00 cm^2. When the current is 12.0 A, the energy stored is 0.390 J. How many turns does the winding have?

30.18 • An air-filled toroidal solenoid has 300 turns of wire, a mean radius of 12.0 cm, and a cross-sectional area of 4.00 cm^2. If the current is 5.00 A, calculate: (a) the magnetic field in the solenoid; (b) the self-inductance of the solenoid; (c) the energy stored in the magnetic field; (d) the energy density in the magnetic field. (e) Check your answer for part (d) by dividing your answer to part (c) by the volume of the solenoid.

30.19 •• A solenoid 25.0 cm long and with a cross-sectional area of 0.500 cm^2 contains 400 turns of wire and carries a current of 80.0 A. Calculate: (a) the magnetic field in the solenoid; (b) the energy density in the magnetic field if the solenoid is filled with air; (c) the total energy contained in the coil's magnetic field (assume the field is uniform); (d) the inductance of the solenoid.

30.20 • It has been proposed to use large inductors as energy storage devices. (a) How much electrical energy is converted to light and thermal energy by a 150-W light bulb in one day? (b) If the amount of energy calculated in part (a) is stored in an inductor in which the current is 80.0 A, what is the inductance?

30.21 •• In a proton accelerator used in elementary particle physics experiments, the trajectories of protons are controlled by bending magnets that produce a magnetic field of 4.80 T. What is the magnetic-field energy in a 10.0-cm^3 volume of space where $B = 4.80$ T?

30.22 • It is proposed to store $1.00 \text{ kW} \cdot \text{h} = 3.60 \times 10^6$ J of electrical energy in a uniform magnetic field with magnitude 0.600 T. (a) What volume (in vacuum) must the magnetic field occupy to store this amount of energy? (b) If instead this amount of energy is to be stored in a volume (in vacuum) equivalent to a cube 40.0 cm on a side, what magnetic field is required?

Section 30.4 The R-L Circuit

30.23 • An inductor with an inductance of 2.50 H and a resistance of 8.00 Ω is connected to the terminals of a battery with an emf of 6.00 V and negligible internal resistance. Find (a) the initial rate of increase of current in the circuit; (b) the rate of increase of current at the instant when the current is 0.500 A; (c) the current 0.250 s after the circuit is closed; (d) the final steady-state current.

30.24 • In Fig. 30.11, $R = 15.0 \ \Omega$ and the battery emf is 6.30 V. With switch S_2 open, switch S_1 is closed. After several minutes, S_1 is opened and S_2 is closed. (a) At 2.00 ms after S_1 is opened, the current has decayed to 0.280 A. Calculate the inductance of the coil. (b) How long after S_1 is opened will the current reach 1.00% of its original value?

30.25 • A 35.0-V battery with negligible internal resistance, a 50.0-Ω resistor, and a 1.25-mH inductor with negligible resistance are all connected in series with an open switch. The switch is suddenly closed. (a) How long after closing the switch will the current through the inductor reach one-half of its maximum value? (b) How long after closing the switch will the energy stored in the inductor reach one-half of its maximum value?

30.26 • In Fig. 30.11, switch S_1 is closed while switch S_2 is kept open. The inductance is $L = 0.115$ H, and the resistance is $R = 120 \ \Omega$. (a) When the current has reached its final value, the energy stored in the inductor is 0.260 J. What is the emf $\mathcal{E}$ of the battery? (b) After the current has reached its final value, S_1 is opened and S_2 is closed. How much time does it take for the energy stored in the inductor to decrease to 0.130 J, half of the original value?

30.27 • In Fig. 30.11, suppose that $\mathcal{E} = 60.0$ V, $R = 240 \ \Omega$, and $L = 0.160$ H. With switch S_2 open, switch S_1 is left closed until a constant current is established. Then S_2 is closed and S_1 opened, taking the battery out of the circuit. (a) What is the initial current in the resistor, just after S_2 is closed and S_1 is opened? (b) What is the current in the resistor at $t = 4.00 \times 10^{-4}$ s? (c) What is the potential difference between points b and c at $t = 4.00 \times 10^{-4}$ s? Which point is at a higher potential? (d) How long does it take the current to decrease to half its initial value?

30.28 • In Fig. 30.11, suppose that $\mathcal{E} = 60.0$ V, $R = 240 \ \Omega$, and $L = 0.160$ H. Initially there is no current in the circuit. Switch S_2 is left open, and switch S_1 is closed. (a) Just after S_1 is closed, what are the potential differences v_{ab} and v_{bc}? (b) A long time (many time constants) after S_1 is closed, what are v_{ab} and v_{bc}? (c) What are v_{ab} and v_{bc} at an intermediate time when $i = 0.150$ A?

30.29 • In Fig. 30.11 switch S_1 is closed while switch S_2 is kept open. The inductance is $L = 0.380$ H, the resistance is $R = 48.0 \ \Omega$, and the emf of the battery is 18.0 V. At time t after S_1 is closed, the current in the circuit is increasing at a rate of $di/dt = 7.20$ A/s. At this instant what is v_{ab}, the voltage across the resistor?

30.30 •• Consider the circuit in Exercise 30.23. (a) Just after the circuit is completed, at what rate is the battery supplying electrical energy to the circuit? (b) When the current has reached its final steady-state value, how much energy is stored in the inductor? What is the rate at which electrical energy is being dissipated in the resistance of the inductor? What is the rate at which the battery is supplying electrical energy to the circuit?

Section 30.5 The L-C Circuit

30.31 • In an L-C circuit, $L = 85.0$ mH and $C = 3.20 \ \mu$F. During the oscillations the maximum current in the inductor is 0.850 mA. (a) What is the maximum charge on the capacitor? (b) What is the magnitude of the charge on the capacitor at an instant when the current in the inductor has magnitude 0.500 mA?

30.32 •• A 15.0-μF capacitor is charged by a 150.0-V power supply, then disconnected from the power and connected in series with a 0.280-mH inductor. Calculate: (a) the oscillation frequency of the circuit; (b) the energy stored in the capacitor at time $t = 0$ ms (the moment of connection with the inductor); (c) the energy stored in the inductor at $t = 1.30$ ms.

30.33 • A 7.50-nF capacitor is charged up to 12.0 V, then disconnected from the power supply and connected in series through a coil. The period of oscillation of the circuit is then measured to be 8.60×10^{-5} s. Calculate: (a) the inductance of the coil; (b) the maximum charge on the capacitor; (c) the total energy of the circuit; (d) the maximum current in the circuit.

30.34 •• A 18.0-μF capacitor is placed across a 22.5-V battery for several seconds and is then connected across a 12.0-mH inductor that has no appreciable resistance. (a) After the capacitor and inductor are connected together, find the maximum current in the circuit. When the current is a maximum, what is the charge on the capacitor? (b) How long after the capacitor and inductor are connected together does it take for the capacitor to be completely discharged for the first time? For the second time? (c) Sketch graphs of the charge on the capacitor plates and the current through the inductor as functions of time.

30.35 • **L-C Oscillations.** A capacitor with capacitance 6.00×10^{-5} F is charged by connecting it to a 12.0-V battery. The capacitor is disconnected from the battery and connected across an inductor with $L = 1.50$ H. (a) What are the angular frequency ω of the electrical oscillations and the period of these oscillations (the time for one oscillation)? (b) What is the initial charge on the capacitor? (c) How much energy is initially stored in the capacitor? (d) What is the charge on the capacitor 0.0230 s after the connection to the inductor is made? Interpret the sign of your answer. (e) At the time given in part (d), what is the current in the inductor? Interpret the sign of your answer. (f) At the time given in part (d), how much electrical energy is stored in the capacitor and how much is stored in the inductor?

30.36 • **A Radio Tuning Circuit.** The minimum capacitance of a variable capacitor in a radio is 4.18 pF. (a) What is the inductance of a coil connected to this capacitor if the oscillation frequency of the L-C circuit is 1600×10^3 Hz, corresponding to one end of the AM radio broadcast band, when the capacitor is set to its minimum capacitance? (b) The frequency at the other end of the broadcast band is 540×10^3 Hz. What is the maximum capacitance of the capacitor if the oscillation frequency is adjustable over the range of the broadcast band?

30.37 •• An L-C circuit containing an 80.0-mH inductor and a 1.25-nF capacitor oscillates with a maximum current of 0.750 A. Calculate: (a) the maximum charge on the capacitor and (b) the oscillation frequency of the circuit. (c) Assuming the capacitor had its maximum charge at time $t = 0$, calculate the energy stored in the inductor after 2.50 ms of oscillation.

30.38 • An L-R-C series circuit has $L = 0.600$ H and $C = 3.00 \mu$F. (a) Calculate the angular frequency of oscillation for the circuit when $R = 0$. (b) What value of R gives critical damping? (c) What is the oscillation frequency ω' when R has half of the value that produces critical damping?

30.39 • An L-R-C series circuit has $L = 0.450$ H, $C = 2.50 \times 10^{-5}$ F, and resistance R. (a) What is the angular frequency of the circuit when $R = 0$? (b) What value must R have to give a 5.0% decrease in angular frequency compared to the value calculated in part (a)?

30.40 • An L-R-C series circuit has $L = 0.400$ H, $C = 7.00 \mu$F, and $R = 320 \Omega$. At $t = 0$ the current is zero and the initial charge on the capacitor is 2.80×10^{-4} C. (a) What are the values of the constants A and ϕ in Eq. (30.28)? (b) How much time does it

take for each complete current oscillation after the switch in this circuit is closed? (c) What is the charge on the capacitor after the first complete current oscillation?

30.41 • For the circuit of Fig. 30.17, let $C = 15.0$ nF, $L = 22$ mH, and $R = 75.0 \Omega$. (a) Calculate the oscillation frequency of the circuit once the capacitor has been charged and the switch has been connected to point a. (b) How long will it take for the amplitude of the oscillation to decay to 10.0% of its original value? (c) What value of R would result in a critically damped circuit?

PROBLEMS

30.42 • An inductor is connected to the terminals of a battery that has an emf of 16.0 V and negligible internal resistance. The current is 4.86 mA at 0.940 ms after the connection is completed. After a long time, the current is 6.45 mA. What are (a) the resistance R of the inductor and (b) the inductance L of the inductor?

30.43 • One solenoid is centered inside another. The outer one has a length of 50.0 cm and contains 6750 coils, while the coaxial inner solenoid is 3.0 cm long and 0.120 cm in diameter and contains 15 coils. The current in the outer solenoid is changing at 49.2 A/s. (a) What is the mutual inductance of these solenoids? (b) Find the emf induced in the inner solenoid.

30.44 •• CALC A coil has 400 turns and self-inductance 7.50 mH. The current in the coil varies with time according to $i = (680 \text{ mA}) \cos(\pi t/0.0250 \text{ s})$. (a) What is the maximum emf induced in the coil? (b) What is the maximum average flux through each turn of the coil? (c) At $t = 0.0180$ s, what is the magnitude of the induced emf?

30.45 •• **Solar Magnetic Energy.** Magnetic fields within a sunspot can be as strong as 0.4 T. (By comparison, the earth's magnetic field is about 1/10,000 as strong.) Sunspots can be as large as 25,000 km in radius. The material in a sunspot has a density of about 3×10^{-4} kg/m^3. Assume μ for the sunspot material is μ_0. If 100% of the magnetic-field energy stored in a sunspot could be used to eject the sunspot's material away from the sun's surface, at what speed would that material be ejected? Compare to the sun's escape speed, which is about 6×10^5 m/s. (*Hint:* Calculate the kinetic energy the magnetic field could supply to 1 m^3 of sunspot material.)

30.46 •• CP CALC **A Coaxial Cable.** A small solid conductor with radius a is supported by insulating, nonmagnetic disks on the axis of a thin-walled tube with inner radius b. The inner and outer conductors carry equal currents i in opposite directions. (a) Use Ampere's law to find the magnetic field at any point in the volume between the conductors. (b) Write the expression for the flux $d\Phi_B$ through a narrow strip of length l parallel to the axis, of width dr, at a distance r from the axis of the cable and lying in a plane containing the axis. (c) Integrate your expression from part (b) over the volume between the two conductors to find the total flux produced by a current i in the central conductor. (d) Show that the inductance of a length l of the cable is

$$L = l\frac{\mu_0}{2\pi}\ln\left(\frac{b}{a}\right)$$

(e) Use Eq. (30.9) to calculate the energy stored in the magnetic field for a length l of the cable.

30.47 •• CP CALC Consider the coaxial cable of Problem 30.46. The conductors carry equal currents i in opposite directions. (a) Use Ampere's law to find the magnetic field at any point in the volume between the conductors. (b) Use the energy density for a magnetic field, Eq. (30.10), to calculate the energy stored in a thin, cylindrical shell between the two conductors. Let the cylindrical shell have inner radius r, outer radius $r + dr$, and length l. (c) Integrate your result in part (b) over the volume between the two conductors to find the total energy stored in the magnetic field for a length l of the cable. (d) Use your result in part (c) and Eq. (30.9) to calculate the inductance L of a length l of the cable. Compare your result to L calculated in part (d) of Problem 30.46.

30.48 •• CALC Consider the circuit in Fig. 30.11 with both switches open. At $t = 0$ switch S_1 is closed while switch S_2 is left open. (a) Use Eq. (30.14) to derive an equation for the rate P_R at which electrical energy is being consumed in the resistor. In terms of $\mathcal{E}$, R, and L, at what value of t is P_R a maximum? What is that maximum value? (b) Use Eqs. (30.14) and (30.15) to derive an equation for P_L, the rate at which energy is being stored in the inductor. (c) What is P_L at $t = 0$ and as $t \rightarrow \infty$? (d) In terms of $\mathcal{E}$, R, and L, at what value of t is P_L a maximum? What is that maximum value? (e) Obtain an expression for $P_\mathcal{E}$, the rate at which the battery is supplying electrical energy to the circuit. In terms of $\mathcal{E}$, R, and L, at what value of t is $P_\mathcal{E}$ a maximum? What is that maximum value?

30.49 • (a) What would have to be the self-inductance of a solenoid for it to store 10.0 J of energy when a 2.00-A current runs through it? (b) If this solenoid's cross-sectional diameter is 4.00 cm, and if you could wrap its coils to a density of 10 coils/mm, how long would the solenoid be? (See Exercise 30.15.) Is this a realistic length for ordinary laboratory use?

30.50 •• CALC An inductor with inductance $L = 0.300$ H and negligible resistance is connected to a battery, a switch S, and two resistors, $R_1 = 12.0\ \Omega$ and $R_2 = 16.0\ \Omega$ (**Fig. P30.50**). The battery has emf 96.0 V and negligible internal resistance. S is closed at $t = 0$. (a) What are the currents i_1, i_2, and i_3 just after S is closed? (b) What are i_1, i_2, and i_3 after S has been closed a long time? (c) What is the value of t for which i_3 has half of the final value that you calculated in part (b)? (d) When i_3 has half of its final value, what are i_1 and i_2?

Figure **P30.50**

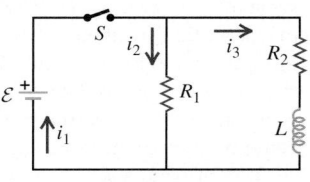

30.51 • **An Electromagnetic Car Alarm.** Your latest invention is a car alarm that produces sound at a particularly annoying frequency of 3500 Hz. To do this, the car-alarm circuitry must produce an alternating electric current of the same frequency. That's why your design includes an inductor and a capacitor in series. The maximum voltage across the capacitor is to be 12.0 V. To produce a sufficiently loud sound, the capacitor must store 0.0160 J of energy. What values of capacitance and inductance should you choose for your car-alarm circuit?

30.52 ••• CALC An inductor with inductance $L = 0.200$ H and negligible resistance is connected to a battery, a switch S, and two resistors, $R_1 = 8.00\ \Omega$ and $R_2 = 6.00\ \Omega$ (**Fig. P30.52**). The battery has emf 48.0 V and negligible internal resistance.

Figure **P30.52**

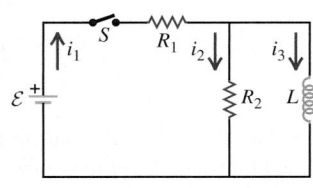

S is closed at $t = 0$. (a) What are the currents i_1, i_2, and i_3 just after S is closed? (b) What are i_1, i_2, and i_3 after S has been closed a long time? (c) Apply Kirchhoff's rules to the circuit and obtain a differential equation for $i_3(t)$. Integrate this equation to obtain an equation for i_3 as a function of the time t that has elapsed since S was closed. (d) Use the equation that you derived in part (c) to calculate the value of t for which i_3 has half of the final value that you calculated in part (b). (e) When i_3 has half of its final value, what are i_1 and i_2?

30.53 • A 7.00-μF capacitor is initially charged to a potential of 16.0 V. It is then connected in series with a 3.75-mH inductor. (a) What is the total energy stored in this circuit? (b) What is the maximum current in the inductor? What is the charge on the capacitor plates at the instant the current in the inductor is maximal?

30.54 •• A 6.40-nF capacitor is charged to 24.0 V and then disconnected from the battery in the circuit and connected in series with a coil that has $L = 0.0660$ H and negligible resistance. After the circuit has been completed, there are current oscillations. (a) At an instant when the charge of the capacitor is 0.0800 μC, how much energy is stored in the capacitor and in the inductor, and what is the current in the inductor? (b) At the instant when the charge on the capacitor is 0.0800 μC, what are the voltages across the capacitor and across the inductor, and what is the rate at which current in the inductor is changing?

30.55 • An L-C circuit consists of a 60.0-mH inductor and a 250-μF capacitor. The initial charge on the capacitor is 6.00 μC, and the initial current in the inductor is zero. (a) What is the maximum voltage across the capacitor? (b) What is the maximum current in the inductor? (c) What is the maximum energy stored in the inductor? (d) When the current in the inductor has half its maximum value, what is the charge on the capacitor and what is the energy stored in the inductor?

30.56 •• A charged capacitor with $C = 590\ \mu$F is connected in series to an inductor that has $L = 0.330$ H and negligible resistance. At an instant when the current in the inductor is $i = 2.50$ A, the current is increasing at a rate of $di/dt = 73.0$ A/s. During the current oscillations, what is the maximum voltage across the capacitor?

30.57 ••• CP In the circuit shown in Fig. P30.57, the switch S has been open for a long time and is suddenly closed. Neither the battery nor the inductors have any appreciable resistance. What do the ammeter and the voltmeter read (a) just after S is closed; (b) after S has been closed a very long time; (c) 0.115 ms after S is closed?

Figure **P30.57**

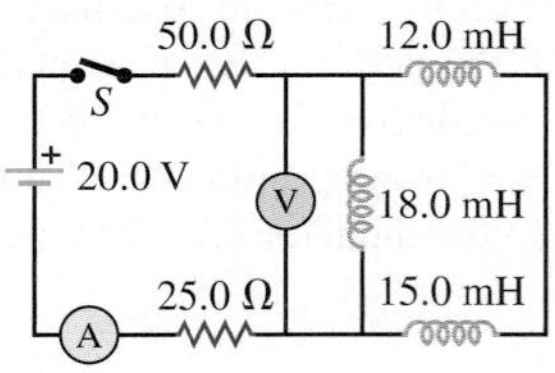

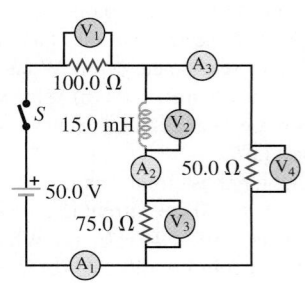

30.58 •• CP In the circuit shown in **Fig. P30.58**, find the reading in each ammeter and voltmeter (a) just after switch S is closed and (b) after S has been closed a very long time.

Figure **P30.58**

30.59 •• CP In the circuit shown in **Fig. P30.59**, switch S is closed at time $t = 0$ with no charge initially on the capacitor. (a) Find the reading of each ammeter and each voltmeter just after S is closed. (b) Find the reading of each meter after a long time has elapsed. (c) Find the maximum charge on the capacitor. (d) Draw a qualitative graph of the reading of voltmeter V_2 as a function of time.

Figure **P30.59**

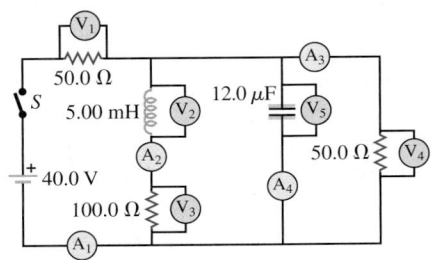

30.60 •• In the circuit shown in **Fig. P30.60**, switch S_1 has been closed for a long enough time so that the current reads a steady 3.50 A. Suddenly, switch S_2 is closed and S_1 is opened at the same instant. (a) What is the maximum charge that the capacitor will receive? (b) What is the current in the inductor at this time?

Figure **P30.60**

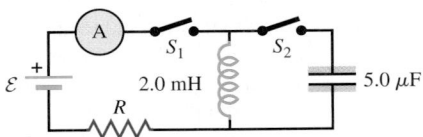

30.61 •• CP In the circuit shown in **Fig. P30.61**, $\mathcal{E} = 60.0$ V, $R_1 = 40.0\ \Omega$, $R_2 = 25.0\ \Omega$, and $L = 0.300$ H. Switch S is closed at $t = 0$. Just after the switch is closed, (a) what is the potential difference v_{ab} across the resistor R_1; (b) which point, a or b, is at a higher potential; (c) what is the potential difference v_{cd} across the inductor L; (d) which point, c or d, is at a higher potential? The switch is left closed a long time and then opened. Just after the switch is opened, (e) what is the potential difference v_{ab} across the resistor R_1; (f) which point, a or b, is at a higher potential; (g) what is the potential difference v_{cd} across the inductor L; (h) which point, c or d, is at a higher potential?

Figure **P30.61**

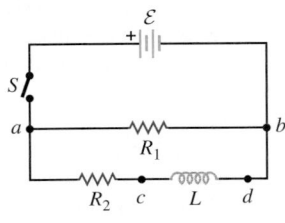

30.62 •• CP In the circuit shown in Fig. P30.61, $\mathcal{E} = 60.0$ V, $R_1 = 40.0\ \Omega$, $R_2 = 25.0\ \Omega$, and $L = 0.300$ H. (a) Switch S is closed. At some time t afterward, the current in the inductor is increasing at a rate of $di/dt = 50.0$ A/s. At this instant, what are the current i_1 through R_1 and the current i_2 through R_2? (*Hint:* Analyze two separate loops: one containing $\mathcal{E}$ and R_1 and the other containing $\mathcal{E}$, R_2, and L.) (b) After the switch has been closed a long time, it is opened again. Just after it is opened, what is the current through R_1?

30.63 •• CALC Consider the circuit shown in **Fig. P30.63**. Let $\mathcal{E} = 36.0$ V, $R_0 = 50.0\ \Omega$, $R = 150\ \Omega$, and $L = 4.00$ H. (a) Switch S_1 is closed and switch S_2 is left open. Just after S_1 is closed, what are the current i_0 through R_0 and the potential differences v_{ac} and v_{cb}? (b) After S_1 has been closed a long time (S_2 is still open) so that the current has reached its final, steady value, what are i_0, v_{ac}, and v_{cb}? (c) Find the expressions for i_0, v_{ac}, and v_{cb} as functions of the time t since S_1 was closed. Your results should agree with part (a) when $t = 0$ and with part (b) when $t \rightarrow \infty$. Graph i_0, v_{ac}, and v_{cb} versus time.

Figure **P30.63**

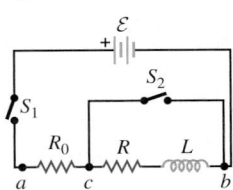

30.64 •• After the current in the circuit of Fig. P30.63 has reached its final, steady value with switch S_1 closed and S_2 open, switch S_2 is closed, thus short-circuiting the inductor. (Switch S_1 remains closed. See Problem 30.63 for numerical values of the circuit elements.) (a) Just after S_2 is closed, what are v_{ac} and v_{cb}, and what are the currents through R_0, R, and S_2? (b) A long time after S_2 is closed, what are v_{ac} and v_{cb}, and what are the currents through R_0, R, and S_2? (c) Derive expressions for the currents through R_0, R, and S_2 as functions of the time t that has elapsed since S_2 was closed. Your results should agree with part (a) when $t = 0$ and with part (b) when $t \rightarrow \infty$. Graph these three currents versus time.

30.65 •• CP In the circuit shown in **Fig. P30.65**, switch S is closed at time $t = 0$. (a) Find the reading of each meter just after S is closed. (b) What does each meter read long after S is closed?

Figure **P30.65**

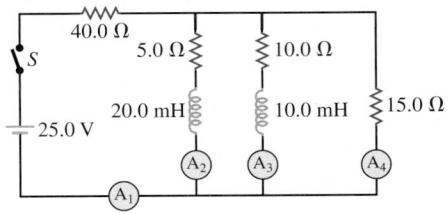

30.66 ••• CP In the circuit shown in **Fig. P30.66**, neither the battery nor the inductors have any appreciable resistance, the capacitors are initially uncharged, and the switch S has been in position 1 for a very long time. (a) What is the current in the circuit? (b) The switch is now suddenly flipped to position 2. Find the maximum charge that each capacitor will receive, and how much time after the switch is flipped it will take them to acquire this charge.

Figure **P30.66**

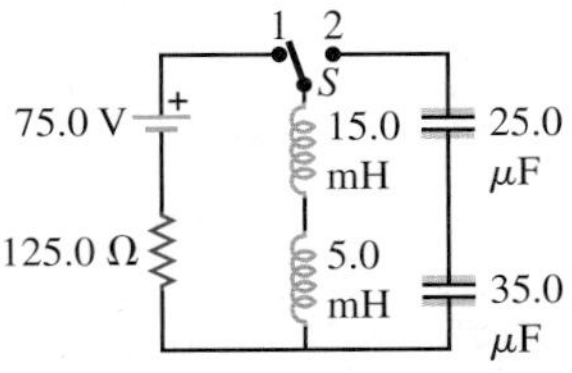

30.67 •• DATA During a summer internship as an electronics technician, you are asked to measure the self-inductance L of a solenoid. You connect the solenoid in series with a 10.0-Ω resistor, a battery that has negligible internal resistance, and a switch. Using an ideal voltmeter, you measure and digitally record the voltage v_L across the solenoid as a function of the time t that has elapsed since the switch is closed. Your measured values are shown in **Fig. P30.67**, where v_L is plotted versus t. In addition, you measure that $v_L = 50.0$ V just after the switch is closed and $v_L = 20.0$ V a long time after it is closed. (a) Apply the loop rule to the circuit and obtain an equation for v_L as a function of t. [*Hint:* Use an analysis similar to that used to derive Eq. (30.15).]

(b) What is the emf $\mathcal{E}$ of the battery? (c) According to your measurements, what is the voltage amplitude across the 10.0-Ω resistor as $t \to \infty$? Use this result to calculate the current in the circuit as $t \to \infty$. (d) What is the resistance R_L of the solenoid? (e) Use the theoretical equation from part (a), Fig. P30.67, and the values of $\mathcal{E}$ and R_L from parts (b) and (d) to calculate L. (*Hint:* According to the equation, what is v_L when $t = \tau$, one time constant? Use Fig. P30.67 to estimate the value of $t = \tau$.)

Figure **P30.67**

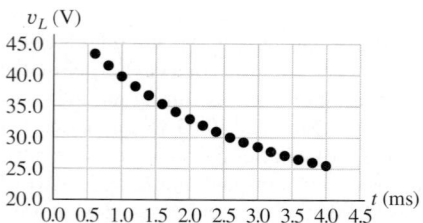

30.68 •• **DATA** You are studying a solenoid of unknown resistance and inductance. You connect it in series with a 50.0-Ω resistor, a 25.0-V battery that has negligible internal resistance, and a switch. Using an ideal voltmeter, you measure and digitally record the voltage v_R across the resistor as a function of the time t that has elapsed after the switch is closed. Your measured values are shown in **Fig. P30.68**, where v_R is plotted versus t. In addition, you measure that $v_R = 0$ just after the switch is closed and $v_R = 25.0$ V a long time after it is closed. (a) What is the resistance R_L of the solenoid? (b) Apply the loop rule to the circuit and obtain an equation for v_R as a function of t. (c) According to the equation that you derived in part (b), what is v_R when $t = \tau$, one time constant? Use Fig. P30.68 to estimate the value of $t = \tau$. What is the inductance of the solenoid? (d) How much energy is stored in the inductor a long time after the switch is closed?

Figure **P30.68**

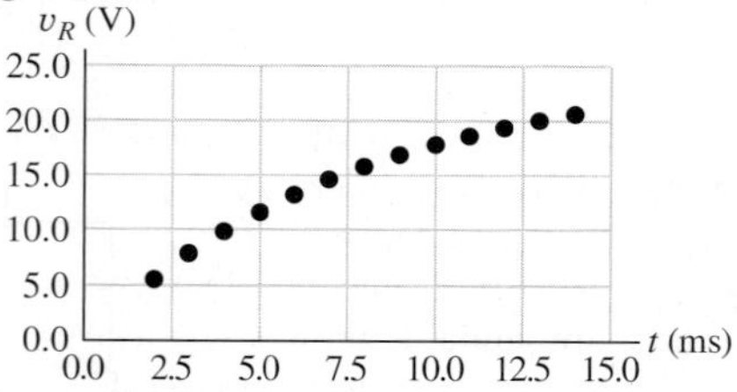

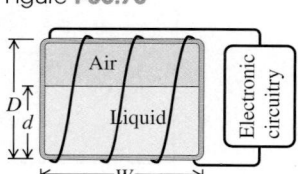

30.69 •• **DATA** To investigate the properties of a large industrial solenoid, you connect the solenoid and a resistor in series with a battery. Switches allow the battery to be replaced by a short circuit across the solenoid and resistor. Therefore Fig. 30.11 applies, with $R = R_{\text{ext}} + R_L$, where R_L is the resistance of the solenoid and R_{ext} is the resistance of the series resistor. With switch S_2 open, you close switch S_1 and keep it closed until the current i in the solenoid is constant (Fig. 30.11). Then you close S_2 and open S_1 simultaneously, using a rapid-response switching mechanism. With high-speed electronics you measure the time t_{half} that it takes for the current to decrease to half of its initial value. You repeat this measurement for several values of R_{ext} and obtain these results:

R_{ext} (Ω)	3.0	4.0	5.0	6.0	7.0	8.0	10.0	12.0
t_{half} (s)	0.735	0.654	0.589	0.536	0.491	0.453	0.393	0.347

(a) Graph your data in the form of $1/t_{\text{half}}$ versus R_{ext}. Explain why the data points plotted this way fall close to a straight line. (b) Use

your graph from part (a) to calculate the resistance R_L and inductance L of the solenoid. (c) If the current in the solenoid is 20.0 A, how much energy is stored there? At what rate is electrical energy being dissipated in the resistance of the solenoid?

CHALLENGE PROBLEMS

30.70 ••• **CP A Volume Gauge.** A tank containing a liquid has turns of wire wrapped around it, causing it to act like an inductor. The liquid content of the tank can be measured by using its inductance to determine the height of the liquid in the tank. The inductance of the tank changes from a value of L_0 corresponding to a relative permeability of 1 when the tank is empty to a value of L_f corresponding to a relative permeability of K_m (the relative permeability of the liquid) when the tank is full. The appropriate electronic circuitry can determine the inductance to five significant figures and thus the effective relative permeability of the combined air and liquid within the rectangular cavity of the tank. The four sides of the tank each have width W and height D (**Fig. P30.70**). The height of the liquid in the tank is d. You can ignore any fringing effects and assume that the relative permeability of the material of which the tank is made can be ignored. (a) Derive an expression for d as a function of L, the inductance corresponding to a certain fluid height, L_0, L_f, and D. (b) What is the inductance (to five significant figures) for a tank $\frac{1}{4}$ full, $\frac{1}{2}$ full, $\frac{3}{4}$ full, and completely full if the tank contains liquid oxygen? Take $L_0 = 0.63000$ H. The magnetic susceptibility of liquid oxygen is $\chi_m = 1.52 \times 10^{-3}$. (c) Repeat part (b) for mercury. The magnetic susceptibility of mercury is given in Table 28.1. (d) For which material is this volume gauge more practical?

Figure **P30.70**

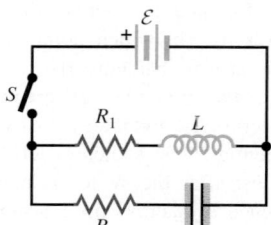

30.71 ••• **CP CALC** Consider the circuit shown in **Fig. P30.71**. Switch S is closed at time $t = 0$, causing a current i_1 through the inductive branch and a current i_2 through the capacitive branch. The initial charge on the capacitor is zero, and the charge at time t is q_2. (a) Derive expressions for i_1, i_2, and q_2 as functions of time. Express your answers in terms of $\mathcal{E}$, L, C, R_1, R_2, and t. For the remainder of the problem let the circuit elements have the following values: $\mathcal{E} = 48$ V, $L = 8.0$ H, $C = 20$ μF, $R_1 = 25$ Ω, and $R_2 = 5000$ Ω. (b) What is the initial current through the inductive branch? What is the initial current through the capacitive branch? (c) What are the currents through the inductive and capacitive branches a long time after the switch has been closed? How long is a "long time"? Explain. (d) At what time t_1 (accurate to two significant figures) will the currents i_1 and i_2 be equal? (*Hint:* You might consider using series expansions for the exponentials.) (e) For the conditions given in part (d), determine i_1. (f) The total current through the battery is $i = i_1 + i_2$. At what time t_2 (accurate to two significant figures) will i equal one-half of its final value? (*Hint:* The numerical work is greatly simplified if one makes suitable approximations. A sketch of i_1 and i_2 versus t may help you decide what approximations are valid.)

Figure **P30.71**

BIO **QUENCHING AN MRI MAGNET.** Magnets carrying very large currents are used to produce the uniform, large-magnitude magnetic fields that are required for *magnetic resonance imaging* (MRI). A typical MRI magnet may be a solenoid that is 2.0 m long and 1.0 m in diameter, has a self-inductance of 4.4 H, and carries a current of 750 A. A normal wire carrying that much current would dissipate a great deal of electrical power as heat, so most MRI magnets are made with coils of superconducting wire cooled by liquid helium at a temperature just under its boiling point (4.2 K). After a current is established in the wire, the power supply is disconnected and the magnet leads are shorted together through a piece of superconductor so that the current flows without resistance as long as the liquid helium keeps the magnet cold.

Under rare circumstances, a small segment of the magnet's wire may lose its superconducting properties and develop resistance. In this segment, electrical energy is converted to thermal energy, which can boil off some of the liquid helium. More of the wire then warms up and loses its superconducting properties, thus dissipating even more energy as heat. Because the latent heat of vaporization of liquid helium is quite low (20.9 kJ/kg), once the wire begins to warm up, all of the liquid helium may boil off rapidly. This event, called a *quench,* can damage the magnet. Also, a large volume of helium gas is generated as the liquid helium boils off, causing an asphyxiation hazard, and the resulting rapid pressure buildup can lead to an explosion. You can see how important it is to keep the wire resistance in an MRI magnet at zero and to have devices that detect a quench and shut down the current immediately.

30.72 How many turns does this typical MRI magnet have? (a) 1100; (b) 3000; (c) 4000; (d) 22,000.

30.73 If a small part of this magnet loses its superconducting properties and the resistance of the magnet wire suddenly rises from 0 to a constant 0.005 Ω, how much time will it take for the current to decrease to half of its initial value? (a) 4.7 min; (b) 10 min; (c) 15 min; (d) 30 min.

30.74 If part of the magnet develops resistance and liquid helium boils away, rendering more and more of the magnet nonsuperconducting, how will this quench affect the time for the current to drop to half of its initial value? (a) The time will be shorter because the resistance will increase; (b) the time will be longer because the resistance will increase; (c) the time will be the same; (d) not enough information is given.

30.75 If all of the magnetic energy stored in this MRI magnet is converted to thermal energy, how much liquid helium will boil off? (a) 27 kg; (b) 38 kg; (c) 60 kg; (d) 110 kg.

Answers

Chapter Opening Question ?

(iii) As explained in Section 30.2, traffic light sensors work by measuring the change in inductance of a coil embedded under the road surface when a car (which contains ferromagnetic material) drives over it.

Test Your Understanding Questions

30.1 (iii) Doubling both the length of the solenoid (l) and the number of turns of wire in the solenoid (N_1) would have *no* effect on the mutual inductance M. Example 30.1 shows that M depends on the ratio of these quantities, which would remain unchanged. This is because the magnetic field produced by the solenoid depends on the number of turns *per unit length,* and the proposed change has no effect on this quantity.

30.2 (iv), (i), (iii), (ii) From Eq. (30.8), the potential difference across the inductor is $V_{ab} = L\,di/dt$. For the four cases we find (i) $V_{ab} = (2.0\,\mu\text{H})(2.0\,\text{A} - 1.0\,\text{A})/(0.50\,\text{s}) = 4.0\,\mu\text{V}$; (ii) $V_{ab} = (4.0\,\mu\text{H})(0 - 3.0\,\text{A})/(2.0\,\text{s}) = -6.0\,\mu\text{V}$; (iii) $V_{ab} = 0$ because the rate of change of current is zero; and (iv) $V_{ab} = (1.0\,\mu\text{H})(4.0\,\text{A} - 0)/(0.25\,\text{s}) = 16\,\mu\text{V}$.

30.3 (a) yes, (b) no Reversing the direction of the current has no effect on the magnetic-field magnitude, but it causes the direction of the magnetic field to reverse. It has no effect on the magnetic-field energy density, which is proportional to the square of the *magnitude* of the magnetic field.

30.4 (a) (i), (b) (ii) Recall that v_{ab} is the potential at a minus the potential at b, and similarly for v_{bc}. For either arrangement of the switches, current flows through the resistor from a to b. The upstream end of the resistor is always at the higher potential, so v_{ab} is positive. With S_1 closed and S_2 open, the current through the inductor flows from b to c and is increasing. The self-induced emf opposes this increase and is therefore directed from c toward b, which means that b is at the higher potential. Hence v_{bc} is positive. With S_1 open and S_2 closed, the inductor current again flows from b to c but is now decreasing. The self-induced emf is directed from b to c in an effort to sustain the decaying current, so c is at the higher potential and v_{bc} is negative.

30.5 (a) positive, (b) electric, (c) negative, (d) electric The capacitor loses energy between stages (a) and (b), so it does positive work on the charges. It does this by exerting an electric force that pushes current away from the positively charged left-hand capacitor plate and toward the negatively charged right-hand plate. At the same time, the inductor gains energy and does negative work on the moving charges. Although the inductor stores magnetic energy, the force that the inductor exerts is *electric*. This force comes about from the inductor's self-induced emf (see Section 30.2).

30.6 (i) and (iii) There are no oscillations if $R^2 \geq 4L/C$. In each case $R^2 = (2.0\,\Omega)^2 = 4.0\,\Omega^2$. In case (i) $4L/C = 4(3.0\,\mu\text{H})/(6.0\,\mu\text{F}) = 2.0\,\Omega^2$, so there are no oscillations (the system is overdamped); in case (ii) $4L/C = 4(6.0\,\mu\text{H})/(3.0\,\mu\text{F}) = 8.0\,\Omega^2$, so there are oscillations (the system is underdamped); and in case (iii) $4L/C = 4(3.0\,\mu\text{H})/(3.0\,\mu\text{F}) = 4.0\,\Omega^2$, so there are no oscillations (the system is critically damped).

Bridging Problem

(a) $7.68 \times 10^{-8}\,\text{J}$ (b) $1.60\,\text{mA}$ (c) $24.8\,\text{mV}$
(d) $1.92 \times 10^{-8}\,\text{J}$, $21.5\,\text{mV}$

Waves from a broadcasting station produce an alternating current in the circuits of a radio (like the one in this classic car). If a radio is tuned to a station at a frequency of 1000 kHz, it will also detect the transmissions from a station broadcasting at (i) 600 kHz; (ii) 800 kHz; (iii) 1200 kHz; (iv) all of these; (v) none of these.

31

ALTERNATING CURRENT

LEARNING GOALS

Looking forward at ...

31.1 How phasors make it easy to describe sinusoidally varying quantities.

31.2 How to use reactance to describe the voltage across a circuit element that carries an alternating current.

31.3 How to analyze an *L-R-C* series circuit with a sinusoidal emf.

31.4 What determines the amount of power flowing into or out of an alternating-current circuit.

31.5 How an *L-R-C* series circuit responds to sinusoidal emfs of different frequencies.

31.6 Why transformers are useful, and how they work.

Looking back at ...

14.2, 14.8 Simple harmonic motion, resonance.

16.5 Resonance and sound.

18.3 Root-mean-square (rms) values.

25.3 Diodes.

26.3 Galvanometers.

28.8 Hysteresis in magnetic materials.

29.2, 29.6, 29.7 Alternating-current generators; eddy currents; displacement current.

30.1, 30.2, 30.5, 30.6 Mutual inductance; voltage across an inductor; *L-C* circuits; *L-R-C* series circuits.

During the 1880s in the United States there was a heated and acrimonious debate between two inventors over the best method of electric-power distribution. Thomas Edison favored direct current (dc)—that is, steady current that does not vary with time. George Westinghouse favored **alternating current (ac),** with sinusoidally varying voltages and currents. He argued that transformers (which we will study in this chapter) can be used to step the voltage up and down with ac but not with dc; low voltages are safer for consumer use, but high voltages and correspondingly low currents are best for long-distance power transmission to minimize i^2R losses in the cables.

Eventually, Westinghouse prevailed, and most present-day household and industrial power distribution systems operate with alternating current. Any appliance that you plug into a wall outlet uses ac. Circuits in modern communication equipment also make extensive use of ac.

In this chapter we will learn how resistors, inductors, and capacitors behave in circuits with sinusoidally varying voltages and currents. Many of the principles that we found useful in Chapter 30 are applicable, along with several new concepts related to the circuit behavior of inductors and capacitors. A key concept in this discussion is *resonance*, which we studied in Chapter 14 for mechanical systems.

31.1 PHASORS AND ALTERNATING CURRENTS

To supply an alternating current to a circuit, a source of alternating emf or voltage is required. An example of such a source is a coil of wire rotating with constant angular velocity in a magnetic field, which we discussed in Example 29.3 (Section 29.2). This develops a sinusoidal alternating emf and is the prototype of the commercial alternating-current generator or *alternator* (see Fig. 29.8).

We use the term **ac source** for any device that supplies a sinusoidally varying voltage (potential difference) v or current i. The usual circuit-diagram symbol for an ac source is

A sinusoidal voltage might be described by a function such as

$$v = V\cos\omega t \qquad (31.1)$$

In this expression, (lowercase) v is the *instantaneous* potential difference; (uppercase) V is the maximum potential difference, which we call the **voltage amplitude;** and ω is the *angular frequency,* equal to 2π times the frequency f (**Fig. 31.1**).

In the United States and Canada, commercial electric-power distribution systems use a frequency $f = 60$ Hz, corresponding to $\omega = (2\pi\ \text{rad})(60\ \text{s}^{-1}) = 377$ rad/s; in much of the rest of the world, $f = 50$ Hz ($\omega = 314$ rad/s) is used. Similarly, a sinusoidal current with a maximum value, or **current amplitude,** of I might be described as

Sinusoidal alternating current:

$$i = I\cos\omega t \qquad (31.2)$$

Instantaneous current · Angular frequency · Time · Current amplitude (maximum current)

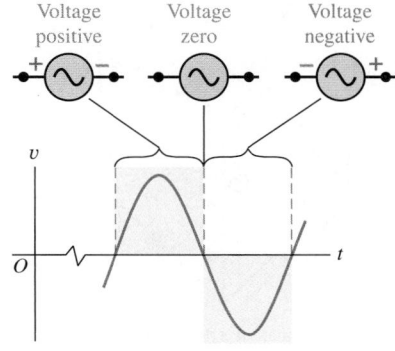

31.1 The voltage across a sinusoidal ac source.

Phasor Diagrams

To represent sinusoidally varying voltages and currents, we will use rotating vector diagrams similar to those we used in the study of simple harmonic motion in Section 14.2 (see Figs. 14.5b and 14.6). In these diagrams the instantaneous value of a quantity that varies sinusoidally with time is represented by the *projection* onto a horizontal axis of a vector with a length equal to the amplitude of the quantity. The vector rotates counterclockwise with constant angular speed ω. These rotating vectors are called **phasors,** and diagrams containing them are called **phasor diagrams. Figure 31.2** shows a phasor diagram for the sinusoidal current described by Eq. (31.2). The projection of the phasor onto the horizontal axis at time t is $I\cos\omega t$; this is why we chose to use the cosine function rather than the sine in Eq. (31.2).

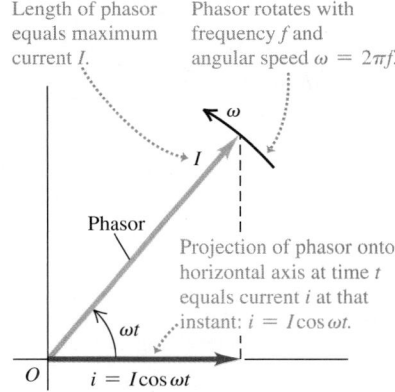

31.2 A phasor diagram.

CAUTION **Just what is a phasor?** A phasor isn't a real physical quantity with a direction in space, such as velocity or electric field. Rather, it's a *geometric* entity that helps us describe physical quantities that vary sinusoidally with time. In Section 14.2 we used a single phasor to represent the position of a particle undergoing simple harmonic motion. Here we'll use phasors to *add* sinusoidal voltages and currents. Combining sinusoidal quantities with phase differences then involves vector addition. We'll use phasors in a similar way in Chapters 35 and 36 in our study of interference effects with light. |

Rectified Alternating Current

How do we measure a sinusoidally varying current? In Section 26.3 we used a d'Arsonval galvanometer to measure steady currents. But if we pass a *sinusoidal* current through a d'Arsonval meter, the torque on the moving coil varies sinusoidally, with one direction half the time and the opposite direction the other half. The needle may wiggle a little if the frequency is low enough, but its average deflection is zero. Hence a d'Arsonval meter by itself isn't very useful for measuring alternating currents.

To get a measurable one-way current through the meter, we can use *diodes,* which we described in Section 25.3. A diode is a device that conducts better in one direction than in the other; an ideal diode has zero resistance for one

31.3 (a) A full-wave rectifier circuit. (b) Graph of the resulting current through the galvanometer G.

(a) A full-wave rectifier circuit

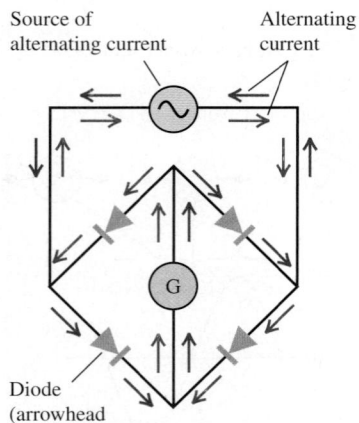

Diode
(arrowhead
and bar indicate the directions in
which current can and cannot pass)

(b) Graph of the full-wave rectified current and its average value, the rectified average current I_{rav}

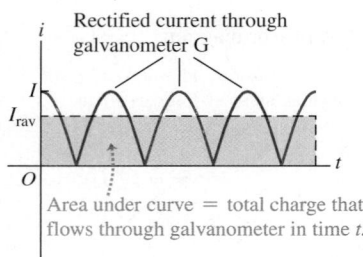

Rectified current through galvanometer G

Area under curve = total charge that flows through galvanometer in time t.

31.4 Calculating the root-mean-square (rms) value of an alternating current.

Meaning of the rms value of a sinusoidal quantity (here, ac current with $I = 3$ A):

① Graph current i versus time.

② *Square* the instantaneous current i.

③ Take the *average* (mean) value of i^2.

④ Take the *square root* of that average.

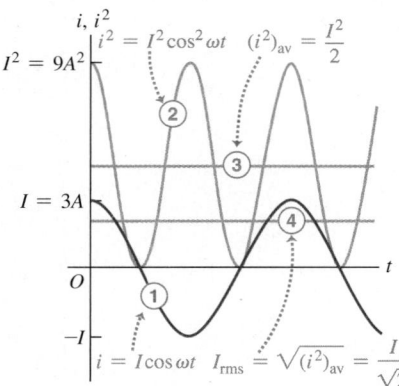

$I^2 = 9A^2$
$i^2 = I^2 \cos^2 \omega t$ $(i^2)_{av} = \dfrac{I^2}{2}$

$I = 3A$

$i = I \cos \omega t$ $I_{rms} = \sqrt{(i^2)_{av}} = \dfrac{I}{\sqrt{2}}$

direction of current and infinite resistance for the other. **Figure 31.3a** shows one possible arrangement, called a *full-wave rectifier circuit*. The current through the galvanometer G is always upward, regardless of the direction of the current from the ac source (i.e., which part of the cycle the source is in). The graph in Fig. 31.3b shows the current through G: It pulsates but always has the same direction, and the average meter deflection is *not* zero.

The **rectified average current** I_{rav} is defined so that during any whole number of cycles, the total charge that flows is the same as though the current were constant with a value equal to I_{rav}. The notation I_{rav} and the name *rectified average current* emphasize that this is *not* the average of the original sinusoidal current. In Fig. 31.3b the total charge that flows in time t corresponds to the area under the curve of i versus t (recall that $i = dq/dt$, so q is the integral of t); this area must equal the rectangular area with height I_{rav}. We see that I_{rav} is less than the maximum current I; the two are related by

> Rectified average value of a sinusoidal current ⋯⋯► $I_{rav} = \dfrac{2}{\pi} I = 0.637I$ ◄⋯⋯ Current amplitude   (31.3)

(The factor of $2/\pi$ is the average value of $|\cos \omega t|$ or of $|\sin \omega t|$; see Example 29.4 in Section 29.2.) The galvanometer deflection is proportional to I_{rav}. The galvanometer scale can be calibrated to read I, I_{rav}, or, most commonly, I_{rms} (discussed below).

Root-Mean-Square (rms) Values

A more useful way to describe a quantity that can be either positive or negative is the *root-mean-square (rms) value*. We used rms values in Section 18.3 in connection with the speeds of molecules in a gas. We *square* the instantaneous current i, take the *average* (mean) value of i^2, and finally take the *square root* of that average. This procedure defines the **root-mean-square current,** denoted as I_{rms} (**Fig. 31.4**). Even when i is negative, i^2 is always positive, so I_{rms} is never zero (unless i is zero at every instant).

Here's how we obtain I_{rms} for a sinusoidal current, like that shown in Fig. 31.4. If the instantaneous current is given by $i = I \cos \omega t$, then

$$i^2 = I^2 \cos^2 \omega t$$

Using a double-angle formula from trigonometry,

$$\cos^2 A = \tfrac{1}{2}(1 + \cos 2A)$$

we find

$$i^2 = I^2 \tfrac{1}{2}(1 + \cos 2\omega t) = \tfrac{1}{2}I^2 + \tfrac{1}{2}I^2 \cos 2\omega t$$

The average of $\cos 2\omega t$ is zero because it is positive half the time and negative half the time. Thus the average of i^2 is simply $I^2/2$. The square root of this is I_{rms}:

> Root-mean-square (rms) value ⋯⋯► $I_{rms} = \dfrac{I}{\sqrt{2}}$ ◄⋯⋯ Current amplitude   (31.4)

In the same way, the root-mean-square value of a sinusoidal voltage is

> Root-mean-square (rms) value of a sinusoidal voltage ⋯⋯► $V_{rms} = \dfrac{V}{\sqrt{2}}$ ◄⋯⋯ Voltage amplitude (maximum value)   (31.5)

We can convert a rectifying ammeter into a voltmeter by adding a series resistor, just as for the dc case discussed in Section 26.3. Meters used for ac voltage and current measurements are nearly always calibrated to read rms values, not maximum or rectified average. Voltages and currents in power distribution systems are always described in terms of their rms values. The usual household power supply, "120-volt ac," has an rms voltage of 120 V (**Fig. 31.5**). The voltage amplitude is

$$V = \sqrt{2}\,V_{\text{rms}}$$
$$= \sqrt{2}\,(120\text{ V}) = 170\text{ V}$$

31.5 This wall socket delivers a root-mean-square voltage of 120 V. Sixty times per second, the instantaneous voltage across its terminals varies from $(\sqrt{2})(120\text{ V}) = 170$ V to -170 V and back again.

EXAMPLE 31.1 CURRENT IN A PERSONAL COMPUTER

The plate on the back of a personal computer says that it draws 2.7 A from a 120-V, 60-Hz line. For this computer, what are (a) the average current, (b) the average of the square of the current, and (c) the current amplitude?

SOLUTION

IDENTIFY and SET UP: This example is about alternating current. In part (a) we find the average, over a complete cycle, of the alternating current. In part (b) we recognize that the 2.7-A current draw of the computer is the rms value I_{rms} —that is, the *square root* of the *mean* (average) of the *square* of the current, $(i^2)_{\text{av}}$. In part (c) we use Eq. (31.4) to relate I_{rms} to the current amplitude.

EXECUTE: (a) The average of *any* sinusoidally varying quantity, over any whole number of cycles, is zero.

(b) We are given $I_{\text{rms}} = 2.7$ A. From the definition of rms value,

$$I_{\text{rms}} = \sqrt{(i^2)_{\text{av}}} \text{ so } (i^2)_{\text{av}} = (I_{\text{rms}})^2 = (2.7\text{ A})^2 = 7.3\text{ A}^2$$

(c) From Eq. (31.4), the current amplitude I is

$$I = \sqrt{2}\,I_{\text{rms}} = \sqrt{2}\,(2.7\text{ A}) = 3.8\text{ A}$$

Figure 31.6 shows graphs of i and i^2 versus time t.

31.6 Our graphs of the current i and the square of the current i^2 versus time t.

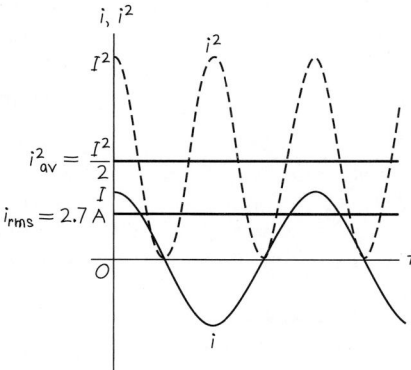

EVALUATE: Why would we be interested in the average of the square of the current? Recall that the rate at which energy is dissipated in a resistor R equals i^2R. This rate varies if the current is alternating, so it is best described by its average value $(i^2)_{\text{av}}R = I_{\text{rms}}^2R$. We'll use this idea in Section 31.4.

TEST YOUR UNDERSTANDING OF SECTION 31.1 The accompanying figure shows four different current phasors with the same angular frequency ω. At the time shown, which phasor corresponds to (a) a positive current that is becoming more positive; (b) a positive current that is decreasing toward zero; (c) a negative current that is becoming more negative; (d) a negative current that is decreasing in magnitude toward zero? ∎

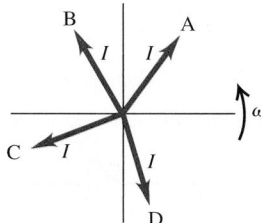

31.2 RESISTANCE AND REACTANCE

In this section we will derive voltage–current relationships for individual circuit elements—resistors, inductors, and capacitors—carrying a sinusoidal current.

Resistor in an ac Circuit

First let's consider a resistor with resistance R through which there is a sinusoidal current given by Eq. (31.2): $i = I\cos\omega t$. The positive direction of current is

31.7 Resistance R connected across an ac source.

(a) Circuit with ac source and resistor

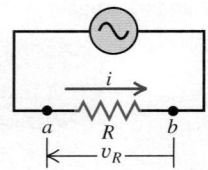

(b) Graphs of current and voltage versus time

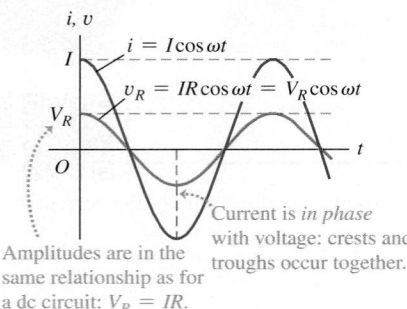

Amplitudes are in the same relationship as for a dc circuit: $V_R = IR$.

Current is in phase with voltage: crests and troughs occur together.

(c) Phasor diagram

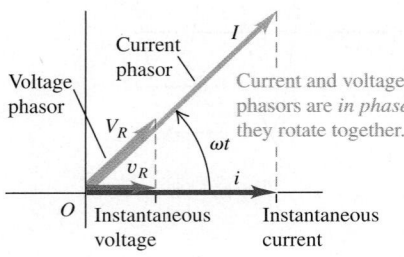

Current and voltage phasors are *in phase*: they rotate together.

counterclockwise around the circuit (**Fig. 31.7a**). The current amplitude (maximum current) is I. From Ohm's law the instantaneous potential v_R of point a with respect to point b (that is, the instantaneous voltage across the resistor) is

$$v_R = iR = (IR)\cos\omega t \tag{31.6}$$

The maximum value of the voltage v_R is V_R, the *voltage amplitude*:

> Amplitude of voltage across····· ·······Current amplitude
> a resistor, ac circuit $\qquad V_R = IR$ ····Resistance $\tag{31.7}$

Hence we can also write

$$v_R = V_R\cos\omega t \tag{31.8}$$

Both the current i and the voltage v_R are proportional to $\cos\omega t$, so the current is *in phase* with the voltage. Equation (31.7) shows that the current and voltage amplitudes are related in the same way as in a dc circuit.

Figure 31.7b shows graphs of i and v_R as functions of time. The vertical scales for current and voltage are different, so the relative heights of the two curves are not significant. The corresponding phasor diagram is given in Fig. 31.7c. Because i and v_R are *in phase* and have the same frequency, the current and voltage phasors rotate together; they are parallel at each instant. Their projections on the horizontal axis represent the instantaneous current and voltage, respectively.

Inductor in an ac Circuit

Now we replace the resistor in Fig. 31.7 with a pure inductor with self-inductance L and zero resistance (**Fig. 31.8a**). Again the current is $i = I\cos\omega t$, and the positive direction of current is counterclockwise around the circuit.

Although there is no resistance, there is a potential difference v_L between the inductor terminals a and b because the current varies with time, giving rise to a self-induced emf. The induced emf in the direction of i is given by Eq. (30.7), $\mathcal{E} = -L\,di/dt$; however, the voltage v_L is *not* simply equal to $\mathcal{E}$. To see why, notice that if the current in the inductor is in the positive (counterclockwise) direction from a to b and is increasing, then di/dt is positive and the induced emf is directed to the left to oppose the increase in current; hence point a is at higher potential than is point b. Thus the potential of point a with respect to point b is positive and is given by $v_L = +L\,di/dt$, the *negative* of the induced emf. (Convince yourself that this expression gives the correct sign of v_L in *all* cases, including i counterclockwise and decreasing, i clockwise and increasing, and i clockwise and decreasing; also review Section 30.2.) So

$$v_L = L\frac{di}{dt} = L\frac{d}{dt}(I\cos\omega t) = -I\omega L\sin\omega t \tag{31.9}$$

31.8 Inductance L connected across an ac source.

(a) Circuit with ac source and inductor

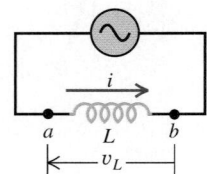

(b) Graphs of current and voltage versus time

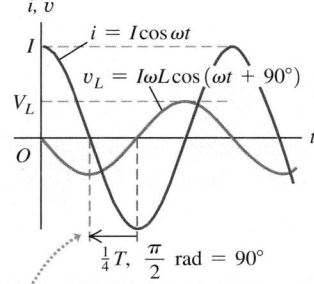

$\frac{1}{4}T$, $\frac{\pi}{2}$ rad = 90°

Voltage curve *leads* current curve by a quarter-cycle (corresponding to $\phi = \pi/2$ rad = 90°).

(c) Phasor diagram

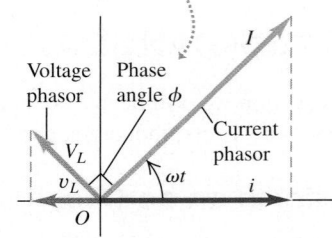

Voltage phasor *leads* current phasor by $\phi = \pi/2$ rad = 90°.

The voltage v_L across the inductor at any instant is proportional to the *rate of change* of the current. The points of maximum voltage on the graph correspond to maximum steepness of the current curve, and the points of zero voltage are the points where the current curve has its maximum and minimum values (Fig. 31.8b). The voltage and current are *out of phase* by a quarter-cycle. Since the voltage peaks occur a quarter-cycle earlier than the current peaks, we say that the voltage *leads* the current by 90°. The phasor diagram in Fig. 31.8c also shows this relationship; the voltage phasor is ahead of the current phasor by 90°.

We can also obtain this phase relationship by rewriting Eq. (31.9) with the identity $\cos(A + 90°) = -\sin A$:

$$v_L = I\omega L\cos(\omega t + 90°) \qquad (31.10)$$

This result shows that the voltage can be viewed as a cosine function with a "head start" of 90° relative to the current.

As we have done in Eq. (31.10), we will usually describe the phase of the *voltage* relative to the *current,* not the reverse. Thus if the current i in a circuit is

$$i = I\cos\omega t$$

and the voltage v of one point with respect to another is

$$v = V\cos(\omega t + \phi)$$

we call ϕ the **phase angle;** it gives the phase of the *voltage* relative to the *current.* For a pure resistor, $\phi = 0$, and for a pure inductor, $\phi = 90°$.

From Eq. (31.9) or (31.10) the amplitude V_L of the inductor voltage is

$$V_L = I\omega L \qquad (31.11)$$

We define the **inductive reactance** X_L of an inductor as

$$X_L = \omega L \qquad \text{(inductive reactance)} \qquad (31.12)$$

Using X_L, we can write Eq. (31.11) in a form similar to Eq. (31.7) for a resistor:

Amplitude of voltage across ⋯ an inductor, ac circuit

⋯ Current amplitude

$$V_L = IX_L \qquad (31.13)$$

Inductive reactance

Because X_L is the ratio of a voltage and a current, its SI unit is the ohm, the same as for resistance.

CAUTION **Inductor voltage and current are not in phase** Equation (31.13) relates the *amplitudes* of the oscillating voltage and current for the inductor in Fig. 31.8a. It does *not* say that the voltage at any instant is equal to the current at that instant multiplied by X_L. As Fig. 31.8b shows, the voltage and current are 90° out of phase. Voltage and current are in phase only for resistors, as in Eq. (31.6). ▮

The Meaning of Inductive Reactance

The inductive reactance X_L is really a description of the self-induced emf that opposes any change in the current through the inductor. From Eq. (31.13), for a given current amplitude I the voltage $v_L = +L\,di/dt$ across the inductor and the self-induced emf $\mathcal{E} = -L\,di/dt$ both have an amplitude V_L that is directly proportional to X_L. According to Eq. (31.12), the inductive reactance and self-induced emf increase with more rapid variation in current (that is, increasing angular frequency ω) and increasing inductance L.

If an oscillating voltage of a given amplitude V_L is applied across the inductor terminals, the resulting current will have a smaller amplitude I for larger values of X_L. Since X_L is proportional to frequency, a high-frequency voltage applied to the inductor gives only a small current, while a lower-frequency voltage of the same amplitude gives rise to a larger current. Inductors are used in some circuit applications, such as power supplies and radio-interference filters, to block high frequencies while permitting lower frequencies or dc to pass through. A circuit device that uses an inductor for this purpose is called a *low-pass filter* (see Problem 31.48).

EXAMPLE 31.2   AN INDUCTOR IN AN AC CIRCUIT

The current amplitude in a pure inductor in a radio receiver is to be 250 μA when the voltage amplitude is 3.60 V at a frequency of 1.60 MHz (at the upper end of the AM broadcast band). (a) What inductive reactance is needed? What inductance? (b) If the voltage amplitude is kept constant, what will be the current amplitude through this inductor at 16.0 MHz? At 160 kHz?

SOLUTION

IDENTIFY and SET UP: The circuit may have other elements, but in this example we don't care: All they do is provide the inductor with an oscillating voltage, so the other elements are lumped into the ac source shown in Fig. 31.8a. We are given the current amplitude I and the voltage amplitude V. Our target variables in part (a) are the inductive reactance X_L at 1.60 MHz and the inductance L, which we find from Eqs. (31.13) and (31.12). Knowing L, we use these equations in part (b) to find X_L and I at any frequency.

EXECUTE: (a) From Eq. (31.13),

$$X_L = \frac{V_L}{I} = \frac{3.60\ \text{V}}{250 \times 10^{-6}\ \text{A}} = 1.44 \times 10^4\ \Omega = 14.4\ \text{k}\Omega$$

From Eq. (31.12), with $\omega = 2\pi f$,

$$L = \frac{X_L}{2\pi f} = \frac{1.44 \times 10^4\ \Omega}{2\pi(1.60 \times 10^6\ \text{Hz})} = 1.43 \times 10^{-3}\ \text{H} = 1.43\ \text{mH}$$

(b) Combining Eqs. (31.12) and (31.13), we find $I = V_L/X_L = V_L/\omega L = V_L/2\pi fL$. Thus the current amplitude is inversely proportional to the frequency f. Since $I = 250\ \mu$A at $f = 1.60$ MHz, the current amplitudes at 16.0 MHz (10f) and 160 kHz = 0.160 MHz (f/10) will be, respectively, one-tenth as great (25.0 μA) and ten times as great (2500 μA = 2.50 mA).

EVALUATE: In general, the lower the frequency of an oscillating voltage applied across an inductor, the greater the amplitude of the resulting oscillating current.

Capacitor in an ac Circuit

CAUTION **Alternating current through a capacitor** Charge can't really move through the capacitor because its two plates are insulated from each other. But as the capacitor charges and discharges, there is at each instant a current i into one plate, an equal current out of the other plate, and an equal *displacement* current between the plates. (You should review the discussion of displacement current in Section 29.7.) Thus we often speak about alternating current *through* a capacitor. ▮

Finally, we connect a capacitor with capacitance C to the source, as in **Fig. 31.9a**, producing a current $i = I\cos\omega t$ through the capacitor. Again, the positive direction of current is counterclockwise around the circuit.

To find the instantaneous voltage v_C across the capacitor—that is, the potential of point a with respect to point b —we first let q denote the charge on the left-hand plate of the capacitor in Fig. 31.9a (so $-q$ is the charge on the right-hand plate). The current i is related to q by $i = dq/dt$; with this definition, positive current corresponds to an increasing charge on the left-hand capacitor plate. Then

$$i = \frac{dq}{dt} = I\cos\omega t$$

Integrating this, we get

$$q = \frac{I}{\omega}\sin\omega t \qquad (31.14)$$

Also, from Eq. (24.1) the charge q equals the voltage v_C multiplied by the capacitance, $q = Cv_C$. Using this in Eq. (31.14), we find

$$v_C = \frac{I}{\omega C}\sin\omega t \qquad (31.15)$$

The instantaneous current i is equal to the rate of change dq/dt of the capacitor charge q; since $q = Cv_C$, i is also proportional to the rate of change of voltage. (Compare to an inductor, for which the situation is reversed and v_L is proportional to the rate of change of i.) Figure 31.9b shows v_C and i as functions of t. Because $i = dq/dt = C\,dv_C/dt$, the current has its greatest magnitude when the v_C curve is rising or falling most steeply and is zero when the v_C curve instantaneously levels off at its maximum and minimum values.

The peaks of capacitor voltage occur a quarter-cycle *after* the corresponding current peaks, and we say that the voltage *lags* the current by 90°. The phasor diagram in Fig. 31.9c shows this relationship; the voltage phasor is behind the current phasor by a quarter-cycle, or 90°.

We can also derive this phase difference by rewriting Eq. (31.15) with the identity $\cos(A - 90°) = \sin A$:

$$v_C = \frac{I}{\omega C}\cos(\omega t - 90°) \qquad (31.16)$$

This corresponds to a phase angle $\phi = -90°$. This cosine function has a "late start" of 90° compared with the current $i = I\cos\omega t$.

Equations (31.15) and (31.16) show that the voltage *amplitude* V_C is

$$V_C = \frac{I}{\omega C} \qquad (31.17)$$

To put this expression in a form similar to Eq. (31.7) for a resistor, $V_R = IR$, we define a quantity X_C, called the **capacitive reactance** of the capacitor, as

$$X_C = \frac{1}{\omega C} \qquad \text{(capacitive reactance)} \qquad (31.18)$$

Then

Amplitude of voltage across ⋯⋯ a capacitor, ac circuit $\quad V_C = IX_C \overset{\text{Current amplitude}}{\underset{\text{Capacitive reactance}}{}}$ $\qquad (31.19)$

The SI unit of X_C is the ohm, the same as for resistance and inductive reactance, because X_C is the ratio of a voltage and a current.

⎯⎯⎯ **CAUTION** **Capacitor voltage and current are not in phase** Remember that Eq. (31.19) for a capacitor, like Eq. (31.13) for an inductor, is *not* a statement about the instantaneous values of voltage and current. The instantaneous values are 90° out of phase, as Fig. 31.9b shows. Rather, Eq. (31.19) relates the *amplitudes* of voltage and current. ▌

The Meaning of Capacitive Reactance

The capacitive reactance of a capacitor is inversely proportional both to the capacitance C and to the angular frequency ω; the greater the capacitance and the higher the frequency, the *smaller* the capacitive reactance X_C. Capacitors tend to pass high-frequency current and to block low-frequency currents and dc, just the opposite of inductors. A device that preferentially passes signals of high frequency is called a *high-pass filter* (see Problem 31.47).

31.9 Capacitor C connected across an ac source.

(a) Circuit with ac source and capacitor

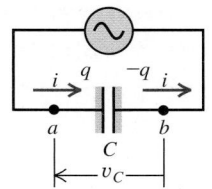

(b) Graphs of current and voltage versus time

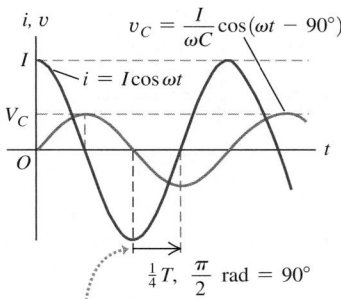

Voltage curve *lags* current curve by a quarter-cycle (corresponding to $\phi = -\pi/2$ rad $= -90°$).

(c) Phasor diagram

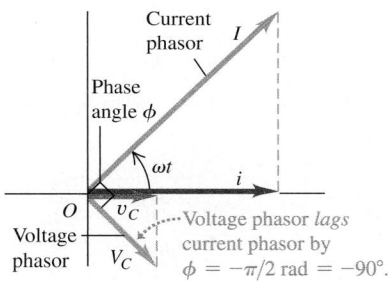

EXAMPLE 31.3   A RESISTOR AND A CAPACITOR IN AN AC CIRCUIT

A 200-Ω resistor is connected in series with a 5.0-μF capacitor. The voltage across the resistor is $v_R = (1.20 \text{ V})\cos(2500 \text{ rad/s})t$ (**Fig. 31.10**). (a) Derive an expression for the circuit current. (b) Determine the capacitive reactance of the capacitor. (c) Derive an expression for the voltage across the capacitor.

SOLUTION

IDENTIFY and SET UP: Since this is a series circuit, the current is the same through the capacitor as through the resistor. Our target variables are the current i, the capacitive reactance X_C, and the

capacitor voltage v_C. We use Eq. (31.6) to find an expression for i in terms of the angular frequency $\omega = 2500$ rad/s, Eq. (31.18) to find X_C, Eq. (31.19) to find the capacitor voltage amplitude V_C, and Eq. (31.16) to write an expression for v_C.

EXECUTE: (a) From Eq. (31.6), $v_R = iR$, we find

$$i = \frac{v_R}{R} = \frac{(1.20 \text{ V})\cos(2500 \text{ rad/s})t}{200 \,\Omega}$$

$$= (6.0 \times 10^{-3} \text{ A})\cos(2500 \text{ rad/s})t$$

(b) From Eq. (31.18), the capacitive reactance at $\omega = 2500$ rad/s is

$$X_C = \frac{1}{\omega C} = \frac{1}{(2500 \text{ rad/s})(5.0 \times 10^{-6} \text{ F})} = 80 \,\Omega$$

(c) From Eq. (31.19), the capacitor voltage amplitude is

$$V_C = IX_C = (6.0 \times 10^{-3} \text{ A})(80 \,\Omega) = 0.48 \text{ V}$$

31.10 Our sketch for this problem.

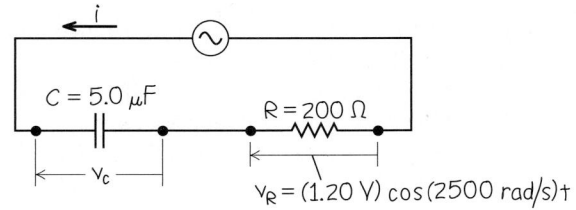

Continued

(The 80-Ω reactance of the capacitor is 40% of the resistor's 200-Ω resistance, so V_C is 40% of V_R.) The instantaneous capacitor voltage is given by Eq. (31.16):

$$v_C = V_C \cos(\omega t - 90°)$$
$$= (0.48 \text{ V})\cos\left[(2500 \text{ rad/s})t - \pi/2 \text{ rad}\right]$$

EVALUATE: Although the same *current* passes through both the capacitor and the resistor, the *voltages* across them are different in both amplitude and phase. Note that in the expression for v_C we converted the 90° to $\pi/2$ rad so that all the angular quantities have the same units. In ac circuit analysis, phase angles are often given in degrees, so be careful to convert to radians when necessary.

31.11 Graphs of R, X_L, and X_C as functions of angular frequency ω.

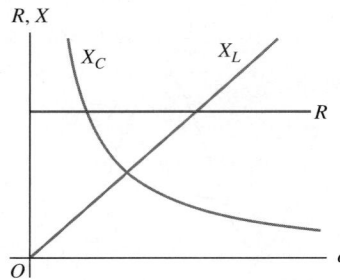

31.12 (a) The two speakers in this loudspeaker system are connected in parallel to the amplifier. (b) Graphs of current amplitude in the tweeter and woofer as functions of frequency for a given amplifier voltage amplitude.

(a) A crossover network in a loudspeaker system

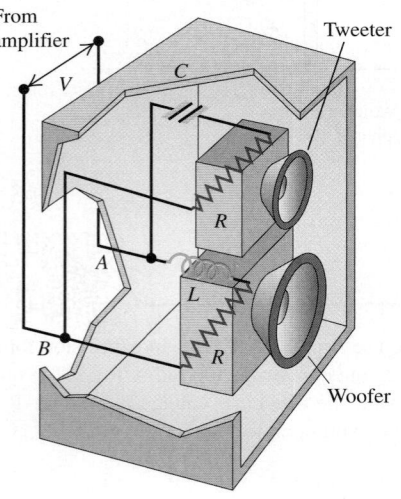

(b) Graphs of rms current as functions of frequency for a given amplifier voltage

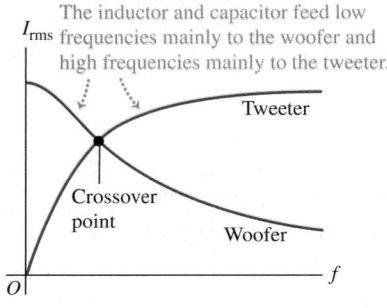

Comparing ac Circuit Elements

Table 31.1 summarizes the relationships of voltage and current amplitudes for the three circuit elements we have discussed. Note again that *instantaneous* voltage and current are proportional in a resistor, where there is zero phase difference between v_R and i (see Fig. 31.7b). The instantaneous voltage and current are *not* proportional in an inductor or capacitor, because there is a 90° phase difference in both cases (see Figs. 31.8b and 39.9b).

TABLE 31.1 Circuit Elements with Alternating Current

Circuit Element	Amplitude Relationship	Circuit Quantity	Phase of v
Resistor	$V_R = IR$	R	In phase with i
Inductor	$V_L = IX_L$	$X_L = \omega L$	Leads i by 90°
Capacitor	$V_C = IX_C$	$X_C = 1/\omega C$	Lags i by 90°

Figure 31.11 shows how the resistance of a resistor and the reactances of an inductor and a capacitor vary with angular frequency ω. Resistance R is independent of frequency, while the reactances X_L and X_C are not. If $\omega = 0$, corresponding to a dc circuit, there is *no* current through a capacitor because $X_C \rightarrow \infty$, and there is no inductive effect because $X_L = 0$. In the limit $\omega \rightarrow \infty$, X_L also approaches infinity, and the current through an inductor becomes vanishingly small; recall that the self-induced emf opposes rapid changes in current. In this same limit, X_C and the voltage across a capacitor both approach zero; the current changes direction so rapidly that no charge can build up on either plate.

Figure 31.12 shows an application of the above discussion to a loudspeaker system. Low-frequency sounds are produced by the *woofer,* which is a speaker with large diameter; the *tweeter,* a speaker with smaller diameter, produces high-frequency sounds. In order to route signals of different frequency to the appropriate speaker, the woofer and tweeter are connected in parallel across the amplifier output. The capacitor in the tweeter branch blocks the low-frequency components of sound but passes the higher frequencies; the inductor in the woofer branch does the opposite.

TEST YOUR UNDERSTANDING OF SECTION 31.2 An oscillating voltage of fixed amplitude is applied across a circuit element. If the frequency of this voltage is increased, will the amplitude of the current through the element (i) increase, (ii) decrease, or (iii) remain the same if it is (a) a resistor, (b) an inductor, or (c) a capacitor? ∎

31.3 THE *L-R-C* SERIES CIRCUIT

Many ac circuits used in practical electronic systems involve resistance, inductive reactance, and capacitive reactance. **Figure 31.13a** shows a simple example: a series circuit containing a resistor, an inductor, a capacitor, and an ac source. (In Section 30.6 we studied an *L-R-C* series circuit *without* a source.)

To analyze this circuit, we'll use a phasor diagram that includes the voltage and current phasors for each of the components. Because of Kirchhoff's loop rule, the instantaneous *total* voltage v_{ad} across all three components is equal to the source voltage at that instant. We will show that the phasor representing this total voltage is the *vector sum* of the phasors for the individual voltages.

Figures 31.13b and 31.13c show complete phasor diagrams for the circuit of Fig. 31.13a. We assume that the source supplies a current i given by $i = I\cos\omega t$. Because the circuit elements are connected in series, the current at any instant is the same at every point in the circuit. Thus a *single phasor I*, with length proportional to the current amplitude, represents the current in *all* circuit elements.

As in Section 31.2, we use the symbols v_R, v_L, and v_C for the instantaneous voltages across R, L, and C, and the symbols V_R, V_L, and V_C for the maximum voltages. We denote the instantaneous and maximum *source* voltages by v and V. Then, in Fig. 31.13a, $v = v_{ad}$, $v_R = v_{ab}$, $v_L = v_{bc}$, and $v_C = v_{cd}$.

The potential difference between the terminals of a resistor is *in phase* with the current in the resistor. Its maximum value V_R is given by Eq. (31.7):

$$V_R = IR$$

The phasor V_R in Fig. 31.13b, in phase with the current phasor I, represents the voltage across the resistor. Its projection onto the horizontal axis at any instant gives the instantaneous potential difference v_R.

The voltage across an inductor *leads* the current by 90°. Its voltage amplitude is given by Eq. (31.13):

$$V_L = IX_L$$

The phasor V_L in Fig. 31.13b represents the voltage across the inductor, and its projection onto the horizontal axis at any instant equals v_L.

The voltage across a capacitor *lags* the current by 90°. Its voltage amplitude is given by Eq. (31.19):

$$V_C = IX_C$$

The phasor V_C in Fig. 31.13b represents the voltage across the capacitor, and its projection onto the horizontal axis at any instant equals v_C.

The instantaneous potential difference v between terminals a and d is equal at every instant to the (algebraic) sum of the potential differences v_R, v_L, and v_C. That is, it equals the sum of the *projections* of the phasors V_R, V_L, and V_C. But the sum of the projections of these phasors is equal to the *projection* of their *vector sum*. So the vector sum V must be the phasor that represents the source voltage v and the instantaneous total voltage v_{ad} across the series of elements.

To form this vector sum, we first subtract the phasor V_C from the phasor V_L. (These two phasors always lie along the same line, with opposite directions.) This gives the phasor $V_L - V_C$. This is always at right angles to the phasor V_R, so from the Pythagorean theorem the magnitude of the phasor V is

$$V = \sqrt{V_R^2 + (V_L - V_C)^2} = \sqrt{(IR)^2 + (IX_L - IX_C)^2} \quad \text{or}$$

$$V = I\sqrt{R^2 + (X_L - X_C)^2} \tag{31.20}$$

We define the **impedance** Z of an ac circuit as the ratio of the voltage amplitude across the circuit to the current amplitude in the circuit. From Eq. (31.20) the impedance of the *L-R-C* series circuit is

$$Z = \sqrt{R^2 + (X_L - X_C)^2} \tag{31.21}$$

so we can rewrite Eq. (31.20) as

31.13 An *L-R-C* series circuit with an ac source.

(a) *L-R-C* series circuit

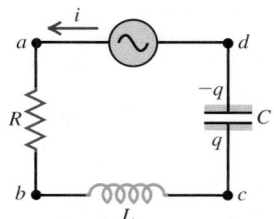

(b) Phasor diagram for the case $X_L > X_C$

Source voltage phasor is the vector sum of the V_R, V_L, and V_C phasors.

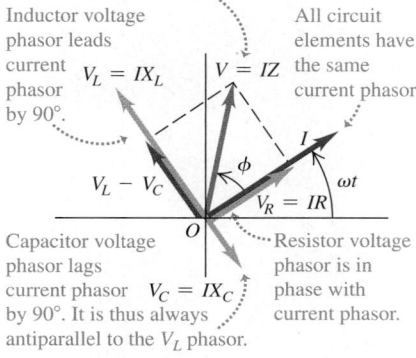

Inductor voltage phasor leads current phasor by 90°.

All circuit elements have the same current phasor.

Capacitor voltage phasor lags current phasor by 90°. It is thus always antiparallel to the V_L phasor.

Resistor voltage phasor is in phase with current phasor.

(c) Phasor diagram for the case $X_L < X_C$

If $X_L < X_C$, the source voltage phasor lags the current phasor, $X < 0$, and ϕ is a negative angle between 0 and −90°.

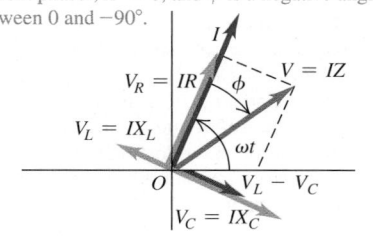

PhET: Circuit Construction Kit (AC+DC)
PhET: Faraday's Electromagnetic Lab

Amplitude of voltage across an ac circuit	$V = IZ$ ← Current amplitude ← Impedance of circuit

$$\begin{array}{c} \text{Amplitude of voltage} \cdots \\ \text{across an ac circuit} \end{array} \quad V = IZ \quad \begin{array}{c} \cdots \text{Current amplitude} \\ \cdots \text{Impedance of circuit} \end{array} \tag{31.22}$$

31.14 This gas-filled glass sphere has an alternating voltage between its surface and the electrode at its center. The glowing streamers show the resulting alternating current that passes through the gas. When you touch the outside of the sphere, your fingertips and the inner surface of the sphere act as the plates of a capacitor, and the sphere and your body together form an *L-R-C* series circuit. The current (which is low enough to be harmless) is drawn to your fingers because the path through your body has a low impedance.

BIO Application **Measuring Body Fat by Bioelectric Impedance Analysis** The electrodes attached to this overweight patient's chest are applying a small ac voltage of frequency 50 kHz. The attached instrumentation measures the amplitude and phase angle of the resulting current through the patient's body. These depend on the relative amounts of water and fat along the path followed by the current, and so provide a sensitive measure of body composition.

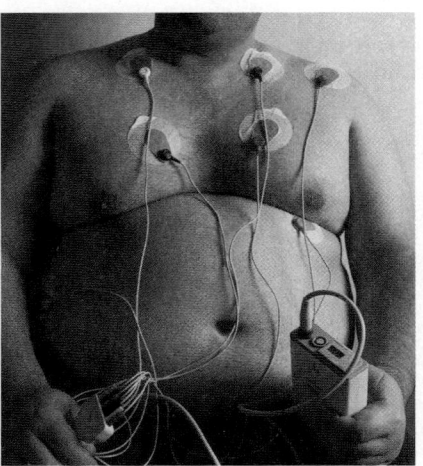

While Eq. (31.21) is valid only for an *L-R-C* series circuit, we can use Eq. (31.22) to define the impedance of *any* network of resistors, inductors, and capacitors as the ratio of the amplitude of the voltage across the network to the current amplitude. The SI unit of impedance is the ohm.

The Meaning of Impedance and Phase Angle

Equation (31.22) has a form similar to $V = IR$, with impedance Z in an ac circuit playing the role of resistance R in a dc circuit. Just as direct current tends to follow the path of least resistance, so alternating current tends to follow the path of lowest impedance (**Fig. 31.14**). Note, however, that impedance is actually a function of R, L, and C, as well as of the angular frequency ω. We can see this by substituting Eq. (31.12) for X_L and Eq. (31.18) for X_C into Eq. (31.21), giving the following complete expression for Z for a series circuit:

$$
\underset{\substack{\text{Impedance of an} \\ \text{\textit{L-R-C} series circuit}}}{} Z = \sqrt{R^2 + [\omega L - (1/\omega C)]^2} \tag{31.23}
$$

Resistance   Inductance   Capacitance   Angular frequency

Hence for a given amplitude V of the source voltage applied to the circuit, the amplitude $I = V/Z$ of the resulting current will be different at different frequencies. We'll explore this frequency dependence in detail in Section 31.5.

In Fig. 31.13b, the angle ϕ between the voltage and current phasors is the phase angle of the source voltage v with respect to the current i; that is, it is the angle by which the source voltage leads the current. From the diagram,

$$
\tan \phi = \frac{V_L - V_C}{V_R} = \frac{I(X_L - X_C)}{IR} = \frac{X_L - X_C}{R}
$$

$$
\underset{\substack{\text{Phase angle of voltage} \\ \text{with respect to current} \\ \text{in an \textit{L-R-C} series circuit}}}{} \tan \phi = \frac{\omega L - 1/\omega C}{R} \tag{31.24}
$$

Inductance   Angular frequency   Capacitance   Resistance

If the current is $i = I\cos \omega t$, then the source voltage v is

$$
v = V\cos(\omega t + \phi) \tag{31.25}
$$

Figure 31.13b shows the behavior of an *L-R-C* series circuit in which $X_L > X_C$. Figure 31.13c shows the behavior when $X_L < X_C$; the voltage phasor V lies on the opposite side of the current phasor I and the voltage *lags* the current. In this case, $X_L - X_C$ is *negative*, $\tan \phi$ is negative, and ϕ is a negative angle between 0° and −90°. Since X_L and X_C depend on frequency, the phase angle ϕ depends on frequency as well. We'll examine the consequences of this in Section 31.5.

All of the expressions that we've developed for an *L-R-C* series circuit are still valid if one of the circuit elements is missing. If the resistor is missing, we set $R = 0$; if the inductor is missing, we set $L = 0$. But if the capacitor is missing, we set $C = \infty$, corresponding to the absence of any potential difference ($v_C = q/C = 0$) or any capacitive reactance ($X_C = 1/\omega C = 0$).

In this entire discussion we have described magnitudes of voltages and currents in terms of their *maximum* values, the voltage and current *amplitudes*. But we remarked at the end of Section 31.1 that these quantities are usually described in terms of rms values, not amplitudes. For any sinusoidally varying quantity, the rms value is always $1/\sqrt{2}$ times the amplitude. All the relationships between voltage and current that we have derived in this and the preceding sections are still valid if we use rms quantities throughout instead of amplitudes. For example, if we divide Eq. (31.22) by $\sqrt{2}$, we get

$$\frac{V}{\sqrt{2}} = \frac{I}{\sqrt{2}}Z$$

which we can rewrite as

$$V_{\text{rms}} = I_{\text{rms}}Z \qquad (31.26)$$

We can translate Eqs. (31.7), (31.13), and (31.19) in exactly the same way.

We have considered only ac circuits in which an inductor. a resistor, and a capacitor are in series. You can do a similar analysis for an *L-R-C parallel* circuit; see Problem 31.54.

PROBLEM-SOLVING STRATEGY 31.1 ALTERNATING-CURRENT CIRCUITS

IDENTIFY *the relevant concepts:* In analyzing ac circuits, we can apply all of the concepts used to analyze direct-current circuits, particularly those in Problem-Solving Strategies 26.1 and 26.2. But now we must distinguish between the amplitudes of alternating currents and voltages and their instantaneous values, and among resistance (for resistors), reactance (for inductors or capacitors), and impedance (for composite circuits).

SET UP *the problem* using the following steps:
1. Draw a diagram of the circuit and label all known and unknown quantities.
2. Identify the target variables.

EXECUTE *the solution* as follows:
1. Use the relationships derived in Sections 31.2 and 31.3 to solve for the target variables, using the following hints.
2. It's almost always easiest to work with angular frequency $\omega = 2\pi f$ rather than ordinary frequency f.
3. Keep in mind the following phase relationships: For a resistor, voltage and current are *in phase,* so the corresponding phasors always point in the same direction. For an inductor, the voltage *leads* the current by 90° (i.e., $\phi = +90° = \pi/2$ radians), so the voltage phasor points 90° counterclockwise from the current phasor. For a capacitor, the voltage *lags* the current by 90° (i.e., $\phi = -90° = -\pi/2$ radians), so the voltage phasor points 90° clockwise from the current phasor.

4. Kirchhoff's rules hold *at each instant.* For example, in a series circuit, the instantaneous current is the same in all circuit elements; in a parallel circuit, the instantaneous potential difference is the same across all circuit elements.
5. Inductive reactance, capacitive reactance, and impedance are analogous to resistance; each represents the ratio of voltage amplitude V to current amplitude I in a circuit element or combination of elements. However, phase relationships are crucial. In applying Kirchhoff's loop rule, you must combine the effects of resistance and reactance by *vector* addition of the corresponding voltage phasors, as in Figs. 31.13b and 31.13c. When several circuit elements are in series, for example, you can't *add* all the numerical values of resistance and reactance to get the impedance; that would ignore the phase relationships.

EVALUATE *your answer:* When working with an *L-R-C series* circuit, you can check your results by comparing the values of the inductive and capacitive reactances X_L and X_C. If $X_L > X_C$, then the voltage amplitude across the inductor is greater than that across the capacitor and the phase angle ϕ is positive (between 0° and 90°). If $X_L < X_C$, then the voltage amplitude across the inductor is less than that across the capacitor and the phase angle ϕ is negative (between 0° and -90°).

EXAMPLE 31.4 AN *L-R-C* SERIES CIRCUIT I

In the series circuit of Fig. 31.13a, suppose $R = 300\ \Omega$, $L = 60\ \text{mH}$, $C = 0.50\ \mu\text{F}$, $V = 50\ \text{V}$, and $\omega = 10{,}000\ \text{rad/s}$. Find the reactances X_L and X_C, the impedance Z, the current amplitude I, the phase angle ϕ, and the voltage amplitude across each circuit element.

SOLUTION

IDENTIFY and SET UP: This problem uses the ideas developed in Section 31.2 and this section about the behavior of circuit elements in an ac circuit. We use Eqs. (31.12) and (31.18) to determine X_L and X_C, and Eq. (31.23) to find Z. We then use Eq. (31.22) to find

the current amplitude and Eq. (31.24) to find the phase angle. The relationships in Table 31.1 then yield the voltage amplitudes.

EXECUTE: The inductive and capacitive reactances are

$$X_L = \omega L = (10{,}000\ \text{rad/s})(60\ \text{mH}) = 600\ \Omega$$

$$X_C = \frac{1}{\omega C} = \frac{1}{(10{,}000\ \text{rad/s})(0.50 \times 10^{-6}\ \text{F})} = 200\ \Omega$$

The impedance Z of the circuit is then

$$Z = \sqrt{R^2 + (X_L - X_C)^2} = \sqrt{(300\ \Omega)^2 + (600\ \Omega - 200\ \Omega)^2}$$
$$= 500\ \Omega$$

Continued

With source voltage amplitude $V = 50$ V, the current amplitude I and phase angle ϕ are

$$I = \frac{V}{Z} = \frac{50 \text{ V}}{500 \text{ }\Omega} = 0.10 \text{ A}$$

$$\phi = \arctan \frac{X_L - X_C}{R} = \arctan \frac{400 \text{ }\Omega}{300 \text{ }\Omega} = 53°$$

From Table 31.1, the voltage amplitudes V_R, V_L, and V_C across the resistor, inductor, and capacitor, respectively, are

$$V_R = IR = (0.10 \text{ A})(300 \text{ }\Omega) = 30 \text{ V}$$

$$V_L = IX_L = (0.10 \text{ A})(600 \text{ }\Omega) = 60 \text{ V}$$

$$V_C = IX_C = (0.10 \text{ A})(200 \text{ }\Omega) = 20 \text{ V}$$

EVALUATE: As in Fig. 31.13b, $X_L > X_C$; hence the voltage amplitude across the inductor is greater than that across the capacitor and ϕ is positive. The value $\phi = 53°$ means that the voltage *leads* the current by 53°.

Note that the source voltage amplitude $V = 50$ V is *not* equal to the sum of the voltage amplitudes across the separate circuit elements: 50 V $\neq$ 30 V + 60 V + 20 V. Instead, V is the *vector sum* of the V_R, V_L, and V_C phasors. If you draw the phasor diagram like Fig. 31.13b for this particular situation, you'll see that V_R, $V_L - V_C$, and V constitute a 3-4-5 right triangle.

EXAMPLE 31.5 AN *L-R-C* SERIES CIRCUIT II

For the *L-R-C* series circuit of Example 31.4, find expressions for the time dependence of the instantaneous current i and the instantaneous voltages across the resistor (v_R), inductor (v_L), capacitor (v_C), and ac source (v).

SOLUTION

IDENTIFY and SET UP: We describe the current by using Eq. (31.2), which assumes that the current is maximum at $t = 0$. The voltages are then given by Eq. (31.8) for the resistor, Eq. (31.10) for the inductor, Eq. (31.16) for the capacitor, and Eq. (31.25) for the source.

EXECUTE: The current and the voltages all oscillate with the same angular frequency, $\omega = 10,000$ rad/s, and hence with the same period, $2\pi/\omega = 2\pi/(10,000 \text{ rad/s}) = 6.3 \times 10^{-4} \text{ s} = 0.63$ ms. From Eq. (31.2), the current is

$$i = I \cos \omega t = (0.10 \text{ A}) \cos(10,000 \text{ rad/s})t$$

The resistor voltage is *in phase* with the current, so

$$v_R = V_R \cos \omega t = (30 \text{ V}) \cos(10,000 \text{ rad/s})t$$

The inductor voltage *leads* the current by 90°, so

$$v_L = V_L \cos(\omega t + 90°) = -V_L \sin \omega t$$
$$= -(60 \text{ V}) \sin(10,000 \text{ rad/s})t$$

The capacitor voltage *lags* the current by 90°, so

$$v_C = V_C \cos(\omega t - 90°) = V_C \sin \omega t$$
$$= (20 \text{ V}) \sin(10,000 \text{ rad/s})t$$

We found in Example 31.4 that the source voltage (equal to the voltage across the entire combination of resistor, inductor, and capacitor) *leads* the current by $\phi = 53°$, so

$$v = V \cos(\omega t + \phi)$$
$$= (50 \text{ V}) \cos\left[(10,000 \text{ rad/s})t + \left(\frac{2\pi \text{ rad}}{360°}\right)(53°)\right]$$
$$= (50 \text{ V}) \cos[(10,000 \text{ rad/s})t + 0.93 \text{ rad}]$$

EVALUATE: Figure 31.15 graphs the four voltages versus time. The inductor voltage has a larger amplitude than the capacitor voltage because $X_L > X_C$. The *instantaneous* source voltage v is always equal to the sum of the instantaneous voltages v_R, v_L, and v_C. You should verify this by measuring the values of the voltages shown in the graph at different values of the time t.

31.15 Graphs of the source voltage v, resistor voltage v_R, inductor voltage v_L, and capacitor voltage v_C as functions of time for the situation of Example 31.4. The current, which is not shown, is in phase with the resistor voltage.

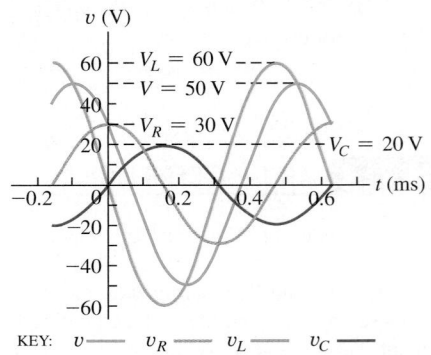

TEST YOUR UNDERSTANDING OF SECTION 31.3 Rank the following ac circuits in order of their current amplitude, from highest to lowest value. (i) The circuit in Example 31.4; (ii) the circuit in Example 31.4 with both the capacitor and inductor removed; (iii) the circuit in Example 31.4 with both the resistor and capacitor removed; (iv) the circuit in Example 31.4 with both the resistor and inductor removed. ❚

31.4 POWER IN ALTERNATING-CURRENT CIRCUITS

Alternating currents play a central role in systems for distributing, converting, and using electrical energy, so it's important to look at power relationships in ac circuits. For an ac circuit with instantaneous current i and current amplitude I, we'll consider an element of that circuit across which the instantaneous potential difference is v with voltage amplitude V. The instantaneous power p delivered to this circuit element is

$$p = vi$$

Let's first see what this means for individual circuit elements. We'll assume in each case that $i = I\cos\omega t$.

Power in a Resistor

Suppose first that the circuit element is a *pure resistor* R, as in Fig. 31.7a; then $v = v_R$ and i are *in phase*. We obtain the graph representing p by multiplying the heights of the graphs of v and i in Fig. 31.7b at each instant. The result is the black curve in **Fig. 31.16a**. The product vi is always positive because v and i are always either both positive or both negative. Hence energy is supplied *to* the resistor at every instant for both directions of i, although the power is not constant.

The power curve for a pure resistor is symmetric about a value equal to one-half its maximum value VI, so the *average power* P_{av} is

$$P_{av} = \tfrac{1}{2}VI \qquad \text{(for a pure resistor)} \qquad (31.27)$$

An equivalent expression is

$$P_{av} = \frac{V}{\sqrt{2}}\frac{I}{\sqrt{2}} = V_{rms}I_{rms} \qquad \text{(for a pure resistor)} \qquad (31.28)$$

Also, $V_{rms} = I_{rms}R$, so we can express P_{av} by any of the equivalent forms

$$P_{av} = I_{rms}{}^2R = \frac{V_{rms}{}^2}{R} = V_{rms}I_{rms} \qquad \text{(for a pure resistor)} \qquad (31.29)$$

Note that the expressions in Eq. (31.29) have the same form as the corresponding relationships for a dc circuit, Eq. (25.18). Also note that they are valid only for pure resistors, not for more complicated combinations of circuit elements.

31.16 Graphs of current, voltage, and power as functions of time for (a) a pure resistor, (b) a pure inductor, (c) a pure capacitor, and (d) an arbitrary ac circuit that can have resistance, inductance, and capacitance.

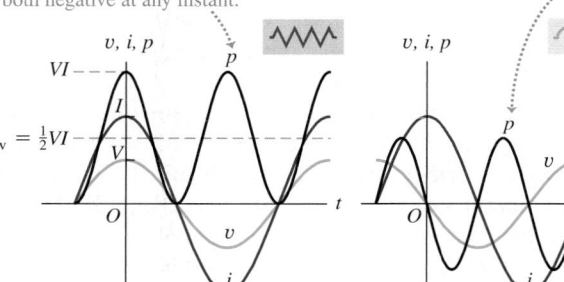
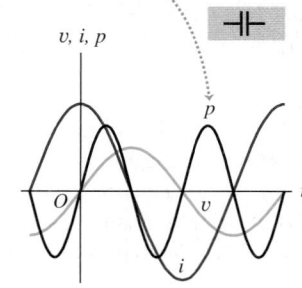

(a) Pure resistor

For a resistor, $p = vi$ is always positive because v and i are either both positive or both negative at any instant.

(b) Pure inductor

For an inductor or capacitor, $p = vi$ is alternately positive and negative, and the average power is zero.

(c) Pure capacitor

(d) Arbitrary ac circuit

For an arbitrary combination of resistors, inductors, and capacitors, the average power is positive.

KEY: Instantaneous current, i —— Instantaneous voltage across device, v —— Instantaneous power input to device, p ——

Power in an Inductor

Next we connect the source to a pure inductor L, as in Fig. 31.8a. The voltage $v = v_L$ leads the current i by 90°. When we multiply the curves of v and i, the product vi is *negative* during the half of the cycle when v and i have *opposite* signs. The power curve, shown in Fig. 31.16b, is symmetric about the horizontal axis; it is positive half the time and negative the other half, and the average power is zero. When p is positive, energy is being supplied to set up the magnetic field in the inductor; when p is negative, the field is collapsing and the inductor is returning energy to the source. The net energy transfer over one cycle is zero.

Power in a Capacitor

Finally, we connect the source to a pure capacitor C, as in Fig. 31.9a. The voltage $v = v_C$ lags the current i by 90°. Figure 31.16c shows the power curve; the average power is again zero. Energy is supplied to charge the capacitor and is returned to the source when the capacitor discharges. The net energy transfer over one cycle is again zero.

Power in a General ac Circuit

In *any* ac circuit, with any combination of resistors, capacitors, and inductors, the voltage v across the entire circuit has some phase angle ϕ with respect to the current i. Then the instantaneous power p is given by

$$p = vi = [V\cos(\omega t + \phi)][I\cos\omega t] \tag{31.30}$$

The instantaneous power curve has the form shown in Fig. 31.16d. The area between the positive loops and the horizontal axis is greater than the area between the negative loops and the horizontal axis, and the average power is positive.

We can derive from Eq. (31.30) an expression for the *average* power P_{av} by using the identity for the cosine of the sum of two angles:

$$p = [V(\cos\omega t\cos\phi - \sin\omega t\sin\phi)][I\cos\omega t]$$
$$= VI\cos\phi\cos^2\omega t - VI\sin\phi\cos\omega t\sin\omega t$$

From the discussion in Section 31.1 that led to Eq. (31.4), the average value of $\cos^2\omega t$ (over one cycle) is $\frac{1}{2}$. Furthermore, $\cos\omega t\sin\omega t$ is equal to $\frac{1}{2}\sin 2\omega t$, whose average over a cycle is zero. So the average power P_{av} is

Average power into a general ac circuit ·······→

Phase angle of voltage with respect to current

$$P_{av} = \tfrac{1}{2}VI\cos\phi = V_{rms}I_{rms}\cos\phi \tag{31.31}$$

Voltage amplitude   Current amplitude   rms voltage   rms current

31.17 Using phasors to calculate the average power for an arbitrary ac circuit.

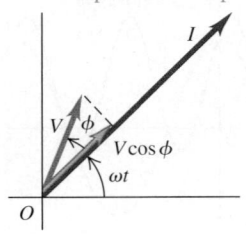

Average power = $\frac{1}{2}I(V\cos\phi)$, where $V\cos\phi$ is the component of V in phase with I.

Figure 31.17 shows the general relationship of the current and voltage phasors. When v and i are in phase, so $\phi = 0$, the average power equals $\frac{1}{2}VI = V_{rms}I_{rms}$; when v and i are 90° out of phase, the average power is zero. In the general case, when v has a phase angle ϕ with respect to i, the average power equals $\frac{1}{2}I$ multiplied by $V\cos\phi$, the component of the voltage phasor that is *in phase* with the current phasor. For the L-R-C series circuit, Figs. 31.13b and 31.13c show that $V\cos\phi$ equals the voltage amplitude V_R for the resistor; hence Eq. (31.31) is the average power dissipated in the resistor. On average there is no energy flow into or out of the inductor or capacitor, so none of P_{av} goes into either of these circuit elements.

The factor $\cos\phi$ is called the **power factor** of the circuit. For a pure resistance, $\phi = 0$, $\cos\phi = 1$, and $P_{av} = V_{rms}I_{rms}$. For a pure inductor or capacitor, $\phi = \pm 90°$, $\cos\phi = 0$, and $P_{av} = 0$. For an *L-R-C* series circuit the power factor is equal to R/Z; we leave the proof of this statement to you (see Exercise 31.21).

A low power factor (large angle ϕ of lag or lead) is usually undesirable in power circuits. The reason is that for a given potential difference, a large current is needed to supply a given amount of power. This results in large i^2R losses in the transmission lines. Many types of ac machinery draw a *lagging* current; that is, the current drawn by the machinery lags the applied voltage. Hence the voltage leads the current, so $\phi > 0$ and $\cos\phi < 1$. The power factor can be corrected toward the ideal value of 1 by connecting a capacitor in parallel with the load. The current drawn by the capacitor *leads* the voltage (that is, the voltage across the capacitor lags the current), which compensates for the lagging current in the other branch of the circuit. The capacitor itself absorbs no net power from the line.

EXAMPLE 31.6   POWER IN A HAIR DRYER

An electric hair dryer is rated at 1500 W (the *average* power) at 120 V (the *rms* voltage). Calculate (a) the resistance, (b) the rms current, and (c) the maximum instantaneous power. Assume that the dryer is a pure resistor. (The heating element acts as a resistor.)

SOLUTION

IDENTIFY and SET UP: We are given $P_{av} = 1500$ W and $V_{rms} = 120$ V. Our target variables are the resistance R, the rms current I_{rms}, and the maximum value p_{max} of the instantaneous power p. We solve Eq. (31.29) to find R, Eq. (31.28) to find I_{rms} from V_{rms} and P_{av}, and Eq. (31.30) to find p_{max}.

EXECUTE: (a) From Eq. (31.29), the resistance is

$$R = \frac{V_{rms}{}^2}{P_{av}} = \frac{(120\ \text{V})^2}{1500\ \text{W}} = 9.6\ \Omega$$

(b) From Eq. (31.28),

$$I_{rms} = \frac{P_{av}}{V_{rms}} = \frac{1500\ \text{W}}{120\ \text{V}} = 12.5\ \text{A}$$

(c) For a pure resistor, the voltage and current are in phase and the phase angle ϕ is zero. Hence from Eq. (31.30), the instantaneous power is $p = VI\cos^2\omega t$ and the maximum instantaneous power is $p_{max} = VI$. From Eq. (31.27), this is twice the average power P_{av}, so

$$p_{max} = VI = 2P_{av} = 2(1500\ \text{W}) = 3000\ \text{W}$$

EVALUATE: We can use Eq. (31.7) to confirm our result in part (b): $I_{rms} = V_{rms}/R = (120\ \text{V})/(9.6\ \Omega) = 12.5$ A. Note that some unscrupulous manufacturers of stereo amplifiers advertise the *peak* power output rather than the lower average value.

EXAMPLE 31.7   POWER IN AN *L-R-C* SERIES CIRCUIT

For the *L-R-C* series circuit of Example 31.4, (a) calculate the power factor and (b) calculate the average power delivered to the entire circuit and to each circuit element.

SOLUTION

IDENTIFY and SET UP: We can use the results of Example 31.4. The power factor is the cosine of the phase angle ϕ, and we use Eq. (31.31) to find the average power delivered in terms of ϕ and the amplitudes of voltage and current.

EXECUTE: (a) The power factor is $\cos\phi = \cos 53° = 0.60$.
(b) From Eq. (31.31),

$$P_{av} = \tfrac{1}{2}VI\cos\phi = \tfrac{1}{2}(50\ \text{V})(0.10\ \text{A})(0.60) = 1.5\ \text{W}$$

EVALUATE: Although P_{av} is the average power delivered to the *L-R-C* combination, all of this power is dissipated in the *resistor*. As Figs. 31.16b and 31.16c show, the average power delivered to a pure inductor or pure capacitor is always zero.

TEST YOUR UNDERSTANDING OF SECTION 31.4   Figure 31.16d shows that during part of a cycle of oscillation, the instantaneous power delivered to the circuit is negative. This means that energy is being extracted from the circuit. (a) Where is the energy extracted from? (i) The resistor; (ii) the inductor; (iii) the capacitor; (iv) the ac source; (v) more than one of these. (b) Where does the energy go? (i) The resistor; (ii) the inductor; (iii) the capacitor; (iv) the ac source; (v) more than one of these. ∎

31.5 RESONANCE IN ALTERNATING-CURRENT CIRCUITS

31.18 How variations in the angular frequency of an ac circuit affect (a) reactance, resistance, and impedance, and (b) impedance, current amplitude, and phase angle.

(a) Reactance, resistance, and impedance as functions of angular frequency

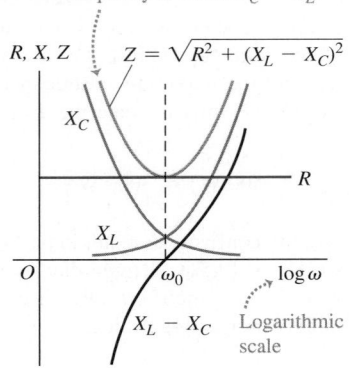

(b) Impedance, current, and phase angle as functions of angular frequency

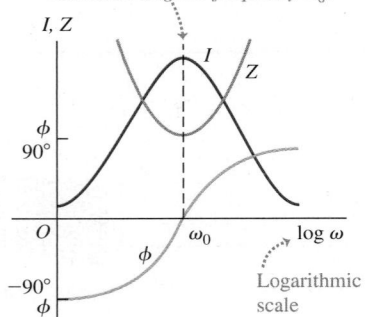

Much of the practical importance of *L-R-C* series circuits arises from the way in which such circuits respond to sources of different angular frequency ω. For example, one type of tuning circuit used in radio receivers is simply an *L-R-C* series circuit. A radio signal of any given frequency produces a current of the same frequency in the receiver circuit, but the amplitude of the current is *greatest* if the signal frequency equals the particular frequency to which the receiver circuit is "tuned." This effect is called *resonance*. The circuit is designed so that signals at other than the tuned frequency produce currents that are too small to make an audible sound come out of the radio's speakers.

To see how an *L-R-C* series circuit can be used in this way, suppose we connect an ac source with constant voltage amplitude V but adjustable angular frequency ω across an *L-R-C* series circuit. The current that appears in the circuit has the same angular frequency as the source and a current amplitude $I = V/Z$, where Z is the impedance of the *L-R-C* series circuit. This impedance depends on the frequency, as Eq. (31.23) shows. **Figure 31.18a** shows graphs of R, X_L, X_C, and Z as functions of ω. We have used a logarithmic angular frequency scale so that we can cover a wide range of frequencies. As the frequency increases, X_L increases and X_C decreases; hence there is always one frequency at which X_L and X_C are equal and $X_L - X_C$ is zero. At this frequency the impedance $Z = \sqrt{R^2 + (X_L - X_C)^2}$ has its *smallest* value, equal simply to the resistance R.

Circuit Behavior at Resonance

As we vary the angular frequency ω of the source, the current amplitude $I = V/Z$ varies as shown in Fig. 31.18b; the *maximum* value of I occurs at the frequency at which the impedance Z is *minimum*. This peaking of the current amplitude at a certain frequency is called **resonance.** The angular frequency ω_0 at which the resonance peak occurs is called the **resonance angular frequency.** At $\omega = \omega_0$ the inductive reactance X_L and capacitive reactance X_C are equal, so $\omega_0 L = 1/\omega_0 C$ and

$$\text{Resonance angular frequency of an } L\text{-}R\text{-}C \text{ series circuit} \qquad \omega_0 = \frac{1}{\sqrt{LC}} \qquad (31.32)$$

Inductance ⤴   Capacitance ⤴

This is equal to the natural angular frequency of oscillation of an *L-C* circuit, which we derived in Section 30.5, Eq. (30.22). The **resonance frequency** f_0 is $\omega_0/2\pi$. At this frequency, the greatest current appears in the circuit for a given source voltage amplitude; f_0 is the frequency to which the circuit is "tuned."

It's instructive to look at what happens to the *voltages* in an *L-R-C* series circuit at resonance. The current at any instant is the same in L and C. The voltage across an inductor always *leads* the current by 90°, or $\frac{1}{4}$ cycle, and the voltage across a capacitor always *lags* the current by 90°. Therefore the instantaneous voltages across L and C always differ in phase by $180°$, or $\frac{1}{2}$ cycle; they have opposite signs at each instant. At the resonance frequency, and *only* at the resonance frequency, $X_L = X_C$ and the voltage amplitudes $V_L = IX_L$ and $V_C = IX_C$ are *equal;* then the instantaneous voltages across L and C add to zero at each instant, and the *total* voltage v_{bd} across the L-C combination in Fig. 31.13a is exactly zero. The voltage across the resistor is then equal to the source voltage. So at the resonance frequency the circuit behaves as if the inductor and capacitor weren't there at all!

The *phase* of the voltage relative to the current is given by Eq. (31.24). At frequencies below resonance, X_C is greater than X_L; the capacitive reactance dominates, the voltage *lags* the current, and the phase angle ϕ is between $0°$ and $-90°$. Above resonance, the inductive reactance dominates, the voltage *leads* the current, and the phase angle ϕ is between zero and $+90°$. Figure 31.18b shows this variation of ϕ with angular frequency.

Tailoring an ac Circuit

If we can vary the inductance L or the capacitance C of a circuit, we can also vary the resonance frequency. This is exactly how a radio is "tuned" to receive a particular station. In the early days of radio this was accomplished by the use of capacitors with movable metal plates whose overlap could be varied to change C. (This is what is being done with the radio tuning knob shown in the photograph that opens this chapter.) Another approach is to vary L by using a coil with a ferrite core that slides in or out.

In an L-R-C series circuit the impedance is a minimum and the current is a maximum at the resonance frequency. The middle curve in **Fig. 31.19** is a graph of current as a function of frequency for such a circuit, with source voltage amplitude $V = 100$ V, $L = 2.0$ H, $C = 0.50\ \mu$F, and $R = 500\ \Omega$. This curve, called a *response curve* or *resonance curve,* has a peak at the resonance angular frequency $\omega_0 = \sqrt{LC} = 1000$ rad/s.

The resonance frequency is determined by L and C; what happens when we change R? Figure 31.19 also shows graphs of I as a function of ω for $R = 200\ \Omega$ and for $R = 2000\ \Omega$. The curves are similar for frequencies far away from resonance, where the impedance is dominated by X_L or X_C. But near resonance, where X_L and X_C nearly cancel each other, the curve is higher and more sharply peaked for small values of R and broader and flatter for large values of R. At resonance, $Z = R$ and $I = V/R$, so the maximum height of the curve is inversely proportional to R.

The shape of the response curve is important in the design of radio receiving circuits. The sharply peaked curve is what makes it possible to discriminate between two stations broadcasting on adjacent frequency bands. But if the peak is *too* sharp, some of the information in the received signal is lost, such as the high-frequency sounds in music. The shape of the resonance curve is also related to the overdamped and underdamped oscillations that we described in Section 30.6. A sharply peaked resonance curve corresponds to a small value of R and a lightly damped oscillating system; a broad, flat curve goes with a large value of R and a heavily damped system.

In this section we have discussed resonance in an L-R-C *series* circuit. Resonance can also occur in an ac circuit in which the inductor, resistor, and capacitor are connected in *parallel*. We leave the details to you (see Problem 31.55).

Resonance phenomena occur not just in ac circuits, but in all areas of physics. We discussed examples of resonance in *mechanical* systems in Sections 14.8 and 16.5. The amplitude of a mechanical oscillation peaks when the driving-force frequency is close to a natural frequency of the system; this is analogous to the peaking of the current in an L-R-C series circuit.

31.19 Graph of current amplitude I as a function of angular frequency ω for an L-R-C series circuit with $V = 100$ V, $L = 2.0$ H, $C = 0.50\ \mu$F, and three different values of the resistance R.

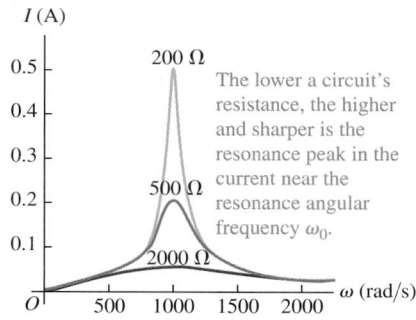

The lower a circuit's resistance, the higher and sharper is the resonance peak in the current near the resonance angular frequency ω_0.

EXAMPLE 31.8   TUNING A RADIO

The series circuit in **Fig. 31.20** is similar to some radio tuning circuits. It is connected to a variable-frequency ac source with an rms terminal voltage of 1.0 V. (a) Find the resonance frequency. At the resonance frequency, find (b) the inductive reactance X_L, the capacitive reactance X_C, and the impedance Z; (c) the rms current I_{rms}; (d) the rms voltage across each circuit element.

SOLUTION

IDENTIFY and SET UP: Figure 31.20 shows an L-R-C series circuit, with ideal meters inserted to measure the rms current and voltages, our target variables. Equation (31.32) gives the formula for the resonance angular frequency ω_0, from which we find the resonance frequency f_0. We use Eqs. (31.12) and (31.18) to find X_L and X_C, which are equal at resonance; at resonance, from Eq. (31.23), we have $Z = R$. We use Eqs. (31.7), (31.13), and (31.19) to find the voltages across the circuit elements.

31.20 A radio tuning circuit at resonance. The circles denote rms current and voltages.

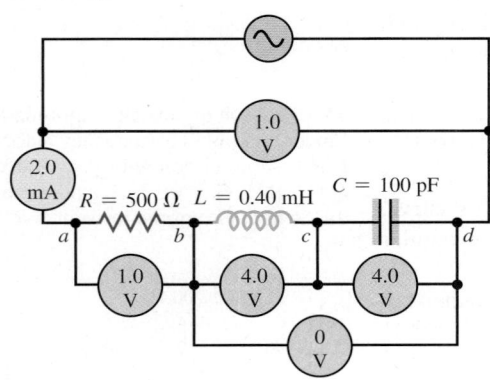

EXECUTE: (a) The values of ω_0 and f_0 are

$$\omega_0 = \frac{1}{\sqrt{LC}} = \frac{1}{\sqrt{(0.40 \times 10^{-3}\ \text{H})(100 \times 10^{-12}\ \text{F})}}$$

$$= 5.0 \times 10^6\ \text{rad/s}$$

$$f_0 = 8.0 \times 10^5\ \text{Hz} = 800\ \text{kHz}$$

This frequency is in the lower part of the AM radio band.
   (b) At this frequency,

$$X_L = \omega L = (5.0 \times 10^6\ \text{rad/s})(0.40 \times 10^{-3}\ \text{H}) = 2000\ \Omega$$

$$X_C = \frac{1}{\omega C} = \frac{1}{(5.0 \times 10^6\ \text{rad/s})(100 \times 10^{-12}\ \text{F})} = 2000\ \Omega$$

Since $X_L = X_C$ at resonance as stated above, $Z = R = 500\ \Omega$.
   (c) From Eq. (31.26) the rms current at resonance is

$$I_{rms} = \frac{V_{rms}}{Z} = \frac{V_{rms}}{R} = \frac{1.0\ \text{V}}{500\ \Omega} = 0.0020\ \text{A} = 2.0\ \text{mA}$$

(d) The rms potential difference across the resistor is

$$V_{R\text{-rms}} = I_{rms}R = (0.0020\ \text{A})(500\ \Omega) = 1.0\ \text{V}$$

The rms potential differences across the inductor and capacitor are

$$V_{L\text{-rms}} = I_{rms}X_L = (0.0020\ \text{A})(2000\ \Omega) = 4.0\ \text{V}$$

$$V_{C\text{-rms}} = I_{rms}X_C = (0.0020\ \text{A})(2000\ \Omega) = 4.0\ \text{V}$$

EVALUATE: The potential differences across the inductor and the capacitor have equal rms values and amplitudes, but are 180° out of phase and so add to zero at each instant. Note also that at resonance, $V_{R\text{-rms}}$ is equal to the source voltage V_{rms}, while in this example, $V_{L\text{-rms}}$ and $V_{C\text{-rms}}$ are both considerably *larger* than V_{rms}.

TEST YOUR UNDERSTANDING OF SECTION 31.5   How does the resonance frequency of an L-R-C series circuit change if the plates of the capacitor are brought closer together? (i) It increases; (ii) it decreases; (iii) it is unaffected. ∎

31.6 TRANSFORMERS

One great advantage of ac over dc for electric-power distribution is that it is much easier to step voltage levels up and down with ac than with dc. For long-distance power transmission it is desirable to use as high a voltage and as small a current as possible; this reduces i^2R losses in the transmission lines, and smaller wires can be used, saving on material costs. Present-day transmission lines operate at rms voltages of the order of 500 kV. However, safety considerations dictate relatively low voltages in generating equipment and in household and industrial power distribution. The standard voltage for household wiring is 120 V in the United States and Canada and 240 V in many other countries. The necessary voltage conversion is accomplished by using **transformers.**

How Transformers Work

Figure 31.21 shows an idealized transformer. Its key components are two coils or *windings*, electrically insulated from each other but wound on the same core. The core is typically made of a material, such as iron, with a very large relative permeability K_m. This keeps the magnetic field lines due to a current in one

31.21 Schematic diagram of an idealized step-up transformer. The primary is connected to an ac source; the secondary is connected to a device with resistance R.

The induced emf *per turn* is the same in both coils, so we adjust the ratio of terminal voltages by adjusting the ratio of turns:

$$\frac{V_2}{V_1} = \frac{N_2}{N_1}$$

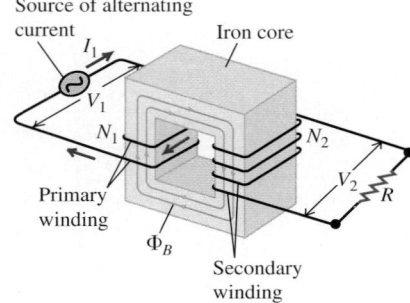

winding almost completely within the core. Hence almost all of these field lines pass through the other winding, maximizing the *mutual inductance* of the two windings (see Section 30.1). The winding to which power is supplied is called the **primary;** the winding from which power is delivered is called the **secondary.** The circuit symbol for a transformer with an iron core is

Here's how a transformer works. The ac source causes an alternating current in the primary, which sets up an alternating flux in the core; this induces an emf in each winding, in accordance with Faraday's law. The induced emf in the secondary gives rise to an alternating current in the secondary, and this delivers energy to the device to which the secondary is connected. All currents and emfs have the same frequency as the ac source.

Let's see how the voltage across the secondary can be made larger or smaller in amplitude than the voltage across the primary. We ignore the resistance of the windings and assume that all the magnetic field lines are confined to the iron core, so at any instant the magnetic flux Φ_B is the same in each turn of the primary and secondary windings. The primary winding has N_1 turns and the secondary winding has N_2 turns. When the magnetic flux changes because of changing currents in the two coils, the resulting induced emfs are

$$\mathcal{E}_1 = -N_1 \frac{d\Phi_B}{dt} \quad \text{and} \quad \mathcal{E}_2 = -N_2 \frac{d\Phi_B}{dt} \qquad (31.33)$$

The flux *per turn* Φ_B is the same in both the primary and the secondary, so Eqs. (31.33) show that the induced emf *per turn* is the same in each. The ratio of the secondary emf $\mathcal{E}_2$ to the primary emf $\mathcal{E}_1$ is therefore equal at any instant to the ratio of secondary to primary turns:

$$\frac{\mathcal{E}_2}{\mathcal{E}_1} = \frac{N_2}{N_1} \qquad (31.34)$$

Since $\mathcal{E}_1$ and $\mathcal{E}_2$ both oscillate with the same frequency as the ac source, Eq. (31.34) also gives the ratio of the amplitudes or of the rms values of the induced emfs. If the windings have zero resistance, the induced emfs $\mathcal{E}_1$ and $\mathcal{E}_2$ are equal to the terminal voltages across the primary and the secondary, respectively; hence

Terminal voltages in a transformer:

Secondary voltage amplitude or rms value

$$\frac{V_2}{V_1} = \frac{N_2}{N_1} \qquad (31.35)$$

Number of turns in secondary

Primary voltage amplitude or rms value

Number of turns in primary

By choosing the appropriate turns ratio N_2/N_1, we may obtain any desired secondary voltage from a given primary voltage. If $N_2 > N_1$, as in Fig. 31.21, then $V_2 > V_1$ and we have a *step-up* transformer; if $N_2 < N_1$, then $V_2 < V_1$ and we have a *step-down* transformer. At a power-generating station, step-up transformers are used; the primary is connected to the power source and the secondary is connected to the transmission lines, giving the desired high voltage for transmission. Near the consumer, step-down transformers lower the voltage to a value suitable for use in home or industry (**Fig. 31.22**).

Even the relatively low voltage provided by a household wall socket is too high for many electronic devices, so a further step-down transformer is necessary.

31.22 The cylindrical can near the top of this power pole is a step-down transformer. It converts the high-voltage ac in the power lines to low-voltage (120 V) ac, which is then distributed to the surrounding homes and businesses.

31.23 An ac adapter like this one converts household ac into low-voltage dc for use in electronic devices. It contains a step-down transformer to change the line voltage to a lower value, typically 3 to 12 V, as well as diodes to convert alternating current to the direct current that small electronic devices require (see Fig. 31.3).

DATA *SPEAKS*

Transformers

When students were given a problem involving transformers, more than 40% gave an incorrect response. Common errors:

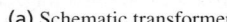

- Forgetting that transformers work for *alternating* current only. A transformer works on the principle that a varying current in a primary coil induces a varying current in a secondary coil. It does not work with constant direct current.

- Confusion over voltage and current. The voltage in a transformer coil is *proportional* to the number of turns in that coil. The current in a transformer coil is *inversely proportional* to the number of turns.

This is the role of an "ac adapter" such as those used to recharge a mobile phone or laptop computer from line voltage (**Fig. 31.23**).

Energy Considerations for Transformers

If the secondary circuit is completed by a resistance R, then the amplitude or rms value of the current in the secondary circuit is $I_2 = V_2/R$. From energy considerations, the power delivered to the primary equals that taken out of the secondary (since there is no resistance in the windings), so

Terminal voltages and currents in a transformer:	Primary voltage amplitude or rms value	Secondary voltage amplitude or rms value	
	$V_1 I_1 = V_2 I_2$		(31.36)
	Current in primary	Current in secondary	

We can combine Eqs. (31.35) and (31.36) and the relationship $I_2 = V_2/R$ to eliminate V_2 and I_2; we obtain

$$\frac{V_1}{I_1} = \frac{R}{(N_2/N_1)^2} \qquad (31.37)$$

This shows that when the secondary circuit is completed through a resistance R, the result is the same as if the *source* had been connected directly to a resistance equal to R divided by the square of the turns ratio, $(N_2/N_1)^2$. In other words, the transformer "transforms" not only voltages and currents, but resistances as well. More generally, we can regard a transformer as "transforming" the *impedance* of the network to which the secondary circuit is completed.

Equation (31.37) has many practical consequences. The power supplied by a source to a resistor depends on the resistances of both the resistor and the source. It can be shown that the power transfer is greatest when the two resistances are *equal*. The same principle applies in both dc and ac circuits. When a high-impedance ac source must be connected to a low-impedance circuit, such as an audio amplifier connected to a loudspeaker, the source impedance can be *matched* to that of the circuit by the use of a transformer with an appropriate turns ratio N_2/N_1.

Real transformers always have some energy losses. (That's why an ac adapter like the one shown in Fig. 31.23 feels warm to the touch after it's been in use for a while; the transformer is heated by the dissipated energy.) The windings have some resistance, leading to i^2R losses. There are also energy losses through hysteresis in the core (see Section 28.8). Hysteresis losses are minimized by the use of soft iron with a narrow hysteresis loop.

Eddy currents (Section 29.6) also cause energy loss in transformers. Consider a section AA through an iron transformer core (**Fig. 31.24a**). Since iron is a conductor, any such section can be pictured as several conducting circuits, one

31.24 (a) Primary and secondary windings in a transformer. (b) Eddy currents in the iron core, shown in the cross section at AA. (c) Using a laminated core reduces the eddy currents.

(a) Schematic transformer

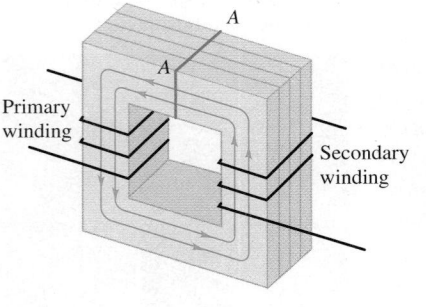

(b) Large eddy currents in solid core

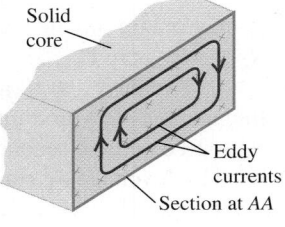

(c) Smaller eddy currents in laminated core

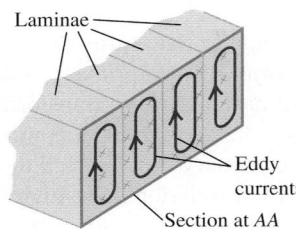

within the other (Fig. 31.24b). The flux through each of these circuits is continually changing, so eddy currents circulate in the entire volume of the core, with lines of flow that form planes perpendicular to the flux. These eddy currents waste energy through i^2R heating and themselves set up an opposing flux.

The effects of eddy currents can be minimized by the use of a *laminated* core—that is, one built up of thin sheets, or laminae. The large electrical surface resistance of each lamina, due either to a natural coating of oxide or to an insulating varnish, effectively confines the eddy currents to individual laminae (Fig. 31.24c). The possible eddy-current paths are narrower, the induced emf in each path is smaller, and the eddy currents are greatly reduced. The alternating magnetic field exerts forces on the current-carrying laminae that cause them to vibrate back and forth; this vibration causes the characteristic "hum" of an operating transformer. You can hear this same "hum" from the magnetic ballast of a fluorescent light fixture (see Section 30.2).

Thanks to the use of soft iron cores and lamination, transformer efficiencies are usually well over 90%; in large installations they may reach 99%.

EXAMPLE 31.9   "WAKE UP AND SMELL THE (TRANSFORMER)!"

A friend returns to the United States from Europe with a 960-W coffeemaker, designed to operate from a 240-V line. (a) What can she do to operate it at the USA-standard 120 V? (b) What current will the coffeemaker draw from the 120-V line? (c) What is the resistance of the coffeemaker? (The voltages are rms values.)

SOLUTION

IDENTIFY and SET UP: Our friend needs a step-up transformer to convert 120-V ac to the 240-V ac that the coffeemaker requires. We use Eq. (31.35) to determine the transformer turns ratio N_2/N_1, $P_{av} = V_{rms}I_{rms}$ for a resistor to find the current draw, and Eq. (31.37) to calculate the resistance.

EXECUTE: (a) To get $V_2 = 240$ V from $V_1 = 120$ V, the required turns ratio is $N_2/N_1 = V_2/V_1 = (240\text{ V})/(120\text{ V}) = 2$. That is, the secondary coil (connected to the coffeemaker) should have twice as many turns as the primary coil (connected to the 120-V line).

(b) We find the rms current I_1 in the 120-V primary by using $P_{av} = V_1I_1$, where P_{av} is the average power drawn by the coffeemaker and hence the power supplied by the 120-V line. (We're assuming that no energy is lost in the transformer.) Hence $I_1 = P_{av}/V_1 = (960\text{ W})/(120\text{ V}) = 8.0$ A. The secondary current is then $I_2 = P_{av}/V_2 = (960\text{ W})/(240\text{ V}) = 4.0$ A.

(c) We have $V_1 = 120$ V, $I_1 = 8.0$ A, and $N_2/N_1 = 2$, so

$$\frac{V_1}{I_1} = \frac{120\text{ V}}{8.0\text{ A}} = 15\ \Omega$$

From Eq. (31.37),

$$R = 2^2(15\ \Omega) = 60\ \Omega$$

EVALUATE: As a check, $V_2/R = (240\text{ V})/(60\ \Omega) = 4.0\text{ A} = I_2$, the same value obtained previously. You can also check this result for R by using the expression $P_{av} = V_2^2/R$ for the power drawn by the coffeemaker.

TEST YOUR UNDERSTANDING OF SECTION 31.6 Each of the following four transformers has 1000 turns in its primary. Rank the transformers from largest to smallest number of turns in the secondary. (i) Converts 120-V ac into 6.0-V ac; (ii) converts 120-V ac into 240-V ac; (iii) converts 240-V ac into 6.0-V ac; (iv) converts 240-V ac into 120-V ac. ▮

Phasors and alternating current: An alternator or ac source produces an emf that varies sinusoidally with time. A sinusoidal voltage or current can be represented by a phasor, a vector that rotates counter-clockwise with constant angular velocity ω equal to the angular frequency of the sinusoidal quantity. Its projection on the horizontal axis at any instant represents the instantaneous value of the quantity.

For a sinusoidal current, the rectified average and rms (root-mean-square) currents are proportional to the current amplitude I. Similarly, the rms value of a sinusoidal voltage is proportional to the voltage amplitude V. (See Example 31.1.)

$$I_{\text{rav}} = \frac{2}{\pi}I = 0.637I \qquad (31.3)$$

$$I_{\text{rms}} = \frac{I}{\sqrt{2}} \qquad (31.4)$$

$$V_{\text{rms}} = \frac{V}{\sqrt{2}} \qquad (31.5)$$

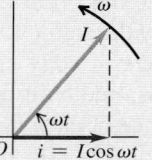

Voltage, current, and phase angle: In general, the instantaneous voltage $v = V\cos(\omega t + \phi)$ between two points in an ac circuit is not in phase with the instantaneous current passing through those points. The quantity ϕ is called the phase angle of the voltage relative to the current.

$$i = I\cos\omega t \qquad (31.2)$$

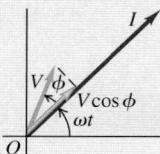

Resistance and reactance: The voltage across a resistor R is in phase with the current. The voltage across an inductor L leads the current by 90° ($\phi = +90°$), while the voltage across a capacitor C lags the current by 90° ($\phi = -90°$). The voltage amplitude across each type of device is proportional to the current amplitude I. An inductor has inductive reactance $X_L = \omega L$, and a capacitor has capacitive reactance $X_C = 1/\omega C$. (See Examples 31.2 and 31.3.)

$$V_R = IR \qquad (31.7)$$

$$V_L = IX_L \qquad (31.13)$$

$$V_C = IX_C \qquad (31.19)$$

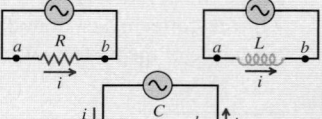

Impedance and the *L-R-C* series circuit: In a general ac circuit, the voltage and current amplitudes are related by the circuit impedance Z. In an L-R-C series circuit, the values of L, R, C, and the angular frequency ω determine the impedance and the phase angle ϕ of the voltage relative to the current. (See Examples 31.4 and 31.5.)

$$V = IZ \qquad (31.22)$$

$$Z = \sqrt{R^2 + [\omega L - (1/\omega C)]^2} \qquad (31.23)$$

$$\tan\phi = \frac{\omega L - 1/\omega C}{R} \qquad (31.24)$$

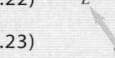

Power in ac circuits: The average power input P_{av} to an ac circuit depends on the voltage and current amplitudes (or, equivalently, their rms values) and the phase angle ϕ of the voltage relative to the current. The quantity $\cos\phi$ is called the power factor. (See Examples 31.6 and 31.7.)

$$P_{\text{av}} = \tfrac{1}{2}VI\cos\phi$$
$$= V_{\text{rms}}I_{\text{rms}}\cos\phi \qquad (31.31)$$

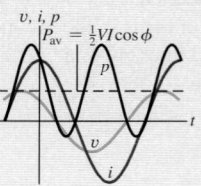

Resonance in ac circuits: In an *L-R-C* series circuit, the current becomes maximum and the impedance becomes minimum at an angular frequency called the resonance angular frequency. This phenomenon is called resonance. At resonance the voltage and current are in phase, and the impedance Z is equal to the resistance R. (See Example 31.8.)

$$\omega_0 = \frac{1}{\sqrt{LC}} \qquad (31.32)$$

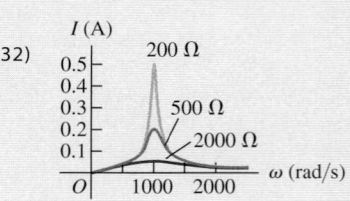

Transformers: A transformer is used to transform the voltage and current levels in an ac circuit. In an ideal transformer with no energy losses, if the primary winding has N_1 turns and the secondary winding has N_2 turns, the amplitudes (or rms values) of the two voltages are related by Eq. (31.35). The amplitudes (or rms values) of the primary and secondary voltages and currents are related by Eq. (31.36). (See Example 31.9.)

$$\frac{V_2}{V_1} = \frac{N_2}{N_1} \quad (31.35)$$

$$V_1 I_1 = V_2 I_2 \quad (31.36)$$

BRIDGING PROBLEM AN ALTERNATING-CURRENT CIRCUIT

A series circuit like the circuit in Fig. 31.13a consists of a 1.50-mH inductor, a 125-Ω resistor, and a 25.0-nF capacitor connected across an ac source having an rms voltage of 35.0 V and variable frequency. (a) At what angular frequencies will the current amplitude be equal to $\frac{1}{3}$ of its maximum possible value? (b) At the frequencies in part (a), what are the current amplitude and the voltage amplitude across each circuit element (including the ac source)?

SOLUTION GUIDE

IDENTIFY and SET UP

1. The maximum current amplitude occurs at the resonance angular frequency. This problem concerns the angular frequencies at which the current amplitude is one-third of that maximum.
2. Choose the equation that will allow you to find the angular frequencies in question, and choose the equations that you will

then use to find the current and voltage amplitudes at each angular frequency.

EXECUTE

3. Find the impedance at the angular frequencies in part (a); then solve for the values of angular frequency.
4. Find the voltage amplitude across the source and the current amplitude for each of the angular frequencies in part (a). (*Hint:* Be careful to distinguish between *amplitude* and *rms value*.)
5. Use the results of steps 3 and 4 to find the reactances at each angular frequency. Then calculate the voltage amplitudes for the resistor, inductor, and capacitor.

EVALUATE

6. Are any voltage amplitudes greater than the voltage amplitude of the source? If so, does this mean your results are in error?

Problems

For assigned homework and other learning materials, go to MasteringPhysics®. (MP)

•, ••, •••: Difficulty levels. **CP**: Cumulative problems incorporating material from earlier chapters. **CALC**: Problems requiring calculus. **DATA**: Problems involving real data, scientific evidence, experimental design, and/or statistical reasoning. **BIO**: Biosciences problems.

DISCUSSION QUESTIONS

Q31.1 Household electric power in most of western Europe is supplied at 240 V, rather than the 120 V that is standard in the United States and Canada. What are the advantages and disadvantages of each system?

Q31.2 The current in an ac power line changes direction 120 times per second, and its average value is zero. Explain how it is possible for power to be transmitted in such a system.

Q31.3 In an ac circuit, why is the average power for an inductor and a capacitor zero, but not for a resistor?

Q31.4 Equation (31.14) was derived by using the relationship $i = dq/dt$ between the current and the charge on the capacitor. In Fig. 31.9a the positive counterclockwise current increases the charge on the capacitor. When the charge on the left plate is positive but decreasing in time, is $i = dq/dt$ still correct or should it be $i = -dq/dt$? Is $i = dq/dt$ still correct when the right-hand plate has positive charge that is increasing or decreasing in magnitude? Explain.

Q31.5 Fluorescent lights often use an inductor, called a ballast, to limit the current through the tubes. Why is it better to use an inductor rather than a resistor for this purpose?

Q31.6 Equation (31.9) says that $v_{ab} = L\, di/dt$ (see Fig. 31.8a). Using Faraday's law, explain why point a is at higher potential than point b when i is in the direction shown in Fig. 31.8a and is increasing in magnitude. When i is counterclockwise and decreasing in magnitude, is $v_{ab} = L\, di/dt$ still correct, or should it be $v_{ab} = -L\, di/dt$? Is $v_{ab} = L\, di/dt$ still correct when i is clockwise and increasing or decreasing in magnitude? Explain.

Q31.7 Is it possible for the power factor of an *L-R-C* series ac circuit to be zero? Justify your answer on *physical* grounds.

Q31.8 In an *L-R-C* series circuit, can the instantaneous voltage across the capacitor exceed the source voltage at that same instant? Can this be true for the instantaneous voltage across the inductor? Across the resistor? Explain.

Q31.9 In an *L-R-C* series circuit, what are the phase angle ϕ and power factor $\cos\phi$ when the resistance is much smaller than the inductive or capacitive reactance and the circuit is operated far from resonance? Explain.

Q31.10 When an *L-R-C* series circuit is connected across a 120-V ac line, the voltage rating of the capacitor may be exceeded even if it is rated at 200 or 400 V. How can this be?

Q31.11 In Example 31.6 (Section 31.4), a hair dryer is treated as a pure resistor. But because there are coils in the heating element and in the motor that drives the blower fan, a hair dryer also has inductance. Qualitatively, does including an inductance increase or decrease the values of R, I_{rms}, and P?

Q31.12 A light bulb and a parallel-plate capacitor with air between the plates are connected in series to an ac source. What happens to the brightness of the bulb when a dielectric is inserted between the plates of the capacitor? Explain.

Q31.13 A coil of wire wrapped on a hollow tube and a light bulb are connected in series to an ac source. What happens to the brightness of the bulb when an iron rod is inserted in the tube?

Q31.14 A circuit consists of a light bulb, a capacitor, and an inductor connected in series to an ac source. What happens to the brightness of the bulb when the inductor is omitted? When the inductor is left in the circuit but the capacitor is omitted? Explain.

Q31.15 A circuit consists of a light bulb, a capacitor, and an inductor connected in series to an ac source. Is it possible for both the capacitor and the inductor to be removed and the brightness of the bulb to remain the same? Explain.

Q31.16 Can a transformer be used with dc? Explain. What happens if a transformer designed for 120-V ac is connected to a 120-V dc line?

Q31.17 An ideal transformer has N_1 windings in the primary and N_2 windings in its secondary. If you double only the number of secondary windings, by what factor does (a) the voltage amplitude in the secondary change, and (b) the effective resistance of the secondary circuit change?

Q31.18 An inductor, a capacitor, and a resistor are all connected in series across an ac source. If the resistance, inductance, and capacitance are all doubled, by what factor does each of the following quantities change? Indicate whether they increase or decrease: (a) the resonance angular frequency; (b) the inductive reactance; (c) the capacitive reactance. (d) Does the impedance double?

Q31.19 You want to double the resonance angular frequency of an *L-R-C* series circuit by changing only the *pertinent* circuit elements all by the same factor. (a) Which ones should you change? (b) By what factor should you change them?

EXERCISES

Section 31.1 Phasors and Alternating Currents

31.1 • You have a special light bulb with a *very* delicate wire filament. The wire will break if the current in it ever exceeds 1.50 A, even for an instant. What is the largest root-mean-square current you can run through this bulb?

31.2 • A sinusoidal current $i = I \cos \omega t$ has an rms value $I_{rms} = 2.10$ A. (a) What is the current amplitude? (b) The current is passed through a full-wave rectifier circuit. What is the rectified average current? (c) Which is larger: I_{rms} or I_{rav}? Explain, using graphs of i^2 and of the rectified current.

31.3 • The voltage across the terminals of an ac power supply varies with time according to Eq. (31.1). The voltage amplitude is $V = 45.0$ V. What are (a) the root-mean-square potential difference V_{rms} and (b) the average potential difference V_{av} between the two terminals of the power supply?

Section 31.2 Resistance and Reactance

31.4 • A capacitor is connected across an ac source that has voltage amplitude 60.0 V and frequency 80.0 Hz. (a) What is the phase angle ϕ for the source voltage relative to the current? Does the source voltage lag or lead the current? (b) What is the capacitance C of the capacitor if the current amplitude is 5.30 A?

31.5 • An inductor with $L = 9.50$ mH is connected across an ac source that has voltage amplitude 45.0 V. (a) What is the phase angle ϕ for the source voltage relative to the current? Does the source voltage lag or lead the current? (b) What value for the frequency of the source results in a current amplitude of 3.90 A?

31.6 • A capacitance C and an inductance L are operated at the same angular frequency. (a) At what angular frequency will they have the same reactance? (b) If $L = 5.00$ mH and $C = 3.50$ μF, what is the numerical value of the angular frequency in part (a), and what is the reactance of each element?

31.7 • **Kitchen Capacitance.** The wiring for a refrigerator contains a starter capacitor. A voltage of amplitude 170 V and frequency 60.0 Hz applied across the capacitor is to produce a current amplitude of 0.850 A through the capacitor. What capacitance C is required?

31.8 • (a) Compute the reactance of a 0.450-H inductor at frequencies of 60.0 Hz and 600 Hz. (b) Compute the reactance of a 2.50-μF capacitor at the same frequencies. (c) At what frequency is the reactance of a 0.450-H inductor equal to that of a 2.50-μF capacitor?

31.9 • (a) What is the reactance of a 3.00-H inductor at a frequency of 80.0 Hz? (b) What is the inductance of an inductor whose reactance is 120 Ω at 80.0 Hz? (c) What is the reactance of a 4.00-μF capacitor at a frequency of 80.0 Hz? (d) What is the capacitance of a capacitor whose reactance is 120 Ω at 80.0 Hz?

31.10 • **A Radio Inductor.** You want the current amplitude through a 0.450-mH inductor (part of the circuitry for a radio receiver) to be 1.80 mA when a sinusoidal voltage with amplitude 12.0 V is applied across the inductor. What frequency is required?

31.11 •• A 0.180-H inductor is connected in series with a 90.0-Ω resistor and an ac source. The voltage across the inductor is $v_L = -(12.0 \text{ V})\sin[(480 \text{ rad/s})t]$. (a) Derive an expression for the voltage v_R across the resistor. (b) What is v_R at $t = 2.00$ ms?

31.12 •• A 250-Ω resistor is connected in series with a 4.80-μF capacitor and an ac source. The voltage across the capacitor is $v_C = (7.60 \text{ V})\sin[(120 \text{ rad/s})t]$. (a) Determine the capacitive reactance of the capacitor. (b) Derive an expression for the voltage v_R across the resistor.

31.13 •• A 150-Ω resistor is connected in series with a 0.250-H inductor and an ac source. The voltage across the resistor is $v_R = (3.80 \text{ V})\cos[(720 \text{ rad/s})t]$. (a) Derive an expression for the circuit current. (b) Determine the inductive reactance of the inductor. (c) Derive an expression for the voltage v_L across the inductor.

Section 31.3 The *L-R-C* Series Circuit

31.14 • You have a 200-Ω resistor, a 0.400-H inductor, and a 6.00-μF capacitor. Suppose you take the resistor and inductor and make a series circuit with a voltage source that has voltage amplitude 30.0 V and an angular frequency of 250 rad/s. (a) What is the impedance of the circuit? (b) What is the current amplitude? (c) What are the voltage amplitudes across the resistor and across the inductor? (d) What is the phase angle ϕ of the source voltage with respect to the current? Does the source voltage lag or lead the current? (e) Construct the phasor diagram.

31.15 • The resistor, inductor, capacitor, and voltage source described in Exercise 31.14 are connected to form an *L-R-C* series circuit. (a) What is the impedance of the circuit? (b) What is the current amplitude? (c) What is the phase angle of the source voltage with respect to the current? Does the source voltage lag or lead the current? (d) What are the voltage amplitudes across the resistor, inductor, and capacitor? (e) Explain how it is possible for the voltage amplitude across the capacitor to be greater than the voltage amplitude across the source.

31.16 •• A 200-Ω resistor, 0.900-H inductor, and 6.00-μF capacitor are connected in series across a voltage source that has voltage amplitude 30.0 V and an angular frequency of 250 rad/s. (a) What are v, v_R, v_L, and v_C at $t = 20.0$ ms? Compare $v_R + v_L + v_C$ to v at this instant. (b) What are V_R, V_L, and V_C? Compare V to $V_R + V_L + V_C$. Explain why these two quantities are not equal.

31.17 • In an *L-R-C* series circuit, the rms voltage across the resistor is 30.0 V, across the capacitor it is 90.0 V, and across the inductor it is 50.0 V. What is the rms voltage of the source?

Section 31.4 Power in Alternating-Current Circuits

31.18 •• A resistor with $R = 300 \ \Omega$ and an inductor are connected in series across an ac source that has voltage amplitude 500 V. The rate at which electrical energy is dissipated in the resistor is 286 W. What is (a) the impedance Z of the circuit; (b) the amplitude of the voltage across the inductor; (c) the power factor?

31.19 • The power of a certain CD player operating at 120 V rms is 20.0 W. Assuming that the CD player behaves like a pure resistor, find (a) the maximum instantaneous power; (b) the rms current; (c) the resistance of this player.

31.20 •• In an *L-R-C* series circuit, the components have the following values: $L = 20.0$ mH, $C = 140$ nF, and $R = 350 \ \Omega$. The generator has an rms voltage of 120 V and a frequency of 1.25 kHz. Determine (a) the power supplied by the generator and (b) the power dissipated in the resistor.

31.21 • (a) Show that for an *L-R-C* series circuit the power factor is equal to R/Z. (b) An *L-R-C* series circuit has phase angle $-31.5°$. The voltage amplitude of the source is 90.0 V. What is the voltage amplitude across the resistor?

31.22 • (a) Use the results of part (a) of Exercise 31.21 to show that the average power delivered by the source in an *L-R-C* series circuit is given by $P_{av} = I_{rms}^2 R$. (b) An *L-R-C* series circuit has $R = 96.0 \ \Omega$, and the amplitude of the voltage across the resistor is 36.0 V. What is the average power delivered by the source?

31.23 • An *L-R-C* series circuit with $L = 0.120$ H, $R = 240 \ \Omega$, and $C = 7.30 \ \mu$F carries an rms current of 0.450 A with a frequency of 400 Hz. (a) What are the phase angle and power factor for this circuit? (b) What is the impedance of the circuit? (c) What is the rms voltage of the source? (d) What average power is delivered by the source? (e) What is the average rate at which electrical energy is converted to thermal energy in the resistor? (f) What is the average rate at which electrical energy is dissipated (converted to other forms) in the capacitor? (g) In the inductor?

31.24 •• An *L-R-C* series circuit is connected to a 120-Hz ac source that has $V_{rms} = 80.0$ V. The circuit has a resistance of 75.0 Ω and an impedance at this frequency of 105 Ω. What average power is delivered to the circuit by the source?

31.25 •• A series ac circuit contains a 250-Ω resistor, a 15-mH inductor, a 3.5-μF capacitor, and an ac power source of voltage amplitude 45 V operating at an angular frequency of 360 rad/s. (a) What is the power factor of this circuit? (b) Find the average power delivered to the entire circuit. (c) What is the average power delivered to the resistor, to the capacitor, and to the inductor?

Section 31.5 Resonance in Alternating-Current Circuits

31.26 •• In an *L-R-C* series circuit the source is operated at its resonant angular frequency. At this frequency, the reactance X_C of the capacitor is 200 Ω and the voltage amplitude across the capacitor is 600 V. The circuit has $R = 300 \ \Omega$. What is the voltage amplitude of the source?

31.27 • **Analyzing an *L-R-C* Circuit.** You have a 200-Ω resistor, a 0.400-H inductor, a 5.00-μF capacitor, and a variable-frequency ac source with an amplitude of 3.00 V. You connect all four elements together to form a series circuit. (a) At what frequency will the current in the circuit be greatest? What will be the current amplitude at this frequency? (b) What will be the current amplitude at an angular frequency of 400 rad/s? At this frequency, will the source voltage lead or lag the current?

31.28 • An *L-R-C* series circuit is constructed using a 175-Ω resistor, a 12.5-μF capacitor, and an 8.00-mH inductor, all connected across an ac source having a variable frequency and a voltage amplitude of 25.0 V. (a) At what angular frequency will the impedance be smallest, and what is the impedance at this frequency? (b) At the angular frequency in part (a), what is the maximum current through the inductor? (c) At the angular frequency in part (a), find the potential difference across the ac source, the resistor, the capacitor, and the inductor at the instant that the current is equal to one-half its greatest positive value. (d) In part (c), how are the potential differences across the resistor, inductor, and capacitor related to the potential difference across the ac source?

31.29 • In an *L-R-C* series circuit, $R = 300 \ \Omega$, $L = 0.400$ H, and $C = 6.00 \times 10^{-8}$ F. When the ac source operates at the resonance frequency of the circuit, the current amplitude is 0.500 A. (a) What is the voltage amplitude of the source? (b) What is the amplitude of the voltage across the resistor, across the inductor, and across the capacitor? (c) What is the average power supplied by the source?

31.30 • An *L-R-C* series circuit consists of a source with voltage amplitude 120 V and angular frequency 50.0 rad/s, a resistor with $R = 400 \ \Omega$, an inductor with $L = 3.00$ H, and a capacitor with capacitance C. (a) For what value of C will the current amplitude in the circuit be a maximum? (b) When C has the value calculated in part (a), what is the amplitude of the voltage across the inductor?

31.31 • In an *L-R-C* series circuit, $R = 150 \ \Omega$, $L = 0.750$ H, and $C = 0.0180 \ \mu$F. The source has voltage amplitude $V = 150$ V and a frequency equal to the resonance frequency of the circuit. (a) What is the power factor? (b) What is the average power delivered by the source? (c) The capacitor is replaced by one with $C = 0.0360 \ \mu$F and the source frequency is adjusted to the new resonance value. Then what is the average power delivered by the source?

31.32 • In an *L-R-C* series circuit, $R = 400 \ \Omega$, $L = 0.350$ H, and $C = 0.0120 \ \mu$F. (a) What is the resonance angular frequency of the circuit? (b) The capacitor can withstand a peak voltage of 670 V. If the voltage source operates at the resonance frequency, what maximum voltage amplitude can it have if the maximum capacitor voltage is not exceeded?

31.33 •• In an *L-R-C* series circuit, $L = 0.280$ H and $C = 4.00 \ \mu$F. The voltage amplitude of the source is 120 V. (a) What is the resonance angular frequency of the circuit? (b) When the source operates at the resonance angular frequency, the current amplitude in the circuit is 1.70 A. What is the resistance R of the resistor? (c) At the resonance angular frequency, what are the peak voltages across the inductor, the capacitor, and the resistor?

Section 31.6 Transformers

31.34 • **Off to Europe!** You plan to take your hair dryer to Europe, where the electrical outlets put out 240 V instead of the 120 V seen in the United States. The dryer puts out 1600 W at 120 V. (a) What could you do to operate your dryer via the 240-V line in Europe? (b) What current will your dryer draw from a European outlet? (c) What resistance will your dryer appear to have when operated at 240 V?

31.35 • **A Step-Down Transformer.** A transformer connected to a 120-V (rms) ac line is to supply 12.0 V (rms) to a portable electronic device. The load resistance in the secondary is 5.00 Ω. (a) What should the ratio of primary to secondary turns of the transformer be? (b) What rms current must the secondary supply? (c) What average power is delivered to the load? (d) What resistance connected directly across the 120-V line would draw the same power as the transformer? Show that this is equal to 5.00 Ω times the square of the ratio of primary to secondary turns.

31.36 • **A Step-Up Transformer.** A transformer connected to a 120-V (rms) ac line is to supply 13,000 V (rms) for a neon sign. To reduce shock hazard, a fuse is to be inserted in the primary circuit; the fuse is to blow when the rms current in the secondary circuit exceeds 8.50 mA. (a) What is the ratio of secondary to primary turns of the transformer? (b) What power must be supplied to the transformer when the rms secondary current is 8.50 mA? (c) What current rating should the fuse in the primary circuit have?

PROBLEMS

31.37 • A coil has a resistance of 48.0 Ω. At a frequency of 80.0 Hz the voltage across the coil leads the current in it by 52.3°. Determine the inductance of the coil.

31.38 •• When a solenoid is connected to a 48.0-V dc battery that has negligible internal resistance, the current in the solenoid is 5.50 A. When this solenoid is connected to an ac source that has voltage amplitude 48.0 V and angular frequency 20.0 rad/s, the current in the solenoid is 3.60 A. What is the inductance of this solenoid?

31.39 •• An L-R-C series circuit has $C = 4.80$ μF, $L = 0.520$ H, and source voltage amplitude $V = 56.0$ V. The source is operated at the resonance frequency of the circuit. If the voltage across the capacitor has amplitude 80.0 V, what is the value of R for the resistor in the circuit?

31.40 •• Five infinite-impedance voltmeters, calibrated to read rms values, are connected as shown in **Fig. P31.40**. Let $R = 200$ Ω, $L = 0.400$ H, $C = 6.00$ μF, and $V = 30.0$ V. What is the reading of each voltmeter if (a) $\omega = 200$ rad/s and (b) $\omega = 1000$ rad/s?

Figure **P31.40**

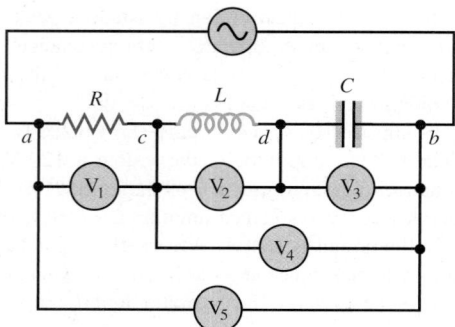

31.41 •• CP A parallel-plate capacitor having square plates 4.50 cm on each side and 8.00 mm apart is placed in series with the following: an ac source of angular frequency 650 rad/s and voltage amplitude 22.5 V; a 75.0-Ω resistor; and an ideal solenoid that is 9.00 cm long, has a circular cross section 0.500 cm in diameter, and carries 125 coils per centimeter. What is the resonance angular frequency of this circuit? (See Exercise 30.15.)

31.42 •• CP A toroidal solenoid has 2900 closely wound turns, cross-sectional area 0.450 cm^2, mean radius 9.00 cm, and resistance $R = 2.80$ Ω. Ignore the variation of the magnetic field across the cross section of the solenoid. What is the amplitude of the current in the solenoid if it is connected to an ac source that has voltage amplitude 24.0 V and frequency 495 Hz?

31.43 •• A series circuit has an impedance of 60.0 Ω and a power factor of 0.720 at 50.0 Hz. The source voltage lags the current. (a) What circuit element, an inductor or a capacitor, should be placed in series with the circuit to raise its power factor? (b) What size element will raise the power factor to unity?

31.44 • A large electromagnetic coil is connected to a 120-Hz ac source. The coil has resistance 400 Ω, and at this source frequency the coil has inductive reactance 250 Ω. (a) What is the inductance of the coil? (b) What must the rms voltage of the source be if the coil is to consume an average electrical power of 450 W?

31.45 •• In an L-R-C series circuit, $R = 300$ Ω, $X_C = 300$ Ω, and $X_L = 500$ Ω. The average electrical power consumed in the resistor is 60.0 W. (a) What is the power factor of the circuit? (b) What is the rms voltage of the source?

31.46 • At a frequency ω_1 the reactance of a certain capacitor equals that of a certain inductor. (a) If the frequency is changed to $\omega_2 = 2\omega_1$, what is the ratio of the reactance of the inductor to that of the capacitor? Which reactance is larger? (b) If the frequency is changed to $\omega_3 = \omega_1/3$, what is the ratio of the reactance of the inductor to that of the capacitor? Which reactance is larger? (c) If the capacitor and inductor are placed in series with a resistor of resistance R to form an L-R-C series circuit, what will be the resonance angular frequency of the circuit?

31.47 •• **A High-Pass Filter.** One application of L-R-C series circuits is to high-pass or low-pass filters, which filter out either the low- or high-frequency components of a signal. A high-pass filter is shown in **Fig. P31.47**, where the output voltage is taken across the L-R combination. (The L-R combination represents an inductive coil that also has resistance due to the large length of wire in the coil.) Derive an expression for V_{out}/V_s, the ratio of the output and source voltage amplitudes, as a function of the angular frequency ω of the source. Show that when ω is small, this ratio is proportional to ω and thus is small, and show that the ratio approaches unity in the limit of large frequency.

Figure **P31.47**

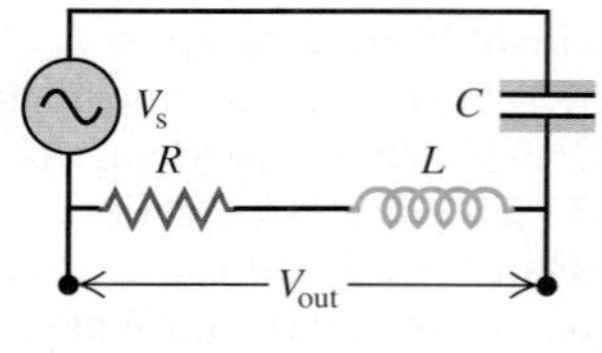

31.48 •• **A Low-Pass Filter.** **Figure P31.48** shows a low-pass filter (see Problem 31.47); the output voltage is taken across the capacitor in an L-R-C series circuit. Derive an expression for

Figure **P31.48**

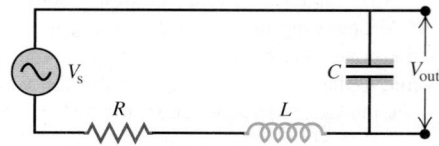

V_{out}/V_s, the ratio of the output and source voltage amplitudes, as a function of the angular frequency ω of the source. Show that when ω is large, this ratio is proportional to ω^{-2} and thus is very small, and show that the ratio approaches unity in the limit of small frequency.

31.49 ••• An *L-R-C* series circuit is connected to an ac source of constant voltage amplitude V and variable angular frequency ω. (a) Show that the current amplitude, as a function of ω, is

$$I = \frac{V}{\sqrt{R^2 + (\omega L - 1/\omega C)^2}}$$

(b) Show that the average power dissipated in the resistor is

$$P = \frac{V^2 R/2}{R^2 + (\omega L - 1/\omega C)^2}$$

(c) Show that I and P are *both* maximum when $\omega = 1/\sqrt{LC}$, the resonance frequency of the circuit. (d) Graph P as a function of ω for $V = 100$ V, $R = 200\ \Omega$, $L = 2.0$ H, and $C = 0.50\ \mu$F. Compare to the light purple curve in Fig. 31.19. Discuss the behavior of I and P in the limits $\omega = 0$ and $\omega \to \infty$.

31.50 ••• An *L-R-C* series circuit is connected to an ac source of constant voltage amplitude V and variable angular frequency ω. Using the results of Problem 31.49, find an expression for (a) the amplitude V_L of the voltage across the inductor as a function of ω and (b) the amplitude V_C of the voltage across the capacitor as a function of ω. (c) Graph V_L and V_C as functions of ω for $V = 100$ V, $R = 200\ \Omega$, $L = 2.0$ H, and $C = 0.50\ \mu$F. (d) Discuss the behavior of V_L and V_C in the limits $\omega = 0$ and $\omega \to \infty$. For what value of ω is $V_L = V_C$? What is the significance of this value of ω?

31.51 •• In an *L-R-C* series circuit the magnitude of the phase angle is 54.0°, with the source voltage lagging the current. The reactance of the capacitor is 350 Ω, and the resistor resistance is 180 Ω. The average power delivered by the source is 140 W. Find (a) the reactance of the inductor; (b) the rms current; (c) the rms voltage of the source.

31.52 •• In an *L-R-C* series circuit, the phase angle is 40.0°, with the source voltage leading the current. The reactance of the capacitor is 400 Ω, and the resistance of the resistor is 200 Ω. The average power delivered by the source is 150 W. Find (a) the reactance of the inductor, (b) the rms current, (c) the rms voltage of the source.

31.53 • An *L-R-C* series circuit has $R = 500\ \Omega$, $L = 2.00$ H, $C = 0.500\ \mu$F, and $V = 100$ V. (a) For $\omega = 800$ rad/s, calculate V_R, V_L, V_C, and ϕ. Using a single set of axes, graph v, v_R, v_L, and v_C as functions of time. Include two cycles of v on your graph. (b) Repeat part (a) for $\omega = 1000$ rad/s. (c) Repeat part (a) for $\omega = 1250$ rad/s.

31.54 •• **The *L-R-C* Parallel Circuit.** A resistor, an inductor, and a capacitor are connected in parallel to an ac source with voltage amplitude V and angular frequency ω. Let the source voltage be given by $v = V\cos \omega t$. (a) Show that each of the instantaneous voltages v_R, v_L, and v_C at any instant is equal to v and that $i = i_R + i_L + i_C$, where i is the current through the source and i_R, i_L, and i_C are the currents through the resistor, inductor, and capacitor, respectively. (b) What are the phases of i_R, i_L, and i_C with respect to v? Use current phasors to represent i, i_R, i_L, and i_C. In a phasor diagram, show the phases of these four currents with respect to v. (c) Use the phasor diagram of part (b) to show that the current amplitude I for the current i through the source is $I = \sqrt{I_R^2 + (I_C - I_L)^2}$. (d) Show that the result of part (c) can be written as $I = V/Z$, with $1/Z = \sqrt{(1/R^2) + [\omega C - (1/\omega L)]^2}$.

31.55 •• The impedance of an *L-R-C* parallel circuit was derived in Problem 31.54. (a) Show that at the resonance angular frequency $\omega_0 = 1/\sqrt{LC}$, the impedance Z is a maximum and therefore the current through the ac source is a minimum. (b) A 100-Ω resistor, a 0.100-μF capacitor, and a 0.300-H inductor are connected in parallel to a voltage source with amplitude 240 V. What is the resonance angular frequency? For this circuit, what is (c) the maximum current through the source at the resonance frequency; (d) the maximum current in the resistor at resonance; (e) the maximum current in the inductor at resonance; (f) the maximum current in the branch containing the capacitor at resonance?

31.56 •• A 400-Ω resistor and a 6.00-μF capacitor are connected in parallel to an ac generator that supplies an rms voltage of 180 V at an angular frequency of 360 rad/s. Use the results of Problem 31.54. Note that since there is no inductor in this circuit, the $1/\omega L$ term is not present in the expression for $1/Z$. Find (a) the current amplitude in the resistor; (b) the current amplitude in the capacitor; (c) the phase angle of the source current with respect to the source voltage; (d) the amplitude of the current through the generator. (e) Does the source current lag or lead the source voltage?

31.57 ••• An *L-R-C* series circuit consists of a 2.50-μF capacitor, a 5.00-mH inductor, and a 75.0-Ω resistor connected across an ac source of voltage amplitude 15.0 V having variable frequency. (a) Under what circumstances is the average power delivered to the circuit equal to $\frac{1}{2}V_{rms}I_{rms}$? (b) Under the conditions of part (a), what is the average power delivered to each circuit element and what is the maximum current through the capacitor?

31.58 •• An *L-R-C* series circuit has $R = 60.0\ \Omega$, $L = 0.800$ H, and $C = 3.00 \times 10^{-4}$ F. The ac source has voltage amplitude 90.0 V and angular frequency 120 rad/s. (a) What is the maximum energy stored in the inductor? (b) When the energy stored in the inductor is a maximum, how much energy is stored in the capacitor? (c) What is the maximum energy stored in the capacitor?

31.59 • In an *L-R-C* series circuit, the source has a voltage amplitude of 120 V, $R = 80.0\ \Omega$, and the reactance of the capacitor is 480 Ω. The voltage amplitude across the capacitor is 360 V. (a) What is the current amplitude in the circuit? (b) What is the impedance? (c) What two values can the reactance of the inductor have? (d) For which of the two values found in part (c) is the angular frequency less than the resonance angular frequency? Explain.

31.60 •• In an *L-R-C* series ac circuit, the source has a voltage amplitude of 240 V, $R = 90.0\ \Omega$, and the reactance of the inductor is 320 Ω. The voltage amplitude across the resistor is 135 V. (a) What is the current amplitude in the circuit? (b) What is the voltage amplitude across the inductor? (c) What two values can the reactance of the capacitor have? (d) For which of the two values found in part (c) is the angular frequency less than the resonance angular frequency? Explain.

31.61 • A resistance R, capacitance C, and inductance L are connected in series to a voltage source with amplitude V and variable angular frequency ω. If $\omega = \omega_0$, the resonance angular frequency, find (a) the maximum current in the resistor; (b) the maximum voltage across the capacitor; (c) the maximum voltage across the inductor; (d) the maximum energy stored in the capacitor; (e) the maximum energy stored in the inductor. Give your answers in terms of R, C, L, and V.

31.62 •• **The Resonance Width.** Consider an *L-R-C* series circuit with a 1.80-H inductor, a 0.900-μF capacitor, and a 300-Ω resistor. The source has terminal rms voltage $V_{rms} = 60.0$ V and variable angular frequency ω. (a) What is the resonance angular frequency ω_0 of the circuit? (b) What is the rms current through the circuit at resonance, I_{rms-0}? (c) For what two values of the angular frequency, ω_1 and ω_2, is the rms current half the resonance value? (d) The quantity $|\omega_1 - \omega_2|$ defines the *resonance width*. Calculate $I_{\text{rms-0}}$ and the resonance width for $R = 300\ \Omega$, $30.0\ \Omega$, and $3.00\ \Omega$. Describe how your results compare to the discussion in Section 31.5.

31.63 •• An *L-R-C* series circuit draws 220 W from a 120-V (rms), 50.0-Hz ac line. The power factor is 0.560, and the source voltage leads the current. (a) What is the net resistance R of the circuit? (b) Find the capacitance of the series capacitor that will result in a power factor of unity when it is added to the original circuit. (c) What power will then be drawn from the supply line?

31.64 •• **DATA** A coworker of yours was making measurements of a large solenoid that is connected to an ac voltage source. Unfortunately, she left for vacation before she completed the analysis, and your boss has asked you to finish it. You are given a graph of $1/I^2$ versus ω^2 (**Fig. P31.64**), where I is the current in the circuit and ω is the angular frequency of the source. A note attached to the graph says that the voltage amplitude of the source was kept constant at 12.0 V. Calculate the resistance and inductance of the solenoid.

Figure **P31.64**

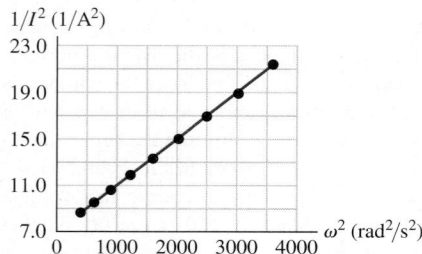

31.65 •• **DATA** You are analyzing an ac circuit that contains a solenoid and a capacitor in series with an ac source that has voltage amplitude 90.0 V and angular frequency ω. For different capacitors in the circuit, each with known capacitance, you measure the value of the frequency ω_{res} for which the current in the circuit is a maximum. You plot your measured values on a graph of ω_{res}^2 versus $1/C$ (**Fig. P31.65**). The maximum current for each value of C is the same, you note, and equal to 4.50 A. Calculate the resistance and inductance of the solenoid.

Figure **P31.65**

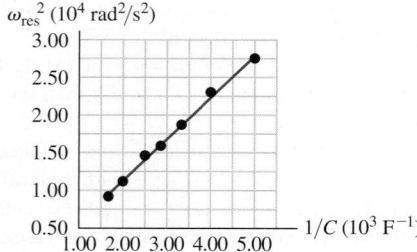

31.66 •• **DATA** You are given this table of data recorded for a circuit that has a resistor, an inductor with negligible resistance, and a capacitor, all in series with an ac voltage source:

f (Hz)	80	160
Z (Ω)	15	13
ϕ (°)	−71	67

Here f is the frequency of the voltage source, Z is the impedance of the circuit, and ϕ is the phase angle. (a) Use the data at both frequencies to calculate the resistance of the resistor. (*Hint*: Use the results of Exercise 31.21.) Calculate the average of these two values of the resistance, and use the result as the value of R in the rest of the analysis. (b) Use the data at 80 Hz and 160 Hz to calculate the inductance L and capacitance C of the circuit. (c) What is the resonance frequency for the circuit, and what are the impedance and phase angle at the resonance frequency?

CHALLENGE PROBLEMS

31.67 •• **CALC** In an *L-R-C* series circuit the current is given by $i = I\cos\omega t$. The voltage amplitudes for the resistor, inductor, and capacitor are V_R, V_L, and V_C. (a) Show that the instantaneous power into the resistor is $p_R = V_R I\cos^2\omega t = \frac{1}{2}V_R I(1 + \cos 2\omega t)$. What does this expression give for the average power into the resistor? (b) Show that the instantaneous power into the inductor is $p_L = -V_L I\sin\omega t\cos\omega t = -\frac{1}{2}V_L I\sin 2\omega t$. What does this expression give for the average power into the inductor? (c) Show that the instantaneous power into the capacitor is $p_C = V_C I\sin\omega t\cos\omega t = \frac{1}{2}V_C I\sin 2\omega t$. What does this expression give for the average power into the capacitor? (d) The instantaneous power delivered by the source is shown in Section 31.4 to be $p = VI\cos\omega t(\cos\phi\cos\omega t - \sin\phi\sin\omega t)$. Show that $p_R + p_L + p_C$ equals p at each instant of time.

31.68 ••• **CALC** (a) At what angular frequency is the voltage amplitude across the *resistor* in an *L-R-C* series circuit at maximum value? (b) At what angular frequency is the voltage amplitude across the *inductor* at maximum value? (c) At what angular frequency is the voltage amplitude across the *capacitor* at maximum value? (You may want to refer to the results of Problem 31.49.)

PASSAGE PROBLEMS

BIO CONVERTING DC TO AC. An individual cell such as an egg cell (an ovum, produced in the ovaries) is commonly organized spatially, as manifested in part by asymmetries in the cell membrane. These asymmetries include nonuniform distributions of ion transport mechanisms, which result in a net electric current entering one region of the membrane and leaving another. These steady cellular currents may regulate cell polarity, leading (in the case of eggs) to embryonic polarity; therefore scientists are interested in measuring them.

These cellular currents move in loops through extracellular fluid. Ohm's law requires that there be voltage differences between any two points in this current-carrying fluid surrounding cells. Although the currents may be significant, the extracellular voltage differences are tiny—on the order of nanovolts. If we can map the voltage differences in the fluid outside a cell, we can calculate the current density by using Ohm's law, assuming that the resistivity of the fluid is known. We cannot measure these voltage differences by spacing two electrodes 10 or 20 μm apart, because the dc impedance (the resistance) of such electrodes is high and the inherent noise in signals detected at the electrodes far exceeds the cellular voltages.

One successful method of measurement uses an electrode with a ball-shaped end made of platinum that is moved sinusoidally between two points in the fluid outside a cell. The electric potential that the electrode measures, with respect to a distant reference electrode, also varies sinusoidally. The dc potential difference between the two extremes (the two points in the fluid) is then converted to a sine-wave ac potential difference. The platinum electrode behaves as a capacitor in series with the resistance of the extracellular fluid. This resistance, called the *access resistance* (R_A), has a value of about $\rho/10a$, where ρ is the resistivity of the fluid (usually expressed in $\Omega \cdot$ cm) and a is the radius of the ball electrode. The platinum ball typically has a diameter of 20 μm and a capacitance of 10 nF; the resistivity of many biological fluids is 100 $\Omega \cdot$ cm.

31.69 What is the dc impedance of the electrode, assuming that it behaves as an ideal capacitor? (a) 0; (b) infinite; (c) $\sqrt{2} \times 10^4$ Ω; (d) $\sqrt{2} \times 10^6$ Ω.

31.70 If the electrode oscillates between two points 20 μm apart at a frequency of $(5000/\pi)$Hz, what is the electrode's impedance? (a) 0; (b) infinite; (c) $\sqrt{2} \times 10^4$ Ω; (d) $\sqrt{2} \times 10^6$ Ω.

31.71 The signal from the oscillating electrode is fed into an amplifier, which reports the measured voltage as an rms value, 1.5 nV. What is the potential difference between the two extremes? (a) 1.5 nV; (b) 3.0 nV; (c) 2.1 nV; (d) 4.2 nV.

31.72 If the frequency at which the electrode is oscillated is increased to a very large value, the electrode's impedance (a) approaches infinity; (b) approaches zero; (c) approaches a constant but nonzero value; (d) does not change.

Answers

Chapter Opening Question ?

(iv) A radio simultaneously detects transmissions at *all* frequencies. However, a radio is an *L-R-C* series circuit, and at any given time it is tuned to have a resonance at just one frequency. Hence the response of the radio to that frequency is much greater than its response to any other frequency, which is why you hear only one broadcasting station through the radio's speaker. (You can sometimes hear a second station if its frequency is sufficiently close to the tuned frequency.)

Test Your Understanding Questions

31.1 (a) D; (b) A; (c) B; (d) C For each phasor, the actual current is represented by the projection of that phasor onto the horizontal axis. The phasors all rotate counterclockwise around the origin with angular speed ω, so at the instant shown the projection of phasor A is positive but trending toward zero; the projection of phasor B is negative and becoming more negative; the projection of phasor C is negative but trending toward zero; and the projection of phasor D is positive and becoming more positive.

31.2 (a) (iii); (b) (ii); (c) (i) For a resistor, $V_R = IR$, so $I = V_R/R$. The voltage amplitude V_R and resistance R do not change with frequency, so the current amplitude I remains constant. For an inductor, $V_L = IX_L = I\omega L$, so $I = V_L/\omega L$. The voltage amplitude V_L and inductance L are constant, so the current amplitude I decreases as the frequency increases. For a capacitor, $V_C = IX_C = I/\omega C$, so $I = V_C\omega C$. The voltage amplitude V_C and capacitance C are constant, so the current amplitude I increases as the frequency increases.

31.3 (iv), (ii), (i), (iii) For the circuit in Example 31.4, $I = V/Z = (50 \text{ V})/(500 \text{ }\Omega) = 0.10$ A. If the capacitor and inductor are removed so that only the ac source and resistor remain, the circuit is like that shown in Fig. 31.7a; then $I = V/R = (50 \text{ V})/(300 \text{ }\Omega) = 0.17$ A. If the resistor and capacitor are removed so that only the

ac source and inductor remain, the circuit is like that shown in Fig. 31.8a; then $I = V/X_L = (50 \text{ V})/(600 \text{ }\Omega) = 0.083$ A. Finally, if the resistor and inductor are removed so that only the ac source and capacitor remain, the circuit is like that shown in Fig. 31.9a; then $I = V/X_C = (50 \text{ V})/(200 \text{ }\Omega) = 0.25$ A.

31.4 (a) (v); (b) (iv) The energy cannot be extracted from the resistor, since energy is dissipated in a resistor and cannot be recovered. Instead, the energy must be extracted from either the inductor (which stores magnetic-field energy) or the capacitor (which stores electric-field energy). Positive power means that energy is being transferred from the ac source to the circuit, so *negative* power implies that energy is being transferred back into the source.

31.5 (ii) The capacitance C increases if the plate spacing is decreased (see Section 24.1). Hence the resonance frequency $f_0 = \omega_0/2\pi = 1/2\pi \sqrt{LC}$ decreases.

31.6 (ii), (iv), (i), (iii) From Eq. (31.35) the turns ratio is $N_2/N_1 = V_2/V_1$, so the number of turns in the secondary is $N_2 = N_1V_2/V_1$. Hence for the four cases we have (i) $N_2 = (1000)(6.0 \text{ V})/(120 \text{ V}) = 50$ turns; (ii) $N_2 = (1000)(240 \text{ V})/(120 \text{ V}) = 2000$ turns; (iii) $N_2 = (1000)(6.0 \text{ V})/(240 \text{ V}) = 25$ turns; and (iv) $N_2 = (1000)(120 \text{ V})/(240 \text{ V}) = 500$ turns. Note that (i), (iii), and (iv) are step-down transformers with fewer turns in the secondary than in the primary, while (ii) is a step-up transformer with more turns in the secondary than in the primary.

Bridging Problem

(a) 8.35×10^4 rad/s and 3.19×10^5 rad/s
(b) At 8.35×10^4 rad/s: $V_{\text{source}} = 49.5$ V, $I = 0.132$ A, $V_R = 16.5$ V, $V_L = 16.5$ V, $V_C = 63.2$ V.
At 3.19×10^5 rad/s: $V_{\text{source}} = 49.5$ V, $I = 0.132$ A, $V_R = 16.5$ V, $V_L = 63.2$ V, $V_C = 16.5$ V.

32 ELECTROMAGNETIC WAVES

What is light? This question has been asked by humans for centuries, but there was no answer until electricity and magnetism were unified into *electromagnetism,* as described by Maxwell's equations. These equations show that a time-varying magnetic field acts as a source of electric field and that a time-varying electric field acts as a source of magnetic field. These $\vec{E}$ and $\vec{B}$ fields can sustain each other, forming an *electromagnetic wave* that propagates through space. Visible light emitted by the glowing filament of a light bulb is one example of an electromagnetic wave; other kinds of electromagnetic waves are produced by wi-fi base stations, x-ray machines, and radioactive nuclei.

In this chapter we'll use Maxwell's equations as the theoretical basis for understanding electromagnetic waves. We'll find that these waves carry both energy and momentum. In sinusoidal electromagnetic waves, the $\vec{E}$ and $\vec{B}$ fields are sinusoidal functions of time and position, with a definite frequency and wavelength. Visible light, radio, x rays, and other types of electromagnetic waves differ only in their frequency and wavelength. Our study of optics in the following chapters will be based in part on the electromagnetic nature of light.

Unlike waves on a string or sound waves in a fluid, electromagnetic waves do not require a material medium; the light that you see coming from the stars at night has traveled without difficulty across tens or hundreds of light-years of (nearly) empty space. Nonetheless, electromagnetic waves and mechanical waves have much in common and are described in much the same language. Before reading further in this chapter, you should review the properties of mechanical waves as discussed in Chapters 15 and 16.

32.1 MAXWELL'S EQUATIONS AND ELECTROMAGNETIC WAVES

In the last several chapters we studied various aspects of electric and magnetic fields. We learned that when the fields don't vary with time, such as an electric field produced by charges at rest or the magnetic field of a steady current, we can analyze the electric and magnetic fields independently without considering interactions between them. But when the fields vary with time, they are no longer independent. Faraday's law (see Section 29.2) tells us that a time-varying magnetic field acts as a source of electric field, as shown by induced emfs in inductors and transformers. Ampere's law, including the displacement current discovered by James Clerk Maxwell (see Section 29.7), shows that a time-varying electric field acts as a source of magnetic field. This mutual interaction between the two fields is summarized in Maxwell's equations, presented in Section 29.7.

Thus, when *either* an electric or a magnetic field is changing with time, a field of the other kind is induced in adjacent regions of space. We are led (as Maxwell was) to consider the possibility of an electromagnetic disturbance, consisting of time-varying electric and magnetic fields, that can propagate through space from one region to another, even when there is no matter in the intervening region. Such a disturbance, if it exists, will have the properties of a *wave,* and an appropriate term is **electromagnetic wave.**

Such waves do exist; radio and television transmission, light, x rays, and many other kinds of radiation are examples of electromagnetic waves. Our goal in this chapter is to see how such waves are explained by the principles of electromagnetism that we have studied thus far and to examine the properties of these waves.

Electricity, Magnetism, and Light

The theoretical understanding of electromagnetic waves actually evolved along a considerably more devious path than the one just outlined. In the early days of electromagnetic theory (the early 19th century), two different units of electric charge were used: one for electrostatics and the other for magnetic phenomena involving currents. In the system of units used at that time, these two units of charge had different physical dimensions. Their *ratio* had units of velocity, and measurements showed that the ratio had a numerical value that was precisely equal to the speed of light, 3.00×10^8 m/s. At the time, physicists regarded this as an extraordinary coincidence and had no idea how to explain it.

In searching to understand this result, Maxwell (**Fig. 32.1**) proved in 1865 that an electromagnetic disturbance should propagate in free space with a speed equal to that of light and hence that light waves were likely to be electromagnetic in nature. At the same time, he discovered that the basic principles of electromagnetism can be expressed in terms of the four equations that we now call **Maxwell's equations,** which we discussed in Section 29.7. These four equations are (1) Gauss's law for electric fields; (2) Gauss's law for magnetic fields, showing the absence of magnetic monopoles; (3) Faraday's law; and (4) Ampere's law, including displacement current:

$$\oint \vec{E} \cdot d\vec{A} = \frac{Q_{\text{encl}}}{\epsilon_0} \qquad \text{(Gauss's law)} \tag{29.18}$$

$$\oint \vec{B} \cdot d\vec{A} = 0 \qquad \text{(Gauss's law for magnetism)} \tag{29.19}$$

$$\oint \vec{E} \cdot d\vec{l} = -\frac{d\Phi_B}{dt} \qquad \text{(Faraday's law)} \tag{29.20}$$

$$\oint \vec{B} \cdot d\vec{l} = \mu_0 \left(i_C + \epsilon_0 \frac{d\Phi_E}{dt} \right)_{\text{encl}} \qquad \text{(Ampere's law)} \tag{29.21}$$

32.1 The Scottish physicist James Clerk Maxwell (1831–1879) was the first person to truly understand the fundamental nature of light. He also made major contributions to thermodynamics, optics, astronomy, and color photography. Albert Einstein described Maxwell's accomplishments as "the most profound and the most fruitful that physics has experienced since the time of Newton."

32.2 (a) Every mobile phone emits signals in the form of electromagnetic waves that are made by accelerating charges. (b) Power lines carry a strong alternating current, which means that a substantial amount of charge is accelerating back and forth and generating electromagnetic waves. These waves can produce a buzzing sound from your car radio when you drive near the lines.

(a)

(b)

These equations apply to electric and magnetic fields *in vacuum*. If a material is present, the electric constant ϵ_0 and magnetic constant μ_0 are replaced by the permittivity ϵ and permeability μ of the material. If the values of ϵ and μ are different at different points in the regions of integration, then ϵ and μ have to be transferred to the left sides of Eqs. (29.18) and (29.21), respectively, and placed inside the integrals. The ϵ in Eq. (29.21) also has to be included in the integral that gives $d\Phi_E/dt$.

According to Maxwell's equations, a point charge at rest produces a static $\vec{E}$ field but no $\vec{B}$ field, whereas a point charge moving with a constant velocity (see Section 28.1) produces both $\vec{E}$ and $\vec{B}$ fields. Maxwell's equations can also be used to show that in order for a point charge to produce electromagnetic waves, the charge must *accelerate*. In fact, in *every* situation where electromagnetic energy is radiated, the source is accelerated charges (**Fig. 32.2**).

Generating Electromagnetic Radiation

One way in which a point charge can be made to emit electromagnetic waves is by making it oscillate in simple harmonic motion, so that it has an acceleration at almost every instant (the exception is when the charge is passing through its equilibrium position). **Figure 32.3** shows some of the electric field lines produced by such an oscillating point charge. Field lines are *not* material objects, but you may nonetheless find it helpful to think of them as behaving somewhat like strings that extend from the point charge off to infinity. Oscillating the charge up and down makes waves that propagate outward from the charge along these "strings." Note that the charge does not emit waves equally in all directions; the waves are strongest at 90° to the axis of motion of the charge, while there are *no* waves along this axis. This is just what the "string" picture would lead you to conclude. There is also a *magnetic* disturbance that spreads outward from the charge; this is not shown in Fig. 32.3. Because the electric and magnetic disturbances spread or radiate away from the source, the name **electromagnetic radiation** is used interchangeably with the phrase "electromagnetic waves."

Electromagnetic waves with macroscopic wavelengths were first produced in the laboratory in 1887 by the German physicist Heinrich Hertz (for whom the SI unit of frequency is named). As a source of waves, he used charges oscillating in *L-C* circuits (see Section 30.5); he detected the resulting electromagnetic waves with other circuits tuned to the same frequency. Hertz also produced electromagnetic *standing waves* and measured the distance between adjacent nodes (one half-wavelength) to determine the wavelength. Knowing the resonant frequency of his circuits, he then found the speed of the waves from the wavelength–frequency relationship $v = \lambda f$. He established that their speed was the same as that of light; this verified Maxwell's theoretical prediction directly.

32.3 Electric field lines of a point charge oscillating in simple harmonic motion, seen at five instants during an oscillation period *T*. The charge's trajectory is in the plane of the drawings. At $t = 0$ the point charge is at its maximum upward displacement. The arrow shows one "kink" in the lines of $\vec{E}$ as it propagates outward from the point charge. The magnetic field (not shown) contains circles that lie in planes perpendicular to these figures and concentric with the axis of oscillation.

(a) $t = 0$ (b) $t = T/4$ (c) $t = T/2$ (d) $t = 3T/4$ (e) $t = T$

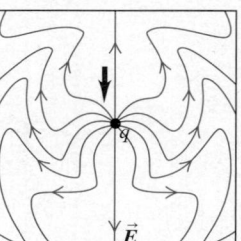 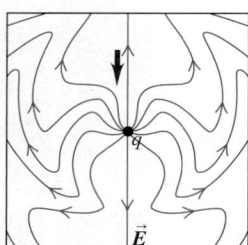 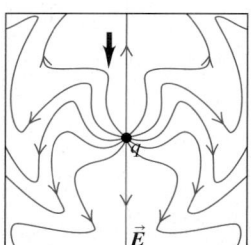 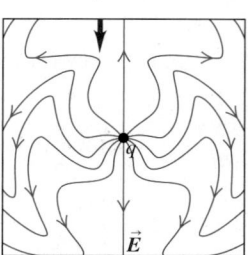 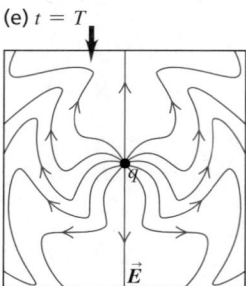

The modern value of the speed of light, c, is 299,792,458 m/s. (Recall from Section 1.3 that this value is the basis of our standard of length: One meter is defined to be the distance that light travels in 1/299,792,458 second.) For our purposes, $c = 3.00 \times 10^8$ m/s is sufficiently accurate.

In the wake of Hertz's discovery, Guglielmo Marconi and others made radio communication a familiar household experience. In a radio *transmitter*, electric charges are made to oscillate along the length of the conducting antenna, producing oscillating field disturbances like those shown in Fig. 32.3. Since many charges oscillate together in the antenna, the disturbances are much stronger than those of a single oscillating charge and can be detected at a much greater distance. In a radio *receiver* the antenna is also a conductor; the fields of the wave emanating from a distant transmitter exert forces on free charges within the receiver antenna, producing an oscillating current that is detected and amplified by the receiver circuitry.

For the remainder of this chapter our concern will be with electromagnetic waves themselves, not with the rather complex problem of how they are produced.

The Electromagnetic Spectrum

The **electromagnetic spectrum** encompasses electromagnetic waves of all frequencies and wavelengths. **Figure 32.4** shows approximate wavelength and frequency ranges for the most commonly encountered portion of the spectrum. Despite vast differences in their uses and means of production, these are all electromagnetic waves with the same propagation speed (in vacuum) $c = 299{,}792{,}458$ m/s. Electromagnetic waves may differ in frequency f and wavelength λ, but the relationship $c = \lambda f$ in vacuum holds for each.

We can detect only a very small segment of this spectrum directly through our sense of sight. We call this range **visible light.** Its wavelengths range from about 380 to 750 nm (380 to 750×10^{-9} m), with corresponding frequencies from about 790 to 400 THz (7.9 to 4.0×10^{14} Hz). Different parts of the visible spectrum evoke in humans the sensations of different colors. **Table 32.1** gives the approximate wavelengths for colors in the visible spectrum.

Ordinary white light includes all visible wavelengths. However, by using special sources or filters, we can select a narrow band of wavelengths within a range of a few nm. Such light is approximately *monochromatic* (single-color) light. Absolutely monochromatic light with only a single wavelength is an unattainable idealization. When we say "monochromatic light with $\lambda = 550$ nm" with reference to a laboratory experiment, we really mean a small band of

TABLE 32.1	Wavelengths of Visible Light
380–450 nm	Violet
450–495 nm	Blue
495–570 nm	Green
570–590 nm	Yellow
590–620 nm	Orange
620–750 nm	Red

32.4 The electromagnetic spectrum. The frequencies and wavelengths found in nature extend over such a wide range that we have to use a logarithmic scale to show all important bands. The boundaries between bands are somewhat arbitrary.

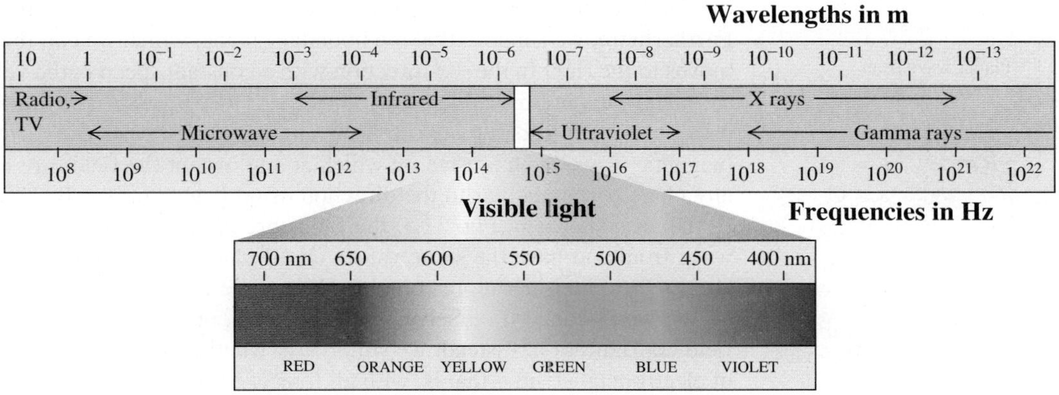

wavelengths *around* 550 nm. Light from a *laser* is much more nearly monochromatic than is light obtainable in any other way.

Invisible forms of electromagnetic radiation are no less important than visible light. Our system of global communication, for example, depends on radio waves: AM radio uses waves with frequencies from 5.4×10^5 Hz to 1.6×10^6 Hz, and FM radio broadcasts are at frequencies from 8.8×10^7 Hz to 1.08×10^8 Hz. Microwaves are also used for communication (for example, by mobile phones and wireless networks) and for weather radar (at frequencies near 3×10^9 Hz). Many cameras have a device that emits a beam of infrared radiation; by analyzing the properties of the infrared radiation reflected from the subject, the camera determines the distance to the subject and automatically adjusts the focus. X rays are able to penetrate through flesh, which makes them invaluable in dentistry and medicine. Gamma rays, the shortest-wavelength type of electromagnetic radiation, are used in medicine to destroy cancer cells.

TEST YOUR UNDERSTANDING OF SECTION 32.1   (a) Is it possible to have a purely electric wave propagate through empty space—that is, a wave made up of an electric field but no magnetic field? (b) What about a purely magnetic wave, with a magnetic field but no electric field? ∎

32.2 PLANE ELECTROMAGNETIC WAVES AND THE SPEED OF LIGHT

We are now ready to develop the basic ideas of electromagnetic waves and their relationship to the principles of electromagnetism. Our procedure will be to postulate a simple field configuration that has wavelike behavior. We'll assume an electric field $\vec{E}$ that has only a y-component and a magnetic field $\vec{B}$ with only a z-component, and we'll assume that both fields move together in the $+x$-direction with a speed c that is initially unknown. (As we go along, it will become clear why we choose $\vec{E}$ and $\vec{B}$ to be perpendicular to the direction of propagation as well as to each other.) Then we will test whether these fields are physically possible by asking whether they are consistent with Maxwell's equations, particularly Ampere's law and Faraday's law. We'll find that the answer is yes, provided that c has a particular value. We'll also show that the *wave equation,* which we encountered during our study of mechanical waves in Chapter 15, can be derived from Maxwell's equations.

A Simple Plane Electromagnetic Wave

32.5 An electromagnetic wave front. The plane representing the wave front moves to the right (in the positive x-direction) with speed c.

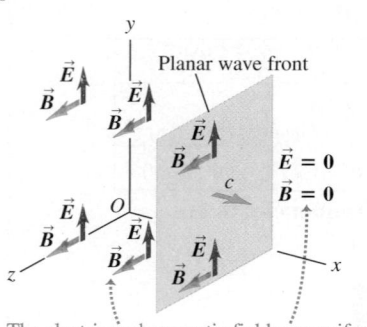

The electric and magnetic fields are uniform behind the advancing wave front and zero in front of it.

Using an *xyz*-coordinate system (**Fig. 32.5**), we imagine that all space is divided into two regions by a plane perpendicular to the x-axis (parallel to the yz-plane). At every point to the left of this plane there are a uniform electric field $\vec{E}$ in the $+y$-direction and a uniform magnetic field $\vec{B}$ in the $+z$-direction, as shown. Furthermore, we suppose that the boundary plane, which we call the *wave front*, moves to the right in the $+x$-direction with a constant speed c, the value of which we'll leave undetermined for now. Thus the $\vec{E}$ and $\vec{B}$ fields travel to the right into previously field-free regions with a definite speed. This is a rudimentary electromagnetic wave. Such a wave, in which at any instant the fields are uniform over any plane perpendicular to the direction of propagation, is called a **plane wave.** In the case shown in Fig. 32.5, the fields are zero for planes to the right of the wave front and have the same values on all planes to the left of the wave front; later we will consider more complex plane waves.

We won't concern ourselves with the problem of actually *producing* such a field configuration. Instead, we simply ask whether it is consistent with the laws of electromagnetism—that is, with all four of Maxwell's equations.

Let us first verify that our wave satisfies Maxwell's first and second equations—that is, Gauss's laws for electric and magnetic fields. To do this, we take as our Gaussian surface a rectangular box with sides parallel to the *xy*-, *xz*-, and *yz*-coordinate planes (**Fig. 32.6**). The box encloses no electric charge. The total electric flux and magnetic flux through the box are both zero, even if part of the box is in the region where $E = B = 0$. This would *not* be the case if $\vec{E}$ or $\vec{B}$ had an *x*-component, parallel to the direction of propagation; if the wave front were inside the box, there would be flux through the left-hand side of the box (at $x = 0$) but not the right-hand side (at $x > 0$). Thus to satisfy Maxwell's first and second equations, the electric and magnetic fields must be perpendicular to the direction of propagation; that is, the wave must be **transverse.**

The next of Maxwell's equations that we'll consider is Faraday's law:

$$\oint \vec{E} \cdot d\vec{l} = -\frac{d\Phi_B}{dt} \tag{32.1}$$

To test whether our wave satisfies Faraday's law, we apply this law to a rectangle *efgh* that is parallel to the *xy*-plane (**Fig. 32.7a**). As shown in Fig. 32.7b, a cross section in the *xy*-plane, this rectangle has height *a* and width Δx. At the time shown, the wave front has progressed partway through the rectangle, and $\vec{E}$ is zero along the side *ef*. In applying Faraday's law we take the vector area $d\vec{A}$ of rectangle *efgh* to be in the $+z$-direction. With this choice the right-hand rule requires that we integrate $\vec{E} \cdot d\vec{l}$ *counterclockwise* around the rectangle. At every point on side *ef*, $\vec{E}$ is zero. At every point on sides *fg* and *he*, $\vec{E}$ is either zero or perpendicular to $d\vec{l}$. Only side *gh* contributes to the integral. On this side, $\vec{E}$ and $d\vec{l}$ are opposite, and we find that the left-hand side of Eq. (32.1) is nonzero:

$$\oint \vec{E} \cdot d\vec{l} = -Ea \tag{32.2}$$

To satisfy Faraday's law, Eq. (32.1), there must be a component of $\vec{B}$ in the *z*-direction (perpendicular to $\vec{E}$) so that there can be a nonzero magnetic flux Φ_B through the rectangle *efgh* and a nonzero derivative $d\Phi_B/dt$. Indeed, in our wave, $\vec{B}$ has *only* a *z*-component. We have assumed that this component is in the *positive z*-direction; let's see whether this assumption is consistent with Faraday's law. During a time interval *dt* the wave front (traveling at speed *c*) moves a distance *c dt* to the right in Fig. 32.7b, sweeping out an area *ac dt* of the rectangle *efgh*. During this interval the magnetic flux Φ_B through the rectangle *efgh* increases by $d\Phi_B = B(ac\,dt)$, so the rate of change of magnetic flux is

$$\frac{d\Phi_B}{dt} = Bac \tag{32.3}$$

Now we substitute Eqs. (32.2) and (32.3) into Faraday's law, Eq. (32.1); we get $-Ea = -Bac$, so

Electromagnetic wave in vacuum:	Electric-field magnitude	Magnetic-field magnitude	
	$E = cB$	Speed of light in vacuum	(32.4)

Our wave is consistent with Faraday's law only if the wave speed *c* and the magnitudes of $\vec{E}$ and $\vec{B}$ are related as in Eq. (32.4). If we had assumed that $\vec{B}$ was in the *negative z*-direction, there would have been an additional minus sign in Eq. (32.4); since *E*, *c*, and *B* are all positive magnitudes, no solution would then have been possible. Furthermore, any component of $\vec{B}$ in the *y*-direction (parallel to $\vec{E}$) would not contribute to the changing magnetic flux Φ_B through the rectangle *efgh* (which is parallel to the *xy*-plane) and so would not be part of the wave.

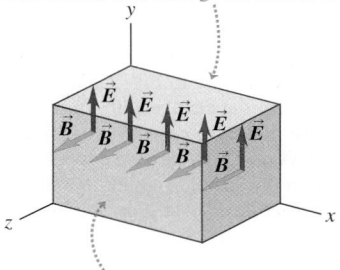

32.6 Gaussian surface for a transverse plane electromagnetic wave.

The electric field is the same on the top and bottom sides of the Gaussian surface, so the total electric flux through the surface is zero.

The magnetic field is the same on the left and right sides of the Gaussian surface, so the total magnetic flux through the surface is zero.

32.7 (a) Applying Faraday's law to a plane wave. (b) In a time *dt*, the magnetic flux through the rectangle in the *xy*-plane increases by an amount $d\Phi_B$. This increase equals the flux through the shaded rectangle with area *ac dt*; that is, $d\Phi_B = Bac\,dt$. Thus $d\Phi_B/dt = Bac$.

(a) In time *dt*, the wave front moves a distance *c dt* in the $+x$-direction.

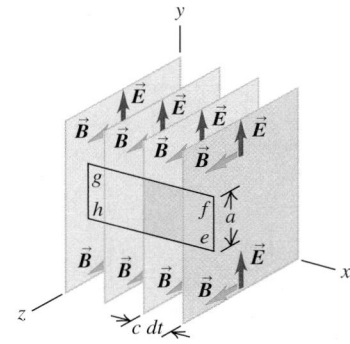

(b) Side view of situation in **(a)**

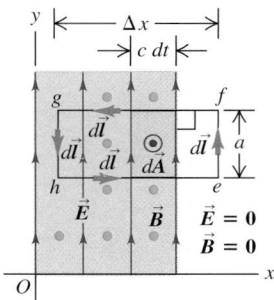

Finally, let's do a similar calculation with Ampere's law, the last of Maxwell's equations. There is no conduction current ($i_C = 0$), so Ampere's law is

$$\oint \vec{B} \cdot d\vec{l} = \mu_0 \epsilon_0 \frac{d\Phi_E}{dt} \tag{32.5}$$

32.8 (a) Applying Ampere's law to a plane wave. (Compare to Fig. 32.7a.) (b) In a time dt, the electric flux through the rectangle in the xz-plane increases by an amount $d\Phi_E$. This increase equals the flux through the shaded rectangle with area $ac\,dt$; that is, $d\Phi_E = Eac\,dt$. Thus $d\Phi_E/dt = Eac$.

(a) In time dt, the wave front moves a distance $c\,dt$ in the +x-direction.

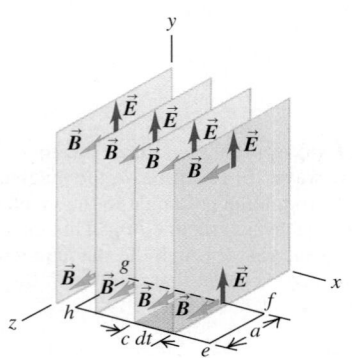

(b) Top view of situation in **(a)**

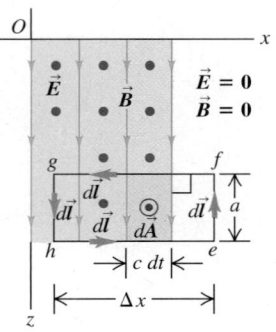

To check whether our wave is consistent with Ampere's law, we move our rectangle so that it lies in the xz-plane (**Fig. 32.8**), and we again look at the situation at a time when the wave front has traveled partway through the rectangle. We take the vector area $d\vec{A}$ in the +y-direction, and so the right-hand rule requires that we integrate $\vec{B} \cdot d\vec{l}$ counterclockwise around the rectangle. The $\vec{B}$ field is zero at every point along side ef, and at each point on sides fg and he it is either zero or perpendicular to $d\vec{l}$. Only side gh, where $\vec{B}$ and $d\vec{l}$ are parallel, contributes to the integral, and

$$\oint \vec{B} \cdot d\vec{l} = Ba \tag{32.6}$$

Hence the left-hand side of Eq. (32.5) is nonzero; the right-hand side must be nonzero as well. Thus $\vec{E}$ must have a y-component (perpendicular to $\vec{B}$) so that the electric flux Φ_E through the rectangle and the time derivative $d\Phi_E/dt$ can be nonzero. Just as we inferred from Faraday's law, we conclude that in an electromagnetic wave, $\vec{E}$ and $\vec{B}$ must be mutually perpendicular.

In a time interval dt the electric flux Φ_E through the rectangle increases by $d\Phi_E = E(ac\,dt)$. Since we chose $d\vec{A}$ to be in the +y-direction, this flux change is positive; the rate of change of electric flux is

$$\frac{d\Phi_E}{dt} = Eac \tag{32.7}$$

Substituting Eqs. (32.6) and (32.7) into Ampere's law, Eq. (32.5), we find $Ba = \epsilon_0 \mu_0 Eac$, so

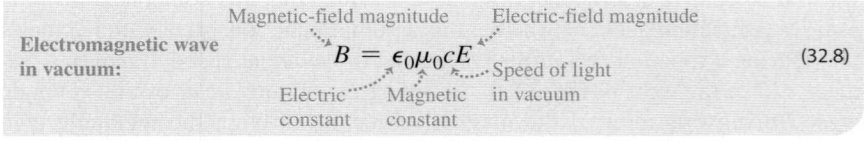

Electromagnetic wave in vacuum:

Magnetic-field magnitude   Electric-field magnitude

$$B = \epsilon_0 \mu_0 cE \tag{32.8}$$

Electric constant   Magnetic constant   Speed of light in vacuum

Our assumed wave obeys Ampere's law only if B, c, and E are related as in Eq. (32.8). The wave must also obey Faraday's law, so Eq. (32.4) must be satisfied as well. This can happen only if $\epsilon_0 \mu_0 c = 1/c$, or

Speed of electromagnetic waves in vacuum

$$c = \frac{1}{\sqrt{\epsilon_0 \mu_0}} \tag{32.9}$$

Electric constant   Magnetic constant

Inserting the numerical values of these quantities, we find

$$c = \frac{1}{\sqrt{(8.85 \times 10^{-12}\ \text{C}^2/\text{N} \cdot \text{m}^2)(4\pi \times 10^{-7}\ \text{N}/\text{A}^2)}}$$

$$= 3.00 \times 10^8\ \text{m/s}$$

Our assumed wave is consistent with all of Maxwell's equations, provided that the wave front moves with the speed given above, which is the speed of light! Recall that the *exact* value of c is defined to be 299,792,458 m/s; the modern value of ϵ_0 is defined to agree with this when used in Eq. (32.9) (see Section 21.3).

Key Properties of Electromagnetic Waves

We chose a simple wave for our study in order to avoid mathematical complications, but this special case illustrates several important features of *all* electromagnetic waves:

1. The wave is *transverse*; both $\vec{E}$ and $\vec{B}$ are perpendicular to the direction of propagation of the wave. The electric and magnetic fields are also perpendicular to each other. The direction of propagation is the direction of the vector product $\vec{E} \times \vec{B}$ (**Fig. 32.9**).

2. There is a definite ratio between the magnitudes of $\vec{E}$ and $\vec{B}$: $E = cB$.

3. The wave travels in vacuum with a definite and unchanging speed.

4. Unlike mechanical waves, which need the particles of a medium such as air to transmit a wave, electromagnetic waves require no medium.

We can generalize this discussion to a more realistic situation. Suppose we have several wave fronts in the form of parallel planes perpendicular to the x-axis, all of which are moving to the right with speed c. Suppose that the $\vec{E}$ and $\vec{B}$ fields are the same at all points within a single region between two planes, but that the fields differ from region to region. The overall wave is a plane wave, but one in which the fields vary in steps along the x-axis. Such a wave could be constructed by superposing several of the simple step waves we have just discussed (shown in Fig. 32.5). This is possible because the $\vec{E}$ and $\vec{B}$ fields obey the superposition principle in waves just as in static situations: When two waves are superposed, the total $\vec{E}$ field at each point is the vector sum of the $\vec{E}$ fields of the individual waves, and similarly for the total $\vec{B}$ field.

We can extend the above development to show that a wave with fields that vary in steps is also consistent with Ampere's and Faraday's laws, provided that the wave fronts all move with the speed c given by Eq. (32.9). In the limit that we make the individual steps infinitesimally small, we have a wave in which the $\vec{E}$ and $\vec{B}$ fields at any instant vary *continuously* along the x-axis. The entire field pattern moves to the right with speed c. In Section 32.3 we will consider waves in which $\vec{E}$ and $\vec{B}$ are *sinusoidal* functions of x and t. Because at each point the magnitudes of $\vec{E}$ and $\vec{B}$ are related by $E = cB$, the periodic variations of the two fields in any periodic traveling wave must be *in phase*.

Electromagnetic waves have the property of **polarization**. In the above discussion the choice of the y-direction for $\vec{E}$ was arbitrary. We could instead have specified the z-axis for $\vec{E}$; then $\vec{B}$ would have been in the $-y$-direction. A wave in which $\vec{E}$ is always parallel to a certain axis is said to be **linearly polarized** along that axis. More generally, *any* wave traveling in the x-direction can be represented as a superposition of waves linearly polarized in the y- and z-directions. We will study polarization in greater detail in Chapter 33.

Derivation of the Electromagnetic Wave Equation

Here is an alternative derivation of Eq. (32.9) for the speed of electromagnetic waves. It is more mathematical than our other treatment, but it includes a derivation of the wave equation for electromagnetic waves. This part of the section can be omitted without loss of continuity in the chapter.

During our discussion of mechanical waves in Section 15.3, we showed that a function $y(x, t)$ that represents the displacement of any point in a mechanical wave traveling along the x-axis must satisfy a differential equation, Eq. (15.12):

$$\frac{\partial^2 y(x, t)}{\partial x^2} = \frac{1}{v^2} \frac{\partial^2 y(x, t)}{\partial t^2} \qquad (32.10)$$

This equation is called the **wave equation,** and v is the speed of propagation of the wave.

32.9 A right-hand rule for electromagnetic waves relates the directions of $\vec{E}$ and $\vec{B}$ and the direction of propagation.

Right-hand rule for an electromagnetic wave:

① Point the thumb of your right hand in the wave's direction of propagation.

② Imagine rotating the $\vec{E}$-field vector 90° in the sense your fingers curl. That is the direction of the $\vec{B}$ field.

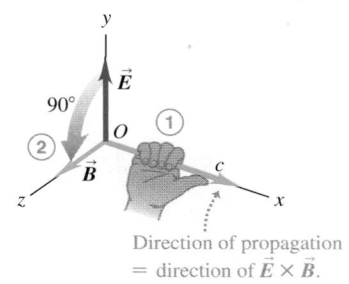

Direction of propagation = direction of $\vec{E} \times \vec{B}$.

DATA *SPEAKS*

Electromagnetic Waves

When students were given a problem involving electromagnetic waves, more than 29% gave an incorrect response. Common errors:

- Forgetting that, in vacuum, all electromagnetic waves travel at the same speed: c. Since $c = \lambda f$, waves with high frequency f have short wavelength λ but travel at the same speed as waves with low f and long λ.

- Confusion about the directions of $\vec{E}$ and $\vec{B}$ and the propagation direction. The electric and magnetic fields in an electromagnetic wave are always perpendicular to each other. The wave propagates in the direction of $\vec{E} \times \vec{B}$, which is perpendicular to both $\vec{E}$ and $\vec{B}$.

To derive the corresponding equation for an electromagnetic wave, we again consider a plane wave. That is, we assume that at each instant, E_y and B_z are uniform over any plane perpendicular to the x-axis, the direction of propagation. But now we let E_y and B_z vary continuously as we go along the x-axis; then each is a function of x and t. We consider the values of E_y and B_z on two planes perpendicular to the x-axis, one at x and one at $x + \Delta x$.

Following the same procedure as previously, we apply Faraday's law to a rectangle lying parallel to the xy-plane, as in **Fig. 32.10**. This figure is similar to Fig. 32.7. Let the left end gh of the rectangle be at position x, and let the right end ef be at position $(x + \Delta x)$. At time t, the values of E_y on these two sides are $E_y(x, t)$ and $E_y(x + \Delta x, t)$, respectively. When we apply Faraday's law to this rectangle, we find that instead of $\oint \vec{E} \cdot d\vec{l} = -Ea$ as before, we have

$$\oint \vec{E} \cdot d\vec{l} = -E_y(x, t)a + E_y(x + \Delta x, t)a$$

$$= a[E_y(x + \Delta x, t) - E_y(x, t)] \tag{32.11}$$

To find the magnetic flux Φ_B through this rectangle, we assume that Δx is small enough that B_z is nearly uniform over the rectangle. In that case, $\Phi_B = B_z(x, t)A = B_z(x, t)a \, \Delta x$, and

$$\frac{d\Phi_B}{dt} = \frac{\partial B_z(x, t)}{\partial t} a \, \Delta x$$

We use partial-derivative notation because B_z is a function of both x and t. When we substitute this expression and Eq. (32.11) into Faraday's law, Eq. (32.1), we get

$$a[E_y(x + \Delta x, t) - E_y(x, t)] = -\frac{\partial B_z}{\partial t} a \, \Delta x$$

$$\frac{E_y(x + \Delta x, t) - E_y(x, t)}{\Delta x} = -\frac{\partial B_z}{\partial t}$$

Finally, imagine shrinking the rectangle down to a sliver so that Δx approaches zero. When we take the limit of this equation as $\Delta x \rightarrow 0$, we get

$$\frac{\partial E_y(x, t)}{\partial x} = -\frac{\partial B_z(x, t)}{\partial t} \tag{32.12}$$

This equation shows that if there is a time-varying component B_z of magnetic field, there must also be a component E_y of electric field that varies with x, and conversely. We put this relationship on the shelf for now; we'll return to it soon.

Next we apply Ampere's law to the rectangle shown in **Fig. 32.11**. The line integral $\oint \vec{B} \cdot d\vec{l}$ becomes

$$\oint \vec{B} \cdot d\vec{l} = -B_z(x + \Delta x, t)a + B_z(x, t)a \tag{32.13}$$

Again assuming that the rectangle is narrow, we approximate the electric flux Φ_E through it as $\Phi_E = E_y(x, t)A = E_y(x, t)a \, \Delta x$. The rate of change of Φ_E, which we need for Ampere's law, is then

$$\frac{d\Phi_E}{dt} = \frac{\partial E_y(x, t)}{\partial t} a \, \Delta x$$

Now we substitute this expression and Eq. (32.13) into Ampere's law, Eq. (32.5):

$$-B_z(x + \Delta x, t)a + B_z(x, t)a = \epsilon_0 \mu_0 \frac{\partial E_y(x, t)}{\partial t} a \, \Delta x$$

32.10 Faraday's law applied to a rectangle with height a and width Δx parallel to the xy-plane.

(a)

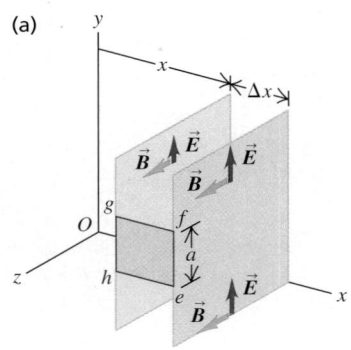

(b) Side view of the situation in **(a)**

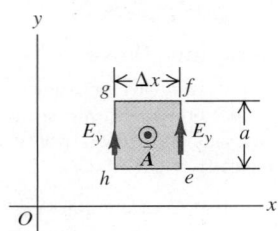

32.11 Ampere's law applied to a rectangle with height a and width Δx parallel to the xz-plane.

(a)

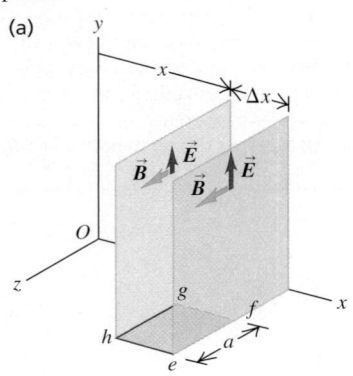

(b) Top view of the situation in **(a)**

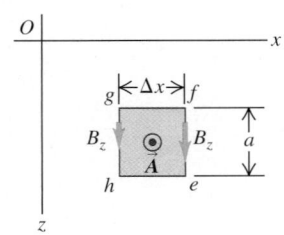

Again we divide both sides by $a\,\Delta x$ and take the limit as $\Delta x \to 0$. We find

$$-\frac{\partial B_z(x, t)}{\partial x} = \epsilon_0\mu_0 \frac{\partial E_y(x, t)}{\partial t} \qquad (32.14)$$

Now comes the final step. We take the partial derivatives of both sides of Eq. (32.12) with respect to x, and we take the partial derivatives of both sides of Eq. (32.14) with respect to t. The results are

$$-\frac{\partial^2 E_y(x, t)}{\partial x^2} = \frac{\partial^2 B_z(x, t)}{\partial x\,\partial t}$$

$$-\frac{\partial^2 B_z(x, t)}{\partial x\,\partial t} = \epsilon_0\mu_0 \frac{\partial^2 E_y(x, t)}{\partial t^2}$$

Combining these two equations to eliminate B_z, we finally find

$$\frac{\partial^2 E_y(x, t)}{\partial x^2} = \epsilon_0\mu_0 \frac{\partial^2 E_y(x, t)}{\partial t^2} \qquad \begin{array}{l}\text{(electromagnetic wave}\\ \text{equation in vacuum)}\end{array} \qquad (32.15)$$

This expression has the same form as the general wave equation, Eq. (32.10). Because the electric field E_y must satisfy this equation, it behaves as a wave with a pattern that travels through space with a definite speed. Furthermore, comparison of Eqs. (32.15) and (32.10) shows that the wave speed v is given by

$$\frac{1}{v^2} = \epsilon_0\mu_0 \qquad \text{or} \qquad v = \frac{1}{\sqrt{\epsilon_0\mu_0}}$$

This agrees with Eq. (32.9) for the speed c of electromagnetic waves.

We can show that B_z also must satisfy the same wave equation as E_y, Eq. (32.15). To prove this, we take the partial derivative of Eq. (32.12) with respect to t and the partial derivative of Eq. (32.14) with respect to x and combine the results. We leave this derivation for you to carry out.

TEST YOUR UNDERSTANDING OF SECTION 32.2 For each of the following electromagnetic waves, state the direction of the magnetic field. (a) The wave is propagating in the positive z-direction, and $\vec{E}$ is in the positive x-direction; (b) the wave is propagating in the positive y-direction, and $\vec{E}$ is in the negative z-direction; (c) the wave is propagating in the negative x-direction, and $\vec{E}$ is in the positive z-direction. ▌

32.3 SINUSOIDAL ELECTROMAGNETIC WAVES

Sinusoidal electromagnetic waves are directly analogous to sinusoidal transverse mechanical waves on a stretched string, which we studied in Section 15.3. In a sinusoidal electromagnetic wave, $\vec{E}$ and $\vec{B}$ at any point in space are sinusoidal functions of time, and at any instant of time the *spatial* variation of the fields is also sinusoidal.

Some sinusoidal electromagnetic waves are *plane waves;* they share with the waves described in Section 32.2 the property that at any instant the fields are uniform over any plane perpendicular to the direction of propagation. The entire pattern travels in the direction of propagation with speed c. The directions of $\vec{E}$ and $\vec{B}$ are perpendicular to the direction of propagation (and to each other), so the wave is *transverse*. Electromagnetic waves produced by an oscillating point charge, shown in Fig. 32.3, are an example of sinusoidal waves that are *not* plane waves. But if we restrict our observations to a relatively small region of space at a sufficiently great distance from the source, even these waves are well approximated by plane waves (**Fig. 32.12**). In the same way, the curved surface of the (nearly) spherical earth appears flat to us because of our small size relative to the earth's radius. In this section we'll restrict our discussion to plane waves.

32.12 Waves passing through a small area at a sufficiently great distance from a source can be treated as plane waves.

Waves that pass through a large area propagate in different directions ...

Source of electromagnetic waves

... but waves that pass through a small area all propagate in nearly the same direction, so we can treat them as plane waves.

The frequency f, the wavelength λ, and the speed of propagation c of any periodic wave are related by the usual wavelength–frequency relationship $c = \lambda f$. If the frequency f is 10^8 Hz (100 MHz), typical of commercial FM radio broadcasts, the wavelength is

$$\lambda = \frac{3 \times 10^8 \text{ m/s}}{10^8 \text{ Hz}} = 3 \text{ m}$$

Figure 32.4 shows the inverse proportionality between wavelength and frequency.

Fields of a Sinusoidal Wave

Figure 32.13 shows a linearly polarized sinusoidal electromagnetic wave traveling in the $+x$-direction. The electric and magnetic fields oscillate in phase: $\vec{E}$ is maximum where $\vec{B}$ is maximum and $\vec{E}$ is zero where $\vec{B}$ is zero. Where $\vec{E}$ is in the $+y$-direction, $\vec{B}$ is in the $+z$-direction; where $\vec{E}$ is in the $-y$-direction, $\vec{B}$ is in the $-z$-direction. At all points the vector product $\vec{E} \times \vec{B}$ is in the direction in which the wave is propagating (the $+x$-direction). We mentioned this in Section 32.2 in the list of characteristics of electromagnetic waves.

> **CAUTION** **In a plane wave, $\vec{E}$ and $\vec{B}$ are everywhere** Figure 32.13 shows $\vec{E}$ and $\vec{B}$ at points on the x-axis only. But, in fact, in a sinusoidal plane wave there are electric and magnetic fields at *all* points in space. Imagine a plane perpendicular to the x-axis (that is, parallel to the yz-plane) at a particular point and time; the fields have the same values at all points in that plane. The values are different on different planes. ▮

We can describe electromagnetic waves by means of *wave functions,* just as we did in Section 15.3 for waves on a string. One form of the wave function for a transverse wave traveling in the $+x$-direction along a stretched string is Eq. (15.7):

$$y(x, t) = A\cos(kx - \omega t)$$

where $y(x, t)$ is the transverse displacement from equilibrium at time t of a point with coordinate x on the string. Here A is the maximum displacement, or *amplitude,* of the wave; ω is its *angular frequency,* equal to 2π times the frequency f; and $k = 2\pi/\lambda$ is the *wave number,* where λ is the wavelength.

Let $E_y(x, t)$ and $B_z(x, t)$ represent the instantaneous values of the y-component of $\vec{E}$ and the z-component of $\vec{B}$, respectively, in Fig. 32.13, and let E_{max} and B_{max} represent the maximum values, or *amplitudes,* of these fields. The wave functions for the wave are then

$$E_y(x, t) = E_{\text{max}}\cos(kx - \omega t) \qquad B_z(x, t) = B_{\text{max}}\cos(kx - \omega t) \quad \text{(32.16)}$$

We can also write the wave functions in vector form:

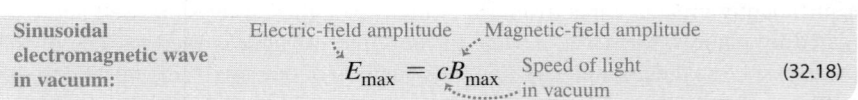

Sinusoidal electromagnetic plane wave, propagating in $+x$-direction:

$$\vec{E}(x, t) = \hat{\jmath}E_{\text{max}}\cos(kx - \omega t)$$
$$\vec{B}(x, t) = \hat{k}B_{\text{max}}\cos(kx - \omega t) \quad \text{(32.17)}$$

Electric field · Electric-field magnitude · Wave number · Angular frequency · Magnetic field · Magnetic-field magnitude

The sine curves in Fig. 32.13 represent the fields as functions of x at time $t = 0$—that is, $\vec{E}(x, t = 0)$ and $\vec{B}(x, t = 0)$. As the wave travels to the right with speed c, Eqs. (32.16) and (32.17) show that at any point the oscillations of $\vec{E}$ and $\vec{B}$ are *in phase.* From Eq. (32.4) the amplitudes must be related by

Sinusoidal electromagnetic wave in vacuum:

$$E_{\text{max}} = cB_{\text{max}} \quad \text{(32.18)}$$

Electric-field amplitude · Magnetic-field amplitude · Speed of light in vacuum

---

32.13 Representation of the electric and magnetic fields as functions of x for a linearly polarized sinusoidal plane electromagnetic wave. One wavelength of the wave is shown at time $t = 0$. The fields are shown for only a few points along the x-axis.

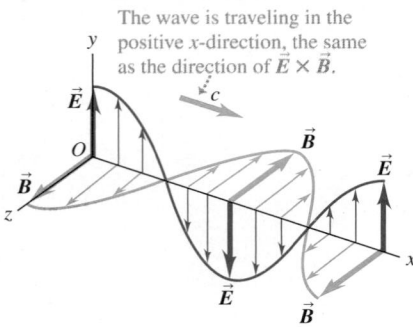

The wave is traveling in the positive x-direction, the same as the direction of $\vec{E} \times \vec{B}$.

$\vec{E}$: y-component only
$\vec{B}$: z-component only

PhET: Radio Waves & Electromagnetic Fields

> **CAUTION** **The symbol k has two meanings** Note the two different k's: the unit vector $\hat{k}$ in the z-direction and the wave number k. Don't get these confused! ▮

These amplitude and phase relationships are also required for $E(x, t)$ and $B(x, t)$ to satisfy Eqs. (32.12) and (32.14), which came from Faraday's law and Ampere's law, respectively. Can you verify this statement? (See Problem 32.34.)

Figure 32.14 shows the $\vec{E}$ and $\vec{B}$ fields of a wave traveling in the *negative* x-direction. At points where $\vec{E}$ is in the positive y-direction, $\vec{B}$ is in the *negative* z-direction; where $\vec{E}$ is in the negative y-direction, $\vec{B}$ is in the *positive* z-direction. As with waves traveling in the $+x$-direction, at any point the oscillations of the $\vec{E}$ and $\vec{B}$ fields of this wave are in phase, and the vector product $\vec{E} \times \vec{B}$ points in the propagation direction. The wave functions for this wave are

$$\vec{E}(x, t) = \hat{j}E_{max}\cos(kx + \omega t)$$

$$\vec{B}(x, t) = -\hat{k}B_{max}\cos(kx + \omega t)$$

(32.19)

(sinusoidal electromagnetic plane wave, propagating in $-x$-direction)

The sinusoidal waves shown in both Figs. 32.13 and 32.14 are linearly polarized in the y-direction; the $\vec{E}$ field is always parallel to the y-axis. Example 32.1 concerns a wave that is linearly polarized in the z-direction.

32.14 Representation of one wavelength of a linearly polarized sinusoidal plane electromagnetic wave traveling in the negative x-direction at $t = 0$. The fields are shown only for points along the x-axis. (Compare with Fig. 32.13.)

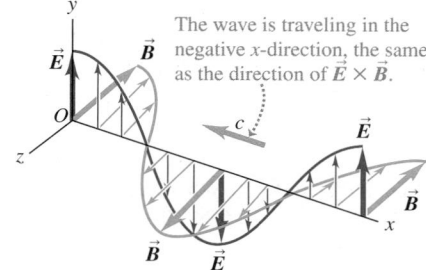

$\vec{E}$: y-component only
$\vec{B}$: z-component only

PROBLEM-SOLVING STRATEGY 32.1 | **ELECTROMAGNETIC WAVES**

IDENTIFY *the relevant concepts:* Many of the same ideas that apply to mechanical waves apply to electromagnetic waves. One difference is that electromagnetic waves are described by two quantities (in this case, electric field $\vec{E}$ and magnetic field $\vec{B}$), rather than by a single quantity, such as the displacement of a string.

SET UP *the problem* using the following steps:
1. Draw a diagram showing the direction of wave propagation and the directions of $\vec{E}$ and $\vec{B}$.
2. Identify the target variables.

EXECUTE *the solution* as follows:
1. Review the treatment of sinusoidal mechanical waves in Chapters 15 and 16, and particularly the four problem-solving strategies suggested there.
2. Keep in mind the basic relationships for periodic waves: $v = \lambda f$ and $\omega = vk$. For electromagnetic waves in vacuum,

$v = c$. Distinguish between ordinary frequency f, usually expressed in hertz, and angular frequency $\omega = 2\pi f$, expressed in rad/s. Remember that the wave number is $k = 2\pi/\lambda$.
3. Concentrate on basic relationships, such as those between $\vec{E}$ and $\vec{B}$ (magnitude, direction, and relative phase), how the wave speed is determined, and the transverse nature of the waves.

EVALUATE *your answer:* Check that your result is reasonable. For electromagnetic waves in vacuum, the magnitude of the magnetic field in teslas is much smaller (by a factor of 3.00×10^8) than the magnitude of the electric field in volts per meter. If your answer suggests otherwise, you probably made an error in using the relationship $E = cB$. (We'll see later in this section that this relationship is different for electromagnetic waves in a material medium.)

EXAMPLE 32.1 **ELECTRIC AND MAGNETIC FIELDS OF A LASER BEAM**

A carbon dioxide laser emits a sinusoidal electromagnetic wave that travels in vacuum in the negative x-direction. The wavelength is 10.6 μm (in the infrared; see Fig. 32.4) and the $\vec{E}$ field is parallel to the z-axis, with $E_{max} = 1.5$ MV/m. Write vector equations for $\vec{E}$ and $\vec{B}$ as functions of time and position.

SOLUTION

IDENTIFY and SET UP: Equations (32.19) describe a wave traveling in the negative x-direction with $\vec{E}$ along the y-axis—that is, a wave that is linearly polarized along the y-axis. By contrast, the wave in this example is linearly polarized along the z-axis. At points where $\vec{E}$ is in the positive z-direction, $\vec{B}$ must be in the positive y-direction for the vector product $\vec{E} \times \vec{B}$ to be in the negative x-direction (the direction of propagation). **Figure 32.15** shows a wave that satisfies these requirements.

32.15 Our sketch for this problem.

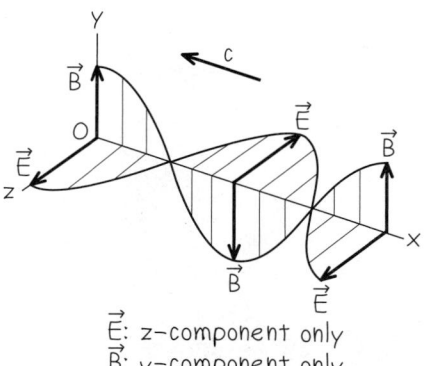

$\vec{E}$: z-component only
$\vec{B}$: y-component only

(Continued)

EXECUTE: A possible pair of wave functions that describe the wave shown in Fig. 32.15 are

$$\vec{E}(x, t) = \hat{k}E_{max}\cos(kx + \omega t)$$

$$\vec{B}(x, t) = \hat{j}B_{max}\cos(kx + \omega t)$$

The plus sign in the arguments of the cosine functions indicates that the wave is propagating in the negative x-direction, as it should. Faraday's law requires that $E_{max} = cB_{max}$ [Eq. (32.18)], so

$$B_{max} = \frac{E_{max}}{c} = \frac{1.5 \times 10^6 \text{ V/m}}{3.0 \times 10^8 \text{ m/s}} = 5.0 \times 10^{-3} \text{ T}$$

(Recall that 1 V = 1 Wb/s and 1 Wb/m² = 1 T.)
We have $\lambda = 10.6 \times 10^{-6}$ m, so the wave number and angular frequency are

$$k = \frac{2\pi}{\lambda} = \frac{2\pi \text{ rad}}{10.6 \times 10^{-6} \text{ m}} = 5.93 \times 10^5 \text{ rad/m}$$

$$\omega = ck = (3.00 \times 10^8 \text{ m/s})(5.93 \times 10^5 \text{ rad/m})$$
$$= 1.78 \times 10^{14} \text{ rad/s}$$

Substituting these values into the above wave functions, we get

$$\vec{E}(x, t) = \hat{k}(1.5 \times 10^6 \text{ V/m})$$
$$\times \cos[(5.93 \times 10^5 \text{ rad/m})x + (1.78 \times 10^{14} \text{ rad/s})t]$$

$$\vec{B}(x, t) = \hat{j}(5.0 \times 10^{-3} \text{ T})$$
$$\times \cos[(5.93 \times 10^5 \text{ rad/m})x + (1.78 \times 10^{14} \text{ rad/s})t]$$

EVALUATE: As we expect, the magnitude B_{max} in teslas is much smaller than the magnitude E_{max} in volts per meter. To check the directions of $\vec{E}$ and $\vec{B}$, note that $\vec{E} \times \vec{B}$ is in the direction of $\hat{k} \times \hat{j} = -\hat{i}$. This is as it should be for a wave that propagates in the negative x-direction.

Our expressions for $\vec{E}(x, t)$ and $\vec{B}(x, t)$ are not the only possible solutions. We could always add a phase angle ϕ to the arguments of the cosine function, so that $kx + \omega t$ would become $kx + \omega t + \phi$. To determine the value of ϕ we would need to know $\vec{E}$ and $\vec{B}$ either as functions of x at a given time t or as functions of t at a given coordinate x. However, the statement of the problem doesn't include this information.

Electromagnetic Waves in Matter

So far, our discussion of electromagnetic waves has been restricted to waves in *vacuum.* But electromagnetic waves can also travel in *matter;* think of light traveling through air, water, or glass. In this subsection we extend our analysis to electromagnetic waves in nonconducting materials—that is, *dielectrics.*

In a dielectric the wave speed is not the same as in vacuum, and we denote it by v instead of c. Faraday's law is unaltered, but in Eq. (32.4), derived from Faraday's law, the speed c is replaced by v. In Ampere's law the displacement current is given not by $\epsilon_0 \, d\Phi_E/dt$, where Φ_E is the flux of $\vec{E}$ through a surface, but by $\epsilon \, d\Phi_E/dt = K\epsilon_0 \, d\Phi_E/dt$, where K is the dielectric constant and ϵ is the permittivity of the dielectric. (We introduced these quantities in Section 24.4.) Also, the constant μ_0 in Ampere's law must be replaced by $\mu = K_m\mu_0$, where K_m is the relative permeability of the dielectric and μ is its permeability (see Section 28.8). Hence Eqs. (32.4) and (32.8) are replaced by

$$E = vB \qquad \text{and} \qquad B = \epsilon\mu vE \qquad (32.20)$$

Following the same procedure as for waves in vacuum, we find that

32.16 The dielectric constant K of water is about 1.8 for visible light, so the speed of visible light in water is slower than in vacuum by a factor of $1/\sqrt{K} = 1/\sqrt{1.8} = 0.75$.

Speed of electromagnetic waves in a dielectric	Permeability	Speed of light in vacuum
$v = \dfrac{1}{\sqrt{\epsilon\mu}} = \dfrac{1}{\sqrt{KK_m}}\dfrac{1}{\sqrt{\epsilon_0\mu_0}} = \dfrac{c}{\sqrt{KK_m}}$		(32.21)
Permittivity · Dielectric constant · Relative permeability · Electric constant · Magnetic constant		

For most dielectrics the relative permeability K_m is nearly equal to unity (except for insulating ferromagnetic materials). When $K_m \cong 1$, $v = c/\sqrt{K}$. Because K is always greater than unity, the speed v of electromagnetic waves in a nonmagnetic dielectric is always *less* than the speed c in vacuum by a factor of $1/\sqrt{K}$ (**Fig. 32.16**). The ratio of the speed c in vacuum to the speed v in a material is known in optics as the **index of refraction** n of the material. When $K_m \cong 1$,

$$\frac{c}{v} = n = \sqrt{KK_m} \cong \sqrt{K} \qquad (32.22)$$

Usually, we can't use the values of K in Table 24.1 in this equation because those values are measured in *constant* electric fields. When the fields oscillate rapidly, there is usually not time for the reorientation of electric dipoles that occurs with steady fields. Values of K with rapidly varying fields are usually much *smaller* than the values in the table. For example, K for water is 80.4 for steady fields but only about 1.8 in the frequency range of visible light. Thus the dielectric "constant" K is actually a function of frequency (the *dielectric function*).

EXAMPLE 32.2 ELECTROMAGNETIC WAVES IN DIFFERENT MATERIALS

(a) Visiting a jewelry store one evening, you hold a diamond up to the light of a sodium-vapor street lamp. The heated sodium vapor emits yellow light with a frequency of 5.09×10^{14} Hz. Find the wavelength in vacuum and the wave speed and wavelength in diamond, for which $K = 5.84$ and $K_m = 1.00$ at this frequency. (b) A 90.0-MHz radio wave (in the FM radio band) passes from vacuum into an insulating ferrite (a ferromagnetic material used in computer cables to suppress radio interference). Find the wavelength in vacuum and the wave speed and wavelength in the ferrite, for which $K = 10.0$ and $K_m = 1000$ at this frequency.

SOLUTION

IDENTIFY and SET UP: In each case we find the wavelength in vacuum by using $c = \lambda f$. To use the corresponding equation $v = \lambda f$ to find the wavelength in a material medium, we find the speed v of electromagnetic waves in the medium from Eq. (32.21), which relates v to the values of dielectric constant K and relative permeability K_m for the medium.

EXECUTE: (a) The wavelength in vacuum of the sodium light is

$$\lambda_{\text{vacuum}} = \frac{c}{f} = \frac{3.00 \times 10^8 \text{ m/s}}{5.09 \times 10^{14} \text{ Hz}} = 5.89 \times 10^{-7} \text{ m} = 589 \text{ nm}$$

The wave speed and wavelength in diamond are

$$v_{\text{diamond}} = \frac{c}{\sqrt{KK_m}} = \frac{3.00 \times 10^8 \text{ m/s}}{\sqrt{(5.84)(1.00)}} = 1.24 \times 10^8 \text{ m/s}$$

$$\lambda_{\text{diamond}} = \frac{v_{\text{diamond}}}{f} = \frac{1.24 \times 10^8 \text{ m/s}}{5.09 \times 10^{14} \text{ Hz}}$$

$$= 2.44 \times 10^{-7} \text{ m} = 244 \text{ nm}$$

(b) Following the same steps as in part (a), we find

$$\lambda_{\text{vacuum}} = \frac{c}{f} = \frac{3.00 \times 10^8 \text{ m/s}}{90.0 \times 10^6 \text{ Hz}} = 3.33 \text{ m}$$

$$v_{\text{ferrite}} = \frac{c}{\sqrt{KK_m}} = \frac{3.00 \times 10^8 \text{ m/s}}{\sqrt{(10.0)(1000)}} = 3.00 \times 10^6 \text{ m/s}$$

$$\lambda_{\text{ferrite}} = \frac{v_{\text{ferrite}}}{f} = \frac{3.00 \times 10^6 \text{ m/s}}{90.0 \times 10^6 \text{ Hz}} = 3.33 \times 10^{-2} \text{ m} = 3.33 \text{ cm}$$

EVALUATE: The speed of light in transparent materials is typically between $0.2c$ and c; our result in part (a) shows that $v_{\text{diamond}} = 0.414c$. As our results in part (b) show, the speed of electromagnetic waves in dense materials like ferrite (for which $v_{\text{ferrite}} = 0.010c$) can be *far* slower than in vacuum.

TEST YOUR UNDERSTANDING OF SECTION 32.3 The first of Eqs. (32.17) gives the electric field for a plane wave as measured at points along the x-axis. For this plane wave, how does the electric field at points *off* the x-axis differ from the expression in Eqs. (32.17)? (i) The amplitude is different; (ii) the phase is different; (iii) both the amplitude and phase are different; (iv) none of these. ∎

32.4 ENERGY AND MOMENTUM IN ELECTROMAGNETIC WAVES

Electromagnetic waves carry energy; the energy in sunlight is a familiar example. Microwave ovens, radio transmitters, and lasers for eye surgery all make use of this wave energy. To understand how to utilize this energy, it's helpful to derive detailed relationships for the energy in an electromagnetic wave.

We begin with the expressions derived in Sections 24.3 and 30.3 for the **energy densities** in electric and magnetic fields; we suggest that you review those derivations now. Equations (24.11) and (30.10) show that in a region of empty space where $\vec{E}$ and $\vec{B}$ fields are present, the total energy density u is

$$u = \tfrac{1}{2}\epsilon_0 E^2 + \frac{1}{2\mu_0}B^2 \qquad (32.23)$$

For electromagnetic waves in vacuum, the magnitudes E and B are related by

$$B = \frac{E}{c} = \sqrt{\epsilon_0\mu_0}\,E \qquad (32.24)$$

Combining Eqs. (32.23) and (32.24), we can also express the energy density u in a simple electromagnetic wave in vacuum as

$$u = \tfrac{1}{2}\epsilon_0 E^2 + \frac{1}{2\mu_0}(\sqrt{\epsilon_0\mu_0}\,E)^2 = \epsilon_0 E^2 \qquad (32.25)$$

This shows that in vacuum, the energy density associated with the $\vec{E}$ field in our simple wave is equal to the energy density of the $\vec{B}$ field. In general, the electric-field magnitude E is a function of position and time, as for the sinusoidal wave described by Eqs. (32.16); thus the energy density u of an electromagnetic wave, given by Eq. (32.25), also depends in general on position and time.

Electromagnetic Energy Flow and the Poynting Vector

Electromagnetic waves such as those we have described are *traveling* waves that transport energy from one region to another. We can describe this energy transfer in terms of energy transferred *per unit time per unit cross-sectional area,* or *power per unit area,* for an area perpendicular to the direction of wave travel.

To see how the energy flow is related to the fields, consider a stationary plane, perpendicular to the x-axis, that coincides with the wave front at a certain time. In a time dt after this, the wave front moves a distance $dx = c\,dt$ to the right of the plane. Consider an area A on this stationary plane (**Fig. 32.17**). The energy in the space to the right of this area had to pass through the area to reach the new location. The volume dV of the relevant region is the base area A times the length $c\,dt$, and the energy dU in this region is the energy density u times this volume:

$$dU = u\,dV = (\epsilon_0 E^2)(Ac\,dt)$$

This energy passes through the area A in time dt. The energy flow per unit time per unit area, which we will call S, is

$$S = \frac{1}{A}\frac{dU}{dt} = \epsilon_0 c E^2 \qquad \text{(in vacuum)} \qquad (32.26)$$

Using Eqs. (32.4) and (32.9), you can derive the alternative forms

$$S = \frac{\epsilon_0}{\sqrt{\epsilon_0\mu_0}}E^2 = \sqrt{\frac{\epsilon_0}{\mu_0}}E^2 = \frac{EB}{\mu_0} \qquad \text{(in vacuum)} \qquad (32.27)$$

The units of S are energy per unit time per unit area, or power per unit area. The SI unit of S is $1\ \text{J/s}\cdot\text{m}^2$ or $1\ \text{W/m}^2$.

We can define a *vector* quantity that describes both the magnitude and direction of the energy flow rate. Introduced by the British physicist John Poynting (1852–1914), this quantity is called the **Poynting vector:**

Poynting vector in vacuum $\qquad \vec{S} = \dfrac{1}{\mu_0}\vec{E}\times\vec{B}$

- Electric field
- Magnetic field
- Magnetic constant

$$(32.28)$$

The vector $\vec{S}$ points in the direction of propagation of the wave (**Fig. 32.18**). Since $\vec{E}$ and $\vec{B}$ are perpendicular, the magnitude of $\vec{S}$ is $S = EB/\mu_0$; from Eqs. (32.26) and (32.27) this is the energy flow per unit area and per unit time through a cross-sectional area perpendicular to the propagation direction. The total energy flow per unit time (power, P) out of any closed surface is the integral of $\vec{S}$ over the surface:

$$P = \oint \vec{S}\cdot d\vec{A}$$

For the sinusoidal waves studied in Section 32.3, as well as for other more complex waves, the electric and magnetic fields at any point vary with time, so the Poynting vector at any point is also a function of time. Because the frequencies of typical electromagnetic waves are very high, the time variation of the Poynting vector is so rapid that it's most appropriate to look at its *average* value. The magnitude of the average value of $\vec{S}$ at a point is called the **intensity** of the radiation at that point. The SI unit of intensity is the same as for S, $1\ \text{W/m}^2$.

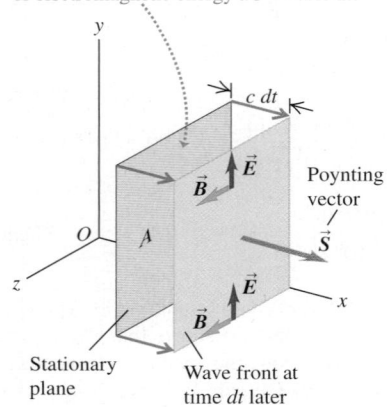

32.17 A wave front at a time dt after it passes through the stationary plane with area A.

At time dt, the volume between the stationary plane and the wave front contains an amount of electromagnetic energy $dU = uAc\,dt$.

32.18 These rooftop solar panels are tilted to be face-on to the sun—that is, face-on to the Poynting vector of electromagnetic waves from the sun, so that the panels can absorb the maximum amount of wave energy.

DEMO

Let's work out the intensity of the sinusoidal wave described by Eqs. (32.17). We first substitute $\vec{E}$ and $\vec{B}$ into Eq. (32.28):

$$\vec{S}(x, t) = \frac{1}{\mu_0} \vec{E}(x, t) \times \vec{B}(x, t)$$

$$= \frac{1}{\mu_0} [\hat{j} E_{max} \cos(kx - \omega t)] \times [\hat{k} B_{max} \cos(kx - \omega t)]$$

The vector product of the unit vectors is $\hat{j} \times \hat{k} = \hat{i}$ and $\cos^2(kx - \omega t)$ is never negative, so $\vec{S}(x, t)$ always points in the positive x-direction (the direction of wave propagation). The x-component of the Poynting vector is

$$S_x(x, t) = \frac{E_{max} B_{max}}{\mu_0} \cos^2(kx - \omega t) = \frac{E_{max} B_{max}}{2\mu_0} [1 + \cos 2(kx - \omega t)]$$

The time average value of $\cos 2(kx - \omega t)$ is zero because at any point, it is positive during one half-cycle and negative during the other half. So the average value of the Poynting vector over a full cycle is $\vec{S}_{av} = \hat{i} S_{av}$, where

$$S_{av} = \frac{E_{max} B_{max}}{2\mu_0}$$

That is, the magnitude of the average value of $\vec{S}$ for a sinusoidal wave (the intensity I of the wave) is $\frac{1}{2}$ the maximum value. You can verify that by using the relationships $E_{max} = B_{max} c$ and $\epsilon_0 \mu_0 = 1/c^2$, we can express the intensity in several equivalent forms:

Intensity of a sinusoidal electromagnetic wave in vacuum

Electric-field amplitude · · · · Magnetic-field amplitude · · · · Electric constant

$$I = S_{av} = \frac{E_{max} B_{max}}{2\mu_0} = \frac{E_{max}^2}{2\mu_0 c} = \frac{1}{2} \sqrt{\frac{\epsilon_0}{\mu_0}} E_{max}^2 = \frac{1}{2} \epsilon_0 c E_{max}^2 \qquad (32.29)$$

Magnitude of average Poynting vector · · · Magnetic constant · · · Speed of light in vacuum

For a wave traveling in the $-x$-direction, represented by Eqs. (32.19), the Poynting vector is in the $-x$-direction at every point, but its magnitude is the same as for a wave traveling in the $+x$-direction. Verifying these statements is left to you.

Throughout this discussion we have considered only electromagnetic waves propagating in vacuum. If the waves are traveling in a dielectric medium, however, the expressions for energy density [Eq. (32.23)], the Poynting vector [Eq. (32.28)], and the intensity of a sinusoidal wave [Eq. (32.29)] must be modified. It turns out that the required modifications are quite simple: Just replace ϵ_0 with the permittivity ϵ of the dielectric, replace μ_0 with the permeability μ of the dielectric, and replace c with the speed v of electromagnetic waves in the dielectric. Remarkably, the energy densities in the $\vec{E}$ and $\vec{B}$ fields are equal even in a dielectric.

BIO Application Laser Surgery Lasers are used widely in medicine as ultra-precise, bloodless "scalpels." They can reach and remove tumors with minimal damage to neighboring healthy tissues, as in the brain surgery shown here. The power output of the laser is typically below 40 W, less than that of a typical light bulb. However, this power is concentrated into a spot from 0.1 to 2.0 mm in diameter, so the intensity of the light (equal to the average value of the Poynting vector) can be as high as 5×10^9 W/m².

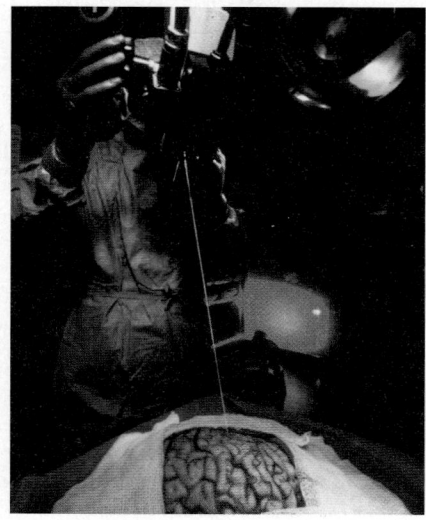

CAUTION Poynting vector vs. intensity At any point x, the magnitude of the Poynting vector varies with time. Hence, the *instantaneous* rate at which electromagnetic energy in a sinusoidal plane wave arrives at a surface is not constant. This may seem to contradict everyday experience; the light from the sun, a light bulb, or the laser in a grocery-store scanner appears steady and unvarying in strength. In fact the Poynting vector from these sources *does* vary in time, but the variation isn't noticeable because the oscillation frequency is so high (around 5×10^{14} Hz for visible light). All that you sense is the *average* rate at which energy reaches your eye, which is why we commonly use intensity (the average value of S) to describe the strength of electromagnetic radiation. ▌

EXAMPLE 32.3 ENERGY IN A NONSINUSOIDAL WAVE

For the nonsinusoidal wave described in Section 32.2, suppose that $E = 100$ V/m $= 100$ N/C. Find the value of B, the energy density u, and the rate of energy flow per unit area S.

SOLUTION

IDENTIFY and SET UP: In this wave $\vec{E}$ and $\vec{B}$ are uniform behind the wave front (and zero ahead of it). Hence the target variables B, u, and S must also be uniform behind the wave front. Given the magnitude E, we use Eq. (32.4) to find B, Eq. (32.25) to find u,

and Eq. (32.27) to find S. (We cannot use Eq. (32.29), which applies to sinusoidal waves only.)

EXECUTE: From Eq. (32.4),

$$B = \frac{E}{c} = \frac{100 \text{ V/m}}{3.00 \times 10^8 \text{ m/s}} = 3.33 \times 10^{-7} \text{ T}$$

From Eq. (32.25),

$$u = \epsilon_0 E^2 = (8.85 \times 10^{-12} \text{ C}^2/\text{N} \cdot \text{m}^2)(100 \text{ N/C})^2$$
$$= 8.85 \times 10^{-8} \text{ N/m}^2 = 8.85 \times 10^{-8} \text{ J/m}^3$$

(Continued)

The magnitude of the Poynting vector is

$$S = \frac{EB}{\mu_0} = \frac{(100 \text{ V/m})(3.33 \times 10^{-7} \text{ T})}{4\pi \times 10^{-7} \text{ T} \cdot \text{m/A}}$$

$$= 26.5 \text{ V} \cdot \text{A/m}^2 = 26.5 \text{ W/m}^2$$

EVALUATE: We can check our result for S by using Eq. (32.26):

$$S = \epsilon_0 c E^2 = (8.85 \times 10^{-12} \text{ C}^2/\text{N} \cdot \text{m}^2)(3.00 \times 10^8 \text{ m/s})$$

$$\times (100 \text{ N/C})^2 = 26.5 \text{ W/m}^2$$

Since $\vec{E}$ and $\vec{B}$ have the same values at all points behind the wave front, u and S likewise have the same value everywhere behind the wave front. In front of the wave front, $\vec{E} = 0$ and $\vec{B} = 0$, and so $u = 0$ and $S = 0$; where there are no fields, there is no field energy.

EXAMPLE 32.4 ENERGY IN A SINUSOIDAL WAVE

A radio station on the earth's surface emits a sinusoidal wave with average total power 50 kW (**Fig. 32.19**). Assuming that the transmitter radiates equally in all directions above the ground (which is unlikely in real situations), find the electric-field and magnetic-field amplitudes E_{max} and B_{max} detected by a satellite 100 km from the antenna.

SOLUTION

IDENTIFY and SET UP: We are given the transmitter's average total power P. The intensity I is the average power per unit area; to find I at 100 km from the transmitter we divide P by the surface area of the hemisphere in Fig. 32.19. For a sinusoidal wave, I is also equal to the magnitude of the average value S_{av} of the Poynting vector, so we can use Eq. (32.29) to find E_{max}; Eq. (32.4) yields B_{max}.

32.19 A radio station radiates waves into the hemisphere shown.

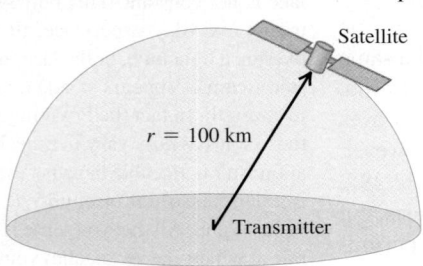

EXECUTE: The surface area of a hemisphere of radius $r = 100 \text{ km} = 1.00 \times 10^5 \text{ m}$ is

$$A = 2\pi R^2 = 2\pi(1.00 \times 10^5 \text{ m})^2 = 6.28 \times 10^{10} \text{ m}^2$$

All the radiated power passes through this surface, so the average power per unit area (that is, the intensity) is

$$I = \frac{P}{A} = \frac{P}{2\pi R^2} = \frac{5.00 \times 10^4 \text{ W}}{6.28 \times 10^{10} \text{ m}^2} = 7.96 \times 10^{-7} \text{ W/m}^2$$

From Eq. (32.29), $I = S_{av} = E_{max}^2/2\mu_0 c$, so

$$E_{max} = \sqrt{2\mu_0 c S_{av}}$$

$$= \sqrt{2(4\pi \times 10^{-7} \text{ T} \cdot \text{m/A})(3.00 \times 10^8 \text{ m/s})(7.96 \times 10^{-7} \text{ W/m}^2)}$$

$$= 2.45 \times 10^{-2} \text{ V/m}$$

Then from Eq. (32.4),

$$B_{max} = \frac{E_{max}}{c} = 8.17 \times 10^{-11} \text{ T}$$

EVALUATE: Note that E_{max} is comparable to fields commonly seen in the laboratory, but B_{max} is extremely small in comparison to $\vec{B}$ fields we saw in previous chapters. For this reason, most detectors of electromagnetic radiation respond to the effect of the electric field, not the magnetic field. Loop radio antennas are an exception (see the Bridging Problem at the end of this chapter).

Electromagnetic Momentum Flow and Radiation Pressure

We've shown that electromagnetic waves transport energy. It can also be shown that electromagnetic waves carry *momentum p*, with a corresponding momentum density (momentum dp per volume dV) of magnitude

$$\frac{dp}{dV} = \frac{EB}{\mu_0 c^2} = \frac{S}{c^2} \tag{32.30}$$

This momentum is a property of the field; it is not associated with the mass of a moving particle in the usual sense.

There is also a corresponding momentum flow rate. The volume dV occupied by an electromagnetic wave (speed c) that passes through an area A in time dt is

$dV = Ac\,dt$. When we substitute this into Eq. (32.30) and rearrange, we find that the momentum flow rate per unit area is

Poynting vector magnitude

Electric-field magnitude

Flow rate of electromagnetic momentum

$$\frac{1}{A}\frac{dp}{dt} = \frac{S}{c} = \frac{EB}{\mu_0 c} \qquad (32.31)$$

Magnetic-field magnitude

Speed of light in vacuum

Momentum transferred per unit surface area per unit time

Magnetic constant

We obtain the *average* rate of momentum transfer per unit area by replacing S in Eq. (32.31) by $S_{av} = I$.

This momentum is responsible for **radiation pressure.** When an electromagnetic wave is completely absorbed by a surface, the wave's momentum is also transferred to the surface. For simplicity we'll consider a surface perpendicular to the propagation direction. Using the ideas developed in Section 8.1, we see that the rate dp/dt at which momentum is transferred to the absorbing surface equals the *force* on the surface. The average force per unit area due to the wave, or *radiation pressure* p_{rad}, is the average value of dp/dt divided by the absorbing area A. (We use the subscript "rad" to distinguish pressure from momentum, for which the symbol p is also used.) From Eq. (32.31) the radiation pressure is

$$p_{rad} = \frac{S_{av}}{c} = \frac{I}{c} \qquad \text{(radiation pressure, wave totally absorbed)} \qquad (32.32)$$

If the wave is totally reflected, the momentum change is twice as great, and

$$p_{rad} = \frac{2S_{av}}{c} = \frac{2I}{c} \qquad \text{(radiation pressure, wave totally reflected)} \qquad (32.33)$$

For example, the value of I (or S_{av}) for direct sunlight, before it passes through the earth's atmosphere, is approximately $1.4\ \text{kW/m}^2$. From Eq. (32.32) the corresponding average pressure on a completely absorbing surface is

$$p_{rad} = \frac{I}{c} = \frac{1.4 \times 10^3\ \text{W/m}^2}{3.0 \times 10^8\ \text{m/s}} = 4.7 \times 10^{-6}\ \text{Pa}$$

From Eq. (32.33) the average pressure on a totally *reflecting* surface is twice this: $2I/c$ or $9.4 \times 10^{-6}\ \text{Pa}$. These are very small pressures, of the order of 10^{-10} atm, but they can be measured with sensitive instruments.

The radiation pressure of sunlight is much greater *inside* the sun than at the earth (see Problem 32.37). Inside stars that are much more massive and luminous than the sun, radiation pressure is so great that it substantially augments the gas pressure within the star and so helps to prevent the star from collapsing under its own gravity. In some cases the radiation pressure of stars can have dramatic effects on the material surrounding them (**Fig. 32.20**).

32.20 At the center of this interstellar gas cloud is a group of intensely luminous stars that exert tremendous radiation pressure on their surroundings. Aided by a "wind" of particles emanating from the stars, over the past million years the radiation pressure has carved out a bubble within the cloud 70 light-years across.

EXAMPLE 32.5   POWER AND PRESSURE FROM SUNLIGHT

An earth-orbiting satellite has solar energy–collecting panels with a total area of $4.0\ \text{m}^2$ (**Fig. 32.21**). If the sun's radiation is perpendicular to the panels and is completely absorbed, find the average solar power absorbed and the average radiation-pressure force.

32.21 Solar panels on a satellite.

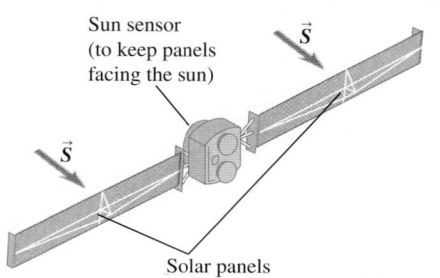

SOLUTION

IDENTIFY and SET UP: This problem uses the relationships among intensity, power, radiation pressure, and force. In the previous discussion, we used the intensity I (average power per unit area) of sunlight to find the radiation pressure p_{rad} (force per unit area) of sunlight on a completely absorbing surface. (These values are for

(Continued)

points above the atmosphere, which is where the satellite orbits.) Multiplying each value by the area of the solar panels gives the average power absorbed and the net radiation force on the panels.

EXECUTE: The intensity I (power per unit area) is 1.4×10^3 W/m^2. Although the light from the sun is not a simple sinusoidal wave, we can still use the relationship that the average power P is the intensity I times the area A:

$$P = IA = (1.4 \times 10^3 \text{ W/m}^2)(4.0 \text{ m}^2)$$

$$= 5.6 \times 10^3 \text{ W} = 5.6 \text{ kW}$$

The radiation pressure of sunlight on an absorbing surface is $p_{rad} = 4.7 \times 10^{-6}$ Pa $= 4.7 \times 10^{-6}$ N/m^2. The total force F is the pressure p_{rad} times the area A:

$$F = p_{rad}A = (4.7 \times 10^{-6} \text{ N/m}^2)(4.0 \text{ m}^2) = 1.9 \times 10^{-5} \text{ N}$$

EVALUATE: The absorbed power is quite substantial. Part of it can be used to power the equipment aboard the satellite; the rest goes into heating the panels, either directly or due to inefficiencies in the photocells contained in the panels.

The total radiation force is comparable to the weight (on the earth) of a single grain of salt. Over time, however, this small force can noticeably affect the orbit of a satellite like that in Fig. 32.21, and so radiation pressure must be taken into account.

TEST YOUR UNDERSTANDING OF SECTION 32.4 Figure 32.13 shows one wavelength of a sinusoidal electromagnetic wave at time $t = 0$. For which of the following four values of x is (a) the energy density a maximum; (b) the energy density a minimum; (c) the magnitude of the instantaneous (not average) Poynting vector a maximum; (d) the magnitude of the instantaneous (not average) Poynting vector a minimum? (i) $x = 0$; (ii) $x = \lambda/4$; (iii) $x = \lambda/2$; (iv) $x = 3\lambda/4$. ∎

32.5 STANDING ELECTROMAGNETIC WAVES

Electromagnetic waves can be *reflected* by the surface of a conductor (like a polished sheet of metal) or of a dielectric (such as a sheet of glass). The superposition of an incident wave and a reflected wave forms a **standing wave.** The situation is analogous to standing waves on a stretched string, discussed in Section 15.7.

Suppose a sheet of a perfect conductor (zero resistivity) is placed in the yz-plane of **Fig. 32.22** and a linearly polarized electromagnetic wave, traveling in the negative x-direction, strikes it. As we discussed in Section 23.4, $\vec{E}$ cannot have a component parallel to the surface of a perfect conductor. Therefore in the present situation, $\vec{E}$ must be zero everywhere in the yz-plane. The electric field of the *incident* electromagnetic wave is *not* zero at all times in the yz-plane. But this incident wave induces oscillating currents on the surface of the conductor, and these currents give rise to an additional electric field. The *net* electric field, which is the vector sum of this field and the incident $\vec{E}$, *is* zero everywhere inside and on the surface of the conductor.

The currents induced on the surface of the conductor also produce a *reflected* wave that travels out from the plane in the $+x$-direction. Suppose the incident wave is described by the wave functions of Eqs. (32.19) (a sinusoidal wave traveling in the $-x$-direction) and the reflected wave by the negative of Eqs. (32.16) (a sinusoidal wave traveling in the $+x$-direction). We take the *negative* of the wave given by Eqs. (32.16) so that the incident and reflected electric fields cancel at $x = 0$ (the plane of the conductor, where the total electric field must be zero). The superposition principle states that the total $\vec{E}$ field at any point is the vector sum of the $\vec{E}$ fields of the incident and reflected waves, and similarly for the $\vec{B}$ field. Therefore the wave functions for the superposition of the two waves are

$$E_y(x, t) = E_{max}[\cos(kx + \omega t) - \cos(kx - \omega t)]$$

$$B_z(x, t) = B_{max}[-\cos(kx + \omega t) - \cos(kx - \omega t)]$$

We can expand and simplify these expressions by using the identities

$$\cos(A \pm B) = \cos A \cos B \mp \sin A \sin B$$

32.22 Representation of the electric and magnetic fields of a linearly polarized electromagnetic standing wave when $\omega t = 3\pi/4$ rad. In any plane perpendicular to the x-axis, E is maximum (an antinode) where B is zero (a node), and vice versa. As time elapses, the pattern does *not* move along the x-axis; instead, at every point the $\vec{E}$ and $\vec{B}$ vectors simply oscillate.

Perfect conductor

$x = \lambda$: nodal plane of $\vec{E}$ antinodal plane of $\vec{B}$

$x = 3\lambda/4$: antinodal plane of $\vec{E}$ nodal plane of $\vec{B}$

The results are

$$E_y(x, t) = -2E_{max} \sin kx \sin \omega t \qquad (32.34)$$

$$B_z(x, t) = -2B_{max} \cos kx \cos \omega t \qquad (32.35)$$

Equation (32.34) is analogous to Eq. (15.28) for a stretched string. We see that at $x = 0$ the electric field $E_y(x = 0, t)$ is *always* zero; this is required by the nature of the ideal conductor, which plays the same role as a fixed point at the end of a string. Furthermore, $E_y(x, t)$ is zero at *all* times at points in those planes perpendicular to the x-axis for which $\sin kx = 0$—that is, $kx = 0, \pi, 2\pi, \ldots$. Since $k = 2\pi/\lambda$, the positions of these planes are

$$x = 0, \frac{\lambda}{2}, \lambda, \frac{3\lambda}{2}, \ldots \qquad \text{(nodal planes of } \vec{E}) \qquad (32.36)$$

These planes are called the **nodal planes** of the $\vec{E}$ field; they are the equivalent of the nodes, or nodal points, of a standing wave on a string. Midway between any two adjacent nodal planes is a plane on which $\sin kx = \pm 1$; on each such plane, the magnitude of $E(x, t)$ equals the maximum possible value of $2E_{max}$ twice per oscillation cycle. These are the **antinodal planes** of $\vec{E}$, corresponding to the antinodes of waves on a string.

The total magnetic field is zero at all times at points in planes on which $\cos kx = 0$. These are the nodal planes of $\vec{B}$, and they occur where

$$x = \frac{\lambda}{4}, \frac{3\lambda}{4}, \frac{5\lambda}{4}, \ldots \qquad \text{(nodal planes of } \vec{B}) \qquad (32.37)$$

There is an antinodal plane of $\vec{B}$ midway between any two adjacent nodal planes.

Figure 32.22 shows a standing-wave pattern at one instant of time. The magnetic field is *not* zero at the conducting surface ($x = 0$). The surface currents that must be present to make $\vec{E}$ exactly zero at the surface cause magnetic fields at the surface. The nodal planes of each field are separated by one half-wavelength. The nodal planes of $\vec{E}$ are midway between those of $\vec{B}$, and vice versa; hence the nodes of $\vec{E}$ coincide with the antinodes of $\vec{B}$, and conversely. Compare this situation to the distinction between pressure nodes and displacement nodes in Section 16.4.

The total electric field is a *sine* function of t, and the total magnetic field is a *cosine* function of t. The sinusoidal variations of the two fields are therefore 90° out of phase at each point. At times when $\sin \omega t = 0$, the electric field is zero *everywhere,* and the magnetic field is maximum. When $\cos \omega t = 0$, the magnetic field is zero everywhere, and the electric field is maximum. This is in contrast to a wave traveling in one direction, as described by Eqs. (32.16) or (32.19) separately, in which the sinusoidal variations of $\vec{E}$ and $\vec{B}$ at any particular point are *in phase*. You can show that Eqs. (32.34) and (32.35) satisfy the wave equation, Eq. (32.15). You can also show that they satisfy Eqs. (32.12) and (32.14), the equivalents of Faraday's and Ampere's laws.

Standing Waves in a Cavity

Let's now insert a second conducting plane, parallel to the first and a distance L from it, along the $+x$-axis. The cavity between the two planes is analogous to a stretched string held at the points $x = 0$ and $x = L$. Both conducting planes must be nodal planes for $\vec{E}$; a standing wave can exist only when the second plane is placed at one of the positions where $E(x, t) = 0$, so L must be an integer multiple of $\lambda/2$. The wavelengths that satisfy this condition are

$$\lambda_n = \frac{2L}{n} \qquad (n = 1, 2, 3, \ldots) \qquad (32.38)$$

32.23 A typical microwave oven sets up a standing electromagnetic wave with $\lambda = 12.2$ cm, a wavelength that is strongly absorbed by the water in food. Because the wave has nodes spaced $\lambda/2 = 6.1$ cm apart, the food must be rotated while cooking. Otherwise, the portion that lies at a node—where the electric-field amplitude is zero—will remain cold.

The corresponding frequencies are

$$f_n = \frac{c}{\lambda_n} = n\frac{c}{2L} \qquad (n = 1, 2, 3, \dots) \qquad (32.39)$$

Thus there is a set of *normal modes,* each with a characteristic frequency, wave shape, and node pattern (**Fig. 32.23**). By measuring the node positions, we can measure the wavelength. If the frequency is known, the wave speed can be determined. This technique was first used by Hertz in the 1880s in his pioneering investigations of electromagnetic waves.

Conducting surfaces are not the only reflectors of electromagnetic waves. Reflections also occur at an interface between two insulating materials with different dielectric or magnetic properties. The mechanical analog is a junction of two strings with equal tension but different linear mass density. In general, a wave incident on such a boundary surface is partly transmitted into the second material and partly reflected back into the first. For example, light is transmitted through a glass window, but its surfaces also reflect light.

EXAMPLE 32.6   INTENSITY IN A STANDING WAVE

Calculate the intensity of the standing wave represented by Eqs. (32.34) and (32.35).

SOLUTION

IDENTIFY and SET UP: The intensity I of the wave is the time-averaged value S_{av} of the magnitude of the Poynting vector $\vec{S}$. To find S_{av}, we first use Eq. (32.28) to find the instantaneous value of $\vec{S}$ and then average it over a whole number of cycles of the wave.

EXECUTE: Using the wave functions of Eqs. (32.34) and (32.35) in Eq. (32.28) for the Poynting vector $\vec{S}$, we find

$$\vec{S}(x, t) = \frac{1}{\mu_0}\vec{E}(x, t) \times \vec{B}(x, t)$$

$$= \frac{1}{\mu_0}\left[-2\hat{\jmath}E_{max}\sin kx \sin \omega t\right] \times \left[-2\hat{k}B_{max}\cos kx \cos \omega t\right]$$

$$= \hat{\imath}\frac{E_{max}B_{max}}{\mu_0}(2\sin kx \cos kx)(2\sin \omega t \cos \omega t) = \hat{\imath}S_x(x, t)$$

Using the identity $\sin 2A = 2\sin A \cos A$, we can rewrite $S_x(x, t)$ as

$$S_x(x, t) = \frac{E_{max}B_{max}\sin 2kx \sin 2\omega t}{\mu_0}$$

The average value of a sine function over any whole number of cycles is zero. Thus *the time average of $\vec{S}$ at any point is zero;* $I = S_{av} = 0$.

EVALUATE: This result is what we should expect. The standing wave is a superposition of two waves with the same frequency and amplitude, traveling in opposite directions. All the energy transferred by one wave is cancelled by an equal amount transferred in the opposite direction by the other wave. When we use electromagnetic waves to transmit power, it is important to avoid reflections that give rise to standing waves.

EXAMPLE 32.7   STANDING WAVES IN A CAVITY

Electromagnetic standing waves are set up in a cavity with two parallel, highly conducting walls 1.50 cm apart. (a) Calculate the longest wavelength λ and lowest frequency f of these standing waves. (b) For a standing wave of this wavelength, where in the cavity does $\vec{E}$ have maximum magnitude? Where is $\vec{E}$ zero? Where does $\vec{B}$ have maximum magnitude? Where is $\vec{B}$ zero?

SOLUTION

IDENTIFY and SET UP: Only certain normal modes are possible for electromagnetic waves in a cavity, just as only certain normal modes are possible for standing waves on a string. The longest possible wavelength and lowest possible frequency correspond to the $n = 1$ mode in Eqs. (32.38) and (32.39); we use these to find λ and f. Equations (32.36) and (32.37) then give the locations of the nodal planes of $\vec{E}$ and $\vec{B}$. The antinodal planes of each field are midway between adjacent nodal planes.

EXECUTE: (a) From Eqs. (32.38) and (32.39), the $n = 1$ wavelength and frequency are

$$\lambda_1 = 2L = 2(1.50 \text{ cm}) = 3.00 \text{ cm}$$

$$f_1 = \frac{c}{2L} = \frac{3.00 \times 10^8 \text{ m/s}}{2(1.50 \times 10^{-2} \text{ m})} = 1.00 \times 10^{10} \text{ Hz} = 10 \text{ GHz}$$

(b) With $n = 1$ there is a single half-wavelength between the walls. The electric field has nodal planes ($\vec{E} = 0$) at the walls and an antinodal plane (where $\vec{E}$ has its maximum magnitude) midway between them. The magnetic field has *antinodal* planes at the walls and a nodal plane midway between them.

EVALUATE: One application of such standing waves is to produce an oscillating $\vec{E}$ field of definite frequency, which is used to probe the behavior of a small sample of material placed in the cavity. To subject the sample to the strongest possible field, it should be placed near the center of the cavity, at the antinode of $\vec{E}$.

TEST YOUR UNDERSTANDING OF SECTION 32.5   In the standing wave described in Example 32.7, is there any point in the cavity where the energy density is zero at all times? If so, where? If not, why not? ▌

CHAPTER 32 SUMMARY

SOLUTIONS TO ALL EXAMPLES

Maxwell's equations and electromagnetic waves: Maxwell's equations predict the existence of electromagnetic waves that propagate in vacuum at the speed of light, c. The electromagnetic spectrum covers frequencies from at least 1 to 10^{24} Hz and a correspondingly broad range of wavelengths. Visible light, with wavelengths from 380 to 750 nm, is a very small part of this spectrum. In a plane wave, $\vec{E}$ and $\vec{B}$ are uniform over any plane perpendicular to the propagation direction. Faraday's law and Ampere's law give relationships between the magnitudes of $\vec{E}$ and $\vec{B}$; requiring that both relationships are satisfied gives an expression for c in terms of ϵ_0 and μ_0. Electromagnetic waves are transverse; the $\vec{E}$ and $\vec{B}$ fields are perpendicular to the direction of propagation and to each other. The direction of propagation is the direction of $\vec{E} \times \vec{B}$.

$$E = cB \tag{32.4}$$
$$B = \epsilon_0 \mu_0 c E \tag{32.8}$$
$$c = \frac{1}{\sqrt{\epsilon_0 \mu_0}} \tag{32.9}$$

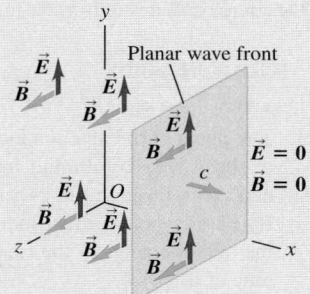

Sinusoidal electromagnetic waves: Equations (32.17) and (32.18) describe a sinusoidal plane electromagnetic wave traveling in vacuum in the $+x$-direction. If the wave is propagating in the $-x$-direction, replace $kx - \omega t$ by $kx + \omega t$. (See Example 32.1.)

$$\vec{E}(x, t) = \hat{j} E_{max} \cos(kx - \omega t)$$
$$\vec{B}(x, t) = \hat{k} B_{max} \cos(kx - \omega t) \tag{32.17}$$
$$E_{max} = c B_{max} \tag{32.18}$$

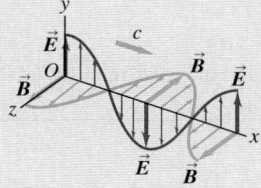

Electromagnetic waves in matter: When an electromagnetic wave travels through a dielectric, the wave speed v is less than the speed of light in vacuum c. (See Example 32.2.)

$$v = \frac{1}{\sqrt{\epsilon \mu}} = \frac{1}{\sqrt{KK_m}} \frac{1}{\sqrt{\epsilon_0 \mu_0}}$$
$$= \frac{c}{\sqrt{KK_m}} \tag{32.21}$$

Energy and momentum in electromagnetic waves: The energy flow rate (power per unit area) in an electromagnetic wave in vacuum is given by the Poynting vector $\vec{S}$. The magnitude of the time-averaged value of the Poynting vector is called the intensity I of the wave. Electromagnetic waves also carry momentum. When an electromagnetic wave strikes a surface, it exerts a radiation pressure p_{rad}. If the surface is perpendicular to the wave propagation direction and is totally absorbing, $p_{rad} = I/c$; if the surface is a perfect reflector, $p_{rad} = 2I/c$. (See Examples 32.3–32.5.)

$$\vec{S} = \frac{1}{\mu_0} \vec{E} \times \vec{B} \tag{32.28}$$

$$I = S_{av} = \frac{E_{max} B_{max}}{2\mu_0} = \frac{E_{max}^2}{2\mu_0 c}$$
$$= \frac{1}{2}\sqrt{\frac{\epsilon_0}{\mu_0}} E_{max}^2$$
$$= \frac{1}{2}\epsilon_0 c E_{max}^2 \tag{32.29}$$

$$\frac{1}{A}\frac{dp}{dt} = \frac{S}{c} = \frac{EB}{\mu_0 c} \tag{32.31}$$
(flow rate of electromagnetic momentum)

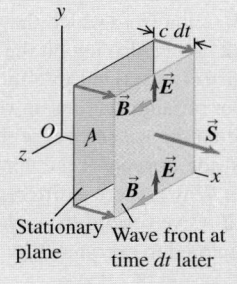

Standing electromagnetic waves: If a perfect reflecting surface is placed at $x = 0$, the incident and reflected waves form a standing wave. Nodal planes for $\vec{E}$ occur at $kx = 0, \pi, 2\pi, \ldots$, and nodal planes for $\vec{B}$ at $kx = \pi/2, 3\pi/2, 5\pi/2, \ldots$. At each point, the sinusoidal variations of $\vec{E}$ and $\vec{B}$ with time are 90° out of phase. (See Examples 32.6 and 32.7.)

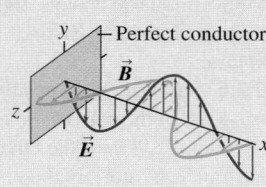

BRIDGING PROBLEM DETECTING ELECTROMAGNETIC WAVES

A circular loop of wire can be used as a radio antenna. If an 18.0-cm-diameter antenna is located 2.50 km from a 95.0-MHz source with a total power of 55.0 kW, what is the maximum emf induced in the loop? The orientation of the antenna loop and the polarization of the wave are as shown in **Fig. 32.24**. Assume that the source radiates uniformly in all directions.

SOLUTION GUIDE

IDENTIFY and SET UP

1. The plane of the antenna loop is perpendicular to the direction of the wave's oscillating magnetic field. This causes a magnetic flux through the loop that varies sinusoidally with time. By Faraday's law, this produces an emf equal in magnitude to the rate of change of the flux. The target variable is the magnitude of this emf.

2. Select the equations that you will need to find (i) the intensity of the wave at the position of the loop, a distance $r = 2.50$ km from the source of power $P = 55.0$ kW; (ii) the amplitude of the sinusoidally varying magnetic field at that position; (iii) the magnetic flux through the loop as a function of time; and (iv) the emf produced by the flux.

EXECUTE

3. Find the wave intensity at the position of the loop.

32.24 Using a circular loop antenna to detect radio waves.

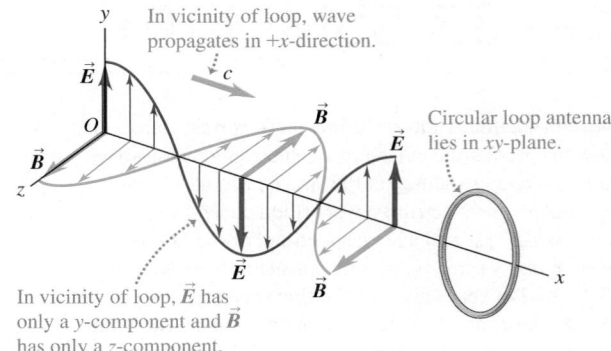

In vicinity of loop, wave propagates in +x-direction.

Circular loop antenna lies in xy-plane.

In vicinity of loop, $\vec{E}$ has only a y-component and $\vec{B}$ has only a z-component.

4. Use your result from step 3 to write expressions for the time-dependent magnetic field at this position and the time-dependent magnetic flux through the loop.

5. Use the results of step 4 to find the time-dependent induced emf in the loop. The amplitude of this emf is your target variable.

EVALUATE

6. Is the induced emf large enough to detect? (If it is, a receiver connected to this antenna will pick up signals from the source.)

Problems

For assigned homework and other learning materials, go to MasteringPhysics®. **MP**

•, ••, •••: Difficulty levels. **CP**: Cumulative problems incorporating material from earlier chapters. **CALC**: Problems requiring calculus. **DATA**: Problems involving real data, scientific evidence, experimental design, and/or statistical reasoning. **BIO**: Biosciences problems.

DISCUSSION QUESTIONS

Q32.1 By measuring the electric and magnetic fields at a point in space where there is an electromagnetic wave, can you determine the direction from which the wave came? Explain.

Q32.2 When driving on the upper level of the Bay Bridge, westbound from Oakland to San Francisco, you can easily pick up a number of radio stations on your car radio. But when driving eastbound on the lower level of the bridge, which has steel girders on either side to support the upper level, the radio reception is much worse. Why is there a difference?

Q32.3 Give several examples of electromagnetic waves that are encountered in everyday life. How are they all alike? How do they differ?

Q32.4 Sometimes neon signs located near a powerful radio station are seen to glow faintly at night, even though they are not turned on. What is happening?

Q32.5 Is polarization a property of all electromagnetic waves, or is it unique to visible light? Can sound waves be polarized? What fundamental distinction in wave properties is involved? Explain.

Q32.6 Suppose that a positive point charge q is initially at rest on the x-axis, in the path of the electromagnetic plane wave described in Section 32.2. Will the charge move after the wave front reaches it? If not, why not? If the charge does move, describe its motion

qualitatively. (Remember that $\vec{E}$ and $\vec{B}$ have the same value at all points behind the wave front.)

Q32.7 The light beam from a searchlight may have an electric-field magnitude of 1000 V/m, corresponding to a potential difference of 1500 V between the head and feet of a 1.5-m-tall person on whom the light shines. Does this cause the person to feel a strong electric shock? Why or why not?

Q32.8 For a certain sinusoidal wave of intensity I, the amplitude of the magnetic field is B. What would be the amplitude (in terms of B) in a similar wave of twice the intensity?

Q32.9 The magnetic-field amplitude of the electromagnetic wave from the laser described in Example 32.1 (Section 32.3) is about 100 times greater than the earth's magnetic field. If you illuminate a compass with the light from this laser, would you expect the compass to deflect? Why or why not?

Q32.10 Most automobiles have vertical antennas for receiving radio broadcasts. Explain what this tells you about the direction of polarization of $\vec{E}$ in the radio waves used in broadcasting.

Q32.11 If a light beam carries momentum, should a person holding a flashlight feel a recoil analogous to the recoil of a rifle when it is fired? Why is this recoil not actually observed?

Q32.12 A light source radiates a sinusoidal electromagnetic wave uniformly in all directions. This wave exerts an average pressure p on a perfectly reflecting surface a distance R away from it. What average pressure (in terms of p) would this wave exert on a perfectly absorbing surface that was twice as far from the source?

Q32.13 Does an electromagnetic *standing* wave have energy? Does it have momentum? Are your answers to these questions the same as for a *traveling* wave? Why or why not?

EXERCISES

Section 32.2 Plane Electromagnetic Waves and the Speed of Light

32.1 • (a) How much time does it take light to travel from the moon to the earth, a distance of 384,000 km? (b) Light from the star Sirius takes 8.61 years to reach the earth. What is the distance from earth to Sirius in kilometers?

32.2 • Consider each of the electric- and magnetic-field orientations given next. In each case, what is the direction of propagation of the wave? (a) $\vec{E}$ in the $+x$-direction, $\vec{B}$ in the $+y$-direction; (b) $\vec{E}$ in the $-y$-direction, $\vec{B}$ in the $+x$-direction; (c) $\vec{E}$ in the $+z$-direction, $\vec{B}$ in the $-x$-direction; (d) $\vec{E}$ in the $+y$-direction, $\vec{B}$ in the $-z$-direction.

32.3 • A sinusoidal electromagnetic wave is propagating in vacuum in the $+z$-direction. If at a particular instant and at a certain point in space the electric field is in the $+x$-direction and has magnitude 4.00 V/m, what are the magnitude and direction of the magnetic field of the wave at this same point in space and instant in time?

32.4 • Consider each of the following electric- and magnetic-field orientations. In each case, what is the direction of propagation of the wave? (a) $\vec{E} = E\hat{i}$, $\vec{B} = -B\hat{j}$; (b) $\vec{E} = E\hat{j}$, $\vec{B} = B\hat{i}$; (c) $\vec{E} = -E\hat{k}$, $\vec{B} = -B\hat{i}$; (d) $\vec{E} = E\hat{i}$, $\vec{B} = -B\hat{k}$.

Section 32.3 Sinusoidal Electromagnetic Waves

32.5 • BIO **Medical X rays.** Medical x rays are taken with electromagnetic waves having a wavelength of around 0.10 nm in air. What are the frequency, period, and wave number of such waves?

32.6 • BIO **Ultraviolet Radiation.** There are two categories of ultraviolet light. Ultraviolet A (UVA) has a wavelength ranging from 320 nm to 400 nm. It is necessary for the production of vitamin D. UVB, with a wavelength in vacuum between 280 nm and 320 nm, is more dangerous because it is much more likely to cause skin cancer. (a) Find the frequency ranges of UVA and UVB. (b) What are the ranges of the wave numbers for UVA and UVB?

32.7 • A sinusoidal electromagnetic wave having a magnetic field of amplitude 1.25 μT and a wavelength of 432 nm is traveling in the $+x$-direction through empty space. (a) What is the frequency of this wave? (b) What is the amplitude of the associated electric field? (c) Write the equations for the electric and magnetic fields as functions of x and t in the form of Eqs. (32.17).

32.8 • An electromagnetic wave of wavelength 435 nm is traveling in vacuum in the $-z$-direction. The electric field has amplitude 2.70×10^{-3} V/m and is parallel to the x-axis. What are (a) the frequency and (b) the magnetic-field amplitude? (c) Write the vector equations for $\vec{E}(z, t)$ and $\vec{B}(z, t)$.

32.9 • Consider electromagnetic waves propagating in air. (a) Determine the frequency of a wave with a wavelength of (i) 5.0 km, (ii) 5.0 μm, (iii) 5.0 nm. (b) What is the wavelength (in meters and nanometers) of (i) gamma rays of frequency 6.50×10^{21} Hz and (ii) an AM station radio wave of frequency 590 kHz?

32.10 • The electric field of a sinusoidal electromagnetic wave obeys the equation $E = (375 \text{ V/m}) \cos[(1.99 \times 10^7 \text{ rad/m})x + (5.97 \times 10^{15} \text{ rad/s})t]$. (a) What is the speed of the wave? (b) What are the amplitudes of the electric and magnetic fields of this wave? (c) What are the frequency, wavelength, and period of the wave? Is this light visible to humans?

32.11 • An electromagnetic wave has an electric field given by $\vec{E}(y, t) = (3.10 \times 10^5 \text{ V/m})\hat{k} \cos[ky - (12.65 \times 10^{12} \text{ rad/s})t]$. (a) In which direction is the wave traveling? (b) What is the wavelength of the wave? (c) Write the vector equation for $\vec{B}(y, t)$.

32.12 • An electromagnetic wave has a magnetic field given by $\vec{B}(x, t) = -(8.25 \times 10^{-9} \text{ T})\hat{j} \cos[(1.38 \times 10^4 \text{ rad/m})x + \omega t]$. (a) In which direction is the wave traveling? (b) What is the frequency f of the wave? (c) Write the vector equation for $\vec{E}(x, t)$.

32.13 • Radio station WCCO in Minneapolis broadcasts at a frequency of 830 kHz. At a point some distance from the transmitter, the magnetic-field amplitude of the electromagnetic wave from WCCO is 4.82×10^{-11} T. Calculate (a) the wavelength; (b) the wave number; (c) the angular frequency; (d) the electric-field amplitude.

32.14 • An electromagnetic wave with frequency 65.0 Hz travels in an insulating magnetic material that has dielectric constant 3.64 and relative permeability 5.18 at this frequency. The electric field has amplitude 7.20×10^{-3} V/m. (a) What is the speed of propagation of the wave? (b) What is the wavelength of the wave? (c) What is the amplitude of the magnetic field?

32.15 • An electromagnetic wave with frequency 5.70×10^{14} Hz propagates with a speed of 2.17×10^8 m/s in a certain piece of glass. Find (a) the wavelength of the wave in the glass; (b) the wavelength of a wave of the same frequency propagating in air; (c) the index of refraction n of the glass for an electromagnetic wave with this frequency; (d) the dielectric constant for glass at this frequency, assuming that the relative permeability is unity.

Section 32.4 Energy and Momentum in Electromagnetic Waves

32.16 • BIO **High-Energy Cancer Treatment.** Scientists are working on a new technique to kill cancer cells by zapping them with ultrahigh-energy (in the range of 10^{12} W) pulses of light that last for an extremely short time (a few nanoseconds). These short pulses scramble the interior of a cell without causing it to explode, as long pulses would do. We can model a typical such cell as a disk 5.0 μm in diameter, with the pulse lasting for 4.0 ns with an average power of 2.0×10^{12} W. We shall assume that the energy is spread uniformly over the faces of 100 cells for each pulse. (a) How much energy is given to the cell during this pulse? (b) What is the intensity (in W/m^2) delivered to the cell? (c) What are the maximum values of the electric and magnetic fields in the pulse?

32.17 •• **Fields from a Light Bulb.** We can reasonably model a 75-W incandescent light bulb as a sphere 6.0 cm in diameter. Typically, only about 5% of the energy goes to visible light; the rest goes largely to nonvisible infrared radiation. (a) What is the visible-light intensity (in W/m^2) at the surface of the bulb? (b) What are the amplitudes of the electric and magnetic fields at this surface, for a sinusoidal wave with this intensity?

32.18 •• A sinusoidal electromagnetic wave from a radio station passes perpendicularly through an open window that has area 0.500 m^2. At the window, the electric field of the wave has rms value 0.0400 V/m. How much energy does this wave carry through the window during a 30.0-s commercial?

32.19 • A space probe 2.0×10^{10} m from a star measures the total intensity of electromagnetic radiation from the star to be 5.0×10^3 W/m². If the star radiates uniformly in all directions, what is its total average power output?

32.20 •• The energy flow to the earth from sunlight is about 1.4 kW/m². (a) Find the maximum values of the electric and magnetic fields for a sinusoidal wave of this intensity. (b) The distance from the earth to the sun is about 1.5×10^{11} m. Find the total power radiated by the sun.

32.21 • The intensity of a cylindrical laser beam is 0.800 W/m². The cross-sectional area of the beam is 3.0×10^{-4} m² and the intensity is uniform across the cross section of the beam. (a) What is the average power output of the laser? (b) What is the rms value of the electric field in the beam?

32.22 • A sinusoidal electromagnetic wave emitted by a cellular phone has a wavelength of 35.4 cm and an electric-field amplitude of 5.40×10^{-2} V/m at a distance of 250 m from the phone. Calculate (a) the frequency of the wave; (b) the magnetic-field amplitude; (c) the intensity of the wave.

32.23 • A monochromatic light source with power output 60.0 W radiates light of wavelength 700 nm uniformly in all directions. Calculate E_{max} and B_{max} for the 700-nm light at a distance of 5.00 m from the source.

32.24 • **Television Broadcasting.** Public television station KQED in San Francisco broadcasts a sinusoidal radio signal at a power of 777 kW. Assume that the wave spreads out uniformly into a hemisphere above the ground. At a home 5.00 km away from the antenna, (a) what average pressure does this wave exert on a totally reflecting surface, (b) what are the amplitudes of the electric and magnetic fields of the wave, and (c) what is the average density of the energy this wave carries? (d) For the energy density in part (c), what percentage is due to the electric field and what percentage is due to the magnetic field?

32.25 •• An intense light source radiates uniformly in all directions. At a distance of 5.0 m from the source, the radiation pressure on a perfectly absorbing surface is 9.0×10^{-6} Pa. What is the total average power output of the source?

32.26 • In the 25-ft Space Simulator facility at NASA's Jet Propulsion Laboratory, a bank of overhead arc lamps can produce light of intensity 2500 W/m² at the floor of the facility. (This simulates the intensity of sunlight near the planet Venus.) Find the average radiation pressure (in pascals and in atmospheres) on (a) a totally absorbing section of the floor and (b) a totally reflecting section of the floor. (c) Find the average momentum density (momentum per unit volume) in the light at the floor.

32.27 •• **BIO Laser Safety.** If the eye receives an average intensity greater than 1.0×10^2 W/m², damage to the retina can occur. This quantity is called the *damage threshold* of the retina. (a) What is the largest average power (in mW) that a laser beam 1.5 mm in diameter can have and still be considered safe to view head-on? (b) What are the maximum values of the electric and magnetic fields for the beam in part (a)? (c) How much energy would the beam in part (a) deliver per second to the retina? (d) Express the damage threshold in W/cm².

32.28 •• A laser beam has diameter 1.20 mm. What is the amplitude of the electric field of the electromagnetic radiation in this beam if the beam exerts a force of 3.8×10^{-9} N on a totally reflecting surface?

32.29 • **Laboratory Lasers.** He–Ne lasers are often used in physics demonstrations. They produce light of wavelength 633 nm and a power of 0.500 mW spread over a cylindrical beam 1.00 mm in diameter (although these quantities can vary). (a) What is the intensity of this laser beam? (b) What are the maximum values of the electric and magnetic fields? (c) What is the average energy density in the laser beam?

Section 32.5 Standing Electromagnetic Waves

32.30 • A standing electromagnetic wave in a certain material has frequency 2.20×10^{10} Hz. The nodal planes of $\vec{B}$ are 4.65 mm apart. Find (a) the wavelength of the wave in this material; (b) the distance between adjacent nodal planes of the $\vec{E}$ field; (c) the speed of propagation of the wave.

32.31 • **Microwave Oven.** The microwaves in a certain microwave oven have a wavelength of 12.2 cm. (a) How wide must this oven be so that it will contain five antinodal planes of the electric field along its width in the standing-wave pattern? (b) What is the frequency of these microwaves? (c) Suppose a manufacturing error occurred and the oven was made 5.0 cm longer than specified in part (a). In this case, what would have to be the frequency of the microwaves for there still to be five antinodal planes of the electric field along the width of the oven?

32.32 • An electromagnetic standing wave in air has frequency 75.0 MHz. (a) What is the distance between nodal planes of the $\vec{E}$ field? (b) What is the distance between a nodal plane of $\vec{E}$ and the closest nodal plane of $\vec{B}$?

PROBLEMS

32.33 •• BIO **Laser Surgery.** Very short pulses of high-intensity laser beams are used to repair detached portions of the retina of the eye. The brief pulses of energy absorbed by the retina weld the detached portions back into place. In one such procedure, a laser beam has a wavelength of 810 nm and delivers 250 mW of power spread over a circular spot 510 μm in diameter. The vitreous humor (the transparent fluid that fills most of the eye) has an index of refraction of 1.34. (a) If the laser pulses are each 1.50 ms long, how much energy is delivered to the retina with each pulse? (b) What average pressure would the pulse of the laser beam exert at normal incidence on a surface in air if the beam is fully absorbed? (c) What are the wavelength and frequency of the laser light inside the vitreous humor of the eye? (d) What are the maximum values of the electric and magnetic fields in the laser beam?

32.34 •• CALC Consider a sinusoidal electromagnetic wave with fields $\vec{E} = E_{max}\hat{j}\cos(kx - \omega t)$ and $\vec{B} = B_{max}\hat{k}\cos(kx - \omega t + \phi)$, with $-\pi \leq \phi \leq \pi$. Show that if $\vec{E}$ and $\vec{B}$ are to satisfy Eqs. (32.12) and (32.14), then $E_{max} = cB_{max}$ and $\phi = 0$. (The result $\phi = 0$ means the $\vec{E}$ and $\vec{B}$ fields oscillate in phase.)

32.35 • A satellite 575 km above the earth's surface transmits sinusoidal electromagnetic waves of frequency 92.4 MHz uniformly in all directions, with a power of 25.0 kW. (a) What is the intensity of these waves as they reach a receiver at the surface of the earth directly below the satellite? (b) What are the amplitudes of the electric and magnetic fields at the receiver? (c) If the receiver has a totally absorbing panel measuring 15.0 cm by 40.0 cm oriented with its plane perpendicular to the direction the waves travel, what average force do these waves exert on the panel? Is this force large enough to cause significant effects?

32.36 •• For a sinusoidal electromagnetic wave in vacuum, such as that described by Eq. (32.16), show that the *average* energy density in the electric field is the same as that in the magnetic field.

32.37 • The sun emits energy in the form of electromagnetic waves at a rate of 3.9×10^{26} W. This energy is produced by nuclear reactions deep in the sun's interior. (a) Find the intensity of electromagnetic radiation and the radiation pressure on an absorbing object at the surface of the sun (radius $r = R = 6.96 \times 10^5$ km) and at $r = R/2$, in the sun's interior. Ignore any scattering of the waves as they move radially outward from the center of the sun. Compare to the values given in Section 32.4 for sunlight just before it enters the earth's atmosphere. (b) The gas pressure at the sun's surface is about 1.0×10^4 Pa; at $r = R/2$, the gas pressure is calculated from solar models to be about 4.7×10^{13} Pa. Comparing with your results in part (a), would you expect that radiation pressure is an important factor in determining the structure of the sun? Why or why not?

32.38 • A small helium–neon laser emits red visible light with a power of 5.80 mW in a beam of diameter 2.50 mm. (a) What are the amplitudes of the electric and magnetic fields of this light? (b) What are the average energy densities associated with the electric field and with the magnetic field? (c) What is the total energy contained in a 1.00-m length of the beam?

32.39 •• CP Two square reflectors, each 1.50 cm on a side and of mass 4.00 g, are located at opposite ends of a thin, extremely light, 1.00-m rod that can rotate without friction and in vacuum about an axle perpendicular to it through its center (**Fig. P32.39**).

Figure **P32.39**

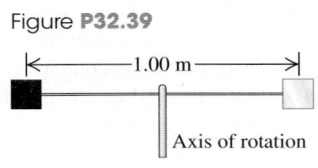

These reflectors are small enough to be treated as point masses in moment-of-inertia calculations. Both reflectors are illuminated on one face by a sinusoidal light wave having an electric field of amplitude 1.25 N/C that falls uniformly on both surfaces and always strikes them perpendicular to the plane of their surfaces. One reflector is covered with a perfectly absorbing coating, and the other is covered with a perfectly reflecting coating. What is the angular acceleration of this device?

32.40 •• A source of sinusoidal electromagnetic waves radiates uniformly in all directions. At a distance of 10.0 m from this source, the amplitude of the electric field is measured to be 3.50 N/C. What is the electric-field amplitude 20.0 cm from the source?

32.41 • CALC CP A cylindrical conductor with a circular cross section has a radius a and a resistivity ρ and carries a constant current I. (a) What are the magnitude and direction of the electric-field vector $\vec{E}$ at a point just inside the wire at a distance a from the axis? (b) What are the magnitude and direction of the magnetic-field vector $\vec{B}$ at the same point? (c) What are the magnitude and direction of the Poynting vector $\vec{S}$ at the same point? (The direction of $\vec{S}$ is the direction in which electromagnetic energy flows into or out of the conductor.) (d) Use the result in part (c) to find the rate of flow of energy into the volume occupied by a length l of the conductor. (*Hint:* Integrate $\vec{S}$ over the surface of this volume.) Compare your result to the rate of generation of thermal energy in the same volume. Discuss why the energy dissipated in a current-carrying conductor, due to its resistance, can be thought of as entering through the cylindrical sides of the conductor.

32.42 •• CP A circular wire loop has a radius of 7.50 cm. A sinusoidal electromagnetic plane wave traveling in air passes through the loop, with the direction of the magnetic field of the wave perpendicular to the plane of the loop. The intensity of the wave at the location of the loop is 0.0275 W/m², and the wavelength of the wave is 6.90 m. What is the maximum emf induced in the loop?

32.43 • In a certain experiment, a radio transmitter emits sinusoidal electromagnetic waves of frequency 110.0 MHz in opposite directions inside a narrow cavity with reflectors at both ends, causing a standing-wave pattern to occur. (a) How far apart are the nodal planes of the magnetic field? (b) If the standing-wave pattern is determined to be in its eighth harmonic, how long is the cavity?

32.44 • The 19th-century inventor Nikola Tesla proposed to transmit electric power via sinusoidal electromagnetic waves. Suppose power is to be transmitted in a beam of cross-sectional area 100 m². What electric- and magnetic-field amplitudes are required to transmit an amount of power comparable to that handled by modern transmission lines (that carry voltages and currents of the order of 500 kV and 1000 A)?

32.45 •• CP **Global Positioning System (GPS).** The GPS network consists of 24 satellites, each of which makes two orbits around the earth per day. Each satellite transmits a 50.0-W (or even less) sinusoidal electromagnetic signal at two frequencies, one of which is 1575.42 MHz. Assume that a satellite transmits half of its power at each frequency and that the waves travel uniformly in a downward hemisphere. (a) What average intensity does a GPS receiver on the ground, directly below the satellite, receive? (*Hint:* First use Newton's laws to find the altitude of the satellite.) (b) What are the amplitudes of the electric and magnetic fields at the GPS receiver in part (a), and how long does it take the signal to reach the receiver? (c) If the receiver is a square panel 1.50 cm on a side that absorbs all of the beam, what average pressure does the signal exert on it? (d) What wavelength must the receiver be tuned to?

32.46 •• CP **Solar Sail.** NASA is giving serious consideration to the concept of *solar sailing*. A solar sailcraft uses a large, low-mass sail and the energy and momentum of sunlight for propulsion. (a) Should the sail be absorbing or reflective? Why? (b) The total power output of the sun is 3.9×10^{26} W. How large a sail is necessary to propel a 10,000-kg spacecraft against the gravitational force of the sun? Express your result in square kilometers. (c) Explain why your answer to part (b) is independent of the distance from the sun.

32.47 •• CP Interplanetary space contains many small particles referred to as *interplanetary dust*. Radiation pressure from the sun sets a lower limit on the size of such dust particles. To see the origin of this limit, consider a spherical dust particle of radius R and mass density ρ. (a) Write an expression for the gravitational force exerted on this particle by the sun (mass M) when the particle is a distance r from the sun. (b) Let L represent the luminosity of the sun, equal to the rate at which it emits energy in electromagnetic radiation. Find the force exerted on the (totally absorbing) particle due to solar radiation pressure, remembering that the intensity of the sun's radiation also depends on the distance r. The relevant area is the cross-sectional area of the particle, *not* the total surface area of the particle. As part of your answer, explain why this is so. (c) The mass density of a typical interplanetary dust particle is about 3000 kg/m³. Find the particle radius R such that the gravitational and radiation forces acting on the particle are equal in magnitude. The luminosity of the sun is 3.9×10^{26} W. Does your answer depend on the distance of the particle from the sun? Why or why not? (d) Explain why dust particles with a radius less than that found in part (c) are unlikely to be found in the solar system. [*Hint:* Construct the ratio of the two force expressions found in parts (a) and (b).]

32.48 •• **DATA** The company where you work has obtained and stored five lasers in a supply room. You have been asked to determine the intensity of the electromagnetic radiation produced by each laser. The lasers are marked with specifications, but unfortunately different information is given for each laser:

Laser A: power = 2.6 W; diameter of cylindrical beam = 2.6 mm
Laser B: amplitude of electric field = 480 V/m
Laser C: amplitude of magnetic field = 8.7×10^{-6} T
Laser D: diameter of cylindrical beam = 1.8 mm; force on totally reflecting surface = 6.0×10^{-8} N
Laser E: average energy density in beam = 3.0×10^{-7} J/m^3

Calculate the intensity for each laser, and rank the lasers in order of increasing intensity. Assume that the laser beams have uniform intensity distributions over their cross sections.

32.49 •• **DATA** Because the speed of light in vacuum (or air) has such a large value, it is very difficult to measure directly. To measure this speed, you conduct an experiment in which you measure the amplitude of the electric field in a laser beam as you change the intensity of the beam. **Figure P32.49** is a graph of the intensity I that you measured versus the square of the amplitude E_{max} of the electric field. The best-fit straight line for your data has a slope of 1.33×10^{-3} J/(V$^2 \cdot$ s). (a) Explain why the data points plotted this way lie close to a straight line. (b) Use this graph to calculate the speed of light in air.

Figure **P32.49**

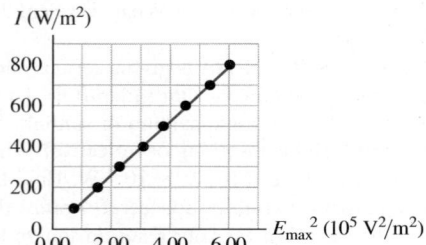

32.50 •• **DATA** As a physics lab instructor, you conduct an experiment on standing waves of microwaves, similar to the standing waves produced in a microwave oven. A transmitter emits microwaves of frequency f. The waves are reflected by a flat metal reflector, and a receiver measures the waves' electric-field amplitude as a function of position in the standing-wave pattern that is produced between the transmitter and reflector (**Fig. P32.50**). You measure the distance d between points of maximum amplitude (antinodes) of the electric field as a function of the frequency of the waves emitted by the transmitter. You obtain the data given in the table.

f (10^9 Hz)	1.0	1.5	2.0	2.5	3.0	3.5	4.0	5.0	6.0	8.0
d (cm)	15.2	9.7	7.7	5.8	5.2	4.1	3.8	3.1	2.3	1.7

Use the data to calculate c, the speed of the electromagnetic waves in air. Because each measured value has some experimental error, plot the data in such a way that the data points will lie close to a straight line, and use the slope of that straight line to calculate c.

Figure **P32.50**

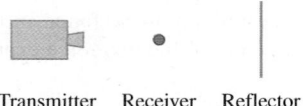

Transmitter   Receiver   Reflector

CHALLENGE PROBLEMS

32.51 ••• **CP** Electromagnetic radiation is emitted by accelerating charges. The rate at which energy is emitted from an accelerating charge that has charge q and acceleration a is given by

$$\frac{dE}{dt} = \frac{q^2 a^2}{6 \pi \epsilon_0 c^3}$$

where c is the speed of light. (a) Verify that this equation is dimensionally correct. (b) If a proton with a kinetic energy of 6.0 MeV is traveling in a particle accelerator in a circular orbit of radius 0.750 m, what fraction of its energy does it radiate per second? (c) Consider an electron orbiting with the same speed and radius. What fraction of its energy does it radiate per second?

32.52 ••• **CP The Classical Hydrogen Atom.** The electron in a hydrogen atom can be considered to be in a circular orbit with a radius of 0.0529 nm and a kinetic energy of 13.6 eV. If the electron behaved classically, how much energy would it radiate per second (see Challenge Problem 32.51)? What does this tell you about the use of classical physics in describing the atom?

32.53 ••• **CALC** Electromagnetic waves propagate much differently in *conductors* than they do in dielectrics or in vacuum. If the resistivity of the conductor is sufficiently low (that is, if it is a sufficiently good conductor), the oscillating electric field of the wave gives rise to an oscillating conduction current that is much larger than the displacement current. In this case, the wave equation for an electric field $\vec{E}(x, t) = E_y(x, t)\hat{\jmath}$ propagating in the $+x$-direction within a conductor is

$$\frac{\partial^2 E_y(x, t)}{\partial x^2} = \frac{\mu}{\rho} \frac{\partial E_y(x, t)}{\partial t}$$

where μ is the permeability of the conductor and ρ is its resistivity. (a) A solution to this wave equation is $E_y(x, t) = E_{max} e^{-k_C x} \cos(k_C x - \omega t)$, where $k_C = \sqrt{\omega\mu/2\rho}$. Verify this by substituting $E_y(x, t)$ into the above wave equation. (b) The exponential term shows that the electric field decreases in amplitude as it propagates. Explain why this happens. (*Hint:* The field does work to move charges within the conductor. The current of these moving charges causes $i^2 R$ heating within the conductor, raising its temperature. Where does the energy to do this come from?) (c) Show that the electric-field amplitude decreases by a factor of $1/e$ in a distance $1/k_C = \sqrt{2\rho/\omega\mu}$, and calculate this distance for a radio wave with frequency $f = 1.0$ MHz in copper (resistivity 1.72×10^{-8} $\Omega \cdot$ m; permeability $\mu = \mu_0$). Since this distance is so short, electromagnetic waves of this frequency can hardly propagate at all into copper. Instead, they are reflected at the surface of the metal. This is why radio waves cannot penetrate through copper or other metals, and why radio reception is poor inside a metal structure.

PASSAGE PROBLEMS

BIO SAFE EXPOSURE TO ELECTROMAGNETIC WAVES.
There have been many studies of the effects on humans of electromagnetic waves of various frequencies. Using these studies, the International Commission on Non-Ionizing Radiation Protection (ICNIRP) produced guidelines for limiting exposure to electromagnetic fields, with the goal of protecting against known adverse health effects. At frequencies of 1 Hz to 25 Hz, the maximum exposure level of electric-field amplitude E_{max} for the general

public is 14 kV/m. (Different guidelines were created for people who have occupational exposure to radiation.) At frequencies of 25 Hz to 3 kHz, the corresponding E_{max} is $350/f$ V/m, where f is the frequency in kHz. (Source: "ICNIRP Statement on the 'Guidelines for Limiting Exposure to Time-Varying Electric, Magnetic, and Electromagnetic Fields (up to 300 GHz)'," 2009; *Health Physics* 97(3): 257–258.)

32.54 In the United States, household electrical power is provided at a frequency of 60 Hz, so electromagnetic radiation at that frequency is of particular interest. On the basis of the ICNIRP guidelines, what is the maximum intensity of an electromagnetic wave at this frequency to which the general public should be exposed? (a) 7.7 W/m^2; (b) 160 W/m^2; (c) 45 kW/m^2; (d) 260 kW/m^2.

32.55 Doubling the frequency of a wave in the range of 25 Hz to 3 kHz represents what change in the maximum allowed electromagnetic-wave intensity? (a) A factor of 2; (b) a factor of $1/\sqrt{2}$; (c) a factor of $\frac{1}{2}$; (d) a factor of $\frac{1}{4}$.

32.56 The ICNIRP also has guidelines for magnetic-field exposure for the general public. In the frequency range of 25 Hz to 3 kHz, this guideline states that the maximum allowed magnetic-field amplitude is $5/f$ T, where f is the frequency in kHz. Which is a more stringent limit on allowable electromagnetic-wave intensity in this frequency range: the electric-field guideline or the magnetic-field guideline? (a) The magnetic-field guideline, because at a given frequency the allowed magnetic field is smaller than the allowed electric field. (b) The electric field guideline, because at a given frequency the allowed intensity calculated from the electric-field guideline is smaller. (c) It depends on the particular frequency chosen (both guidelines are frequency dependent). (d) Neither—for any given frequency, the guidelines represent the same electromagnetic-wave intensity.

Answers

Chapter Opening Question **?**

(i) Metals are reflective because they are good conductors of electricity. When an electromagnetic wave strikes a conductor, the electric field of the wave sets up currents on the conductor surface that generate a reflected wave. For a perfect conductor, the requirement that the electric-field component parallel to the surface must be zero implies that this reflected wave is just as intense as the incident wave. Tarnished metals are less shiny because their surface is oxidized and less conductive; polishing the metal removes the oxide and exposes the conducting metal.

Test Your Understanding Questions

32.1 (a) no, (b) no   A purely electric wave would have a varying electric field. Such a field necessarily generates a magnetic field through Ampere's law, Eq. (29.21), so a purely electric wave is impossible. In the same way, a purely magnetic wave is impossible: The varying magnetic field in such a wave would automatically give rise to an electric field through Faraday's law, Eq. (29.20).

32.2 (a) positive *y*-direction, (b) negative *x*-direction, (c) positive *y*-direction   You can verify these answers by using the right-hand rule to show that $\vec{E} \times \vec{B}$ in each case is in the direction of propagation, or by using the rule shown in Fig. 32.9.

32.3 (iv)   In an ideal electromagnetic plane wave, at any instant the fields are the same anywhere in a plane perpendicular to the direction of propagation. The plane wave described by Eqs. (32.17) is propagating in the *x*-direction, so the fields depend on the coordinate *x* and time *t* but do *not* depend on the coordinates *y* and *z*.

32.4 (a) (i) and (iii), (b) (ii) and (iv), (c) (i) and (iii), (d) (ii) and (iv)   Both the energy density *u* and the Poynting vector magnitude *S* are maximum where the $\vec{E}$ and $\vec{B}$ fields have their maximum magnitudes. (The directions of the fields don't matter.) From Fig. 32.13, this occurs at $x = 0$ and $x = \lambda/2$. Both *u* and *S* have a minimum value of zero; that occurs where $\vec{E}$ and $\vec{B}$ are both zero. From Fig. 32.13, this occurs at $x = \lambda/4$ and $x = 3\lambda/4$.

32.5 no   There are places where $\vec{E} = 0$ at all times (at the walls) and the electric energy density $\frac{1}{2}\epsilon_0 E^2$ is always zero. There are also places where $\vec{B} = 0$ at all times (on the plane midway between the walls) and the magnetic energy density $B^2/2\mu_0$ is always zero. However, there are *no* locations where both $\vec{E}$ and $\vec{B}$ are always zero. Hence the energy density at any point in the standing wave is always nonzero.

Bridging Problem

0.0368 V

? When a cut diamond is
illuminated with white
light, it sparkles brilliantly with
a spectrum of vivid colors.
These distinctive visual features
are a result of (i) light traveling
much slower in diamond than
in air; (ii) light of different colors
traveling at different speeds in
diamond; (iii) diamond absorbing light of certain colors;
(iv) both (i) and (ii); (v) all of (i),
(ii), and (iii).

33

THE NATURE AND PROPAGATION OF LIGHT

Blue lakes, ochre deserts, green forests, and multicolored rainbows can be enjoyed by anyone who has eyes with which to see them. But by study-ing the branch of physics called **optics,** which deals with the behavior of light and other electromagnetic waves, we can reach a deeper appreciation of the visible world. A knowledge of the properties of light allows us to understand the blue color of the sky and the design of optical devices such as telescopes, microscopes, cameras, eyeglasses, and the human eye. The same basic principles of optics also lie at the heart of modern developments such as the laser, optical fibers, holograms, and new techniques in medical imaging.

The importance of optics to physics, and to science and engineering in general, is so great that we will devote the next four chapters to its study. In this chapter we begin with a study of the laws of reflection and refraction and the concepts of dispersion, polarization, and scattering of light. Along the way we compare the various possible descriptions of light in terms of particles, rays, or waves, and we introduce Huygens's principle, an important link that connects the ray and wave viewpoints. In Chapter 34 we'll use the ray description of light to understand how mirrors and lenses work, and we'll see how mirrors and lenses are used in optical instruments such as cameras, microscopes, and telescopes. We'll explore the wave characteristics of light further in Chapters 35 and 36.

33.1 THE NATURE OF LIGHT

Until the time of Isaac Newton (1642–1727), most scientists thought that light consisted of streams of particles (called *corpuscles*) emitted by light sources. Galileo and others tried (unsuccessfully) to measure the speed of light. Around 1665, evidence of *wave* properties of light began to be discovered. By the early 19th century, evidence that light is a wave had grown very persuasive.

In 1873, James Clerk Maxwell predicted the existence of electromagnetic waves and calculated their speed of propagation, as we learned in Section 32.2. This development, along with the experimental work of Heinrich Hertz starting in 1887, showed conclusively that light is indeed an electromagnetic wave.

The Two Personalities of Light

The wave picture of light is not the whole story, however. Several effects associated with emission and absorption of light reveal a particle aspect, in that the energy carried by light waves is packaged in discrete bundles called *photons* or *quanta*. These apparently contradictory wave and particle properties have been reconciled since 1930 with the development of quantum electrodynamics, a comprehensive theory that includes *both* wave and particle properties. The *propagation* of light is best described by a wave model, but understanding emission and absorption requires a particle approach.

The fundamental sources of all electromagnetic radiation are electric charges in accelerated motion. All bodies emit electromagnetic radiation as a result of thermal motion of their molecules; this radiation, called *thermal radiation,* is a mixture of different wavelengths. At sufficiently high temperatures, all matter emits enough visible light to be self-luminous; a very hot body appears "red-hot" (**Fig. 33.1**) or "white-hot." Thus hot matter in any form is a light source. Familiar examples are a candle flame, hot coals in a campfire, and the coils in an electric toaster oven or room heater.

Light is also produced during electrical discharges through ionized gases. The bluish light of mercury-arc lamps, the orange-yellow of sodium-vapor lamps, and the various colors of "neon" signs are familiar. A variation of the mercury-arc lamp is the *fluorescent* lamp (see Fig. 30.7). This light source uses a material called a *phosphor* to convert the ultraviolet radiation from a mercury arc into visible light. This conversion makes fluorescent lamps more efficient than incandescent lamps in transforming electrical energy into light.

In most light sources, light is emitted independently by different atoms within the source; in a *laser,* by contrast, atoms are induced to emit light in a cooperative, coherent fashion. The result is a very narrow beam of radiation that can be enormously intense and that is much more nearly *monochromatic,* or single-frequency, than light from any other source. Lasers are used by physicians for microsurgery, in a DVD or Blu-ray player to scan the information recorded on a video disc, in industry to cut through steel and to fuse high-melting-point materials, and in many other applications (**Fig. 33.2**).

No matter what its source, electromagnetic radiation travels in vacuum at the same speed c. As we saw in Sections 1.3 and 32.1, this speed is defined to be

$$c = 2.99792458 \times 10^8 \text{ m/s}$$

or 3.00×10^8 m/s to three significant figures. The duration of one second is defined by the cesium clock (see Section 1.3), so one meter is defined to be the distance that light travels in $1/299{,}792{,}458$ s.

Waves, Wave Fronts, and Rays

We often use the concept of a **wave front** to describe wave propagation. We introduced this concept in Section 32.2 to describe the leading edge of a wave. More generally, we define a wave front as *the locus of all adjacent points at which the phase of vibration of a physical quantity associated with the wave is the same.* That is, at any instant, all points on a wave front are at the same part of the cycle of their variation.

When we drop a pebble into a calm pool, the expanding circles formed by the wave crests, as well as the circles formed by the wave troughs between them, are wave fronts. Similarly, when sound waves spread out in still air from a

33.1 An electric heating element emits primarily infrared radiation. But if its temperature is high enough, it also emits a discernible amount of visible light.

33.2 Ophthalmic surgeons use lasers for repairing detached retinas and for cauterizing blood vessels in retinopathy. Pulses of blue-green light from an argon laser are ideal for this purpose, since they pass harmlessly through the transparent part of the eye but are absorbed by red pigments in the retina.

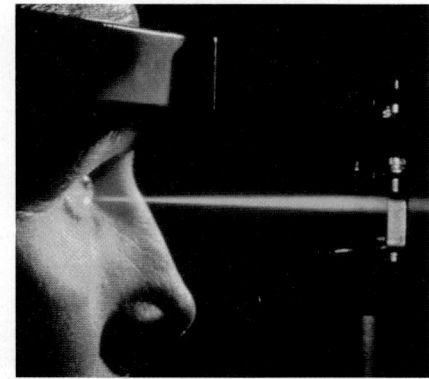

33.3 Spherical wave fronts of sound spread out uniformly in all directions from a point source in a motionless medium, such as still air, that has the same properties in all regions and in all directions. Electromagnetic waves in vacuum also spread out as shown here.

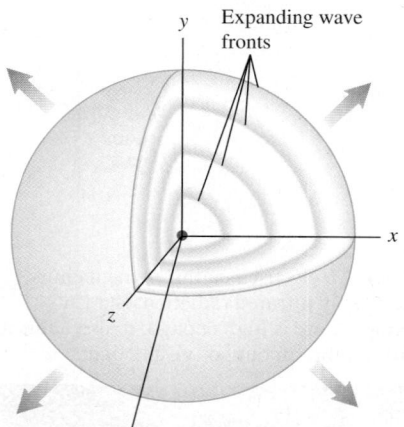

Point sound source producing spherical sound waves (alternating compressions and rarefactions of air)

33.4 Wave fronts (blue) and rays (purple).

(a)

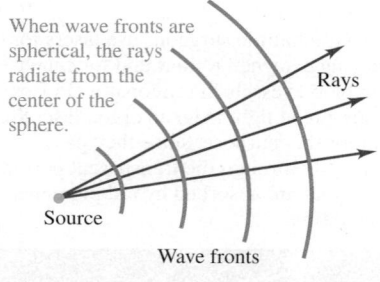

When wave fronts are spherical, the rays radiate from the center of the sphere.

Rays

Source

Wave fronts

(b)

When wave fronts are planar, the rays are perpendicular to the wave fronts and parallel to each other.

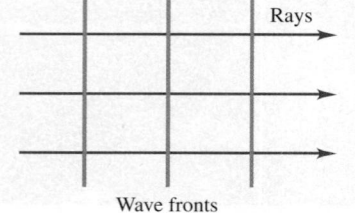

Rays

Wave fronts

pointlike source, or when electromagnetic radiation spreads out from a pointlike emitter, any spherical surface that is concentric with the source is a wave front, as shown in **Fig. 33.3**. In diagrams of wave motion we usually draw only parts of a few wave fronts, often choosing consecutive wave fronts that have the same phase and thus are one wavelength apart, such as crests of water waves. Similarly, a diagram for sound waves might show only the "pressure crests," the surfaces over which the pressure is maximum, and a diagram for electromagnetic waves might show only the "crests" on which the electric or magnetic field is maximum.

We will often use diagrams that show the shapes of the wave fronts or their cross sections in some reference plane. For example, when electromagnetic waves are radiated by a small light source, we can represent the wave fronts as spherical surfaces concentric with the source or, as in **Fig. 33.4a**, by the circular intersections of these surfaces with the plane of the diagram. Far away from the source, where the radii of the spheres have become very large, a section of a spherical surface can be considered as a plane, and we have a *plane* wave like those discussed in Sections 32.2 and 32.3 (Fig. 33.4b).

To describe the directions in which light propagates, it's often convenient to represent a light wave by **rays** rather than by wave fronts. In a particle theory of light, rays are the paths of the particles. From the wave viewpoint *a ray is an imaginary line along the direction of travel of the wave.* In Fig. 33.4a the rays are the radii of the spherical wave fronts, and in Fig. 33.4b they are straight lines perpendicular to the wave fronts. When waves travel in a homogeneous isotropic material (a material with the same properties in all regions and in all directions), the rays are always straight lines normal to the wave fronts. At a boundary surface between two materials, such as the surface of a glass plate in air, the wave speed and the direction of a ray may change, but the ray segments in the air and in the glass are straight lines.

The next several chapters will give you many opportunities to see the interplay of the ray, wave, and particle descriptions of light. The branch of optics for which the ray description is adequate is called **geometric optics;** the branch dealing specifically with wave behavior is called **physical optics.** This chapter and the following one are concerned mostly with geometric optics. In Chapters 35 and 36 we will study wave phenomena and physical optics.

TEST YOUR UNDERSTANDING OF SECTION 33.1 Some crystals are *not* isotropic: Light travels through the crystal at a higher speed in some directions than in others. In a crystal in which light travels at the same speed in the *x*- and *z*-directions but faster in the *y*-direction, what would be the shape of the wave fronts produced by a light source at the origin? (i) Spherical, like those shown in Fig. 33.3; (ii) ellipsoidal, flattened along the *y*-axis; (iii) ellipsoidal, stretched out along the *y*-axis. ▮

33.2 REFLECTION AND REFRACTION

In this section we'll use the *ray* model of light to explore two of the most important aspects of light propagation: **reflection** and **refraction.** When a light wave strikes a smooth interface separating two transparent materials (such as air and glass or water and glass), the wave is in general partly *reflected* and partly *refracted* (transmitted) into the second material, as shown in **Fig. 33.5a**. For example, when you look into a restaurant window from the street, you see a reflection of the street scene, but a person inside the restaurant can look out through the window at the same scene as light reaches him by refraction.

The segments of plane waves shown in Fig. 33.5a can be represented by bundles of rays forming *beams* of light (Fig. 33.5b). For simplicity we often draw only one ray in each beam (Fig. 33.5c). Representing these waves in terms of rays is the basis of geometric optics. We begin our study with the behavior of an individual ray.

33.5 (a) A plane wave is in part reflected and in part refracted at the boundary between two media (in this case, air and glass). The light that reaches the inside of the coffee shop is refracted twice, once entering the glass and once exiting the glass. (b), (c) How light behaves at the interface between the air outside the coffee shop (material a) and the glass (material b). For the case shown here, material b has a larger index of refraction than material a ($n_b > n_a$) and the angle θ_b is smaller than θ_a.

(a) Plane waves reflected and refracted from a window

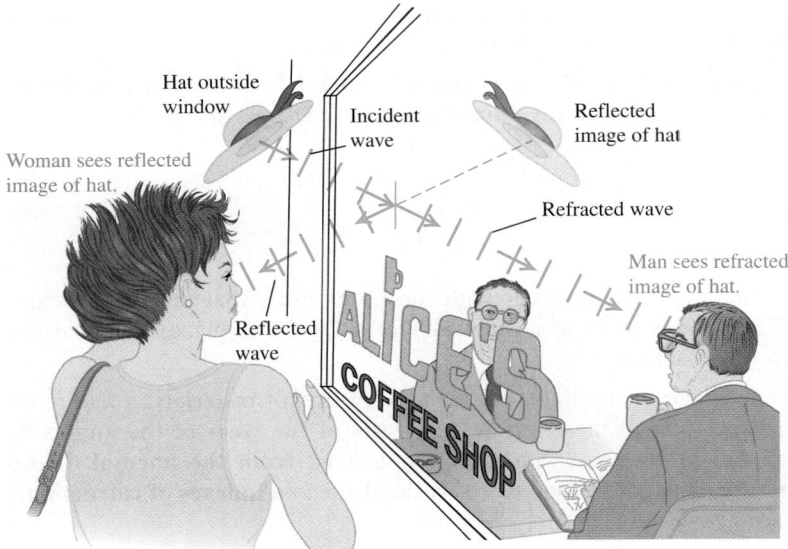

(b) The waves in the outside air and glass represented by rays

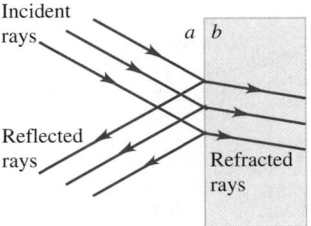

(c) The representation simplified to show just one set of rays

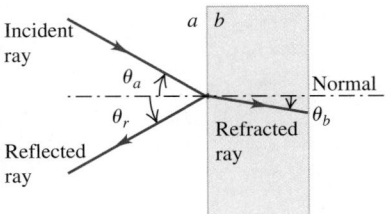

We describe the directions of the incident, reflected, and refracted (transmitted) rays at a smooth interface between two optical materials in terms of the angles they make with the *normal* (perpendicular) to the surface at the point of incidence, as shown in Fig. 33.5c. If the interface is rough, both the transmitted light and the reflected light are scattered in various directions, and there is no single angle of transmission or reflection. Reflection at a definite angle from a very smooth surface is called **specular reflection** (from the Latin word for "mirror"); scattered reflection from a rough surface is called **diffuse reflection** (**Fig. 33.6**). Both kinds of reflection can occur with either transparent materials or *opaque* materials that do not transmit light. The vast majority of objects in your environment (including plants, other people, and this book) are visible to you because they reflect light in a diffuse manner from their surfaces. Our primary concern, however, will be with specular reflection from a very smooth surface such as highly polished glass or metal. Unless stated otherwise, when referring to "reflection" we will always mean *specular* reflection.

The **index of refraction** of an optical material (also called the **refractive index**), denoted by n, plays a central role in geometric optics:

$$\text{Index of refraction}\cdots n = \frac{c}{v} \cdots \text{Speed of light in vacuum} \atop \text{of an optical material} \qquad\qquad\quad \cdots \text{Speed of light in the material} \tag{33.1}$$

Light always travels *more slowly* in a material than in vacuum, so the value of n in anything other than vacuum is always greater than unity. For vacuum, $n = 1$. Since n is a ratio of two speeds, it is a pure number without units. (In Section 32.3 we described the relationship of the value of n to the electric and magnetic properties of a material.)

CAUTION **Wave speed and index of refraction** Keep in mind that the wave speed v is *inversely* proportional to the index of refraction n. The greater the index of refraction in a material, the *slower* the wave speed in that material. ∎

33.6 Two types of reflection.

(a) Specular reflection

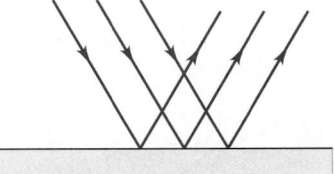

(b) Diffuse reflection

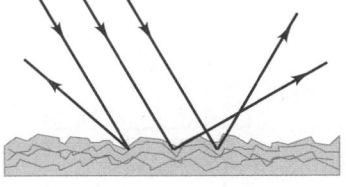

33.7 The laws of reflection and refraction.

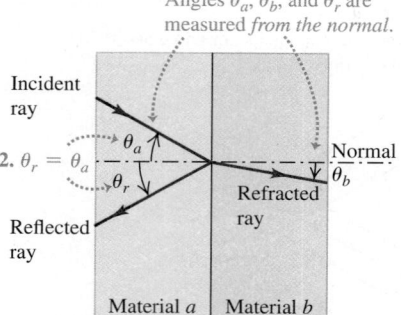

1. The incident, reflected, and refracted rays and the normal to the surface all lie in the same plane.

Angles θ_a, θ_b, and θ_r are measured *from the normal*.

Incident ray

2. $\theta_r = \theta_a$

Normal

Refracted ray

Reflected ray

Material *a* | Material *b*

3. When a monochromatic light ray crosses the interface between two given materials *a* and *b*, the angles θ_a and θ_b are related to the indexes of refraction of *a* and *b* by

$$\frac{\sin\theta_a}{\sin\theta_b} = \frac{n_b}{n_a}$$

33.8 Refraction and reflection in three cases. (a) Material *b* has a larger index of refraction than material *a*. (b) Material *b* has a smaller index of refraction than material *a*. (c) The incident light ray is normal to the interface between the materials.

(a) A ray entering a material of *larger* index of refraction bends *toward* the normal.

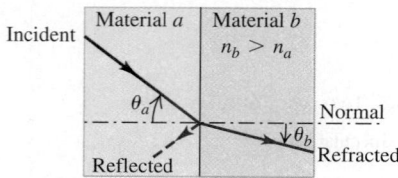

Material *a* | Material *b* $n_b > n_a$

Incident

θ_a

Normal

Reflected

θ_b

Refracted

(b) A ray entering a material of *smaller* index of refraction bends *away from* the normal.

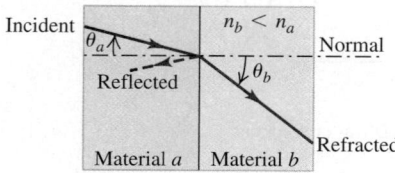

Incident

$n_b < n_a$

θ_a

Normal

Reflected

θ_b

Refracted

Material *a* | Material *b*

(c) A ray oriented along the normal does not bend, regardless of the materials.

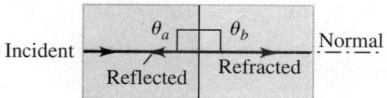

θ_a | θ_b

Incident

Normal

Reflected | Refracted

The Laws of Reflection and Refraction

Experimental studies of reflection and refraction at a smooth interface between two optical materials lead to the following conclusions (**Fig. 33.7**):

1. **The incident, reflected, and refracted rays and the normal to the surface all lie in the same plane.** This plane, called the **plane of incidence,** is perpendicular to the plane of the boundary surface between the two materials. We always draw ray diagrams so that the incident, reflected, and refracted rays are in the plane of the diagram.

2. **The angle of reflection θ_r is equal to the angle of incidence θ_a for all wavelengths and for any pair of materials.** That is, in Fig. 33.5c,

Law of reflection:

Angle of reflection (measured from normal)

$$\theta_r = \theta_a \quad \text{Angle of incidence (measured from normal)}$$

(33.2)

This relationship, together with the observation that the incident and reflected rays and the normal all lie in the same plane, is called the **law of reflection.**

3. For monochromatic light and for a given pair of materials, *a* and *b*, on opposite sides of the interface, **the ratio of the sines of the angles θ_a and θ_b, where both angles are measured from the normal to the surface, is equal to the inverse ratio of the two indexes of refraction:**

$$\frac{\sin\theta_a}{\sin\theta_b} = \frac{n_b}{n_a}$$

(33.3)

or

Law of refraction:

Angle of incidence (measured from normal)

$$n_a\sin\theta_a = n_b\sin\theta_b \quad \text{Angle of refraction (measured from normal)}$$

Index of refraction for material with incident light   Index of refraction for material with refracted light

(33.4)

This result, together with the observation that the incident and refracted rays and the normal all lie in the same plane, is called the **law of refraction** or **Snell's law,** after the Dutch scientist Willebrord Snell (1591–1626). This law was actually first discovered in the 10th century by the Persian scientist Ibn Sahl. The discovery that $n = c/v$ came much later.

While these results were first observed experimentally, they can be derived theoretically from a wave description of light. We do this in Section 33.7.

Equations (33.3) and (33.4) show that when a ray passes from one material (*a*) into another material (*b*) having a larger index of refraction ($n_b > n_a$) and hence a slower wave speed, the angle θ_b with the normal is *smaller* in the second material than the angle θ_a in the first; hence the ray is bent *toward* the normal (**Fig. 33.8a**). When the second material has a *smaller* index of refraction than the first material ($n_b < n_a$) and hence a faster wave speed, the ray is bent *away from* the normal (Fig. 33.8b).

No matter what the materials on either side of the interface, in the case of *normal* incidence the transmitted ray is not bent at all (Fig. 33.8c). In this case $\theta_a = 0$ and $\sin\theta_a = 0$, so from Eq. (33.4) θ_b is also equal to zero; the transmitted ray is also normal to the interface. From Eq. (33.2), θ_r is also equal to zero, so the reflected ray travels back along the same path as the incident ray.

The law of refraction explains why a partially submerged ruler or drinking straw appears bent; light rays coming from below the surface change in direction at the air–water interface, so the rays appear to be coming from a position above

33.9 (a) This ruler is actually straight, but it appears to bend at the surface of the water. (b) Light rays from any submerged object bend away from the normal when they emerge into the air. As seen by an observer above the surface of the water, the object appears to be much closer to the surface than it actually is.

(a) A straight ruler half-immersed in water

(b) Why the ruler appears bent

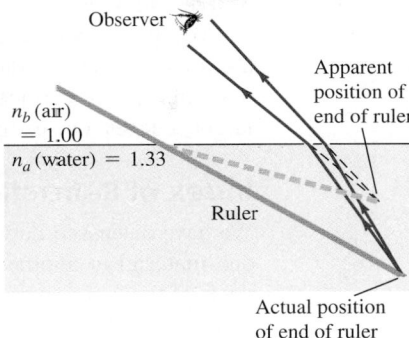

their actual point of origin (**Fig. 33.9**). A similar effect explains the appearance of the setting sun (**Fig. 33.10**).

An important special case is refraction that occurs at an interface between vacuum, for which the index of refraction is unity by definition, and a material. When a ray passes from vacuum into a material (b), so that $n_a = 1$ and $n_b > 1$, the ray is always bent *toward* the normal. When a ray passes from a material into vacuum, so that $n_a > 1$ and $n_b = 1$, the ray is always bent *away from* the normal.

The laws of reflection and refraction apply regardless of which side of the interface the incident ray comes from. If a ray of light approaches the interface in Fig. 33.8a or 33.8b from the right rather than from the left, there are again reflected and refracted rays, and they lie in the same plane as the incident ray and the normal to the surface. Furthermore, the path of a refracted ray is *reversible;* it follows the same path when going from b to a as when going from a to b. [You can verify this by using Eq. (33.4).] Since the reflected and incident angles are the same, the path of a reflected ray is also reversible. That's why when you see someone's eyes in a mirror, they can also see you.

The *intensities* of the reflected and refracted rays depend on the angle of incidence, the two indexes of refraction, and the polarization (that is, the direction of the electric-field vector) of the incident ray. The fraction reflected is smallest at normal incidence ($\theta_a = 0°$), where it is about 4% for an air–glass interface. This fraction increases with increasing angle of incidence to 100% at grazing incidence, when $\theta_a = 90°$. (It's possible to use Maxwell's equations to predict the amplitude, intensity, phase, and polarization states of the reflected and refracted waves. Such an analysis is beyond our scope, however.)

BIO **Application Transparency and Index of Refraction** An eel in its larval stage is nearly as transparent as the seawater in which it swims. The larva in this photo is nonetheless easy to see because its index of refraction is higher than that of seawater, so that some of the light striking it is reflected instead of transmitted. The larva appears particularly shiny around its edges because the light reaching the camera from those points struck the larva at near-grazing incidence ($\theta_a = 90°$), resulting in almost 100% reflection.

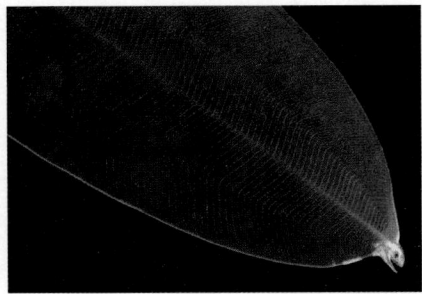

(a) Atmosphere (not to scale)

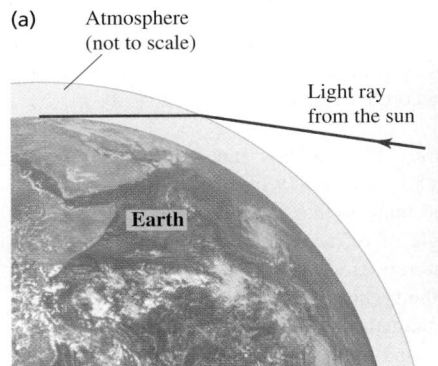

(b)

33.10 (a) The index of refraction of air is slightly greater than 1, so light rays from the setting sun bend downward when they enter our atmosphere. (The effect is exaggerated in this figure.) (b) Stronger refraction occurs for light coming from the lower limb of the sun (the part that appears closest to the horizon), which passes through denser air in the lower atmosphere. As a result, the setting sun appears flattened vertically. (See Problem 33.51.)

Index of Refraction for Yellow Sodium Light,

| TABLE 33.1 | $\lambda_0 = 589$ nm |

Substance	Index of Refraction, n
Solids	
Ice (H_2O)	1.309
Fluorite (CaF_2)	1.434
Polystyrene	1.49
Rock salt (NaCl)	1.544
Quartz (SiO_2)	1.544
Zircon ($ZrO_2 \cdot SiO_2$)	1.923
Diamond (C)	2.417
Fabulite ($SrTiO_3$)	2.409
Rutile (TiO_2)	2.62
Glasses (typical values)	
Crown	1.52
Light flint	1.58
Medium flint	1.62
Dense flint	1.66
Lanthanum flint	1.80
Liquids at 20°C	
Methanol (CH_3OH)	1.329
Water (H_2O)	1.333
Ethanol (C_2H_5OH)	1.36
Carbon tetrachloride (CCl_4)	1.460
Turpentine	1.472
Glycerine	1.473
Benzene	1.501
Carbon disulfide (CS_2)	1.628

The index of refraction depends not only on the substance but also on the wavelength of the light. The dependence on wavelength is called *dispersion;* we will consider it in Section 33.4. Indexes of refraction for several solids and liquids are given in **Table 33.1** for a particular wavelength of yellow light.

The index of refraction of air at standard temperature and pressure is about 1.0003, and we will usually take it to be exactly unity. The index of refraction of a gas increases as its density increases. Most glasses used in optical instruments have indexes of refraction between about 1.5 and 2.0. A few substances have larger indexes; one example is diamond, with 2.417 (see Table 33.1).

Index of Refraction and the Wave Aspects of Light

We have discussed how the direction of a light ray changes when it passes from one material to another material with a different index of refraction. What aspects of the *wave* characteristics of the light change when this happens?

First, the frequency f of the wave does *not* change when passing from one material to another. That is, the number of wave cycles arriving per unit time must equal the number leaving per unit time; this is a statement that the boundary surface cannot create or destroy waves.

Second, the wavelength λ of the wave *is* different in general in different materials. This is because in any material, $v = \lambda f$; since f is the same in any material as in vacuum and v is always less than the wave speed c in vacuum, λ is also correspondingly reduced. Thus the wavelength λ of light in a material is *less than* the wavelength λ_0 of the same light in vacuum. From the above discussion, $f = c/\lambda_0 = v/\lambda$. Combining this with Eq. (33.1), $n = c/v$, we find

$$\text{Wavelength of light} \cdots\!\!\rightarrow \lambda = \frac{\lambda_0}{n} \cdots\!\!\begin{matrix}\text{Wavelength of light in vacuum}\\ \text{Index of refraction}\\ \text{of the material}\end{matrix} \qquad (33.5)$$

When a wave passes from one material into a second material with larger index of refraction, so $n_b > n_a$, the wave speed decreases. The wavelength $\lambda_b = \lambda_0/n_b$ in the second material is then shorter than the wavelength $\lambda_a = \lambda_0/n_a$ in the first material. If instead the second material has a smaller index of refraction than the first material, so $n_b < n_a$, then the wave speed increases. Then the wavelength λ_b in the second material is longer than the wavelength λ_a in the first material. This makes intuitive sense; the waves get "squeezed" (the wavelength gets shorter) if the wave speed decreases and get "stretched" (the wavelength gets longer) if the wave speed increases.

PROBLEM-SOLVING STRATEGY 33.1 | REFLECTION AND REFRACTION

IDENTIFY *the relevant concepts:* Use geometric optics, discussed in this section, whenever light (or electromagnetic radiation of *any* frequency and wavelength) encounters a boundary between materials. In general, part of the light is reflected back into the first material and part is refracted into the second material.

SET UP *the problem* using the following steps:
1. In problems involving rays and angles, start by drawing a large, neat diagram. Label all known angles and indexes of refraction.
2. Identify the target variables.

EXECUTE *the solution* as follows:
1. Apply the laws of reflection, Eq. (33.2), and refraction, Eq. (33.4). Measure angles of incidence, reflection, and refraction with respect to the *normal* to the surface, *never* from the surface itself.

2. Apply geometry or trigonometry in working out angular relationships. Remember that the sum of the acute angles of a right triangle is 90° (they are *complementary*) and the sum of the interior angles in any triangle is 180°.
3. The frequency of the electromagnetic radiation does not change when it moves from one material to another; the wavelength changes in accordance with Eq. (33.5), $\lambda = \lambda_0/n$.

EVALUATE *your answer:* In problems that involve refraction, check that your results are consistent with Snell's law ($n_a \sin\theta_a = n_b \sin\theta_b$). If the second material has a higher index of refraction than the first, the angle of refraction must be *smaller* than the angle of incidence: The refracted ray bends toward the normal. If the first material has the higher index of refraction, the refracted angle must be larger than the incident angle: The refracted ray bends away from the normal.

EXAMPLE 33.1   REFLECTION AND REFRACTION

In **Fig. 33.11**, material a is water and material b is glass with index of refraction 1.52. The incident ray makes an angle of 60.0° with the normal; find the directions of the reflected and refracted rays.

SOLUTION

IDENTIFY and SET UP: This is a problem in geometric optics. We are given the angle of incidence $\theta_a = 60.0°$ and the indexes of

33.11 Reflection and refraction of light passing from water to glass.

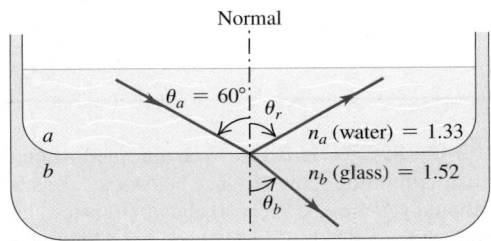

refraction $n_a = 1.33$ and $n_b = 1.52$. We must find the angles of reflection and refraction θ_r and θ_b; to do this we use Eqs. (33.2) and (33.4), respectively. Figure 33.11 shows the rays and angles; n_b is slightly greater than n_a, so by Snell's law [Eq. (33.4)] θ_b is slightly smaller than θ_a.

EXECUTE: According to Eq. (33.2), the angle the reflected ray makes with the normal is the same as that of the incident ray, so $\theta_r = \theta_a = 60.0°$.

To find the direction of the refracted ray we use Snell's law, Eq. (33.4):

$$n_a \sin\theta_a = n_b \sin\theta_b$$

$$\sin\theta_b = \frac{n_a}{n_b}\sin\theta_a = \frac{1.33}{1.52}\sin 60.0° = 0.758$$

$$\theta_b = \arcsin(0.758) = 49.3°$$

EVALUATE: The second material has a larger refractive index than the first, as in Fig. 33.8a. Hence the refracted ray is bent toward the normal and $\theta_b < \theta_a$.

EXAMPLE 33.2   INDEX OF REFRACTION IN THE EYE

The wavelength of the red light from a helium-neon laser is 633 nm in air but 474 nm in the aqueous humor inside your eyeball. Calculate the index of refraction of the aqueous humor and the speed and frequency of the light in it.

SOLUTION

IDENTIFY and SET UP: The key ideas here are (i) the definition of index of refraction n in terms of the wave speed v in a medium and the speed c in vacuum, and (ii) the relationship between wavelength λ_0 in vacuum and wavelength λ in a medium of index n. We use Eq. (33.1), $n = c/v$; Eq. (33.5), $\lambda = \lambda_0/n$; and $v = \lambda f$.

EXECUTE: The index of refraction of air is very close to unity, so we assume that the wavelength λ_0 in vacuum is the same as that in air, 633 nm. Then from Eq. (33.5),

$$\lambda = \frac{\lambda_0}{n} \qquad n = \frac{\lambda_0}{\lambda} = \frac{633 \text{ nm}}{474 \text{ nm}} = 1.34$$

This is about the same index of refraction as for water. Then, using $n = c/v$ and $v = \lambda f$, we find

$$v = \frac{c}{n} = \frac{3.00 \times 10^8 \text{ m/s}}{1.34} = 2.25 \times 10^8 \text{ m/s}$$

$$f = \frac{v}{\lambda} = \frac{2.25 \times 10^8 \text{ m/s}}{474 \times 10^{-9} \text{ m}} = 4.74 \times 10^{14} \text{ Hz}$$

EVALUATE: Although the speed and wavelength have different values in air and in the aqueous humor, the *frequency* in air, f_0, is the same as the frequency f in the aqueous humor:

$$f_0 = \frac{c}{\lambda_0} = \frac{3.00 \times 10^8 \text{ m/s}}{633 \times 10^{-9} \text{ m}} = 4.74 \times 10^{14} \text{ Hz}$$

When a light wave passes from one material into another, both the wave speed and wavelength change but the wave frequency is unchanged.

EXAMPLE 33.3   A TWICE-REFLECTED RAY

Two mirrors are perpendicular to each other. A ray traveling in a plane perpendicular to both mirrors is reflected from one mirror at P, then the other at Q, as shown in **Fig. 33.12** (next page). What is the ray's final direction relative to its original direction?

SOLUTION

IDENTIFY and SET UP: This problem involves the law of reflection, which we must apply twice (once for each mirror).

EXECUTE: For mirror 1 the angle of incidence is θ_1, and this equals the angle of reflection. The sum of interior angles in the triangle PQR is 180°, so we see that the angles of both incidence and reflection for mirror 2 are $90° - \theta_1$. The total change in direction of the ray after both reflections is therefore $2(90° - \theta_1) + 2\theta_1 = 180°$. That is, the ray's final direction is opposite to its original direction.

Continued

33.12 A ray moving in the *xy*-plane. The first reflection changes the sign of the *y*-component of its velocity, and the second reflection changes the sign of the *x*-component.

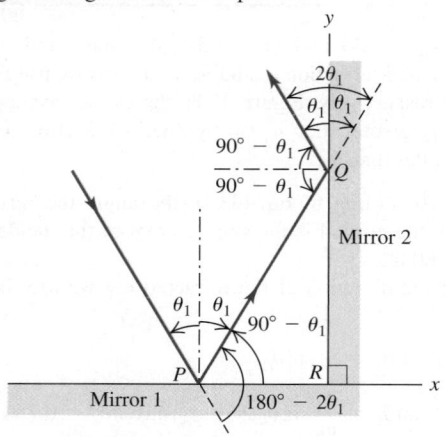

EVALUATE: An alternative viewpoint is that reflection reverses the sign of the component of light velocity perpendicular to the surface but leaves the other components unchanged. We invite you to verify this in detail. You should also be able to use this result to show that when a ray of light is successively reflected by three mirrors forming a corner of a cube (a "corner reflector"), its final direction is again opposite to its original direction. This principle is widely used in tail-light lenses and bicycle reflectors to improve their night-time visibility. *Apollo* astronauts placed arrays of corner reflectors on the moon. By use of laser beams reflected from these arrays, the earth–moon distance has been measured to within 0.15 m.

TEST YOUR UNDERSTANDING OF SECTION 33.2   You are standing on the shore of a lake. You spot a tasty fish swimming some distance below the lake surface. (a) If you want to spear the fish, should you aim the spear (i) above, (ii) below, or (iii) directly at the apparent position of the fish? (b) If instead you use a high-power laser to simultaneously kill and cook the fish, should you aim the laser (i) above, (ii) below, or (iii) directly at the apparent position of the fish? ❚

33.3 TOTAL INTERNAL REFLECTION

We have described how light is partially reflected and partially transmitted at an interface between two materials with different indexes of refraction. Under certain circumstances, however, *all* of the light can be reflected back from the interface, with none of it being transmitted, even though the second material is transparent. **Figure 33.13a** shows how this can occur. Several rays are shown radiating from a point source in material *a* with index of refraction n_a. The rays strike the surface of a second material *b* with index n_b, where $n_a > n_b$. (Materials *a* and *b* could be water and air, respectively.) From Snell's law of refraction,

$$\sin\theta_b = \frac{n_a}{n_b}\sin\theta_a$$

Because n_a/n_b is greater than unity, $\sin\theta_b$ is larger than $\sin\theta_a$; the ray is bent *away from* the normal. Thus there must be some value of θ_a *less than* 90° for which $\sin\theta_b = 1$ and $\theta_b = 90°$. This is shown by ray 3 in the diagram, which emerges just grazing the surface at an angle of refraction of 90°. Compare Fig. 33.13a to the photograph of light rays in Fig. 33.13b.

33.13 (a) Total internal reflection. The angle of incidence for which the angle of refraction is 90° is called the critical angle: This is the case for ray 3. The reflected portions of rays 1, 2, and 3 are omitted for clarity. (b) Rays of laser light enter the water in the fishbowl from above; they are reflected at the bottom by mirrors tilted at slightly different angles. One ray undergoes total internal reflection at the air–water interface.

(a) Total internal reflection

Total internal reflection occurs only if $n_b < n_a$.

At the critical angle of incidence, θ_{crit}, the angle of refraction $\theta_b = 90°$.

Any ray with $\theta_a > \theta_{crit}$ shows total internal reflection.

(b) A light beam enters the top left of the tank, then reflects at the bottom from mirrors tilted at different angles. One beam undergoes total internal reflection at the air–water interface.

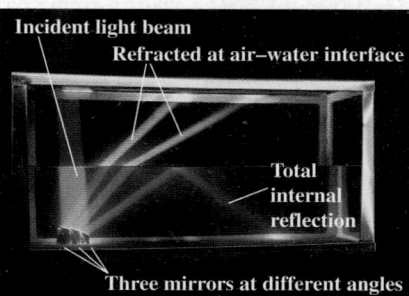

Incident light beam
Refracted at air–water interface
Total internal reflection
Three mirrors at different angles

The angle of incidence for which the refracted ray emerges tangent to the surface is called the **critical angle,** denoted by θ_{crit}. (A more detailed analysis using Maxwell's equations shows that as the incident angle approaches the critical angle, the transmitted intensity approaches zero.) If the angle of incidence is *larger* than the critical angle, $\sin\theta_b$ would have to be greater than unity, which is impossible. Beyond the critical angle, the ray *cannot* pass into the upper material; it is trapped in the lower material and is completely reflected at the boundary surface. This situation, called **total internal reflection,** occurs only when a ray in material a is incident on a second material b whose index of refraction is *smaller* than that of material a (that is, $n_b < n_a$).

We can find the critical angle for two given materials a and b by setting $\theta_b = 90°$ ($\sin\theta_b = 1$) in Snell's law. We then have

$$\text{Critical angle for total internal reflection} \quad \sin\theta_{crit} = \frac{n_b \cdots \text{Index of refraction of second material}}{n_a \cdots \text{Index of refraction of first material}} \qquad (33.6)$$

Total internal reflection will occur if the angle of incidence θ_a is larger than or equal to θ_{crit}.

Applications of Total Internal Reflection

Total internal reflection finds numerous uses in optical technology. As an example, consider glass with index of refraction $n = 1.52$. If light propagating within this glass encounters a glass–air interface, the critical angle is

$$\sin\theta_{crit} = \frac{1}{1.52} = 0.658 \qquad \theta_{crit} = 41.1°$$

The light will be *totally reflected* if it strikes the glass–air surface at an angle of 41.1° or larger. Because the critical angle is slightly smaller than 45°, it is possible to use a prism with angles of 45°−45°−90° as a totally reflecting surface. As reflectors, totally reflecting prisms have some advantages over metallic surfaces such as ordinary coated-glass mirrors. While no metallic surface reflects 100% of the light incident on it, light can be *totally* reflected by a prism. These reflecting properties of a prism are unaffected by tarnishing.

A 45°−45°−90° prism, used as in **Fig. 33.14a**, is called a *Porro* prism. Light enters and leaves at right angles to the hypotenuse and is totally reflected at each of the shorter faces. The total change of direction of the rays is 180°. Binoculars often use combinations of two Porro prisms, as in Fig. 33.14b.

(a) Total internal reflection in a Porro prism

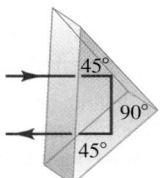

45°

90°

45°

If the incident beam is oriented as shown, total internal reflection occurs on the 45° faces (because, for a glass–air interface, $\theta_{crit} = 41.1$).

(b) Binoculars use Porro prisms to reflect the light to each eyepiece.

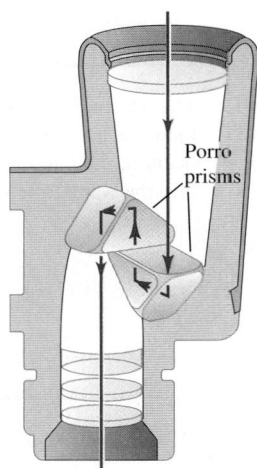

Porro prisms

33.14 (a) Total internal reflection in a Porro prism. (b) A combination of two Porro prisms in binoculars.

33.15 A transparent rod with refractive index greater than that of the surrounding material.

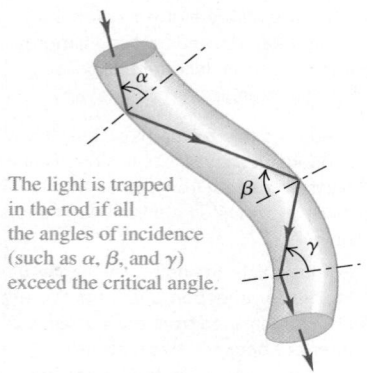

The light is trapped in the rod if all the angles of incidence (such as α, β, and γ) exceed the critical angle.

33.16 This colored x-ray image of a patient's abdomen shows an endoscope winding through the colon.

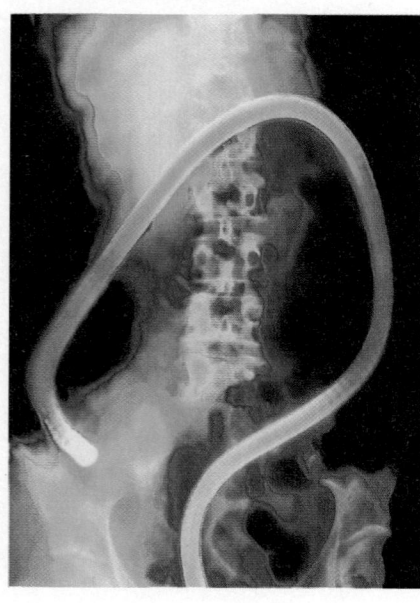

When a beam of light enters at one end of a transparent rod (**Fig. 33.15**), the light can be totally reflected internally if the index of refraction of the rod is greater than that of the surrounding material. The light is "trapped" within even a curved rod, provided that the curvature is not too great. A bundle of fine glass or plastic fibers behaves in the same way and has the advantage of being flexible. A bundle may consist of thousands of individual fibers, each of the order of 0.002 to 0.01 mm in diameter. If the fibers are assembled in the bundle so that the relative positions of the ends are the same (or mirror images) at both ends, the bundle can transmit an image.

Fiber-optic devices have found a wide range of medical applications in instruments called *endoscopes,* which can be inserted directly into the bronchial tubes, the bladder, the colon, and other organs for direct visual examination (**Fig. 33.16**). A bundle of fibers can even be enclosed in a hypodermic needle for studying tissues and blood vessels far beneath the skin.

Fiber optics also have applications in communication systems. The rate at which information can be transmitted by a wave (light, radio, or whatever) is proportional to the frequency. To see qualitatively why this is so, consider modulating (modifying) the wave by chopping off some of the wave crests. Suppose each crest represents a binary digit, with a chopped-off crest representing a zero and an unmodified crest representing a one. The number of binary digits we can transmit per unit time is thus proportional to the frequency of the wave. Infrared and visible-light waves have much higher frequency than do radio waves, so a modulated laser beam can transmit an enormous amount of information through a single fiber-optic cable.

Another advantage of optical fibers is that they can be made thinner than conventional copper wire, so more fibers can be bundled together in a cable of a given diameter. Hence more distinct signals (for instance, different phone lines) can be sent over the same cable. Because fiber-optic cables are electrical insulators, they are immune to electrical interference from lightning and other sources, and they don't allow unwanted currents between source and receiver. For these and other reasons, fiber-optic cables play an important role in long-distance telephone, television, and Internet communication.

Total internal reflection also plays an important role in the design of jewelry. The brilliance of diamond is due in large measure to its very high index of refraction ($n = 2.417$) and correspondingly small critical angle. Light entering a cut diamond is totally internally reflected from facets on its back surface and then emerges from its front surface (see the photograph that opens this chapter). "Imitation diamond" gems, such as cubic zirconia, are made from less expensive crystalline materials with comparable indexes of refraction.

CONCEPTUAL EXAMPLE 33.4 **A LEAKY PERISCOPE**

SOLUTION

A submarine periscope uses two totally reflecting 45°–45°–90° prisms with total internal reflection on the sides adjacent to the 45° angles. Explain why the periscope will no longer work if it springs a leak and the bottom prism is covered with water.

SOLUTION

The critical angle for water ($n_b = 1.33$) on glass ($n_a = 1.52$) is

$$\theta_{\text{crit}} = \arcsin\frac{1.33}{1.52} = 61.0°$$

The 45° angle of incidence for a totally reflecting prism is *smaller* than this new 61° critical angle, so total internal reflection does not occur at the glass–water interface. Most of the light is transmitted into the water, and very little is reflected back into the prism.

TEST YOUR UNDERSTANDING OF SECTION 33.3 In which of the following situations is there total internal reflection? (i) Light propagating in water ($n = 1.33$) strikes a water–air interface at an incident angle of 70°; (ii) light propagating in glass ($n = 1.52$) strikes a glass–water interface at an incident angle of 70°; (iii) light propagating in water strikes a water–glass interface at an incident angle of 70°. ▮

33.4 DISPERSION

Ordinary white light is a superposition of waves with all visible wavelengths. The speed of light *in vacuum* is the same for all wavelengths, but the speed in a material substance is different for different wavelengths. Therefore the index of refraction of a material depends on wavelength. The dependence of wave speed and index of refraction on wavelength is called **dispersion.**

Figure 33.17 shows the variation of index of refraction n with wavelength for some common optical materials. Note that the horizontal axis of this figure is the wavelength of the light *in vacuum,* λ_0; the wavelength in the material is given by Eq. (33.5), $\lambda = \lambda_0/n$. In most materials the value of n *decreases* with increasing wavelength and decreasing frequency, and thus n *increases* with decreasing wavelength and increasing frequency. In such a material, light of longer wavelength has greater speed than light of shorter wavelength.

Figure 33.18 shows a ray of white light incident on a prism. The deviation (change of direction) produced by the prism increases with increasing index of refraction and frequency and decreasing wavelength. So violet light is deviated most, and red is deviated least. When it comes out of the prism, the light is spread out into a fan-shaped beam, as shown. The light is said to be *dispersed* into a spectrum. The amount of dispersion depends on the *difference* between the refractive indexes for violet light and for red light. From Fig. 33.17 we can see that for fluorite, the difference between the indexes for red and violet is small, and the dispersion will also be small. A better choice of material for a prism whose purpose is to produce a spectrum would be silicate flint glass, for which there is a larger difference in the value of n between red and violet.

As we mentioned in Section 33.3, the brilliance of diamond is due in **?** part to its unusually large refractive index; another important factor is its large dispersion, which causes white light entering a diamond to emerge as a multicolored spectrum. Crystals of rutile and of strontium titanate, which can be produced synthetically, have about eight times the dispersion of diamond.

33.17 Variation of index of refraction n with wavelength for different transparent materials. The horizontal axis shows the wavelength λ_0 of the light *in vacuum;* the wavelength in the material is equal to $\lambda = \lambda_0/n$.

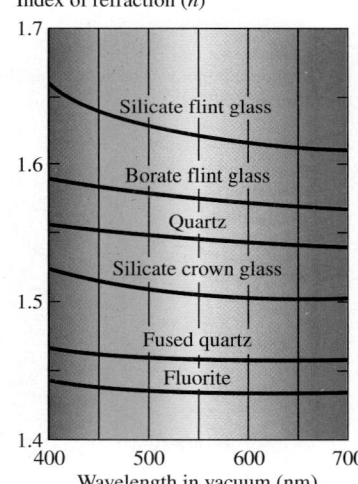

33.18 Dispersion of light by a prism. The band of colors is called a spectrum.

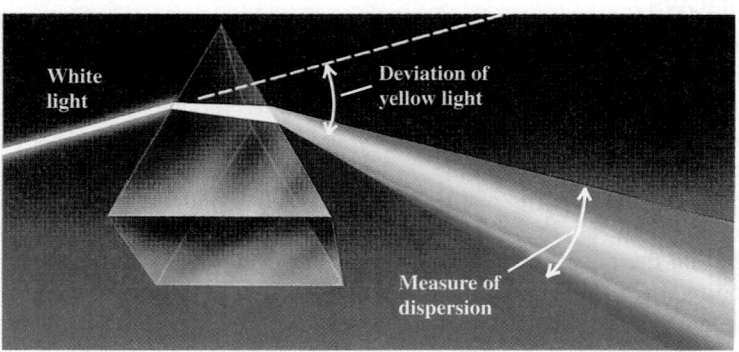

33.19 How rainbows form.

(a) A double rainbow

Secondary rainbow
(note reversed colors)

Primary rainbow

(b) The paths of light rays entering the upper half of a raindrop

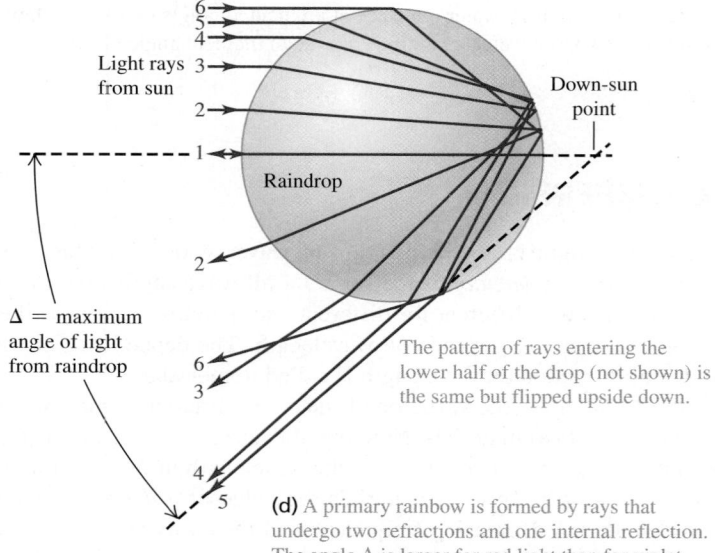

Light rays from sun

Down-sun point

Raindrop

Δ = maximum angle of light from raindrop

The pattern of rays entering the lower half of the drop (not shown) is the same but flipped upside down.

(c) Forming a rainbow. The sun in this illustration is directly behind the observer at *P*.

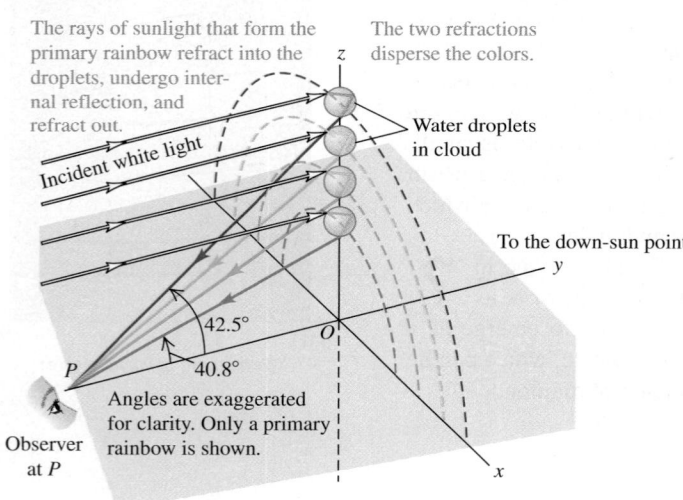

The rays of sunlight that form the primary rainbow refract into the droplets, undergo internal reflection, and refract out.

The two refractions disperse the colors.

Incident white light

Water droplets in cloud

To the down-sun point

42.5°

40.8°

Angles are exaggerated for clarity. Only a primary rainbow is shown.

Observer at *P*

(d) A primary rainbow is formed by rays that undergo two refractions and one internal reflection. The angle Δ is larger for red light than for violet.

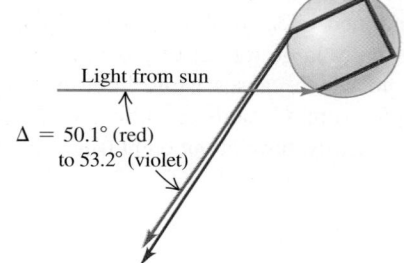

Light from sun

Δ = 40.8° (violet) to 42.5° (red)

(e) A secondary rainbow is formed by rays that undergo two refractions and *two* internal reflections. The angle Δ is larger for violet light than for red.

Light from sun

Δ = 50.1° (red) to 53.2° (violet)

Rainbows

When you experience the beauty of a rainbow, as in **Fig. 33.19a**, you are seeing the combined effects of dispersion, refraction, and reflection. Sunlight comes from behind you, enters a water droplet, is (partially) reflected from the back surface of the droplet, and is refracted again upon exiting the droplet (Fig. 33.19b). A light ray that enters the middle of the raindrop is reflected straight back. All other rays exit the raindrop within an angle Δ of that middle ray, with many rays "piling up" at the angle Δ. What you see is a disk of light of angular radius Δ centered on the down-sun point (the point in the sky opposite the sun); due to the "piling up" of light rays, the disk is brightest around its rim, which we see as a rainbow (Fig. 33.19c). Because no light reaches your eye from angles larger than Δ, the sky looks dark outside the rainbow (see Fig. 33.19a). The value of the angle Δ depends on the index of refraction of the water that makes up the rain-drops, which in turn depends on the wavelength (Fig. 33.19d). The bright disk of red light is slightly larger than that for orange light, which in turn is slightly larger than that for yellow light, and so on. As a result, you see the rainbow as a band of colors.

In many cases you can see a second, larger rainbow. It is the result of dispersion, refraction, and *two* reflections from the back surface of the droplet (Fig. 33.19e). Each time a light ray hits the back surface, part of the light is refracted out of the drop (not shown in Fig. 33.19); after two such hits, relatively little light is left inside the drop, which is why the secondary rainbow is noticeably fainter than the primary rainbow. Just as a mirror held up to a book reverses the printed letters, so the second reflection reverses the sequence of colors in the secondary rainbow. You can see this effect in Fig. 33.19a.

33.5 POLARIZATION

Polarization is a characteristic of all transverse waves. This chapter is about light, but to introduce some basic polarization concepts, let's go back to the transverse waves on a string that we studied in Chapter 15. For a string that in equilibrium lies along the x-axis, the displacements may be along the y-direction, as in **Fig. 33.20a**. In this case the string always lies in the xy-plane. But the displacements might instead be along the z-axis, as in Fig. 33.20b; then the string always lies in the xz-plane.

When a wave has only y-displacements, we say that it is **linearly polarized** in the y-direction; a wave with only z-displacements is linearly polarized in the z-direction. For mechanical waves we can build a **polarizing filter,** or **polarizer,** that permits only waves with a certain polarization direction to pass. In Fig. 33.20c the string can slide vertically in the slot without friction, but no horizontal motion is possible. This filter passes waves that are polarized in the y-direction but blocks those that are polarized in the z-direction.

This same language can be applied to electromagnetic waves, which also have polarization. As we learned in Chapter 32, an electromagnetic wave is a transverse wave; the fluctuating electric and magnetic fields are perpendicular to each other and to the direction of propagation. We always define the direction of polarization of an electromagnetic wave to be the direction of the *electric*-field vector $\vec{E}$, not the magnetic field, because many common electromagnetic-wave detectors respond to the electric forces on electrons in materials, not the magnetic forces. Thus the electromagnetic wave described by Eq. (32.17),

$$\vec{E}(x, t) = \hat{j}E_{\max}\cos(kx - \omega t)$$

$$\vec{B}(x, t) = \hat{k}B_{\max}\cos(kx - \omega t)$$

is said to be polarized in the y-direction because the electric field has only a y-component.

CAUTION The meaning of "polarization" It's unfortunate that the same word "polarization" that is used to describe the direction of $\vec{E}$ in an electromagnetic wave is also used to describe the shifting of electric charge within a body, such as in response to a nearby charged body; we described this latter kind of polarization in Section 21.2 (see Fig. 21.7). Don't confuse these two concepts! ▮

33.20 (a), (b) Polarized waves on a string. (c) Making a polarized wave on a string from an unpolarized one using a polarizing filter.

(a) Transverse wave linearly polarized in the y-direction

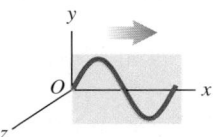

(b) Transverse wave linearly polarized in the z-direction

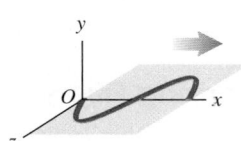

(c) The slot functions as a polarizing filter, passing only components polarized in the y-direction.

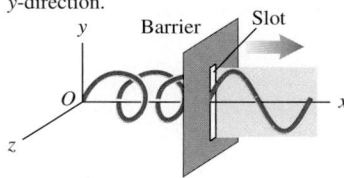

33.21 (a) Electrons in the red and white broadcast antenna oscillate vertically, producing vertically polarized electromagnetic waves that propagate away from the antenna in the horizontal direction. (The small gray antennas are for relaying cellular phone signals.) (b) No matter how this light bulb is oriented, the random motion of electrons in the filament produces unpolarized light waves.

(a)

(b)

Polarizing Filters

Waves emitted by a radio transmitter are usually linearly polarized. The vertical antennas that are used for radio broadcasting emit waves that, in a horizontal plane around the antenna, are polarized in the vertical direction (parallel to the antenna) (**Fig. 33.21a**).

The situation is different for visible light. Light from incandescent light bulbs and fluorescent light fixtures is *not* polarized (Fig. 33.21b). The "antennas" that radiate light waves are the molecules that make up the sources. The waves emitted by any one molecule may be linearly polarized, like those from a radio antenna. But any actual light source contains a tremendous number of molecules with random orientations, so the emitted light is a random mixture of waves linearly polarized in all possible transverse directions. Such light is called **unpolarized light** or **natural light.** To create polarized light from unpolarized natural light requires a filter that is analogous to the slot for mechanical waves in Fig. 33.20c.

Polarizing filters for electromagnetic waves have different details of construction, depending on the wavelength. For microwaves with a wavelength of a few centimeters, a good polarizer is an array of closely spaced, parallel conducting wires that are insulated from each other. (Think of a barbecue grill with the outer metal ring replaced by an insulating one.) Electrons are free to move along the length of the conducting wires and will do so in response to a wave whose $\vec{E}$ field is parallel to the wires. The resulting currents in the wires dissipate energy by I^2R heating; the dissipated energy comes from the wave, so whatever wave passes through the grid is greatly reduced in amplitude. Waves with $\vec{E}$ oriented perpendicular to the wires pass through almost unaffected, since electrons cannot move through the air between the wires. Hence a wave that passes through such a filter will be predominantly polarized in the direction perpendicular to the wires.

The most common polarizing filter for visible light is a material known by the trade name Polaroid, widely used for sunglasses and polarizing filters for camera lenses. This material incorporates substances that have **dichroism,** a selective absorption in which one of the polarized components is absorbed much more strongly than the other (**Fig. 33.22**). A Polaroid filter transmits 80% or more of the intensity of a wave that is polarized parallel to the **polarizing axis** of the material, but only 1% or less for waves that are polarized perpendicular to this axis. In one type of Polaroid filter, long-chain molecules within the filter are oriented with their axis perpendicular to the polarizing axis; these molecules preferentially absorb light that is polarized along their length, much like the conducting wires in a polarizing filter for microwaves.

33.22 A Polaroid filter is illuminated by unpolarized natural light (shown by $\vec{E}$ vectors that point in all directions perpendicular to the direction of propagation). The transmitted light is linearly polarized along the polarizing axis (shown by $\vec{E}$ vectors along the polarization direction only).

Filter only partially absorbs vertically polarized component of light.

Incident unpolarized light

Polarizing axis

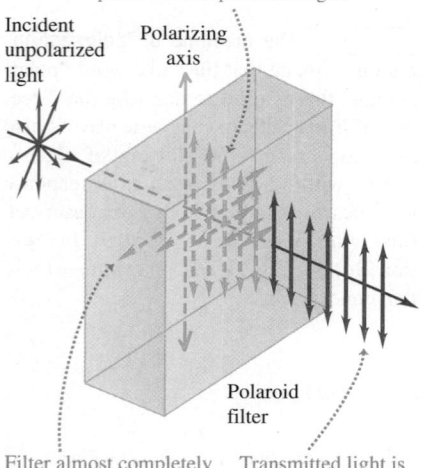

Polaroid filter

Filter almost completely absorbs horizontally polarized component of light.

Transmitted light is linearly polarized in the vertical direction.

Using Polarizing Filters

An *ideal* polarizing filter (polarizer) passes 100% of the incident light that is polarized parallel to the filter's polarizing axis but completely blocks all light

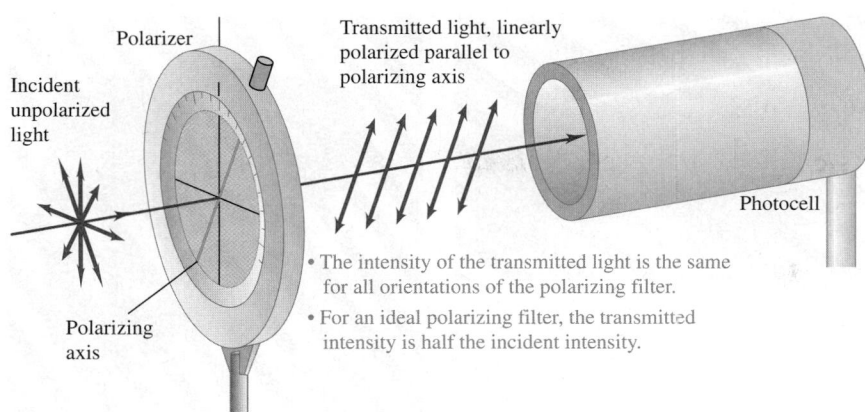

that is polarized perpendicular to this axis. Such a device is an unattainable idealization, but the concept is useful in clarifying the basic ideas. In the following discussion we will assume that all polarizing filters are ideal. In **Fig. 33.23** unpolarized light is incident on a flat polarizing filter. The $\vec{E}$ vector of the incident wave can be represented in terms of components parallel and perpendicular to the polarizer axis (shown in blue); only the component of $\vec{E}$ parallel to the polarizing axis is transmitted. Hence the light emerging from the polarizer is linearly polarized parallel to the polarizing axis.

When unpolarized light is incident on an ideal polarizer as in Fig. 33.23, the intensity of the transmitted light is *exactly half* that of the incident unpolarized light, no matter how the polarizing axis is oriented. Here's why: We can resolve the $\vec{E}$ field of the incident wave into a component parallel to the polarizing axis and a component perpendicular to it. Because the incident light is a random mixture of all states of polarization, these two components are, on average, equal. The ideal polarizer transmits only the component that is parallel to the polarizing axis, so half the incident intensity is transmitted.

What happens when the linearly polarized light emerging from a polarizer passes through a second polarizer, or *analyzer,* as in **Fig. 33.24**? Suppose the polarizing axis of the analyzer makes an angle ϕ with the polarizing axis of the first polarizer. We can resolve the linearly polarized light that is transmitted by the first polarizer into two components, as shown in Fig. 33.24, one parallel and the other perpendicular to the axis of the analyzer. Only the parallel component, with amplitude $E\cos\phi$, is transmitted by the analyzer. The transmitted intensity is greatest when $\phi = 0$, and it is zero when the polarizer and analyzer are *crossed*

DEMO

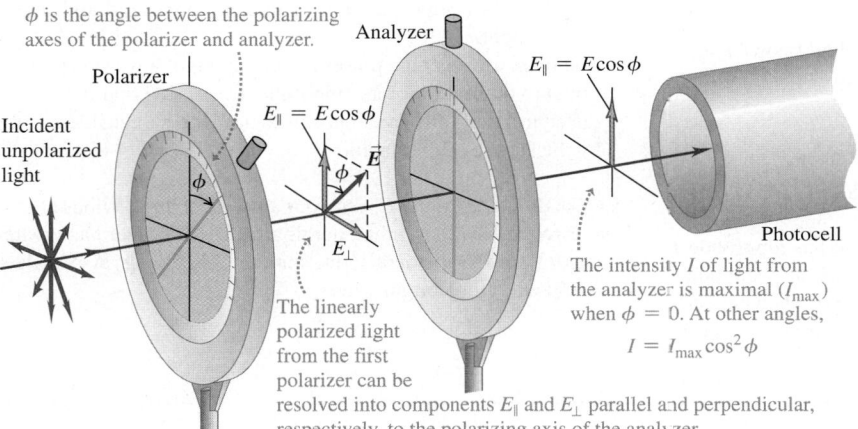

33.25 These photos show the view through Polaroid sunglasses whose polarizing axes are (left) aligned ($\phi = 0$) and (right) perpendicular ($\phi = 90°$). The transmitted intensity is greatest when the axes are aligned; it is zero when the axes are perpendicular.

so that $\phi = 90°$ (**Fig. 33.25**). To determine the direction of polarization of the light transmitted by the first polarizer, rotate the analyzer until the photocell in Fig. 33.24 measures zero intensity; the polarization axis of the first polarizer is then perpendicular to that of the analyzer.

To find the transmitted intensity at intermediate values of the angle ϕ, we recall from Section 32.4 that the intensity of an electromagnetic wave is proportional to the *square* of the amplitude of the wave [see Eq. (32.29)]. The ratio of transmitted to incident *amplitude* is $\cos\phi$, so the ratio of transmitted to incident *intensity* is $\cos^2\phi$. Thus the intensity transmitted is

Intensity of polarized light passed through an analyzer

Malus's law:
$$I = I_{\text{max}} \cos^2\phi \quad \text{(33.7)}$$
Angle between polarization axis of light and polarizing axis of analyzer

Maximum transmitted intensity

This relationship, discovered experimentally by Étienne-Louis Malus in 1809, is called **Malus's law.** Malus's law applies *only* if the incident light passing through the analyzer is already linearly polarized.

PROBLEM-SOLVING STRATEGY 33.2 | LINEAR POLARIZATION

IDENTIFY *the relevant concepts:* In all electromagnetic waves, including light waves, the direction of polarization is the direction of the $\vec{E}$ field and is perpendicular to the propagation direction. Problems about polarizers are therefore about the components of $\vec{E}$ parallel and perpendicular to the polarizing axis.

SET UP *the problem* using the following steps:
1. Start by drawing a large, neat diagram. Label all known angles, including the angles of all polarizing axes.
2. Identify the target variables.

EXECUTE *the solution* as follows:
1. Remember that a polarizer lets pass only electric-field components parallel to its polarizing axis.
2. If the incident light is linearly polarized and has amplitude E and intensity I_{max}, the light that passes through an ideal polarizer has amplitude $E\cos\phi$ and intensity $I_{\text{max}}\cos^2\phi$, where ϕ is the angle between the incident polarization direction and the filter's polarizing axis.

3. Unpolarized light is a random mixture of all possible polarization states, so on the average it has equal components in any two perpendicular directions. When passed through an ideal polarizer, unpolarized light becomes linearly polarized light with half the incident intensity. Partially linearly polarized light is a superposition of linearly polarized and unpolarized light.
4. The intensity (average power per unit area) of a wave is proportional to the *square* of its amplitude. If you find that two waves differ in amplitude by a certain factor, their intensities differ by the square of that factor.

EVALUATE *your answer:* Check your answer for any obvious errors. If your results say that light emerging from a polarizer has greater intensity than the incident light, something's wrong: A polarizer can't add energy to a light wave.

EXAMPLE 33.5   TWO POLARIZERS IN COMBINATION

In Fig. 33.24 the incident unpolarized light has intensity I_0. Find the intensities transmitted by the first and second polarizers if the angle between the axes of the two filters is 30°.

SOLUTION

IDENTIFY and SET UP: This problem involves a polarizer (a polarizing filter on which unpolarized light shines, producing polarized light) and an analyzer (a second polarizing filter on which the polarized light shines). We are given the intensity I_0 of the incident light and the angle $\phi = 30°$ between the axes of the polarizers. We use Malus's law, Eq. (33.7), to solve for the intensities of the light emerging from each polarizer.

EXECUTE: The incident light is unpolarized, so the intensity of the linearly polarized light transmitted by the first polarizer is $I_0/2$. From Eq. (33.7) with $\phi = 30°$, the second polarizer reduces the intensity by a further factor of $\cos^2 30° = \frac{3}{4}$. Thus the intensity transmitted by the second polarizer is

$$\left(\frac{I_0}{2}\right)\left(\tfrac{3}{4}\right) = \tfrac{3}{8}I_0$$

EVALUATE: Note that the intensity decreases after each passage through a polarizer. The only situation in which the transmitted intensity does *not* decrease is if the polarizer is ideal (so it absorbs none of the light that passes through it) and if the incident light is linearly polarized along the polarizing axis, so $\phi = 0$.

Polarization by Reflection

Unpolarized light can be polarized, either partially or totally, by *reflection*. In **Fig. 33.26**, unpolarized natural light is incident on a reflecting surface between two transparent optical materials. For most angles of incidence, waves for which the electric-field vector $\vec{E}$ is perpendicular to the plane of incidence (that is, parallel to the reflecting surface) are reflected more strongly than those for which $\vec{E}$ lies in this plane. In this case the reflected light is *partially polarized* in the direction perpendicular to the plane of incidence.

But at one particular angle of incidence, called the **polarizing angle** θ_p, the light for which $\vec{E}$ lies in the plane of incidence is *not reflected at all* but is completely refracted. At this same angle of incidence the light for which $\vec{E}$ is perpendicular to the plane of incidence is partially reflected and partially refracted. The *reflected* light is therefore *completely* polarized perpendicular to the plane of incidence, as shown in Fig. 33.26. The *refracted* (transmitted) light is *partially* polarized parallel to this plane; the refracted light is a mixture of the component parallel to the plane of incidence, all of which is refracted, and the remainder of the perpendicular component.

In 1812 the British scientist Sir David Brewster discovered that when the angle of incidence is equal to the polarizing angle θ_p, the reflected ray and the refracted

33.26 When light is incident on a reflecting surface at the polarizing angle, the reflected light is linearly polarized.

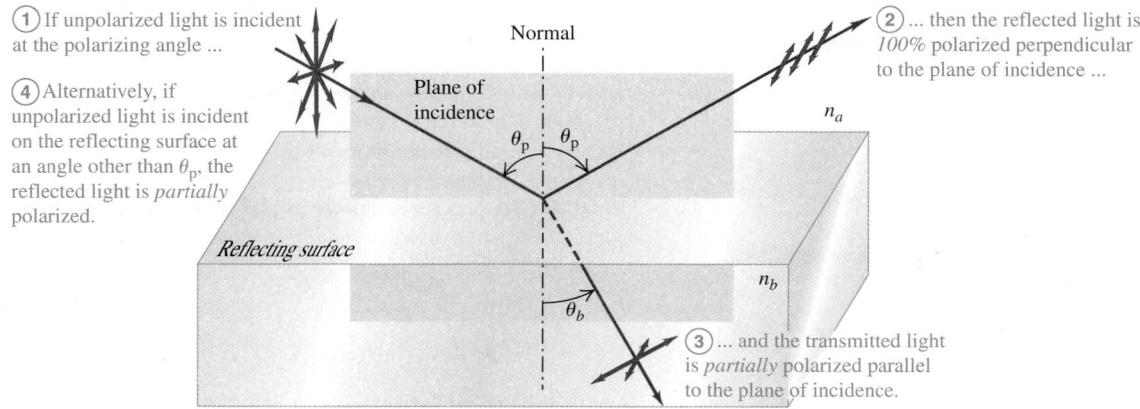

33.27 The significance of the polarizing angle. The open circles represent a component of $\vec{E}$ that is perpendicular to the plane of the figure (the plane of incidence) and parallel to the surface between the two materials.

Note: This is a side view of the situation shown in Fig. 33.26.

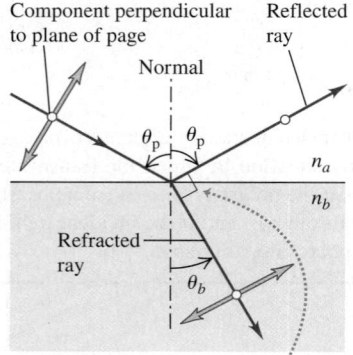

Component perpendicular to plane of page | Reflected ray

When light strikes a surface at the polarizing angle, the reflected and refracted rays are perpendicular to each other and
$$\tan\theta_p = \frac{n_b}{n_a}$$

ray are perpendicular to each other (**Fig. 33.27**). In this case the angle of refraction θ_b equals $90° - \theta_p$. From the law of refraction,

$$n_a\sin\theta_p = n_b\sin\theta_b = n_b\sin(90° - \theta_p) = n_b\cos\theta_p$$

Since $(\sin\theta_p)/(\cos\theta_p) = \tan\theta_p$, we can rewrite this equation as

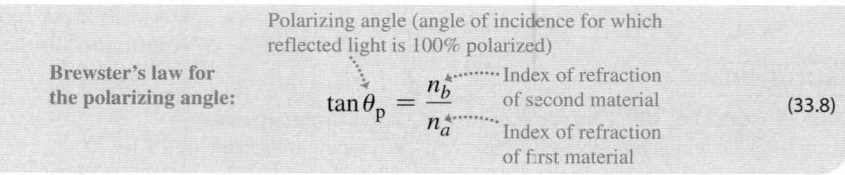

Polarizing angle (angle of incidence for which reflected light is 100% polarized)

Brewster's law for the polarizing angle:
$$\tan\theta_p = \frac{n_b}{n_a} \qquad (33.8)$$

Index of refraction of second material
Index of refraction of first material

This relationship is known as **Brewster's law.** Although discovered experimentally, it can also be *derived* from a wave model by using Maxwell's equations.

Polarization by reflection is the reason polarizing filters are widely used in sunglasses (Fig. 33.25). When sunlight is reflected from a horizontal surface, the plane of incidence is vertical, and the reflected light contains a preponderance of light that is polarized in the horizontal direction. When the reflection occurs at a smooth asphalt road surface, it causes unwanted glare. To eliminate this glare, the polarizing axis of the lens material is made vertical, so very little of the horizontally polarized light reflected from the road is transmitted to the eyes. The glasses also reduce the overall intensity of the transmitted light to somewhat less than 50% of the intensity of the unpolarized incident light.

EXAMPLE 33.6 **REFLECTION FROM A SWIMMING POOL'S SURFACE**

Sunlight reflects off the smooth surface of a swimming pool. (a) For what angle of reflection is the reflected light completely polarized? (b) What is the corresponding angle of refraction? (c) At night, an underwater floodlight is turned on in the pool. Repeat parts (a) and (b) for rays from the floodlight that strike the surface from below.

SOLUTION

IDENTIFY and SET UP: This problem involves polarization by reflection at an air–water interface in parts (a) and (b) and at a water–air interface in part (c). **Figure 33.28** shows our sketches.

33.28 Our sketches for this problem.

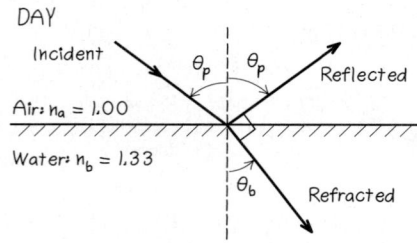

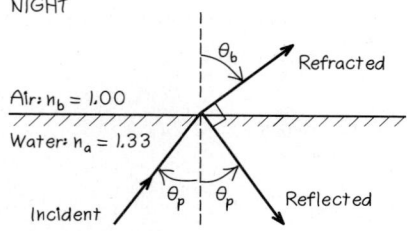

For both cases our first target variable is the polarizing angle θ_p, which we find from Brewster's law, Eq. (33.8). For this angle of reflection, the angle of refraction θ_b is the complement of θ_p (that is, $\theta_b = 90° - \theta_p$).

EXECUTE: (a) During the day (shown in the upper part of Fig. 33.28) the light moves in air toward water, so $n_a = 1.00$ (air) and $n_b = 1.33$ (water). From Eq. (33.8),

$$\theta_p = \arctan\frac{n_b}{n_a} = \arctan\frac{1.33}{1.00} = 53.1°$$

(b) The incident light is at the polarizing angle, so the reflected and refracted rays are perpendicular; hence

$$\theta_b = 90° - \theta_p = 90° - 53.1° = 36.9°$$

(c) At night (shown in the lower part of Fig. 33.28) the light moves in water toward air, so now $n_a = 1.33$ and $n_b = 1.00$. Again using Eq. (33.8), we have

$$\theta_p = \arctan\frac{1.00}{1.33} = 36.9°$$

$$\theta_b = 90° - 36.9° = 53.1°$$

EVALUATE: We check our answer in part (b) by using Snell's law, $n_a\sin\theta_a = n_b\sin\theta_b$, to solve for θ_b:

$$\sin\theta_b = \frac{n_a\sin\theta_p}{n_b} = \frac{1.00\sin 53.1°}{1.33} = 0.600$$

$$\theta_b = \arcsin(0.600) = 36.9°$$

Note that the two polarizing angles found in parts (a) and (c) add to 90°. This is *not* an accident; can you see why?

Circular and Elliptical Polarization

Light and other electromagnetic radiation can also have *circular* or *elliptical* polarization. To introduce these concepts, let's return once more to mechanical waves on a stretched string. In Fig. 33.20, suppose the two linearly polarized waves in parts (a) and (b) are in phase and have equal amplitude. When they are superposed, each point in the string has simultaneous y- and z-displacements of equal magnitude. A little thought shows that the resultant wave lies in a plane oriented at 45° to the y- and z-axes (i.e., in a plane making a 45° angle with the xy- and xz-planes). The amplitude of the resultant wave is larger by a factor of $\sqrt{2}$ than that of either component wave, and the resultant wave is linearly polarized.

But now suppose the two equal-amplitude waves differ in phase by a quarter-cycle. Then the resultant motion of each point corresponds to a superposition of two simple harmonic motions at right angles, with a quarter-cycle phase difference. The y-displacement at a point is greatest at times when the z-displacement is zero, and vice versa. The string as a whole then no longer moves in a single plane. It can be shown that each point on the rope moves in a *circle* in a plane parallel to the yz-plane. Successive points on the rope have successive phase differences, and the overall motion of the string has the appearance of a rotating helix, as shown to the left of the polarizing filter in Fig. 33.20c. Such a superposition of two linearly polarized waves is called **circular polarization.**

Figure 33.29 shows the analogous situation for an electromagnetic wave. Two sinusoidal waves of equal amplitude, polarized in the y- and z-directions and with a quarter-cycle phase difference, are superposed. The result is a wave in which the $\vec{E}$ vector at each point has a constant magnitude but *rotates* around the direction of propagation. The wave in Fig. 33.29 is propagating toward you and the $\vec{E}$ vector appears to be rotating clockwise, so it is called a *right circularly polarized* electromagnetic wave. If instead the $\vec{E}$ vector of a wave coming toward you appears to be rotating counterclockwise, it is called a *left circularly polarized* electromagnetic wave.

If the phase difference between the two component waves is something other than a quarter-cycle, or if the two component waves have different amplitudes, then each point on the string traces out not a circle but an *ellipse*. The resulting wave is said to be **elliptically polarized.**

For electromagnetic waves with radio frequencies, circular or elliptical polarization can be produced by using two antennas at right angles, fed from the same transmitter but with a phase-shifting network that introduces the appropriate phase difference. For light, the phase shift can be introduced by use of a material

33.29 Circular polarization of an electromagnetic wave moving toward you parallel to the x-axis. The y-component of $\vec{E}$ lags the z-component by a quarter-cycle. This phase difference results in right circular polarization.

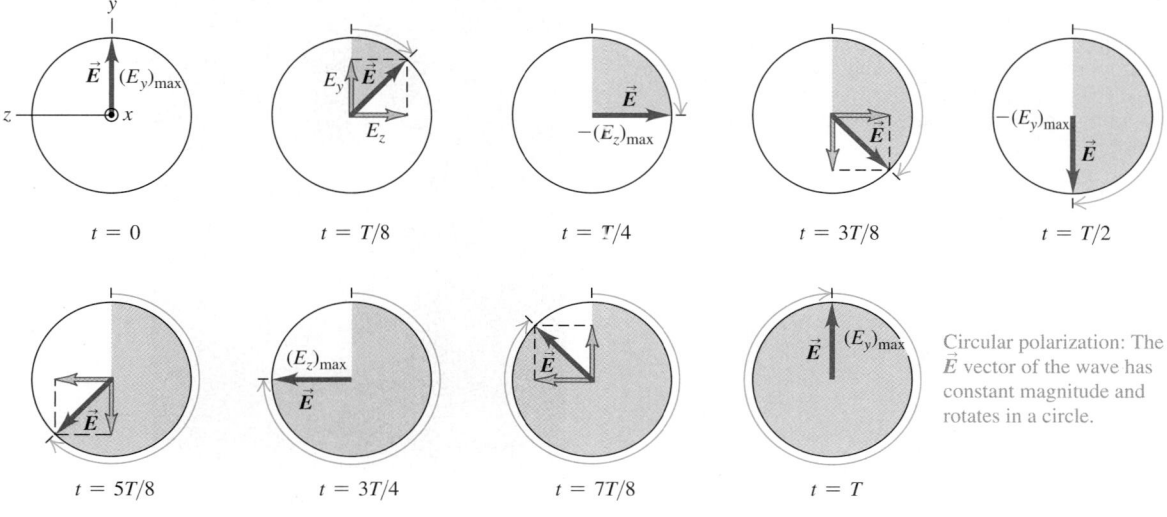

Circular polarization: The $\vec{E}$ vector of the wave has constant magnitude and rotates in a circle.

Application Birefringence and Liquid Crystal Displays In each pixel of an LCD computer screen is a birefringent material called a liquid crystal. This material is composed of rod-shaped molecules that align to produce a fluid with two different indexes of refraction. The liquid crystal is placed between linear polarizing filters with perpendicular polarizing axes, and the sandwich of filters and liquid crystal is backlighted. The two polarizers by themselves would not transmit light, but like the birefringent object in Fig. 33.30, the liquid crystal allows light to pass through. Varying the voltage across a pixel turns the birefringence effect on and off, changing the pixel from bright to dark and back again.

Microscope image of a liquid crystal

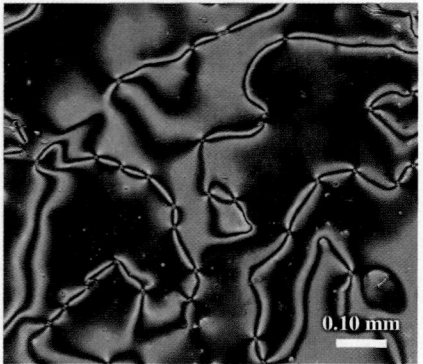

0.10 mm

Liquid crystal display

that exhibits *birefringence*—that is, has different indexes of refraction for different directions of polarization. A common example is calcite ($CaCO_3$). When a calcite crystal is oriented appropriately in a beam of unpolarized light, its refractive index (for a wavelength in vacuum of 589 nm) is 1.658 for one direction of polarization and 1.486 for the perpendicular direction. When two waves with equal amplitude and with perpendicular directions of polarization enter such a material, they travel with different speeds. If they are in phase when they enter the material, then in general they are no longer in phase when they emerge. If the crystal is just thick enough to introduce a quarter-cycle phase difference, then the crystal converts linearly polarized light to circularly polarized light. Such a crystal is called a *quarter-wave plate*. Such a plate also converts circularly polarized light to linearly polarized light. Can you prove this?

Photoelasticity

Some optical materials that are not normally birefringent become so when they are subjected to mechanical stress. This is the basis of the science of *photoelasticity*. Stresses in girders, boiler plates, gear teeth, and cathedral pillars can be analyzed by constructing a transparent model of the object, usually of a plastic material, subjecting it to stress, and examining it between a polarizer and an analyzer in the crossed position. Very complicated stress distributions can be studied by these optical methods.

Figure 33.30 is a photograph of a photoelastic model under stress. The polarized light that enters the model can be thought of as having a component along each of the two directions of the birefringent plastic. Since these two components travel through the plastic at different speeds, the light that emerges from the other side of the model can have a different overall direction of polarization. Hence some of this transmitted light will be able to pass through the analyzer even though its polarization axis is at a 90° angle to the polarizer's axis, and the stressed areas in the plastic will appear as bright spots. The amount of birefringence is different for different wavelengths and hence different colors of light; the color that appears at each location in Fig. 33.30 is that for which the transmitted light is most nearly polarized along the analyzer's polarization axis.

TEST YOUR UNDERSTANDING OF SECTION 33.5 You are taking a photograph of a sunlit office building at sunrise, so the plane of incidence is nearly horizontal. In order to minimize the reflections from the building's windows, you place a polarizing filter on the camera lens. How should you orient the filter? (i) With the polarizing axis vertical; (ii) with the polarizing axis horizontal; (iii) either orientation will minimize the reflections just as well; (iv) neither orientation will have any effect. ▮

33.30 This plastic model of an artificial hip joint was photographed between two polarizing filters (a polarizer and an analyzer) with perpendicular polarizing axes. The colored interference pattern reveals the direction and magnitude of stresses in the model. Engineers use these results to help design the actual hip joint (used in hip replacement surgery), which is made of metal.

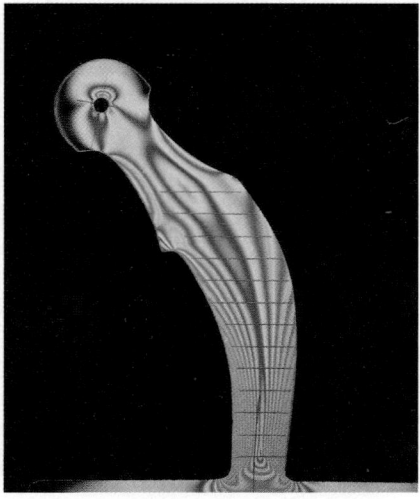

33.6 SCATTERING OF LIGHT

The sky is blue. Sunsets are red. Skylight is partially polarized; that's why the sky looks darker from some angles than from others when it is viewed through Polaroid sunglasses. As we will see, a single phenomenon is responsible for all of these effects.

When you look at the daytime sky, the light that you see is sunlight that has been absorbed and then re-radiated in a variety of directions. This process is called **scattering.** (If the earth had no atmosphere, the sky would appear as black in the daytime as it does at night, just as it does to an astronaut in space or on the moon.) **Figure 33.31** shows some of the details of the scattering process. Sunlight, which is unpolarized, comes from the left along the x-axis and passes over an observer looking vertically upward along the y-axis. (We are viewing the situation from the side.) Consider the molecules of the earth's atmosphere located at point O. The electric field in the beam of sunlight sets the electric charges in these molecules into vibration. Since light is a transverse wave, the direction of the electric field in any component of the sunlight lies in the yz-plane, and the motion of the charges takes place in this plane. There is no field, and hence no motion of charges, in the direction of the x-axis.

An incident light wave sets the electric charges in the molecules at point O vibrating along the line of $\vec{E}$. We can resolve this vibration into two components, one along the y-axis and the other along the z-axis. Each component in the incident light produces the equivalent of two molecular "antennas," oscillating with the same frequency as the incident light and lying along the y- and z-axes.

We mentioned in Chapter 32 that an oscillating charge, like those in an antenna, does not radiate in the direction of its oscillation. (See Fig. 32.3 in Section 32.1.) Thus the "antenna" along the y-axis does not send any light to the observer directly below it, although it does emit light in other directions. Therefore the only light reaching this observer comes from the other molecular "antenna," corresponding to the oscillation of charge along the z-axis. This light is linearly polarized, with its electric field along the z-axis (parallel to the "antenna"). The red vectors on the y-axis below point O in Fig. 33.31 show the direction of polarization of the light reaching the observer.

As the original beam of sunlight passes though the atmosphere, its intensity decreases as its energy goes into the scattered light. Detailed analysis of the scattering process shows that the intensity of the light scattered from air molecules increases in proportion to the fourth power of the frequency (inversely to the fourth power of the wavelength). Thus the intensity ratio for the two ends of the visible spectrum is $(750 \text{ nm}/380 \text{ nm})^4 = 15$. Roughly speaking, scattered light contains 15 times as much blue light as red, and that's why the sky is blue.

BIO Application Bee Vision and Polarized Light from the Sky The eyes of a bee can detect the polarization of light. Bees use this ability when they navigate between the hive and food sources. As Fig. 33.31 would suggest, a bee sees unpolarized light if it looks directly toward the sun and sees completely polarized light if it looks 90° away from the sun. These polarizations are unaffected by the presence of clouds, so a bee can navigate relative to the sun even on an overcast day.

Eyes

33.31 When the sunbathing observer on the left looks up, he sees blue, polarized sunlight that has been scattered by air molecules. The observer on the right sees reddened, unpolarized light when he looks at the sun.

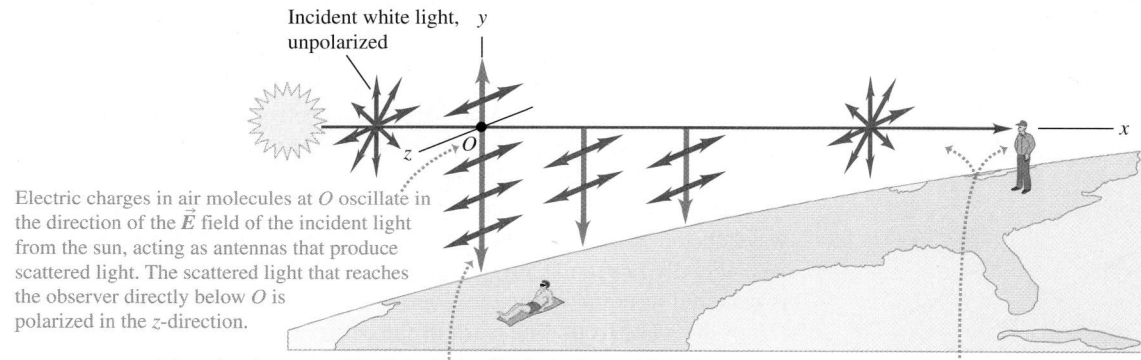

Incident white light, y unpolarized

Electric charges in air molecules at O oscillate in the direction of the $\vec{E}$ field of the incident light from the sun, acting as antennas that produce scattered light. The scattered light that reaches the observer directly below O is polarized in the z-direction.

Air molecules scatter blue light more effectively than red light; we see the sky overhead by scattered light, so it looks blue.

This observer sees reddened sunlight because most of the blue light has been scattered out.

33.32 Clouds are white because they efficiently scatter sunlight of all wavelengths.

Clouds contain a high concentration of suspended water droplets or ice crystals, which also scatter light. Because of this high concentration, light passing through the cloud has many more opportunities for scattering than does light passing through a clear sky. Thus light of *all* wavelengths is eventually scattered out of the cloud, so the cloud looks white (**Fig. 33.32**). Milk looks white for the same reason; the scattering is due to fat globules suspended in the milk.

Near sunset, when sunlight has to travel a long distance through the earth's atmosphere, a substantial fraction of the blue light is removed by scattering. White light minus blue light looks yellow or red. This explains the yellow or red hue that we so often see from the setting sun (and that is seen by the observer at the far right of Fig. 33.31).

33.7 HUYGENS'S PRINCIPLE

The laws of reflection and refraction of light rays, as introduced in Section 33.2, were discovered experimentally long before the wave nature of light was firmly established. However, we can *derive* these laws from wave considerations and show that they are consistent with the wave nature of light.

We begin with a principle called **Huygens's principle.** This principle, stated originally by the Dutch scientist Christiaan Huygens in 1678, is a geometrical method for finding, from the known shape of a wave front at some instant, the shape of the wave front at some later time. Huygens assumed that **every point of a wave front may be considered the source of secondary wavelets that spread out in all directions with a speed equal to the speed of propagation of the wave.** The new wave front at a later time is then found by constructing a surface *tangent* to the secondary wavelets or, as it is called, the *envelope* of the wavelets. All the results that we obtain from Huygens's principle can also be obtained from Maxwell's equations, but Huygens's simple model is easier to use.

Figure 33.33 illustrates Huygens's principle. The original wave front AA' is traveling outward from a source, as indicated by the arrows. We want to find the shape of the wave front after a time interval t. We assume that v, the speed of propagation of the wave, is the same at all points. Then in time t the wave front travels a distance vt. We construct several circles (traces of spherical wavelets) with radius $r = vt$, centered at points along AA'. The trace of the envelope of these wavelets, which is the new wave front, is the curve BB'.

33.33 Applying Huygens's principle to wave front AA' to construct a new wave front BB'.

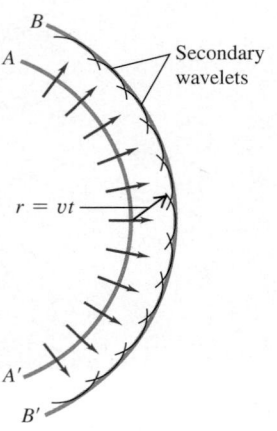

Reflection and Huygens's Principle

To derive the law of reflection from Huygens's principle, we consider a plane wave approaching a plane reflecting surface. In **Fig. 33.34a** the lines AA', OB', and NC' represent successive positions of a wave front approaching the surface MM'. Point A on the wave front AA' has just arrived at the reflecting surface. We can use Huygens's principle to find the position of the wave front after a time interval t. With points on AA' as centers, we draw several secondary wavelets with radius vt. The wavelets that originate near the upper end of AA' spread out unhindered, and their envelope gives the portion OB' of the new wave front. If the reflecting surface were not there, the wavelets originating near the lower end of AA' would similarly reach the positions shown by the broken circular arcs. Instead, these wavelets strike the reflecting surface.

The effect of the reflecting surface is to *change the direction* of travel of those wavelets that strike it, so the part of a wavelet that would have penetrated the surface actually lies to the left of it, as shown by the full lines. The first such wavelet is centered at point A; the envelope of all such reflected wavelets is the portion OB of the wave front. The trace of the entire wave front at this instant is the bent line BOB'. A similar construction gives the line CNC' for the wave front after another interval t.

33.34 Using Huygens's principle to derive the law of reflection.

(a) Successive positions of a plane wave AA' as it is reflected from a plane surface

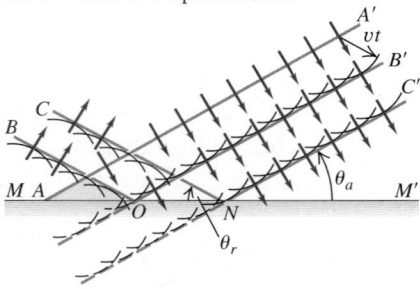

(b) Magnified portion of **(a)**

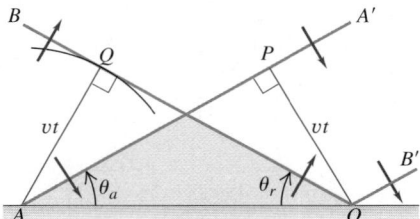

From plane geometry the angle θ_a between the incident *wave front* and the *surface* is the same as that between the incident *ray* and the *normal* to the surface and is therefore the angle of incidence. Similarly, θ_r is the angle of reflection. To find the relationship between these angles, we consider Fig. 33.34b. From O we draw $OP = vt$, perpendicular to AA'. Now OB, by construction, is tangent to a circle of radius vt with center at A. If we draw AQ from A to the point of tangency, the triangles APO and OQA are congruent because they are right triangles with the side AO in common and with $AQ = OP = vt$. The angle θ_a therefore equals the angle θ_r, and we have the law of reflection.

Refraction and Huygens's Principle

We can derive the law of *refraction* by a similar procedure. In **Fig. 33.35a** we consider a wave front, represented by line AA', for which point A has just arrived at the boundary surface SS' between two transparent materials a and b, with indexes of refraction n_a and n_b and wave speeds v_a and v_b. (The *reflected* waves are not shown; they proceed as in Fig. 33.34.) We can apply Huygens's principle to find the position of the refracted wave fronts after a time t.

With points on AA' as centers, we draw several secondary wavelets. Those originating near the upper end of AA' travel with speed v_a and, after a time interval t, are spherical surfaces of radius $v_a t$. The wavelet originating at point A, however, is traveling in the second material b with speed v_b and at time t is a spherical surface of radius $v_b t$. The envelope of the wavelets from the original wave front is the plane whose trace is the bent line BOB'. A similar construction leads to the trace CPC' after a second interval t.

The angles θ_a and θ_b between the surface and the incident and refracted wave fronts are the angle of incidence and the angle of refraction, respectively. To find the relationship between these angles, refer to Fig. 33.35b. We draw $OQ = v_a t$, perpendicular to AQ, and we draw $AB = v_b t$, perpendicular to BO. From the right triangle AOQ,

$$\sin \theta_a = \frac{v_a t}{AO}$$

and from the right triangle AOB,

$$\sin \theta_b = \frac{v_b t}{AO}$$

Combining these, we find

$$\frac{\sin \theta_a}{\sin \theta_b} = \frac{v_a}{v_b} \qquad (33.9)$$

We have defined the index of refraction n of a material as the ratio of the speed of light c in vacuum to its speed v in the material: $n_a = c/v_a$ and $n_b = c/v_b$. Thus

$$\frac{n_b}{n_a} = \frac{c/v_b}{c/v_a} = \frac{v_a}{v_b}$$

and we can rewrite Eq. (33.9) as

$$\frac{\sin \theta_a}{\sin \theta_b} = \frac{n_b}{n_a} \quad \text{or}$$

$$n_a \sin \theta_a = n_b \sin \theta_b$$

which we recognize as Snell's law, Eq. (33.4). So we have derived Snell's law from a wave theory. Alternatively, we can regard Snell's law as an experimental result that defines the index of refraction of a material; in that case this analysis helps confirm the relationship $v = c/n$ for the speed of light in a material.

33.35 Using Huygens's principle to derive the law of refraction. The case $v_b < v_a$ ($n_b > n_a$) is shown.

(a) Successive positions of a plane wave AA' as it is refracted by a plane surface

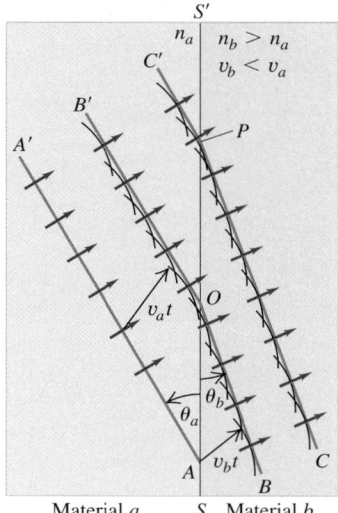

(b) Magnified portion of **(a)**

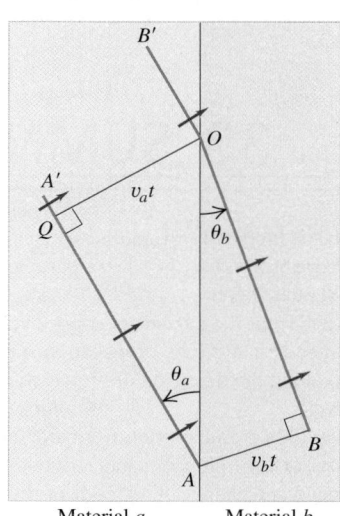

33.36 How mirages are formed.

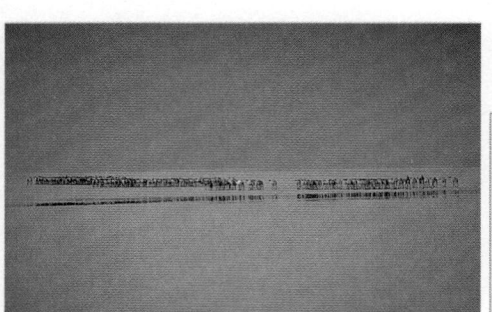

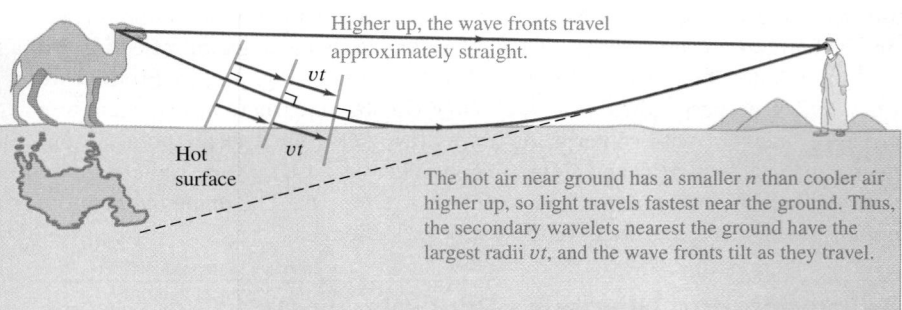

Higher up, the wave fronts travel approximately straight.

vt

Hot surface

vt

The hot air near ground has a smaller n than cooler air higher up, so light travels fastest near the ground. Thus, the secondary wavelets nearest the ground have the largest radii vt, and the wave fronts tilt as they travel.

Mirages are an example of Huygens's principle in action. When the surface of pavement or desert sand is heated intensely by the sun, a hot, less dense, smaller-n layer of air forms near the surface. The speed of light is slightly greater in the hotter air near the ground, the Huygens wavelets have slightly larger radii, the wave fronts tilt slightly, and rays that were headed toward the surface with a large angle of incidence (near 90°) can be bent up as shown in **Fig. 33.36**. Light farther from the ground is bent less and travels nearly in a straight line. The observer sees the object in its natural position, with an inverted image below it, as though seen in a horizontal reflecting surface. A thirsty traveler can interpret the apparent reflecting surface as a sheet of water.

It is important to keep in mind that Maxwell's equations are the fundamental relationships for electromagnetic wave propagation. But Huygens's principle provides a convenient way to visualize this propagation.

TEST YOUR UNDERSTANDING OF SECTION 33.7   Sound travels faster in warm air than in cold air. Imagine a weather front that runs north-south, with warm air to the west of the front and cold air to the east. A sound wave traveling in a northeast direction in the warm air encounters this front. How will the direction of this sound wave change when it passes into the cold air? (i) The wave direction will deflect toward the north; (ii) the wave direction will deflect toward the east; (iii) the wave direction will be unchanged. ▌

| CHAPTER **33 SUMMARY**

SOLUTIONS TO ALL EXAMPLES

Light and its properties: Light is an electromagnetic wave. When emitted or absorbed, it also shows particle properties. It is emitted by accelerated electric charges.

A wave front is a surface of constant phase; wave fronts move with a speed equal to the propagation speed of the wave. A ray is a line along the direction of propagation, perpendicular to the wave fronts.

When light is transmitted from one material to another, the frequency of the light is unchanged, but the wavelength and wave speed can change. The index of refraction n of a material is the ratio of the speed of light in vacuum c to the speed v in the material. If λ_0 is the wavelength in vacuum, the same wave has a shorter wavelength λ in a medium with index of refraction n. (See Example 33.2.)

$$n = \frac{c}{v} \tag{33.1}$$

$$\lambda = \frac{\lambda_0}{n} \tag{33.5}$$

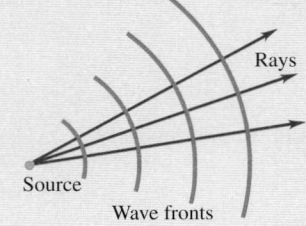

Reflection and refraction: At a smooth interface between two optical materials, the incident, reflected, and refracted rays and the normal to the interface all lie in a single plane called the plane of incidence. The law of reflection states that the angles of incidence and reflection are equal. The law of refraction relates the angles of incidence and refraction to the indexes of refraction of the materials. (See Examples 33.1 and 33.3.)

$$\theta_r = \theta_a \quad (33.2)$$
(law of reflection)

$$n_a \sin \theta_a = n_b \sin \theta_b \quad (33.4)$$
(law of refraction)

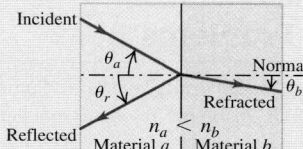

Total internal reflection: When a ray travels in a material of index of refraction n_a toward a material of index $n_b < n_a$, total internal reflection occurs at the interface when the angle of incidence equals or exceeds a critical angle θ_{crit}. (See Example 33.4.)

$$\sin \theta_{crit} = \frac{n_b}{n_a} \quad (33.6)$$

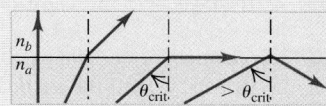

Polarization of light: The direction of polarization of a linearly polarized electromagnetic wave is the direction of the $\vec{E}$ field. A polarizing filter passes waves that are linearly polarized along its polarizing axis and blocks waves polarized perpendicularly to that axis. When polarized light of intensity I_{max} is incident on a polarizing filter used as an analyzer, the intensity I of the light transmitted through the analyzer depends on the angle ϕ between the polarization direction of the incident light and the polarizing axis of the analyzer. (See Example 33.5.)

$$I = I_{max} \cos^2 \phi \quad (33.7)$$
(Malus's law)

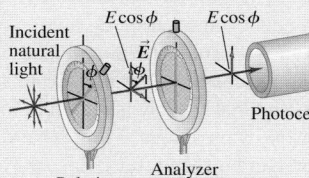

Polarization by reflection: When unpolarized light strikes an interface between two materials, Brewster's law states that the reflected light is completely polarized perpendicular to the plane of incidence (parallel to the interface) if the angle of incidence equals the polarizing angle θ_p. (See Example 33.6.)

$$\tan \theta_p = \frac{n_b}{n_a} \quad (33.8)$$
(Brewster's law)

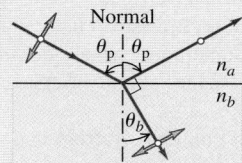

Huygens's principle: Huygens's principle states that if the position of a wave front at one instant is known, then the position of the front at a later time can be constructed by imagining the front as a source of secondary wavelets. Huygens's principle can be used to derive the laws of reflection and refraction.

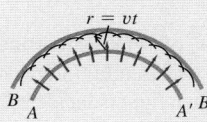

BRIDGING PROBLEM REFLECTION AND REFRACTION

Figure 33.37 shows a rectangular glass block that has a metal reflector on one face and water on an adjoining face. A light beam strikes the reflector as shown. You gradually increase the angle θ of the light beam. If $\theta \geq 59.2°$, no light enters the water. What is the speed of light in this glass?

33.37 Glass bounded by water and a metal reflector.

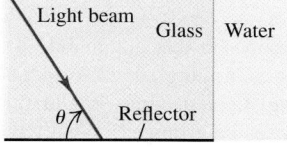

SOLUTION GUIDE

IDENTIFY and SET UP

1. Specular reflection occurs where the light ray in the glass strikes the reflector. If no light is to enter the water, we require that there be reflection only and no refraction where this ray strikes the glass–water interface—that is, there must be total internal reflection.

2. The target variable is the speed of light v in the glass, which you can determine from the index of refraction n of the glass. (Table 33.1 gives the index of refraction of water.) Write down the equations you will use to find n and v.

EXECUTE

3. Use the figure to find the angle of incidence of the ray at the glass–water interface.

4. Use the result of step 3 to find n.

5. Use the result of step 4 to find v.

EVALUATE

6. How does the speed of light in the glass compare to the speed in water? Does this make sense?

Problems

•, ••, •••: Difficulty levels. **CP**: Cumulative problems incorporating material from earlier chapters. **CALC**: Problems requiring calculus.
DATA: Problems involving real data, scientific evidence, experimental design, and/or statistical reasoning. **BIO**: Biosciences problems.

DISCUSSION QUESTIONS

Q33.1 Light requires about 8 minutes to travel from the sun to the earth. Is it delayed appreciably by the earth's atmosphere? Explain.

Q33.2 Sunlight or starlight passing through the earth's atmosphere is always bent toward the vertical. Why? Does this mean that a star is not really where it appears to be? Explain.

Q33.3 A beam of light goes from one material into another. On *physical* grounds, explain *why* the wavelength changes but the frequency and period do not.

Q33.4 A student claimed that, because of atmospheric refraction (see Discussion Question Q33.2), the sun can be seen after it has set and that the day is therefore longer than it would be if the earth had no atmosphere. First, what does she mean by saying that the sun can be seen after it has set? Second, comment on the validity of her conclusion.

Q33.5 When hot air rises from a radiator or heating duct, objects behind it appear to shimmer or waver. What causes this?

Q33.6 Devise straightforward experiments to measure the speed of light in a given glass using (a) Snell's law; (b) total internal reflection; (c) Brewster's law.

Q33.7 Sometimes when looking at a window, you see two reflected images slightly displaced from each other. What causes this?

Q33.8 If you look up from underneath toward the surface of the water in your aquarium, you may see an upside-down reflection of your pet fish in the surface of the water. Explain how this can happen.

Q33.9 A ray of light in air strikes a glass surface. Is there a range of angles for which total internal reflection occurs? Explain.

Q33.10 When light is incident on an interface between two materials, the angle of the refracted ray depends on the wavelength, but the angle of the reflected ray does not. Why should this be?

Q33.11 A salesperson at a bargain counter claims that a certain pair of sunglasses has Polaroid filters; you suspect that the glasses are just tinted plastic. How could you find out for sure?

Q33.12 Does it make sense to talk about the polarization of a *longitudinal* wave, such as a sound wave? Why or why not?

Q33.13 How can you determine the direction of the polarizing axis of a single polarizer?

Q33.14 It has been proposed that automobile windshields and headlights should have polarizing filters to reduce the glare of oncoming lights during night driving. Would this work? How should the polarizing axes be arranged? What advantages would this scheme have? What disadvantages?

Q33.15 When a sheet of plastic food wrap is placed between two crossed polarizers, no light is transmitted. When the sheet is stretched in one direction, some light passes through the crossed polarizers. What is happening?

Q33.16 If you sit on the beach and look at the ocean through Polaroid sunglasses, the glasses help to reduce the glare from sunlight reflecting off the water. But if you lie on your side on the beach, there is little reduction in the glare. Explain why there is a difference.

Q33.17 When unpolarized light is incident on two crossed polarizers, no light is transmitted. A student asserted that if a third polarizer is inserted between the other two, some transmission will occur. Does this make sense? How can adding a third filter *increase* transmission?

Q33.18 For the old "rabbit-ear" style TV antennas, it's possible to alter the quality of reception considerably simply by changing the orientation of the antenna. Why?

Q33.19 In Fig. 33.31, since the light that is scattered out of the incident beam is polarized, why is the transmitted beam not also partially polarized?

Q33.20 You are sunbathing in the late afternoon when the sun is relatively low in the western sky. You are lying flat on your back, looking straight up through Polaroid sunglasses. To minimize the amount of sky light reaching your eyes, how should you lie: with your feet pointing north, east, south, west, or in some other direction? Explain your reasoning.

Q33.21 Light scattered from blue sky is strongly polarized because of the nature of the scattering process described in Section 33.6. But light scattered from white clouds is usually *not* polarized. Why not?

Q33.22 Atmospheric haze is due to water droplets or smoke particles ("smog"). Such haze reduces visibility by scattering light, so that the light from distant objects becomes randomized and images become indistinct. Explain why visibility through haze can be improved by wearing red-tinted sunglasses, which filter out blue light.

Q33.23 The explanation given in Section 33.6 for the color of the setting sun should apply equally well to the *rising* sun, since sunlight travels the same distance through the atmosphere to reach your eyes at either sunrise or sunset. Typically, however, sunsets are redder than sunrises. Why? (*Hint:* Particles of all kinds in the atmosphere contribute to scattering.)

Q33.24 Huygens's principle also applies to sound waves. During the day, the temperature of the atmosphere decreases with increasing altitude above the ground. But at night, when the ground cools, there is a layer of air just above the surface in which the temperature *increases* with altitude. Use this to explain why sound waves from distant sources can be heard more clearly at night than in the daytime. (*Hint:* The speed of sound increases with increasing temperature. Use the ideas displayed in Fig. 33.36 for light.)

Q33.25 Can water waves be reflected and refracted? Give examples. Does Huygens's principle apply to water waves? Explain.

EXERCISES

Section 33.2 Reflection and Refraction

33.1 • Two plane mirrors intersect at right angles. A laser beam strikes the first of them at a point 11.5 cm from their point of intersection, as shown in **Fig. E33.1**. For what angle of incidence at the first mirror will this ray strike the midpoint of the second mirror (which is 28.0 cm long) after reflecting from the first mirror?

Figure **E33.1**

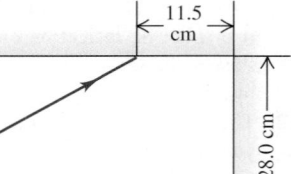

33.2 • BIO **Light Inside the Eye.** The vitreous humor, a transparent, gelatinous fluid that fills most of the eyeball, has an index of refraction of 1.34. Visible light ranges in wavelength from 380 nm (violet) to 750 nm (red), as measured in air. This light travels through the vitreous humor and strikes the rods and cones at the surface of the retina. What are the ranges of (a) the wavelength, (b) the frequency, and (c) the speed of the light just as it approaches the retina within the vitreous humor?

33.3 • A beam of light has a wavelength of 650 nm in vacuum. (a) What is the speed of this light in a liquid whose index of refraction at this wavelength is 1.47? (b) What is the wavelength of these waves in the liquid?

33.4 • Light with a frequency of 5.80×10^{14} Hz travels in a block of glass that has an index of refraction of 1.52. What is the wavelength of the light (a) in vacuum and (b) in the glass?

33.5 • A light beam travels at 1.94×10^8 m/s in quartz. The wavelength of the light in quartz is 355 nm. (a) What is the index of refraction of quartz at this wavelength? (b) If this same light travels through air, what is its wavelength there?

33.6 •• Light of a certain frequency has a wavelength of 526 nm in water. What is the wavelength of this light in benzene?

33.7 •• A parallel beam of light in air makes an angle of 47.5° with the surface of a glass plate having a refractive index of 1.66. (a) What is the angle between the reflected part of the beam and the surface of the glass? (b) What is the angle between the refracted beam and the surface of the glass?

33.8 •• A laser beam shines along the surface of a block of transparent material (see **Fig. E33.8**). Half of the beam goes straight to a detector, while the other half travels through the block and then hits the detector. The time delay between the arrival of the two light beams at the detector is 6.25 ns. What is the index of refraction of this material?

Figure **E33.8**

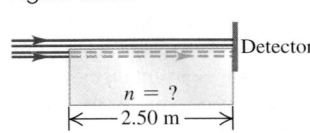

33.9 • Light traveling in air is incident on the surface of a block of plastic at an angle of 62.7° to the normal and is bent so that it makes a 48.1° angle with the normal in the plastic. Find the speed of light in the plastic.

33.10 • (a) A tank containing methanol has walls 2.50 cm thick made of glass of refractive index 1.550. Light from the outside air strikes the glass at a 41.3° angle with the normal to the glass. Find the angle the light makes with the normal in the methanol. (b) The tank is emptied and refilled with an unknown liquid. If light incident at the same angle as in part (a) enters the liquid in the tank at an angle of 20.2° from the normal, what is the refractive index of the unknown liquid?

33.11 •• As shown in **Fig. E33.11**, a layer of water covers a slab of material X in a beaker. A ray of light traveling upward follows the path indicated. Using the information on the figure, find (a) the index of refraction of material X and (b) the angle the light makes with the normal in the *air*.

Figure **E33.11**

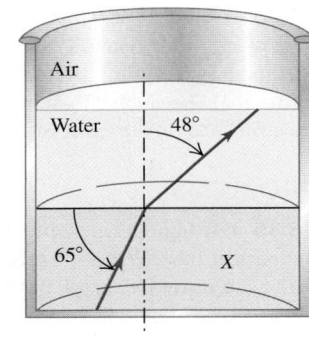

33.12 •• A horizontal, parallel-sided plate of glass having a refractive index of 1.52 is in contact with the surface of water in a tank. A ray coming from above in air makes an angle of

incidence of 35.0° with the normal to the top surface of the glass. (a) What angle does the ray refracted into the water make with the normal to the surface? (b) What is the dependence of this angle on the refractive index of the glass?

33.13 • A ray of light is incident on a plane surface separating two sheets of glass with refractive indexes 1.70 and 1.58. The angle of incidence is 62.0°, and the ray originates in the glass with $n = 1.70$. Compute the angle of refraction.

33.14 • A ray of light traveling in water is incident on an interface with a flat piece of glass. The wavelength of the light in the water is 726 nm, and its wavelength in the glass is 544 nm. If the ray in water makes an angle of 56.0° with respect to the normal to the interface, what angle does the refracted ray in the glass make with respect to the normal?

Section 33.3 Total Internal Reflection

33.15 • **Light Pipe.** Light enters a solid pipe made of plastic having an index of refraction of 1.60. The light travels parallel to the upper part of the pipe (**Fig. E33.15**). You want to cut the face AB so that all the light will reflect back into the pipe after it first strikes that face. (a) What is the largest that θ can be if the pipe is in air? (b) If the pipe is immersed in water of refractive index 1.33, what is the largest that θ can be?

Figure **E33.15**

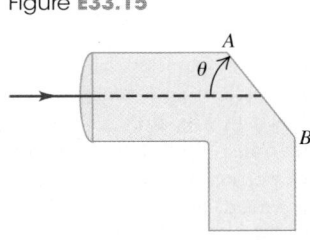

33.16 • A flat piece of glass covers the top of a vertical cylinder that is completely filled with water. If a ray of light traveling in the glass is incident on the interface with the water at an angle of $\theta_a = 36.2°$, the ray refracted into the water makes an angle of 49.8° with the normal to the interface. What is the smallest value of the incident angle θ_a for which none of the ray refracts into the water?

33.17 •• The critical angle for total internal reflection at a liquid–air interface is 42.5°. (a) If a ray of light traveling in the liquid has an angle of incidence at the interface of 35.0°, what angle does the refracted ray in the air make with the normal? (b) If a ray of light traveling in air has an angle of incidence at the interface of 35.0°, what angle does the refracted ray in the liquid make with the normal?

33.18 • A beam of light is traveling inside a solid glass cube that has index of refraction 1.62. It strikes the surface of the cube from the inside. (a) If the cube is in air, at what minimum angle with the normal inside the glass will this light *not* enter the air at this surface? (b) What would be the minimum angle in part (a) if the cube were immersed in water?

33.19 • A ray of light is traveling in a glass cube that is totally immersed in water. You find that if the ray is incident on the glass–water interface at an angle to the normal larger than 48.7°, no light is refracted into the water. What is the refractive index of the glass?

33.20 • At the very end of Wagner's series of operas *Ring of the Nibelung*, Brünnhilde takes the golden ring from the finger of the dead Siegfried and throws it into the Rhine, where it sinks to the bottom of the river. Assuming that the ring is small enough compared to the depth of the river to be treated as a point and that the Rhine is 10.0 m deep where the ring goes in, what is the area of the largest circle at the surface of the water over which light from the ring could escape from the water?

33.21 • Light is incident along the normal on face AB of a glass prism of refractive index 1.52, as shown in **Fig. E33.21**. Find the largest value the angle α can have without any light refracted out of the prism at face AC if (a) the prism is immersed in air and (b) the prism is immersed in water.

Figure **E33.21**

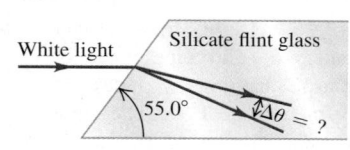

Section 33.4 Dispersion

33.22 • The indexes of refraction for violet light ($\lambda = 400$ nm) and red light ($\lambda = 700$ nm) in diamond are 2.46 and 2.41, respectively. A ray of light traveling through air strikes the diamond surface at an angle of 53.5° to the normal. Calculate the angular separation between these two colors of light in the refracted ray.

33.23 •• A narrow beam of white light strikes one face of a slab of silicate flint glass. The light is traveling parallel to the two adjoining faces, as shown in **Fig. E33.23**. For the transmitted light inside the glass, through what angle $\Delta\theta$ is the portion of the visible spectrum between 400 nm and 700 nm dispersed? (Consult the graph in Fig. 33.17.)

Figure **E33.23**

White light | Silicate flint glass
55.0° | $\Delta\theta = ?$

33.24 • A beam of light strikes a sheet of glass at an angle of 57.0° with the normal in air. You observe that red light makes an angle of 38.1° with the normal in the glass, while violet light makes a 36.7° angle. (a) What are the indexes of refraction of this glass for these colors of light? (b) What are the speeds of red and violet light in the glass?

Section 33.5 Polarization

33.25 • Unpolarized light with intensity I_0 is incident on two polarizing filters. The axis of the first filter makes an angle of 60.0° with the vertical, and the axis of the second filter is horizontal. What is the intensity of the light after it has passed through the second filter?

33.26 •• (a) At what angle above the horizontal is the sun if sunlight reflected from the surface of a calm lake is completely polarized? (b) What is the plane of the electric-field vector in the reflected light?

33.27 •• A beam of unpolarized light of intensity I_0 passes through a series of ideal polarizing filters with their polarizing axes turned to various angles as shown in **Fig. E33.27**. (a) What is the light intensity (in terms of I_0) at points A, B, and C? (b) If we remove the middle filter, what will be the light intensity at point C?

Figure **E33.27**

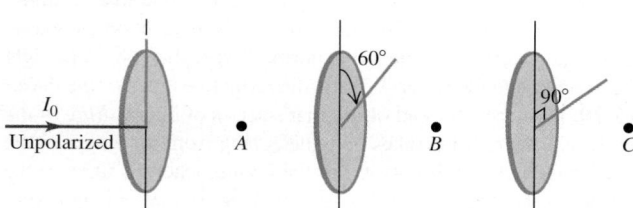

33.28 •• Light of original intensity I_0 passes through two ideal polarizing filters having their polarizing axes oriented as shown in **Fig. E33.28**. You want to adjust the angle ϕ so that the intensity

at point P is equal to $I_0/10$. (a) If the original light is unpolarized, what should ϕ be? (b) If the original light is linearly polarized in the same direction as the polarizing axis of the first polarizer the light reaches, what should ϕ be?

Figure **E33.28**

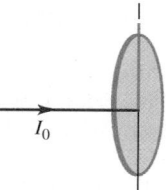
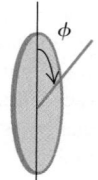

33.29 • A parallel beam of unpolarized light in air is incident at an angle of 54.5° (with respect to the normal) on a plane glass surface. The reflected beam is completely linearly polarized. (a) What is the refractive index of the glass? (b) What is the angle of refraction of the transmitted beam?

33.30 • The refractive index of a certain glass is 1.66. For what incident angle is light reflected from the surface of this glass completely polarized if the glass is immersed in (a) air and (b) water?

33.31 •• A beam of polarized light passes through a polarizing filter. When the angle between the polarizing axis of the filter and the direction of polarization of the light is θ, the intensity of the emerging beam is I. If you now want the intensity to be $I/2$, what should be the angle (in terms of θ) between the polarizing angle of the filter and the original direction of polarization of the light?

33.32 ••• Three polarizing filters are stacked, with the polarizing axis of the second and third filters at 23.0° and 62.0°, respectively, to that of the first. If unpolarized light is incident on the stack, the light has intensity 55.0 W/cm² after it passes through the stack. If the incident intensity is kept constant but the second polarizer is removed, what is the intensity of the light after it has passed through the stack?

33.33 •• Unpolarized light of intensity 20.0 W/cm² is incident on two polarizing filters. The axis of the first filter is at an angle of 25.0° counterclockwise from the vertical (viewed in the direction the light is traveling), and the axis of the second filter is at 62.0° counterclockwise from the vertical. What is the intensity of the light after it has passed through the second polarizer?

33.34 • **Three Polarizing Filters.** Three polarizing filters are stacked with the polarizing axes of the second and third at 45.0° and 90.0°, respectively, with that of the first. (a) If unpolarized light of intensity I_0 is incident on the stack, find the intensity and state of polarization of light emerging from each filter. (b) If the second filter is removed, what is the intensity of the light emerging from each remaining filter?

Section 33.6 Scattering of Light

33.35 • A beam of white light passes through a uniform thickness of air. If the intensity of the scattered light in the middle of the green part of the visible spectrum is I, find the intensity (in terms of I) of scattered light in the middle of (a) the red part of the spectrum and (b) the violet part of the spectrum. Consult Table 32.1.

PROBLEMS

33.36 • A light beam is directed parallel to the axis of a hollow cylindrical tube. When the tube contains only air, the light takes 8.72 ns to travel the length of the tube, but when the tube is filled with a transparent jelly, the light takes 1.82 ns longer to travel its length. What is the refractive index of this jelly?

33.37 •• **BIO** **Heart Sonogram.** Physicians use high-frequency ($f = 1$–5 MHz) sound waves, called ultrasound, to image internal organs. The speed of these ultrasound waves is 1480 m/s in muscle and 344 m/s in air. We define the index of refraction of a material for sound waves to be the ratio of the speed of sound in air to the speed of sound in the material. Snell's law then applies to the refraction of sound waves. (a) At what angle from the normal does an ultrasound beam enter the heart if it leaves the lungs at an angle of 9.73° from the normal to the heart wall? (Assume that the speed of sound in the lungs is 344 m/s.) (b) What is the critical angle for sound waves in air incident on muscle?

33.38 ••• In a physics lab, light with wavelength 490 nm travels in air from a laser to a photocell in 17.0 ns. When a slab of glass 0.840 m thick is placed in the light beam, with the beam incident along the normal to the parallel faces of the slab, it takes the light 21.2 ns to travel from the laser to the photocell. What is the wavelength of the light in the glass?

33.39 •• A ray of light is incident in air on a block of a transparent solid whose index of refraction is n. If $n = 1.38$, what is the *largest* angle of incidence θ_a for which total internal reflection will occur at the vertical face (point A shown in **Fig. P33.39**)?

Figure **P33.39**

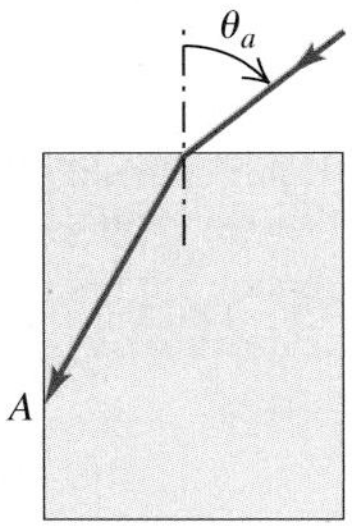

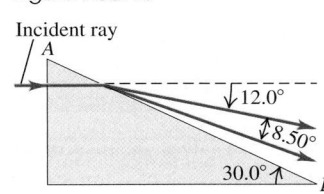

33.40 • A light ray in air strikes the right-angle prism shown in **Fig. P33.40**. The prism angle at B is 30.0°. This ray consists of two different wavelengths. When it emerges at face AB, it has been split into two different rays that diverge from each other by 8.50°. Find the index of refraction of the prism for each of the two wavelengths.

Figure **P33.40**

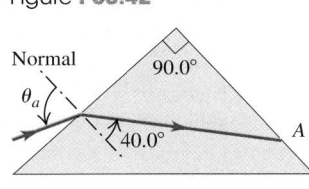

33.41 •• A ray of light traveling *in* a block of glass ($n = 1.52$) is incident on the top surface at an angle of 57.2° with respect to the normal in the glass. If a layer of oil is placed on the top surface of the glass, the ray is totally reflected. What is the maximum possible index of refraction of the oil?

33.42 •• A ray of light traveling in air is incident at angle θ_a on one face of a 90.0° prism made of glass. Part of the light refracts into the prism and strikes the opposite face at point A (**Fig. P33.42**). If the ray at A is at the critical angle, what is the value of θ_a?

Figure **P33.42**

33.43 ••• A glass plate 2.50 mm thick, with an index of refraction of 1.40, is placed between a point source of light with wavelength 540 nm (in vacuum) and a screen. The distance from source to screen is 1.80 cm. How many wavelengths are there between the source and the screen?

33.44 • After a long day of driving you take a late-night swim in a motel swimming pool. When you go to your room, you realize that you have lost your room key in the pool. You borrow a powerful flashlight and walk around the pool, shining the light into it. The light shines on the key, which is lying on the bottom of the pool, when the flashlight is held 1.2 m above the water surface and is directed at the surface a horizontal distance of 1.5 m from the

edge (**Fig. P33.44**). If the water here is 4.0 m deep, how far is the key from the edge of the pool?

Figure **P33.44**

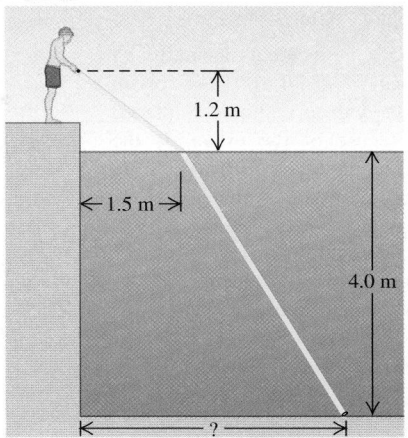

33.45 • You sight along the rim of a glass with vertical sides so that the top rim is lined up with the opposite edge of the bottom (**Fig. P33.45a**). The glass is a thin-walled, hollow cylinder 16.0 cm high. The diameter of the top and bottom of the glass is 8.0 cm. While you keep your eye in the same position, a friend fills the glass with a transparent liquid, and you then see a dime that is lying at the center of the bottom of the glass (Fig. P33.45b). What is the index of refraction of the liquid?

Figure **P33.45**

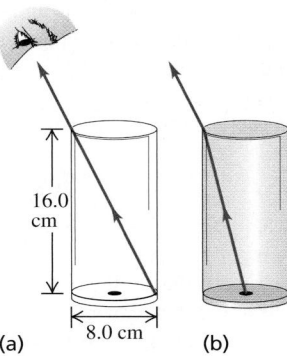

33.46 •• Optical fibers are constructed with a cylindrical core surrounded by a sheath of cladding material. Common materials used are pure silica ($n_2 = 1.450$) for the cladding and silica doped with germanium ($n_1 = 1.465$) for the core. (a) What is the critical angle θ_{crit} for light traveling in the core and reflecting at the interface with the cladding material? (b) The numerical aperture (NA) is defined as the angle of incidence θ_i at the flat end of the cable for which light is incident on the core–cladding interface at angle θ_{crit} (**Fig. P33.46**). Show that $\sin\theta_i = \sqrt{n_1^2 - n_2^2}$. (c) What is the value of θ_i for $n_1 = 1.465$ and $n_2 = 1.450$?

Figure **P33.46**

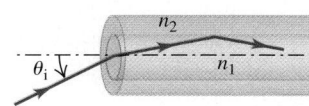

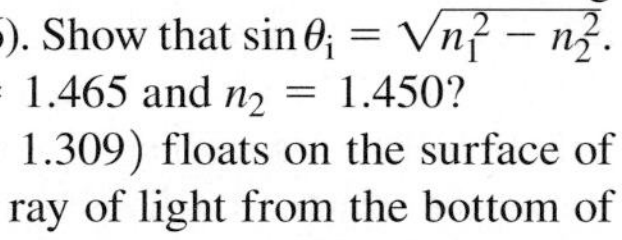

33.47 • A thin layer of ice ($n = 1.309$) floats on the surface of water ($n = 1.333$) in a bucket. A ray of light from the bottom of the bucket travels upward through the water. (a) What is the largest angle with respect to the normal that the ray can make at the ice–water interface and still pass out into the air above the ice? (b) What is this angle after the ice melts?

33.48 •• A 45°–45°–90° prism is immersed in water. A ray of light is incident normally on one of its shorter faces. What is the minimum index of refraction that the prism must have if this ray is to be totally reflected within the glass at the long face of the prism?

33.49 •• The prism shown in **Fig. P33.49** has a refractive index of 1.66, and the angles A are 25.0°. Two light rays m and n are parallel as they enter the prism. What is the angle between them after they emerge?

Figure **P33.49**

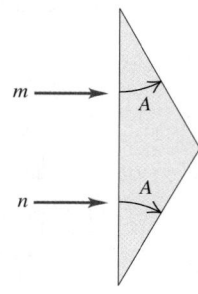

33.50 •• Light is incident normally on the short face of a 30°–60°–90° prism (**Fig. P33.50**). A drop of liquid is placed on the hypotenuse of the prism. If the index of refraction of the prism is 1.56, find the maximum index that the liquid may have for the light to be totally reflected.

Figure **P33.50**

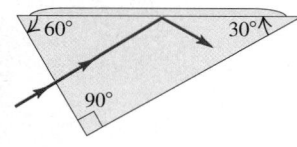

33.51 •• When the sun is either rising or setting and appears to be just on the horizon, it is in fact *below* the horizon. The explanation for this seeming paradox is that light from the sun bends slightly when entering the earth's atmosphere, as shown in **Fig. P33.51**. Since our perception is based on the idea that light travels in straight lines, we perceive the light to be coming from an apparent position that is an angle δ above the sun's true position. (a) Make the simplifying assumptions that the atmosphere has uniform density, and hence uniform index of refraction n, and extends to a height h above the earth's surface, at which point it abruptly stops. Show that the angle δ is given by

$$\delta = \arcsin\left(\frac{nR}{R+h}\right) - \arcsin\left(\frac{R}{R+h}\right)$$

where $R = 6378$ km is the radius of the earth. (b) Calculate δ using $n = 1.0003$ and $h = 20$ km. How does this compare to the angular radius of the sun, which is about one quarter of a degree? (In actuality a light ray from the sun bends gradually, not abruptly, since the density and refractive index of the atmosphere change gradually with altitude.)

Figure **P33.51**

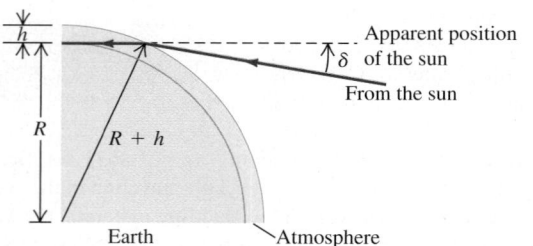

33.52 •• A horizontal cylindrical tank 2.20 m in diameter is half full of water. The space above the water is filled with a pressurized gas of unknown refractive index. A small laser can move along the curved bottom of the water and aims a light beam toward the center of the water surface (**Fig. P33.52**). You observe that when the laser has moved a distance $S = 1.09$ m or more (measured along the

Figure **P33.52**

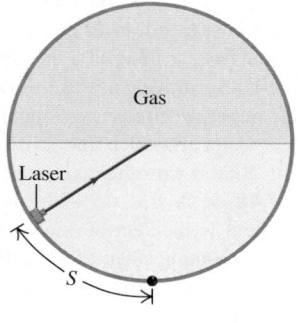

curved surface) from the lowest point in the water, no light enters the gas. (a) What is the index of refraction of the gas? (b) What minimum time does it take the light beam to travel from the laser to the rim of the tank when (i) $S > 1.09$ m and (ii) $S < 1.09$ m?

33.53 •• **Angle of Deviation.** The incident angle θ_a shown in **Fig. P33.53** is chosen so that the light passes symmetrically through the prism, which has refractive index n and apex angle A. (a) Show that the angle of deviation δ (the angle between the initial and final directions of the ray) is given by

$$\sin\frac{A+\delta}{2} = n\sin\frac{A}{2}$$

(When the light passes through symmetrically, as shown, the angle of deviation is a minimum.) (b) Use the result of part (a) to find the angle of deviation for a ray of light passing symmetrically through a prism having three equal angles ($A = 60.0°$) and $n = 1.52$. (c) A certain glass has a refractive index of 1.61 for red light (700 nm) and 1.66 for violet light (400 nm). If both colors pass through symmetrically, as described in part (a), and if $A = 60.0°$, find the difference between the angles of deviation for the two colors.

Figure **P33.53**

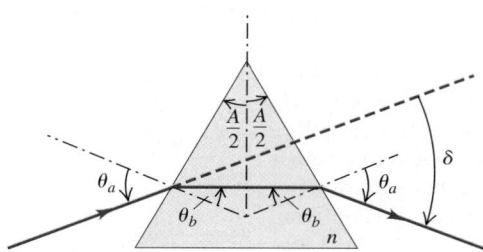

33.54 •• Light is incident in air at an angle θ_a (**Fig. P33.54**) on the upper surface of a transparent plate, the surfaces of the plate being plane and parallel to each other. (a) Prove that $\theta_a = \theta'_a$. (b) Show that this is true for any number of different parallel plates. (c) Prove that the lateral displacement d of the emergent beam is given by the relationship

$$d = t\frac{\sin(\theta_a - \theta'_b)}{\cos\theta'}$$

where t is the thickness of the plate. (d) A ray of light is incident at an angle of 66.0° on one surface of a glass plate 2.40 cm thick with an index of refraction of 1.80. The medium on either side of the plate is air. Find the lateral displacement between the incident and emergent rays.

Figure **P33.54**

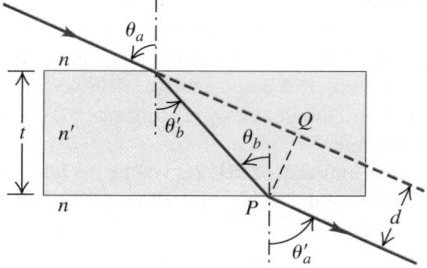

33.55 •• A beam of unpolarized sunlight strikes the vertical plastic wall of a water tank at an unknown angle. Some of the light reflects from the wall and enters the water (**Fig. P33.55**). The refractive index of the plastic wall is 1.61. If the light that has been reflected from the wall into the water is observed to be completely polarized, what angle does this beam make with the normal inside the water?

33.56 • A thin beam of white light is directed at a flat sheet of silicate flint glass at an angle of 20.0° to the surface of the sheet. Due to dispersion in the glass, the beam is spread out in a spectrum as shown in **Fig. P33.56**. The refractive index of silicate flint glass versus wavelength is graphed in Fig. 33.17. (a) The rays a and b shown in Fig. P33.56 correspond to the extreme wavelengths shown in Fig. 33.17. Which corresponds to red and which to violet? Explain your reasoning. (b) For what thickness d of the glass sheet will the spectrum be 1.0 mm wide, as shown (see Problem 33.54)?

33.57 •• DATA In physics lab, you are studying the properties of four transparent liquids. You shine a ray of light (in air) onto the surface of each liquid—A, B, C, and D—one at a time, at a 60.0° angle of incidence; you then measure the angle of refraction. The table gives your data:

Liquid	A	B	C	D
θ_a (°)	36.4	40.5	32.1	35.2

The wavelength of the light when it is traveling in air is 589 nm. (a) Find the refractive index of each liquid at this wavelength. Use Table 33.1 to identify each liquid, assuming that all four are listed in the table. (b) For each liquid, what is the dielectric constant K at the frequency of the 589-nm light? For each liquid, the relative permeability (K_m) is very close to unity. (c) What is the frequency of the light in air and in each liquid?

33.58 •• DATA Given small samples of three liquids, you are asked to determine their refractive indexes. However, you do not have enough of each liquid to measure the angle of refraction for light refracting from air into the liquid. Instead, for each liquid, you take a rectangular block of glass ($n = 1.52$) and place a drop of the liquid on the top surface of the block. You shine a laser beam with wavelength 638 nm in vacuum at one side of the block and measure the largest angle of incidence θ_a for which there is total internal reflection at the interface between the glass and the liquid (**Fig. P33.58**). Your results are given in the table:

Liquid	A	B	C
θ_a (°)	52.0	44.3	36.3

What is the refractive index of each liquid at this wavelength?

Figure **P33.55**

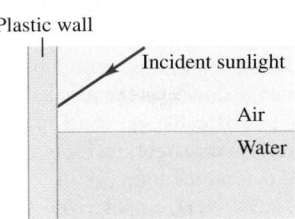

Plastic wall

Incident sunlight

Air

Water

Figure **P33.56**

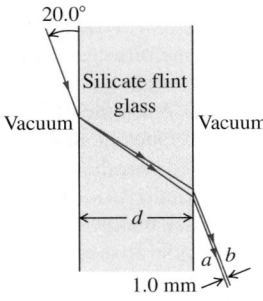

20.0°

Silicate flint glass

Vacuum

Vacuum

d

1.0 mm

a b

Figure **P33.58**

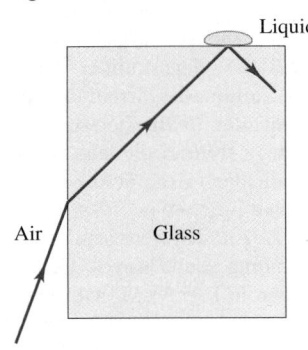

Liquid

Air

Glass

33.59 •• DATA A beam of light traveling horizontally is made of an unpolarized component with intensity I_0 and a polarized component with intensity I_p. The plane of polarization of the polarized component is oriented at an angle θ with respect to the vertical. **Figure P33.59** is a graph of the total intensity I_{total} after the light passes through a polarizer versus the angle α that the polarizer's axis makes with respect to the vertical. (a) What is the orientation of the polarized component? (That is, what is θ?) (b) What are the values of I_0 and I_p?

Figure **P33.59**

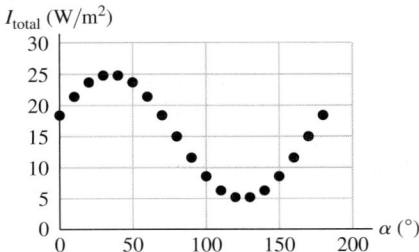

CHALLENGE PROBLEMS

33.60 ••• CALC A rainbow is produced by the reflection of sunlight by spherical drops of water in the air. **Figure P33.60** shows a ray that refracts into a drop at point A, is reflected from the back surface of the drop at point B, and refracts back into the air at point C. The angles of incidence and refraction, θ_a and θ_b, are shown at points A and C, and the angles of incidence and reflection, θ_a and θ_r, are shown at point B. (a) Show that $\theta_a^B = \theta_b^A$, $\theta_a^C = \theta_b^A$, and $\theta_b^C = \theta_a^A$. (b) Show that the angle in radians between the ray before it enters the drop at A and after it exits at C (the total angular deflection of the ray) is $\Delta = 2\theta_a^A - 4\theta_b^A + \pi$. (Hint: Find the angular deflections that occur at A, B, and C, and add them to get Δ.) (c) Use Snell's law to write Δ in terms of θ_a^A and n, the refractive index of the water in the drop. (d) A rainbow will form when the angular deflection Δ is *stationary* in the incident angle θ_a^A—that is, when $d\Delta/d\theta_a^A = 0$. If this condition is satisfied, all the rays with incident angles close to θ_a^A will be sent back in the same direction, producing a bright zone in the sky. Let θ_1 be the value of θ_a^A for which this occurs. Show that $\cos^2 \theta_1 = \frac{1}{3}(n^2 - 1)$. (Hint: You may find the derivative formula $d(\arcsin u(x))/dx = (1 - u^2)^{-1/2}(du/dx)$ helpful.) (e) The index

Figure **P33.60**

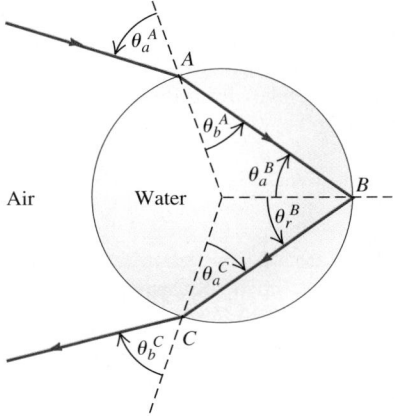

Air

Water

θ_a^A

A

θ_b^A

θ_a^B

θ_r^B

B

θ_a^C

θ_b^C C

of refraction in water is 1.342 for violet light and 1.330 for red light. Use the results of parts (c) and (d) to find θ_1 and Δ for violet and red light. Do your results agree with the angles shown in Fig. 33.19d? When you view the rainbow, which color, red or violet, is higher above the horizon?

33.61 ••• **CALC** A *secondary rainbow* is formed when the incident light undergoes two internal reflections in a spherical drop of water as shown in Fig. 33.19e. (See Challenge Problem 33.60.) (a) In terms of the incident angle θ_a^A and the refractive index n of the drop, what is the angular deflection Δ of the ray? That is, what is the angle between the ray before it enters the drop and after it exits? (b) What is the incident angle θ_2 for which the derivative of Δ with respect to the incident angle θ_a^A is zero? (c) The indexes of refraction for red and violet light in water are given in part (e) of Challenge Problem 33.60. Use the results of parts (a) and (b) to find θ_2 and Δ for violet and red light. Do your results agree with the angles shown in Fig. 33.19e? When you view a secondary rainbow, is red or violet higher above the horizon? Explain.

PASSAGE PROBLEMS

BIO SEEING POLARIZED LIGHT. Some insect eyes have two types of cells that are sensitive to the plane of polarization of light. In a simple model, one cell type (type H) is sensitive to horizontally polarized light only, and the other cell type (type V) is sensitive to vertically polarized light only. To study the responses of these cells, researchers fix the insect in a normal, upright position so that one eye is illuminated by a light source. Then several experiments are carried out.

33.62 First, light with a plane of polarization at 45° to the horizontal shines on the insect. Which statement is true about the two types of cells? (a) Both types detect this light. (b) Neither type detects this light. (c) Only type H detects the light. (d) Only type V detects the light.

33.63 Next, unpolarized light is reflected off a smooth horizontal piece of glass, and the reflected light shines on the insect. Which statement is true about the two types of cells? (a) When the light is directly above the glass, only type V detects the reflected light. (b) When the light is directly above the glass, only type H detects the reflected light. (c) When the light is about 35° above the horizontal, type V responds much more strongly than type H does. (d) When the light is about 35° above the horizontal, type H responds much more strongly than type V does.

33.64 To vary the angle as well as the intensity of polarized light, ordinary unpolarized light is passed through one polarizer with its transmission axis vertical, and then a second polarizer is placed between the first polarizer and the insect. When the light leaving the second polarizer has half the intensity of the original unpolarized light, which statement is true about the two types of cells? (a) Only type H detects this light. (b) Only type V detects this light. (c) Both types detect this light, but type H detects more light. (d) Both types detect this light, but type V detects more light.

Answers

Chapter Opening Question ?

(iv) The brilliance and color of a diamond are due to total internal reflection from its surfaces (Section 33.3) and to dispersion, which spreads this light into a spectrum (Section 33.4).

Test Your Understanding Questions

33.1 (iii) The waves go farther in the y-direction in a given amount of time than in the other directions, so the wave fronts are elongated in the y-direction.

33.2 (a) (ii), (b) (iii) As shown in the figure, light rays coming from the fish bend away from the normal when they pass from the water ($n = 1.33$) into the air ($n = 1.00$). As a result, the fish appears to be higher in the water than it actually is. Hence you should aim a spear *below* the apparent position of the fish. If you use a laser beam, you should aim *at* the apparent position of the fish: The beam of laser light takes the same path from you to the fish as ordinary light takes from the fish to you (though in the opposite direction).

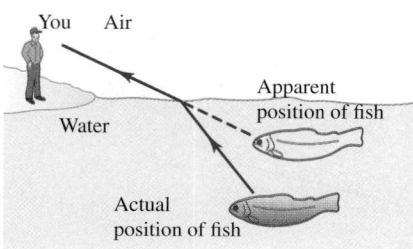

33.3 (i), (ii) Total internal reflection can occur only if two conditions are met: n_b must be less than n_a, and the critical angle θ_{crit} (where $\sin\theta_{crit} = n_b/n_a$) must be smaller than the angle of incidence θ_a. In the first two cases both conditions are met: The critical angles are (i) $\theta_{crit} = \sin^{-1}(1/1.33) = 48.8°$ and (ii) $\theta_{crit} = \sin^{-1}(1.33/1.52) = 61.0°$, both of which are smaller than $\theta_a = 70°$. In the third case $n_b = 1.52$ is greater than $n_a = 1.33$, so total internal reflection cannot occur for any incident angle.

33.5 (ii) The sunlight reflected from the windows of the high-rise building is partially polarized in the vertical direction, perpendicular to the horizontal plane of incidence. The Polaroid filter in front of the lens is oriented with its polarizing axis perpendicular to the dominant direction of polarization of the reflected light.

33.7 (ii) Huygens's principle applies to waves of all kinds, including sound waves. Hence this situation is exactly like that shown in Fig. 33.35, with material a representing the warm air, material b representing the cold air in which the waves travel more slowly, and the interface between the materials representing the weather front. North is toward the top of the figure and east is toward the right, so Fig. 33.35 shows that the rays (which indicate the direction of propagation) deflect toward the east.

Bridging Problem

1.93×10^8 m/s

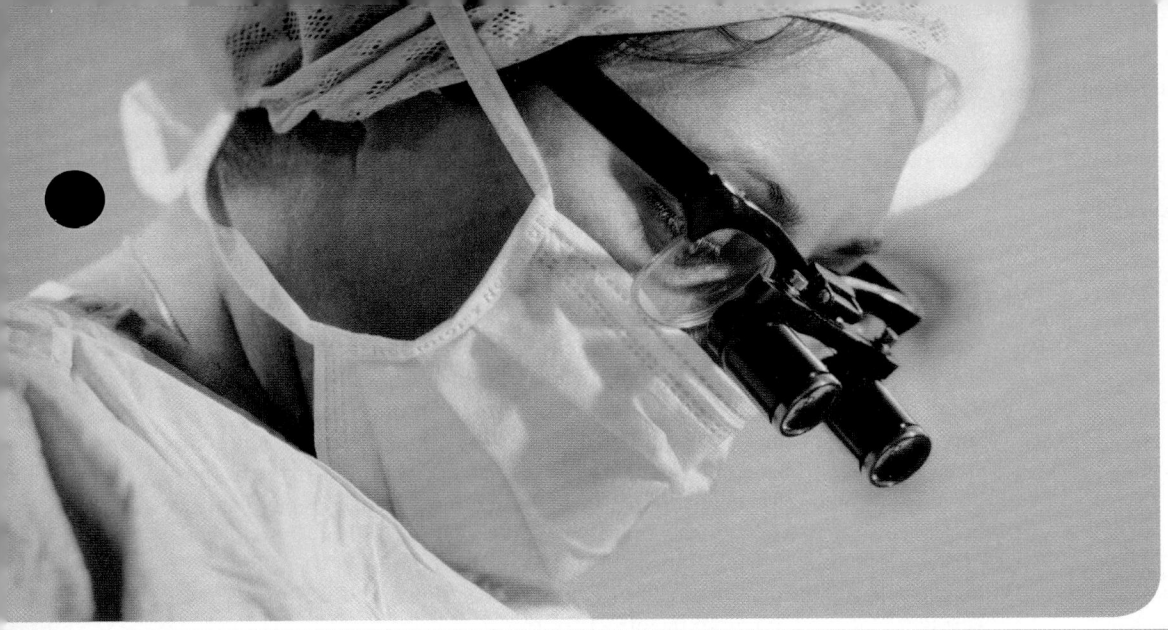

? This surgeon performing microsurgery needs a sharp, magnified view of the surgical site. To obtain this, she's wearing glasses with magnifying lenses that must be (i) at a particular distance from her eye; (ii) at a particular distance from the object being magnified; (iii) both (i) and (ii); (iv) neither (i) nor (ii).

34 GEOMETRIC OPTICS

LEARNING GOALS

Looking forward at …

34.1 How a plane mirror forms an image.

34.2 Why concave and convex mirrors form images of different kinds.

34.3 How images can be formed by a curved interface between two transparent materials.

34.4 What aspects of a lens determine the type of image that it produces.

34.5 What determines the field of view of a camera lens.

34.6 What causes various defects in human vision, and how they can be corrected.

34.7 The principle of the simple magnifier.

34.8 How microscopes and telescopes work.

Looking back at …

33.2 Reflection and refraction.

Your reflection in the bathroom mirror, the view of the moon through a telescope, an insect seen through a magnifying lens—all of these are examples of *images*. In each case the object that you're looking at appears to be in a different place than its actual position: Your reflection is on the other side of the mirror, the moon appears to be much closer when seen through a telescope, and an insect seen through a magnifying lens appears *more distant* (so your eye can focus on it easily). In each case, light rays that come from a point on an object are deflected by reflection or refraction (or a combination of the two), so they converge toward or appear to diverge from a point called an *image point*. Our goal in this chapter is to see how this is done and to explore the different kinds of images that can be made with simple optical devices.

To understand images and image formation, all we need are the ray model of light, the laws of reflection and refraction (Section 33.2), and some simple geometry and trigonometry. The key role played by geometry in our analysis explains why we give the name *geometric optics* to the study of how light rays form images. We'll begin our analysis with one of the simplest image-forming optical devices, a plane mirror. We'll go on to study how images are formed by curved mirrors, by refracting surfaces, and by thin lenses. Our results will lay the foundation for understanding many familiar optical instruments, including camera lenses, magnifiers, the human eye, microscopes, and telescopes.

34.1 REFLECTION AND REFRACTION AT A PLANE SURFACE

Before discussing what is meant by an image, we first need the concept of **object** as it is used in optics. By an *object* we mean anything from which light rays radiate. This light could be emitted by the object itself if it is *self-luminous*, like the glowing filament of a light bulb. Alternatively, the light could be emitted by another source (such as a lamp or the sun) and then reflected from the object;

34.1 Light rays radiate from a point object P in all directions. For an observer to see this object directly, there must be no obstruction between the object and the observer's eyes.

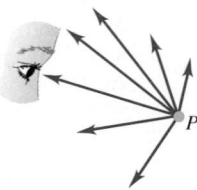

34.2 Light rays from the object at point P are reflected from a plane mirror. The reflected rays entering the eye look as though they had come from image point P'.

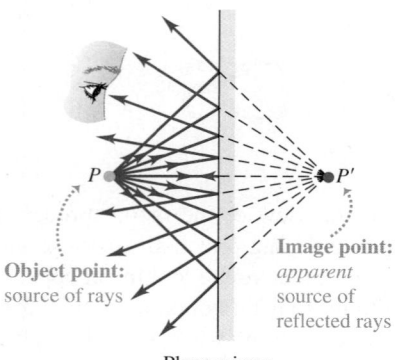

Object point: source of rays

Image point: *apparent* source of reflected rays

Plane mirror

34.3 Light rays from the object at point P are refracted at the plane interface. The refracted rays entering the eye look as though they had come from image point P'.

When $n_a > n_b$, P' is closer to the surface than P; for $n_a < n_b$, the reverse is true.

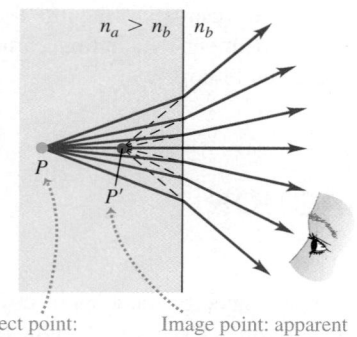

$n_a > n_b$ | n_b

Object point: source of rays

Image point: apparent source of refracted rays

an example is the light you see coming from the pages of this book. **Figure 34.1** shows light rays radiating in all directions from an object at a point P. Note that light rays from the object reach the observer's left and right eyes at different angles; these differences are processed by the observer's brain to infer the *distance* from the observer to the object.

The object in Fig. 34.1 is a **point object** that has no physical extent. Real objects with length, width, and height are called **extended objects.** To start with, we'll consider only an idealized point object, since we can always think of an extended object as being made up of a very large number of point objects.

Suppose some of the rays from the object strike a smooth, plane reflecting surface (**Fig. 34.2**). This could be the surface of a material with a different index of refraction, which reflects part of the incident light, or a polished metal surface that reflects almost 100% of the light that strikes it. We will always draw the reflecting surface as a black line with a shaded area behind it, as in Fig. 34.2. Bathroom mirrors have a thin sheet of glass that lies in front of and protects the reflecting surface; we'll ignore the effects of this thin sheet.

According to the law of reflection, all rays striking the surface are reflected at an angle from the normal equal to the angle of incidence. Since the surface is plane, the normal is in the same direction at all points on the surface, and we have *specular* reflection. After the rays are reflected, their directions are the same as though they had come from point P'. We call point P an *object point* and point P' the corresponding *image point,* and we say that the reflecting surface forms an **image** of point P. An observer who can see only the rays reflected from the surface, and who doesn't know that he's seeing a reflection, *thinks* that the rays originate from the image point P'. The image point is therefore a convenient way to describe the directions of the various reflected rays, just as the object point P describes the directions of the rays arriving at the surface *before* reflection.

If the surface in Fig. 34.2 were *not* smooth, the reflection would be *diffuse.* Rays reflected from different parts of the surface would go in uncorrelated directions (see Fig. 33.6b), and there would be no definite image point P' from which all reflected rays seem to emanate. You can't see your reflection in a tarnished piece of metal because its surface is rough; polishing the metal smoothes the surface so that specular reflection occurs and a reflected image becomes visible.

A plane *refracting* surface also forms an image (**Fig. 34.3**). Rays coming from point P are refracted at the interface between two optical materials. When the angles of incidence are small, the final directions of the rays after refraction are the same as though they had come from an *image point P'* as shown. In Section 33.2 we described how this effect makes underwater objects appear closer to the surface than they really are (see Fig. 33.9).

In both Figs. 34.2 and 34.3 the rays do not actually pass through the image point P'. Indeed, if the mirror in Fig. 34.2 is opaque, there is no light at all on its right side. If the outgoing rays don't actually pass through the image point, we call the image a **virtual image.** Later we will see cases in which the outgoing rays really *do* pass through an image point, and we will call the resulting image a **real image.** The images that are formed on a projection screen, on the electronic sensor in a camera, and on the retina of your eye are real images.

Image Formation by a Plane Mirror

Let's concentrate for now on images produced by *reflection;* we'll return to refraction later in the chapter. **Figure 34.4** shows how to find the precise location of the virtual image P' that a plane mirror forms of an object at P. The diagram shows two rays diverging from an object point P at a distance s to the left of a plane mirror. We call s the **object distance.** The ray PV is perpendicular to the mirror surface, and it returns along its original path.

The ray PB makes an angle θ with PV. It strikes the mirror at an angle of incidence θ and is reflected at an equal angle with the normal. When we extend the two reflected rays backward, they intersect at point P', at a distance s' behind the mirror. We call s' the **image distance.** The line between P and P' is perpendicular to the mirror. The two triangles PVB and $P'VB$ are congruent, so P and P' are at equal distances from the mirror, and s and s' have equal magnitudes. The image point P' is located exactly opposite the object point P as far *behind* the mirror as the object point is from the front of the mirror.

We can repeat the construction of Fig. 34.4 for each ray diverging from P. The directions of *all* the outgoing reflected rays are the same as though they had originated at point P', confirming that P' is the *image* of P. No matter where the observer is located, she will always see the image at the point P'.

Sign Rules

Before we go further, let's introduce some general sign rules. These may seem unnecessarily complicated for the simple case of an image formed by a plane mirror, but we want to state the rules in a form that will be applicable to *all* the situations we will encounter later. These will include image formation by a plane or spherical reflecting or refracting surface, or by a pair of refracting surfaces forming a lens. Here are the rules:

1. **Sign rule for the object distance:** When the object is on the same side of the reflecting or refracting surface as the incoming light, object distance s is positive; otherwise, it is negative.

2. **Sign rule for the image distance:** When the image is on the same side of the reflecting or refracting surface as the outgoing light, image distance s' is positive; otherwise, it is negative.

3. **Sign rule for the radius of curvature of a spherical surface:** When the center of curvature C is on the same side as the outgoing light, the radius of curvature is positive; otherwise, it is negative.

Figure 34.5 illustrates rules 1 and 2 for two different situations. For a mirror the incoming and outgoing sides are always the same; for example, in Figs. 34.2, 34.4, and 34.5a they are both on the left side. For the refracting surfaces in Figs. 34.3 and 34.5b the incoming and outgoing sides are on the left and right sides, respectively, of the interface between the two materials. (Note that other textbooks may use different rules.)

In Figs. 34.4 and 34.5a the object distance s is *positive* because the object point P is on the incoming side (the left side) of the reflecting surface. The image distance s' is *negative* because the image point P' is *not* on the outgoing side (the left side) of the surface. The object and image distances s and s' are related by

$$s = -s' \qquad \text{(plane mirror)} \qquad (34.1)$$

For a plane reflecting or refracting surface, the radius of curvature is infinite and not a particularly interesting or useful quantity; in these cases we really don't need sign rule 3. But this rule will be of great importance when we study image formation by *curved* reflecting and refracting surfaces later in the chapter.

Image of an Extended Object: Plane Mirror

Next we consider an *extended* object with finite size. For simplicity we often consider an object that has only one dimension, like a slender arrow, oriented parallel to the reflecting surface; an example is the arrow PQ in **Fig. 34.6**. The distance from the head to the tail of an arrow oriented in this way is called its *height*. In Fig. 34.6 the height is y. The image formed by such an extended object is an extended image; to each point on the object, there corresponds a point on

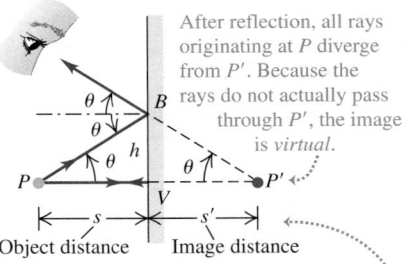

34.4 Construction for determining the location of the image formed by a plane mirror. The image point P' is as far behind the mirror as the object point P is in front of it.

After reflection, all rays originating at P diverge from P'. Because the rays do not actually pass through P', the image is *virtual*.

Object distance   Image distance

Triangles PVB and $P'VB$ are congruent, so $|s| = |s'|$.

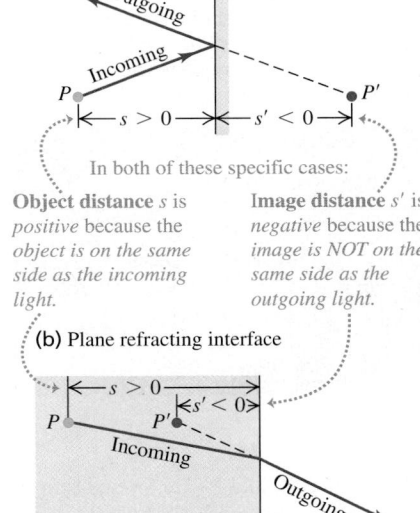

34.5 For both of these situations, the object distance s is positive (rule 1) and the image distance s' is negative (rule 2).

(a) Plane mirror

In both of these specific cases:

| Object distance s is *positive* because the object is on the same side as the incoming light. | Image distance s' is *negative* because the image is NOT on the same side as the outgoing light. |

(b) Plane refracting interface

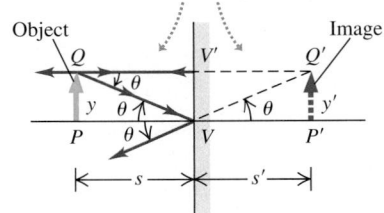

34.6 Construction for determining the height of an image formed by reflection at a plane reflecting surface.

For a plane mirror, PQV and $P'Q'V$ are congruent, so $y = y'$ and the object and image are the same size (the lateral magnification is 1).

the image. Two of the rays from Q are shown; *all* the rays from Q appear to diverge from its image point Q' after reflection. The image of the arrow is the line $P'Q'$, with height y'. Other points of the object PQ have image points between P' and Q'. The triangles PQV and $P'Q'V$ are congruent, so the object PQ and image $P'Q'$ have the same size and orientation, and $y = y'$.

The ratio of image height to object height, y'/y, in *any* image-forming situation is called the **lateral magnification** m; that is,

$$\underset{\substack{\text{Lateral magnification} \\ \text{in an image-forming situation}}}{} m = \frac{y' \cdots\text{Image height}}{y \cdots\text{Object height}} \qquad (34.2)$$

For a plane mirror $y = y'$, so the lateral magnification m is unity. When you look at yourself in a plane mirror, your image is the same size as the real you.

In Fig. 34.6 the image arrow points in the *same* direction as the object arrow; we say that the image is **erect.** In this case, y and y' have the same sign, and the lateral magnification m is positive. The image formed by a plane mirror is always erect, so y and y' have both the same magnitude and the same sign; from Eq. (34.2) the lateral magnification of a plane mirror is always $m = +1$. Later we will encounter situations in which the image is **inverted;** that is, the image arrow points in the direction *opposite* to that of the object arrow. For an inverted image, y and y' have *opposite* signs, and the lateral magnification m is *negative*.

The object in Fig. 34.6 has only one dimension. **Figure 34.7** shows a *three-dimensional* object and its three-dimensional virtual image formed by a plane mirror. The object and image are related in the same way as a left hand and a right hand.

CAUTION **Reflections in a plane mirror** At this point, you may be asking, "Why does a plane mirror reverse images left and right but not top and bottom?" This question is quite misleading! As Fig. 34.7 shows, the up-down image $P'Q'$ and the left-right image $P'S'$ are parallel to their objects and are not reversed at all. Only the front-back image $P'R'$ is reversed relative to PR. Hence it's most correct to say that a plane mirror reverses *back to front*. When an object and its image are related in this way, the image is said to be **reversed;** this means that *only* the front-back dimension is reversed. ▮

The reversed image of a three-dimensional object formed by a plane mirror is the same *size* as the object in all its dimensions. When the transverse dimensions of object and image are in the same direction, the image is erect. Thus a plane mirror always forms an erect but reversed image (**Fig. 34.8**).

An important property of all images formed by reflecting or refracting surfaces is that an *image* formed by one surface or optical device can serve as the *object* for a second surface or device. **Figure 34.9** shows a simple example. Mirror 1 forms an image P'_1 of the object point P, and mirror 2 forms another image P'_2, each in the way we have just discussed. But in addition, the image P'_1 formed by mirror 1 serves as an object for mirror 2, which then forms an image of this object at point P'_3 as shown. Similarly, mirror 1 uses the image P'_2 formed by mirror 2 as an object and forms an image of it. We leave it to you to show that this image point is also at P'_3. The idea that an image formed by one device can act as the object for a second device is of great importance. We'll use it later in this chapter to locate the image formed by two successive curved-surface refractions in a lens. We'll also use it to understand image formation by combinations of lenses, as in a microscope or a refracting telescope.

TEST YOUR UNDERSTANDING OF SECTION 34.1 If you walk directly toward a plane mirror at a speed v, at what speed does your image approach you? (i) Slower than v; (ii) v; (iii) faster than v but slower than $2v$; (iv) $2v$; (v) faster than $2v$. ▮

34.7 The image formed by a plane mirror is virtual, erect, and reversed. It is the same size as the object.

An image made by a plane mirror is reversed back to front: the image thumb $P'R'$ and object thumb PR point in opposite directions (toward each other).

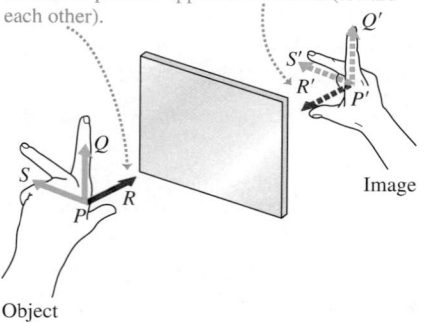

34.8 The image formed by a plane mirror is reversed; the image of a right hand is a left hand, and so on. (The hand is resting on a horizontal mirror.) Are images of the letters I, H, and T reversed?

34.9 Images P'_1 and P'_2 are formed by a single reflection of each ray from the object at P. Image P'_3, located by treating either of the other images as an object, is formed by a double reflection of each ray.

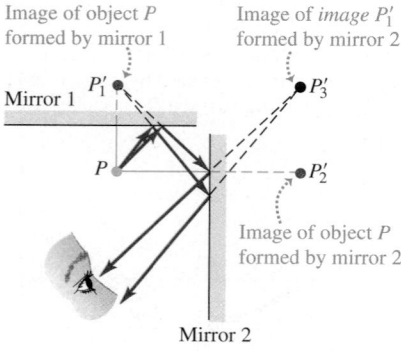

34.2 REFLECTION AT A SPHERICAL SURFACE

A plane mirror produces an image that is the same size as the object. But there are many applications for mirrors in which the image and object must be of different sizes. A magnifying mirror used when applying makeup gives an image that is *larger* than the object, and surveillance mirrors (used in stores to help spot shoplifters) give an image that is *smaller* than the object. There are also applications of mirrors in which a *real* image is desired, so light rays do indeed pass through the image point P'. A plane mirror by itself cannot perform any of these tasks. Instead, *curved* mirrors are used.

Image of a Point Object: Spherical Mirror

We'll consider the special (and easily analyzed) case of image formation by a *spherical* mirror. **Figure 34.10a** shows a spherical mirror with radius of curvature R, with its concave side facing the incident light. The **center of curvature** of the surface (the center of the sphere of which the surface is a part) is at C, and the **vertex** of the mirror (the center of the mirror surface) is at V. The line CV is called the **optic axis.** Point P is an object point that lies on the optic axis; for the moment, we assume that the distance from P to V is greater than R.

Ray PV, passing through C, strikes the mirror normally and is reflected back on itself. Ray PB, at an angle α with the axis, strikes the mirror at B, where the angles of incidence and reflection are θ. The reflected ray intersects the axis at point P'. We will show shortly that *all* rays from P intersect the axis at the *same* point P', as in Fig. 34.10b, provided that the angle α is small. Point P' is therefore the *image* of object point P. Unlike the reflected rays in Fig. 34.1, the reflected rays in Fig. 34.10b actually do intersect at point P', then diverge from P' *as if* they had originated at this point. Thus P' is a *real* image.

To see the usefulness of having a real image, suppose that the mirror is in a darkened room in which the only source of light is a self-luminous object at P. If you place a small piece of photographic film at P', all the rays of light coming from point P that reflect off the mirror will strike the same point P' on the film; when developed, the film will show a single bright spot, representing a sharply focused image of the object at point P. This principle is at the heart of most astronomical telescopes, which use large concave mirrors to make photographs of celestial objects. With a *plane* mirror like that in Fig. 34.2, the light rays never actually pass through the image point, and the image can't be recorded on film. Real images are *essential* for photography.

Let's now locate the real image point P' in Fig. 34.10a and prove that all rays from P intersect at P' (provided that their angle with the optic axis is small). The object distance, measured from the vertex V, is s; the image distance, also measured from V, is s'. The signs of s, s', and the radius of curvature R are determined by the sign rules given in Section 34.1. The object point P is on the same side as the incident light, so according to sign rule 1, s is positive. The image point P' is on the same side as the reflected light, so according to sign rule 2, the image distance s' is also positive. The center of curvature C is on the same side as the reflected light, so according to sign rule 3, R, too, is positive; R is always positive when reflection occurs at the *concave* side of a surface (**Fig. 34.11**).

We now use the following theorem from plane geometry: An exterior angle of a triangle equals the sum of the two opposite interior angles. Applying this theorem to triangles PBC and $P'BC$ in Fig. 34.10a, we have

$$\phi = \alpha + \theta \qquad \beta = \phi + \theta$$

Eliminating θ between these equations gives

$$\alpha + \beta = 2\phi \qquad (34.3)$$

34.10 (a) A concave spherical mirror forms a real image of a point object P on the mirror's optic axis. (b) The eye sees some of the outgoing rays and perceives them as having come from P'.

(a) Construction for finding the position P' of an image formed by a concave spherical mirror

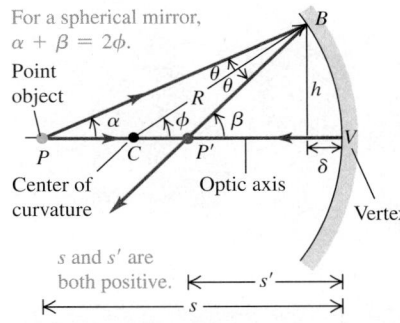

For a spherical mirror, $\alpha + \beta = 2\phi$.

s and s' are both positive.

(b) The paraxial approximation, which holds for rays with small α

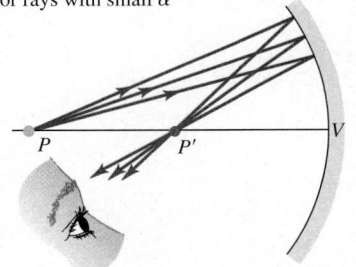

All rays from P that have a small angle α pass through P', forming a real image.

34.11 The sign rule for the radius of a spherical mirror.

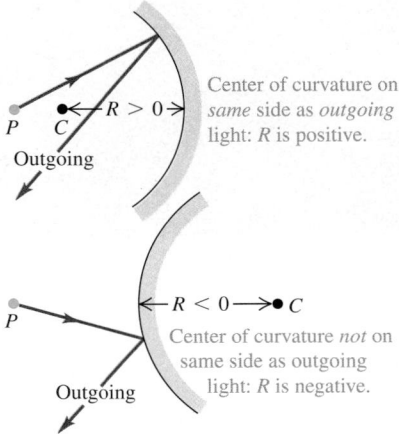

Center of curvature on *same* side as *outgoing* light: R is positive.

Center of curvature *not* on same side as outgoing light: R is negative.

34.12 (a), (b) Soon after the Hubble Space Telescope (HST) was placed in orbit in 1990, it was discovered that the concave primary mirror (also called the *objective mirror*) was too shallow by about $\frac{1}{50}$ the width of a human hair, leading to spherical aberration of the star's image. (c) After corrective optics were installed in 1993, the effects of spherical aberration were almost completely eliminated.

(a) The 2.4-m-diameter primary mirror of the Hubble Space Telescope

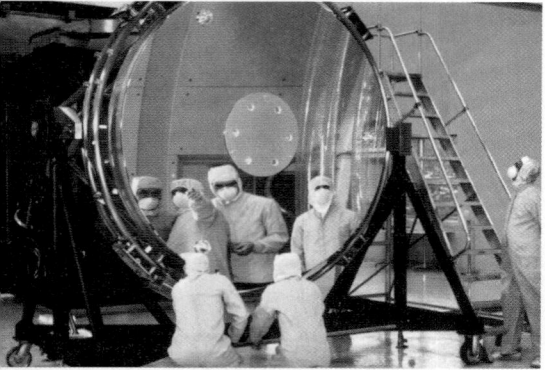

(b) A star seen with the original mirror

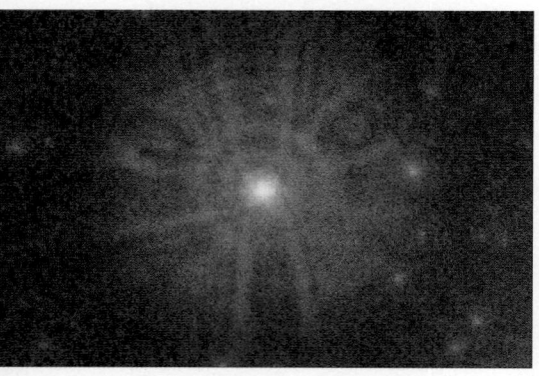

(c) The same star with corrective optics

We may now compute the image distance s'. Let h represent the height of point B above the optic axis, and let δ represent the short distance from V to the foot of this vertical line. We now write expressions for the tangents of α, β, and ϕ, remembering that s, s', and R are all positive quantities:

$$\tan\alpha = \frac{h}{s - \delta} \qquad \tan\beta = \frac{h}{s' - \delta} \qquad \tan\phi = \frac{h}{R - \delta}$$

These trigonometric equations cannot be solved as simply as the corresponding algebraic equations for a plane mirror. However, *if the angle α is small,* the angles β and ϕ are also small. The tangent of an angle that is much less than one radian is nearly equal to the angle itself (measured in radians), so we can replace $\tan\alpha$ by α, and so on, in the equations above. Also, if α is small, we can ignore the distance δ compared with s', s, and R. So for small angles we have the following approximate relationships:

$$\alpha = \frac{h}{s} \qquad \beta = \frac{h}{s'} \qquad \phi = \frac{h}{R}$$

Substituting these into Eq. (34.3) and dividing by h, we obtain a general relationship among s, s', and R:

$$\frac{1}{s} + \frac{1}{s'} = \frac{2}{R} \qquad \text{(object–image relationship, spherical mirror)} \quad (34.4)$$

This equation does not contain the angle α. Hence *all* rays from P that make sufficiently small angles with the axis intersect at P' after they are reflected; this verifies our earlier assertion. Such rays, nearly parallel to the axis and close to it, are called **paraxial rays.** (The term **paraxial approximation** is often used for the approximations we have just described.) Since all such reflected light rays converge on the image point, a concave mirror is also called a *converging mirror.*

Be sure you understand that Eq. (34.4), as well as many similar relationships that we will derive later in this chapter and the next, is only *approximately* correct. It results from a calculation containing approximations, and it is valid only for paraxial rays. If we increase the angle α that a ray makes with the optic axis, the point P' where the ray intersects the optic axis moves somewhat closer to the vertex than for a paraxial ray. As a result, a spherical mirror, unlike a plane mirror, does not form a precise point image of a point object; the image is "smeared out." This property of a spherical mirror is called **spherical aberration.** When the primary mirror of the Hubble Space Telescope (**Fig. 34.12a**) was manufactured, tiny errors were made in its shape that led to an unacceptable amount of spherical aberration (Fig. 34.12b). The performance of the telescope improved dramatically after the installation of corrective optics (Fig. 34.12c).

If the radius of curvature becomes infinite ($R = \infty$), the mirror becomes *plane,* and Eq. (34.4) reduces to Eq. (34.1) for a plane reflecting surface.

Focal Point and Focal Length

When the object point P is very far from the spherical mirror ($s = \infty$), the incoming rays are parallel. (The star shown in Fig. 34.12c is an example of such a distant object.) From Eq. (34.4) the image distance s' in this case is given by

$$\frac{1}{\infty} + \frac{1}{s'} = \frac{2}{R} \qquad s' = \frac{R}{2}$$

The situation is shown in **Fig. 34.13a**. The beam of incident parallel rays converges, after reflection from the mirror, to a point F at a distance $R/2$ from the vertex of the mirror. The point F at which the incident parallel rays converge is called the **focal point;** we say that these rays are brought to a focus. The distance from the vertex to the focal point, denoted by f, is called the **focal length.** We see that f is related to the radius of curvature R by

$$f = \frac{R}{2} \qquad \text{(focal length of a spherical mirror)} \qquad (34.5)$$

Figure 34.13b shows the opposite situation. Now the *object* is placed at the focal point F, so the object distance is $s = f = R/2$. The image distance s' is again given by Eq. (34.4):

$$\frac{2}{R} + \frac{1}{s'} = \frac{2}{R} \qquad \frac{1}{s'} = 0 \qquad s' = \infty$$

With the object at the focal point, the reflected rays in Fig. 34.13b are parallel to the optic axis; they meet only at a point infinitely far from the mirror, so the image is at infinity.

Thus the focal point F of a spherical mirror has the properties that (1) any incoming ray parallel to the optic axis is reflected through the focal point and (2) any incoming ray that passes through the focal point is reflected parallel to the optic axis. For spherical mirrors these statements are true only for paraxial rays. For parabolic mirrors these statements are *exactly* true. Spherical or parabolic mirrors are used in flashlights and headlights to form the light from the bulb into a parallel beam. Some solar-power plants use an array of plane mirrors to simulate an approximately spherical concave mirror; sunlight is collected by the mirrors and directed to the focal point, where a steam boiler is placed. (The concepts of focal point and focal length also apply to lenses, as we'll see in Section 34.4.)

We will usually express the relationship between object and image distances for a mirror, Eq. (34.4), in terms of the focal length f:

$$\frac{1}{s} + \frac{1}{s'} = \frac{1}{f} \qquad \text{(object–image relationship, spherical mirror)} \qquad (34.6)$$

Image of an Extended Object: Spherical Mirror

Now suppose we have an object with *finite* size, represented by the arrow PQ in **Fig. 34.14**, perpendicular to the optic axis CV. The image of P formed by paraxial rays is at P'. The object distance for point Q is very nearly equal to that for point P, so the image $P'Q'$ is nearly straight and perpendicular to the axis.

34.13 The focal point and focal length of a concave mirror.

(a) All parallel rays incident on a spherical mirror reflect through the focal point.

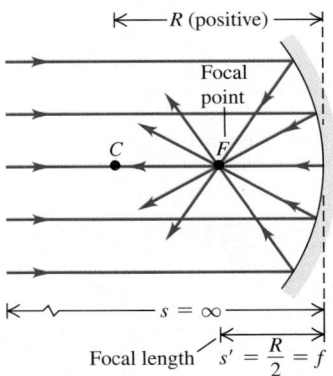

(b) Rays diverging from the focal point reflect to form parallel outgoing rays.

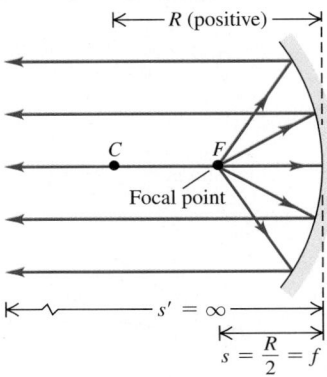

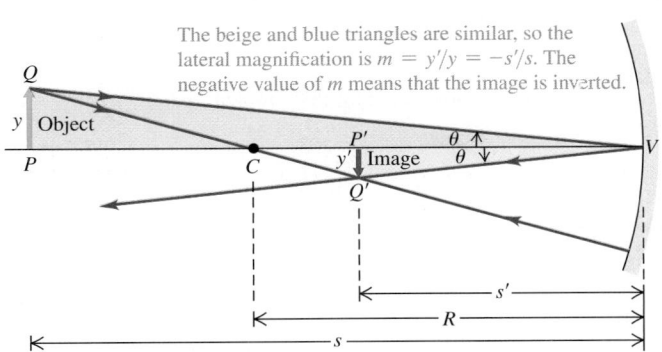

The beige and blue triangles are similar, so the lateral magnification is $m = y'/y = -s'/s$. The negative value of m means that the image is inverted.

34.14 Construction for determining the position, orientation, and height of an image formed by a concave spherical mirror.

Application Satellite Television Dishes A dish antenna used to receive satellite TV broadcasts is actually a concave parabolic mirror. The waves are of much lower frequency than visible light (1.2 to 1.8 × 10^{10} Hz compared with 4.0 to 7.9 × 10^{14} Hz), but the laws of reflection are the same. The transmitter in orbit is so far away that the arriving waves have essentially parallel rays, as in Fig. 34.13a. The dish reflects the waves and brings them to a focus at a feed horn, from which they are "piped" to a decoder that extracts the signal.

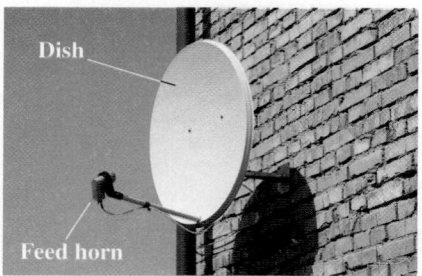
Dish

Feed horn

Dish = segment of a curved mirror. Only a segment away from the optic axis is used so that the feed horn does not block incoming waves.

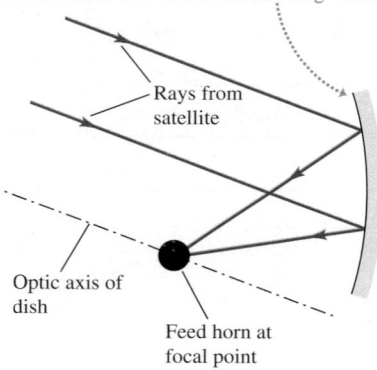
Rays from satellite

Optic axis of dish

Feed horn at focal point

Note that the object and image arrows have different sizes, y and y', respectively, and that they have opposite orientation. In Eq. (34.2) we defined the *lateral magnification m* as the ratio of image size y' to object size y:

$$m = \frac{y'}{y}$$

Because triangles PVQ and $P'VQ'$ in Fig. 34.14 are *similar*, we also have the relationship $y/s = -y'/s'$. The negative sign is needed because object and image are on opposite sides of the optic axis; if y is positive, y' is negative. Therefore

$$m = \frac{y'}{y} = -\frac{s'}{s} \qquad \text{(lateral magnification, spherical mirror)} \qquad (34.7)$$

If m is positive, the image is erect in comparison to the object; if m is negative, the image is *inverted* relative to the object, as in Fig. 34.14. For a *plane* mirror, $s = -s'$, so $y' = y$ and $m = +1$; since m is positive, the image is erect, and since $|m| = 1$, the image is the same size as the object.

CAUTION **Lateral magnification can be less than 1** Although the ratio of image size to object size is called the *lateral magnification*, the image formed by a mirror or lens may be larger than, smaller than, or the same size as the object. If it is smaller, then the lateral magnification is less than unity in absolute value: $|m| < 1$. The image formed by an astronomical telescope mirror or a camera lens is usually *much* smaller than the object. For example, the image of the bright star shown in Fig. 34.12c is just a few millimeters across, while the star itself is hundreds of thousands of kilometers in diameter. ▌

In our discussion of concave mirrors we have so far considered only objects that lie *outside* or at the focal point, so that the object distance s is greater than or equal to the (positive) focal length f. In this case the image point is on the same side of the mirror as the outgoing rays, and the image is real and inverted. If an object is *inside* the focal point of a concave mirror, so that $s < f$, the resulting image is *virtual* (that is, the image point is on the opposite side of the mirror from the object), *erect,* and *larger* than the object. Mirrors used when you apply makeup (referred to at the beginning of this section) are concave mirrors; in use, the distance from the face to the mirror is less than the focal length, and you see an enlarged, erect image. You can prove these statements about concave mirrors by applying Eqs. (34.6) and (34.7). We'll also be able to verify these results later in this section, after we've learned some graphical methods for relating the positions and sizes of the object and the image.

EXAMPLE 34.1   IMAGE FORMATION BY A CONCAVE MIRROR I

A concave mirror forms an image, on a wall 3.00 m in front of the mirror, of a headlamp filament 10.0 cm in front of the mirror. (a) What are the radius of curvature and focal length of the mirror? (b) What is the lateral magnification? What is the image height if the object height is 5.00 mm?

SOLUTION

IDENTIFY and SET UP: Figure 34.15 shows our sketch. Our target variables are the radius of curvature R, focal length f, lateral magnification m, and image height y'. We are given the distances from the mirror to the object (s) and from the mirror to the image (s'). We solve Eq. (34.4) for R, and then use Eq. (34.5) to find f. Equation (34.7) yields both m and y'.

34.15 Our sketch for this problem.

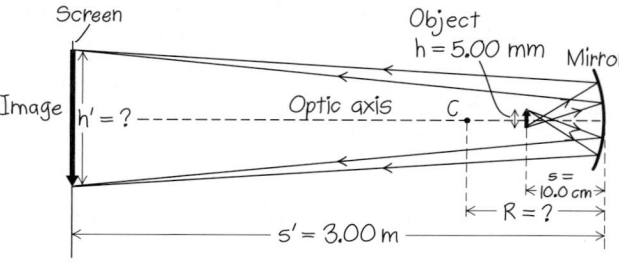

EXECUTE: (a) Both the object and the image are on the concave (reflective) side of the mirror, so both s and s' are positive; we have $s = 10.0$ cm and $s' = 300$ cm. We solve Eq. (34.4) for R:

$$\frac{1}{10.0\text{ cm}} + \frac{1}{300\text{ cm}} = \frac{2}{R}$$

$$R = 2\left(\frac{1}{10.0\text{ cm}} + \frac{1}{300\text{ cm}}\right)^{-1} = 19.4\text{ cm}$$

The focal length of the mirror is $f = R/2 = 9.7$ cm.

(b) From Eq. (34.7) the lateral magnification is

$$m = -\frac{s'}{s} = -\frac{300\text{ cm}}{10.0\text{ cm}} = -30.0$$

Because m is negative, the image is inverted. The height of the image is 30.0 times the height of the object, or $(30.0)(5.00\text{ mm}) = 150$ mm.

EVALUATE: Our sketch indicates that the image is inverted; our calculations agree. Note that the object (at $s = 10.0$ cm) is just outside the focal point ($f = 9.7$ cm). This is very similar to what is done in automobile headlights. With the filament close to the focal point, the concave mirror produces a beam of nearly parallel rays.

CONCEPTUAL EXAMPLE 34.2   IMAGE FORMATION BY A CONCAVE MIRROR II

In Example 34.1, suppose that the lower half of the mirror's reflecting surface is covered with nonreflective soot. What effect will this have on the image of the filament?

SOLUTION

It would be natural to guess that the image would now show only half of the filament. But in fact the image will still show the *entire* filament. You can see why by examining Fig. 34.10b. Light rays coming from any object point P are reflected from *all* parts of the mirror and converge on the corresponding image point P'. If part of

the mirror surface is made nonreflective (or is removed altogether), rays from the remaining reflective surface still form an image of every part of the object.

Reducing the reflecting area reduces the light energy reaching the image point, however: The image becomes *dimmer*. If the area is reduced by one-half, the image will be one-half as bright. Conversely, *increasing* the reflective area makes the image brighter. To make reasonably bright images of faint stars, astronomical telescopes use mirrors that are up to several meters in diameter (see Fig. 34.12a).

Convex Mirrors

In **Fig. 34.16a** the *convex* side of a spherical mirror faces the incident light. The center of curvature is on the side opposite to the outgoing rays; according to sign rule 3 in Section 34.1, R is negative (see Fig. 34.11). Ray PB is reflected, with the angles of incidence and reflection both equal to θ. The reflected ray, projected backward, intersects the axis at P'. As with a concave mirror, all rays from P that are reflected by the mirror diverge from the same point P', provided that the angle α is small. Therefore P' is the image of P. The object distance s is positive, the image distance s' is negative, and the radius of curvature R is *negative* for a *convex* mirror.

Figure 34.16b shows two rays diverging from the head of the arrow PQ and the virtual image $P'Q'$ of this arrow. The same procedure that we used for a concave mirror can be used to show that for a convex mirror, the expressions for the object–image relationship and the lateral magnification are

$$\frac{1}{s} + \frac{1}{s'} = \frac{2}{R} \quad \text{and} \quad m = \frac{y'}{y} = -\frac{s'}{s}$$

34.16 Image formation by a convex mirror.

(a) Construction for finding the position of an image formed by a convex mirror

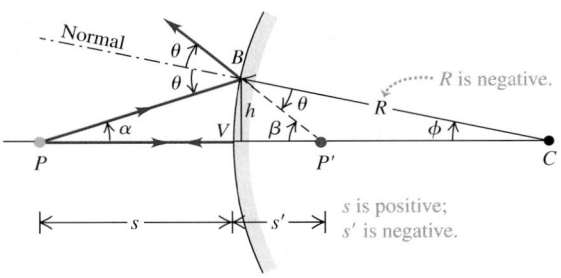

(b) Construction for finding the magnification of an image formed by a convex mirror

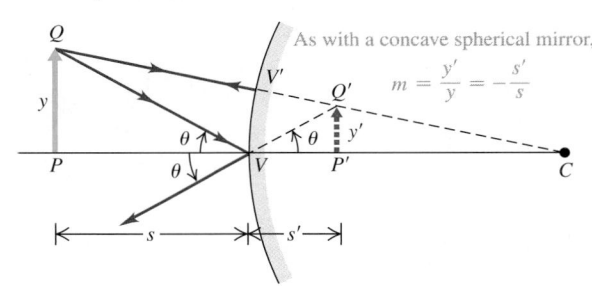

34.17 The focal point and focal length of a convex mirror.

(a) Paraxial rays incident on a convex spherical mirror diverge from a virtual focal point.

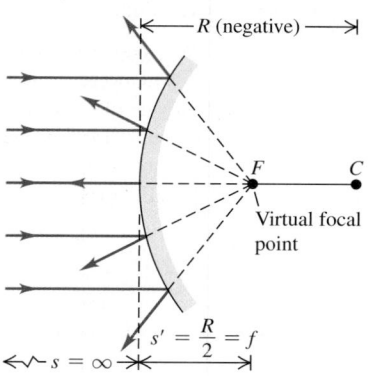

(b) Rays aimed at the virtual focal point are parallel to the axis after reflection.

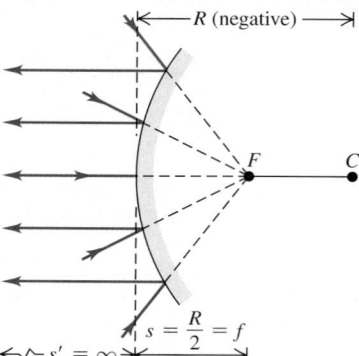

DATA *SPEAKS*

Image Formation by Mirrors

When students were given a problem involving image formation by mirrors, more than 59% gave an incorrect response. Common errors:

- Not using the law of reflection properly. For a mirror (plane or curved), the incident and reflected rays make the same angle with the normal to the mirror. The reflected ray originates at the point where the incident ray hits the mirror.

- Confusion about lateral magnification. The lateral magnification m depends on only the ratio of image distance s' to object distance s. If s' and s have different values but have the same ratio in two situations, then the value of m is the same.

These expressions are exactly the same as Eqs. (34.4) and (34.7) for a concave mirror. Thus when we use our sign rules consistently, Eqs. (34.4) and (34.7) are valid for both concave and convex mirrors.

When R is negative (convex mirror), incoming rays that are parallel to the optic axis are not reflected through the focal point F. Instead, they diverge as though they had come from the point F at a distance f *behind* the mirror, as shown in **Fig. 34.17a**. In this case, f is the focal length, and F is called a *virtual focal point*. The corresponding image distance s' is negative, so both f and R are negative, and Eq. (34.5), $f = R/2$, holds for convex as well as concave mirrors. In Fig. 34.17b the incoming rays are converging as though they would meet at the virtual focal point F, and they are reflected parallel to the optic axis.

In summary, Eqs. (34.4) through (34.7), the basic relationships for image formation by a spherical mirror, are valid for *both* concave and convex mirrors, provided that we use the sign rules consistently.

EXAMPLE 34.3 **SANTA'S IMAGE PROBLEM**

Santa checks himself for soot, using his reflection in a silvered Christmas tree ornament 0.750 m away (**Fig. 34.18a**). The diameter of the ornament is 7.20 cm. Standard reference texts state that he is a "right jolly old elf," so we estimate his height to be 1.6 m. Where and how tall is the image of Santa formed by the ornament? Is it erect or inverted?

SOLUTION

IDENTIFY and SET UP: Figure 34.18b shows the situation. Santa is the object, and the surface of the ornament closest to him acts as a convex mirror. The relationships among object distance, image distance, focal length, and magnification are the same as

34.18 (a) The ornament forms a virtual, reduced, erect image of Santa. **(b)** Our sketch of two of the rays forming the image.

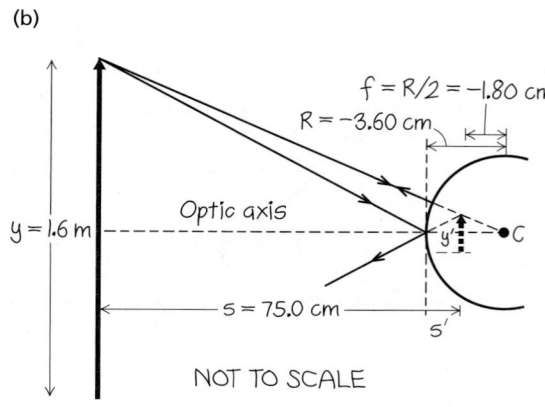

NOT TO SCALE

for concave mirrors, provided we use the sign rules consistently. The radius of curvature and the focal length of a convex mirror are *negative*. The object distance is $s = 0.750 \text{ m} = 75.0 \text{ cm}$, and Santa's height is $y = 1.6 \text{ m}$. We solve Eq. (34.6) to find the image distance s', and then use Eq. (34.7) to find the lateral magnification m and the image height y'. The sign of m tells us whether the image is erect or inverted.

EXECUTE: The radius of the mirror (half the diameter) is $R = -(7.20 \text{ cm})/2 = -3.60 \text{ cm}$, and the focal length is $f = R/2 = -1.80 \text{ cm}$. From Eq. (34.6),

$$\frac{1}{s'} = \frac{1}{f} - \frac{1}{s} = \frac{1}{-1.80 \text{ cm}} - \frac{1}{75.0 \text{ cm}}$$

$$s' = -1.76 \text{ cm}$$

Because s' is negative, the image is behind the mirror—that is, on the side opposite to the outgoing light (Fig. 34.18b)—and it is

virtual. The image is about halfway between the front surface of the ornament and its center.

From Eq. (34.7), the lateral magnification and the image height are

$$m = \frac{y'}{y} = -\frac{s'}{s} = -\frac{-1.76 \text{ cm}}{75.0 \text{ cm}} = 0.0234$$

$$y' = my = (0.0234)(1.6 \text{ m}) = 3.8 \times 10^{-2} \text{ m} = 3.8 \text{ cm}$$

EVALUATE: Our sketch indicates that the image is erect so both m and y' are positive; our calculations agree. When the object distance s is positive, a convex mirror *always* forms an erect, virtual, reduced, reversed image. For this reason, convex mirrors are used at blind intersections, for surveillance in stores, and as wide-angle rear-view mirrors for cars and trucks. (Many such mirrors read "Objects in mirror are closer than they appear.")

Graphical Methods for Mirrors

In Examples 34.1 and 34.3, we used Eqs. (34.6) and (34.7) to find the position and size of the image formed by a mirror. We can also determine the properties of the image by a simple *graphical* method. This method consists of finding the point of intersection of a few particular rays that diverge from a point of the object (such as point Q in **Fig. 34.19**) and are reflected by the mirror. Then (ignoring aberrations) *all* rays from this object point that strike the mirror will intersect at the same point. For this construction we always choose an object point that is *not* on the optic axis. Four rays that we can usually draw easily are shown in Fig. 34.19. These are called **principal rays.**

1. *A ray parallel to the axis*, after reflection, passes through the focal point F of a concave mirror or appears to come from the (virtual) focal point of a convex mirror.

2. *A ray through (or proceeding toward) the focal point F* is reflected parallel to the axis.

3. *A ray along the radius* through or away from the center of curvature C intersects the surface normally and is reflected back along its original path.

4. *A ray to the vertex V* is reflected forming equal angles with the optic axis.

34.19 The graphical method of locating an image formed by a spherical mirror. The colors of the rays are for identification only; they do not refer to specific colors of light.

(a) Principal rays for concave mirror

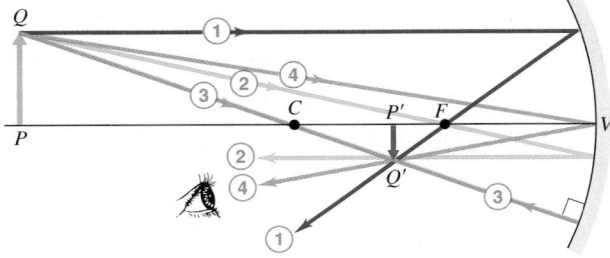

(b) Principal rays for convex mirror

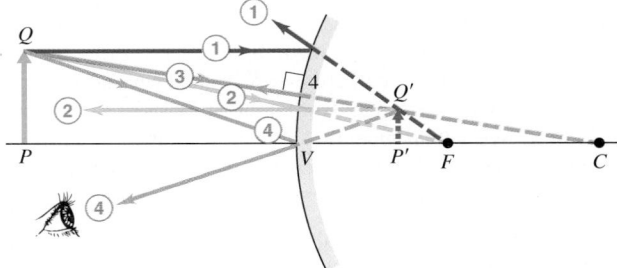

① Ray parallel to axis reflects through focal point.
② Ray through focal point reflects parallel to axis.
③ Ray through center of curvature intersects the surface normally and reflects along its original path.
④ Ray to vertex reflects symmetrically around optic axis.

① Reflected parallel ray appears to come from focal point.
② Ray toward focal point reflects parallel to axis.
③ As with concave mirror: Ray radial to center of curvature intersects the surface normally and reflects along its original path.
④ As with concave mirror: Ray to vertex reflects symmetrically around optic axis.

Once we have found the position of the image point by means of the intersection of any two of these principal rays (1, 2, 3, 4), we can draw the path of any other ray from the object point to the same image point.

CAUTION **Principal rays are not the only rays** Although we've emphasized the principal rays, in fact *any* ray from the object that strikes the mirror will pass through the image point (for a real image) or appear to originate from the image point (for a virtual image). Usually, you need to draw only the principal rays in order to locate the image. ∎

PROBLEM-SOLVING STRATEGY 34.1) **IMAGE FORMATION BY MIRRORS**

IDENTIFY *the relevant concepts:* Problems involving image formation by mirrors can be solved in two ways: using principal-ray diagrams and using equations. A successful problem solution uses *both* approaches.

SET UP *the problem:* Identify the target variables. One of them is likely to be the focal length, the object distance, or the image distance, with the other two quantities given.

EXECUTE *the solution* as follows:
1. Draw a large, clear principal-ray diagram if you have enough information.
2. Orient your diagram so that incoming rays go from left to right. Draw only the principal rays; color-code them as in Fig. 34.19. If possible, use graph paper or quadrille-ruled paper. Use a ruler and measure distances carefully! A freehand sketch will *not* give good results.
3. If your principal rays don't converge at a real image point, you may have to extend them straight backward to locate a virtual

image point, as in Fig. 34.19b. We recommend drawing the extensions with broken lines.
4. Measure the resulting diagram to obtain the magnitudes of the target variables.
5. Solve for the target variables by using Eq. (34.6), $1/s + 1/s' = 1/f$, and the lateral magnification equation, Eq. (34.7), as appropriate. Apply the sign rules given in Section 34.1 to object and image distances, radii of curvature, and object and image heights.
6. Use the sign rules to interpret the results that you deduced from your ray diagram and calculations. Note that the *same* sign rules (given in Section 34.1) work for all four cases in this chapter: reflection and refraction from plane and spherical surfaces.

EVALUATE *your answer:* Check that the results of your calculations agree with your ray-diagram results for image position, image size, and whether the image is real or virtual.

EXAMPLE 34.4 **CONCAVE MIRROR WITH VARIOUS OBJECT DISTANCES**

A concave mirror has a radius of curvature with absolute value 20 cm. Find graphically the image of an object in the form of an arrow perpendicular to the axis of the mirror at object distances of (a) 30 cm, (b) 20 cm, (c) 10 cm, and (d) 5 cm. Check the construction by *computing* the size and lateral magnification of each image.

SOLUTION

IDENTIFY and SET UP: We must use graphical methods *and* calculations to analyze the image made by a mirror. The mirror is concave, so its radius of curvature is $R = +20$ cm and its focal length is $f = R/2 = +10$ cm. Our target variables are the image distances s' and lateral magnifications m corresponding to four cases with successively smaller object distances s. In each case we solve Eq. (34.6) for s' and use $m = -s'/s$ to find m.

EXECUTE: Figure 34.20 shows the principal-ray diagrams for the four cases. Study each of these diagrams carefully and confirm that each numbered ray is drawn in accordance with the rules given earlier (under "Graphical Methods for Mirrors"). Several points are worth noting. First, in case (b) the object and image distances are equal. Ray 3 cannot be drawn in this case because a ray from Q through the center of curvature C does not strike the mirror. In case (c), ray 2 cannot be drawn because a ray from Q through

F does not strike the mirror. In this case the outgoing rays are parallel, corresponding to an infinite image distance. In case (d), the outgoing rays diverge; they have been extended backward to the *virtual image point Q'*, from which they appear to diverge. Case (d) illustrates the general observation that an object placed inside the focal point of a concave mirror produces a virtual image.

Measurements of the figures, with appropriate scaling, give the following approximate image distances: (a) 15 cm; (b) 20 cm; (c) ∞ or −∞ (because the outgoing rays are parallel and do not converge at any finite distance); (d) −10 cm. To *compute* these distances, we solve Eq. (34.6) for s' and insert $f = 10$ cm:

$$\text{(a)} \quad \frac{1}{30 \text{ cm}} + \frac{1}{s'} = \frac{1}{10 \text{ cm}} \qquad s' = 15 \text{ cm}$$

$$\text{(b)} \quad \frac{1}{20 \text{ cm}} + \frac{1}{s'} = \frac{1}{10 \text{ cm}} \qquad s' = 20 \text{ cm}$$

$$\text{(c)} \quad \frac{1}{10 \text{ cm}} + \frac{1}{s'} = \frac{1}{10 \text{ cm}} \qquad s' = \infty \text{ (or } -\infty)$$

$$\text{(d)} \quad \frac{1}{5 \text{ cm}} + \frac{1}{s'} = \frac{1}{10 \text{ cm}} \qquad s' = -10 \text{ cm}$$

The signs of s' tell us that the image is real in cases (a) and (b) and virtual in case (d).

The lateral magnifications measured from the figures are approximately (a) $-\frac{1}{2}$; (b) -1; (c) ∞ or $-\infty$; (d) $+2$. From Eq. (34.7),

(a) $m = -\dfrac{15 \text{ cm}}{30 \text{ cm}} = -\dfrac{1}{2}$

(b) $m = -\dfrac{20 \text{ cm}}{20 \text{ cm}} = -1$

(c) $m = -\dfrac{\infty \text{ cm}}{10 \text{ cm}} = -\infty \text{ (or } +\infty)$

(d) $m = -\dfrac{-10 \text{ cm}}{5 \text{ cm}} = +2$

The signs of m tell us that the image is inverted in cases (a) and (b) and erect in case (d).

EVALUATE: Notice the trend of the results in the four cases. When the object is far from the mirror, as in Fig. 34.20a, the image is smaller than the object, inverted, and real. As the object distance s decreases, the image moves farther from the mirror and gets larger (Fig. 34.20b). When the object is at the focal point, the image is at infinity (Fig. 34.20c). When the object is inside the focal point, the image becomes larger than the object, erect, and virtual (Fig. 34.20d). You can confirm these conclusions by looking at objects reflected in the concave bowl of a shiny metal spoon.

34.20 Using principal-ray diagrams to locate the image $P'Q'$ made by a concave mirror.

(a) Construction for $s = 30$ cm

All principal rays can be drawn. The image is inverted.

(b) Construction for $s = 20$ cm

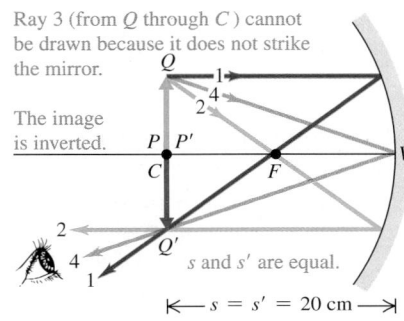

Ray 3 (from Q through C) cannot be drawn because it does not strike the mirror.

The image is inverted.

s and s' are equal.

$\longleftarrow s = s' = 20 \text{ cm} \longrightarrow$

(c) Construction for $s = 10$ cm

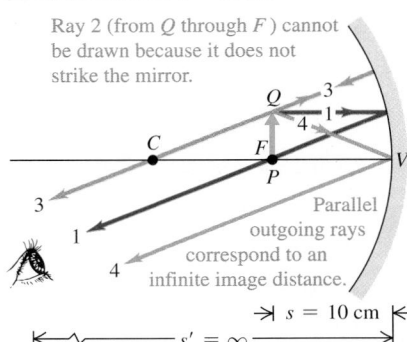

Ray 2 (from Q through F) cannot be drawn because it does not strike the mirror.

Parallel outgoing rays correspond to an infinite image distance.

$s = 10$ cm

$s' = \infty$

(d) Construction for $s = 5$ cm

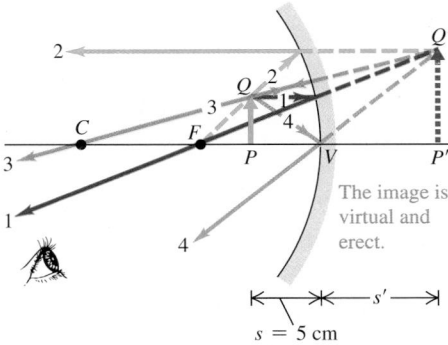

The image is virtual and erect.

$s = 5$ cm

TEST YOUR UNDERSTANDING OF SECTION 34.2 A cosmetics mirror is designed so that your reflection appears right-side up and enlarged. (a) Is the mirror concave or convex? (b) To see an enlarged image, what should be the distance from the mirror (of focal length f) to your face? (i) $|f|$; (ii) less than $|f|$; (iii) greater than $|f|$. ∎

34.3 REFRACTION AT A SPHERICAL SURFACE

As we mentioned in Section 34.1, images can be formed by refraction as well as by reflection. To begin with, let's consider refraction at a spherical surface—that is, at a spherical interface between two optical materials with different indexes of refraction. This analysis is directly applicable to some real optical systems, such as the human eye. It also provides a stepping-stone for the analysis of lenses, which usually have *two* spherical (or nearly spherical) surfaces.

34.21 Construction for finding the position of the image point P' of a point object P formed by refraction at a spherical surface. The materials to the left and right of the interface have indexes of refraction n_a and n_b, respectively. In the case shown here, $n_a < n_b$.

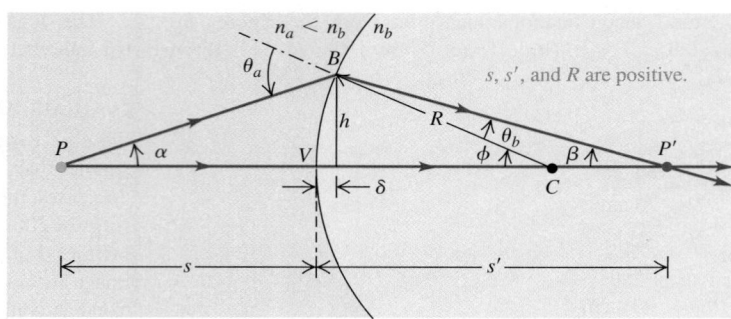

Image of a Point Object: Spherical Refracting Surface

In **Fig. 34.21** a spherical surface with radius R forms an interface between two materials with different indexes of refraction n_a and n_b. The surface forms an image P' of an object point P; we want to find how the object and image distances (s and s') are related. We will use the same sign rules that we used for spherical mirrors. The center of curvature C is on the outgoing side of the surface, so R is positive. Ray PV strikes the vertex V and is perpendicular to the surface (that is, to the plane that is tangent to the surface at the point of incidence V). It passes into the second material without deviation. Ray PB, making an angle α with the axis, is incident at an angle θ_a with the normal and is refracted at an angle θ_b. These rays intersect at P', a distance s' to the right of the vertex. The figure is drawn for the case $n_a < n_b$. Both the object and image distances are positive.

We are going to prove that if the angle α is small, *all* rays from P intersect at the same point P', so P' is the *real image* of P. We use much the same approach as we did for spherical mirrors in Section 34.2. We again use the theorem that an exterior angle of a triangle equals the sum of the two opposite interior angles; applying this to the triangles PBC and $P'BC$ gives

$$\theta_a = \alpha + \phi \qquad \phi = \beta + \theta_b \tag{34.8}$$

From the law of refraction,

$$n_a \sin\theta_a = n_b \sin\theta_b$$

Also, the tangents of α, β, and ϕ are

$$\tan\alpha = \frac{h}{s+\delta} \qquad \tan\beta = \frac{h}{s'-\delta} \qquad \tan\phi = \frac{h}{R-\delta} \tag{34.9}$$

For paraxial rays, θ_a and θ_b are both small in comparison to a radian, and we may approximate both the sine and tangent of either of these angles by the angle itself (measured in radians). The law of refraction then gives

$$n_a\theta_a = n_b\theta_b$$

Combining this with the first of Eqs. (34.8), we obtain

$$\theta_b = \frac{n_a}{n_b}(\alpha + \phi)$$

When we substitute this into the second of Eqs. (34.8), we get

$$n_a\alpha + n_b\beta = (n_b - n_a)\phi \tag{34.10}$$

Now we use the approximations $\tan\alpha = \alpha$, and so on, in Eqs. (34.9) and also ignore the small distance δ; those equations then become

$$\alpha = \frac{h}{s} \qquad \beta = \frac{h}{s'} \qquad \phi = \frac{h}{R}$$

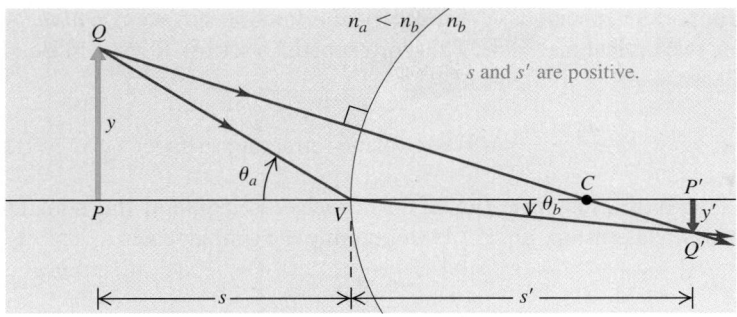

Finally, we substitute these into Eq. (34.10) and divide out the common factor h. We obtain

$$\frac{n_a}{s} + \frac{n_b}{s'} = \frac{n_b - n_a}{R} \qquad \text{(object–image relationship,}$$
$$\text{spherical refracting surface)} \qquad (34.11)$$

This equation does not contain the angle α, so the image distance is the same for *all* paraxial rays emanating from P; this proves that P' is the image of P.

To obtain the lateral magnification m for this situation, we use the construction in **Fig. 34.22**. We draw two rays from point Q, one through the center of curvature C and the other incident at the vertex V. From the triangles PQV and $P'Q'V$,

$$\tan \theta_a = \frac{y}{s} \qquad \tan \theta_b = \frac{-y'}{s'}$$

and from the law of refraction,

$$n_a \sin \theta_a = n_b \sin \theta_b$$

For small angles,

$$\tan \theta_a = \sin \theta_a \qquad \tan \theta_b = \sin \theta_b$$

so finally

$$\frac{n_a y}{s} = -\frac{n_b y'}{s'} \qquad \text{or}$$

$$m = \frac{y'}{y} = -\frac{n_a s'}{n_b s} \qquad \text{(lateral magnification,}$$
$$\text{spherical refracting surface)} \qquad (34.12)$$

Equations (34.11) and (34.12) apply to both convex and concave refracting surfaces, provided that you use the sign rules consistently. It doesn't matter whether n_b is greater or less than n_a. To verify these statements, construct diagrams like Figs. 34.21 and 34.22 for the following three cases: (i) $R > 0$ and $n_a > n_b$, (ii) $R < 0$ and $n_a < n_b$, and (iii) $R < 0$ and $n_a > n_b$. Then in each case, use your diagram to again derive Eqs. (34.11) and (34.12).

Here's a final note on the sign rule for the radius of curvature R of a surface. For the convex reflecting surface in Fig. 34.16, we considered R negative, but the convex *refracting* surface in Fig. 34.21 has a *positive* value of R. This may seem inconsistent, but it isn't. The rule is that R is positive if the center of curvature C is on the outgoing side of the surface and negative if C is on the other side. For the convex reflecting surface in Fig. 34.16, R is negative because point C is to the right of the surface but outgoing rays are to the left. For the convex refracting surface in Fig. 34.21, R is positive because both C and the outgoing rays are to the right of the surface.

Refraction at a curved surface is one reason gardeners avoid watering plants at midday. As sunlight enters a water drop resting on a leaf (**Fig. 34.23**), the light rays are refracted toward each other as in Figs. 34.21 and 34.22. The sunlight that strikes the leaf is therefore more concentrated and able to cause damage.

34.23 Light rays refract as they pass through the curved surfaces of these water droplets.

An important special case of a spherical refracting surface is a *plane* surface between two optical materials. This corresponds to setting $R = \infty$ in Eq. (34.11). In this case,

$$\frac{n_a}{s} + \frac{n_b}{s'} = 0 \qquad \text{(plane refracting surface)} \qquad (34.13)$$

To find the lateral magnification m for this case, we combine this equation with the general relationship, Eq. (34.12), obtaining the simple result

$$m = 1$$

That is, the image formed by a *plane* refracting surface always has the same lateral size as the object, and it is always erect.

An example of image formation by a plane refracting surface is the appearance of a partly submerged drinking straw or canoe paddle. When viewed from some angles, the submerged part appears to be only about three-quarters of its actual distance below the surface. (We commented on the appearance of a submerged object in Section 33.2; see Fig. 33.9.)

EXAMPLE 34.5   IMAGE FORMATION BY REFRACTION I

A cylindrical glass rod (**Fig. 34.24**) has index of refraction 1.52. It is surrounded by air. One end is ground to a hemispherical surface with radius $R = 2.00$ cm. A small object is placed on the axis of the rod, 8.00 cm to the left of the vertex. Find (a) the image distance and (b) the lateral magnification.

SOLUTION

IDENTIFY and SET UP: This problem uses the ideas of refraction at a curved surface. Our target variables are the image distance s' and the lateral magnification m. Here material a is air ($n_a = 1.00$) and material b is the glass of which the rod is made ($n_b = 1.52$). We are given $s = 8.00$ cm. The center of curvature of the spherical surface is on the outgoing side of the surface, so the radius is positive: $R = +2.00$ cm. We solve Eq. (34.11) for s', and we use Eq. (34.12) to find m.

EXECUTE: (a) From Eq. (34.11),

$$\frac{1.00}{8.00 \text{ cm}} + \frac{1.52}{s'} = \frac{1.52 - 1.00}{+2.00 \text{ cm}}$$

$$s' = +11.3 \text{ cm}$$

34.24 The glass rod in air forms a real image.

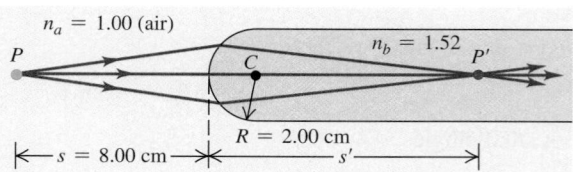

(b) From Eq. (34.12),

$$m = -\frac{n_a s'}{n_b s} = -\frac{(1.00)(11.3 \text{ cm})}{(1.52)(8.00 \text{ cm})} = -0.929$$

EVALUATE: Because the image distance s' is positive, the image is formed 11.3 cm to the *right* of the vertex (on the outgoing side), as Fig. 34.24 shows. The value of m tells us that the image is somewhat smaller than the object and that it is inverted. If the object is an arrow 1.000 mm high, pointing upward, the image is an arrow 0.929 mm high, pointing downward.

EXAMPLE 34.6   IMAGE FORMATION BY REFRACTION II

The glass rod of Example 34.5 is immersed in water, which has index of refraction $n = 1.33$ (**Fig. 34.25**). The object distance is again 8.00 cm. Find the image distance and lateral magnification.

SOLUTION

IDENTIFY and SET UP: The situation is the same as in Example 34.5 except that now $n_a = 1.33$. We again use Eqs. (34.11) and (34.12) to determine s' and m, respectively.

EXECUTE: Our solution of Eq. (34.11) in Example 34.5 yields

$$\frac{1.33}{8.00 \text{ cm}} + \frac{1.52}{s'} = \frac{1.52 - 1.33}{+2.00 \text{ cm}}$$

$$s' = -21.3 \text{ cm}$$

The lateral magnification in this case is

$$m = \frac{n_a s'}{n_b s} = -\frac{(1.33)(-21.3 \text{ cm})}{(1.52)(8.00 \text{ cm})} = +2.33$$

EVALUATE: The negative value of s' means that the refracted rays do not converge, but appear to diverge from a point 21.3 cm to the *left* of the vertex. We saw a similar case in the reflection of light from a convex mirror; in both cases we call the result a *virtual image*. The vertical image is erect (because m is positive) and 2.33 times as large as the object.

34.25 When immersed in water, the glass rod forms a virtual image.

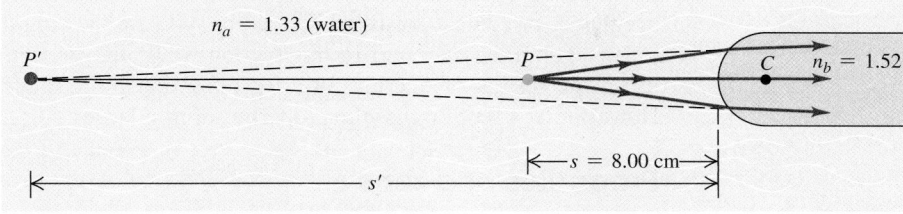

EXAMPLE 34.7   APPARENT DEPTH OF A SWIMMING POOL

If you look straight down into a swimming pool where it is 2.00 m deep, how deep does it appear to be?

SOLUTION

IDENTIFY and SET UP: Figure 34.26 shows the situation. The surface of the water acts as a plane refracting surface. To determine the pool's apparent depth, we imagine an arrow PQ painted on the bottom. The pool's refracting surface forms a virtual image $P'Q'$ of this arrow. We solve Eq. (34.13) to find the image depth s'; that's the pool's apparent depth.

EXECUTE: The object distance is the actual depth of the pool, $s = 2.00$ m. Material a is water ($n_a = 1.33$) and material b is

air ($n_b = 1.00$). From Eq. (34.13),

$$\frac{n_a}{s} + \frac{n_b}{s'} = \frac{1.33}{2.00 \text{ m}} + \frac{1.00}{s'} = 0$$

$$s' = -1.50 \text{ m}$$

The image distance is negative. By the sign rules in Section 34.1, this means that the image is virtual and on the incoming side of the refracting surface—that is, on the same side as the object, namely underwater. The pool's apparent depth is 1.50 m, or just 75% of its true depth.

EVALUATE: Recall that the lateral magnification for a plane refracting surface is $m = 1$. Hence the image $P'Q'$ of the arrow has the same *horizontal length* as the actual arrow PQ (**Fig. 34.27**). Only its depth is different from that of PQ.

34.26 Arrow $P'Q'$ is the virtual image of underwater arrow PQ. The angles of the ray with the vertical are exaggerated for clarity.

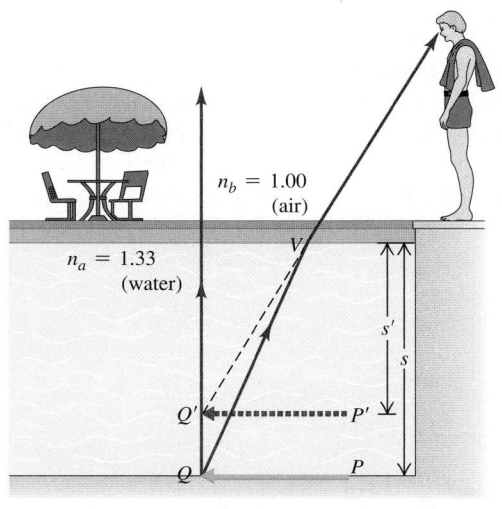

34.27 The submerged portion of this straw appears to be at a shallower depth (closer to the surface) than it actually is.

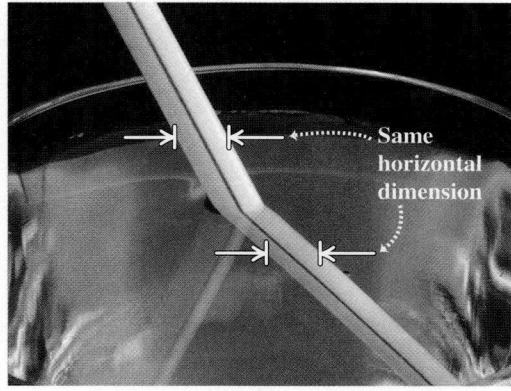

TEST YOUR UNDERSTANDING OF SECTION 34.3 The water droplets in Fig. 34.23 have radius of curvature R and index of refraction $n = 1.33$. Can they form an image of the sun on the leaf? ∎

34.4 THIN LENSES

The most familiar and widely used optical device (after the plane mirror) is the *lens*. A lens is an optical system with two refracting surfaces. The simplest lens has two *spherical* surfaces close enough together that we can ignore the distance between them (the thickness of the lens); we call this a **thin lens.** If you wear eyeglasses or contact lenses while reading, you are viewing these words through a pair of thin lenses. Later in this section we'll analyze thin lenses in detail by using the results of Section 34.3 for refraction by a single spherical surface. However, let's first discuss the properties of thin lenses.

Properties of a Lens

A lens of the shape shown in **Fig. 34.28** has an important property: When a beam of rays parallel to the axis passes through the lens, the rays converge to a point F_2 (Fig. 34.28a) and form a real image at that point. Such a lens is called a **converging lens.** Similarly, rays passing through point F_1 emerge from the lens as a beam of parallel rays (Fig. 34.28b). The points F_1 and F_2 are called the first and second *focal points,* and the distance f (measured from the center of the lens) is called the *focal length.* Note the similarities between the two focal points of a converging lens and the single focal point of a concave mirror (see Fig. 34.13). As for a concave mirror, the focal length of a converging lens is defined to be a *positive* quantity, and such a lens is also called a *positive lens.*

The central horizontal line in Fig. 34.28 is called the *optic axis,* as with spherical mirrors. The centers of curvature of the two spherical surfaces lie on and define the optic axis. The two focal lengths in Fig. 34.28, both labeled f, *are always equal* for a thin lens, even when the two sides have different curvatures. We will show this result later in the section, when we derive the relationship of f to the index of refraction of the lens and the radii of curvature of its surfaces.

Image of an Extended Object: Converging Lens

Like a concave mirror, a converging lens can form an image of an extended object. **Figure 34.29** shows how to find the position and lateral magnification of an image made by a thin converging lens. Using the same notation and sign rules as before, we let s and s' be the object and image distances, respectively, and let y and y' be the object and image heights. Ray QA, parallel to the optic axis before refraction, passes through the second focal point F_2 after refraction. Ray QOQ' passes undeflected straight through the center of the lens because at the center the two surfaces are parallel and (we have assumed) very close together. There is refraction where the ray enters and leaves the material but no net change in direction.

The two angles labeled α in Fig. 34.29 are equal, so the two right triangles PQO and $P'Q'O$ are *similar* and ratios of corresponding sides are equal. Thus

$$\frac{y}{s} = -\frac{y'}{s'} \quad \text{or} \quad \frac{y'}{y} = -\frac{s'}{s} \tag{34.14}$$

34.28 F_1 and F_2 are the first and second focal points of a converging thin lens. The numerical value of f is positive.

(a)

Optic axis (passes through centers of curvature of both lens surfaces)

Second focal point: the point to which incoming parallel rays converge

F_1 F_2

$\overset{\longleftarrow}{f} \overset{\longleftarrow}{f}$

Focal length
• Measured from lens center
• Always the same on both sides of the lens
• Positive for a converging thin lens

(b)

First focal point: Rays diverging from this point emerge from the lens parallel to the axis.

F_1 F_2

$\overset{\longleftarrow}{f} \overset{\longleftarrow}{f}$

34.29 Construction used to find image position for a thin lens. To emphasize that the lens is assumed to be very thin, the ray QAQ' is shown as bent at the midplane of the lens rather than at the two surfaces and ray QOQ' is shown as a straight line.

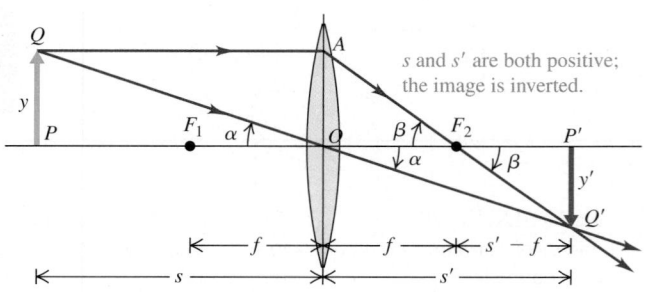

(The reason for the negative sign is that the image is below the optic axis and y' is negative.) Also, the two angles labeled β are equal, and the two right triangles OAF_2 and $P'Q'F_2$ are similar, so

$$\frac{y}{f} = -\frac{y'}{s' - f} \quad \text{or}$$

$$\frac{y'}{y} = -\frac{s' - f}{f} \tag{34.15}$$

We now equate Eqs. (34.14) and (34.15), divide by s', and rearrange to obtain

Object–image relationship, thin lens:	$\dfrac{1}{s} + \dfrac{1}{s'} = \dfrac{1}{f}$ ⋯⋯Focal length of lens	(34.16)

Object distance   Image distance

Equation (34.14) also gives the lateral magnification $m = y'/y$ for the lens:

$$m = -\frac{s'}{s} \quad \text{(lateral magnification, thin lens)} \tag{34.17}$$

The negative sign tells us that when s and s' are both positive, as in Fig. 34.29, the image is *inverted,* and y and y' have opposite signs.

Equations (34.16) and (34.17) are the basic equations for thin lenses. They are *exactly* the same as the corresponding equations for spherical mirrors, Eqs. (34.6) and (34.7). As we will see, the same sign rules that we used for spherical mirrors are also applicable to lenses. In particular, consider a lens with a positive focal length (a converging lens). When an object is outside the first focal point F_1 of this lens (that is, when $s > f$), the image distance s' is positive (that is, the image is on the same side as the outgoing rays); this image is real and inverted, as in Fig. 34.29. An object placed inside the first focal point of a converging lens, so that $s < f$, produces an image with a negative value of s'; this image is located on the same side of the lens as the object and is virtual, erect, and larger than the object. You can verify these statements algebraically by using Eqs. (34.16) and (34.17); we'll also verify them in the next section, using graphical methods analogous to those introduced for mirrors in Section 34.2.

Figure 34.30 shows how a lens forms a three-dimensional image of a three-dimensional object. Point R is nearer the lens than point P. From Eq. (34.16), image point R' is farther from the lens than is image point P', and the image $P'R'$ points in the same direction as the object PR. Arrows $P'S'$ and $P'Q'$ are reversed relative to PS and PQ.

Let's compare Fig. 34.30 with Fig. 34.7, which shows the image formed by a plane *mirror.* We note that the image formed by the lens is inverted, but it is *not* reversed front to back along the optic axis. That is, if the object is a left hand, its image is also a left hand. You can verify this by pointing your left thumb along PR,

A real image made by a converging lens is inverted but *not* reversed back to front: the image thumb $P'R'$ and object thumb PR point in the same direction.

34.30 The image $S'P'Q'R'$ of a three-dimensional object $SPQR$ is not reversed by a lens.

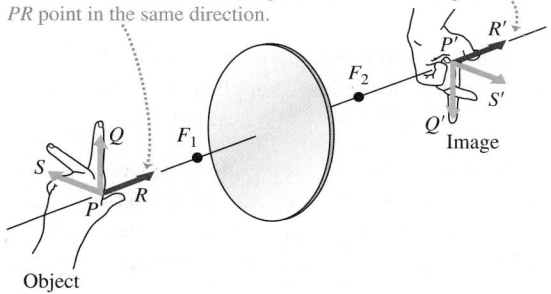

34.31 F_2 and F_1 are the second and first focal points of a diverging thin lens, respectively. The numerical value of f is negative.

(a)

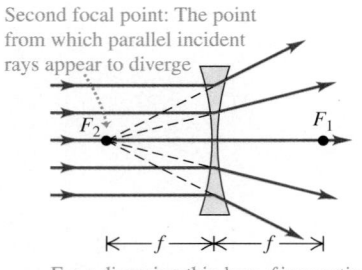

Second focal point: The point from which parallel incident rays appear to diverge

F_2 F_1

For a diverging thin lens, f is negative.

(b)

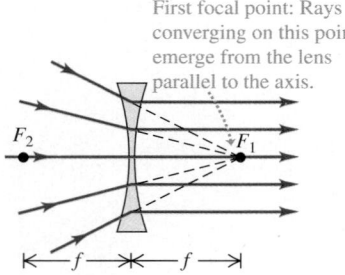

First focal point: Rays converging on this point emerge from the lens parallel to the axis.

F_2 F_1

34.32 Various types of lenses.

(a)   **Converging lenses**

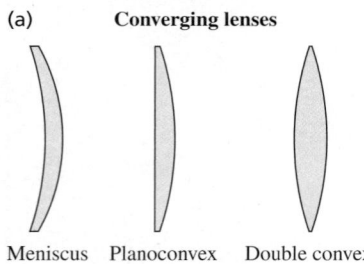

Meniscus   Planoconvex   Double convex

(b)   **Diverging lenses**

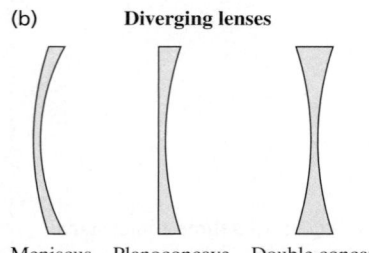

Meniscus   Planoconcave   Double concave

your left forefinger along PQ, and your left middle finger along PS. Then rotate your hand 180°, using your thumb as an axis; this brings the fingers into coincidence with $P'Q'$ and $P'S'$. In other words, an *inverted* image is equivalent to an image that has been rotated by 180° about the lens axis.

Diverging Lenses

So far we have been discussing *converging* lenses. **Figure 34.31** shows a **diverging lens;** the beam of parallel rays incident on this lens *diverges* after refraction. The focal length of a diverging lens is a negative quantity, and the lens is also called a *negative lens*. The focal points of a negative lens are reversed, relative to those of a positive lens. The second focal point, F_2, of a negative lens is the point from which rays that are originally parallel to the axis *appear to diverge* after refraction, as in Fig. 34.31a. Incident rays converging toward the first focal point F_1, as in Fig. 34.31b, emerge from the lens parallel to its axis. Comparing with Section 34.2, you can see that a diverging lens has the same relationship to a converging lens as a convex mirror has to a concave mirror.

Equations (34.16) and (34.17) apply to *both* positive and negative lenses. **Figure 34.32** shows various types of lenses, both converging and diverging. Here's an important observation: *Any lens that is thicker at its center than at its edges is a converging lens with positive f; and any lens that is thicker at its edges than at its center is a diverging lens with negative f* (provided that the lens has a greater index of refraction than the surrounding material). We can prove this by using the *lensmaker's equation,* which it is our next task to derive.

The Lensmaker's Equation

We'll now derive Eq. (34.16) in more detail and at the same time derive the *lensmaker's equation,* which is a relationship among the focal length f, the index of refraction n of the lens, and the radii of curvature R_1 and R_2 of the lens surfaces. We use the principle that an image formed by one reflecting or refracting surface can serve as the object for a second reflecting or refracting surface.

We begin with the somewhat more general problem of two spherical interfaces separating three materials with indexes of refraction n_a, n_b, and n_c, as shown in **Fig. 34.33**. The object and image distances for the first surface are s_1 and s_1', and those for the second surface are s_2 and s_2'. We assume that the lens is thin, so that the distance t between the two surfaces is small in comparison with the object and image distances and can therefore be ignored. This is usually the case with eyeglass lenses (**Fig. 34.34**). Then s_2 and s_1' have the same magnitude but opposite sign. For example, if the first image is on the outgoing side of the first surface, s_1' is positive. But when viewed as an object for the second surface, the first image is *not* on the incoming side of that surface. So we can say that $s_2 = -s_1'$.

We need to use the single-surface equation, Eq. (34.11), twice, once for each surface. The two resulting equations are

$$\frac{n_a}{s_1} + \frac{n_b}{s_1'} = \frac{n_b - n_a}{R_1}$$

$$\frac{n_b}{s_2} + \frac{n_c}{s_2'} = \frac{n_c - n_b}{R_2}$$

Ordinarily, the first and third materials are air or vacuum, so we set $n_a = n_c = 1$. The second index n_b is that of the lens, which we can call simply n. Substituting these values and the relationship $s_2 = -s_1'$, we get

$$\frac{1}{s_1} + \frac{n}{s_1'} = \frac{n - 1}{R_1}$$

$$-\frac{n}{s_1'} + \frac{1}{s_2'} = \frac{1 - n}{R_2}$$

34.33 The image formed by the first surface of a lens serves as the object for the second surface. The distances s_1' and s_2 are taken to be equal; this is a good approximation if the lens thickness t is small.

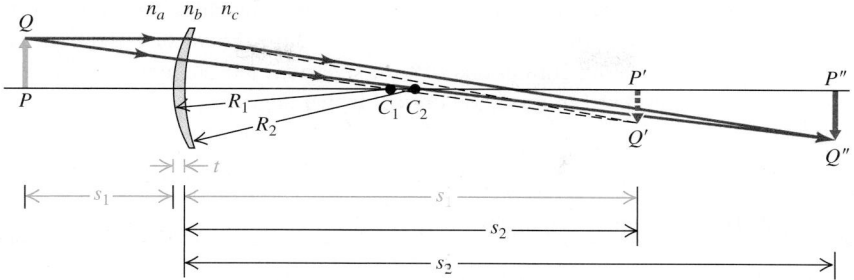

To get a relationship between the initial object position s_1 and the final image position s_2', we add these two equations. This eliminates the term n/s_1':

$$\frac{1}{s_1} + \frac{1}{s_2'} = (n - 1)\left(\frac{1}{R_1} - \frac{1}{R_2}\right)$$

Finally, thinking of the lens as a single unit, we rename the object distance simply s instead of s_1, and we rename the final image distance s' instead of s_2':

$$\frac{1}{s} + \frac{1}{s'} = (n - 1)\left(\frac{1}{R_1} - \frac{1}{R_2}\right) \tag{34.18}$$

Now we compare this with the other thin-lens equation, Eq. (34.16). We see that the object and image distances s and s' appear in exactly the same places in both equations and that the focal length f is given by the **lensmaker's equation:**

34.34 These eyeglass lenses satisfy the thin-lens approximation: Their thickness is small compared to the object and image distances.

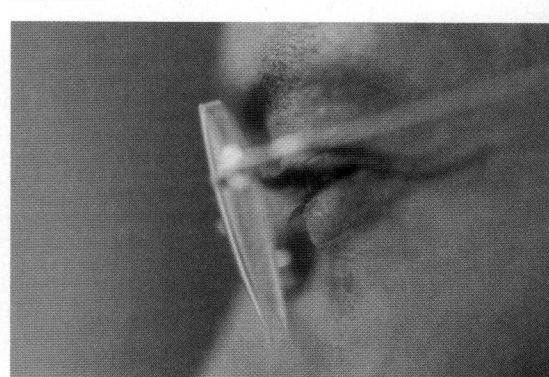

> **Lensmaker's equation for a thin lens:**
>
> Index of refraction of lens material
> $$\frac{1}{f} = (n - 1)\left(\frac{1}{R_1} - \frac{1}{R_2}\right) \tag{34.19}$$
> Focal length   Radius of curvature of first surface   Radius of curvature of second surface

In the process of rederiving the relationship among object distance, image distance, and focal length for a thin lens, we have also derived Eq. (34.19), an expression for the focal length f of a lens in terms of its index of refraction n and the radii of curvature R_1 and R_2 of its surfaces. This can be used to show that all the lenses in Fig. 34.32a are converging lenses with $f > 0$ and that all the lenses in Fig. 34.32b are diverging lenses with $f < 0$.

We use all our sign rules from Section 34.1 with Eqs. (34.18) and (34.19). For example, in **Fig. 34.35**, s, s', and R_1 are positive, but R_2 is negative.

It is not hard to generalize Eq. (34.19) to the situation in which the lens is immersed in a material with an index of refraction greater than unity. We invite you to work out the lensmaker's equation for this more general situation.

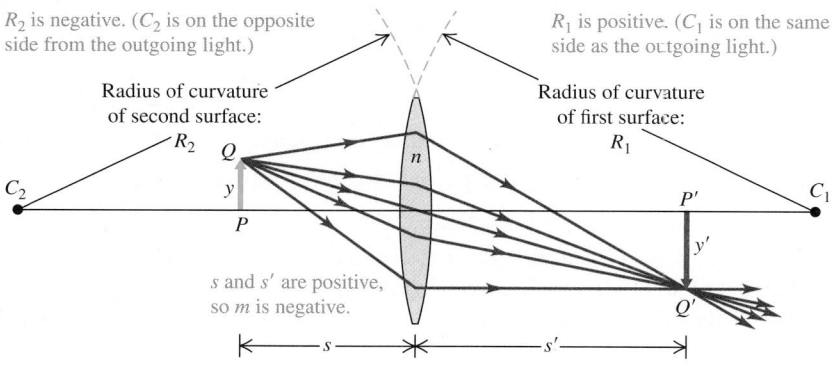

R_2 is negative. (C_2 is on the opposite side from the outgoing light.)

Radius of curvature of second surface: R_2

R_1 is positive. (C_1 is on the same side as the outgoing light.)

Radius of curvature of first surface: R_1

s and s' are positive, so m is negative.

34.35 A converging thin lens with a positive focal length f.

We stress that the paraxial approximation is indeed an approximation! Rays that are at sufficiently large angles to the optic axis of a spherical lens will not be brought to the same focus as paraxial rays; this is the same problem of spherical aberration that plagues spherical *mirrors* (see Section 34.2). To avoid this and other limitations of thin spherical lenses, lenses of more complicated shape are used in precision optical instruments.

EXAMPLE 34.8 DETERMINING THE FOCAL LENGTH OF A LENS

(a) Suppose the absolute values of the radii of curvature of the lens surfaces in Fig. 34.35 are both equal to 10 cm and the index of refraction of the glass is $n = 1.52$. What is the focal length f of the lens? (b) Suppose the lens in Fig. 34.31 also has $n = 1.52$ and the absolute values of the radii of curvature of its lens surfaces are also both equal to 10 cm. What is the focal length of this lens?

SOLUTION

IDENTIFY and SET UP: We are asked to find the focal length f of (a) a lens that is convex on both sides (Fig. 34.35) and (b) a lens that is concave on both sides (Fig. 34.31). In both cases we solve the lensmaker's equation, Eq. (34.19), to determine f. We apply the sign rules given in Section 34.1 to the radii of curvature R_1 and R_2 to take account of whether the surfaces are convex or concave.

EXECUTE: (a) The lens in Fig. 34.35 is *double convex:* The center of curvature of the first surface (C_1) is on the outgoing side of the lens, so R_1 is positive, and the center of curvature of the second surface (C_2) is on the *incoming* side. Hence $R_1 = +10$ cm and $R_2 = -10$ cm. Then from Eq. (34.19),

$$\frac{1}{f} = (1.52 - 1)\left(\frac{1}{+10 \text{ cm}} - \frac{1}{-10 \text{ cm}}\right)$$

$$f = 9.6 \text{ cm}$$

(b) The lens in Fig. 34.31 is *double concave:* The center of curvature of the first surface is on the *incoming* side, so R_1 is negative, and the center of curvature of the second surface is on the outgoing side, so R_2 is positive. Hence in this case $R_1 = -10$ cm and $R_2 = +10$ cm. Again using Eq. (34.19), we get

$$\frac{1}{f} = (1.52 - 1)\left(\frac{1}{-10 \text{ cm}} - \frac{1}{+10 \text{ cm}}\right)$$

$$f = -9.6 \text{ cm}$$

EVALUATE: In part (a) the focal length is *positive,* so this is a converging lens; this makes sense, since the lens is thicker at its center than at its edges. In part (b) the focal length is *negative,* so this is a diverging lens; this also makes sense, since the lens is thicker at its edges than at its center.

Graphical Methods for Lenses

We can determine the position and size of an image formed by a thin lens by using a graphical method very similar to the one we used in Section 34.2 for spherical mirrors. Again we draw a few special rays called *principal rays* that diverge from a point of the object that is *not* on the optic axis. The intersection of these rays, after they pass through the lens, determines the position and size of the image. In using this graphical method, we will consider the entire deviation of a ray as occurring at the midplane of the lens, as shown in **Fig. 34.36**. This is consistent with the assumption that the distance between the lens surfaces is negligible.

The three principal rays whose paths are usually easy to trace for lenses are shown in Fig. 34.36:

1. *A ray parallel to the axis* emerges from the lens in a direction that passes through the second focal point F_2 of a converging lens, or appears to come from the second focal point of a diverging lens.

2. *A ray through the center of the lens* is not appreciably deviated; at the center of the lens the two surfaces are parallel, so this ray emerges at essentially the same angle at which it enters and along essentially the same line.

3. *A ray through (or proceeding toward) the first focal point F_1* emerges parallel to the axis.

When the image is real, the position of the image point is determined by the intersection of any two rays 1, 2, and 3 (Fig. 34.36a). When the image is virtual, we extend the diverging outgoing rays backward to their intersection point to find the image point (Fig. 34.36b).

34.36 The graphical method of locating an image formed by a thin lens. The colors of the rays are for identification only; they do not refer to specific colors of light. (Compare Fig. 34.19 for spherical mirrors.)

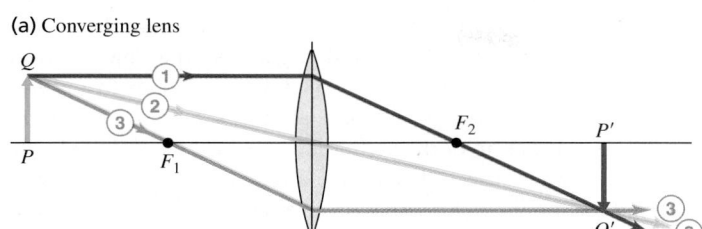

(a) Converging lens

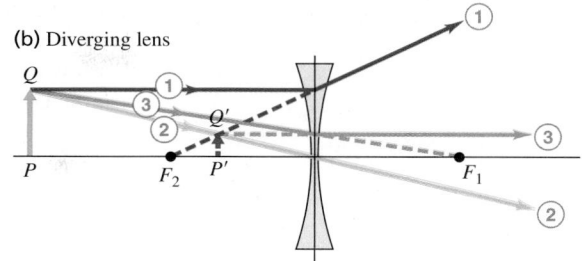

(b) Diverging lens

① Parallel incident ray refracts to pass through second focal point F_2.

② Ray through center of lens does not deviate appreciably.

③ Ray through the first focal point F_1 emerges parallel to the axis.

① Parallel incident ray appears after refraction to have come from the second focal point F_2.

② Ray through center of lens does not deviate appreciably.

③ Ray aimed at the first focal point F_1 emerges parallel to the axis.

CAUTION **Principal rays are not the only rays** Keep in mind that *any* ray from the object that strikes the lens will pass through the image point (for a real image) or appear to originate from the image point (for a virtual image). (We made a similar comment about image formation by mirrors in Section 34.2.) We've emphasized the principal rays because they're the only ones you need to draw to locate the image. ▮

Figure 34.37 shows principal-ray diagrams for a converging lens for several object distances. We suggest you study each of these diagrams very carefully, comparing each numbered ray with the above description.

Parts (a), (b), and (c) of Fig. 34.37 help explain what happens in focusing a camera. For a photograph to be in sharp focus, the electronic sensor or film must be at the position of the real image made by the camera's lens. The image distance increases as the object is brought closer, so the sensor is moved farther behind the lens (i.e., the lens is moved farther in front of the sensor).

DEMO

34.37 Formation of images by a thin converging lens for various object distances. The principal rays are numbered. (Compare Fig. 34.20 for a concave spherical mirror.)

(a) Object O is outside focal point; image I is real.

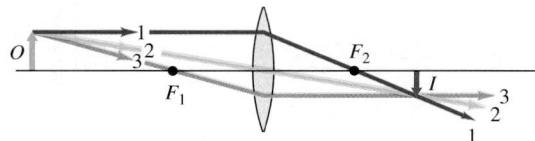

(b) Object O is closer to focal point; image I is real and farther away.

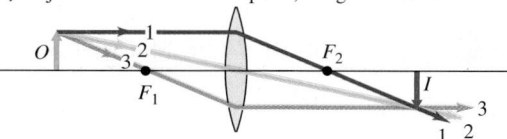

(c) Object O is even closer to focal point; image I is real and even farther away.

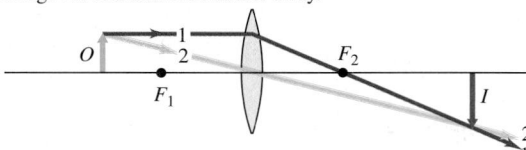

(d) Object O is at focal point; image I is at infinity.

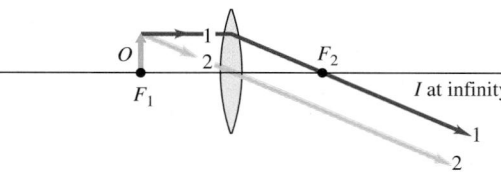

(e) Object O is inside focal point; image I is virtual and larger than object.

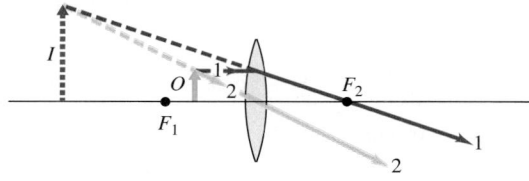

(f) A virtual object O (light rays are *converging* on lens)

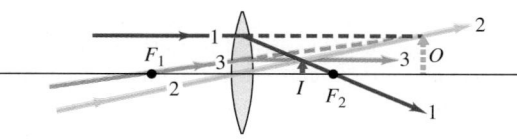

In Fig. 34.37d the object is at the focal point; ray 3 can't be drawn because it doesn't pass through the lens. In Fig. 34.37e the object distance is less than the focal length. The outgoing rays are divergent, and the image is *virtual;* its position is located by extending the outgoing rays backward, so the image distance s' is negative. Note also that the image is erect and larger than the object. (We'll see the usefulness of this in Section 34.6.) Figure 34.37f corresponds to a *virtual object*. The incoming rays do not diverge from a real object, but are *converging* as though they would meet at the tip of the virtual object O on the right side; the object distance s is negative in this case. The image is real and is located between the lens and the second focal point. This situation can arise if the rays that strike the lens in Fig. 34.37f emerge from another converging lens (not shown) to the left of the figure.

PROBLEM-SOLVING STRATEGY 34.2) **IMAGE FORMATION BY THIN LENSES**

IDENTIFY *the relevant concepts:* Review Problem-Solving Strategy 34.1 (Section 34.2) for mirrors, which is equally applicable here. As for mirrors, you should use *both* principal-ray diagrams and equations to solve problems that involve image formation by lenses.

SET UP *the problem:* Identify the target variables.

EXECUTE *the solution* as follows:
1. Draw a large principal-ray diagram if you have enough information, using graph paper or quadrille-ruled paper. Orient your diagram so that incoming rays go from left to right. Draw the rays with a ruler, and measure distances carefully.
2. Draw the principal rays so they change direction at the midplane of the lens, as in Fig. 34.36. For a lens there are only three principal rays (compared to four for a mirror). Draw all three whenever possible; the intersection of any two rays determines the image location, but the third ray should pass through the same point.

3. If the outgoing principal rays diverge, extend them backward to find the virtual image point on the *incoming* side of the lens, as in Fig. 34.27e.
4. Solve Eqs. (34.16) and (34.17), as appropriate, for the target variables. Carefully use the sign rules given in Section 34.1.
5. The *image* from a first lens or mirror may serve as the *object* for a second lens or mirror. In finding the object and image distances for this intermediate image, be sure you include the distance between the two elements (lenses and/or mirrors) correctly.

EVALUATE *your answer:* Your calculated results must be consistent with your ray-diagram results. Check that they give the same image position and image size, and that they agree on whether the image is real or virtual.

EXAMPLE 34.9 **IMAGE POSITION AND MAGNIFICATION WITH A CONVERGING LENS**

Use ray diagrams to find the image position and magnification for an object at each of the following distances from a converging lens with a focal length of 20 cm: (a) 50 cm; (b) 20 cm; (c) 15 cm; (d) −40 cm. Check your results by calculating the image position and lateral magnification by using Eqs. (34.16) and (34.17), respectively.

SOLUTION

IDENTIFY and SET UP: We are given the focal length $f = 20$ cm and four object distances s. Our target variables are the corresponding image distances s' and lateral magnifications m. We solve Eq. (34.16) for s', and find m from Eq. (34.17), $m = -s'/s$.

EXECUTE: Figures 34.37a, d, e, and f, respectively, show the appropriate principal-ray diagrams. You should be able to reproduce these without referring to the figures. Measuring these diagrams yields the approximate results: $s' = 35$ cm, $-\infty$, -40 cm, and 15 cm, and $m = -\frac{2}{3}$, $+\infty$, $+3$, and $+\frac{1}{3}$, respectively.

Calculating the image distances from Eq. (34.16), we find

(a) $\dfrac{1}{50 \text{ cm}} + \dfrac{1}{s'} = \dfrac{1}{20 \text{ cm}}$ $s' = 33.3$ cm

(b) $\dfrac{1}{20 \text{ cm}} + \dfrac{1}{s'} = \dfrac{1}{20 \text{ cm}}$ $s' = \pm \infty$

(c) $\dfrac{1}{15 \text{ cm}} + \dfrac{1}{s'} = \dfrac{1}{20 \text{ cm}}$ $s' = -60$ cm

(d) $\dfrac{1}{-40 \text{ cm}} + \dfrac{1}{s'} = \dfrac{1}{20 \text{ cm}}$ $s' = 13.3$ cm

The graphical results are fairly close to these except for part (c); the accuracy of the diagram in Fig. 34.37e is limited because the rays extended backward have nearly the same direction.

From Eq. (34.17),

(a) $m = -\dfrac{33.3 \text{ cm}}{50 \text{ cm}} = -\frac{2}{3}$ (b) $m = -\dfrac{\pm \infty \text{ cm}}{20 \text{ cm}} = \pm \infty$

(c) $m = -\dfrac{-60 \text{ cm}}{15 \text{ cm}} = +4$ (d) $m = -\dfrac{13.3 \text{ cm}}{-40 \text{ cm}} = +\frac{1}{3}$

EVALUATE: Note that the image distance s' is positive in parts (a) and (d) but negative in part (c). This makes sense: The image is real in parts (a) and (d) but virtual in part (c). The light rays that emerge from the lens in part (b) are parallel and never converge, so the image can be regarded as being at either $+\infty$ or $-\infty$.

The values of magnification m tell us that the image is inverted in part (a) and erect in parts (c) and (d), in agreement with the principal-ray diagrams. The infinite value of magnification in part (b) is another way of saying that the image is formed infinitely far away.

EXAMPLE 34.10 **IMAGE FORMATION BY A DIVERGING LENS**

A beam of parallel rays spreads out after passing through a thin diverging lens, as if the rays all came from a point 20.0 cm from the center of the lens. You want to use this lens to form an erect, virtual image that is $\frac{1}{3}$ the height of the object. (a) Where should the object be placed? Where will the image be? (b) Draw a principal-ray diagram.

SOLUTION

IDENTIFY and SET UP: The result with parallel rays shows that the focal length is $f = -20$ cm. We want the lateral magnification to be $m = +\frac{1}{3}$ (positive because the image is to be erect). Our target variables are the object distance s and the image distance s'. In part (a), we solve the magnification equation, Eq. (34.17), for s' in terms of s; we then use the object–image relationship, Eq. (34.16), to find s and s' individually.

EXECUTE: (a) From Eq. (34.17), $m = +\frac{1}{3} = -s'/s$, so $s' = -s/3$. We insert this result into Eq. (34.16) and solve for the object distance s:

$$\frac{1}{s} + \frac{1}{-s/3} = \frac{1}{s} - \frac{3}{s} = -\frac{2}{s} = \frac{1}{f}$$

$$s = -2f = -2(-20.0 \text{ cm}) = 40.0 \text{ cm}$$

The object should be 40.0 cm from the lens. The image distance will be

$$s' = -\frac{s}{3} = -\frac{40.0 \text{ cm}}{3} = -13.3 \text{ cm}$$

The image distance is negative, so the object and image are on the same side of the lens.

(b) **Figure 34.38** is a principal-ray diagram for this problem, with the rays numbered as in Fig. 34.36b.

EVALUATE: You should be able to draw a principal-ray diagram like Fig. 34.38 without referring to the figure. From your diagram, you can confirm our results in part (a) for the object and image distances. You can also check our results for s and s' by substituting them back into Eq. (34.16).

34.38 Principal-ray diagram for an image formed by a thin diverging lens.

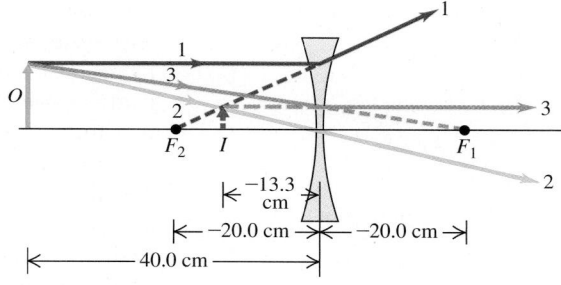

EXAMPLE 34.11 **AN IMAGE OF AN IMAGE**

Converging lenses A and B, of focal lengths 8.0 cm and 6.0 cm, respectively, are placed 36.0 cm apart. Both lenses have the same optic axis. An object 8.0 cm high is placed 12.0 cm to the left of lens A. Find the position, size, and orientation of the image produced by the lenses in combination. (Such combinations are used in telescopes and microscopes, to be discussed in Section 34.7.)

34.39 Principal-ray diagram for a combination of two converging lenses. The first lens (A) makes a real image of the object. This real image acts as an object for the second lens (B).

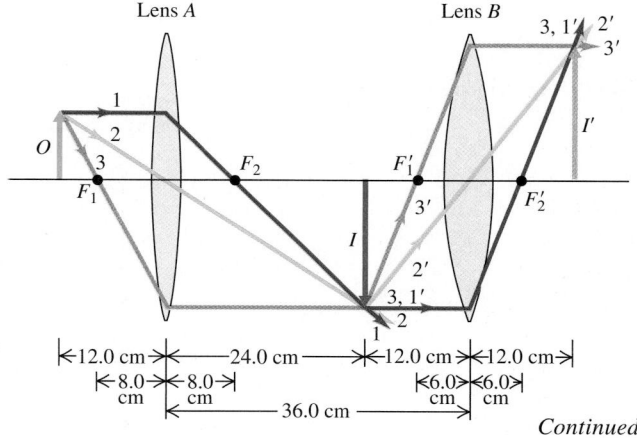

SOLUTION

IDENTIFY and SET UP: Figure 34.39 shows the situation. The object O lies outside the first focal point F_1 of lens A, which therefore produces a real image I. The light rays that strike lens B diverge from this real image just as if I was a material object; image I therefore acts as an *object* for lens B. Our goal is to determine the properties of the image I' made by lens B. We use both ray-diagram and computational methods to do this.

Continued

EXECUTE: In Fig. 34.39 we have drawn principal rays 1, 2, and 3 from the head of the object arrow O to find the position of the image I made by lens A, and principal rays 1′, 2′, and 3′ from the head of I to find the position of the image I' made by lens B (even though rays 2′ and 3′ don't actually exist in this case). The image is inverted *twice*, once by each lens, so the second image I' has the same orientation as the original object.

We first find the position and size of the first image I. Applying Eq. (34.16), $1/s + 1/s' = 1/f$, to lens A gives

$$\frac{1}{12.0 \text{ cm}} + \frac{1}{s'_{I,A}} = \frac{1}{8.0 \text{ cm}} \qquad s'_{I,A} = +24.0 \text{ cm}$$

Image I is 24.0 cm to the right of lens A. The lateral magnification is $m_A = -(24.0 \text{ cm})/(12.0 \text{ cm}) = -2.00$, so image I is inverted and twice as tall as object O.

Image I is 36.0 cm − 24.0 cm = 12.0 cm to the left of lens B, so the object distance for lens B is +12.0 cm. Applying Eq. (34.16) to lens B then gives

$$\frac{1}{12.0 \text{ cm}} + \frac{1}{s'_{I',B}} = \frac{1}{6.0 \text{ cm}} \qquad s'_{I',B} = +12.0 \text{ cm}$$

The final image I' is 12.0 cm to the right of lens B. The magnification produced by lens B is $m_B = -(12.0 \text{ cm})/(12.0 \text{ cm}) = -1.00$.

EVALUATE: The value of m_B means that the final image I' is just as large as the first image I but has the opposite orientation. The *overall* magnification is $m_A m_B = (-2.00)(-1.00) = +2.00$. Hence the final image I' is $(2.00)(8.0 \text{ cm}) = 16$ cm tall and has the same orientation as the original object O, just as Fig. 34.39 shows.

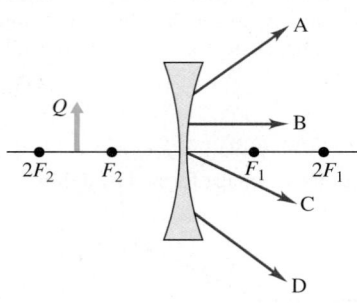

TEST YOUR UNDERSTANDING OF SECTION 34.4 A diverging lens and an object are positioned as shown in the accompanying figure. Which of the rays A, B, C, and D could emanate from point Q at the top of the object? ▮

34.5 CAMERAS

The concept of *image,* which is so central to understanding simple mirror and lens systems, plays an equally important role in the analysis of optical instruments (also called *optical devices*). Among the most common optical devices are cameras, which make an image of an object and record it either electronically or on film.

The basic elements of a **camera** are a light-tight box ("camera" is a Latin word meaning "a room or enclosure"), a converging lens, a shutter to open the lens for a prescribed length of time, and a light-sensitive recording medium (**Fig. 34.40**). In digital cameras (including mobile-phone cameras), this is an electronic sensor; in older cameras, it is photographic film. The lens forms an inverted real image on the recording medium of the object being photographed. High-quality camera lenses have several elements, permitting partial correction of various *aberrations,* including the dependence of index of refraction on wavelength and the limitations imposed by the paraxial approximation.

When the camera is in proper *focus,* the position of the recording medium coincides with the position of the real image formed by the lens. The resulting photograph will then be as sharp as possible. With a converging lens, the image distance increases as the object distance decreases (see **Figs. 34.41a**, 34.41b, and 34.41c, and the discussion in Section 34.4). Hence in "focusing" the camera, we move the lens closer to the sensor or film for a distant object and farther from the sensor or film for a nearby object.

34.40 Key elements of a digital camera.

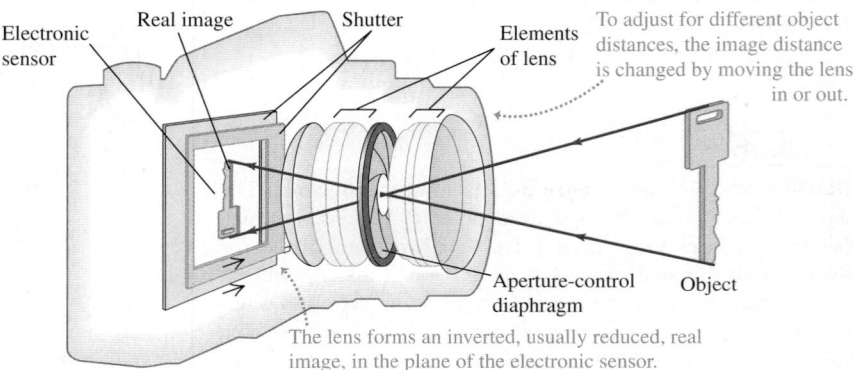

Electronic sensor

Real image

Shutter

Elements of lens

To adjust for different object distances, the image distance is changed by moving the lens in or out.

Aperture-control diaphragm

Object

The lens forms an inverted, usually reduced, real image, in the plane of the electronic sensor.

34.41 (a), (b), (c) Three photographs taken with the same camera from the same position, using lenses with focal lengths $f = 28$ mm, 70 mm, and 135 mm. Increasing the focal length increases the image size proportionately. (d) The larger the value of f, the narrower the angle of view. The angles shown here are for a camera with image area 24 mm $\times$ 36 mm (corresponding to 35-mm film) and refer to the angle of view along the 36-mm width of the film.

(a) $f = 28$ mm

(b) $f = 70$ mm

(c) $f = 135$ mm

(d) The angles of view for the photos in (a)–(c)

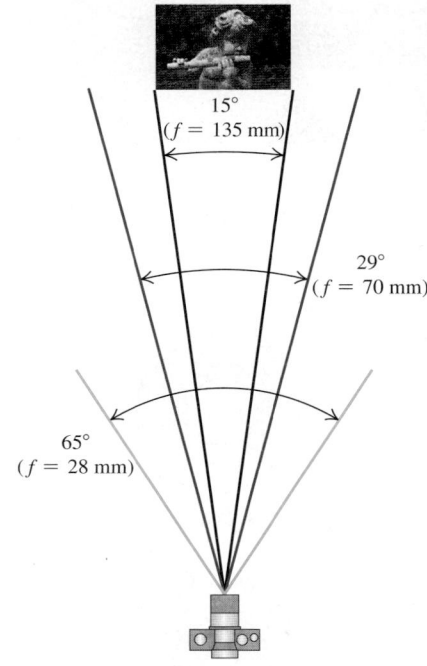

Camera Lenses: Focal Length

The choice of the focal length f for a camera lens depends on the size of the electronic sensor or film and the desired angle of view. Figure 34.41 shows three photographs taken on 35-mm film with the same camera at the same position, but with lenses of different focal lengths. A lens of long focal length, called a *telephoto* lens, gives a narrow angle of view and a large image of a distant object (such as the statue in Fig. 34.41c); a lens of short focal length gives a small image and a wide angle of view (as in Fig. 34.41a) and is called a *wide-angle* lens. To understand this behavior, recall that the focal length is the distance from the lens to the image when the object is infinitely far away. In general, for *any* object distance, using a lens of longer focal length gives a greater image distance. This also increases the height of the image; as discussed in Section 34.4, the ratio of the image height y' to the object height y (the *lateral magnification*) is equal in absolute value to the ratio of the image distance s' to the object distance s [Eq. (34.17)]:

$$m = \frac{y'}{y} = -\frac{s'}{s}$$

With a lens of short focal length, the ratio s'/s is small, and a distant object gives only a small image. When a lens with a long focal length is used, the image of this same object may entirely cover the area of the electronic sensor or film. Hence the longer the focal length, the narrower the angle of view (Fig. 34.41d).

Camera Lenses: *f*-Number

For a camera to record the image properly, the total light energy per unit area reaching the electronic sensor or film (the "exposure") must fall within certain limits. This is controlled by the *shutter* and the *lens aperture*. The shutter controls the time interval during which light enters the lens. This is usually adjustable in steps corresponding to factors of about 2, often from 1 s to $\frac{1}{1000}$ s.

The intensity of light reaching the sensor or film is proportional to the area viewed by the camera lens and to the effective area of the lens. The size of the area that the lens "sees" is proportional to the square of the angle of view of the lens, and so is roughly proportional to $1/f^2$. The effective area of the lens is controlled by means of an adjustable lens aperture, or *diaphragm,* a nearly circular hole with variable diameter D; hence the effective area is proportional to D^2. Putting these factors together, we see that the intensity of light reaching the sensor or film with a particular lens is proportional to D^2/f^2.

Photographers commonly express the light-gathering capability of a lens in terms of the ratio f/D, called the **f-number** of the lens:

$$f\text{-number of a lens} = \frac{f \;\cdots\text{Focal length of lens}}{D \;\cdots\text{Aperture diameter}} \qquad (34.20)$$

For example, a lens with a focal length $f = 50$ mm and an aperture diameter $D = 25$ mm has an f-number of 2, or "an aperture of $f/2$." The light intensity reaching the sensor or film is *inversely* proportional to the square of the f-number.

For a lens with a variable-diameter aperture, increasing the diameter by a factor of $\sqrt{2}$ changes the f-number by $1/\sqrt{2}$ and increases the intensity at the sensor or film by a factor of 2. Adjustable apertures usually have scales labeled with successive numbers (often called *f-stops*) related by factors of $\sqrt{2}$, such as

$$f/2 \qquad f/2.8 \qquad f/4 \qquad f/5.6 \qquad f/8 \qquad f/11 \qquad f/16$$

and so on. The larger numbers represent smaller apertures and exposures, and each step corresponds to a factor of 2 in intensity (**Fig. 34.42**). The actual *exposure* (total amount of light reaching the sensor or film) is proportional to both the aperture area and the time of exposure. Thus $f/4$ and $\frac{1}{500}$ s, $f/5.6$ and $\frac{1}{250}$ s, and $f/8$ and $\frac{1}{125}$ s all correspond to the same exposure.

34.42 A camera lens with an adjustable diaphragm.

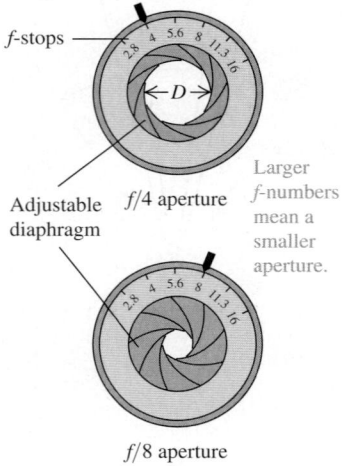

Changing the diameter by a factor of $\sqrt{2}$ changes the intensity by a factor of 2.

f-stops

$\leftarrow D \rightarrow$

Adjustable diaphragm

$f/4$ aperture

Larger f-numbers mean a smaller aperture.

$f/8$ aperture

Zoom Lenses and Projectors

Many photographers use a *zoom lens,* which is not a single lens but a complex collection of several lens elements that give a continuously variable focal length, often over a range as great as 10 to 1. **Figures 34.43a** and 34.43b show a simple system with variable focal length, and Fig. 34.43c shows a typical zoom lens for a digital single-lens reflex camera. Zoom lenses give a range of image sizes of a given object. It is an enormously complex problem in optical design to keep the image in focus and maintain a constant f-number while the focal length changes. When you vary the focal length of a typical zoom lens, two groups of elements move within the lens and a diaphragm opens and closes.

A *digital projector* for viewing lecture slides, photos, or movies operates very much like a digital camera in reverse. In the most common type of digital projector, the pixels of data to be projected are shown on a small, transparent liquid-crystal-display (LCD) screen inside the projector and behind the projection lens. A lamp illuminates the LCD screen, which acts as an object for the lens. The lens forms a real, enlarged, inverted image of the LCD screen. Because the image is inverted, the pixels shown on the LCD screen are upside down so that the image on the projection screen appears right-side up.

34.43 A simple zoom lens uses a converging lens and a diverging lens in tandem. (a) When the two lenses are close together, the combination behaves like a single lens of long focal length. (b) If the two lenses are moved farther apart, the combination behaves like a short-focal-length lens. (c) This zoom lens contains twelve elements arranged in four groups.

(a) Zoom lens set for long focal length

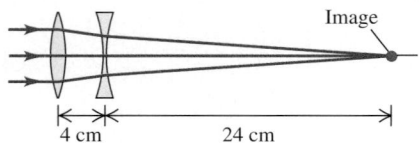

4 cm — 24 cm — Image

(b) Zoom lens set for short focal length

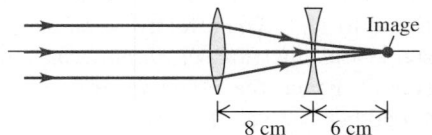

8 cm — 6 cm — Image

(c) A practical zoom lens

EXAMPLE 34.12 PHOTOGRAPHIC EXPOSURES

A common telephoto lens for a 35-mm film camera has a focal length of 200 mm; its f-stops range from $f/2.8$ to $f/22$. (a) What is the corresponding range of aperture diameters? (b) What is the corresponding range of image intensities on the film?

SOLUTION

IDENTIFY and SET UP: Part (a) of this problem uses the relationship among lens focal length f, aperture diameter D, and f-number. Part (b) uses the relationship between intensity and aperture diameter. We use Eq. (34.20) to relate D (the target variable) to the f-number and the focal length $f = 200$ mm. The intensity of the light reaching the film is proportional to D^2/f^2; since f is the same in each case, we conclude that the intensity in this case is proportional to D^2, the square of the aperture diameter.

EXECUTE: (a) From Eq. (34.20), the diameter ranges from

$$D = \frac{f}{f\text{-number}} = \frac{200 \text{ mm}}{2.8} = 71 \text{ mm}$$

to

$$D = \frac{200 \text{ mm}}{22} = 9.1 \text{ mm}$$

(b) Because the intensity is proportional to D^2, the ratio of the intensity at $f/2.8$ to the intensity at $f/22$ is

$$\left(\frac{71 \text{ mm}}{9.1 \text{ mm}}\right)^2 = \left(\frac{22}{2.8}\right)^2 = 62 \qquad \text{(about } 2^6\text{)}$$

EVALUATE: If the correct exposure time at $f/2.8$ is $\frac{1}{1000}$ s, then the exposure at $f/22$ is $(62)\left(\frac{1}{1000} \text{ s}\right) = \frac{1}{16}$ s to compensate for the lower intensity. In general, the smaller the aperture and the larger the f-number, the longer the required exposure. Nevertheless, many photographers prefer to use small apertures so that only the central part of the lens is used to make the image. This minimizes aberrations that occur near the edges of the lens and gives the sharpest possible image.

TEST YOUR UNDERSTANDING OF SECTION 34.5   When used with 35-mm film (image area 24 mm × 36 mm), a lens with $f = 50$ mm gives a 45° angle of view and is called a "normal lens." When used with an electronic sensor that measures 5 mm × 5 mm, this same lens is (i) a wide-angle lens; (ii) a normal lens; (iii) a telephoto lens. ∎

34.6 THE EYE

MP **PhET:** Color Vision

The optical behavior of the eye is similar to that of a camera. **Figure 34.44** shows the essential parts of the human eye, considered as an optical system. The eye is nearly spherical and about 2.5 cm in diameter. The front portion is somewhat more sharply curved and is covered by a tough, transparent membrane called the *cornea*. The region behind the cornea contains a liquid called the *aqueous humor*. Next comes the *crystalline lens,* a capsule containing a fibrous jelly, hard

34.44 (a) The eye. (b) There are two types of light-sensitive cells on the retina. Rods are more sensitive to light than cones, but only the cones are sensitive to differences in color. A typical human eye contains about 1.3×10^8 rods and about 7×10^6 cones.

(a) Diagram of the eye

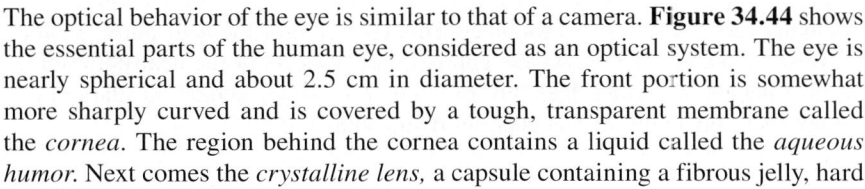

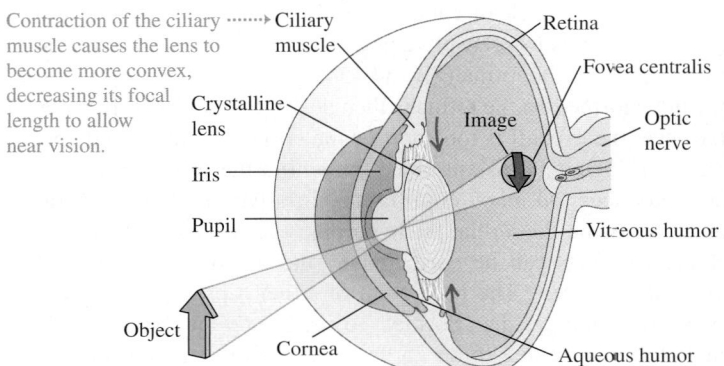

Contraction of the ciliary muscle causes the lens to become more convex, decreasing its focal length to allow near vision.

Ciliary muscle · Crystalline lens · Iris · Pupil · Object · Cornea · Retina · Fovea centralis · Image · Optic nerve · Vitreous humor · Aqueous humor

(b) Scanning electron micrograph showing retinal rods and cones in different colors

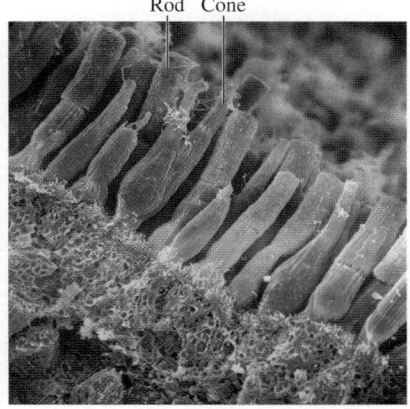

Rod   Cone

BIO Application Focusing in the Animal Kingdom The crystalline lens and ciliary muscle found in humans and other mammals are among a number of focusing mechanisms used by animals. Birds can change the shape not only of their lens but also of the corneal surface. In aquatic animals the corneal surface is not very useful for focusing because its index of refraction is close to that of water. Thus, focusing is accomplished entirely by the lens, which is nearly spherical. Fish focus by using a muscle to move the lens either inward or outward. Whales and dolphins achieve the same effect by filling or emptying a fluid chamber behind the lens to move the lens in or out.

34.45 Refractive errors for (a) a normal eye, (b) a myopic (nearsighted) eye, and (c) a hyperopic (farsighted) eye viewing a very distant object. In each case, the eye is shown with the ciliary muscle relaxed. The dashed blue curve indicates the required position of the retina.

(a) Normal eye

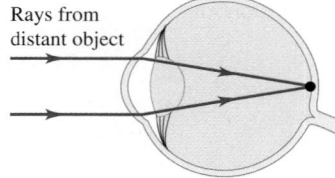

Rays from distant object

(b) Myopic (nearsighted) eye

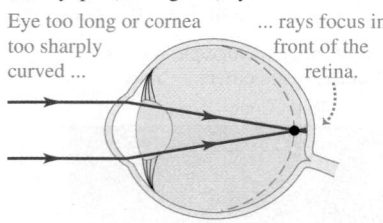

Eye too long or cornea too sharply curved ...

... rays focus in front of the retina.

(c) Hyperopic (farsighted) eye

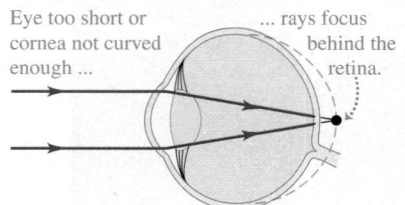

Eye too short or cornea not curved enough ...

... rays focus behind the retina.

at the center and progressively softer at the outer portions. The crystalline lens is held in place by ligaments that attach it to the ciliary muscle, which encircles it. Behind the lens, the eye is filled with a thin watery jelly called the *vitreous humor*. The indexes of refraction of both the aqueous humor and the vitreous humor are about 1.336, nearly equal to that of water. The crystalline lens, while not homogeneous, has an average index of 1.437. This is not very different from the indexes of the aqueous and vitreous humors. As a result, most of the refraction of light entering the eye occurs at the outer surface of the cornea.

Refraction at the cornea and the surfaces of the lens produces a *real image* of the object being viewed. This image is formed on the light-sensitive *retina*, lining the rear inner surface of the eye. The retina plays the same role as the electronic sensor in a digital camera. The *rods* and *cones* in the retina act like an array of miniature photocells (Fig. 34.44b); they sense the image and transmit it via the *optic nerve* to the brain. Vision is most acute in a small central region called the *fovea centralis,* about 0.25 mm in diameter.

In front of the lens is the *iris*. It contains an aperture with variable diameter called the *pupil,* which opens and closes to adapt to changing light intensity. The receptors of the retina also have intensity adaptation mechanisms.

For an object to be seen sharply, the image must be formed exactly at the location of the retina. The eye adjusts to different object distances s by changing the focal length f of its lens; the lens-to-retina distance, corresponding to s', does not change. (Contrast this with focusing a camera, in which the focal length is fixed and the lens-to-sensor distance is changed.) For the normal eye, an object at infinity is sharply focused when the ciliary muscle is relaxed. To focus sharply on a closer object, the tension in the ciliary muscle surrounding the lens increases, the ciliary muscle contracts, the lens bulges, and the radii of curvature of its surfaces decrease; this decreases f. This process is called *accommodation*.

The extremes of the range over which distinct vision is possible are known as the *far point* and the *near point* of the eye. The far point of a normal eye is at infinity. The position of the near point depends on the amount by which the ciliary muscle can increase the curvature of the crystalline lens. The range of accommodation gradually diminishes with age because the crystalline lens grows throughout a person's life (it is about 50% larger at age 60 than at age 20) and the ciliary muscles are less able to distort a larger lens. For this reason, the near point gradually recedes as one grows older. This recession of the near point is called *presbyopia*. **Table 34.1** shows the approximate position of the near point for an average person at various ages. For example, an average person 50 years of age cannot focus on an object that is closer than about 40 cm.

Defects of Vision

Several common defects of vision result from incorrect distance relationships in the eye. A normal eye forms an image on the retina of an object at infinity when the eye is relaxed (**Fig. 34.45a**). In the *myopic* (nearsighted) eye, the eyeball is too long from front to back in comparison with the radius of curvature of the cornea (or the cornea is too sharply curved), and rays from an object at infinity are focused in front of the retina (Fig. 34.45b). The most distant object for which an image can be formed on the retina is then nearer than infinity. In the *hyperopic* (farsighted) eye, the eyeball is too short or the cornea is not curved enough, and the image of an infinitely distant object is behind the retina (Fig. 34.45c). The myopic eye produces *too much* convergence in a parallel bundle of rays for an image to be formed on the retina; the hyperopic eye, *not enough* convergence.

All of these defects can be corrected by the use of corrective lenses (eyeglasses or contact lenses). The near point of either a presbyopic or a hyperopic eye is *farther* from the eye than normal. To see clearly an object at normal reading distance (often assumed to be 25 cm), we need a lens that forms a virtual image of the object at or beyond the near point. This can be accomplished by a converging (positive) lens (**Fig. 34.46**). In effect the lens moves the object farther

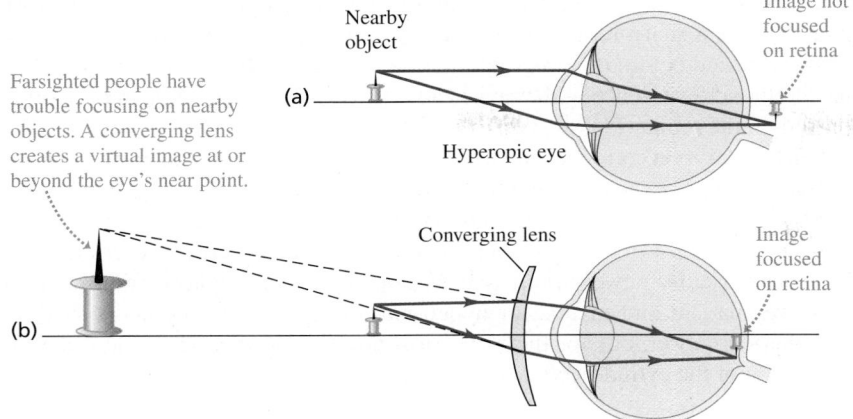

34.46 (a) An uncorrected hyperopic (farsighted) eye. (b) A positive (converging) lens gives the extra convergence needed for a hyperopic eye to focus the image on the retina.

Nearby object

Image not focused on retina

Farsighted people have trouble focusing on nearby objects. A converging lens creates a virtual image at or beyond the eye's near point.

(a) Hyperopic eye

Converging lens

Image focused on retina

(b)

34.47 (a) An uncorrected myopic (nearsighted) eye. (b) A negative (diverging) lens spreads the rays farther apart to compensate for the excessive convergence of the myopic eye.

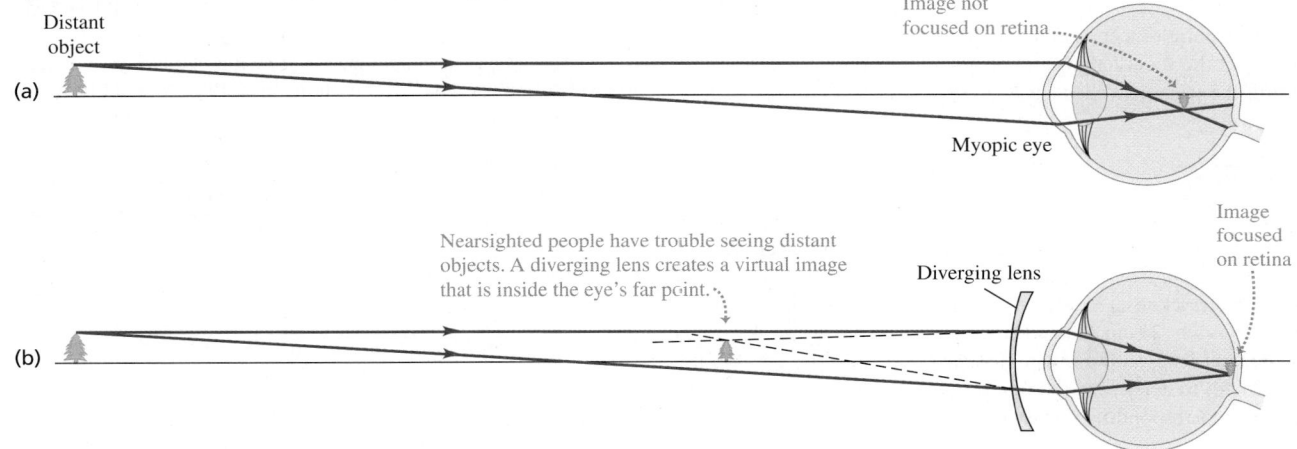

Distant object

Image not focused on retina

(a) Myopic eye

Image focused on retina

Nearsighted people have trouble seeing distant objects. A diverging lens creates a virtual image that is inside the eye's far point.

Diverging lens

(b)

away from the eye to a point where a sharp retinal image can form. Similarly, correcting the myopic eye involves the use of a diverging (negative) lens to move the image closer to the eye than the actual object is (**Fig. 34.47**).

Astigmatism is a different type of defect in which the surface of the cornea is not spherical but rather more sharply curved in one plane than in another. As a result, horizontal lines may be imaged in a different plane from vertical lines (**Fig. 34.48a**). Astigmatism may make it impossible, for example, to focus clearly on both the horizontal and vertical bars of a window at the same time.

Astigmatism can be corrected by use of a lens with a *cylindrical* surface. For example, suppose the curvature of the cornea in a horizontal plane is correct to

TABLE 34.1	Receding of Near Point with Age
Age (years)	**Near Point (cm)**
10	7
20	10
30	14
40	22
50	40
60	200

(a) Vertical lines are imaged in front of the retina.

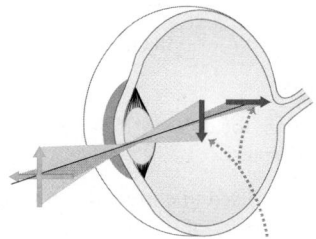

Shape of eyeball or lens causes vertical and horizontal elements to focus at different distances.

(b) A cylindrical lens corrects for astigmatism.

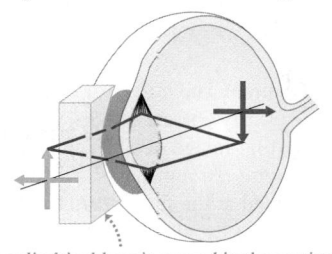

This cylindrical lens is curved in the vertical, but not the horizontal, direction; it changes the focal length of vertical elements.

34.48 One type of astigmatism and how it is corrected.

focus rays from infinity on the retina but the curvature in the vertical plane is too great to form a sharp retinal image. When a cylindrical lens with its axis horizontal is placed before the eye, the rays in a horizontal plane are unaffected, but the additional divergence of the rays in a vertical plane causes these to be sharply imaged on the retina (Fig. 34.48b).

Lenses for vision correction are usually described in terms of the **power,** defined as the reciprocal of the focal length expressed in meters. The unit of power is the **diopter.** Thus a lens with $f = 0.50$ m has a power of 2.0 diopters, $f = -0.25$ m corresponds to -4.0 diopters, and so on. The numbers on a prescription for glasses are usually powers expressed in diopters. When the correction involves both astigmatism and myopia or hyperopia, there are three numbers: one for the spherical power, one for the cylindrical power, and an angle to describe the orientation of the cylinder axis.

EXAMPLE 34.13 **CORRECTING FOR FARSIGHTEDNESS**

The near point of a certain hyperopic eye is 100 cm in front of the eye. Find the focal length and power of the contact lens that will permit the wearer to see clearly an object that is 25 cm in front of the eye.

SOLUTION

IDENTIFY and SET UP: Figure 34.49 shows the situation. We want the lens to form a virtual image of the object at the near point of the eye, 100 cm from it. The contact lens (which we treat as having negligible thickness) is at the surface of the cornea, so the object distance is $s = 25$ cm. The virtual image is on the incoming side of the contact lens, so the image distance is $s' = -100$ cm. We use Eq. (34.16) to determine the required focal length f of the contact lens; the corresponding power is $1/f$.

EXECUTE: From Eq. (34.16),

$$\frac{1}{f} = \frac{1}{s} + \frac{1}{s'} = \frac{1}{+25 \text{ cm}} + \frac{1}{-100 \text{ cm}}$$

$$f = +33 \text{ cm}$$

34.49 Using a contact lens to correct for farsightedness. For clarity, the eye and contact lens are shown much larger than the scale of the figure; the 2.5-cm diameter of the eye is actually much smaller than the focal length f of the contact lens.

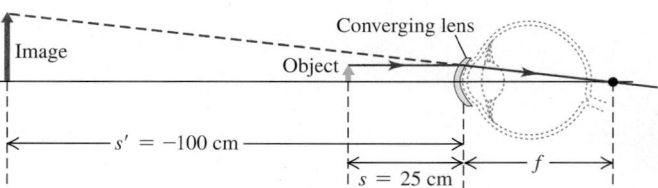

We need a converging lens with focal length $f = 33$ cm and power $1/(0.33 \text{ m}) = +3.0$ diopters.

EVALUATE: In this example we used a contact lens to correct hyperopia. Had we used eyeglasses, we would have had to account for the separation between the eye and the eyeglass lens, and a somewhat different power would have been required (see Example 34.14).

EXAMPLE 34.14 **CORRECTING FOR NEARSIGHTEDNESS**

The far point of a certain myopic eye is 50 cm in front of the eye. Find the focal length and power of the eyeglass lens that will permit the wearer to see clearly an object at infinity. Assume that the lens is worn 2 cm in front of the eye.

SOLUTION

IDENTIFY and SET UP: Figure 34.50 shows the situation. The far point of a myopic eye is nearer than infinity. To see clearly objects beyond the far point, we need a lens that forms a virtual image of such objects no farther from the eye than the far point. Assume that the virtual image of the object at infinity is formed at the far point, 50 cm in front of the eye (48 cm in front of the eyeglass lens). Then when the object distance is $s = \infty$, we want the image

distance to be $s' = -48$ cm. As in Example 34.13, we use the values of s and s' to calculate the required focal length.

EXECUTE: From Eq. (34.16),

$$\frac{1}{f} = \frac{1}{s} + \frac{1}{s'} = \frac{1}{\infty} + \frac{1}{-48 \text{ cm}}$$

$$f = -48 \text{ cm}$$

We need a *diverging* lens with focal length $f = -48$ cm and power $1/(-0.48 \text{ m}) = -2.1$ diopters.

EVALUATE: If a *contact* lens were used to correct this myopia, we would need $f = -50$ cm and a power of -2.0 diopters. Can you see why?

34.50 Using an eyeglass lens to correct for nearsightedness. For clarity, the eye and eyeglass lens are shown much larger than the scale of the figure.

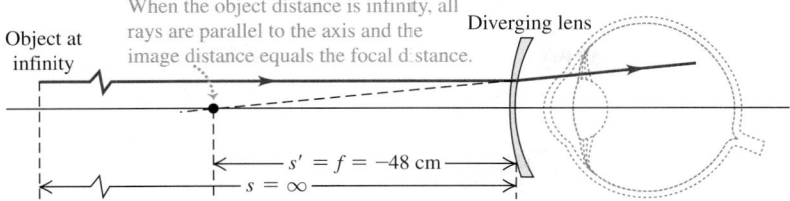

Object at infinity

When the object distance is infinity, all rays are parallel to the axis and the image distance equals the focal distance.

Diverging lens

$s' = f = -48$ cm

$s = \infty$

TEST YOUR UNDERSTANDING OF SECTION 34.6   A certain eyeglass lens is thin at its center, even thinner at its top and bottom edges, and relatively thick at its left and right edges. What defects of vision is this lens intended to correct? (i) Hyperopia for objects oriented both vertically and horizontally; (ii) myopia for objects oriented both vertically and horizontally; (iii) hyperopia for objects oriented vertically and myopia for objects oriented horizontally; (iv) hyperopia for objects oriented horizontally and myopia for objects oriented vertically. ∎

34.7 THE MAGNIFIER

The apparent size of an object is determined by the size of its image on the retina. If the eye is unaided, this size depends on the *angle* θ subtended by the object at the eye, called its **angular size** (**Fig. 34.51a**).

To look closely at a small object such as an insect or a crystal, you bring it close to your eye, making the subtended angle and the retinal image as large as possible. But your eye cannot focus sharply on objects that are closer than the near point, so an object subtends the largest possible viewing angle when it is placed at the near point. In the following discussion we will assume an average viewer for whom the near point is 25 cm from the eye.

A converging lens can be used to form a virtual image that is larger and farther from the eye than the object itself, as shown in Fig. 34.51b. Then the object can be moved closer to the eye, and the angular size of the image may be substantially larger than the angular size of the object at 25 cm without the lens. A lens used in this way is called a **magnifier,** otherwise known as a *magnifying glass* or a *simple magnifier*. The virtual image is most comfortable to view when it is placed at infinity, so that the ciliary muscle of the eye is relaxed; this means that the object is placed at the focal point F_1 of the magnifier. In the following discussion we assume that this is done.

In Fig. 34.51a the object is at the near point, where it subtends an angle θ at the eye. In Fig. 34.51b a magnifier in front of the eye forms an image at infinity,

BIO Application The Telephoto Eyes of Chameleons The crystalline lens of a human eye can change shape but is always a converging (positive) lens. The lens in the eye of a chameleon lizard (family Chamaeleonidae) is different: It can change shape to be either a converging or a *diverging* (negative) lens. When it acts as a diverging lens just behind the cornea (which acts as a converging lens), the combination is like the long-focal-length zoom lens shown in Fig. 34.43a. This "telephoto vision" gives the chameleon a sharp view of potential insect prey.

34.51 (a) The angular size θ is largest when the object is at the near point. (b) The magnifier gives a virtual image at infinity. This virtual image appears to the eye to be a real object subtending a larger angle θ' at the eye.

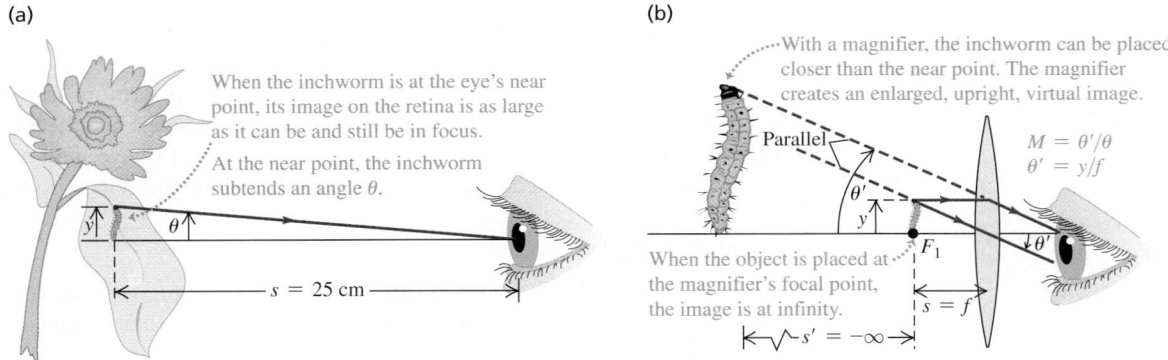

(a)

When the inchworm is at the eye's near point, its image on the retina is as large as it can be and still be in focus.

At the near point, the inchworm subtends an angle θ.

y θ

$s = 25$ cm

(b)

With a magnifier, the inchworm can be placed closer than the near point. The magnifier creates an enlarged, upright, virtual image.

Parallel

θ y

F_1 θ'

When the object is placed at the magnifier's focal point, the image is at infinity.

$s = f$

$s' = -\infty$

$M = \theta'/\theta$
$\theta' = y/f$

and the angle subtended at the magnifier is θ'. The usefulness of the magnifier is given by the ratio of the angle θ' (with the magnifier) to the angle θ (without the magnifier). This ratio is called the **angular magnification** M:

$$M = \frac{\theta'}{\theta} \qquad \text{(angular magnification)} \tag{34.21}$$

CAUTION **Angular magnification vs. lateral magnification** Don't confuse *angular* magnification M with *lateral* magnification m. Angular magnification is the ratio of the *angular* size of an image to the angular size of the corresponding object; lateral magnification refers to the ratio of the *height* of an image to the height of the corresponding object. For the situation shown in Fig. 34.51b, the angular magnification is about 3×, since the inchworm subtends an angle about three times larger than that in Fig. 34.51a; hence the inchworm will look about three times larger to the eye. The *lateral* magnification $m = -s'/s$ in Fig. 34.51b is *infinite* because the virtual image is at infinity, but that doesn't mean that the inchworm looks infinitely large through the magnifier! When dealing with a magnifier, M is useful but m is not. ▮

To find the value of M, we first assume that the angles are small enough that each angle (in radians) is equal to its sine and its tangent. Using Fig. 34.51a and drawing the ray in Fig. 34.51b that passes undeviated through the center of the lens, we find that θ and θ' (in radians) are

$$\theta = \frac{y}{25\ \text{cm}} \qquad \theta' = \frac{y}{f}$$

Combining these expressions with Eq. (34.21), we find

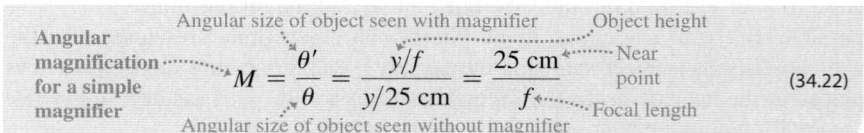

It may seem that we can make the angular magnification as large as we like by decreasing the focal length f. In fact, the aberrations of a simple double-convex lens set a limit to M of about 3× to 4×. If these aberrations are corrected, the angular magnification may be made as great as 20×. A compound microscope, discussed in the next section, provides even greater magnification.

TEST YOUR UNDERSTANDING OF SECTION 34.7   You are using a magnifier to examine a gem. If you change to a different magnifier with twice the focal length of the first one, you will have to hold the object at (i) twice the distance and the angular magnification will be twice as great; (ii) twice the distance and the angular magnification will be $\frac{1}{2}$ as great; (iii) $\frac{1}{2}$ the distance and the angular magnification will be twice as great; (iv) $\frac{1}{2}$ the distance and the angular magnification will be $\frac{1}{2}$ as great. ▮

34.8 MICROSCOPES AND TELESCOPES

Cameras, eyeglasses, and magnifiers use a single lens to form an image. Two important optical devices that use *two* lenses are the microscope and the telescope. In each device a primary lens, or *objective,* forms a real image, and a second lens, or *eyepiece,* is used as a magnifier to make an enlarged, virtual image.

Microscopes

Figure 34.52a shows the essential features of a **microscope,** sometimes called a *compound microscope.* To analyze this system, we use the principle that an image formed by one optical element such as a lens or mirror can serve as the object

34.52 (a) Elements of a microscope. (b) The object O is placed just outside the first focal point of the objective (the distance s_1 has been exaggerated for clarity). (c) This microscope image shows single-celled organisms about 2×10^{-4} m (0.2 mm) across. Typical light microscopes can resolve features as small as 2×10^{-7} m, comparable to the wavelength of light.

(a) Elements of a microscope

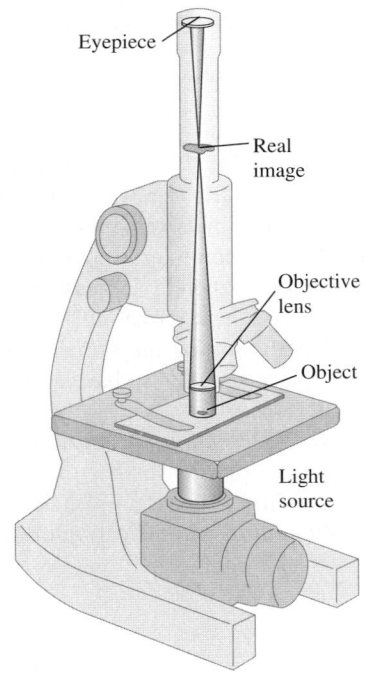

(b) Microscope optics

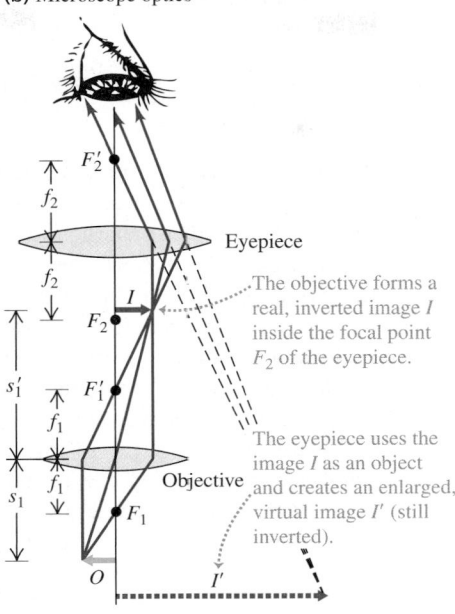

The objective forms a real, inverted image I inside the focal point F_2 of the eyepiece.

The eyepiece uses the image I as an object and creates an enlarged, virtual image I' (still inverted).

(c) Single-celled freshwater algae (*Micrasterias denticulata*)

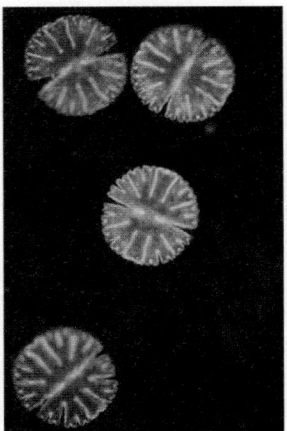

for a second element. We used this principle in Section 34.4 when we derived the lensmaker's equation by repeated application of the single-surface refraction equation; we used this principle again in Example 34.11 (Section 34.4), in which the image formed by a lens was used as the object of a second lens.

The object O to be viewed is placed just beyond the first focal point F_1 of the **objective,** a converging lens that forms a real and enlarged image I (Fig. 34.52b). In a properly designed instrument this image lies just inside the first focal point F_2 of a second converging lens called the **eyepiece** or *ocular.* (The reason the image should lie just *inside* F_2 is left for you to discover.) The eyepiece acts as a simple magnifier, as discussed in Section 34.7, and forms a final virtual image I' of I. The position of I' may be anywhere between the near and far points of the eye. Both the objective and the eyepiece of an actual microscope are highly corrected compound lenses with several optical elements, but for simplicity we show them here as simple thin lenses.

As for a simple magnifier, what matters when viewing through a microscope is the *angular* magnification M. The overall angular magnification of the compound microscope is the product of two factors. The first factor is the *lateral* magnification m_1 of the objective, which determines the linear size of the real image I; the second factor is the *angular* magnification M_2 of the eyepiece, which relates the angular size of the virtual image seen through the eyepiece to the angular size that the real image I would have if you viewed it *without* the eyepiece. The first of these factors is given by

$$m_1 = -\frac{s_1'}{s_1} \tag{34.23}$$

where s_1 and s_1' are the object and image distances, respectively, for the objective lens. Ordinarily, the object is very close to the focal point, and the resulting image distance s_1' is very great in comparison to the focal length f_1 of the objective lens. Thus s_1 is approximately equal to f_1, and we can write $m_1 = -s_1'/f_1$.

The real image I is close to the focal point F_2 of the eyepiece, so to find the eyepiece angular magnification, we can use Eq. (34.22): $M_2 = (25 \text{ cm})/f_2$, where f_2 is the focal length of the eyepiece (considered as a simple lens). The overall angular magnification M of the compound microscope (apart from a negative sign, which is customarily ignored) is the product of the two magnifications:

$$M = m_1 M_2 = \frac{(25 \text{ cm}) s_1'}{f_1 f_2} \qquad \text{(angular magnification for a microscope)} \qquad (34.24)$$

where s_1', f_1, and f_2 are measured in centimeters. The final image is inverted with respect to the object. Microscope manufacturers usually specify the values of m_1 and M_2 rather than the focal lengths of the objective and eyepiece.

Equation (34.24) shows that the angular magnification of a microscope can be increased by using an objective of shorter focal length f_1, thereby increasing m_1 and the size of the real image I. Most optical microscopes have a rotating "turret" with three or more objectives of different focal lengths so that the same object can be viewed at different magnifications. The eyepiece should also have a short focal length f_2 to help to maximize the value of M.

To use a microscope to take a photograph (called a *photomicrograph* or *micrograph*), the eyepiece is removed and a camera placed so that the real image I falls on the camera's electronic sensor or film. Figure 34.52c shows such a photograph. In this case what matters is the *lateral* magnification of the microscope as given by Eq. (34.23).

Telescopes

The optical system of a **telescope** is similar to that of a compound microscope. In both instruments the image formed by an objective is viewed through an eyepiece. The key difference is that the telescope is used to view large objects at large distances and the microscope is used to view small objects close at hand. Another difference is that many telescopes use a curved mirror, not a lens, as an objective.

Figure 34.53 shows an *astronomical telescope*. Because this telescope uses a lens as an objective, it is called a *refracting telescope* or *refractor*. The objective

34.53 Optical system of an astronomical refracting telescope.

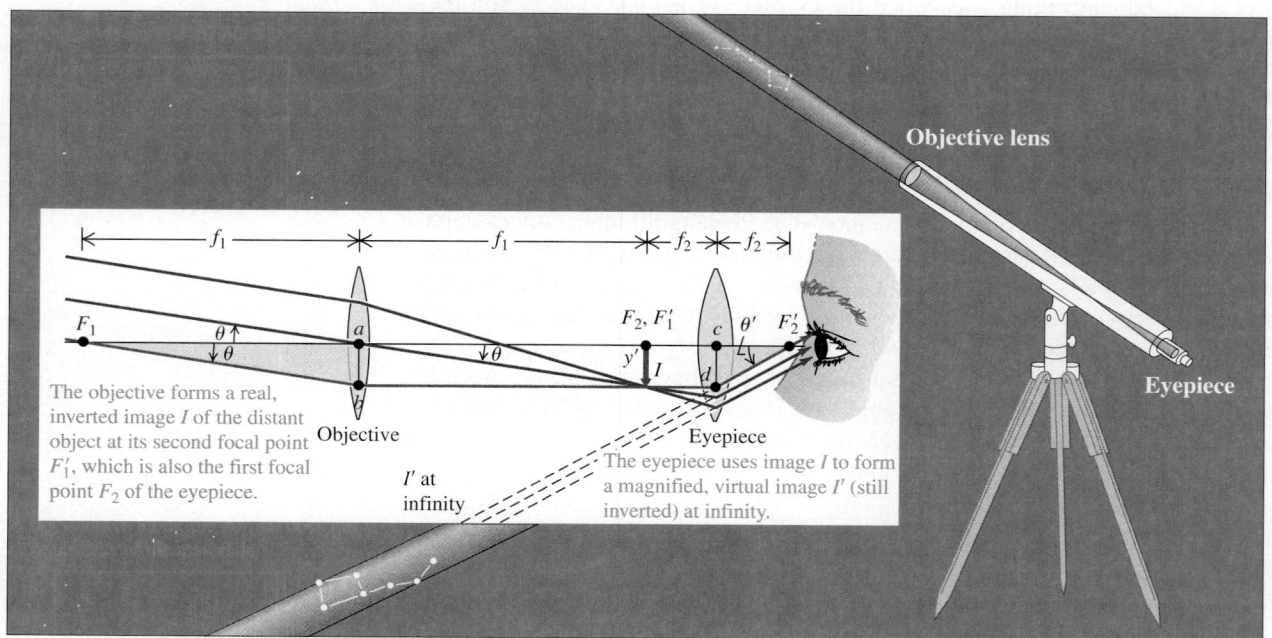

lens forms a real, reduced image I of the object. This image is the object for the eyepiece lens, which forms an enlarged, virtual image of I. Objects that are viewed with a telescope are usually so far away from the instrument that the first image I is formed very nearly at the second focal point of the objective lens. If the final image I' formed by the eyepiece is at infinity (for most comfortable viewing by a normal eye), the first image must also be at the first focal point of the eyepiece. The distance between objective and eyepiece, which is the length of the telescope, is therefore the *sum* of the focal lengths of objective and eyepiece, $f_1 + f_2$.

The angular magnification M of a telescope is defined as the ratio of the angle subtended at the eye by the final image I' to the angle subtended at the (unaided) eye by the object. We can express this ratio in terms of the focal lengths of objective and eyepiece. In Fig. 34.53 the ray passing through F_1, the first focal point of the objective, and through F'_2, the second focal point of the eyepiece, is shown in red. The object (not shown) subtends an angle θ at the objective and would subtend essentially the same angle at the unaided eye. Also, since the observer's eye is placed just to the right of the focal point F'_2, the angle subtended at the eye by the final image is very nearly equal to the angle θ'. Because bd is parallel to the optic axis, the distances ab and cd are equal to each other and also to the height y' of the real image I. Because the angles θ and θ' are small, they may be approximated by their tangents. From the right triangles F_1ab and F'_2cd,

$$\theta = \frac{-y'}{f_1}$$

$$\theta' = \frac{y'}{f_2}$$

and the angular magnification M is

$$M = \frac{\theta'}{\theta} = -\frac{y'/f_2}{y'/f_1} = -\frac{f_1}{f_2} \qquad \text{(angular magnification for a telescope)} \qquad (34.25)$$

The angular magnification M of a telescope is equal to the ratio of the focal length of the objective to that of the eyepiece. The negative sign shows that the final image is inverted. Equation (34.25) shows that to achieve good angular magnification, a *telescope* should have a *long* objective focal length f_1. By contrast, Eq. (34.24) shows that a *microscope* should have a *short* objective focal length. However, a telescope objective with a long focal length should also have a large diameter D so that the f-number f_1/D will not be too large; as described in Section 34.5, a large f-number means a dim, low-intensity image. Telescopes typically do not have interchangeable objectives; instead, the magnification is varied by using different eyepieces with different focal lengths f_2. Just as for a microscope, smaller values of f_2 give larger angular magnifications.

An inverted image is no particular disadvantage for astronomical observations. When we use a telescope or binoculars—essentially a pair of telescopes mounted side by side—to view objects on the earth, though, we want the image to be right-side up. In prism binoculars, this is accomplished by reflecting the light several times along the path from the objective to the eyepiece. The combined effect of the reflections is to flip the image both horizontally and vertically. Binoculars are usually described by two numbers separated by a multiplication sign, such as 7×50. The first number is the angular magnification M, and the second is the diameter of the objective lenses (in millimeters). The diameter helps to determine the light-gathering capacity of the objective lenses and thus the brightness of the image.

34.54 (a), (b), (c) Three designs for reflecting telescopes. (d) This photo shows the interior of the Gemini North telescope, which uses the design shown in (c). The objective mirror is 8 meters in diameter.

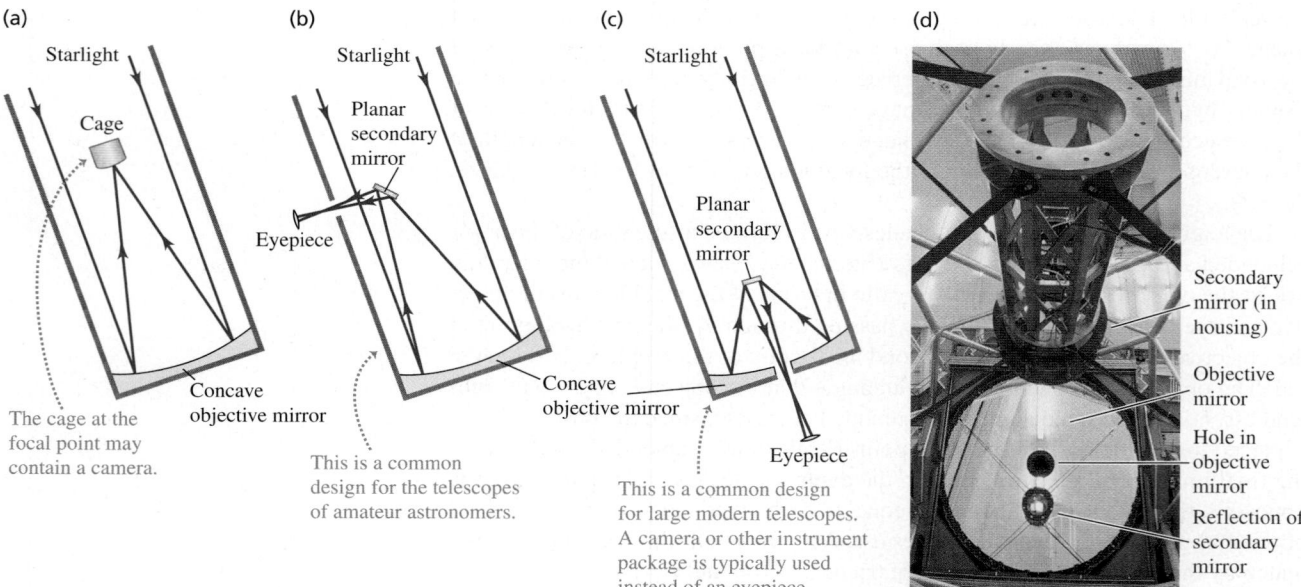

(a)

Starlight

Cage

The cage at the focal point may contain a camera.

Concave objective mirror

(b)

Starlight

Planar secondary mirror

Eyepiece

This is a common design for the telescopes of amateur astronomers.

Concave objective mirror

(c)

Starlight

Planar secondary mirror

Eyepiece

This is a common design for large modern telescopes. A camera or other instrument package is typically used instead of an eyepiece.

(d)

Secondary mirror (in housing)

Objective mirror

Hole in objective mirror

Reflection of secondary mirror

In the *reflecting telescope* (**Fig. 34.54a**) the objective lens is replaced by a concave mirror. In large telescopes this scheme has many advantages. Mirrors are inherently free of chromatic aberrations (dependence of focal length on wavelength), and spherical aberrations (associated with the paraxial approximation) are easier to correct than with a lens. The reflecting surface is sometimes nonspherical. The material of the mirror need not be transparent, and it can be made more rigid than a lens, which has to be supported only at its edges.

The largest reflecting telescope in the world, the Gran Telescopio Canarias in the Canary Islands, has an objective mirror of overall diameter 10.4 m made up of 36 separate hexagonal reflecting elements.

One challenge in designing reflecting telescopes is that the image is formed in front of the objective mirror, in a region traversed by incoming rays. Isaac Newton devised one solution to this problem. A flat secondary mirror oriented at 45° to the optic axis causes the image to be formed in a hole on the side of the telescope, where it can be magnified with an eyepiece (Fig. 34.54b). Another solution uses a secondary mirror that causes the focused light to pass through a hole in the objective mirror (Fig. 34.54c). Large research telescopes, as well as many amateur telescopes, use this design (Fig. 34.54d).

Like a microscope, when a telescope is used for photography the eyepiece is removed and an electronic sensor is placed at the position of the real image formed by the objective. (Some long-focal-length "lenses" for photography are actually reflecting telescopes used in this way.) Most telescopes used for astronomical research are never used with an eyepiece.

TEST YOUR UNDERSTANDING OF SECTION 34.8 Which gives a lateral magnification of greater absolute value? (i) The objective lens in a microscope (Fig. 34.52); (ii) the objective lens in a refracting telescope (Fig. 34.53); or (iii) not enough information is given to decide. ▌

Reflection or refraction at a plane surface: When rays diverge from an object point P and are reflected or refracted, the directions of the outgoing rays are the same as though they had diverged from a point P' called the image point. If they actually converge at P' and diverge again beyond it, P' is a real image of P; if they only appear to have diverged from P', it is a virtual image. Images can be either erect or inverted.

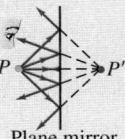

Plane mirror

Lateral magnification: The lateral magnification m in any reflecting or refracting situation is defined as the ratio of image height y' to object height y. When m is positive, the image is erect; when m is negative, the image is inverted.

$$m = \frac{y'}{y} \qquad (34.2)$$

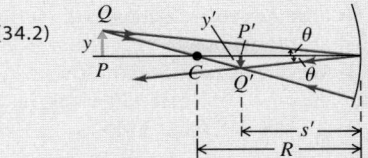

Focal point and focal length: The focal point of a mirror is the point where parallel rays converge after reflection from a concave mirror, or the point from which they appear to diverge after reflection from a convex mirror. Rays diverging from the focal point of a concave mirror are parallel after reflection; rays converging toward the focal point of a convex mirror are parallel after reflection. The distance from the focal point to the vertex is called the focal length, denoted as f. The focal points of a lens are defined similarly.

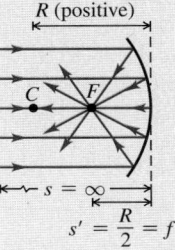

$$s' = \frac{R}{2} = f$$

Relating object and image distances: The formulas for object distance s and image distance s' for plane and spherical mirrors and single refracting surfaces are summarized in the table. The equation for a plane surface can be obtained from the corresponding equation for a spherical surface by setting $R = \infty$. (See Examples 34.1–34.7.)

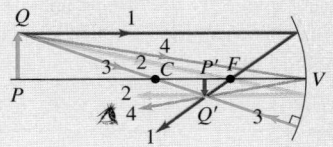

	Plane Mirror	Spherical Mirror	Plane Refracting Surface	Spherical Refracting Surface
Object and image distances	$\dfrac{1}{s} + \dfrac{1}{s'} = 0$	$\dfrac{1}{s} + \dfrac{1}{s'} = \dfrac{2}{R} = \dfrac{1}{f}$	$\dfrac{n_a}{s} + \dfrac{n_b}{s'} = 0$	$\dfrac{n_a}{s} + \dfrac{n_b}{s'} = \dfrac{n_b - n_a}{R}$
Lateral magnification	$m = -\dfrac{s'}{s} = 1$	$m = -\dfrac{s'}{s}$	$m = -\dfrac{n_a s'}{n_b s} = 1$	$m = -\dfrac{n_a s'}{n_b s}$

Object–image relationships derived in this chapter are valid for only rays close to and nearly parallel to the optic axis; these are called paraxial rays. Nonparaxial rays do not converge precisely to an image point. This effect is called spherical aberration.

Thin lenses: The object–image relationship, given by Eq. (34.16), is the same for a thin lens as for a spherical mirror. Equation (34.19), the lensmaker's equation, relates the focal length of a lens to its index of refraction and the radii of curvature of its surfaces. (See Examples 34.8–34.11.)

$$\frac{1}{s} + \frac{1}{s'} = \frac{1}{f} \qquad (34.16)$$

$$\frac{1}{f} = (n - 1)\left(\frac{1}{R_1} - \frac{1}{R_2}\right) \qquad (34.19)$$

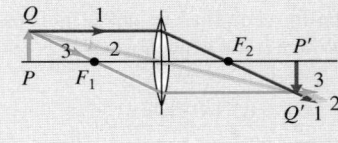

Sign rules: The following sign rules are used with all plane and spherical reflecting and refracting surfaces:

- $s > 0$ when the object is on the incoming side of the surface (a real object); $s < 0$ otherwise.

- $s' > 0$ when the image is on the outgoing side of the surface (a real image); $s' < 0$ otherwise.
- $R > 0$ when the center of curvature is on the outgoing side of the surface; $R < 0$ otherwise.
- $m > 0$ when the image is erect; $m < 0$ when inverted.

Cameras: A camera forms a real, inverted, reduced image of the object being photographed on a light-sensitive surface. The amount of light striking this surface is controlled by the shutter speed and the aperture. The intensity of this light is inversely proportional to the square of the *f*-number of the lens. (See Example 34.12.)

$$f\text{-number} = \frac{\text{Focal length}}{\text{Aperture diameter}} \qquad (34.20)$$
$$= \frac{f}{D}$$

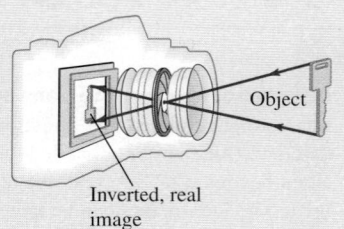

Inverted, real
image

The eye: In the eye, refraction at the surface of the cornea forms a real image on the retina. Adjustment for various object distances is made by squeezing the lens, thereby making it bulge and decreasing its focal length. A nearsighted eye is too long for its lens; a farsighted eye is too short. The power of a corrective lens, in diopters, is the reciprocal of the focal length in meters. (See Examples 34.13 and 34.14.)

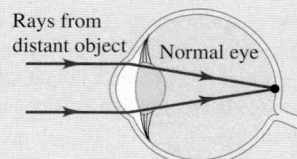

Rays from
distant object   Normal eye

The simple magnifier: The simple magnifier creates a virtual image whose angular size θ' is larger than the angular size θ of the object itself at a distance of 25 cm, the nominal closest distance for comfortable viewing. The angular magnification M of a simple magnifier is the ratio of the angular size of the virtual image to that of the object at this distance.

$$M = \frac{\theta'}{\theta} = \frac{25\ \text{cm}}{f} \qquad (34.22)$$

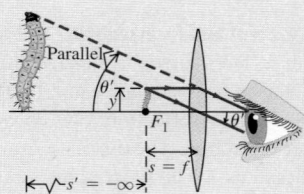

Microscopes and telescopes: In a compound microscope, the objective lens forms a first image in the barrel of the instrument, and the eyepiece forms a final virtual image, often at infinity, of the first image. The telescope operates on the same principle, but the object is far away. In a reflecting telescope, the objective lens is replaced by a concave mirror, which eliminates chromatic aberrations.

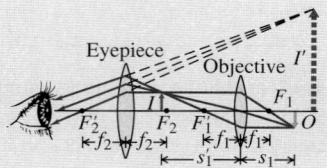

BRIDGING PROBLEM   IMAGE FORMATION BY A WINE GOBLET

A thick-walled wine goblet can be considered to be a hollow glass sphere with an outer radius of 4.00 cm and an inner radius of 3.40 cm. The index of refraction of the goblet glass is 1.50. (a) A beam of parallel light rays enters the side of the empty goblet along a horizontal radius. Where, if anywhere, will an image be formed? (b) The goblet is filled with white wine ($n = 1.37$). Where is the image formed?

SOLUTION GUIDE

IDENTIFY and SET UP

1. The goblet is *not* a thin lens, so you cannot use the thin-lens formula. Instead, you must think of the inner and outer surfaces of the goblet walls as spherical refracting surfaces. The image formed by one surface serves as the object for the next surface. Draw a diagram that shows the goblet and the light rays that enter it.
2. Choose the appropriate equation that relates the image and object distances for a spherical refracting surface.

EXECUTE

3. For the empty goblet, each refracting surface has glass on one side and air on the other. Find the position of the image formed by the first surface, the outer wall of the goblet. Use this as the object for the second surface (the inner wall of the same side of the goblet) and find the position of the second image. (*Hint*: Be sure to account for the thickness of the goblet wall.)
4. Continue the process of step 3. Consider the refractions at the inner and outer surfaces of the glass on the opposite side of the goblet, and find the position of the final image. (*Hint*: Be sure to account for the distance between the two sides of the goblet.)
5. Repeat steps 3 and 4 for the case in which the goblet is filled with wine.

EVALUATE

6. Are the images real or virtual? How can you tell?

Problems

For assigned homework and other learning materials, go to MasteringPhysics®. (MP)

•, ••, •••: Difficulty levels. CP: Cumulative problems incorporating material from earlier chapters. CALC: Problems requiring calculus.
DATA: Problems involving real data, scientific evidence, experimental design, and/or statistical reasoning. BIO: Biosciences problems.

DISCUSSION QUESTIONS

Q34.1 A spherical mirror is cut in half horizontally. Will an image be formed by the bottom half of the mirror? If so, where will the image be formed?

Q34.2 For the situation shown in Fig. 34.3, is the image distance s' positive or negative? Is the image real or virtual? Explain your answers.

Q34.3 The laws of optics also apply to electromagnetic waves invisible to the eye. A satellite TV dish is used to detect radio waves coming from orbiting satellites. Why is a curved reflecting surface (a "dish") used? The dish is always concave, never convex; why? The actual radio receiver is placed on an arm and suspended in front of the dish. How far in front of the dish should it be placed?

Q34.4 Explain why the focal length of a *plane* mirror is infinite, and explain what it means for the focal point to be at infinity.

Q34.5 If a spherical mirror is immersed in water, does its focal length change? Explain.

Q34.6 For what range of object positions does a concave spherical mirror form a real image? What about a convex spherical mirror?

Q34.7 When a room has mirrors on two opposite walls, an infinite series of reflections can be seen. Discuss this phenomenon in terms of images. Why do the distant images appear fainter?

Q34.8 For a spherical mirror, if $s = f$, then $s' = \infty$, and the lateral magnification m is infinite. Does this make sense? If so, what does it mean?

Q34.9 You may have noticed a small convex mirror next to your bank's ATM. Why is this mirror convex, as opposed to flat or concave? What considerations determine its radius of curvature?

Q34.10 A student claims that she can start a fire on a sunny day using just the sun's rays and a concave mirror. How is this done? Is the concept of image relevant? Can she do the same thing with a convex mirror? Explain.

Q34.11 A person looks at his reflection in the concave side of a shiny spoon. Is it right side up or inverted? Does it matter how far his face is from the spoon? What if he looks in the convex side? (Try this yourself!)

Q34.12 In Example 34.4 (Section 34.2), there appears to be an ambiguity for the case $s = 10$ cm as to whether s' is $+\infty$ or $-\infty$ and whether the image is erect or inverted. How is this resolved? Or is it?

Q34.13 Suppose that in the situation of Example 34.7 of Section 34.3 (see Fig. 34.26) a vertical arrow 2.00 m tall is painted on the side of the pool beneath the water line. According to the calculations in the example, this arrow would appear to the person shown in Fig. 34.26 to be 1.50 m long. But the discussion following Eq. (34.13) states that the magnification for a plane refracting surface is $m = 1$, which suggests that the arrow would appear to the person to be 2.00 m long. How can you resolve this apparent contradiction?

Q34.14 The bottom of the passenger-side mirror on your car notes, "Objects in mirror are closer than they appear." Is this true? Why?

Q34.15 How could you very quickly make an approximate measurement of the focal length of a converging lens? Could the same method be applied if you wished to use a diverging lens? Explain.

Q34.16 The focal length of a simple lens depends on the color (wavelength) of light passing through it. Why? Is it possible for a lens to have a positive focal length for some colors and negative for others? Explain.

Q34.17 When a converging lens is immersed in water, does its focal length increase or decrease in comparison with the value in air? Explain.

Q34.18 A spherical air bubble in water can function as a lens. Is it a converging or diverging lens? How is its focal length related to its radius?

Q34.19 Can an image formed by one reflecting or refracting surface serve as an object for a second reflection or refraction? Does it matter whether the first image is real or virtual? Explain.

Q34.20 If a piece of photographic film is placed at the location of a real image, the film will record the image. Can this be done with a virtual image? How might one record a virtual image?

Q34.21 According to the discussion in Section 34.2, light rays are reversible. Are the formulas in the table in this chapter's Summary still valid if object and image are interchanged? What does reversibility imply with respect to the *forms* of the various formulas?

Q34.22 You've entered a survival contest that will include building a crude telescope. You are given a large box of lenses. Which two lenses do you pick? How do you quickly identify them?

Q34.23 BIO You can't see clearly underwater with the naked eye, but you *can* if you wear a face mask or goggles (with air between your eyes and the mask or goggles). Why is there a difference? Could you instead wear eyeglasses (with water between your eyes and the eyeglasses) in order to see underwater? If so, should the lenses be converging or diverging? Explain.

Q34.24 You take a lens and mask it so that light can pass through only the bottom half of the lens. How does the image formed by the masked lens compare to the image formed before masking?

EXERCISES

Section 34.1 Reflection and Refraction at a Plane Surface

34.1 • A candle 4.85 cm tall is 39.2 cm to the left of a plane mirror. Where is the image formed by the mirror, and what is the height of this image?

34.2 • The image of a tree just covers the length of a plane mirror 4.00 cm tall when the mirror is held 35.0 cm from the eye. The tree is 28.0 m from the mirror. What is its height?

34.3 • A pencil that is 9.0 cm long is held perpendicular to the surface of a plane mirror with the tip of the pencil lead 12.0 cm from the mirror surface and the end of the eraser 21.0 cm from the mirror surface. What is the length of the image of the pencil that is formed by the mirror? Which end of the image is closer to the mirror surface: the tip of the lead or the end of the eraser?

Section 34.2 Reflection at a Spherical Surface

34.4 • A concave mirror has a radius of curvature of 34.0 cm. (a) What is its focal length? (b) If the mirror is immersed in water (refractive index 1.33), what is its focal length?

34.5 • An object 0.600 cm tall is placed 16.5 cm to the left of the vertex of a concave spherical mirror having a radius of curvature of 22.0 cm. (a) Draw a principal-ray diagram showing the formation of the image. (b) Determine the position, size, orientation, and nature (real or virtual) of the image.

34.6 • Repeat Exercise 34.5 for the case in which the mirror is convex.

34.7 •• The diameter of Mars is 6794 km, and its minimum distance from the earth is 5.58×10^7 km. When Mars is at this distance, find the diameter of the image of Mars formed by a spherical, concave telescope mirror with a focal length of 1.75 m.

34.8 •• An object is 18.0 cm from the center of a spherical silvered-glass Christmas tree ornament 6.00 cm in diameter. What are the position and magnification of its image?

34.9 • A coin is placed next to the convex side of a thin spherical glass shell having a radius of curvature of 18.0 cm. Reflection from the surface of the shell forms an image of the 1.5-cm-tall coin that is 6.00 cm behind the glass shell. Where is the coin located? Determine the size, orientation, and nature (real or virtual) of the image.

34.10 • You hold a spherical salad bowl 60 cm in front of your face with the bottom of the bowl facing you. The bowl is made of polished metal with a 35-cm radius of curvature. (a) Where is the image of your 5.0-cm-tall nose located? (b) What are the image's size, orientation, and nature (real or virtual)?

34.11 • A spherical, concave shaving mirror has a radius of curvature of 32.0 cm. (a) What is the magnification of a person's face when it is 12.0 cm to the left of the vertex of the mirror? (b) Where is the image? Is the image real or virtual? (c) Draw a principal-ray diagram showing the formation of the image.

34.12 • For a concave spherical mirror that has focal length $f = +18.0$ cm, what is the distance of an object from the mirror's vertex if the image is real and has the same height as the object?

34.13 • **Dental Mirror.** A dentist uses a curved mirror to view teeth on the upper side of the mouth. Suppose she wants an erect image with a magnification of 2.00 when the mirror is 1.25 cm from a tooth. (Treat this problem as though the object and image lie along a straight line.) (a) What kind of mirror (concave or convex) is needed? Use a ray diagram to decide, without performing any calculations. (b) What must be the focal length and radius of curvature of this mirror? (c) Draw a principal-ray diagram to check your answer in part (b).

34.14 • For a convex spherical mirror that has focal length $f = -12.0$ cm, what is the distance of an object from the mirror's vertex if the height of the image is half the height of the object?

34.15 • The thin glass shell shown in **Fig. E34.15** has a spherical shape with a radius of curvature of 12.0 cm, and both of its surfaces can act as mirrors. A seed 3.30 mm high is placed 15.0 cm from the center of the mirror along the optic axis, as shown in the figure. (a) Calculate the location and height of the image of this seed. (b) Suppose now that the shell is reversed. Find the location and height of the seed's image.

Figure **E34.15**

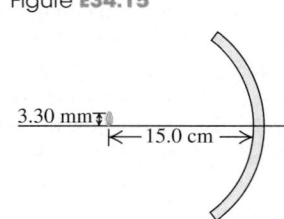

3.30 mm ↕

← 15.0 cm →

Section 34.3 Refraction at a Spherical Surface

34.16 •• A tank whose bottom is a mirror is filled with water to a depth of 20.0 cm. A small fish floats motionless 7.0 cm under the surface of the water. (a) What is the apparent depth of the fish when viewed at normal incidence? (b) What is the apparent depth of the image of the fish when viewed at normal incidence?

34.17 • A speck of dirt is embedded 3.50 cm below the surface of a sheet of ice ($n = 1.309$). What is its apparent depth when viewed at normal incidence?

34.18 • A transparent liquid fills a cylindrical tank to a depth of 3.60 m. There is air above the liquid. You look at normal incidence at a small pebble at the bottom of the tank. The apparent depth of the pebble below the liquid's surface is 2.45 m. What is the refractive index of this liquid?

34.19 • A person swimming 0.80 m below the surface of the water in a swimming pool looks at the diving board that is directly overhead and sees the image of the board that is formed by refraction at the surface of the water. This image is a height of 5.20 m above the swimmer. What is the actual height of the diving board above the surface of the water?

34.20 • A person is lying on a diving board 3.00 m above the surface of the water in a swimming pool. She looks at a penny that is on the bottom of the pool directly below her. To her, the penny appears to be a distance of 7.00 m from her. What is the depth of the water at this point?

34.21 •• **A Spherical Fish Bowl.** A small tropical fish is at the center of a water-filled, spherical fish bowl 28.0 cm in diameter. (a) Find the apparent position and magnification of the fish to an observer outside the bowl. The effect of the thin walls of the bowl may be ignored. (b) A friend advised the owner of the bowl to keep it out of direct sunlight to avoid blinding the fish, which might swim into the focal point of the parallel rays from the sun. Is the focal point actually within the bowl?

34.22 • The left end of a long glass rod 6.00 cm in diameter has a convex hemispherical surface 3.00 cm in radius. The refractive index of the glass is 1.60. Determine the position of the image if an object is placed in air on the axis of the rod at the following distances to the left of the vertex of the curved end: (a) infinitely far, (b) 12.0 cm; (c) 2.00 cm.

34.23 •• The glass rod of Exercise 34.22 is immersed in oil ($n = 1.45$). An object placed to the left of the rod on the rod's axis is to be imaged 1.20 m inside the rod. How far from the left end of the rod must the object be located to form the image?

34.24 •• The left end of a long glass rod 8.00 cm in diameter, with an index of refraction of 1.60, is ground and polished to a convex hemispherical surface with a radius of 4.00 cm. An object in the form of an arrow 1.50 mm tall, at right angles to the axis of the rod, is located on the axis 24.0 cm to the left of the vertex of the convex surface. Find the position and height of the image of the arrow formed by paraxial rays incident on the convex surface. Is the image erect or inverted?

34.25 •• Repeat Exercise 34.24 for the case in which the end of the rod is ground to a *concave* hemispherical surface with radius 4.00 cm.

34.26 •• The glass rod of Exercise 34.25 is immersed in a liquid. An object 14.0 cm from the vertex of the left end of the rod and on its axis is imaged at a point 9.00 cm from the vertex inside the liquid. What is the index of refraction of the liquid?

Section 34.4 Thin Lenses

34.27 • An insect 3.75 mm tall is placed 22.5 cm to the left of a thin planoconvex lens. The left surface of this lens is flat, the right surface has a radius of curvature of magnitude 13.0 cm, and the index of refraction of the lens material is 1.70. (a) Calculate the location and size of the image this lens forms of the insect.

Is it real or virtual? Erect or inverted? (b) Repeat part (a) if the lens is reversed.

34.28 • A lens forms an image of an object. The object is 16.0 cm from the lens. The image is 12.0 cm from the lens on the same side as the object. (a) What is the focal length of the lens? Is the lens converging or diverging? (b) If the object is 8.50 mm tall, how tall is the image? Is it erect or inverted? (c) Draw a principal-ray diagram.

34.29 • A converging meniscus lens (see Fig. 34.32a) with a refractive index of 1.52 has spherical surfaces whose radii are 7.00 cm and 4.00 cm. What is the position of the image if an object is placed 24.0 cm to the left of the lens? What is the magnification?

34.30 • A converging lens with a focal length of 70.0 cm forms an image of a 3.20-cm-tall real object that is to the left of the lens. The image is 4.50 cm tall and inverted. Where are the object and image located in relation to the lens? Is the image real or virtual?

34.31 •• A converging lens forms an image of an 8.00-mm-tall real object. The image is 12.0 cm to the left of the lens, 3.40 cm tall, and erect. What is the focal length of the lens? Where is the object located?

34.32 • A photographic slide is to the left of a lens. The lens projects an image of the slide onto a wall 6.00 m to the right of the slide. The image is 80.0 times the size of the slide. (a) How far is the slide from the lens? (b) Is the image erect or inverted? (c) What is the focal length of the lens? (d) Is the lens converging or diverging?

34.33 •• A double-convex thin lens has surfaces with equal radii of curvature of magnitude 2.50 cm. Using this lens, you observe that it forms an image of a very distant tree at a distance of 1.87 cm from the lens. What is the index of refraction of the lens?

34.34 • A converging lens with a focal length of 9.00 cm forms an image of a 4.00-mm-tall real object that is to the left of the lens. The image is 1.30 cm tall and erect. Where are the object and image located? Is the image real or virtual?

34.35 • BIO **The Cornea As a Simple Lens.** The cornea behaves as a thin lens of focal length approximately 1.8 cm, although this varies a bit. The material of which it is made has an index of refraction of 1.38, and its front surface is convex, with a radius of curvature of 5.0 mm. (a) If this focal length is in air, what is the radius of curvature of the back side of the cornea? (b) The closest distance at which a typical person can focus on an object (called the near point) is about 25 cm, although this varies considerably with age. Where would the cornea focus the image of an 8.0-mm-tall object at the near point? (c) What is the height of the image in part (b)? Is this image real or virtual? Is it erect or inverted? (*Note:* The results obtained here are not strictly accurate because, on one side, the cornea has a fluid with a refractive index different from that of air.)

34.36 •• A lensmaker wants to make a magnifying glass from glass that has an index of refraction $n = 1.55$ and a focal length of 20.0 cm. If the two surfaces of the lens are to have equal radii, what should that radius be?

34.37 • For each thin lens shown in **Fig. E34.37**, calculate the location of the image of an object that is 18.0 cm to the left of

the lens. The lens material has a refractive index of 1.50, and the radii of curvature shown are only the magnitudes.

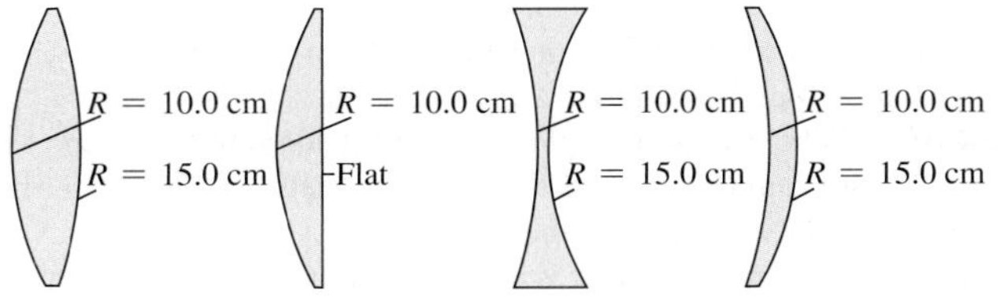

34.38 • A converging lens with a focal length of 12.0 cm forms a virtual image 8.00 mm tall, 17.0 cm to the right of the lens. Determine the position and size of the object. Is the image erect or inverted? Are the object and image on the same side or opposite sides of the lens? Draw a principal-ray diagram for this situation.

34.39 • Repeat Exercise 34.38 for the case in which the lens is diverging, with a focal length of −48.0 cm.

34.40 • An object is 16.0 cm to the left of a lens. The lens forms an image 36.0 cm to the right of the lens. (a) What is the focal length of the lens? Is the lens converging or diverging? (b) If the object is 8.00 mm tall, how tall is the image? Is it erect or inverted? (c) Draw a principal-ray diagram.

34.41 •• **Combination of Lenses I.** A 1.20-cm-tall object is 50.0 cm to the left of a converging lens of focal length 40.0 cm. A second converging lens, this one having a focal length of 60.0 cm, is located 300.0 cm to the right of the first lens along the same optic axis. (a) Find the location and height of the image (call it I_1) formed by the lens with a focal length of 40.0 cm. (b) I_1 is now the object for the second lens. Find the location and height of the image produced by the second lens. This is the final image produced by the combination of lenses.

34.42 •• **Combination of Lenses II.** Repeat Exercise 34.41 using the same lenses except for the following changes: (a) The second lens is a *diverging* lens having a focal length of magnitude 60.0 cm. (b) The first lens is a *diverging* lens having a focal length of magnitude 40.0 cm. (c) Both lenses are diverging lenses having focal lengths of the same *magnitudes* as in Exercise 34.41.

34.43 •• **Combination of Lenses III.** Two thin lenses with a focal length of magnitude 12.0 cm, the first diverging and the second converging, are located 9.00 cm apart. An object 2.50 mm tall is placed 20.0 cm to the left of the first (diverging) lens. (a) How far from this first lens is the final image formed? (b) Is the final image real or virtual? (c) What is the height of the final image? Is it erect or inverted? (*Hint:* See the preceding two problems.)

34.44 • BIO **The Lens of the Eye.** The crystalline lens of the human eye is a double-convex lens made of material having an index of refraction of 1.44 (although this varies). Its focal length in air is about 8.0 mm, which also varies. We shall assume that the radii of curvature of its two surfaces have the same magnitude. (a) Find the radii of curvature of this lens. (b) If an object 16 cm tall is placed 30.0 cm from the eye lens, where would the lens focus it and how tall would the image be? Is this image real or virtual? Is it erect or inverted? (*Note:* The results obtained here are not strictly accurate because the lens is embedded in fluids having refractive indexes different from that of air.)

Section 34.5 Cameras

34.45 •• A camera lens has a focal length of 200 mm. How far from the lens should the subject for the photo be if the lens is 20.4 cm from the sensor?

34.46 • You wish to project the image of a slide on a screen 9.00 m from the lens of a slide projector. (a) If the slide is placed 15.0 cm from the lens, what focal length lens is required? (b) If the dimensions of the picture on a 35-mm color slide are 24 mm × 36 mm, what is the minimum size of the projector screen required to accommodate the image?

34.47 • When a camera is focused, the lens is moved away from or toward the digital image sensor. If you take a picture of your friend, who is standing 3.90 m from the lens, using a camera with

Figure **E34.37**

(a) $R = 10.0$ cm / $R = 15.0$ cm (b) $R = 10.0$ cm / Flat (c) $R = 10.0$ cm / $R = 15.0$ cm (d) $R = 10.0$ cm / $R = 15.0$ cm

a lens with an 85-mm focal length, how far from the sensor is the lens? Will the whole image of your friend, who is 175 cm tall, fit on a sensor that is 24 mm × 36 mm?

34.48 • **Zoom Lens.** Consider the simple model of the zoom lens shown in Fig. 34.43a. The converging lens has focal length $f_1 = 12$ cm, and the diverging lens has focal length $f_2 = -12$ cm. The lenses are separated by 4 cm as shown in Fig. 34.43a. (a) For a distant object, where is the image of the converging lens? (b) The image of the converging lens serves as the object for the diverging lens. What is the object distance for the diverging lens? (c) Where is the final image? Compare your answer to Fig. 34.43a. (d) Repeat parts (a), (b), and (c) for the situation shown in Fig. 34.43b, in which the lenses are separated by 8 cm.

34.49 •• A camera lens has a focal length of 180.0 mm and an aperture diameter of 16.36 mm. (a) What is the f-number of the lens? (b) If the correct exposure of a certain scene is $\frac{1}{30}$ s at $f/11$, what is the correct exposure at $f/2.8$?

Section 34.6 The Eye

34.50 •• BIO **Curvature of the Cornea.** In a simplified model of the human eye, the aqueous and vitreous humors and the lens all have a refractive index of 1.40, and all the bending occurs at the cornea, whose vertex is 2.60 cm from the retina. What should be the radius of curvature of the cornea such that the image of an object 40.0 cm from the cornea's vertex is focused on the retina?

34.51 •• BIO (a) Where is the near point of an eye for which a contact lens with a power of +2.75 diopters is prescribed? (b) Where is the far point of an eye for which a contact lens with a power of −1.30 diopters is prescribed for distant vision?

34.52 • BIO **Contact Lenses.** Contact lenses are placed right on the eyeball, so the distance from the eye to an object (or image) is the same as the distance from the lens to that object (or image). A certain person can see distant objects well, but his near point is 45.0 cm from his eyes instead of the usual 25.0 cm. (a) Is this person nearsighted or farsighted? (b) What type of lens (converging or diverging) is needed to correct his vision? (c) If the correcting lenses will be contact lenses, what focal length lens is needed and what is its power in diopters?

34.53 •• BIO **Ordinary Glasses.** Ordinary glasses are worn in front of the eye and usually 2.0 cm in front of the eyeball. Suppose that the person in Exercise 34.52 prefers ordinary glasses to contact lenses. What focal length lenses are needed to correct his vision, and what is their power in diopters?

34.54 • BIO A person can see clearly up close but cannot focus on objects beyond 75.0 cm. She opts for contact lenses to correct her vision. (a) Is she nearsighted or farsighted? (b) What type of lens (converging or diverging) is needed to correct her vision? (c) What focal length contact lens is needed, and what is its power in diopters?

34.55 •• BIO If the person in Exercise 34.54 chooses ordinary glasses over contact lenses, what power lens (in diopters) does she need to correct her vision if the lenses are 2.0 cm in front of the eye?

Section 34.7 The Magnifier

34.56 •• A thin lens with a focal length of 6.00 cm is used as a simple magnifier. (a) What angular magnification is obtainable with the lens if the object is at the focal point? (b) When an object is examined through the lens, how close can it be brought to the lens? Assume that the image viewed by the eye is at the near point, 25.0 cm from the eye, and that the lens is very close to the eye.

34.57 • The focal length of a simple magnifier is 8.00 cm. Assume the magnifier is a thin lens placed very close to the eye. (a) How far in front of the magnifier should an object be placed if the image is formed at the observer's near point, 25.0 cm in front of her eye? (b) If the object is 1.00 mm high, what is the height of its image formed by the magnifier?

34.58 • You want to view through a magnifier an insect that is 2.00 mm long. If the insect is to be at the focal point of the magnifier, what focal length will give the image of the insect an angular size of 0.032 radian?

Section 34.8 Microscopes and Telescopes

34.59 •• The focal length of the eyepiece of a certain microscope is 18.0 mm. The focal length of the objective is 8.00 mm. The distance between objective and eyepiece is 19.7 cm. The final image formed by the eyepiece is at infinity. Treat all lenses as thin. (a) What is the distance from the objective to the object being viewed? (b) What is the magnitude of the linear magnification produced by the objective? (c) What is the overall angular magnification of the microscope?

34.60 •• **Resolution of a Microscope.** The image formed by a microscope objective with a focal length of 5.00 mm is 160 mm from its second focal point. The eyepiece has a focal length of 26.0 mm. (a) What is the angular magnification of the microscope? (b) The unaided eye can distinguish two points at its near point as separate if they are about 0.10 mm apart. What is the minimum separation between two points that can be observed (or resolved) through this microscope?

34.61 •• A telescope is constructed from two lenses with focal lengths of 95.0 cm and 15.0 cm, the 95.0-cm lens being used as the objective. Both the object being viewed and the final image are at infinity. (a) Find the angular magnification for the telescope. (b) Find the height of the image formed by the objective of a building 60.0 m tall, 3.00 km away. (c) What is the angular size of the final image as viewed by an eye very close to the eyepiece?

34.62 •• The eyepiece of a refracting telescope (see Fig. 34.53) has a focal length of 9.00 cm. The distance between objective and eyepiece is 1.20 m, and the final image is at infinity. What is the angular magnification of the telescope?

34.63 •• A reflecting telescope (**Fig. E34.63**) is to be made by using a spherical mirror with a radius of curvature of 1.30 m and an eyepiece with a focal length of 1.10 cm. The final image is at infinity. (a) What should the distance between the eyepiece and the mirror vertex be if the object is taken to be at infinity? (b) What will the angular magnification be?

Figure **E34.63**

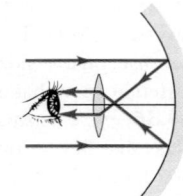

PROBLEMS

34.64 •• What is the size of the smallest vertical plane mirror in which a woman of height h can see her full-length image?

34.65 • If you run away from a plane mirror at 3.60 m/s, at what speed does your image move away from you?

34.66 • Where must you place an object in front of a concave mirror with radius R so that the image is erect and $2\frac{1}{2}$ times the size of the object? Where is the image?

34.67 •• A concave mirror is to form an image of the filament of a headlight lamp on a screen 8.00 m from the mirror. The filament is 6.00 mm tall, and the image is to be 24.0 cm tall. (a) How far in front of the vertex of the mirror should the filament be placed? (b) What should be the radius of curvature of the mirror?

34.68 • A light bulb is 3.00 m from a wall. You are to use a concave mirror to project an image of the bulb on the wall, with the image 3.50 times the size of the object. How far should the mirror be from the wall? What should its radius of curvature be?

34.69 •• CP CALC You are in your car driving on a highway at 25 m/s when you glance in the passenger-side mirror (a convex mirror with radius of curvature 150 cm) and notice a truck approaching. If the image of the truck is approaching the vertex of the mirror at a speed of 1.9 m/s when the truck is 2.0 m from the mirror, what is the speed of the truck relative to the highway?

34.70 •• A layer of benzene ($n = 1.50$) that is 4.20 cm deep floats on water ($n = 1.33$) that is 5.70 cm deep. What is the apparent distance from the upper benzene surface to the bottom of the water when you view these layers at normal incidence?

34.71 •• **Rear-View Mirror.** A mirror on the passenger side of your car is convex and has a radius of curvature with magnitude 18.0 cm. (a) Another car is behind your car, 9.00 m from the mirror, and this car is viewed in the mirror by your passenger. If this car is 1.5 m tall, what is the height of the image? (b) The mirror has a warning attached that objects viewed in it are closer than they appear. Why is this so?

34.72 •• **Figure P34.72** shows a small plant near a thin lens. The ray shown is one of the principal rays for the lens. Each square is 2.0 cm along the horizontal direction, but the vertical direction is not to the same scale.

Figure **P34.72**

Use information from the diagram for the following: (a) Using only the ray shown, decide what type of lens (converging or diverging) this is. (b) What is the focal length of the

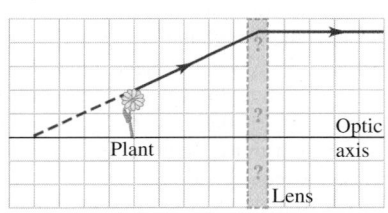

lens? (c) Locate the image by drawing the other two principal rays. (d) Calculate where the image should be, and compare this result with the graphical solution in part (c).

34.73 •• **Pinhole Camera.** A pinhole camera is just a rectangular box with a tiny hole in one face. The film is on the face opposite this hole, and that is where the image is formed. The camera forms an image *without* a lens. (a) Make a clear ray diagram to show how a pinhole camera can form an image on the film without using a lens. (*Hint:* Put an object outside the hole, and then draw rays passing through the hole to the opposite side of the box.) (b) A certain pinhole camera is a box that is 25 cm square and 20.0 cm deep, with the hole in the middle of one of the 25 cm × 25 cm faces. If this camera is used to photograph a fierce chicken that is 18 cm high and 1.5 m in front of the camera, how large is the image of this bird on the film? What is the lateral magnification of this camera?

34.74 ••• A microscope is focused on the upper surface of a glass plate. A second plate is then placed over the first. To focus on the bottom surface of the second plate, the microscope must be raised 0.780 mm. To focus on the upper surface, it must be raised another 2.10 mm. Find the index of refraction of the second plate.

34.75 •• What should be the index of refraction of a transparent sphere in order for paraxial rays from an infinitely distant object to be brought to a focus at the vertex of the surface opposite the point of incidence?

34.76 •• **A Glass Rod.** Both ends of a glass rod with index of refraction 1.60 are ground and polished to convex hemispherical surfaces. The radius of curvature at the left end is 6.00 cm, and the radius of curvature at the right end is 12.0 cm. The length of the rod between vertices is 40.0 cm. The object for the surface at the left end is an arrow that lies 23.0 cm to the left of the vertex of this surface. The arrow is 1.50 mm tall and at right angles to the axis. (a) What constitutes the object for the surface at the right end of the rod? (b) What is the object distance for this surface? (c) Is the object for this surface real or virtual? (d) What is the position of the final image? (e) Is the final image real or virtual? Is it erect or inverted with respect to the original object? (f) What is the height of the final image?

34.77 •• (a) You want to use a lens with a focal length of 35.0 cm to produce a real image of an object, with the height of the image twice the height of the object. What kind of lens do you need, and where should the object be placed? (b) Suppose you want a virtual image of the same object, with the same magnification—what kind of lens do you need, and where should the object be placed?

34.78 •• **Autocollimation.** You place an object alongside a white screen, and a plane mirror is 60.0 cm to the right of the object and the screen, with the surface of the mirror tilted slightly from the perpendicular to the line from object to mirror. You then place a converging lens between the object and the mirror. Light from the object passes through the lens, reflects from the mirror, and passes back through the lens; the image is projected onto the screen. You adjust the distance of the lens from the object until a sharp image of the object is focused on the screen. The lens is then 22.0 cm from the object. Because the screen is alongside the object, the distance from object to lens is the same as the distance from screen to lens. (a) Draw a sketch that shows the locations of the object, lens, plane mirror, and screen. (b) What is the focal length of the lens?

34.79 •• A lens forms a real image that is 214 cm away from the object and $1\frac{2}{3}$ times its height. What kind of lens is this, and what is its focal length?

34.80 • **Figure P34.80** shows an object and its image formed by a thin lens. (a) What is the focal length of the lens, and what type of lens (converging or diverging) is it? (b) What is the height of the image? Is it real or virtual?

Figure **P34.80**

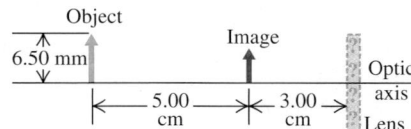

34.81 • **Figure P34.81** shows an object and its image formed by a thin lens. (a) What is the focal length of the lens, and what type of lens (converging or diverging) is it? (b) What is the height of the image? Is it real or virtual?

Figure **P34.81**

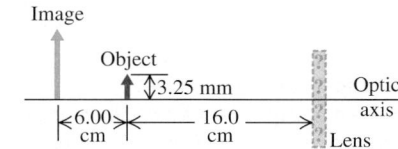

34.82 ••• A transparent rod 30.0 cm long is cut flat at one end and rounded to a hemispherical surface of radius 10.0 cm at the other end. A small object is embedded within the rod along its axis and halfway between its ends, 15.0 cm from the flat end and 15.0 cm from the vertex of the curved end. When the rod is viewed from its flat end, the apparent depth of the object is 8.20 cm from the flat end. What is its apparent depth when the rod is viewed from its curved end?

34.83 • BIO **Focus of the Eye.** The cornea of the eye has a radius of curvature of approximately 0.50 cm, and the aqueous humor behind it has an index of refraction of 1.35. The thickness of the cornea itself is small enough that we shall neglect it. The depth of a typical human eye is around 25 mm. (a) What would have to be the radius of curvature of the cornea so that it alone would focus the image of a distant mountain on the retina, which is at the back of the eye opposite the cornea? (b) If the cornea focused the mountain correctly on the retina as described in part (a), would it also focus the text from a computer screen on the retina if that screen were 25 cm in front of the eye? If not, where would it focus that text: in front of or behind the retina? (c) Given that the cornea has a radius of curvature of about 5.0 mm, where does it actually focus the mountain? Is this in front of or behind the retina? Does this help you see why the eye needs help from a lens to complete the task of focusing?

34.84 • The radii of curvature of the surfaces of a thin converging meniscus lens are $R_1 = +12.0$ cm and $R_2 = +28.0$ cm. The index of refraction is 1.60. (a) Compute the position and size of the image of an object in the form of an arrow 5.00 mm tall, perpendicular to the lens axis, 45.0 cm to the left of the lens. (b) A second converging lens with the same focal length is placed 3.15 m to the right of the first. Find the position and size of the final image. Is the final image erect or inverted with respect to the original object? (c) Repeat part (b) except with the second lens 45.0 cm to the right of the first.

34.85 • An object to the left of a lens is imaged by the lens on a screen 30.0 cm to the right of the lens. When the lens is moved 4.00 cm to the right, the screen must be moved 4.00 cm to the left to refocus the image. Determine the focal length of the lens.

34.86 • An object is placed 22.0 cm from a screen. (a) At what two points between object and screen may a converging lens with a 3.00-cm focal length be placed to obtain an image on the screen? (b) What is the magnification of the image for each position of the lens?

34.87 •• A convex mirror and a concave mirror are placed on the same optic axis, separated by a distance $L = 0.600$ m. The radius of curvature of each mirror has a magnitude of 0.360 m. A light source is located a distance x from the concave mirror, as shown in **Fig. P34.87**. (a) What distance x will result in the rays from the source returning to the source after reflecting first from the convex mirror and then from the concave mirror? (b) Repeat part (a), but now let the rays reflect first from the concave mirror and then from the convex one.

Figure **P34.87**

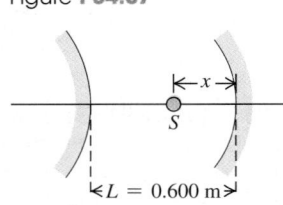

34.88 •• A screen is placed a distance d to the right of an object. A converging lens with focal length f is placed between the object and the screen. In terms of f, what is the smallest value d can have for an image to be in focus on the screen?

34.89 •• As shown in **Fig. P34.89**, the candle is at the center of curvature of the concave mirror, whose focal length is 10.0 cm.

The converging lens has a focal length of 32.0 cm and is 85.0 cm to the right of the candle. The candle is viewed looking through the lens from the right. The lens forms two images of the candle. The first is formed by light passing directly through the lens. The second image is formed from the light that goes from the candle to the mirror, is reflected, and then passes through the lens. (a) For each of these two images, draw a principal-ray diagram that locates the image. (b) For each image, answer the following questions: (i) Where is the image? (ii) Is the image real or virtual? (iii) Is the image erect or inverted with respect to the original object?

Figure **P34.89**

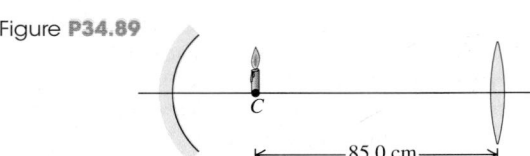

34.90 •• **Two Lenses in Contact.** (a) Prove that when two thin lenses with focal lengths f_1 and f_2 are placed *in contact*, the focal length f of the combination is given by the relationship

$$\frac{1}{f} = \frac{1}{f_1} + \frac{1}{f_2}$$

(b) A converging meniscus lens (see Fig. 34.32a) has an index of refraction of 1.55 and radii of curvature for its surfaces of magnitudes 4.50 cm and 9.00 cm. The concave surface is placed upward and filled with carbon tetrachloride (CCl_4), which has $n = 1.46$. What is the focal length of the CCl_4–glass combination?

34.91 ••• When an object is placed at the proper distance to the left of a converging lens, the image is focused on a screen 30.0 cm to the right of the lens. A diverging lens is now placed 15.0 cm to the right of the converging lens, and it is found that the screen must be moved 19.2 cm farther to the right to obtain a sharp image. What is the focal length of the diverging lens?

34.92 •• (a) Repeat the derivation of Eq. (34.19) for the case in which the lens is totally immersed in a liquid of refractive index n_{liq}. (b) A lens is made of glass that has refractive index 1.60. In air, the lens has focal length $+18.0$ cm. What is the focal length of this lens if it is totally immersed in a liquid that has refractive index 1.42?

34.93 ••• A convex spherical mirror with a focal length of magnitude 24.0 cm is placed 20.0 cm to the left of a plane mirror. An object 0.250 cm tall is placed midway between the surface of the plane mirror and the vertex of the spherical mirror. The spherical mirror forms multiple images of the object. Where are the two images of the object formed by the spherical mirror that are closest to the spherical mirror, and how tall is each image?

34.94 •• BIO **What Is the Smallest Thing We Can See?** The smallest object we can resolve with our eye is limited by the size of the light receptor cells in the retina. In order for us to distinguish any detail in an object, its image cannot be any smaller than a single retinal cell. Although the size depends on the type of cell (rod or cone), a diameter of a few microns (μm) is typical near the center of the eye. We shall model the eye as a sphere 2.50 cm in diameter with a single thin lens at the front and the retina at the rear, with light receptor cells 5.0 μm in diameter. (a) What is the smallest object you can resolve at a near point of 25 cm? (b) What angle is subtended by this object at the eye? Express your answer in units of minutes ($1° = 60$ min), and compare it with the typical experimental value of about 1.0 min. (*Note:* There are other limitations, such as the bending of light as it passes through the pupil, but we shall ignore them here.)

34.95 • Three thin lenses, each with a focal length of 40.0 cm, are aligned on a common axis; adjacent lenses are separated by 52.0 cm. Find the position of the image of a small object on the axis, 80.0 cm to the left of the first lens.

34.96 •• A camera with a 90-mm-focal-length lens is focused on an object 1.30 m from the lens. To refocus on an object 6.50 m from the lens, by how much must the distance between the lens and the sensor be changed? To refocus on the more distant object, is the lens moved toward or away from the sensor?

34.97 •• BIO In one form of cataract surgery the person's natural lens, which has become cloudy, is replaced by an artificial lens. The refracting properties of the replacement lens can be chosen so that the person's eye focuses on distant objects. But there is no accommodation, and glasses or contact lenses are needed for close vision. What is the power, in diopters, of the corrective contact lenses that will enable a person who has had such surgery to focus on the page of a book at a distance of 24 cm?

34.98 •• BIO **A Nearsighted Eye.** A certain very nearsighted person cannot focus on anything farther than 36.0 cm from the eye. Consider the simplified model of the eye described in Exercise 34.50. If the radius of curvature of the cornea is 0.75 cm when the eye is focusing on an object 36.0 cm from the cornea vertex and the indexes of refraction are as described in Exercise 34.50, what is the distance from the cornea vertex to the retina? What does this tell you about the shape of the nearsighted eye?

34.99 •• BIO A person with a near point of 85 cm, but excellent distant vision, normally wears corrective glasses. But he loses them while traveling. Fortunately, he has his old pair as a spare. (a) If the lenses of the old pair have a power of +2.25 diopters, what is his near point (measured from his eye) when he is wearing the old glasses if they rest 2.0 cm in front of his eye? (b) What would his near point be if his old glasses were contact lenses instead?

34.100 •• **The Galilean Telescope.** Figure P34.100 is a diagram of a *Galilean telescope*, or *opera glass*, with both the object and its final image at infinity. The image I serves as a virtual object for the eyepiece. The final image is virtual and erect. (a) Prove that the angular magnification is $M = -f_1/f_2$. (b) A Galilean telescope is to be constructed with the same objective lens as in Exercise 34.61. What focal length should the eyepiece have if this telescope is to have the same magnitude of angular magnification as the one in Exercise 34.61? (c) Compare the lengths of the telescopes.

Figure **P34.100**

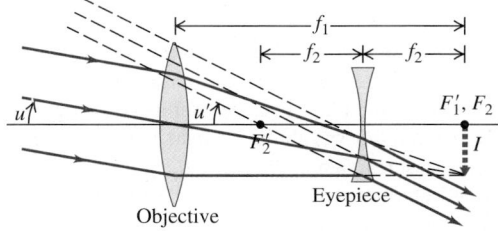

34.101 ••• **Focal Length of a Zoom Lens.** Figure P34.101 shows a simple version of a zoom lens. The converging lens has focal length f_1 and the diverging lens has focal length $f_2 = -|f_2|$. The two lenses are separated by a variable distance d that is always less than f_1 Also, the magnitude of the focal length of the diverging

lens satisfies the inequality $|f_2| > (f_1 - d)$. To determine the effective focal length of the combination lens, consider a bundle of parallel rays of radius r_0 entering the converging lens. (a) Show that the radius of the ray bundle decreases to $r_0' = r_0(f_1 - d)/f_1$ at the point that it enters the diverging lens. (b) Show that the final image I' is formed a distance $s_2' = |f_2|(f_1 - d)/(|f_2| - f_1 + d)$ to the right of the diverging lens. (c) If the rays that emerge from the diverging lens and reach the final image point are extended backward to the left of the diverging lens, they will eventually expand to the original radius r_0 at some point Q. The distance from the final image I' to the point Q is the *effective focal length f* of the lens combination; if the combination were replaced by a single lens of focal length f placed at Q, parallel rays would still be brought to a focus at I'. Show that the effective focal length is given by $f = f_1|f_2|/(|f_2| - f_1 + d)$. (d) If $f_1 = 12.0$ cm, $f_2 = -18.0$ cm, and the separation d is adjustable between 0 and 4.0 cm, find the maximum and minimum focal lengths of the combination. What value of d gives $f = 30.0$ cm?

Figure **P34.101**

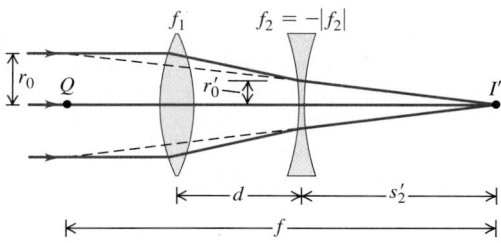

34.102 •• DATA In setting up an experiment for a high school biology lab, you use a concave spherical mirror to produce real images of a 4.00-mm-tall firefly. The firefly is to the right of the mirror, on the mirror's optic axis, and serves as a real object for the mirror. You want to determine how far the object must be from the mirror's vertex (that is, object distance s) to produce an image of a specified height. First you place a square of white cardboard to the right of the object and find what its distance from the vertex needs to be so that the image is sharply focused on it. Next you measure the height of the sharply focused images for five values of s. For each s value, you then calculate the lateral magnification m. You find that if you graph your data with s on the vertical axis and $1/m$ on the horizontal axis, then your measured points fall close to a straight line. (a) Explain why the data plotted this way should fall close to a straight line. (b) Use the graph in **Fig. P34.102** to calculate the focal length of the mirror. (c) How far from the mirror's vertex should you place the object in order for the image to be real, 8.00 mm tall, and inverted? (d) According to Fig. P34.102, starting from the position that you calculated in part (c), should you move the object closer to the mirror or farther from it to increase the height of the inverted, real image? What distance should you move the object in order to increase the image height from 8.00 mm to 12.00 mm? (e) Explain why $1/m$ approaches zero as s approaches 25 cm. Can you produce a sharp image on the cardboard when $s = 25$ cm? (f) Explain why you can't see sharp images on the cardboard when $s < 25$ cm (and m is positive).

Figure **P34.102**

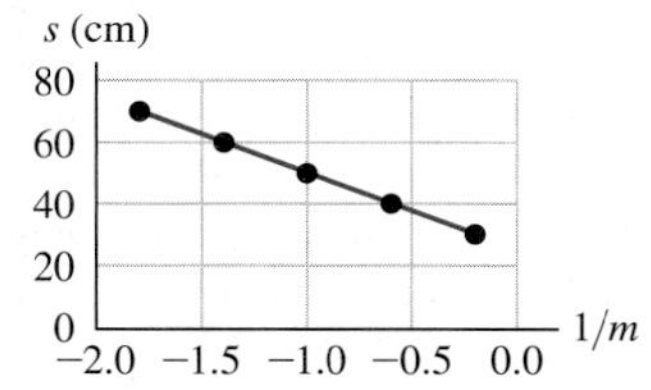

34.103 •• **DATA** It is your first day at work as a summer intern at an optics company. Your supervisor hands you a diverging lens and asks you to measure its focal length. You know that with a *converging* lens, you can measure the focal length by placing an object a distance s to the left of the lens, far enough from the lens for the image to be real, and viewing the image on a screen that is to the right of the lens. By adjusting the position of the screen until the image is in sharp focus, you can determine the image distance s' and then use Eq. (34.16) to calculate the focal length f of the lens. But this procedure won't work with a diverging lens—by itself, a diverging lens produces only virtual images, which can't be projected onto a screen. Therefore, to determine the focal length of a diverging lens, you do the following: First you take a *converging* lens and measure that, for an object 20.0 cm to the left of the lens, the image is 29.7 cm to the right of the lens. You then place a *diverging* lens 20.0 cm to the right of the converging lens and measure the final image to be 42.8 cm to the right of the converging lens. Suspecting some inaccuracy in measurement, you repeat the lens-combination measurement with the same object distance for the converging lens but with the diverging lens 25.0 cm to the right of the converging lens. You measure the final image to be 31.6 cm to the right of the converging lens. (a) Use both lens-combination measurements to calculate the focal length of the diverging lens. Take as your best experimental value for the focal length the average of the two values. (b) Which position of the diverging lens, 20.0 cm to the right or 25.0 cm to the right of the converging lens, gives the tallest image?

34.104 •• **DATA** The science museum where you work is constructing a new display. You are given a glass rod that is surrounded by air and was ground on its left-hand end to form a hemispherical surface there. You must determine the radius of curvature of that surface and the index of refraction of the glass. Remembering the optics portion of your physics course, you place a small object to the left of the rod, on the rod's optic axis, at a distance s from the vertex of the hemispherical surface. You measure the distance s' of the image from the vertex of the surface, with the image being to the right of the vertex. Your measurements are as follows:

s (cm)	22.5	25.0	30.0	35.0	40.0	45.0
s' (cm)	271.6	148.3	89.4	71.1	60.8	53.2

Recalling that the object–image relationships for thin lenses and spherical mirrors include reciprocals of distances, you plot your data as $1/s'$ versus $1/s$. (a) Explain why your data points plotted this way lie close to a straight line. (b) Use the slope and y-intercept of the best-fit straight line to your data to calculate the index of refraction of the glass and the radius of curvature of the hemispherical surface of the rod. (c) Where is the image if the object distance is 15.0 cm?

CHALLENGE PROBLEMS

34.105 ••• **CALC** (a) For a lens with focal length f, find the smallest distance possible between the object and its real image. (b) Graph the distance between the object and the real image as a function of the distance of the object from the lens. Does your graph agree with the result you found in part (a)?

34.106 ••• **An Object at an Angle.** A 16.0-cm-long pencil is placed at a 45.0° angle, with its center 15.0 cm above the optic axis and 45.0 cm from a lens with a 20.0-cm focal length as shown in **Fig. P34.106**. (Note that the figure is not drawn to scale.) Assume that the diameter of the lens is large enough for the paraxial approximation to be valid. (a) Where is the image of the pencil? (Give the location of the images of the points A, B, and C on the object, which are located at the eraser, point, and center of the pencil, respectively.) (b) What is the length of the image (that is, the distance between the images of points A and B)? (c) Show the orientation of the image in a sketch.

Figure **P34.106**

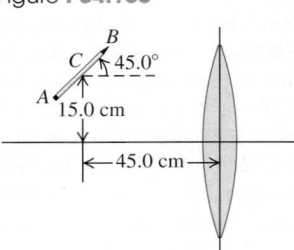

34.107 ••• **BIO** People with normal vision cannot focus their eyes underwater if they aren't wearing a face mask or goggles and there is water in contact with their eyes (see Discussion Question Q34.23). (a) Why not? (b) With the simplified model of the eye described in Exercise 34.50, what corrective lens (specified by focal length as measured in air) would be needed to enable a person underwater to focus an infinitely distant object? (Be careful—the focal length of a lens underwater is *not* the same as in air! See Problem 34.92. Assume that the corrective lens has a refractive index of 1.62 and that the lens is used in eyeglasses, not goggles, so there is water on both sides of the lens. Assume that the eyeglasses are 2.00 cm in front of the eye.)

PASSAGE PROBLEMS

BIO AMPHIBIAN VISION. The eyes of amphibians such as frogs have a much flatter cornea but a more strongly curved (almost spherical) lens than do the eyes of air-dwelling mammals. In mammalian eyes, the shape (and therefore the focal length) of the lens changes to enable the eye to focus at different distances. In amphibian eyes, the shape of the lens doesn't change. Amphibians focus on objects at different distances by using specialized muscles to move the lens closer to or farther from the retina, like the focusing mechanism of a camera. In air, most frogs are nearsighted; correcting the distance vision of a typical frog in air would require contact lenses with a power of about –6.0 D.

34.108 A frog can see an insect clearly at a distance of 10 cm. At that point the effective distance from the lens to the retina is 8 mm. If the insect moves 5 cm farther from the frog, by how much and in which direction does the lens of the frog's eye have to move to keep the insect in focus? (a) 0.02 cm, toward the retina; (b) 0.02 cm, away from the retina; (c) 0.06 cm, toward the retina; (d) 0.06 cm, away from the retina.

34.109 What is the farthest distance at which a typical "nearsighted" frog can see clearly in air? (a) 12 m; (b) 6.0 m; (c) 80 cm; (d) 17 cm.

34.110 Given that frogs are nearsighted in air, which statement is most likely to be true about their vision in water? (a) They are even more nearsighted; because water has a higher index of refraction than air, a frog's ability to focus light increases in water. (b) They are less nearsighted, because the cornea is less effective at refracting light in water than in air. (c) Their vision is no different, because only structures that are internal to the eye can affect the eye's ability to focus. (d) The images projected on the retina are no longer inverted, because the eye in water functions as a diverging lens rather than a converging lens.

34.111 To determine whether a frog can judge distance by means of the amount its lens must move to focus on an object, researchers covered one eye with an opaque material. An insect was placed in front of the frog, and the distance that the frog snapped its tongue out to catch the insect was measured with high-speed video. The experiment was repeated with a contact lens over the eye to determine whether the frog could correctly judge the distance under these conditions. If such an experiment is performed twice, once with a lens of power –9 D and once with a lens of power –15 D, in which case does the frog have to focus at a shorter distance, and why? (a) With the –9-D lens; because the lenses are diverging, the lens with the longer focal length creates an image that is closer to the frog. (b) With the –15-D lens; because the lenses are diverging, the lens with the shorter focal length creates an image that is closer to the frog. (c) With the –9-D lens; because the lenses are converging, the lens with the longer focal length creates a larger real image. (d) With the –15-D lens; because the lenses are converging, the lens with the shorter focal length creates a larger real image.

Answers

Chapter Opening Question ?

(ii) A magnifying lens (simple magnifier) produces a virtual image with a large angular size that is infinitely far away, so you can see it in sharp focus with your eyes relaxed. (A surgeon would not appreciate having to strain her eyes while working.) The object should be at the focal point of the lens, so the object and lens are separated by one focal length. The distance from magnifier to eye is not crucial.

Test Your Understanding Questions

34.1 (iv) When you are a distance s from the mirror, your image is a distance s on the other side of the mirror and the distance from you to your image is $2s$. As you move toward the mirror, the distance $2s$ changes at twice the rate of the distance s, so your image moves toward you at speed $2v$.

34.2 (a) concave, (b) (ii) A convex mirror always produces an erect image, but that image is smaller than the object (see Fig. 34.16b). Hence a concave mirror must be used. The image will be erect and enlarged only if the distance from the object (your face) to the mirror is less than the focal length of the mirror, as in Fig. 34.20d.

34.3 no The sun is very far away, so the object distance is essentially infinite: $s = \infty$ and $1/s = 0$. Material a is air ($n_a = 1.00$) and material b is water ($n_b = 1.33$), so the image position s' is

$$\frac{n_a}{s} + \frac{n_b}{s'} = \frac{n_b - n_a}{R} \quad \text{or} \quad 0 + \frac{1.33}{s'} = \frac{1.33 - 1.00}{R}$$

$$s' = \frac{1.33}{0.33}R = 4.0R$$

The image would be formed 4.0 drop radii from the front surface of the drop. But since each drop is only a part of a complete sphere, the distance from the front to the back of the drop is less than $2R$. Thus the rays of sunlight never reach the image point, and the drops do not form an image of the sun on the leaf. The rays are nonetheless concentrated and can cause damage to the leaf.

34.4 A and C When rays A and D are extended backward, they pass through focal point F_2; thus, before they passed through the lens, they were parallel to the optic axis. The figures show that ray A emanated from point Q, but ray D did not. Ray B is parallel to the optic axis, so before it passed through the lens, it was directed toward focal point F_1. Hence it cannot have come from point Q. Ray C passes through the center of the lens and hence is not deflected by its passage; tracing the ray backward shows that it emanates from point Q.

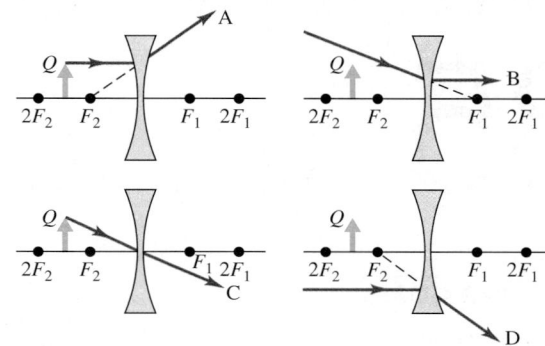

34.5 (iii) The smaller image area of the electronic sensor means that the angle of view is decreased for a given focal length. Individual objects make images of the same size in either case; when a smaller light-sensitive area is used, fewer images fit into the area and the field of view is narrower.

34.6 (iii) This lens is designed to correct for a type of astigmatism. Along the vertical axis, the lens is configured as a converging lens; along the horizontal axis, the lens is configured as a diverging lens. Hence the eye is hyperopic (see Fig. 34.46) for objects that are oriented vertically but myopic for objects that are oriented horizontally (see Fig. 34.47). Without correction, the eye focuses vertical objects behind the retina but horizontal objects in front of the retina.

34.7 (ii) The object must be held at the focal point, which is twice as far away if the focal length f is twice as great. Equation (34.24) shows that the angular magnification M is inversely proportional to f, so doubling the focal length makes M $\frac{1}{2}$ as great. To improve the magnification, you should use a magnifier with a *shorter* focal length.

34.8 (i) The objective lens of a microscope makes enlarged images of small objects, so the absolute value of its lateral magnification m is greater than 1. By contrast, the objective lens of a refracting telescope makes *reduced* images. For example, the moon is thousands of kilometers in diameter, but its image may fit on an electronic sensor a few centimeters across. Thus $|m|$ is much less than 1 for a refracting telescope. (In both cases m is negative because the objective makes an inverted image, which is why the question asks about the absolute value of m.)

Bridging Problem

(a) 29.9 cm to the left of the goblet
(b) 3.73 cm to the right of the goblet

? When white light shines downward on a thin, horizontal layer of oil, light waves reflected from the upper and lower surfaces of the film of oil interfere, producing vibrant colors. The color that you see reflected from a certain spot on the film depends on (i) the film thickness at that spot; (ii) the index of refraction of the oil; (iii) the index of refraction of the material below the oil; (iv) both (i) and (ii); (v) all of (i), (ii), and (iii).

35 INTERFERENCE

LEARNING GOALS

Looking forward at …

35.1 What happens when two waves combine, or interfere, in space.

35.2 How to understand the interference pattern formed by the interference of two coherent light waves.

35.3 How to calculate the intensity at various points in an interference pattern.

35.4 How interference occurs when light reflects from the two surfaces of a thin film.

35.5 How interference makes it possible to measure extremely small distances.

Looking back at …

14.2, 31.1 Phasors.

15.3, 15.6, 15.7 Wave number, wave superposition, standing waves on a string.

16.4 Standing sound waves.

32.1, 32.4, 32.5 Electromagnetic spectrum, wave intensity, standing electromagnetic waves.

An ugly black oil spot on the pavement can become a thing of beauty after a rain, when the oil reflects a rainbow of colors. Multicolored reflections can also be seen from the surfaces of soap bubbles and DVDs. How is it possible for colorless objects to produce these remarkable colors?

In our discussion of lenses, mirrors, and optical instruments we used the model of *geometric optics,* in which we represent light as *rays,* straight lines that are bent at a reflecting or refracting surface. But light is fundamentally a *wave,* and in some situations we have to consider its wave properties explicitly. If two or more light waves of the same frequency overlap at a point, the total effect depends on the *phases* of the waves as well as their amplitudes. The resulting patterns of light are a result of the *wave* nature of light and cannot be understood on the basis of rays. Optical effects that depend on the wave nature of light are grouped under the heading **physical optics.**

In this chapter we'll look at *interference* phenomena that occur when two waves combine. Effects that occur when *many* sources of waves are present are called *diffraction* phenomena; we'll study these in Chapter 36. In that chapter we'll see that diffraction effects occur whenever a wave passes through an aperture or around an obstacle. They are important in practical applications of physical optics such as diffraction gratings, x-ray diffraction, and holography.

While our primary concern is with light, interference and diffraction can occur with waves of *any* kind. As we go along, we'll point out applications to other types of waves such as sound and water waves.

35.1 INTERFERENCE AND COHERENT SOURCES

As we discussed in Chapter 15, the term **interference** refers to any situation in which two or more waves overlap in space. When this occurs, the total wave at any point at any instant of time is governed by the **principle of superposition,** which we introduced in Section 15.6 in the context of waves on a string. This principle also applies to electromagnetic waves and is the most important principle in all of physical optics. The principle of superposition states:

> When two or more waves overlap, the resultant displacement at any point and at any instant is found by adding the instantaneous displacements that would be produced at the point by the individual waves if each were present alone.

(In some special situations, such as electromagnetic waves propagating in a crystal, this principle may not apply. A discussion of these is beyond our scope.)

We use the term "displacement" in a general sense. With waves on the surface of a liquid, we mean the actual displacement of the surface above or below its normal level. With sound waves, the term refers to the excess or deficiency of pressure. For electromagnetic waves, we usually mean a specific component of electric or magnetic field.

Interference in Two or Three Dimensions

We have already discussed one important case of interference, in which two identical waves propagating in opposite directions combine to produce a *standing wave*. We saw this in Section 15.7 for transverse waves on a string and in Section 16.4 for longitudinal waves in a fluid filling a pipe; we described the same phenomenon for electromagnetic waves in Section 32.5. In all of these cases the waves propagated along only a single axis: along a string, along the length of a fluid-filled pipe, or along the propagation direction of an electromagnetic plane wave. But light waves can (and do) travel in *two* or *three* dimensions, as can any kind of wave that propagates in a two- or three-dimensional medium. In this section we'll see what happens when we combine waves that spread out in two or three dimensions from a pair of identical wave sources.

Interference effects are most easily seen when we combine *sinusoidal* waves with a single frequency f and wavelength λ. **Figure 35.1** shows a "snapshot" of a *single* source S_1 of sinusoidal waves and some of the wave fronts produced by this source. The figure shows only the wave fronts corresponding to wave *crests,* so the spacing between successive wave fronts is one wavelength. The material surrounding S_1 is uniform, so the wave speed is the same in all directions, and there is no refraction (and hence no bending of the wave fronts). If the waves are two-dimensional, like waves on the surface of a liquid, the circles in Fig. 35.1 represent circular wave fronts; if the waves propagate in three dimensions, the circles represent spherical wave fronts spreading away from S_1.

In optics, sinusoidal waves are characteristic of **monochromatic light** (light of a single color). While it's fairly easy to make water waves or sound waves of a single frequency, common sources of light *do not* emit monochromatic (single-frequency) light. For example, incandescent light bulbs and flames emit a continuous distribution of wavelengths. By far the most nearly monochromatic light source is the *laser*. An example is the helium–neon laser, which emits red light at 632.8 nm with a wavelength range of the order of ± 0.000001 nm, or about one part in 10^9. In this chapter and the next, we will assume that we are working with monochromatic waves (unless we explicitly state otherwise).

Constructive and Destructive Interference

Figure 35.2a (next page) shows two identical sources of monochromatic waves, S_1 and S_2. The two sources produce waves of the same amplitude and the same wavelength λ. In addition, the two sources are permanently *in phase;* they vibrate in unison. They might be two loudspeakers driven by the same amplifier, two radio antennas powered by the same transmitter, or two small slits in an opaque screen, illuminated by the same monochromatic light source. We will see that if

35.1 A "snapshot" of sinusoidal waves of frequency f and wavelength λ spreading out from source S_1 in all directions.

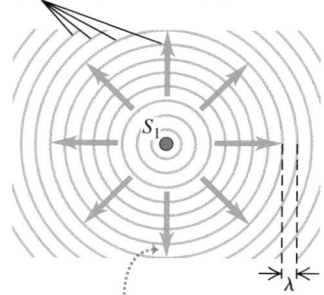

Wave fronts: crests of the wave (frequency f) separated by one wavelength λ

The wave fronts move outward from source S_1 at the wave speed $v = f\lambda$.

35.2 (a) A "snapshot" of sinusoidal waves spreading out from two coherent sources S_1 and S_2. Constructive interference occurs at point a (equidistant from the two sources) and (b) at point b. (c) Destructive interference occurs at point c.

(a) Two coherent wave sources separated by a distance 4λ

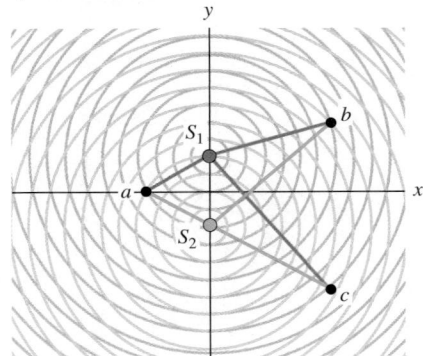

(b) Conditions for constructive interference: Waves interfere constructively if their path lengths differ by an integral number of wavelengths: $r_2 - r_1 = m\lambda$.

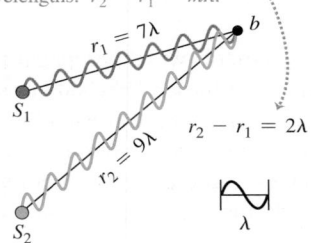

(c) Conditions for destructive interference: Waves interfere destructively if their path lengths differ by a half-integral number of wavelengths: $r_2 - r_1 = (m + \frac{1}{2})\lambda$.

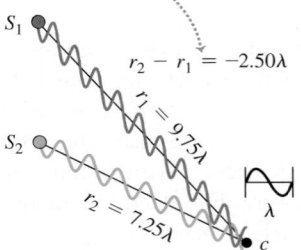

there were not a constant phase relationship between the two sources, the phenomena we are about to discuss would not occur. Two monochromatic sources of the same frequency and with a constant phase relationship (not necessarily in phase) are said to be **coherent.** We also use the term *coherent waves* (or, for light waves, *coherent light*) to refer to the waves emitted by two such sources.

If the waves emitted by the two coherent sources are *transverse,* like electromagnetic waves, then we will also assume that the wave disturbances produced by both sources have the same *polarization* (that is, they lie along the same line). For example, the sources S_1 and S_2 in Fig. 35.2a could be two radio antennas in the form of long rods oriented parallel to the z-axis (perpendicular to the plane of the figure); at any point in the xy-plane the waves produced by both antennas have $\vec{E}$ fields with only a z-component. Then we need only a single scalar function to describe each wave; this makes the analysis much easier.

We position the two sources of equal amplitude, equal wavelength, and (if the waves are transverse) the same polarization along the y-axis in Fig. 35.2a, equidistant from the origin. Consider a point a on the x-axis. From symmetry the two distances from S_1 to a and from S_2 to a are *equal*; waves from the two sources thus require equal times to travel to a. Hence waves that leave S_1 and S_2 in phase arrive at a in phase, and the total amplitude at a is *twice* the amplitude of each individual wave. This is true for *any* point on the x-axis.

Similarly, the distance from S_2 to point b is exactly two wavelengths *greater* than the distance from S_1 to b. A wave crest from S_1 arrives at b exactly two cycles earlier than a crest emitted at the same time from S_2, and again the two waves arrive in phase. As at point a, the total amplitude is the sum of the amplitudes of the waves from S_1 and S_2.

In general, when waves from two or more sources arrive at a point *in phase,* they reinforce each other: The amplitude of the resultant wave is the *sum* of the amplitudes of the individual waves. This is called **constructive interference** (Fig. 35.2b). Let the distance from S_1 to any point P be r_1, and let the distance from S_2 to P be r_2. For constructive interference to occur at P, the path difference $r_2 - r_1$ for the two sources must be an integral multiple of the wavelength λ:

$$r_2 - r_1 = m\lambda \qquad (m = 0, \pm 1, \pm 2, \pm 3, \ldots) \qquad \begin{array}{l}\text{(constructive} \\ \text{interference,} \\ \text{sources in phase)}\end{array} \qquad (35.1)$$

In Fig. 35.2a, points a and b satisfy Eq. (35.1) with $m = 0$ and $m = +2$, respectively.

Something different occurs at point c in Fig. 35.2a. At this point, the path difference $r_2 - r_1 = -2.50\lambda$, which is a *half-integral* number of wavelengths. Waves from the two sources arrive at point c exactly a half-cycle out of phase. A crest of one wave arrives at the same time as a crest in the opposite direction (a "trough") of the other wave (Fig. 35.2c). The resultant amplitude is the *difference* between the two individual amplitudes. If the individual amplitudes are equal, then the total amplitude is *zero*! This cancellation or partial cancellation of the individual waves is called **destructive interference.** The condition for destructive interference in the situation shown in Fig. 35.2a is

$$r_2 - r_1 = \left(m + \tfrac{1}{2}\right)\lambda \quad (m = 0, \pm 1, \pm 2, \pm 3, \dots)$$

(destructive interference, sources in phase) (35.2)

The path difference at point *c* in Fig. 35.2a satisfies Eq. (35.2) with $m = -3$.

Figure 35.3 shows the same situation as in Fig. 35.2a, but with red curves that show all positions where *constructive* interference occurs. On each curve, the path difference $r_2 - r_1$ is equal to an integer *m* times the wavelength, as in Eq. (35.1). These curves are called **antinodal curves.** They are directly analogous to *antinodes* in the standing-wave patterns described in Chapters 15 and 16 and Section 32.5. In a standing wave formed by interference between waves propagating in opposite directions, the antinodes are points at which the amplitude is maximum; likewise, the wave amplitude in the situation of Fig. 35.3 is maximum along the antinodal curves. Not shown in Fig. 35.3 are the **nodal curves,** which are the curves that show where *destructive* interference occurs in accordance with Eq. (35.2); these are analogous to the *nodes* in a standing-wave pattern. A nodal curve lies between each two adjacent antinodal curves in Fig. 35.3; one such curve, corresponding to $r_2 - r_1 = -2.50\lambda$, passes through point *c*.

In some cases, such as two loudspeakers or two radio-transmitter antennas, the interference pattern is three-dimensional. Think of rotating the color curves of Fig. 35.3 around the *y*-axis; then maximum constructive interference occurs at all points on the resulting surfaces of revolution.

CAUTION **Interference patterns are not standing waves** In the standing waves described in Sections 15.7, 16.4, and 32.5, the interference is between two waves propagating in opposite directions; there is *no* net energy flow in either direction (the energy in the wave is left "standing"). In the situations shown in Figs. 35.2a and 35.3, there is likewise a stationary pattern of antinodal and nodal curves, but there is a net flow of energy *outward* from the two sources. All that interference does is to "channel" the energy flow so that it is greatest along the antinodal curves and least along the nodal curves. ▌

For Eqs. (35.1) and (35.2) to hold, the two sources must have the same wavelength and must *always* be in phase. These conditions are rather easy to satisfy for sound waves. But with *light* waves there is no practical way to achieve a constant phase relationship (coherence) with two independent sources. This is because of the way light is emitted. In ordinary light sources, atoms gain excess energy by thermal agitation or by impact with accelerated electrons. Such an "excited" atom begins to radiate energy and continues until it has lost all the energy it can, typically in a time of the order of 10^{-8} s. The many atoms in a source ordinarily radiate in an unsynchronized and random phase relationship, and the light that is emitted from *two* such sources has no definite phase relationship.

However, the light from a single source can be split so that parts of it emerge from two or more regions of space, forming two or more *secondary sources.* Then any random phase change in the source affects these secondary sources equally and does not change their *relative* phase.

The distinguishing feature of light from a *laser* is that the emission of light from many atoms is synchronized in frequency and phase. As a result, the random phase changes mentioned above occur much less frequently. Definite phase relationships are preserved over correspondingly much greater lengths in the beam, and laser light is much more coherent than ordinary light.

TEST YOUR UNDERSTANDING OF SECTION 35.1 Consider a point in Fig. 35.3 on the positive *y*-axis above S_1. Does this point lie on (i) an antinodal curve; (ii) a nodal curve; or (iii) neither? (*Hint:* The distance between S_1 and S_2 is 4λ.) ▌

35.3 The same as Fig. 35.2a, but with red antinodal curves (curves of maximum amplitude) superimposed. All points on each curve satisfy Eq. (35.1) with the value of *m* shown. The nodal curves (not shown) lie between each adjacent pair of antinodal curves.

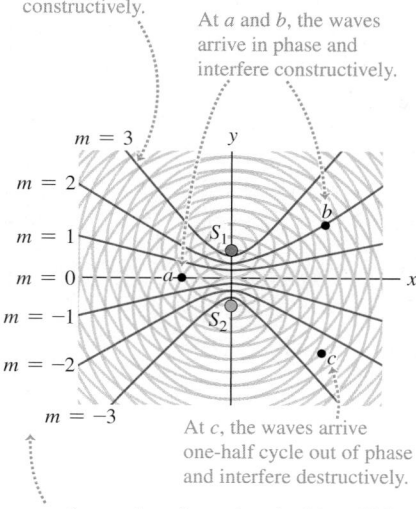

Antinodal curves (red) mark positions where the waves from S_1 and S_2 interfere constructively.

At *a* and *b*, the waves arrive in phase and interfere constructively.

At *c*, the waves arrive one-half cycle out of phase and interfere destructively.

$m =$ the number of wavelengths λ by which the path lengths from S_1 and S_2 differ.

BIO Application Phase Difference, Path Difference, and Localization in Human Hearing Your auditory system uses the phase differences between sounds received by your left and your right ears for *localization*—determining the direction from which the sounds are coming. For sound waves with frequencies lower than about 800 Hz (which are important in speech and music), the distance between your ears is less than a half-wavelength and the phase difference between the waves detected by each ear is less than a half cycle. Remarkably, your brain can detect this phase difference, determine the corresponding path difference, and use this information to localize the direction of the sound source.

35.4 The concepts of constructive interference and destructive interference apply to these water waves as well as to light waves and sound waves.

35.2 TWO-SOURCE INTERFERENCE OF LIGHT

The interference pattern produced by two coherent sources of *water* waves of the same wavelength can be readily seen in a ripple tank with a shallow layer of water (**Fig. 35.4**). This pattern is not directly visible when the interference is between *light* waves, since light traveling in a uniform medium cannot be seen. (Sunlight in a room is made visible by scattering from airborne dust.)

Figure 35.5a shows one of the earliest quantitative experiments to reveal the interference of light from two sources, first performed in 1800 by the English scientist Thomas Young. Let's examine this important experiment in detail. A light source (not shown) emits monochromatic light; however, this light is not suitable for use in an interference experiment because emissions from different parts of an ordinary source are not synchronized. To remedy this, the light is directed at a screen with a narrow slit S_0, 1 μm or so wide. The light emerging from the slit originated from only a small region of the light source; thus slit S_0 behaves more nearly like the idealized source shown in Fig. 35.1. (In modern versions of the experiment, a laser is used as a source of coherent light, and the slit S_0 isn't needed.) The light from slit S_0 falls on a screen with two other narrow slits S_1 and S_2, each 1 μm or so wide and a few tens or hundreds of micrometers apart. Cylindrical wave fronts spread out from slit S_0 and reach slits S_1 and S_2 *in phase* because they travel equal distances from S_0. The waves *emerging* from slits S_1 and S_2 are therefore also always in phase, so S_1 and S_2 are *coherent* sources. The interference of waves from S_1 and S_2 produces a pattern in space like that to the right of the sources in Figs. 35.2a and 35.3.

To visualize the interference pattern, a screen is placed so that the light from S_1 and S_2 falls on it (Fig. 35.5b). The screen will be most brightly illuminated at points P, where the light waves from the slits interfere constructively, and will be darkest at points where the interference is destructive.

To simplify the analysis of Young's experiment, we assume that the distance R from the slits to the screen is so large in comparison to the distance d between the slits that the lines from S_1 and S_2 to P are very nearly parallel, as in Fig. 35.5c.

35.5 (a) Young's experiment to show interference of light passing through two slits. A pattern of bright and dark areas appears on the screen (see Fig. 35.6). (b) Geometrical analysis of Young's experiment. For the case shown, $r_2 > r_1$ and both y and θ are positive. If point P is on the other side of the screen's center, $r_2 < r_1$ and both y and θ are negative. (c) Approximate geometry when the distance R to the screen is much greater than the distance d between the slits.

(a) Interference of light waves passing through two slits

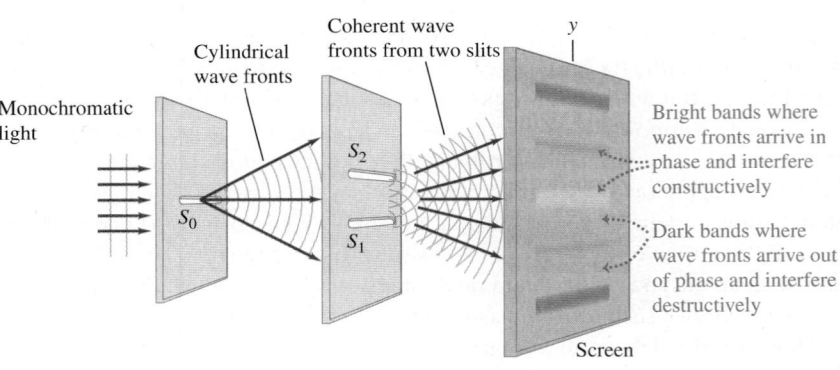

Bright bands where wave fronts arrive in phase and interfere constructively

Dark bands where wave fronts arrive out of phase and interfere destructively

(b) Actual geometry (seen from the side)

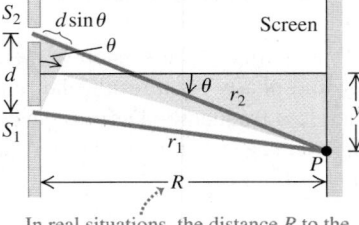

In real situations, the distance R to the screen is usually very much greater than the distance d between the slits ...

(c) Approximate geometry

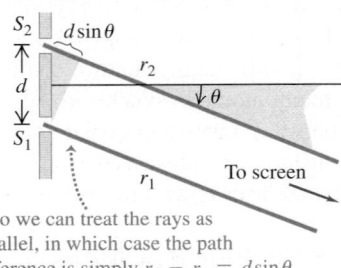

... so we can treat the rays as parallel, in which case the path difference is simply $r_2 - r_1 = d\sin\theta$.

This is usually the case for experiments with light; the slit separation is typically a few millimeters, while the screen may be a meter or more away. The difference in path length is then given by

PhET: Wave Interference

$$r_2 - r_1 = d \sin \theta \qquad (35.3)$$

where θ is the angle between a line from slits to screen (shown in blue in Fig. 35.5c) and the normal to the plane of the slits (shown as a thin black line).

Constructive and Destructive Two-Slit Interference

We found in Section 35.1 that constructive interference (reinforcement) occurs at points where the path difference is an integral number of wavelengths, $m\lambda$, where $m = 0, \pm 1, \pm 2, \pm 3, \ldots$. So the bright regions on the screen in Fig. 35.5a occur at angles θ for which

Constructive interference, two slits:

Distance between slits   Wavelength

$$d \sin \theta = m\lambda \qquad (m = 0, \pm 1, \pm 2, \ldots) \qquad (35.4)$$

Angle of line from slits to mth bright region on screen

Similarly, destructive interference (cancellation) occurs, forming dark regions on the screen, at points for which the path difference is a half-integral number of wavelengths, $(m + \frac{1}{2})\lambda$:

Destructive interference, two slits:

Distance between slits   Wavelength

$$d \sin \theta = (m + \tfrac{1}{2})\lambda \qquad (m = 0, \pm 1, \pm 2, \ldots) \qquad (35.5)$$

Angle of line from slits to mth dark region on screen

Thus the pattern on the screen of Figs. 35.5a and 35.5b is a succession of bright and dark bands, or **interference fringes,** parallel to the slits S_1 and S_2. A photograph of such a pattern is shown in **Fig. 35.6**.

We can derive an expression for the positions of the centers of the bright bands on the screen. In Fig. 35.5b, y is measured from the center of the pattern, corresponding to the distance from the center of Fig. 35.6. Let y_m be the distance from the center of the pattern ($\theta = 0$) to the center of the mth bright band. Let θ_m be the corresponding value of θ; then

$$y_m = R \tan \theta_m$$

In such experiments, the distances y_m are often much smaller than the distance R from the slits to the screen. Hence θ_m is very small, $\tan \theta_m \approx \sin \theta_m$, and

$$y_m = R \sin \theta_m$$

Combining this with Eq. (35.4), we find that *for small angles only,*

Constructive interference, Young's experiment (small angles only):

Position of mth bright band   Wavelength

$$y_m = R \frac{m\lambda}{d} \qquad (m = 0, \pm 1, \pm 2, \ldots) \qquad (35.6)$$

Distance from slits to screen   Distance between slits

We can measure R and d, as well as the positions y_m of the bright fringes, so this experiment provides a direct measurement of the wavelength λ. Young's experiment was in fact the first direct measurement of wavelengths of light.

The distance between adjacent bright bands in the pattern is *inversely* proportional to the distance d between the slits. The closer together the slits are, the more the pattern spreads out. When the slits are far apart, the bands in the pattern are closer together.

35.6 Photograph of interference fringes produced on a screen in Young's double-slit experiment. The center of the pattern is a bright band corresponding to $m = 0$ in Eq. (35.4); this point on the screen is equidistant from the two slits.

m (constructive interference, bright regions)	$m + 1/2$ (destructive interference, dark regions)
	$\leftarrow 11/2$
$5 \rightarrow$	$\leftarrow 9/2$
$4 \rightarrow$	$\leftarrow 7/2$
$3 \rightarrow$	$\leftarrow 5/2$
$2 \rightarrow$	$\leftarrow 3/2$
$1 \rightarrow$	$\leftarrow 1/2$
$0 \rightarrow$	
$-1 \rightarrow$	$\leftarrow -1/2$
$-2 \rightarrow$	$\leftarrow -3/2$
$-3 \rightarrow$	$\leftarrow -5/2$
$-4 \rightarrow$	$\leftarrow -7/2$
$-5 \rightarrow$	$\leftarrow -9/2$
	$\leftarrow -11/2$

CAUTION **Equation (35.6) is for small angles only** While Eqs. (35.4) and (35.5) are valid for any angle, Eq. (35.6) is valid for *small* angles only. It can be used *only* if the distance R from slits to screen is much greater than the slit separation d and if R is much greater than the distance y_m from the center of the interference pattern to the mth bright band. █

While we have described the experiment that Young performed with visible light, the results given in Eqs. (35.4) and (35.5) are valid for *any* type of wave, provided that the resultant wave from two coherent sources is detected at a point that is far away in comparison to the separation d.

EXAMPLE 35.1   TWO-SLIT INTERFERENCE

Figure 35.7 shows a two-slit interference experiment in which the slits are 0.200 mm apart and the screen is 1.00 m from the slits. The $m = 3$ bright fringe in the figure is 9.49 mm from the central fringe. Find the wavelength of the light.

35.7 Using a two-slit interference experiment to measure the wavelength of light.

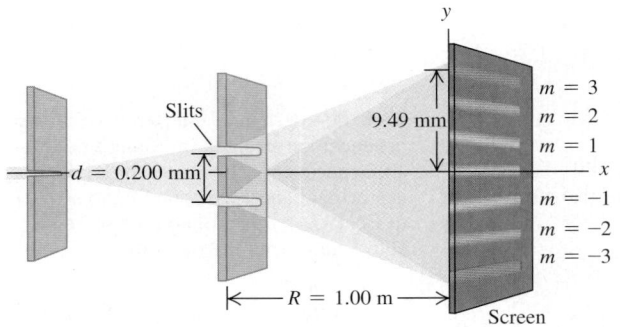

SOLUTION

IDENTIFY and SET UP: Our target variable in this two-slit interference problem is the wavelength λ. We are given the slit separation $d = 0.200$ mm, the distance from slits to screen $R = 1.00$ m, and the distance $y_3 = 9.49$ mm on the screen from the center of the interference pattern to the $m = 3$ bright fringe. We may use Eq. (35.6) to find λ, since the value of R is so much greater than the values of d or y_3.

EXECUTE: We solve Eq. (35.6) for λ for the case $m = 3$:

$$\lambda = \frac{y_m d}{mR} = \frac{(9.49 \times 10^{-3} \text{ m})(0.200 \times 10^{-3} \text{ m})}{(3)(1.00 \text{ m})}$$

$$= 633 \times 10^{-9} \text{ m} = 633 \text{ nm}$$

EVALUATE: This bright fringe could also correspond to $m = -3$. Can you show that this gives the same result for λ?

EXAMPLE 35.2   BROADCAST PATTERN OF A RADIO STATION

It is often desirable to radiate most of the energy from a radio transmitter in particular directions rather than uniformly in all directions. Pairs or rows of antennas are often used to produce the desired radiation pattern. As an example, consider two identical vertical antennas 400 m apart, operating at 1500 kHz = 1.5×10^6 Hz (near the top end of the AM broadcast band) and oscillating in phase. At distances much greater than 400 m, in what directions is the intensity from the two antennas greatest?

SOLUTION

IDENTIFY and SET UP: The antennas, shown in **Fig. 35.8**, correspond to sources S_1 and S_2 in Fig. 35.5. Hence we can apply the ideas of two-slit interference to this problem. Since the resultant wave is detected at distances much greater than $d = 400$ m, we may use Eq. (35.4) to give the directions of the intensity maxima, the values of θ for which the path difference is zero or a whole number of wavelengths.

EXECUTE: The wavelength is $\lambda = c/f = 200$ m. From Eq. (35.4) with $m = 0$, ± 1, and ± 2, the intensity maxima are given by

$$\sin \theta = \frac{m\lambda}{d} = \frac{m(200 \text{ m})}{400 \text{ m}} = \frac{m}{2} \qquad \theta = 0, \pm 30°, \pm 90°$$

In this example, values of m greater than 2 or less than -2 give values of $\sin \theta$ greater than 1 or less than -1, which is impossible. There is *no* direction for which the path difference is three or more wavelengths, so values of m of ± 3 or beyond have no meaning in this example.

35.8 Two radio antennas broadcasting in phase. The purple arrows indicate the directions of maximum intensity. The waves that are emitted toward the lower half of the figure are not shown.

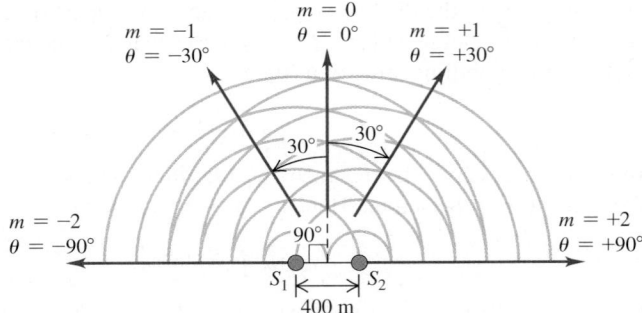

EVALUATE: We can check our result by calculating the angles for *minimum* intensity, using Eq. (35.5). There should be one intensity minimum between each pair of intensity maxima, just as in Fig. 35.6. From Eq. (35.5), with $m = -2$, -1, 0, and 1,

$$\sin \theta = \frac{(m + \frac{1}{2})\lambda}{d} = \frac{m + \frac{1}{2}}{2} \qquad \theta = \pm 14.5°, \pm 48.6°$$

These angles fall between the angles for intensity maxima, as they should. The angles are not small, so the angles for the minima are *not* exactly halfway between the angles for the maxima.

TEST YOUR UNDERSTANDING OF SECTION 35.2   You shine a tunable laser (whose wavelength can be adjusted by turning a knob) on a pair of closely spaced slits. The light emerging from the two slits produces an interference pattern on a screen like that shown in Fig. 35.6. If you adjust the wavelength so that the laser light changes from red to blue, how will the spacing between bright fringes change? (i) The spacing increases; (ii) the spacing decreases; (iii) the spacing is unchanged; (iv) not enough information is given to decide. ▮

35.3 INTENSITY IN INTERFERENCE PATTERNS

In Section 35.2 we found the positions of maximum and minimum intensity in a two-source interference pattern. Let's now see how to find the intensity at *any* point in the pattern. To do this, we have to combine the two sinusoidally varying fields (from the two sources) at a point P in the radiation pattern, taking proper account of the phase difference of the two waves at point P, which results from the path difference. The intensity is then proportional to the square of the resultant electric-field amplitude, as we learned in Section 32.4.

To calculate the intensity, we will assume, as in Section 35.2, that the waves from the two sources have equal amplitude E and the same polarization. This assumes that the sources are identical and ignores the slight amplitude difference caused by the unequal path lengths (the amplitude decreases with increasing distance from the source). From Eq. (32.29), each source by itself would give an intensity $\frac{1}{2}\epsilon_0 c E^2$ at point P. If the two sources are in phase, then the waves that arrive at P differ in phase by an amount ϕ that is proportional to the difference in their path lengths, $(r_2 - r_1)$. Then we can use the following expressions for the two electric fields superposed at P:

$$E_1(t) = E\cos(\omega t + \phi)$$
$$E_2(t) = E\cos\omega t$$

The superposition of the two fields at P is a sinusoidal function with some amplitude E_P that depends on E and the phase difference ϕ. First we'll work on finding the amplitude E_P if E and ϕ are known. Then we'll find the intensity I of the resultant wave, which is proportional to E_P^2. Finally, we'll relate the phase difference ϕ to the path difference, which is determined by the geometry of the situation.

Amplitude in Two-Source Interference

To add the two sinusoidal functions with a phase difference, we use the same *phasor* representation that we used for simple harmonic motion (see Section 14.2) and for voltages and currents in ac circuits (see Section 31.1). We suggest that you review these sections now. Each sinusoidal function is represented by a rotating vector (phasor) whose projection on the horizontal axis at any instant represents the instantaneous value of the sinusoidal function.

In **Fig. 35.9**, E_1 is the horizontal component of the phasor representing the wave from source S_1, and E_2 is the horizontal component of the phasor for the wave from S_2. As shown in the diagram, both phasors have the same magnitude E, but E_1 is *ahead* of E_2 in phase by an angle ϕ. Both phasors rotate counterclockwise with constant angular speed ω, and the sum of the projections on the horizontal axis at any time gives the instantaneous value of the total E field at point P. Thus the amplitude E_P of the resultant sinusoidal wave at P is the magnitude of the dark red phasor in the diagram (labeled E_P); this is the *vector sum* of the other two phasors. To find E_P, we use the law of cosines and the trigonometric identity $\cos(\pi - \phi) = -\cos\phi$:

$$E_P^2 = E^2 + E^2 - 2E^2\cos(\pi - \phi)$$
$$= E^2 + E^2 + 2E^2\cos\phi$$

DATA *SPEAKS*

Two-Source Interference

When students were given a problem involving interference of waves from two sources, more than 34% gave an incorrect response. Common errors:

- Confusion about the kind of sources required to cause interference. For there to be a steady interference pattern from two wave sources, the two sources must be monochromatic, emit waves of the same frequency, and have a fixed phase relationship.

- Confusion about constructive and destructive interference. Constructive interference occurs at points where waves from two sources arrive in phase (the crest of one wave aligns with a crest of the other wave). Destructive interference occurs at points where waves from two sources arrive out of phase (the crest of one wave aligns with a trough of the other wave).

35.9 Phasor diagram for the superposition at a point P of two waves of equal amplitude E with a phase difference ϕ.

Then, using the identity $1 + \cos\phi = 2\cos^2(\phi/2)$, we obtain

$$E_P^2 = 2E^2(1 + \cos\phi) = 4E^2\cos^2\left(\frac{\phi}{2}\right)$$

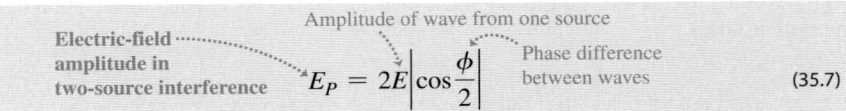

Electric-field amplitude in two-source interference $E_P = 2E\left|\cos\dfrac{\phi}{2}\right|$ — Amplitude of wave from one source — Phase difference between waves (35.7)

You can also obtain this result without using phasors.

When the two waves are in phase, $\phi = 0$ and $E_P = 2E$. When they are exactly a half-cycle out of phase, $\phi = \pi$ rad $= 180°$, $\cos(\phi/2) = \cos(\pi/2) = 0$, and $E_P = 0$. Thus the superposition of two sinusoidal waves with the same frequency and amplitude but with a phase difference yields a sinusoidal wave with the same frequency and an amplitude between zero and twice the individual amplitudes, depending on the phase difference.

Intensity in Two-Source Interference

To obtain the intensity I at point P, we recall from Section 32.4 that I is equal to the average magnitude of the Poynting vector, S_{av}. For a sinusoidal wave with electric-field amplitude E_P, this is given by Eq. (32.29) with E_{max} replaced by E_P. Thus we can express the intensity in several equivalent forms:

$$I = S_{av} = \frac{E_P^2}{2\mu_0 c} = \frac{1}{2}\sqrt{\frac{\epsilon_0}{\mu_0}}E_P^2 = \frac{1}{2}\epsilon_0 c E_P^2 \qquad (35.8)$$

The essential content of these expressions is that I is proportional to E_P^2. When we substitute Eq. (35.7) into the last expression in Eq. (35.8), we get

$$I = \frac{1}{2}\epsilon_0 c E_P^2 = 2\epsilon_0 c E^2 \cos^2\frac{\phi}{2} \qquad (35.9)$$

In particular, the *maximum* intensity I_0, which occurs at points where the phase difference is zero ($\phi = 0$), is

$$I_0 = 2\epsilon_0 c E^2$$

Note that the maximum intensity I_0 is *four times* (not twice) as great as the intensity $\frac{1}{2}\epsilon_0 c E^2$ from each individual source. Substituting the expression for I_0 into Eq. (35.9), we find

Intensity in two-source interference $I = I_0\cos^2\dfrac{\phi}{2}$ — Maximum intensity — Phase difference between waves (35.10)

The intensity depends on the phase difference ϕ and varies between I_0 and zero. If we average Eq. (35.10) over all possible phase differences, the result is $I_0/2 = \epsilon_0 c E^2$ [the average of $\cos^2(\phi/2)$ is $\frac{1}{2}$]. This is just twice the intensity from each individual source, as we should expect. The total energy output from the two sources isn't changed by the interference effects, but the energy is redistributed (see Section 35.1).

Phase Difference and Path Difference

Our next task is to find the phase difference ϕ between the two fields at any point P. We know that ϕ is proportional to the difference in path length from the two sources to point P. When the path difference is one wavelength, the phase difference is one cycle, and $\phi = 2\pi$ rad $= 360°$. When the path difference is $\lambda/2$, $\phi = \pi$ rad $= 180°$, and so on. That is, the ratio of the phase difference ϕ to 2π is equal to the ratio of the path difference $r_2 - r_1$ to λ:

$$\frac{\phi}{2\pi} = \frac{r_2 - r_1}{\lambda}$$

Phase difference in two-source interference

Path difference   Wave number $= 2\pi/\lambda$

$$\phi = \frac{2\pi}{\lambda}(r_2 - r_1) = k(r_2 - r_1) \qquad (35.11)$$

Wavelength   Distance from source 2   Distance from source 1

We introduced the wave number $k = 2\pi/\lambda$ in Section 15.3.

If the material in the space between the sources and P is anything other than vacuum, we must use the wavelength *in the material* in Eq. (35.11). If λ_0 and k_0 are the wavelength and wave number, respectively, in vacuum and the material has index of refraction n, then

$$\lambda = \frac{\lambda_0}{n} \qquad \text{and} \qquad k = nk_0 \qquad (35.12)$$

Finally, if the point P is far away from the sources in comparison to their separation d, the path difference is given by Eq. (35.3):

$$r_2 - r_1 = d\sin\theta$$

Combining this with Eq. (35.11), we find

$$\phi = k(r_2 - r_1) = kd\sin\theta = \frac{2\pi d}{\lambda}\sin\theta \qquad (35.13)$$

When we substitute this into Eq. (35.10), we find

$$I = I_0\cos^2\left(\tfrac{1}{2}kd\sin\theta\right) = I_0\cos^2\left(\frac{\pi d}{\lambda}\sin\theta\right) \qquad \begin{array}{l}\text{(intensity far from} \\ \text{two sources)}\end{array} \qquad (35.14)$$

Maximum intensity occurs when the cosine has the values ± 1—that is, when

$$\frac{\pi d}{\lambda}\sin\theta = m\pi \qquad (m = 0, \pm 1, \pm 2, \dots)$$

or

$$d\sin\theta = m\lambda$$

in agreement with Eq. (35.4). You can also derive Eq. (35.5) for the zero-intensity directions from Eq. (35.14).

As we noted in Section 35.2, in experiments with light we visualize the interference pattern due to two slits by using a screen placed at a distance R from the slits. We can describe positions on the screen with the coordinate y; the positions of the bright fringes are given by Eq. (35.6), where ordinarily $y \ll R$. In that case, $\sin\theta$ is approximately equal to y/R, and we obtain the following expressions for the intensity at *any* point on the screen as a function of y:

$$I = I_0\cos^2\left(\frac{kdy}{2R}\right) = I_0\cos^2\left(\frac{\pi dy}{\lambda R}\right) \qquad \begin{array}{l}\text{(intensity in two-slit} \\ \text{interference)}\end{array} \qquad (35.15)$$

35.10 Intensity distribution in the interference pattern from two identical slits.

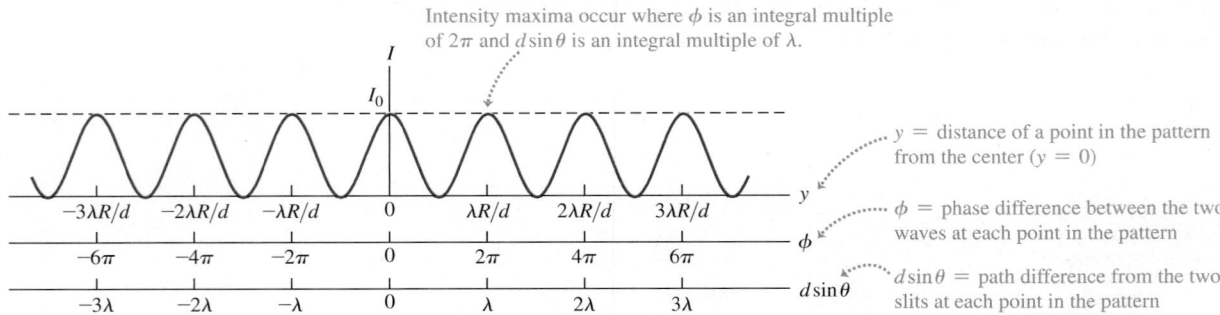

Intensity maxima occur where ϕ is an integral multiple of 2π and $d\sin\theta$ is an integral multiple of λ.

y = distance of a point in the pattern from the center ($y = 0$)

ϕ = phase difference between the two waves at each point in the pattern

$d\sin\theta$ = path difference from the two slits at each point in the pattern

Figure 35.10 shows a graph of Eq. (35.15); we can compare this with the photographically recorded pattern of Fig. 35.6. All peaks in Fig. 35.10 have the same intensity, while those in Fig. 35.6 fade off as we go away from the center. We'll explore the reasons for this variation in peak intensity in Chapter 36.

EXAMPLE 35.3 **A DIRECTIONAL TRANSMITTING ANTENNA ARRAY**

Suppose the two identical radio antennas of Fig. 35.8 are moved to be only 10.0 m apart and the broadcast frequency is increased to $f = 60.0$ MHz. At a distance of 700 m from the point midway between the antennas and in the direction $\theta = 0$ (see Fig. 35.8), the intensity is $I_0 = 0.020$ W/m². At this same distance, find (a) the intensity in the direction $\theta = 4.0°$; (b) the direction near $\theta = 0$ for which the intensity is $I_0/2$; and (c) the directions in which the intensity is zero.

SOLUTION

IDENTIFY and SET UP: This problem involves the intensity distribution as a function of angle. Because the 700-m distance from the antennas to the point at which the intensity is measured is much greater than the distance $d = 10.0$ m between the antennas, the amplitudes of the waves from the two antennas are very nearly equal. Hence we can use Eq. (35.14) to relate intensity I and angle θ.

EXECUTE: The wavelength is $\lambda = c/f = 5.00$ m. The spacing $d = 10.0$ m between the antennas is just twice the wavelength (as was the case in Example 35.2), so $d/\lambda = 2.00$ and Eq. (35.14) becomes

$$I = I_0\cos^2\left(\frac{\pi d}{\lambda}\sin\theta\right) = I_0\cos^2[(2.00\pi \text{ rad})\sin\theta]$$

(a) When $\theta = 4.0°$,

$$I = I_0\cos^2[(2.00\pi \text{ rad})\sin 4.0°] = 0.82I_0$$

$$= (0.82)(0.020 \text{ W/m}^2) = 0.016 \text{ W/m}^2$$

(b) The intensity I equals $I_0/2$ when the cosine in Eq. (35.14) has the value $\pm 1/\sqrt{2}$. The smallest angles at which this occurs correspond to $2.00\pi\sin\theta = \pm\pi/4$ rad, so $\sin\theta = \pm(1/8.00) = \pm 0.125$ and $\theta = \pm 7.2°$.

(c) The intensity is zero when $\cos[(2.00\pi \text{ rad})\sin\theta] = 0$. This occurs for $2.00\pi\sin\theta = \pm\pi/2, \pm 3\pi/2, \pm 5\pi/2, \ldots$, or $\sin\theta = \pm 0.250, \pm 0.750, \pm 1.25, \ldots$. Values of $\sin\theta$ greater than 1 have no meaning, so the answers are

$$\theta = \pm 14.5°, \pm 48.6°$$

EVALUATE: The condition in part (b) that $I = I_0/2$, so that $(2.00\pi \text{ rad})\sin\theta = \pm\pi/4$ rad, is also satisfied when $\sin\theta = \pm 0.375, \pm 0.625$, or ± 0.875 so that $\theta = \pm 22.0°, \pm 38.7°$, or $\pm 61.0°$. (Can you verify this?) It would be incorrect to include these angles in the solution, however, because the problem asked for the angle *near* $\theta = 0$ at which $I = I_0/2$. These additional values of θ aren't the ones we're looking for.

TEST YOUR UNDERSTANDING OF SECTION 35.3 A two-slit interference experiment uses coherent light of wavelength 5.00×10^{-7} m. Rank the following points in the interference pattern according to the intensity at each point, from highest to lowest. (i) A point that is closer to one slit than the other by 4.00×10^{-7} m; (ii) a point where the light waves received from the two slits are out of phase by 4.00 rad; (iii) a point that is closer to one slit than the other by 7.50×10^{-7} m; (iv) a point where the light waves received by the two slits are out of phase by 2.00 rad. ∎

35.4 INTERFERENCE IN THIN FILMS

You often see bright bands of color when light reflects from a thin layer of oil floating on water or from a soap bubble (see the photograph that opens this chapter). These are the results of interference. Light waves are reflected from the front and back surfaces of such thin films, and constructive interference between the two reflected waves (with different path lengths) occurs in different places for different wavelengths. **Figure 35.11a** shows the situation. Light shining on the upper surface of a thin film with thickness t is partly reflected at the upper surface (path abc). Light *transmitted* through the upper surface is partly reflected at the lower surface (path $abdef$). The two reflected waves come together at point P on the retina of the eye. Depending on the phase relationship, they may interfere constructively or destructively. Different colors have different wavelengths, so the interference may be constructive for some colors and destructive for others. That's why we see colored patterns in the photograph that opens this chapter (which shows a thin film of oil floating on water) and in Fig. 35.11b (which shows thin films of soap solution that make up the bubble walls). The complex shapes of the colored patterns result from variations in the thickness of the film.

Thin-Film Interference and Phase Shifts During Reflection

Let's look at a simplified situation in which *monochromatic* light reflects from two nearly parallel surfaces at nearly normal incidence. **Figure 35.12** shows two plates of glass separated by a thin wedge, or film, of air. We want to consider interference between the two light waves reflected from the surfaces adjacent to the air wedge. (Reflections also occur at the top surface of the upper plate and the bottom surface of the lower plate; to keep our discussion simple, we won't include these.) The situation is the same as in Fig. 35.11a except that the film (wedge) thickness is not uniform. The path difference between the two waves is just twice the thickness t of the air wedge at each point. At points where $2t$ is an integer number of wavelengths, we expect to see constructive interference and a bright area; where it is a half-integer number of wavelengths, we expect to see destructive interference and a dark area. Where the plates are in contact, there is practically *no* path difference, and we expect a bright area.

When we carry out the experiment, the bright and dark fringes appear, but they are interchanged! Along the line where the plates are in contact, we find a *dark* fringe, not a bright one. This suggests that one or the other of the reflected waves has undergone a half-cycle phase shift during its reflection. In that case the two waves that are reflected at the line of contact are a half-cycle out of phase even though they have the same path length.

In fact, this phase shift can be predicted from Maxwell's equations and the electromagnetic nature of light. The details of the derivation are beyond our scope, but here is the result. Suppose a light wave with electric-field amplitude E_i is traveling in an optical material with index of refraction n_a. It strikes, at normal incidence, an interface with another optical material with index n_b. The amplitude E_r of the wave reflected from the interface is given by

$$E_r = \frac{n_a - n_b}{n_a + n_b} E_i \qquad \text{(normal incidence)} \qquad (35.16)$$

This result shows that the incident and reflected amplitudes have the same sign when n_a is larger than n_b and opposite signs when n_b is larger than n_a. Because amplitudes must always be positive or zero, a *negative* value means that the

35.11 (a) A diagram and (b) a photograph showing interference of light reflected from a thin film.

(a) Interference between rays reflected from the two surfaces of a thin film

Light reflected from the upper and lower surfaces of the film comes together in the eye at P and undergoes interference.

Some colors interfere constructively and others destructively, creating the color bands we see.

(b) Colorful reflections from a soap bubble

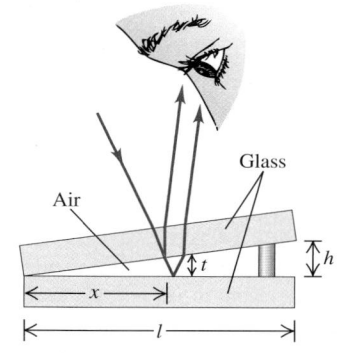

35.12 Interference between light waves reflected from the two sides of an air wedge separating two glass plates. The angles and the thickness of the air wedge have been exaggerated for clarity; in the text we assume that the light strikes the upper plate at normal incidence and that the distances h and t are much less than l.

35.13 Upper figures: electromagnetic waves striking an interface between optical materials at normal incidence (shown as a small angle for clarity). Lower figures: mechanical wave pulses on ropes.

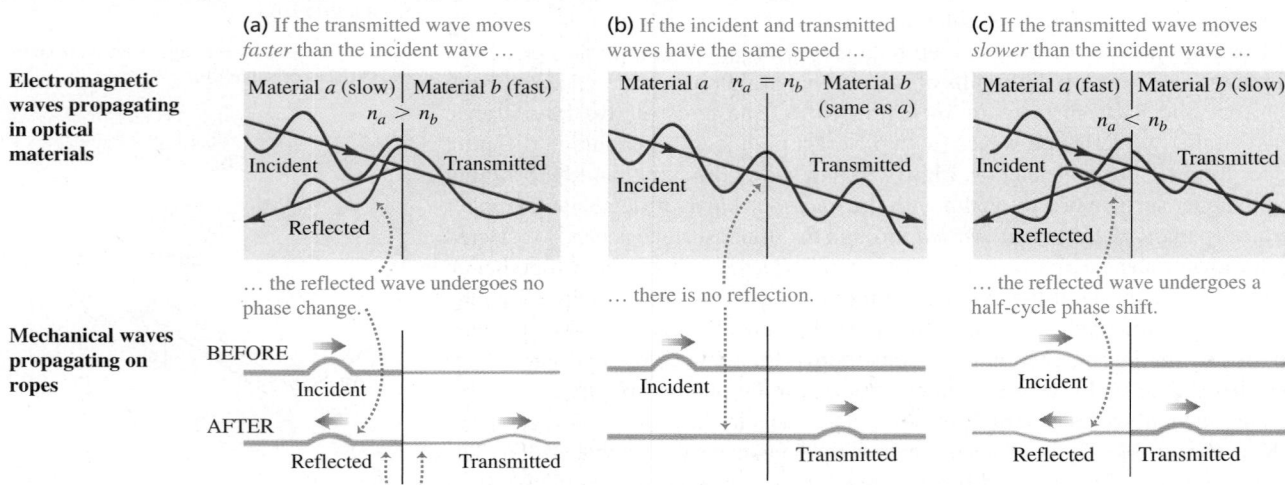

wave actually undergoes a half-cycle (180°) phase shift. **Figure 35.13** shows three possibilities:

Figure 35.13a: When $n_a > n_b$, light travels more slowly in the first material than in the second. In this case, E_r and E_i have the same sign, and the phase shift of the reflected wave relative to the incident wave is zero. This is analogous to reflection of a transverse mechanical wave on a heavy rope at a point where it is tied to a lighter rope.

Figure 35.13b: When $n_a = n_b$, the amplitude E_r of the reflected wave is zero. In effect there is *no* interface, so there is *no* reflected wave.

Figure 35.13c: When $n_a < n_b$, light travels more slowly in the second material than in the first. In this case, E_r and E_i have opposite signs, and the phase shift of the reflected wave relative to the incident wave is π rad (a half-cycle). This is analogous to reflection (with inversion) of a transverse mechanical wave on a light rope at a point where it is tied to a heavier rope.

Let's compare with the situation of Fig. 35.12. For the wave reflected from the upper surface of the air wedge, n_a (glass) is greater than n_b, so this wave has zero phase shift. For the wave reflected from the lower surface, n_a (air) is less than n_b (glass), so this wave has a half-cycle phase shift. Waves that are reflected from the line of contact have no path difference to give additional phase shifts, and they interfere destructively; this is what we observe. You can use the above principle to show that for normal incidence, the wave reflected at point b in Fig. 35.11a is shifted by a half-cycle, while the wave reflected at d is not (if there is air below the film).

We can summarize this discussion mathematically. If the film has thickness t, the light is at normal incidence and has wavelength λ in the film; if neither or both of the reflected waves from the two surfaces have a half-cycle reflection phase shift, the conditions for constructive and destructive interference are

Constructive reflection	$2t = m\lambda \quad (m = 0, 1, 2, \dots)$	(35.17a)
(From thin film, no relative phase shift)	Thickness of film — Wavelength	
Destructive reflection	$2t = \left(m + \tfrac{1}{2}\right)\lambda \quad (m = 0, 1, 2, \dots)$	(35.17b)

If *one* of the two waves has a half-cycle reflection phase shift, the conditions for constructive and destructive interference are reversed:

Constructive reflection	$2t = \left(m + \frac{1}{2}\right)\lambda \quad (m = 0, 1, 2, \dots)$	(35.18a)
(From thin film, half-cycle phase shift)	Thickness of film $\qquad$ Wavelength	
Destructive reflection	$2t = m\lambda \quad (m = 0, 1, 2, \dots)$	(35.18b)

Thin and Thick Films

We have emphasized *thin* films in our discussion because of a principle we introduced in Section 35.1: In order for two waves to cause a steady interference pattern, the waves must be *coherent*, with a definite and constant phase relationship. The sun and light bulbs emit light in a stream of short bursts, each of which is only a few micrometers long (1 micrometer = $1\ \mu m = 10^{-6}$ m). If light reflects from the two surfaces of a thin film, the two reflected waves are part of the same burst (**Fig. 35.14a**). Hence these waves are coherent and interference occurs as we have described. If the film is too thick, however, the two reflected waves will belong to different bursts (Fig. 35.14b). There is no definite phase relationship between different light bursts, so the two waves are incoherent and there is no fixed interference pattern. That's why you see interference colors in light reflected from a soap bubble a few micrometers thick (see Fig. 35.11b), but you do *not* see such colors in the light reflected from a pane of window glass with a thickness of a few millimeters (a thousand times greater).

35.14 (a) Light reflecting from a thin film produces a steady interference pattern, but (b) light reflecting from a thick film does not.

(a) Light reflecting from a thin film

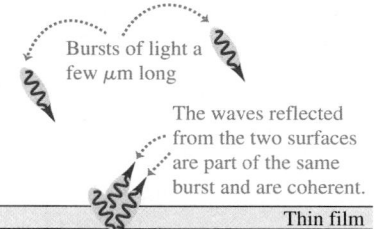

Bursts of light a few μm long

The waves reflected from the two surfaces are part of the same burst and are coherent.

Thin film

(b) Light reflecting from a thick film

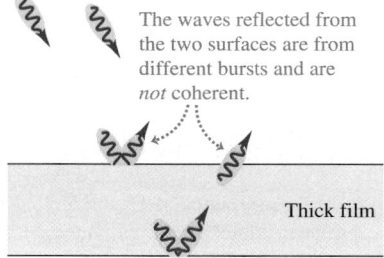

The waves reflected from the two surfaces are from different bursts and are *not* coherent.

Thick film

PROBLEM-SOLVING STRATEGY 35.1   INTERFERENCE IN THIN FILMS

IDENTIFY *the relevant concepts:* Problems with thin films involve interference of two waves, one reflected from the film's front surface and one reflected from the back surface. Typically you will be asked to relate the wavelength, the film thickness, and the index of refraction of the film.

SET UP *the problem* using the following steps:
1. Make a drawing showing the geometry of the film. Identify the materials that adjoin the film; their properties determine whether one or both of the reflected waves have a half-cycle phase shift.
2. Identify the target variable.

EXECUTE *the solution* as follows:
1. Apply the rule for phase changes to each reflected wave: There is a half-cycle phase shift when $n_b > n_a$ and none when $n_b < n_a$.

2. If *neither* reflected wave undergoes a phase shift, or if *both* do, use Eqs. (35.17). If only one reflected wave undergoes a phase shift, use Eqs. (35.18).
3. Solve the resulting equation for the target variable. Use the wavelength $\lambda = \lambda_0/n$ of light *in the film* in your calculations, where n is the index of refraction of the film. (For air, $n = 1.000$ to four-figure precision.)
4. If you are asked about the wave that is transmitted through the film, remember that minimum intensity in the reflected wave corresponds to maximum *transmitted* intensity, and vice versa.

EVALUATE *your answer:* Interpret your results by examining what would happen if the wavelength were changed or if the film had a different thickness.

EXAMPLE 35.4   THIN-FILM INTERFERENCE I

Suppose the two glass plates in Fig. 35.12 are two microscope slides 10.0 cm long. At one end they are in contact; at the other end they are separated by a piece of paper 0.0200 mm thick. What is the spacing of the interference fringes seen by reflection? Is the fringe at the line of contact bright or dark? Assume monochromatic light with a wavelength in air of $\lambda = \lambda_0 = 500$ nm.

35.15 Our sketch for this problem.

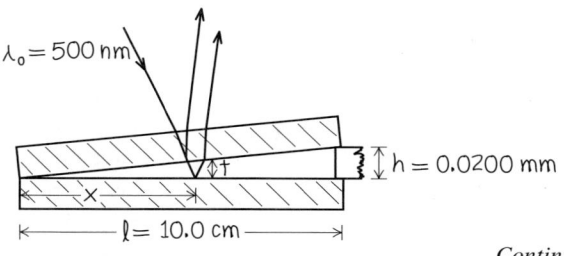

$\lambda_0 = 500$ nm

$h = 0.0200$ mm

$\ell = 10.0$ cm

SOLUTION

IDENTIFY and SET UP: Figure 35.15 depicts the situation. We'll consider only interference between the light reflected from the

Continued

upper and lower surfaces of the air wedge between the microscope slides. [The top slide has a relatively great thickness, about 1 mm, so we can ignore interference between the light reflected from its upper and lower surfaces (see Fig. 35.14b).] Light travels more slowly in the glass of the slides than it does in air. Hence the wave reflected from the upper surface of the air wedge has no phase shift (see Fig. 35.13a), while the wave reflected from the lower surface has a half-cycle phase shift (see Fig. 35.13c).

EXECUTE: Since only one of the reflected waves undergoes a phase shift, the condition for *destructive* interference (a dark fringe) is Eq. (35.18b):

$$2t = m\lambda_0 \qquad (m = 0, 1, 2, \ldots)$$

From similar triangles in Fig. 35.15 the thickness t of the air wedge at each point is proportional to the distance x from the line of contact:

$$\frac{t}{x} = \frac{h}{l}$$

Combining this with Eq. (35.18b), we find

$$\frac{2xh}{l} = m\lambda_0$$

$$x = m\frac{l\lambda_0}{2h} = m\frac{(0.100 \text{ m})(500 \times 10^{-9} \text{ m})}{(2)(0.0200 \times 10^{-3} \text{ m})} = m(1.25 \text{ mm})$$

Successive dark fringes, corresponding to $m = 1, 2, 3, \ldots$, are spaced 1.25 mm apart. Substituting $m = 0$ into this equation gives $x = 0$, which is where the two slides touch (at the left-hand side of Fig. 35.15). Hence there is a dark fringe at the line of contact.

EVALUATE: Our result shows that the fringe spacing is proportional to the wavelength of the light used; the fringes would be farther apart with red light (larger λ_0) than with blue light (smaller λ_0). If we use white light, the reflected light at any point is a mixture of wavelengths for which constructive interference occurs; the wavelengths that interfere destructively are weak or absent in the reflected light. (This same effect explains the colors seen when a soap bubble is illuminated by white light, as in Fig. 35.11b).

EXAMPLE 35.5 | **THIN-FILM INTERFERENCE II**

Suppose the glass plates of Example 35.4 have $n = 1.52$ and the space between plates contains water ($n = 1.33$) instead of air. What happens now?

SOLUTION

IDENTIFY and SET UP: The index of refraction of the water wedge is still less than that of the glass on either side of it, so the phase shifts are the same as in Example 35.4. Once again we use Eq. (35.18b) to find the positions of the dark fringes; the only difference is that the wavelength λ in this equation is now the wavelength in water instead of in air.

EXECUTE: In the film of water ($n = 1.33$), the wavelength is $\lambda = \lambda_0/n = (500 \text{ nm})/(1.33) = 376 \text{ nm}$. When we replace λ_0 by λ in the expression from Example 35.4 for the position x of the mth dark fringe, we find that the fringe spacing is reduced by the same factor of 1.33 and is equal to 0.940 mm. There is still a dark fringe at the line of contact.

EVALUATE: Can you see that to obtain the same fringe spacing as in Example 35.4, the dimension h in Fig. 35.15 would have to be reduced to $(0.0200 \text{ mm})/1.33 = 0.0150 \text{ mm}$? This shows that what matters in thin-film interference is the *ratio* t/λ between film thickness and wavelength. [Consider Eqs. (35.17) and (35.18).]

EXAMPLE 35.6 | **THIN-FILM INTERFERENCE III**

Suppose the upper of the two plates of Example 35.4 is a plastic with $n = 1.40$, the wedge is filled with a silicone grease with $n = 1.50$, and the bottom plate is a dense flint glass with $n = 1.60$. What happens now?

SOLUTION

IDENTIFY and SET UP: The geometry is again the same as shown in Fig. 35.15, but now half-cycle phase shifts occur at *both* surfaces of the grease wedge (see Fig. 35.13c). Hence there is no *relative* phase shift and we must use Eq. (35.17b) to find the positions of the dark fringes.

EXECUTE: The value of λ to use in Eq. (35.17b) is the wavelength in the silicone grease, $\lambda = \lambda_0/n = (500 \text{ nm})/1.50 = 333$ nm. You can readily show that the fringe spacing is 0.833 mm. Note that the two reflected waves from the line of contact are in phase (they both undergo the same phase shift), so the line of contact is at a *bright* fringe.

EVALUATE: What would happen if you carefully removed the upper microscope slide so that the grease wedge retained its shape? There would still be half-cycle phase changes at the upper and lower surfaces of the wedge, so the pattern of fringes would be the same as with the upper slide present.

Newton's Rings

Figure 35.16a shows the convex surface of a lens in contact with a plane glass plate. A thin film of air is formed between the two surfaces. When you view the setup with monochromatic light, you see circular interference fringes (Fig. 35.16b). These were studied by Newton and are called **Newton's rings**.

(a) A convex lens in contact with a glass plane

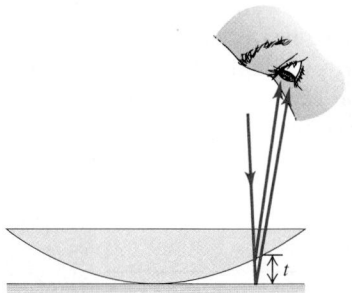

(b) Newton's rings: circular interference fringes

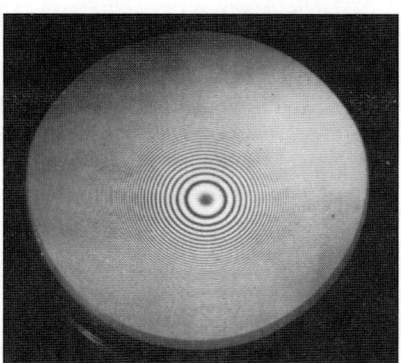

35.16 (a) Air film between a convex lens and a plane surface. The thickness of the film t increases from zero as we move out from the center, giving (b) a series of alternating dark and bright rings for monochromatic light.

We can use interference fringes to compare the surfaces of two optical parts by placing the two in contact and observing the interference fringes. **Figure 35.17** is a photograph made during the grinding of a telescope objective lens. The lower, larger-diameter, thicker disk is the correctly shaped master, and the smaller, upper disk is the lens under test. The "contour lines" are Newton's interference fringes; each one indicates an additional distance between the specimen and the master of one half-wavelength. At 10 lines from the center spot the distance between the two surfaces is five wavelengths, or about 0.003 mm. This isn't very good; high-quality lenses are routinely ground with a precision of less than one wavelength. The surface of the primary mirror of the Hubble Space Telescope was ground to a precision of better than $\frac{1}{50}$ wavelength. Unfortunately, it was ground to incorrect specifications, creating one of the most precise errors in the history of optical technology (see Section 34.2).

Nonreflective and Reflective Coatings

Nonreflective coatings for lens surfaces make use of thin-film interference. A thin layer or film of hard transparent material with an index of refraction smaller than that of the glass is deposited on the lens surface, as in **Fig. 35.18**. Light is reflected from both surfaces of the layer. In both reflections the light is reflected from a medium of greater index than that in which it is traveling, so the same phase change occurs in both reflections. If the film thickness is a quarter (one-fourth) of the wavelength *in the film* (assuming normal incidence), the total path difference is a half-wavelength. Light reflected from the first surface is then a half-cycle out of phase with light reflected from the second, and there is destructive interference.

The thickness of the nonreflective coating can be a quarter-wavelength for only one particular wavelength. This is usually chosen in the central yellow-green portion of the spectrum ($\lambda = 550$ nm), where the eye is most sensitive. Then there is somewhat more reflection at both longer (red) and shorter (blue) wavelengths, and the reflected light has a purple hue. The overall reflection from a lens or prism surface can be reduced in this way from 4–5% to less than 1%. This also increases the net amount of light that is *transmitted* through the lens, since light that is not reflected will be transmitted. The same principle is used to minimize reflection from silicon photovoltaic solar cells ($n = 3.5$) by use of a thin surface layer of silicon monoxide (SiO, $n = 1.45$); this helps to increase the amount of light that actually reaches the solar cells.

If a quarter-wavelength thickness of a material with an index of refraction *greater* than that of glass is deposited on glass, then the reflectivity is *increased*, and the deposited material is called a **reflective coating.** In this case there is a half-cycle phase shift at the air–film interface but none at the film–glass interface, and reflections from the two sides of the film interfere constructively.

35.17 The surface of a telescope objective lens under inspection during manufacture.

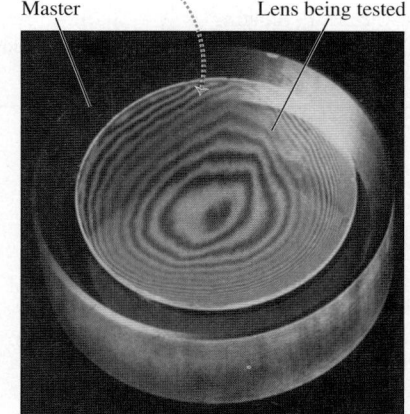

Fringes map lack of fit between lens and master.

Master   Lens being tested

35.18 A nonreflective coating has an index of refraction intermediate between those of glass and air.

Destructive interference occurs when
• the film is about $\frac{1}{4}\lambda$ thick and
• the light undergoes a phase change at both reflecting surfaces,
so that the two reflected waves emerge from the film about $\frac{1}{2}$ cycle out of phase.

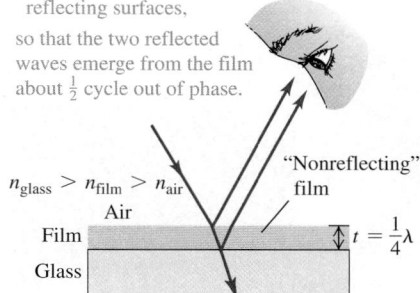

$n_{glass} > n_{film} > n_{air}$

"Nonreflecting" film

Air
Film
Glass

$t = \frac{1}{4}\lambda$

For example, a coating with refractive index 2.5 causes 38% of the incident energy to be reflected, compared with 4% or so with no coating. By use of multiple-layer coatings, it is possible to achieve nearly 100% transmission or reflection for particular wavelengths. Some practical applications of these coatings are for color separation in television cameras and for infrared "heat reflectors" in motion-picture projectors, solar cells, and astronauts' visors.

EXAMPLE 35.7   A NONREFLECTIVE COATING

A common lens coating material is magnesium fluoride (MgF_2), with $n = 1.38$. What thickness should a nonreflective coating have for 550-nm light if it is applied to glass with $n = 1.52$?

SOLUTION

IDENTIFY and SET UP: This coating is of the sort shown in Fig. 35.18. The thickness must be one-quarter of the wavelength of this light *in the coating*.

EXECUTE: The wavelength in air is $\lambda_0 = 550$ nm, so its wavelength in the MgF_2 coating is $\lambda = \lambda_0/n = (550 \text{ nm})/1.38 = 400$ nm. The coating thickness should be one-quarter of this, or $\lambda/4 = 100$ nm.

EVALUATE: This is a very thin film, no more than a few hundred molecules thick. Note that this coating is *reflective* for light whose wavelength is *twice* the coating thickness; light of that wavelength reflected from the coating's lower surface travels one wavelength farther than light reflected from the upper surface, so the two waves are in phase and interfere constructively. This occurs for light with a wavelength in MgF_2 of 200 nm and a wavelength in air of $(200 \text{ nm})(1.38) = 276$ nm. This is an ultraviolet wavelength (see Section 32.1), so designers of optical lenses with nonreflective coatings need not worry about such enhanced reflection.

TEST YOUR UNDERSTANDING OF SECTION 35.4   A thin layer of benzene ($n = 1.501$) lies on top of a sheet of fluorite ($n = 1.434$). It is illuminated from above with light whose wavelength in benzene is 400 nm. Which of the following possible thicknesses of the benzene layer will maximize the brightness of the reflected light? (i) 100 nm; (ii) 200 nm; (iii) 300 nm; (iv) 400 nm. ∎

35.5 THE MICHELSON INTERFEROMETER

An important experimental device that uses interference is the **Michelson interferometer.** Michelson interferometers are used to make precise measurements of wavelengths and of very small distances, such as the minute changes in thickness of an axon when a nerve impulse propagates along its length. Like the Young two-slit experiment, a Michelson interferometer takes monochromatic light from a single source and divides it into two waves that follow different paths. In Young's experiment, this is done by sending part of the light through one slit and part through another; in a Michelson interferometer a device called a *beam splitter* is used. Interference occurs in both experiments when the two light waves are recombined.

How a Michelson Interferometer Works

Figure 35.19 shows the principal components of a Michelson interferometer. A ray of light from a monochromatic source A strikes the beam splitter C, which is a glass plate with a thin coating of silver on its right side. Part of the light (ray 1) passes through the silvered surface and the compensator plate D and is reflected from mirror M_1. It then returns through D and is reflected from the silvered surface of C to the observer. The remainder of the light (ray 2) is reflected from the silvered surface at point P to the mirror M_2 and back through C to the observer's eye. The purpose of the compensator plate D is to ensure that rays 1 and 2 pass through the same thickness of glass; plate D is cut from the same piece of glass as plate C, so their thicknesses are identical to within a fraction of a wavelength.

BIO Application Interference and Butterfly Wings Many of the most brilliant colors in the animal world are created by *interference* rather than by pigments. These photos show the butterfly *Morpho rhetenor* and the microscopic scales that cover the upper surfaces of its wings. The scales have a profusion of tiny ridges (middle photo); these carry regularly spaced flanges (bottom photo) that function as reflectors. These are spaced so that the reflections interfere constructively for blue light. The multilayered structure reflects 70% of the blue light that strikes it, giving the wings a mirrorlike brilliance. (The undersides of the wings do not have these structures and are a dull brown.)

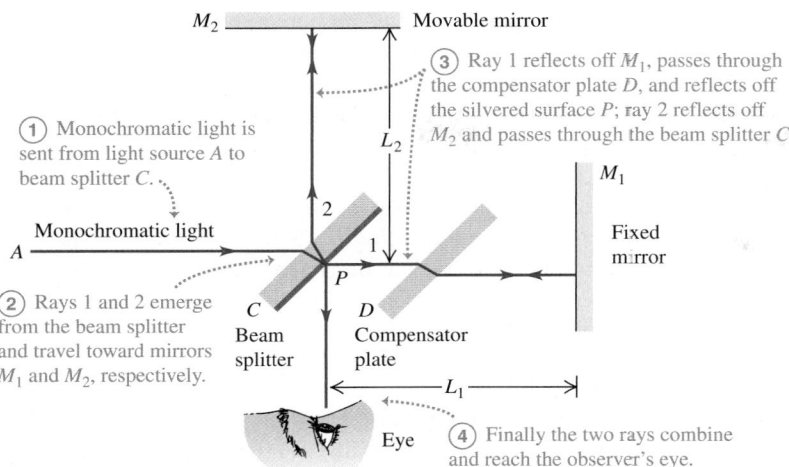

① Monochromatic light is sent from light source A to beam splitter C.

② Rays 1 and 2 emerge from the beam splitter and travel toward mirrors M_1 and M_2, respectively.

③ Ray 1 reflects off M_1, passes through the compensator plate D, and reflects off the silvered surface P; ray 2 reflects off M_2 and passes through the beam splitter C.

④ Finally the two rays combine and reach the observer's eye.

35.19 A schematic Michelson interferometer. The observer sees an interference pattern that results from the difference in path lengths for rays 1 and 2.

The whole apparatus in Fig. 35.19 is mounted on a very rigid frame, and the position of mirror M_2 can be adjusted with a fine, very accurate micrometer screw. If the distances L_1 and L_2 are exactly equal and the mirrors M_1 and M_2 are exactly at right angles, the virtual image of M_1 formed by reflection at the silvered surface of plate C coincides with mirror M_2. If L_1 and L_2 are *not* exactly equal, the image of M_1 is displaced slightly from M_2; and if the mirrors are not exactly perpendicular, the image of M_1 makes a slight angle with M_2. Then the mirror M_2 and the virtual image of M_1 play the same roles as the two surfaces of a wedge-shaped thin film (see Section 35.4), and light reflected from these surfaces forms the same sort of interference fringes.

Suppose the angle between mirror M_2 and the virtual image of M_1 is just large enough that five or six vertical fringes are present in the field of view. If we now move the mirror M_2 slowly either backward or forward a distance $\lambda/2$, the difference in path length between rays 1 and 2 changes by λ, and each fringe moves to the left or right a distance equal to the fringe spacing. If we observe the fringe positions through a telescope with a crosshair eyepiece and m fringes cross the crosshairs when we move the mirror a distance y, then

$$y = m\frac{\lambda}{2} \quad \text{or} \quad \lambda = \frac{2y}{m} \qquad (35.19)$$

If m is several thousand, the distance y is large enough that it can be measured with good accuracy, and we can obtain an accurate value for the wavelength λ. Alternatively, if the wavelength is known, a distance y can be measured by simply counting fringes when M_2 is moved by this distance. In this way, distances that are comparable to a wavelength of light can be measured with relative ease.

The Michelson-Morley Experiment

The original application of the Michelson interferometer was to the historic **Michelson-Morley experiment.** Before the electromagnetic theory of light became established, most physicists thought that the propagation of light waves occurred in a medium called the **ether,** which was believed to permeate all space. In 1887 the American scientists Albert Michelson and Edward Morley used the Michelson interferometer in an attempt to detect the motion of the earth through the ether. Suppose the interferometer in Fig. 35.19 is moving from left to right relative to the ether. According to the ether theory, this would lead to changes in the speed of light in the portions of the path shown as horizontal lines in the figure. There would be fringe shifts relative to the positions that the fringes would have if the instrument were at rest in the ether. Then when the entire instrument was rotated 90°, the other portions of the paths would be similarly affected, giving a fringe shift in the opposite direction.

BIO Application Imaging Cells with a Michelson Interferometer This false-color image of a human colon cancer cell was made by using a microscope that was mated to a Michelson interferometer. The cell is in one arm of the interferometer, and light passing through the cell undergoes a phase shift that depends on the cell thickness and the organelles within the cell. The fringe pattern can then be used to construct a three-dimensional view of the cell. Scientists have used this technique to observe how different types of cells behave when prodded by microscopic probes. Cancer cells turn out to be "softer" than normal cells, a distinction that may make cancer stem cells easier to identify.

Michelson and Morley expected that the motion of the earth through the ether would cause a shift of about four-tenths of a fringe when the instrument was rotated. The shift that was actually observed was less than a hundredth of a fringe and, within the limits of experimental uncertainty, appeared to be exactly zero. Despite its orbital motion around the sun, the earth appeared to be *at rest* relative to the ether. This negative result baffled physicists until 1905, when Albert Einstein developed the special theory of relativity (which we will study in detail in Chapter 37). Einstein postulated that the speed of a light wave in vacuum has the same magnitude *c* relative to *all* inertial reference frames, no matter what their velocity may be relative to each other. The presumed ether then plays no role, and the concept of an ether has been abandoned.

TEST YOUR UNDERSTANDING OF SECTION 35.5 You are observing the pattern of fringes in a Michelson interferometer like that shown in Fig. 35.19. If you change the index of refraction (but not the thickness) of the compensator plate, will the pattern change? ❚

CHAPTER 35 SUMMARY

SOLUTIONS TO ALL EXAMPLES

Interference and coherent sources: Monochromatic light is light with a single frequency. Coherence is a definite, unchanging phase relationship between two waves. The overlap of waves from two coherent sources of monochromatic light forms an interference pattern. The principle of superposition states that the total wave disturbance at any point is the sum of the disturbances from the separate waves.

Two-source interference of light: When two sources are in phase, constructive interference occurs where the difference in path length from the two sources is zero or an integer number of wavelengths; destructive interference occurs where the path difference is a half-integer number of wavelengths. If two sources separated by a distance *d* are both very far from a point *P*, and the line from the sources to *P* makes an angle θ with the line perpendicular to the line of the sources, then the condition for constructive interference at *P* is Eq. (35.4). The condition for destructive interference is Eq. (35.5). When θ is very small, the position y_m of the *m*th bright fringe on a screen located a distance *R* from the sources is given by Eq. (35.6). (See Examples 35.1 and 35.2.)

$$d\sin\theta = m\lambda \quad (m = 0, \pm1, \pm2, \ldots)$$
(constructive interference) $\quad$ (35.4)

$$d\sin\theta = \left(m + \tfrac{1}{2}\right)\lambda$$
$$(m = 0, \pm1, \pm2, \ldots)$$ $\quad$ (35.5)
(destructive interference)

$$y_m = R\frac{m\lambda}{d} \quad (m = 0, \pm1, \pm2, \ldots)$$ $\quad$ (35.6)
(bright fringes)

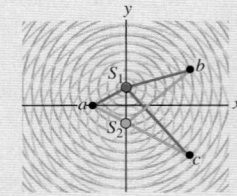

Intensity in interference patterns: When two sinusoidal waves with equal amplitude *E* and phase difference ϕ are superimposed, the resultant amplitude E_P and intensity *I* are given by Eqs. (35.7) and (35.10), respectively. If the two sources emit in phase, the phase difference ϕ at a point *P* (located a distance r_1 from source 1 and a distance r_2 from source 2) is directly proportional to the path difference $r_2 - r_1$. (See Example 35.3.)

$$E_P = 2E\left|\cos\frac{\phi}{2}\right|$$ $\quad$ (35.7)

$$I = I_0\cos^2\frac{\phi}{2}$$ $\quad$ (35.10)

$$\phi = \frac{2\pi}{\lambda}(r_2 - r_1) = k(r_2 - r_1)$$ $\quad$ (35.11)

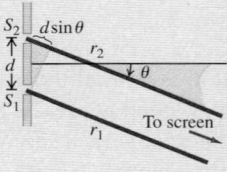

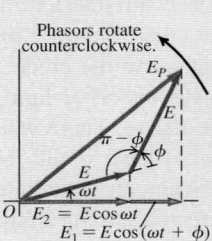

Interference in thin films: When light is reflected from both sides of a thin film of thickness t and no phase shift occurs at either surface, constructive interference between the reflected waves occurs when $2t$ is equal to an integral number of wavelengths. If a half-cycle phase shift occurs at one surface, this is the condition for destructive interference. A half-cycle phase shift occurs during reflection whenever the index of refraction in the second material is greater than that in the first. (See Examples 35.4–35.7.)

$2t = m\lambda \quad (m = 0, 1, 2, \ldots)$
(constructive reflection from   (35.17a)
thin film, no relative phase shift)

$2t = \left(m + \frac{1}{2}\right)\lambda \quad (m = 0, 1, 2, \ldots)$
(destructive reflection from   (35.17b)
thin film, no relative phase shift)

$2t = \left(m + \frac{1}{2}\right)\lambda \quad (m = 0, 1, 2, \ldots)$
(constructive reflection from   (35.18a)
thin film, half-cycle phase shift)

$2t = m\lambda \quad (m = 0, 1, 2, \ldots)$
(destructive reflection from   (35.18b)
thin film, half-cycle phase shift)

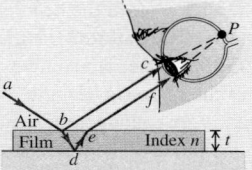

Michelson interferometer: The Michelson interferometer uses a monochromatic light source and can be used for high-precision measurements of wavelengths. Its original purpose was to detect motion of the earth relative to a hypothetical ether, the supposed medium for electromagnetic waves. The ether has never been detected, and the concept has been abandoned; the speed of light is the same relative to all observers. This is part of the foundation of the special theory of relativity.

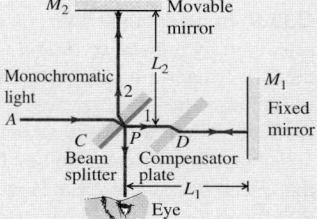

BRIDGING PROBLEM | **THIN-FILM INTERFERENCE IN AN OIL SLICK**

An oil tanker spills a large amount of oil ($n = 1.45$) into the sea ($n = 1.33$). (a) If you look down onto the oil spill from overhead, what predominant wavelength of light do you see at a point where the oil is 380 nm thick? What color is the light? (*Hint:* See Table 32.1.) (b) In the water under the slick, what visible wavelength (as measured in air) is predominant in the transmitted light at the same place in the slick as in part (a)?

SOLUTION GUIDE

IDENTIFY and SET UP

1. The oil layer acts as a thin film, so we must consider interference between light that reflects from the top and bottom surfaces of the oil. If a wavelength is prominent in the *transmitted* light, there is destructive interference for that wavelength in the *reflected* light.

2. Choose the appropriate interference equations that relate the thickness of the oil film and the wavelength of light. Take account of the indexes of refraction of the air, oil, and water.

EXECUTE

3. For part (a), find the wavelengths for which there is constructive interference as seen from above the oil film. Which of these are in the visible spectrum?

4. For part (b), find the visible wavelength for which there is destructive interference as seen from above the film. (This will ensure that there is substantial transmitted light at the wavelength.)

EVALUATE

5. If a diver below the water's surface shines a light up at the bottom of the oil film, at what wavelengths would there be constructive interference in the light that reflects back downward?

Problems | For assigned homework and other learning materials, go to MasteringPhysics®. (**MP**)

•, ••, •••: Difficulty levels. CP: Cumulative problems incorporating material from earlier chapters. CALC: Problems requiring calculus.
DATA: Problems involving real data, scientific evidence, experimental design, and/or statistical reasoning. BIO: Biosciences problems.

DISCUSSION QUESTIONS

Q35.1 A two-slit interference experiment is set up, and the fringes are displayed on a screen. Then the whole apparatus is immersed in the nearest swimming pool. How does the fringe pattern change?

Q35.2 Could an experiment similar to Young's two-slit experiment be performed with sound? How might this be carried out? Does it matter that sound waves are longitudinal and electromagnetic waves are transverse? Explain.

Q35.3 Monochromatic coherent light passing through two thin slits is viewed on a distant screen. Are the bright fringes equally spaced on the screen? If so, why? If not, which ones are closest to being equally spaced?

Q35.4 In a two-slit interference pattern on a distant screen, are the bright fringes midway between the dark fringes? Is this ever a good approximation?

Q35.5 Would the headlights of a distant car form a two-source interference pattern? If so, how might it be observed? If not, why not?

Q35.6 The two sources S_1 and S_2 shown in Fig. 35.3 emit waves of the same wavelength λ and are in phase with each other. Suppose S_1 is a weaker source, so that the waves emitted by S_1 have half the amplitude of the waves emitted by S_2. How would this affect the positions of the antinodal lines and nodal lines? Would there be total reinforcement at points on the antinodal curves? Would there be total cancellation at points on the nodal curves? Explain your answers.

Q35.7 Could the Young two-slit interference experiment be performed with gamma rays? If not, why not? If so, discuss differences in the experimental design compared to the experiment with visible light.

Q35.8 Coherent red light illuminates two narrow slits that are 25 cm apart. Will a two-slit interference pattern be observed when the light from the slits falls on a screen? Explain.

Q35.9 Coherent light with wavelength λ falls on two narrow slits separated by a distance d. If d is less than some minimum value, no dark fringes are observed. Explain. In terms of λ, what is this minimum value of d?

Q35.10 A fellow student, who values memorizing equations above understanding them, combines Eqs. (35.4) and (35.13) to "prove" that ϕ can *only* equal $2\pi m$. How would you explain to this student that ϕ can have values other than $2\pi m$?

Q35.11 If the monochromatic light shown in Fig. 35.5a were replaced by white light, would a two-slit interference pattern be seen on the screen? Explain.

Q35.12 In using the superposition principle to calculate intensities in interference patterns, could you add the intensities of the waves instead of their amplitudes? Explain.

Q35.13 A glass windowpane with a thin film of water on it reflects less than when it is perfectly dry. Why?

Q35.14 A *very* thin soap film ($n = 1.33$), whose thickness is much less than a wavelength of visible light, looks black; it appears to reflect no light at all. Why? By contrast, an equally thin layer of soapy water ($n = 1.33$) on glass ($n = 1.50$) appears quite shiny. Why is there a difference?

Q35.15 Interference can occur in thin films. Why is it important that the films be *thin*? Why don't you get these effects with a relatively *thick* film? Where should you put the dividing line between "thin" and "thick"? Explain your reasoning.

Q35.16 If we shine white light on an air wedge like that shown in Fig. 35.12, the colors that are weak in the light *reflected* from any point along the wedge are strong in the light *transmitted* through the wedge. Explain why this should be so.

Q35.17 Monochromatic light is directed at normal incidence on a thin film. There is destructive interference for the reflected light, so the intensity of the reflected light is very low. What happened to the energy of the incident light?

Q35.18 When a thin oil film spreads out on a puddle of water, the thinnest part of the film looks dark in the resulting interference pattern. What does this tell you about the relative magnitudes of the refractive indexes of oil and water?

EXERCISES

Section 35.1 Interference and Coherent Sources

35.1 • Two small stereo speakers A and B that are 1.40 m apart are sending out sound of wavelength 34 cm in all directions and all in phase. A person at point P starts out equidistant from both speakers and walks so that he is always 1.50 m from speaker B (**Fig. E35.1**). For what values of x will the sound this person hears

Figure **E35.1**

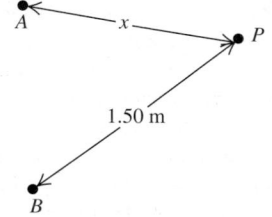

be (a) maximally reinforced, (b) cancelled? Limit your solution to the cases where $x \leq 1.50$ m.

35.2 •• Two speakers that are 15.0 m apart produce in-phase sound waves of frequency 250.0 Hz in a room where the speed of sound is 340.0 m/s. A woman starts out at the midpoint between the two speakers. The room's walls and ceiling are covered with absorbers to eliminate reflections, and she listens with only one ear for best precision. (a) What does she hear: constructive or destructive interference? Why? (b) She now walks slowly toward one of the speakers. How far from the center must she walk before she first hears the sound reach a minimum intensity? (c) How far from the center must she walk before she first hears the sound maximally enhanced?

35.3 •• A radio transmitting station operating at a frequency of 120 MHz has two identical antennas that radiate in phase. Antenna B is 9.00 m to the right of antenna A. Consider point P between the antennas and along the line connecting them, a horizontal distance x to the right of antenna A. For what values of x will constructive interference occur at point P?

35.4 • **Radio Interference.** Two radio antennas A and B radiate in phase. Antenna B is 120 m to the right of antenna A. Consider point Q along the extension of the line connecting the antennas, a horizontal distance of 40 m to the right of antenna B. The frequency, and hence the wavelength, of the emitted waves can be varied. (a) What is the longest wavelength for which there will be destructive interference at point Q? (b) What is the longest wavelength for which there will be constructive interference at point Q?

35.5 • Two speakers, emitting identical sound waves of wavelength 2.0 m in phase with each other, and an observer are located as shown in **Fig. E35.5**. (a) At the observer's location, what is the path difference for waves from the two speakers? (b) Will the sound waves interfere constructively or destructively at the observer's location—or something in between constructive and destructive? (c) Suppose the observer now increases her distance from the closest speaker to 17.0 m, staying directly in front of the same speaker as initially. Answer the questions of parts (a) and (b) for this new situation.

Figure **E35.5**

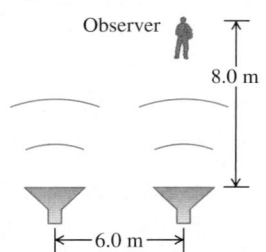

35.6 • Two light sources can be adjusted to emit monochromatic light of any visible wavelength. The two sources are coherent, 2.04 μm apart, and in line with an observer, so that one source is 2.04 μm farther from the observer than the other. (a) For what visible wavelengths (380 to 750 nm) will the observer see the brightest light, owing to constructive interference? (b) How would your answers to part (a) be affected if the two sources were not in line with the observer, but were still arranged so that one source is 2.04 μm farther away from the observer than the other? (c) For what visible wavelengths will there be *destructive* interference at the location of the observer?

Section 35.2 Two-Source Interference of Light

35.7 • Young's experiment is performed with light from excited helium atoms ($\lambda = 502$ nm). Fringes are measured carefully on a screen 1.20 m away from the double slit, and the center of the 20th fringe (not counting the central bright fringe) is found to be 10.6 mm from the center of the central bright fringe. What is the separation of the two slits?

35.8 •• Coherent light with wavelength 450 nm falls on a pair of slits. On a screen 1.80 m away, the distance between dark fringes is 3.90 mm. What is the slit separation?

35.9 •• Two slits spaced 0.450 mm apart are placed 75.0 cm from a screen. What is the distance between the second and third dark lines of the interference pattern on the screen when the slits are illuminated with coherent light with a wavelength of 500 nm?

35.10 •• If the entire apparatus of Exercise 35.9 (slits, screen, and space in between) is immersed in water, what then is the distance between the second and third dark lines?

35.11 •• Two thin parallel slits that are 0.0116 mm apart are illuminated by a laser beam of wavelength 585 nm. (a) On a very large distant screen, what is the *total* number of bright fringes (those indicating complete constructive interference), including the central fringe and those on both sides of it? Solve this problem without calculating all the angles! (*Hint:* What is the largest that $\sin\theta$ can be? What does this tell you is the largest value of m?) (b) At what angle, relative to the original direction of the beam, will the fringe that is most distant from the central bright fringe occur?

35.12 • Coherent light with wavelength 400 nm passes through two very narrow slits that are separated by 0.200 mm, and the interference pattern is observed on a screen 4.00 m from the slits. (a) What is the width (in mm) of the central interference maximum? (b) What is the width of the first-order bright fringe?

35.13 •• Two very narrow slits are spaced 1.80 μm apart and are placed 35.0 cm from a screen. What is the distance between the first and second dark lines of the interference pattern when the slits are illuminated with coherent light with $\lambda = 550$ nm? (*Hint:* The angle θ in Eq. (35.5) is *not* small.)

35.14 •• Coherent light that contains two wavelengths, 660 nm (red) and 470 nm (blue), passes through two narrow slits that are separated by 0.300 mm. Their interference pattern is observed on a screen 4.00 m from the slits. What is the distance on the screen between the first-order bright fringes for the two wavelengths?

35.15 •• Coherent light with wavelength 600 nm passes through two very narrow slits and the interference pattern is observed on a screen 3.00 m from the slits. The first-order bright fringe is at 4.84 mm from the center of the central bright fringe. For what wavelength of light will the first-order dark fringe be observed at this same point on the screen?

35.16 •• Coherent light of frequency 6.32×10^{14} Hz passes through two thin slits and falls on a screen 85.0 cm away. You observe that the third bright fringe occurs at ± 3.11 cm on either side of the central bright fringe. (a) How far apart are the two slits? (b) At what distance from the central bright fringe will the third dark fringe occur?

Section 35.3 Intensity in Interference Patterns

35.17 •• In a two-slit interference pattern, the intensity at the peak of the central maximum is I_0. (a) At a point in the pattern where the phase difference between the waves from the two slits is 60.0°, what is the intensity? (b) What is the path difference for 480-nm light from the two slits at a point where the phase difference is 60.0°?

35.18 • Coherent sources A and B emit electromagnetic waves with wavelength 2.00 cm. Point P is 4.86 m from A and 5.24 m from B. What is the phase difference at P between these two waves?

35.19 • Coherent light with wavelength 500 nm passes through narrow slits separated by 0.340 mm. At a distance from the slits large compared to their separation, what is the phase difference (in radians) in the light from the two slits at an angle of 23.0° from the centerline?

35.20 • Two slits spaced 0.260 mm apart are 0.900 m from a screen and illuminated by coherent light of wavelength 660 nm. The intensity at the center of the central maximum ($\theta = 0°$) is I_0. What is the distance on the screen from the center of the central maximum (a) to the first minimum; (b) to the point where the intensity has fallen to $I_0/2$?

35.21 • Consider two antennas separated by 9.00 m that radiate in phase at 120 MHz, as described in Exercise 35.3. A receiver placed 150 m from both antennas measures an intensity I_0. The receiver is moved so that it is 1.8 m closer to one antenna than to the other. (a) What is the phase difference ϕ between the two radio waves produced by this path difference? (b) In terms of I_0, what is the intensity measured by the receiver at its new position?

35.22 •• Two slits spaced 0.0720 mm apart are 0.800 m from a screen. Coherent light of wavelength λ passes through the two slits. In their interference pattern on the screen, the distance from the center of the central maximum to the first minimum is 3.00 mm. If the intensity at the peak of the central maximum is 0.0600 W/m^2, what is the intensity at points on the screen that are (a) 2.00 mm and (b) 1.50 mm from the center of the central maximum?

Section 35.4 Interference in Thin Films

35.23 • What is the thinnest film of a coating with $n = 1.42$ on glass ($n = 1.52$) for which destructive interference of the red component (650 nm) of an incident white light beam in air can take place by reflection?

35.24 •• **Nonglare Glass.** When viewing a piece of art that is behind glass, one often is affected by the light that is reflected off the front of the glass (called *glare*), which can make it difficult to see the art clearly. One solution is to coat the outer surface of the glass with a film to cancel part of the glare. (a) If the glass has a refractive index of 1.62 and you use TiO$_2$, which has an index of refraction of 2.62, as the coating, what is the minimum film thickness that will cancel light of wavelength 505 nm? (b) If this coating is too thin to stand up to wear, what other thickness would also work? Find only the three thinnest ones.

35.25 •• Two rectangular pieces of plane glass are laid one upon the other on a table. A thin strip of paper is placed between them at one edge so that a very thin wedge of air is formed. The plates are illuminated at normal incidence by 546-nm light from a mercury-vapor lamp. Interference fringes are formed, with 15.0 fringes per centimeter. Find the angle of the wedge.

35.26 •• A plate of glass 9.00 cm long is placed in contact with a second plate and is held at a small angle with it by a metal strip 0.0800 mm thick placed under one end. The space between the plates is filled with air. The glass is illuminated from above with light having a wavelength in air of 656 nm. How many interference fringes are observed per centimeter in the reflected light?

35.27 •• A uniform film of TiO_2, 1036 nm thick and having index of refraction 2.62, is spread uniformly over the surface of crown glass of refractive index 1.52. Light of wavelength 520.0 nm falls at normal incidence onto the film from air. You want to increase the thickness of this film so that the reflected light cancels. (a) What is the *minimum* thickness of TiO_2 that you must *add* so the reflected light cancels as desired? (b) After you make the adjustment in part (a), what is the path difference between the light reflected off the top of the film and the light that cancels it after traveling through the film? Express your answer in (i) nanometers and (ii) wavelengths of the light in the TiO_2 film.

35.28 • A plastic film with index of refraction 1.70 is applied to the surface of a car window to increase the reflectivity and thus to keep the car's interior cooler. The window glass has index of refraction 1.52. (a) What minimum thickness is required if light of wavelength 550 nm in air reflected from the two sides of the film is to interfere constructively? (b) Coatings as thin as that calculated in part (a) are difficult to manufacture and install. What is the next greater thickness for which constructive interference will also occur?

35.29 • The walls of a soap bubble have about the same index of refraction as that of plain water, $n = 1.33$. There is air both inside and outside the bubble. (a) What wavelength (in air) of visible light is most strongly reflected from a point on a soap bubble where its wall is 290 nm thick? To what color does this correspond (see Fig. 32.4 and Table 32.1)? (b) Repeat part (a) for a wall thickness of 340 nm.

35.30 •• A researcher measures the thickness of a layer of benzene ($n = 1.50$) floating on water by shining monochromatic light onto the film and varying the wavelength of the light. She finds that light of wavelength 575 nm is reflected most strongly from the film. What does she calculate for the minimum thickness of the film?

35.31 •• **Compact Disc Player.** A compact disc (CD) is read from the bottom by a semiconductor laser with wavelength 790 nm passing through a plastic substrate of refractive index 1.8. When the beam encounters a pit, part of the beam is reflected from the pit and part from the flat region between the pits, so these two beams interfere with each other (**Fig. E35.31**). What must the minimum pit depth be so that the part of the beam reflected from a pit cancels the part of the beam reflected from the flat region? (It is this cancellation that allows the player to recognize the beginning and end of a pit.)

Figure **E35.31**

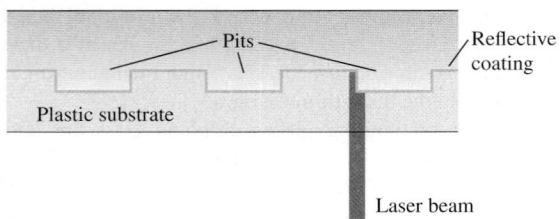

35.32 • What is the thinnest soap film (excluding the case of zero thickness) that appears black when illuminated with light with wavelength 480 nm? The index of refraction of the film is 1.33, and there is air on both sides of the film.

Section 35.5 The Michelson Interferometer

35.33 • How far must the mirror M_2 (see Fig. 35.19) of the Michelson interferometer be moved so that 1800 fringes of He-Ne laser light ($\lambda = 633$ nm) move across a line in the field of view?

35.34 • Jan first uses a Michelson interferometer with the 606-nm light from a krypton-86 lamp. He displaces the movable mirror away from him, counting 818 fringes moving across a line in his field of view. Then Linda replaces the krypton lamp with filtered 502-nm light from a helium lamp and displaces the movable mirror toward her. She also counts 818 fringes, but they move across the line in her field of view opposite to the direction they moved for Jan. Assume that both Jan and Linda counted to 818 correctly. (a) What distance did each person move the mirror? (b) What is the resultant displacement of the mirror?

PROBLEMS

35.35 •• One round face of a 3.25-m, solid, cylindrical plastic pipe is covered with a thin black coating that completely blocks light. The opposite face is covered with a fluorescent coating that glows when it is struck by light. Two straight, thin, parallel scratches, 0.225 mm apart, are made in the center of the black face. When laser light of wavelength 632.8 nm shines through the slits perpendicular to the black face, you find that the central bright fringe on the opposite face is 5.82 mm wide, measured between the dark fringes that border it on either side. What is the index of refraction of the plastic?

35.36 ••• Newton's rings are visible when a planoconvex lens is placed on a flat glass surface. For a particular lens with an index of refraction of $n = 1.50$ and a glass plate with an index of $n = 1.80$, the diameter of the third bright ring is 0.640 mm. If water ($n = 1.33$) now fills the space between the lens and the glass plate, what is the new diameter of this ring? Assume the radius of curvature of the lens is much greater than the wavelength of the light.

35.37 • **BIO Coating Eyeglass Lenses.** Eyeglass lenses can be coated on the *inner* surfaces to reduce the reflection of stray light to the eye. If the lenses are medium flint glass of refractive index 1.62 and the coating is fluorite of refractive index 1.432, (a) what minimum thickness of film is needed on the lenses to cancel light of wavelength 550 nm reflected toward the eye at normal incidence? (b) Will any other wavelengths of visible light be cancelled or enhanced in the reflected light?

35.38 •• **BIO Sensitive Eyes.** After an eye examination, you put some eyedrops on your sensitive eyes. The cornea (the front part of the eye) has an index of refraction of 1.38, while the eyedrops have a refractive index of 1.45. After you put in the drops, your friends notice that your eyes look red, because red light of wavelength 600 nm has been reinforced in the reflected light. (a) What is the minimum thickness of the film of eyedrops on your cornea? (b) Will any other wavelengths of visible light be reinforced in the reflected light? Will any be cancelled? (c) Suppose you had contact lenses, so that the eyedrops went on them instead of on your corneas. If the refractive index of the lens material is 1.50 and the layer of eyedrops has the same thickness as in part (a), what wavelengths of visible light will be reinforced? What wavelengths will be cancelled?

35.39 •• Two flat plates of glass with parallel faces are on a table, one plate on the other. Each plate is 11.0 cm long and has a refractive index of 1.55. A very thin sheet of metal foil is inserted under the end of the upper plate to raise it slightly at that end, in a manner similar to that discussed in Example 35.4. When you view the glass plates from above with reflected white light, you observe that, at 1.15 mm from the line where the sheets are in contact, the violet light of wavelength 400.0 nm is enhanced in this reflected light, but no visible light is enhanced closer to the line of contact. (a) How far from the line of contact will green light (of wavelength 550.0 nm) and orange light (of wavelength 600.0 nm) first be enhanced? (b) How far from the line of contact will the violet, green, and orange light again be enhanced in the reflected light? (c) How thick is the metal foil holding the ends of the plates apart?

35.40 •• In a setup similar to that of Problem 35.39, the glass has an index of refraction of 1.53, the plates are each 8.00 cm long, and the metal foil is 0.015 mm thick. The space between the plates is filled with a jelly whose refractive index is not known precisely, but is known to be greater than that of the glass. When you illuminate these plates from above with light of wavelength 525 nm, you observe a series of equally spaced dark fringes in the reflected light. You measure the spacing of these fringes and find that there are 10 of them every 6.33 mm. What is the index of refraction of the jelly?

35.41 ••• Suppose you illuminate two thin slits by monochromatic coherent light in air and find that they produce their first interference *minima* at ±35.20° on either side of the central bright spot. You then immerse these slits in a transparent liquid and illuminate them with the same light. Now you find that the first minima occur at ±19.46° instead. What is the index of refraction of this liquid?

35.42 •• CP CALC A very thin sheet of brass contains two thin parallel slits. When a laser beam shines on these slits at normal incidence and room temperature (20.0°C), the first interference dark fringes occur at ±26.6° from the original direction of the laser beam when viewed from some distance. If this sheet is now slowly heated to 135°C, by how many degrees do these dark fringes change position? Do they move closer together or farther apart? See Table 17.1 for pertinent information, and ignore any effects that might occur due to a change in the thickness of the slits. (*Hint:* Thermal expansion normally produces very small changes in length, so you can use differentials to find the change in the angle.)

35.43 •• Two radio antennas radiating in phase are located at points A and B, 200 m apart (**Fig. P35.43**). The radio waves have a frequency of 5.80 MHz. A radio receiver is moved out from point B along a line perpendicular to the line connecting A and B (line BC shown in Fig. P35.43). At what distances from B will there be *destructive* interference? (*Note:* The distance of the receiver from the sources is not large in comparison to the separation of the sources, so Eq. (35.5) does not apply.)

Figure **P35.43**

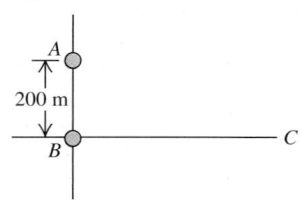

35.44 •• Two speakers A and B are 3.50 m apart, and each one is emitting a frequency of 444 Hz. However, because of signal delays in the cables, speaker A is one-fourth of a period *ahead* of speaker B. For points far from the speakers, find all the angles relative to the centerline (**Fig. P35.44**) at which the sound from these speakers cancels. Include angles on *both* sides of the centerline. The speed of sound is 340 m/s.

Figure **P35.44**

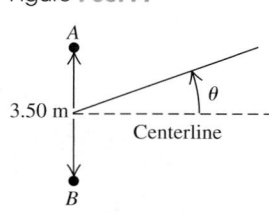

35.45 •• CP A thin uniform film of refractive index 1.750 is placed on a sheet of glass of refractive index 1.50. At room temperature (20.0°C), this film is just thick enough for light with wavelength 582.4 nm reflected off the top of the film to be cancelled by light reflected from the top of the glass. After the glass is placed in an oven and slowly heated to 170°C, you find that the film cancels reflected light with wavelength 588.5 nm. What is the coefficient of linear expansion of the film? (Ignore any changes in the refractive index of the film due to the temperature change.)

35.46 ••• **GPS Transmission.** The GPS (Global Positioning System) satellites are approximately 5.18 m across and transmit two low-power signals, one of which is at 1575.42 MHz (in the UHF band). In a series of laboratory tests on the satellite, you put two 1575.42-MHz UHF transmitters at opposite ends of the satellite. These broadcast in phase uniformly in all directions. You measure the intensity at points on a circle that is several hundred meters in radius and centered on the satellite. You measure angles on this circle relative to a point that lies along the centerline of the satellite (that is, the perpendicular bisector of a line that extends from one transmitter to the other). At this point on the circle, the measured intensity is 2.00 W/m². (a) At how many other angles in the range 0° < θ < 90° is the intensity also 2.00 W/m²? (b) Find the four smallest angles in the range 0° < θ < 90° for which the intensity is 2.00 W/m². (c) What is the intensity at a point on the circle at an angle of 4.65° from the centerline?

35.47 •• White light reflects at normal incidence from the top and bottom surfaces of a glass plate (n = 1.52). There is air above and below the plate. Constructive interference is observed for light whose wavelength in air is 477.0 nm. What is the thickness of the plate if the next longer wavelength for which there is constructive interference is 540.6 nm?

35.48 •• Laser light of wavelength 510 nm is traveling in air and shines at normal incidence onto the flat end of a transparent plastic rod that has n = 1.30. The end of the rod has a thin coating of a transparent material that has refractive index 1.65. What is the minimum (nonzero) thickness of the coating (a) for which there is maximum transmission of the light into the rod; (b) for which transmission into the rod is minimized?

35.49 •• Red light with wavelength 700 nm is passed through a two-slit apparatus. At the same time, monochromatic visible light with another wavelength passes through the same apparatus. As a result, most of the pattern that appears on the screen is a mixture of two colors; however, the center of the third bright fringe (m = 3) of the red light appears pure red, with none of the other color. What are the possible wavelengths of the second type of visible light? Do you need to know the slit spacing to answer this question? Why or why not?

35.50 •• BIO **Reflective Coatings and Herring.** Herring and related fish have a brilliant silvery appearance that camouflages them while they are swimming in a sunlit ocean. The silveriness is due to *platelets* attached to the surfaces of these fish. Each platelet is made up of several alternating layers of crystalline guanine (n = 1.80) and of cytoplasm (n = 1.333, the same as water), with a guanine layer on the outside in contact with the surrounding water (**Fig. P35.50**). In one typical platelet, the guanine

Figure **P35.50**

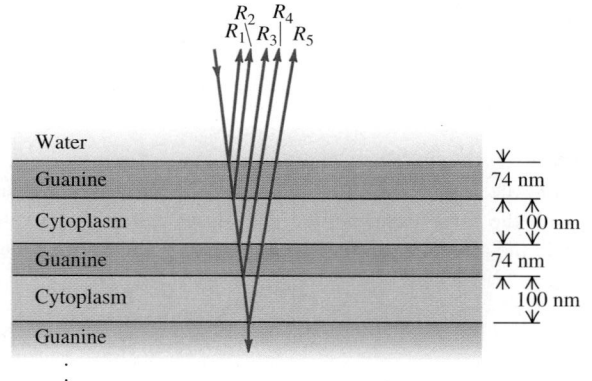

layers are 74 nm thick and the cytoplasm layers are 100 nm thick. (a) For light striking the platelet surface at normal incidence, for which vacuum wavelengths of visible light will all of the reflections R_1, R_2, R_3, R_4, and R_5, shown in Fig. P35.50, be approximately in phase? If white light is shone on this platelet, what color will be most strongly reflected (see Fig. 32.4)? The surface of a herring has very many platelets side by side with layers of different thickness, so that *all* visible wavelengths are reflected. (b) Explain why such a "stack" of layers is more reflective than a single layer of guanine with cytoplasm underneath. (A stack of five guanine layers separated by cytoplasm layers reflects more than 80% of incident light at the wavelength for which it is "tuned.") (c) The color that is most strongly reflected from a platelet depends on the angle at which it is viewed. Explain why this should be so. (You can see these changes in color by examining a herring from different angles. Most of the platelets on these fish are oriented in the same way, so that they are vertical when the fish is swimming.)

35.51 •• After a laser beam passes through two thin parallel slits, the first completely dark fringes occur at $\pm 19.0°$ with the original direction of the beam, as viewed on a screen far from the slits. (a) What is the ratio of the distance between the slits to the wavelength of the light illuminating the slits? (b) What is the smallest angle, relative to the original direction of the laser beam, at which the intensity of the light is $\frac{1}{10}$ the maximum intensity on the screen?

35.52 •• DATA In your summer job at an optics company, you are asked to measure the wavelength λ of the light that is produced by a laser. To do so, you pass the laser light through two narrow slits that are separated by a distance d. You observe the interference pattern on a screen that is 0.900 m from the slits and measure the separation Δy between adjacent bright fringes in the portion of the pattern that is near the center of the screen. Using a microscope, you measure d. But both Δy and d are small and difficult to measure accurately, so you repeat the measurements for several pairs of slits, each with a different value of d. Your results are shown in **Fig. P35.52**, where you have plotted Δy versus $1/d$. The line in the graph is the best-fit straight line for the data. (a) Explain why the data points plotted this way fall close to a straight line. (b) Use Fig. P35.52 to calculate λ.

Figure **P35.52**

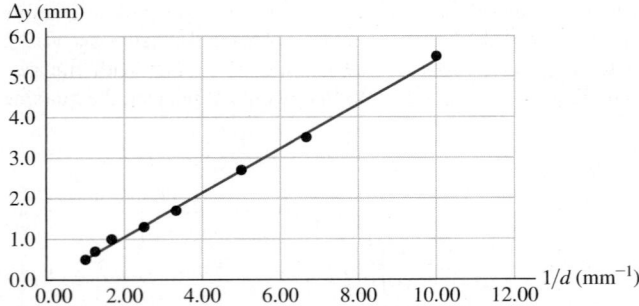

35.53 •• DATA Short-wave radio antennas A and B are connected to the same transmitter and emit coherent waves in phase and with the same frequency f. You must determine the value of f and the placement of the antennas that produce a maximum

intensity through constructive interference at a receiving antenna that is located at point P, which is at the corner of your garage. First you place antenna A at a point 240.0 m due east of P. Next you place antenna B on the line that connects A and P, a distance x due east of P, where $x < 240.0$ m. Then you measure that a maximum in the total intensity from the two antennas occurs when $x = 210.0$ m, 216.0 m, and 222.0 m. You don't investigate smaller or larger values of x. (Treat the antennas as point sources.) (a) What is the frequency f of the waves that are emitted by the antennas? (b) What is the greatest value of x, with $x < 240.0$ m, for which the interference at P is destructive?

35.54 •• DATA In your research lab, a very thin, flat piece of glass with refractive index 1.40 and uniform thickness covers the opening of a chamber that holds a gas sample. The refractive indexes of the gases on either side of the glass are very close to unity. To determine the thickness of the glass, you shine coherent light of wavelength λ_0 in vacuum at normal incidence onto the surface of the glass. When $\lambda_0 = 496$ nm, constructive interference occurs for light that is reflected at the two surfaces of the glass. You find that the next shorter wavelength in vacuum for which there is constructive interference is 386 nm. (a) Use these measurements to calculate the thickness of the glass. (b) What is the longest wavelength in vacuum for which there is constructive interference for the reflected light?

CHALLENGE PROBLEMS

35.55 ••• CP The index of refraction of a glass rod is 1.48 at $T = 20.0°C$ and varies linearly with temperature, with a coefficient of $2.50 \times 10^{-5}/C°$. The coefficient of linear expansion of the glass is $5.00 \times 10^{-6}/C°$. At 20.0°C the length of the rod is 3.00 cm. A Michelson interferometer has this glass rod in one arm, and the rod is being heated so that its temperature increases at a rate of 5.00 C°/min. The light source has wavelength $\lambda = 589$ nm, and the rod initially is at $T = 20.0°C$. How many fringes cross the field of view each minute?

35.56 ••• CP **Figure P35.56** shows an interferometer known as *Fresnel's biprism*. The magnitude of the prism angle A is extremely small. (a) If S_0 is a very narrow source slit, show that the separation of the two virtual coherent sources S_1 and S_2 is given by $d = 2aA(n - 1)$, where n is the index of refraction of the material of the prism. (b) Calculate the spacing of the fringes of green light with wavelength 500 nm on a screen 2.00 m from the biprism. Take $a = 0.200$ m, $A = 3.50$ mrad, and $n = 1.50$.

Figure **P35.56**

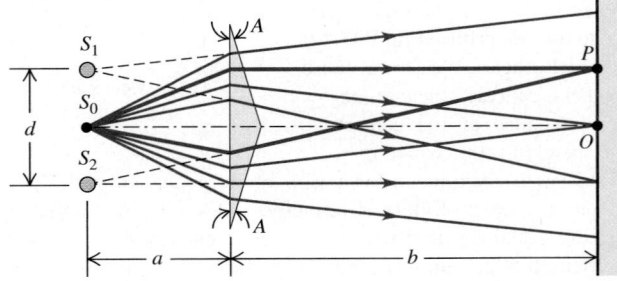

INTERFERENCE AND SOUND WAVES. Interference occurs with not only light waves but also all frequencies of electromagnetic waves and all other types of waves, such as sound and water waves. Suppose that your physics professor sets up two sound speakers in the front of your classroom and uses an electronic oscillator to produce sound waves of a single frequency. When she turns the oscillator on (take this to be its original setting), you and many students hear a loud tone while other students hear nothing. (The speed of sound in air is 340 m/s.)

35.57 The professor then adjusts the apparatus. The frequency that you hear does not change, but the loudness decreases. Now all of your fellow students can hear the tone. What did the professor do? (a) She turned off the oscillator. (b) She turned down the volume of the speakers. (c) She changed the phase relationship of the speakers. (d) She disconnected one speaker.

35.58 The professor returns the apparatus to the original setting. She then adjusts the speakers again. All of the students who had heard nothing originally now hear a loud tone, while you and the others who had originally heard the loud tone hear nothing.

What did the professor do? (a) She turned off the oscillator. (b) She turned down the volume of the speakers. (c) She changed the phase relationship of the speakers. (d) She disconnected one speaker.

35.59 The professor again returns the apparatus to its original setting, so you again hear the original loud tone. She then slowly moves one speaker away from you until it reaches a point at which you can no longer hear the tone. If she has moved the speaker by 0.34 m (farther from you), what is the frequency of the tone? (a) 1000 Hz; (b) 2000 Hz; (c) 500 Hz; (d) 250 Hz.

35.60 The professor once again returns the apparatus to its original setting, but now she adjusts the oscillator to produce sound waves of half the original frequency. What happens? (a) The students who originally heard a loud tone again hear a loud tone, and the students who originally heard nothing still hear nothing. (b) The students who originally heard a loud tone now hear nothing, and the students who originally heard nothing now hear a loud tone. (c) Some of the students who originally heard a loud tone again hear a loud tone, but others in that group now hear nothing. (d) Among the students who originally heard nothing, some still hear nothing but others now hear a loud tone.

Answers

Chapter Opening Question ?

(v) The colors appear as a result of constructive interference between light waves reflected from the upper and lower surfaces of the oil film. The wavelength of light for which the most constructive interference occurs at a point, and hence the color that appears the brightest at that point, depends on (1) the thickness of the film (which determines the path difference between light waves that reflect off the two surfaces), (2) the oil's index of refraction (which gives the wavelength of light in the oil a different value than in air), and (3) the index of refraction of the material below the oil (which determines whether the wave that reflects from the lower surface undergoes a half-cycle phase shift). (See Examples 35.4, 35.5, and 35.6 in Section 35.4.)

Test Your Understanding Questions

35.1 (i) At any point P on the positive y-axis above S_1, the distance r_2 from S_2 to P is greater than the distance r_1 from S_1 to P by 4λ. This corresponds to $m = 4$ in Eq. (35.1), the equation for constructive interference. Hence all such points make up an antinodal curve.

35.2 (ii) Blue light has a shorter wavelength than red light (see Section 32.1). Equation (35.6) tells us that the distance y_m from the center of the pattern to the mth bright fringe is proportional to the wavelength λ. Hence all of the fringes will move toward the center of the pattern as the wavelength decreases, and the spacing between fringes will decrease.

35.3 (i), (iv), (ii), (iii) In cases (i) and (iii) we are given the wavelength λ and path difference $d\sin\theta$. Hence we use Eq. (35.14),

$I = I_0\cos^2[(\pi d\sin\theta)/\lambda]$. In parts (ii) and (iii) we are given the phase difference ϕ and we use Eq. (35.10), $I = I_0\cos^2(\phi/2)$. We find:
(i) $I = I_0\cos^2[\pi(4.00 \times 10^{-7}\ \text{m})/(5.00 \times 10^{-7}\ \text{m})] = I_0\cos^2(0.800\pi\ \text{rad}) = 0.655I_0$;
(ii) $I = I_0\cos^2[(4.00\ \text{rad})/2] = I_0\cos^2(2.00\ \text{rad}) = 0.173I_0$;
(iii) $I = I_0\cos^2[\pi(7.50 \times 10^{-7}\ \text{m})/(5.00 \times 10^{-7}\ \text{m})] = I_0\cos^2(1.50\pi\ \text{rad}) = 0$;
(iv) $I = I_0\cos^2[(2.00\ \text{rad})/2] = I_0\cos^2(1.00\ \text{rad}) = 0.292I_0$.

35.4 (i) and (iii) Benzene has a larger index of refraction than air, so light that reflects off the upper surface of the benzene undergoes a half-cycle phase shift. Fluorite has a *smaller* index of refraction than benzene, so light that reflects off the benzene–fluorite interface does not undergo a phase shift. Hence the equation for constructive reflection is Eq. (35.18a), $2t = (m + \frac{1}{2})\lambda$, which we can rewrite as $t = (m + \frac{1}{2})\lambda/2 = (m + \frac{1}{2})(400\ \text{mm})/2 = 100\ \text{nm},\ 300\ \text{nm},\ 500\ \text{nm},\ \ldots$.

35.5 yes Changing the index of refraction changes the wavelength of the light inside the compensator plate, and so changes the number of wavelengths within the thickness of the plate. Hence this has the same effect as changing the distance L_1 from the beam splitter to mirror M_1, which would change the interference pattern.

Bridging Problem

(a) 441 nm
(b) 551 nm

? Flies have *compound* eyes with thousands of miniature lenses. The overall diameter of the eye is about 1 mm, but each lens is only about 20 μm in diameter and produces an individual image of a small region in the fly's field of view. Compared to the resolving power of the human eye (in which the light-gathering region is about 16 mm across), the ability of a fly's eye to resolve small details is (i) worse because the lenses are so small; (ii) worse because the eye as a whole is so small; (iii) better because the lenses are so small; (iv) better because the eye as a whole is so small; (v) about the same.

36 DIFFRACTION

LEARNING GOALS

Looking forward at …

36.1 What happens when coherent light shines on an object with an edge or aperture.

36.2 How to understand the diffraction pattern formed when coherent light passes through a narrow slit.

36.3 How to calculate the intensity at various points in a single-slit diffraction pattern.

36.4 What happens when coherent light shines on an array of narrow, closely spaced slits.

36.5 How scientists use diffraction gratings for precise measurements of wavelength.

36.6 How x-ray diffraction reveals the arrangement of atoms in a crystal.

36.7 How diffraction sets limits on the smallest details that can be seen with an optical system.

36.8 How holograms work.

Looking back at …

33.4, 33.7 Prisms and dispersion; Huygens's principle.

34.4, 34.5 Image formation by a lens; *f*-number.

35.1–35.3 Coherent light, two-slit interference, and phasors.

Everyone is used to the idea that sound bends around corners. If sound didn't behave this way, you couldn't hear a police siren that's out of sight or the speech of a person whose back is turned to you. But *light* can bend around corners as well. When light from a point source falls on a straightedge and casts a shadow, the edge of the shadow is never perfectly sharp. Some light appears in the area that we expect to be in the shadow, and we find alternating bright and dark fringes in the illuminated area. In general, light emerging from apertures doesn't precisely follow the predictions of the straight-line ray model of geometric optics.

The reason for these effects is that light, like sound, has wave characteristics. In Chapter 35 we studied the interference patterns that can arise when two light waves are combined. In this chapter we'll investigate interference effects due to combining *many* light waves. Such effects are referred to as *diffraction*. The behavior of waves after they pass through an aperture is an example of diffraction; each infinitesimal part of the aperture acts as a source of waves, and these waves interfere, producing a pattern of bright and dark fringes.

Similar patterns appear when light emerges from *arrays* of apertures. The nature of these patterns depends on the color of the light and the size and spacing of the apertures. Examples of this effect include the colors of iridescent butterflies and the "rainbow" you see reflected from the surface of a compact disc. We'll explore similar effects with x rays that are used to study the atomic structure of solids and liquids. Finally, we'll look at the physics of a *hologram*, a special kind of interference pattern used to form three-dimensional images.

36.1 FRESNEL AND FRAUNHOFER DIFFRACTION

According to geometric optics, when an opaque object is placed between a point light source and a screen, as in **Fig. 36.1**, the shadow of the object forms a perfectly sharp line. No light at all strikes the screen at points within the shadow, and the area outside the shadow is illuminated nearly uniformly. But as we saw in Chapter 35, the *wave* nature of light causes effects that can't be understood with geometric optics. An important class of such effects occurs when light strikes a barrier that has an aperture or an edge. The interference patterns formed in such a situation are grouped under the heading **diffraction.**

Figure 36.2 shows an example of diffraction. The photograph in Fig. 36.2a was made by placing a razor blade halfway between a pinhole, illuminated by monochromatic light, and a photographic film. The film recorded the shadow cast by the blade. Figure 36.2b is an enlargement of a region near the shadow of the right edge of the blade. The position of the *geometric* shadow line is indicated by arrows. The area outside the geometric shadow is bordered by alternating bright and dark bands. There is some light in the shadow region, although this is not very visible in the photograph. The first bright band in Fig. 36.2b, just to the right of the geometric shadow, is considerably brighter than in the region of uniform illumination to the extreme right. This simple experiment gives us some idea of the richness and complexity of diffraction.

We don't often observe diffraction patterns such as Fig. 36.2 in everyday life because most ordinary light sources are neither monochromatic nor point sources. If we use a white frosted light bulb instead of a point source to illuminate the razor blade in Fig. 36.2, each wavelength of the light from every point of the bulb forms its own diffraction pattern, but the patterns overlap so much that we can't see any individual pattern.

Diffraction and Huygens's Principle

We can use Huygens's principle (see Section 33.7) to analyze diffraction patterns. This principle states that we can consider every point of a wave front as a source of secondary wavelets. These spread out in all directions with a speed equal to the speed of propagation of the wave. The position of the wave front at any later time is the *envelope* of the secondary wavelets at that time. To find the resultant displacement at any point, we use the superposition principle to combine all the individual displacements produced by these secondary waves.

In Fig. 36.1, both the point source and the screen are relatively close to the obstacle forming the diffraction pattern. This situation is described as *near-field diffraction* or **Fresnel diffraction,** pronounced "Freh-nell" (after the French scientist Augustin Jean Fresnel, 1788–1827). By contrast, we use the term **Fraunhofer diffraction** (after the German physicist Joseph von Fraunhofer, 1787–1826) for situations in which the source, obstacle, and screen are far enough apart that we can consider all lines from the source to the obstacle to be parallel, and can likewise consider all lines from the obstacle to a given point on the screen to be parallel. We will restrict the following discussion

36.1 A point source of light illuminates a straightedge.

Geometric optics predicts that this situation should produce a sharp boundary between illumination and solid shadow.

That's NOT what really happens!

DOESN'T HAPPEN

Point source

Straightedge

Area of illumination

Geometric shadow

Screen

36.2 An example of diffraction.

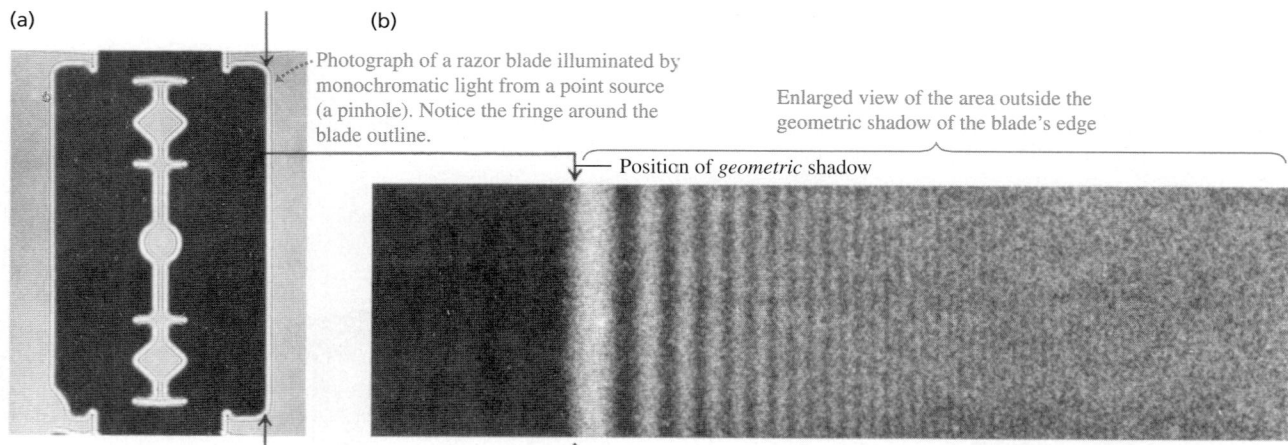

(a)

(b)

Photograph of a razor blade illuminated by monochromatic light from a point source (a pinhole). Notice the fringe around the blade outline.

Enlarged view of the area outside the geometric shadow of the blade's edge

Position of *geometric* shadow

to Fraunhofer diffraction, which is usually simpler to analyze in detail than Fresnel diffraction.

Diffraction is sometimes described as "the bending of light around an obstacle." But the process that causes diffraction is present in the propagation of *every* wave. When part of the wave is cut off by some obstacle, we observe diffraction effects that result from interference of the remaining parts of the wave fronts. Optical instruments typically use only a limited portion of a wave; for example, a telescope uses only the part of a wave that is admitted by its objective lens or mirror. Thus diffraction plays a role in nearly all optical phenomena.

Finally, we emphasize that there is no fundamental distinction between *interference* and *diffraction*. In Chapter 35 we used the term *interference* for effects involving waves from a small number of sources, usually two. *Diffraction* usually involves a *continuous* distribution of Huygens's wavelets across the area of an aperture, or a very large number of sources or apertures. But both interference and diffraction are consequences of superposition and Huygens's principle.

TEST YOUR UNDERSTANDING OF SECTION 36.1   Can *sound* waves undergo diffraction around an edge? ▮

PhET: Wave Interference

36.2 DIFFRACTION FROM A SINGLE SLIT

In this section we'll discuss the diffraction pattern formed by plane-wave (parallel-ray) monochromatic light when it emerges from a long, narrow slit, as shown in **Fig. 36.3**. We call the narrow dimension the *width,* even though in this figure it is a vertical dimension.

According to geometric optics, the transmitted beam should have the same cross section as the slit, as in Fig. 36.3a. What is *actually* observed is the pattern shown in Fig. 36.3b. The beam spreads out vertically after passing through the slit. The diffraction pattern consists of a central bright band, which may be much broader than the width of the slit, bordered by alternating dark and bright bands with rapidly decreasing intensity. About 85% of the power in the transmitted beam is in the central bright band, whose width is *inversely* proportional to the slit width. In general, the narrower the slit, the broader the entire diffraction pattern. (The *horizontal* spreading of the beam in Fig. 36.3b is negligible because the horizontal dimension of the slit is relatively large.) You can observe a similar diffraction pattern by looking at a point source, such as a distant street light, through a narrow slit formed between your two thumbs held in front of your eye; the retina of your eye acts as the screen.

36.3 (a) The "shadow" of a horizontal slit as incorrectly predicted by geometric optics. (b) A horizontal slit actually produces a diffraction pattern. The slit width has been greatly exaggerated.

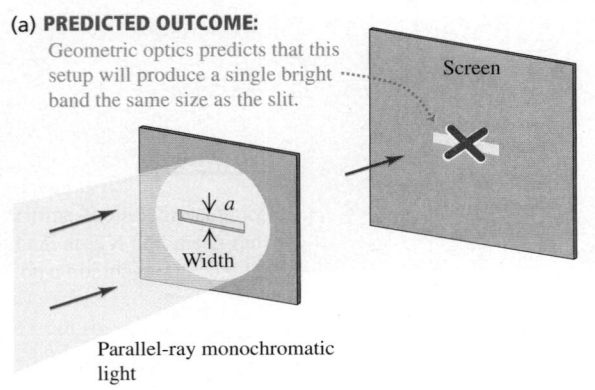

(a) PREDICTED OUTCOME:

Geometric optics predicts that this setup will produce a single bright band the same size as the slit.

Screen

Parallel-ray monochromatic light

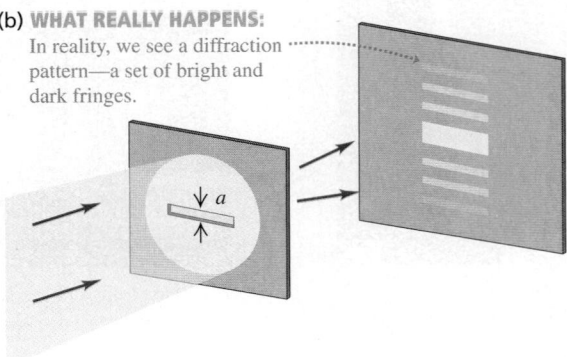

(b) WHAT REALLY HAPPENS:

In reality, we see a diffraction pattern—a set of bright and dark fringes.

36.4 Diffraction by a single rectangular slit. The long sides of the slit are perpendicular to the figure.

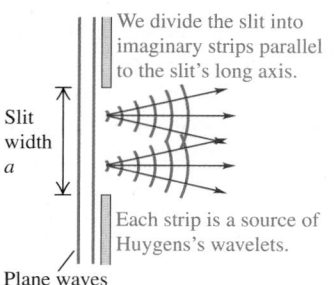

(a) A slit as a source of wavelets

We divide the slit into imaginary strips parallel to the slit's long axis.

Slit width *a*

Each strip is a source of Huygens's wavelets.

Plane waves incident on the slit

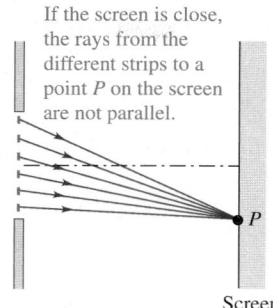

(b) Fresnel (near-field) diffraction

If the screen is close, the rays from the different strips to a point *P* on the screen are not parallel.

Screen

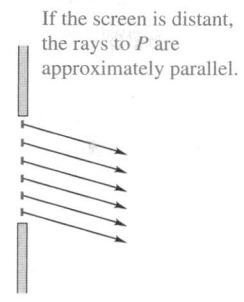

(c) Fraunhofer (far-field) diffraction

If the screen is distant, the rays to *P* are approximately parallel.

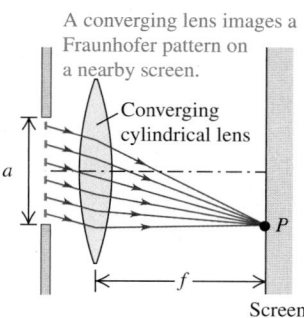

(d) Imaging Fraunhofer diffraction

A converging lens images a Fraunhofer pattern on a nearby screen.

Converging cylindrical lens

a

f

Screen

Single-Slit Diffraction: Locating the Dark Fringes

Figure 36.4 shows a side view of the same setup; the long sides of the slit are perpendicular to the figure, and plane waves are incident on the slit from the left. According to Huygens's principle, each element of area of the slit opening can be considered as a source of secondary waves. In particular, imagine dividing the slit into several narrow strips of equal width, parallel to the long edges and perpendicular to the page. Figure 36.4a shows two such strips. Cylindrical secondary wavelets, shown in cross section, spread out from each strip.

In Fig. 36.4b a screen is placed to the right of the slit. We can calculate the resultant intensity at a point *P* on the screen by adding the contributions from the individual wavelets, taking proper account of their various phases and amplitudes. It's easiest to do this calculation if we assume that the screen is far enough away that all the rays from various parts of the slit to a particular point *P* on the screen are parallel, as in Fig. 36.4c. An equivalent situation is Fig. 36.4d, in which the rays to the lens are parallel and the lens forms a reduced image of the same pattern that would be formed on an infinitely distant screen without the lens. We might expect that the various light paths through the lens would introduce additional phase shifts, but in fact it can be shown that all the paths have *equal* phase shifts, so this is not a problem.

The situation of Fig. 36.4b is Fresnel diffraction; those in Figs. 36.4c and 36.4d, where the outgoing rays are considered parallel, are Fraunhofer diffraction. We can derive quite simply the most important characteristics of the Fraunhofer diffraction pattern from a single slit. First consider two narrow strips, one just below the top edge of the drawing of the slit and one at its center, shown in end view in **Fig. 36.5**. The difference in path length to point *P* is $(a/2) \sin\theta$, where *a* is the slit width and θ is the angle between the perpendicular to the slit

(a)

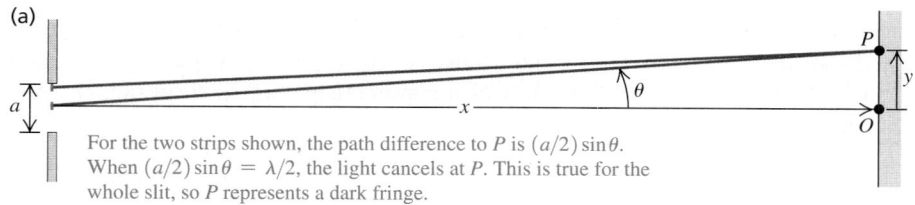

For the two strips shown, the path difference to *P* is $(a/2) \sin\theta$. When $(a/2) \sin\theta = \lambda/2$, the light cancels at *P*. This is true for the whole slit, so *P* represents a dark fringe.

(b) Enlarged view of the top half of the slit

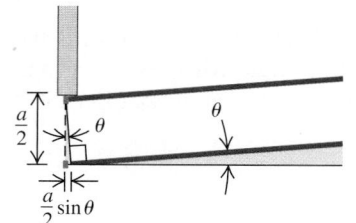

$\frac{a}{2} \sin\theta$

θ is usually very small, so we can use the approximations $\sin\theta = \theta$ and $\tan\theta = \theta$. Then the condition for a dark band is

$$y_m = x \frac{m\lambda}{a}$$

36.5 Side view of a horizontal slit. When the distance *x* to the screen is much greater than the slit width *a*, the rays from a distance $a/2$ apart may be considered parallel.

and a line from the center of the slit to P. Suppose this path difference happens to be equal to $\lambda/2$; then light from these two strips arrives at point P with a half-cycle phase difference, and cancellation occurs.

Similarly, light from two strips immediately *below* the two in the figure also arrives at P a half-cycle out of phase. In fact, the light from *every* strip in the top half of the slit cancels out the light from a corresponding strip in the bottom half. Hence the combined light from the entire slit completely cancels at P, giving a dark fringe in the interference pattern. A dark fringe occurs whenever

$$\frac{a}{2}\sin\theta = \pm\frac{\lambda}{2} \quad \text{or} \quad \sin\theta = \pm\frac{\lambda}{a} \tag{36.1}$$

The plus-or-minus ($\pm$) sign in Eq. (36.1) says that there are symmetric dark fringes above and below point O in Fig. 36.5a. The upper fringe ($\theta > 0$) occurs at a point P where light from the bottom half of the slit travels $\lambda/2$ farther to P than does light from the top half; the lower fringe ($\theta < 0$) occurs where light from the *top* half travels $\lambda/2$ farther than light from the *bottom* half.

We may also divide the slit into quarters, sixths, and so on, and use the above argument to show that a dark fringe occurs whenever $\sin\theta = \pm 2\lambda/a$, $\pm 3\lambda/a$, and so on. Thus the condition for a *dark* fringe is

Dark fringes, single-slit diffraction:
$$\sin\theta = \frac{m\lambda}{a} \quad (m = \pm 1, \pm 2, \pm 3, \dots) \tag{36.2}$$

Angle of line from center of slit to mth dark fringe on screen / Slit width / Wavelength

For example, if the slit width is equal to ten wavelengths ($a = 10\lambda$), dark fringes occur at $\sin\theta = \pm\frac{1}{10}, \pm\frac{2}{10}, \pm\frac{3}{10}, \dots$. Between the dark fringes are bright fringes. Note that $\sin\theta = 0$ corresponds to a *bright* band; in this case, light from the entire slit arrives at P in phase. Thus it would be wrong to put $m = 0$ in Eq. (36.2).

With light, the wavelength λ is of the order of 500 nm $= 5 \times 10^{-7}$ m. This is often much smaller than the slit width a; a typical slit width is 10^{-2} cm $= 10^{-4}$ m. Therefore the values of θ in Eq. (36.2) are often so small that the approximation $\sin\theta \approx \theta$ (where θ is in radians) is a very good one. In that case we can rewrite this equation as

$$\theta = \frac{m\lambda}{a} \quad (m = \pm 1, \pm 2, \pm 3, \dots) \quad \text{(for small angles } \theta \text{ in radians)}$$

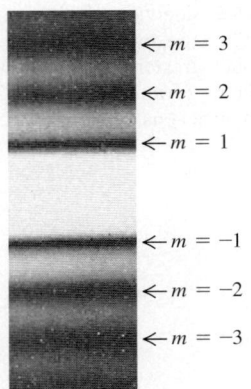

36.6 Photograph of the Fraunhofer diffraction pattern of a single horizontal slit.

$\leftarrow m = 3$
$\leftarrow m = 2$
$\leftarrow m = 1$
$\leftarrow m = -1$
$\leftarrow m = -2$
$\leftarrow m = -3$

Also, if the distance from slit to screen is x, as in Fig. 36.5a, and the vertical distance of the mth dark band from the center of the pattern is y_m, then $\tan\theta = y_m/x$. For small θ we may also approximate $\tan\theta$ by θ (in radians). We then find

$$y_m = x\frac{m\lambda}{a} \quad \text{(for } y_m \ll x) \tag{36.3}$$

Figure 36.6 is a photograph of a single-slit diffraction pattern with the $m = \pm 1, \pm 2$, and ± 3 minima labeled. The central bright fringe is wider than the other bright fringes; in the small-angle approximation used in Eq. (36.3), it is exactly twice as wide.

CAUTION Single-slit diffraction vs. two-slit interference Equation (36.3) has the same form as the equation for the two-slit pattern, Eq. (35.6), except that in Eq. (36.3) we use x rather than R for the distance to the screen. But Eq. (36.3) gives the positions of the *dark* fringes in a *single-slit* pattern rather than the *bright* fringes in a *double-slit* pattern. Also, $m = 0$ in Eq. (36.2) is *not* a dark fringe. Be careful! ▮

EXAMPLE 36.1 SINGLE-SLIT DIFFRACTION

You pass 633-nm laser light through a narrow slit and observe the diffraction pattern on a screen 6.0 m away. The distance on the screen between the centers of the first minima on either side of the central bright fringe is 32 mm (**Fig. 36.7**). How wide is the slit?

36.7 A single-slit diffraction experiment.

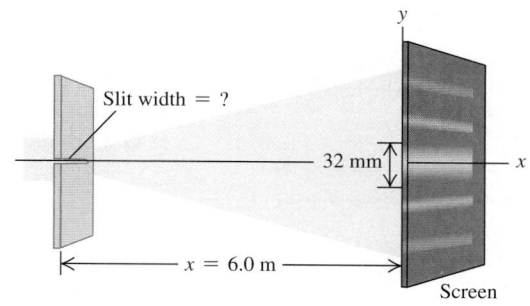

SOLUTION

IDENTIFY and SET UP: This problem involves the relationship between the positions of dark fringes in a single-slit diffraction pattern and the slit width a (our target variable). The distances between fringes on the screen are much smaller than the slit-to-screen distance, so the angle θ shown in Fig. 36.5a is very small and we can use Eq. (36.3) to solve for a.

EXECUTE: The first minimum corresponds to $m = 1$ in Eq. (36.3). The distance y_1 from the central maximum to the first minimum on either side is half the distance between the two first minima, so $y_1 = (32 \text{ mm})/2 = 16$ mm. Solving Eq. (36.3) for a, we find

$$a = \frac{x\lambda}{y_1} = \frac{(6.0 \text{ m})(633 \times 10^{-9} \text{ m})}{16 \times 10^{-3} \text{ m}} = 2.4 \times 10^{-4} \text{ m} = 0.24 \text{ mm}$$

EVALUATE: The angle θ is small only if the wavelength is small compared to the slit width. Since $\lambda = 633$ nm $= 6.33 \times 10^{-7}$ m and we have found $a = 0.24$ mm $= 2.4 \times 10^{-4}$ m, our result is consistent with this: The wavelength is $(6.33 \times 10^{-7} \text{ m})/(2.4 \times 10^{-4} \text{ m}) = 0.0026$ as large as the slit width. Can you show that the distance between the *second* minima on either side is $2(32 \text{ mm}) = 64$ mm, and so on?

TEST YOUR UNDERSTANDING OF SECTION 36.2 Rank the following single-slit diffraction experiments in order of the size of the angle from the center of the diffraction pattern to the first dark fringe, from largest to smallest: (i) Wavelength 400 nm, slit width 0.20 mm; (ii) wavelength 600 nm, slit width 0.20 mm; (iii) wavelength 400 nm, slit width 0.30 mm; (iv) wavelength 600 nm, slit width 0.30 mm. ▮

36.3 INTENSITY IN THE SINGLE-SLIT PATTERN

We can derive an expression for the intensity distribution for the single-slit diffraction pattern by the same phasor-addition method that we used in Section 35.3 for the two-slit interference pattern. We again imagine a plane wave front at the slit subdivided into a large number of strips. We superpose the contributions of the Huygens wavelets from all the strips at a point P on a distant screen at an angle θ from the normal to the slit plane (**Fig. 36.8a**, next page). To do this, we use a phasor to represent the sinusoidally varying $\vec{E}$ field from each strip. The magnitude of the vector sum of the phasors at each point P is the amplitude E_P of the total $\vec{E}$ field at that point. The intensity at P is proportional to E_P^2.

At the point O shown in Fig. 36.8a, corresponding to the center of the pattern where $\theta = 0$, there are negligible path differences for $x \gg a$; the phasors are all essentially *in phase* (that is, have the same direction). In Fig. 36.8b we draw the phasors at time $t = 0$ and denote the resultant amplitude at O by E_0. In this illustration we have divided the slit into 14 strips.

36.8 Using phasor diagrams to find the amplitude of the $\vec{E}$ field in single-slit diffraction. Each phasor represents the $\vec{E}$ field from a single strip within the slit.

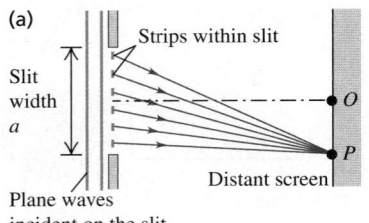

(a)

Strips within slit

Slit width a

O

P

Plane waves incident on the slit

Distant screen

(b) At the center of the diffraction pattern (point O), the phasors from all strips within the slit are in phase.

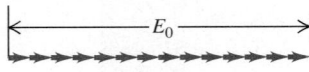

E_0

(c) Phasor diagram at a point slightly off the center of the pattern; β = total phase difference between the first and last phasors.

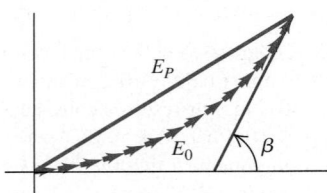

E_P

E_0

β

(d) As in (c), but in the limit that the slit is subdivided into infinitely many strips

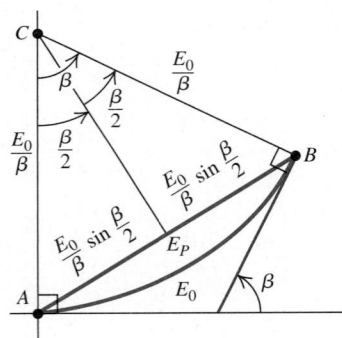

C

β

$\dfrac{E_0}{\beta}$

$\dfrac{\beta}{2}$

$\dfrac{E_0}{\beta}$

$\dfrac{\beta}{2}$

$\dfrac{E_0}{\beta}\sin\dfrac{\beta}{2}$

B

$\dfrac{E_0}{\beta}\sin\dfrac{\beta}{2}$

E_P

A

E_0

β

Now consider wavelets arriving from different strips at point P in Fig. 36.8a, at an angle θ from point O. Because of the differences in path length, there are now phase differences between wavelets coming from adjacent strips; the corresponding phasor diagram is shown in Fig. 36.8c. The vector sum of the phasors is now part of the perimeter of a many-sided polygon, and E_P, the amplitude of the resultant electric field at P, is the *chord*. The angle β is the total phase difference between the wave received at P from the top strip of Fig. 36.8a and the wave received at P from the bottom strip.

We may imagine dividing the slit into narrower and narrower strips. In the limit that there is an infinite number of infinitesimally narrow strips, the curved trail of phasors becomes an *arc of a circle* (Fig. 36.8d), with arc length equal to the length E_0 in Fig. 36.8b. The center C of this arc is found by constructing perpendiculars at A and B. From the relationship among arc length, radius, and angle, the radius of the arc is E_0/β; the amplitude E_P of the resultant electric field at P is equal to the chord AB, which is $2(E_0/\beta)\sin(\beta/2)$. (Note that β *must* be in radians!) We then have

$$E_P = E_0 \frac{\sin(\beta/2)}{\beta/2} \qquad \text{(amplitude in single-slit diffraction)} \qquad (36.4)$$

The intensity at each point on the screen is proportional to the square of the amplitude given by Eq. (36.4). If I_0 is the intensity in the straight-ahead direction where $\theta = 0$ and $\beta = 0$, then the intensity I at any point is

$$I = I_0 \left[\frac{\sin(\beta/2)}{\beta/2}\right]^2 \qquad \text{(intensity in single-slit diffraction)} \qquad (36.5)$$

We can express the phase difference β in terms of geometric quantities, as we did for the two-slit pattern. From Eq. (35.11) the phase difference is $2\pi/\lambda$ times the path difference. Figure 36.5 shows that the path difference between the ray from the top of the slit and the ray from the middle of the slit is $(a/2)\sin\theta$. The path difference between the rays from the top of the slit and the bottom of the slit is twice this, so

$$\beta = \frac{2\pi}{\lambda}a\sin\theta \qquad (36.6)$$

and Eq. (36.5) becomes

Angle of line from center of slit to position on screen

Intensity in single-slit diffraction

$$I = I_0 \left\{\frac{\sin[\pi a(\sin\theta)/\lambda]}{\pi a(\sin\theta)/\lambda}\right\}^2 \qquad (36.7)$$

Intensity at $\theta = 0$ Slit width   Wavelength

This equation expresses the intensity directly in terms of the angle θ. In many calculations it is easier first to calculate the phase angle β, from Eq. (36.6), and then to use Eq. (36.5).

Equation (36.7) is plotted in **Fig. 36.9a**. Note that the central intensity peak is much larger than any of the others. This means that most of the power in the wave remains within an angle θ from the perpendicular to the slit, where $\sin\theta = \lambda/a$ (the first diffraction minimum). You can see this easily in Fig. 36.9b, which is a photograph of water waves undergoing single-slit diffraction. Note also that the peak intensities in Fig. 36.9a decrease rapidly as we go away from the center of the pattern. (Compare Fig. 36.6, which shows a single-slit diffraction pattern for light.)

The dark fringes in the pattern are the places where $I = 0$. These occur at points for which the numerator of Eq. (36.5) is zero so that β is a multiple of 2π. From Eq. (36.6) this corresponds to

$$\frac{a\sin\theta}{\lambda} = m \quad (m = \pm 1, \pm 2, \dots)$$

$$\sin\theta = \frac{m\lambda}{a} \quad (m = \pm 1, \pm 2, \dots) \tag{36.8}$$

This agrees with our previous result, Eq. (36.2). Note again that $\beta = 0$ (corresponding to $\theta = 0$) is *not* a minimum. Equation (36.5) is indeterminate at $\beta = 0$, but we can evaluate the limit as $\beta \to 0$ by using L'Hôpital's rule. We find that at $\beta = 0$, $I = I_0$, as we should expect.

Intensity Maxima in the Single-Slit Pattern

We can also use Eq. (36.5) to calculate the positions of the peaks, or *intensity maxima*, and the intensities at these peaks. This is not quite as simple as it may appear. We might expect the peaks to occur where the sine function reaches the value ± 1—namely, where $\beta = \pm\pi, \pm 3\pi, \pm 5\pi$, or in general,

$$\beta \approx \pm(2m + 1)\pi \quad (m = 0, 1, 2, \dots) \tag{36.9}$$

This is *approximately* correct, but because of the factor $(\beta/2)^2$ in the denominator of Eq. (36.5), the maxima don't occur precisely at these points. When we take the derivative of Eq. (36.5) with respect to β and set it equal to zero to try to find the maxima and minima, we get a transcendental equation that has to be solved numerically. In fact there is *no* maximum near $\beta = \pm\pi$. The first maxima on either side of the central maximum, near $\beta = \pm 3\pi$, actually occur at $\pm 2.860\pi$. The second side maxima, near $\beta = \pm 5\pi$, are actually at $\pm 4.918\pi$, and so on. The error in Eq. (36.9) vanishes in the limit of large m—that is, for intensity maxima far from the center of the pattern.

To find the intensities at the side maxima, we substitute these values of β back into Eq. (36.5). Using the approximate expression in Eq. (36.9), we get

$$I_m \approx \frac{I_0}{\left(m + \frac{1}{2}\right)^2 \pi^2} \tag{36.10}$$

where I_m is the intensity of the mth side maximum and I_0 is the intensity of the central maximum. Equation (36.10) gives the series of intensities

$$0.0450I_0 \qquad 0.0162I_0 \qquad 0.0083I_0$$

and so on. As we have pointed out, this equation is only approximately correct. The actual intensities of the side maxima turn out to be

$$0.0472I_0 \qquad 0.0165I_0 \qquad 0.0083I_0 \qquad \cdots$$

These intensities decrease very rapidly, as Fig. 36.9a also shows. Even the first side maxima have less than 5% of the intensity of the central maximum.

36.9 (a) Intensity versus angle in single-slit diffraction. The values of m label intensity minima given by Eq. (36.8). Most of the wave power goes into the central intensity peak (between the $m = 1$ and $m = -1$ intensity minima). (b) These water waves passing through a small aperture behave exactly like light waves in single-slit diffraction. Only the diffracted waves within the central intensity peak are visible; the waves at larger angles are too faint to see.

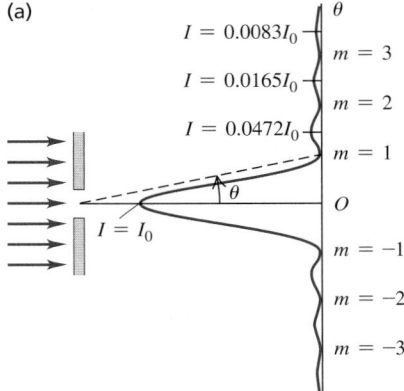

(a)

(b)

DATA *SPEAKS*

Single-Slit Diffraction

When students were given a problem involving diffraction of waves through a slit, more than 30% gave an incorrect response. Common errors:

- Confusion about the positions of dark fringes. Equation (36.2) gives the angle from the mth dark fringe to the center of the diffraction pattern—*not* the angle from a dark fringe on one side of the pattern to the corresponding dark fringe on the other side.

- Confusion about how the slit width a and wavelength λ affect the width of the diffraction pattern. Decreasing a or increasing λ makes the pattern broader; increasing a or decreasing λ makes the pattern narrower.

36.10 The single-slit diffraction pattern depends on the ratio of the slit width a to the wavelength λ.

(a) $a = \lambda$

If the slit width is equal to or narrower than the wavelength, only one broad maximum forms.

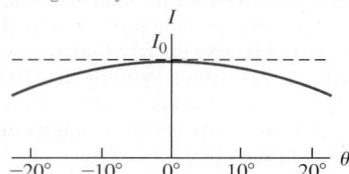

(b) $a = 5\lambda$

(c) $a = 8\lambda$

The wider the slit (or the shorter the wavelength), the narrower and sharper is the central peak.

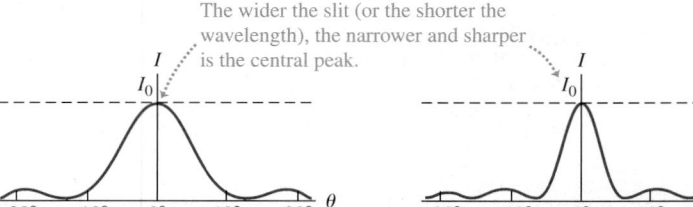

Width of the Single-Slit Pattern

For small angles the angular spread of the diffraction pattern is inversely proportional to the ratio of the slit width a to the wavelength λ. **Figure 36.10** shows graphs of intensity I as a function of the angle θ for three values of the ratio a/λ.

With light waves, the wavelength λ is often much smaller than the slit width a, and the values of θ in Eqs. (36.6) and (36.7) are so small that the approximation $\sin\theta = \theta$ is very good. With this approximation the position θ_1 of the first ($m = 1$) minimum, corresponding to $\beta/2 = \pi$, is, from Eq. (36.7),

$$\theta_1 = \frac{\lambda}{a} \tag{36.11}$$

This characterizes the width (angular spread) of the central maximum, and we see that it is *inversely* proportional to the slit width a. When the small-angle approximation is valid, the central maximum is exactly twice as wide as each side maximum. When a is of the order of a centimeter or more, θ_1 is so small that we can consider practically all the light to be concentrated at the geometrical focus. But when a is less than λ, the central maximum spreads over 180°, and the fringe pattern is not seen at all.

It's important to keep in mind that diffraction occurs for *all* kinds of waves, not just light. Sound waves undergo diffraction when they pass through a slit or aperture such as an ordinary doorway. The sound waves used in speech have wavelengths of about a meter or greater, and a typical doorway is less than 1 m wide; in this situation, a is less than λ, and the central intensity maximum extends over 180°. This is why the sounds coming through an open doorway can easily be heard by an eavesdropper hiding out of sight around the corner. In the same way, sound waves can bend around the head of an instructor who faces the blackboard while lecturing (**Fig. 36.11**). By contrast, there is essentially no diffraction of visible light through a doorway because the width a is very much greater than the wavelength λ (of order 5×10^{-7} m). You can *hear* around corners because typical sound waves have relatively long wavelengths; you cannot *see* around corners because the wavelength of visible light is very short.

36.11 The sound waves used in speech have a long wavelength (about 1 m) and can easily bend around this instructor's head. By contrast, light waves have very short wavelengths and undergo very little diffraction. Hence you can't *see* around his head!

EXAMPLE 36.2 **SINGLE-SLIT DIFFRACTION: INTENSITY I**

(a) The intensity at the center of a single-slit diffraction pattern is I_0. What is the intensity at a point in the pattern where there is a 66-radian phase difference between wavelets from the two edges of the slit? (b) If this point is 7.0° away from the central maximum, how many wavelengths wide is the slit?

SOLUTION

IDENTIFY and SET UP: In our analysis of Fig. 36.8 we used the symbol β for the phase difference between wavelets from the two

edges of the slit. In part (a) we use Eq. (36.5) to find the intensity I at the point in the pattern where $\beta = 66$ rad. In part (b) we need to find the slit width a as a multiple of the wavelength λ so our target variable is a/λ. We are given the angular position θ of the point where $\beta = 66$ rad, so we can use Eq. (36.6) to solve for a/λ.

EXECUTE: (a) We have $\beta/2 = 33$ rad, so from Eq. (36.5),

$$I = I_0\left[\frac{\sin(33 \text{ rad})}{33 \text{ rad}}\right]^2 = (9.2 \times 10^{-4})I_0$$

(b) From Eq. (36.6),

$$\frac{a}{\lambda} = \frac{\beta}{2\pi \sin\theta} = \frac{66 \text{ rad}}{(2\pi \text{ rad})\sin 7.0°} = 86$$

For example, for 550-nm light the slit width is $a = (86)(550 \text{ nm}) = 4.7 \times 10^{-5}$ m $= 0.047$ mm, or roughly $\frac{1}{20}$ mm.

EVALUATE: To what point in the diffraction pattern does this value of β correspond? To find out, note that $\beta = 66$ rad is approximately equal to 21π. This is an odd multiple of π, corresponding to the form $(2m + 1)\pi$ found in Eq. (36.9) for the intensity *maxima*. Hence $\beta = 66$ rad corresponds to a point near the tenth $(m = 10)$ maximum. This is well beyond the range shown in Fig. 36.9a, which shows only maxima out to $m = \pm 3$.

EXAMPLE 36.3   SINGLE-SLIT DIFFRACTION: INTENSITY II

In the experiment described in Example 36.1 (Section 36.2), the intensity at the center of the pattern is I_0. What is the intensity at a point on the screen 3.0 mm from the center of the pattern?

SOLUTION

IDENTIFY and SET UP: This is similar to Example 36.2, except that we are not given the value of the phase difference β at the point in question. We use geometry to determine the angle θ for our point and then use Eq. (36.7) to find the intensity I (the target variable).

EXECUTE: Referring to Fig. 36.5a, we have $y = 3.0$ mm and $x = 6.0$ m, so $\tan\theta = y/x = (3.0 \times 10^{-3} \text{ m})/(6.0 \text{ m}) = 5.0 \times 10^{-4}$.

This is so small that the values of $\tan\theta$, $\sin\theta$, and θ (in radians) are all nearly the same. Then, using Eq. (36.7),

$$\frac{\pi a \sin\theta}{\lambda} = \frac{\pi(2.4 \times 10^{-4} \text{ m})(5.0 \times 10^{-4})}{6.33 \times 10^{-7} \text{ m}}$$

$$= 0.60$$

$$I = I_0 \left(\frac{\sin 0.60}{0.60} \right)^2 = 0.89 I_0$$

EVALUATE: Figure 36.9a shows that an intensity this high can occur only within the central intensity maximum. This checks out; from Example 36.1, the first intensity minimum ($m = 1$ in Fig. 36.9a) is $(32 \text{ mm})/2 = 16$ mm from the center of the pattern, so the point in question here at $y = 3$ mm does, indeed, lie within the central maximum.

TEST YOUR UNDERSTANDING OF SECTION 36.3 Coherent electromagnetic radiation is sent through a slit of width 0.0100 mm. For which of the following wavelengths will there be *no* points in the diffraction pattern where the intensity is zero? (i) Blue light of wavelength 500 nm; (ii) infrared light of wavelength 10.6 μm; (iii) microwaves of wavelength 1.00 mm; (iv) ultraviolet light of wavelength 50.0 nm. ▮

36.4 MULTIPLE SLITS

In Sections 35.2 and 35.3 we analyzed interference from two point sources and from two very narrow slits; in this analysis we ignored effects due to the finite (that is, nonzero) slit width. In Sections 36.2 and 36.3 we considered the diffraction effects that occur when light passes through a single slit of finite width. Additional interesting effects occur when we have two slits with finite width or when there are several very narrow slits.

Two Slits of Finite Width

Let's take another look at the two-slit pattern in the more realistic case in which the slits have finite width. If the slits are narrow in comparison to the wavelength, we can assume that light from each slit spreads out uniformly in all directions to the right of the slit. We used this assumption in Section 35.3 to calculate the interference pattern described by Eq. (35.10) or (35.15), consisting of a series of equally spaced, equally intense maxima. However, when the slits have finite width, the peaks in the two-slit interference pattern are modulated by the single-slit diffraction pattern characteristic of the width of each slit.

36.12 Finding the intensity pattern for two slits of finite width.

(a) Single-slit diffraction pattern for a slit width a

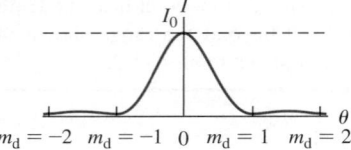

$m_d = -2$ $m_d = -1$ 0 $m_d = 1$ $m_d = 2$

(b) Two-slit interference pattern for narrow slits whose separation d is four times the width of the slit in **(a)**

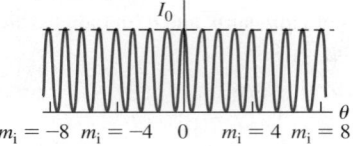

$m_i = -8$ $m_i = -4$ 0 $m_i = 4$ $m_i = 8$

(c) Calculated intensity pattern for two slits of width a and separation $d = 4a$, including both interference and diffraction effects

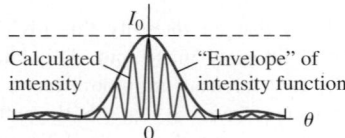

Calculated intensity   "Envelope" of intensity function

0

(d) Photograph of the pattern calculated in **(c)**

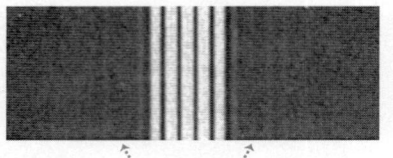

For $d = 4a$, every fourth interference maximum at the sides ($m_i = \pm 4, \pm 8, ...$) is missing.

36.13 Multiple-slit diffraction. Here a lens is used to give a Fraunhofer pattern on a nearby screen, as in Fig. 36.4d.

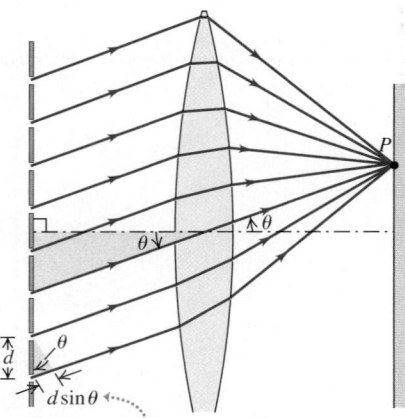

Maxima occur where the path difference for adjacent slits is a whole number of wavelengths: $d \sin\theta = m\lambda$.

Figure 36.12a shows the intensity in a single-slit diffraction pattern with slit width a. The *diffraction minima* are labeled by the integer $m_d = \pm 1, \pm 2, ...$ ("d" for "diffraction"). Figure 36.12b shows the pattern formed by two very narrow slits with distance d between slits, where d is four times as great as the single-slit width a in Fig. 36.12a; that is, $d = 4a$. The *interference maxima* are labeled by the integer $m_i = 0, \pm 1, \pm 2, ...$ ("i" for "interference"). We note that the spacing between adjacent minima in the single-slit pattern is four times as great as in the two-slit pattern. Now suppose we widen each of the narrow slits to the same width a as that of the single slit in Fig. 36.12a. Figure 36.12c shows the pattern from two slits with width a, separated by a distance (between centers) $d = 4a$. The effect of the finite width of the slits is to superimpose the two patterns—that is, to multiply the two intensities at each point. The two-slit peaks are in the same positions as before, but their intensities are modulated by the single-slit pattern, which acts as an "envelope" for the intensity function. The expression for the intensity shown in Fig. 36.12c is proportional to the product of the two-slit and single-slit expressions, Eqs. (35.10) and (36.5):

$$I = I_0 \cos^2 \frac{\phi}{2} \left[\frac{\sin(\beta/2)}{\beta/2} \right]^2 \quad \text{(two slits of finite width)} \quad (36.12)$$

where, as before,

$$\phi = \frac{2\pi d}{\lambda} \sin\theta \qquad \beta = \frac{2\pi a}{\lambda} \sin\theta$$

In Fig. 36.12c, every fourth interference maximum at the sides is *missing* because these interference maxima ($m_i = \pm 4, \pm 8, ...$) coincide with diffraction minima ($m_d = \pm 1, \pm 2, ...$). This can also be seen in Fig. 36.12d, which is a photograph of an actual pattern with $d = 4a$. You should be able to convince yourself that there will be "missing" maxima whenever d is an integer multiple of a.

Figures 36.12c and 36.12d show that as you move away from the central bright maximum of the two-slit pattern, the intensity of the maxima decreases. This is a result of the single-slit modulating pattern shown in Fig. 36.12a; mathematically, the decrease in intensity arises from the factor $(\beta/2)^2$ in the denominator of Eq. (36.12). You can also see this decrease in Fig. 35.6 (Section 35.2). The narrower the slits, the broader the single-slit pattern (as in Fig. 36.10) and the slower the decrease in intensity from one interference maximum to the next.

Shall we call the pattern in Fig. 36.12d *interference* or *diffraction*? It's really both, since it results from the superposition of waves coming from various parts of the two apertures.

Several Slits

Next let's consider patterns produced by *several* very narrow slits. As we will see, systems of narrow slits are of tremendous practical importance in *spectroscopy,* the determination of the particular wavelengths of light coming from a source. Assume that each slit is narrow in comparison to the wavelength, so its diffraction pattern spreads out nearly uniformly. **Figure 36.13** shows an array of eight narrow slits, with distance d between adjacent slits. Constructive interference occurs for rays at angle θ to the normal that arrive at point P with a path difference between adjacent slits equal to an integer number of wavelengths:

$$d \sin\theta = m\lambda \qquad (m = 0, \pm 1, \pm 2, ...)$$

This means that reinforcement occurs when the phase difference ϕ at P for light from adjacent slits is an integer multiple of 2π. That is, the maxima in the pattern occur at the *same* positions as for *two* slits with the same spacing.

What happens *between* the maxima is different with multiple slits, however. In the two-slit pattern, there is exactly one intensity minimum located midway between each pair of maxima, corresponding to angles for which the phase

36.14 Phasor diagrams for light passing through eight narrow slits. Intensity maxima occur when the phase difference $\phi = 0, 2\pi, 4\pi, \ldots$. Between the maxima at $\phi = 0$ and $\phi = 2\pi$ are seven minima, corresponding to $\phi = \pi/4, \pi/2, 3\pi/4, \pi, 5\pi/4, 3\pi/2$, and $7\pi/4$. Can you draw phasor diagrams for the other minima?

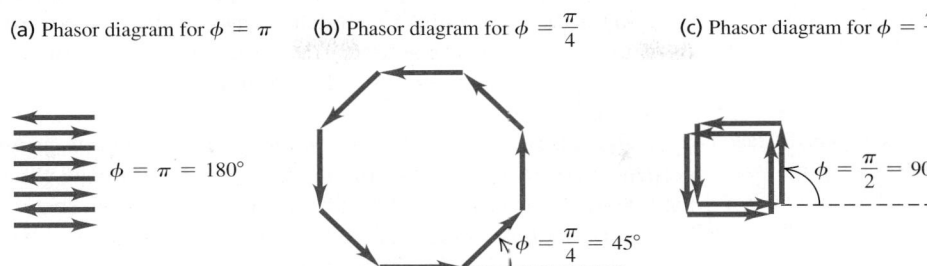

(a) Phasor diagram for $\phi = \pi$ **(b)** Phasor diagram for $\phi = \dfrac{\pi}{4}$ **(c)** Phasor diagram for $\phi = \dfrac{\pi}{2}$

$\phi = \pi = 180°$

$\phi = \dfrac{\pi}{4} = 45°$

$\phi = \dfrac{\pi}{2} = 90°$

difference between waves from the two sources is $\pi, 3\pi, 5\pi$, and so on. In the eight-slit pattern these are also minima because the light from adjacent slits cancels out in pairs, corresponding to the phasor diagram in **Fig. 36.14a**. But these are not the only minima in the eight-slit pattern. For example, when the phase difference ϕ from adjacent sources is $\pi/4$, the phasor diagram is as shown in Fig. 36.14b; the total (resultant) phasor is zero, and the intensity is zero. When $\phi = \pi/2$, we get the phasor diagram of Fig. 36.14c, and again both the total phasor and the intensity are zero. More generally, the intensity with eight slits is zero whenever ϕ is an integer multiple of $\pi/4$, *except* when ϕ is a multiple of 2π. Thus there are seven minima for every maximum.

Figure 36.15b shows the result of a detailed calculation of the eight-slit pattern. The large maxima, called *principal maxima,* are in the same positions as for the two-slit pattern of Fig. 36.15a but are much narrower. If the phase difference ϕ between adjacent slits is slightly different from a multiple of 2π, the waves from slits 1 and 2 will be only a little out of phase; however, the phase difference between slits 1 and 3 will be greater, that between slits 1 and 4 will be greater still, and so on. This leads to a partial cancellation for angles that are only slightly different from the angle for a maximum, giving the narrow maxima in Fig. 36.15b. The maxima are even narrower with 16 slits (Fig. 36.15c).

You should show that when there are N slits, there are $(N - 1)$ minima between each pair of principal maxima and a minimum occurs whenever ϕ is an integral multiple of $2\pi/N$ (except when ϕ is an integral multiple of 2π, which gives a principal maximum). There are small *secondary* intensity maxima between the minima; these become smaller in comparison to the principal maxima as N increases. The greater the value of N, the narrower the principal maxima become. From an energy standpoint the total power in the entire pattern is proportional to N. The height of each principal maximum is proportional to N^2, so from energy conservation the width of each principal maximum must be proportional to $1/N$. In the next section we'll see why the details of the multiple-slit pattern are of great practical importance.

TEST YOUR UNDERSTANDING OF SECTION 36.4   Suppose two slits, each of width a, are separated by a distance $d = 2.5a$. Are there any missing maxima in the interference pattern produced by these slits? If so, which are missing? If not, why not? ❚

36.5 THE DIFFRACTION GRATING

We have just seen that increasing the number of slits in an interference experiment (while keeping the spacing of adjacent slits constant) gives interference patterns in which the maxima are in the same positions, but progressively narrower, than with two slits. Because these maxima are so narrow, their angular position, and hence the wavelength, can be measured to very high precision. As we will see, this effect has many important applications.

36.15 Interference patterns for N equally spaced, very narrow slits. (a) Two slits. (b) Eight slits. (c) Sixteen slits. The vertical scales are different for each graph; the maximum intensity is I_0 for a single slit and $N^2 I_0$ for N slits. The width of each peak is proportional to $1/N$.

(a) $N = 2$: two slits produce one minimum between adjacent maxima.

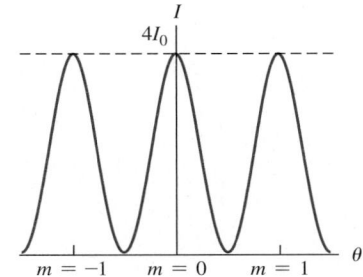

$4I_0$

$m = -1$ $m = 0$ $m = 1$

(b) $N = 8$: eight slits produce taller, narrower maxima in the same locations, separated by seven minima.

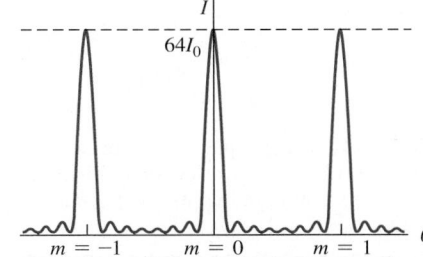

$64I_0$

$m = -1$ $m = 0$ $m = 1$

(c) $N = 16$: with 16 slits, the maxima are even taller and narrower, with more intervening minima.

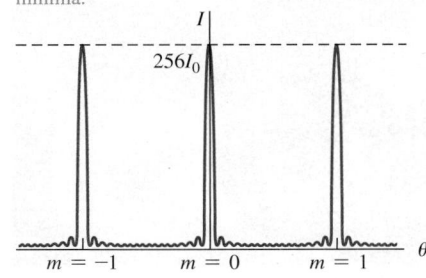

$256I_0$

$m = -1$ $m = 0$ $m = 1$

An array of a large number of parallel slits, all with the same width a and spaced equal distances d between centers, is called a **diffraction grating.** The first one was constructed by Fraunhofer using fine wires. Gratings can be made by using a diamond point to scratch many equally spaced grooves on a glass or metal surface, or by photographic reduction of a pattern of black and white stripes on paper. For a grating, what we have been calling *slits* are often called *rulings* or *lines.*

In **Fig. 36.16,** GG' is a cross section of a *transmission grating;* the slits are perpendicular to the plane of the page, and an interference pattern is formed by the light that is transmitted through the slits. The diagram shows only six slits; an actual grating may contain several thousand. The spacing d between centers of adjacent slits is called the *grating spacing.* A plane monochromatic wave is incident normally on the grating from the left side. We assume far-field (Fraunhofer) conditions; that is, the pattern is formed on a screen that is far enough away that all rays emerging from the grating and going to a particular point on the screen can be considered to be parallel.

We found in Section 36.4 that the principal intensity maxima with multiple slits occur in the same directions as for the two-slit pattern. These are the directions for which the path difference for adjacent slits is an integer number of wavelengths. So the positions of the maxima are once again

36.16 A portion of a transmission diffraction grating. The separation between the centers of adjacent slits is d.

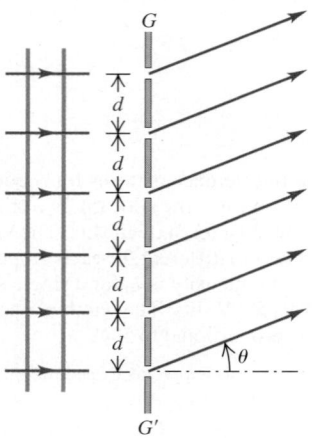

	Distance between slits ····· Wavelength
Intensity maxima, multiple slits:	$d\sin\theta = m\lambda$ $(m = 0, \pm1, \pm2, \dots)$ (36.13)
	Angle of line from center of slit array to mth bright region on screen

The intensity patterns for two, eight, and 16 slits displayed in Fig. 36.15 show the progressive increase in sharpness of the maxima as the number of slits increases.

When a grating containing hundreds or thousands of slits is illuminated by a beam of parallel rays of monochromatic light, the pattern is a series of very sharp lines at angles determined by Eq. (36.13). The $m = \pm1$ lines are called the *first-order lines,* the $m = \pm2$ lines the *second-order lines,* and so on. If the grating is illuminated by white light with a continuous distribution of wavelengths, each value of m corresponds to a continuous spectrum in the pattern. The angle for each wavelength is determined by Eq. (36.13); for a given value of m, long wavelengths (the red end of the spectrum) lie at larger angles (that is, are deviated more from the straight-ahead direction) than do the shorter wavelengths at the violet end of the spectrum.

As Eq. (36.13) shows, the sines of the deviation angles of the maxima are proportional to the ratio λ/d. For substantial deviation to occur, the grating spacing d should be of the same order of magnitude as the wavelength λ. Gratings for use with visible light (λ from 400 to 700 nm) usually have about 1000 slits per millimeter; the value of d is the *reciprocal* of the number of slits per unit length, so d is of the order of $\frac{1}{1000}$ mm = 1000 nm.

In a *reflection grating,* the array of equally spaced slits shown in Fig. 36.16 is replaced by an array of equally spaced ridges or grooves on a reflective screen. The reflected light has maximum intensity at angles where the phase difference between light waves reflected from adjacent ridges or grooves is an integral multiple of 2π. If light of wavelength λ is incident normally on a reflection grating with a spacing d between adjacent ridges or grooves, the *reflected* angles at which intensity maxima occur are given by Eq. (36.13).

The rainbow-colored reflections from the surface of a DVD are a reflection-grating effect (**Fig. 36.17**). The "grooves" are tiny pits 0.12 μm deep in the surface of the disc, with a uniform radial spacing of 0.74 μm = 740 nm. Information is coded on the DVD by varying the *length* of the pits. The reflection-grating aspect of the disc is merely an aesthetic side benefit.

36.17 Microscopic pits on the surface of this DVD act as a reflection grating, splitting white light into its component colors.

EXAMPLE 36.4 WIDTH OF A GRATING SPECTRUM

The wavelengths of the visible spectrum are approximately 380 nm (violet) to 750 nm (red). (a) Find the angular limits of the first-order visible spectrum produced by a plane grating with 600 slits per millimeter when white light falls normally on the grating. (b) Do the first-order and second-order spectra overlap? What about the second-order and third-order spectra? Do your answers depend on the grating spacing?

SOLUTION

IDENTIFY and SET UP: We must find the angles spanned by the visible spectrum in the first-, second-, and third-order spectra. These correspond to $m = 1$, 2, and 3 in Eq. (36.13).

EXECUTE: (a) The grating spacing is

$$d = \frac{1}{600 \text{ slits/mm}} = 1.67 \times 10^{-6} \text{ m}$$

We solve Eq. (36.13) for θ:

$$\theta = \arcsin \frac{m\lambda}{d}$$

Then for $m = 1$, the angular deviations θ_{v1} and θ_{r1} for violet and red light, respectively, are

$$\theta_{v1} = \arcsin\left(\frac{380 \times 10^{-9} \text{ m}}{1.67 \times 10^{-6} \text{ m}}\right) = 13.2°$$

$$\theta_{r1} = \arcsin\left(\frac{750 \times 10^{-9} \text{ m}}{1.67 \times 10^{-6} \text{ m}}\right) = 26.7°$$

That is, the first-order visible spectrum appears with deflection angles from $\theta_{v1} = 13.2°$ (violet) to $\theta_{r1} = 26.7°$ (red).

(b) With $m = 2$ and $m = 3$, our equation $\theta = \arcsin(m\lambda/d)$ for 380-mm violet light yields

$$\theta_{v2} = \arcsin\left(\frac{2(380 \times 10^{-9} \text{ m})}{1.67 \times 10^{-6} \text{ m}}\right) = 27.1°$$

$$\theta_{v3} = \arcsin\left(\frac{3(380 \times 10^{-9} \text{ m})}{1.67 \times 10^{-6} \text{ m}}\right) = 43.0°$$

For 750-nm red light, this same equation gives

$$\theta_{r2} = \arcsin\left(\frac{2(750 \times 10^{-9} \text{ m})}{1.67 \times 10^{-6} \text{ m}}\right) = 63.9°$$

$$\theta_{r3} = \arcsin\left(\frac{3(750 \times 10^{-9} \text{ m})}{1.67 \times 10^{-6} \text{ m}}\right) = \arcsin(1.35) = \text{undefined}$$

Hence the second-order spectrum extends from 27.1° to 63.9° and the third-order spectrum extends from 43.0° to 90° (the largest possible value of θ). The undefined value of θ_{r3} means that the third-order spectrum reaches $\theta = 90° = \arcsin(1)$ at a wavelength shorter than 750 nm; you should be able to show that this happens for $\lambda = 557$ nm. Hence the first-order spectrum (from 13.2° to 26.7°) does not overlap with the second-order spectrum, but the second- and third-order spectra do overlap. You can convince yourself that this is true for any value of the grating spacing d.

EVALUATE: The fundamental reason the first-order and second-order visible spectra don't overlap is that the human eye is sensitive to only a narrow range of wavelengths. Can you show that if the eye could detect wavelengths from 380 nm to 900 nm (in the near-infrared range), the first and second orders *would* overlap?

Grating Spectrographs

Diffraction gratings are widely used to measure the spectrum of light emitted by a source, a process called *spectroscopy* or *spectrometry*. Light incident on a grating of known spacing is dispersed into a spectrum. The angles of deviation of the maxima are then measured, and Eq. (36.13) is used to compute the wavelength. With a grating that has many slits, very sharp maxima are produced, and the angle of deviation (and hence the wavelength) can be measured very precisely.

An important application of this technique is to astronomy. As light generated within the sun passes through the sun's atmosphere, certain wavelengths are selectively absorbed. The result is that the spectrum of sunlight produced by a diffraction grating has dark *absorption lines* (**Fig. 36.18**). Experiments in the laboratory show that different types of atoms and ions absorb light of different wavelengths. By comparing these laboratory results with the wavelengths of absorption lines in the spectrum of sunlight, astronomers can deduce the chemical composition of the sun's atmosphere. The same technique is used to make chemical assays of galaxies that are millions of light-years away.

Figure 36.19, next page, shows one design for a *grating spectrograph* used in astronomy. A transmission grating is used in the figure; in other setups, a reflection grating is used. In older designs a prism was used rather than a grating, and a spectrum was formed by dispersion (see Section 33.4) rather than diffraction. However, there is no simple relationship between wavelength and angle of deviation for a prism, prisms absorb some of the light that passes through them, and they are less effective for many nonvisible wavelengths that are important in astronomy. For these and other reasons, gratings are preferred in precision applications.

36.18 (a) A visible-light photograph of the sun. (b) Sunlight is dispersed into a spectrum by a diffraction grating. Specific wavelengths are absorbed as sunlight passes through the sun's atmosphere, leaving dark lines in the spectrum.

(a)

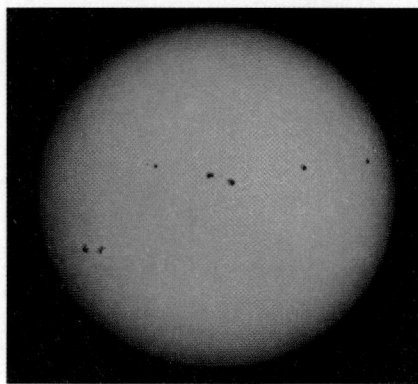

(b)

36.19 A schematic diagram of a diffraction-grating spectrograph for use in astronomy. Note that the light does not strike the grating normal to its surface, so the intensity maxima are given by a somewhat different expression than Eq. (36.13).

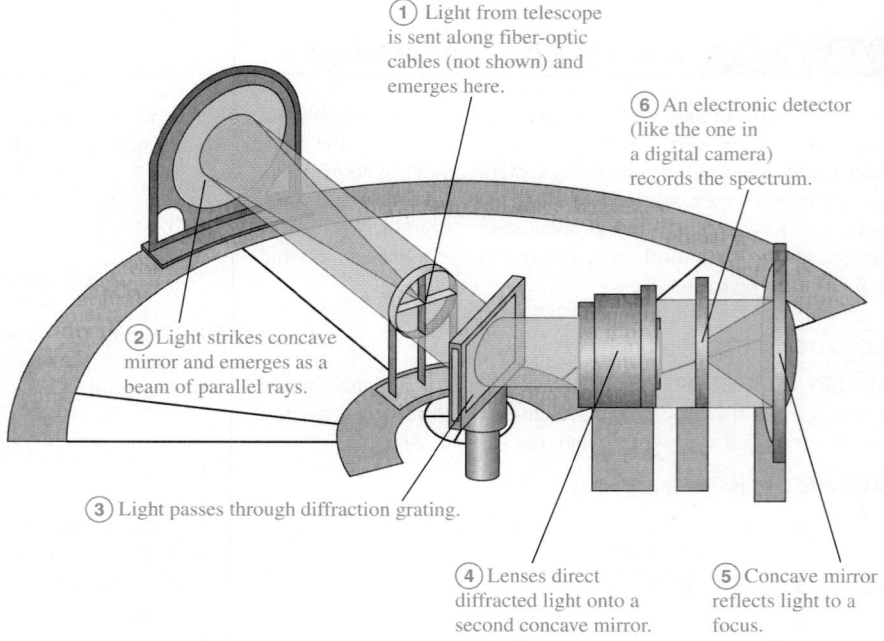

① Light from telescope is sent along fiber-optic cables (not shown) and emerges here.

⑥ An electronic detector (like the one in a digital camera) records the spectrum.

② Light strikes concave mirror and emerges as a beam of parallel rays.

③ Light passes through diffraction grating.

④ Lenses direct diffracted light onto a second concave mirror.

⑤ Concave mirror reflects light to a focus.

BIO Application Detecting DNA with Diffraction Diffraction gratings are used in a common piece of laboratory equipment known as a spectrophotometer. Light shining across a diffraction grating is dispersed into its component wavelengths. A slit is used to block all but a very narrow range of wavelengths, producing a beam of almost perfectly monochromatic light. The instrument then measures how much of that light is absorbed by a solution of biological molecules. For example, the sample tube shown here contains a solution of DNA, which is transparent to visible light but which strongly absorbs ultraviolet light with a wavelength of exactly 260 nm. Therefore, by illuminating the sample with 260-nm light and measuring the amount absorbed, we can determine the concentration of DNA in the solution.

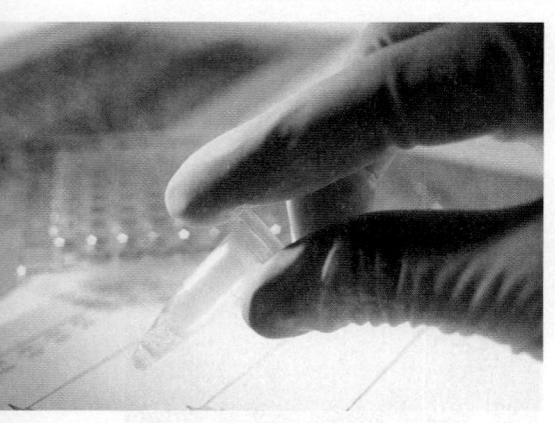

CAUTION **Watch out for different uses of the symbol** d Don't confuse the slit spacing d with the differential "d" in the angular interval $d\theta$ or in the phase shift increment $d\phi$! ▮

Resolution of a Grating Spectrograph

In spectroscopy it is often important to distinguish slightly differing wavelengths. The minimum wavelength difference $\Delta\lambda$ that can be distinguished by a spectrograph is described by the **chromatic resolving power** R, defined as

$$R = \frac{\lambda}{\Delta\lambda} \qquad \text{(chromatic resolving power)} \qquad (36.14)$$

As an example, when sodium atoms are heated, they emit strongly at the yellow wavelengths 589.00 nm and 589.59 nm. A spectrograph that can barely distinguish these two lines in the spectrum (called the *sodium doublet*) has a chromatic resolving power $R = (589.00 \text{ nm})/(0.59 \text{ nm}) = 1000$. (You can see these wavelengths when boiling water on a gas range. If the water boils over onto the flame, dissolved sodium from table salt emits a burst of yellow light.)

We can derive an expression for the resolving power of a diffraction grating used in a spectrograph. Two different wavelengths give diffraction maxima at slightly different angles. As a reasonable (though arbitrary) criterion, let's assume that we can distinguish them as two separate peaks if the maximum of one coincides with the first minimum of the other.

From our discussion in Section 36.4 the mth-order maximum occurs when the phase difference ϕ for adjacent slits is $\phi = 2\pi m$. The first minimum beside that maximum occurs when $\phi = 2\pi m + 2\pi/N$, where N is the number of slits. The phase difference is also given by $\phi = (2\pi d\sin\theta)/\lambda$, so the angular interval $d\theta$ corresponding to a small increment $d\phi$ in the phase shift can be obtained from the differential of this equation:

$$d\phi = \frac{2\pi d\cos\theta \, d\theta}{\lambda}$$

When $d\phi = 2\pi/N$, this corresponds to the angular interval $d\theta$ between a maximum and the first adjacent minimum. Thus $d\theta$ is given by

$$\frac{2\pi}{N} = \frac{2\pi d\cos\theta \, d\theta}{\lambda} \qquad \text{or} \qquad d\cos\theta \, d\theta = \frac{\lambda}{N}$$

Now we need to find the angular spacing $d\theta$ between maxima for two slightly different wavelengths. The positions of these maxima are given by $d\sin\theta = m\lambda$, and the differential of this equation gives

$$d\cos\theta\, d\theta = m\, d\lambda$$

According to our criterion, the limit or resolution is reached when these two angular spacings are equal. Equating the two expressions for the quantity $(d\cos\theta\, d\theta)$, we find

$$\frac{\lambda}{N} = m\, d\lambda \qquad \text{and} \qquad \frac{\lambda}{d\lambda} = Nm$$

If $\Delta\lambda$ is small, we can replace $d\lambda$ by $\Delta\lambda$, and the resolving power R is

$$R = \frac{\lambda}{\Delta\lambda} = Nm \qquad\qquad (36.15)$$

The greater the number of slits N, the better the resolution; also, the higher the order m of the diffraction-pattern maximum that we use, the better the resolution.

TEST YOUR UNDERSTANDING OF SECTION 36.5 What minimum number of slits would be required in a grating to resolve the sodium doublet in the fourth order? (i) 250; (ii) 400; (iii) 1000; (iv) 4000. ∎

36.6 X-RAY DIFFRACTION

X rays were discovered by Wilhelm Röntgen (1845–1923) in 1895, and early experiments suggested that they were electromagnetic waves with wavelengths of the order of 10^{-10} m. At about the same time, the idea began to emerge that in a crystalline solid the atoms are arranged in a regular repeating pattern, with spacing between adjacent atoms also of the order of 10^{-10} m. Putting these two ideas together, Max von Laue (1879–1960) proposed in 1912 that a crystal might serve as a kind of three-dimensional diffraction grating for x rays. That is, a beam of x rays might be scattered (that is, absorbed and re-emitted) by the individual atoms in a crystal, and the scattered waves might interfere just like waves from a diffraction grating.

The first **x-ray diffraction** experiments were performed in 1912 by Friedrich, Knipping, and von Laue, using the experimental setup shown in **Fig. 36.20a**. The scattered x rays *did* form an interference pattern, which they recorded on photographic film. Figure 36.20b is a photograph of such a pattern. These experiments

36.20 (a) An x-ray diffraction experiment. (b) Diffraction pattern (or *Laue pattern*) formed by directing a beam of x rays at a thin section of quartz crystal.

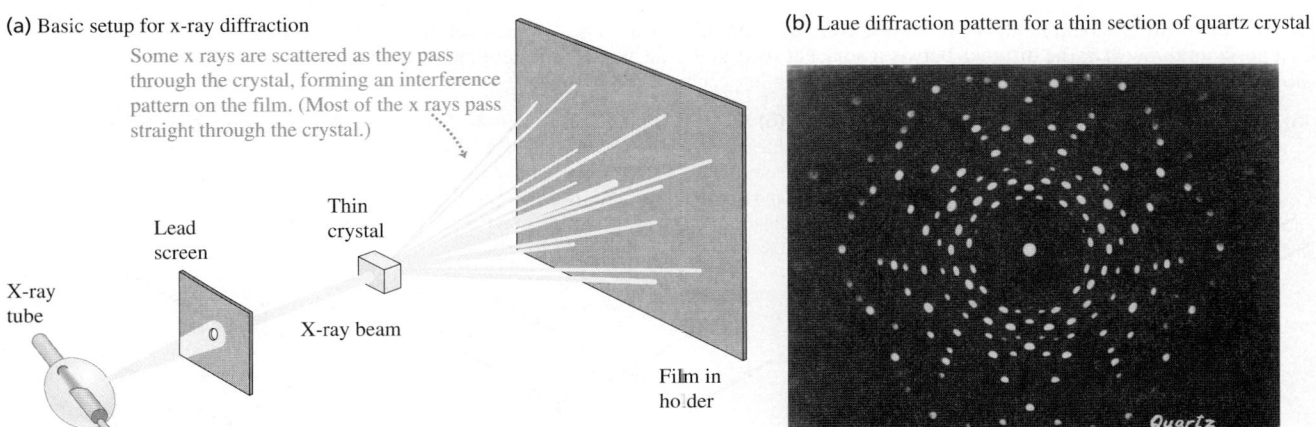

(a) Basic setup for x-ray diffraction

Some x rays are scattered as they pass through the crystal, forming an interference pattern on the film. (Most of the x rays pass straight through the crystal.)

X-ray tube

Lead screen

Thin crystal

X-ray beam

Film in holder

(b) Laue diffraction pattern for a thin section of quartz crystal

Quartz

36.21 Model of the arrangement of ions in a crystal of NaCl (table salt). The spacing of adjacent atoms is 0.282 nm. (The electron clouds of the atoms actually overlap slightly.)

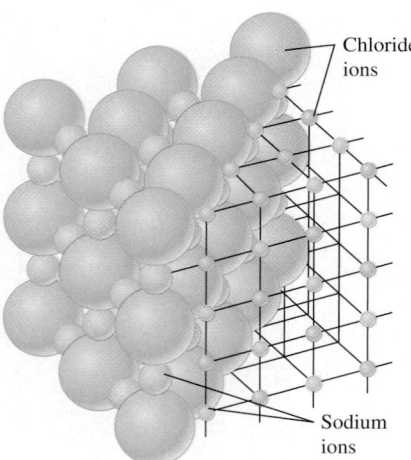

Chloride ions

Sodium ions

verified that x rays *are* waves, or at least have wavelike properties, and also that the atoms in a crystal *are* arranged in a regular pattern (**Fig. 36.21**). Since that time, x-ray diffraction has proved to be an invaluable research tool, both for measuring x-ray wavelengths and for studying the structure of crystals and complex molecules.

A Simple Model of X-Ray Diffraction

To better understand x-ray diffraction, we consider first a two-dimensional scattering situation, as shown in **Fig. 36.22a**, in which a plane wave is incident on a rectangular array of scattering centers. The situation might be a ripple tank with an array of small posts or x rays incident on an array of atoms. In the case of electromagnetic waves, the wave induces an oscillating electric dipole moment in each scatterer. These dipoles act like little antennas, emitting scattered waves. The resulting interference pattern is the superposition of all these scattered waves. The situation is different from that with a diffraction grating, in which the waves from all the slits are emitted *in phase* (for a plane wave at normal incidence). Here the scattered waves are *not* all in phase because their distances from the *source* are different. To compute the interference pattern, we have to consider the *total* path differences for the scattered waves, including the distances from source to scatterer and from scatterer to observer.

As Fig. 36.22b shows, the path length from source to observer is the same for all the scatterers in a single row if the two angles θ_a and θ_r are equal. Scattered radiation from *adjacent* rows is *also* in phase if the path difference for adjacent rows is an integer number of wavelengths. Figure 36.22c shows that this path difference is $2d\sin\theta$, where θ is the common value of θ_a and θ_r. Therefore the conditions for radiation from the *entire array* to reach the observer in phase are (1) the angle of incidence must equal the angle of scattering and (2) the path difference for adjacent rows must equal $m\lambda$, where m is an integer. We can express the second condition, called the **Bragg condition** in honor of x-ray diffraction pioneers Sir William Bragg and his son Laurence Bragg, as

Bragg condition for constructive interference from an array:	Distance between adjacent rows in array········Wavelength $$2d\sin\theta = m\lambda \quad (m = 1, 2, 3, \dots)$$ Angle of line from surface of array to *m*th bright region on screen	(36.16)

CAUTION **Scattering from an array** In Eq. (36.16) the angle θ is measured with respect to the *surface* of the crystal rather than with respect to the *normal* to the plane of an array of slits or a grating. Also, note that the path difference in Eq. (36.16) is $2d\sin\theta$, not $d\sin\theta$ as in Eq. (36.13) for a diffraction grating. ▮

36.22 A two-dimensional model of scattering from a rectangular array. The distance between adjacent atoms in a horizontal row is a; the distance between adjacent rows is d. The angles in (b) are measured from the *surface* of the array, not from its normal.

(a) Scattering of waves from a rectangular array

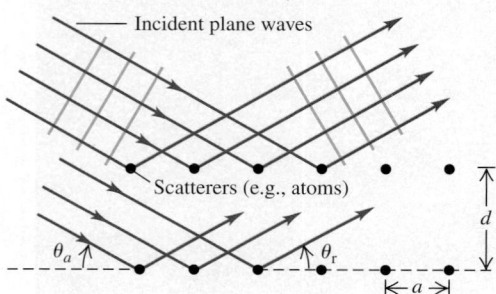

Incident plane waves

Scatterers (e.g., atoms)

(b) Scattering from adjacent atoms in a row
Interference from adjacent atoms in a row is constructive when the path lengths $a\cos\theta_a$ and $a\cos\theta_r$ are equal, so that the angle of incidence θ_a equals the angle of reflection (scattering) θ_r.

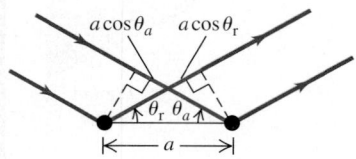

(c) Scattering from atoms in adjacent rows
Interference from atoms in adjacent rows is constructive when the path difference $2d\sin\theta$ is an integral number of wavelengths, as in Eq. (36.16).

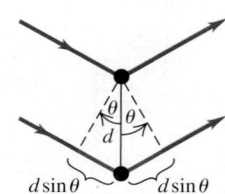

(a) Spacing of planes is $d = a/\sqrt{2}$.

(b) Spacing of planes is $d = a/\sqrt{3}$.

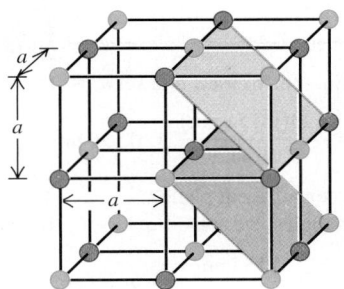

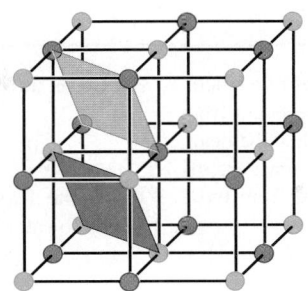

36.23 A cubic crystal and two different families of crystal planes. There are also three sets of planes parallel to the cube faces, with spacing a.

In directions for which Eq. (36.16) is satisfied, we see a strong maximum in the interference pattern. We can describe this interference in terms of *reflections* of the wave from the horizontal rows of scatterers in Fig. 36.22a. Strong reflection (constructive interference) occurs at angles such that the incident and scattered angles are equal and Eq. (36.16) is satisfied. Since $\sin\theta$ can never be greater than 1, Eq. (36.16) says that to have constructive interference the quantity $m\lambda$ must be less than $2d$ and so λ must be less than $2d/m$. For example, the value of d in an NaCl crystal (see Fig. 36.21) is only 0.282 nm. Hence to have the mth-order maximum present in the diffraction pattern, λ must be less than $2(0.282\ \text{nm})/m$; that is, $\lambda < 0.564$ nm for $m = 1$, $\lambda < 0.282$ nm for $m = 2$, $\lambda < 0.188$ nm for $m = 3$, and so on. These are all x-ray wavelengths (see Fig. 32.4), which is why x rays are used for studying crystal structure.

We can extend this discussion to a three-dimensional array by considering *planes* of scatterers instead of *rows*. **Figure 36.23** shows two different sets of parallel planes that pass through all the scatterers. Waves from all the scatterers in a given plane interfere constructively if the angles of incidence and scattering are equal. There is also constructive interference between planes when Eq. (36.16) is satisfied, where d is now the distance between adjacent planes. Because there are many different sets of parallel planes, there are also many values of d and many sets of angles that give constructive interference for the whole crystal lattice. This phenomenon is called **Bragg reflection.**

CAUTION **Bragg *reflection* is really Bragg *interference*** While we are using the term *reflection*, remember that we are dealing with an *interference* effect. The reflections from various planes are closely analogous to interference effects in thin films (see Section 35.4). ▌

As Fig. 36.20b shows, in x-ray diffraction there is nearly complete cancellation in all but certain very specific directions in which constructive interference occurs and forms bright spots. Such a pattern is usually called an x-ray *diffraction* pattern, although *interference* pattern might be more appropriate.

We can determine the wavelength of x rays by examining the diffraction pattern for a crystal of known structure and known spacing between atoms, just as we determined wavelengths of visible light by measuring patterns from slits or gratings. (The spacing between atoms in simple crystals of known structure, such as sodium chloride, can be found from the density of the crystal and Avogadro's number.) Then, once we know the x-ray wavelength, we can use x-ray diffraction to explore the structure and determine the spacing between atoms in crystals with unknown structure.

X-ray diffraction is by far the most important experimental tool in the investigation of crystal structure of solids. X-ray diffraction also plays an important role in studies of the structures of liquids and of organic molecules. It has been one of the chief experimental techniques in working out the double-helix structure of DNA (**Fig. 36.24**) and subsequent advances in molecular genetics.

36.24 The British scientist Rosalind Franklin made this groundbreaking x-ray diffraction image of DNA in 1953. The dark bands arranged in a cross provided the first evidence of the helical structure of the DNA molecule.

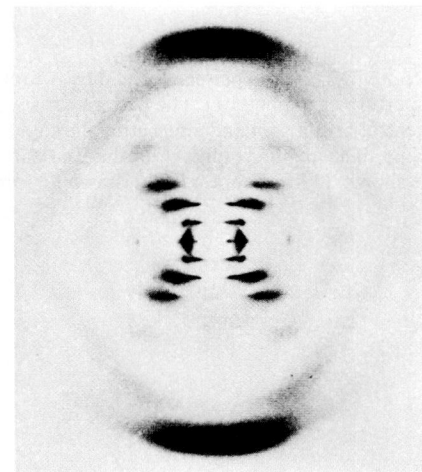

EXAMPLE 36.5 X-RAY DIFFRACTION

You direct a beam of 0.154-nm x rays at certain planes of a silicon crystal. As you increase the angle of incidence of the beam from zero, the first strong interference maximum occurs when the beam makes an angle of 34.5° with the planes. (a) How far apart are the planes? (b) Will you find other interference maxima from these planes at greater angles of incidence?

SOLUTION

IDENTIFY and SET UP: This problem involves Bragg reflection of x rays from the planes of a crystal. In part (a) we use the Bragg condition, Eq. (36.16), to find the distance d between adjacent planes from the known wavelength $\lambda = 0.154$ nm and angle of incidence $\theta = 34.5°$ for the $m = 1$ interference maximum. Given the value of d, we use the Bragg condition again in part (b) to find the values of θ for interference maxima corresponding to other values of m.

EXECUTE: (a) We solve Eq. (36.16) for d and set $m = 1$:

$$d = \frac{m\lambda}{2\sin\theta} = \frac{(1)(0.154 \text{ nm})}{2\sin 34.5°} = 0.136 \text{ nm}$$

This is the distance between adjacent planes.

(b) To calculate other angles, we solve Eq. (36.16) for $\sin\theta$:

$$\sin\theta = \frac{m\lambda}{2d} = m\frac{0.154 \text{ nm}}{2(0.136 \text{ nm})} = m(0.566)$$

Values of m of 2 or greater give values of $\sin\theta$ greater than unity, which is impossible. Hence there are *no* other angles for interference maxima for this particular set of crystal planes.

EVALUATE: Our result in part (b) shows that there *would* be a second interference maximum if the quantity $2\lambda/2d = \lambda/d$ were less than 1. This would be the case if the wavelength of the x rays were less than $d = 0.136$ nm. How short would the wavelength need to be to have *three* interference maxima?

TEST YOUR UNDERSTANDING OF SECTION 36.6 You are doing an x-ray diffraction experiment with a crystal in which the atomic planes are 0.200 nm apart. You are using x rays of wavelength 0.0900 nm. What is the highest-order maximum present in the diffraction pattern? (i) Third; (ii) fourth; (iii) fifth; (iv) sixth; (v) seventh. ∎

36.7 CIRCULAR APERTURES AND RESOLVING POWER

We have studied in detail the diffraction patterns formed by long, thin slits or arrays of slits. But an aperture of *any* shape forms a diffraction pattern. The diffraction pattern formed by a *circular* aperture is of special interest because of its role in limiting how well an optical instrument can resolve fine details. In principle, we could compute the intensity at any point P in the diffraction pattern by dividing the area of the aperture into small elements, finding the resulting wave amplitude and phase at P, and then integrating over the aperture area to find the resultant amplitude and intensity at P. In practice, the integration cannot be carried out in terms of elementary functions. We will simply *describe* the pattern and quote a few relevant numbers.

The diffraction pattern formed by a circular aperture consists of a central bright spot surrounded by a series of bright and dark rings, as **Fig. 36.25** shows.

36.25 Diffraction pattern formed by a circular aperture of diameter D. The pattern consists of a central bright spot and alternating dark and bright rings. The angular radius θ_1 of the first dark ring is shown. (This diagram is not drawn to scale.)

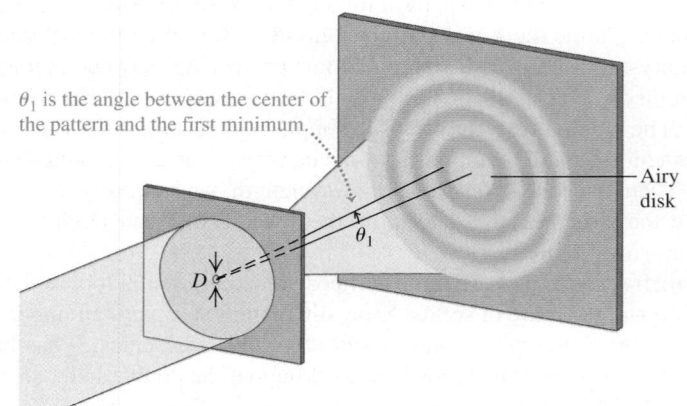

We can describe the pattern in terms of the angle θ, representing the angular radius of each ring. The angular radius θ_1 of the first *dark* ring is given by

	Angular radius of first dark ring = angular radius of Airy disk	
Diffraction by a circular aperture:	$$\sin\theta_1 = 1.22\frac{\lambda}{D}$$ Wavelength Aperture diameter	(36.17)

The angular radii of the next two dark rings are given by

$$\sin\theta_2 = 2.23\frac{\lambda}{D} \qquad \sin\theta_3 = 3.24\frac{\lambda}{D} \qquad (36.18)$$

The central bright spot is called the **Airy disk,** in honor of Sir George Airy (1801–1892), who first derived the expression for the intensity in the pattern. The angular radius of the Airy disk is that of the first dark ring, given by Eq. (36.17). The angular radii of the first three *bright* rings outside the Airy disk are

$$\sin\theta = 1.63\frac{\lambda}{D}, \qquad 2.68\frac{\lambda}{D}, \qquad 3.70\frac{\lambda}{D} \qquad (36.19)$$

The intensities in the bright rings drop off very quickly with increasing angle. When D is much larger than the wavelength λ, the usual case for optical instruments, the peak intensity in the first ring is only 1.7% of the value at the center of the Airy disk, and the peak intensity of the second ring is only 0.4%. Most (85%) of the light energy falls within the Airy disk. **Figure 36.26** shows a diffraction pattern from a circular aperture.

Diffraction and Image Formation

Diffraction has far-reaching implications for image formation by lenses and mirrors. In our study of optical instruments in Chapter 34 we assumed that a lens with focal length f focuses a parallel beam (plane wave) to a *point* at a distance f from the lens. We now see that what we get is *not* a point but the diffraction pattern just described. If we have two point objects, their images are not two points but two diffraction patterns. When the objects are close together, their diffraction patterns overlap; if they are close enough, their patterns overlap almost completely and cannot be distinguished. The effect is shown in **Fig. 36.27**, which presents the patterns for four very small "point" sources of light. In Fig. 36.27a the image of source 1 is well separated from the others, but the images of the sources 3 and 4 have merged. In Fig. 36.27b, with a larger aperture diameter and hence smaller Airy disks, images 3 and 4 are better resolved. In Fig. 36.27c, with a still larger aperture, they are well resolved.

A widely used criterion for resolution of two point objects, proposed by the English physicist Lord Rayleigh (1842–1919) and called **Rayleigh's criterion,** is that the objects are just barely resolved (that is, distinguishable) if the center of one diffraction pattern coincides with the first minimum of the other. In that case the angular separation of the image centers is given by Eq. (36.17). The angular separation of the *objects* is the same as that of the *images* made by a telescope, microscope, or other optical device. So two point objects are barely resolved when their angular separation is given by Eq. (36.17).

The minimum separation of two objects that can just be resolved by an optical instrument is called the **limit of resolution** of the instrument. The smaller the limit of resolution, the greater the *resolution,* or **resolving power,** of the instrument. Diffraction sets the ultimate limits on resolution of lenses. *Geometric optics* may make it seem that we can make images as large as we like. Eventually, though, we always reach a point at which the image becomes larger but does not gain in detail. The images in Fig. 36.27 would not become sharper if enlarged.

36.26 Photograph of the diffraction pattern formed by a circular aperture.

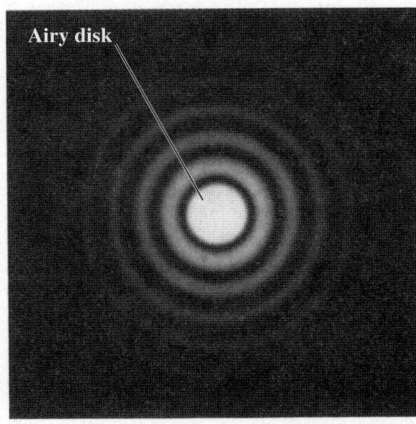

Airy disk

36.27 Diffraction patterns of four very small ("point") sources of light. The photographs were made with a circular aperture in front of the lens. (a) The aperture is so small that the patterns of sources 3 and 4 overlap and are barely resolved by Rayleigh's criterion. Increasing the size of the aperture decreases the size of the diffraction patterns, as shown in (b) and (c).

(a) Small aperture

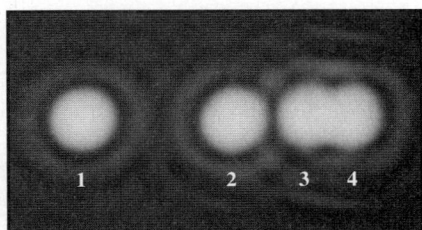

(b) Medium aperture

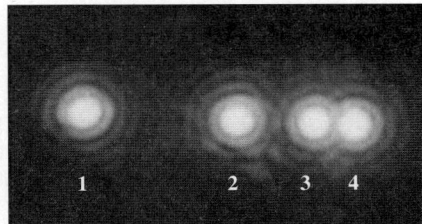

(c) Large aperture

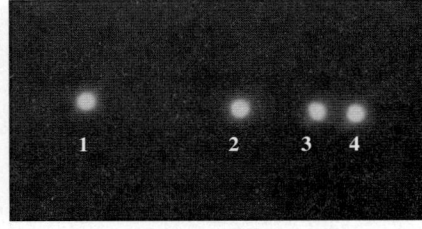

Application Bigger Telescope, Better Resolution The large aperture diameter of very large telescopes minimizes diffraction effects. The effective diameter of a telescope can be increased by using arrays of smaller telescopes. The Very Large Array (VLA) in New Mexico is a collection of 27 radio telescopes, each 25 m in diameter, that can be spread out in a Y-shaped arrangement 36 km across. Hence the effective aperture diameter is 36 km, giving the VLA a limit of resolution of 5×10^{-8} rad at a radio wavelength of 1.5 cm. If your eye had this angular resolution, you could read the "20/20" line on an eye chart more than 30 km away!

CAUTION Resolving power vs. chromatic resolving power Don't confuse the resolving power of an optical instrument with the *chromatic* resolving power of a grating (Section 36.5). Resolving power refers to the ability to distinguish the images of objects that appear close to each other, when looking either through an optical instrument or at a photograph made with the instrument. Chromatic resolving power describes how well different wavelengths can be distinguished in a spectrum formed by a diffraction grating. ▌

Rayleigh's criterion combined with Eq. (36.17) shows that resolution (resolving power) improves with larger diameter; it also improves with shorter wavelengths. Ultraviolet microscopes have higher resolution than visible-light microscopes. In electron microscopes the resolution is limited by the wavelengths associated with the electrons, which have wavelike aspects (to be discussed further in Chapter 39). These wavelengths can be made 100,000 times smaller than wavelengths of visible light, with a corresponding gain in resolution. Resolving power also explains the difference in storage capacity between DVDs (introduced in 1995) and Blu-ray discs (introduced in 2003). Information is stored in both of these in a series of tiny pits. In order not to lose information in the scanning process, the scanning optics must be able to resolve two adjacent pits so that they do not seem to blend into a single pit (see sources 3 and 4 in Fig. 36.27). The blue scanning laser used in a Blu-ray player has a shorter wavelength (405 nm) and hence better resolving power than the 650-nm red laser in a DVD player. Hence pits can be spaced closer together in a Blu-ray disc than in a DVD, and more information can be stored on a disc of the same size (50 gigabytes on a Blu-ray disc versus 4.7 gigabytes on a DVD).

EXAMPLE 36.6 RESOLVING POWER OF A CAMERA LENS

A camera lens with focal length $f = 50$ mm and maximum aperture $f/2$ forms an image of an object 9.0 m away. (a) If the resolution is limited by diffraction, what is the minimum distance between two points on the object that are barely resolved? What is the corresponding distance between image points? (b) How does the situation change if the lens is "stopped down" to $f/16$? Use $\lambda = 500$ nm in both cases.

SOLUTION

IDENTIFY and SET UP: This example uses the ideas about resolving power, image formation by a lens (Section 34.4), and f-number (Section 34.5). From Eq. (34.20), the f-number of a lens is its focal length f divided by the aperture diameter D. We use this equation to determine D and then use Eq. (36.17) (the Rayleigh criterion) to find the angular separation θ between two barely resolved points on the object. We then use the geometry of image formation by a lens to determine the distance y between those points and the distance y' between the corresponding image points.

EXECUTE: (a) The aperture diameter is $D = f/(f\text{-number}) = (50 \text{ mm})/2 = 25 \text{ mm} = 25 \times 10^{-3}$ m. From Eq. (36.17) the angular separation θ of two object points that are barely resolved is

$$\theta \approx \sin\theta = 1.22\frac{\lambda}{D} = 1.22\frac{500 \times 10^{-9} \text{ m}}{25 \times 10^{-3} \text{ m}} = 2.4 \times 10^{-5} \text{ rad}$$

We know from our thin-lens analysis in Section 34.4 that, apart from sign, $y/s = y'/s'$ [see Eq. (34.14)]. Thus the angular separations of the object points and the corresponding image points are both equal to θ. Because the object distance s is much greater than the focal length $f = 50$ mm, the image distance s' is approximately equal to f. Thus

$$\frac{y}{9.0 \text{ m}} = 2.4 \times 10^{-5} \qquad y = 2.2 \times 10^{-4} \text{ m} = 0.22 \text{ mm}$$

$$\frac{y'}{50 \text{ mm}} = 2.4 \times 10^{-5} \qquad y' = 1.2 \times 10^{-3} \text{ mm}$$

$$= 0.0012 \text{ mm} \approx \tfrac{1}{800} \text{ mm}$$

(b) The aperture diameter is now $(50 \text{ mm})/16$, or one-eighth as large as before. The angular separation between barely resolved points is eight times as great, and the values of y and y' are also eight times as great as before:

$$y = 1.8 \text{ mm} \qquad y' = 0.0096 \text{ mm} = \tfrac{1}{100} \text{ mm}$$

Only the best camera lenses can approach this resolving power.

EVALUATE: Many photographers use the smallest possible aperture for maximum sharpness, since lens aberrations cause light rays that are far from the optic axis to converge to a different image point than do rays near the axis. But as this example shows, diffraction effects become more significant at small apertures. One cause of fuzzy images has to be balanced against another.

TEST YOUR UNDERSTANDING OF SECTION 36.7 You have been asked to compare four proposals for telescopes to be placed in orbit above the blurring effects of the earth's atmosphere. Rank the proposed telescopes in order of their ability to resolve small details, from best to worst. (i) A radio telescope 100 m in diameter observing at a wavelength of 21 cm; (ii) an optical telescope 2.0 m in diameter observing at a wavelength of 500 nm; (iii) an ultraviolet telescope 1.0 m in diameter observing at a wavelength of 100 nm; (iv) an infrared telescope 2.0 m in diameter observing at a wavelength of 10 μm. ❚

36.8 HOLOGRAPHY

Holography is a technique for recording and reproducing an image of an object through the use of interference effects. Unlike the two-dimensional images recorded by an ordinary photograph or television system, a holographic image is truly three-dimensional. Such an image can be viewed from different directions to reveal different sides and from various distances to reveal changing perspective. If you had never seen a hologram, you wouldn't believe it was possible!

Figure 36.28a shows the basic procedure for making a hologram. We illuminate the object to be holographed with monochromatic light, and we place a photographic film so that it is struck by scattered light from the object and also by direct light from the source. In practice, the light source must be a laser, for reasons we will discuss later. Interference between the direct and scattered light forms a complex interference pattern that is recorded on the film.

To form the images, we simply project light through the developed film (Fig. 36.28b). Two images are formed: a virtual image on the side of the film nearer the source and a real image on the opposite side.

Holography and Interference Patterns

A complete analysis of holography is beyond our scope, but we can gain some insight into the process by looking at how a single point is holographed and imaged. Consider the interference pattern that is formed on a sheet of photographic negative film by the superposition of an incident plane wave and a spherical wave, as shown in **Fig. 36.29a**, next page. The spherical wave originates at a point source P at a distance b_0 from the film; P may in fact be a small object that scatters part of the incident plane wave. We assume that the two waves are monochromatic and coherent and that the phase relationship is such that constructive interference occurs at point O on the diagram. Then constructive interference will *also* occur at any point Q on the film that is farther from P than O is by an integer number of wavelengths. That is, if $b_m - b_0 = m\lambda$, where m is an integer, then constructive interference occurs. The points where this condition is satisfied

36.28 (a) A hologram is the record on film of the interference pattern formed with light from the coherent source and light scattered from the object. (b) Images are formed when light is projected through the hologram. The observer sees the virtual image formed behind the hologram.

(a) Recording a hologram

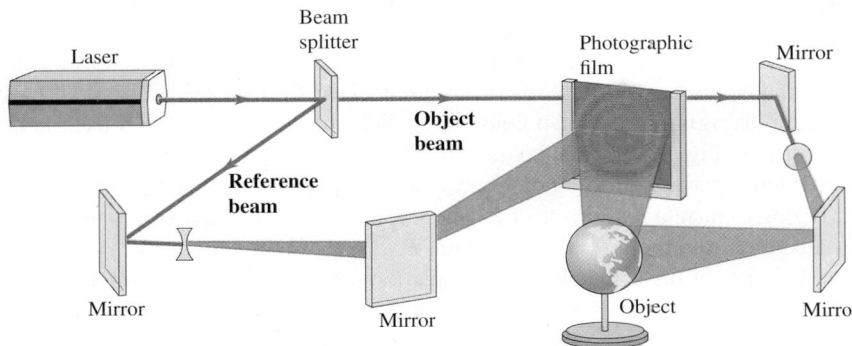

(b) Viewing the hologram

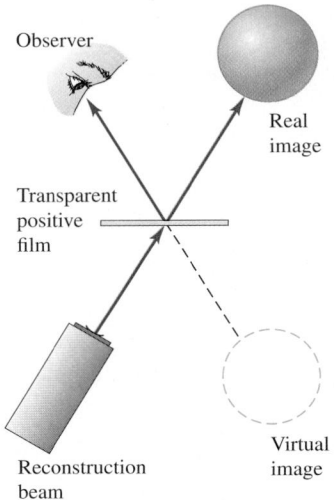

36.29 (a) Constructive interference of the plane and spherical waves occurs in the plane of the film at every point Q for which the distance b_m from P is greater than the distance b_0 from P to O by an integral number of wavelengths $m\lambda$. For the point Q shown, $m = 2$. (b) When a plane wave strikes a transparent positive print of the developed film, the diffracted wave consists of a wave converging to P' and then diverging again and a diverging wave that appears to originate at P. These waves form the real and virtual images, respectively.

(a)

(b)

form circles on the film centered at O, with radii r_m given by

$$b_m - b_0 = \sqrt{b_0^2 + r_m^2} - b_0 = m\lambda \qquad (m = 1, 2, 3, \ldots) \qquad (36.20)$$

Solving this for r_m^2, we find

$$r_m^2 = \lambda(2mb_0 + m^2\lambda)$$

Ordinarily, b_0 is very much larger than λ, so we ignore the second term in parentheses and obtain

$$r_m = \sqrt{2m\lambda b_0} \qquad (m = 1, 2, 3, \ldots) \qquad (36.21)$$

The interference pattern consists of a series of concentric bright circular fringes with radii given by Eq. (36.21). Between these bright fringes are dark fringes.

Now we develop the film and make a transparent positive print, so the bright-fringe areas have the greatest transparency on the film. Then we illuminate it with monochromatic plane-wave light of the same wavelength λ that we used initially. In Fig. 36.29b, consider a point P' at a distance b_0 along the axis from the film. The centers of successive bright fringes differ in their distances from P' by an integer number of wavelengths, and therefore a strong *maximum* in the diffracted wave occurs at P'. That is, light converges to P' and then diverges from it on the opposite side. Therefore P' is a *real image* of point P.

This is not the entire diffracted wave, however. The interference of the wave-lets that spread out from all the transparent areas forms a second spherical wave that is diverging rather than converging. When this wave is traced back behind the film in Fig. 36.29b, it appears to be spreading out from point P. Thus the total diffracted wave from the hologram is a superposition of a spherical wave converging to form a real image at P' and a spherical wave that diverges as though it had come from the virtual image point P.

Because of the principle of superposition for waves, what is true for the imaging of a single point is also true for the imaging of any number of points. The film records the superposed interference pattern from the various points, and when light is projected through the film, the various image points are reproduced simultaneously. Thus the images of an extended object can be recorded and reproduced just as for a single point object. **Figure 36.30** shows photographs of a holographic image from two different angles, showing the changing perspective in this three-dimensional image.

36.30 Two views of the same hologram seen from different angles.

In making a hologram, we have to overcome two practical problems. First, the light used must be *coherent* over distances that are large in comparison to the dimensions of the object and its distance from the film. Ordinary light sources *do not* satisfy this requirement, for reasons that we discussed in Section 35.1. Therefore laser light is essential for making a hologram. (Ordinary white light can be used for *viewing* certain types of hologram, such as those used on credit cards.) Second, extreme mechanical stability is needed. If any relative motion of source, object, or film occurs during exposure, even by as much as a quarter of a wavelength, the interference pattern on the film is blurred enough to prevent satisfactory image formation. These obstacles are not insurmountable, however, and holography has become important in research, entertainment, and a wide variety of technological applications.

CHAPTER 36 SUMMARY

SOLUTIONS TO ALL EXAMPLES

Fresnel and Fraunhofer diffraction: Diffraction occurs when light passes through an aperture or around an edge. When the source and the observer are so far away from the obstructing surface that the outgoing rays can be considered parallel, it is called Fraunhofer diffraction. When the source or the observer is relatively close to the obstructing surface, it is Fresnel diffraction.

Single-slit diffraction: Monochromatic light sent through a narrow slit of width a produces a diffraction pattern on a distant screen. Equation (36.2) gives the condition for destructive interference (a dark fringe) at a point P in the pattern at angle θ. Equation (36.7) gives the intensity in the pattern as a function of θ. (See Examples 36.1–36.3.)

$$\sin\theta = \frac{m\lambda}{a} \tag{36.2}$$
$$(m = \pm 1, \pm 2, \pm 3, \dots)$$

$$I = I_0 \left\{ \frac{\sin[\pi a(\sin\theta)/\lambda]}{\pi a(\sin\theta)/\lambda} \right\}^2 \tag{36.7}$$

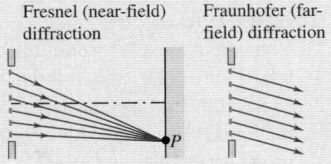

Diffraction gratings: A diffraction grating consists of a large number of thin parallel slits, spaced a distance d apart. The condition for maximum intensity in the interference pattern is the same as for the two-source pattern, but the maxima for the grating are very sharp and narrow. (See Example 36.4.)

$$d\sin\theta = m\lambda \tag{36.13}$$
$$(m = 0, \pm 1, \pm 2, \pm 3, \dots)$$

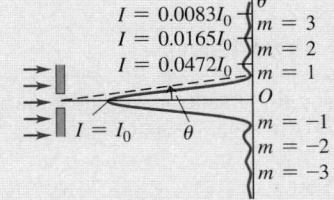

X-ray diffraction: A crystal serves as a three-dimensional diffraction grating for x rays with wavelengths of the same order of magnitude as the spacing between atoms in the crystal. For a set of crystal planes spaced a distance d apart, constructive interference occurs when the angles of incidence and scattering (measured from the crystal planes) are equal and when the Bragg condition [Eq. (36.16)] is satisfied. (See Example 36.5.)

$$2d\sin\theta = m\lambda$$
$$(m = 1, 2, 3, \ldots)$$

(36.16)

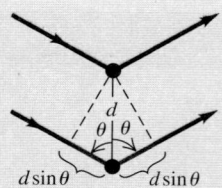

$d\sin\theta \qquad d\sin\theta$

Circular apertures and resolving power: The diffraction pattern from a circular aperture of diameter D consists of a central bright spot, called the Airy disk, and a series of concentric dark and bright rings. Equation (36.17) gives the angular radius θ_1 of the first dark ring, equal to the angular size of the Airy disk. Diffraction sets the ultimate limit on resolution (image sharpness) of optical instruments. According to Rayleigh's criterion, two point objects are just barely resolved when their angular separation θ is given by Eq. (36.17). (See Example 36.6.)

$$\sin\theta_1 = 1.22\frac{\lambda}{D}$$

(36.17)

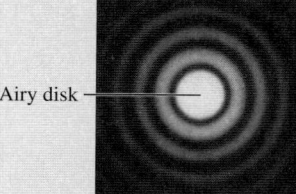

Airy disk

BRIDGING PROBLEM   OBSERVING THE EXPANDING UNIVERSE

An astronomer who is studying the light from a galaxy has identified the spectrum of hydrogen but finds that the wavelengths are somewhat shifted from those found in the laboratory. In the lab, the H_α line in the hydrogen spectrum has a wavelength of 656.3 nm. The astronomer is using a transmission diffraction grating having 5758 lines/cm in the first order and finds that the first bright fringe for the H_α line occurs at $\pm 23.41°$ from the central spot. How fast is the galaxy moving? Express your answer in m/s and as a percentage of the speed of light. Is the galaxy moving toward us or away from us?

SOLUTION GUIDE

IDENTIFY and SET UP

1. You can use the information about the grating to find the wavelength of the H_α line in the galaxy's spectrum.
2. In Section 16.8 we learned about the Doppler effect for electromagnetic radiation: The frequency that we receive from a moving source, such as the galaxy, is different from the frequency

that is emitted. Equation (16.30) relates the emitted frequency, the received frequency, and the velocity of the source (the target variable). The equation $c = f\lambda$ relates the frequency f and wavelength λ through the speed of light c.

EXECUTE

3. Find the wavelength of the H_α spectral line in the received light.
4. Rewrite Eq. (16.30) as a formula for the velocity v of the galaxy in terms of the received wavelength and the wavelength emitted by the source.
5. Solve for v. Express it in m/s and as a percentage of c, and decide whether the galaxy is moving toward us or moving away.

EVALUATE

6. Is your answer consistent with the relative sizes of the received wavelength and the emitted wavelength?

Problems

For assigned homework and other learning materials, go to MasteringPhysics®.

°, °°, °°°: Difficulty levels. **CP:** Cumulative problems incorporating material from earlier chapters. **CALC:** Problems requiring calculus. **DATA:** Problems involving real data, scientific evidence, experimental design, and/or statistical reasoning. **BIO:** Biosciences problems.

DISCUSSION QUESTIONS

Q36.1 Why can we readily observe diffraction effects for sound waves and water waves, but not for light? Is this because light travels so much faster than these other waves? Explain.

Q36.2 What is the difference between Fresnel and Fraunhofer diffraction? Are they different *physical* processes? Explain.

Q36.3 You use a lens of diameter D and light of wavelength λ and frequency f to form an image of two closely spaced and distant objects. Which of the following will increase the resolving power? (a) Use a lens with a smaller diameter; (b) use light of higher frequency; (c) use light of longer wavelength. In each case justify your answer.

Q36.4 Light of wavelength λ and frequency f passes through a single slit of width a. The diffraction pattern is observed on a screen a distance x from the slit. Which of the following will *decrease* the width of the central maximum? (a) Decrease the slit width; (b) decrease the frequency f of the light; (c) decrease the wavelength λ of the light; (d) decrease the distance x of the screen from the slit. In each case justify your answer.

Q36.5 In a diffraction experiment with waves of wavelength λ, there will be *no* intensity minima (that is, no dark fringes) if the slit width is small enough. What is the maximum slit width for which this occurs? Explain your answer.

Q36.6 An interference pattern is produced by four parallel and equally spaced narrow slits. By drawing appropriate phasor diagrams, explain why there is an interference minimum when the phase difference ϕ from adjacent slits is (a) $\pi/2$; (b) π; (c) $3\pi/2$. In each case, for which pairs of slits is there totally destructive interference?

Q36.7 Phasor Diagram for Eight Slits. An interference pattern is produced by eight equally spaced narrow slits. The caption for Fig. 36.14 claims that minima occur for $\phi = 3\pi/4, \pi/4, 3\pi/2$, and $7\pi/4$. Draw the phasor diagram for each of these four cases, and explain why each diagram proves that there is in fact a minimum. In each case, for which pairs of slits is there totally destructive interference?

Q36.8 A rainbow ordinarily shows a range of colors (see Section 33.4). But if the water droplets that form the rainbow are small enough, the rainbow will appear white. Explain why, using diffraction ideas. How small do you think the raindrops would have to be for this to occur?

Q36.9 Some loudspeaker horns for outdoor concerts (at which the entire audience is seated on the ground) are wider vertically than horizontally. Use diffraction ideas to explain why this is more efficient at spreading the sound uniformly over the audience than either a square speaker horn or a horn that is wider horizontally than vertically. Would this still be the case if the audience were seated at different elevations, as in an amphitheater? Why or why not?

Q36.10 Figure 31.12 (Section 31.2) shows a loudspeaker system. Low-frequency sounds are produced by the *woofer*, which is a speaker with large diameter; the *tweeter*, a speaker with smaller diameter, produces high-frequency sounds. Use diffraction ideas to explain why the tweeter is more effective for distributing high-frequency sounds uniformly over a room than is the woofer.

Q36.11 Information is stored on an audio compact disc, CD-ROM, or DVD disc in a series of pits on the disc. These pits are scanned by a laser beam. An important limitation on the amount of information that can be stored on such a disc is the width of the laser beam. Explain why this should be, and explain how using a shorter-wavelength laser allows more information to be stored on a disc of the same size.

Q36.12 With which color of light can the Hubble Space Telescope see finer detail in a distant astronomical object: red, blue, or ultraviolet? Explain your answer.

Q36.13 At the end of Section 36.4, the following statements were made about an array of N slits. Explain, using phasor diagrams, why each statement is true. (a) A minimum occurs whenever ϕ is an integral multiple of $2\pi/N$, except when ϕ is an integral multiple of 2π (which gives a principal maximum). (b) There are $(N-1)$ minima between each pair of principal maxima.

Q36.14 Could x-ray diffraction effects with crystals be observed by using visible light instead of x rays? Why or why not?

Q36.15 Why is a diffraction grating better than a two-slit setup for measuring wavelengths of light?

Q36.16 One sometimes sees rows of evenly spaced radio antenna towers. A student remarked that these act like diffraction gratings. What did she mean? Why would one want them to act like a diffraction grating?

Q36.17 If a hologram is made using 600-nm light and then viewed with 500-nm light, how will the images look compared to those observed when viewed with 600-nm light? Explain.

Q36.18 A hologram is made using 600-nm light and then viewed by using white light from an incandescent bulb. What will be seen? Explain.

Q36.19 Ordinary photographic film reverses black and white, in the sense that the most brightly illuminated areas become blackest upon development (hence the term *negative*). Suppose a hologram negative is viewed directly, without making a positive transparency. How will the resulting images differ from those obtained with the positive? Explain.

EXERCISES

Section 36.2 Diffraction from a Single Slit

36.1 •• Monochromatic light from a distant source is incident on a slit 0.750 mm wide. On a screen 2.00 m away, the distance from the central maximum of the diffraction pattern to the first minimum is measured to be 1.35 mm. Calculate the wavelength of the light.

36.2 • Parallel rays of green mercury light with a wavelength of 546 nm pass through a slit covering a lens with a focal length of 60.0 cm. In the focal plane of the lens, the distance from the central maximum to the first minimum is 8.65 mm. What is the width of the slit?

36.3 •• Light of wavelength 585 nm falls on a slit 0.0666 mm wide. (a) On a very large and distant screen, how many *totally* dark fringes (indicating complete cancellation) will there be, including both sides of the central bright spot? Solve this problem *without* calculating all the angles! (*Hint:* What is the largest that $\sin\theta$ can be? What does this tell you is the largest that m can be?) (b) At what angle will the dark fringe that is most distant from the central bright fringe occur?

36.4 • Light of wavelength 633 nm from a distant source is incident on a slit 0.750 mm wide, and the resulting diffraction pattern is observed on a screen 3.50 m away. What is the distance between the two dark fringes on either side of the central bright fringe?

36.5 •• Diffraction occurs for all types of waves, including sound waves. High-frequency sound from a distant source with wavelength 9.00 cm passes through a slit 12.0 cm wide. A microphone is placed 8.00 m directly in front of the center of the slit, corresponding to point O in Fig. 36.5a. The microphone is then moved in a direction perpendicular to the line from the center of the slit to point O. At what distances from O will the intensity detected by the microphone be zero?

36.6 • **CP Tsunami!** On December 26, 2004, a violent earthquake of magnitude 9.1 occurred off the coast of Sumatra. This quake triggered a huge tsunami (similar to a tidal wave) that killed more than 150,000 people. Scientists observing the wave on the open ocean measured the time between crests to be 1.0 h and the speed of the wave to be 800 km/h. Computer models of the evolution of this enormous wave showed that it bent around the continents and spread to all the oceans of the earth. When the wave reached the gaps between continents, it diffracted between them as through a slit. (a) What was the wavelength of this tsunami? (b) The distance between the southern tip of Africa and northern Antarctica is about 4500 km, while the distance between the southern end of Australia and Antarctica is about 3700 km. As an approximation, we can model this wave's behavior by using Fraunhofer diffraction. Find the smallest angle away from the central maximum for which the waves would cancel after going through each of these continental gaps.

36.7 •• CP A series of parallel linear water wave fronts are traveling directly toward the shore at 15.0 cm/s on an otherwise placid lake. A long concrete barrier that runs parallel to the shore at a distance of 3.20 m away has a hole in it. You count the wave crests and observe that 75.0 of them pass by each minute, and you also observe that no waves reach the shore at ±61.3 cm from the point directly opposite the hole, but waves do reach the shore everywhere within this distance. (a) How wide is the hole in the barrier? (b) At what other angles do you find no waves hitting the shore?

36.8 • Monochromatic electromagnetic radiation with wavelength λ from a distant source passes through a slit. The diffraction pattern is observed on a screen 2.50 m from the slit. If the width of the central maximum is 6.00 mm, what is the slit width a if the wavelength is (a) 500 nm (visible light); (b) 50.0 μm (infrared radiation); (c) 0.500 nm (x rays)?

36.9 •• **Doorway Diffraction.** Sound of frequency 1250 Hz leaves a room through a 1.00-m-wide doorway (see Exercise 36.5). At which angles relative to the centerline perpendicular to the doorway will someone outside the room hear no sound? Use 344 m/s for the speed of sound in air and assume that the source and listener are both far enough from the doorway for Fraunhofer diffraction to apply. You can ignore effects of reflections.

36.10 • CP Light waves, for which the electric field is given by $E_y(x, t) = E_{\max} \sin[(1.40 \times 10^7 \text{ m}^{-1})x - \omega t]$, pass through a slit and produce the first dark bands at ±28.6° from the center of the diffraction pattern. (a) What is the frequency of this light? (b) How wide is the slit? (c) At which angles will other dark bands occur?

36.11 •• Red light of wavelength 633 nm from a helium–neon laser passes through a slit 0.350 mm wide. The diffraction pattern is observed on a screen 3.00 m away. Define the width of a bright fringe as the distance between the minima on either side. (a) What is the width of the central bright fringe? (b) What is the width of the first bright fringe on either side of the central one?

Section 36.3 Intensity in the Single-Slit Pattern

36.12 •• Public Radio station KXPR-FM in Sacramento broadcasts at 88.9 MHz. The radio waves pass between two tall skyscrapers that are 15.0 m apart along their closest walls. (a) At what horizontal angles, relative to the original direction of the waves, will a distant antenna not receive any signal from this station? (b) If the maximum intensity is 3.50 W/m² at the antenna, what is the intensity at ±5.00° from the center of the central maximum at the distant antenna?

36.13 •• Monochromatic light of wavelength 580 nm passes through a single slit and the diffraction pattern is observed on a screen. Both the source and screen are far enough from the slit for Fraunhofer diffraction to apply. (a) If the first diffraction minima are at ±90.0°, so the central maximum completely fills the screen, what is the width of the slit? (b) For the width of the slit as calculated in part (a), what is the ratio of the intensity at $\theta = 45.0°$ to the intensity at $\theta = 0$?

36.14 •• Monochromatic light of wavelength $\lambda = 620$ nm from a distant source passes through a slit 0.450 mm wide. The diffraction pattern is observed on a screen 3.00 m from the slit. In terms of the intensity I_0 at the peak of the central maximum, what is the intensity of the light at the screen the following distances from the center of the central maximum: (a) 1.00 mm; (b) 3.00 mm; (c) 5.00 mm?

36.15 •• A slit 0.240 mm wide is illuminated by parallel light rays of wavelength 540 nm. The diffraction pattern is observed on a screen that is 3.00 m from the slit. The intensity at the center of the central maximum ($\theta = 0°$) is 6.00×10^{-6} W/m². (a) What is the distance on the screen from the center of the central maximum to the first minimum? (b) What is the intensity at a point on the screen midway between the center of the central maximum and the first minimum?

36.16 • Monochromatic light of wavelength 592 nm from a distant source passes through a slit that is 0.0290 mm wide. In the resulting diffraction pattern, the intensity at the center of the central maximum ($\theta = 0°$) is 4.00×10^{-5} W/m². What is the intensity at a point on the screen that corresponds to $\theta = 1.20°$?

36.17 •• A single-slit diffraction pattern is formed by monochromatic electromagnetic radiation from a distant source passing through a slit 0.105 mm wide. At the point in the pattern 3.25° from the center of the central maximum, the total phase difference between wavelets from the top and bottom of the slit is 56.0 rad. (a) What is the wavelength of the radiation? (b) What is the intensity at this point, if the intensity at the center of the central maximum is I_0?

Section 36.4 Multiple Slits

36.18 • Parallel rays of monochromatic light with wavelength 568 nm illuminate two identical slits and produce an interference pattern on a screen that is 75.0 cm from the slits. The centers of the slits are 0.640 mm apart and the width of each slit is 0.434 mm. If the intensity at the center of the central maximum is 5.00×10^{-4} W/m², what is the intensity at a point on the screen that is 0.900 mm from the center of the central maximum?

36.19 • **Number of Fringes in a Diffraction Maximum.** In Fig. 36.12c the central diffraction maximum contains exactly seven interference fringes, and in this case $d/a = 4$. (a) What must the ratio d/a be if the central maximum contains exactly five fringes? (b) In the case considered in part (a), how many fringes are contained within the first diffraction maximum on one side of the central maximum?

36.20 •• **Diffraction and Interference Combined.** Consider the interference pattern produced by two parallel slits of width a and separation d, in which $d = 3a$. The slits are illuminated by normally incident light of wavelength λ. (a) First we ignore diffraction effects due to the slit width. At what angles θ from the central maximum will the next four maxima in the two-slit interference pattern occur? Your answer will be in terms of d and λ. (b) Now we include the effects of diffraction. If the intensity at $\theta = 0°$ is I_0, what is the intensity at each of the angles in part (a)? (c) Which double-slit interference maxima are missing in the pattern? (d) Compare your results to those illustrated in Fig. 36.12c. In what ways are your results different?

36.21 •• An interference pattern is produced by light of wavelength 580 nm from a distant source incident on two identical parallel slits separated by a distance (between centers) of 0.530 mm. (a) If the slits are very narrow, what would be the angular positions of the first-order and second-order, two-slit interference maxima? (b) Let the slits have width 0.320 mm. In terms of the intensity I_0 at the center of the central maximum, what is the intensity at each of the angular positions in part (a)?

36.22 •• Laser light of wavelength 500.0 nm illuminates two identical slits, producing an interference pattern on a screen 90.0 cm from the slits. The bright bands are 1.00 cm apart, and the third bright bands on either side of the central maximum are missing in the pattern. Find the width and the separation of the two slits.

Section 36.5 The Diffraction Grating

36.23 • When laser light of wavelength 632.8 nm passes through a diffraction grating, the first bright spots occur at $\pm 17.8°$ from the central maximum. (a) What is the line density (in lines/cm) of this grating? (b) How many additional bright spots are there beyond the first bright spots, and at what angles do they occur?

36.24 •• Monochromatic light is at normal incidence on a plane transmission grating. The first-order maximum in the interference pattern is at an angle of $11.3°$. What is the angular position of the fourth-order maximum?

36.25 • If a diffraction grating produces its third-order bright band at an angle of $78.4°$ for light of wavelength 681 nm, find (a) the number of slits per centimeter for the grating and (b) the angular location of the first-order and second-order bright bands. (c) Will there be a fourth-order bright band? Explain.

36.26 • If a diffraction grating produces a third-order bright spot for red light (of wavelength 700 nm) at $65.0°$ from the central maximum, at what angle will the second-order bright spot be for violet light (of wavelength 400 nm)?

36.27 • Visible light passes through a diffraction grating that has 900 slits/cm, and the interference pattern is observed on a screen that is 2.50 m from the grating. (a) Is the angular position of the first-order spectrum small enough for $\sin\theta \approx \theta$ to be a good approximation? (b) In the first-order spectrum, the maxima for two different wavelengths are separated on the screen by 3.00 mm. What is the difference in these wavelengths?

36.28 • The wavelength range of the visible spectrum is approximately 380–750 nm. White light falls at normal incidence on a diffraction grating that has 350 slits/mm. Find the angular width of the visible spectrum in (a) the first order and (b) the third order. (*Note:* An advantage of working in higher orders is the greater angular spread and better resolution. A disadvantage is the overlapping of different orders, as shown in Example 36.4.)

36.29 • (a) What is the wavelength of light that is deviated in the first order through an angle of $13.5°$ by a transmission grating having 5000 slits/cm? (b) What is the second-order deviation of this wavelength? Assume normal incidence.

36.30 •• **CDs and DVDs as Diffraction Gratings.** A laser beam of wavelength $\lambda = 632.8$ nm shines at normal incidence on the reflective side of a compact disc. (a) The tracks of tiny pits in which information is coded onto the CD are 1.60 μm apart. For what angles of reflection (measured from the normal) will the intensity of light be maximum? (b) On a DVD, the tracks are only 0.740 μm apart. Repeat the calculation of part (a) for the DVD.

36.31 • A typical laboratory diffraction grating has 5.00×10^3 lines/cm, and these lines are contained in a 3.50-cm width of grating. (a) What is the chromatic resolving power of such a grating in the first order? (b) Could this grating resolve the lines of the sodium doublet (see Section 36.5) in the first order? (c) While doing spectral analysis of a star, you are using this grating in the *second* order to resolve spectral lines that are very close to the 587.8002-nm spectral line of iron. (i) For wavelengths longer than the iron line, what is the shortest wavelength you could distinguish from the iron line? (ii) For wavelengths shorter than the iron line, what is the longest wavelength you could distinguish from the iron line? (iii) What is the range of wavelengths you could *not* distinguish from the iron line?

36.32 • **Identifying Isotopes by Spectra.** Different isotopes of the same element emit light at slightly different wavelengths. A wavelength in the emission spectrum of a hydrogen atom is 656.45 nm; for deuterium, the corresponding wavelength is 656.27 nm. (a) What minimum number of slits is required to resolve these two wavelengths in second order? (b) If the grating has 500.00 slits/mm, find the angles and angular separation of these two wavelengths in the second order.

36.33 • The light from an iron arc includes many different wavelengths. Two of these are at $\lambda = 587.9782$ nm and $\lambda = 587.8002$ nm. You wish to resolve these spectral lines in first order using a grating 1.20 cm in length. What minimum number of slits per centimeter must the grating have?

Section 36.6 X-Ray Diffraction

36.34 • If the planes of a crystal are 3.50 Å (1 Å $= 10^{-10}$ m $= 1$ Ångstrom unit) apart, (a) what wavelength of electromagnetic waves is needed so that the first strong interference maximum in the Bragg reflection occurs when the waves strike the planes at an angle of $22.0°$, and in what part of the electromagnetic spectrum do these waves lie? (See Fig. 32.4.) (b) At what other angles will strong interference maxima occur?

36.35 • X rays of wavelength 0.0850 nm are scattered from the atoms of a crystal. The second-order maximum in the Bragg reflection occurs when the angle θ in Fig. 36.22 is $21.5°$. What is the spacing between adjacent atomic planes in the crystal?

36.36 • Monochromatic x rays are incident on a crystal for which the spacing of the atomic planes is 0.440 nm. The first-order maximum in the Bragg reflection occurs when the incident and reflected x rays make an angle of $39.4°$ with the crystal planes. What is the wavelength of the x rays?

Section 36.7 Circular Apertures and Resolving Power

36.37 •• Monochromatic light with wavelength 620 nm passes through a circular aperture with diameter 7.4 μm. The resulting diffraction pattern is observed on a screen that is 4.5 m from the aperture. What is the diameter of the Airy disk on the screen?

36.38 •• Monochromatic light with wavelength 490 nm passes through a circular aperture, and a diffraction pattern is observed on a screen that is 1.20 m from the aperture. If the distance on the screen between the first and second dark rings is 1.65 mm, what is the diameter of the aperture?

36.39 • Two satellites at an altitude of 1200 km are separated by 28 km. If they broadcast 3.6-cm microwaves, what minimum receiving-dish diameter is needed to resolve (by Rayleigh's criterion) the two transmissions?

36.40 • **BIO** If you can read the bottom row of your doctor's eye chart, your eye has a resolving power of 1 arcminute, equal to $\frac{1}{60}$ degree. If this resolving power is diffraction limited, to what effective diameter of your eye's optical system does this correspond? Use Rayleigh's criterion and assume $\lambda = 550$ nm.

36.41 •• The VLBA (Very Long Baseline Array) uses a number of individual radio telescopes to make one unit having an equivalent diameter of about 8000 km. When this radio telescope is focusing radio waves of wavelength 2.0 cm, what would have to be the diameter of the mirror of a visible-light telescope focusing light of wavelength 550 nm so that the visible-light telescope has the same resolution as the radio telescope?

36.42 •• **Searching for Planets Around Other Stars.** If an optical telescope focusing light of wavelength 550 nm has a perfectly ground mirror, what would the minimum mirror diameter have to be so that the telescope could resolve a Jupiter-size planet around our nearest star, Alpha Centauri, which is about 4.3 light-years from earth? (Consult Appendix F.)

36.43 •• **Hubble Versus Arecibo.** The Hubble Space Telescope has an aperture of 2.4 m and focuses visible light (380–750 nm). The Arecibo radio telescope in Puerto Rico is 305 m (1000 ft) in diameter (it is built in a mountain valley) and focuses radio waves of wavelength 75 cm. (a) Under optimal viewing conditions, what is the smallest crater that each of these telescopes could resolve on our moon? (b) If the Hubble Space Telescope were to be converted to surveillance use, what is the highest orbit above the surface of the earth it could have and still be able to resolve the license plate (not the letters, just the plate) of a car on the ground? Assume optimal viewing conditions, so that the resolution is diffraction limited.

36.44 • **Photography.** A wildlife photographer uses a moderate telephoto lens of focal length 135 mm and maximum aperture $f/4.00$ to photograph a bear that is 11.5 m away. Assume the wavelength is 550 nm. (a) What is the width of the smallest feature on the bear that this lens can resolve if it is opened to its maximum aperture? (b) If, to gain depth of field, the photographer stops the lens down to $f/22.0$, what would be the width of the smallest resolvable feature on the bear?

36.45 • **Observing Jupiter.** You are asked to design a space telescope for earth orbit. When Jupiter is 5.93×10^8 km away (its closest approach to the earth), the telescope is to resolve, by Rayleigh's criterion, features on Jupiter that are 250 km apart. What minimum-diameter mirror is required? Assume a wavelength of 500 nm.

PROBLEMS

36.46 •• Coherent monochromatic light of wavelength λ passes through a narrow slit of width a, and a diffraction pattern is observed on a screen that is a distance x from the slit. On the screen, the width w of the central diffraction maximum is twice the distance x. What is the ratio a/λ of the width of the slit to the wavelength of the light?

36.47 •• BIO **Thickness of Human Hair.** Although we have discussed single-slit diffraction only for a slit, a similar result holds when light bends around a straight, thin object, such as a strand of hair. In that case, a is the width of the strand. From actual laboratory measurements on a human hair, it was found that when a beam of light of wavelength 632.8 nm was shone on a single strand of hair, and the diffracted light was viewed on a screen 1.25 m away, the first dark fringes on either side of the central bright spot were 5.22 cm apart. How thick was this strand of hair?

36.48 •• CP A loudspeaker with a diaphragm that vibrates at 960 Hz is traveling at 80.0 m/s directly toward a pair of holes in a very large wall. The speed of sound in the region is 344 m/s. Far from the wall, you observe that the sound coming through the openings first cancels at $\pm 11.4°$ with respect to the direction in which the speaker is moving. (a) How far apart are the two openings? (b) At what angles would the sound first cancel if the source stopped moving?

36.49 ••• Laser light of wavelength 632.8 nm falls normally on a slit that is 0.0250 mm wide. The transmitted light is viewed on a distant screen where the intensity at the center of the central bright fringe is 8.50 W/m². (a) Find the maximum number of totally dark fringes on the screen, assuming the screen is large enough to show them all. (b) At what angle does the dark fringe that is most distant from the center occur? (c) What is the maximum intensity of the bright fringe that occurs immediately before the dark fringe in part (b)? Approximate the angle at which this fringe occurs by assuming it is midway between the angles to the dark fringes on either side of it.

36.50 • **Grating Design.** Your boss asks you to design a diffraction grating that will disperse the first-order visible spectrum through an angular range of 27.0°. (See Example 36.4 in Section 36.5.) (a) What must be the number of slits per centimeter for this grating? (b) At what angles will the first-order visible spectrum begin and end?

36.51 • **Measuring Refractive Index.** A thin slit illuminated by light of frequency f produces its first dark band at $\pm 38.2°$ in air. When the entire apparatus (slit, screen, and space in between) is immersed in an unknown transparent liquid, the slit's first dark bands occur instead at $\pm 21.6°$. Find the refractive index of the liquid.

36.52 •• **Underwater Photography.** An underwater camera has a lens with focal length in air of 35.0 mm and a maximum aperture of $f/2.80$. The film it uses has an emulsion that is sensitive to light of frequency 6.00×10^{14} Hz. If the photographer takes a picture of an object 2.75 m in front of the camera with the lens wide open, what is the width of the smallest resolvable detail on the subject if the object is (a) a fish underwater with the camera in the water and (b) a person on the beach with the camera out of the water?

36.53 ••• CALC The intensity of light in the Fraunhofer diffraction pattern of a single slit is given by Eq. (36.5). Let $\gamma = \beta/2$. (a) Show that the equation for the values of γ at which I is a maximum is $\tan \gamma = \gamma$. (b) Determine the two smallest positive values of γ that are solutions of this equation. (*Hint:* You can use a trial-and-error procedure. Guess a value of γ and adjust your guess to bring $\tan \gamma$ closer to γ. A graphical solution of the equation is very helpful in locating the solutions approximately, to get good initial guesses.) (c) What are the positive values of γ for the first, second, and third minima on one side of the central maximum? Are the γ values in part (b) precisely halfway between the γ values for adjacent minima? (d) If $a = 12\lambda$, what are the angles θ (in degrees) that locate the first minimum, the first maximum beyond the central maximum, and the second minimum?

36.54 •• A slit 0.360 mm wide is illuminated by parallel rays of light that have a wavelength of 540 nm. The diffraction pattern is observed on a screen that is 1.20 m from the slit. The intensity at the center of the central maximum ($\theta = 0°$) is I_0. (a) What is the distance on the screen from the center of the central maximum to the first minimum? (b) What is the distance on the screen from the center of the central maximum to the point where the intensity has fallen to $I_0/2$?

36.55 •• CP CALC In a large vacuum chamber, monochromatic laser light passes through a narrow slit in a thin aluminum plate and forms a diffraction pattern on a screen that is 0.620 m from the slit. When the aluminum plate has a temperature of 20.0°C, the width of the central maximum in the diffraction pattern is 2.75 mm. What is the change in the width of the central maximum when the temperature of the plate is raised to 520.0°C? Does the width of the central diffraction maximum increase or decrease when the temperature is increased?

36.56 •• CP In a laboratory, light from a particular spectrum line of helium passes through a diffraction grating and the second-order maximum is at 18.9° from the center of the central bright fringe. The same grating is then used for light from a distant galaxy that is moving away from the earth with a speed of 2.65×10^7 m/s. For the light from the galaxy, what is the angular location of the second-order maximum for the same spectral line as was observed in the lab? (See Section 16.8.)

36.57 • What is the longest wavelength that can be observed in the third order for a transmission grating having 9200 slits/cm? Assume normal incidence.

36.58 •• It has been proposed to use an array of infrared telescopes spread over thousands of kilometers of space to observe planets orbiting other stars. Consider such an array that has an effective diameter of 6000 km and observes infrared radiation at a wavelength of 10 μm. If it is used to observe a planet orbiting the star 70 Virginis, which is 59 light-years from our solar system, what is the size of the smallest details that the array might resolve on the planet? How does this compare to the diameter of the planet, which is assumed to be similar to that of Jupiter $(1.40 \times 10^5$ km)? (Although the planet of 70 Virginis is thought to be at least 6.6 times more massive than Jupiter, its radius is probably not too different from that of Jupiter. Such large planets are thought to be composed primarily of gases, not rocky material, and hence can be greatly compressed by the mutual gravitational attraction of different parts of the planet.)

36.59 • A diffraction grating has 650 slits/mm. What is the highest order that contains the entire visible spectrum? (The wavelength range of the visible spectrum is approximately 380–750 nm.)

36.60 •• *Quasars,* an abbreviation for *quasi-stellar radio sources,* are distant objects that look like stars through a telescope but that emit far more electromagnetic radiation than an entire normal galaxy of stars. An example is the bright object below and to the left of center in **Fig. P36.60**; the other elongated objects in this image are normal galaxies. The leading model for the structure of a quasar is a galaxy with a supermassive black hole at its center. In this model, the radiation is emitted by interstellar gas and dust within the galaxy as this material falls toward the black hole. The radiation is thought to emanate from a region just a few light-years in diameter. (The diffuse glow surrounding the bright quasar shown in Fig. P36.60 is thought to be this quasar's host galaxy.) To investigate this model of quasars and to study other exotic astronomical objects, the Russian Space Agency plans to place a radio telescope in an orbit that extends to 77,000 km from the earth. When the signals from this telescope are combined with signals from the ground-based telescopes of the VLBA, the resolution will be that of a single radio telescope 77,000 km in diameter. What is the size of the smallest detail that this arrangement could resolve in quasar 3C 405, which is 7.2×10^8 light-years from earth, using radio waves at a frequency of 1665 MHz? (*Hint:* Use Rayleigh's criterion.) Give your answer in light-years and in kilometers.

Figure **P36.60**

36.61 •• A glass sheet is covered by a very thin opaque coating. In the middle of this sheet there is a thin scratch 0.00125 mm thick. The sheet is totally immersed beneath the surface of a liquid. Parallel rays of monochromatic coherent light with wavelength 612 nm in air strike the sheet perpendicular to its surface and pass through the scratch. A screen is placed in the liquid a distance of 30.0 cm away from the sheet and parallel to it. You observe that the first dark fringes on either side of the central bright fringe on the screen are 22.4 cm apart. What is the refractive index of the liquid?

36.62 •• BIO **Resolution of the Eye.** The maximum resolution of the eye depends on the diameter of the opening of the pupil (a diffraction effect) and the size of the retinal cells. The size of the retinal cells (about 5.0 μm in diameter) limits the size of an object at the near point (25 cm) of the eye to a height of about 50 μm. (To get a reasonable estimate without having to go through complicated calculations, we shall ignore the effect of the fluid in the eye.) (a) Given that the diameter of the human pupil is about 2.0 mm, does the Rayleigh criterion allow us to resolve a 50-μm-tall object at 25 cm from the eye with light of wavelength 550 nm? (b) According to the Rayleigh criterion, what is the shortest object we could resolve at the 25-cm near point with light of wavelength 550 nm? (c) What angle would the object in part (b) subtend at the eye? Express your answer in minutes (60 min $= 1°$), and compare it with the experimental value of about 1 min. (d) Which effect is more important in limiting the resolution of our eyes: diffraction or the size of the retinal cells?

36.63 •• DATA While researching the use of laser pointers, you conduct a diffraction experiment with two thin parallel slits. Your result is the pattern of closely spaced bright and dark fringes shown in **Fig. P36.63**. (Only the central portion of the pattern is shown.) You measure that the bright spots are equally spaced at 1.53 mm center to center (except for the missing spots) on a screen that is 2.50 m from the slits. The light source was a helium–neon laser producing a wavelength of 632.8 nm. (a) How far apart are the two slits? (b) How wide is each one?

Figure **P36.63**

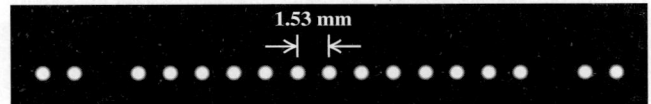

1.53 mm

36.64 •• DATA Your physics study partner tells you that the width of the central bright band in a single-slit diffraction pattern is inversely proportional to the width of the slit. This means that the width of the central maximum increases when the width of the slit decreases. The claim seems counterintuitive to you, so you make measurements to test it. You shine monochromatic laser light with wavelength λ onto a very narrow slit of width a and measure the width w of the central maximum in the diffraction pattern that is produced on a screen 1.50 m from the slit. (By "width," you mean the distance on the screen between the two minima on either side of the central maximum.) Your measurements are given in the table.

a (μm)	0.78	0.91	1.04	1.82	3.12	5.20	7.80	10.40	15.60
w (m)	2.68	2.09	1.73	0.89	0.51	0.30	0.20	0.15	0.10

(a) If w is inversely proportional to a, then the product aw is constant, independent of a. For the data in the table, graph aw versus a. Explain why aw is not constant for smaller values of a. (b) Use your graph in part (a) to calculate the wavelength λ of the laser light. (c) What is the angular position of the first minimum in the diffraction pattern for (i) $a = 0.78$ μm and (ii) $a = 15.60$ μm?

36.65 •• DATA At the metal fabrication company where you work, you are asked to measure the diameter D of a very small circular hole in a thin, vertical metal plate. To do so, you pass coherent monochromatic light with wavelength 562 nm through the hole and observe the diffraction pattern on a screen that is a distance x from the hole. You measure the radius r of the first dark ring in the diffraction pattern (see Fig. 36.26). You make the measurements for four values of x. Your results are given in the table.

x (m)	1.00	1.50	2.00	2.50
r (cm)	5.6	8.5	11.6	14.1

(a) Use each set of measurements to calculate D. Because the measurements contain some error, calculate the average of the four values of D and take that to be your reported result. (b) For $x = 1.00$ m, what are the radii of the second and third dark rings in the diffraction pattern?

CHALLENGE PROBLEMS

36.66 ••• **Intensity Pattern of N Slits.** (a) Consider an arrangement of N slits with a distance d between adjacent slits. The slits emit coherently and in phase at wavelength λ. Show that at a time t, the electric field at a distant point P is

$$E_P(t) = E_0\cos(kR - \omega t) + E_0\cos(kR - \omega t + \phi)$$
$$+ E_0\cos(kR - \omega t + 2\phi) + \cdots$$
$$+ E_0\cos(kR - \omega t + (N-1)\phi)$$

where E_0 is the amplitude at P of the electric field due to an individual slit, $\phi = (2\pi d\sin\theta)/\lambda$, θ is the angle of the rays reaching P (as measured from the perpendicular bisector of the slit arrangement), and R is the distance from P to the most distant slit. In this problem, assume that R is much larger than d. (b) To carry out the sum in part (a), it is convenient to use the complex-number relationship $e^{iz} = \cos z + i\sin z$, where $i = \sqrt{-1}$. In this expression, $\cos z$ is the *real part* of the complex number e^{iz}, and $\sin z$ is its *imaginary part*. Show that the electric field $E_P(t)$ is equal to the real part of the complex quantity

$$\sum_{n=0}^{N-1} E_0 e^{i(kR - \omega t + n\phi)}$$

(c) Using the properties of the exponential function that $e^A e^B = e^{(A+B)}$ and $(e^A)^n = e^{nA}$, show that the sum in part (b) can be written as

$$E_0\left(\frac{e^{iN\phi} - 1}{e^{i\phi} - 1}\right)e^{i(kR-\omega t)}$$
$$= E_0\left(\frac{e^{iN\phi/2} - e^{-iN\phi/2}}{e^{i\phi/2} - e^{-i\phi/2}}\right)e^{i[kR-\omega t+(N-1)\phi/2]}$$

Then, using the relationship $e^{iz} = \cos z + i\sin z$, show that the (real) electric field at point P is

$$E_P(t) = \left[E_0\frac{\sin(N\phi/2)}{\sin(\phi/2)}\right]\cos[kR - \omega t + (N-1)\phi/2]$$

The quantity in the first square brackets in this expression is the amplitude of the electric field at P. (d) Use the result for the electric-field amplitude in part (c) to show that the intensity at an angle θ is

$$I = I_0\left[\frac{\sin(N\phi/2)}{\sin(\phi/2)}\right]^2$$

where I_0 is the maximum intensity for an individual slit. (e) Check the result in part (d) for the case $N = 2$. It will help to recall

that $\sin 2A = 2\sin A\cos A$. Explain why your result differs from Eq. (35.10), the expression for the intensity in two-source interference, by a factor of 4. (*Hint:* Is I_0 defined in the same way in both expressions?)

36.67 ••• **CALC Intensity Pattern of N Slits, Continued.** Part (d) of Challenge Problem 36.66 gives an expression for the intensity in the interference pattern of N identical slits. Use this result to verify the following statements. (a) The maximum intensity in the pattern is $N^2 I_0$. (b) The principal maximum at the center of the pattern extends from $\phi = -2\pi/N$ to $\phi = 2\pi/N$, so its width is inversely proportional to $1/N$. (c) A minimum occurs whenever ϕ is an integral multiple of $2\pi/N$, except when ϕ is an integral multiple of 2π (which gives a principal maximum). (d) There are $(N-1)$ minima between each pair of principal maxima. (e) Halfway between two principal maxima, the intensity can be no greater than I_0; that is, it can be no greater than $1/N^2$ times the intensity at a principal maximum.

36.68 ••• **CALC** It is possible to calculate the intensity in the single-slit Fraunhofer diffraction pattern *without* using the phasor method of Section 36.3. Let y' represent the position of a point within the slit of width a in Fig. 36.5a, with $y' = 0$ at the center of the slit so that the slit extends from $y' = -a/2$ to $y' = a/2$. We imagine dividing the slit up into infinitesimal strips of width dy', each of which acts as a source of secondary wavelets. (a) The amplitude of the total wave at the point O on the distant screen in Fig. 36.5a is E_0. Explain why the amplitude of the wavelet from each infinitesimal strip within the slit is $E_0(dy'/a)$, so that the electric field of the wavelet a distance x from the infinitesimal strip is $dE = E_0(dy'/a)\sin(kx - \omega t)$. (b) Explain why the wavelet from each strip as detected at point P in Fig. 36.5a can be expressed as

$$dE = E_0\frac{dy'}{a}\sin[k(D - y'\sin\theta) - \omega t]$$

where D is the distance from the center of the slit to point P and $k = 2\pi/\lambda$. (c) By integrating the contributions dE from all parts of the slit, show that the total wave detected at point P is

$$E = E_0\sin(kD - \omega t)\frac{\sin[ka(\sin\theta)/2]}{ka(\sin\theta)/2}$$
$$= E_0\sin(kD - \omega t)\frac{\sin[\pi a(\sin\theta)/\lambda]}{\pi a(\sin\theta)/\lambda}$$

(The trigonometric identities in Appendix B will be useful.) Show that at $\theta = 0°$, corresponding to point O in Fig. 36.5a, the wave is $E = E_0\sin(kD - \omega t)$ and has amplitude E_0, as stated in part (a). (d) Use the result of part (c) to show that if the intensity at point O is I_0, then the intensity at a point P is given by Eq. (36.7).

PASSAGE PROBLEMS

BRAGG REFLECTION ON A DIFFERENT SCALE. A *colloid* consists of particles of one type of substance dispersed in another substance. Suspensions of electrically charged microspheres (microscopic spheres, such as polystyrene) in a liquid such as water can form a colloidal crystal when the microspheres arrange themselves in a regular repeating pattern under the influence of the electrostatic force. Colloidal crystals can selectively manipulate different wavelengths of visible light. Just as we can study crystalline solids by using Bragg reflection of x rays, we can study colloidal crystals through Bragg scattering of visible light from the regular arrangement of charged microspheres. Because the light is traveling through a liquid when it experiences the path differences that lead to constructive interference, it is the wavelength in the liquid that determines the angles at which Bragg reflections are seen. In one experiment, laser light with a wavelength in vacuum of 650 nm is passed through a sample of charged polystyrene spheres in water. A strong interference maximum is then observed when the incident and reflected beams make an angle of 39° with the colloidal crystal planes.

36.69 Why is visible light, which has much longer wavelengths than x rays do, used for Bragg reflection experiments on colloidal crystals? (a) The microspheres are suspended in a liquid, and it is more difficult for x rays to penetrate liquid than it is for visible light. (b) The irregular spacing of the microspheres allows the longer-wavelength visible light to produce more destructive interference than can x rays. (c) The microspheres are much larger than atoms in a crystalline solid, and in order to get interference maxima at reasonably large angles, the wavelength must be much longer

than the size of the individual scatterers. (d) The microspheres are spaced more widely than atoms in a crystalline solid, and in order to get interference maxima at reasonably large angles, the wavelength must be comparable to the spacing between scattering planes.

36.70 What plane spacing in the colloidal crystal could produce the maximum in this experiment? (a) 390 nm; (b) 520 nm; (c) 650 nm; (d) 780 nm.

36.71 When the light is passed through the bottom of the sample container, the interference maximum is observed to be at 41°; when it is passed through the top, the corresponding maximum is at 37°. What is the best explanation for this observation? (a) The microspheres are more tightly packed at the bottom, because they tend to settle in the suspension. (b) The microspheres are more tightly packed at the top, because they tend to float to the top of the suspension. (c) The increased pressure at the bottom makes the microspheres smaller there. (d) The maximum at the bottom corresponds to $m = 2$, whereas the maximum at the top corresponds to $m = 1$.

Answers

Chapter Opening Question ?

(i) For an optical system that uses a lens, the ability to resolve fine details—its resolving power, or resolution—improves as the lens diameter D increases (Section 36.7). Each miniature lens in a fly's eye produces its own image, so these images have very poor resolution, compared to those produced by a human eye, because the lens is so small. However, a fly's eye is much better than a human eye at detecting movement.

Test Your Understanding Questions

36.1 yes When you hear the voice of someone standing around a corner, you are hearing sound waves that underwent diffraction. If there were no diffraction or reflection of sound, you could hear sounds only from objects that were in plain view.

36.2 (ii), (i) and (iv) (tie), (iii) The angle θ of the first dark fringe is given by Eq. (36.2) with $m = 1$, or $\sin\theta = \lambda/a$. The larger the value of the ratio λ/a, the larger the value of $\sin\theta$ and hence the value of θ. The ratio λ/a in each case is
(i) $(400 \text{ nm})/(0.20 \text{ mm}) = (4.0 \times 10^{-7} \text{ m})/(2.0 \times 10^{-4} \text{ m}) = 2.0 \times 10^{-3}$;
(ii) $(600 \text{ nm})/(0.20 \text{ mm}) = (6.0 \times 10^{-7} \text{ m})/(2.0 \times 10^{-4} \text{ m}) = 3.0 \times 10^{-3}$;
(iii) $(400 \text{ nm})/(0.30 \text{ mm}) = (4.0 \times 10^{-7} \text{ m})/(3.0 \times 10^{-4} \text{ m}) = 1.3 \times 10^{-3}$;
(iv) $(600 \text{ nm})/(0.30 \text{ mm}) = (6.0 \times 10^{-7} \text{ m})/(3.0 \times 10^{-4} \text{ m}) = 2.0 \times 10^{-3}$.

36.3 (ii) and (iii) If the slit width a is less than the wavelength λ, there are no points in the diffraction pattern at which the intensity is zero (see Fig. 36.10a). The slit width is 0.0100 mm $= 1.00 \times 10^{-5}$ m, so this condition is satisfied for (ii) ($\lambda = 10.6 \ \mu\text{m} = 1.06 \times 10^{-5}$ m) and (iii) ($\lambda = 1.00 \text{ mm} = 1.00 \times 10^{-3}$ m) but not for (i) ($\lambda = 500 \text{ nm} = 5.00 \times 10^{-7}$ m) or (iv) ($\lambda = 50.0 \text{ nm} = 5.00 \times 10^{-8}$ m).

36.4 yes; $m_i = \pm 5, \pm 10, \ldots$ A "missing maximum" satisfies both $d\sin\theta = m_i\lambda$ (the condition for an interference maximum)

and $a\sin\theta = m_d\lambda$ (the condition for a diffraction minimum). Substituting $d = 2.5a$, we can combine these two conditions into the relationship $m_i = 2.5m_d$. This is satisfied for $m_i = \pm 5$ and $m_d = \pm 2$ (the fifth interference maximum is missing because it coincides with the second diffraction minimum), $m_i = \pm 10$ and $m_d = \pm 4$ (the tenth interference maximum is missing because it coincides with the fourth diffraction minimum), and so on.

36.5 (i) As described in the text, the resolving power needed is $R = Nm = 1000$. In the first order ($m = 1$) we need $N = 1000$ slits, but in the fourth order ($m = 4$) we need only $N = R/m = 1000/4 = 250$ slits. (These numbers are only approximate because of the arbitrary nature of our criterion for resolution and because real gratings always have slight imperfections in the shapes and spacings of the slits.)

36.6 (ii) The angular position of the mth maximum is given by Eq. (36.16), $2d\sin\theta = m\lambda$. This gives $m = (2d\sin\theta)/\lambda$. The sine function can never be greater than 1, so the largest value of m in the pattern can be no greater than $2d/\lambda = 2(0.200 \text{ nm})/(0.0900 \text{ nm}) = 4.44$. Since m must be an integer, the highest-order maximum in the pattern is $m = 4$ (fourth order). The $m = 5, 6, 7, \ldots$ maxima do not appear.

36.7 (iii), (ii), (iv), (i) Rayleigh's criterion combined with Eq. (36.17) shows that the smaller the value of the ratio λ/D, the better the resolving power of a telescope of diameter D. For the four telescopes, this ratio is equal to (i) $(21 \text{ cm})/(100 \text{ m}) = (0.21 \text{ m})/(100 \text{ m}) = 2.1 \times 10^{-3}$; (ii) $(500 \text{ nm})/(2.0 \text{ m}) = (5.0 \times 10^{-7} \text{ m})/(2.0 \text{ m}) = 2.5 \times 10^{-7}$; (iii) $(100 \text{ nm})/(1.0 \text{ m}) = (1.0 \times 10^{-7} \text{ m})/(1.0 \text{ m}) = 1.0 \times 10^{-7}$; (iv) $(10 \ \mu\text{m})/(2.0 \text{ m}) = (1.0 \times 10^{-5} \text{ m})/(2.0 \text{ m}) = 5.0 \times 10^{-6}$.

Bridging Problem

1.501×10^7 m/s or 5.00% of c; away from us

? At Brookhaven National Laboratory in New York, atomic nuclei are accelerated to 99.995% of the ultimate speed limit of the universe—the speed of light, *c*. Compared to the kinetic energy of a nucleus moving at 99.000% of *c*, the kinetic energy of the same nucleus moving at 99.995% of *c* is about (i) 0.001% greater; (ii) 0.% greater; (iii) 1% greater; (iv) 2% greater; (v) 16 times greater.

37 RELATIVITY

In 1905, Albert Einstein—then an unknown 25-year-old assistant in the Swiss patent office—published four papers of extraordinary importance. One was an analysis of Brownian motion; a second (for which he was awarded the Nobel Prize) was on the photoelectric effect. In the last two, Einstein introduced his **special theory of relativity,** proposing drastic revisions in the Newtonian concepts of space and time.

Einstein based the special theory of relativity on two postulates. One states that the laws of physics are the same in all inertial frames of reference; the other states that the speed of light in vacuum is the same in all inertial frames. These innocent-sounding propositions have far-reaching implications. Here are three: (1) Events that are simultaneous for one observer may not be simultaneous for another. (2) When two observers moving relative to each other measure a time interval or a length, they may not get the same results. (3) For the conservation principles for momentum and energy to be valid in all inertial systems, Newton's second law and the equations for momentum and kinetic energy have to be revised.

Relativity has important consequences in *all* areas of physics, including electromagnetism, atomic and nuclear physics, and high-energy physics. Although many of the results derived in this chapter may run counter to your intuition, the theory is in solid agreement with experimental observations.

37.1 INVARIANCE OF PHYSICAL LAWS

Let's take a look at the two postulates that make up the special theory of relativity. Both postulates describe what is seen by an observer in an *inertial frame of reference*, which we introduced in Section 4.2. The theory is "special" in the sense that it applies to observers in such special reference frames.

Einstein's First Postulate

Einstein's first postulate, called the **principle of relativity,** states: **The laws of physics are the same in every inertial frame of reference.** If the laws differed, that difference could distinguish one inertial frame from the others or make one frame more "correct" than another. As an example, suppose you watch two children playing catch with a ball while the three of you are aboard a train moving with constant velocity. Your observations of the motion *of the ball,* no matter how carefully done, can't tell you how fast (or whether) the train is moving. This is because Newton's laws of motion are the same in every inertial frame.

Another example is the electromotive force (emf) induced in a coil of wire by a nearby moving permanent magnet. In the frame of reference in which the *coil* is stationary (**Fig. 37.1a**), the moving magnet causes a change of magnetic flux through the coil, and this induces an emf. In a different frame of reference in which the *magnet* is stationary (Fig. 37.1b), the motion of the coil through a magnetic field induces the emf. According to the principle of relativity, both of these frames of reference are equally valid. Hence the same emf must be induced in both situations shown in Fig. 37.1. As we saw in Section 29.1, this is indeed the case, so Faraday's law is consistent with the principle of relativity. Indeed, *all* of the laws of electromagnetism are the same in every inertial frame of reference.

Equally significant is the prediction of the speed of electromagnetic radiation, derived from Maxwell's equations (see Section 32.2). According to this analysis, light and all other electromagnetic waves travel in vacuum with a constant speed, now defined to equal exactly 299,792,458 m/s. (We often use the approximate value $c = 3.00 \times 10^8$ m/s.) As we will see, the speed of light in vacuum plays a central role in the theory of relativity.

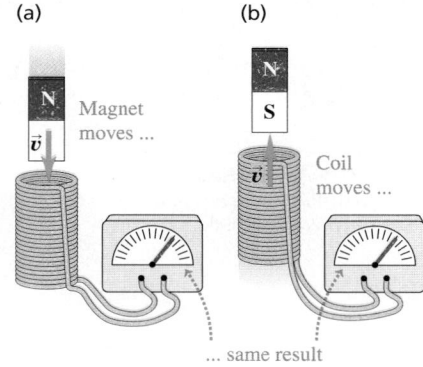

37.1 The same emf is induced in the coil whether (a) the magnet moves relative to the coil or (b) the coil moves relative to the magnet.

Einstein's Second Postulate

During the 19th century, most physicists believed that light traveled through a hypothetical medium called the *ether,* just as sound waves travel through air. If so, the speed of light measured by observers would depend on their motion relative to the ether and would therefore be different in different directions. The Michelson-Morley experiment, described in Section 35.5, was an effort to detect motion of the earth relative to the ether.

Einstein's conceptual leap was to recognize that if Maxwell's equations are valid in all inertial frames, then the speed of light in vacuum should also be the same in all frames and in all directions. In fact, Michelson and Morley detected *no* ether motion across the earth, and the ether concept has been discarded. Although Einstein may not have known about this negative result, it supported his bold hypothesis. We call this **Einstein's second postulate: The speed of light in vacuum is the same in all inertial frames of reference and is independent of the motion of the source.**

Let's think about what this means. Suppose two observers measure the speed of light in vacuum. One is at rest with respect to the light source, and the other is moving away from it. Both are in inertial frames of reference. According to the principle of relativity, the two observers must obtain the same result, despite the fact that one is moving with respect to the other.

If this seems too easy, consider the following situation. A spacecraft moving past the earth at 1000 m/s fires a missile straight ahead with a speed of 2000 m/s (relative to the spacecraft) (**Fig. 37.2**, next page). What is the missile's speed relative to the earth? Simple, you say; this is an elementary problem in relative velocity (see Section 3.5). The correct answer, according to Newtonian mechanics, is 3000 m/s. But now suppose the spacecraft turns on a searchlight, pointing in the same direction in which the missile was fired. An observer on the spacecraft measures the speed of light emitted by the searchlight and obtains the value c. According to Einstein's second postulate, the motion of the light after it has left the source cannot depend on the motion of the source. So the observer on earth who measures the speed of this same light must also obtain the value c, *not* $c + 1000$ m/s. This result contradicts our elementary notion of relative velocities, and it may not appear to agree with common sense. But "common sense" is intuition based on everyday experience, and this does not usually include measurements of the speed of light.

37.2 (a) Newtonian mechanics makes correct predictions about relatively slow-moving objects; (b) it makes incorrect predictions about the behavior of light.

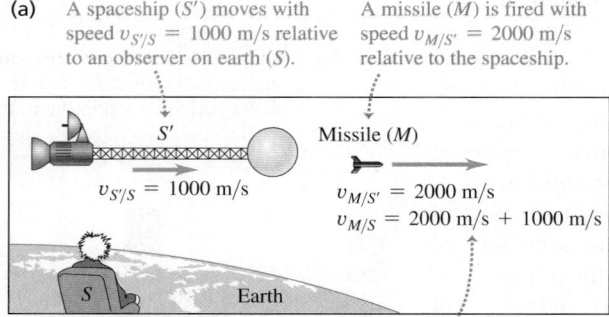

(a) A spaceship (S') moves with speed $v_{S'/S} = 1000$ m/s relative to an observer on earth (S).

A missile (M) is fired with speed $v_{M/S'} = 2000$ m/s relative to the spaceship.

$v_{S'/S} = 1000$ m/s

$v_{M/S'} = 2000$ m/s
$v_{M/S} = 2000$ m/s + 1000 m/s

NEWTONIAN MECHANICS HOLDS: Newtonian mechanics tells us correctly that the missile moves with speed $v_{M/S} = 3000$ m/s relative to the observer on earth.

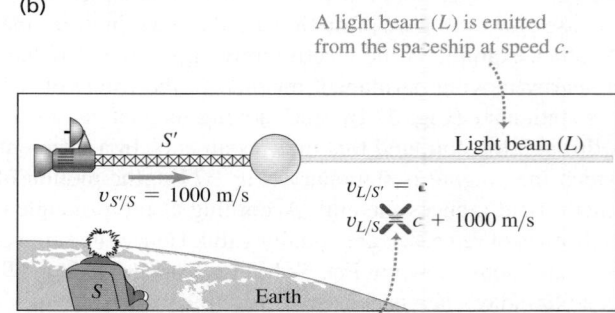

(b) A light beam (L) is emitted from the spaceship at speed c.

Light beam (L)

$v_{S'/S} = 1000$ m/s

$v_{L/S'} = c$
$v_{L/S} \ne c + 1000$ m/s

NEWTONIAN MECHANICS FAILS: Newtonian mechanics tells us *incorrectly* that the light moves at a speed greater than c relative to the observer on earth ... which would contradict Einstein's second postulate.

The Ultimate Speed Limit

Einstein's second postulate immediately implies the following result: **It is impossible for an inertial observer to travel at c, the speed of light in vacuum.**

We can prove this by showing that travel at c implies a logical contradiction. Suppose that the spacecraft S' in Fig. 37.2b is moving at the speed of light relative to an observer on the earth, so that $v_{S'/S} = c$. If the spacecraft turns on a headlight, the second postulate now asserts that the earth observer S measures the headlight beam to be also moving at c. Thus this observer measures that the headlight beam and the spacecraft move together and are always at the same point in space. But Einstein's second postulate also asserts that the headlight beam moves at a speed c relative to the spacecraft, so they *cannot* be at the same point in space. This contradictory result can be avoided only if it is impossible for an inertial observer, such as a passenger on the spacecraft, to move at c. As we go through our discussion of relativity, you may find yourself asking the question Einstein asked himself as a 16-year-old student, "What would I see if I were traveling at the speed of light?" Einstein realized only years later that his question's basic flaw was that he could *not* travel at c.

The Galilean Coordinate Transformation

37.3 The position of particle P can be described by the coordinates x and y in frame of reference S or by x' and y' in frame S'.

Frame S' moves relative to frame S with constant velocity u along the common x-x'-axis.

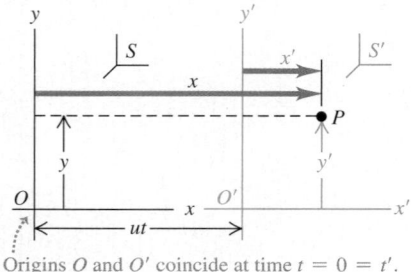

Origins O and O' coincide at time $t = 0 = t'$.

Let's restate this argument symbolically, using two inertial frames of reference, labeled S for the observer on earth and S' for the moving spacecraft, as shown in **Fig. 37.3**. To keep things as simple as possible, we have omitted the z-axes. The x-axes of the two frames lie along the same line, but the origin O' of frame S' moves relative to the origin O of frame S with constant velocity u along the common x-x'-axis. We on earth set our clocks so that the two origins coincide at time $t = 0$, so their separation at a later time t is ut.

CAUTION Choose your inertial frame coordinates wisely Many of the equations derived in this chapter are true *only* if you define your inertial reference frames as stated in the preceding paragraph. The positive x-direction must be the direction in which the origin O' moves relative to the origin O. In Fig. 37.3 this direction is to the right; if instead O' moves to the left relative to O, you must define the positive x-direction to be to the left. ▮

Now think about how we describe the motion of a particle P. This might be an exploratory vehicle launched from the spacecraft or a pulse of light from a laser. We can describe the *position* of this particle by using the earth coordinates

(x, y, z) in S or the spacecraft coordinates (x', y', z') in S'. Figure 37.3 shows that these are simply related by

$$x = x' + ut \qquad y = y' \qquad z = z' \qquad \text{(Galilean coordinate transformation)} \qquad (37.1)$$

These equations, based on the familiar Newtonian notions of space and time, are called the **Galilean coordinate transformation.**

If particle P moves in the x-direction, its instantaneous velocity v_x as measured by an observer stationary in S is $v_x = dx/dt$. Its velocity v'_x as measured by an observer stationary in S' is $v'_x = dx'/dt$. We can derive a relationship between v_x and v'_x by taking the derivative with respect to t of the first of Eqs. (37.1):

$$\frac{dx}{dt} = \frac{dx'}{dt} + u$$

Now dx/dt is the velocity v_x measured in S, and dx'/dt is the velocity v'_x measured in S', so we get the *Galilean velocity transformation* for one-dimensional motion:

$$v_x = v'_x + u \qquad \text{(Galilean velocity transformation)} \qquad (37.2)$$

Although the notation differs, this result agrees with our discussion of relative velocities in Section 3.5.

Now here's the fundamental problem. Applied to the speed of light in vacuum, Eq. (37.2) says that $c = c' + u$. Einstein's second postulate, supported subsequently by a wealth of experimental evidence, says that $c = c'$. This is a genuine inconsistency, not an illusion, and it demands resolution. If we accept this postulate, we are forced to conclude that Eqs. (37.1) and (37.2) *cannot* be precisely correct, despite our convincing derivation. These equations have to be modified to bring them into harmony with this principle.

The resolution involves some very fundamental modifications in our kinematic concepts. The first idea to be changed is the seemingly obvious assumption that the observers in frames S and S' use the same *time scale*, formally stated as $t = t'$. Alas, we are about to show that this everyday assumption cannot be correct; the two observers *must* have different time scales. We must define the velocity v' in frame S' as $v' = dx'/dt'$, not as dx'/dt; the two quantities are not the same. The difficulty lies in the concept of *simultaneity*, which is our next topic. A careful analysis of simultaneity will help us develop the appropriate modifications of our notions about space and time.

TEST YOUR UNDERSTANDING OF SECTION 37.1 As a high-speed spaceship flies past you, it fires a strobe light that sends out a pulse of light in all directions. An observer aboard the spaceship measures a spherical wave front that spreads away from the spaceship with the same speed c in all directions. (a) What is the shape of the wave front that *you* measure? (i) Spherical; (ii) ellipsoidal, with the longest axis of the ellipsoid along the direction of the spaceship's motion; (iii) ellipsoidal, with the shortest axis of the ellipsoid along the direction of the spaceship's motion; (iv) not enough information is given to decide. (b) As measured by you, does the wave front remain centered on the spaceship? ▌

37.2 RELATIVITY OF SIMULTANEITY

Measuring times and time intervals involves the concept of **simultaneity.** In a given frame of reference, an **event** is an occurrence that has a definite position and time (**Fig. 37.4**). When you say that you awoke at seven o'clock, you mean that two events (your awakening and your clock showing 7:00) occurred *simultaneously*. The fundamental problem in measuring time intervals is this: In general, two events that are simultaneous in one frame of reference are *not* simultaneous in a second frame that is moving relative to the first, even if both are inertial frames.

37.4 An event has a definite position and time—for instance, on the pavement directly below the center of the Eiffel Tower at midnight on New Year's Eve.

A Thought Experiment in Simultaneity

This may seem to be contrary to common sense. To illustrate the point, here is a version of one of Einstein's *thought experiments*—mental experiments that follow concepts to their logical conclusions. Imagine a train moving with a speed comparable to c, with uniform velocity (**Fig. 37.5**). Two lightning bolts strike a passenger car, one near each end. Each bolt leaves a mark on the car and one on the ground at the instant the bolt hits. The points on the ground are labeled A and B in the figure, and the corresponding points on the car are A' and B'. Stanley is stationary on the ground at O, midway between A and B. Mavis is moving with the train at O' in the middle of the passenger car, midway between A' and B'. Both Stanley and Mavis see both light flashes emitted from the points where the lightning strikes.

Suppose the two wave fronts from the lightning strikes reach Stanley at O simultaneously. He knows that he is the same distance from B and A, so Stanley concludes that the two bolts struck B and A simultaneously. Mavis agrees that the two wave fronts reached Stanley at the same time, but she disagrees that the flashes were emitted simultaneously.

Stanley and Mavis agree that the two wave fronts do not reach Mavis at the same time. Mavis at O' is moving to the right with the train, so she runs into

37.5 A thought experiment in simultaneity.

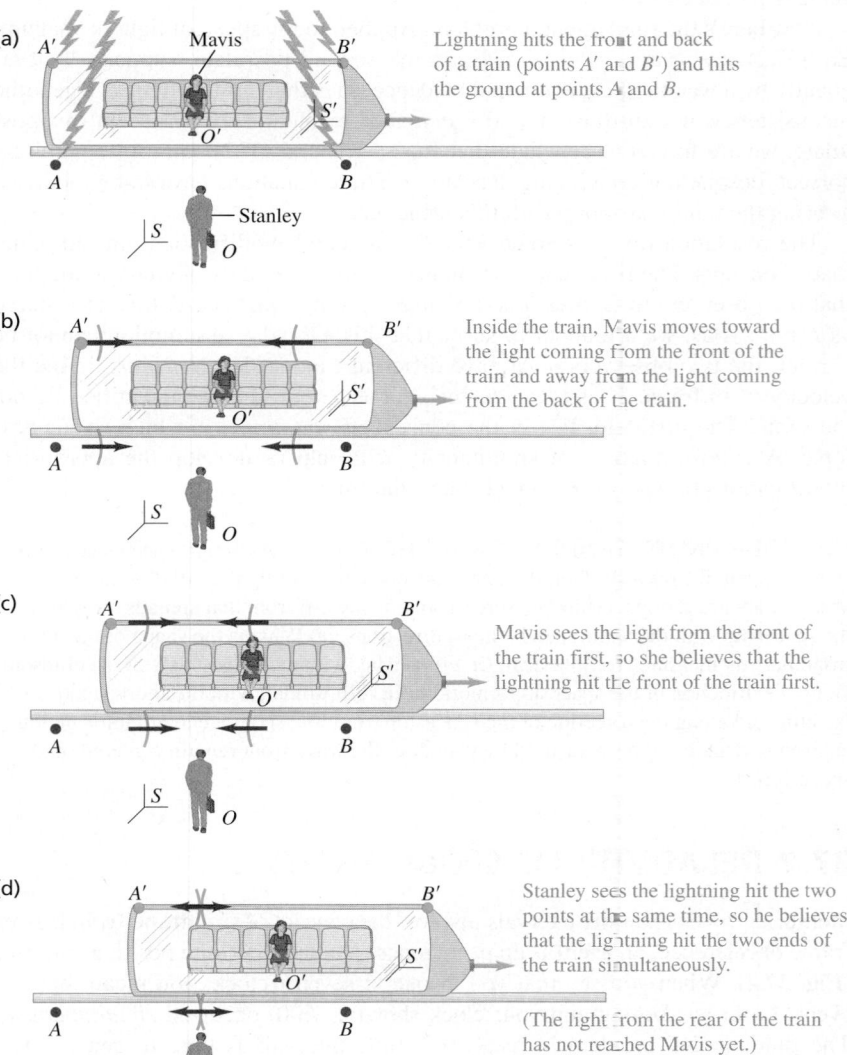

(a) Lightning hits the front and back of a train (points A' and B') and hits the ground at points A and B.

(b) Inside the train, Mavis moves toward the light coming from the front of the train and away from the light coming from the back of the train.

(c) Mavis sees the light from the front of the train first, so she believes that the lightning hit the front of the train first.

(d) Stanley sees the lightning hit the two points at the same time, so he believes that the lightning hit the two ends of the train simultaneously.

(The light from the rear of the train has not reached Mavis yet.)

the wave front from B' *before* the wave front from A' catches up to her. However, because she is in the middle of the passenger car equidistant from A' and B', her observation is that both wave fronts took the same time to reach her because both moved the same distance at the same speed c. (Recall that the speed of each wave front with respect to *either* observer is c.) Thus she concludes that the lightning bolt at B' struck *before* the one at A'. Stanley at O measures the two events to be simultaneous, but Mavis at O' does not! *Whether or not two events at different x-axis locations are simultaneous depends on the state of motion of the observer.*

You may want to argue that in this example the lightning bolts really *are* simultaneous and that if Mavis at O' could communicate with the distant points without the time delay caused by the finite speed of light, she would realize this. But that would be erroneous; the finite speed of information transmission is not the real issue. If O' is midway between A' and B', then in her frame of reference the time for a signal to travel from A' to O' is the same as that from B' to O'. Two signals arrive simultaneously at O' only if they were emitted simultaneously at A' and B'. In this example they *do not* arrive simultaneously at O', and so Mavis must conclude that the events at A' and B' were *not* simultaneous.

Furthermore, there is no basis for saying that Stanley is right and Mavis is wrong, or vice versa. According to the principle of relativity, no inertial frame of reference is more correct than any other in the formulation of physical laws. Each observer is correct *in his or her own frame of reference*. In other words, simultaneity is not an absolute concept. Whether two events are simultaneous depends on the frame of reference. As we mentioned at the beginning of this section, simultaneity plays an essential role in measuring time intervals. It follows that *the time interval between two events may be different in different frames of reference.* So our next task is to learn how to compare time intervals in different frames of reference.

TEST YOUR UNDERSTANDING OF SECTION 37.2 Stanley, who works for the rail system shown in Fig. 37.5, has carefully synchronized the clocks at all of the rail stations. At the moment that Stanley measures all of the clocks striking noon, Mavis is on a high-speed passenger car traveling from Ogdenville toward North Haverbrook. According to Mavis, when the Ogdenville clock strikes noon, what time is it in North Haverbrook? (i) Noon; (ii) before noon; (iii) after noon. ∎

37.3 RELATIVITY OF TIME INTERVALS

We can derive a quantitative relationship between time intervals in different coordinate systems. To do this, let's consider another thought experiment. As before, a frame of reference S' moves along the common x-x'-axis with constant speed u relative to a frame S. As discussed in Section 37.1, u must be less than the speed of light c. Mavis, who is riding along with frame S', measures the time interval between two events that occur at the *same* point in space. Event 1 is when a flash of light from a light source leaves O'. Event 2 is when the flash returns to O', having been reflected from a mirror a distance d away, as shown in **Fig. 37.6a** (next page). We label the time interval Δt_0, using the subscript zero as a reminder that the apparatus is at rest, with zero velocity, in frame S'. The flash of light moves a total distance $2d$, so the time interval is

$$\Delta t_0 = \frac{2d}{c} \tag{37.3}$$

The round-trip time measured by Stanley in frame S is a different interval Δt; in his frame of reference the two events occur at *different* points in space.

37.6 (a) Mavis, in frame of reference S', observes a light pulse emitted from a source at O' and reflected back along the same line. (b) How Stanley (in frame of reference S) and Mavis observe the same light pulse. The positions of O' at the times of departure and return of the pulse are shown.

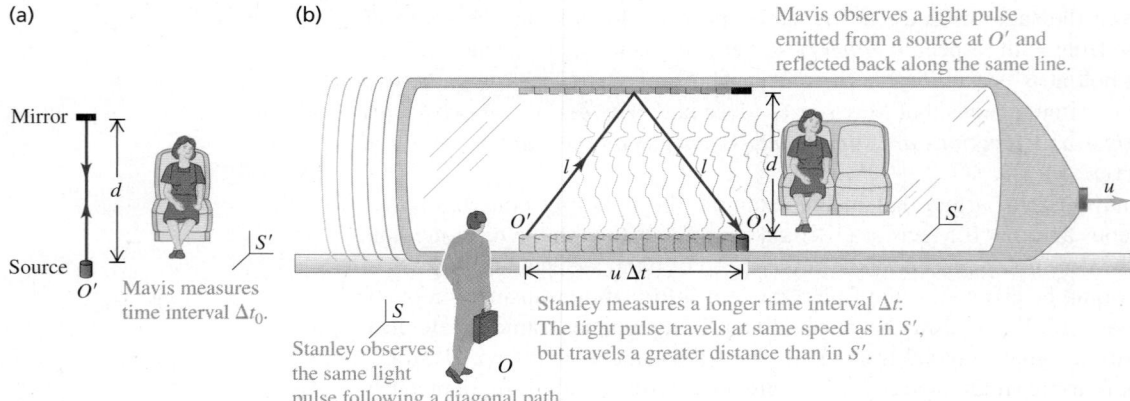

(a)

Mirror

d

Source

O' Mavis measures time interval Δt_0.

S'

(b)

Mavis observes a light pulse emitted from a source at O' and reflected back along the same line.

l l d

O' O'

$u \Delta t$

u

S'

S

Stanley observes the same light pulse following a diagonal path.

O

Stanley measures a longer time interval Δt: The light pulse travels at same speed as in S', but travels a greater distance than in S'.

During the time Δt, the source moves relative to S a distance $u\,\Delta t$ (Fig. 37.6b). In S' the round-trip distance is $2d$ perpendicular to the relative velocity, but the round-trip distance in S is the longer distance $2l$, where

$$l = \sqrt{d^2 + \left(\frac{u\,\Delta t}{2}\right)^2}$$

In writing this expression, we have assumed that both observers measure the same distance d. We will justify this assumption in the next section. The speed of light is the same for both observers, so the round-trip time measured in S is

$$\Delta t = \frac{2l}{c} = \frac{2}{c}\sqrt{d^2 + \left(\frac{u\,\Delta t}{2}\right)^2} \tag{37.4}$$

We would like to have a relationship between Δt and Δt_0 that is independent of d. To get this, we solve Eq. (37.3) for d and substitute the result into Eq. (37.4):

$$\Delta t = \frac{2}{c}\sqrt{\left(\frac{c\,\Delta t_0}{2}\right)^2 + \left(\frac{u\,\Delta t}{2}\right)^2} \tag{37.5}$$

Now we square this and solve for Δt; the result is

$$\Delta t = \frac{\Delta t_0}{\sqrt{1 - u^2/c^2}}$$

Since the quantity $\sqrt{1 - u^2/c^2}$ is less than 1, Δt is greater than Δt_0: Thus Stanley measures a *longer* round-trip time for the light pulse than does Mavis.

Time Dilation and Proper Time

We may generalize this important result. Suppose that in a particular frame of reference, two events occur at the same point in space. If these events are two ticks of a clock, then this is the frame of reference at which the clock is at rest. We call this the *rest frame* of the clock. There is only one frame of reference in which a clock is at rest, and there are infinitely many in which it is moving. Therefore the time interval measured between two events (such as two ticks of the clock) that occur at the same point in a particular frame is a more fundamental quantity than the interval between events at different points. We use the term **proper time** to describe the time interval between two events that occur *at the same point*.

Let Δt_0 be the proper time between the two events—that is, the time as measured by an observer at rest in the frame in which the events occur at the same point. Then our above result says that an observer in a second frame moving with constant speed u relative to the rest frame will measure the time interval to be Δt, where

Time dilation:

$$\Delta t = \frac{\Delta t_0}{\sqrt{1 - u^2/c^2}} \quad (37.6)$$

Proper time between two events (measured in rest frame)

Speed of light in vacuum

Speed of second frame relative to rest frame

Time interval between same events measured in second frame of reference

We recall that no inertial observer can travel at $u = c$ and we note that $\sqrt{1 - u^2/c^2}$ is imaginary for $u > c$. Thus Eq. (37.6) gives sensible results only when $u < c$. The denominator of Eq. (37.6) is always smaller than 1, so Δt is always *larger* than Δt_0. Thus we call this effect **time dilation.**

Think of an old-fashioned pendulum clock that has one second between ticks, as measured by Mavis in the clock's rest frame; this is Δt_0. If the clock's rest frame is moving relative to Stanley, he measures a time between ticks Δt that is longer than one second. In brief, *observers measure any clock to run slow if it moves relative to them* (**Fig. 37.7**). Note that this conclusion is a direct result of the fact that the speed of light in vacuum is the same in both frames of reference.

The quantity $1/\sqrt{1 - u^2/c^2}$ in Eq. (37.6) is called the **Lorentz factor.** It appears often in relativity and is denoted by the symbol γ (the Greek letter gamma):

Lorentz factor

$$\gamma = \frac{1}{\sqrt{1 - u^2/c^2}} \quad (37.7)$$

Speed of light in vacuum

Speed of one frame of reference relative to another

In terms of this symbol, we can express the time dilation formula, Eq. (37.6), as

Time dilation:

$$\Delta t = \gamma \Delta t_0 \quad (37.8)$$

Proper time between two events (measured in rest frame)

Lorentz factor relating the two frames

Time interval between same events measured in second frame of reference

As a further simplification, u/c is sometimes given the symbol β (the Greek letter beta); then $\gamma = 1/\sqrt{1 - \beta^2}$.

Figure 37.8 shows a graph of γ as a function of the relative speed u of two frames of reference. When u is very small compared to c, u^2/c^2 is much smaller than 1 and γ is very nearly *equal* to 1. In that limit, Eqs. (37.6) and (37.8) approach the Newtonian relationship $\Delta t = \Delta t_0$, corresponding to the same time interval in all frames of reference.

If the relative speed u is great enough that γ is appreciably greater than 1, the speed is said to be *relativistic;* if the difference between γ and 1 is negligibly small, the speed u is called *nonrelativistic.* Thus $u = 6.00 \times 10^7$ m/s $= 0.200c$ (for which $\gamma = 1.02$) is a relativistic speed, but $u = 6.00 \times 10^4$ m/s $= 0.000200c$ (for which $\gamma = 1.00000002$) is a nonrelativistic speed.

CAUTION **Measuring time intervals** It is important to note that the time interval Δt in Eq. (37.6) involves events that occur *at different space points* in the frame of reference S. Note also that any differences between Δt and the proper time Δt_0 are *not* caused by differences in the times required for light to travel from those space points to an observer at rest in S. We assume that our observer can correct for differences in light transit times, just as an astronomer who's observing the sun understands that an event seen now on earth actually occurred 500 s ago on the sun's surface. Alternatively, we can use *two* observers, one stationary at the location of the first event and the other at the second, each with his or her own clock. We can synchronize these two clocks without difficulty, as long as they are at rest in the same frame of reference. For example, we could send a light pulse simultaneously to the two clocks from a point midway between them. When the pulses arrive, the observers set their clocks to a prearranged time. (But clocks that are synchronized in one frame of reference *are not* in general synchronized in any other frame.) ▌

37.7 This image shows an exploding star, called a *supernova,* within a distant galaxy. The brightness of a typical supernova decays at a certain rate. But supernovae that are moving away from us at a substantial fraction of the speed of light decay more slowly, in accordance with Eq. (37.6). The decaying supernova is a moving "clock" that runs slow.

Galaxy

Supernova

37.8 The Lorentz factor $\gamma = 1/\sqrt{1 - u^2/c^2}$ as a function of the relative speed u of two frames of reference.

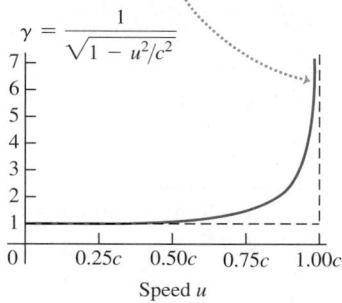

As speed u approaches the speed of light c, γ approaches infinity.

$$\gamma = \frac{1}{\sqrt{1 - u^2/c^2}}$$

Speed u

37.9 A frame of reference pictured as a coordinate system with a grid of synchronized clocks.

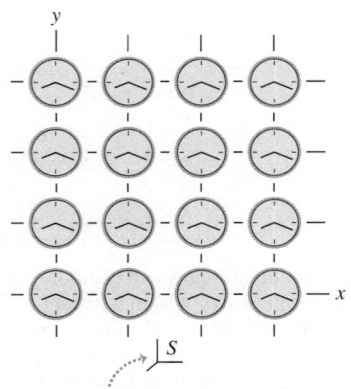

The grid is three dimensional; identical planes of clocks lie in front of and behind the page, connected by grid lines perpendicular to the page.

In thought experiments, it's often helpful to imagine many observers with synchronized clocks at rest at various points in a particular frame of reference. We can picture a frame of reference as a coordinate grid with lots of synchronized clocks distributed around it, as suggested by **Fig. 37.9**. Only when a clock is moving relative to a given frame of reference do we have to watch for ambiguities of synchronization or simultaneity.

Throughout this chapter we will frequently use phrases like "Stanley *observes* that Mavis passes the point $x = 5.00$ m, $y = 0$, $z = 0$ at time 2.00 s." This means that Stanley is using a grid of clocks in his frame of reference, like the grid shown in Fig. 37.9, to record the time of an event. We could restate the phrase as "When Mavis passes the point at $x = 5.00$ m, $y = 0$, $z = 0$, the clock at that location in Stanley's frame of reference reads 2.00 s." We will avoid using phrases like "Stanley *sees* that Mavis is at a certain point at a certain time," because there is a time delay for light to travel to Stanley's eye from the position of an event.

PROBLEM-SOLVING STRATEGY 37.1 | TIME DILATION

IDENTIFY *the relevant concepts:* The concept of time dilation is used whenever we compare the time intervals between events as measured by observers in different inertial frames of reference.

SET UP *the problem* using the following steps:
1. First decide what two events define the beginning and the end of the time interval. Then identify the two frames of reference in which the time interval is measured.
2. Identify the target variable.

EXECUTE *the solution* as follows:
1. In many problems, the time interval as measured in one frame of reference is the *proper* time Δt_0. This is the time interval between two events in a frame of reference in which the two events occur at the same point in space. In a second frame of reference that has a speed u relative to that first frame, there is a longer time interval Δt between the same two events. In this second frame the two events occur at different points. You will need to decide in which frame the time interval is Δt_0 and in which frame it is Δt.
2. Use Eq. (37.6) or (37.8) to relate Δt_0 and Δt, and then solve for the target variable.

EVALUATE *your answer:* Note that Δt is never smaller than Δt_0, and u is never greater than c. If your results suggest otherwise, you need to rethink your calculation.

EXAMPLE 37.1 | TIME DILATION AT 0.990*c*

High-energy subatomic particles coming from space interact with atoms in the earth's upper atmosphere, in some cases producing unstable particles called *muons*. A muon decays into other particles with a mean lifetime of 2.20 μs $= 2.20 \times 10^{-6}$ s as measured in a reference frame in which it is at rest. If a muon is moving at 0.990*c* relative to the earth, what will an observer on earth measure its mean lifetime to be?

SOLUTION

IDENTIFY and SET UP: The muon's lifetime is the time interval between two events: the production of the muon and its subsequent decay. Our target variable is the lifetime in your frame of reference on earth, which we call frame S. We are given the lifetime in a frame S' in which the muon is at rest; this is its *proper* lifetime, $\Delta t_0 = 2.20 \ \mu$s. The relative speed of these two frames is $u = 0.990c$. We use Eq. (37.6) to relate the lifetimes in the two frames.

EXECUTE: The muon moves relative to the earth between the two events, so the two events occur at different positions as measured in S and the time interval in that frame is Δt (the target variable). From Eq. (37.6),

$$\Delta t = \frac{\Delta t_0}{\sqrt{1 - u^2/c^2}} = \frac{2.20 \ \mu\text{s}}{\sqrt{1 - (0.990)^2}} = 15.6 \ \mu\text{s}$$

EVALUATE: Our result predicts that the mean lifetime of the muon in the earth frame (Δt) is about seven times longer than in the muon's frame (Δt_0). This prediction has been verified experimentally; indeed, this was the first experimental confirmation of the time dilation formula, Eq. (37.6).

EXAMPLE 37.2 **TIME DILATION AT AIRLINER SPEEDS**

An airplane flies from San Francisco to New York (about 4800 km, or 4.80×10^6 m) at a steady speed of 300 m/s (about 670 mi/h). How much time does the trip take, as measured by an observer on the ground? By an observer in the plane?

SOLUTION

IDENTIFY and SET UP: Here we're interested in the time interval between the airplane departing from San Francisco and landing in New York. The target variables are the time intervals as measured in the frame of reference of the ground S and in the frame of reference of the airplane S'.

EXECUTE: As measured in S the two events occur at different positions (San Francisco and New York), so the time interval measured by ground observers corresponds to Δt in Eq. (37.6). To find it, we simply divide the distance by the speed $u = 300$ m/s:

$$\Delta t = \frac{4.80 \times 10^6 \text{ m}}{300 \text{ m/s}} = 1.60 \times 10^4 \text{ s} \quad \text{(about } 4\tfrac{1}{2} \text{ hours)}$$

In the airplane's frame S', San Francisco and New York passing under the plane occur at the same point (the position of the plane). Hence the time interval in the airplane is a proper time, corresponding to Δt_0 in Eq. (37.6). We have

$$\frac{u^2}{c^2} = \frac{(300 \text{ m/s})^2}{(3.00 \times 10^8 \text{ m/s})^2} = 1.00 \times 10^{-12}$$

From Eq. (37.6),

$$\Delta t_0 = (1.60 \times 10^4 \text{ s})\sqrt{1 - 1.00 \times 10^{-12}}$$

The square root can't be evaluated with adequate precision with an ordinary calculator. But we can approximate it using the binomial theorem (see Appendix B):

$$(1 - 1.00 \times 10^{-12})^{1/2} = 1 - \left(\tfrac{1}{2}\right)(1.00 \times 10^{-12}) + \cdots$$

The remaining terms are of the order of 10^{-24} or smaller and can be discarded. The approximate result for Δt_0 is

$$\Delta t_0 = (1.60 \times 10^4 \text{ s})(1 - 0.50 \times 10^{-12})$$

The proper time Δt_0, measured in the airplane, is very slightly less (by less than one part in 10^{12}) than the time measured on the ground.

EVALUATE: We don't notice such effects in everyday life. But present-day atomic clocks (see Section 1.3) can attain a precision of about one part in 10^{13}. A cesium clock traveling a long distance in an airliner has been used to measure this effect and thereby verify Eq. (37.6) even at speeds much less than c.

EXAMPLE 37.3 **JUST WHEN IS IT PROPER?**

Mavis boards a spaceship and then zips past Stanley on earth at a relative speed of $0.600c$. At the instant she passes him, they both start timers. (a) A short time later Stanley measures that Mavis has traveled 9.00×10^7 m beyond him and is passing a space station. What does Stanley's timer read as she passes the space station? What does Mavis's timer read? (b) Stanley starts to blink just as Mavis flies past him, and Mavis measures that the blink takes 0.400 s from beginning to end. According to Stanley, what is the duration of his blink?

SOLUTION

IDENTIFY and SET UP: This problem involves time dilation for two *different* sets of events measured in Stanley's frame of reference (which we call S) and in Mavis's frame of reference (which we call S'). The two events of interest in part (a) are when Mavis passes Stanley and when Mavis passes the space station; the target variables are the time intervals between these two events as measured in S and in S'. The two events in part (b) are the start and finish of Stanley's blink; the target variable is the time interval between these two events as measured in S.

EXECUTE: (a) The two events, Mavis passing the earth and Mavis passing the space station, occur at different positions in Stanley's frame but at the same position in Mavis's frame. Hence Stanley

measures time interval Δt, while Mavis measures the *proper* time Δt_0. As measured by Stanley, Mavis moves at $0.600c = 0.600(3.00 \times 10^8 \text{ m/s}) = 1.80 \times 10^8$ m/s and travels 9.00×10^7 m in time $\Delta t = (9.00 \times 10^7 \text{ m})/(1.80 \times 10^8 \text{ m/s}) = 0.500$ s. From Eq. (37.6), Mavis's timer reads an elapsed time of

$$\Delta t_0 = \Delta t \sqrt{1 - u^2/c^2} = 0.500 \text{ s} \sqrt{1 - (0.600)^2} = 0.400 \text{ s}$$

(b) It is tempting to answer that Stanley's blink lasts 0.500 s in his frame. But this is wrong, because we are now considering a *different* pair of events than in part (a). The start and finish of Stanley's blink occur at the same point in his frame S but at different positions in Mavis's frame S', so the time interval of 0.400 s that she measures between these events is equal to Δt. The duration of the blink measured on Stanley's timer is the proper time Δt_0:

$$\Delta t_0 = \Delta t \sqrt{1 - u^2/c^2} = 0.400 \text{ s} \sqrt{1 - (0.600)^2} = 0.320 \text{ s}$$

EVALUATE: This example illustrates the relativity of simultaneity. In Mavis's frame she passes the space station at the same instant that Stanley finishes his blink, 0.400 s after she passed Stanley. Hence these two events are simultaneous to Mavis in frame S'. But these two events are *not* simultaneous to Stanley in his frame S: According to his timer, he finishes his blink after 0.320 s and Mavis passes the space station after 0.500 s.

The Twin Paradox

Equations (37.6) and (37.8) for time dilation suggest an apparent paradox called the **twin paradox.** Consider identical twin astronauts named Eartha and Astrid. Eartha remains on earth while her twin Astrid takes off on a high-speed trip through the galaxy. Because of time dilation, Eartha observes Astrid's heartbeat and all other life processes proceeding more slowly than her own. Thus to Eartha, Astrid ages more slowly; when Astrid returns to earth she is younger (has aged less) than Eartha.

Here is the paradox: All inertial frames are equivalent. Can't Astrid make exactly the same arguments to conclude that Eartha is in fact the younger? Then each twin measures the other to be younger when they're back together, and that's a paradox.

To resolve the paradox, note that the twins are *not* identical in all respects. While Eartha remains in an approximately inertial frame at all times, Astrid must *accelerate* with respect to that frame during parts of her trip in order to leave, turn around, and return to earth. Eartha's reference frame is always approximately inertial; Astrid's is often far from inertial. Thus there is a real physical difference between the circumstances of the two twins. Careful analysis shows that Eartha is correct; when Astrid returns, she *is* younger than Eartha.

TEST YOUR UNDERSTANDING OF SECTION 37.3   Samir (who is standing on the ground) starts his stopwatch at the instant that Maria flies past him in her spaceship at a speed of $0.600c$. At the same instant, Maria starts her stopwatch. (a) As measured in Samir's frame of reference, what is the reading on Maria's stopwatch at the instant that Samir's stopwatch reads 10.0 s? (i) 10.0 s; (ii) less than 10.0 s; (iii) more than 10.0 s. (b) As measured in Maria's frame of reference, what is the reading on Samir's stopwatch at the instant that Maria's stopwatch reads 10.0 s? (i) 10.0 s; (ii) less than 10.0 s; (iii) more than 10.0 s. ▌

37.4 RELATIVITY OF LENGTH

Not only does the time interval between two events depend on the observer's frame of reference, but the *distance* between two points may also depend on the observer's frame of reference. The concept of simultaneity is involved. Suppose you want to measure the length of a moving car. One way is to have two assistants make marks on the pavement at the positions of the front and rear bumpers. Then you measure the distance between the marks. But your assistants have to make their marks *at the same time*. If one marks the position of the front bumper at one time and the other marks the position of the rear bumper half a second later, you won't get the car's true length. Since we've learned that simultaneity isn't an absolute concept, we have to proceed with caution.

Lengths Parallel to the Relative Motion

To develop a relationship between lengths that are measured parallel to the direction of motion in various coordinate systems, we consider another thought experiment. We attach a light source to one end of a ruler and a mirror to the other end. The ruler is at rest in reference frame S', and its length in this frame is l_0 (**Fig. 37.10a**). Then the time Δt_0 required for a light pulse to make the round trip from source to mirror and back is

$$\Delta t_0 = \frac{2l_0}{c} \tag{37.9}$$

This is a *proper* time interval because departure and return occur at the same point in S'.

In reference frame S the ruler is moving to the right with speed u during this travel of the light pulse (Fig. 37.10b). The length of the ruler in S is l, and the time of travel from source to mirror, as measured in S, is Δt_1. During this interval the ruler, with source and mirror attached, moves a distance $u\,\Delta t_1$. The total length of path d from source to mirror is not l, but rather

$$d = l + u\,\Delta t_1 \tag{37.10}$$

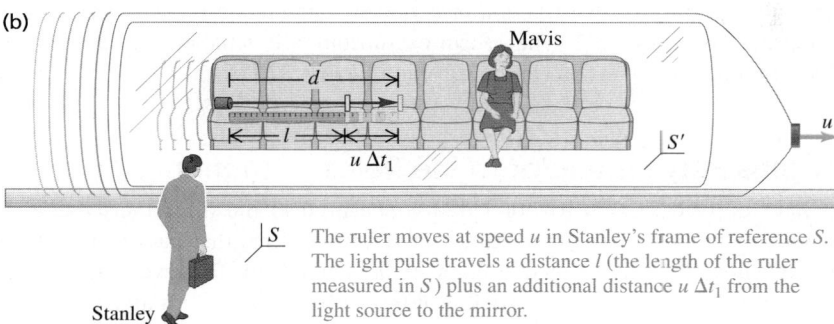

(a)

Source

Mirror

Mavis

l_0

S'

The ruler is stationary in Mavis's frame of reference S'. The light pulse travels a distance l_0 from the light source to the mirror.

(b)

Mavis

d

l

$u \Delta t_1$

S'

u

S

Stanley

The ruler moves at speed u in Stanley's frame of reference S. The light pulse travels a distance l (the length of the ruler measured in S) plus an additional distance $u \Delta t_1$ from the light source to the mirror.

37.10 (a) A ruler is at rest in Mavis's frame S'. A light pulse is emitted from a source at one end of the ruler, reflected by a mirror at the other end, and returned to the source position. (b) Motion of the light pulse as measured in Stanley's frame S.

The light pulse travels with speed c, so it is also true that

$$d = c \, \Delta t_1 \qquad (37.11)$$

Combining Eqs. (37.10) and (37.11) to eliminate d, we find

$$c \, \Delta t_1 = l + u \, \Delta t_1 \quad \text{or}$$

$$\Delta t_1 = \frac{l}{c - u} \qquad (37.12)$$

(Dividing the distance l by $c - u$ does *not* mean that light travels with speed $c - u$, but rather that the distance the pulse travels in S is greater than l.)

In the same way we can show that the time Δt_2 for the return trip from mirror to source is

$$\Delta t_2 = \frac{l}{c + u} \qquad (37.13)$$

The *total* time $\Delta t = \Delta t_1 + \Delta t_2$ for the round trip, as measured in S, is

$$\Delta t = \frac{l}{c - u} + \frac{l}{c + u} = \frac{2l}{c(1 - u^2/c^2)} \qquad (37.14)$$

We also know that Δt and Δt_0 are related by Eq. (37.6) because Δt_0 is a proper time in S'. Thus Eq. (37.9) for the round-trip time in the rest frame S' of the ruler becomes

$$\Delta t \sqrt{1 - \frac{u^2}{c^2}} = \frac{2l_0}{c} \qquad (37.15)$$

Finally, we combine Eqs. (37.14) and (37.15) to eliminate Δt and simplify:

Proper length of object (measured in rest frame)

Length contraction:

$$l = l_0 \sqrt{1 - \frac{u^2}{c^2}} = \frac{l_0}{\gamma} \qquad (37.16)$$

Speed of second frame relative to rest frame

Lorentz factor relating the two frames

Length in second frame of reference moving parallel to object's length

Speed of light in vacuum

[We have used the Lorentz factor γ defined in Eq. (37.7).] Thus the length l measured in S, in which the ruler is moving, is *shorter* than the length l_0 measured in its rest frame S'.

CAUTION **Length contraction is real** This is *not* an optical illusion! The ruler really is shorter in reference frame S than it is in S'. ▮

37.11 The speed at which electrons traverse the 3-km beam line of the SLAC National Accelerator Laboratory is slower than c by less than 1 cm/s. As measured in the reference frame of such an electron, the beam line (which extends from the top to the bottom of this photograph) is only about 15 cm long!

Beam line

A length measured in the frame in which the body is at rest (the rest frame of the body) is called a **proper length;** thus l_0 is a proper length in S', and the length measured in any other frame moving relative to S' is *less than* l_0. This effect is called **length contraction.**

When u is very small in comparison to c, γ approaches 1. Thus in the limit of small speeds we approach the Newtonian relationship $l = l_0$. This and the corresponding result for time dilation show that Eqs. (37.1), the Galilean coordinate transformation, are usually sufficiently accurate for relative speeds much smaller than c. If u is a reasonable fraction of c, however, the quantity $\sqrt{1 - u^2/c^2}$ can be appreciably less than 1. Then l can be substantially smaller than l_0, and the effects of length contraction can be substantial (**Fig. 37.11**).

Lengths Perpendicular to the Relative Motion

We have derived Eq. (37.16) for lengths measured in the direction *parallel* to the relative motion of the two frames of reference. Lengths that are measured *perpendicular* to the direction of motion are *not* contracted. To prove this, consider two identical meter sticks. One stick is at rest in frame S and lies along the positive y-axis with one end at O, the origin of S. The other is at rest in frame S' and lies along the positive y'-axis with one end at O', the origin of S'. Frame S' moves in the positive x-direction relative to frame S. Observers Stanley and Mavis, at rest in S and S' respectively, station themselves at the 50-cm mark of their sticks. At the instant the two origins coincide, the two sticks lie along the same line. At this instant, Mavis makes a mark on Stanley's stick at the point that coincides with her own 50-cm mark, and Stanley does the same to Mavis's stick.

Suppose for the sake of argument that Stanley observes Mavis's stick as longer than his own. Then the mark Stanley makes on her stick is *below* its center. In that case, Mavis will think Stanley's stick has become shorter, since half of its length coincides with *less* than half her stick's length. So Mavis observes moving sticks getting shorter and Stanley observes them getting longer. But this implies an asymmetry between the two frames that contradicts the basic postulate of relativity that tells us all inertial frames are equivalent. We conclude that consistency with the postulates of relativity requires that both observers measure the rulers as having the *same* length, even though to each observer one of them is stationary and the other is moving (**Fig. 37.12**). So *there is no length contraction perpendicular to the direction of relative motion of the coordinate systems.* We used this result in our derivation of Eq. (37.6) in assuming that the distance d is the same in both frames of reference.

For example, suppose a moving rod of length l_0 makes an angle θ_0 with the direction of relative motion (the x-axis) as measured in its rest frame. Its length component in that frame parallel to the motion, $l_0 \cos\theta_0$, is contracted to $(l_0 \cos\theta_0)/\gamma$. However, its length component perpendicular to the motion, $l_0 \sin\theta_0$, remains the same.

37.12 The meter sticks are perpendicular to the relative velocity. For any value of u, both Stanley and Mavis measure either meter stick to have a length of 1 meter.

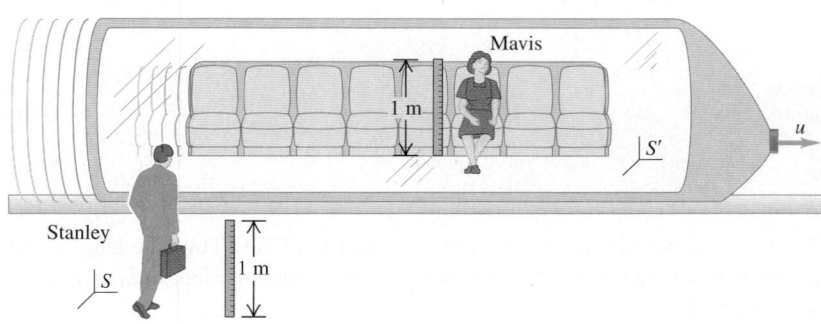

PROBLEM-SOLVING STRATEGY 37.2) LENGTH CONTRACTION

IDENTIFY *the relevant concepts:* The concept of length contraction is used whenever we compare the length of an object as measured by observers in different inertial frames of reference.

SET UP *the problem* using the following steps:
1. Decide what defines the length in question. If the problem describes an object such as a ruler, it is just the distance between the ends of the object. If the problem is about a distance between two points in space, it helps to envision an object like a ruler that extends from one point to the other.
2. Identify the target variable.

EXECUTE *the solution* as follows:
1. Determine the reference frame in which the object in question is at rest. In this frame, the length of the object is its proper

length l_0. In a second reference frame moving at speed u relative to the first frame, the object has contracted length l.
2. Keep in mind that length contraction occurs only for lengths parallel to the direction of relative motion of the two frames. Any length that is perpendicular to the relative motion is the same in both frames.
3. Use Eq. (37.16) to relate l and l_0, and then solve for the target variable.

EVALUATE *your answer:* Check that your answers make sense: l is never larger than l_0, and u is never greater than c.

EXAMPLE 37.4   HOW LONG IS THE SPACESHIP?

A spaceship flies past earth at a speed of $0.990c$. A crew member on board the spaceship measures its length, obtaining the value 400 m. What length do observers measure on earth?

SOLUTION

IDENTIFY and SET UP: This problem is about the nose-to-tail length of the spaceship as measured on the spaceship and on earth. This length is along the direction of relative motion (**Fig. 37.13**), so there will be length contraction. The spaceship's 400-m length is the *proper* length l_0 because it is measured in the frame in which the spaceship is at rest. Our target variable is the length l measured in the earth frame, relative to which the spaceship is moving at $u = 0.990c$.

EXECUTE: From Eq. (37.16), the length in the earth frame is

$$l = l_0 \sqrt{1 - \frac{u^2}{c^2}} = (400 \text{ m}) \sqrt{1 - (0.990)^2} = 56.4 \text{ m}$$

EVALUATE: The spaceship is shorter in a frame in which it is in motion than in a frame in which it is at rest. To measure the length l, two earth observers with synchronized clocks could measure the

37.13 Measuring the length of a moving spaceship.

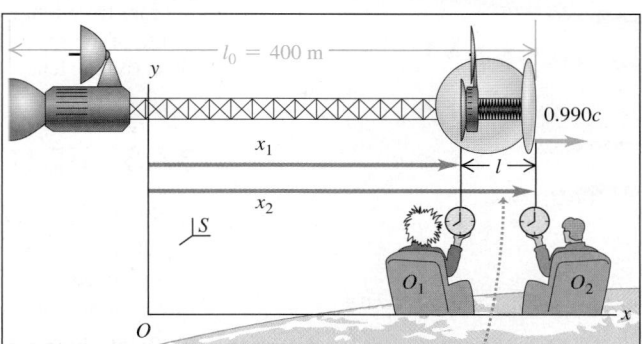

The two observers on earth (S) must measure x_2 and x_1 simultaneously to obtain the correct length $l = x_2 - x_1$ in their frame of reference.

positions of the two ends of the spaceship simultaneously in the earth's reference frame, as shown in Fig. 37.13. (These two measurements will *not* appear simultaneous to an observer in the spaceship.)

EXAMPLE 37.5   HOW FAR APART ARE THE OBSERVERS?

Observers O_1 and O_2 in Fig. 37.13 are 56.4 m apart on the earth. How far apart does the spaceship crew measure them to be?

SOLUTION

IDENTIFY and SET UP: In this example the 56.4-m distance is the *proper* length l_0. It represents the length of a ruler that extends from O_1 to O_2 and is at rest in the earth frame in which the observers are at rest. Our target variable is the length l of this ruler measured in the spaceship frame, in which the earth and ruler are moving at $u = 0.990c$.

EXECUTE: As in Example 37.4, but with $l_0 = 56.4$ m,

$$l = l_0 \sqrt{1 - \frac{u^2}{c^2}} = (56.4 \text{ m}) \sqrt{1 - (0.990)^2} = 7.96 \text{ m}$$

EVALUATE: This answer does *not* say that the crew measures their spaceship to be both 400 m long and 7.96 m long. As measured on earth, the tail of the spacecraft is at the position of O_1 at the same instant that the nose of the spacecraft is at the position of O_2. Hence the length of the spaceship measured on earth equals the 56.4-m distance between O_1 and O_2. But in the spaceship frame O_1 and O_2 are only 7.96 m apart, and the nose (which is 400 m in front of the tail) passes O_2 before the tail passes O_1.

37.14 Computer simulation of the appearance of an array of 25 rods with square cross section. The center rod is viewed end-on. The simulation ignores color changes in the array caused by the Doppler effect (see Section 37.6).

(a) Array at rest

(b) Array moving to the right at $0.2c$

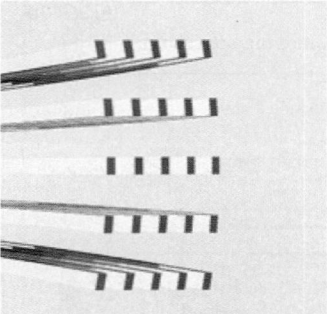

(c) Array moving to the right at $0.9c$

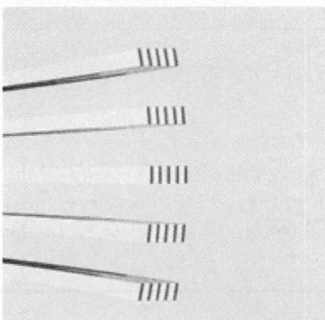

How an Object Moving Near *c* Would Appear

Let's think a little about the visual appearance of a moving three-dimensional body. If we could see the positions of all points of the body simultaneously, it would appear to shrink only in the direction of motion. But we *don't* see all the points simultaneously; light from points farther from us takes longer to reach us than does light from points near to us, so we see the farther points at the positions they had at earlier times.

Suppose we have a rectangular rod with its faces parallel to the coordinate planes. When we look end-on at the center of the closest face of such a rod at rest, we see only that face. (See the center rod in computer-generated **Fig. 37.14a**.) But when that rod is moving past us toward the right at an appreciable fraction of the speed of light, we may also see its left side because of the earlier-time effect just described. That is, we can see some points that we couldn't see when the rod was at rest because the rod moves out of the way of the light rays from those points to us. Conversely, some light that can get to us when the rod is at rest is blocked by the moving rod. Because of all this, the rods in Figs. 37.14b and 37.14c appear rotated and distorted.

TEST YOUR UNDERSTANDING OF SECTION 37.4 A miniature spaceship is flying past you, moving horizontally at a substantial fraction of the speed of light. At a certain instant, you observe that the nose and tail of the spaceship align exactly with the two ends of a meter stick that you hold in your hands. Rank the following distances in order from longest to shortest: (i) The proper length of the meter stick; (ii) the proper length of the spaceship; (iii) the length of the spaceship measured in your frame of reference; (iv) the length of the meter stick measured in the spaceship's frame of reference. ∎

37.5 THE LORENTZ TRANSFORMATIONS

In Section 37.1 we discussed the Galilean coordinate transformation equations, Eqs. (37.1). They relate the coordinates (x, y, z) of a point in frame of reference S to the coordinates (x', y', z') of the point in a second frame S'. The second frame moves with constant speed u relative to S in the positive direction along the common x-x'-axis. This transformation also assumes that the time scale is the same in the two frames of reference, so $t = t'$. This Galilean transformation, as we have seen, is valid only in the limit when u approaches zero. We are now ready to derive more general transformations that are consistent with the principle of relativity. The more general relationships are called the **Lorentz transformations.**

The Lorentz Coordinate Transformation

Our first question is this: When an event occurs at point (x, y, z) at time t, as observed in a frame of reference S, what are the coordinates (x', y', z') and time t' of the event as observed in a second frame S' moving relative to S with constant speed u in the $+x$-direction?

To derive the coordinate transformation, we refer to **Fig. 37.15**, which is the same as Fig. 37.3. As before, we assume that the origins coincide at the initial time $t = 0 = t'$. Then in S the distance from O to O' at time t is still ut. The coordinate x' is a *proper length* in S', so in S it is contracted by the factor $1/\gamma = \sqrt{1 - u^2/c^2}$, as in Eq. (37.16). Thus the distance x from O to P, as measured in S, is not simply $x = ut + x'$, as in the Galilean coordinate transformation, but

$$x = ut + x'\sqrt{1 - \frac{u^2}{c^2}} \tag{37.17}$$

Solving this equation for x', we obtain

$$x' = \frac{x - ut}{\sqrt{1 - u^2/c^2}} \tag{37.18}$$

Equation (37.18) is part of the Lorentz coordinate transformation; another part is the equation giving t' in terms of x and t. To obtain this, we note that the principle of relativity requires that the *form* of the transformation from S to S' be identical to that from S' to S. The only difference is a change in the sign of the relative velocity component u. Thus from Eq. (37.17) it must be true that

$$x' = -ut' + x\sqrt{1 - \frac{u^2}{c^2}} \qquad (37.19)$$

We now equate Eqs. (37.18) and (37.19) to eliminate x'. This gives us an equation for t' in terms of x and t. You can do the algebra to show that

$$t' = \frac{t - ux/c^2}{\sqrt{1 - u^2/c^2}} \qquad (37.20)$$

As we discussed previously, lengths perpendicular to the direction of relative motion are not affected by the motion, so $y' = y$ and $z' = z$.

Collecting our results, we have the *Lorentz coordinate transformation:*

Velocity of S' relative to S in positive direction along x-x'-axis

Lorentz coordinate transformation:
Spacetime coordinates of an event are x, y, z, t in frame S and x', y', z', t' in frame S'.

$$x' = \frac{x - ut}{\sqrt{1 - u^2/c^2}} = \gamma(x - ut)$$

Lorentz factor relating the two frames

$$y' = y$$

Speed of light in vacuum

$$z' = z \qquad (37.21)$$

$$t' = \frac{t - ux/c^2}{\sqrt{1 - u^2/c^2}} = \gamma(t - ux/c^2)$$

These equations are the relativistic generalization of the Galilean coordinate transformation, Eqs. (37.1) and $t = t'$. For values of u that approach zero, $\sqrt{1 - u^2/c^2}$ and γ approach 1, and the ux/c^2 term approaches zero. In this limit, Eqs. (37.21) become identical to Eqs. (37.1) along with $t = t'$. In general, though, both the coordinates and time of an event in one frame depend on its coordinates and time in another frame. *Space and time have become intertwined; we can no longer say that length and time have absolute meanings independent of the frame of reference.* For this reason, we refer to time and the three dimensions of space collectively as a four-dimensional entity called **spacetime,** and we call (x, y, z, t) together the **spacetime coordinates** of an event.

The Lorentz Velocity Transformation

We can use Eqs. (37.21) to derive the relativistic generalization of the Galilean velocity transformation, Eq. (37.2). We consider only one-dimensional motion along the x-axis and use the term "velocity" as being short for the "x-component of the velocity." Suppose that in a time dt a particle moves a distance dx, as measured in frame S. We obtain the corresponding distance dx' and time dt' in S' by taking differentials of Eqs. (37.21):

$$dx' = \gamma(dx - u\,dt)$$
$$dt' = \gamma(dt - u\,dx/c^2)$$

We divide the first equation by the second and then divide the numerator and denominator of the result by dt to obtain

$$\frac{dx'}{dt'} = \frac{\dfrac{dx}{dt} - u}{1 - \dfrac{u}{c^2}\dfrac{dx}{dt}}$$

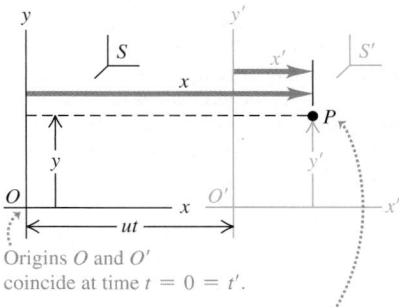

37.15 As measured in frame of reference S, x' is contracted to x'/γ, so $x = ut + (x'/\gamma)$ and $x' = \gamma(x - ut)$.

Frame S' moves relative to frame S with constant velocity u along the common x-x'-axis.

Origins O and O' coincide at time $t = 0 = t'$.

The Lorentz coordinate transformation relates the spacetime coordinates of an event as measured in the two frames: (x, y, z, t) in frame S and (x', y', z', t') in frame S'.

DATA *SPEAKS*

The Lorentz Transformations

When students were given a problem involving the Lorentz transformation equations, more than 25% gave an incorrect response. Common errors:

- Confusion about which reference frame is which. It's essential in every problem to draw a diagram showing which reference frame goes with which observer.

- Confusion about length contraction. Remember that lengths perpendicular to the direction of relative motion do not contract.

Application Relative Velocity and Reference Frames A relay race illustrates the frames of reference used in Eqs. (37.22) and (37.23). The runner in purple is the particle, and the two frames of reference in which the particle's motion is observed are our rest frame S (we are spectators standing next to the track) and the rest frame S' of the runner in red, who has speed u relative to us. Since the runner in red is moving to the left relative to us, we must take the positive x-direction to the left. The runner in purple has positive velocity v_x relative to us (she is moving to the left); if the runner in red is moving faster ($u > v_x$), then from Eq. (37.22) the runner in purple has a *negative* velocity v'_x relative to the runner in red.

Reference frame S: Our rest frame as we stand next to the track

Reference frame S': Rest frame of runner in red. She moves with speed u relative to us.

Runner in purple has velocity v_x in S and velocity v'_x in S'.

CAUTION **Use the correct reference frame coordinates** The Lorentz transformation equations given by Eqs. (37.21), (37.22), and (37.23) assume that frame S' is moving in the positive x-direction with velocity u relative to frame S. Always set up your coordinate system to follow this convention. ∎

Now dx/dt is the velocity v_x in S, and dx'/dt' is the velocity v'_x in S', so

Lorentz velocity transformation (velocity in S' in terms of velocity in S):

x-velocity of object in frame S' x-velocity of object in frame S

$$v'_x = \frac{v_x - u}{1 - uv_x/c^2}$$ (37.22)

Velocity of S' relative to S in positive direction along x-x'-axis

Speed of light in vacuum

When u and v_x are much smaller than c, the denominator in Eq. (37.22) approaches 1, and we approach the nonrelativistic result $v'_x = v_x - u$. The opposite extreme is the case $v_x = c$; then we find

$$v'_x = \frac{c - u}{1 - uc/c^2} = \frac{c(1 - u/c)}{1 - u/c} = c$$

This says that anything moving with velocity $v_x = c$ measured in S also has velocity $v'_x = c$ measured in S', despite the relative motion of the two frames. So Eq. (37.22) is consistent with Einstein's postulate that the speed of light in vacuum is the same in all inertial frames of reference.

The principle of relativity tells us there is no fundamental distinction between the two frames S and S'. Thus the expression for v_x in terms of v'_x must have the same form as Eq. (37.22), with v_x changed to v'_x, and vice versa, and the sign of u reversed. Carrying out these operations with Eq. (37.22), we find

Lorentz velocity transformation (velocity in S in terms of velocity in S'):

x-velocity of object in frame S x-velocity of object in frame S'

$$v_x = \frac{v'_x + u}{1 + uv'_x/c^2}$$ (37.23)

Velocity of S' relative to S in positive direction along x-x'-axis

Speed of light in vacuum

You can also obtain this equation by solving Eq. (37.22) for v_x. Both Eqs. (37.22) and (37.23) are *Lorentz velocity transformations* for one-dimensional motion.

When u is less than c, the Lorentz velocity transformations show us that a body moving with a speed less than c in one frame of reference always has a speed less than c in *every other* frame of reference. This is one reason for concluding that no material body may travel with a speed equal to or greater than c relative to *any* inertial frame of reference. Later we'll see that the relativistic generalizations of energy and momentum give further support to this hypothesis.

PROBLEM-SOLVING STRATEGY 37.3 | LORENTZ TRANSFORMATIONS

IDENTIFY *the relevant concepts:* The Lorentz *coordinate* transformation equations relate the spacetime coordinates of an event in one inertial reference frame to the coordinates of the same event in a second inertial frame. The Lorentz *velocity* transformation equations relate the velocity of an object in one inertial reference frame to its velocity in a second inertial frame.

SET UP *the problem* using the following steps:
1. Identify the target variable.
2. Define the two inertial frames S and S'. Remember that S' moves relative to S at a constant velocity u in the $+x$-direction.
3. If the coordinate transformation equations are needed, make a list of spacetime coordinates in the two frames, such as x_1, x'_1, t_1, t'_1, and so on. Label carefully which of these you know and which you don't.
4. In velocity-transformation problems, clearly identify u (the relative velocity of the two frames of reference), v_x (the velocity of the object relative to S), and v'_x (the velocity of the object relative to S').

EXECUTE *the solution* as follows:
1. In a coordinate-transformation problem, use Eqs. (37.21) to solve for the spacetime coordinates of the event as measured in S' in terms of the corresponding values in S. (If you need to solve for the spacetime coordinates in S in terms of the corresponding values in S', you can easily convert the expressions in Eqs. (37.21): Replace all of the primed quantities with unprimed ones, and vice versa, and replace u with $-u$.)
2. In a velocity-transformation problem, use either Eq. (37.22) or Eq. (37.23), as appropriate, to solve for the target variable.

EVALUATE *your answer:* Don't be discouraged if some of your results don't seem to make sense or if they disagree with "common sense." It takes time to develop intuition about relativity; you'll gain it with experience.

EXAMPLE 37.6 WAS IT RECEIVED BEFORE IT WAS SENT?

Winning an interstellar race, Mavis pilots her spaceship across a finish line in space at a speed of $0.600c$ relative to that line. A "hooray" message is sent from the back of her ship (event 2) at the instant (in her frame of reference) that the front of her ship crosses the line (event 1). She measures the length of her ship to be 300 m. Stanley is at the finish line and is at rest relative to it. When and where does he measure events 1 and 2 to occur?

SOLUTION

IDENTIFY and SET UP: This example involves the Lorentz coordinate transformation. Our derivation of this transformation assumes that the origins of frames S and S' coincide at $t = 0 = t'$. Thus for simplicity we fix the origin of S at the finish line and the origin of S' at the front of the spaceship so that Stanley and Mavis measure event 1 to be at $x = 0 = x'$ and $t = 0 = t'$.

Mavis in S' measures her spaceship to be 300 m long, so she has the "hooray" sent from 300 m behind her spaceship's front at the instant she measures the front to cross the finish line. That is, she measures event 2 at $x' = -300$ m and $t' = 0$.

Our target variables are the coordinate x and time t of event 2 that Stanley measures in S.

EXECUTE: To solve for the target variables, we modify the first and last of Eqs. (37.21) to give x and t as functions of x' and t'. We do so in the same way that we obtained Eq. (37.23) from Eq. (37.22). We remove the primes from x' and t', add primes to x and t, and replace each u with $-u$. The results are

$$x = \gamma(x' + ut') \quad \text{and} \quad t = \gamma(t' + ux'/c^2)$$

From Eq. (37.7), $\gamma = 1.25$ for $u = 0.600c = 1.80 \times 10^8$ m/s. We also substitute $x' = -300$ m, $t' = 0$, $c = 3.00 \times 10^8$ m/s, and $u = 1.80 \times 10^8$ m/s in the equations for x and t to find $x = -375$ m at $t = -7.50 \times 10^{-7}$ s $= -0.750$ μs for event 2.

EVALUATE: Mavis says that the events are simultaneous, but Stanley says that the "hooray" was sent *before* Mavis crossed the finish line. This does not mean that the effect preceded the cause. The fastest that Mavis can send a signal the length of her ship is 300 m$/(3.00 \times 10^8$ m/s$) = 1.00$ μs. She cannot send a signal from the front at the instant it crosses the finish line that would cause a "hooray" to be broadcast from the back at the same instant. She would have to send that signal from the front at least 1.00 μs before then, so she had to slightly anticipate her success.

EXAMPLE 37.7 RELATIVE VELOCITIES

(a) A spaceship moving away from the earth at $0.900c$ fires a robot space probe in the same direction as its motion at $0.700c$ relative to the spaceship. What is the probe's velocity relative to the earth? (b) A scoutship is sent to catch up with the spaceship by traveling at $0.950c$ relative to the earth. What is the velocity of the scoutship relative to the spaceship?

SOLUTION

IDENTIFY and SET UP: This example uses the Lorentz velocity transformation. Let the earth and spaceship reference frames be S and S', respectively (**Fig. 37.16**); their relative velocity is $u = 0.900c$. In part (a) we are given the probe velocity $v_x' = 0.700c$ with respect to S', and the target variable is the velocity v_x of the probe relative to S. In part (b) we are given the velocity $v_x = 0.950c$ of the scoutship relative to S, and the target variable is its velocity v_x' relative to S'.

EXECUTE: (a) We use Eq. (37.23) to find the probe velocity relative to the earth:

$$v_x = \frac{v_x' + u}{1 + uv_x'/c^2} = \frac{0.700c + 0.900c}{1 + (0.900c)(0.700c)/c^2} = 0.982c$$

(b) We use Eq. (37.22) to find the scoutship velocity relative to the spaceship:

$$v_x' = \frac{v_x - u}{1 - uv_x/c^2} = \frac{0.950c - 0.900c}{1 - (0.900c)(0.950c)/c^2} = 0.345c$$

EVALUATE: What would the Galilean velocity transformation formula, Eq. (37.2), say? In part (a) we would have found the probe's velocity relative to the earth to be $v_x = v_x' + u = 0.700c + 0.900c = 1.600c$, which is greater than c and hence impossible. In part (b), we would have found the scoutship's velocity relative to the spaceship to be $v_x' = v_x - u = 0.950c - 0.900c = 0.050c$; the relativistically correct value, $v_x' = 0.345c$, is almost seven times greater than the incorrect Galilean value.

37.16 The spaceship, robot space probe, and scoutship.

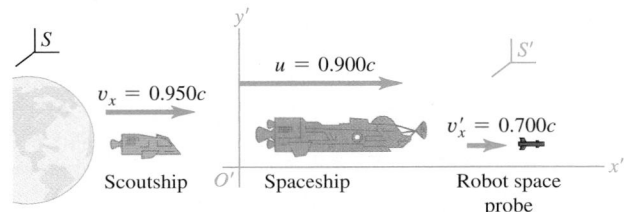

S y' $u = 0.900c$ S'

$v_x = 0.950c$

$v_x' = 0.700c$

x'

Scoutship O' Spaceship Robot space probe

TEST YOUR UNDERSTANDING OF SECTION 37.5 (a) In frame S events P_1 and P_2 occur at the same x-, y-, and z-coordinates, but event P_1 occurs before event P_2. In frame S', which event occurs first? (b) In frame S events P_3 and P_4 occur at the same time t and the same y- and z-coordinates, but event P_3 occurs at a less positive x-coordinate than event P_4. In frame S', which event occurs first? ∎

37.6 THE DOPPLER EFFECT FOR ELECTROMAGNETIC WAVES

An additional important consequence of relativistic kinematics is the Doppler effect for electromagnetic waves. In Section 16.8 we quoted without proof the formula, Eq. (16.30), for the frequency shift that results from motion of a source of electromagnetic waves relative to an observer. We can now derive that result.

Here's a statement of the problem. A source of light is moving with constant speed u toward Stanley, who is stationary in an inertial frame (**Fig. 37.17**). As measured in its rest frame, the source emits light waves with frequency f_0 and period $T_0 = 1/f_0$. What is the frequency f of these waves as received by Stanley?

Let T be the time interval between *emission* of successive wave crests as observed in Stanley's reference frame. Note that this is *not* the interval between the *arrival* of successive crests at his position, because the crests are emitted at different points in Stanley's frame. In measuring only the frequency f he receives, he does not take into account the difference in transit times for successive crests. Therefore the frequency he receives is *not* $1/T$. What is the equation for f?

During a time T the crests ahead of the source move a distance cT, and the source moves a shorter distance uT in the same direction. The distance λ between successive crests—that is, the wavelength—is thus $\lambda = (c - u)T$, as measured in Stanley's frame. The frequency that he measures is c/λ. Therefore

$$f = \frac{c}{(c - u)T} \tag{37.24}$$

So far we have followed a pattern similar to that for the Doppler effect for sound from a moving source (see Section 16.8). In that discussion our next step was to equate T to the time T_0 between emissions of successive wave crests by the source. However, due to time dilation it is *not* relativistically correct to equate T to T_0. The time T_0 is measured in the rest frame of the source, so it is a proper time. From Eq. (37.6), T_0 and T are related by

$$T = \frac{T_0}{\sqrt{1 - u^2/c^2}} = \frac{cT_0}{\sqrt{c^2 - u^2}}$$

or, since $T_0 = 1/f_0$,

$$\frac{1}{T} = \frac{\sqrt{c^2 - u^2}}{cT_0} = \frac{\sqrt{c^2 - u^2}}{c}f_0$$

Remember, $1/T$ is not equal to f. We must substitute this expression for $1/T$ into Eq. (37.24) to find f:

$$f = \frac{c}{c - u}\frac{\sqrt{c^2 - u^2}}{c}f_0$$

37.17 The Doppler effect for light. A light source moving at speed u relative to Stanley emits a wave crest, then travels a distance uT toward an observer and emits the next crest. In Stanley's reference frame S, the second crest is a distance λ behind the first crest.

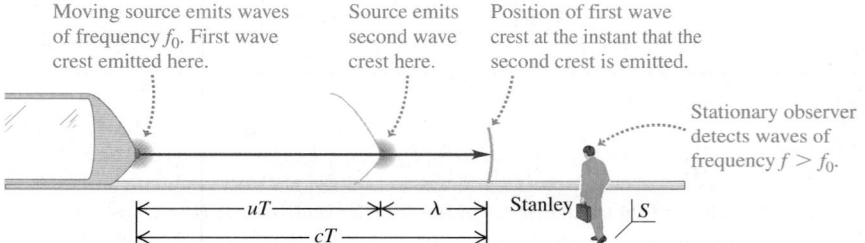

Moving source emits waves of frequency f_0. First wave crest emitted here.

Source emits second wave crest here.

Position of first wave crest at the instant that the second crest is emitted.

Stationary observer detects waves of frequency $f > f_0$.

Stanley

S

Using $c^2 - u^2 = (c - u)(c + u)$ gives

> **Doppler effect, electromagnetic waves, source approaching observer:**
>
> Frequency measured by observer ⋯⋯⋯⋯⋯⋯⋯⋯ Frequency measured in rest frame of source
>
> $$f = \sqrt{\frac{c + u}{c - u}}\, f_0 \qquad (37.25)$$
>
> Speed of light in vacuum ⋯⋯⋯ Speed of source relative to observer

This shows that when the source moves *toward* the observer, the observed frequency f is *greater* than the emitted frequency f_0. The difference $f - f_0 = \Delta f$ is called the Doppler frequency shift. When u/c is much smaller than 1, the fractional shift $\Delta f/f$ is also small and is approximately equal to u/c:

$$\frac{\Delta f}{f} = \frac{u}{c}$$

When the source moves *away from* the observer, we change the sign of u in Eq. (37.25) to get

$$f = \sqrt{\frac{c - u}{c + u}}\, f_0 \qquad \text{(Doppler effect, electromagnetic waves, source moving away from observer)} \qquad (37.26)$$

This agrees with Eq. (16.30) with minor notation changes.

With light, unlike sound, there is no distinction between motion of source and motion of observer; only the *relative* velocity of the two is significant. The last four paragraphs of Section 16.8 discuss several practical applications of the Doppler effect with light and other electromagnetic radiation; we suggest you review those paragraphs now. **Figure 37.18** shows one common application.

37.18 This handheld radar gun emits a radio beam of frequency f_0, which in the frame of reference of an approaching car has a higher frequency f given by Eq. (37.25). The reflected beam also has frequency f in the car's frame, but has an even higher frequency f' in the police officer's frame. The radar gun calculates the car's speed by comparing the frequencies of the emitted beam and the doubly Doppler-shifted reflected beam. (Compare Example 16.18 in Section 16.8.)

EXAMPLE 37.8   A JET FROM A BLACK HOLE

Many galaxies have supermassive black holes at their centers (see Section 13.8). As material swirls around such a black hole, it is heated, becomes ionized, and generates strong magnetic fields. The resulting magnetic forces steer some of the material into high-speed jets that blast out of the galaxy and into intergalactic space (**Fig. 37.19**). The light we observe from the jet in Fig. 37.19 has a frequency of 6.66×10^{14} Hz (in the far ultraviolet; see Fig. 32.4), but in the reference frame of the jet material the light has a frequency of 5.55×10^{13} Hz (in the infrared). What is the speed of the jet material with respect to us?

SOLUTION

IDENTIFY and SET UP: This problem involves the Doppler effect for electromagnetic waves. The frequency we observe is $f = 6.66 \times 10^{14}$ Hz, and the frequency in the frame of the source is $f_0 = 5.55 \times 10^{13}$ Hz. Since $f > f_0$, the jet is approaching us and we use Eq. (37.25) to find the target variable u.

EXECUTE: We need to solve Eq. (37.25) for u. We'll leave it as an exercise for you to show that the result is

$$u = \frac{(f/f_0)^2 - 1}{(f/f_0)^2 + 1}\, c$$

We have $f/f_0 = (6.66 \times 10^{14}\ \text{Hz})/(5.55 \times 10^{13}\ \text{Hz}) = 12.0$, so

$$u = \frac{(12.0)^2 - 1}{(12.0)^2 + 1}\, c = 0.986c$$

37.19 This image shows a fast-moving jet 5000 light-years in length emanating from the center of the galaxy M87. The light from the jet is emitted by fast-moving electrons spiraling around magnetic field lines (see Fig. 27.18).

EVALUATE: Because the frequency shift is quite substantial, it would have been erroneous to use the approximate expression $\Delta f/f = u/c$. Had you done so, you would have found $u = c(\Delta f/f_0) = c(6.66 \times 10^{14}\ \text{Hz} - 5.55 \times 10^{13}\ \text{Hz})/(5.55 \times 10^{13}\ \text{Hz}) = 11.0c$. This result cannot be correct because the jet material cannot travel faster than light.

37.7 RELATIVISTIC MOMENTUM

Newton's laws of motion have the same form in all inertial frames of reference. When we use transformations to change from one inertial frame to another, the laws should be *invariant* (unchanging). But we have just learned that the principle of relativity forces us to replace the Galilean transformations with the more general Lorentz transformations. As we will see, this requires corresponding generalizations in the laws of motion and the definitions of momentum and energy.

The principle of conservation of momentum states that *when two bodies interact, the total momentum is constant,* provided that the net external force acting on the bodies in an inertial reference frame is zero (for example, if they form an isolated system, interacting only with each other). If conservation of momentum is a valid physical law, it must be valid in *all* inertial frames of reference. Now, here's the problem: Suppose we look at a collision in one inertial coordinate system S and find that momentum is conserved. Then we use the Lorentz transformation to obtain the velocities in a second inertial system S'. We find that if we use the Newtonian definition of momentum ($\vec{p} = m\vec{v}$), momentum is *not* conserved in the second system! The only way to make momentum conservation consistent with relativity is to generalize the *definition* of momentum.

We won't derive the correct relativistic generalization of momentum, but here is the result. Suppose we measure the mass of a particle to be m when it is at rest relative to us: We often call m the **rest mass.** We will use the term *material particle* for a particle that has a nonzero rest mass. When such a particle has a velocity $\vec{v}$, its **relativistic momentum** $\vec{p}$ is

$$\text{Relativistic momentum} \cdots\cdots\blacktriangleright \vec{p} = \frac{m\vec{v}}{\sqrt{1 - v^2/c^2}} \tag{37.27}$$

Rest mass of particle
Velocity of particle
Speed of particle
Speed of light in vacuum

When the particle's speed v is much less than c, this is approximately equal to the Newtonian expression $\vec{p} = m\vec{v}$, but in general the momentum is greater in magnitude than mv (**Fig. 37.20**). In fact, as v approaches c, the momentum approaches infinity.

37.20 Graph of the magnitude of the momentum of a particle of rest mass m as a function of speed v. Also shown is the Newtonian prediction, which gives correct results only at speeds much less than c.

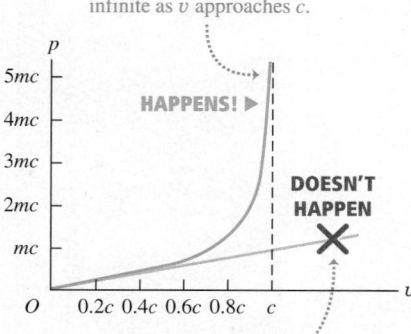

Relativistic momentum becomes infinite as v approaches c.

HAPPENS! ▶

DOESN'T HAPPEN

Newtonian mechanics incorrectly predicts that momentum becomes infinite only if v becomes infinite.

Relativity, Newton's Second Law, and Relativistic Mass

What about the relativistic generalization of Newton's second law? In Newtonian mechanics the most general form of the second law is

$$\vec{F} = \frac{d\vec{p}}{dt} \tag{37.28}$$

That is, the net force $\vec{F}$ on a particle equals the time rate of change of its momentum. Experiments show that this result is still valid in relativistic mechanics, provided that we use the relativistic momentum given by Eq. (37.27). That is, the relativistically correct generalization of Newton's second law is

$$\vec{F} = \frac{d}{dt} \frac{m\vec{v}}{\sqrt{1 - v^2/c^2}} \tag{37.29}$$

Because momentum is no longer directly proportional to velocity, the rate of change of momentum is no longer directly proportional to the acceleration. As a result, *constant force does not cause constant acceleration.* For example, when the net force and the velocity are both along the x-axis, Eq. 37.29 gives

$$F = \frac{m}{(1 - v^2/c^2)^{3/2}} a \qquad (\vec{F} \text{ and } \vec{v} \text{ along the same line}) \tag{37.30}$$

where a is the acceleration, also along the x-axis. Solving Eq. (37.30) for the acceleration a gives

$$a = \frac{F}{m}\left(1 - \frac{v^2}{c^2}\right)^{3/2}$$

We see that as a particle's speed increases, the acceleration caused by a given force continuously *decreases*. As the speed approaches c, the acceleration approaches zero, no matter how great a force is applied. Thus it is impossible to accelerate a particle with nonzero rest mass to a speed equal to or greater than c. We again see that the speed of light in vacuum represents an ultimate speed limit.

Equation (37.27) for relativistic momentum is sometimes interpreted to mean that a rapidly moving particle undergoes an increase in mass. If the mass at zero velocity (the rest mass) is denoted by m, then the "relativistic mass" m_{rel} is

$$m_{rel} = \frac{m}{\sqrt{1 - v^2/c^2}}$$

Indeed, when we consider the motion of a system of particles (such as rapidly moving ideal-gas molecules in a stationary container), the total rest mass of the system is the sum of the relativistic masses of the particles, not the sum of their rest masses.

However, if blindly applied, the concept of relativistic mass has its pitfalls. As Eq. (37.29) shows, the relativistic generalization of Newton's second law is *not* $\vec{F} = m_{rel}\vec{a}$, and we will show in Section 37.8 that the relativistic kinetic energy of a particle is *not* $K = \frac{1}{2}m_{rel}v^2$. The use of relativistic mass has its supporters and detractors, some quite strong in their opinions. We will mostly deal with individual particles, so we will sidestep the controversy and use Eq. (37.27) as the generalized definition of momentum with m as a constant for each particle, independent of its state of motion.

The quantity $1/\sqrt{1 - v^2/c^2}$ in Eqs. (37.27) and (37.29) is the Lorentz factor γ from Eq. (37.7) (Section 37.3), but with a difference: We've replaced u, the relative speed of two coordinate systems, by v, the speed of a particle in a particular coordinate system—that is, the speed of the particle's *rest frame* with respect to that system. In terms of γ, Eqs. (37.27) and (37.30) become

Rest mass of particle ╌ Velocity of particle

Relativistic momentum ┄┄┄▸ $\vec{p} = \gamma m\vec{v}$ Lorentz factor relating rest frame of particle and frame of observer   (37.31)

$$F = \gamma^3 ma \qquad (\vec{F} \text{ and } \vec{v} \text{ along the same line}) \qquad (37.32)$$

In linear accelerators (used in medicine as well as nuclear and elementary-particle physics; see Fig. 37.11) the net force $\vec{F}$ and the velocity $\vec{v}$ of the accelerated particle are along the same straight line. But for much of the path in most *circular* accelerators the particle moves in uniform circular motion at constant speed v. Then the net force and velocity are perpendicular, so the force can do no work on the particle and the kinetic energy and speed remain constant. Thus the denominator in Eq. (37.29) is constant, and we obtain

$$F = \frac{m}{(1 - v^2/c^2)^{1/2}}a = \gamma ma \qquad (\vec{F} \text{ and } \vec{v} \text{ perpendicular}) \qquad (37.33)$$

Recall from Section 3.4 that if the particle moves in a circle, the net force and acceleration are directed inward along the radius r, and $a = v^2/r$.

What about the general case in which $\vec{F}$ and $\vec{v}$ are neither along the same line nor perpendicular? Then we can resolve the net force $\vec{F}$ at any instant into components parallel to and perpendicular to $\vec{v}$. The resulting acceleration will have corresponding components obtained from Eqs. (37.32) and (37.33). Because of the different γ^3 and γ factors, the acceleration components will not be proportional to the net force components. That is, *unless the net force on a relativistic particle is either along the same line as the particle's velocity or perpendicular to it, the net force and acceleration vectors are not parallel.*

EXAMPLE 37.9   **RELATIVISTIC DYNAMICS OF AN ELECTRON**

An electron (rest mass 9.11×10^{-31} kg, charge -1.60×10^{-19} C) is moving opposite to an electric field of magnitude $E = 5.00 \times 10^5$ N/C. All other forces are negligible in comparison to the electric-field force. (a) Find the magnitudes of momentum and of acceleration at the instants when $v = 0.010c$, $0.90c$, and $0.99c$. (b) Find the corresponding accelerations if a net force of the same magnitude is perpendicular to the velocity.

SOLUTION

IDENTIFY and SET UP: In addition to the expressions from this section for relativistic momentum and acceleration, we need the relationship between electric force and electric field from Chapter 21. In part (a) we use Eq. (37.31) to determine the magnitude of momentum; the force acts along the same line as the velocity, so we use Eq. (37.32) to determine the magnitude of acceleration. In part (b) the force is perpendicular to the velocity, so we use Eq. (37.33) rather than Eq. (37.32).

EXECUTE: (a) For $v = 0.010c$, $0.90c$, and $0.99c$ we have $\gamma = \sqrt{1 - v^2/c^2} = 1.00$, 2.29, and 7.09, respectively. The values of the momentum magnitude $p = \gamma m v$ are

$$p_1 = (1.00)(9.11 \times 10^{-31} \text{ kg})(0.010)(3.00 \times 10^8 \text{ m/s})$$
$$= 2.7 \times 10^{-24} \text{ kg} \cdot \text{m/s at } v_1 = 0.010c$$

$$p_2 = (2.29)(9.11 \times 10^{-31} \text{ kg})(0.90)(3.00 \times 10^8 \text{ m/s})$$
$$= 5.6 \times 10^{-22} \text{ kg} \cdot \text{m/s at } v_2 = 0.90c$$

$$p_3 = (7.09)(9.11 \times 10^{-31} \text{ kg})(0.99)(3.00 \times 10^8 \text{ m/s})$$
$$= 1.9 \times 10^{-21} \text{ kg} \cdot \text{m/s at } v_3 = 0.99c$$

From Eq. (21.4), the magnitude of the force on the electron is

$$F = |q|E = (1.60 \times 10^{-19} \text{ C})(5.00 \times 10^5 \text{ N/C})$$
$$= 8.00 \times 10^{-14} \text{ N}$$

From Eq. (37.32), $a = F/\gamma^3 m$. For $v = 0.010c$ and $\gamma = 1.00$,

$$a_1 = \frac{8.00 \times 10^{-14} \text{ N}}{(1.00)^3 (9.11 \times 10^{-31} \text{ kg})} = 8.8 \times 10^{16} \text{ m/s}^2$$

The accelerations at the two higher speeds are smaller than the nonrelativistic value by factors of $\gamma^3 = 12.0$ and 356, respectively:

$$a_2 = 7.3 \times 10^{15} \text{ m/s}^2 \qquad a_3 = 2.5 \times 10^{14} \text{ m/s}^2$$

(b) From Eq. (37.33), $a = F/\gamma m$ if $\vec{F}$ and $\vec{v}$ are perpendicular. When $v = 0.010c$ and $\gamma = 1.00$,

$$a_1 = \frac{8.00 \times 10^{-14} \text{ N}}{(1.00)(9.11 \times 10^{-31} \text{ kg})} = 8.8 \times 10^{16} \text{ m/s}^2$$

Now the accelerations at the two higher speeds are smaller by factors of $\gamma = 2.29$ and 7.09, respectively:

$$a_2 = 3.8 \times 10^{16} \text{ m/s}^2 \qquad a_3 = 1.2 \times 10^{16} \text{ m/s}^2$$

These accelerations are larger than the corresponding ones in part (a) by factors of γ^2.

EVALUATE: Our results in part (a) show that at higher speeds, the relativistic values of momentum differ more and more from the nonrelativistic values calculated from $p = mv$. The momentum at $0.99c$ is more than three times as great as at $0.90c$ because of the increase in the factor γ. Our results also show that the acceleration drops off very quickly as v approaches c.

TEST YOUR UNDERSTANDING OF SECTION 37.7   According to relativistic mechanics, when you double the speed of a particle, the magnitude of its momentum increases by (i) a factor of 2; (ii) a factor greater than 2; (iii) a factor between 1 and 2 that depends on the mass of the particle. ∎

37.8 RELATIVISTIC WORK AND ENERGY

When we developed the relationship between work and kinetic energy in Chapter 6, we used Newton's laws of motion. Since we have generalized these laws according to the principle of relativity, we need a corresponding generalization of the equation for kinetic energy.

Relativistic Kinetic Energy

We use the work–energy theorem, beginning with the definition of work. When the net force and displacement are in the same direction, the work done by that force is $W = \int F\, dx$. We substitute the expression for F from Eq. (37.30), the relativistic version of Newton's second law for straight-line motion. In moving a particle of rest mass m from point x_1 to point x_2,

$$W = \int_{x_1}^{x_2} F\, dx = \int_{x_1}^{x_2} \frac{ma\, dx}{(1 - v_x^2/c^2)^{3/2}} \tag{37.34}$$

We've replaced v in Eq. (37.34) with v_x because the motion is along the x-axis only. So v_x is the varying x-component of the particle's velocity as the net force accelerates it. To derive the generalized expression for kinetic energy K, first remember that the kinetic energy of a particle equals the net work done on it in moving it from rest to speed v: $K = W$. Thus we let the speeds be zero at point x_1 and v at point x_2. It's useful to convert Eq. (37.34) to an integral on v_x. To do this, note that dx and dv_x are the infinitesimal changes in x and v_x, respectively, in the time interval dt. Because $v_x = dx/dt$ and $a = dv_x/dt$, we can rewrite $a\, dx$ in Eq. (37.34) as

$$a\, dx = \frac{dv_x}{dt}\, dx = dx \frac{dv_x}{dt} = \frac{dx}{dt}\, dv_x = v_x dv_x$$

Making these substitutions gives us

$$K = W = \int_0^v \frac{m v_x dv_x}{(1 - v_x^2/c^2)^{3/2}} \tag{37.35}$$

We can evaluate this integral by a simple change of variable; the final result is

Rest mass of particle Speed of light in vacuum

Relativistic kinetic energy $\quad K = \dfrac{mc^2}{\sqrt{1 - v^2/c^2}} - mc^2 = (\gamma - 1)mc^2 \tag{37.36}$

Speed of particle   Lorentz factor relating rest frame of particle and frame of observer

As v approaches c, the kinetic energy approaches infinity. If Eq. (37.36) is correct, it must also approach the Newtonian expression $K = \frac{1}{2}mv^2$ when v is much smaller than c (**Fig. 37.21**). To verify this, we expand the radical, using the binomial theorem in the form

$$(1 + x)^n = 1 + nx + n(n - 1)x^2/2 + \cdots$$

In our case, $n = -\frac{1}{2}$ and $x = -v^2/c^2$, and we get

$$\gamma = \left(1 - \frac{v^2}{c^2}\right)^{-1/2} = 1 + \frac{1}{2}\frac{v^2}{c^2} + \frac{3}{8}\frac{v^4}{c^4} + \cdots$$

Combining this with $K = (\gamma - 1)mc^2$, we find

$$K = \left(1 + \frac{1}{2}\frac{v^2}{c^2} + \frac{3}{8}\frac{v^4}{c^4} + \cdots - 1\right)mc^2 = \frac{1}{2}mv^2 + \frac{3}{8}\frac{mv^4}{c^2} + \cdots \tag{37.37}$$

When v is much smaller than c, all the terms in the series in Eq. (37.37) except the first are negligibly small, and we obtain the Newtonian expression $\frac{1}{2}mv^2$.

Rest Energy and $E = mc^2$

Equation (37.36) for the kinetic energy of a moving particle includes a term $mc^2/\sqrt{1 - v^2/c^2}$ that depends on the motion and a second energy term mc^2 that is independent of the motion. It seems that the kinetic energy of a particle is the difference between some **total energy** E and an energy mc^2 that it has even when it is at rest. Thus we can rewrite Eq. (37.36) as

Kinetic Rest   Rest mass   Speed of light
energy energy of particle in vacuum

Total energy of a particle $\quad E = K + mc^2 = \dfrac{mc^2}{\sqrt{1 - v^2/c^2}} = \gamma mc^2 \tag{37.38}$

Speed of particle   Lorentz factor relating rest frame of particle and frame of observer

For a particle at rest ($K = 0$), we see that $E = mc^2$. The energy mc^2 associated with rest mass m rather than motion is called the **rest energy** of the particle.

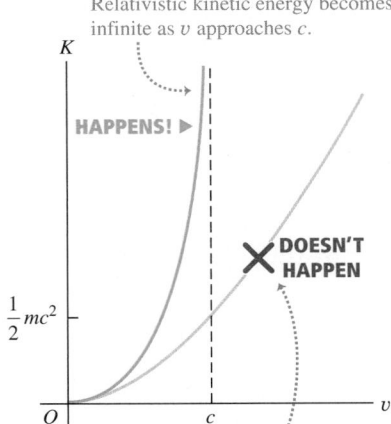

37.21 Graph of the kinetic energy of a particle of rest mass m as a function of speed v. Also shown is the Newtonian prediction, which gives correct results only at speeds much less than c.

Relativistic kinetic energy becomes infinite as v approaches c.

HAPPENS! ▶

✗ **DOESN'T HAPPEN**

Newtonian mechanics incorrectly predicts that kinetic energy becomes infinite only if v becomes infinite.

Application **Monitoring Mass-Energy Conversion** Although the control room of a nuclear power plant is very complex, the physical principle on which such a plant operates is a simple one: Part of the rest energy of atomic nuclei is converted to thermal energy, which in turn is used to produce steam to drive electric generators.

There is abundant experimental evidence that rest energy really does exist. The simplest example is the decay of a neutral *pion*. This is an unstable subatomic particle of rest mass m_π; when it decays, it disappears and electromagnetic radiation appears. If a neutral pion has no kinetic energy before its decay, the total energy of the radiation after its decay is found to equal exactly $m_\pi c^2$. In many other fundamental particle transformations the sum of the rest masses of the particles changes. In every case there is a corresponding energy change, consistent with the assumption of a rest energy mc^2 associated with a rest mass m.

Historically, the principles of conservation of mass and of energy developed quite independently. The theory of relativity shows that they are actually two special cases of a single broader conservation principle, the *principle of conservation of mass and energy*. In some physical phenomena, neither the sum of the rest masses of the particles nor the total energy other than rest energy is separately conserved, but there is a more general conservation principle: In an isolated system, when the sum of the rest masses changes, there is always a change in $1/c^2$ times the total energy other than the rest energy. This change is equal in magnitude but opposite in sign to the change in the sum of the rest masses.

This more general mass-energy conservation law is the fundamental principle involved in the generation of nuclear power. When a uranium nucleus undergoes fission in a nuclear reactor, the sum of the rest masses of the resulting fragments is *less than* the rest mass of the parent nucleus. An amount of energy is released that equals the mass decrease multiplied by c^2. Most of this energy can be used to produce steam to operate turbines for electric power generators.

We can also relate the total energy E of a particle (kinetic energy plus rest energy) directly to its momentum by combining Eq. (37.27) for relativistic momentum and Eq. (37.38) for total energy to eliminate the particle's velocity. The simplest procedure is to rewrite these equations in the following forms:

$$\left(\frac{E}{mc^2}\right)^2 = \frac{1}{1 - v^2/c^2} \quad \text{and} \quad \left(\frac{p}{mc}\right)^2 = \frac{v^2/c^2}{1 - v^2/c^2}$$

Subtracting the second of these from the first and rearranging, we find

	Total energy	Rest energy	Magnitude of momentum	
Total energy, rest energy, and momentum:		$E^2 = (mc^2)^2 + (pc)^2$		(37.39)
		Rest mass	Speed of light in vacuum	

Again we see that for a particle at rest ($p = 0$), $E = mc^2$.

Equation (37.39) also suggests that a particle may have energy and momentum even when it has no rest mass. In such a case, $m = 0$ and

$$E = pc \quad \text{(zero rest mass)} \tag{37.40}$$

In fact, zero rest mass particles do exist. Such particles always travel at the speed of light in vacuum. One example is the *photon*, the quantum of electromagnetic radiation (to be discussed in Chapter 38). Photons are emitted and absorbed during changes of state of an atomic or nuclear system when the energy and momentum of the system change.

EXAMPLE 37.10 | **ENERGETIC ELECTRONS**

SOLUTION

(a) Find the rest energy of an electron ($m = 9.109 \times 10^{-31}$ kg, $q = -e = -1.602 \times 10^{-19}$ C) in joules and in electron volts. (b) Find the speed of an electron that has been accelerated by an electric field, from rest, through a potential increase of 20.0 kV or of 5.00 MV (typical of a high-voltage x-ray machine).

SOLUTION

IDENTIFY and SET UP: This problem uses the ideas of rest energy, relativistic kinetic energy, and (from Chapter 23) electric potential energy. We use $E = mc^2$ to find the rest energy and Eqs. (37.7) and (37.38) to find the speed that gives the stated total energy.

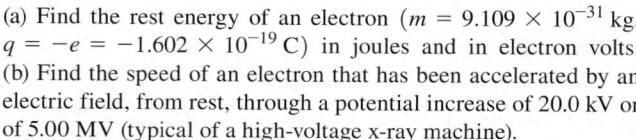

EXECUTE: (a) The rest energy is

$$mc^2 = (9.109 \times 10^{-31} \text{ kg})(2.998 \times 10^8 \text{ m/s})^2$$

$$= 8.187 \times 10^{-14} \text{ J}$$

From the definition of the electron volt in Section 23.2, $1 \text{ eV} = 1.602 \times 10^{-19}$ J. Using this, we find

$$mc^2 = (8.187 \times 10^{-14} \text{ J}) \frac{1 \text{ eV}}{1.602 \times 10^{-19} \text{ J}}$$

$$= 5.11 \times 10^5 \text{ eV} = 0.511 \text{ MeV}$$

(b) In calculations such as this, it is often convenient to work with the quantity $\gamma = 1/\sqrt{1 - v^2/c^2}$ from Eq. (37.38). Solving this for v, we find

$$v = c\sqrt{1 - (1/\gamma)^2}$$

The total energy E of the accelerated electron is the sum of its rest energy mc^2 and the kinetic energy eV_{ba} that it gains from the work done on it by the electric field in moving from point a to point b:

$$E = \gamma mc^2 = mc^2 + eV_{ba} \qquad \text{or}$$

$$\gamma = 1 + \frac{eV_{ba}}{mc^2}$$

An electron accelerated through a potential increase of $V_{ba} = 20.0$ kV gains 20.0 keV of energy, so for this electron

$$\gamma = 1 + \frac{20.0 \times 10^3 \text{ eV}}{0.511 \times 10^6 \text{ eV}} = 1.039$$

and

$$v = c\sqrt{1 - (1/1.039)^2} = 0.272c = 8.15 \times 10^7 \text{ m/s}$$

Repeating the calculation for $V_{ba} = 5.00$ MV, we find $eV_{ba}/mc^2 = 9.78$, $\gamma = 10.78$, and $v = 0.996c$.

EVALUATE: With $V_{ba} = 20.0$ kV, the added kinetic energy of 20.0 keV is less than 4% of the rest energy of 0.511 MeV, and the final speed is about one-fourth the speed of light. With $V_{ba} = 5.00$ MV, the added kinetic energy of 5.00 MeV is much greater than the rest energy and the speed is close to c.

> **CAUTION** **Three electron energies** All electrons have *rest* energy 0.511 MeV. An electron accelerated from rest through a 5.00-MeV potential increase has *kinetic* energy 5.00 MeV (we call it a "5.00-MeV electron") and *total* energy 5.51 MeV. Be careful to distinguish these energies from one another. ▌

EXAMPLE 37.11 A RELATIVISTIC COLLISION

Two protons (each with mass $m_p = 1.67 \times 10^{-27}$ kg) are initially moving with equal speeds in opposite directions. They continue to exist after a head-on collision that also produces a neutral pion of mass $m_\pi = 2.40 \times 10^{-28}$ kg (**Fig. 37.22**). If all three particles are at rest after the collision, find the initial speed of the protons. Energy is conserved in the collision.

37.22 In this collision the kinetic energy of two protons is transformed into the rest energy of a new particle, a pion.

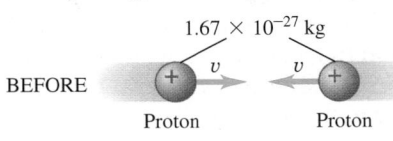
1.67 × 10⁻²⁷ kg

BEFORE

Proton   Proton

AFTER

Pion (2.40 × 10⁻²⁸ kg)

SOLUTION

IDENTIFY and SET UP: Relativistic total energy is conserved in the collision, so we can equate the (unknown) total energy of the two protons before the collision to the combined rest energies of the two protons and the pion after the collision. We then use Eq. (37.38) to find the speed of each proton.

EXECUTE: The total energy of each proton before the collision is $\gamma m_p c^2$. By conservation of energy,

$$2(\gamma m_p c^2) = 2(m_p c^2) + m_\pi c^2$$

$$\gamma = 1 + \frac{m_\pi}{2m_p} = 1 + \frac{2.40 \times 10^{-28} \text{ kg}}{2(1.67 \times 10^{-27} \text{ kg})} = 1.072$$

From Eq. (37.38), the initial proton speed is

$$v = c\sqrt{1 - (1/\gamma)^2} = 0.360c$$

EVALUATE: The proton rest energy is 938 MeV, so the initial kinetic energy of each proton is $(\gamma - 1)m_p c^2 = 0.072 m_p c^2 = (0.072)(938 \text{ MeV}) = 67.5$ MeV. You can verify that the rest energy $m_\pi c^2$ of the pion is twice this, or 135 MeV. All the kinetic energy "lost" in this completely inelastic collision is transformed into the rest energy of the pion.

TEST YOUR UNDERSTANDING OF SECTION 37.8 A proton is accelerated from rest by a constant force that always points in the direction of the particle's motion. Compared to the amount of kinetic energy that the proton gains during the first meter of its travel, how much kinetic energy does the proton gain during one meter of travel while it is moving at 99% of the speed of light? (i) The same amount; (ii) a greater amount; (iii) a smaller amount. ▌

37.9 NEWTONIAN MECHANICS AND RELATIVITY

The sweeping changes required by the principle of relativity go to the very roots of Newtonian mechanics, including the concepts of length and time, the equations of motion, and the conservation principles. Thus it may appear that we have destroyed the foundations on which Newtonian mechanics is built. In one sense this is true, yet the Newtonian formulation is still accurate whenever speeds are small in comparison with the speed of light in vacuum. In such cases, time dilation, length contraction, and the modifications of the laws of motion are so small that they are unobservable. In fact, every one of the principles of Newtonian mechanics survives as a special case of the more general relativistic formulation.

The laws of Newtonian mechanics are not *wrong;* they are *incomplete.* They are a limiting case of relativistic mechanics. They are *approximately* correct when all speeds are small in comparison to c, and they become exactly correct in the limit when all speeds approach zero. Thus relativity does not completely destroy the laws of Newtonian mechanics but *generalizes* them. This is a common pattern in the development of physical theory. Whenever a new theory is in partial conflict with an older, established theory, the new must yield the same predictions as the old in areas in which the old theory is supported by experimental evidence. Every new physical theory must pass this test, called the **correspondence principle.**

The General Theory of Relativity

At this point we may ask whether the special theory of relativity gives the final word on mechanics or whether *further* generalizations are possible or necessary. For example, inertial frames have occupied a privileged position in our discussion. Can the principle of relativity be extended to noninertial frames as well?

Here's an example that illustrates some implications of this question. A student decides to go over Niagara Falls while enclosed in a large wooden box. During her free fall she doesn't fall to the floor of the box because both she and the box are in free fall with a downward acceleration of 9.8 m/s^2. But an alternative interpretation, from her point of view, is that she doesn't fall to the floor because her gravitational interaction with the earth has suddenly been turned off. As long as she remains in the box and it remains in free fall, she cannot tell whether she is indeed in free fall or whether the gravitational interaction has vanished.

A similar problem occurs in a space station in orbit around the earth. Objects in the space station *seem* to be weightless, but without looking outside the station there is no way to determine whether gravity has been turned off or whether the station and all its contents are accelerating toward the center of the earth. **Figure 37.23** makes a similar point for a spaceship that is not in free fall but may be accelerating relative to an inertial frame or be at rest on the earth's surface.

These considerations form the basis of Einstein's **general theory of relativity.** If we cannot distinguish experimentally between a uniform gravitational field at a particular location and a uniformly accelerated reference frame, then there cannot be any real distinction between the two. Pursuing this concept, we may try to represent *any* gravitational field in terms of special characteristics of the coordinate system. This turns out to require even more sweeping revisions of our space-time concepts than did the special theory of relativity. In the general theory of relativity the geometric properties of space are affected by the presence of matter (**Fig. 37.24**).

The general theory of relativity has passed several experimental tests, including three proposed by Einstein. One test has to do with understanding the rotation of the axes of the planet Mercury's elliptical orbit, called the *precession of the perihelion.* (The perihelion is the point of closest approach to the sun.) A second test concerns the apparent bending of light rays from distant stars when they pass near the sun. The third test is the *gravitational red shift,* the increase in wavelength of light proceeding outward from a massive source. Some details of the general theory are more

37.23 Without information from outside the spaceship, the astronaut cannot distinguish situation (b) from situation (c).

(a) An astronaut is about to drop her watch in a spaceship.

(b) In gravity-free space, the floor accelerates upward at $a = g$ and hits the watch.

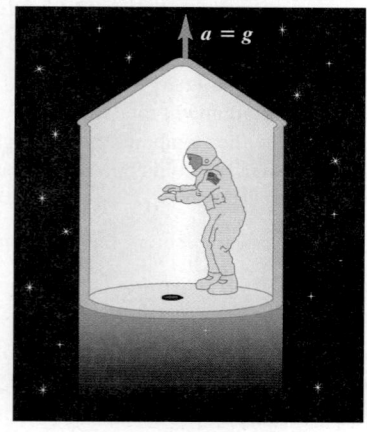

(c) On the earth's surface, the watch accelerates downward at $a = g$ and hits the floor.

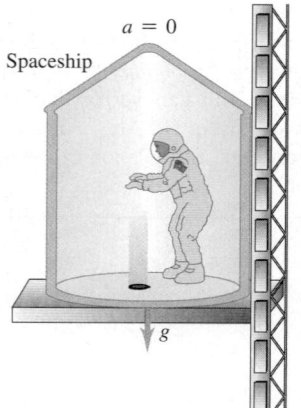

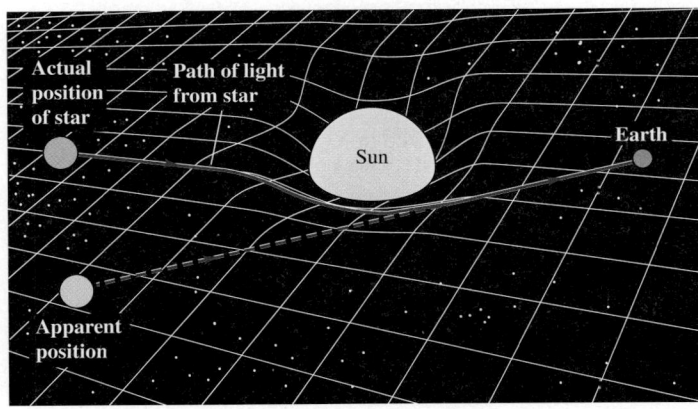

Actual position of star

Path of light from star

Sun

Earth

Apparent position

37.24 A two-dimensional representation of curved space. We imagine the space (a plane) as being distorted as shown by a massive object (the sun). Light from a distant star (solid line) follows the distorted surface on its way to the earth. The dashed line shows the direction from which the light *appears* to be coming. The effect is greatly exaggerated; for the sun, the maximum deviation is only 0.00048°.

37.25 A GPS receiver uses radio signals from the orbiting GPS satellites to determine its position. To account for the effects of relativity, the receiver must be tuned to a slightly higher frequency (10.23 MHz) than the frequency emitted by the satellites (10.22999999543 MHz).

difficult to test, but this theory has played a central role in investigations of the formation and evolution of stars, black holes, and studies of the evolution of the universe.

The general theory of relativity may seem to be an exotic bit of knowledge with little practical application. In fact, this theory plays an essential role in the global positioning system (GPS), which makes it possible to determine your position on the earth's surface to within a few meters using a handheld receiver (**Fig. 37.25**). The heart of the GPS system is a collection of more than two dozen satellites in very precise orbits. Each satellite emits carefully timed radio signals, and a GPS receiver simultaneously detects the signals from several satellites. The receiver then calculates the time delay between when each signal was emitted and when it was received, and uses this information to calculate the receiver's position. To ensure the proper timing of the signals, it's necessary to include corrections due to the special theory of relativity (because the satellites are moving relative to the receiver on earth) as well as the general theory (because the satellites are higher in the earth's gravitational field than the receiver). The corrections due to relativity are small—less than one part in 10^9—but are crucial to the superb precision of the GPS system.

| CHAPTER **37 SUMMARY** | **SOLUTIONS TO ALL EXAMPLES** |

Invariance of physical laws, simultaneity: All of the fundamental laws of physics have the same form in all inertial frames of reference. The speed of light in vacuum is the same in all inertial frames and is independent of the motion of the source. Simultaneity is not an absolute concept; events that are simultaneous in one frame are not necessarily simultaneous in a second frame moving relative to the first.

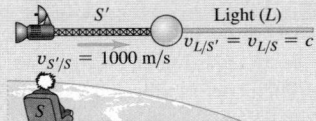

S' Light (L)
$v_{L/S'} = v_{L/S} = c$
$v_{S'/S} = 1000$ m/s
S

Time dilation: If two events occur at the same space point in a particular frame of reference, the time interval Δt_0 between the events as measured in that frame is called a proper time interval. If this frame moves with constant velocity u relative to a second frame, the time interval Δt between the events as observed in the second frame is longer than Δt_0. (See Examples 37.1–37.3.)

$$\Delta t = \frac{\Delta t_0}{\sqrt{1 - u^2/c^2}} = \gamma \, \Delta t_0$$

$$(37.6), (37.8)$$

$$\gamma = \frac{1}{\sqrt{1 - u^2/c^2}} \qquad (37.7)$$

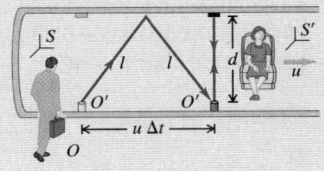

Length contraction: If two points are at rest in a particular frame of reference, the distance l_0 between the points as measured in that frame is called a proper length. If this frame moves with constant velocity u relative to a second frame and the distances are measured parallel to the motion, the distance l between the points as measured in the second frame is shorter than l_0. (See Examples 37.4 and 37.5.)

$$l = l_0 \sqrt{1 - \frac{u^2}{c^2}} = \frac{l_0}{\gamma} \qquad (37.16)$$

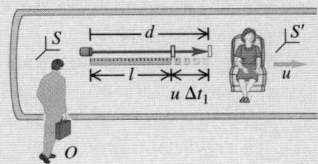

The Lorentz transformations: The Lorentz coordinate transformations relate the coordinates and time of an event in an inertial frame S to the coordinates and time of the same event as observed in a second inertial frame S' moving at velocity u relative to the first. For one-dimensional motion, a particle's velocities v_x in S and v'_x in S' are related by the Lorentz velocity transformation. (See Examples 37.6 and 37.7.)

$$x' = \frac{x - ut}{\sqrt{1 - u^2/c^2}} = \gamma(x - ut)$$

$$y' = y \qquad z' = z \qquad (37.21)$$

$$t' = \frac{t - ux/c^2}{\sqrt{1 - u^2/c^2}} = \gamma(t - ux/c^2)$$

$$v'_x = \frac{v_x - u}{1 - uv_x/c^2} \qquad (37.22)$$

$$v_x = \frac{v'_x + u}{1 + uv'_x/c^2} \qquad (37.23)$$

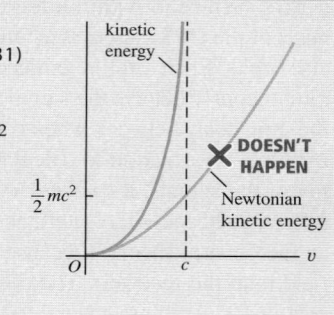

The Doppler effect for electromagnetic waves: The Doppler effect is the frequency shift in light from a source due to the relative motion of source and observer. For a source moving toward the observer with speed u, Eq. (37.25) gives the received frequency f in terms of the emitted frequency f_0. (See Example 37.8.)

$$f = \sqrt{\frac{c + u}{c - u}} f_0 \qquad (37.25)$$

Moving source emits light of frequency f_0.

Stationary observer detects light of frequency $f > f_0$.

Relativistic momentum and energy: For a particle of rest mass m moving with velocity $\vec{v}$, the relativistic momentum $\vec{p}$ is given by Eq. (37.27) or (37.31) and the relativistic kinetic energy K is given by Eq. (37.36). The total energy E is the sum of the kinetic energy and the rest energy mc^2. The total energy can also be expressed in terms of the magnitude of momentum p and rest mass m. (See Examples 37.9–37.11.)

$$\vec{p} = \frac{m\vec{v}}{\sqrt{1 - v^2/c^2}} = \gamma m\vec{v} \quad (37.27), (37.31)$$

$$K = \frac{mc^2}{\sqrt{1 - v^2/c^2}} - mc^2 = (\gamma - 1)mc^2 \qquad (37.36)$$

$$E = K + mc^2 = \frac{mc^2}{\sqrt{1 - v^2/c^2}} = \gamma mc^2 \qquad (37.38)$$

$$E^2 = (mc^2)^2 + (pc)^2 \qquad (37.39)$$

BRIDGING PROBLEM COLLIDING PROTONS

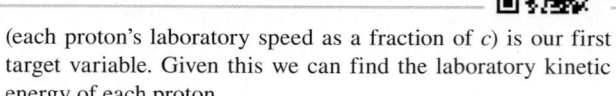

In an experiment, two protons are shot directly toward each other. Their speeds are such that in the frame of reference of each proton, the other proton is moving at $0.500c$. (a) What does an observer in the laboratory measure for the speed of each proton? (b) What is the kinetic energy of each proton as measured by an observer in the laboratory? (c) What is the kinetic energy of each proton as measured by the other proton?

SOLUTION GUIDE

IDENTIFY and SET UP

1. This problem uses the Lorentz velocity transformation, which allows us to relate the velocity v_x of a proton in one frame to its velocity v'_x in a different frame. It also uses the idea of relativistic kinetic energy.

2. Draw a sketch of the situation. Take the x-axis to be the line of motion of the protons, and take the $+x$-direction to be to the right. In the frame in which the left-hand proton is at rest, the right-hand proton has velocity $-0.500c$. In the laboratory frame the two protons have velocities $-\alpha c$ and $+\alpha c$, where α

(each proton's laboratory speed as a fraction of c) is our first target variable. Given this we can find the laboratory kinetic energy of each proton.

EXECUTE

3. Write a Lorentz velocity-transformation equation that relates the velocity of the right-hand proton in the laboratory frame to its velocity in the frame of the left-hand proton. Solve this equation for α. (*Hint:* Remember that α cannot be greater than 1. Why?)

4. Use your result from step 3 to find the laboratory kinetic energy of each proton.

5. Find the kinetic energy of the right-hand proton as measured in the frame of the left-hand proton.

EVALUATE

6. How much total kinetic energy must be imparted to the protons by a scientist in the laboratory? If the experiment were to be repeated with one proton stationary, what kinetic energy would have to be given to the other proton for the collision to be equivalent?

Problems

•, ••, •••: Difficulty levels. CP: Cumulative problems incorporating material from earlier chapters. CALC: Problems requiring calculus.
DATA: Problems involving real data, scientific evidence, experimental design, and/or statistical reasoning. BIO: Biosciences problems.

DISCUSSION QUESTIONS

Q37.1 You are standing on a train platform watching a high-speed train pass by. A light inside one of the train cars is turned on and then a little later it is turned off. (a) Who can measure the proper time interval for the duration of the light: you or a passenger on the train? (b) Who can measure the proper length of the train car: you or a passenger on the train? (c) Who can measure the proper length of a sign attached to a post on the train platform: you or a passenger on the train? In each case explain your answer.

Q37.2 If simultaneity is not an absolute concept, does that mean that we must discard the concept of causality? If event A is to *cause* event B, A must occur first. Is it possible that in some frames A appears to be the cause of B, and in others B appears to be the cause of A? Explain.

Q37.3 A rocket is moving to the right at $\frac{1}{2}$ the speed of light relative to the earth. A light bulb in the center of a room inside the rocket suddenly turns on. Call the light hitting the front end of the room event A and the light hitting the back of the room event B (**Fig. Q37.3**). Which event occurs first, A or B, or are they simultaneous, as viewed by (a) an astronaut riding in the rocket and (b) a person at rest on the earth?

Figure **Q37.3**

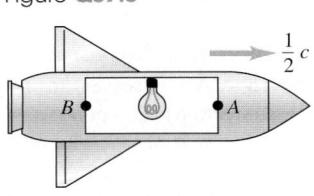

Q37.4 A spaceship is traveling toward the earth from the space colony on Asteroid 1040A. The ship is at the halfway point of the trip, passing Mars at a speed of $0.9c$ relative to the Mars frame of reference. At the same instant, a passenger on the spaceship receives a radio message from her boyfriend on 1040A and another from her sister on earth. According to the passenger on the ship, were these messages sent simultaneously or at different times? If at different times, which one was sent first? Explain your reasoning.

Q37.5 The average life span in the United States is about 70 years. Does this mean that it is impossible for an average person to travel a distance greater than 70 light-years away from the earth? (A light-year is the distance light travels in a year.) Explain.

Q37.6 You are holding an elliptical serving platter. How would you need to travel for the serving platter to appear round to another observer?

Q37.7 Two events occur at the same space point in a particular inertial frame of reference and are simultaneous in that frame. Is it possible that they may not be simultaneous in a different inertial frame? Explain.

Q37.8 A high-speed train passes a train platform. Larry is a passenger on the train, Adam is standing on the train platform, and David is riding a bicycle toward the platform in the same direction as the train is traveling. Compare the length of a train car as measured by Larry, Adam, and David.

Q37.9 The theory of relativity sets an upper limit on the speed that a particle can have. Are there also limits on the energy and momentum of a particle? Explain.

Q37.10 A student asserts that a material particle must always have a speed slower than that of light, and a massless particle must always move at exactly the speed of light. Is she correct? If so, how do massless particles such as photons and neutrinos acquire this speed? Can't they start from rest and accelerate? Explain.

Q37.11 The speed of light relative to still water is 2.25×10^8 m/s. If the water is moving past us, the speed of light we measure depends on the speed of the water. Do these facts violate Einstein's second postulate? Explain.

Q37.12 When a monochromatic light source moves toward an observer, its wavelength appears to be shorter than the value measured when the source is at rest. Does this contradict the hypothesis that the speed of light is the same for all observers? Explain.

Q37.13 In principle, does a hot gas have more mass than the same gas when it is cold? Explain. In practice, would this be a measurable effect? Explain.

Q37.14 Why do you think the development of Newtonian mechanics preceded the more refined relativistic mechanics by so many years?

Q37.15 What do you think would be different in everyday life if the speed of light were 10 m/s instead of 3.00×10^8 m/s?

EXERCISES

Section 37.2 Relativity of Simultaneity

37.1 • Suppose the two lightning bolts shown in Fig. 37.5a are simultaneous to an observer on the train. Show that they are *not* simultaneous to an observer on the ground. Which lightning strike does the ground observer measure to come first?

Section 37.3 Relativity of Time Intervals

37.2 • The positive muon (μ^+), an unstable particle, lives on average 2.20×10^{-6} s (measured in its own frame of reference) before decaying. (a) If such a particle is moving, with respect to the laboratory, with a speed of $0.900c$, what average lifetime is measured in the laboratory? (b) What average distance, measured in the laboratory, does the particle move before decaying?

37.3 • How fast must a rocket travel relative to the earth so that time in the rocket "slows down" to half its rate as measured by earth-based observers? Do present-day jet planes approach such speeds?

37.4 • A spaceship flies past Mars with a speed of $0.985c$ relative to the surface of the planet. When the spaceship is directly overhead, a signal light on the Martian surface blinks on and then off. An observer on Mars measures that the signal light was on for $75.0 \ \mu$s. (a) Does the observer on Mars or the pilot on the spaceship measure the proper time? (b) What is the duration of the light pulse measured by the pilot of the spaceship?

37.5 • The negative pion (π^-) is an unstable particle with an average lifetime of 2.60×10^{-8} s (measured in the rest frame of the pion). (a) If the pion is made to travel at very high speed relative to a laboratory, its average lifetime is measured in the laboratory to be 4.20×10^{-7} s. Calculate the speed of the pion expressed as a fraction of c. (b) What distance, measured in the laboratory, does the pion travel during its average lifetime?

37.6 •• As you pilot your space utility vehicle at a constant speed toward the moon, a race pilot flies past you in her spaceracer at a constant speed of $0.800c$ relative to you. At the instant the spaceracer passes you, both of you start timers at zero. (a) At the instant when you measure that the spaceracer has traveled 1.20×10^8 m past you, what does the race pilot read on her timer? (b) When the race pilot reads the value calculated in part (a) on her timer, what does she measure to be your distance from her? (c) At the instant when the race pilot reads the value calculated in part (a) on her timer, what do you read on yours?

37.7 •• A spacecraft flies away from the earth with a speed of 4.80×10^6 m/s relative to the earth and then returns at the same speed. The spacecraft carries an atomic clock that has been carefully synchronized with an identical clock that remains at rest on earth. The spacecraft returns to its starting point 365 days (1 year) later, as measured by the clock that remained on earth. What is the difference in the elapsed times on the two clocks, measured in hours? Which clock, the one in the spacecraft or the one on earth, shows the shorter elapsed time?

37.8 • An alien spacecraft is flying overhead at a great distance as you stand in your backyard. You see its searchlight blink on for 0.150 s. The first officer on the spacecraft measures that the searchlight is on for 12.0 ms. (a) Which of these two measured times is the proper time? (b) What is the speed of the spacecraft relative to the earth, expressed as a fraction of the speed of light c?

Section 37.4 Relativity of Length

37.9 • A spacecraft of the Trade Federation flies past the planet Coruscant at a speed of $0.600c$. A scientist on Coruscant measures the length of the moving spacecraft to be 74.0 m. The spacecraft later lands on Coruscant, and the same scientist measures the length of the now stationary spacecraft. What value does she get?

37.10 • A meter stick moves past you at great speed. Its motion relative to you is parallel to its long axis. If you measure the length of the moving meter stick to be 1.00 ft (1 ft = 0.3048 m)—for example, by comparing it to a 1-foot ruler that is at rest relative to you—at what speed is the meter stick moving relative to you?

37.11 •• **Why Are We Bombarded by Muons?** Muons are unstable subatomic particles that decay to electrons with a mean lifetime of 2.2 μs. They are produced when cosmic rays bombard the upper atmosphere about 10 km above the earth's surface, and they travel very close to the speed of light. The problem we want to address is why we see any of them at the earth's surface. (a) What is the greatest distance a muon could travel during its 2.2-μs lifetime? (b) According to your answer in part (a), it would seem that muons could never make it to the ground. But the 2.2-μs lifetime is measured in the frame of the muon, and muons are moving very fast. At a speed of $0.999c$, what is the mean lifetime of a muon as measured by an observer at rest on the earth? How far would the muon travel in this time? Does this result explain why we find muons in cosmic rays? (c) From the point of view of the muon, it still lives for only 2.2 μs, so how does it make it to the ground? What is the thickness of the 10 km of atmosphere through which the muon must travel, as measured by the muon? Is it now clear how the muon is able to reach the ground?

37.12 • An unstable particle is created in the upper atmosphere from a cosmic ray and travels straight down toward the surface of the earth with a speed of $0.99540c$ relative to the earth. A scientist at rest on the earth's surface measures that the particle is created at an altitude of 45.0 km. (a) As measured by the scientist, how much time does it take the particle to travel the 45.0 km to the surface of the earth? (b) Use the length-contraction formula to calculate the distance from where the particle is created to the surface of the earth as measured in the particle's frame. (c) In the particle's frame, how much time does it take the particle to travel from where it is created to the surface of the earth? Calculate this time both by the time dilation formula and from the distance calculated in part (b). Do the two results agree?

37.13 • As measured by an observer on the earth, a spacecraft runway on earth has a length of 3600 m. (a) What is the length of the runway as measured by a pilot of a spacecraft flying past at a speed of 4.00×10^7 m/s relative to the earth? (b) An observer on earth measures the time interval from when the spacecraft is directly over one end of the runway until it is directly over the other end. What result does she get? (c) The pilot of the spacecraft measures the time it takes him to travel from one end of the runway to the other end. What value does he get?

37.14 • A rocket ship flies past the earth at 91.0% of the speed of light. Inside, an astronaut who is undergoing a physical examination is having his height measured while he is lying down parallel to the direction in which the ship is moving. (a) If his height is measured to be 2.00 m by his doctor inside the ship, what height would a person watching this from the earth measure? (b) If the earth-based person had measured 2.00 m, what would the doctor in the spaceship have measured for the astronaut's height? Is this a reasonable height? (c) Suppose the astronaut in part (a) gets up after the examination and stands with his body perpendicular to the direction of motion. What would the doctor in the rocket and the observer on earth measure for his height now?

Section 37.5 The Lorentz Transformations

37.15 • An observer in frame S' is moving to the right $(+x$-direction$)$ at speed $u = 0.600c$ away from a stationary observer in frame S. The observer in S' measures the speed v' of a particle moving to the right away from her. What speed v does the observer in S measure for the particle if (a) $v' = 0.400c$; (b) $v' = 0.900c$; (c) $v' = 0.990c$?

37.16 • Space pilot Mavis zips past Stanley at a constant speed relative to him of $0.800c$. Mavis and Stanley start timers at zero when the front of Mavis's ship is directly above Stanley. When Mavis reads 5.00 s on her timer, she turns on a bright light under the front of her spaceship. (a) Use the Lorentz coordinate transformation derived in Example 37.6 to calculate x and t as measured by Stanley for the event of turning on the light. (b) Use the time dilation formula, Eq. (37.6), to calculate the time interval between the two events (the front of the spaceship passing overhead and turning on the light) as measured by Stanley. Compare to the value of t you calculated in part (a). (c) Multiply the time interval by Mavis's speed, both as measured by Stanley, to calculate the distance she has traveled as measured by him when the light turns on. Compare to the value of x you calculated in part (a).

37.17 •• A pursuit spacecraft from the planet Tatooine is attempting to catch up with a Trade Federation cruiser. As measured by an observer on Tatooine, the cruiser is traveling away from the planet with a speed of $0.600c$. The pursuit ship is traveling at a speed of $0.800c$ relative to Tatooine, in the same direction as the cruiser. (a) For the pursuit ship to catch the cruiser, should the velocity of the cruiser relative to the pursuit ship be directed toward or away from the pursuit ship? (b) What is the speed of the cruiser relative to the pursuit ship?

37.18 •• An enemy spaceship is moving toward your starfighter with a speed, as measured in your frame, of $0.400c$. The enemy ship fires a missile toward you at a speed of $0.700c$ relative to the enemy ship (**Fig. E37.18**). (a) What is the speed of the missile

Figure **E37.18**

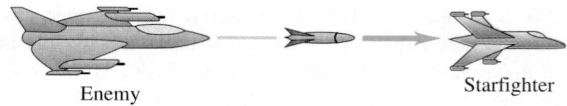

Enemy   Starfighter

relative to you? Express your answer in terms of the speed of light. (b) If you measure that the enemy ship is 8.00×10^6 km away from you when the missile is fired, how much time, measured in your frame, will it take the missile to reach you?

37.19 •• Two particles are created in a high-energy accelerator and move off in opposite directions. The speed of one particle, as measured in the laboratory, is $0.650c$, and the speed of each particle relative to the other is $0.950c$. What is the speed of the second particle, as measured in the laboratory?

37.20 •• Two particles in a high-energy accelerator experiment are approaching each other head-on, each with a speed of $0.9380c$ as measured in the laboratory. What is the magnitude of the velocity of one particle relative to the other?

37.21 •• Two particles in a high-energy accelerator experiment approach each other head-on with a relative speed of $0.890c$. Both particles travel at the same speed as measured in the laboratory. What is the speed of each particle, as measured in the laboratory?

37.22 • An imperial spaceship, moving at high speed relative to the planet Arrakis, fires a rocket toward the planet with a speed of $0.920c$ relative to the spaceship. An observer on Arrakis measures that the rocket is approaching with a speed of $0.360c$. What is the speed of the spaceship relative to Arrakis? Is the spaceship moving toward or away from Arrakis?

Section 37.6 The Doppler Effect for Electromagnetic Waves

37.23 • **Tell It to the Judge.** (a) How fast must you be approaching a red traffic light ($\lambda = 675$ nm) for it to appear yellow ($\lambda = 575$ nm)? Express your answer in terms of the speed of light. (b) If you used this as a reason not to get a ticket for running a red light, how much of a fine would you get for speeding? Assume that the fine is $1.00 for each kilometer per hour that your speed exceeds the posted limit of 90 km/h.

37.24 • Electromagnetic radiation from a star is observed with an earth-based telescope. The star is moving away from the earth at a speed of $0.520c$. If the radiation has a frequency of 8.64×10^{14} Hz in the rest frame of the star, what is the frequency measured by an observer on earth?

37.25 • A source of electromagnetic radiation is moving in a radial direction relative to you. The frequency you measure is 1.25 times the frequency measured in the rest frame of the source. What is the speed of the source relative to you? Is the source moving toward you or away from you?

Section 37.7 Relativistic Momentum

37.26 • **Relativistic Baseball.** Calculate the magnitude of the force required to give a 0.145-kg baseball an acceleration $a = 1.00$ m/s^2 in the direction of the baseball's initial velocity when this velocity has a magnitude of (a) 10.0 m/s; (b) $0.900c$; (c) $0.990c$. (d) Repeat parts (a), (b), and (c) if the force and acceleration are perpendicular to the velocity.

37.27 • A proton has momentum with magnitude p_0 when its speed is $0.400c$. In terms of p_0, what is the magnitude of the proton's momentum when its speed is doubled to $0.800c$?

37.28 • **When Should You Use Relativity?** As you have seen, relativistic calculations usually involve the quantity γ. When γ is appreciably greater than 1, we must use relativistic formulas instead of Newtonian ones. For what speed v (in terms of c) is the value of γ (a) 1.0% greater than 1; (b) 10% greater than 1; (c) 100% greater than 1?

37.29 • (a) At what speed is the momentum of a particle twice as great as the result obtained from the nonrelativistic expression mv? Express your answer in terms of the speed of light. (b) A force is applied to a particle along its direction of motion. At what speed is the magnitude of force required to produce a given acceleration twice as great as the force required to produce the same acceleration when the particle is at rest? Express your answer in terms of the speed of light.

37.30 • An electron is acted upon by a force of 5.00×10^{-15} N due to an electric field. Find the acceleration this force produces in each case: (a) The electron's speed is 1.00 km/s. (b) The electron's speed is 2.50×10^8 m/s and the force is parallel to the velocity.

Section 37.8 Relativistic Work and Energy

37.31 •• What is the speed of a particle whose kinetic energy is equal to (a) its rest energy and (b) five times its rest energy?

37.32 • If a muon is traveling at $0.999c$, what are its momentum and kinetic energy? (The mass of such a muon at rest in the laboratory is 207 times the electron mass.)

37.33 • A proton (rest mass 1.67×10^{-27} kg) has total energy that is 4.00 times its rest energy. What are (a) the kinetic energy of the proton; (b) the magnitude of the momentum of the proton; (c) the speed of the proton?

37.34 •• (a) How much work must be done on a particle with mass m to accelerate it (a) from rest to a speed of $0.090c$ and (b) from a speed of $0.900c$ to a speed of $0.990c$? (Express the answers in terms of mc^2.) (c) How do your answers in parts (a) and (b) compare?

37.35 • **An Antimatter Reactor.** When a particle meets its antiparticle, they annihilate each other and their mass is converted to light energy. The United States uses approximately 1.0×10^{20} J of energy per year. (a) If all this energy came from a futuristic antimatter reactor, how much mass of matter and antimatter fuel would be consumed yearly? (b) If this fuel had the density of iron (7.86 g/cm^3) and were stacked in bricks to form a cubical pile, how high would it be? (Before you get your hopes up, antimatter reactors are a *long* way in the future—if they ever will be feasible.)

37.36 •• Electrons are accelerated through a potential difference of 750 kV, so that their kinetic energy is 7.50×10^5 eV. (a) What is the ratio of the speed v of an electron having this energy to the speed of light, c? (b) What would the speed be if it were computed from the principles of classical mechanics?

37.37 • A particle has rest mass 6.64×10^{-27} kg and momentum 2.10×10^{-18} kg·m/s. (a) What is the total energy (kinetic plus rest energy) of the particle? (b) What is the kinetic energy of the particle? (c) What is the ratio of the kinetic energy to the rest energy of the particle?

37.38 • **Creating a Particle.** Two protons (each with rest mass $M = 1.67 \times 10^{-27}$ kg) are initially moving with equal speeds in opposite directions. The protons continue to exist after a collision that also produces an η^0 particle (see Chapter 44). The rest mass of the η^0 is $m = 9.75 \times 10^{-28}$ kg. (a) If the two protons and the η^0 are all at rest after the collision, find the initial speed of the protons, expressed as a fraction of the speed of light. (b) What is the kinetic energy of each proton? Express your answer in MeV. (c) What is the rest energy of the η^0, expressed in MeV? (d) Discuss the relationship between the answers to parts (b) and (c).

37.39 • Compute the kinetic energy of a proton (mass 1.67×10^{-27} kg) using both the nonrelativistic and relativistic expressions, and compute the ratio of the two results (relativistic divided by nonrelativistic) for speeds of (a) 8.00×10^7 m/s and (b) 2.85×10^8 m/s.

37.40 • What is the kinetic energy of a proton moving at (a) $0.100c$; (b) $0.500c$; (c) $0.900c$? How much work must be done to (d) increase the proton's speed from $0.100c$ to $0.500c$ and (e) increase the proton's speed from $0.500c$ to $0.900c$? (f) How do the last two results compare to results obtained in the nonrelativistic limit?

37.41 • (a) Through what potential difference does an electron have to be accelerated, starting from rest, to achieve a speed of $0.980c$? (b) What is the kinetic energy of the electron at this speed? Express your answer in joules and in electron volts.

37.42 • The sun produces energy by nuclear fusion reactions, in which matter is converted into energy. By measuring the amount of energy we receive from the sun, we know that it is producing energy at a rate of 3.8×10^{26} W. (a) How many kilograms of matter does the sun lose each second? Approximately how many tons of matter is this (1 ton = 2000 lb)? (b) At this rate, how long would it take the sun to use up all its mass?

PROBLEMS

37.43 • After being produced in a collision between elementary particles, a positive pion (π^+) must travel down a 1.90-km-long tube to reach an experimental area. A π^+ particle has an average lifetime (measured in its rest frame) of 2.60×10^{-8} s; the π^+ we are considering has this lifetime. (a) How fast must the π^+ travel if it is not to decay before it reaches the end of the tube? (Since u will be very close to c, write $u = (1 - \Delta)c$ and give your answer in terms of Δ rather than u.) (b) The π^+ has a rest energy of 139.6 MeV. What is the total energy of the π^+ at the speed calculated in part (a)?

37.44 • Inside a spaceship flying past the earth at three-fourths the speed of light, a pendulum is swinging. (a) If each swing takes 1.80 s as measured by an astronaut performing an experiment inside the spaceship, how long will the swing take as measured by a person at mission control (on earth) who is watching the experiment? (b) If each swing takes 1.80 s as measured by a person at mission control, how long will it take as measured by the astronaut in the spaceship?

37.45 ••• The starships of the Solar Federation are marked with the symbol of the federation, a circle, while starships of the Denebian Empire are marked with the empire's symbol, an ellipse whose major axis is 1.40 times longer than its minor axis ($a = 1.40b$ in **Fig. P37.45**). How fast, relative to an observer, does an empire ship have to travel for its marking to be confused with the marking of a federation ship?

Figure **P37.45**

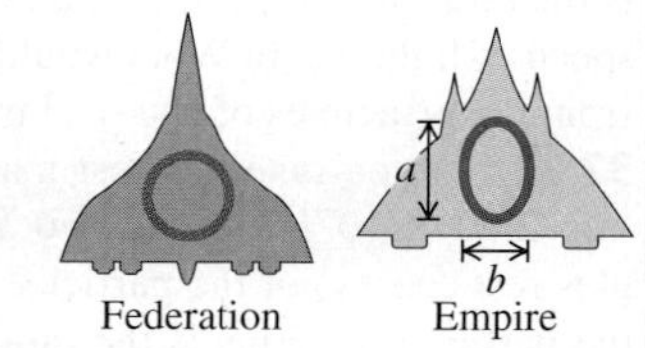

Federation   Empire

37.46 •• A cube of metal with sides of length a sits at rest in a frame S with one edge parallel to the x-axis. Therefore, in S the cube has volume a^3. Frame S' moves along the x-axis with a speed u. As measured by an observer in frame S', what is the volume of the metal cube?

37.47 •• A space probe is sent to the vicinity of the star Capella, which is 42.2 light-years from the earth. (A light-year is the distance light travels in a year.) The probe travels with a speed of $0.9930c$. An astronaut recruit on board is 19 years old when the probe leaves the earth. What is her biological age when the probe reaches Capella?

37.48 •• A muon is created 55.0 km above the surface of the earth (as measured in the earth's frame). The average lifetime of a muon, measured in its own rest frame, is 2.20 μs, and the muon we are considering has this lifetime. In the frame of the muon, the earth is moving toward the muon with a speed of $0.9860c$. (a) In the muon's frame, what is its initial height above the surface of the earth? (b) In the muon's frame, how much closer does the earth get during the lifetime of the muon? What fraction is this of the muon's original height, as measured in the muon's frame? (c) In the earth's frame, what is the lifetime of the muon? In the earth's frame, how far does the muon travel during its lifetime? What fraction is this of the muon's original height in the earth's frame?

37.49 • **The Large Hadron Collider (LHC).** Physicists and engineers from around the world came together to build the largest accelerator in the world, the Large Hadron Collider (LHC) at the CERN Laboratory in Geneva, Switzerland. The machine accelerates protons to high kinetic energies in an underground ring 27 km in circumference. (a) What is the speed v of a proton in the LHC if the proton's kinetic energy is 7.0 TeV? (Because v is very close to c, write $v = (1 - \Delta)c$ and give your answer in terms of Δ.) (b) Find the relativistic mass, m_{rel}, of the accelerated proton in terms of its rest mass.

37.50 •• The net force $\vec{F}$ on a particle of mass m is directed at $30.0°$ counterclockwise from the $+x$-axis. At one instant of time, the particle is traveling in the $+x$-direction with a speed (measured relative to the earth) of $0.700c$. At this instant, what is the direction of the particle's acceleration?

37.51 •• **Everyday Time Dilation.** Two atomic clocks are carefully synchronized. One remains in New York, and the other is loaded on an airliner that travels at an average speed of 250 m/s and then returns to New York. When the plane returns, the elapsed time on the clock that stayed behind is 4.00 h. By how much will the readings of the two clocks differ, and which clock will show the shorter elapsed time? (*Hint:* Since $u \ll c$, you can simplify $\sqrt{1 - u^2/c^2}$ by a binomial expansion.)

37.52 •• The distance to a particular star, as measured in the earth's frame of reference, is 7.11 light-years (1 light-year is the distance that light travels in 1 y). A spaceship leaves the earth and takes 3.35 y to arrive at the star, as measured by passengers on the ship. (a) How long does the trip take, according to observers on earth? (b) What distance for the trip do passengers on the spacecraft measure?

37.53 • CP **Čerenkov Radiation.** The Russian physicist P. A. Čerenkov discovered that a charged particle traveling in a solid with a speed exceeding the speed of light in that material radiates electromagnetic radiation. (This is analogous to the sonic boom produced by an aircraft moving faster than the speed of sound in air; see Section 16.9. Čerenkov shared the 1958 Nobel Prize for this discovery.) What is the minimum kinetic energy (in electron volts) that an electron must have while traveling inside a slab of crown glass ($n = 1.52$) in order to create this Čerenkov radiation?

37.54 •• Scientists working with a particle accelerator determine that an unknown particle has a speed of 1.35×10^8 m/s and a momentum of 2.52×10^{-19} kg·m/s. From the curvature of the particle's path in a magnetic field, they also deduce that it has a positive charge. Using this information, identify the particle.

37.55 • CP A nuclear bomb containing 12.0 kg of plutonium explodes. The sum of the rest masses of the products of the explosion is less than the original rest mass by one part in 10^4. (a) How much energy is released in the explosion? (b) If the explosion takes place in 4.00 μs, what is the average power developed by the bomb? (c) What mass of water could the released energy lift to a height of 1.00 km?

37.56 •• In the earth's rest frame, two protons are moving away from each other at equal speed. In the frame of each proton, the other proton has a speed of 0.700c. What does an observer in the rest frame of the earth measure for the speed of each proton?

37.57 • In certain radioactive beta decay processes, the beta particle (an electron) leaves the atomic nucleus with a speed of 99.95% the speed of light relative to the decaying nucleus. If this nucleus is moving at 75.00% the speed of light in the laboratory reference frame, find the speed of the emitted electron relative to the laboratory reference frame if the electron is emitted (a) in the same direction that the nucleus is moving and (b) in the opposite direction from the nucleus's velocity. (c) In each case in parts (a) and (b), find the kinetic energy of the electron as measured in (i) the laboratory frame and (ii) the reference frame of the decaying nucleus.

37.58 •• Two events are observed in a frame of reference S to occur at the same space point, the second occurring 1.80 s after the first. In a frame S′ moving relative to S, the second event is observed to occur 2.15 s after the first. What is the difference between the positions of the two events as measured in S′?

37.59 • One of the wavelengths of light emitted by hydrogen atoms under normal laboratory conditions is $\lambda = 656.3$ nm, in the red portion of the electromagnetic spectrum. In the light emitted from a distant galaxy this same spectral line is observed to be Doppler-shifted to $\lambda = 953.4$ nm, in the infrared portion of the spectrum. How fast are the emitting atoms moving relative to the earth? Are they approaching the earth or receding from it?

37.60 •• **Albert in Wonderland.** Einstein and Lorentz, being avid tennis players, play a fast-paced game on a court where they stand 20.0 m from each other. Being very skilled players, they play without a net. The tennis ball has mass 0.0580 kg. You can ignore gravity and assume that the ball travels parallel to the ground as it travels between the two players. Unless otherwise specified, all measurements are made by the two men. (a) Lorentz serves the ball at 80.0 m/s. What is the ball's kinetic energy? (b) Einstein slams a return at 1.80×10^8 m/s. What is the ball's kinetic energy? (c) During Einstein's return of the ball in part (a), a white rabbit runs beside the court in the direction from Einstein to Lorentz. The rabbit has a speed of 2.20×10^8 m/s relative to the two men. What is the speed of the rabbit relative to the ball? (d) What does the rabbit measure as the distance from Einstein to Lorentz? (e) How much time does it take for the rabbit to run 20.0 m, according to the players? (f) The white rabbit carries a pocket watch. He uses this watch to measure the time (as he sees it) for the distance from Einstein to Lorentz to pass by under him. What time does he measure?

37.61 •• **Measuring Speed by Radar.** A baseball coach uses a radar device to measure the speed of an approaching pitched baseball. This device sends out electromagnetic waves with frequency f_0 and then measures the shift in frequency Δf of the waves reflected from the moving baseball. If the fractional frequency shift produced by a baseball is $\Delta f / f_0 = 2.86 \times 10^{-7}$, what is the baseball's speed in km/h? (*Hint:* Are the waves Doppler-shifted a second time when reflected off the ball?)

37.62 •• A spaceship moving at constant speed u relative to us broadcasts a radio signal at constant frequency f_0. As the spaceship approaches us, we receive a higher frequency f; after it has passed, we receive a lower frequency. (a) As the spaceship passes by, so it is instantaneously moving neither toward nor away from us, show that the frequency we receive is not f_0, and derive an expression for the frequency we do receive. Is the frequency we receive higher or lower than f_0? (*Hint:* In this case, successive wave crests move the same distance to the observer and so they have the same transit time. Thus f equals 1/T. Use the time dilation formula to relate the periods in the stationary and moving frames.) (b) A spaceship emits electromagnetic waves of frequency $f_0 = 345$ MHz as measured in a frame moving with the ship. The spaceship is moving at a constant speed 0.758c relative to us. What frequency f do we receive when the spaceship is approaching us? When it is moving away? In each case what is the shift in frequency, $f - f_0$? (c) Use the result of part (a) to calculate the frequency f and the frequency shift $(f - f_0)$ we receive at the instant that the ship passes by us. How does the shift in frequency calculated here compare to the shifts calculated in part (b)?

37.63 • CP In a particle accelerator a proton moves with constant speed 0.750c in a circle of radius 628 m. What is the net force on the proton?

37.64 •• CP The French physicist Armand Fizeau was the first to measure the speed of light accurately. He also found experimentally that the speed, relative to the lab frame, of light traveling in a tank of water that is itself moving at a speed V relative to the lab frame is

$$v = \frac{c}{n} + kV$$

where $n = 1.333$ is the index of refraction of water. Fizeau called k the dragging coefficient and obtained an experimental value of $k = 0.44$. What value of k do you calculate from relativistic transformations?

37.65 •• DATA As a research scientist at a linear accelerator, you are studying an unstable particle. You measure its mean lifetime Δt as a function of the particle's speed relative to your laboratory equipment. You record the speed of the particle u as a fraction of the speed of light in vacuum c. The table gives the results of your measurements.

u/c	0.70	0.80	0.85	0.88	0.90	0.92	0.94
$\Delta t\ (10^{-8}\ \text{s})$	3.57	4.41	5.02	5.47	6.05	6.58	7.62

(a) Your team leader suggests that if you plot your data as $(\Delta t)^2$ versus $(1 - u^2/c^2)^{-1}$, the data points will be fit well by a straight line. Construct this graph and verify the team leader's prediction. Use the best-fit straight line to your data to calculate the mean lifetime of the particle in its rest frame. (b) What is the speed of the particle relative to your lab equipment (expressed as u/c) if the lifetime that you measure is four times its rest-frame lifetime?

37.66 •• DATA You are an astronomer investigating four astronomical sources of infrared radiation. You have identified the nature of each source, so you know the frequency f_0 of each when it is at rest relative to you. Your detector, which is at rest relative to the earth, measures the frequency f of the moving source. Your results are given in the table.

Source	A	B	C	D
f (THz)	7.1	5.4	6.1	8.1
f_0 (THz)	9.2	8.6	7.9	8.9

(a) Which source is moving at the highest speed relative to your detector? What is its speed? Is that source moving toward or away from the detector? (b) Which source is moving at the lowest speed relative to your detector? What is its speed? Is that source moving toward or away from the detector? (c) For source B, what frequency would your detector measure if the source were moving at the same speed relative to the detector but toward it rather than away from it?

37.67 •• DATA You are a scientist studying small aerosol parti-cles that are contained in a vacuum chamber. The particles carry a net charge, and you use a uniform electric field to exert a constant force of 8.00×10^{-14} N on one of them. That particle moves in the direction of the exerted force. Your instruments measure the acceleration of the particle as a function of its speed v. The table gives the results of your measurements for this particular particle.

v/c	0.60	0.65	0.70	0.75	0.80	0.85
a (10^3 m/s^2)	20.3	17.9	14.8	11.2	8.5	5.9

(a) Graph your data so that the data points are well fit by a straight line. Use the slope of this line to calculate the mass m of the particle. (b) What magnitude of acceleration does the exerted force produce if the speed of the particle is 100 m/s?

CHALLENGE PROBLEMS

37.68 ••• CP **Determining the Masses of Stars.** Many of the stars in the sky are actually *binary stars,* in which two stars orbit about their common center of mass. If the orbital speeds of the stars are high enough, the motion of the stars can be detected by the Doppler shifts of the light they emit. Stars for which this is the case are called *spectroscopic binary stars.* **Figure P37.68** shows the simplest case of a spectroscopic binary star: two iden-tical stars, each with mass m, orbiting their center of mass in a circle of radius R. The plane of the stars' orbits is edge-on to the line of sight of an observer on the earth. (a) The light produced by heated hydrogen gas in a laboratory on the earth has a frequency of 4.568110×10^{14} Hz. In the light received from the stars by a telescope on the earth, hydrogen light is observed to vary in fre-quency between 4.567710×10^{14} Hz and 4.568910×10^{14} Hz. Determine whether the binary star system as a whole is moving toward or away from the earth, the speed of this motion, and the orbital speeds of the stars. (*Hint:* The speeds involved are much less than c, so you may use the approximate result $\Delta f/f = u/c$ given in Section 37.6.) (b) The light from each star in the binary system varies from its maximum frequency to its minimum fre-quency and back again in 11.0 days. Determine the orbital radius R and the mass m of each star. Give your answer for m in kilo-grams and as a multiple of the mass of the sun, 1.99×10^{30} kg. Compare the value of R to the distance from the earth to the sun, 1.50×10^{11} m. (This technique is actually used in astronomy to determine the masses of stars. In practice, the problem is more complicated because the two stars in a binary system are usually not identical, the orbits are usually not circular, and the plane of the orbits is usually tilted with respect to the line of sight from the earth.)

Figure **P37.68**

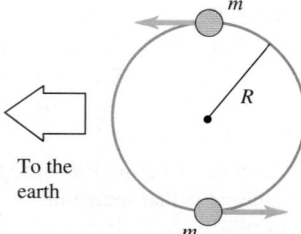

To the earth

37.69 ••• CP **Kaon Production.** In high-energy physics, new particles can be created by collisions of fast-moving projectile particles with stationary particles. Some of the kinetic energy of the incident particle is used to create the mass of the new particle. A proton–proton collision can result in the creation of a negative kaon (K^-) and a positive kaon (K^+):

$$p + p \rightarrow p + p + K^- + K^+$$

(a) Calculate the minimum kinetic energy of the incident proton that will allow this reaction to occur if the second (target) proton is initially at rest. The rest energy of each kaon is 493.7 MeV, and the rest energy of each proton is 938.3 MeV. (*Hint:* It is useful here to work in the frame in which the total momentum is zero. But note that the Lorentz transformation must be used to relate the velocities in the laboratory frame to those in the zero-total-momentum frame.) (b) How does this calculated minimum kinetic energy compare with the total rest mass energy of the created kaons? (c) Suppose that instead the two protons are both in motion with velocities of equal magnitude and opposite direction. Find the minimum combined kinetic energy of the two protons that will allow the reaction to occur. How does this calculated minimum kinetic energy compare with the total rest mass energy of the created kaons? (This example shows that when colliding beams of particles are used instead of a stationary target, the energy require-ments for producing new particles are reduced substantially.)

37.70 ••• CP CALC **Relativity and the Wave Equation.** (a) Consider the Galilean transformation along the x-direction: $x' = x - vt$ and $t' = t$. In frame S the wave equation for electro-magnetic waves in a vacuum is

$$\frac{\partial^2 E(x, t)}{\partial x^2} - \frac{1}{c^2} \frac{\partial^2 E(x, t)}{\partial t^2} = 0$$

where E represents the electric field in the wave. Show that by using the Galilean transformation the wave equation in frame S' is found to be

$$\left(1 - \frac{v^2}{c^2}\right) \frac{\partial^2 E(x', t')}{\partial x'^2} + \frac{2v}{c^2} \frac{\partial^2 E(x', t')}{\partial x' \partial t'} - \frac{1}{c^2} \frac{\partial^2 E(x', t')}{\partial t'^2} = 0$$

This has a different form than the wave equation in S. Hence the Galilean transformation *violates* the first relativity postulate that all physical laws have the same form in all inertial reference frames. (*Hint:* Express the derivatives $\partial/\partial x$ and $\partial/\partial t$ in terms of $\partial/\partial x'$ and $\partial/\partial t'$ by use of the chain rule.) (b) Repeat the analysis of part (a), but use the Lorentz coordinate transformations, Eqs. (37.21), and show that in frame S' the wave equation has the same form as in frame S:

$$\frac{\partial^2 E(x', t')}{\partial x'^2} - \frac{1}{c^2} \frac{\partial^2 E(x', t')}{\partial t'^2} = 0$$

Explain why this shows that the speed of light in vacuum is c in both frames S and S'.

PASSAGE PROBLEMS

SPEED OF LIGHT. Our universe has properties that are deter-mined by the values of the fundamental physical constants, and it would be a much different place if the charge of the electron, the mass of the proton, or the speed of light was substantially differ-ent from its actual value. For instance, the speed of light is so great that the effects of relativity usually go unnoticed in everyday events. Let's imagine an alternate universe where the speed of light is 1,000,000 times less than it is in our universe to see what would happen.

37.71 An airplane has a length of 60 m when measured at rest. When the airplane is moving at 180 m/s (400 mph) in the alternate universe, how long would the plane appear to be to a stationary observer? (a) 24 m; (b) 36 m; (c) 48 m; (d) 60 m; (e) 75 m.

37.72 If the airplane of Passage Problem 37.71 has a rest mass of 20,000 kg, what is its relativistic mass when the plane is moving at 180 m/s? (a) 8000 kg; (b) 12,000 kg; (c) 16,000 kg; (d) 25,000 kg; (e) 33,300 kg.

37.73 In our universe, the rest energy of an electron is approximately 8.2×10^{-14} J. What would it be in the alternate universe? (a) 8.2×10^{-8} J; (b) 8.2×10^{-26} J; (c) 8.2×10^{-2} J; (d) 0.82 J.

37.74 In the alternate universe, how fast must an object be moving for it to have a kinetic energy equal to its rest mass? (a) 225 m/s; (b) 260 m/s; (c) 300 m/s; (d) The kinetic energy could not be equal to the rest mass.

Answers

Chapter Opening Question ?

(v) From Eq. (37.36), the relativistic expression for the kinetic energy of a particle of mass m moving at speed v is $K = (\gamma - 1)mc^2$, where $\gamma = 1/\sqrt{1 - v^2/c^2}$. If $v = 0.99000c$, $\gamma - 1 = 6.08881$; if $v = 0.99995c$, $\gamma - 1 = 99.001$, which is 16.260 times greater than the value at $v = 0.99000c$. As the speed approaches c, a relatively small increase in v corresponds to a large increase in kinetic energy (see Fig. 37.21).

Test Your Understanding Questions

37.1 (a) (i), (b) no You, too, will measure a spherical wave front that expands at the same speed c in all directions. This is a consequence of Einstein's second postulate. The wave front that you measure does *not* stay centered on the position of the moving spaceship; rather, it is centered on the point P where the spaceship was located at the instant that it emitted the light pulse. For example, suppose the spaceship is moving at speed $c/2$. When your watch shows that a time t has elapsed since the pulse of light was emitted, your measurements will show that the wave front is a sphere of radius ct centered on P and that the spaceship is a distance $ct/2$ from P.

37.2 (iii) In Mavis's frame of reference, the two events (the Ogdenville clock striking noon and the North Haverbrook clock striking noon) are not simultaneous. Figure 37.5 shows that the event toward the front of the rail car occurs first. Since the rail car is moving toward North Haverbrook, that clock struck noon before the one on Ogdenville. So, according to Mavis, it is after noon in North Haverbrook.

37.3 (a) (ii), (b) (ii) The statement that moving clocks run slow refers to any clock that is moving relative to an observer. Maria and her stopwatch are moving relative to Samir, so Samir measures Maria's stopwatch to be running slow and to have ticked off fewer seconds than his own stopwatch. Samir and his stopwatch are moving relative to Maria, so she likewise measures Samir's stopwatch to be running slow. Each observer's measurement is correct for his or her own frame of reference. *Both* observers conclude that a moving stopwatch runs slow. This is consistent with the principle of relativity (see Section 37.1), which states that the laws of physics are the same in all inertial frames of reference.

37.4 (ii), (i) and (iii) (tie), (iv) You measure both the rest length of the stationary meter stick and the contracted length of the moving spaceship to be 1 meter. The rest length of the spaceship is greater than the contracted length that you measure, and so must be greater than 1 meter. A miniature observer on board the spaceship would measure a contracted length for the meter stick of less than 1 meter. Note that in your frame of reference the nose and tail of the spaceship can simultaneously align with the two ends of the meter stick, since in your frame of reference they have the same length of 1 meter. In the spaceship's frame these two alignments cannot happen simultaneously because the meter stick is shorter than the spaceship. Section 37.2 tells us that this shouldn't be a surprise; two events that are simultaneous to one observer may not be simultaneous to a second observer moving relative to the first one.

37.5 (a) P_1, (b) P_4 (a) The last of Eqs. (37.21) tells us the times of the two events in S': $t_1' = \gamma(t_1 - ux_1/c^2)$ and $t_2' = \gamma(t_2 - ux_2/c^2)$. In frame S the two events occur at the same x-coordinate, so $x_1 = x_2$, and event P_1 occurs before event P_2, so $t_1 < t_2$. Hence you can see that $t_1' < t_2'$ and event P_1 happens before P_2 in frame S', too. This says that if event P_1 happens before P_2 in a frame of reference S where the two events occur at the same position, then P_1 happens before P_2 in any other frame moving relative to S. (b) In frame S the two events occur at different x-coordinates such that $x_3 < x_4$, and events P_3 and P_4 occur at the same time, so $t_3 = t_4$. Hence you can see that $t_3' = \gamma(t_3 - ux_3/c^2)$ is greater than $t_4' = \gamma(t_4 - ux_4/c^2)$, so event P_4 happens before P_3 in frame S'. This says that even though the two events are simultaneous in frame S, they need not be simultaneous in a frame moving relative to S.

37.7 (ii) Equation (37.27) tells us that the magnitude of momentum of a particle with mass m and speed v is $p = mv/\sqrt{1 - v^2/c^2}$. If v increases by a factor of 2, the numerator mv increases by a factor of 2 *and* the denominator $\sqrt{1 - v^2/c^2}$ decreases. Hence p increases by a factor greater than 2. (Note that in order to double the speed, the initial speed must be less than $c/2$. That's because the speed of light is the ultimate speed limit.)

37.8 (i) As the proton moves a distance s, the constant force of magnitude F does work $W = Fs$ and increases the kinetic energy by an amount $\Delta K = W = Fs$. This is true no matter what the speed of the proton before moving this distance. Thus the constant force increases the proton's kinetic energy by the same amount during the first meter of travel as during any subsequent meter of travel. (It's true that as the proton approaches the ultimate speed limit of c, the increase in the proton's *speed* is less and less with each subsequent meter of travel. That's not what the question is asking, however.)

Bridging Problem

(a) $0.268c$ **(b)** 35.6 MeV   **(c)** 145 MeV

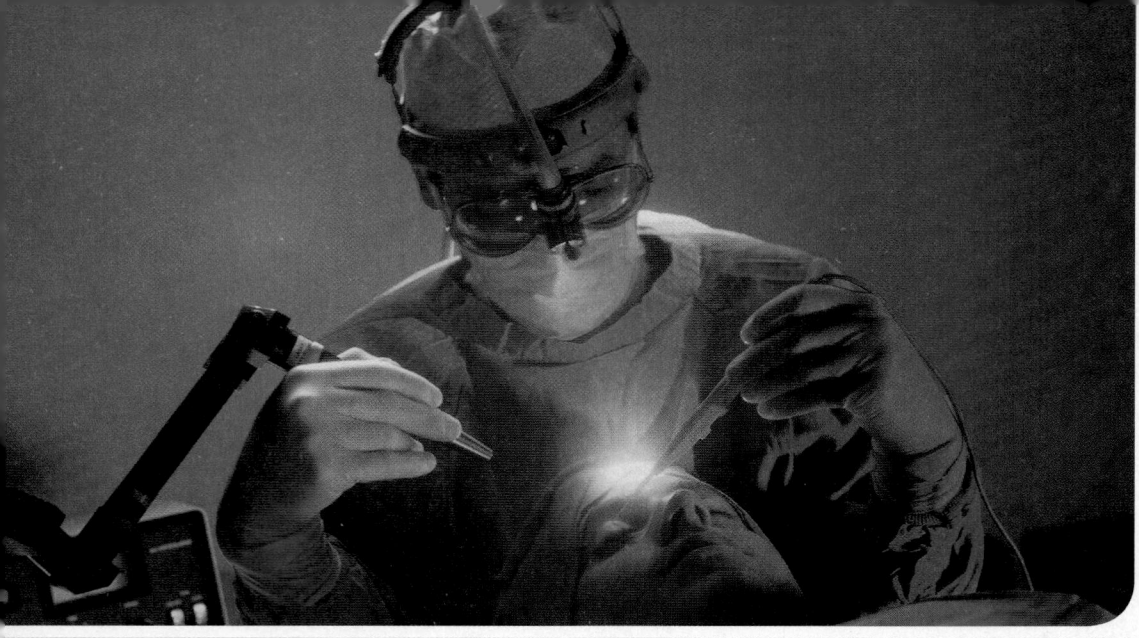

38 PHOTONS: LIGHT WAVES BEHAVING AS PARTICLES

PhET: Photoelectric Effect

In Chapter 32 we saw how Maxwell, Hertz, and others established firmly that light is an electromagnetic wave. Interference, diffraction, and polarization, discussed in Chapters 35 and 36, further demonstrate this *wave nature* of light.

When we look more closely at the emission, absorption, and scattering of electromagnetic radiation, however, we discover a completely different aspect of light. We find that the energy of an electromagnetic wave is *quantized;* it is emitted and absorbed in particle-like packages of definite energy, called *photons.* The energy of a single photon is proportional to the frequency of the radiation.

We'll find that light and other electromagnetic radiation exhibits *wave–particle duality:* Light acts sometimes like waves and sometimes like particles. Interference and diffraction demonstrate wave behavior, while emission and absorption of photons demonstrate the particle behavior. This radical reinterpretation of light will lead us in the next chapter to no less radical changes in our views of the nature of matter.

38.1 LIGHT ABSORBED AS PHOTONS: THE PHOTOELECTRIC EFFECT

A phenomenon that gives insight into the nature of light is the **photoelectric effect,** in which a material emits electrons from its surface when illuminated (**Fig. 38.1**). To escape from the surface, an electron must absorb enough energy from the incident light to overcome the attraction of positive ions in the material. These attractions constitute a potential-energy barrier; the light supplies the "kick" that enables the electron to escape.

The photoelectric effect has a number of applications. Digital cameras and night-vision scopes use it to convert light energy into an electric signal that is reconstructed into an image (**Fig. 38.2**). Sunlight striking the moon causes surface dust to eject electrons, leaving the dust particles with a positive charge. The mutual electric repulsion of these charged dust particles causes them to rise above the moon's surface, a phenomenon that was observed from lunar orbit by the *Apollo* astronauts.

Threshold Frequency and Stopping Potential

In Section 32.1 we explored the wave model of light, which Maxwell formulated two decades before the photoelectric effect was observed. Is the photoelectric effect consistent with this model? **Figure 38.3a** (next page) shows a modern version of one of the experiments that explored this question. Two conducting electrodes are enclosed in an evacuated glass tube and connected by a battery, and the cathode is illuminated. Depending on the potential difference V_{AC} between the two electrodes, electrons emitted by the illuminated cathode (called *photoelectrons*) may travel across to the anode, producing a *photocurrent* in the external circuit. (The tube is evacuated to a pressure of 0.01 Pa or less to minimize collisions between the electrons and gas molecules.)

The illuminated cathode emits photoelectrons with various kinetic energies. If the electric field points toward the cathode, as in Fig. 38.3a, all the electrons are accelerated toward the anode and contribute to the photocurrent. But by reversing the field and adjusting its strength as in Fig. 38.3b, we can prevent the less energetic electrons from reaching the anode. In fact, we can determine the *maximum* kinetic energy K_{max} of the emitted electrons by making the potential of the anode relative to the cathode, V_{AC}, just negative enough so that the current stops. This occurs for $V_{AC} = -V_0$, where V_0 is called the **stopping potential.** As an electron moves from the cathode to the anode, the potential decreases by V_0 and negative work $-eV_0$ is done on the (negatively charged) electron. The most energetic electron leaves the cathode with kinetic energy $K_{max} = \frac{1}{2}mv_{max}^2$ and has zero kinetic energy at the anode. Using the work–energy theorem, we have

$$W_{tot} = -eV_0 = \Delta K = 0 - K_{max}$$
$$K_{max} = \tfrac{1}{2}mv_{max}^2 = eV_0$$

(maximum kinetic energy of photoelectrons) (38.1)

Hence by measuring the stopping potential V_0, we can determine the maximum kinetic energy with which electrons leave the cathode. (We are ignoring any effects due to differences in the materials of the cathode and anode.)

In this experiment, how does the photocurrent depend on the voltage across the electrodes and on the frequency and intensity of the light? Based on Maxwell's picture of light as an electromagnetic wave, here is what we would predict:

Wave-Model Prediction 1: We saw in Section 32.4 that the intensity of an electromagnetic wave depends on its amplitude but not on its frequency. So the photoelectric effect should occur for light of any frequency, and *the magnitude of the photocurrent should not depend on the frequency of the light.*

Wave-Model Prediction 2: It takes a certain minimum amount of energy, called the **work function,** to eject a single electron from a particular surface (see Fig. 38.1). If the light falling on the surface is very faint, some time may elapse before the total energy absorbed by the surface equals the work function. Hence, for faint illumination, *we expect a time delay* between when we switch on the light and when photoelectrons appear.

Wave-Model Prediction 3: Because the energy delivered to the cathode surface depends on the intensity of illumination, *we expect the stopping potential to increase with increasing light intensity.* Since intensity does not depend on frequency, we further expect that *the stopping potential should not depend on the frequency of the light.*

38.1 The photoelectric effect.

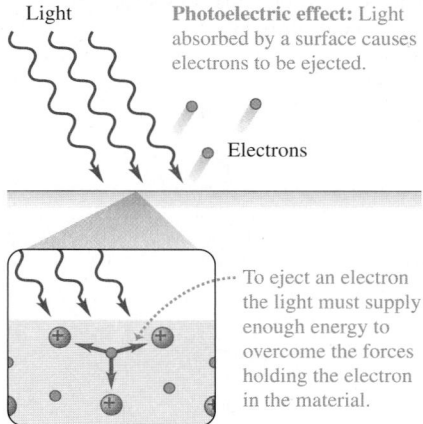

Light

Photoelectric effect: Light absorbed by a surface causes electrons to be ejected.

Electrons

To eject an electron the light must supply enough energy to overcome the forces holding the electron in the material.

38.2 (a) A night-vision scope makes use of the photoelectric effect. Photons entering the scope strike a plate, ejecting electrons that pass through a thin disk in which there are millions of tiny channels. The current through each channel is amplified electronically and then directed toward a screen that glows when hit by electrons. (b) The image formed on the screen, which is a combination of these millions of glowing spots, is thousands of times brighter than the naked-eye view.

(a)

(b)

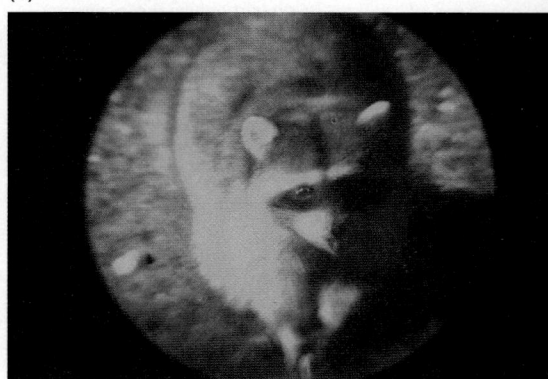

38.3 An experiment testing whether the photoelectric effect is consistent with the wave model of light.

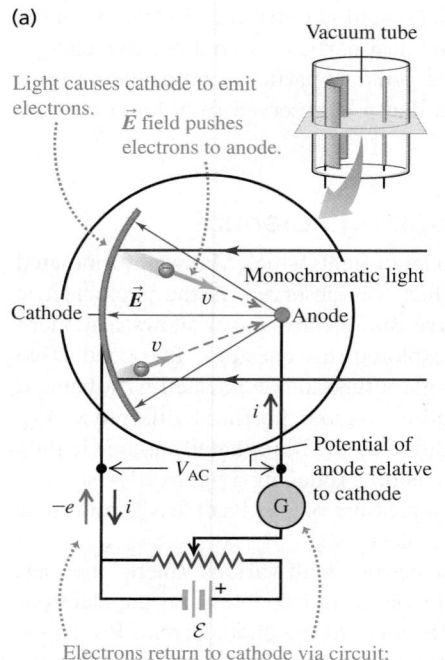

(a)

Vacuum tube

Light causes cathode to emit electrons.

$\vec{E}$ field pushes electrons to anode.

Cathode

$\vec{E}$

v

Monochromatic light

Anode

v

i

Potential of anode relative to cathode

V_{AC}

$-e$ i

G

$\mathcal{E}$

Electrons return to cathode via circuit; galvanometer measures current.

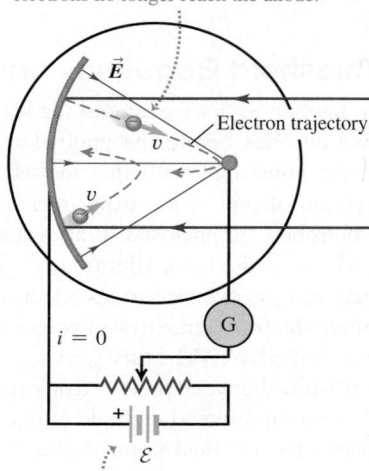

(b)

We now reverse the electric field so that it tends to repel electrons from the anode. Above a certain field strength, electrons no longer reach the anode.

$\vec{E}$

v

Electron trajectory

v

$i = 0$

G

$\mathcal{E}$

The **stopping potential** at which the current ceases has absolute value V_0.

The experimental results proved to be *very* different from these predictions. Here is what was found in the years between 1877 and 1905:

Experimental Result 1: *The photocurrent depends on the light frequency.* For a given material, monochromatic light with a frequency below a minimum **threshold frequency** produces *no* photocurrent, regardless of intensity. For most metals the threshold frequency is in the ultraviolet (corresponding to wavelengths λ between 200 and 300 nm), but for other materials like potassium oxide and cesium oxide it is in the visible spectrum (λ between 380 and 750 nm).

Experimental Result 2: There is *no measurable time delay* between when the light is turned on and when the cathode emits photoelectrons (assuming the frequency of the light exceeds the threshold frequency). This is true no matter how faint the light is.

Experimental Result 3: *The stopping potential does not depend on intensity, but does depend on frequency.* **Figure 38.4** shows graphs of photocurrent as a function of potential difference V_{AC} for light of a given frequency and two different intensities. The reverse potential difference $-V_0$ needed to reduce the current to zero is the same for both intensities. The only effect of increasing the intensity is to increase the number of electrons per second and hence the photocurrent i. (The curves level off when V_{AC} is large and positive because at that point all the emitted electrons are being collected by the anode.) If the intensity is held constant but the frequency is increased, the stopping potential also increases. In other words, the greater the light frequency, the higher the energy of the ejected photoelectrons.

These results directly contradict Maxwell's description of light as an electromagnetic wave. A solution to this dilemma was provided by Albert Einstein in 1905. His proposal involved nothing less than a new picture of the nature of light.

38.4 Photocurrent i for light frequency f as a function of the potential V_{AC} of the anode with respect to the cathode.

The stopping potential V_0 is independent of the light intensity ...

... but the photocurrent i for large positive V_{AC} is directly proportional to the intensity.

i

f is constant.

Constant intensity $2I$

Constant intensity I

$-V_0$ 0 V_{AC}

Einstein's Photon Explanation

Einstein made the radical postulate that a beam of light consists of small packages of energy called **photons** or *quanta*. This postulate was an extension of an idea developed five years earlier by Max Planck to explain the properties of blackbody radiation, which we discussed in Section 17.7. (We'll explore

Planck's ideas in Section 39.5.) In Einstein's picture, the energy E of an individual photon is equal to a constant times the photon frequency f. From the relationship $f = c/\lambda$ for electromagnetic waves in vacuum, we have

Planck's constant

Energy of a photon $\cdots\blacktriangleright E = hf = \dfrac{hc}{\lambda}$ Speed of light in vacuum (38.2)

Frequency   Wavelength

Here **Planck's constant,** h, is a universal constant. Its numerical value, to the accuracy known at present, is

$$h = 6.62606957(29) \times 10^{-34}\ \text{J} \cdot \text{s}$$

In Einstein's picture, an individual photon arriving at the surface in Fig. 38.1a or 38.2 is absorbed by a single electron. This energy transfer is an all-or-nothing process, in contrast to the continuous transfer of energy in the wave theory of light; the electron gets all of the photon's energy or none at all. The electron can escape from the surface only if the energy it acquires is greater than the work function ϕ. Thus photoelectrons will be ejected only if $hf > \phi$, or $f > \phi/h$. Einstein's postulate therefore explains why the photoelectric effect occurs only for frequencies greater than a minimum threshold frequency. This postulate is also consistent with the observation that greater intensity causes a greater photocurrent (Fig. 38.4). Greater intensity at a particular frequency means a greater number of photons per second absorbed, and thus a greater number of electrons emitted per second and a greater photocurrent.

Einstein's postulate also explains why there is no delay between illumination and the emission of photoelectrons. As soon as photons of sufficient energy strike the surface, electrons can absorb them and be ejected.

Finally, Einstein's postulate explains why the stopping potential for a given surface depends only on the light frequency. Recall that ϕ is the *minimum* energy needed to remove an electron from the surface. Einstein applied conservation of energy to find that the *maximum* kinetic energy $K_{\max} = \frac{1}{2}mv_{\max}^2$ for an emitted electron is the energy hf gained from a photon minus the work function ϕ:

$$K_{\max} = \tfrac{1}{2}mv_{\max}^2 = hf - \phi \qquad (38.3)$$

Substituting $K_{\max} = eV_0$ from Eq. (38.1), we find

Maximum kinetic energy of photoelectron   Energy of absorbed photon

Photoelectric effect: $eV_0 = hf - \phi$ Work function (38.4)

Light frequency

Magnitude of   Stopping   Planck's
electron charge   potential   constant

Equation (38.4) shows that the stopping potential V_0 increases with increasing frequency f. The intensity doesn't appear in Eq. (38.4), so V_0 is independent of intensity. As a check of Eq. (38.4), we can measure the stopping potential V_0 for each of several values of frequency f for a given cathode material (**Fig. 38.5**). A graph of V_0 as a function of f turns out to be a straight line, verifying Eq. (38.4), and from such a graph we can determine both the work function ϕ for the material and the value of the quantity h/e. After the electron charge $-e$ was measured by Robert Millikan in 1909, Planck's constant h could also be determined from these measurements.

Electron energies and work functions are usually expressed in electron volts (eV), defined in Section 23.2. To four significant figures,

$$1\ \text{eV} = 1.602 \times 10^{-19}\ \text{J}$$

CAUTION Photons are not "particles" in the usual sense It's common, but inaccurate, to envision photons as miniature billiard balls. Billiard balls have a rest mass and travel slower than the speed of light c, while photons travel at the speed of light and have *zero* rest mass. Furthermore, photons have wave aspects (frequency and wavelength) that are easy to observe. The photon concept is a very strange one, and the true nature of photons is difficult to visualize in a simple way. We'll discuss this in more detail in Section 38.4.

DATA *SPEAKS*

Photons

When students were given a problem involving photons and their properties, more than 20% gave an incorrect response. Common errors:

- Confusion about photon energy, frequency, and wavelength. The greater the frequency of a photon, the greater the photon energy and the shorter its wavelength; the longer the wavelength of a photon, the smaller the photon energy and the lower its frequency [see Eq. (38.2)].

- Confusion about the photoelectric effect. The *greater* the work function of the material, the *smaller* the kinetic energy of the electrons emitted when photons of a given frequency shine on the material [see Eq. (38.3)].

38.5 Stopping potential as a function of frequency for a particular cathode material.

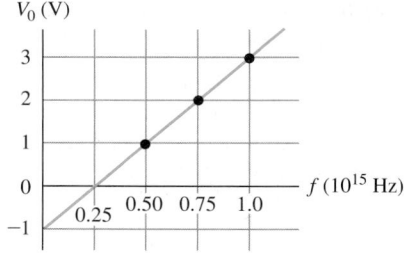

TABLE 38.1	Work Functions of Several Elements
Element	**Work Function (eV)**
Aluminum	4.3
Carbon	5.0
Copper	4.7
Gold	5.1
Nickel	5.1
Silicon	4.8
Silver	4.3
Sodium	2.7

38.6 Stopping potential as a function of frequency for two cathode materials having different work functions ϕ.

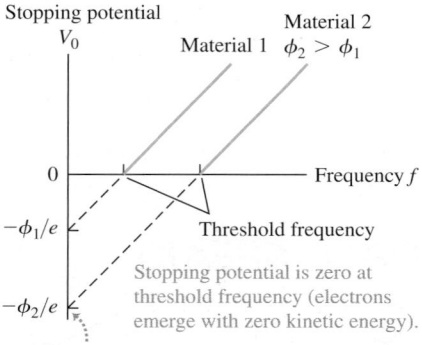

For each material,

$$eV_0 = hf - \phi \quad \text{or} \quad V_0 = \frac{hf}{e} - \frac{\phi}{e}$$

so the plots have same slope h/e but different intercepts $-\phi/e$ on the vertical axis.

To this accuracy, Planck's constant is

$$h = 6.626 \times 10^{-34}\,\text{J}\cdot\text{s} = 4.136 \times 10^{-15}\,\text{eV}\cdot\text{s}$$

Table 38.1 lists the work functions of several elements. These values are approximate because they are very sensitive to surface impurities. The greater the work function, the higher the minimum frequency needed to emit photoelectrons (**Fig. 38.6**).

The photon picture also explains other phenomena in which light is absorbed. A *suntan* is caused when the energy in sunlight triggers a chemical reaction in skin cells that leads to increased production of the pigment melanin. This reaction can occur only if a specific molecule in the cell absorbs a certain minimum amount of energy. A short-wavelength ultraviolet photon has enough energy to trigger the reaction, but a longer-wavelength visible-light photon does not. Hence ultraviolet light causes tanning, while visible light cannot.

Photon Momentum

Einstein's photon concept applies to *all* regions of the electromagnetic spectrum, including radio waves, x rays, and so on. A photon of any frequency f and wavelength λ has energy E given by Eq. (38.2). Furthermore, according to the special theory of relativity, every particle that has energy must have momentum. Photons have zero rest mass, and a particle with zero rest mass and energy E has momentum with magnitude p given by $E = pc$ [Section 37.8; see (Eq. 37.40)]. Thus the magnitude p of the momentum of a photon is

$$\text{Momentum of a photon} \cdots\!\blacktriangleright p = \frac{E}{c} = \frac{hf}{c} = \frac{h}{\lambda} \cdots\!\text{Wavelength} \qquad (38.5)$$

Photon energy — Planck's constant — Speed of light in vacuum — Frequency

The direction of the photon's momentum is simply the direction in which the electromagnetic wave is moving.

IDENTIFY *the relevant concepts:* The energy and momentum of an individual photon are proportional to the frequency and inversely proportional to the wavelength. Einstein's interpretation of the photoelectric effect is that energy is conserved as a photon ejects an electron from a material surface.

SET UP *the problem:* Identify the target variable. It could be the photon's wavelength λ, frequency f, energy E, or momentum p. If the problem involves the photoelectric effect, the target variable could be the maximum kinetic energy of photoelectrons K_{max}, the stopping potential V_0, or the work function ϕ.

EXECUTE *the solution* as follows:
1. Use Eqs. (38.2) and (38.5) to relate the energy and momentum of a photon to its wavelength and frequency. If the problem involves the photoelectric effect, use Eqs. (38.1), (38.3), and (38.4) to relate

the photon frequency, stopping potential, work function, and maximum photoelectron kinetic energy.
2. The electron volt (eV), which we introduced in Section 23.2, is a convenient unit. It is the kinetic energy gained by an electron when it moves freely through an increase of potential of one volt: $1\,\text{eV} = 1.602 \times 10^{-19}\,\text{J}$. If the photon energy E is given in electron volts, use $h = 4.136 \times 10^{-15}\,\text{eV}\cdot\text{s}$; if E is in joules, use $h = 6.626 \times 10^{-34}\,\text{J}\cdot\text{s}$.

EVALUATE *your answer:* In problems involving photons, at first the numbers will be unfamiliar to you and errors will not be obvious. It helps to remember that a visible-light photon with $\lambda = 600$ nm and $f = 5 \times 10^{14}$ Hz has an energy E of about 2 eV, or about 3×10^{-19} J.

EXAMPLE 38.1 LASER-POINTER PHOTONS

A laser pointer with a power output of 5.00 mW emits red light (λ = 650 nm). (a) What is the magnitude of the momentum of each photon? (b) How many photons does the laser pointer emit each second?

SOLUTION

IDENTIFY and SET UP: This problem involves the ideas of (a) photon momentum and (b) photon energy. In part (a) we'll use Eq. (38.5) and the given wavelength to find the magnitude of each photon's momentum. In part (b), Eq. (38.5) gives the energy per photon, and the power output tells us the energy emitted per second. We can combine these quantities to calculate the number of photons emitted per second.

EXECUTE: (a) We have λ = 650 nm = 6.50×10^{-7} m, so from Eq. (38.5) the photon momentum is

$$p = \frac{h}{\lambda} = \frac{6.626 \times 10^{-34}\,\text{J} \cdot \text{s}}{6.50 \times 10^{-7}\,\text{m}}$$
$$= 1.02 \times 10^{-27}\,\text{kg} \cdot \text{m/s}$$

(Recall that 1 J = 1 kg $\cdot$ m²/s².)

(b) From Eq. (38.5), the energy of a single photon is

$$E = pc = (1.02 \times 10^{-27}\,\text{kg} \cdot \text{m/s})(3.00 \times 10^{8}\,\text{m/s})$$
$$= 3.06 \times 10^{-19}\,\text{J} = 1.91\,\text{eV}$$

The laser pointer emits energy at the rate of 5.00×10^{-3} J/s, so it emits photons at the rate of

$$\frac{5.00 \times 10^{-3}\,\text{J/s}}{3.06 \times 10^{-19}\,\text{J/photon}} = 1.63 \times 10^{16}\,\text{photons/s}$$

EVALUATE: The result in part (a) is very small; a typical oxygen molecule in room-temperature air has 2500 times more momentum. As a check on part (b), we can calculate the photon energy from Eq. (38.2):

$$E = hf = \frac{hc}{\lambda} = \frac{(6.626 \times 10^{-34}\,\text{J} \cdot \text{s})(3.00 \times 10^{8}\,\text{m/s})}{6.50 \times 10^{-7}\,\text{m}}$$
$$= 3.06 \times 10^{-19}\,\text{J} = 1.91\,\text{eV}$$

Our result in part (b) shows that a huge number of photons leave the laser pointer each second, each of which has an infinitesimal amount of energy. Hence the discreteness of the photons isn't noticed, and the radiated energy appears to be a continuous flow.

EXAMPLE 38.2 A PHOTOELECTRIC-EFFECT EXPERIMENT

While conducting a photoelectric-effect experiment with light of a certain frequency, you find that a reverse potential difference of 1.25 V is required to reduce the current to zero. Find (a) the maximum kinetic energy and (b) the maximum speed of the emitted photoelectrons.

SOLUTION

IDENTIFY and SET UP: The value of 1.25 V is the stopping potential V_0 for this experiment. We'll use this in Eq. (38.1) to find the maximum photoelectron kinetic energy K_{max}, and from this we'll find the maximum photoelectron speed.

EXECUTE: (a) From Eq. (38.1),

$$K_{max} = eV_0 = (1.60 \times 10^{-19}\,\text{C})(1.25\,\text{V}) = 2.00 \times 10^{-19}\,\text{J}$$

(Recall that 1 V = 1 J/C.) In terms of electron volts,

$$K_{max} = eV_0 = e(1.25\,\text{V}) = 1.25\,\text{eV}$$

because the electron volt (eV) is the magnitude of the electron charge e times one volt (1 V).

(b) From $K_{max} = \frac{1}{2}mv_{max}^{2}$ we get

$$v_{max} = \sqrt{\frac{2K_{max}}{m}} = \sqrt{\frac{2(2.00 \times 10^{-19}\,\text{J})}{9.11 \times 10^{-31}\,\text{kg}}}$$
$$= 6.63 \times 10^{5}\,\text{m/s}$$

EVALUATE: The value of v_{max} is about 0.2% of the speed of light, so we are justified in using the nonrelativistic expression for kinetic energy. (An equivalent justification is that the electron's 1.25-eV kinetic energy is much less than its rest energy mc^2 = 0.511 MeV = 5.11×10^{5} eV.)

EXAMPLE 38.3 DETERMINING ϕ AND h EXPERIMENTALLY

For a particular cathode material in a photoelectric-effect experiment, you measure stopping potentials $V_0 = 1.0$ V for light of wavelength $\lambda = 600$ nm, 2.0 V for 400 nm, and 3.0 V for 300 nm. Determine the work function ϕ for this material and the implied value of Planck's constant h.

SOLUTION

IDENTIFY and SET UP: This example uses the relationship among stopping potential V_0, frequency f, and work function ϕ in the photoelectric effect. According to Eq. (38.4), a graph of V_0 versus f should be a straight line as in Fig. 38.5 or 38.6. Such a graph is completely determined by its slope and the value at which it intercepts the vertical axis; we will use these to determine the values of the target variables ϕ and h.

EXECUTE: We rewrite Eq. (38.4) as

$$V_0 = \frac{h}{e}f - \frac{\phi}{e}$$

In this form we see that the slope of the line is h/e and the vertical-axis intercept (corresponding to $f = 0$) is $-\phi/e$. The frequencies, obtained from $f = c/\lambda$ and $c = 3.00 \times 10^8$ m/s, are 0.50×10^{15} Hz, 0.75×10^{15} Hz, and 1.0×10^{15} Hz, respectively. From a graph of these data (see Fig. 38.6), we find

$$-\frac{\phi}{e} = \text{vertical intercept} = -1.0 \text{ V}$$

$$\phi = 1.0 \text{ eV} = 1.6 \times 10^{-19} \text{ J}$$

and

$$\text{Slope} = \frac{\Delta V_0}{\Delta f} - \frac{3.0 \text{ V} - (-1.0 \text{ V})}{1.00 \times 10^{15} \text{ s}^{-1} - 0} = 4.0 \times 10^{-15} \text{ J} \cdot \text{s/C}$$

$$h = \text{slope} \times e = (4.0 \times 10^{-15} \text{ J} \cdot \text{s/C})(1.60 \times 10^{-19} \text{ C})$$

$$= 6.4 \times 10^{-34} \text{ J} \cdot \text{s}$$

EVALUATE: The value of Planck's constant h determined from your experiment differs from the accepted value by only about 3%. The small value $\phi = 1.0$ eV tells us that the cathode surface is not composed solely of one of the elements in Table 38.1.

BIO Application Sterilizing with High-Energy Photons One technique for killing harmful microorganisms is to illuminate them with ultraviolet light with a wavelength shorter than 254 nm. If a photon of such short wavelength strikes a DNA molecule within a microorganism, the energy of the photon is great enough to break the bonds within the molecule. This renders the microorganism unable to grow or reproduce. Such ultraviolet germicidal irradiation is used for medical sanitation, to keep laboratories sterile (as shown here), and to treat both drinking water and wastewater.

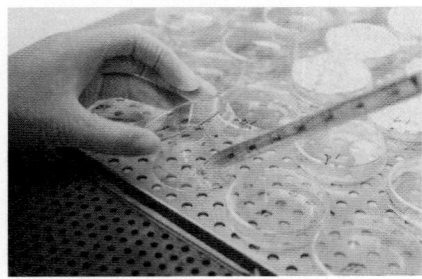

TEST YOUR UNDERSTANDING OF SECTION 38.1 Silicon films become better electrical conductors when illuminated by photons with energies of 1.14 eV or greater, an effect called *photoconductivity*. Which of the following wavelengths of electromagnetic radiation can cause photoconductivity in silicon films? (i) Ultraviolet light with $\lambda = 300$ nm; (ii) red light with $\lambda = 600$ nm; (iii) infrared light with $\lambda = 1200$ nm; (iv) both (i) and (ii); (v) all of (i), (ii), and (iii). ▌

38.2 LIGHT EMITTED AS PHOTONS: X-RAY PRODUCTION

The photoelectric effect provides convincing evidence that light is *absorbed* in the form of photons. For physicists to accept Einstein's radical photon concept, however, it was also necessary to show that light is *emitted* as photons. An experiment that demonstrates this convincingly is the inverse of the photoelectric effect: Instead of releasing electrons from a surface by shining electromagnetic radiation on it, we cause a surface to emit radiation—specifically, *x rays*—by bombarding it with fast-moving electrons.

X-Ray Photons

X rays were first produced in 1895 by the German physicist Wilhelm Röntgen, using an apparatus similar in principle to the setup shown in **Fig. 38.7**. When the cathode is heated to a very high temperature, it releases electrons in a process called *thermionic emission*. (As in the photoelectric effect, the minimum energy that an individual electron must be given to escape from the cathode's surface is equal to the work function for the surface. In this case the energy is provided to the electrons by heat rather than by light.) The electrons are then accelerated toward the anode by a potential difference V_{AC}. The bulb is evacuated (residual pressure 10^{-7} atm or less), so the electrons can travel from the cathode to the anode without colliding with air molecules. When V_{AC} is a few thousand volts or more, x rays are emitted from the anode surface.

The anode produces x rays in part simply by slowing the electrons abruptly. (Recall from Section 32.1 that accelerated charges emit electromagnetic waves.) This process is called *bremsstrahlung* (German for "braking radiation"). Because the electrons undergo accelerations of very great magnitude, they emit much of their radiation at short wavelengths in the x-ray range, about 10^{-9} to 10^{-12} m (1 nm to 1 pm). (X-ray wavelengths can be measured quite precisely by crystal diffraction techniques, which we discussed in Section 36.6.) Most electrons are braked by a series of collisions and interactions with anode atoms, so bremsstrahlung produces a continuous spectrum of electromagnetic radiation.

Just as we did for the photoelectric effect in Section 38.1, let's compare what Maxwell's wave theory of electromagnetic radiation would predict about this radiation to what is observed experimentally.

Wave-Model Prediction: The electromagnetic waves produced when an electron slams into the anode should be analogous to the sound waves produced by crashing cymbals together. These waves include sounds of all frequencies. By analogy, the x rays produced by bremsstrahlung should have a spectrum that includes *all* frequencies and hence *all* wavelengths.

Experimental Result: **Figure 38.8** shows bremsstrahlung spectra obtained when the same cathode and anode are used with four different accelerating voltages V_{AC}. *Not all x-ray frequencies and wavelengths are emitted:* Each spectrum has a maximum frequency f_{max} and a corresponding minimum wavelength λ_{min}. The greater the value of V_{AC}, the higher the maximum frequency and the shorter the minimum wavelength.

The wave model of electromagnetic radiation cannot explain these experimental results. But we can readily understand them by using the photon model. An electron has charge $-e$ and gains kinetic energy eV_{AC} when accelerated through a potential increase V_{AC}. The most energetic photon (highest frequency and shortest wavelength) is produced if the electron is braked to a stop all at once when it hits the anode, so that all of its kinetic energy goes to produce one photon; that is,

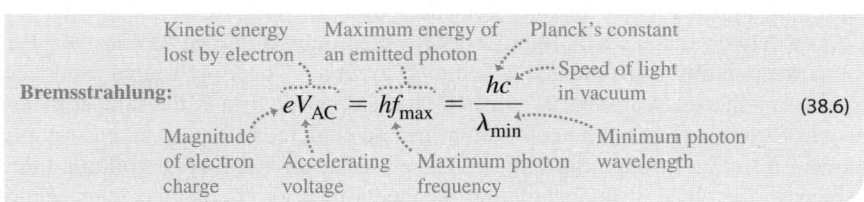

Kinetic energy lost by electron — Maximum energy of an emitted photon — Planck's constant — Speed of light in vacuum

Bremsstrahlung:
$$eV_{AC} = hf_{max} = \frac{hc}{\lambda_{min}} \qquad (38.6)$$

Magnitude of electron charge — Accelerating voltage — Maximum photon frequency — Minimum photon wavelength

(In this equation we ignore the work function of the target anode and the initial kinetic energy of the electrons "boiled off" from the cathode. These energies are very small compared to the kinetic energy eV_{AC} gained due to the potential difference.) If only a portion of an electron's kinetic energy goes into producing a photon, the photon energy will be less than eV_{AC} and the wavelength will be greater than λ_{min}. Experiment shows that the measured values for λ_{min} for different values of eV_{AC} (see Fig. 38.8) agree with Eq. (38.6). Note that according to Eq. (38.6), the maximum frequency and minimum wavelength in the bremsstrahlung process do not depend on the target material; this also agrees with experiment. So we can conclude that the photon picture of electromagnetic radiation is valid for the *emission* as well as the absorption of radiation.

The apparatus shown in Fig. 38.7 can also produce x rays by a second process in which electrons transfer their kinetic energy partly or completely to individual atoms within the target. It turns out that this process not only is consistent with the photon model of electromagnetic radiation, but also provides insight into the structure of atoms. We'll return to this process in Section 41.5.

38.7 An apparatus used to produce x rays, similar to Röntgen's 1895 apparatus.

Electrons are emitted thermionically from the heated cathode and are accelerated toward the anode; when they strike it, x rays are produced.

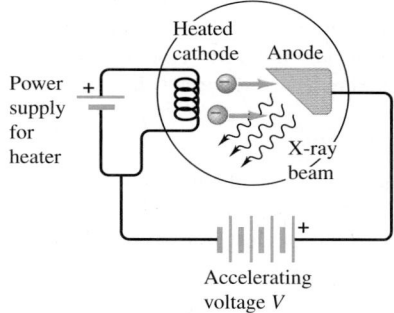

38.8 The continuous spectrum of x rays produced when a tungsten target is struck by electrons accelerated through a voltage V_{AC}. The curves represent different values of V_{AC}; points a, b, c, and d show the minimum wavelength for each voltage.

Vertical axis: x-ray intensity per unit wavelength

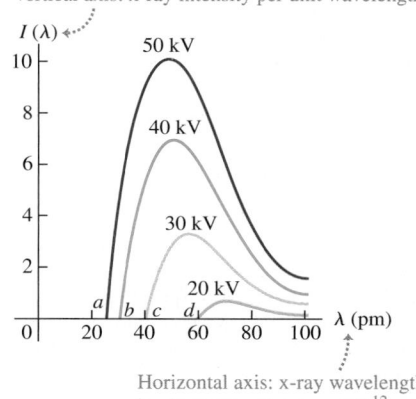

Horizontal axis: x-ray wavelength in picometers (1 pm = 10^{-12} m)

EXAMPLE 38.4 PRODUCING X RAYS

Electrons in an x-ray tube accelerate through a potential difference of 10.0 kV before striking a target. If an electron produces one photon on impact with the target, what is the minimum wavelength of the resulting x rays? Find the answer by expressing energies in both SI units and electron volts.

SOLUTION

IDENTIFY and SET UP: To produce an x-ray photon with minimum wavelength and hence maximum energy, all of the electron's kinetic energy must go into producing a single x-ray photon. We'll use Eq. (38.6) to determine the wavelength.

EXECUTE: From Eq. (38.6), using SI units we have

$$\lambda_{min} = \frac{hc}{eV_{AC}} = \frac{(6.626 \times 10^{-34}\ \text{J} \cdot \text{s})(3.00 \times 10^{8}\ \text{m/s})}{(1.602 \times 10^{-19}\ \text{C})(10.0 \times 10^{3}\ \text{V})}$$

$$= 1.24 \times 10^{-10}\ \text{m} = 0.124\ \text{nm}$$

Using electron volts, we have

$$\lambda_{min} = \frac{hc}{eV_{AC}} = \frac{(4.136 \times 10^{-15}\ \text{eV} \cdot \text{s})(3.00 \times 10^{8}\ \text{m/s})}{e(10.0 \times 10^{3}\ \text{V})}$$

$$= 1.24 \times 10^{-10}\ \text{m} = 0.124\ \text{nm}$$

In the second calculation, the "e" for the magnitude of the electron charge cancels the "e" in the unit "eV," because the electron volt (eV) is the magnitude of the electron charge e times one volt (1 V).

EVALUATE: To check our result, recall from Example 38.1 that a 1.91-eV photon has a wavelength of 650 nm. Here the electron energy, and therefore the x-ray photon energy, is 10.0×10^{3} eV = 10.0 keV, about 5000 times greater than in Example 38.1, and the wavelength is about $\frac{1}{5000}$ as great as in Example 38.1. This makes sense, since wavelength and photon energy are inversely proportional.

Applications of X Rays

X rays have many practical applications in medicine and industry. Because x-ray photons are of such high energy, they can penetrate several centimeters of solid matter. Hence they can be used to visualize the interiors of materials that are opaque to ordinary light, such as broken bones or defects in structural steel. The object to be visualized is placed between an x-ray source and an electronic detector (like that used in a digital camera). The darker an area in the image recorded by such a detector, the greater the radiation exposure. Bones are much more effective x-ray absorbers than soft tissue, so bones appear as light areas. A crack or air bubble allows greater transmission and shows as a dark area.

A widely used and vastly improved x-ray technique is *computed tomography*; the corresponding instrument is called a *CT scanner*. The x-ray source produces a thin, fan-shaped beam that is detected on the opposite side of the subject by an array of several hundred detectors in a line. Each detector measures absorption along a thin line through the subject. The entire apparatus is rotated around the subject in the plane of the beam, and the changing photon-counting rates of the detectors are recorded digitally. A computer processes this information and reconstructs a picture of absorption over an entire cross section of the subject (see **Fig. 38.9**). Differences in absorption as small as 1% or less can be detected with CT scans, and tumors and other anomalies that are much too small to be seen with older x-ray techniques can be detected.

X rays cause damage to living tissues. As x-ray photons are absorbed in tissues, their energy breaks molecular bonds and creates highly reactive free radicals (such as neutral H and OH), which in turn can disturb the molecular structure of proteins and especially genetic material. Young and rapidly growing cells are particularly susceptible, which is why x rays are useful for selective destruction of cancer cells. Conversely, however, a cell may be damaged by radiation but survive, continue dividing, and produce generations of defective cells; thus x rays can *cause* cancer.

Even when the organism itself shows no apparent damage, excessive exposure to x rays can cause changes in the organism's reproductive system that will affect its offspring. A careful assessment of the balance between risks and benefits of radiation exposure is essential in each individual case.

38.9 This radiologist is operating a CT scanner (seen through the window) from a separate room to avoid repeated exposure to x rays.

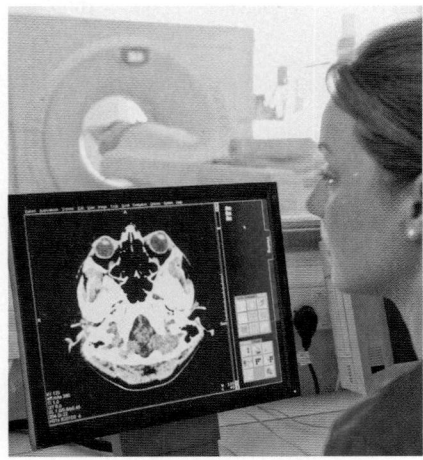

TEST YOUR UNDERSTANDING OF SECTION 38.2   In the apparatus shown in Fig. 38.7, suppose you increase the number of electrons that are emitted from the cathode per second while keeping the potential difference V_{AC} the same. How will this affect the intensity I and minimum wavelength λ_{min} of the emitted x rays? (i) I and λ_{min} will both increase; (ii) I will increase but λ_{min} will be unchanged; (iii) I will increase but λ_{min} will decrease; (iv) I will remain the same but λ_{min} will decrease; (v) none of these. ∎

38.3 LIGHT SCATTERED AS PHOTONS: COMPTON SCATTERING AND PAIR PRODUCTION

The final aspect of light that we must test against Einstein's photon model is its behavior after the light is produced and before it is eventually absorbed. We can do this by considering the *scattering* of light. As we discussed in Section 33.6, scattering is what happens when light bounces off particles such as molecules in the air.

Compton Scattering

Let's see what Maxwell's wave model and Einstein's photon model predict for how light behaves when it undergoes scattering by a single electron, such as an individual electron within an atom.

> ***Wave-Model Prediction:*** In the wave description, scattering would be a process of absorption and re-radiation. Part of the energy of the light wave would be absorbed by the electron, which would oscillate in response to the oscillating electric field of the wave. The oscillating electron would act like a miniature antenna (see Section 32.1), re-radiating its acquired energy as *scattered* waves in a variety of directions. The frequency at which the electron oscillates would be the same as the frequency of the incident light, and the re-radiated light would have the same frequency as the oscillations of the electron. So, *in the wave model, the scattered light and incident light have the same frequency and same wavelength.*

> ***Photon-Model Prediction:*** In the photon model we imagine the scattering process as a collision of two *particles*, the incident photon and an electron that is initially at rest (**Fig. 38.10a**). The incident photon would give up part of its energy and momentum to the electron, which recoils as a result of this impact. The scattered photon that remains can fly off at a variety of angles ϕ with respect to the incident direction, but it has less energy and less momentum than the incident photon (Fig. 38.10b). The energy and momentum of a photon are given by $E = hf = hc/\lambda$ (Eq. 38.2) and $p = hf/c = h/\lambda$ (Eq. 38.5). Therefore, *in the photon model, the scattered light has a lower frequency f and longer wavelength λ than the incident light.*

The definitive experiment that tested these predictions was carried out in 1922 by the American physicist Arthur H. Compton. He aimed a beam of x rays at a solid target and measured the wavelength of the radiation scattered from the target (**Fig. 38.11**). Compton discovered that some of the scattered radiation has

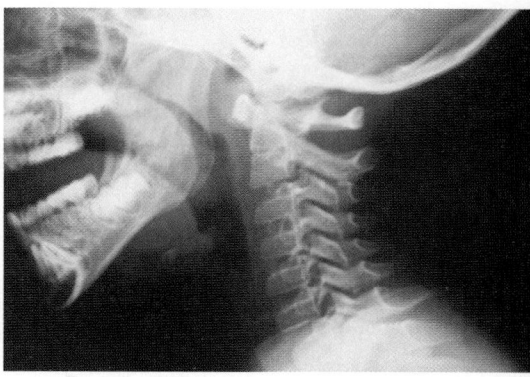

38.10 The photon model of light scattering by an electron.

(a) Before collision: The target electron is at rest.

Incident photon: wavelength λ, momentum $\vec{p}$ Target electron (at rest)

(b) After collision: The angle between the directions of the scattered photon and the incident photon is ϕ.

Scattered photon: wavelength λ', momentum $\vec{p}'$

ϕ

Recoiling electron: momentum $\vec{P}_e$

38.11 A Compton-effect experiment.

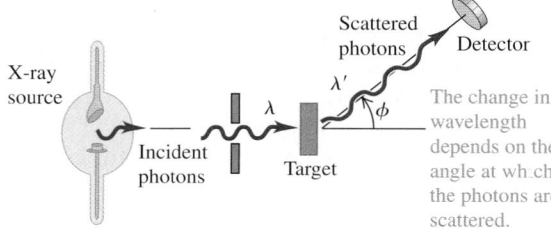

X-ray source

Incident photons

Target

λ

Scattered photons

λ'

ϕ

Detector

The change in wavelength depends on the angle at which the photons are scattered.

smaller frequency (longer wavelength) than the incident radiation and that the change in wavelength depends on the angle through which the radiation is scattered. This is precisely what the photon model predicts for light scattered from electrons in the target, a process that is now called **Compton scattering.**

Specifically, Compton found that if the scattered radiation emerges at an angle ϕ with respect to the incident direction, as shown in Fig. 38.11, then

Compton scattering:
$$\underset{\text{Wavelength of scattered radiation}}{\lambda'} - \underset{\text{Wavelength of incident radiation}}{\lambda} = \frac{\overset{\text{Planck's constant}}{h}}{\underset{\text{Electron rest mass}}{mc} \underset{\text{Speed of light in vacuum}}{}}(1 - \cos\phi) \overset{\text{Scattering angle}}{} \tag{38.7}$$

In other words, λ' is greater than λ. The quantity h/mc that appears in Eq. (38.7) has units of length. Its numerical value is

$$\frac{h}{mc} = \frac{6.626 \times 10^{-34}\,\text{J}\cdot\text{s}}{(9.109 \times 10^{-31}\,\text{kg})(2.998 \times 10^8\,\text{m/s})} = 2.426 \times 10^{-12}\,\text{m}$$

Compton showed that Einstein's photon theory, combined with the principles of conservation of energy and conservation of momentum, provides a beautifully clear explanation of his experimental results. We outline the derivation below. The electron recoil energy may be in the relativistic range, so we have to use the relativistic energy–momentum relationships, Eqs. (37.39) and (37.40). The incident photon has momentum $\vec{p}$, with magnitude p and energy pc. The scattered photon has momentum $\vec{p}\,'$, with magnitude p' and energy $p'c$. The electron is initially at rest, so its initial momentum is zero and its initial energy is its rest energy mc^2. The final electron momentum $\vec{P}_\text{e}$ has magnitude P_e, and the final electron energy is $E_\text{e}^2 = (mc^2)^2 + (P_\text{e}c)^2$. Then energy conservation gives us the relationship

$$pc + mc^2 = p'c + E_\text{e}$$

Rearranging, we find

$$(pc - p'c + mc^2)^2 = E_\text{e}^2 = (mc^2)^2 + (P_\text{e}c)^2 \tag{38.8}$$

We can eliminate the electron momentum $\vec{P}_\text{e}$ from Eq. (38.8) by using momentum conservation. From **Fig. 38.12** we see that $\vec{p} = \vec{p}\,' + \vec{P}_\text{e}$, or

$$\vec{P}_\text{e} = \vec{p} - \vec{p}\,' \tag{38.9}$$

By taking the scalar product of each side of Eq. (38.9) with itself, we find

$$P_\text{e}^2 = p^2 + p'^2 - 2pp'\cos\phi \tag{38.10}$$

We now substitute this expression for P_e^2 into Eq. (38.8) and multiply out the left side. We divide out a common factor c^2; several terms cancel, and when the resulting equation is divided through by (pp'), the result is

$$\frac{mc}{p'} - \frac{mc}{p} = 1 - \cos\phi \tag{38.11}$$

Finally, we substitute $p' = h/\lambda'$ and $p = h/\lambda$, then multiply by h/mc to obtain Eq. (38.7).

When the wavelengths of x rays scattered at a certain angle are measured, the curve of intensity per unit wavelength as a function of wavelength has two peaks (**Fig. 38.13**). The longer-wavelength peak represents Compton scattering. The shorter-wavelength peak, λ_0, is at the wavelength of the incident x rays and corresponds to x-ray scattering from tightly bound electrons. In such scattering processes the entire atom must recoil, so m in Eq. (38.7) is the mass of the entire atom rather than of a single electron. The resulting wavelength shifts are negligible.

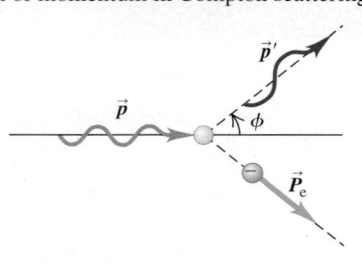
38.12 Vector diagram showing conservation of momentum in Compton scattering.

Conservation of momentum during Compton scattering

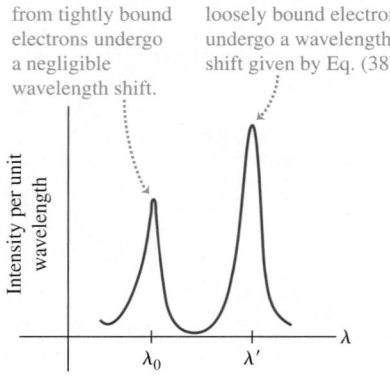
38.13 Intensity as a function of wavelength for photons scattered at an angle of 135° in a Compton-scattering experiment.

Photons scattered from tightly bound electrons undergo a negligible wavelength shift.

Photons scattered from loosely bound electrons undergo a wavelength shift given by Eq. (38.7).

EXAMPLE 38.5 | COMPTON SCATTERING

You use 0.124-nm x-ray photons in a Compton-scattering experiment. (a) At what angle is the wavelength of the scattered x rays 1.0% longer than that of the incident x rays? (b) At what angle is it 0.050% longer?

SOLUTION

IDENTIFY and SET UP: We'll use the relationship between scattering angle and wavelength shift in the Compton effect. In each case our target variable is the angle ϕ (see Fig. 38.10b). We solve for ϕ by using Eq. (38.7).

EXECUTE: (a) In Eq. (38.7) we want $\Delta\lambda = \lambda' - \lambda$ to be 1.0% of 0.124 nm, so $\Delta\lambda = 0.00124$ nm $= 1.24 \times 10^{-12}$ m. Using the value $h/mc = 2.426 \times 10^{-12}$ m, we find

$$\Delta\lambda = \frac{h}{mc}(1 - \cos\phi)$$

$$\cos\phi = 1 - \frac{\Delta\lambda}{h/mc} = 1 - \frac{1.24 \times 10^{-12} \text{ m}}{2.426 \times 10^{-12} \text{ m}} = 0.4889$$

$$\phi = 60.7°$$

(b) For $\Delta\lambda$ to be 0.050% of 0.124 nm, or 6.2×10^{-14} m,

$$\cos\phi = 1 - \frac{6.2 \times 10^{-14} \text{ m}}{2.426 \times 10^{-12} \text{ m}} = 0.9744$$

$$\phi = 13.0°$$

EVALUATE: Our results show that smaller scattering angles give smaller wavelength shifts. Thus in a grazing collision the photon energy loss and the electron recoil energy are smaller than when the scattering angle is larger. This is just what we would expect for an elastic collision, whether between a photon and an electron or between two billiard balls.

Pair Production

Another effect that can be explained only with the photon picture involves *gamma rays,* the shortest-wavelength and highest-frequency variety of electromagnetic radiation. If a gamma-ray photon of sufficiently short wavelength is fired at a target, it may not scatter. Instead, as depicted in **Fig. 38.14**, it may disappear completely and be replaced by two new particles: an electron and a **positron** (a particle that has the same rest mass m as an electron but has a positive charge $+e$ rather than the negative charge $-e$ of the electron). This process, called **pair production,** was first observed by the physicists Patrick Blackett and Giuseppe Occhialini in 1933. The electron and positron have to be produced in pairs in order to conserve electric charge: The incident photon has zero charge, and the electron–positron pair has net charge $(-e) + (+e) = 0$. Enough energy must be available to account for the rest energy $2mc^2$ of the two particles. To four significant figures, this minimum energy is

$$E_{min} = 2mc^2 = 2(9.109 \times 10^{-31} \text{ kg})(2.998 \times 10^8 \text{ m/s})^2$$

$$= 1.637 \times 10^{-13} \text{ J} = 1.022 \text{ MeV}$$

Thus the photon must have at least this much energy to produce an electron–positron pair. From Eq. (38.2), $E = hc/\lambda$, the photon wavelength has to be shorter than

$$\lambda_{max} = \frac{hc}{E_{min}} = \frac{(6.626 \times 10^{-34} \text{ J} \cdot \text{s})(2.998 \times 10^8 \text{ m/s})}{1.637 \times 10^{-13} \text{ J}}$$

$$= 1.213 \times 10^{-12} \text{ m} = 1.213 \times 10^{-3} \text{ nm} = 1.213 \text{ pm}$$

This is a very short wavelength, about $\frac{1}{1000}$ as large as the x-ray wavelengths that Compton used in his scattering experiments. (The requisite minimum photon energy is actually a bit higher than 1.022 MeV, so the photon wavelength must be a bit shorter than 1.213 pm. The reason is that when the incident photon encounters an atomic nucleus in the target, some of the photon energy goes into the kinetic energy of the recoiling nucleus.) Just as for the photoelectric effect, the wave model of electromagnetic radiation cannot explain why pair production occurs only when very short wavelengths are used.

38.14 (a) Photograph of bubble-chamber tracks of electron–positron pairs that are produced when 300-MeV photons strike a lead sheet. A magnetic field directed out of the photograph made the electrons (e^-) and positrons (e^+) curve in opposite directions. (b) Diagram showing the pair-production process for two of the gamma-ray photons (γ).

(a)

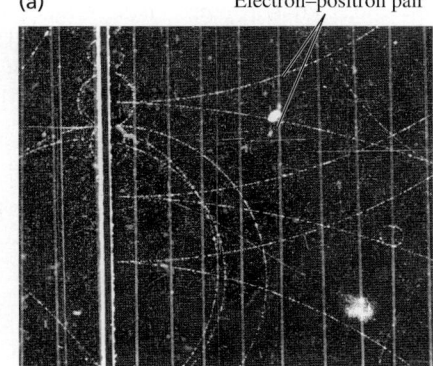

(b)

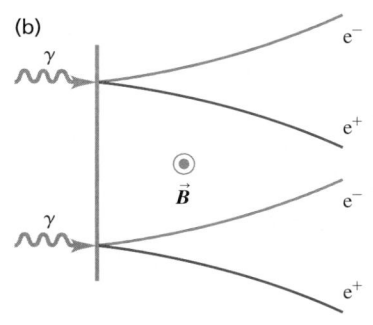

The inverse process, *electron–positron pair annihilation,* occurs when a positron and an electron collide. Both particles disappear, and two (or occasionally three) photons can appear, with total energy of at least $2m_ec^2 = 1.022$ MeV. Decay into a *single* photon is impossible because such a process could not conserve both energy and momentum. It's easiest to analyze this annihilation process in the frame of reference called the *center-of-momentum system,* in which the total momentum is zero. It is the relativistic generalization of the center-of-mass system that we discussed in Section 8.5.

EXAMPLE 38.6 PAIR ANNIHILATION

An electron and a positron, initially far apart, move toward each other with the same speed. They collide head-on, annihilating each other and producing two photons. Find the energies, wavelengths, and frequencies of the photons if the initial kinetic energies of the electron and positron are (a) both negligible and (b) both 5.000 MeV. The electron rest energy is 0.511 MeV.

SOLUTION

IDENTIFY and SET UP: Just as in the elastic collisions we studied in Chapter 8, both momentum and energy are conserved in pair annihilation. The electron and positron are initially far apart, so the initial electric potential energy is zero and the initial energy is the sum of the particle kinetic and rest energies. The final energy is the sum of the photon energies. The total initial momentum is zero; the total momentum of the two photons must likewise be zero. We find the photon energy E by using conservation of energy, conservation of momentum, and the relationship $E = pc$ (see Section 38.1). We then calculate the wavelengths and frequencies from $E = hc/\lambda = hf$.

EXECUTE: If the total momentum of the two photons is to be zero, their momenta must have equal magnitudes p and opposite directions. From $E = pc = hc/\lambda = hf$, the two photons must also have the same energy E, wavelength λ, and frequency f.

Before the collision the energy of each electron is $K + mc^2$, where K is its kinetic energy and $mc^2 = 0.511$ MeV. Conservation of energy then gives

$$(K + mc^2) + (K + mc^2) = E + E$$

Hence the energy of each photon is $E = K + mc^2$.

(a) In this case the electron kinetic energy K is negligible compared to its rest energy mc^2, so each photon has energy $E = mc^2 = 0.511$ MeV. The corresponding photon wavelength and frequency are

$$\lambda = \frac{hc}{E} = \frac{(4.136 \times 10^{-15} \text{ eV} \cdot \text{s})(3.00 \times 10^8 \text{ m/s})}{0.511 \times 10^6 \text{ eV}}$$

$$= 2.43 \times 10^{-12} \text{ m} = 2.43 \text{ pm}$$

$$f = \frac{E}{h} = \frac{0.511 \times 10^6 \text{ eV}}{4.136 \times 10^{-15} \text{ eV} \cdot \text{s}} = 1.24 \times 10^{20} \text{ Hz}$$

(b) In this case $K = 5.000$ MeV, so each photon has energy $E = 5.000$ MeV $+ 0.511$ MeV $= 5.511$ MeV. Proceeding as in part (a), you can show that the photon wavelength is 0.2250 pm and the frequency is 1.333×10^{21} Hz.

EVALUATE: As a check, recall from Example 38.1 that a 650-nm visible-light photon has energy 1.91 eV and frequency 4.62×10^{14} Hz. The photon energy in part (a) is about 2.5×10^5 times greater. As expected, the photon's wavelength is shorter and its frequency higher than those for a visible-light photon by the same factor. You can check the results for part (b) in the same way.

TEST YOUR UNDERSTANDING OF SECTION 38.3 If you used visible-light photons in the experiment shown in Fig. 38.11, would the photons undergo a wavelength shift due to the scattering? If so, is it possible to detect the shift with the human eye? ▮

38.4 WAVE–PARTICLE DUALITY, PROBABILITY, AND UNCERTAINTY

We have studied many examples of the behavior of light and other electromagnetic radiation. Some, including the interference and diffraction effects described in Chapters 35 and 36, demonstrate conclusively the *wave* nature of light. Others, the subject of the present chapter, point with equal force to the *particle* nature of light. This *wave–particle duality* means that light has two aspects that seem to be in direct conflict. How can light be a wave and a particle at the same time?

We can find the answer to this apparent wave–particle conflict in the **principle of complementarity,** first stated by the Danish physicist Niels Bohr in 1928. The wave descriptions and the particle descriptions are complementary. That is, we need both to complete our model of nature, but we will never need to use both at the same time to describe a single part of an occurrence.

Diffraction and Interference in the Photon Picture

Let's start by considering again the diffraction pattern for a single slit, which we analyzed in Sections 36.2 and 36.3. Instead of recording the pattern on a digital camera chip or photographic film, we use a detector called a *photomultiplier* that can actually detect individual photons. Using the setup shown in **Fig. 38.15**, we place the photomultiplier at various positions for equal time intervals, count the photons at each position, and plot out the intensity distribution.

We find that, on average, the distribution of photons agrees with our predictions from Section 36.3. At points corresponding to the maxima of the pattern, we count many photons; at minimum points, we count almost none; and so on. The graph of the counts at various points gives the same diffraction pattern that we predicted with Eq. (36.7).

But suppose we now reduce the intensity to such a low level that only a few photons per second pass through the slit. We now record a series of discrete strikes, each representing a single photon. While we *cannot predict* where any given photon will strike, over time the accumulating strikes build up the familiar diffraction pattern we expect for a wave. To reconcile the wave and particle aspects of this pattern, we have to regard the pattern as a *statistical* distribution that tells us how many photons, on average, go to each spot. Equivalently, the pattern tells us the *probability* that any individual photon will land at a given spot. If we shine our faint light beam on a two-slit apparatus, we get an analogous result (**Fig. 38.16**). Again we can't predict exactly where an individual photon will go; the interference pattern is a statistical distribution.

How does the principle of complementarity apply to these diffraction and interference experiments? The wave description, not the particle description, explains the single- and double-slit patterns. But the particle description, not the wave description, explains why the photomultiplier records discrete packages of energy. The two descriptions complete our understanding of the results. For instance, suppose we consider an individual photon and ask how it knows "which way to go" when passing through the slit. This question seems like a conundrum, but that is because it is framed in terms of a *particle* description—whereas it is the *wave* nature of light that determines the distribution of photons. Conversely, the fact that the photomultiplier detects faint light as a sequence of individual "spots" can't be explained in wave terms.

Probability and Uncertainty

Although photons have energy and momentum, they are nonetheless very different from the particle model we used for Newtonian mechanics in Chapters 4 through 8. The Newtonian particle model treats an object as a point mass. We can describe the location and state of motion of such a particle at any instant with three spatial coordinates and three components of momentum, and we can then predict the particle's future motion. This model doesn't work at all for photons, however: We *cannot* treat a photon as a point object. This is because there are fundamental limitations on the precision with which we can simultaneously determine the position and momentum of a photon. Many aspects of a photon's behavior can be stated only in terms of *probabilities*. (In Chapter 39 we will find that the non-Newtonian ideas we develop for photons in this section also apply to particles such as electrons.)

To get more insight into the problem of measuring a photon's position and momentum simultaneously, let's look again at the single-slit diffraction of light.

38.15 Single-slit diffraction pattern of light observed with a movable photomultiplier. The curve shows the intensity distribution predicted by the wave picture. The photon distribution is shown by the numbers of photons counted at various positions.

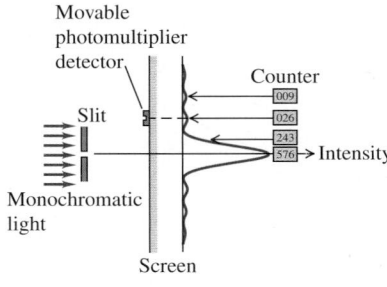

38.16 These images record the positions where individual photons in a two-slit interference experiment strike the screen. As more photons reach the screen, a recognizable interference pattern appears.

After 21 photons reach the screen

After 1000 photons reach the screen

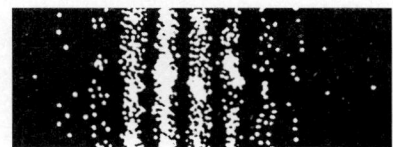

After 10,000 photons reach the screen

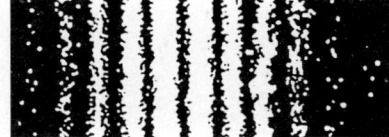

PhET: Fourier: Making Waves
PhET: Quantum Wave Interference

38.17 Interpreting single-slit diffraction in terms of photon momentum.

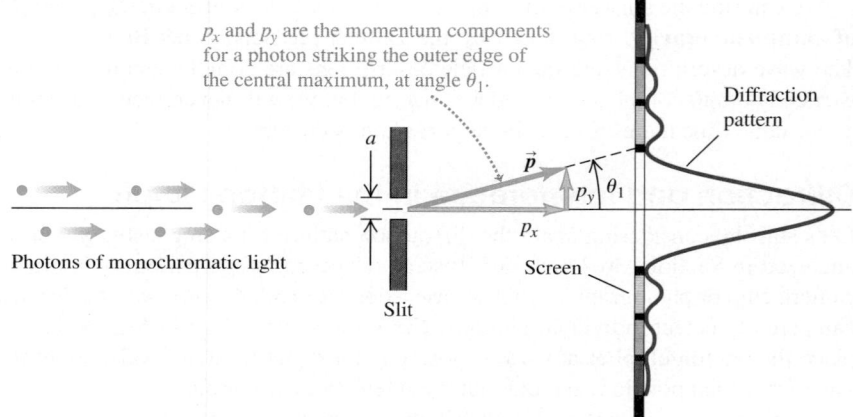

Suppose the wavelength λ is much less than the slit width a (**Fig. 38.17**). Then most (85%) of the photons go into the central maximum of the diffraction pattern, and the remainder go into other parts of the pattern. We use θ_1 to denote the angle between the central maximum and the first minimum. Using Eq. (36.2) with $m = 1$, we find that θ_1 is given by $\sin\theta_1 = \lambda/a$. Since we assume $\lambda \ll a$, it follows that θ_1 is very small, $\sin\theta_1$ is very nearly equal to θ_1 (in radians), and

$$\theta_1 = \frac{\lambda}{a} \tag{38.12}$$

Even though the photons all have the same initial state of motion, they don't all follow the same path. We can't predict the exact trajectory of any individual photon from knowledge of its initial state; we can only describe the *probability* that an individual photon will strike a given spot on the screen. This fundamental indeterminacy has no counterpart in Newtonian mechanics.

Furthermore, there are fundamental *uncertainties* in both the position and the momentum of an individual particle, and these uncertainties are related inseparably. To clarify this point, let's go back to Fig. 38.17. A photon that strikes the screen at the outer edge of the central maximum, at angle θ_1, must have a component of momentum p_y in the y-direction, as well as a component p_x in the x-direction, despite the fact that initially the beam was directed along the x-axis. From the geometry of the situation the two components are related by $p_y/p_x = \tan\theta_1$. Since θ_1 is small, we may use the approximation $\tan\theta_1 = \theta_1$, and

$$p_y = p_x\theta_1 \tag{38.13}$$

Substituting Eq. (38.12), $\theta_1 = \lambda/a$, into Eq. (38.13) gives

$$p_y = p_x\frac{\lambda}{a} \tag{38.14}$$

Equation (38.14) says that for the 85% of the photons that strike the detector within the central maximum (that is, at angles between $-\lambda/a$ and $+\lambda/a$), the y-component of momentum is spread out over a range from $-p_x\lambda/a$ to $+p_x\lambda/a$. Now let's consider *all* the photons that pass through the slit and strike the screen. Again, they may hit above or below the center of the pattern, so their component p_y may be positive or negative. However the symmetry of the diffraction pattern shows us the average value $(p_y)_{av} = 0$. There will be an *uncertainty* Δp_y in the y-component of momentum at least as great as $p_x\lambda/a$. That is,

$$\Delta p_y \geq p_x\frac{\lambda}{a} \tag{38.15}$$

The narrower the slit width a, the broader is the diffraction pattern and the greater is the uncertainty in the y-component of momentum p_y.

The photon wavelength λ is related to the momentum p_x by Eq. (38.5), which we can rewrite as $\lambda = h/p_x$. Using this relationship in Eq. (38.15) and simplifying, we find

$$\Delta p_y \geq p_x \frac{h}{p_x a} = \frac{h}{a}$$

$$\Delta p_y a \geq h \qquad (38.16)$$

What does Eq. (38.16) mean? The slit width a represents an uncertainty in the y-component of the *position* of a photon as it passes through the slit. We don't know exactly *where* in the slit each photon passes through. So both the y-position and the y-component of momentum have uncertainties, and the two uncertainties are related by Eq. (38.16). We can reduce the *momentum* uncertainty Δp_y only by reducing the width of the diffraction pattern. To do this, we have to increase the slit width a, which increases the *position* uncertainty. Conversely, when we *decrease* the position uncertainty by narrowing the slit, the diffraction pattern broadens and the corresponding momentum uncertainty *increases*.

You may protest that it doesn't seem to be consistent with common sense for a photon not to have a definite position and momentum. But what we call *common sense* is based on familiarity gained through experience. Our usual experience includes very little contact with the microscopic behavior of particles like photons. Sometimes we have to accept conclusions that violate our intuition when we are dealing with areas that are far removed from everyday experience.

The Uncertainty Principle

In more general discussions of uncertainty relationships, the uncertainty of a quantity is usually described in terms of the statistical concept of *standard deviation,* which is a measure of the spread or dispersion of a set of numbers around their average value. Suppose we now begin to describe uncertainties in this way [neither Δp_y nor a in Eq. (38.16) is a standard deviation]. If a coordinate x has an uncertainty Δx and if the corresponding momentum component p_x has an uncertainty Δp_x, then we find that in general

| Heisenberg uncertainty principle for position and momentum: | Uncertainty in coordinate x ⋯⋯ Planck's constant divided by 2π ⋯⋯ $\Delta x \Delta p_x \geq \hbar/2$ Uncertainty in corresponding momentum component p_x | (38.17) |

The quantity $\hbar$ (pronounced "h-bar") is Planck's constant divided by 2π:

$$\hbar = \frac{h}{2\pi} = 1.054571628(53) \times 10^{-34} \text{ J} \cdot \text{s}$$

We will use $\hbar$ frequently to avoid writing a lot of factors of 2π in later equations.

Equation (38.17) is one form of the **Heisenberg uncertainty principle,** first discovered by the German physicist Werner Heisenberg (1901–1976). It states that, in general, it is impossible to simultaneously determine both the position and the momentum of a particle with arbitrarily great precision, as classical physics would predict. Instead, the uncertainties in the two quantities play complementary roles, as we have described. **Figure 38.18** shows the relationship between the two uncertainties. Our derivation of Eq. (38.16), a less refined form of the uncertainty principle given by Eq. (38.17), shows that this principle has its roots in the wave aspect of photons. We will see in Chapter 39 that electrons and other subatomic particles also have a wave aspect, and the same uncertainty principle applies to them as well.

It is tempting to suppose that we could get greater precision by using more sophisticated detectors of position and momentum. This turns out not to be possible. To detect a particle, the detector must *interact* with it, and this interaction

CAUTION h versus h-bar It's common for students to plug in the value of h when what they wanted was $\hbar = h/2\pi$, or vice versa. Don't make the same mistake, or your answer will be off by a factor of 2π! ▌

38.18 The Heisenberg uncertainty principle for position and momentum components. It is impossible for the product $\Delta x \Delta p_x$ to be less than $\hbar/2 = h/4\pi$.

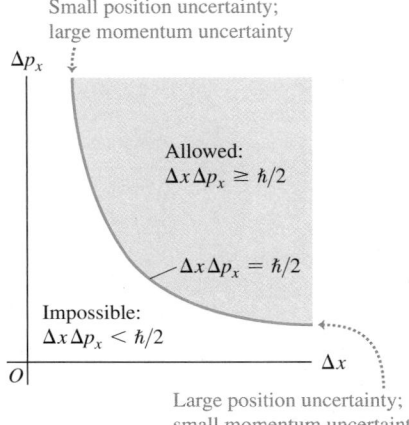

Small position uncertainty; large momentum uncertainty

Δp_x

Allowed: $\Delta x \Delta p_x \geq \hbar/2$

$\Delta x \Delta p_x = \hbar/2$

Impossible: $\Delta x \Delta p_x < \hbar/2$

O $\qquad \Delta x$

Large position uncertainty; small momentum uncertainty

Application Butterfly Hunting with Heisenberg Because $\hbar$ has such a small value, the Heisenberg uncertainty principle comes into play only for objects on the scale of atoms or smaller. To visualize what this principle means, imagine that we could make the value of $\hbar$ larger by a factor of 10^{34} so that $\hbar = 1.05$ J·s. If you trap a butterfly in a butterfly net, you know the butterfly's position to within the 0.25-m diameter of the net. Then the uncertainty in the butterfly's position is approximately $\Delta x = 0.25$ m. The minimum uncertainty in its momentum is then $\Delta p_x = (\hbar/2\,\Delta x) = (1.05$ J·s$)/2(0.25$ m$) = 2.1$ kg·m/s, so just by trapping the butterfly you could impart this much momentum to it. A typical butterfly has a mass of about 3×10^{-4} kg. With this much momentum, the butterfly's speed would be about 7000 m/s (about 20 times the speed of sound!) and its kinetic energy about 7000 J (that of a baseball traveling at about 300 m/s, just under the speed of sound). By confining the butterfly in the net, you could give it so much momentum and kinetic energy that it could burst out of the net!

unavoidably changes the state of motion of the particle, introducing uncertainty about its original state. For example, we could imagine placing an electron at a certain point in the middle of the slit in Fig. 38.17. If the photon passes through the middle, we would see the electron recoil. We would then know that the photon passed through that point in the slit, and we would be much more certain about the x-coordinate of the photon. However, the collision between the photon and the electron would change the photon momentum, giving us greater uncertainty in the value of that momentum. A more detailed analysis of such hypothetical experiments shows that the uncertainties we have described are fundamental and intrinsic. They *cannot* be circumvented *even in principle* by any experimental technique, no matter how sophisticated.

There is nothing special about the x-axis. In a three-dimensional situation with coordinates (x, y, z) there is an uncertainty relationship for each coordinate and its corresponding momentum component: $\Delta x\Delta p_x \geq \hbar/2$, $\Delta y\Delta p_y \geq \hbar/2$, and $\Delta z\Delta p_z \geq \hbar/2$. However, the uncertainty in one coordinate is *not* related to the uncertainty in a different component of momentum. For example, Δx is not related directly to Δp_y.

Waves and Uncertainty

Here's an alternative way to understand the Heisenberg uncertainty principle in terms of the properties of waves. Consider a sinusoidal electromagnetic wave propagating in the positive x-direction with its electric field polarized in the y-direction. If the wave has wavelength λ, frequency f, and amplitude A, we can write the wave function as

$$E_y(x, t) = A\sin(kx - \omega t) \qquad (38.18)$$

In this expression the wave number is $k = 2\pi/\lambda$ and the angular frequency is $\omega = 2\pi f$. We can think of the wave function in Eq. (38.18) as a description of a photon with a definite wavelength and a definite frequency. In terms of k and ω we can express the momentum and energy of the photon as

$$p_x = \frac{h}{\lambda} = \frac{h}{2\pi}\frac{2\pi}{\lambda} = \hbar k \qquad \begin{array}{l}\text{(photon momentum in}\\ \text{terms of wave number)}\end{array} \qquad (38.19a)$$

$$E = hf = \frac{h}{2\pi}2\pi f = \hbar\omega \qquad \begin{array}{l}\text{(photon energy in terms}\\ \text{of angular frequency)}\end{array} \qquad (38.19b)$$

Using Eqs. (38.19) in Eq. (38.18), we can rewrite our photon wave equation as

$$E_y(x, t) = A\sin\left[(p_x x - Et)/\hbar\right] \qquad \begin{array}{l}\text{(wave function for a}\\ \text{photon with } x\text{-momentum}\\ p_x \text{ and energy } E)\end{array} \qquad (38.20)$$

Since this wave function has a definite value of x-momentum p_x, there is *no* uncertainty in the value of this quantity: $\Delta p_x = 0$. The Heisenberg uncertainty principle, Eq. (38.17), says that $\Delta x\Delta p_x \geq \hbar/2$. If Δp_x is zero, then Δx must be infinite. Indeed, the wave described by Eq. (38.20) extends along the entire x-axis and has the same amplitude everywhere. The price we pay for knowing the photon's momentum precisely is that we have no idea *where* the photon is!

In practical situations we always have *some* idea where a photon is. To describe this situation, we need a wave function that is more localized in space. We can create one by superimposing two or more sinusoidal functions. To keep things simple, we'll consider only waves propagating in the positive x-direction. For example, let's add together two sinusoidal wave functions like those in Eqs. (38.18) and (38.20), but with slightly different wavelengths and frequencies and hence slightly different values p_{x1} and p_{x2} of x-momentum and slightly

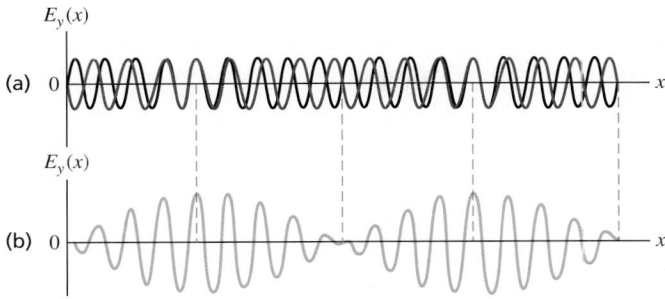

$E_y(x)$

(a) 0 ———————————— x

$E_y(x)$

(b) 0 ———————————— x

38.19 (a) Two sinusoidal waves with slightly different wave numbers k and hence slightly different values of momentum $p_x = \hbar k$ shown at one instant of time. (b) The superposition of these waves has a momentum equal to the average of the two individual values of momentum. The amplitude varies, giving the total wave a lumpy character not possessed by either individual wave.

different values E_1 and E_2 of energy. The total wave function is

$$E_y(x, t) = A_1 \sin[(p_{1x}x - E_1 t)/\hbar] + A_2 \sin[(p_{2x}x - E_2 t)/\hbar] \quad (38.21)$$

Consider what this wave function looks like at a particular instant of time, say, $t = 0$. At this instant Eq. (38.21) becomes

$$E_y(x, t = 0) = A_1 \sin(p_{1x}x/\hbar) + A_2 \sin(p_{2x}x/\hbar) \quad (38.22)$$

Figure 38.19a is a graph of the individual wave functions at $t = 0$ for the case $A_2 = -A_1$, and Fig. 38.19b graphs the combined wave function $E_y(x, t = 0)$ given by Eq. (38.22). We saw something very similar to Fig. 38.19b in our discussion of beats in Section 16.7: When we superimposed two sinusoidal waves with slightly different frequencies (see Fig. 16.25), the resulting wave exhibited amplitude variations not present in the original waves. In the same way, a photon represented by the wave function in Eq. (38.21) is most likely to be found in the regions where the wave function's amplitude is greatest. That is, the photon is *localized*. However, the photon's momentum no longer has a definite value because we began with two different x-momentum values, p_{x1} and p_{x2}. This agrees with the Heisenberg uncertainty principle: By decreasing the uncertainty in the photon's position, we have increased the uncertainty in its momentum.

Uncertainty in Energy

Our discussion of combining waves also shows that there is an uncertainty principle that involves *energy* and *time*. To see why this is so, imagine measuring the combined wave function described by Eq. (38.21) at a certain position, say $x = 0$, over a period of time. At $x = 0$, the wave function from Eq. (38.21) becomes

$$E_y(x, t) = A_1 \sin(-E_1 t/\hbar) + A_2 \sin(-E_2 t/\hbar)$$
$$= -A_1 \sin(E_1 t/\hbar) - A_2 \sin(E_2 t/\hbar) \quad (38.23)$$

What we measure at $x = 0$ is a combination of two oscillating electric fields with slightly different angular frequencies $\omega_1 = E_1/\hbar$ and $\omega_2 = E_2/\hbar$. This is exactly the phenomenon of beats that we discussed in Section 16.7 (compare Fig. 16.25). The amplitude of the combined field rises and falls, so the photon described by this field is localized in *time* as well as in position. The photon is most likely to be found at the times when the amplitude is large. The price we pay for localizing the photon in time is that the wave does not have a definite energy. By contrast, if the photon is described by a sinusoidal wave like that in Eq. (38.20) that *does* have a definite energy E but that has the same amplitude at all times, we have no idea when the photon will appear at $x = 0$. So the better we know the photon's energy, the less certain we are of when we will observe the photon.

Just as for the momentum–position uncertainty principle, we can write a mathematical expression for the uncertainty principle that relates energy and time. In fact, except for an overall minus sign, Eq. (38.23) is identical to Eq. (38.22) if we replace the x-momentum p_x by energy E and the position x by time t.

This tells us that in the momentum–position uncertainty relation, Eq. (38.17), we can replace the momentum uncertainty Δp_x with the energy uncertainty ΔE and replace the position uncertainty Δx with the time uncertainty Δt. The result is

Heisenberg uncertainty principle for energy and time:

Time uncertainty of a phenomenon ⋯ Planck's constant divided by 2π

$$\Delta t\,\Delta E \geq \hbar/2 \qquad (38.24)$$

Energy uncertainty of same phenomenon

In practice, any real photon has a limited spatial extent and hence passes any point in a limited amount of time. The following example illustrates how this affects the momentum and energy of the photon.

EXAMPLE 38.7 ULTRASHORT LASER PULSES AND THE UNCERTAINTY PRINCIPLE

Many varieties of lasers emit light in the form of pulses rather than a steady beam. A tellurium–sapphire laser can produce light at a wavelength of 800 nm in ultrashort pulses that last only 4.00×10^{-15} s (4.00 femtoseconds, or 4.00 fs). The energy in a single pulse produced by one such laser is $2.00 \ \mu\text{J} = 2.00 \times 10^{-6}$ J, and the pulses propagate in the positive x-direction. Find (a) the frequency of the light; (b) the energy and minimum energy uncertainty of a single photon in the pulse; (c) the minimum frequency uncertainty of the light in the pulse; (d) the spatial length of the pulse, in meters and as a multiple of the wavelength; (e) the momentum and minimum momentum uncertainty of a single photon in the pulse; and (f) the approximate number of photons in the pulse.

SOLUTION

IDENTIFY and SET UP: It's important to distinguish between the light pulse as a whole (which contains a very large number of photons) and an individual photon within the pulse. The 4.00-fs pulse duration represents the time it takes the pulse to emerge from the laser; it is also the time *uncertainty* for an individual photon within the pulse, since we don't know when during the pulse that photon emerges. Similarly, the position uncertainty of a photon is the spatial length of the pulse, since a given photon could be found anywhere within the pulse. To find our target variables, we'll use the relationships for photon energy and momentum from Section 38.1 and the two Heisenberg uncertainty principles, Eqs. (38.17) and (38.24).

EXECUTE: (a) From the relationship $c = \lambda f$, the frequency of 800-nm light is

$$f = \frac{c}{\lambda} = \frac{3.00 \times 10^8 \ \text{m/s}}{8.00 \times 10^{-7} \ \text{m}} = 3.75 \times 10^{14} \ \text{Hz}$$

(b) From Eq. (38.2) the energy of a single 800-nm photon is

$$E = hf = (6.626 \times 10^{-34} \ \text{J} \cdot \text{s})(3.75 \times 10^{14} \ \text{Hz})$$
$$= 2.48 \times 10^{-19} \ \text{J}$$

The time uncertainty equals the pulse duration, $\Delta t = 4.00 \times 10^{-15}$ s. From Eq. (38.24) the minimum uncertainty in energy corresponds to the case $\Delta t\,\Delta E = \hbar/2$, so

$$\Delta E = \frac{\hbar}{2\Delta t} = \frac{1.055 \times 10^{-34} \ \text{J} \cdot \text{s}}{2(4.00 \times 10^{-15} \ \text{s})} = 1.32 \times 10^{-20} \ \text{J}$$

This is 5.3% of the photon energy $E = 2.48 \times 10^{-19}$ J, so the energy of a given photon is uncertain by at least 5.3%. The uncertainty could be greater, depending on the shape of the pulse.

(c) From the relationship $f = E/h$, the minimum frequency uncertainty is

$$\Delta f = \frac{\Delta E}{h} = \frac{1.32 \times 10^{-20} \ \text{J}}{6.626 \times 10^{-34} \ \text{J} \cdot \text{s}} = 1.99 \times 10^{13} \ \text{Hz}$$

This is 5.3% of the frequency $f = 3.75 \times 10^{14}$ Hz we found in part (a). Hence these ultrashort pulses do not have a definite frequency; the average frequency of many such pulses will be 3.75×10^{14} Hz, but the frequency of any individual pulse can be anywhere from 5.3% higher to 5.3% lower.

(d) The spatial length Δx of the pulse is the distance that the front of the pulse travels during the time $\Delta t = 4.00 \times 10^{-15}$ s it takes the pulse to emerge from the laser:

$$\Delta x = c\Delta t = (3.00 \times 10^8 \ \text{m/s})(4.00 \times 10^{-15} \ \text{s})$$
$$= 1.20 \times 10^{-6} \ \text{m}$$

$$\Delta x = \frac{1.20 \times 10^{-6} \ \text{m}}{8.00 \times 10^{-7} \ \text{m/wavelength}} = 1.50 \ \text{wavelengths}$$

This justifies the term *ultrashort*. The pulse is less than two wavelengths long!

(e) From Eq. (38.5), the momentum of an average photon in the pulse is

$$p_x = \frac{E}{c} = \frac{2.48 \times 10^{-19} \ \text{J}}{3.00 \times 10^8 \ \text{m/s}} = 8.28 \times 10^{-28} \ \text{kg} \cdot \text{m/s}$$

The spatial uncertainty is $\Delta x = 1.20 \times 10^{-6}$ m. From Eq. (38.17) minimum momentum uncertainty corresponds to $\Delta x\,\Delta p_x = \hbar/2$, so

$$\Delta p_x = \frac{\hbar}{2\Delta x} = \frac{1.055 \times 10^{-34} \ \text{J} \cdot \text{s}}{2(1.20 \times 10^{-6} \ \text{m})} = 4.40 \times 10^{-29} \ \text{kg} \cdot \text{m/s}$$

This is 5.3% of the average photon momentum p_x. An individual photon within the pulse can have a momentum that is 5.3% greater or less than the average.

(f) To estimate the number of photons in the pulse, we divide the total pulse energy by the average photon energy:

$$\frac{2.00 \times 10^{-6} \ \text{J/pulse}}{2.48 \times 10^{-19} \ \text{J/photon}} = 8.06 \times 10^{12} \ \text{photons/pulse}$$

The energy of an individual photon is uncertain, so this is the *average* number of photons per pulse.

EVALUATE: The percentage uncertainties in energy and momentum are large because this laser pulse is so short. If the pulse were longer, both Δt and Δx would be greater and the corresponding uncertainties in photon energy and photon momentum would be smaller.

Our calculation in part (f) shows an important distinction between photons and other kinds of particles. In principle it is possible to make an exact count of the number of electrons, protons, and neutrons in an object such as this book. If you repeated the count, you would get the same answer as the first time. By contrast, if you counted the number of photons in a laser pulse you would *not* necessarily get the same answer every time! The uncertainty in photon energy means that on each count there could be a different number of photons whose individual energies sum to 2.00×10^{-6} J. That's yet another of the many strange properties of photons.

TEST YOUR UNDERSTANDING OF SECTION 38.4 Through which of the following angles is a photon of wavelength λ most likely to be deflected after passing through a slit of width a? Assume that λ is much less than a. (i) $\theta = \lambda/a$; (ii) $\theta = 3\lambda/2a$; (iii) $\theta = 2\lambda/a$; (iv) $\theta = 3\lambda/a$; (v) not enough information given to decide. ∎

CHAPTER **38 SUMMARY**

SOLUTIONS TO ALL EXAMPLES

Photons: Electromagnetic radiation behaves as both waves and particles. The energy in an electromagnetic wave is carried in units called photons. The energy E of one photon is proportional to the wave frequency f and inversely proportional to the wavelength λ, and is proportional to a universal quantity h called Planck's constant. The momentum of a photon has magnitude E/c. (See Example 38.1.)

$$E = hf = \frac{hc}{\lambda} \qquad (38.2)$$

$$p = \frac{E}{c} = \frac{hf}{c} = \frac{h}{\lambda} \qquad (38.5)$$

The photoelectric effect: In the photoelectric effect, a surface can eject an electron by absorbing a photon whose energy hf is greater than or equal to the work function ϕ of the material. The stopping potential V_0 is the voltage required to stop a current of ejected electrons from reaching an anode. (See Examples 38.2 and 38.3.)

$$eV_0 = hf - \phi \qquad (38.4)$$

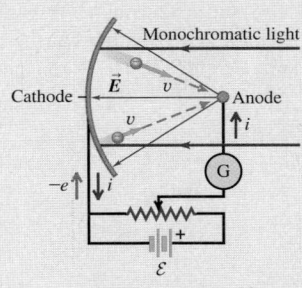

Photon production, photon scattering, and pair production: X rays can be produced when electrons accelerated to high kinetic energy across a potential increase V_{AC} strike a target. The photon model explains why the maximum frequency and minimum wavelength produced are given by Eq. (38.6). (See Example 38.4.) In Compton scattering a photon transfers some of its energy and momentum to an electron with which it collides. For free electrons (mass m), the wavelengths of incident and scattered photons are related to the photon scattering angle ϕ by Eq. (38.7). (See Example 38.5.) In pair production a photon of sufficient energy can disappear and be replaced by an electron–positron pair. In the inverse process, an electron and a positron can annihilate and be replaced by a pair of photons. (See Example 38.6.)

$$eV_{AC} = hf_{max} = \frac{hc}{\lambda_{min}} \qquad (38.6)$$
(bremsstrahlung)

$$\lambda' - \lambda = \frac{h}{mc}(1 - \cos\phi) \qquad (38.7)$$
(Compton scattering)

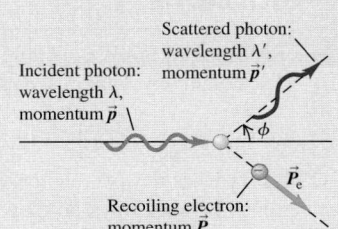

The Heisenberg uncertainty principle: It is impossible to determine both a photon's position and its momentum at the same time to arbitrarily high precision. The precision of such measurements for the *x*-components is limited by the Heisenberg uncertainty principle, Eq. (38.17); there are corresponding relationships for the *y*- and *z*-components. The uncertainty ΔE in the energy of a state that is occupied for a time Δt is given by Eq. (38.24). In these expressions, $\hbar = h/2\pi$. (See Example 38.7.)

$$\Delta x \, \Delta p_x \geq \hbar/2 \qquad (38.17)$$
(Heisenberg uncertainty principle for position and momentum)

$$\Delta t \, \Delta E \geq \hbar/2 \qquad (38.24)$$
(Heisenberg uncertainty principle for energy and time)

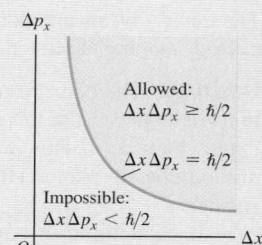

BRIDGING PROBLEM   COMPTON SCATTERING AND ELECTRON RECOIL

An incident x-ray photon is scattered from a free electron that is initially at rest. The photon is scattered straight back at an angle of 180° from its initial direction. The wavelength of the scattered photon is 0.0830 nm. (a) What is the wavelength of the incident photon? (b) What are the magnitude of the momentum and the speed of the electron after the collision? (c) What is the kinetic energy of the electron after the collision?

SOLUTION GUIDE

IDENTIFY and SET UP

1. In this problem a photon is scattered by an electron initially at rest. In Section 38.3 you learned how to relate the wavelengths of the incident and scattered photons; in this problem you must also find the momentum, speed, and kinetic energy of the recoiling electron. You can find these because momentum and energy are conserved in the collision. Draw a diagram showing the momentum vectors of the photon and electron before and after the scattering.

2. Which key equation can be used to find the incident photon wavelength? What is the photon scattering angle ϕ in this problem?

EXECUTE

3. Use the equation you selected in step 2 to find the wavelength of the incident photon.

4. Use momentum conservation and your result from step 3 to find the momentum of the recoiling electron. (*Hint:* All of the momentum vectors are along the same line, but not all point in the same direction. Be careful with signs.)

5. Find the speed of the recoiling electron from your result in step 4. (*Hint:* Assume that the electron is nonrelativistic, so you can use the relationship between momentum and speed from Chapter 8. This is acceptable if the speed of the electron is less than about 0.1*c*. Is it?)

6. Use your result from step 4 or step 5 to find the electron kinetic energy.

EVALUATE

7. You can check your answer in step 6 by finding the difference between the energies of the incident and scattered photons. Is your result consistent with conservation of energy?

Problems

For assigned homework and other learning materials, go to MasteringPhysics®.

•, ••, •••: Difficulty levels. **CP**: Cumulative problems incorporating material from earlier chapters. **CALC**: Problems requiring calculus. **DATA**: Problems involving real data, scientific evidence, experimental design, and/or statistical reasoning. **BIO**: Biosciences problems.

DISCUSSION QUESTIONS

Q38.1 In what ways do photons resemble other particles such as electrons? In what ways do they differ? Do photons have mass? Do they have electric charge? Can they be accelerated? What mechanical properties do they have?

Q38.2 There is a certain probability that a single electron may simultaneously absorb *two* identical photons from a high-intensity laser. How would such an occurrence affect the threshold frequency and the equations of Section 38.1? Explain.

Q38.3 According to the photon model, light carries its energy in packets called quanta or photons. Why then don't we see a series of flashes when we look at things?

Q38.4 Would you expect effects due to the photon nature of light to be generally more important at the low-frequency end of the electromagnetic spectrum (radio waves) or at the high-frequency end (x rays and gamma rays)? Why?

Q38.5 During the photoelectric effect, light knocks electrons out of metals. So why don't the metals in your home lose their electrons when you turn on the lights?

Q38.6 Most black-and-white photographic film (with the exception of some special-purpose films) is less sensitive to red light than blue light and has almost no sensitivity to infrared. How can these properties be understood on the basis of photons?

Q38.7 Human skin is relatively insensitive to visible light, but ultraviolet radiation can cause severe burns. Does this have anything to do with photon energies? Explain.

Q38.8 Explain why Fig. 38.4 shows that most photoelectrons have kinetic energies less than $hf - \phi$, and also explain how these smaller kinetic energies occur.

Q38.9 In a photoelectric-effect experiment, the photocurrent i for large positive values of V_{AC} has the same value no matter what the light frequency f (provided that f is higher than the threshold frequency f_0). Explain why.

Q38.10 In an experiment involving the photoelectric effect, if the intensity of the incident light (having frequency higher than the threshold frequency) is reduced by a factor of 10 without changing anything else, which (if any) of the following statements about this process will be true? (a) The number of photoelectrons will most likely be reduced by a factor of 10. (b) The maximum kinetic energy of the ejected photoelectrons will most likely be reduced by a factor of 10. (c) The maximum speed of the ejected photoelectrons will most likely be reduced by a factor of 10. (d) The maximum speed of the ejected photoelectrons will most likely be reduced by a factor of $\sqrt{10}$. (e) The time for the first photoelectron to be ejected will be increased by a factor of 10.

Q38.11 The materials called *phosphors* that coat the inside of a fluorescent lamp convert ultraviolet radiation (from the mercury-vapor discharge inside the tube) into visible light. Could one also make a phosphor that converts visible light to ultraviolet? Explain.

Q38.12 In a photoelectric-effect experiment, which of the following will increase the maximum kinetic energy of the photoelectrons? (a) Use light of greater intensity; (b) use light of higher frequency; (c) use light of longer wavelength; (d) use a metal surface with a larger work function. In each case justify your answer.

Q38.13 A photon of frequency f undergoes Compton scattering from an electron at rest and scatters through an angle ϕ. The frequency of the scattered photon is f'. How is f' related to f? Does your answer depend on ϕ? Explain.

Q38.14 Can Compton scattering occur with protons as well as electrons? For example, suppose a beam of x rays is directed at a target of liquid hydrogen. (Recall that the nucleus of hydrogen consists of a single proton.) Compared to Compton scattering with electrons, what similarities and differences would you expect? Explain.

Q38.15 Why must engineers and scientists shield against x-ray production in high-voltage equipment?

Q38.16 In attempting to reconcile the wave and particle models of light, some people have suggested that the photon rides up and down on the crests and troughs of the electromagnetic wave. What things are *wrong* with this description?

Q38.17 Some lasers emit light in pulses that are only 10^{-12} s in duration. The length of such a pulse is $(3 \times 10^8 \text{ m/s})(10^{-12} \text{ s}) = 3 \times 10^{-4} \text{ m} = 0.3 \text{ mm}$. Can pulsed laser light be as monochromatic as light from a laser that emits a steady, continuous beam? Explain.

EXERCISES

Section 38.1 Light Absorbed as Photons: The Photoelectric Effect

38.1 • A photon of green light has a wavelength of 520 nm. Find the photon's frequency, magnitude of momentum, and energy. Express the energy in both joules and electron volts.

38.2 • BIO **Response of the Eye.** The human eye is most sensitive to green light of wavelength 505 nm. Experiments have found that when people are kept in a dark room until their eyes adapt to the darkness, a *single* photon of green light will trigger receptor cells in the rods of the retina. (a) What is the frequency of this photon? (b) How much energy (in joules and electron volts) does it deliver to the receptor cells? (c) To appreciate what a small amount of energy this is, calculate how fast a typical bacterium of mass 9.5×10^{-12} g would move if it had that much energy.

38.3 • A 75-W light source consumes 75 W of electrical power. Assume all this energy goes into emitted light of wavelength 600 nm. (a) Calculate the frequency of the emitted light. (b) How many photons per second does the source emit? (c) Are the answers to parts (a) and (b) the same? Is the frequency of the light the same thing as the number of photons emitted per second? Explain.

38.4 • BIO A laser used to weld detached retinas emits light with a wavelength of 652 nm in pulses that are 20.0 ms in duration. The average power during each pulse is 0.600 W. (a) How much energy is in each pulse in joules? In electron volts? (b) What is the energy of one photon in joules? In electron volts? (c) How many photons are in each pulse?

38.5 • A photon has momentum of magnitude 8.24×10^{-28} kg $\cdot$ m/s. (a) What is the energy of this photon? Give your answer in joules and in electron volts. (b) What is the wavelength of this photon? In what region of the electromagnetic spectrum does it lie?

38.6 • The photoelectric threshold wavelength of a tungsten surface is 272 nm. Calculate the maximum kinetic energy of the electrons ejected from this tungsten surface by ultraviolet radiation of frequency 1.45×10^{15} Hz. Express the answer in electron volts.

38.7 •• A clean nickel surface is exposed to light of wavelength 235 nm. What is the maximum speed of the photoelectrons emitted from this surface? Use Table 38.1.

38.8 •• What would the minimum work function for a metal have to be for visible light (380–750 nm) to eject photoelectrons?

38.9 •• When ultraviolet light with a wavelength of 400.0 nm falls on a certain metal surface, the maximum kinetic energy of the emitted photoelectrons is measured to be 1.10 eV. What is the maximum kinetic energy of the photoelectrons when light of wavelength 300.0 nm falls on the same surface?

38.10 •• The photoelectric work function of potassium is 2.3 eV. If light that has a wavelength of 190 nm falls on potassium, find (a) the stopping potential in volts; (b) the kinetic energy, in electron volts, of the most energetic electrons ejected; (c) the speed of these electrons.

38.11 • When ultraviolet light with a wavelength of 254 nm falls on a clean copper surface, the stopping potential necessary to stop emission of photoelectrons is 0.181 V. (a) What is the photoelectric threshold wavelength for this copper surface? (b) What is the work function for this surface, and how does your calculated value compare with that given in Table 38.1?

Section 38.2 Light Emitted as Photons: X-Ray Production

38.12 • The cathode-ray tubes that generated the picture in early color televisions were sources of x rays. If the acceleration voltage in a television tube is 15.0 kV, what are the shortest-wavelength x rays produced by the television?

38.13 • Protons are accelerated from rest by a potential difference of 4.00 kV and strike a metal target. If a proton produces one photon on impact, what is the minimum wavelength of the resulting x rays? How does your answer compare to the minimum wavelength if 4.00-keV electrons are used instead? Why do x-ray tubes use electrons rather than protons to produce x rays?

38.14 •• (a) What is the minimum potential difference between the filament and the target of an x-ray tube if the tube is to produce x rays with a wavelength of 0.150 nm? (b) What is the shortest wavelength produced in an x-ray tube operated at 30.0 kV?

Section 38.3 Light Scattered as Photons: Compton Scattering and Pair Production

38.15 • An x ray with a wavelength of 0.100 nm collides with an electron that is initially at rest. The x ray's final wavelength is 0.110 nm. What is the final kinetic energy of the electron?

38.16 • X rays are produced in a tube operating at 24.0 kV. After emerging from the tube, x rays with the minimum wavelength produced strike a target and undergo Compton scattering through an angle of 45.0°. (a) What is the original x-ray wavelength? (b) What is the wavelength of the scattered x rays? (c) What is the energy of the scattered x rays (in electron volts)?

38.17 •• X rays with initial wavelength 0.0665 nm undergo Compton scattering. What is the longest wavelength found in the scattered x rays? At which scattering angle is this wavelength observed?

38.18 •• A photon with wavelength $\lambda = 0.1385$ nm scatters from an electron that is initially at rest. What must be the angle between the direction of propagation of the incident and scattered photons if the speed of the electron immediately after the collision is 8.90×10^6 m/s?

38.19 •• If a photon of wavelength 0.04250 nm strikes a free electron and is scattered at an angle of 35.0° from its original direction, find (a) the change in the wavelength of this photon; (b) the wavelength of the scattered light; (c) the change in energy of the photon (is it a loss or a gain?); (d) the energy gained by the electron.

38.20 •• A photon scatters in the backward direction ($\phi = 180°$) from a free proton that is initially at rest. What must the wavelength of the incident photon be if it is to undergo a 10.0% change in wavelength as a result of the scattering?

38.21 •• X rays with an initial wavelength of 0.900×10^{-10} m undergo Compton scattering. For what scattering angle is the wavelength of the scattered x rays greater by 1.0% than that of the incident x rays?

38.22 • An electron and a positron are moving toward each other and each has speed $0.500c$ in the lab frame. (a) What is the kinetic energy of each particle? (b) The e^+ and e^- meet head-on and annihilate. What is the energy of each photon that is produced? (c) What is the wavelength of each photon? How does the wavelength compare to the photon wavelength when the initial kinetic energy of the e^+ and e^- is negligibly small (see Example 38.6)?

Section 38.4 Wave–Particle Duality, Probability, and Uncertainty

38.23 • An ultrashort pulse has a duration of 9.00 fs and produces light at a wavelength of 556 nm. What are the momentum and momentum uncertainty of a single photon in the pulse?

38.24 • A horizontal beam of laser light of wavelength 585 nm passes through a narrow slit that has width 0.0620 mm. The intensity of the light is measured on a vertical screen that is 2.00 m from the slit. (a) What is the minimum uncertainty in the vertical component of the momentum of each photon in the beam after the photon has passed through the slit? (b) Use the result of part (a) to estimate the width of the central diffraction maximum that is observed on the screen.

38.25 • A laser produces light of wavelength 625 nm in an ultrashort pulse. What is the minimum duration of the pulse if the minimum uncertainty in the energy of the photons is 1.0%?

PROBLEMS

38.26 •• (a) If the average frequency emitted by a 120-W light bulb is 5.00×10^{14} Hz and 10.0% of the input power is emitted as visible light, approximately how many visible-light photons are emitted per second? (b) At what distance would this correspond to 1.00×10^{11} visible-light photons per cm^2 per second if the light is emitted uniformly in all directions?

38.27 •• CP BIO **Removing Vascular Lesions.** A pulsed dye laser emits light of wavelength 585 nm in 450-μs pulses. Because this wavelength is strongly absorbed by the hemoglobin in the blood, the method is especially effective for removing various types of blemishes due to blood, such as port-wine–colored birthmarks. To get a reasonable estimate of the power required for such laser surgery, we can model the blood as having the same specific heat and heat of vaporization as water (4190 J/kg·K, 2.256×10^6 J/kg). Suppose that each pulse must remove 2.0 μg of blood by evaporating it, starting at 33°C. (a) How much energy must each pulse deliver to the blemish? (b) What must be the power output of this laser? (c) How many photons does each pulse deliver to the blemish?

38.28 • A 2.50-W beam of light of wavelength 124 nm falls on a metal surface. You observe that the maximum kinetic energy of the ejected electrons is 4.16 eV. Assume that each photon in the beam ejects a photoelectron. (a) What is the work function (in electron volts) of this metal? (b) How many photoelectrons are ejected each second from this metal? (c) If the power of the light beam, but not its wavelength, were reduced by half, what would the answer to part (b)? (d) If the wavelength of the beam, but not its power, were reduced by half, what would be the answer to part (b)?

38.29 •• An incident x-ray photon of wavelength 0.0900 nm is scattered in the backward direction from a free electron that is initially at rest. (a) What is the magnitude of the momentum of the scattered photon? (b) What is the kinetic energy of the electron after the photon is scattered?

38.30 •• CP A photon with wavelength $\lambda = 0.0980$ nm is incident on an electron that is initially at rest. If the photon scatters in the backward direction, what is the magnitude of the linear momentum of the electron just after the collision with the photon?

38.31 •• CP A photon with wavelength $\lambda = 0.1050$ nm is incident on an electron that is initially at rest. If the photon scatters at an angle of 60.0° from its original direction, what are the magnitude and direction of the linear momentum of the electron just after it collides with the photon?

38.32 •• CP A photon of wavelength 4.50 pm scatters from a free electron that is initially at rest. (a) For $\phi = 90.0°$, what is the kinetic energy of the electron immediately after the collision with the photon? What is the ratio of this kinetic energy to the rest energy of the electron? (b) What is the speed of the electron immediately after the collision? (c) What is the magnitude of the momentum of the electron immediately after the collision? What is the ratio of this momentum value to the nonrelativistic expression mv?

38.33 •• Nuclear fusion reactions at the center of the sun produce gamma-ray photons with energies of about 1 MeV (10^6 eV). By contrast, what we see emanating from the sun's surface are visible-light photons with wavelengths of about 500 nm. A simple model that explains this difference in wavelength is that a photon undergoes Compton scattering many times—in fact, about 10^{26} times, as suggested by models of the solar interior—as it travels from the center of the sun to its surface. (a) Estimate the increase in wavelength of a photon in an average Compton-scattering event. (b) Find the angle in degrees through which the photon is scattered in the scattering event described in part (a). (*Hint:* A useful approximation is $\cos\phi \approx 1 - \phi^2/2$, which is valid for $\phi \ll 1$. Note that ϕ is in radians in this expression.) (c) It is estimated that a photon takes about 10^6 years to travel from the core to the surface of the sun. Find the average distance that light can travel within the interior of the sun without being scattered. (This distance is roughly equivalent to how far you could see if you were inside the sun and could survive the extreme temperatures there. As your answer shows, the interior of the sun is *very* opaque.)

38.34 •• CP An x-ray tube is operating at voltage V and current I. (a) If only a fraction p of the electric power supplied is converted into x rays, at what rate is energy being delivered to the target? (b) If the target has mass m and specific heat c (in J/kg $\cdot$ K), at what average rate would its temperature rise if there were no thermal losses? (c) Evaluate your results from parts (a) and (b) for an x-ray tube operating at 18.0 kV and 60.0 mA that converts 1.0% of the electric power into x rays. Assume that the 0.250-kg target is made of lead ($c = 130$ J/kg $\cdot$ K). (d) What must the physical properties of a practical target material be? What would be some suitable target elements?

38.35 •• A photon with wavelength 0.1100 nm collides with a free electron that is initially at rest. After the collision the wavelength is 0.1132 nm. (a) What is the kinetic energy of the electron after the collision? What is its speed? (b) If the electron is suddenly stopped (for example, in a solid target), all of its kinetic energy is used to create a photon. What is the wavelength of this photon?

38.36 •• An x-ray photon is scattered from a free electron (mass m) at rest. The wavelength of the scattered photon is λ', and the final speed of the struck electron is v. (a) What was the initial wavelength λ of the photon? Express your answer in terms of λ', v, and m. (*Hint:* Use the relativistic expression for the electron kinetic energy.) (b) Through what angle ϕ is the photon scattered? Express your answer in terms of λ, λ', and m. (c) Evaluate your results in parts (a) and (b) for a wavelength of 5.10×10^{-3} nm for the scattered photon and a final electron speed of 1.80×10^8 m/s. Give ϕ in degrees.

38.37 •• DATA In developing night-vision equipment, you need to measure the work function for a metal surface, so you perform a photoelectric-effect experiment. You measure the stopping potential V_0 as a function of the wavelength λ of the light that is incident on the surface. You get the results in the table.

λ (nm)	100	120	140	160	180	200
V_0 (V)	7.53	5.59	3.98	2.92	2.06	1.43

In your analysis, you use $c = 2.998 \times 10^8$ m/s and $e = 1.602 \times 10^{-19}$ C, which are values obtained in other experiments. (a) Select a way to plot your results so that the data points fall close to a straight line. Using that plot, find the slope and y-intercept of the best-fit straight line to the data. (b) Use the results of part (a) to calculate Planck's constant h (as a test of your data) and the work function (in eV) of the surface. (c) What is the longest wavelength of light that will produce photoelectrons from this surface? (d) What wavelength of light is required to produce photoelectrons with kinetic energy 10.0 eV?

38.38 •• DATA While analyzing smoke detector designs that rely on the photoelectric effect, you are evaluating surfaces made from each of the materials listed in Table 38.1. One particular application uses ultraviolet light with wavelength 270 nm. (a) For which of the materials in Table 38.1 will this light produce photoelectrons?

(b) Which material will result in photoelectrons of the greatest kinetic energy? What will be the maximum speed of the photoelectrons produced as they leave this material's surface? (c) What is the longest wavelength that will produce photoelectrons from a gold surface, if the surface has a work function equal to the value given for gold in Table 38.1? (d) For the wavelength calculated in part (c), what will be the maximum kinetic energy of the photoelectrons produced from a sodium surface that has a work function equal to the value given in Table 38.1 for sodium?

38.39 •• DATA To test the photon concept, you perform a Compton-scattering experiment in a research lab. Using photons of very short wavelength, you measure the wavelength λ' of scattered photons as a function of the scattering angle ϕ, the angle between the direction of a scattered photon and the incident photon. You obtain these results.

ϕ (deg)	30.6	58.7	90.2	119.2	151.3
λ' (pm)	5.52	6.40	7.60	8.84	9.69

Your analysis assumes that the target is a free electron at rest. (a) Graph your data as λ' versus $1 - \cos\phi$. What are the slope and y-intercept of the best-fit straight line to your data? (b) The Compton wavelength λ_C is defined as $\lambda_C = h/mc$, where m is the mass of an electron. Use the results of part (a) to calculate λ_C. (c) Use the results of part (a) to calculate the wavelength λ of the incident light.

CHALLENGE PROBLEM

38.40 ••• Consider Compton scattering of a photon by a *moving* electron. Before the collision the photon has wavelength λ and is moving in the $+x$-direction, and the electron is moving in the $-x$-direction with total energy E (including its rest energy mc^2). The photon and electron collide head-on. After the collision, both are moving in the $-x$-direction (that is, the photon has been scattered by 180°). (a) Derive an expression for the wavelength λ' of the scattered photon. Show that if $E \gg mc^2$, where m is the rest mass of the electron, your result reduces to

$$\lambda' = \frac{hc}{E}\left(1 + \frac{m^2c^4\lambda}{4hcE}\right)$$

(b) A beam of infrared radiation from a CO_2 laser ($\lambda = 10.6$ μm) collides head-on with a beam of electrons, each of total energy $E = 10.0$ GeV (1 GeV $= 10^9$ eV). Calculate the wavelength λ' of the scattered photons, assuming a 180° scattering angle. (c) What kind of scattered photons are these (infrared, microwave, ultraviolet, etc.)? Can you think of an application of this effect?

PASSAGE PROBLEMS

BIO **RADIATION THERAPY FOR TUMORS.** Malignant tumors are commonly treated with targeted x-ray radiation therapy. To generate these medical x rays, a linear accelerator directs a high-energy beam of electrons toward a metal target—typically tungsten. As they near the tungsten nuclei, the electrons are deflected and accelerated, emitting high-energy photons via bremsstrahlung. The resulting x rays are collimated into a beam that is directed at the tumor. The photons can deposit energy in the tumor through Compton and photoelectric interactions. A typical tumor has 10^8 cells/cm^3, and in a full treatment, 4-MeV photons may produce a dose of 70 Gy in 35 fractional exposures on different days. The *gray* (Gy) is a measure of the absorbed energy dose of radiation per unit mass of tissue: 1 Gy $= 1$ J/kg.

38.41 How much energy is imparted to one cell during one day's treatment? Assume that the specific gravity of the tumor is 1 and that $1 \text{ J} = 6 \times 10^{18} \text{ eV}$. (a) 120 keV; (b) 12 MeV; (c) 120 MeV; (d) 120×10^3 MeV.

38.42 While interacting with molecules (mainly water) in the tumor tissue, each Compton electron or photoelectron causes a series of ionizations, each of which takes about 40 eV. Estimate the maximum number of ionizations that one photon generated by this linear accelerator can produce in tissue. (a) 100; (b) 1000; (c) 10^4; (d) 10^5.

38.43 The high-energy photons can undergo Compton scattering off electrons in the tumor. The energy imparted by a photon is a maximum when the photon scatters straight back from the electron. In this process, what is the maximum energy that a photon with the energy described in the passage can give to an electron? (a) 3.8 MeV; (b) 2.0 MeV; (c) 0.40 MeV; (d) 0.23 MeV.

38.44 The probability of a photon interacting with tissue via the photoelectric effect or the Compton effect depends on the photon energy. Use **Fig. P38.44** to determine the best description of how the photons from the linear accelerator described in the passage interact with a tumor. (a) Via the Compton effect only; (b) mostly via the photoelectric effect until they have lost most of their energy,

Figure **P38.44**

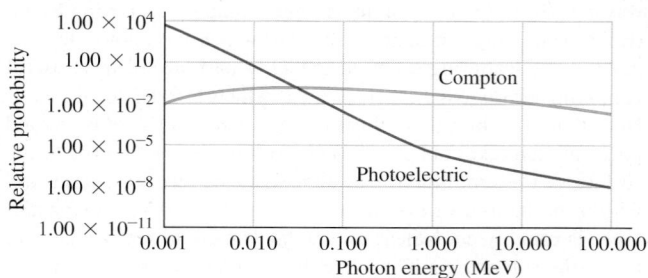

and then mostly via the Compton effect; (c) mostly via the Compton effect until they have lost most of their energy, and then mostly via the photoelectric effect; (d) via the Compton effect and the photoelectric effect equally.

38.45 Higher-energy photons might be desirable for the treatment of certain tumors. Which of these actions would generate higher-energy photons in this linear accelerator? (a) Increasing the number of electrons that hit the tungsten target; (b) accelerating the electrons through a higher potential difference; (c) both (a) and (b); (d) none of these.

Answers

Chapter Opening Question ?

(i) The energy of a photon E is inversely proportional to its wavelength λ: The shorter the wavelength, the more energetic is the photon. Since visible light has shorter wavelengths than infrared light, the headlamp emits photons of greater energy. However, the light from the infrared laser is far more *intense* (delivers much more energy per second per unit area to the patient's skin) because it emits many more photons per second than does the headlamp and concentrates them onto a very small spot.

Test Your Understanding Questions

38.1 (iv) From Eq. (38.2), a photon of energy $E = 1.14 \text{ eV}$ has wavelength

$$\lambda = hc/E$$
$$= (4.136 \times 10^{-15} \text{ eV} \cdot \text{s})(3.00 \times 10^8 \text{ m/s})/(1.14 \text{ eV})$$
$$= 1.09 \times 10^{-6} \text{ m} = 1090 \text{ nm}$$

This is in the infrared part of the spectrum. Since wavelength is inversely proportional to photon energy, the *minimum* photon energy of 1.14 eV corresponds to the *maximum* wavelength that causes photoconductivity in silicon. Thus the wavelength must be 1090 nm or less.

38.2 (ii) Equation (38.6) shows that the minimum wavelength of x rays produced by bremsstrahlung depends on the potential difference V_{AC} but does *not* depend on the rate at which electrons strike the anode. Increasing the number of electrons per second will only cause an increase in the number of x-ray photons emitted per second (that is, the x-ray intensity I).

38.3 yes, no Equation (38.7) shows that the wavelength shift $\Delta\lambda = \lambda' - \lambda$ depends only on the photon scattering angle ϕ, not on the wavelength of the incident photon. So a visible-light photon scattered through an angle ϕ undergoes the same wavelength shift as an x-ray photon. Equation (38.7) also shows that this shift is of the order of $h/mc = 2.426 \times 10^{-12} \text{ m} = 0.002426 \text{ nm}$. This is a few percent of the wavelength of x rays (see Example 38.5), so the effect is noticeable in x-ray scattering. However, h/mc is a tiny fraction of the wavelength of visible light (between 380 and 750 nm). The human eye cannot distinguish such minuscule differences in wavelength (that is, differences in color).

38.4 (ii) There is *zero* probability that a photon will be deflected by one of the angles where the diffraction pattern has zero intensity. These angles are given by $a \sin \theta = m\lambda$ with $m = \pm 1$, $\pm 2, \pm 3, \ldots$. Since λ is much less than a, we can write these angles as $\theta = m\lambda/a = \pm\lambda/a, \pm 2\lambda/a, \pm 3\lambda/a, \ldots$. These values include answers (i), (iii), and (iv), so it is impossible for a photon to be deflected through any of these angles. The intensity is not zero at $\theta = 3\lambda/2a$ (located between two zeros in the diffraction pattern), so there is some probability that a photon will be deflected through this angle.

Bridging Problem

(a) 0.0781 nm   **(b)** $1.65 \times 10^{-23} \text{ kg} \cdot \text{m/s}$, $1.81 \times 10^7 \text{ m/s}$
(c) $1.49 \times 10^{-16} \text{ J}$

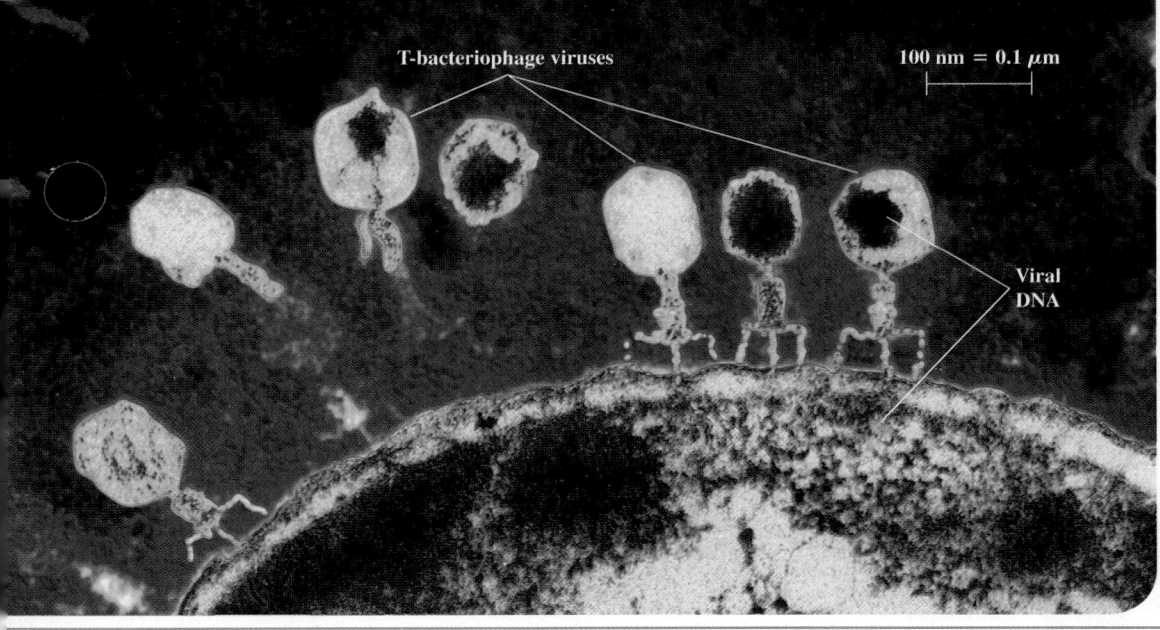

T-bacteriophage viruses

100 nm = 0.1 μm

Viral DNA

? Viruses (shown in blue) have landed on an *E. coli* bacterium and injected their DNA, converting the bacterium into a virus factory. This false-color image was made by using a beam of electrons rather than a light beam. Electrons are used for imaging such fine details because, compared to visible-light photons, (i) electrons can have much shorter wavelengths; (ii) electrons can have much longer wavelengths; (iii) electrons can have much less momentum; (iv) electrons have more total energy for the same momentum; (v) more than one of these.

39 PARTICLES BEHAVING AS WAVES

I n Chapter 38 we discovered one aspect of nature's wave–particle duality: Light and other electromagnetic radiation act sometimes like waves and sometimes like particles. Interference and diffraction demonstrate wave behavior, while emission and absorption of photons demonstrate particle behavior.

If light waves can behave like particles, can the particles of matter behave like waves? The answer is a resounding yes. Electrons can interfere and diffract just like other kinds of waves. The wave nature of electrons is not merely a laboratory curiosity: It is the fundamental reason why atoms, which according to classical physics should be unstable, are able to exist. In this chapter the wave nature of matter will help us understand the structure of atoms, the operating principles of a laser, and the curious properties of the light emitted by a heated, glowing object. Without the wave picture of matter, there would be no way to explain these phenomena.

In Chapter 40 we'll introduce an even more complete wave picture of matter called *quantum mechanics*. Through the remainder of this book we'll use the ideas of quantum mechanics to understand the nature of molecules, solids, atomic nuclei, and the fundamental particles that are the building blocks of our universe.

39.1 ELECTRON WAVES

In 1924 a French physicist, Louis de Broglie (pronounced "de broy"; **Fig. 39.1**, next page), made a remarkable proposal about the nature of matter. His reasoning, freely paraphrased, went like this: Nature loves symmetry. Light is dualistic in nature, behaving in some situations like waves and in others like particles. If nature is symmetric, this duality should also hold for matter. Electrons, which we usually think of as *particles*, may in some situations behave like *waves*.

If a particle acts like a wave, it should have a wavelength and a frequency. De Broglie postulated that a free particle with rest mass m, moving with non-relativistic speed v, should have a wavelength λ related to its momentum

39.1 Louis-Victor de Broglie, the seventh Duke de Broglie (1892–1987), broke with family tradition by choosing a career in physics rather than as a diplomat. His revolutionary proposal that particles have wave characteristics—for which de Broglie won the 1929 Nobel Prize in physics—was published in his doctoral thesis.

(MP)

PhET: Davisson-Germer: Electron Diffraction

$p = mv$ in exactly the same way as for a photon, as expressed by Eq. (38.5) from Section 38.1: $\lambda = h/p$. The **de Broglie wavelength** of a particle is then

De Broglie wavelength of a particle ......

Planck's constant

$$\lambda = \frac{h}{p} = \frac{h}{mv} \qquad (39.1)$$

Particle's momentum ·· ·Particle's mass
Particle's speed

If the particle's speed is an appreciable fraction of the speed of light c, we replace mv in Eq. (39.1) with $\gamma mv = mv/\sqrt{1 - v^2/c^2}$ [Eq. (37.27) from Section 37.7]. The frequency f, according to de Broglie, is also related to the particle's energy E in exactly the same way as for a photon:

Energy of a particle ····

Planck's constant

$$E = hf \qquad (39.2)$$

Frequency

CAUTION **Not all photon equations apply to particles with mass** Be careful when applying $E = hf$ to particles with nonzero rest mass, such as electrons. Unlike a photon, they do *not* travel at speed c, so the equations $f = c/\lambda$ and $E = pc$ do *not* apply to them! ▮

Observing the Wave Nature of Electrons

De Broglie's proposal was a bold one, made at a time when there was no direct experimental evidence that particles have wave characteristics. But within a few years, his ideas were resoundingly verified by a diffraction experiment with electrons. This experiment was analogous to those we described in Section 36.6, in which atoms in a crystal act as a three-dimensional diffraction grating for x rays. An x-ray beam is strongly reflected when it strikes a crystal at an angle that gives constructive interference among the waves scattered from the various atoms in the crystal. These interference effects demonstrate the *wave* nature of x rays.

In 1927 the American physicists Clinton Davisson and Lester Germer, working at the Bell Telephone Laboratories, were studying the surface of a piece of nickel by directing a beam of *electrons* at the surface and observing how many electrons bounced off at various angles. **Figure 39.2** shows an experimental setup like theirs. Like many ordinary metals, the sample was *polycrystalline:* It consisted of many randomly oriented microscopic crystals bonded together. As a result, the electron beam reflected diffusely, like light bouncing off a rough surface (see Fig. 33.6b), with a smooth distribution of intensity as a function of the angle θ.

During the experiment an accident occurred that permitted air to enter the vacuum chamber, and an oxide film formed on the metal surface. To remove this film, Davisson and Germer baked the sample in a high-temperature oven. Unknown to them, this had the effect of creating large regions within the nickel with crystal

39.2 An apparatus similar to that used by Davisson and Germer to discover electron diffraction.

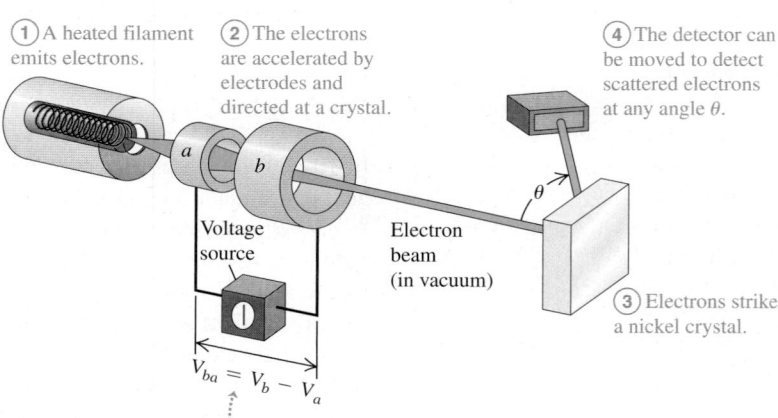

1 A heated filament emits electrons.

2 The electrons are accelerated by electrodes and directed at a crystal.

4 The detector can be moved to detect scattered electrons at any angle θ.

3 Electrons strike a nickel crystal.

Voltage source

Electron beam (in vacuum)

$V_{ba} = V_b - V_a$

$V_{ba} > 0$, so electrons speed up in moving from a to b.

planes that were continuous over the width of the electron beam. From the perspective of the electrons, the sample looked like a *single* crystal of nickel.

When the observations were repeated with this sample, the results were quite different. Now strong maxima in the intensity of the reflected electron beam occurred at specific angles (**Fig. 39.3a**), in contrast to the smooth variation of intensity with angle that Davisson and Germer had observed before the accident. The angular positions of the maxima depended on the accelerating voltage V_{ba} used to produce the electron beam. Davisson and Germer were familiar with de Broglie's hypothesis, and they noticed the similarity of this behavior to x-ray diffraction. This was not the effect they had been looking for, but they immediately recognized that the electron beam was being *diffracted*. They had discovered a very direct experimental confirmation of the wave hypothesis.

Davisson and Germer could determine the speeds of the electrons from the accelerating voltage, so they could compute the de Broglie wavelength from Eq. (39.1). If an electron is accelerated from rest at point a to point b through a potential increase $V_{ba} = V_b - V_a$ as shown in Fig. 39.2, the work done on the electron eV_{ba} equals its kinetic energy K. Using $K = \left(\frac{1}{2}\right)m\upsilon^2 = p^2/2m$ for a nonrelativistic particle, we have

$$eV_{ba} = \frac{p^2}{2m} \qquad p = \sqrt{2meV_{ba}}$$

We substitute this into Eq. (39.1) for the de Broglie wavelength of the electron:

De Broglie wavelength of an electron
$$\lambda = \frac{h}{p} = \frac{h}{\sqrt{2meV_{ba}}} \qquad (39.3)$$

Planck's constant
Accelerating voltage
Electron momentum
Electron mass
Magnitude of electron charge

The greater the accelerating voltage V_{ba}, the shorter the wavelength of the electron.

To predict the angles at which strong reflection occurs, note that the electrons were scattered primarily by the planes of atoms near the surface of the crystal. Atoms in a surface plane are arranged in rows, with a distance d that can be measured by x-ray diffraction techniques. These rows act like a reflecting diffraction grating; the angles at which strong reflection occurs are the same as for a grating with center-to-center distance d between its slits (Fig. 39.3b). From Eq. (36.13) the angles of maximum reflection are given by

$$d\sin\theta = m\lambda \qquad (m = 1, 2, 3, \ldots) \qquad (39.4)$$

where θ is the angle shown in Fig. 39.2. (Note that the geometry in Fig. 39.3b is different from that for Fig. 36.22, so Eq. (39.4) is different from Eq. (36.16).) Davisson and Germer found that the angles predicted by this equation, with the de Broglie wavelength given by Eq. (39.3), agreed with the observed values (Fig. 39.3a). Thus the accidental discovery of **electron diffraction** was the first direct evidence confirming de Broglie's hypothesis.

In 1928, just a year after the Davisson–Germer discovery, the English physicist G. P. Thomson carried out electron-diffraction experiments using a thin, polycrystalline, metallic foil as a target. Debye and Sherrer had used a similar technique several years earlier to study x-ray diffraction from polycrystalline specimens. In these experiments the beam passes *through* the target rather than being reflected from it. Because of the random orientations of the individual microscopic crystals in the foil, the diffraction pattern consists of intensity maxima forming rings around the direction of the incident beam. Thomson's results again confirmed the de Broglie relationship. **Figure 39.4** shows both x-ray and electron diffraction patterns for a polycrystalline aluminum foil. (G. P. Thomson was the son of J. J. Thomson, who 31 years earlier discovered the electron. Davisson and the younger Thomson shared the 1937 Nobel Prize in physics for their discoveries.)

39.3 (a) Intensity of the scattered electron beam in Fig. 39.2 as a function of the scattering angle θ. (b) Electron waves scattered from two adjacent atoms interfere constructively when $d\sin\theta = m\lambda$. In the case shown here, $\theta = 50°$ and $m = 1$.

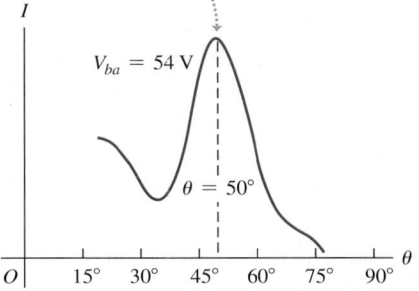

(a) This peak in the intensity of scattered electrons is due to constructive interference between electron waves scattered by different surface atoms.

$V_{ba} = 54$ V

$\theta = 50°$

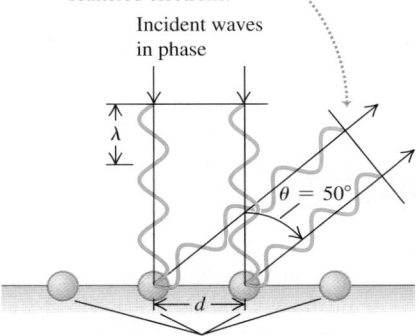

(b) If the scattered waves are in phase, there is a peak in the intensity of scattered electrons.

Incident waves in phase

λ

$\theta = 50°$

d

Atoms on surface of crystal

39.4 X-ray and electron diffraction. The upper half of the photo shows the diffraction pattern for 71-pm x rays passing through aluminum foil. The lower half, with a different scale, shows the diffraction pattern for 600-eV electrons from aluminum. The similarity shows that electrons undergo the same kind of diffraction as x rays.

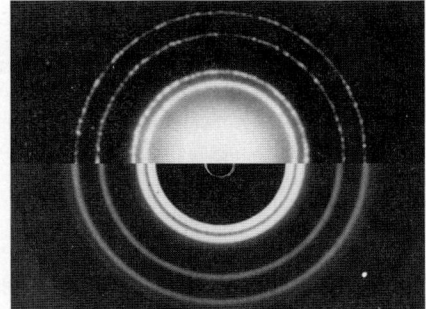

Top: x-ray diffraction

Bottom: electron diffraction

Additional diffraction experiments were soon carried out in many laboratories using not only electrons but also various ions and low-energy neutrons. All of these are in agreement with de Broglie's bold predictions. Thus the wave nature of particles, so strange in 1924, became firmly established in the years that followed.

PROBLEM-SOLVING STRATEGY 39.1 | **WAVELIKE PROPERTIES OF PARTICLES**

IDENTIFY *the relevant concepts:* Particles have wavelike properties. A particle's (de Broglie) wavelength is inversely proportional to its momentum, and its frequency is proportional to its energy.

SET UP *the problem:* Identify the target variables and decide which equations you will use to calculate them.

EXECUTE *the solution* as follows:
1. Use Eq. (39.1) to relate a particle's momentum p to its wavelength λ; use Eq. (39.2) to relate its energy E to its frequency f.
2. Nonrelativistic kinetic energy may be expressed as either $K = \frac{1}{2}mv^2$ or (because $p = mv$) $K = p^2/2m$. The latter form is useful in calculations involving the de Broglie wavelength.
3. You may express energies in either joules or electron volts, using $h = 6.626 \times 10^{-34}$ J·s or $h = 4.136 \times 10^{-15}$ eV·s as appropriate.

EVALUATE *your answer:* To check numerical results, it helps to remember some approximate orders of magnitude. Here's a partial list:

Size of an atom: 10^{-10} m $= 0.1$ nm

Mass of an atom: 10^{-26} kg

Mass of an electron: $m = 10^{-30}$ kg; $mc^2 = 0.511$ MeV

Electron charge magnitude: 10^{-19} C

kT at room temperature: $\frac{1}{40}$ eV

Difference between energy levels of an atom (to be discussed in Section 39.3): 1 to 10 eV

Speed of an electron in the Bohr model of a hydrogen atom (to be discussed in Section 39.3): 10^6 m/s

---

EXAMPLE 39.1 **AN ELECTRON-DIFFRACTION EXPERIMENT**

In an electron-diffraction experiment using an accelerating voltage of 54 V, an intensity maximum occurs for $\theta = 50°$ (see Fig. 39.3a). X-ray diffraction indicates that the atomic spacing in the target is $d = 2.18 \times 10^{-10}$ m $= 0.218$ nm. The electrons have negligible kinetic energy before being accelerated. Find the electron wavelength.

SOLUTION

IDENTIFY, SET UP, and EXECUTE: We'll determine λ from both de Broglie's equation, Eq. (39.3), and the diffraction equation, Eq. (39.4). From Eq. (39.3),

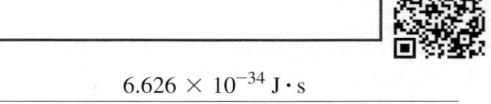

$$\lambda = \frac{6.626 \times 10^{-34} \text{ J·s}}{\sqrt{2(9.109 \times 10^{-31} \text{ kg})(1.602 \times 10^{-19} \text{ C})(54 \text{ V})}}$$
$$= 1.7 \times 10^{-10} \text{ m} = 0.17 \text{ nm}$$

Alternatively, using Eq. (39.4) and assuming $m = 1$, we get

$$\lambda = d\sin\theta = (2.18 \times 10^{-10} \text{ m})\sin 50° = 1.7 \times 10^{-10} \text{ m}$$

EVALUATE: The two numbers agree within the accuracy of the experimental results, which gives us an excellent check on our calculations. Note that this electron wavelength is less than the spacing between the atoms.

EXAMPLE 39.2 **ENERGY OF A THERMAL NEUTRON**

Find the speed and kinetic energy of a neutron ($m = 1.675 \times 10^{-27}$ kg) with de Broglie wavelength $\lambda = 0.200$ nm, a typical interatomic spacing in crystals. Compare this energy with the average translational kinetic energy of an ideal-gas molecule at room temperature ($T = 20°C = 293$ K).

SOLUTION

IDENTIFY and SET UP: This problem uses the relationships between particle speed and wavelength, between particle speed and kinetic energy, and between gas temperature and the average

kinetic energy of a gas molecule. We'll find the neutron speed v by using Eq. (39.1) and from that calculate the neutron kinetic energy $K = \frac{1}{2}mv^2$. We'll use Eq. (18.16) to find the average kinetic energy of a gas molecule.

EXECUTE: From Eq. (39.1), the neutron speed is

$$v = \frac{h}{\lambda m} = \frac{6.626 \times 10^{-34} \text{ J·s}}{(0.200 \times 10^{-9} \text{ m})(1.675 \times 10^{-27} \text{ kg})}$$
$$= 1.98 \times 10^3 \text{ m/s}$$

The neutron kinetic energy is

$$K = \tfrac{1}{2}mv^2 = \tfrac{1}{2}(1.675 \times 10^{-27}\,\text{kg})(1.98 \times 10^3\,\text{m/s})^2$$
$$= 3.28 \times 10^{-21}\,\text{J} = 0.0205\,\text{eV}$$

From Eq. (18.16), the average translational kinetic energy of an ideal-gas molecule at $T = 293$ K is

$$\tfrac{1}{2}m(v^2)_{av} = \tfrac{3}{2}kT = \tfrac{3}{2}(1.38 \times 10^{-23}\,\text{J/K})(293\,\text{K})$$
$$= 6.07 \times 10^{-21}\,\text{J} = 0.0379\,\text{eV}$$

The two energies are comparable in magnitude, which is why a neutron with kinetic energy in this range is called a *thermal neutron*. Diffraction of thermal neutrons is used to study crystal and molecular structure in the same way as x-ray diffraction. Neutron diffraction has proved to be especially useful in the study of large organic molecules.

EVALUATE: Note that the calculated neutron speed is much less than the speed of light. This justifies our use of the nonrelativistic form of Eq. (39.1).

De Broglie Waves and the Macroscopic World

If the de Broglie picture is correct and matter has wave aspects, you might wonder why we don't see these aspects in everyday life. As an example, we know that waves diffract when sent through a single slit. Yet when we walk through a doorway (a kind of single slit), we don't worry about our body diffracting!

The main reason we don't see these effects on human scales is that Planck's constant h has such a minuscule value. As a result, the de Broglie wavelengths of even the smallest ordinary objects that you can see are extremely small, and the wave effects are unimportant. For instance, what is the wavelength of a falling grain of sand? If the grain's mass is 5×10^{-10} kg and its diameter is 0.07 mm $= 7 \times 10^{-5}$ m, it will fall in air with a terminal speed of about 0.4 m/s. The magnitude of its momentum is $p = mv = (5 \times 10^{-10}\,\text{kg})(0.4\,\text{m/s}) = 2 \times 10^{-10}\,\text{kg} \cdot \text{m/s}$. The de Broglie wavelength of this falling sand grain is then

$$\lambda = \frac{h}{p} = \frac{6.626 \times 10^{-34}\,\text{J} \cdot \text{s}}{2 \times 10^{-10}\,\text{kg} \cdot \text{m/s}} = 3 \times 10^{-24}\,\text{m}$$

Not only is this wavelength far smaller than the diameter of the sand grain, but it's also far smaller than the size of a typical atom (about 10^{-10} m). A more massive, faster-moving object would have an even larger momentum and an even smaller de Broglie wavelength. The effects of such tiny wavelengths are so small that they are never noticed in daily life.

The Electron Microscope

The **electron microscope** offers an important and interesting example of the interplay of wave and particle properties of electrons. An electron beam can be used to form an image of an object in much the same way as a light beam. A ray of light can be bent by reflection or refraction, and an electron trajectory can be bent by an electric or magnetic field. Rays of light diverging from a point on an object can be brought to convergence by a converging lens or concave mirror, and electrons diverging from a small region can be brought to convergence by electric and/or magnetic fields.

The analogy between light rays and electrons goes deeper. The *ray* model of geometric optics is an approximate representation of the more general *wave* model. Geometric optics (ray optics) is valid whenever interference and diffraction effects can be ignored. Similarly, the model of an electron as a point particle following a line trajectory is an approximate description of the actual behavior of the electron; this model is useful when we can ignore effects associated with the wave nature of electrons.

How is an electron microscope superior to an optical microscope? The *resolution* of an optical microscope is limited by diffraction effects, as we discussed in Section 36.7. Since an optical microscope uses wavelengths around 500 nm, it can't resolve objects smaller than a few hundred nanometers, no matter how carefully its lenses are made. The resolution of an electron microscope is similarly limited by the wavelengths of the electrons, but these wavelengths may be many thousands of times smaller than wavelengths of visible light. As a result, the useful magnification of an electron microscope can be thousands of times greater than that of an optical microscope.

Note that the ability of the electron microscope to form a magnified image *does not* depend on the wave properties of electrons. Within the limitations of the Heisenberg uncertainty principle (which we'll discuss in Section 39.6), we can compute the electron trajectories by treating them as classical charged particles under the action of electric and magnetic forces. Only when we talk about *resolution* do the wave properties become important.

EXAMPLE 39.3 AN ELECTRON MICROSCOPE

In an electron microscope, the nonrelativistic electron beam is formed by a setup similar to the electron gun used in the Davisson–Germer experiment (see Fig. 39.2). The electrons have negligible kinetic energy before they are accelerated. What accelerating voltage is needed to produce electrons with wavelength 10 pm = 0.010 nm (roughly 50,000 times smaller than typical visible-light wavelengths)?

SOLUTION

IDENTIFY, SET UP, and EXECUTE: We can use the same concepts we used to understand the Davisson–Germer experiment. The accelerating voltage is the quantity V_{ba} in Eq. (39.3). Rewrite this equation to solve for V_{ba}:

$$V_{ba} = \frac{h^2}{2me\lambda^2}$$

$$= \frac{(6.626 \times 10^{-34} \text{ J·s})^2}{2(9.109 \times 10^{-31} \text{ kg})(1.602 \times 10^{-19} \text{ C})(10 \times 10^{-12} \text{ m})^2}$$

$$= 1.5 \times 10^4 \text{ V} = 15,000 \text{ V}$$

EVALUATE: It is easy to attain 15-kV accelerating voltages from 120-V or 240-V line voltage by using a step-up transformer (Section 31.6) and a rectifier (Section 31.1). The accelerated electrons have kinetic energy 15 keV; since the electron rest energy is 0.511 MeV = 511 keV, these electrons are indeed nonrelativistic.

Types of Electron Microscope

39.5 Schematic diagram of a transmission electron microscope (TEM).

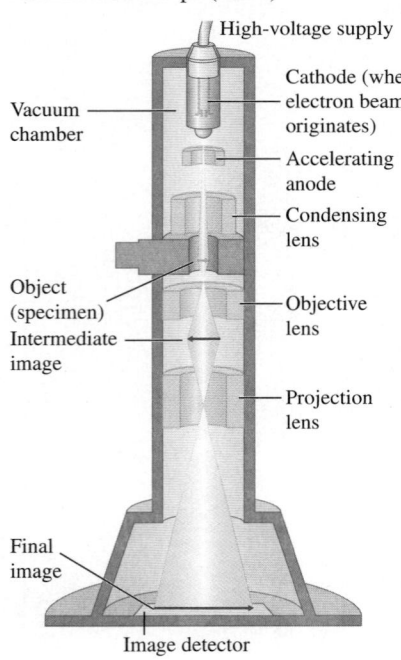

Figure 39.5 shows the design of a *transmission electron microscope,* in which electrons actually pass through the specimen being studied. The specimen to be viewed can be no more than 10 to 100 nm thick so the electrons are not slowed appreciably as they pass through. The electrons used in a transmission electron microscope are emitted from a hot cathode and accelerated by a potential difference, typically 40 to 400 kV. They then pass through a condensing "lens" that uses magnetic fields to focus the electrons into a parallel beam before they pass through the specimen. The beam then passes through two more magnetic lenses: an objective lens that forms an intermediate image of the specimen and a projection lens that produces a final real image of the intermediate image. The objective and projection lenses play the roles of the objective and eyepiece lenses, respectively, of a compound optical microscope (see Section 34.8). The final image is projected onto a fluorescent screen for viewing or photographing. The entire apparatus, including the specimen, must be enclosed in a vacuum container; otherwise, electrons would scatter off air molecules and muddle the image. The image that opens this chapter was made with a transmission electron microscope.

We might think that when the electron wavelength is 0.01 nm (as it is in Example 39.3), the resolution would also be about 0.01 nm. In fact, it is seldom better than 0.1 nm, in part because the focal length of a magnetic lens depends on the electron speed, which is never exactly the same for all electrons in the beam.

An important variation is the *scanning electron microscope.* The electron beam is focused to a very fine line and scanned across the specimen. The beam knocks additional electrons off the specimen wherever it hits. These ejected electrons are collected by an anode that is kept at a potential a few hundred volts positive with respect to the specimen. The current of ejected electrons flowing to the collecting anode varies as the microscope beam sweeps across the specimen. The varying strength of the current is then used to create a "map" of the scanned specimen, and this map forms a greatly magnified image of the specimen.

This scheme has several advantages. The specimen can be thick because the beam does not need to pass through it. Also, the knock-off electron production depends on the *angle* at which the beam strikes the surface. Thus scanning electron micrographs have an appearance that is much more three-dimensional than conventional visible-light micrographs (**Fig. 39.6**). The resolution is typically of the order of 10 nm, not as good as a transmission electron microscope but still much finer than the best optical microscopes.

TEST YOUR UNDERSTANDING OF SECTION 39.1 (a) A proton has a slightly smaller mass than a neutron. Compared to the neutron described in Example 39.2, would a proton of the same wavelength have (i) more kinetic energy; (ii) less kinetic energy; or (iii) the same kinetic energy? (b) Example 39.1 shows that to give electrons a wavelength of 1.7×10^{-10} m, they must be accelerated from rest through a voltage of 54 V and so acquire a kinetic energy of 54 eV. Does a photon of this same energy also have a wavelength of 1.7×10^{-10} m? ▌

39.2 THE NUCLEAR ATOM AND ATOMIC SPECTRA

Every neutral atom contains at least one electron. How does the wave aspect of electrons affect atomic structure? As we will see, it is crucial for understanding not only the structure of atoms but also how they interact with light. Historically, the quest to understand the nature of the atom was intimately linked with both the idea that electrons have wave characteristics and the notion that light has particle characteristics. Before we explore how these ideas shaped atomic theory, it's useful to look at what was known about atoms—as well as what remained mysterious—by the first decade of the 20th century.

Line Spectra

Heated materials emit light, and different materials emit different kinds of light. The coils of a toaster glow red when in operation, the flame of a match has a characteristic yellow color, and the flame from a gas range is a distinct blue. To analyze these different types of light, we can use a prism or a diffraction grating to separate the various wavelengths in a beam of light into a spectrum. If the light source is a hot solid (such as the filament of an incandescent light bulb) or liquid, the spectrum is *continuous;* light of all wavelengths is present (**Fig. 39.7a**). But if the source is a heated *gas,* such as the neon in a sign or the sodium vapor formed when table salt is thrown into a campfire, the spectrum includes only a few colors in the form of isolated sharp parallel lines (Fig. 39.7b). (Each "line" is

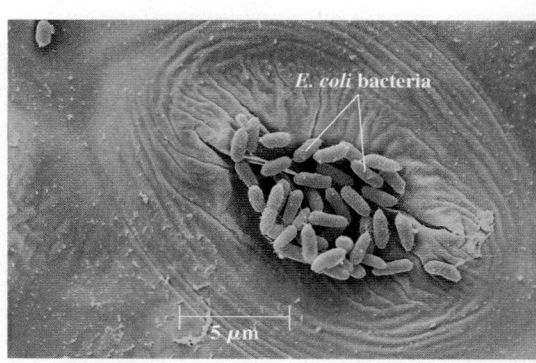

39.6 This scanning electron microscope image shows *Escherichia coli* bacteria crowded into a stoma, or respiration opening, on the surface of a lettuce leaf. (False color has been added.) If not washed off before the lettuce is eaten, these bacteria can be a health hazard. The *transmission* electron micrograph that opens this chapter shows a greatly magnified view of the surface of an *E. coli* bacterium.

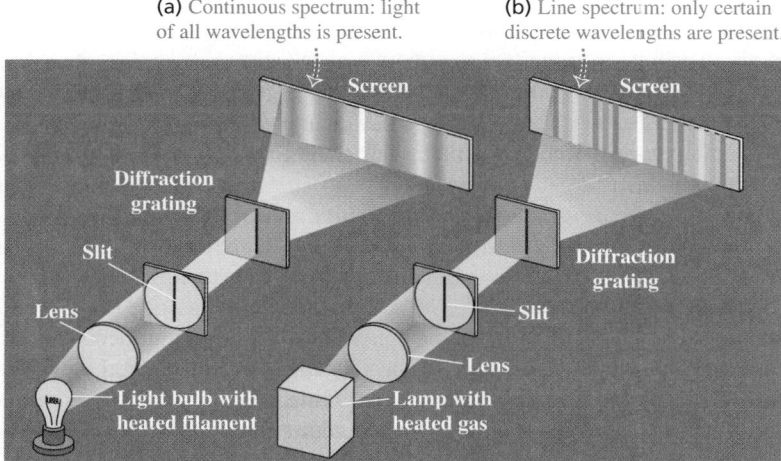

(a) Continuous spectrum: light of all wavelengths is present.

(b) Line spectrum: only certain discrete wavelengths are present.

39.7 (a) Continuous spectrum produced by a glowing light bulb filament. (b) Emission line spectrum emitted by a lamp containing a heated gas.

Application Using Spectra to Analyze an Interstellar Gas Cloud The light from this glowing gas cloud—located in the Small Magellanic Cloud, a small satellite galaxy of the Milky Way some 200,000 light-years (1.9×10^{18} km) from earth—has an emission line spectrum. Despite its immense distance, astronomers can tell that this cloud is composed mostly of hydrogen because its spectrum is dominated by red light at a wavelength of 656.3 nm, a wavelength emitted by hydrogen and no other element.

39.9 The absorption line spectrum of the sun. (The spectrum "lines" read from left to right and from top to bottom, like text on a page.) The spectrum is produced by the sun's relatively cool atmosphere, which absorbs photons from deeper, hotter layers. The absorption lines thus indicate what kinds of atoms are present in the solar atmosphere.

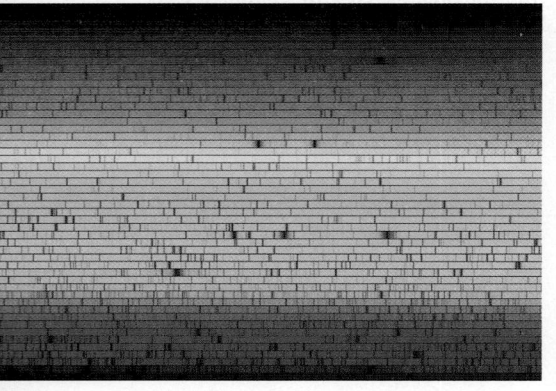

39.8 The emission line spectra of several kinds of atoms and molecules. No two are alike. Note that the spectrum of water vapor (H_2O) is similar to that of hydrogen (H_2), but there are important differences that make it straightforward to distinguish these two spectra.

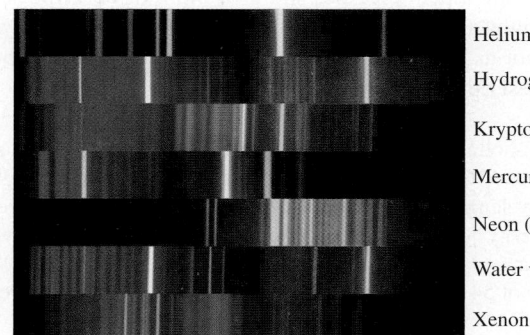

Helium (He)

Hydrogen (H_2)

Krypton (Kr)

Mercury (Hg)

Neon (Ne)

Water vapor (H_2O)

Xenon

an image of the spectrograph slit, deviated through an angle that depends on the wavelength of the light forming that image; see Section 36.5.) A spectrum of this sort is called an **emission line spectrum,** and the lines are called **spectral lines.** Each spectral line corresponds to a definite wavelength and frequency.

It was discovered early in the 19th century that each element in its gaseous state has a unique set of wavelengths in its line spectrum. The spectrum of hydrogen always contains a certain set of wavelengths; mercury produces a different set, neon still another, and so on (**Fig. 39.8**). Scientists find the use of spectra to identify elements and compounds to be an invaluable tool. For instance, astronomers have detected the spectra from more than 100 different molecules in interstellar space, including some that are not found naturally on earth.

While a *heated* gas selectively *emits* only certain wavelengths, a *cool* gas selectively *absorbs* certain wavelengths. If we pass white (continuous-spectrum) light through a gas and look at the *transmitted* light with a spectrometer, we find a series of dark lines corresponding to the wavelengths that have been absorbed (**Fig. 39.9**). This is called an **absorption line spectrum.** What's more, a given kind of atom or molecule absorbs the *same* characteristic set of wavelengths when it's cool as it emits when heated. Hence scientists can use absorption line spectra to identify substances in the same manner that they use emission line spectra.

As useful as emission line spectra and absorption line spectra are, they presented a quandary to scientists: *Why* does a given kind of atom emit and absorb only certain very specific wavelengths? To answer this question, we need to have a better idea of what the inside of an atom is like. We know that atoms are much smaller than the wavelengths of visible light, so there is no hope of actually using that light to *see* an atom. But we can still describe how the mass and electric charge are distributed throughout the volume of the atom.

Here's where things stood in 1910. In 1897 the English physicist J. J. Thomson had discovered the electron and measured its charge-to-mass ratio e/m. By 1909, the American physicist Robert Millikan had made the first measurements of the electron charge $-e$. These and other experiments showed that almost all the mass of an atom had to be associated with the *positive* charge, not with the electrons. It was also known that the overall size of atoms is of the order of 10^{-10} m and that all atoms except hydrogen contain more than one electron.

In 1910 the best available model of atomic structure was one developed by Thomson. He envisioned the atom as a sphere of some as yet unidentified positively charged substance, within which the electrons were embedded like raisins in cake. This model offered an explanation for line spectra. If the atom collided with another atom, as in a heated gas, each electron would oscillate around its equilibrium position with a characteristic frequency and emit electromagnetic radiation with that frequency. If the atom were illuminated with light of many frequencies, each electron would selectively absorb only light whose frequency matched the electron's natural oscillation frequency. (This is the phenomenon of resonance that we discussed in Section 14.8.)

Rutherford's Exploration of the Atom

The first experiments designed to test Thomson's model by probing the interior structure of the atom were carried out in 1910–1911 by Ernest Rutherford (**Fig. 39.10**) and two of his students, Hans Geiger and Ernest Marsden, at the University of Manchester in England. These experiments consisted of shooting a beam of charged particles at thin foils of various elements and observing how the foil deflected the particles.

The particle accelerators now in common use in laboratories had not yet been invented, and Rutherford's projectiles were *alpha particles* emitted from naturally radioactive elements. The nature of these alpha particles was not completely understood, but it was known that they are ejected from unstable nuclei with speeds of the order of 10^7 m/s, are positively charged, and can travel several centimeters through air or 0.1 mm or so through solid matter before they are brought to rest by collisions.

Figure 39.11 is a schematic view of Rutherford's experimental setup. A radioactive substance at the left emits alpha particles. Thick lead screens stop all particles except those in a narrow beam. The beam passes through the foil target (consisting of gold, silver, or copper) and strikes screens coated with zinc sulfide, creating a momentary flash, or *scintillation*. Rutherford and his students counted the numbers of particles deflected through various angles.

The atoms in a metal foil are packed together like marbles in a box (not spaced apart). Because the particle beam passes through the foil, the alpha particles must pass through the interior of atoms. Within an atom, the charged alpha particle will interact with the electrons and the positive charge. (Because the *total* charge of the atom is zero, alpha particles feel little electric force outside an atom.) An electron has about 7300 times less mass than an alpha particle, so momentum considerations indicate that the atom's electrons cannot appreciably deflect the alpha particle—any more than a swarm of gnats deflects a tossed pebble. Any deflection will be due to the positively charged material that makes up almost all of the atom's mass.

39.10 Born in New Zealand, Ernest Rutherford (1871–1937) spent his professional life in England and Canada. Before carrying out the experiments that established the existence of atomic nuclei, he shared (with Frederick Soddy) the 1908 Nobel Prize in chemistry for showing that radioactivity results from the disintegration of atoms.

PhET: Rutherford Scattering

39.11 The Rutherford scattering experiments investigated what happens to alpha particles fired at a thin gold foil. The results of this experiment helped reveal the structure of atoms.

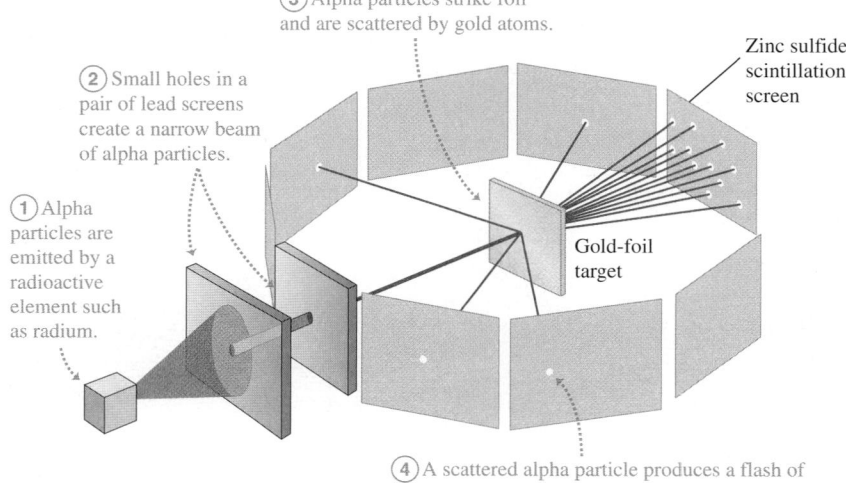

③ Alpha particles strike foil and are scattered by gold atoms.

② Small holes in a pair of lead screens create a narrow beam of alpha particles.

① Alpha particles are emitted by a radioactive element such as radium.

Zinc sulfide scintillation screen

Gold-foil target

④ A scattered alpha particle produces a flash of light when it hits a scintillation screen, showing the direction in which it was scattered.

39.12 A comparison of Thomson's and Rutherford's models of the atom.

(a) Thomson's model of the atom: An alpha particle is scattered through only a small angle.

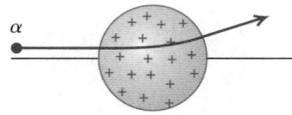

(b) Rutherford's model of the atom: An alpha particle can be scattered through a large angle by the compact, positively charged nucleus (not drawn to scale).

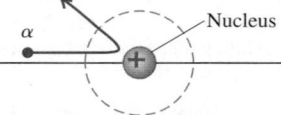

In the Thomson model, the positive charge and the negative electrons are distributed through the whole atom. Hence the electric field inside the atom should be quite small, and the electric force on an alpha particle that enters the atom should be quite weak. The maximum deflection to be expected is then only a few degrees (**Fig. 39.12a**). The results of the Rutherford experiments were *very* different from the Thomson prediction. Some alpha particles were scattered by nearly 180°—that is, almost straight backward (Fig. 39.12b). Rutherford later wrote:

> It was quite the most incredible event that ever happened to me in my life. It was almost as incredible as if you had fired a 15-inch shell at a piece of tissue paper and it came back and hit you.

Clearly the Thomson model was wrong and a new model was needed. Suppose the positive charge, instead of being distributed through a sphere with atomic dimensions (of the order of 10^{-10} m), is all concentrated in a much *smaller* volume. Then it would act like a point charge down to much smaller distances. The maximum electric field repelling the alpha particle would be much larger, and the amazing large-angle scattering that Rutherford observed could occur. Rutherford developed this model and called the concentration of positive charge the **nucleus.** He again computed the numbers of particles expected to be scattered through various angles. Within the accuracy of his experiments, the computed and measured results agreed, down to distances of the order of 10^{-14} m. His experiments therefore established that the atom does have a nucleus—a very small, very dense structure, no larger than 10^{-14} m in diameter. The nucleus occupies only about 10^{-12} of the total volume of the atom or less, but it contains *all* the positive charge and at least 99.95% of the total mass of the atom.

Figure 39.13 shows a computer simulation of alpha particles with a kinetic energy of 5.0 MeV being scattered from a gold nucleus of radius 7.0×10^{-15} m (the actual value) and from a nucleus with a hypothetical radius ten times larger. In the second case there is *no* large-angle scattering. The presence of large-angle scattering in Rutherford's experiments thus attested to the small size of the nucleus.

Later experiments showed that all nuclei are composed of positively charged protons (discovered in 1918) and electrically neutral neutrons (discovered in 1930). For example, the gold atoms in Rutherford's experiments have 79 protons and 118 neutrons. In fact, an alpha particle is itself the nucleus of a helium atom, with two protons and two neutrons. It is much more massive than an electron but only about 2% as massive as a gold nucleus, which helps explain why alpha particles are scattered by gold nuclei but not by electrons.

39.13 Computer simulation of scattering of 5.0-MeV alpha particles from a gold nucleus. Each curve shows a possible alpha-particle trajectory. (a) The scattering curves match Rutherford's experimental data if a radius of 7.0×10^{-15} m is assumed for a gold nucleus. (b) A model with a much larger radius for the gold nucleus does not match the data.

(a) A gold nucleus with radius 7.0×10^{-15} m gives large-angle scattering.

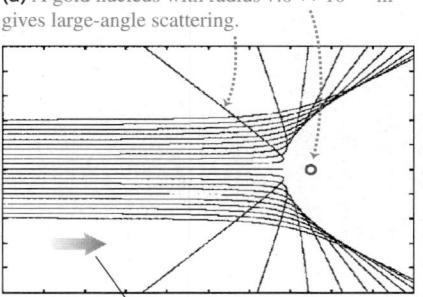

(b) A nucleus with 10 times the radius of the nucleus in (a) shows *no* large-scale scattering.

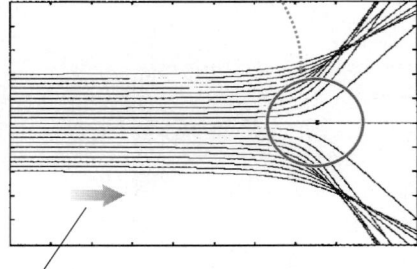

Motion of incident 5.0-MeV alpha particles

EXAMPLE 39.4 A RUTHERFORD EXPERIMENT

An alpha particle (charge $2e$) is aimed directly at a gold nucleus (charge $79e$). What minimum initial kinetic energy must the alpha particle have to approach within 5.0×10^{-14} m of the center of the gold nucleus before reversing direction? Assume that the gold nucleus, which has about 50 times the mass of an alpha particle, remains at rest.

SOLUTION

IDENTIFY: The repulsive electric force exerted by the gold nucleus makes the alpha particle slow to a halt as it approaches, then reverse direction. This force is conservative, so the total mechanical energy (kinetic energy of the alpha particle plus electric potential energy of the system) is conserved.

SET UP: Let point 1 be the initial position of the alpha particle, very far from the gold nucleus, and let point 2 be 5.0×10^{-14} m from the center of the gold nucleus. Our target variable is the kinetic energy K_1 of the alpha particle at point 1 that allows it to reach point 2 with $K_2 = 0$. To find this we'll use the law of conservation of energy and Eq. (23.9) for electric potential energy, $U = qq_0/4\pi\epsilon_0 r$.

EXECUTE: At point 1 the separation r of the alpha particle and gold nucleus is effectively infinite, so from Eq. (23.9) $U_1 = 0$. At point 2 the potential energy is

$$U_2 = \frac{1}{4\pi\epsilon_0} \frac{qq_0}{r}$$

$$= (9.0 \times 10^9 \text{ N} \cdot \text{m}^2/\text{C}^2) \frac{(2)(79)(1.60 \times 10^{-19} \text{ C})^2}{5.0 \times 10^{-14} \text{ m}}$$

$$= 7.3 \times 10^{-13} \text{ J} = 4.6 \times 10^6 \text{ eV} = 4.6 \text{ MeV}$$

In accordance with energy conservation, $K_1 + U_1 = K_2 + U_2$, so $K_1 = K_2 + U_2 - U_1 = 0 + 4.6 \text{ MeV} - 0 = 4.6 \text{ MeV}$. Thus, to approach within 5.0×10^{-14} m, the alpha particle must have initial kinetic energy $K_1 = 4.6 \text{ MeV}$.

EVALUATE: Alpha particles emitted from naturally occurring radioactive elements typically have energies in the range 4 to 6 MeV. For example, the common isotope of radium, ^{226}Ra, emits an alpha particle with energy 4.78 MeV.

Was it valid to assume that the gold nucleus remains at rest? To find out, note that when the alpha particle stops momentarily, all of its initial momentum has been transferred to the gold nucleus. An alpha particle has a mass $m_\alpha = 6.64 \times 10^{-27}$ kg; if its initial kinetic energy $K_1 = \frac{1}{2}mv_1^2$ is 7.3×10^{-13} J, you can show that its initial speed is $v_1 = 1.5 \times 10^7$ m/s and its initial momentum is $p_1 = m_\alpha v_1 = 9.8 \times 10^{-20}$ kg $\cdot$ m/s. A gold nucleus (mass $m_{\text{Au}} = 3.27 \times 10^{-25}$ kg) with this much momentum has a much slower speed $v_{\text{Au}} = 3.0 \times 10^5$ m/s and kinetic energy $K_{\text{Au}} = \frac{1}{2}mv_{\text{Au}}^2 = 1.5 \times 10^{-14}$ J $= 0.092$ MeV. This *recoil kinetic energy* of the gold nucleus is only 2% of the total energy in this situation, so we are justified in ignoring it.

The Failure of Classical Physics

Rutherford's discovery of the atomic nucleus raised a serious question: What prevented the negatively charged electrons from falling into the positively charged nucleus due to the strong electrostatic attraction? Rutherford suggested that perhaps the electrons *revolve* in orbits about the nucleus, just as the planets revolve around the sun.

But according to classical electromagnetic theory, any accelerating electric charge (either oscillating or revolving) radiates electromagnetic waves. An example is the radiation from an oscillating point charge that we depicted in Fig. 32.3 (Section 32.1). An electron orbiting inside an atom would always have a centripetal acceleration toward the nucleus, and so should be emitting radiation *at all times*. The energy of an orbiting electron should therefore decrease continuously, its orbit should become smaller and smaller, and it should spiral into the nucleus within a fraction of a second (**Fig. 39.14**). Even worse, according to classical theory the *frequency* of the electromagnetic waves emitted should equal the frequency of revolution. As the electrons radiated energy, their angular speeds would change continuously, and they would emit a *continuous* spectrum (a mixture of all frequencies), not the *line* spectrum actually observed.

Thus Rutherford's model of electrons orbiting the nucleus, which is based on Newtonian mechanics and classical electromagnetic theory, makes three entirely *wrong* predictions about atoms: They should emit light continuously, they should be unstable, and the light they emit should have a continuous spectrum. Clearly a radical reappraisal of physics on the scale of the atom was needed. In the next section we will see the bold idea that led to a new understanding of the atom, and see how this idea meshes with de Broglie's no less bold notion that electrons have wave attributes.

39.14 Classical physics makes predictions about the behavior of atoms that do not match reality.

ACCORDING TO CLASSICAL PHYSICS:

- An orbiting electron is accelerating, so it should radiate electromagnetic waves.
- The waves would carry away energy, so the electron should lose energy and spiral inward.
- The electron's angular speed would increase as its orbit shrank, so the frequency of the radiated waves should increase.

Thus, classical physics says that atoms should collapse within a fraction of a second and should emit light with a continuous spectrum as they do so.

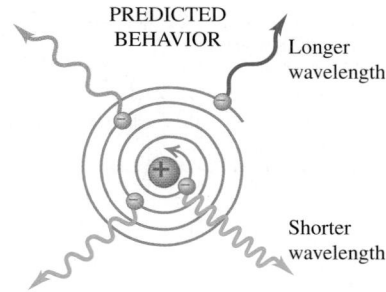

PREDICTED BEHAVIOR

Longer wavelength

Shorter wavelength

IN FACT:

- Atoms are stable.
- They emit light only when excited, and only at specific frequencies (as a line spectrum).

TEST YOUR UNDERSTANDING OF SECTION 39.2   Suppose you repeated Rutherford's scattering experiment with a thin sheet of solid hydrogen in place of the gold foil. (Hydrogen is a solid at temperatures below 14.0 K.) The nucleus of a hydrogen atom is a single proton, with about one-fourth the mass of an alpha particle. Compared to the original experiment with gold foil, would you expect the alpha particles in this experiment to undergo (i) more large-angle scattering; (ii) the same amount of large-angle scattering; or (iii) less large-angle scattering? ▌

39.3 ENERGY LEVELS AND THE BOHR MODEL OF THE ATOM

In 1913 a young Danish physicist working with Ernest Rutherford at the University of Manchester made a revolutionary proposal to explain both the stability of atoms and their emission and absorption line spectra. The physicist was Niels Bohr (**Fig. 39.15**), and his innovation was to combine the photon concept that we introduced in Chapter 38 with a fundamentally new idea: The energy of an atom can have only certain particular values. His hypothesis represented a clean break from 19th-century ideas.

Photon Emission and Absorption by Atoms

Bohr's reasoning went like this. The emission line spectrum of an element tells us that atoms of that element emit photons with only certain specific frequencies f and hence certain specific energies $E = hf$. During the emission of a photon, the internal energy of the atom changes by an amount equal to the energy of the photon. Therefore, said Bohr, each atom must be able to exist with only certain specific values of internal energy. Each atom has a set of possible **energy levels.** An atom can have an amount of internal energy equal to any one of these levels, but it *cannot* have an energy *intermediate* between two levels. All isolated atoms of a given element have the same set of energy levels, but atoms of different elements have different sets.

Suppose an atom is raised, or *excited*, to a high energy level. (In a hot gas this happens when fast-moving atoms undergo inelastic collisions with each other or with the walls of the gas container. In an electric discharge tube, such as those used in a neon light fixture, atoms are excited by collisions with fast-moving electrons.) According to Bohr, an excited atom can make a *transition* from one energy level to a lower level by emitting a photon with energy equal to the energy *difference* between the initial and final levels (**Fig. 39.16**):

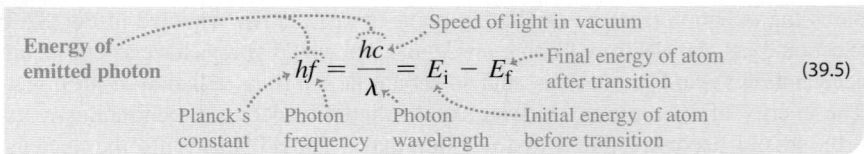

Energy of emitted photon ···
Speed of light in vacuum
$$hf = \frac{hc}{\lambda} = E_i - E_f \qquad (39.5)$$
Final energy of atom after transition
Planck's constant · Photon frequency · Photon wavelength · Initial energy of atom before transition

For example, an excited lithium atom emits red light with wavelength $\lambda = 671$ nm. The corresponding photon energy is

$$E = \frac{hc}{\lambda} = \frac{(6.63 \times 10^{-34}\ \text{J} \cdot \text{s})(3.00 \times 10^8\ \text{m/s})}{671 \times 10^{-9}\ \text{m}}$$
$$= 2.96 \times 10^{-19}\ \text{J} = 1.85\ \text{eV}$$

This photon is emitted during a transition like that shown in Fig. 39.16 between two levels of the atom that differ in energy by $E_i - E_f = 1.85$ eV.

Emission line spectra (Fig. 39.8) show that many different wavelengths are emitted by each atom. Hence each kind of atom must have a number of energy levels, with different spacings in energy between them. Each wavelength in the spectrum corresponds to a transition between two specific atomic energy levels.

39.15 Niels Bohr (1885–1962) was a young postdoctoral researcher when he proposed the novel idea that the energy of an atom could have only certain discrete values. He won the 1922 Nobel Prize in physics for these ideas. Bohr went on to make seminal contributions to nuclear physics and to become a passionate advocate for the free exchange of scientific ideas among all nations.

39.16 An excited atom emitting a photon.

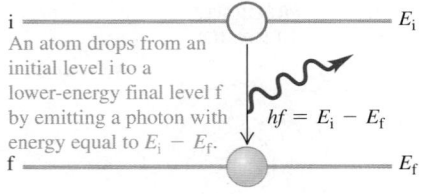

i ———————○——————— E_i
An atom drops from an initial level i to a lower-energy final level f by emitting a photon with energy equal to $E_i - E_f$. $hf = E_i - E_f$
f ———————●——————— E_f

PhET: Models of the Hydrogen Atom

CAUTION **Producing a line spectrum** The lines of an emission line spectrum, such as the helium spectrum shown at the top of Fig. 39.8, are *not* all produced by a single atom. The sample of helium gas that produced the spectrum in Fig. 39.8 contained a large number of helium atoms; these were excited in an electric discharge tube to various energy levels. The spectrum of the gas shows the light emitted from all the different transitions that occurred in different atoms of the sample. ▮

The observation that atoms are stable means that each atom has a *lowest* energy level, called the **ground level.** Levels with energies greater than the ground level are called **excited levels.** An atom in an excited level, called an *excited atom,* can make a transition into the ground level by emitting a photon as in Fig. 39.16. But since there are no levels below the ground level, an atom in the ground level cannot lose energy and so cannot emit a photon.

Collisions are not the only way that an atom's energy can be raised from one level to a higher level. If an atom initially in the lower energy level in Fig. 39.16 is struck by a photon with just the right amount of energy, the photon can be *absorbed* and the atom will end up in the higher level (**Fig. 39.17**). As an example, we previously mentioned two levels in the lithium atom with an energy difference of 1.85 eV. For a photon to be absorbed and excite the atom from the lower level to the higher one, the photon must have an energy of 1.85 eV and a wavelength of 671 nm. In other words, an atom *absorbs* the same wavelengths that it *emits.* This explains the correspondence between an element's emission line spectrum and its absorption line spectrum that we described in Section 39.2.

Note that a lithium atom *cannot* absorb a photon with a slightly longer wavelength (say, 672 nm) or one with a slightly shorter wavelength (say, 670 nm). That's because these photons have, respectively, slightly too little or slightly too much energy to raise the atom's energy from one level to the next, and an atom cannot have an energy that's intermediate between levels. This explains why absorption line spectra have distinct dark lines (see Fig. 39.9): Atoms can absorb only photons with specific wavelengths.

An atom that's been excited into a high energy level, either by photon absorption or by collisions, does not stay there for long. After a short time, called the *lifetime* of the level (typically around 10^{-8} s), the excited atom will emit a photon and make a transition into a lower excited level or the ground level. A cool gas that's illuminated by white light to make an *absorption* line spectrum thus also produces an *emission* line spectrum when viewed from the side, since when the atoms de-excite they emit photons in all directions (**Fig. 39.18**). To keep a gas of

39.17 An atom absorbing a photon. (Compare with Fig. 39.16.)

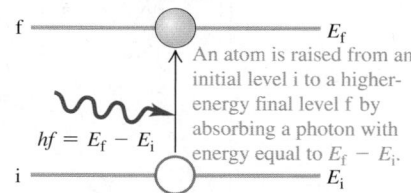

f ———————●——————— E_f

An atom is raised from an initial level i to a higher-energy final level f by absorbing a photon with energy equal to $E_f - E_i$.

$hf = E_f - E_i$

i ———————○——————— E_i

39.18 When a beam of white light with a continuous spectrum passes through a cool gas, the transmitted light has an absorption spectrum. The absorbed light energy excites the gas and causes it to emit light of its own, which has an emission spectrum.

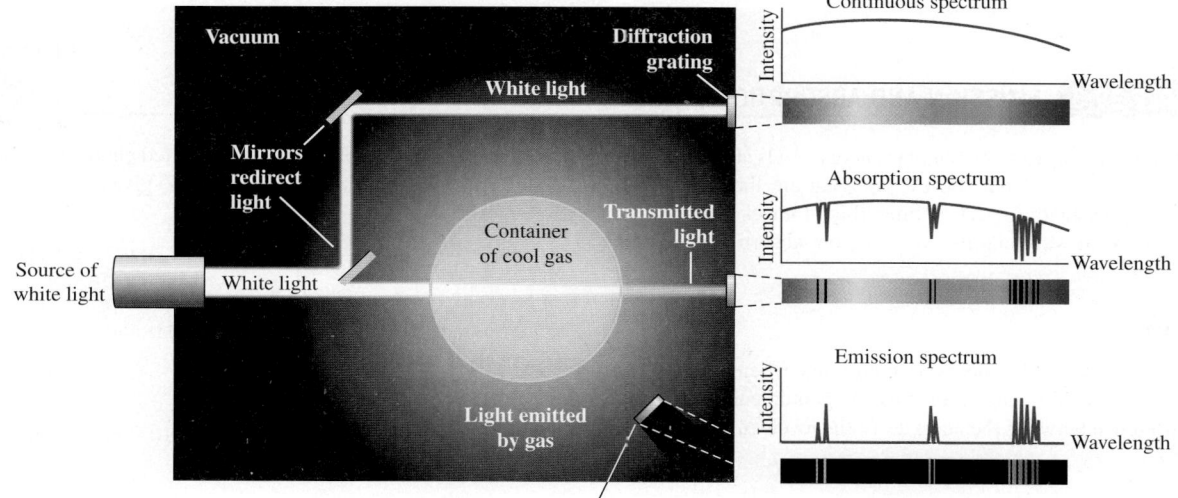

Diffraction grating for light emitted by gas

39.19 (a) Energy levels of the sodium atom relative to the ground level. Numbers on the lines between levels are wavelengths of the light emitted or absorbed during transitions between those levels. The column labels, such as $^2S_{1/2}$, refer to certain quantum states of the atom. (b) When a sodium compound is placed in a flame, sodium atoms are excited into the lowest excited levels. As they drop back to the ground level, the atoms emit photons of yellow-orange light with wavelengths 589.0 and 589.6 nm.

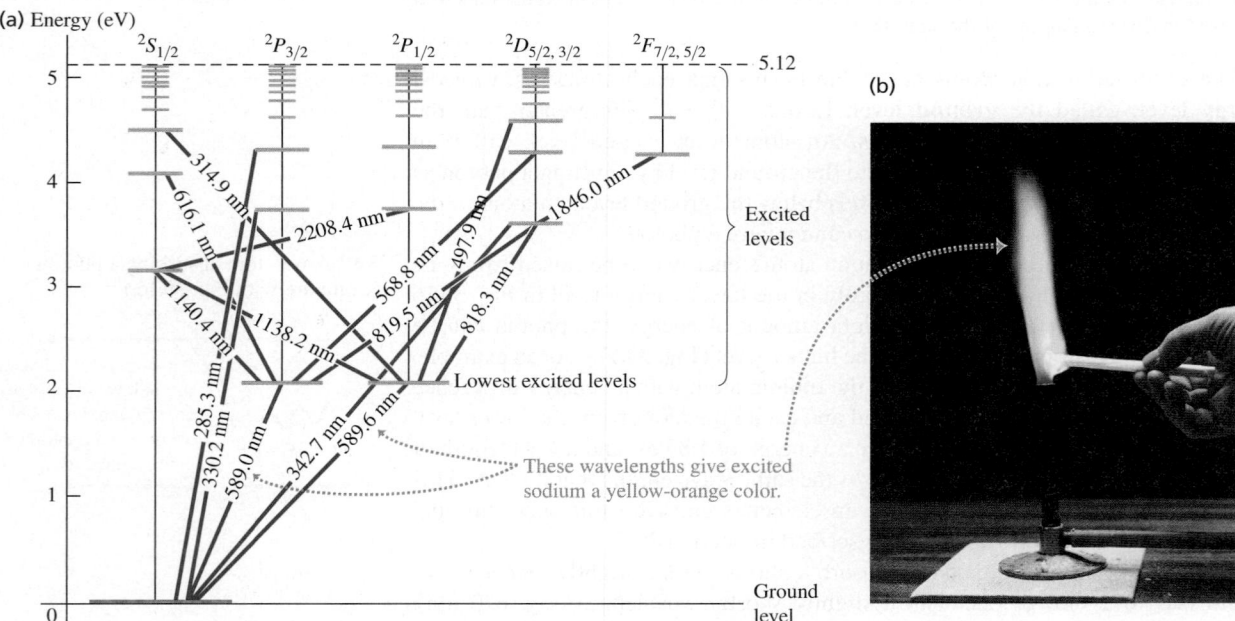

(a) Energy (eV)

(b)

These wavelengths give excited sodium a yellow-orange color.

DEMO

atoms glowing, you have to continually provide energy to the gas in order to re-excite atoms so that they can emit more photons. If you turn off the energy supply (for example, by turning off the electric current through a neon light fixture, or by shutting off the light source in Fig. 39.18), the atoms drop back into their ground levels and cease to emit light.

By working backward from the observed emission line spectrum of an element, physicists can deduce the arrangement of energy levels in an atom of that element. As an example, **Fig. 39.19a** shows some of the energy levels for a sodium atom. You may have noticed the yellow-orange light emitted by sodium vapor street lights. Sodium atoms emit this characteristic yellow-orange light with wavelengths 589.0 and 589.6 nm when they make transitions from the two closely spaced levels labeled *lowest excited levels* to the ground level. A standard test for the presence of sodium compounds is to look for this yellow-orange light from a sample placed in a flame (Fig. 39.19b).

EXAMPLE 39.5 | **EMISSION AND ABSORPTION SPECTRA**

A hypothetical atom (**Fig. 39.20a**) has energy levels at 0.00 eV (the ground level), 1.00 eV, and 3.00 eV. (a) What are the frequencies and wavelengths of the spectral lines this atom can emit when excited? (b) What wavelengths can this atom absorb if it is in its ground level?

SOLUTION

IDENTIFY and SET UP: Energy is conserved when a photon is emitted or absorbed. In each transition the photon energy is equal to the difference between the energies of the levels involved in the transition.

EXECUTE: (a) The possible energies of emitted photons are 1.00 eV, 2.00 eV, and 3.00 eV. For 1.00 eV, Eq. (39.2) gives

$$f = \frac{E}{h} = \frac{1.00 \text{ eV}}{4.136 \times 10^{-15} \text{ eV} \cdot \text{s}} = 2.42 \times 10^{14} \text{ Hz}$$

For 2.00 eV and 3.00 eV, $f = 4.84 \times 10^{14}$ Hz and 7.25×10^{14} Hz, respectively. For 1.00-eV photons,

$$\lambda = \frac{c}{f} = \frac{3.00 \times 10^8 \text{ m/s}}{2.42 \times 10^{14} \text{ Hz}} = 1.24 \times 10^{-6} \text{ m} = 1240 \text{ nm}$$

39.20 (a) Energy-level diagram for the hypothetical atom, showing the possible transitions for emission from excited levels and for absorption from the ground level. (b) Emission spectrum of this hypothetical atom.

(a)

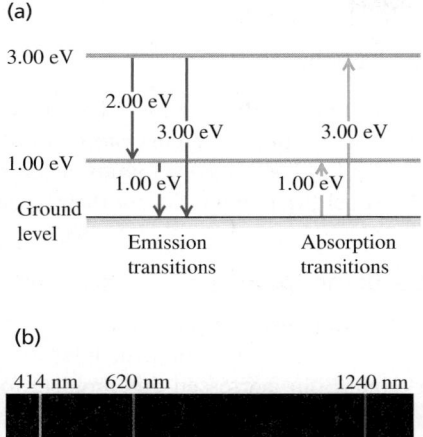

(b)

This is in the infrared region of the spectrum (Fig. 39.20b). For 2.00 eV and 3.00 eV, the wavelengths are 620 nm (red) and 414 nm (violet), respectively.

(b) From the ground level, only a 1.00-eV or a 3.00-eV photon can be absorbed (Fig. 39.20a); a 2.00-eV photon cannot be absorbed because the atom has no energy level 2.00 eV above the ground level. Passing light from a hot solid through a gas of these hypothetical atoms (almost all of which would be in the ground level if the gas were cool) would yield a continuous spectrum with dark absorption lines at 1240 nm and 414 nm.

EVALUATE: Note that if a gas of these atoms were at a sufficiently high temperature, collisions would excite a number of atoms into the 1.00-eV energy level. Such excited atoms *can* absorb 2.00-eV photons, as Fig. 39.20a shows, and an absorption line at 620 nm would appear in the spectrum. Thus the observed spectrum of a given substance depends on its energy levels and its temperature.

Suppose we take a gas of the hypothetical atoms in Example 39.5 and illuminate it with violet light of wavelength 414 nm. Atoms in the ground level can absorb this photon and make a transition to the 3.00-eV level. Some of these atoms will make a transition back to the ground level by emitting a 414-nm photon. But other atoms will return to the ground level in two steps, first emitting a 620-nm photon to transition to the 1.00-eV level, then a 1240-nm photon to transition back to the ground level. Thus this gas will emit longer-wavelength radiation than it absorbs, a phenomenon called *fluorescence*. For example, the electric discharge in a fluorescent lamp causes the mercury vapor in the tube to emit ultraviolet radiation. This radiation is absorbed by the atoms of the coating on the inside of the tube. The coating atoms then re-emit light in the longer-wavelength, visible portion of the spectrum. Fluorescent lamps are more efficient than incandescent lamps in converting electrical energy to visible light because they do not waste as much energy producing (invisible) infrared photons.

Our discussion of energy levels and spectra has concentrated on *atoms,* but the same ideas apply to *molecules.* Figure 39.8 shows the emission line spectra of two molecules, hydrogen (H_2) and water (H_2O). Just as for sodium or other atoms, physicists can work backward from these molecular spectra and deduce the arrangement of energy levels for each kind of molecule. We'll return to molecules and molecular structure in Chapter 42.

BIO Application Fish Fluorescence
When illuminated by blue light, this tropical lizardfish (family Synodontidae) fluoresces and emits longer-wavelength green light. The fluorescence may be a sexual signal or a way for the fish to camouflage itself among coral (which also have a green fluorescence).

The Franck–Hertz Experiment: Are Energy Levels Real?

Are atomic energy levels real, or just a convenient fiction that helps us to explain spectra? In 1914, the German physicists James Franck and Gustav Hertz answered this question when they found direct experimental evidence for the existence of atomic energy levels.

Franck and Hertz studied the motion of electrons through mercury vapor under the action of an electric field. They found that when the electron kinetic energy was 4.9 eV or greater, the vapor emitted ultraviolet light of wavelength 250 nm. Suppose mercury atoms have an excited energy level 4.9 eV above the ground level. An atom can be raised to this level by collision with an electron;

it later decays back to the ground level by emitting a photon. From the photon formula $E = hc/\lambda$, the wavelength of the photon should be

$$\lambda = \frac{hc}{E} = \frac{(4.136 \times 10^{-15}\ \text{eV} \cdot \text{s})(3.00 \times 10^8\ \text{m/s})}{4.9\ \text{eV}}$$

$$= 2.5 \times 10^{-7}\ \text{m} = 250\ \text{nm}$$

This is equal to the wavelength that Franck and Hertz measured, which demonstrates that this energy level actually exists in the mercury atom. Similar experiments with other atoms yield the same kind of evidence for atomic energy levels. Franck and Hertz shared the 1925 Nobel Prize in physics for their research.

Electron Waves and the Bohr Model of Hydrogen

Bohr's hypothesis established the relationship between atomic spectra and energy levels. By itself, however, it provided no general principles for *predicting* the energy levels of a particular atom. Bohr addressed this problem for the case of the simplest atom, hydrogen, which has just one electron.

In the **Bohr model,** Bohr postulated that each energy level of a hydrogen atom corresponds to a specific *stable* circular orbit of the electron around the nucleus. In a break with classical physics, Bohr further postulated that an electron in such an orbit does *not* radiate. Instead, an atom radiates energy only when an electron makes a transition from an orbit of energy E_i to a different orbit with lower energy E_f, emitting a photon of energy $hf = E_i - E_f$ in the process.

As a result of a rather complicated argument that related the angular frequency of the light emitted to the angular speed of the electron in highly excited energy levels, Bohr found that the magnitude of the electron's angular momentum is *quantized;* that is, this magnitude must be an integral multiple of $h/2\pi$. (Because $1\ \text{J} = 1\ \text{kg} \cdot \text{m}^2/\text{s}^2$, the SI units of Planck's constant h, J·s, are the same as the SI units of angular momentum, usually written as $\text{kg} \cdot \text{m}^2/\text{s}$.) Let's number the orbits by an integer n, where $n = 1, 2, 3, \ldots$, and call the radius of orbit $n\ r_n$ and the speed of the electron in that orbit v_n. The value of n for each orbit is called the **principal quantum number** for the orbit. From Section 10.5, Eq. (10.25), the magnitude of the angular momentum of an electron of mass m in such an orbit is $L_n = mv_n r_n$ (**Fig. 39.21**). So Bohr's argument led to

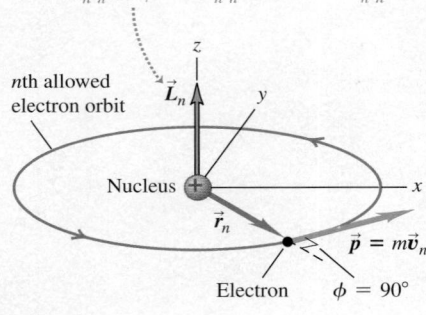

39.21 Calculating the angular momentum of an electron in a circular orbit around an atomic nucleus.

Angular momentum $\vec{L}_n$ of orbiting electron is perpendicular to plane of orbit (since we take origin to be at nucleus) and has magnitude $L = mv_n r_n \sin\phi = mv_n r_n \sin 90° = mv_n r_n$.

*n*th allowed electron orbit

Nucleus

$\vec{r}_n$

$\vec{p} = m\vec{v}_n$

Electron $\phi = 90°$

	Orbital angular momentum	Principal quantum number ($n = 1, 2, 3, \ldots$)	
Quantization of angular momentum:	$L_n = mv_n r_n = n\dfrac{h}{2\pi}$	Planck's constant	(39.6)
	Electron mass Electron speed Electron orbital radius		

Instead of going through Bohr's argument to justify Eq. (39.6), we can use de Broglie's picture of electron waves. Rather than visualizing the orbiting electron as a particle moving around the nucleus in a circular path, think of it as a sinusoidal *standing wave* with wavelength λ that extends around the circle. A standing wave on a string transmits no energy (see Section 15.7), and electrons in Bohr's orbits radiate no energy. For the wave to "come out even" and join onto itself smoothly, the circumference of this circle must include some *whole number* of wavelengths, as **Fig. 39.22** suggests. Hence for an orbit with radius r_n and circumference $2\pi r_n$, we must have $2\pi r_n = n\lambda_n$, where λ_n is the wavelength and $n = 1, 2, 3, \ldots$. According to the de Broglie relationship, Eq. (39.1), the wavelength of a particle with rest mass m moving with nonrelativistic speed v_n is $\lambda_n = h/mv_n$. Combining $2\pi r_n = n\lambda_n$ and $\lambda_n = h/mv_n$, we find $2\pi r_n = nh/mv_n$ or

$$mv_n r_n = n\frac{h}{2\pi}$$

This is the same as Bohr's result, Eq. (39.6). Thus a wave picture of the electron leads naturally to the quantization of the electron's angular momentum.

Now let's consider a model of the hydrogen atom that is Newtonian in spirit but incorporates this quantization assumption (**Fig. 39.23**). This atom consists of a single electron with mass m and charge $-e$ in a circular orbit around a single proton with charge $+e$. The proton is nearly 2000 times as massive as the electron, so we can assume that the proton does not move. We learned in Section 5.4 that when a particle with mass m moves with speed v_n in a circular orbit with radius r_n, its centripetal (inward) acceleration is v_n^2/r_n. According to Newton's second law, a radially inward net force with magnitude $F = mv_n^2/r_n$ is needed to cause this acceleration. We discussed in Section 13.4 how the gravitational attraction provides that inward force for satellite orbits. In hydrogen the force is provided by the electrical attraction between the proton (charge $+e$) and the electron (charge $-e$). From Coulomb's law, Eq. (21.2),

$$F = \frac{1}{4\pi\epsilon_0}\frac{|(+e)(-e)|}{r_n^2} = \frac{1}{4\pi\epsilon_0}\frac{e^2}{r_n^2}$$

Hence Newton's second law states that

$$\frac{1}{4\pi\epsilon_0}\frac{e^2}{r_n^2} = \frac{mv_n^2}{r_n} \tag{39.7}$$

When we solve Eqs. (39.6) and (39.7) simultaneously for r_n and v_n, we get

Radius of nth orbit in the Bohr model ⋯⋯
$$r_n = \epsilon_0\frac{n^2h^2}{\pi me^2} \tag{39.8}$$
Principal quantum number $(n = 1, 2, 3, \dots)$
Planck's constant
Magnitude of electron charge
Electric constant   Electron mass

Orbital speed in nth orbit in the Bohr model ⋯⋯
$$v_n = \frac{1}{\epsilon_0}\frac{e^2}{2nh} \tag{39.9}$$
Magnitude of electron charge
Planck's constant
Electric constant   Principal quantum number $(n = 1, 2, 3, \dots)$

Equation (39.8) shows that the orbit radius r_n is proportional to n^2, so the smallest orbit radius corresponds to $n = 1$. We'll denote this minimum radius, called the *Bohr radius*, as a_0:

$$a_0 = \epsilon_0\frac{h^2}{\pi me^2} \quad \text{(Bohr radius)} \tag{39.10}$$

Then we can rewrite Eq. (39.8) as

Radius of nth orbit in the Bohr model ⋯⋯
$$r_n = n^2a_0 \tag{39.11}$$
Bohr radius
Principal quantum number $(n = 1, 2, 3, \dots)$

The permitted orbits have radii a_0, $4a_0$, $9a_0$, and so on.

You can find the numerical values of the quantities on the right-hand side of Eq. (39.10) in Appendix F. Using these values, we find that the radius a_0 of the smallest Bohr orbit is

$$a_0 = \frac{(8.854 \times 10^{-12}\,\text{C}^2/\text{N}\cdot\text{m}^2)(6.626 \times 10^{-34}\,\text{J}\cdot\text{s})^2}{\pi(9.109 \times 10^{-31}\,\text{kg})(1.602 \times 10^{-19}\,\text{C})^2}$$

$$= 5.29 \times 10^{-11}\,\text{m}$$

This gives an atomic diameter of about 10^{-10} m $= 0.1$ nm, which is consistent with atomic dimensions estimated by other methods.

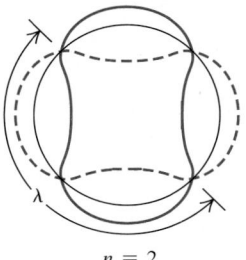

39.22 The idea of fitting a standing electron wave around a circular orbit. For the wave to join onto itself smoothly, the circumference of the orbit must be an integral number n of wavelengths.

$n = 2$

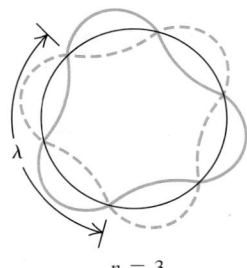

$n = 3$

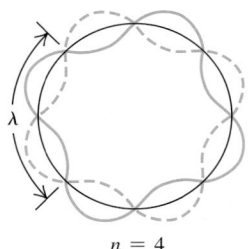

$n = 4$

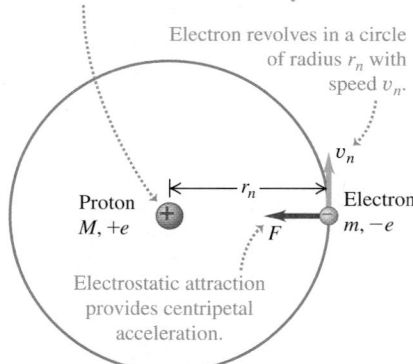

39.23 The Bohr model of the hydrogen atom.

Proton is assumed to be stationary.

Electron revolves in a circle of radius r_n with speed v_n.

v_n

Proton $M, +e$ r_n Electron $m, -e$

F

Electrostatic attraction provides centripetal acceleration.

Equation (39.9) shows that the orbital speed v_n is proportional to $1/n$. Hence the greater the value of n, the larger the orbital radius of the electron and the slower its orbital speed. (We saw the same relationship between orbital radius and speed for satellite orbits in Section 13.4.) We leave it to you to calculate the speed in the $n = 1$ orbit, which is the greatest possible speed of the electron in the hydrogen atom (see Exercise 39.23); the result is $v_1 = 2.19 \times 10^6$ m/s. This is less than 1% of the speed of light, so relativistic considerations aren't significant.

Hydrogen Energy Levels in the Bohr Model

We can now use Eqs. (39.8) and (39.9) to find the kinetic and potential energies K_n and U_n when the electron is in the orbit with quantum number n:

$$K_n = \tfrac{1}{2}mv_n{}^2 = \frac{1}{\epsilon_0{}^2}\frac{me^4}{8n^2h^2} \qquad \text{(kinetic energies in the Bohr model)} \tag{39.12}$$

$$U_n = -\frac{1}{4\pi\epsilon_0}\frac{e^2}{r_n} = -\frac{1}{\epsilon_0{}^2}\frac{me^4}{4n^2h^2} \qquad \text{(potential energies in the Bohr model)} \tag{39.13}$$

The electric potential energy is negative because we have taken its value to be zero when the electron is infinitely far from the nucleus. We are interested only in the *differences* in energy between orbits, so the reference position doesn't matter. The total energy E_n is the sum of the kinetic and potential energies:

$$E_n = K_n + U_n = -\frac{1}{\epsilon_0{}^2}\frac{me^4}{8n^2h^2} \qquad \text{(total energies in the Bohr model)} \tag{39.14}$$

Since E_n in Eq. (39.14) has a different value for each n, you can see that this equation gives the *energy levels* of the hydrogen atom in the Bohr model. Each distinct orbit corresponds to a distinct energy level.

Figure 39.24 depicts the orbits and energy levels. We label the possible energy levels of the atom by values of the quantum number n. For each value of n there are corresponding values of orbit radius r_n, speed v_n, angular momentum

39.24 Two ways to represent the energy levels of the hydrogen atom and the transitions between them. Note that the radius of the nth permitted orbit is actually n^2 times the radius of the $n = 1$ orbit.

(a) Permitted orbits of an electron in the Bohr model of a hydrogen atom (not to scale). Arrows indicate the transitions responsible for some of the lines of various series.

(b) Energy-level diagram for hydrogen, showing some transitions corresponding to the various series

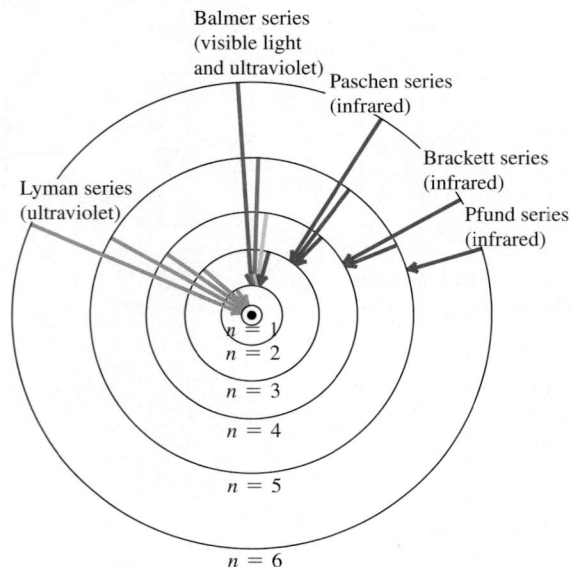

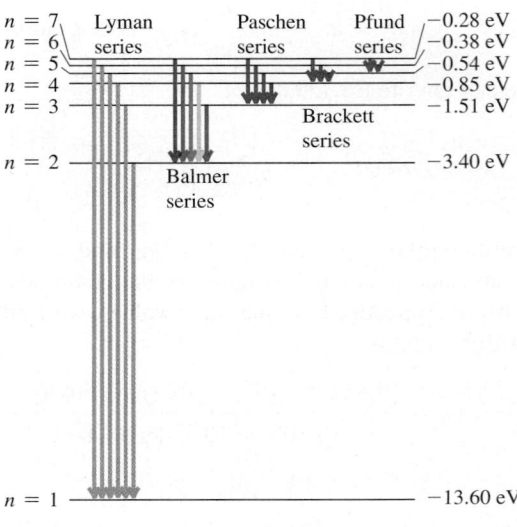

$L_n = nh/2\pi$, and total energy E_n. The energy of the atom is least when $n = 1$ and E_n has its most negative value. This is the *ground level* of the hydrogen atom; it is the level with the smallest orbit, of radius a_0. For $n = 2, 3, \ldots$, the absolute value of E_n is smaller and the energy is progressively larger (less negative).

Figure 39.24 also shows some of the possible transitions from one electron orbit to an orbit of lower energy. Consider a transition from orbit n_U (for "upper") to a smaller orbit n_L (for "lower"), with $n_L < n_U$—or, equivalently, from *level* n_U to a lower *level* n_L. Then the energy hc/λ of the emitted photon of wavelength λ is equal to $E_{n_U} - E_{n_L}$. Before we use this relationship to solve for λ, it's convenient to rewrite Eq. (39.14) for the energies as

Total energy for *n*th orbit in the Bohr model

$$E_n = -\frac{hcR}{n^2}, \quad \text{where} \quad R = \frac{me^4}{8\epsilon_0^2 h^3 c} \tag{39.15}$$

Planck's constant · Speed of light in vacuum · Electron mass · Magnitude of electron charge · Principal quantum number $(n = 1, 2, 3, \ldots)$ · Rydberg constant · Electric constant

The quantity R in Eq. (39.15) is called the **Rydberg constant** (named for the Swedish physicist Johannes Rydberg, who did pioneering work on the hydrogen spectrum). When we substitute the numerical values of the fundamental physical constants m, c, e, h, and ϵ_0, all of which can be determined quite independently of the Bohr theory, we find that $R = 1.097 \times 10^7 \text{ m}^{-1}$. Now we solve for the wavelength of the photon emitted in a transition from level n_U to level n_L:

$$\frac{hc}{\lambda} = E_{n_U} - E_{n_L} = \left(-\frac{hcR}{n_U^2}\right) - \left(-\frac{hcR}{n_L^2}\right) = hcR\left(\frac{1}{n_L^2} - \frac{1}{n_U^2}\right)$$

$$\frac{1}{\lambda} = R\left(\frac{1}{n_L^2} - \frac{1}{n_U^2}\right) \qquad \begin{array}{l}\text{(hydrogen wavelengths in the}\\\text{Bohr model, } n_L < n_U)\end{array} \tag{39.16}$$

Equation (39.16) is a *theoretical prediction* of the wavelengths found in the *emission* line spectrum of hydrogen atoms. When a hydrogen atom *absorbs* a photon, an electron makes a transition from a level n_L to a *higher* level n_U. This can happen only if the photon energy hc/λ is equal to $E_{n_U} - E_{n_L}$, which is the same condition expressed by Eq. (39.16). So this equation also predicts the wavelengths found in the *absorption* line spectrum of hydrogen.

How does this prediction compare with experiment? If $n_L = 2$, corresponding to transitions to the second energy level in Fig. 39.24, the wavelengths predicted by Eq. (39.16)—collectively called the *Balmer series* (**Fig. 39.25**)—are all in the visible and ultraviolet parts of the electromagnetic spectrum. If we let $n_L = 2$ and $n_U = 3$ in Eq. (39.16) we obtain the wavelength of the H_α line:

$$\frac{1}{\lambda} = (1.097 \times 10^7 \text{ m}^{-1})\left(\tfrac{1}{4} - \tfrac{1}{9}\right) \qquad \text{or} \qquad \lambda = 656.3 \text{ nm}$$

PhET: Neon Lights and Other Discharge Lamps

39.25 The Balmer series of spectral lines for atomic hydrogen. You can see these same lines in the spectrum of *molecular* hydrogen (H_2) shown in Fig. 39.8, as well as additional lines that are present only when two hydrogen atoms are combined to make a molecule.

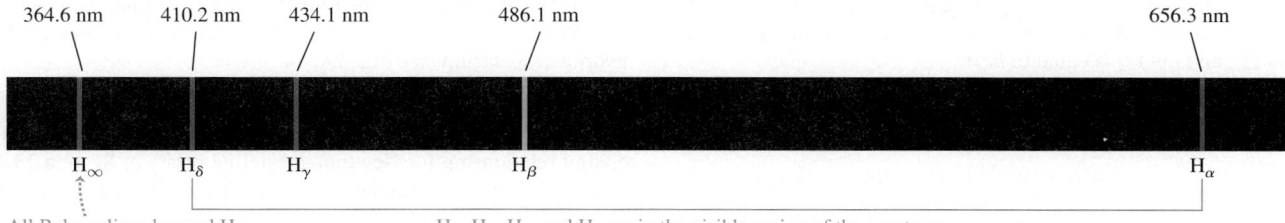

364.6 nm   410.2 nm   434.1 nm   486.1 nm   656.3 nm

H_∞ H_δ H_γ H_β H_α

All Balmer lines beyond H_δ are in the ultraviolet spectrum.

H_α, H_β, H_γ, and H_δ are in the visible region of the spectrum.

DATA *SPEAKS*

The Hydrogen Spectrum

When students were given a problem involving the spectrum of atomic hydrogen, more than 36% gave an incorrect response. Common errors:

- Confusion about energy levels, photon energy, and wavelength. The difference in energy between two energy levels of an atom equals the energy of a photon emitted or absorbed in a transition between these levels. Hence the *greater* the energy difference, the *shorter* the wavelength of the photon.

- Confusion about transitions between energy levels. A transition can "skip over" levels, so the quantum number *n* can change by more than 1 (for example, when an atom starts in the $n = 5$ level, emits a photon, and ends up in the $n = 2$ level).

With $n_L = 2$ and $n_U = 4$ we obtain the wavelength of the H$_\beta$ line, and so on. With $n_L = 2$ and $n_U = \infty$ we obtain the shortest wavelength in the series, $\lambda = 364.6$ nm. These theoretical predictions are within 0.1% of the observed hydrogen wavelengths! This close agreement provides very strong and direct confirmation of Bohr's theory.

The Bohr model also predicts many other wavelengths in the hydrogen spectrum, as Fig. 39.24 shows. The observed wavelengths of all of these series, each of which is named for its discoverer, match the predicted values with the same percent accuracy as for the Balmer series. The *Lyman series* of spectral lines is caused by transitions between the ground level and the excited levels, corresponding to $n_L = 1$ and $n_U = 2, 3, 4, \ldots$ in Eq. (39.16). The energy difference between the ground level and any of the excited levels is large, so the emitted photons have wavelengths in the ultraviolet part of the electromagnetic spectrum. Transitions among the higher energy levels involve a much smaller energy difference, so the photons emitted in these transitions have little energy and long, infrared wavelengths. That's the case for both the *Brackett series* ($n_L = 3$ and $n_U = 4, 5, 6, \ldots$, corresponding to transitions between the third and higher energy levels) and the *Pfund series* ($n_L = 4$ and $n_U = 5, 6, 7, \ldots$, with transitions between the fourth and higher energy levels).

Figure 39.24 shows only transitions in which a hydrogen atom emits a photon. But as we discussed previously, the wavelengths of those photons that an atom can *absorb* are the same as those that it can emit. For example, a hydrogen atom in the $n = 2$ level can absorb a 656.3-nm photon and end up in the $n = 3$ level.

One additional test of the Bohr model is its predicted value of the *ionization energy* of the hydrogen atom. This is the energy required to remove the electron completely from the atom. Ionization corresponds to a transition from the ground level ($n = 1$) to an infinitely large orbit radius ($n = \infty$), so the energy that must be added to the atom is $E_\infty - E_1 = 0 - E_1 = -E_1$ (recall that E_1 is negative). Substituting the constants from Appendix F into Eq. (39.15) gives an ionization energy of 13.606 eV. The ionization energy can also be measured directly; the result is 13.60 eV. These two values agree within 0.1%.

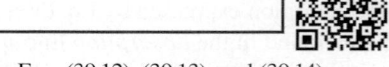

EXAMPLE 39.6 EXPLORING THE BOHR MODEL

Find the kinetic, potential, and total energies of the hydrogen atom in the first excited level, and find the wavelength of the photon emitted in a transition from that level to the ground level.

SOLUTION

IDENTIFY and SET UP: This problem uses the ideas of the Bohr model. We use simplified versions of Eqs. (39.12), (39.13), and (39.14) to find the energies of the atom, and Eq. (39.16), $hc/\lambda = E_{n_U} - E_{n_L}$, to find the photon wavelength λ in a transition from $n_U = 2$ (the first excited level) to $n_L = 1$ (the ground level).

EXECUTE: We could evaluate Eqs. (39.12), (39.13), and (39.14) for the nth level by substituting the values of m, e, ϵ_0, and h. But we can simplify the calculation by comparing with Eq. (39.15), which shows that the constant $me^4/8\epsilon_0^2 h^2$ that appears in Eqs. (39.12), (39.13), and (39.14) is equal to hcR:

$$\frac{me^4}{8\epsilon_0^2 h^2} = hcR$$

$$= (6.626 \times 10^{-34} \text{ J} \cdot \text{s})(2.998 \times 10^8 \text{ m/s})$$

$$\times (1.097 \times 10^7 \text{ m}^{-1})$$

$$= 2.179 \times 10^{-18} \text{ J} = 13.60 \text{ eV}$$

This allows us to rewrite Eqs. (39.12), (39.13), and (39.14) as

$$K_n = \frac{13.60 \text{ eV}}{n^2} \qquad U_n = \frac{-27.20 \text{ eV}}{n^2} \qquad E_n = \frac{-13.60 \text{ eV}}{n^2}$$

For the first excited level ($n = 2$), we have $K_2 = 3.40$ eV, $U_2 = -6.80$ eV, and $E_2 = -3.40$ eV. For the ground level ($n = 1$), $E_1 = -13.60$ eV. The energy of the emitted photon is then $E_2 - E_1 = -3.40$ eV $- (-13.60$ eV$) = 10.20$ eV, and

$$\lambda = \frac{hc}{E_2 - E_1} = \frac{(4.136 \times 10^{-15} \text{ eV} \cdot \text{s})(3.00 \times 10^8 \text{ m/s})}{10.20 \text{ eV}}$$

$$= 1.22 \times 10^{-7} \text{ m} = 122 \text{ nm}$$

This is the wavelength of the Lyman-alpha (L_α) line, the longest-wavelength line in the Lyman series of ultraviolet lines in the hydrogen spectrum (see Fig. 39.24).

EVALUATE: The total mechanical energy for any level is negative and is equal to one-half the potential energy. We found the same energy relationship for Newtonian satellite orbits in Section 13.4. The situations are similar because both the electrostatic and gravitational forces are inversely proportional to $1/r^2$.

Nuclear Motion and the Reduced Mass of an Atom

The Bohr model is so successful that we can justifiably ask why its predictions for the wavelengths and ionization energy of hydrogen differ from the measured values by about 0.1%. The explanation is that we assumed that the nucleus (a proton) remains at rest. However, as **Fig. 39.26** shows, the proton and electron *both* orbit about their common center of mass (see Section 8.5). It turns out that we can take this motion into account by using in Bohr's equations not the electron rest mass m but a quantity called the **reduced mass** m_r of the system. For a system composed of two bodies of masses m_1 and m_2, the reduced mass is

$$m_r = \frac{m_1 m_2}{m_1 + m_2} \tag{39.17}$$

For ordinary hydrogen we let m_1 equal m and m_2 equal the proton mass, $m_p = 1836.2m$. Thus ordinary hydrogen has a reduced mass of

$$m_r = \frac{m(1836.2m)}{m + 1836.2m} = 0.99946m$$

When this value is used instead of the electron mass m in the Bohr equations, the predicted values agree very well with the measured values.

In an atom of deuterium, also called *heavy hydrogen,* the nucleus is not a single proton but a proton and a neutron bound together to form a composite body called the *deuteron.* The reduced mass of the deuterium atom turns out to be $0.99973m$. Equations (39.15) and (39.16) (with m replaced by m_r) show that all wavelengths are inversely proportional to m_r. Thus the wavelengths of the deuterium spectrum should be those of hydrogen divided by $(0.99973m)/(0.99946m) = 1.00027$. This is a small effect but well within the precision of modern spectrometers. This small wavelength shift led the American scientist Harold Urey to the discovery of deuterium in 1932, an achievement that earned him the 1934 Nobel Prize in chemistry.

Hydrogenlike Atoms

We can extend the Bohr model to other one-electron atoms, such as singly ionized helium (He^+), doubly ionized lithium (Li^{2+}), and so on. Such atoms are called *hydrogenlike* atoms. In such atoms, the nuclear charge is not e but Ze, where Z is the *atomic number,* equal to the number of protons in the nucleus. The effect in the previous analysis is to replace e^2 everywhere by Ze^2. You should verify that the orbital radii r_n given by Eq. (39.8) become smaller by a factor of Z, and the energy levels E_n given by Eq. (39.14) are multiplied by Z^2. The reduced-mass correction in these cases is even less than 0.1% because the nuclei are more massive than the single proton of ordinary hydrogen. **Figure 39.27** compares the energy levels for H and for He^+, which has $Z = 2$.

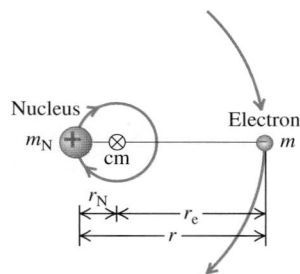

39.26 The nucleus and the electron both orbit around their common center of mass. The distance r_N has been exaggerated for clarity; for ordinary hydrogen it actually equals $r_e/1836.2$.

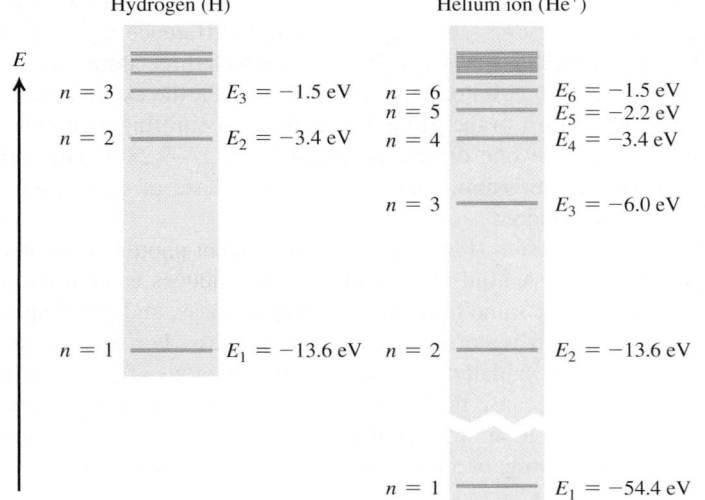

39.27 Energy levels of H and He^+. The energy expression, Eq. (39.14), is multiplied by $Z^2 = 4$ for He^+, so the energy of an He^+ ion with a given n is almost exactly four times that of an H atom with the same n. (There are small differences of the order of 0.05% because the reduced masses are slightly different.)

Atoms of the alkali metals (at the far left-hand side of the periodic table; see Appendix D) have one electron outside a core consisting of the nucleus and the inner electrons, with net core charge $+e$. These atoms are approximately hydrogenlike, especially in excited levels. Physicists have studied alkali atoms in which the outer electron has been excited into a very large orbit with $n = 1000$. From Eq. (39.8), the radius of such a *Rydberg atom* with $n = 1000$ is $n^2 = 10^6$ times the Bohr radius, or about 0.05 mm—about the size of a small grain of sand.

Although the Bohr model predicted the energy levels of the hydrogen atom correctly, it raised as many questions as it answered. It combined elements of classical physics with new postulates that were inconsistent with classical ideas. The model provided no insight into what happens during a transition from one orbit to another; the angular speeds of the electron motion were not in general the angular frequencies of the emitted radiation, a result that is contrary to classical electrodynamics. Attempts to extend the model to atoms with two or more electrons were not successful. An electron moving in one of Bohr's circular orbits forms a current loop and should produce a magnetic dipole moment (see Section 27.7). However, a hydrogen atom in its ground level has *no* magnetic moment due to orbital motion. In Chapters 40 and 41 we will find that an even more radical departure from classical concepts was needed before the understanding of atomic structure could progress further.

TEST YOUR UNDERSTANDING OF SECTION 39.3 Consider the possible transitions between energy levels in a He$^+$ ion. For which of these transitions in He$^+$ will the wavelength of the emitted photon be nearly the same as one of the wavelengths emitted by excited H atoms? (i) $n = 2$ to $n = 1$; (ii) $n = 3$ to $n = 2$; (iii) $n = 4$ to $n = 3$; (iv) $n = 4$ to $n = 2$; (v) more than one of these; (vi) none of these. ▮

39.4 THE LASER

The **laser** is a light source that produces a beam of highly coherent and very nearly monochromatic light as a result of cooperative emission from many atoms. The name "laser" is an acronym for "light amplification by stimulated emission of radiation." We can understand the principles of laser operation from what we have learned about atomic energy levels and photons. To do this we'll have to introduce two new concepts: *stimulated emission* and *population inversion*.

Spontaneous and Stimulated Emission

Consider a gas of atoms in a transparent container. Each atom is initially in its ground level of energy E_g and also has an excited level of energy E_{ex}. If we shine light of frequency f on the container, an atom can absorb one of the photons provided the photon energy $E = hf$ equals the energy difference $E_{ex} - E_g$ between the levels. **Figure 39.28a** shows this process, in which three atoms A each absorb a photon and go into the excited level. Some time later, the excited atoms (which we denote as A*) return to the ground level by each emitting a photon with the same frequency as the one originally absorbed (Fig. 39.28b). This process is called **spontaneous emission.** The direction and phase of each spontaneously emitted photon are random.

In **stimulated emission** (Fig. 39.28c), each incident photon encounters a previously excited atom. A kind of resonance effect induces each atom to emit a second photon with the same frequency, direction, phase, and polarization as the incident photon, which is not changed by the process. For each atom there is one photon before a stimulated emission and two photons after—thus the name *light amplification*. Because the two photons have the same phase, they emerge together as *coherent* radiation. The laser makes use of stimulated emission to produce a beam consisting of a large number of such coherent photons.

39.28 Three processes in which atoms interact with light.

(a) Absorption

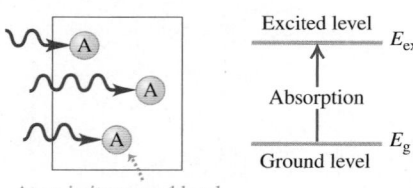

Atom in its ground level

(b) Spontaneous emission

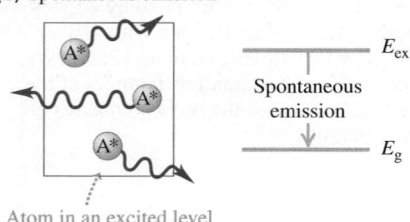

Atom in an excited level

(c) Stimulated emission

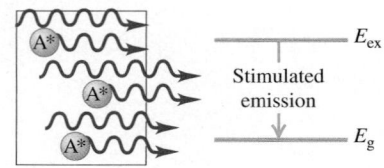

To discuss stimulated emission from atoms in excited levels. we need to know something about how many atoms are in each of the various energy levels. First, we need to make the distinction between the terms *energy level* and *state*. A system may have more than one way to attain a given energy level; each different way is a different **state.** For instance, there are two ways of putting an ideal unstretched spring in a given energy level. Remembering that the spring potential energy is $U = \frac{1}{2}kx^2$, we could compress the spring by $x = -b$ or we could stretch it by $x = +b$ to get the same $U = \frac{1}{2}kb^2$. The Bohr model had only one state in each energy level, but we will find in Chapter 41 that the hydrogen atom (Fig. 39.24b) actually has two *ground states* in its -13.60-eV ground level, eight *excited states* in its -3.40-eV first excited level, and so on.

The Maxwell–Boltzmann distribution function (see Section 18.5) determines the number of atoms in a given state in a gas. The function tells us that when the gas is in thermal equilibrium at absolute temperature T, the number n_i of atoms in a state with energy E_i equals $Ae^{-E_i/kT}$, where k is the Boltzmann constant and A is another constant determined by the total number of atoms in the gas. (In Section 18.5, E was the kinetic energy $\frac{1}{2}mv^2$ of a gas molecule; here we're talking about the internal energy of an atom.) Because of the negative exponent, fewer atoms are in higher-energy states. If E_g is a ground-state energy and E_{ex} is the energy of an excited state, then the ratio of numbers of atoms in the two states is

$$\frac{n_{ex}}{n_g} = \frac{Ae^{-E_{ex}/kT}}{Ae^{-E_g/kT}} = e^{-(E_{ex}-E_g)/kT} \tag{39.18}$$

For example, suppose $E_{ex} - E_g = 2.0\,\text{eV} = 3.2 \times 10^{-19}\,\text{J}$, the energy of a 620-nm visible-light photon. At $T = 3000\,\text{K}$ (roughly the temperature of the filament in an incandescent light bulb or restaurant heat lamp),

$$\frac{E_{ex} - E_g}{kT} = \frac{3.2 \times 10^{-19}\,\text{J}}{(1.38 \times 10^{-23}\,\text{J/K})(3000\,\text{K})} = 7.73$$

and

$$e^{-(E_{ex}-E_g)/kT} = e^{-7.73} = 0.00044$$

That is, the fraction of atoms in a state 2.0 eV above a ground state is extremely small, even at this high temperature. The point is that at any reasonable temperature there aren't enough atoms in excited states for any appreciable amount of stimulated emission from these states to occur. Rather, a photon emitted by one of the rare excited atoms will almost certainly be absorbed by an atom in the ground state rather than encountering another excited atom.

Enhancing Stimulated Emission: Population Inversions

To make a laser, we need to promote stimulated emission by increasing the number of atoms in excited states. Can we do that simply by illuminating the container with radiation of frequency $f = E/h$ corresponding to the energy difference $E = E_{ex} - E_g$, as in Fig. 39.28a? Some of the atoms absorb photons of energy E and are raised to the excited state, and the population ratio n_{ex}/n_g momentarily increases. But because n_g is originally so much larger than n_{ex}, an enormously intense beam of light would be required to momentarily increase n_{ex} to a value comparable to n_g. The rate at which energy is *absorbed* from the beam by the n_g ground-state atoms far exceeds the rate at which energy is added to the beam by stimulated emission from the relatively rare (n_{ex}) excited atoms.

We need to create a *nonequilibrium* situation in which there are more atoms in a higher-energy state than in a lower-energy state. Such a situation is called a **population inversion.** Then the rate of energy radiation by stimulated emission can *exceed* the rate of absorption, and the system will act as a net *source* of radiation with photon energy E. We can achieve a population inversion by

39.29 (a), (b), (c) Stages in the operation of a four-level laser. (d) The light emitted by atoms making spontaneous transitions from state E_2 to state E_1 is reflected between mirrors, so it continues to stimulate emission and gives rise to coherent light. One mirror is partially transmitting and allows the high-intensity light beam to escape.

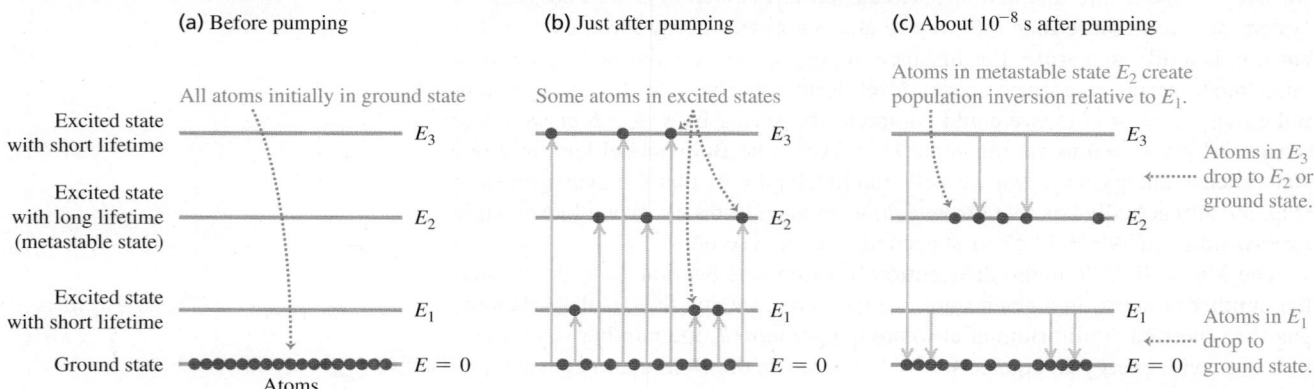

(a) Before pumping

(b) Just after pumping

(c) About 10^{-8} s after pumping

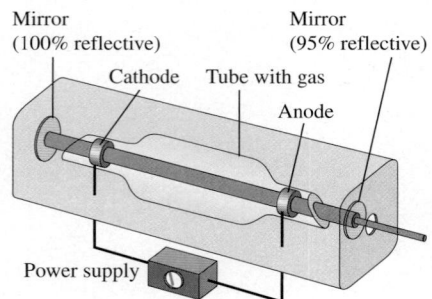

(d) Schematic of gas laser

starting with atoms that have the right kinds of excited states. **Figure 39.29a** shows an energy-level diagram for such an atom with a ground state and *three* excited states of energies E_1, E_2, and E_3. A laser that uses a material with energy levels like these is called a *four-level laser*. For the laser action to work, the states of energies E_1 and E_3 must have ordinary short lifetimes of about 10^{-8} s, while the state of energy E_2 must have an unusually long lifetime of 10^{-3} s or so. Such a long-lived **metastable state** can occur if, for instance, there are restrictions imposed by conservation of angular momentum that hinder photon emission from this state. (We'll discuss these restrictions in Chapter 41.) The metastable state is the one that we want to populate.

To produce a population inversion, we *pump* the material to excite the atoms out of the ground state into the states of energies E_1, E_2, and E_3 (Fig. 39.29b). If the atoms are in a gas, we can do this by inserting two electrodes into the gas container. When a burst of sufficiently high voltage is applied to the electrodes, an electric discharge occurs. Collisions between ionized atoms and electrons carrying the discharge current then excite the atoms to various energy states. Within about 10^{-8} s the atoms that are excited to states E_1 and E_3 undergo spontaneous photon emission, so these states end up depopulated. But atoms "pile up" in the metastable state with energy E_2. The number of atoms in the metastable state is *less* than the number in the ground state, but is *much greater* than in the nearly unoccupied state of energy E_1. Hence there is a population inversion of state E_2 relative to state E_1 (Fig. 39.29c). You can see why we need the two levels E_1 and E_3: Atoms that undergo spontaneous emission from the E_3 level help to populate the E_2 level, and the presence of the E_1 level makes a population inversion possible.

Over the next 10^{-3} s, some of the atoms in the long-lived metastable state E_2 transition to state E_1 by spontaneous emission. The emitted photons of energy $hf = E_2 - E_1$ are sent back and forth through the gas many times by a pair of parallel mirrors (Fig. 39.29d), so that they can *stimulate* emission from as many of the atoms in state E_2 as possible. The net result of all these processes is a beam of light of frequency f that can be quite intense, has parallel rays, is highly monochromatic, and is spatially *coherent* at all points within a given cross section—that is, a laser beam. One of the mirrors is partially transparent, so a portion of the beam emerges.

What we've described is a *pulsed* laser that produces a burst of coherent light every time the atoms are pumped. Pulsed lasers are used in LASIK eye surgery (an acronym for *laser-assisted in situ keratomileusis*) to reshape the cornea and correct for nearsightedness, farsightedness, or astigmatism. In a *continuous* laser, such as those found in the barcode scanners used at retail checkout counters, energy is supplied to the atoms continuously (for instance, by having the power supply in Fig. 39.29d provide a steady voltage to the electrodes) and a steady beam of light emerges from the laser. For such a laser the pumping must be intense enough to sustain the population inversion, so that the rate at which atoms are added to level E_2 through pumping equals the rate at which atoms in this level emit a photon and transition to level E_1.

Since a special arrangement of energy levels is needed for laser action, it's not surprising that only certain materials can be used to make a laser. Some types of laser use a solid, transparent material such as neodymium glass rather than a gas. The most common kind of laser—used in laser printers (Section 21.1), laser pointers, and to read the data on the disc in a DVD player or Blu-ray player—is a *semiconductor laser,* which doesn't use atomic energy levels at all. As we'll discuss in Chapter 42, these lasers instead use the energy levels of electrons that are free to roam throughout the volume of the semiconductors.

TEST YOUR UNDERSTANDING OF SECTION 39.4 An ordinary neon light fixture like those used in advertising signs emits red light of wavelength 632.8 nm. Neon atoms are also used in a helium–neon laser (a type of gas laser). The light emitted by a neon light fixture is (i) spontaneous emission; (ii) stimulated emission; (iii) both spontaneous and stimulated emission. ▌

39.5 CONTINUOUS SPECTRA

PhET: Blackbody Spectrum
PhET: The Greenhouse Effect

Emission line spectra come from matter in the gaseous state, in which the atoms are so far apart that interactions between them are negligible and each atom behaves as an isolated system. By contrast, a heated solid or liquid (in which atoms are close to each other) nearly always emits radiation with a *continuous* distribution of wavelengths rather than a line spectrum.

Here's an analogy that suggests why there is a difference. A tuning fork emits sound waves of a single definite frequency (a pure tone) when struck. But if you tightly pack a suitcase full of tuning forks and then shake the suitcase, the proximity of the tuning forks to each other affects the sound that they produce. What you hear is mostly noise, which is sound with a continuous distribution of all frequencies. In the same manner, isolated atoms in a gas emit light of certain distinct frequencies when excited, but if the same atoms are crowded together in a solid or liquid they produce a continuous spectrum of light.

In this section we'll study an idealized case of continuous-spectrum radiation from a hot, dense object. Just as was the case for the emission line spectrum of light from an atom, we'll find that we can understand the continuous spectrum only if we use the ideas of energy levels and photons.

In the same way that an atom's emission spectrum has the same lines as its absorption spectrum, the ideal surface for *emitting* light with a continuous spectrum is one that also *absorbs* all wavelengths of electromagnetic radiation. Such an ideal surface is called a *blackbody* because it would appear perfectly black when illuminated; it would reflect no light at all. The continuous-spectrum radiation that a blackbody emits is called **blackbody radiation.** Like a perfectly frictionless incline or a massless rope, a perfect blackbody does not exist but is nonetheless a useful idealization.

A good approximation to a blackbody is a hollow box with a small aperture in one wall (**Fig. 39.30**). Light that enters the aperture will eventually be absorbed by the walls of the box, so the box is a nearly perfect absorber. Conversely, when we heat the box, the light that emanates from the aperture is nearly ideal blackbody radiation with a continuous spectrum.

By 1900 blackbody radiation had been studied extensively, and three characteristics had been established. First, the total intensity I (the average rate of radiation of energy per unit surface area or average power per area) emitted from

39.30 A hollow box with a small aperture behaves like a blackbody. When the box is heated, the electromagnetic radiation that emerges from the aperture has a blackbody spectrum.

Hollow box with small aperture (cross section)

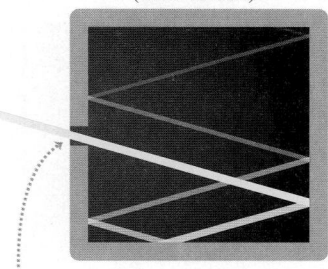

Light that enters box is eventually absorbed. Hence box approximates a perfect blackbody.

39.31 This close-up view of the sun's surface shows two dark sunspots. Their temperature is about 4000 K, while the surrounding solar material is at $T = 5800$ K. From the Stefan–Boltzmann law, the intensity from a given area of sunspot is only $(4000 \text{ K}/5800 \text{ K})^4 = 0.23$ as great as the intensity from the same area of the surrounding material—which is why sunspots appear dark.

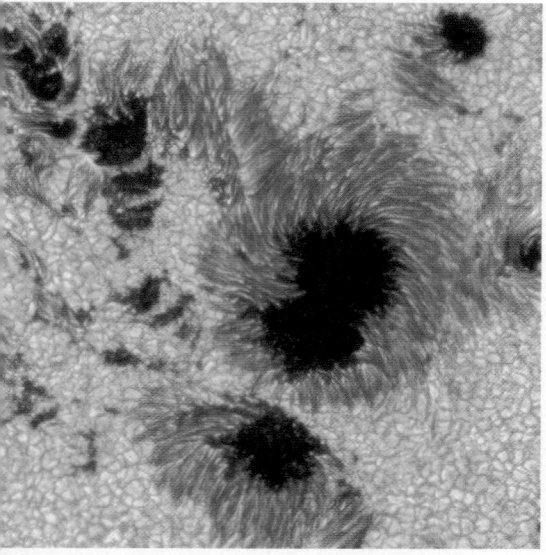

39.32 These graphs show the spectral emittance $I(\lambda)$ for radiation from a blackbody at three different temperatures.

As the temperature increases, the peak of the spectral emittance curve becomes higher and shifts to shorter wavelengths.

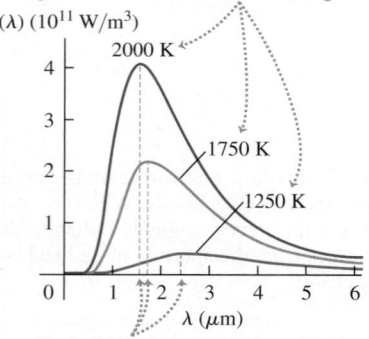

$I(\lambda)$ (10^{11} W/m^3)

2000 K
1750 K
1250 K

λ (μm)

Dashed blue lines are values of λ_{m} in Eq. (39.21) for each temperature.

the surface of an ideal radiator is proportional to the fourth power of the absolute temperature (**Fig. 39.31**). This is the **Stefan–Boltzmann law:**

Stefan–Boltzmann law for a blackbody:

Intensity of radiation from blackbody

$$I = \sigma T^4$$

Absolute temperature of blackbody

Stefan–Boltzmann constant

(39.19)

We encountered a version of this relationship in Section 17.7 during our study of heat transfer. In SI units, the value of the Stefan–Boltzmann constant σ is

$$\sigma = 5.670373(21) \times 10^{-8} \text{ W/m}^2 \cdot \text{K}^4$$

Second, the intensity is not uniformly distributed over all wavelengths. Its distribution can be measured and described by the intensity per wavelength interval $I(\lambda)$, called the *spectral emittance*. Thus $I(\lambda) \, d\lambda$ is the intensity corresponding to wavelengths in the interval from λ to $\lambda + d\lambda$. The *total* intensity I, given by Eq. (39.19), is the *integral* of the distribution function $I(\lambda)$ over all wavelengths, which equals the area under the $I(\lambda)$-versus-λ curve:

$$I = \int_0^\infty I(\lambda) \, d\lambda \qquad (39.20)$$

CAUTION *Spectral emittance vs. intensity* Although we use the symbol $I(\lambda)$ for spectral emittance, keep in mind that spectral emittance is *not* the same thing as intensity I. Intensity is power per unit area, with units W/m^2. Spectral emittance is power per unit area *per unit wavelength interval*, with units W/m^3. ▌

Figure 39.32 shows the measured spectral emittances $I(\lambda)$ for blackbody radiation at three different temperatures. Each has a peak wavelength λ_{m} at which the emitted intensity per wavelength interval is largest. Experiment shows that λ_{m} is inversely proportional to T, so their product is constant and equal to 2.90×10^{-3} m $\cdot$ K. This observation is called the **Wien displacement law:**

Wien displacement law for a blackbody:

Peak wavelength in spectral emittance curve

$$\lambda_{\text{m}} T = 2.90 \times 10^{-3} \text{ m} \cdot \text{K}$$

Absolute temperature of blackbody

(39.21)

As the temperature rises, the peak of $I(\lambda)$ becomes higher and shifts to shorter wavelengths. Yellow light has shorter wavelengths than red light, so a body that glows yellow is hotter and brighter than one of the same size that glows red.

Third, experiments show that the *shape* of the distribution function is the same for all temperatures. We can make a curve for one temperature fit any other temperature by simply changing the scales on the graph.

Rayleigh and the "Ultraviolet Catastrophe"

During the last decade of the 19th century, many attempts were made to derive these empirical results about blackbody radiation from basic principles. In one attempt, the English physicist Lord Rayleigh considered the light enclosed within a rectangular box like that shown in Fig. 39.30. Such a box, he reasoned, has a series of possible *normal modes* for electromagnetic waves, as we discussed in Section 32.5. It also seemed reasonable to assume that the distribution of energy among the various modes would be given by the equipartition principle (see Section 18.4), which had been used successfully in the analysis of heat capacities.

Including both the electric- and magnetic-field energies, Rayleigh assumed that the total energy of each normal mode was equal to kT. Then by computing the *number* of normal modes corresponding to a wavelength interval $d\lambda$, Rayleigh calculated the expected distribution of wavelengths in the radiation within the box. Finally, he computed the predicted intensity distribution $I(\lambda)$ for the radiation emerging from the hole. His result was quite simple:

$$I(\lambda) = \frac{2\pi ckT}{\lambda^4} \quad \text{(Rayleigh's calculation)} \quad (39.22)$$

At large wavelengths this formula agrees quite well with the experimental results shown in Fig. 39.32, but there is serious disagreement at small wavelengths. The experimental curves in Fig. 39.32 fall toward zero at small λ. By contrast, Rayleigh's prediction in Eq. (39.22) goes in the opposite direction, approaching infinity as $1/\lambda^4$, a result that was called in Rayleigh's time the "ultraviolet catastrophe." Even worse, the integral of Eq. (39.22) over all λ is infinite, indicating an infinitely large *total* radiated intensity. Clearly, something is wrong.

Planck and the Quantum Hypothesis

Finally, in 1900, the German physicist Max Planck succeeded in deriving a function, now called the **Planck radiation law,** that agreed very well with experimental intensity distribution curves. In his derivation he made what seemed at the time to be a crazy assumption: that electromagnetic oscillators (electrons) in the walls of Rayleigh's box vibrating at a frequency f could have only certain values of energy equal to nhf, where $n = 0, 1, 2, 3, \ldots$ and h is the constant that now bears Planck's name. These oscillators were in equilibrium with the electromagnetic waves in the box, so they both emitted and absorbed light. His assumption gave quantized energy levels and said that the energy in each normal mode was also a multiple of hf. This was in sharp contrast to Rayleigh's point of view that each normal mode could have any amount of energy.

Planck was not comfortable with this quantum hypothesis; he regarded it as a calculational trick rather than a fundamental principle. In a letter to a friend, he called it "an act of desperation" into which he was forced because "a theoretical explanation had to be found at any cost, whatever the price." But five years later, Einstein identified the energy change hf between levels as the energy of a photon (see Section 38.1), and other evidence quickly mounted. By 1915 there was little doubt about the validity of the quantum concept and the existence of photons. By discussing atomic spectra *before* continuous spectra, we have departed from the historical order of things. The credit for inventing the concept of quantization of energy levels goes to Planck, even though he didn't believe it at first. He received the 1918 Nobel Prize in physics for his achievements.

Figure 39.33 shows energy-level diagrams for two of the oscillators that Planck envisioned in the walls of the rectangular box, one with a low frequency

39.33 Energy levels for two of the oscillators that Planck envisioned in the walls of a blackbody like that shown in Fig. 39.30. The spacing between adjacent energy levels for each oscillator is hf, which is smaller for the low-frequency oscillator.

Low-frequency oscillator	High-frequency oscillator
$12hf$	$2hf$
$11hf$	
$10hf$	
$9hf$	
$8hf$	
$7hf$	
$6hf$	hf
$5hf$	
$4hf$	
$3hf$	
$2hf$	
hf	
0	0

(a)

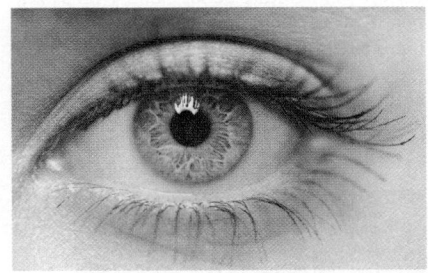

(b)

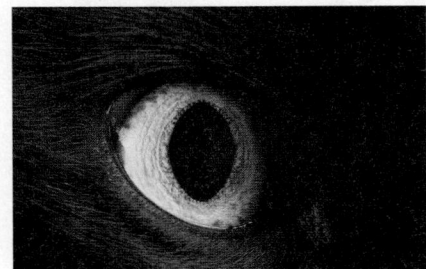

(c)

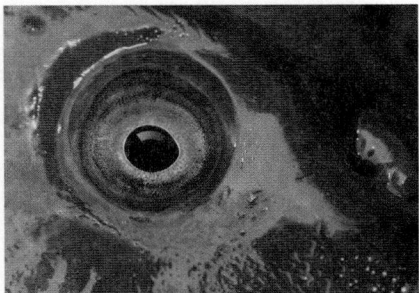

Application **Star Colors and the Planck Radiation Law** Stars (with radiation very similar to that of a blackbody) have a broad range of surface temperatures, from lower than 2500 K to higher than 30,000 K. The Wien displacement law and the shape of the Planck spectral emittance curve explain why these stars have different colors. From Eq. (39.21), a star with a high surface temperature of, say, 12,000 K has a short peak wavelength λ_m in the ultraviolet. Hence such a star emits more blue light than red light and appears blue to the eye. A star with a low surface temperature of, say, 3000 K has a long peak wavelength λ_m in the infrared, emits more red light than blue light, and appears red to the eye. For a star like the sun, which has a surface temperature of 5800 K, λ_m lies in the visible spectrum and the star appears white.

High-temperature stars appear blue.

Low-temperature stars appear red.

Visible spectrum

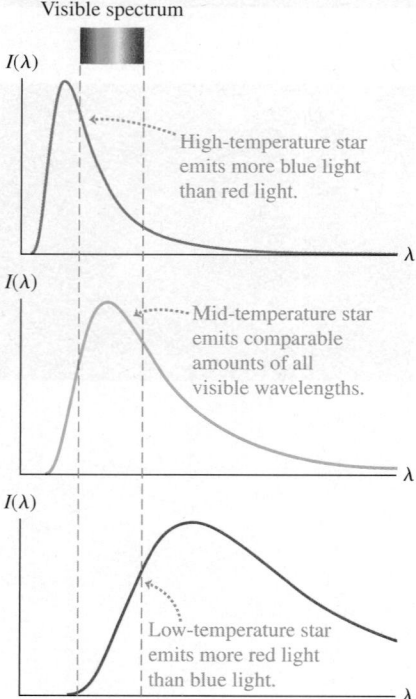

High-temperature star emits more blue light than red light.

Mid-temperature star emits comparable amounts of all visible wavelengths.

Low-temperature star emits more red light than blue light.

and the other with a high frequency. The spacing in energy between adjacent levels is hf. This spacing is small for the low-frequency oscillator that emits and absorbs photons of low frequency f and long wavelength $\lambda = c/f$. The energy spacing is greater for the high-frequency oscillator, which emits high-frequency photons of short wavelength.

According to Rayleigh's picture, both of these oscillators have the same amount of energy kT and are equally effective at emitting radiation. In Planck's model, however, the high-frequency oscillator is very ineffective as a source of light. To see why, we can use the ideas from Section 39.4 about the populations of various energy states. If we consider all the oscillators of a given frequency f in a box at temperature T, the number of oscillators that have energy nhf is $Ae^{-nhf/kT}$. The ratio of the number of oscillators in the first excited state ($n = 1$, energy hf) to the number of oscillators in the ground state ($n = 0$, energy zero) is

$$\frac{n_1}{n_0} = \frac{Ae^{-hf/kT}}{Ae^{-(0)/kT}} = e^{-hf/kT} \tag{39.23}$$

Let's evaluate Eq. (39.23) for $T = 2000$ K, one of the temperatures shown in Fig. 39.32. At this temperature $kT = 2.76 \times 10^{-20}$ J $= 0.172$ eV. For an oscillator that emits photons of wavelength $\lambda = 3.00\ \mu$m, $hf = hc/\lambda = 0.413$ eV; for a higher-frequency oscillator that emits photons of wavelength $\lambda = 0.500\ \mu$m, $hf = hc/\lambda = 2.48$ eV. For these two cases Eq. (39.23) gives

$$\frac{n_1}{n_0} = e^{-hf/kT} = 0.0909 \text{ for } \lambda = 3.00\ \mu\text{m}$$

$$\frac{n_1}{n_0} = e^{-hf/kT} = 5.64 \times 10^{-7} \text{ for } \lambda = 0.500\ \mu\text{m}$$

The value for $\lambda = 3.00\ \mu$m means that of all the oscillators that can emit light at this wavelength, 0.0909 of them—about one in 11—are in the first excited state. These excited oscillators can each emit a 3.00-μm photon and contribute it to the radiation inside the box. Hence we would expect that this radiation would be rather plentiful in the spectrum of radiation from a 2000 K blackbody. By contrast, the value for $\lambda = 0.500\ \mu$m means that only 5.64×10^{-7} (about one in two million) of the oscillators that can emit this wavelength are in the first excited state. An oscillator can't emit if it's in the ground state, so the amount of radiation in the box at this wavelength is *tremendously* suppressed compared to Rayleigh's prediction. That's why the spectral emittance curve for 2000 K in Fig. 39.32 has such a low value at $\lambda = 0.500\ \mu$m and shorter wavelengths. So Planck's quantum hypothesis provided a natural way to suppress the spectral emittance of a blackbody at short wavelengths, and hence averted the ultraviolet catastrophe that plagued Rayleigh's calculations.

We won't go into all the details of Planck's derivation of the spectral emittance. Here is his result:

Spectral emittance of blackbody · · · Planck's constant · · · Speed of light in vacuum

Planck radiation law:

$$I(\lambda) = \frac{2\pi hc^2}{\lambda^5 (e^{hc/\lambda kT} - 1)} \tag{39.24}$$

Absolute temperature of blackbody

Wavelength · · · Boltzmann constant

This function turns out to agree well with experimental emittance curves such as those in Fig. 39.32.

The Planck radiation law also contains the Wien displacement law and the Stefan–Boltzmann law as consequences. To derive the Wien law, we find the value of λ at which $I(\lambda)$ is maximum by taking the derivative of Eq. (39.24) and setting it equal to zero. We leave it to you to fill in the details; the result is

$$\lambda_m = \frac{hc}{4.965kT} \qquad (39.25)$$

To obtain this result, you have to solve the equation

$$5 - x = 5e^{-x} \qquad (39.26)$$

The root of this equation, found by trial and error or more sophisticated means, is 4.965 to four significant figures. You should evaluate the constant $hc/4.965k$ and show that it agrees with the experimental value of 2.90×10^{-3} m $\cdot$ K given in Eq. (39.21).

We can obtain the Stefan–Boltzmann law for a blackbody by integrating Eq. (39.24) over all λ to find the *total* radiated intensity (see Problem 39.61). This is not a simple integral; the result is

$$I = \int_0^\infty I(\lambda)\, d\lambda = \frac{2\pi^5 k^4}{15c^2 h^3}T^4 = \sigma T^4 \qquad (39.27)$$

in agreement with Eq. (39.19). Our result in Eq. (39.27) also shows that the constant σ in that law can be expressed in terms of other fundamental constants:

$$\sigma = \frac{2\pi^5 k^4}{15c^2 h^3} \qquad (39.28)$$

Substitute the values of k, c, and h from Appendix F and verify that you obtain the value $\sigma = 5.6704 \times 10^{-8}$ W/m$^2 \cdot$ K^4 for the Stefan–Boltzmann constant.

The Planck radiation law, Eq. (39.24), looks so different from the unsuccessful Rayleigh expression, Eq. (39.22), that it may seem unlikely that they would agree for any value of λ. But when λ is large, the exponent in the denominator of Eq. (39.24) is very small. We can then use the approximation $e^x \approx 1 + x$ (for x much less than 1). You should verify that when this is done, the result approaches Eq. (39.22), showing that the two expressions do agree in the limit of very large λ. We also note that the Rayleigh expression does not contain h. At very long wavelengths (very small photon energies), quantum effects become unimportant.

EXAMPLE 39.7   **LIGHT FROM THE SUN**

To a good approximation, the sun's surface is a blackbody with a surface temperature of 5800 K. (We are ignoring the absorption produced by the sun's atmosphere, shown in Fig. 39.9.) (a) At what wavelength does the sun emit most strongly? (b) What is the total radiated power per unit surface area?

SOLUTION

IDENTIFY and SET UP: Our target variables are the peak-intensity wavelength λ_m and the radiated power per area I. Hence we'll use the Wien displacement law, Eq. (39.21) (which relates λ_m to the blackbody temperature T), and the Stefan–Boltzmann law, Eq. (39.19) (which relates I to T).

EXECUTE: (a) From Eq. (39.21),

$$\lambda_m = \frac{2.90 \times 10^{-3} \text{ m} \cdot \text{K}}{T} = \frac{2.90 \times 10^{-3} \text{ m} \cdot \text{K}}{5800 \text{ K}}$$

$$= 0.500 \times 10^{-6} \text{ m} = 500 \text{ nm}$$

(b) From Eq. (39.19),

$$I = \sigma T^4 = (5.67 \times 10^{-8} \text{ W/m}^2 \cdot \text{K}^4)(5800 \text{ K})^4$$

$$= 6.42 \times 10^7 \text{ W/m}^2 = 64.2 \text{ MW/m}^2$$

EVALUATE: The 500-nm wavelength found in part (a) is near the middle of the visible spectrum. This should not be a surprise: The human eye evolved to take maximum advantage of natural light.

The enormous value $I = 64.2$ MW/m^2 that we obtained in part (b) is the intensity at the *surface* of the sun, which is a sphere of radius 6.96×10^8 m. When this radiated energy reaches the earth, 1.50×10^{11} m away, the intensity has decreased by the factor $[(6.96 \times 10^8 \text{ m})/(1.50 \times 10^{11} \text{ m})]^2 = 2.15 \times 10^{-5}$ to the still-impressive 1.4 kW/m^2.

EXAMPLE 39.8 **A SLICE OF SUNLIGHT**

Find the power per unit area radiated from the sun's surface in the wavelength range 600.0 to 605.0 nm.

SOLUTION

IDENTIFY and SET UP: This question concerns the power emitted by a blackbody over a narrow range of wavelengths, and so involves the spectral emittance $I(\lambda)$ given by the Planck radiation law, Eq. (39.24). This requires that we find the area under the $I(\lambda)$ curve between 600.0 and 605.0 nm. We'll *approximate* this area as the product of the height of the curve at the median wavelength $\lambda = 602.5$ nm and the width of the interval, $\Delta\lambda = 5.0$ nm. From Example 39.7, $T = 5800$ K.

EXECUTE: To obtain the height of the $I(\lambda)$ curve at $\lambda = 602.5$ nm $= 6.025 \times 10^{-7}$ m, we first evaluate the quantity $hc/\lambda kT$ in Eq. (39.24) and then substitute the result into Eq. (39.24):

$$\frac{hc}{\lambda kT} = \frac{(6.626 \times 10^{-34}\,\text{J}\cdot\text{s})(2.998 \times 10^8\,\text{m/s})}{(6.025 \times 10^{-7}\,\text{m})(1.381 \times 10^{-23}\,\text{J/K})(5800\,\text{K})} = 4.116$$

$$I(\lambda) = \frac{2\pi(6.626 \times 10^{-34}\,\text{J}\cdot\text{s})(2.998 \times 10^8\,\text{m/s})^2}{(6.025 \times 10^{-7}\,\text{m})^5(e^{4.116} - 1)}$$
$$= 7.81 \times 10^{13}\,\text{W/m}^3$$

The intensity in the 5.0-nm range from 600.0 to 605.0 nm is then approximately

$$I(\lambda)\Delta\lambda = (7.81 \times 10^{13}\,\text{W/m}^3)(5.0 \times 10^{-9}\,\text{m})$$
$$= 3.9 \times 10^5\,\text{W/m}^2 = 0.39\,\text{MW/m}^2$$

EVALUATE: In part (b) of Example 39.7, we found the power radiated per unit area by the sun at *all* wavelengths to be $I = 64.2$ MW/m²; here we have found that the power radiated per unit area in the wavelength range from 600 to 605 nm is $I(\lambda)\Delta\lambda = 0.39$ MW/m², about 0.6% of the total.

TEST YOUR UNDERSTANDING OF SECTION 39.5 (a) Does a blackbody at 2000 K emit x rays? (b) Does it emit radio waves?

39.6 THE UNCERTAINTY PRINCIPLE REVISITED

The discovery of the dual wave–particle nature of matter forces us to reevaluate the kinematic language we use to describe the position and motion of a particle. In classical Newtonian mechanics we think of a particle as a point. We can describe its location and state of motion at any instant with three spatial coordinates and three components of velocity. But because matter also has a wave aspect, when we look at the behavior on a small enough scale—comparable to the de Broglie wavelength of the particle—we can no longer use the Newtonian description. Certainly no Newtonian particle would undergo diffraction like electrons do (Section 39.1).

To demonstrate just how non-Newtonian the behavior of matter can be, let's look at an experiment involving the two-slit interference of electrons (**Fig. 39.34**).

39.34 (a) A two-slit interference experiment for electrons. (b) The interference pattern after 28, 1000, and 10,000 electrons.

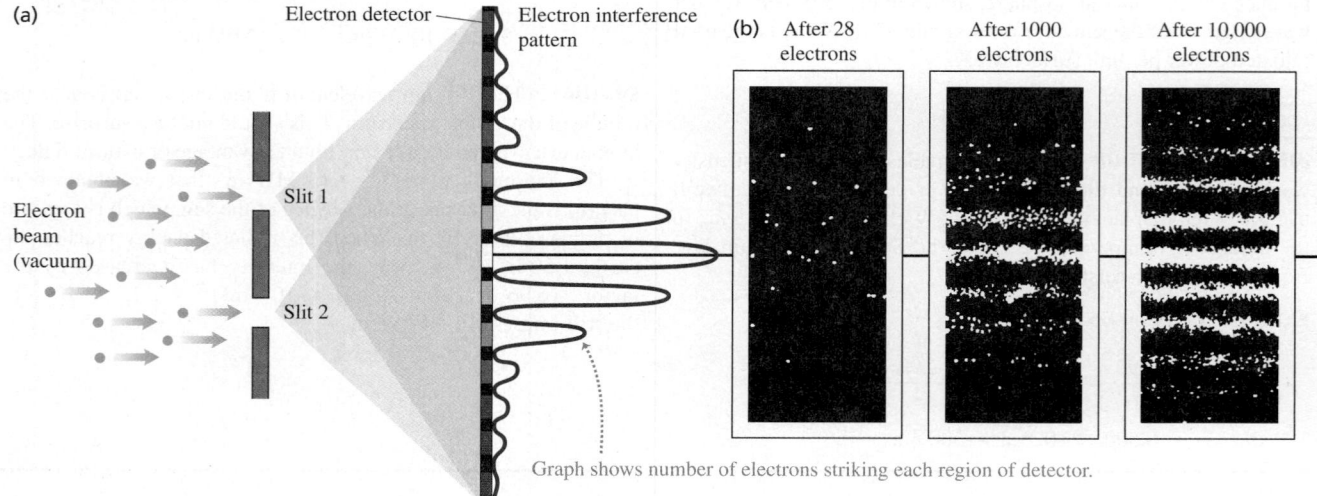

Graph shows number of electrons striking each region of detector.

We aim an electron beam at two parallel slits, as we did for light in Section 38.4. (The electron experiment has to be done in vacuum so that the electrons don't collide with air molecules.) What kind of pattern appears on the detector on the other side of the slits? The answer is: *exactly the same* kind of interference pattern we saw for photons in Section 38.4! Moreover, the principle of complementarity, which we introduced in Section 38.4, tells us that we cannot apply the wave and particle models simultaneously to describe any single element of this experiment. Thus we *cannot* predict exactly where in the pattern (a wave phenomenon) any individual electron (a particle) will land. We can't even ask which slit an individual electron passes through. If we tried to look at where the electrons were going by shining a light on them—that is, by scattering photons off them—the electrons would recoil, which would modify their motions so that the two-slit interference pattern would not appear.

CAUTION Electron two-slit interference is not interference between two electrons It's a common misconception that the pattern in Fig. 39.34b is due to the interference between *two* electron waves, each representing an electron passing through one slit. To show that this cannot be the case, we can send just one electron at a time through the apparatus. It makes no difference; we end up with the same interference pattern. In a sense, each electron wave interferes with itself. ∎

The Heisenberg Uncertainty Principles for Matter

Just as electrons and photons show the same behavior in a two-slit interference experiment, electrons and other forms of matter obey the same Heisenberg uncertainty principles as photons do:

$$\Delta x \Delta p_x \geq \hbar/2$$
$$\Delta y \Delta p_y \geq \hbar/2$$
$$\Delta z \Delta p_z \geq \hbar/2$$

(Heisenberg uncertainty principle for position and momentum) (39.29)

$$\Delta t \, \Delta E \geq \hbar/2$$

(Heisenberg uncertainty principle for energy and time interval) (39.30)

In these equations $\hbar = h/2\pi = 1.055 \times 10^{-34}$ J·s. The uncertainty principle for energy and time interval has a direct application to energy levels. We have assumed that each energy level in an atom has a very definite energy. However, Eq. (39.30) says that this is not true for all energy levels. A system that remains in a metastable state for a very long time (large Δt) can have a very well-defined energy (small ΔE), but if it remains in a state for only a short time (small Δt) the uncertainty in energy must be correspondingly greater (large ΔE). **Figure 39.35** illustrates this idea.

39.35 The longer the lifetime Δt of a state, the smaller is its spread in energy (shown by the width of the energy levels).

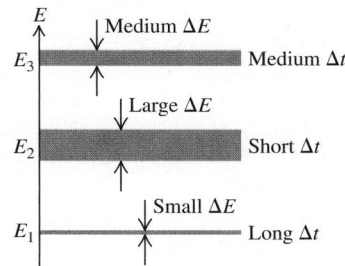

EXAMPLE 39.9 **THE UNCERTAINTY PRINCIPLE: POSITION AND MOMENTUM**

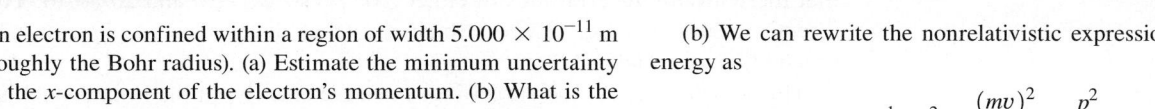

An electron is confined within a region of width 5.000×10^{-11} m (roughly the Bohr radius). (a) Estimate the minimum uncertainty in the x-component of the electron's momentum. (b) What is the kinetic energy of an electron with this magnitude of momentum? Express your answer in both joules and electron volts.

SOLUTION

IDENTIFY and SET UP: This problem uses the Heisenberg uncertainty principle for position and momentum and the relationship between a particle's momentum and its kinetic energy. The electron could be anywhere within the region, so we take $\Delta x = 5.000 \times 10^{-11}$ m as its position uncertainty. We then find the momentum uncertainty Δp_x from Eq. (39.29) and the kinetic energy from the relationships $p = mv$ and $K = \frac{1}{2}mv^2$.

EXECUTE: (a) From Eqs. (39.29), for a given value of Δx, the uncertainty in momentum is minimum when the product $\Delta x \, \Delta p_x$ equals $\hbar/2$. Hence

$$\Delta p_x = \frac{\hbar}{2\Delta x} = \frac{1.055 \times 10^{-34}\ \text{J} \cdot \text{s}}{2(5.000 \times 10^{-11}\ \text{m})} = 1.055 \times 10^{-24}\ \text{J} \cdot \text{s/m}$$

$$= 1.055 \times 10^{-24}\ \text{kg} \cdot \text{m/s}$$

(b) We can rewrite the nonrelativistic expression for kinetic energy as

$$K = \tfrac{1}{2}mv^2 = \frac{(mv)^2}{2m} = \frac{p^2}{2m}$$

Hence an electron with a magnitude of momentum equal to Δp_x from part (a) has kinetic energy

$$K = \frac{p^2}{2m} = \frac{(1.055 \times 10^{-24}\ \text{kg} \cdot \text{m/s})^2}{2(9.11 \times 10^{-31}\ \text{kg})}$$

$$= 6.11 \times 10^{-19}\ \text{J} = 3.81\ \text{eV}$$

EVALUATE: This energy is typical of electron energies in atoms. This agreement suggests that the uncertainty principle is deeply involved in atomic structure.

A similar calculation explains why electrons in atoms do not fall into the nucleus. If an electron were confined to the interior of a nucleus, its position uncertainty would be $\Delta x \approx 10^{-14}$ m. This would give the electron a momentum uncertainty about 5000 times greater than that of the electron in this example, and a kinetic energy so great that the electron would immediately be ejected from the nucleus.

EXAMPLE 39.10 THE UNCERTAINTY PRINCIPLE: ENERGY AND TIME

A sodium atom in one of the states labeled "Lowest excited levels" in Fig. 39.19a remains in that state, on average, for 1.6×10^{-8} s before it makes a transition to the ground state, emitting a photon with wavelength 589.0 nm and energy 2.105 eV. What is the uncertainty in energy of that excited state? What is the wavelength spread of the corresponding spectral line?

SOLUTION

IDENTIFY and SET UP: We use the Heisenberg uncertainty principle for energy and time interval and the relationship between photon energy and wavelength. The average time that the atom spends in this excited state is equal to Δt in Eq. (39.30). We find the minimum uncertainty in the energy of the excited state by replacing the $\geq$ sign in Eq. (39.30) with an equals sign and solving for ΔE.

EXECUTE: From Eq. (39.30),

$$\Delta E = \frac{\hbar}{2\Delta t} = \frac{1.055 \times 10^{-34} \text{ J} \cdot \text{s}}{2(1.6 \times 10^{-8} \text{ s})}$$

$$= 3.3 \times 10^{-27} \text{ J} = 2.1 \times 10^{-8} \text{ eV}$$

The atom remains in the ground state indefinitely, so that state has *no* associated energy uncertainty. The fractional uncertainty of the *photon* energy is therefore

$$\frac{\Delta E}{E} = \frac{2.1 \times 10^{-8} \text{ eV}}{2.105 \text{ eV}} = 1.0 \times 10^{-8}$$

You can use some simple calculus and the relationship $E = hc/\lambda$ to show that $\Delta\lambda/\lambda \approx \Delta E/E$, so that the corresponding spread in wavelength, or "width," of the spectral line is approximately

$$\Delta\lambda = \lambda \frac{\Delta E}{E} = (589.0 \text{ nm})(1.0 \times 10^{-8}) = 0.0000059 \text{ nm}$$

EVALUATE: This irreducible uncertainty $\Delta\lambda$ is called the *natural line width* of this particular spectral line. Though very small, it is within the limits of resolution of present-day spectrometers. Ordinarily, the natural line width is much smaller than the line width arising from other causes such as the Doppler effect and collisions among the rapidly moving atoms.

The Uncertainty Principle and the Limits of the Bohr Model

We saw in Section 39.3 that the Bohr model of the hydrogen atom was tremendously successful. However, the Heisenberg uncertainty principle for position and momentum shows that this model *cannot* be a correct description of how an electron in an atom behaves. Figure 39.22 shows that in the Bohr model as interpreted by de Broglie, an electron wave moves in a plane around the nucleus. Let's call this the xy-plane, so the z-axis is perpendicular to the plane. Hence the Bohr model says that an electron is always found at $z = 0$, and its z-momentum p_z is always zero (the electron does not move out of the xy-plane). But this implies that there are *no* uncertainties in either z or p_z, so $\Delta z = 0$ and $\Delta p_z = 0$. This directly contradicts Eq. (39.29), which says that the product $\Delta z \Delta p_z$ must be greater than or equal to $\hbar/2$.

This conclusion isn't too surprising, since the electron in the Bohr model is a mix of particle and wave ideas (the electron moves in an orbit like a miniature planet, but has a wavelength). To get an accurate picture of how electrons behave inside an atom and elsewhere, we need a description that is based *entirely* on the electron's wave properties. Our goal in Chapter 40 will be to develop this description, which we call *quantum mechanics*. To do this we'll introduce the *Schrödinger equation*, the fundamental equation that describes the dynamics of matter waves. This equation, as we will see, is as fundamental to quantum mechanics as Newton's laws are to classical mechanics or as Maxwell's equations are to electromagnetism.

TEST YOUR UNDERSTANDING OF SECTION 39.6 Rank the following situations according to the uncertainty in x-momentum, from largest to smallest. The mass of the proton is 1836 times the mass of the electron. (i) An electron whose x-coordinate is known to within 2×10^{-15} m; (ii) an electron whose x-coordinate is known to within 4×10^{-15} m; (iii) a proton whose x-coordinate is known to within 2×10^{-15} m; (iv) a proton whose x-coordinate is known to within 4×10^{-15} m. ∎

De Broglie waves and electron diffraction: Electrons and other particles have wave properties. A particle's wavelength depends on its momentum in the same way as for photons. A nonrelativistic electron accelerated from rest through a potential difference V_{ba} has a wavelength given by Eq. (39.3). Electron microscopes use the very small wavelengths of fast-moving electrons to make images with resolution thousands of times finer than is possible with visible light. (See Examples 39.1–39.3.)

$$\lambda = \frac{h}{p} = \frac{h}{mv} \quad (39.1)$$

$$E = hf \quad (39.2)$$

$$\lambda = \frac{h}{p} = \frac{h}{\sqrt{2meV_{ba}}} \quad (39.3)$$

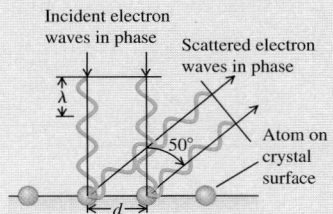

The nuclear atom: The Rutherford scattering experiments show that most of an atom's mass and all of its positive charge are concentrated in a tiny, dense nucleus at the center of the atom. (See Example 39.4.)

Atomic line spectra and energy levels: The energies of atoms are quantized: They can have only certain definite values, called energy levels. When an atom makes a transition from an energy level E_i to a lower level E_f, it emits a photon of energy $E_i - E_f$. The same photon can be absorbed by an atom in the lower energy level, which excites the atom to the upper level. (See Example 39.5.)

$$hf = \frac{hc}{\lambda} = E_i - E_f \quad (39.5)$$

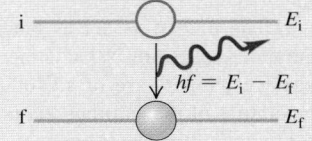

The Bohr model: In the Bohr model of the hydrogen atom, the permitted values of angular momentum are integral multiples of $h/2\pi$. The integer multiplier n is called the principal quantum number for the level. The orbital radii are proportional to n^2. The energy levels of the hydrogen atom are given by Eq. (39.15), where R is the Rydberg constant. (See Example 39.6.)

$$L_n = mv_n r_n = n\frac{h}{2\pi} \quad (39.6)$$
$$(n = 1, 2, 3, \dots)$$

$$r_n = \epsilon_0 \frac{n^2 h^2}{\pi m e^2} = n^2 a_0 \quad (39.8),\ (39.11)$$

$$v_n = \frac{1}{\epsilon_0} \frac{e^2}{2nh} \quad (39.9)$$

$$E_n = -\frac{hcR}{n^2} = -\frac{13.60\ \text{eV}}{n^2} \quad (39.15)$$
$$(n = 1, 2, 3, \dots)$$

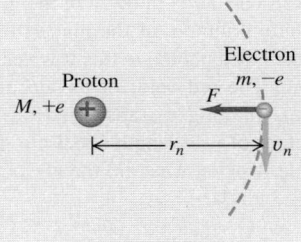

The laser: The laser operates on the principle of stimulated emission, by which many photons with identical wavelength and phase are emitted. Laser operation requires a nonequilibrium condition called a population inversion, in which more atoms are in a higher-energy state than are in a lower-energy state.

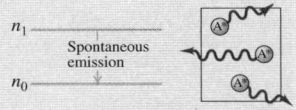

Blackbody radiation: The total radiated intensity (average power radiated per area) from a blackbody surface is proportional to the fourth power of the absolute temperature T. The quantity $\sigma = 5.67 \times 10^{-8}\ \text{W/m}^2 \cdot \text{K}^4$ is called the Stefan–Boltzmann constant. The wavelength λ_m at which a blackbody radiates most strongly is inversely proportional to T. The Planck radiation law gives the spectral emittance $I(\lambda)$ (intensity per wavelength interval in blackbody radiation). (See Examples 39.7 and 39.8.)

$$I = \sigma T^4$$
(Stefan–Boltzmann law) $\quad (39.19)$

$$\lambda_m T = 2.90 \times 10^{-3}\ \text{m} \cdot \text{K}$$
(Wien displacement law) $\quad (39.21)$

$$I(\lambda) = \frac{2\pi hc^2}{\lambda^5 (e^{hc/\lambda kT} - 1)}$$
(Planck radiation law) $\quad (39.24)$

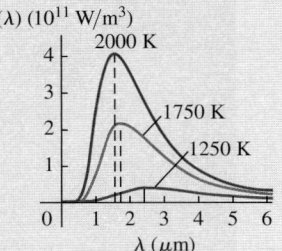

The Heisenberg uncertainty principle for particles: The same uncertainty considerations that apply to photons also apply to particles such as electrons. The uncertainty ΔE in the energy of a state that is occupied for a time Δt is given by Eq. (39.30), $\Delta t \, \Delta E \geq \hbar/2$. (See Examples 39.9 and 39.10.)

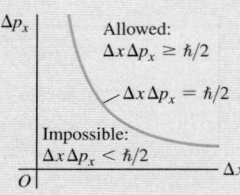

BRIDGING PROBLEM   HOT STARS AND HYDROGEN CLOUDS

Figure 39.36 shows a cloud, or *nebula,* of glowing hydrogen in interstellar space. The atoms in this cloud are excited by short-wavelength radiation emitted by the bright blue stars at the center of the nebula.

(a) The blue stars act as blackbodies and emit light with a continuous spectrum. What is the wavelength at which a star with a surface temperature of 15,100 K (about $2\frac{1}{2}$ times the surface temperature of the sun) has the maximum spectral emittance? In what region of the electromagnetic spectrum is this?

(b) Figure 39.32 shows that most of the energy radiated by a blackbody is at wavelengths between about one half and three times the wavelength of maximum emittance. If a hydrogen atom near the star in part (a) is initially in the ground level, what is the principal quantum number of the highest energy level to which it could be excited by a photon in this wavelength range?

(c) The red color of the nebula is primarily due to hydrogen atoms making a transition from $n = 3$ to $n = 2$ and emitting photons of wavelength 656.3 nm. In the Bohr model as interpreted by de Broglie, what are the *electron* wavelengths in the $n = 2$ and $n = 3$ levels?

39.36 The Rosette Nebula.

SOLUTION GUIDE

IDENTIFY and SET UP

1. To solve this problem you need to use your knowledge of both blackbody radiation (Section 39.5) and the Bohr model of the hydrogen atom (Section 39.3).

2. In part (a) the target variable is the wavelength at which the star emits most strongly; in part (b) the target variable is a principal quantum number, and in part (c) it is the de Broglie wavelength of an electron in the $n = 2$ and $n = 3$ Bohr orbits (see Fig. 39.24). Select the equations you will need to find the target variables. (*Hint:* In Section 39.5 you learned how to find the energy change involved in a transition between two given levels of a hydrogen atom. Part (b) is a variation on this: You are to find the final level in a transition that starts in the $n = 1$ level and involves the absorption of a photon of a given wavelength and hence a given energy.)

EXECUTE

3. Use the Wien displacement law to find the wavelength at which the star has maximum spectral emittance. In what part of the electromagnetic spectrum is this wavelength?

4. Use your result from step 3 to find the range of wavelengths in which the star radiates most of its energy. Which end of this range corresponds to a photon with the greatest energy?

5. Write an expression for the wavelength of a photon that must be absorbed to cause an electron transition from the ground level ($n = 1$) to a higher level n. Solve for the value of n that corresponds to the highest-energy photon in the range you calculated in step 4. (*Hint:* Remember that n must be an integer.)

6. Find the electron wavelengths that correspond to the $n = 2$ and $n = 3$ orbits shown in Fig. 39.22.

EVALUATE

7. Check your result in step 5 by calculating the wavelength needed to excite a hydrogen atom from the ground level into the level *above* the highest-energy level that you found in step 5. Is it possible for light in the range of wavelengths you found in step 4 to excite hydrogen atoms from the ground level into this level?

8. How do the electron wavelengths you found in step 6 compare to the wavelength of a *photon* emitted in a transition from the $n = 3$ level to the $n = 2$ level?

Problems

•, ••, •••: Difficulty levels. CP: Cumulative problems incorporating material from earlier chapters. CALC: Problems requiring calculus. DATA: Problems involving real data, scientific evidence, experimental design, and/or statistical reasoning. BIO: Biosciences problems.

DISCUSSION QUESTIONS

Q39.1 If a proton and an electron have the same speed, which has the longer de Broglie wavelength? Explain.

Q39.2 If a proton and an electron have the same kinetic energy, which has the longer de Broglie wavelength? Explain.

Q39.3 Does a photon have a de Broglie wavelength? If so, how is it related to the wavelength of the associated electromagnetic wave? Explain.

Q39.4 When an electron beam goes through a very small hole, it produces a diffraction pattern on a screen, just like that of light. Does this mean that an electron spreads out as it goes through the hole? What does this pattern mean?

Q39.5 Galaxies tend to be strong emitters of Lyman-α photons (from the $n = 2$ to $n = 1$ transition in atomic hydrogen). But the intergalactic medium—the very thin gas between the galaxies—tends to *absorb* Lyman-α photons. What can you infer from these observations about the temperature in these two environments? Explain.

Q39.6 A doubly ionized lithium atom (Li^{++}) is one that has had two of its three electrons removed. The energy levels of the remaining single-electron ion are closely related to those of the hydrogen atom. The nuclear charge for lithium is $+3e$ instead of just $+e$. How are the energy levels related to those of hydrogen? How is the *radius* of the ion in the ground level related to that of the hydrogen atom? Explain.

Q39.7 The emission of a photon by an isolated atom is a recoil process in which momentum is conserved. Thus Eq. (39.5) should include a recoil kinetic energy K_r for the atom. Why is this energy negligible in that equation?

Q39.8 How might the energy levels of an atom be measured directly—that is, without recourse to analysis of spectra?

Q39.9 Elements in the gaseous state emit line spectra with well-defined wavelengths. But hot solid bodies always emit a continuous spectrum—that is, a continuous smear of wavelengths. Can you account for this difference?

Q39.10 As a body is heated to a very high temperature and becomes self-luminous, the apparent color of the emitted radiation shifts from red to yellow and finally to blue as the temperature increases. Why does the color shift? What other changes in the character of the radiation occur?

Q39.11 Do the planets of the solar system obey a distance law ($r_n = n^2 r_1$) as the electrons of the Bohr atom do? Should they? Why (or why not)? (Consult Appendix F for the appropriate distances.)

Q39.12 You have been asked to design a magnet system to steer a beam of 54-eV electrons like those described in Example 39.1 (Section 39.1). The goal is to be able to direct the electron beam to a specific target location with an accuracy of ± 1.0 mm. In your design, do you need to take the wave nature of electrons into account? Explain.

Q39.13 Why go through the expense of building an electron microscope for studying very small objects such as organic molecules? Why not just use extremely short electromagnetic waves, which are much cheaper to generate?

Q39.14 Which has more total energy: a hydrogen atom with an electron in a high shell (large n) or in a low shell (small n)? Which is moving faster: the high-shell electron or the low-shell electron? Is there a contradiction here? Explain.

Q39.15 Does the uncertainty principle have anything to do with marksmanship? That is, is the accuracy with which a bullet can be aimed at a target limited by the uncertainty principle? Explain.

Q39.16 Suppose a two-slit interference experiment is carried out using an electron beam. Would the same interference pattern result if one slit at a time is uncovered instead of both at once? If not, why not? Doesn't each electron go through one slit or the other? Or does every electron go through both slits? Discuss the latter possibility in light of the principle of complementarity.

Q39.17 Equation (39.30) states that the energy of a system can have uncertainty. Does this mean that the principle of conservation of energy is no longer valid? Explain.

Q39.18 Laser light results from transitions from long-lived metastable states. Why is it more monochromatic than ordinary light?

Q39.19 Could an electron-diffraction experiment be carried out using three or four slits? Using a grating with many slits? What sort of results would you expect with a grating? Would the uncertainty principle be violated? Explain.

Q39.20 As the lower half of Fig. 39.4 shows, the diffraction pattern made by electrons that pass through aluminum foil is a series of concentric rings. But if the aluminum foil is replaced by a single crystal of aluminum, only certain points on these rings appear in the pattern. Explain.

Q39.21 Why can an electron microscope have greater magnification than an ordinary microscope?

Q39.22 When you check the air pressure in a tire, a little air always escapes; the process of making the measurement changes the quantity being measured. Think of other examples of measurements that change or disturb the quantity being measured.

EXERCISES

Section 39.1 Electron Waves

39.1 • (a) An electron moves with a speed of 4.70×10^6 m/s. What is its de Broglie wavelength? (b) A proton moves with the same speed. Determine its de Broglie wavelength.

39.2 •• For crystal diffraction experiments (discussed in Section 39.1), wavelengths on the order of 0.20 nm are often appropriate. Find the energy in electron volts for a particle with this wavelength if the particle is (a) a photon; (b) an electron; (c) an alpha particle ($m = 6.64 \times 10^{-27}$ kg).

39.3 • An electron has a de Broglie wavelength of 2.80×10^{-10} m. Determine (a) the magnitude of its momentum and (b) its kinetic energy (in joules and in electron volts).

39.4 •• **Wavelength of an Alpha Particle.** An alpha particle ($m = 6.64 \times 10^{-27}$ kg) emitted in the radioactive decay of uranium-238 has an energy of 4.20 MeV. What is its de Broglie wavelength?

39.5 • An electron is moving with a speed of 8.00×10^6 m/s. What is the speed of a proton that has the same de Broglie wavelength as this electron?

39.6 • (a) A nonrelativistic free particle with mass m has kinetic energy K. Derive an expression for the de Broglie wavelength of the particle in terms of m and K. (b) What is the de Broglie wavelength of an 800-eV electron?

39.7 • (a) If a photon and an electron each have the same energy of 20.0 eV, find the wavelength of each. (b) If a photon and an electron each have the same wavelength of 250 nm, find the energy of each. (c) You want to study an organic molecule that is about 250 nm long using either a photon or an electron microscope. Approximately what wavelength should you use, and which probe, the electron or the photon, is likely to damage the molecule the least?

39.8 •• What is the de Broglie wavelength for an electron with speed (a) $v = 0.480c$ and (b) $v = 0.960c$? (*Hint:* Use the correct relativistic expression for linear momentum if necessary.)

39.9 • **Wavelength of a Bullet.** Calculate the de Broglie wavelength of a 5.00-g bullet that is moving at 340 m/s. Will the bullet exhibit wavelike properties?

39.10 •• Through what potential difference must electrons be accelerated if they are to have (a) the same wavelength as an x ray of wavelength 0.220 nm and (b) the same energy as the x ray in part (a)?

39.11 •• (a) What accelerating potential is needed to produce electrons of wavelength 5.00 nm? (b) What would be the energy of photons having the same wavelength as these electrons? (c) What would be the wavelength of photons having the same energy as the electrons in part (a)?

39.12 •• CP A beam of electrons is accelerated from rest through a potential difference of 0.100 kV and then passes through a thin slit. When viewed far from the slit, the diffracted beam shows its first diffraction minima at $\pm 14.6°$ from the original direction of the beam. (a) Do we need to use relativity formulas? How do you know? (b) How wide is the slit?

39.13 •• A beam of neutrons that all have the same energy scatters from atoms that have a spacing of 0.0910 nm in the surface plane of a crystal. The $m = 1$ intensity maximum occurs when the angle θ in Fig. 39.2 is 28.6°. What is the kinetic energy (in electron volts) of each neutron in the beam?

39.14 • (a) In an electron microscope, what accelerating voltage is needed to produce electrons with wavelength 0.0600 nm? (b) If protons are used instead of electrons, what accelerating voltage is needed to produce protons with wavelength 0.0600 nm? (*Hint:* In each case the initial kinetic energy is negligible.)

39.15 • A CD-ROM is used instead of a crystal in an electron-diffraction experiment. The surface of the CD-ROM has tracks of tiny pits with a uniform spacing of 1.60 μm. (a) If the speed of the electrons is 1.26×10^4 m/s, at which values of θ will the $m = 1$ and $m = 2$ intensity maxima appear? (b) The scattered electrons in these maxima strike at normal incidence a piece of photographic film that is 50.0 cm from the CD-ROM. What is the spacing on the film between these maxima?

Section 39.2 The Nuclear Atom and Atomic Spectra

39.16 •• CP A 4.78-MeV alpha particle from a ^{226}Ra decay makes a head-on collision with a uranium nucleus. A uranium nucleus has 92 protons. (a) What is the distance of closest approach of the alpha particle to the center of the nucleus? Assume that the uranium nucleus remains at rest and that the distance of closest

approach is much greater than the radius of the uranium nucleus. (b) What is the force on the alpha particle at the instant when it is at the distance of closest approach?

39.17 • A beam of alpha particles is incident on a target of lead. A particular alpha particle comes in "head-on" to a particular lead nucleus and stops 6.50×10^{-14} m away from the center of the nucleus. (This point is well outside the nucleus.) Assume that the lead nucleus, which has 82 protons, remains at rest. The mass of the alpha particle is 6.64×10^{-27} kg. (a) Calculate the electrostatic potential energy at the instant that the alpha particle stops. Express your result in joules and in MeV. (b) What initial kinetic energy (in joules and in MeV) did the alpha particle have? (c) What was the initial speed of the alpha particle?

Section 39.3 Energy Levels and the Bohr Model of the Atom

39.18 • The silicon–silicon single bond that forms the basis of the mythical silicon-based creature the Horta has a bond strength of 3.80 eV. What wavelength of photon would you need in a (mythical) phasor disintegration gun to destroy the Horta?

39.19 •• A hydrogen atom is in a state with energy -1.51 eV. In the Bohr model, what is the angular momentum of the electron in the atom, with respect to an axis at the nucleus?

39.20 • A hydrogen atom initially in its ground level absorbs a photon, which excites the atom to the $n = 3$ level. Determine the wavelength and frequency of the photon.

39.21 • A triply ionized beryllium ion, Be^{3+} (a beryllium atom with three electrons removed), behaves very much like a hydrogen atom except that the nuclear charge is four times as great. (a) What is the ground-level energy of Be^{3+}? How does this compare to the ground-level energy of the hydrogen atom? (b) What is the ionization energy of Be^{3+}? How does this compare to the ionization energy of the hydrogen atom? (c) For the hydrogen atom, the wavelength of the photon emitted in the $n = 2$ to $n = 1$ transition is 122 nm (see Example 39.6). What is the wavelength of the photon emitted when a Be^{3+} ion undergoes this transition? (d) For a given value of n, how does the radius of an orbit in Be^{3+} compare to that for hydrogen?

39.22 •• Consider the Bohr-model description of a hydrogen atom. (a) Calculate $E_2 - E_1$ and $E_{10} - E_9$. As n increases, does the energy separation between adjacent energy levels increase, decrease, or stay the same? (b) Show that $E_{n+1} - E_n$ approaches $(27.2 \text{ eV})/n^3$ as n becomes large. (c) How does $r_{n+1} - r_n$ depend on n? Does the radial distance between adjacent orbits increase, decrease, or stay the same as n increases?

39.23 • (a) Using the Bohr model, calculate the speed of the electron in a hydrogen atom in the $n = 1$, 2, and 3 levels. (b) Calculate the orbital period in each of these levels. (c) The average lifetime of the first excited level of a hydrogen atom is 1.0×10^{-8} s. In the Bohr model, how many orbits does an electron in the $n = 2$ level complete before returning to the ground level?

39.24 • Consider the Bohr-model description of a hydrogen atom. (a) Calculate K_1, U_1, and E_1 for the $n = 1$ energy level. How are K_1 and U_1 related? (b) Show that for any value of n, both $U_n = -2K_n$ and $K_n = -E_n$.

39.25 • CP The energy-level scheme for the hypothetical one-electron element Searsium is shown in **Fig. E39.25**. The potential energy is taken to be zero for an electron at an infinite distance from the nucleus. (a) How much energy (in electron volts) does it take to ionize an electron from the ground level? (b) An 18-eV photon is absorbed by a Searsium atom in its ground level. As the

Figure **E39.25**

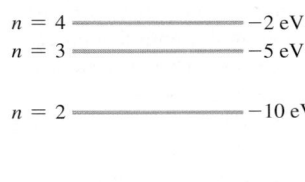

$n = 4$ ———————— -2 eV
$n = 3$ ———————— -5 eV

$n = 2$ ———————— -10 eV

$n = 1$ ———◯——— -20 eV

atom returns to its ground level, what possible energies can the emitted photons have? Assume that there can be transitions between all pairs of levels. (c) What will happen if a photon with an energy of 8 eV strikes a Searsium atom in its ground level? Why? (d) Photons emitted in the Searsium transitions $n = 3 \rightarrow n = 2$ and $n = 3 \rightarrow n = 1$ will eject photoelectrons from an unknown metal, but the photon emitted from the transition $n = 4 \rightarrow n = 3$ will not. What are the limits (maximum and minimum possible values) of the work function of the metal?

39.26 • (a) For one-electron ions with nuclear charge Z, what is the speed of the electron in a Bohr-model orbit labeled with n? Give your answer in terms of v_1, the orbital speed for the $n = 1$ Bohr orbit in hydrogen. (b) What is the largest value of Z for which the $n = 1$ orbital speed is less than 10% of the speed of light in vacuum?

39.27 • In a set of experiments on a hypothetical one-electron atom, you measure the wavelengths of the photons emitted from transitions ending in the ground level ($n = 1$), as shown in the energy-level diagram in **Fig. E39.27**. You also observe that it takes 17.50 eV to ionize this atom. (a) What is the energy of the atom in each of the levels ($n = 1$, $n = 2$, etc.) shown in the figure? (b) If an electron made a transition from the $n = 4$ to the $n = 2$ level, what wavelength of light would it emit?

Figure **E39.27**

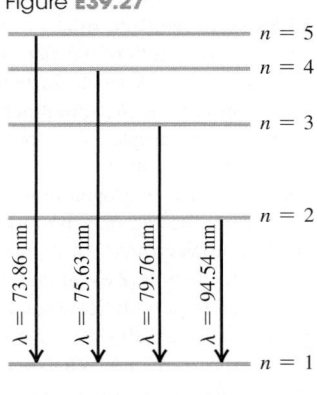

$n = 5$
$n = 4$
$n = 3$
$n = 2$

$\lambda = 73.86$ nm
$\lambda = 75.63$ nm
$\lambda = 79.76$ nm
$\lambda = 94.54$ nm

$n = 1$

39.28 • Find the longest and shortest wavelengths in the Lyman and Paschen series for hydrogen. In what region of the electromagnetic spectrum does each series lie?

39.29 • (a) An atom initially in an energy level with $E = -6.52$ eV absorbs a photon that has wavelength 860 nm. What is the internal energy of the atom after it absorbs the photon? (b) An atom initially in an energy level with $E = -2.68$ eV emits a photon that has wavelength 420 nm. What is the internal energy of the atom after it emits the photon?

39.30 •• Use Balmer's formula to calculate (a) the wavelength, (b) the frequency, and (c) the photon energy for the H_γ line of the Balmer series for hydrogen.

Section 39.4 The Laser

39.31 • BIO **Laser Surgery.** Using a mixture of CO_2, N_2, and sometimes He, CO_2 lasers emit a wavelength of 10.6 μm. At power outputs of 0.100 kW, such lasers are used for surgery. How many photons per second does a CO_2 laser deliver to the tissue during its use in an operation?

39.32 • BIO **Removing Birthmarks.** Pulsed dye lasers emit light of wavelength 585 nm in 0.45-ms pulses to remove skin blemishes such as birthmarks. The beam is usually focused onto a circular spot 5.0 mm in diameter. Suppose that the output of one such laser is 20.0 W. (a) What is the energy of each photon, in eV? (b) How many photons per square millimeter are delivered to the blemish during each pulse?

39.33 • How many photons per second are emitted by a 7.50-mW CO_2 laser that has a wavelength of 10.6 μm?

39.34 • BIO **PRK Surgery.** Photorefractive keratectomy (PRK) is a laser-based surgical procedure that corrects near- and farsightedness by removing part of the lens of the eye to change its curvature and hence focal length. This procedure can remove layers 0.25 μm thick using pulses lasting 12.0 ns from a laser beam of wavelength 193 nm. Low-intensity beams can be used because each individual photon has enough energy to break the covalent bonds of the tissue. (a) In what part of the electromagnetic spectrum does this light lie? (b) What is the energy of a single photon? (c) If a 1.50-mW beam is used, how many photons are delivered to the lens in each pulse?

39.35 • A large number of neon atoms are in thermal equilibrium. What is the ratio of the number of atoms in a $5s$ state to the number in a $3p$ state at (a) 300 K; (b) 600 K; (c) 1200 K? The energies of these states, relative to the ground state, are $E_{5s} = 20.66$ eV and $E_{3p} = 18.70$ eV. (d) At any of these temperatures, the rate at which a neon gas will spontaneously emit 632.8-nm radiation is quite low. Explain why.

39.36 • Figure 39.19a shows the energy levels of the sodium atom. The two lowest excited levels are shown in columns labeled $^2P_{3/2}$ and $^2P_{1/2}$. Find the ratio of the number of atoms in a $^2P_{3/2}$ state to the number in a $^2P_{1/2}$ state for a sodium gas in thermal equilibrium at 500 K. In which state are more atoms found?

Section 39.5 Continuous Spectra

39.37 •• A 100-W incandescent light bulb has a cylindrical tungsten filament 30.0 cm long, 0.40 mm in diameter, and with an emissivity of 0.26. (a) What is the temperature of the filament? (b) For what wavelength does the spectral emittance of the bulb peak? (c) Incandescent light bulbs are not very efficient sources of visible light. Explain why this is so.

39.38 • Determine λ_m, the wavelength at the peak of the Planck distribution, and the corresponding frequency f, at these temperatures: (a) 3.00 K; (b) 300 K; (c) 3000 K.

39.39 • Radiation has been detected from space that is characteristic of an ideal radiator at $T = 2.728$ K. (This radiation is a relic of the Big Bang at the beginning of the universe.) For this temperature, at what wavelength does the Planck distribution peak? In what part of the electromagnetic spectrum is this wavelength?

39.40 • The shortest visible wavelength is about 400 nm. What is the temperature of an ideal radiator whose spectral emittance peaks at this wavelength?

39.41 •• Two stars, both of which behave like ideal blackbodies, radiate the same total energy per second. The cooler one has a surface temperature T and a diameter 3.0 times that of the hotter star. (a) What is the temperature of the hotter star in terms of T? (b) What is the ratio of the peak-intensity wavelength of the hot star to the peak-intensity wavelength of the cool star?

39.42 • The wavelength 10.0 μm is in the infrared region of the electromagnetic spectrum, whereas 600 nm is in the visible region and 100 nm is in the ultraviolet. What is the temperature of an ideal blackbody for which the peak wavelength λ_m is equal to each of these wavelengths?

39.43 • **Sirius B.** The brightest star in the sky is Sirius, the Dog Star. It is actually a binary system of two stars, the smaller one (Sirius B) being a white dwarf. Spectral analysis of Sirius B indicates that its surface temperature is 24,000 K and that it radiates energy at a total rate of 1.0×10^{25} W. Assume that it behaves like an ideal blackbody. (a) What is the total radiated intensity of Sirius B? (b) What is the peak-intensity wavelength? Is this wavelength visible to humans? (c) What is the radius of Sirius B? Express your answer in kilometers and as a fraction of our sun's radius. (d) Which star radiates more *total* energy per second, the hot Sirius B or the (relatively) cool sun with a surface temperature of 5800 K? To find out, calculate the ratio of the total power radiated by our sun to the power radiated by Sirius B.

Section 39.6 The Uncertainty Principle Revisited

39.44 • A pesky 1.5-mg mosquito is annoying you as you attempt to study physics in your room, which is 5.0 m wide and 2.5 m high. You decide to swat the bothersome insect as it flies toward you, but you need to estimate its speed to make a successful hit. (a) What is the maximum uncertainty in the horizontal position of the mosquito? (b) What limit does the Heisenberg uncertainty principle place on your ability to know the horizontal velocity of this mosquito? Is this limitation a serious impediment to your attempt to swat it?

39.45 • (a) The uncertainty in the y-component of a proton's position is 2.0×10^{-12} m. What is the minimum uncertainty in a simultaneous measurement of the y-component of the proton's velocity? (b) The uncertainty in the z-component of an electron's velocity is 0.250 m/s. What is the minimum uncertainty in a simultaneous measurement of the z-coordinate of the electron?

39.46 • A 10.0-g marble is gently placed on a horizontal tabletop that is 1.75 m wide. (a) What is the maximum uncertainty in the horizontal position of the marble? (b) According to the Heisenberg uncertainty principle, what is the minimum uncertainty in the horizontal velocity of the marble? (c) In light of your answer to part (b), what is the longest time the marble could remain on the table? Compare this time to the age of the universe, which is approximately 14 billion years. (*Hint:* Can you know that the horizontal velocity of the marble is *exactly* zero?)

39.47 • A scientist has devised a new method of isolating individual particles. He claims that this method enables him to detect simultaneously the position of a particle along an axis with a standard deviation of 0.12 nm and its momentum component along this axis with a standard deviation of 3.0×10^{-25} kg · m/s. Use the Heisenberg uncertainty principle to evaluate the validity of this claim.

39.48 • (a) The x-coordinate of an electron is measured with an uncertainty of 0.30 mm. What is the x-component of the electron's velocity, v_x, if the minimum percent uncertainty in a simultaneous measurement of v_x is 1.0%? (b) Repeat part (a) for a proton.

39.49 • An atom in a metastable state has a lifetime of 5.2 ms. What is the uncertainty in energy of the metastable state?

PROBLEMS

39.50 • An atom with mass m emits a photon of wavelength λ. (a) What is the recoil speed of the atom? (b) What is the kinetic energy K of the recoiling atom? (c) Find the ratio K/E, where E is the energy of the emitted photon. If this ratio is much less than unity, the recoil of the atom can be neglected in the emission process. Is the recoil of the atom more important for small or large atomic masses? For long or short wavelengths? (d) Calculate K (in

electron volts) and K/E for a hydrogen atom (mass 1.67×10^{-27} kg) that emits an ultraviolet photon of energy 10.2 eV. Is recoil an important consideration in this emission process?

39.51 •• The negative muon has a charge equal to that of an electron but a mass that is 207 times as great. Consider a hydrogenlike atom consisting of a proton and a muon. (a) What is the reduced mass of the atom? (b) What is the ground-level energy (in electron volts)? (c) What is the wavelength of the radiation emitted in the transition from the $n = 2$ level to the $n = 1$ level?

39.52 • A large number of hydrogen atoms are in thermal equilibrium. Let n_2/n_1 be the ratio of the number of atoms in an $n = 2$ excited state to the number of atoms in an $n = 1$ ground state. At what temperature is n_2/n_1 equal to (a) 10^{-12}; (b) 10^{-8}; (c) 10^{-4}? (d) Like the sun, other stars have continuous spectra with dark absorption lines (see Fig. 39.9). The absorption takes place in the star's atmosphere, which in all stars is composed primarily of hydrogen. Explain why the Balmer absorption lines are relatively weak in stars with low atmospheric temperatures such as the sun (atmosphere temperature 5800 K) but strong in stars with higher atmospheric temperatures.

39.53 • (a) What is the smallest amount of energy in electron volts that must be given to a hydrogen atom initially in its ground level so that it can emit the H_α line in the Balmer series? (b) How many different possibilities of spectral-line emissions are there for this atom when the electron starts in the $n = 3$ level and eventually ends up in the ground level? Calculate the wavelength of the emitted photon in each case.

39.54 •• In the Bohr model of the hydrogen atom, what is the de Broglie wavelength of the electron when it is in (a) the $n = 1$ level and (b) the $n = 4$ level? In both cases, compare the de Broglie wavelength to the circumference $2\pi r_n$ of the orbit.

39.55 ••• A sample of hydrogen atoms is irradiated with light with wavelength 85.5 nm, and electrons are observed leaving the gas. (a) If each hydrogen atom were initially in its ground level, what would be the maximum kinetic energy in electron volts of these photoelectrons? (b) A few electrons are detected with energies as much as 10.2 eV greater than the maximum kinetic energy calculated in part (a). How can this be?

39.56 •• Take 380–750 nm to be the wavelength range of the visible spectrum. (a) What are the largest and smallest photon energies for visible light? (b) The lowest six energy levels of the one-electron He^+ ion are given in Fig. 39.27. For these levels, what transitions give absorption or emission of visible-light photons?

39.57 •• **The Red Supergiant Betelgeuse.** The star Betelgeuse has a surface temperature of 3000 K and is 600 times the diameter of our sun. (If our sun were that large, we would be inside it!) Assume that it radiates like an ideal blackbody. (a) If Betelgeuse were to radiate all of its energy at the peak-intensity wavelength, how many photons per second would it radiate? (b) Find the ratio of the power radiated by Betelgeuse to the power radiated by our sun (at 5800 K).

39.58 •• **CP** Light from an ideal spherical blackbody 15.0 cm in diameter is analyzed by using a diffraction grating that has 3850 lines/cm. When you shine this light through the grating, you observe that the peak-intensity wavelength forms a first-order bright fringe at $\pm 14.4°$ from the central bright fringe. (a) What is the temperature of the blackbody? (b) How long will it take this sphere to radiate 12.0 MJ of energy at constant temperature?

39.59 • What must be the temperature of an ideal blackbody so that photons of its radiated light having the peak-intensity wavelength can excite the electron in the Bohr-model hydrogen atom from the ground level to the $n = 4$ energy level?

39.60 ·· **An Ideal Blackbody.**   A large cavity that has a very small hole and is maintained at a temperature T is a good approximation to an ideal radiator or blackbody. Radiation can pass into or out of the cavity only through the hole. The cavity is a perfect absorber, since any radiation incident on the hole becomes trapped inside the cavity. Such a cavity at 400°C has a hole with area 4.00 mm². How long does it take for the cavity to radiate 100 J of energy through the hole?

39.61 ·· **CALC** (a) Write the Planck distribution law in terms of the frequency f, rather than the wavelength λ, to obtain $I(f)$. (b) Show that

$$\int_0^\infty I(\lambda)\, d\lambda = \frac{2\pi^5 k^4}{15c^2 h^3} T^4$$

where $I(\lambda)$ is the Planck distribution formula of Eq. (39.24). *Hint:* Change the integration variable from λ to f. You will need to use the following tabulated integral:

$$\int_0^\infty \frac{x^3}{e^{\alpha x} - 1}\, dx = \frac{1}{240}\left(\frac{2\pi}{\alpha}\right)^4$$

(c) The result of part (b) is I and has the form of the Stefan–Boltzmann law, $I = \sigma T^4$ (Eq. 39.19). Evaluate the constants in part (b) to show that σ has the value given in Section 39.5.

39.62 ·· **CP** A beam of 40-eV electrons traveling in the $+x$-direction passes through a slit that is parallel to the y-axis and 5.0 μm wide. The diffraction pattern is recorded on a screen 2.5 m from the slit. (a) What is the de Broglie wavelength of the electrons? (b) How much time does it take the electrons to travel from the slit to the screen? (c) Use the width of the central diffraction pattern to calculate the uncertainty in the y-component of momentum of an electron just after it has passed through the slit. (d) Use the result of part (c) and the Heisenberg uncertainty principle (Eq. 39.29 for y) to estimate the minimum uncertainty in the y-coordinate of an electron just after it has passed through the slit. Compare your result to the width of the slit.

39.63 · (a) What is the energy of a photon that has wavelength 0.10 μm? (b) Through approximately what potential difference must electrons be accelerated so that they will exhibit wave nature in passing through a pinhole 0.10 μm in diameter? What is the speed of these electrons? (c) If protons rather than electrons were used, through what potential difference would protons have to be accelerated so they would exhibit wave nature in passing through this pinhole? What would be the speed of these protons?

39.64 · **CP** Electrons go through a single slit 300 nm wide and strike a screen 24.0 cm away. At angles of $\pm 20.0°$ from the center of the diffraction pattern, no electrons hit the screen, but electrons hit at all points closer to the center. (a) How fast were these electrons moving when they went through the slit? (b) What will be the next pair of larger angles at which no electrons hit the screen?

39.65 ·· **CP** A beam of electrons is accelerated from rest and then passes through a pair of identical thin slits that are 1.25 nm apart. You observe that the first double-slit interference dark fringe occurs at $\pm 18.0°$ from the original direction of the beam when viewed on a distant screen. (a) Are these electrons relativistic? How do you know? (b) Through what potential difference were the electrons accelerated?

39.66 · **CP** Coherent light is passed through two narrow slits whose separation is 20.0 μm. The second-order bright fringe in the interference pattern is located at an angle of 0.0300 rad. If electrons are used instead of light, what must the kinetic energy (in electron volts) of the electrons be if they are to produce an interference pattern for which the second-order maximum is also at 0.0300 rad?

39.67 ·· **CP** An electron beam and a photon beam pass through identical slits. On a distant screen, the first dark fringe occurs at the same angle for both of the beams. The electron speeds are much slower than that of light. (a) Express the energy of a photon in terms of the kinetic energy K of one of the electrons. (b) Which is greater, the energy of a photon or the kinetic energy of an electron?

39.68 · **BIO** What is the de Broglie wavelength of a red blood cell, with mass 1.00×10^{-11} g, that is moving with a speed of 0.400 cm/s? Do we need to be concerned with the wave nature of the blood cells when we describe the flow of blood in the body?

39.69 · High-speed electrons are used to probe the interior structure of the atomic nucleus. For such electrons the expression $\lambda = h/p$ still holds, but we must use the relativistic expression for momentum, $p = mv/\sqrt{1 - v^2/c^2}$. (a) Show that the speed of an electron that has de Broglie wavelength λ is

$$v = \frac{c}{\sqrt{1 + (mc\lambda/h)^2}}$$

(b) The quantity h/mc equals 2.426×10^{-12} m. (As we saw in Section 38.3, this same quantity appears in Eq. (38.7), the expression for Compton scattering of photons by electrons.) If λ is small compared to h/mc, the denominator in the expression found in part (a) is close to unity and the speed v is very close to c. In this case it is convenient to write $v = (1 - \Delta)c$ and express the speed of the electron in terms of Δ rather than v. Find an expression for Δ valid when $\lambda \ll h/mc$. [*Hint:* Use the binomial expansion $(1 + z)^n = 1 + nz + [n(n-1)z^2/2] + \cdots$, valid for the case $|z| < 1$.] (c) How fast must an electron move for its de Broglie wavelength to be 1.00×10^{-15} m, comparable to the size of a proton? Express your answer in the form $v = (1 - \Delta)c$, and state the value of Δ.

39.70 · Suppose that the uncertainty of position of an electron is equal to the radius of the $n = 1$ Bohr orbit for hydrogen. Calculate the simultaneous minimum uncertainty of the corresponding momentum component, and compare this with the magnitude of the momentum of the electron in the $n = 1$ Bohr orbit. Discuss your results.

39.71 · **CP** (a) A particle with mass m has kinetic energy equal to three times its rest energy. What is the de Broglie wavelength of this particle? (*Hint:* You must use the relativistic expressions for momentum and kinetic energy: $E^2 = (pc)^2 + (mc^2)^2$ and $K = E - mc^2$.) (b) Determine the numerical value of the kinetic energy (in MeV) and the wavelength (in meters) if the particle in part (a) is (i) an electron and (ii) a proton.

39.72 · **Proton Energy in a Nucleus.** The radii of atomic nuclei are of the order of 5.0×10^{-15} m. (a) Estimate the minimum uncertainty in the momentum of a proton if it is confined within a nucleus. (b) Take this uncertainty in momentum to be an estimate of the magnitude of the momentum. Use the relativistic relationship between energy and momentum, Eq. (37.39), to obtain an estimate of the kinetic energy of a proton confined within a nucleus. (c) For a proton to remain bound within a nucleus, what must the magnitude of the (negative) potential energy for a proton be within the nucleus? Give your answer in eV and in MeV. Compare to the potential energy for an electron in a hydrogen atom, which has a magnitude of a few tens of eV. (This shows why the interaction that binds the nucleus together is called the "strong nuclear force.")

39.73 • **Electron Energy in a Nucleus.** The radii of atomic nuclei are of the order of 5.0×10^{-15} m. (a) Estimate the minimum uncertainty in the momentum of an electron if it is confined within a nucleus. (b) Take this uncertainty in momentum to be an estimate of the magnitude of the momentum. Use the relativistic relationship between energy and momentum, Eq. (37.39), to obtain an estimate of the kinetic energy of an electron confined within a nucleus. (c) Compare the energy calculated in part (b) to the magnitude of the Coulomb potential energy of a proton and an electron separated by 5.0×10^{-15} m. On the basis of your result, could there be electrons within the nucleus? (*Note:* It is interesting to compare this result to that of Problem 39.72.)

39.74 • The neutral pion (π^0) is an unstable particle produced in high-energy particle collisions. Its mass is about 264 times that of the electron, and it exists for an average lifetime of 8.4×10^{-17} s before decaying into two gamma-ray photons. Using the relationship $E = mc^2$ between rest mass and energy, find the uncertainty in the mass of the particle and express it as a fraction of the mass.

39.75 • **Doorway Diffraction.** If your wavelength were 1.0 m, you would undergo considerable diffraction in moving through a doorway. (a) What must your speed be for you to have this wavelength? (Assume that your mass is 60.0 kg.) (b) At the speed calculated in part (a), how many years would it take you to move 0.80 m (one step)? Will you notice diffraction effects as you walk through doorways?

39.76 • **Atomic Spectra Uncertainties.** A certain atom has an energy level 2.58 eV above the ground level. Once excited to this level, the atom remains in this level for 1.64×10^{-7} s (on average) before emitting a photon and returning to the ground level. (a) What is the energy of the photon (in electron volts)? What is its wavelength (in nanometers)? (b) What is the smallest possible uncertainty in energy of the photon? Give your answer in electron volts. (c) Show that $|\Delta E/E| = |\Delta\lambda/\lambda|$ if $|\Delta\lambda/\lambda| \ll 1$. Use this to calculate the magnitude of the smallest possible uncertainty in the wavelength of the photon. Give your answer in nanometers.

39.77 •• For x rays with wavelength 0.0300 nm, the $m = 1$ intensity maximum for a crystal occurs when the angle θ in Fig. 39.2 is 35.8°. At what angle θ does the $m = 1$ maximum occur when a beam of 4.50-keV electrons is used instead? Assume that the electrons also scatter from the atoms in the surface plane of this same crystal.

39.78 •• A certain atom has an energy state 3.50 eV above the ground state. When excited to this state, the atom remains for 2.0 μs, on average, before it emits a photon and returns to the ground state. (a) What are the energy and wavelength of the photon? (b) What is the smallest possible uncertainty in energy of the photon?

39.79 •• BIO **Structure of a Virus.** To investigate the structure of extremely small objects, such as viruses, the wavelength of the probing wave should be about one-tenth the size of the object for sharp images. But as the wavelength gets shorter, the energy of a photon of light gets greater and could damage or destroy the object being studied. One alternative is to use electron matter waves instead of light. Viruses vary considerably in size, but 50 nm is not unusual. Suppose you want to study such a virus, using a wave of wavelength 5.00 nm. (a) If you use light of this wavelength, what would be the energy (in eV) of a single photon? (b) If you use an electron of this wavelength, what would be its kinetic energy (in eV)? Is it now clear why matter waves (such as in the electron microscope) are often preferable to electromagnetic waves for studying microscopic objects?

39.80 •• CALC **Zero-Point Energy.** Consider a particle with mass m moving in a potential $U = \frac{1}{2}kx^2$, as in a mass–spring system. The total energy of the particle is $E = (p^2/2m) + \frac{1}{2}kx^2$. Assume that p and x are approximately related by the Heisenberg uncertainty principle, so $px \approx h$. (a) Calculate the minimum possible value of the energy E, and the value of x that gives this minimum E. This lowest possible energy, which is not zero, is called the *zero-point energy*. (b) For the x calculated in part (a), what is the ratio of the kinetic to the potential energy of the particle?

39.81 •• CALC A particle with mass m moves in a potential energy $U(x) = A|x|$, where A is a positive constant. In a simplified picture, quarks (the constituents of protons, neutrons, and other particles, as will be described in Chapter 44) have a potential energy of interaction of approximately this form, where x represents the separation between a pair of quarks. Because $U(x) \to \infty$ as $x \to \infty$, it's not possible to separate quarks from each other (a phenomenon called *quark confinement*). (a) Classically, what is the force acting on this particle as a function of x? (b) Using the uncertainty principle as in Problem 39.80, determine approximately the zero-point energy of the particle.

39.82 •• Imagine another universe in which the value of Planck's constant is 0.0663 J · s, but in which the physical laws and all other physical constants are the same as in our universe. In this universe, two physics students are playing catch. They are 12 m apart, and one throws a 0.25-kg ball directly toward the other with a speed of 6.0 m/s. (a) What is the uncertainty in the ball's horizontal momentum, in a direction perpendicular to that in which it is being thrown, if the student throwing the ball knows that it is located within a cube with volume 125 cm³ at the time she throws it? (b) By what horizontal distance could the ball miss the second student?

39.83 •• DATA For your work in a mass spectrometry lab, you are investigating the absorption spectrum of one-electron ions. To maintain the atoms in an ionized state, you hold them at low density in an ion trap, a device that uses a configuration of electric fields to confine ions. The majority of the ions are in their ground state, so that is the initial state for the absorption transitions that you observe. (a) If the longest wavelength that you observe in the absorption spectrum is 13.56 nm, what is the atomic number Z for the ions? (b) What is the next shorter wavelength that the ions will absorb? (c) When one of the ions absorbs a photon of wavelength 6.78 nm, a free electron is produced. What is the kinetic energy (in electron volts) of the electron?

39.84 •• DATA In the crystallography lab where you work, you are given a single crystal of an unknown substance to identify. To obtain one piece of information about the substance, you repeat the Davisson–Germer experiment to determine the spacing of the atoms in the surface planes of the crystal. You start with electrons that are essentially stationary and accelerate them through a potential difference of magnitude V_{ac}. The electrons then scatter off the atoms on the surface of the crystal (as in Fig. 39.3b). Next you measure the angle θ that locates the first-order diffraction peak. Finally, you repeat the measurement for different values of V_{ac}. Your results are given in the table.

V_{ac} (V)	106.3	69.1	49.9	25.2	16.9	13.6
θ (°)	20.4	24.8	30.2	45.5	59.1	73.1

(a) Graph your data in the form $\sin\theta$ versus $1/\sqrt{V_{ac}}$. What is the slope of the straight line that best fits the data points when plotted in this way? (b) Use your results from part (a) to calculate the value of d for this crystal.

39.85 •• DATA As an amateur astronomer, you are studying the apparent brightness of stars. You know that a star's apparent brightness depends on its distance from the earth and also on the fraction of its radiated energy that is in the visible region of the electromagnetic spectrum. But, as a first step, you search the Internet for information on the surface temperatures and radii of some selected stars so that you can calculate their total radiated power. You find the data given in the table.

Star	Polaris	Vega	Antares	α Centauri B
Surface temperature (K)	6015	9602	3400	5260
Radius relative to that of the sun (R_{sun})	46	2.73	883	0.865

The radius is given in units of the radius of the sun, $R_{sun} = 6.96 \times 10^8$ m. The surface temperature is the effective temperature that gives the measured photon luminosity of the star if the star is assumed to radiate as an ideal blackbody. The photon luminosity is the power emitted in the form of photons. (a) Which star in the table has the greatest radiated power? (b) For which of these stars, if any, is the peak wavelength λ_m in the visible range (380–750 nm)? (c) The sun has a total radiated power of 3.85×10^{26} W. Which of these stars, if any, have a total radiated power less than that of our sun?

CHALLENGE PROBLEMS

39.86 ••• CP CALC You have entered a contest in which the contestants drop a marble with mass 20.0 g from the roof of a building onto a small target 25.0 m below. From uncertainty considerations, what is the typical distance by which you will miss the target, given that you aim with the highest possible precision? (*Hint:* The uncertainty Δx_f in the x-coordinate of the marble when it reaches the ground comes in part from the uncertainty Δx_i in the x-coordinate initially and in part from the initial uncertainty in v_x. The latter gives rise to an uncertainty Δv_x in the horizontal motion of the marble as it falls. The values of Δx_i and Δv_x are related by the uncertainty principle. A small Δx_i gives rise to a large Δv_x, and vice versa. Find the value of Δx_i that gives the smallest total uncertainty in x at the ground. Ignore any effects of air resistance.)

39.87 ••• (a) Show that in the Bohr model, the frequency of revolution of an electron in its circular orbit around a stationary hydrogen nucleus is $f = me^4/4\epsilon_0^2 n^3 h^3$. (b) In classical physics, the frequency of revolution of the electron is equal to the frequency of the radiation that it emits. Show that when n is very large, the frequency of revolution does indeed equal the radiated frequency calculated from Eq. (39.5) for a transition from $n_1 = n + 1$ to $n_2 = n$. (This illustrates Bohr's *correspondence principle,* which is often used as a check on quantum calculations. When n is small, quantum physics gives results that are very different from those of classical physics. When n is large, the differences are not

significant, and the two methods then "correspond." In fact, when Bohr first tackled the hydrogen atom problem, he sought to determine f as a function of n such that it would correspond to classical results for large n.)

PASSAGE PROBLEMS

BIO **ION MICROSCOPES.** Whereas electron microscopes make use of the wave properties of electrons, ion microscopes make use of the wave properties of atomic ions, such as helium ions (He^+), to image materials. A helium ion has a mass 7300 times that of an electron. In a typical helium-ion microscope, helium ions are accelerated by a high voltage of 10–50 kV and focused into a beam onto the sample to be imaged. At these energies, the ions don't travel very far into the sample, so this type of microscope is used primarily for the surface imaging of biological structures. The use of helium ions with much greater energies (in the MeV range) has been proposed as a way to image the entire thickness of a sample, because these faster helium ions can pass all the way through biological samples such as cells. In this type of ion microscope, the energy lost as the ion beam passes through different parts of a cell can be measured and related to the distribution of material in the cell, with thicker parts of the cell causing greater energy loss. [Source: "Whole-Cell Imaging at Nanometer Resolutions Using Fast and Slow Focused Helium Ions," by Xiao Chen et al., *Biophysical Journal* 101(7): 1788–1793, Oct. 5, 2011.]

39.88 How does the wavelength of a helium ion compare to that of an electron accelerated through the same potential difference? (a) The helium ion has a longer wavelength, because it has greater mass. (b) The helium ion has a shorter wavelength, because it has greater mass. (c) The wavelengths are the same, because the kinetic energy is the same. (d) The wavelengths are the same, because the electric charge is the same.

39.89 Can the first type of helium-ion microscope, used for surface imaging, produce helium ions with a wavelength of 0.1 pm? (a) Yes; the voltage required is 21 kV. (b) Yes; the voltage required is 42 kV. (c) No; a voltage higher than 50 kV is required. (d) No; a voltage lower than 10 kV is required.

39.90 Why is it easier to use helium ions rather than neutral helium atoms in such a microscope? (a) Helium atoms are not electrically charged, and only electrically charged particles have wave properties. (b) Helium atoms form molecules, which are too large to have wave properties. (c) Neutral helium atoms are more difficult to focus with electric and magnetic fields. (d) Helium atoms have much larger mass than helium ions do and thus are more difficult to accelerate.

39.91 In the second type of helium-ion microscope, a 1.2-MeV ion passing through a cell loses 0.2 MeV per μm of cell thickness. If the energy of the ion can be measured to 6 keV, what is the smallest difference in thickness that can be discerned? (a) 0.03 μm; (b) 0.06 μm; (c) 3 μm; (d) 6 μm.

Answers

Chapter Opening Question ?

(i) The smallest detail visible in an image is comparable to the wavelength used to make the image. Electrons can easily be given a large momentum p and hence a short wavelength $\lambda = h/p$, and so can be used to resolve extremely fine details. (See Section 39.1.)

Test Your Understanding Questions

39.1 (a) (i), (b) no From Example 39.2, the speed of a particle is $v = h/\lambda m$ and the kinetic energy is $K = \frac{1}{2}mv^2 = (m/2)(h/\lambda m)^2 = h^2/2\lambda^2 m$. This shows that for a given wavelength, the kinetic energy is inversely proportional to the mass. Hence the proton, with a smaller mass, has more kinetic energy than the neutron. For part (b), the energy of a photon is $E = hf$, and the frequency of a photon is $f = c/\lambda$. Hence $E = hc/\lambda$ and $\lambda = hc/E = (4.136 \times 10^{-15} \text{ eV} \cdot \text{s})(2.998 \times 10^8 \text{ m/s})/(54 \text{ eV}) = 2.3 \times 10^{-8}$ m. This is more than 100 times greater than the wavelength of an electron of the same energy. While both photons and electrons have wavelike properties, they have different relationships between their energy and momentum and hence between their frequency and wavelength.

39.2 (iii) Because the alpha particle is more massive, it won't bounce back in even a head-on collision with a proton that's initially at rest, any more than a bowling ball would when colliding with a Ping-Pong ball at rest (see Fig. 8.23b). Thus there would be *no* large-angle scattering in this case. Rutherford saw large-angle scattering in his experiment because gold nuclei are more massive than alpha particles (see Fig. 8.23a).

39.3 (iv) Figure 39.27 shows that many (though *not* all) of the energy levels of He$^+$ are the same as those of H. Hence photons emitted during transitions between corresponding pairs of levels in He$^+$ and H have the same energy E and the same wavelength $\lambda = hc/E$. An H atom that drops from the $n = 2$ level to the $n = 1$ level emits a photon of energy 10.20 eV and wavelength 122 nm (see Example 39.6); a He$^+$ ion emits a photon of the same energy and wavelength when it drops from the $n = 4$ level to the $n = 2$ level. Inspecting Fig. 39.27 will show you that every even-numbered level in He$^+$ matches a level in H, while none of the odd-numbered He$^+$ levels do. The first three He$^+$ transitions given in the question ($n = 2$ to $n = 1$, $n = 3$ to $n = 2$, and $n = 4$ to $n = 3$) all involve an odd-numbered level, so none of their wavelengths match a wavelength emitted by H atoms.

39.4 (i) In a neon light fixture, a large potential difference is applied between the ends of a neon-filled glass tube. This ionizes some of the neon atoms, allowing a current of electrons to flow through the gas. Some of the neon atoms are struck by fast-moving electrons, making them transition to an excited level. From this level the atoms undergo *spontaneous* emission, as depicted in Fig. 39.28b, and emit 632.8-nm photons in the process. No population inversion occurs and the photons are not trapped by mirrors as shown in Fig. 39.29d, so there is no stimulated emission. Hence there is no laser action.

39.5 (a) yes, (b) yes The Planck radiation law, Eq. (39.24), shows that an ideal blackbody emits radiation at *all* wavelengths: The spectral emittance $I(\lambda)$ is equal to zero only for $\lambda = 0$ and in the limit $\lambda \to \infty$. So a blackbody at 2000 K does indeed emit both x rays and radio waves. However, Fig. 39.32 shows that the spectral emittance for this temperature is very low for wavelengths much shorter than 1 μm (including x rays) and for wavelengths much longer than a few μm (including radio waves). Hence such a blackbody emits very little in the way of x rays or radio waves.

39.6 (i) and (iii) (tie), (ii) and (iv) (tie) According to the Heisenberg uncertainty principle, the smaller the uncertainty Δx in the x-coordinate, the greater the uncertainty Δp_x in the x-momentum. The relationship between Δx and Δp_x does not depend on the mass of the particle, and so is the same for a proton as for an electron.

Bridging Problem

(a) 192 nm; ultraviolet   **(b)** $n = 4$
(c) $\lambda_2 = 0.665$ nm, $\lambda_3 = 0.997$ nm

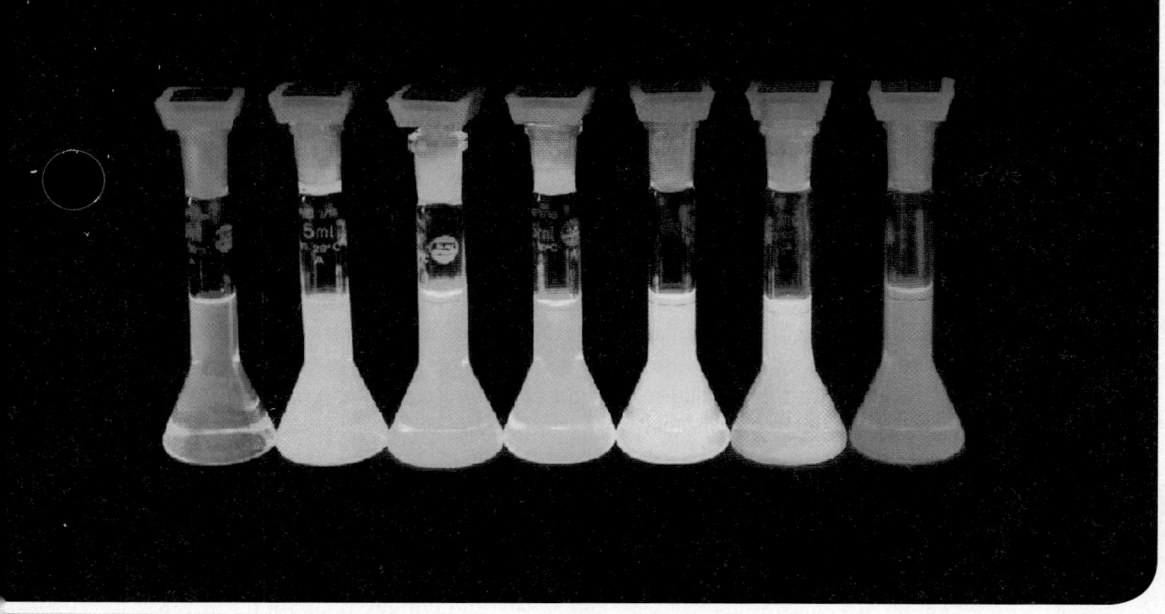

40 QUANTUM MECHANICS I: WAVE FUNCTIONS

LEARNING GOALS

Looking forward at …

40.1 The wave function that describes the behavior of a particle and the Schrödinger equation that this function must satisfy.

40.2 How to calculate the wave functions and energy levels for a particle confined to a box.

40.3 How to analyze the quantum-mechanical behavior of a particle in a potential well.

40.4 How quantum mechanics makes it possible for particles to go where Newtonian mechanics says they cannot.

40.5 How to use quantum mechanics to analyze a harmonic oscillator.

40.6 How measuring a quantum-mechanical system can change that system's state.

Looking back at …

I n Chapter 39 we found that particles can behave like waves. In fact, it turns out that we can use the wave picture to completely describe the behavior of a particle. This approach, called *quantum mechanics,* is the key to understanding the behavior of matter on the molecular, atomic, and nuclear scales. In this chapter we'll see how to find the *wave function* of a particle by solving the *Schrödinger equation,* which is as fundamental to quantum mechanics as Newton's laws are to mechanics or as Maxwell's equations are to electromagnetism.

We'll begin with a quantum-mechanical analysis of a *free particle* that moves along a straight line without being acted on by forces of any kind. We'll then consider particles that are acted on by forces and are trapped in *bound states,* just as electrons are bound within an atom. We'll see that solving the Schrödinger equation automatically gives the possible energy levels for the system.

Besides energies, solving the Schrödinger equation gives us the probabilities of finding a particle in various regions. One surprising result is that there is a nonzero probability that microscopic particles will pass through thin barriers, even though such a process is forbidden by Newtonian mechanics.

In this chapter we'll consider the Schrödinger equation for one-dimensional motion only. In Chapter 41 we'll see how to extend this equation to three-dimensional problems such as the hydrogen atom. The hydrogen-atom wave functions will in turn form the foundation for our analysis of more complex atoms, of the periodic table of the elements, of x-ray energy levels, and of other properties of atoms.

40.1 WAVE FUNCTIONS AND THE ONE-DIMENSIONAL SCHRÖDINGER EQUATION

We have now seen compelling evidence that on an atomic or subatomic scale, an object such as an electron cannot be described simply as a classical, Newtonian point particle. Instead, we must take into account its *wave* characteristics. In the Bohr model of the hydrogen atom (Section 39.3) we tried to have it both ways: We

40.1 These children are talking over a cup-and-string "telephone." The displacement of the string is completely described by a wave function $y(x, t)$. In an analogous way, a particle is completely described by a quantum-mechanical wave function $\Psi(x, y, z, t)$.

CAUTION **Particle waves vs. mechanical waves** Unlike waves on a string or sound waves in air, the wave function for a particle is *not* a mechanical wave that needs a material medium in order to propagate. The wave function describes the particle, but we cannot define the function itself in terms of anything material. We can describe only how it is related to physically observable effects. ∎

pictured the electron as a classical particle in a circular orbit around the nucleus, and used the de Broglie relationship between particle momentum and wavelength to explain why only orbits of certain radii are allowed. As we saw in Section 39.6, however, the Heisenberg uncertainty principle tells us that a hybrid description of this kind can't be wholly correct. In this section we'll explore how to describe the state of a particle by using *only* the language of waves. This new description, called **quantum mechanics,** replaces the classical scheme of describing the state of a particle by its coordinates and velocity components.

Our new quantum-mechanical scheme for describing a particle has a lot in common with the language of classical wave motion. In Section 15.3 of Chapter 15, we described transverse waves on a string by specifying the position of each point in the string at each instant of time by means of a *wave function* $y(x, t)$ that represents the displacement from equilibrium, at time t, of a point on the string at a distance x from the origin (**Fig. 40.1**). Once we know the wave function for a particular wave motion, we know everything there is to know about the motion. For example, we can find the velocity and acceleration of any point on the string at any time. We worked out specific forms for these functions for *sinusoidal* waves, in which each particle undergoes simple harmonic motion.

We followed a similar pattern for sound waves in Chapter 16. The wave function $p(x, t)$ for a wave traveling along the x-direction represented the pressure variation at any point x and any time t. In Section 32.3 we used *two* wave functions to describe the $\vec{E}$ and $\vec{B}$ fields in an electromagnetic wave.

Thus it's natural to use a wave function as the central element of our new language of quantum mechanics. The customary symbol for this wave function is the Greek letter psi, Ψ or ψ. In general, we'll use an uppercase Ψ to denote a function of all the space coordinates and time, and a lowercase ψ for a function of the space coordinates only—*not* of time. Just as the wave function $y(x, t)$ for mechanical waves on a string provides a complete description of the motion, so the wave function $\Psi(x, y, z, t)$ for a particle contains all the information that can be known about the particle.

Waves in One Dimension: Waves on a String

The wave function of a particle depends in general on all three dimensions of space. For simplicity, however, we'll begin our study of these functions by considering *one-dimensional* motion, in which a particle of mass m moves parallel to the x-axis and the wave function Ψ depends on the coordinate x and the time t only. (In the same way, we studied one-dimensional kinematics in Chapter 2 before going on to study two- and three-dimensional motion in Chapter 3.)

What does a one-dimensional quantum-mechanical wave look like, and what determines its properties? We can answer this question by first recalling the properties of a wave on a string. We saw in Section 15.3 that any wave function $y(x, t)$ that describes a wave on a string must satisfy the *wave equation:*

$$\frac{\partial^2 y(x, t)}{\partial x^2} = \frac{1}{v^2} \frac{\partial^2 y(x, t)}{\partial t^2} \qquad \text{(wave equation for waves on a string)} \qquad (40.1)$$

In Eq. (40.1) v is the speed of the wave, which is the same no matter what the wavelength. As an example, consider the following wave function for a wave of wavelength λ and frequency f moving in the positive x-direction along a string:

$$y(x, t) = A\cos(kx - \omega t) + B\sin(kx - \omega t) \qquad \begin{array}{l}\text{(sinusoidal wave} \\ \text{on a string)}\end{array} \qquad (40.2)$$

Here $k = 2\pi/\lambda$ is the *wave number* and $\omega = 2\pi f$ is the *angular frequency.* (We used these same quantities for mechanical waves in Chapter 15 and electromagnetic waves in Chapter 32.) The quantities A and B are constants that determine the amplitude and phase of the wave. The expression in Eq. (40.2) is a valid wave function if and only if it satisfies the wave equation, Eq. (40.1). To check this, take the first and second derivatives of $y(x, t)$ with respect to x and take the first and second derivatives with respect to t:

$$\frac{\partial y(x, t)}{\partial x} = -kA \sin(kx - \omega t) + kB \cos(kx - \omega t) \qquad (40.3a)$$

$$\frac{\partial^2 y(x, t)}{\partial x^2} = -k^2 A \cos(kx - \omega t) - k^2 B \sin(kx - \omega t) \qquad (40.3b)$$

$$\frac{\partial y(x, t)}{\partial t} = \omega A \sin(kx - \omega t) - \omega B \cos(kx - \omega t) \qquad (40.3c)$$

$$\frac{\partial^2 y(x, t)}{\partial t^2} = -\omega^2 A \cos(kx - \omega t) - \omega^2 B \sin(kx - \omega t) \qquad (40.3d)$$

If we substitute Eqs. (40.3b) and (40.3d) into the wave equation, Eq. (40.1), we get

$$-k^2 A \cos(kx - \omega t) - k^2 B \sin(kx - \omega t)$$
$$= \frac{1}{v^2} [-\omega^2 A \cos(kx - \omega t) - \omega^2 B \sin(kx - \omega t)] \qquad (40.4)$$

For Eq. 40.4 to be satisfied at all coordinates x and all times t, the coefficients of $\cos(kx - \omega t)$ must be the same on both sides of the equation, and likewise for the coefficients of $\sin(kx - \omega t)$. Both of these conditions will be satisfied if

$$k^2 = \frac{\omega^2}{v^2} \qquad \text{or} \qquad \omega = vk \qquad \text{(waves on a string)} \qquad (40.5)$$

Since $\omega = 2\pi f$ and $k = 2\pi/\lambda$, Eq. (40.5) is equivalent to

$$2\pi f = v\frac{2\pi}{\lambda} \qquad \text{or} \qquad v = \lambda f \qquad \text{(waves on a string)}$$

This equation is just the familiar relationship among wave speed, wavelength, and frequency for waves on a string. So our calculation shows that Eq. (40.2) is a valid wave function for waves on a string for any values of A and B, provided that ω and k are related by Eq. (40.5).

Waves in One Dimension: Particle Waves

What we need is a quantum-mechanical version of the wave equation, Eq. (40.1), valid for particle waves. We expect this equation to involve partial derivatives of the wave function $\Psi(x, t)$ with respect to x and with respect to t. However, this new equation *cannot* be the same as Eq. (40.1) for waves on a string because the relationship between ω and k is different. We can show this by considering a **free particle,** one that experiences no force at all as it moves along the x-axis. For such a particle the potential energy $U(x)$ has the same value for all x (recall from Chapter 7 that $F_x = -dU(x)/dx$, so zero force means the potential energy has zero derivative). For simplicity let $U = 0$ for all x. Then the energy of the free particle is equal to its kinetic energy, which we can express in terms of its momentum p:

$$E = \tfrac{1}{2}mv^2 = \frac{m^2 v^2}{2m} = \frac{(mv)^2}{2m} = \frac{p^2}{2m} \qquad \text{(energy of a free particle)} \qquad (40.6)$$

The de Broglie relationships (Section 39.1) tell us that the energy E is proportional to the angular frequency ω and the momentum p is proportional to the wave number:

$$E = hf = \frac{h}{2\pi} 2\pi f = \hbar\omega \qquad (40.7a)$$

$$p = \frac{h}{\lambda} = \frac{h}{2\pi} \frac{2\pi}{\lambda} = \hbar k \qquad (40.7b)$$

Remember that $\hbar = h/2\pi$. If we substitute Eqs. (40.7) into Eq. (40.6), we find that the relationship between ω and k for a free particle is

$$\hbar\omega = \frac{\hbar^2 k^2}{2m} \qquad \text{(free particle)} \tag{40.8}$$

Equation (40.8) is *very* different from the corresponding relationship for waves on a string, Eq. (40.5): The angular frequency ω for particle waves is proportional to the *square* of the wave number, while for waves on a string ω is directly proportional to k. Our task is therefore to construct a quantum-mechanical version of the wave equation whose free-particle solutions satisfy Eq. (40.8).

We'll attack this problem by assuming a sinusoidal wave function $\Psi(x, t)$ of the same form as Eq. (40.2) for a sinusoidal wave on a string. For a wave on a string, Eq. (40.2) represents a wave of wavelength $\lambda = 2\pi/k$ and frequency $f = \omega/2\pi$ propagating in the positive x-direction. By analogy, our sinsuoidal wave function $\Psi(x, t)$ represents a free particle of mass m, momentum $p = \hbar k$, and energy $E = \hbar\omega$ moving in the positive x-direction:

$$\Psi(x, t) = A\cos(kx - \omega t) + B\sin(kx - \omega t) \qquad \substack{\text{(sinusoidal wave} \\ \text{function representing} \\ \text{a free particle)}} \tag{40.9}$$

The wave number k and angular frequency ω in Eq. (40.9) must satisfy Eq. (40.8). If you look at Eq. (40.3b), you'll see that taking the second derivative of $\Psi(x, t)$ in Eq. (40.9) with respect to x gives us $\Psi(x, t)$ multiplied by $-k^2$. Hence if we multiply $\partial^2\Psi(x, t)/\partial x^2$ by $-\hbar^2/2m$, we get

$$-\frac{\hbar^2}{2m}\frac{\partial^2\Psi(x, t)}{\partial x^2} = -\frac{\hbar^2}{2m}\left[-k^2 A\cos(kx - \omega t) - k^2 B\sin(kx - \omega t)\right]$$

$$= \frac{\hbar^2 k^2}{2m}\left[A\cos(kx - \omega t) + B\sin(kx - \omega t)\right] \tag{40.10}$$

$$= \frac{\hbar^2 k^2}{2m}\Psi(x, t)$$

Equation (40.10) suggests that $(-\hbar^2/2m)\partial^2\Psi(x, t)/\partial x^2$ should be one side of our quantum-mechanical wave equation, with the other side equal to $\hbar\omega\Psi(x, t)$ in order to satisfy Eq. (40.8). If you look at Eq. (40.3c), you'll see that taking the *first* time derivative of $\Psi(x, t)$ in Eq. (40.9) brings out a factor of ω. So we'll make the educated guess that the right-hand side of our quantum-mechanical wave equation involves $\hbar = h/2\pi$ times $\partial\Psi(x, t)/\partial t$. So our tentative equation is

$$-\frac{\hbar^2}{2m}\frac{\partial^2\Psi(x, t)}{\partial x^2} = C\hbar\frac{\partial\Psi(x, t)}{\partial t} \tag{40.11}$$

At this point we include a constant C as a "fudge factor" to make sure that everything turns out right. Now let's substitute the wave function from Eq. (40.9) into Eq. (40.11). From Eq. (40.10) and Eq. (40.3c), we get

$$\frac{\hbar^2 k^2}{2m}\left[A\cos(kx - \omega t) + B\sin(kx - \omega t)\right]$$

$$= C\hbar\omega\left[A\sin(kx - \omega t) - B\cos(kx - \omega t)\right] \tag{40.12}$$

From Eq. (40.8), $\hbar\omega = \hbar^2 k^2/2m$, so we can cancel these factors on the two sides of Eq. (40.12). What remains is

$$A\cos(kx - \omega t) + B\sin(kx - \omega t)$$

$$= CA\sin(kx - \omega t) - CB\cos(kx - \omega t) \tag{40.13}$$

As in our discussion above of the wave equation for waves on a string, in order for Eq. (40.13) to be satisfied for all values of x and all values of t, the coefficients of $\cos(kx - \omega t)$ must be the same on both sides of the equation, and likewise for the coefficients of $\sin(kx - \omega t)$. Hence we have the following relationships among the coefficients A, B, and C in Eqs. (40.9) and (40.11):

$$A = -CB \tag{40.14a}$$

$$B = CA \tag{40.14b}$$

If we use Eq. (40.14b) to eliminate B from Eq. (40.14a), we get $A = -C^2A$, which means that $C^2 = -1$. Thus C is equal to the *imaginary* number $i = \sqrt{-1}$, and Eq. (40.11) becomes

$$-\frac{\hbar^2}{2m}\frac{\partial^2\Psi(x,t)}{\partial x^2} = i\hbar\frac{\partial\Psi(x,t)}{\partial t} \qquad \text{(one-dimensional Schrödinger equation for a free particle)} \tag{40.15}$$

Equation (40.15) is the one-dimensional **Schrödinger equation** for a free particle, developed in 1926 by the Austrian physicist Erwin Schrödinger (**Fig. 40.2**). The presence of the imaginary number i in Eq. (40.15) means that the solutions to the Schrödinger equation are complex quantities, with a real part and an imaginary part. (The imaginary part of $\Psi(x,t)$ is a real function multiplied by the imaginary number $i = \sqrt{-1}$.) An example is our free-particle wave function from Eq. (40.9). Since we found that $C = i$ in Eqs. (40.14), it follows from Eq. (40.14b) that $B = iA$. Then Eq. (40.9) becomes

$$\Psi(x,t) = A[\cos(kx - \omega t) + i\sin(kx - \omega t)] \qquad \text{(sinusoidal wave function representing a free particle)} \tag{40.16}$$

The real part of $\Psi(x,t)$ is $\mathrm{Re}\,\Psi(x,t) = A\cos(kx - \omega t)$ and the imaginary part is $\mathrm{Im}\,\Psi(x,t) = A\sin(kx - \omega t)$. **Figure 40.3** graphs the real and imaginary parts of $\Psi(x,t)$ at $t = 0$, so $\Psi(x,0) = A\cos kx + iA\sin kx$.

We can rewrite Eq. (40.16) with *Euler's formula,* which states that for any angle θ,

$$e^{i\theta} = \cos\theta + i\sin\theta$$
$$e^{-i\theta} = \cos(-\theta) + i\sin(-\theta) = \cos\theta - i\sin\theta \tag{40.17}$$

Thus our sinusoidal free-particle wave function becomes

$$\Psi(x,t) = Ae^{i(kx - \omega t)} = Ae^{ikx}e^{-i\omega t} \qquad \text{(sinusoidal wave function representing a free particle)} \tag{40.18}$$

If k is positive in Eq. (40.16), the wave function represents a free particle moving in the positive x-direction with momentum $p = \hbar k$ and energy $E = \hbar\omega = \hbar^2 k^2/2m$. If k is negative, the momentum and hence the motion are in the negative x-direction. (With a negative value of k, the wavelength is $\lambda = 2\pi/|k|$).

Interpreting the Wave Function

The complex nature of the wave function for a free particle makes this function challenging to interpret. (We certainly haven't needed imaginary numbers before this point to describe real physical phenomena.) Here's how to think about this function: $\Psi(x,t)$ describes the *distribution* of a particle in space, just as the wave functions for an electromagnetic wave describe the distribution of the electric and magnetic fields. When we worked out interference and diffraction patterns in Chapters 35 and 36, we found that the intensity I of the radiation at any point in a pattern is proportional to the square of the electric-field magnitude—that is, to E^2. In the photon interpretation of interference and diffraction (see Section 38.4), the intensity at each point is proportional to the number of photons striking around that point or, alternatively, to the *probability* that any individual photon will

40.2 Erwin Schrödinger (1887–1961) developed the equation that bears his name in 1926, an accomplishment for which he shared (with the British physicist P. A. M. Dirac) the 1933 Nobel Prize in physics. His grave marker is adorned with a version of Eq. (40.15).

40.3 The spatial wave function $\Psi(x,t) = Ae^{i(kx - \omega t)}$ for a free particle of definite momentum $p = \hbar k$ is a complex function: It has both a real part and an imaginary part. These are graphed here as functions of x for $t = 0$.

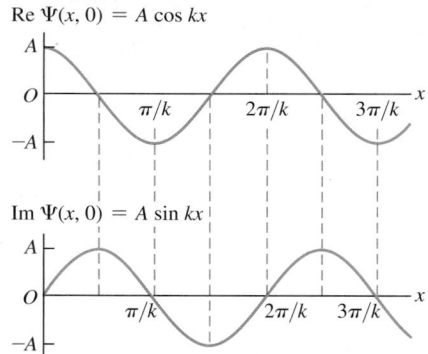

strike around the point. Thus the square of the electric-field magnitude at each point is proportional to the probability of finding a photon around that point.

In exactly the same way, the square of the wave function of a particle at each point tells us about the probability of finding the particle around that point. More precisely, we should say the square of the *absolute value* of the wave function, $|\Psi|^2$. This is necessary because, as we have seen, the wave function is a complex quantity with real and imaginary parts.

For a particle that can move only along the x-direction, the quantity $|\Psi(x, t)|^2 dx$ is the probability that the particle will be found at time t at a coordinate in the range from x to $x + dx$. The particle is most likely to be found in regions where $|\Psi|^2$ is large, and so on. This interpretation, first made by the German physicist Max Born (**Fig. 40.4**), requires that the wave function Ψ be *normalized*. That is, the integral of $|\Psi(x, t)|^2 dx$ over all possible values of x must equal exactly 1. In other words, the probability is exactly 1, or 100%, that the particle is *somewhere*.

40.4 In 1926, the German physicist Max Born (1882–1970) devised the interpretation that $|\Psi|^2$ is the probability distribution function for a particle that is described by the wave function Ψ. He also coined the term "quantum mechanics" (in the original German, *Quantenmechanik*). For his contributions, Born shared (with Walther Bothe) the 1954 Nobel Prize in physics.

CAUTION **Interpreting $|\Psi|^2$** Note that $|\Psi(x, t)|^2$ itself is *not* a probability. Rather, $|\Psi(x, t)|^2 dx$ is the probability of finding the particle between position x and position $x + dx$ at time t. If the length dx is made smaller, it becomes less likely that the particle will be found within that length, so the probability decreases. A better name for $|\Psi(x, t)|^2$ is the **probability distribution function,** since it describes how the probability of finding the particle at different locations is distributed over space. Another common name for $|\Psi(x, t)|^2$ is the *probability density.*

We can use the probability interpretation of $|\Psi|^2$ to get a better understanding of Eq. (40.18), the wave function for a free particle. This function describes a particle that has a definite momentum $p = \hbar k$ in the x-direction and *no* uncertainty in momentum: $\Delta p_x = 0$. The Heisenberg uncertainty principle for position and momentum, Eqs. (39.29), says that $\Delta x \Delta p_x \geq \hbar/2$. If Δp_x is zero, then Δx must be infinite, and we have no idea where along the x-axis the particle can be found. (We saw a similar result for photons in Section 38.4.) We can show this by calculating the probability distribution function $|\Psi(x, t)|^2$: the product of Ψ and its *complex conjugate* Ψ^*. To find the complex conjugate of a complex number, we replace all i with $-i$. For example, the complex conjugate of $c = a + ib$, where a and b are real, is $c^* = a - ib$, so $|c|^2 = c^*c = (a + ib)(a - ib) = a^2 + b^2$ (recall that $i^2 = -1$). The complex conjugate of Eq. (40.18) is

$$\Psi^*(x, t) = A^* e^{-i(kx-\omega t)} = A^* e^{-ikx} e^{i\omega t}$$

(We have to allow for the possibility that the coefficient A is itself a complex number.) Hence the probability distribution function is

$$|\Psi(x, t)|^2 = \Psi^*(x, t)\Psi(x, t) = (A^* e^{-ikx} e^{i\omega t})(A e^{ikx} e^{-i\omega t})$$

$$= A^* A e^0 = |A|^2$$

The probability distribution function doesn't depend on position, which says that we are equally likely to find the particle *anywhere* along the x-axis! Mathematically, this is because the sinusoidal wave function $\Psi(x, t) = A e^{i(kx-\omega t)} = A[\cos(kx - \omega t) + i\sin(kx - \omega t)]$ extends all the way from $x = -\infty$ to $x = +\infty$ with the same amplitude A. This also means that the wave function can't be normalized: The integral of $|\Psi(x, t)|^2$ over all space is infinite for any value of A.

Note also that the wave function in Eq. (40.18) describes a particle with a definite energy $E = \hbar\omega$, so there is zero uncertainty in energy: $\Delta E = 0$. The Heisenberg uncertainty principle for energy and time interval, $\Delta t \Delta E \geq \hbar/2$ [Eq. (39.30)], tells us that the time uncertainty Δt for this particle is infinite. In other words, we can have no idea *when* the particle will pass a given point on the x-axis. That also agrees with our result $|\Psi(x, t)|^2 = |A|^2$; the probability distribution function has the same value at all times.

Since we always have some idea of where a particle is, the wave function given in Eq. (40.18) isn't a realistic description. In our study of light in Section 38.4, we saw that we can make a wave function that's more *localized* in space by superimposing two or more sinusoidal functions. (This would be a good time to review that section.) As an illustration, let's calculate $\left|\Psi(x, t)\right|^2$ for a wave function of this kind.

EXAMPLE 40.1   A LOCALIZED FREE-PARTICLE WAVE FUNCTION

The wave function $\Psi(x, t) = Ae^{i(k_1 x - \omega_1 t)} + Ae^{i(k_2 x - \omega_2 t)}$ is a superposition of *two* free-particle wave functions of the form given by Eq. (40.18). Both k_1 and k_2 are positive. (a) Show that this wave function satisfies the Schrödinger equation for a free particle of mass m. (b) Find the probability distribution function for $\Psi(x, t)$.

SOLUTION

IDENTIFY and SET UP: Both wave functions $Ae^{i(k_1 x - \omega_1 t)}$ and $Ae^{i(k_2 x - \omega_2 t)}$ represent a particle moving in the positive x-direction, but with different momenta and kinetic energies: $p_1 = \hbar k_1$ and $E_1 = \hbar\omega_1 = \hbar^2 k_1^2/2m$ for the first function, $p_2 = \hbar k_2$ and $E_2 = \hbar\omega_2 = \hbar^2 k_2^2/2m$ for the second function. To test whether a superposition of these is also a valid wave function for a free particle, we'll see whether our function $\Psi(x, t)$ satisfies the free-particle Schrödinger equation, Eq. (40.15). We'll use the derivatives of the exponential function: $(d/du)e^{au} = ae^{au}$ and $(d^2/du^2)e^{au} = a^2 e^{au}$. The probability distribution function $\left|\Psi(x, t)\right|^2$ is the product of $\Psi(x, t)$ and its complex conjugate.

EXECUTE: (a) If we substitute $\Psi(x, t)$ into Eq. (40.15), the left-hand side of the equation is

$$-\frac{\hbar^2}{2m}\frac{\partial^2 \Psi(x, t)}{\partial x^2} = -\frac{\hbar^2}{2m}\frac{\partial^2 (Ae^{i(k_1 x - \omega_1 t)} + Ae^{i(k_2 x - \omega_2 t)})}{\partial x^2}$$

$$= -\frac{\hbar^2}{2m}\left[(ik_1)^2 Ae^{i(k_1 x - \omega_1 t)} + (ik_2)^2 Ae^{i(k_2 x - \omega_2 t)}\right]$$

$$= \frac{\hbar^2 k_1^2}{2m}Ae^{i(k_1 x - \omega_1 t)} + \frac{\hbar^2 k_2^2}{2m}Ae^{i(k_2 x - \omega_2 t)}$$

The right-hand side is

$$i\hbar\frac{\partial \Psi(x, t)}{\partial t} = i\hbar\frac{\partial (Ae^{i(k_1 x - \omega_1 t)} + Ae^{i(k_2 x - \omega_2 t)})}{\partial t}$$

$$= i\hbar\left[(-i\omega_1)Ae^{i(k_1 x - \omega_1 t)} + (-i\omega_2)Ae^{i(k_2 x - \omega_2 t)}\right]$$

$$= \hbar\omega_1 Ae^{i(k_1 x - \omega_1 t)} + \hbar\omega_2 Ae^{i(k_2 x - \omega_2 t)}$$

The two sides *are* equal, provided that $\hbar\omega_1 = \hbar^2 k_1^2/2m$ and $\hbar\omega_2 = \hbar^2 k_2^2/2m$. These are just the relationships that we noted above. So we conclude that $\Psi(x, t) = Ae^{i(k_1 x - \omega_1 t)} + Ae^{i(k_2 x - \omega_2 t)}$ is a valid free-particle wave function. In general, if we take any two wave functions that are solutions of the Schrödinger equation and then make a superposition of these to create a third wave function $\Psi(x, t)$, then $\Psi(x, t)$ is also a solution of the Schrödinger equation.

(b) The complex conjugate of $\Psi(x, t)$ is

$$\Psi^*(x, t) = A^* e^{-i(k_1 x - \omega_1 t)} + A^* e^{-i(k_2 x - \omega_2 t)}$$

Hence

$$\left|\Psi(x, t)\right|^2$$

$$= \Psi^*(x, t)\Psi(x, t)$$

$$= \left(A^* e^{-i(k_1 x - \omega_1 t)} + A^* e^{-i(k_2 x - \omega_2 t)}\right)\left(Ae^{i(k_1 x - \omega_1 t)} + Ae^{i(k_2 x - \omega_2 t)}\right)$$

$$= A^* A\left[\begin{array}{c} e^{-i(k_1 x - \omega_1 t)}e^{i(k_1 x - \omega_1 t)} + e^{-i(k_2 x - \omega_2 t)}e^{i(k_2 x - \omega_2 t)} \\ + e^{-i(k_1 x - \omega_1 t)}e^{i(k_2 x - \omega_2 t)} + e^{-i(k_2 x - \omega_2 t)}e^{i(k_1 x - \omega_1 t)} \end{array}\right]$$

$$= |A|^2\left[e^0 + e^0 + e^{i[(k_2 - k_1)x - (\omega_2 - \omega_1)t]} + e^{-i[(k_2 - k_1)x - (\omega_2 - \omega_1)t]}\right]$$

To simplify this expression, recall that $e^0 = 1$. From Euler's formula, $e^{i\theta} = \cos\theta + i\sin\theta$ and $e^{-i\theta} = \cos\theta - i\sin\theta$, so $e^{i\theta} + e^{-i\theta} = 2\cos\theta$. Hence

$$\left|\Psi(x, t)\right|^2 = |A|^2\{2 + 2\cos[(k_2 - k_1)x - (\omega_2 - \omega_1)t]\}$$

$$= 2|A|^2\{1 + \cos[(k_2 - k_1)x - (\omega_2 - \omega_1)t]\}$$

EVALUATE: Figure 40.5 is a graph of the probability distribution function $\left|\Psi(x, t)\right|^2$ at $t = 0$. The value of $\left|\Psi(x, t)\right|^2$ varies between 0 and $4|A|^2$; probabilities can never be negative! The particle has become *somewhat* localized: The particle is most likely to be found near a point where $\left|\Psi(x, t)\right|^2$ is maximum (where the functions $Ae^{i(k_1 x - \omega_1 t)}$ and $Ae^{i(k_2 x - \omega_2 t)}$ interfere constructively) and is very unlikely to be found near a point where $\left|\Psi(x, t)\right|^2 = 0$ (where $Ae^{i(k_1 x - \omega_1 t)}$ and $Ae^{i(k_2 x - \omega_2 t)}$ interfere destructively).

40.5 The probability distribution function at $t = 0$ for $\Psi(x, t) = Ae^{i(k_1 x - \omega_1 t)} + Ae^{i(k_2 x - \omega_2 t)}$.

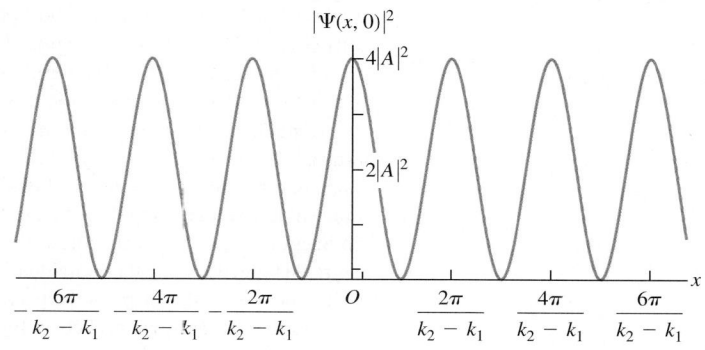

Continued

Note also that the probability distribution function is not stationary: It moves in the positive x-direction like the particle that it represents. To see this, recall from Section 15.3 that a sinusoidal wave given by $y(x, t) = A\cos(kx - \omega t)$ moves in the positive x-direction with speed $v = \omega/k$; since $|\Psi(x, t)|^2$ includes a term $\cos[(k_2 - k_1)x - (\omega_2 - \omega_1)t]$, the probability distribution moves at a speed $v_{av} = (\omega_2 - \omega_1)/(k_2 - k_1)$. The subscript "av" reminds us that v_{av} represents the *average* value of the particle's speed.

The price we pay for localizing the particle somewhat is that, unlike a particle represented by Eq. (40.18), it no longer has either a definite momentum or a definite energy. That's consistent with the Heisenberg uncertainty principles: If we decrease the uncertainties about where a particle is and when it passes a certain point, the uncertainties in its momentum and energy must increase.

The average momentum of the particle is $p_{av} = (\hbar k_2 + \hbar k_1)/2$, which is the average of the momenta associated with the free-particle wave functions we added to create $\Psi(x, t)$. This corresponds to the particle having an average speed $v_{av} = p_{av}/m = (\hbar k_2 + \hbar k_1)/2m$. Can you show that this is equal to the expression $v_{av} = (\omega_2 - \omega_1)/(k_2 - k_1)$ that we found above?

Wave Packets

40.6 Superposing a large number of sinusoidal waves with different wave numbers and appropriate amplitudes can produce a wave pulse that has a wavelength $\lambda_{av} = 2\pi/k_{av}$ and is localized within a region of space of length Δx. This localized pulse has aspects of both particle and wave.

(a) Real part of the wave function at time t

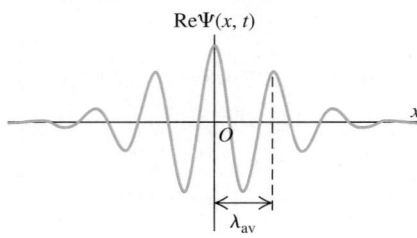

(b) Imaginary part of the wave function at time t

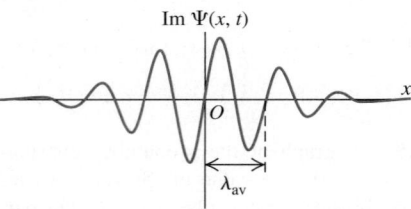

(c) Probability distribution function at time t

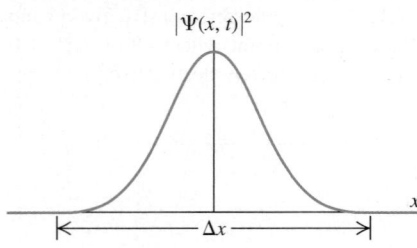

The wave function that we examined in Example 40.1 is not very well localized: The probability distribution function still extends from $x = -\infty$ to $x = +\infty$. Hence this wave function can't be normalized. To make a wave function that's more highly localized, imagine superposing two additional sinusoidal waves with different wave numbers and amplitudes so as to reinforce alternate maxima of $|\Psi(x, t)|^2$ in Fig. 40.5 and cancel out the in-between ones. Finally, if we superpose waves with a very large number of different wave numbers, we can construct a wave with only *one* maximum of $|\Psi(x, t)|^2$ (**Fig. 40.6**). Then we have something that begins to look like both a particle and a wave. It is a particle in the sense that it is localized in space; if we look from a distance, it may look like a point. But it also has a periodic structure that is characteristic of a wave.

A localized wave pulse like that shown in Fig. 40.6 is called a **wave packet.** We can represent a wave packet by an expression such as

$$\Psi(x, t) = \int_{-\infty}^{\infty} A(k)e^{i(kx - \omega t)}\, dk \qquad (40.19)$$

This integral represents a superposition of a very large number of waves, each with a different wave number k and angular frequency $\omega = \hbar k^2/2m$, and each with an amplitude $A(k)$ that depends on k.

There is an important relationship between the two functions $\Psi(x, t)$ and $A(k)$, which we show qualitatively in **Fig. 40.7**. If the function $A(k)$ is sharply peaked, as in Fig. 40.7a, we are superposing only a narrow range of wave numbers. The resulting wave pulse is then relatively broad (Fig. 40.7b). But if we use a wider range of wave numbers, so that the function $A(k)$ is broader (Fig. 40.7c), then the wave pulse is more narrowly localized (Fig. 40.7d). This is simply the uncertainty principle in action. A narrow range of k means a narrow range of $p_x = \hbar k$ and thus a small Δp_x; the result is a relatively large Δx. A broad range of k corresponds to a large Δp_x, and the resulting Δx is smaller. You can see that the uncertainty principle for position and momentum, $\Delta x \Delta p_x \geq \hbar/2$, is really just a consequence of the properties of integrals like Eq. (40.19).

CAUTION **Matter waves versus light waves in vacuum** We can regard both a wave packet that represents a particle and a short pulse of light from a laser as superpositions of waves of different wave numbers and angular frequencies. An important difference is that the speed of light in vacuum is the same for all wavelengths λ and hence all wave numbers $k = 2\pi/\lambda$, but the speed of a matter wave is *different* for different wavelengths. You can see this from the formula for the speed of the wave crests in a periodic wave, $v = \lambda f = \omega/k$. For a matter wave, $\omega = \hbar k^2/2m$, so $v = \hbar k/2m = h/2m\lambda$. Hence matter waves with longer wavelengths and smaller wave numbers travel more slowly than those with short wavelengths and large wave numbers. (This shouldn't be too surprising. The de Broglie relationships that we learned in Section 39.1 tell us that shorter wavelength corresponds to greater momentum and hence a greater speed.) Since the individual sinusoidal waves that make up a wave packet travel at different speeds, the shape of the packet changes as it moves. That's why we've specified the time for which the wave packets in Figs. 40.6 and 40.7 are drawn; at later times, the packets become more spread out. By contrast, a pulse of light waves in vacuum retains the same shape at all times because all of its constituent sinusoidal waves travel together at the same speed.

The One-Dimensional Schrödinger Equation with Potential Energy

The one-dimensional Schrödinger equation that we presented in Eq. (40.15) is valid only for free particles, for which the potential energy function is zero: $U(x) = 0$. But for an electron within an atom, a proton within an atomic nucleus, and many other real situations, the potential energy plays an important role. To study the behavior of matter waves in these situations, we need a version of the Schrödinger equation that describes a particle moving in the presence of a non-zero potential energy function $U(x)$. This equation is

General one-dimensional Schrödinger equation:

Planck's constant divided by 2π — Particle's wave function

$$-\frac{\hbar^2}{2m}\frac{\partial^2\Psi(x,t)}{\partial x^2} + U(x)\Psi(x,t) = i\hbar\frac{\partial\Psi(x,t)}{\partial t} \qquad (40.20)$$

Particle's mass — Potential-energy function

Note that if $U(x) = 0$, Eq. (40.20) reduces to the free-particle Schrödinger equation given in Eq. (40.15).

Here's the motivation behind Eq. (40.20). If $\Psi(x,t)$ is a sinusoidal wave function for a free particle, $\Psi(x,t) = Ae^{i(kx-\omega t)} = Ae^{ikx}e^{-i\omega t}$, the derivative terms in Eq. (40.20) become

$$-\frac{\hbar^2}{2m}\frac{\partial^2\Psi(x,t)}{\partial x^2} = -\frac{\hbar^2}{2m}\frac{\partial^2}{\partial x^2}(Ae^{ikx}e^{-i\omega t}) = -\frac{\hbar^2}{2m}(ik)^2(Ae^{ikx}e^{-i\omega t})$$

$$= \frac{\hbar^2 k^2}{2m}\Psi(x,t)$$

$$i\hbar\frac{\partial\Psi(x,t)}{\partial t} = i\hbar\frac{\partial}{\partial t}(Ae^{ikx}e^{-i\omega t}) = i\hbar(-i\omega)(Ae^{ikx}e^{-i\omega t}) = \hbar\omega\Psi(x,t)$$

In these expressions $(\hbar^2 k^2/2m)\Psi(x,t)$ is just the kinetic energy $K = p^2/2m = \hbar^2 k^2/2m$ multiplied by the wave function, and $\hbar\omega\Psi(x,t)$ is the total energy $E = \hbar\omega$ multiplied by the wave function. So for a wave function of this kind, Eq. (40.20) says that kinetic energy times $\Psi(x,t)$ plus potential energy times $\Psi(x,t)$ equals total energy times $\Psi(x,t)$. That's equivalent to the statement in classical physics that the sum of kinetic energy and potential energy equals total mechanical energy: $K + U = E$.

The observations we've just made certainly aren't a *proof* that Eq. (40.20) is correct. The real reason we know this equation *is* correct is that it works: Predictions made with this equation agree with experimental results. Later in this chapter we'll apply Eq. (40.20) to several physical situations, each with a different form of the function $U(x)$.

Stationary States

We saw in our discussion of wave packets that any free-particle wave function can be built up as a superposition of sinusoidal wave functions of the form $\Psi(x,t) = Ae^{ikx}e^{-i\omega t}$. Each such sinusoidal wave function corresponds to a state of definite energy $E = \hbar\omega = \hbar^2 k^2/2m$ and definite angular frequency $\omega = E/\hbar$, so we can rewrite these functions as $\Psi(x,t) = Ae^{ikx}e^{-iEt/\hbar}$. If the potential-energy function $U(x)$ is nonzero, these sinusoidal wave functions do not satisfy the Schrödinger equation, Eq. (40.20), and so these functions cannot be the basic "building blocks" of more complicated wave functions. However, we can still write the wave function for a state of definite energy E in the form

Time-dependent wave function for a state of definite energy

Time-independent wave function

$$\Psi(x,t) = \psi(x)e^{-iEt/\hbar} \qquad (40.21)$$

Energy of state — Planck's constant divided by 2π

40.7 How varying the function $A(k)$ in the wave-packet expression, Eq. (40.19), changes the character of the wave function $\Psi(x,t)$ (shown here at a specific time $t = 0$).

(a) $A(k)$

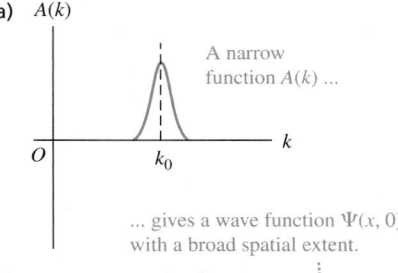

A narrow function $A(k)$ ...

... gives a wave function $\Psi(x,0)$ with a broad spatial extent.

(b) Re $\Psi(x,0)$

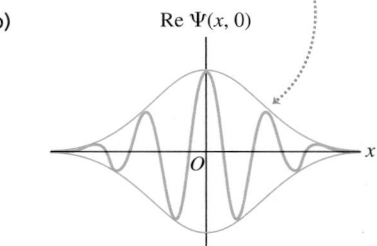

(c) $A(k)$

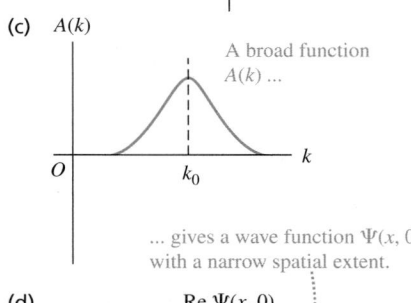

A broad function $A(k)$ ...

... gives a wave function $\Psi(x,0)$ with a narrow spatial extent.

(d) Re $\Psi(x,0)$

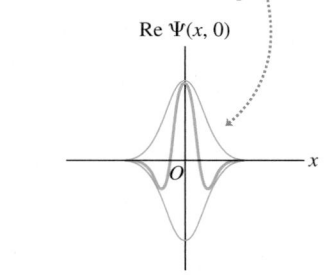

That is, the wave function $\Psi(x, t)$ for a state of definite energy is the product of a *time-independent* wave function $\psi(x)$ and a factor $e^{-iEt/\hbar}$. (For the free-particle sinusoidal wave function, $\psi(x) = Ae^{ikx}$.) States of definite energy are of tremendous importance in quantum mechanics. For example, for each energy level in a hydrogen atom (Section 39.3) there is a specific wave function. It is possible for an atom to be in a state that does not have a definite energy. The wave function for any such state can be written as a combination of definite-energy wave functions, in precisely the same way that a free-particle wave packet can be written as a superposition of sinusoidal wave functions of definite energy as in Eq. (40.19).

A state of definite energy is commonly called a **stationary state.** To see where this name comes from, let's multiply Eq. (40.21) by its complex conjugate to find the probability distribution function $|\Psi|^2$:

$$|\Psi(x, t)|^2 = \Psi^*(x, t)\Psi(x, t) = [\psi^*(x)e^{+iEt/\hbar}][\psi(x)e^{-iEt/\hbar}]$$

$$= \psi^*(x)\psi(x)e^{(+iEt/\hbar)+(-iEt/\hbar)} = |\psi(x)|^2 e^0 \qquad (40.22)$$

$$= |\psi(x)|^2$$

Since $|\psi(x)|^2$ does not depend on time, Eq. (40.22) shows that the same must be true for the probability distribution function $|\Psi(x, t)|^2$. This justifies the term "stationary state" for a state of definite energy.

PhET: Quantum Tunneling and Wave Packets

CAUTION **A stationary state does not mean a stationary particle** Saying that a particle is in a stationary state does *not* mean that the particle is at rest. It's the *probability distribution* (that is, the relative likelihood of finding the particle at various positions), not the particle itself, that's stationary. ▮

The Schrödinger equation, Eq. (40.20), becomes quite a bit simpler for stationary states. To see this, we substitute Eq. (40.21) into Eq. (40.20):

$$-\frac{\hbar^2}{2m}\frac{\partial^2[\psi(x)e^{-iEt/\hbar}]}{\partial x^2} + U(x)\psi(x)e^{-iEt/\hbar} = i\hbar\frac{\partial[\psi(x)e^{-iEt/\hbar}]}{\partial t}$$

The derivative in the first term on the left-hand side is with respect to x, so the factor $e^{-iEt/\hbar}$ comes outside of the derivative. Now we take the derivative with respect to t on the right-hand side of the equation:

$$-\frac{\hbar^2}{2m}\frac{d^2\psi(x)}{dx^2}e^{-iEt/\hbar} + U(x)\psi(x)e^{-iEt/\hbar} = i\hbar\left(\frac{-iE}{\hbar}\right)[\psi(x)e^{-iEt/\hbar}]$$

$$= E\psi(x)e^{-iEt/\hbar}$$

If we divide both sides of this equation by $e^{-iEt/\hbar}$, we get

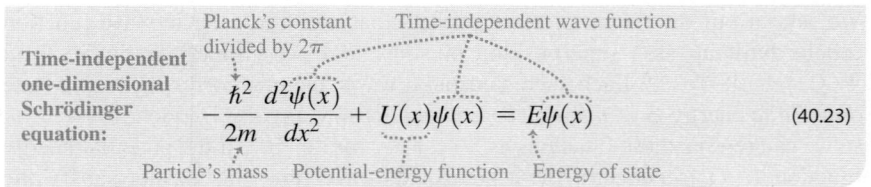

Time-independent one-dimensional Schrödinger equation:
$$-\frac{\hbar^2}{2m}\frac{d^2\psi(x)}{dx^2} + U(x)\psi(x) = E\psi(x) \qquad (40.23)$$

This is called the **time-independent one-dimensional Schrödinger equation.** The time-dependent factor $e^{-iEt/\hbar}$ does not appear, and Eq. (40.23) involves only the time-independent wave function $\psi(x)$. We'll devote much of this chapter to solving this equation to find the definite-energy, stationary-state wave functions $\psi(x)$ and the corresponding values of E—that is, the energies of the allowed levels—for different physical situations.

EXAMPLE 40.2 A STATIONARY STATE

Consider the wave function $\psi(x) = A_1 e^{ikx} + A_2 e^{-ikx}$, where k is positive. Is this a valid time-independent wave function for a free particle in a stationary state? What is the energy corresponding to this wave function?

SOLUTION

IDENTIFY and SET UP: A valid stationary-state wave function for a free particle must satisfy the time-independent Schrödinger equation, Eq. (40.23), with $U(x) = 0$. To test the given function $\psi(x)$, we simply substitute it into the left-hand side of the equation. If the result is a constant times $\psi(x)$, then the wave function is indeed a solution and the constant is equal to the particle energy E.

EXECUTE: Substituting $\psi(x) = A_1 e^{ikx} + A_2 e^{-ikx}$ and $U(x) = 0$ into Eq. (40.23), we obtain

$$-\frac{\hbar^2}{2m}\frac{d^2\psi(x)}{dx^2} = -\frac{\hbar^2}{2m}\frac{d^2(A_1 e^{ikx} + A_2 e^{-ikx})}{dx^2}$$

$$= -\frac{\hbar^2}{2m}\left[(ik)^2 A_1 e^{ikx} + (-ik)^2 A_2 e^{-ikx}\right]$$

$$= \frac{\hbar^2 k^2}{2m}(A_1 e^{ikx} + A_2 e^{-ikx}) = \frac{\hbar^2 k^2}{2m}\psi(x)$$

The result is a constant times $\psi(x)$, so this $\psi(x)$ is indeed a valid stationary-state wave function for a free particle. Comparing with Eq. (40.23) shows that the constant on the right-hand side is the particle energy: $E = \hbar^2 k^2/2m$.

EVALUATE: Note that $\psi(x)$ is a *superposition* of two different wave functions: one function $(A_1 e^{ikx})$ that represents a particle with magnitude of momentum $p = \hbar k$ moving in the positive x-direction, and one function $(A_2 e^{-ikx})$ that represents a particle with the same magnitude of momentum moving in the negative x-direction. So while the combined wave function $\psi(x)$ represents a stationary state with a definite energy, this state does *not* have a definite momentum. We'll see in Section 40.2 that such a wave function can represent a *standing wave,* and we'll explore situations in which such standing matter waves can arise.

TEST YOUR UNDERSTANDING OF SECTION 40.1 Does a wave packet given by Eq. (40.19) represent a stationary state? ∎

40.2 PARTICLE IN A BOX

An important problem in quantum mechanics is how to use the time-independent Schrödinger equation, Eq. (40.23), to determine the possible energy levels and the corresponding wave functions for various systems. That is, for a given potential energy function $U(x)$, what are the possible stationary-state wave functions $\psi(x)$, and what are the corresponding energies E?

In Section 40.1 we solved this problem for the case $U(x) = 0$, corresponding to a *free* particle. The allowed wave functions and corresponding energies are

$$\psi(x) = Ae^{ikx} \qquad E = \frac{\hbar^2 k^2}{2m} \qquad \text{(free particle)} \qquad (40.24)$$

The wave number k is equal to $2\pi/\lambda$, where λ is the wavelength. We found that k can have any real value, so the energy E of a free particle can have any value from zero to infinity. Furthermore, the particle can be found with equal probability at any value of x from $-\infty$ to $+\infty$.

Now let's look at a simple model in which a particle is *bound* so that it cannot escape to infinity, but rather is confined to a restricted region of space. Our system consists of a particle confined between two rigid walls separated by a distance L (**Fig. 40.8**). The motion is purely one dimensional, with the particle moving along the x-axis only and the walls at $x = 0$ and $x = L$. The potential energy corresponding to the rigid walls is infinite, so the particle cannot escape; between the walls, the potential energy is zero (**Fig. 40.9**). This situation is often described as a **"particle in a box."** This model might represent an electron that is free to move within a long, straight molecule or along a very thin wire.

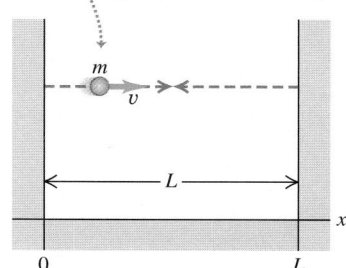

40.8 The Newtonian view of a particle in a box.

A particle with mass m moves along a straight line at constant speed, bouncing between two rigid walls a distance L apart.

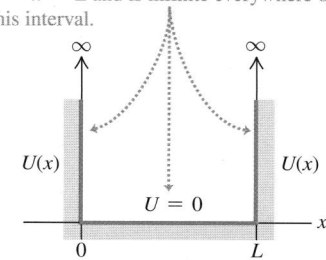

40.9 The potential-energy function for a particle in a box.

The potential energy U is zero in the interval $0 < x < L$ and is infinite everywhere outside this interval.

Wave Functions for a Particle in a Box

To solve the Schrödinger equation for this system, we begin with some restrictions on the particle's stationary-state wave function $\psi(x)$. Because the particle is confined to the region $0 \leq x \leq L$, we expect the probability distribution function $|\Psi(x, t)|^2 = |\psi(x)|^2$ and the wave function $\psi(x)$ to be zero outside that region. This agrees with the Schrödinger equation: If the term $U(x)\psi(x)$ in Eq. (40.23) is to be finite, then $\psi(x)$ must be zero where $U(x)$ is infinite.

Furthermore, $\psi(x)$ must be a *continuous* function to be a mathematically well-behaved solution to the Schrödinger equation. This implies that $\psi(x)$ must be zero at the region's boundary, $x = 0$ and $x = L$. These two conditions serve as *boundary conditions* for the problem. They should look familiar, because they are the same conditions that we used to find the normal modes of a vibrating string in Section 15.8 (**Fig. 40.10**); you should review that discussion.

An additional condition is that to calculate the second derivative $d^2\psi(x)/dx^2$ in Eq. (40.23), the *first* derivative $d\psi(x)/dx$ must also be continuous except at points where the potential energy becomes infinite (as it does at the walls of the box). This is analogous to the requirement that a vibrating string, like those shown in Fig. 40.10, can't have any kinks in it (which would correspond to a discontinuity in the first derivative of the wave function) except at the ends of the string.

We now solve for the wave functions in the region $0 \leq x \leq L$ subject to the above conditions. In this region $U(x) = 0$, so $\psi(x)$ in this region must satisfy

$$-\frac{\hbar^2}{2m}\frac{d^2\psi(x)}{dx^2} = E\psi(x) \qquad \text{(particle in a box)} \qquad (40.25)$$

Equation (40.25) is the *same* Schrödinger equation as for a free particle, so it is tempting to conclude that the wave functions and energies are given by Eq. (40.24). It is true that $\psi(x) = Ae^{ikx}$ satisfies the Schrödinger equation with $U(x) = 0$, is continuous, and has a continuous first derivative $d\psi(x)/dx = ikAe^{ikx}$. However, this wave function does *not* satisfy the boundary conditions that $\psi(x)$ must be zero at $x = 0$ and $x = L$: At $x = 0$ the wave function in Eq. (40.24) is equal to $Ae^0 = A$, and at $x = L$ it is equal to Ae^{ikL}. (These would be equal to zero if $A = 0$, but then the wave function would be zero and there would be no particle at all!)

The way out of this dilemma is to recall Example 40.2 (Section 40.1), in which we found that a more general stationary-state solution to the time-independent Schrödinger equation with $U(x) = 0$ is

$$\psi(x) = A_1 e^{ikx} + A_2 e^{-ikx} \qquad (40.26)$$

This wave function is a superposition of two waves: one traveling in the $+x$-direction of amplitude A_1, and one traveling in the $-x$-direction with the same wave number but amplitude A_2. This is analogous to a standing wave on a string (Fig. 40.10), which we can regard as the superposition of two sinusoidal waves propagating in opposite directions (see Section 15.7). The energy that corresponds to Eq. (40.26) is $E = \hbar^2 k^2/2m$, just as for a single wave.

To see whether the wave function given by Eq. (40.26) can satisfy the boundary conditions, let's first rewrite it in terms of sines and cosines by using Euler's formula, Eq. (40.17):

$$\psi(x) = A_1(\cos kx + i \sin kx) + A_2[\cos(-kx) + i \sin(-kx)]$$
$$= A_1(\cos kx + i \sin kx) + A_2(\cos kx - i \sin kx) \qquad (40.27)$$
$$= (A_1 + A_2)\cos kx + i(A_1 - A_2)\sin kx$$

At $x = 0$ this is equal to $\psi(0) = A_1 + A_2$, which must equal zero to satisfy the boundary condition at that point. Hence $A_2 = -A_1$, and Eq. (40.27) becomes

$$\psi(x) = 2iA_1 \sin kx = C \sin kx \qquad (40.28)$$

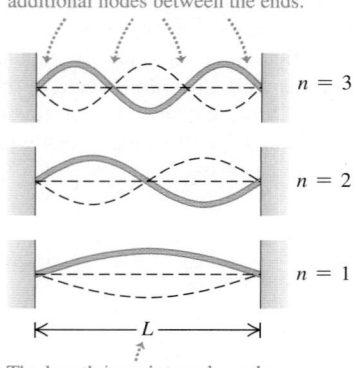

40.10 Normal modes of vibration for a string with length L, held at both ends.

Each end is a node, and there are $n - 1$ additional nodes between the ends.

$n = 3$

$n = 2$

$n = 1$

The length is an integral number of half-wavelengths: $L = n\lambda_n/2$.

We have simplified the expression by introducing the constant $C = 2iA_1$. (We'll come back to this constant later.) We can also satisfy the second boundary condition that $\psi = 0$ at $x = L$ by choosing values of k such that $kL = n\pi$ ($n = 1, 2, 3, \ldots$). Hence Eq. (40.28) does indeed give the stationary-state wave functions for a particle in a box in the region $0 \le x \le L$. (Outside this region, $\psi(x) = 0$.) The possible values of k and the wavelength $\lambda = 2\pi/k$ are

$$k = \frac{n\pi}{L} \quad \text{and} \quad \lambda = \frac{2\pi}{k} = \frac{2L}{n} \quad (n = 1, 2, 3, \ldots) \tag{40.29}$$

Just as for the string in Fig. 40.10, the length L of the region is an integral number of half-wavelengths.

Energy Levels for a Particle in a Box

The possible energy levels for a particle in a box are given by $E = \hbar^2 k^2/2m = p^2/2m$, where $p = \hbar k = (h/2\pi)(2\pi/\lambda) = h/\lambda$ is the magnitude of momentum of a free particle with wave number k and wavelength λ. This makes sense, since inside the region $0 \le x \le L$ the potential energy is zero and the energy is all kinetic. For each value of n, there are corresponding values of p, λ, and E; let's call them p_n, λ_n, and E_n. Putting the pieces together, we get

$$p_n = \frac{h}{\lambda_n} = \frac{nh}{2L} \tag{40.30}$$

and so the energy levels for a particle in a box are

Energy levels for a particle in a box

$$E_n = \frac{p_n{}^2}{2m} = \frac{n^2 h^2}{8mL^2} = \frac{n^2 \pi^2 \hbar^2}{2mL^2} \quad \left(n = 1, 2, 3, \ldots\right) \tag{40.31}$$

Magnitude of momentum — p_n. Planck's constant — h. Planck's constant divided by 2π — $\hbar$. Particle's mass — m. Width of box — L. Quantum number — n.

Each energy level has its own value of the quantum number n and a corresponding wave function, which we denote by ψ_n. When we replace k in Eq. (40.28) by $n\pi/L$ from Eq. (40.29), we find

$$\psi_n(x) = C \sin \frac{n\pi x}{L} \quad (n = 1, 2, 3, \ldots) \tag{40.32}$$

The energy-level diagram in **Fig. 40.11a** shows the five lowest levels for a particle in a box. The energy levels are proportional to n^2, so successively higher levels are spaced farther and farther apart. There are an infinite number of levels because the walls are perfectly rigid; even a particle of infinitely great kinetic

Application Particles in a Polymer "Box" Polyacetylene is one of a class of long-chain organic molecules that conduct electricity along their length. The molecule is made up of a large number of (C_2H_2) units, called *monomers* (only three monomers are shown here). Electrons can move freely along the length of the molecule but not perpendicular to the length, so the molecule is like a one-dimensional "box" for electrons. The length L of the molecule depends on the number of monomers. Experiment shows that the allowed energy levels agree well with Eq. (40.31): The greater the number of monomers and the greater the length L, the lower the energy levels and the smaller the spacing between these levels.

Polyacetylene

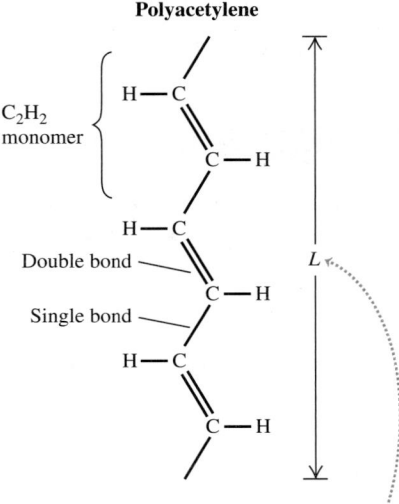

C_2H_2 monomer

Double bond

Single bond

L

Electrons are confined to the length L of the molecule (like particles in a box).

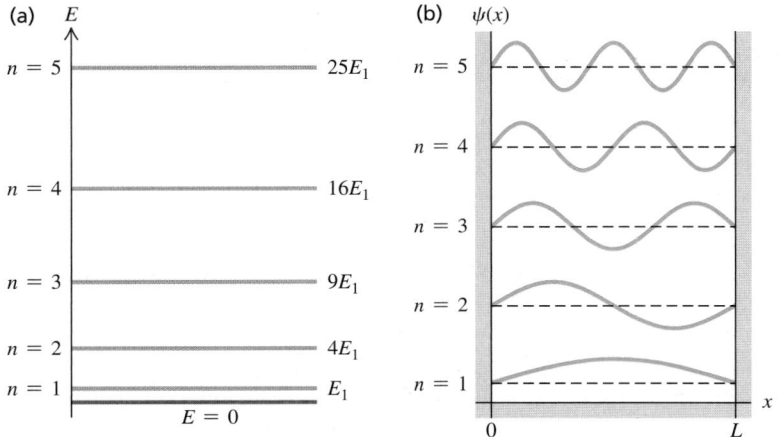

(a) Energy-level diagram

$n = 5$ — $25E_1$
$n = 4$ — $16E_1$
$n = 3$ — $9E_1$
$n = 2$ — $4E_1$
$n = 1$ — E_1
$E = 0$

(b) $\psi(x)$

$n = 5$
$n = 4$
$n = 3$
$n = 2$
$n = 1$

0 — L

40.11 (a) Energy-level diagram for a particle in a box. Each energy is $n^2 E_1$, where E_1 is the ground-level energy. (b) Wave functions for a particle in a box, with $n = 1, 2, 3, 4,$ and 5. **CAUTION:** The five graphs have been displaced vertically for clarity, as in Fig. 40.10. Each of the horizontal dashed lines represents $\psi = 0$ for the respective wave function.

energy is confined within the box. Figure 40.11b shows graphs of the wave functions $\psi_n(x)$ for $n = 1, 2, 3, 4,$ and 5. Note that these functions look identical to those for a standing wave on a string (see Fig. 40.10).

> CAUTION **A particle in a box cannot have zero energy** The energy of a particle in a box *cannot* be zero. Equation (40.31) shows that $E = 0$ would require $n = 0$, but substituting $n = 0$ into Eq. (40.32) gives a zero wave function. Since a particle is described by a *nonzero* wave function, there cannot be a particle with $E = 0$. This is a consequence of the Heisenberg uncertainty principle: A particle in a zero-energy state would have a definite value of momentum (precisely zero), so its position uncertainty would be infinite and the particle could be found anywhere along the x-axis. But this is impossible, since a particle in a box can be found only between $x = 0$ and $x = L$. Hence $E = 0$ is not allowed. By contrast, the allowed stationary-state wave functions with $n = 1, 2, 3, \ldots$ do not represent states of definite momentum (each is an equal mixture of a state of x-momentum $+p_n = nh/2L$ and a state of x-momentum $-p_n = -nh/2L$). Hence each stationary state has a nonzero momentum uncertainty, consistent with having a finite position uncertainty. ∎

EXAMPLE 40.3 ELECTRON IN AN ATOM-SIZE BOX

Find the first two energy levels for an electron confined to a one-dimensional box 5.0×10^{-10} m across (about the diameter of an atom).

SOLUTION

IDENTIFY AND SET UP: This problem uses what we have learned in this section about a particle in a box. The first two energy levels correspond to $n = 1$ and $n = 2$ in Eq. (40.31).

EXECUTE: From Eq. (40.31),

$$E_1 = \frac{h^2}{8mL^2} = \frac{(6.626 \times 10^{-34}\,\text{J}\cdot\text{s})^2}{8(9.109 \times 10^{-31}\,\text{kg})(5.0 \times 10^{-10}\,\text{m})^2}$$

$$= 2.4 \times 10^{-19}\,\text{J} = 1.5\,\text{eV}$$

$$E_2 = \frac{2^2 h^2}{8mL^2} = 4E_1 = 9.6 \times 10^{-19}\,\text{J} = 6.0\,\text{eV}$$

EVALUATE: The difference between the first two energy levels is $E_2 - E_1 = 4.5$ eV. An electron confined to a box is different from an electron bound in an atom, but it is reassuring that this result is of the same order of magnitude as the difference between actual atomic energy levels.

You can show that for a proton or neutron ($m = 1.67 \times 10^{-27}$ kg) confined to a box 1.1×10^{-14} m across (the width of a medium-sized atomic nucleus), the energies of the first two levels are about a million times larger: $E_1 = 1.7 \times 10^6$ eV $= 1.7$ MeV, $E_2 = 4E_1 = 6.8$ MeV, $E_2 - E_1 = 5.1$ MeV. This suggests why nuclear reactions (which involve transitions between energy levels in nuclei) release so much more energy than chemical reactions (which involve transitions between energy levels of electrons in atoms).

Finally, you can show (see Exercise 40.9) that the energy levels of a billiard ball ($m = 0.2$ kg) confined to a box 1.3 m across—the width of a billiard table—are separated by about 5×10^{-67} J. Quantum effects won't disturb a game of billiards.

Probability and Normalization

Let's look a bit more closely at the wave functions for a particle in a box, keeping in mind the *probability* interpretation of the wave function ψ that we discussed in Section 40.1. In our one-dimensional situation the quantity $|\psi(x)|^2\, dx$ is proportional to the probability that the particle will be found within a small interval dx about x. For a particle in a box,

$$|\psi(x)|^2\, dx = C^2 \sin^2 \frac{n\pi x}{L}\, dx$$

Figure 40.12 shows graphs of both $\psi(x)$ and $|\psi(x)|^2$ for $n = 1, 2,$ and 3. Note that not all positions are equally likely. By contrast, in classical mechanics the particle is equally likely to be found at any position between $x = 0$ and $x = L$. We see from Fig. 40.12b that $|\psi(x)|^2 = 0$ at some points, so there is zero probability of finding the particle at exactly these points. Don't let that bother you; the uncertainty principle has already shown us that we can't measure position exactly. The particle is localized only to be somewhere between $x = 0$ and $x = L$.

The particle must be *somewhere* on the *x*-axis—that is, somewhere between $x = -\infty$ and $x = +\infty$. So the *sum* of the probabilities for all the *dx*'s everywhere (the *total* probability of finding the particle) must equal 1. That's the normalization condition that we discussed in Section 40.1:

Normalization condition, time-independent wave function:

Integral over all *x*

$$\int_{-\infty}^{\infty} |\psi(x)|^2 \, dx = 1 \qquad (40.33)$$

Probability distribution function

Time-independent wave function

Probability that particle is *somewhere* on *x*-axis

A wave function is said to be *normalized* if it has a constant such as *C* in Eq. (40.32) that is calculated to make the total probability equal 1 in Eq. (40.33). For a normalized wave function, $|\psi(x)|^2 \, dx$ is not merely proportional to, but *equals,* the probability of finding the particle between the coordinates *x* and *x* + *dx*. That's why we call $|\psi(x)|^2$ the probability distribution function. (In Section 40.1 we called $|\Psi(x, t)|^2$ the probability distribution function. For the case of a stationary-state wave function, however, $|\Psi(x, t)|^2$ is equal to $|\psi(x)|^2$.)

Let's normalize the particle-in-a-box wave functions $\psi_n(x)$ given by Eq. (40.32). Since $\psi_n(x)$ is zero except between $x = 0$ and $x = L$, Eq. (40.33) becomes

$$\int_0^L C^2 \sin^2 \frac{n\pi x}{L} \, dx = 1 \qquad (40.34)$$

You can evaluate this integral by using the trigonometric identity $\sin^2 \theta = \frac{1}{2}(1 - \cos 2\theta)$; the result is $C^2 L/2$. Thus our probability interpretation of the wave function demands that $C^2 L/2 = 1$, or $C = (2/L)^{1/2}$; the constant *C* is *not* arbitrary. (This is in contrast to the classical vibrating string problem, in which *C* represents an amplitude that depends on initial conditions.) Thus the normalized stationary-state wave functions for a particle in a box are

Stationary-state wave functions for a particle in a box

Quantum number

$$\psi_n(x) = \sqrt{\frac{2}{L}} \sin \frac{n\pi x}{L} \quad (n = 1, 2, 3, \ldots) \qquad (40.35)$$

Width of box

40.12 Graphs of (a) $\psi(x)$ and (b) $|\psi(x)|^2$ for the first three wave functions ($n = 1, 2, 3$) for a particle in a box. The horizontal dashed lines represent $\psi(x) = 0$ and $|\psi(x)|^2 = 0$ for each of the three levels. The value of $|\psi(x)|^2 \, dx$ at each point is the probability of finding the particle in a small interval *dx* about the point. As in Fig. 40.11b, the three graphs in each part have been displaced vertically for clarity.

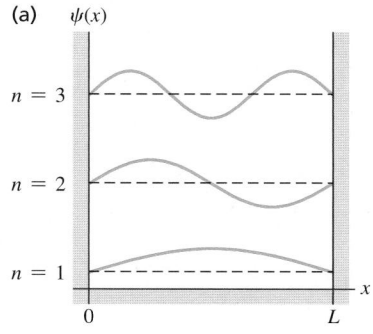

(a) $\psi(x)$

$n = 3$

$n = 2$

$n = 1$

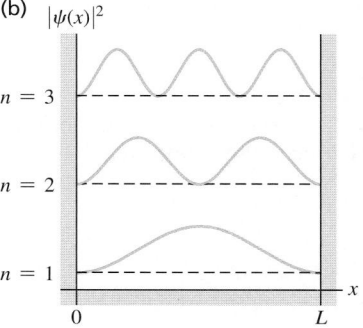

(b) $|\psi(x)|^2$

$n = 3$

$n = 2$

$n = 1$

EXAMPLE 40.4 A NONSINUSOIDAL WAVE FUNCTION?

(a) Show that $\psi(x) = Ax + B$, where *A* and *B* are constants, is a solution of the Schrödinger equation for an $E = 0$ energy level of a particle in a box. (b) What constraints do the boundary conditions at $x = 0$ and $x = L$ place on the constants *A* and *B*?

SOLUTION

IDENTIFY and SET UP: To be physically reasonable, a wave function must satisfy both the Schrödinger equation and the appropriate boundary conditions. In part (a) we'll substitute $\psi(x)$ into the Schrödinger equation for a particle in a box, Eq. (40.25), to determine whether it is a solution. In part (b) we'll see what restrictions on $\psi(x)$ arise from applying the boundary conditions that $\psi(x) = 0$ at $x = 0$ and $x = L$.

EXECUTE: (a) From Eq. (40.25), the Schrödinger equation for an $E = 0$ energy level of a particle in a box is

$$-\frac{\hbar^2}{2m} \frac{d^2\psi(x)}{dx^2} = E\psi(x) = 0$$

in the region $0 \le x \le L$. Differentiating $\psi(x) = Ax + B$ twice with respect to *x* gives $d^2\psi(x)/dx^2 = 0$, so the left side of the equation is zero, and so $\psi(x) = Ax + B$ is a solution of this Schrödinger equation for $E = 0$. (Note that both $\psi(x)$ and its derivative $d\psi(x)/dx = A$ are continuous functions, as they must be.)

(b) Applying the boundary condition at $x = 0$ gives $\psi(0) = B = 0$, and so $\psi(x) = Ax$. Applying the boundary condition at $x = L$ gives $\psi(L) = AL = 0$, so $A = 0$. Hence $\psi(x) = 0$ both inside the box ($0 \le x \le L$) *and outside:* There is *zero* probability of finding the particle anywhere with this wave function, and so $\psi(x) = Ax + B$ is *not* a physically valid wave function.

EVALUATE: The moral is that there are many functions that satisfy the Schrödinger equation for a given physical situation, but most of these—including the function considered here—have to be rejected because they don't satisfy the appropriate boundary conditions.

Time Dependence

The wave functions $\psi_n(x)$ in Eq. (40.35) depend only on the *spatial* coordinate x. Equation (40.21) shows that if $\psi(x)$ is the wave function for a state of definite energy E, the full time-dependent wave function is $\Psi(x, t) = \psi(x)e^{-iEt/\hbar}$. Hence the *time-dependent* stationary-state wave functions for a particle in a box are

$$\Psi_n(x, t) = \sqrt{\frac{2}{L}} \sin\left(\frac{n\pi x}{L}\right)e^{-iE_nt/\hbar} \qquad (n = 1, 2, 3, \ldots) \qquad (40.36)$$

In this expression the energies E_n are given by Eq. (40.31). The higher the quantum number n, the greater the angular frequency $\omega_n = E_n/\hbar$ at which the wave function oscillates. Note that since $|e^{-iE_nt/\hbar}|^2 = e^{+iE_nt/\hbar}e^{-iE_nt/\hbar} = e^0 = 1$, the probability distribution function $|\Psi_n(x, t)|^2 = (2/L)\sin^2(n\pi x/L)$ is independent of time and does *not* oscillate. (Remember, this is why we say that these states of definite energy are *stationary*.)

TEST YOUR UNDERSTANDING OF SECTION 40.2 If a particle in a box is in the nth energy level, what is the average value of its x-component of momentum p_x? (i) $nh/2L$; (ii) $(\sqrt{2}/2)nh/L$; (iii) $(1/\sqrt{2})nh/L$; (iv) $[1/(2\sqrt{2})]nh/L$; (v) zero. ∎

40.3 POTENTIAL WELLS

PhET: Double Wells and Covalent Bonds
PhET: Quantum Bound States

A **potential well** is a potential-energy function $U(x)$ that has a minimum. We introduced this term in Section 7.5, and we also used it in our discussion of periodic motion in Chapter 14. In Newtonian mechanics a particle trapped in a potential well can vibrate back and forth with periodic motion. Our first application of the Schrödinger equation, the particle in a box, involved a rudimentary potential well with a function $U(x)$ that is zero within a certain interval and infinite everywhere else. As we mentioned in Section 40.2, this function corresponds to a few situations found in nature, but the correspondence is only approximate.

A better approximation to several physical situations is a **finite well,** which is a potential well that has straight sides but *finite* height. **Figure 40.13** shows a potential-energy function that is zero in the interval $0 \le x \le L$ and has the value U_0 outside this interval. This function is often called a **square-well potential.** It could serve as a simple model of an electron within a metallic sheet with thickness L, moving perpendicular to the surfaces of the sheet. The electron can move freely inside the metal but has to climb a potential-energy barrier with height U_0 to escape from either surface of the metal. The energy U_0 is related to the *work function* that we discussed in Section 38.1 in connection with the photoelectric effect. In three dimensions, a spherical version of a finite well gives an approximate description of the motions of protons and neutrons within a nucleus.

40.13 A square-well potential.

The potential energy U is zero within the potential well (in the interval $0 \le x \le L$) and has the constant value U_0 outside this interval.

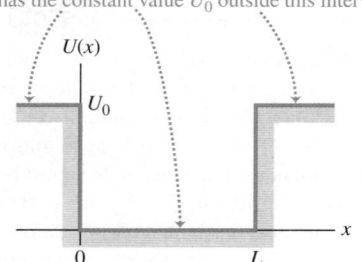

Bound States of a Square-Well Potential

In Newtonian mechanics, the particle is trapped (localized) in a well if the total mechanical energy E is less than U_0. In quantum mechanics, such a trapped state is often called a **bound state.** All states are bound for an infinitely deep well like the one we described in Section 40.2. For a finite well like that shown in Fig. 40.13, if E is greater than U_0, the particle is *not* bound.

Let's see how to solve the Schrödinger equation for the bound states of a square-well potential. Our goal is to find the energies and wave functions for which $E < U_0$. It's easiest to consider separately the regions where $U = 0$ and where $U = U_0$. Inside the square well $(0 \le x \le L)$, where $U = 0$, the time-independent Schrödinger equation is

$$-\frac{\hbar^2}{2m}\frac{d^2\psi(x)}{dx^2} = E\psi(x) \quad \text{or} \quad \frac{d^2\psi(x)}{dx^2} = -\frac{2mE}{\hbar^2}\psi(x) \qquad (40.37)$$

This is the same as Eq. (40.25) from Section 40.2, which describes a particle in a box. As in Section 40.2, we can express the solutions of this equation as combinations of $\cos kx$ and $\sin kx$, where $E = \hbar^2 k^2/2m$. We can rewrite the relationship between E and k as $k = \sqrt{2mE}/\hbar$. Hence inside the square well we have

$$\psi(x) = A\cos\left(\frac{\sqrt{2mE}}{\hbar}x\right) + B\sin\left(\frac{\sqrt{2mE}}{\hbar}x\right) \quad \text{(inside the well)} \qquad (40.38)$$

where A and B are constants. So far, this looks a lot like the particle-in-a-box analysis in Section 40.2. The difference is that for the square-well potential, the potential energy outside the well is not infinite, so the wave function $\psi(x)$ outside the well is *not* zero.

For the regions outside the well $(x < 0$ and $x > L)$ the potential-energy function in the time-independent Schrödinger equation is $U = U_0$:

$$-\frac{\hbar^2}{2m}\frac{d^2\psi(x)}{dx^2} + U_0\psi(x) = E\psi(x) \quad \text{or} \quad \frac{d^2\psi(x)}{dx^2} = \frac{2m(U_0 - E)}{\hbar^2}\psi(x) \qquad (40.39)$$

The quantity $U_0 - E$ is positive, so the solutions of this equation are exponential functions rather than sines or cosines. Using κ (the Greek letter kappa) to represent the quantity $[2m(U_0 - E)]^{1/2}/\hbar$ and taking κ as positive, we can write the solutions as

$$\psi(x) = Ce^{\kappa x} + De^{-\kappa x} \quad \text{(outside the well)} \qquad (40.40)$$

where C and D are constants with different values in the two regions $x < 0$ and $x > L$. Note that ψ can't be allowed to approach infinity as $x \to +\infty$ or $x \to -\infty$. [If it did, we wouldn't be able to satisfy the normalization condition, Eq. (40.33).] This means that in Eq. (40.40), we must have $D = 0$ for $x < 0$ and $C = 0$ for $x > L$.

Our calculations so far show that the bound-state wave functions for a finite well are sinusoidal inside the well [Eq. (40.38)] and exponential outside it [Eq. (40.40)]. We have to *match* the wave functions inside and outside the well so that they satisfy the boundary conditions that we mentioned in Section 40.2: $\psi(x)$ and $d\psi(x)/dx$ must be continuous at the boundary points $x = 0$ and $x = L$. If the wave function $\psi(x)$ or the slope $d\psi(x)/dx$ were to change discontinuously at a point, the second derivative $d^2\psi(x)/dx^2$ would be *infinite* at that point. That would violate the time-independent Schrödinger equation, Eq. (40.23), which says that at every point $d^2\psi(x)/dx^2$ is proportional to $U - E$. For a finite well $U - E$ is finite everywhere, so $d^2\psi(x)/dx^2$ must also be finite everywhere.

Matching the sinusoidal and exponential functions at the boundary points so that they join smoothly is possible only for certain specific values of the total energy E, so this requirement determines the possible energy levels of the finite square well. There is no simple formula for the energy levels as there was for the infinitely deep well. Finding the levels is a fairly complex mathematical problem that requires solving a transcendental equation by numerical approximation; we won't go into the details. **Figure 40.14** shows the general shape of a possible wave function. The most striking features of this wave function are the "exponential tails" that extend outside the well into regions that are forbidden by Newtonian mechanics (because in those regions the particle would have negative kinetic energy). We see that there is some probability for finding the particle *outside* the potential well, which would be impossible in classical mechanics. In Section 40.4 we'll discuss an amazing result of this effect.

40.14 A possible wave function for a particle in a finite potential well. The function is sinusoidal inside the well $(0 \le x \le L)$ and exponential outside it. It approaches zero asymptotically at large $|x|$. The functions must join smoothly at $x = 0$ and $x = L$; the wave function and its derivative must be continuous.

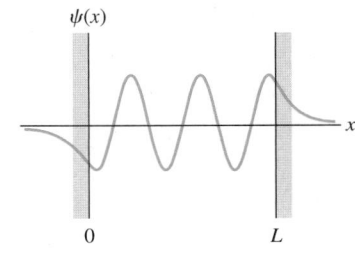

EXAMPLE 40.5 OUTSIDE A FINITE WELL

(a) Show that Eq. (40.40), $\psi(x) = Ce^{\kappa x} + De^{-\kappa x}$, is indeed a solution of the time-independent Schrödinger equation outside a finite well of height U_0. (b) What happens to $\psi(x)$ in the limit $U_0 \to \infty$?

SOLUTION

IDENTIFY and SET UP: In part (a), we try the given function $\psi(x)$ in the time-independent Schrödinger equation for $x < 0$ and for $x > L$, Eq. (40.39). In part (b), we note that in the limit $U_0 \to \infty$ the finite well becomes an *infinite* well, like that for a particle in a box (Section 40.2). So in this limit the wave functions outside a finite well must reduce to the wave functions outside the box.

EXECUTE: (a) We must show that $\psi(x) = Ce^{\kappa x} + De^{-\kappa x}$ satisfies $d^2\psi(x)/dx^2 = [2m(U_0 - E)/\hbar^2]\psi(x)$. We recall that $(d/du)e^{au} = ae^{au}$ and $(d^2/du^2)e^{au} = a^2e^{au}$; the left-hand side of the Schrödinger equation is then

$$\frac{d^2\psi(x)}{dx^2} = \frac{d^2}{dx^2}(Ce^{\kappa x}) + \frac{d^2}{dx^2}(De^{-\kappa x})$$

$$= C\kappa^2 e^{\kappa x} + D(-\kappa)^2 e^{-\kappa x}$$

$$= \kappa^2(Ce^{\kappa x} + De^{-\kappa x}) = \kappa^2\psi(x)$$

Since from Eq. (40.40) $\kappa^2 = 2m(U_0 - E)/\hbar^2$, this is equal to the right-hand side of the equation. The equation is satisfied, and $\psi(x)$ is a solution.

(b) As U_0 approaches infinity, κ also approaches infinity. In the region $x < 0$, $\psi(x) = Ce^{\kappa x}$; as $\kappa \to \infty$, $\kappa x \to -\infty$ (since x is negative) and $e^{\kappa x} \to 0$, so the wave function approaches zero for all $x < 0$. Likewise, we can show that the wave function also approaches zero for all $x > L$. This is just what we found in Section 40.2; the wave function for a particle in a box must be zero outside the box.

EVALUATE: Our result in part (b) shows that the infinite square well is a *limiting case* of the finite well. We've seen many cases in Newtonian mechanics where it's important to consider limiting cases (such as Examples 5.11 and 5.13 in Section 5.2). Limiting cases are no less important in quantum mechanics.

DATA *SPEAKS*

The Square-Well Potential

When students were given a problem involving a particle in a square-well potential, more than 41% gave an incorrect response. Common errors:

- Confusion about energy levels. The energy of a particle in a well is given relative to the *bottom* of the well (taken to be $E = 0$), not the *top* of the well (see Fig. 40.13). If the depth of the well is U_0, then for a bound state, $E < U_0$.

- Confusion about wave functions. The narrower the width L of the well, the farther outside the well the "exponential tails" of the wave function of a bound state extend.

Comparing Finite and Infinite Square Wells

Let's continue the comparison of the finite-depth potential well with the infinitely deep well, which we began in Example 40.5. First, because the wave functions for the finite well don't go to zero at $x = 0$ and $x = L$, the wavelength of the sinusoidal part of each wave function is *longer* than it would be with an infinite well. This increase in λ corresponds to a reduced magnitude of momentum $p = h/\lambda$ and therefore a reduced energy. Thus each energy level, including the ground level, is *lower* for a finite well than for an infinitely deep well with the same width.

Second, a well with finite depth U_0 has only a *finite* number of bound states and corresponding energy levels, compared to the *infinite* number for an infinitely deep well. How many levels there are depends on the magnitude of U_0 in comparison with the ground-level energy for the infinitely deep well (IDW), which we call $E_{1-\text{IDW}}$. From Eq. (40.31),

$$E_{1-\text{IDW}} = \frac{\pi^2\hbar^2}{2mL^2} \qquad \text{(ground-level energy, infinitely deep well)} \qquad (40.41)$$

When the well is very deep so U_0 is much larger than $E_{1-\text{IDW}}$, there are many bound states and the energies of the lowest few are nearly the same as the energies for the infinitely deep well. When U_0 is only a few times as large as $E_{1-\text{IDW}}$ there are only a few bound states. (There is always at least *one* bound state, no matter how shallow the well.) As with the infinitely deep well, there is no state with $E = 0$; such a state would violate the uncertainty principle.

Figure 40.15 shows the case $U_0 = 6E_{1-\text{IDW}}$, for which there are three bound states. In the figure, we express the energy levels both as fractions of the well depth U_0 and as multiples of $E_{1-\text{IDW}}$. If the well were infinitely deep, the lowest three levels, as given by Eq. (40.31), would be $E_{1-\text{IDW}}$, $4E_{1-\text{IDW}}$, and $9E_{1-\text{IDW}}$. Figure 40.15 also shows the wave functions for the three bound states.

40.15 (a) Wave functions for the three bound states for a particle in a finite potential well of depth $U_0 = 6E_{1-\text{IDW}}$. (Here $E_{1-\text{IDW}}$ is the ground-level energy for an infinite well of the same width.) The horizontal brown line for each wave function corresponds to $\psi = 0$; the vertical placement of these lines indicates the energy of each bound state (compare Fig. 40.11). (b) Energy-level diagram for this system. All energies greater than U_0 are possible; states with $E > U_0$ form a continuum.

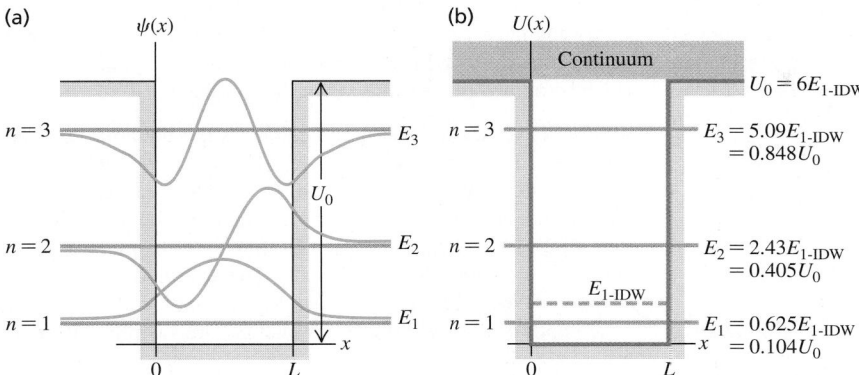

It turns out that when U_0 is less than $E_{1-\text{IDW}}$, there is only one bound state. In the limit when U_0 is *much smaller* than $E_{1-\text{IDW}}$ (a very shallow well), the energy of this single state is approximately $E = 0.68U_0$.

Figure 40.16 shows graphs of the probability distributions—that is, the values of $|\psi|^2$—for the wave functions shown in Fig. 40.15a. As with the infinite well, not all positions are equally likely. Unlike the infinite well, there is some probability of finding the particle outside the well in the classically forbidden regions.

There are also states for which E is *greater* than U_0. In these *free-particle states* the particle is not bound but is free to move through all values of x. *Any* energy E greater than U_0 is possible, so the free-particle states form a *continuum* rather than a discrete set of states with definite energy levels. The free-particle wave functions are sinusoidal both inside and outside the well. The wavelength is shorter inside the well than outside, corresponding to greater kinetic energy inside the well than outside it.

Figure 40.17 shows particles in a *two*-dimensional finite potential well. Example 40.6 describes another application of the square-well potential.

40.16 Probability distribution functions $|\psi(x)|^2$ for the square-well wave functions shown in Fig. 40.15. The horizontal brown line for each wave function corresponds to $|\psi|^2 = 0$.

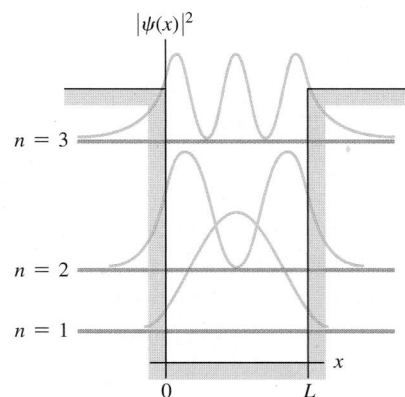

40.17 To make this image, 48 iron atoms (shown as yellow peaks) were placed in a circle on a copper surface. The "elevation" at each point inside the circle indicates the electron density within the circle. The standing-wave pattern is very similar to the probability distribution function for a particle in a one-dimensional finite potential well. (This image was made with a scanning tunneling microscope, discussed in Section 40.4.)

EXAMPLE 40.6 AN ELECTRON IN A FINITE WELL

An electron is trapped in a square well 0.50 nm across (roughly five times a typical atomic diameter). (a) Find the ground-level energy $E_{1-\text{IDW}}$ if the well is infinitely deep. (b) Find the energy levels if the actual well depth U_0 is six times the ground-level energy found in part (a). (c) Find the wavelength of the photon emitted when the electron makes a transition from the $n = 2$ level to the $n = 1$ level. In what region of the electromagnetic spectrum does the photon wavelength lie? (d) If the electron is in the $n = 1$ (ground) level and absorbs a photon, what is the minimum photon energy that will free the electron from the well? In what region of the spectrum does the wavelength of this photon lie?

SOLUTION

IDENTIFY and SET UP: Equation (40.41) gives the ground-level energy $E_{1-\text{IDW}}$ for an infinitely deep well, and Fig. 40.15b shows the energies for a square well with $U_0 = 6E_{1-\text{IDW}}$. The energy of the photon emitted or absorbed in a transition is equal to the difference in energy between two levels involved in the transition; the photon wavelength is given by $E = hc/\lambda$ (see Chapter 38).

EXECUTE: (a) From Eq. (40.41),

$$E_{1-\text{IDW}} = \frac{\pi^2 \hbar^2}{2mL^2} = \frac{\pi^2 (1.055 \times 10^{-34} \text{ J} \cdot \text{s})^2}{2(9.11 \times 10^{-31} \text{ kg})(0.50 \times 10^{-9} \text{ m})^2}$$

$$= 2.4 \times 10^{-19} \text{ J} = 1.5 \text{ eV}$$

(b) We have $U_0 = 6E_{1-\text{IDW}} = 6(1.5 \text{ eV}) = 9.0 \text{ eV}$. We can read off the energy levels from Fig. 40.15b:

$$E_1 = 0.625E_{1-\text{IDW}} = 0.625(1.5 \text{ eV}) = 0.94 \text{ eV}$$
$$E_2 = 2.43E_{1-\text{IDW}} = 2.43(1.5 \text{ eV}) = 3.6 \text{ eV}$$
$$E_3 = 5.09E_{1-\text{IDW}} = 5.09(1.5 \text{ eV}) = 7.6 \text{ eV}$$

(c) The photon energy and wavelength for the $n = 2$ to $n = 1$ transition are

$$E_2 - E_1 = 3.6 \text{ eV} - 0.94 \text{ eV} = 2.7 \text{ eV}$$

$$\lambda = \frac{hc}{E} = \frac{(4.136 \times 10^{-15} \text{ eV} \cdot \text{s})(3.00 \times 10^8 \text{ m/s})}{2.7 \text{ eV}}$$

$$= 460 \text{ nm}$$

in the blue region of the visible spectrum.

(d) We see from Fig. 40.15b that the minimum energy needed to free the electron from the well from the $n = 1$ level is $U_0 - E_1 = 9.0 \text{ eV} - 0.94 \text{ eV} = 8.1 \text{ eV}$, which is three times the 2.7-eV photon energy found in part (c). Hence the corresponding photon wavelength is one-third of 460 nm, or (to two significant figures) 150 nm, which is in the ultraviolet region of the spectrum.

EVALUATE: As a check, you can calculate the bound-state energies by using the formulas $E_1 = 0.104U_0$, $E_2 = 0.405U_0$, and $E_3 = 0.848U_0$ given in Fig. 40.15b. As an additional check, note that the first three energy levels of an infinitely deep well of the same width are $E_{1-\text{IDW}} = 1.5 \text{ eV}$, $E_{2-\text{IDW}} = 4E_{1-\text{IDW}} = 6.0 \text{ eV}$, and $E_{3-\text{IDW}} = 9E_{1-\text{IDW}} = 13.5 \text{ eV}$. The energies we found in part (b) are less than these values: As we mentioned earlier, the finite depth of the well lowers the energy levels compared to the levels for an infinitely deep well.

One application of these ideas is *quantum dots,* which are nanometer-sized particles of a semiconductor such as cadmium selenide (CdSe). An electron within a quantum dot behaves much like a particle in a finite potential well of width L equal to the size of the dot. When quantum dots are illuminated with ultraviolet light, the electrons absorb the ultraviolet photons and are excited into high energy levels, such as the $n = 3$ level described in this example. If the electron returns to the ground level ($n = 1$) in two or more steps (for example, from $n = 3$ to $n = 2$ and from $n = 2$ to $n = 1$), one step will involve emitting a visible-light photon, as we have calculated here. (We described this process of *fluorescence* in Section 39.3.) Increasing the value of L decreases the energies of the levels and hence the spacing between them, and thus decreases the energy and increases the wavelength of the emitted photons. The photograph that opens this chapter shows quantum dots of different sizes in solution: Each emits a characteristic wavelength that depends on the dot size. Quantum dots can be injected into living tissue and their fluorescent glow used as a tracer for biological research and for medicine. They may also be the key to a new generation of lasers and ultrafast computers.

40.18 A potential-energy barrier. According to Newtonian mechanics, if the total energy of the system is E_1, a particle to the left of the barrier can go no farther than $x = a$. If the total energy is greater than E_2, the particle can pass over the barrier.

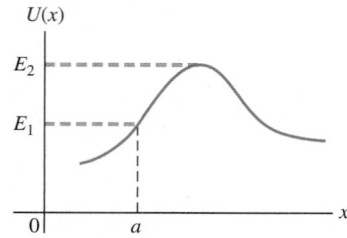

TEST YOUR UNDERSTANDING OF SECTION 40.3 Suppose that the width of the finite potential well shown in Fig. 40.15 is reduced by one-half. How must the value of U_0 change so that there are still just three bound energy levels whose energies are the fractions of U_0 shown in Fig. 40.15b? U_0 must (i) increase by a factor of four; (ii) increase by a factor of two; (iii) remain the same; (iv) decrease by a factor of one-half; (v) decrease by a factor of one-fourth. ∎

40.4 POTENTIAL BARRIERS AND TUNNELING

A **potential barrier** is the opposite of a potential well; it is a potential-energy function with a *maximum*. **Figure 40.18** shows an example. In classical Newtonian mechanics, if a particle (such as a roller coaster) is located to the left of the barrier (which might be a hill), and if the total mechanical energy of the system is E_1, the particle cannot move farther to the right than $x = a$. If it did, the potential energy U would be greater than the total energy E and the kinetic energy $K = E - U$ would be negative. This is impossible in classical mechanics since $K = \frac{1}{2}mv^2$ can never be negative.

A quantum-mechanical particle behaves differently: If it encounters a barrier like the one in Fig. 40.18 and has energy less than E_2, it *may* appear on the other side. This phenomenon is called *tunneling*. In quantum-mechanical tunneling, unlike macroscopic, mechanical tunneling, the particle does not actually push through the barrier and loses no energy in the process.

Tunneling Through a Rectangular Barrier

To understand how tunneling can occur, let's look at the potential-energy function $U(x)$ shown in **Fig. 40.19**. It's like Fig. 40.13 turned upside-down; the potential energy is zero everywhere except in the range $0 \leq x \leq L$, where it has the value U_0. This might represent a simple model for the potential energy of an electron in the presence of two slabs of metal separated by an air gap of thickness L. The potential energy is lower within either slab than in the gap between them.

Let's consider solutions of the Schrödinger equation for this potential-energy function for the case in which E is less than U_0. We can use our results from Section 40.3. In the regions $x < 0$ and $x > L$, where $U = 0$, the solution is sinusoidal and is given by Eq. (40.38). Within the barrier ($0 \leq x \leq L$), $U = U_0$ and the solution is exponential as in Eq. (40.40). Just as with the finite potential well, the functions have to join smoothly at the boundary points $x = 0$ and $x = L$, which means that both $\psi(x)$ and $d\psi(x)/dx$ have to be continuous at these points.

These requirements lead to a wave function like the one shown in **Fig. 40.20**. The function is *not* zero inside the barrier (the region forbidden by Newtonian mechanics). Even more remarkable, a particle that is initially to the *left* of the barrier has some probability of being found to the *right* of the barrier. How great this probability is depends on the width L of the barrier and the particle's energy E in comparison with the barrier height U_0. The **tunneling probability** T that the particle gets through the barrier is proportional to the square of the ratio of the amplitudes of the sinusoidal wave functions on the two sides of the barrier. These amplitudes are determined by matching wave functions and their derivatives at the boundary points, a fairly involved mathematical problem. When T is much smaller than unity, it is given approximately by

$$T = Ge^{-2\kappa L} \text{ where } G = 16\frac{E}{U_0}\left(1 - \frac{E}{U_0}\right) \text{ and } \kappa = \frac{\sqrt{2m(U_0 - E)}}{\hbar} \quad (40.42)$$

(probability of tunneling)

The probability decreases rapidly with increasing barrier width L. It also depends critically on the energy difference $U_0 - E$, which in Newtonian physics is the additional kinetic energy the particle would need to be able to climb over the barrier.

40.19 A rectangular potential-energy barrier with width L and height U_0. According to Newtonian mechanics, if the total energy E is less than U_0, a particle cannot pass over this barrier but is confined to the side where it starts.

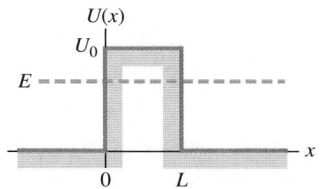

PhET: Quantum Tunneling and Wave Packets

The wave function is exponential within the barrier ($0 \leq x \leq L$) ...

... and sinusoidal outside the barrier.

The function and its derivative (slope) are continuous at $x = 0$ and $x = L$, so the sinusoidal and exponential functions join smoothly.

40.20 A possible wave function for a particle tunneling through the potential-energy barrier shown in Fig. 40.19.

EXAMPLE 40.7 TUNNELING THROUGH A BARRIER

A 2.0-eV electron encounters a barrier 5.0 eV high. What is the probability that it will tunnel through the barrier if the barrier width is (a) 1.00 nm and (b) 0.50 nm?

SOLUTION

IDENTIFY and SET UP: This problem uses the ideas of tunneling through a rectangular barrier, as in Figs. 40.19 and 40.20. Our target variable is the tunneling probability T in Eq. (40.42), which we evaluate for the given values $E = 2.0$ eV (electron energy), $U = 5.0$ eV (barrier height), $m = 9.11 \times 10^{-31}$ kg (mass of the electron), and $L = 1.00$ nm or 0.50 nm (barrier width).

EXECUTE: First we evaluate G and κ in Eq. (40.42), using $E = 2.0$ eV:

$$G = 16 \left(\frac{2.0 \text{ eV}}{5.0 \text{ eV}} \right) \left(1 - \frac{2.0 \text{ eV}}{5.0 \text{ eV}} \right) = 3.8$$

$$U_0 - E = 5.0 \text{ eV} - 2.0 \text{ eV} = 3.0 \text{ eV} = 4.8 \times 10^{-19} \text{ J}$$

$$\kappa = \frac{\sqrt{2(9.11 \times 10^{-31} \text{ kg})(4.8 \times 10^{-19} \text{ J})}}{1.055 \times 10^{-34} \text{ J} \cdot \text{s}} = 8.9 \times 10^{9} \text{ m}^{-1}$$

(a) When $L = 1.00$ nm $= 1.00 \times 10^{-9}$ m, we have $2\kappa L = 2(8.9 \times 10^{9} \text{ m}^{-1})(1.00 \times 10^{-9} \text{ m}) = 17.8$ and $T = Ge^{-2\kappa L} = 3.8 e^{-17.8} = 7.1 \times 10^{-8}$.

(b) When $L = 0.50$ nm, one-half of 1.00 nm, $2\kappa L$ is one-half of 17.8, or 8.9. Hence $T = 3.8 e^{-8.9} = 5.2 \times 10^{-4}$.

EVALUATE: Halving the width of this barrier increases the tunneling probability T by a factor of $(5.2 \times 10^{-4})/(7.1 \times 10^{-8}) = 7.3 \times 10^{3}$, or nearly ten thousand. The tunneling probability is an *extremely* sensitive function of the barrier width.

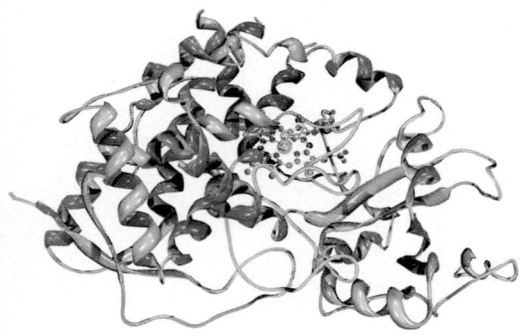

BIO Application Electron Tunneling in Enzymes Protein molecules play essential roles as enzymes in living organisms. Enzymes like the one shown here are large molecules, and in many cases their function depends on the ability of electrons to tunnel across the space that separates one part of the molecule from another. Without tunneling, life as we know it would be impossible!

Applications of Tunneling

Tunneling has a number of practical applications, some of considerable importance. When you twist two copper wires together or close the contacts of a switch, current passes from one conductor to the other despite a thin layer of nonconducting copper oxide between them. The electrons tunnel through this thin insulating layer. A *tunnel diode* is a semiconductor device in which electrons tunnel through a potential barrier. The current can be switched on and off very quickly (within a few picoseconds) by varying the height of the barrier. A *Josephson junction* consists of two superconductors separated by an oxide layer a few atoms (1 to 2 nm) thick. Electron pairs in the superconductors can tunnel through the barrier layer, giving such a device unusual circuit properties. Josephson junctions are useful for establishing precise voltage standards and measuring tiny magnetic fields, and they play a crucial role in the developing field of quantum computing.

The *scanning tunneling microscope* (STM) uses electron tunneling to create images of surfaces down to the scale of individual atoms. An extremely sharp conducting needle is brought very close to the surface, within 1 nm or so (**Fig. 40.21a**). When the needle is at a positive potential with respect to the surface, electrons can tunnel through the surface potential-energy barrier and reach the needle. As Example 40.7 shows, the tunneling probability and hence the tunneling current are very sensitive to changes in the width L of the barrier (the distance between the surface and the needle tip). In one mode of operation the needle is scanned across the surface while being moved perpendicular to the surface to

---

40.21 (a) Schematic diagram of the probe of a scanning tunneling microscope (STM). As the sharp conducting probe is scanned across the surface in the x- and y-directions, it is also moved in the z-direction to maintain a constant tunneling current. The changing position of the probe is recorded and used to construct an image of the surface. (b) This colored STM image shows "quantum wires": thin strips, just 10 atoms wide, of a conductive rare-earth silicide atop a silicon surface. Such quantum wires may one day be the basis of ultraminiaturized circuits.

(a)

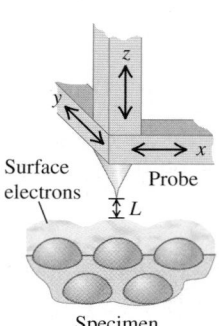

Surface electrons

Probe

L

Specimen

(b)

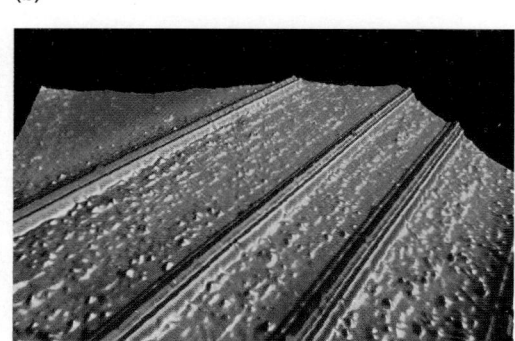

maintain a constant tunneling current. The needle motion is recorded; after many parallel scans, an image of the surface can be reconstructed. Extremely precise control of needle motion, including isolation from vibration, is essential. Figure 40.21b shows an STM image. (Figure 40.17 is also an STM image.)

Tunneling is also of great importance in nuclear physics. A fusion reaction can occur when two nuclei tunnel through the barrier caused by their electrical repulsion and approach each other closely enough for the attractive nuclear force to cause them to fuse. Fusion reactions occur in the cores of stars, including the sun; without tunneling, the sun wouldn't shine. The emission of alpha particles from unstable nuclei such as radium also involves tunneling. An alpha particle is a cluster of two protons and two neutrons (the same as a nucleus of the most common form of helium). Such clusters form naturally within larger atomic nuclei. An alpha particle trying to escape from a nucleus encounters a potential barrier that results from the combined effect of the attractive nuclear force and the electrical repulsion of the remaining part of the nucleus (**Fig. 40.22**). To escape, the alpha particle must tunnel through this barrier. Depending on the barrier height and width for a given kind of alpha-emitting nucleus, the tunneling probability can be low or high, and the alpha-emitting material will have low or high radioactivity. Recall from Section 39.2 that Ernest Rutherford used alpha particles from a radioactive source to discover the atomic nucleus. Although Rutherford did not know it, tunneling allowed these alpha particles to escape from their parent nuclei, which made his experiments possible! We'll learn more about alpha decay in Chapter 43.

TEST YOUR UNDERSTANDING OF SECTION 40.4   Is it possible for a particle undergoing tunneling to be found *within* the barrier rather than on either side of it? ▌

40.5 THE HARMONIC OSCILLATOR

Systems that *oscillate* are of tremendous importance in the physical world, from the oscillations of your eardrums in response to a sound wave to the vibrations of the ground caused by an earthquake. Oscillations are equally important on the microscopic scale where quantum effects dominate. The molecules of the air around you can be set into vibration when they collide with each other, the protons and neutrons in an excited atomic nucleus can oscillate in opposite directions, and a microwave oven transfers energy to food by making water molecules in the food flip back and forth. In this section we'll look at the solutions of the Schrödinger equation for the simplest kind of vibrating system, the quantum-mechanical harmonic oscillator.

As we learned in Section 14.2, a **harmonic oscillator** is a particle with mass m that moves along the x-axis under the influence of a conservative force $F_x = -k'x$. The constant k' is called the *force constant*. (In Section 14.2 we used the symbol k for the force constant. In this section we'll use the symbol k' instead to minimize confusion with the wave number $k = 2\pi/\lambda$.) The force is proportional to the particle's displacement x from its equilibrium position, $x = 0$. The corresponding potential-energy function is $U = \frac{1}{2}k'x^2$ (**Fig. 40.23**). In Newtonian mechanics, when the particle is displaced from equilibrium, it undergoes sinusoidal motion with frequency $f = (1/2\pi)(k'/m)^{1/2}$ and angular frequency $\omega = 2\pi f = (k'/m)^{1/2}$. The amplitude (that is, the maximum displacement from equilibrium) of these Newtonian oscillations is A, which is related to the energy E of the oscillator by $E = \frac{1}{2}k'A^2$.

Let's make an enlightened guess about the energy levels of a quantum-mechanical harmonic oscillator. In classical physics an electron oscillating with angular frequency ω emits electromagnetic radiation with that same angular frequency. It's reasonable to guess that when an excited quantum-mechanical harmonic oscillator with angular frequency $\omega = (k'/m)^{1/2}$ (according to Newtonian mechanics, at least) makes a transition from one energy level to a

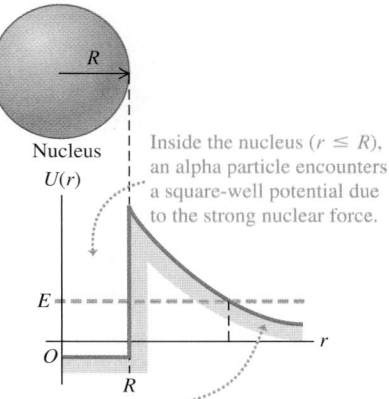

40.22 Approximate potential-energy function for an alpha particle interacting with a nucleus of radius R. If an alpha particle inside the nucleus has energy E greater than zero, it can tunnel through the barrier and escape from the nucleus.

Nucleus
$U(r)$

Inside the nucleus ($r \leq R$), an alpha particle encounters a square-well potential due to the strong nuclear force.

E

O

R

r

Outside the nucleus ($r > R$), an alpha particle experiences a $1/r$ potential due to electrostatic repulsion.

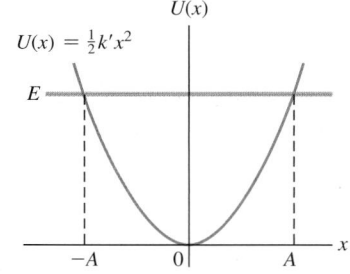

40.23 Potential-energy function for the harmonic oscillator. In Newtonian mechanics the amplitude A is related to the total energy E by $E = \frac{1}{2}k'A^2$, and the particle is restricted to the range from $x = -A$ to $x = A$. In quantum mechanics the particle can be found at $x > A$ or $x < -A$.

$U(x)$

$U(x) = \frac{1}{2}k'x^2$

E

$-A$ 0 A x

lower level, it would emit a photon with this same angular frequency ω. The energy of such a photon is $hf = (2\pi\hbar)(\omega/2\pi) = \hbar\omega$. So we would expect that the spacing between adjacent energy levels of the harmonic oscillator would be

$$hf = \hbar\omega = \hbar\sqrt{\frac{k'}{m}} \qquad (40.43)$$

That's the same spacing between energy levels that Planck assumed in deriving his radiation law (see Section 39.5). It was a good assumption; as we'll see, the energy levels are in fact half-integer $\left(\frac{1}{2}, \frac{3}{2}, \frac{5}{2}, \dots\right)$ multiples of $\hbar\omega$.

Wave Functions, Boundary Conditions, and Energy Levels

We'll begin our quantum-mechanical analysis of the harmonic oscillator by writing down the one-dimensional time-independent Schrödinger equation, Eq. (40.23), with $\frac{1}{2}k'x^2$ in place of U:

$$-\frac{\hbar^2}{2m}\frac{d^2\psi(x)}{dx^2} + \frac{1}{2}k'x^2\psi(x) = E\psi(x) \qquad \text{(Schrödinger equation for the harmonic oscillator)} \qquad (40.44)$$

The solutions of this equation are wave functions for the physically possible states of the system.

In the discussion of square-well potentials in Section 40.2 we found that the energy levels are determined by boundary conditions at the walls of the well. However, the harmonic-oscillator potential has no walls as such; what, then, are the appropriate boundary conditions? Classically, $|x|$ cannot be greater than the amplitude A given by $E = \frac{1}{2}k'A^2$. Quantum mechanics does allow some penetration into classically forbidden regions, but the probability decreases as that penetration increases. Thus the wave functions must approach zero as $|x|$ grows large.

Satisfying the requirement that $\psi(x) \rightarrow 0$ as $|x| \rightarrow \infty$ is not as trivial as it may seem. To see why this is, let's rewrite Eq. (40.44) in the form

$$\frac{d^2\psi(x)}{dx^2} = \frac{2m}{\hbar^2}\left(\frac{1}{2}k'x^2 - E\right)\psi(x) \qquad (40.45)$$

40.24 Possible behaviors of harmonic-oscillator wave functions in the region $\frac{1}{2}k'x^2 > E$. In this region, $\psi(x)$ and $d^2\psi(x)/dx^2$ have the same sign. The curve is concave upward when $d^2\psi(x)/dx^2$ is positive and concave downward when $d^2\psi(x)/dx^2$ is negative.

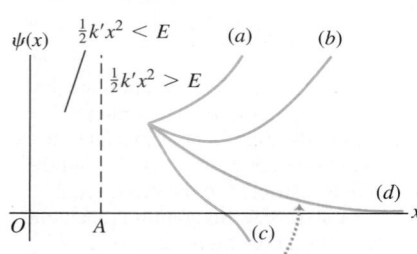

Only curve d, which approaches the x-axis asymptotically for large x, is an acceptable wave function for this system.

Equation (40.45) shows that when x is large enough (either positive or negative) to make the quantity $\left(\frac{1}{2}k'x^2 - E\right)$ positive, the function $\psi(x)$ and its second derivative $d^2\psi(x)/dx^2$ have the same sign. **Figure 40.24** shows four possible kinds of behavior of $\psi(x)$ beginning at a point where x is greater than the classical amplitude A, so that $\frac{1}{2}k'x^2 - \frac{1}{2}k'A^2 = \frac{1}{2}k'x^2 - E > 0$. Let's look at these four cases more closely. If $\psi(x)$ is positive as shown in Fig. 40.24, Eq. (40.45) tells us that $d^2\psi(x)/dx^2$ is also positive and the function is *concave upward*. Note also that $d^2\psi(x)/dx^2$ is the rate of change of the *slope* of $\psi(x)$; this will help us understand how our four possible wave functions behave.

- *Curve a:* The slope of $\psi(x)$ is positive at point x. Since $d^2\psi(x)/dx^2 > 0$, the function curves upward increasingly steeply and goes to infinity. This violates the boundary condition that $\psi(x) \rightarrow 0$ as $|x| \rightarrow \infty$, so this isn't a viable wave function.
- *Curve b:* The slope of $\psi(x)$ is negative at point x, and $d^2\psi(x)/dx^2$ has a large positive value. Hence the slope changes rapidly from negative to positive and keeps on increasing—so, again, the wave function goes to infinity. This wave function isn't viable either.
- *Curve c:* As for curve b, the slope is negative at point x. However, $d^2\psi(x)/dx^2$ now has a *small* positive value, so the slope increases only gradually as $\psi(x)$ decreases to zero and crosses over to negative values. Equation (40.45) tells us that once $\psi(x)$ becomes negative, $d^2\psi(x)/dx^2$ also becomes negative. Hence the curve becomes concave *downward* and heads for *negative* infinity. This wave function, too, fails to satisfy the requirement that $\psi(x) \rightarrow 0$ as $|x| \rightarrow \infty$ and thus isn't viable.

- *Curve d:* If the slope of $\psi(x)$ at point x is negative, and the positive value of $d^2\psi(x)/dx^2$ at this point is neither too large nor too small, the curve bends just enough to glide in asymptotically to the x-axis. In this case $\psi(x)$, $d\psi(x)/dx$, and $d^2\psi(x)/dx^2$ all approach zero at large x. This case offers the only hope of satisfying the boundary condition that $\psi(x) \to 0$ as $|x| \to \infty$, and it occurs only for certain very special values of the energy E.

This qualitative discussion suggests how the boundary conditions as $|x| \to \infty$ determine the possible energy levels for the quantum-mechanical harmonic oscillator. It turns out that these boundary conditions are satisfied only if the energy E is equal to one of the values E_n:

Energy levels for a harmonic oscillator · · · · Quantum number

$$E_n = \left(n + \tfrac{1}{2}\right)\hbar\sqrt{\frac{k'}{m}} = \left(n + \tfrac{1}{2}\right)\hbar\omega \quad (n = 0, 1, 2, \dots) \quad (40.46)$$

Planck's constant divided by 2π · · Particle's mass · · Force constant · · Oscillation angular frequency

Note that the ground level of energy $E_0 = \tfrac{1}{2}\hbar\omega$ is denoted by the quantum number $n = 0$, *not* $n = 1$.

Equation (40.46) confirms our guess [(Eq. 40.43)] that adjacent energy levels are separated by a constant interval of $\hbar\omega = hf$, as Planck assumed in 1900. There are infinitely many levels; this shouldn't be surprising because we are dealing with an infinitely deep potential well. As $|x|$ increases, $U = \tfrac{1}{2}k'x^2$ increases without bound.

Figure 40.25 shows the lowest six energy levels and the potential-energy function $U(x)$. For each level n, the value of $|x|$ at which the horizontal line representing the total energy E_n intersects $U(x)$ gives the amplitude A_n of the corresponding Newtonian oscillator.

40.25 Energy levels for the harmonic oscillator. The spacing between any two adjacent levels is $\Delta E = \hbar\omega$. The energy of the ground level is $E_0 = \tfrac{1}{2}\hbar\omega$.

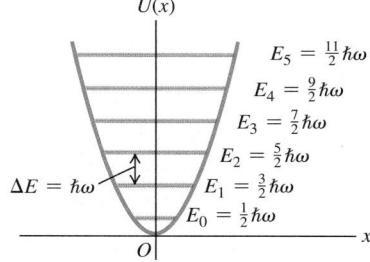

EXAMPLE 40.8 VIBRATION IN A CRYSTAL

A sodium atom of mass 3.82×10^{-26} kg vibrates within a crystal. The potential energy increases by 0.0075 eV when the atom is displaced 0.014 nm from its equilibrium position. Treat the atom as a harmonic oscillator. (a) Find the angular frequency of the oscillations according to Newtonian mechanics. (b) Find the spacing (in electron volts) of adjacent vibrational energy levels according to quantum mechanics. (c) What is the wavelength of a photon emitted as the result of a transition from one level to the next lower level? In what region of the electromagnetic spectrum does this lie?

SOLUTION

IDENTIFY and SET UP: We'll find the force constant k' from the expression $U = \tfrac{1}{2}k'x^2$ for potential energy. We'll then find the angular frequency $\omega = (k'/m)^{1/2}$ and use this in Eq. (40.46) to find the spacing between adjacent energy levels. We'll calculate the wavelength of the emitted photon as in Example 40.6.

EXECUTE: We are given that $U = 0.0075$ eV $= 1.2 \times 10^{-21}$ J when $x = 0.014 \times 10^{-9}$ m, so we can solve $U = \tfrac{1}{2}k'x^2$ for k':

$$k' = \frac{2U}{x^2} = \frac{2(1.2 \times 10^{-21}\text{ J})}{(0.014 \times 10^{-9}\text{ m})^2} = 12.2\text{ N/m}$$

(a) The Newtonian angular frequency is

$$\omega = \sqrt{\frac{k'}{m}} = \sqrt{\frac{12.2\text{ N/m}}{3.82 \times 10^{-26}\text{ kg}}} = 1.79 \times 10^{13}\text{ rad/s}$$

(b) From Eq. (40.46) and Fig. 40.25, the spacing between adjacent energy levels is

$$\hbar\omega = (1.055 \times 10^{-34}\text{ J}\cdot\text{s})(1.79 \times 10^{13}\text{ s}^{-1})$$

$$= 1.89 \times 10^{-21}\text{ J}\left(\frac{1\text{ eV}}{1.602 \times 10^{-19}\text{ J}}\right) = 0.0118\text{ eV}$$

(c) The energy E of the emitted photon is equal to the energy lost by the oscillator in the transition, 0.0118 eV. Then

$$\lambda = \frac{hc}{E} = \frac{(4.136 \times 10^{-15}\text{ eV}\cdot\text{s})(3.00 \times 10^8\text{ m/s})}{0.0118\text{ eV}}$$

$$= 1.05 \times 10^{-4}\text{ m} = 105\ \mu\text{m}$$

This photon wavelength is in the infrared region of the spectrum.

EVALUATE: This example shows us that interatomic force constants are a few newtons per meter, about the same as those of household springs or spring-based toys such as the Slinky®. It also suggests that we can learn about the vibrations of molecules by measuring the radiation that they emit in transitioning to a lower vibrational state. We will explore this idea further in Chapter 42.

40.26 The first four wave functions for the harmonic oscillator. The amplitude A of a Newtonian oscillator with the same total energy is shown for each. Each wave function penetrates somewhat into the classically forbidden regions $|x| > A$. The total number of finite maxima and minima for each function is $n + 1$, one more than the quantum number.

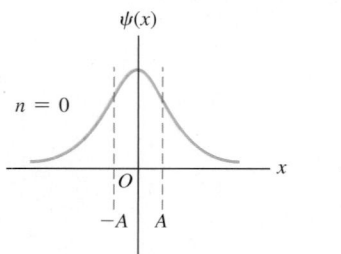
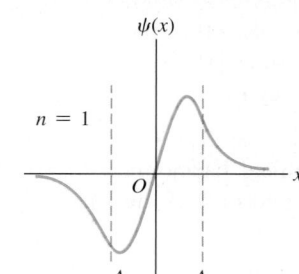
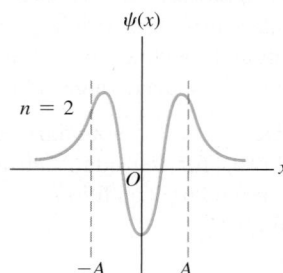
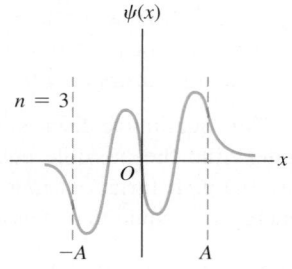

Comparing Quantum and Newtonian Oscillators

The wave functions for the levels $n = 0, 1, 2, \ldots$ of the harmonic oscillator are called *Hermite functions;* they aren't encountered in elementary calculus courses but are well known to mathematicians. Each Hermite function is an exponential function multiplied by a polynomial in x. The harmonic-oscillator wave function corresponding to $n = 0$ and $E = E_0$ (the ground level) is

$$\psi(x) = Ce^{-\sqrt{mk'}\, x^2/2\hbar} \tag{40.47}$$

The constant C is chosen to normalize the function—that is, to make $\int_{-\infty}^{\infty} |\psi|^2 \, dx = 1$. (We're using C rather than A as a normalization constant in this section, since we've already appropriated the symbol A to denote the Newtonian amplitude of a harmonic oscillator.) You can find C by using the following result from integral tables:

$$\int_{-\infty}^{\infty} e^{-a^2 x^2} dx = \frac{\sqrt{\pi}}{a}$$

To confirm that $\psi(x)$ as given by Eq. (40.47) really *is* a solution of the Schrödinger equation for the harmonic oscillator, you can calculate the second derivative of this wave function, substitute it into Eq. (40.44), and verify that the equation is satisfied if the energy E is equal to $E_0 = \frac{1}{2}\hbar\omega$ (see Exercise 40.34). It's a little messy, but the result is satisfying and worth the effort.

Figure 40.26 shows the the first four harmonic-oscillator wave functions. Each graph also shows the amplitude A of a Newtonian harmonic oscillator with the same energy—that is, the value of A determined from

$$\tfrac{1}{2}k'A^2 = \left(n + \tfrac{1}{2}\right)\hbar\omega \tag{40.48}$$

In each case there is some penetration of the wave function into the regions $|x| > A$ that are forbidden by Newtonian mechanics. This is similar to the effect that we noted in Section 40.3 for a particle in a finite square well.

Figure 40.27 shows the probability distributions $|\psi(x)|^2$ for these states. Each graph also shows the probability distribution determined from Newtonian physics, in which the probability of finding the particle near a randomly chosen

40.27 Probability distribution functions $|\psi(x)|^2$ for the harmonic-oscillator wave functions shown in Fig. 40.26. The amplitude A of the Newtonian motion with the same energy is shown for each. The blue lines show the corresponding probability distributions for the Newtonian motion. As n increases, the averaged-out quantum-mechanical functions resemble the Newtonian curves more and more.

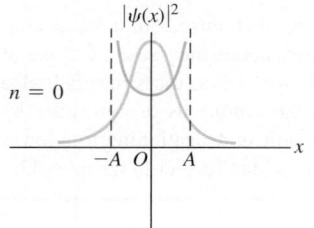
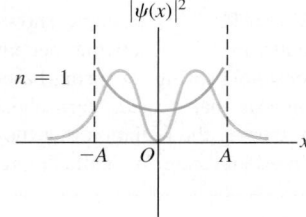
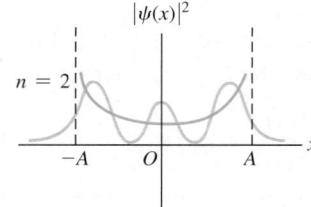
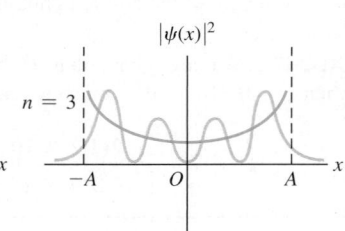

point is inversely proportional to the particle's speed at that point. If we average out the wiggles in the quantum-mechanical probability curves, the results for $n > 0$ resemble the Newtonian predictions. This agreement improves with increasing n; **Fig. 40.28** shows the classical and quantum-mechanical probability functions for $n = 10$. Notice that the spacing between zeros of $|\psi(x)|^2$ in Fig. 40.28 increases with increasing distance from $x = 0$. This makes sense from the Newtonian perspective: As a particle moves away from $x = 0$, its kinetic energy K and the magnitude p of its momentum both decrease. Thinking quantum-mechanically, this means that the wavelength $\lambda = h/p$ increases, so the spacing between zeros of $\psi(x)$ (and hence of $|\psi(x)|^2$) also increases.

In the Newtonian analysis of the harmonic oscillator the minimum energy is zero, with the particle at rest at its equilibrium position $x = 0$. This is not possible in quantum mechanics; no solution of the Schrödinger equation has $E = 0$ and satisfies the boundary conditions. Furthermore, such a state would violate the Heisenberg uncertainty principle because there would be no uncertainty in either position or momentum. The energy must be at least $\frac{1}{2}\hbar\omega$ for the system to conform to the uncertainty principle. To see qualitatively why this is so, consider a Newtonian oscillator with total energy $\frac{1}{2}\hbar\omega$. We can find the amplitude A and the maximum velocity just as we did in Section 14.3. When the particle is at its maximum displacement $(x = \pm A)$ and instantaneously at rest, $K = 0$ and $E = U = \frac{1}{2}k'A^2$. When the particle is at equilibrium $(x = 0)$ and moving at its maximum speed, $U = 0$ and $E = K = \frac{1}{2}mv_{max}^2$. Setting $E = \frac{1}{2}\hbar\omega$, we find

$$E = \tfrac{1}{2}k'A^2 = \tfrac{1}{2}\hbar\omega = \tfrac{1}{2}\hbar\left(\frac{k'}{m}\right)^{1/2} \quad \text{so} \quad A = \frac{\hbar^{1/2}}{k'^{1/4}m^{1/4}}$$

$$E = \tfrac{1}{2}mv_{max}^2 = \tfrac{1}{2}k'A^2 \quad \text{so} \quad v_{max} = A\left(\frac{k'}{m}\right)^{1/2} = \frac{\hbar^{1/2}k'^{1/4}}{m^{3/4}}$$

The maximum *momentum* of the particle is

$$p_{max} = mv_{max} = \hbar^{1/2}k'^{1/4}m^{1/4}$$

Here's where the Heisenberg uncertainty principle comes in. It turns out that the uncertainties in the particle's position and momentum (calculated as standard deviations) are, respectively, $\Delta x = A/\sqrt{2} = A/2^{1/2}$ and $\Delta p_x = p_{max}/\sqrt{2} = p_{max}/2^{1/2}$. Then the product of the two uncertainties is

$$\Delta x\,\Delta p_x = \left(\frac{\hbar^{1/2}}{2^{1/2}\,k'^{1/4}\,m^{1/4}}\right)\left(\frac{\hbar^{1/2}\,k'^{1/4}\,m^{1/4}}{2^{1/2}}\right) = \frac{\hbar}{2}$$

This product equals the minimum value allowed by Eq. (39.29), $\Delta x\,\Delta p_x \geq \hbar/2$, and thus satisfies the uncertainty principle. If the energy had been less than $\frac{1}{2}\hbar\omega$, the product $\Delta x\,\Delta p_x$ would have been less than $\hbar/2$, and the uncertainty principle would have been violated.

Even when a potential-energy function isn't precisely parabolic in shape, we may be able to approximate it by the harmonic-oscillator potential for sufficiently small displacements from equilibrium. **Figure 40.29** shows a typical potential-energy function for an interatomic force in a molecule. At large separations the curve of $U(r)$ versus r levels off, corresponding to the absence of force at great distances. But the curve is approximately parabolic near the minimum of $U(r)$ (the equilibrium separation of the atoms). Near equilibrium the molecular vibration is approximately simple harmonic with energy levels given by Eq. (40.46), as we assumed in Example 40.8.

40.28 Newtonian and quantum-mechanical probability distribution functions for a harmonic oscillator for the state $n = 10$. The Newtonian amplitude A is also shown.

The larger the value of n, the more closely the quantum-mechanical probability distribution (green) matches the Newtonian probability distribution (blue).

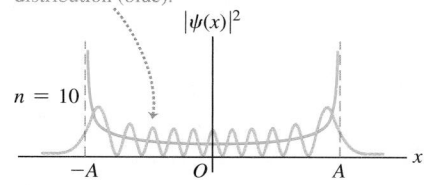

40.29 A potential-energy function describing the interaction of two atoms in a diatomic molecule. The distance r is the separation between the centers of the atoms, and the equilibrium separation is $r = r_0$. The energy needed to dissociate the molecule is U_∞.

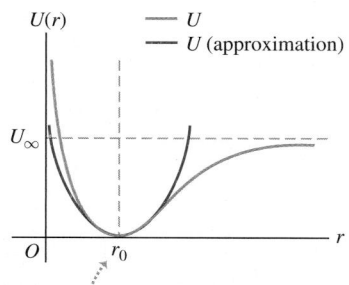

When r is near r_0, the potential-energy curve is approximately parabolic (as shown by the red curve) and the system behaves approximately like a harmonic oscillator.

TEST YOUR UNDERSTANDING OF SECTION 40.5 A quantum-mechanical system initially in its ground level absorbs a photon and ends up in the first excited state. The system then absorbs a second photon and ends up in the second excited state. For which of the following systems does the second photon have a longer wavelength than the first one? (i) A harmonic oscillator; (ii) a hydrogen atom; (iii) a particle in a box. ∎

40.6 MEASUREMENT IN QUANTUM MECHANICS

We've seen how to use the Schrödinger equation to calculate the stationary-state wave functions and energy levels for various potential-energy functions $U(x)$. We've also seen how to interpret the wave function $\Psi(x, t)$ of a particle in terms of the probability distribution function $|\Psi(x, t)|^2$. We'll conclude with a brief discussion of what happens when we try to *measure* the properties of a quantum-mechanical particle. As we will see, the consequences of such a measurement can be startlingly different from what happens when we measure the properties of a familiar Newtonian particle, such as a marble or billiard ball.

Let's consider a "particle in a box"—that is, a particle in an infinite square well of width L, as described in Section 40.2. This particle of mass m is free to move along the x-axis in the region $0 \leq x \leq L$ but cannot move beyond this region. Let's suppose the particle is in a stationary state with definite energy E, equal to one of the energy levels E_n given by Eq. (40.31). If we measure the x-component of momentum of this particle, what is the result?

First let's consider the answer to that question for a Newtonian particle in a box (see Fig. 40.8). This could be a hockey puck sliding on frictionless ice and bouncing back and forth between two parallel walls. The energy E of the puck is equal to its kinetic energy $p^2/2m$, so the magnitude of its momentum is $p = \sqrt{2mE}$. The x-component of its momentum p_x is therefore

$$p_x = +\sqrt{2mE} \quad \text{or} \quad p_x = -\sqrt{2mE} \tag{40.49}$$

Whether p_x is positive or negative depends on whether the hockey puck is moving in the $+x$-direction (then $p_x = +\sqrt{2mE}$) or the $-x$-direction (then $p_x = -\sqrt{2mE}$). To determine which value of p_x is correct at a given time, we need only look at the puck to see in which direction it's moving.

We can't make such an observation in the dark; we need to shine some light on the hockey puck. We know from Section 38.1 that light comes in the form of photons and that a photon of wavelength λ has momentum $p = h/\lambda$. When we shine light on the puck to observe it, the photons collide with the puck and *change* its momentum. The mere act of measuring the puck's momentum can affect the quantity that we're trying to measure! The good news is that this change is minuscule: A hockey puck of mass $m = 0.165$ kg moving at speed $v = 1.00$ m/s has momentum $p = mv = 0.165$ kg $\cdot$ m/s, while a visible-light photon of wavelength 500 nm has momentum $p = h/\lambda = 1.33 \times 10^{-27}$ kg $\cdot$ m/s. Even if we directed all of the photons from a 100-W light source onto the puck for a 1.00-s burst of light, the total momentum in this burst would be only 3.33×10^{-7} kg $\cdot$ m/s, and the resulting change in the momentum of the puck would be negligible. In general, we can measure any of the properties of a Newtonian particle—its momentum, position, energy, and so on—without appreciably changing the quantity that we are measuring.

The situation is very different for a quantum-mechanical particle in a box. From Eq. (40.21) the state of such a particle with energy $E = E_n$ is described by the wave function

$$\Psi(x, t) = \psi_n(x)e^{-iE_n t/\hbar} = \psi_n(x)e^{-i\omega_n t} \tag{40.50}$$

In Eq. (40.50) the angular frequency is $\omega_n = E_n/\hbar$ and the time-independent, stationary-state wave function $\psi_n(x)$ is given by Eq. (40.35):

$$\psi_n(x) = \sqrt{\frac{2}{L}} \sin \frac{n\pi x}{L} \quad (n = 1, 2, 3, \ldots) \tag{40.51}$$

This is a state of definite *energy,* but it is *not* a state of definite momentum: It represents a standing wave with equal amounts of momentum in the $+x$-direction and the $-x$-direction. To make this more explicit, recall from

Eqs. (40.30) and (40.31) that the magnitude of momentum in a state of energy E_n is $p_n = \sqrt{2mE_n} = nh/2L = n\pi\hbar/L$, and the corresponding wave number is $k_n = p_n/\hbar = n\pi/L$. So we can replace $n\pi/L$ in Eq. (40.51) with k_n:

$$\psi_n(x) = \sqrt{\frac{2}{L}} \sin k_n x$$

Recall also Euler's formula from Eq. (40.17): $e^{i\theta} = \cos\theta + i\sin\theta$ and $e^{-i\theta} = \cos\theta - i\sin\theta$. Hence $\sin\theta = (e^{i\theta} - e^{-i\theta})/2i$, and we can write

$$\psi_n(x) = \sqrt{\frac{2}{L}}\left(\frac{e^{ik_n x} - e^{-ik_n x}}{2i}\right) = \frac{1}{i\sqrt{2L}}(e^{ik_n x} - e^{-ik_n x}) \qquad (40.52)$$

Now we substitute Eq. (40.52) into Eq. (40.50) and distribute the factors $1/i\sqrt{2L}$ and $e^{-i\omega_n t}$:

$$\Psi(x, t) = \frac{1}{i\sqrt{2L}}(e^{ik_n x} - e^{-ik_n x})e^{-i\omega_n t}$$

$$= \frac{1}{i\sqrt{2L}}e^{ik_n x}e^{-i\omega_n t} - \frac{1}{i\sqrt{2L}}e^{-ik_n x}e^{-i\omega_n t} \qquad (40.53)$$

In Eq. (40.53) the $e^{ik_n x}e^{-i\omega_n t}$ term is a wave function for a free particle with energy $E_n = \hbar\omega_n$ and a *positive* x-component of momentum $p_x = p_n = \hbar k_n$. In the $e^{-ik_n x}e^{-i\omega_n t}$ term, k_n is replaced by $-k_n$, so this term is a wave function for a free particle with the same energy $E_n = \hbar\omega_n$ but a *negative* x-component of momentum $p_x = -p_n = -\hbar k_n$. These two possible values of p_x are the same as for a Newtonian particle in a box, Eq. (40.49). The difference is that as the Newtonian particle bounces back and forth between the walls of the box, it has positive p_x half of the time and negative p_x half of the time. Only its time-averaged value of p_x is zero. But because both terms for positive p_x and negative p_x are present in Eq. (40.53), the quantum-mechanical particle has *both* signs of the x-component of momentum present at *all* times. As we stated earlier, this stationary state for a quantum-mechanical particle in a box has a definite energy [both terms in Eq. (40.53) have the same value of ω_n and hence the same value of $E_n = \hbar\omega_n$] but does not have a definite momentum. Because the $e^{ik_n x}e^{-i\omega_n t}$ and $e^{-ik_n x}e^{-i\omega_n t}$ terms in Eq. (40.53) have coefficients of the same magnitude, $1/\sqrt{2L}$, the *instantaneous* average value of p_x for the quantum-mechanical particle is zero (the average of $\hbar k_n$ and $-\hbar k_n$) at *all* times.

What value do we get if we *measure* the momentum of the quantum-mechanical particle in a box? As for the Newtonian particle, we can measure the momentum by shining a light on it. Let's fire a single photon, moving in the $-y$-direction, at the particle and allow the photon and particle to collide (**Fig. 40.30**). Before the collision the total x-component of momentum of the system of photon and particle is zero. Momentum is conserved in the collision, so the same is true after the collision. After the collision, whichever sign of p_x the photon has, the x-component of momentum of the particle will have the opposite sign. If the photon is detected in detector A, we conclude that the particle has $p_x = \hbar k_n$; if instead the photon is detected in detector B, we conclude that the particle has $p_x = -\hbar k_n$.

In this experiment, we need to be even more concerned about how the photon changes the momentum of the particle than in the Newtonian case. For an electron in a box of width $L = 1.00 \times 10^{-6}$ m $= 1.00 \, \mu$m, the electron momentum has a minimum magnitude of $p = 3.31 \times 10^{-28}$ kg·m/s (corresponding to the $n = 1$ energy level), which is only about one-quarter that of a visible-light photon of wavelength 500 nm. To minimize the change in the electron's magnitude of momentum due to the collision, we should use a photon of much longer wavelength (say, a radio-wave photon) and hence much smaller momentum.

40.30 Using photon scattering to measure the x-component of momentum of a particle in a box.

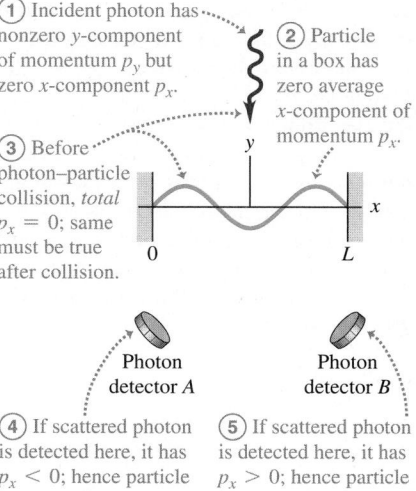

① Incident photon has nonzero y-component of momentum p_y but zero x-component p_x.

② Particle in a box has zero average x-component of momentum p_x.

③ Before photon–particle collision, *total* $p_x = 0$; same must be true after collision.

④ If scattered photon is detected here, it has $p_x < 0$; hence particle has $p_x > 0$.

⑤ If scattered photon is detected here, it has $p_x > 0$; hence particle has $p_x < 0$.

Photon detector A

Photon detector B

Even when we use a photon with the lowest possible momentum, however, we find that the state of the particle in the box *must* change as a result of the experiment. Here's a summary of the results:

1. If the measurement shows that the particle has positive $p_x = \hbar k_n$, the wave function *changes* from that given in Eq. (40.53) to one with an $e^{ik_nx}e^{-i\omega_nt}$ term *only*. The other term, which corresponds to $p_x = -\hbar k_n$, disappears. We say that the wave function, which was a combination of two terms with different values of p_x, has undergone *wave-function collapse*—it has collapsed to one term with $p_x = \hbar k_n$ as a consequence of measuring the value of p_x. To test this result, we fire a second photon immediately after the first. The second photon scatters from the particle as we would expect if the particle had the value $p_x = \hbar k_n$.

2. If the measurement shows that the particle has negative $p_x = -\hbar k_n$, the wave function collapses in the opposite way: It changes to one with an $e^{-ik_nx}e^{-i\omega_nt}$ term only. The $p_x = \hbar k_n$ term disappears.

3. If we repeat the experiment many times, each time starting with the particle described by the wave function in Eq. (40.53), 50% of the time we measure the particle to have $p_x = \hbar k_n$ and 50% of the time we measure the particle to have $p_x = -\hbar k_n$. For any given time that we try the experiment, there is no way to predict which outcome will occur. We can state only that there is equal probability of either outcome.

These results reveal a fact of quantum-mechanical life: *Measuring a physical property of a system can change the wave function of that system.* By measuring the value of p_x for a particle in a box, we changed the wave function from one that was a combination of two wave functions, one for $p_x = \hbar k_n$ and one for $p_x = -\hbar k_n$, to one with a definite value of p_x. This change in the wave function is not described by the time-dependent Schrödinger equation [Eq. (40.20)] but is a consequence of the measurement process. It is also independent of how the measurement is carried out: No matter how small the momentum of the incident photon shown in Fig. 40.30, the same collapse of the wave function takes place. Indeed, *any* experiment to measure p_x for a particle in a box in a steady state, no matter how the experiment is designed, will have the results that we described earlier.

(*After* the measurement, the wave function will undergo further change that *is* described by the Schrödinger equation. Neither $e^{ik_nx}e^{-i\omega_nt}$ nor $e^{-ik_nx}e^{-i\omega_nt}$ by itself satisfies the boundary conditions for a particle in a box—namely, that the wave function vanishes at $x = 0$ and $x = L$. The wave function must evolve to satisfy these conditions.)

> **CAUTION** Quantum measurement misconceptions If we measure the particle to have $p_x = \hbar k_n$, does that mean it had $p_x = \hbar k_n$ before the measurement? No; the particle acquired that value as a result of the measurement. If we measure the particle to have $p_x = \hbar k_n$ instead of $p_x = -\hbar k_n$, does that mean there was some bias in the way we did the measurement? Again, no; the result of any given experiment is random. All quantum mechanics can do is predict the probability that this experiment will give us a certain result. ∎

Note that not every measurement of a quantum-mechanical system causes a change in the wave function. If we perform an experiment that measures only the *energy* of a particle given by the wave function in Eq. (40.53), the wave function does *not* change. That's because the wave function already corresponds to a state of definite energy $E_n = \hbar\omega_n$, so there is a 100% probability that we will measure that value of energy.

You may ask, Does the wave function really collapse? Most physicists would answer yes, but some theorists have devised alternative models of what happens in a quantum-mechanical measurement. One model, called the *many-worlds interpretation,* asserts that there is a *universal* wave function that describes all particles in the universe. Whenever a measurement of any sort takes place, whether of human origin (like our experiment) or natural origin (for example, a photon of sunlight scattering from an electron in an atom in the atmosphere), this universal wave function does not collapse. Instead, every measurement causes the universe to branch into alternative timelines. So, when we carry out the experiment depicted in Fig. 40.30, the universe splits into one timeline in which the photon goes into detector *A* and a second timeline in which the photon goes into detector *B*. These two timelines then no longer communicate.

As weird as these aspects of quantum mechanics are, others are far weirder. We will investigate these in Chapter 41 after we have learned more about the nature of the electron.

TEST YOUR UNDERSTANDING OF SECTION 40.6 A particle in a box is described by a wave function that is a combination of the $n = 1$ and $n = 2$ stationary states: $\Psi(x, t) = C\psi_1(x)e^{-iE_1t/\hbar} + D\psi_2(x)e^{-iE_2t/\hbar}$, where $\psi_1(x)$ and $\psi_2(x)$ are given by Eq. (40.35), E_1 and E_2 are given by Eq. (40.31), and C and D are nonzero constants. If you carry out an experiment to measure the energy of this particle, the result is *guaranteed* to be (i) E_1; (ii) E_2; (iii) $(E_1 + E_2)/2$; (iv) intermediate between E_1 and E_2, with a value that depends on the values of C and D; (v) none of these. ∎

CHAPTER 40 SUMMARY

SOLUTIONS TO ALL EXAMPLES

Wave functions: The wave function for a particle contains all of the information about that particle. If the particle moves in one dimension in the presence of a potential energy function $U(x)$, the wave function $\Psi(x, t)$ obeys the one-dimensional Schrödinger equation. (For a *free* particle on which no forces act, $U(x) = 0$.) The quantity $|\Psi(x, t)|^2$, called the probability distribution function, determines the relative probability of finding a particle near a given position at a given time. If the particle is in a state of definite energy, called a stationary state, $\Psi(x, t)$ is a product of a function $\psi(x)$ that depends on only spatial coordinates and a function $e^{-iEt/\hbar}$ that depends on only time. For a stationary state, the probability distribution function is independent of time.

A spatial stationary-state wave function $\psi(x)$ for a particle that moves in one dimension in the presence of a potential-energy function $U(x)$ satisfies the time-independent Schrödinger equation. More complex wave functions can be constructed by superposing stationary-state wave functions. These can represent particles that are localized in a certain region, thus representing both particle and wave aspects. (See Examples 40.1 and 40.2.)

$$-\frac{\hbar^2}{2m}\frac{\partial^2\Psi(x, t)}{\partial x^2} + U(x)\Psi(x, t)$$
$$= i\hbar\frac{\partial\Psi(x, t)}{\partial t} \qquad (40.20)$$
(general 1-D Schrödinger equation)

$$\Psi(x, t) = \psi(x)e^{-iEt/\hbar} \qquad (40.21)$$
(time-dependent wave function for a state of definite energy)

$$-\frac{\hbar^2}{2m}\frac{d^2\psi(x)}{dx^2} + U(x)\psi(x) = E\psi(x)$$
(time-independent 1-D Schrödinger equation) $\qquad (40.23)$

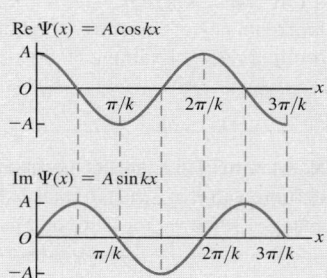

Particle in a box: The energy levels for a particle of mass m in a box (an infinitely deep square potential well) with width L are given by Eq. (40.31). The corresponding normalized stationary-state wave functions of the particle are given by Eq. (40.35). (See Examples 40.3 and 40.4.)

$$E_n = \frac{p_n^2}{2m} = \frac{n^2h^2}{8mL^2} = \frac{n^2\pi^2\hbar^2}{2mL^2}$$
$$(n = 1, 2, 3, \dots) \qquad (40.31)$$

$$\psi_n(x) = \sqrt{\frac{2}{L}}\sin\frac{n\pi x}{L}$$
$$(n = 1, 2, 3, \dots) \qquad (40.35)$$

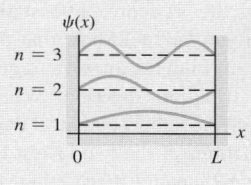

Wave functions and normalization: To be a solution of the Schrödinger equation, the wave function $\psi(x)$ and its derivative $d\psi(x)/dx$ must be continuous everywhere except where the potential-energy function $U(x)$ has an infinite discontinuity. Wave functions are usually normalized so that the total probability of finding the particle somewhere is unity.

$$\int_{-\infty}^{\infty}|\psi(x)|^2\,dx = 1 \qquad (40.33)$$
(normalization condition)

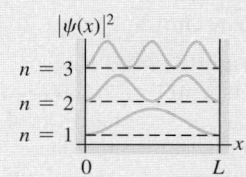

Finite potential well: In a potential well with finite depth U_0, the energy levels are lower than those for an infinitely deep well with the same width, and the number of energy levels corresponding to bound states is finite. The levels are obtained by matching wave functions at the well walls to satisfy the continuity of $\psi(x)$ and $d\psi(x)/dx$. (See Examples 40.5 and 40.6.)

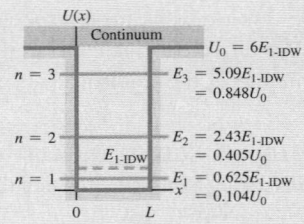

Potential barriers and tunneling: There is a certain probability that a particle will penetrate a potential-energy barrier even though its initial energy is less than the barrier height. This process is called tunneling. (See Example 40.7.)

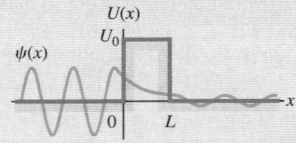

Quantum harmonic oscillator: The energy levels for the harmonic oscillator (for which $U(x) = \frac{1}{2}k'x^2$) are given by Eq. (40.46). The spacing between any two adjacent levels is $\hbar\omega$, where $\omega = \sqrt{k'/m}$ is the oscillation angular frequency of the corresponding Newtonian harmonic oscillator. (See Example 40.8.)

$$E_n = \left(n + \tfrac{1}{2}\right)\hbar\sqrt{\frac{k'}{m}} = \left(n + \tfrac{1}{2}\right)\hbar\omega$$

$$(n = 0, 1, 2, 3, \dots) \qquad (40.46)$$

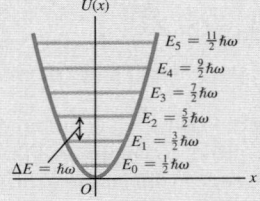

Measurement in quantum mechanics: If the wave function of a particle does not correspond to a definite value of a certain physical property (such as momentum or energy), the wave function changes when we measure that property. This phenomenon is called wave-function collapse.

BRIDGING PROBLEM   A PACKET IN A BOX

A particle of mass m in an infinitely deep well (see Fig. 40.9) has the following wave function in the region from $x = 0$ to $x = L$:

$$\Psi(x, t) = \frac{1}{\sqrt{2}}\psi_1(x)e^{-iE_1t/\hbar} + \frac{1}{\sqrt{2}}\psi_2(x)e^{-iE_2t/\hbar}$$

Here $\psi_1(x)$ and $\psi_2(x)$ are the normalized stationary-state wave functions for the first two levels ($n = 1$ and $n = 2$), given by Eq. (40.35). E_1 and E_2, given by Eq. (40.31), are the energies of these levels. The wave function is zero for $x < 0$ and for $x > L$. (a) Find the probability distribution function for this wave function. (b) Does $\Psi(x, t)$ represent a stationary state of definite energy? How can you tell? (c) Show that the wave function $\Psi(x, t)$ is normalized. (d) Find the angular frequency of oscillation of the probability distribution function. What is the interpretation of this oscillation? (e) Suppose instead that $\Psi(x, t)$ is a combination of the wave functions of the two lowest levels of a finite well of length L and height U_0 equal to six times the energy of the lowest-energy bound state of an infinite well of length L. What would be the angular frequency of the probability distribution function in this case?

SOLUTION GUIDE

IDENTIFY and SET UP

1. In Section 40.1 we saw how to interpret a combination of two free-particle wave functions of different energies. In this problem you need to apply these same ideas to a combination of wave functions for the infinite well (Section 40.2) and the finite well (Section 40.3).

EXECUTE

2. Write down the full time-dependent wave function $\Psi(x, t)$ and its complex conjugate $\Psi^*(x, t)$ by using the functions $\psi_1(x)$ and $\psi_2(x)$ from Eq. (40.35). Use these to calculate the probability distribution function, and decide whether or not this function depends on time.

3. To check for normalization, you'll need to verify that when you integrate the probability distribution function from step 2 over all values of x, the integral is equal to 1. [*Hint:* The trigonometric identities $\sin^2\theta = \frac{1}{2}(1 - \cos 2\theta)$ and $\sin\theta \sin\phi = \cos(\theta - \phi) - \cos(\theta + \phi)$ may be helpful.]

4. To find the answer to part (d) you'll need to identify the oscillation angular frequency ω_{osc} in your expression from step 2 for the probability distribution function. To interpret the oscillations, draw graphs of the probability distribution functions at times $t = 0$, $t = T/4$, $t = T/2$, and $t = 3T/4$, where $T = 2\pi/\omega_{osc}$ is the oscillation period of the probability distribution function.

5. For the finite well you do not have simple expressions for the first two stationary-state wave functions $\psi_1(x)$ and $\psi_2(x)$. However, you can still find the oscillation angular frequency ω_{osc}, which is related to the energies E_1 and E_2 in the same way as for the infinite-well case. (Can you see why?)

EVALUATE

6. Why are the factors of $1/\sqrt{2}$ in the wave function $\Psi(x, t)$ important?

7. Why do you suppose the oscillation angular frequency for a finite well is lower than for an infinite well of the same width?

Problems

•, ••, •••: Difficulty levels. **CP**: Cumulative problems incorporating material from earlier chapters. **CALC**: Problems requiring calculus.
DATA: Problems involving real data, scientific evidence, experimental design, and/or statistical reasoning. **BIO**: Biosciences problems.

DISCUSSION QUESTIONS

Q40.1 If quantum mechanics replaces the language of Newtonian mechanics, why don't we have to use wave functions to describe the motion of macroscopic bodies such as baseballs and cars?

Q40.2 A student remarks that the relationship of ray optics to the more general wave picture is analogous to the relationship of Newtonian mechanics, with well-defined particle trajectories, to quantum mechanics. Comment on this remark.

Q40.3 As Eq. (40.21) indicates, the time-dependent wave function for a stationary state is a complex number having a real part and an imaginary part. How can this function have any physical meaning, since part of it is *imaginary*?

Q40.4 Why must the wave function of a particle be normalized?

Q40.5 If a particle is in a stationary state, does that mean that the particle is not moving? If a particle moves in empty space with constant momentum $\vec{p}$ and hence constant energy $E = p^2/2m$, is it in a stationary state? Explain your answers.

Q40.6 For the particle in a box, we chose $k = n\pi/L$ with $n = 1, 2, 3, \ldots$ to fit the boundary condition that $\psi = 0$ at $x = L$. However, $n = 0, -1, -2, -3, \ldots$ also satisfy that boundary condition. Why didn't we also choose those values of n?

Q40.7 If ψ is normalized, what is the physical significance of the area under a graph of $|\psi|^2$ versus x between x_1 and x_2? What is the total area under the graph of $|\psi|^2$ when all x are included? Explain.

Q40.8 For a particle in a box, what would the probability distribution function $|\psi|^2$ look like if the particle behaved like a classical (Newtonian) particle? Do the actual probability distributions approach this classical form when n is very large? Explain.

Q40.9 In Chapter 15 we represented a standing wave as a superposition of two waves traveling in opposite directions. Can the wave functions for a particle in a box also be thought of as a combination of two traveling waves? Why or why not? What physical interpretation does this representation have? Explain.

Q40.10 A particle in a box is in the ground level. What is the probability of finding the particle in the right half of the box? (Refer to Fig. 40.12, but don't evaluate an integral.) Is the answer the same if the particle is in an excited level? Explain.

Q40.11 The wave functions for a particle in a box (see Fig. 40.12a) are zero at certain points. Does this mean that the particle can't move past one of these points? Explain.

Q40.12 For a particle confined to an infinite square well, is it correct to say that each state of definite energy is also a state of definite wavelength? Is it also a state of definite momentum? Explain. (*Hint:* Remember that momentum is a vector.)

Q40.13 For a particle in a finite potential well, is it correct to say that each bound state of definite energy is also a state of definite wavelength? Is it a state of definite momentum? Explain.

Q40.14 In Fig. 40.12b, the probability function is zero at the points $x = 0$ and $x = L$, the "walls" of the box. Does this mean that the particle never strikes the walls? Explain.

Q40.15 A particle is confined to a finite potential well in the region $0 < x < L$. How does the area under the graph of $|\psi|^2$ in the region $0 < x < L$ compare to the total area under the graph of $|\psi|^2$ when including all possible x?

Q40.16 Compare the wave functions for the first three energy levels for a particle in a box of width L (see Fig. 40.12a) to the corresponding wave functions for a finite potential well of the same width (see Fig. 40.15a). How does the wavelength in the interval $0 \le x \le L$ for the $n = 1$ level of the particle in a box compare to the corresponding wavelength for the $n = 1$ level of the finite potential well? Use this to explain why E_1 is less than $E_{1-\text{IDW}}$ in the situation depicted in Fig. 40.15b.

Q40.17 It is stated in Section 40.3 that a finite potential well always has at least one bound level, no matter how shallow the well. Does this mean that as $U_0 \to 0$, $E_1 \to 0$? Does this violate the Heisenberg uncertainty principle? Explain.

Q40.18 Figure 40.15a shows that the higher the energy of a bound state for a finite potential well, the more the wave function extends outside the well (into the intervals $x < 0$ and $x > L$). Explain why this happens.

Q40.19 In classical (Newtonian) mechanics, the total energy E of a particle can never be less than the potential energy U because the kinetic energy K cannot be negative. Yet in barrier tunneling (see Section 40.4) a particle passes through regions where E is less than U. Is this a contradiction? Explain.

Q40.20 Figure 40.17 shows the scanning tunneling microscope image of 48 iron atoms placed on a copper surface, the pattern indicating the density of electrons on the copper surface. What can you infer about the potential-energy function inside the circle of iron atoms?

Q40.21 Qualitatively, how would you expect the probability for a particle to tunnel through a potential barrier to depend on the height of the barrier? Explain.

Q40.22 The wave function shown in Fig. 40.20 is nonzero for both $x < 0$ and $x > L$. Does this mean that the particle splits into two parts when it strikes the barrier, with one part tunneling through the barrier and the other part bouncing off the barrier? Explain.

Q40.23 The probability distributions for the harmonic-oscillator wave functions (see Figs. 40.27 and 40.28) begin to resemble the classical (Newtonian) probability distribution when the quantum number n becomes large. Would the distributions become the same as in the classical case in the limit of very large n? Explain.

Q40.24 In Fig. 40.28, how does the probability of finding a particle in the center half of the region $-A < x < A$ compare to the probability of finding the particle in the outer half of the region? Is this consistent with the physical interpretation of the situation?

Q40.25 Compare the allowed energy levels for the hydrogen atom, the particle in a box, and the harmonic oscillator. What are the values of the quantum number n for the ground level and the second excited level of each system?

Q40.26 Sketch the wave function for the potential-energy well shown in **Fig. Q40.26** when E_1 is less than U_0 and when E_3 is greater than U_0.

Figure **Q40.26**

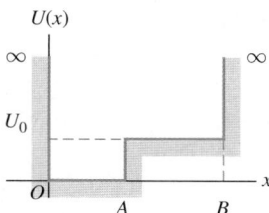

Q40.27 (a) A particle in a box has wave function $\Psi(x, t) = \psi_2(x)e^{-iE_2t/\hbar}$, where ψ_n and E_n are given by Eqs. (40.35) and (40.31), respectively. If the energy of the particle is measured, what is the result? (b) If instead the particle has wave function $\Psi(x, t) = (1/\sqrt{2})(\psi_1(x)e^{-iE_1t/\hbar} + \psi_2(x)e^{-iE_2t/\hbar})$ and the energy of the particle is measured, what is the result? (c) If we had many identical particles with the wave function of part (b) and measured the energy of each, what would be the average value of all of the measurements? Can we say that, before the measurement was made, each particle had this average energy? Explain.

EXERCISES

Section 40.1 Wave Functions and the One-Dimensional Schrödinger Equation

40.1 • An electron is moving as a free particle in the $-x$-direction with momentum that has magnitude 4.50×10^{-24} kg $\cdot$ m/s. What is the one-dimensional time-dependent wave function of the electron?

40.2 • A free particle moving in one dimension has wave function

$$\Psi(x, t) = A\left[e^{i(kx-\omega t)} - e^{i(2kx-4\omega t)}\right]$$

where k and ω are positive real constants. (a) At $t = 0$ what are the two smallest positive values of x for which the probability function $|\Psi(x, t)|^2$ is a maximum? (b) Repeat part (a) for time $t = 2\pi/\omega$. (c) Calculate v_{av} as the distance the maxima have moved divided by the elapsed time. Compare your result to the expression $v_{av} = (\omega_2 - \omega_1)/(k_2 - k_1)$ from Example 40.1.

40.3 • Consider the free-particle wave function of Example 40.1. Let $k_2 = 3k_1 = 3k$. At $t = 0$ the probability distribution function $|\Psi(x, t)|^2$ has a maximum at $x = 0$. (a) What is the smallest positive value of x for which the probability distribution function has a maximum at time $t = 2\pi/\omega$, where $\omega = \hbar k^2/2m$? (b) From your result in part (a), what is the average speed with which the probability distribution is moving in the $+x$-direction? Compare your result to the expression $v_{av} = (\omega_2 - \omega_1)/(k_2 - k_1)$ from Example 40.1.

40.4 • A particle is described by a wave function $\psi(x) = Ae^{-\alpha x^2}$, where A and α are real, positive constants. If the value of α is increased, what effect does this have on (a) the particle's uncertainty in position and (b) the particle's uncertainty in momentum? Explain your answers.

40.5 • Consider a wave function given by $\psi(x) = A\sin kx$, where $k = 2\pi/\lambda$ and A is a real constant. (a) For what values of x is there the highest probability of finding the particle described by this wave function? Explain. (b) For which values of x is the probability zero? Explain.

40.6 •• Compute $|\Psi|^2$ for $\Psi = \psi\sin\omega t$, where ψ is time independent and ω is a real constant. Is this a wave function for a stationary state? Why or why not?

40.7 • CALC Let ψ_1 and ψ_2 be two solutions of Eq. (40.23) with energies E_1 and E_2, respectively, where $E_1 \neq E_2$. Is $\psi = A\psi_1 + B\psi_2$, where A and B are nonzero constants, a solution to Eq. (40.23)? Explain your answer.

Section 40.2 Particle in a Box

40.8 •• CALC A particle moving in one dimension (the x-axis) is described by the wave function

$$\psi(x) = \begin{cases} Ae^{-bx}, & \text{for } x \geq 0 \\ Ae^{bx}, & \text{for } x < 0 \end{cases}$$

where $b = 2.00$ m^{-1}, $A > 0$, and the $+x$-axis points toward the right. (a) Determine A so that the wave function is normalized. (b) Sketch the graph of the wave function. (c) Find the probability of finding this particle in each of the following regions: (i) within 50.0 cm of the origin, (ii) on the left side of the origin (can you first guess the answer by looking at the graph of the wave function?), (iii) between $x = 0.500$ m and $x = 1.00$ m.

40.9 • **Ground-Level Billiards.** (a) Find the lowest energy level for a particle in a box if the particle is a billiard ball ($m = 0.20$ kg) and the box has a width of 1.3 m, the size of a billiard table. (Assume that the billiard ball slides without friction rather than rolls; that is, ignore the *rotational* kinetic energy.) (b) Since the energy in part (a) is all kinetic, to what speed does this correspond? How much time would it take at this speed for the ball to move from one side of the table to the other? (c) What is the difference in energy between the $n = 2$ and $n = 1$ levels? (d) Are quantum-mechanical effects important for the game of billiards?

40.10 • A proton is in a box of width L. What must the width of the box be for the ground-level energy to be 5.0 MeV, a typical value for the energy with which the particles in a nucleus are bound? Compare your result to the size of a nucleus—that is, on the order of 10^{-14} m.

40.11 •• Find the width L of a one-dimensional box for which the ground-state energy of an electron in the box equals the absolute value of the ground state of a hydrogen atom.

40.12 •• When a hydrogen atom undergoes a transition from the $n = 2$ to the $n = 1$ level, a photon with $\lambda = 122$ nm is emitted. (a) If the atom is modeled as an electron in a one-dimensional box, what is the width of the box in order for the $n = 2$ to $n = 1$ transition to correspond to emission of a photon of this energy? (b) For a box with the width calculated in part (a), what is the ground-state energy? How does this correspond to the ground-state energy of a hydrogen atom? (c) Do you think a one-dimensional box is a good model for a hydrogen atom? Explain. (*Hint:* Compare the spacing between adjacent energy levels as a function of n.)

40.13 •• A certain atom requires 3.0 eV of energy to excite an electron from the ground level to the first excited level. Model the atom as an electron in a box and find the width L of the box.

40.14 • An electron in a one-dimensional box has ground-state energy 2.00 eV. What is the wavelength of the photon absorbed when the electron makes a transition to the second excited state?

40.15 •• CALC **Normalization of the Wave Function.** Consider a particle moving in one dimension, which we shall call the x-axis. (a) What does it mean for the wave function of this particle to be *normalized*? (b) Is the wave function $\psi(x) = e^{ax}$, where a is a positive real number, normalized? Could this be a valid wave function? (c) If the particle described by the wave function $\psi(x) = Ae^{-bx}$, where A and b are positive real numbers, is confined to the range $x \geq 0$, determine A (including its units) so that the wave function is normalized.

40.16 • Recall that $|\psi|^2 dx$ is the probability of finding the particle that has normalized wave function $\psi(x)$ in the interval x to $x + dx$. Consider a particle in a box with rigid walls at $x = 0$ and $x = L$. Let the particle be in the ground level and use ψ_n as given in Eq. (40.35). (a) For which values of x, if any, in the range from 0 to L is the probability of finding the particle zero? (b) For which values of x is the probability highest? (c) In parts (a) and (b) are your answers consistent with Fig. 40.12? Explain.

40.17 • Repeat Exercise 40.16 for the particle in the first excited level.

40.18 • (a) Find the excitation energy from the ground level to the third excited level for an electron confined to a box of width 0.360 nm. (b) The electron makes a transition from the $n = 1$ to $n = 4$ level by absorbing a photon. Calculate the wavelength of this photon.

40.19 • An electron is in a box of width 3.0×10^{-10} m. What are the de Broglie wavelength and the magnitude of the momentum of the electron if it is in (a) the $n = 1$ level; (b) the $n = 2$ level; (c) the $n = 3$ level? In each case how does the wavelength compare to the width of the box?

40.20 •• When an electron in a one-dimensional box makes a transition from the $n = 1$ energy level to the $n = 2$ level, it absorbs a photon of wavelength 426 nm. What is the wavelength of that photon when the electron undergoes a transition (a) from the $n = 2$ to the $n = 3$ energy level and (b) from the $n = 1$ to the $n = 3$ energy level? (c) What is the width L of the box?

Section 40.3 Potential Wells

40.21 • An electron is bound in a square well of depth $U_0 = 6E_{1-\text{IDW}}$. What is the width of the well if its ground-state energy is 2.00 eV?

40.22 •• An electron is moving past the square well shown in Fig. 40.13. The electron has energy $E = 3U_0$. What is the ratio of the de Broglie wavelength of the electron in the region $x > L$ to the wavelength for $0 < x < L$?

40.23 •• An electron is bound in a square well of width 1.50 nm and depth $U_0 = 6E_{1-\text{IDW}}$. If the electron is initially in the ground level and absorbs a photon, what maximum wavelength can the photon have and still liberate the electron from the well?

40.24 •• An electron is in the ground state of a square well of width $L = 4.00 \times 10^{-10}$ m. The depth of the well is six times the ground-state energy of an electron in an infinite well of the same width. What is the kinetic energy of this electron after it has absorbed a photon of wavelength 72 nm and moved away from the well?

40.25 •• A proton is bound in a square well of width 4.0 fm $= 4.0 \times 10^{-15}$ m. The depth of the well is six times the ground-level energy $E_{1-\text{IDW}}$ of the corresponding infinite well. If the proton makes a transition from the level with energy E_1 to the level with energy E_3 by absorbing a photon, find the wavelength of the photon.

40.26 •• An electron is bound in a square well that has a depth equal to six times the ground-level energy $E_{1-\text{IDW}}$ of an infinite well of the same width. The longest-wavelength photon that is absorbed by this electron has a wavelength of 582 nm. Determine the width of the well.

Section 40.4 Potential Barriers and Tunneling

40.27 •• (a) An electron with initial kinetic energy 32 eV encounters a square barrier with height 41 eV and width 0.25 nm. What is the probability that the electron will tunnel through the barrier? (b) A proton with the same kinetic energy encounters the same barrier. What is the probability that the proton will tunnel through the barrier?

40.28 •• **Alpha Decay.** In a simple model for a radioactive nucleus, an alpha particle ($m = 6.64 \times 10^{-27}$ kg) is trapped by a square barrier that has width 2.0 fm and height 30.0 MeV. (a) What is the tunneling probability when the alpha particle encounters the barrier if its kinetic energy is 1.0 MeV below the top of the barrier (**Fig. E40.28**)?

Figure **E40.28**

(b) What is the tunneling probability if the energy of the alpha particle is 10.0 MeV below the top of the barrier?

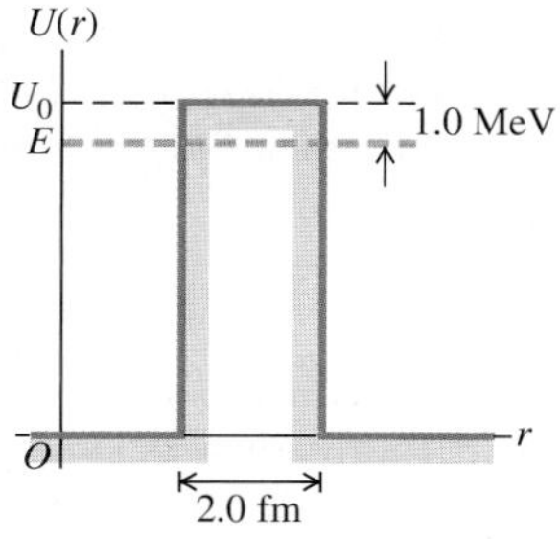

40.29 • An electron with initial kinetic energy 6.0 eV encounters a barrier with height 11.0 eV. What is the probability of tunneling if the width of the barrier is (a) 0.80 nm and (b) 0.40 nm?

40.30 • An electron with initial kinetic energy 5.0 eV encounters a barrier with height U_0 and width 0.60 nm. What is the transmission coefficient if (a) $U_0 = 7.0$ eV; (b) $U_0 = 9.0$ eV; (c) $U_0 = 13.0$ eV?

40.31 •• An electron is moving past the square barrier shown in Fig. 40.19, but the energy of the electron is *greater* than the barrier height. If $E = 2U_0$, what is the ratio of the de Broglie wavelength of the electron in the region $x > L$ to the wavelength for $0 < x < L$?

40.32 • A proton with initial kinetic energy 50.0 eV encounters a barrier of height 70.0 eV. What is the width of the barrier if the probability of tunneling is 8.0×10^{-3}? How does this compare with the barrier width for an electron with the same energy tunneling through a barrier of the same height with the same probability?

Section 40.5 The Harmonic Oscillator

40.33 • A wooden block with mass 0.250 kg is oscillating on the end of a spring that has force constant 110 N/m. Calculate the ground-level energy and the energy separation between adjacent levels. Express your results in joules and in electron volts. Are quantum effects important?

40.34 • CALC Show that $\psi(x)$ given by Eq. (40.47) is a solution to Eq. (40.44) with energy $E_0 = \hbar\omega/2$.

40.35 • Chemists use infrared absorption spectra to identify chemicals in a sample. In one sample, a chemist finds that light of wavelength 5.8 μm is absorbed when a molecule makes a transition from its ground harmonic oscillator level to its first excited level. (a) Find the energy of this transition. (b) If the molecule can be treated as a harmonic oscillator with mass 5.6×10^{-26} kg, find the force constant.

40.36 • A harmonic oscillator absorbs a photon of wavelength 6.35 μm when it undergoes a transition from the ground state to the first excited state. What is the ground-state energy, in electron volts, of the oscillator?

40.37 •• The ground-state energy of a harmonic oscillator is 5.60 eV. If the oscillator undergoes a transition from its $n = 3$ to $n = 2$ level by emitting a photon, what is the wavelength of the photon?

40.38 •• While undergoing a transition from the $n = 1$ to the $n = 2$ energy level, a harmonic oscillator absorbs a photon of wavelength 6.50 μm. What is the wavelength of the absorbed photon when this oscillator undergoes a transition (a) from the $n = 2$ to the $n = 3$ energy level and (b) from the $n = 1$ to the $n = 3$ energy level? (c) What is the value of $\sqrt{k'/m}$, the angular oscillation frequency of the corresponding Newtonian oscillator?

40.39 • In Section 40.5 it is shown that for the ground level of a harmonic oscillator, $\Delta x \Delta p_x = \hbar/2$. Do a similar analysis for an excited level that has quantum number n. How does the uncertainty product $\Delta x \Delta p_x$ depend on n?

40.40 •• For the ground-level harmonic oscillator wave function $\psi(x)$ given in Eq. (40.47), $|\psi|^2$ has a maximum at $x = 0$. (a) Compute the ratio of $|\psi|^2$ at $x = +A$ to $|\psi|^2$ at $x = 0$, where A is given by Eq. (40.48) with $n = 0$ for the ground level. (b) Compute the ratio of $|\psi|^2$ at $x = +2A$ to $|\psi|^2$ at $x = 0$. In each case is your result consistent with what is shown in Fig. 40.27?

40.41 •• For the sodium atom of Example 40.8, find (a) the ground-state energy; (b) the wavelength of a photon emitted when the $n = 4$ to $n = 3$ transition occurs; (c) the energy difference for any $\Delta n = 1$ transition.

PROBLEMS

40.42 •• CALC Consider the wave packet defined by

$$\psi(x) = \int_0^\infty B(k)\cos kx\, dk$$

Let $B(k) = e^{-\alpha^2 k^2}$. (a) The function $B(k)$ has its maximum value at $k = 0$. Let k_h be the value of k at which $B(k)$ has fallen to half its maximum value, and define the width of $B(k)$ as $w_k = k_h$. In terms of α, what is w_k? (b) Use integral tables to evaluate the integral that gives $\psi(x)$. For what value of x is $\psi(x)$ maximum? (c) Define the width of $\psi(x)$ as $w_x = x_h$, where x_h is the positive value of x at which $\psi(x)$ has fallen to half its maximum value. Calculate w_x in terms of α. (d) The momentum p is equal to $hk/2\pi$, so the width of B in momentum is $w_p = hw_k/2\pi$. Calculate the product $w_p w_x$ and compare to the Heisenberg uncertainty principle.

40.43 •• A particle of mass m in a one-dimensional box has the following wave function in the region $x = 0$ to $x = L$:

$$\Psi(x, t) = \frac{1}{\sqrt{2}}\psi_1(x)e^{-iE_1 t/\hbar} + \frac{1}{\sqrt{2}}\psi_3(x)e^{-iE_3 t/\hbar}$$

Here $\psi_1(x)$ and $\psi_3(x)$ are the normalized stationary-state wave functions for the $n = 1$ and $n = 3$ levels, and E_1 and E_3 are the energies of these levels. The wave function is zero for $x < 0$ and for $x > L$. (a) Find the value of the probability distribution function at $x = L/2$ as a function of time. (b) Find the angular frequency at which the probability distribution function oscillates.

40.44 •• CALC (a) Using the integral in Problem 40.42, determine the wave function $\psi(x)$ for a function $B(k)$ given by

$$B(k) = \begin{cases} 0 & k < 0 \\ 1/k_0, & 0 \le k \le k_0 \\ 0, & k > k_0 \end{cases}$$

This represents an equal combination of all wave numbers between 0 and k_0. Thus $\psi(x)$ represents a particle with average wave number $k_0/2$, with a total spread or uncertainty in wave number of k_0. We will call this spread the *width* w_k of $B(k)$, so $w_k = k_0$. (b) Graph $B(k)$ versus k and $\psi(x)$ versus x for the case $k_0 = 2\pi/L$, where L is a length. Locate the point where $\psi(x)$ has its maximum value and label this point on your graph. Locate the two points closest to this maximum (one on each side of it) where $\psi(x) = 0$, and define the distance along the x-axis between these two points as w_x, the width of $\psi(x)$. Indicate the distance w_x on your graph. What is the value of w_x if $k_0 = 2\pi/L$? (c) Repeat part (b) for the case $k_0 = \pi/L$. (d) The momentum p is equal to $hk/2\pi$, so the width of B in momentum is $w_p = hw_k/2\pi$. Calculate the product $w_p w_x$ for each of the cases $k_0 = 2\pi/L$ and $k_0 = \pi/L$. Discuss your results in light of the Heisenberg uncertainty principle.

40.45 •• CALC Consider a beam of free particles that move with velocity $v = p/m$ in the x-direction and are incident on a potential-energy step $U(x) = 0$, for $x < 0$, and $U(x) = U_0 < E$, for $x > 0$. The wave function for $x < 0$ is $\psi(x) = Ae^{ik_1 x} + Be^{-ik_1 x}$, representing incident and reflected particles, and for $x > 0$ is

$\psi(x) = Ce^{ik_2 x}$, representing transmitted particles. Use the conditions that both ψ and its first derivative must be continuous at $x = 0$ to find the constants B and C in terms of k_1, k_2, and A.

40.46 • CALC A particle is in the ground level of a box that extends from $x = 0$ to $x = L$. (a) What is the probability of finding the particle in the region between 0 and $L/4$? Calculate this by integrating $|\psi(x)|^2\, dx$, where ψ is normalized, from $x = 0$ to $x = L/4$. (b) What is the probability of finding the particle in the region $x = L/4$ to $x = L/2$? (c) How do the results of parts (a) and (b) compare? Explain. (d) Add the probabilities calculated in parts (a) and (b). (e) Are your results in parts (a), (b), and (d) consistent with Fig. 40.12b? Explain.

40.47 • **Photon in a Dye Laser.** An electron in a long, organic molecule used in a dye laser behaves approximately like a particle in a box with width 4.18 nm. What is the wavelength of the photon emitted when the electron undergoes a transition (a) from the first excited level to the ground level and (b) from the second excited level to the first excited level?

40.48 •• Consider a particle in a box with rigid walls at $x = 0$ and $x = L$. Let the particle be in the ground level. Calculate the probability $|\psi|^2 dx$ that the particle will be found in the interval x to $x + dx$ for (a) $x = L/4$; (b) $x = L/2$; (c) $x = 3L/4$.

40.49 •• Repeat Problem 40.48 for a particle in the first excited level.

40.50 •• CP A particle is confined within a box with perfectly rigid walls at $x = 0$ and $x = L$. Although the magnitude of the instantaneous force exerted on the particle by the walls is infinite and the time over which it acts is zero, the impulse (that involves a product of force and time) is both finite and quantized. Show that the impulse exerted by the wall at $x = 0$ is $(nh/L)\hat{\imath}$ and that the impulse exerted by the wall at $x = L$ is $-(nh/L)\hat{\imath}$. (*Hint:* You may wish to review Section 8.1.)

40.51 •• CALC What is the probability of finding a particle in a box of length L in the region between $x = L/4$ and $x = 3L/4$ when the particle is in (a) the ground level and (b) the first excited level? (*Hint:* Integrate $|\psi(x)|^2\, dx$, where ψ is normalized, between $L/4$ and $3L/4$.) (c) Are your results in parts (a) and (b) consistent with Fig. 40.12b? Explain.

40.52 •• The *penetration distance* η in a finite potential well is the distance at which the wave function has decreased to $1/e$ of the wave function at the classical turning point:

$$\psi(x = L + \eta) = \frac{1}{e}\psi(L)$$

The penetration distance can be shown to be

$$\eta = \frac{\hbar}{\sqrt{2m(U_0 - E)}}$$

The probability of finding the particle beyond the penetration distance is nearly zero. (a) Find η for an electron having a kinetic energy of 13 eV in a potential well with $U_0 = 20$ eV. (b) Find η for a 20.0-MeV proton trapped in a 30.0-MeV-deep potential well.

40.53 •• CALC A fellow student proposes that a possible wave function for a free particle with mass m (one for which the potential-energy function $U(x)$ is zero) is

$$\psi(x) = \begin{cases} e^{+\kappa x}, & x < 0 \\ e^{-\kappa x}, & x \ge 0 \end{cases}$$

where κ is a positive constant. (a) Graph this proposed wave function. (b) Show that the proposed wave function satisfies the Schrödinger equation for $x < 0$ if the energy is $E = -\hbar^2\kappa^2/2m$—

that is, if the energy of the particle is *negative*. (c) Show that the proposed wave function also satisfies the Schrödinger equation for $x \geq 0$ with the same energy as in part (b). (d) Explain why the proposed wave function is nonetheless *not* an acceptable solution of the Schrödinger equation for a free particle. (*Hint:* What is the behavior of the function at $x = 0$?) It is in fact impossible for a free particle (one for which $U(x) = 0$) to have an energy less than zero.

40.54 • An electron with initial kinetic energy 5.5 eV encounters a square potential barrier of height 10.0 eV. What is the width of the barrier if the electron has a 0.50% probability of tunneling through the barrier?

40.55 • CALC (a) For the finite potential well of Fig. 40.13, what relationships among the constants A and B of Eq. (40.38) and C and D of Eq. (40.40) are obtained by applying the boundary condition that ψ be continuous at $x = 0$ and at $x = L$? (b) What relationships among A, B, C, and D are obtained by applying the boundary condition that $d\psi/dx$ be continuous at $x = 0$ and at $x = L$?

40.56 • CP A harmonic oscillator consists of a 0.020-kg mass on a spring. The oscillation frequency is 1.50 Hz, and the mass has a speed of 0.480 m/s as it passes the equilibrium position. (a) What is the value of the quantum number n for its energy level? (b) What is the difference in energy between the levels E_n and E_{n+1}? Is this difference detectable?

40.57 • For small amplitudes of oscillation the motion of a pendulum is simple harmonic. For a pendulum with a period of 0.500 s, find the ground-level energy and the energy difference between adjacent energy levels. Express your results in joules and in electron volts. Are these values detectable?

40.58 •• CALC (a) Show by direct substitution in the Schrödinger equation for the one-dimensional harmonic oscillator that the wave function $\psi_1(x) = A_1 x e^{-\alpha^2 x^2/2}$, where $\alpha^2 = m\omega/\hbar$, is a solution with energy corresponding to $n = 1$ in Eq. (40.46). (b) Find the normalization constant A_1. (c) Show that the probability density has a minimum at $x = 0$ and maxima at $x = \pm 1/\alpha$, corresponding to the classical turning points for the ground state $n = 0$.

40.59 •• CP (a) The wave nature of particles results in the quantum-mechanical situation that a particle confined in a box can assume only wavelengths that result in standing waves in the box, with nodes at the box walls. Use this to show that an electron confined in a one-dimensional box of length L will have energy levels given by

$$E_n = \frac{n^2 h^2}{8mL^2}$$

(*Hint:* Recall that the relationship between the de Broglie wavelength and the speed of a nonrelativistic particle is $mv = h/\lambda$. The energy of the particle is $\frac{1}{2}mv^2$.) (b) If a hydrogen atom is modeled as a one-dimensional box with length equal to the Bohr radius, what is the energy (in electron volts) of the lowest energy level of the electron?

40.60 ••• Consider a potential well defined as $U(x) = \infty$ for $x < 0$, $U(x) = 0$ for $0 < x < L$, and $U(x) = U_0 > 0$ for $x > L$ (**Fig. P40.60**). Consider a particle with mass m and kinetic energy $E < U_0$ that is trapped in the well. (a) The boundary condition at the infinite wall ($x = 0$) is $\psi(0) = 0$. What must the form of the function $\psi(x)$ for $0 < x < L$ be in

Figure **P40.60**

order to satisfy both the Schrödinger equation and this boundary condition? (b) The wave function must remain finite as $x \rightarrow \infty$. What must the form of the function $\psi(x)$ for $x > L$ be in order to satisfy both the Schrödinger equation and this boundary condition at infinity? (c) Impose the boundary conditions that ψ and $d\psi/dx$ are continuous at $x = L$. Show that the energies of the allowed levels are obtained from solutions of the equation $k \cot kL = -\kappa$, where $k = \sqrt{2mE}/\hbar$ and $\kappa = \sqrt{2m(U_0 - E)}/\hbar$.

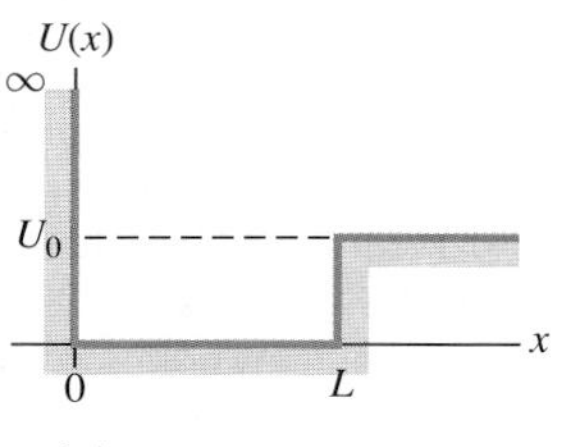

40.61 •• DATA In your research on new solid-state devices, you are studying a solid-state structure that can be modeled accurately as an electron in a one-dimensional infinite potential well (box) of width L. In one of your experiments, electromagnetic radiation is absorbed in transitions in which the initial state is the $n = 1$ ground state. You measure that light of frequency $f = 9.0 \times 10^{14}$ Hz is absorbed and that the next higher absorbed frequency is 16.9×10^{14} Hz. (a) What is quantum number n for the final state in each of the transitions that leads to the absorption of photons of these frequencies? (b) What is the width L of the potential well? (c) What is the longest wavelength in air of light that can be absorbed by an electron if it is initially in the $n = 1$ state?

40.62 •• DATA As an intern at a research lab, you study the transmission of electrons through a potential barrier. You know the height of the barrier, 8.0 eV, but must measure the width L of the barrier. When you measure the tunneling probability T as a function of the energy E of the electron, you get the results shown in the table.

E (eV)	4.0	5.0	6.0	7.0	7.6
T	2.4×10^{-6}	1.5×10^{-5}	1.2×10^{-4}	1.3×10^{-3}	8.1×10^{-3}

(a) For each value of E, calculate the quantities G and κ that appear in Eq. (40.42). Graph $\ln(T/G)$ versus κ. Explain why your data points, when plotted this way, fall close to a straight line. (b) Use the slope of the best-fit straight line to the data in part (a) to calculate L.

40.63 •• DATA When low-energy electrons pass through an ionized gas, electrons of certain energies pass through the gas as if the gas atoms weren't there and thus have transmission coefficients (tunneling probabilities) T equal to unity. The gas ions can be modeled approximately as a rectangular barrier. The value of $T = 1$ occurs when an integral or half-integral number of de Broglie wavelengths of the electron as it passes over the barrier equal the width L of the barrier. You are planning an experiment to measure this effect. To assist you in designing the necessary apparatus, you estimate the electron energies E that will result in $T = 1$. You assume a barrier height of 10 eV and a width of 1.8×10^{-10} m. Calculate the three lowest values of E for which $T = 1$.

CHALLENGE PROBLEMS

40.64 ••• CALC **The WKB Approximation.** It can be a challenge to solve the Schrödinger equation for the bound-state energy levels of an arbitrary potential well. An alternative approach that can yield good approximate results for the energy levels is the *WKB approximation* (named for the physicists Gregor Wentzel, Hendrik Kramers, and Léon Brillouin, who pioneered its application to quantum mechanics). The WKB approximation begins from three physical statements: (i) According to de Broglie, the magnitude of momentum p of a quantum-mechanical particle is $p = h/\lambda$. (ii) The magnitude of momentum is related to the kinetic energy K by the relationship $K = p^2/2m$. (iii) If there are no nonconservative forces, then in Newtonian mechanics the energy E for a particle is constant and equal at each point to the sum of the kinetic and potential energies at that point: $E = K + U(x)$, where x is the coordinate. (a) Combine these three relationships to show that the wavelength of the particle at a coordinate x can be written as

$$\lambda(x) = \frac{h}{\sqrt{2m[E - U(x)]}}$$

Thus we envision a quantum-mechanical particle in a potential well $U(x)$ as being like a free particle, but with a wavelength $\lambda(x)$ that is a function of position. (b) When the particle moves into a region of increasing potential energy, what happens to its wavelength? (c) At a point where $E = U(x)$, Newtonian mechanics says that the particle has zero kinetic energy and must be instantaneously at rest. Such a point is called a *classical turning point,* since this is where a Newtonian particle must stop its motion and reverse direction. As an example, an object oscillating in simple harmonic motion with amplitude A moves back and forth between the points $x = -A$ and $x = +A$; each of these is a classical turning point, since there the potential energy $\frac{1}{2}k'x^2$ equals the total energy $\frac{1}{2}k'A^2$. In the WKB expression for $\lambda(x)$, what is the wavelength at a classical turning point? (d) For a particle in a box with length L, the walls of the box are classical turning points (see Fig. 40.8). Furthermore, the number of wavelengths that fit within the box must be a half-integer (see Fig. 40.10), so that $L = (n/2)\lambda$ and hence $L/\lambda = n/2$, where $n = 1, 2, 3, \ldots$. [Note that this is a restatement of Eq. (40.29).] The WKB scheme for finding the allowed bound-state energy levels of an *arbitrary* potential well is an extension of these observations. It demands that for an allowed energy E, there must be a half-integer number of wavelengths between the classical turning points for that energy. Since the wavelength in the WKB approximation is not a constant but depends on x, the number of wavelengths between the classical turning points a and b for a given value of the energy is the integral of $1/\lambda(x)$ between those points:

$$\int_a^b \frac{dx}{\lambda(x)} = \frac{n}{2} \quad (n = 1, 2, 3, \ldots)$$

Using the expression for $\lambda(x)$ you found in part (a), show that the *WKB condition for an allowed bound-state energy* can be written as

$$\int_a^b \sqrt{2m[E - U(x)]} \, dx = \frac{nh}{2} \quad (n = 1, 2, 3, \ldots)$$

(e) As a check on the expression in part (d), apply it to a particle in a box with walls at $x = 0$ and $x = L$. Evaluate the integral and show that the allowed energy levels according to the WKB approximation are the same as those given by Eq. (40.31). (*Hint:* Since the walls of the box are infinitely high, the points $x = 0$ and $x = L$ are classical turning points for *any* energy E. Inside the box, the potential energy is zero.) (f) For the finite square well shown in Fig. 40.13, show that the WKB expression given in part (d) predicts the *same* bound-state energies as for an infinite square well of the same width. (*Hint:* Assume $E < U_0$. Then the classical turning points are at $x = 0$ and $x = L$.) This shows that the WKB approximation does a poor job when the potential-energy function changes discontinuously, as for a finite potential well. In the next two problems we consider situations in which the potential-energy function changes gradually and the WKB approximation is much more useful.

40.65 ••• CALC The WKB approximation (see Challenge Problem 40.64) can be used to calculate the energy levels for a harmonic oscillator. In this approximation, the energy levels are the solutions to the equation

$$\int_a^b \sqrt{2m[E - U(x)]} \, dx = \frac{nh}{2} \quad n = 1, 2, 3, \ldots$$

Here E is the energy, $U(x)$ is the potential-energy function, and $x = a$ and $x = b$ are the classical turning points (the points at which E is equal to the potential energy, so the Newtonian kinetic energy would be zero). (a) Determine the classical turning points for a harmonic oscillator with energy E and force constant k'. (b) Carry out the integral in the WKB approximation and show that the energy levels in this approximation are $E_n = \hbar\omega$, where $\omega = \sqrt{k'/m}$ and $n = 1, 2, 3, \ldots$. (*Hint:* Recall that $\hbar = h/2\pi$. A useful standard integral is

$$\int \sqrt{A^2 - x^2} \, dx = \frac{1}{2}\left[x\sqrt{A^2 - x^2} + A^2 \arcsin\left(\frac{x}{|A|}\right) \right]$$

where arcsin denotes the inverse sine function. Note that the integrand is even, so the integral from $-x$ to x is equal to twice the integral from 0 to x.) (c) How do the approximate energy levels found in part (b) compare with the true energy levels given by Eq. (40.46)? Does the WKB approximation give an underestimate or an overestimate of the energy levels?

40.66 ••• CALC Protons, neutrons, and many other particles are made of more fundamental particles called *quarks* and *antiquarks* (the antimatter equivalent of quarks). A quark and an antiquark can form a bound state with a variety of different energy levels, each of which corresponds to a different particle observed in the laboratory. As an example, the ψ particle is a low-energy bound state of a so-called charm quark and its antiquark, with a rest energy of 3097 MeV; the $\psi(2S)$ particle is an excited state of this same quark–antiquark combination, with a rest energy of 3686 MeV. A simplified representation of the potential energy of interaction between a quark and an antiquark is $U(x) = A|x|$, where A is a positive constant and x represents the distance between the quark and the antiquark. You can use the WKB approximation (see Challenge Problem 40.64) to determine the bound-state energy levels for this potential-energy function. In the WKB approximation, the energy levels are the solutions to the equation

$$\int_a^b \sqrt{2m[E - U(x)]} \, dx = \frac{nh}{2} \quad (n = 1, 2, 3, \ldots)$$

Here E is the energy, $U(x)$ is the potential-energy function, and $x = a$ and $x = b$ are the classical turning points (the points at which E is equal to the potential energy, so the Newtonian kinetic energy would be zero). (a) Determine the classical turning points for the potential $U(x) = A|x|$ and for an energy E. (b) Carry out the above integral and show that the allowed energy levels in the WKB approximation are given by

$$E_n = \frac{1}{2m}\left(\frac{3mAh}{4}\right)^{2/3} n^{2/3} \quad (n = 1, 2, 3, \ldots)$$

(*Hint:* The integrand is even, so the integral from $-x$ to x is equal to twice the integral from 0 to x.) (c) Does the difference in energy between successive levels increase, decrease, or remain the same as n increases? How does this compare to the behavior of the energy levels for the harmonic oscillator? For the particle in a box? Can you suggest a simple rule that relates the difference in energy between successive levels to the shape of the potential-energy function?

PASSAGE PROBLEMS

QUANTUM DOTS. A *quantum dot* is a type of crystal so small that quantum effects are significant. One application of quantum dots is in fluorescence imaging, in which a quantum dot is bound to a molecule or structure of interest. When the quantum dot is illuminated with light, it absorbs photons and then re-emits photons at a different wavelength. This phenomenon is called *fluorescence*. The wavelength that a quantum dot emits when stimulated with light depends on the dot's size, so the synthesis of quantum dots with different photon absorption and emission properties may be possible. We can understand many quantum-dot properties via a model in which a particle of mass M (roughly the mass of the electron) is confined to a two-dimensional rigid square box of sides L. In this model, the quantum-dot energy levels are given by $E_{m,n} = (m^2 + n^2)(\pi^2\hbar^2)/2ML^2$, where m and n are integers 1, 2, 3, ...

40.67 According to this model, which statement is true about the energy-level spacing of dots of different sizes? (a) Smaller dots have equally spaced levels, but larger dots have energy levels that get farther apart as the energy increases. (b) Larger dots have greater spacing between energy levels than do smaller dots. (c) Smaller dots have greater spacing between energy levels than do larger dots. (d) The spacing between energy levels is independent of the dot size.

40.68 When a given dot with side length L makes a transition from its first excited state to its ground state, the dot emits green (550 nm) light. If a dot with side length $1.1L$ is used instead, what wavelength is emitted in the same transition, according to this model? (a) 600 nm; (b) 670 nm; (c) 500 nm; (d) 460 nm.

40.69 Dots that are the same size but made from different materials are compared. In the same transition, a dot of material 1 emits a photon of longer wavelength than the dot of material 2 does. Based on this model, what is a possible explanation? (a) The mass of the confined particle in material 1 is greater. (b) The mass of the confined particle in material 2 is greater. (c) The confined particles make more transitions per second in material 1. (d) The confined particles make more transitions per second in material 2.

40.70 One advantage of the quantum dot is that, compared to many other fluorescent materials, excited states have relatively long lifetimes (10 ns). What does this mean for the spread in the energy of the photons emitted by quantum dots? (a) Quantum dots emit photons of more well-defined energies than do other fluorescent materials. (b) Quantum dots emit photons of less well-defined energies than do other fluorescent materials. (c) The spread in the energy is affected by the size of the dot, not by the lifetime. (d) There is no spread in the energy of the emitted photons, regardless of the lifetime.

Answers

Chapter Opening Question ?

(i) When an electron in one of these particles—called *quantum dots*—makes a transition from an excited level to a lower level, it emits a photon whose energy is equal to the difference in energy between the levels. The smaller the quantum dot, the larger the energy spacing between levels and hence the shorter (bluer) the wavelength of the emitted photons. See Example 40.6 (Section 40.3) for more details.

Test Your Understanding Questions

40.1 no Equation (40.19) represents a superposition of wave functions with different values of wave number k and hence different values of energy $E = \hbar^2 k^2/2m$. The state that this combined wave function represents is not a state of definite energy, and therefore not a stationary state. Another way to see this is to note that there is a factor $e^{-iEt/\hbar}$ inside the integral in Eq. (40.19), with a different value of E for each value of k. This wave function therefore has a very complicated time dependence, and the probability distribution function $|\Psi(x, t)|^2$ does depend on time.

40.2 (v) Our derivation of the stationary-state wave functions for a particle in a box shows that they are superpositions of waves propagating in opposite directions, just like a standing wave on a string. One wave has momentum in the positive x-direction, while the other wave has an equal magnitude of momentum in the negative x-direction. The *total* x-component of momentum is zero.

40.3 (i) The energy levels are arranged as shown in Fig. 40.15b if $U_0 = 6E_{1-\text{IDW}}$, where $E_{1-\text{IDW}} = \pi^2\hbar^2/2mL^2$ is the ground-level energy of an infinite well. If the well width L is reduced to one-half of its initial value, $E_{1-\text{IDW}}$ increases by a factor of four and so U_0 must also increase by a factor of four. The energies E_1, E_2, and E_3 shown in Fig. 40.15b are all specific fractions of U_0, so they will also increase by a factor of four.

40.4 yes Figure 40.20 shows a possible wave function $\psi(x)$ for tunneling. Since $\psi(x)$ is not zero within the barrier $(0 \leq x \leq L)$, there is some probability that the particle can be found there.

40.5 (ii) If the second photon has a longer wavelength and hence lower energy than the first photon, the difference in energy between the first and second excited levels must be less than the difference between the ground level and the first excited level. This is the case for the hydrogen atom, for which the energy difference between levels decreases as the energy increases (see Fig. 39.24). By contrast, the energy difference between successive levels increases for a particle in a box (see Fig. 40.11b) and is constant for a harmonic oscillator (see Fig. 40.25).

40.6 (v) The value of the energy of a particle in a box must be equal to one of the allowed energy levels, so the measured value will be either E_1 or E_2. *Neither* of these results is guaranteed. If $|C| = |D|$, then E_1 and E_2 are of equal probability. E_1 is the more likely result if $|C| > |D|$, and E_2 is the more likely result if $|C| < |D|$.

Bridging Problem

(a) $|\Psi(x, t)|^2 = \dfrac{1}{L}\left[\sin^2\dfrac{\pi x}{L} + \sin^2\dfrac{2\pi x}{L} \right.$

$\left. + 2\sin\dfrac{\pi x}{L}\sin\dfrac{2\pi x}{L}\cos\left(\dfrac{(E_2 - E_1)t}{\hbar}\right) \right]$

(b) no   (d) $\dfrac{3\pi^2\hbar}{2mL^2}$ (e) $\dfrac{0.903\pi^2\hbar}{mL^2}$

? Lithium (with three electrons per atom) is a metal that burns spontaneously in water, while helium (with two electrons per atom) is a gas that undergoes almost no chemical reactions. The additional electron makes lithium behave very differently from helium primarily because (i) the third electron is strongly repelled by electric forces from the other two electrons; (ii) the third electron and larger nucleus make the lithium atom more massive than the helium atom; (iii) there is a limit on the number of electrons that can occupy a given quantum-mechanical state; (iv) the lithium nucleus has more positive charge than a helium nucleus has.

41 QUANTUM MECHANICS II: ATOMIC STRUCTURE

Some physicists claim that all of chemistry is contained in the Schrödinger equation. This is somewhat of an exaggeration, but this equation can teach us a great deal about the chemical behavior of elements, the periodic table, and the nature of chemical bonds.

In order to learn about the quantum-mechanical structure of atoms, we'll first construct a three-dimensional version of the Schrödinger equation. We'll try this equation out by looking at a three-dimensional version of a particle in a box: a particle confined to a cubical volume.

We'll then see that we can learn a great deal about the structure and properties of *all* atoms from the solutions to the Schrödinger equation for the hydrogen atom. These solutions have quantized values of orbital angular momentum; we don't need to impose quantization as we did with the Bohr model. We label the states with a set of quantum numbers, which we'll use later with many-electron atoms as well. We'll find that the electron also has an intrinsic *spin* angular momentum with its own set of quantized values.

We'll also encounter the exclusion principle, a kind of microscopic zoning ordinance that is the key to understanding many-electron atoms. This principle says that no two electrons in an atom can have the same quantum-mechanical state. We'll then use the principles of this chapter to explain the characteristic x-ray spectra of atoms. Finally, we'll end our discussion of quantum mechanics with a look at the curious concept of quantum entanglement and its application to the new science of quantum computing.

41.1 THE SCHRÖDINGER EQUATION IN THREE DIMENSIONS

We have discussed the Schrödinger equation and its applications only for *one-dimensional* problems, the analog of a Newtonian particle moving along a straight line. The straight-line model is adequate for some applications, but to understand atomic structure, we need a three-dimensional generalization.

It's not difficult to guess what the three-dimensional Schrödinger equation should look like. First, the wave function Ψ is a function of time and all three space coordinates (x, y, z). In general, the potential-energy function also depends on all three coordinates and can be written as $U(x, y, z)$. Next, recall from Section 40.1 that the term $-(\hbar^2/2m)\partial^2\Psi/\partial x^2$ in the one-dimensional Schrödinger equation, Eq. (40.20), is related to the kinetic energy of the particle in the state described by the wave function Ψ. For example, if we insert into this term the wave function $\Psi(x, t) = Ae^{ikx}e^{-i\omega t}$ for a free particle with magnitude of momentum $p = \hbar k$ and kinetic energy $K = p^2/2m$, we obtain $-(\hbar^2/2m)(ik)^2Ae^{ikx}e^{-i\omega t} = (\hbar^2 k^2/2m)Ae^{ikx}e^{-i\omega t} = (p^2/2m)\Psi(x, t) = K\Psi(x, t)$. If the particle can move in three dimensions, its momentum has three components (p_x, p_y, p_z) and its kinetic energy is

$$K = \frac{p_x^2}{2m} + \frac{p_y^2}{2m} + \frac{p_z^2}{2m} \tag{41.1}$$

These observations, taken together, suggest that the correct generalization of the Schrödinger equation to three dimensions is

$$-\frac{\hbar^2}{2m}\left(\frac{\partial^2\Psi(x, y, z, t)}{\partial x^2} + \frac{\partial^2\Psi(x, y, z, t)}{\partial y^2} + \frac{\partial^2\Psi(x, y, z, t)}{\partial z^2}\right)$$

$$+ U(x, y, z)\Psi(x, y, z, t) = i\hbar\frac{\partial\Psi(x, y, z, t)}{\partial t} \tag{41.2}$$

(general three-dimensional Schrödinger equation)

The three-dimensional wave function $\Psi(x, y, z, t)$ has a similar interpretation as in one dimension. The wave function itself is a complex quantity with both a real part and an imaginary part, but $|\Psi(x, y, z, t)|^2$—the square of its absolute value, equal to the product of $\Psi(x, y, z, t)$ and its complex conjugate $\Psi^*(x, y, z, t)$—is real and either positive or zero at every point in space. We interpret $|\Psi(x, y, z, t)|^2\, dV$ as the *probability* of finding the particle within a small volume dV centered on the point (x, y, z) at time t, so $|\Psi(x, y, z, t)|^2$ is the *probability distribution function* in three dimensions. The *normalization condition* on the wave function is that the probability that the particle is *somewhere* in space is exactly 1. Hence the integral of $|\Psi(x, y, z, t)|^2$ over all space must equal 1:

$$\int |\Psi(x, y, z, t)|^2\, dV = 1 \qquad \begin{array}{l}\text{(normalization condition}\\ \text{in three dimensions)}\end{array} \tag{41.3}$$

If the wave function $\Psi(x, y, z, t)$ represents a state of a definite energy E—that is, a stationary state—we can write it as the product of a spatial wave function $\psi(x, y, z)$ and a function of time $e^{-iEt/\hbar}$:

$$\Psi(x, y, z, t) = \psi(x, y, z)e^{-iEt/\hbar} \qquad \begin{array}{l}\text{(time-dependent wave function}\\ \text{for a state of definite energy)}\end{array} \tag{41.4}$$

(Compare this to Eq. (40.21) for a one-dimensional state of definite energy.) If we substitute Eq. (41.4) into Eq. (41.2), the right-hand side of the equation becomes $i\hbar\psi(x, y, z)(-iE/\hbar)e^{-iEt/\hbar} = E\psi(x, y, z)e^{-iEt/\hbar}$. We can then divide both sides by the factor $e^{-iEt/\hbar}$, leaving the *time-independent* Schrödinger equation in three dimensions for a stationary state:

Time-independent three-dimensional Schrödinger equation:

Planck's constant divided by 2π

Time-independent wave function

$$-\frac{\hbar^2}{2m}\left(\frac{\partial^2\psi(x, y, z)}{\partial x^2} + \frac{\partial^2\psi(x, y, z)}{\partial y^2} + \frac{\partial^2\psi(x, y, z)}{\partial z^2}\right) \tag{41.5}$$

$$+ U(x, y, z)\psi(x, y, z) = E\psi(x, y, z)$$

Particle's mass

Potential-energy function

Energy of state

The probability distribution function for a stationary state is just the square of the absolute value of the spatial wave function: $|\psi(x, y, z)e^{-iEt/\hbar}|^2 = \psi^*(x, y, z)e^{+iEt/\hbar}\psi(x, y, z)e^{-iEt/\hbar} = |\psi(x, y, z)|^2$. Note that this doesn't depend on time. (As we discussed in Section 40.1, that's why we call these states *stationary*.) Hence for a stationary state the wave function normalization condition, Eq. (41.3), becomes

$$\int |\psi(x, y, z)|^2 \, dV = 1 \qquad \text{(normalization condition for a stationary state in three dimensions)} \qquad (41.6)$$

We won't pretend that we have *derived* Eqs. (41.2) and (41.5). Like their one-dimensional versions, these equations have to be tested by comparison of their predictions with experimental results. Happily, Eqs. (41.2) and (41.5) both pass this test with flying colors, so we are confident that they *are* the correct equations.

An important topic that we will address in this chapter is the solutions for Eq. (41.5) for the stationary states of the hydrogen atom. The potential-energy function for an electron in a hydrogen atom is *spherically symmetric*; it depends only on the distance $r = (x^2 + y^2 + z^2)^{1/2}$ from the origin of coordinates. To take advantage of this symmetry, it's best to use *spherical coordinates* rather than the Cartesian coordinates (x, y, z) to solve the Schrödinger equation for the hydrogen atom. Before introducing these new coordinates and investigating the hydrogen atom, it's useful to look at the three-dimensional version of the particle in a box that we considered in Section 40.2. Solving this simpler problem will give us insight into the more complicated stationary states found in atomic physics.

TEST YOUR UNDERSTANDING OF SECTION 41.1 In a certain region of space the potential-energy function for a quantum-mechanical particle is zero. In this region the wave function $\psi(x, y, z)$ for a certain stationary state is real and satisfies $\partial^2\psi/\partial x^2 > 0$, $\partial^2\psi/\partial y^2 > 0$, and $\partial^2\psi/\partial z^2 > 0$. The particle has a definite energy E that is positive. What can you conclude about $\psi(x, y, z)$ in this region? (i) It must be positive; (ii) it must be negative; (iii) it must be zero; (iv) not enough information given to decide. ∎

41.2 PARTICLE IN A THREE-DIMENSIONAL BOX

41.1 A particle is confined in a cubical box with walls at $x = 0$, $x = L$, $y = 0$, $y = L$, $z = 0$, and $z = L$.

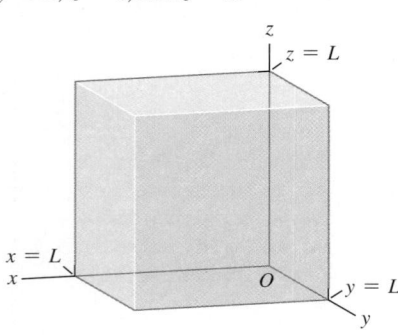

Consider a particle enclosed within a cubical box of side L. This could represent an electron that's free to move anywhere within the interior of a solid metal cube but cannot escape the cube. We'll choose the origin to be at one corner of the box, with the x-, y-, and z-axes along edges of the box. Then the particle is confined to the region $0 \le x \le L$, $0 \le y \le L$, $0 \le z \le L$ (**Fig. 41.1**). What are the stationary states of this system?

As for the model of a particle in a one-dimensional box that we considered in Section 40.2, we'll say that the potential energy is zero inside the box but infinite outside. Hence the spatial wave function $\psi(x, y, z)$ must be zero outside the box in order that the term $U(x, y, z)\psi(x, y, z)$ in the time-independent Schrödinger equation, Eq. (41.5), not be infinite. Consequently the probability distribution function $|\psi(x, y, z)|^2$ is zero outside the box, and the probability that the particle will be found there is zero. Inside the box, the spatial wave function for a stationary state obeys the time-independent Schrödinger equation, Eq. (41.5), with $U(x, y, z) = 0$:

$$-\frac{\hbar^2}{2m}\left(\frac{\partial^2\psi(x, y, z)}{\partial x^2} + \frac{\partial^2\psi(x, y, z)}{\partial y^2} + \frac{\partial^2\psi(x, y, z)}{\partial z^2}\right) = E\psi(x, y, z)$$

(particle in a three-dimensional box) (41.7)

In order for the wave function to be continuous from the inside to the outside of the box, $\psi(x, y, z)$ must equal zero on the walls. Hence our boundary conditions are that $\psi(x, y, z) = 0$ at $x = 0$, $x = L$, $y = 0$, $y = L$, $z = 0$, and $z = L$.

Guessing a solution to a complicated partial differential equation like Eq. (41.7) seems like quite a challenge. To make progress, recall that we wrote the time-*dependent* wave function for a stationary state as the product of one function that depends on only the spatial coordinates x, y, and z and a second function that depends on only the time t: $\Psi(x, y, z, t) = \psi(x, y, z)e^{-iEt/\hbar}$. In the same way, let's try a technique called *separation of variables:* We'll write the spatial wave function $\psi(x, y, z)$ as a product of one function X that depends on only x, a second function Y that depends on only y, and a third function Z that depends on only z:

$$\psi(x, y, z) = X(x)Y(y)Z(z) \tag{41.8}$$

If we substitute Eq. (41.8) into Eq. (41.7), we get

$$-\frac{\hbar^2}{2m}\left(Y(y)Z(z)\frac{d^2X(x)}{dx^2} + X(x)Z(z)\frac{d^2Y(y)}{dy^2} + X(x)Y(y)\frac{d^2Z(z)}{dz^2}\right)$$
$$= EX(x)Y(y)Z(z) \tag{41.9}$$

The partial derivatives in Eq. (41.7) have become ordinary derivatives since they act on functions of a single variable. Now we divide both sides of Eq. (41.9) by the product $X(x)Y(y)Z(z)$:

$$\left(-\frac{\hbar^2}{2m}\frac{1}{X(x)}\frac{d^2X(x)}{dx^2}\right) + \left(-\frac{\hbar^2}{2m}\frac{1}{Y(y)}\frac{d^2Y(y)}{dy^2}\right) + \left(-\frac{\hbar^2}{2m}\frac{1}{Z(z)}\frac{d^2Z(z)}{dz^2}\right) = E$$

$$\tag{41.10}$$

The right-hand side of Eq. (41.10) is the energy of the stationary state. Since E is a constant that does not depend on the values of x, y, and z, the left-hand side of the equation must also be independent of the values of x, y, and z. Hence the first term in parentheses on the left-hand side of Eq. (41.10) must equal a constant that doesn't depend on x, the second term in parentheses must equal another constant that doesn't depend on y, and the third term in parentheses must equal a third constant that doesn't depend on z. Let's call these constants E_X, E_Y, and E_Z, respectively. We then have a separate equation for each of the three functions $X(x)$, $Y(y)$, and $Z(z)$:

$$-\frac{\hbar^2}{2m}\frac{d^2X(x)}{dx^2} = E_XX(x) \tag{41.11a}$$

$$-\frac{\hbar^2}{2m}\frac{d^2Y(y)}{dy^2} = E_YY(y) \tag{41.11b}$$

$$-\frac{\hbar^2}{2m}\frac{d^2Z(z)}{dz^2} = E_ZZ(z) \tag{41.11c}$$

To satisfy the boundary conditions that $\psi(x, y, z) = X(x)Y(y)Z(z)$ be equal to zero on the walls of the box, we demand that $X(x) = 0$ at $x = 0$ and $x = L$, $Y(y) = 0$ at $y = 0$ and $y = L$, and $Z(z) = 0$ at $z = 0$ and $z = L$.

How can we interpret the three constants E_X, E_Y, and E_Z in Eqs. (41.11)? From Eq. (41.10), they are related to the energy E by

$$E_X + E_Y + E_Z = E \tag{41.12}$$

Equation (41.12) should remind you of Eq. (41.1) in Section 41.1, which states that the kinetic energy of a particle is the sum of contributions coming from its x-, y-, and z-components of momentum. Hence the constants E_X, E_Y, and E_Z tell us how much of the particle's energy is due to motion along each of the three coordinate axes. (Inside the box the potential energy is zero, so the particle's energy is purely kinetic.)

Equations (41.11) represent an enormous simplification; we've reduced the problem of solving a fairly complex *partial* differential equation with three independent variables to the much simpler problem of solving three separate *ordinary* differential equations with one independent variable each. What's more, each of these ordinary differential equations is the same as the time-independent Schrödinger equation for a particle in a *one-dimensional* box, Eq. (40.25), and with exactly the same boundary conditions at 0 and L. (The only differences are that some of the quantities are labeled by different symbols.) By comparing with our work in Section 40.2, you can see that the solutions to Eqs. (41.11) are

$$X_{n_X}(x) = C_X \sin\frac{n_X \pi x}{L} \quad (n_X = 1, 2, 3, \ldots) \tag{41.13a}$$

$$Y_{n_Y}(y) = C_Y \sin\frac{n_Y \pi y}{L} \quad (n_Y = 1, 2, 3, \ldots) \tag{41.13b}$$

$$Z_{n_Z}(z) = C_Z \sin\frac{n_Z \pi z}{L} \quad (n_Z = 1, 2, 3, \ldots) \tag{41.13c}$$

where C_X, C_Y, and C_Z are constants. The corresponding values of E_X, E_Y, and E_Z are

$$E_X = \frac{n_X{}^2 \pi^2 \hbar^2}{2mL^2} \quad (n_X = 1, 2, 3, \ldots) \tag{41.14a}$$

$$E_Y = \frac{n_Y{}^2 \pi^2 \hbar^2}{2mL^2} \quad (n_Y = 1, 2, 3, \ldots) \tag{41.14b}$$

$$E_Z = \frac{n_Z{}^2 \pi^2 \hbar^2}{2mL^2} \quad (n_Z = 1, 2, 3, \ldots) \tag{41.14c}$$

There is only one quantum number n for the one-dimensional particle in a box, but *three* quantum numbers n_X, n_Y, and n_Z for the three-dimensional box. If we substitute Eqs. (41.13) back into Eq. (41.8) for the total spatial wave function, $\psi(x, y, z) = X(x)Y(y)Z(z)$, we get the following stationary-state wave functions for a particle in a three-dimensional cubical box:

$$\psi_{n_X,n_Y,n_Z}(x, y, z) = C\sin\frac{n_X \pi x}{L}\sin\frac{n_Y \pi y}{L}\sin\frac{n_Z \pi z}{L}$$

$$(n_X = 1, 2, 3, \ldots; n_Y = 1, 2, 3, \ldots; n_Z = 1, 2, 3, \ldots) \tag{41.15}$$

where $C = C_X C_Y C_Z$. The value of the constant C is determined by the normalization condition, Eq. (41.6).

In Section 40.2 we saw that the stationary-state wave functions for a particle in a one-dimensional box were analogous to standing waves on a string. In a similar way, the *three*-dimensional wave functions given by Eq. (41.15) are analogous to standing electromagnetic waves in a cubical cavity like the interior of a microwave oven (see Section 32.5). In a microwave oven there are "dead spots" where the wave intensity is zero, corresponding to the nodes of the standing wave. (The moving platform in a microwave oven ensures even cooking by making sure that no part of the food sits at any "dead spot.") In a similar fashion, the probability distribution function corresponding to Eq. (41.15) can have "dead spots" where there is zero probability of finding the particle. As an example, consider the case $(n_X, n_Y, n_Z) = (2, 1, 1)$. From Eq. (41.15), the probability distribution function for this case is

$$|\psi_{2,1,1}(x, y, z)|^2 = |C|^2 \sin^2\frac{2\pi x}{L}\sin^2\frac{\pi y}{L}\sin^2\frac{\pi z}{L}$$

As **Fig. 41.2a** shows, this probability distribution function is zero on the plane $x = L/2$, where $\sin^2(2\pi x/L) = \sin^2 \pi = 0$. The particle is most likely

41.2 Probability distribution function $|\psi_{n_X,n_Y,n_Z}(x, y, z)|^2$ for (n_X, n_Y, n_Z) equal to (a) (2, 1, 1), (b) (1, 2, 1), and (c) (1, 1, 2). The value of $|\psi|^2$ is proportional to the density of dots. The wave function is zero on the walls of the box and on a midplane of the box, so $|\psi|^2 = 0$ at these locations.

(a) $|\psi_{2,1,1}|^2$

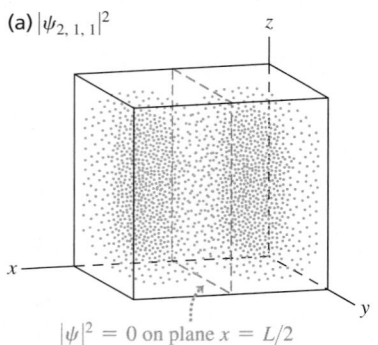

$|\psi|^2 = 0$ on plane $x = L/2$

(b) $|\psi_{1,2,1}|^2$

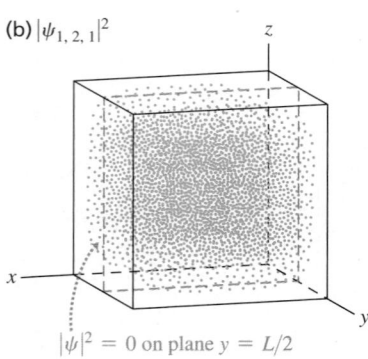

$|\psi|^2 = 0$ on plane $y = L/2$

(c) $|\psi_{1,1,2}|^2$

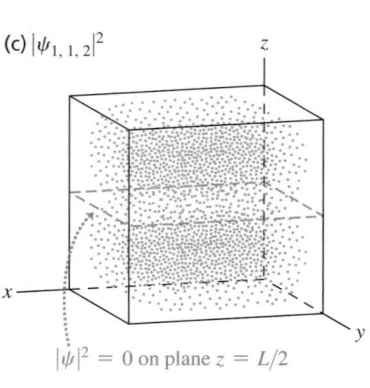

$|\psi|^2 = 0$ on plane $z = L/2$

to be found near where all three of the sine-squared functions are greatest, at $(x, y, z) = (L/4, L/2, L/2)$ or $(x, y, z) = (3L/4, L/2, L/2)$. Figures 41.2b and 41.2c show the similar cases $(n_X, n_Y, n_Z) = (1, 2, 1)$ and $(n_X, n_Y, n_Z) = (1, 1, 2)$. For higher values of the quantum numbers n_X, n_Y, and n_Z there are additional planes on which the probability distribution function equals zero, just as the probability distribution function $|\psi(x)|^2$ for a one-dimensional box has more zeros for higher values of n (see Fig. 40.12).

EXAMPLE 41.1   PROBABILITY IN A THREE-DIMENSIONAL BOX

(a) Find the value of the constant C that normalizes the wave function of Eq. (41.15). (b) Find the probability that the particle will be found somewhere in the region $0 \leq x \leq L/4$ (**Fig. 41.3**) for the cases (i) $(n_X, n_Y, n_Z) = (1, 2, 1)$, (ii) $(n_X, n_Y, n_Z) = (2, 1, 1)$, and (iii) $(n_X, n_Y, n_Z) = (3, 1, 1)$.

SOLUTION

IDENTIFY and SET UP: Equation (41.6) tells us that to normalize the wave function, we have to choose the value of C so that the integral of the probability distribution function $|\psi_{n_X, n_Y, n_Z}(x, y, z)|^2$ over the volume within the box equals 1. (The integral is actually over *all* space, but the particle-in-a-box wave functions are zero outside the box.)

The probability of finding the particle within a certain volume within the box equals the integral of the probability distribution function over that volume. Hence in part (b) we'll integrate $|\psi_{n_X, n_Y, n_Z}(x, y, z)|^2$ for the given values of (n_X, n_Y, n_Z) over the volume $0 \leq x \leq L/4, 0 \leq y \leq L, 0 \leq z \leq L$.

EXECUTE: (a) From Eq. (41.15),

$$|\psi_{n_X, n_Y, n_Z}(x, y, z)|^2 = |C|^2 \sin^2 \frac{n_X \pi x}{L} \sin^2 \frac{n_Y \pi y}{L} \sin^2 \frac{n_Z \pi z}{L}$$

Hence the normalization condition is

$$\int |\psi_{n_X, n_Y, n_Z}(x, y, z)|^2 \, dV$$

$$= |C|^2 \int_{x=0}^{x=L} \int_{y=0}^{y=L} \int_{z=0}^{z=L} \sin^2 \frac{n_X \pi x}{L} \sin^2 \frac{n_Y \pi y}{L} \sin^2 \frac{n_Z \pi z}{L} \, dx \, dy \, dz$$

$$= |C|^2 \left(\int_{x=0}^{x=L} \sin^2 \frac{n_X \pi x}{L} \, dx \right) \left(\int_{y=0}^{y=L} \sin^2 \frac{n_Y \pi y}{L} \, dy \right)$$

$$\times \left(\int_{z=0}^{z=L} \sin^2 \frac{n_Z \pi z}{L} \, dz \right)$$

$$= 1$$

We can use the identity $\sin^2 \theta = \frac{1}{2}(1 - \cos 2\theta)$ and the variable substitution $\theta = n_X \pi x/L$ to show that

$$\int \sin^2 \frac{n_X \pi x}{L} \, dx = \frac{L}{2n_X \pi} \left[\frac{n_X \pi x}{L} - \frac{1}{2} \sin \left(\frac{2n_X \pi x}{L} \right) \right]$$

$$= \frac{x}{2} - \frac{L}{4n_X \pi} \sin \left(\frac{2n_X \pi x}{L} \right)$$

If we evaluate this integral between $x = 0$ and $x = L$, the result is $L/2$ (recall that $\sin 0 = 0$ and $\sin 2n_X \pi = 0$ for any integer n_X).

41.3 What is the probability that the particle is in the dark-colored quarter of the box?

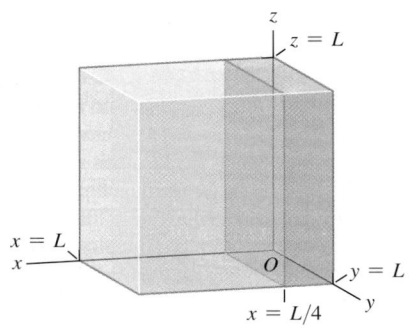

The y- and z-integrals each yield the same result, so the normalization condition is

$$|C|^2 \left(\frac{L}{2} \right) \left(\frac{L}{2} \right) \left(\frac{L}{2} \right) = |C|^2 \left(\frac{L}{2} \right)^3 = 1$$

or $|C|^2 = (2/L)^3$. If we choose C to be real and positive, then $C = (2/L)^{3/2}$.

(b) We have the same y- and z-integrals as in part (a), but now the limits of integration on the x-integral are $x = 0$ and $x = L/4$:

$$P = \int_{0 \leq x \leq L/4} |\psi_{n_X, n_Y, n_Z}|^2 \, dV$$

$$= |C|^2 \left(\int_{x=0}^{x=L/4} \sin^2 \frac{n_X \pi x}{L} \, dx \right) \left(\int_{y=0}^{y=L} \sin^2 \frac{n_Y \pi y}{L} \, dy \right)$$

$$\times \left(\int_{z=0}^{z=L} \sin^2 \frac{n_Z \pi z}{L} \, dz \right)$$

The x-integral is

$$\int_{x=0}^{x=L/4} \sin^2 \frac{n_X \pi x}{L} \, dx = \left[\frac{x}{2} - \frac{L}{4n_X \pi} \sin \left(\frac{2n_X \pi x}{L} \right) \right] \Bigg|_{x=0}^{x=L/4}$$

$$= \frac{L}{8} - \frac{L}{4n_X \pi} \sin \left(\frac{n_X \pi}{2} \right)$$

Hence the probability of finding the particle somewhere in the region $0 \leq x \leq L/4$ is

$$P = \left(\frac{2}{L} \right)^3 \left[\frac{L}{8} - \frac{L}{4n_X \pi} \sin \left(\frac{n_X \pi}{2} \right) \right] \left(\frac{L}{2} \right) \left(\frac{L}{2} \right)$$

$$= \frac{1}{4} - \frac{1}{2n_X \pi} \sin \left(\frac{n_X \pi}{2} \right)$$

Continued

This depends only on the value of n_X, not on n_Y or n_Z. Hence for the three cases we have

(i) $n_X = 1$: $P = \frac{1}{4} - \dfrac{1}{2(1)\pi} \sin\left(\dfrac{\pi}{2}\right) = \frac{1}{4} - \dfrac{1}{2\pi}(1)$

$$= \frac{1}{4} - \frac{1}{2\pi} = 0.091$$

(ii) $n_X = 2$: $P = \frac{1}{4} - \dfrac{1}{2(2)\pi} \sin\left(\dfrac{2\pi}{2}\right) = \frac{1}{4} - \dfrac{1}{4\pi} \sin\pi$

$$= \frac{1}{4} - 0 = 0.250$$

(iii) $n_X = 3$: $P = \frac{1}{4} - \dfrac{1}{2(3)\pi} \sin\left(\dfrac{3\pi}{2}\right) = \frac{1}{4} - \dfrac{1}{6\pi}(-1)$

$$= \frac{1}{4} + \frac{1}{6\pi} = 0.303$$

EVALUATE: You can see why the probabilities in part (b) are different by looking at part (b) of Fig. 40.12, which shows $\sin^2 n_X\pi x/L$ for $n_X = 1$, 2, and 3. For $n_X = 2$ the area under the curve between $x = 0$ and $x = L/4$ (equal to the integral between these two points) is exactly $\frac{1}{4}$ of the total area between $x = 0$ and $x = L$. For $n_X = 1$ the area between $x = 0$ and $x = L/4$ is less than $\frac{1}{4}$ of the total area, and for $n_X = 3$ it is greater than $\frac{1}{4}$ of the total area.

Energy Levels, Degeneracy, and Symmetry

From Eqs. (41.12) and (41.14), the allowed energies for a particle of mass m in a cubical box of side L are

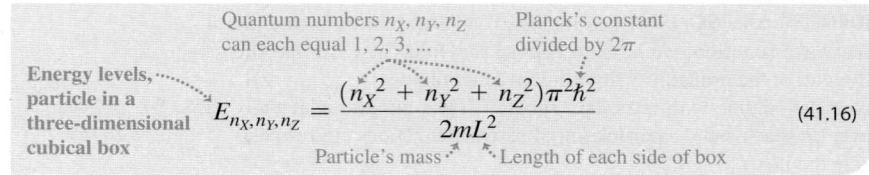

Quantum numbers n_X, n_Y, n_Z can each equal 1, 2, 3, ...   Planck's constant divided by 2π

Energy levels, particle in a three-dimensional cubical box

$$E_{n_X,n_Y,n_Z} = \frac{(n_X^2 + n_Y^2 + n_Z^2)\pi^2\hbar^2}{2mL^2} \tag{41.16}$$

Particle's mass   Length of each side of box

Figure 41.4 shows the six lowest energy levels given by Eq. (41.16). Note that most energy levels correspond to more than one set of quantum numbers (n_X, n_Y, n_Z) and hence to more than one quantum state. Having two or more distinct quantum states with the same energy is called **degeneracy,** and states with the same energy are said to be **degenerate.** For example, Fig. 41.4 shows that the states $(n_X, n_Y, n_Z) = (2, 1, 1)$, $(1, 2, 1)$, and $(1, 1, 2)$ are degenerate. By comparison, for a particle in a one-dimensional box there is just one state for each energy level (see Fig. 40.11a) and no degeneracy.

The reason the cubical box exhibits degeneracy is that it is *symmetric:* All sides of the box have the same dimensions. As an illustration, Fig. 41.2 shows the probability distribution functions for the three states $(n_X, n_Y, n_Z) = (2, 1, 1)$,

41.4 Energy-level diagram for a particle in a three-dimensional cubical box. We label each level with the quantum numbers of the states (n_X, n_Y, n_Z) with that energy. Several of the levels are degenerate (more than one state has the same energy). The lowest (ground) level, $(n_X, n_Y, n_Z) = (1, 1, 1)$, has energy $E_{1,1,1} = (1^2 + 1^2 + 1^2)\pi^2\hbar^2/2mL^2 = 3\pi^2\hbar^2/2mL^2$; we show the energies of the other levels as multiples of $E_{1,1,1}$.

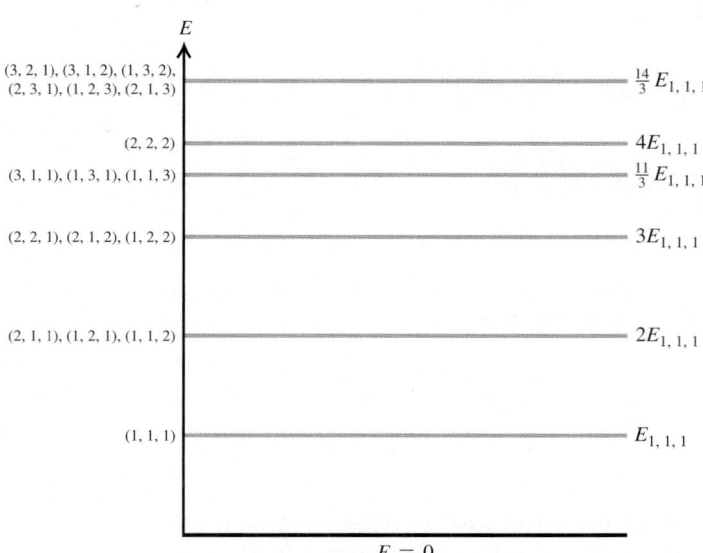

$(1, 2, 1)$, and $(1, 1, 2)$. You can transform any one of these three states into a different one by simply rotating the cubical box by $90°$. This rotation doesn't change the energy, so the three states are degenerate.

Since degeneracy is a consequence of symmetry, we can remove the degeneracy by making the box asymmetric. We do this by giving the three sides of the box different lengths L_X, L_Y, and L_Z. If we repeat the steps that we followed to solve the time-independent Schrödinger equation, we find that the energy levels are given by

$$E_{n_X, n_Y, n_Z} = \left(\frac{n_X^2}{L_X^2} + \frac{n_Y^2}{L_Y^2} + \frac{n_Z^2}{L_Z^2} \right) \frac{\pi^2 \hbar^2}{2m} \quad \begin{array}{l} (n_X = 1, 2, 3, \ldots; \\ n_Y = 1, 2, 3, \ldots; \\ n_Z = 1, 2, 3, \ldots) \end{array} \quad (41.17)$$

(energy levels, particle in a three-dimensional box with sides of length L_X, L_Y, and L_Z)

If L_X, L_Y, and L_Z are all different, the states $(n_X, n_Y, n_Z) = (2, 1, 1)$, $(1, 2, 1)$, and $(1, 1, 2)$ have different energies and hence are no longer degenerate. Note that Eq. (41.17) reduces to Eq. (41.16) if $L_X = L_Y = L_Z = L$.

Returning to a particle in a three-dimensional cubical box, let's summarize the differences from the one-dimensional case that we examined in Section 40.2:

- We can write the wave function for a three-dimensional stationary state as a product of three functions, one for each spatial coordinate. Only a single function of the coordinate x is needed in one dimension.
- In the three-dimensional case, three quantum numbers are needed to describe each stationary state. Only one quantum number is needed in the one-dimensional case.
- Most of the energy levels for the three-dimensional case are degenerate: More than one stationary state has this energy. There is no degeneracy in the one-dimensional case.
- For a stationary state of the three-dimensional case, there are surfaces on which the probability distribution function $|\psi|^2$ is zero. In the one-dimensional case there are positions on the x-axis where $|\psi|^2$ is zero.

We'll see these same features in the following section for a three-dimensional situation that's more realistic than a particle in a cubical box: a hydrogen atom in which a negatively charged electron orbits a positively charged nucleus.

TEST YOUR UNDERSTANDING OF SECTION 41.2 Rank the following states of a particle in a cubical box of side L in order from highest to lowest energy: (i) $(n_X, n_Y, n_Z) = (2, 3, 2)$; (ii) $(n_X, n_Y, n_Z) = (4, 1, 1)$; (iii) $(n_X, n_Y, n_Z) = (2, 2, 3)$; (iv) $(n_X, n_Y, n_Z) = (1, 3, 3)$. ❙

41.3 THE HYDROGEN ATOM

Let's continue the discussion of the hydrogen atom that we began in Chapter 39. In the Bohr model, electrons move in circular orbits like Newtonian particles, but with quantized values of angular momentum. While this model gave the correct energy levels of the hydrogen atom, as deduced from spectra, it had many conceptual difficulties. It mixed classical physics with new and seemingly contradictory concepts. It provided no insight into the process by which photons are emitted and absorbed. It could not be generalized to atoms with more than one electron. It predicted the wrong magnetic properties for the hydrogen atom. And perhaps most important, its picture of the electron as a localized point particle was inconsistent with the more general view we developed in Chapters 39 and 40. To go beyond the Bohr model, let's apply the Schrödinger equation to find the wave functions for stationary states (states of definite energy) of the hydrogen atom. As in Section 39.3, we include the motion of the nucleus by simply replacing the electron mass m with the reduced mass m_r.

The Schrödinger Equation for the Hydrogen Atom

41.5 The Schrödinger equation for the hydrogen atom can be solved most readily by using spherical coordinates.

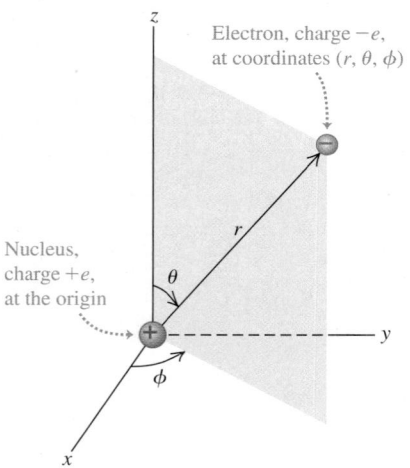

We discussed the three-dimensional version of the Schrödinger equation in Section 41.1. The potential-energy function is *spherically symmetric:* It depends only on the distance $r = (x^2 + y^2 + z^2)^{1/2}$ from the origin of coordinates:

$$U(r) = -\frac{1}{4\pi\epsilon_0}\frac{e^2}{r} \qquad (41.18)$$

The hydrogen-atom problem is best formulated in spherical coordinates (r, θ, ϕ), shown in **Fig. 41.5**; the spherically symmetric potential-energy function depends only on r, not on θ or ϕ. The Schrödinger equation with this potential-energy function can be solved exactly; the solutions are combinations of familiar functions. Without going into a lot of detail, we can describe the most important features of the procedure and the results.

First, we find the solutions by using the same method of separation of variables that we employed for a particle in a cubical box in Section 41.2. We express the wave function $\psi(r, \theta, \phi)$ as a product of three functions, each one a function of only one of the three coordinates:

$$\psi(r, \theta, \phi) = R(r)\Theta(\theta)\Phi(\phi) \qquad (41.19)$$

That is, the function $R(r)$ depends on only r, $\Theta(\theta)$ depends on only θ, and $\Phi(\phi)$ depends on only ϕ. Just as for a particle in a three-dimensional box, when we substitute Eq. (41.19) into the Schrödinger equation, we get three separate ordinary differential equations. One equation involves only r and $R(r)$, a second involves only θ and $\Theta(\theta)$, and a third involves only ϕ and $\Phi(\phi)$:

$$-\frac{\hbar^2}{2m_r r^2}\frac{d}{dr}\left(r^2\frac{dR(r)}{dr}\right) + \left(\frac{\hbar^2 l(l+1)}{2m_r r^2} + U(r)\right)R(r) = ER(r) \qquad (41.20a)$$

$$\frac{1}{\sin\theta}\frac{d}{d\theta}\left(\sin\theta\frac{d\Theta(\theta)}{d\theta}\right) + \left(l(l+1) - \frac{m_l^2}{\sin^2\theta}\right)\Theta(\theta) = 0 \qquad (41.20b)$$

$$\frac{d^2\Phi(\phi)}{d\phi^2} + m_l^2\Phi(\phi) = 0 \qquad (41.20c)$$

CAUTION Two uses of the symbol *m* Don't confuse the constant m_l in Eqs. (41.20b) and (41.20c) with the similar symbol m_r for the reduced mass of the electron and nucleus (see Section 39.3). The constant m_l is a dimensionless number; the reduced mass m_r has units of kilograms.

In Eqs. (41.20) E is the energy of the stationary state and l and m_l are constants that we'll discuss later.

We won't attempt to solve this set of three equations, but we can describe how it's done. As for the particle in a cubical box, the physically acceptable solutions of these three equations are determined by boundary conditions. The radial function $R(r)$ in Eq. (41.20a) must approach zero at large r, because we are describing *bound states* of the electron that are localized near the nucleus. This is analogous to the requirement that the harmonic-oscillator wave functions (see Section 40.5) must approach zero at large x. The angular functions $\Theta(\theta)$ and $\Phi(\phi)$ in Eqs. (41.20b) and (41.20c) must be *finite* for all relevant values of the angles. For example, there are solutions of the Θ equation that become infinite at $\theta = 0$ and $\theta = \pi$; these are unacceptable, since $\psi(r, \theta, \phi)$ must be normalizable. Furthermore, the angular function $\Phi(\phi)$ in Eq. (41.20c) must be *periodic*. For example, (r, θ, ϕ) and $(r, \theta, \phi + 2\pi)$ describe the same point, so $\Phi(\phi + 2\pi)$ must equal $\Phi(\phi)$.

The allowed radial functions $R(r)$ turn out to be an exponential function $e^{-\alpha r}$ (where α is positive) multiplied by a polynomial in r. The functions $\Theta(\theta)$ are polynomials containing various powers of $\sin\theta$ and $\cos\theta$, and the functions $\Phi(\phi)$ are simply proportional to $e^{im_l\phi}$, where $i = \sqrt{-1}$ and m_l is an integer that may be positive, zero, or negative.

In the process of finding solutions that satisfy the boundary conditions, we also find the corresponding energy levels. We denote the energies of these levels [E in

Eq. (41.20a)] by E_n ($n = 1, 2, 3, \ldots$). These turn out to be *identical* to those from the Bohr model, as given by Eq. (39.15), with the electron rest mass m replaced by the reduced mass m_r. Rewriting that equation with $\hbar = h/2\pi$, we have

Reduced mass ⸱⸱⸱ ⸱⸱Magnitude of electron charge

Energy levels ⸱⸱⸱
of hydrogen
$$E_n = -\frac{1}{(4\pi\epsilon_0)^2}\frac{m_r e^4}{2n^2\hbar^2} = -\frac{13.60 \text{ eV}}{n^2} \qquad (41.21)$$

Electric   Principal quantum number   Planck's constant
constant   ($n = 1, 2, 3, \ldots$)   divided by 2π

As in Section 39.3, we call n the **principal quantum number.**

Equation (41.21) is an important validation of our Schrödinger-equation analysis of the hydrogen atom. The Schrödinger analysis is quite different from the Bohr model, both mathematically and conceptually, yet both yield the same energy-level scheme—a scheme that agrees with the energies determined from spectra. As we will see, the Schrödinger analysis can explain many more aspects of the hydrogen atom than can the Bohr model.

Quantization of Orbital Angular Momentum

The solutions to Eqs. (41.20) that satisfy the boundary conditions mentioned above also have quantized values of *orbital angular momentum*. That is, only certain discrete values of the magnitude and components of orbital angular momentum are permitted. In discussing the Bohr model in Section 39.3, we mentioned that quantization of angular momentum was a result with little fundamental justification. With the Schrödinger equation it appears automatically.

The possible values of the magnitude L of orbital angular momentum $\vec{L}$ are determined by the requirement that the $\Theta(\theta)$ function in Eq. (41.20b) must be finite at $\theta = 0$ and $\theta = \pi$. In a level with energy E_n and principal quantum number n, the possible values of L are

Orbital quantum number

Magnitude of
orbital angular ⸱⸱⸱⸱
momentum,
hydrogen atom
$$L = \sqrt{l(l + 1)}\,\hbar \quad (l = 0, 1, 2, \ldots, n - 1) \qquad (41.22)$$

Planck's constant   Principal quantum number
divided by 2π ($n = 1, 2, 3, \ldots$)

The **orbital quantum number** l in Eq. (41.22) is the same l that appears in Eqs. (41.20a) and (41.20b). In the Bohr model, each energy level corresponded to a single value of angular momentum. Equation (41.22) shows that in fact there are n different possible values of L for the nth energy level.

An interesting feature of Eq. (41.22) is that the orbital angular momentum is *zero* for $l = 0$ states. This result disagrees with the Bohr model, in which the electron always moved in a circle of definite radius and L was never zero. The $l = 0$ wave functions ψ depend only on r; for these states, the functions $\Theta(\theta)$ and $\Phi(\phi)$ are constants. Thus the wave functions for $l = 0$ states are spherically symmetric. There is nothing in their probability distribution $|\psi|^2$ to favor one direction over any other, and there is no orbital angular momentum.

The permitted values of the *component* of $\vec{L}$ in a given direction, say the z-component L_z, are determined by the requirement that the $\Phi(\phi)$ function must equal $\Phi(\phi + 2\pi)$. The possible values of L_z are

Orbital magnetic quantum number

z-component of
orbital angular ⸱⸱⸱⸱
momentum,
hydrogen atom
$$L_z = m_l\hbar \qquad (m_l = 0, \pm 1, \pm 2, \ldots, \pm l) \qquad (41.23)$$

Planck's constant   Orbital quantum number
divided by 2π

41.6 (a) When $l = 2$, the magnitude of the angular momentum vector $\vec{L}$ is $\sqrt{6}\hbar = 2.45\hbar$, but $\vec{L}$ does not have a definite direction. In this semiclassical vector picture, $\vec{L}$ makes an angle of 35.3° with the z-axis when the z-component has its maximum value of $2\hbar$. (b) These cones show the possible directions of $\vec{L}$ for different values of L_z.

(a)

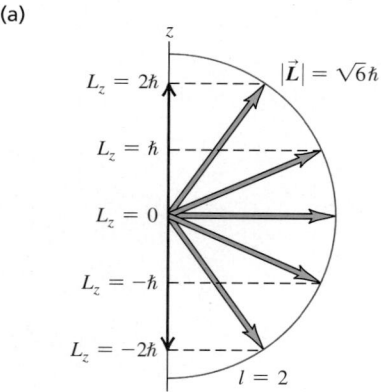

(b)

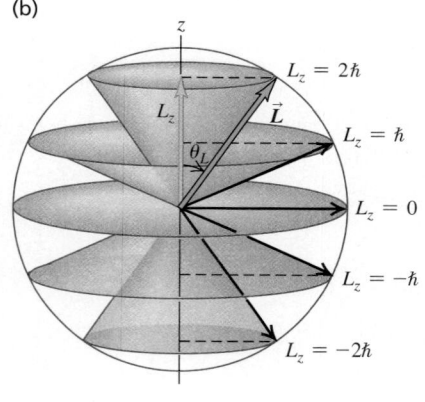

The quantum number m_l is the same as that in Eqs. (41.20b) and (41.20c). We see that m_l can be zero or a positive or negative integer up to, but no larger in magnitude than, l. That is, $|m_l| \leq l$. For example, if $l = 1$, m_l can equal 1, 0, or -1. For reasons that will emerge later, we call m_l the *orbital magnetic quantum number,* or **magnetic quantum number** for short.

The component L_z can never be quite as large as L (unless both are zero). For example, when $l = 2$, the largest possible value of m_l is also 2; then Eqs. (41.22) and (41.23) give

$$L = \sqrt{2(2+1)}\,\hbar = \sqrt{6}\hbar = 2.45\hbar$$

$$L_z = 2\hbar$$

Figure 41.6 shows the situation. The minimum value of the angle θ_L between the vector $\vec{L}$ and the z-axis is

$$\theta_L = \arccos \frac{L_z}{L}$$

$$= \arccos \frac{2}{2.45} = 35.3°$$

That $|L_z|$ is always less than L is also required by the uncertainty principle. Suppose we could know the precise *direction* of the orbital angular momentum vector. Then we could let that be the direction of the z-axis, and L_z would equal L. This corresponds to a particle moving in the xy-plane only, in which case the z-component of the linear momentum $\vec{p}$ would be zero with no uncertainty Δp_z. Then the uncertainty principle $\Delta z \Delta p_z \geq \hbar$ requires infinite uncertainty Δz in the coordinate z. This is impossible for a localized state; we conclude that we can't know the direction of $\vec{L}$ precisely. Thus, as we've already stated, the component of $\vec{L}$ in a given direction can never be quite as large as its magnitude L. Also, if we can't know the direction of $\vec{L}$ precisely, we can't determine the components L_x and L_y precisely. Thus we show *cones* of possible directions for $\vec{L}$ in Fig. 41.6b.

You may wonder why we have singled out the z-axis. We can't determine all three components of orbital angular momentum with certainty, so we arbitrarily pick one as the component we want to measure. When we discuss interactions of the atom with a magnetic field, we will consistently choose the positive z-axis to be in the direction of $\vec{B}$.

Quantum Number Notation

The wave functions for the hydrogen atom are determined by the values of three quantum numbers n, l, and m_l. (Compare this to the particle in a three-dimensional box that we considered in Section 41.2. There, too, three quantum numbers were needed to describe each stationary state.) The energy E_n is determined by the principal quantum number n according to Eq. (41.21). The magnitude of orbital angular momentum is determined by the orbital quantum number l, as in Eq. (41.22). The component of orbital angular momentum in a specified axis direction (customarily the z-axis) is determined by the magnetic quantum number m_l, as in Eq. (41.23). The energy does not depend on the values of l or m_l (**Fig. 41.7**), so for each energy level E_n given by Eq. (41.21), there is more than one distinct state having the same energy but different quantum numbers. That is, these states are *degenerate,* just like most of the states of a particle in a three-dimensional box. As for the three-dimensional box, degeneracy arises because the hydrogen atom is symmetric: If you rotate the atom through any angle, the potential-energy function at a distance r from the nucleus has the same value.

States with various values of the orbital quantum number l are often labeled with letters, according to the following scheme:

$l = 0$: s states $l = 3$: f states
$l = 1$: p states $l = 4$: g states
$l = 2$: d states $l = 5$: h states

and so on alphabetically. This seemingly irrational choice of the letters s, p, d, and f originated in the early days of spectroscopy and has no fundamental significance. In an important form of *spectroscopic notation* that we'll use often, a state with $n = 2$ and $l = 1$ is called a $2p$ state; a state with $n = 4$ and $l = 0$ is a $4s$ state; and so on. Only s states $(l = 0)$ are spherically symmetric.

Here's another bit of notation. The radial extent of the wave functions increases with the principal quantum number n, and we can speak of a region of space associated with a particular value of n as a **shell**. Especially in discussions of many-electron atoms, these shells are denoted by capital letters:

$n = 1$: K shell
$n = 2$: L shell
$n = 3$: M shell
$n = 4$: N shell

and so on alphabetically. For each n, different values of l correspond to different *subshells*. For example, the L shell $(n = 2)$ contains the $2s$ and $2p$ subshells.

Table 41.1 shows some of the possible combinations of the quantum numbers n, l, and m_l for hydrogen-atom wave functions. The spectroscopic notation and the shell notation for each are also shown.

41.7 The energy for an orbiting satellite such as the Hubble Space Telescope depends on the average distance between the satellite and the center of the earth. It does *not* depend on whether the orbit is circular (with a large orbital angular momentum L) or elliptical (in which case L is smaller). In the same way, the energy of a hydrogen atom does not depend on the orbital angular momentum.

| TABLE 41.1 | Quantum States of the Hydrogen Atom |

n	l	m_l	Spectroscopic Notation	Shell
1	0	0	$1s$	K
2	0	0	$2s$	
2	1	$-1, 0, 1$	$2p$	L
3	0	0	$3s$	
3	1	$-1, 0, 1$	$3p$	M
3	2	$-2, -1, 0, 1, 2$	$3d$	
4	0	0	$4s$	N

and so on

PROBLEM-SOLVING STRATEGY 41.1 ATOMIC STRUCTURE

IDENTIFY *the relevant concepts:* Many problems in atomic structure can be solved simply by reference to the quantum numbers n, l, and m_l that describe the total energy E, the magnitude of the orbital angular momentum $\vec{L}$, the z-component of $\vec{L}$, and other properties of an atom.

SET UP *the problem:* Identify the target variables and choose the appropriate equations, which may include Eqs. (41.21), (41.22), and (41.23).

EXECUTE *the solution* as follows:
1. Be sure you understand the possible values of the quantum numbers n, l, and m_l for the hydrogen atom. They are all

integers; n is always greater than zero, l can be zero or positive up to $n - 1$, and m_l can range from $-l$ to l. You should know how to count the number of (n, l, m_l) states in each shell (K, L, M, and so on) and subshell ($3s$, $3p$, $3d$, and so on). Be able to *construct* Table 41.1, not just to write it from memory.
2. Solve for the target variables.

EVALUATE *your answer:* It helps to be familiar with typical magnitudes in atomic physics. For example, the electric potential energy of a proton and electron 0.10 nm apart (typical of atomic dimensions) is about -15 eV. Visible light has wavelengths around 500 nm and frequencies around 5×10^{14} Hz. Problem-Solving Strategy 39.1 (Section 39.1) gives other typical magnitudes.

EXAMPLE 41.2 COUNTING HYDROGEN STATES

How many distinct (n, l, m_l) states of the hydrogen atom with $n = 3$ are there? What are their energies?

SOLUTION

IDENTIFY and SET UP: This problem uses the relationships among the principal quantum number n, orbital quantum number l, magnetic quantum number m_l, and energy of a state for the hydrogen atom. We use the rule that l can have n integer values, from 0 to $n - 1$, and that m_l can have $2l + 1$ values, from $-l$ to l. Equation (41.21) gives the energy of any particular state.

EXECUTE: When $n = 3$, l can be 0, 1, or 2. When $l = 0$, m_l can be only 0 (1 state). When $l = 1$, m_l can be -1, 0, or 1 (3 states). When $l = 2$, m_l can be -2, -1, 0, 1, or 2 (5 states). The total number

of (n, l, m_l) states with $n = 3$ is therefore $1 + 3 + 5 = 9$. (In Section 41.5 we'll find that the total number of $n = 3$ states is in fact twice this, or 18, because of electron spin.)

The energy of a hydrogen-atom state depends only on n, so all 9 of these states have the same energy. From Eq. (41.21),

$$E_3 = \frac{-13.60 \text{ eV}}{3^2} = -1.51 \text{ eV}$$

EVALUATE: For a given value of n, the total number of (n, l, m_l) states turns out to be n^2. In this case $n = 3$ and there are $3^2 = 9$ states. Remember that the ground level of hydrogen has $n = 1$ and $E_1 = -13.6$ eV; the $n = 3$ excited states have a higher (less negative) energy.

EXAMPLE 41.3 ANGULAR MOMENTUM IN AN EXCITED LEVEL OF HYDROGEN

Consider the $n = 4$ states of hydrogen. (a) What is the maximum magnitude L of the orbital angular momentum? (b) What is the maximum value of L_z? (c) What is the minimum angle between $\vec{L}$ and the z-axis? Give your answers to parts (a) and (b) in terms of $\hbar$.

SOLUTION

IDENTIFY and SET UP: We again need to relate the principal quantum number n and the orbital quantum number l for a hydrogen atom. We also need to relate the value of l and the magnitude and possible directions of the orbital angular momentum vector. We'll use Eq. (41.22) in part (a) to determine the maximum value of L; then we'll use Eq. (41.23) in part (b) to determine the maximum value of L_z. The angle between $\vec{L}$ and the z-axis is minimum when L_z is maximum (so that $\vec{L}$ is most nearly aligned with the positive z-axis).

EXECUTE: (a) When $n = 4$, the maximum value of the orbital quantum number l is $(n - 1) = (4 - 1) = 3$; from Eq. (41.22),

$$L_{\text{max}} = \sqrt{3(3 + 1)}\, \hbar = \sqrt{12}\, \hbar = 3.464\hbar$$

(b) For $l = 3$ the maximum value of m_l is 3. From Eq. (41.23),

$$(L_z)_{\text{max}} = 3\hbar$$

(c) The *minimum* allowed angle between $\vec{L}$ and the z-axis corresponds to the *maximum* allowed values of L_z and m_l (Fig. 41.6b shows an $l = 2$ example). For the state with $l = 3$ and $m_l = 3$,

$$\theta_{\text{min}} = \arccos \frac{(L_z)_{\text{max}}}{L} = \arccos \frac{3\hbar}{3.464\hbar} = 30.0°$$

EVALUATE: As a check, you can verify that θ is greater than 30.0° for all states with smaller values of l.

Electron Probability Distributions

Rather than picturing the electron as a point particle moving in a precise circle, the Schrödinger equation gives an electron *probability distribution* surrounding the nucleus. The hydrogen-atom probability distributions are three-dimensional, so they are harder to visualize than the two-dimensional circular orbits of the Bohr model. It's helpful to look at the *radial probability distribution function* $P(r)$—that is, the probability per radial length for the electron to be found at various distances from the proton. From Section 41.1 the probability for finding the electron in a small volume element dV is $|\psi|^2 \, dV$. (We assume that ψ is normalized in accordance with Eq. (41.6)—that is, that the integral of $|\psi|^2 \, dV$ over all space equals unity so that there is 100% probability of finding the electron somewhere in the universe.) Let's take as our volume element a thin spherical shell with inner radius r and outer radius $r + dr$. The volume dV of this shell is approximately its area $4\pi r^2$ multiplied by its thickness dr:

$$dV = 4\pi r^2 \, dr \tag{41.24}$$

We denote by $P(r)\,dr$ the probability of finding the particle within the radial range dr; then, using Eq. (41.24), we get

Probability that the electron is between r and $r + dr$	Radial probability distribution function	Probability distribution function

$$P(r)\,dr = |\psi|^2\,dV = |\psi|^2 4\pi r^2\,dr \qquad (41.25)$$

Wave function　Volume of spherical shell with inner radius r, outer radius $r + dr$

For wave functions that depend on θ and ϕ as well as r, we use the value of $|\psi|^2$ averaged over all angles in Eq. (41.25).

Figure 41.8 shows graphs of $P(r)$ for several hydrogen-atom wave functions. The r scales are labeled in multiples of a, the smallest distance between the electron and the nucleus in the Bohr model:

Radius of smallest orbit in Bohr model	Electric constant	Planck's constant	Planck's constant divided by 2π

$$a = \frac{\epsilon_0 h^2}{\pi m_r e^2} = \frac{4\pi\epsilon_0 \hbar^2}{m_r e^2} = 5.29 \times 10^{-11}\text{ m} \qquad (41.26)$$

Reduced mass　Magnitude of electron charge

As for a particle in a cubical box (see Section 41.2), there are some locations where the probability is zero. These surfaces are planes for a particle in a box; for a hydrogen atom these are spherical surfaces (that is, surfaces of constant r). Note that for the states that have the largest possible l for each n (such as $1s$, $2p$, $3d$, and $4f$ states), $P(r)$ has a single maximum at $n^2 a$. For these states, the electron is most likely to be found at the distance from the nucleus that is predicted by the Bohr model, $r = n^2 a$.

Figure 41.8 shows *radial* probability distribution functions $P(r) = 4\pi r^2 |\psi|^2$, which indicate the relative probability of finding the electron within a thin spherical shell of radius r. By contrast, **Figs. 41.9** and **41.10** (next page) show the *three-dimensional* probability distribution functions $|\psi|^2$, which indicate the relative probability of finding the electron within a small box at a given position. The darker the blue "cloud," the greater the value of $|\psi|^2$. (These are similar to the "clouds" shown in Fig. 41.2.) Figure 41.9 shows cross sections of the spherically symmetric probability clouds for the lowest three s subshells, for which $|\psi|^2$ depends only on the radial coordinate r. Figure 41.10 shows cross sections of the clouds for other electron states for which $|\psi|^2$ depends on both r and θ. For these states the probability distribution function is zero for certain values of θ as well as for certain values of r. In *any* stationary state of the hydrogen atom, $|\psi|^2$ is independent of ϕ.

41.8 Radial probability distribution functions $P(r)$ for several hydrogen-atom wave functions, plotted as functions of the ratio r/a [see Eq. (41.26)]. For each function, the number of maxima is $(n - l)$. The curves for which $l = n - 1$ ($1s$, $2p$, $3d$, ...) have only one maximum, located at $r = n^2 a$.

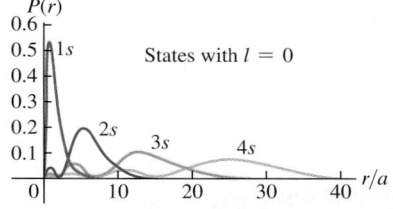

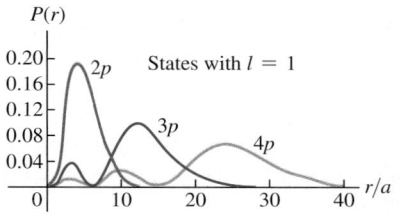

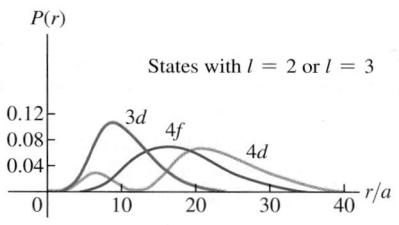

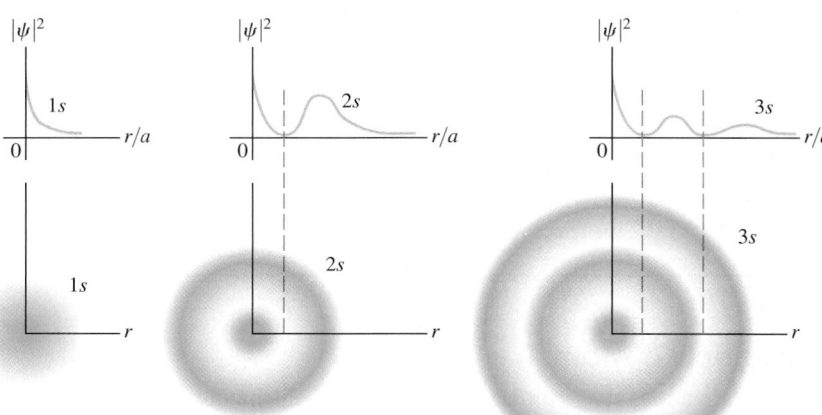

41.9 Three-dimensional probability distribution functions $|\psi|^2$ for the spherically symmetric $1s$, $2s$, and $3s$ hydrogen-atom wave functions.

41.10 Cross sections of three-dimensional probability distributions for a few quantum states of the hydrogen atom. They are not to the same scale. Mentally rotate each drawing about the z-axis to obtain the three-dimensional representation of $|\psi|^2$. For example, the $2p$, $m_l = \pm 1$ probability distribution looks like a fuzzy donut.

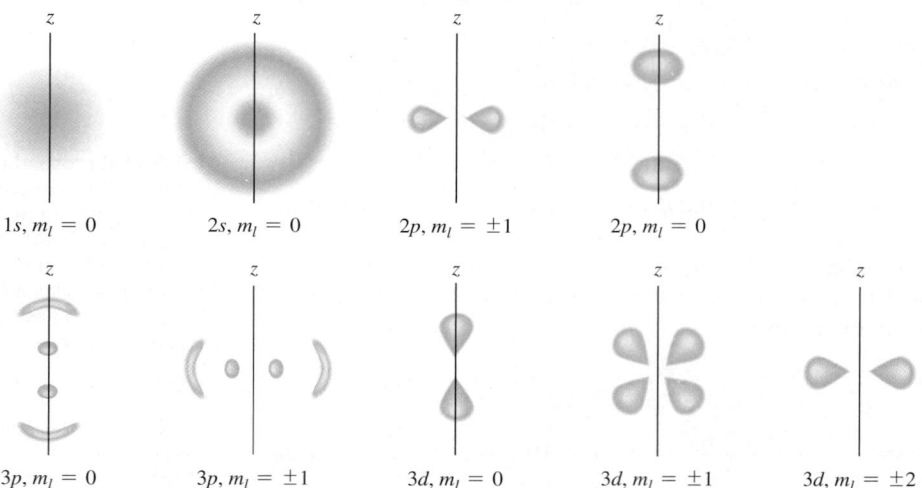

$1s$, $m_l = 0$ $2s$, $m_l = 0$ $2p$, $m_l = \pm 1$ $2p$, $m_l = 0$

$3p$, $m_l = 0$ $3p$, $m_l = \pm 1$ $3d$, $m_l = 0$ $3d$, $m_l = \pm 1$ $3d$, $m_l = \pm 2$

EXAMPLE 41.4 A HYDROGEN WAVE FUNCTION

The ground-state wave function for hydrogen (a $1s$ state) is

$$\psi_{1s}(r) = \frac{1}{\sqrt{\pi a^3}} e^{-r/a}$$

(a) Verify that this function is normalized. (b) What is the probability that the electron will be found at a distance less than a from the nucleus?

SOLUTION

IDENTIFY and SET UP: This example is similar to Example 41.1 in Section 41.2. We need to show that this wave function satisfies the condition that the probability of finding the electron *somewhere* is 1. We then need to find the probability that it will be found in the region $r < a$. In part (a) we'll carry out the integral $\int |\psi|^2 \, dV$ over all space; if it is equal to 1, the wave function is normalized. In part (b) we'll carry out the same integral over a spherical volume that extends from the origin (the nucleus) out to a distance a from the nucleus.

EXECUTE: (a) Since the wave function depends only on the radial coordinate r, we can choose our volume elements to be spherical shells of radius r, thickness dr, and volume dV given by Eq. (41.24). We then have

$$\int_{\text{all space}} |\psi_{1s}|^2 \, dV = \int_0^\infty \frac{1}{\pi a^3} e^{-2r/a} (4\pi r^2 \, dr)$$

$$= \frac{4}{a^3} \int_0^\infty r^2 e^{-2r/a} \, dr$$

You can find the following indefinite integral in a table of integrals or by integrating by parts:

$$\int r^2 e^{-2r/a} \, dr = \left(-\frac{ar^2}{2} - \frac{a^2 r}{2} - \frac{a^3}{4} \right) e^{-2r/a}$$

Evaluating this between the limits $r = 0$ and $r = \infty$ is simple; it is zero at $r = \infty$ because of the exponential factor, and at $r = 0$ only the last term in the parentheses survives. Thus the value of the definite integral is $a^3/4$. Putting it all together, we find

$$\int_0^\infty |\psi_{1s}|^2 \, dV = \frac{4}{a^3} \int_0^\infty r^2 e^{-2r/a} \, dr = \frac{4}{a^3} \frac{a^3}{4} = 1$$

The wave function *is* normalized.

(b) To find the probability P that the electron is found within $r < a$, we carry out the same integration but with the limits 0 and a. We'll leave the details to you. From the upper limit we get $-5e^{-2}a^3/4$; the final result is

$$P = \int_0^a |\psi_{1s}|^2 \, 4\pi r^2 \, dr = \frac{4}{a^3} \left(-\frac{5a^3 e^{-2}}{4} + \frac{a^3}{4} \right)$$

$$= (-5e^{-2} + 1) = 1 - 5e^{-2} = 0.323$$

EVALUATE: Our results tell us that in a ground state we expect to find the electron at a distance from the nucleus less than a about $\frac{1}{3}$ of the time and at a greater distance about $\frac{2}{3}$ of the time. It's hard to tell, but in Fig. 41.8, about $\frac{2}{3}$ of the area under the $1s$ curve is at distances greater than a (that is, $r/a > 1$).

Hydrogenlike Atoms

Two generalizations that we discussed with the Bohr model in Section 39.3 are equally valid in the Schrödinger analysis. First, if the "atom" is not composed of a single proton and a single electron, using the reduced mass m_r of the system in Eqs. (41.21) and (41.26) will lead to changes that are substantial for some exotic systems. One example is *positronium,* in which a positron and an electron orbit each other; another is a *muonic atom,* in which the electron is replaced by an unstable particle called a muon that has the same charge as an electron but is 207 times more massive. Second, our analysis is applicable to single-electron ions, such as He$^+$, Li^{2+}, and so on. For such ions we replace e^2 by Ze^2 in Eqs. (41.21) and (41.26), where Z is the number of protons (the **atomic number**).

TEST YOUR UNDERSTANDING OF SECTION 41.3   Rank the following states of the hydrogen atom in order from highest to lowest probability of finding the electron in the vicinity of $r = 5a$: (i) $n = 1, l = 0, m_l = 0$; (ii) $n = 2, l = 1, m_l = +1$; (iii) $n = 2, l = 1, m_l = 0$. ∎

41.4 THE ZEEMAN EFFECT

The **Zeeman effect** is the splitting of atomic energy levels and the associated spectral lines when the atoms are placed in a magnetic field (**Fig. 41.11**). This effect confirms experimentally the quantization of angular momentum. In this section we'll assume that the only angular momentum is the *orbital* angular momentum of a single electron and learn why we call m_l the magnetic quantum number.

Atoms contain charges in motion, so it should not be surprising that magnetic forces cause changes in that motion and in the energy levels. In 1896 the Dutch physicist Pieter Zeeman was the first to show that in the presence of a magnetic field, some spectral lines were split into groups of closely spaced lines (**Fig. 41.12**). This effect now bears his name.

Magnetic Moment of an Orbiting Electron

Let's begin our analysis of the Zeeman effect by reviewing the concept of *magnetic dipole moment* or *magnetic moment,* introduced in Section 27.7. A plane current loop with vector area $\vec{A}$ carrying current I has a magnetic moment $\vec{\mu}$ given by

$$\vec{\mu} = I\vec{A} \tag{41.27}$$

When a magnetic dipole of moment $\vec{\mu}$ is placed in a magnetic field $\vec{B}$, the field exerts a torque $\vec{\tau} = \vec{\mu} \times \vec{B}$ on the dipole. The potential energy U associated with this interaction is given by Eq. (27.27):

$$U = -\vec{\mu} \cdot \vec{B} \tag{41.28}$$

Now let's use Eqs. (41.27) and (41.28) and the Bohr model to look at the interaction of a hydrogen atom with a magnetic field. The orbiting electron with speed v is equivalent to a current loop with radius r and area πr^2. The average current I is the average charge per unit time that passes a given point of the orbit. This is equal to the charge magnitude e divided by the time T for one revolution, given by $T = 2\pi r/v$. Thus $I = ev/2\pi r$, and from Eq. (41.27) the magnitude μ of the magnetic moment is

$$\mu = IA = \frac{ev}{2\pi r}\pi r^2 = \frac{evr}{2} \tag{41.29}$$

We can also express this in terms of the magnitude L of the orbital angular momentum. From Eq. (10.28) the angular momentum of a particle in a circular orbit is $L = mvr$, so Eq. (41.29) becomes

$$\mu = \frac{e}{2m}L \tag{41.30}$$

The ratio of the magnitude of $\vec{\mu}$ to the magnitude of $\vec{L}$ is $\mu/L = e/2m$ and is called the *gyromagnetic ratio.*

41.11 Magnetic effects on the spectrum of sunlight. **(a)** The slit of a spectrograph is positioned along the black line crossing a portion of a sunspot. **(b)** The 0.4-T magnetic field in the sunspot (a thousand times greater than the earth's field) splits the middle spectral line into three lines.

(a)

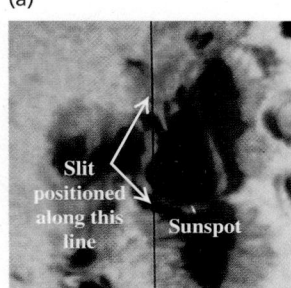

Slit
positioned
along this
line   Sunspot

(b)

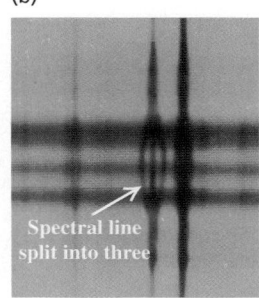

Spectral line
split into three

41.12 The normal Zeeman effect. Compare this to the magnetic splitting in the solar spectrum shown in Fig. 41.11b.

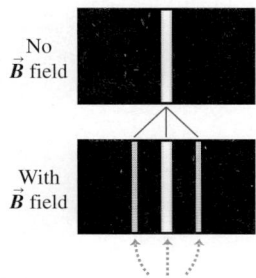

No
$\vec{B}$ field

With
$\vec{B}$ field

When an excited gas is placed in a magnetic field, the interaction of orbital magnetic moments with the field splits individual spectral lines of the gas into sets of three lines.

In the Bohr model, $L = nh/2\pi = n\hbar$, where $n = 1, 2, \ldots$. For an $n = 1$ state (a ground state), Eq. (41.30) becomes $\mu = (e/2m)\hbar$. This quantity is a natural unit for magnetic moment; it is called one **Bohr magneton,** denoted by μ_B:

$$\mu_B = \frac{e\hbar}{2m} \qquad \text{(definition of the Bohr magneton)} \qquad (41.31)$$

(We defined this quantity in Section 28.8.) Evaluating Eq. (41.31) gives

$$\mu_B = 5.788 \times 10^{-5} \text{ eV/T} = 9.274 \times 10^{-24} \text{ J/T or A} \cdot \text{m}^2$$

Note that the units J/T and A $\cdot$ m^2 are equivalent.

While the Bohr model suggests that the orbital motion of an atomic electron gives rise to a magnetic moment, this model does *not* give correct predictions about magnetic interactions. As an example, the Bohr model predicts that an electron in a hydrogen-atom ground state has an orbital magnetic moment of magnitude μ_B. But the Schrödinger picture tells us that such a ground-state electron is in an s state with zero angular momentum, so the orbital magnetic moment must be *zero!* To get the correct results, we must describe the states by using Schrödinger wave functions.

It turns out that in the Schrödinger formulation, electrons have the same ratio of μ to L (gyromagnetic ratio) as in the Bohr model—namely, $e/2m$. Suppose the magnetic field $\vec{B}$ is directed along the $+z$-axis. From Eq. (41.28) the interaction energy U of the atom's magnetic moment with the field is

$$U = -\mu_z B \qquad (41.32)$$

where μ_z is the z-component of the vector $\vec{\mu}$.

Now we use Eq. (41.30) to find μ_z, recalling that e is the *magnitude* of the electron charge and that the actual charge is $-e$. Because the electron charge is negative, the orbital angular momentum and magnetic moment vectors are opposite. We find

$$\mu_z = -\frac{e}{2m} L_z \qquad (41.33)$$

For the Schrödinger wave functions, $L_z = m_l \hbar$, with $m_l = 0, \pm 1, \pm 2, \ldots, \pm l$, so Eq. (41.33) becomes

$$\mu_z = -\frac{e}{2m} L_z = -m_l \frac{e\hbar}{2m} \qquad (41.34)$$

CAUTION Again, two uses of the symbol m As in Section 41.3, the symbol m is used in two ways in Eq. (41.34). Don't confuse the electron mass m with the magnetic quantum number m_l.

Finally, using Eq. (41.31) for the Bohr magneton, we can express the interaction energy from Eq. (41.32) as

Orbital magnetic interaction energy

Magnetic dipole component in direction of $\vec{B}$ — Magnitude of electron charge — Planck's constant divided by 2π

$$U = -\mu_z B = m_l \frac{e\hbar}{2m} B = m_l \mu_B B \qquad (41.35)$$

Magnetic-field magnitude — Electron mass — Bohr magneton — Orbital magnetic quantum number $= 0, \pm 1, \pm 2, \ldots, \pm l$

The magnetic field shifts the energy of each orbital state by an amount U. The interaction energy U depends on the value of m_l because m_l determines the orientation of the orbital magnetic moment relative to the magnetic field. This dependence is the reason m_l is called the magnetic quantum number.

The values of m_l range from $-l$ to $+l$ in steps of one, so an energy level with a particular value of the orbital quantum number l contains $(2l + 1)$ different orbital states. Without a magnetic field these states all have the same energy; that is, they are degenerate. The magnetic field removes this degeneracy. In the presence of a magnetic field they are split into $2l + 1$ distinct energy levels;

m_l m_l m_l

——————————————————————— $E = 0$
$n = 4$ ——— 0- - - - - - - - - - - - - - 2 - - - - - - -0.85 eV
$n = 3$ ——— 0- - - - 0 $\begin{smallmatrix}1\\-1\end{smallmatrix}$ - - - 0 $\begin{smallmatrix}1\\-1\\-2\end{smallmatrix}$ - - - - - - -1.51 eV

$l = 2$

$n = 2$ ——— 0- - - - 0 $\begin{smallmatrix}1\\-1\end{smallmatrix}$ - - - - - - - - - - - - - -3.40 eV
$l = 1$

$n = 1$ ——— 0- -13.60 eV
$l = 0$

41.13 This energy-level diagram for hydrogen shows how the levels are split when the electron's orbital magnetic moment interacts with an external magnetic field. The values of m_l are shown adjacent to the various levels. The relative magnitudes of the level splittings are exaggerated for clarity. The $n = 4$ splittings are not shown; can you draw them in?

adjacent levels differ in energy by $(e\hbar/2m)B = \mu_B B$. We can understand this in terms of the connection between degeneracy and symmetry. With a magnetic field applied along the z-axis, the atom is no longer completely symmetric under rotation: There is a preferred direction in space. By removing the symmetry, we remove the degeneracy of states.

Figure 41.13 shows the effect on the energy levels of hydrogen. Spectral lines corresponding to transitions from one set of levels to another set are correspondingly split and appear as a series of three closely spaced spectral lines replacing a single line. As the following example shows, the splitting of spectral lines is quite small because the value of $\mu_B B$ is small even for substantial magnetic fields.

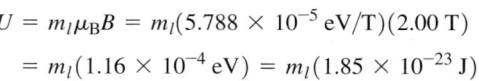

SOLUTION

EXAMPLE 41.5 AN ATOM IN A MAGNETIC FIELD

An atom in a state with $l = 1$ emits a photon with wavelength 600.000 nm as it decays to a state with $l = 0$. If the atom is placed in a magnetic field with magnitude $B = 2.00$ T, what are the shifts in the energy levels and in the wavelength that result from the interaction between the atom's orbital magnetic moment and the magnetic field?

SOLUTION

IDENTIFY and SET UP: This problem concerns the splitting of atomic energy levels by a magnetic field (the Zeeman effect). We use Eq. (41.35) to determine the energy-level shifts. The relationship $E = hc/\lambda$ between the energy and wavelength of a photon then lets us calculate the wavelengths emitted during transitions from the $l = 1$ states to the $l = 0$ state.

EXECUTE: The energy of a 600-nm photon is

$$E = \frac{hc}{\lambda} = \frac{(4.14 \times 10^{-15} \text{ eV} \cdot \text{s})(3.00 \times 10^8 \text{ m/s})}{600 \times 10^{-9} \text{ m}}$$

$$= 2.07 \text{ eV}$$

If there is no external magnetic field, that is the difference in energy between the $l = 0$ and $l = 1$ levels.

With a 2.00-T field present, Eq. (41.35) shows that there is no shift of the $l = 0$ state (which has $m_l = 0$). For the $l = 1$ states, the splitting of levels is given by

$$U = m_l \mu_B B = m_l (5.788 \times 10^{-5} \text{ eV/T})(2.00 \text{ T})$$

$$= m_l (1.16 \times 10^{-4} \text{ eV}) = m_l (1.85 \times 10^{-23} \text{ J})$$

The possible values of m_l for $l = 1$ are -1, 0, and $+1$, and the three corresponding levels are separated by equal intervals of 1.16×10^{-4} eV. This is a small fraction of the 2.07-eV photon energy:

$$\frac{\Delta E}{E} = \frac{1.16 \times 10^{-4} \text{ eV}}{2.07 \text{ eV}} = 5.60 \times 10^{-5}$$

The corresponding *wavelength* shifts are about $(5.60 \times 10^{-5}) \times (600 \text{ nm}) = 0.034$ nm. The original 600.000-nm line is split into a triplet with wavelengths 599.966, 600.000, and 600.034 nm.

EVALUATE: Even though 2.00 T would be a strong field in most laboratories, the wavelength splittings are extremely small. Nonetheless, modern spectrographs have more than enough chromatic resolving power to measure these splittings (see Section 36.5).

41.14 This figure shows how the splitting of the energy levels of a d state ($l = 2$) depends on the magnitude B of an external magnetic field, assuming only an orbital magnetic moment.

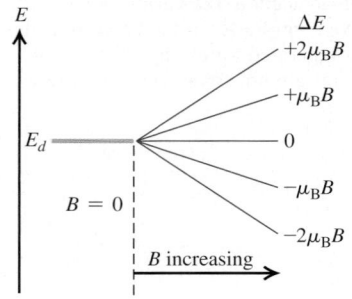

41.15 The cause of the normal Zeeman effect. The magnetic field splits the levels, but selection rules allow transitions with only three different energy changes, giving three different photon frequencies and wavelengths.

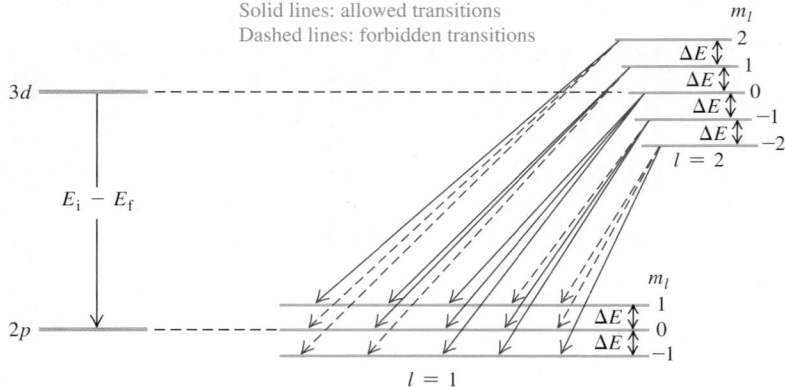

Selection Rules

Figure 41.14 shows what happens to a set of d states ($l = 2$) as the magnetic field increases. With zero field the five states $m_l = -2$, -1, 0, 1, and 2 are degenerate (have the same energy), but the applied field spreads the states out. **Figure 41.15** shows the splittings of both the $3d$ and $2p$ states. Equal energy differences $(e\hbar/2m)B = \mu_B B$ separate adjacent levels. In the absence of a magnetic field, a transition from a $3d$ to a $2p$ state would yield a single spectral line with photon energy $E_i - E_f$. With the levels split as shown, it might seem that there are five possible photon energies.

In fact, there are only three possibilities. Not all combinations of initial and final levels are possible because of a restriction associated with conservation of angular momentum. The photon ordinarily carries off one unit ($\hbar$) of angular momentum, which leads to the requirements that in a transition l must change by 1 and m_l must change by 0 or ± 1. These requirements are called **selection rules.** Transitions that obey these rules are called *allowed transitions*; those that don't are *forbidden transitions*. In Fig. 41.15 we show the allowed transitions by solid arrows. You should count the possible transition energies to convince yourself that the nine solid arrows give only three possible energies; the zero-field value $E_i - E_f$, and that value plus or minus $\Delta E = (e\hbar/2m)B = \mu_B B$. Figure 41.12 shows the corresponding spectral lines.

What we have described is called the *normal* Zeeman effect. It is based entirely on the orbital angular momentum of the electron. However, it leaves out a very important consideration: the electron *spin* angular momentum, the subject of the next section.

TEST YOUR UNDERSTANDING OF SECTION 41.4 In this section we assumed that the magnetic field points in the positive z-direction. Would the results be different if the magnetic field pointed in the positive x-direction? ❚

41.5 ELECTRON SPIN

Despite the success of the Schrödinger equation in predicting the energy levels of the hydrogen atom, experimental observations indicate that it doesn't tell the whole story of the behavior of electrons in atoms. First, spectroscopists have found magnetic-field splitting into other than the three equally spaced lines that we explained in Section 41.4 (see Fig. 41.12). Before this effect was understood, it was called the *anomalous* Zeeman effect to distinguish it from the "normal"

41.16 Illustrations of the normal and anomalous Zeeman effects for two elements, zinc and sodium. The brackets under each illustration show the "normal" splitting predicted by ignoring the effect of electron spin.

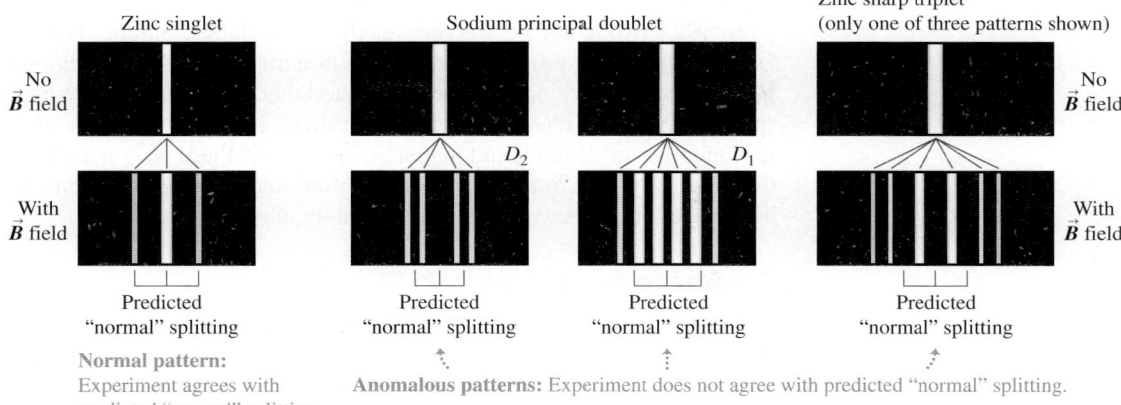

effect discussed in the preceding section. **Figure 41.16** shows both kinds of splittings.

Second, some energy levels show splittings that resemble the Zeeman effect even when there is *no* external magnetic field. For example, when the lines in the hydrogen spectrum are examined with a high-resolution spectrograph, some lines are found to consist of sets of closely spaced lines called *multiplets*. Similarly, the orange-yellow line of sodium, corresponding to the transition $4p \rightarrow 3s$ of the outer electron, is found to be a doublet ($\lambda = 589.0, 589.6$ nm), suggesting that the $4p$ level might in fact be two closely spaced levels. The Schrödinger equation in its original form didn't predict any of this.

The Stern–Gerlach Experiment

Similar anomalies appeared in 1922 in atomic-beam experiments performed in Germany by Otto Stern and Walter Gerlach. When they passed a beam of neutral atoms through a nonuniform magnetic field (**Fig. 41.17**), atoms were deflected according to the orientation of their magnetic moments with respect to the field. These experiments demonstrated the quantization of angular momentum in a very direct way. If there were only orbital angular momentum, the deflections would split the beam into an odd number ($2l + 1$) of different components. However, some atomic beams were split into an *even* number of components. If we use a different symbol j for an angular momentum quantum number,

PhET: Stern–Gerlach Experiment

41.17 The Stern–Gerlach experiment.

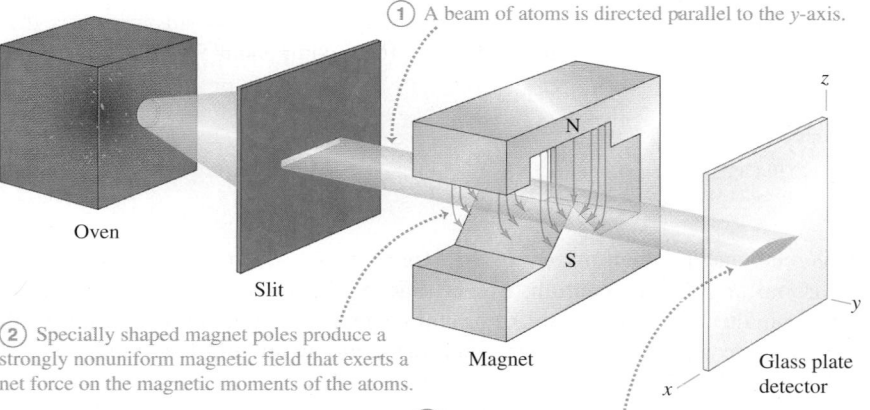

① A beam of atoms is directed parallel to the y-axis.

Oven

Slit

② Specially shaped magnet poles produce a strongly nonuniform magnetic field that exerts a net force on the magnetic moments of the atoms.

Magnet

③ Each atom is deflected upward or downward according to the orientation of its magnetic moment.

Glass plate detector

setting $2j + 1$ equal to an even number gives $j = \frac{1}{2}, \frac{3}{2}, \frac{5}{2}, \ldots$, suggesting a half-integer angular momentum. This can't be understood on the basis of the Bohr model and similar pictures of atomic structure.

In 1925 two graduate students in the Netherlands, Samuel Goudsmidt and George Uhlenbeck, proposed that the electron might have some additional motion. Using a semiclassical model, they suggested that the electron might behave like a spinning sphere of charge instead of a particle. If so, it would have an additional *spin* angular momentum and magnetic moment. If these were quantized in much the same way as *orbital* angular momentum and magnetic moment, they might help explain the observed energy-level anomalies.

An Analogy for Electron Spin

To introduce the concept of **electron spin,** let's start with an analogy. The earth travels in a nearly circular orbit around the sun, and at the same time it *rotates* on its axis. Each motion has its associated angular momentum, which we call the *orbital* and *spin* angular momentum, respectively. The total angular momentum of the earth is the vector sum of the two. If we were to model the earth as a single point, it would have no moment of inertia about its spin axis and thus no spin angular momentum. But when our model includes the finite size of the earth, spin angular momentum becomes possible.

In the Bohr model, suppose the electron is not just a point charge but a small spinning sphere that orbits the nucleus. Then the electron has not only orbital angular momentum but also spin angular momentum associated with the rotation of its mass about its axis. The sphere carries an electric charge, so the spinning motion leads to current loops and to a magnetic moment, as we discussed in Section 27.7. In a magnetic field, the *spin* magnetic moment has an interaction energy in addition to that of the *orbital* magnetic moment (the normal Zeeman-effect interaction that we discussed in Section 41.4). We should see additional Zeeman shifts due to the spin magnetic moment.

As we mentioned, such shifts *are* indeed observed in precise spectroscopic analysis. This and a variety of other experimental evidence have shown conclusively that the electron *does* have a spin angular momentum and a spin magnetic moment that do not depend on its orbital motion but are intrinsic to the electron itself. The origin of this spin angular momentum is fundamentally quantum-mechanical, so it's not correct to model the electron as a spinning charged sphere. But just as the Bohr model can be a useful conceptual picture for the motion of an electron in an atom, the spinning-sphere analogy can help you visualize the intrinsic spin angular momentum of an electron.

Spin Quantum Numbers

Like orbital angular momentum, the spin angular momentum of an electron (denoted by $\vec{S}$) is found to be quantized. Suppose we have an apparatus that measures a particular component of $\vec{S}$, say the z-component S_z. We find that the only possible values are

$$
\underset{\substack{\text{z-component of}\\ \text{spin angular momentum}\\ \text{of electron}}}{} S_z = m_s \hbar \qquad (41.36)
$$

Spin magnetic quantum number $= \pm\frac{1}{2}$ ······· Planck's constant divided by 2π

This relationship is reminiscent of the expression $L_z = m_l \hbar$ for the z-component of orbital angular momentum, except that $|S_z|$ is *one-half* of $\hbar$ instead of an *integer* multiple. In analogy to the orbital magnetic quantum number m_l, we call the quantum number m_s the **spin magnetic quantum number.** Since m_s has only two possible values, $+\frac{1}{2}$ and $-\frac{1}{2}$, it follows that the spin angular momentum vector $\vec{S}$ can have only two orientations in space relative to the z-axis: "*spin up*" with a z-component of $+\frac{1}{2}\hbar$ and "*spin down*" with a z-component of $-\frac{1}{2}\hbar$.

Equation (41.36) also suggests that the magnitude S of the spin angular momentum is given by an expression analogous to Eq. (41.22) with the orbital quantum number l replaced by the **spin quantum number** $s = \frac{1}{2}$:

> Magnitude of spin angular momentum of electron ⋯⋯➤ $S = \sqrt{\frac{1}{2}\left(\frac{1}{2} + 1\right)}\hbar = \sqrt{\frac{3}{4}}\hbar$ (41.37)
>
> Maximum value of spin magnetic quantum number $= \frac{1}{2}$
>
> Planck's constant divided by 2π

The electron is often called a "spin-one-half particle" or "spin-$\frac{1}{2}$ particle."

We see that to label the state of the electron in a hydrogen atom completely, we need *four* quantum numbers: n, l, and m_l (described in Section 41.3) to specify the electron's motion relative to the nucleus, plus the spin magnetic quantum number m_s to specify the electron spin orientation.

To visualize the quantized spin of an electron in a hydrogen atom, think of the electron probability distribution function $|\psi|^2$ as a cloud surrounding the nucleus like those shown in Figs. 41.9 and 41.10. Then imagine many tiny spin arrows distributed throughout the cloud, either all with components in the $+z$-direction or all with components in the $-z$-direction. But don't take this picture too seriously.

Just as the orbital magnetic moment of the electron is proportional to its orbital angular momentum $\vec{L}$ (see Section 41.4), the electron's spin magnetic moment is proportional to its spin angular momentum $\vec{S}$. The z-component of the spin magnetic moment (μ_z) turns out to be related to S_z by

$$\mu_z = -(2.00232)\frac{e}{2m}S_z \qquad (41.38)$$

where $-e$ and m are (as usual) the charge and mass of the electron. When the atom is placed in a magnetic field, the interaction energy $-\vec{\mu} \cdot \vec{B}$ of the spin magnetic dipole moment with the field causes further splittings in energy levels and in the corresponding spectral lines.

Equation (41.38) shows that the gyromagnetic ratio for electron spin is approximately *twice* as great as the value $e/2m$ for *orbital* angular momentum and magnetic dipole moment. This result has no classical analog. But in 1928 Paul Dirac developed a relativistic generalization of the Schrödinger equation for electrons. His equation gave a spin gyromagnetic ratio of exactly $2(e/2m)$. It took another two decades to develop the area of physics called *quantum electrodynamics*, abbreviated QED, which predicts the value we've given to "only" six significant figures as 2.00232. QED now predicts a value that agrees with the currently accepted experimental value of 2.00231930436153(53), making QED the most precise theory in all science.

BIO Application Electron Spins and Dating Human Origins In many atoms, the net spin of all of the electrons is zero (as many electrons are "spin up" as are "spin down"). If these atoms are ionized and lose an electron, however, the net spin of the ion that remains is nonzero. This happens naturally in tooth enamel, where ionization is caused by radioactivity in the environment. The longer a tooth is exposed, the more ions are present. To find the age of fossil teeth, such as those in this skull of *Homo neanderthalensis*, a sample of the enamel is placed in a strong magnetic field. The ion spins align opposite to this field (become "spin down"). The sample is then illuminated with microwave photons of just the right energy to flip the spins to the higher-energy configuration aligned with the field ("spin up"). The amount of microwave energy absorbed in this process (called *electron spin resonance*) indicates the number of ions present and hence the age of the enamel.

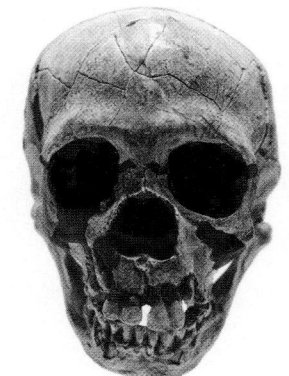

EXAMPLE 41.6 | **ENERGY OF ELECTRON SPIN IN A MAGNETIC FIELD**

Calculate the interaction energy for an electron in an $l = 0$ state in a magnetic field with magnitude 2.00 T.

SOLUTION

IDENTIFY and SET UP: For $l = 0$ the electron has zero orbital angular momentum and zero orbital magnetic moment. Hence the only magnetic interaction is that between the $\vec{B}$ field and the spin magnetic moment $\vec{\mu}$. From Eq. (41.28), the interaction energy is $U = -\vec{\mu} \cdot \vec{B}$. As in Section 41.4, we take $\vec{B}$ to be in the positive z-direction so that $U = -\mu_z B$ [Eq. (41.32)]. Equation (41.38) gives μ_z in terms of S_z, and Eq. (41.36) gives S_z.

EXECUTE: Combining Eqs. (41.36) and (41.38), we have

$$\mu_z = -(2.00232)\left(\frac{e}{2m}\right)\left(\pm\frac{1}{2}\hbar\right)$$

$$= \mp\frac{1}{2}(2.00232)\left(\frac{e\hbar}{2m}\right) = \mp(1.00116)\mu_B$$

$$= \mp(1.00116)(9.274 \times 10^{-24}\text{ J/T})$$

$$= \mp 9.285 \times 10^{-24}\text{ J/T} = \mp 5.795 \times 10^{-5}\text{ eV/T}$$

Continued

Then from Eq. (41.32),

$$U = -\mu_z B = \pm(9.285 \times 10^{-24} \text{ J/T})(2.00 \text{ T})$$

$$= \pm 1.86 \times 10^{-23} \text{ J} = \pm 1.16 \times 10^{-4} \text{ eV}$$

The positive value of U and the negative value of μ_z correspond to $S_z = +\frac{1}{2}\hbar$ (spin up); the negative value of U and the positive value of μ_z correspond to $S_z = -\frac{1}{2}\hbar$ (spin down).

EVALUATE: Let's check the *signs* of our results. If the electron is spin down, $\vec{S}$ points generally opposite to $\vec{B}$. Then the magnetic moment $\vec{\mu}$ (which is opposite to $\vec{S}$ because the electron charge is negative) points generally parallel to $\vec{B}$, and μ_z is positive. From Eq. (41.28), $U = -\vec{\mu} \cdot \vec{B}$, the interaction energy is negative if $\vec{\mu}$ and $\vec{B}$ are parallel. Our results show that U is indeed negative in this case. We can similarly confirm that U must be positive and μ_z negative for a spin-up electron.

The red lines in **Fig. 41.18** show how the interaction energies for the two spin states vary with the magnetic-field magnitude B. The graphs are straight lines because, from Eq. (41.32), U is proportional to B.

41.18 An $l = 0$ level of a single electron is split by interaction of the spin magnetic moment with an external magnetic field. The greater the magnitude B of the magnetic field, the greater the splitting. The quantity 5.795×10^{-5} eV/T is just $(1.00116)\mu_B$.

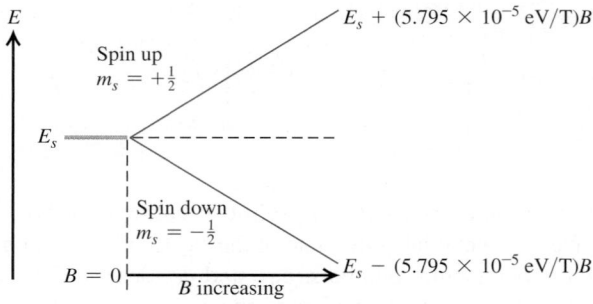

Spin-Orbit Coupling

We mentioned earlier that the spin magnetic dipole moment also gives splitting of energy levels even when there is *no* external field. One cause involves the orbital motion of the electron. In the Bohr model, observers moving with the electron would see the positively charged nucleus revolving around them (just as to earthbound observers the sun seems to be orbiting the earth). This apparent motion of charge causes a magnetic field at the location of the electron, as measured in the electron's moving frame of reference. The resulting interaction with the spin magnetic moment causes a twofold splitting of this level, corresponding to the two possible orientations of electron spin.

Discussions based on the Bohr model can't be taken too seriously, but a similar result can be derived from the Schrödinger equation. The interaction energy U can be expressed in terms of the scalar product of the angular momentum vectors $\vec{L}$ and $\vec{S}$. This effect is called **spin-orbit coupling;** it is responsible for the small energy difference between the two closely spaced, lowest excited levels of sodium shown in Fig. 39.19a and for the corresponding doublet (589.0, 589.6 nm) in the spectrum of sodium.

EXAMPLE 41.7 AN EFFECTIVE MAGNETIC FIELD

To six significant figures, the wavelengths of the two spectral lines that make up the sodium doublet are $\lambda_1 = 588.995$ nm and $\lambda_2 = 589.592$ nm. Calculate the effective magnetic field experienced by the electron in the $3p$ levels of the sodium atom.

SOLUTION

IDENTIFY and SET UP: The two lines in the sodium doublet result from transitions from the two $3p$ levels, which are split by spin-orbit coupling, to the $3s$ level, which is *not* split because it has $L = 0$. We picture the spin-orbit coupling as an interaction between the electron spin magnetic moment and an effective magnetic field due to the nucleus. This example is like Example 41.6 in reverse: There we were given B and found the difference between the energies of the two spin states, while here we use the energy difference to find the target variable B. The difference in energy

between the two $3p$ levels is equal to the difference in energy between the two photons of the sodium doublet. We use this relationship and the results of Example 41.6 to determine B.

EXECUTE: The energies of the two photons are $E_1 = hc/\lambda_1$ and $E_2 = hc/\lambda_2$. Here $E_1 > E_2$ because $\lambda_1 < \lambda_2$, so the difference in their energies is

$$\Delta E = \frac{hc}{\lambda_1} - \frac{hc}{\lambda_2} = hc\left(\frac{\lambda_2 - \lambda_1}{\lambda_2 \lambda_1}\right)$$

$$= (4.136 \times 10^{-15} \text{ eV} \cdot \text{s})(2.998 \times 10^8 \text{ m/s})$$

$$\times \frac{(589.592 \times 10^{-9} \text{ m}) - (588.995 \times 10^{-9} \text{ m})}{(589.592 \times 10^{-9} \text{ m})(588.995 \times 10^{-9} \text{ m})}$$

$$= 0.00213 \text{ eV} = 3.41 \times 10^{-22} \text{ J}$$

This equals the energy difference between the two $3p$ levels. The spin-orbit interaction raises one level by 1.70×10^{-22} J (one-half of 3.41×10^{-22} J) and lowers the other by 1.70×10^{-22} J. From Example 41.6, the amount each state is raised or lowered is $|U| = (1.00116)\mu_{\mathrm{B}}B$, so

$$B = \left| \frac{U}{(1.00116)\mu_{\mathrm{B}}} \right| = \frac{1.70 \times 10^{-22}\,\mathrm{J}}{9.28 \times 10^{-24}\,\mathrm{J/T}} = 18.0\,\mathrm{T}$$

EVALUATE: The electron experiences a *very* strong effective magnetic field. To produce a steady, macroscopic field of this magnitude in the laboratory requires state-of-the-art electromagnets.

Combining Orbital and Spin Angular Momenta

The orbital and spin angular momenta ($\vec{L}$ and $\vec{S}$, respectively) can combine in various ways. The vector sum of $\vec{L}$ and $\vec{S}$ is the *total* angular momentum $\vec{J}$:

$$\vec{J} = \vec{L} + \vec{S} \tag{41.39}$$

The possible values of the magnitude J are given in terms of a quantum number j, called the **total angular momentum quantum number:**

$$J = \sqrt{j(j+1)}\,\hbar \tag{41.40}$$

We can then have states in which $j = |l \pm \frac{1}{2}|$. The $l + \frac{1}{2}$ states correspond to the case in which the vectors $\vec{L}$ and $\vec{S}$ have parallel z-components; for the $l - \frac{1}{2}$ states, $\vec{L}$ and $\vec{S}$ have antiparallel z-components. For example, when $l = 1$, j can be $\frac{1}{2}$ or $\frac{3}{2}$. In another spectroscopic notation these p states are labeled $^2P_{1/2}$ and $^2P_{3/2}$, respectively. The superscript is the number of possible spin orientations, the letter P (now capitalized) indicates states with $l = 1$, and the subscript is the value of j. We used this scheme to label the energy levels of the sodium atom in Fig. 39.19a.

In addition to shifts in energy levels due to magnetic effects within the atom, there are shifts of the same magnitude due to relativistic corrections to the kinetic energy of the electron. (In the Bohr model, an electron in the $n = 1$ orbit of hydrogen moves at about 1% of the speed of light.) The term "fine structure" refers to the energy-level shifts caused by magnetic and relativistic effects together, as well as to the line splittings that result from these shifts. Including these effects, the energy levels of the hydrogen atom are

Energy levels of hydrogen, including fine structure

$$E_{n,j} = -\frac{13.60\,\mathrm{eV}}{n^2}\left[1 + \frac{\alpha^2}{n^2}\left(\frac{n}{j+\frac{1}{2}} - \frac{3}{4}\right)\right] \tag{41.41}$$

Fine-structure constant

Principal quantum number ($n = 1, 2, 3, \dots$)

Total angular momentum quantum number

The *fine-structure constant* α that appears in Eq. (41.41) is a dimensionless number:

$$\alpha = \frac{1}{4\pi\epsilon_0}\frac{e^2}{\hbar c} = 7.2973525698(24) \times 10^{-3} \quad \text{(fine-structure constant)} \tag{41.42}$$

To five significant figures, $\alpha = 7.2974 \times 10^{-3} = 1/137.04$.

In Section 41.3 we found that the energy levels of the hydrogen atom are degenerate: All states that have the same principal quantum number n have the same energy. Our more complete treatment including fine structure shows that this degeneracy is removed: States with the same n but different values of the total angular momentum quantum number j have different energies. Example 41.8 illustrates this for the $n = 2$ levels of hydrogen.

EXAMPLE 41.8 FINE STRUCTURE AND SPECTRAL-LINE SPLITTING

For an electron with orbital quantum number $l = 0$, the angular momentum is due to spin alone and the only possible value of the total angular momentum quantum number is $j = \frac{1}{2}$. If $l = 1$, two values are possible: $j = \frac{3}{2}$ (the spin and orbital angular momentum vectors are in roughly the same direction and so add together) and $j = \frac{1}{2}$ (the spin and orbital angular momentum vectors are in roughly opposite directions and so partially cancel). (a) Find the energies of a state of the electron in a hydrogen atom with $n = 2$, $l = 1$, $j = \frac{3}{2}$ (a $^2P_{3/2}$ state) and a state with $n = 2$, $l = 1$, $j = \frac{1}{2}$ (a $^2P_{1/2}$ state), and calculate the difference between the two energies. Which state has the higher energy? (b) Find the difference in wavelengths between (i) a photon emitted in a transition from a state with $n = 2$, $l = 1$, $j = \frac{3}{2}$ to a state with $n = 1$, $l = 0$, $j = \frac{1}{2}$ and (ii) a photon emitted in a transition from a state with $n = 2$, $l = 1$, $j = \frac{1}{2}$ to a state with $n = 1$, $l = 0$, $j = \frac{1}{2}$. Which photon has the longer wavelength?

SOLUTION

IDENTIFY and SET UP: In part (a) we use Eq. (41.41) to find the difference in energy between these two states, which have the same n value but different j values. The difference between the two energies is due to fine structure, so we expect this difference to be small. In part (b) both transitions end in the same state with $n = 1$, so we recognize from Section 39.3 that both are members of the Lyman series. If there were no fine structure, the two initial states would have the same energies and both photons would have the same energy E and hence the same wavelength $\lambda = hc/E$. But because the two initial states differ slightly in energy, the photons in the two transitions will have slightly different wavelengths.

EXECUTE: (a) From Eq. (41.41), the energies of the two states are

$$E_{n=2,\,j=3/2} = -\frac{13.60\text{ eV}}{2^2}\left[1 + \frac{\alpha^2}{2^2}\left(\frac{2}{\frac{3}{2} + \frac{1}{2}} - \frac{3}{4}\right)\right]$$

$$= -3.40\text{ eV}\left(1 + \frac{\alpha^2}{16}\right)$$

$$E_{n=2,\,j=1/2} = -\frac{13.60\text{ eV}}{2^2}\left[1 + \frac{\alpha^2}{2^2}\left(\frac{2}{\frac{1}{2} + \frac{1}{2}} - \frac{3}{4}\right)\right]$$

$$= -3.40\text{ eV}\left(1 + \frac{5\alpha^2}{16}\right)$$

The fine-structure terms involving α^2 cause both states to have lower (more negative) energies than in the Bohr model, in which both states would have energy $E_2 = -3.40$ eV. The fine-structure term is five times greater for the $j = \frac{1}{2}$ state, so the $j = \frac{3}{2}$ state has the higher (less negative) energy. Using the value of the fine-structure constant α from Eq. (41.42), we get the difference in energy between the two states:

$$E_{n=2,\,j=3/2} - E_{n=2,\,j=1/2}$$

$$= \left[-3.40\text{ eV}\left(1 + \frac{\alpha^2}{16}\right)\right] - \left[-3.40\text{ eV}\left(1 + \frac{5\alpha^2}{16}\right)\right]$$

$$= 3.40\text{ eV}\left(\frac{4\alpha^2}{16}\right) = (3.40\text{ eV})\left(\frac{4}{16}\right)\left(\frac{1}{137.04}\right)^2$$

$$= 4.53 \times 10^{-5}\text{ eV}$$

(b) The photon energy in each case equals the difference between the energies of the initial and final states of the electron. The final electron state for both transitions has $n = 1$ and $j = \frac{1}{2}$, which from Eq. (41.41) has energy

$$E_{n=1,\,j=1/2} = -\frac{13.60\text{ eV}}{1^2}\left[1 + \frac{\alpha^2}{1^2}\left(\frac{1}{\frac{1}{2} + \frac{1}{2}} - \frac{3}{4}\right)\right]$$

$$= -13.60\text{ eV}\left(1 + \frac{\alpha^2}{4}\right)$$

Note that as for the two $n = 2$ states in part (a), the fine-structure correction to the $n = 1$ state makes the energy more negative. The photon energies for the two transitions are then

$$E_{\text{photon}}\left(n = 2, l = 1, j = \tfrac{3}{2} \text{ to } n = 1, l = 0, j = \tfrac{1}{2}\right)$$

$$= E_{n=2,\,j=3/2} - E_{n=1,\,j=1/2}$$

$$= \left[-3.40\text{ eV}\left(1 + \frac{\alpha^2}{16}\right)\right] - \left[-13.60\text{ eV}\left(1 + \frac{\alpha^2}{4}\right)\right]$$

$$= 10.20\text{ eV} + (3.40\text{ eV})\left(\frac{15\alpha^2}{16}\right)$$

$$= 10.20\text{ eV} + 1.70 \times 10^{-4}\text{ eV}$$

$$E_{\text{photon}}\left(n = 2, l = 1, j = \tfrac{1}{2} \text{ to } n = 1, l = 0, j = \tfrac{1}{2}\right)$$

$$= E_{n=2,\,j=1/2} - E_{n=1,\,j=1/2}$$

$$= \left[-3.40\text{ eV}\left(1 + \frac{5\alpha^2}{16}\right)\right] - \left[-13.60\text{ eV}\left(1 + \frac{\alpha^2}{4}\right)\right]$$

$$= 10.20\text{ eV} + (3.40\text{ eV})\left(\frac{11\alpha^2}{16}\right)$$

$$= 10.20\text{ eV} + 1.24 \times 10^{-4}\text{ eV}$$

The photon emitted when the initial state is $n = 2$, $l = 1$, $j = \frac{1}{2}$ has a lower energy E_{photon} and hence will have a longer wavelength, as given by the equation $\lambda = hc/E_{\text{photon}}$. If you plug the two photon energies into this equation, your calculator will tell you that $\lambda = 1.216 \times 10^{-7}$ m $= 121.6$ nm in both cases because the energy difference is so small. To find the wavelength difference $\Delta\lambda$, we instead take the differential of both sides of $\lambda = hc/E_{\text{photon}}$:

$$d\lambda = d\left(\frac{hc}{E_{\text{photon}}}\right) = -\frac{hc}{(E_{\text{photon}})^2}dE_{\text{photon}}$$

$$= -\left(\frac{hc}{E_{\text{photon}}}\right)\left(\frac{1}{E_{\text{photon}}}\right)dE_{\text{photon}} = -\frac{\lambda}{E_{\text{photon}}}dE_{\text{photon}}$$

The minus sign means that a *decrease* in photon energy corresponds to an *increase* in photon wavelength. Replacing $d\lambda$ with $\Delta\lambda$ (the wavelength difference that we seek) and dE_{photon} with ΔE_{photon}, we get the difference between the two photon wavelengths:

$$\Delta\lambda = -\frac{\lambda}{E_{\text{photon}}}\Delta E_{\text{photon}}$$

To four significant digits, we have $\lambda = 121.6$ nm and $E_{\text{photon}} = 10.20$ eV. We find the photon energy difference ΔE_{photon} from the two expressions above, subtracting the larger energy from the smaller one so that $\Delta\lambda$ is positive:

$$\Delta\lambda = -\frac{121.6 \text{ nm}}{10.20 \text{ eV}}\left\{\left[10.20 \text{ eV} + (3.40 \text{ eV})\left(\frac{11\alpha^2}{16}\right)\right]\right.$$

$$\left. - \left[10.20 \text{ eV} + (3.40 \text{ eV})\left(\frac{15\alpha^2}{16}\right)\right]\right\}$$

$$= -\frac{121.6 \text{ nm}}{10.20 \text{ eV}}(3.40 \text{ eV})\left(-\frac{4\alpha^2}{16}\right)$$

$$= \frac{121.6 \text{ nm}}{10.20 \text{ eV}}(3.40 \text{ eV})\left(\frac{4}{16}\right)\left(\frac{1}{137.04}\right)^2$$

$$= 5.40 \times 10^{-4} \text{ nm}$$

EVALUATE: This line splitting is very small, as predicted. Fine structure is fine indeed! It is nonetheless observable with a diffraction grating that has a sufficient number of lines (see Section 36.5). The measured wavelengths are 121.567364 nm for the transition that begins in the $j = \frac{1}{2}$ state and 121.566824 nm for the transition that begins in the $j = \frac{3}{2}$ state. These are ultraviolet wavelengths.

There are also states of the hydrogen atom with $n = 2$, $l = 0$, $j = \frac{1}{2}$. [From Eq. (41.41), these states have the same energy as those with $n = 2$, $l = 1$, $j = \frac{1}{2}$; the energy $E_{n,j}$ depends on n and j but not on l.] However, an electron in an $n = 2$, $l = 0$, $j = \frac{1}{2}$ state *cannot* emit a photon and transition to an $n = 1$, $l = 0$, $j = \frac{1}{2}$ state. Such a transition is forbidden by the selection rule that l must change by 1 when a photon is emitted (see Section 41.4).

Additional, much smaller splittings are associated with the fact that the *nucleus* of the atom has a magnetic dipole moment that interacts with the orbital and/or spin magnetic dipole moments of the electrons. These effects are called *hyperfine structure*. For example, the ground level of hydrogen is split into two states, separated by only 5.9×10^{-6} eV. The photon that is emitted in the transitions between these states has a wavelength of 21 cm. Radio astronomers use this wavelength to map clouds of interstellar hydrogen gas that are too cold to emit visible light (**Fig. 41.19**).

TEST YOUR UNDERSTANDING OF SECTION 41.5 In which of the following situations is the magnetic moment of an electron perfectly aligned with a magnetic field that points in the positive z-direction? (i) $m_s = +\frac{1}{2}$; (ii) $m_s = -\frac{1}{2}$; (iii) both (i) and (ii); (iv) neither (i) nor (ii). ▮

41.6 MANY-ELECTRON ATOMS AND THE EXCLUSION PRINCIPLE

So far our analysis of atomic structure has concentrated on the hydrogen atom. That's natural; neutral hydrogen, with only one electron, is the simplest atom. Let's now take what we've learned about the hydrogen atom and apply that knowledge to the more complicated case of many-electron atoms.

An atom in its normal (electrically neutral) state has Z electrons and Z protons. Recall from Section 41.3 that we call Z the *atomic number*. The total electric charge of such an atom is exactly zero because the neutron has no charge while the proton and electron charges have the same magnitude but opposite sign.

A complete understanding of such a general atom requires that we know the wave function that describes the behavior of all Z of its electrons. This wave function depends on $3Z$ coordinates (three for each electron), so its complexity increases very rapidly with increasing Z. What's more, each of the Z electrons interacts not only with the nucleus but also with every other electron. The potential energy is therefore a complicated function of all $3Z$ coordinates, and the Schrödinger equation contains second derivatives with respect to all of them. Finding exact solutions to such equations is such a complex task that it has not been successfully achieved even for the neutral helium atom, which has only two electrons.

Fortunately, various approximation schemes are available. The simplest approximation is to ignore all interactions between electrons and consider each electron as moving under the action only of the nucleus (considered to be a point charge). In this approximation, we write a separate wave function for each *individual* electron. Each such function is like that for the hydrogen atom, specified by four quantum numbers (n, l, m_l, m_s). The nuclear charge is Ze

41.19 (a) In a visible-light image, these three distant galaxies appear to be unrelated. But in fact these galaxies are connected by immense streamers of hydrogen gas. This is revealed by (b) the false-color image made with a radio telescope tuned to the 21-cm wavelength emitted by hydrogen atoms.

(a) Galaxies in visible light (negative image; galaxies appear dark)

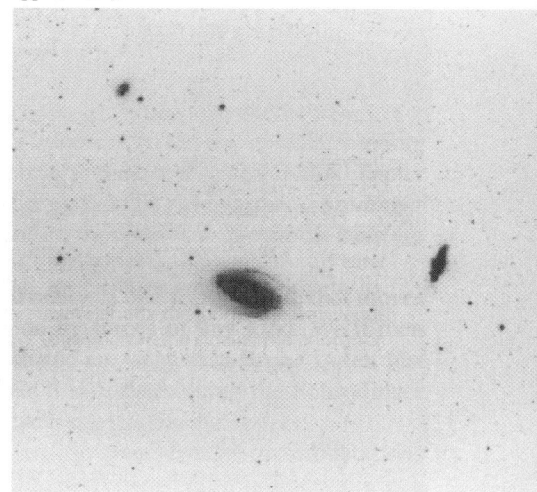

(b) Radio image of the same galaxies at wavelength 21 cm

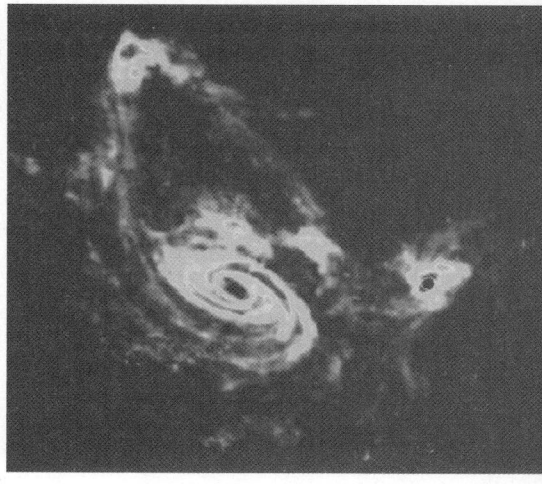

instead of e, so we replace every factor of e^2 in the wave functions and the energy levels by Ze^2. In particular, the energy levels are given by Eq. (41.21) with e^4 replaced by Z^2e^4:

$$E_n = -\frac{1}{(4\pi\epsilon_0)^2}\frac{m_r Z^2 e^4}{2n^2\hbar^2} = -\frac{Z^2}{n^2}(13.6 \text{ eV}) \qquad (41.43)$$

This approximation is fairly drastic; when there are many electrons, their interactions with each other are as important as the interaction of each with the nucleus. So this model isn't very useful for quantitative predictions.

The Central-Field Approximation

A less drastic and more useful approximation is to think of all the electrons together as making up a charge cloud that is, on average, *spherically symmetric*. We can then think of each individual electron as moving in the total electric field due to the nucleus and this averaged-out cloud of all the other electrons. There is a corresponding spherically symmetric potential-energy function $U(r)$. This picture is called the **central-field approximation;** it provides a useful starting point for understanding atomic structure.

In the central-field approximation we can again deal with one-electron wave functions. The Schrödinger equation for these functions differs from the equation for hydrogen, which we discussed in Section 41.3, only in that the $1/r$ potential-energy function is replaced by a different function $U(r)$. Now, Eqs. (41.20) show that $U(r)$ does not appear in the differential equations for $\Theta(\theta)$ and $\Phi(\phi)$. So those angular functions are exactly the same as for hydrogen, and the orbital angular momentum *states* are also the same as before. The quantum numbers l, m_l, and m_s have the same meanings as before, and Eqs. (41.22) and (41.23) again give the magnitude and z-component of the orbital angular momentum.

The radial wave functions and probabilities are different than for hydrogen because of the change in $U(r)$, so the energy levels are no longer given by Eq. (41.21). We can still label a state by using the four quantum numbers (n, l, m_l, m_s). In general, the energy of a state now depends on both n and l, rather than just on n as with hydrogen. (Due to fine-structure effects, the energy can also depend on the total angular momentum quantum number j. These effects are generally small, however, so we ignore them for this discussion.) The restrictions on the values of the quantum numbers are the same as before:

Allowed values of quantum numbers for one-electron wave functions:	Principal quantum number	Orbital magnetic quantum number	Spin magnetic quantum number	
	$n \geq 1$	$0 \leq l \leq n-1$ $\quad$ $|m_l| \leq l$	$m_s = \pm\frac{1}{2}$	(41.44)
		Orbital quantum number		

The Exclusion Principle

To understand the structure of many-electron atoms, we need an additional principle, the *exclusion principle*. To see why this principle is needed, let's consider the lowest-energy state, or *ground state,* of a many-electron atom. In the one-electron states of the central-field model, there is a lowest-energy state (corresponding to an $n = 1$ state of hydrogen). We might expect that in the ground state of a complex atom, *all* the electrons should be in this lowest state. If so, then we should see only gradual changes in physical and chemical properties when we look at the behavior of atoms with increasing numbers of electrons (Z).

Such gradual changes are *not* what is observed. Instead, properties of elements vary widely from one to the next, with each element having its own distinct personality. For example, the elements fluorine, neon, and sodium have 9, 10, and 11 electrons, respectively, per atom. Fluorine ($Z = 9$) is a *halogen;* it tends strongly to form compounds in which each fluorine atom acquires an extra electron.

Sodium ($Z = 11$) is an *alkali metal;* it forms compounds in which each sodium atom *loses* an electron. Neon ($Z = 10$) is a *noble gas,* forming no compounds at all. Such observations show that in the ground state of a complex atom the electrons *cannot* all be in the lowest-energy states. But why not?

The key to this puzzle, discovered by the Austrian physicist Wolfgang Pauli (**Fig. 41.20**) in 1925, is called the **exclusion principle.** This principle states that **no two electrons can occupy the same quantum-mechanical state** in a given system. That is, **no two electrons in an atom can have the same values of all four quantum numbers** (n, l, m_l, m_s). Each quantum state corresponds to a certain distribution of the electron "cloud" in space. Therefore the principle also says, in effect, that no more than two electrons with opposite values of the quantum number m_s can occupy the same region of space. We shouldn't take this last statement too seriously because the electron probability functions don't have sharp, definite boundaries. But the exclusion principle limits the amount by which electron wave functions can overlap. Think of it as the quantum-mechanical analog of a university rule that allows only one student per desk. This same exclusion principle applies to all spin-$\frac{1}{2}$ particles, not just electrons. (We'll see in Chapter 43 that protons and neutrons are also spin-$\frac{1}{2}$ particles. As a result, the exclusion principle plays an important role in the structure of atomic nuclei.)

41.20 The key to understanding the periodic table of the elements was the discovery by Wolfgang Pauli (1900–1958) of the exclusion principle. Pauli received the 1945 Nobel Prize in physics for his accomplishment. This photo shows Pauli (on the left) and Niels Bohr watching the physics of a toy top spinning on the floor—a macroscopic analog of a microscopic electron with spin.

CAUTION **The meaning of the exclusion principle** Don't confuse the exclusion principle with the electrical repulsion between electrons. While both effects tend to keep electrons within an atom separated from each other, they are very different in character. Two electrons can always be pushed closer together by adding energy to combat electrical repulsion, but *nothing* can overcome the exclusion principle and force two electrons into the same quantum-mechanical state. ▮

Table 41.2 lists some of the sets of quantum numbers for electron states in an atom. It's similar to Table 41.1 (Section 41.3), but we've added the number of states in each subshell and shell. Because of the exclusion principle, the "number of states" is the same as the *maximum* number of electrons that can be found in those states. For each state, m_s can be either $+\frac{1}{2}$ or $-\frac{1}{2}$.

As with the hydrogen wave functions, different states correspond to different spatial distributions; electrons with larger values of n are concentrated at larger distances from the nucleus. Figure 41.8 (Section 41.3) shows this effect. When an atom has more than two electrons, they can't all huddle down in the low-energy $n = 1$ states nearest to the nucleus because there are only two of these states; the exclusion principle forbids multiple occupancy of a state. Some electrons are forced into states farther away, with higher energies. Each value of n corresponds roughly to a region of space around the nucleus in the form of a spherical *shell.* Hence we speak of the K shell as the region that is occupied by the electrons in the $n = 1$ states, the L shell as the region of the $n = 2$ states, and so on. States with the same n but different l form *subshells,* such as the $3p$ subshell.

TABLE 41.2	Quantum States of Electrons in the First Four Shells				
n	l	m_l	Spectroscopic Notation	Number of States	Shell
1	0	0	$1s$	2	K
2	0	0	$2s$	2	L
2	1	$-1, 0, 1$	$2p$	6	
3	0	0	$3s$	2	M
3	1	$-1, 0, 1$	$3p$	6	
3	2	$-2, -1, 0, 1, 2$	$3d$	10	
4	0	0	$4s$	2	N
4	1	$-1, 0, 1$	$4p$	6	
4	2	$-2, -1, 0, 1, 2$	$4d$	10	
4	3	$-3, -2, -1, 0, 1, 2, 3$	$4f$	14	

The Periodic Table

We can use the exclusion principle to derive the most important features of the structure and chemical behavior of multielectron atoms, including the periodic table of the elements. Let's imagine constructing a neutral atom by starting with a bare nucleus with Z protons and then adding Z electrons, one by one. To obtain the ground state of the atom as a whole, we fill the lowest-energy electron states (those closest to the nucleus, with the smallest values of n and l) first, and we use successively higher states until all the electrons are in place. The chemical properties of an atom are determined principally by interactions involving the outermost, or *valence,* electrons, so we particularly want to learn how these electrons are arranged.

Let's look at the ground-state electron configurations for the first few atoms (in order of increasing Z). For hydrogen the ground state is $1s$; the single electron is in a state $n = 1$, $l = 0$, $m_l = 0$, and $m_s = \pm\frac{1}{2}$. In the helium atom ($Z = 2$), *both* electrons are in $1s$ states, with opposite spins; one has $m_s = -\frac{1}{2}$ and the other has $m_s = +\frac{1}{2}$. We denote the helium ground state as $1s^2$. (The superscript 2 is not an exponent; the notation $1s^2$ tells us that there are two electrons in the $1s$ subshell. Also, the superscript 1 is understood, as in $2s$.) For helium the K shell is completely filled, and all others are empty. Helium is a noble gas; it has no tendency to gain or lose an electron, and it forms no compounds.

Lithium ($Z = 3$) has three electrons. In its ground state, two are in $1s$ states and one is in a $2s$ state, so we denote the lithium ground state as $1s^2 2s$. On average, the $2s$ electron is considerably farther from the nucleus than are the $1s$ electrons (**Fig. 41.21**). According to Gauss's law, the *net* charge Q_{encl} attracting the $2s$ electron is nearer to $+e$ than to the value $+3e$ it would have without the two $1s$ electrons present. As a result, the $2s$ electron is loosely bound; only 5.4 eV is required to remove it, compared with the 30.6 eV given by Eq. (41.43) with $Z = 3$ and $n = 2$. Chemically, lithium is an *alkali metal*. It forms ionic compounds in which each lithium atom loses an electron and has a valence of $+1$.

Next is beryllium ($Z = 4$); its ground-state configuration is $1s^2 2s^2$, with its two valence electrons filling the s subshell of the L shell. Beryllium is the first of the *alkaline earth* elements, forming ionic compounds in which the valence of the atoms is $+2$.

Table 41.3 shows the ground-state electron configurations of the first 30 elements. The L shell can hold eight electrons. At $Z = 10$, both the K and L shells are filled, and there are no electrons in the M shell. We expect this to be a particularly stable configuration, with little tendency to gain or lose electrons. This element is neon, a noble gas with no known compounds. The next element after neon is sodium ($Z = 11$), with filled K and L shells and one electron in the M shell. Its "noble-gas-plus-one-electron" structure resembles that of lithium; both are alkali metals. The element *before* neon is fluorine, with $Z = 9$. It has a vacancy in the L shell and has an affinity for an extra electron to fill the shell. Fluorine forms ionic compounds in which it has a valence of -1. This behavior is characteristic of the *halogens* (fluorine, chlorine, bromine, iodine, and astatine), all of which have "noble-gas-minus-one" configurations (**Fig. 41.22**).

Proceeding down the list, we can understand the regularities in chemical behavior displayed by the **periodic table of the elements** (Appendix D) on the basis of electron configurations. The similarity of elements in each *group* (vertical column) of the periodic table is the result of similarity in outer-electron configuration. All the noble gases (helium, neon, argon, krypton, xenon, and radon) have filled-shell or filled-shell plus filled p subshell configurations. All the alkali metals (lithium, sodium, potassium, rubidium, cesium, and francium) have "noble-gas-plus-one" configurations. All the alkaline earth metals (beryllium, magnesium, calcium, strontium, barium, and radium) have "noble-gas-plus-two" configurations, and, as we just mentioned, all the halogens (fluorine, chlorine, bromine, iodine, and astatine) have "noble-gas-minus-one" structures.

41.21 Schematic representation of the charge distribution in a lithium atom. The nucleus has a charge of $+3e$.

On average, the $2s$ electron is considerably farther from the nucleus than the $1s$ electrons. Therefore, it experiences a net nuclear charge of approximately $+3e - 2e = +e$ (rather than $+3e$).

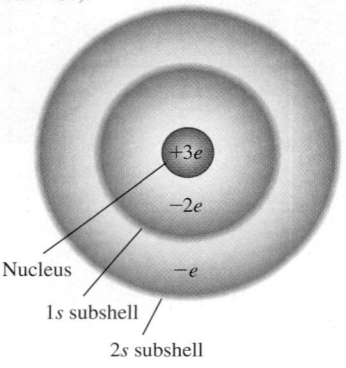

Nucleus

$1s$ subshell

$2s$ subshell

$+3e$

$-2e$

$-e$

41.22 Salt (sodium chloride, NaCl) dissolves readily in water, making seawater salty. This is due to the electron configurations of sodium and chlorine: Sodium can easily lose an electron to form an Na^+ ion, and chlorine can easily gain an electron to form a Cl^- ion. These ions are held in solution because they are attracted to the polar ends of water molecules (see Fig. 21.30a).

TABLE 41.3 | **Ground-State Electron Configurations**

Element	Symbol	Atomic Number (Z)	Electron Configuration
Hydrogen	H	1	$1s$
Helium	He	2	$1s^2$
Lithium	Li	3	$1s^2 2s$
Beryllium	Be	4	$1s^2 2s^2$
Boron	B	5	$1s^2 2s^2 2p$
Carbon	C	6	$1s^2 2s^2 2p^2$
Nitrogen	N	7	$1s^2 2s^2 2p^3$
Oxygen	O	8	$1s^2 2s^2 2p^4$
Fluorine	F	9	$1s^2 2s^2 2p^5$
Neon	Ne	10	$1s^2 2s^2 2p^6$
Sodium	Na	11	$1s^2 2s^2 2p^6 3s$
Magnesium	Mg	12	$1s^2 2s^2 2p^6 3s^2$
Aluminum	Al	13	$1s^2 2s^2 2p^6 3s^2 3p$
Silicon	Si	14	$1s^2 2s^2 2p^6 3s^2 3p^2$
Phosphorus	P	15	$1s^2 2s^2 2p^6 3s^2 3p^3$
Sulfur	S	16	$1s^2 2s^2 2p^6 3s^2 3p^4$
Chlorine	Cl	17	$1s^2 2s^2 2p^6 3s^2 3p^5$
Argon	Ar	18	$1s^2 2s^2 2p^6 3s^2 3p^6$
Potassium	K	19	$1s^2 2s^2 2p^6 3s^2 3p^6 4s$
Calcium	Ca	20	$1s^2 2s^2 2p^6 3s^2 3p^6 4s^2$
Scandium	Sc	21	$1s^2 2s^2 2p^6 3s^2 3p^6 4s^2 3d$
Titanium	Ti	22	$1s^2 2s^2 2p^6 3s^2 3p^6 4s^2 3d^2$
Vanadium	V	23	$1s^2 2s^2 2p^6 3s^2 3p^6 4s^2 3d^3$
Chromium	Cr	24	$1s^2 2s^2 2p^6 3s^2 3p^6 4s3d^5$
Manganese	Mn	25	$1s^2 2s^2 2p^6 3s^2 3p^6 4s^2 3d^5$
Iron	Fe	26	$1s^2 2s^2 2p^6 3s^2 3p^6 4s^2 3d^6$
Cobalt	Co	27	$1s^2 2s^2 2p^6 3s^2 3p^6 4s^2 3d^7$
Nickel	Ni	28	$1s^2 2s^2 2p^6 3s^2 3p^6 4s^2 3d^8$
Copper	Cu	29	$1s^2 2s^2 2p^6 3s^2 3p^6 4s3d^{10}$
Zinc	Zn	30	$1s^2 2s^2 2p^6 3s^2 3p^6 4s^2 3d^{10}$

A slight complication occurs with the M and N shells because the $3d$ and $4s$ subshell levels ($n = 3$, $l = 2$, and $n = 4$, $l = 0$, respectively) have similar energies. (We'll discuss in the next subsection why this happens.) Argon ($Z = 18$) has all the $1s$, $2s$, $2p$, $3s$, and $3p$ subshells filled, but in potassium ($Z = 19$) the additional electron goes into a $4s$ energy state rather than a $3d$ state (because the $4s$ state has slightly lower energy).

The next several elements have one or two electrons in the $4s$ subshell and increasing numbers in the $3d$ subshell. These elements are all metals with rather similar chemical and physical properties; they form the first *transition series,* starting with scandium ($Z = 21$) and ending with zinc ($Z = 30$), for which all the $3d$ and $4s$ subshells are filled.

Something similar happens with $Z = 57$ through $Z = 71$, which have one or two electrons in the $6s$ subshell but only partially filled $4f$ and $5d$ subshells. These *rare earth* elements all have very similar physical and chemical properties. Another such series, called the *actinide* series, starts with $Z = 91$.

Screening

We have mentioned that in the central-field model, the energy levels depend on l as well as n. Let's take sodium ($Z = 11$) as an example. If 10 of its electrons fill its K and L shells, the energies of some of the states for the remaining electron are found experimentally to be

$3s$ states: -5.138 eV

$3p$ states: -3.035 eV

$3d$ states: -1.521 eV

$4s$ states: -1.947 eV

BIO Application **Electron Configurations and Bone Cancer Radiotherapy** The orange spots in this colored x-ray image are bone cancer tumors. One method of treating bone cancer is to inject a radioactive isotope of strontium (^{89}Sr) into a patient's vein. Strontium is chemically similar to calcium because in both atoms the two outer electrons are in an s state (the structures are $1s^2 2s^2 2p^6 3s^2 3p^6 4s^2 3d^{10} 4p^6 5s^2$ for strontium and $1s^2 2s^2 2p^6 3s^2 3p^6 4s^2$ for calcium). Hence the strontium is readily taken up by the tumors, where calcium turnover is more rapid than in healthy bone. Radiation from the strontium helps destroy the tumors.

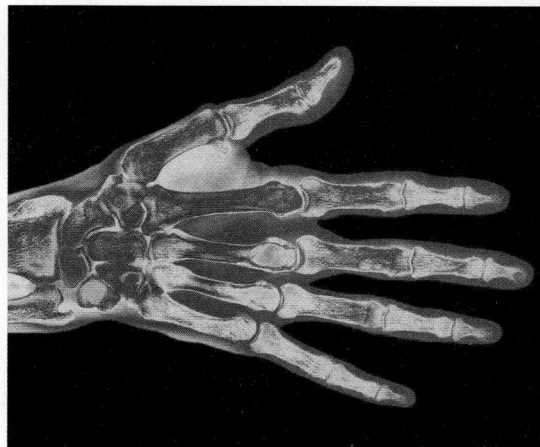

The 3s states are the lowest (most negative); one is the ground state for the 11th electron in sodium. The energy of the 3d states is quite close to the energy of the $n = 3$ state in hydrogen. The surprise is that the 4s state energy is 0.426 eV *below* the 3d state, even though the 4s state has larger n.

We can understand these results by using Gauss's law (Section 22.3). For any spherically symmetric charge distribution, the electric-field magnitude at a distance r from the center is $Q_{encl}/4\pi\epsilon_0 r^2$, where Q_{encl} is the total charge enclosed within a sphere with radius r. Mentally remove the outer (valence) electron atom from a sodium atom. What you have left is a spherically symmetric collection of 10 electrons (filling the K and L shells) and 11 protons, so $Q_{encl} = -10e + 11e = +e$. If the 11th electron is completely outside this collection of charges, it is attracted by an effective charge of $+e$, not $+11e$. This is a more extreme example of the effect depicted in Fig. 41.21.

This effect is called **screening;** the 10 electrons *screen* 10 of the 11 protons in the sodium nucleus, leaving an effective net charge of $+e$. From the viewpoint of the 11th electron, this is equivalent to reducing the number of protons in the nucleus from $Z = 11$ to a smaller *effective atomic number* Z_{eff}. If the 11th electron is *completely* outside the charge distribution of the other 10 electrons, then $Z_{eff} = 1$. Since the probability distribution of the 11th electron does extend somewhat into those of the other electrons, in fact Z_{eff} is greater than 1 (but still much less than 11). In general, an electron that spends all its time completely outside a positive charge $Z_{eff}e$ has energy levels given by the hydrogen expression with e^2 replaced by $Z_{eff}e^2$. From Eq. (41.43) this is

Energy levels of an electron with screening $\cdots\cdots\blacktriangleright$ Effective (screened) atomic number

$$E_n = -\frac{Z_{eff}^2}{n^2}(13.6\text{ eV}) \qquad (41.45)$$

$\cdots\cdots$ Principal quantum number

CAUTION **Different equations for different atoms** Equations (41.21), (41.43), and (41.45) all give values of E_n in terms of $(13.6\text{ eV})/n^2$, but they don't apply in general to the same atoms. Equation (41.21) is for hydrogen *only*. Equation (41.43) is for only the case in which there is no interaction with any other electron (and is thus accurate only when the atom has just one electron). Equation (41.45) is useful when one electron is screened from the nucleus by other electrons. ▮

DATA *SPEAKS*

Many-Electron Atoms and Electron States

When students were given a problem involving quantum-mechanical states in many-electron atoms, more than 32% gave an incorrect response. Common errors:

- **Confusion about quantum numbers.** There are limits on the values of the four quantum numbers n, l, m_l, and m_s. For a given n value, l can be no greater than $n - 1$; for a given l value, m_l can be no greater than l and no less than $-l$; and m_s has only two possible values, $+\frac{1}{2}$ and $-\frac{1}{2}$.

- **Confusion about electron subshells.** A *subshell* corresponds to a given value of n and l. The total number of electrons that can be present in a given subshell is $2(2l + 1)$ (that is, two possible values of m_s multiplied by $2l + 1$ possible values of m_l, from l through 0 to $-l$).

Now let's use the radial probability functions shown in Fig. 41.8 to explain why the energy of a sodium 3d state is approximately the same as the $n = 3$ value of hydrogen, -1.51 eV. The distribution for the 3d state (for which l has the maximum value $n - 1$) has one peak, and its most probable radius is *outside* the positions of the electrons with $n = 1$ or 2. (Those electrons also are pulled closer to the nucleus than in hydrogen because they are less effectively screened from the positive charge $11e$ of the nucleus.) Thus in sodium a 3d electron spends most of its time well outside the $n = 1$ and $n = 2$ states (the K and L shells). The 10 electrons in these shells screen about ten-elevenths of the charge of the 11 protons, leaving a net charge of about $Z_{eff}e = (1)e$. Then, from Eq. (41.45), the corresponding energy is approximately $-(1)^2(13.6\text{ eV})/3^2 = -1.51$ eV. This approximation is very close to the experimental value of -1.521 eV.

Looking again at Fig. 41.8, we see that the radial probability density for the 3p state (for which $l = n - 2$) has two peaks and that for the 3s state ($l = n - 3$) has three peaks. For sodium the first small peak in the 3p distribution gives a 3p electron a higher probability (compared to the 3d state) of being *inside* the charge distributions for the electrons in the $n = 2$ states. That is, a 3p electron is less completely screened from the nucleus than is a 3d electron because it spends some of its time within the filled K and L shells. Thus for the 3p electrons, Z_{eff} is greater than unity. From Eq. (41.45) the 3p energy is lower (more negative) than the 3d energy of −1.521 eV. The actual value is −3.035 eV. A 3s electron spends even more time within the inner electron shells than a 3p electron does, giving an even larger Z_{eff} and an even more negative energy.

This discussion shows that the energy levels given by Eq. (41.45) depend on both the principal quantum number n and the orbital quantum number l. That's because the value of Z_{eff} is different for the 3s state ($n = 3, l = 0$), the 3p state ($n = 3, l = 1$), and the 3d state ($n = 3, l = 2$).

EXAMPLE 41.9   DETERMINING Z_{eff} EXPERIMENTALLY

The measured energy of a 3s state of sodium is −5.138 eV. Calculate the value of Z_{eff}.

SOLUTION

IDENTIFY and SET UP: Sodium has a single electron in the M shell outside filled K and L shells. The ten K and L electrons partially screen the single M electron from the $+11e$ charge of the nucleus; our goal is to determine the extent of this screening. We are given $n = 3$ and $E_n = -5.138$ eV, so we can use Eq. (41.45) to determine Z_{eff}.

EXECUTE: Solving Eq. (41.45) for Z_{eff}, we have

$$Z_{eff}^2 = -\frac{n^2 E_n}{13.6\ \text{eV}} = -\frac{3^2(-5.138\ \text{eV})}{13.6\ \text{eV}} = 3.40$$

$$Z_{eff} = 1.84$$

EVALUATE: The effective charge attracting a 3s electron is 1.84e. Sodium's 11 protons are screened by an average of 11 − 1.84 = 9.16 electrons instead of 10 electrons because the 3s electron spends some time within the inner (K and L) shells.

Each alkali metal (lithium, sodium, potassium, rubidium, and cesium) has one more electron than the corresponding noble gas (helium, neon, argon, krypton, and xenon). This extra electron is mostly outside the other electrons in the filled shells and subshells. Therefore all the alkali metals behave similarly to sodium.

EXAMPLE 41.10   ENERGIES FOR A VALENCE ELECTRON

The valence electron in potassium has a 4s ground state. Calculate the approximate energy of the $n = 4$ state having the smallest Z_{eff}, and discuss the relative energies of the 4s, 4p, 4d, and 4f states.

SOLUTION

IDENTIFY and SET UP: The state with the smallest Z_{eff} is the one in which the valence electron spends the most time outside the inner filled shells and subshells, so that it is most effectively screened from the charge of the nucleus. Once we have determined which state has the smallest Z_{eff}, we can use Eq. (41.45) to determine the energy of this state.

EXECUTE: A 4f state has $n = 4$ and $l = 3 = 4 - 1$. Thus it is the state of greatest orbital angular momentum for $n = 4$, and thus the state in which the electron spends the most time outside the electron charge clouds of the inner filled shells and subshells. This makes Z_{eff} for a 4f state close to unity. Equation (41.45) then gives

$$E_4 = -\frac{Z_{eff}^2}{n^2}(13.6\ \text{eV}) = -\frac{1}{4^2}(13.6\ \text{eV}) = -0.85\ \text{eV}$$

This approximation agrees with the measured energy of the sodium 4f state to the precision given.

An electron in a 4d state spends a bit more time within the inner shells, and its energy is therefore a bit more negative (measured to be −0.94 eV). For the same reason, a 4p state has an even lower energy (measured to be −2.73 eV) and a 4s state has the lowest energy (measured to be −4.339 eV).

EVALUATE: We can extend this analysis to the *singly ionized alkaline earth elements:* Be^+, Mg^+, Ca^+, Sr^+, and Ba^+. For any allowed value of n, the highest-l state ($l = n - 1$) of the one remaining outer electron sees an effective charge of almost $+2e$, so for these states, $Z_{eff} = 2$. A 3d state for Mg^+, for example, has an energy of about $-2^2(13.6\ \text{eV})/3^2 = -6.0$ eV.

TEST YOUR UNDERSTANDING OF SECTION 41.6 If electrons did *not* obey the exclusion principle, would it be easier or more difficult to remove the first electron from sodium? ▮

41.7 X-RAY SPECTRA

X-ray spectra provide an example of the richness and power of the model of atomic structure that we derived in the preceding section. In Section 38.2 we discussed how x-ray photons are produced when electrons strike a metal target (see Fig. 38.7). In this section we'll see how the spectrum of x rays produced in this way depends on the type of metal used in the target and how the ideas of atomic energy levels and screening help us understand this dependence.

Characteristic X Rays and Atomic Energy Levels

X-ray diffraction techniques (see Section 36.6) make it possible to measure x-ray wavelengths quite precisely (to within 0.1% or less). **Figure 41.23** shows the spectrum of x rays produced when fast-moving electrons strike a target of the metal molybdenum. This spectrum has two features:

1. There is a *continuous* spectrum of wavelengths (see Fig. 38.8 in Section 38.2), with a minimum wavelength (corresponding to a maximum frequency and a maximum photon energy) that is determined by the voltage V_{AC} used to accelerate the electrons. As we saw in Section 38.2, this continuous spectrum is due to *bremsstrahlung*, in which the electrons slow down as they interact with the metal atoms in the target and convert their kinetic energy into photons. The minimum wavelength λ_{min} corresponds to all of the kinetic energy eV_{AC} of the electron being converted into the energy of a single photon of energy hc/λ_{min}, so

$$\lambda_{min} = \frac{hc}{eV_{AC}} \qquad (41.46)$$

This continuous-spectrum radiation is nearly independent of the target material in the x-ray tube.

2. Depending on the accelerating voltage, sharp peaks may be superimposed on the continuous bremsstrahlung spectrum, as in Fig. 41.23. These peaks are caused when the target atoms are struck by high-energy electrons and emit x rays of very definite wavelengths. Unlike the continuous spectrum, the wavelengths of the peaks are *different* for different target elements; they form what is called a *characteristic x-ray spectrum* for each target element.

In 1913 the English physicist Henry G. J. Moseley undertook a careful experimental study of characteristic x-ray spectra. He found that the most intense short-wavelength line in the characteristic x-ray spectrum from a particular target element, called the K_α line, varied smoothly with that element's atomic number Z (**Fig. 41.24**). This is in sharp contrast to optical spectra, in which elements with adjacent Z-values have spectra that often bear no resemblance to each other.

41.23 Graph of intensity per unit wavelength as a function of wavelength for x rays produced with an accelerating voltage of 35 kV and a molybdenum target. The curve is a smooth function similar to the bremsstrahlung spectra in Fig. 38.8 (Section 38.2), but with two sharp spikes corresponding to part of the characteristic x-ray spectrum for molybdenum.

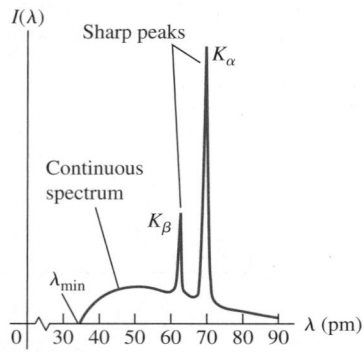

41.24 The square root of Moseley's measured frequencies of the K_α line for 14 elements.

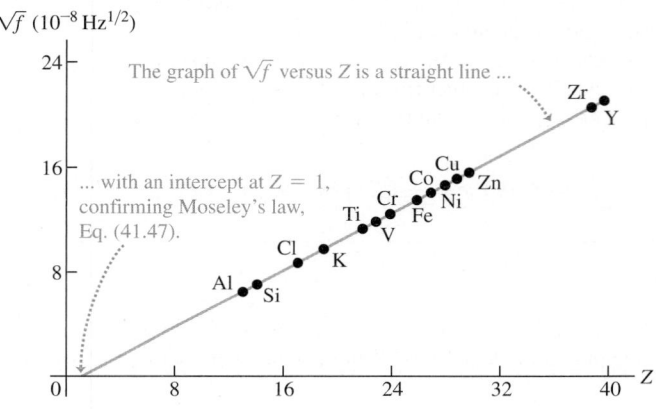

Moseley found that the relationship could be expressed in terms of x-ray frequencies f by a simple formula called *Moseley's law:*

Moseley's law:

Frequency of K_α line in characteristic x-ray spectrum of an element

$$f = (2.48 \times 10^{15}\text{ Hz})(Z - 1)^2 \qquad (41.47)$$

Atomic number of element

Moseley went far beyond this empirical relationship; he showed how characteristic x-ray spectra could be understood on the basis of energy levels of atoms in the target. His analysis was based on the Bohr model, published in the same watershed year of 1913. We will recast it somewhat, using the ideas of atomic structure that we discussed in Section 41.6. First recall that the *outer* electrons of an atom are responsible for optical spectra. Their excited states are usually only a few electron volts above their ground state. In transitions from excited states to the ground state, they usually emit photons in or near the visible region.

Characteristic x rays, by contrast, are emitted in transitions involving the *inner* shells of a complex atom. In an x-ray tube the electrons may strike the target with enough energy to knock electrons out of the inner shells of the target atoms. These inner electrons are much closer to the nucleus than are the electrons in the outer shells; they are much more tightly bound, and hundreds or thousands of electron volts may be required to remove them.

Suppose one electron is knocked out of the K shell. This process leaves a vacancy, which we'll call a *hole.* (One electron remains in the K shell.) The hole can then be filled by an electron falling in from one of the outer shells, such as the $L, M, N, \ldots$ shell. This transition is accompanied by a decrease in the energy of the atom (because *less* energy would be needed to remove an electron from an $L, M, N, \ldots$ shell), and an x-ray photon is emitted with energy equal to this decrease. Each state has definite energy, so the emitted x rays have definite wavelengths; the emitted spectrum is a *line* spectrum.

We can estimate the energy and frequency of K_α x-ray photons by using the concept of screening from Section 41.6. A K_α x-ray photon is emitted when an electron in the L shell $(n = 2)$ drops down to fill a hole in the K shell $(n = 1)$. As the electron drops down, it is attracted by the Z protons in the nucleus screened by the one remaining electron in the K shell. We therefore approximate the energy by Eq. (41.45), with $Z_\text{eff} = Z - 1$, $n_\text{i} = 2$, and n_f. The energy before the transition is

$$E_i \approx -\frac{(Z - 1)^2}{2^2}(13.6\text{ eV}) = -(Z - 1)^2(3.4\text{ eV})$$

and the energy after the transition is

$$E_\text{f} \approx -\frac{(Z - 1)^2}{1^2}(13.6\text{ eV}) = -(Z - 1)^2(13.6\text{ eV})$$

$E_{K_\alpha} = E_\text{i} - E_\text{f} \approx (Z - 1)^2(-3.4\text{ eV} + 13.6\text{ eV})$ is the energy of the K_α x-ray photon. That is,

$$E_{K_\alpha} \approx (Z - 1)^2(10.2\text{ eV}) \qquad (41.48)$$

The frequency of the photon is its energy divided by Planck's constant:

$$f = \frac{E}{h} \approx \frac{(Z - 1)^2(10.2\text{ eV})}{4.136 \times 10^{-15}\text{ eV}\cdot\text{s}} = (2.47 \times 10^{15}\text{ Hz})(Z - 1)^2$$

This relationship agrees almost exactly with Moseley's experimental law, Eq. (41.47). Indeed, considering the approximations we have made, the agreement is better than we have a right to expect. But our calculation does show how Moseley's law can be understood on the bases of screening and transitions between energy levels.

Application X Rays in Forensic Science When a handgun is fired, a cloud of gunshot residue (GSR) is ejected from the barrel. The x-ray emission spectrum of GSR includes characteristic peaks from lead (Pb), antimony (Sb), and barium (Ba). If a sample taken from a suspect's skin or clothing has an x-ray emission spectrum with these characteristics, it indicates that the suspect recently fired a gun.

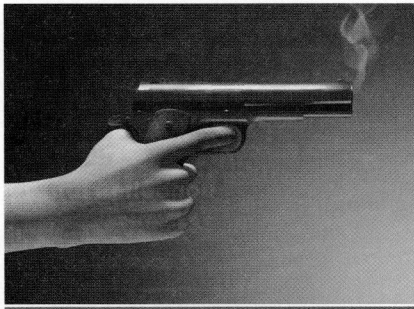

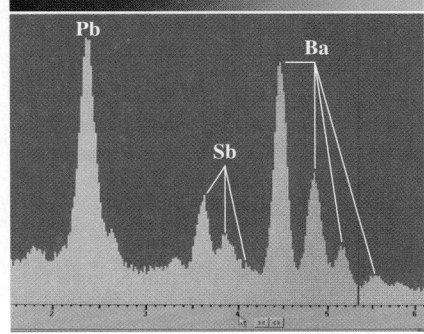

41.25 Wavelengths of the K_α, K_β, and K_γ lines of tungsten (W), molybdenum (Mo), and copper (Cu).

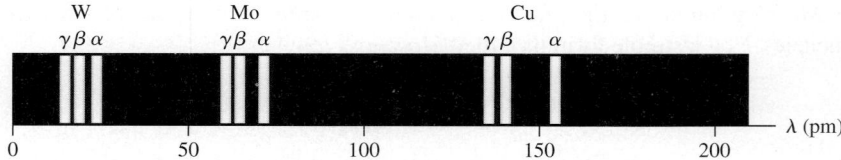

The three lines in each series are called the K_α, K_β, and K_γ lines. The K_α line is produced by the transition of an L electron to the vacancy in the K shell, the K_β line by an M electron, and the K_γ line by an N electron.

The hole in the K shell may also be filled by an electron falling from the M or N shell, assuming that these are occupied. If so, the x-ray spectrum of a large group of atoms of a single element shows a series, named the K series, of three lines, called the K_α, K_β, and K_γ lines. These three lines result from transitions in which the K-shell hole is filled by an L, M, or N electron, respectively. **Figure 41.25** shows the K series for tungsten ($Z = 74$), molybdenum ($Z = 42$), and copper ($Z = 29$).

There are other series of x-ray lines, called the L, M, and N series, that are produced after the ejection of electrons from the L, M, and N shells rather than the K shell. Electrons in these outer shells are farther away from the nucleus and are not held as tightly as are those in the K shell, so removing these outer electrons requires less energy. Hence the x-ray photons that are emitted when these vacancies are filled have lower energy than those in the K series.

EXAMPLE 41.11 **CHEMICAL ANALYSIS BY X-RAY EMISSION**

You measure the K_α wavelength for an unknown element, obtaining the value 0.0709 nm. What is the element?

SOLUTION

IDENTIFY and SET UP: To determine which element this is, we need to know its atomic number Z. We can find this by using Moseley's law, which relates the frequency of an element's K_α x-ray emission line to that element's atomic number Z. We'll use the relationship $f = c/\lambda$ to calculate the frequency for the K_α line, and then use Eq. (41.47) to find the corresponding value of the atomic number Z. We'll then consult the periodic table (Appendix D) to determine which element has this atomic number.

EXECUTE: The frequency is

$$f = \frac{c}{\lambda} = \frac{3.00 \times 10^8 \text{ m/s}}{0.0709 \times 10^{-9} \text{ m}} = 4.23 \times 10^{18} \text{ Hz}$$

Solving Moseley's law for Z, we get

$$Z = 1 + \sqrt{\frac{f}{2.48 \times 10^{15} \text{ Hz}}} = 1 + \sqrt{\frac{4.23 \times 10^{18} \text{ Hz}}{2.48 \times 10^{15} \text{ Hz}}} = 42.3$$

We know that Z has to be an integer; we conclude that $Z = 42$, corresponding to the element molybdenum.

EVALUATE: If you're worried that our calculation did not give an integer for Z, remember that Moseley's law is an empirical relationship. There are slight variations from one atom to another due to differences in the structure of the electron shells. Nonetheless, this example suggests the power of Moseley's law.

Niels Bohr commented that it was Moseley's observations, not the alpha-particle scattering experiments of Rutherford, Geiger, and Marsden (see Section 39.2), that truly convinced physicists that the atom consists of a positive nucleus surrounded by electrons in motion. Unlike Bohr or Rutherford, Moseley did not receive a Nobel Prize for his important work; these awards are given to living scientists only, and Moseley (at age 27) was killed in combat during the First World War.

X-Ray Absorption Spectra

We can also observe x-ray *absorption* spectra. Unlike optical spectra, the absorption wavelengths are usually not the same as those for emission, especially in many-electron atoms, and do not give simple line spectra. For example, the K_α emission line results from a transition from the L shell to a hole in the K shell. The reverse transition doesn't occur in atoms with $Z \geq 10$ because in the atom's ground state, there is no vacancy in the L shell. To be absorbed, a photon must have enough energy to move an electron to an empty state. Since empty states are only a few electron volts in energy below the free-electron continuum, the minimum absorption energies in many-electron atoms are about the same as the minimum energies that are needed to remove an electron from its shell. Experimentally, if we gradually increase the accelerating voltage and hence the maximum photon energy, we observe sudden increases in absorption when we reach these minimum energies. These sudden jumps of absorption are called *absorption edges* (**Fig. 41.26**).

Characteristic x-ray spectra provide a very useful analytical tool. Satellite-borne x-ray spectrometers are used to study x-ray emission lines from highly excited atoms in distant astronomical sources. X-ray spectra are also used in air-pollution monitoring and in studies of the abundance of various elements in rocks.

TEST YOUR UNDERSTANDING OF SECTION 41.7 A beam of photons is passed through a sample of high-temperature atomic hydrogen. At what photon energy would you expect there to be an absorption edge like that shown in Fig. 41.26? (i) 13.60 eV; (ii) 3.40 eV; (iii) 1.51 eV; (iv) all of these; (v) none of these. ∎

41.26 When a beam of x rays is passed through a slab of molybdenum, the extent to which the beam is absorbed depends on the energy E of the x-ray photons. A sharp increase in absorption occurs at the K absorption edge at 20 keV. The increase occurs because photons with energies above this value can excite an electron from the K shell of a molybdenum atom into an empty state.

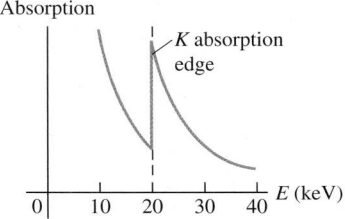

41.8 QUANTUM ENTANGLEMENT

We've seen that quantum mechanics is very successful at correctly predicting the results of experiments. As we will see in the remaining chapters, quantum mechanics is the basis of all electronic devices and is essential to our understanding of atomic nuclei and subatomic particles. But even though the central ideas of quantum mechanics have been established for decades, some aspects of the theory continue to baffle physicists and remain topics of ongoing research. We close this chapter with a discussion of one of these topics, *quantum entanglement*.

The Wave Function for Two Identical Particles

To understand what is meant by "quantum entanglement," let's consider how to write the wave function for two identical particles, such as two electrons (the electrons in the neutral helium atom, for instance). We'll use the subscripts 1 and 2 to refer to these particles.

As we discussed in Section 41.6, a wave function that describes both particles is a function of both the coordinates (x_1, y_1, z_1) of particle 1 and the coordinates (x_2, y_2, z_2) of particle 2. As a shorthand, we'll use the position vectors $\vec{r}_1 = x_1\hat{i} + y_1\hat{j} + z_1\hat{k}$ and $\vec{r}_2 = x_2\hat{i} + y_2\hat{j} + z_2\hat{k}$. In terms of these vectors, we can write the time-dependent two-particle wave function as $\Psi(\vec{r}_1, \vec{r}_2, t)$. Just as for a single particle, if the system of two particles has total energy E, we can write this time-dependent wave function as the product of a time-*independent* two-particle wave function $\psi(\vec{r}_1, \vec{r}_2)$ and a factor that depends on time only:

$$\Psi(\vec{r}_1, \vec{r}_2, t) = \psi(\vec{r}_1, \vec{r}_2)e^{-iEt/\hbar} \qquad (41.49)$$

We saw in Section 41.2 that a technique called *separation of variables* is useful for expressing a wave function that depends on several variables as a product of functions of the individual variables. Let's see whether we can express the time-independent two-particle wave function $\psi(\vec{r}_1, \vec{r}_2)$ in Eq. (41.49) as a product of a function of $\vec{r}_1$ and a function of $\vec{r}_2$. We interpret the function of $\vec{r}_1$ to be the *single*-particle wave function for particle 1 and the function of $\vec{r}_2$ to be the single-particle wave function for particle 2. Suppose particle 1 is in a state A, for which the single-particle wave function is ψ_A, and particle 2 is in a state B, for which the single-particle wave function is ψ_B. [In the hydrogen atom, with two electrons, one electron could be in the spin-up state ($n = 1$, $l = 0$, $m_l = 0$, and $m_s = +\frac{1}{2}$) and the other could be in the spin-down state ($n = 1$, $l = 0$, $m_l = 0$, and $m_s = -\frac{1}{2}$).] Using separation of variables, we would write

$$\psi(\vec{r}_1, \vec{r}_2) = \psi_A(\vec{r}_1)\psi_B(\vec{r}_2) \qquad (41.50)$$

(first guess at the two-particle wave function)

However, Eq. (41.50) *cannot* be correct, because particles 1 and 2 are *identical* and *indistinguishable*. We may be able to state with confidence that one particle is in state A and the other is in state B, but it's impossible to specify which particle is in which state. (There's no way even in principle to "tag" the particles.)

To account for this, let's make an improved guess for the two-particle wave function: a combination of two terms like Eq. (41.50)—one term for which particle 1 is in state A and particle 2 is in state B, and one term for which particle 1 is in state B and particle 2 is in state A. Our improved guess is then

$$\psi(\vec{r}_1, \vec{r}_2) = \frac{1}{\sqrt{2}}[\psi_A(\vec{r}_1)\psi_B(\vec{r}_2) \pm \psi_B(\vec{r}_1)\psi_A(\vec{r}_2)] \qquad (41.51)$$

(second guess at the two-particle wave function)

The factor $1/\sqrt{2}$ in Eq. (41.51) ensures that if ψ_A and ψ_B are normalized, then $\psi(\vec{r}_1, \vec{r}_2)$ will be normalized as well. Note that the terms $\psi_A(\vec{r}_1)\psi_B(\vec{r}_2)$ and $\psi_B(\vec{r}_1)\psi_A(\vec{r}_2)$ appear with equal magnitudes, so that the two possibilities (particle 1 in A and particle 2 in B, or particle 1 in B and particle 2 in A) are equally probable.

How can we decide whether the $\pm$ sign in Eq. (41.51) should be a plus or a minus? If the particles are two electrons, or two of any other type of spin-$\frac{1}{2}$ particle, the Pauli exclusion principle (Section 41.6) tells us that we must use the minus sign:

$$\psi(\vec{r}_1, \vec{r}_2) = \frac{1}{\sqrt{2}}[\psi_A(\vec{r}_1)\psi_B(\vec{r}_2) - \psi_B(\vec{r}_1)\psi_A(\vec{r}_2)] \qquad (41.52)$$

(two-particle wave function, spin-$\frac{1}{2}$ particles)

To check this, suppose we demand that *both* particles be in the same state A. Then we would replace ψ_B in Eq. (41.52) with ψ_A:

$$\psi(\vec{r}_1, \vec{r}_2) = \frac{1}{\sqrt{2}}[\psi_A(\vec{r}_1)\psi_A(\vec{r}_2) - \psi_A(\vec{r}_1)\psi_A(\vec{r}_2)] = 0$$

The zero value of the wave function says that our demand cannot be met. This is in agreement with the Pauli exclusion principle: No two electrons, and indeed no two identical spin-$\frac{1}{2}$ particles of any kind, can occupy the same quantum-mechanical state.

Measurement and "Spooky Action at a Distance"

The wave function in Eq. (41.52) shares features with that of the quantum-mechanical particle in a box that we discussed in Section 40.6. That particle's wave function is also a combination of two terms of equal magnitude representing different situations: one in which the momentum of the particle is in the $+x$-direction, so $p_x > 0$, and one in which its momentum is in the $-x$-direction, so $p_x < 0$. We saw in Section 40.6 that if we *measured* the momentum of the particle, we would get either $p_x > 0$ or $p_x < 0$. Making such a measurement causes the wave function to *collapse*, and only the term corresponding to the measured value of p_x survives. The other term disappears from the wave function.

The same sort of wave-function collapse happens in our two-particle system. Suppose we make a measurement of one particle—call it particle 1—and determine that it is in state A. The measurement causes the wave function in Eq. (41.52) to collapse to $\psi(\vec{r}_1, \vec{r}_2) = \psi_A(\vec{r}_1)\psi_B(\vec{r}_2)$. This wave equation corresponds to particle 1 being in state A but also corresponds to particle 2 being in state B. It follows that, after the measurement, particle 2 *must* be in state B, even though we have not directly measured the state of particle 2. In other words, making a measurement on *one* particle affects the state of the *other* particle. Schrödinger described this situation by saying that the two particles are *entangled*.

If the two entangled particles are the two electrons within a helium atom, the idea that their states are entangled may not seem troubling. After all, these electrons are in very close proximity (a helium atom is only about 0.1 nm in diameter) and exert substantial electric forces on each other. You might imagine that when we measure electron 1 to be in state A (say, spin up with $m_s = +\frac{1}{2}$), it exerts forces on electron 2 that require electron 2 to be in state B (say, spin down with $m_s = -\frac{1}{2}$).

But suppose we arrange for two identical particles to be in an entangled state in which the particles are *not* close to each other, so they cannot exert forces on each other. When the same kind of measurement experiment is done on such a distant pair of entangled particles, the result is the same as if they are close together: If we measure particle 1 to be in state A and subsequently make a measurement on particle 2, we always find that particle 2 is in state B. If instead we measure particle 1 to be in state B and then make a measurement on particle 2, we always find that particle 2 is in state A. So measuring the state of one particle affects the state of the other particle, even when the two particles *cannot* exert forces on each other (**Fig. 41.27**). This finding has been confirmed with entangled particles that are *more than 300 km apart!* (These experiments with very large distances are done with photons rather than electrons. Like electrons, photons have spin, and the "spin-up" and "spin-down" states correspond to left and right circular polarization. The only difference is that photons are spin-1 particles, not spin-$\frac{1}{2}$, and do not obey the Pauli exclusion principle. As a result, we must use the plus sign rather than the minus sign in Eq. (41.51) to describe two entangled photons. The rest of the physics is identical, however.)

These results contradict the idea of *locality*—the notion that a particle responds to forces or fields that act at its position only, not at some other point in space. We used locality in Chapters 4 and 13 when we expressed the gravitational force on a particle of mass m as $\vec{F}_g = m\vec{g}$, where $\vec{g}$ is the acceleration due to gravity at the point in space where the particle is located. We used locality again in Chapters 21 and 27 when we wrote the electric force $\vec{F}_E$ and the magnetic force $\vec{F}_B$ on a particle of charge q moving with velocity $\vec{v}$ as $\vec{F}_E = q\vec{E}$ and $\vec{F}_B = q\vec{v} \times \vec{B}$, where $\vec{E}$ and $\vec{B}$, respectively, are the electric and magnetic fields at the position of the particle. But the interaction between two entangled, widely separated particles seems *not* to obey locality. For this reason, Albert Einstein

41.27 If two particles are in an entangled state, making a measurement of one particle determines the result of a subsequent measurement of the other particle.

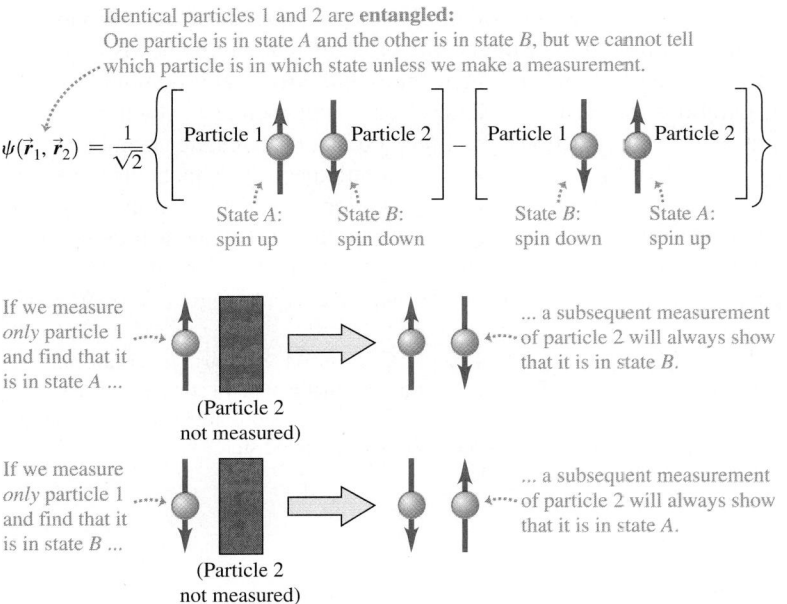

Identical particles 1 and 2 are **entangled:**
One particle is in state A and the other is in state B, but we cannot tell which particle is in which state unless we make a measurement.

$$\psi(\vec{r}_1, \vec{r}_2) = \frac{1}{\sqrt{2}}\left\{\left[\text{Particle 1} \quad \text{Particle 2}\right] - \left[\text{Particle 1} \quad \text{Particle 2}\right]\right\}$$

State A: spin up   State B: spin down   State B: spin down   State A: spin up

If we measure *only* particle 1 and find that it is in state A ...

(Particle 2 not measured)

... a subsequent measurement of particle 2 will always show that it is in state B.

If we measure *only* particle 1 and find that it is in state B ...

(Particle 2 not measured)

... a subsequent measurement of particle 2 will always show that it is in state A.

Because the two particles are entangled, measuring the state of particle 1 determines the state of particle 2. This is true even if the two particles do not interact.

referred to the results of an experiment like that shown in Fig. 41.27 as "spooky action at a distance." Spooky or not, quantum mechanics appears to be intrinsically nonlocal.

What makes these results even more striking is that no matter how far apart the two entangled particles are, there appears to be *zero delay* between the time that we make a measurement on one particle and the time that the state of the other particle changes as a result. At first glance this seems to violate a key idea of Einstein's special theory of relativity: that signals of any kind—radio waves, light signals, or beams of particles—cannot travel faster than the speed of light in vacuum, *c*. If measuring the state of particle 1 in Fig. 41.27 causes the state of particle 2 to change instantaneously, couldn't we make a "quantum radio" that sends signals faster than *c*, with particle 1 as the transmitter and particle 2 as the receiver?

The answer is no. The "message" in our quantum radio would be the result of a measurement of particle 1 by a physicist (call her Primo) at that particle's position. Primo's measurement collapses the wave function of the two particles, and her result would be that particle 1 is in either state *A* or state *B*. Another physicist (call him Secondo) at the position of particle 2 would measure particle 2 to be in state *B* if Primo measured particle 1 to be in state *A*, and to be in state *A* if Primo measured particle 1 to be in state *B*. But Secondo would have no way of knowing whether his result was caused by Primo making a measurement first or by *Secondo* himself making an independent measurement of particle 2 *without* Primo having made any measurement. (Secondo could determine this later by, for instance, sending text messages back and forth with Primo. But that method of communication involves signals that travel at the speed of light, not instantaneously.) So our quantum radio would transmit no information at all and would not allow us to communicate at speeds faster than *c*.

A remarkable practical application of quantum entanglement is *quantum computing*. In a conventional ("classical") computer, the memory is made up of *bits*. Each bit has only two possible values (say, 0 or 1), so a computer memory with *N* bits can have any of 2^N different configurations. (This is analogous to coins that can be either heads up or tails up. Figure 20.21 in Section 20.8 shows the possible configurations of four coins; the number of possibilities is $2^4 = 16$.) In a quantum computer, bits are replaced with *qubits* (short for "quantum bits"). An example is a spin-$\frac{1}{2}$ electron that can be in a spin-up state $\left(m_s = +\frac{1}{2}\right)$ or a spin-down state $\left(m_s = -\frac{1}{2}\right)$, as usual, but can also be in *any combination* of these states. The wave function of *N* entangled qubits can correspond to any of 2^N configurations (like ordinary bits or coins that can be heads up or tails up) or to an entangled state in which the qubits are in any combination of these configurations. So, unlike a classical computer memory, which can be in only one of its 2^N configurations at a time, a quantum computer memory can essentially be in *all* of these configurations simultaneously. This holds the promise of the ability to do certain types of computations, such as those involved in breaking codes in cryptography, much more rapidly than a classical computer could. As of this writing, the quest to build a fully quantum computer is still in its early stages, but intensive research is under way and rapid progress is being made.

TEST YOUR UNDERSTANDING OF SECTION 41.8 Particle 1 is an electron that can be in state *C* or *D*. Particle 2 is a proton that can be in state *E* or *F*. Is $\psi(\vec{r}_1, \vec{r}_2) = (1/\sqrt{2})[\psi_C(\vec{r}_1)\psi_E(\vec{r}_2) + \psi_C(\vec{r}_1)\psi_F(\vec{r}_2)]$ a possible wave function for this two-particle system? If so, does it represent an entangled state? ▮

Three-dimensional problems: The time-independent Schrödinger equation for three-dimensional problems is given by Eq. (41.5).

$$-\frac{\hbar^2}{2m}\left(\frac{\partial^2\psi(x,y,z)}{\partial x^2} + \frac{\partial^2\psi(x,y,z)}{\partial y^2} + \frac{\partial^2\psi(x,y,z)}{\partial z^2}\right) + U(x,y,z)\psi(x,y,z)$$
$$= E\psi(x,y,z)$$

(three-dimensional time-independent Schrödinger equation)　　(41.5)

Particle in a three-dimensional box: The wave function for a particle in a cubical box is the product of a function of x only, a function of y only, and a function of z only. Each stationary state is described by three quantum numbers (n_X, n_Y, n_Z). Most of the energy levels given by Eq. (41.16) exhibit degeneracy: More than one quantum state has the same energy. (See Example 41.1.)

$$E_{n_X,n_Y,n_Z} = \frac{(n_X{}^2 + n_Y{}^2 + n_Z{}^2)\pi^2\hbar^2}{2mL^2}$$

$(n_X = 1, 2, 3, \ldots;$
$n_Y = 1, 2, 3, \ldots;$
$n_Z = 1, 2, 3, \ldots)$
(energy levels, particle in a three-dimensional cubical box)　　(41.16)

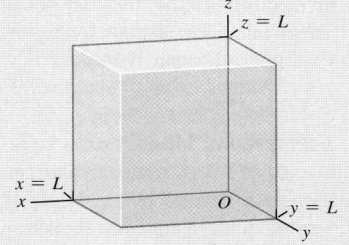

The hydrogen atom: The Schrödinger equation for the hydrogen atom gives the same energy levels as the Bohr model. If the nucleus has charge Ze, there is an additional factor of Z^2 in the numerator of Eq. (41.21). The possible magnitudes L of orbital angular momentum are given by Eq. (41.22), and the possible values of the z-component of orbital angular momentum are given by Eq. (41.23). (See Examples 41.2 and 41.3.)

The probability that an atomic electron is between r and $r + dr$ from the nucleus is $P(r)\,dr$, given by Eq. (41.25). Atomic distances are often measured in units of a, the smallest distance between the electron and the nucleus in the Bohr model. (See Example 41.4.)

$$E_n = -\frac{1}{(4\pi\epsilon_0)^2}\frac{m_re^4}{2n^2\hbar^2} = -\frac{13.60\text{ eV}}{n^2}$$

(energy levels of hydrogen)　　(41.21)

$$L = \sqrt{l(l+1)}\,\hbar$$
$(l = 0, 1, 2, \ldots, n-1)$　　(41.22)

$$L_z = m_l\hbar$$
$(m_l = 0, \pm1, \pm2, \ldots, \pm l)$　　(41.23)

$$P(r)\,dr = |\psi|^2\,dV = |\psi|^2\,4\pi r^2\,dr$$
　　(41.25)

$$a = \frac{\epsilon_0h^2}{\pi m_re^2} = \frac{4\pi\epsilon_0\hbar^2}{m_re^2}$$
$$= 5.29 \times 10^{-11}\text{ m}$$　　(41.26)

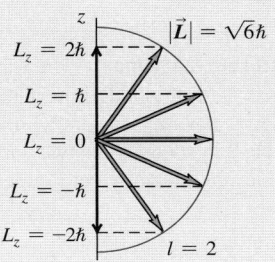

The Zeeman effect: The interaction energy of an electron (mass m) with magnetic quantum number m_l in a magnetic field $\vec{B}$ along the $+z$-direction is given by Eq. (41.35), where $\mu_B = e\hbar/2m$ is called the Bohr magneton. (See Example 41.5.)

$$U = -\mu_z B = m_l\frac{e\hbar}{2m}B = m_l\mu_B B$$
$(m_l = 0, \pm1, \pm2, \ldots, \pm l)$　　(41.35)

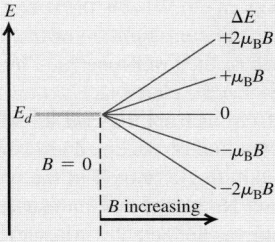

Electron spin: An electron has an intrinsic spin angular momentum of magnitude S, given by Eq. (41.37). The possible values of the z-component of the spin angular momentum are $S_z = m_s\hbar$, where $m_s = \pm\frac{1}{2}$. (See Examples 41.6 and 41.7.)

An orbiting electron experiences an interaction between its spin and the effective magnetic field produced by the relative motions of electron and nucleus. This spin-orbit coupling, along with relativistic effects, splits the energy levels according to their total angular momentum quantum number j. (See Example 41.8.)

$$S = \sqrt{\tfrac{1}{2}\left(\tfrac{1}{2}+1\right)}\,\hbar = \sqrt{\tfrac{3}{4}}\,\hbar$$　　(41.37)

$$S_z = m_s\hbar \quad \left(m_s = \pm\tfrac{1}{2}\right)$$　　(41.36)

$$E_{n,j} = -\frac{13.60\text{ eV}}{n^2}\left[1 + \frac{\alpha^2}{n^2}\left(\frac{n}{j+\frac{1}{2}} - \frac{3}{4}\right)\right]$$
　　(41.41)

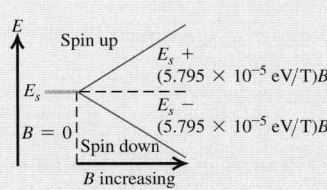

Many-electron atoms: In a hydrogen atom, the quantum numbers n, l, m_l, and m_s of the electron have certain allowed values given by Eq. (41.44). In a many-electron atom, the allowed quantum numbers for each electron are the same as in hydrogen, but the energy levels depend on both n and l because of screening, the partial cancellation of the field of the nucleus by the inner electrons. If the effective (screened) charge attracting an electron is $Z_{\text{eff}}e$, the energies of the levels are given approximately by Eq. (41.45). (See Examples 41.9 and 41.10.)

$$n \geq 1 \quad 0 \leq l \leq n - 1$$
$$|m_l| \leq l \quad m_s = \pm\tfrac{1}{2} \qquad (41.44)$$

$$E_n = -\frac{Z_{\text{eff}}^2}{n^2}(13.6 \text{ eV}) \qquad (41.45)$$

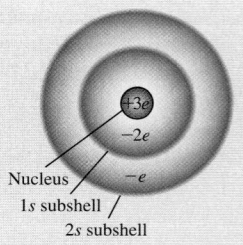

Nucleus

$1s$ subshell
$2s$ subshell

X-ray spectra: Moseley's law states that the frequency of a K_α x ray from a target with atomic number Z is given by Eq. (41.47). Characteristic x-ray spectra result from transitions to a hole in an inner energy level of an atom. (See Example 41.11.)

$$f = (2.48 \times 10^{15} \text{ Hz})(Z - 1)^2 \qquad (41.47)$$

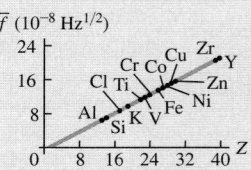

Quantum entanglement: The wave function of two identical particles can be such that neither particle is itself in a definite state. For example, the wave function could be a combination of one term with particle 1 in state A and particle 2 in state B and one term with particle 1 in state B and particle 2 in state A. The two particles are said to be entangled, since measuring the state of one particle automatically determines the results of subsequent measurements of the other particle.

$$\psi(\vec{r}_1, \vec{r}_2) = \frac{1}{\sqrt{2}}\left\{\begin{bmatrix} 1 & 2 \end{bmatrix} - \begin{bmatrix} 1 & 2 \end{bmatrix}\right\}$$

BRIDGING PROBLEM   A MANY-ELECTRON ATOM IN A BOX

An atom of titanium (Ti) has 22 electrons and has a radius of 1.47×10^{-10} m. As a simple model of this atom, imagine putting 22 electrons into a cubical box that has the same volume as a titanium atom. (a) What is the length of each side of the box? (b) What will be the configuration of the 22 electrons? (c) Find the energies of each of the levels occupied by the electrons. (Ignore the electric forces that the electrons exert on each other.) (d) You remove one of the electrons from the lowest level. As a result, one of the electrons from the highest occupied level drops into the lowest level to fill the hole, emitting a photon in the process. What is the energy of this photon? How does this compare to the energy of the K_α photon for titanium as predicted by Moseley's law?

SOLUTION GUIDE

IDENTIFY and SET UP

1. In this problem you'll use ideas from Section 41.2 about a particle in a cubical box. You'll also apply the exclusion principle from Section 41.6 to find the electron configuration of this cubical "atom." The ideas about x-ray spectra from Section 41.7 are also important.

2. The target variables are (a) the dimensions of the box, (b) the electron configurations (like those given in Table 41.3 for real atoms), (c) the occupied energy levels of the cubical box, and (d) the energy of the emitted photon.

EXECUTE

3. Use your knowledge of geometry to find the length of each side of the box.

4. Each electron state is described by four quantum numbers: n_X, n_Y, and n_Z as described in Section 41.2 and the spin magnetic quantum number m_s described in Section 41.5. Use the exclusion principle to determine the quantum numbers of each of the 22 electrons in the "atom." (*Hint:* Figure 41.4 in Section 41.2 shows the first several energy levels of a cubical box relative to the ground level $E_{1,1,1}$.)

5. Use your results from steps 3 and 4 to find the energies of each of the occupied levels.

6. Use your result from step 5 to find the energy of the photon emitted when an electron makes a transition from the highest occupied level to the ground level. Compare this to the energy that we calculated for titanium by using Moseley's law.

EVALUATE

7. Is this cubical "atom" a useful model for titanium? Why or why not?

8. In this problem you ignored the electrical interactions between electrons. To estimate how large these are, find the electrostatic potential energy of two electrons separated by half the length of the box. How does this compare to the energy levels you calculated in step 5? Is it a good approximation to ignore these interactions?

Problems

•, ••, •••: Difficulty levels. **CP**: Cumulative problems incorporating material from earlier chapters. **CALC**: Problems requiring calculus. **DATA**: Problems involving real data, scientific evidence, experimental design, and/or statistical reasoning. **BIO**: Biosciences problems.

DISCUSSION QUESTIONS

Q41.1 Particle A is described by the wave function $\psi(x, y, z)$. Particle B is described by the wave function $\psi(x, y, z)e^{i\phi}$, where ϕ is a real constant. How does the probability of finding particle A within a volume dV around a certain point in space compare with the probability of finding particle B within this same volume?

Q41.2 What are the most significant differences between the Bohr model of the hydrogen atom and the Schrödinger analysis? What are the similarities?

Q41.3 For a body orbiting the sun, such as a planet, comet, or asteroid, is there any restriction on the z-component of its orbital angular momentum such as there is with the z-component of the electron's orbital angular momentum in hydrogen? Explain.

Q41.4 Why is the analysis of the helium atom much more complex than that of the hydrogen atom, either in a Bohr type of model or using the Schrödinger equation?

Q41.5 The Stern–Gerlach experiment is always performed with beams of *neutral* atoms. Wouldn't it be easier to form beams using *ionized* atoms? Why won't this work?

Q41.6 (a) If two electrons in hydrogen atoms have the same principal quantum number, can they have different orbital angular momenta? How? (b) If two electrons in hydrogen atoms have the same orbital quantum number, can they have different principal quantum numbers? How?

Q41.7 In the Stern–Gerlach experiment, why is it essential for the magnetic field to be *inhomogeneous* (that is, nonuniform)?

Q41.8 In the ground state of the helium atom one electron must have "spin down" and the other "spin up." Why?

Q41.9 An electron in a hydrogen atom is in an s level, and the atom is in a magnetic field $\vec{B} = B\hat{k}$. Explain why the "spin up" state $\left(m_s = +\frac{1}{2}\right)$ has a higher energy than the "spin down" state $\left(m_s = -\frac{1}{2}\right)$.

Q41.10 The central-field approximation is more accurate for alkali metals than for transition metals such as iron, nickel, or copper. Why?

Q41.11 Table 41.3 shows that for the ground state of the potassium atom, the outermost electron is in a $4s$ state. What does this tell you about the relative energies of the $3d$ and $4s$ levels for this atom? Explain.

Q41.12 Do gravitational forces play a significant role in atomic structure? Explain.

Q41.13 Why do the transition elements ($Z = 21$ to 30) all have similar chemical properties?

Q41.14 Use Table 41.3 to help determine the ground-state electron configuration of the neutral gallium atom (Ga) as well as the ions Ga$^+$ and Ga$^-$. Gallium has an atomic number of 31.

Q41.15 On the basis of the Pauli exclusion principle, the structure of the periodic table of the elements shows that there must be a fourth quantum number in addition to n, l, and m_l. Explain.

Q41.16 A small amount of magnetic-field splitting of spectral lines occurs even when the atoms are not in a magnetic field. What causes this?

Q41.17 The ionization energies of the alkali metals (that is, the lowest energy required to remove one outer electron when the atom is in its ground state) are about 4 or 5 eV, while those of the noble gases are in the range from 11 to 25 eV. Why is there a difference?

Q41.18 For magnesium, the first ionization potential is 7.6 eV. The second ionization potential (additional energy required to remove a second electron) is almost twice this, 15 eV, and the third ionization potential is much larger, about 80 eV. How can these numbers be understood?

Q41.19 What is the "central-field approximation" and why is it only an approximation?

Q41.20 The nucleus of a gold atom contains 79 protons. How does the energy required to remove a $1s$ electron completely from a gold atom compare with the energy required to remove the electron from the ground level in a hydrogen atom? In what region of the electromagnetic spectrum would a photon with this energy for each of these two atoms lie?

Q41.21 (a) Can you show that the orbital angular momentum of an electron in any given direction (e.g., along the z-axis) is *always* less than or equal to its total orbital angular momentum? In which cases would the two be equal to each other? (b) Is the result in part (a) true for a classical object, such as a spinning top or planet?

Q41.22 An atom in its ground level absorbs a photon with energy equal to the K absorption edge. Does absorbing this photon ionize this atom? Explain.

Q41.23 Can a hydrogen atom emit x rays? If so, how? If not, why not?

Q41.24 A system of two electrons has the wave function $\psi(\vec{r}_1, \vec{r}_2) = (1/\sqrt{2})[\psi_\alpha(\vec{r}_1)\psi_\beta(\vec{r}_2) - \psi_\beta(\vec{r}_1)\psi_\alpha(\vec{r}_2)]$, where ψ_α is a normalized wave function for a state with $S_z = +\frac{1}{2}\hbar$ and ψ_β is a normalized wave function for a state with $S_z = -\frac{1}{2}\hbar$. (a) If S_z for electron 1 is measured, what are the possible results? What is the probability of each result? (b) If S_z for electron 2 is measured, what are the possible results? What is the probability of each result? (c) If measurement of S_z for electron 1 yields the value $\frac{1}{2}\hbar$, what are the possible results of a subsequent measurement of S_z for electron 2? What is the probability of each result being obtained? Explain.

Q41.25 Repeat Discussion Question Q41.24 for the wave function $\psi(\vec{r}_1, \vec{r}_2) = \psi_\alpha(\vec{r}_1)\psi_\alpha(\vec{r}_2)$.

EXERCISES

Section 41.2 Particle in a Three-Dimensional Box

41.1 • For a particle in a three-dimensional cubical box, what is the degeneracy (number of different quantum states with the same energy) of the energy levels (a) $3\pi^2\hbar^2/2mL^2$ and (b) $9\pi^2\hbar^2/2mL^2$?

41.2 • **CP** Model a hydrogen atom as an electron in a cubical box with side length L. Set the value of L so that the volume of the box equals the volume of a sphere of radius $a = 5.29 \times 10^{-11}$ m, the Bohr radius. Calculate the energy separation between the ground and first excited levels, and compare the result to this energy separation calculated from the Bohr model.

41.3 • **CP** A photon is emitted when an electron in a three-dimensional cubical box of side length 8.00×10^{-11} m makes a transition from the $n_X = 2$, $n_Y = 2$, $n_Z = 1$ state to the $n_X = 1$, $n_Y = 1$, $n_Z = 1$ state. What is the wavelength of this photon?

41.4 • For each of the following states of a particle in a three-dimensional cubical box, at what points is the probability distribution function a maximum: (a) $n_X = 1$, $n_Y = 1$, $n_Z = 1$ and (b) $n_X = 2$, $n_Y = 2$, $n_Z = 1$?

41.5 •• A particle is in the three-dimensional cubical box of Section 41.1. For the state $n_X = 2, n_Y = 2, n_Z = 1$, for what planes (in addition to the walls of the box) is the probability distribution function zero? Compare this number of planes to the corresponding number of planes where $|\psi|^2$ is zero for the lower-energy state $n_X = 2, n_Y = 1, n_Z = 1$ and for the ground state $n_X = 1, n_Y = 1, n_Z = 1$.

41.6 • What is the energy difference between the two lowest energy levels for a proton in a cubical box with side length 1.00×10^{-14} m, the approximate diameter of a nucleus?

Section 41.3 The Hydrogen Atom

41.7 •• Consider an electron in the N shell. (a) What is the smallest orbital angular momentum it could have? (b) What is the largest orbital angular momentum it could have? Express your answers in terms of $\hbar$ and in SI units. (c) What is the largest orbital angular momentum this electron could have in any chosen direction? Express your answers in terms of $\hbar$ and in SI units. (d) What is the largest spin angular momentum this electron could have in any chosen direction? Express your answers in terms of $\hbar$ and in SI units. (e) For the electron in part (c), what is the ratio of its spin angular momentum in the z-direction to its orbital angular momentum in the z-direction?

41.8 • An electron is in the hydrogen atom with $n = 5$. (a) Find the possible values of L and L_z for this electron, in units of $\hbar$. (b) For each value of L, find all the possible angles between $\vec{L}$ and the z-axis. (c) What are the maximum and minimum values of the magnitude of the angle between $\vec{L}$ and the z-axis?

41.9 • The orbital angular momentum of an electron has a magnitude of 4.716×10^{-34} kg·m²/s. What is the angular momentum quantum number l for this electron?

41.10 • Consider states with angular momentum quantum number $l = 2$. (a) In units of $\hbar$, what is the largest possible value of L_z? (b) In units of $\hbar$, what is the value of L? Which is larger: L or the maximum possible L_z? (c) For each allowed value of L_z, what angle does the vector $\vec{L}$ make with the $+z$-axis? How does the minimum angle for $l = 2$ compare to the minimum angle for $l = 3$ calculated in Example 41.3?

41.11 •• In a particular state of the hydrogen atom, the angle between the angular momentum vector $\vec{L}$ and the z-axis is $\theta = 26.6°$. If this is the smallest angle for this particular value of the orbital quantum number l, what is l?

41.12 •• A hydrogen atom is in a state that has $L_z = 2\hbar$. In the semiclassical vector model, the angular momentum vector $\vec{L}$ for this state makes an angle $\theta_L = 63.4°$ with the $+z$-axis. (a) What is the l quantum number for this state? (b) What is the smallest possible n quantum number for this state?

41.13 • Calculate, in units of $\hbar$, the magnitude of the maximum orbital angular momentum for an electron in a hydrogen atom for states with a principal quantum number of 2, 20, and 200. Compare each with the value of $n\hbar$ postulated in the Bohr model. What trend do you see?

41.14 • (a) Make a chart showing all possible sets of quantum numbers l and m_l for the states of the electron in the hydrogen atom when $n = 4$. How many combinations are there? (b) What are the energies of these states?

41.15 •• (a) How many different $5g$ states does hydrogen have? (b) Which of the states in part (a) has the largest angle between $\vec{L}$ and the z-axis, and what is that angle? (c) Which of the states in part (a) has the smallest angle between $\vec{L}$ and the z-axis, and what is that angle?

41.16 •• CALC (a) What is the probability that an electron in the $1s$ state of a hydrogen atom will be found at a distance less than $a/2$ from the nucleus? (b) Use the results of part (a) and of Example 41.4 to calculate the probability that the electron will be found at distances between $a/2$ and a from the nucleus.

41.17 • Show that $\Phi(\phi) = e^{im_l\phi} = \Phi(\phi + 2\pi)$ (that is, show that $\Phi(\phi)$ is periodic with period 2π) if and only if m_l is restricted to the values $0, \pm 1, \pm 2, \ldots$ (*Hint:* Euler's formula states that $e^{i\phi} = \cos\phi + i\sin\phi$.)

Section 41.4 The Zeeman Effect

41.18 • A hydrogen atom is in a d state. In the absence of an external magnetic field, the states with different m_l values have (approximately) the same energy. Consider the interaction of the magnetic field with the atom's orbital magnetic dipole moment. (a) Calculate the splitting (in electron volts) of the m_l levels when the atom is put in a 0.800-T magnetic field that is in the $+z$-direction. (b) Which m_l level will have the lowest energy? (c) Draw an energy-level diagram that shows the d levels with and without the external magnetic field.

41.19 • A hydrogen atom in a $3p$ state is placed in a uniform external magnetic field $\vec{B}$. Consider the interaction of the magnetic field with the atom's orbital magnetic dipole moment. (a) What field magnitude B is required to split the $3p$ state into multiple levels with an energy difference of 2.71×10^{-5} eV between adjacent levels? (b) How many levels will there be?

41.20 •• CP A hydrogen atom undergoes a transition from a $2p$ state to the $1s$ ground state. In the absence of a magnetic field, the energy of the photon emitted is 122 nm. The atom is then placed in a strong magnetic field in the z-direction. Ignore spin effects; consider only the interaction of the magnetic field with the atom's orbital magnetic moment. (a) How many different photon wavelengths are observed for the $2p \rightarrow 1s$ transition? What are the m_l values for the initial and final states for the transition that leads to each photon wavelength? (b) One observed wavelength is exactly the same with the magnetic field as without. What are the initial and final m_l values for the transition that produces a photon of this wavelength? (c) One observed wavelength with the field is longer than the wavelength without the field. What are the initial and final m_l values for the transition that produces a photon of this wavelength? (d) Repeat part (c) for the wavelength that is shorter than the wavelength in the absence of the field.

41.21 • A hydrogen atom in the $5g$ state is placed in a magnetic field of 0.600 T that is in the z-direction. (a) Into how many levels is this state split by the interaction of the atom's orbital magnetic dipole moment with the magnetic field? (b) What is the energy separation between adjacent levels? (c) What is the energy separation between the level of lowest energy and the level of highest energy?

Section 41.5 Electron Spin

41.22 •• A hydrogen atom in the $n = 1, m_s = -\frac{1}{2}$ state is placed in a magnetic field with a magnitude of 1.60 T in the $+z$-direction. (a) Find the magnetic interaction energy (in electron volts) of the electron with the field. (b) Is there any orbital magnetic dipole moment interaction for this state? Explain. Can there be an orbital magnetic dipole moment interaction for $n \neq 1$?

41.23 •• CP **Classical Electron Spin.** (a) If you treat an electron as a classical spherical object with a radius of 1.0×10^{-17} m, what angular speed is necessary to produce a spin angular momentum of magnitude $\sqrt{\frac{3}{4}}\hbar$? (b) Use $v = r\omega$ and the result of part (a) to calculate the speed v of a point at the electron's equator. What does your result suggest about the validity of this model?

41.24 • CP The hyperfine interaction in a hydrogen atom between the magnetic dipole moment of the proton and the spin magnetic dipole moment of the electron splits the ground level into two levels separated by 5.9×10^{-6} eV. (a) Calculate the wavelength and frequency of the photon emitted when the atom makes a transition between these states, and compare your answer to the value given at the end of Section 41.5. In what part of the electromagnetic spectrum does this lie? Such photons are emitted by cold hydrogen clouds in interstellar space; by detecting these photons, astronomers can learn about the number and density of such clouds. (b) Calculate the effective magnetic field experienced by the electron in these states (see Fig. 41.18). Compare your result to the effective magnetic field due to the spin-orbit coupling calculated in Example 41.7.

41.25 • Calculate the energy difference between the $m_s = \frac{1}{2}$ ("spin up") and $m_s = -\frac{1}{2}$ ("spin down") levels of a hydrogen atom in the 1s state when it is placed in a 1.45-T magnetic field in the *negative z*-direction. Which level, $m_s = \frac{1}{2}$ or $m_s = -\frac{1}{2}$, has the lower energy?

41.26 • A hydrogen atom in a particular orbital angular momentum state is found to have j quantum numbers $\frac{7}{2}$ and $\frac{9}{2}$. (a) What is the letter that labels the value of l for the state? (b) If $n = 5$, what is the energy difference between the $j = \frac{7}{2}$ and $j = \frac{9}{2}$ levels?

Section 41.6 Many-Electron Atoms and the Exclusion Principle

41.27 • Make a list of the four quantum numbers n, l, m_l, and m_s for each of the 10 electrons in the ground state of the neon atom. Do *not* refer to Table 41.2 or 41.3.

41.28 • For germanium (Ge, $Z = 32$), make a list of the number of electrons in each subshell (1s, 2s, 2p, . . .). Use the allowed values of the quantum numbers along with the exclusion principle; do *not* refer to Table 41.3.

41.29 •• (a) Write out the ground-state electron configuration (1s^2, 2s^2, . . .) for the beryllium atom. (b) What element of next-larger Z has chemical properties similar to those of beryllium? Give the ground-state electron configuration of this element. (c) Use the procedure of part (b) to predict what element of next-larger Z than in (b) will have chemical properties similar to those of the element you found in part (b), and give its ground-state electron configuration.

41.30 •• (a) Write out the ground-state electron configuration (1s^2, 2s^2, . . .) for the carbon atom. (b) What element of next-larger Z has chemical properties similar to those of carbon? Give the ground-state electron configuration for this element.

41.31 • The 5s electron in rubidium (Rb) sees an effective charge of 2.771e. Calculate the ionization energy of this electron.

41.32 • The energies of the 4s, 4p, and 4d states of potassium are given in Example 41.10. Calculate Z_{eff} for each state. What trend do your results show? How can you explain this trend?

41.33 • (a) The doubly charged ion N^{2+} is formed by removing two electrons from a nitrogen atom. What is the ground-state electron configuration for the N^{2+} ion? (b) Estimate the energy of the least strongly bound level in the L shell of N^{2+}. (c) The doubly charged ion P^{2+} is formed by removing two electrons from a phosphorus atom. What is the ground-state electron configuration for the P^{2+} ion? (d) Estimate the energy of the least strongly bound level in the M shell of P^{2+}.

41.34 • (a) The energy of the 2s state of lithium is -5.391 eV. Calculate the value of Z_{eff} for this state. (b) The energy of the 4s state of potassium is -4.339 eV. Calculate the value of Z_{eff} for

this state. (c) Compare Z_{eff} for the 2s state of lithium, the 3s state of sodium (see Example 41.9), and the 4s state of potassium. What trend do you see? How can you explain this trend?

41.35 • Estimate the energy of the highest-l state for (a) the L shell of Be$^+$ and (b) the N shell of Ca$^+$.

Section 41.7 X-Ray Spectra

41.36 • A K_α x ray emitted from a sample has an energy of 7.46 keV. Of which element is the sample made?

41.37 • Calculate the frequency, energy (in keV), and wavelength of the K_α x ray for the elements (a) calcium (Ca, $Z = 20$); (b) cobalt (Co, $Z = 27$); (c) cadmium (Cd, $Z = 48$).

41.38 •• The energies for an electron in the K, L, and M shells of the tungsten atom are $-69,500$ eV, $-12,000$ eV, and -2200 eV, respectively. Calculate the wavelengths of the K_α and K_β x rays of tungsten.

PROBLEMS

41.39 • In terms of the ground-state energy $E_{1,1,1}$, what is the energy of the highest level occupied by an electron when 10 electrons are placed into a cubical box?

41.40 •• An electron is in a three-dimensional box with side lengths $L_X = 0.600$ nm and $L_Y = L_Z = 2L_X$. What are the quantum numbers n_X, n_Y, and n_Z and the energies, in eV, for the four lowest energy levels? What is the degeneracy of each (including the degeneracy due to spin)?

41.41 •• CALC A particle is in the three-dimensional cubical box of Section 41.2. (a) Consider the cubical volume defined by $0 \le x \le L/4$, $0 \le y \le L/4$, and $0 \le z \le L/4$. What fraction of the total volume of the box is this cubical volume? (b) If the particle is in the ground state ($n_X = 1$, $n_Y = 1$, $n_Z = 1$), calculate the probability that the particle will be found in the cubical volume defined in part (a). (c) Repeat the calculation of part (b) when the particle is in the state $n_X = 2$, $n_Y = 1$, $n_Z = 1$.

41.42 ••• An electron is in a three-dimensional box. The x- and z-sides of the box have the same length, but the y-side has a different length. The two lowest energy levels are 2.24 eV and 3.47 eV, and the degeneracy of each of these levels (including the degeneracy due to the electron spin) is two. (a) What are the n_X, n_Y, and n_Z quantum numbers for each of these two levels? (b) What are the lengths L_X, L_Y, and L_Z for each side of the box? (c) What are the energy, the quantum numbers, and the degeneracy (including the spin degeneracy) for the next higher energy state?

41.43 •• CALC A particle in the three-dimensional cubical box of Section 41.2 is in the ground state, where $n_X = n_Y = n_Z = 1$. (a) Calculate the probability that the particle will be found somewhere between $x = 0$ and $x = L/2$. (b) Calculate the probability that the particle will be found somewhere between $x = L/4$ and $x = L/2$. Compare your results to the result of Example 41.1 for the probability of finding the particle in the region $x = 0$ to $x = L/4$.

41.44 •• CP CALC **A Three-Dimensional Isotropic Harmonic Oscillator.** An isotropic harmonic oscillator has the potential-energy function $U(x, y, z) = \frac{1}{2}k'(x^2 + y^2 + z^2)$. (*Isotropic* means that the force constant k' is the same in all three coordinate directions.) (a) Show that for this potential, a solution to Eq. (41.5) is given by $\psi = \psi_{n_x}(x)\psi_{n_y}(y)\psi_{n_z}(z)$. In this expression, $\psi_{n_x}(x)$ is a solution to the one-dimensional harmonic-oscillator Schrödinger equation, Eq. (40.44), with energy $E_{n_x} = \left(n_x + \frac{1}{2}\right)\hbar\omega$. The functions $\psi_{n_y}(y)$ and $\psi_{n_z}(z)$ are analogous one-dimensional wave functions for oscillations in the y- and z-directions. Find the energy associated with this ψ. (b) From your results in part (a) what are the ground-level and first-excited-level energies of the three-dimensional isotropic oscillator? (c) Show that there is only one state (one set of quantum numbers n_x, n_y, and n_z) for the ground level but three states for the first excited level.

41.45 •• **CP CALC Three-Dimensional Anisotropic Harmonic Oscillator.** An oscillator has the potential-energy function $U(x, y, z) = \frac{1}{2}k_1'(x^2 + y^2) + \frac{1}{2}k_2'z^2$, where $k_1' > k_2'$. This oscillator is called *anisotropic* because the force constant is not the same in all three coordinate directions. (a) Find a general expression for the energy levels of the oscillator (see Problem 41.44). (b) From your results in part (a), what are the ground-level and first-excited-level energies of this oscillator? (c) How many states (different sets of quantum numbers n_x, n_y, and n_z) are there for the ground level and for the first excited level? Compare to part (c) of Problem 41.44.

41.46 •• **CALC** A particle is described by the normalized wave function $\psi(x, y, z) = Axe^{-\alpha x^2}e^{-\beta y^2}e^{-\gamma z^2}$, where A, α, β, and γ are all real, positive constants. The probability that the particle will be found in the infinitesimal volume $dx\,dy\,dz$ centered at the point (x_0, y_0, z_0) is $|\psi(x_0, y_0, z_0)|^2\,dx\,dy\,dz$. (a) At what value of x_0 is the particle most likely to be found? (b) Are there values of x_0 for which the probability of the particle being found is zero? If so, at what x_0?

41.47 •• (a) Show that the total number of atomic states (including different spin states) in a shell of principal quantum number n is $2n^2$. [*Hint:* The sum of the first N integers $1 + 2 + 3 + \cdots + N$ is equal to $N(N + 1)/2$.] (b) Which shell has 50 states?

41.48 •• (a) What is the lowest possible energy (in electron volts) of an electron in hydrogen if its orbital angular momentum is $\sqrt{20}\,\hbar$? (b) What are the largest and smallest values of the z-component of the orbital angular momentum (in terms of $\hbar$) for the electron in part (a)? (c) What are the largest and smallest values of the spin angular momentum (in terms of $\hbar$) for the electron in part (a)? (d) What are the largest and smallest values of the orbital angular momentum (in terms of $\hbar$) for an electron in the M shell of hydrogen?

41.49 •• **CALC** Consider a hydrogen atom in the $1s$ state. (a) For what value of r is the potential energy $U(r)$ equal to the total energy E? Express your answer in terms of a. This value of r is called the *classical turning point,* since this is where a Newtonian particle would stop its motion and reverse direction. (b) For r greater than the classical turning point, $U(r) > E$. Classically, the particle cannot be in this region, since the kinetic energy cannot be negative. Calculate the probability of the electron being found in this classically forbidden region.

41.50 • **CALC** For a hydrogen atom, the probability $P(r)$ of finding the electron within a spherical shell with inner radius r and outer radius $r + dr$ is given by Eq. (41.25). For a hydrogen atom in the $1s$ ground state, at what value of r does $P(r)$ have its maximum value? How does your result compare to the distance between the electron and the nucleus for the $n = 1$ state in the Bohr model, Eq. (41.26)?

41.51 •• **CALC** The normalized radial wave function for the $2p$ state of the hydrogen atom is $R_{2p} = (1/\sqrt{24a^5})re^{-r/2a}$. After we average over the angular variables, the radial probability function becomes $P(r)\,dr = (R_{2p})^2r^2\,dr$. At what value of r is $P(r)$ for the $2p$ state a maximum? Compare your results to the radius of the $n = 2$ state in the Bohr model.

41.52 • **CP Rydberg Atoms.** *Rydberg atoms* are atoms whose outermost electron is in an excited state with a *very* large principal

quantum number. Rydberg atoms have been produced in the laboratory and detected in interstellar space. (a) Why do all neutral Rydberg atoms with the same n value have essentially the same ionization energy, independent of the total number of electrons in the atom? (b) What is the ionization energy for a Rydberg atom with a principal quantum number of 300? By the Bohr model, what is the radius of the Rydberg electron's orbit? (c) Repeat part (b) for $n = 600$.

41.53 •• (a) For an excited state of hydrogen, show that the smallest angle that the orbital angular momentum vector $\vec{L}$ can have with the z-axis is

$$(\theta_L)_{\min} = \arccos\left(\frac{n - 1}{\sqrt{n(n - 1)}}\right)$$

(b) What is the corresponding expression for $(\theta_L)_{\max}$, the largest possible angle between $\vec{L}$ and the z-axis?

41.54 •• An atom in a $3d$ state emits a photon of wavelength 475.082 nm when it decays to a $2p$ state. (a) What is the energy (in electron volts) of the photon emitted in this transition? (b) Use the selection rules described in Section 41.4 to find the allowed transitions if the atom is now in an external magnetic field of 3.500 T. Ignore the effects of the electron's spin. (c) For the case in part (b), if the energy of the $3d$ state was originally -8.50000 eV with no magnetic field present, what will be the energies of the states into which it splits in the magnetic field? (d) What are the allowed wavelengths of the light emitted during transition in part (b)?

41.55 •• **CALC Spectral Analysis.** While studying the spectrum of a gas cloud in space, an astronomer magnifies a spectral line that results from a transition from a p state to an s state. She finds that the line at 575.050 nm has actually split into three lines, with adjacent lines 0.0462 nm apart, indicating that the gas is in an external magnetic field. (Ignore effects due to electron spin.) What is the strength of the external magnetic field?

41.56 •• **CP Stern–Gerlach Experiment.** In a Stern–Gerlach experiment, the deflecting force on the atom is $F_z = -\mu_z(dB_z/dz)$, where μ_z is given by Eq. (41.38) and dB_z/dz is the magnetic-field gradient. In a particular experiment, the magnetic-field region is 50.0 cm long; assume the magnetic-field gradient is constant in that region. A beam of silver atoms enters the magnetic field with a speed of 375 m/s. What value of dB_z/dz is required to give a separation of 1.0 mm between the two spin components as they exit the field? (*Note:* The magnetic dipole moment of silver is the same as that for hydrogen, since its valence electron is in an $l = 0$ state.)

41.57 • **CP** A large number of hydrogen atoms in $1s$ states are placed in an external magnetic field that is in the $+z$-direction. Assume that the atoms are in thermal equilibrium at room temperature, $T = 300$ K. According to the Maxwell–Boltzmann distribution (see Section 39.4), what is the ratio of the number of atoms in the $m_s = \frac{1}{2}$ state to the number in the $m_s = -\frac{1}{2}$ state when the magnetic-field magnitude is (a) 5.00×10^{-5} T (approximately the earth's field); (b) 0.500 T; (c) 5.00 T?

41.58 •• **Effective Magnetic Field.** An electron in a hydrogen atom is in the $2p$ state. In a simple model of the atom, assume that the electron circles the proton in an orbit with radius r equal to the Bohr-model radius for $n = 2$. Assume that the speed v of the orbiting electron can be calculated by setting $L = mvr$ and taking L to have the quantum-mechanical value for a $2p$ state. In the frame of the electron, the proton orbits with radius r and speed v. Model the orbiting proton as a circular current loop, and calculate the magnetic field it produces at the location of the electron.

41.59 •• **Weird Universe.** In another universe, the electron is a spin-$\frac{3}{2}$ rather than a spin-$\frac{1}{2}$ particle, but all other physics are the same as in our universe. In this universe, (a) what are the atomic numbers of the lightest two inert gases? (b) What is the ground-state electron configuration of sodium?

41.60 •• A lithium atom has three electrons, and the $^2S_{1/2}$ ground-state electron configuration is $1s^2 2s$. The $1s^2 2p$ excited state is split into two closely spaced levels, $^2P_{3/2}$ and $^2P_{1/2}$, by the spin-orbit interaction (see Example 41.7 in Section 41.5). A photon with wavelength $67.09608 \ \mu m$ is emitted in the $^2P_{3/2} \rightarrow {}^2S_{1/2}$ transition, and a photon with wavelength $67.09761 \ \mu m$ is emitted in the $^2P_{1/2} \rightarrow {}^2S_{1/2}$ transition. Calculate the effective magnetic field seen by the electron in the $1s^2 2p$ state of the lithium atom. How does your result compare to that for the $3p$ level of sodium found in Example 41.7?

41.61 •• A hydrogen atom in an $n = 2$, $l = 1$, $m_l = -1$ state emits a photon when it decays to an $n = 1$, $l = 0$, $m_l = 0$ ground state. (a) In the absence of an external magnetic field, what is the wavelength of this photon? (b) If the atom is in a magnetic field in the $+z$-direction and with a magnitude of 2.20 T, what is the shift in the wavelength of the photon from the zero-field value? Does the magnetic field increase or decrease the wavelength? Disregard the effect of electron spin. [*Hint:* Use the result of Problem 39.76(c).]

41.62 •• **CP** **Electron Spin Resonance.** Electrons in the lower of two spin states in a magnetic field can absorb a photon of the right frequency and move to the higher state. (a) Find the magnetic-field magnitude B required for this transition in a hydrogen atom with $n = 1$ and $l = 0$ to be induced by microwaves with wavelength λ. (b) Calculate the value of B for a wavelength of 4.20 cm.

41.63 • Estimate the minimum and maximum wavelengths of the characteristic x rays emitted by (a) vanadium ($Z = 23$) and (b) rhenium ($Z = 45$). Discuss any approximations that you make.

41.64 •• A hydrogen atom initially in an $n = 3$, $l = 1$ state makes a transition to the $n = 2$, $l = 0$, $j = \frac{1}{2}$ state. Find the difference in wavelength between the following two photons: one emitted in a transition that starts in the $n = 3$, $l = 1$, $j = \frac{3}{2}$ state and one that starts instead in the $n = 3$, $l = 1$, $j = \frac{1}{2}$ state. Which photon has the longer wavelength?

41.65 •• **DATA** In studying electron screening in multielectron atoms, you begin with the alkali metals. You look up experimental data and find the results given in the table.

Element	Li	Na	K	Rb	Cs	Fr
Ionization energy (kJ/mol)	520.2	495.8	418.8	403.0	375.7	380

The ionization energy is the minimum energy required to remove the least-bound electron from a ground-state atom. (a) The units kJ/mol given in the table are the minimum energy in kJ required to ionize 1 mol of atoms. Convert the given values for ionization energy to the energy in eV required to ionize one atom. (b) What is the value of the nuclear charge Z for each element in the table? What is the n quantum number for the least-bound electron in the ground state? (c) Calculate Z_{eff} for this electron in each alkali-metal atom. (d) The ionization energies decrease as Z increases. Does Z_{eff} increase or decrease as Z increases? Why does Z_{eff} have this behavior?

41.66 •• **DATA** You are studying the absorption of electromagnetic radiation by electrons in a crystal structure. The situation is well described by an electron in a cubical box of side length L. The electron is initially in the ground state. (a) You observe that the longest-wavelength photon that is absorbed has a wavelength in air of $\lambda = 624 \ nm$. What is L? (b) You find that $\lambda = 234 \ nm$ is also absorbed when the initial state is still the ground state. What is the value of n^2 for the final state in the transition for which this wavelength is absorbed, where $n^2 = n_X^2 + n_Y^2 + n_Z^2$? What is the degeneracy of this energy level (including the degeneracy due to electron spin)?

41.67 •• **DATA** While working in a magnetics lab, you conduct an experiment in which a hydrogen atom in the $n = 1$ state is in a magnetic field of magnitude B. A photon of wavelength λ (in air) is absorbed in a transition from the $m_s = -\frac{1}{2}$ to the $m_s = +\frac{1}{2}$ state. The wavelengths λ as a function of B are given in the table.

B (T)	0.51	0.74	1.03	1.52	2.02	2.48	2.97
λ (mm)	21.4	14.3	10.7	7.14	5.35	4.28	3.57

(a) Graph the data in the table as photon frequency f versus B, where $f = c/\lambda$. Find the slope of the straight line that gives the best fit to the data. (b) Use your results of part (a) to calculate $|\mu_z|$, the magnitude of the spin magnetic moment. (c) Let $\gamma = |\mu_z|/|S_z|$ denote the gyromagnetic ratio for electron spin. Use your result of part (b) to calculate γ. What is the value of $\gamma/(e/2m)$ given by your experimental data?

CHALLENGE PROBLEMS

41.68 ••• Each of $2N$ electrons (mass m) is free to move along the x-axis. The potential-energy function for each electron is $U(x) = \frac{1}{2}k'x^2$, where k' is a positive constant. The electric and magnetic interactions between electrons can be ignored. Use the exclusion principle to show that the minimum energy of the system of $2N$ electrons is $\hbar N^2 \sqrt{k'/m}$. (*Hint:* See Section 40.5 and the hint given in Problem 41.47.)

41.69 ••• **CP** Consider a simple model of the helium atom in which two electrons, each with mass m, move around the nucleus (charge $+2e$) in the same circular orbit. Each electron has orbital angular momentum $\hbar$ (that is, the orbit is the smallest-radius Bohr orbit), and the two electrons are always on opposite sides of the nucleus. Ignore the effects of spin. (a) Determine the radius of the orbit and the orbital speed of each electron. [*Hint:* Follow the procedure used in Section 39.3 to derive Eqs. (39.8) and (39.9). Each electron experiences an attractive force from the nucleus and a repulsive force from the other electron.] (b) What is the total kinetic energy of the electrons? (c) What is the potential energy of the system (the nucleus and the two electrons)? (d) In this model, how much energy is required to remove both electrons to infinity? How does this compare to the experimental value of 79.0 eV?

PASSAGE PROBLEMS

BIO **ATOMS OF UNUSUAL SIZE.** In photosynthesis in plants, light is absorbed in light-harvesting complexes that consist of protein and pigment molecules. The absorbed energy is then transported to a specialized complex called the *reaction center*. Quantum-mechanical effects may play an important role in this energy transfer. In a recent experiment, researchers cooled rubidium atoms to a very low temperature to study a similar energy-transfer process in the lab. Laser light was used to excite an electron in each atom to a state with large n. This highly excited electron behaves much like the single electron in a hydrogen atom, with an effective (screened) atomic number $Z_{\text{eff}} = 1$. Because n is so large, though, the excited electron is quite far from the atomic nucleus, with an orbital radius of approximately $1 \ \mu m$, and is weakly bound. Using these so-called *Rydberg atoms,* the researchers were able to study the way energy is transported from one atom to the next. This process may be a model for understanding energy transport in photosynthesis. (Source: "Observing the Dynamics of Dipole-Mediated Energy Transport by Interaction Enhanced Imaging," by G. Günter et al., *Science* 342(6161): 954–956, Nov. 2013.)

41.70 In the Bohr model, what is the principal quantum number n at which the excited electron is at a radius of 1 μm? (a) 140; (b) 400; (c) 20; (d) 81.

41.71 Take the size of a Rydberg atom to be the diameter of the orbit of the excited electron. If the researchers want to perform this experiment with the rubidium atoms in a gas, with atoms separated by a distance 10 times their size, the density of atoms per cubic centimeter should be about (a) 10^5 atoms/cm^3; (b) 10^8 atoms/cm^3; (c) 10^{11} atoms/cm^3; (d) 10^{21} atoms/cm^3.

41.72 Assume that the researchers place an atom in a state with $n = 100$, $l = 2$. What is the magnitude of the orbital angular momentum $\vec{L}$ associated with this state? (a) $\sqrt{2}\,\hbar$; (b) $\sqrt{6}\,\hbar$; (c) $\sqrt{200}\,\hbar$; (d) $\sqrt{10,100}\,\hbar$.

41.73 How many different possible electron states are there in the $n = 100$, $l = 2$ subshell? (a) 2; (b) 100; (c) 10,000; (d) 10.

Answers

Chapter Opening Question ?

(iii) The Pauli exclusion principle is responsible. Helium is inert because its two electrons fill the K shell; lithium is very reactive because its third electron must go into the L shell and is loosely bound. See Section 41.6 for more details.

Test Your Understanding Questions

41.1 (iv) If $U(x, y, z) = 0$ in a certain region of space, we can rewrite the time-independent Schrödinger equation [Eq. (41.5)] for that region as $\partial^2\psi/\partial x^2 + \partial^2\psi/\partial y^2 + \partial^2\psi/\partial z^2 = (-2mE/\hbar^2)\psi$. We are told that all of the second derivatives of $\psi(x, y, z)$ are positive in this region, so the left-hand side of this equation is positive. Hence the right-hand side $(-2mE/\hbar^2)\psi$ must also be positive. Since $E > 0$, the quantity $-2mE/\hbar^2$ is negative, and so $\psi(x, y, z)$ must be negative.

41.2 (iv), (ii), (i) and (iii) (tie) Equation (41.16) shows that the energy levels for a cubical box are proportional to the quantity $n_X^2 + n_Y^2 + n_Z^2$. Hence ranking in order of this quantity is the same as ranking in order of energy. For the four cases we are given, we have (i) $n_X^2 + n_Y^2 + n_Z^2 = 2^2 + 3^2 + 2^2 = 17$; (ii) $n_X^2 + n_Y^2 + n_Z^2 = 4^2 + 1^2 + 1^2 = 18$; (iii) $n_X^2 + n_Y^2 + n_Z^2 = 2^2 + 2^2 + 3^2 = 17$; and (iv) $n_X^2 + n_Y^2 + n_Z^2 = 1^2 + 3^2 + 3^2 = 19$. The states $(n_X, n_Y, n_Z) = (2, 3, 2)$ and $(n_X, n_Y, n_Z) = (2, 2, 3)$ have the same energy (they are degenerate).

41.3 (ii) and (iii) (tie), (i) An electron in a state with principal quantum number n is most likely to be found at $r = n^2 a$. This result is independent of the values of the quantum numbers l and m_l. Hence an electron with $n = 2$ (most likely to be found at $r = 4a$) is more likely to be found near $r = 5a$ than an electron with $n = 1$ (most likely to be found at $r = a$).

41.4 no All that matters is the component of the electron's orbital magnetic moment along the direction of $\vec{B}$. We called this quantity μ_z in Eq. (41.32) because we *defined* the positive z-axis to be in the direction of $\vec{B}$. In reality, the names of the axes are arbitrary.

41.5 (iv) For the magnetic moment to be perfectly aligned with the z-direction, the z-component of the spin vector $\vec{S}$ would have to have the same absolute value as $\vec{S}$. However, the possible values of S_z are $\pm\frac{1}{2}\hbar$ [Eq. (41.36)], while the magnitude of the spin vector is $S = \sqrt{\frac{3}{4}}\,\hbar$ [Eq. (41.37)]. Hence $\vec{S}$ can never be perfectly aligned with any one direction in space.

41.6 more difficult If there were no exclusion principle, all 11 electrons in the sodium atom would be in the level of lowest energy (the $1s$ level) and the configuration would be $1s^{11}$. Consequently, it would be more difficult to remove the first electron. (In a real sodium atom the valence electron is in a screened $3s$ state, which has a comparatively high energy.)

41.7 (iv) An absorption edge appears if the photon energy is just high enough to remove an electron in a given energy level from the atom. In a sample of high-temperature hydrogen we expect to find atoms whose electron is in the ground level ($n = 1$), the first excited level ($n = 2$), and the second excited level ($n = 3$). From Eq. (41.21) these levels have energies $E_n = (-13.60 \text{ eV})/n^2 = -13.60 \text{ eV}, -3.40 \text{ eV},$ and -1.51 eV (see Fig. 39.24b).

41.8 yes; no This wave function says that it is equally possible that the electron (particle 1) is in state C and the proton (particle 2) is in state E or that particle 1 is in state C and particle 2 is in state F. Since particles 1 and 2 are not identical and are distinguishable, there is no reason we can't know that particle 1 is in state C independent of which state particle 2 is in. So this is a valid wave function for the system, and the two particles are not entangled. If we measure the state of the electron alone, we are guaranteed to get C as a result; a subsequent measurement of the proton's state will give either E or F with equal probability, the same as if we had not first measured the state of the electron. Similarly, if we first measure the state of the proton, the results of that measurement will not affect a subsequent measurement of the state of the electron (for which the result is guaranteed to be C).

Bridging Problem

(a) 2.37×10^{-10} m

(b) Values of (n_X, n_Y, n_Z, m_s) for the 22 electrons: $(1, 1, 1, +\frac{1}{2})$, $(1, 1, 1, -\frac{1}{2})$, $(2, 1, 1, +\frac{1}{2})$, $(2, 1, 1, -\frac{1}{2})$, $(1, 2, 1, +\frac{1}{2})$, $(1, 2, 1, -\frac{1}{2})$, $(1, 1, 2, +\frac{1}{2})$, $(1, 1, 2, -\frac{1}{2})$, $(2, 2, 1, +\frac{1}{2})$, $(2, 2, 1, -\frac{1}{2})$, $(2, 1, 2, +\frac{1}{2})$, $(2, 1, 2, -\frac{1}{2})$, $(1, 2, 2, +\frac{1}{2})$, $(1, 2, 2, -\frac{1}{2})$, $(3, 1, 1, +\frac{1}{2})$, $(3, 1, 1, -\frac{1}{2})$, $(1, 3, 1, +\frac{1}{2})$, $(1, 3, 1, -\frac{1}{2})$, $(1, 1, 3, +\frac{1}{2})$, $(1, 1, 3, -\frac{1}{2})$, $(2, 2, 2, +\frac{1}{2})$, $(2, 2, 2, -\frac{1}{2})$

(c) 20.1 eV, 40.2 eV, 60.3 eV, 73.7 eV, and 80.4 eV

(d) 60.3 eV versus 4.52×10^3 eV

? Although Venus is almost twice as far as Mercury is from the sun, it has a higher surface temperature: 735 K (462°C = 863°F). The reason is that Venus has a thick, cloud-shrouded atmosphere (shown here in false color) that is 96.5% carbon dioxide (CO_2). Molecules of CO_2 are a potent agent for raising Venus's temperature because (i) they absorb infrared radiation in vibrational transitions; (ii) they absorb infrared radiation in electronic transitions; (iii) they absorb ultraviolet radiation in vibrational transitions; (iv) they absorb ultraviolet radiation in electronic transitions; (v) more than one of these.

42 MOLECULES AND CONDENSED MATTER

solated atoms, which we studied in Chapter 41, are the exception; usually we find atoms combined to form molecules or more extended structures we call condensed matter (liquid or solid). In this chapter we'll study the attractive forces, called molecular bonds, that cause atoms to combine into molecules. We will see that just as atoms have quantized energies determined by the quantum-mechanical state of their electrons, so molecules have quantized energies determined by their rotational and vibrational states.

The same physical principles behind molecular bonds also apply to the study of condensed matter, in which various types of bonding occur. We'll explore the concept of energy bands and see how it helps us understand the properties of solids. Then we'll look more closely at the properties of a special class of solids called semiconductors. Devices using semiconductors are found in every mobile phone, TV, and computer used today.

42.1 TYPES OF MOLECULAR BONDS

We can use our discussion of atomic structure in Chapter 41 as a basis for exploring the nature of *molecular bonds*, the interactions that hold atoms together to form stable structures such as molecules and solids.

Ionic Bonds

The **ionic bond** is an interaction between oppositely charged *ionized* atoms. The most familiar example is sodium chloride (NaCl), in which the sodium (Na) atom loses its one $3s$ electron, which fills the vacancy in the $3p$ subshell of the chlorine (Cl) atom.

Let's look at the energy balance in this transaction. Removing the $3s$ electron from a neutral Na atom requires 5.138 eV of energy; this is called the *ionization energy* of Na. The neutral Cl atom can attract an extra electron into the vacancy

42.1 When the separation r between two oppositely charged ions is large, the potential energy $U(r)$ is proportional to $1/r$ as for point charges and the force is attractive. As r decreases, the charge clouds of the two atoms overlap and the force becomes less attractive. If r is less than the equilibrium separation r_0, the force is repulsive.

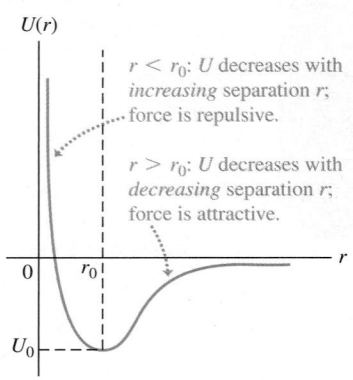

$r < r_0$: U decreases with *increasing* separation r; force is repulsive.

$r > r_0$: U decreases with *decreasing* separation r; force is attractive.

in the $3p$ subshell, where it is incompletely screened by the other electrons and therefore is attracted to the nucleus. This state has 3.613 eV lower energy than a state with a neutral Cl atom and a distant free electron; 3.613 eV is the magnitude of the *electron affinity* of chlorine. Thus creating the well-separated Na^+ and Cl^- ions requires a net investment of only 5.138 eV − 3.613 eV = 1.525 eV.

When their mutual attraction brings the Na^+ and Cl^- ions together, the magnitude of their negative potential energy is determined by their separation r (**Fig. 42.1**). The exclusion principle (Section 41.6), which states that only one electron can occupy a given quantum-mechanical state, limits how small this separation can be. As r decreases, the exclusion principle distorts the charge clouds, so the ions no longer interact like point charges and the interaction eventually becomes repulsive.

The minimum electric potential energy for NaCl turns out to be −5.7 eV at a separation of 0.24 nm. The net energy released in creating the ions and letting them come together to the equilibrium separation of 0.24 nm is 5.7 eV − 1.525 eV = 4.2 eV. Thus, if the kinetic energy of the ions is ignored, 4.2 eV is the *binding energy* of the NaCl molecule, the energy that is needed to dissociate the molecule into separate neutral atoms.

Ionic bonds can involve more than one electron per atom. For instance, alkaline earth elements form ionic compounds in which an atom loses *two* electrons; an example is magnesium chloride, or $Mg^{2+}(Cl^-)_2$. Ionic bonds that involve a loss of more than two electrons are relatively rare. Instead, a different kind of bond, the *covalent* bond, comes into operation. We'll discuss this type of bond below.

EXAMPLE 42.1 **ELECTRIC POTENTIAL ENERGY OF THE NaCl MOLECULE**

Find the electric potential energy of an Na^+ ion and a Cl^- ion separated by 0.24 nm. Consider the ions as point charges.

SOLUTION

IDENTIFY and SET UP: Equation (23.9) in Section 23.1 tells us that the electric potential energy of two point charges q and q_0 separated by a distance r is $U = qq_0/4\pi\epsilon_0 r$.

EXECUTE: We have $q = +e$ (for Na^+), $q_0 = -e$ (for Cl^-), and $r = 0.24$ nm $= 0.24 \times 10^{-9}$ m. From Eq. (23.9),

$$U = -\frac{1}{4\pi\epsilon_0} \frac{e^2}{r_0} = -(9.0 \times 10^9 \text{ N}\cdot\text{m}^2/\text{C}^2)\frac{(1.6 \times 10^{-19} \text{ C})^2}{0.24 \times 10^{-9} \text{ m}}$$

$$= -9.6 \times 10^{-19} \text{J} = -6.0 \text{ eV}$$

EVALUATE: This result agrees fairly well with the observed value of −5.7 eV. The reason for the difference is that when the two ions are at their equilibrium separation of 0.24 nm, the outer regions of their electron clouds overlap. Hence the two ions don't behave exactly like point charges.

PhET: Double Wells and Covalent Bonds

Covalent Bonds

Unlike the transaction that occurs in an ionic bond, in a **covalent bond** there is no net transfer of electrons from one atom to another. The simplest covalent bond is found in the hydrogen molecule, a structure containing two protons and two electrons. As the separate atoms (**Fig. 42.2a**) come together, the electron wave functions are distorted and become more concentrated in the region between the two protons (Fig. 42.2b). The net attraction of the electrons for each proton more than balances the repulsion of the two protons and of the two electrons.

The attractive interaction is then supplied by a *pair* of electrons, one contributed by each atom, with charge clouds that are concentrated primarily in the region between the two atoms. The energy of the covalent bond in the hydrogen molecule H_2 is −4.48 eV.

As we saw in Section 41.6, the exclusion principle permits two electrons to occupy the same region of space (that is, to be in the same spatial quantum state) only when they have opposite spins. Hence the two electrons in the H_2 covalent bond (Fig. 42.2b) must have opposite spins, since both occupy the same region between the two nuclei. Opposite spins are an essential requirement for a covalent bond, and no more than two electrons can participate in such a bond.

However, an atom with several electrons in its outermost shell can form several covalent bonds. The bonding of carbon and hydrogen atoms, of central importance in organic chemistry, is an example. In the *methane* molecule (CH_4) the carbon atom is at the center of a regular tetrahedron, with a hydrogen atom at each corner. The carbon atom has four electrons in its L shell, and each of these four electrons forms a covalent bond with one of the four hydrogen atoms (**Fig. 42.3**). Similar patterns occur in more complex organic molecules.

Covalent bonds are highly directional. In the methane molecule the wave function for each of carbon's four valence electrons is a combination of the $2s$ and $2p$ wave functions called a *hybrid wave function*. The probability distribution for each one has a lobe protruding toward a corner of a tetrahedron. This symmetric arrangement minimizes the overlap of wave functions for the electron pairs, which in turn minimizes the positive potential energy associated with repulsion between the pairs.

Ionic and covalent bonds represent two extremes in molecular bonding, but there is no sharp division between the two types. Often there is a *partial* transfer of one or more electrons from one atom to another. As a result, many molecules that have dissimilar atoms have electric dipole moments—that is, a preponderance of positive charge at one end and of negative charge at the other. Such molecules are called *polar* molecules. Water molecules have large electric dipole moments; these are responsible for the exceptionally large dielectric constant of liquid water (see Sections 24.4 and 24.5).

Van der Waals Bonds

Ionic and covalent bonds, with typical bond energies of 1 to 5 eV, are called *strong bonds*. There are also two types of weaker bonds. One of these, the **van der Waals bond**, is an interaction between the electric dipole moments of atoms or molecules; typical energies are 0.1 eV or less. The bonding of water molecules in the liquid and solid states results partly from dipole–dipole interactions.

No atom has a permanent electric dipole moment, nor do many molecules. However, fluctuating charge distributions can lead to fluctuating dipole moments; these in turn can induce dipole moments in neighboring structures. Overall, the resulting dipole–dipole interaction is attractive, giving a weak bonding of atoms or molecules. The interaction potential energy drops off very quickly with distance r between molecules, usually as $1/r^6$. The liquefaction and solidification of the inert gases and of molecules such as H_2, O_2, and N_2 are due to induced-dipole van der Waals interactions. Not much thermal-agitation energy is needed to break these weak bonds, so such substances usually exist in the liquid and solid states only at very low temperatures.

Hydrogen Bonds

In the other type of weak bond, the **hydrogen bond**, a proton (H^+ ion) gets between two atoms, polarizing them and attracting them by means of the induced dipoles. This bond is unique to hydrogen-containing compounds because only hydrogen has a singly ionized state with no remaining electron cloud; the hydrogen ion is a bare proton, much smaller than any other singly ionized atom. The bond energy is usually less than 0.5 eV. The hydrogen bond is responsible for the cross-linking of long-chain organic molecules such as polyethylene (used in plastic bags). Hydrogen bonding also plays a role in the structure of ice.

All these bond types hold the atoms together in *solids* as well as in molecules. Indeed, a solid is in many respects a giant molecule. Still another type of

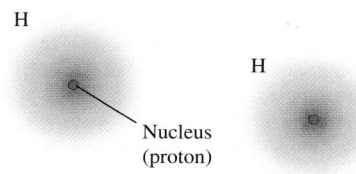

42.2 Covalent bond in a hydrogen molecule.

(a) Separate hydrogen atoms

H

H

Nucleus
(proton)

Individual H atoms are usually widely separated and do not interact.

(b) H_2 molecule

H_2

Covalent bond: the charge clouds for the two electrons with opposite spins are concentrated in the region between the nuclei.

42.3 Schematic diagram of the methane (CH_4) molecule. The carbon atom is at the center of a regular tetrahedron and forms four covalent bonds with the hydrogen atoms at the corners. Each covalent bond includes two electrons with opposite spins, forming a charge cloud that is concentrated between the carbon atom and a hydrogen atom.

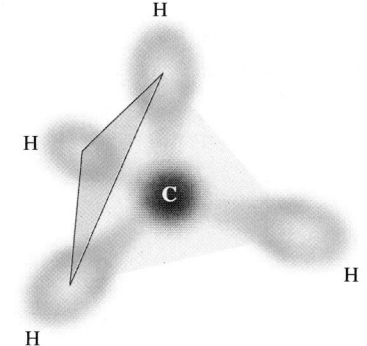

H

H

C

H

H

BIO Application Molecular Zipper
A DNA molecule functions like a twisted zipper. Each of the two strands of the "zipper" consists of an outer backbone and inward-facing nucleotide "teeth"; hydrogen bonds between facing teeth "zip" the strands together. The covalent bonds that hold together the atoms of each strand are strong, whereas the hydrogen bonds are relatively weak, so that the cell's biochemical machinery can easily separate the strands for reading or copying.

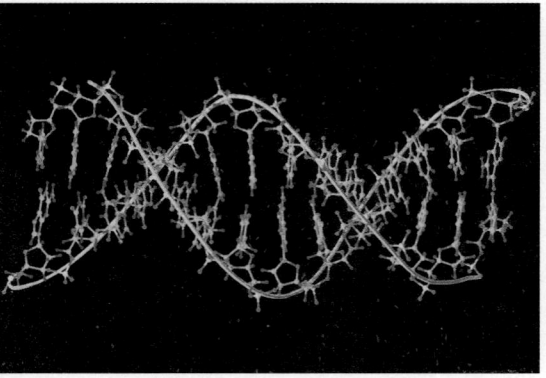

42.4 A diatomic molecule modeled as two point masses m_1 and m_2 separated by a distance r_0. The distances of the masses from the center of mass are r_1 and r_2, where $r_1 + r_2 = r_0$.

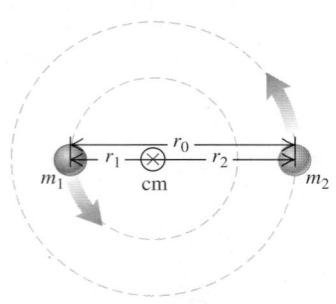

42.5 The ground level and first four excited rotational energy levels for a diatomic molecule. The levels are not equally spaced.

E

$10\hbar^2/I$ ———— $l = 4$

$6\hbar^2/I$ ———— $l = 3$

$3\hbar^2/I$ ———— $l = 2$

$\hbar^2/I$ ———— $l = 1$
0 ———— $l = 0$

bonding, the *metallic bond*, comes into play in the structure of metallic solids. We'll return to this subject in Section 42.3.

TEST YOUR UNDERSTANDING OF SECTION 42.1   If electrons obeyed the exclusion principle but did *not* have spin, how many electrons could participate in a covalent bond? (i) One; (ii) two; (iii) three; (iv) more than three. ∎

42.2 MOLECULAR SPECTRA

Molecules have energy levels that are associated with rotation of a molecule as a whole and with vibration of the atoms relative to each other. Just as transitions between energy levels in atoms lead to atomic spectra, transitions between rotational and vibrational levels in molecules lead to *molecular spectra*.

Rotational Energy Levels

In this discussion we'll concentrate mostly on *diatomic* molecules, to keep things as simple as possible. In **Fig. 42.4** we picture a diatomic molecule as a rigid dumbbell (two point masses m_1 and m_2 separated by a constant distance r_0) that can *rotate* about axes through its center of mass, perpendicular to the line joining them. What are the energy levels associated with this motion?

We showed in Section 10.5 that when a rigid body rotates with angular speed ω about a perpendicular axis through its center of mass, the magnitude L of its angular momentum is given by Eq. (10.28), $L = I\omega$, where I is its moment of inertia for that axis. Its kinetic energy is given by Eq. (9.17), $K = \frac{1}{2}I\omega^2$. Combining these two equations, we find $K = L^2/2I$. There is no potential energy U, so the kinetic energy K is equal to the total mechanical energy E:

$$E = \frac{L^2}{2I} \tag{42.1}$$

Zero potential energy means that U does not depend on the angular coordinate of the molecule. But the potential-energy function U for the hydrogen atom (see Section 41.3) also has no dependence on angular coordinates. Thus the angular solutions to the Schrödinger equation for rigid-body rotation are the same as for the hydrogen atom, and the angular momentum is quantized in the same way. As in Eq. (41.22),

$$L = \sqrt{l(l + 1)}\,\hbar \qquad (l = 0, 1, 2, \dots) \tag{42.2}$$

Combining Eqs. (42.1) and (42.2), we obtain the *rotational energy levels:*

Rotational quantum number ($l = 0, 1, 2, \dots$)

Rotational energy levels of a diatomic molecule $\cdots\!\rightarrow E_l = l(l + 1)\dfrac{\hbar^2}{2I}$ $\cdots$ Planck's constant divided by 2π $\tag{42.3}$

Moment of inertia for axis through molecule's cm

Figure 42.5 is an energy-level diagram showing these rotational levels. The $l = 0$ ground level has zero angular momentum (no rotation and zero rotational energy E). The spacing of adjacent levels increases with increasing l.

We can express the moment of inertia I in Eqs. (42.1) and (42.3) in terms of the *reduced mass* m_r of the molecule:

Mass of atom 1   Mass of atom 2

Reduced mass of a $\cdots\!\rightarrow m_r = \dfrac{m_1 m_2}{m_1 + m_2}$ $\tag{42.4}$
diatomic molecule

We introduced the reduced mass in Section 39.3 to accommodate the finite nuclear mass of the hydrogen atom. In Fig. 42.4 the distances r_1 and r_2 are the distances from the center of mass to the centers of the atoms. By the definition of the center of mass, $m_1 r_1 = m_2 r_2$, and the figure also shows that $r_0 = r_1 + r_2$. Solving these equations for r_1 and r_2, we find

PhET: The Greenhouse Effect

$$r_1 = \frac{m_2}{m_1 + m_2} r_0 \qquad r_2 = \frac{m_1}{m_1 + m_2} r_0 \qquad (42.5)$$

The moment of inertia is $I = m_1 r_1^2 + m_2 r_2^2$; substituting Eq. (42.5), we find

$$I = m_1 \frac{m_2^2}{(m_1 + m_2)^2} r_0^2 + m_2 \frac{m_1^2}{(m_1 + m_2)^2} r_0^2 = \frac{m_1 m_2}{m_1 + m_2} r_0^2 \quad \text{or}$$

Moment of inertia ⋯⋯
of a diatomic molecule, $I = m_r r_0^2$ Reduced mass   Distance between centers
axis through molecule's cm   of molecule's two atoms   (42.6)

The moment of inertia is the same as that of an equivalent *single* point mass m_r that orbits the axis in a circle of radius r_0.

To conserve angular momentum and account for the angular momentum of the emitted or absorbed photon, the allowed transitions between rotational states must satisfy the same selection rule that we discussed in Section 41.4 for allowed transitions between the states of an atom: l must change by exactly one unit; that is, $\Delta l = \pm 1$.

EXAMPLE 42.2   ROTATIONAL SPECTRUM OF CARBON MONOXIDE

The two nuclei in the carbon monoxide (CO) molecule are 0.1128 nm apart. The mass of the most common carbon atom is 1.993×10^{-26} kg; that of the most common oxygen atom is 2.656×10^{-26} kg. (a) Find the energies of the lowest three rotational energy levels of CO. Express your results in meV (1 meV = 10^{-3} eV). (b) Find the wavelength of the photon emitted in the transition from the $l = 2$ to the $l = 1$ level.

SOLUTION

IDENTIFY and SET UP: This problem uses the ideas developed in this section about the rotational energy levels of molecules. We are given the distance r_0 between the atoms and their masses m_1 and m_2. We find the reduced mass m_r from Eq. (42.4), the moment of inertia I from Eq. (42.6), and the energies E_l from Eq. (42.3). The energy E of the emitted photon is equal to the difference in energy between the $l = 2$ and $l = 1$ levels. (This transition obeys the $\Delta l = \pm 1$ selection rule, since $\Delta l = 1 - 2 = -1$.) We determine the photon wavelength by using $E = hc/\lambda$.

EXECUTE: (a) From Eqs. (42.4) and (42.6), the reduced mass and moment of inertia of the CO molecule are:

$$m_r = \frac{m_1 m_2}{m_1 + m_2}$$

$$= \frac{(1.993 \times 10^{-26} \text{ kg})(2.656 \times 10^{-26} \text{ kg})}{(1.993 \times 10^{-26} \text{ kg}) + (2.656 \times 10^{-26} \text{ kg})}$$

$$= 1.139 \times 10^{-26} \text{ kg}$$

$$I = m_r r_0^2$$

$$= (1.139 \times 10^{-26} \text{ kg})(0.1128 \times 10^{-9} \text{ m})^2$$

$$= 1.449 \times 10^{-46} \text{ kg} \cdot \text{m}^2$$

The rotational levels are given by Eq. (42.3):

$$E_l = l(l + 1)\frac{\hbar^2}{2I} = l(l + 1)\frac{(1.0546 \times 10^{-34} \text{ J} \cdot \text{s})^2}{2(1.449 \times 10^{-46} \text{ kg} \cdot \text{m}^2)}$$

$$= l(l + 1)(3.838 \times 10^{-23} \text{ J}) = l(l + 1)0.2395 \text{ meV}$$

(1 meV = 10^{-3} eV.) Substituting $l = 0, 1, 2$, we find

$$E_0 = 0 \qquad E_1 = 0.479 \text{ meV} \qquad E_2 = 1.437 \text{ meV}$$

(b) The photon energy and wavelength are

$$E = E_2 - E_1 = 0.958 \text{ meV}$$

$$\lambda = \frac{hc}{E} = \frac{(4.136 \times 10^{-15} \text{ eV} \cdot \text{s})(3.00 \times 10^8 \text{ m/s})}{0.958 \times 10^{-3} \text{ eV}}$$

$$= 1.29 \times 10^{-3} \text{ m} = 1.29 \text{ mm}$$

EVALUATE: The differences between the first few rotational energy levels of CO are very small (about 1 meV = 10^{-3} eV) compared to the differences between atomic energy levels (typically a few eV). Hence a photon emitted by a CO molecule in a transition from the $l = 2$ to the $l = 1$ level has very low energy and a very long wavelength compared to the visible light emitted by excited atoms. Photon wavelengths for rotational transitions in molecules are typically in the microwave and far infrared regions of the spectrum.

In this example we were given the equilibrium separation between the atoms, also called the *bond length*, and we used it to calculate one of the wavelengths emitted by excited CO molecules. In experiments, scientists work this problem backward: By measuring the long-wavelength emissions of a sample of diatomic molecules, they determine the moment of inertia of the molecule and hence the bond length.

42.6 A diatomic molecule modeled as two point masses m_1 and m_2 connected by a spring with force constant k'.

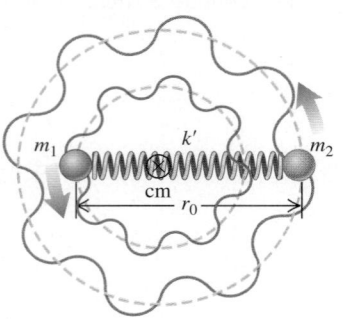

42.7 The ground level and first three excited vibrational levels for a diatomic molecule, assuming small displacements from equilibrium so we can treat the oscillations as simple harmonic. (Compare Fig. 40.25.)

E

$\frac{7}{2}\hbar\omega$ ——————— $n = 3$

$\frac{5}{2}\hbar\omega$ ——————— $n = 2$

$\frac{3}{2}\hbar\omega$ ——————— $n = 1$

$\frac{1}{2}\hbar\omega$ ——————— $n = 0$

0 - - - - - -

42.8 Energy-level diagram for vibrational and rotational energy levels of a diatomic molecule. For each vibrational level (n) there is a series of more closely spaced rotational levels (l). Several transitions corresponding to a single band in a band spectrum are shown. These transitions obey the selection rule $\Delta l = \pm 1$.

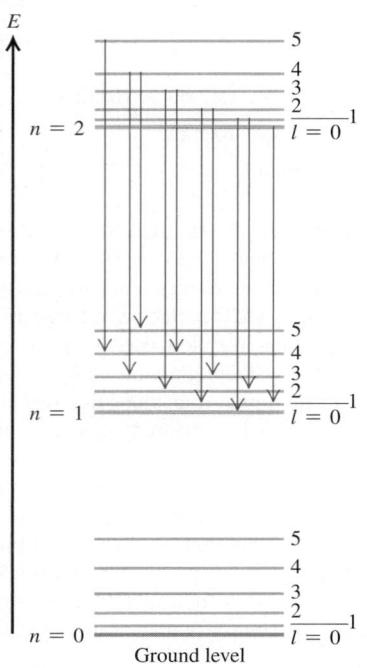

Vibrational Energy Levels

Molecules are never completely rigid. In a more realistic model of a diatomic molecule we represent the connection between atoms not as a rigid rod but as a spring (**Fig. 42.6**). Then, in addition to rotating, the atoms of the molecule can *vibrate* about their equilibrium positions along the line joining them. For small oscillations the restoring force can be taken as proportional to the displacement from the equilibrium separation r_0 (like a spring that obeys Hooke's law with a force constant k'), and the system is a harmonic oscillator. We discussed the quantum-mechanical harmonic oscillator in Section 40.5. The energy levels are given by Eq. (40.46) with the mass m replaced by the reduced mass m_r:

Vibrational energy levels of a diatomic molecule
$$E_n = (n + \tfrac{1}{2})\hbar\omega = (n + \tfrac{1}{2})\hbar\sqrt{\frac{k'}{m_r}} \tag{42.7}$$
Vibrational quantum number ($n = 0, 1, 2, \dots$); Planck's constant divided by 2π; Oscillation angular frequency; Reduced mass; Force constant

The spacing in energy between any two adjacent vibrational levels is

$$\Delta E = \hbar\omega = \hbar\sqrt{\frac{k'}{m_r}} \tag{42.8}$$

Figure 42.7 is an energy-level diagram showing these vibrational levels. As an example, for the CO molecule of Example 42.2 the spacing $\hbar\omega$ between levels is 0.2690 eV. From Eq. (42.8) this corresponds to a force constant of 1.90×10^3 N/m, which is a fairly loose spring. (To stretch a macroscopic spring with this value of k' by 1.0 cm would require a force of only 19 N, or about 4 lb.) Force constants for diatomic molecules are typically about 100 to 2000 N/m.

CAUTION **Watch out for k, k', and K** As in Section 40.5 we're using k' for the force constant, this time to minimize confusion with Boltzmann's constant k, the gas constant per molecule (introduced in Section 18.3). Besides the quantities k and k', we also use the absolute temperature unit $1\,\text{K} = 1$ kelvin. ∎

Rotation and Vibration Combined

Visible-light photons have energies between 1.65 eV and 3.26 eV. The 0.2690-eV energy difference between vibrational levels for carbon monoxide (CO) corresponds to a photon of wavelength 4.613 μm, in the infrared region of the spectrum. This is much closer to the visible region than is the photon in the rotational transition in Example 42.2. In general the energy differences for molecular *vibration* are much smaller than those that produce atomic spectra, but much larger than the energy differences for molecular *rotation*.

When we include *both* rotational and vibrational energies, the energy levels for our diatomic molecule are

$$E_{nl} = l(l + 1)\frac{\hbar^2}{2I} + (n + \tfrac{1}{2})\hbar\sqrt{\frac{k'}{m_r}} \tag{42.9}$$

Figure 42.8 shows the energy-level diagram. For each value of n there are many values of l, forming a series of closely spaced levels.

The red arrows in Fig. 42.8 show several possible transitions in which a molecule goes from a level with $n = 2$ to a level with $n = 1$ by emitting a photon. As we mentioned, these transitions must obey the selection rule $\Delta l = \pm 1$ to conserve angular momentum. Another selection rule states that if the vibrational level changes, the vibrational quantum number n in Eq. (42.9) must increase by 1 ($\Delta n = 1$) if a photon is absorbed or decrease by 1 ($\Delta n = -1$) if a photon is emitted.

42.9 A typical molecular band spectrum.

Illustrating these selection rules, Fig. 42.8 shows that a molecule in the $n = 2$, $l = 4$ level can emit a photon and drop into the $n = 1$, $l = 5$ level ($\Delta n = -1$, $\Delta l = +1$) or the $n = 1$, $l = 3$ level ($\Delta n = -1$, $\Delta l = -1$), but is forbidden from making a $\Delta n = -1$, $\Delta l = 0$ transition into the $n = 1$, $l = 4$ level.

Transitions between states with various pairs of n-values give different series of spectrum lines, and the resulting spectrum has a series of *bands*. Each band corresponds to a particular vibrational transition, and each individual line in a band represents a particular rotational transition, with the selection rule $\Delta l = \pm 1$. **Figure 42.9** shows a typical *band spectrum*.

All molecules can have excited states of the *electrons* in addition to the rotational and vibrational states that we have described. In general, these lie at higher energies than the rotational and vibrational states, and there is no simple rule relating them. When there is a transition between electronic states, the $\Delta n = \pm 1$ selection rule for the vibrational levels no longer holds.

EXAMPLE 42.3 **VIBRATION-ROTATION SPECTRUM OF CARBON MONOXIDE**

Consider again the CO molecule of Example 42.2. Find the wavelength of the photon emitted by a CO molecule when its vibrational energy changes and its rotational energy is (a) initially zero and (b) finally zero.

SOLUTION

IDENTIFY and SET UP: We need to use the selection rules for the vibrational and rotational transitions of a diatomic molecule. Since a photon is emitted as the vibrational energy changes, the selection rule $\Delta n = -1$ tells us that the vibrational quantum number n decreases by 1 in both parts (a) and (b). In part (a) the initial value of l is zero; the selection rule $\Delta l = \pm 1$ tells us that the *final* value of l is 1, so the rotational energy increases in this case. In part (b) the *final* value of l is zero; $\Delta l = \pm 1$ then tells us that the *initial* value of l is 1, and the rotational energy decreases.

The energy E of the emitted photon is the difference between the initial and final energies of the molecule, accounting for the change in both vibrational and rotational energies. In part (a) E equals the difference $\hbar\omega$ between adjacent vibrational energy levels *minus* the rotational energy that the molecule *gains*; in part (b) E equals $\hbar\omega$ *plus* the rotational energy that the molecule *loses*. Example 42.2 tells us that the difference between the $l = 0$ and $l = 1$ rotational energy levels is 0.479 meV = 0.000479 eV, and we learned above that the vibrational energy-level separation

for CO is $\hbar\omega = 0.2690$ eV. We use $E = hc/\lambda$ to determine the corresponding wavelengths (our target variables).

EXECUTE: (a) The CO molecule loses $\hbar\omega = 0.2690$ eV of vibrational energy and gains 0.000479 eV of rotational energy. Hence the energy E that goes into the emitted photon equals 0.2690 eV *less* 0.000479 eV, or 0.2685 eV. The photon wavelength is

$$\lambda = \frac{hc}{E} = \frac{(4.136 \times 10^{-15}\ \text{eV} \cdot \text{s})(2.998 \times 10^8\ \text{m/s})}{0.2685\ \text{eV}}$$

$$= 4.618 \times 10^{-6}\ \text{m} = 4.618\ \mu\text{m}$$

(b) Now the CO molecule loses $\hbar\omega = 0.2690$ eV of vibrational energy and also loses 0.000479 eV of rotational energy, so the energy that goes into the photon is $E = 0.2690$ eV $+ 0.000479$ eV $= 0.2695$ eV. The wavelength is

$$\lambda = \frac{hc}{E} = \frac{(4.136 \times 10^{-15}\ \text{eV} \cdot \text{s})(2.998 \times 10^8\ \text{m/s})}{0.2695\ \text{eV}}$$

$$= 4.601 \times 10^{-6}\ \text{m} = 4.601\ \mu\text{m}$$

EVALUATE: In part (b) the molecule loses more energy than it does in part (a), so the emitted photon must have greater energy and a shorter wavelength. That's just what our results show.

Complex Molecules

We can apply these same principles to more complex molecules. A molecule with three or more atoms has several different kinds or *modes* of vibratory motion. Each mode has its own set of energy levels, related to its frequency by Eq. (42.7). In nearly all cases the associated radiation lies in the infrared region of the electromagnetic spectrum.

42.10 The carbon dioxide molecule can vibrate in three different modes. For clarity, the atoms are not shown to scale: The separation between atoms is actually comparable to their diameters.

(a) Bending mode

When carbon atom moves upward ...

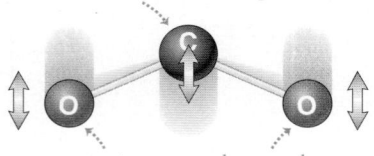

... oxygen atoms move downward, and vice versa.

(b) Symmetric stretching mode

While carbon atom remains at rest ...

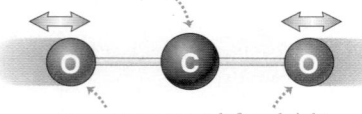

... oxygen atoms move left and right in opposite directions.

(c) Asymmetric stretching mode

When carbon atom moves right ...

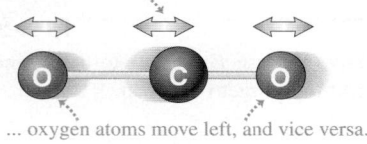

... oxygen atoms move left, and vice versa.

Infrared spectroscopy has proved to be an extremely valuable analytical tool. It provides information about the strength, rigidity, and length of molecular bonds and the structure of complex molecules. Also, because every molecule (like every atom) has its characteristic spectrum, infrared spectroscopy can be used to identify unknown compounds.

One molecule that can readily absorb and emit infrared radiation is carbon dioxide (CO_2). **Figure 42.10** shows the three possible modes of vibration of a CO_2 molecule. A number of transitions are possible between excited levels of the same vibrational mode as well as between levels of different vibrational modes. The energy differences are less than 1 eV in all of these transitions, and so involve infrared photons of wavelength longer than 1 μm. Hence a gas of CO_2 can readily absorb light at a number of different infrared wavelengths. This makes CO_2 primarily responsible for the greenhouse effect (Section 17.7) on the earth, even though CO_2 is only 0.04% of our atmosphere by volume. On Venus, however, the atmosphere has more than 90 times the total mass of our atmosphere and is almost entirely CO_2. The resulting greenhouse effect is tremendous: The surface temperature on Venus is more than 400 kelvins higher than what it would be if the planet had no atmosphere at all.

TEST YOUR UNDERSTANDING OF SECTION 42.2 A rotating diatomic molecule emits a photon when it makes a transition from level (n, l) to level $(n - 1, l - 1)$. If the value of l increases but n is unchanged, does the wavelength of the emitted photon (i) increase, (ii) decrease, or (iii) remain unchanged? ❚

42.3 STRUCTURE OF SOLIDS

The term *condensed matter* includes both solids and liquids. In both states, the interactions between atoms or molecules are strong enough to give the material a definite volume that changes relatively little with applied stress. In condensed matter, adjacent atoms attract one another until their outer electron charge clouds begin to overlap significantly. Thus the distances between adjacent atoms in condensed matter are about the same as the diameters of the atoms themselves, typically 0.1 to 0.5 nm. Also, when we speak of the distances between atoms, we mean the center-to-center (nucleus-to-nucleus) distances.

Ordinarily, we think of a liquid as a material that can flow and of a solid as a material with a definite shape. However, if you heat a horizontal glass rod in the flame of a burner, you'll find that the rod begins to sag (flow) more and more easily as its temperature rises. Glass has no definite transition from solid to liquid, and no definite melting point. On this basis, we can consider glass at room temperature as being an extremely viscous liquid. Tar and butter show similar behavior.

What is the microscopic difference between materials like glass or butter and solids like ice or copper, which do have definite melting points? Ice and copper are examples of *crystalline solids* in which the atoms have *long-range order*, a recurring pattern of atomic positions that extends over many atoms. This pattern is called the *crystal structure*. In contrast, glass at room temperature is an example of an *amorphous* solid, one that has no long-range order, but only *short-range order* (correlations between neighboring atoms or molecules). Liquids also have only short-range order. The boundaries between crystalline solid, amorphous solid, and liquid may be sometimes blurred. Some solids, crystalline when perfect, can form with so many imperfections in their structure that they have almost no long-range order. (Yet another kind of order is found in a *liquid crystal*, which is made up of rod-shaped or disc-shaped molecules of an organic compound. The positions of the molecules in the liquid are not fixed, but there is *orientational* order; the axes of the molecules tend to align with each other. This ordering can extend over a distance of many molecules.)

BIO Application Using Crystals to Determine Protein Structure Protein molecules can form crystals, such as these crystals of insulin (a protein composed of 51 amino acids). All of the molecules within a single crystal of a protein have the same orientation; how the crystal diffracts x rays or neutrons depends on the shape and size of the molecules. By analyzing these diffraction patterns, scientists have deduced the molecular structures of more than 100,000 types of proteins.

42.11 Portions of some common types of crystal lattices.

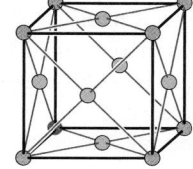
(a) Simple cubic (sc)

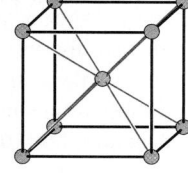
(b) Face-centered cubic (fcc)

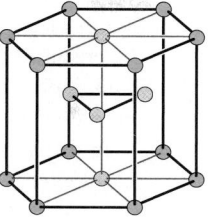
(c) Body-centered cubic (bcc)

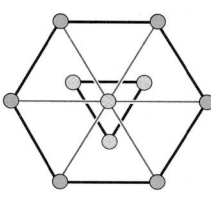
(d) Hexagonal close packed (hcp)

(e) Top view, hexagonal close packed

Nearly everything we know about crystal structure was learned from diffraction experiments with x rays, electrons, or neutrons. A typical distance between atoms is of the order of 0.1 nm. You can show that 12.4-keV x rays, 150-eV electrons, and 0.0818-eV neutrons all have wavelengths $\lambda = 0.1$ nm.

Crystal Lattices and Structures

An essential part of understanding crystals is the idea of a *crystal lattice*, which is a repeating pattern of mathematical points that extends throughout space. There are 14 general types of such patterns; **Fig. 42.11** shows small portions of some common examples. The *simple cubic lattice* (sc) has a lattice point at each corner of a cubic array (Fig. 42.11a). The *face-centered cubic lattice* (fcc) is like the simple cubic but with an additional lattice point at the center of each cube face (Fig. 42.11b). The *body-centered cubic lattice* (bcc) is like the simple cubic but with an additional point at the center of each cube (Fig. 42.11c). The *hexagonal close-packed lattice* has layers of lattice points in hexagonal patterns, each hexagon made up of six equilateral triangles (Figs. 42.11d and 42.11e).

> CAUTION **A perfect crystal lattice is infinitely large** Figure 42.11 shows just enough lattice points so that you can easily visualize the pattern; the lattice, a mathematical abstraction, extends throughout space. Thus the lattice points shown repeat endlessly in all directions. ▮

In a crystal structure, a single atom or a group of atoms is associated with each lattice point. The group may contain the same or different kinds of atoms. This atom or group of atoms is called a *basis*. Thus a complete description of a crystal structure includes both the lattice and the basis. We initially consider *perfect crystals*, or *ideal single crystals*, in which the crystal structure extends uninterrupted throughout space.

The bcc and fcc structures are two common simple crystal structures. The alkali metals have a bcc structure—that is, a bcc lattice with a basis of one atom at each lattice point. Each atom in a bcc structure has eight nearest neighbors (**Fig. 42.12a**). The elements Al, Ca, Cu, Ag, and Au have an fcc structure—that is, an fcc lattice with a basis of one atom at each lattice point. Each atom in an fcc structure has 12 nearest neighbors (Fig. 42.12b).

Figure 42.13 shows a representation of the structure of sodium chloride (NaCl, ordinary salt). It may look like a simple cubic structure, but it isn't. The sodium and chloride ions each form an fcc structure, so we can think roughly of the sodium chloride structure as being composed of two interpenetrating fcc structures. More correctly, the sodium chloride crystal structure of Fig. 42.13 has an fcc lattice with one chloride ion at each lattice point and one sodium ion half a cube length above it. That is, its basis consists of one chloride and one sodium ion.

Another example is the *diamond structure*; it's called that because it is the crystal structure of carbon in the diamond form. It's also the crystal structure of

42.12 (a) The bcc *structure* is composed of a bcc *lattice* with a basis of one atom for each lattice point. **(b)** The fcc *structure* is composed of an fcc *lattice* with a basis of one atom for each lattice point. These structures repeat precisely to make up perfect crystals.

(a) The bcc structure   **(b)** The fcc structure

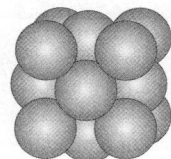

42.13 Representation of part of the sodium chloride crystal structure. The distances between ions are exaggerated.

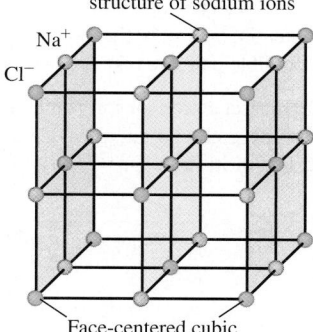

Face-centered cubic structure of sodium ions

Na^+
Cl^-

Face-centered cubic structure of chloride ions

42.14 The diamond structure, shown as two interpenetrating face-centered cubic structures with distances between atoms exaggerated. Relative to the corresponding green atom, each purple atom is shifted up, back, and to the left by a distance $a/4$.

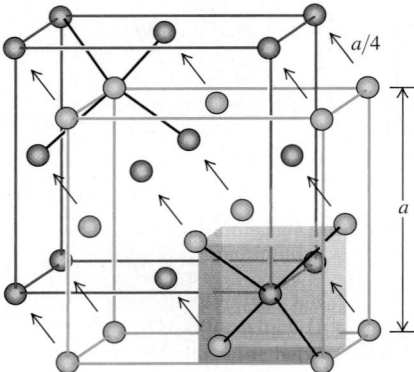

silicon, germanium, and gray tin. The diamond lattice is fcc; the basis consists of one atom at each lattice point and a second *identical* atom displaced a quarter of a cube length in each of the three cube-edge directions. **Figure 42.14** will help you visualize this. The shaded volume in Fig. 42.14 shows the bottom right front eighth of the basic cube; the four atoms at alternate corners of this cube are at the corners of a regular tetrahedron, and there is an additional atom at the center. Thus each atom in the diamond structure is at the center of a regular tetrahedron with four nearest-neighbor atoms at the corners.

In the diamond structure, both the purple and green spheres in Fig. 42.14 represent *identical* atoms—for example, both carbon or both silicon. In the cubic zinc sulfide structure, the purple spheres represent one type of atom and the green spheres represent a *different* type. For example, in zinc sulfide (ZnS) each zinc atom (purple in Fig. 42.14) is at the center of a regular tetrahedron with four sulfur atoms (green in Fig. 42.14) at its corners, and vice versa. Gallium arsenide (GaAs) and similar compounds have this same structure.

Bonding in Solids

The forces that are responsible for the regular arrangement of atoms in a crystal are the same as those involved in molecular bonds, plus one additional type. Not surprisingly, *ionic* and *covalent* molecular bonds are found in ionic and covalent crystals, respectively. The most familiar *ionic crystals* are the alkali halides, such as ordinary salt (NaCl). The positive sodium ions and the negative chloride ions occupy adjacent positions in the crystal (see Fig. 42.13). The attractive forces are the familiar Coulomb's-law forces between charged particles. These forces have no preferred direction, and the arrangement in which the material crystallizes is partly determined by the relative sizes of the two ions. Such a structure is *stable* in the sense that it has lower total energy than the separated ions (see the following example). The negative potential energies of pairs of opposite charges are greater in absolute value than the positive energies of pairs of like charges because the pairs of unlike charges are closer together, on average.

EXAMPLE 42.4 **POTENTIAL ENERGY OF AN IONIC CRYSTAL**

Imagine a one-dimensional ionic crystal consisting of a very large number of alternating positive and negative ions with charges e and $-e$, with equal spacing a along a line. Show that the total interaction potential energy is negative, which means that such a "crystal" is stable.

SOLUTION

IDENTIFY and SET UP: We treat each ion as a point charge and use our results from Section 23.1 for the electric potential energy of a collection of point charges. Equations (23.10) and (23.11) tell us to consider the electric potential energy U of each pair of charges. The total potential energy of the system is the sum of the values of U for every possible pair; we take the number of pairs to be infinite.

EXECUTE: Let's pick an ion somewhere in the middle of the line and add the potential energies of its interactions with all the ions to one side of it. From Eq. (23.11), that sum is

$$\Sigma U = -\frac{e^2}{4\pi\epsilon_0}\frac{1}{a} + \frac{e^2}{4\pi\epsilon_0}\frac{1}{2a} - \frac{e^2}{4\pi\epsilon_0}\frac{1}{3a} + \cdots$$

$$= -\frac{e^2}{4\pi\epsilon_0 a}\left(1 - \tfrac{1}{2} + \tfrac{1}{3} - \tfrac{1}{4} + \cdots\right)$$

You may notice that the series in parentheses resembles the Taylor series for the function $\ln(1 + x)$:

$$\ln(1 + x) = x - \frac{x^2}{2} + \frac{x^3}{3} - \frac{x^4}{4} + \cdots$$

When $x = 1$ this becomes the series in parentheses above, so

$$\Sigma U = -\frac{e^2}{4\pi\epsilon_0 a}\ln 2$$

This is certainly a negative quantity. The atoms on the other side of the ion we're considering make an equal contribution to the potential energy. And if we include the potential energies of all pairs of atoms, the sum is certainly negative.

EVALUATE: We conclude that this one-dimensional ionic "crystal" is stable: It has lower energy than the zero electric potential energy that is obtained when all the ions are infinitely far apart from each other.

Types of Crystals

Carbon, silicon, germanium, and tin in the diamond structure are simple examples of *covalent crystals*. These elements are in Group IV of the periodic table, meaning that each atom has four electrons in its outermost shell. Each atom forms a covalent bond with each of four adjacent atoms at the corners of a tetrahedron (Fig. 42.14). These bonds are strongly directional because of the asymmetric electron distributions dictated by the exclusion principle (see Fig. 42.3), and the result is the tetrahedral diamond structure.

A third crystal type, less directly related to the chemical bond than are ionic or covalent crystals, is the **metallic crystal.** In this structure, one or more of the outermost electrons in each atom become detached from the parent atom (leaving a positive ion). The detached electrons are free to move through the crystal and are not localized near the ion from which they originated. So we can picture a metallic crystal as an array of positive ions immersed in a sea of freed electrons whose attraction for the positive ions holds the crystal together (**Fig. 42.15**).

This sea of electrons, which gives metals their high electrical and thermal conductivities, has many of the properties of a gas. Indeed, we speak of the *electron-gas model* of metallic solids. The simplest version of this model is the *free-electron model*, which ignores interactions with the ions completely (except at the surface). We'll return to this model in Section 42.5.

In a metallic crystal the freed electrons are shared among *many* atoms. This gives a bonding that is neither localized nor strongly directional. The crystal structure is determined primarily by considerations of *close packing*—that is, the maximum number of atoms that can fit into a given volume. The two most common metallic crystal lattices are the face-centered cubic and hexagonal close-packed (see Figs. 42.11b, 42.11d, and 42.11e). In structures composed of these lattices with a basis of one atom, each atom has 12 nearest neighbors.

Hydrogen bonds and van der Waals forces also play a role in the structure of some solids. In polyethylene and similar polymers, covalent bonding of atoms forms long-chain molecules, and hydrogen bonding forms cross-links between adjacent chains. In solid water, both hydrogen bonds and van der Waals forces are significant in determining the crystal structures of ice.

Our discussion has centered on perfect crystals. Real crystals show a variety of departures from this idealized structure. Materials are often *polycrystalline*, composed of many small single crystals bonded together at *grain boundaries*. There may be *point defects* within a crystal: *Interstitial* atoms may occur in places where they do not belong, and there may be *vacancies*, positions that should be occupied by an atom but are not. A point defect of interest in semiconductors, which we will discuss in Section 42.6, is the *substitutional impurity*, a foreign atom replacing a regular atom (for example, arsenic in a silicon crystal).

There are several basic types of extended defects called *dislocations*. One type is the *edge dislocation*, shown schematically in **Fig. 42.16**, in which one plane of atoms slips relative to another. The mechanical properties of metallic crystals are influenced strongly by the presence of dislocations. The ductility and malleability of some metals depend on the presence of dislocations that can move through the crystal during plastic deformations. The biggest extended defect of all, present in *all* real crystals, is the surface of the material with its dangling bonds and abrupt change in potential energy.

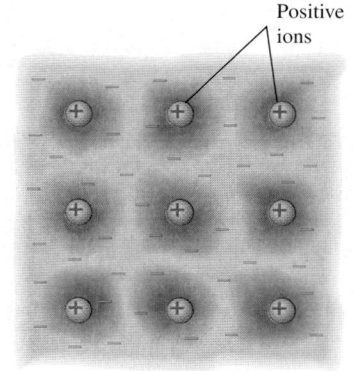

42.15 In a metallic solid, one or more electrons are detached from each atom and are free to wander around the crystal, forming an "electron gas." The wave functions for these electrons extend over many atoms. The positive ions vibrate around fixed locations in the crystal.

Positive ions

42.16 An edge dislocation in two dimensions. In three dimensions an edge dislocation would look like an extra plane of atoms slipped partway into the crystal.

You can see the irregularity most easily by viewing the figure from various directions at a grazing angle with the page.

TEST YOUR UNDERSTANDING OF SECTION 42.3   If a is the distance in an NaCl crystal from an Na^+ ion to one of its nearest-neighbor Cl^- ions, what is the distance from an Na^+ ion to one of its *next-to-nearest*-neighbor Cl^- ions? (i) $a\sqrt{2}$; (ii) $a\sqrt{3}$; (iii) $2a$; (iv) none of these. ∎

42.17 The concept of energy bands was first developed by the Swiss-American physicist Felix Bloch (1905–1983) in his doctoral thesis. Our modern understanding of electrical conductivity stems from that landmark work. Bloch's work in nuclear physics brought him (along with Edward Purcell) the 1952 Nobel Prize in physics.

42.18 Origin of energy bands in a solid. (a) As the distance r between atoms decreases, the energy levels spread into bands. The vertical line at r_0 shows the actual atomic spacing in the crystal. (b) Symbolic representation of energy bands.

(a)   (b)

Actual separation of atoms in the crystal

E

O r_0 r

42.4 ENERGY BANDS

The **energy-band** concept, introduced in 1928 (**Fig. 42.17**), is a great help in understanding several properties of solids. To introduce the idea, suppose we have a large number N of identical atoms, far enough apart that their interactions are negligible. Every atom has the same energy-level diagram. We can draw an energy-level diagram for the *entire system*. It looks just like the diagram for a single atom, but the exclusion principle, applied to the entire system, permits each state to be occupied by N electrons (one per atom) instead of just one.

Now we begin to push the atoms uniformly closer together. Because of the electrical interactions and the exclusion principle, the wave functions begin to distort, especially those of the outer, or *valence*, electrons. The corresponding energies also shift, some upward and some downward, by varying amounts, as the valence electron wave functions become less localized and extend over more and more atoms. (The inner electrons in an atom are affected much less by nearby atoms than are the valence electrons, and their energy levels remain relatively sharp.) Thus the valence states that formerly gave the *system* a state with a sharp energy level that could accommodate N electrons now give a *band* containing N closely spaced levels (**Fig. 42.18**). Ordinarily, N is somewhere near the order of Avogadro's number (10^{24}), so we can accurately treat the levels as forming a *continuous* distribution of energies within a band. Between adjacent energy bands are gaps where there are *no* allowed energy levels.

Insulators, Semiconductors, and Conductors

The nature of the energy bands determines whether the material is an electrical insulator, a semiconductor, or a conductor. In particular, what matters are the extent to which the states in each band are occupied and the spacing, called the *band gap* or *energy gap*, between adjacent bands.

In an *insulator* at absolute zero temperature, the highest band that is completely filled, called the **valence band,** is also the highest band that has *any* electrons in it. The next higher band, called the **conduction band,** is completely empty; there are no electrons in its states (**Fig. 42.19a**). Imagine what happens if an electric field is applied to a material of this kind. To move in response to the field, an electron would have to go into a different quantum state with a slightly different energy. It can't do that, however, because all the neighboring states are already occupied. The only way such an electron can move is to jump across the energy gap into the conduction band, where there are plenty of nearby unoccupied states. At any temperature above absolute zero there is some probability this jump can happen, because an electron can gain energy from thermal motion. In an insulator, however, the energy gap between the valence and conduction bands can be 5 eV or more, and that much thermal energy is not ordinarily available. Hence little or no current flows in response to an applied electric field, and the electrical

42.19 Three types of energy-band structure.

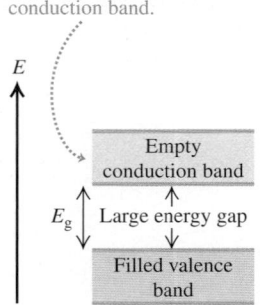

(a) In an insulator at absolute zero, there are no electrons in the conduction band.

E

Empty conduction band

E_g Large energy gap

Filled valence band

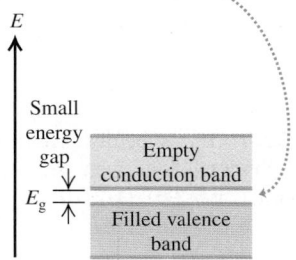

(b) A semiconductor has the same band structure as an insulator but a smaller gap between the valence and conduction bands.

E

Small energy gap

E_g

Empty conduction band

Filled valence band

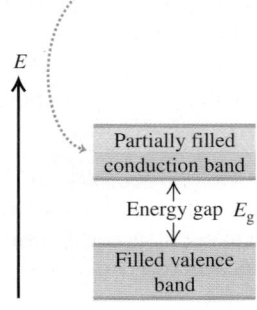

(c) A conductor has a partially filled conduction band.

E

Partially filled conduction band

Energy gap E_g

Filled valence band

conductivity (Section 25.2) is low. The thermal conductivity (Section 17.7), which also depends on mobile electrons, is likewise low.

We saw in Section 24.4 that an insulator becomes a conductor if it is subjected to a large enough electric field; this is called *dielectric breakdown*. If the electric field is of order 10^{10} V/m, there is a potential difference of a few volts over a distance comparable to atomic sizes. In this case the field can do enough work on a valence electron to boost it across the energy gap and into the conduction band. (In practice dielectric breakdown occurs for fields much less than 10^{10} V/m, because imperfections in the structure of an insulator provide some more accessible energy states *within* the energy gap.)

As in an insulator, a *semiconductor* at absolute zero has an empty conduction band above the full valence band. The difference is that in a semiconductor the energy gap between these bands is relatively small and electrons can more readily jump into the conduction band (Fig. 42.19b). As the temperature of a semiconductor increases, the population in the conduction band increases very rapidly, as does the electrical conductivity. For example, in a semiconductor near room temperature with an energy gap of 1 eV, the number of conduction electrons doubles when the temperature rises by just 10°C. We will use the concept of energy bands to explore semiconductors in more depth in Section 42.6.

In a *conductor* such as a metal, there are electrons in the conduction band even at absolute zero (Fig. 42.19c). The metal sodium is an example. An analysis of the atomic energy-level diagram for sodium (see Fig. 39.19a) shows that for an isolated sodium atom, the six lowest excited states (all 3*p* states) are about 2.1 eV above the two 3*s* ground states. In solid sodium, however, the atoms are so close together that the 3*s* and 3*p* *bands* spread out and overlap into a single band. Each sodium atom contributes one electron to the band, leaving an Na$^+$ ion behind. Each atom also contributes eight *states* to that band (two 3*s*, six 3*p*), so the band is only one-eighth occupied. We call this structure a *conduction* band because it is only partially occupied. Electrons near the top of the filled portion of the band have many adjacent unoccupied states available, and they can easily gain or lose small amounts of energy in response to an applied electric field. Therefore these electrons are mobile, giving solid sodium its high electrical and thermal conductivity. A similar description applies to other conducting materials.

PhET: Band Structure
PhET: Conductivity

EXAMPLE 42.5   **PHOTOCONDUCTIVITY IN GERMANIUM**

At room temperature, pure germanium has an almost completely filled valence band separated by a 0.67-eV gap from an almost completely empty conduction band. It is a poor electrical conductor, but its conductivity increases greatly when it is irradiated with electromagnetic waves of a certain maximum wavelength. What is that wavelength?

SOLUTION

IDENTIFY and SET UP: The conductivity of a semiconductor increases greatly when electrons are excited from the valence band into the conduction band. In germanium, the excitation occurs when an electron absorbs a photon with an energy of at least $E_{min} = 0.67$ eV. From $E = hc/\lambda$, the *maximum* wavelength λ_{max} (our target variable) corresponds to this *minimum* photon energy.

EXECUTE: The wavelength of a photon with energy $E_{min} = 0.67$ eV is

$$\lambda_{max} = \frac{hc}{E_{min}} = \frac{(4.136 \times 10^{-15} \text{ eV} \cdot \text{s})(3.00 \times 10^8 \text{ m/s})}{0.67 \text{ eV}}$$

$$= 1.9 \times 10^{-6} \text{ m} = 1.9 \ \mu\text{m} = 1900 \text{ nm}$$

EVALUATE: This wavelength is in the infrared part of the spectrum, so visible-light photons (which have shorter wavelength and higher energy) will also induce conductivity in germanium. As we'll see in Section 42.7, semiconductor crystals are widely used as photocells and for many other applications.

TEST YOUR UNDERSTANDING OF SECTION 42.4   One type of thermometer works by measuring the temperature-dependent electrical resistivity of a sample. Which of the following types of material displays the greatest change in resistivity for a given temperature change? (i) Insulator; (ii) semiconductor; (iii) conductor. ∎

42.5 FREE-ELECTRON MODEL OF METALS

Studying the energy states of electrons in metals can give us a lot of insight into their electrical and magnetic properties, the electron contributions to heat capacities, and other behavior. As we discussed in Section 42.3, one of the distinguishing features of a metal is that one or more valence electrons are detached from their home atom and can move freely within the metal, with wave functions that extend over many atoms.

The **free-electron model** assumes that these electrons don't interact at all with the ions or with each other, but that there are infinite potential-energy barriers at the surfaces. The idea is that a typical electron moves so rapidly within the metal that it "sees" the effect of the ions and other electrons as a uniform potential-energy function, whose value we can choose to be zero.

We can represent the surfaces of the metal by the same cubical box that we analyzed in Section 41.2 (the three-dimensional version of the particle in a box studied in Section 40.2). If the box has sides of length L (**Fig. 42.20**), the energies of the stationary states (quantum states of definite energy) are

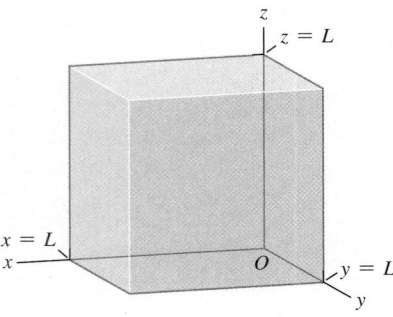

42.20 A cubical box with side length L. We studied this three-dimensional version of the infinite square well in Section 41.2. The energy levels for a particle in this box are given by Eq. (42.10).

$$E_{n_X, n_Y, n_Z} = \frac{(n_X{}^2 + n_Y{}^2 + n_Z{}^2)\pi^2\hbar^2}{2mL^2} \qquad \begin{array}{l} (n_X = 1, 2, 3, \ldots; \\ \; n_Y = 1, 2, 3, \ldots; \\ \; n_Z = 1, 2, 3, \ldots) \end{array} \quad (42.10)$$

Each state is labeled by the three positive-integer quantum numbers (n_X, n_Y, n_Z).

Density of States

Later we'll need to know the *number dn* of quantum states that have energies in a given range dE. The number of states per unit energy range dn/dE is called the **density of states,** denoted by $g(E)$. We'll begin by working out an expression for $g(E)$. Think of a three-dimensional space with coordinates (n_X, n_Y, n_Z) (**Fig. 42.21**). The radius n_{rs} of a sphere centered at the origin in that space is $n_{rs} = (n_X{}^2 + n_Y{}^2 + n_Z{}^2)^{1/2}$. Each point with integer coordinates in that space represents one spatial quantum state. Thus each point corresponds to one unit of volume in the space, and the total number of points with integer coordinates inside a sphere equals the volume of the sphere, $\frac{4}{3}\pi n_{rs}{}^3$. Because all our n's are positive, we must take only one *octant* of the sphere, with $\frac{1}{8}$ the total volume, or $(\frac{1}{8})(\frac{4}{3}\pi n_{rs}{}^3) = \frac{1}{6}\pi n_{rs}{}^3$. The particles are electrons, so each point corresponds to *two* states with opposite spin components $(m_s = \pm\frac{1}{2})$, and the total number n of electron states corresponding to points inside the octant is twice $\frac{1}{6}\pi n_{rs}{}^3$, or

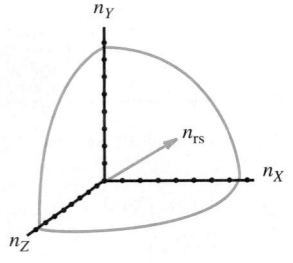

42.21 The allowed values of n_X, n_Y, and n_Z are positive integers for the electron states in the free-electron gas model. Including spin, there are two states for each unit volume in n space.

$$n = \frac{\pi n_{rs}{}^3}{3} \qquad (42.11)$$

The energy E of states at the surface of the sphere can be expressed in terms of n_{rs}. Equation (42.10) becomes

$$E = \frac{n_{rs}{}^2\pi^2\hbar^2}{2mL^2} \qquad (42.12)$$

We can combine Eqs. (42.11) and (42.12) to get a relationship between E and n that doesn't contain n_{rs}. We'll leave the details for you to work out; the total number of states with energies of E or less is

$$n = \frac{(2m)^{3/2}VE^{3/2}}{3\pi^2\hbar^3} \qquad (42.13)$$

where $V = L^3$ is the volume of the box.

To get the number of states dn in an energy interval dE, we treat n and E as continuous variables and take differentials of both sides of Eq. (42.13):

$$dn = \frac{(2m)^{3/2}VE^{1/2}}{2\pi^2\hbar^3}\,dE \qquad (42.14)$$

The density of states $g(E)$ is equal to dn/dE, so from Eq. (42.14) we get

Density of states, free-electron model:

Number of states per unit energy range near E ⋯⋯ Electron mass

$$g(E) = \frac{(2m)^{3/2}V}{2\pi^2\hbar^3}E^{1/2} \qquad (42.15)$$

Volume

Planck's constant divided by 2π ⋯ Electron energy

Fermi–Dirac Distribution

Let's now see how the electrons are distributed among the various quantum states at any given temperature. The Maxwell–Boltzmann distribution states that the average number of particles in a state of energy E is proportional to $e^{-E/kT}$ (see Sections 18.5 and 39.4). However, there are two reasons why it wouldn't be right to use the Maxwell–Boltzmann distribution. The first reason is the exclusion principle. At absolute zero the Maxwell–Boltzmann function predicts that *all* the electrons would go into the two ground states of the system, with $n_X = n_Y = n_Z = 1$ and $m_s = \pm\frac{1}{2}$. But the exclusion principle allows only one electron in each state. At absolute zero the electrons can fill up the lowest *available* states, but they cannot *all* go into the lowest states. Thus at absolute zero the distribution function is as shown in **Fig. 42.22**. All states with energies E less than some value E_{F0} are occupied, so the occupation probability $f(E) = 1$; all states with energies greater than this value are unoccupied, so $f(E) = 0$.

The second reason we can't use the Maxwell–Boltzmann distribution is more subtle. That distribution assumes that the particles are *distinguishable*. But, as we discussed in Section 41.8, electrons are *indistinguishable*; it's impossible to "tag" electrons to know which is which. If one electron is in state A and the other is in state B, there's no way to tell whether electron 1 is in state A and electron 2 is in state B, or electron 1 is in state B and electron 2 is in state A.

The statistical distribution function that emerges from the exclusion principle and the indistinguishability requirement is called (after its inventors) the **Fermi–Dirac distribution.** This distribution gives the probability $f(E)$ that at temperature T, a particular state of energy E is occupied by an electron:

Fermi–Dirac distribution:

Probability that a given state is occupied by an electron

$$f(E) = \frac{1}{e^{(E-E_F)/kT} + 1} \qquad (42.16)$$

Absolute temperature

Energy of state   Fermi energy   Boltzmann constant

The energy E_F is called the **Fermi energy** or the *Fermi level*; we'll discuss its significance below. We use E_{F0} for its value at absolute zero ($T = 0$) and E_F for other temperatures. We can accurately let $E_F = E_{F0}$ for metals because the Fermi energy does not change much with temperature for solid conductors. However, it is not safe to assume that $E_F = E_{F0}$ for semiconductors, in which the Fermi energy usually does change with temperature.

Figure 42.23 shows graphs of Eq. (42.16) for three temperatures. The $kT = \frac{1}{40}E_F$ curve shows that for $kT \ll E_F$, $f(E)$ approaches the shape of the absolute-zero distribution function shown in Fig. 42.22.

42.22 The probability distribution for occupation of free-electron energy states at absolute zero.

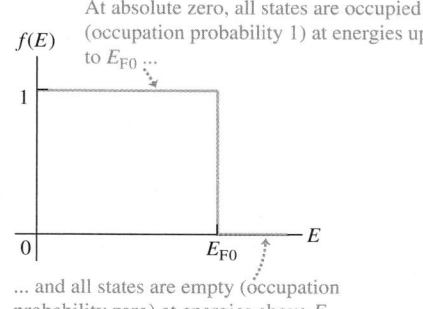

At absolute zero, all states are occupied (occupation probability 1) at energies up to E_{F0}

... and all states are empty (occupation probability zero) at energies above E_{F0}.

42.23 Graphs of the Fermi–Dirac distribution function for various values of kT, assuming that the Fermi energy E_F is independent of the temperature T.

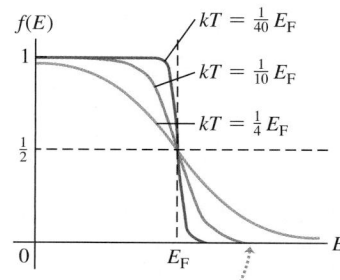

As T increases, more and more of the electrons are excited to states with energy $E > E_F$.

Note that when $E = E_F$, the exponent $(E - E_F)/kT$ in Eq. (42.16) is zero, so $f(E) = f(E_F) = \frac{1}{2}$ for any temperature T. That is, the probability is $\frac{1}{2}$ that a state at the Fermi energy contains an electron. This means that at $E = E_F$, half the states are filled (and half are empty). For states with $E < E_F$, the exponent is negative and the occupation probability $f(E)$ is greater than $\frac{1}{2}$; for states with $E > E_F$, $f(E)$ is less than $\frac{1}{2}$ and approaches zero for E much larger than kT.

EXAMPLE 42.6   PROBABILITIES IN THE FREE-ELECTRON MODEL

For free electrons in a solid, at what energy is the probability that a particular state is occupied equal to (a) 0.01 and (b) 0.99?

SOLUTION

IDENTIFY and SET UP: This problem asks us to explore the Fermi–Dirac distribution. Equation (42.16) gives the occupation probability $f(E)$ for a given energy E. If we solve this equation for E, we get an expression for the energy that corresponds to a given occupation probability—which is just what we need.

EXECUTE: Using Eq. (42.16), you can show that

$$E = E_F + kT \ln\left(\frac{1}{f(E)} - 1\right)$$

(a) When $f(E) = 0.01$,

$$E = E_F + kT \ln\left(\frac{1}{0.01} - 1\right) = E_F + 4.6kT$$

The probability that a state $4.6kT$ above the Fermi level is occupied is only 0.01, or 1%.

(b) When $f(E) = 0.99$,

$$E = E_F + kT \ln\left(\frac{1}{0.99} - 1\right) = E_F - 4.6kT$$

The probability that a state $4.6kT$ below the Fermi level is occupied is 0.99, or 99%.

EVALUATE: At very low temperatures, $4.6kT$ is much less than E_F. Then the occupation probability of levels even slightly below E_F is nearly 1 (100%), and that for levels even slightly above E_F is nearly zero (see Fig. 42.23). In general, if the probability is P that a state with an energy ΔE *above* E_F is occupied, then the probability is $1 - P$ that a state ΔE *below* E_F is occupied. We leave the proof to you.

Electron Concentration and Fermi Energy

Equation (42.16) gives the probability that any specific state with energy E is occupied at a temperature T. To get the actual number of electrons in any energy range dE, we have to multiply this probability by the number dn of states in that range $g(E)\,dE$. Thus the number dN of electrons with energies in the range dE is

$$dN = g(E)f(E)\,dE = \frac{(2m)^{3/2}VE^{1/2}}{2\pi^2\hbar^3} \frac{1}{e^{(E-E_F)/kT} + 1}\,dE \qquad (42.17)$$

The Fermi energy E_F is determined by the total number N of electrons; at any temperature the electron states are filled up to a point at which all electrons are accommodated. At absolute zero there is a simple relationship between E_{F0} and N. *All* states below E_{F0} are filled; in Eq. (42.13) we set n equal to the total number of electrons N and E to the Fermi energy at absolute zero E_{F0}:

$$N = \frac{(2m)^{3/2}VE_{F0}^{3/2}}{3\pi^2\hbar^3} \qquad (42.18)$$

Solving Eq. (42.18) for E_{F0}, we get

$$E_{F0} = \frac{3^{2/3}\pi^{4/3}\hbar^2}{2m}\left(\frac{N}{V}\right)^{2/3} \qquad (42.19)$$

CAUTION **Electron concentration and number of electrons** Don't confuse the electron concentration n with any quantum number n. Furthermore, the number of states is *not* in general the same as the total number of electrons N. ▮

The quantity N/V is the number of free electrons per unit volume. It is called the *electron concentration* and is usually denoted by n.

If we replace N/V with n, Eq. (42.19) becomes

$$E_{F0} = \frac{3^{2/3}\pi^{4/3}\hbar^2 n^{2/3}}{2m} \qquad (42.20)$$

EXAMPLE 42.7 THE FERMI ENERGY IN COPPER

At low temperatures, copper has a free-electron concentration $n = 8.45 \times 10^{28}$ m^{-3}. Using the free-electron model, find the Fermi energy for solid copper, and find the speed of an electron with a kinetic energy equal to the Fermi energy.

SOLUTION

IDENTIFY and SET UP: This problem uses the relationship between Fermi energy and free-electron concentration. Because copper is a solid conductor, its Fermi energy changes very little with temperature and we can use the expression for the Fermi energy at absolute zero, Eq. (42.20). We'll use the nonrelativistic formula $E_F = \frac{1}{2}mv_F^2$ to find the *Fermi speed* v_F that corresponds to kinetic energy E_F.

EXECUTE: Using the given value of n, we solve for E_F and v_F:

$$E_F = \frac{3^{2/3}\pi^{4/3}(1.055 \times 10^{-34} \text{ J} \cdot \text{s})^2(8.45 \times 10^{28} \text{ m}^{-3})^{2/3}}{2(9.11 \times 10^{-31} \text{ kg})}$$

$$= 1.126 \times 10^{-18} \text{ J} = 7.03 \text{ eV}$$

$$v_F = \sqrt{\frac{2E_F}{m}} = \sqrt{\frac{2(1.126 \times 10^{-18} \text{ J})}{9.11 \times 10^{-31} \text{ kg}}} = 1.57 \times 10^6 \text{ m/s}$$

EVALUATE: Our values of E_F and v_F are within the ranges of typical values for metals, 1.6–14 eV and 0.8–2.2 $\times 10^6$ m/s, respectively. Note that the calculated Fermi speed is far less than the speed of light $c = 3.00 \times 10^8$ m/s, which justifies our use of the nonrelativistic formula $\frac{1}{2}mv_F^2 = E_F$.

Our calculated Fermi energy is much larger than kT at ordinary temperatures. (At room temperature $T = 20°C = 293$ K, the quantity kT equals $(1.381 \times 10^{-23}$ J/K$)(293$ K$) = 4.04 \times 10^{-21}$ J $= 0.0254$ eV.) So it is a good approximation to take almost all the states below E_F as completely full and almost all those above E_F as completely empty (see Fig. 42.22).

We can also use Eq. (42.15) to find $g(E)$ if E and V are known. You can show that if $E = 7.03$ eV and $V = 1$ cm^3, $g(E)$ is about 2×10^{22} states/eV. This huge number shows why we were justified in treating n and E as continuous variables in our density-of-states derivation.

Average Free-Electron Energy

We can calculate the *average* free-electron energy in a metal at absolute zero by using the same ideas that we used to find E_{F0}. From Eq. (42.17) the number dN of electrons with energies in the range dE is $g(E)f(E)\,dE$. The energy of these electrons is $E\,dN = Eg(E)f(E)\,dE$. At absolute zero we substitute $f(E) = 1$ from $E = 0$ to $E = E_{F0}$ and $f(E) = 0$ for all other energies. Therefore the total energy E_{tot} of all the N electrons is

$$E_{\text{tot}} = \int_0^{E_{F0}} Eg(E)(1)\,dE + \int_{E_{F0}}^{\infty} Eg(E)(0)\,dE = \int_0^{E_{F0}} Eg(E)\,dE$$

The simplest way to evaluate this expression is to compare Eqs. (42.15) and (42.19). You'll see that

$$g(E) = \frac{3NE^{1/2}}{2E_{F0}^{3/2}}$$

Substituting this expression into the integral and using $E_{\text{av}} = E_{\text{tot}}/N$, we get

$$E_{\text{av}} = \frac{3}{2E_{F0}^{3/2}} \int_0^{E_{F0}} E^{3/2}\,dE = \tfrac{3}{5}E_{F0} \qquad (42.21)$$

At absolute zero the average free-electron energy equals $\frac{3}{5}$ of the Fermi energy.

EXAMPLE 42.8 FREE-ELECTRON GAS VERSUS IDEAL GAS

(a) Find the average energy of the free electrons in copper at absolute zero (see Example 42.7). (b) What would be the average kinetic energy of electrons if they behaved like an ideal gas at room temperature, 20°C (see Section 18.3)? What would be the speed of an electron with this kinetic energy? Compare these ideal-gas values with the (correct) free-electron values.

SOLUTION

IDENTIFY and SET UP: Free electrons in a metal behave like a kind of gas. In part (a) we use Eq. (42.21) to determine the average kinetic energy of free electrons in terms of the Fermi energy

Continued

at absolute zero, which we know for copper from Example 42.7. In part (b) we treat electrons as an ideal gas at room temperature: Eq. (18.16) then gives the average kinetic energy per electron as $E_{av} = \frac{3}{2}kT$, and $E_{av} = \frac{1}{2}mv^2$ gives the corresponding electron speed v.

EXECUTE: (a) From Example 42.7, the Fermi energy in copper at absolute zero is 1.126×10^{-18} J = 7.03 eV. According to Eq. (42.21), the average energy is $\frac{3}{5}$ of this, or 6.76×10^{-19} J = 4.22 eV.

(b) In Example 42.7 we found that $kT = 4.04 \times 10^{-21}$ J = 0.0254 eV at room temperature $T = 20°C = 293$ K. If electrons behaved like an ideal gas at this temperature, the average kinetic energy per electron would be $\frac{3}{2}$ of this, or 6.07×10^{-21} J = 0.0379 eV. The speed of an electron with this kinetic energy is

$$v = \sqrt{\frac{2E_{av}}{m}} = \sqrt{\frac{2(6.07 \times 10^{-21}\,\text{J})}{9.11 \times 10^{-31}\,\text{kg}}}$$
$$= 1.15 \times 10^5 \text{ m/s}$$

EVALUATE: The ideal-gas model predicts an average energy that is about 1% of the value given by the free-electron model, and a speed that is about 7% of the free-electron Fermi speed

$v_F = 1.57 \times 10^6$ m/s that we found in Example 42.7. Thus temperature plays a *very* small role in determining the properties of electrons in metals; their average energies are determined almost entirely by the exclusion principle.

A similar analysis allows us to determine the contributions of electrons to the heat capacities of a solid metal. If there is one conduction electron per atom, the principle of equipartition of energy (see Section 18.4) would predict that the kinetic energies of these electrons contribute $3R/2$ to the molar heat capacity at constant volume C_V. But when kT is much smaller than E_F, which is usually the situation in metals, only those few electrons near the Fermi level can find empty states and change energy appreciably when the temperature changes. The number of such electrons is proportional to kT/E_F, so we expect that the electron molar heat capacity at constant volume is proportional to $(kT/E_F)(3R/2) = (3kT/2E_F)R$. A more detailed analysis shows that the actual electron contribution to C_V for a solid metal is $(\pi^2 kT/2E_F)R$, not far from our prediction. You can verify that if $T = 293$ K and $E_F = 7.03$ eV, the electron contribution to C_V is $0.018R$, which is only 1.2% of the (incorrect) $3R/2$ prediction of the equipartition principle. Because the electron contribution is so small, the overall heat capacity of most solid metals is due primarily to vibration of the atoms in the crystal structure (see Fig. 18.18 in Section 18.4).

TEST YOUR UNDERSTANDING OF SECTION 42.5 An ideal gas obeys the relationship $pV = nRT$ (see Section 18.1): For a given volume V and a number of moles n, as the temperature T decreases, the pressure p decreases proportionately and tends to zero as T approaches absolute zero. Is this also true of the free-electron gas in a solid metal? ∎

42.6 SEMICONDUCTORS

PhET: Semiconductors
PhET: Conductivity

A **semiconductor** has an electrical resistivity that is intermediate between those of good conductors and of good insulators. The tremendous importance of semiconductors in present-day electronics stems in part from the fact that their electrical properties are very sensitive to very small concentrations of impurities. We'll discuss the basic concepts, using the semiconductor elements silicon (Si) and germanium (Ge) as examples.

Silicon and germanium are in Group IV of the periodic table. Both have four electrons in the outermost atomic subshells ($3s^2 3p^2$ for silicon, $4s^2 4p^2$ for germanium), and both crystallize in the covalently bonded diamond structure discussed in Section 42.3 (see Fig. 42.14). Because all four of the outer electrons are involved in the bonding, at absolute zero the band structure (see Section 42.4) has a completely empty conduction band (see Fig. 42.19b). As we discussed in Section 42.4, at very low temperatures electrons cannot jump from the filled valence band into the conduction band. This property makes these materials insulators at very low temperatures; their electrons have no nearby states available into which they can move in response to an applied electric field.

However, in semiconductors the energy gap E_g between the valence and conduction bands is small in comparison to the gap of 5 eV or more for many insulators; room-temperature values are 1.12 eV for silicon and only 0.67 eV for germanium. Thus even at room temperature a substantial number of electrons can gain enough energy to jump the gap to the conduction band, where they are dissociated from their parent atoms and are free to move about the crystal. The number of these electrons increases rapidly with temperature.

EXAMPLE 42.9 JUMPING A BAND GAP

Consider a material with the band structure described above, with its Fermi energy in the middle of the gap (**Fig. 42.24**). Find the probability that a state at the bottom of the conduction band is occupied at $T = 300$ K, and compare that with the probability at $T = 310$ K, for band gaps of (a) 0.200 eV; (b) 1.00 eV; (c) 5.00 eV.

SOLUTION

IDENTIFY and SET UP: The Fermi–Dirac distribution function gives the probability that a state of energy E is occupied at temperature T. Figure 42.24 shows that the state of interest at the bottom of the conduction band has an energy $E = E_F + E_g/2$ that is greater than the Fermi energy E_F, with $E - E_F = E_g/2$. Figure 42.23 shows that

42.24 Band structure of a semiconductor. At absolute zero a completely filled valence band is separated by a narrow energy gap E_g of 1 eV or so from a completely empty conduction band. At ordinary temperatures, a number of electrons are excited to the conduction band.

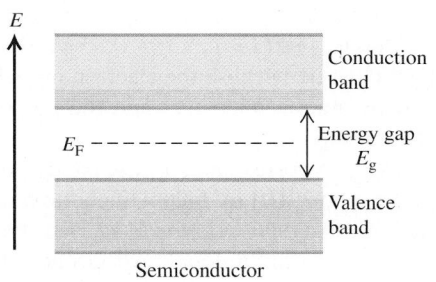

the higher the temperature, the larger the fraction of electrons with energies greater than the Fermi energy.

EXECUTE: (a) When $E_g = 0.200$ eV,

$$\frac{E - E_F}{kT} = \frac{E_g}{2kT} = \frac{0.100 \text{ eV}}{(8.617 \times 10^{-5} \text{ eV/K})(300 \text{ K})} = 3.87$$

$$f(E) = \frac{1}{e^{3.87} + 1} = 0.0205$$

For $T = 310$ K, the exponent is 3.74 and $f(E) = 0.0231$, a 13% increase in probability for a temperature rise of 10 K.

(b) For $E_g = 1.00$ eV, both exponents are five times as large as in part (a), namely 19.3 and 18.7; the values of $f(E)$ are 4.0×10^{-9} and 7.4×10^{-9}. In this case the (low) probability nearly doubles with a temperature rise of 10-K.

(c) For $E_g = 5.0$ eV, the exponents are 96.7 and 93.6; the values of $f(E)$ are 1.0×10^{-42} and 2.3×10^{-41}. The (extremely low) probability increases by a factor of 23 for a 10-K temperature rise.

EVALUATE: This example illustrates two important points. First, the probability of finding an electron in a state at the bottom of the conduction band is extremely sensitive to the width of the band gap. At room temperature, the probability is about 2% for a 0.200-eV gap, a few in a thousand million for a 1.00-eV gap, and essentially zero for a 5.00-eV gap. (Pure diamond, with a 5.47-eV band gap, has essentially no electrons in the conduction band and is an excellent insulator.) Second, for any given band gap the probability depends strongly on temperature, and even more strongly for large gaps than for small ones.

In principle, we could continue the calculation in Example 42.9 to find the actual density $n = N/V$ of electrons in the conduction band at any temperature. To do this, we would have to evaluate the integral $\int g(E)f(E) \, dE$ from the bottom of the conduction band to its top. [To do this we would need to know the density of states function $g(E)$. This function for a semiconductor is more complicated than the free-electron model expression given by Eq. (42.15).] Once we know n, we can *begin* to determine the resistivity of the material (and its temperature dependence) by using the analysis of Section 25.2, which you may want to review. But next we'll see that the electrons in the conduction band don't tell the whole story about conduction in semiconductors.

Holes

When an electron is removed from a covalent bond, it leaves a vacancy behind. An electron from a neighboring atom can move into this vacancy, leaving the neighbor with the vacancy. In this way the vacancy, called a **hole,** can travel through the material and serve as an additional current carrier. It's like describing the motion of a bubble in a liquid. In a pure, or *intrinsic*, semiconductor, valence-band holes and conduction-band electrons are always present in equal numbers. When an electric field is applied, they move in opposite directions (**Fig. 42.25**). Thus a hole in the valence band behaves like a positively charged particle, even though the moving charges in that band are electrons. The conductivity that we just described for a pure semiconductor is called *intrinsic conductivity*. Another kind of conductivity, to be discussed in the next subsection, is due to impurities.

An analogy helps to picture conduction in an intrinsic semiconductor. The valence band at absolute zero is like a floor of a parking garage that's filled bumper to bumper with cars (which represent electrons). No cars can move

42.25 Motion of electrons in the conduction band and of holes in the valence band of a semiconductor under the action of an applied electric field $\vec{E}$.

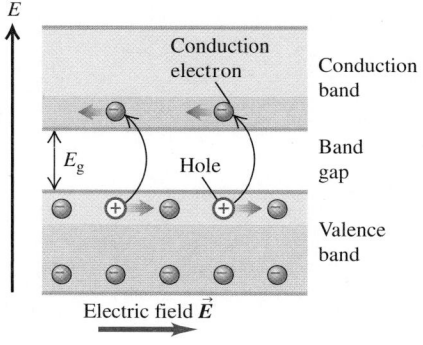

because there is nowhere for them to go. But if one car is moved to the vacant floor above, it can move freely, just as electrons can move freely in the conduction band. Also, the empty space that it leaves permits cars to move on the nearly filled floor, thereby moving the empty space just as holes move in the normally filled valence band.

Impurities

Suppose we mix into melted germanium ($Z = 32$) a small amount of arsenic ($Z = 33$), the next element after germanium in the periodic table. This deliberate addition of impurity elements is called *doping*. Arsenic is in Group V; it has *five* valence electrons. When one of these electrons is removed, the remaining electron structure is essentially identical to that of germanium. The only difference is that it is smaller; the arsenic nucleus has a charge of $+33e$ rather than $+32e$, and it pulls the electrons in a little more. An arsenic atom can comfortably take the place of a germanium atom as a substitutional impurity. Four of its five valence electrons form the necessary nearest-neighbor covalent bonds.

The fifth valence electron is very loosely bound (**Fig. 42.26a**); it doesn't participate in the covalent bonds, and it is screened from the nuclear charge of $+33e$ by the 32 electrons, leaving a net effective charge of about $+e$. We might guess that the binding energy would be of the same order of magnitude as the energy of the $n = 4$ level in hydrogen—that is, $(\frac{1}{4})^2(13.6 \text{ eV}) = 0.85 \text{ eV}$. In fact, it is much smaller than this, only about 0.01 eV, because the electron probability distribution actually extends over many atomic diameters and the polarization of intervening atoms provides additional screening.

The energy level of this fifth electron corresponds in the band picture to an isolated energy level lying in the gap, about 0.01 eV below the bottom of the conduction band (Fig. 42.26b). This level is called a *donor level*, and the impurity atom that is responsible for it is simply called a *donor*. All Group V elements, including N, P, As, Sb, and Bi, can serve as donors. At room temperature, kT is about 0.025 eV. This is substantially greater than 0.01 eV, so at ordinary temperatures, most electrons can gain enough energy to jump from donor levels into the conduction band, where they are free to wander through the material. The remaining ionized donor stays at its site in the structure and does not participate in conduction.

Example 42.9 shows that at ordinary temperatures and with a band gap of 1.0 eV, only a very small fraction (of the order of 10^{-9}) of the states at the bottom of the conduction band in a pure semiconductor contain electrons to participate in intrinsic conductivity. Thus we expect the conductivity of such a semiconductor to be about 10^{-9} as great as that of good metallic conductors, and measurements bear out this prediction. However, a concentration of donors as small as one part in 10^8 can increase the conductivity so drastically that conduction due to impurities becomes by far the dominant mechanism. In this case the conductivity is due almost entirely to *negative* charge (electron) motion. We call the material an ***n*-type semiconductor,** with *n*-type impurities.

Adding atoms of an element in Group III (B, Al, Ga, In, Tl), with only *three* valence electrons, has an analogous effect. An example is gallium ($Z = 31$); as a substitutional impurity in germanium, the gallium atom would like to form four covalent bonds, but it has only three outer electrons. It can, however, steal an electron from a neighboring germanium atom to complete the required four covalent bonds (**Fig. 42.27a**). The resulting atom has the same electron configuration as Ge but is somewhat larger because gallium's nuclear charge is smaller, $+31e$ instead of $+32e$.

This theft leaves the neighboring atom with a *hole*, or missing electron. The hole acts as a positive charge that can move through the crystal just as with intrinsic conductivity. The stolen electron is bound to the gallium atom in a level called an *acceptor level* about 0.01 eV above the top of the valence band

42.26 An *n*-type semiconductor.

(a) A donor (*n*-type) impurity atom has a fifth valence electron that does not participate in the covalent bonding and is very loosely bound.

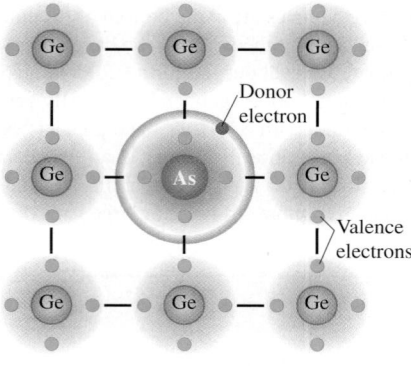

(b) Energy-band diagram for an *n*-type semiconductor at a low temperature. One donor electron has been excited from the donor levels into the conduction band.

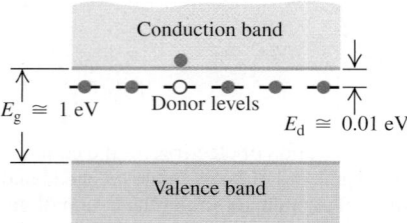

(Fig. 42.27b). The gallium atom, called an *acceptor*, thus accepts an electron to complete its desire for four covalent bonds. This extra electron gives the previously neutral gallium atom a net charge of $-e$. The resulting gallium ion is *not* free to move. In a semiconductor that is doped with acceptors, we consider the conductivity to be almost entirely due to *positive* charge (hole) motion. We call the material a ***p*-type semiconductor,** with *p*-type impurities. Some semiconductors are doped with *both* n- and p-type impurities. Such materials are called *compensated* semiconductors.

CAUTION The meaning of "*p*-type" and "*n*-type" Saying that a material is a *p*-type semiconductor does *not* mean that the material has a positive charge; ordinarily, it would be neutral. Rather, it means that its *majority carriers* of current are positive holes (and therefore its *minority carriers* are negative electrons). The same idea holds for an *n*-type semiconductor; ordinarily, it will *not* have a negative charge, but its majority carriers are negative electrons. ▌

We can verify the assertion that the current in *n*- and *p*-type semiconductors really *is* carried by electrons and holes, respectively, by using the Hall effect (see Section 27.9). The sign of the Hall emf is opposite in the two cases. Hall-effect devices constructed from semiconductor materials are used in probes to measure magnetic fields and the currents that cause those fields.

TEST YOUR UNDERSTANDING OF SECTION 42.6   Would there be any advantage to adding *n*-type or *p*-type impurities to copper? ▌

42.7 SEMICONDUCTOR DEVICES

Semiconductor devices play an indispensable role in contemporary electronics. In the early days of radio and television, transmitting and receiving equipment relied on vacuum tubes, but these have been replaced by solid-state devices, including transistors, diodes, integrated circuits, and other semiconductor devices. All modern consumer electronic devices use semiconductor devices of various kinds.

One simple semiconductor device is the *photocell* (**Fig. 42.28**). When a thin slab of semiconductor is irradiated with an electromagnetic wave whose photons have at least as much energy as the band gap between the valence and conduction bands, an electron in the valence band can absorb a photon and jump to the conduction band, where it and the hole it left behind contribute to the conductivity (see Example 42.5 in Section 42.4). The conductivity therefore increases with wave intensity, thus increasing the current *I* in the photocell circuit of Fig. 42.28. Hence the ammeter reading indicates the intensity of the light.

Detectors for charged particles operate on the same principle. An external circuit applies a voltage across a semiconductor. An energetic charged particle passing through the semiconductor collides inelastically with valence electrons, exciting them from the valence to the conduction band and creating pairs of holes and conduction electrons. The conductivity increases momentarily, causing a pulse of current in the external circuit. Solid-state detectors are widely used in nuclear and high-energy physics research.

The *p-n* Junction

In many semiconductor devices the essential principle is the fact that the conductivity of the material is controlled by impurity concentrations, which can be varied within wide limits from one region of a device to another. An example is the ***p-n* junction** at the boundary between one region of a semiconductor with *p*-type impurities and another region containing *n*-type impurities. One way of fabricating a *p-n* junction is to deposit some *n*-type material on the *very* clean surface of some *p*-type material. (We can't just stick p- and n-type pieces together and expect the junction to work properly because of the impossibility of matching their surfaces at the atomic level.)

42.27 A *p*-type semiconductor.

(a) An acceptor (*p*-type) impurity atom has only three valence electrons, so it can borrow an electron from a neighboring atom. The resulting hole is free to move about the crystal.

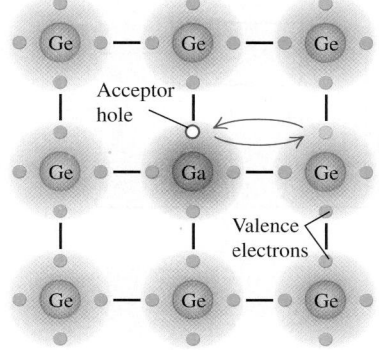

(b) Energy-band diagram for a *p*-type semiconductor at a low temperature. One acceptor level has accepted an electron from the valence band, leaving a hole behind.

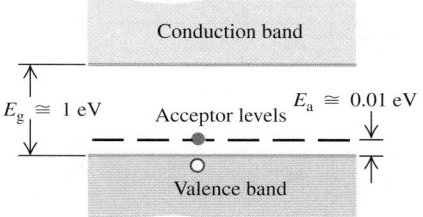

42.28 A semiconductor photocell in a circuit. The more intense the light falling on the photocell, the greater the conductivity of the photocell and the greater the current measured by the ammeter (A).

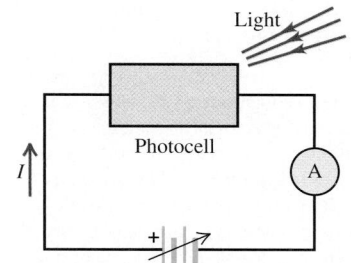

42.29 (a) A semiconductor *p-n* junction in a circuit. (b) Graph showing the asymmetric current–voltage relationship. The curve is described by Eq. (42.22).

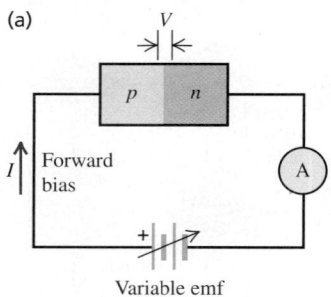

(a)

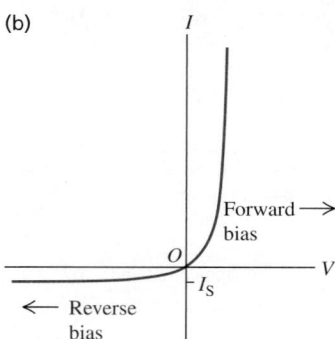

(b)

DATA *SPEAKS*

Semiconductors

When students were given a problem involving semiconductors, more than 24% gave an incorrect response. Common errors:

- Confusion about *p*-type and *n*-type semiconductors. The mobile charges (holes) in a *p*-type semiconductor are positive, but the semiconductor as a whole does not have a net positive charge. Similarly, the mobile charges (electrons) in an *n*-type semiconductor are negative, but the semiconductor does not have a net negative charge.

- Confusion about band gaps. An electron can absorb a photon and be promoted from the valence band to the conduction band, but only if it is given enough energy to bridge the gap between these bands. Hence a photon with an energy less than that of the band gap cannot be absorbed.

When a *p-n* junction is connected to an external circuit, as in **Fig. 42.29a**, and the potential difference $V_p - V_n = V$ across the junction is varied, the current I varies as shown in Fig. 42.29b. In striking contrast to the symmetric behavior of resistors that obey Ohm's law and give a straight line on an *I–V* graph, a *p-n* junction conducts much more readily in the direction from *p* to *n* than the reverse. Such a (mostly) one-way device is called a **diode rectifier.** Later we'll discuss a simple model of *p-n* junction behavior that predicts a current–voltage relationship in the form

$$
\begin{array}{ll}
\text{Current through} & I = I_S(e^{eV/kT} - 1) \\
\text{a } p\text{-}n \text{ junction} & e = 2.71828\ldots
\end{array}
\tag{42.22}
$$

Saturation current · · · Voltage · · · · · Absolute temperature · · · Boltzmann constant · · · Magnitude of electron charge

CAUTION **Two different uses of e** In $e^{eV/kT}$ the base of the exponent also uses the symbol *e*, standing for the base of the natural logarithms, 2.71828 . . . . This *e* is quite different from $e = 1.602 \times 10^{-19}$ C in the exponent. ▌

Equation (42.22) is valid for both positive and negative values of *V*; note that *V* and *I* always have the same sign. As *V* becomes very negative, *I* approaches the value $-I_S$. The magnitude I_S (always positive) is called the *saturation current*.

Currents Through a *p-n* Junction

We can understand the behavior of a *p-n* junction diode qualitatively on the basis of the mechanisms for conductivity in the two regions. Suppose, as in Fig. 42.29a, you connect the positive terminal of the battery to the *p* region and the negative terminal to the *n* region. Then the *p* region is at higher potential than the *n* region, corresponding to positive *V* in Eq. (42.22), and the resulting electric field is in the direction *p* to *n*. This is called the *forward* direction, and the positive potential difference is called *forward bias*. Holes, plentiful in the *p* region, flow easily across the junction into the *n* region, and free electrons, plentiful in the *n* region, easily flow into the *p* region; these movements of charge constitute a *forward* current. Connecting the battery with the opposite polarity gives *reverse bias*, and the field tends to push electrons from *p* to *n* and holes from *n* to *p*. But there are very few free electrons in the *p* region and very few holes in the *n* region. As a result, the current in the *reverse* direction is much smaller than that with the same potential difference in the forward direction.

Suppose you have a box with a barrier separating the left and right sides: You fill the left side with oxygen gas and the right side with nitrogen gas. What happens if the barrier leaks? Oxygen diffuses to the right, and nitrogen diffuses to the left. A similar diffusion occurs across a *p-n* junction. First consider the equilibrium situation with no applied voltage (**Fig. 42.30**). The many holes in the *p* region act like a hole gas that diffuses across the junction into the *n* region. Once there, the holes recombine with some of the many free electrons. Similarly, electrons diffuse from the *n* region to the *p* region and fall into some of the many holes there. The hole and electron diffusion currents lead to a net positive charge in the *n* region and a net negative charge in the *p* region, causing an electric field in the direction from *n* to *p* at the junction. The potential energy associated with this field raises the electron energy levels in the *p* region relative to the same levels in the *n* region.

There are four currents across the junction, as shown. The diffusion processes lead to *recombination currents* of holes and electrons, labeled i_{pr} and i_{nr} in Fig. 42.30. At the same time, electron–hole pairs are generated in the junction region by thermal excitation. The electric field described above sweeps these electrons and holes out of the junction; electrons are swept opposite the field to the *n* side, and holes are swept in the same direction as the field to the *p* side. The corresponding currents, called *generation currents*, are labeled i_{pg} and i_{ng}. At equilibrium the magnitudes of the generation and recombination currents are equal:

$$
|i_{pg}| = |i_{pr}| \quad \text{and} \quad |i_{ng}| = |i_{nr}|
\tag{42.23}
$$

42.30 A p-n junction in equilibrium, with no externally applied field or potential difference. The generation currents (subscript g) and recombination currents (subscript r) exactly balance. The Fermi energy E_F is the same on both sides of the junction. The excess positive and negative charges on the n and p sides produce an electric field $\vec{E}$ in the direction shown.

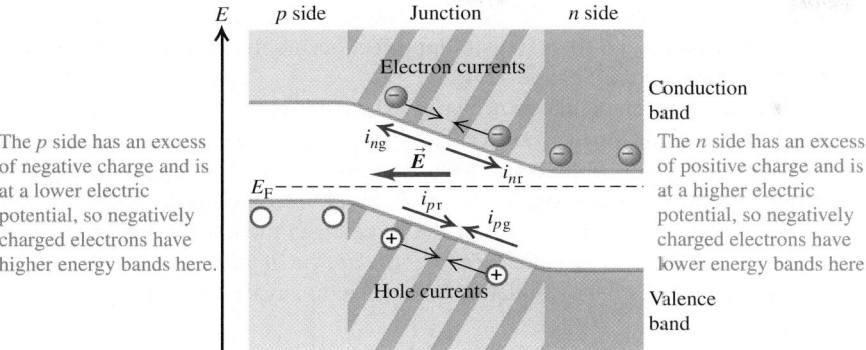

The p side has an excess of negative charge and is at a lower electric potential, so negatively charged electrons have higher energy bands here.

The n side has an excess of positive charge and is at a higher electric potential, so negatively charged electrons have lower energy bands here.

In thermal equilibrium the Fermi energy is the same at each point across the junction.

Now we apply a forward bias—that is, a positive potential difference V across the junction. A forward bias *decreases* the electric field in the junction region. It also decreases the difference between the energy levels on the p and n sides (**Fig. 42.31**) by an amount $\Delta E = -eV$. It becomes easier for the electrons in the n region to climb the potential-energy hill and diffuse into the p region and for the holes in the p region to diffuse into the n region. This effect increases both recombination currents by the Maxwell–Boltzmann factor $e^{-\Delta E/kT} = e^{eV/kT}$. (We don't have to use the Fermi–Dirac distribution because most of the available states for the diffusing electrons and holes are empty, so the exclusion principle has little effect.) The generation currents don't change appreciably, so the net hole current is

$$
\begin{aligned}
i_{p\text{tot}} &= i_{pr} - |i_{pg}| \\
&= |i_{pg}|e^{eV/kT} - |i_{pg}| \\
&= |i_{pg}|(e^{eV/kT} - 1)
\end{aligned}
\tag{42.24}
$$

The net electron current $i_{n\text{tot}}$ is given by a similar expression, so the total current $I = i_{p\text{tot}} + i_{n\text{tot}}$ is

$$
I = I_S(e^{eV/kT} - 1)
\tag{42.25}
$$

in agreement with Eq. (42.22). We can repeat this entire discussion for reverse bias (negative V and I) with the same result. Therefore Eq. (42.22) is valid for both positive and negative values.

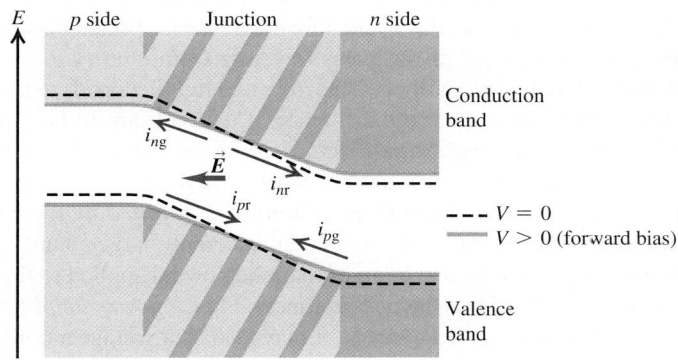

--- $V = 0$
—— $V > 0$ (forward bias)

42.31 A p-n junction under forward-bias conditions. The potential difference between p and n regions is reduced, as is the electric field within the junction. The recombination currents increase but the generation currents are nearly constant, causing a net current from left to right. (Compare Fig. 42.30.)

42.32 Under reverse-bias conditions the potential-energy difference between the p and n sides of a junction is greater than at equilibrium. If this difference is great enough, the bottom of the conduction band on the n side may actually be below the top of the valence band on the p side.

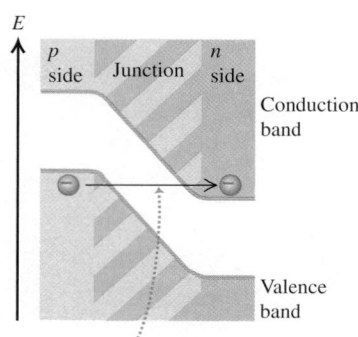

If a p-n junction under reverse bias is thin enough, electrons can tunnel from the valence band to the conduction band (a process called Zener breakdown).

BIO Application Swallow This Semiconductor Device This tiny capsule—designed to be swallowed by a patient—contains a miniature camera with a CCD light detector, plus six LEDs to illuminate the subject. The capsule radios high-resolution images to an external recording unit as it passes painlessly through the patient's stomach and intestines. This technique makes it possible to examine the small intestine, which is not readily accessible with conventional endoscopy.

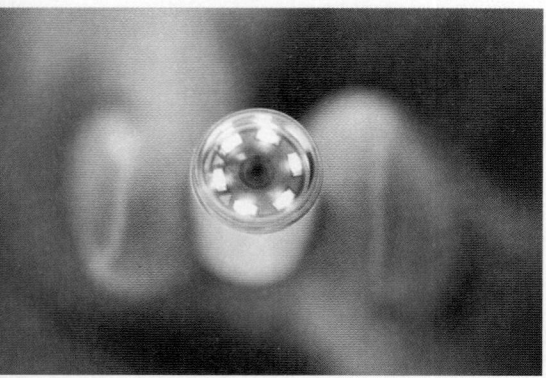

Several effects make the behavior of practical p-n junction diodes more complex than this simple analysis predicts. One effect, *avalanche breakdown*, occurs under large reverse bias. The electric field in the junction is so great that the carriers can gain enough energy between collisions to create electron–hole pairs during inelastic collisions. The electrons and holes then gain energy and collide to form more pairs, and so on. (A similar effect occurs in dielectric breakdown in insulators, discussed in Section 42.4.)

A second type of breakdown begins when the reverse bias becomes large enough that the top of the valence band in the p region is just higher in energy than the bottom of the conduction band in the n region (**Fig. 42.32**). If the junction region is thin enough, the probability becomes large that electrons can *tunnel* from the valence band of the p region to the conduction band of the n region. This process is called *Zener breakdown*. It occurs in Zener diodes, which are used for voltage regulation and protection against voltage surges.

Semiconductor Devices and Light

A *light-emitting diode (LED)* is a p-n junction diode that emits light. When the junction is forward biased, many holes are pushed from their p region to the junction region, and many electrons are pushed from their n region to the junction region. In the junction region the electrons fall into holes (recombine). In recombining, the electron can emit a photon with energy approximately equal to the band gap. This energy (and therefore the photon wavelength and the color of the light) can be varied by using materials with different band gaps. Light-emitting diodes are very energy-efficient light sources and have many applications, including automobile lamps, traffic signals, and flat-screen displays.

The reverse process is called the *photovoltaic effect*. Here the material absorbs photons, and electron–hole pairs are created. Pairs that are created in the p-n junction, or close enough to migrate to it without recombining, are separated by the electric field we described above that sweeps the electrons to the n side and the holes to the p side. We can connect this device to an external circuit, where it becomes a source of emf and power. Such a device is often called a *solar cell*, although sunlight isn't required. *Any* light with photon energies greater than the band gap will do. You might have a calculator powered by such cells. Production of low-cost photovoltaic cells for large-scale solar energy conversion is a very active field of research. The same basic physics is used in charge-coupled device (CCD) image detectors, digital cameras, and video cameras.

Transistors

A *bipolar junction transistor* includes two p-n junctions in a "sandwich" configuration, which may be either p-n-p or n-p-n. **Figure 42.33** shows such a p-n-p transistor. The three regions are called the emitter, base, and collector, as shown. When there is no current in the left loop of the circuit, there is only a very small current through the resistor R because the voltage across the base–collector junction is in the reverse direction. But when a forward bias is applied between emitter and base, as shown, most of the holes traveling from emitter to base travel *through* the base (which is typically both narrow and lightly doped) to the second junction, where they come under the influence of the collector-to-base potential difference and flow on through the collector to give an increased current to the resistor.

In this way the current in the collector circuit is *controlled* by the current in the emitter circuit. Furthermore, V_c may be considerably larger than V_e, so the *power* dissipated in R may be much larger than the power supplied to the emitter circuit by the battery V_e. Thus the device functions as a *power amplifier*. If the potential drop across R is greater than V_e, it may also be a voltage amplifier.

In this configuration the *base* is the common element between the "input" and "output" sides of the circuit. Another widely used arrangement is the *common-emitter* circuit, shown in **Fig. 42.34**. In this circuit the current in the collector side of the circuit is much larger than that in the base side, and the result is current amplification.

The *field-effect transistor* (**Fig. 42.35**) is an important type. In one variation a slab of *p*-type silicon is made with two *n*-type regions on the top, called the *source* and the *drain*; a metallic conductor is fastened to each. A third electrode called the *gate* is separated from the slab, source, and drain by an insulating layer of SiO_2. When there is no charge on the gate and a potential difference of either polarity is applied between the source and the drain, there is very little current because one of the *p-n* junctions is reverse biased.

Now we place a positive charge on the gate. With dimensions of the order of 10^{-6} m, it takes little charge to provide a substantial electric field. Thus there is very little current into or out of the gate. There aren't many free electrons in the *p*-type material, but there are some, and the effect of the field is to attract them toward the positive gate. The resulting greatly enhanced concentration of electrons near the gate (and between the two junctions) permits current to flow between the source and the drain. The current is very sensitive to the gate charge and potential, and the device functions as an amplifier. The device just described is called an *enhancement-type MOSFET* (metal-oxide-semiconductor field-effect transistor).

Integrated Circuits

A further refinement in semiconductor technology is the *integrated circuit*. By successively depositing layers of material and etching patterns to define current paths, we can combine the functions of several MOSFETs, capacitors, and resistors on a single square of semiconductor material that may be only a few millimeters on a side. An elaboration of this idea leads to *large-scale integrated circuits*. The resulting integrated circuit chips are the heart of all pocket calculators and present-day computers, large and small (**Fig. 42.36**).

The first semiconductor devices were invented in 1947. Since then, they have completely revolutionized the electronics industry through miniaturization, reliability, speed, energy usage, and cost. They have found applications in communications, computer systems, control systems, and many other areas. In transforming these areas, they have changed, and continue to change, human civilization itself.

TEST YOUR UNDERSTANDING OF SECTION 42.7 Suppose a negative charge is placed on the gate of the MOSFET shown in Fig. 42.35. Will a substantial current flow between the source and the drain? ∎

42.33 Schematic diagram of a *p-n-p* transistor and circuit.

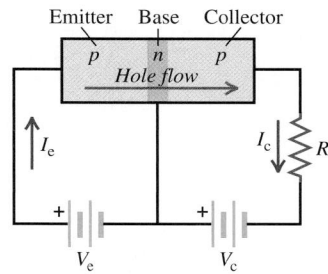

- When $V_e = 0$, the current is very small.
- When a potential V_e is applied between emitter and base, holes travel from the emitter to the base.
- When V_c is sufficiently large, most of the holes continue into the collector.

42.34 A common-emitter circuit.

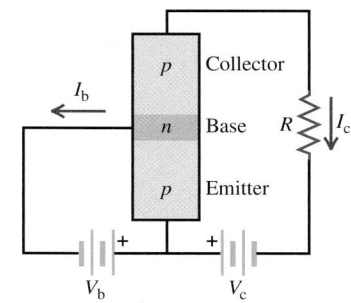

- When $V_b = 0$, I_c is very small, and most of the voltage V_c appears across the base–collector junction.
- As V_b increases, the base–collector potential decreases, and more holes can diffuse into the collector; thus, I_c increases. Ordinarily, I_c is much larger than I_b.

42.35 A field-effect transistor. The current from source to drain is controlled by the potential difference between the source and the drain and by the charge on the gate; no current flows through the gate.

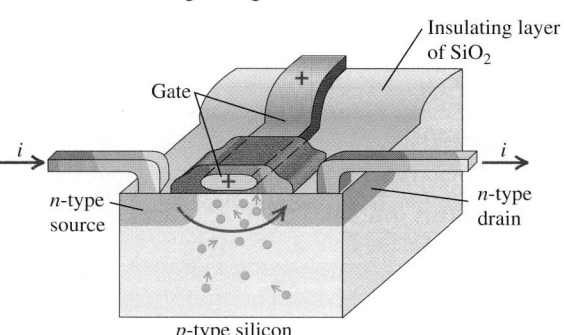

42.36 An integrated circuit chip smaller than your fingertip can contain millions of transistors.

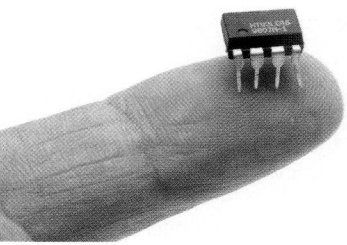

42.8 SUPERCONDUCTIVITY

Superconductivity is the complete disappearance of all electrical resistance at low temperatures. We described this property at the end of Section 25.2 and the magnetic properties of type-I and type-II superconductors in Section 29.8. In this section we'll relate superconductivity to the structure and energy-band model of a solid.

Although superconductivity was discovered in 1911, it was not well understood on a theoretical basis until 1957. That year, the American physicists John Bardeen, Leon Cooper, and Robert Schrieffer published the theory of superconductivity, now called the BCS theory, that earned them the Nobel Prize in physics in 1972. (It was Bardeen's second; he shared his first for his work on the development of the transistor.) The key to the BCS theory is an interaction between *pairs* of conduction electrons, called *Cooper pairs*, caused by an interaction with the positive ions of the crystal. Here's a rough picture of what happens. A free electron exerts attractive forces on nearby positive ions, pulling them slightly closer together. The resulting slight concentration of positive charge then exerts an attractive force on another free electron with momentum opposite to the first. At ordinary temperatures this electron-pair interaction is very small in comparison to energies of thermal motion, but at very low temperatures it is significant.

Bound together this way, the pairs of electrons cannot *individually* gain or lose very small amounts of energy, as they would ordinarily be able to do in a partly filled conduction band. Their pairing gives an energy gap in the allowed electron quantum levels, and at low temperatures there is not enough collision energy to jump this gap. Therefore the electrons can move freely through the crystal without any energy exchange through collisions—that is, with zero resistance.

Since 1987 physicists have discovered a number of compounds that remain superconducting at temperatures above 77 K (the boiling point of liquid nitrogen). The original pairing mechanism of the BCS theory cannot explain the properties of these *high-temperature superconductors*. Instead, it appears that electrons in these materials form pairs due to magnetic interactions between their spins.

| CHAPTER 42 **SUMMARY**

SOLUTIONS TO ALL EXAMPLES

Molecular bonds and molecular spectra: The principal types of molecular bonds are ionic, covalent, van der Waals, and hydrogen bonds. In a diatomic molecule the rotational energy levels are given by Eq. (42.3), where I is the moment of inertia of the molecule, m_r is its reduced mass, and r_0 is the distance between the two atoms. The vibrational energy levels are given by Eq. (42.7), where k' is the effective force constant of the interatomic force. (See Examples 42.1–42.3.)

$$E_l = l(l+1)\frac{\hbar^2}{2I} \quad (l = 0, 1, 2, \dots)$$
(42.3)

$$I = m_r r_0^2$$
(42.6)

$$m_r = \frac{m_1 m_2}{m_1 + m_2}$$
(42.4)

$$E_n = (n + \tfrac{1}{2})\hbar\omega = (n + \tfrac{1}{2})\hbar\sqrt{\frac{k'}{m_r}}$$
$$(n = 0, 1, 2, \dots)$$
(42.7)

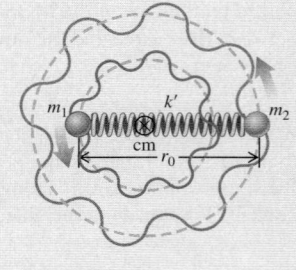

Solids and energy bands: Interatomic bonds in solids are of the same types as in molecules plus one additional type, the metallic bond. Associating the basis with each lattice point gives the crystal structure. (See Example 42.4.)

When atoms are bound together in condensed matter, their outer energy levels spread out into bands. At absolute zero, insulators and conductors have a completely filled valence band separated by an energy gap from an empty conduction band. Conductors, including metals, have partially filled conduction bands. (See Example 42.5.)

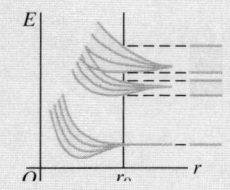

Free-electron model of metals: In the free-electron model of the behavior of conductors, the electrons are treated as completely free particles within the conductor. In this model the density of states is given by Eq. (42.15). The probability that an energy state of energy E is occupied is given by the Fermi–Dirac distribution, Eq. (42.16), which is a consequence of the exclusion principle. In Eq. (42.16), E_F is the Fermi energy. (See Examples 42.6–42.8.)

$$g(E) = \frac{(2m)^{3/2}V}{2\pi^2\hbar^3}E^{1/2} \quad (42.15)$$

$$f(E) = \frac{1}{e^{(E-E_F)/kT} + 1} \quad (42.16)$$

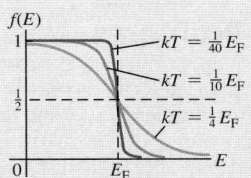

Semiconductors: A semiconductor has an energy gap of about 1 eV between its valence and conduction bands. Its electrical properties may be drastically changed by the addition of small concentrations of donor impurities, giving an n-type semiconductor, or acceptor impurities, giving a p-type semiconductor. (See Example 42.9.)

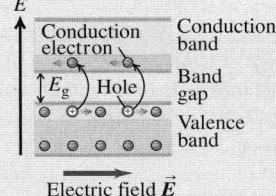

Semiconductor devices: Many semiconductor devices, including diodes, transistors, and integrated circuits, use one or more p-n junctions. The current–voltage relationship for an ideal p-n junction diode is given by Eq. (42.22).

$$I = I_S(e^{eV/kT} - 1) \quad (42.22)$$

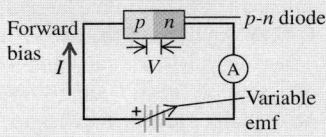

BRIDGING PROBLEM   MOLECULAR VIBRATION AND SEMICONDUCTOR BAND GAP

At 80 K, the band gap in the semiconductor indium antimonide (InSb) is 0.230 eV. A photon emitted by a hydrogen fluoride (HF) molecule undergoing a vibration-rotation transition from ($n = 1$, $l = 0$) to ($n = 0, l = 1$) is absorbed by an electron at the top of the valence band of InSb. (a) How far above the top of the band gap (in eV) is the final state of the electron? (b) What is the probability that the final state was already occupied? The vibration frequency for HF is 1.24×10^{14} Hz, the mass of a hydrogen atom is 1.67×10^{-27} kg, the mass of a fluorine atom is 3.15×10^{-26} kg, and the equilibrium distance between the two nuclei is 0.092 nm. Assume that the Fermi energy for InSb is in the middle of the gap.

SOLUTION GUIDE

IDENTIFY and SET UP

1. This problem involves what you learned about molecular transitions in Section 42.2, about the Fermi–Dirac distribution in Section 42.5, and about semiconductors in Section 42.6.
2. Equation (42.9) gives the combined vibrational-rotational energy in the initial and final molecular states. The difference between the initial and final molecular energies equals the energy E of the emitted photon, which is in turn equal to the

energy gained by the InSb valence electron when it absorbs that photon. The probability that the final state is occupied is given by the Fermi–Dirac distribution, Eq. (42.16).

EXECUTE

3. Before you can use Eq. (42.9), you'll first need to use the data given to calculate the moment of inertia I and the quantity $\hbar\omega$ for the HF molecule. (*Hint:* Be careful not to confuse frequency f and angular frequency ω.)
4. Use your results from step 3 to calculate the initial and final energies of the HF molecule. (*Hint:* Does the vibrational energy increase or decrease? What about the rotational energy?)
5. Use your result from step 4 to find the energy imparted to the InSb electron. Determine the final energy of this electron relative to the bottom of the conduction band.
6. Use your result from step 5 to determine the probability that the InSb final state is already occupied.

EVALUATE

7. Is the molecular transition of the HF molecule allowed? Which is larger: the vibrational energy change or the rotational energy change?
8. Is it likely that the excited InSb electron will be blocked from entering a state in the conduction band?

Problems

For assigned homework and other learning materials, go to MasteringPhysics®. **MP**

•, ••, •••: Difficulty levels. **CP**: Cumulative problems incorporating material from earlier chapters. **CALC**: Problems requiring calculus.
DATA: Problems involving real data, scientific evidence, experimental design, and/or statistical reasoning. **BIO**: Biosciences problems.

DISCUSSION QUESTIONS

Q42.1 Van der Waals bonds occur in many molecules, but hydrogen bonds occur only with materials that contain hydrogen. Why is this type of bond unique to hydrogen?

Q42.2 The bonding of gallium arsenide (GaAs) is said to be 31% ionic and 69% covalent. Explain.

Q42.3 The H_2^+ molecule consists of two hydrogen nuclei and a single electron. What kind of molecular bond do you think holds this molecule together? Explain.

Q42.4 The moment of inertia for an axis through the center of mass of a diatomic molecule calculated from the wavelength emitted in an $l = 19 \rightarrow l = 18$ transition is different from the moment of inertia calculated from the wavelength of the photon emitted in an $l = 1 \rightarrow l = 0$ transition. Explain this difference. Which transition corresponds to the larger moment of inertia?

Q42.5 Analysis of the photon absorption spectrum of a diatomic molecule shows that the vibrational energy levels for small values of n are very nearly equally spaced but the levels for large n are not equally spaced. Discuss the reason for this observation. Do you expect the adjacent levels to move closer together or farther apart as n increases? Explain.

Q42.6 Discuss the differences between the rotational and vibrational energy levels of the deuterium ("heavy hydrogen") molecule D_2 and those of the ordinary hydrogen molecule H_2. A deuterium atom has twice the mass of an ordinary hydrogen atom.

Q42.7 Various organic molecules have been discovered in interstellar space. Why were these discoveries made with radio telescopes rather than optical telescopes?

Q42.8 The air you are breathing contains primarily nitrogen (N_2) and oxygen (O_2). Many of these molecules are in excited rotational energy levels $(l = 1, 2, 3, \ldots)$, but almost all of them are in the vibrational ground level $(n = 0)$. Explain this difference between the rotational and vibrational behaviors of the molecules.

Q42.9 In what ways do atoms in a diatomic molecule behave as though they were held together by a spring? In what ways is this a poor description of the interaction between the atoms?

Q42.10 Individual atoms have discrete energy levels, but certain solids (which are made up of only individual atoms) show energy bands and gaps. What causes the solids to behave so differently from the atoms of which they are composed?

Q42.11 What factors determine whether a material is a conductor of electricity or an insulator? Explain.

Q42.12 Ionic crystals are often transparent, whereas metallic crystals are always opaque. Why?

Q42.13 Speeds of molecules in a gas vary with temperature, whereas speeds of electrons in the conduction band of a metal are nearly independent of temperature. Why are these behaviors so different?

Q42.14 Use the band model to explain how it is possible for some materials to undergo a semiconductor-to-metal transition as the temperature or pressure varies.

Q42.15 An isolated zinc atom has a ground-state electron configuration of filled $1s$, $2s$, $2p$, $3s$, $3p$, and $4s$ subshells. How can zinc be a conductor if its valence subshell is full?

Q42.16 The assumptions of the *free-electron model* of metals may seem contrary to reason, since electrons exert powerful electric forces on each other. Give some reasons why these assumptions actually make physical sense.

Q42.17 Why are materials that are good thermal conductors also good electrical conductors? What kinds of problems does this pose for the design of appliances such as clothes irons and electric heaters? Are there materials that do not follow this general rule?

Q42.18 What is the essential characteristic for an element to serve as a donor impurity in a semiconductor such as Si or Ge? For it to serve as an acceptor impurity? Explain.

Q42.19 There are several methods for removing electrons from the surface of a semiconductor. Can holes be removed from the surface? Explain.

Q42.20 A student asserts that silicon and germanium become good insulators at very low temperatures and good conductors at very high temperatures. Do you agree? Explain your reasoning.

Q42.21 The electrical conductivities of most metals decrease gradually with increasing temperature, but the intrinsic conductivity of semiconductors always *increases* rapidly with increasing temperature. What causes the difference?

Q42.22 How could you make compensated silicon that has twice as many acceptors as donors?

Q42.23 The saturation current I_S for a p-n junction, Eq. (42.22), depends strongly on temperature. Explain why.

Q42.24 Why does tunneling limit the miniaturization of MOSFETs?

EXERCISES

Section 42.1 Types of Molecular Bonds

42.1 • If the energy of the H_2 covalent bond is -4.48 eV, what wavelength of light is needed to break that molecule apart? In what part of the electromagnetic spectrum does this light lie?

42.2 • **An Ionic Bond.** (a) Calculate the electric potential energy for a K^+ ion and a Br^- ion separated by a distance of 0.29 nm, the equilibrium separation in the KBr molecule. Treat the ions as point charges. (b) The ionization energy of the potassium atom is 4.3 eV. Atomic bromine has an electron affinity of 3.5 eV. Use these data and the results of part (a) to estimate the binding energy of the KBr molecule. Do you expect the actual binding energy to be higher or lower than your estimate? Explain your reasoning.

42.3 • For the H_2 molecule the equilibrium spacing of the two protons is 0.074 nm. The mass of a hydrogen atom is 1.67×10^{-27} kg. Calculate the wavelength of the photon emitted in the rotational transition $l = 2$ to $l = 1$.

42.4 • During each of these processes, a photon of light is given up. In each process, what wavelength of light is given up, and in what part of the electromagnetic spectrum is that wavelength? (a) A molecule decreases its vibrational energy by 0.198 eV; (b) an atom decreases its energy by 7.80 eV; (c) a molecule decreases its rotational energy by 4.80×10^{-3} eV.

Section 42.2 Molecular Spectra

42.5 • A hypothetical NH molecule makes a rotational-level transition from $l = 3$ to $l = 1$ and gives off a photon of wavelength 1.780 nm in doing so. What is the separation between the two atoms in this molecule if we model them as point masses? The mass of hydrogen is 1.67×10^{-27} kg, and the mass of nitrogen is 2.33×10^{-26} kg.

42.6 •• The H_2 molecule has a moment of inertia of 4.6×10^{-48} kg $\cdot$ m^2. What is the wavelength λ of the photon absorbed when H_2 makes a transition from the $l = 3$ to the $l = 4$ rotational level?

42.7 • The water molecule has an $l = 1$ rotational level 1.01×10^{-5} eV above the $l = 0$ ground level. Calculate the wavelength and frequency of the photon absorbed by water when it undergoes a rotational-level transition from $l = 0$ to $l = 1$. The magnetron oscillator in a microwave oven generates microwaves with a frequency of 2450 MHz. Does this make sense, in view of the frequency you calculated in this problem? Explain.

42.8 • Two atoms of cesium (Cs) can form a Cs_2 molecule. The equilibrium distance between the nuclei in a Cs_2 molecule is 0.447 nm. Calculate the moment of inertia about an axis through the center of mass of the two nuclei and perpendicular to the line joining them. The mass of a cesium atom is 2.21×10^{-25} kg.

42.9 •• CP The rotational energy levels of CO are calculated in Example 42.2. If the energy of the rotating molecule is described by the classical expression $K = \frac{1}{2}I\omega^2$, for the $l = 1$ level what are (a) the angular speed of the rotating molecule; (b) the linear speed of each atom; (c) the rotational period (the time for one rotation)?

42.10 •• The average kinetic energy of an ideal-gas atom or molecule is $\frac{3}{2}kT$, where T is the Kelvin temperature (Chapter 18). The rotational inertia of the H_2 molecule is 4.6×10^{-48} kg $\cdot$ m^2. What is the value of T for which $\frac{3}{2}kT$ equals the energy separation between the $l = 0$ and $l = 1$ energy levels of H_2? What does this tell you about the number of H_2 molecules in the $l = 1$ level at room temperature?

42.11 • A lithium atom has mass 1.17×10^{-26} kg, and a hydrogen atom has mass 1.67×10^{-27} kg. The equilibrium separation between the two nuclei in the LiH molecule is 0.159 nm. (a) What is the difference in energy between the $l = 3$ and $l = 4$ rotational levels? (b) What is the wavelength of the photon emitted in a transition from the $l = 4$ to the $l = 3$ level?

42.12 • If a sodium chloride (NaCl) molecule could undergo an $n \rightarrow n - 1$ vibrational transition with no change in rotational quantum number, a photon with wavelength 20.0 μm would be emitted. The mass of a sodium atom is 3.82×10^{-26} kg, and the mass of a chlorine atom is 5.81×10^{-26} kg. Calculate the force constant k' for the interatomic force in NaCl.

42.13 •• When a hypothetical diatomic molecule having atoms 0.8860 nm apart undergoes a rotational transition from the $l = 2$ state to the next lower state, it gives up a photon having energy 8.841×10^{-4} eV. When the molecule undergoes a vibrational transition from one energy state to the next lower energy state, it gives up 0.2560 eV. Find the force constant of this molecule.

42.14 • The vibrational and rotational energies of the CO molecule are given by Eq. (42.9). Calculate the wavelength of the photon absorbed by CO in each of these vibration-rotation transitions: (a) $n = 0$, $l = 2 \rightarrow n = 1$, $l = 3$; (b) $n = 0$, $l = 3 \rightarrow n = 1$, $l = 2$; (c) $n = 0$, $l = 4 \rightarrow n = 1$, $l = 3$.

Section 42.3 Structure of Solids

42.15 • **Density of NaCl.** The spacing of adjacent atoms in a crystal of sodium chloride is 0.282 nm. The mass of a sodium atom is 3.82×10^{-26} kg, and the mass of a chlorine atom is 5.89×10^{-26} kg. Calculate the density of sodium chloride.

42.16 • Potassium bromide (KBr) has a density of 2.75×10^3 kg/m^3 and the same crystal structure as NaCl. The mass of a potassium atom is 6.49×10^{-26} kg, and the mass of a bromine

atom is 1.33×10^{-25} kg. (a) Calculate the average spacing between adjacent atoms in a KBr crystal. (b) How does the value calculated in part (a) compare with the spacing in NaCl (see Exercise 42.15)? Is the relationship between the two values qualitatively what you would expect? Explain.

Section 42.4 Energy Bands

42.17 • The maximum wavelength of light that a certain silicon photocell can detect is 1.11 μm. (a) What is the energy gap (in electron volts) between the valence and conduction bands for this photocell? (b) Explain why pure silicon is opaque.

42.18 • The gap between valence and conduction bands in diamond is 5.47 eV. (a) What is the maximum wavelength of a photon that can excite an electron from the top of the valence band into the conduction band? In what region of the electromagnetic spectrum does this photon lie? (b) Explain why pure diamond is transparent and colorless. (c) Most gem diamonds have a yellow color. Explain how impurities in the diamond can cause this color.

42.19 • The gap between valence and conduction bands in silicon is 1.12 eV. A nickel nucleus in an excited state emits a gamma-ray photon with wavelength 9.31×10^{-4} nm. How many electrons can be excited from the top of the valence band to the bottom of the conduction band by the absorption of this gamma ray?

Section 42.5 Free-Electron Model of Metals

42.20 • Calculate v_{rms} for free electrons with average kinetic energy $\frac{3}{2}kT$ at a temperature of 300 K. How does your result compare to the speed of an electron with a kinetic energy equal to the Fermi energy of copper, calculated in Example 42.7? Why is there such a difference between these speeds?

42.21 • Calculate the density of states $g(E)$ for the free-electron model of a metal if $E = 7.0$ eV and $V = 1.0$ cm^3. Express your answer in units of states per electron volt.

42.22 • The Fermi energy of sodium is 3.23 eV. (a) Find the average energy E_{av} of the electrons at absolute zero. (b) What is the speed of an electron that has energy E_{av}? (c) At what Kelvin temperature T is kT equal to E_F? (This is called the *Fermi temperature* for the metal. It is approximately the temperature at which molecules in a classical ideal gas would have the same kinetic energy as the fastest-moving electron in the metal.)

42.23 • CP Silver has a Fermi energy of 5.48 eV. Calculate the electron contribution to the molar heat capacity at constant volume of silver, C_V, at 300 K. Express your result (a) as a multiple of R and (b) as a fraction of the actual value for silver, $C_V = 25.3$ J/mol $\cdot$ K. (c) Is the value of C_V due principally to the electrons? If not, to what is it due? (*Hint:* See Section 18.4.)

42.24 •• At the Fermi temperature T_F, $E_F = kT_F$ (see Exercise 42.22). When $T = T_F$, what is the probability that a state with energy $E = 2E_F$ is occupied?

42.25 •• For a solid metal having a Fermi energy of 8.500 eV, what is the probability, at room temperature, that a state having an energy of 8.520 eV is occupied by an electron?

Section 42.6 Semiconductors

42.26 • Pure germanium has a band gap of 0.67 eV. The Fermi energy is in the middle of the gap. (a) For temperatures of 250 K, 300 K, and 350 K, calculate the probability $f(E)$ that a state at the bottom of the conduction band is occupied. (b) For each temperature in part (a), calculate the probability that a state at the top of the valence band is empty.

42.27 • Germanium has a band gap of 0.67 eV. Doping with arsenic adds donor levels in the gap 0.01 eV below the bottom of the conduction band. At a temperature of 300 K, the probability is 4.4×10^{-4} that an electron state is occupied at the bottom of the conduction band. Where is the Fermi level relative to the conduction band in this case?

Section 42.7 Semiconductor Devices

42.28 •• (a) Suppose a piece of very pure germanium is to be used as a light detector by observing, through the absorption of photons, the increase in conductivity resulting from generation of electron–hole pairs. If each pair requires 0.67 eV of energy, what is the maximum wavelength that can be detected? In what portion of the spectrum does it lie? (b) What are the answers to part (a) if the material is silicon, with an energy requirement of 1.12 eV per pair, corresponding to the gap between valence and conduction bands in that element?

42.29 • CP At a temperature of 290 K, a certain *p-n* junction has a saturation current $I_S = 0.500$ mA. (a) Find the current at this temperature when the voltage is (i) 1.00 mV, (ii) −1.00 mV, (iii) 100 mV, and (iv) −100 mV. (b) Is there a region of applied voltage where the diode obeys Ohm's law?

42.30 • For a certain *p-n* junction diode, the saturation current at room temperature (20°C) is 0.950 mA. What is the resistance of this diode when the voltage across it is (a) 85.0 mV and (b) −50.0 mV?

42.31 •• (a) A forward-bias voltage of 15.0 mV produces a positive current of 9.25 mA through a *p-n* junction at 300 K. What does the positive current become if the forward-bias voltage is reduced to 10.0 mV? (b) For reverse-bias voltages of −15.0 mV and −10.0 mV, what is the reverse-bias negative current?

42.32 •• A *p-n* junction has a saturation current of 6.40 mA. (a) At a temperature of 300 K, what voltage is needed to produce a positive current of 40.0 mA? (b) For a voltage equal to the negative of the value calculated in part (a), what is the negative current?

PROBLEMS

42.33 •• A hypothetical diatomic molecule of oxygen (mass = 2.656×10^{-26} kg) and hydrogen (mass = 1.67×10^{-27} kg) emits a photon of wavelength 2.39 μm when it makes a transition from one vibrational state to the next lower state. If we model this molecule as two point masses at opposite ends of a massless spring, (a) what is the force constant of this spring, and (b) how many vibrations per second is the molecule making?

42.34 • When a diatomic molecule undergoes a transition from the $l = 2$ to the $l = 1$ rotational state, a photon with wavelength 54.3 μm is emitted. What is the moment of inertia of the molecule for an axis through its center of mass and perpendicular to the line connecting the nuclei?

42.35 •• CP (a) The equilibrium separation of the two nuclei in an NaCl molecule is 0.24 nm. If the molecule is modeled as charges $+e$ and $-e$ separated by 0.24 nm, what is the electric dipole moment of the molecule (see Section 21.7)? (b) The measured electric dipole moment of an NaCl molecule is 3.0×10^{-29} C·m. If this dipole moment arises from point charges $+q$ and $-q$ separated by 0.24 nm, what is q? (c) A definition of the *fractional ionic character* of the bond is q/e. If the sodium atom has charge $+e$ and the chlorine atom has charge $-e$, the fractional ionic character would be equal to 1. What is the actual fractional ionic character for the bond in NaCl? (d) The

equilibrium distance between nuclei in the hydrogen iodide (HI) molecule is 0.16 nm, and the measured electric dipole moment of the molecule is 1.5×10^{-30} C·m. What is the fractional ionic character for the bond in HI? How does your answer compare to that for NaCl calculated in part (c)? Discuss reasons for the difference in these results.

42.36 • The binding energy of a potassium chloride molecule (KCl) is 4.43 eV. The ionization energy of a potassium atom is 4.3 eV, and the electron affinity of chlorine is 3.6 eV. Use these data to estimate the equilibrium separation between the two atoms in the KCl molecule. Explain why your result is only an estimate and not a precise value.

42.37 • (a) For the sodium chloride molecule (NaCl) discussed at the beginning of Section 42.1, what is the maximum separation of the ions for stability if they may be regarded as point charges? That is, what is the largest separation for which the energy of an Na^+ ion and a Cl^- ion, calculated in this model, is lower than the energy of the two separate atoms Na and Cl? (b) Calculate this distance for the potassium bromide molecule, described in Exercise 42.2.

42.38 • When a NaF molecule makes a transition from the $l = 3$ to the $l = 2$ rotational level with no change in vibrational quantum number or electronic state, a photon with wavelength 3.83 mm is emitted. A sodium atom has mass 3.82×10^{-26} kg, and a fluorine atom has mass 3.15×10^{-26} kg. Calculate the equilibrium separation between the nuclei in a NaF molecule. How does your answer compare with the value for NaCl given in Section 42.1? Is this result reasonable? Explain.

42.39 •• CP Consider a gas of diatomic molecules (moment of inertia I) at an absolute temperature T. If E_g is a ground-state energy and E_{ex} is the energy of an excited state, then the Maxwell–Boltzmann distribution (see Section 39.4) predicts that the ratio of the numbers of molecules in the two states is $n_{ex}/n_g = e^{-(E_{ex} - E_g)/kT}$. (a) Explain why the ratio of the number of molecules in the lth rotational energy *level* to the number of molecules in the ground-state ($l = 0$) rotational level is

$$\frac{n_l}{n_0} = (2l + 1)e^{-[l(l+1)\hbar^2]/2IkT}$$

(*Hint:* For each value of l, how many states are there with different values of m_l?) (b) Determine the ratio n_l/n_0 for a gas of CO molecules at 300 K for (i) $l = 1$; (ii) $l = 2$; (iii) $l = 10$; (iv) $l = 20$; (v) $l = 50$. The moment of inertia of the CO molecule is given in Example 42.2 (Section 42.2). (c) Your results in part (b) show that as l is increased, the ratio n_l/n_0 first increases and then decreases. Explain why.

42.40 ••• CALC Part (a) of Problem 42.39 gives an equation for the number of diatomic molecules in the lth rotational level to the number in the ground-state rotational level. (a) Derive an expression for the value of l for which this ratio is the largest. (b) For the CO molecule at $T = 300$ K, for what value of l is this ratio a maximum? (The moment of inertia of the CO molecule is given in Example 42.2.)

42.41 • **Spectral Lines from Isotopes.** The equilibrium separation for NaCl is 0.2361 nm. The mass of a sodium atom is 3.8176×10^{-26} kg. Chlorine has two stable isotopes, ^{35}Cl and ^{37}Cl, that have different masses but identical chemical properties. The atomic mass of ^{35}Cl is 5.8068×10^{-26} kg, and the atomic mass of ^{37}Cl is 6.1384×10^{-26} kg. (a) Calculate the wavelength of the photon emitted in the $l = 2 \rightarrow l = 1$ and $l = 1 \rightarrow l = 0$ transitions for Na^{35}Cl. (b) Repeat part (a) for Na^{37}Cl. What are the differences in the wavelengths for the two isotopes?

42.42 •• Our galaxy contains numerous *molecular clouds*, regions many light-years in extent in which the density is high enough and the temperature low enough for atoms to form into molecules. Most of the molecules are H_2, but a small fraction of the molecules are carbon monoxide (CO). Such a molecular cloud in the constellation Orion is shown in **Fig. P42.42**. The upper image was made with an ordinary visible-light telescope; the lower image shows the molecular cloud in Orion as imaged with a radio telescope tuned to a wavelength emitted by CO in a rotational transition. The different colors in the radio image indicate regions of the cloud that are moving either toward us (blue) or away from us (red) relative to the motion of the cloud as a whole, as determined by the Doppler shift of the radiation. (Since a molecular cloud has about 10,000 hydrogen molecules for each CO molecule, it might seem more reasonable to tune a radio telescope to emissions from H_2 than to emissions from CO. Unfortunately, it turns out that the H_2 molecules in molecular clouds do not radiate in either the radio or visible portions of the electromagnetic spectrum.) (a) Using the data in Example 42.2 (Section 42.2), calculate the energy and wavelength of the photon emitted by a CO molecule in an $l = 1 \rightarrow l = 0$ rotational transition. (b) As a rule, molecules in a gas at temperature T will be found in a certain excited rotational energy level, provided the energy of that level is no higher than kT (see Problem 42.39). Use this rule to explain why astronomers can detect radiation from CO in molecular clouds even though the typical temperature of a molecular cloud is a very low 20 K.

Figure **P42.42**

42.43 • The force constant for the internuclear force in a hydrogen molecule (H_2) is $k' = 576$ N/m. A hydrogen atom has mass 1.67×10^{-27} kg. Calculate the zero-point vibrational energy for H_2 (that is, the vibrational energy the molecule has in the $n = 0$ ground vibrational level). How does this energy compare in magnitude with the H_2 bond energy of -4.48 eV?

42.44 • When an OH molecule undergoes a transition from the $n = 0$ to the $n = 1$ vibrational level, its internal vibrational energy increases by 0.463 eV. Calculate the frequency of vibration and the force constant for the interatomic force. (The mass of an oxygen atom is 2.66×10^{-26} kg, and the mass of a hydrogen atom is 1.67×10^{-27} kg.)

42.45 • The hydrogen iodide (HI) molecule has equilibrium separation 0.160 nm and vibrational frequency 6.93×10^{13} Hz. The mass of a hydrogen atom is 1.67×10^{-27} kg, and the mass of an iodine atom is 2.11×10^{-25} kg. (a) Calculate the moment of inertia of HI about a perpendicular axis through its center of mass. (b) Calculate the wavelength of the photon emitted

in each of the following vibration-rotation transitions: (i) $n = 1$, $l = 1 \rightarrow n = 0$, $l = 0$; (ii) $n = 1$, $l = 2 \rightarrow n = 0$, $l = 1$; (iii) $n = 2, l = 2 \rightarrow n = 1, l = 3$.

42.46 • Suppose the hydrogen atom in HF (see the Bridging Problem for this chapter) is replaced by an atom of deuterium, an isotope of hydrogen with a mass of 3.34×10^{-27} kg. The force constant is determined by the electron configuration, so it is the same as for the normal HF molecule. (a) What is the vibrational frequency of this molecule? (b) What wavelength of light corresponds to the energy difference between the $n = 1$ and $n = 0$ levels? In what region of the spectrum does this wavelength lie?

42.47 •• Compute the Fermi energy of potassium by making the simple approximation that each atom contributes one free electron. The density of potassium is 851 kg/m³, and the mass of a single potassium atom is 6.49×10^{-26} kg.

42.48 •• CALC The one-dimensional calculation of Example 42.4 (Section 42.3) can be extended to three dimensions. For the three-dimensional fcc NaCl lattice, the result for the potential energy of a pair of Na^+ and Cl^- ions due to the electrostatic interaction with all of the ions in the crystal is $U = -\alpha e^2/4\pi\epsilon_0 r$, where $\alpha = 1.75$ is the *Madelung constant*. Another contribution to the potential energy is a repulsive interaction at small ionic separation r due to overlap of the electron clouds. This contribution can be represented by A/r^8, where A is a positive constant, so the expression for the total potential energy is

$$U_{\text{tot}} = -\frac{\alpha e^2}{4\pi\epsilon_0 r} + \frac{A}{r^8}$$

(a) Let r_0 be the value of the ionic separation r for which U_{tot} is a minimum. Use this definition to find an equation that relates r_0 and A, and use this to write U_{tot} in terms of r_0. For NaCl, $r_0 = 0.281$ nm. Obtain a numerical value (in electron volts) of U_{tot} for NaCl. (b) The quantity $-U_{\text{tot}}$ is the energy required to remove a Na^+ ion and a Cl^- ion from the crystal. Forming a pair of neutral atoms from this pair of ions involves the release of 5.14 eV (the ionization energy of Na) and the expenditure of 3.61 eV (the electron affinity of Cl). Use the result of part (a) to calculate the energy required to remove a pair of neutral Na and Cl atoms from the crystal. The experimental value for this quantity is 6.39 eV; how well does your calculation agree?

42.49 ••• Metallic lithium has a bcc crystal structure. Each unit cell is a cube of side length $a = 0.35$ nm. (a) For a bcc lattice, what is the number of atoms per unit volume? Give your answer in terms of a. (*Hint:* How many atoms are there per unit cell?) (b) Use the result of part (a) to calculate the zero-temperature Fermi energy E_{F0} for metallic lithium. Assume there is one free electron per atom.

42.50 •• DATA To determine the equilibrium separation of the atoms in the HCl molecule, you measure the rotational spectrum of HCl. You find that the spectrum contains these wavelengths (among others): 60.4 μm, 69.0 μm, 80.4 μm, 96.4 μm, and 120.4 μm. (a) Use your measured wavelengths to find the moment of inertia of the HCl molecule about an axis through the center of mass and perpendicular to the line joining the two nuclei. (b) The value of l changes by ± 1 in rotational transitions. What value of l for the upper level of the transition gives rise to each of these wavelengths? (c) Use your result of part (a) to calculate the equilibrium separation of the atoms in the HCl molecule. The mass of a chlorine atom is 5.81×10^{-26} kg, and the mass of a hydrogen atom is 1.67×10^{-27} kg. (d) What is the longest-wavelength line in the rotational spectrum of HCl?

42.51 •• DATA The table gives the occupation probabilities $f(E)$ as a function of the energy E for a solid conductor at a fixed temperature T.

$f(E)$	0.064	0.173	0.390	0.661	0.856	0.950
E (eV)	3.0	2.5	2.0	1.5	1.0	0.5

To determine the Fermi energy of the solid material, you are asked to analyze this information in terms of the Fermi–Dirac distribution. (a) Graph the values in the table as E versus $\ln\{[1/f(E)] - 1\}$. Find the slope and y-intercept of the best-fit straight line for the data points when they are plotted this way. (b) Use your results of part (a) to calculate the temperature T and the Fermi energy of the material.

42.52 •• DATA A p-n junction is part of the control mechanism for a wind turbine that is used to generate electricity. The turbine has been malfunctioning, so you are running diagnostics. You can remotely change the bias voltage V applied to the junction and measure the current through the junction. With a forward bias voltage of $+5.00$ mV, the current is $I_f = 0.407$ mA. With a reverse bias voltage of -5.00 mV, the current is $I_r = -0.338$ mA. Assume that Eq. (42.22) accurately represents the current–voltage relationship for the junction, and use these two results to calculate the temperature T and saturation current I_S for the junction. [*Hint:* In your analysis, let $x = e^{eV/kT}$. Apply Eq. (42.22) to each measurement and obtain a quadratic equation for x.]

CHALLENGE PROBLEMS

42.53 •• CALC Consider a system of N free electrons within a volume V. Even at absolute zero, such a system exerts a pressure p on its surroundings due to the motion of the electrons. To calculate this pressure, imagine that the volume increases by a small amount dV. The electrons will do an amount of work $p\,dV$ on their surroundings, which means that the total energy E_{tot} of the electrons will change by an amount $dE_{tot} = -p\,dV$. Hence $p = -dE_{tot}/dV$. (a) Show that the pressure of the electrons at absolute zero is

$$p = \frac{3^{2/3}\pi^{4/3}\hbar^2}{5m}\left(\frac{N}{V}\right)^{5/3}$$

(b) Evaluate this pressure for copper, which has a free-electron concentration of 8.45×10^{28} m^{-3}. Express your result in pascals and in atmospheres. (c) The pressure you found in part (b) is extremely high. Why, then, don't the electrons in a piece of copper simply explode out of the metal?

42.54 •• CALC When the pressure p on a material increases by an amount Δp, the volume of the material will change from V to $V + \Delta V$, where ΔV is negative. The *bulk modulus B* of the material is defined to be the ratio of the pressure change Δp to the absolute value $|\Delta V/V|$ of the fractional volume change. The greater the bulk modulus, the greater the pressure increase required for a given fractional volume change, and the more incompressible the material (see Section 11.4). Since $\Delta V < 0$, the bulk modulus can be written as $B = -\Delta p/(\Delta V/V_0)$. In the limit that the pressure and volume changes are very small, this becomes

$$B = -V\frac{dp}{dV}$$

(a) Use the result of Problem 42.53 to show that the bulk modulus for a system of N free electrons in a volume V at low temperatures is $B = \frac{5}{3}p$. (*Hint:* The quantity p in the expression $B = -V(dp/dV)$ is the *external* pressure on the system. Can you explain why this is equal to the *internal* pressure of the system itself, as found in Problem 42.53?) (b) Evaluate the bulk modulus for the electrons in copper, which has a free-electron concentration of 8.45×10^{28} m^{-3}. Express your result in pascals. (c) The actual bulk modulus of copper is 1.4×10^{11} Pa. Based on your result in part (b), what fraction of this is due to the free electrons in copper? (This result shows that the free electrons in a metal play a major role in making the metal resistant to compression.) What do you think is responsible for the remaining fraction of the bulk modulus?

42.55 •• In the discussion of free electrons in Section 42.5, we assumed that we could ignore the effects of relativity. This is not a safe assumption if the Fermi energy is greater than about $\frac{1}{100}mc^2$ (that is, more than about 1% of the rest energy of an electron). (a) Assume that the Fermi energy at absolute zero, as given by Eq. (42.19), is equal to $\frac{1}{100}mc^2$. Show that the electron concentration is

$$\frac{N}{V} = \frac{2^{3/2}m^3c^3}{3000\pi^2\hbar^3}$$

and determine the numerical value of N/V. (b) Is it a good approximation to ignore relativistic effects for electrons in a metal such as copper, for which the electron concentration is 8.45×10^{28} m^{-3}? Explain. (c) A *white dwarf star* is what is left behind by a star like the sun after it has ceased to produce energy by nuclear reactions. (Our own sun will become a white dwarf star in another 6×10^9 years or so.) A typical white dwarf has mass 2×10^{30} kg (comparable to the sun) and radius 6000 km (comparable to that of the earth). The gravitational attraction of different parts of the white dwarf for each other tends to compress the star; what prevents it from compressing is the pressure of free electrons within the star (see Problem 42.53). Use both of the following assumptions to estimate the electron concentration within a typical white dwarf star: (i) the white dwarf star is made of carbon, which has a mass per atom of 1.99×10^{-26} kg; and (ii) all six of the electrons from each carbon atom are able to move freely throughout the star. (d) Is it a good approximation to ignore relativistic effects in the structure of a white dwarf star? Explain.

PASSAGE PROBLEMS

DIODE TEMPERATURE SENSOR. The current–voltage characteristics of a forward-biased p-n junction diode depend strongly on temperature, as shown in the figure. As a result, diodes can be used as temperature sensors. In actual operation, the voltage is adjusted to keep the current through the diode constant at a specified value, such as 100 mA, and the temperature is determined from a measurement of the voltage at that current.

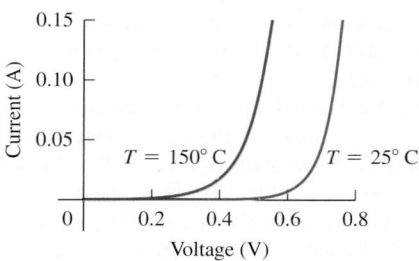

42.56 The sensitivity of a diode thermometer depends on how much the voltage changes for a given temperature change, with the current remaining constant. What is the sensitivity for this diode thermometer, operated at 100 mA, for a temperature change from 25°C to 150°C? (a) +0.2 mV/°C; (b) +2.0 mV/°C; (c) −0.2 mV/°C; (d) −2.0 mV/°C.

42.57 Which statement best explains the temperature dependence of the current–voltage characteristics that the graph shows? At higher temperatures: (a) The band gap is larger, so the electron–hole pairs have more energy, which causes the current at a given voltage to be larger. (b) More electrons can move to the conduction band, which causes the current at a given voltage to be larger. (c) All of the electrons in the valence band move to the conduction band, and the diode behaves like a metal and follows Ohm's law. (d) The acceptor and donor impurity atoms are free to move through the material, which causes the current at a given voltage to be larger.

42.58 If the voltage rather than the current is kept constant, what happens as the temperature increases from 25°C to 150°C? (a) At first the current increases, then it decreases. (b) The current increases. (c) The current decreases, eventually approaching zero. (d) The current does not change unless the voltage also changes.

Answers

Chapter Opening Question ?

(i) Venus must radiate energy into space at the same rate that it receives energy in the form of sunlight. However, carbon dioxide (CO_2) molecules in the atmosphere absorb infrared radiation emitted by the surface of Venus and re-emit it toward the ground. This involves a transition between vibrational states of the CO_2 molecule (see Section 42.2). To compensate for this and to maintain the balance between emitted and received energy, the surface temperature of Venus and hence the rate of blackbody radiation from the surface increase.

Test Your Understanding Questions

42.1 (i) The exclusion principle states that only one electron can be in a given state. Real electrons have spin, so two electrons (one spin up, one spin down) can be in a given *spatial* state and hence two can participate in a given covalent bond between two atoms. If electrons obeyed the exclusion principle but did not have spin, that state of an electron would be completely described by its spatial distribution and only *one* electron could participate in a covalent bond. (We will learn in Chapter 44 that this situation is wholly imaginary: There are subatomic particles without spin, but they do *not* obey the exclusion principle.)

42.2 (ii) Figure 42.5 shows that the difference in energy between adjacent rotational levels increases with increasing l. Hence, as l increases, the energy E of the emitted photon increases and the wavelength $\lambda = hc/E$ decreases.

42.3 (ii) In Fig. 42.13 let a be the distance between adjacent Na^+ and Cl^- ions. This figure shows that the Cl^- ion that is the next nearest neighbor to a Na^+ ion is on the opposite corner of a cube of side a. The distance between these two ions is $\sqrt{a^2 + a^2 + a^2} = \sqrt{3a^2} = a\sqrt{3}$.

42.4 (ii) A small temperature change causes a substantial increase in the population of electrons in a semiconductor's conduction band and a comparably substantial increase in conductivity. The conductivity of conductors and insulators varies more gradually with temperature.

42.5 no The kinetic-molecular model of an ideal gas (Section 18.3) shows that the gas pressure is proportional to the average translational kinetic energy E_{av} of the particles that make up the gas. In a classical ideal gas, E_{av} is directly proportional to the average temperature T, so the pressure decreases as T decreases. In a free-electron gas, the average kinetic energy per electron is *not* related simply to T; as Example 42.8 shows, for the free-electron gas in a metal, E_{av} is almost completely a consequence of the exclusion principle at room temperature and colder. Hence the pressure of a free-electron gas in a solid metal does *not* change appreciably between room temperature and absolute zero.

42.6 no Pure copper is already an excellent conductor since it has a partially filled conduction band (Fig. 42.19c). Furthermore, copper forms a metallic crystal (Fig. 42.15) as opposed to the covalent crystals of silicon or germanium, so the scheme of using an impurity to donate or accept an electron does not work for copper. In fact, adding impurities to copper *decreases* the conductivity because an impurity tends to scatter electrons, impeding the flow of current.

42.7 no A negative charge on the gate will repel, not attract, electrons in the *p*-type silicon. Hence the electron concentration in the region between the two *p-n* junctions will be made even smaller. With so few charge carriers present in this region, very little current will flow between the source and the drain.

Bridging Problem

(a) 0.278 eV

(b) 1.74×10^{-25}

43 NUCLEAR PHYSICS

Every atom contains at its center an extremely dense, positively charged *nucleus,* which is much smaller than the overall size of the atom but contains most of its total mass. In this chapter we'll look at several important general properties of nuclei and of the nuclear force that holds protons and neutrons together within a nucleus. The stability or instability of a particular nucleus is determined by the competition between the attractive nuclear force among the protons and neutrons and the repulsive electrical interactions among the protons. Unstable nuclei *decay,* transforming themselves spontaneously into other nuclei by a variety of processes. Nuclear reactions can also be induced by impact on a nucleus of a particle or another nucleus. Two classes of reactions of special interest are *fission* and *fusion.* Fission is the process that takes place within a nuclear reactor used for generating power. We could not survive without the energy released by one nearby fusion reactor, our sun.

43.1 PROPERTIES OF NUCLEI

As we described in Section 39.2, Rutherford found that the nucleus is tens of thousands of times smaller in radius than the atom itself. Since Rutherford's initial experiments, many additional scattering experiments have been performed with high-energy protons, electrons, neutrons, and alpha particles. These experiments show that we can model a nucleus as a sphere with a radius R that depends on the total number of *nucleons* (neutrons and protons) in the nucleus. This number is called the **nucleon number** A. The radii of most nuclei are represented quite well by the equation

Experimentally determined constant $= 1.2 \times 10^{-15} \text{ m} = 1.2 \text{ fm}$

Radius of an atomic nucleus $\rightarrow R = R_0 A^{1/3}$ Nucleon number $=$ total number of protons and neutrons   (43.1)

The nucleon number A in Eq. (43.1) is also called the **mass number** because it is the nearest whole number to the mass of the nucleus measured in unified atomic mass units (u). (The proton mass and the neutron mass are both approximately 1 u.) The best current conversion factor is

$$1\ u = 1.660538921(73) \times 10^{-27}\ kg$$

In Section 43.2 we'll discuss the masses of nuclei in more detail. Note that when we speak of the masses of nuclei and particles, we mean their *rest* masses.

Nuclear Density

The volume V of a sphere is equal to $4\pi R^3/3$, so Eq. (43.1) shows that the *volume* of a nucleus is proportional to A. Dividing A (the approximate mass in u) by the volume gives us the approximate density and cancels out A. Thus *all nuclei have approximately the same density*. This fact is of crucial importance in understanding nuclear structure.

EXAMPLE 43.1   **CALCULATING NUCLEAR PROPERTIES**

The most common kind of iron nucleus has mass number $A = 56$. Find the radius, approximate mass, and approximate density of the nucleus.

SOLUTION

IDENTIFY and SET UP: Equation (43.1) tells us how the nuclear radius R depends on the mass number A. The mass of the nucleus in atomic mass units is approximately equal to the value of A, and the density ρ is mass divided by volume.

EXECUTE: The radius and approximate mass are

$$R = R_0 A^{1/3} = (1.2 \times 10^{-15}\ m)(56)^{1/3}$$
$$= 4.6 \times 10^{-15}\ m = 4.6\ fm$$
$$m \approx (56\ u)(1.66 \times 10^{-27}\ kg/u) = 9.3 \times 10^{-26}\ kg$$

The volume V of the nucleus (which we treat as a sphere of radius R) and its density ρ are

$$V = \tfrac{4}{3}\pi R^3 = \tfrac{4}{3}\pi R_0^{\,3} A = \tfrac{4}{3}\pi(4.6 \times 10^{-15}\ m)^3$$
$$= 4.1 \times 10^{-43}\ m^3$$
$$\rho = \frac{m}{V} \approx \frac{9.3 \times 10^{-26}\ kg}{4.1 \times 10^{-43}\ m^3} = 2.3 \times 10^{17}\ kg/m^3$$

EVALUATE: As we mentioned above, *all* nuclei have approximately this same density. The density of solid iron is about $7000\ kg/m^3$; the iron nucleus is more than 10^{13} times as dense as iron in bulk. Such densities are also found in *neutron stars,* which are similar to gigantic nuclei made almost entirely of neutrons. A 1-cm cube of material with this density would have a mass of $2.3 \times 10^{11}\ kg$, or 230 million metric tons!

Nuclides and Isotopes

The building blocks of the nucleus are the proton and the neutron. In a neutral atom, there is one electron for every proton in the nucleus. We introduced these particles in Section 21.1; we'll recount the discovery of the neutron and proton in Chapter 44. The masses of these particles are

Proton: $m_p = 1.007276\ u = 1.672622 \times 10^{-27}\ kg$

Neutron: $m_n = 1.008665\ u = 1.674927 \times 10^{-27}\ kg$

Electron: $m_e = 0.000548580\ u = 9.10938 \times 10^{-31}\ kg$

The number of protons in a nucleus is the **atomic number** Z. The number of neutrons is the **neutron number** N. The nucleon number or mass number A is the sum of the number of protons Z and the number of neutrons N:

$$A = Z + N \qquad\qquad (43.2)$$

TABLE 43.1 Compositions of Some Common Nuclides

Z = atomic number (number of protons)
N = neutron number
$A = Z + N$ = mass number (total number of nucleons)

Nucleus	Z	N	$A = Z + N$
^1_1H	1	0	1
^2_1H	1	1	2
^4_2He	2	2	4
^6_3Li	3	3	6
^7_3Li	3	4	7
^9_4Be	4	5	9
$^{10}_5\text{B}$	5	5	10
$^{11}_5\text{B}$	5	6	11
$^{12}_6\text{C}$	6	6	12
$^{13}_6\text{C}$	6	7	13
$^{14}_7\text{N}$	7	7	14
$^{16}_8\text{O}$	8	8	16
$^{23}_{11}\text{Na}$	11	12	23
$^{65}_{29}\text{Cu}$	29	36	65
$^{200}_{80}\text{Hg}$	80	120	200
$^{235}_{92}\text{U}$	92	143	235
$^{238}_{92}\text{U}$	92	146	238

Application Using Isotopes to Measure Ancient Climate This sample of ice from Antarctica was deposited tens of thousands of years ago. The deeper the sample, the further in the past the ice was deposited. Most of the water molecules (H_2O) in the ice contain the oxygen isotope $^{16}_8\text{O}$, but a small percentage contain the heavier isotope $^{18}_8\text{O}$. Water molecules that contain the lighter isotope evaporate more readily, but condense less readily, than water molecules that include the heavier isotope, and these processes vary with temperature. Measuring the ratio of $^{18}_8\text{O}$ to $^{16}_8\text{O}$ in an ancient ice sample thus allows scientists to determine the average ocean temperature at the time the sample was deposited. Scientists also measure the amount of atmospheric carbon dioxide (CO_2) that was trapped in the ice when it was deposited. These observations have helped confirm the idea that high atmospheric CO_2 concentrations go hand in hand with high temperatures, a key principle for understanding 21st-century climate change (see Section 17.7).

A single nuclear species having specific values of both Z and N is called a **nuclide.** Table 43.1 lists values of A, Z, and N for some nuclides. The electron structure of an atom, which is responsible for its chemical properties, is determined by the charge Ze of the nucleus. The table shows some nuclides that have the same number of protons Z but a different number of neutrons N. These nuclides are called **isotopes** of that element. A familiar example is chlorine (Cl, $Z = 17$). About 76% of chlorine nuclei have $N = 18$; the other 24% have $N = 20$. Different isotopes of an element usually have slightly different physical properties such as melting and boiling temperatures and diffusion rates. The two common isotopes of uranium with $A = 235$ and 238 are usually separated industrially by taking advantage of the different diffusion rates of gaseous uranium hexafluoride (UF_6) containing the two isotopes.

Table 43.1 also shows the usual notation for individual nuclides: the symbol of the element, with a pre-subscript equal to Z and a pre-superscript equal to the mass number A. The general format for an element El is ^A_ZEl. The isotopes of chlorine mentioned above, with $A = 35$ and 37, are written $^{35}_{17}\text{Cl}$ and $^{37}_{17}\text{Cl}$ and pronounced "chlorine-35" and "chlorine-37," respectively. This name of the element determines the atomic number Z, so the pre-subscript Z is sometimes omitted, as in ^{35}Cl.

Table 43.2 gives the masses of some common atoms, including their electrons. Note that this table gives masses of *neutral* atoms (with Z electrons) rather than masses of *bare* nuclei, because it is much more difficult to measure masses of bare nuclei with high precision. The mass of a neutral carbon-12 atom is exactly 12 u; that's how the unified atomic mass unit is defined. The masses of other atoms are *approximately* equal to A atomic mass units, as we stated earlier. In fact, the atomic masses are *less* than the sum of the masses of their parts (the Z protons, the Z electrons, and the N neutrons). We'll explain this very important mass difference in the next section.

| TABLE 43.2 | Neutral Atomic Masses for Some Light Nuclides |

Element and Isotope	Atomic Number, Z	Neutron Number, N	Atomic Mass (u)	Mass Number, A
Hydrogen ($^1_1$H)	1	0	1.007825	1
Deuterium ($^2_1$H)	1	1	2.014102	2
Tritium ($^3_1$H)	1	2	3.016049	3
Helium ($^3_2$He)	2	1	3.016029	3
Helium ($^4_2$He)	2	2	4.002603	4
Lithium ($^6_3$Li)	3	3	6.015123	6
Lithium ($^7_3$Li)	3	4	7.016005	7
Beryllium ($^9_4$Be)	4	5	9.012182	9
Boron ($^{10}_5$B)	5	5	10.012937	10
Boron ($^{11}_5$B)	5	6	11.009305	11
Carbon ($^{12}_6$C)	6	6	12.000000	12
Carbon ($^{13}_6$C)	6	7	13.003355	13
Nitrogen ($^{14}_7$N)	7	7	14.003074	14
Nitrogen ($^{15}_7$N)	7	8	15.000109	15
Oxygen ($^{16}_8$O)	8	8	15.994915	16
Oxygen ($^{17}_8$O)	8	9	16.999132	17
Oxygen ($^{18}_8$O)	8	10	17.999161	18

Source: G. Audi, A. H. Wapstra, and C. Thibault, *Nuclear Physics* **A729**, 337 (2003).

Nuclear Spins and Magnetic Moments

Like electrons, nucleons (protons and neutrons) are spin-$\frac{1}{2}$ particles with spin angular momenta given by the same equations as in Section 41.5. The magnitude of the spin angular momentum $\vec{S}$ of a nucleon is

$$S = \sqrt{\tfrac{1}{2}(\tfrac{1}{2} + 1)}\,\hbar = \sqrt{\tfrac{3}{4}}\,\hbar \qquad (43.3)$$

and the z-component is

$$S_z = \pm\tfrac{1}{2}\hbar \qquad (43.4)$$

In addition to its spin angular momentum, a nucleon may have *orbital* angular momentum $\vec{L}$ associated with its motion within the nucleus. The values of $\vec{L}$ and of its z-component L_z for a nucleon are quantized in the same way as for an electron in an atom.

The *total* angular momentum $\vec{J}$ of the nucleus is the vector sum of the individual spin and orbital angular momenta of all the nucleons. It has magnitude

$$J = \sqrt{j(j + 1)}\,\hbar \qquad (43.5)$$

and z-component

$$J_z = m_j\hbar \qquad (m_j = -j, -j + 1, \ldots, j - 1, j) \qquad (43.6)$$

The total nuclear angular momentum quantum number j is usually called the *nuclear spin,* even though in general it refers to a combination of the orbital and spin angular momenta of the nucleons that make up the nucleus. When the total number of nucleons A is *even, j* is an integer; when it is *odd, j* is a half-integer. All nuclides for which both Z and N are even have $J = 0$. As we will see, this happens because nucleons tend to form pairs with opposite spin components.

Associated with nuclear angular momentum is a *magnetic moment.* When we discussed *electron* magnetic moments in Section 41.4, we introduced the Bohr magneton $\mu_B = e\hbar/2m_e$ as a natural unit of magnetic moment. We found that the magnitude of the z-component of the electron spin magnetic moment is almost exactly equal to μ_B; that is, $|\mu_{sz}|_{\text{electron}} \approx \mu_B$. In discussing *nuclear* magnetic moments, we can define an analogous quantity, the **nuclear magneton** μ_n:

$$\mu_n = \frac{e\hbar}{2m_p} = 5.05078 \times 10^{-27} \text{ J/T} = 3.15245 \times 10^{-8} \text{ eV/T} \tag{43.7}$$

(nuclear magneton)

The proton mass m_p is 1836 times larger than the electron mass m_e, so the nuclear magneton μ_n is 1836 times smaller than the Bohr magneton μ_B.

We might expect the magnitude of the z-component of the spin magnetic moment of the proton to be approximately μ_n. Instead, it turns out to be

$$|\mu_{sz}|_{\text{proton}} = 2.7928\mu_n \tag{43.8}$$

Even more surprising, the neutron, which has zero charge, has a spin magnetic moment; its z-component has magnitude

$$|\mu_{sz}|_{\text{neutron}} = 1.9130\mu_n \tag{43.9}$$

The proton has a positive charge; as expected, its spin magnetic moment $\vec{\mu}$ is parallel to its spin angular momentum $\vec{S}$. However, $\vec{\mu}$ and $\vec{S}$ are opposite for a neutron, as would be expected for a *negative* charge distribution. These *anomalous* magnetic moments arise because the proton and neutron aren't really fundamental particles but are made of simpler particles called *quarks*. We'll discuss quarks in some detail in Chapter 44.

The magnetic moment of an entire nucleus is typically a few nuclear magnetons. When a nucleus is placed in an external magnetic field $\vec{B}$, there is an interaction energy $U = -\vec{\mu} \cdot \vec{B} = -\mu_z B$ just as with atomic magnetic moments. The components of the magnetic moment in the direction of the field μ_z are quantized, so a series of energy levels results from this interaction.

EXAMPLE 43.2 PROTON SPIN FLIPS

Protons are placed in a 2.30-T magnetic field that points in the positive z-direction. (a) What is the energy difference between states with the z-component of proton spin angular momentum parallel and antiparallel to the field? (b) A proton can make a transition from one of these states to the other by emitting or absorbing a photon with the appropriate energy. Find the frequency and wavelength of such a photon.

SOLUTION

IDENTIFY and SET UP: The proton is a spin-$\frac{1}{2}$ particle with a magnetic moment $\vec{\mu}$ in the same direction as its spin $\vec{S}$, so its energy depends on the orientation of its spin relative to an applied magnetic field $\vec{B}$. If the z-component of $\vec{S}$ is aligned with $\vec{B}$, then μ_z is equal to the positive value given in Eq. (43.8). If the z-component of $\vec{S}$ is opposite $\vec{B}$, then μ_z is the negative of this value. The interaction energy in either case is $U = -\mu_z B$; the difference between these energies is our target variable in part (a). We find the photon frequency and wavelength by using $E = hf = hc/\lambda$.

EXECUTE: (a) When the z-components of $\vec{S}$ and $\vec{\mu}$ are parallel to $\vec{B}$, the interaction energy is

$$U = -|\mu_z|B = -(2.7928)(3.152 \times 10^{-8} \text{ eV/T})(2.30 \text{ T})$$
$$= -2.025 \times 10^{-7} \text{ eV}$$

When the z-components of $\vec{S}$ and $\vec{\mu}$ are antiparallel to the field, the energy is $+2.025 \times 10^{-7}$ eV. Hence the energy *difference* between the states is

$$\Delta E = 2(2.025 \times 10^{-7} \text{ eV}) = 4.05 \times 10^{-7} \text{ eV}$$

(b) The corresponding photon frequency and wavelength are

$$f = \frac{\Delta E}{h} = \frac{4.05 \times 10^{-7} \text{ eV}}{4.136 \times 10^{-15} \text{ eV} \cdot \text{s}} = 9.79 \times 10^7 \text{ Hz} = 97.9 \text{ MHz}$$

$$\lambda = \frac{c}{f} = \frac{3.00 \times 10^8 \text{ m/s}}{9.79 \times 10^7 \text{ s}^{-1}} = 3.06 \text{ m}$$

EVALUATE: This frequency is in the middle of the FM radio band. When a hydrogen specimen is placed in a 2.30-T magnetic field and irradiated with radio waves of this frequency, proton *spin flips* can be detected by the absorption of energy from the radiation.

Nuclear Magnetic Resonance and MRI

PhET: Simplified MRI

Spin-flip experiments of the sort referred to in Example 43.2 are called *nuclear magnetic resonance* (NMR). They have been carried out with many different nuclides. Frequencies and magnetic fields can be measured very precisely, so this technique permits precise measurements of nuclear magnetic moments. An elaboration of this basic idea leads to *magnetic resonance imaging* (MRI), a noninvasive medical imaging technique that discriminates among various types of body tissues on the basis of the differing environments of protons in the tissues (**Fig. 43.1**).

The magnetic moment of a nucleus is also the *source* of a magnetic field. In an atom the interaction of an electron's magnetic moment with the field of the nucleus's magnetic moment causes additional splittings in atomic energy levels and spectra. We called this effect *hyperfine structure* in Section 41.5. Measurements of the hyperfine structure may be used to directly determine the nuclear spin.

TEST YOUR UNDERSTANDING OF SECTION 43.1 (a) By what factor must the mass number of a nucleus increase to double its volume? (i) $\sqrt[3]{2}$; (ii) $\sqrt{2}$; (iii) 2; (iv) 4; (v) 8. (b) By what factor must the mass number increase to double the radius of the nucleus? (i) $\sqrt[3]{2}$; (ii) $\sqrt{2}$; (iii) 2; (iv) 4; (v) 8. ❙

43.1 Magnetic resonance imaging (MRI).

(a)

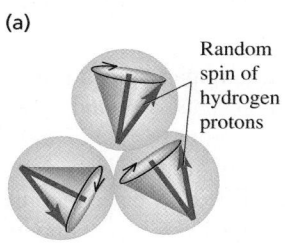

Protons, the nuclei of hydrogen atoms in the tissue under study, normally have random spin orientations.

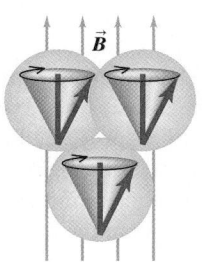

In the presence of a strong magnetic field, the spins become aligned with a component parallel to $\vec{B}$.

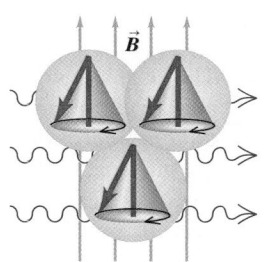

A brief radio signal causes the spins to flip orientation.

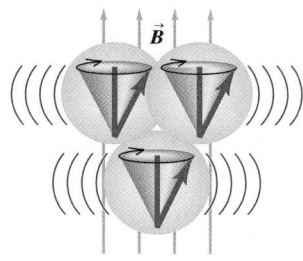

As the protons realign with the $\vec{B}$ field, they emit radio waves that are picked up by sensitive detectors.

(b) Since $\vec{B}$ has a different value at different locations in the tissue, the radio waves from different locations have different frequencies. This makes it possible to construct an image.

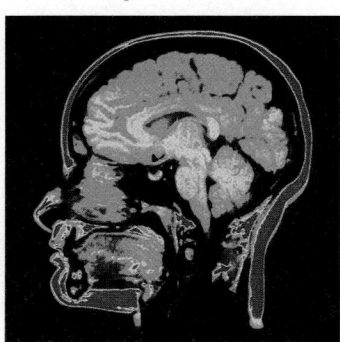

(c) An electromagnet used for MRI

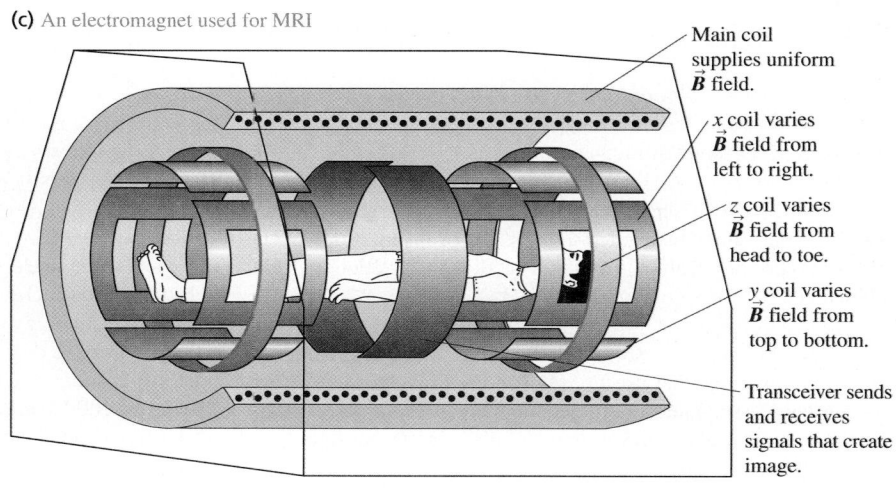

Main coil supplies uniform $\vec{B}$ field.

x coil varies $\vec{B}$ field from left to right.

z coil varies $\vec{B}$ field from head to toe.

y coil varies $\vec{B}$ field from top to bottom.

Transceiver sends and receives signals that create image.

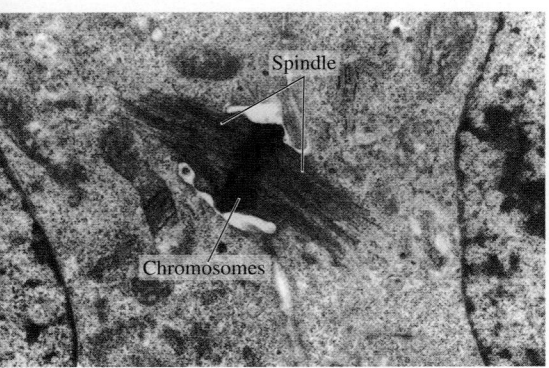

43.2 NUCLEAR BINDING AND NUCLEAR STRUCTURE

Because energy must be added to a nucleus to separate it into its individual protons and neutrons, the total rest energy E_0 of the separated nucleons is greater than the rest energy of the nucleus. The energy that must be added to separate the nucleons is called the **binding energy** E_B; it is the magnitude of the energy by which the nucleons are bound together. Thus the rest energy of the nucleus is $E_0 - E_B$. The binding energy is defined as

Binding energy of a nucleus with Z protons, N neutrons

Atomic number → Z
Neutron number → N
(Speed of light in vacuum)^2 = 931.5 MeV/u

$$E_B = (ZM_H + Nm_n - {}^A_Z M)c^2 \tag{43.10}$$

Mass of hydrogen atom
Neutron mass
Mass of neutral atom containing nucleus

Note that Eq. (43.10) does not include Zm_p, the mass of Z protons. Rather, it contains ZM_H, the mass of Z protons and Z electrons combined as Z neutral ${}^1_1H$ atoms, to balance the Z electrons included in ${}^A_Z M$, the mass of the neutral atom.

The simplest nucleus is that of hydrogen, a single proton. Next comes the nucleus of ${}^2_1H$, the isotope of hydrogen with mass number 2, usually called *deuterium*. Its nucleus consists of a proton and a neutron bound together to form a particle called the *deuteron*. By using values from Table 43.2 in Eq. (43.10), we find that the binding energy of the deuteron is

$$E_B = (1.007825 \text{ u} + 1.008665 \text{ u} - 2.014102 \text{ u})(931.5 \text{ MeV/u})$$
$$= 2.224 \text{ MeV}$$

This much energy would be required to pull the deuteron apart into a proton and a neutron. An important measure of how tightly a nucleus is bound is the *binding energy per nucleon*, E_B/A. At $(2.224 \text{ MeV})/(2 \text{ nucleons}) = 1.112 \text{ MeV}$ per nucleon, ${}^2_1H$ has the lowest binding energy per nucleon of all nuclides.

Using the equivalence of rest mass and energy (Section 37.8), we see that the mass of a nucleus is always *less* than the total mass of its nucleons by an amount $\Delta M = E_B/c^2$, called the *mass defect*. For example, the mass defect of ${}^2_1H$ is $\Delta M = E_B/c^2 = (2.224 \text{ MeV})/(931.5 \text{ MeV/u}) = 0.002388 \text{ u}$.

PROBLEM-SOLVING STRATEGY 43.1 | **NUCLEAR PROPERTIES**

IDENTIFY *the relevant concepts:* The key properties of a nucleus are its mass, radius, binding energy, mass defect, binding energy per nucleon, and angular momentum.

SET UP *the problem:* Once you have identified the target variables, assemble the equations needed to solve the problem. A relatively small number of equations from this section and Section 43.1 are all you need.

EXECUTE *the solution:* Solve for the target variables. Binding-energy calculations that use Eq. (43.10) often involve subtracting two nearly equal quantities. To get enough precision in the difference, you may need to carry as many as nine significant figures, if that many are available.

EVALUATE *your answer:* It's useful to be familiar with the following benchmark magnitudes. Protons and neutrons are about 1840 times as massive as electrons. Nuclear radii are of the order of 10^{-15} m. The electric potential energy of two protons in a nucleus is roughly 10^{-13} J or 1 MeV, so nuclear interaction energies are typically a few MeV rather than a few eV as with atoms. The binding energy per nucleon is about 1% of the nucleon rest energy. (The ionization energy of the hydrogen atom is only 0.003% of the electron's rest energy.) Angular momenta are determined only by the value of $\hbar$, so they are of the same order of magnitude in both nuclei and atoms. Nuclear magnetic moments, however, are about a factor of 1000 *smaller* than those of electrons in atoms because nuclei are so much more massive than electrons.

EXAMPLE 43.3 THE MOST STRONGLY BOUND NUCLIDE

Find the mass defect, the total binding energy, and the binding energy per nucleon of $^{62}_{28}$Ni, which has the highest binding energy per nucleon of all nuclides (**Fig. 43.2**). The neutral atomic mass of $^{62}_{28}$Ni is 61.928345 u.

SOLUTION

IDENTIFY and SET UP: The mass defect ΔM is the difference between the mass of the nucleus and the combined mass of its constituent nucleons. The binding energy E_B is this quantity multiplied by c^2, and the binding energy per nucleon is E_B divided by the mass number A. We use Eq. (43.10), $\Delta M = ZM_H + Nm_n - {}^A_Z M$, to determine both the mass defect and the binding energy.

EXECUTE: With $Z = 28$, $M_H = 1.007825$ u, $N = A - Z = 62 - 28 = 34$, $m_n = 1.008665$ u, and ${}^A_Z M = 61.928345$ u,

Eq. (43.10) gives $\Delta M = 0.585365$ u. The binding energy is then

$$E_B = (0.585365 \text{ u})(931.5 \text{ MeV/u}) = 545.3 \text{ MeV}$$

The binding energy *per nucleon* is $E_B/A = (545.3 \text{ MeV})/62$, or 8.795 MeV per nucleon.

EVALUATE: Our result means that it would take a minimum of 545.3 MeV to pull a $^{62}_{28}$Ni completely apart into 28 protons and 34 neutrons. The mass defect of $^{62}_{28}$Ni is about 1% of the atomic (or the nuclear) mass. The binding energy is therefore about 1% of the rest energy of the nucleus, and the binding energy per nucleon is about 1% of the rest energy of a nucleon. Note that the mass defect is more than half the mass of a nucleon, which suggests how tightly bound nuclei are.

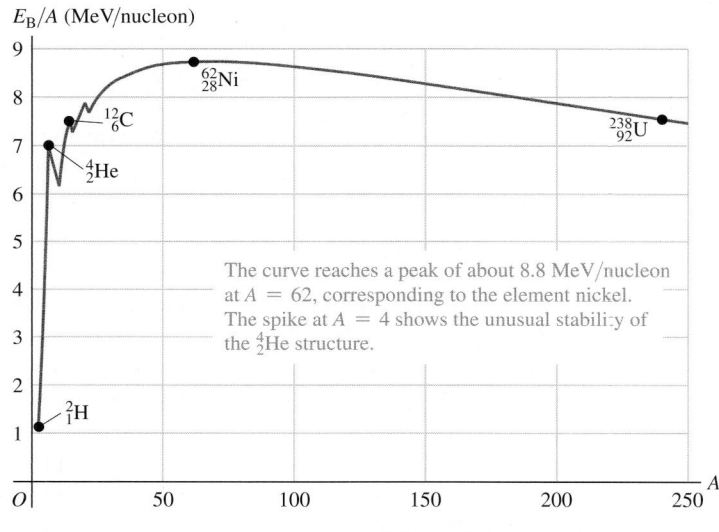

43.2 Approximate binding energy per nucleon as a function of mass number A (the total number of nucleons) for stable nuclides.

The curve reaches a peak of about 8.8 MeV/nucleon at $A = 62$, corresponding to the element nickel. The spike at $A = 4$ shows the unusual stability of the $^4_2$He structure.

Nearly all stable nuclides, from the lightest to the most massive, have binding energies in the range of 7–9 MeV per nucleon. Figure 43.2 is a graph of binding energy per nucleon as a function of the mass number A. Note the spike at $A = 4$, showing the unusually large binding energy per nucleon of the $^4_2$He nucleus (alpha particle) relative to its neighbors. To explain this curve, we must consider the interactions among the nucleons.

The Nuclear Force

The force that binds protons and neutrons together in the nucleus, despite the electrical repulsion of the protons, is an example of the *strong interaction* that we mentioned in Section 5.5. In the context of nuclear structure, this interaction is called the *nuclear force*. Here are some of its characteristics. First, it does not depend on charge; neutrons as well as protons are bound, and the binding is the same for both. Second, it has short range, of the order of nuclear dimensions—that is, 10^{-15} m. (Otherwise, the nucleus would grow by pulling in additional protons and neutrons.) But within its range, the nuclear force is much stronger than electric forces; otherwise, the nucleus could never be stable. It would be nice if we could write a simple equation like Newton's law of gravitation or Coulomb's law for this force, but physicists have yet to fully determine its dependence on the separation r. Third, the nearly constant density of nuclear matter and the nearly constant binding energy per nucleon of larger nuclides show that a particular nucleon cannot interact simultaneously with *all* the other nucleons in a nucleus, but only with those few in its immediate vicinity. This is different from electric forces; *every* proton in the nucleus repels every other one. This limited number of interactions is called *saturation;* it is analogous to covalent bonding in molecules and solids. Finally, the nuclear force favors binding of *pairs* of protons or neutrons with opposite spins and of *pairs of pairs*—that is, a pair of protons and a pair of neutrons, each pair having opposite spins. Hence the alpha particle (two protons and two neutrons) is an exceptionally stable nucleus for its mass number. We'll see other evidence for pairing effects in nuclei in the next subsection. (In Section 42.8 we described an analogous pairing that binds opposite-spin electrons in Cooper pairs in the BCS theory of superconductivity.)

The analysis of nuclear structure is more complex than the analysis of many-electron atoms. Two different kinds of interactions are involved (electrical and nuclear). Even so, we can gain some insight into nuclear structure by the use of simple models. We'll discuss briefly two rather different but successful models, the *liquid-drop model* and the *shell model.*

The Liquid-Drop Model

The **liquid-drop model,** first proposed in 1928 by the Russian physicist George Gamow and later expanded on by Niels Bohr, is suggested by the observation that all nuclei have nearly the same density. The individual nucleons are analogous to molecules of a liquid, held together by short-range interactions and surface-tension effects. We can use this simple picture to derive a formula for the estimated total binding energy of a nucleus. We'll include five contributions:

1. We've remarked that nuclear forces show *saturation;* an individual nucleon interacts only with a few of its nearest neighbors. This effect gives a binding-energy term that is proportional to the number of nucleons. We write this term as C_1A, where C_1 is an experimentally determined constant.

2. The nucleons on the surface of the nucleus are less tightly bound than those in the interior because they have no neighbors outside the surface. This decrease in the binding energy gives a *negative* energy term proportional to the surface area $4\pi R^2$. Because R is proportional to $A^{1/3}$, this term is proportional to $A^{2/3}$; we write it as $-C_2A^{2/3}$, where C_2 is another constant.

3. Every one of the Z protons repels every one of the $(Z - 1)$ other protons. The total repulsive electric potential energy is proportional to $Z(Z - 1)$ and inversely proportional to the radius R and thus to $A^{1/3}$. This energy term is negative because the nucleons are less tightly bound than they would be without the electrical repulsion. We write this correction as $-C_3Z(Z - 1)/A^{1/3}$.

4. Observations show that nuclei are most tightly bound if N is close to Z for small A and N is greater than Z (but not too much greater) for larger A. We need a negative energy term corresponding to the difference $|N - Z|$. The best agreement with observed binding energies is obtained if this term is proportional to $(N - Z)^2/A$. If we use $N = A - Z$ to express this energy in terms of A and Z, this correction is $-C_4(A - 2Z)^2/A$.

5. Finally, the nuclear force favors *pairing* of protons and of neutrons. This energy term is positive (more binding) if both Z and N are even, negative (less binding) if both Z and N are odd, and zero otherwise. The best fit to the data occurs with the form $\pm C_5A^{-4/3}$ for this term.

The total estimated binding energy E_B is the sum of these five terms:

$$E_B = C_1A - C_2A^{2/3} - C_3\frac{Z(Z - 1)}{A^{1/3}} - C_4\frac{(A - 2Z)^2}{A} \pm C_5A^{-4/3} \quad \text{(43.11)}$$

(nuclear binding energy)

The constants C_1, C_2, C_3, C_4, and C_5, chosen to make this formula best fit the observed binding energies of nuclides, are

$$C_1 = 15.75 \text{ MeV}$$
$$C_2 = 17.80 \text{ MeV}$$
$$C_3 = 0.7100 \text{ MeV}$$
$$C_4 = 23.69 \text{ MeV}$$
$$C_5 = 39 \text{ MeV}$$

The constant C_1 is the binding energy per nucleon due to the saturated nuclear force. This energy is almost 16 MeV per nucleon, about double the *total* binding energy per nucleon in most nuclides.

If we use Eq. (43.11) to estimate the binding energy E_B, we can solve Eq. (43.10) to use it to estimate the mass of any neutral atom:

$$^A_Z M = Z M_H + N m_n - \frac{E_B}{c^2} \qquad \text{(semiempirical mass formula)} \qquad (43.12)$$

Equation (43.12) is called the *semiempirical mass formula*. The name is apt; the equation is *empirical* in the sense that the C's have to be determined empirically (experimentally), yet it does have a sound theoretical basis.

EXAMPLE 43.4 ESTIMATING BINDING ENERGY AND MASS

For the nuclide $^{62}_{28}$Ni of Example 43.3, (a) calculate the five terms in the binding energy and the total estimated binding energy, and (b) find the neutral atomic mass using the semiempirical mass formula.

SOLUTION

IDENTIFY and SET UP: We use the liquid-drop model of the nucleus and its five contributions to the binding energy, as given by Eq. (43.11), to calculate the total binding energy E_B. We then use Eq. (43.12) to find the neutral atomic mass $^{62}_{28}M$.

EXECUTE: (a) With $Z = 28$, $A = 62$, and $N = 34$, the five terms in Eq. (43.11) are

1. $C_1 A = (15.75 \text{ MeV})(62) = 976.5 \text{ MeV}$
2. $-C_2 A^{2/3} = -(17.80 \text{ MeV})(62)^{2/3} = -278.8 \text{ MeV}$
3. $-C_3 \dfrac{Z(Z-1)}{A^{1/3}} = -(0.7100 \text{ MeV}) \dfrac{(28)(27)}{(62)^{1/3}}$
   $= -135.6 \text{ MeV}$

4. $-C_4 \dfrac{(A-2Z)^2}{A} = -(23.69 \text{ MeV}) \dfrac{(62-56)^2}{62}$
   $= -13.8 \text{ MeV}$
5. $+C_5 A^{-4/3} = (39 \text{ MeV})(62)^{-4/3} = 0.2 \text{ MeV}$

The pairing correction (term 5) is by far the smallest of all the terms; it is positive because both Z and N are even. The sum of all five terms is the total estimated binding energy, $E_B = 548.5$ MeV.

(b) We use $E_B = 548.5$ MeV in Eq. (43.12):

$$^{62}_{28}M = 28(1.007825 \text{ u}) + 34(1.008665 \text{ u}) - \frac{548.5 \text{ MeV}}{931.5 \text{ MeV/u}}$$

$$= 61.925 \text{ u}$$

EVALUATE: The binding energy of $^{62}_{28}$Ni calculated in part (a) is only about 0.6% larger than the true value of 545.3 MeV found in Example 43.3, and the mass calculated in part (b) is only about 0.005% smaller than the measured value of 61.928345 u. The semiempirical mass formula can be quite accurate!

The liquid-drop model and the mass formula derived from it are quite successful in correlating nuclear masses, and we will see later that they help in understanding decay processes of unstable nuclides. Other aspects of nuclei, such as angular momentum and excited states, are better approached with different models.

The Shell Model

The **shell model** of nuclear structure is analogous to the central-field approximation in atomic physics (see Section 41.6). We picture each nucleon as moving in a potential that represents the averaged-out effect of all the other nucleons. Although this is a very simplified model, in several respects it works out quite well.

43.3 Approximate potential-energy functions for a nucleon in a nucleus. The approximate nuclear radius is R.

(a) The potential energy U_{nuc} due to the nuclear force is the same for protons and neutrons. For neutrons, it is the *total* potential energy.

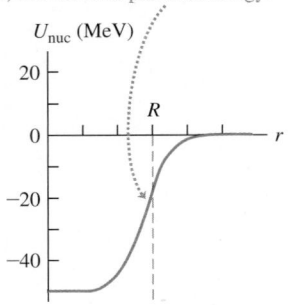

(b) For protons, the total potential energy U_{tot} is the sum of the nuclear (U_{nuc}) and electric (U_{el}) potential energies.

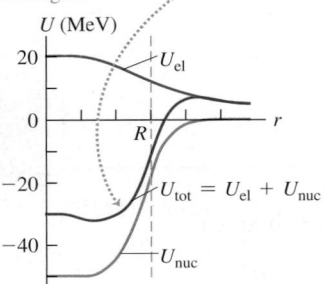

The potential-energy function for the nuclear force is the same for protons as for neutrons. **Figure 43.3a** shows a reasonable assumption for the shape of this function: a spherical version of the square-well potential we discussed in Section 40.3. The corners are somewhat rounded because the nucleus doesn't have a sharply defined surface. For protons there is an additional potential energy associated with electrical repulsion. We consider each proton to interact with a sphere of uniform charge density, with radius R and total charge $(Z - 1)e$. Figure 43.3b shows the nuclear, electric, and total potential energies for a proton as functions of the distance r from the center of the nucleus.

In principle, we could solve the Schrödinger equation for a proton or neutron moving in such a potential. For any spherically symmetric potential energy, the angular-momentum states are the same as for the electrons in the central-field approximation in atomic physics. In particular, we can use the concept of *filled shells and subshells* and their relationship to stability. As we saw in Section 41.6, the exclusion principle forbids more than one electron from occupying any given quantum-mechanical state in a multi-electron atom. This explains why the values $Z = 2, 10, 18, 36, 54,$ and 86 (the atomic numbers of the noble gases) correspond to atoms with particularly stable electron arrangements.

A comparable effect occurs in nuclear structure. Like electrons, protons and neutrons are spin-$\frac{1}{2}$ particles. So no more than one proton can be in any given quantum-mechanical state in a nucleus, and likewise for neutrons. Just as for electrons in atoms, there are certain numbers of protons *or* of neutrons, called *magic numbers,* that correspond to particularly stable nuclei—that is, nuclei with particularly high binding energies. The magic numbers are 2, 8, 20, 28, 50, 82, and 126. These numbers are different from those for electrons in atoms because the potential-energy function is different and the nuclear spin-orbit interaction is much stronger and of opposite sign than in atoms. So nuclear subshells fill up in a different order from those for electrons in an atom. Nuclides in which Z is a magic number tend to have an above-average number of stable isotopes. (Nuclides with $Z = 126$ have not been observed in nature.) There are several *doubly magic* nuclides for which both Z and N are magic, including

$$^{4}_{2}\text{He} \qquad ^{16}_{8}\text{O} \qquad ^{40}_{20}\text{Ca} \qquad ^{48}_{20}\text{Ca} \qquad ^{208}_{82}\text{Pb}$$

All these nuclides have substantially higher binding energy per nucleon than do nuclides with neighboring values of N or Z. They also all have zero nuclear spin. The magic numbers correspond to filled-shell or -subshell configurations of nucleon energy levels with a relatively large jump in energy to the next allowed level.

TEST YOUR UNDERSTANDING OF SECTION 43.2 Rank the following nuclei in order from largest to smallest value of the binding energy per nucleon. (i) $^{4}_{2}\text{He}$; (ii) $^{52}_{24}\text{Cr}$; (iii) $^{152}_{62}\text{Sm}$; (iv) $^{200}_{80}\text{Hg}$; (v) $^{252}_{92}\text{Cf}$. ∎

43.3 NUCLEAR STABILITY AND RADIOACTIVITY

Among about 2500 known nuclides, fewer than 300 are stable. The others are unstable structures that decay to form other nuclides by emitting particles and electromagnetic radiation, a process called **radioactivity.** The time scale of these decay processes ranges from a small fraction of a microsecond to billions of years. The *stable* nuclides are shown by dots on the graph in **Fig. 43.4,** where the neutron number N and proton number (or atomic number) Z for each nuclide are plotted. Such a chart is called a *Segrè chart,* after its inventor, the Italian-American physicist Emilio Segrè (1905–1989).

Each blue line in Fig. 43.4 represents a specific value of the mass number $A = Z + N$. Most lines of constant A pass through only one or two stable nuclides; that is, there is usually a very narrow range of stability for a given mass

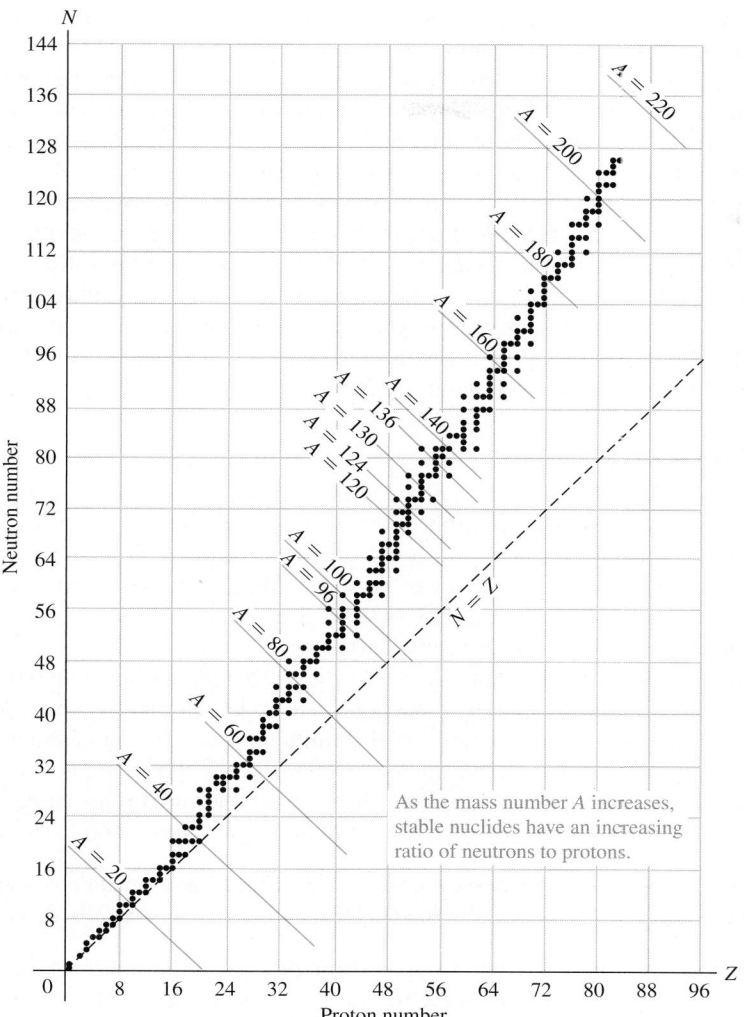

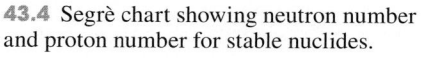

43.4 Segrè chart showing neutron number and proton number for stable nuclides.

number. The lines at $A = 20$, 40, 60, and 80 are examples. In four exceptional cases ($A = 94$, 124, 130, and 136), these lines pass through *three* stable nuclides.

Only four stable nuclides have both odd Z and odd N:

$$^2_1\text{H} \qquad ^6_3\text{Li} \qquad ^{10}_5\text{B} \qquad ^{14}_7\text{N}$$

These are called *odd-odd nuclides*. The absence of other odd-odd nuclides shows the importance of pairing in adding to nuclear stability. Also, there is *no* stable nuclide with $A = 5$ or $A = 8$. The doubly magic ^4_2He nucleus, with a pair of protons and a pair of neutrons, has no interest in accepting a fifth particle into its structure. Collections of eight nucleons decay to smaller nuclides, with a ^8_4Be nucleus immediately splitting into two ^4_2He nuclei.

The stable nuclides define a rather narrow region on the Segrè chart. For low mass numbers, the numbers of protons and neutrons are approximately equal, $N \approx Z$. The ratio N/Z increases gradually with A, up to about 1.6 at large mass numbers, because of the increasing influence of the electrical repulsion of the protons. Points to the right of the stability region represent nuclides that have too many protons relative to neutrons. In these cases, repulsion wins, and the nucleus comes apart. To the left are nuclides with too many neutrons relative to protons. In these cases the energy associated with the neutrons is out of balance with that associated with the protons, and the nuclides decay in a process that converts neutrons to protons. The graph also shows that no nuclide with $A > 209$ or

$Z > 83$ is stable. A nucleus is unstable if it is too big. Note that there is no stable nuclide with $Z = 43$ (technetium) or 61 (promethium).

Nearly 90% of the 2500 known nuclides are *radioactive*; they are not stable but decay into other nuclides. Many of these radioactive nuclides occur in nature. For example, you are very slightly radioactive because of unstable nuclides such as carbon-14 ($^{14}_{6}C$) and potassium-40 ($^{40}_{19}K$) that are present throughout your body. The study of radioactivity began in 1896, one year after Wilhelm Röntgen discovered x rays (Section 36.6). Henri Becquerel discovered a radiation from uranium salts that seemed similar to x rays. Investigation in the following two decades by Marie and Pierre Curie, Ernest Rutherford, and many others revealed that the emissions consist of positively and negatively charged particles and neutral rays. These particles were given the names *alpha, beta,* and *gamma* because of their differing penetration characteristics.

PhET: Alpha Decay

Alpha Decay

When unstable nuclides decay into different nuclides, they usually emit alpha (α) or beta (β) particles. An **alpha particle** is a ^{4}He nucleus, two protons and two neutrons bound together, with total spin zero. Alpha decay occurs principally with nuclei that are too large to be stable. When a nucleus emits an alpha particle, its N and Z values each decrease by 2 and A decreases by 4, moving it closer to stable territory on the Segrè chart.

Figure 43.5a shows the alpha decay of radium-226 ($^{226}_{88}Ra$). Spontaneous alpha decay of this kind can occur only if energy is released in the process; this released energy goes into the kinetic energy of the emitted α particle and of the nucleus that remains, called the *daughter nucleus.* (For the decay shown in Fig. 43.5a, the daughter nucleus is radon-222, $^{222}_{86}Rn$.) The original nucleus (in this case, $^{226}_{88}Ra$) is called the *parent nucleus.* You can use mass-energy conservation to show that

> alpha decay is possible whenever the mass of the original neutral atom is greater than the sum of the masses of the final neutral atom and a neutral $^{4}_{2}He$ atom.

In alpha decay, the α particle tunnels through a potential-energy barrier, as Fig. 43.5b shows. You may want to review the discussion of tunneling in Section 40.4.

Alpha particles are always emitted with definite kinetic energies, determined by conservation of momentum and energy in the alpha-decay process. As Fig. 43.5c shows, an α particle emitted in the decay of $^{226}_{88}Ra$ can have either of *two* possible energies, depending on the energy level of the $^{222}_{86}Rn$ daughter nucleus just after the decay. (Later in this section we'll discuss the photon-emission process shown in Fig. 43.5c.)

43.5 Alpha decay of the unstable radium nuclide $^{226}_{88}Ra$. The alpha particles used in the Rutherford scattering experiment (Section 39.2) were emitted by this nuclide.

(a)

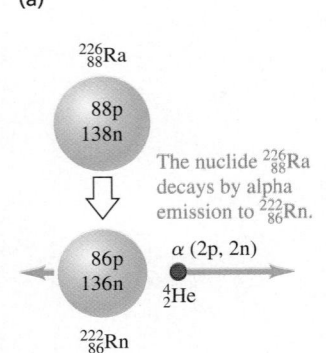

(b) Potential-energy curve for an α particle and a $^{222}_{86}Rn$ nucleus

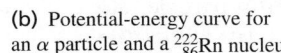

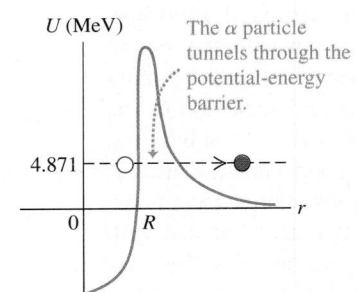

(c) Energy-level diagram for the system

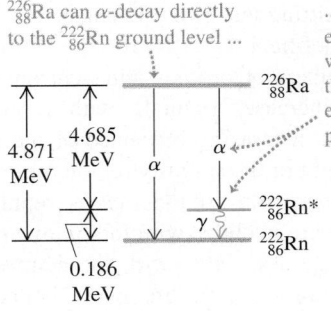

Alpha particles are emitted at high speeds, typically a few percent of the speed of light (see the following example). Nonetheless, because of their charge and mass, alpha particles can travel only several centimeters in air, or a few tenths or hundredths of a millimeter through solids, before they are brought to rest by collisions.

EXAMPLE 43.5   ALPHA DECAY OF RADIUM

Show that the α-emission process $^{226}_{88}\text{Ra} \rightarrow ^{222}_{86}\text{Rn} + ^{4}_{2}\text{He}$ (Fig. 43.5a) is energetically possible, and calculate the kinetic energy of the emitted α particle. The neutral atomic masses are 226.025410 u for $^{226}_{88}\text{Ra}$, 222.017578 u for $^{222}_{86}\text{Rn}$, and 4.002603 u for $^{4}_{2}\text{He}$.

SOLUTION

IDENTIFY and SET UP: Alpha emission is possible if the mass of the $^{226}_{88}\text{Ra}$ atom is greater than the sum of the atomic masses of $^{222}_{86}\text{Rn}$ and $^{4}_{2}\text{He}$. The mass difference between the initial radium atom and the final radon and helium atoms corresponds (through $E = mc^2$) to the energy E released in the decay. Because momentum is conserved as well as energy, *both* the alpha particle and the $^{222}_{86}\text{Rn}$ atom are in motion after the decay; we will have to account for this in determining the kinetic energy of the alpha particle.

EXECUTE: The difference in mass between the original nucleus and the decay products is

226.025410 u $-$ (222.017578 u $+$ 4.002603 u) $= +0.005229$ u

Since this is positive, α decay is energetically possible. The energy equivalent of this mass difference is

$$E = (0.005229 \text{ u})(931.5 \text{ MeV/u}) = 4.871 \text{ MeV}$$

In this process the $^{222}_{86}\text{Rn}$ nucleus is produced in its ground level (Fig. 43.5c). Thus we expect the decay products to emerge with total kinetic energy 4.871 MeV. Momentum is also conserved; if the parent $^{226}_{88}\text{Ra}$ nucleus is at rest, the daughter $^{222}_{86}\text{Rn}$ nucleus and the α particle have momenta of equal magnitude p but opposite direction. Kinetic energy is $K = \frac{1}{2}mv^2 = p^2/2m$: Since p is the same for the two particles, the kinetic energy divides inversely as their masses. Hence the α particle gets $222/(222 + 4)$ of the total, or 4.78 MeV.

EVALUATE: Experiment shows that $^{226}_{88}\text{Ra}$ does emit α particles with a kinetic energy of 4.78 MeV. Check your results by verifying that the alpha particle and the $^{222}_{86}\text{Rn}$ nucleus produced in the decay have the same magnitude of momentum $p = mv$. You can calculate the speed v of each of the decay products from its respective kinetic energy [note that the $^{222}_{86}\text{Rn}$ nucleus gets $4/(222 + 4)$ of the 4.871 MeV released]. You'll find that the alpha particle moves at $0.0506c = 1.52 \times 10^7 \text{ m/s}$; if momentum is conserved, you should find that the $^{222}_{86}\text{Rn}$ nucleus moves $\frac{4}{222}$ as fast. Does it?

Beta Decay

There are three different simple types of *beta decay: beta-minus, beta-plus,* and *electron capture.* A **beta-minus particle** (β^-) is an electron. There are no electrons in the nucleus waiting to be emitted; instead, emission of a β^- involves *transformation* of a neutron into a proton, an electron, and a third particle called an *antineutrino.* In fact, if you freed a neutron from a nucleus, it would decay into a proton, an electron, and an antineutrino in an average time of about 15 minutes.

Beta particles can be identified and their speeds can be measured with techniques that are similar to the Thomson e/m experiment we described in Section 27.5. The speeds of beta particles range up to 0.9995 of the speed of light, so their motion is highly relativistic. They are emitted with a continuous spectrum of energies. This would not be possible if the only two particles were the β^- and the recoiling nucleus, since energy and momentum conservation would then require a definite speed for the β^-. Thus there must be a *third* particle involved. From conservation of charge, it must be neutral, and from conservation of angular momentum, it must be a spin-$\frac{1}{2}$ particle.

This third particle is an antineutrino, the *antiparticle* of a **neutrino.** The symbol for a neutrino is ν_e (the Greek letter nu). Both the neutrino and the antineutrino have zero charge and very small mass and therefore produce very little observable effect when passing through matter. Both evaded detection until 1953, when Frederick Reines and Clyde Cowan succeeded in observing the antineutrino directly. We now know that there are at least three varieties of neutrinos, each with its corresponding antineutrino; one is associated with beta decay and the other two are associated with the decay of two unstable particles, the muon and the tau particle. We'll discuss these particles in more detail in Chapter 44. The antineutrino that is emitted in β^- decay is denoted as $\bar{\nu}_e$. The basic process of β^- decay is

$$n \longrightarrow p + \beta^- + \bar{\nu}_e \qquad (43.13)$$

Beta-minus decay usually occurs with nuclides for which the neutron-to-proton ratio N/Z is too large for stability. In β^- decay, N decreases by 1, Z increases by 1, and A doesn't change. You can use mass-energy conservation to show that

beta-minus decay can occur whenever the mass of the original neutral atom is larger than that of the final atom.

EXAMPLE 43.6 WHY COBALT-60 IS A BETA-MINUS EMITTER

The nuclide $^{60}_{27}\text{Co}$, an odd-odd unstable nucleus, is used in medical and industrial applications of radiation. Show that it is unstable relative to β^- decay. The atomic masses you need are 59.933817 u for $^{60}_{27}\text{Co}$ and 59.930786 u for $^{60}_{28}\text{Ni}$.

SOLUTION

IDENTIFY and SET UP: Beta-minus decay is possible if the mass of the original neutral atom is greater than that of the final atom. We must first identify the nuclide that will result if $^{60}_{27}\text{Co}$ undergoes β^- decay and then compare its neutral atomic mass to that of $^{60}_{27}\text{Co}$.

EXECUTE: In the presumed β^- decay of $^{60}_{27}\text{Co}$, Z increases by 1 from 27 to 28 and A remains at 60, so the final nuclide is $^{60}_{28}\text{Ni}$.

The neutral atomic mass of $^{60}_{27}\text{Co}$ is greater than that of $^{60}_{28}\text{Ni}$ by 0.003031 u, so β^- decay *can* occur.

EVALUATE: With three decay products in β^- decay—the $^{60}_{28}\text{Ni}$ nucleus, the electron, and the antineutrino—the energy can be shared in many different ways that are consistent with conservation of energy and momentum. It's impossible to predict precisely how the energy will be shared for the decay of a particular $^{60}_{27}\text{Co}$ nucleus. By contrast, in alpha decay there are just two decay products, and their energies and momenta are determined uniquely (see Example 43.5).

We have noted that β^- decay occurs with nuclides that have too large a neutron-to-proton ratio N/Z. Nuclides for which N/Z is too *small* for stability can emit a *positron,* the electron's antiparticle, which is identical to the electron but with positive charge. (We'll discuss the positron in more detail in Chapter 44.) The basic process, called *beta-plus decay* (β^+), is

$$p \longrightarrow n + \beta^+ + \nu_e \qquad (43.14)$$

where β^+ is a positron and ν_e is the electron neutrino.

Beta-plus decay can occur whenever the mass of the original neutral atom is at least two electron masses larger than that of the final atom.

You can show this by using mass-energy conservation.

The third type of beta decay is *electron capture.* There are a few nuclides for which β^+ emission is not energetically possible but in which an orbital electron (usually in the innermost K shell) can combine with a proton in the nucleus to form a neutron and a neutrino. The neutron remains in the nucleus and the neutrino is emitted. The basic process is

$$p + \beta^- \longrightarrow n + \nu_e \qquad (43.15)$$

You can use mass-energy conservation to show that

> **electron capture can occur whenever the mass of the original neutral atom is larger than that of the final atom.**

In all types of beta decay, A remains constant. However, in beta-plus decay and electron capture, N increases by 1 and Z decreases by 1 as the neutron-to-proton ratio increases toward a more stable value. The reaction of Eq. (43.15) also helps explain the formation of a neutron star, mentioned in Example 43.1.

CAUTION Beta decay inside and outside nuclei The beta-decay reactions given by Eqs. (43.13), (43.14), and (43.15) occur *within* a nucleus. Although the decay of a neutron outside the nucleus proceeds through the reaction of Eq. (43.13), the reaction of Eq. (43.14) is forbidden by mass-energy conservation for a proton outside the nucleus. The reaction of Eq. (43.15) can occur outside the nucleus only with the addition of some extra energy, as in a collision.

EXAMPLE 43.7   WHY COBALT-57 IS NOT A BETA-PLUS EMITTER

The nuclide $^{57}_{27}\text{Co}$ is an odd-even unstable nucleus. Show that it cannot undergo β^+ decay, but that it *can* decay by electron capture. The atomic masses you need are 56.936291 u for $^{57}_{27}\text{Co}$ and 56.935394 u for $^{57}_{26}\text{Fe}$.

SOLUTION

IDENTIFY and SET UP: Beta-plus decay is possible if the mass of the original neutral atom is greater than that of the final atom plus two electron masses (0.001097 u). Electron capture is possible if the mass of the original atom is greater than that of the final atom. We must first identify the nuclide that will result if $^{57}_{27}\text{Co}$ undergoes β^+ decay or electron capture and then find the corresponding mass difference.

EXECUTE: The original nuclide is $^{57}_{27}\text{Co}$. In both the presumed β^+ decay and electron capture, Z decreases by 1 from 27 to 26, and A remains at 57, so the final nuclide is $^{57}_{26}\text{Fe}$. Its mass is less than that of $^{57}_{27}\text{Co}$ by 0.000897 u, a value smaller than 0.001097 u (two electron masses), so β^+ decay *cannot* occur. However, the mass of the original atom is greater than the mass of the final atom, so electron capture *can* occur.

EVALUATE: In electron capture there are just two decay products, the final nucleus and the emitted neutrino. As in alpha decay (Example 43.5) but unlike in β^- decay (Example 43.6), the decay products of electron capture have unique energies and momenta. In Section 43.4 we'll see how to relate the probability that electron capture will occur to the *half-life* of this nuclide.

Gamma Decay

The energy of internal motion of a nucleus is quantized. A typical nucleus has a set of allowed energy levels, including a *ground state* (state of lowest energy) and several *excited states*. Because of the great strength of nuclear interactions, excitation energies of nuclei are typically of the order of 1 MeV, compared with a few eV for atomic energy levels. In ordinary physical and chemical transformations the nucleus always remains in its ground state. When a nucleus is placed in an excited state, either by bombardment with high-energy particles or by a radioactive transformation, it can decay to the ground state by emission of one or more photons called **gamma rays** or *gamma-ray photons,* with typical energies of 10 keV to 5 MeV. This process is called *gamma (γ) decay.* For example, alpha particles emitted from ^{226}Ra have two possible kinetic energies, either 4.784 MeV or 4.602 MeV. Including the recoil energy of the resulting ^{222}Rn nucleus, these correspond to a total released energy of 4.871 MeV or 4.685 MeV, respectively (see Fig. 43.5c). When an alpha particle with the smaller energy is emitted, the ^{222}Rn nucleus is left in an excited state. It then decays to its ground state by emitting a gamma-ray photon with energy

$$(4.871 - 4.685)\ \text{MeV} = 0.186\ \text{MeV}$$

CAUTION γ decay vs. α and β decay In both α and β decay, the Z value of a nucleus changes and the nucleus of one element becomes the nucleus of a different element. In γ decay, the element does *not* change; the nucleus merely goes from an excited state to a less excited state.

43.6 Earthquakes are caused in part by the radioactive decay of ^{238}U in the earth's interior. These decays release energy that helps produce convection currents in the earth's interior. Such currents drive the motions of the earth's crust, including the sudden sharp motions that we call earthquakes (like the one that caused this damage).

Radioactive Decay Series

When a radioactive nucleus decays, the resulting (daughter) nucleus may also be unstable. In this case a *series* of successive decays occurs until a stable configuration is reached. Several such series are found in nature. The most abundant radioactive nuclide found on earth is the uranium isotope ^{238}U, which undergoes a series of 14 decays, including eight α emissions and six β^- emissions, terminating at a stable isotope of lead, ^{206}Pb (**Fig. 43.6**).

Radioactive decay series can be represented on a Segrè chart, as in **Fig. 43.7**. The neutron number N is plotted vertically, and the atomic number Z is plotted horizontally. In alpha emission, both N and Z decrease by 2. In β^- emission, N decreases by 1 and Z increases by 1. The decays can also be represented in equation form; the first two decays in the series are written as

$$^{238}\text{U} \rightarrow {}^{234}\text{Th} + \alpha \qquad \text{and} \qquad {}^{234}\text{Th} \rightarrow {}^{234}\text{Pa} + \beta^- + \bar{\nu}_e$$

43.7 Segrè chart showing the uranium ^{238}U decay series, terminating with the stable nuclide ^{206}Pb. The times are half-lives (discussed in the next section), given in years (y), days (d), hours (h), minutes (m), or seconds (s).

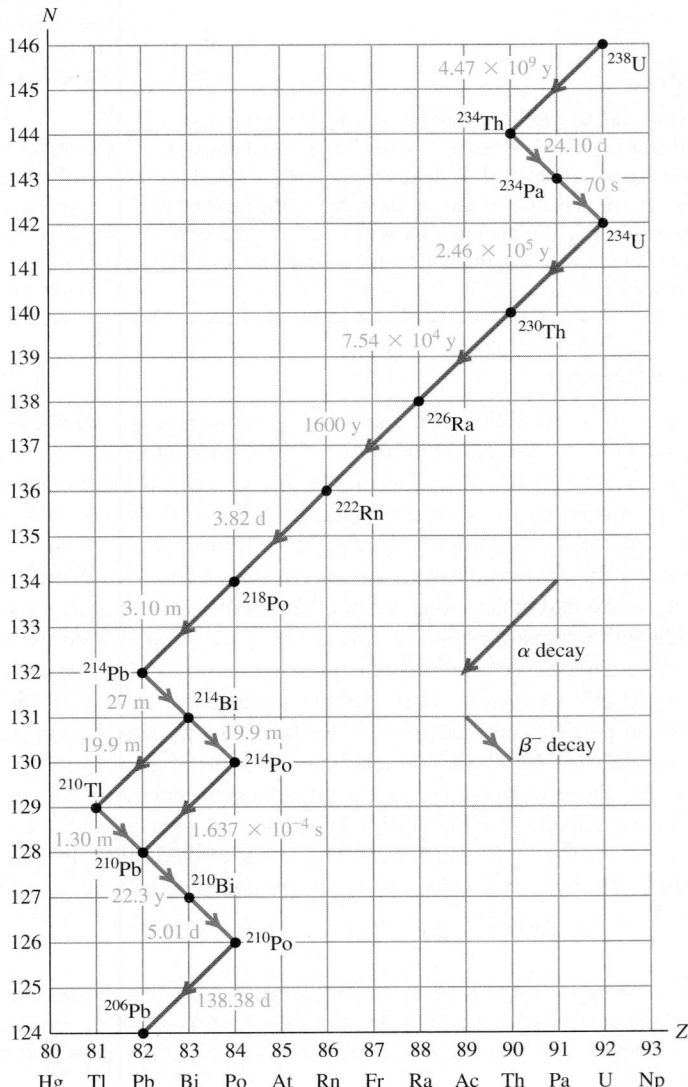

DATA *SPEAKS*

Nuclear Decays

When students were given a problem involving radioactive decay, more than 24% gave an incorrect response. Common errors:

- Confusing alpha, beta, and gamma decays. In alpha decay, atomic number Z decreases by 2 and mass number A decreases by 4. In beta-minus decay, Z increases by 1 and A is unchanged; in beta-plus decay or electron capture, Z decreases by 1 and A is unchanged. In gamma decay, both Z and A remain unchanged; the final nucleus is simply a less excited state of the initial nucleus.

- Confusion about the fate of radioactive atoms. In alpha, beta, and gamma decays, atoms do not disappear; they transform into other atoms.

or more briefly as

$$^{238}\text{U} \xrightarrow{\alpha} {}^{234}\text{Th} \qquad \text{and} \qquad {}^{234}\text{Th} \xrightarrow{\beta^-} {}^{234}\text{Pa}$$

In the second process, the beta decay leaves the daughter nucleus ^{234}Pa in an excited state, from which it decays to the ground state by emitting a gamma-ray photon. An excited state is denoted by an asterisk, so we can represent the γ emission as

$$^{234}\text{Pa}^* \rightarrow {}^{234}\text{Pa} + \gamma \qquad \text{or} \qquad {}^{234}\text{Pa}^* \xrightarrow{\gamma} {}^{234}\text{Pa}$$

An interesting feature of the ^{238}U decay series is the branching that occurs at ^{214}Bi. This nuclide decays to ^{210}Pb by emission of an α and a β^-, which can occur in either order. We also note that the series includes unstable isotopes of several elements that also have stable isotopes, including thallium (Tl), lead (Pb), and bismuth (Bi). The unstable isotopes of these elements that occur in the ^{238}U series all have too many neutrons to be stable.

Many other decay series are known. Two of these occur in nature, one starting with the uncommon isotope ^{235}U and ending with ^{207}Pb, the other starting with thorium (^{232}Th) and ending with ^{208}Pb.

TEST YOUR UNDERSTANDING OF SECTION 43.3   A nucleus with atomic number Z and neutron number N undergoes two decay processes. The result is a nucleus with atomic number $Z - 3$ and neutron number $N - 1$. Which decay processes may have taken place? (i) Two β^- decays; (ii) two β^+ decays; (iii) two α decays; (iv) an α decay and a β^- decay; (v) an α decay and a β^+ decay. ❚

43.4 ACTIVITIES AND HALF-LIVES

Suppose you have a certain number of nuclei of a particular radioactive nuclide. If no more are produced, that number decreases in a simple manner as the nuclei decay. This decrease is a statistical process; there is no way to predict when any individual nucleus will decay. No change in physical or chemical environment, such as chemical reactions or heating or cooling, greatly affects most decay rates. The rate varies over an extremely wide range for different nuclides.

Radioactive Decay Rates

Let $N(t)$ be the (very large) number of radioactive nuclei in a sample at time t, and let $dN(t)$ be the (negative) change in that number during a short time interval dt. (We'll use $N(t)$ to minimize confusion with the neutron number N.) The number of decays during the interval dt is $-dN(t)$. The rate of change of $N(t)$ is the negative quantity $dN(t)/dt$; thus $-dN(t)/dt$ is called the *decay rate* or the **activity** of the specimen. The larger the number of nuclei in the specimen, the more nuclei decay during any time interval. That is, the activity is directly proportional to $N(t)$; it equals a constant λ multiplied by $N(t)$:

$$-\frac{dN(t)}{dt} = \lambda N(t) \qquad (43.16)$$

The constant λ is called the **decay constant,** and it has different values for different nuclides. A large value of λ corresponds to rapid decay; a small value corresponds to slower decay. Solving Eq. (43.16) for λ shows us that λ is the ratio of the number of decays per time to the number of remaining radioactive nuclei; λ can then be interpreted as the *probability per unit time* that any individual nucleus will decay.

43.8 The number of nuclei in a sample of a radioactive element as a function of time. The sample's activity has an exponential decay curve with the same shape.

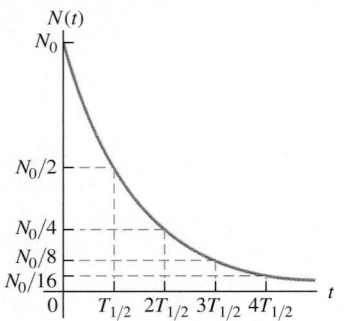

The situation is reminiscent of a discharging capacitor, which we studied in Section 26.4. Equation (43.16) has the same form as the negative of Eq. (26.15), with q and $1/RC$ replaced by $N(t)$ and λ. Then we can make the same substitutions in Eq. (26.16), with the initial number of nuclei $N(0) = N_0$, to find

Number of remaining nuclei at time t in sample of radioactive element ⋯⋯⋯⋯⋯⋯ Number of nuclei at $t = 0$

$$N(t) = N_0 e^{-\lambda t} \qquad \text{Time} \qquad (43.17)$$

Decay constant

Figure 43.8 is a graph of this function.

The **half-life** $T_{1/2}$ is the time required for the number of radioactive nuclei to decrease to one-half the original number N_0. Then half of the remaining radioactive nuclei decay during a second interval $T_{1/2}$, and so on. The numbers remaining after successive half-lives are $N_0/2, N_0/4, N_0/8, \ldots$.

To get the relationship between the half-life $T_{1/2}$ and the decay constant λ, we set $N(t)/N_0 = \frac{1}{2}$ and $t = T_{1/2}$ in Eq. (43.17), obtaining

$$\frac{1}{2} = e^{-\lambda T_{1/2}}$$

We take logarithms of both sides and solve for $T_{1/2}$:

$$T_{1/2} = \frac{\ln 2}{\lambda} = \frac{0.693}{\lambda} \qquad (43.18)$$

The mean lifetime T_{mean}, generally called the *lifetime,* of an unstable nucleus or particle is proportional to the half-life $T_{1/2}$:

Half-life of nucleus or particle

Lifetime of unstable ⋯⋯⋯ nucleus or particle

$$T_{\text{mean}} = \frac{1}{\lambda} = \frac{T_{1/2}}{\ln 2} = \frac{T_{1/2}}{0.693} \qquad (43.19)$$

Decay constant of nucleus or particle

In particle physics the life of an unstable particle is usually described by the lifetime, not the half-life.

Because the activity $-dN(t)/dt$ at any time equals $\lambda N(t)$, Eq. (43.17) tells us that the activity also depends on time as $e^{-\lambda t}$. Thus the graph of activity versus time has the same shape as Fig. 43.8. Also, after successive half-lives, the activity is one-half, one-fourth, one-eighth, and so on of the original activity.

CAUTION **A half-life may not be enough** It is sometimes implied that any radioactive sample will be safe after a half-life has passed. That's wrong. If your radioactive waste initially has ten times too much activity for safety, it is not safe after one half-life, when it still has five times too much. Even after three half-lives it still has 25% more activity than is safe. The number of radioactive nuclei and the activity approach zero only as t approaches infinity. ▌

A common unit of activity is the **curie**, abbreviated Ci, which is defined to be 3.70×10^{10} decays per second. This is approximately equal to the activity of one gram of radium-226. The SI unit of activity is the *becquerel,* abbreviated Bq. One becquerel is one decay per second, so

$$1\ \text{Ci} = 3.70 \times 10^{10}\ \text{Bq} = 3.70 \times 10^{10}\ \text{decays/s}$$

EXAMPLE 43.8   ACTIVITY OF ^{57}Co

The isotope ^{57}Co decays by electron capture to ^{57}Fe with a half-life of 272 d. The ^{57}Fe nucleus is produced in an excited state, and it almost instantaneously emits gamma rays that we can detect. (a) Find the mean lifetime and decay constant for ^{57}Co. (b) If the activity of a ^{57}Co radiation source is now 2.00 μCi, how many ^{57}Co nuclei does the source contain? (c) What will be the activity after one year?

SOLUTION

IDENTIFY and SET UP: This problem uses the relationships among decay constant λ, lifetime T_{mean}, and activity $-dN(t)/dt$. In part (a) we use Eq. (43.19) to find λ and T_{mean} from $T_{1/2}$. In part (b), we use Eq. (43.16) to calculate the number of nuclei $N(t)$ from the activity. Finally, in part (c) we use Eqs. (43.16) and (43.17) to find the activity after one year.

EXECUTE: (a) It's convenient to convert the half-life from days to seconds:

$$T_{1/2} = (272 \text{ d})(86,400 \text{ s/d})$$

$$= 2.35 \times 10^7 \text{ s}$$

From Eq. (43.19), we find that the mean lifetime and the decay constant are

$$T_{mean} = \frac{T_{1/2}}{\ln 2} = \frac{2.35 \times 10^7 \text{ s}}{0.693}$$

$$= 3.39 \times 10^7 \text{ s} = 392 \text{ days}$$

$$\lambda = \frac{1}{T_{mean}} = 2.95 \times 10^{-8} \text{ s}^{-1}$$

(b) The activity $-dN(t)/dt$ is given as 2.00 μCi, so

$$-\frac{dN(t)}{dt} = 2.00 \text{ } \mu\text{Ci} = (2.00 \times 10^{-6})(3.70 \times 10^{10} \text{ s}^{-1})$$

$$= 7.40 \times 10^4 \text{ decays/s}$$

From Eq. (43.16) this is equal to $\lambda N(t)$, so we find

$$N(t) = -\frac{dN(t)/dt}{\lambda} = \frac{7.40 \times 10^4 \text{ s}^{-1}}{2.95 \times 10^{-8} \text{ s}^{-1}}$$

$$= 2.51 \times 10^{12} \text{ nuclei}$$

If you feel we're being too cavalier about the "units" decays and nuclei, you can use decays/(nucleus $\cdot$ s) as the unit for λ.

(c) From Eq. (43.17) the number $N(t)$ of nuclei remaining after one year (3.156 $\times$ 10^7 s) is

$$N(t) = N_0 e^{-\lambda t} = N_0 e^{-(2.95 \times 10^{-8} \text{s}^{-1})(3.156 \times 10^7 \text{s})}$$

$$= 0.394 N_0$$

The number of nuclei has decreased to 0.394 of the original number. Equation (43.16) says that the activity is proportional to the number of nuclei, so the activity has decreased by this same factor to $(0.394)(2.00 \text{ } \mu\text{Ci}) = 0.788 \text{ } \mu\text{Ci}$.

EVALUATE: The number of nuclei found in part (b) is equivalent to 4.17×10^{-12} mol, with a mass of 2.38×10^{-10} g. This is a far smaller mass than even the most sensitive balance can measure.

After one 272-day half-life, the number of ^{57}Co nuclei has decreased to $N_0/2$; after $2(272 \text{ d}) = 544$ d, it has decreased to $N_0/2^2 = N_0/4$. This result agrees with our answer to part (c), which says that after 365 d the number of nuclei is between $N_0/2$ and $N_0/4$.

Radioactive Dating

An important application of radioactivity is the dating of archaeological and geological specimens by measuring the concentration of radioactive isotopes. The most familiar example is *carbon dating*. The unstable isotope ^{14}C, produced during nuclear reactions in the atmosphere that result from cosmic-ray bombardment, gives a small proportion of ^{14}C in the CO_2 in the atmosphere. Plants that obtain their carbon from this source contain the same proportion of ^{14}C as the atmosphere. When a plant dies, it stops taking in carbon, and its ^{14}C β^- decays to ^{14}N with a half-life of 5730 years. By measuring the proportion of ^{14}C in the remains, we can determine how long ago the organism died.

One difficulty with radiocarbon dating is that the ^{14}C concentration in the atmosphere changes over long time intervals. Corrections can be made on the basis of other data such as measurements of tree rings that show annual growth cycles. Similar radioactive techniques are used with other isotopes for dating geological specimens. Some rocks, for example, contain the unstable potassium isotope ^{40}K, a beta emitter that decays to the stable nuclide ^{40}Ar with a half-life of 2.4 $\times$ 10^8 y. The age of the rock can be determined by comparing the concentrations of ^{40}K and ^{40}Ar.

EXAMPLE 43.9 RADIOCARBON DATING

Before 1900 the activity per unit mass of atmospheric carbon due to the presence of ^{14}C averaged about 0.255 Bq per gram of carbon. (a) What fraction of carbon atoms were ^{14}C? (b) In analyzing an archaeological specimen containing 500 mg of carbon, you observe 174 decays in one hour. What is the age of the specimen, assuming that its activity per unit mass of carbon when it died was that average value of the air?

SOLUTION

IDENTIFY and SET UP: The key idea is that the present-day activity of a biological sample containing ^{14}C is related to both the elapsed time since it stopped taking in atmospheric carbon and its activity at that time. We use Eqs. (43.16) and (43.17) to solve for the age t of the specimen. In part (a) we determine the number of ^{14}C atoms $N(t)$ from the activity $-dN(t)/dt$ by using Eq. (43.16). We find the total number of carbon atoms in 500 mg by using the molar mass of carbon (12.011 g/mol, given in Appendix D), and we use the result to calculate the fraction of carbon atoms that are ^{14}C. The activity decays at the same rate as the number of ^{14}C nuclei; we use this and Eq. (43.17) to solve for the age t of the specimen.

EXECUTE: (a) To use Eq. (43.16), we must first find the decay constant λ from Eq. (43.18):

$$T_{1/2} = 5730 \text{ y} = (5730 \text{ y})(3.156 \times 10^7 \text{ s/y}) = 1.808 \times 10^{11} \text{ s}$$

$$\lambda = \frac{\ln 2}{T_{1/2}} = \frac{0.693}{1.808 \times 10^{11} \text{ s}} = 3.83 \times 10^{-12} \text{ s}^{-1}$$

Then, from Eq. (43.16),

$$N(t) = \frac{-dN/dt}{\lambda} = \frac{0.255 \text{ s}^{-1}}{3.83 \times 10^{-12} \text{ s}^{-1}} = 6.65 \times 10^{10} \text{ atoms}$$

The *total* number of C atoms in 1 gram (1/12.011 mol) is $(1/12.011)(6.022 \times 10^{23}) = 5.01 \times 10^{22}$. The ratio of ^{14}C atoms to all C atoms is

$$\frac{6.65 \times 10^{10}}{5.01 \times 10^{22}} = 1.33 \times 10^{-12}$$

Only four carbon atoms in every 3×10^{12} are ^{14}C.

(b) Assuming that the activity per gram of carbon in the specimen when it died ($t = 0$) was 0.255 Bq/g = $(0.255 \text{ s}^{-1} \cdot \text{g}^{-1}) \times (3600 \text{ s/h}) = 918 \text{ h}^{-1} \cdot \text{g}^{-1}$, the activity of 500 mg of carbon then was $(0.500 \text{ g})(918 \text{ h}^{-1} \cdot \text{g}^{-1}) = 459 \text{ h}^{-1}$. The observed activity now, at time t, is 174 h^{-1}. Since the activity is proportional to the number of radioactive nuclei, the activity ratio $174/459 = 0.379$ equals the number ratio $N(t)/N_0$.

Now we solve Eq. (43.17) for t and insert values for $N(t)/N_0$ and λ:

$$t = \frac{\ln(N(t)/N_0)}{-\lambda} = \frac{\ln 0.379}{-3.83 \times 10^{-12} \text{ s}^{-1}} = 2.53 \times 10^{11} \text{ s} = 8020 \text{ y}$$

EVALUATE: After 8020 y the ^{14}C activity has decreased from 459 to 174 decays per hour. The specimen died and stopped taking CO_2 out of the air about 8000 years ago.

Radiation in the Home

A serious health hazard in some areas is the accumulation in houses of ^{222}Rn, an inert, colorless, odorless radioactive gas. Looking at the ^{238}U decay chain in Fig. 43.7, we see that the half-life of ^{222}Rn is 3.82 days. If so, why not just move out of the house for a while and let it decay away? The answer is that ^{222}Rn is continuously being *produced* by the decay of ^{226}Ra, which is found in minute quantities in the rocks and soil on which some houses are built. It's a dynamic equilibrium situation, in which the rate of production equals the rate of decay. The reason ^{222}Rn is a bigger hazard than the other elements in the ^{238}U decay series is that it's a gas. During its short half-life of 3.82 days it can migrate from the soil into your house. If a ^{222}Rn nucleus decays in your lungs, it emits a damaging α particle and its daughter nucleus ^{218}Po, which is *not* chemically inert and is likely to stay in your lungs until it decays, emits another damaging α particle and so on down the ^{238}U decay series.

How much of a hazard is radon? Although reports indicate values as high as 3500 pCi/L, the average activity per volume in the air inside American homes due to ^{222}Rn is about 1.5 pCi/L (over a thousand decays each second in an average-sized room). *If* your environment has this level of activity, it has been estimated that a lifetime exposure would reduce your life expectancy by about 40 days. For comparison, smoking one pack of cigarettes per day reduces life expectancy by 6 years, and it is estimated that the average emission from all the nuclear power plants in the world reduces life expectancy by anywhere from 0.01 day to 5 days. These figures include catastrophes such as the nuclear reactor disasters at Chernobyl, Ukraine (1986), and Fukushima, Japan (2011), for which the *local* effect on life expectancy is much greater.

TEST YOUR UNDERSTANDING OF SECTION 43.4 Which sample contains a greater number of nuclei: a 5.00-μCi sample of ^{240}Pu (half-life 6560 y) or a 4.45-μCi sample of ^{243}Am (half-life 7370 y)? (i) The ^{240}Pu sample; (ii) the ^{243}Am sample; (iii) both have the same number of nuclei. ∎

43.5 BIOLOGICAL EFFECTS OF RADIATION

The above discussion of radon introduced the interaction of radiation with living organisms, a topic of vital interest and importance. Under *radiation* we include radioactivity (alpha, beta, gamma, and neutrons) and electromagnetic radiation such as x rays. As these particles pass through matter, they lose energy, breaking molecular bonds and creating ions—hence the term *ionizing radiation*. Charged particles interact directly with the electrons in the material. X rays and γ rays interact by the photoelectric effect, in which an electron absorbs a photon and breaks loose from its site, or by Compton scattering (see Section 38.3). Neutrons cause ionization indirectly through collisions with nuclei or absorption by nuclei with subsequent radioactive decay of the resulting nuclei.

These interactions are extremely complex. It is well known that excessive exposure to radiation, including sunlight, x rays, and all the nuclear radiations, can destroy tissues. In mild cases it results in a burn, as with common sunburn. Greater exposure can cause very severe illness or death by a variety of mechanisms, including massive destruction of tissue cells, alterations of genetic material, and destruction of the components in bone marrow that produce red blood cells.

Calculating Radiation Doses

Radiation dosimetry is the quantitative description of the effect of radiation on living tissue. The *absorbed dose* of radiation is defined as the energy delivered to the tissue per unit mass. The SI unit of absorbed dose, the joule per kilogram, is called the *gray* (Gy); 1 Gy = 1 J/kg. Another unit is the *rad,* defined as

$$1 \text{ rad} = 0.01 \text{ J/kg} = 0.01 \text{ Gy}$$

Absorbed dose by itself is not an adequate measure of biological effect because equal energies of different kinds of radiation cause different extents of biological effect. This variation is described by a numerical factor called the **relative biological effectiveness (RBE)**, also called the *quality factor* (QF), of each specific radiation. X rays with 200 keV of energy are defined to have an RBE of unity, and the effects of other radiations can be compared experimentally. **Table 43.3** shows approximate values of RBE for several radiations. All these values depend somewhat on the kind of tissue in which the radiation is absorbed and on the energy of the radiation.

The biological effect is described by the product of the absorbed dose and the RBE of the radiation; this quantity is called the *biologically equivalent dose,* or simply the equivalent dose. The SI unit of equivalent dose for humans is the sievert (Sv):

$$\text{Equivalent dose (Sv)} = \text{RBE} \times \text{Absorbed dose (Gy)} \qquad (43.20)$$

A more common unit, corresponding to the rad, is the rem (an abbreviation of *röntgen equivalent for man*):

$$\text{Equivalent dose (rem)} = \text{RBE} \times \text{Absorbed dose (rad)} \qquad (43.21)$$

Thus the unit of the RBE is 1 Sv/Gy or 1 rem/rad, and 1 rem = 0.01 Sv.

TABLE 43.3 Relative Biological Effectiveness (RBE) for Several Types of Radiation

Radiation	RBE (Sv/Gy or rem/rad)
X rays and γ rays	1
Electrons	1.0–1.5
Slow neutrons	3–5
Protons	10
α particles	20
Heavy ions	20

EXAMPLE 43.10 DOSE FROM A MEDICAL X RAY

During a diagnostic x-ray examination a 1.2-kg portion of a broken leg receives an equivalent dose of 0.40 mSv. (a) What is the equivalent dose in mrem? (b) What is the absorbed dose in mrad and in mGy? (c) If the x-ray energy is 50 keV, how many x-ray photons are absorbed?

SOLUTION

IDENTIFY and SET UP: We are asked to relate the equivalent dose (the biological effect of the radiation, measured in sieverts or rems) to the absorbed dose (the energy absorbed per mass, measured in grays or rads). In part (a) we use the conversion factor 1 rem = 0.01 Sv for equivalent dose. Table 43.3 gives the RBE for x rays; we use this value in part (b) to determine the absorbed dose from Eqs. (43.20) and (43.21). Finally, in part (c) we use the mass and the definition of absorbed dose to find the total energy absorbed and the total number of photons absorbed.

EXECUTE: (a) The equivalent dose in mrem is

$$\frac{0.40 \text{ mSv}}{0.01 \text{ Sv/rem}} = 40 \text{ mrem}$$

(b) For x rays, RBE = 1 rem/rad or 1 Sv/Gy, so the absorbed dose is

$$\frac{40 \text{ mrem}}{1 \text{ rem/rad}} = 40 \text{ mrad}$$

$$\frac{0.40 \text{ mSv}}{1 \text{ Sv/Gy}} = 0.40 \text{ mGy} = 4.0 \times 10^{-4} \text{ J/kg}$$

(c) The total energy absorbed is

$$(4.0 \times 10^{-4} \text{ J/kg})(1.2 \text{ kg}) = 4.8 \times 10^{-4} \text{ J} = 3.0 \times 10^{15} \text{ eV}$$

The number of x-ray photons is

$$\frac{3.0 \times 10^{15} \text{ eV}}{5.0 \times 10^4 \text{ eV/photon}} = 6.0 \times 10^{10} \text{ photons}$$

EVALUATE: The absorbed dose is relatively large because x rays have a low RBE. If the ionizing radiation had been a beam of α particles, for which RBE = 20, the absorbed dose needed for an equivalent dose of 0.40 mSv would be only 0.020 mGy, corresponding to a smaller total absorbed energy of 2.4×10^{-5} J.

Radiation Hazards

Here are a few numbers for perspective. To convert from Sv to rem, simply multiply by 100. An ordinary chest x-ray exam delivers about 0.20–0.40 mSv to about 5 kg of tissue. Radiation exposure from cosmic rays and natural radioactivity in soil, building materials, and so on is of the order of 2–3 mSv per year at sea level and twice that at an elevation of 1500 m (5000 ft). A whole-body dose of up to about 0.20 Sv causes no immediately detectable effect. A short-term whole-body dose of 5 Sv or more usually causes death within a few days or weeks. A localized dose of 100 Sv causes complete destruction of the exposed tissues.

The long-term hazards of radiation exposure in causing various cancers and genetic defects have been widely publicized, and the question of whether there is any "safe" level of radiation exposure has been hotly debated. U.S. government regulations are based on a maximum *yearly* exposure, from all except natural resources, of 2 to 5 mSv. Workers with occupational exposure to radiation are permitted an average of 20 mSv per year. Recent studies suggest that these limits are too high and that even extremely small exposures carry hazards, but it is very difficult to gather reliable statistics on the effects of low doses. It has become clear that any use of x rays for medical diagnosis should be preceded by a very careful estimation of the relationship of risk to possible benefit.

Another sharply debated question is that of radiation hazards from nuclear power plants. The radiation level from these plants is *not* negligible. However, to make a meaningful evaluation of hazards, we must compare these levels with the alternatives, such as coal-powered plants. The health hazards of coal smoke are serious and well documented, and the natural radioactivity in the smoke from a coal-fired power plant is believed to be roughly 100 times as great as that from a properly operating nuclear plant with equal capacity. But the comparison is not this simple; the possibility of a nuclear accident and the very serious problem of safe disposal of radioactive waste from nuclear plants must also be considered. **Figure 43.9** shows one estimate of the various sources of radiation exposure for the U.S. population. Ionizing radiation is a two-edged sword; it poses very

43.9 Contribution of various sources to the total average radiation exposure in the U.S. population, expressed as percentages of the total.

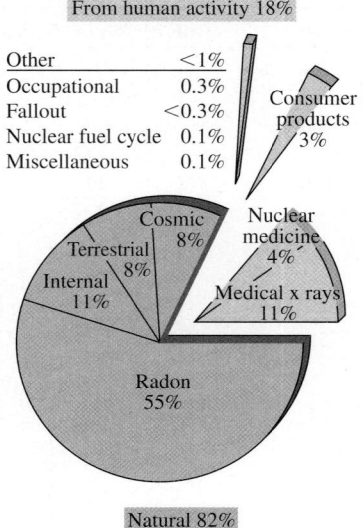

serious health hazards, yet it also provides many benefits to humanity, including the diagnosis and treatments of disease and a wide variety of analytical techniques.

Beneficial Uses of Radiation

Radiation is widely used in medicine for intentional selective destruction of tissue such as tumors. The hazards are considerable, but if the disease would be fatal without treatment, any hazard may be preferable. Artificially produced isotopes are often used as radiation sources. Such isotopes have several advantages over naturally radioactive isotopes. They may have shorter half-lives and correspondingly greater activity. Isotopes can be chosen that emit the type and energy of radiation desired. Some artificial isotopes have been replaced by photon, proton, and electron beams from linear accelerators.

Nuclear medicine is an expanding field of application. Radioactive isotopes have virtually the same electron configurations and resulting chemical behavior as stable isotopes of the same element. But the location and concentration of radioactive isotopes can easily be detected by measurements of the radiation they emit. A familiar example is the use of radioactive iodine for thyroid studies. Nearly all the iodine ingested is either eliminated or stored in the thyroid, and the body's chemical reactions do not discriminate between the unstable isotope ^{131}I and the stable isotope ^{127}I. A minute quantity of ^{131}I is fed or injected into the patient, and the speed with which it becomes concentrated in the thyroid provides a measure of thyroid function. The half-life is 8.02 days, so there are no long-lasting radiation hazards. By use of more sophisticated scanning detectors, one can also obtain a "picture" of the thyroid, which shows enlargement and other abnormalities. This procedure, a type of *autoradiography,* is comparable to photographing the glowing filament of an incandescent light bulb by using the light emitted by the filament itself. If this process discovers cancerous thyroid nodules, they can be destroyed by much larger quantities of ^{131}I.

Another useful nuclide for nuclear medicine is technetium-99 (^{99}Tc), which is formed in an excited state by the β^- decay of molybdenum (^{99}Mo). The technetium then decays to its ground state by emitting a γ-ray photon with energy 143 keV. The half-life is 6.01 hours, unusually long for γ emission. (The ground state of ^{99}Tc is also unstable, with a half-life of 2.11×10^5 y; it decays by β^- emission to the stable ruthenium nuclide ^{99}Ru.) The chemistry of technetium is such that it can readily be attached to organic molecules that are taken up by various organs of the body. A small quantity of such technetium-bearing molecules is injected into a patient, and a scanning detector or *gamma camera* is used to produce an image, or *scintigram,* that reveals which parts of the body take up these γ-emitting molecules. This technique, in which ^{99}Tc acts as a radioactive *tracer,* plays an important role in locating cancers, embolisms, and other pathologies (**Fig. 43.10**).

Tracer techniques have many other applications. Tritium (^3H), a radioactive hydrogen isotope, is used to tag molecules in complex organic reactions; radioactive tags on pesticide molecules, for example, can be used to trace their passage through food chains. In the world of machinery, radioactive iron can be used to study piston-ring wear. Laundry detergent manufacturers have even used radioactive dirt to test the effectiveness of their products.

Many direct effects of radiation are also useful, such as strengthening polymers by cross-linking, sterilizing surgical tools, dispersing unwanted static electricity in the air, and intentionally ionizing the air in smoke detectors. Gamma rays are also used to sterilize and preserve some food products.

BIO Application A Radioactive Building The United States Capitol building in Washington, DC, is made of granite that contains a small amount of naturally radioactive uranium. As a result, the radiation exposure to someone working in the Capitol is 0.85 mSv per year. The health effects of this are negligible; it is estimated that a person who spent 20 years inside the Capitol would have an extra 0.1% chance of getting cancer, above the 10% chance due to all other causes during the same 20-year period.

43.10 This colored scintigram shows where a chemical containing radioactive ^{99}Tc was taken up by a patient's lungs. The orange color in the lung on the left indicates strong γ-ray emission by the ^{99}Tc, which shows that the chemical was able to pass into this lung through the bloodstream. The lung on the right shows weaker emission, indicating the presence of an embolism (a blood clot or other obstruction in an artery) that is restricting the flow of blood to this lung.

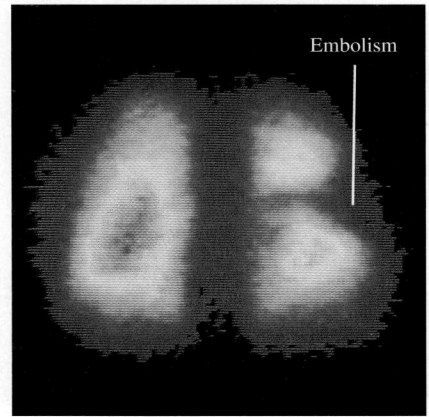

Embolism

TEST YOUR UNDERSTANDING OF SECTION 43.5 Alpha particles have 20 times the relative biological effectiveness of 200-keV x rays. Which would be better to use to radiate tissue deep inside the body? (i) A beam of alpha particles; (ii) a beam of 200-keV x rays; (iii) both are equally effective. ▮

43.6 NUCLEAR REACTIONS

In the preceding sections we studied the decay of unstable nuclei, especially spontaneous emission of an α or β particle, sometimes followed by γ emission. Nothing needs to be done to initiate this decay, and nothing can be done to control it. This section examines some *nuclear reactions,* rearrangements of nuclear components that result from a bombardment by a particle rather than a spontaneous natural process. Rutherford suggested in 1919 that a massive particle with sufficient kinetic energy might be able to penetrate a nucleus. The result would be either a new nucleus with greater atomic number and mass number or a decay of the original nucleus. Rutherford verified this when he bombarded nitrogen (^{14}N) with α particles and obtained an oxygen (^{17}O) nucleus and a proton:

$$\,^4_2\text{He} + \,^{14}_7\text{N} \rightarrow \,^{17}_8\text{O} + \,^1_1\text{H} \qquad (43.22)$$

Rutherford used alpha particles from naturally radioactive sources. In Chapter 44 we'll describe some of the particle accelerators that are now used to initiate nuclear reactions.

Nuclear reactions are subject to several *conservation laws.* The classical conservation principles for charge, momentum, angular momentum, and energy (including rest energies) are obeyed in all nuclear reactions. An additional conservation law, not anticipated by classical physics, is conservation of the total number of nucleons. The numbers of protons and neutrons need not be conserved separately; in β decay, neutrons and protons change into one another. We'll study the basis of the conservation of nucleon number in Chapter 44.

When two nuclei interact, charge conservation requires that the sum of the initial atomic numbers must equal the sum of the final atomic numbers. Because of conservation of nucleon number, the sum of the initial mass numbers must also equal the sum of the final mass numbers. In general, these are *not* elastic collisions, and the total initial mass does *not* equal the total final mass.

Reaction Energy

The difference between the masses before and after the reaction corresponds to the **reaction energy,** according to the mass–energy relationship $E = mc^2$. If initial particles A and B interact to produce final particles C and D, the reaction energy Q is defined as

$$Q = (M_A + M_B - M_C - M_D)c^2 \quad \text{(reaction energy)} \qquad (43.23)$$

To balance the electrons, we use the neutral atomic masses in Eq. (43.23). That is, we use the mass of $^1_1$H for a proton, $^2_1$H for a deuteron, $^4_2$He for an α particle, and so on. When Q is positive, the total mass decreases and the total kinetic energy increases. Such a reaction is called an *exoergic reaction.* When Q is negative, the mass increases and the kinetic energy decreases, and the reaction is called an *endoergic reaction.* The terms *exothermal* and *endothermal,* borrowed from chemistry, are also used. In an endoergic reaction the reaction cannot occur at all unless the initial kinetic energy in the center-of-mass reference frame is at least as great as $|Q|$. That is, there is a **threshold energy,** the minimum kinetic energy to make an endoergic reaction go.

EXAMPLE 43.11 | **EXOERGIC AND ENDOERGIC REACTIONS**

(a) When a lithium-7 nucleus is bombarded by a proton, two alpha particles (^{4}He) are produced. Find the reaction energy. (b) Calculate the reaction energy for the reaction $^4_2\text{He} + \,^{14}_7\text{N} \rightarrow \,^{17}_8\text{O} + \,^1_1\text{H}$.

SOLUTION

IDENTIFY and SET UP: The reaction energy Q for any nuclear reaction equals c^2 times the difference between the total initial mass

and the total final mass, as in Eq. (43.23). Table 43.2 gives the required masses.

EXECUTE: (a) The reaction is $^1_1\text{H} + \,^7_3\text{Li} \rightarrow \,^4_2\text{He} + \,^4_2\text{He}$. The initial and final masses and their respective sums are

A:	$^1_1$H	1.007825 u	C:	$^4_2$He	4.002603 u
B:	$^7_3$Li	7.016005 u	D:	$^4_2$He	4.002603 u
		8.023830 u			8.005206 u

The mass decreases by 0.018624 u. From Eq. (43.23), the reaction energy is

$$Q = (0.018624 \text{ u})(931.5 \text{ MeV/u}) = +17.35 \text{ MeV}$$

(b) The initial and final masses are

A: $^4_2$He	4.002603 u	C: $^{17}_8$O	16.999132 u
B: $^{14}_7$N	14.003074 u	D: $^1_1$H	1.007825 u
	18.005677 u		18.006957 u

The mass increases by 0.001280 u, and the corresponding reaction energy is

$$Q = (-0.001280 \text{ u})(931.5 \text{ MeV/u}) = -1.192 \text{ MeV}$$

EVALUATE: The reaction in part (a) is *exoergic:* The final total kinetic energy of the two separating alpha particles is 17.35 MeV greater than the initial total kinetic energy of the proton and the lithium nucleus. The reaction in part (b) is *endoergic:* In the center-of-mass system—that is, in a head-on collision with zero total momentum—the minimum total initial kinetic energy required for this reaction to occur is 1.192 MeV.

Ordinarily, the endoergic reaction of part (b) of Example 43.11 would be produced by bombarding stationary ^{14}N nuclei with alpha particles from an accelerator. In this case an alpha's kinetic energy must be *greater than* 1.192 MeV. If all the alpha's kinetic energy went solely to increasing the rest energy, the final kinetic energy would be zero, and momentum would not be conserved. When a particle with mass m and kinetic energy K collides with a stationary particle with mass M, the total kinetic energy K_{cm} in the center-of-mass coordinate system (the energy available to cause reactions) is

$$K_{cm} = \frac{M}{M + m} K \qquad (43.24)$$

This expression assumes that the kinetic energies of the particles and nuclei are much less than their rest energies. We leave the derivation of Eq. (43.24) to you. In part (b) of Example 43.11, $M = 14.003074$ u and $m = 4.002603$ u, so $M/(M + m) = (14.003074 \text{ u})/(18.005677 \text{ u}) = 0.7777$ and $K_{cm} = 0.7777K$. Since K_{cm} must be at least 1.192 MeV, the α particle's kinetic energy K must be at least $(1.192 \text{ MeV})/0.7777 = 1.533$ MeV.

For a charged particle such as a proton or an α particle to penetrate the nucleus of another atom and cause a reaction, it must usually have enough initial kinetic energy to overcome the potential-energy barrier caused by the repulsive electrostatic forces. In the reaction of part (a) of Example 43.11, if we treat the proton and the ^{7}Li nucleus as spherically symmetric charges with radii given by Eq. (43.1), their centers will be 3.5×10^{-15} m apart when they touch. The repulsive potential energy of the proton (charge $+e$) and the ^{7}Li nucleus (charge $+3e$) at this separation r is

$$U = \frac{1}{4\pi\epsilon_0} \frac{(e)(3e)}{r} = (9.0 \times 10^9 \text{ N} \cdot \text{m}^2/\text{C}^2) \frac{(3)(1.6 \times 10^{-19} \text{ C})^2}{3.5 \times 10^{-15} \text{ m}}$$

$$= 2.0 \times 10^{-13} \text{ J} = 1.2 \text{ MeV}$$

Even though the reaction is exoergic, the proton must have a minimum kinetic energy of about 1.2 MeV for the reaction to occur, unless the proton *tunnels* through the barrier (see Section 40.4).

Neutron Absorption

Absorption of *neutrons* by nuclei forms an important class of nuclear reactions. Heavy nuclei bombarded by neutrons can undergo a series of neutron absorptions alternating with beta decays, in which the mass number A increases by as much as 25. Some of the *transuranic elements,* elements having Z larger than 92, are produced in this way. These elements have not been found in nature. Many transuranic elements, having Z possibly as high as 118, have been identified.

The analytical technique of *neutron activation analysis* uses similar reactions. When bombarded by neutrons, many stable nuclides absorb a neutron to become unstable and then undergo β^- decay. The energies of the β^- and associated γ emissions depend on the unstable nuclide and provide a means of identifying it and the original stable nuclide. Quantities of elements that are far too small for conventional chemical analysis can be detected in this way.

TEST YOUR UNDERSTANDING OF SECTION 43.6   The reaction described in part (a) of Example 43.11 is exoergic. Can it happen naturally when a sample of solid lithium is placed in a flask of hydrogen gas? ▌

43.7 NUCLEAR FISSION

PhET: Nuclear Fission

Nuclear fission is a decay process in which an unstable nucleus splits into two fragments of comparable mass. Fission was discovered in 1938 through the experiments of Otto Hahn and Fritz Strassman in Germany. Pursuing earlier work by Enrico Fermi, they bombarded uranium ($Z = 92$) with neutrons. The resulting radiation did not coincide with that of any known radioactive nuclide. Urged on by their colleague Lise Meitner, they used meticulous chemical analysis to reach the astonishing but inescapable conclusion that they had found a radioactive isotope of barium ($Z = 56$). Later, radioactive krypton ($Z = 36$) was also found. Meitner and Otto Frisch correctly interpreted these results as showing that uranium nuclei were splitting into two massive fragments called *fission fragments*. Two or three free neutrons usually appear along with the fission fragments and, very occasionally, a light nuclide such as ^{3}H.

43.11 Mass distribution of fission fragments from the fission of ^{236}U* (an excited state of ^{236}U), which is produced when ^{235}U absorbs a neutron. The vertical scale is logarithmic.

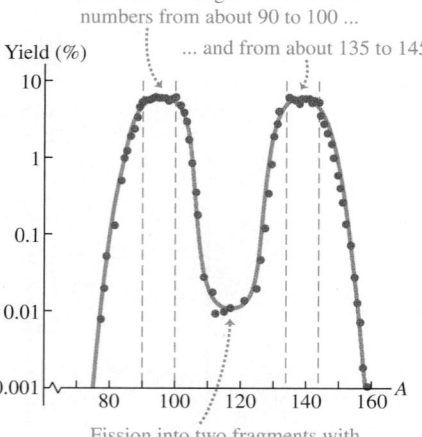

Both the common isotope (99.3%) ^{238}U and the uncommon isotope (0.7%) ^{235}U (as well as several other nuclides) can be easily split by neutron bombardment: ^{235}U by slow neutrons (kinetic energy less than 1 eV) but ^{238}U only by fast neutrons with a minimum of about 1 MeV of kinetic energy. Fission resulting from neutron absorption is called *induced fission*. Some nuclides can also undergo *spontaneous fission* without initial neutron absorption, but this is quite rare. When ^{235}U absorbs a neutron, the resulting nuclide ^{236}U* is in a highly excited state and splits into two fragments almost instantaneously. Strictly speaking, it is ^{236}U*, not ^{235}U, that undergoes fission, but it's usual to speak of the fission of ^{235}U.

Over 100 different nuclides, representing more than 20 different elements, have been found among the fission products. **Figure 43.11** shows the distribution of mass numbers for fission fragments from the fission of ^{235}U.

Fission Reactions

You should check the following two typical fission reactions for conservation of nucleon number and charge:

$$^{235}_{92}\text{U} + {}^1_0\text{n} \rightarrow {}^{236}_{92}\text{U}^* \rightarrow {}^{144}_{56}\text{Ba} + {}^{89}_{36}\text{Kr} + 3{}^1_0\text{n}$$

$$^{235}_{92}\text{U} + {}^1_0\text{n} \rightarrow {}^{236}_{92}\text{U}^* \rightarrow {}^{140}_{54}\text{Xe} + {}^{94}_{38}\text{Sr} + 2{}^1_0\text{n}$$

The total kinetic energy of the fission fragments is enormous, about 200 MeV (compared to typical α and β energies of a few MeV). The reason for this is that nuclides at the high end of the mass spectrum (near $A = 240$) are less tightly bound than those nearer the middle ($A = 90$ to 145). Referring to Fig. 43.2, we see that the average binding energy per nucleon is about 7.6 MeV at $A = 240$ but about 8.5 MeV at $A = 120$. Therefore a rough estimate of the expected *increase* in binding energy during fission is about 8.5 MeV − 7.6 MeV = 0.9 MeV per nucleon, or a total of $(235)(0.9 \text{ MeV}) \approx 200$ MeV.

CAUTION   **Binding energy and rest energy**   It may seem to be a violation of conservation of energy to have an increase in both the binding energy and the kinetic energy during a fission reaction. But relative to the total rest energy E_0 of the separated nucleons, the rest energy of the nucleus is E_0 *minus* E_B. Thus an *increase* in binding energy corresponds to a *decrease* in rest energy as rest energy is converted to the kinetic energy of the fission fragments. ▌

43.12 A liquid-drop model of fission.

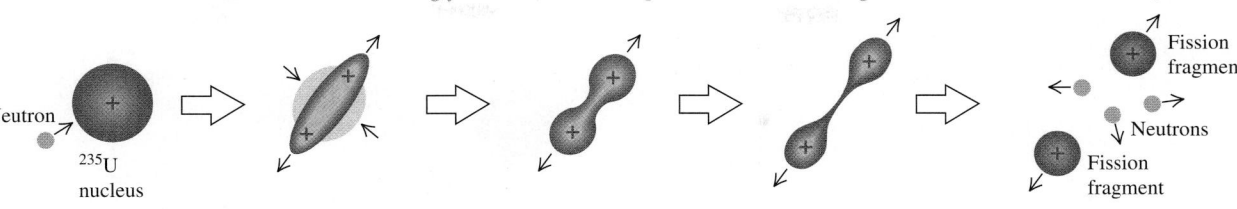

(a) A ^{235}U nucleus absorbs a neutron.

(b) The resulting ^{236}U* nucleus is in a highly excited state and oscillates strongly.

(c) A neck develops, and electrical repulsion pushes the two lobes apart.

(d) The two lobes separate, forming fission fragments.

(e) The fragments emit neutrons at the time of fission (or occasionally a few seconds later).

Neutron

^{235}U nucleus

Fission fragment

Neutrons

Fission fragment

Fission fragments always have too many neutrons to be stable. We noted in Section 43.3 that the neutron-to-proton ratio (N/Z) for stable nuclides is about 1 for light nuclides but almost 1.6 for the heaviest nuclides because of the increasing influence of the electrical repulsion of the protons. The N/Z value for stable nuclides is about 1.3 at $A = 100$ and 1.4 at $A = 150$. The fragments have about the same N/Z as ^{235}U, about 1.55. They usually respond to this surplus of neutrons by undergoing a series of β^- decays (each of which increases Z by 1 and decreases N by 1) until a stable value of N/Z is reached. A typical example is

$$^{140}_{54}\text{Xe} \xrightarrow{\beta^-} {}^{140}_{55}\text{Cs} \xrightarrow{\beta^-} {}^{140}_{56}\text{Ba} \xrightarrow{\beta^-} {}^{140}_{57}\text{La} \xrightarrow{\beta^-} {}^{140}_{58}\text{Ce}$$

The nuclide ^{140}Ce is stable. This series of β^- decays produces, on average, about 15 MeV of additional kinetic energy. The neutron excess of fission fragments also explains why two or three free neutrons are released during the fission.

Fission appears to set an upper limit on the production of transuranic nuclei, mentioned in Section 43.6, that are relatively stable. There are theoretical reasons to expect that nuclei near $Z = 114$, $N = 184$ or 196, might be stable with respect to spontaneous fission. In the shell model (see Section 43.2), these numbers correspond to filled shells and subshells in the nuclear energy-level structure. Such *superheavy nuclei* would still be unstable with respect to alpha emission. In 2009 it was confirmed that there are at least four isotopes with $Z = 114$, the longest-lived of which has a half-life due to alpha decay of about 2.6 s.

Liquid-Drop Model

We can understand fission qualitatively on the basis of the liquid-drop model of the nucleus (see Section 43.2). The process is shown in **Fig. 43.12** in terms of an electrically charged liquid drop. These sketches shouldn't be taken too literally, but they may help to develop your intuition about fission. A ^{235}U nucleus absorbs a neutron (Fig. 43.12a), becoming a ^{236}U* nucleus with excess energy (Fig. 43.12b). This excess energy causes violent oscillations, during which a neck between two lobes develops (Fig. 43.12c). The electrical repulsion of these two lobes stretches the neck farther (Fig. 43.12d), and finally two smaller fragments are formed (Fig. 43.12e) that move rapidly apart.

This qualitative picture has been developed into a more quantitative theory to explain why some nuclei undergo fission and others don't. **Figure 43.13** shows a hypothetical potential-energy function for two possible fission fragments. If neutron absorption results in an excitation energy greater than the energy barrier height U_B, fission occurs immediately. Even when there isn't quite enough energy to surmount the barrier, fission can take place by quantum-mechanical *tunneling,* discussed in Section 40.4. In principle, many stable heavy nuclei can fission by tunneling. But the probability depends very critically on the height and width of the barrier. For most nuclei this process is so unlikely that it is never observed.

43.13 Hypothetical potential-energy function for two fission fragments in a fissionable nucleus. At distances r beyond the range of the nuclear force, the potential energy varies approximately as $1/r$. Fission occurs if there is an excitation energy greater than U_B or an appreciable probability for tunneling through the potential-energy barrier.

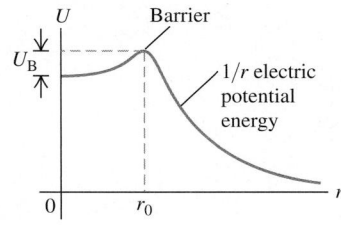

43.14 Schematic diagram of a nuclear fission chain reaction.

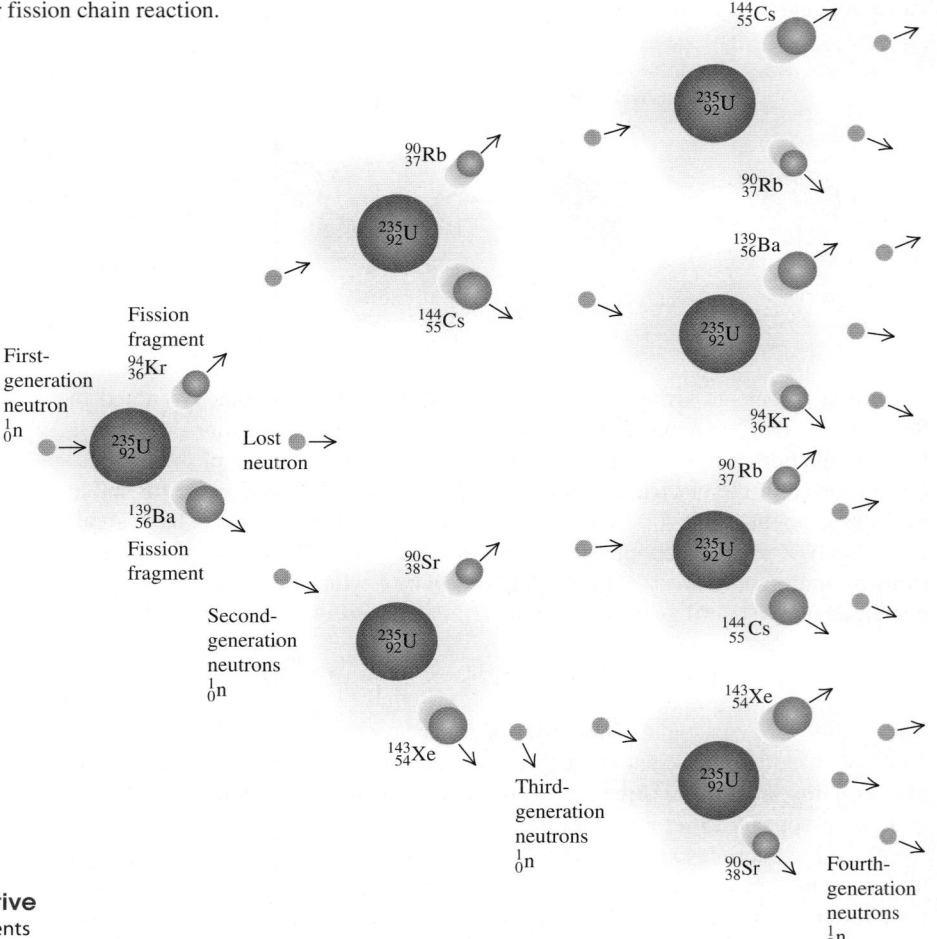

Chain Reactions

Fission of a uranium nucleus, triggered by neutron bombardment, releases other neutrons that can trigger more fissions, suggesting the possibility of a **chain reaction** (**Fig. 43.14**). The chain reaction may be made to proceed slowly and in a controlled manner in a nuclear reactor or explosively in a bomb. The energy release in a nuclear chain reaction is enormous, far greater than that in any chemical reaction. (In a sense, *fire* is a chemical chain reaction.) For example, when uranium is "burned" to uranium dioxide in the chemical reaction

$$U + O_2 \rightarrow UO_2$$

the heat of combustion is about 4500 J/g. Expressed as energy per atom, this is about 11 eV per atom. By contrast, fission liberates about 200 MeV per atom, nearly 20 million times as much energy.

Nuclear Reactors

A *nuclear reactor* is a system in which a controlled nuclear chain reaction is used to liberate energy. In a nuclear power plant, this energy is used to generate steam, which operates a turbine and turns an electrical generator.

On average, each fission of a ^{235}U nucleus produces about 2.5 free neutrons, so 40% of the neutrons are needed to sustain a chain reaction. A ^{235}U nucleus is much more likely to absorb a low-energy neutron (less than 1 eV) than one of the higher-energy neutrons (1 MeV or so) that are liberated during fission. In a nuclear reactor the higher-energy neutrons are slowed down by collisions with nuclei in the surrounding material, called the *moderator,* so they are much more likely to cause further fissions. In nuclear power plants, the moderator is often

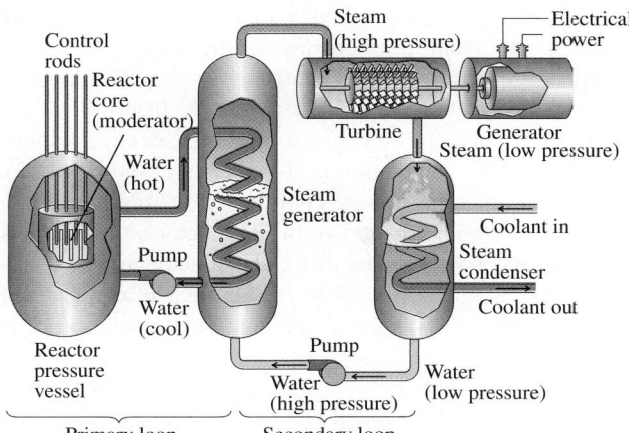

43.15 Schematic diagram of a nuclear power plant.

water, occasionally graphite. The *rate* of the reaction is controlled by inserting or withdrawing *control rods* made of elements (such as boron or cadmium) whose nuclei *absorb* neutrons without undergoing any additional reaction. The isotope ^{238}U can also absorb neutrons, leading to $^{239}U^*$, but not with high enough probability for it to sustain a chain reaction by itself. Thus uranium that is used in reactors is often "enriched" by increasing the proportion of ^{235}U above the natural value of 0.7%, typically to 3% or so, by isotope-separation processing.

The most familiar application of nuclear reactors is for the generation of electrical power. As was noted above, the fission energy appears as kinetic energy of the fission fragments, and its immediate result is to increase the internal energy of the fuel elements and the surrounding moderator. This increase in internal energy is transferred as heat to generate steam to drive turbines, which spin the electrical generators. **Figure 43.15** is a schematic diagram of a nuclear power plant. The energetic fission fragments heat the water surrounding the reactor core. The steam generator is a heat exchanger that takes heat from this highly radioactive water and generates nonradioactive steam to run the turbines.

A typical nuclear plant has an electric-generating capacity of 1000 MW (or 10^9 W). The turbines are heat engines and are subject to the efficiency limitations imposed by the second law of thermodynamics, discussed in Chapter 20. In modern nuclear plants the overall efficiency is about one-third, so 3000 MW of thermal power from the fission reaction is needed to generate 1000 MW of electrical power.

EXAMPLE 43.12   URANIUM CONSUMPTION IN A NUCLEAR REACTOR

What mass of ^{235}U must undergo fission each day to provide 3000 MW of thermal power?

SOLUTION

IDENTIFY and SET UP: Fission of ^{235}U liberates about 200 MeV per atom. We use this and the mass of the ^{235}U atom to determine the required amount of uranium.

EXECUTE: Each second, we need 3000 MJ or 3000×10^6 J. Each fission provides 200 MeV, or

$$(200 \text{ MeV/fission})(1.6 \times 10^{-13} \text{ J/MeV}) = 3.2 \times 10^{-11} \text{ J/fission}$$

The number of fissions needed each second is

$$\frac{3000 \times 10^6 \text{ J}}{3.2 \times 10^{-11} \text{ J/fission}} = 9.4 \times 10^{19} \text{ fissions}$$

Each ^{235}U atom has a mass of $(235 \text{ u})(1.66 \times 10^{-27} \text{ kg/u}) = 3.9 \times 10^{-25}$ kg, so the mass of ^{235}U that undergoes fission each second is

$$(9.4 \times 10^{19})(3.9 \times 10^{-25} \text{ kg}) = 3.7 \times 10^{-5} \text{ kg} = 37 \text{ } \mu g$$

In one day (86,400 s), the total consumption of ^{235}U is

$$(3.7 \times 10^{-5} \text{ kg/s})(86,400 \text{ s}) = 3.2 \text{ kg}$$

EVALUATE: For comparison, a 1000-MW coal-fired power plant burns 10,600 tons (about 10 million kg) of coal per day!

We mentioned above that about 15 MeV of the energy released after fission of a ^{235}U nucleus comes from the β^- decays of the fission fragments. This fact poses a serious problem with respect to control and safety of reactors. Even after the chain reaction has been completely stopped by insertion of control rods into the core, heat continues to be evolved by the β^- decays, which cannot be stopped. For a 3000-MW reactor this heat power is initially very large, about 200 MW. In the event of total loss of cooling water, this power is more than enough to cause a catastrophic meltdown of the reactor core and possible penetration of the containment vessel. The difficulty in achieving a "cold shutdown" following an accident at the Three Mile Island nuclear power plant in Pennsylvania in March 1979 was a result of the continued evolution of heat due to β^- decays.

The catastrophe of April 26, 1986, at Chernobyl reactor No. 4 in Ukraine resulted from a combination of an inherently unstable design and several human errors committed during a test of the emergency core cooling system. Too many control rods were withdrawn to compensate for a decrease in power caused by a buildup of neutron absorbers such as ^{135}Xe. The power level rose from 1% of normal to 100 times normal in 4 seconds; a steam explosion ruptured pipes in the core cooling system and blew the heavy concrete cover off the reactor. The graphite moderator caught fire and burned for several days, and there was a meltdown of the core. The total activity of the radioactive material released into the atmosphere has been estimated as about 10^8 Ci.

TEST YOUR UNDERSTANDING OF SECTION 43.7 The fission of ^{235}U can be triggered by the absorption of a slow neutron by a nucleus. Can a slow *proton* be used to trigger ^{235}U fission? ∎

43.8 NUCLEAR FUSION

In a **nuclear fusion** reaction, two or more small light nuclei come together, or *fuse,* to form a larger nucleus. Fusion reactions release energy for the same reason as fission reactions: The binding energy per nucleon after the reaction is greater than before. Referring to Fig. 43.2, we see that the binding energy per nucleon increases with A up to about $A = 60$, so fusion of nearly any two light nuclei to make a nucleus with A less than 60 is likely to be an exoergic reaction. In comparison to fission, we are moving toward the peak of this curve from the opposite side. Another way to express the energy relationships is that the total mass of the products is less than that of the initial particles.

Here are three examples of energy-liberating fusion reactions, written in terms of the neutral atoms:

$$^1_1\text{H} + {}^1_1\text{H} \rightarrow {}^2_1\text{H} + \beta^+ + \nu_e$$

$$^2_1\text{H} + {}^1_1\text{H} \rightarrow {}^3_2\text{He} + \gamma$$

$$^3_2\text{He} + {}^3_2\text{He} \rightarrow {}^4_2\text{He} + {}^1_1\text{H} + {}^1_1\text{H}$$

In the first reaction, two protons combine to form a deuteron (^{2}H), with the emission of a positron (β^+) and an electron neutrino. In the second, a proton and a deuteron combine to form the nucleus of the light isotope of helium, ^{3}He, with the emission of a gamma ray. Now double the first two reactions to provide the two ^{3}He nuclei that fuse in the third reaction to form an alpha particle (^{4}He) and two protons. Together the reactions make up the process called the *proton-proton chain* (**Fig. 43.16**).

① Two protons combine to form a deuteron (^{2}H) ...

③ A third proton combines with the deuteron, forming a helium nucleus (^{3}He) and emitting a gamma-ray photon.

④ Two ^{3}He nuclei fuse, forming a ^{4}He nucleus and releasing two protons.

43.16 The proton-proton chain.

② ... as well as a positron (β^+) and an electron neutrino (ν_e).

The net effect of the chain is the conversion of four protons into one α particle, two positrons, two electron neutrinos, and two γ's. We can calculate the energy release from this part of the process: The mass of an α particle plus two positrons is the mass of neutral ^{4}He, the neutrinos have zero (or negligible) mass, and the gammas have zero mass.

Mass of four protons	4.029106 u
Mass of ^{4}He	4.002603 u
Mass difference and energy release	0.026503 u and 24.69 MeV

The two positrons that are produced during the first step of the proton-proton chain collide with two electrons; mutual annihilation of the four particles takes place, and their rest energy is converted into $4(0.511 \text{ MeV}) = 2.044 \text{ MeV}$ of gamma radiation. Thus the total energy released is $(24.69 + 2.044) \text{ MeV} = 26.73 \text{ MeV}$. The proton-proton chain takes place in the interior of the sun and other stars (**Fig. 43.17**). Each gram of the sun's mass contains about 4.5×10^{23} protons. If all of these protons were fused into helium, the energy released would be about 130,000 kWh. If the sun were to continue to radiate at its present rate, it would take about 75×10^9 years to exhaust its supply of protons. As we will soon see, fusion reactions can occur only at extremely high temperatures; in the sun, these temperatures are found only deep within the interior. Hence the sun cannot fuse *all* of its protons and can sustain fusion for a total of only about 10×10^9 years in total. The present age of the solar system (including the sun) is 4.54×10^9 years, so the sun is about halfway through its available store of protons.

43.17 The energy released as starlight comes from fusion reactions deep within a star's interior. When a star is first formed and for most of its life, it converts the hydrogen in its core into helium. As a star ages, the core temperature can become high enough for additional fusion reactions that convert helium into carbon, oxygen, and other elements.

EXAMPLE 43.13 **A FUSION REACTION**

Two deuterons fuse to form a *triton* (a nucleus of tritium, or ^{3}H) and a proton. How much energy is liberated?

SOLUTION

IDENTIFY and SET UP: This is a nuclear reaction of the type discussed in Section 43.6. We use Eq. (43.23) to find the energy released.

EXECUTE: Adding one electron to each nucleus makes each a neutral atom; we find their masses in Table 43.2. Substituting into

Eq. (43.23), we find

$$Q = [2(2.014102 \text{ u}) - 3.016049 \text{ u} - 1.007825 \text{ u}] \times (931.5 \text{ MeV/u}) = 4.03 \text{ MeV}$$

EVALUATE: Thus 4.03 MeV is released in the reaction; the triton and proton together have 4.03 MeV more kinetic energy than the two deuterons had together.

Achieving Fusion

For two nuclei to undergo fusion, they must come together to within the range of the nuclear force, typically of the order of 2×10^{-15} m. To do this, they must overcome the electrical repulsion of their positive charges. For two protons at this distance, the corresponding potential energy is about 1.2×10^{-13} J or 0.7 MeV; this represents the total initial *kinetic* energy that the fusion nuclei must have—for example, 0.6×10^{-13} J each in a head-on collision.

Atoms have this much energy only at extremely high temperatures. The discussion of Section 18.3 showed that the average translational kinetic energy of a gas molecule at temperature T is $\frac{3}{2}kT$, where k is Boltzmann's constant. The temperature at which this is equal to $E = 0.6 \times 10^{-13}$ J is determined by the relationship

$$E = \tfrac{3}{2}kT$$

$$T = \frac{2E}{3k} = \frac{2(0.6 \times 10^{-13} \text{ J})}{3(1.38 \times 10^{-23} \text{ J/K})} = 3 \times 10^9 \text{ K}$$

Fusion reactions are possible at lower temperatures because the Maxwell–Boltzmann distribution function (see Section 18.5) gives a small fraction of protons with kinetic energies much higher than the average value. The proton-proton reaction occurs at "only" 1.5×10^7 K at the center of the sun, making it an extremely low-probability process; but that's why the sun is expected to last so long. At these temperatures the fusion reactions are called *thermonuclear* reactions.

Intensive efforts are under way to achieve controlled fusion reactions, which potentially represent an enormous new resource of energy (see Fig. 24.11). At the temperatures mentioned, light atoms are fully ionized, and the resulting state of matter is called a *plasma*. In one kind of experiment using *magnetic confinement,* a plasma is heated to extremely high temperature by an electrical discharge, while being contained by appropriately shaped magnetic fields. In another, through *inertial confinement,* pellets of the material to be fused are heated by a high-intensity laser beam (see **Fig. 43.18**). Some of the reactions being studied are

$$^{2}_{1}\text{H} + {}^{2}_{1}\text{H} \rightarrow {}^{3}_{1}\text{H} + {}^{1}_{1}\text{H} + 4.0 \text{ MeV} \qquad (1)$$

$$^{3}_{1}\text{H} + {}^{2}_{1}\text{H} \rightarrow {}^{4}_{2}\text{He} + {}^{1}_{0}\text{n} + 17.6 \text{ MeV} \qquad (2)$$

$$^{2}_{1}\text{H} + {}^{2}_{1}\text{H} \rightarrow {}^{3}_{2}\text{He} + {}^{1}_{0}\text{n} + 3.3 \text{ MeV} \qquad (3)$$

$$^{3}_{2}\text{He} + {}^{2}_{1}\text{H} \rightarrow {}^{4}_{2}\text{He} + {}^{1}_{1}\text{H} + 18.3 \text{ MeV} \qquad (4)$$

We considered reaction (1) in Example 43.13; two deuterons fuse to form a triton and a proton. In reaction (2) a triton combines with another deuteron to form an alpha particle and a neutron. The result of these two reactions together is the conversion of three deuterons into an alpha particle, a proton, and a neutron, with 21.6 MeV of energy liberated. Reactions (3) and (4) together achieve the same conversion. In a plasma that contains deuterons, the two pairs of reactions occur with roughly equal probability. As yet, no one has succeeded in producing these reactions under controlled conditions in such a way as to yield a net surplus of usable energy.

Methods of achieving fusion that don't require high temperatures are also being studied; these are called *cold fusion.* One successful scheme uses an unusual hydrogen molecule ion. The usual H_2^+ ion consists of two protons bound by one shared electron; the nuclear spacing is about 0.1 nm. If the protons are replaced by a deuteron (^2H) and a triton (^3H) and the electron by a *muon,* which is 208 times as massive as the electron, the spacing is reduced by a factor of 208. The probability then becomes appreciable for the two nuclei to tunnel through the narrow repulsive potential-energy barrier and fuse in reaction (2) above. The prospect of making this process, called *muon-catalyzed fusion,* into a practical energy source is still distant.

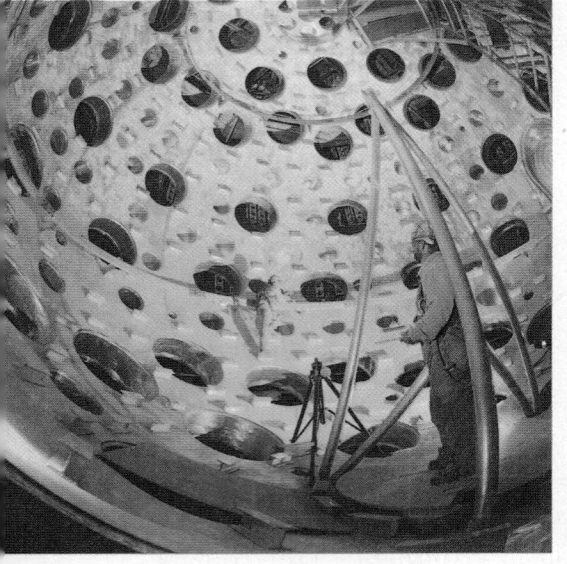

43.18 This target chamber at the National Ignition Facility in California has apertures for 192 powerful laser beams. The lasers deliver 5×10^{14} W of power for a few nanoseconds to a millimeter-sized pellet of deuterium and tritium at the center of the chamber, thus triggering thermonuclear fusion.

TEST YOUR UNDERSTANDING OF SECTION 43.8 Are *all* fusion reactions exoergic? ∎

Nuclear properties: A nucleus is composed of A nucleons (Z protons and N neutrons). All nuclei have about the same density. The radius of a nucleus with mass number A is given approximately by Eq. (43.1). A single nuclear species of a given Z and N is called a nuclide. Isotopes are nuclides of the same element (same Z) that have different numbers of neutrons. Nuclear masses are measured in atomic mass units. Nucleons have angular momentum and a magnetic moment. (See Examples 43.1 and 43.2.)

$$R = R_0 A^{1/3}$$
$$(R_0 = 1.2 \times 10^{-15} \text{ m})$$
(43.1)

Nuclear binding and structure: The mass of a nucleus is always less than the mass of the protons and neutrons within it. The mass difference multiplied by c^2 gives the binding energy E_B. The binding energy for a given nuclide is determined by the nuclear force, which is short range and favors pairs of particles, and by the electrical repulsion between protons. A nucleus is unstable if A or Z is too large or if the ratio N/Z is wrong. Two widely used models of the nucleus are the liquid-drop model and the shell model; the latter is analogous to the central-field approximation for atomic structure. (See Examples 43.3 and 43.4.)

$$E_B = (ZM_H + Nm_n - {}_Z^A M)c^2 \quad (43.10)$$

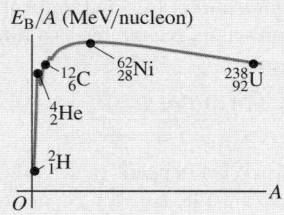

Radioactive decay: Unstable nuclides usually emit an alpha particle (a ${}_2^4$He nucleus) or a beta particle (an electron) in the process of changing to another nuclide, sometimes followed by a gamma-ray photon. The rate of decay of an unstable nucleus is described by the decay constant λ, the half-life $T_{1/2}$, or the lifetime T_{mean}. If the number of nuclei at time $t = 0$ is N_0 and no more are produced, the number at time t is given by Eq. (43.17). (See Examples 43.5–43.9.)

$$N(t) = N_0 e^{-\lambda t} \quad (43.17)$$
$$T_{\text{mean}} = \frac{1}{\lambda} = \frac{T_{1/2}}{\ln 2} = \frac{T_{1/2}}{0.693} \quad (43.19)$$

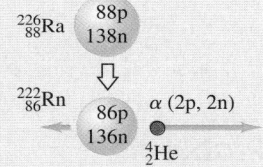

Biological effects of radiation: The biological effect of any radiation depends on the product of the energy absorbed per unit mass and the relative biological effectiveness (RBE), which is different for different radiations. (See Example 43.10.)

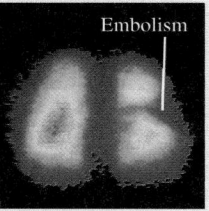

Nuclear reactions: In a nuclear reaction, two nuclei or particles collide to produce two new nuclei or particles. Reactions can be exoergic or endoergic. Several conservation laws, including charge, energy, momentum, angular momentum, and nucleon number, are obeyed. Energy is released by the fission of a heavy nucleus into two lighter, always unstable, nuclei. Energy is also released by the fusion of two light nuclei into a heavier nucleus. (See Examples 43.11–43.13.)

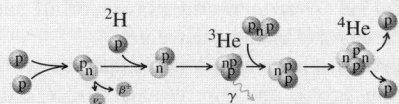

BRIDGING PROBLEM SATURATION OF ^{128}I PRODUCTION

In an experiment, the iodine isotope ^{128}I is created by irradiating a sample of ^{127}I with a beam of neutrons, yielding 1.50×10^6 ^{128}I nuclei per second. Initially no ^{128}I nuclei are present. A ^{128}I nucleus decays by β^- emission with a half-life of 25.0 min. (a) To what nuclide does ^{128}I decay? (b) Could that nuclide decay back to ^{128}I by β^+ emission? Why or why not? (c) After the sample has been irradiated for a long time, what is the maximum number of ^{128}I atoms that can be present in the sample? What is the maximum activity that can be produced? (This steady-state situation is called *saturation*.) (d) Find an expression for the number of ^{128}I atoms present in the sample as a function of time.

SOLUTION GUIDE

IDENTIFY and SET UP

1. What happens to the values of Z, N, and A in β^- decay? What must be true for β^- decay to be possible? For β^+ decay to be possible?
2. You'll need to write an equation for the rate of change dN/dt of the number N of ^{128}I atoms in the sample, taking account of both the creation of ^{128}I by the neutron irradiation and the

decay of any ^{128}I present. In the steady state, how do the rates of these two processes compare?
3. List the unknown quantities for each part of the problem and identify your target variables.

EXECUTE

4. Find the values of Z and N of the nuclide produced by the decay of ^{128}I. What element is this?
5. Decide whether this nuclide can decay back to ^{128}I.
6. Inspect your equation for dN/dt. What is the value of dN/dt in the steady state? Use this to solve for the steady-state values of N and the activity.
7. Solve your dN/dt equation for the function $N(t)$. (*Hint:* See Section 26.4.)

EVALUATE

8. Your result from step 6 tells you the value of N after a long time (that is, for large values of t). Is this consistent with your result from step 7? What would constitute a "long time" under these conditions?

Problems

For assigned homework and other learning materials, go to MasteringPhysics®.

°, °°, °°°: Difficulty levels. CP: Cumulative problems incorporating material from earlier chapters. CALC: Problems requiring calculus. DATA: Problems involving real data, scientific evidence, experimental design, and/or statistical reasoning. BIO: Biosciences problems.

DISCUSSION QUESTIONS

Q43.1 BIO Neutrons have a magnetic dipole moment and can undergo spin flips by absorbing electromagnetic radiation. Why, then, are protons rather than neutrons used in MRI of body tissues? (See Fig. 43.1.)

Q43.2 In Eq. (43.11), as the total number of nucleons becomes larger, the importance of the second term in the equation decreases relative to that of the first term. Does this make physical sense? Explain.

Q43.3 Why aren't the masses of all nuclei integer multiples of the mass of a single nucleon?

Q43.4 The only two stable nuclides with more protons than neutrons are $^1_1$H and $^3_2$He. Why is $Z > N$ so uncommon?

Q43.5 What are the six known elements for which Z is a magic number? Discuss what properties these elements have as a consequence of their special values of Z.

Q43.6 The binding energy per nucleon for most nuclides doesn't vary much (see Fig. 43.2). Is there similar consistency in the *atomic* energy of atoms, on an "energy per electron" basis? If so, why? If not, why not?

Q43.7 Heavy, unstable nuclei usually decay by emitting an α or a β particle. Why don't they usually emit a single proton or neutron?

Q43.8 As stars age, they use up their supply of hydrogen and eventually begin producing energy by a reaction that involves the fusion of three helium nuclei to form a carbon nucleus. Would you

expect the interiors of these old stars to be hotter or cooler than the interiors of younger stars? Explain.

Q43.9 Since lead is a stable element, why doesn't the ^{238}U decay series shown in Fig. 43.7 stop at lead, ^{214}Pb?

Q43.10 In the ^{238}U decay series shown in Fig. 43.7, some nuclides in the series are found much more abundantly in nature than others, even though every ^{238}U nucleus goes through every step in the series before finally becoming ^{206}Pb. Why don't the intermediate nuclides all have the same abundance?

Q43.11 Compared to α particles with the same energy, β particles can much more easily penetrate through matter. Why is this?

Q43.12 If ^A_ZEl$_i$ represents the initial nuclide, what is the decay process or processes if the final nuclide is (a) $^A_{Z+1}$El$_f$; (b) $^{A-4}_{Z-2}$El$_f$; (c) $^A_{Z-1}$El$_f$?

Q43.13 In a nuclear decay equation, why can we represent an electron as $^0_{-1}\beta^-$? What are the equivalent representations for a positron, a neutrino, and an antineutrino?

Q43.14 Why is the alpha, beta, or gamma decay of an unstable nucleus unaffected by the *chemical* situation of the atom, such as the nature of the molecule or solid in which it is bound? The chemical situation of the atom can, however, have an effect on the half-life in electron capture. Why is this?

Q43.15 In the process of *internal conversion*, a nucleus decays from an excited state to a ground state by giving the excitation energy directly to an atomic electron rather than emitting a gamma-ray photon. Why can this process also produce x-ray photons?

Q43.16 In Example 43.9 (Section 43.4), the activity of atmospheric carbon *before* 1900 was given. Discuss why this activity may have changed since 1900.

Q43.17 BIO One problem in radiocarbon dating of biological samples, especially very old ones, is that they can easily be contaminated with modern biological material during the measurement process. What effect would such contamination have on the estimated age? Why is such contamination a more serious problem for samples of older material than for samples of younger material?

Q43.18 The most common radium isotope found on earth, ^{226}Ra, has a half-life of about 1600 years. If the earth was formed well over 10^9 years ago, why is there any radium left now?

Q43.19 Fission reactions occur only for nuclei with large nucleon numbers, while exoergic fusion reactions occur only for nuclei with small nucleon numbers. Why is this?

Q43.20 When a large nucleus splits during nuclear fission, the daughter nuclei of the fission fly apart with enormous kinetic energy. Why does this happen?

EXERCISES

Section 43.1 Properties of Nuclei

43.1 • How many protons and how many neutrons are there in a nucleus of the most common isotope of (a) silicon, $^{28}_{14}$Si; (b) rubidium, $^{85}_{37}$Rb; (c) thallium, $^{205}_{81}$Tl?

43.2 •• Neutrons are placed in a magnetic field with magnitude 2.30 T. (a) What is the energy difference between the states with the nuclear spin angular momentum components parallel and antiparallel to the field? Which state is lower in energy: the one with its spin component parallel to the field or the one with its spin component antiparallel to the field? How do your results compare with the energy states for a proton in the same field (see Example 43.2)? (b) The neutrons can make transitions from one of these states to the other by emitting or absorbing a photon with energy equal to the energy difference of the two states. Find the frequency and wavelength of such a photon.

43.3 • Hydrogen atoms are placed in an external magnetic field. The protons can make transitions between states in which the nuclear spin component is parallel and antiparallel to the field by absorbing or emitting a photon. What magnetic-field magnitude is required for this transition to be induced by photons with frequency 22.7 MHz?

Section 43.2 Nuclear Binding and Nuclear Structure

43.4 •• The nuclei $^{11}_5$B and $^{11}_6$C are called *mirror nuclei,* because the number of protons in $^{11}_5$B equals the number of neutrons in $^{11}_6$C and the number of neutrons in $^{11}_5$B equals the number of protons in $^{11}_6$C. The atomic mass of $^{11}_6$C is 11.011434 u, and the atomic mass of $^{11}_5$B is given in Table 43.2. (a) Calculate the binding energies of $^{11}_5$B and $^{11}_6$C. (b) Which of the two nuclei has the larger binding energy? Why is this so?

43.5 • The most common isotope of boron is $^{11}_5$B. (a) Determine the total binding energy of $^{11}_5$B from Table 43.2 in Section 43.1. (b) Calculate this binding energy from Eq. (43.11). (Why is the fifth term zero?) Compare to the result you obtained in part (a). What is the percent difference? Compare the accuracy of Eq. (43.11) for $^{11}_5$B to its accuracy for $^{62}_{28}$Ni (see Example 43.4).

43.6 • The most common isotope of uranium, $^{238}_{92}$U, has atomic mass 238.050788 u. Calculate (a) the mass defect; (b) the binding energy (in MeV); (c) the binding energy per nucleon.

43.7 • Calculate (a) the total binding energy and (b) the binding energy per nucleon of ^{12}C. (c) What percent of the rest mass of this nucleus is its total binding energy?

43.8 •• An alpha particle is strongly bound. The $^{12}_6$C nucleus might be modeled as a composite of three alpha particles. Compare the binding energy of $^{12}_6$C with three times the binding energy of an alpha particle. Which of these quantities is larger, and why might this be so?

43.9 • CP A photon with a wavelength of 3.50×10^{-13} m strikes a deuteron, splitting it into a proton and a neutron. (a) Calculate the kinetic energy released in this interaction. (b) Assuming the two particles share the energy equally, and taking their masses to be 1.00 u, calculate their speeds after the photodisintegration.

43.10 • Calculate the mass defect, the binding energy (in MeV), and the binding energy per nucleon of (a) the nitrogen nucleus, $^{14}_7$N, and (b) the helium nucleus, $^4_2$He. (c) How does the binding energy per nucleon compare for these two nuclei?

43.11 • Use Eq. (43.11) to calculate the binding energy per nucleon for the nuclei $^{86}_{36}$Kr and $^{180}_{73}$Ta. Do your results confirm what is shown in Fig. 43.2—that for *A* greater than 62 the binding energy per nucleon decreases as *A* increases?

Section 43.3 Nuclear Stability and Radioactivity

43.12 • (a) Is the decay $n \rightarrow p + \beta^- + \bar{v}_e$ energetically possible? If not, explain why not. If so, calculate the total energy released. (b) Is the decay $p \rightarrow n + \beta^+ + v_e$ energetically possible? If not, explain why not. If so, calculate the total energy released.

43.13 • What nuclide is produced in the following radioactive decays? (a) α decay of $^{239}_{94}$Pu; (b) β^- decay of $^{24}_{11}$Na; (c) β^+ decay of $^{15}_8$O.

43.14 •• CP ^{238}U decays spontaneously by α emission to ^{234}Th. Calculate (a) the total energy released by this process and (b) the recoil velocity of the ^{234}Th nucleus. The atomic masses are 238.050788 u for ^{238}U and 234.043601 u for ^{234}Th.

43.15 •• The atomic mass of ^{14}C is 14.003242 u. Show that the β^- decay of ^{14}C is energetically possible, and calculate the energy released in the decay.

43.16 • What particle (α particle, electron, or positron) is emitted in the following radioactive decays? (a) $^{27}_{14}$Si $\rightarrow$ $^{27}_{13}$Al; (b) $^{238}_{92}$U $\rightarrow$ $^{234}_{90}$Th; (c) $^{74}_{33}$As $\rightarrow$ $^{74}_{34}$Se.

43.17 •• (a) Calculate the energy released by the electron-capture decay of $^{57}_{27}$Co (see Example 43.7). (b) A negligible amount of this energy goes to the resulting $^{57}_{26}$Fe atom as kinetic energy. About 90% of the time, the $^{57}_{26}$Fe nucleus emits two successive gamma-ray photons after the electron-capture process, of energies 0.122 MeV and 0.014 MeV, respectively, in decaying to its ground state. What is the energy of the neutrino emitted in this case?

43.18 • Tritium ($^3_1$H) is an unstable isotope of hydrogen; its mass, including one electron, is 3.016049 u. (a) Show that tritium must be unstable with respect to beta decay because the decay products ($^3_2$He plus an emitted electron) have less total mass than the tritium. (b) Determine the total kinetic energy (in MeV) of the decay products, taking care to account for the electron masses correctly.

Section 43.4 Activities and Half-Lives

43.19 • If a 6.13-g sample of an isotope having a mass number of 124 decays at a rate of 0.350 Ci, what is its half-life?

43.20 • **BIO** Radioactive isotopes used in cancer therapy have a "shelf-life," like pharmaceuticals used in chemotherapy. Just after it has been manufactured in a nuclear reactor, the activity of a sample of ^{60}Co is 5000 Ci. When its activity falls below 3500 Ci, it is considered too weak a source to use in treatment. You work in the radiology department of a large hospital. One of these ^{60}Co sources in your inventory was manufactured on October 6, 2011. It is now April 6, 2014. Is the source still usable? The half-life of ^{60}Co is 5.271 years.

43.21 •• The common isotope of uranium, ^{238}U, has a half-life of 4.47×10^9 years, decaying to ^{234}Th by alpha emission. (a) What is the decay constant? (b) What mass of uranium is required for an activity of 1.00 curie? (c) How many alpha particles are emitted per second by 10.0 g of uranium?

43.22 •• **BIO** **Radiation Treatment of Prostate Cancer.** In many cases, prostate cancer is treated by implanting 60 to 100 small seeds of radioactive material into the tumor. The energy released from the decays kills the tumor. One isotope that is used (there are others) is palladium (^{103}Pd)), with a half-life of 17 days. If a typical grain contains 0.250 g of ^{103}Pd, (a) what is its initial activity rate in Bq, and (b) what is the rate 68 days later?

43.23 •• A 12.0-g sample of carbon from living matter decays at the rate of 184 decays/minute due to the radioactive ^{14}C in it. What will be the decay rate of this sample in (a) 1000 years and (b) 50,000 years?

43.24 •• **BIO** **Radioactive Tracers.** Radioactive isotopes are often introduced into the body through the bloodstream. Their spread through the body can then be monitored by detecting the appearance of radiation in different organs. One such tracer is ^{131}I, a β^- emitter with a half-life of 8.0 d. Suppose a scientist introduces a sample with an activity of 325 Bq and watches it spread to the organs. (a) Assuming that all of the sample went to the thyroid gland, what will be the decay rate in that gland 24 d (about $3\frac{1}{2}$ weeks) later? (b) If the decay rate in the thyroid 24 d later is measured to be 17.0 Bq, what percentage of the tracer went to that gland? (c) What isotope remains after the I-131 decays?

43.25 •• The unstable isotope ^{40}K is used for dating rock samples. Its half-life is 1.28×10^9 y. (a) How many decays occur per second in a sample containing 1.63×10^{-6} g of ^{40}K? (b) What is the activity of the sample in curies?

43.26 • As a health physicist, you are being consulted about a spill in a radiochemistry lab. The isotope spilled was 400 μCi of ^{131}Ba, which has a half-life of 12 days. (a) What mass of ^{131}Ba was spilled? (b) Your recommendation is to clear the lab until the radiation level has fallen 1.00 μCi. How long will the lab have to be closed?

43.27 • Measurements on a certain isotope tell you that the decay rate decreases from 8318 decays/min to 3091 decays/min in 4.00 days. What is the half-life of this isotope?

43.28 • A radioactive isotope has a half-life of 43.0 min. At $t = 0$ its activity is 0.376 Ci. What is its activity at $t = 2.00$ h?

43.29 • The radioactive nuclide ^{199}Pt has a half-life of 30.8 minutes. A sample is prepared that has an initial activity of 7.56×10^{11} Bq. (a) How many ^{199}Pt nuclei are initially present in the sample? (b) How many are present after 30.8 minutes? What is the activity at this time? (c) Repeat part (b) for a time 92.4 minutes after the sample is first prepared.

43.30 •• **Radiocarbon Dating.** At an archeological site, a sample from timbers containing 500 g of carbon provides 2690 decays/min. What is the age of the sample?

Section 43.5 Biological Effects of Radiation

43.31 •• **BIO** (a) If a chest x ray delivers 0.25 mSv to 5.0 kg of tissue, how many *total* joules of energy does this tissue receive? (b) Natural radiation and cosmic rays deliver about 0.10 mSv per year at sea level. Assuming an RBE of 1, how many rem and rads is this dose, and how many joules of energy does a 75-kg person receive in a year? (c) How many chest x rays like the one in part (a) would it take to deliver the same *total* amount of energy to a 75-kg person as she receives from natural radiation in a year at sea level, as described in part (b)?

43.32 • **BIO** **Radiation Overdose.** If a person's entire body is exposed to 5.0 J/kg of x rays, death usually follows within a few days. (a) Express this lethal radiation dose in Gy, rad, Sv, and rem. (b) How much total energy does a 70.0-kg person absorb from such a dose? (c) If the 5.0 J/kg came from a beam of protons instead of x rays, what would be the answers to parts (a) and (b)?

43.33 •• **BIO** A nuclear chemist receives an accidental radiation dose of 5.0 Gy from slow neutrons (RBE = 4.0). What does she receive in rad, rem, and J/kg?

43.34 •• **BIO** A person exposed to fast neutrons receives a radiation dose of 300 rem on part of his hand, affecting 25 g of tissue. The RBE of these neutrons is 10. (a) How many rad did he receive? (b) How many joules of energy did he receive? (c) Suppose the person received the same rad dosage, but from beta rays with an RBE of 1.0 instead of neutrons. How many rem would he have received?

43.35 • **BIO** **Food Irradiation.** Food is often irradiated with either x rays or electron beams to help prevent spoilage. A low dose of 5–75 kilorads (krad) helps to reduce and kill inactive parasites, a medium dose of 100–400 krad kills microorganisms and pathogens such as salmonella, and a high dose of 2300–5700 krad sterilizes food so that it can be stored without refrigeration. (a) A dose of 175 krad kills spoilage microorganisms in fish. If x rays are used, what would be the dose in Gy, Sv, and rem, and how much energy would a 220-g portion of fish absorb? (See Table 43.3.) (b) Repeat part (a) if electrons of RBE 1.50 are used instead of x rays.

43.36 • **BIO** **To Scan or Not to Scan?** It has become popular for some people to have yearly whole-body scans (CT scans, formerly called CAT scans) using x rays, just to see if they detect anything suspicious. A number of medical people have recently questioned the advisability of such scans, due in part to the radiation they impart. Typically, one such scan gives a dose of 12 mSv, applied to the *whole body*. By contrast, a chest x ray typically administers 0.20 mSv to only 5.0 kg of tissue. How many chest x rays would deliver the same *total* amount of energy to the body of a 75-kg person as one whole-body scan?

43.37 •• **BIO** A 67-kg person accidentally ingests 0.35 Ci of tritium. (a) Assume that the tritium spreads uniformly throughout the body and that each decay leads on the average to the absorption of 5.0 keV of energy from the electrons emitted in the decay. The half-life of tritium is 12.3 y, and the RBE of the electrons is 1.0. Calculate the absorbed dose in rad and the equivalent dose in rem during one week. (b) The β^- decay of tritium releases more than 5.0 keV of energy. Why is the average energy absorbed less than the total energy released in the decay?

43.38 • **BIO** In an industrial accident a 65-kg person receives a lethal whole-body equivalent dose of 5.4 Sv from x rays. (a) What is the equivalent dose in rem? (b) What is the absorbed dose in rad? (c) What is the total energy absorbed by the person's body? How does this amount of energy compare to the amount of energy required to raise the temperature of 65 kg of water 0.010 C°?

43.39 •• CP BIO In a diagnostic x-ray procedure, 5.00×10^{10} photons are absorbed by tissue with a mass of 0.600 kg. The x-ray wavelength is 0.0200 nm. (a) What is the total energy absorbed by the tissue? (b) What is the equivalent dose in rem?

Section 43.6 Nuclear Reactions
Section 43.7 Nuclear Fission
Section 43.8 Nuclear Fusion

43.40 •• Calculate the reaction energy Q for the reaction $p + {}_{1}^{3}H \rightarrow {}_{1}^{2}H + {}_{1}^{2}H$. Is this reaction exoergic or endoergic?

43.41 • Consider the nuclear reaction

$$ {}_{1}^{2}H + {}_{4}^{9}Be \rightarrow X + {}_{2}^{4}He $$

where X is a nuclide. (a) What are the values of Z and A for the nuclide X? (b) How much energy is liberated? (c) Estimate the threshold energy for this reaction.

43.42 • **Energy from Nuclear Fusion.** Calculate the energy released in the fusion reaction

$$ {}_{2}^{3}He + {}_{1}^{2}H \rightarrow {}_{2}^{4}He + {}_{1}^{1}H $$

43.43 •• At the beginning of Section 43.7 the equation of a fission process is given in which ^{235}U is struck by a neutron and undergoes fission to produce ^{144}Ba, ^{89}Kr, and three neutrons. The measured masses of these isotopes are 235.043930 u (^{235}U), 143.922953 u (^{144}Ba), 88.917631 u (^{89}Kr), and 1.0086649 u (neutron). (a) Calculate the energy (in MeV) released by each fission reaction. (b) Calculate the energy released per gram of ^{235}U, in MeV/g.

43.44 • The United States uses about 1.4×10^{19} J of electrical energy per year. If all this energy came from the fission of ^{235}U, which releases 200 MeV per fission event, (a) how many kilograms of ^{235}U would be used per year, and (b) how many kilograms of uranium would have to be mined per year to provide that much ^{235}U? (Recall that only 0.70% of naturally occurring uranium is ^{235}U.)

43.45 • Consider the nuclear reaction

$$ {}_{2}^{4}He + {}_{3}^{7}Li \rightarrow X + {}_{0}^{1}n $$

where X is a nuclide. (a) What are Z and A for the nuclide X? (b) Is energy absorbed or liberated? How much?

43.46 •• Consider the nuclear reaction

$$ {}_{14}^{28}Si + \gamma \rightarrow {}_{12}^{24}Mg + X $$

where X is a nuclide. (a) What are Z and A for the nuclide X? (b) Ignoring the effects of recoil, what minimum energy must the photon have for this reaction to occur? The mass of a ${}_{14}^{28}Si$ atom is 27.976927 u, and the mass of a ${}_{12}^{24}Mg$ atom is 23.985042 u.

PROBLEMS

43.47 • **Comparison of Energy Released per Gram of Fuel.** (a) When gasoline is burned, it releases 1.3×10^{8} J of energy per gallon (3.788 L). Given that the density of gasoline is 737 kg/m³, express the quantity of energy released in J/g of fuel. (b) During fission, when a neutron is absorbed by a ^{235}U nucleus, about 200 MeV of energy is released for each nucleus that undergoes fission. Express this quantity in J/g of fuel. (c) In the proton-proton chain that takes place in stars like our sun, the overall fusion reaction can be summarized as six protons fusing to form one ^{4}He nucleus with two leftover protons and the liberation of 26.7 MeV of

energy. The fuel is the six protons. Express the energy produced here in units of J/g of fuel. Notice the huge difference between the two forms of nuclear energy, on the one hand, and the chemical energy from gasoline, on the other. (d) Our sun produces energy at a measured rate of 3.86×10^{26} W. If its mass of 1.99×10^{30} kg were all gasoline, how long could it last before consuming all its fuel? (*Historical note:* Before the discovery of nuclear fusion and the vast amounts of energy it releases, scientists were confused. They knew that the earth was at least many millions of years old, but could not explain how the sun could survive that long if its energy came from chemical burning.)

43.48 •• (a) Calculate the minimum energy required to remove one proton from the nucleus $^{12}_{6}C$. This is called the proton-removal energy. (*Hint:* Find the difference between the mass of a $^{12}_{6}C$ nucleus and the mass of a proton plus the mass of the nucleus formed when a proton is removed from $^{12}_{6}C$.) (b) How does the proton-removal energy for $^{12}_{6}C$ compare to the binding energy per nucleon for $^{12}_{6}C$, calculated using Eq. (43.10)?

43.49 •• (a) Calculate the minimum energy required to remove one neutron from the nucleus $^{17}_{8}O$. This is called the neutron-removal energy. (See Problem 43.48.) (b) How does the neutron-removal energy for $^{17}_{8}O$ compare to the binding energy per nucleon for $^{17}_{8}O$, calculated using Eq. (43.10)?

43.50 •• The isotope $^{110}_{47}Ag$ is created by irradiation of a sample of $^{109}_{47}Ag$ nuclei with neutrons. The $^{109}_{47}Ag$ nucleus is stable and the number of $^{109}_{47}Ag$ nuclei in the sample is large, so we take the number of $^{109}_{47}Ag$ nuclei to be constant. Therefore, with a constant flux of neutrons, the $^{110}_{47}Ag$ nuclei are produced at a constant rate of 8.40×10^{3} nuclei per second. The $^{110}_{47}Ag$ isotope decays by β^{-} emission to the stable nucleus $^{110}_{48}Cd$, with a half-life of 24.6 s. Initially, at $t = 0$, only $^{109}_{47}Ag$ nuclei are present. (a) After steady state has been reached and the number of $^{110}_{47}Ag$ nuclei in the sample is constant, how many $^{110}_{47}Ag$ nuclei are there? (b) How many $^{110}_{47}Ag$ are there in the sample at $t = 24.6$ s? (*Hint:* Refer to the Bridging Problem for this chapter.)

43.51 • BIO **Radioactive Fallout.** One of the problems of in-air testing of nuclear weapons (or, even worse, the *use* of such weapons!) is the danger of radioactive fallout. One of the most problematic nuclides in such fallout is strontium-90 (^{90}Sr), which breaks down by β^{-} decay with a half-life of 28 years. It is chemically similar to calcium and therefore can be incorporated into bones and teeth, where, due to its rather long half-life, it remains for years as an internal source of radiation. (a) What is the daughter nucleus of the ^{90}Sr decay? (b) What percentage of the original level of ^{90}Sr is left after 56 years? (c) How long would you have to wait for the original level to be reduced to 6.25% of its original value?

43.52 •• CP Thorium $^{230}_{90}Th$ decays to radium $^{226}_{88}Ra$ by α emission. The masses of the neutral atoms are 230.033134 u for $^{230}_{90}Th$ and 226.025410 u for $^{226}_{88}Ra$. If the parent thorium nucleus is at rest, what is the kinetic energy of the emitted α particle? (Be sure to account for the recoil of the daughter nucleus.)

43.53 •• The atomic mass of $^{25}_{12}Mg$ is 24.985837 u, and the atomic mass of $^{25}_{13}Al$ is 24.990428 u. (a) Which of these nuclei will decay into the other? (b) What type of decay will occur? Explain how you determined this. (c) How much energy (in MeV) is released in the decay?

43.54 •• The polonium isotope $^{210}_{84}Po$ has atomic mass 209.982874 u. Other atomic masses are $^{206}_{82}Pb$, 205.974465 u; $^{209}_{83}Bi$, 208.980399 u; $^{210}_{83}Bi$, 209.984120 u; $^{209}_{84}Po$, 208.982430 u; and $^{210}_{85}At$, 209.987148 u. (a) Show that the alpha decay of $^{210}_{84}Po$ is energetically possible, and find the energy of the emitted

α particle. (b) Is $^{210}_{84}$Po energetically stable with respect to emission of a proton? Why or why not? (c) Is $^{210}_{84}$Po energetically stable with respect to emission of a neutron? Why or why not? (d) Is $^{210}_{84}$Po energetically stable with respect to β^- decay? Why or why not? (e) Is $^{210}_{84}$Po energetically stable with respect to β^+ decay? Why or why not?

43.55 •• **BIO** **Irradiating Ourselves!** The radiocarbon in our bodies is one of the naturally occurring sources of radiation. Let's see how large a dose we receive. ^{14}C decays via β^- emission, and 18% of our body's mass is carbon. (a) Write out the decay scheme of carbon-14 and show the end product. (A neutrino is also produced.) (b) Neglecting the effects of the neutrino, how much kinetic energy (in MeV) is released per decay? The atomic mass of ^{14}C is 14.003242 u. (c) How many grams of carbon are there in a 75-kg person? How many decays per second does this carbon produce? (*Hint:* Use data from Example 43.9.) (d) Assuming that all the energy released in these decays is absorbed by the body, how many MeV/s and J/s does the ^{14}C release in this person's body? (e) Consult Table 43.3 and use the largest appropriate RBE for the particles involved. What radiation dose does the person give himself in a year, in Gy, rad, Sv, and rem?

43.56 •• **BIO** **Pion Radiation Therapy.** A neutral pion (π^0) has a mass of 264 times the electron mass and decays with a lifetime of 8.4×10^{-17} s to two photons. Such pions are used in the radiation treatment of some cancers. (a) Find the energy and wavelength of these photons. In which part of the electromagnetic spectrum do they lie? What is the RBE for these photons? (b) If you want to deliver a dose of 200 rem (which is typical) in a single treatment to 25 g of tumor tissue, how many π^0 mesons are needed?

43.57 •• Calculate the mass defect for the β^+ decay of $^{11}_6$C. Is this decay energetically possible? Why or why not? The atomic mass of $^{11}_6$C is 11.011434 u.

43.58 •• **BIO** A person ingests an amount of a radioactive source that has a very long lifetime and activity 0.52 μCi. The radioactive material lodges in her lungs, where all of the emitted 4.0-MeV α particles are absorbed within a 0.50-kg mass of tissue. Calculate the absorbed dose and the equivalent dose for one year.

43.59 • **We Are Stardust.** In 1952 spectral lines of the element technetium-99 (^{99}Tc) were discovered in a red giant star. Red giants are very old stars, often around 10 billion years old, and near the end of their lives. Technetium has *no* stable isotopes, and the half-life of ^{99}Tc is 200,000 years. (a) For how many half-lives has the ^{99}Tc been in the red giant star if its age is 10 billion years? (b) What fraction of the original ^{99}Tc would be left at the end of that time? This discovery was extremely important because it provided convincing evidence for the theory (now essentially known to be true) that most of the atoms heavier than hydrogen and helium were made inside stars by thermonuclear fusion and other nuclear processes. If the ^{99}Tc had been part of the star since it was born, the amount remaining after 10 billion years would have been so minute that it would not have been detectable. This knowledge is what led the late astronomer Carl Sagan to proclaim that "we are stardust."

43.60 • **BIO** A 70.0-kg person experiences a whole-body exposure to α radiation with energy 4.77 MeV. A total of 7.75×10^{12} α particles are absorbed. (a) What is the absorbed dose in rad? (b) What is the equivalent dose in rem? (c) If the source is 0.0320 g of ^{226}Ra (half-life 1600 y) somewhere in the body, what is the activity of this source? (d) If all of the alpha particles produced are absorbed, what time is required for this dose to be delivered?

43.61 •• Measurements indicate that 27.83% of all rubidium atoms currently on the earth are the radioactive ^{87}Rb isotope. The rest are the stable ^{85}Rb isotope. The half-life of ^{87}Rb is 4.75×10^{10} y. Assuming that no rubidium atoms have been formed since, what percentage of rubidium atoms were ^{87}Rb when our solar system was formed 4.6×10^9 y ago?

43.62 • The nucleus $^{15}_8$O has a half-life of 122.2 s; $^{19}_8$O has a half-life of 26.9 s. If at some time a sample contains equal amounts of $^{15}_8$O and $^{19}_8$O, what is the ratio of $^{15}_8$O to $^{19}_8$O (a) after 3.0 min and (b) after 12.0 min?

43.63 •• **BIO** A ^{60}Co source with activity 2.6×10^{-4} Ci is embedded in a tumor that has mass 0.200 kg. The source emits γ photons with average energy 1.25 MeV. Half the photons are absorbed in the tumor, and half escape. (a) What energy is delivered to the tumor per second? (b) What absorbed dose (in rad) is delivered per second? (c) What equivalent dose (in rem) is delivered per second if the RBE for these γ rays is 0.70? (d) What exposure time is required for an equivalent dose of 200 rem?

43.64 •• **An Oceanographic Tracer.** Nuclear weapons tests in the 1950s and 1960s released significant amounts of radioactive tritium ($^3_1$H, half-life 12.3 years) into the atmosphere. The tritium atoms were quickly bound into water molecules and rained out of the air, most of them ending up in the ocean. For any of this tritium-tagged water that sinks below the surface, the amount of time during which it has been isolated from the surface can be calculated by measuring the ratio of the decay product, $^3_2$He, to the remaining tritium in the water. For example, if the ratio of $^3_2$He to $^3_1$H in a sample of water is 1:1, the water has been below the surface for one half-life, or approximately 12 years. This method has provided oceanographers with a convenient way to trace the movements of subsurface currents in parts of the ocean. Suppose that in a particular sample of water, the ratio of $^3_2$He to $^3_1$H is 4.3 to 1.0. How many years ago did this water sink below the surface?

43.65 • A bone fragment found in a cave believed to have been inhabited by early humans contains 0.29 times as much ^{14}C as an equal amount of carbon in the atmosphere when the organism containing the bone died. (See Example 43.9 in Section 43.4.) Find the approximate age of the fragment.

43.66 •• **BIO** In the 1986 disaster at the Chernobyl reactor in eastern Europe, about $\frac{1}{8}$ of the ^{137}Cs present in the reactor was released. The isotope ^{137}Cs has a half-life of 30.07 y for β decay, with the emission of a total of 1.17 MeV of energy per decay. Of this, 0.51 MeV goes to the emitted electron; the remaining 0.66 MeV goes to a γ ray. The radioactive ^{137}Cs is absorbed by plants, which are eaten by livestock and humans. How many ^{137}Cs atoms would need to be present in each kilogram of body tissue if an equivalent dose for one week is 3.5 Sv? Assume that all of the energy from the decay is deposited in 1.0 kg of tissue and that the RBE of the electrons is 1.5.

43.67 •• Consider the fusion reaction $^2_1\text{H} + ^2_1\text{H} \rightarrow ^3_2\text{He} + ^1_0\text{n}$. (a) Estimate the barrier energy by calculating the repulsive electrostatic potential energy of the two $^2_1$H nuclei when they touch. (b) Compute the energy liberated in this reaction in MeV and in joules. (c) Compute the energy liberated *per mole* of deuterium, remembering that the gas is diatomic, and compare with the heat of combustion of hydrogen, about 2.9×10^5 J/mol.

43.68 •• **DATA** As a scientist in a nuclear physics research lab, you are conducting a photodisintegration experiment to verify the binding energy of a deuteron. A photon with wavelength λ in air is absorbed by a deuteron, which breaks apart into a neutron and a proton. The two fragments share the released kinetic energy equally, and the deuteron can be assumed to be initially at rest. You measure the speed of the proton after the disintegration as a function of the wavelength λ of the photon. Your experimental results are given in the table.

λ (10^{-13} m)	3.50	3.75	4.00	4.25	4.50	4.75	5.00
v (10^6 m/s)	11.3	10.2	9.1	8.1	7.2	6.1	4.9

(a) Graph the data as v^2 versus $1/\lambda$. Explain why the data points, when graphed this way, should follow close to a straight line. Find the slope and y-intercept of the straight line that gives the best fit to the data. (b) Assume that h and c have their accepted values. Use your results from part (a) to calculate the mass of the proton and the binding energy (in MeV) of the deuteron.

43.69 •• DATA Your company develops radioactive isotopes for medical applications. In your work there, you measure the activity of a radioactive sample. Your results are given in the table.

Time (h)	Decays/s
0	20,000
0.5	14,800
1.0	11,000
1.5	8130
2.0	6020
2.5	4460
3.0	3300
4.0	1810
5.0	1000
6.0	550
7.0	300

(a) Find the half-life of the sample. (b) How many radioactive nuclei were present in the sample at $t = 0$? (c) How many were present after 7.0 h?

43.70 ••• DATA In your job as a health physicist, you measure the activity of a mixed sample of radioactive elements. Your results are given in the table.

Time (h)	Decays/s
0	7500
0.5	4120
1.0	2570
1.5	1790
2.0	1350
2.5	1070
3.0	872
4.0	596
5.0	404
6.0	288
7.0	201
8.0	140
9.0	98
10.0	68
12.0	33

(a) What minimum number of different nuclides are present in the mixture? (b) What are their half-lives? (c) How many nuclei of each type are initially present in the sample? (d) How many of each type are present at $t = 5.0$ h?

CHALLENGE PROBLEMS

43.71 ••• **Industrial Radioactivity.** Radioisotopes are used in a variety of manufacturing and testing techniques. Wear measurements can be made using the following method. An automobile engine is produced using piston rings with a total mass of 100 g, which includes 9.4 μCi of ^{59}Fe whose half-life is 45 days. The engine is test-run for 1000 hours, after which the oil is drained and its activity is measured. If the activity of the engine oil is 84 decays/s, how much mass was worn from the piston rings per hour of operation?

43.72 ••• Many radioactive decays occur within a sequence of decays—for example, $^{234}_{92}\text{U} \rightarrow {}^{230}_{88}\text{Th} \rightarrow {}^{226}_{84}\text{Ra}$. The half-life for the $^{234}_{92}\text{U} \rightarrow {}^{230}_{88}\text{Th}$ decay is 2.46×10^5 y, and the half-life for the $^{230}_{88}\text{Th} \rightarrow {}^{226}_{84}\text{Ra}$ decay is 7.54×10^4 y. Let 1 refer to $^{234}_{92}\text{U}$, 2 to $^{230}_{88}\text{Th}$, and 3 to $^{226}_{84}\text{Ra}$; let λ_1 be the decay constant for the $^{234}_{92}\text{U} \rightarrow {}^{230}_{88}\text{Th}$ decay and λ_2 be the decay constant for the $^{230}_{88}\text{Th} \rightarrow {}^{226}_{84}\text{Ra}$ decay. The amount of $^{230}_{88}\text{Th}$ present at any time depends on the rate at which it is produced by the decay of $^{234}_{92}\text{U}$ and the rate by which it is depleted by its decay to $^{226}_{84}\text{Ra}$. Therefore, $dN_2(t)/dt = \lambda_1 N_1(t) - \lambda_2 N_2(t)$. If we start with a sample that contains N_{10} nuclei of $^{234}_{92}\text{U}$ and nothing else, then $N(t) = N_{01}e^{-\lambda_1 t}$. Thus $dN_2(t)/dt = \lambda_1 N_{10}e^{-\lambda_1 t} - \lambda_2 N_2(t)$. This differential equation for $N_2(t)$ can be solved as follows. Assume a trial solution of the form $N_2(t) = N_{10}[h_1 e^{-\lambda_1 t} + h_2 e^{-\lambda_2 t}]$, where h_1 and h_2 are constants. (a) Since $N_2(0) = 0$, what must be the relationship between h_1 and h_2? (b) Use the trial solution to calculate $dN_2(t)/dt$, and substitute that into the differential equation for $N_2(t)$. Collect the coefficients of $e^{-\lambda_1 t}$ and $e^{-\lambda_2 t}$. Since the equation must hold at all t, each of these coefficients must be zero. Use this requirement to solve for h_1 and thereby complete the determination of $N_2(t)$. (c) At time $t = 0$, you have a pure sample containing 30.0 g of $^{234}_{92}\text{U}$ and nothing else. What mass of $^{230}_{88}\text{Th}$ is present at time $t = 2.46 \times 10^5$ y, the half-life for the $^{234}_{92}\text{U}$ decay?

PASSAGE PROBLEMS

BIO **RADIOACTIVE IODINE IN MEDICINE.** Iodine in the body is preferentially taken up by the thyroid gland. Therefore, radioactive iodine in small doses is used to image the thyroid and in large doses is used to kill thyroid cells to treat some types of cancer or thyroid disease. The iodine isotopes used have relatively short half-lives, so they must be produced in a nuclear reactor or accelerator. One isotope frequently used for imaging is ^{123}I; it has a half-life of 13.2 h and emits a 0.16-MeV gamma-ray photon. One method of producing ^{123}I is in the nuclear reaction $^{123}\text{Te} + \text{p} \rightarrow {}^{123}\text{I} + \text{n}$. The atomic masses relevant to this reaction are ^{123}Te, 122.904270 u; ^{123}I, 122.905589 u; n, 1.008665 u; and ^{1}H, 1.007825 u.

The iodine isotope commonly used for treatment of disease is ^{131}I, which is produced by irradiating ^{130}Te in a nuclear reactor to form ^{131}Te. The ^{131}Te then decays to ^{131}I. ^{131}I undergoes β^- decay with a half-life of 8.04 d, emitting electrons with energies up to 0.61 MeV and gamma-ray photons of energy 0.36 MeV. A typical thyroid cancer treatment might involve administration of 3.7 GBq of ^{131}I.

43.73 Which reaction produces ^{131}Te in the nuclear reactor? (a) $^{130}\text{Te} + \text{n} \rightarrow {}^{131}\text{Te}$; (b) $^{130}\text{I} + \text{n} \rightarrow {}^{131}\text{Te}$; (c) $^{132}\text{Te} + \text{n} \rightarrow {}^{131}\text{Te}$; (d) $^{132}\text{I} + \text{n} \rightarrow {}^{131}\text{Te}$.

43.74 Which type of radioactive decay produces ^{131}I from ^{131}Te? (a) Alpha decay; (b) β^- decay; (c) β^+ decay; (d) gamma decay.

43.75 How many ^{131}I atoms are administered in a typical thyroid cancer treatment? (a) 4.2×10^{10}; (b) 1.0×10^{12}; (c) 2.5×10^{14}; (d) 3.7×10^{15}.

43.76 In the reaction that produces ^{123}I, is there a minimum kinetic energy the protons need to make the reaction go? (a) No, because the proton has a smaller mass than the neutron. (b) No, because the total initial mass is smaller than the total final mass.

(c) Yes, because the proton has a smaller mass than the neutron.
(d) Yes, because the total initial mass is smaller than the total final mass.

43.77 Why might ^{123}I be preferred for imaging over ^{131}I? (a) The atomic mass of ^{123}I is smaller, so the ^{123}I particles travel farther through tissue. (b) Because ^{123}I emits only gamma-ray photons, the radiation dose to the body is lower with that isotope. (c) The beta particles emitted by ^{131}I can leave the body, whereas the gamma-ray photons emitted by ^{123}I cannot. (d) ^{123}I is radioactive, whereas ^{131}I is not.

Answers

Chapter Opening Question ?

(iv) When an organism dies, it stops taking in carbon from atmospheric CO_2. Some of this carbon is radioactive ^{14}C, which decays with a half-life of 5730 years. By measuring the proportion of ^{14}C that remains in the specimen, scientists can determine how long ago the organism died. (See Section 43.4.)

Test Your Understanding Questions

43.1 (a) (iii), (b) (v) The radius R is proportional to the cube root of the mass number A, while the volume is proportional to R^3 and hence to $(A^{1/3})^3 = A$. Therefore, doubling the volume requires increasing the mass number by a factor of 2; doubling the radius implies increasing both the volume and the mass number by a factor of $2^3 = 8$.

43.2 (ii), (iii), (iv), (v), (i) You can find the answers by inspecting Fig. 43.2. The binding energy per nucleon is lowest for very light nuclei such as $^4_2$He, is greatest around $A = 60$, and then decreases with increasing A.

43.3 (v) Two protons and two neutrons are lost in an α decay, so Z and N each decrease by 2. A β^+ decay changes a proton to a neutron, so Z decreases by 1 and N increases by 1. The net result is that Z decreases by 3 and N decreases by 1.

43.4 (iii) The activity $-dN(t)/dt$ of a sample is the product of the number of nuclei in the sample $N(t)$ and the decay constant $\lambda = (\ln 2)/T_{1/2}$. Hence $N(t) = [-dN(t)/dt]T_{1/2}/(\ln 2)$. Taking the ratio of this expression for ^{240}Pu to this same expression for ^{243}Am, the factors of $\ln 2$ cancel and we get

$$\frac{N_{\text{Pu}}}{N_{\text{Am}}} = \frac{(-dN_{\text{Pu}}/dt)T_{1/2-\text{Pu}}}{(-dN_{\text{Am}}/dt)T_{1/2-\text{Am}}} = \frac{(5.00 \ \mu\text{Ci})(6560 \ \text{y})}{(4.45 \ \mu\text{Ci})(7370 \ \text{y})} = 1.00$$

The two samples contain *equal* numbers of nuclei. The ^{243}Am sample has a longer half-life and hence a slower decay rate, so it has a lower activity than the ^{240}Pu sample.

43.5 (ii) We saw in Section 43.3 that alpha particles can travel only a very short distance before they are stopped. By contrast,

x-ray photons are very penetrating, so they can easily pass into the body.

43.6 no The reaction $^1_1\text{H} + ^7_3\text{Li} \rightarrow ^4_2\text{He} + ^4_2\text{He}$ is a *nuclear* reaction that can take place only if a proton (a hydrogen nucleus) comes into contact with a lithium nucleus. If the hydrogen is in atomic form, the interaction between its electron cloud and the electron cloud of a lithium atom keeps the two nuclei from getting close to each other. Even if isolated protons are used, they must be fired at the lithium atoms with enough kinetic energy to overcome the electrical repulsion between the protons and the lithium nuclei. The statement that the reaction is exoergic means that more energy is released by the reaction than had to be put in to make the reaction occur.

43.7 no Because the neutron has no electric charge, it experiences no electrical repulsion from a ^{235}U nucleus. Hence a slow-moving neutron can approach and enter a ^{235}U nucleus, thereby providing the excitation needed to trigger fission. By contrast, a slow-moving *proton* (charge $+e$) feels a strong electrical repulsion from a ^{235}U nucleus (charge $+92e$). It never gets close to the nucleus, so it cannot trigger fission.

43.8 no Fusion reactions between sufficiently light nuclei are exoergic because the binding energy per nucleon E_{B}/A increases. If the nuclei are too massive, however, E_{B}/A decreases and fusion is *endoergic* (i.e., it takes in energy rather than releasing it). As an example, imagine fusing together two nuclei of $A = 100$ to make a single nucleus with $A = 200$. From Fig. 43.2, E_{B}/A is more than 8.5 MeV for the $A = 100$ nuclei but is less than 8 MeV for the $A = 200$ nucleus. Such a fusion reaction is possible, but requires a substantial input of energy.

Bridging Problem

(a) ^{128}Xe

(b) no; β^+ emission would be endoergic

(c) 3.25×10^9 atoms, 1.50×10^6 Bq

(d) $N(t) = (3.25 \times 10^9 \ \text{atoms})(1 - e^{-(4.62 \times 10^{-4} \ \text{s}^{-1})t})$

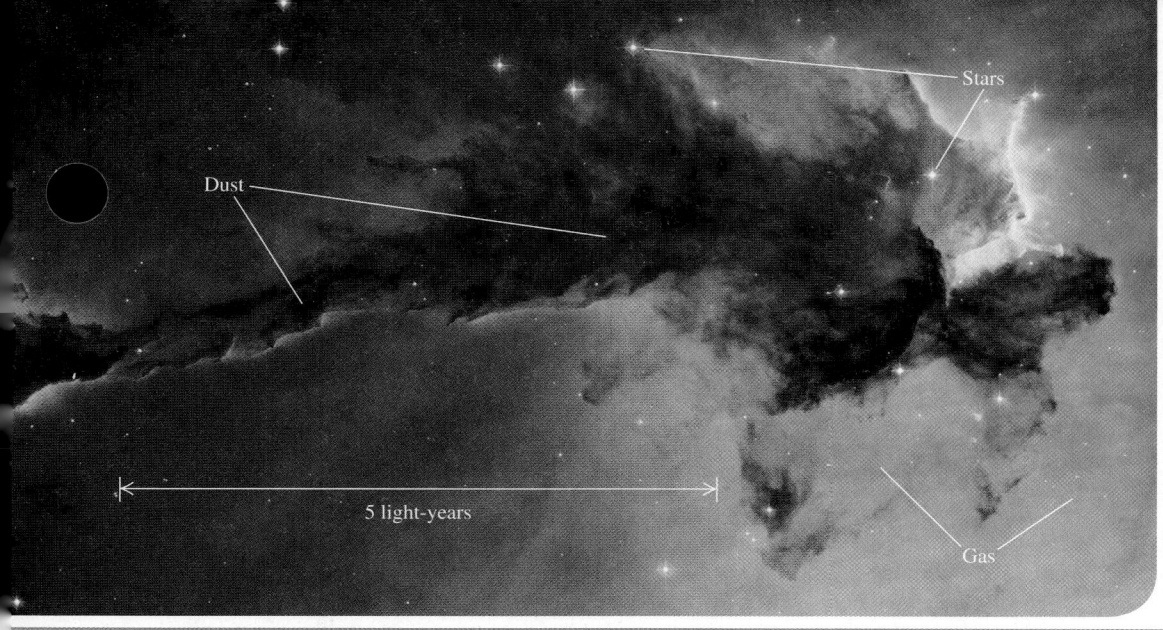

? This image shows a portion of the Eagle Nebula, a region some 6500 light-years away where new stars are forming. The luminous stars, glowing gas, and opaque dust clouds are all made of "normal" matter—that is, atoms and their constituents. What percentage of the mass and energy in the universe is composed of "normal" matter? (i) 75% to 100%; (ii) 50% to 75%; (iii) 25% to 50%; (iv) 5% to 25%; (v) less than 5%.

Dust

Stars

5 light-years

Gas

44 PARTICLE PHYSICS AND COSMOLOGY

What are the most fundamental constituents of matter? How did the universe begin? And what is the fate of our universe? In this chapter we will explore what physicists and astronomers have learned in their quest to answer these questions.

The chapter title, "Particle Physics and Cosmology," may seem strange. Fundamental particles are the *smallest* things in the universe, and cosmology deals with the *biggest* thing there is—the universe itself. Nonetheless, we'll see in this chapter that physics on the most microscopic scale plays an essential role in determining the nature of the universe on the largest scale.

The development of high-energy accelerators and associated detectors has been crucial in our emerging understanding of particles. We can classify particles and their interactions in several ways in terms of conservation laws and symmetries, some of which are absolute and others of which are obeyed only in certain kinds of interactions. We'll conclude by discussing our present understanding of the nature and evolution of the universe as a whole.

44.1 FUNDAMENTAL PARTICLES—A HISTORY

The idea that the world is made of fundamental particles has a long history. In about 400 B.C. the Greek philosophers Democritus and Leucippus suggested that matter is made of indivisible particles that they called *atoms,* a word derived from *a-* (not) and *tomos* (cut or divided). This idea lay dormant until about 1804, when the English scientist John Dalton (1766–1844), often called the father of modern chemistry, discovered that many chemical phenomena could be explained if atoms of each element are the basic, indivisible building blocks of matter.

The Electron and the Proton

Toward the end of the 19th century it became clear that atoms are *not* indivisible. The characteristic spectra of elements suggested that atoms have internal structure, and J. J. Thomson's discovery of the negatively charged *electron* in 1897 showed that atoms could be taken apart into charged particles. Rutherford's experiments in 1910–11 (see Section 39.2) revealed that an atom's positive charge resides in a small, dense nucleus. In 1919 Rutherford made an additional discovery: When alpha particles are fired into nitrogen, one product is hydrogen gas. He reasoned that the hydrogen nucleus is a constituent of the nuclei of heavier atoms such as nitrogen, and that a collision with a fast-moving alpha particle can dislodge one of those hydrogen nuclei. Thus the hydrogen nucleus is an elementary particle that Rutherford named the *proton*. The following decade saw the blossoming of quantum mechanics, including the Schrödinger equation. Physicists were on their way to understanding the principles that underlie atomic structure.

The Photon

Einstein explained the photoelectric effect in 1905 by assuming that the energy of electromagnetic waves is quantized; that is, it comes in little bundles called *photons* with energy $E = hf$. Atoms and nuclei can emit (create) and absorb (destroy) photons (see Section 38.1). Considered as particles, photons have zero charge and zero rest mass. (Note that any discussions of a particle's mass in this chapter will refer to its rest mass.) In particle physics, a photon is denoted by the symbol γ (the Greek letter gamma).

The Neutron

In 1930 the German physicists Walther Bothe and Herbert Becker observed that when beryllium, boron, or lithium was bombarded by alpha particles, the target material emitted a radiation that had much greater penetrating power than the original alpha particles. Experiments by the English physicist James Chadwick in 1932 showed that the emitted particles were electrically neutral, with mass approximately equal to that of the proton. Chadwick christened these particles *neutrons* (symbol n or $^1_0$n). A typical reaction of the type studied by Bothe and Becker, with a beryllium target, is

$$^4_2\text{He} + {}^9_4\text{Be} \rightarrow {}^{12}_6\text{C} + {}^1_0\text{n} \tag{44.1}$$

Elementary particles are usually detected by their electromagnetic effects—for instance, by the ionization that they cause when they pass through matter. (This is the principle of the cloud chamber, described below.) Because neutrons have no charge, they are difficult to detect directly; they interact hardly at all with electrons and produce little ionization when they pass through matter. However, neutrons can be slowed down by scattering from nuclei, and they can penetrate a nucleus. Hence slow neutrons can be detected by means of a nuclear reaction in which a neutron is absorbed and an alpha particle is emitted. An example is

$$^1_0\text{n} + {}^{10}_5\text{B} \rightarrow {}^7_3\text{Li} + {}^4_2\text{He} \tag{44.2}$$

The ejected alpha particle is easy to detect because it is charged. Later experiments showed that neutrons and protons, like electrons, are spin-$\frac{1}{2}$ particles (see Section 43.1).

The discovery of the neutron cleared up a mystery about the composition of the nucleus. Before 1930 the mass of a nucleus was thought to be due only to protons, but no one understood why the charge-to-mass ratio was not the same for all nuclides. It soon became clear that all nuclides (except $^1_1$H) contain both protons and neutrons. Hence the proton, the neutron, and the electron are the building blocks of atoms. However, that is not the end of the particle story; these are not the only particles, and particles can do more than build atoms.

The Positron

The positive electron, or positron, was discovered by the American physicist Carl D. Anderson in 1932, during an investigation of particles bombarding the earth from space. **Figure 44.1** shows a historic photograph made with a *cloud chamber,* an instrument used to visualize the tracks of charged particles. The chamber contained a supercooled vapor; a charged particle passing through the vapor causes ionization, and the ions trigger the condensation of liquid droplets from the vapor. The droplets make a visible track showing the charged particle's path.

The cloud chamber in Fig. 44.1 is in a magnetic field directed into the plane of the photograph. The particle has passed through a thin lead plate (which extends from left to right in the figure) that lies within the chamber. The track is more tightly curved above the plate than below it, showing that the speed was less above the plate than below it. Therefore the particle had to be moving upward; it could not have gained energy passing through the lead. The thickness and curvature of the track suggested that its mass and the magnitude of its charge equaled those of the electron. But the directions of the magnetic field and the velocity in the magnetic force equation $\vec{F} = q\vec{v} \times \vec{B}$ showed that the particle had *positive* charge. Anderson christened this particle the *positron.*

To theorists, the appearance of the positron was a welcome development. In 1928 the English physicist Paul Dirac had developed a relativistic generalization of the Schrödinger equation for the electron. In Section 41.5 we discussed how Dirac's ideas helped explain the spin magnetic moment of the electron.

One of the puzzling features of the Dirac equation was that for a *free* electron it predicted not only a continuum of energy states greater than its rest energy $m_e c^2$, as expected, but also a continuum of *negative* energy states *less than* $-m_e c^2$ (**Fig. 44.2a**). That posed a problem. What was to prevent an electron from emitting a photon with energy $2m_e c^2$ or greater and hopping from a positive state to a negative state? It wasn't clear what these negative-energy states meant, and there was no obvious way to get rid of them. Dirac's ingenious interpretation was that all the negative-energy states were filled with electrons, and that these electrons were for some reason unobservable. The exclusion principle (see Section 41.6) would forbid a transition to a state that was already occupied.

A vacancy in a negative-energy state would act like a positive charge, just as a hole in the valence band of a semiconductor (see Section 42.6) acts like a positive charge. Initially, Dirac tried to argue that such vacancies were protons. But after Anderson's discovery it became clear that the vacancies were observed physically as *positrons.* Furthermore, the Dirac energy-state picture provides a mechanism for the *creation* of positrons. When an electron in a negative-energy state absorbs a photon with energy greater than $2m_e c^2$, it goes to a positive state (Fig. 44.2b), in which it becomes observable. The vacancy that it leaves behind is observed as a positron; the result is the creation of an electron–positron pair. Similarly, when an electron in a positive-energy state falls into a vacancy, both the electron and the vacancy (that is, the positron) disappear, and photons are emitted (Fig. 44.2c). Thus the Dirac theory leads naturally to the conclusion that, like photons, *electrons can be created and destroyed.* While photons can be created and destroyed singly, electrons can be produced or destroyed only in electron–positron pairs or in association with other particles. (Creating or destroying an electron alone would mean creating or destroying an amount of charge $-e$, which would violate the conservation of electric charge.)

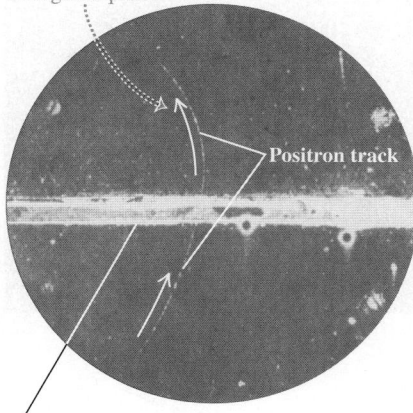

44.1 Photograph of the cloud-chamber track made by the first positron ever identified. The photograph was made by Carl D. Anderson in 1932.

The positron follows a curved path owing to the presence of a magnetic field.

The track is more strongly curved above the lead plate, showing that the positron was traveling upward and lost energy and speed as it passed through the plate.

Positron track

Lead plate (6 mm thick)

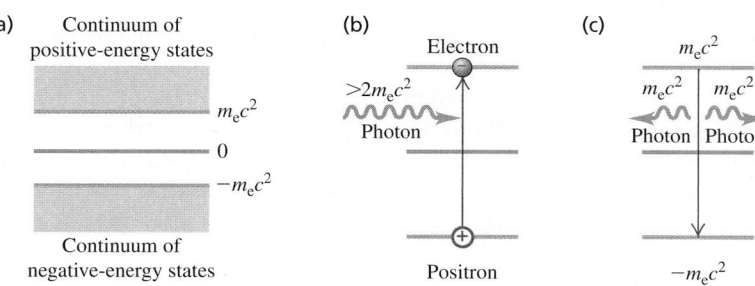

(a) Continuum of positive-energy states

$m_e c^2$

0

$-m_e c^2$

Continuum of negative-energy states

(b) Electron

$>2m_e c^2$

Photon

Positron

(c) $m_e c^2$

$m_e c^2$ | $m_e c^2$

Photon | Photon

$-m_e c^2$

44.2 (a) Energy states for a free electron predicted by the Dirac equation. (b) Raising an electron from an $E < 0$ state to an $E > 0$ state corresponds to electron–positron pair production. (c) An electron dropping from an $E > 0$ state to a vacant $E < 0$ state corresponds to electron–positron pair annihilation.

44.3 (a) Photograph of bubble-chamber tracks of electron–positron pairs that are produced when 300-MeV photons strike a lead sheet. A magnetic field directed out of the photograph made the electrons and positrons curve in opposite directions. (b) Diagram showing the pair-production process for two of the photons.

(a)

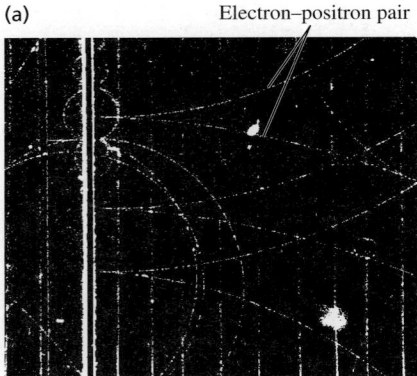

Electron–positron pair

(b)

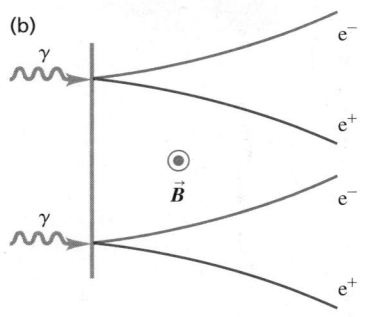

A technique called positron emission tomography (PET) can be used to identify the early stages of Alzheimer's disease. A patient is administered a glucose-like compound called FDG in which one oxygen atom is replaced by radioactive ^{18}F. FDG accumulates in active areas of the brain, where glucose metabolism is high. The ^{18}F undergoes β^+ decay (positron emission) with a half-life of 110 minutes, and the emitted positron immediately annihilates with an atomic electron to produce two gamma-ray photons. A scanner detects both photons, then calculates where the annihilation took place—the site of FDG accumulation. These PET images—which show areas of strongest emission, and hence greatest glucose metabolism, in red—reveal changes in the brains of patients.

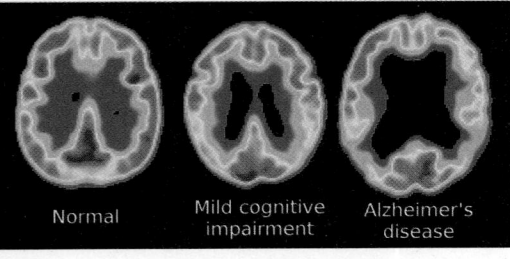

In 1949 the American physicist Richard Feynman showed that a positron could be described mathematically as an electron traveling backward in time. His reformulation of the Dirac theory eliminated difficult calculations involving the infinite sea of negative-energy states and put electrons and positrons on the same footing. But the creation and destruction of electron–positron pairs remain. The Dirac theory provides the beginning of a theoretical framework for creation and destruction of all fundamental particles.

Experiment and theory tell us that the masses of the positron and electron are identical and that their charges are equal in magnitude but opposite in sign. The positron's spin angular momentum $\vec{S}$ and magnetic moment $\vec{\mu}$ are parallel; they are opposite for the electron. However, $\vec{S}$ and $\vec{\mu}$ have the same magnitude for both particles because they have the same spin. We use the term **antiparticle** for a particle that is related to another particle as the positron is to the electron. Each kind of particle has a corresponding antiparticle. For a few kinds of particles (necessarily all neutral) the particle and antiparticle are identical, and we can say that they are their own antiparticles. The photon is an example; there is no way to distinguish a photon from an antiphoton. We'll use the standard symbols e^- for the electron and e^+ for the positron. The generic term "electron" often includes both electrons and positrons. Other antiparticles may be denoted by a bar over the particle's symbol; for example, an antiproton is $\bar{p}$. We'll see other examples of antiparticles later.

Positrons do not occur in ordinary matter. Electron–positron pairs are produced during high-energy collisions of charged particles or γ rays with matter. This process is called e^+e^- *pair production* (**Fig. 44.3**). The minimum available energy required for electron–positron pair production equals the rest energy $2m_ec^2$ of the two particles:

$$E_{min} = 2m_ec^2 = 2(9.109 \times 10^{-31}\text{ kg})(2.998 \times 10^8\text{ m/s})^2$$
$$= 1.637 \times 10^{-13}\text{ J} = 1.022\text{ MeV}$$

The inverse process, e^+e^- *pair annihilation*, occurs when a positron and an electron collide (see Example 38.6 in Section 38.3). Both particles disappear, and two (or occasionally three) photons can appear, with total energy of at least $2m_ec^2 = 1.022$ MeV. Decay into a *single* photon is impossible: Such a process could not conserve both energy and momentum.

Positrons also occur in the decay of some unstable nuclei, in which they are called beta-plus particles (β^+). We discussed β^+ decay in Section 43.3.

We'll frequently represent particle masses in terms of the equivalent rest energy by using $m = E/c^2$. Then typical mass units are MeV/c^2; for example, $m = 0.511$ MeV/c^2 for an electron or positron.

Particles As Force Mediators

In classical physics we describe the interaction of charged particles in terms of electric and magnetic forces. In quantum mechanics we can describe this interaction in terms of emission and absorption of photons. Two electrons repel each other as one emits a photon and the other absorbs it, just as two skaters can push each other apart by tossing a heavy ball back and forth between them (**Fig. 44.4a**). For an electron and a proton, in which the charges are opposite and the force is attractive, we imagine the skaters trying to grab the ball away from each other (Fig. 44.4b). The electromagnetic interaction between two charged particles is *mediated* or transmitted by photons.

If charged-particle interactions are mediated by photons, where does the energy to create the photons come from? Recall from our discussion of the uncertainty principle (see Sections 38.4 and 39.6) that a state that exists for a short time Δt has an uncertainty ΔE in its energy such that

$$\Delta E\,\Delta t \geq \frac{\hbar}{2} \tag{44.3}$$

This uncertainty permits the creation of a photon with energy ΔE, provided that it lives no longer than the time Δt given by Eq. (44.3). A photon that can exist for a short time because of this energy uncertainty is called a *virtual photon*. It's as though there were an energy bank; you can borrow energy, provided that you pay it back within the time limit. According to Eq. (44.3), the more you borrow, the sooner you have to pay it back.

Mesons

Is there a particle that mediates the *nuclear* force? By the mid-1930s the nuclear force between two nucleons (neutrons or protons) appeared to be described by a potential energy $U(r)$ with the general form

$$U(r) = -f^2\left(\frac{e^{-r/r_0}}{r}\right) \qquad \text{(nuclear potential energy)} \qquad (44.4)$$

The constant f characterizes the strength of the interaction, and r_0 describes its range. **Figure 44.5** compares the absolute value of this function with the function f^2/r, which would be analogous to the *electric* interaction of two protons:

$$U(r) = \frac{1}{4\pi\epsilon_0}\frac{e^2}{r} \qquad \text{(electric potential energy)} \qquad (44.5)$$

In 1935 the Japanese physicist Hideki Yukawa suggested that a hypothetical particle that he called a **meson** might mediate the nuclear force. He showed that the range of the force was related to the mass of the particle. Yukawa argued that the particle must live for a time Δt long enough to travel a distance comparable to the range r_0 of the nuclear force. This range was known from the sizes of nuclei and other information to be about 1.5×10^{-15} m = 1.5 fm. If we assume that an average particle's speed is comparable to c and travels about half the range, its lifetime Δt must be about

$$\Delta t = \frac{r_0}{2c} = \frac{1.5 \times 10^{-15}\text{ m}}{2(3.0 \times 10^8\text{ m/s})} = 2.5 \times 10^{-24}\text{ s}$$

From Eq. (44.3), the minimum necessary uncertainty ΔE in energy is

$$\Delta E = \frac{\hbar}{2\Delta t} = \frac{1.05 \times 10^{-34}\text{ J}\cdot\text{s}}{2(2.5 \times 10^{-24}\text{ s})} = 2.1 \times 10^{-11}\text{ J} = 130\text{ MeV}$$

The mass equivalent Δm of this energy is about 250 times the electron mass:

$$\Delta m = \frac{\Delta E}{c^2} = \frac{2.1 \times 10^{-11}\text{ J}}{(3.00 \times 10^8\text{ m/s})^2} = 2.3 \times 10^{-28}\text{ kg} = 130\text{ MeV}/c^2$$

Yukawa postulated that an as yet undiscovered particle with this mass serves as the messenger for the nuclear force.

A year later, Carl Anderson and his colleague Seth Neddermeyer discovered in cosmic radiation two new particles, now called **muons.** The μ^- has charge equal to that of the electron, and its antiparticle the μ^+ has a positive charge with equal magnitude. The two particles have equal mass, about 207 times the electron mass. But it soon became clear that muons were *not* Yukawa's particles because they interacted with nuclei only very weakly.

In 1947 a family of three particles, called π *mesons* or **pions,** were discovered. Their charges are $+e$, $-e$, and zero, and their masses are about 270 times the electron mass. The pions interact strongly with nuclei, and they *are* the particles predicted by Yukawa. Other, heavier mesons, the ω and ρ, evidently also act as shorter-range messengers of the nuclear force. The complexity of this explanation suggests that the nuclear force has simpler underpinnings; these involve the quarks

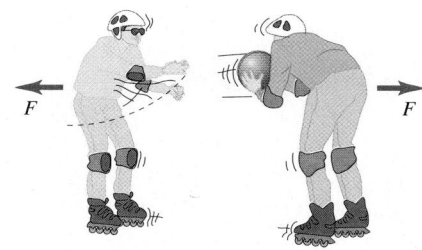

44.4 An analogy for how particles act as force mediators.

(a) Two skaters exert repulsive forces on each other by tossing a ball back and forth.

(b) Two skaters exert attractive forces on each other when one tries to grab the ball out of the other's hands.

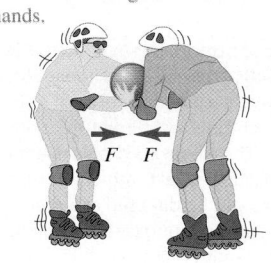

44.5 Graph of the magnitude of the Yukawa potential-energy function for nuclear forces, $|U(r)| = f^2 e^{-r/r_0}/r$. The function $U(r) = f^2/r$, proportional to the potential energy for Coulomb's law, is also shown. The two functions are similar at small r, but the Yukawa potential energy drops off much more quickly at large r.

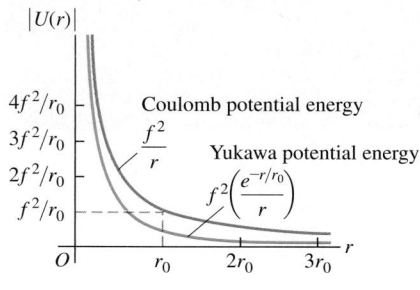

and gluons that we'll discuss in Section 44.4. Before discussing mesons further, we'll describe some particle accelerators and detectors to see how mesons and other particles are created in a controlled fashion and observed.

TEST YOUR UNDERSTANDING OF SECTION 44.1   Each of the following particles can be exchanged between two protons, two neutrons, or a neutron and a proton as part of the nuclear force. Rank the particles in order of the range of the interaction that they mediate, from largest to smallest range. (i) The π^+ (pi-plus) meson, mass $140 \text{ MeV}/c^2$; (ii) the ρ^+ (rho-plus) meson, mass $776 \text{ MeV}/c^2$; (iii) the η^0 (eta-zero) meson, mass $548 \text{ MeV}/c^2$; (iv) the ω^0 (omega-zero) meson, mass $783 \text{ MeV}/c^2$. ∎

44.2 PARTICLE ACCELERATORS AND DETECTORS

Early nuclear physicists used alpha and beta particles from naturally occurring radioactive elements for their experiments, but they were restricted in energy to the few MeV that are available in such random decays. Present-day particle accelerators can produce precisely controlled beams of particles, from electrons and positrons up to heavy ions, with a wide range of energies. These beams have three main uses. First, high-energy particles can collide to produce new particles, just as a collision of an electron and a positron can produce photons. Second, a high-energy particle has a short de Broglie wavelength and so can probe the small-scale interior structure of other particles, just as electron microscopes (see Section 39.1) can give better resolution than optical microscopes. Third, they can be used to produce nuclear reactions of scientific or medical use.

Linear Accelerators

Particle accelerators use electric and magnetic fields to accelerate and guide beams of charged particles. A *linear accelerator* (linac) accelerates particles in a straight line. J. J. Thomson's cathode-ray tubes were early examples of linacs. Modern linacs use a series of electrodes with gaps to give the particles a series of boosts. Most present-day high-energy linear accelerators use a traveling electromagnetic wave; the charged particles "ride" the wave in more or less the way that a surfer rides an incoming ocean wave. In the highest-energy linac in the world today, at the SLAC National Accelerator Laboratory, electrons and positrons can be accelerated to 50 GeV in a tube 3 km long. At this energy their de Broglie wavelengths are 0.025 fm, much smaller than the size of a proton or a neutron.

The Cyclotron

Many accelerators use magnets to deflect the charged particles into circular paths. The first was the *cyclotron,* invented in 1931 by E. O. Lawrence and M. Stanley Livingston at the University of California (**Fig. 44.6a**). Particles with mass m and

BIO Application Linear Accelerators in Medicine Electron linear accelerators that provide a kinetic energy of 4–20 MeV are important tools in the treatment of many cancers. The electrons themselves are used to irradiate superficial tumors. Alternatively, the electrons can be directed at a metal target; then bremsstrahlung (see Section 38.2) produces x rays that are used to irradiate tumors that lie deeper inside the patient.

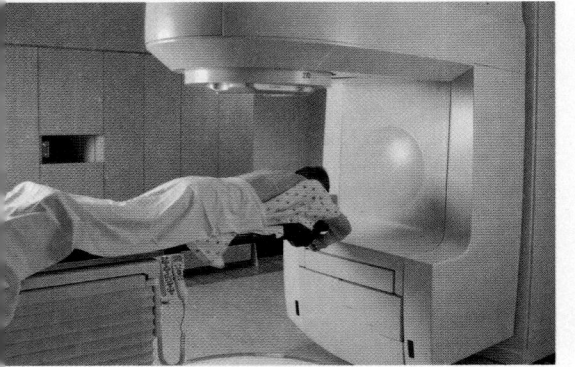

44.6 Layout and operation of a cyclotron.

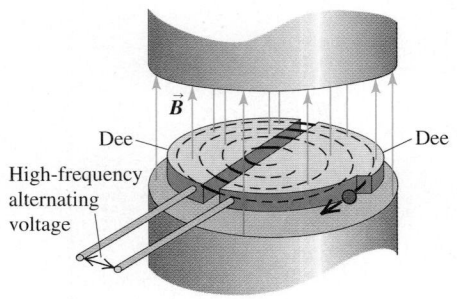

(a) Schematic diagram of a cyclotron

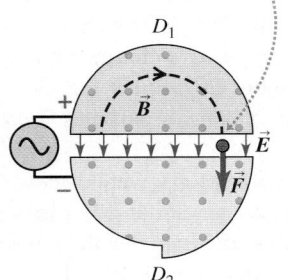

(b) As the positive particle reaches the gap, it is accelerated by the electric-field force ...

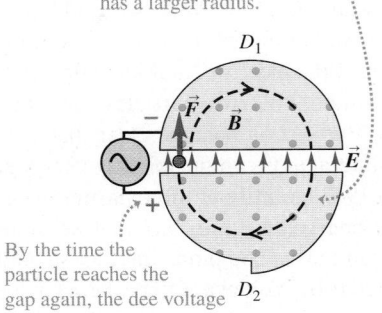

(c) ... and the next semicircular orbit has a larger radius.

By the time the particle reaches the gap again, the dee voltage has reversed and the particle is again accelerated.

charge q move inside a vacuum chamber in a uniform magnetic field $\vec{B}$ that is perpendicular to the plane of their paths. In Section 27.4 we showed that in such a field, a particle with speed v moves in a circular path with radius r given by

$$r = \frac{mv}{|q|B} \qquad (44.6)$$

and with angular speed (angular frequency) ω given by

$$\omega = \frac{v}{r} = \frac{|q|B}{m} \qquad (44.7)$$

An alternating potential difference is applied between the two hollow electrodes D_1 and D_2 (called *dees*), creating an electric field in the gap between them. The polarity of the potential difference and electric field is changed precisely twice each revolution (Figs. 44.6b and 44.6c), so that the particles get a push each time they cross the gap. The pushes increase their speed and kinetic energy, boosting them into paths of larger radius. The maximum speed v_{max} and kinetic energy K_{max} are determined by the radius R of the largest possible path. Solving Eq. (44.6) for v, we find $v = |q|Br/m$ and $v_{max} = |q|BR/m$. Assuming nonrelativistic speeds, we have

$$K_{max} = \tfrac{1}{2}mv_{max}^2 = \frac{q^2B^2R^2}{2m} \qquad (44.8)$$

EXAMPLE 44.1 FREQUENCY AND ENERGY IN A PROTON CYCLOTRON

One cyclotron built during the 1930s has a path of maximum radius 0.500 m and a magnetic field of magnitude 1.50 T. If it is used to accelerate protons, find (a) the frequency of the alternating voltage applied to the dees and (b) the maximum particle energy.

SOLUTION

IDENTIFY and SET UP: The frequency f of the applied voltage must equal the frequency of the proton orbital motion. Equation (44.7) gives the *angular* frequency ω of the proton orbital motion; we find f from $f = \omega/2\pi$. The proton reaches its maximum energy K_{max}, given by Eq. (44.8), when the radius of its orbit equals the radius of the dees.

EXECUTE: (a) For protons, $q = 1.60 \times 10^{-19}$ C and $m = 1.67 \times 10^{-27}$ kg. From Eq. (44.7),

$$f = \frac{\omega}{2\pi} = \frac{|q|B}{2\pi m} = \frac{(1.60 \times 10^{-19}\ \text{C})(1.50\ \text{T})}{2\pi(1.67 \times 10^{-27}\ \text{kg})}$$

$$= 2.3 \times 10^7\ \text{Hz} = 23\ \text{MHz}$$

(b) From Eq. (44.8) the maximum kinetic energy is

$$K_{max} = \frac{(1.60 \times 10^{-19}\ \text{C})^2(1.50\ \text{T})^2(0.50\ \text{m})^2}{2(1.67 \times 10^{-27}\ \text{kg})}$$

$$= 4.3 \times 10^{-12}\ \text{J} = 2.7 \times 10^7\ \text{eV} = 27\ \text{MeV}$$

This proton kinetic energy is much larger than that available from natural radioactive sources.

EVALUATE: From Eq. (44.6) or Eq. (44.7), the proton speed is $v = 7.2 \times 10^7$ m/s, which is about 25% of the speed of light. At such speeds, relativistic effects are beginning to become important. Since we ignored these effects in our calculation, the results for f and K_{max} are in error by a few percent; this is why we kept only two significant figures.

The maximum energy that can be attained with a cyclotron is limited by relativistic effects. The relativistic version of Eq. (44.7) is

$$\omega = \frac{|q|B}{m}\sqrt{1 - v^2/c^2}$$

As the particles speed up, their angular frequency ω *decreases,* and their motion gets out of phase with the alternating dee voltage. In the *synchrocyclotron* the particles are accelerated in bursts. For each burst, the frequency of the alternating voltage is decreased as the particles speed up, maintaining the correct phase relationship with the particles' motion.

Another limitation of the cyclotron is the difficulty of building very large electromagnets. The largest synchrocyclotron ever built has a vacuum chamber that is about 8 m in diameter and accelerates protons to energies of about 600 MeV.

The Synchrotron

To attain higher energies, a type of machine called the *synchrotron* is more practical. Particles move in a vacuum chamber in the form of a thin doughnut called the *accelerating ring*. The particle beam is bent to follow the ring by a series of electromagnets placed around the ring. As the particles speed up, the magnetic field is increased so that the particles retrace the same trajectory over and over. The Large Hadron Collider (LHC) near Geneva, Switzerland, is the highest-energy accelerator in the world (**Fig. 44.7**). It is designed to accelerate protons to a maximum energy of 7 TeV, or 7×10^{12} eV. (As we'll see in Section 44.3, *hadrons* are a class of elementary particles that includes protons and neutrons.)

As we pointed out in Section 32.1, accelerated charges radiate electromagnetic energy. In an accelerator in which the particles move in curved paths, this radiation is often called *synchrotron radiation*. High-energy accelerators are typically constructed underground to provide protection from this radiation. From the accelerator standpoint, synchrotron radiation is undesirable, since the energy given to an accelerated particle is radiated right back out. It can be minimized by making the accelerator radius r large so that the centripetal acceleration v^2/r is small. On the positive side, synchrotron radiation is used as a source of well-controlled high-frequency electromagnetic waves.

Available Energy

When a beam of high-energy particles collides with a stationary target, not all the kinetic energy of the incident particles is *available* to form new particle states. Because momentum must be conserved, the particles emerging from the collision must have some net motion and thus some kinetic energy. The discussion following Example 43.11 (Section 43.6) presented a nonrelativistic example of this principle. The maximum available energy is the kinetic energy in the frame of reference in which the total momentum is zero. We call this the *center-of-momentum system;* it is the relativistic generalization of the center-of-mass system that we discussed in Section 8.5. In this system the total kinetic energy after the collision can be zero, so that the maximum amount of the initial kinetic energy becomes available to cause the reaction being studied.

Consider the *laboratory system,* in which a target particle with mass M is initially at rest and is bombarded by a particle with mass m and total energy (including rest energy) E_m. The total available energy E_a in the center-of-momentum system (including rest energies of all the particles) can be shown to be

$$E_a^2 = 2Mc^2 E_m + (Mc^2)^2 + (mc^2)^2 \qquad \text{(available energy)} \qquad (44.9)$$

When the masses of the target and projectile particles are equal, we can simplify:

$$E_a^2 = 2mc^2(E_m + mc^2) \qquad \text{(available energy, equal masses)} \qquad (44.10)$$

If in addition E_m is much greater than mc^2, we can ignore the second term in the parentheses in Eq (44.10). Then E_a is

$$E_a = \sqrt{2mc^2 E_m} \qquad \text{(available energy, equal masses, } E_m \gg mc^2) \qquad (44.11)$$

The square root in Eq. (44.11) is a disappointing result for an accelerator designer: Doubling the energy E_m of the bombarding particle increases the available energy E_a by only a factor of $\sqrt{2} = 1.414$. Examples 44.2 and 44.3 explore the limitations of having a stationary target particle.

44.7 (a) The Large Hadron Collider at the European Organization for Nuclear Research (CERN). The underground accelerating ring (shown by the red circle) is 100 m underground and 8.5 km in diameter, so large that it spans the border between Switzerland and France. (Note the Alps in the background.) When accelerated to 7 TeV, protons travel around the ring more than 11,000 times per second. (b) An engineer working on one of the 9593 superconducting electromagnets around the LHC ring.

(a)

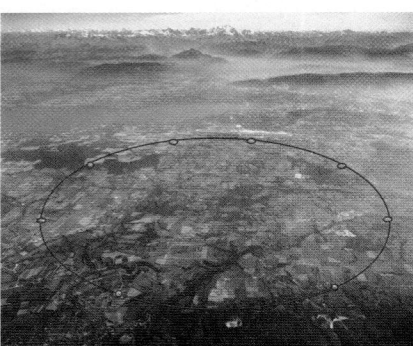

(b)

EXAMPLE 44.2 THRESHOLD ENERGY FOR PION PRODUCTION

A proton (rest energy 938 MeV) with kinetic energy K collides with a proton at rest. Both protons survive the collision, and a neutral pion (π^0, rest energy 135 MeV) is produced. What is the threshold energy (minimum value of K) for this process?

SOLUTION

IDENTIFY and SET UP: The final state includes the two original protons (mass m) and the pion (mass m_π). The threshold energy corresponds to the minimum-energy case in which all three particles are at rest in the center-of-momentum system. The total available energy E_a in that system must be at least the total rest energy, $2mc^2 + m_\pi c^2$. We use this to solve Eq. (44.10) for the total energy E_m of the bombarding proton; the kinetic energy K (our target variable) is then E_m minus the proton rest energy mc^2.

EXECUTE: We substitute $E_a = 2mc^2 + m_\pi c^2$ into Eq. (44.10), simplify, and solve for E_m:

$$4m^2c^4 + 4mm_\pi c^4 + m_\pi^2 c^4 = 2mc^2 E_m + 2(mc^2)^2$$

$$E_m = mc^2 + m_\pi c^2\left(2 + \frac{m_\pi}{2m}\right) = mc^2 + K$$

$$K = m_\pi c^2\left(2 + \frac{m_\pi}{2m}\right)$$

We see that the bombarding proton's kinetic energy K must be somewhat greater than twice the pion rest energy $m_\pi c^2$. With $mc^2 = 938$ MeV and $m_\pi c^2 = 135$ MeV, we have $m_\pi/2m = 0.072$ and

$$K = (135\ \text{MeV})(2 + 0.072) = 280\ \text{MeV}$$

EVALUATE: Compare this result with the result of Example 37.11 (Section 37.8), where we found that a pion can be produced in a head-on collision of two protons, each with only 67.5 MeV of kinetic energy. We discuss the energy advantage of such collisions in the next subsection.

EXAMPLE 44.3 INCREASING THE AVAILABLE ENERGY

The Fermilab accelerator in Illinois was designed to bombard stationary targets with 800-GeV protons. (a) What is the available energy E_a in a proton-proton collision? (b) What is E_a if the beam energy is increased to 980 GeV?

SOLUTION

IDENTIFY and SET UP: Our target variable is the available energy E_a in a stationary-target collision between identical particles. In both parts (a) and (b) the beam energy E_m is much larger than the proton rest energy $mc^2 = 938$ MeV $= 0.938$ GeV, so we can safely use the approximation of Eq. (44.11).

EXECUTE: (a) For $E_m = 800$ GeV, Eq. (44.11) gives

$$E_a = \sqrt{2(0.938\ \text{GeV})(800\ \text{GeV})} = 38.7\ \text{GeV}$$

(b) For $E_m = 980$ GeV,

$$E_a = \sqrt{2(0.938\ \text{GeV})(980\ \text{GeV})} = 42.9\ \text{GeV}$$

EVALUATE: With a stationary-proton target, increasing the proton beam energy by 180 GeV increases the available energy by only 4.2 GeV! This shows a major limitation of experiments in which one of the colliding particles is initially at rest. Below we describe how physicists can overcome this limitation.

Colliding Beams

The limitation illustrated by Example 44.3 is circumvented in *colliding-beam* experiments. In these experiments there is no stationary target; instead, beams of particles moving in opposite directions are tightly focused onto one another so that head-on collisions can occur. Usually the two colliding particles have momenta of equal magnitude and opposite direction, so the total momentum is zero. Hence the laboratory system is also the center-of-momentum system, and the available energy is maximized.

The highest-energy colliding beams available are those at the Large Hadron Collider (see Fig. 44.7). In operation, 2808 bunches of 7-TeV protons circulate around the ring, half in one direction and half in the opposite direction. Each bunch contains about 10^{11} protons. Magnets steer the oppositely moving bunches to collide at interaction points. The available energy E_a in the resulting head-on collisions is the *total* energy of the two colliding particles: $E_a = 2 \times 7\ \text{TeV} = 14\ \text{TeV}$. (Strictly, E_a is 14 TeV minus the rest energy of the two colliding protons. But this rest energy is only $2mc^2 = 2(938\ \text{MeV}) = 1.876 \times 10^{-3}$ TeV, which is so small compared to 14 TeV that it can be ignored.) The very large available energy at the Large Hadron Collider makes it possible to produce particles that have never been seen before (see Section 44.5).

Detectors

A wide variety of devices have been designed to measure the properties of subatomic particles. Many detectors use the ionization caused by charged particles as they move through a gas, liquid, or solid. The ions along the particle's path give rise to droplets of liquid in the supersaturated vapor of a cloud chamber (Fig. 44.1) or cause small volumes of vapor in the superheated liquid of a bubble chamber (Fig. 44.3a). In a semiconducting solid the ionization can take the form of electron–hole pairs. We discussed their detection in Section 42.7. *Wire chambers* contain arrays of closely spaced wires that detect the ions. The charge collected and time information from each wire are processed by using computers to reconstruct the particle trajectories. The detectors at the Large Hadron Collider use an array of devices to follow the tracks of particles produced by collisions between protons (**Fig. 44.8**). The giant solenoid in the photo that opens Chapter 28 is at the heart of one these detector arrays. The intense magnetic field of the solenoid helps identify newly produced particles, which curve in different directions and along paths of different radii depending on their charge and energy.

Cosmic-Ray Experiments

Large numbers of particles called *cosmic rays* continually bombard the earth from sources both within and beyond our galaxy. These particles consist mostly of neutrinos, protons, and heavier nuclei, with energies ranging from less than 1 MeV to more than 10^{20} eV. The earth's atmosphere and magnetic field protect us from much of this radiation. This means that cosmic-ray experimentation often must be carried out above all or most of the atmosphere by means of rockets or high-altitude balloons.

In contrast, neutrino detectors are buried below the earth's surface in tunnels or mines or submerged deep in the ocean. This is done to screen out all other types of particles so that only neutrinos, which interact only very weakly with matter, reach the detector. Because neutrino interactions with matter are so weak, neutrino detectors must consist of huge amounts of matter: The Super-Kamiokande detector looks for flashes of light produced when a neutrino interacts in a tank containing 5×10^7 kg of water (see Section 44.5).

Cosmic rays were important in early particle physics, and their study currently brings us important information about the rest of the universe. Although cosmic rays provide a source of high-energy particles that does not depend on expensive accelerators, most particle physicists use accelerators because the high-energy cosmic-ray particles they want are too few and too random.

TEST YOUR UNDERSTANDING OF SECTION 44.2 In a colliding-beam experiment, a 90-GeV electron collides head-on with a 90-GeV positron. The electron and the positron annihilate each other, forming a single virtual photon that then transforms into other particles. Does the virtual photon obey the same relationship $E = pc$ as real photons do? ▮

44.3 PARTICLES AND INTERACTIONS

We have mentioned the array of subatomic particles that were known as of 1947: photons, electrons, positrons, protons, neutrons, muons, and pions. Since then, literally hundreds of additional particles have been discovered in accelerator experiments. The vast majority of known particles are *unstable* and decay spontaneously into other particles. Particles of all kinds, whether stable or unstable, can be created or destroyed in interactions between particles. Each such

44.8 This computer-generated image shows the result of a simulated collision between two protons (not shown) in one of the interaction regions at the Large Hadron Collider. The view is along the beampipe. The different color tracks show different types of particles emerging from the collision. A variety of different detectors surround the collision region. (Note the woman in a red dress, drawn for scale.)

DATA *SPEAKS*

Particle Collisions

When students were given a problem involving collisions between elementary particles, more than 46% gave an incorrect response. Common errors:

- Confusion about the energy released in particle–antiparticle annihilation. If a particle of mass m collides with and annihilates its antiparticle, the released energy is greater than or equal to the *combined* rest energy $2mc^2$ of the particle and antiparticle.

- Confusion about available energy. To produce a new particle of mass m in a collision, the *available* energy must be at least mc^2. If the target particle is at rest, the available energy can be far less than the kinetic energy of the bombarding particle.

interaction involves the exchange of virtual particles, which exist on borrowed energy allowed by the uncertainty principle.

Although the world of subatomic particles and their interactions is complex, some key results bring order and simplicity to the seeming chaos. One key simplification is that there are only four fundamental types of interactions, each mediated or transmitted by the exchange of certain characteristic virtual particles. Furthermore, not all particles respond to all four kinds of interaction. In this section we will examine the fundamental interactions more closely and see how physicists classify particles in terms of the ways in which they interact.

Four Forces and Their Mediating Particles

In Section 5.5 we first described the four fundamental types of forces or interactions (**Fig. 44.9**). They are, in order of decreasing strength:

1. The strong interaction
2. The electromagnetic interaction
3. The weak interaction
4. The gravitational interaction

The *electromagnetic* and *gravitational* interactions are familiar from classical physics. Both are characterized by a $1/r^2$ dependence on distance. In this scheme, the mediating particles for both interactions have mass zero and are stable as ordinary particles. The mediating particle for the electromagnetic interaction is the familiar photon, which has spin 1. (That means its spin quantum number is $s = 1$, so the magnitude of its spin angular momentum is $S = \sqrt{s(s + 1)}\,\hbar = \sqrt{2}\,\hbar$.) The mediating particle for the gravitational force is the spin-2 *graviton* ($s = 2$, $S = \sqrt{s(s + 1)}\,\hbar = \sqrt{6}\,\hbar$). The graviton has not yet been observed experimentally because the gravitational force is very much weaker than the electromagnetic force. For example, the gravitational attraction of two protons is smaller than their electrical repulsion by a factor of about 10^{36}. The gravitational force is of primary importance in the structure of stars and the large-scale behavior of the universe, but it is not believed to play a significant role in particle interactions at the energies that are currently attainable.

The other two forces are less familiar. One, usually called the *strong interaction,* is responsible for the nuclear force and also for the production of pions and several other particles in high-energy collisions. At the most fundamental level, the mediating particle for the strong interaction is called a *gluon.* However, the force between nucleons is more easily described in terms of mesons as the mediating particles. We'll discuss the spin-1, massless gluon in Section 44.4.

Equation (44.4) is a possible potential-energy function for the nuclear force. The strength of the interaction is described by the constant f^2, which has units of energy times distance. A better basis for comparison with other forces is the dimensionless ratio $f^2/\hbar c$, called the *coupling constant* for the interaction. (We invite you to verify that this ratio is a pure number and so must have the same value in all systems of units.) The observed behavior of nuclear forces suggests that $f^2/\hbar c \approx 1$. The dimensionless coupling constant for *electromagnetic* interactions is the fine-structure constant, which we introduced in Section 41.5:

$$\frac{1}{4\pi\epsilon_0}\frac{e^2}{\hbar c} = 7.2974 \times 10^{-3} = \frac{1}{137.04} \qquad (44.12)$$

Thus the strong interaction is roughly 100 times as strong as the electromagnetic interaction; however, it drops off with distance more quickly than $1/r^2$.

The fourth interaction is called the *weak* interaction. It is responsible for beta decay, such as the conversion of a neutron into a proton, an electron, and an antineutrino. It is also responsible for the decay of many unstable particles (pions into muons, muons into electrons, and so on). Its mediating particles are the short-lived particles W^+, W^-, and Z^0. The existence of these particles

44.9 The ties that bind us together originate in the fundamental interactions of nature. The nuclei within our bodies are held together by the strong interaction. The electromagnetic interaction binds nuclei and electrons together to form atoms, binds atoms together to form molecules, and binds molecules together to form us.

TABLE 44.1	**Four Fundamental Interactions**					
	Relative			**Mediating Particle**		
Interaction	**Strength**	**Range**	**Name**	**Mass**	**Charge**	**Spin**
Strong	1	Short ($\sim$1 fm)	Gluon	0	0	1
Electromagnetic	$\frac{1}{137.04}$	Long ($1/r^2$)	Photon	0	0	1
Weak	10^{-9}	Short ($\sim$0.001 fm)	$W^\pm, Z^0$	80.4, 91.2 GeV/c^2	$\pm e, 0$	1
Gravitational	10^{-38}	Long ($1/r^2$)	Graviton	0	0	2

was confirmed in 1983 in experiments at CERN, for which Carlo Rubbia and Simon van der Meer were awarded the Nobel Prize in 1984. The $W^\pm$ and Z^0 have spin 1 like the photon and the gluon, but they are *not* massless. In fact, they have enormous masses, 80.4 GeV/c^2 for the W's and 91.2 GeV/c^2 for the Z^0. With such massive mediating particles the weak interaction has a much shorter range than the strong interaction. It also lives up to its name by being weaker than the strong interaction by a factor of about 10^9.

Table 44.1 compares the main features of these four fundamental interactions.

More Particles

In Section 44.1 we mentioned the discoveries of muons in 1937 and of pions in 1947. The electric charges of the muons and the charged pions have the same magnitude e as the electron charge. The positive muon μ^+ is the antiparticle of the negative muon μ^-. Each has spin $\frac{1}{2}$, like the electron, and a mass of about $207m_e = 106$ MeV/c^2. Muons are unstable; each decays with a lifetime of 2.2×10^{-6} s into an electron of the same sign, a neutrino, and an antineutrino.

There are three kinds of pions, all with spin 0; they have *no* spin angular momentum. The π^+ and π^- have masses of $273m_e = 140$ MeV/c^2. They are unstable; each $\pi^\pm$ decays with a lifetime of 2.6×10^{-8} s into a muon of the same sign along with a neutrino for the π^+ and an antineutrino for the π^-. The π^0 is somewhat less massive, $264m_e = 135$ MeV/c^2, and it decays with a lifetime of 8.4×10^{-17} s into two photons. The π^+ and π^- are antiparticles of one another, while the π^0 is its own antiparticle. (That is, there is no distinction between particle and antiparticle for the π^0.)

The existence of the *antiproton* $\bar{p}$ had been suspected ever since the discovery of the positron. The $\bar{p}$ was found in 1955, when proton–antiproton ($p\bar{p}$) pairs were created by use of a beam of 6-GeV protons from the Bevatron at the University of California, Berkeley. The *antineutron* $\bar{n}$ was found soon afterward. After 1960, as higher-energy accelerators and more sophisticated detectors were developed, a veritable blizzard of new unstable particles were identified. To describe and classify them, we need a small blizzard of new terms.

Initially, particles were classified by mass into three categories: (1) leptons ("light ones" such as electrons); (2) mesons ("intermediate ones" such as pions); and (3) baryons ("heavy ones" such as nucleons and more massive particles). But this scheme has been superseded by a more useful one in which particles are classified in terms of their *interactions*. For instance, *hadrons* (which include mesons and baryons) have strong interactions, and *leptons* do not.

In the following discussion we will also distinguish between **fermions,** which have half-integer spins, and **bosons,** which have zero or integer spins. Fermions obey the exclusion principle, on which the Fermi-Dirac distribution function (see Section 42.5) is based. Bosons do not obey the exclusion principle (there is no limit on how many bosons can occupy the same quantum state) and have a different distribution function, the Bose-Einstein distribution.

TABLE 44.2 The Six Leptons

Particle Name	Symbol	Anti-particle	Mass (MeV/c^2)	L_e	L_μ	L_τ	Lifetime (s)	Principal Decay Modes
Electron	e^-	e^+	0.511	+1	0	0	Stable	
Electron neutrino	ν_e	$\bar{\nu}_e$	$<2 \times 10^{-6}$	+1	0	0	Stable	
Muon	μ^-	μ^+	105.7	0	+1	0	2.20×10^{-6}	$e^- \bar{\nu}_e \nu_\mu$
Muon neutrino	ν_μ	$\bar{\nu}_\mu$	<0.19	0	+1	0	Stable	
Tau	τ^-	τ^+	1777	0	0	+1	2.9×10^{-13}	$\mu^- \bar{\nu}_\mu \nu_\tau$ or $e^- \bar{\nu}_e \nu_\tau$
Tau neutrino	ν_τ	$\bar{\nu}_\tau$	<18.2	0	0	+1	Stable	

Note: In addition to the limits on the individual neutrino masses, there is a much more stringent limit on the *sum* of the masses of the three types of neutrinos. Evidence suggests that this sum is less than about 3×10^{-7} MeV/$c^2 = 0.3$ eV/c^2.

Leptons

The **leptons,** which do not have strong interactions, include six particles: the electron (e^-) and its neutrino (ν_e), the muon (μ^-) and its neutrino (ν_μ), and the tau particle (τ^-) and its neutrino (ν_τ). Each of these has a distinct antiparticle. All leptons have spin $\frac{1}{2}$ and thus are fermions. **Table 44.2** shows the family of leptons. The taus have mass $3478m_e = 1777$ MeV/c^2. Taus and muons are unstable; a τ^- decays into a μ^- plus a tau neutrino and a muon antineutrino, or an electron plus a tau neutrino and an electron antineutrino. A μ^- decays into an electron plus a muon neutrino and an electron antineutrino. They have relatively long lifetimes because their decays are mediated by the weak interaction. Despite their zero charge, a neutrino is distinct from an antineutrino; the spin angular momentum of a neutrino has a component that is opposite its linear momentum, while for an antineutrino that component is parallel to its linear momentum. Because neutrinos are so elusive, physicists have only been able to place upper limits on the rest masses of the ν_e, the ν_μ, and the ν_τ. It was thought that the rest masses of the neutrinos were zero; compelling recent evidence indicates that they have small but nonzero masses. We'll return to this point and its implications later.

Leptons obey a *conservation principle.* Corresponding to the three pairs of leptons are three lepton numbers L_e, L_μ, and L_τ. The electron e^- and the electron neutrino ν_e are assigned $L_e = 1$, and their antiparticles e^+ and $\bar{\nu}_e$ are given $L_e = -1$. Corresponding assignments of L_μ and L_τ are made for the μ and τ particles and their neutrinos. **In all interactions, each lepton number is separately conserved.** For example, in the decay of the μ^-, the lepton numbers are

$$\mu^- \rightarrow e^- + \bar{\nu}_e + \nu_\mu$$
$$L_\mu = 1 \quad L_e = 1 \quad L_e = -1 \quad L_\mu = 1$$

These conservation principles have no counterpart in classical physics.

EXAMPLE 44.4 **LEPTON NUMBER CONSERVATION**

Check conservation of lepton numbers for these decay schemes:

(a) $\mu^+ \rightarrow e^+ + \nu_e + \bar{\nu}_\mu$

(b) $\pi^- \rightarrow \mu^- + \bar{\nu}_\mu$

(c) $\pi^0 \rightarrow \mu^- + e^+ + \nu_e$

SOLUTION

IDENTIFY and SET UP: Lepton number conservation requires that L_e, L_μ, and L_τ (given in Table 44.2) separately have the same sums after the decay as before.

EXECUTE: We tabulate L_e and L_μ for each decay scheme. An antiparticle has the opposite lepton number from its corresponding particle listed in Table 44.2. No τ particles or τ neutrinos appear in any of the schemes, so $L_\tau = 0$ both before and after each decay and L_τ is conserved.

(a) $\mu^+ \rightarrow e^+ + \nu_e + \bar{\nu}_\mu$

$L_e: 0 = -1 + 1 + 0$

$L_\mu: -1 = 0 + 0 + (-1)$

Continued

(b) $\pi^- \rightarrow \mu^- + \bar{\nu}_\mu$

$L_e: 0 = 0 + 0$

$L_\mu: 0 = 1 + (-1)$

(c) $\pi^0 \rightarrow \mu^- + e^+ + \nu_e$

$L_e: 0 = 0 + (-1) + 1$

$L_\mu: 0 \neq 1 + 0 + 0$

EVALUATE: Decays (a) and (b) are consistent with lepton number conservation and are observed. Decay (c) violates the conservation of L_μ and has *never* been observed. Physicists used these and other experimental results to deduce the principle that all three lepton numbers must separately be conserved.

Hadrons

Hadrons, the strongly interacting particles, are a more complex family than leptons. Each hadron has an antiparticle, often denoted with an overbar, as with the antiproton $\bar{p}$. There are two subclasses of hadrons: *mesons* and *baryons*. **Table 44.3** shows some of the many hadrons that are currently known. (We'll explain *strangeness* and *quark content* later in this section and in the next one.)

Mesons include the pions that have already been mentioned, K mesons or *kaons,* η mesons, and others that we will discuss later. Mesons have spin 0 or 1 and therefore are all bosons. There are no stable mesons; all mesons decay to less massive particles, obeying all the conservation laws for such decays.

Baryons include the nucleons and several particles called *hyperons,* including the Λ, Σ, Ξ, and Ω. These resemble nucleons but are more massive. Baryons have half-integer spin, and therefore all are fermions. The only stable baryon is the proton; a free neutron decays to a proton, and hyperons decay to other hyperons or to nucleons by various processes. Baryons obey the *conservation of baryon number,* analogous to conservation of lepton numbers, again with no counterpart in classical physics. We assign a baryon number $B = 1$ to each baryon (p, n, Λ, Σ, and so on) and $B = -1$ to each antibaryon ($\bar{p}$, $\bar{n}$, $\bar{\Lambda}$, $\bar{\Sigma}$, and so on).

In all interactions, the total baryon number is conserved.

This principle is the reason the mass number A was conserved in all of the nuclear reactions that we studied in Chapter 43.

TABLE 44.3 **Some Hadrons and Their Properties**

Particle	Mass (MeV/c^2)	Charge Ratio, Q/e	Spin	Baryon Number, B	Strangeness, S	Mean Lifetime (s)	Typical Decay Modes	Quark Content
Mesons								
π^0	135.0	0	0	0	0	8.4×10^{-17}	$\gamma\,\gamma$	$u\bar{u}, d\bar{d}$
π^+	139.6	+1	0	0	0	2.60×10^{-8}	$\mu^+\nu_\mu$	$u\bar{d}$
π^-	139.6	-1	0	0	0	2.60×10^{-8}	$\mu^-\bar{\nu}_\mu$	$\bar{u}d$
K^+	493.7	+1	0	0	+1	1.24×10^{-8}	$\mu^+\nu_\mu$	$u\bar{s}$
K^-	493.7	-1	0	0	-1	1.24×10^{-8}	$\mu^-\bar{\nu}_\mu$	$\bar{u}s$
η^0	547.3	0	0	0	0	$\approx 10^{-18}$	$\gamma\,\gamma$	$u\bar{u}, d\bar{d}, s\bar{s}$
Baryons								
p	938.3	+1	$\frac{1}{2}$	1	0	Stable	—	uud
n	939.6	0	$\frac{1}{2}$	1	0	886	$pe^-\bar{\nu}_e$	udd
Λ^0	1116	0	$\frac{1}{2}$	1	-1	2.63×10^{-10}	$p\pi^-$ or $n\pi^0$	uds
Σ^+	1189	+1	$\frac{1}{2}$	1	-1	8.02×10^{-11}	$p\pi^0$ or $n\pi^+$	uus
Σ^0	1193	0	$\frac{1}{2}$	1	-1	7.4×10^{-20}	$\Lambda^0\gamma$	uds
Σ^-	1197	-1	$\frac{1}{2}$	1	-1	1.48×10^{-10}	$n\pi^-$	dds
Ξ^0	1315	0	$\frac{1}{2}$	1	-2	2.90×10^{-10}	$\Lambda^0\pi^0$	uss
Ξ^-	1321	-1	$\frac{1}{2}$	1	-2	1.64×10^{-10}	$\Lambda^0\pi^-$	dss
Δ^{++}	1232	+2	$\frac{3}{2}$	1	0	$\approx 10^{-23}$	$p\pi^+$	uuu
Ω^-	1672	-1	$\frac{3}{2}$	1	-3	8.2×10^{-11}	$\Lambda^0 K^-$	sss
Λ_c^+	2285	+1	$\frac{1}{2}$	1	0	2.0×10^{-13}	$pK^-\pi^+$	udc

EXAMPLE 44.5 BARYON NUMBER CONSERVATION

Check conservation of baryon number for these reactions:

(a) $n + p \rightarrow n + p + p + \bar{p}$

(b) $n + p \rightarrow n + p + \bar{n}$

SOLUTION

IDENTIFY and SET UP: This example is similar to Example 44.4. We compare the total baryon number before and after each reaction, using data from Table 44.3.

EXECUTE: We tabulate the baryon numbers, noting that a baryon has $B = 1$ and an antibaryon has $B = -1$:

(a) $n + p \rightarrow n + p + p + \bar{p}$: $1 + 1 = 1 + 1 + 1 + (-1)$

(b) $n + p \rightarrow n + p + \bar{n}$: $1 + 1 \neq 1 + 1 + (-1)$

EVALUATE: Reaction (a) is consistent with baryon number conservation. It can occur if enough energy is available in the $n + p$ collision. Reaction (b) violates baryon number conservation and has never been observed.

EXAMPLE 44.6 ANTIPROTON CREATION

What is the minimum proton energy required to produce an antiproton in a collision with a stationary proton?

SOLUTION

IDENTIFY and SET UP: The reaction must conserve baryon number, charge, and energy. Since the target and bombarding protons are of equal mass and the target is at rest, we determine the minimum energy E_m of the bombarding proton from Eq. (44.10).

EXECUTE: Conservation of charge and conservation of baryon number forbid the creation of an antiproton by itself; it must be created as part of a proton–antiproton pair. The complete reaction is

$$p + p \rightarrow p + p + p + \bar{p}$$

For this reaction to occur, the minimum available energy E_a in Eq. (44.10) is the final rest energy $4mc^2$ of three protons and an antiproton. Equation (44.10) then gives

$$(4mc^2)^2 = 2mc^2(E_m + mc^2)$$
$$E_m = 7mc^2$$

EVALUATE: The energy E_m of the bombarding proton includes its rest energy mc^2, so its minimum *kinetic* energy must be $6mc^2 = 6(938 \text{ MeV}) = 5.63 \text{ GeV}$.

The search for the antiproton was a principal reason for the construction of the Bevatron at the University of California, Berkeley, with beam energy of 6 GeV. The search succeeded in 1955, and Emilio Segrè and Owen Chamberlain were later awarded the Nobel Prize for this discovery.

Strangeness

The K mesons and the Λ and Σ hyperons were discovered during the late 1950s. Because of their unusual behavior they were called *strange particles.* They were produced in high-energy collisions such as $\pi^- + p$, and a K meson and a hyperon were always produced *together.* The relatively high rate of production of these particles suggested that it was a *strong*-interaction process, but their relatively long lifetimes suggested that their decay was a *weak*-interaction process. The K^0 appeared to have *two* lifetimes, one about 9×10^{-11} s and another nearly 600 times longer. Were the K mesons strongly interacting hadrons or not?

The search for the answer to this question led physicists to introduce a new quantity called **strangeness.** The hyperons Λ^0 and $\Sigma^{\pm,0}$ were assigned a strangeness quantum number $S = -1$, and the associated K^0 and K^+ mesons were assigned $S = +1$. The corresponding antiparticles had opposite strangeness, $S = +1$ for $\overline{\Lambda}^0$ and $\overline{\Sigma}^{\pm,0}$ and $S = -1$ for $\overline{K}^0$ and K^-. Then strangeness was *conserved* in production processes such as

$$p + \pi^- \rightarrow \Sigma^- + K^+$$
$$p + \pi^- \rightarrow \Lambda^0 + K^0$$

The process

$$p + \pi^- \rightarrow p + K^-$$

does not conserve strangeness and it does not occur.

When strange particles decay individually, strangeness is usually *not* conserved. Typical processes include

$$\Sigma^+ \rightarrow n + \pi^+$$
$$\Lambda^0 \rightarrow p + \pi^-$$
$$K^- \rightarrow \pi^+ + \pi^- + \pi^-$$

CAUTION **Strangeness vs. spin** Take care not to confuse the symbol S for strangeness with the identical symbol for the magnitude of the spin angular momentum. ∎

In each of these decays, the initial strangeness is 1 or -1, and the final value is zero. All observations of these particles are consistent with the conclusion that *strangeness is conserved in strong interactions but it can change by zero or one unit in weak interactions.* There is no counterpart to the strangeness quantum number in classical physics.

Conservation Laws

The decay of strange particles provides our first example of a *conditional conservation law,* one that is obeyed in some interactions and not in others. By contrast, several conservation laws are obeyed in *all* interactions. These include the familiar conservation laws; energy, momentum, angular momentum, and electric charge. These are called *absolute conservation laws.* Baryon number and the three lepton numbers are also conserved in all interactions. Strangeness is conserved in strong and electromagnetic interactions but *not* in all weak interactions.

Two other quantities, which are conserved in some but not all interactions, are useful in classifying particles and their interactions. One is *isospin,* a quantity that is used to describe the charge independence of the strong interactions. The other is *parity,* which describes the comparative behavior of two systems that are mirror images of each other. Isospin is conserved in strong interactions, which are charge independent, but not in electromagnetic or weak interactions. (The electromagnetic interaction is certainly *not* charge independent.) Parity is conserved in strong and electromagnetic interactions but not in weak ones. The Chinese-American physicists T. D. Lee and C. N. Yang received the Nobel Prize in 1957 for laying the theoretical foundations for nonconservation of parity in weak interactions.

This discussion shows that conservation laws provide another basis for classifying particles and their interactions. Each conservation law is also associated with a *symmetry* property of the system. A familiar example is angular momentum. If a system is in an environment that has spherical symmetry, no torque can act on it because the direction of the torque would violate the symmetry. In such a system, total angular momentum is *conserved.* When a conservation law is violated, the interaction may be described as a *symmetry-breaking interaction.*

TEST YOUR UNDERSTANDING OF SECTION 44.3 From conservation of energy, a particle of mass m and rest energy mc^2 can decay only if the decay products have a total mass less than m. (The remaining energy goes into the kinetic energy of the decay products.) Can a proton decay into less massive mesons? ∎

44.4 QUARKS AND GLUONS

The leptons form a fairly neat package: three particles and three neutrinos, each with its antiparticle, and a conservation law relating their numbers. Physicists believe that leptons are genuinely fundamental particles. The hadron family, by comparison, is a mess. Table 44.3 (in Section 44.3) contains only a sample of well over 100 hadrons that have been discovered since 1960, and it has become clear that these particles *do not* represent the most fundamental level of the structure of matter.

Our present understanding of the structure of hadrons is based on a proposal made initially in 1964 by the American physicist Murray Gell-Mann and his collaborators. In this proposal, hadrons are not fundamental particles but are composite structures whose constituents are spin-$\frac{1}{2}$ fermions called **quarks.** (The name is from the line "Three quarks for Muster Mark!" from *Finnegans Wake,* by James Joyce.) Each baryon is composed of three quarks (qqq), each antibaryon of three antiquarks ($\bar{q}\bar{q}\bar{q}$), and each meson of a quark–antiquark pair ($q\bar{q}$). Table 44.3 gives the quark content of many hadrons. No other compositions seem to be necessary. This scheme requires that quarks have electric charges with magnitudes $\frac{1}{3}$ and $\frac{2}{3}$ of the electron charge e, which had been thought to be the smallest unit of charge. Each quark also has a fractional value $\frac{1}{3}$ for its baryon number B, and each antiquark has a baryon-number value $-\frac{1}{3}$. In a meson, a quark and antiquark combine with net baryon number 0 and can have their spin angular momentum components parallel to form a spin-1 meson or antiparallel to form a spin-0 meson. Similarly, the three quarks in a baryon combine with net baryon number 1 and can form a spin-$\frac{1}{2}$ baryon or a spin-$\frac{3}{2}$ baryon.

The Three Original Quarks

The first (1964) quark theory included three types (called *flavors*) of quarks, labeled u (up), d (down), and s (strange). Their principal properties are listed in **Table 44.4**. The corresponding antiquarks $\bar{u}$, $\bar{d}$, and $\bar{s}$ have opposite values of charge Q, B, and S. Protons, neutrons, π and K mesons, and several hyperons can be constructed from these three quarks. For example, the proton quark content is uud. Checking Table 44.4, we see that the values of Q/e add to 1 and that the values of the baryon number B also add to 1, as we should expect. The neutron is udd, with total $Q = 0$ and $B = 1$. The π^+ meson is $u\bar{d}$, with $Q/e = 1$ and $B = 0$, and the K$^+$ meson is $u\bar{s}$. Checking the values of S for the quark content, we see that the proton, neutron, and π^+ have strangeness 0 and that the K$^+$ has strangeness 1, in agreement with Table 44.3. The antiproton is $\bar{p} = \bar{u}\bar{u}\bar{d}$, the negative pion is $\pi^- = \bar{u}d$, and so on. The quark content can also be used to explain hadron excited states and magnetic moments. **Figure 44.10** shows the quark content of two baryons and two mesons.

44.10 Quark content of four hadrons. The various color combinations that are needed for color neutrality are not shown.

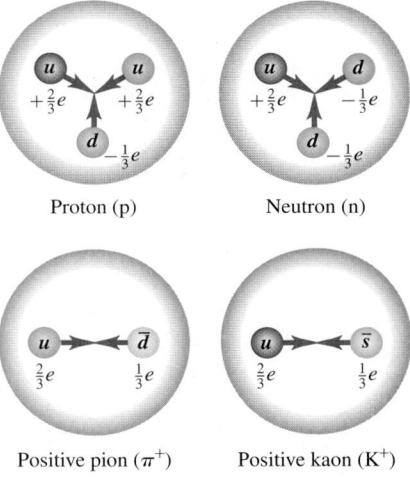

Proton (p) Neutron (n)

Positive pion (π^+) Positive kaon (K$^+$)

TABLE 44.4	Properties of the Three Original Quarks						
Symbol	Q/e	Spin	Baryon Number, B	Strangeness, S	Charm, C	Bottomness, B'	Topness, T
u	$\frac{2}{3}$	$\frac{1}{2}$	$\frac{1}{3}$	0	0	0	0
d	$-\frac{1}{3}$	$\frac{1}{2}$	$\frac{1}{3}$	0	0	0	0
s	$-\frac{1}{3}$	$\frac{1}{2}$	$\frac{1}{3}$	-1	0	0	0

EXAMPLE 44.7 DETERMINING THE QUARK CONTENT OF BARYONS

Given that they contain only u, d, s, $\bar{u}$, $\bar{d}$, and/or $\bar{s}$, find the quark content of (a) Σ^+ and (b) $\bar{\Lambda}^0$. The Σ^+ and Λ^0 (the antiparticle of the $\bar{\Lambda}^0$) are both baryons with strangeness $S = -1$.

SOLUTION

IDENTIFY and SET UP: We use the idea that the total charge of each baryon is the sum of the individual quark charges, and similarly for the baryon number and strangeness. We use the quark properties given in Table 44.4.

EXECUTE: Baryons contain three quarks. If $S = -1$, exactly *one* of the three must be an s quark, which has $S = -1$ and $Q/e = -\frac{1}{3}$.

(a) The Σ^+ has $Q/e = +1$, so the other two quarks must both be u quarks (each of which has $Q/e = +\frac{2}{3}$). Hence the quark content of Σ^+ is uus.

(b) First we find the quark content of the Λ^0. To yield zero total charge, the other two quarks must be u ($Q/e = +\frac{2}{3}$) and d ($Q/e = -\frac{1}{3}$), so the quark content of the Λ^0 is uds. The quark content of the $\bar{\Lambda}^0$ is therefore $\bar{u}\,\bar{d}\,\bar{s}$.

EVALUATE: Although the Λ^0 and $\bar{\Lambda}^0$ are both electrically neutral and have the same mass, they are different particles: Λ^0 has $B = 1$ and $S = -1$, while $\bar{\Lambda}^0$ has $B = -1$ and $S = 1$.

Motivating the Quark Model

What caused physicists to suspect that hadrons were made up of something smaller? The magnetic moment of the neutron (see Section 43.1) was one of the first reasons. In Section 27.7 we learned that a magnetic moment results from a circulating current (a motion of electric charge). But the neutron has *no* charge, or, to be more accurate, no *total* charge. It could be made up of smaller particles whose charges add to zero. The quantum motion of these particles within the neutron would then give its surprising nonzero magnetic moment. To verify this hypothesis by "seeing" inside a neutron, we need a probe with a wavelength that is much less than the neutron's size of about a femtometer. This probe should not be affected by the strong interaction, so that it won't interact with the neutron as a whole but will penetrate into it and interact electromagnetically with these supposed smaller charged particles. A probe with these properties is an electron with energy above 10 GeV. In experiments carried out at SLAC, such electrons were scattered from neutrons and protons to help show that nucleons are indeed made up of fractionally charged, spin-$\frac{1}{2}$ pointlike particles.

The Eightfold Way

Symmetry considerations play a very prominent role in particle theory. Here are two examples. Consider the eight spin-$\frac{1}{2}$ baryons we've mentioned: the familiar p and n; the strange Λ^0, Σ^+, Σ^0, and Σ^-; and the doubly strange Ξ^0 and Ξ^-. For each we plot the value of strangeness S versus the value of charge Q in **Fig. 44.11**. The result is a hexagonal pattern. A similar plot for the nine spin-0 mesons (six shown in Table 44.3 plus three others not included in that table) is shown in **Fig. 44.12**; the particles fall in exactly the same hexagonal pattern! In each plot, all the particles have masses that are within about $\pm 200 \text{ MeV}/c^2$ of the median mass value of that plot, with variations due to differences in quark masses and internal potential energies.

The symmetries that lead to these and similar patterns are collectively called the **eightfold way.** They were discovered in 1961 by Murray Gell-Mann and independently by Yu'val Ne'eman. (The name is a slightly irreverent reference to the Noble Eightfold Path, a set of principles for right living in Buddhism.) A similar pattern for the spin-$\frac{3}{2}$ baryons contains *ten* particles, arranged in a triangular

44.11 (a) Plot of S and Q values for spin-$\frac{1}{2}$ baryons, showing the symmetry pattern of the eightfold way. (b) Quark content of each spin-$\frac{1}{2}$ baryon. The quark contents of the Σ^0 and Λ^0 are the same; the Σ^0 is an excited state of the Λ^0 and can decay into it by photon emission.

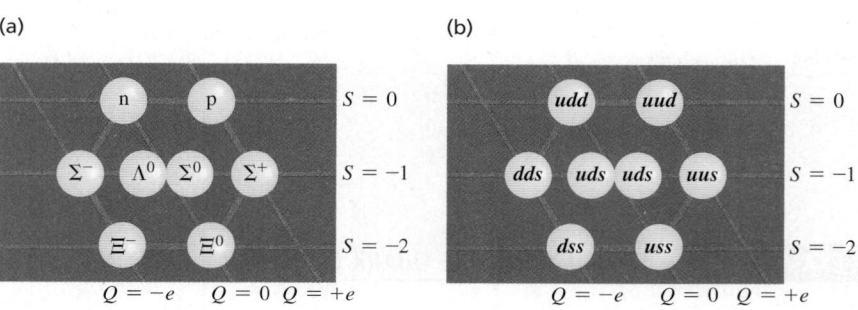

44.12 (a) Plot of S and Q values for nine spin-0 mesons, showing the symmetry pattern of the eightfold way. Each particle is on the opposite side of the hexagon from its antiparticle; each of the three particles in the center is its own antiparticle. (b) Quark content of each spin-0 meson. The particles in the center are different mixtures of the three quark–antiquark pairs shown.

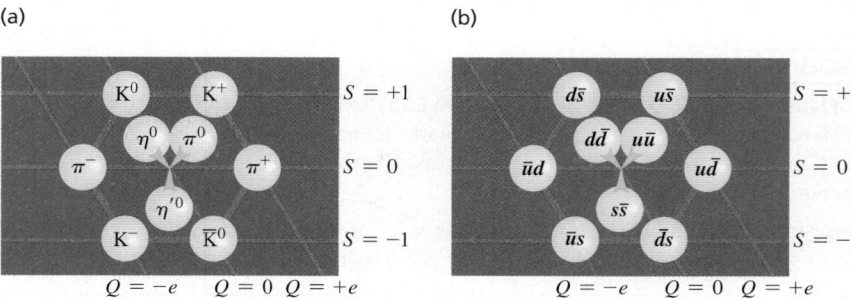

pattern like pins in a bowling alley. When this pattern was first discovered, one of the particles was missing. But Gell-Mann gave it a name anyway (Ω^-), predicted the properties it should have, and told experimenters what they should look for. Three years later, the particle was found during an experiment at Brookhaven National Laboratory, a spectacular success for Gell-Mann's theory. The whole series of events is reminiscent of the way in which Mendeleev used gaps in the periodic table of the elements to predict properties of undiscovered elements and to guide chemists in their search for these elements.

What binds quarks to one another? The attractive interactions among quarks are mediated by massless spin-1 bosons called **gluons** in much the same way that photons mediate the electromagnetic interaction or that pions mediated the nucleon–nucleon force in the old Yukawa theory.

Color

Quarks, having spin $\frac{1}{2}$, are fermions and so are subject to the exclusion principle. This would seem to forbid a baryon having two or three quarks with the same flavor and same spin component. To avoid this difficulty, it is assumed that each quark comes in three varieties, which are whimsically called *colors*. Red, green, and blue are the usual choices. The exclusion principle applies separately to each color. A baryon always contains one red, one green, and one blue quark, so the baryon itself has no net color. Each gluon has a color–anticolor combination (for example, blue–antired) that allows it to transmit color when exchanged, and color is conserved during emission and absorption of a gluon by a quark. The gluon-exchange process changes the colors of the quarks in such a way that there is always one quark of each color in every baryon. The color of an individual quark changes continually as gluons are exchanged.

Similar processes occur in mesons such as pions. The quark–antiquark pairs of mesons have canceling color and anticolor (for example, blue and antiblue), so mesons also have no net color. Suppose a pion initially consists of a blue quark and an antiblue antiquark. The blue quark can become a red quark by emitting a blue–antired virtual gluon. The gluon is then absorbed by the antiblue antiquark, converting it to an antired antiquark (**Fig. 44.13**). Color is conserved in each emission and absorption, but a blue–antiblue pair has become a red–antired pair. Such changes occur continually, so we have to think of a pion as a superposition of three quantum states: blue–antiblue, green–antigreen, and red–antired. On a larger scale, the strong interaction between nucleons was described in Section 44.3 as due to the exchange of virtual mesons. In terms of quarks and gluons, these mediating virtual mesons are quark–antiquark systems bound together by the exchange of gluons.

The theory of strong interactions is known as *quantum chromodynamics* (QCD). No one has been able to isolate an individual quark, and indeed QCD predicts that quarks are bound in such a way that it is impossible to obtain a free quark. An impressive body of experimental evidence supports the correctness of the quark model and the idea that quantum chromodynamics is the key to understanding the strong interactions.

Three More Quarks

Before the tau particles were discovered, there were four known leptons. This fact, together with some puzzling decay rates, led to the speculation that there might be a fourth quark flavor. This quark is labeled c (the *charmed* quark); it has $Q/e = \frac{2}{3}$, $B = \frac{1}{3}$, $S = 0$, and a new quantum number **charm** $C = +1$. This was confirmed in 1974 by the observation at both SLAC and the Brookhaven National Laboratory of a meson, now named ψ, with mass 3097 MeV/c^2. This meson was found to have several decay modes, decaying into e^+e^-, $\mu^+\mu^-$, or hadrons. The mean lifetime was found to be about 10^{-20} s. These results are consistent with ψ being a spin-1 $c\bar{c}$ system. Almost immediately after this,

44.13 (a) A pion containing a blue quark and an antiblue antiquark. (b) The blue quark emits a blue–antired gluon, changing to a red quark. (c) The gluon is absorbed by the antiblue antiquark, which becomes an antired antiquark. The pion now consists of a red–antired quark–antiquark pair. The actual quantum state of the pion is an equal superposition of red–antired, green–antigreen, and blue–antiblue pairs.

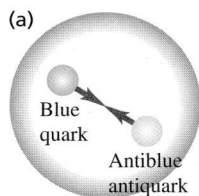

(a)

Blue quark

Antiblue antiquark

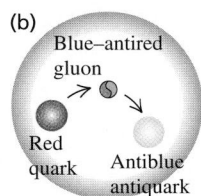

(b)

Blue–antired gluon

Red quark

Antiblue antiquark

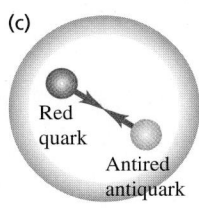

(c)

Red quark

Antired antiquark

TABLE 44.5 Properties of the Six Quarks

Symbol	Q/e	Spin	Baryon Number, B	Strangeness, S	Charm, C	Bottomness, B'	Topness, T
u	$\frac{2}{3}$	$\frac{1}{2}$	$\frac{1}{3}$	0	0	0	0
d	$-\frac{1}{3}$	$\frac{1}{2}$	$\frac{1}{3}$	0	0	0	0
s	$-\frac{1}{3}$	$\frac{1}{2}$	$\frac{1}{3}$	-1	0	0	0
c	$\frac{2}{3}$	$\frac{1}{2}$	$\frac{1}{3}$	0	$+1$	0	0
b	$-\frac{1}{3}$	$\frac{1}{2}$	$\frac{1}{3}$	0	0	-1	0
t	$\frac{2}{3}$	$\frac{1}{2}$	$\frac{1}{3}$	0	0	0	$+1$

similar mesons of greater mass were observed and identified as excited states of the $c\bar{c}$ system. A few years later, individual mesons with a nonzero net charm quantum number, D^0 ($c\bar{u}$) and D^+ ($c\bar{d}$), and a charmed baryon, Λ_c^+ (udc), were also observed.

In 1977 a meson with mass 9460 MeV/c^2, called upsilon (Υ), was discovered at Brookhaven. Because it had properties similar to ψ, it was conjectured that the meson was really the bound system of a new quark, b (the *bottom* quark), and its antiquark, $\bar{b}$. The bottom quark has the value -1 of a new quantum number B' called *bottomness*. Excited states of the Υ were soon observed, as were the B^+ ($\bar{b}u$) and B^0 ($\bar{b}d$) mesons.

With the five flavors of quarks (u, d, s, c, and b) and the six flavors of leptons (e, μ, τ, ν_e, ν_μ, and ν_τ) it was an appealing conjecture that nature is symmetric in its building blocks and that therefore there should be a *sixth* quark. This quark, labeled t (top), would have $Q/e = \frac{2}{3}$, $B = \frac{1}{3}$, and a new quantum number, $T = 1$. In 1995, groups using two different detectors at Fermilab's Tevatron announced the discovery of the top quark. The groups collided 0.9-TeV protons with 0.9-TeV antiprotons, but even with 1.8 TeV of available energy, a top–antitop ($t\bar{t}$) pair was detected in fewer than two of every 10^{11} collisions! **Table 44.5** lists some properties of the six quarks. Each has a corresponding antiquark with opposite values of Q, B, S, C, B', and T.

CAUTION Bottomness vs. baryon number Don't confuse the bottomness quantum number B' with baryon number B. For example, the proton (which has zero bottomness and is a baryon) has $B' = 0$ and $B = +1$; the B^+ meson (which includes an antibottom quark but is not a baryon) has $B' = +1$ and $B = 0$.

TEST YOUR UNDERSTANDING OF SECTION 44.4 Is it possible to have a baryon with charge $Q = +e$ and strangeness $S = -2$? ∎

44.5 THE STANDARD MODEL AND BEYOND

The particles and interactions that we've discussed in this chapter provide a reasonably comprehensive picture of the fundamental building blocks of nature. There is enough confidence in the basic correctness of this picture that it is called the **standard model.**

The standard model includes three families of particles: (1) the six leptons, which have no strong interactions; (2) the six quarks, from which all hadrons are made; and (3) the particles that mediate the various interactions. These mediators are gluons for the strong interaction among quarks, photons for the electromagnetic interaction, the $W^\pm$ and Z^0 particles for the weak interaction, and the graviton for the gravitational interaction.

Electroweak Unification

Theoretical physicists have long dreamed of combining all the interactions of nature into a single unified theory. As a first step, Einstein spent much of his later life trying to develop a field theory that would unify gravitation and electromagnetism. He was only partly successful.

Between 1961 and 1967, Sheldon Glashow, Abdus Salam, and Steven Weinberg developed a theory that unifies the weak and electromagnetic forces. One outcome of their **electroweak theory** is a prediction of the weak-force mediator particles, the $W^{\pm}$ and Z^0 bosons, including their masses. The basic idea is that the mass difference between photons (zero mass) and the weak bosons ($\approx 100 \text{ GeV}/c^2$) makes the electromagnetic and weak interactions behave quite differently at low energies. At sufficiently high energies (well above 100 GeV), however, the distinction disappears, and the two merge into a single interaction. This prediction was verified in 1983 in experiments with proton-antiproton collisions at CERN. The weak bosons were found, again with the help provided by the theoretical description, and their observed masses agreed with the predictions of the electroweak theory, a wonderful convergence of theory and experiment. The electroweak theory and quantum chromodynamics form the backbone of the standard model. Glashow, Salam, and Weinberg received the Nobel Prize in 1979.

In the electroweak theory photons are massless but the weak bosons are very massive. To account for the broken symmetry among these interaction mediators, a field called the *Higgs field* was proposed by theoretical physicists in the 1960s. We use the symbol $\phi(\vec{r}, t)$ to denote the value of this field at position $\vec{r}$ and time t. (Unlike the electric and magnetic fields, which are vectors, the Higgs field is a scalar quantity.) According to the theory, the mass of the weak bosons is proportional to the absolute value of ϕ_{av}, where ϕ_{av} is the average value of $\phi(\vec{r}, t)$ over space. **Figure 44.14** shows a simplified model of how ϕ_{av} depends on energy. At very high energies, the value of the Higgs field $\phi(\vec{r}, t)$ oscillates between positive and negative values, so its average value is $\phi_{av} = 0$ (Fig. 44.14a). But at low energies, $\phi(\vec{r}, t)$ oscillates around either a positive average value $\phi_{av} = +\phi_0$ or a negative average value $\phi_{av} = -\phi_0$ (Fig. 44.14b). The oscillation is no longer symmetric around $\phi = 0$, so the symmetry has been broken. Hence at low energies, the weak bosons acquire a nonzero mass proportional to $|\phi_{av}|$, which is equal to ϕ_0 for either of the cases shown in Fig. 44.14b.

This theory also predicts that there should be a particle called the *Higgs boson* associated with the Higgs field itself. (In an analogous way, the photon is the particle associated with the electromagnetic field.) The Higgs boson was predicted to be unstable, have zero charge and spin 0, and have a large mass. An important mission of the Large Hadron Collider at CERN was to produce the Higgs boson from the available energy in proton-proton collisions and thereby verify the existence of the Higgs field. In 2012 the first Higgs bosons were detected in such collisions, with the predicted properties. This suggests that the concept of the Higgs field—that all of space is filled with a field that gives mass to the weak bosons—is indeed correct. The Nobel Prize was awarded in 2013 to François Englert and Peter Higgs, two of the theorists who first proposed the idea of the Higgs field in 1964. Current experiments show that the mass of the Higgs boson is about $125 \text{ GeV}/c^2$, even greater than the masses of the $W^{\pm}$ and Z^0 weak bosons.

Grand Unified Theories

Perhaps at sufficiently high energies the strong interaction and the electroweak interaction have a convergence similar to that between the electromagnetic and weak interactions. If so, they can be unified to give a comprehensive theory of strong, weak, and electromagnetic interactions. Such schemes, called **grand unified theories** (GUTs), are still speculative.

Some grand unified theories predict the decay of the proton (in violation of conservation of baryon number), with an estimated lifetime of more than 10^{28} years. (For comparison the age of the universe is known to be 1.38×10^{10} years.) With a lifetime of 10^{28} years, six metric tons of protons would be expected to have only one decay per day, so huge amounts of material must be examined. Some of the neutrino detectors that we mentioned in Section 44.2 originally looked for, and failed to find, evidence of proton decay. Current estimates set the proton lifetime well over 10^{33} years. Some GUTs also predict the existence of magnetic monopoles, which we mentioned in Chapter 27. At present there is no confirmed experimental evidence that magnetic monopoles exist.

44.14 The Higgs-field value ϕ can oscillate, much like the value of the coordinate of a ball rolling within a trough with two minima. The average value of ϕ determines the masses of the $W^{\pm}$ and Z^0 weak bosons. **(a)** At high energies, the average value of ϕ is zero and the $W^{\pm}$ and Z^0 are massless (like the photon). **(b)** At low energies, the symmetry is broken. The average value of ϕ is nonzero, and the $W^{\pm}$ and Z^0 acquire nonzero masses.

(a) When the energy E of the system is high, the ball can oscillate between these two limits ...

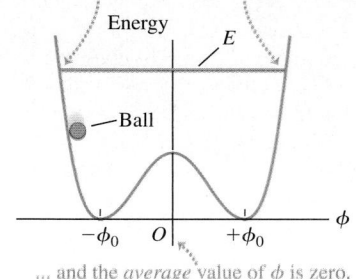

... and the *average* value of ϕ is zero.

(b) When the energy E of the system is low, the ball is trapped near one of the two minima ...

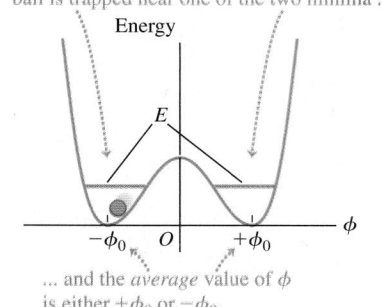

... and the *average* value of ϕ is either $+\phi_0$ or $-\phi_0$.

44.15 This photo shows the interior of the Super-Kamiokande neutrino detector in Japan. When in operation, the detector is filled with 5×10^7 kg of water. A neutrino passing through the detector can produce a faint flash of light, which is detected by the 13,000 photomultiplier tubes lining the detector walls. Data from this detector were the first to indicate that neutrinos have mass.

In the standard model, the neutrinos have zero mass. Nonzero values are controversial because experiments to determine neutrino masses are difficult both to perform and to analyze. In most GUTs the neutrinos *must* have nonzero masses. If neutrinos do have mass, transitions called *neutrino oscillations* can occur, in which one type of neutrino (ν_e, ν_μ, or ν_τ) changes into another type. In 1998, scientists using the Super-Kamiokande neutrino detector in Japan (**Fig. 44.15**) reported the discovery of oscillations between muon neutrinos and tau neutrinos. Subsequent measurements at the Sudbury Neutrino Observatory in Canada have confirmed the existence of neutrino oscillations. This discovery is evidence for exciting physics beyond that predicted by the standard model.

The discovery of neutrino oscillations cleared up a long-standing mystery. Since the 1960s, physicists have been using sensitive detectors to look for electron neutrinos produced by nuclear fusion reactions in the sun's core (see Section 43.8). However, the observed flux of solar electron neutrinos is only one-third of the predicted value. The explanation was provided in 2002 by the Sudbury Neutrino Observatory, which can detect neutrinos of all three flavors. The results showed that the combined flux of solar neutrinos of *all* flavors is equal to the theoretical prediction for the flux of *electron* neutrinos. The explanation is that the sun is producing electron neutrinos at the predicted rate, but that two-thirds of these electron neutrinos are transformed into muon or tau neutrinos during their flight from the sun's core to a detector on earth.

Supersymmetric Theories and TOEs

The ultimate dream of theorists is to unify all four fundamental interactions, adding gravitation to the strong and electroweak interactions that are included in GUTs. Such a unified theory is whimsically called a Theory of Everything (TOE). It turns out that an essential ingredient of such theories is a space-time continuum with more than four dimensions. The additional dimensions are "rolled up" into extremely tiny structures that we ordinarily do not notice. Depending on the scale of these structures, it may be possible for the next generation of particle accelerators to reveal the presence of extra dimensions.

Another ingredient of many theories is *supersymmetry,* which gives every boson and fermion a "superpartner" of the other spin type. For example, the proposed supersymmetric partner of the spin-$\frac{1}{2}$ electron is a spin-0 particle called the *selectron,* and that of the spin-1 photon is a spin-$\frac{1}{2}$ *photino.* As yet, no superpartner particles have been discovered, perhaps because they are too massive to be produced by the present generation of accelerators. Within a few years, new data from the Large Hadron Collider will help us decide whether these intriguing theories have merit.

TEST YOUR UNDERSTANDING OF SECTION 44.5 One aspect of the standard model is that a *d* quark can transform into a *u* quark, an electron, and an antineutrino by means of the weak interaction. If this happens to a *d* quark inside a neutron, what kind of particle remains afterward in addition to the electron and antineutrino? (i) A proton; (ii) a Σ^-; (iii) a Σ^+; (iv) a Λ^0 or a Σ^0; (v) any of these. ▌

44.6 THE EXPANDING UNIVERSE

In the last two sections of this chapter we'll explore briefly the connections between the early history of the universe and the interactions of fundamental particles. It is remarkable that there are such close ties between physics on the smallest scale that we've explored experimentally (the range of the weak interaction, of the order of 10^{-18} m) and physics on the largest scale (the universe itself, of the order of at least 10^{26} m).

Gravitational interactions play an essential role in the large-scale behavior of the universe. We saw in Chapter 13 how the law of gravitation explains the motions of planets in the solar system. Astronomical evidence shows that gravitational forces also dominate in larger systems such as galaxies and clusters of galaxies (**Fig. 44.16**).

Until early in the 20th century it was usually assumed that the universe was *static;* stars might move relative to each other, but there was not thought to be any overall expansion or contraction. But measurements that were begun in 1912 by Vesto Slipher at Lowell Observatory in Arizona, and continued in the 1920s by Edwin Hubble with the help of Milton Humason at Mount Wilson in California, indicated that the universe is *not* static. The motions of galaxies relative to the earth can be measured by observing the shifts in the wavelengths of their spectra. For distant galaxies these shifts are always toward longer wavelength, so they appear to be receding from us and from each other. Astronomers first assumed that these were Doppler shifts and used a relationship between the wavelength λ_0 of light measured now on earth from a source receding at speed v and the wavelength λ_S measured in the rest frame of the source when it was emitted. We can derive this relationship by inverting Eq. (37.25) for the Doppler effect, making subscript changes, and using $\lambda = c/f$; the result is

$$\lambda_0 = \lambda_S \sqrt{\frac{c + v}{c - v}} \qquad (44.13)$$

Wavelengths from receding sources are always shifted toward longer wavelengths; this increase in λ is called the **redshift.** We can solve Eq. (44.13) for v:

$$v = \frac{(\lambda_0/\lambda_S)^2 - 1}{(\lambda_0/\lambda_S)^2 + 1}c \qquad (44.14)$$

CAUTION **Redshift, not Doppler shift** Equations (44.13) and (44.14) are from the *special* theory of relativity and refer to the Doppler effect. As we'll see, the redshift from *distant* galaxies is caused by an effect that is explained by the *general* theory of relativity and is *not* a Doppler shift. However, as the ratio v/c and the fractional wavelength change $(\lambda_0 - \lambda_S)/\lambda_S$ become small, the general theory's equations approach Eqs. (44.13) and (44.14), and those equations may be used. ▌

44.16 (a) The galaxy M101 is a larger version of the Milky Way galaxy of which our solar system is a part. Like all galaxies, M101 is held together by the mutual gravitational attraction of its stars, gas, dust, and other matter, all of which orbit around the galaxy's center of mass. M101 is 25 million light-years away. (b) This image shows part of the Coma cluster, an immense grouping of over 1000 galaxies that lies 300 million light-years from us. The galaxies within the cluster are all in motion. Gravitational forces between the galaxies prevent them from escaping from the cluster.

(a)

(b)

EXAMPLE 44.8   **RECESSION SPEED OF A GALAXY**

The spectral lines of various elements are detected in light from a galaxy in the constellation Ursa Major. An ultraviolet line from singly ionized calcium ($\lambda_S = 393$ nm) is observed at wavelength $\lambda_0 = 414$ nm, redshifted into the visible portion of the spectrum. At what speed is this galaxy receding from us?

SOLUTION

IDENTIFY and SET UP: This example uses the relationship between redshift and recession speed for a distant galaxy. We can use the wavelengths λ_S at which the light is emitted and λ_0 that we detect on earth in Eq. (44.14) to determine the galaxy's recession speed v if the fractional wavelength shift is not too great.

EXECUTE: The fractional wavelength redshift for this galaxy is $\lambda_0/\lambda_S = (414 \text{ nm})/(393 \text{ nm}) = 1.053$. This is only a 5.3% increase, so we can use Eq. (44.14) with reasonable accuracy:

$$v = \frac{(1.053)^2 - 1}{(1.053)^2 + 1}c = 0.0516c = 1.55 \times 10^7 \text{ m/s}$$

EVALUATE: The galaxy is receding from the earth at 5.16% of the speed of light. Rather than going through this calculation, astronomers often just state the *redshift* $z = (\lambda_0 - \lambda_S)/\lambda_S = (\lambda_0/\lambda_S) - 1$. This galaxy has redshift $z = 0.053$.

44.17 Graph of recession speed versus distance for several galaxies. The best-fit straight line illustrates Hubble's law. The slope of the line is the Hubble constant, H_0.

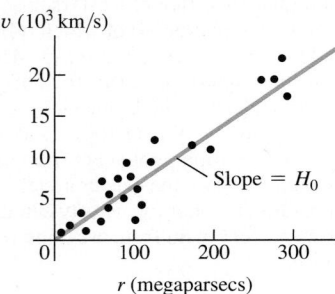

The Hubble Law

Analysis of redshifts from many distant galaxies led Edwin Hubble to a remarkable conclusion: The speed of recession v of a galaxy is proportional to its distance r from us (**Fig. 44.17**). This relationship is now called the **Hubble law;** expressed as an equation,

$$v = H_0 r \qquad (44.15)$$

where H_0 is an experimental quantity commonly called the *Hubble constant,* since at any given time it is constant over space. Determining H_0 has been a key goal of the Hubble Space Telescope, which can measure distances to galaxies with unprecedented accuracy. The current best value is $2.18 \times 10^{-18}\ \text{s}^{-1}$, with an uncertainty of 2%.

Astronomical distances are often measured in *parsecs* (pc); one parsec is the distance at which there is a one-arcsecond ($1/3600°$) angular separation between two objects 1.50×10^{11} m apart (the average distance from the earth to the sun). A distance of 1 pc is equal to 3.26 *light-years* (ly), where 1 ly $= 9.46 \times 10^{12}$ km is the distance that light travels in one year. The Hubble constant is then commonly expressed in the mixed units (km/s)/Mpc (kilometers per second per megaparsec), where 1 Mpc $= 10^6$ pc:

$$H_0 = (2.18 \times 10^{-18}\ \text{s}^{-1})\left(\frac{9.46 \times 10^{12}\ \text{km}}{1\ \text{ly}}\right)\left(\frac{3.26\ \text{ly}}{1\ \text{pc}}\right)\left(\frac{10^6\ \text{pc}}{1\ \text{Mpc}}\right) = 67.3\ \frac{\text{km/s}}{\text{Mpc}}$$

EXAMPLE 44.9 **DETERMINING DISTANCE WITH THE HUBBLE LAW**

Use the Hubble law to find the distance from earth to the galaxy in Ursa Major described in Example 44.8.

SOLUTION

IDENTIFY and SET UP: The Hubble law relates the redshift of a distant galaxy to its distance r from earth. We solve Eq. (44.15) for r and substitute the recession speed v from Example 44.8.

EXECUTE: Using $H_0 = 67.3\ (\text{km/s})/\text{Mpc} = 6.73 \times 10^4\ (\text{m/s})/\text{Mpc}$,

$$r = \frac{v}{H_0} = \frac{1.55 \times 10^7\ \text{m/s}}{6.73 \times 10^4\ (\text{m/s})/\text{Mpc}} = 230\ \text{Mpc}$$

$$= 2.3 \times 10^8\ \text{pc} = 7.5 \times 10^8\ \text{ly} = 7.1 \times 10^{24}\ \text{m}$$

EVALUATE: A distance of 230 million parsecs (750 million light-years) is truly stupendous, but many galaxies lie much farther away. To appreciate the immensity of this distance, consider that our farthest-ranging unmanned spacecraft have traveled only about 0.002 ly from our planet.

Another aspect of Hubble's observations was that, *in all directions,* distant galaxies appeared to be receding from us. There is no particular reason to think that our galaxy is at the very center of the universe; if we lived in some other galaxy, every distant galaxy would still seem to be moving away. That is, at any given time, *the universe looks more or less the same, no matter where in the universe we are.* This important idea is called the **cosmological principle.** There are local fluctuations in density, but on average, the universe looks the same from all locations. Thus the Hubble constant is constant in space although not necessarily constant in time, and the laws of physics are the same everywhere.

The Big Bang

The Hubble law suggests that at some time in the past, all the matter in the universe was far more concentrated than it is today. It was then blown apart in a rapid expansion called the **Big Bang,** giving all observable matter more or less the velocities that we observe today. When did this happen? According to the

Hubble law, matter at a distance r away from us is traveling with speed $v = H_0 r$. The time t needed to travel a distance r is

$$t = \frac{r}{v} = \frac{r}{H_0 r} = \frac{1}{H_0} = 4.59 \times 10^{17} \text{ s} = 1.45 \times 10^{10} \text{ y}$$

By this hypothesis the Big Bang occurred about 14 billion years ago. It assumes that all speeds are *constant* after the Big Bang; that is, it ignores any change in the expansion rate due to gravitational attraction or other effects. We'll return to this point later. For now, however, notice that the age of the earth determined from radioactive dating (see Section 43.4) is 4.54 billion (4.54×10^9) years. It's encouraging that our hypothesis tells us that the universe is older than the earth!

Expanding Space

The general theory of relativity takes a radically different view of the expansion just described. According to this theory, the increased wavelength is *not* caused by a Doppler shift as the universe expands into a previously empty void. Rather, the increase comes from *the expansion of space itself* and everything in intergalactic space, including the wavelengths of light traveling to us from distant sources. This is not an easy concept to grasp, and if you haven't encountered it before, it may sound like doubletalk.

An analogy may help you develop some intuition on this point. Imagine we are all bugs crawling around on a horizontal surface. We can't leave the surface, and we can see in any direction along the surface, but not up or down. We are then living in a two-dimensional world; some writers have called such a world *flatland*. If the surface is a plane, we can locate our position with two Cartesian coordinates (x, y). If the plane extends indefinitely in both the x- and y-directions, we described our space as having *infinite* extent, or as being *unbounded*. No matter how far we go, we never reach an edge or a boundary.

An alternative habitat for us bugs would be the surface of a sphere of radius R. The space would still seem infinite—we could crawl forever and never reach an edge or a boundary. Yet in this case the space is *finite* or *bounded*. To describe the location of a point in this space, we could still use two coordinates: latitude and longitude, or the spherical coordinates θ and ϕ shown in Fig. 41.5.

Now suppose the spherical surface is that of a balloon (**Fig. 44.18**). As we inflate the balloon more and more, increasing the radius R, the coordinates of a point don't change, yet the distance between any two points gets larger and larger. Furthermore, as R increases, the *rate of change* of distance between two points (their recession speed) is proportional to their distance apart. *The recession speed is proportional to the distance,* just as with the Hubble law. For example, the distance from Pittsburgh to Miami is twice as great as the distance from Pittsburgh to Boston. If the earth were to begin to swell, Miami would recede from Pittsburgh twice as fast as Boston would.

Although the quantity R isn't one of the two coordinates giving the position of a point on the balloon's surface, it nevertheless plays an essential role in any discussion of distance. It is the radius of curvature of our two-dimensional universe, and it is also a varying *scale factor* that changes as this universe expands.

Generalizing this picture to three dimensions isn't so easy. We have to think of our three-dimensional space as being embedded in a space with four or more dimensions, just as we visualized the two-dimensional spherical flatland as being embedded in a three-dimensional Cartesian space. Our real three-space is *not* Cartesian; to describe its characteristics in any small region requires at least one additional parameter, the curvature of space, which is analogous to the radius of the sphere. In a sense, this scale factor, which we'll continue to call R, describes the *size* of the universe, just as the radius of the sphere described the size of our two-dimensional spherical universe. We'll return later to the question of whether the universe is bounded or unbounded.

44.18 An inflating balloon as an analogy for an expanding universe.

(a) Points (representing galaxies) on the surface of a balloon are described by their latitude and longitude coordinates.

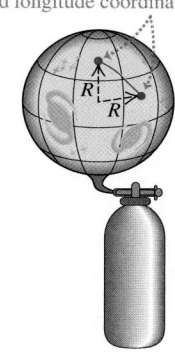

(b) The radius R of the balloon has increased. The coordinates of the points are the same, but the distance between them has increased.

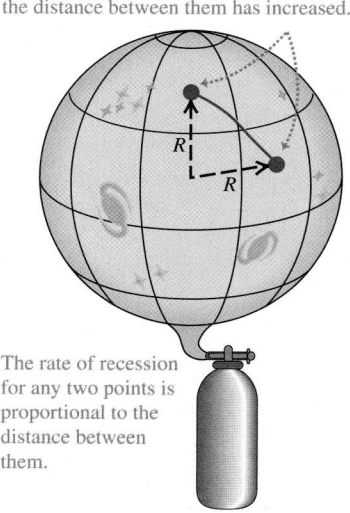

The rate of recession for any two points is proportional to the distance between them.

Any length that is measured in intergalactic space is proportional to R, so the wavelength of light traveling to us from a distant galaxy increases along with every other dimension as the universe expands. That is,

$$\frac{\lambda_0}{\lambda} = \frac{R_0}{R} \qquad (44.16)$$

The zero subscripts refer to the values of the wavelength and scale factor *now*, just as H_0 is the current value of the Hubble constant. The quantities λ and R without subscripts are the values at *any* time—past, present, or future. For the galaxy described in Examples 44.8 and 44.9, we have $\lambda_0 = 414$ nm and $\lambda = \lambda_S = 393$ nm, so Eq. (44.16) gives $R_0/R = 1.053$. That is, the scale factor *now* (R_0) is 5.3% larger than it was 750 million years ago when the light was emitted from that galaxy in Ursa Major. This increase of wavelength with time as the scale factor increases in our expanding universe is called the *cosmological redshift*. The farther away an object is, the longer its light takes to get to us and the greater the change in R and λ. The current largest measured wavelength ratio for galaxies is about 8.6, meaning that the volume of space itself is about $(8.6)^3 \approx 640$ times larger than it was when the light was emitted. Do *not* attempt to substitute $\lambda_0/\lambda_S = 8.6$ into Eq. (44.14) to find the recession speed; that equation is accurate only for small cosmological redshifts and $v \ll c$. The actual value of v depends on the density of the universe, the value of H_0, and the expansion history of the universe.

Here's a surprise: If the distance from us in the Hubble law is large enough, then the speed of recession is greater than the speed of light! This does *not* violate the special theory of relativity because the recession speed is *not* caused by the motion of the astronomical object relative to some coordinates in its region of space. Rather, $v > c$ when two sets of coordinates move apart fast enough as space itself expands. In other words, there are objects whose coordinates have been moving away from our coordinates so fast that light from them hasn't had enough time in the entire history of the universe to reach us. What we see is just the *observable* universe; we have no direct evidence about what lies beyond its horizon.

CAUTION **The universe isn't expanding into emptiness** The balloon shown in Fig. 44.18 is expanding into the empty space around it. It's a common misconception to picture the universe in the same way as a large but finite collection of galaxies that's expanding into unoccupied space. The reality is quite different! All evidence shows that our universe is *infinite:* It has no edges, so there is nothing "outside" it and it isn't "expanding into" anything. The expansion of the universe simply means that the scale factor of the universe is increasing. A good two-dimensional analogy is to think of the universe as a flat, infinitely large rubber sheet that's stretching and expanding much like the surface of the balloon in Fig. 44.18. In a sense, the infinite universe is becoming more infinite! ▌

Critical Density

In an expanding universe, gravitational attractions between galaxies should slow the initial expansion. But by how much? If these attractions are strong enough, the universe should expand more and more slowly, eventually stop, and then begin to contract, perhaps all the way down to what's been called a *Big Crunch.* On the other hand, if gravitational forces are much weaker, they slow the expansion only a little, and the universe should continue to expand forever.

The situation is analogous to the problem of escape speed of a projectile launched from the earth. We studied this problem in Example 13.5 (Section 13.3). The total energy $E = K + U$ when a projectile of mass m and speed v is at a distance r from the center of the earth (mass m_E) is

$$E = \tfrac{1}{2}mv^2 - \frac{Gmm_E}{r}$$

If E is positive, the projectile has enough kinetic energy to move infinitely far from the earth $(r \to \infty)$ and have some kinetic energy left over. If E is negative, the kinetic energy $K = \frac{1}{2}mv^2$ becomes zero and the projectile stops when $r = -Gmm_\mathrm{E}/E$. In that case, no greater value of r is possible, and the projectile can't escape the earth's gravity.

We can carry out a similar analysis for the universe. Whether the universe continues to expand indefinitely should depend on the average *density* of matter. If matter is relatively dense, there is a lot of gravitational attraction to slow and eventually stop the expansion and make the universe contract again. If not, the expansion should continue indefinitely. We can derive an expression for the *critical density* ρ_c needed to just barely stop the expansion.

Here's a calculation based on Newtonian mechanics; it isn't relativistically correct, but it illustrates the idea. Consider a large sphere with radius R, containing many galaxies (**Fig. 44.19**), with total mass M. Suppose our own galaxy has mass m and is located at the surface of this sphere. According to the cosmological principle, the average distribution of matter within the sphere is uniform. The total gravitational force on our galaxy is just the force due to the mass M inside the sphere. The force on our galaxy and potential energy U due to this spherically symmetric distribution are the same as though m and M were both points, so $U = -GmM/R$, just as in Section 13.3. The net force from all the uniform distribution of mass *outside* the sphere is zero, so we'll ignore it.

The total energy E (kinetic plus potential) for our galaxy is

$$E = \tfrac{1}{2}mv^2 - \frac{GmM}{R} \tag{44.17}$$

If E is *positive*, our galaxy has enough energy to escape from the gravitational attraction of the mass M inside the sphere; in this case the universe should keep expanding forever. If E is negative, our galaxy cannot escape and the universe should eventually pull back together. The crossover between these two cases occurs when $E = 0$, so

$$\tfrac{1}{2}mv^2 = \frac{GmM}{R} \tag{44.18}$$

The total mass M inside the sphere is the volume $4\pi R^3/3$ times the density ρ_c:

$$M = \tfrac{4}{3}\pi R^3 \rho_\mathrm{c}$$

We'll assume that the speed v of our galaxy relative to the center of the sphere is given by the Hubble law: $v = H_0 R$. Substituting these expressions for m and v into Eq. (44.18), we get

$$\tfrac{1}{2}m(H_0 R)^2 = \frac{Gm}{R}\left(\tfrac{4}{3}\pi R^3 \rho_\mathrm{c}\right) \qquad \text{or}$$

$$\rho_\mathrm{c} = \frac{3H_0^2}{8\pi G} \qquad \text{(critical density of the universe)} \tag{44.19}$$

This is the *critical density*. If the average density is less than ρ_c, the universe should continue to expand indefinitely; if it is greater, the universe should eventually stop expanding and begin to contract.

Putting numbers into Eq. (44.19), we find

$$\rho_\mathrm{c} = \frac{3(2.18 \times 10^{-18}\ \mathrm{s}^{-1})^2}{8\pi(6.67 \times 10^{-11}\ \mathrm{N \cdot m^2/kg^2})} = 8.50 \times 10^{-27}\ \mathrm{kg/m^3}$$

The mass of a hydrogen atom is 1.67×10^{-27} kg, so this density is equivalent to about five hydrogen atoms per cubic meter.

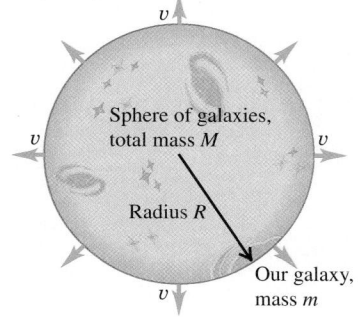

44.19 An imaginary sphere of galaxies. The net gravitational force exerted on our galaxy (at the surface of the sphere) by the other galaxies is the same as if all of their mass were concentrated at the center of the sphere. (Since the universe is infinite, there's also an infinity of galaxies outside this sphere.)

Sphere of galaxies, total mass M

Radius R

Our galaxy, mass m

Dark Matter, Dark Energy, and the Accelerating Universe

Astronomers have made extensive studies of the average density of matter in the universe. One way to do so is to count the number of galaxies in a patch of sky. Based on the mass of an average star and the number of stars in an average galaxy, this effort gives an estimate of the average density of *luminous* matter in the universe—that is, matter that emits electromagnetic radiation. (You are made of luminous matter because you emit infrared radiation as a consequence of your temperature; see Sections 17.7 and 39.5.) It's also necessary to take into account other luminous matter within a galaxy, including the tenuous gas and dust between the stars.

Another technique is to study the motions of galaxies within clusters of galaxies (see Fig. 44.16b). The motions are so slow that we can't actually see galaxies changing positions within a cluster. However, observations show that different galaxies within a cluster have somewhat different redshifts, which indicates that the galaxies are moving relative to the center of mass of the cluster. The speeds of these motions are related to the gravitational force exerted on each galaxy by the other members of the cluster, which in turn depends on the total mass of the cluster. By measuring these speeds, astronomers can determine the average density of *all* kinds of matter within the cluster, whether or not the matter emits electromagnetic radiation.

Observations using these and other techniques show that the average density of *all* matter in the universe is 31.5% of the critical density, but the average density of *luminous* matter is only 4.9% of the critical density. In other words, most of the matter in the universe is not luminous: It does not emit electromagnetic radiation of *any* kind. At present, the nature of this **dark matter** remains an outstanding mystery. Some proposed candidates for dark matter are WIMPs (weakly interacting massive particles, which are hypothetical subatomic particles far more massive than those produced in accelerator experiments) and MACHOs (massive compact halo objects, which include objects such as black holes that might form "halos" around galaxies). Whatever the true nature of dark matter, it is by far the dominant form of matter in the universe. For every kilogram of the conventional matter that has been our subject for most of this book—including electrons, protons, atoms, molecules, blocks on inclined planes, planets, and stars— there are about *five and a half* kilograms of dark matter.

Since the average density of matter in the universe is less than the critical density, it might seem fair to conclude that the universe will continue to expand indefinitely, and that gravitational attraction between matter in different parts of the universe should slow the expansion down (albeit not enough to stop it). One way to test this prediction is to examine the redshifts of extremely distant objects. The more distant a galaxy is, the more time it takes that light to reach us from that galaxy, so the further back in time we look when we observe that galaxy. If the expansion of the universe has been slowing down, the expansion must have been more rapid in the distant past. Thus we would expect very distant galaxies to have *greater* redshifts than predicted by the Hubble law, Eq. (44.15).

Only since the 1990s has it become possible to measure accurately both the distances and the redshifts of extremely distant galaxies. The results have been totally surprising: Very distant galaxies, seen as they were when the universe was a small fraction of its present age (**Fig. 44.20**), have *smaller* redshifts than predicted by the Hubble law! The implication is that the expansion of the universe was slower in the past than it is now, so the expansion has been *speeding up* rather than slowing down.

If gravitational attraction should make the expansion slow down, why is it speeding up instead? Our best explanation is that space is suffused with a kind of energy that has no gravitational effect and emits no electromagnetic radiation,

44.20 The bright spots in this image are not stars but entire galaxies. We see the most distant of these, magnified in the inset, as it was 13.1 billion years ago, when the universe was just 700 million years old. At that time the scale factor of the universe was only about 12% as large as it is now. (The red color of this galaxy is due to its very large redshift.) By comparison, we see the relatively nearby Coma cluster (see Fig. 44.16b) as it was 300 million years ago, when the scale factor was 98% of the present-day value.

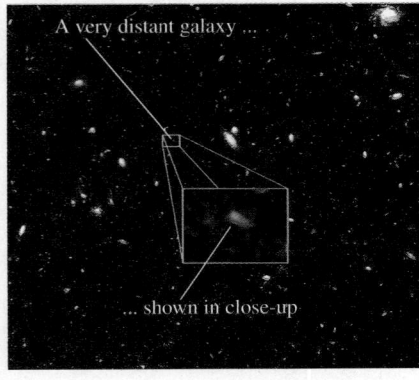

A very distant galaxy ...

... shown in close-up

but rather acts as a kind of "antigravity" that produces a universal *repulsion*. This invisible, immaterial energy is called **dark energy.** As the name suggests, the nature of dark energy is poorly understood but is the subject of very active research.

Observations show that the *energy* density of dark energy (measured in, say, joules per cubic meter) is 68.5% of the critical density times c^2; that is, it is equal to $0.685\rho_c c^2$. As described above, the average density of matter of all kinds is 31.5% of the critical density. From the Einstein relationship $E = mc^2$, the average *energy* density of matter in the universe is therefore $0.315\rho_c c^2$. Because the energy density of dark energy is nearly three times greater than that of matter, the expansion of the universe will continue to accelerate. This expansion will never stop, and the universe will never contract.

If we account for energy of *all* kinds, the average energy density of the universe is equal to $0.685\rho_c c^2 + 0.315\rho_c c^2 = 1.00\rho_c c^2$. Of this, 68.5% is the mysterious dark energy, 26.6% is the no less mysterious dark matter, and a mere 4.9% is well-understood conventional matter. How little we know about the contents of our universe (**Fig. 44.21**)! When we take account of the density of matter in the universe (which tends to slow the expansion of space) and the density of dark energy (which tends to speed up the expansion), the age of the universe turns out to be 13.8 billion (1.38×10^{10}) years.

What is the significance of the result that within observational error, the average energy density of the universe is equal to $\rho_c c^2$? It tells us that the universe is infinite and unbounded, but just barely so. If the average energy density were even slightly larger than $\rho_c c^2$, the universe would be finite like the surface of the balloon depicted in Fig. 44.18. As of this writing, the observational error in the average energy density is less than 1%, but we can't be totally sure that the universe *is* unbounded. Improving these measurements will be an important task for physicists and astronomers in the years ahead.

TEST YOUR UNDERSTANDING OF SECTION 44.6 Is it accurate to say that your body is made of "ordinary" matter? ▌

44.7 THE BEGINNING OF TIME

What an odd title for the very last section of a book! We will describe in general terms some of the current theories about the very early history of the universe and their relationship to fundamental particle interactions. We'll find that an astonishing amount happened in the very first second.

Temperatures

The early universe was extremely dense and extremely hot. and the average particle energies were extremely large, all many orders of magnitude beyond anything that exists in the present universe. We can compare particle energy E and absolute temperature T by using the equipartition principle (see Section 18.4):

$$E = \tfrac{3}{2}kT \qquad (44.20)$$

In this equation k is Boltzmann's constant, which we'll often express in eV/K:

$$k = 8.617 \times 10^{-5} \text{ eV/K}$$

Thus we can replace Eq. (44.20) by $E \approx (10^{-4} \text{ eV/K})T = (10^{-13} \text{ GeV/K})T$ when we're discussing orders of magnitude.

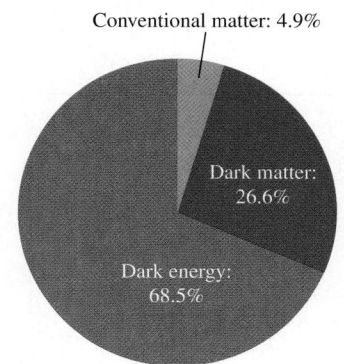

44.21 The composition of our universe. Conventional matter includes all of the familiar sorts of matter that you see around you, including your body, our planet, and the sun and stars.

Conventional matter: 4.9%

Dark matter: 26.6%

Dark energy: 68.5%

BIO Application A Fossil Both Ancient and Recent This fossil trilobite is an example of a group of marine arthropods that flourished in earth's oceans from 540 to 250 million years ago. (By comparison, the first dinosaurs did not appear until 230 million years ago.) From our perspective, this makes trilobites almost unfathomably ancient. But compared to the time that has elapsed since the Big Bang, 13.8 billion years, even trilobites are a very recent phenomenon: They first appeared when the universe was already 96% of its present age.

EXAMPLE 44.10 TEMPERATURE AND ENERGY

(a) What is the average kinetic energy E (in eV) of particles at room temperature ($T = 290$ K) and at the surface of the sun ($T = 5800$ K)? (b) What approximate temperature corresponds to the ionization energy of the hydrogen atom and to the rest energies of the electron and the proton?

SOLUTION

IDENTIFY and SET UP: In this example we are to apply the equipartition principle. We use Eq. (44.20) to relate the target variables E and T.

EXECUTE: (a) At room temperature, from Eq. (44.20),

$$E = \tfrac{3}{2}kT = \tfrac{3}{2}(8.617 \times 10^{-5} \text{ eV/K})(290 \text{ K}) = 0.0375 \text{ eV}$$

The temperature at the sun's surface is higher than room temperature by a factor of $(5800 \text{ K})/(290 \text{ K}) = 20$, so the average kinetic energy there is $20(0.0375 \text{ eV}) = 0.75$ eV.

(b) The ionization energy of hydrogen is 13.6 eV. Using the approximation $E \approx (10^{-4} \text{ eV/K})T$, we have

$$T \approx \frac{E}{10^{-4} \text{ eV/K}} = \frac{13.6 \text{ eV}}{10^{-4} \text{ eV/K}} \approx 10^{5} \text{ K}$$

Repeating this calculation for the rest energies of the electron ($E = 0.511$ MeV) and proton ($E = 938$ MeV) gives temperatures of 10^{10} K and 10^{13} K, respectively.

EVALUATE: Temperatures in excess of 10^{5} K are found in the sun's interior, so most of the hydrogen there is ionized. Temperatures of 10^{10} K or 10^{13} K are not found anywhere in the solar system; as we will see, temperatures were this high in the very early universe.

Uncoupling of Interactions

We've characterized the expansion of the universe by a continual increase of the scale factor R, which we can think of very roughly as characterizing the *size* of the universe, and by a corresponding decrease in average density. As the total gravitational potential energy increased during expansion, there were corresponding *decreases* in temperature and average particle energy. As this happened, the basic interactions became progressively uncoupled.

To understand the uncouplings, recall that the unification of the electromagnetic and weak interactions occurs at energies that are large enough that the differences in mass among the various spin-1 bosons that mediate the interactions become insignificant by comparison. The electromagnetic interaction is mediated by the massless photon, and the weak interaction is mediated by the weak bosons $W^{\pm}$ and Z^{0} with masses of the order of $100 \text{ GeV}/c^2$. At energies much *less* than 100 GeV, the two interactions seem quite different. But at energies much *greater* than 100 GeV, they become part of a single interaction, because the $W^{\pm}$ and Z^{0} weak bosons become massless like the photon. (This occurs because the average value ϕ_{av} of the Higgs field is zero at high energy, as in Fig. 44.14a.)

The grand unified theories (GUTs) provide a similar behavior for the strong interaction. It becomes unified with the electroweak interaction at energies of the order of 10^{14} GeV, but at lower energies the two appear quite distinct. One of the reasons GUTs are still very speculative is that there is no way to do controlled experiments in this energy range, which is larger by a factor of 10^{11} than energies available with any current accelerator.

Finally, at sufficiently high energies and short distances, it is assumed that gravitation becomes unified with the other three interactions. The distance at which this happens is thought to be of the order of 10^{-35} m. This distance, called the *Planck length* l_P, is determined by the speed of light c and the fundamental constants of quantum mechanics and gravitation, h and G, respectively:

$$l_P = \sqrt{\frac{\hbar G}{c^3}} = 1.616 \times 10^{-35} \text{ m} \tag{44.21}$$

You should verify that this combination of constants has units of length. The *Planck time* $t_P = l_P/c$ is the time required for light to travel a distance l_P:

$$t_P = \frac{l_P}{c} = \sqrt{\frac{\hbar G}{c^5}} = 0.539 \times 10^{-43} \text{ s} \tag{44.22}$$

If we mentally go backward in time, we have to stop when we reach $t = 10^{-43}$ s because we have no adequate theory that unifies all four interactions. So as yet we have no way of knowing what happened or how the universe behaved at times earlier than the Planck time or when its size was less than the Planck length.

The Standard Model of the History of the Universe

The description that follows is called the *standard model* of the history of the universe. The title indicates that there are substantial areas of theory that rest on solid experimental foundations and are quite generally accepted. The figure on pages 1512–1513 is a graphical description of this history, with the characteristic temperature, particle energy, and scale factor at various times. Referring to this figure frequently will help you to understand the following discussion.

In this standard model, the temperature of the universe at time $t = 10^{-43}$ s (the Planck time) was about 10^{32} K, and the average energy per particle was approximately

$$E \approx (10^{-13} \text{ GeV/K})(10^{32} \text{ K}) = 10^{19} \text{ GeV}$$

In a totally unified theory this is about the energy below which gravity begins to behave as a separate interaction. This time therefore marked the transition from any proposed TOE to the GUT period.

During the GUT period, roughly $t = 10^{-43}$ to 10^{-35} s, the strong and electroweak forces were still unified, and the universe consisted of a soup of quarks and leptons transforming into each other so freely that there was no distinction between the two families of particles. Other, much more massive particles may also have been freely created and destroyed. One important characteristic of GUTs is that at sufficiently high energies, baryon number is not conserved. (We mentioned earlier the proposed decay of the proton, which has not yet been observed.) Thus by the end of the GUT period the numbers of quarks and antiquarks may have been unequal. This point has important implications; we'll return to it at the end of the section.

By $t = 10^{-35}$ s the temperature had decreased to about 10^{27} K and the average energy to about 10^{14} GeV. At this energy the strong force separated from the electroweak force (**Fig. 44.22**), and baryon number and lepton numbers began to be separately conserved. This separation of the strong force was analogous to a phase change such as boiling a liquid, with an associated heat of vaporization. Think of it as being similar to boiling a heavy nucleus, pulling the particles apart

44.22 Schematic diagram showing the times and energies at which the various interactions are thought to have uncoupled. The energy scale is backward because the average energy decreased as the age of the universe increased.

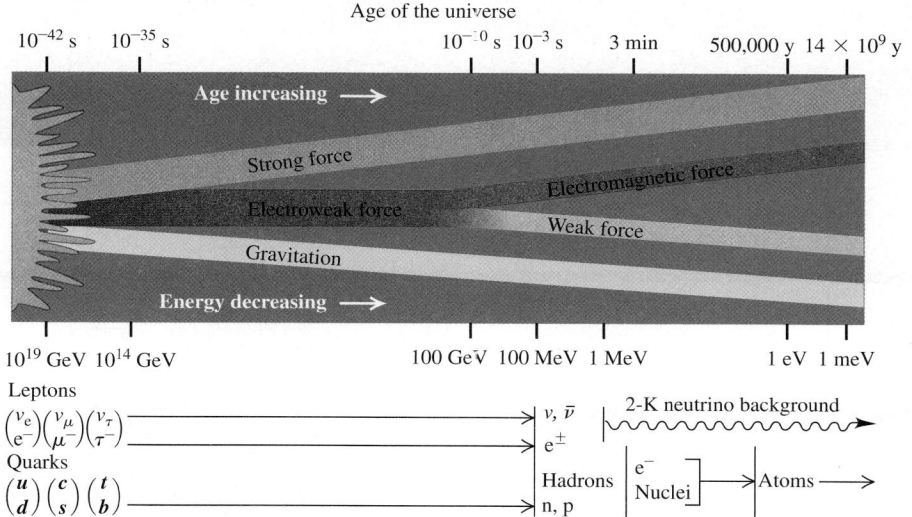

**AGE OF QUARKS AND
GLUONS (GUT Period)**
Dense concentration of matter and
antimatter; gravity a separate force;
more quarks than antiquarks.
Inflationary period (10^{-35} s): rapid expansion,
strong force separates from
electroweak force.

**AGE OF NUCLEONS AND
ANTINUCLEONS**
Quarks bind together to form
nucleons and antinucleons; energy
too low for nucleon–antinucleon
pair production at 10^{-2} s.

**AGE OF
NUCLEOSYNTHESIS**
Stable deuterons; matter
74% H, 25% He, 1%
heavier nuclei.

AGE OF LEPTONS
Leptons distinct from quarks;
$W^{\pm}$ and Z^0 bosons mediate
weak force (10^{-12} s).

**BIG
BANG** 10^{-43} s 10^{-32} s 10^{-6} s   225 s 10^3 s

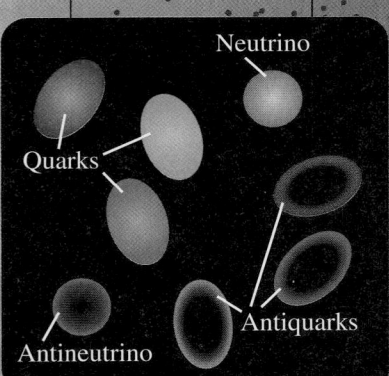

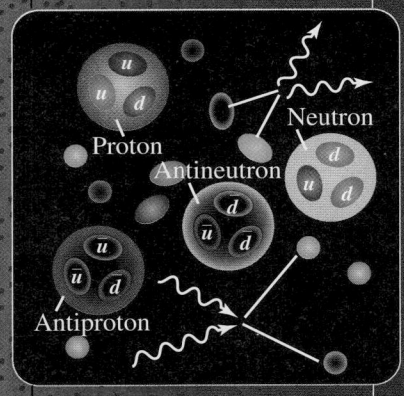

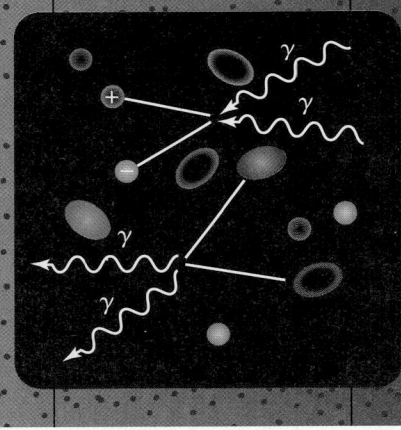

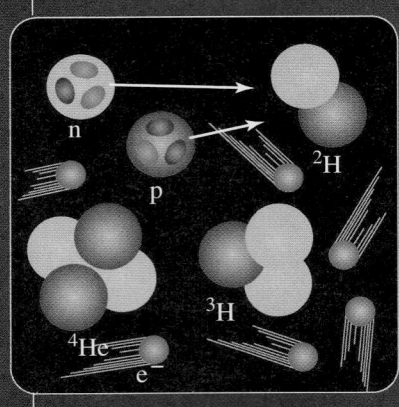

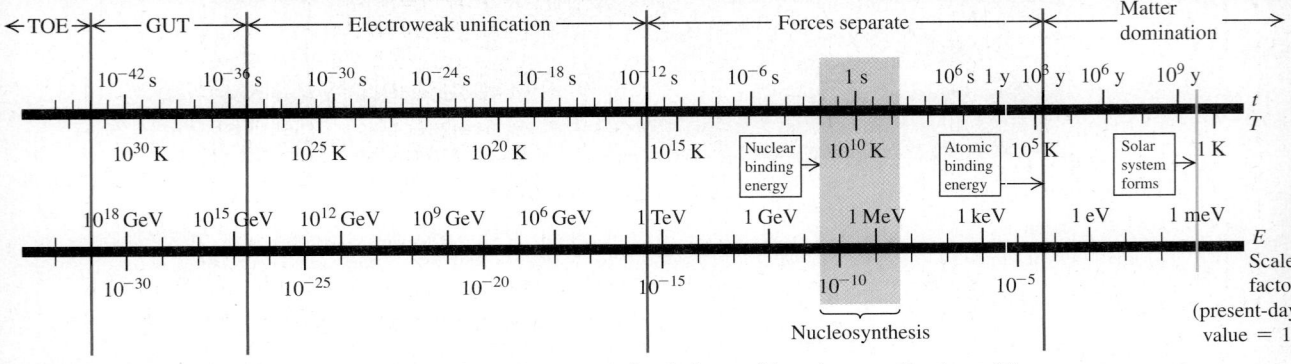

Logarithmic scales show characteristic temperature, particle energy, and scale factor of the universe as functions of time.

A Brief History of the Universe

OF IONS
Expanding, cooling gas of ionized H and He.

AGE OF ATOMS
Neutral atoms form; universe becomes transparent to most light.

AGE OF STARS AND GALAXIES
Thermonuclear fusion begins in stars, forming heavier nuclei.

10^{13} s

10^{15} s

NOW

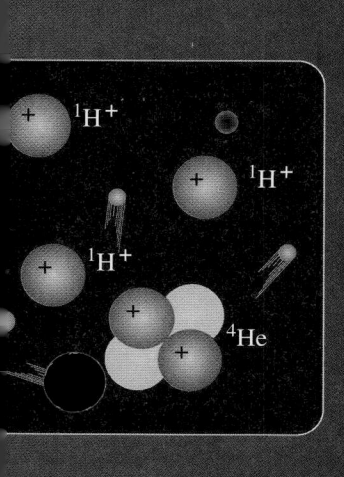

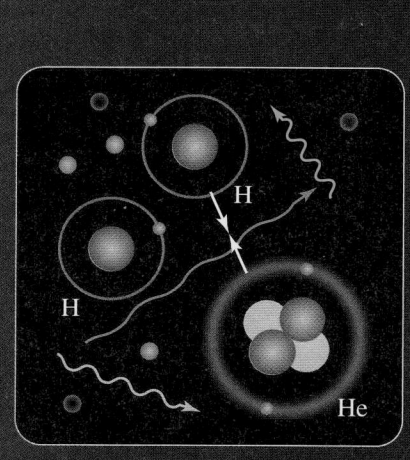

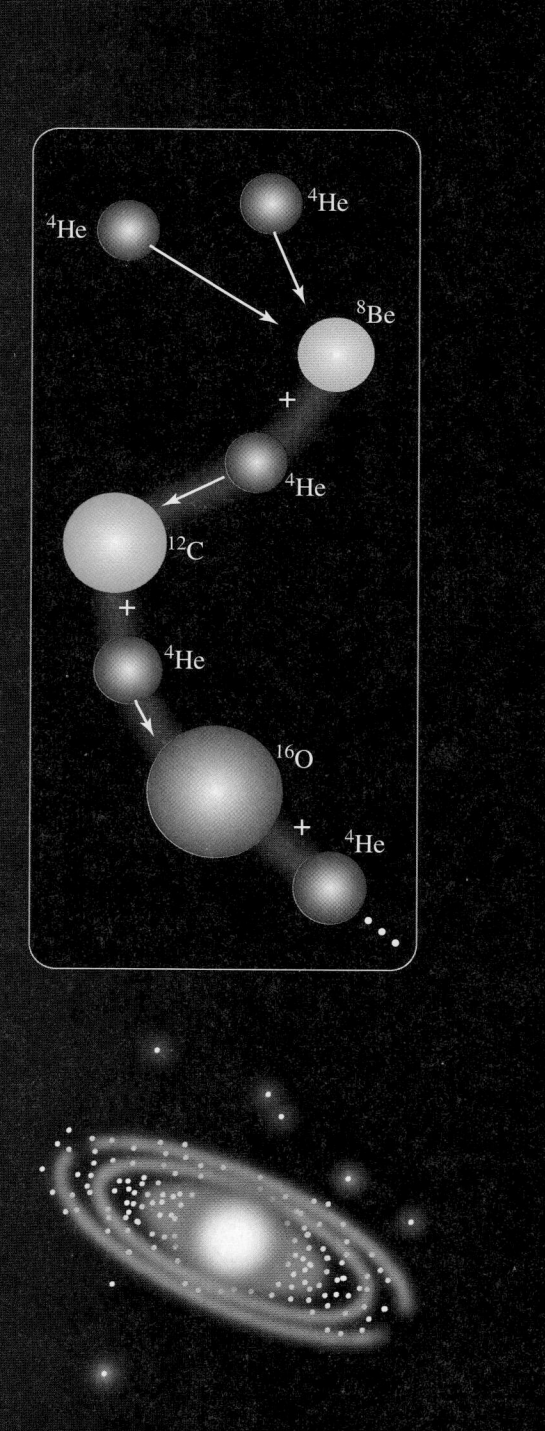

beyond the short range of the nuclear force. As a result, the universe underwent a dramatic expansion (far more rapid than the present-day expansion rate) called *cosmic inflation.* In one model, the scale factor R increased by a factor of 10^{50} in 10^{-32} s.

At $t = 10^{-32}$ s the universe was a mixture of quarks, leptons, and the mediating bosons (gluons, photons, and the weak bosons $W^{\pm}$ and Z^0). It continued to expand and cool from the inflationary period to $t = 10^{-6}$ s, when the temperature was about 10^{13} K and typical energies were about 1 GeV (comparable to the rest energy of a nucleon; see Example 44.11). The quarks then began to bind together, forming nucleons and antinucleons. There were still enough photons of sufficient energy to produce nucleon–antinucleon pairs to balance the process of nucleon–antinucleon annihilation. However, by about $t = 10^{-2}$ s, most photon energies fell well below the threshold energy for such pair production. There was a slight excess of nucleons over antinucleons; as a result, virtually all of the antinucleons and most of the nucleons annihilated one another. A similar equilibrium occurred later between the production of electron–positron pairs from photons and the annihilation of such pairs. At about $t = 14$ s the average energy dropped to around 1 MeV, below the threshold for e^+e^- pair production. After pair production ceased, virtually all of the remaining positrons were annihilated, leaving the universe with many more protons and electrons than the antiparticles of each.

Up until about $t = 1$ s, neutrons and neutrinos could be produced in the endoergic reaction

$$e^- + p \rightarrow n + \nu_e$$

After this time, most electrons no longer had enough energy for this reaction. The average neutrino energy also decreased, and as the universe expanded, equilibrium reactions that involved *absorption* of neutrinos (which occurred with decreasing probability) became inoperative. At this time, in effect, the flux of neutrinos and antineutrinos throughout the universe uncoupled from the rest of the universe. Because of the extraordinarily low probability for neutrino absorption, most of this flux is still present today, although cooled greatly by expansion. The standard model of the universe predicts a present neutrino temperature of about 2 K, but no experiment has yet been able to test this prediction.

Nucleosynthesis

At about $t = 1$ s, the ratio of protons to neutrons was determined by the Boltzmann distribution factor $e^{-\Delta E/kT}$, where ΔE is the difference between the neutron and proton rest energies: $\Delta E = 1.294$ MeV. At a temperature of about 10^{10} K, this distribution factor gives about 4.5 times as many protons as neutrons. However, as we have discussed, free neutrons (with a half-life of 887 s) decay spontaneously to protons. This decay caused the proton-to-neutron ratio to increase until about $t = 225$ s. At this time, the temperature was about 10^9 K, and the average energy was well below 2 MeV.

This energy distribution was critical because the binding energy of the *deuteron* (a neutron and a proton bound together) is 2.22 MeV (see Section 43.2). A neutron bound in a deuteron does not decay spontaneously. As the average energy decreased, a proton and a neutron could combine to form a deuteron, and there were fewer and fewer photons with 2.22 MeV or more of energy to dissociate the deuterons again. Therefore the combining of protons and neutrons into deuterons halted the decay of free neutrons.

The formation of deuterons starting at about $t = 225$ s marked the beginning of the period of formation of nuclei, or *nucleosynthesis.* At this time, there were about seven protons for each neutron. The deuteron (^{2}H) can absorb a neutron and form a triton (^{3}H), or it can absorb a proton and form ^{3}He. Then ^{3}H can absorb a proton and ^{3}He can absorb a neutron, each yielding ^{4}He (the alpha particle). A few ^{7}Li nuclei may also have formed by fusion of ^{3}H and ^{4}He nuclei. According to the theory, essentially all the ^{1}H and ^{4}He in the present universe formed at this time. But then the building of nuclei almost ground to a halt. The reason is that *no* nuclide with mass number $A = 5$ has a half-life greater than 10^{-21} s. Alpha particles simply do not permanently absorb neutrons or protons. The nuclide ^{8}Be that is formed by fusion of two ^{4}He nuclei is unstable, with an extremely short half-life, about 7×10^{-17} s. At this time, the average energy was still much too large for electrons to be bound to nuclei; there were not yet any atoms.

CONCEPTUAL EXAMPLE 44.11   **THE RELATIVE ABUNDANCE OF HYDROGEN AND HELIUM IN THE UNIVERSE**

SOLUTION

Nearly all of the protons and neutrons in the seven-to-one ratio at $t = 225$ s either formed ^{4}He or remained as ^{1}H. After this time, what was the resulting relative abundance of ^{1}H and ^{4}He, by mass?

SOLUTION

The ^{4}He nucleus contains two protons and two neutrons. For every two neutrons present at $t = 225$ s there were 14 protons. The two neutrons and two of the 14 protons make up one ^{4}He nucleus,

leaving 12 protons (^{1}H nuclei). So there were eventually 12 ^{1}H nuclei for every ^{4}He nucleus. The masses of ^{1}H and ^{4}He are about 1 u and 4 u, respectively, so there were 12 u of ^{1}H for every 4 u of ^{2}He. Therefore the relative abundance, by mass, was 75% ^{1}H and 25% ^{4}He. This result agrees very well with estimates of the present H–He ratio in the universe, an important confirmation of this part of the theory.

Further nucleosynthesis did not occur until very much later, well after $t = 10^{13}$ s (about 380,000 y). At that time, the temperature was about 3000 K, and the average energy was a few tenths of an electron volt. Because the ionization energies of hydrogen and helium atoms are 13.6 eV and 24.5 eV, respectively, almost all the hydrogen and helium was electrically neutral (not ionized). With the electrical repulsions of the nuclei canceled out, gravitational attraction could slowly pull the neutral atoms together to form clouds of gas and eventually stars. Thermonuclear reactions in stars then produced all of the more massive nuclei. In Section 43.8 we discussed one cycle of thermonuclear reactions in which ^{1}H becomes ^{4}He.

For stars whose mass is 40% of the sun's mass or greater, as the hydrogen is consumed the star's core begins to contract as the inward gravitational pressure exceeds the outward gas and radiation pressure. The gravitational potential energy decreases as the core contracts, so the kinetic energy of nuclei in the core increases. Eventually the core temperature becomes high enough to begin another process, *helium fusion.* First two ^{4}He nuclei fuse to form ^{8}Be, which is highly unstable. But because a star's core is so dense and collisions among nuclei are so frequent, there is a nonzero probability that a third ^{4}He nucleus will fuse with the ^{8}Be nucleus before it can decay. The result is the stable nuclide ^{12}C. This is called the *triple-alpha process,* since three ^{4}He nuclei (that is, alpha particles) fuse to form one carbon nucleus. Then successive fusions with ^{4}He give ^{16}O, ^{20}Ne, and ^{24}Mg. All these reactions are exoergic. They release energy to heat up the star, and ^{12}C and ^{16}O can fuse to form elements with higher and higher atomic number.

For nuclides that can be created in this manner, the binding energy per nucleon peaks at mass number $A = 56$ with the nuclide ^{56}Fe, so exoergic fusion reactions stop with Fe. But successive neutron captures followed by beta decays can continue the synthesis of more massive nuclei. If the star is massive enough, it may eventually explode as a *supernova,* sending out into space the heavy elements that were produced by the earlier processes (**Fig. 44.23**; see also Fig. 37.7). In space, the debris and other interstellar matter can gravitationally bunch together to form a new generation of stars and planets. Our sun is one such "second-generation" star. The sun's planets and everything on them (including you) contain matter that was long ago blasted into space by an exploding supernova.

44.23 The Veil Nebula in the constellation Cygnus is a remnant of a supernova explosion that occurred more than 20,000 years ago. The gas ejected from the supernova is still moving very rapidly. Collisions between this fast-moving gas and the tenuous material of interstellar space excite the gas and cause it to glow. The portion of the nebula shown here is about 40 ly (12 pc) in length.

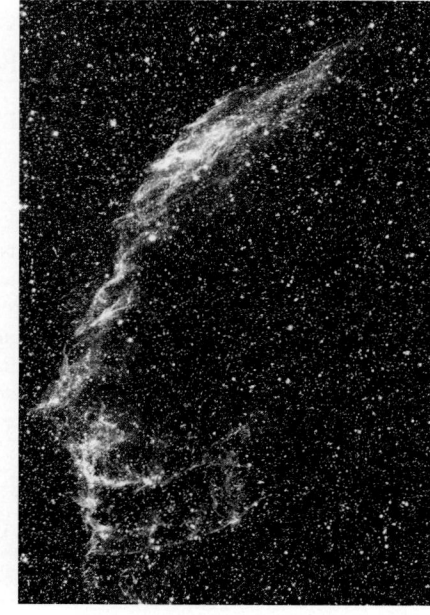

44.24 This false-color map shows microwave radiation from the entire sky mapped onto an oval. When this radiation was emitted 380,000 years after the Big Bang, the regions shown in blue were slightly cooler and denser than average. Within these cool, dense regions formed galaxies, including the Milky Way galaxy of which our solar system, our earth, and our selves are part.

Background Radiation

In 1965 Arno Penzias and Robert Wilson, working at Bell Telephone Laboratories in New Jersey on satellite communications, turned a microwave antenna skyward and found a background signal that had no apparent preferred direction. (This signal produces about 1% of the "hash" you see on an analog TV that's tuned to an unused channel.) Further research has shown that the radiation that is received has a frequency spectrum that fits Planck's blackbody radiation law, Eq. (39.24) (Section 39.5). The wavelength of peak intensity is 1.063 mm (in the microwave region of the spectrum), with a corresponding absolute temperature $T = 2.725$ K. Penzias and Wilson contacted physicists at Princeton University who had begun the design of an antenna to search for radiation that was a remnant from the early evolution of the universe. We mentioned above that neutral atoms began to form at about $t = 380,000$ years when the temperature was 3000 K. With far fewer charged particles present than previously, the universe became transparent at this time to electromagnetic radiation of long wavelength. The 3000-K blackbody radiation therefore survived, cooling to its present 2.725-K temperature as the universe expanded. The *cosmic background radiation* is among the most clear-cut experimental confirmations of the Big Bang theory. **Figure 44.24** shows a modern map of the cosmic background radiation.

EXAMPLE 44.12 **EXPANSION OF THE UNIVERSE**

By approximately what factor has the universe expanded since $t = 380,000$ y?

SOLUTION

IDENTIFY and SET UP: We use the idea that as the universe has expanded, all intergalactic wavelengths have expanded with it. The Wien displacement law, Eq. (39.21), relates the peak wavelength λ_m in blackbody radiation to the temperature T. Given the temperatures of the cosmic background radiation today (2.725 K) and at $t = 380,000$ y (3000 K) we can determine the factor by which wavelengths have changed and hence determine the factor by which the universe has expanded.

EXECUTE: We rewrite Eq. (39.21) as

$$\lambda_m = \frac{2.90 \times 10^{-3} \text{ m} \cdot \text{K}}{T}$$

Hence the peak wavelength λ_m is inversely proportional to T. As the universe expands, all intergalactic wavelengths (including λ_m) increase in proportion to the scale factor R. The temperature has decreased by the factor $(3000 \text{ K})/(2.725 \text{ K}) \approx 1100$, so λ_m and the scale factor must both have *increased* by this factor. Thus, between $t = 380,000$ y and the present, the universe has expanded by a factor of about 1100.

EVALUATE: Our results show that since $t = 380,000$ y, any particular intergalactic *volume* has increased by a factor of about $(1100)^3 = 1.3 \times 10^9$. They also show that when the cosmic background radiation was emitted, its peak wavelength was $\frac{1}{1100}$ of the present-day value of 1.063 mm, or 967 nm. This is in the infrared region of the spectrum.

Matter and Antimatter

One of the most remarkable features of our universe is the asymmetry between matter and antimatter. You might think that the universe should have equal numbers of protons and antiprotons and of electrons and positrons, but this doesn't appear to be the case. Theories of the early universe must explain this imbalance.

We've mentioned that most GUTs include violation of conservation of baryon number at energies at which the strong and electroweak interactions have converged. If particle–antiparticle symmetry is also violated, we have a mechanism for making more quarks than antiquarks, more leptons than antileptons, and eventually more matter than antimatter. One serious problem is that any asymmetry that is created in this way during the GUT era might be wiped out by the electroweak interaction after the end of the GUT era. If so, there must be some mechanism that creates particle–antiparticle asymmetry at a much *later* time. The problem of the matter–antimatter asymmetry is still very much an open one.

There are still many unanswered questions at the intersection of particle physics and cosmology. Is the energy density of the universe precisely equal to $\rho_c c^2$, or are there small but important differences? What is dark energy? Has the density of dark energy remained constant over the history of the universe, or has the density changed? What is dark matter? What happened during the first 10^{-43} s after the Big Bang? Can we see evidence that the strong and electroweak interactions undergo a grand unification at high energies? The search for the answers to these and many other questions about our physical world continues to be one of the most exciting adventures of the human mind.

TEST YOUR UNDERSTANDING OF SECTION 44.7   Given a sufficiently powerful telescope, could we detect photons emitted earlier than $t = 380{,}000$ y? ▌

CHAPTER **44 SUMMARY**

SOLUTIONS TO ALL EXAMPLES

Fundamental particles: Each particle has an antiparticle; some particles are their own antiparticles. Particles can be created and destroyed, some of them (including electrons and positrons) only in pairs or in conjunction with other particles and antiparticles.

Particles serve as mediators for the fundamental interactions. The photon is the mediator of the electromagnetic interaction. Yukawa proposed the existence of mesons to mediate the nuclear interaction. Mediating particles that can exist only because of the uncertainty principle for energy are called virtual particles.

Particle accelerators and detectors: Cyclotrons, synchrotrons, and linear accelerators are used to accelerate charged particles to high energies for experiments with particle interactions. Only part of the beam energy is available to cause reactions with targets at rest. This problem is avoided in colliding-beam experiments. (See Examples 44.1–44.3.)

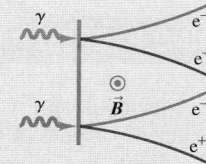

High-frequency alternating voltage

Particles and interactions: Four fundamental interactions are found in nature: the strong, electromagnetic, weak, and gravitational interactions. Particles can be described in terms of their interactions and of quantities that are conserved in all or some of the interactions.

Fermions have half-integer spins; bosons have integer spins. Leptons, which are fermions, have no strong interactions. Strongly interacting particles are called hadrons. They include mesons, which are always bosons, and baryons, which are always fermions. There are conservation laws for three different lepton numbers and for baryon number. Additional quantum numbers, including strangeness and charm, are conserved in some interactions. (See Examples 44.4–44.6.)

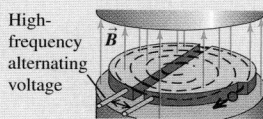

Quarks: Hadrons are composed of quarks. There are thought to be six types of quarks. The interaction between quarks is mediated by gluons. Quarks and gluons have an additional attribute called color. (See Example 44.7.)

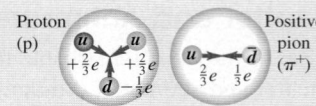

Symmetry and the unification of interactions: Symmetry considerations play a central role in all fundamental-particle theories. The electromagnetic and weak interactions become unified at high energies into the electroweak interaction. In grand unified theories the strong interaction is also unified with these interactions, but at much higher energies.

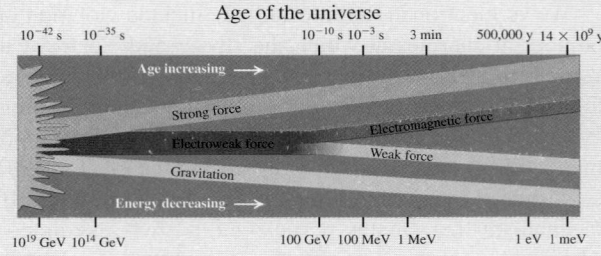

The expanding universe and its composition: The Hubble law shows that galaxies are receding from each other and that the universe is expanding. Observations show that the rate of expansion is accelerating due to the presence of dark energy, which makes up 68.5% of the energy in the universe. Only 4.9% of the energy in the universe is in the form of conventional matter; the remaining 26.6% is dark matter, whose nature is poorly understood. (See Examples 44.8 and 44.9.)

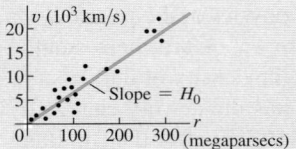

The history of the universe: In the standard model of the universe, a Big Bang gave rise to the first fundamental particles. They eventually formed into the lightest atoms as the universe expanded and cooled. The cosmic background radiation is a relic of the time when these atoms formed. The heavier elements were manufactured much later by fusion reactions inside stars. (See Examples 44.10–44.12.)

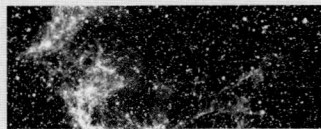

BRIDGING PROBLEM   HYPERONS, PIONS, AND THE EXPANDING UNIVERSE

(a) A Λ^0 hyperon decays into a neutron and a neutral pion (π^0). Find the kinetic energies of the decay products and the fraction of the kinetic energy that is carried off by each particle. (b) A π^0 is at rest in the galaxy shown in Fig. 44.20. If a physicist on earth detects one of the two photons emitted in the decay of this π^0, what is the energy of this detected photon?

SOLUTION GUIDE

IDENTIFY and SET UP

1. Which quantities are conserved in the Λ^0 decay? In the π^0 decay?
2. The universe expanded during the time that the photon traveled from the cluster to earth. How does this affect the wavelength and energy of the photon that the physicist detects?
3. List the unknown quantities for each part of the problem and identify the target variables.
4. Select the equations that will allow you to solve for the target variables.

EXECUTE

5. Write the conservation equations for the decay of the Λ^0. [*Hint:* It's useful to write the energy E of a particle in terms of its momentum p and mass m with $E = (p^2c^2 + m^2c^4)^{1/2}$.]
6. Solve the conservation equations for the energy of one of the decay products. (*Hint:* Rearrange the energy conservation equation so that one of the $(p^2c^2 + m^2c^4)^{1/2}$ terms is on one side of the equation. Then square both sides.) Then use $K = E - mc^2$.
7. Find the fraction of the total kinetic energy that goes into the neutron and into the pion.
8. Write the conservation equations for the decay of the π^0 at rest and find the energy of each emitted photon. By what factor does the wavelength of this photon change as it travels from the galaxy to earth? By what factor does the photon *energy* change? (*Hint:* See Fig. 44.20.)

EVALUATE

9. Which of the Λ^0 decay products should have the greater kinetic energy? Should the detected π^0 decay photon have more or less energy than when it was emitted?

Problems

For assigned homework and other learning materials, go to MasteringPhysics®.

•, ••, •••: Difficulty levels. CP: Cumulative problems incorporating material from earlier chapters. CALC: Problems requiring calculus. DATA: Problems involving real data, scientific evidence, experimental design, and/or statistical reasoning. BIO: Biosciences problems.

DISCUSSION QUESTIONS

Q44.1 Is it possible that some parts of the universe contain anti-matter whose atoms have nuclei made of antiprotons and antineutrons, surrounded by positrons? How could we detect this condition without actually going there? Can we detect these antiatoms by identifying the light they emit as composed of antiphotons? Explain. What problems might arise if we actually *did* go there?

Q44.2 Given the Heisenberg uncertainty principle, is it possible to create particle–antiparticle pairs that exist for extremely short periods of time before annihilating? Does this mean that empty space is really empty?

Q44.3 When they were first discovered during the 1930s and 1940s, there was confusion as to the identities of pions and muons. What are the similarities and most significant differences?

Q44.4 The gravitational force between two electrons is weaker than the electric force by the order of 10^{-40}. Yet the gravitational interactions of matter were observed and analyzed long before electrical interactions were understood. Why?

Q44.5 When a π^0 decays to two photons, what happens to the quarks of which it was made?

Q44.6 Why can't an electron decay to two photons? To two neutrinos?

Q44.7 According to the standard model of the fundamental particles, what are the similarities between baryons and leptons? What are the most important differences?

Q44.8 According to the standard model of the fundamental particles, what are the similarities between quarks and leptons? What are the most important differences?

Q44.9 The quark content of the neutron is **udd**. (a) What is the quark content of the antineutron? Explain your reasoning. (b) Is the neutron its own antiparticle? Why or why not? (c) The quark content of the ψ is $c\bar{c}$. Is the ψ its own antiparticle? Explain your reasoning.

Q44.10 Does the universe have a center? Explain.

Q44.11 Does it make sense to ask, "If the universe is expanding, what is it expanding into?"

Q44.12 Assume that the universe has an edge. Placing yourself at that edge in a thought experiment, explain why this assumption violates the cosmological principle.

Q44.13 Explain why the cosmological principle requires that H_0 must have the same value everywhere in space, but does not require that it be constant in time.

EXERCISES

Section 44.1 Fundamental Particles—A History

44.1 • A neutral pion at rest decays into two photons. Find the energy, frequency, and wavelength of each photon. In which part of the electromagnetic spectrum does each photon lie? (Use the pion mass given in terms of the electron mass in Section 44.1.)

44.2 •• CP Two equal-energy photons collide head-on and annihilate each other, producing a $\mu^+\mu^-$ pair. The muon mass is given in terms of the electron mass in Section 44.1. (a) Calculate the maximum wavelength of the photons for this to occur. If the photons have this wavelength, describe the motion of the μ^+ and μ^- immediately after they are produced. (b) If the wavelength of each

photon is half the value calculated in part (a), what is the speed of each muon after they have moved apart? Use correct relativistic expressions for momentum and energy.

44.3 •• A positive pion at rest decays into a positive muon and a neutrino. (a) Approximately how much energy is released in the decay? (Assume the neutrino has zero rest mass. Use the muon and pion masses given in terms of the electron mass in Section 44.1.) (b) Why can't a positive muon decay into a positive pion?

44.4 • A proton and an antiproton annihilate, producing two photons. Find the energy, frequency, and wavelength of each photon (a) if the p and $\bar{p}$ are initially at rest and (b) if the p and $\bar{p}$ collide head-on, each with an initial kinetic energy of 620 MeV.

44.5 •• CP For the nuclear reaction given in Eq. (44.2) assume that the initial kinetic energy and momentum of the reacting particles are negligible. Calculate the speed of the α particle immediately after it leaves the reaction region.

44.6 •• Estimate the range of the force mediated by an ω^0 meson that has mass 783 MeV/c^2.

44.7 • The starship *Enterprise*, of television and movie fame, is powered by combining matter and antimatter. If the entire 400-kg antimatter fuel supply of the *Enterprise* combines with matter, how much energy is released? How does this compare to the U.S. yearly energy use, which is roughly 1.0×10^{20} J?

Section 44.2 Particle Accelerators and Detectors

44.8 • An electron with a total energy of 30.0 GeV collides with a stationary positron. (a) What is the available energy? (b) If the electron and positron are accelerated in a collider, what total energy corresponds to the same available energy as in part (a)?

44.9 • Deuterons in a cyclotron travel in a circle with radius 32.0 cm just before emerging from the dees. The frequency of the applied alternating voltage is 9.00 MHz. Find (a) the magnetic field and (b) the kinetic energy and speed of the deuterons upon emergence.

44.10 • The magnetic field in a cyclotron that accelerates protons is 1.70 T. (a) How many times per second should the potential across the dees reverse? (This is twice the frequency of the circulating protons.) (b) The maximum radius of the cyclotron is 0.250 m. What is the maximum speed of the proton? (c) Through what potential difference must the proton be accelerated from rest to give it the speed that you calculated in part (b)?

44.11 • (a) A high-energy beam of alpha particles collides with a stationary helium gas target. What must the total energy of a beam particle be if the available energy in the collision is 16.0 GeV? (b) If the alpha particles instead interact in a colliding-beam experiment, what must the energy of each beam be to produce the same available energy?

44.12 •• Let ω_{nr} be the nonrelativistic cyclotron angular frequency given by Eq. (44.7), and let ω_r be the corresponding relativistic value, $\omega_r = (|q|B/m)\sqrt{1 - v^2/c^2}$. (a) What is the speed v of a proton for which $\omega_r = 0.90\omega_{nr}$ so that the two expressions differ by 10%? (b) What is the kinetic energy (in MeV) of a proton with the speed calculated in part (a)? Use the nonrelativistic expression for kinetic energy.

44.13 •• (a) What is the speed of a proton that has total energy 1000 GeV? (b) What is the angular frequency ω of a proton with the speed calculated in part (a) in a magnetic field of 4.00 T? Use both the nonrelativistic Eq. (44.7) and the correct relativistic expression, and compare the results.

44.14 •• Calculate the minimum beam energy in a proton-proton collider to initiate the $p + p \rightarrow p + p + \eta^0$ reaction. The rest energy of the η^0 is 547.3 MeV (see Table 44.3).

44.15 • In Example 44.3 it was shown that a proton beam with an 800-GeV beam energy gives an available energy of 38.7 GeV for collisions with a stationary proton target. (a) You are asked to design an upgrade of the accelerator that will double the available energy in stationary-target collisions. What beam energy is required? (b) In a colliding-beam experiment, what total energy of each beam is needed to give an available energy of $2(38.7 \text{ GeV}) = 77.4 \text{ GeV}$?

44.16 •• You work for a start-up company that is planning to use antiproton annihilation to produce radioactive isotopes for medical applications. One way to produce antiprotons is by the reaction $p + p \rightarrow p + p + p + \bar{p}$ in proton-proton collisions. (a) You first consider a colliding-beam experiment in which the two proton beams have equal kinetic energies. To produce an antiproton via this reaction, what is the required minimum kinetic energy of the protons in each beam? (b) You then consider the collision of a proton beam with a stationary proton target. For this experiment, what is the required minimum kinetic energy of the protons in the beam?

Section 44.3 Particles and Interactions

44.17 • A K^+ meson at rest decays into two π mesons. (a) What are the allowed combinations of π^0, π^+, and π^- as decay products? (b) Find the total kinetic energy of the π mesons.

44.18 • How much energy is released when a μ^- muon at rest decays into an electron and two neutrinos? Neglect the small masses of the neutrinos.

44.19 • What is the mass (in kg) of the Z^0? What is the ratio of the mass of the Z^0 to the mass of the proton?

44.20 • Table 44.3 shows that a Σ^0 decays into a Λ^0 and a photon. (a) Calculate the energy of the photon emitted in this decay, if the Λ^0 is at rest. (b) What is the magnitude of the momentum of the photon? Is it reasonable to ignore the final momentum and kinetic energy of the Λ^0? Explain.

44.21 • If a Σ^+ at rest decays into a proton and a π^0, what is the total kinetic energy of the decay products?

44.22 • The discovery of the Ω^- particle helped confirm Gell-Mann's eightfold way. If an Ω^- decays into a Λ^0 and a K^-, what is the total kinetic energy of the decay products?

44.23 • In which of the following decays are the three lepton numbers conserved? In each case, explain your reasoning. (a) $\mu^- \rightarrow e^- + \nu_e + \bar{\nu}_\mu$; (b) $\tau^- \rightarrow e^- + \bar{\nu}_e + \nu_\tau$; (c) $\pi^+ \rightarrow e^+ + \gamma$; (d) $n \rightarrow p + e^- + \bar{\nu}_e$.

44.24 • Which of the following reactions obey the conservation of baryon number? (a) $p + p \rightarrow p + e^+$; (b) $p + n \rightarrow 2e^+ + e^-$; (c) $p \rightarrow n + e^- + \bar{\nu}_e$; (d) $p + \bar{p} \rightarrow 2\gamma$.

44.25 • In which of the following reactions or decays is strangeness conserved? In each case, explain your reasoning. (a) $K^+ \rightarrow \mu^+ + \nu_\mu$; (b) $n + K^+ \rightarrow p + \pi^0$; (c) $K^+ + K^- \rightarrow \pi^0 + \pi^0$; (d) $p + K^- \rightarrow \Lambda^0 + \pi^0$.

Section 44.4 Quarks and Gluons

44.26 •• Determine the electric charge, baryon number, strangeness quantum number, and charm quantum number for the following quark combinations: (a) uus, (b) $c\bar{s}$, (c) $\overline{dd}\,\overline{u}$, and (d) $\bar{c}b$.

44.27 • Determine the electric charge, baryon number, strangeness quantum number, and charm quantum number for the following quark combinations: (a) uds; (b) $c\bar{u}$; (c) ddd; and (d) $d\bar{c}$. Explain your reasoning.

44.28 • What is the total kinetic energy of the decay products when an upsilon particle at rest decays to $\tau^+ + \tau^-$?

44.29 • Given that each particle contains only combinations of $u, d, s, \bar{u}, \bar{d}$, and $\bar{s}$, use the method of Example 44.7 to deduce the quark content of (a) a particle with charge $+e$, baryon number 0, and strangeness $+1$; (b) a particle with charge $+e$, baryon number -1, and strangeness $+1$; (c) a particle with charge 0, baryon number $+1$, and strangeness -2.

Section 44.5 The Standard Model and Beyond

44.30 • Section 44.5 states that current experiments show that the mass of the Higgs boson is about $125 \text{ GeV}/c^2$. What is the ratio of the mass of the Higgs boson to the mass of a proton?

Section 44.6 The Expanding Universe

44.31 • The spectrum of the sodium atom is detected in the light from a distant galaxy. (a) If the 590.0-nm line is redshifted to 658.5 nm, at what speed is the galaxy receding from the earth? (b) Use the Hubble law to calculate the distance of the galaxy from the earth.

44.32 •• In an experiment done in a laboratory on the earth, the wavelength of light emitted by a hydrogen atom in the $n = 4$ to $n = 2$ transition is 486.1 nm. In the light emitted by the quasar 3C273 (see Problem 36.60), this spectral line is redshifted to 563.9 nm. Assume the redshift is described by Eq. (44.14) and use the Hubble law to calculate the distance in light-years of this quasar from the earth.

44.33 • A galaxy in the constellation Pisces is 5210 Mly from the earth. (a) Use the Hubble law to calculate the speed at which this galaxy is receding from earth. (b) What redshifted ratio λ_0/λ_S is expected for light from this galaxy?

44.34 • **Redshift.** The definition of the redshift z is given in Example 44.8. (a) Show that Eq. (44.13) can be written as $1 + z = ([1 + \beta]/[1 - \beta])^{1/2}$, where $\beta = v/c$. (b) The observed redshift for a certain galaxy is $z = 0.700$. Find the speed of the galaxy relative to the earth; assume the redshift is described by Eq. (44.14). (c) Use the Hubble law to find the distance of this galaxy from the earth.

Section 44.7 The Beginning of Time

44.35 • Calculate the reaction energy Q (in MeV) for the reaction $e^- + p \rightarrow n + \nu_e$. Is this reaction endoergic or exoergic?

44.36 • Calculate the energy (in MeV) released in the triple-alpha process $3\,^4\text{He} \rightarrow {}^{12}\text{C}$.

44.37 • CP The 2.728-K blackbody radiation has its peak wavelength at 1.062 mm. What was the peak wavelength at $t = 700{,}000$ y when the temperature was 3000 K?

44.38 • Calculate the reaction energy Q (in MeV) for the nucleosynthesis reaction

$$^{12}_6\text{C} + {}^4_2\text{He} \rightarrow {}^{16}_8\text{O}$$

Is this reaction endoergic or exoergic?

PROBLEMS

44.39 •• CP BIO **Radiation Therapy with π^- Mesons.** Beams of π^- mesons are used in radiation therapy for certain cancers. The energy comes from the complete decay of the π^- to *stable* particles. (a) Write out the complete decay of a π^- meson to stable particles. What are these particles? (b) How much energy is released from the complete decay of a single π^- meson to stable particles? (You can ignore the very small masses of the neutrinos.) (c) How many π^- mesons need to decay to give a dose of 50.0 Gy to 10.0 g of tissue? (d) What would be the equivalent dose in part (c) in Sv and in rem? Consult Table 43.3 and use the largest appropriate RBE for the particles involved in this decay.

44.40 •• A proton and an antiproton collide head-on with equal kinetic energies. Two γ rays with wavelengths of 0.720 fm are produced. Calculate the kinetic energy of the incident proton.

44.41 •• Calculate the threshold kinetic energy for the reaction $p + p \rightarrow p + p + K^+ + K^-$ if a proton beam is incident on a stationary proton target.

44.42 •• Calculate the threshold kinetic energy for the reaction $\pi^- + p \rightarrow \Sigma^0 + K^0$ if a π^- beam is incident on a stationary proton target. The K^0 has a mass of 497.7 MeV/c^2.

44.43 • Each of the following reactions is missing a single particle. Calculate the baryon number, charge, strangeness, and the three lepton numbers (where appropriate) of the missing particle, and from this identify the particle. (a) $p + p \rightarrow p + \Lambda^0 + ?$; (b) $K^- + n \rightarrow \Lambda^0 + ?$; (c) $p + \bar{p} \rightarrow n + ?$; (d) $\bar{\nu}_\mu + p \rightarrow n + ?$

44.44 •• An η^0 meson at rest decays into three π mesons. (a) What are the allowed combinations of π^0, π^+, and π^- as decay products? (b) Find the total kinetic energy of the π mesons.

44.45 • The ϕ meson has mass 1019.4 MeV/c^2 and a measured energy width of 4.4 MeV/c^2. Using the uncertainty principle, estimate the lifetime of the ϕ meson.

44.46 • Estimate the energy width (energy uncertainty) of the ψ if its mean lifetime is 7.6×10^{-21} s. What fraction is this of its rest energy?

44.47 •• CP BIO One proposed proton decay is $p^+ \rightarrow e^+ + \pi^0$, which violates both baryon and lepton number conservation, so the proton lifetime is expected to be very long. Suppose the proton half-life were 1.0×10^{18} y. (a) Calculate the energy deposited per kilogram of body tissue (in rad) due to the decay of the protons in your body in one year. Model your body as consisting entirely of water. Only the two protons in the hydrogen atoms in each H_2O molecule would decay in the manner shown; do you see why? Assume that the π^0 decays to two γ rays, that the positron annihilates with an electron, and that all the energy produced in the primary decay and these secondary decays remains in your body. (b) Calculate the equivalent dose (in rem) assuming an RBE of 1.0 for all the radiation products, and compare with the 0.1 rem due to the natural background and the 5.0-rem guideline for industrial workers. Based on your calculation, can the proton lifetime be as short as 1.0×10^{18} y?

44.48 •• A ϕ meson (see Problem 44.45) at rest decays via $\phi \rightarrow K^+ + K^-$. It has strangeness 0. (a) Find the kinetic energy of the K^+ meson. (Assume that the two decay products share kinetic energy equally, since their masses are equal.) (b) Suggest a reason the decay $\phi \rightarrow K^+ + K^- + \pi^0$ has not been observed. (c) Suggest reasons the decays $\phi \rightarrow K^+ + \pi^-$ and $\phi \rightarrow K^+ + \mu^-$ have not been observed.

44.49 •• **Cosmic Jerk.** The densities of ordinary matter and dark matter have decreased as the universe has expanded, since the same amount of mass occupies an ever-increasing volume. Yet observations suggest that the density of dark energy has remained constant over the entire history of the universe. (a) Explain why the expansion of the universe actually slowed down in its early history but is speeding up today. "Jerk" is the term for a change in acceleration, so the change in cosmic expansion from slowing down to speeding up is called *cosmic jerk*. (b) Calculations show that the change in acceleration took place when the combined density of matter of all kinds was equal to twice the density of dark energy. Compared to today's value of the scale factor, what was the scale factor at that time? (c) We see the galaxies in Figs. 44.16b and 44.20 as they were 300 million years ago and 13.1 billion years ago. Was the expansion of the universe slowing down or speeding up at these times? (*Hint:* See the caption for Fig. 44.20.)

44.50 ••• CP A Ξ^- particle at rest decays to a Λ^0 and a π^-. (a) Find the total kinetic energy of the decay products. (b) What fraction of the energy is carried off by each particle? (Use relativistic expressions for momentum and energy.)

44.51 ••• CP A Σ^- particle moving in the $+x$-direction with kinetic energy 180 MeV decays into a π^- and a neutron. The π^- moves in the $+y$-direction. What is the kinetic energy of the neutron, and what is the direction of its velocity? Use relativistic expressions for energy and momentum.

44.52 ••• CP The K^0 meson has rest energy 497.7 MeV. A K^0 meson moving in the $+x$-direction with kinetic energy 225 MeV decays into a π^+ and a π^-, which move off at equal angles above and below the $+x$-axis. Calculate the kinetic energy of the π^+ and the angle it makes with the $+x$-axis. Use relativistic expressions for energy and momentum.

44.53 •• DATA While tuning up a medical cyclotron for use in isotope production, you obtain the data given in the table.

B (T)	0.10	0.20	0.30	0.40
K_{max} (MeV)	0.068	0.270	0.608	1.080

B is the uniform magnetic field in the cyclotron, and K_{max} is the maximum kinetic energy of the particle being accelerated, which is a proton. The radius R of the proton path at maximum kinetic energy has the same value for each magnetic-field value. (a) Compare the kinetic energy values in the table to the rest energy mc^2 of a proton. Is it necessary to use relativistic expressions in your analysis? Explain. (b) Graph your data as K_{max} versus B^2. Use the slope of the best-fit straight line to your data to find R. (c) What is the maximum kinetic energy for a 0.25-T magnetic field? (d) What is the angular frequency ω of the proton when $B = 0.40$ T?

44.54 •• DATA The decay products from the decay of short-lived unstable particles can provide evidence that these particles have been produced in a collision experiment. As an initial step in designing an experiment to detect short-lived hadrons, you make a literature study of their decays. Table 44.3 gives experimental data for the mass and typical decay modes of the particles Σ^-, Ξ^0, Δ^{++}, and Ω^-. (a) Which of these four particles has the largest mass? The smallest? (b) By the decay modes shown in the table, for which of these particles do the decay products have the greatest total kinetic energy? The least?

44.55 •• DATA You have entered a graduate program in particle physics and are learning about the use of symmetry. You begin by repeating the analysis that led to the prediction of the Ω^- particle. Nine of the spin-$\frac{3}{2}$ baryons are four Δ particles, each with mass 1232 MeV/c^2, strangeness 0, and charges $+2e$, $+e$, 0, and $-e$; three Σ^* particles, each with mass 1385 MeV/c^2, strangeness -1, and charges $+e$, 0, and $-e$; and two Ξ^* particles, each with mass 1530 MeV/c^2, strangeness -2, and charges 0 and $-e$.

(a) Place these particles on a plot of S versus Q. Deduce the Q and S values of the tenth spin-$\frac{3}{2}$ baryon, the Ω^- particle, and place it on your diagram. Also label the particles with their masses. The mass of the Ω^- is 1672 MeV/c^2; is this value consistent with your diagram? (b) Deduce the three-quark combinations (of u, d, and s) that make up each of these ten particles. Redraw the plot of S versus Q from part (a) with each particle labeled by its quark content. What regularities do you see?

CHALLENGE PROBLEM

44.56 ••• CP Consider a collision in which a stationary particle with mass M is bombarded by a particle with mass m, speed v_0, and total energy (including rest energy) E_m. (a) Use the Lorentz transformation to write the velocities v_m and v_M of particles m and M in terms of the speed v_{cm} of the center of momentum. (b) Use the fact that the total momentum in the center-of-momentum frame is zero to obtain an expression for v_{cm} in terms of m, M, and v_0. (c) Combine the results of parts (a) and (b) to obtain Eq. (44.9) for the total energy in the center-of-momentum frame.

PASSAGE PROBLEMS

BIO **LOOKING UNDER THE HOOD OF PET.** In the imaging method called positron emission tomography (PET), a patient is injected with molecules containing atoms that have nuclei with an excess of protons. As they decay into neutrons, these protons emit positrons. An emitted positron travels a short distance and slows to near-zero velocity; when it encounters an electron, they may annihilate each other and emit two photons in opposite directions. The patient is enclosed in a circular array of detectors, with the tissue to be imaged centered in the array. If two photons of the proper energy strike two detectors simultaneously (within 10 ns), we can conclude that the photons were produced by positron–electron annihilation somewhere along a line connecting the detectors. By observing many such simultaneous events, we can create a map of the distribution of positron-emitting atoms in the tissue. However, photons can be absorbed or scattered as they pass through tissue. The number of photons remaining after they travel a distance x through tissue is given by $N = N_0 e^{-\mu x}$, where N_0 is the initial number of photons and μ is the attenuation coefficient, which is approximately 0.1 cm^{-1} for photons of this energy. The index of refraction of biological tissue for x rays is 1.

44.57 What is the energy of each photon produced by positron–electron annihilation? (a) $\frac{1}{2}m_e v^2$, where v is the speed of the emitted positron; (b) $m_e v^2$; (c) $\frac{1}{2}m_e c^2$; (d) $m_e c^2$.

44.58 Suppose that positron–electron annihilations occur on the line 3 cm from the center of the line connecting two detectors. Will the resultant photons be counted as having arrived at these detectors simultaneously? (a) No, because the time difference between their arrivals is 100 ms; (b) no, because the time difference is 200 ms; (c) yes, because the time difference is 0.1 ns; (d) yes, because the time difference is 0.2 ns.

44.59 If the annihilation photons come from a part of the body that is separated from the detector by 20 cm of tissue, what percentage of the photons that originally travelled toward the detector remains after they have passed through the tissue? (a) 1.4%; (b) 8.6%; (c) 14%; (d) 86%.

Answers

Chapter Opening Question ?

(v) Only 4.9% of the mass and energy of the universe is in the form of "normal" matter. Of the rest, 26.6% is poorly understood dark matter and 68.5% is even more mysterious dark energy.

Test Your Understanding Questions

44.1 (i), (iii), (ii), (iv) The more massive the virtual particle, the shorter its lifetime and the shorter the distance that it can travel during its lifetime.

44.2 no In a head-on collision between an electron and a positron of equal energy, the net momentum is zero. Since both momentum and energy are conserved in the collision, the virtual photon also has momentum $p = 0$ but has energy $E = 90$ GeV + 90 GeV = 180 GeV. Hence the relationship $E = pc$ is definitely *not* true for this virtual photon.

44.3 no Mesons all have baryon number $B = 0$, while a proton has $B = 1$. The decay of a proton into one or more mesons would require that baryon number *not* be conserved. No violation of this conservation principle has ever been observed, so the proposed decay is impossible.

44.4 no Only the s quark, with $S = -1$, has nonzero strangeness. For a baryon to have $S = -2$, it must have two s quarks and one quark of a different flavor. Since each s quark has charge $-\frac{1}{3}e$, the nonstrange quark must have charge $+\frac{5}{3}e$ to make the net charge equal to $+e$. But *no* quark has charge $+\frac{5}{3}e$, so the proposed baryon is impossible.

44.5 (i) If a d quark in a neutron (quark content udd) undergoes the process $d \rightarrow u + e^- + \bar{\nu}_e$, the remaining baryon has quark content uud and hence is a proton (see Fig. 44.11). An electron is the same as a β^- particle, so the net result is beta-minus decay: $n \rightarrow p + \beta^- + \bar{\nu}_e$.

44.6 yes . . . and no The material of which your body is made is ordinary to us on earth. But from a cosmic perspective your material is quite *extraordinary:* Only 4.9% of the mass and energy in the universe is in the form of atoms.

44.7 no Prior to $t = 380,000$ y the temperature was so high that atoms could not form, so free electrons and protons were plentiful. These charged particles are very effective at scattering photons, so light could not propagate very far and the universe was opaque. The oldest photons that we can detect date from the time $t = 380,000$ y when atoms formed and the universe became transparent.

Bridging Problem

(a) Neutron: 5.78 MeV (0.140 of total);
    pion: 35.62 MeV (0.860 of total)
(b) 8.1 MeV

APPENDIX A

THE INTERNATIONAL SYSTEM OF UNITS

The Système International d'Unités, abbreviated SI, is the system developed by the General Conference on Weights and Measures and adopted by nearly all the industrial nations of the world. The following material is adapted from the National Institute of Standards and Technology (**http://physics.nist.gov/cuu**).

Quantity	Name of unit	Symbol	
SI base units			
length	meter	m	
mass	kilogram	kg	
time	second	s	
electric current	ampere	A	
thermodynamic temperature	kelvin	K	
amount of substance	mole	mol	
luminous intensity	candela	cd	
SI derived units			**Equivalent units**
area	square meter	m^2	
volume	cubic meter	m^3	
frequency	hertz	Hz	s^{-1}
mass density (density)	kilogram per cubic meter	kg/m^3	
speed, velocity	meter per second	m/s	
angular velocity	radian per second	rad/s	
acceleration	meter per second squared	m/s^2	
angular acceleration	radian per second squared	rad/s^2	
force	newton	N	$kg \cdot m/s^2$
pressure (mechanical stress)	pascal	Pa	N/m^2
kinematic viscosity	square meter per second	m^2/s	
dynamic viscosity	newton-second per square meter	$N \cdot s/m^2$	
work, energy, quantity of heat	joule	J	$N \cdot m$
power	watt	W	J/s
quantity of electricity	coulomb	C	$A \cdot s$
potential difference, electromotive force	volt	V	J/C, W/A
electric field strength	volt per meter	V/m	N/C
electrical resistance	ohm	Ω	V/A
capacitance	farad	F	$A \cdot s/V$
magnetic flux	weber	Wb	$V \cdot s$
inductance	henry	H	$V \cdot s/A$
magnetic flux density	tesla	T	Wb/m^2
magnetic field strength	ampere per meter	A/m	
magnetomotive force	ampere	A	
luminous flux	lumen	lm	$cd \cdot sr$
luminance	candela per square meter	cd/m^2	
illuminance	lux	lx	lm/m^2
wave number	1 per meter	m^{-1}	
entropy	joule per kelvin	J/K	
specific heat capacity	joule per kilogram-kelvin	$J/kg \cdot K$	
thermal conductivity	watt per meter-kelvin	$W/m \cdot K$	

Quantity	Name of unit	Symbol	Equivalent units
radiant intensity	watt per steradian	W/sr	
activity (of a radioactive source)	becquerel	Bq	s^{-1}
radiation dose	gray	Gy	J/kg
radiation dose equivalent	sievert	Sv	J/kg
SI supplementary units			
plane angle	radian	rad	
solid angle	steradian	sr	

Definitions of SI Units

meter (m)   The *meter* is the length equal to the distance traveled by light, in vacuum, in a time of 1/299,792,458 second.

kilogram (kg)   The *kilogram* is the unit of mass; it is equal to the mass of the international prototype of the kilogram. (The international prototype of the kilogram is a particular cylinder of platinum-iridium alloy that is preserved in a vault at Sévres, France, by the International Bureau of Weights and Measures.)

second (s)   The *second* is the duration of 9,192,631,770 periods of the radiation corresponding to the transition between the two hyperfine levels of the ground state of the cesium-133 atom.

ampere (A)   The *ampere* is that constant current that, if maintained in two straight parallel conductors of infinite length, of negligible circular cross section, and placed 1 meter apart in vacuum, would produce between these conductors a force equal to 2×10^{-7} newton per meter of length.

kelvin (K)   The *kelvin,* unit of thermodynamic temperature, is the fraction 1/273.16 of the thermodynamic temperature of the triple point of water.

ohm (Ω)   The *ohm* is the electric resistance between two points of a conductor when a constant difference of potential of 1 volt, applied between these two points, produces in this conductor a current of 1 ampere, this conductor not being the source of any electromotive force.

coulomb (C)   The *coulomb* is the quantity of electricity transported in 1 second by a current of 1 ampere.

candela (cd)   The *candela* is the luminous intensity, in a given direction, of a source that emits monochromatic radiation of frequency 540×10^{12} hertz and that has a radiant intensity in that direction of 1/683 watt per steradian.

mole (mol)   The *mole* is the amount of substance of a system that contains as many elementary entities as there are carbon atoms in 0.012 kg of carbon 12. The elementary entities must be specified and may be atoms, molecules, ions, electrons, other particles, or specified groups of such particles.

newton (N)   The *newton* is that force that gives to a mass of 1 kilogram an acceleration of 1 meter per second per second.

joule (J)   The *joule* is the work done when the point of application of a constant force of 1 newton is displaced a distance of 1 meter in the direction of the force.

watt (W)   The *watt* is the power that gives rise to the production of energy at the rate of 1 joule per second.

volt (V)   The *volt* is the difference of electric potential between two points of a conducting wire carrying a constant current of 1 ampere, when the power dissipated between these points is equal to 1 watt.

weber (Wb)   The *weber* is the magnetic flux that, linking a circuit of one turn, produces in it an electromotive force of 1 volt as it is reduced to zero at a uniform rate in 1 second.

lumen (lm)   The *lumen* is the luminous flux emitted in a solid angle of 1 steradian by a uniform point source having an intensity of 1 candela.

farad (F)   The *farad* is the capacitance of a capacitor between the plates of which there appears a difference of potential of 1 volt when it is charged by a quantity of electricity equal to 1 coulomb.

henry (H)   The *henry* is the inductance of a closed circuit in which an electromotive force of 1 volt is produced when the electric current in the circuit varies uniformly at a rate of 1 ampere per second.

radian (rad)   The *radian* is the plane angle between two radii of a circle that cut off on the circumference an arc equal in length to the radius.

steradian (sr)   The *steradian* is the solid angle that, having its vertex in the center of a sphere, cuts off an area of the surface of the sphere equal to that of a square with sides of length equal to the radius of the sphere.

SI Prefixes   To form the names of multiples and submultiples of SI units, apply the prefixes listed in Appendix F.

APPENDIX B

USEFUL MATHEMATICAL RELATIONS

Algebra

$$a^{-x} = \frac{1}{a^x} \qquad a^{(x+y)} = a^x a^y \qquad a^{(x-y)} = \frac{a^x}{a^y}$$

Logarithms: If $\log a = x$, then $a = 10^x$. $\log a + \log b = \log(ab)$ $\log a - \log b = \log(a/b)$ $\log(a^n) = n \log a$

If $\ln a = x$, then $a = e^x$. $\ln a + \ln b = \ln(ab)$ $\ln a - \ln b = \ln(a/b)$ $\ln(a^n) = n \ln a$

Quadratic formula: If $ax^2 + bx + c = 0$, $\qquad x = \dfrac{-b \pm \sqrt{b^2 - 4ac}}{2a}$.

Binomial Theorem

$$(a + b)^n = a^n + na^{n-1}b + \frac{n(n-1)a^{n-2}b^2}{2!} + \frac{n(n-1)(n-2)a^{n-3}b^3}{3!} + \cdots$$

Trigonometry

In the right triangle ABC, $x^2 + y^2 = r^2$.

Definitions of the trigonometric functions:
$$\sin\alpha = y/r \qquad \cos\alpha = x/r \qquad \tan\alpha = y/x$$

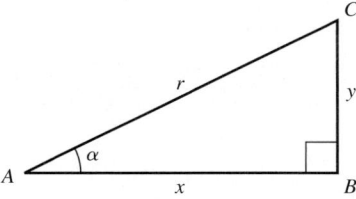

Identities:

$$\sin^2\alpha + \cos^2\alpha = 1 \qquad\qquad \tan\alpha = \frac{\sin\alpha}{\cos\alpha}$$

$$\sin 2\alpha = 2\sin\alpha\cos\alpha \qquad\qquad \cos 2\alpha = \cos^2\alpha - \sin^2\alpha = 2\cos^2\alpha - 1$$
$$= 1 - 2\sin^2\alpha$$

$$\sin\tfrac{1}{2}\alpha = \sqrt{\frac{1 - \cos\alpha}{2}} \qquad\qquad \cos\tfrac{1}{2}\alpha = \sqrt{\frac{1 + \cos\alpha}{2}}$$

$$\sin(-\alpha) = -\sin\alpha \qquad\qquad \sin(\alpha \pm \beta) = \sin\alpha\cos\beta \pm \cos\alpha\sin\beta$$

$$\cos(-\alpha) = \cos\alpha \qquad\qquad \cos(\alpha \pm \beta) = \cos\alpha\cos\beta \mp \sin\alpha\sin\beta$$

$$\sin(\alpha \pm \pi/2) = \pm\cos\alpha \qquad\qquad \sin\alpha + \sin\beta = 2\sin\tfrac{1}{2}(\alpha + \beta)\cos\tfrac{1}{2}(\alpha - \beta)$$

$$\cos(\alpha \pm \pi/2) = \mp\sin\alpha \qquad\qquad \cos\alpha + \cos\beta = 2\cos\tfrac{1}{2}(\alpha + \beta)\cos\tfrac{1}{2}(\alpha - \beta)$$

For *any* triangle $A'B'C'$ (not necessarily a right triangle) with sides a, b, and c and angles α, β, and γ:

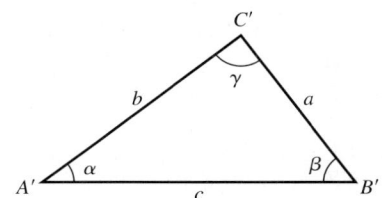

Law of sines: $\dfrac{\sin\alpha}{a} = \dfrac{\sin\beta}{b} = \dfrac{\sin\gamma}{c}$

Law of cosines: $c^2 = a^2 + b^2 - 2ab\cos\gamma$

Geometry

Circumference of circle of radius r:	$C = 2\pi r$	Surface area of sphere of radius r:	$A = 4\pi r^2$
Area of circle of radius r:	$A = \pi r^2$	Volume of cylinder of radius r and height h:	$V = \pi r^2 h$
Volume of sphere of radius r:	$V = 4\pi r^3/3$		

Calculus

Derivatives:

$$\frac{d}{dx}x^n = nx^{n-1} \qquad\qquad \frac{d}{dx}\ln ax = \frac{1}{x} \qquad\qquad \frac{d}{dx}e^{ax} = ae^{ax}$$

$$\frac{d}{dx}\sin ax = a\cos ax \qquad\qquad \frac{d}{dx}\cos ax = -a\sin ax$$

Integrals:

$$\int x^n\,dx = \frac{x^{n+1}}{n+1} \quad (n \neq -1) \qquad \int \frac{dx}{x} = \ln x \qquad \int e^{ax}\,dx = \frac{1}{a}e^{ax}$$

$$\int \sin ax\,dx = -\frac{1}{a}\cos ax \qquad \int \cos ax\,dx = \frac{1}{a}\sin ax \qquad \int \frac{dx}{\sqrt{a^2 - x^2}} = \arcsin\frac{x}{a}$$

$$\int \frac{dx}{\sqrt{x^2 + a^2}} = \ln\left(x + \sqrt{x^2 + a^2}\right) \qquad \int \frac{dx}{x^2 + a^2} = \frac{1}{a}\arctan\frac{x}{a} \qquad \int \frac{dx}{(x^2 + a^2)^{3/2}} = \frac{1}{a^2}\frac{x}{\sqrt{x^2 + a^2}}$$

$$\int \frac{x\,dx}{(x^2 + a^2)^{3/2}} = -\frac{1}{\sqrt{x^2 + a^2}}$$

Power series (convergent for range of x shown):

$$(1 + x)^n = 1 + nx + \frac{n(n-1)x^2}{2!} + \frac{n(n-1)(n-2)}{3!}x^3 \qquad \tan x = x + \frac{x^3}{3} + \frac{2x^5}{15} + \frac{17x^7}{315} + \cdots (|x| < \pi/2)$$

$$+ \cdots (|x| < 1) \qquad\qquad e^x = 1 + x + \frac{x^2}{2!} + \frac{x^3}{3!} + \cdots (\text{all } x)$$

$$\sin x = x - \frac{x^3}{3!} + \frac{x^5}{5!} - \frac{x^7}{7!} + \cdots (\text{all } x) \qquad \ln(1 + x) = x - \frac{x^2}{2} + \frac{x^3}{3} - \frac{x^4}{4} + \cdots (|x| < 1)$$

$$\cos x = 1 - \frac{x^2}{2!} + \frac{x^4}{4!} - \frac{x^6}{6!} + \cdots (\text{all } x)$$

APPENDIX C

THE GREEK ALPHABET

Name	Capital	Lowercase	Name	Capital	Lowercase	Name	Capital	Lowercase
Alpha	A	α	Iota	I	ι	Rho	P	ρ
Beta	B	β	Kappa	K	κ	Sigma	Σ	σ
Gamma	Γ	γ	Lambda	Λ	λ	Tau	T	τ
Delta	Δ	δ	Mu	M	μ	Upsilon	Y	υ
Epsilon	E	ϵ	Nu	N	ν	Phi	Φ	ϕ
Zeta	Z	ζ	Xi	Ξ	ξ	Chi	X	χ
Eta	H	η	Omicron	O	o	Psi	Ψ	ψ
Theta	Θ	θ	Pi	Π	π	Omega	Ω	ω

APPENDIX D

PERIODIC TABLE OF THE ELEMENTS

Group → Period ↓	1	2	3	4	5	6	7	8	9	10	11	12	13	14	15	16	17	18
1	1 **H** 1.008																	2 **He** 4.003
2	3 **Li** 6.941	4 **Be** 9.012											5 **B** 10.811	6 **C** 12.011	7 **N** 14.007	8 **O** 15.999	9 **F** 18.998	10 **Ne** 20.180
3	11 **Na** 22.990	12 **Mg** 24.305											13 **Al** 26.982	14 **Si** 28.086	15 **P** 30.974	16 **S** 32.065	17 **Cl** 35.453	18 **Ar** 39.948
4	19 **K** 39.098	20 **Ca** 40.078	21 **Sc** 44.956	22 **Ti** 47.867	23 **V** 50.942	24 **Cr** 51.996	25 **Mn** 54.938	26 **Fe** 55.845	27 **Co** 58.933	28 **Ni** 58.693	29 **Cu** 63.546	30 **Zn** 65.409	31 **Ga** 69.723	32 **Ge** 72.64	33 **As** 74.922	34 **Se** 78.96	35 **Br** 79.904	36 **Kr** 83.798
5	37 **Rb** 85.468	38 **Sr** 87.62	39 **Y** 88.906	40 **Zr** 91.224	41 **Nb** 92.906	42 **Mo** 95.94	43 **Tc** (98)	44 **Ru** 101.07	45 **Rh** 102.905	46 **Pd** 106.42	47 **Ag** 107.868	48 **Cd** 112.411	49 **In** 114.818	50 **Sn** 118.710	51 **Sb** 121.760	52 **Te** 127.60	53 **I** 126.904	54 **Xe** 131.293
6	55 **Cs** 132.905	56 **Ba** 137.327	71 **Lu** 174.967	72 **Hf** 178.49	73 **Ta** 180.948	74 **W** 183.84	75 **Re** 186.207	76 **Os** 190.23	77 **Ir** 192.217	78 **Pt** 195.078	79 **Au** 196.967	80 **Hg** 200.59	81 **Tl** 204.383	82 **Pb** 207.2	83 **Bi** 208.980	84 **Po** (209)	85 **At** (210)	86 **Rn** (222)
7	87 **Fr** (223)	88 **Ra** (226)	103 **Lr** (262)	104 **Rf** (261)	105 **Db** (262)	106 **Sg** (266)	107 **Bh** (270)	108 **Hs** (269)	109 **Mt** (278)	110 **Ds** (281)	111 **Rg** (281)	112 **Cn** (285)	113 **Uut** (284)	114 **Fl** (289)	115 **Uup** (288)	116 **Lv** (292)	117 **Uus** (294)	118 **Uuo** (294)

Lanthanoids

57 **La** 138.905	58 **Ce** 140.116	59 **Pr** 140.908	60 **Nd** 144.24	61 **Pm** (145)	62 **Sm** 150.36	63 **Eu** 151.96	64 **Gd** 157.25	65 **Tb** 158.925	66 **Dy** 162.500	67 **Ho** 164.930	68 **Er** 167.259	69 **Tm** 168.934	70 **Yb** 173.04

Actinoids

89 **Ac** (227)	90 **Th** (232)	91 **Pa** (231)	92 **U** (238)	93 **Np** (237)	94 **Pu** (244)	95 **Am** (243)	96 **Cm** (247)	97 **Bk** (247)	98 **Cf** (251)	99 **Es** (252)	100 **Fm** (257)	101 **Md** (258)	102 **No** (259)

For each element, the average atomic mass of the mixture of isotopes occurring in nature is shown. For elements having no stable isotope, the approximate atomic mass of the longest-lived isotope is shown in parentheses. All atomic masses are expressed in atomic mass units $(1\ u = 1.660538921(73) \times 10^{-27}\ kg)$, equivalent to grams per mole (g/mol).

APPENDIX E

UNIT CONVERSION FACTORS

Length
1 m = 100 cm = 1000 mm = $10^6 \, \mu$m = 10^9 nm
1 km = 1000 m = 0.6214 mi
1 m = 3.281 ft = 39.37 in.
1 cm = 0.3937 in.
1 in. = 2.540 cm
1 ft = 30.48 cm
1 yd = 91.44 cm
1 mi = 5280 ft = 1.609 km
1 Å = 10^{-10} m = 10^{-8} cm = 10^{-1} nm
1 nautical mile = 6080 ft
1 light-year = 9.461×10^{15} m

Area
1 cm^2 = 0.155 $in.^2$
1 m^2 = $10^4 \, cm^2$ = 10.76 ft^2
1 $in.^2$ = 6.452 cm^2
1 ft^2 = 144 $in.^2$ = 0.0929 m^2

Volume
1 liter = 1000 cm^3 = $10^{-3} \, m^3$ = 0.03531 ft^3 = 61.02 $in.^3$
1 ft^3 = 0.02832 m^3 = 28.32 liters = 7.477 gallons
1 gallon = 3.788 liters

Time
1 min = 60 s
1 h = 3600 s
1 d = 86,400 s
1 y = 365.24 d = 3.156×10^7 s

Angle
1 rad = 57.30° = 180°/π
1° = 0.01745 rad = $\pi/180$ rad
1 revolution = 360° = 2π rad
1 rev/min (rpm) = 0.1047 rad/s

Speed
1 m/s = 3.281 ft/s
1 ft/s = 0.3048 m/s
1 mi/min = 60 mi/h = 88 ft/s
1 km/h = 0.2778 m/s = 0.6214 mi/h
1 mi/h = 1.466 ft/s = 0.4470 m/s = 1.609 km/h
1 furlong/fortnight = 1.662×10^{-4} m/s

Acceleration
1 m/s^2 = 100 cm/s^2 = 3.281 ft/s^2
1 cm/s^2 = 0.01 m/s^2 = 0.03281 ft/s^2
1 ft/s^2 = 0.3048 m/s^2 = 30.48 cm/s^2
1 mi/h $\cdot$ s = 1.467 ft/s^2

Mass
1 kg = 10^3 g = 0.0685 slug
1 g = 6.85×10^{-5} slug
1 slug = 14.59 kg
1 u = 1.661×10^{-27} kg
1 kg has a weight of 2.205 lb when g = 9.80 m/s^2

Force
1 N = 10^5 dyn = 0.2248 lb
1 lb = 4.448 N = 4.448×10^5 dyn

Pressure
1 Pa = 1 N/m^2 = 1.450×10^{-4} $lb/in.^2$ = 0.0209 lb/ft^2
1 bar = 10^5 Pa
1 $lb/in.^2$ = 6895 Pa
1 lb/ft^2 = 47.88 Pa
1 atm = 1.013×10^5 Pa = 1.013 bar
   = 14.7 $lb/in.^2$ = 2117 lb/ft^2
1 mm Hg = 1 torr = 133.3 Pa

Energy
1 J = 10^7 ergs = 0.239 cal
1 cal = 4.186 J (based on 15° calorie)
1 ft $\cdot$ lb = 1.356 J
1 Btu = 1055 J = 252 cal = 778 ft $\cdot$ lb
1 eV = 1.602×10^{-19} J
1 kWh = 3.600×10^6 J

Mass–Energy Equivalence
1 kg $\leftrightarrow$ 8.988×10^{16} J
1 u $\leftrightarrow$ 931.5 MeV
1 eV $\leftrightarrow$ 1.074×10^{-9} u

Power
1 W = 1 J/s
1 hp = 746 W = 550 ft $\cdot$ lb/s
1 Btu/h = 0.293 W

APPENDIX F

NUMERICAL CONSTANTS

Fundamental Physical Constants*

Name	Symbol	Value
Speed of light in vacuum	c	2.99792458×10^8 m/s
Magnitude of charge of electron	e	$1.602176565(35) \times 10^{-19}$ C
Gravitational constant	G	$6.67384(80) \times 10^{-11}$ N·m^2/kg^2
Planck's constant	h	$6.62606957(29) \times 10^{-34}$ J·s
Boltzmann constant	k	$1.3806488(13) \times 10^{-23}$ J/K
Avogadro's number	N_A	$6.02214129(27) \times 10^{23}$ molecules/mol
Gas constant	R	$8.3144621(75)$ J/mol·K
Mass of electron	m_e	$9.10938291(40) \times 10^{-31}$ kg
Mass of proton	m_p	$1.672621777(74) \times 10^{-27}$ kg
Mass of neutron	m_n	$1.674927351(74) \times 10^{-27}$ kg
Magnetic constant	μ_0	$4\pi \times 10^{-7}$ Wb/A·m
Electric constant	$\epsilon_0 = 1/\mu_0 c^2$	$8.854187817\ldots \times 10^{-12}$ C^2/N·m^2
	$1/4\pi\epsilon_0$	$8.987551787\ldots \times 10^9$ N·m^2/C^2

Other Useful Constants*

Mechanical equivalent of heat		4.186 J/cal ($15°$ calorie)
Standard atmospheric pressure	1 atm	1.01325×10^5 Pa
Absolute zero	0 K	$-273.15°$C
Electron volt	1 eV	$1.602176565(35) \times 10^{-19}$ J
Atomic mass unit	1 u	$1.660538921(73) \times 10^{-27}$ kg
Electron rest energy	$m_e c^2$	$0.510998928(11)$ MeV
Volume of ideal gas ($0°$C and 1 atm)		$22.413968(20)$ liter/mol
Acceleration due to gravity (standard)	g	9.80665 m/s^2

*Source: National Institute of Standards and Technology (**http://physics.nist.gov/cuu**). Numbers in parentheses show the uncertainty in the final digits of the main number; for example, the number 1.6454(21) means 1.6454 ± 0.0021. Values shown without uncertainties are exact.

Astronomical Data[†]

Body	Mass (kg)	Radius (m)	Orbit radius (m)	Orbital period
Sun	1.99×10^{30}	6.96×10^8	—	—
Moon	7.35×10^{22}	1.74×10^6	3.84×10^8	27.3 d
Mercury	3.30×10^{23}	2.44×10^6	5.79×10^{10}	88.0 d
Venus	4.87×10^{24}	6.05×10^6	1.08×10^{11}	224.7 d
Earth	5.97×10^{24}	6.37×10^6	1.50×10^{11}	365.3 d
Mars	6.42×10^{23}	3.39×10^6	2.28×10^{11}	687.0 d
Jupiter	1.90×10^{27}	6.99×10^7	7.78×10^{11}	11.86 y
Saturn	5.68×10^{26}	5.82×10^7	1.43×10^{12}	29.45 y
Uranus	8.68×10^{25}	2.54×10^7	2.87×10^{12}	84.02 y
Neptune	1.02×10^{26}	2.46×10^7	4.50×10^{12}	164.8 y
Pluto‡	1.31×10^{22}	1.15×10^6	5.91×10^{12}	247.9 y

[†]Source: NASA (**http://solarsystem.nasa.gov/planets/**). For each body, "radius" is its average radius and "orbit radius" is its average distance from the sun or (for the moon) from the earth.

‡In August 2006, the International Astronomical Union reclassified Pluto and similar small objects that orbit the sun as "dwarf planets."

Prefixes for Powers of 10

Power of ten	Prefix	Abbreviation	Pronunciation
10^{-24}	yocto-	y	*yoc*-toe
10^{-21}	zepto-	z	*zep*-toe
10^{-18}	atto-	a	*at*-toe
10^{-15}	femto-	f	*fem*-toe
10^{-12}	pico-	p	*pee*-koe
10^{-9}	nano-	n	*nan*-oe
10^{-6}	micro-	μ	*my*-crow
10^{-3}	milli-	m	*mil*-i
10^{-2}	centi-	c	*cen*-ti
10^3	kilo-	k	*kil*-oe
10^6	mega-	M	*meg*-a
10^9	giga-	G	*jig*-a or *gig*-a
10^{12}	tera-	T	*ter*-a
10^{15}	peta-	P	*pet*-a
10^{18}	exa-	E	*ex*-a
10^{21}	zetta-	Z	*zet*-a
10^{24}	yotta-	Y	*yot*-a

Examples:

1 femtometer = 1 fm = 10^{-15} m

1 picosecond = 1 ps = 10^{-12} s

1 nanocoulomb = 1 nC = 10^{-9} C

1 microkelvin = 1 μK = 10^{-6} K

1 millivolt = 1 mV = 10^{-3} V

1 kilopascal = 1 kPa = 10^3 Pa

1 megawatt = 1 MW = 10^6 W

1 gigahertz = 1 GHz = 10^9 Hz

ANSWERS TO ODD-NUMBERED PROBLEMS

Chapter 1

1.1 a) 1.61 km b) 3.28×10^3 ft
1.3 1.02 ns
1.5 5.36 L
1.7 31.7 y
1.9 a) 23.4 km/L b) 1.4 tanks
1.11 9.0 cm
1.13 4.2×10^{-12} cm^3, 1.3×10^{-5} mm^2
1.15 0.45%
1.17 a) no b) no c) no d) no e) no
1.19 $\approx 4 \times 10^8$
1.21 $\approx$ \$70 million
1.23 2×10^5
1.25 7.8 km, 38° north of east
1.27 $A_x = 0$, $A_y = -8.00$ m, $B_x = 7.50$ m,
$B_y = 13.0$ cm, $C_x = -10.9$ cm,
$C_y = -5.07$ m, $D_x = -7.99$ m, $D_y = 6.02$ m
1.29 a) -6.00 m b) 11.3 m
1.31 a) 9.01 m, 33.7° b) 9.01 m, 33.7°
c) 22.3 m, 250° d) 22.3 m, 70.3°
1.33 2.81 km, 38.5° north of west
1.35 a) 2.48 cm, 18.4° b) 4.09 cm, 83.7°
c) 4.09 cm, 264°
1.37 $\vec{A} = -(8.00\text{ m})\hat{\jmath}$,
$\vec{B} = (7.50\text{ m})\hat{\imath} + (+13.0\text{ m})\hat{\jmath}$,
$\vec{C} = (-10.9\text{ m})\hat{\imath} + (-5.07\text{ m})\hat{\jmath}$,
$\vec{D} = (-7.99\text{ m})\hat{\imath} + (6.02\text{ m})\hat{\jmath}$
1.39 a) $\vec{A} = (1.23\text{ m})\hat{\imath} + (3.38\text{ m})\hat{\jmath}$,
$\vec{B} = (-2.08\text{ m})\hat{\imath} + (-1.20\text{ m})\hat{\jmath}$
b) $\vec{C} = (12.0\text{ m})\hat{\imath} + (14.9\text{ m})\hat{\jmath}$
c) 19.2 m, 51.2°
1.41 a) $A = 5.38$, $B = 4.36$
b) $-5.00\hat{\imath} + 2.00\hat{\jmath} + 7.00\hat{k}$
c) 8.83, yes
1.43 a) -104 m^2 b) -148 m^2 c) 40.6 m^2
1.45 a) 165° b) 28° c) 90°
1.47 a) $(-63.9\text{ m}^2)\hat{k}$ b) $(63.9\text{ m}^2)\hat{k}$
1.49 a) 5.51 g/cm^3
b) 1.1×10^6 g/cm^3
c) 4.7×10^{14} g/cm^3
1.51 a) 1.64×10^4 km b) $2.57r_E$
1.53 a) 2200 g b) 2.1 m
1.55 a) (2.8 ± 0.3) cm^3 b) 170 ± 20
1.57 $\approx 6 \times 10^{27}$
1.59 179 N, 358 N, 45.8° east of north, or 393 N, 786 N, 45.8° south of east
1.61 144 m, 41° south of west
1.63 7.55 N
1.65 60.9 km, 33.0° south of west
1.67 28.8 m, 11.4° north of east
1.69 71.9 m, 64.1° north of west
1.71 160 N, 13° below horizontal
1.73 a) 818 m, 15.8° west of south
1.75 18.6° east of south, 29.6 m
1.77 28.2 m
1.79 124°
1.81 156 m^2
1.83 28.0 m
1.85 $C_x = -8.0$, $C_y = -6.1$
1.87 D, F, B, C, A, E
1.89 b) (i) 0.9857 AU (ii) 1.3820 AU
(iii) 1.695 AU c) 54.6°
1.91 a) 76.2 ly b) 129°
1.93 choice (a)

Chapter 2

2.1 25.0 m
2.3 55 min
2.5 a) 0.312 m/s b) 1.56 m/s
2.7 a) 12.0 m/s b) (i) 0 (ii) 15.0 m/s
(iii) 12.0 m/s c) 13.3 m/s
2.9 a) 2.33 m/s, 2.33 m/s
b) 2.33 m/s, 0.33 m/s
2.11 6.7 m/s, 6.7 m/s, 0, -40.0 m/s, -40.0 m/s, -40.0 m/s, 0
2.13 a) no b) (i) 12.8 m/s^2 (ii) 3.50 m/s^2
(iii) 0.718 m/s^2
2.15 a) 2.00 cm/s, 50.0 cm, -0.125 cm/s^2
b) 16.0 s c) 32.0 s
d) 6.20 s, 1.23 cm/s;
25.8 s, -1.23 cm/s; 36.4 s, -2.55 cm/s
2.17 a) 0.500 m/s^2 b) 0, 1.00 m/s^2
2.19 a) 8.33 m/s b) 1.11 m/s^2
2.21 a) 675 m/s^2 b) 0.0667 s
2.23 1.70 m
2.25 0.38 m
2.27 a) 3.1×10^6 m/s^2 = 3.2×10^5 g
b) 1.6 ms c) no
2.29 a) (i) 5.59 m/s^2 (ii) 7.74 m/s^2
b) (i) 179 m (ii) 1.23×10^4 m
2.31 a) 0, 6.3 m/s^2, -11.2 m/s^2
b) 100 m, 230 m, 320 m
2.33 2.69 m/s
2.35 a) 2.94 m/s b) 0.600 s
2.37 1.67 s
2.39 a) 33.5 m b) 15.8 m/s
2.41 a) $t = \sqrt{2d/g}$ b) 0.190 s
2.43 a) 646 m b) 16.4 s, 112 m/s
2.45 a) 249 m/s^2 b) 25.4 c) 101 m
d) no (if a is constant)
2.47 0.0868 m/s^2
2.49 37.6 m/s
2.51 a) 467 m b) 110 m/s
2.53 a) $x = (0.25\text{ m/s}^3)t^3 - (0.010\text{ m/s}^4)t^4$,
$v_x = (0.75\text{ m/s}^3)t^2 - (0.040\text{ m/s}^4)t^3$
b) 39.1 m/s
2.55 a) 10.0 m b) (i) 8.33 m/s (ii) 9.09 m/s
(iii) 9.52 m/s
2.57 250 km
2.59 a) 197 m/s b) 169 m/s
2.61 a) 92.0 m/s b) 92.0 m/s
2.63 67 m
2.65 a) 7.56 s b) 37.2 m
c) 25.7 m/s (car), 15.9 m/s (truck)
2.67 a) 15.9 s b) 393 m c) 29.5 m/s
2.69 a) -4.00 m/s b) 12.0 m/s
2.71 a) $2.64H$ b) $2.64T$
2.73 a) 6.69 m/s b) 4.49 m c) 1.42 s
2.75 a) 3.3 s b) $9H$
2.77 6.75 s
2.79 a) 380 m b) 184 m
2.81 a) 0.625 m/s^3 b) 107 m
2.83 a) car A b) 2.27 s, 5.73 s c) 1.00 s, 4.33 s
d) 2.67 s
2.85 a) 0.0510 s^2/m b) lower than c) no
2.87 4.8
2.89 a) 8.3 m/s b) (i) 0.411 m (ii) 1.15 km
c) 9.8 m/s d) 4.9 m/s
2.91 choice (b)

Chapter 3

3.1 a) 1.4 m/s, -1.3 m/s b) 1.9 m/s, 317°
3.3 a) 7.1 cm/s, 45°
b) 5.0 cm/s, 90°; 7.1 cm/s, 45°; 11 cm/s, 27°
3.5 b) -8.67 m/s^2, -2.33 m/s^2
c) 8.98 m/s^2, 195°
3.7 b) $\vec{v} = \alpha\hat{\imath} - 2\beta t\hat{\jmath}$, $\vec{a} = -2\beta\hat{\jmath}$
c) 5.4 m/s, 297°; 2.4 m/s^2, 270°
d) speeding up and turning right
3.9 a) 1.13 m b) 0.528 m
c) $v_x = 1.10$ m/s, $v_y = -4.70$ m/s, 4.83 m/s, 76.8° below the horizontal
3.11 2.57 m
3.13 a) 24.1 m/s b) 31.0 m/s
3.15 1.28 m/s^2
3.17 a) 0.683 s, 2.99 s
b) 24.0 m/s, 11.3 m/s; 24.0 m/s, -11.3 m/s
c) 30.0 m/s, 36.9° below the horizontal
3.19 a) 1.5 m b) -0.89 m/s
3.21 a) 13.6 m/s b) 34.6 m/s c) 103 m
3.23 a) 0.034 m/s^2 = $0.0034g$ b) 1.4 h
3.25 120 m/s^2, 270 mph
3.27 a) 2.57 m/s^2 upward
b) 2.57 m/s^2 downward
c) 14.7 s
3.29 a) 32.9 m/s b) 27.7 m/s c) 35.5 rpm
3.31 a) 14 s b) 70 s
3.33 0.36 m/s, 52.5° south of west
3.35 a) 4.7 m/s, 25° south of east b) 120 s
c) 240 m
3.37 a) 24° west of south b) 5.5 h
3.39 a) $A = 0$, $B = 2.00$ m/s^2,
$C = 50.0$ m, $D = 0.500$ m/s^3
b) $\vec{a} = (4.00\text{ m/s}^2)\hat{\imath}$, $\vec{v} = 0$
c) $v_x = 40.0$ m/s, $v_y = 150$ m/s, 155 m/s
d) $\vec{r} = (200\text{ m})\hat{\imath} + (550\text{ m})\hat{\jmath}$
3.41 $2b/3c$
3.43 a) 128 m b) 315 m
3.45 31 m/s
3.47 274 m
3.49 795 m
3.51 33.7 m
3.53 a) 42.8 m/s b) 42.0 m
3.55 a) 16.6 m/s
b) 10.9 m/s, 40.5° below the horizontal
3.57 a) 1.50 m/s b) 4.66 m
3.59 a) 6.91 m c) no
3.61 a) 4.25 m/s b) 10.6 m
3.63 a) 17.8 m/s b) in the river, 28.4 m horizontally from his launch point
3.65 a) 49.5 m/s b) 50 m
3.67 a) 81.6 m b) 245 m
c) in the cart
3.69 a) 13.3 m/s b) 3.8 m
3.71 a) 44.7 km/h, 26.6° west of south
b) 10.5° north of west
3.73 7.39 m/s, 12.4° north of east
3.75 3.01 m/s, 33.7° north of east
3.77 a) graph R^2 versus h b) 16.4 m/s
c) 23.8 m
3.79 70.5°
3.81 5.15 s
3.83 choice (b)
3.85 choice (c)

Chapter 4

4.1   494 N, 31.8°
4.3   3.15 N
4.5   a) −8.10 N, 3.00 N   b) 8.64 N
4.7   46.7 N, opposite to the motion of the skater
4.9   21.8 kg
4.11   a) 3.12 m, 3.12 m/s   b) 21.9 m, 6.24 m/s
4.13   a) 45.0 N, between 2.0 s and 4.0 s
       b) between 2.0 s and 4.0 s   c) 0 s, 6.0 s
4.15   a) $A = 100$ N, $B = 12.5$ N/s^2
       b) (i) 21.6 N, 2.70 m/s^2 (ii) 134 N, 16.8 m/s^2
       c) 26.6 m/s^2
4.17   2940 N
4.19   a) 4.49 kg   b) 4.49 kg, 8.13 N
4.21   825 N, blocks
4.23   50 N
4.25   b) yes
4.27   a) yes   b) no
4.29   b) 142 N
4.31   2.58 s
4.33   a) 17 N, 90° clockwise from the +x-axis
       b) 840 N
4.35   a) 4.85 m/s   b) 16.2 m/s^2 upward
       c) 1470 N upward (on him), 2360 N downward
       (on ground)
4.37   a) 153 N
4.39   a) 2.50 m/s^2 b) 10.0 N
       c) to the right, $F > T$ d) 25.0 N
4.41   a) 4.4 m   b) 300 m/s
       c) (i) 2.7×10^4 N   (ii) 9.0×10^3 N
4.43   b) 0.049 N   c) 410mg
4.45   a) 0.603 m/s^2, upward
       b) 1.26 m/s^2, downward
4.47   a) 7.79 m/s   b) 50.6 m/s^2 upward
       c) $F_{\text{ground}} - mg$ upward, 4530 N upward,
       6.16mg
4.49   a) 4.34 kg   b) 5.30 kg
4.51   7.78 m
4.53   a) largest: Ferrari; smallest: Alpha Romeo and
       Honda Civic   b) largest: Ferrari; smallest:
       Volvo   c) 7.5 kN, smaller   d) zero
4.55   b) 26 kg, 8.3 m/s^2
4.57   choice (d)
4.59   choice (a)

Chapter 5

5.1   a) 25.0 N   b) 50.0 N
5.3   a) 990 N, 735 N   b) 926 N
5.5   48°
5.7   a) $T_A = 0.732w$, $T_B = 0.897w$, $T_C = w$
       b) $T_A = 2.73w$, $T_B = 3.35w$, $T_C = w$
5.9   a) 574 N   b) 607 N
5.11   a) 1.10×10^8 N   b) 5w c) 8.4 s
5.13   a) 4610 m/s$^2 = 470g$
       b) 9.70×10^5 N $= 471w$ c) 0.0187 s
5.15   b) 2.96 m/s^2 c) 191 N; greater than; less than
5.17   b) 3.75 m/s^2 c) 2.48 kg
       d) $T <$ weight of the hanging block
5.19   a) 0.832 m/s^2 b) 17.3 s
5.21   a) 3.4 m/s   c) 2.2w
5.23   a) 14.0 m   b) 18.0 m/s
5.25   50°
5.27   a) 33 N   b) 3.1 m
5.29   a) μ_s: 0.710; μ_k: 0.472   b) 258 N
       c) (i) 51.8 N   (ii) 4.97 m/s^2
5.31   a) 18.3 m/s^2 b) 2.29 m/s^2
5.33   a) 57.1 N   b) 146 N up the ramp
5.35   a) 52.5 m   b) 16.0 m/s
5.37   a) $\mu_k(m_A + m_B)g$ b) $\mu_k m_A g$

5.39   a) 0.218 m/s   b) 11.7 N
5.41   a) $\dfrac{\mu_k mg}{\cos\theta - \mu_k \sin\theta}$ b) $1/\tan\theta$
5.43   b) 8.2 m/s
5.45   a) 61.8 N   b) 30.4 N
5.47   3.66 s
5.49   a) 21.0°, no   b) 11,800 N (car), 23,600 N (truck)
5.51   6200 N (horizontal cable), 1410 N (upper cable)
5.53   a) 1.5 rev/min   b) 0.92 rev/min
5.55   a) 38.3 m/s = 138 km/h   b) 3580 N
5.57   2.42 m/s
5.59   a) 1.73 m/s^2 c) 0.0115 N upward
       d) 0.0098 N
5.61   a) rope making 60° angle   b) 6400 N
5.63 $T_B = 4960$ N, $T_C = 1200$ N
5.65   a) 470 N   b) 163 N
5.67   762 N
5.69   a) (i) −3.80 m/s   (ii) 24.6 m/s   b) 4.36 m
       c) 2.45 s
5.71   a) 11.4 N   b) 2.57 kg
5.73   12.3 m/s
5.75   1.78 m/s
5.77   a) $m_1(\sin\alpha + \mu_k \cos\alpha)$
       b) $m_1(\sin\alpha - \mu_k \cos\alpha)$
       c) $m_1(\sin\alpha - \mu_s \cos\alpha) \le m_2 \le$
       $m_1(\sin\alpha + \mu_s \cos\alpha)$
5.79   a) 1.44 N   b) 1.80 N
5.81   920 N
5.83   a) 88.0 N northward   b) 78 N southward
5.85   a) 294 N (18.0-cm wire), 152 N, 152 N
       b) 40.0 N
5.87   3.0 N
5.89   a) 12.9 kg   b) $T_{AB} = 47.2$ N, $T_{BC} = 101$ N
5.91 $a_1 = \dfrac{2m_2 g}{4m_1 + m_2}$, $a_2 = \dfrac{m_2 g}{4m_1 + m_2}$
5.93   1.46 m above the floor
5.95 g/μ_s
5.97   b) 0.452
5.99   0.34
5.101 b) 8.8 N   c) 31.0 N   d) 1.54 m/s^2
5.103 $v = (2mg/k)\left[\frac{1}{2} + e^{-(k/m)t}\right]$
5.105 b) 0.28   c) no
5.107 a) 81.1°   b) no
       c) The bead rides at the bottom of the hoop.
5.109 a) 0.371   b) 0.290
       c) yes, same slope, less-negative intercept
5.111 a) 5/8 in.   b) 23.9 kN   c) 3.57 kN, smaller
       d) larger; accurate
5.113 $F = (M + m)g \tan\alpha$
5.115 $\cos^2\beta$
5.117 choice (b)

Chapter 6

6.1   a) 3.60 J   b) −0.900 J   c) 0   d) 0   e) 2.70 J
6.3   a) 74 N   b) 333 J   c) −330 J   d) 0,0   e) 0
6.5   a) −1750 J   b) no
6.7   a) (i) 9.00 J   (ii) −9.00 J
       b) (i) 0   (ii) 9.00 J   (iii) −9.00 J   (iv) 0
       c) zero for each block
6.9   a) (i) 0   (ii) 0   b) (i) 0   (ii) −25.1 J
6.11   a) 374 J   b) −333 J   c) 0   d) 41 J
       e) 352 J
6.13   −572 J
6.15   a) 120 J   b) −108 J   c) 24.3 J
6.17   a) 36,000 J   b) 4
6.19   a) 1.0×10^{16} J   b) 2.4 times
6.21   a) 43.2 m/s   b) 101 m/s
6.23 $\sqrt{2gh(1 + \mu_k/\tan\alpha)}$

6.25   48.0 N
6.27   a) 4.48 m/s   b) 3.61 m/s
6.29   a) 4.96 m/s   b) 1.43 m/s^2; 4.96 m/s, same
6.31   a) $v_0^2/2\mu_k g$ b) (i) $\frac{1}{2}$ (ii) 4   (iii) 2
6.33   a) 40.0 N/m   b) 0.456 N
6.35   b) 13.1 cm (bottom), 14.1 cm (middle),
       15.2 cm (top)
6.37   a) 2.83 m/s   b) 3.46 m/s
6.39   8.5 cm
6.41   a) 1.76   b) 0.666 m/s
6.43   a) 4.0 J   b) 0   c) −1.0 J   d) 3.0 J
       e) −1.0 J
6.45   a) 2.83 m/s   b) 2.40 m/s
6.47   a) 0.0565 m   b) no, 0.57 J
6.49   8.17 m/s
6.51   360,000 J; 100 m/s
6.53 $(3.9 \times 10^{13})P$
6.55   745 W $\approx$ 1 hp
6.57   a) 84.6/min   b) 22.7/min
6.59   29.6 kW
6.61   0.20 W
6.63   a) 608 J   b) −395 J   c) 0   d) −189 J
       e) 24 J   f) 1.5 m/s
6.65   a) 5.62 J (20.0-N block), 3.38 J (12.0-N
       block)   b) 2.58 J (20.0-N block), 1.54 J
       (12.0-N block)
6.67   a) 1.8 m/s = 4.0 mi/h
       b) 180 m/s$^2 \approx 18g$, 900 N
6.69   a) 5.11 m   b) 0.304   c) 10.3 m
6.71   a) 0.074 N   b) 4.7 N   c) 0.22 J
6.73 6.3×10^4 N/m
6.75   1.1 m
6.77   a) 2.39 m/s   b) 9.42 m/s, away from the wall
6.79   a) 0.600 m   b) 1.50 m/s
6.81   0.786
6.83   1.3 m
6.85   a) 1.10×10^5 J   b) 1.30×10^5 J
       c) 3.99 kW
6.87   3.6 h
6.89   a) 1.26×10^5 J   b) 1.46 W
6.91   b) $v^2 = -\dfrac{k}{m}d^2 + 2d\left[\dfrac{k}{m}(0.400\text{ m}) - \mu_k g\right]$
       c) 1.29 m/s, 0.204 m   d) 12.0 N/m, 0.800
6.93   a) $Mv^2/6$ b) 6.1 m/s   c) 3.9 m/s
       d) 0.40 J, 0.60 J
6.95   choice (a)
6.97   choice (d)

Chapter 7

7.1   a) 6.6×10^5 J   b) -7.7×10^5 J
7.3   a) 610 N   b) (i) 0   (ii) 550 J
7.5   a) 24.0 m/s   b) 24.0 m/s   c) part (b)
7.7   a) 2.0 m/s   b) $9.8/10^{-7}$ J, 2.0 J/kg   c) 200 m,
       63 m/s   d) 5.9 J/kg   e) in its tensed legs
7.9   a) (i) 0   (ii) 0.98 J   b) 2.8 m/s
       c) Only gravity is constant.   d) 5.1 N
7.11   −5400 J
7.13   a) 660 J   b) −118 J   c) 353 J   d) 190 J
       e) 3.16 m/s^2, 6.16 m/s, 190 J
7.15   a) 52.0 J   b) 3.25 J
7.17   a) (i) 4U_0 (ii) $U_0/4$
       b) (i) $x_0\sqrt{2}$ (ii) $x_0/\sqrt{2}$
7.19   a) 5.48 cm   b) 3.92 cm
7.21   a) 6.32 cm   b) 12 cm
7.23   a) 3.03 m/s, as it leaves the spring
       b) 95.9 m/s^2, when the spring has its maximum
       compression
7.25   a) 4.46×10^5 N/m   b) 0.128 m
7.27   a) −5.4 J   b) −5.4 J   c) −10.8 J
       d) nonconservative

7.29   a) 8.16 m/s   b) 766 J
7.31   1.29 N, +x-direction
7.33   130 m/s^2, 132° counterclockwise from the +x-axis
7.35   a) $F(r) = (12a/r^{13}) - (6b/r^7)$
       b) $(2a/b)^{1/6}$, yes   c) $b^2/4a$
       d) $a = 6.67 \times 10^{-138}$ J·m^{12}, $b = 6.41 \times 10^{-78}$ J·m^6
7.37   a) zero (gravel), 637 N (box)   b) 2.99 m/s
7.39   0.41
7.41   a) 16.0 m/s   b) 11,500 N
7.43   a) 20.0 m along the rough bottom   b) −78.4 J
7.45   a) 22.2 m/s   b) 16.4 m   c) no
7.47   0.602 m
7.49   15.5 m/s
7.51   4.4 m/s
7.53   a) 7.00 m/s   b) 8.82 N
7.55   48.2°
7.57   a) 0.392   b) −0.83 J
7.59   a) $U(x) = \frac{1}{2}\alpha x^2 + \frac{1}{3}\beta x^3$ b) 7.85 m/s
7.61   a) $\alpha/(x + x_0)$ b) 3.27 m/s
7.63   7.01 m/s
7.65   a) 0.747 m/s   b) 0.931 m/s
7.67   a) 0.480 m/s   b) 0.566 m/s
7.69   a) 3.87 m/s   b) 0.10 m
7.71   0.456 N
7.73   119 J
7.75   a) −50.6 J   b) −67.5 J   c) nonconservative
7.77   a) 57.0 m   b) 16.5 m
       c) negative work done by air resistance
7.79   a) yes   b) 0.14 J   d) −1.0 m, 0, 1.0 m
       e) positive: −1.5 m < x < −1.0 m and 0 < x < 1.0 m; negative: −1.0 m < x < 0 and 1.0 m < x < 1.5 m   f) −0.55 m, 0.12 J
7.81   choice (c)
7.83   choice (b)

Chapter 8

8.1   a) 1.20×10^5 kg·m/s
       b) (i) 60.0 m/s   (ii) 26.8 m/s
8.3   a) −30 kg·m/s, −55 kg·m/s
       b) 0, 52 kg·m/s   c) 0, −3.0 kg·m/s
8.5   a) 22.5 kg·m/s, to the left   b) 838 J
8.7   562 N, not significant
8.9   a) 10.8 m/s, to the right
       b) 0.750 m/s, to the left
8.11   a) 500 N/s^2 b) 5810 N·s   c) 2.70 m/s
8.13   a) 2.50 N·s, in the direction of the force
       b) (i) 6.25 m/s, to the right   (ii) 3.75 m/s, to the right
8.15   0.593 kg·m/s
8.17   0.87 kg·m/s, in the same direction as the bullet is traveling
8.19   a) 6.79 m/s   b) 55.2 J
8.21   a) 0.790 m/s   b) −0.0023 J
8.23   1.97 m/s
8.25   a) 0.0559 m/s   b) 0.0313 m/s
8.27   a) 7.20 m/s, 38.0° from Rebecca's original direction   b) −680 J
8.29   a) 4.3 m/s   c) 4.3 m/s
8.31   a) A: 29.3 m/s; B: 20.7 m/s   b) 19.6%
8.33   a) 0.846 m/s   b) 2.10 J
8.35   a) $−1.4 \times 10^{-6}$ km/h, no
       b) $−6.7 \times 10^{-8}$ km/h, no
8.37   5.9 m/s, 58° north of east
8.39   5.46 m/s, 36.0° south of east
8.41   19.5 m/s (car), 21.9 m/s (truck)
8.43   a) 2.93 cm   b) 866 J   c) 1.73 J
8.45   13.6 N

8.47   a) 3.00 J; 0.500 m/s for both
       b) A: −1.00 m/s; B: 1.00 m/s
8.49   a) $v_1/3$ b) $K_1/9$ c) 10
8.51   (0.0444 m, 0.0556 m)
8.53   2520 km
8.55   0.700 m to the right and 0.700 m upward
8.57   0.73 m/s
8.59 $F_x = −(1.50$ N/s$)t$, $F_y = 0.25$ N, $F_z = 0$
8.61   a) 0.053 kg   b) 5.19 N
8.63   a) 7.2×10^{-66} b) 0.223
8.65   a) −1.14 N·s, 0.330 N·s
       b) 0.04 m/s, 1.8 m/s
8.67   a) 5.21 J, −0.0833 m/s
       b) −2.17 m/s (A), 0.333 m/s (B)
8.69   a) 1.75 m/s, 0.260 m/s   b) −0.092 J
8.71   0.946 m
8.73   1.8 m
8.75   a) $a_A = 162$ m/s^2, $a_B = 54.0$ m/s^2
       b) $v_A = 5.23$ m/s, $v_E = 1.74$ m/s
8.77   12 m/s (SUV), 21 m/s (sedan)
8.79   a) 2.60 m/s   b) 325 m/s
8.81   a) 5.3 m/s   b) 5.7 m
8.83   53.7°
8.85   a) 0.0781   b) 248 J   c) 0.441 J
8.87   a) 9.35 m/s   b) 3.29 m/s
8.89 1.61×10^{-22} kg·m/s, to the left
8.91   1.33 m
8.93   0.400 m/s
8.95   250 J
8.97   a) 71.6 m/s (0.28-kg piece), 14.3 m/s (1.40-kg piece)   b) 347 m
8.99   a) yes   b) no, decreases by 4800 J
8.101 a) maximum: C, minimum: B
       b) 69 N/m   c) 0.12 m
8.103 a) $g/3$ b) 14.7 m   c) 29.4 g
8.105 a) 0, $4a/3\pi$
8.107 choice (b)
8.109 choice (b)

Chapter 9

9.1   a) 0.600 rad, 34.4°   b) 6.27 cm   c) 1.05 m
9.3   a) rad/s, rad/s^3 b) (i) 0   (ii) 15.0 rad/s^2
       c) 9.50 rad
9.5   a) $\omega_z = \gamma + 3\beta t^2$ b) 0.400 rad/s
       c) 1.30 rad/s, 0.700 rad/s
9.7   a) $\pi/4$ rad, 2.00 rad/s, −0.139 rad/s^3
       b) 0   c) 19.5 rad, 9.36 rad/s
9.9   a) 2.00 rad/s   b) 4.58 rad
9.11   a) 24.0 s   b) 68.8 rev
9.13   3.00 rad/s
9.15   a) 300 rpm   b) 75.0 s, 312 rev
9.17   9.00 rev
9.19   a) 1.99×10^{-7} rad/s
       b) 7.27×10^{-5} rad/s   c) 2.98×10^4 m/s
       d) 463 m/s   e) 0.0337 m/s^2, 0
9.21   a) 15.1 m/s^2 b) 15.1 m/s^2
9.23   a) 0.180 m/s^2, 0, 0.180 m/s^2
       b) 0.180 m/s^2, 0.377 m/s^2, 0.418 m/s^2
       c) 0.180 m/s^2, 0.754 m/s^2, 0.775 m/s^2
9.25   0.107 m, no
9.27   a) 0.831 m/s   b) 109 m/s^2
9.29   a) (i) 0.469 kg·m^2 (ii) 0.117 kg·m^2 (iii) 0
       b) (i) 0.0433 kg·m^2 (ii) 0.0722 kg·m^2
       c) (i) 0.0288 kg·m^2 (ii) 0.0144 kg·m^2
9.31   a) 1.93 kg·m^2 b) 6.53 kg·m^2
       c) 1.15 kg·m^2
9.33   0.193 kg·m^2
9.35   8.52 kg·m^2
9.37   6.49 m/s
9.39   0.600 kg·m^2

9.41 7.35×10^4 J
9.43   a) 0.673 m   b) 45.5%
9.45   46.5 kg
9.47   a) f^5 b) 6.37×10^8 J
9.49   an axis that is parallel to a diameter and is 0.516R from the center
9.51 $M(a^2 + b^2)/3$
9.53 $\frac{1}{2}MR^2$
9.55   a) $\gamma L^2/2$ b) $ML^2/2$ c) $ML^2/6$
9.57   7.68 m
9.59   a) 0.600 m/s^3 b) $\alpha = (2.40$ rad/s$^3)t$
       c) 3.54 s   d) 17.7 rad
9.61   13.8 rad/s^2
9.63   a) 1.70 m/s   b) 94.2 rad/s
9.65   2.99 cm
9.67   a) 7.36 m   b) 327 m/s^2
9.69   4.65 kg·m^2
9.71   a) −0.882 J   b) 5.42 rad/s   c) 5.42 m/s
       d) 5.42 m/s compared to 4.43 m/s
9.73   1.46 m/s
9.75 $\sqrt{\dfrac{2gd(m_B - \mu_k m_A)}{m_A + m_B + I/R^2}}$
9.77   a) 2.25×10^{-3} kg·m^2 b) 3.40 m/s
       c) 4.95 m/s
9.79   13.9 m
9.81   a) 1.05 rad/s   b) 5.0 J   c) 78.5 J   d) 6.4%
9.85   a) 55.3 kg   b) 0.804 kg·m^2
9.87   a) 4.0 rev, no   b) 15 rad/s   c) 9.5 rad/s
9.89   a) yes   b) 3.15 m/s   c) 0.348 kg·m^2
       d) 36.4 N
9.91   a) $s(\theta) = r_0\theta + \dfrac{\beta}{2}\theta^2$
       b) $\theta(t) = \dfrac{1}{\beta}\left(\sqrt{r_0^2 + 2\beta vt} - r_0\right)$
       c) $\omega_z(t) = \dfrac{v}{\sqrt{r_0^2 + 2\beta vt}}$,
       $\alpha_z(t) = -\dfrac{\beta v^2}{(r_0^2 + 2\beta vt)^{3/2}}$, no
       d) 25.0 mm, 0.247 µm/rad, 2.13×10^4 rev
9.93   choice (d)
9.95   choice (d)

Chapter 10

10.1   a) 40.0 N·m, out of the page
       b) 34.6 N·m, out of the page
       c) 20.0 N·m, out of the page
       d) 17.3 N·m, into the page   e) 0   f) 0
10.3   2.50 N·m, out of the page
10.5   b) $-\hat{k}$ c) $(-1.05$ N·m$)\hat{k}$
10.7   a) 2.56 N·m
       b) 4.25 N·m, perpendicular to handle
10.9   8.38 N·m
10.11 a) 14.8 rad/s^2 b) 1.52 s
10.13 a) 7.5 N (at book on table), 18.2 N (at hanging book)   b) 0.16 kg·m^2
10.15 0.255 kg·m^2
10.17 a) 1.56 m/s   b) 5.35 J
       c) (i) 3.12 m/s to the right   (ii) 0
       (iii) 2.21 m/s at 45° below the horizontal
       d) (i) 1.56 m/s to the right   (ii) 1.56 m/s to the left   (iii) 1.56 m/s downward
10.19 a) $\frac{1}{3}$ b) $\frac{2}{7}$ c) $\frac{2}{5}$ d) $\frac{5}{13}$
10.21 a) 0.613   b) no   c) no slipping
10.23 14.0 m
10.25 a) 3.76 m   b) 8.58 m/s
10.27 a) 67.9 rad/s   b) 8.35 J
10.29 a) 0.309 rad/s   b) 100 J   c) 6.67 W
10.31 a) 0.704 N·m   b) 157 rad   c) 111 J
       d) 111 J

10.33 a) 358 N·m   b) 1790 N   c) 83.8 m/s

10.35 a) 115 kg·m^2/s into the page
  b) 125 kg·m^2/s^2 out of the page

10.37 4.71×10^{-6} kg·m^2/s

10.39 a) A: rad/s^2; B: rad/s^4
  b) (i) 59.0 kg·m^2/s   (ii) 56.1 N·m

10.41 4600 rad/s

10.43 1.14 rev/s

10.45 a) 1.38 rad/s   b) 1080 J, 495 J

10.47 a) 0.120 rad/s   b) 3.20×10^{-4} J
  c) work done by the bug

10.49 a) 5.88 rad/s

10.51 a) 1.62 N   b) 1800 rev/min

10.53 2.4×10^{-12} N·m

10.55 0.483

10.57 a) 16.3 rad/s^2 b) no, decrease
  c) 5.70 rad/s

10.59 0.921 m/s^2, 7.68 rad/s^2, 35.5 N (at A),
  21.4 N (at B)

10.61 a) 293 N   b) 16.2 rad/s^2

10.63 a) 2.88 m/s^2 b) 6.13 m/s^2

10.65 270 N

10.67 $a = \dfrac{2g}{2 + (R/b)^2}$, $\alpha = \dfrac{2g}{2b + R^2/b}$,

  $T = \dfrac{2mg}{2(b/R)^2 + 1}$

10.69 a) $3H_0/5$

10.71 29.0 m/s

10.73 a) 26.0 m/s   b) no change

10.75 $g/3$

10.77 1.87 m

10.79 a) $\frac{6}{19}v/L$ b) $\frac{3}{19}$

10.81 3200 J

10.83 5.41 m

10.85 a) 2.00 rad/s   b) 6.58 rad/s

10.87 0.776 rad/s

10.89 a) A: solid sphere, B: solid cylinder,
  C: hollow sphere, D: hollow cylinder
  b) same   c) D d) 0.350

10.91 a) $mv_1^2 r_1^2/r^3$ b) $\dfrac{mv_1^2}{2}r_1^2\left(\dfrac{1}{r_2^2} - \dfrac{1}{r_1^2}\right)$
  c) same

10.93 a) 39.2 N upward, 39.2 N upward
  b) 60.0 N upward, 18.4 N upward   c) 165 N
  upward, 86.2 N downward   d) 0.0940 rev/s

10.95 choice (c)

10.97 choice (a)

Chapter 11

11.1 29.8 cm

11.3 1.35 m

11.5 6.6 kN

11.7 a) 1000 N, 0.800 m from end where 600-N
  force is applied   b) 800 N, 0.75 m from end
  where 600-N force is applied

11.9 a) 550 N   b) 0.614 m from A

11.11 a) 1920 N   b) 1140 N

11.13 a) $T = 2.60w$; 3.28w, 37.6°
  b) $T = 4.10w$; 5.39w, 48.8°

11.15 a) 3410 N   b) 3410 N, 7600 N

11.17 b) 533 N   c) 600 N, 267 N; downward

11.19 220 N (left), 255 N (right), 42°

11.21 a) 0.800 m   b) clockwise
  c) 0.800 m, clockwise

11.23 b) 208 N

11.25 1.9 mm

11.27 2.0×10^{11} Pa

11.29 a) 3.1×10^{-3} (upper), 2.0×10^{-3} (lower)
  b) 1.6 mm (upper), 1.0 mm (lower)

11.31 a) 150 atm   b) 1.5 km, no

11.33 4.8×10^9 Pa, 2.1×10^{-10} Pa^{-1}

11.35 b) 6.6×10^5 N   c) 1.8 mm

11.37 7.36×10^6 Pa

11.39 3.41×10^7 Pa

11.41 10.2 m/s^2

11.43 20.0 kg

11.45 a) 525 N   b) 222 N, 328 N   c) 1.48

11.47 a) 140 N   b) 6 cm to the right

11.49 a) 409 N   b) 161 N

11.51 49.9 cm

11.53 a) 370 N   b) when he starts to raise his leg
  c) no

11.55 a) 3 cm   b) lean backward

11.57 5500 N

11.59 b) 2000 N $= 2.72mg$ c) 4.4 mm

11.61 a) 4.90 m   b) 60 N

11.63 a) 175 N at each hand, 200 N at each foot
  b) 91 N at each hand and at each foot

11.65 a) 1150 N   b) 1940 N   c) 918 N   d) 0.473

11.67 590 N (person above), 1370 N (person below);
  person above

11.69 a) $\dfrac{T_{\max}hD}{L\sqrt{h^2 + D^2}}$

  b) $\dfrac{T_{\max}h}{L\sqrt{h^2 + D^2}}\left(1 - \dfrac{D^2}{h^2 + D^2}\right)$, positive

11.71 a) 71.5 kg
  b) 380 N, 25.2° above the horizontal

11.73 a) 375 N   b) 325 N   c) 512 N

11.75 a) 0.424 N (A), 1.47 N (B), 0.424 N (C)
  b) 0.848 N

11.77 a) 27° to tip, 31° to slip, tips first
  b) 27° to tip, 22° to slip, slips first

11.79 a) 80 N (A), 870 N (B)   b) 1.92 m

11.81 a) 1.0 cm   b) 0.86 cm

11.83 a) 0.70 m from A b) 0.60 m from A

11.85 a) 4.2×10^4 N   b) 65 m

11.87 b) $x = 1.50 \text{ m} + \dfrac{(1.30 \text{ m})m_1 - (0.38 \text{ m})M}{m_2}$
  c) 1.59 kg   d) 1.50 m

11.89 a) 391 N (4.00-m ladder), 449 N (3.00-m ladder)
  b) 322 N   c) 334 N   d) 937 N

11.91 a) 0.66 mm   b) 0.022 J   c) 8.35×10^{-3} J
  d) -3.04×10^{-2} J   e) 3.04×10^{-2} J

11.93 choice (a)

11.95 choice (d)

Chapter 12

12.1 no (41.8 N)

12.3 7020 kg/m^3; yes

12.5 1.6

12.7 61.6 N

12.9 a) 1.86×10^6 Pa   b) 184 m

12.11 0.581 m

12.13 a) 1.90×10^4 Pa
  b) causes additional force on their walls

12.15 2.8 m

12.17 6.0×10^4 Pa

12.19 2.27×10^5 Pa

12.21 a) 636 Pa   b) (i) 1170 Pa   (ii) 1170 Pa

12.23 10.9

12.25 a) 2.19×10^7 N   b) 2.17×10^7 N
  c) 5.79×10^8 N

12.27 0.122 m

12.29 6.43×10^{-4} m^3, 2.78×10^3 kg/m^3

12.31 10.5 N

12.33 a) 116 Pa   b) 921 Pa
  c) 0.822 kg, 822 kg/m^3

12.35 1640 kg/m^3

12.37 9.6 m/s

12.39 a) 17.0 m/s   b) 0.317 m

12.41 28.4 m/s

12.43 1.47×10^5 Pa

12.45 2.03×10^4 Pa

12.47 2.25×10^5 Pa

12.49 1.19D

12.51 a) $(p_0 - p)\pi D^2/4$ b) 776 N

12.53 a) 5.9×10^5 N   b) 1.8×10^5 N

12.55 2.61×10^4 N·m

12.57 0.964 cm, rise

12.59 a) 1470 Pa   b) 13.9 cm

12.61 a) 0.0500 m^3 b) 10.0 kg

12.63 9.8×10^6 kg, yes

12.65 a) 0.30   b) 0.70

12.67 a) 8.27×10^3 m^3 b) 83.8 kN

12.69 a) 16.5 cm   b) 1.75 m

12.71 a) 5.07 m/s, 1.28   b) 32.4 min, 2.08

12.73 a) 53.9 N   b) 31.0 m/s^2

12.75 a) $1 - \dfrac{\rho_B}{\rho_L}$ b) $\left(\dfrac{\rho_L - \rho_B}{\rho_L - \rho_w}\right)L$ c) 4.60 cm

12.77 a) $2\sqrt{h(H - h)}$ b) h

12.79 5.47 m

12.81 a) 0.200 m^3/s   b) 6.97×10^4 Pa

12.83 $3h_1$

12.85 b) no

12.87 a) 2.5×10^{-4} m^2/Pa (slope), 16 m^2
  (intercept)   b) 8.2 m, 800 kg/m^3

12.89 choice (b)

12.91 choice (a)

Chapter 13

13.1 a) 2.18

13.3 a) 1.2×10^{-11} m/s^2 b) 15 days
  c) no, increase

13.5 2.1×10^{-9} m/s^2, downward

13.7 a) 2.4×10^{-3} N
  b) $F_{\text{moon}}/F_{\text{earth}} = 3.5 \times 10^{-6}$

13.9 a) 0.634 m from 3m
  b) (i) unstable   (ii) stable

13.11 1.38×10^7 m

13.13 a) 0.37 m/s^2 b) 1700 kg/m^3

13.15 610 N, 735 N (on earth), astronaut and satellite
  have same acceleration; no

13.17 a) 5030 m/s   b) 60,200 m/s

13.19 9.03 m/s^2

13.21 a) 7410 m/s   b) 1.71 h

13.23 7330 m/s

13.25 a) 4.1 m/s = 9.1 mph, yes   b) 2.6 h

13.27 a) 82,700 m/s   b) 14.5 days

13.29 a) 7.84×10^9 s = 248 y
  b) 4.44×10^{12} m, 7.38×10^{12} m

13.31 2.3×10^{30} kg = 1.2M_S

13.33 a) (i) 5.31×10^{-9} N   (ii) 2.67×10^{-9} N

13.35 a) $-\dfrac{GmM}{\sqrt{x^2 + a^2}}$ b) $-GmM/x$
  c) $\dfrac{GmMx}{(x^2 + a^2)^{3/2}}$, toward the ring   d) GmM/x^2
  e) $U = -GmM/a$, $F_x = 0$

13.37 a) 33.7 N   b) 32.8 N

13.39 a) 4.3×10^{37} kg = $(2.1 \times 10^7)M_S$ b) no
  c) 6.32×10^{10} m, yes

13.41 9.16×10^{13} N

13.43 a) 9.67×10^{-12} N, at 45° above $+x$-axis
  b) 3.02×10^{-5} m/s

13.45 a) 2.00×10^{-10} N, 161° above $+x$-axis
  b) $x = 0$, $y = 1.32$ m

13.47 b) (i) 1.49×10^{-5} m/s (50.0-kg sphere),
  7.46×10^{-6} m/s (100.0-kg sphere)
  (ii) 2.24×10^{-5} m/s
  c) 26.6 m

13.49 a) 3.59×10^7 m

13.51 177 m/s

13.53 a) 7.36 h b) 2.47 h

13.55 1.83×10^{27} kg

13.57 22.8 m

13.59 6060 km/h

13.61 $v = \sqrt{\dfrac{2Gm_E h}{R_E(R_E + h)}}$

13.63 a) $GM^2/4R^2$
   b) $v = \sqrt{GM/4R}$, $T = 4\pi\sqrt{R^3/GM}$
   c) $GM^2/4R$

13.65 6.8×10^4 m/s

13.67 a) 7900 s b) 1.53
   c) 8430 m/s (perigee), 5510 m/s (apogee)
   d) 2420 m/s; 3250 m/s; perigee

13.69 5.38×10^9 J

13.71 9.34 m/s²

13.73 $GmMx/(a^2 + x^2)^{3/2}$

13.75 a) $U(r) = \dfrac{Gm_E m}{2R_E^3}r^2$ b) 7.91×10^3 m/s

13.77 a) It is considerable and shows no apparent pattern.
   b) Earth (5500 kg/m³), Mercury (5400 kg/m³), Venus (5300 kg/m³), Mars (3900 kg/m³), Neptune (1600 kg/m³), Uranus (1200 kg/m³), Jupiter (1200 kg/m³), Saturn (530 kg/m³)
   c) no effect d) 93 m/s²

13.79 a) opposite; opposite b) 259 days
   c) 44.1°

13.81 $\dfrac{2GMm}{a^2}\left(1 - \dfrac{x}{\sqrt{a^2 + x^2}}\right)$

13.83 choice (c)

Chapter 14

14.1 a) 2.15 ms, 2930 rad/s
   b) 2.00×10^4 Hz, 1.26×10^5 rad/s
   c) 1.3×10^{-15} s $\leq T \leq 2.3 \times 10^{-15}$ s, 4.3×10^{14} Hz $\leq f \leq 7.5 \times 10^{14}$ Hz
   d) 2.0×10^{-7} s, 3.1×10^7 rad/s

14.3 5530 rad/s, 1.14 ms

14.5 0.0625 s

14.7 a) 0.80 s b) 1.25 Hz c) 7.85 rad/s
   d) 3.0 cm e) 148 N/m

14.9 a) 0.167 s b) 37.7 rad/s c) 0.0844 kg

14.11 a) 0.150 s b) 0.0750 s

14.13 a) 0.98 m b) $\pi/2$ rad
   c) $x = (-0.98 \text{ m}) \sin[(12.2 \text{ rad/s})t]$

14.15 a) -2.71 m/s² b) $x = (1.46 \text{ cm}) \times \cos[(15.7 \text{ rad/s})t + 0.715 \text{ rad}]$,
   $v_x = (-22.9 \text{ cm/s}) \times \sin[(15.7 \text{ rad/s})t + 0.715 \text{ rad}]$,
   $a_x = (-359 \text{ cm/s}^2) \times \cos[(15.7 \text{ rad/s})t + 0.715 \text{ rad}]$

14.17 120 kg

14.19 a) 0.253 kg b) 1.21 cm c) 3.03 N

14.21 a) 1.51 s b) 26.0 N/m
   c) 30.8 cm/s d) 1.92 N
   e) -0.0125 m, 30.4 cm/s, 0.216 m/s²
   f) 0.324 N

14.23 a) $x = (0.0030 \text{ m}) \cos[(2760 \text{ rad/s})t]$
   b) 8.3 m/s, 2.3×10^4 m/s²
   c) $da_x/dt = (6.3 \times 10^7 \text{ m/s}^3) \times \sin[(2760 \text{ rad/s})t]$, 6.3×10^7 m/s³

14.25 92.2 m/s²

14.27 a) 0.0336 J b) 0.0150 m c) 0.669 m/s

14.29 a) 1.20 m/s b) 1.11 m/s c) 36 m/s²
   d) 13.5 m/s² e) 0.36 J

14.31 $3M; \frac{3}{4}$

14.33 0.240 m

14.35 a) 0.376 m b) 59.3 m/s² c) 119 N

14.37 a) 4.06 cm b) 1.21 m/s c) 29.8 rad/s

14.39 a) 0, 0, 3.92 J, 3.92 J b) 3.92 J, 0, 0, 3.92 J
   c) 0.98 J, 0.98 J, 1.96 J, 3.92 J

14.41 a) 2.7×10^{-8} kg·m²
   b) 4.3×10^{-6} N·m/rad

14.43 0.0294 kg·m²

14.45 a) 0.25 s b) 0.25 s

14.47 0.407 swing per second

14.49 10.7 m/s²

14.51 a) 2.84 s b) 2.89 s c) 2.89 s; −2%

14.53 A: $2\pi\sqrt{\dfrac{L}{g}}$, B: $\dfrac{2\sqrt{2}}{3}\left(2\pi\sqrt{\dfrac{L}{g}}\right)$; pendulum A

14.55 0.129 kg·m²

14.57 A: $2\pi\sqrt{\dfrac{L}{g}}$, B: $\sqrt{\dfrac{11}{10}}\left(2\pi\sqrt{\dfrac{L}{g}}\right)$, pendulum B

14.59 a) 0.30 J

14.61 a) 0.393 Hz b) 1.73 kg/s

14.63 a) $A_1/3$ b) $2A_1$

14.65 0.353 m

14.67 a) 1.34 m/s b) 1.90 m/s²

14.69 a) 24.4 cm b) 0.221 s c) 1.19 m/s

14.71 2.00 m

14.73 $0.921\left(\dfrac{1}{2\pi}\sqrt{\dfrac{g}{L}}\right)$

14.75 a) 0.784 s b) -1.12×10^{-4} s per s; shorter
   c) 0.419 s

14.77 a) 0.150 m/s b) 0.112 m/s² downward
   c) 0.700 s d) 4.38 m

14.79 a) 2.6 m/s b) 0.21 m c) 0.49 s

14.81 1.17 s

14.83 0.421 s

14.85 0.705 Hz, 14.5°

14.87 $2\pi\sqrt{\dfrac{M}{3k}}$

14.89 a) 1.60 s b) 0.625 Hz c) 3.93 rad/s
   d) 5.1 cm; 0.4 s, 1.2 s, 1.8 s
   e) 79 cm/s²; 0.4 s, 1.2 s, 1.8 s f) 4.9 kg

14.91 b) The angular amplitude increases as L decreases. c) about 53°

14.93 a) $Mv^2/6$ c) $\omega = \sqrt{3k/M}$, $M' = M/3$

14.95 choice (a)

Chapter 15

15.1 a) 0.439 m, 1.28 ms b) 0.219 m

15.3 220 m/s = 800 km/h

15.5 a) 1.7 cm to 17 m
   b) 4.3×10^{14} Hz to 7.5×10^{14} Hz
   c) 1.5 cm d) 6.4 cm

15.7 a) 25.0 Hz, 0.0400 s, 19.6 rad/m
   b) $y(x, t) = (0.0700 \text{ m}) \times \cos[(19.6 \text{ m}^{-1})x + (157 \text{ rad/s})t]$
   c) 4.95 cm d) 0.0050 s

15.9 a) yes b) yes c) no
   d) $v_y = \omega A \cos(kx + \omega t)$, $a_y = -\omega^2 A \sin(kx + \omega t)$

15.11 a) 4 mm b) 0.040 s c) 0.14 m, 3.6 m/s
   d) 0.24 m, 6.0 m/s e) no

15.13 b) $+x$-direction

15.15 a) 17.5 m/s b) 0.146 m
   c) both would increase by a factor of $\sqrt{2}$

15.17 0.337 kg

15.19 a) 9.53 N b) 20.8 m/s

15.21 a) 10.0 m/s b) 0.250 m
   c) $y(x, t) = (3.00 \text{ cm}) \times \cos[(8.00\pi \text{ rad/m})x - (80.0\pi \text{ rad/s})t]$
   d) 1890 m/s² e) yes

15.23 4.10 mm

15.25 a) 95 km b) 0.25 μW/m² c) 110 kW

15.27 a) 0.050 W/m² b) 22 kJ

15.29 9.48×10^{27} W

15.37 a) $(1.33 \text{ m})n$, $n = 0, 1, 2, \ldots$
   b) $(1.33 \text{ m})\left(n + \frac{1}{2}\right)$, $n = 0, 1, 2, \ldots$

15.39 a) 96.0 m/s b) 461 N
   c) 1.13 m/s, 4.26 m/s²

15.41 b) 2.80 cm c) 277 cm
   d) 185 cm, 7.96 m/s, 0.126 s, 1470 cm/s
   e) 280 cm/s
   f) $y(x, t) = (5.60 \text{ cm}) \times \sin[(0.0906 \text{ rad/cm})x] \sin[(133 \text{ rad/s})t]$

15.43 4.0 m, 2.0 m, 1.33 m

15.45 a) 45.0 cm b) no

15.47 a) 311 m/s b) 246 Hz c) 245 Hz, 1.40 m

15.49 a) 20.0 Hz, 126 rad/s, 3.49 rad/m
   b) $y(x, t) = (2.50 \times 10^{-3} \text{ m}) \times \cos[(3.49 \text{ rad/m})x - (126 \text{ rad/s})t]$
   c) $y(0, t) = (2.50 \times 10^{-3} \text{ m}) \cos[(126 \text{ rad/s})t]$
   d) $y(1.35 \text{ m}, t) = (2.50 \times 10^{-3} \text{ m}) \times \cos[(126 \text{ rad/s})t - 3\pi/2 \text{ rad}]$
   e) 0.315 m/s f) -2.50×10^{-3} m, 0

15.51 a) $\dfrac{7L}{2}\sqrt{\dfrac{\mu_1}{F}}$ b) no

15.53 a) 62.1 m

15.55 13.7 Hz, 25.0 m

15.57 1.83 m

15.59 361 Hz (copper), 488 Hz (aluminum)

15.61 a) 18.8 cm b) 0.0169 kg

15.63 a) 7.07 cm b) 0.400 kW

15.65 $(0.800 \text{ Hz})n$, $n = 1, 2, 3, \ldots$

15.67 a) 2.22 g b) 2.24×10^4 m/s²

15.69 233 N

15.71 1780 kg/m³

15.73 a) 148 N b) 26%

15.75 c) 47.5 Hz d) 138 g

15.77 a) 392 N b) 392 N + (7.70 N/m)x
   c) 3.89 s

15.79 choice (b)

Chapter 16

16.1 a) 0.344 m b) 1.2×10^{-5} m
   c) 6.9 m, 50 Hz

16.3 a) 7.78 Pa b) 77.8 Pa c) 778 Pa

16.5 a) 90 m b) 102 kHz c) 1.4 cm
   d) 4.4 mm to 8.8 mm e) 6.2 MHz

16.7 90.8 m

16.9 81.4°C

16.11 0.16 s

16.13 a) 5.5×10^{-15} J b) 0.074 mm/s

16.15 15.0 cm

16.17 a) 4.14 Pa b) 0.0208 W/m² c) 103 dB

16.19 a) 4.4×10^{-12} W/m² b) 6.4 dB
   c) 5.8×10^{-11} m

16.21 14.0 dB

16.23 a) 2.0×10^{-7} W/m² b) 6.0 m
   c) 290 m d) yes, no

16.25 a) fundamental: 0.60 m; 0, 1.20 m; first overtone: 0.30 m, 0.90 m; 0, 0.60 m, 1.20 m; second overtone: 0.20 m, 0.60 m, 1.00 m; 0, 0.40 m, 0.80 m, 1.20 m
   b) fundamental: 0; 1.20 m; first overtone: 0, 0.80 m; 0.40 m, 1.20 m; second overtone: 0, 0.48 m, 0.96 m; 0.24 m, 0.72 m, 1.20 m

16.27 506 Hz, 1517 Hz, 2529 Hz

16.29 a) 35.2 Hz b) 17.6 Hz

16.31 a) 614 Hz b) 1230 Hz

16.33 a) 137 Hz, 0.50 m b) 137 Hz, 2.51 m

16.35 a) 172 Hz b) 86 Hz

16.37 0.125 m

16.39 a) $(820 \text{ Hz})n$, $n = 1, 2, 3, \ldots$
   b) $(410 \text{ Hz})(2n + 1)$, $n = 0, 1, 2, \ldots$

16.41 a) 433 Hz b) loosen

16.43 1.3 Hz

16.45 780 m/s
16.47 a) 375 Hz b) 371 Hz c) 4 Hz
16.49 a) 0.25 m/s b) 0.91 m
16.51 19.8 m/s
16.53 a) 1910 Hz b) 0.188 m
16.55 a) 7.02 m/s, toward b) 1404 Hz
16.57 a) 36.0° b) 2.94 s
16.59 a) 1.00 b) 8.00
    c) 4.73×10^{-8} m = 47.3 nm
16.61 flute harmonic $3n$ resonates with string
    harmonic $4n$, $n = 1, 3, 5, \ldots$
16.63 a) stopped b) 7th and 9th c) 0.439 m
16.65 a) 0.026 m, 0.53 m, 1.27 m, 2.71 m, 9.01 m
    b) 0.26 m, 0.86 m, 1.84 m, 4.34 m c) 86 Hz
16.67 a) 0.0823 m b) 120 Hz
16.69 b) 2.0 m/s
16.71 a) 38 Hz b) no
16.73 a) 375 m/s b) 1.39 c) 0.8 cm
16.75 d) 9.69 cm/s, 667 m/s^2
16.77 choice (b)
16.79 choice (a)
16.81 choice (b)

Chapter 17

17.1 a) −81.0°F b) 134.1°F c) 88.0°F
17.3 a) 27.2 C° b) −55.6 C°
17.5 a) −18.0 F° b) −10.0 C°
17.7 0.964 atm
17.9 a) −282°C b) no, 47,600 Pa
17.11 0.39 m
17.13 1.9014 cm; 1.8964 cm
17.15 49.4°C
17.17 1.7×10^{-5} (C°)$^{-1}$
17.19 a) 1.431 cm^2 b) 1.436 cm^2
17.21 a) 6.0 mm b) -1.0×10^8 Pa
17.23 555 kJ
17.25 23 min
17.27 240 J/kg · K
17.29 0.526 C°
17.31 45.2 C°
17.33 0.0613 C°
17.35 a) 215 J/kg · K b) water c) too small
17.37 0.114 kg
17.39 27.5°C
17.41 150°C
17.43 7.6 min
17.45 54.5 kJ, 13.0 kcal, 51.7 Btu
17.47 357 m/s
17.49 3.45 L
17.51 5.05×10^{15} kg
17.53 0.0674 kg
17.55 190 g
17.57 a) 222 K/m b) 10.7 W c) 73.3°C
17.59 a) −0.86°C b) 24 W/m^2
17.61 4.0×10^{-3} W/m · C°
17.63 105.5°C
17.65 a) 21 kW b) 6.4 kW
17.67 15 W
17.69 2.1 cm^2
17.71 35.0°C
17.73 a) 35.1°M b) 39.6 C°
17.75 69.4°C
17.77 23.0 cm (first rod), 7.0 cm (second rod)
17.79 b) 1.9×10^8 Pa
17.81 a) 87°C b) −80°C
17.83 460 s
17.85 a) 83.6 J b) 1.86 J/mol · K
    c) 5.60 J/mol · K
17.87 a) 4.20×10^7 J b) 10.7 C° c) 30.0 C°
17.89 a) 0.60 kg b) 0.80 bottle/h
17.91 3.4×10^5 J/kg

17.93 a) no b) 0.0°C, 0.156 kg
17.95 a) 86.1°C
    b) no ice, 0.130 kg liquid water, no steam
17.97 a) 100°C
    b) 0.0214 kg steam, 0.219 kg liquid water
17.99 a) 93.9 W b) 1.35
17.101 2.9
17.103 a) 59.8°C b) 42.7°C c) 8.40 W
17.105 c) 170 h d) 1.5×10^{10} s ≈ 500 y, no
17.107 5.82 g
17.109 a) 1.04 kW b) 87.1 W c) 1.13 kW
    d) 28 g e) 1.1 bottles
17.111 a) 3.00×10^4 J/kg b) 1.00×10^3 J/kg · K
    (liquid), 1.33×10^3 J/kg · K (solid)
17.113 A: 216 W/m · K, B: 130 W/m · K
17.115 a) $H = \dfrac{(T_2 - T_1)2\pi kL}{\ln(b/a)}$
    b) $T = T_2 - \dfrac{(T_2 - T_1)\ln(r/a)}{\ln(b/a)}$
    d) 73°C e) 49 W
17.117 choice (a)
17.119 choice (a)

Chapter 18

18.1 a) 0.122 mol b) 14,700 Pa, 0.145 atm
18.3 0.100 atm
18.5 a) 0.0136 kg/m^3, 67.6 kg/m^3, 5.39 kg/m^3
    b) $0.011\rho_E$, $56\rho_E$, $4.5\rho_E$
18.7 503°C
18.9 19.7 kPa
18.11 0.159 L
18.13 $0.0508V$
18.15 a) 70.2°C b) yes
18.17 850 m
18.19 a) 6.95×10^{-16} kg b) 2.32×10^{-13} kg/m^3
18.21 55.6 mol, 3.35×10^{25} molecules
18.23 a) 2.20×10^6 molecules
    b) 2.44×10^{19} molecules
18.25 6.4×10^{-6} m
18.27 a) 5.83×10^7 J b) 242 m/s
18.29 (d) must be true; the others could be true.
18.31 a) 1.93×10^6 m/s, no b) 7.3×10^{10} K
18.33 a) 6.21×10^{-21} J b) 2.34×10^5 m^2/s^2
    c) 484 m/s d) 2.57×10^{-23} kg · m/s
    e) 1.24×10^{-19} N f) 1.24×10^{-17} Pa
    g) 8.17×10^{21} molecules
    h) 2.45×10^{22} molecules
18.35 3800°C
18.37 a) 1870 J b) 1120 J
18.39 a) 741 J/kg · K, $c_w = 5.65c_{N_2}$
    b) 5.65 kg; 4850 L
18.41 a) 337 m/s b) 380 m/s c) 412 m/s
18.43 a) 610 Pa b) 22.12 MPa
18.45 18.0 cm^3, $V_{20°C} = 0.32V_{cp}$
18.47 a) 11.8 kPa b) 0.566 L
18.49 272°C
18.51 0.195 kg
18.53 a) −179°C b) 1.2×10^{26} molecules/m^3
    c) $\rho_T = 4.8\rho_e$
18.55 1.92 atm
18.57 a) 30.7 cylinders b) 8420 N c) 7800 N
18.59 a) 26.2 m/s b) 16.1 m/s, 5.44 m/s c) 1.74 m
18.61 ≈5×10^{27} atoms
18.63 a) A b) B c) 4250°C d) B
18.65 a) 6.00 $\times 10^3$ Pa b) 32.8 m/s
18.67 a) 4.65×10^{-26} kg b) 6.11×10^{-21} J
    c) 2.04×10^{24} molecules d) 12.5 kJ
18.69 b) r_2 c) $r_1 = \dfrac{R_0}{2^{1/6}}$, $r_2 = R_0$, $2^{-1/6}$ d) U_0
18.71 a) $2R = 16.6$ J/mol · K b) less than

18.73 b) 1.40×10^5 K (N$_2$), 1.01×10^4 K (H$_2$)
    c) 6370 K (N$_2$), 459 K (H$_2$)
18.75 $3kT/m$, same
18.77 b) $0.0421N$ c) $(2.94 \times 10^{-21})N$
    d) $0.0297N$, $(2.08 \times 10^{-21})N$
    e) $0.0595N$, $(4.15 \times 10^{-21})N$
18.79 a) $p_0 + \dfrac{mg}{\pi r^2}$ b) $-\left(\dfrac{y}{h}\right)(p_0\pi r^2 + mg)$
    c) $\dfrac{1}{2\pi}\sqrt{\dfrac{g}{h}\left(1 + \dfrac{p_0\pi r^2}{mg}\right)}$, no
18.81 a) 42.6% b) 3 km c) 1 km
18.83 a) 4.5×10^{11} m
    b) 703 m/s, 6.4×10^8 s (≈20 y)
    c) 1.4×10^{-14} Pa
    d) 650 m/s, $v_H > v_{esc}$, evaporate
    f) 2×10^5 K, $>3T_{sun}$, no
18.85 choice (a)
18.87 choice (c)

Chapter 19

19.1 b) 1330 J
19.3 b) −6180 J
19.5 a) 1.04 atm
19.7 a) $(p_1 - p_2)(V_2 - V_1)$
    b) negative of work done in reverse direction
19.9 a) 34.7 kJ b) 80.4 kJ c) no
19.11 a) 278 K, at a b) 0; 162 J c) 53 J
19.13 a) $T_a = 535$ K, $T_b = 9350$ K, $T_c = 15,000$ K
    b) 21 kJ done by gas c) 36 kJ
19.15 a) 0 b) $T_b = 2T_a$ c) $U_b = U_a + 700$ J
19.17 b) 208 J c) on the piston d) 712 J
    e) 920 J f) 208 J
19.19 a) 948 K b) 900 K
19.21 $\frac{2}{5}$
19.23 a) 747 J b) 1.30
19.25 a) −605 J b) 0 c) yes, 605 J, liberate
19.27 a) 476 kPa b) −10.6 kJ c) 1.59, heated
19.29 b) 314 J c) −314 J
19.31 11.6°C
19.33 a) increase b) 4.8 kJ
19.35 a) 0.681 mol b) 0.0333 m^3
    c) 2.23 kJ d) 0
19.37 a) 45.0 J b) liberate, 65.0 J c) 23.0 J, 22.0 J
19.39 a) the same b) 4.0 kJ, absorb c) 8.0 kJ
19.41 b) −2460 J
19.43 a) 0.80 L b) 305 K, 1220 K, 1220 K
    c) ab: 76 J, into the gas
       ca: −107 J, out of the gas
       bc: 56 J, into the gas
    d) ab: 76 J, increased
       bc: 0, no change
       ca: −76 J, decreased
19.45 a) 837°C b) 11.5 kJ c) 40.3 kJ d) 42.4 kJ
19.47 b) 6.00 L, 2.5×10^4 Pa, 75.0 K
    c) 95 J d) heat it at constant volume
19.49 b) 11.9 C°
19.51 a) 0.168 m b) 196°C c) 70.1 kJ
19.53 a) $Q = 450$ J, $\Delta U = 0$
    b) $Q = 0$, $\Delta U = -450$ J
    c) $Q = 1125$ J, $\Delta U = 675$ J
19.55 a) $W = 738$ J, $Q = 2590$ J, $\Delta U = 1850$ J
    b) $W = 0$, $Q = -1850$ J, $\Delta U = -1850$ J
    c) $\Delta U = 0$
19.57 a) $W = -187$ J, $Q = -654$ J, $\Delta U = -467$ J
    b) $W = 113$ J, $Q = 0$, $\Delta U = -113$ J
    c) $W = 0$, $Q = 580$ J, $\Delta U = 580$ J
19.59 a) a: adiabatic, b: isochoric, c: isobaric
    b) 28.0°C c) a: −30.0 J, a: 0, a: 20.0 J
    d) a
    e) a: decrease, b: stay the same, c: increase

19.61 b) -300 J, out of the gas

19.63 choice (c)

19.65 choice (d)

Chapter 20

20.1 a) 6500 J b) 34%

20.3 a) 23% b) 12,400 J c) 0.350 g

d) 222 kW = 298 hp

20.5 a) 12.3 atm b) 5470 J, ca c) 3723 J, bc

d) 1747 J e) 31.9%

20.7 a) 58% b) 1.4%

20.9 a) 14.8 kJ b) 45.8 kJ

20.11 1.2 h

20.13 a) 215 J b) 378 K c) 39.0%

20.15 a) 38 kJ b) 590°C

20.17 a) 492 J b) 212 W c) 5.4

20.19 44.5 hp

20.21 a) 429 J/K b) -393 J/K c) 36 J/K

20.23 a) irreversible b) 1250 J/K

20.25 -6.31 J/K

20.27 a) 6.05×10^3 J/K

b) about five times greater for vaporization

20.29 a) no b) 18.3 J/K c) 18.3 J/K

20.31 10.0 J/K

20.33 a) 121 J b) 3800 cycles

20.35 a) 90.2 J b) 320 J c) 45°C d) 0

e) 263 g

20.37 -5.8 J/K, decrease

20.39 b) absorbed: bc; rejected: ab and ca

c) $T_a = T_b = 241$ K, $T_c = 481$ K

d) 610 J, 610 J e) 8.7%

20.41 a) 21.0 kJ (enters), 16.6 kJ (leaves)

b) 4.4 kJ, 21% c) $e = 0.31e_{max}$

20.43 a) 7.0% b) 3.0 MW; 2.8 MW

c) 6×10^5 kg/h = 6×10^5 L/h

20.45 a) 1: 2.00 atm, 4.00 L; 2: 2.00 atm, 6.00 L;

3: 1.11 atm, 6.00 L; 4: 1.67 atm, 4.00 L

b) $1 \to 2$: 1422 J, 405 J; $2 \to 3$: -1355 J, 0;

$3 \to 4$: -274 J, -274 J; $4 \to 1$: 339 J, 0

c) 131 J d) 7.44%; $e = 0.168e_C$

20.47 $1 - T_C/T_H$, same

20.49 a) 122 J, -78 J b) 5.10×10^{-4} m³

c) b: 2.32 MPa, 4.81×10^{-5} m³, 771 K

c: 4.01 MPa, 4.81×10^{-5} m³, 1332 K

d: 0.147 MPa, 5.10×10^{-4} m³, 518 K

d) 61.1%, 77.5%

20.51 6.23

20.55 a) A: 28.9%, B: 38.3%, C: 53.8%, D: 24.4%

b) C c) B > D > A

20.57 a) 4.83% b) 4.83% c) 6.25%

d) $e = \dfrac{0.80T_d - 200}{12T_d - 2700}$, 6.67%

20.59 choice (b)

20.61 choice (d)

Chapter 21

21.1 a) 2.00×10^{10} b) 8.59×10^{-13}

21.3 3.4×10^{18} m/s² (proton),

6.3×10^{21} m/s² (electron)

21.5 1.3 nC

21.7 3.7 km

21.9 a) 0.742 μC b) 0.371 μC, 1.48 μC

21.11 a) 2.21×10^4 m/s²

21.13 $+0.750$ nC

21.15 1.8×10^{-4} N, in the $+x$-direction

21.17 $x = -0.144$ m

21.19 2.58 μN, in the $-y$-direction

21.21 a) 8.80×10^{-9} N, attractive

b) 8.22×10^{-8} N; about 10 times larger than

the bonding force

21.23 a) 4.40×10^{-16} N b) 2.63×10^{11} m/s²

c) 2.63×10^5 m/s

21.25 a) 3.30×10^6 N/C, to the left b) 14.2 ns

c) 1.80×10^3 N/C, to the right

21.27 a) -21.9 μC b) 1.02×10^{-7} N/C

21.29 a) 364 N/C b) no; 2.73 μm, downward

21.31 1.79×10^6 m/s

21.33 a) $-\hat{j}$ b) $\dfrac{\sqrt{2}}{2}\hat{i} + \dfrac{\sqrt{2}}{2}\hat{j}$

c) $-0.39\hat{i} + 0.92\hat{j}$

21.35 a) 633 km/s b) 15.9 km/s

21.37 a) 0

b) for $|x| < a$: $E_x = -\dfrac{q}{\pi\epsilon_0}\dfrac{ax}{(x^2 - a^2)^2}$;

for $x > a$: $E_x = \dfrac{q}{2\pi\epsilon_0}\dfrac{x^2 + a^2}{(x^2 - a^2)^2}$;

for $x < -a$: $E_x = -\dfrac{q}{2\pi\epsilon_0}\dfrac{x^2 + a^2}{(x^2 - a^2)^2}$

21.39 a) (i) 574 N/C, $+x$-direction (ii) 268 N/C,

$-x$-direction (iii) 404 N/C, $-x$-direction

b) (i) 9.20×10^{-17} N, $-x$-direction

(ii) 4.30×10^{-17} N, $+x$-direction

(iii) 6.48×10^{-17} N, $+x$-direction

21.41 1.04×10^7 N/C, toward the -2.00-μC charge

21.43 a) 8740 N/C, to the right

b) 6540 N/C, to the right

c) 1.40×10^{-15} N, to the right

21.45 1.73×10^{-8} N, toward the point midway

between the electrons

21.47 a) $E_x = E_y = E = 0$ b) $E_x = 2660$ N/C,

$E_y = 0$, $E = 2660$ N/C, $+x$-direction

c) $E_x = 129$ N/C, $E_y = -510$ N/C,

$E = 526$ N/C, 284° counterclockwise from

the $+x$-axis d) $E_x = 0$, $E_y = 1380$ N/C,

$E = 1380$ N/C, $+y$-direction

21.49 a) 1.14×10^5 N/C, toward the center of the

disk b) 8.92×10^4 N/C, toward the center

of the disk

c) 1.46×10^5 N/C, toward the charge

21.51 a) $(7.0 \text{ N/C})\hat{i}$ b) $(1.75 \times 10^{-5} \text{ N})\hat{i}$

21.53 a) 1.4×10^{-11} C·m, from q_1 toward q_2

b) 860 N/C

21.55 a) $\vec{p}$ aligned in the same or the opposite

direction as $\vec{E}$

b) stable: $\vec{p}$ aligned in the same direction as $\vec{E}$;

unstable: $\vec{p}$ aligned in the opposite direction

21.57 a) 1680 N, from the -5.00-μC charge toward

the -5.00-μC charge b) 22.3 N·m, clockwise

21.59 b) $\dfrac{Q^2}{8\pi\epsilon_0 L^2}(1 + 2\sqrt{2})$, away from the center

of the square

21.61 a) 8.63×10^{-5} N, -5.52×10^{-5} N

b) 1.02×10^{-4} N, 32.6° below the $+x$-axis

21.63 b) 2.80 μC c) 39.5°

21.65 3.41×10^4 N/C, to the left

21.67 between the charges, 0.24 m from the 0.500-nC

charge

21.69 at $x = d/3$, $q = -4Q/9$

21.71 a) $\dfrac{6q^2}{4\pi\epsilon_0 L^2}$, away from the vacant corner

b) $\dfrac{3q^2}{4\pi\epsilon_0 L^2}\left(\sqrt{2} + \dfrac{1}{2}\right)$, toward the center of

the square

21.73 a) 6.0×10^{23} b) 4.1×10^{-31} N (gravita-

tional), 510 kN (electric)

c) yes (electric), no (gravitational)

21.75 2190 km/s

21.77 a) $\dfrac{mv_0^2 \sin^2\alpha}{2eE}$ b) $\dfrac{mv_0^2 \sin^2 2\alpha}{eE}$

d) h_{max}: 0.418 m, d: 2.89 m

21.79 a) $E_x = \dfrac{Q}{4\pi\epsilon_0 a}\left(\dfrac{1}{x - a} - \dfrac{1}{x}\right)$, $E_y = 0$

b) $\dfrac{qQ}{4\pi\epsilon_0 a}\left(\dfrac{1}{r} - \dfrac{1}{r + a}\right)\hat{i}$

21.81 a) -7.99 nC b) -24.0 nC

21.83 a) 1.56 N/C, $+x$-direction c) smaller

d) 4.7%

21.85 $E_x = E_y = \dfrac{Q}{2\pi^2\epsilon_0 a^2}$

21.87 a) 6.25×10^4 N/C, 225° counterclockwise

from an axis pointing to the right at point P

b) 1.00×10^{-14} N, opposite the electric field

direction

21.89 a) 1.15×10^6 N/C, to the left

b) 1.58×10^5 N/C, to the left

c) 1.58×10^5 N/C, to the right

21.91 a) $\pi(R_2^2 - R_1^2)\sigma$

b) $\dfrac{\sigma}{2\epsilon_0}\left(\dfrac{1}{\sqrt{(R_1/x)^2 + 1}} - \dfrac{1}{\sqrt{(R_2/x)^2 + 1}}\right)\dfrac{|x|}{x}\hat{i}$

c) $\dfrac{\sigma}{2\epsilon_0}\left(\dfrac{1}{R_1} - \dfrac{1}{R_2}\right)x\hat{i}$; $x \ll R_1$

d) $\dfrac{1}{2\pi}\sqrt{\dfrac{q\sigma}{2\epsilon_0 m}\left(\dfrac{1}{R_1} - \dfrac{1}{R_2}\right)}$

21.93 a) $q_1 = 8.00$ μC, $q_2 = 3.00$ μC b) 7.49 N,

in the $-x$-direction c) $x = 0.248$ m

21.95 b) $q_1 < 0$, $q_2 > 0$ c) 0.843 μC d) 56.2 N

21.97 a) $\dfrac{Q}{2\pi\epsilon_0 L}\left(\dfrac{1}{2x + a} - \dfrac{1}{2L + 2x + a}\right)$

21.99 choice (c)

21.101 choice (b)

Chapter 22

22.1 a) 1.8 N·m²/C b) no c) (i) 0° (ii) 90°

22.3 a) 3.53×10^5 N·m²/C b) 3.13 μC

22.5 $\pi r^2 E$

22.7 0.977 N·m²/C, inward

22.9 a) 0 b) 1.22×10^8 N/C, radially inward

c) 3.64×10^7 N/C, radially inward

22.11 a) 1.17×10^5 N·m²/C b) no change

22.13 0.0810 N

22.15 1.35×10^{10}

22.17 a) 6.47×10^5 N/C, $+y$-direction

b) 7.2×10^4 N/C, $-y$-direction

22.19 a) 5.73 μC/m² b) 6.47×10^5 N/C

c) -5.65×10^4 N·m²/C

22.21 a) 0.260 μC/m³ b) 1960 N/C

22.23 a) 6.56×10^{-21} J b) 1.20×10^5 m/s

22.25 a) 6.00 nC b) -1.00 nC

22.27 σ/ϵ_0 (between), 0 (outside)

22.29 a) $2\pi R\sigma$ b) $\dfrac{\sigma R}{\epsilon_0 r}$ c) $\dfrac{\lambda}{2\pi\epsilon_0 r}$

22.31 1.16 km/s

22.33 10.2°

22.35 a) 750 N·m²/C b) 0

c) 577 N/C, $+x$-direction

d) within and outside

22.37 a) -0.598 nC b) within and outside

22.39 a) $\dfrac{\lambda}{2\pi\epsilon_0 r}$, radially outward

b) $\dfrac{\lambda}{2\pi\epsilon_0 r}$, radially outward

d) $-\lambda$ (inner), $+\lambda$ (outer)

22.41 a) $\dfrac{\rho r}{2\epsilon_0}$ b) $\dfrac{\lambda}{2\pi\epsilon_0 r}$

c) They are equal.

22.43 a) $r < R$: 0; $R < r < 2R$: $\dfrac{1}{4\pi\epsilon_0}\dfrac{Q}{r^2}$, radially

outward; $r > 2R$: $\dfrac{1}{4\pi\epsilon_0}\dfrac{2Q}{r^2}$, radially outward

22.45 a) (i) 0   (ii) 0   (iii) $\dfrac{q}{2\pi\epsilon_0 r^2}$, radially outward

(iv) 0   (v) $\dfrac{3q}{2\pi\epsilon_0 r^2}$, radially outward

b) (i) 0   (ii) $+2q$ (iii) $-2q$ (iv) $+6q$

22.47 a) $\dfrac{qQ}{4\pi\epsilon_0 r^2}$, toward the center of the shell   b) 0

22.49 a) $\dfrac{\alpha}{2\epsilon_0}\left(1 - \dfrac{a^2}{r^2}\right)$ b) $q = +2\pi\alpha a^2$, $E = \dfrac{\alpha}{2\epsilon_0}$

22.51 $R/2$

22.53 c) $\dfrac{Qr}{4\pi\epsilon_0 R^3}\left(4 - \dfrac{3r}{R}\right)$ e) $2R/3$, $\dfrac{Q}{3\pi\epsilon_0 R^2}$

22.55 b) $|x| > d$ (outside the slab): $\dfrac{\rho_0 d}{3\epsilon_0}\dfrac{x}{|x|}\hat{\imath}$;

$|x| < d$ (inside the slab): $\dfrac{\rho_0 x^3}{3\epsilon_0 d^2}\hat{\imath}$

22.57 b) $\dfrac{\rho\vec{b}}{3\epsilon_0}$

22.59 a) uniform line of charge: A; uniformly
charged sphere: B   b) $\lambda = 1.50 \times 10^{-7}$ C/m,
$\rho = 2.81 \times 10^{-3}$ C/m^3

22.61 (i) 377 N/C   (ii) 653 N/C   (iii) 274 N/C
(iv) 0

22.63 choice (a)

22.65 choice (b)

Chapter 23

23.1 -0.356 J

23.3 3.46×10^{-13} J $= 2.16$ MeV

23.5 a) 12.5 m/s   b) 0.323 m

23.7 1.94×10^{-5} N

23.9 a) 13.6 km/s; very long after release
b) 2.45×10^{17} m/s^2; just after release

23.11 $-q/2$

23.13 7.42 m/s, faster

23.15 a) 0   b) 0.750 mJ   c) -2.06 mJ

23.17 a) 0   b) -175 kV   c) -0.875 J

23.19 a) -737 V   b) -704 V   c) 8.2×10^{-8} J

23.21 b) $V = \dfrac{q}{4\pi\epsilon_0}\left(\dfrac{1}{|x|} - \dfrac{2}{|x-a|}\right)$

c) $x = -a, a/3$ e) $V = \dfrac{q}{4\pi\epsilon_0 x}$

23.23 a) b b) 800 V/m   c) -48.0 μJ

23.25 a) (i) 180 V   (ii) -270 V   (iii) -450 V
b) 719 V, inner shell

23.27 a) oscillatory   b) 1.67×10^7 m/s

23.29 150 m/s

23.31 a) 94.9 nC/m   b) no, less   c) 0

23.33 a) 78.2 kV   b) 0

23.35 0.474 J

23.37 a) 8.00 kV/m   b) 19.2 μN   c) 0.864 μJ
d) -0.864 μJ

23.39 -760 V

23.41 a) (i) $V = \dfrac{q}{4\pi\epsilon_0}\left(\dfrac{1}{r_a} - \dfrac{1}{r_b}\right)$

(ii) $V = \dfrac{q}{4\pi\epsilon_0}\left(\dfrac{1}{r} - \dfrac{1}{r_b}\right)$ (iii) $V = 0$

d) 0   e) $E = \dfrac{q-Q}{4\pi\epsilon_0 r^2}$

23.43 a) $E_x = -Ay + 2Bx$, $E_y = -Ax - C$, $E_z = 0$
b) $x = -C/A$, $y = -2BC/A^2$, any value of z

23.45 a) 0.762 nC

23.47 a) -0.360 μJ   b) $x = 0.074$ m

23.49 4.2×10^6 V

23.51 a) 4.79 MeV, 7.66×10^{-13} J
b) 5.17×10^{-14} m

23.53 a) -21.5 μJ   b) -2.83 kV   c) 35.4 kV/m

23.55 a) 7.85×10^4 V/m$^{4/3}$
b) $E_x(x) = -(1.05 \times 10^5$ V/m$^{4/3})x^{1/3}$
c) 3.13×10^{-15} N, toward the anode

23.57 a) $-\dfrac{1.46q^2}{\pi\epsilon_0 d}$

23.59 47.8 V

23.61 a) (i) $V = (\lambda/2\pi\epsilon_0)\ln(b/a)$
(ii) $V = (\lambda/2\pi\epsilon_0)\ln(b/r)$ (iii) $V = 0$
d) $(\lambda/2\pi\epsilon_0)\ln(b/a)$

23.63 a) 1.76×10^{-16} N, downward
b) 1.93×10^{14} m/s^2, downward   c) 0.822 cm
d) 15.3°   e) 3.29 cm

23.65 a) 97.1 kV/m   b) 30.3 pC

23.67 $\dfrac{3}{5}\left(\dfrac{Q^2}{4\pi\epsilon_0 R}\right)$

23.69 360 kV

23.71 a) 50.0 g: 216 m/s^2, 12.7 m/s; 150.0 g:
7.20 m/s^2, 4.24 m/s

23.73 a) $\dfrac{Q}{4\pi\epsilon_0 a}\ln\left(\dfrac{x+a}{x}\right)$

b) $\dfrac{Q}{4\pi\epsilon_0 a}\ln\left(\dfrac{a + \sqrt{a^2 + y^2}}{y}\right)$

c) (a): $\dfrac{Q}{4\pi\epsilon_0 x}$, (b): $\dfrac{Q}{4\pi\epsilon_0 y}$

23.75 a) $\frac{1}{3}$ b) 3

23.77 a) 7580 km/s   b) 7260 km/s
c) 2.3×10^9 K; 6.4×10^9 K

23.79 a) $A = -6.0$ V/m^2, $B = -4.0$ V/m^3,
$C = -2.0$ V/m^6, $D = 10$ V, $l = 2.0$, $m = 3.0$,
$n = 6.0$
b) (0, 0, 0): 10.0 V, 0; (0.50 m, 0.50 m, 0.50 m):
8.0 V, 6.7 V/m; (1.00 m, 1.00 m, 1.00 m):
-2.0 V, 21 V/m

23.81 c) 4.79×10^{-19} C (drop 1), 1.59×10^{-19} C
(drop 2), 8.09×10^{-19} C (drop 3),
3.23×10^{-19} C (drop 4)
d) 3 (drop 1), 5 (drop 3), 2 (drop 4)
e) 1.60×10^{-19} C (drop 1), 1.59×10^{-19} C
(drop 2), 1.62×10^{-19} C (drops 3 and 4);
1.61×10^{-19} C

23.83 1.01×10^{-12} m, 1.11×10^{-13} m,
2.54×10^{-14} m

23.85 choice (b)

Chapter 24

24.1 a) 10.0 kV   b) 22.6 cm^2 c) 8.00 pF

24.3 a) 604 V   b) 90.8 cm^2 c) 1840 kV/m
d) 16.3 μC/m^2

24.5 a) 120 μC   b) 60 μC   c) 480 μC

24.7 a) 1.05 mm   b) 84.0 V

24.9 a) 4.35 pF   b) 2.30 V

24.11 a) 15.0 pF   b) 3.09 cm   c) 31.2 kN/C

24.13 a) 17.5 cm   b) 25.5 nC

24.15 a) series   b) 5000

24.17 a) $Q_1 = Q_2 = 22.4$ μC, $Q_3 = 44.8$ μC,
$Q_4 = 67.2$ μC   b) $V_1 = V_2 = 5.6$ V,
$V_3 = 11.2$ V, $V_4 = 16.8$ V   c) 11.2 V

24.19 a) $Q_1 = 156$ μC, $Q_2 = 260$ μC
b) $V_1 = V_2 = 52.0$ V

24.21 a) 19.3 nF   b) 482 nC   c) 162 nC
d) 25 V

24.23 0.0283 J/m^3

24.25 a) 90.0 pF   b) 0.0152 m^3 c) 4.5 kV
d) 1.80 μJ

24.27 a) $U_p = 4U_s$ b) $Q_p = 2Q_s$ c) $E_p = 2E_s$

24.29 a) 24.2 μC   b) $Q_{35} = 7.7$ μC,
$Q_{75} = 16.5$ μC   c) 2.66 μJ
d) $U_{35} = 0.85$ mJ, $U_{75} = 1.81$ mJ   e) 220 V

24.31 a) 1.60 nC   b) 8.05

24.33 a) 3.60 mJ (before), 13.5 mJ (after)
b) 9.9 mJ, increase

24.35 a) 0.620 μC/m^2 b) 1.28

24.37 0.0135 m^2

24.39 a) 6.3 μC   b) 6.3 μC   c) none

24.41 a) 10.1 V   b) 2.25

24.43 a) $\dfrac{Q}{\epsilon_0 AK}$ b) $\dfrac{Qd}{\epsilon_0 AK}$ c) $K\dfrac{\epsilon_0 A}{d} = KC_0$

24.45 a) 421 J   b) 0.054 F

24.47 a) 0.531 pF   b) 0.224 mm

24.49 a) 0.0160 C   b) 533 V   c) 4.26 J   d) 2.14 J

24.51 a) 158 μJ   b) 72.1 μJ

24.53 a) 2.5 μF
b) $Q_1 = 550$ μC, $Q_2 = 370$ μC, $Q_3 = Q_4 =$
180 μC, $Q_5 = 550$ μC; $V_1 = 65$ V, $V_2 = 87$ V,
$V_3 = V_4 = 43$ V, $V_5 = 65$ V

24.55 $C_2 = 6.00$ μF, $C_3 = 4.50$ μF

24.57 a) 76 μC   b) 1.4 mJ   c) 11 V   d) 1.3 mJ

24.59 a) 2.3 μF
b) $Q_1 = 970$ μC, $Q_2 = 640$ μC   c) 47 V

24.61 a) 3.91   b) 22.8 V

24.63 1.67 μF

24.65 0.185 μJ

24.67 b) 2.38 nF

24.69 a) $C_1 = 6.00$ μF, $C_2 = 3.00$ μF
b) same charge; C_2 stores more energy
c) C_1 stores more charge and energy

24.71 a) first (connected)   b) 144 cm^2
c) disconnected

24.73 choice (c)

24.75 choice (a)

Chapter 25

25.1 1.0 C

25.3 a) 3.12×10^{19} b) 1.51×10^6 A/m^2
c) 0.111 mm/s
d) both (b) and (c) would increase

25.5 a) 110 min   b) 440 min   c) $v_d \propto 1/d^2$

25.7 a) 330 C   b) 41 A

25.9 9.0 μA

25.11 a) 1.06×10^{-5} $\Omega \cdot$m   b) 0.00105 (C°)$^{-1}$

25.13 a) 0.206 mV   b) 0.176 mV

25.15 a) 1.21 V/m   b) 0.0145 Ω c) 0.182 V

25.17 0.125 Ω

25.19 a) 4.67×10^{-8} Ω b) 6.72×10^{-4} Ω

25.21 a) 11 A   b) 3.1 V   c) 0.28 Ω

25.23 a) 99.54 Ω b) 0.0158 Ω

25.25 a) 27.4 V   b) 12.3 MJ

25.27 a) 0   b) 5.0 V   c) 5.0 V

25.29 3.08 V, 0.067 Ω, 1.80 Ω

25.31 a) 1.41 A, clockwise   b) 13.7 V   c) -1.0 V

25.33 a) 0.471 A, counterclockwise   b) 15.2 V

25.35 a) 144 Ω b) 240 Ω
c) 100 W: 0.833 A; 60 W: 0.500 A

25.37 a) 29.8 W   b) 0.248 A

25.39 a) 3.1 W   b) 7.2 W   c) 4.1 W

25.41 a) 300 W   b) 0.90 J

25.43 a) 2.6 MJ   b) 0.063 L   c) 1.6 h

25.45 12.3%

25.47 a) 24.0 W   b) 4.0 W   c) 20.0 W

25.49 a) 1.55×10^{-12} s

25.51 a) 3.65×10^{-8} $\Omega \cdot$m   b) 172 A
c) 2.58 mm/s

25.53 0.060 Ω

25.55 a) 2.5 mA b) 21.4 μV/m c) 85.5 μV/m
d) 0.180 mV

25.57 a) 80 C° b) no

25.59 $\dfrac{\rho h}{\pi r_1 r_2}$

25.61 a) 0.36 Ω b) 8.94 V

25.63 a) 1.0 kΩ b) 100 V c) 10 W

25.65 a) \$78.90 b) \$140.27

25.67 a) $I_A\left(1 + \dfrac{R_A}{r + R}\right)$ b) 0.0429 Ω

25.69 a) 171 μΩ b) 176 μV/m
c) left: 54.7 μΩ; right: 116 μΩ

25.71 a) 204 V b) 199 J

25.73 6.67 V

25.75 b) no c) yes d) 9.40 W e) 4.12 W

25.77 a) $R = \dfrac{\rho_0 L}{A}\left(1 - \dfrac{1}{e}\right)$, $I = \dfrac{V_0 A}{\rho_0 L\left(1 - \dfrac{1}{e}\right)}$

b) $E(x) = \dfrac{V_0 e^{-x/L}}{L\left(1 - \dfrac{1}{e}\right)}$

c) $V(x) = \dfrac{V_0\left(e^{-x/L} - \dfrac{1}{e}\right)}{1 - \dfrac{1}{e}}$

25.79 choice (c)

25.81 choice (d)

Chapter 26

26.1 $3R/4$

26.3 22.5 W

26.5 a) 3.50 A b) 4.50 A c) 3.15 A
d) 3.25 A

26.7 0.769 A

26.9 a) 8.80 Ω b) 3.18 A c) 3.18 A
d) $V_{1.60} = 5.09$ V, $V_{2.40} = 7.63$ V,
$V_{4.80} = 15.3$ V e) $P_{1.60} = 16.2$ W,
$P_{2.40} = 24.3$ W, $P_{4.80} = 48.5$ W f) greatest

26.11 a) $I_1 = 8.00$ A, $I_3 = 12.0$ A b) 84.0 V

26.13 5.00 Ω; $I_{3.00} = 8.00$ A, $I_{4.00} = 9.00$ A,
$I_{6.00} = 4.00$ A, $I_{12.0} = 3.00$ A

26.15 a) $I_1 = 1.50$ A, $I_2 = I_3 = I_4 = 0.500$ A
b) $P_1 = 10.1$ W, $P_2 = P_3 = P_4 = 1.12$ W;
bulb R_1 c) $I_1 = 1.33$ A, $I_2 = I_3 = 0.667$ A
d) $P_1 = 8.00$ W, $P_2 = P_3 = 2.00$ W
e) brighter: R_2 and R_3; less bright: R_1

26.17 18.0 V, 3.00 A

26.19 1010 s

26.21 a) 0.100 A b) $P_{400} = 4.0$ W, $P_{800} = 8.0$ W
c) 12.0 W d) $I_{400} = 0.300$ A, $I_{800} = 0.150$ A
e) $P_{400} = 36.0$ W, $P_{800} = 18.0$ W
f) 54.0 W g) series: 800-Ω bulb;
parallel: 400-Ω bulb h) parallel

26.23 a) 20.0 Ω b) A_2: 4.00 A; A_3: 12.0 A;
A_4: 14.0 A; A_5: 8.00 A

26.25 a) 2.00 A b) 5.00 Ω c) 42.0 V d) 3.50 A

26.27 a) 8.00 A b) $\mathcal{E}_1 = 36.0$ V, $\mathcal{E}_2 = 54.0$ V
c) 9.00 Ω

26.29 a) 1.60 A (top), 1.40 A (middle),
0.20 A (bottom) b) 10.4 V

26.31 a) 36.4 V b) 0.500 A

26.33 a) 2.14 V, a b) 0.050 A, 0; down

26.35 a) 0.641 Ω b) 975 Ω

26.37 a) 17.9 V b) 22.7 V c) 21.4%

26.39 a) 0.849 μF b) 2.89 s

26.41 a) 0 b) 245 V c) 0 d) 32.7 mA
e) (a): 245 V; (b): 0; (c): 1.13 mC; (d): 0

26.43 a) 4.21 ms b) 0.125 A

26.45 192 μC

26.47 13.6 A

26.49 a) 0.937 A b) 0.606 A

26.51 a) 165 μC b) 463 Ω c) 12.6 ms

26.53 900 W

26.55 a) 2.2 A, 4.4 V, 9.7 W
b) 16.3 W; more brightly

26.57 a) +0.22 V b) 0.464 A

26.59 $I_1 = 0.848$ A, $I_2 = 2.14$ A, $I_3 = 0.171$ A

26.61 $I_{2.00} = 5.21$ A, $I_{4.00} = 1.11$ A,
$I_{5.00} = 6.32$ A

26.63 a) 109 V; no b) 13.5 s

26.65 a) 186 V, upper terminal positive
b) 3.00 A, upward c) 20.0 Ω

26.67 a) −12.0 V b) 1.71 A c) 4.21 Ω

26.69 a) $P_1 + P_2$ b) $\dfrac{P_1 P_2}{P_1 + P_2}$

26.71 a) 1.35 W b) 8.31 ms c) 0.337 W

26.73 a) 114 V b) 263 V c) 266 V

26.75 a) 18.0 V b) a c) 6.00 V
d) both decrease by 36.0 μC

26.77 a) $V_{224} = 24.8$ V, $V_{589} = 65.2$ V
b) 3840 Ω c) 62.6 V d) no

26.79 1.7 MΩ, 3.1 μF

26.81 a) −1.23 ms (slope), 79.5 μC (y-intercept)
b) 247 Ω, 15.9 V c) 1.22 ms d) 11.9 V

26.85 b) 4 c) 3.2 MΩ, 4.0 × 10^{-3}
d) 3.4 × 10^{-4} e) 0.88

26.87 choice (d)

Chapter 27

27.1 a) $(-6.68 \times 10^{-4} \text{ N})\hat{k}$
b) $(6.68 \times 10^{-4} \text{ N})\hat{i} + (7.27 \times 10^{-4} \text{ N})\hat{j}$

27.3 a) positive b) 0.0505 N

27.5 9490 km/s

27.7 a) $B_x = -0.175$ T, $B_z = -0.256$ T b) B_y
c) 0; 90°

27.9 a) 1.46 T, in the xz-plane at 40° from the
$+x$-axis toward the $-z$-axis
b) 7.47×10^{-16} N, in the xz-plane at 50° from
the $-x$-axis toward the $-z$-axis

27.11 a) 3.05 mWb b) 1.83 mWb c) 0

27.13 −0.78 mWb

27.15 a) 0.160 mT, into the page b) 0.111 μs

27.17 7.93×10^{-10} N, toward the south

27.19 a) 2.84×10^6 m/s, negative b) yes c) same

27.21 a) 835 km/s b) 26.2 ns c) 7.27 kV

27.23 0.838 mT

27.25 a) $(1.60 \times 10^{-14} \text{ N})\hat{j}$ b) yes
c) helix; no d) 1.40 cm

27.27 a) 7900 N/C, $\hat{i}$ b) 7900 N/C, $\hat{i}$

27.29 0.0445 T, out of the page

27.31 a) 4.92 km/s b) 9.96×10^{-26} kg

27.33 2.0 cm

27.35 0.724 N, 63.4° below the current direction in
the upper wire segment

27.37 a) 817 V b) 113 m/s^2

27.39 a) a b) 3.21 kg

27.41 b) $F_{cd} = 1.20$ N c) 0.420 N·m

27.43 a) A_2 b) 290 rad/s^2

27.45 a) $-NIAB\hat{i}$, 0 b) 0, $-NIAB$ c) $+NIAB\hat{i}$, 0
d) 0, $+NIAB$

27.47 a) 1.13 A, 3.69 A c) 98.2 V d) 362 W

27.49 a) 4.7 mm/s
b) $+4.5 \times 10^{-3}$ V/m, $+z$-direction
c) 53 μV

27.51 a) $-\dfrac{F_2}{qv_1}\hat{j}$ b) $F_2/\sqrt{2}$

27.53 a) 8.3×10^6 m/s b) 0.14 T

27.55 3.45 T, perpendicular to the coin's initial
velocity

27.57 a) −3.89 μC
b) $(7.60 \times 10^{14} \text{ m/s}^2)\hat{i} + (5.70 \times 10^{14} \text{ m/s}^2)\hat{j}$
c) 2.90 cm d) 2.88×10^7 Hz
e) (0.0290 m, 0, 0.874 m)

27.59 1.6 mm

27.61 $\dfrac{Mg \tan \theta}{LB}$, right to left

27.63 a) 8.46 mT b) 27.2 cm c) 2.2 cm; yes

27.65 a) ILB, to the right b) $\dfrac{v^2 m}{2ILB}$ c) 1960 km

27.67 1.97 N, 68.3° clockwise from the left-hand
segment

27.69 0.024 T, $+y$-direction

27.71 a) $F_{PQ} = 0$; $F_{RP} = 12.0$ N, into the page;
$F_{QR} = 12.0$ N, out of the page
b) 0 c) $\tau_{PQ} = \tau_{RP} = 0$; $\tau_{QR} = 3.60$ N·m
d) 3.60 N·m; yes e) out

27.73 $-(0.444 \text{ N})\hat{j}$

27.75 b) left: $(B_0 LI/2)\hat{i}$; top: $-IB_0 L\hat{j}$; right:
$-(B_0 LI/2)\hat{i}$; bottom: 0 c) $-IB_0 L\hat{j}$

27.77 a) $-IA\,\hat{k}$ b) $B_x = \dfrac{3D}{IA}, B_y = \dfrac{4D}{IA}, B_z = -\dfrac{12D}{IA}$

27.79 b) 1.85×10^{-28} kg c) 1.20 kV
d) 8.32×10^5 m/s

27.81 a) 5.14 m b) 1.72 μs c) 6.08 mm
d) 3.05 cm

27.83 choice (c)

27.85 choice (a)

Chapter 28

28.1 a) $-(19.2 \mu\text{T})\hat{k}$ b) 0 c) $(19.2 \mu\text{T})\hat{i}$
d) $(6.79 \mu\text{T})\hat{i}$

28.3 a) 60.0 nT, out of the page at A and B
b) 0.120 μT, out of the page c) 0

28.5 a) 0 b) $-(1.31 \mu\text{T})\hat{k}$ c) $-(0.462 \mu\text{T})\hat{k}$
d) $(1.31 \mu\text{T})\hat{j}$

28.7 $(97.5 \text{ nT})\hat{k}$

28.9 a) 0.440 μT, out of the page
b) 16.7 nT, out of the page c) 0

28.11 a) $(50.0 \text{ pT})\hat{j}$ b) $-(50.0 \text{ pT})\hat{i}$
c) $-(17.7 \text{ pT})(\hat{i} - \hat{j})$ d) 0

28.13 17.6 μT, into the page

28.15 a) 0.8 mT b) 40 μT (20 times larger)

28.17 250 μA

28.19 a) 10.0 A b) at all points directly above the
wire c) at all points directly east of the wire

28.21 a) $-(0.10 \mu\text{T})\hat{i}$ b) 2.19 μT, at 46.8° from
the $+x$-axis to the $+z$-axis c) $(7.9 \mu\text{T})\hat{i}$

28.23 a) 0 b) 6.67 μT, toward the top of the page
c) 7.54 μT, to the left

28.25 a) 0 b) 0 c) 0.40 mT, to the left

28.27 a) P: 41 μT, into the page; Q: 25 μT, out of the
page b) P: 9.0 μT, out of the page;
Q: 9.0 μT, into the page

28.29 a) 6.00 μN; repulsive b) 24.0 μN

28.31 46 μN/m; repulsive; no

28.33 0.38 μA

28.35 $\dfrac{\mu_0 |I_1 - I_2|}{4R}$; 0

28.37 a) 25.1 μT b) 503 μT; no

28.39 18.0 A, counterclockwise

28.41 a) 305 A b) -3.83×10^{-4} T·m

28.43 a) $\mu_0 I/2\pi r$ b) 0

28.45 a) 2.83 mT b) 35.0 μT; no

28.47 a) 1790 turns per meter b) 63.0 m

28.49 a) 3.72 MA b) 249 kA c) 237 A

28.51 1.11 mT

28.53 a) (i) 1.13 mT   (ii) 4.68 MA/m   (iii) 5.88 T

28.55 a) 1.00 μT, into the page   b) $(74.9 \text{ nN})\hat{\jmath}$

28.57 a) in the plane of the wires, between them, 0.300 m from the 75.0-A wire
b) in the plane of the wires, 0.200 m from the 25.0-A wire and 0.600 m from the 75.0-A wire

28.59 a) 5.7×10^{12} m/s^2, away from the wire
b) 32.5 N/C, away from the wire   c) no

28.61 a) 81 A   b) 2.4×10^{-3} N/m

28.63 a) 2.00 A, out of the page
b) 2.13 μT, upward   c) 2.06 μT

28.65 23.2 A

28.67 a) $\dfrac{\mu_0 N I a^2}{2} \times$

$\left\{ \dfrac{1}{[(x + a/2)^2 + a^2]^{3/2}} + \dfrac{1}{[(x - a/2)^2 + a^2]^{3/2}} \right\}$

c) $\left(\dfrac{4}{5}\right)^{3/2} \dfrac{\mu_0 N I}{a}$ d) 20.2 mT   e) 0, 0

28.69 a) $\dfrac{3I}{2\pi R^3}$ b) (i) $B = \dfrac{\mu_0 I r^2}{2\pi R^3}$ (ii) $B = \dfrac{\mu_0 I}{2\pi r}$

28.71 b) $B = \dfrac{\mu_0 I_0}{2\pi r}$ c) $\dfrac{I_0 r^2}{a^2}\left(2 - \dfrac{r^2}{a^2}\right)$

d) $B = \dfrac{\mu_0 I_0 r}{2\pi a^2}\left(2 - \dfrac{r^2}{a^2}\right)$

28.73 a) $B = \mu_0 I n/2$, $+x$-direction
b) $B = \mu_0 I n/2$, $-x$-direction

28.75 a) $I_0 = 2\pi b \delta (1 - e^{-a/\delta})$, 81.5 A   b) $\dfrac{\mu_0 I_0}{2\pi r}$

c) $\left(\dfrac{e^{r/\delta} - 1}{e^{a/\delta} - 1}\right) I_0$ d) $\dfrac{\mu_0 I}{2\pi r}\left(\dfrac{e^{r/\delta} - 1}{e^{a/\delta} - 1}\right)$

e) $r = \delta$: 175 μT; $r = a$: 326 μT; $r = 2a$: 163 μT

28.77 a) no   c) 65 A, 1.2 cm

28.79 b) $\dfrac{1}{2g}\left(\dfrac{\mu_0 Q_0^2}{4\pi \lambda R C d}\right)^2$

28.81 choice (b)

28.83 choice (c)

Chapter 29

29.1 a) 17.1 mV   b) 28.5 mA

29.3 a) $Q = NBA/R$ c) no

29.5 a) 34 V   b) counterclockwise

29.7 a) $\mu_0 i/2\pi r$, into the page   b) $\dfrac{\mu_0 i}{2\pi r} L dr$

c) $\dfrac{\mu_0 i L}{2\pi} \ln(b/a)$ d) $\dfrac{\mu_0 L}{2\pi} \ln(b/a)\dfrac{di}{dt}$

e) 0.506 μV

29.9 a) 5.44 mV   b) clockwise

29.11 a) bAv b) clockwise
c) bAv, counterclockwise

29.13 10.4 rad/s

29.15 a) counterclockwise   b) clockwise
c) no induced current

29.17 a) C: counterclockwise; A: clockwise
b) toward the wire

29.19 a) a to b b) b to a c) b to a

29.21 a) clockwise   b) no induced current
c) counterclockwise

29.23 13.2 mA, counterclockwise

29.25 a) 0.675 V   b) b c) 2.25 V/m, b to a
d) b e) (i) 0   (ii) 0

29.27 46.2 m/s = 103 mph; no

29.29 a) 3.00 V   b) b to a
c) 0.800 N, to the right   d) 6.00 W for each

29.31 a) counterclockwise   b) 42.4 mW

29.33 35.0 m/s, to the right

29.35 a) 0.225 A, clockwise   b) 0
c) 0.225 A, counterclockwise

29.37 a) $\pi r_1^2 \dfrac{dB}{dt}$ b) $\dfrac{r_1}{2}\dfrac{dB}{dt}$ c) $\dfrac{R^2}{2r_2}\dfrac{dB}{dt}$ e) $\dfrac{\pi R^2}{4}\dfrac{dB}{dt}$

f) $\pi R^2 \dfrac{dB}{dt}$ g) $\pi R^2 \dfrac{dB}{dt}$

29.39 9.21 A/s

29.41 0.950 mV

29.43 a) 0.599 nC   b) 6.00 mA   c) 6.00 mA

29.45 a) inside: $B = 0$, $\vec{M} = -(0.103 \text{ MA/m})\hat{\imath}$;
outside: $\vec{B} = (0.130 \text{ T})\hat{\imath}$, $M = 0$
b) inside and outside: $\vec{B} = (0.260 \text{ T})\hat{\imath}$, $M = 0$

29.47 a) 3.7 A   b) 1.33 mA   c) counterclockwise

29.49 16.2 μV

29.51 a) $\dfrac{\mu_0 I a b v}{2\pi r(r + a)}$ b) clockwise

29.53 a) 17.9 mV   b) a to b

29.55 $\mu_0 I W/4\pi$

29.57 a) $\dfrac{\mu_0 I v}{2\pi} \ln(1 + L/d)$ b) a c) 0

29.59 a) 0.165 V   b) 0.165 V   c) 0; 0.0412 V

29.61 a) $B^2 L^2 v/R$

29.63 a: $\dfrac{qr}{2}\dfrac{dB}{dt}$, to the left; b: $\dfrac{qr}{2}\dfrac{dB}{dt}$, toward the top of the page; c: 0

29.65 5.0 s

29.67 a) 0.3071 s^{-1} b) 3.69 T   c) a d) 2.26 s

29.69 a) a to b b) $\dfrac{Rmg \tan\phi}{L^2 B^2 \cos\phi}$ c) $\dfrac{mg \tan\phi}{LB}$

d) $\dfrac{Rm^2 g^2 \tan^2\phi}{L^2 B^2}$ e) $\dfrac{Rm^2 g^2 \tan^2\phi}{L^2 B^2}$; same

29.71 choice (c)

29.73 choice (c)

Chapter 30

30.1 a) 0.270 V; yes   b) 0.270 V

30.3 6.32 μH

30.5 a) 1.96 H   b) 7.11 mWb

30.7 a) 1940   b) 800 A/s

30.9 a) 0.250 H   b) 0.450 mWb

30.11 a) 4.68 mV   b) a

30.13 a) 1000   b) 2.09 Ω

30.15 b) 0.111 μH

30.17 2850

30.19 a) 0.161 T   b) 10.3 kJ/m^3 c) 0.129 J
d) 40.2 μH

30.21 91.7 J

30.23 a) 2.40 A/s   b) 0.800 A/s   c) 0.413 A
d) 0.750 A

30.25 a) 17.3 μs   b) 30.7 μs

30.27 a) 0.250 A   b) 0.137 A   c) 32.9 V; c
d) 0.462 ms

30.29 15.3 V

30.31 a) 443 nC   b) 358 nC

30.33 a) 25.0 mH   b) 90.0 nC   c) 0.540 μJ
d) 6.58 mA

30.35 a) 105 rad/s, 59.6 ms   b) 0.720 mC
c) 4.32 mJ   d) -0.542 mC
e) -0.050 A, counterclockwise
f) $U_C = 2.45$ mJ, $U_L = 1.87$ mJ

30.37 a) 7.50 μC   b) 15.9 kHz   c) 21.2 mJ

30.39 a) 298 rad/s   b) 83.8 Ω

30.41 a) 8.76 kHz   b) 1.35 ms   c) 2420 Ω

30.43 a) 0.288 μH   b) 14.2 μV

30.45 20 km/s; about 30 times smaller

30.47 a) $\dfrac{\mu_0 i}{2\pi r}$ b) $\dfrac{\mu_0 i^2 l}{4\pi r} dr$ c) $\dfrac{\mu_0 i^2 l}{4\pi} \ln(b/a)$

d) $\dfrac{\mu_0 I}{2\pi} \ln(b/a)$

30.49 a) 5.00 H   b) 31.7 m; no

30.51 222 μF, 9.31 μH

30.53 a) 0.896 mJ   b) 0.691 A; 0

30.55 a) 24.0 mV   b) 1.55 mA   c) 72.1 nJ
d) 5.20 μC, 18.0 nJ

30.57 a) 0, 20.0 V   b) 0.267 A, 0   c) 0.147 A, 9.0 V

30.59 a) $A_1 = A_4 = 0.800$ A, $A_2 = A_3 = 0$;
$V_1 = 40.0$ V, $V_2 = V_3 = V_4 = V_5 = 0$
b) $A_1 = 0.480$ A, $A_2 = 0.160$ A,
$A_3 = 0.320$ A, $A_4 = 0$; $V_1 = 24.0$ V, $V_2 = 0$,
$V_3 = V_4 = V_5 = 16.0$ V   c) 192 μC

30.61 a) 60.0 V   b) a c) 60.0 V   d) c
e) -96.0 V   f) b g) 156 V   h) d

30.63 a) 0; $v_{ac} = 0$, $v_{cb} = 36.0$ V
b) 0.180 A, $v_{ac} = 9.0$ V, $v_{cb} = 27.0$ V
c) $i_0 = (0.180 \text{ A})(1 - e^{-t/(0.020 \text{ s})})$,
$v_{ac} = (9.0 \text{ V})(1 - e^{-t/(0.020 \text{ s})})$,
$v_{cb} = (9.0 \text{ V})(3.00 + e^{-t/(0.020 \text{ s})})$

30.65 a) $A_1 = A_4 = 0.455$ A, $A_2 = A_3 = 0$
b) $A_1 = 0.585$ A, $A_2 = 0.320$ A,
$A_3 = 0.160$ A, $A_4 = 0.107$ A

30.67 a) $v_L = \dfrac{\epsilon}{R + R_L}(R_L + R e^{-(R+R_L)t/L})$

b) 50.0 V   c) 30.0 V; 3.00 A   d) 6.67 Ω
e) 40.0 mH

30.69 b) 5.0 Ω, 8.5 H   c) 1.7 kJ; 2.0 kW

30.71 a) $i_1 = \dfrac{\epsilon}{R_1}(1 - e^{-R_1 t/L})$, $i_2 = \dfrac{\epsilon}{R_2} e^{-t/R_2 C}$,

$q_2 = \epsilon C(1 - e^{-t/R_2 C})$
b) $i_1 = 0$, $i_2 = 9.60$ mA
c) $i_1 = 1.92$ A, $i_2 = 0$; $t \gg L/R_1$ and
$t \gg R_2 C$
d) 1.6 ms   e) 9.4 mA   f) 0.22 s

30.73 choice (b)

30.75 choice (c)

Chapter 31

31.1 1.06 A

31.3 a) 31.8 V   b) 0

31.5 a) 90°; lead   b) 193 Hz

31.7 13.3 μF

31.9 a) 1510 Ω b) 0.239 H   c) 497 Ω
d) 16.6 μF

31.11 a) (12.5 V) cos[(480 rad/s)t]   b) 7.17 V

31.13 a) $i = (0.0253 \text{ A}) \cos[(720 \text{ rad/s})t]$
b) 180 Ω
c) $v_L = -(4.56 \text{ V}) \sin[(720 \text{ rad/s})t]$

31.15 a) 601 Ω b) 49.9 mA   c) $-70.6°$; lag
d) $V_R = 9.98$ V, $V_L = 4.99$ V, $V_C = 33.3$ V

31.17 50.0 V

31.19 a) 40.0 W   b) 0.167 A   c) 720 Ω

31.21 b) 76.7 V

31.23 a) 45.8°, 0.697   b) 344 Ω c) 155 V
d) 48.6 W   e) 48.6 W   f) 0   g) 0

31.25 a) 0.302   b) 0.370 W
c) 0.370 W (resistor), 0, 0

31.27 a) 113 Hz; 15.0 mA   b) 7.61 mA; lag

31.29 a) 150 V   b) $V_R = 150$ V,
$V_L = V_C = 1290$ V   c) 37.5 W

31.31 a) 1.00   b) 75.0 W   c) 75.0 W

31.33 a) 945 rad/s   b) 70.6 Ω
c) $V_L = V_C = 450$ V, $V_R = 120$ V

31.35 a) 10   b) 2.40 A   c) 28.8 W   d) 500 Ω

31.37 0.124 H

31.39 230 Ω

31.41 3.59×10^7 rad/s

31.43 a) inductor b) 0.133 H
31.45 a) 0.831 b) 161 W
31.47 $\dfrac{V_{out}}{V_s} = \sqrt{\dfrac{R^2 + \omega^2 L^2}{R^2 + \left(\omega L - \dfrac{1}{\omega C}\right)^2}}$
31.51 a) 102 Ω b) 0.882 A c) 270 V
31.53 a) $V_R = 48.6$ V, $V_L = 155$ V, $V_C = 243$ V, $-60.9°$
   b) $V_R = 100$ V, $V_L = V_C = 400$ V, 0°
   c) $V_R = 48.6$ V, $V_L = 243$ V, $V_C = 155$ V, $+60.9°$
31.55 b) 5770 rad/s c) 2.40 A d) 2.40 A
   e) 0.139 A f) 0.139 A
31.57 a) $\omega = 28{,}800$ rad/s so $\phi = 60°$
   b) $P_R = 0.375$ W, $P_L = P_C = 0$; 0.100 A
31.59 a) 0.750 A b) 160 Ω c) 341 Ω, 619 Ω
   d) 341 Ω
31.61 a) $\dfrac{V}{R}$ b) $\dfrac{V}{R}\sqrt{\dfrac{L}{C}}$ c) $\dfrac{V}{R}\sqrt{\dfrac{L}{C}}$ d) $\dfrac{1}{2}L\dfrac{V^2}{R^2}$
   e) $\dfrac{1}{2}L\dfrac{V^2}{R^2}$
31.63 a) 20.6 Ω b) 105 μF c) 699 W
31.65 20.0 Ω, 0.18 H
31.67 a) $\frac{1}{2}V_R I$ b) 0 c) 0
31.69 choice (b)
31.71 choice (d)

Chapter 32

32.1 a) 1.28 s b) 8.15×10^{13} km
32.3 13.3 nT, $+y$-direction
32.5 3.0×10^{18} Hz, 3.3×10^{-19} s, 6.3×10^{10} rad/s
32.7 a) 6.94×10^{14} Hz b) 375 V/m
   c) $E(x, t) = (375$ V/m$) \times$
   $\cos[(1.45 \times 10^7 $ rad/m$)x$
   $\quad - (4.36 \times 10^{15} $ rad/s$)t]$,
   $B(x, t) = (1.25 \, \mu$T$) \times$
   $\cos[(1.45 \times 10^7 $ rad/m$)x$
   $\quad - (4.36 \times 10^{15} $ rad/s$)t]$
32.9 a) (i) 60 kHz (ii) 6.0×10^{13} Hz
   (iii) 6.0×10^{16} Hz
   b) (i) 4.62×10^{-14} m $= 4.62 \times 10^{-5}$ nm
   (ii) 508 m $= 5.08 \times 10^{11}$ nm
32.11 a) $+y$-direction b) 0.149 mm
   c) $\vec{B} = (1.03$ mT$) \cos[(4.22 \times 10^4 $ rad/m$)y$
   $\quad - (1.265 \times 10^{13} $ rad/s$)t]\hat{\imath}$
32.13 a) 361 m b) 0.0174 rad/m
   c) 5.22×10^6 rad/s d) 0.0144 V/m
32.15 a) 0.381 μm b) 0.526 μm
   c) 1.38 d) 1.90
32.17 a) 330 W/m^2 b) 500 V/m, 1.7 μT
32.19 2.5×10^{25} W
32.21 a) 0.24 mW b) 17.4 V/m
32.23 12.0 V/m, 40.0 nT
32.25 850 kW
32.27 a) 0.18 mW b) 274 V/m, 0.913 μT
   c) 0.18 mJ/s d) 0.010 W/cm^2
32.29 a) 637 W/m^2 b) 693 V/m, 2.31 μT
   c) 2.12 μJ/m^3
32.31 a) 30.5 cm b) 2.46 GHz c) 2.11 GHz
32.33 a) 0.375 mJ b) 4.08 mPa c) 604 nm, 3.70×10^{14} Hz d) 30.3 kV/m, 101 μT
32.35 b) 6.02×10^{-9} W/m^2 c) 2.13×10^{-3} N/C, 7.10×10^{-12} T c) 1.20×10^{-18} N; no
32.37 a) at $r = R$: 64 MW/m^2, 0.21 Pa; at $r = R/2$: 260 MW/m^2, 0.85 Pa b) no
32.39 3.89×10^{-13} rad/s^2
32.41 a) $\rho I/\pi a^2$, in the direction of the current
   b) $\mu_0 I/2\pi a$, counterclockwise if the current is out of the page

c) $\dfrac{\rho I^2}{2\pi^2 a^3}$, radially inward d) $\dfrac{\rho I^2}{\pi a^2} = I^2 R$
32.43 a) 1.363 m b) 10.90 m
32.45 a) 9.75×10^{-15} W/m^2
   b) 2.71 μV/m, 9.03×10^{-15} T, 67.3 ms
   c) 3.25×10^{-23} Pa d) 0.190 m
32.47 a) $\dfrac{4\rho G\pi M R^3}{3r^2}$ b) $\dfrac{LR^2}{4cr^2}$ c) 0.19 μm; no
32.49 b) 3.00×10^8 m/s
32.51 b) 1.39×10^{-11} c) 2.54×10^{-8}
32.53 c) 66.0 μm
32.55 choice (d)

Chapter 33

33.1 39.4°
33.3 a) 2.04×10^8 m/s b) 442 nm
33.5 a) 1.55 b) 550 nm
33.7 a) 47.5° b) 66.0°
33.9 2.51×10^8 m/s
33.11 a) 2.34 b) 82°
33.13 71.8°
33.15 a) 51.3° b) 33.6°
33.17 a) 58.1° b) 22.8°
33.19 1.77
33.21 a) 48.9° b) 28.7°
33.23 0.6°
33.25 $0.375 I_0$
33.27 a) A: $I_0/2$, B: $0.125 I_0$, C: $0.0938 I_0$ b) 0
33.29 a) 1.40 b) 35.5°
33.31 $\arccos\left(\dfrac{\cos\theta}{\sqrt{2}}\right)$
33.33 6.38 W/cm^2
33.35 a) $0.364 I$ b) $2.70 I$
33.37 a) 46.7° b) 13.4°
33.39 72.1°
33.41 1.28
33.43 3.52×10^4
33.45 1.84
33.47 a) 48.6° b) 48.6°
33.49 39.1°
33.51 b) 0.23°; about the same
33.53 b) 38.9° c) 5.0°
33.55 23.3°
33.57 a) A: 1.46, carbon tetrachloride; B: 1.33, water; C: 1.63, carbon disulfide; D: 1.50, benzene
   b) A: 2.13, B: 1.77, C: 2.66, D: 2.25
   c) all: 5.09×10^{14} Hz
33.59 a) 35° b) $I_0 = 10$ W/m^2, $I_p = 20$ W/m^2
33.61 a) $\Delta = 2\theta_a^A - 6\arcsin\left(\dfrac{\sin\theta_a^A}{n}\right) + 2\pi$
   b) $\theta_2 = \arccos\sqrt{\dfrac{n^2 - 1}{8}}$
   c) violet: $\theta_2 = 71.55°$, $\Delta = 233.2°$
   red: $\theta_2 = 71.94°$, $\Delta = 230.1°$; violet
33.63 choice (d)

Chapter 34

34.1 39.2 cm to the right of the mirror, 4.85 cm
34.3 9.0 cm; tip of the lead
34.5 b) 33.0 cm to the left of the vertex, 1.20 cm, inverted, real
34.7 0.213 mm
34.9 18.0 cm from the vertex; 0.50 cm, erect, virtual
34.11 a) $+4.00$
   b) 48.0 cm to the right of the mirror; virtual
34.13 a) concave b) $f = 2.50$ cm, $R = 5.00$ cm
34.15 a) 10.0 cm to the left of the shell vertex, 2.20 mm

b) 4.29 cm to the right of the shell vertex, 0.944 mm
34.17 2.67 cm
34.19 3.30 m
34.21 a) at the center of the bowl, 1.33 b) no
34.23 39.5 cm
34.25 8.35 cm to the left of the vertex, 0.326 mm; erect
34.27 a) 107 cm to the right of the lens, 17.8 mm; real; inverted b) the same
34.29 71.2 cm to the right of the lens; -2.97
34.31 3.69 cm; 2.82 cm to the left of the lens
34.33 1.67
34.35 a) 18.6 mm b) 19 mm from the cornea
   c) 0.61 mm; real; inverted
34.37 a) 36.0 cm to the right of the lens b) 180 cm to the left of the lens c) 7.20 cm to the left of the lens d) 13.8 cm to the left of the lens
34.39 26.3 cm from the lens, 12.4 mm; erect; same side
34.41 a) 200 cm to the right of the first lens, 4.80 cm
   b) 150 cm to the right of the second lens, 7.20 cm
34.43 a) 53.0 cm b) real c) 2.50 mm; inverted
34.45 10.2 m
34.47 8.69 cm; no
34.49 a) $f/11$ b) $1/480$ s $= 2.1$ ms
34.51 a) 80.0 cm b) 76.9 cm
34.53 49.4 cm, 2.02 diopters
34.55 -1.37 diopters
34.57 a) 6.06 cm b) 4.12 mm
34.59 a) 8.37 mm b) 21.4 c) -297
34.61 a) -6.33 b) 1.90 cm c) 0.127 rad
34.63 a) 0.661 m b) 59.1
34.65 7.20 m/s
34.67 a) 20.0 cm b) 39.0 cm
34.69 51 m/s
34.71 a) 1.49 cm
34.73 b) 2.4 cm; -0.133
34.75 2.00
34.77 a) converging, 52.5 cm from the lens
   b) converging, 17.5 cm from the lens
34.79 converging, $+50.2$ cm
34.81 a) 58.7 cm, converging b) 4.48 mm; virtual
34.83 a) 6.48 mm b) no, behind the retina
   c) 19.3 mm from the cornea; in front of the retina
34.85 10.6 cm
34.87 a) 0.24 m b) 0.24 m
34.89 b) first image: (i) 51.3 cm to the right of the lens (ii) real (iii) inverted
   second image: (i) 51.3 cm to the right of the lens (ii) real (iii) erect
34.91 -26.7 cm
34.93 7.06 cm to the left of the spherical mirror vertex, 0.177 cm tall; 13.3 cm to the left of the spherical mirror vertex, 0.111 cm tall
34.95 134 cm to the left of the object
34.97 4.17 diopters
34.99 a) 30.9 cm b) 29.2 cm
34.101 d) 36.0 cm, 21.6 cm; $d = 1.2$ cm
34.103 a) -16.6 cm b) 20.0 cm to the right
34.105 a) $4f$
34.107 b) 1.74 cm
34.109 choice (d)
34.111 choice (b)

Chapter 35

35.1 a) 14 cm, 48 cm, 82 cm, 116 cm, 150 cm
   b) 31 cm, 65 cm, 99 cm, 133 cm
35.3 0.75 m, 2.00 m, 3.25 m, 4.50 m, 5.75 m, 7.00 m, 8.25 m

35.5 a) 2.0 m b) constructively
c) 1.0 m, destructively
35.7 1.14 mm
35.9 0.83 mm
35.11 a) 39 b) $\pm 73.3°$
35.13 12.6 cm
35.15 1200 nm
35.17 a) $0.750 I_0$ b) 80 nm
35.19 1670 rad
35.21 a) 4.52 rad b) $0.404 I_0$
35.23 114 nm
35.25 0.0234°
35.27 a) 55.6 nm
b) (i) 2180 nm (ii) 11.0 wavelengths
35.29 a) 514 nm; green b) 603 nm; orange
35.31 0.11 μm
35.33 0.570 mm
35.35 1.57
35.37 a) 96.0 nm b) no, no
35.39 a) 1.58 mm (green), 1.72 mm (orange)
b) 3.45 mm (violet), 4.74 mm (green),
5.16 mm (orange) c) 9.57 μm
35.41 1.730
35.43 761 m, 219 m, 90.1 m, 20.0 m
35.45 6.8×10^{-5} (C°)$^{-1}$
35.47 1.33 μm
35.49 600 nm, 467 nm; no
35.51 a) 1.54 b) $\pm 15.0°$
35.53 a) 50 MHz b) 237.0 m
35.55 14.0
35.57 choice (d)
35.59 choice (c)

Chapter 36

36.1 506 nm
36.3 a) 226 b) $\pm 83.0°$
36.5 9.07 m
36.7 a) 63.8 cm
b) $\pm 22.1°$, $\pm 34.3°$, $\pm 48.8°$, $\pm 70.1°$
36.9 $\pm 16.0°$, $\pm 33.4°$, $\pm 55.6°$
36.11 a) 10.9 mm b) 5.4 mm
36.13 a) 580 nm b) 0.128
36.15 a) 6.75 mm b) 2.43 μW/m^2
36.17 a) 668 nm b) $(9.36 \times 10^{-5}) I_0$
36.19 a) 3 b) 2
36.21 a) 0.0627°, 0.125° b) $0.249 I_0$, $0.0256 I_0$
36.23 a) 4830 lines/cm b) 4; $\pm 37.7°$, $\pm 66.5°$
36.25 a) 4790 slits/cm b) 19.1°, 40.8° c) no
36.27 a) yes b) 13.3 nm
36.29 a) 467 nm b) 27.8°
36.31 a) 17,500 b) yes
c) (i) 587.8170 nm (ii) 587.7834 nm
(iii) 587.7834 nm $< \lambda <$ 587.8170 nm
36.33 2752 slits/cm
36.35 0.232 nm
36.37 92 cm
36.39 1.88 m
36.41 220 m
36.43 a) 73 m (Hubble), 1100 km (Arecibo)
b) 1600 km
36.45 1.45 m
36.47 30.2 μm
36.49 a) 78 b) $\pm 80.8°$ c) 555 μW/m^2
36.51 1.68
36.53 a) 4.49 rad, 7.73 rad c) 3.14 rad, 6.28 rad,
9.42 rad; no d) 4.78°, 6.84°, 9.59°
36.55 -0.033 mm; decrease
36.57 360 nm
36.59 second
36.61 1.40
36.63 a) 1.03 mm b) 0.148 mm

36.65 a) 12.1 μm b) 10.4 cm, 15.2 cm
36.69 choice (d)
36.71 choice (a)

Chapter 37

37.1 bolt A
37.3 $0.867c$; no
37.5 a) $0.998c$ b) 126 m
37.7 1.12 h, in the spacecraft
37.9 92.5 m
37.11 a) 0.66 km b) 49 μs; 15 km c) 0.45 km
37.13 a) 3570 m b) 90.0 μs c) 89.2 μs
37.15 a) $0.806c$ b) $0.974c$ c) $0.997c$
37.17 a) toward b) $0.385c$
37.19 $0.784c$
37.21 $0.611c$
37.23 a) $0.159c$ b) \$172 million
37.25 $0.220c$; toward you
37.27 $3.06 p_0$
37.29 a) $0.866c$ b) $0.608c$
37.31 a) $0.866c$ b) $0.986c$
37.33 a) 0.450 nJ b) 1.94×10^{-18} kg $\cdot$ m/s
c) $0.968c$
37.35 a) 1110 kg b) 52.1 cm
37.37 a) 0.867 nJ b) 0.270 nJ c) 0.452
37.39 a) 5.34 pJ (nonrel), 5.65 pJ (rel), 1.06
b) 67.8 pJ (nonrel), 331 pJ (rel), 4.88
37.41 a) 2.06 MV b) 0.330 pJ = 2.06 MeV
37.43 a) $\Delta = 8.42 \times 10^{-6}$ b) 34.0 GeV
37.45 $0.700c$
37.47 42.5 y
37.49 a) $\Delta = 9 \times 10^{-9}$ b) $7000 m$
37.51 5.01 ns, clock on plane
37.53 0.168 MeV
37.55 a) 1.08×10^{14} J b) 2.70×10^{19} W
c) 1.10×10^{10} kg
37.57 a) $0.999929c$ b) $-0.9965c$ c) (i) 42.4 MeV
(a), 5.60 MeV (b) (ii) 15.7 MeV (a and b)
37.59 $0.357c$; receding
37.61 154 km/h
37.63 2.04×10^{-13} N
37.65 a) 2.6×10^{-8} s b) 0.97
37.67 a) 2.0×10^{-18} kg b) 4.0×10^4 m/s^2
37.69 a) 2494 MeV b) 2.526 times
c) 987.4 MeV, twice as much
37.71 choice (c)
37.73 choice (b)

Chapter 38

38.1 5.77×10^{14} Hz, 1.28×10^{-27} kg $\cdot$ m/s,
3.84×10^{-19} J = 2.40 eV
38.3 a) 5.0×10^{14} Hz b) 2.3×10^{20} photons/s
c) no
38.5 a) 2.47×10^{-19} J = 1.54 eV
b) 804 nm; infrared
38.7 249 km/s
38.9 2.14 eV
38.11 a) 264 nm b) 4.70 eV, same
38.13 0.311 nm; same
38.15 1.13 keV
38.17 0.0714 nm; 180°
38.19 a) 4.39×10^{-4} nm b) 0.04294 nm
c) 300 eV, loss d) 300 eV
38.21 51.0°
38.23 1.19×10^{-27} kg $\cdot$ m/s, 1.96×10^{-29} kg $\cdot$ m/s
38.25 16.6 fs
38.27 a) 5.07 mJ b) 11.3 W
c) 1.49×10^{16} photons/s
38.29 a) 6.99×10^{-24} kg $\cdot$ m/s b) 705 eV
38.31 6.28×10^{-24} kg $\cdot$ m/s, 59.4°

38.33 a) 5×10^{-33} m b) $(4 \times 10^{-9})°$ c) 0.1 mm
38.35 a) 319 eV; 1.06×10^7 m/s b) 3.89 nm
38.37 a) V_0 versus $1/\lambda$; 1.23×10^{-6} V $\cdot$ m (slope),
-4.76 V (y-intercept) b) 6.58×10^{-34} J $\cdot$ s,
4.76 eV c) 260 nm d) 84.0 nm
38.39 a) 2.40 pm (slope), 5.21 pm (y-intercept)
b) 2.40 pm c) 5.21 pm
38.41 choice (c)
38.43 choice (a)
38.45 choice (b)

Chapter 39

39.1 a) 0.155 nm b) 8.46×10^{-14} m
39.3 a) 2.37×10^{-24} kg $\cdot$ m/s
b) 3.08×10^{-18} J = 19.3 eV
39.5 4.36 km/s
39.7 a) 62.0 nm (photon), 0.274 nm (electron)
b) 4.96 eV (photon), 2.41×10^{-5} eV (electron)
c) ≈ 250 nm, electron
39.9 3.90×10^{-34} m, no
39.11 a) 0.0607 V b) 248 eV c) 20.5 μm
39.13 0.432 eV
39.15 a) 2.07°, 4.14° b) 1.81 cm
39.17 a) 5.82×10^{-13} J = 3.63 MeV
b) 5.82×10^{-13} J = 3.63 MeV
c) 1.32×10^7 m/s
39.19 3.16×10^{-34} kg $\cdot$ m^2/s
39.21 a) -218 eV; 16 times b) 218 eV; 16 times
c) 7.60 nm d) $\frac{1}{4}$ hydrogen radius
39.23 a) 2.18×10^6 m/s, 1.09×10^6 m/s,
7.27×10^5 m/s b) 1.53×10^{-16} s,
1.22×10^{-15} s, 4.13×10^{-15} s c) 8.2×10^6
39.25 a) 20 eV b) 3 eV, 5 eV, 8 eV, 10 eV, 15 eV,
18 eV c) photo will not be absorbed
d) 3 eV $< \phi <$ 5 eV
39.27 a) -17.50 eV, -4.38 eV, -1.95 eV, -1.10 eV,
-0.71 eV b) 378 nm
39.29 a) -5.08 eV b) -5.64 eV
39.31 5.32×10^{21} photons/s
39.33 4.00×10^{17} photons/s
39.35 a) 1.2×10^{-33} b) 3.5×10^{-17}
c) 5.9×10^{-9}
39.37 a) 2060 K b) 1410 nm
39.39 1.06 mm; microwave
39.41 a) $1.7 T$ b) 0.58
39.43 a) 1.9×10^{10} W/m^2 b) 20 nm; no
c) 6510 km = $0.0093 R_{sun}$ d) sun; 39
39.45 a) 1.6×10^4 m/s b) 2.3×10^{-4} m
39.47 not valid
39.49 6.34×10^{-14} eV
39.51 a) 1.69×10^{-28} kg b) -2.53 keV
c) 0.655 nm
39.53 a) 12.1 eV b) 3; 103 nm, 122 nm, 657 nm
39.55 a) 0.90 eV
39.57 a) 5×10^{49} photons/s b) 30,000
39.59 29,800 K
39.61 a) $I(f) = \dfrac{2\pi h f^5}{c^3 (e^{hf/kT} - 1)}$
39.63 a) 12 eV b) 0.15 mV; 7.3 km/s
c) 0.082 μV; 4.0 m/s
39.65 a) no b) 2.52 V
39.67 a) $E = c\sqrt{2mK}$ b) photon
39.69 b) $\Delta = \dfrac{m^2 c^2 \lambda^2}{2h^2}$
c) $v = (1 - 8.50 \times 10^{-8})c$, $\Delta = 8.50 \times 10^{-8}$
39.71 a) $\dfrac{h}{mc\sqrt{15}}$
b) (i) 1.53 MeV, 6.26×10^{-13} m
(ii) 2810 MeV, 3.41×10^{-16} m

39.73 a) 1.1×10^{-20} kg·m/s   b) 19 MeV
   c) $|U_{Coul}| = 0.015K$; no

39.75 a) 1.1×10^{-35} m/s   b) 2.3×10^{27} y; no

39.77 20.9°

39.79 a) 248 eV   b) 0.0603 eV

39.81 a) $F = -\dfrac{A|x|}{x}$, where $x \neq 0$

   b) $E = \dfrac{3}{2}\left(\dfrac{h^2 A^2}{m}\right)^{1/3}$

39.83 a) 3   b) 11.40 nm   c) 60.5 eV

39.85 a) Antares   b) Polaris and α Centauri B
   c) α Centauri B

39.89 choice (a)

39.91 choice (a)

Chapter 40

40.1 $\Psi(x, t) = Ae^{-i(4.27 \times 10^{10}\,\mathrm{m}^{-1})x}e^{-i(1.05 \times 10^{17}\,\mathrm{s}^{-1})t}$

40.3 a) $8\pi/k$ b) $4\omega/k$; same

40.5 a) $\lambda/4, 3\lambda/4, 5\lambda/4, \ldots$ b) $0, \lambda/2, 3\lambda/2, \ldots$

40.7 no

40.9 a) 1.6×10^{-67} J   b) 1.3×10^{-33} m/s;
   1.0×10^{33} s   c) 4.9×10^{-67} J   d) no

40.11 0.166 nm

40.13 0.61 nm

40.15 b) no; no   c) $\sqrt{2b}$

40.17 a) $0, L/2, L$ b) $L/4, 3L/4$ c) yes

40.19 a) 6.0×10^{-10} m (twice the width of the box),
   1.1×10^{-24} kg·m/s   b) 3.0×10^{-10} m (same
   as the width of the box), 2.2×10^{-24} kg·m/s
   c) 2.0×10^{-10} m (2/3 the width of the box),
   3.3×10^{-24} kg·m/s

40.21 3.43×10^{-10} m

40.23 1.38 μm

40.25 22 fm

40.27 a) 0.0013   b) 10^{-143}

40.29 a) 4.4×10^{-8} b) 4.2×10^{-4}

40.31 $1/\sqrt{2}$

40.33 1.11×10^{-33} J $= 6.93 \times 10^{-15}$ eV,
   2.22×10^{-33} J $= 1.39 \times 10^{-14}$ eV; no

40.35 a) 0.21 eV   b) 5900 N/m

40.37 111 nm

40.39 $(2n + 1)\dfrac{\hbar}{2}$, increases with n

40.41 a) 5.89×10^{-3} eV   b) 106 μm   c) 0.0118 eV

40.43 a) $|\Psi(x, t)|^2 = \dfrac{2}{L}\left[1 - \cos\left(\dfrac{4\pi^2 \hbar t}{mL^2}\right)\right]$

   b) $\dfrac{4\pi^2 \hbar}{mL^2}$

40.45 $B = \left(\dfrac{k_1 - k_2}{k_1 + k_2}\right)A$, $C = \left(\dfrac{2k_2}{k_1 + k_2}\right)A$

40.47 a) 19.2 μm   b) 11.5 μm

40.49 a) $(2/L)dx$ b) 0   c) $(2/L)dx$

40.51 a) 0.818   b) 0.500   c) yes

40.55 a) $A = C$, $B\sin kL + A\cos kL = De^{-\kappa L}$,
   where $k = \dfrac{\sqrt{2mE}}{\hbar}$

   b) $kB = \kappa C$, $kB\cos kL - kA\sin kL = -\kappa De^{-\kappa L}$

40.57 6.63×10^{-34} J $= 4.14 \times 10^{-15}$ eV,
   1.33×10^{-33} J $= 8.30 \times 10^{-15}$ eV, no

40.59 b) 134 eV

40.61 a) 3, 4   b) 0.90 nm   c) 890 nm

40.63 22 eV, 56 eV, 110 eV

40.65 a) $x = \pm\sqrt{2E/k'}$ c) underestimate

40.67 choice (c)

40.69 choice (a)

Chapter 41

41.1 a) 1   b) 3

41.3 3.51 nm

41.5 $(2, 2, 1)$: $x = L/2$, $y = L/2$; $(2, 1, 1)$: $x = L/2$;
   $(1, 1, 1)$: none

41.7 a) 0   b) $\sqrt{12}\,\hbar$, 3.65×10^{-34} kg·m^2/s
   c) $3\hbar$, 3.16×10^{-34} kg·m^2/s
   d) $\frac{1}{2}\hbar$, 5.27×10^{-35} kg·m^2/s   e) $\frac{1}{6}$

41.9 4

41.11 4

41.13 $1.414\hbar$, $19.49\hbar$, $199.5\hbar$; as n increases, the
   maximum L gets closer to $n\hbar$.

41.15 a) 18   b) $m_l = -4$, 153.4°
   c) $m_l = +4$, 26.6°

41.19 a) 0.468 T   b) 3

41.21 a) 9   b) 3.47×10^{-5} eV   c) 2.78×10^{-4} eV

41.23 a) 2.5×10^{30} rad/s
   b) 2.5×10^{13} m/s; not valid since $v > c$

41.25 1.68×10^{-4} eV; $m_s = +\frac{1}{2}$

41.27 $n = 1, l = 0, m_l = 0, m_s = \pm\frac{1}{2}$: 2 states;
   $n = 2, l = 0, m_l = 0, m_s = \pm\frac{1}{2}$: 2 states;
   $n = 2, l = 1, m_l = 0, \pm 1, m_s = \pm\frac{1}{2}$: 6 states

41.29 a) $1s^2 2s^2$ b) magnesium; $1s^2 2s^2 2p^6 3s^2$
   c) calcium, $1s^2 2s^2 2p^6 3s^2 3p^6 4s^2$

41.31 4.18 eV

41.33 a) $1s^2 2s^2 2p$ b) -30.5 eV
   c) $1s^2 2s^2 2p^6 3s^2 3p$ d) -13.6 eV

41.35 a) -13.6 eV   b) -3.4 eV

41.37 a) 8.95×10^{17} Hz, 3.71 keV, 3.35×10^{-10} m
   b) 1.68×10^{18} Hz, 6.96 keV, 1.79×10^{-10} m
   c) 5.48×10^{18} Hz, 22 7 keV, 5.47×10^{-11} m

41.39 $3E_{1,1,1}$

41.41 a) $\frac{1}{64} = 0.0156$ b) 7.50×10^{-4}
   c) 2.06×10^{-3}

41.43 a) 0.500   b) 0.409

41.45 a) $E = \hbar\left[(n_x + n_y + 1)\omega_1^2 + \left(n_z + \frac{1}{2}\right)\omega_2^2\right]$,
   with n_x, n_y, n_z nonnegative integers
   b) $\hbar\left(\omega_1^2 + \frac{1}{2}\omega_2^2\right)$, $\hbar\left(\omega_1^2 + \frac{3}{2}\omega_2^2\right)$ c) 1

41.47 b) $n = 5$ shell

41.49 a) $2a$ b) 0.238

41.51 $4a$; same

41.53 b) $(\theta_L)_{max} = \arccos\left(-\sqrt{1 - 1/n}\right)$

41.55 3.00 T

41.57 a) $0.99999978 = 1 - 2.2 \times 10^{-7}$
   b) 0.9978   c) 0.978

41.59 a) 4, 20   b) $1s^4 2s^4 2p^3$

41.61 a) 122 nm   b) 1.52 pm; increase

41.63 a) 0.188 nm, 0.250 nm
   b) 0.0471 nm, 0.0624 rm

41.65 a) Li: 5.391 eV; Na: 5.139 eV; K: 4.341 eV;
   Rb: 4.177 eV; Cs: 3.894 eV; Fr: 3.9 eV
   b) Li: 3; 2; Na: 11; 3; K: 19; 4; Rb: 37; 5;
   Cs: 55; 6; Fr: 87; 7
   c) Li: 1.26; Na: 1.84; K: 2.26; Rb: 2.77;
   Cs: 3.21; Fr: 3.8   d) increase

41.67 a) 2.84×10^{10} Hz/T   b) 9.41×10^{-24} J/T
   c) 1.78×10^{11} Hz/T; 2.03

41.69 a) 3.02×10^{-11} m, 3.83×10^6 m/s
   b) 83.5 eV   c) -166.9 eV   d) 83.4 eV

41.71 choice (b)

41.73 choice (d)

Chapter 42

42.1 277 nm; ultraviolet

42.3 40.8 μm

42.5 5.65×10^{-13} m

42.7 2440 MHz, 0.123 m; yes

42.9 a) 1.03×10^{12} rad/s
   b) 66.3 m/s (C), 49.8 m/s (O)   c) 6.10 ps

42.11 a) 7.49×10^{-3} eV   b) 166 μm

42.13 30.27 N/m

42.15 2170 kg/m^3

42.17 a) 1.12 eV

42.19 1.20×10^6

42.21 1.5×10^{22} states per electron volt

42.23 a) $0.0233R$ b) $0.00767 = 0.767\%$
   c) no, motion of the ions

42.25 $0.312 = 31.2\%$

42.27 0.20 eV below the bottom of the conduction
   band

42.29 a) (i) 0.0204 mA   (ii) -0.0196 mA
   (iii) 26.8 mA   (iv) -0.491 mA
   b) yes, where -1.0 mV $< V < +1.0$ mV

42.31 a) 5.56 mA   b) -5.18 mA, -3.77 mA

42.33 a) 977 N/m   b) 1.25×10^{14} Hz

42.35 a) 3.8×10^{-29} C·m   b) 1.3×10^{-19} C
   c) 0.81   d) 0.058, much less

42.37 a) 0.96 nm   b) 1.8 nm

42.39 b) (i) 2.95   (ii) 4.73   (iii) 7.57   (iv) 0.838
   (v) 5.69×10^{-9}

42.41 a) 1.146 cm, 2.291 cm
   b) 1.171 cm, 2.341 cm; 0.025 cm $(2 \rightarrow 1)$,
   0.050 cm $(1 \rightarrow 0)$

42.43 0.274 eV; much less

42.45 a) 4.24×10^{-47} kg·m^2
   b) (i) 4.30 μm   (ii) 4.28 μm   (iii) 4.40 μm

42.47 2.03 eV

42.49 a) $2/a^3$ b) 4.7 eV

42.51 a) 0.445 eV (slope), 1.80 eV (y-intercept)
   b) 5170 K, 1.80 eV

42.53 b) 3.81×10^{10} Pa $= 3.76 \times 10^5$ atm

42.55 a) 1.67×10^{33} m^{-3} b) yes
   c) 6.66×10^{35} m^{-3} d) no

42.57 choice (b)

Chapter 43

43.1 a) 14 p, 14 n   b) 37 p, 48 n   c) 81 p, 124 n

43.3 0.533 T

43.5 a) 76.21 MeV
   b) 76.68 MeV; 0.6%; greater accuracy for $^{62}_{28}$Ni

43.7 a) 92.16 MeV   b) 7.680 MeV/nucleon
   c) 0.8245%

43.9 a) 1.32 MeV   b) 1.13×10^7 m/s

43.11 $^{86}_{36}$K: 8.73 MeV/nucleon;
   $^{180}_{73}$Ta: 8.08 MeV/nucleon; yes

43.13 a) $^{235}_{92}$U   b) $^{24}_{12}$Mg   c) $^{15}_{7}$N

43.15 156 keV

43.17 a) 0.836 MeV   b) 0.700 MeV

43.19 5.01×10^4 y

43.21 a) 4.92×10^{-18} s^{-1} b) 2990 kg
   c) 1.24×10^5 decays/s

43.23 a) 163 decays/min   b) 0.435 decay/min

43.25 a) 0.421 decay/s   b) 11.4 pCi

43.27 2.80 days

43.29 a) 2.02×10^{15}
   b) 1.01×10^{15}; 3.78×10^{11} decays/s
   c) 2.53×10^{14}; 9.45×10^{10} decays/s

43.31 a) 1.2 mJ   b) 10 mrem, 10 mrad, 7.5 mJ
   c) 6.2

43.33 500 rad, 2000 rem, 5.0 J/kg

43.35 a) 1.75 kGy, 1.75 kSv, 175 krem, 385 J
   b) 1.75 kGy, 2.625 kSv, 262.5 krem, 385 J

43.37 a) 9.32 rad, 9.32 rem

43.39 a) 0.497 mJ   b) 0.0828 rem

43.41 a) $Z = 3$, $A = 7$ b) 7.150 MeV
   c) 1.4 MeV

43.43 a) 173.3 MeV   b) 4.42×10^{23} MeV/g

43.45 a) $Z = 5$, $A = 10$ b) absorbed; 2.79 MeV

43.47 a) 4.7×10^4 J/g   b) 8.2×10^{10} J/g
   c) 4.3×10^{11} J/g   d) 7600 y

43.49 a) 4.14 MeV   b) 7.75 MeV/nucleon, about
   half the binding energy per nucleon

43.51 a) $^{90}_{39}$Y   b) 25%   c) 112 y

43.53 a) $^{25}_{13}$Al will decay into $^{25}_{12}$Mg.
b) β^+ or electron capture   c) 3.254 MeV (β^+), 4.277 MeV (electron capture)

43.55 a) $^{14}_{6}$C $\rightarrow$ e$^-$ + $^{14}_{7}$N + $\bar{v}_e$ b) 0.156 MeV
c) 13.5 kg; 3400 decays/s
d) 530 MeV/s = 8.5×10^{-11} J/s
e) 36 μGy, 3.6 mrad, 36 μSv, 3.6 mrem

43.57 1.03×10^{-3} u; yes

43.59 a) 5.0×10^4 b) $10^{-15,000}$

43.61 29.2%

43.63 a) 0.96 μJ/s   b) 0.48 mrad/s   c) 0.34 mrem
d) 6.9 days

43.65 1.0×10^4 y

43.67 a) 0.48 MeV
b) 3.270 MeV = 5.239×10^{-13} J
c) 3.155×10^{-11} J/mol, more than a million times larger

43.69 a) 1.16 h   b) 1.20×10^8 c) 1.81×10^6

43.71 4.59×10^{-5} g/h

43.73 choice (a)

43.75 choice (d)

43.77 choice (b)

Chapter 44

44.1 a) 69 MeV, 1.7×10^{22} Hz, 18 fm; gamma ray

44.3 a) 32 MeV

44.5 9.26×10^6 m/s

44.7 7.2×10^{19} J; 70%

44.9 a) 1.18 T   b) 3.42 MeV, 1.81×10^7 m/s

44.11 a) 30.6 GeV   b) 8.0 GeV

44.13 a) 0.999999559c b) 3.83×10^8 rad/s (nonrel), 3.59×10^5 rad/s (rel)

44.15 a) 3200 GeV   b) 38.7 GeV

44.17 a) π^0, π^+ b) 219.1 MeV

44.19 1.63×10^{-25} kg; 97.2

44.21 116 MeV

44.23 (b) and (d)

44.25 (c) and (d)

44.27 a) 0, 1, -1, 0   b) 0, 0, 0, 1   c) $-e$, 1, 0, 0
d) $-e$, 0, 0, -1

44.29 a) $u\bar{s}$ b) $\bar{d}\,\bar{d}\,\bar{s}$ c) uss

44.31 a) 3.28×10^7 m/s   b) 1590 Mly

44.33 a) 1.08×10^5 km/s   b) 1.46

44.35 -0.783 MeV; endoergic

44.37 966 nm

44.39 a) $\pi^- \rightarrow \mu^- + v \rightarrow e^- + 3v$; an electron and neutrinos   b) 139 MeV   c) 2.24×10^{10}
d) 50 Sv, 5.0 krem

44.41 2.494 GeV

44.43 a) 0, $+e$, 1, $L_e = L_\mu = L_\tau = 0$, K$^+$
b) 0, $-e$, 0, $L_e = L_\mu = L_\tau = 0$, π^-
c) -1, 0, 0, $L_e = L_\mu = L_\tau = 0$, antineutron ($\bar{n}$)
d) 0, $+e$, 0, $L_\mu = -1$, $L_e = L_\tau = 0$, μ^+

44.45 7.5×10^{-23} s

44.47 a) 0.70 rad   b) 0.70 rem, 7 times, 2%; no

44.49 b) $R/R_0 = 0.574$ c) speeding up at 300 My, slowing down at 13.1 Gy

44.51 230 MeV, 12.5° below the $+x$-axis

44.53 a) all are much less; no   b) 37.5 cm
c) 0.42 MeV   d) 3.8×10^7 rad/s

44.55 a) $Q = -1$, $S = -3$; yes
b) Δ: *ddd*, *udd*, *uud*, *uuu*, Σ^*: *dds*, *uds*, *uus*, Ξ^*: *dss*, *uss*, Ω^-: *sss*

44.57 choice (d)

44.59 choice (c)

CREDITS

Chapter 1 Opener: Minerva Studio/Shutterstock; 1.1a: Michele Perbellini/Shutterstock; 1.1b: Studio Bazile/Thales/ESA; 1.4: AFP/Getty Images/Newscom; 1.5ab: NASA; 1.5c: JPL/NASA; 1.5d: Photodisc/Getty Images; 1.5e: Chad Baker/Photodisc/Getty Images; 1.5f: Veeco Instruments, Inc; 1.6: Pearson; 1.7: ND/Roger Viollet/Getty Images; Appl. p. 10: Tyler Olsen/Shutterstock

Chapter 2 Opener: Vibe Images/Fotolia; 2.4: Michael Dalder/Reuters/Landov; 2.5: Wolfgang Rattay/Reuters; Appl. p. 46: NASA; 2.22: Richard Megna/Fundamental Photographs; 2.26: Andreas Stirnberg/Getty Images; 2.27: Guichaoua/Alamy; E2.54 (chart): Source: "The Flying Leap of the Flea" by M. Rothschild, Y. Schlein, K. Parker, C. Neville, and S. Sternberg in the November 1973 *Scientific American*

Chapter 3 Opener: Feng Li/Getty Images; Appl. p. 71: Luca Lozzi/Getty Images; 3.8: Dominique Douieb/PhotoAlto Agency/Getty Images; 3.16: Richard Megna/Fundamental Photographs; 3.19: Fundamental Photographs; Appl. p. 85: David Wall/Alamy; 3.31: Hart Matthews JHM/GAC/Reuters

Chapter 4 Opener: Monkey Business/Fotolia; p. 105 (law): Newton, Isaac. 1845. *Newton's Principia: The Mathematical Principles*. Andrew Mott (trans.) New York: Daniel Adee; Appl. p. 106: Prisca Koller/Fotolia; 4.11: Wayne Eastep/The Image Bank/Getty Images; 4.16: Albert Gea/Reuters; p. 111 (law): Newton, Isaac. 1845. *Newton's Principia: The Mathematical Principles*. Andrew Mott (trans.) New York: Daniel Adee; Appl. p. 111: Kadmy/Fotolia; 4.19: Cheryl A. Meyer/Shutterstock; p. 116 (law): Newton, Isaac. 1845. *Newton's Principia: The Mathematical Principles*. Andrew Mott (trans.) New York: Daniel Adee; 4.28: Maksym Gorpenyuk/Shutterstock; 4.29a: PCN Black Photography/Alamy; 4.29b: John W. McDonough/Sports Illustrated/Getty Images; 4.29c: Roy Pedersen/Fotolia; P4.53 (table): Data from www.autosnout.com

Chapter 5 Opener: Brian A Jackson/Shutterstock; 5.11: NASA; 5.16: Efired/Shutterstock; Appl. p. 144: Alex Kosev/Shutterstock; Appl. p. 148: Eye of Science/Science Source; 5.26b: 2happy/Shutterstock; Appl. p. 155: Suthep/Shutterstock; 5.38a: JPL/Space Science Institute/NASA; 5.38b: Jason Stitt/Shutterstock; 5.38c: FikMik/Shutterstock; 5.38d: Shots Studio/Shutterstock; P5.78 (chart): Source: "The Flying Leap of the Flea" by M. Rothschild, Y. Schlein, K. Parker, C. Neville, and S. Sternberg in the November 1973 *Scientific American;* P5.111 (table): Courtesy of The Engineering Toolbox. www.engineeringtoolbox.com/wire-rope-strength-d_1518.html

Chapter 6 Opener: Brocreative/Shutterstock; 6.1: mariiya/Fotolia; Appl. p. 173: Steve Gschmeissner/Science Source; 6.13: Mikadun/Shutterstock; Appl. p. 186: Steve Gschmeissner/Science Source; 6.26: Fox Photos/Hulton Archive/Getty Images; Appl. p. 190: Gayvoronskaya_Yana/Shutterstock; 6.27a: Keystone/Hulton Archive/Getty Images; 6.27b: Anthony Hall/Fotolia; 6.28: Fandu/Fotolia; P6.90 (chart): Data from Science Buddies. www.sciencebuddies.org/science-fair-projects/project_ideas/Physics_Springs_Tutorial.shtml

Chapter 7 Opener: Nagel Photography/Shutterstock; 7.1: Alistair Michael Thomas/Shutterstock; Appl. p. 205: Erni/Shutterstock; 7.3: Robert F. Bukaty/AP Images; 7.5: hinnamsaisuy/Shutterstock; 7.12: ejwhite/Shutterstock; Appl. p. 214: G. Ronald Austing/Science Source; 7.15: JNP/Shutterstock; Appl. p. 220: fotoedu/Shutterstock; 7.21: Kletr/Shutterstock; Appl. p. 223: LaiQuocAnh/Shutterstock; 7.24: Peter Menzel/Science Source; P7.78 (table): Courtesy of the EngineersHandbook.com

Chapter 8 Opener: Colorful High Speed Photographs/Moment Select/Getty Images; 8.2: Alex Emanuel Koch/Shutterstock; Appl. p. 239: Willie Linn/Shutterstock; 8.4: Vereshchagin Dmitry/Shutterstock; 8.5: Jim Cummins/The Image Bank/Getty Images; 8.7ab: Andrew Davidhazy; 8.17: FPG/Archive Photos/Getty Images; 8.22: David Leah/The Image Bank/Getty Images; 8.30: Richard Megna/Fundamental Photographs; Appl. p. 258: Elliotte Rusty Harold/Shutterstock; 8.34: NASA; P8.101 (table): Courtesy of Chuck Hawks. www.chuckhawks.com/handgun_power_chart.htm

Chapter 9 Opener: Blanscape/Shutterstock; 9.3b: connel/Shutterstock; Appl. p. 278: Hybrid Medical Animation/Science Source; Appl. p. 285 (top): Dan Rodney/Shutterstock; Appl. p. 285 (bottom): Ammit Jack/Shutterstock; 9.18: David J. Phillip/AP Images; 9.19: DenisNata/Shutterstock; 9.22: NASA; P9.86: NASA

Chapter 10 Opener: Lonely Planet Images; 10.7: 68/Ocean/Corbis; Appl. p. 311 (right): Bruce MacQueen/Shutterstock; Appl. p. 311 (left): David Lentink/Science Source; 10.14: Robert Young/Fotolia; 10.17: gorillaimages/Shutterstock; Appl. p. 315: Chris DeRidder/Shutterstock; 10.22: Bjorn Heller/Shutterstock; 10.28: Gerard Lacz/Photoshot; P10.88: Data from Chevrolet

Chapter 11 Opener: nito/Shutterstock; 11.3: Turleyt/Fotolia; 11.8a: Maridav/Shutterstock; 11.12a: Benjamin Marin Rubio/Shutterstock; 11.12b: Richard Carey/Fotolia; 11.12c: Andrew Bret Wallis/Photodisc/Getty Images; 11.13: Djomas/Shutterstock; Appl. p. 350: Dante Fenolio/Science Source

Chapter 12 Opener (left): orlandin/Shutterstock; Opener (right): Isabelle Kuehn/Shutterstock; Appl. p. 370: Jorg Hackemann/Shutterstock; 12.6: SPL/Science Source; p. 373 (law): Pascal, Blaise. 1653. *Treatise on the Equilibrium of Liquids;* 12.9b: Andrey Jitkov/Shutterstock; Appl. p. 375: Lisa F. Young/Shutterstock; p. 376 (principle): Source: Archimedes; 12.14: Pali A/Fotolia; 12.19: Pearson; 12.20: Popov Nikolay/Shutterstock; 12.22: Fribus Ekaterina/Shutterstock; Appl. p. 382: Kairos69/Shutterstock; 12.28: Shutterstock; Appl. p. 386: Edward Lara/Shutterstock; 12.30a: jupeart/Shutterstock; 12.30b: Btrseller/Shutterstock; 12.31f: Harold Edgerton at MIT, copyright 2014. Courtesy of Palm Press, Inc.

Chapter 13 Opener: JPL-Caltech/SSI/Cornell/NASA; p. 398 (law): Source: Newton published the law of gravitation in 1687. *Philosophiae naturalis principia mathematica* (MacLehose, 1726) [1871] Reprinted for Sir William Thomson and Hugh Blackburn; 13.3 (Jupiter): JPL/University of Arizona/NASA; 13.3 (inset): JPL/Cornell University/NASA; 13.6: ESA/Hubble/NASA; Appl. p. 403: NASA; 13.7: swisshippo/Fotolia; 13.13: National Aeronautics and Space Administration/NASA; 13.14: Based on Newton, Isaac. 1728. *A Treatise on the System of the World;* 13.16: John F. Kennedy/Space Center/NASA; 13.17: ESA and M. Showalter/NASA; p. 411 (laws): Kepler, Johannes. 1597. *Mysterium Cosmographicum;* Appl. p. 412: NASA; 13.21b: NASA; p. 419 (quote): Source: John Michell, (1784). "On the Means of Discovering the Distance, Magnitude, & c. of the Fixed Stars, in Consequence of the Diminution of the Velocity of Their Light, in Case Such a Diminution Should be Found to Take Place in any of Them, and Such Other Data Should be Procured from Observations, as Would be Farther Necessary for That Purpose". *Philosophical Transactions of the Royal Society* 74 (0): 35–57; 13.27ab: NASA; 13.28: NASA; 13.29: Andrea Ghez/UCLA; P13.76 (table): Source: ssd.jpl.nasa.gov; P13.77 (table): Source: nssdc.gsfc.nasa.gov/planetary/factsheet

Chapter 14 Opener: blurAZ/Shutterstock; Appl. p. 434: Steve Byland/Shutterstock; 14.7: American Diagnostic Corporation; 14.21a: Zurijeta/Shutterstock; 14.25: Christopher Griffin/Alamy; Appl. p. 455: Symbiot/Shutterstock; Appl. p. 456: Sonya Etchison/Shutterstock

Chapter 15 Opener: Walter D. Mooney/U.S. Geological Survey; Appl. p. 469: Marco PoloCollection/Balan Madhavan/Alamy; 15.2: Charles Platiau/Reuters; 15.5: EpicStockMedia/Shutterstock; 15.12: Kodda/Shutterstock; Appl. p. 482: Christian Delbert/Shutterstock; 15.23a–d: Richard Megna/Fundamental Photographs; 15.25: Shmeliova Natalia/Shutterstock; 15.27: National Optical Astronomy Observatories

Chapter 16 Opener: Eduard Kyslynskyy/Shutterstock; 16.5 (line art): Berg, Richard E.; Stork, David G., *The Physics Of Sound*, 1st Ed., ©1982. Reprinted and electronically reproduced by permission of Pearson Education, Inc., Upper Saddle River, New Jersey; 16.5a: olly/Shutterstock; 16.5b: Lebrecht Music and Arts Photo Library/Alamy; Appl. p. 510: Steve Thorne/Getty Images; 16.6: Geoff Dann/DK Images; 16.9: Kretztechnik/Science Source; 16.10: auremar/Shutterstock; 16.15: Digoarpi/Shutterstock; Appl. p. 523: Piotr Marcinski/Shutterstock; 16.20: Martin Bough/Fundamental

INDEX

For users of the three-volume edition: pages 1–682 are in Volume 1; pages 683–1253 are in Volume 2; pages 1218–1522 are in Volume 3. Pages 1254–1522 are not in the Standard Edition.

NOTE: Page numbers followed by f indicate figures; those followed by t indicate tables.

Astronomical Data[†]

Body	Mass (kg)	Radius (m)	Orbit radius (m)	Orbital period
Sun	1.99×10^{30}	6.96×10^{8}	—	—
Moon	7.35×10^{22}	1.74×10^{6}	3.84×10^{8}	27.3 d
Mercury	3.30×10^{23}	2.44×10^{6}	5.79×10^{10}	88.0 d
Venus	4.87×10^{24}	6.05×10^{6}	1.08×10^{11}	224.7 d
Earth	5.97×10^{24}	6.37×10^{6}	1.50×10^{11}	365.3 d
Mars	6.42×10^{23}	3.39×10^{6}	2.28×10^{11}	687.0 d
Jupiter	1.90×10^{27}	6.99×10^{7}	7.78×10^{11}	11.86 y
Saturn	5.68×10^{26}	5.82×10^{7}	1.43×10^{12}	29.45 y
Uranus	8.68×10^{25}	2.54×10^{7}	2.87×10^{12}	84.02 y
Neptune	1.02×10^{26}	2.46×10^{7}	4.50×10^{12}	164.8 y
Pluto[‡]	1.31×10^{22}	1.15×10^{6}	5.91×10^{12}	247.9 y

[†]Source: NASA (**http://solarsystem.nasa.gov/planets/**). For each body, "radius" is its average radius and "orbit radius" is its average distance from the sun or (for the moon) from the earth.

[‡]In August 2006, the International Astronomical Union reclassified Pluto and similar small objects that orbit the sun as "dwarf planets."

Prefixes for Powers of 10

Power of ten	Prefix	Abbreviation	Pronunciation
10^{-24}	yocto-	y	*yoc*-toe
10^{-21}	zepto-	z	*zep*-toe
10^{-18}	atto-	a	*at*-toe
10^{-15}	femto-	f	*fem*-toe
10^{-12}	pico-	p	*pee*-koe
10^{-9}	nano-	n	*nan*-oe
10^{-6}	micro-	μ	*my*-crow
10^{-3}	milli-	m	*mil*-i
10^{-2}	centi-	c	*cen*-ti
10^{3}	kilo-	k	*kil*-oe
10^{6}	mega-	M	*meg*-a
10^{9}	giga-	G	*jig*-a or *gig*-a
10^{12}	tera-	T	*ter*-a
10^{15}	peta-	P	*pet*-a
10^{18}	exa-	E	*ex*-a
10^{21}	zetta-	Z	*zet*-a
10^{24}	yotta-	Y	*yot*-a

Examples:

1 femtometer = 1 fm = 10^{-15} m

1 picosecond = 1 ps = 10^{-12} s

1 nanocoulomb = 1 nC = 10^{-9} C

1 microkelvin = 1 μK = 10^{-6} K

1 millivolt = 1 mV = 10^{-3} V

1 kilopascal = 1 kPa = 10^{3} Pa

1 megawatt = 1 MW = 10^{6} W

1 gigahertz = 1 GHz = 10^{9} Hz